08

DISCARD

Haz Mat Data

Second Edition

HazMat Data

For First Response, Transportation, Storage, and Security

Second Edition

Richard P. Pohanish

A JOHN WILEY & SONS, INC., PUBLICATION

Copyright © 2004 by John Wiley & Sons, Inc. All rights reserved.

Published by John Wiley & Sons, Inc., Hoboken, New Jersey.
Published simultaneously in Canada.

No part of this publication may be reproduced, stored in a retrieval system or transmitted in any form or by any means, electronic, mechanical, photocopying, recording, scanning or otherwise, except as permitted under Section 107 or 108 of the 1976 United States Copyright Act, without either the prior written permission of the Publisher, or authorization through payment of the appropriate per-copy fee to the Copyright Clearance Center, Inc., 222 Rosewood Drive, Danvers, MA 01923, (978) 750-8400, fax (978) 646-8600, or on the web at www.copyright.com. Requests to the Publisher for permission should be addressed to the Permissions Department, John Wiley & Sons, Inc., 111 River Street, Hoboken, NJ 07030, (201) 748-6011, fax (201) 748-6008.

Limit of Liability/Disclaimer of Warranty: While the publisher and author have used their best efforts in preparing this book, they make no representation or warranties with respect to the accuracy or completeness of the contents of this book and specifically disclaim any implied warranties of merchantability or fitness for a particular purpose. No warranty may be created or extended by sales representatives or written sales materials. The advice and strategies contained herein may not be suitable for your situation. You should consult with a professional where appropriate. Neither the publisher nor author shall be liable for any loss of profit or any other commercial damages, including but not limited to special, incidental, consequential, or other damages.

For general information on our other products and services please contact our Customer Care Department within the U.S. at 877-762-2974, outside the U.S. at 317-572-3993 or fax 317-572-4002.

Wiley also publishes its books in a variety of electronic formats. Some content that appears in print, however, may not be available in electronic format.

Library of Congress Cataloging-in-Publication Data is available.

ISBN 0-471-27328-7

Printed in the United States of America.

10 9 8 7 6 5 4 3 2 1

NOTICE

This reference is intended to provide data about chemical hazards and guidelines for those trained in hazardous materials response. It is not intended as a primary source of research information. **As with any reference, it cannot include all information or discuss all situations that might occur. It cannot be assumed that all necessary warnings and precautionary measures are contained in this work, and that other, or additional, information or assessments may not be required. Most of all, it cannot replace the training and experience of individual responders.** Extreme care has been taken in the preparation of this work and, to the best knowledge of the publisher and the editors, the information presented is accurate and no warranty is expressed or implied. No warranty, express or implied, is made. Information may not be available for some chemicals; consequently, an absence of data does not necessarily mean that a substance is not hazardous. For major incidents it will be necessary to obtain additional detailed information from other resources as well as more expertise from those with more extensive training. Neither the publisher nor the editors assume any liability or responsibility for completeness or accuracy of the information presented or any damages of any kind alleged to result in connection with, or arising from, the use of this book. The publisher and the editors strongly encourage all readers, and users of chemicals, to follow the manufacturers' or suppliers' current instructions, technical bulletins, and material safety data sheets (MSDSs) for specific use, handling, and storage of all chemical materials.

ACKNOWLEDGEMENTS

Thanks are due to those individuals who reviewed and constructively commented on the development of this second edition. Wesley Van Pelt, Ph.D. , in particular, deserves special thanks for his valuable advice and criticism. The assistance of Alan Schneider, D.Sc., Marine Technical and Hazardous Materials Division, Headquarters, U.S. Coast Guard, Washington D.C., and Steve Lawhorne, U.S. Army, Aberdeen Proving Grounds, Maryland, in producing the initial edition of this work is gratefully acknowledged. I also want to acknowledge the good work of the scientists, contract employees, and others who developed various documents and databases that provided so much of the data that were compiled for this work. Thanks are due to my editor, Bob Esposito of John Wiley & Sons, for his encouragement and suggestions.

CONTENTS

Introduction		ix
	How to Use This Book	ix
Conversion Factors		xvii
International Code of Signals Phonetic Alphabet		xix
Guide to Compatibility of Chemicals		xxi
	Exceptions to the Chart	xxxvi
	Reactivity Groups	xxxviii
Key to Abbreviations, Symbols, and Acronyms		xlv
Chemical Records A to Z		1
Appendix A	Emergency Response	1171
Appendix B	List of Marine Pollutants (§172.101 Appendix B)	1185
Appendix C	Synonym and Trade Name Index	1191
Appendix D	CAS Number Index	1257
Appendix E	Chemicals Likely Involved in Terrorist Incidents	1263

INTRODUCTION

This work covers nearly 1450 hazardous chemicals found in the industrial workplace and frequently transported in bulk. Chemical warfare (CW) agents are seen as a threat to civilian populations and many have been included. Biological agents and most drugs have not been included in this edition. Each chemical presented in the record includes data of interest to those who must manage the use of chemicals and those who must respond to accidents and spills. The data is presented in a uniform format to make it easy for users to find critical information quickly or to compare data in related records.

The objectives of this book are (1) to provide critical data for those who must initially respond to fixed facility, transportation, and terrorist incidents and to help them to limit the consequences of these incidents and (2) to present data on heavily used and widely transported chemicals in a portable package. To this end, the author compiled data for this volume by utilizing significant features of several important works originally produced by agencies of federal and state governments and by prominent professional and advisory organizations.

This book is directed at first responders trained at the Awareness and/or Operational levels as defined by the Occupational Safety and Health Administration (OSHA 1910-120) and the National Fire Protection Association (NFPA 472, 19992). First Responders at the *Awareness Level* are individuals who are likely to witness or discover a hazardous substance release and who have been trained to initiate an emergency response sequence by notifying the authorities of the release. At this level, first responders shall have sufficient training or have had sufficient experience to objectively demonstrate competency in the following areas: (a) an understanding of what hazardous substances are, and the risks associated with them in an incident; (b) an understanding of the potential outcome associated with an emergency created when hazardous substances are present; (c) the ability to recognize the presence of hazardous substances in an emergency; (d) the ability to identify the hazardous substances, if possible; and (e) an understanding of the role of the first responder awareness individual in the employers emergency response plan, including site security and control and the U.S. Department of Transportation's Emergency Response Guide (ERG). First Responders at the *Operations Level* are individuals who respond to releases or potential releases of hazardous substances as part of the initial response to the site for the purpose of protecting nearby persons, property, or the environment form the effects of the release. They are trained to respond in a defensive fashion without actually trying to stop the release. Their function is to contain the release from a safe distance, keep it from spreading, and prevent exposures. First responders at the Operations Level shall have received at least eight hours of training or have sufficient experience to objectively demonstrate competency in the following areas in addition to those listed for the Awareness Level and employer shall so certify: (a) knowledge of the basic hazard and risk assessment techniques; (b) knowhow to select and use proper personal protective equipment provided to the first responder operations level; (c) an understanding of basic hazardous materials terms; (d) knowhow to perform basic control, containment, and /or confinement operations within the capabilities of the resources and personal protective equipment available with their unit; (e) knowhow to implement basic decontamination procedures; and (f) an understanding of the relevant standard operating procedures and termination procedures.

Any comments or criticism from readers regarding this volume, as well as, suggestions for future improvements, are welcomed by the author.

How to Use This Book

Each of the chemical records contains 13 sections that include regulatory names, trade names, and chemical synonyms, identification, chemical formula, DOT information and reportable quantity. The description section lists physical description, warning properties, and lists of dangers that may be faced by first responders, a simple four-part hazard classification (based on the NFPA-704 Rating System). The emergency response section includes evacuation recommendations from the US DOT. The exposure and health hazards sections included PPE and respirator recommendations. This is followed by sections listing fire data, chemical reactivity, environmental data, shipping information, hazard classification for bulk water transportation, and physical and chemical properties.

A new feature of this book is a chart entitled "**Chemicals Likely Involved in Terrorist Incidents**" This chart is the last page of the book, and it lists kind of agents, examples of war agents and the record number, examples of industrial chemicals and record number, a brief summary of physical properties (a more complete list is incorporated in the chemical records or profiles), early symptoms of exposure, and field detection using both M8 paper and the M256-A1 Chemical Agent Detection Kit. The record number can be used to immediately access the chemical agent of interest.

Following is a explanation of the individual fields and special terms found in each chemical record. In a few cases where data items were not available, values were estimated by standard, reliable procedures; all such values are labeled "estimate." As accurate values are found, they will be included in subsequent editions of this work.

NAME SECTION

The top line of each hazardous materials record contains an arbitrary alphanumeric record number assigned by the author. The numbering system is sequential and designed to allow for the insertion of additional chemical records in future editions of this work. For ease of use, indices (both CAS and Synonym) are found in the *Appendix* and are keyed to record (Rec.) numbers. Each record is arranged alphabetically by a common chemical name used by regulatory and advisory bodies. In a few cases, product names or trivial names may be used because they are often encountered in the chemical industry. The chemical name section includes alternate scientific, product, trade, and other synonym names that are used for each hazardous substance.

Synonyms: The synonym field for many records also includes identification numbers, including the following: EEC (European Economic Community); RCRA (Resource Conservation and Recovery Act). These identifying numbers may be useful to shippers, environmental technicians, and safety personnel.

IDENTIFICATION SECTION

CAS Registry Number: A unique identifier is assigned to each chemical registered with the Chemical Abstracts Service (CAS) of the American Chemical Society. This number is used to identify chemicals on the basis of their molecular structure. CAS numbers, in the format xxx-xx-x, can be used in conjunction with chemical names for positive identification, as well as for searching computerized databases. To avoid confusion with like-sounding names (e.g, "benzene" and "benzine"), CAS number should always be used in conjunction with substance names. While not commonly used in shipping, the CAS number is nevertheless important. These numbers may be found on product containers and Material Safety Data Sheets (MSDS) and is used as an indexing number in more detailed reference sources.

Formula: A commonly used one-line molecular chemical formula has been provided, and in many cases a more detailed version is also presented.

DOT ID number(s): The hazard identification number is assigned to the substance by the U.S. Department of Transportation (DOT). The DOT ID number format is UNxxxx or NAxxxx. This identification (ID) number identifies substances regulated by DOT and must appear on shipping documents, or the exterior of packages, and on specified containers. ID numbers containing a UN prefix are also known as United Nations numbers and are authorized for use with international shipments of hazardous materials. The "NA" prefix is used for shipments between Canada and the United States and may not be used for international shipments. DOT Guide number: A four-digit number in the format xx xx assigned by the U.S. Department of Transportation (DOT). This number is a guide to initial actions to be taken by professional first responders to stabilize an emergency situation and to protect the general public. These action guides are found in the *2000 Emergency Response Guidebook* (ERG 2000), published by U.S. Department of Transportation, Research and Special Programs Administration. DOT Guide numbers are not unique for specific substances.

Proper shipping name: Names of hazardous materials used on shipping papers, placards, and packaging markings (49 CFR, 171, et al.) It should be noted by the reader that it is the shipper's responsibility to determine the proper shipping name and to provide it where required. In every case, 49 CFR, 172.101, Appendix A should be used to verify the proper shipping name of hazardous substances. In some cases a material may require more than one hazardous materials label. Materials designated as "Forbidden" may not be transported.

Reportable quantity (RQ): Releases of CERCLA and EHS hazardous substances in quantities equal to, or greater than, their RQs are subject to immediate reporting to the U.S. Environmental Protection Agency (EPA): Those labeled CERCLA are reportable to the National Response Center (NRC) and to state and Local Emergency Planning Committees (LEPCs) authorities under section 304 of SARA Title III (EPCRA). CERCLA hazardous substances, along with their reportable quantities, are listed in 40 CFR Part 302, Table 302.4. Substances labeled "EHS" are subject to state and local reporting under SARA Title III (EPCRA), Section 304. The National Response Center is operated by the U.S. Coast Guard. If a material is identified as a marine pollutant, it must by reported immediately to the NRC. When in doubt as to whether the release equals or exceeds the reporting levels, the NRC should be notified 24/7/365 at 1-800-424-8802 (Toll-free in the United States, Canada, and the U.S. Virgin Islands); 202-267-2675 in the District of Columbia. Calling other emergency response numbers such as CHRMTREC®, CHEM-TEL, Inc, INFOTRAC, etc., does NOT constitute compliance with regulatory requirements to call the NRC.

DESCRIPTION SECTION

This is a three-part section that contains the following: (1) A summary of observable characteristics of the substance including state (solid, liquid, or gas), color, odor description and solubility or miscibility in water: Where a compound may be shipped either as a liquid or solid, both designations are given. Color descriptions are listed for pure materials. Occasionally the color of a chemical changes when it dissolves in water or becomes a gas. Odor descriptions are listed for pure materials. The expression "characteristic" is used only when no other reasonable description was found. When the chemical reacts with water, producing heat, and/or another chemical, it is noted here. (2) A summary of important characteristics and dangers that may be encountered by first responders and firefighters. The author has bulleted this section to make it easier to read and included important notes and warnings. Included in this summary is toxic products of combustion. Some chemicals decompose or burn to give off toxic and irritating gases. Such gases may also be given off by chemicals that vaporize in the heat of a fire without either decomposing or burning. The specific combustion products are usually not well known over the wide variety of conditions existing in fires; some are very hazardous. (3) Hazard classification summary information for nearly all of the 1430 materials in this book. This information is based on the NFPA-704 Rating System using NFPA 49 definitions along with available data on the material or on structurally similar materials. The NFPA system provides basic information to firefighting and other first response personnel. The system is usually presented as a four-colored "diamond" and identifies the hazards of materials based on three principal categories: Health (Blue), Flammability (Red), and Reactivity (Yellow). A numerical rating indicates the degree of severity ranging from 4 (indicating a severe hazard) to 0 (indicating no hazard). The bottom of the diamond is white and can contain the symbol OXY (for a strong oxidizer) or W (for materials that react with water).

TABLE 1
National Fire Protection Association (NFPA)
Fire Diamond

Health Hazard (Blue)		Flammability (Red)		Reactivity (Yellow)	
Signal	Type of Possible Injury	Signal	Susceptibility of Materials to Burn	Signal	Susceptibility to Release of Energy
4	Materials which on very short exposure could cause death or major residual injury.	4	Materials which will rapidly or completely vaporize at atmospheric pressure and normal ambient temperature, or which are readily dispersed in air and which will burn readily.	4	Materials that, in themselves, are readily capable of detonation, explosive decomposition or explosive reaction at normal temperatures and pressures.
3	Materials which on short exposure could cause serious temporary or residual injury.	3	Liquids and solids that can be easily ignited under almost all normal temperature conditions.	3	Materials that, in themselves, are capable of detonation or explosive decomposition but require a strong initiating source or which must be heated under confinement before initiation or which react explosively with water.
2	Materials that, on intense or continued (but not chronic) exposure, could cause temporary incapacitation or possible residual injury.	2	Materials that must be moderately heated or exposed to relatively high ambient temperatures before ignition can occur.	2	Materials that readily undergo violent chemical changes at elevated temperatures and pressure or which react violently with water or which may form explosive mixtures with water.
1	Materials which on exposure would cause irritation but only minor residual injury.	1	Materials that must be preheated before ignition can occur.	1	Materials that in themselves are normally stable, but which become unstable at elevated temperatures and pressures.
0	Materials that, on exposure under fire conditions, would offer no hazard beyond that of ordinary combustible material.	0	Materials that will not burn.	0	Materials that in themselves are normally stable, even under fire exposure conditions, and which are not reactive with water.

Special Notice (White): Water Reactive: Avoid the use of Water
OXY: Oxidizer
Radioactive

EMERGENCY RESPONSE SECTION
Emergency response, initial isolation and protective distances, fire, exposure, and response to discharge have been included for first responders, NOT the general public. Contains first response information based on the *2000 Emergency Response Guidebook* (ERG 2000) "Guides" (orange) section. In many cases, the information may be more chemical-specific and less general. For example, Appendix A contains a Guide 119 for chlorosilanes and another Guide 119 for other non-chlorosilane chemicals. The Evacuation section contains three sections: Public Safety, Spill, and Fire for both Toxic to Humans (TIH) and non-TIH materials.

EXPOSURE SECTION
Short term effects: This section contains warnings such as HARMFUL IF ABSORBED THROUGH THE SKIN and brief descriptions of the effects observed in humans when the vapor (gas) is inhaled, when the liquid or solid is ingested (swallowed), and when the liquid or solid comes in contact with the eyes or skin. "First-aid" procedures are recommended. They deal with exposure to the vapor (gas), liquid, or solid and include inhalation, ingestion (swallowing) and contact with eyes or skin. The instruction "Do NOT induce vomiting" is given if an unusual hazard is associated with the chemical being sucked into the lungs (aspiration) while the patient is vomiting. "Seek medical attention" or "Call a doctor" is recommended in those cases where only competent medical personnel can treat the injury properly. In all cases of human exposure, seek medical assistance as soon as possible. Many records contain additional information in "medical notes" such as *Note to physician or authorized medical personnel*. These "notes" contain reminders and treatments that might be considered in the case of exposure.

HEALTH HAZARDS SECTION
Personal protective equipment (PPE): The items listed are those recommended by (a) NIOSH and/or OSHA, (b) manufacturers, either in technical bulletins or on Material Safety Data Sheets (MSDSs), (c) the Chemical Manufacturers Association (CMA), or (d) the National Safety Council (NSC), for use by personnel while responding to fire or accidental discharge of the chemical. They are intended to protect the lungs, eyes, and skin. Although safety showers and eyewash fountains are considered to be important protective equipment for the handling of almost all chemicals, they are not usually listed. Respirator codes found in the "NIOSH Pocket Guide" have been included to ease updating.
Exposure limits: Where available, this field contains legally enforceable airborne Permissible Exposure Limits (PELs) from OSHA, recommended airborne exposure limits from NIOSH or ACGIH, and special warnings when a chemical substance is a Special Health Hazard

Substance. Each are described below. TLVs have not been developed as legal standards, and the ACGIH does not advocate their use as such. The TLV is defined as the time weighted average (TWA) concentration for a normal 8-hour workday and a 40-hour workweek, to which nearly all workers may be repeatedly exposed, day after day, without adverse effects. A ceiling value (TLV-C) is the concentration that should not be exceeded during any part of the working exposure. If instantaneous monitoring is not feasible, then the TLV-C can be assessed by sampling over a 15-minute periods except for those substances that may cause immediate irritation when exposures are short. As some people become ill after exposure to concentrations lower than the exposure limits, this value cannot be used to define exactly what is a "safe" or "dangerous" concentration. ACGIH threshold limit values (TLVs) are reprinted with permission of the American Conference of Governmental Industrial Hygienists, Inc., from the booklet entitled, *Threshold Limit Values for Chemical Substances and Physical Agents and Biological Exposure Indices*. This booklet is revised on an annual basis. No entry appears when the chemical is a mixture; it is possible to calculate the TLV for a mixture only when the TLV for each component of the mixture is known and the composition of the mixture by weight is also known. According to ACGIH, the latest edition of their publication, *Documentation of the Threshold Limit Values and Biological Exposure Indices*, is necessary to fully interpret and implement the TLVs.

OSHA permissible exposure limits (PELs), are found in Tables Z-1, Z-2, and Z-3 of OSHA General Industry Air Contaminants Standard (29 CFR 1910.1000) that were effective on July 1, 1993 and which are currently enforced by OSHA. Unless otherwise noted, PELs are the time weighted average (TWA) concentrations that must not be exceeded during any 8-hour shift of a 40-hour workweek. An OSHA ceiling concentration must not be exceeded during any part of the workday; if instantaneous monitoring is not feasible, the ceiling must be assessed as a 15-minute TWA exposure. In addition there are a number of substances from Table Z-2 that have PEL ceiling values that must not be exceeded except for a maximum peak over a specified period (e.g., a 5-minute maximum peak in any 2 hours).

NIOSH Recommended Exposure Limits (RELs) are time weighted average (TWA) concentrations for up to a 10-hour work day during a 40-hour work week. A ceiling REL should not be exceeded at any time. Exposure limits are usually expressed in units of parts per million (ppm) – that is, the parts of vapor (gas) per million parts of contaminated air by volume at 25°C/77°F and one atmosphere pressure. For a chemical that forms a fine mist or dust, the concentration is given in milligrams per cubic meter (mg/m^3).

Short-term exposure limits (15 minute TWA): This field contains Short-term exposure limits (STELs) from ACGIH, NIOSH and OSHA. The parts of vapor (gas per million parts of contaminated air by volume at 25°C/77°F and one atmosphere pressure is given. The limits are given in milligrams per cubic meter (mg/m^3) for chemicals that can form a fine mist or dust. Unless otherwise specified, the STEL is a 15-minute TWA exposure that should not be exceeded at any time during the workday.

Toxicity by ingestion: The grade and corresponding LD$_{50}$ value are those defined by the National Academy of Sciences, Committee on Hazardous Materials, *Evaluation of the Hazard of Bulk Water Transportation of Industrial Chemicals. A Tentative Guide*, Washington, DC, 1972. Data were also collected from other sources and converted to the appropriate grade. The term LD$_{50}$ signifies that about 50% of the animals given the specified dose by mouth will die. Thus, for a Grade 4 chemical (below 50 mg/kg) the toxic dose for 50% of animals weighing 70 kg (150 lb) is 70 x 50 = 3500 mg = 3.5 g, or less than 1 teaspoonful; it might be as little as a few drops. For a Grade 1 chemical (5 to 15 g/kg), the LD$_{50}$ would be between a pint and a quart for a 150-lb man. All LD$_{50}$ values have been obtained using small laboratory animals such as rodents, cats, and dogs. The substantial risks taken in using these values for estimating human toxicity are the same as those taken when new drugs are administered to humans for the first time.

Long-term health effects: Where there is evidence that the chemical can cause cancer, mutagenic effects, teratogenic effects, or a delayed injury to vital organs such as the liver or kidney, a description of the effect is given.

Vapor (gas) irritant characteristics: This field was updated to reflect irritant characteristics found in the literature.

Liquid or solid irritant characteristics: This field was updated to reflect irritant characteristics found in the literature.

Odor threshold: This is the lowest concentration in air that most humans can detect by smell. Some values ranges are reported. The value cannot be relied on to prevent overexposure, because human sensitivity to odors varies over wide limits, some chemicals cannot be smelled at toxic concentrations, odors can be masked by other odors, and some compounds rapidly deaden the sense of smell.

IDLH (immediately dangerous to life and health) value: This concentration represents a maximum level from which one could escape within 30 minutes without any escape-impairing symptoms or any irreversible health effects. However, the 30-minute period is meant to represent a margin of safety and is NOT meant to imply that any person should stay in the work environments any longer than necessary. In fact, every effort should be made to exit immediately. The concentrations are reported in either parts per million (ppm) or milligrams per cubic meter (mg/m^3).

FIRE DATA SECTION

Flash point: This is defined as the lowest temperature at which vapors above a volatile combustible substance will ignite in air when exposed to a flame. Depending on the test method used, the values given are either Tag Closed Cup (cc) (ASTM D56) or Cleveland Open Cup (oc) (ASTM D93). The values, along with those in "Flammable Limits in Air" and "Autoignition Temperature" below, give an indication of the relative flammability of the chemical. In general, the open cup value is about 10-15°F higher than the closed cup value.

Flammable limits in air: The percent concentration in air (by volume) is given for the LEL [lower explosive (flammable) limit in air, % by volume] and UEL [upper explosive (flammable) limit in air, % by volume], at room temperature, unless otherwise specified. The values, along with those in Flash Point and Autoignition Temperature give an indication of the relative flammability of the chemical.

Fire extinguishing agents not to be used: The agents listed must not be used because they react with the chemical and have the potential to create an additional hazard. In some cases they are listed because they are ineffective in putting out the fire.

Behavior in fire: Any characteristic behavior that might increase significantly the hazard involved in a fire is described. Unusual difficulty in extinguishing the fire is noted.

Autoignition temperature: This is the minimum temperature at which the material will ignite without a spark or flame being present. Values given are only approximate and may change substantially with changes in geometry, gas, or vapor concentrations, presence of catalysts, or other factors. Also called Ignition Temperature.

Electrical hazard: The electrical group and class is based on the explosive characteristics of air mixtures of gases or vapors. This information is available for relatively few chemicals, so an absence of data does not necessarily mean that the substance is not hazardous in the presence of electrical equipment.

Burning rate: The value is the rate (in millimeters per minute) at which the depth of a pool of liquid decreases as the liquid burns.
Adiabatic flame temperature: The value is the temperature in degrees Fahrenheit of the flame when the material is burned under adiabatic conditions.
Stoichiometric air-to-fuel ratio: The value is the ratio of air to the compound in question required for stoichiometric combustion. Since it is a ratio, the value is dimensionless.

CHEMICAL REACTIVITY SECTION
Reactivity with water: Where a hazard does result, it is described here and/or in the Description section.
Binary reactants: This field describes important hazardous chemical incompatibilities or reactions of each substance with another material including structural materials such as metal, wood, plastics, cement, and glass. The nature of the hazard, such as severe corrosion formation of a flammable gas, is described. This list is by no means complete or all-inclusive. In some cases a very small quantity of material can act as a catalyst and produce violent reactions such as polymerization, disassociation, and condensation. Catalysts, when known, are listed.
   The accidental mixing of one chemical group with another can in some cases be expected to result in a vigorous and hazardous chemical reaction. The generation of toxic gases, the heating, overflow, and rupture of containers and fire explosion are possible consequences of such reactions. The purpose of the Compatibility Chart, which is fully explained in 46 CFR 150, is to show chemical combinations believed to be dangerously reactive in the case of accidental mixing.
Neutralizing agents for acids and caustics: In all cases involving accidental discharge, dilution with water may be followed by use of the agent specified, particularly if the material cannot be flushed away; the agent specified need not necessarily be used.
Polymerization: Some chemicals can undergo rapid polymerization to form sticky, resinous materials, with the liberation of much heat. The containers may explode. For these chemicals the conditions under which the reaction can occur are given. See "Heat of polymerization" field under Physical and Chemical Properties section for quantitative data.
Inhibitor of polymerization: The chemical names and concentrations of inhibitors added by the manufacturer to prevent polymerization are given.
Reactivity group: The value is the number of the reactivity group, if any, to which the compound belongs. The reactivity groups include compounds of similar, though not identical, reactive characteristics. General reactivity traits that have been identified for a group will apply more or less to each compound in the group, depending upon the individual compound itself.
Compatibility class : An entry appears when the chemical has been assigned to one of the 43 cargo groups listed in Code of Federal Regulations, Title 46, Part 150, *Compatibility of Cargoes and Operational Requirements for Bulk Liquid Hazardous Waste Cargoes*.

ENVIRONMENTAL DATA SECTION
Food Chain concentration potential: If the chemical is consumed by fish, marine plants, waterfowl, and so on, that are in turn eaten by other species, the substance may accumulate and ultimately be consumed by humans. Where this occurs, an indication of the potential hazard and its significance is given. Using the octanol/water coefficient (Log P_{ow}), a substance's potential for harming the environment can be determined. The higher the value, the greater the chance for accumulation of the substance in living tissues, primarily fats. Values greater than 3.0 have this potential.
Water pollution: This subsection lists substances that are marine pollutants meeting the definition of any hazard class or division as defined in Appendix B to §172.101 *List of Marine Pollutants*. Severe marine pollutants are noted where appropriate. The purpose of the terms used in this section is to describe in a general way the cautionary and corrective responses to substances of concern. Spill or leak information provided is intended to be used only as a guide. *Issue warning* is used when the chemical is a poison, has a high flammability, is a water contaminant, is an air contaminant (so as to be hazardous to life), is an oxidizing material, or is corrosive. *Restrict access* is used for those chemicals that are unusually and immediately hazardous to personnel unless they are protected properly by appropriate protective clothing, eye protection, respiratory protection equipment, and so on. "Evacuate area" is used primarily for unusually poisonous chemicals or these that ignite easily. *Mechanical containment* is used for water-insoluble chemicals that float and do not evaporate readily. *Should be removed* is used for chemicals that cannot be allowed to disperse because of potentially harmful effect on humans or on the ecological system in general. The term is not used unless there is a reasonable chance of preventing dispersal, after a discharge or leak, by chemical and physical treatment. *Chemical and physical treatment* is recommended for chemicals that can be removed by skimming, pumping, dredging, burning, neutralization, absorption, coagulation, or precipitation. The corrective response may also include the use of dispersing agents, sinking agents, and biological treatment. *Disperse and flush* is used for chemicals that can be made nonhazardous to humans by simple dilution with water. In a few cases the response is indicated even when the compound reacts with water because, when proper care is taken, dilution is still the most effective way of removing the primary hazard.

SHIPPING INFORMATION SECTION
Grades or purity: The grades USP (United States Pharmacopoeia) and CP (chemically pure) are quite pure. Where "technical" or "commercial" grades are given, the percent by weight of the pure chemical present is usually indicated. In a few cases the identity of the major impurities is given. If the properties of the less pure grades differ significantly from those of the pure substance, the differences in properties are described in general terms.
Storage temperature: The range of temperatures at which the chemical is normally shipped in bulk by water transport is given. "Ambient" means the temperature of the surroundings.
Inert atmosphere: The terms used are "inerted," "padded," "ventilated (forced)," "ventilated (natural)," and "no requirement." They are given when found in the Code of Federal Regulations (CFR), Title 46, beginning in Part 151.05.
Venting: The terms used are "open," "pressure-vacuum," and "safety relief." They are given when found in the Code of Federal Regulations (CFR), Title 46, beginning in Part 151.05.
Stability during transport: The term "stable" means that the chemical will not decompose in a hazardous manner under the conditions of temperature, pressure, and mechanical shock that are normally encountered during shipment; the term does NOT apply to fire situations. Where there is a possibility of hazardous decomposition, an indication of the conditions and the nature of the hazard is given.

NAS HAZARD CLASSIFICATION FOR BULK WATER TRANSPORTATION SECTION
NAS Hazard Rating for Bulk Water Transportation: The indicated ratings are from *"Evaluation of the Hazard of Bulk Water Transportation of Industrial Chemicals, A Tentative Guide,"* An outline of the rating system is given in Table 2.

TABLE 2
EXPLANATION OF NAS HAZARD RATINGS

Health

Rating	Fire	Vapor Irritant	Liquid or Solid	Poisons irritant
0	No hazard	No effect	No effect	No effect
1	Flash point (cc) above 140°F/60°C.	Slight effect	Causes skin smarting	Slightly toxic
2	Flash point (cc) 100 to 140°/ 38–60°C	Moderate irritation; temporary effect	First-degree burns on short exposure	Intermediate toxicity
3	Flash point (cc) below 100°F/38°C; BP above 100°F/38°C	Irritating; cannot be tolerated	Second-degree burns; few minutes exposure	Moderately toxic
4	Flash point (cc) below 100°F/38°C; BP 100°F/38°C	Severe effect; may do permanent injury	Second- and third-degree burns	Severely toxic

Water Pollution

Rating	Human Toxicity	Aquatic toxicity	Aesthetic effect
0	Nontoxic; LD_{50} >15 g/kg	Acute threshold limits above 10,000 ppm	No significant pollution; gases and odorless liquids
1	Practically nontoxic; LD_{50} 5 to 15 g/kg	Threshold limits 1000 to 10,000	Mold-odored light oils and soluble chemicals
2	Slightly toxic; LD_{50} 0.5 to 5 g/kg	Threshold limits 100 to 1000	Mild-odored, colorless, water insoluble oils; BP 150 to 450°F/66- 232°C
3	Moderately toxic; LD_{50} to 500 mg/kg	Threshold limits 1 to 100	Light-colored, high boiling soluble compound
4	Toxic LD_{50} < 50 mg/kg	Threshold limits below 1 ppm	Heavy oils, colored or bad odors.

Reactivity

Rating	Other chemicals	Water	Self-reaction
0	Inactive; may be attacked by materials rated 4	No reaction	No reaction
1	React only with materials rated 4	Mild reaction; unlikely to be hazardous	Mild self-reaction under some conditions
2	React with materials rated 3 or 4	Moderate reaction	Will undergo self-reaction if contaminated; do not require stabilizer
3	React with each other and with materials rated 2 or 4	More vigorous reaction; may be hazardous	Vigorous self reaction; require stabilizer
4	React with each other and materials rated 0 to 3	Vigorous reaction; likely to be hazardous	Self-oxidizing chemical; capable of explosion or detonation

PHYSICAL AND CHEMICAL PROPERTIES SECTION

Physical state @ 59°F/15°C and 1 atm: The statement indicates whether the chemical is a solid, liquid, or gas after it has reached equilibrium with its surroundings at ordinary conditions of temperature and pressure.

Molecular weight: The value given is the weight of a molecule of the chemical relative to a value of 12 for one atom of carbon. The molecular weight is useful in converting from molecular units to weight units and in calculating the pressure, volume and temperature relationships for gaseous materials. The ratio of the densities of any two gases is approximately equal to the ratio of their molecular weights (see Vapor density). The molecular weights of mixtures can be calculated if both the identity and quantity of each component of the mixture are known. Because the composition of mixtures described in this manual is not known exactly, or because it varies from one shipment to another, no molecular weights are given for such mixtures.

Boiling point @ 1 atm: The boiling point of the anhydrous substance at 1 atm (760 mm Hg).

Melting/Freezing point: The melting/freezing point is the temperature at which a solid changes to liquid or a liquid changes to a solid. For example, liquid water changes to solid ice at 0°C/32°F. Some liquids solidify very slowly even when cooled below their freezing point. When liquids are not pure (for example, salt water) their freezing points are lowered slightly.

BP below 59°F/15°C	gas
BP between 59°F/15°C and 86°F/30°C	gas or liquid
Melting point below 59°F/15°C	liquid
MP between 59°F/15°C and 86°F/30°C	liquid or solid

Critical temperature: The maximum temperature at which a liquid can exist, no matter what the pressure on it, is called the critical temperature. For example, the critical temperature of water is 372°C/705°F. The value can be used to estimate many properties whose values are not immediately available.

Critical pressure: The vapor pressure of a chemical at the critical temperature (see above) is called the critical pressure. For example, the critical pressure of water is 218 atm. Values are given in pounds per square inch absolute, atmospheres, and mega-newtons per square meter. The value can be used for estimating many property values that are not immediately available.

Specific gravity (water = 1): The specific gravity of a chemical is the ratio of the weight of the solid or liquid to the weight of an equal volume of water at 39°F/4°C (or at some other specified temperature). In the case of liquids of limited solubility, the specific gravity will predict whether the product will sink or float on water; for example, If the specific gravity is less than 1.0 (or less than 1.03 in seawater) the chemical will float; if higher than 1.0, it will sink.

Liquid surface tension: This property is a measure of the tensile force at the surface of a liquid that tends to shape liquid fragments into spherical drops. Values are expressed in dynes per centimeter and newtons per meter. Liquids with high surface tensions show less tendency to spread. Water has a surface tension of about 73 dynes/cm; seawater has a slightly higher value.

Liquid-water interfacial tension: The value is a measure of the tensile forces existing at the interface between a liquid and water. Approximately, it is the difference between the individual surface tension of the liquid and that of water. Low values of the interfacial tension indicate that the chemical spreads readily on a water surface. Values are expressed in dynes per centimeter and newtons per meter.

Relative vapor density (air = 1): Also known as "vapor (gas) specific gravity," this is actually a specific gravity rather than a true density because it equals the ratio of the weight of a vapor or gas (with no air present) compared to the weight of an equal volume of air at the same temperature and pressure. Values less than 1 indicate that the vapor or gas tends to rise and values greater than 1 indicate that it tends to settle. However, temperature effects must be considered: For example, although methane at 68°F/20°C has a vapor density of 0.55, it becomes denser at lower temperatures. At –259°F/–162°C, the boiling point, the vapor is heavier than air. Vapors from an open container of boiling methane fall rather than rise.

Ratio of specific heats of vapor (gas): This property is the ratio of the specific heat at constant pressure (C_p) to the specific heat at constant volume (C_v); its value is always greater than one. In most cases it was calculated by use of the expression

$$\frac{C_p}{C_v} = \frac{C_p}{(C_p - R)}$$

where R is the universal gas constant. The ratio varies slightly with temperature; the value given is at 20°C/68°F. The ratio is often of value in estimating temperature changes when gases are compressed or expanded. Higher values of the ratio lead to larger temperature changes for a given pressure change.

Latent heat of vaporization: The value is the heat that must be added to the specified weight of a liquid before it can change to vapor (gas). It varies with temperature; the value given is that at the boiling point at 1 atm (see boiling point). The units used are Btu per pound, calories per gram, and joules per kilogram. No value is given for chemicals with very high boiling points at 1 atm, because such substances are considered essentially nonvolatile.

Heat of combustion: The value is the amount of heat liberated when the specified weight is burned in oxygen at 77°F/25°C. The products of combustion, including water, are assumed to remain as gases; the value given is usually referred to as the "lower heat value." The negative sign before the value indicates that heat is given off when the chemical burns. The units used are Btu per pound, calories per gram, and joules per kilogram.

Heat of decomposition: The value is the amount of heat liberated when the specified weight decomposes to more stable substances. The value is given for very few chemicals, because most are stable and do not decompose under the conditions of temperature and pressure encountered during shipment. The negative sign before the value simply indicates that heat is given off during the decomposition. The value does not include heat given off when the chemical burns. The units used are Btu per pound, calories per gram, and joules per kilogram.

Heat of solution: The value represents the heat liberated when the specified weight of chemical is dissolved in a relatively large amount of water at 77°F/25°C ("infinite dilution"). A negative sign before the value indicates that heat is given off, causing a rise in temperature. (A few chemicals absorb heat when they dissolve, causing the temperature to fall.) The units used are Btu per pound, calories per gram, and joules per kilogram. In those few cases where the chemical reacts with water and the reaction products dissolve, the heat given off during the reaction is included in the heat of solution.

Heat of polymerization: The value is the heat liberated when the specified weight of the compound (usually called the monomer) polymerizes to form the polymer. In some cases the heat liberated is so great that the temperature rises significantly, and the material may burst its container or catch fire. The negative sign before the value indicates that heat is given off during the polymerization reaction. The units used are Btu per pound, calories per gram, and joules per kilogram.

Heat of fusion: The value is the number of Btu needed to change one pound of solid to liquid with no change in temperature.

Limiting value: A chemical-specific concentration in water in mole fraction units below which the contribution to the evolution of toxic or flammable vapor at the water surface can be assumed to be negligible.

Vapor pressure: This field displays both the vapor pressure and the Reid vapor pressure when both are found in the literature. Vapor pressure is defined as the equilibrium pressure of the saturated vapor above the liquid, measured in millimeters of mercury (760 mm Hg = 14.7 psia) at 20°C/68°F unless another temperature is specified. Conversion is done as follows:

$$psi = \frac{mmHg \times 14.7}{760}$$

Reid vapor pressure is defined as the equilibrium pressure exerted by vapor over the liquid at 100°F/38°C, expressed as pounds/in^2 absolute, as defined in 46 CFR30.10-59.

CONVERSION FACTORS (* Denotes Exact Value)

TO CONVERT:	TO:	MULTIPLY BY:
LENGTH		
inches	millimeters	25.4*
inches	feet	0.0833
feet	inches	12*
feet	meters	0.3048*
feet	yards	0.3333
feet	miles (U.S. statute)	0.0001894
yards	feet	3*
yards	miles (U.S. statute)	0.0005682
miles (U.S. statute)	feet	5280*
miles (U.S. statute)	yards	1760*
miles (U.S. statute)	meters	1609*
miles (U.S. statute)	nautical miles	0.868
meters	feet	3.281
meters	yards	1.094
meters	miles (U.S. statute)	0.0006214
nautical miles	miles (U.S. statute)	1.152
AREA		
square inches	square centimeters	6.452
square inches	square feet	0.006944
square feet	square inches	144*
square feet	square meters	0.09290
square meters	square feet	10.76
square miles	square yards	3,097,600*
square yards	square feet	9*
VOLUME		
cubic inches	cubic centimeters	16.39
cubic inches	cubic feet	0.0005787
cubic feet	cubic inches	1728*
cubic feet	cubic meters	0.02832
cubic feet	U.S. gallons	7.481
cubic meters	cubic feet	35.31
liters	quarts (U.S. liquid)	1.057
quarts (U.S. liquid)	liters	0.9463
U.S. gallons	barrels (petroleum)	0.02381
U.S. gallons	cubic feet	0.1337
U.S. gallons	Imperial gallons	0.8327
barrels (petroleum)	U.S. gallons	42*
Imperial gallons	U.S. gallons	1.201
milliliters	cubic centimeters	1*
TIME		
seconds	minutes	0.01667
seconds	hours	0.0002778
seconds	days	0.00001157
minutes	seconds	60*
minutes	hours	0.01667
minutes	days	0.0006944
hours	seconds	3600*
hours	minutes	60*
hours	days	0.04167
MASS OR WEIGHT		
pounds	kilograms	0.4536
pounds	short tons	0.0005*
pounds	long tons	0.0004464
pounds	metric tons	0.0004536
tons (short)	pounds	2000*
tons (metric)	pounds	2205
tons (long)	pounds	2240*
kilograms	pounds	2.205
tons (metric tons)	kilograms	1000*
ENERGY		
calories	Btu	0.003968

calories	joules	4.187
Btu	calories	252.0
Btu	joules	1055
joules	calories	0.2388
joules	Btu	0.0009479

VELOCITY

feet per second	meters per second	0.3048
feet per second	miles per hour	0.6818
feet per second	knots	0.5921
meters per second	feet per second	3.281
meters per second	miles per hour	2.237
miles per hour	meters per second	0.4470
miles per hour	feet per second	1.467
knots	meters per second	0.5148
knots	miles per hour	1.151
knots	feet per second	1.689

DENSITY

pounds per cubic foot	gams per cubic centimeter	0.01602
grams per cubic centimeter	pounds per cubic foot	62.42
grams per cubic centimeter	kilograms per cubic meter	1000*
kilograms per cubic meter	grams per cubic centimeter	0.001*

PRESSURE

pounds per square inch absolute (psia)	kilo-newtons per square meter	kN/m^2
psia	atmospheres	0.0680
psia	inches of water	27.67
psia	millimeters of mercury (torr)	51.72
pounds per square inch gauge (psig)	psia	add 14.70
millimeters of mercury (torr)	psia	0.01934
millimeters of mercury (torr)	kN/m^2	0.1333
inches of water	psia	0.03614
inches of mercury	atmospheres	0.03342
kilograms per square centimeter	millimeters of mercury (torr)	735.6
inches of water	kN/m^2	0.2491
kilograms per square centimeter	atmospheres	0.9678
atmospheres	kN/m^2	101.3
kilograms per square centimeter	psia	14.22
atmospheres	psia	14.70
bars	kN/m^2	100*
kilonewtons per square meter (kN/m^2)	psia	0.1450
bars	atmospheres	0.9869
kilonewtons per square meter (kN/m^2)	atmospheres	0.009869
bars	kilograms per square centimeter	1.020

VISCOSITY

centipoises	pounds per foot per second	0.0006720
pounds per foot per second	centipoises	1488
centipoises	poises	0.01*
centipoises	newton seconds per square meter	0.001*
poises	grams per centimeter	1*
grams per centimeter/second/second	poises	1*
new seconds per square meter	centipoises	1000*

THERMAL CONDUCTIVITY

Btu per hour per foot per °F	watts per meter-kelvin	1.731
Btu per hour per foot per meter per °F	kilocalories per hour per °C	1.488
watts per meter-kelvin	Btu per hour per foot per °F	0.5778
kilocalories per hour per meter per °C	watts per meter-kelvin	1.163
kilocalories per hour per meter per °C	Btu per hour per foot per °F	0.6720

HEAT CAPACITY

Btu per pound per °F	calories per gram per °C	1*
Btu per pound per °F	joules per kilogram-°K	4187
joules per kilogram-°K	Btu per pound per °F	0.0002388
Calories per gram per °C	Btu per pound per °F	1*

CONCENTRATION (in water solution)

parts per million (ppm)	milligrams per liter	1*
milligrams per liter	ppm	1*
milligrams per cubic	grams per cubic centimeter	1 x 10

meter grams per cubic centimeter	milligrams per cubic meter	1 x 10
grams per cubic centimeter	pounds per cubic foot	62.42
pounds per cubic foot	grams per cubic centimeter	0.01602

TEMPERATURE

degrees Kelvin (°K)	degrees Rankine (°R)	1.8*
degrees Rankine (°R)	degrees Kelvin (°K)	0.5556
degrees Centigrade (°C)	degrees Fahrenheit (°F)	first multiply by 1.8, then add 32
degrees Fahrenheit (°F)	degrees Centigrade (°C)	first subtract 32, then multiply by 0.5556
degrees Centigrade (°C)	degrees Kelvin (°K)	add 273.2
degrees Fahrenheit (°F)	degrees Rankine (°R)	add 459.7

FLOW

cubic feet per second	U.S. gallons per minute	448.9
U.S. gallons per minute	cubic feet per second	0.002228

UNIVERSAL GAS CONSTANT (R)

8.314 joules per gram mole-°K
1.987 calories per gram mole-°K
1.987 Btu per pound mole per °F
10.73 psia-cubic feet per pound mole per °F
82.057 atm-cubic centimeters per gram mole-°K
62.361 millimeters mercury liter per gram mole-°K

INTERNATIONAL CODE OF SIGNALS: PHONETIC ALPHABET

A	Alpha (Alfa)	N	November
B	Bravo	O	Oscar
C	Charlie	P	Papa
D	Delta	Q	Quebec
E	Echo	R	Romeo
F	Foxtrot	S	Sierra
G	Golf	T	Tango
H	Hotel	U	Uniform
I	India	V	Victor
J	Julliett	W	Whiskey
K	Kilo	X	X-ray
L	Lima	Y	Yankee
M	Mike	Z	Zulu

GUIDE TO COMPATIBILITY OF CHEMICALS

The accidental mixing of one chemical shipment with another; can, in some cases, be expected to result in a vigorous and hazardous chemical reaction. The generation of toxic gases, the heating, overflow, and rupture of cargo tanks, and fire and explosion are possible consequences of such reactions.

The purpose of the Compatibility Chart is to show chemical combinations believed to be dangerously reactive in the case of accidental mixing. It should be recognized, however, that the Chart provides a broad grouping of chemicals with an extensive variety of possible binary combinations. Although one group, generally speaking, can be considered dangerously reactive with another group where an "X" appears on the Chart, there may exist between the groups some combinations which would not dangerously react. The Chart should therefore NOT be used as an infallible guide. It is offered as an aid in the safe storage of chemicals, with the recommendation that proper safeguards be taken to avoid accidental mixing of binary mixtures for which an "X" appears on the Chart.

The following procedure explains how the Guide should be used in determining compatibility information: (1) Determine the reactivity group of a particular product by referring to the alphabetical list in the incompatibility index. (2) Enter the Chart with the reactivity group. Proceed across the page. An "X" indicates a reactivity group that forms a potentially unsafe combination with the product in question. *For example*, crotonaldehyde is listed in the incompatibility index as belonging in Group 19 (Aldehydes) and also has a notation, (2), which is explained in the footnotes to the incompatibility index. The Compatibility Chart shows that chemicals in group 19 should be segregated from sulfuric and nitric acids, caustics, ammonia, and all types of amines (aliphatic, alkanol, and aromatic). Footnote (2) refers the user to the "exceptions" chart where exceptions to the Compatibility Chart are listed. Here, crotonaldehyde is listed as also being incompatible with Group 1, nonoxidizing acids. It is recognized that there are wide variations in the reaction rates of individual chemicals within the broad groupings shown reactive by the Compatibility Chart. Some individual materials in one group will react violently with some of the materials in another group and cause great hazard; others will react slowly, or not at all. Accordingly, a useful addition to the Guide would be the identification of specific materials which might not follow the characteristic reactivities of the rest of the materials in its Group. A few such combinations are listed in the exceptions table.

INCOMPATIBILITY INDEX

CHEMICAL NAME	Group No.	Record No.	Related record number	Other information
Acetaldehyde	19	A:0150		
Acetic acid	04(2)	A:0200		
Acetic anhydride	11	A:0250		Water
Acetone	18(2)	A:0300		
Acetone cyanohydrin	0(1,2)	A:0350		
Acetonitrile	37	A:0400		
Acetophenone	18	A:0450		
Acetyl chloride		A:0600		Water
Acrolein	19(2)	A:0800		
Acrylamide solution	10	A:0850		
Acrylic acid	4(2)	A:0900		
Acrylonitrile	15(2)	A:0950		
Adiponitrile	37	A:1050		
Alcohols (C_{13} and above)	20		T:3050	
			T:0450	
			P:0400	
Alcoholic beverages	20			
Alkyl(C_9–C_{17}) benzenes	32		D:0750	
			D:8100	
			U:0400	
			T:3150	
			T:0550	
Alkylbenzenesulfonic acid	0(1,2)	A:1150		
Alkylbenzenesulfonic acid, sodium salt solutions	33			
Alkyl phthalates	34			
Allyl alcohol	15(2)	A:1200		
Allyl chloride	15	A:1300		
Allyl trichlorsilane		A:1400		Water
Aluminum Chloride		A:1450		Water
Aluminum phosphide		A:1700		Water
Aluminum sulfate solution	43(2)	A:1800	A:1750	
2-(2-Aminoethoxy)ethanol	8	A:1850		
Aminoethylethanolamine	8	A:1850		

N-Aminoethyl piperazine	7	A:1900		
2-Amino-2-hydroxy methyl-1,3-propanediol solution	43			
Ammonia, anhydrous	6	A:2000		
Ammonium bicarbonate		A:2150		Oxy
Ammonium bisulfite solution	43(2)	A:2250		
Ammonium dichromate		A:2600		Oxy
Ammonium hydrogen phosphate solution	0			
Ammonium hydroxide (<28% aqueous NH_3)	6	A:2850		
Ammonium nitrate		A:3150		Oxy
Ammonium nitrate solution	0(1)	A:3150		Oxy
Ammonium nitrate-urea solution (containing NH_3)	6	U:1000		
Ammonium nitrate-urea solution (not containing NH_3)	43	A:3300		
Ammonium perchlorate		A:3500		Oxy
Ammonium polyphosphate solution	43	A:3250	A:3600	
Ammonium sulfate solution	43	A:3850		
Ammonium sulfide solution	5	A:3900		
Ammonium thiosulfate solution	43	A:4100		
Amyl acetate	34	A:4200	I:0100 A:4200 A:4250 A:4300	
Amyl alcohol	20	A:4350	I:0150 A:4350	
Amylene	30	P:0800		
Amyl methyl ketone	18	A:4500		
Amyl nitrate		A:4550		Oxy
n-Amyl trichlorsilane		A:4700		Water
Aniline	9	A:4750		
Anthracene oil (coal tar fraction)	33	O:3200		
Asphalt	33	A:5600		
Asphalt blending stocks: roofers flux	33	A:5650		
Asphalt blending stocks: straight run residue	33	A:5700		
Aviation alkylates	33	G:0250		
Barium chlorate		B:0150		Oxy
Barium nitrate		B:0250		Oxy
Barium peroxide		B:0400		Oxy
Benzene	32	B:0550		
Benzenesulfonyl chloride	0(1,2)	B:0750		
Benzoyl chloride		B:1100		Water
Benzylacetate	34	B:1150		
Benzyl alcohol	21	B:1200		
Benzyl chloride	36	B:1350		
Boron tribromide		B:2150		Water
Bromine		B:2250		Oxy
Bromine pentafluoride		B:2300		Water/Oxy
Bromine trifluoride		B:2350		Water/Oxy
Butadiene	30	B:2850		
Butane	31	B:2900	I:0250	
Butene	30	B:3850	I:0450	
Butyl acetate	34	B:3100	I:0300 B:3100 B:3150 B:3200	
Butyl acrylate	14	B:3250	B:3300	
Butyl alcohol	20(2)	B:3350	I:0350 B:3400 B:3450	

Butylamine	7	B:3500	I:4100	
			B:3500	
			B:3550	
			B:3600	
Butyl benzyl phthalate	34		B:3650	
tert-Butyl hydroperoxide			B:4200	Oxy
Butylene	30	B:3850	I:0450	
Butylene glycol	20(2)	B:3900	B:3950	
			B:4000	
1,2-Butylene oxide	16	B:4050		
Butyl ether	41	B:5400		
Butyl formate	34	B:4100		
Butyl methacrylate	14	B:4450	B:4500	
			B:4550	
Butyraldehyde	19	B:4800	B:4850	
Butyric acid	4	B:4900	I:0500	
Calcium		C:0500		Water
Calcium carbide		C:0650		Water
Calcium chlorate		C:0700		Oxy
Calcium chloride solution	43	C:0750		
Calcium hypochlorite solution	5	C:1000		Oxy
Camphor oil	18	C:1400		
Caprolactam	22	C:1450		
Carbolic oil	21	P:1200		
Carbon disulfide	38	C:1700		
Carbon tetrachloride	36	C:1850		
Carnauba wax	34	W:0100		
Cashew nut shell oil (untreated)	4	O:3100		
Caustic potash solution	5(2)	P:2950		
Caustic soda solution	05(2)	S:3800		
Chlorine	0(1)	C:2100		Oxy
Chlorine trifluoride		C:2150		Oxy
Chloroacetic acid solution	4	C:2300	M:5750	
Chlorobenzene	36	C:2400		
Chlorodifluoromethane	36	M:5800		
Chloroform	36	C:2500		
Chlorohydrins	17	C:2550	E:0300	
Chloronitrobenzene	0	C:2650		
Chloropropionic acid	4	C:2900	C:2950	
Chlorosulfonic acid	0(1)	C:3000		Water/Oxy
Chlorotoluene	36	C:3100	C:3050	
			C:3150	
Chromic anhydride		C:3300		Oxy
Chromyl chloride		C:3450		Water
Coal tar	33	O:3200		
Corn syrup	43	C:4950		
Cottonseed oil, fatty acid	34	C:1400		
Creosote	21(2)	C:5050		
Cresols	21	C:5100	C:5150	
			C:5200	
			C:5250	
Cresylate spent caustic solution	5	C:5300		
Cresylic acid	21	C:5100		
Cresylic acid, sodium salt solution	5	C:5300		
Crotonaldehyde	19(2)	C:5400		
Cumene	32	C:5450		
Cumene hydroperoxide		C:5500		Oxy
Cupric nitrate		C:4650		Oxy
Cycloheptane	31	C:5800		
Cyclohexane	31	C:5850		
Cyclohexanol	20	C:5900		
Cyclohexanone	18	C:5950		
Cyclohexylamine	7	C:6100		
1,3-Cyclopentadiene dimer	30	D:2150		
Cyclopentadiene, styrene, benzene mix	30			

Cyclopentane	31	C:6200		
Cyclopentene	30	C:6250		
p-Cymene	32	C:6350		
Decahydronaphthalene	33	D:0400		
Decaldehyde	19	D:0450	I:0600	
Decane	31	D:0500		
Decanoic acid	4	D:0550		
Decene	30	D:0600		
Decyl acrylate	14	D:0650	I:0650	
Decyl alcohol	20(2)	D:0700	I:0700	
n-Decylbenzene	32	D:0750		
Dextrose solution	43	D:0850		
Diacetone alcohol	20(2)	D:0900		
Dialkyl(C_7–C_{13}) phthalates	34		D:4000	
			D:5250	
Diammonium salt of zinc (ETDA)	43	D:0950		
Dibutyl amine	7	D:5350		
Dibutyl phthalate	34	D:1250		
Dibenzoyl peroxide		D:1050		Oxy
Dichlorobenzene	36	D:1500	D:1550	
			D:1450	
Dichlorodifluoromethane	36	D:1650		
1,1-Dichloroethane	36	D:1700		
2,2-Dichloroethyl ether	41	D:1800		
2,2'-Dichloroisopropyl ether	36	D:1950		
Dichloromethane	36	D:2050		
2,4-Dichlorophenol	21	D:2200		
Dichloropropane	36	D:2250	D:2300	
			D:2350	
1,3-Dichloropropene	15	D:2400	D:2450	
			D:2500	
Dichloropropene, dichloropropane mixture	15	D:2500		
2,2-Dichloroproprionic acid	4	D:0200		
Dicyclopentadiene	30	D:2150		
Diethanolamine	8	D:2750		
Diethylamine	7	D:2800		
Diethylaminoethanol	8	D:3500		
2,6-Diethylaniline	9	D:2850		
Diethylbenzene	32	D:2900		
Diethylene glycol	40	D:3000		
Diethylene glycol monobutyl ether	40	D:3300		
Diethylene glycol monobutyl ether acetate	34	D:3350		
Diethylene glycol dibutyl ether	40	D:3050		
Diethylene glycol monoethyl ether	40	D:3400		
Diethylene glycol ethyl ether acetate	34	D:3150		
Diethylene glycol methyl ether	40	D:3415		
Diethylene glycol methyl ether acetate	34	D:3250		
Diethylene glycol phthalate	34	D:3425		
Diethylenetriamine	07(2)	D:3450		
Diethylethanolamine	8	D:3500		
Diethyl ether	41	E:5600		
Di-(2-ethylhexyl) adipate	34	D:4050		
Di-(2-ethylhexyl)phosphoric acid	1	D:3650		
Di-(2-ethylhexyl) phthalate	34	D:3600	D:5250	
Diethyl phthalate	34	D:3750		
Diethyl sulfate	34	D:3800		
Diethyl zinc		D:3850		Water
Diglycidyl ether of bisphenol A	41	B:1950		
Diheptyl phthalate	34	D:4000		
Di-n-hexyl adipate	34	D:4050		
Diisobutylamine	7	D:4100		
Diisobutylcarbinol	20	D:4150		
Diisobutylene	30	D:4200		

Diisobutyl ketone	18	D:4250		
Diisobutyl phthalate	34	D:4300		
Diisodecyl phthalate	34	D:4350		
Diisononyl adipate	34	D:4400		
Diisononyl phthalate	34	D:4450		
Diisooctyl phthalate	34	D:4500		
Diisopropanolamine	8	D:4550		
Diisopropylamine	7	D:4600		
Diisopropylbenzene	32	D:4650		
Diisopropyl naphthalene	32	D:4750		
Dimethyl acetamide	10	D:4800		
N,N-Dimethylacetamide solution	10	D:4850		
Dimethyl adipate	34	D:4900		
Dimethylamine	7	D:4950		
Dimethylamine solution	7	D:4950		
2,6-Dimethylaniline	9	D:5000		
Dimethyldichlorosilane		D:5150		Water
Dimethylethanolamine	8	D:5100		
Dimethylformamide	10	D:5550		
Dimethyl glutarate	34	D:5600		
Dimethyl hydrogen phosphite	34	D:5800		
2,2-Dimethyloctanoic acid	4	D:5850		
Dimethyl phthalate	34	D:5900		
Dimethylpolysiloxane	34	D:5950		
2,2-Dimethylpropane-1,3-diol	20	D:6000		
Dimethyl succinate	34	D:6050		
2,4-Dinitrotoluene	42	D:6800		
2,6-Dinitrotoluene	42	D:6850		
3,4-Dinitrotoluene	42	D:6900		
Dinonyl phthalate	34	D:6950		
Dioctyl phthalate	34	D:7000		
1,4-Dioxane	41	D:7100		
Dipentene	30	D:7150		
Diphenyl	32	D:7200		
Diphenyl, diphenyl oxide	33	D:8850		
Diphenyl ether	41	D:7350		
Diphenylmethane diisocyanate	12	D:7400		
Di-n-propylamine	7	D:7450		
Dipropylene glycol	40	D:7500		
Dipropylene glycol dibenzoate	34	D:7550		
Dipropylene glycol methyl ether	40	D:7600		
Distillates: flashed feed stocks	33	D:7750		
Distillates: straight run	33	D:7800		
Ditridecyl phthalate	34	D:7850		
Diundecyl phthalate	34	D:7900		
Dodecane	31			
Dodecanol	20	D:8000	L:2000	
Dodecene	30	D:8050		
Dodecyl alcohol	20	D:8000	L:2000	
Dodecylbenzene	32	D:8100		
Dodecylbenzenesulfonic acid	0(2)	D:8150		
Dodecyl diphenyl oxide disulfonate solution	43	D:8400		
Dodecylmethacrylate	14	D:8450		
Dodecyl, pentadecyl methacrylate	14	D:8500		
Dodecyl phenol	21	D:8550		
Epichlorohydrin	17	E:0300		
Ethane	31	E:0500		
Ethanolamine	8	M:5950		
2-Ethoxyethanol	20	E:0800		
2-Ethoxyethyl acetate	34	E:0900		
Ethoxylated alcohols, (C_{11}–C_{15})	20	E:1000		
		E:1200		
		E:1300		
		E:1350		
		E:1400		

Ethoxy triglycol	40	E:1500	
Ethyl acetate	34	E:1600	
Ethyl acetoacetate	34	E:1700	
Ethyl acrylate	14	E:1800	
Ethyl alcohol	20(2)	E:1900	
Ethyl aluminum dichloride		E:2000	Water
Ethylamine	7(2)	E:2200	
Ethylamine solution	7	E:2200	
Ethyl amyl ketone	18	E:2300	
Ethyl benzene	32	E:2400	
Ethyl butanol	20	E:2500	
N-Ethyl-n-butylamine	7	E:2600	
Ethyl butyrate	34	E:2800	
Ethyl chloride	36	E:2900	
Ethyl chlorothioformate	0(2)	E:3200	
N-Ethylcyclohexylamine	7	E:6400	
Ethylene	30	E:3500	
Ethylene chlorohydrin	20	E:3600	
Ethylene cyanohydrin	20	E:3700	
Ethylenediamine	7	E:3800	
Ethylene dibromide	36	E:4000	
Ethylene dichloride	36(2)	E:4100	
Ethylene glycol	20(2)	E:4200	
Ethylene glycol acetate	34	E:4300	
Ethylene glycol diacetate	34	E:4400	
Ethylene glycol dibutyl ether	40	E:4500	
Ethylene glycol ethyl ether	40	E:0800	
Ethylene glycol ethyl ether acetate	34	E:0900	
Ethylene glycol methyl ether	40	E:4700	
Ethylene glycol methyl ether acetate	34	E:4800	
Ethylene glycol monobutyl ether	40	E:4900	
Ethylene glycol monobutyl ether acetate	34	E:5000	
Ethylene glycol monoethyl ether	40	E:0800	
Ethylene glycol monoethyl ether acetate	34	E:0900	
Ethylene glycol phenyl ether	40	E:5200	
Ethylene glycol propyl ether	40	E:5200	
Ethylene oxide	0(1)	E:5500	
Ethyl ether	41	E:5600	
Ethyl-3-ethoxypropionate	34	E:5700	
Ethylhexaldehyde	19	E:5900	
2-Ethylhexanoic acid	4	E:6000	
2-Ethylhexanol	20	E:6100	
2-Ethylhexyl acrylate	14	E:6300	
2-Ethylhexylamine	7	E:6500	
Ethyl hexyl phthalate	34	E:6550	
Ethylidene norbornene	30(2)	E:6600	
Ethyl methacrylate	14	E:6900	
o-Ethyl phenol	21	E:7200	
Ethyl propionate	34	E:7600	
2-Ethyl-3-propylacrolein	19(2)	E:7700	
Ethyl toluene	32	E:7900	
Ethyl trichlorsilane		E:8000	Water
Ferric chloride solution	1	F:0200	
Fluorine		F:0800	Water/Oxy
Formaldehyde solution	19(2)	F:1250	
Formamide	10	F:1300	
Formic acid	4(2)	F:1350	
Furfural	19	F:1500	
Furfuryl alcohol	20(2)	F:1550	
Gas oil: cracked	33	G:0150	
Gasoline blending stocks: alkylates	33	G:0300	
Gasoline blending stocks: reformates	33	G:0350	

Guide to compatibility of chemicals

Gasolines:				
Automotive (not > 4.23 g Pb/gal)	33	G:0200		
Aviation (not > 4.86 g Pb/gal)	33	G:0250		
Casinghead (natural)	33	G:0400		
Polymer	33	G:0450		
Straight run	33	G:0500		
Glutaraldehyde solution	19	G:0600		
Glycerine	20(2)	G:0650		
Glicidyl methacrylate	14	G:0675		
Glyoxal solutions	19	G:0700		
Heptane	31	H:0150		
Heptanoic acid	4	H:0200		
Heptanol	20	H:0250		
Heptene	30	H:0300		
Heptyl acetate	34	H:0350		
Hexamethylenediamine solution	7	H:0800		
Hexamethylenetetramine	7	H:0900		
Hexamethylenetetramine solutions	7	H:0900		
Hexamethylenimine	7	H:0850		
Hexane	31(2)	H:0950	I:0750	
Hexanoic acid	4	H:1000		
Hexanol	20	H:1050		
Hexene	30	H:1100		
Hexyl acetate	34	H:1150		
Hexylene glycol	20	H:1200		
Hydrochloric acid	1	H:1450		
Hydrochloric acid, spent	1	H:1450		
Hydrofluorosilicic acid	1	H:1300		
Hydrogen peroxide solutions	0(1)	H:1600		Oxy
2-Hydroxyethyl acrylate	0(1,2)	H:1750		
2-Hydroxy-4-(methylthio)butanoic acid	4	H:1900		
Lead nitrate		L:1300		Oxy
iso-Butyl isobutyrate	34	B:4300		
Isophorone	18(2)	I:0950		
Isophorone diamine	7	I:1000		
Isophorone diisocyanate	12	I:1050		Water
Isophthalic acid		I:1100		
Isoprene	30	I:1150		
Isopropylbenzene	32	C:5450		
Jet Fuels:				
JP-1	33	J:0100	K:0150	
JP-3	33	J:0150		
JP-4	33	J:0200		
JP-5	33	J:0250		
Kerosene	33	K:0150	J:0100	
			J:0250	
			O:2400	
			O:4600	
Latex, liquid synthetic	43	L:0200		
Lauric acid	34	L:0300		
Lauroyl peroxide		L:0400		Oxy
Lead nitrate		L:1300		Oxy
Lithium		L:2400		Water
Lithium aluminum hydride		L:2500		Water
Lithium bichromate		L:2600		Oxy
Lithium hydride		L:2800		Water
Magnesium (and alloys)		M:0100		Water
Magnesium nitrate		M:0150		Oxy
Magnesium perchlorate		M:0200		Oxy
Maleic anhydride	11	M:0350		
Mesityl oxide	18(2)	M:1150		
Methacrylic acid	4	M:1200		
Methacrylonitrile	15	M:1300		
Methane	31	M:1400		
3-Methoxy-1-butanol	20			
3-Methoxybutyl acetate	34	M:1550		

1-Methoxy-2-propyl acetate	34			
Methoxy triglycol	40			
Methyl isocyanate		M:4600		Water
Methyl acetate	34	M:1550		
Methyl acetoacetate	34	M:1600		
Methyl acetylene, propadiene mixture	30	M:1650		
Methyl acrylate	14	M:1700		
Methyl alcohol	20(2)	M:1750		
Methylamine	7	M:1850		
Methylamine solution	7	M:1850		
Methyl amyl acetate	34	M:1950		
Methyl amyl alcohol	20	M:2000		
Methyl amyl ketone	18	A:4500		
Methyl bromide	36	M:3100		
Methyl butenol	20	M:3200		
Methyl butyl ketone	18	M:3350		
Methyl *tert*-butyl ether	41(2)	M:3300		
Methylbutynol	20	M:3250		
3-Methyl butyraldehyde	19	I:1700		
Methyl butyrate	34	M:3400		
Methyl chloride	36	M:3450		
Methyl chloroacetate		M:3500		Water
Methylcyclohexane	31	M:3650		
Methylcyclopentadiene dimer	30	M:3750		
Methyl diethanolamine	8	M:4000		
2-Methyl-6-ethyl aniline	9	M:4050		
Methyl ethyl ketone	18(2)	M:4100		
2-Methyl-5-ethyl pyridine		M:4150		
Methyl formal	41	M:4200		
Methyl formate	34	M:4250		
Methyl heptyl ketone	18	M:4300		
2-Methyl-2-hydroxy-3-butyne	20	M:4400		
Methyl isoamyl ketone	18			
Methyl isobutyl carbinol	20	M:4500		
Methyl isobutyl ketone	18(2)	M:4550		
Methyl isocyanate	12	M:4600		Water
Methyl methacrylate	14	M:4800		
3-Methyl-3-methoxy butanol	20			
3-Methyl-3-methoxybutyl acetate	34			
Methyl naphthalene	32	M:4850		
Methylolureas	19			
2-Methyl pentane	31	I:0750		
2-Methyl-1-pentene	30	M:4950		
4-Methyl-1-pentene	30	M:5000		
Methylpyridine	9	M:5250	M:4950	
N-Methyl-2-pyrrolidone	9	M:5300		
Methyl salicylate	34	M:5350		
α-Methylstyrene	30	M:5400		
Methyl trichlorosilane		M:5450		Water
Metolachlor	34	M:5550		
Mineral spirits	33	M:5600	N:0300	
Molasses	20			
Monochlorodifluoro methane	36	M:5800		
Morpholine	7(2)	M:6100		
Motor fuel anti-knock compounds containing lead alkyls	0(1)	M:6150		
Myrcene	30	M:6200		
Naphtha:				
Coal tar	33	N:0350		
Cracking fraction	33(2)			
Petroleum	33	N:0200		
Solvent	33	N:0250		
Stoddard solvent	33	N:0300		
Varnish makers' & painters'	33	N:0200		

Naphthalene	32	N:0400		
Naphthenic acids	4	N:0450		
Naphthenic acid, sodium salt solution	43			
Neodecanoic acid	4	N:0550		
Nickel nitrate hexahydrate		N:1100		Oxy
Nitrating acid	0(1)			
Nitric acid	3	N:1350		Oxy (70% or less)
Nitric acid	0(1)	N:1350		Oxy (Greater than 70%)
Nitrobenzene	42	N:1550		
o-Nitrochlorobenzene	42	C:2650		
Nitroethane	42	N:1600		
Nitrogen tetroxide		N:1700		Oxy
2-Nitrophenol (molten)	0(1,2)	N:1850		
3-Nitrophenol (molten)	0(1,2)	N:1900		
4-Nitrophenol (molten)	0(1,2)	N:1950		
Nitropropane	42	N:2000		
m-Nitrotoluene	42	N:2150		
o-Nitrotoluene	42	N:2200		
p-Nitrotoluene	42	N:2250		
Nonane	31	N:2350		
Nonanoic acid	4			
Nonene	30	N:2450		
Nonyl alcohol	20(2)	N:2400		
Nonyl methacrylate	14			
Nonylphenol	21	N:2550		
Nonyl phenol (ethoxylated)	40			
Nonyl phenol poly(4-12) ethoxylates	40			
Nonyl phenol sulfide solution	33			
1-Octadecene	30			
Octadecenoamide	10			
Octane	31	O:0100		
Octanoic acid	4	O:0200		
Octene	30	O:0400		
Octyl alcohol (octanol)	20(2)	O:0300	I:0850	
Octyl aldehyde	19	O:0500	I:0800	
Octyl decyl adipate	34			
Octyl epoxy tallate	34	O:0600		
Octyl nitrate	34(2)			
Octyl phenol	21			
Oils, edible:				
Castor	34	O:1200		
Coconut	34	O:1300		
Corn	34			
Cottonseed	34	O:1400		
Fish	34(2)	O:1500		
Lard	34	O:1600		
Olive	34	O:1700		
Palm	34(2)	O:1800		
Peanut	34	O:1900		
Safflower	34	O:2000		
Soya bean	34	O:2100		
Tucum	34	O:2200		
Vegetable	34	O:2300		
Oils, fuel:				
No. 1	33	O:2400		
No. 1-D	33	O:0900	O:1000	
No. 2	33	O:2500		
No. 2-D	33	O:0900	O:1100	
No. 4	33	O:2600		
No. 5	33	O:2700		
No. 6	33	O:2800		
Oils, miscellaneous				
Absorption	33	O:2900		
Clarified	33	O:0700		
Coal tar	33	O:3200		

Coconut	34	O:1300			
Cottonseed oil, fatty acid	34	O:1400			
Crude	33	O:0800			
Diesel	33	O:0900			
Linseed	33	O:3400			
Lubricating	33	O:3500			
Mineral	33	O:3600			
Mineral seal	33	O:3700			
Motor	33	O:3800			
Neatsfoot	33	O:3900			
Penetrating	33	O:4000			
Pine	33	O:4100			
Range	33	O:2400			
Residual	33		O:2600		
			O:2700		
			O:2800		
Resin	33	O:4200			
Road	33	O:4300			
Rosin	33	O:4200			
Seal	34				
Soybean	34	O:2100			
Sperm	33	O:4400			
Spindle	33	O:4500			
Spray	33	O:4600			
Tall	34	O:4700			
Tall, fatty acid	34(2)				
Tanner's	33	O:4800			
Transformer	33	O:4900			
Tung	34				
Turbine	33	O:5000			
White (mineral)	33	O:3600			
α-Olefins (C$_{13}$ and above)	30		T:3100		
Oleic acid	04	O:5100			
Oleum	0(1,2)	O:5400			
Oxalic acid		O:5000		Water	
Oxygen		O:5600		Oxy	
Paraffin wax	31	P:0100			
Paraldehyde	19	P:0175			
Pentachloroethane	36	P:0300			
Pentadecanol	20	P:0400			
1,3-Pentadiene	30	P:0450	P:0500		
Pentaethylenehexamine, Tetraethylenepentamine mixture	7				
Pentane	31	P:0650	I:0900		
Pentene	30	P:0800			
Pentene, miscellaneous hydrocarbon mixture	30(2)				
Peracetic acid		P:0900		Oxy	
Perchloric acid		P:0950		Oxy	
Perchloroethylene	36	T:0400			
Petrolatum	33	P:1100			
Phenol	21	P:1200			
1-Phenyl-1-xylyl ethane	32	P:1450			
Phosphoric acid	1	P:1600	P:2350		
Phosphorus	0(1)	P:1650			
		P:1700			
		P:1750		Water	
Phosphorus oxychloride		P:1800		Water	
Phosphorus pentasulfide		P:1850		Water	
Phosphorus tribromide		P:1900		Water	
Phosphorus trichloride		P:1950		Water	
Phthalic anhydride (molten)	11	P:2000			
Pinene	30	P:2050			
Pine oil	33	O:4100			
Polybutene	30	P:2150			

Polyethylene glycols	40	P:2450		
		P:4300		
Polyethylene polyamines	7(2)	P:2250		
Polymethylene polyphenyl isocyanate	12	P:2300		
Polypropylene	30	P:2400		
Polypropylene glycols	40	P:2450		
Polypropylene glycol methyl ether	40	P:4300		
Potassium		P:2500		Water
Potassium chlorate		P:2700		Oxy
Potassium dichloro-s-triazinetrione		P:2850		Oxy
Potassium dichromate		P:2900		Oxy
Potassium hydroxide solution	5(2)	P:2950		
Potassium nitrate		P:3050		Oxy
Potassium oleate	34	P:3100	O:5200	
Potassium permanganate		P:3200		Oxy
Potassium peroxide		P:3250		Water/Oxy
Propane	31	P:3300		
Propanolamine	8	P:3350	M:6000	
Propionaldehyde	19	P:3650		
Propionic acid	4	P:3700		
Propionic anhydride	11	P:3750		
Propionitrile	37	P:3800		
n-Propoxypropanol	40	P:3850		
Propyl acetate	34	P:3900	I:1200	
Propyl alcohol	20(2)	P:3950	I:1250	
Propylamine	7		I:1300	
			P:4000	
			D:7450	
Propylbenzene	32	P:4050		
iso-Propylcyclohexane	31			
Propylene	30	P:4150		
Propylene glycol	20(2)	P:4200		
Propylene glycol ethyl ether	40	P:4250		
Propylene glycol methyl ether	40	P:4300		
Propylene oxide	16	P:4450		
Propylene tetramer	30	P:4500		
Propylene trimer	30	P:4550		
Propyl ether	41		I:1400	
			P:4600	
n-Propyl nitrate		P:4700		Oxy
Pyridine	9	P:4800		
Pyridine bases	9			
Rosin oil	33	O:4200		
Salicylaldehyde	19	S:0100		
Sewage sludge	43			
Silver nitrate		S:1000		Oxy
Sodium		S:1300		Water
Sodium aluminate solution	5	S:1600		
Sodium borohydride, sodium hydroxide solution	5	S:2600		
Sodium chlorate		S:2800		Oxy
Sodium chlorate solution	0(1,2)	S:2900	S:2800	Oxy
Sodium cyanide solution	5	S:3100		
Sodium dichloro-s-triazinetrione dihydrate		S:3200		Oxy
Sodium dichromate solution	0(1,2)	S:3300		
Sodium fluoride		S:3400		Oxy
Sodium hydride		S:3600		Water
Sodium hydro sulfide solution	5(2)	S:3700		
Sodium hydroxide solution	5(2)	S:3800		
Sodium hypochlorite solution	5	S:3900		
Sodium nitrate		S:4300		Oxy
Sodium nitrite		S:4400		Oxy
Sodium nitrite solution	05	S:4500		Oxy
Sodium silicate solution	43(2)	S:5100		
Sodium thiocyanate solution	0(1,2)	S:5500		

Sorbitol solutions	20	S:5700		
Stearic acid	34	S:5900		
Styrene	30	S:6300		
Sulfolane	39	S:6500		
Sulfur	0(1)	S:6600		
Sulfur chloride		S:7000		Water
Sulfuric acid	(2)2	S:6800		Water
Sulfuric acid, spent	2	S:6900		Water
Sulfuryl chloride		S:7100		Water
Tallow	34(2)	T:0150		
Tallow fatty acid	34(2)	T:0100		
Tallow fatty alcohol	20	T:0200		
Tetrachloroethane	36	T:0350		
Tetradecanol	20	T:0450		
Tetradecene	30	T:0500		
Tetradecylbenzene	32	T:0550		
Tetraethylene glycol	40	T:0650		
Tetraethylenepentamine	7	T:0700		
Tetrahydrofuran	41	T:0900		
Tetrahydronaphthalene	32	T:0950		
1,2,3,5-Tetramethyl benzene	32	T:1000		
Thorium nitrate		T:1500		Oxy
Titanium tetrachloride	2	T:1550		Water
Toluene	32	T:1600		
Toluenediamine	9	T:1650		
Toluene diisocyanate	12	T:1700		Water
o-Toluidine	9	T:1850		
Tributyl phosphate	34	T:2000		
1,2,4-Trichlorobenzene	36	T:2150		
1,1,1-Trichloroethane	36(2)	T:2250		
1,1,2-Trichloroethane	36	T:2300		
Trichloroethylene	36(2)	T:2350		
1,2,3-Trichloropropane	36	T:2750		
Trichlorosilane		T:2800		Water
Trichloro-*s*-triazine trione		T:2850		Oxy
Tricresyl phosphate	34	T:2950		
Tridecane	31	T:3000		
Tridecanol	20	T:3050		
Tridecene	30	T:3100		
Tridecylbenzene	32	T:3150		
Triethanolamine	8(2)	T:3200		
Triethylaluminum		T:3250	Water	
Triethylamine	7	T:3300		
Triethylbenzene	32	T:3350		
Triethylene glycol	40	T:3400		
Triethylene glycol di-(2-ethylbutyrate)	34	T:3450		
Triethylene glycol ethyl ether	40	T:3500		
Triethylenetetramine	7(2)	T:3600		
Triethyl phosphate	34	T:3650		
Triethyl phosphite	34(2)	T:3700		
Triisobutylene	30	T:3900		
Triisobutylaluminum		T:3850		Water
Triisopropanolamine	8	T:3950		
Trimethylacetic acid	4	T:4000		
Trimethylbenzene	32	T:4100		
Trimethylchlorosilane		T:4150		Water
Trimethylhexamethylene-diamine (2,2,4- and 2,4,4-)	7	T:4200	T:4250	
Trimethylhexamethylene diisocyanate (2,2,4- and 2,4,4-)	12	T:4300		
2,2,4-Trimethyl pentanediol-1,3-diisobutyrate	34	T:4350		
2,2,4-Trimethyl-1,3-pentane-diol-1-isobutyrate	34	T:4350		
2,2,4-Trimethyl-3-pentanol-1-isobutyrate	34	T:4350		

Trimethyl phosphite	34(2)	T:4400		
Tripropylene	30	T:4500		
Tripropylene glycol	40	T:4550		
Tripropylene glycol methyl ether	40	T:4600		
Trixylenyl phosphate	34	T:4700		
Turpentine	30	T:4750		
Undecanoic acid	4	U:0100		
Undecanol	20	U:0200		
Undecene	30	U:0300		
Undecyl alcohol	20	U:0200		
Undecylbenzene	32	U:0400		
Urea, ammonium nitrate solution (containing NH_3)	6	U:1000		
Urea, ammonium nitrate solution (not containing NH_3)	43	A:3300		
Uranyl nitrate		U:0700		Oxy
Valeraldehyde	19	V:0100	I:1700	
Vanillin black liquor	5	V:0600		
Vinyl acetate	13	V:0700		
Vinyl chloride	35	V:0800		
Vinyl ethyl ether	13	V:1000		
Vinylidene chloride	35	V:1200		
Vinyl neodecanoate	13	V:1400		
Vinyl toluene	13	V:1500		
Vinyl trichlorsilane		V:1600		Water
Waxes:				
Carnauba	34	W:0100		
Paraffin	31	P:0100		
White Spirit [low (15–20%) aromatic]	33	W:0200		
Xylene	32	X:0100	X:0200 X:0300	
Xylenols	21	X:0400		
Zinc bichromate		Z:0500		Oxy
Zirconium Tetrachloride		Z:2800		Water

Footnotes to the incompatibility index: The Guide is based, in part, upon information provided by the National Academy of Sciences, U.S. Coast Guard Advisory Committee on Hazardous Materials and represents the current information available to the Coast Guard on chemical compatibility. *Note 1*: Because of very high reactivity or unusual conditions of carriage or potential compatibility problems, this product is not assigned to a specific group in the Compatibility Chart. For additional compatibility information, contact Commandant (G-MTH), U.S. Coast Guard, 2100 Second Street, SW., Washington, DC 20593-0001. Telephone (202) 267-1577. *Note 2*: See "Exceptions to the Chart."

GUIDE TO THE COMPATIBILITY OF CHEMICALS

CARGO COMPATIBILITY

CARGO GROUPS \ REACTIVE GROUPS	1 NON-OXIDIZING MINERAL ACIDS	2 SULFURIC ACID	3 NITRIC ACID	4 ORGANIC ACIDS	5 CAUSTICS	6 AMMONIA	7 ALIPHATIC AMINES	8 ALKANOLAMINES	9 AROMATIC AMINES	10 AMIDES	11 ORGANIC ANHYDRIDES	12 ISOCYANATES	13 VINYL ACETATE	14 ACRYLATES	15 SUBSTITUTED ALLYLS	16 ALKYLENE OXIDES	17 EPICHLOROHYDRIN	18 KETONES	19 ALDEHYDES	20 ALCOHOLS, GLYCOLS	21 PHENOLS, CRESOLS	22 CAPROLACTAM SOLUTION
1 NON-OXIDIZING MINERAL ACIDS		X			X	X	X	X	X	X	X	X				X	X					
2 SULFURIC ACID	X		X		X	X	X	X	X	X	X	X	X	X	X	X	X	X	X	X	X	X
3 NITRIC ACID		X		X	X	X	X	X	X	X	X	X	X	X	X	X	X	X	X	X	X	
4 ORGANIC ACIDS			X		X	X		X				X				X	X					
5 CAUSTICS	X	X	X	X							X		X			X	X	X	X	X	X	X
6 AMMONIA	X	X	X	X							X	X				X	X	X	X	X	X	X
7 ALIPHATIC AMINES	X	X	X		X	X					X	X	X	X	X	X	X	X	X	X	X	
8 ALKANOLAMINES	X	X	X	X	X	X					X	X	X	X	X	X	X					
9 AROMATIC AMINES	X	X	X									X				X	X					
10 AMIDES	X	X	X													X	X					
11 ORGANIC ANHYDRIDES	X	X	X		X	X	X	X	X							X						
12 ISOCYANATES	X	X	X	X	X	X	X	X	X							X						
13 VINYL ACETATE		X	X	X		X																
14 ACRYLATES		X	X																			
15 SUBSTITUTED ALLYLS		X	X																			
16 ALKYLENE OXIDES	X	X	X	X	X	X	X	X	X													
17 EPICHLOROHYDRIN	X	X	X	X	X	X	X	X														
18 KETONES		X	X																			

Guide to the compatibility of chemicals

CARGO COMPATIBILITY

CARGO GROUPS	1 NON-OXIDIZING MINERAL ACIDS	2 SULFURIC ACID	3 NITRIC ACID	4 ORGANIC ACIDS	5 CAUSTICS	6 AMMONIA	7 ALIPHATIC AMINES	8 ALKANOLAMINES	9 AROMATIC AMINES	10 AMIDES	11 ORGANIC ANHYDRIDES	12 ISOCYANATES	13 VINYL ACETATE	14 ACRYLATES	15 SUBSTITUTED ALLYLS	16 ALKYLENE OXIDES	17 EPICHLOROHYDRIN	18 KETONES	19 ALDEHYDES	20 ALCOHOLS, GLYCOLS	21 PHENOLS, CRESOLS	22 CAPROLACTAM SOLUTION
19 ALDEHYDES		X	X		X	X	X	X	X													
20 ALCOHOLS, GLYCOLS		X	X		X																	
21 PHENOLS, CRESOLS		X	X		X		X															
22 CAPROLACTAM SOLUTION		X			X		X			X		X										
30 OLEFINS		X	X																			
31 PARAFFINS																						
32 AROMATIC HYDROCARBONS			X																			
33 MISC. HYDROCARBON MIXTURES			X																			
34 ESTERS		X	X																			
35 VINYL HALIDES		X	X																			X
36 HALOGENATED HYDROCARBONS																						
37 NITRILES		X					X	X														
38 CARBON DISULFIDE								X														
39 SULFOLANE																						
40 GLYCOL ETHERS		X																				
41 ETHERS		X	X				X	X	X			X										
42 NITROCOMPOUNDS								X														
43 MISC. WATER SOLUTIONS		X										X										

Note: "X" indicates a reactivity group that forms an unsafe combination.

EXCEPTIONS TO THE CHART–Part A

The binary combinations listed below have been tested by the National Academy of Sciences and found not to be dangerously reactive. These combinations are exceptions to the Compatibility Chart and may be stowed in adjacent tanks.

MEMBER OF REACTIVE GROUP	COMPATIBLE WITH
Acetone (18)	Diethylenetriamine (7)
Acetone cyanohydrin (0)	Acetic acid (4)
Acrylonitrile (15)	Triethanolamine (8)
1,3-Butylene glycol (20)	Morpholine (7)
1,4-Butylene glycol (20)	Ethylamine (7)
	Triethanolamine (8)
Caustic potash, 50% or less (5)	Ethyl alcohol (20
	Ethylene glycol (20)
	Isopropyl alcohol (20)
	Methyl alcohol (20)
	iso-Octyl alcohol (20)
Caustic soda, 50% or less (5)	Butyl alcohol (20)
tert-Butyl alcohol–methanol mixtures (7)	Decyl alcohol (20)
	Diacetone alcohol (20)
	Diethylene glycol (40)
	Ethyl alcohol (20)
	Ethyl alcohol (40%, whiskey) (20)
	Ethylene glycol (20)
	Ethylene glycol, Diethylene glycol mixture (20)
	Ethyl hexanol (Octyl alcohol) (20)
	Isotridecanol (20)
	Methyl alcohol (20)
	Nonyl alcohol (20)
	Propyl alcohol (20)
	Propylene glycol (20)
	Sodium chlorate (0)
Dodecyl- and tetradecylamine mixture (7)	Tall oil, fatty acid (34)
Ethylenediamine (7)	Butyl alcohol (20)
	tert-Butyl alcohol (20)
	Butylene glycol (20)
	Creosote (21)
	Diethylene glycol (40)
	Ethyl alcohol (20)
	Ethylene glycol (20)
	Ethyl hexanol (20)
	Glycerine (20)
	Isononyl alcohol (20)
	Isophorone (18)
	Methyl butyl ketone (18)
	Methyl isobutyl ketone (18)
	Methyl ethyl ketone (18)
	Propyl alcohol (20)
	Propylene glycol (20)
Oleum (0)	Hexane (31)
	Dichloromethane (36)
	Perchloroethylene (36)
1,2-Propylene glycol (20)	Diethylenetriamine (7)
	Polyethylene polyamines (7)
	Triethylenetetramine (7)
Sulfuric acid (2)	Coconut oil, coconut oil acid, palm oil, and tallow (34)
Sulfuric acid, 98% or less (2)	Choice white grease tallow (34)

EXCEPTIONS TO THE CHART–Part B

The binary combinations listed below have been determined to be *dangerously reactive*, based on either data obtained in the literature or on laboratory testing which has been carried out by the National Academy of Sciences. These combinations are exceptions to the Compatibility Chart and *should not be stowed next to one another.*

Acetone cyanohydrin (0) is not compatible with Groups 1–12, 16, 17, and 22.

Acrolein (19) is not compatible with Group 1, Nonoxidizing Mineral Acids.

Acrylic acid (4) is not compatible with Group 9, Aromatic Amines.

Alkylbenzenesulfonic acid (0) is not compatible with Groups 1–3, 5–9, 15, 16, 18, 19, 30, 34, 37, and strong oxidizers.

Allyl alcohol (15) is not compatible with Group 12, Isocyanates.

Aluminum sulfate solution (43) is not compatible with Groups 5–11.

Ammonium bisulfite solution (43) is not compatible with Groups 1, 3, 4, and 5.

Benzenesulfonyl chloride (0) is not compatible with Groups 5–7, and 43.

γ-Butyrolactone (0) is not compatible with Groups 1–9.

Crotonaldehyde (19) is not compatible with Group 1, Nonoxidizing Mineral Acids.

Cyclohexanone-cyclohexanol mixture (18) is not compatible with Group 12, Isocyanates.

2,4-Dichlorophenoxyacetic acid–triisopropanolamine salt solution (43) is not compatible with Group 3, Nitric acid.

2,4-Dichlorophenoxyacetic acid–dimethylamine salt solution (0) is not compatible with Groups 1–5, 11, 12, and 16.

Dimethyl hydrogen phosphite (34) is not compatible with Groups 1 and 4.

Dimethyl naphthalene sulfonic acid, sodium salt solution (34) is not compatible with Group 12, Formaldehyde, and strong oxidizing agents.

Dodecylbenzenesulfonic acid (0) is not compatible with oxidizing agents and Groups 1, 2, 3, 5, 6, 7, 8, 9, 15, 16, 18, 19, 30, 34, and 37.

Ethyl chlorothioformate (0) is not compatible with Groups 5, 6, 7, 8, and 9.

Ethylenediamine (7) is not compatible with Ethylene dichloride (36).

Ethylene dichloride (36) is not compatible with Ethylenediamine (7).

Ethylidene norbonene (30) is not compatible with Groups 1–3 and 5–8.

2-Ethyl-3-propylacrolein (19) is not compatible with Group 1, Nonoxidizing mineral acids.

Ferric hydroxyethyylethylenediamine triacetic acid–Sodium salt solution (43) is not compatible with Group 3, Nitric acid.

Fish oil (34) is not compatible with Sulfuric acid (2).

Formaldehyde (over 50%) in Methyl alcohol (over 30%) (19) is not compatible with Group 12, Isocyanates.

Formic acid (4) is not compatible with Furfural alcohol (20).

Furfuryl alcohol (20) is not compatible with Group 1, Nonoxidizing mineral acids and formic acid (4).

2-Hydroxyethyl acrylate is not compatible with Groups 2, 3, 5–8, and 12.

Isophorone (18) is not compatible with Group 8, Alkanolamines.

Magnesium chloride solution (0) is not compatible with Groups 2, 3, 5, 6, and 12.

Mesityl oxide (18) is not compatible with Group 8, Alkanolamines.

Methyl *tert*-butyl ether (41) is not compatible with Group 1, Nonoxidizing mineral acids.

Naphtha, cracking fraction (33) is not compatible with strong acids, caustics or oxidizing agents.

o-Nitrophenol (0) is not compatible with Groups 2, 3, and 5–10.

Octyl nitrates (all isomers) (34) is not compatible with Group 1, Nonoxidizing Mineral Acids.

Oleum (0) is not compatible with sulfuric acid (2) and 1,1,1-trichloroethane (36).

Pentene–miscellaneous hydrocarbon mixtures (30) are not compatible with strong acids or oxidizing agents.

Sodium chlorate solution (50% or less) (0) is not compatible with Groups 1–3, 5, 7, 8, 10, 12, 13, 17, and 20.

Sodium dichromate solution (70% or less) (0) is not compatible with Groups 1–3, 5, 7, 8, 10, 12, 13, 17, and 20.

Sodium dimethyl naphthalene sulfonate solution (34) is not compatible with Group 12, Formaldehyde and strong oxidizing agents.

Sodium hydrogen sulfide, sodium carbonate solution (0) is not compatible with Groups 6 (Ammonia) and 7 (Aliphatic amines).

Sodium hydrosulfide (5) is not compatible with Groups 6 (Ammonia) and 7 (Aliphatic amines).

Sodium hydrosulfide–ammonium sulfide solution (5) is not compatible with Groups 6 (Ammonia) and 7 (Aliphatic amines).

Sodium polyacrylate solution (43) is not compatible with Group 3, Nitric Acid.

Sodium salt of ferric hydroxyethylethylenediamine triacetic acid solution (43) is not compatible with Group 3, Nitric acid.

Sodium silicate solution (43) is not compatible with Group 3, Nitric acid.

Sodium sulfide–hydrosulfide solution (0) is not compatible with Groups 6 (Ammonia) and 7 (Aliphatic amines).

Sodium thiocyanate (56% or less) (0) is not compatible with Groups 1–4.

Sulfuric acid (2) is not compatible with fish oil (34), or oleum (0).

Tallow fatty acid (34) is not compatible with Group 5, Caustics.

1,1,1-Trichloroethane (36) is not compatible with Oleum (0).

Trichlorethylene (36) is not compatible with Group 5, Caustics.

Triethyl phosphite (34) is not compatible with Groups 1 and 4.

Trimethyl phosphite (34) is not compatible with Groups 1 and 4.

REACTIVITY GROUPS

0. UNASSIGNED
Acetone cyanohydrin (1,2)
Alkylbenzenesulfonic acid (1,2)
Aluminum chloride–hydrochloric acid solution
Ammonium hydrogen phosphate solution
Ammonium nitrate solution (1)
Ammonium thiocyanate–ammonium thiosulfate solution
Benzenesulfonyl chloride (1,2)
γ-Butyrolactone (1,2)
Chlorine (1)
Chlorosulfonic acid (1)
2,4-Dichlorophenoxyacetic acid–dimethylamine salt solution (1,2)
Dimethylamine salt of 2,4-dichlorophenoxyacetic acid solution (1,2)
Diphenylol propane-epichlorohydrin resins (1)
Dodecylbenzenesulfonic acid (2)
Ethyl chlorothioformate (2)
Ethylene oxide (1)
2-Hydroxyethyl acrylate (1,2)
Magnesium chloride solution (1,2)
Molasses residue
Motor fuel anti-knock compounds containing lead alkyls (1)
Naphthalene sulfonic acid–formaldehyde copolymer, sodium salt solution
Nitrating acid (1)
Nitric acid (Greater than 70%) (1)
o-Nitrophenol (1,2)
Noxious liquid substance, n.o.s. (NLS's)
Oleum (1,2)
Phosphorus (1)
Sodium chlorate solution (1,2)
Sodium dichromate solution (1,2)
Sodium hydrogen sulfide–sodium carbonate solution (2)
Sodium sulfide, hydrosulfide solution (1,2)
Sodium thiocyanate solution (1,2)
Sulfur (1)
Urea-ammonium mono- and di-hydrogen phosphate–potassium chloride solution

1. NONOXIDIZING MINERAL ACIDS
Di-(2-ethylhexyl)phosphoric acid
Ferric chloride solution
Hydrochloric acid
Hydrochloric acid, spent
Hydrofluorosilicic acid
Phosphoric acid

2. SULFURIC ACIDS
Sulfuric acid (2)
Sulfuric acid, spent
Titanium tetrachloride

3. NITRIC ACID
Ferric nitrate–nitric acid solution
Nitric acid (70% or less)

4. ORGANIC ACIDS
Acetic acid (2)
Acrylic acid (2)
Butyric acid
Cashew nut shell oil (untreated)
Chloroacetic acid solution
Chloropropionic acid
Cyclohexane oxidation product acid water
Decanoic acid
2,2-Dichloropropionic acid
2,2-Dimethyloctanoic acid
2-Ethylhexanoic acid
Formic acid (2)
n-Heptanoic acid
Hexanoic acid
2-Hydroxy-4-(methylthio)butanoic acid
Methacrylic acid
Naphthenic acids
Neodecanoic acid
Nonanoic acid
Octanoic acid
Propionic acid
Trimethylacetic acid
Undecanoic acid

5. CAUSTICS
Ammonium sulfide solution
Calcium hypochlorite solutions
Caustic potash solution (2)
Caustic soda solution (2)
Cresylate spent caustic solution
Cresylic acid, sodium salt solution
Kraft black liquor
Kraft pulping liquors
Mercaptobenzothiazol, sodium salt solution
Potassium hydroxide solution (2)
Sodium aluminate solution
Sodium borohydride–sodium hydroxide solution
Sodium carbonate solutions
Sodium cyanide solution
Sodium hydrosulfide solution (2)
Sodium hydrosulfide–ammonium sulfide solution (2)
Sodium hydroxide solution (2)
Sodium hypochlorite solution
Sodium 2-mercaptobenzothiazol solution
Sodium nitrite solution
Vanillin black liquor

6. AMMONIA
Ammonia, anhydrous
Ammonium hydroxide (28% or less containing NH_3)
Ammonium nitrate–urea solution (containing containing NH_3)
Urea–ammonium nitrate solution (containing NH_3)

7. ALIPHATIC AMINES
N-Aminoethyl piperazine
Butylamine
Cyclohexylamine
Di-n-propylamine
Dibutyl amine
Diethylamine
Diethylenetriamine (2)
Diisobutylamine
Diisopropylamine
Dimethylamine
Dimethylamine solution
N,N-Dimethylcyclohexylamine
Di-n-propylamine
Dodecylamine, tetradecylamine mixture (2)
Ethylamine (2)
Ethylamine solution
N-Ethyl-n-butylamine
N-Ethylcyclohexylamine
Ethylenediamine (2)
2-Ethylhexylamine
Hexamethylenediamine solution
Hexamethylenetetramine
Hexamethylenetetramine solution
Hexamethylenimine
Isophorone diamine
Metam sodium solution
Methylamine

Methylamine solution
Morpholine (2)
Pentaethylenehexamine–tetraethylenepentamine mixture
Polyethylene polyamines (2)
Propyl amine
Tetraethylenepentamine
Triethylamine
Triethylenetetramine (2)
Trimethylhexamethylenediamine (2,2,4- and 2,4,4-)

8. ALKANOLAMINES
2-(2-Aminoethoxy)ethanol
Aminoethyldiethanolamine–aminoethylethanolamine solution
Aminoethylethanolamine
Amino-2-methyl-1-propanol
Diethanolamine
Diethylaminoethanol
Diethylethanolamine
Diisopropanolamine
Dimethylethanolamine
Ethanolamine
Propanolamine
Triethanolamine (2)
Triisopropanolamine

9. AROMATIC AMINES
Aniline
4-Chloro-2-methylphenoxyacetic acid–dimethylamine salt solution
2,6-Diethylaniline
Dimethylamine salt of 4-Chloro-2-methylphenoxyacetic acid solution
2,6-Dimethylaniline
2-Ethyl-6-methyl-N)1-methyl-2-methoxyethyl)aniline
4,4'-Methylenediamine (43% or less)–polymethylene polyphenylamine–o-dichlorobenzene mixtures
2-Methyl-6-ethyl aniline
2-Methyl-5-ethyl pyridine
Methylpyridine
N-Methyl-2-pyrrolidone
3-Methylpyridine
N-Methyl pyrrolidone
Pyridine
Pyridine bases
Toluenediamine
o-Toluidine

10. AMIDES
Acrylamide solution
N,N-Dimethyl acetamide
N,N-Dimethylacetamide solution
Dimethylformamide
Formamide
Octadecenoamide

11. ORGANIC ANHYDRIDES
Acetic anhydride
Maleic anhydride
Phthalic anhydride
Propionic anhydride

12. ISOCYANATES
Diphenylmethane diisocyanate
Isophorone diisocyanate
Polymethylene polyphenyl isocyanate
Toluene diisocyanate
Trimethylhexamethylene diisocyanate (2,2,4- and 2,4,4-)

13. VINYL ACETATE
Vinyl acetate
Vinyl ethyl ether
Vinyl neodecanoate
Vinyl toluene

14. ACRYLATES
Butyl acrylate
Butyl methacrylate
Butyl methacrylate, decyl methacrylate, cetyl eicosyl methacrylate mixture
Cetyl eicosyl methacrylate
Decyl acrylate
Dodecylmethacrylate
Dodecyl, pentadecyl methacrylate
Ethyl acrylate
2-Ethylhexyl acrylate
Ethyl methacrylate
Glycidyl methacrylate
Methyl acrylate
Methyl methacrylate
Nonyl methacrylate
Polyalkyl (C18–C22) acrylate in xylene

15. SUBSTITUTED ALLYLS
Acrylonitrile (2)
Allyl alcohol (2)
Allyl chloride
1,3-Dichloropropene
Dichloropropene–dichloropropane mixture
Methacrylonitrile

16. ALKYLENE OXIDES
Butylene oxide
Ethylene oxide–propylene oxide mixture
Propylene oxide

17. EPICHLOROHYDRIN
Chlorohydrins
Epichlorohydrin

18. KETONES
Acetone (2)
Acetophenone
Amyl methyl ketone
Butyl heptyl ketone
Camphor oil
Cyclohexanone
Cyclohexanone–cyclohexanol mixture (2)
Diisobutyl ketone
Epoxy resin
Ethyl amyl ketone
Isophorone (2)
Ketone residue
Mesityl oxide (2)
Methyl amyl ketone
Methyl butyl ketone
Methyl diethenaolamine
Methyl ethyl ketone (2)
Methyl heptyl ketone
Methyl isoamyl ketone
Methyl isobutyl ketone (2)
Methyl propyl ketone

19. ALDEHYDES
Acetal
Acetaldehyde
Acrolein (2)
Butyraldehyde
Crotonaldehyde (2)
Decaldehyde
Ethylhexaldehyde
2-Ethyl-3-propylacrolein (2)
Formaldehyde solution (2)
Formaldehyde–methanol mixtures (2)
Furfural
Glutaraldehyde solution

Glyoxal solutions
3-Methyl butyraldehyde
Methylolureas
Octyl aldehyde
Paraldehyde
Pentyl aldehyde
Propionaldehyde
Salicylaldehyde
Valeraldehyde

20. ALCOHOLS, GLYCOLS
Acrylonitrile–styrene copolymer dispersion in polyether polyol
Alcoholic beverages
Alcohol polyethoxylates
Alcohol polyethoxylates, secondary
Alcohols (C_{13} and above)
Amyl alcohol
Behenyl alcohol
Brake fluid base mixtures
Butyl alcohol (2)
Butylene glycol (2)
1,3-Butylene glycol
1,4-Butylene glycol
2,3-Butylene glycol
Choline chloride solutions
Cyclohexanol
Decyl alcohol (2)
Diacetone alcohol (2)
Diisobutylcarbinol
2,2-Dimethylpropane-1,3-diol
Dodecanol
Dodecyl alcohol
Ethoxylated alcohols, C_{11}-C_{15}
2-Ethoxyethanol
Ethyl alcohol (2)
Ethyl butanol
Ethylene chlorohydrin
Ethylene cyanohydrin
Ethylene glycol (2)
2-Ethylhexanol
Furfuryl alcohol (2)
Glycerine (2)
Heptanol
Hexanol
Hexylene glycol
3-Methoxy-1-butanol
Methyl alcohol (2)
Methyl amyl alcohol
Methyl butenol
Methylbutynol
2-Methyl-2-hydroxy-3-butyne
Methyl isobutyl carbinol
3-Methyl-3-methoxybutanol
Molasses
Nonyl alcohol (2)
Octyl alcohol (2)
Pentadecanol
Polyalkylene oxide polyol
Polybutadiene, hydroxyl terminated
Polyglycerol
Propyl alcohol (2)
Propylene glycol (2)
Rum
Sorbitol solutions
Tallow fatty alcohol
Tetradecanol
Tridecanol
Trimethylol propane polyethoxylate
Undecanol
Undecyl alcohol

21. PHENOLS, CRESOLS
Benzyl alcohol
Carbolic oil
Creosote (2)
Cresols
Cresylic acid
2,4-Dichlorophenol
Dodecyl phenol
o-Ethyl phenol
Nonylphenol
Octyl phenol
Phenol
Xylenols

22. CAPROLACTAM SOLUTIONS
Caprolactam solution

23–29. UNASSIGNED GROUPS

30. OLEFINS
Amylene
Butadiene
Butadiene–butylene mixtures (containing acetylenes)
Butene
Butene oligomer
Butylene
1,5,9-Cyclododecatriene
1,3-Cyclopentadiene dimer
Cyclopentadiene polymers
Cyclopentadiene– styrene– benzene mixtures
Cyclopentene
Decene
Dicyclopentadiene
Diisobutylene
Dipentene
Dodecene
Ethylene
Ethylidene norbornene (2)
1-Heptene
Hexene
Isoprene
Methyl acetylene– propadiene mixture
Methylcyclopentadiene dimer
2-Methyl-1-pentene
4-Methyl-1-pentene
α-Methyl styrene
Myrcene
Nonene
1-Octadecene
Octene
Olefin mixtures
α-Olefins (C_6–C_{18}) mixtures
α-Olefins (C_{13} and above)
1,3-Pentadiene
Pentene
Pentene–miscellaneous hydrocarbon mixture (2)
Pinene
Polybutene
Polypropylene
Propylene
Propylene butylene polymer
Propylene dimer
Propylene tetramer
Propylene trimer
Styrene
Tetradecene

Tridecene
Triisobutylene
Tripropylene
Turpentine
Undecene

31. PARAFFINS
Butane
Cycloaliphatic resins
Cycloheptane
Cyclohexane
Cyclopentane
Decane
Dodecane
Ethane
Heptane
Hexane (2)
Methane
Methylcyclohexane
2-Methyl pentane
Nonane
Octane
n-Paraffins (C_{10}–C_{20})
Pentane
Propane
iso-Propylcyclohexane
Tridecane
Waxes: Paraffin

32. AROMATIC HYDROCARBONS
Alkyl acrylate–vinyl pyridine copolymer in toluene
Alkyl (C_9–C_{17}) benzenes
Benzene
Benzene–hydrocarbon mixture (10% benzene or more)
Benzene–toluene–xylene mixture
Butylbenzene
Butyl phenol formaldehyde resin in xylene
Butyl toluene
Cumene
Cymene
Decylbenzene
Dialkyl (C_{10}–C_{14}) benzenes
Diethylbenzene
Diisopropylbenzene
Diisopropyl naphthalene
Diphenyl
Dodecylbenzene
Ethyl benzene
Ethyl toluene
Isopropylbenzene
Methyl naphthalene
Naphthalene
1-Phenyl-1-xylyl ethane
Propylbenzene
Pseudocumene
Tetradecylbenzene
Tetrahydronaphthalene
1,2,3,5-Tetramethylbenzene
Toluene
Tridecylbenzene
Triethylbenzene
Trimethylbenzene
Undecylbenzene
Xylene

33. MISCELLANEOUS HYDROCARBON MIXTURES
Alkylbenzenesulfonic acid, sodium salt solutions
Asphalt blending stocks: roofers flux
Asphalt blending stocks: straight run residue
Aviation alkylates
Calcium sulfonate, Calcium carbonate, Hydrocarbon solvent mixture
Carbon black, base
Coal tar
Coal tar pitch
Decahydronaphthalene
Diphenyl–diphenyl ether mixture
Distillates: flashed feed stocks
Distillates: straight run
Drilling mud (low toxicity) (if flammable or combustible)
Fatty acid amides
Gas oil: cracked
Gasoline blending stocks: alkylates
Gasoline blending stocks: reformates
Gasolines:
  Automotive (not over 4.23 gm Pb/gal.)
  Aviation (not over 4.86 gm Pb/gal.)
  Casinghead (natural)
  Polymer
  Straight run
Glycols–resins–solvents mixture
Herbicide (C_{15}-H_{22}-NO_2-Cl)
Jet Fuels:
  JP-1
  JP-3
  JP-4
  JP-5
  JP-8
Kerosene
Magnesium nonyl phenol sulfide
Maleic anhydride copolymer
Mineral spirits
Naphtha:
  Coal tar solvent
  Cracking fraction (2)
  Petroleum
  Solvent
  Stoddard solvent
  Varnish makers' and painters'
Nonyl phenol sulfide solution
Oil, fuel:
  No. 1
  No. 1-D
  No. 2
  No. 2-D
  No. 4
  No. 5
  No. 6
Oil, miscellaneous:
  Absorption
  Aliphatic
  Aromatic
  Clarified
  Coal
  Crude
  Diesel
  Heartcut distillate
  Linseed
  Lubricating
  Mineral
  Mineral seal
  Motor
  Neatsfoot
  Penetrating
  Pine

Range
Resin
Resinous petroleum
Rosin
Soybean (epoxidized)
Sperm
Spindle
Spray
Tanner's
Turbine
White (mineral)
Residual
Road
Transformer
Oxyalkylated alkyl phenol formaldehyde
Petrolatum
Pine
Pine oil
Polyalkenyl succinic anhydride amine
White spirit [low (15–20%) aromatic]

34. ESTERS
Acetyl tributyl citrate
Alkyl phthalates
Amyl acetate
Amyl tallate
Benzene tricarboxylic acid, trioctyl ester
Benzylacetate
Butyl acetate
Butyl benzyl phthalate
n-Butyl butyrate
Butyl formate
n-Butyl butyrate
iso-Butyl isobutyrate
Calcium naphthenate in mineral oil
Calcium nitrate–magnesium nitrate–potassium chloride solution
Coconut oil, fatty acid
Cottonseed oil, fatty acid
Cyclohexyl acetate
Dialkyl (C_7–C_{13}) phthalates
Dibutyl phthalate
Diethylene glycol butyl ether acetate
Diethylene glycol ethyl ether acetate
Diethylene glycol methyl ether acetate
Diethylene glycol phthalate
Di-(2-ethylhexyl)adipate
Di-(2-ethylhexyl)phthalate
Diethyl phthalate
Diethyl sulfate
Diheptyl phthalate
Di-n-hexyl adipate
Diisobutyl phthalate
Diisodecyl phthalate
Diisononyl adipate
Diisononyl phthalate
Diisooctyl phthalate
Dimethyl adipate
Dimethylcyclicsiloxane hydrolyzate
Dimethyl glutarate
Dimethyl hydrogen phosphite (2)
Dimethyl naphthalene sulfonic acid, sodium salt solution (2)
Dimethyl phthalate
Dimethylpolysiloxane
Dimethyl succinate
Dinonyl phthalate
Dioctyl phthalate
Dipropylene glycol dibenzoate
Ditridecyl phthalate
2-Dodecenylsuccinic acid, dipotassium salt solution
Diundecyl phthalate
2-Ethoxyethyl acetate
Ethyl acetate
Ethyl acetoacetate
Ethyl butyrate
Ethylene glycol acetate
Ethylene glycol butyl ether acetate
Ethylene glycol diacetate
Ethylene glycol ethyl ether acetate
Ethylene glycol methyl ether acetate
Ethyl-3-ethoxypropionate
Ethyl hexyl phthalate
Ethyl propionate
Fatty acidsd (saturated, C_{13} and above)
Glycerol polyalkoxylate
Glyceryl triacetate
Glycidyl ester of tridecylacetic acid
Glycidyl ester of Versatic acid
Glycol diacetate
Heptyl acetate
Hexyl acetate
Lauric acid
Magnesium sulfonate
3-Methoxybutyl acetate
1-Methoxy-2-propyl acetate
Methyl acetate
Methyl acetoacetate
Methyl amyl acetate
Methyl butyrate
Methyl formate
3-Methyl-3-methoxybutyl acetate
Methyl salicylate
Metolachlor
Naphthalene sulfonic acid, sodium salt solution
Octyl decyl adipate
Octyl epoxy tallate
Octyl nitrate (2)
Oil, edible:
 Babassu
 Castor
 Coconut
 Corn
 Cottonseed
 Fish (2)
 Lard
 Olive
 Palm (2)
 Palm kernel
 Peanut
 Rapeseed
 Rice bran
 Safflower
 Soya bean
 Sunflower seed
 Tucum
 Vegetable
Oil, miscellaneous:
 Coconut oil, fatty acid methyl ester
 Cotton seed oil, fatty acid
 Palm oil, fatty acid methyl ester
 Palm oil, methyl ester
 Soapstock
 Tall
 Tall, fatty acid (2)

Tung
Oleic acid
Palm kernel oil, fatty acid
Palm kernel oil, fatty acid methyl ester
Palm stearin
n-Pentyl propionate
Polydimethylsiloxane
Polyferric sulfate solution
Polymethylsiloxane
Poly(20)oxyethylene sorbitan monooleate
Polysiloxane
Potassium oleate
Propyl acetate
Sodium acetate solution
Sodium benzoate solution
Sodium dimethyl naphthalene sulfonate solution (2)
Sodium naphthalene sulfonate solution
Stearic acid
Tall oil
Tallow (2)
Tallow fatty acid (2)
Triarylphosphate
Tributyl phosphate
Tricresyl phosphate
Triethylene glycol di-(2-ethylbutyrate)
Triethyl phosphate
Triethyl phosphite (2)
Triisooctyl trimellitate
2,2,4-Trimethyl pentanediol-1,3-diisobutyrate
2,2,4-Trimethyl-1,3-pentanediol-1-isobutyrate
2,2,4-Trimethyl-3-pentanol-1-isobutyrate
Trimethyl phosphite (2)
Trisodium nitrilotriacetate
Trixylenyl phosphate
Vinyl acetate–fumarate copolymer
Waxes: carnauba

35. VINYL HALIDES
Vinyl chloride
Vinylidene chloride

36. HALOGENATED HYDROCARBONS
Benzyl chloride
Carbon tetrachloride
Chlorinated paraffins (C_{10}–C_{13})
Chlorinated paraffins (C_{14}–C_{17})
Chlorobenzene
Chlorodifluoromethane
Chloroform
Chlorotoluene
Dichlorobenzene
Dichlorodifluoromethane
1,1-Dichloroethane
2,2'-Dichloroisopropyl ether
Dichloromethane
Dichloropropane
Ethyl chloride
Ethylene dibromide
Ethylene dichloride (2)
Methyl bromide
Methyl chloride
Monochlorodifluoromethane
Pentachloroethane
Perchloroethylene
1,1,2,2-Tetrachloroethane
1,2,4-Trichlorobenzene
1,1,1-Trichloroethane (2)
1,1,2-Trichloroethane
Trichloroethylene (2)
1,2,3-Trichloropropane
1,1,2-trichloro-1,2,2-trifluoroethane

37. NITRILES
Acetonitrile
Adiponitrile
Lactonitrile solution
3-Pentenenitrile
Propionitrile
Tallow nitrile

38. CARBON DISULFIDE
Carbon disulfide

39. SULFOLANE
Sulfolane

40. GLYCOL ETHERS
Diethylene glycol
Diethylene glycol butyl ether
Diethylene glycol dibutyl ether
Diethylene glycol ethyl ether
Diethylene glycol methyl ether
Diethylene glycol phenyl ether
Dipropylene glycol
Dipropylene glycol methyl ether
Ethoxy triglycol
Ethylene glycol tert-butyl ether
Ethylene glycol butyl ether
Ethylene glycol dibutyl ether
Ethylene glycol ethyl ether
Ethylene glycol isopropyl ether
Ethylene glycol methyl ether
Ethylene glycol phenyl ether
Ethylene glycol phenyl ether–diethylene glycol phenyl ether mixture
Ethylene glycol propyl ether
Methoxy triglycol
Nonyl phenol (ethoxylated)
Nonyl phenol poly(4-12)ethoxylates
Polyalkylene glycols–polyalkylene glycol monoalkyl ethers mixtures
Polyethylene glycol
Polyethylene glycol dimethyl ether
Polyethylene glycol monoalkyl ether
Polypropylene glycol
Polypropylene glycol methyl ether
n-Propoxypropanol
Propylene glycol ethyl ether
Propylene glycol methyl ether
Propylene glycol monoalkyl ether
Tetraethylene glycol
Triethylene glycol
Triethylene glycol butyl ether
Triethylene glycol butyl ether mixture
Triethylene glycol ether mixture
Triethylene glycol ethyl ether
Tripropylene glycol
Tripropylene glycol methyl ether

41. ETHERS
Butyl ether
2,2-Dichloroethyl ether
Diethyl ether
Diglycidyl ether of bisphenol F
Diglycidyl ether of bisphenol A
Dimethyl furan
1,4-Dioxane
Diphenyl ether
Diphenyl ether–diphenyl phenyl ether mixture
Ethyl ether

Methyl tert-butyl ether (2)
Methyl formal
Propyl ether
Tetrahydrofuran

42. NITROCOMPOUNDS
1,1-Dichloro-1-nitroethane
Dinitrotoluene
Nitrobenzene
o-Nitrochlorobenzene
Nitroethane
Nitropropane
Nitropropane,–nitroethane mixture
Nitrotoluene

43. MISCELLANEOUS WATER SOLUTIONS
Aluminum sulfate solution (2)
2-Amino-2-hydroxymethyl-1,3-propanediol solution
Ammonium bisulfite solution (2)
Ammonium nitrate–urea solution (not containing NH_3)
Ammonium polyphosphate solution
Ammonium sulfate solution
Ammonium thiosulfate solution
Calcium bromide solution
Calcium chloride solution
Corn syrup
Dextrose solution
Diammonium salt of zinc EDTA
2,4-Dichlorophenoxyacetic acid, diethanolamine salt solution
2,4-Dichlorophenoxyacetic acid, triisopropanolamine salt solution (2)
Didecyl dimethyl ammonium chloride–ethanol mixture solution
Diethanolamine salt of 2,4-dichlorophenoxyacetic acid solution
Dodecyl diphenyl oxide disulfonate solution
Drilling brine (containing calcium, potassium, or sodium salts)
Drilling brine (containing zinc salts)
Drilling mud (low toxicity) (if nonflammable or noncombustible)
Ethylenediaminetetracetic acid, tetrasodium salt solution
Ethylene–vinyl acetate copolymer emulsion
Ferric hydroxyethylethylenediaminetriacetic acid, trisodium salt solution (2)
Fish solubles (water based fish meal extracts)
Fructose solution
Fumaric adduct of rosin, water dispersion
N-(Hydroxyethyl)ethylenediaminetriacetic acid, trisodium salt solution
Kaolin clay slurry
Latex, liquid synthetic
Lignin liquor
Naphthenic acid, sodium salt solution
Rosin soap (disproportionated) solution
Sewage sludge
Sodium alkyl sulfonate solution
Sodium hydrogen sulfite solution
Sodium polyacrylate solution (2)
Sodium salt of ferric hydroxyethylethylenediamine triacetic acid solution
Sodium silicate solution (2)
Tall oil soap (disproportionated) solution
Tetrasodium salt of EDTA solution
Triisopropanolamine salt of 2,4-dichlorophenoxyacetic acid solution
Urea–ammonium nitrate solution (not containing NH_3)
Urea–ammonium phosphate solution
Vegetable protein solution

Footnotes to tables: (1) Because of very high reactivity or potential compatibility problems, this product is not assigned to a specific group in the Compatibility Chart. For additional compatibility information, contact Commandant (G-MTH), U.S. Coast Guard, 2100 Second Street, SW, Washington, DC 20593-0001. Telephone (202) 267-1577. (2) See Exceptions to the Chart.

KEY TO ABBREVIATIONS, SYMBOLS, AND ACRONYMS

α-	Greek letter alpha; used as a prefix to denote the carbon atom in a straight-chain compound to which the principal group is attached
as-	Prefix for asymmetric
ACGIH	American Conference of Governmental Industrial Hygienists
asym-	Prefix for asymmetric
@	At
atm.	Atmosphere
β	Greek letter beta
BLEVE	Boiling Liquid Expanding Vapor Explosion
C	Centigrade
CAS	Chemical Abstract Service
cc	Closed cup; cubic centimeter
CERCLA	Comprehensive Environmental Response, Compensation, and Liability Act
CFR	Code of Federal Regulations
cis-	(Latin, on this side). Indicating one of two geometrical isomers in which certain atoms or groups are on the same side of a plane
cyclo-	(Greek, circle). Cyclic, ring structure; as cyclohexane
Δ or δ	Greek letter delta
deriv.	Derivative
DOT	U.S. Department of Transportation
DOT ID	U.S. Department of Transportation Identification Numbers
ε	Greek letter epsilon
EEC	European Economic Community
EHS	Extremely hazardous substance
EPA	U.S. Environmental Protection Agency
F	Fahrenheit
FDA	U.S. Food and Drug Administration
FEMA	Federal Emergency Management Agency
FR	*Federal Register*
γ	Greek letter gamma
IARC	International Agency for Research on Cancer
IDLH	Immediately Dangerous to Life or Health
iso-	(Greek, equal, alike). Usually denoting an isomer of a compound
kg	Weight in kilograms (one thousand grams)
K	Kelvin
lb	Weight in pounds
LEL	Lower explosive (flammable) limit in air, % by volume at room temperature or other temperature as noted
m-	Abbreviation for *"meta-,"* a prefix used to distinguish between isomers or nearly related compounds
μ	Micro
μg	Microgram(s)
m^3	Cubic meter
mg	Milligram
MSDS	Material Safety Data Sheets
NIOSH	National Institute for Occupational Safety and Health
n-	Abbreviation for "normal," referring to the arrangement of carbon atoms in a chemical molecule prefix for normal
N-	Symbol used in some chemical names, indicating that the next section of the name refers to a chemical group attached to a nitrogen atom.; the bond to the nitrogen atom
NCI	National Cancer Institute
NTP	National Toxicology Program
o-	Abbreviation for *"ortho"*- a prefix used to distinguish between isomers or nearly related compounds.
ω	Greek letter omega
oc	Open cup
OSHA	Occupational Safety and Health Administration
Oxy	Oxidizer or oxidizing agent
p-	Abbreviation for *"para-, "* a prefix used to distinguish between isomers or nearly related compounds.
PEL	Permissible exposure level (OSHA)
ppb	Parts per billion
PPE	Personal Protective Equipment
ppm	Parts per million
prim-	Prefix for "primary"

®	Symbol for a registered trademark or proprietary product
REL	Recommended exposure limits (NIOSH)
RQ	Reportable quantity
SCBA	Self-Contained Breathing Apparatus
sec-	Prefix for "secondary"
soln.	Solution
STEL	Short-term exposure limit
sym-	Abbreviation for "symmetrical," referring to a particular arrangement of elements within a chemical molecule
t-	Prefix for 'tertiary'
temp.	Temperature
tert-	Abbreviation for "tertiary," referring to a particular arrangement of elements within a chemical molecule
TLV®	Threshold Limit Value; a registered trademark of ACGIH
trans-	(Latin, across). Indicating that one of two geometrical isomers in which certain atoms or groups are on opposite sides of a plane.
TWA	Time weighted average
UEL	Upper explosive (flammable) limit in air, % by volume at room temperature or other temperature as noted
unsym-	Prefix for "asymmetric"
USDA	U.S. Department of Agriculture
>	Symbol for "greater than"
<	Symbol for "less than"
°	Degrees of temperature
%	Percent

ACETAL
REC. A:0100

SYNONYMS: ACETAL DIETHYLIQUE (French); ACETALDEHYDE DIETHYLACETAL; 1,1-DIAETHOXY-AETHAN (German); DIAETHYLACETAL (German); DIETHYL ACETAL; 1,1-DIETHOXYETHANE; DIETHYL ACETAL; EEC No. 605-015-00-1; ETHYLIDENE DIETHYL ETHER

IDENTIFICATION
CAS Number: 105-57-7
Formula: $C_6H_{14}O_2$
DOT ID Number: UN 1088; DOT Guide Number: 127
Proper Shipping Name: Acetal

DESCRIPTION: Colorless liquid. Pungent, woody odor. Floats on water surface; slightly soluble.

Highly flammable • Containers may BLEVE when exposed to fire • Vapors may form explosive mixture with air • Vapors are heavier than air and will collect and stay in low areas • Vapors may travel long distances to ignition sources and flashback • Vapors in confined areas (e.g., tanks, sewers, buildings) may explode when exposed to fire • Highly irritating to the eyes and respiratory tract; this material has anesthetic properties • Toxic products of combustion may include carbon monoxide.

Hazard Classification (based on NFPA-704 Rating System)
Health Hazards (Blue): 2; Flammability (Red): 3; Reactivity (Yellow): 0

EMERGENCY RESPONSE: See Appendix A (127)
Evacuation:
Public safety: Isolate spill area for at least 25 to 50 meters (80 to 160 feet) in all directions.
Spill: Large spill–Consider initial downwind evacuation for at least 300 meters (1000 feet).
Fire: Isolate for 800 meters (½ mile) in all directions, especially if tank, rail car, or tank truck is involved in fire.

EXPOSURE
Short-term effects: *SEEK MEDICAL ATTENTION.* **Vapor:** May irritate the upper respiratory tract. High concentrations act as a central nervous system depressant. Symptoms of exposure include headache, dizziness, drowsiness, abdominal pain, and nausea. May be harmful if inhaled or absorbed through skin. Move victim to fresh air. *IF BREATHING HAS STOPPED,* give artificial respiration. IF breathing is difficult, administer oxygen. **Liquid:** *HARMFUL IF INGESTED OR ABSORBED THROUGH THE SKIN.* May cause irritation to eyes and skin. *IF IN EYES,* flush with plenty of water for at least 15 minutes. Remove contaminated clothing and shoes. Flush affected areas with plenty of running water. *IF SWALLOWED,* do nothing except keep victim warm.

HEALTH HAZARDS
Personal protective equipment (PPE): B-Level. Butyl rubber is generally suitable for aldehydes.
Recommendations for respirator selection: *At concentrations above the NIOSH REL, or where there is no REL, at any detectable concentration:* SCBAF:PD,PP (any self-contained breathing apparatus that has a full facepiece and is operated in a pressure-demand or other positive-pressure mode); or SAF:PD,PP:ASCBA (any supplied-air respirator that has a full facepiece and is operated in a pressure-demand or other positive-pressure mode in combination with an auxiliary self-contained breathing apparatus operated in a pressure-demand or other positive-pressure mode).

Toxicity by ingestion: Grade 2: LD_{50} = 3.5 g/kg (mouse).
Long-term health effects: Central nervous system depressant.
Vapor (gas) irritant characteristics: Vapors cause a smarting of the eyes or respiratory system. The effect may be temporary.
Liquid or solid irritant characteristics: If spilled on clothing and allowed to remain, may cause smarting and reddening of skin.

FIRE DATA
Flash point: –5°F/–21°C (cc)
Flammable limits in air: LEL: 1.65; UEL: 10.4%.
Fire extinguishing agents not to be used: Water may pose a problem.
Autoignition temperature: 446°F/230°C/503°K.
Electrical hazard: Flow or agitation of substance may generate electrostatic charges due to low conductivity.

CHEMICAL REACTIVITY
Binary reactants: Violent reaction with oxidizers.
Polymerization: Forms heat-sensitive explosive peroxides on contact with air. Old samples may explode upon heating. May polymerize on standing.
Reactivity group: 19
Compatibility class: Aldehyde

ENVIRONMENTAL DATA
Food chain concentration potential: Log P_{ow} = 0.82. Unlikely to accumulate.
Water pollution: DOT Appendix B, §172.101–marine pollutants. Effects of low concentrations on aquatic life are not known. May be dangerous if it enters nearby water intakes; notify operators. Notify local health and wildlife officials.

SHIPPING INFORMATION
Grades of purity: 99%; **Inert atmosphere:** None; **Venting:** None; **Stability during transport:** Stable

NAS HAZARD CLASSIFICATION FOR BULK WATER TRANSPORTATION
FIRE: 3
HEALTH: Vapor irritant: 1; Liquid or solid irritant: 1; Poisons: 2
WATER POLLUTION: Human toxicity: 2; Aquatic toxicity: 1; Aesthetic effect: 2
REACTIVITY: Other chemicals: 1; Water: 1; Self-reaction: –

PHYSICAL AND CHEMICAL PROPERTIES
Physical state @ 59°F/15°C and 1 atm: Liquid.
Molecular weight: 118.2
Boiling point @ 1 atm: 216°F/102°C/375°K.
Melting/Freezing point: –148°F/–100°C/173°K.
Specific gravity (water = 1): 0.831 @ 68°F/20°C.
Liquid surface tension: 21.65 dyne/cm = 0.022 N/m @ 68°F/20°C.
Relative vapor density (air = 1): 4.1
Latent heat of vaporization: 119.2 Btu/lb = 66.2 cal/g = 2.8×10^5 J/kg.
Vapor pressure: (Reid) 1.1 psia; 10 mm @ 8°C.

ACETALDEHYDE
REC. A:0150

SYNONYMS: ACETALDEHIDO (Spanish); ACETIC ALDEHYDE; ACETEHYD (German); ACETEHYDE; ACETIC EHYDE; EEC No. 605-003-00-6; EHYDE ACETIQUE (French); ETHANAL; ETHYL ALDEHYDE; ETHYL EHYDE; RCRA No. U001

Acetaldehyde

IDENTIFICATION
CAS Number: 75-07-0
Formula: C_2H_4O
DOT ID Number: UN 1089; DOT Guide Number: 129
Proper Shipping Name: Acetaldehyde
Reportable Quantity (RQ): **(CERCLA)** 1000 lb/454 kg

DESCRIPTION: Colorless, mobile, watery liquid. Sharp, fruity odor. Floats on water surface; soluble. Large amounts of flammable, irritating vapor is produced.

Highly flammable • Corrosive eyes and respiratory tract; skin or eye contact may cause burns or vision impairment; inhalation symptoms may be delayed; this material has anesthetic properties • May explode without warning when exposed to heat, dust, or corrosive or oxidizing agents • Firefighting gear (including SCBA) does not provide adequate protection. If exposure occurs, remove and isolate gear immediately and thoroughly decontaminate personnel • Containers may BLEVE when exposed to fire •Vapors may form explosive mixture with air •Vapors are heavier than air and will collect and stay in low areas • Vapors may travel long distances to ignition sources and flashback • Vapors in confined areas (e.g., tanks, sewers, buildings) may explode when exposed to fire • Toxic products of combustion may include carbon monoxide. *Note:* Oxidizes freely in air, forming unstable peroxides; these peroxides may explode without warning when exposed to heat, dust, air pressure, corrosives, or oxidizers.

Hazard Classification (based on NFPA-704 Rating System)
Health Hazards (Blue): 3; Flammability (Red): 4; Reactivity (Yellow): 2

EMERGENCY RESPONSE: See Appendix A (129)
Evacuation:
Public safety: Isolate spill area for at least 50 to 100 meters (160 to 330 feet) in all directions.
Spill: Large spill–Consider initial downwind evacuation for at least 300 meters (1000 feet).
Fire: Isolate for 800 meters (½ mile) in all directions, especially if tank, rail car, or tank truck is involved in fire.

EXPOSURE
Short-term effects: *SEEK MEDICAL ATTENTION.* **Vapor:** Irritating to eyes, nose, and throat. *IF INHALED*, will, will cause nausea, vomiting, headache, difficult breathing, or loss of consciousness. Move to fresh air. *IF BREATHING HAS STOPPED*, give artificial respiration. IF breathing is difficult, administer oxygen. Lung edema may develop. **Liquid:** Will burn skin and eyes. Harmful if swallowed. Remove contaminated clothing and shoes. Flush affected areas with plenty of water. *IF IN EYES*, hold eyelids open and flush with plenty of water. *IF SWALLOWED* and victim is *CONSCIOUS AND ABLE TO SWALLOW*, have victim drink 4 to 8 ounces of water.
Note to physician or authorized medical personnel: Pulmonary edema may be delayed. Medical observation is recommended for 24 to 48 hours after inhalation overexposure. As first aid for pulmonary edema, a physician or authorized medical personnel may consider administering a corticosteroid spray. Cigarette smoking may exacerbate pulmonary injury and should be discouraged for at least 72 hours following exposure.

HEALTH HAZARDS
Personal protective equipment (PPE): B-Level PPE. Chemical protective material(s) reported to have good to excellent resistance: Butyl rubber, Teflon®, Silvershield®. OSHA Z-1-A air contaminant.

Recommendations for respirator selection: NIOSH
At any concentrations above the NIOSH REL, or where there is no REL, at any detectable concentration: SCBAF:PD,PP (any self-contained breathing apparatus that has a full facepiece and is operated in a pressure-demand or other positive-pressure mode); or SAF:PD,PP:ASCBA (any supplied-air respirator that has a full facepiece and is operated in a pressure-demand or other positive-pressure mode in combination with an auxiliary self-contained breathing apparatus operated in a pressure-demand or other positive pressure mode). *ESCAPE:* GMFOV [any air-purifying, full-facepiece respirator (gas mask) with a chin-style, front-or back-mounted organic vapor canister]; or SCBAE (any appropriate escape-type, self-contained breathing apparatus).
Exposure limits (TWA unless otherwise noted): OSHA PEL 200 ppm (360 mg/m^3); potential human carcinogen; reduce exposure to lowest feasible level.
Short-term exposure limits (15-minute TWA): ACGIH TLVceiling 25 ppm (45 mg/m^3).
Toxicity by ingestion: Grade 2; LD_{50} = 0.5 to 5g/kg (rat).
Long-term health effects: IARC possible carcinogen; rating 2B; sufficient animal evidence; NTP anticipated carcinogen
Vapor (gas) irritant characteristics: Corrosive to the respiratory tract. Vapor is irritating such that personnel will not usually tolerate moderate or high concentrations.
Liquid or solid irritant characteristics: Corrosive to the skin and eyes. If spilled on clothing and allowed to remain, may cause smarting and reddening of the skin.
Odor threshold: 0.067 ppm.
IDLH value: Potential carcinogen; 2000 ppm.

FIRE DATA
Flash point: –36°F/–38°C (cc); –58°F/–50°C (oc).
Flammable limits in air: LEL: 4%; UEL: 60%.
Fire extinguishing agents not to be used: Water may be ineffective.
Behavior in fire: Closed containers my rupture or explode.
Autoignition temperature: 347°F/175°C/448°K.
Electrical hazard: Class I, Group C; Flow or agitation of substance may generate electrostatic charges due to low conductivity.
Burning rate: 3.3 mm/min
Stoichiometric air-to-fuel ratio: 7.800 (estimate).

CHEMICAL REACTIVITY
Binary reactants: A strong reducing agent. Reacts with strong oxidizers, strong acids, strong bases, alcohols, ammonia, amines, cyanides, halogens, phenols, ketones, hydrogen cyanide, hydrogen sulfide, anhydrides. Exposure to heat, dust, corrosives, or oxidizers can cause explosive polymerization. Slightly corrosive to mild steel.
Polymerization: Can form explosive peroxides under air pressure. Slowly polymerizes to paraldehyde (acetic acid also reported).
Reactivity group: 19
Compatibility class: Aldehyde

ENVIRONMENTAL DATA
Food chain concentration potential: Log K_{ow} = 0.42 (estimate). Unlikely to accumulate.
Water pollution: DOT Appendix B, §172.101–marine pollutant. Harmful to aquatic life in very low concentrations. May be dangerous if it enters nearby water intakes; notify operators. Notify local health and wildlife officials. **Response to discharge:** Issue warning–high flammability. Disperse and flush.

SHIPPING INFORMATION
Grades of purity: More than 99%; **Storage temperature:**

Ambient. Bulk quantities should be provided with refrigeration; **Inert atmosphere:** Inerted; **Venting:** Safety relief; **Stability during transport:** Stable.

NAS HAZARD CLASSIFICATION FOR BULK WATER TRANSPORTATION
FIRE: 4
HEALTH: Vapor irritant: 3; Liquid or solid irritant: 1; Poisons: 2
WATER POLLUTION: Human toxicity: 2; Aquatic toxicity: 3; Aesthetic effect: 2
REACTIVITY: Other chemicals: 2; Water: 0; Self-reaction: 1

PHYSICAL AND CHEMICAL PROPERTIES
Physical state @ 59°F/15°C and 1 atm: Liquid.
Molecular weight: 44.05
Boiling point @ 1 atm: 68.7°F/20.4°C/293.6°K.
Melting/Freezing point: −189°F/−123°C/150°K.
Critical temperature: 370°F/188°C/461°K.
Critical pressure: 820 psia = 56 atm = 5.7 MN/m^2.
Specific gravity (water = 1): 0.780 @ 68°F/20°C (liquid).
Relative vapor density (air = 1): 1.5
Ratio of specific heats of vapor (gas): 1.182
Latent heat of vaporization: 245 Btu/lb = 136 cal/g = 5.69 x 10^5 J/kg.
Heat of combustion: −10,600 Btu/lb = −5890 cal/g = −246.4 x 10^5 J/kg.
Vapor pressure: (Reid) 25.6 psia; 740 mm.

ACETIC ACID   **REC. A:0200**

SYNONYMS: ACETIC ACID, GLACIAL; ACIDE ACETIQUE (French); ACIDO ACETICO (Spanish); EEC No. (100%) 607-002-01-3; EEC No. (85% in water) 607-002-00-6; ESSIGSAEURE (German); ETHANOIC ACID; ETHYLIC ACID; GLACIAL ACETIC ACID; METHANE CARBOXYLIC ACID; VINEGAR ACID

IDENTIFICATION
CAS Number: 64-19-7
Formula: C$_2$H$_4$O$_2$; CH$_3$COOH
DOT ID Number: UN 2789 solution greater than 80% acid, by mass; UN 2790 solution, with more than 10% but not more than 80% acid, by mass; DOT Guide Number: 132 (UN2789); 153 (UN 2790)
Proper Shipping Name: Acetic acid, glacial or Acetic acid solution, more than 80% acid, by mass (UN 2789); Acetic acid solution, more than 10% but not more than 80% acid, by mass (UN 2790)
Reportable Quantity (RQ): **(CERCLA)** 5000 lb/2270 kg

DESCRIPTION: Colorless watery liquid. Sour, pungent, vinegar odor. Sinks and mixes with water; produces heat. Irritating vapor is produced.

Corrosive to the skin, eyes, and respiratory tract; prolonged contact with skin or eyes can cause burns and vision impairment; inhalation symptoms may be delayed • Very flammable • Containers may BLEVE when exposed to fire •Vapors may form explosive mixture with air • Vapors are heavier than air and will collect and stay in low areas • Vapors may travel long distances to ignition sources and flashback • Vapors in confined areas (e.g., tanks, sewers, buildings) may explode when exposed to fire • Toxic products of combustion may include carbon monoxide.

Hazard Classification (based on NFPA-704 Rating System)
Health Hazards (Blue): 3; Flammability (Red): 2; Reactivity (Yellow): 0

EMERGENCY RESPONSE, solution greater than 80% acid, by mass: See Appendix A (132)
Evacuation:
Public safety: Isolate spill area for at least 100 to 200 meters (330 to 660 feet) in all directions.
Spill: Increase, as necessary, the isolation distance shown above, in "Public safety."
Fire: Isolate for 800 meters (½ mile) in all directions, especially if tank, rail car, or tank truck is involved in fire.

EMERGENCY RESPONSE, solution, with > 10% but not >80% acid, by mass: See Appendix A (153)
Evacuation:
Public safety: Isolate the area of spill or leak for at least 25 to 50 meters (80 to 160 feet) in all directions.
Spill: Increase, in the downwind direction, as necessary, the distance shown under "Public Safety."
Fire: If tank, rail car, or tank truck is involved in fire, isolate for at least 800 meters (½ mile) in all directions; also, consider initial evacuation for 800 meters (½ mile) in all directions.

EXPOSURE
Short-term effects: *SEEK MEDICAL ATTENTION.* **Vapor:** Irritating to nose and throat. *IF INHALED*, will, will cause coughing, nausea, vomiting, or difficult breathing. May cause lung edema; physical exertion will aggravate this condition. Move to fresh air. *IF BREATHING HAS STOPPED,* give artificial respiration. IF breathing is difficult, administer oxygen. **Liquid or solid:** Will burn skin and eyes. Harmful if swallowed. Remove contaminated clothing and shoes. Flush affected areas with plenty of water. *IF IN EYES*, hold eyelids open and flush with plenty of water. *IF SWALLOWED* and victim is *CONSCIOUS AND ABLE TO SWALLOW*, have victim drink 4 to 8 ounces of water. **Do NOT induce vomiting.**
Note to physician or authorized medical personnel: Medical observation is recommended for 24 to 48 hours after breathing overexposure, as pulmonary edema may be delayed. As first aid for pulmonary edema, consider administering a corticosteroid spray. Cigarette smoking may exacerbate pulmonary injury and should be discouraged for at least 72 hours following exposure.

HEALTH HAZARDS
Personal protective equipment (PPE): B-Level PPE. OSHA Table Z-1-A air contaminant. Protective clothing should be worn when skin contact might occur. Respiratory protection necessary when exposed to vapor. Complete eye protection. Chemical protective material(s) reported to have good to excellent resistance: butyl rubber, Teflon®, Saranex®, natural rubber, neoprene, neoprene+natural rubber, neoprene/natural rubber, nitrile. The following materials offer limited protection: Chlorinated polyethylene, Viton®.
Recommendations for respirator selection: NIOSH/OSHA
50 ppm: SA:CF* (any supplied-air respirator operated in a continuous-flow mode); or PAPROV* [any powered, air-purifying respirator with organic vapor cartridge(s)]; CCRFOV [any air-purifying, full-facepiece respirator (gas mask) with a chin-style, front-or back-mounted acid gas canister]; or GMFOV [any air-purifying, full-facepiece respirator (gas mask) with a chin-style, front-or back-mounted organic vapor canister]; or SCBAF (any self-contained breathing apparatus with a full facepiece); or SAF (any supplied-air respirator with a full facepiece). *EMERGENCY OR*

Acetic anhydride

OR PLANNED ENTRY INTO UNKNOWN CONCENTRATIONS OR IDLH CONDITIONS: SCBAF:PD,PP (any self-contained breathing apparatus that has a full facepiece and is operated in a pressure-demand or other positive-pressure mode); or SAF:PD,PP:ASCBA (any supplied-air respirator that has a full facepiece and is operated in a pressure-demand or other positive-pressure mode in combination with an auxiliary self-contained breathing apparatus operated in a pressure-demand or other positive-pressure mode). *ESCAPE:* GMFOV [any air-purifying, full-facepiece respirator (gas mask) with a chin-style, front-or back-mounted organic vapor canister] or SCBAE (any appropriate escape-type, self-contained breathing apparatus). *Note:* Substance causes eye irritation or damage; eye protection needed.
Exposure limits (TWA unless otherwise noted): ACGIH TLV 10 ppm (25 mg/m^3); OSHA PEL 10 ppm (25 mg/m^3).
Short-term exposure limits (15-minute TWA): ACGIH STEL 15 ppm (37 mg/m^3); NIOSH STEL 15 ppm (37 mg/m^3).
Toxicity by ingestion: Grade 2; LD_{50} = 0.5 to 5.0 g/kg (rat).
Vapor (gas) irritant characteristics: Corrosive to the respiratory tract. Vapors cause irritation such that personnel will find high concentrations unpleasant. The effect is temporary.
Liquid or solid irritant characteristics: Severe skin irritant; may cause pain and second-degree burns after a few minutes of contact.
Odor threshold: 0.037–0.15 ppm.
IDLH value: 50 ppm.

FIRE DATA
Flash point: 112°F/45°C (oc); 103°F/40°C (cc).
Flammable limits in air: LEL 5.4%; UEL 16.0%.
Autoignition temperature: 867°F/463°C/736°K.
Electrical hazard: Class I, Group D.
Burning rate: 1.6 mm/min

CHEMICAL REACTIVITY
Reactivity with water: None reported.
Binary reactants: Strong oxidizers (especially chromic acid, peroxides and nitric acid), chromium trioxide; strong caustics; potassium permanganate. Highly corrosive to metal, especially when diluted; forms flammable hydrogen gas. Excellent solvent for many synthetic resins or rubber.
Neutralizing agents for acids and caustics: Dilute with water, rinse with sodium bicarbonate solution.
Reactivity group: 4
Compatibility class: Organic acids

ENVIRONMENTAL DATA
Food chain concentration potential: Log P_{ow} = –0.28. Unlikely to accumulate.
Water pollution: Harmful to aquatic life in very low concentrations. May be dangerous if it enters nearby water intakes; notify operators. Notify local health and wildlife officials.
Response to discharge: Issue warning–corrosive. Disperse and flush.

SHIPPING INFORMATION
Grades of purity: Commercial; USP; CP; **Storage temperature:** Ambient; **Inert atmosphere:** None; **Venting:** Open

NAS HAZARD CLASSIFICATION FOR BULK WATER TRANSPORTATION
FIRE: 2
HEALTH: Vapor irritant: 2; Liquid or solid irritant: 3; Poisons: 2
WATER POLLUTION: Human toxicity: 1; Aquatic toxicity: 2; Aesthetic effect: 2
REACTIVITY: Other chemicals: 2; Water: 0; Self-reaction: 0

PHYSICAL AND CHEMICAL PROPERTIES
Physical state @ 59°F/15°C and 1 atm: Liquid.
Molecular weight: 60.05
Boiling point @ 1 atm: 244°F/117.9°C/391.1°K.
Melting/Freezing point: 62.1°F/16.7°C/290°K.
Critical temperature: 611°F/321.6°C/594.8°K.
Critical pressure: 839 psia = 57.1 atm = 5.78 MN/m^2.
Specific gravity (water = 1): 1.051 @ 68°F/20°C (liquid).
Relative vapor density (air = 1): 2.1
Ratio of specific heats of vapor (gas): 1.145
Latent heat of vaporization: 17.1 Btu/lb = 96.7 cal/g = 4.05 x 10^5 J/kg.
Heat of combustion: –5645 Btu/lb = –3136 cal/g = –131.3 x 10^5 J/kg.
Heat of fusion: 45.91 cal/g.
Vapor pressure: (Reid) 0.60 psia; 11 mm.

ACETIC ANHYDRIDE REC. A:0250

SYNONYMS: ACETIC ACID, ANHYDRIDE; ACETIC OXIDE; ACETYL ANHYDRIDE; ACETYL ETHER; ACETYL OXIDE; ANHIDRIDO ACETICO (Spanish); ANHYDRIDE ACETIQUE (French); EEC No. 607-008-00-9; ESSIGSAEURE ANHYDRID (German); ETHANOIC ANHYDRATE; ETHANOIC ANHYDRIDE

IDENTIFICATION
CAS Number: 108-24-7
Formula: $C_4H_6O_6$; $CH_3CO \cdot O \cdot COCH_3$
DOT ID Number: UN 1715; DOT Guide Number: 137
Proper Shipping Name: Acetic anhydride
Reportable Quantity (RQ): (CERCLA) 5000 lb/2270 kg

DESCRIPTION: Colorless, watery liquid. Strong, vinegar odor. Sinks and reacts slowly with water forming acetic acid (vinegar) and heat. Irritating vapor is produced.

Flammable • Corrosive to the skin, eyes, and respiratory tract; contact with skin or eyes can cause burns and vision impairment; inhalation symptoms may be delayed • Firefighting gear (including SCBA) does not provide adequate protection. If exposure occurs, remove and isolate gear immediately and thoroughly decontaminate personnel • Containers may BLEVE when exposed to fire •Vapors may form explosive mixture with air •Vapors are heavier than air and will collect and stay in low areas • Vapors in confined areas (e.g., tanks, sewers, buildings) may explode when exposed to fire • Reacts explosively with a large number of chemicals • Toxic products of combustion may include carbon monoxide.

Hazard Classification (based on NFPA-704 Rating System)
Health Hazards (Blue): 3; Flammability (Red): 2; Reactivity (Yellow): 1

EMERGENCY RESPONSE: See Appendix A (137)
Evacuation:
Public safety: Isolate the area of spill or leak for at least 50 to 100 meters (160 to 330 feet) in all directions.
Spill: Increase, in the downwind direction, as necessary, the distance shown under "Public Safety."
Fire: If any large container is involved in fire, isolate for at least 800 meters (½ mile) in all directions; also, consider initial evacuation for 800 meters (½ mile) in all directions.

EXPOSURE

Short-term effects: *SEEK MEDICAL ATTENTION.*
Vapor: Will burn eyes. Irritating to nose and throat. *IF INHALED,* will, will cause nausea, vomiting, or difficult breathing. Move to fresh air. May cause lung edema; physical exertion will aggravate this condition. *IF IN EYES*, hold eyelids open and flush with plenty of water. *IF BREATHING HAS STOPPED,* give artificial respiration; *avoid mouth-to-mouth resuscitation; use bag/mask apparatus.* IF breathing is difficult, administer oxygen. **Liquid:** Will burn skin and eyes. Harmful if swallowed. Remove contaminated clothing and shoes. Flush affected areas with plenty of water. *IF IN EYES*, hold eyelids open and flush with plenty of water. *IF SWALLOWED* and victim is *CONSCIOUS AND ABLE TO SWALLOW*, have victim drink 4 to 8 ounces of water. **Do NOT induce vomiting.**
Note to physician or authorized medical personnel: Medical observation is recommended for 24 to 48 hours after breathing overexposure, as pulmonary edema may be delayed. As first aid for pulmonary edema, consider administering a corticosteroid spray. Cigarette smoking may exacerbate pulmonary injury and should be discouraged for at least 72 hours following exposure.

HEALTH HAZARDS

Personal protective equipment (PPE): A-Level PPE. OSHA Table Z-1-A air contaminant. Chemical protective material(s) reported to have good to excellent resistance: butyl rubber, chlorinated polyethylene, polycarbonate, neoprene, Teflon®.
Recommendations for respirator selection: NIOSH/OSHA
125 ppm: SA:CF* (any supplied-air respirator operated in a continuous-flow mode); or PAPROV* [any powered, air-purifying respirator with organic vapor cartridge(s)]. *200 ppm:* CCRFOV [any air-purifying, full-facepiece respirator (gas mask) with a chin-style, front-or back-mounted acid gas canister]; or GMFOV [any air-purifying, full-facepiece respirator (gas mask) with a chin-style, front-or back-,mounted organic vapor canister]; or PAPRTOV* [any powered, air-purifying respirator with a tight-fitting facepiece and organic vapor cartridge(s)]; or SCBAF (any self-contained breathing apparatus with a full facepiece); or SAF (any supplied-air respirator with a full facepiece).*EMERGENCY OR PLANNED ENTRY INTO UNKNOWN CONCENTRATIONS OR IDLH CONDITIONS:* SCBAF:PD,PP (any self-contained breathing apparatus that has a full facepiece and is operated in a pressure-demand or other positive-pressure mode); or SAF:PD,PP:ASCBA (any supplied-air respirator that has a full facepiece and is operated in a pressure-demand or other positive-pressure mode in combination with an auxiliary self-contained breathing apparatus operated in a pressure-demand or other positive pressure mode).*ESCAPE:* GMFOV[any air-purifying, full-facepiece respirator (gas mask) with a chin-style, front- or back-mounted organic vapor canister]; or SCBAE (any appropriate escape-type, self-contained breathing apparatus). *Note:* Substance causes eye irritation or damage; eye protection needed.
Exposure limits (TWA unless otherwise noted): ACGIH TLV ceiling 5 ppm (21 mg/m^3); NIOSH REL ceiling 5 ppm (21 mg/m^3); OSHA PEL 5 ppm (20 mg/m^3).
Toxicity by ingestion: Grade 2; LD$_{50}$ = 0.5 to 5 g/kg (rat).
Long-term health effects: Dermatitis may result from prolonged or repeated contact.
Vapor (gas) irritant characteristics: Vapor is irritating such that personnel will not usually tolerate moderate or high vapor concentrations.
Liquid or solid irritant characteristics: Severe skin or eye irritant, may cause pain and second-degree burns after a few minutes of contact.

Odor threshold: Less than 0.14 ppm..
IDLH value: 200 ppm..

FIRE DATA

Flash point: 136°F/58°C (oc); 120°F/49°C (cc).
Flammable limits in air: LEL: 2.7%; UEL: 10.3%.
Fire extinguishing agents not to be used: Direct contact with water is not recommended. Water and foam react, but heat liberated may not be enough to create a hazard. Dry chemical forced below the surface can cause foaming and boiling. Use water spray to cool containers exposed to fire.
Autoignition temperature: 601°F/316°C.
Electrical hazard: Class I, Group D. Due to low electric conductivity, this substance may generate electrostatic charges as a result of agitation and flow.
Burning rate: 3.3 mm/min

CHEMICAL REACTIVITY

Reactivity with water: Reacts slowly with cold water to form acetic acid; considerable heat liberated when water spray is used. Corrosive in the presence of water or moisture.
Binary reactants: Reacts with alcohols, strong oxidizers (especially chromic acid), amines, alcohols, strong caustics. Attacks iron, steel, and other metals.
Neutralizing agents for acids and caustics: Dilute with water; use sodium bicarbonate solution to rinse.
Reactivity group: 11
Compatibility class: Organic anhydrides

ENVIRONMENTAL DATA

Water pollution: Harmful to aquatic life in very low concentrations. May be dangerous if it enters nearby water intakes; notify operators. Notify local health and wildlife officials. **Response to discharge:** Issue warning–corrosive. Disperse and flush.

SHIPPING INFORMATION

Grades of purity: Pure: 99% minimum; Technical: 75–98.5%.
Storage temperature: Ambient; **Inert atmosphere:** None;
Venting: Pressure-vacuum; **Stability during transport:** Stable

NAS HAZARD CLASSIFICATION FOR BULK WATER TRANSPORTATION

FIRE: 2
HEALTH: Vapor irritant: 3; Liquid or solid irritant: 3; Poisons: 3
WATER POLLUTION: Human toxicity: 2; Aquatic toxicity: 1; Aesthetic effect: 2
REACTIVITY: Other chemicals: 2; Water: 3; Self-reaction: 2

PHYSICAL AND CHEMICAL PROPERTIES

Physical state @ 59°F/15°C and 1 atm: Liquid.
Molecular weight: 102.09
Boiling point @ 1 atm: 282°F/139°C/412°K.
Melting/Freezing point: –99°F/–73°C/200°K.
Critical temperature: 565°F/296°C/569°K.
Critical pressure: 679 psia = 46.2 atm = 4.68 MN/m^2.
Specific gravity (water = 1): 1.08
Relative vapor density (air = 1): 3.5
Ratio of specific heats of vapor (gas): 1.093
Latent heat of vaporization: 119 Btu/lb = 66.2 cal/g = 2.77 x 10^5 J/kg.
Heat of combustion: –7058 Btu/lb = –3921 cal/g = –164.2 x 10^5 J/kg.
Vapor pressure: (Reid) 0.3 psia; 4 mm.

ACETONE
REC. A:0300

SYNONYMS: ACETONA (Spanish); DIMETHYLFORMALDEHYDE; DIMETHYLFORMEHYDE; DIMETHYLKETAL; DIMETHYL EEC No. 606-001-00-8; KETONE; KETONE, DIMETHYL; KETONE PROPANE; β-KETOPROPANE; METHYL KETONE; PROPANONE; 2-PROPANONE; PYROACETIC ACID; PYROACETIC ETHER; RCRA No. U002

IDENTIFICATION
CAS Number: 67-64-1
Formula: C_3H_6O; CH_3COCH_3
DOT ID Number: UN 1090; UN 1091; DOT Guide Number: 127
Proper Shipping Name: Acetone; Acetone oils
Reportable Quantity (RQ): **(CERCLA)** 5000 lb/2270 kg

DESCRIPTION: Colorless watery liquid. Smells like fingernail polish remover. Floats on water surface; soluble; flammable, large amount of irritating vapor is produced.

Highly flammable • Containers may BLEVE when exposed to fire • Vapors may form explosive mixture with air • Vapors are heavier than air and will collect and stay in low areas • Vapors may travel long distances to ignition sources and flashback • Vapors in confined areas (e.g., tanks, sewers, buildings) may explode when exposed to fire • Highly irritating to the eyes and respiratory tract; this material has anesthetic properties • Runoff from fire control may cause pollution • Toxic products of combustion may include carbon monoxide.

Hazard Classification (based on NFPA-704 Rating System)
Health Hazards (Blue): 1; Flammability (Red): 3; Reactivity (Yellow): 0

EMERGENCY RESPONSE: See Appendix A (127)
Evacuation:
Public safety: Isolate spill area for at least 25 to 50 meters (80 to 160 feet) in all directions.
Spill: Large spill–Consider initial downwind evacuation for at least 300 meters (1000 feet).
Fire: Isolate for 800 meters (½ mile) in all directions, especially if tank, rail car, or tank truck is involved in fire.

EXPOSURE
Short-term effects: *SEEK MEDICAL ATTENTION.* **Vapor:** Irritating to eyes, nose, and throat. *IF INHALED*, will, may cause difficult breathing or loss of consciousness. Move to fresh air. *IF BREATHING HAS STOPPED*, give artificial respiration. IF breathing is difficult, administer oxygen. **Liquid:** Irritating to eyes. Not irritating to skin. *IF IN EYES*, hold eyelids open and flush with plenty of water. The use of alcoholic beverages may enhance the toxic effect.

HEALTH HAZARDS
Personal protective equipment (PPE): B-Level PPE. OSHA Table Z-1-A air contaminant. Organic vapor canister or air–supplied mask; synthetic rubber gloves; chemical safety goggles or face splash shield.
Recommendations for respirator selection: NIOSH
2500 ppm: CCRFOV [any chemical cartridge respirator with a full facepiece and organic vapor cartridge(s)]; or PAPROV [any powered, air-purifying respirator with organic vapor cartridge(s)]; or GMFOV [any air-purifying, full-facepiece respirator (gas mask) with a chin-style, front- or back-mounted acid gas canister]; or SA (any supplied-air respirator); or SCBAF (any self-contained breathing apparatus with a full facepiece). *EMERGENCY OR PLANNED ENTRY INTO UNKNOWN CONCENTRATIONS OR IDLH CONDITIONS:* SCBAF:PD,PP (any self-contained breathing apparatus that has a full facepiece and is operated in a pressure-demand or other positive-pressure mode); or SAF:PD,PP:ASCBA (any supplied-air respirator that has a full facepiece and is operated in a pressure-demand or other positive-pressure mode in combination with an auxiliary self-contained breathing apparatus operated in a pressure-demand or other positive pressure mode). *ESCAPE:* GMFOV [any air-purifying, full-facepiece respirator (gas mask) with a chin-style, front- or back-mounted organic vapor canister]; or, SCBAE (any appropriate escape-type, self-contained breathing apparatus). *Note:* Substance reported to cause eye irritation or damage; may require eye protection.
Exposure limits (TWA unless otherwise noted): ACGIH TLV 500 ppm (1200 mg/m^3); OSHA PEL 1000 ppm (2400 mg/m^3); NIOSH REL 250 ppm (590 mg/m^3).
Short-term exposure limits (15-minute TWA): ACGIH STEL 750 ppm (1770 mg/m^3).
Toxicity by ingestion: Grade 1; LD_{50} = 5 to 15 g/kg (dog).
Long-term health effects: May effect the nervous system.
Vapor (gas) irritant characteristics: If present in high concentrations, vapors cause irritation of the eyes or respiratory system. Effect may be temporary.
Liquid or solid irritant characteristics: May cause irritation to the skin, eyes, and respiratory tract. It is very volatile and evaporates quickly from the skin.
Odor threshold: 3.5–650 ppm.
IDLH value: 2500 ppm.

FIRE DATA
Flash point: 4°F/–16°C (oc); –4°F/–20°C (cc).
Flammable limits in air: LEL: 2.5%; UEL: 12.8%.
Fire extinguishing agents not to be used: Water in straight hose stream will scatter and spread fire and should not be used.
Autoignition temperature: 869°F/465°C/738°K.
Electrical hazard: Class I, Group D. Due to low electric conductivity, this substance may generate electrostatic charges as a result of agitation and flow.
Burning rate: 3.9 mm/min

CHEMICAL REACTIVITY
Binary reactants: Oxidizers, acids, chloroform (a basic environment may cause a violent explosion).
Reactivity group: 18
Compatibility class: Ketone

ENVIRONMENTAL DATA
Food chain concentration potential: Log P_{ow} = –0.25. Unlikely to accumulate.
Water pollution: Dangerous to aquatic life in high concentrations. May be dangerous if it enters nearby water intakes; notify operators. Notify local health and pollution control officials. **Response to discharge:** Issue warning–high flammability. Disperse and flush.

SHIPPING INFORMATION
Grades of purity: Technical: 99.5% plus 0.5% water; Reagent: 99.5% plus 0.5% water; **Storage temperature:** Ambient; **Inert atmosphere:** None; **Venting:** Open (flame arrester) or pressure-vacuum; plastic; **Stability during transport:** Stable.

NAS HAZARD CLASSIFICATION FOR BULK WATER TRANSPORTATION
FIRE: 3

HEALTH: Vapor irritant: 1; Liquid or solid irritant: 0; Poisons: 0
WATER POLLUTION: Human toxicity: 1; Aquatic toxicity: 1; Aesthetic effect: 1
REACTIVITY: Other chemicals: 2; Water: 0; Self-reaction: 1

PHYSICAL AND CHEMICAL PROPERTIES
Physical state @ 59°F/15°C and 1 atm: Liquid.
Molecular weight: 58.08
Boiling point @ 1 atm: 133°F/56°C/329°K.
Melting/Freezing point: −138°F/−94.7°C/178.5°K.
Critical temperature: 455°F/235°C/508°K.
Critical pressure: 682 psia = 46.4 atm = 4.70 MN/m².
Specific gravity (water = 1): 0.791 @ 68°F/20°C (liquid).
Relative vapor density (air = 1): 2.0
Ratio of specific heats of vapor (gas): 1.127
Latent heat of vaporization: 220 Btu/lb = 122 cal/g = 5.11×10^5 J/kg.
Heat of combustion: −12,250 Btu/lb = −6808 cal/g = -285.0×10^5 J/kg.
Heat of fusion: 23.42 cal/g.
Vapor pressure: (Reid) 7.25 psia; 180 mm.

ACETONE CYANOHYDRIN   REC. A:0350

SYNONYMS: ACETONA, CIANHIDRINA de (Spanish); ACETONCYANHYDRIN (German); ACETONECYANHYDRINE (French); ACETONE CYANOHYDRIN; ACETONE CYANOHYDRIN; CYANHYDRINE D'ACETONE (French); 2-CYANO-2-PROPONAL; α-HYDROXY ISOBUTYRONITRILE; EEC No. 608-004-00-X; 2-HYDROXYISOBUTYRONITRILE; HYDROXY ISOBUTYRO NITRITE; 2-HYDROXY-2-METHYLPROPIONITRILE; 2-METHYLACTONITRILE; 2-PROPANE CYANOHYDRIN; PROPANENITRILE, 2-HYDROXY-2-METHYL-; RCRA No. P069

IDENTIFICATION
CAS Number: 75-86-5
Formula: C_4H_7NO; $(CH_3)_2C(OH)CN$
DOT ID Number: UN 1541; DOT Guide Number: 155
Proper Shipping Name: Acetone cyanohydrin, stabilized
Reportable Quantity (RQ): **(CERCLA)** 10 lb/4.54 kg

DESCRIPTION: Colorless watery liquid. Mild, almond odor (poor warning). Floats on water surface; soluble; reacts producing poisonous hydrogen cyanide vapor.

Poison! • Combustible • Exposure to heat produces cyanide gas • Breathing the vapor, skin or eye contact, or swallowing the material can kill you; produces cyanide in the body • Firefighting gear (including SCBA) does not provide adequate protection. If exposure occurs, remove and isolate gear immediately and thoroughly decontaminate personnel • Containers may BLEVE when exposed to fire • Vapors may form explosive mixture with air • Vapors are heavier than air and will collect and stay in low areas • Vapors in confined areas (e.g., tanks, sewers, buildings) may explode when exposed to fire • Toxic products of combustion may include carbon monoxide, nitrogen oxide, and hydrogen cyanide • Do not attempt rescue.

Hazard Classification (based on NFPA-704 Rating System)
Health Hazards (Blue): 4; Flammability (Red): 2; Reactivity (Yellow): 2

EMERGENCY RESPONSE: See Appendix A (155)

Evacuation:
Public safety: Isolate the area of spill or leak for at least 50 to 100 meters (160 to 330 feet) in all directions.
Spill: Small spill–First: Isolate in all directions, 95 meters (300 feet). Then: Protect persons downwind. DAY: 0.3 km (0.2 mile); NIGHT: 1.3 km (0.8 mile). Large spill–First: Isolate in all directions 240 meters (800 feet); Then: Protect persons downwind, DAY: 1.1 km (0.7 mile); NIGHT: 4.8 km (3.0 miles).
Fire: If tank, rail car, or tank truck is involved in fire, isolate for at least 800 meters (½ mile) in all directions; also, consider initial evacuation for 800 meters (½ mile) in all directions.

EXPOSURE
Short-term effects: *SEEK MEDICAL ATTENTION.* **Vapor or liquid:** *POISONOUS IF INHALED OR ABSORBED THROUGH THE SKIN.* Irritating to eyes, nose, and throat. Move to fresh air. *IF BREATHING HAS STOPPED,* give artificial respiration; *avoid mouth-to-mouth resuscitation; use bag/mask apparatus.* IF breathing is difficult, administer oxygen. **Liquid:** *POISONOUS IF SWALLOWED.* Will burn skin and eyes. Remove contaminated clothing and shoes. Flush affected areas with plenty of water. *IF SWALLOWED* and victim is *CONSCIOUS AND ABLE TO SWALLOW*, have victim drink 4 to 8 ounces of water and have victim induce vomiting. *IF SWALLOWED* and victim is *UNCONSCIOUS OR HAVING CONVULSIONS*, do nothing except keep victim warm.

Note to physician or authorized medical personnel: Consider the use of amyl nitrite perles if symptoms of cyanide poisoning develop. If symptoms indicate, initial treatment includes the cyanide antidote kit. In all cases, break an amyl nitrite perle in a gauze pad and hold lightly under victim's nose for 15 seconds, repeating 5 times at about 15-second intervals; if necessary (and if sodium nitrite infusions will be delayed), repeat procedure every 3 minutes with fresh perles until 3 or 4 have been used. Avoid breathing the vapor while administering it to the victim. Administer sodium nitrite IV, ASAP. The usual adult dose is 10 to 20 mL of a 3% solution infused over no less than 5 minutes; the average child dose is 0.15 to 0.20 mL/kg. Monitor blood pressure during administration, and slow the rate of infusion if hypotention develops. Next, infuse sodium thiosulfate IV. The usual adult dose is 50 mL of a 25% solution infused over 10 to 20 minutes; the average child dose is 1.65 mL/kg. Repeat with nitrite and thiosulfate as required.

HEALTH HAZARDS
Personal protective equipment (PPE): A-Level PPE. Air–supplied mask with canister approved for use with acrylonitrile in less than 2% concentrations; rubber or plastic gloves; cover goggles or face mask; rubber boots; slicker suit; safety helmet. Chemical protective material(s) reported to have good to excellent resistance: butyl rubber, Teflon®, polcarbonate.
Recommendations for respirator selection: NIOSH
10 ppm: SA (any supplied-air respirator). *25 ppm:* SA:CF (any supplied-air respirator operated in a continuous-flow mode). *50 ppm:* SCBAF (any self-contained breathing apparatus with a full facepiece); or SAF (any supplied-air respirator with a full facepiece). *250 ppm:* SAF:PD,PP (any supplied-air respirator that has a full facepiece and is operated in a pressure-demand or other positive-pressure mode). *EMERGENCY OR PLANNED ENTRY INTO UNKNOWN CONCENTRATIONS OR IDLH CONDITIONS:* SCBAF:PD,PP (any self-contained breathing apparatus that has a full facepiece and is operated in a pressure-demand or other positive-pressure mode); or SAF:PD,PP:ASCBA (any supplied-air respirator that has a full facepiece and is operated in a pressure-demand or other positive-pressure mode in combination with an

auxiliary self-contained breathing apparatus operated in a pressure-demand or other positive pressure mode). *ESCAPE:* GMFOV [any air-purifying, full-facepiece respirator (gas mask) with a chin-style, front- or back-mounted organic vapor canister]; or, SCBAE (any appropriate escape-type, self-contained breathing apparatus).
Exposure limits (TWA unless otherwise noted): AIHA WEEL 2 ppm; NIOSH REL ceiling 1 ppm (4 mg/m^3)/15 minutes; skin contact contributes significantly in overall exposure.
Short-term exposure limits (15-minute TWA): ACGIH ceiling 4.7 ppm (5 mg/m^3).
Toxicity by ingestion: Grade 4; LD_{50} below 50 mg/kg (mice).
Long-term health effects: Causes liver damage in rats.
Vapor (gas) irritant characteristics: Vapors irritate the eyes and respiratory system if present in high concentrations. The effect is temporary.
Liquid or solid irritant characteristics: Causes smarting of the skin and first-degree burns on short exposure and may cause secondary burns on long exposure.
Odor threshold: 3 ppm.
IDLH value: 50 mg/m^3 as cyanide.

FIRE DATA
Flash point: 165°F/74°C (cc).
Flammable limits in air: LEL: 2.25%; UEL 11%.
Fire extinguishing agents not to be used: Soda-acid extinguisher.
Autoignition temperature: 1270°F/688°C/961°K.
Electrical hazard: Class I, Group D.

CHEMICAL REACTIVITY
Reactivity with water: Decomposition forms hydrogen cyanide.
Binary reactants: Strong oxidizers, sulfuric acid, strong caustics, and reducing agents. Slowly decomposes to acetone and hydrogen cyanide at room temperatures, under normal storage conditions; rate is accelerated by an increase in pH (contact with alkalis), water content or temperature. Causes rubber to swell. Corrosive to carbon steel, stainless steel, Teflon®.
Reactivity group: 0
Compatibility class: Unassigned cargoes

ENVIRONMENTAL DATA
Water pollution: DOT Appendix B, §172.101–marine pollutant. Harmful to aquatic life in very low concentrations. May be dangerous if it enters nearby water intakes; notify operators. Notify local health and wildlife officials. **Response to discharge:** Issue warning–poison. Restrict access. Should be removed–human toxicity.

SHIPPING INFORMATION
Grades of purity: 98-99%; **Storage temperature:** Ambient; **Inert atmosphere:** None; **Venting:** Pressure-vacuum; **Stability during transport:** Stable.

NAS HAZARD CLASSIFICATION FOR BULK WATER TRANSPORTATION
FIRE: 1
HEALTH: Vapor irritant: 1; Liquid or solid irritant: 2; Poisons: 4
WATER POLLUTION: Human toxicity: 4; Aquatic toxicity: 3; Aesthetic effect: 3
REACTIVITY: Other chemicals: 2; Water: 3; Self-reaction: 0

PHYSICAL AND CHEMICAL PROPERTIES
Physical state @ 59°F/15°C and 1 atm: Liquid.
Molecular weight: 85.11
Boiling point @ 1 atm: Decomposes. 248°F/120°C/393°K.
Melting/Freezing point: –4°F/–20°C/253°K.
Specific gravity (water = 1): 0.925 @ 77°F/25°C (liquid).
Relative vapor density (air = 1): 2.9
Ratio of specific heats of vapor (gas): (estimate) 1.074
Vapor pressure: (Reid) 0.3 psia; 0.8 mm.

ACETONITRILE   REC. A:0400

SYNONYMS: ACETONITRILO (Spanish); CYANOMETHANE; CYANURE de METHYL (French); EEC No. 608-001-00-3; ETHANENITRILE; ETHYL NITRIL; ETHYL NITRILE; METHANECARBONITRIL; METHANECARBONITRILE; METHANE, CYANO-; METHYL CYANIDE; RCRA U00308

IDENTIFICATION
CAS Number: 75-05-8
Formula: C_2H_3N; CH_3CN
DOT ID Number: UN 1648; DOT Guide Number: 131
Proper Shipping Name: Acetonitrile
Reportable Quantity (RQ): **(CERCLA)** 5000 lb/2270 kg

DESCRIPTION: Colorless, watery liquid. Sweet, ether-like odor. Floats on water surface; soluble; flammable, irritating vapor is produced.

Poison! • Breathing the vapors, skin contact or swallowing the liquid can kill you; converted to cyanide in the body • Firefighting gear (including SCBA) does not provide adequate protection • If exposure occurs, remove and isolate gear immediately and thoroughly decontaminate personnel • Odor is not a reliable indicator of the presence of toxic amounts of the vapor • Highly flammable • Containers may BLEVE when exposed to fire • Vapors may form explosive mixture with air • Vapors are heavier than air and will collect and stay in low areas • Vapors may travel long distances to ignition sources and flashback • Vapors in confined areas (e.g., tanks, sewers, buildings) may explode when exposed to fire • Irritating to the skin, eyes, and respiratory tract • Toxic products of combustion may include hydrogen cyanide and nitrogen oxides • Do not put yourself in danger by entering a contaminated area to rescue a victim.

Hazard Classification (based on NFPA-704 Rating System)
Health Hazards (Blue): 2; Flammability (Red): 3; Reactivity (Yellow): 0

EMERGENCY RESPONSE: See Appendix A (131)
Evacuation:
Public safety: Isolate spill area for at least 100 to 200 meters (330 to 660 feet) in all directions.
Spill: Increase, as necessary, the isolation distance shown above, in "Public safety."
Fire: Isolate for 800 meters (½ mile) in all directions, especially if tank, rail car, or tank truck is involved in fire.

EXPOSURE
Short-term effects: *SEEK MEDICAL ATTENTION.* **Vapor:** Irritating to eyes, nose, and throat. *IF INHALED*, will, will cause difficult breathing. Move to fresh air. *IF BREATHING HAS STOPPED,* give artificial respiration; *avoid mouth-to-mouth resuscitation; use bag/mask apparatus.* IF breathing is difficult, administer oxygen. **Liquid:** *HARMFUL IF ABSORBED THROUGH THE SKIN.* Irritating to skin and eyes. Harmful if swallowed. Remove contaminated clothing and shoes. Flush affected areas with plenty of water. *IF IN EYES*, hold eyelids open and flush with plenty of water. *IF SWALLOWED* and victim is

CONSCIOUS AND ABLE TO SWALLOW, have victim drink 4 to 8 ounces of water.

Note to physician or authorized medical personnel: Consider the use of amyl nitrite perles if symptoms of cyanide poisoning develop. If symptoms indicate, initial treatment includes the cyanide antidote kit. In all cases, break an amyl nitrite perle in a gauze pad and hold lightly under victim's nose for 15 seconds, repeating 5 times at about 15-second intervals; if necessary (and if sodium nitrite infusions will be delayed), repeat procedure every 3 minutes with fresh perles until 3 or 4 have been used. Avoid breathing the vapor while administering it to the victim. Administer sodium nitrite IV, ASAP. The usual adult dose is 10 to 20 mL of a 3% solution infused over no less than 5 minutes; the average child dose is 0.15 to 0.20 mL/kg. Monitor blood pressure during administration, and slow the rate of infusion if hypotention develops. Next, infuse sodium thiosulfate IV. The usual adult dose is 50 mL of a 25% solution infused over 10 to 20 minutes; the average child dose is 1.65 mL/kg. Repeat with nitrite and thiosulfate as required.

HEALTH HAZARDS
Personal protective equipment (PPE): A-Level PPE. OSHA Table Z-1-A air contaminant. Chemical protective material(s) reported to have good to excellent resistance: butyl rubber, Teflon®
Recommendations for respirator selection: NIOSH
200 ppm: CCRFOV [any air-purifying, full-facepiece respirator (gas mask) with a chin-style, front-or back-mounted acid gas canister]; SA (any supplied-air respirator); or SCBA (any self-contained breathing apparatus); *500 ppm*: SA:CF (any supplied-air respirator operated in a continuous-flow mode); or PAPROV [any powered, air-purifying respirator with organic vapor cartridge(s)]; or CCRFOV [any chemical cartridge respirator with a full facepiece and organic vapor cartridge(s)]; or GMFOV [any air-purifying, full-facepiece respirator (gas mask) with a chin-style, front-or back-mounted acid gas canister]; or SCBAF (any self-contained breathing apparatus with a full facepiece); or SAF (any supplied-air respirator with a full facepiece). *EMERGENCY OR PLANNED ENTRY INTO UNKNOWN CONCENTRATIONS OR IDLH CONDITIONS:* SCBAF:PD,PP (any self-contained breathing apparatus that has a full facepiece and is operated in a pressure-demand or other positive-pressure mode); or SAF:PD,PP:ASCBA (any supplied-air respirator that has a full facepiece and is operated in a pressure-demand or other positive-pressure mode in combination with an auxiliary self-contained breathing apparatus operated in a pressure-demand or other positive pressure mode. *ESCAPE:* GMFOV [any air-purifying, full-facepiece respirator (gas mask) with a chin-style, front- or back-mounted organic vapor canister]; or SCBAE (any appropriate escape-type, self-contained breathing apparatus).
Exposure limits (TWA unless otherwise noted): ACGIH TLV 40 ppm (67 mg/m^3); OSHA PEL 40 ppm (70 mg/m^3); NIOSH 20 ppm (34 mg/m^3).
Short-term exposure limits (15-minute TWA): ACGIH STEL 60 ppm (101 mg/m^3).
Toxicity by ingestion: Grade 3; LD$_{50}$ = 50 to 500 mg/kg (guinea pig).
Vapor (gas) irritant characteristics: Vapors cause smarting of the eyes or respiratory system if present in high concentrations. Effect may be temporary.
Liquid or solid irritant characteristics: Irritates the eyes, skin, and respiratory tract. If spilled on clothing and allowed to remain, may cause smarting and reddening of the skin.
Odor threshold: 40 ppm (70 mg/m^3).
IDLH value: 500 ppm.

FIRE DATA
Flash point: 42°F/6°C (oc).
Flammable limits in air: LEL: 3.0%; UEL: 16.0%.
Fire extinguishing agents not to be used: Water may be ineffective.
Autoignition temperature: 975°F/524°C/797°K.
Electrical hazard: Class I, Group D. Due to low electric conductivity, this substance may generate electrostatic charges as a result of agitation and flow.
Burning rate: 2.7 mm/min

CHEMICAL REACTIVITY
Reactivity with water: Reacts with steam.
Binary reactants: Strong oxidizers; acids.
Reactivity group: 37
Compatibility class: Nitriles

ENVIRONMENTAL DATA
Food chain concentration potential: Log P$_{ow}$ = –0.33. Unlikely to accumulate.
Water pollution: DOT Appendix B, §172.101–marine pollutant. Dangerous to aquatic life in high concentrations. May be dangerous if it enters nearby water intakes; notify operators. Notify local health and wildlife officials. **Response to discharge:** Issue warning–high flammability. Disperse and flush.

SHIPPING INFORMATION
Storage temperature: Ambient; **Inert atmosphere:** None; **Venting:** Pressure-vacuum; **Stability during transport:** Stable

NAS HAZARD CLASSIFICATION FOR BULK WATER TRANSPORTATION
FIRE: 3
HEALTH: Vapor irritant: 1; Liquid or solid irritant: 1; Poisons: 3
WATER POLLUTION: Human toxicity: 2; Aquatic toxicity: 1; Aesthetic effect: 1
REACTIVITY: Other chemicals: 2; Water: 0; Self-reaction: 0

PHYSICAL AND CHEMICAL PROPERTIES
Physical state @ 59°F/15°C and 1 atm: Liquid.
Molecular weight: 41.05
Boiling point @ 1 atm: 179°F/81.6°C/354.8°K.
Melting/Freezing point: –50.3°F/–45.7°C/227.5°K.
Critical temperature: 526.5°F/274.7°C/547.9°K.
Critical pressure: 701 psia = 47.7 atm = 4.83 MN/m^2.
Specific gravity (water = 1): 0.787 @ 68°F/20°C (liquid).
Relative vapor density (air = 1): 1.4
Ratio of specific heats of vapor (gas): 1.192
Latent heat of vaporization: 313 Btu/lb = 174 cal/g = 7.29 x 10^5 J/kg.
Heat of combustion: –13,360 Btu/lb = –7420 cal/g = –310.7 x 10^5 J/kg.
Vapor pressure: (Reid) 0.02 psia; 73 mm.

ACETOPHENONE   **REC. A:0450**

SYNONYMS: ACETOFENONA (Spanish); ACETYL BENZENE; BENZOYL METHIDE HYPNONE; HYPNONE; KETONE METHYL PHENYL; METHYL PHENYL KETONE; 1-PHENYLETHANONE; PHENYL METHYL KETONE; RCRA No. U004

IDENTIFICATION
CAS Number: 98-86-2

10   Acetyl acetone

Formula: C_8H_8O; $C_6H_5COCH_3$
DOT ID Number: NA 1993; DOT Guide Number: 128
Proper Shipping Name: Combustible liquid, n.o.s.
Reportable Quantity (RQ): **(CERCLA)** 5000 lb/2270 kg

DESCRIPTION: Colorless liquid. Flowery, sweet odor, like orange blossoms and jasmine. Sinks slowly in water; slightly soluble.

Highly flammable • Containers may BLEVE when exposed to fire • Vapors may form explosive mixture with air • Vapors are heavier than air and will collect and stay in low areas • Vapors may travel long distances to ignition sources and flashback • Vapors in confined areas (e.g., tanks, sewers, buildings) may explode when exposed to fire • Irritating to the skin, eyes, and respiratory tract • May have a narcotic or hypnotic effect in high concentrations • Toxic products of combustion may include carbon monoxide.

Hazard Classification (based on NFPA-704 Rating System)
Health Hazards (Blue): 1; Flammability (Red): 2; Reactivity (Yellow): 0

EMERGENCY RESPONSE: See Appendix A (128)
Evacuation:
Public safety: Isolate spill area for at least 25 to 50 meters (80 to 160 feet) in all directions.
Spill: Large spill–Consider initial downwind evacuation for at least 300 meters (1000 feet).
Fire: Isolate for 800 meters (½ mile) in all directions, especially if tank, rail car, or tank truck is involved in fire.

EXPOSURE
Short-term effects: *SEEK MEDICAL ATTENTION.* **Liquid or solid:** Irritating to skin or eyes. Harmful if swallowed. Remove contaminated clothing and shoes. Flush affected areas with plenty of water. *IF IN EYES*, hold eyelids open and flush with plenty of water. *IF SWALLOWED* and victim is *CONSCIOUS AND ABLE TO SWALLOW*, have victim drink 4 to 8 ounces of water. The use of alcoholic beverages may enhance the toxic effect.

HEALTH HAZARDS
Personal protective equipment (PPE): B-Level PPE. Protect eyes and skin from direct contact.
Exposure limits (TWA unless otherwise noted): ACGIH TLV 10 ppm (49 mg/m^3).
Toxicity by ingestion: Grade 2; LD_{50} = 0.5 to 5 g/kg.
Long-term health effects: May damage the nervous system.
Vapor (gas) irritant characteristics: Irritates the eyes, skin, and respiratory tract.
Liquid or solid irritant characteristics: If spilled on clothing and allowed to remain, may cause smarting and reddening of the skin. Possible defatting of the skin.
Odor threshold: 0.0039–2.05 ppm.

FIRE DATA
Flash point: 180°F/82°C (oc); 170°F/77°C (cc).
Autoignition temperature: 1058°F/570°C/843°K.

ENVIRONMENTAL DATA
Food chain concentration potential: Log P_{ow} = 1.6.
Water pollution: Effect of low concentrations on aquatic life is unknown. Fouling to shoreline. *May be dangerous if it enters nearby water intakes; notify operators.* Notify local health and pollution control officials. **Response to discharge:** Issue warning–water contaminant. Disperse and flush.

SHIPPING INFORMATION
Grades of purity: Technical: 99+% acetophenone; **Stability during transport:** Stable

PHYSICAL AND CHEMICAL PROPERTIES
Physical state @ 59°F/15°C and 1 atm: Liquid.
Molecular weight: 120.15
Boiling point @ 1 atm: 395.1°F/201.7°C/474.9°K.
Melting/Freezing point: 67.5°F/19.7°C/292.9°K.
Critical temperature: 802°F/428°C/701°K.
Critical pressure: 560 pisa = 38 atm = 3.8 MN/m^2.
Specific gravity (water = 1): 1.028 @ 68°F/20°C (liquid).
Liquid surface tension: 12 dynes/cm = 0.012 N/m at 30°C.
Liquid water interfacial tension: (estimate) 40 dynes/cm = 0.04 N/m at 27°C.
Relative vapor density (air = 1): 4.1
Ratio of specific heats of vapor (gas): (estimate) 1.071
Latent heat of vaporization: 150 Btu/lb = 83.6 cal/g = 3.5 x 10^5 J/kg.
Heat of combustion: –14,850 Btu/lb = –8250 cal/g = –345.4 x 10^5 J/kg.
Vapor pressure: 0.99 mm.

ACETYL ACETONE   **REC. A:0500**

SYNONYMS: ACETILACETONA (Spanish); ACETOACETONE; DIACETYL METHANE; EEC No. 606-029-00-0; PENTANEDIONE; 2,4-PENTANEDIONE; PENTANE-2,4-DIONE

IDENTIFICATION
CAS Number: 123-54-6
Formula: $C_5H_8O_2$
DOT ID Number: UN 2310; DOT Guide Number: 127
Proper Shipping Name: Pentane-2,4-dione

DESCRIPTION: Colorless liquid. Unpleasant odor. Floats and mixes slowly with water. Flammable, irritating vapor is produced.

Highly flammable • Containers may BLEVE when exposed to fire • Vapors may form explosive mixture with air • Vapors are heavier than air and will collect and stay in low areas • Vapors may travel long distances to ignition sources and flashback • Vapors in confined areas (e.g., tanks, sewers, buildings) may explode when exposed to fire • Irritating to the skin, eyes, and respiratory tract • Toxic products of combustion may include carbon monoxide.

Hazard Classification (based on NFPA-704 Rating System)
Health Hazards (Blue): 2; Flammability (Red): 2; Reactivity (Yellow): 0

EMERGENCY RESPONSE: See Appendix A (127)
Evacuation:
Public safety: Isolate spill area for at least 25 to 50 meters (80 to 160 feet) in all directions.
Spill: Large spill–Consider initial downwind evacuation for at least 300 meters (1000 feet).
Fire: Isolate for 800 meters (½ mile) in all directions, especially if tank, rail car, or tank truck is involved in fire.

EXPOSURE
Short-term effects: *SEEK MEDICAL ATTENTION.* **Vapor:** Irritating to eyes. *IF INHALED*, will, will cause dizziness,

coughing, headache, or loss of consciousness. *IF IN EYES*, hold eyelids open and flush with plenty of water. *IF BREATHING HAS STOPPED*, give artificial respiration. IF breathing is difficult, administer oxygen. **Liquid:** Irritating to skin and eyes. Harmful if swallowed. Remove contaminated clothing and shoes. Flush affected areas with plenty of water. *IF IN EYES*, hold eyelids open and flush with plenty of water. *IF SWALLOWED* and victim is *UNCONSCIOUS OR HAVING CONVULSIONS*, do nothing except keep victim warm.

HEALTH HAZARDS
Personal protective equipment (PPE): B-Level PPE. Safety glasses; eye bath and safety shower; air–supplied mask for concentrations above 2%.
Toxicity by ingestion: Grade 2; oral LD_{50} = 1000 mg/kg (rat).
Long-term health effects: May cause damage to the nervous system.
Vapor (gas) irritant characteristics: May irritate the eyes, skin, and respiratory tract.
Liquid or solid irritant characteristics: May cause defatting of the skin.

FIRE DATA
Flash point: 105°F/41°C (oc); 93°F/34°C (cc).
Flammable limits in air: LEL: 2.4%; UEL: 11.6%.
Fire extinguishing agents not to be used: Water may be ineffective on fire.
Autoignition temperature: 644°F/340°C.
Burning rate: 3.6 mm/min

CHEMICAL REACTIVITY
Binary reactants: Reacts with strong oxidizers, organic acids, isocyanates and amines. May dissolve plastics. Forms explosive mixture with air at 93°F/34°C.

ENVIRONMENTAL DATA
Food chain concentration potential: Log P_{ow} = 0.14. Unlikely to accumulate.
Water pollution: Effect of low concentration on aquatic life is unknown. May be dangerous if it enters nearby water intakes; notify operators. Notify local health and wildlife officials. **Response to discharge:** Issue warning–water contaminant. Disperse and flush.

SHIPPING INFORMATION
Grades of purity: Commercial; **Storage temperature:** Ambient; **Inert atmosphere:** None; **Venting:** Open (flame arrester); **Stability during transport:** Stable.

PHYSICAL AND CHEMICAL PROPERTIES
Physical state @ 59°F/15°C and 1 atm: Liquid.
Molecular weight: 100.12
Boiling point @ 1 atm: 284.7°F/140.4°C/413.6°K.
Melting/Freezing point: −10.3°F/−23.5°C/249.7°K.
Specific gravity (water = 1): 0.975 @ 68°F/20°C.
Liquid surface tension: 31.2 dynes/cm = 0.0312 N/m @ 68°F/20°C.
Relative vapor density (air = 1): 3.45
Ratio of specific heats of vapor (gas): (estimate) 1.072
Latent heat of vaporization: 194 Btu/lb = 108 cal/g = 4.52×10^5 J/kg.
Heat of combustion: −11,070 Btu/lb = −6,150 cal/g = -257×10^5 J/kg.
Heat of solution: −11.5 Btu/lb = −6.4 cal/g = -0.27×10^5 J/kg.
Vapor pressure: 7.1 mm.

ACETYL BROMIDE   REC. A:0550

SYNONYMS: ACETIC ACID BROMIDE; ACETIC BROMIDE; ACETILO de BROMURA (Spanish); ETHANOYL BROMIDE

IDENTIFICATION
CAS Number: 506-96-7
Formula: C_2H_3BrO; CH_3COBr
DOT ID Number: UN 1716; DOT Guide Number: 156
Proper Shipping Name: Acetyl bromide
Reportable Quantity (RQ): **(CERCLA)** 5000 lb/2270 kg

DESCRIPTION: Colorless liquid. Sharp, acrid, unpleasant odor. Sinks in water; violent reaction forming acetic acid and toxic hydrogen bromide gas, with release of heat.

Poison! • Corrosive • Highly flammable • Do not use water or foam • Breathing the vapor, skin or eye contact, or swallowing the material can kill you • Firefighting gear (including SCBA) does not provide adequate protection. If exposure occurs, remove and isolate gear immediately and thoroughly decontaminate personnel • Containers may BLEVE when exposed to fire • Vapors may form explosive mixture with air • Vapors are heavier than air and will collect and stay in low areas • Vapors may travel long distances to ignition sources and flashback • Vapors in confined areas (e.g., tanks, sewers, buildings) may explode when exposed to fire • Toxic products of combustion may include nitrogen oxide • Attacks and corrodes most metals and wood in the presence of moisture • Do not put yourself in danger by entering a contaminated area to rescue a victim.

Hazard Classification (based on NFPA-704 Rating System)
Health Hazards (Blue): 3; Flammability (Red): 3; Reactivity (Yellow): 2; Special Notice (White): Water reactive

EMERGENCY RESPONSE: See Appendix A (156)
Evacuation:
Public safety: Isolate the area of spill or leak for at least 50 to 100 meters (160 to 330 feet) in all directions.
Spill: Increase, in the downwind direction, as necessary, the distance shown under "Public Safety."
Fire: If tank, rail car, or tank truck is involved in fire, isolate for at least 800 meters (½ mile) in all directions; also, consider initial evacuation for 800 meters (½ mile) in all directions.

EXPOSURE
Short-term effects: *SEEK MEDICAL ATTENTION*. **Vapor:** Irritating to eyes, nose, and throat. *IF INHALED*, will, will cause difficult breathing. *May cause lung edema; physical exertion will aggravate this condition*. *IF IN EYES*, hold eyelids open and flush with plenty of water. *IF BREATHING HAS STOPPED*, give artificial respiration; *avoid mouth-to-mouth resuscitation; use bag/mask apparatus*. IF breathing is difficult, administer oxygen. **Liquid:** Will burn skin and eyes. Harmful if swallowed. Remove contaminated clothing and shoes. Flush affected areas with plenty of water. *IF IN EYES*, hold eyelids open and flush with plenty of water. *IF SWALLOWED* and victim is *CONSCIOUS AND ABLE TO SWALLOW*, have victim drink 4 to 8 ounces of water. *IF SWALLOWED* and victim is *UNCONSCIOUS OR HAVING CONVULSIONS*, do nothing except keep victim warm.
Note to physician or authorized medical personnel: Medical observation is recommended for 24 to 48 hours after breathing overexposure, as pulmonary edema may be delayed. As first aid for pulmonary edema, consider administering a corticosteroid spray.

Cigarette smoking may exacerbate pulmonary injury and should be discouraged for at least 72 hours following exposure.

HEALTH HAZARDS
Personal protective equipment (PPE): A-Level PPE. Wear corrosive-resistant gloves and clothing to prevent any reasonable probability of skin contact. Contact lenses should not be worn when working with this chemical. Chemical protective material(s) reported to offer protection: Teflon®

Respirator: Where the potential for exposures to acetyl bromide exists, use a MSHA/NIOSH approved supplied-air respirator with a full facepiece operated in the positive pressure mode or with a full facepiece, hood, or helmet in the continuous flow mode, or use a MSHA/NIOSH approved self-contained breathing apparatus with a full facepiece operated in pressure-demand or other positive pressure mode.
Exposure limits (TWA unless otherwise noted): ACGIH TLV 0.1 ppm as bromine; OSHA PEL 0.1 ppm (0.7 mg/m^3) as bromine.
Short-term exposure limits (15-minute TWA): NIOSH STEL 0.3 ppm (2 mg/m^3) as bromine.
Toxicity by ingestion: Grade 2; oral rat LD$_{50}$ = 3,310 mg/kg (acetic acid). Decomposes violently in water, forming bromic acid and acetic acid).
Vapor (gas) irritant characteristics: Corrosive to the eyes, skin, and respiratory tract. Possible lung edema.
Liquid or solid irritant characteristics: Corrosive to the skin.
Odor threshold: 0.0005 ppm.
IDLH value: 3 ppm as bromine.

FIRE DATA
Fire extinguishing agents not to be used: Water; foam (vigorous reaction).

CHEMICAL REACTIVITY
Reactivity with water: Reacts violently, forming corrosive and toxic fumes of hydrogen bromide
Binary reactants: Reacts with air forming corrosive heavier-than-air hydrogen bromide vapors. Reacts violently with many materials including alcohols forming a corrosive mist. Attacks and corrodes wood and most metals in the presence of moisture. Flammable hydrogen gas may collect in enclosed spaces.
Neutralizing agents for acids and caustics: Flood with water, rinse with dilute sodium bicarbonate or soda ash solution.
Reactivity group: 34
Compatibility class: Esters.

ENVIRONMENTAL DATA
Food chain concentration potential: Negative; unlikely to accumulate.
Water pollution: Effect of low concentrations on aquatic life is unknown. May be dangerous if it enters nearby water intakes; notify operators. Notify local health and wildlife officials. **Response to discharge:** Issue warning–corrosive. Restrict access. Evacuate area. Disperse and flush.

SHIPPING INFORMATION
Grades of purity: Analytical; Commercial; **Storage temperature:** Ambient; **Inert atmosphere:** Padded; **Venting:** Pressure-vacuum; **Stability during transport:** Stable if protected from moisture.

PHYSICAL AND CHEMICAL PROPERTIES
Physical state @ 59°F/15°C and 1 atm: Liquid.
Molecular weight: 122.95
Boiling point @ 1 atm: 169°F/76°C/349°K.
Melting/Freezing point: –141.7°F/–96.5°C/176.7°K.
Specific gravity (water = 1): 1.66 at 16°C (liquid).
Relative vapor density (air = 1): 4.24
Ratio of specific heats of vapor (gas): (estimate) 1.144
Latent heat of vaporization: 106 Btu/lb = 59 cal/g = 2.5 x 10^5 J/kg.
Vapor pressure: 101.0 mm.

ACETYL CHLORIDE   REC. A:0600

SYNONYMS: ACETIC ACID CHLORIDE; ACETIC CHLORIDE; CLORURO de ACETILO (Spanish); EEC No. 607-011-00-5; ETHANOYL CHLORIDE; RCRA No. U006

IDENTIFICATION
CAS Number: 75-36-5
Formula: C$_2$H$_3$ClO
DOT ID Number: UN 1717; DOT Guide Number: 132
Proper Shipping Name: Acetyl chloride
Reportable Quantity (RQ): **(CERCLA)** 5000 lb/2270 kg

DESCRIPTION: Colorless to yellow liquid. Sharp, pungent odor. Floats on the surface of water; reacts violently, forming toxic hydrogen chloride and acetic acid; large amounts if irritating vapor is produced.

Poison! • Do not use water or foam directly on material • Breathing the vapor can kill you; skin or eye contact causes severe burns, impaired vision, or blindness • Firefighting gear (including SCBA) does not provide adequate protection. If exposure occurs, remove and isolate gear immediately and thoroughly decontaminate personnel • Containers may BLEVE when exposed to fire • Toxic products of combustion may include hydrogen chloride and phosgene • Attacks and corrodes most metals the presence of moisture, forming flammable hydrogen gas • Do not put yourself in danger by entering a contaminated area to rescue a victim.

Hazard Classification (based on NFPA-704 Rating System)
Health Hazards (Blue): 3; Flammability (Red): 3; Reactivity (Yellow): 2; Special Notice (White): Water reactive

EMERGENCY RESPONSE: See Appendix A (132)
Evacuation:
Public safety: Isolate spill area for at least 100 to 200 meters (330 to 660 feet) in all directions.
Spill: Increase, as necessary, the isolation distance shown above, in "Public safety."
Fire: Isolate for 800 meters (½ mile) in all directions, especially if tank, rail car, or tank truck is involved in fire.

EXPOSURE
Short-term effects: *SEEK MEDICAL ATTENTION*. **Vapor:** Irritating to eyes, nose, and throat. *IF INHALED*, will, will cause difficult breathing. May cause lung edema; physical exertion will aggravate this condition. Move victim to fresh air. If breathing is difficult, administer oxygen. **Liquid:** Will burn skin and eyes. Harmful if swallowed. Remove contaminated clothing and shoes. Flush affected areas with plenty of water. *IF IN EYES*, hold eyelids open and flush with plenty of water. *IF SWALLOWED* and victim is *CONSCIOUS AND ABLE TO SWALLOW*, have victim drink 4 to 8 ounces of water. **Do NOT induce vomiting**.
Note to physician or authorized medical personnel: Medical observation is recommended for 24 to 48 hours after breathing overexposure, as pulmonary edema may be delayed. As first aid for

pulmonary edema, consider administering a corticosteroid spray. Cigarette smoking may exacerbate pulmonary injury and should be discouraged for at least 72 hours following exposure.

HEALTH HAZARDS
Personal protective equipment (PPE): A-Level PPE. Wear sealed clothing, gloves, and approved respirator. Chemical protective material(s) reported to offer protection: Teflon®.

Recommendations for respirator selection: NIOSH/OSHA as hydrogen chloride.
50 ppm: CCRS* [any chemical cartridge respirator with cartridge(s) providing protection against the compound of concern)]; or GMFS [any air-purifying, full-facepiece respirator (gas mask) with a chin-style, front- or back-mounted canister providing protection against the compound of concern]; or PAPRS* [any powered, air-purifying respirator with cartridge(s) providing protection against the compound of concern]; or SA* (any supplied-air respirator); or SCBAF (any self-contained breathing apparatus with a full facepiece). *EMERGENCY OR PLANNED ENTRY INTO UNKNOWN CONCENTRATIONS OR IDLH CONDITIONS:* SCBAF:PD,PP (any self-contained breathing apparatus that has a full facepiece and is operated in a pressure-demand or other positive-pressure mode); or SAF:PD,PP:ASCBA (any supplied-air respirator that has a full facepiece and is operated in a pressure-demand or other positive-pressure mode in combination with an auxiliary self-contained breathing apparatus operated in a pressure-demand or other positive pressure mode). *ESCAPE:* GMFAG [any air-purifying, full-facepiece respirator (gas mask) with a chin-style, front- or back-mounted organic vapor cannister]; or SCBAE (any appropriate escape-type, self-contained breathing apparatus).
Note: Substance reported to cause eye irritation or damage; may require eye protection.
Exposure limits (TWA unless otherwise noted): ACGIH TLV 5 ppm; NIOSH OSHA ceiling 5 ppm (7 mg/m^3) as hydrochloric acid.
Toxicity by ingestion: Readily hydrolyzes to form hydrochloric and acetic acids. Oral human LD_{LO} = 1470 mg/kg (acetic acid). Grade 2; oral rat LD_{50} = 3310 mg/kg (acetic acid).
Long-term health effects: Dermatitis may result from prolonged or repeated skin contact. Possible lung damage.
Vapor (gas) irritant characteristics: Vapors cause severe irritation of eyes and throat and can cause eye and lung injury. They cannot be tolerated even at low concentrations.
Liquid or solid irritant characteristics: Severe skin irritant. Causes second-and third-degree burns on short contact and is very injurious to the eyes.
Odor threshold: Acetic acid: 1 ppm; hydrochloric acid: 1 ppm.
IDLH value: 50 ppm as hydrochloric acid.

FIRE DATA
Flash point: 40°F/4°C (cc).
Flammable limits in air: LEL: 7.3%; UEL: 19%.
Fire extinguishing agents not to be used: Water, foam.
Autoignition temperature: 734°F/390°C/663°K.
Electrical hazard: Class I, Group D.
Burning rate: 2.6 mm/min

CHEMICAL REACTIVITY
Reactivity with water: Violent decomposition, evolving hydrogen chloride and acetic acid.
Binary reactants: Highly corrosive to most metals in the presence of moisture. Reacts with strong oxidizers, bases, dimethylsulfoxide, metal powders, alcohol and amines. Reacts with air to form corrosive acid fumes.

Neutralizing agents for acids and caustics: Following dilution with water, limestone or sodium bicarbonate can be used.

ENVIRONMENTAL DATA
Food chain concentration potential: Negative; unlikely to accumulate.
Water pollution: Effect of low concentrations on aquatic life is unknown. May be dangerous if it enters nearby water intakes; notify operators. Notify local health and wildlife officials.

Response to discharge: Issue warning–high flammability, corrosive. Restrict access. Disperse and flush.

SHIPPING INFORMATION
Grades of purity: Commercial; **Storage temperature:** Ambient; **Inert atmosphere:** None; **Venting:** None; **Stability during transport:** Stable

NAS HAZARD CLASSIFICATION FOR BULK WATER TRANSPORTATION
FIRE: 3
HEALTH: Vapor irritant: 4; Liquid or solid irritant: 4; Poisons: 3
WATER POLLUTION: Human toxicity: 2; Aquatic toxicity: 3; Aesthetic effect: 2
REACTIVITY: Other chemicals: 3; Water: 4; Self-reaction: 0

PHYSICAL AND CHEMICAL PROPERTIES
Physical state @ 59°F/15°C and 1 atm: Liquid.
Molecular weight: 78.5
Boiling point @ 1 atm: 124°F/51°C/324°K.
Melting/Freezing point: –170°F/–112°C/161°K.
Critical temperature: (estimate) 475°F/246°C/519°K.
Critical pressure: (estimate) 845 psia = 57.5 atm = 5.83 MN/m^2.
Specific gravity (water = 1): 1.1039 at 21°C (liquid).
Liquid surface tension: 26 dynes/cm = 0.026 N/m @ 68°F/20°C.
Relative vapor density (air = 1): 2.7
Ratio of specific heats of vapor (gas): 1.1467
Latent heat of vaporization: 160 Btu/lb = 88 cal/g = 3.7 x 10^5 J/kg.
Heat of combustion: –6000 Btu/lb = –3300 cal/g = –140 x 10^5 J/kg.
Heat of solution: (estimate) –54 Btu/lb = –30 cal/g = –1.3 x 10^5 J/kg.
Vapor pressure: 213 mm.

ACETYLENE   REC. A:0650

SYNONYMS: ACETILENO (Spanish); ACETYLEN; EEC No. 601-015-00-0; ETHINE; ETHYNE; NARCYLEN; WELDING GAS

IDENTIFICATION
CAS Number: 74-86-2
Formula: C_2H_2
DOT ID Number: UN 1001; DOT Guide Number: 116
Proper Shipping Name: Acetylene, dissolved

DESCRIPTION: Colorless compressed gas. Not shipped by tank car. Often shipped and stored dissolved in acetone. Mild garlic or ether-like odor. Only shipped in cylinders that cannot be used for any other commodity; cylinders have a *fusible plug* not a relief valve.

Extremely flammable • Forms explosive mixture with air • Gas may cause dizziness or asphyxiation without warning • Contact with liquid may cause frostbite • Reacts explosively with many materials • Gas is slightly lighter than air, and will disperse slowly unless confined • Gas may travel to source of ignition and flashback • Exposure of cylinders to elevated temperatures, fire, and flame may cause cylinders to rupture or cause frangible disk to burst, releasing entire contents of cylinder • Ruptured or venting cylinders may rocket through buildings and/or travel a considerable distance.

Hazard Classification (based on NFPA-704 Rating System)
Health Hazards (Blue): 0; Flammability (Red): 4; Reactivity (Yellow): 3

EMERGENCY RESPONSE: See Appendix A (116)
Evacuation:
Public safety: Isolate spill area for at least 100 meters (330 feet) in all directions.
Spill: Consider initial downwind evacuation for at least 800 meters (½ mile).
Fire: Isolate for 1600 meters (1 mile) in all directions, especially if tank, rail car, or tank truck is involved in fire.

EXPOSURE
Short-term effects: *SEEK MEDICAL ATTENTION.* **Gas:** Not irritating to eyes, nose or throat. *IF INHALED*, will, will cause headache, difficult breathing, or loss of consciousness. Move to fresh air. *IF BREATHING HAS STOPPED*, give artificial respiration. IF breathing is difficult, administer oxygen.

HEALTH HAZARDS
Personal protective equipment (PPE): B-Level PPE. Wear thermal protective clothing.
Exposure limits (TWA unless otherwise noted): NIOSH ceiling 2500 ppm (2,662 mg/m^3); ACGIH TLV simple asphyxiant.
Long-term health effects: May cause damage to the nervous system.
Vapor (gas) irritant characteristics: High concentrations may cause suffocation and death.
Liquid or solid irritant characteristics: Rapid evaporation of liquid may cause frostbite.

FIRE DATA
Flash point: Flammable gas.
Flammable limits in air: LEL: 2.5%; UEL: 82%.
Fire extinguishing agents not to be used: CO$_2$, dry chemical and water.
Fire extinguishing agents not to be used: Spray are not generally recommended because the discharged gas or volatile liquid may create a more serious explosion hazard.
Behavior in fire: May explode in fire
Autoignition temperature: 581°F/305°C/578°K.
Electrical hazard: Class I, Group A. Due to low electric conductivity, this substance may generate electrostatic charges as a result of agitation and flow.
Adiabatic flame temperature: 2907°F/1597°C (estimate).
Stoichiometric air-to-fuel ratio: 13.18 (estimate).

CHEMICAL REACTIVITY
Binary reactants: Forms an explosive mixture with air. Under certain conditions forms spontaneously explosive (and shock-sensitive) acetylide compounds with copper, mercury, silver, and brasses (containing more than 66% copper). Reacts with zinc, oxygen, and other oxidizing agents such as fluorine, chlorine, bromine, and halogens.

ENVIRONMENTAL DATA
Water pollution: Not harmful to aquatic life. **Response to discharge:** Issue warning–high flammability.

SHIPPING INFORMATION
Grades of purity: Commercial grade acetylene is supplied dissolved in acetone under pressure in cylinders; **Stability during transport:** Stable as shipped.

PHYSICAL AND CHEMICAL PROPERTIES
Physical state @ 59°F/15°C and 1 atm: Gas
Molecular weight: 26.04
Boiling point @ 1 atm: –119°F/84.0°C/189.2°K (sublimes).
Melting/Freezing point: –118°F/–83°C/190.2°K (sublimes).
Critical temperature: 95.3°F/35.2°C/308.4°K.
Critical pressure: 890.7 psia = 60.59 atm = 6.138 MN/m^2.
Specific gravity (water = 1): 0.613 at –80°C (liquid).
Relative vapor density (air = 1): 0.9
Ratio of specific heats of vapor (gas): 1.235
Heat of combustion: –20,747 Btu/lb = –11,526 cal/g = –482.57 x 10^5 J/kg.
Vapor pressure: 33.9 mm; 44.2 atm.

ACETYL PEROXIDE SOLUTION   REC. A:0700

SYNONYMS: ACETYL PEROXIDE; DIACETYL PEROXIDE SOLUTION; PEROXIDO de ACETILO (Spanish)

IDENTIFICATION
CAS Number: 110-22-5
Formula: C$_4$H$_6$O$_4$; CH$_3$CO·OOCO·CH$_3$
DOT ID Number: UN 2084; DOT Guide Number: 148
Proper Shipping Name: Acetyl peroxide, solid, or with more than 25% in solution

DESCRIPTION: Colorless liquid. Sharp, pungent odor. Sinks in water.
Note: Do not allow substance to warm up above 86°F/30°C. Material must be kept below 80.6°F/27°C. Obtain liquid nitrogen, dry ice or ice for cooling. If none can be obtained, evacuate area immediately.

Flammable • Thermally unstable; may explode from heat, shock, or contamination •Strong oxidizer which may react spontaneously with low flash point organics or reducing agents. Heat forms oxygen; will increase the activity of an existing fire • May explode on contact with combustibles (wood, paper, oil, clothing, etc.) • Containers may BLEVE when exposed to fire • Vapors are heavier than air and will collect and stay in low areas • Vapors may travel long distances to ignition sources and flashback • Irritating to the skin, eyes, and respiratory tract • Toxic products of combustion may include CO$_2$ and carbon monoxide *Note*: The dry material forms explosive mixtures.

Hazard Classification (based on NFPA-704 Rating System)
Health Hazards (Blue): 1; Flammability (Red): 2; Reactivity (Yellow): 4

EMERGENCY RESPONSE: See Appendix A (148)
Evacuation:
Public safety: Isolate the area of spill or leak for at least 50 to 100 meters (160 to 330 feet) in all directions.
Spill: Consider initial evacuation for at least 250 meters (800 feet).

Fire: If tank, rail car, or tank truck is involved in fire, isolate for at least 800 meters (½ mile) in all directions; also, consider initial evacuation for 800 meters (½ mile) in all directions.

EXPOSURE
Short-term effects: *SEEK MEDICAL ATTENTION*. **Vapor:** Irritating to eyes, nose, and throat. Move victim to fresh air. **Liquid:** Irritating to skin and eyes. Harmful if swallowed. Remove contaminated clothing and shoes. Flush affected areas with plenty of water. *IF IN EYES*, hold eyelids open and flush with plenty of water. *IF SWALLOWED* and victim is *CONSCIOUS AND ABLE TO SWALLOW*, have victim drink 4 to 8 ounces of water and have victim induce vomiting. *IF SWALLOWED* and victim is *UNCONSCIOUS OR HAVING CONVULSIONS*, do nothing except keep victim warm.

HEALTH HAZARDS
Personal protective equipment (PPE): B-Level PPE. Chemical protective material(s) reported to have good to excellent resistance: polycarbonate, chlorinated polyethylene, butyl rubber.
Vapor (gas) irritant characteristics: May cause eye and respiratory irritation.
Liquid or solid irritant characteristics: May cause irritation of the eyes and skin.

FIRE DATA
Flash point: 113°F/45°C (oc).
Behavior in fire: May explode. Burns with accelerating intensity.
Autoignition temperature: Explodes

CHEMICAL REACTIVITY
Binary reactants: May ignite combustible materials such as wood.

ENVIRONMENTAL DATA
Food chain concentration potential: Negative; unlikely to accumulate.
Water pollution: Effect of low concentrations on aquatic life is unknown. May be dangerous if it enters nearby water intakes; notify operators. Notify local health and wildlife officials. **Response to discharge:** Issue warning–oxidizing material, water contaminant. Should be removed. Chemical and physical treatment.

SHIPPING INFORMATION
Grades of purity: 25% acetyl peroxide; 75% dimethyl phthalate; **Storage temperature:** Unstable above 50°F/10°C/283°K; **Inert atmosphere:** None; **Venting:** Pressure-vacuum; **Stability during transport:** Heat- and shock-sensitive crystals may separate at very low temperatures during transport.

PHYSICAL AND CHEMICAL PROPERTIES
Physical state @ 59°F/15°C and 1 atm: Liquid.
Molecular weight: Mixture
Boiling point @ 1 atm: (decomposes) 145°F/63°C/336°K @ 21 mm.
Melting/Freezing point: 18°F/–8°C/265°K.
Specific gravity (water = 1): 1.2 @ 68°F/20°C (liquid).
Liquid surface tension: (estimate) 30 dynes/cm = 0.030 N/m @ 68°F/20°C.
Liquid water interfacial tension: (estimate) 30 dynes/cm = 0.030 N/m @ 68°F/20°C.
Relative vapor density (air = 1): 4.07
Heat of combustion: (estimate) –15,700 Btu/lb = –8750 cal/g = –366 x 10^5 J/kg.
Heat of decomposition: (estimate) –50 Btu/lb = –28 cal/g = –1.2 x 10^5 J/kg.

ACRIDINE   REC. A:0750

SYNONYMS: ACRIDINA (Spanish); 9-AZAANTHRACENE; 10-AZAANTHRACENE; BENZO[*b*]QUINOLINE; 2,3-BENZOQUINOLINE; DIBENZO[*b,e*]PYRIDINE

IDENTIFICATION
CAS Number: 260-94-6
Formula: $C_{13}H_9N$
DOT ID Number: UN 2713; DOT Guide Number: 153
Proper Shipping Name: Acridine

DESCRIPTION: Colorless or slightly yellow crystalline solid. Weak irritating odor. Sinks in water; practically insoluble in cold.

Poison! • Combustible solid • Breathing the dust, skin or eye contact, or swallowing the material can kill you • Firefighting gear (including SCBA) does not provide adequate protection. If exposure occurs, remove and isolate gear immediately and thoroughly decontaminate personnel • Containers may BLEVE when exposed to fire • Concentrated dust in confined areas (e.g., tanks, sewers, buildings) may explode when exposed to fire • Toxic products of combustion may include nitrogen oxide • Do not put yourself in danger by entering a contaminated area to rescue a victim.

Hazard Classification (based on NFPA-704 Rating System)
Health Hazards (Blue): 2; Flammability (Red): 2; Reactivity (Yellow): 0

EMERGENCY RESPONSE: See Appendix A (153)
Evacuation:
Public safety: Isolate the area of spill or leak for at least 25 to 50 meters (80 to 160 feet) in all directions.
Spill: Increase, in the downwind direction, as necessary, the distance shown under "Public Safety."
Fire: If tank, rail car, or tank truck is involved in fire, isolate for at least 800 meters (½ mile) in all directions; also, consider initial evacuation for 800 meters (½ mile) in all directions.

EXPOSURE
Short-term effects: *SEEK MEDICAL ATTENTION*. **Dust:** Irritating to eyes, nose, and throat. *IF INHALED*, will, will cause coughing or difficult breathing. *IF IN EYES*, hold eyelids open and flush with plenty of water. *IF BREATHING HAS STOPPED*, give artificial respiration; *avoid mouth-to-mouth resuscitation; use bag/mask apparatus*. IF breathing is difficult, administer oxygen. **Solid:** Irritating to skin and eyes. Harmful if swallowed. Remove contaminated clothing and shoes. Flush affected areas with plenty of water. *IF IN EYES*, hold eyelids open and flush with plenty of water. *IF SWALLOWED* and victim is *UNCONSCIOUS OR HAVING CONVULSIONS*, do nothing except keep victim warm.

HEALTH HAZARDS
Personal protective equipment (PPE): A-Level PPE.
Exposure limits (TWA unless otherwise noted): OSHA PEL 0.2 mg/m^3.
Toxicity by ingestion: Grade 2; oral rat LD_{50} = 2000 mg/kg.

FIRE DATA
Flash point: Combustible solid.
Fire extinguishing agents not to be used: Water may not be effective.
Behavior in fire: Sublimes before melting.

ENVIRONMENTAL DATA
Food chain concentration potential: Negative; unlikely to accumulate.
Water pollution: Harmful to aquatic life in very low concentrations. May be dangerous if it enters nearby water intakes; notify operators. Notify local health and wildlife officials.
Response to discharge: Issue warning–water contaminant. Mechanical containment. Should be removed. Chemical and physical treatment.

SHIPPING INFORMATION
Grades of purity: Commercial; **Storage temperature:** Ambient; **Inert atmosphere:** None; **Venting:** Open; **Stability during transport:** Stable.

PHYSICAL AND CHEMICAL PROPERTIES
Physical state @ 59°F/15°C and 1 atm: Solid.
Molecular weight: 179.08
Boiling point @ 1 atm: 655°F/346°C/619°K.
Melting/Freezing point: 230°F/110°C/383°K.
Specific gravity (water = 1): (approximate) 1.2 @ 68°F/20°C (solid).
Heat of combustion: $-15,800$ Btu/lb = -8790 cal/g = -368×10^5 J/kg.

ACROLEIN   REC. A:0800

SYNONYMS: ACREHYDE; ACROLEINA (Spanish); ACROLEINE (French); ACRYLEHYDE; ACRYLIC ALDEHYDE; ACRYLALDEHYDE; ACRYLIC EHYDE; ALLYLALDEHYDE; ALLYLEHYDE; EEC No. 607-061-00-8; EHYDE ACRYLIQUE (French); ALLYL EHYDE; AQUALIN; AQUALINE; BIOCIDE; ETHYLENE ALDEHYDE; ETHYLENE EHYDE; PROPENAL; 2-PROPENAL; PROP-2-EN-1-AL; 2-PROPEN-1-ONE; RCRA No. P003; SLIMICIDE

IDENTIFICATION
CAS Number: 107-02-8
Formula: C_3H_4O; $CH_2=CH \cdot CHO$
DOT ID Number: UN 1092; DOT Guide Number: 131P
Proper Shipping Name: Acrolein, inhibited
Reportable Quantity (RQ): **(CERCLA)** 1 lb/0.454 kg

DESCRIPTION: Colorless to light yellow watery liquid. Sharp, irritating odor. Floats on the surface of water; soluble. Poisonous, flammable vapor is produced.

Poison! • Highly flammable • Corrosive to the skin, eyes, and respiratory tract; contact with skin or eyes can cause burns and vision impairment; inhalation or ingestion can kill you; inhalation symptoms may be delayed; this material has anesthetic properties • Firefighting gear (including SCBA) does not provide adequate protection. If exposure occurs, remove and isolate gear immediately and thoroughly decontaminate personnel • Containers may BLEVE when exposed to fire • Vapors may form explosive mixture with air • Vapors are heavier than air and will collect and stay in low areas • Vapors may travel long distances to ignition sources and flashback • Vapors in confined areas (e.g., tanks, sewers, buildings) may explode when exposed to fire • Toxic products of combustion may include carbon monoxide and poisonous acrolein • Do not put yourself in danger by entering a contaminated area to rescue a victim. *Warning*: Inhalation exposures of 100 ppm (10 minutes) can be fatal.

Hazard Classification (based on NFPA-704 Rating System)
Health Hazards (Blue): 4; Flammability (Red): 3; Reactivity (Yellow): 3

EMERGENCY RESPONSE: See Appendix A (131)
Evacuation:
Public safety: Isolate spill area for at least 100 to 200 meters (330 to 660 feet) in all directions.
Spill: Small spill–First: Isolate in all directions 125 meters (400 feet); Then: Protect persons downwind, DAY: 0.5 km (0.3 mile); NIGHT: 2.3 km (1.4 miles). Large spill–First: Isolate in all directions 305 meters (1000 feet); Then: Protect persons downwind, DAY: 1.9 km (1.2 miles); NIGHT: 8.4 km (5.2 miles).
Fire: Isolate for 800 meters (½ mile) in all directions, especially if tank, rail car, or tank truck is involved in fire.
EXPOSURE
Short-term effects: *SEEK MEDICAL ATTENTION.*
Vapors: *VAPOR POISONOUS IF INHALED.* Irritating to eyes, nose, and throat. May cause lung edema; physical exertion will aggravate this condition. Move victim to fresh air. *IF IN EYES*, hold eyelids open and flush with plenty of water. *IF BREATHING HAS STOPPED,* give artificial respiration. IF breathing is difficult, administer oxygen. **Liquids:** *POISONOUS IF SWALLOWED OR ABSORBED THROUGH THE SKIN.* Will burn eyes. Irritating to skin. Remove contaminated clothing and shoes. Flush affected areas with plenty of water. *IF IN EYES*, hold eyelids open and flush with plenty of water. *IF SWALLOWED* and victim is *CONSCIOUS AND ABLE TO SWALLOW*, have victim drink 4 to 8 ounces of water and have victim induce vomiting. *IF SWALLOWED* and victim is *UNCONSCIOUS OR HAVING CONVULSIONS,* do nothing except keep victim warm.
Note to physician or authorized medical personnel: Medical observation is recommended for 24 to 48 hours after breathing overexposure, as pulmonary edema may be delayed. As first aid for pulmonary edema, consider administering a corticosteroid spray. Cigarette smoking may exacerbate pulmonary injury and should be discouraged for at least 72 hours following exposure. No oils or ointments should be used unless ordered by the physician.

HEALTH HAZARDS
Personal protective equipment (PPE): A-Level PPE. OSHA Table Z-1-A air contaminant. Chemical protective material(s) reported to have good to excellent resistance: butyl rubber, polycarbonate, Viton®/chlorobutyl.
Recommendations for respirator selection: OSHA/NIOSH
2 ppm: SA:CF (any supplied-air respirator operated in a continuous-flow mode); or PAPROV [any powered, air-purifying respirator with organic vapor cartridge(s)]; or CCRFOV [any air-purifying, full-facepiece respirator (gas mask) with a chin-style, front-or back-mounted acid gas canister]; or GMFOV [any air-purifying, full-facepiece respirator (gas mask) with a chin-style, front-or back-mounted organic vapor canister]; or SCBAF (any self-contained breathing apparatus with a full facepiece); or SAF (any supplied-air respirator with a full facepiece). *EMERGENCY OR PLANNED ENTRY INTO UNKNOWN CONCENTRATIONS OR IDLH CONDITIONS:* SCBAF:PD,PP (any self-contained breathing apparatus that has a full facepiece and is operated in a pressure-demand or other positive-pressure mode); or SAF:PD,PP:ASCBA (any supplied-air respirator that has a full facepiece and is operated in a pressure-demand or other positive-pressure mode in combination with an auxiliary, self-contained breathing apparatus operated in a pressure-demand or other positive pressure mode. *ESCAPE:* GMFOV [any air-purifying, full-facepiece respirator (gas mask) with a chin-style, front-or back-mounted organic vapor canister]; or SCBAE (any appropriate

escape-type, self-contained breathing apparatus). *Note:* Substance reported to causes eye irritation or damage; may require eye protection.
Exposure limits (TWA unless otherwise noted): OSHA PEL 0.1 ppm (0.25 mg/m^3); NIOSH REL 0.1 ppm (0.25 mg/m^3).
Short-term exposure limits (15-minute TWA): ACGIH STEL ceiling 0.1 ppm (0.23 mg/m^3); NIOSH STEL 0.3 ppm (0.8 mg/m^3).
Toxicity by ingestion: Grade 4; LD$_{50}$ in rats below 50 mg/kg.
Long-term health effects: Grade 4; oral rat LD$_{50}$ = 46 mg/kg Grade 4; oral rabbit LD$_{50}$ = 7 mg/kg; IARC rating 3
Vapor (gas) irritant characteristics: Vapors cause severe irritation of eyes and throat and can cause eye and lung injury. They cannot be tolerated even at low concentrations.
Liquid or solid irritant characteristics: Causes smarting of the skin and first-degree burns on short exposure; may cause second-degree burns on long exposure.
Odor threshold: 0.21 ppm.
IDLH value: 2 ppm.

FIRE DATA
Flash point: < 0°F/< –18°C (oc); –14°F/–26°C (cc).
Flammable limits in air: LEL: 2.8%; UEL 31%.
Fire extinguishing agents not to be used: Water may be ineffective, but water spray can be used to cool containers.
Behavior in fire: distance to a source of ignition and flashback. Polymerization may take place, and containers may rupture and explode in fire.
Autoignition temperature: 428°F/220°C/493.2°K.
Electrical hazard: Class I, Group B (C). Due to low electric conductivity, this substance may generate electrostatic charges as a result of agitation and flow.
Burning rate: 3.8 mm/min

CHEMICAL REACTIVITY
Binary reactants: Oxidizers, acids, alkalis, ammonia, amines, peroxides, zinc and chromium. Shock-sensitive peroxides may be formed over time.
Polymerization: Undergoes uncatalyzed polymerization reaction around 392°F/200°C. Light promotes polymerization; **Inhibitor of polymerization:** Hydroquinone: 0.10 to 0.25%.
Reactivity group: 19
Compatibility class: Aldehyde

ENVIRONMENTAL DATA
Food chain concentration potential: Log P$_{ow}$ = –0.01. Unlikely to accumulate.
Water pollution: DOT Appendix B, §172.101–marine pollutant. Harmful to aquatic life in very low concentrations. May be dangerous if it enters nearby water intakes; notify operators. Notify local health and wildlife officials. **Response to discharge:** Issue warning–high flammability, air contaminant. Restrict access. Evacuate area. Disperse and flush.

SHIPPING INFORMATION
Grades of purity: Industrial, 92+%; **Storage temperature:** Ambient; **Inert atmosphere:** None; **Venting:** Pressure-vacuum; **Stability during transport:** Stable when inhibited.

NAS HAZARD CLASSIFICATION FOR BULK WATER TRANSPORTATION
FIRE: 3
HEALTH: Vapor irritant: 4; Liquid or solid irritant: 3; Poisons: 4
WATER POLLUTION: Human toxicity: 4; Aquatic toxicity: 3; Aesthetic effect: 3
REACTIVITY: Other chemicals: 3; Water: 0; Self-reaction: 3

PHYSICAL AND CHEMICAL PROPERTIES
Physical state @ 59°F/15°C and 1 atm: Liquid.
Molecular weight: 56.1
Boiling point @ 1 atm: 127°F/53°C/326°K.
Melting/Freezing point: –125°F/–87°C/186°K.
Critical temperature: (estimate) 489°F/254°C/527°K.
Critical pressure: (estimate) 737 psia = 50.0 atm = 5.08 MN/m^2.
Specific gravity (water = 1): 0.843 @ 68°F/20°C (liquid).
Liquid surface tension: 24 dynes/cm = 0.024 N/m @ 68°F/20°C.
Liquid water interfacial tension: (estimate) 35 dynes/cm = 0.035 N/m @ 68°F/20°C.
Relative vapor density (air = 1): 1.94
Ratio of specific heats of vapor (gas): 1.1487
Latent heat of vaporization: 216 Btu/lb = 120 cal/g = 5.02 x 10^5 J/kg.
Heat of combustion: –12,500 Btu/lb = –6950 cal/g = –290 x 10^5 J/kg.
Heat of polymerization: (estimate) –50 Btu/lb = –28 cal/g = –1.2 x 10^5 J/kg.
Vapor pressure: (Reid) 8.6 psia; 210 mm.

ACRYLAMIDE   REC. A:0850

SYNONYMS: ACRILAMIDA (Spanish); ACRYLAMIDE MONOMER; ACRYLIC ACID AMIDE (50%); ACRYLIC AMIDE; ACRYLIC AMIDE 50%; EEC No. 616-003-00-0; ETHYLENECARBOXAMIDE; ETHYLENE MONOCLINIC TABLETS CARBOXAMIDE; PROPENAMIDE; 2-PROPENAMIDE; PROPENAMIDE; VINYL AMIDE; RCRA No. U007

IDENTIFICATION
CAS Number: 79-06-1
Formula: C$_3$H$_5$NO; CH$_2$=CHCONH$_2$·H$_2$O
DOT ID Number: UN 2074; DOT Guide Number: 153P
Proper Shipping Name: Acrylamide
Reportable Quantity (RQ): **(EHS/CERCLA)** 5000 lb/2270 kg

DESCRIPTION: White crystals or powder. May be dissolved in a flammable liquid. Odorless. Soluble in water.

Poison! • Combustible • Polymerization hazard • Heat above 120°F/50°C can induce polymerization with rapid release of energy; sealed containers may rupture explosively • May react with itself blocking relief valves; leading to tank explosions • May decompose explosively above 176°F/80°C releasing toxic ammonia gas • Containers may BLEVE when exposed to fire • Vapors in confined areas (e.g., tanks, sewers, buildings) may explode when exposed to fire • Irritating to the skin, eyes, and respiratory tract • Toxic products of combustion may include ammonia, carbon monoxide, and nitrogen oxides.

Hazard Classification (based on NFPA-704 Rating System)
Health Hazards (Blue): 2; Flammability (Red): 2; Reactivity (Yellow): 2

EMERGENCY RESPONSE: See Appendix A (153)
Evacuation:
Public safety: Isolate the area of spill or leak for at least 25 to 50 meters (80 to 160 feet) in all directions.
Spill: Increase, in the downwind direction, as necessary, the distance shown under "Public Safety."
Fire: If tank, rail car, or tank truck is involved in fire, isolate for at

least 800 meters (½ mile) in all directions; also, consider initial evacuation for 800 meters (½ mile) in all directions.

EXPOSURE
Short-term effects: *SEEK MEDICAL ATTENTION.* **Vapor:** Irritating to eyes, nose, and throat. Harmful if inhaled. *IF IN EYES*, hold eyelids open and flush with plenty of water.*IF BREATHING HAS STOPPED*, give artificial respiration; *avoid mouth-to-mouth resuscitation; use bag/mask apparatus.* IF breathing is difficult, administer oxygen. **Liquid:** *HARMFUL IF ABSORBED THROUGH THE SKIN.* Will burn skin and eyes. Harmful if swallowed. Remove contaminated clothing and shoes. Flush affected areas with plenty of water. *IF IN EYES*, hold eyelids open and flush with plenty of water. *IF SWALLOWED* and victim is *CONSCIOUS AND ABLE TO SWALLOW*, have victim drink 4 to 8 ounces of water and have victim induce vomiting. *IF SWALLOWED* and victim is *UNCONSCIOUS OR HAVING CONVULSIONS*, do nothing except keep victim warm. The use of alcoholic beverages may enhance the toxic effect.

HEALTH HAZARDS
Personal protective equipment (PPE): A-Level PPE. OSHA Table Z-1-A air contaminant. Sealed chemical protective materials offers limited protection: polyethylene.
Recommendations for respirator selection: NIOSH
At any concentrations above the NIOSH REL, or where there is no REL, at any detectable concentration: SCBAF:PD,PP (any self-contained breathing apparatus that has a full facepiece and is operated in a pressure-demand or other positive-pressure mode); or SAF:PD,PP:ASCBA (any supplied-air respirator that has a full facepiece and is operated in a pressure-demand or other positive-pressure mode in combination with an auxiliary self-contained breathing apparatus operated in a pressure-demand or other positive pressure mode). *ESCAPE:* GMFOV [any air-purifying, full-facepiece respirator (gas mask) with a chin-style, front-or back-mounted organic vapor canister]; or SCBAE (any appropriate escape-type, self-contained breathing apparatus).
Exposure limits (TWA unless otherwise noted): ACGIH TLV 0.03 mg/m^3; potential human carcinogen; OSHA PEL 0.3 mg/m^3; NIOSH 0.03 mg/m^3; potential human carcinogen; reduce exposure to lowest feasible level; skin contact contributes significantly in overall exposure
Toxicity by ingestion: Grade 3; oral rat LD$_{50}$ = 170 mg/kg.
Long-term health effects: May cause liver, kidney, brain, and/or nervous system damage. Repeated exposure to small amounts may cause essentially reversible neurological effects. OSHA specifically regulated carcinogen. IARC possible carcinogen; rating 2B
Vapor (gas) irritant characteristics: May cause irritation of the eyes and respiratory tract. Nervous system effects.
Liquid or solid irritant characteristics: May cause eye and skin irritation.
IDLH value: Suspected human carcinogen; 60 mg/m^3.

FIRE DATA
Flash point: 280°F/138°C (cc).
Autoignition temperature: 464°F/240°C/513°K.
Electrical hazard: Class II, Group G.

CHEMICAL REACTIVITY reported
Binary reactants: Strong oxidizers, reducing agents, peroxides, acids, bases and vinyl polymerization initiators.
Polymerization: May polymerize violently and explode upon melting or in strong light; may occur above 176°F/80°C; **Inhibitor of polymerization:** Oxygen (air) plus 50 ppm of copper as copper sulfate.

Reactivity group: 10 (solution)
Compatibility class: Amides

ENVIRONMENTAL DATA
Food chain concentration potential: Negative; unlikely to accumulate.
Water pollution: Effect of low concentrations on aquatic life is unknown. May be dangerous if it enters nearby water intakes; notify operators. Notify local health and wildlife officials. **Response to discharge:** Issue warning–water contaminant. Disperse and flush.

SHIPPING INFORMATION
Grades of purity: 15–50% solution in water; **Storage temperature:** Below 50°C (122°F); **Inert atmosphere:** Ventilated (natural); **Venting:** Open; **Stability during transport:** Stable.

NAS HAZARD CLASSIFICATION FOR BULK WATER TRANSPORTATION
FIRE: 0
HEALTH: Vapor irritant: 0; Liquid or solid irritant: 1; Poisons: 3
WATER POLLUTION: Human toxicity: 3; Aquatic toxicity: 3; Aesthetic effect: 0
REACTIVITY: Other chemicals: 2; Water: 0; Self-reaction: 3

PHYSICAL AND CHEMICAL PROPERTIES
Physical state @ 59°F/15°C and 1 atm: Liquid.
Molecular weight: 71.1 (solute only).
Boiling point @ 1 atm: Decomposes. 347–572°F/175–300°C/448–573°K.
Melting/Freezing point: 184°F/85°C/358°K.
Specific gravity (water = 1): 1.05 @ 77°F/25°C (liquid).
Vapor pressure: 0.007 mm.

ACRYLIC ACID REC. A:0900

SYNONYMS: ACROLEIC ACID; ACRYLIC ACID, GLACIAL; EEC No. 607-061-00-8; ETHYLENE CARBOXYLIC ACID; GLACIAL ACRYLIC ACID; PROPANEACID; 2-PROPENOIC ACID; VINYL FORMIC ACID; RCRA No. U008

IDENTIFICATION
CAS Number: 79-10-7
Formula: $C_3H_4O_2$; $CH_2=CHCOOH$
DOT ID Number: UN 2218; DOT Guide Number: 132P
Proper Shipping Name: Acrylic acid, inhibited
Reportable Quantity (RQ): **(CERCLA)** 5000 lb/2270 kg

DESCRIPTION: Colorless watery liquid. Irritating, rancid odor. Sinks and mixes with water. Irritating vapor is produced. Freezes at 54°F/12°C.

Highly flammable • A tear gas • Corrosive to the skin, eyes, and respiratory tract; contact with skin or eyes can cause burns and vision impairment; inhalation symptoms may be delayed • Firefighting gear (including SCBA) does not provide adequate protection. If exposure occurs, remove and isolate gear immediately and thoroughly decontaminate personnel • Containers may BLEVE when exposed to fire •Vapors may form explosive mixture with air • Vapors are heavier than air and will collect and stay in low areas • Vapors may travel long distances to ignition sources and flashback • Vapors in confined areas (e.g., tanks, sewers, buildings) may explode when exposed to fire • Polymerization hazard • Heat can induce polymerization with rapid release of energy; sealed containers may rupture explosively • May react with itself blocking relief valves; leading to tank explosions • Toxic products of combustion may include toxic carbon monoxide and irritating gases • Do not put yourself in danger by entering a contaminated area to rescue a victim.

Hazard Classification (based on NFPA-704 Rating System)
Health Hazards (Blue): 3; Flammability (Red): 2; Reactivity (Yellow): 2

EMERGENCY RESPONSE: See Appendix A (132)
Evacuation:
Public safety: Isolate spill area for at least 100 to 200 meters (330 to 660 feet) in all directions.
Spill: Increase, as necessary, the isolation distance shown above, in "Public safety."
Fire: Isolate for 800 meters (½ mile) in all directions, especially if tank, rail car, or tank truck is involved in fire.

EXPOSURE
Short-term effects: *SEEK MEDICAL ATTENTION.* **Vapor:** Irritating to eyes, nose, and throat. Move to fresh air. *IF BREATHING HAS STOPPED,* give artificial respiration. IF breathing is difficult, administer oxygen. Lung edema may develop. **Liquid or solid:** *HARMFUL IF ABSORBED THROUGH THE SKIN.* Will burn skin and eyes. Harmful if swallowed. Remove contaminated clothing and shoes. Flush affected areas with plenty of water. *IF IN EYES,* hold eyelids open and flush with plenty of water. *IF SWALLOWED* and victim is *CONSCIOUS AND ABLE TO SWALLOW,* have victim drink 4 to 8 ounces of water. **Do NOT induce vomiting.**
Note to physician or authorized medical personnel: Medical observation is recommended for 24 to 48 hours after breathing overexposure, as pulmonary edema may be delayed. As first aid for pulmonary edema, consider administering a corticosteroid spray. Cigarette smoking may exacerbate pulmonary injury and should be discouraged for at least 72 hours following exposure.

HEALTH HAZARDS
Personal protective equipment (PPE): A-Level PPE. OSHA Table Z-1-A air contaminant. Chemical protective material(s) reported to have good to excellent resistance: butyl rubber, natural rubber, nitrile, neoprene, Teflon®, chlorinated polyethylene, Saranex®.
Exposure limits (TWA unless otherwise noted): ACGIH TLV 2 ppm (5.9 mg/m^3); NIOSH 2 ppm (6 mg/m^3); skin contact contributes significantly in overall exposure.
Toxicity by ingestion: Grade 2; LD$_{50}$ = 0.5 to 5 g/kg (rat).
Long-term health effects: May cause kidney damage; IARC rating 3.
Vapor (gas) irritant characteristics: Vapor is irritating such that personnel will not usually tolerate moderate or high vapor concentrations. Corrosive to the eyes, skin, and respiratory tract.
Liquid or solid irritant characteristics: Corrosive skin and eye irritant; may cause pain and second-degree burns after a few minutes of contact.
Odor threshold: 0.092 ppm.

FIRE DATA
Flash point: 118°F/48°C (oc)(glacial); 122°F/50°C (oc).
Flammable limits in air: LEL: 2.4%; UEL: 8.0% (glacial); LEL: 5.3%; UEL: 26.0%.
Autoignition temperature: 835°F/446°C/719°C.
Electrical hazard: Class I, Group D.
Burning rate: 1.6 mm/min

CHEMICAL REACTIVITY
Binary reactants: Oxidizers, amines, alkalis, ammonium hydroxide, acids, chlorosulfonic acid, oleum, ethylene diamine, ethyleneimine; 2-aminoethanol. Corrosive to iron, steel and other metals.
Neutralizing agents for acids and caustics: Wash with water, rinse with sodium bicarbonate solution.
Polymerization: May occur in prolonged heat, on contact with acids, iron salts; releases high energy rapidly; may explode under confinement; **Inhibitor of polymerization:** Monomethyl ether of hydroquinone 180–200 ppm; phenothiazine (for technical grades) 1000 ppm; hydroquinone (0.1%); methylene blue (0.5–1%); *N,N'*-diphenyl-*p*-phenylenediamine (0.05%).
Reactivity group: 4
Compatibility class: Organic acids

ENVIRONMENTAL DATA
Food chain concentration potential: Log P_{ow} = 0.16–0.37. Unlikely to accumulate.
Water pollution: Effect of low concentrations on aquatic life is unknown. May be dangerous if it enters nearby water intakes; notify operators. Notify local health and pollution control officials.
Response to discharge: Issue warning–corrosive. Restrict access. Disperse and flush.

SHIPPING INFORMATION
Grades of purity: Technical: 94.0%; glacial: 98.0–99.5%; **Storage temperature:** 60–75°F; **Stability during transport:** Normally unstable, but may not detonate.

NAS HAZARD CLASSIFICATION FOR BULK WATER TRANSPORTATION
FIRE: 2
HEALTH: Vapor irritant: 3; Liquid or solid irritant: 3; Poisons: 2
WATER POLLUTION: Human toxicity: 1; Aquatic toxicity: 3; Aesthetic effect: 2
REACTIVITY: Other chemicals: 2; Water: 0; Self-reaction: 3

PHYSICAL AND CHEMICAL PROPERTIES
Physical state @ 59°F/15°C and 1 atm: Liquid.
Molecular weight: 72.06
Boiling point @ 1 atm: 286.3°F/141.3°C/414.5°K.
Melting/Freezing point: 54.1°F/12.3°C/285.5°K.
Critical temperature: 648°F/342°C/615°K.
Critical pressure: 840 psia = 57 atm = 5.8 MN/m^2.
Specific gravity (water = 1): 1.0497 @ 68°F/20°C (liquid).
Relative vapor density (air = 1): 2.5
Ratio of specific heats of vapor (gas): 1.121
Latent heat of vaporization: 272.7 Btu/lb = 151.5 cal/g = 6.343 x 10^5 J/kg.
Heat of combustion: –8100 But/lb = –4500 cal/g = –188.4 x 10^5 J/kg.
Heat of polymerization: –463 Btu/lb = –257 cal/g = –10.8 x 10^5 J/kg.
Heat of fusion: 30.03 cal/g.
Vapor pressure: (Reid) 0.2 psia; 3–3.3 mm.

ACRYLONITRILE REC. A:0950

SYNONYMS: ACRILONITRILO (Spanish); ACRYLNITRIL (German); ACRYLON; ACRYLONITRILE MONOMER; AN; CARBACRYL; CIANURO VILILICO (Spanish); CYANOETHYLENE; CYANURE de VINYLE (French); EEC No. 608-00300-4; FUMIGRAIN; MILLER'S FUMIGRAIN; NITRILE

ACRYLIQUE (French); PROPENENITRILE; 2-PROPENENITRILE; TL 314; VENTOX; VINYL CYANIDE; VINYL CYANIDE, PROPENENITRILE; RCRA No. U009

IDENTIFICATION
CAS Number: 107-13-1
Formula: C_3H_3N; $CH_2=CHCN$; C_3H_3N
DOT ID Number: UN 1093; DOT Guide Number: 131P
Proper Shipping Name: Acrylonitrile, inhibited
Reportable Quantity (RQ): **(CERCLA)** 100 lb/45.4 kg

DESCRIPTION: Colorless to light yellow, watery liquid. Strong, pungent, onion or garlic odor. Floats on water surface; moderately soluble; poisonous, flammable vapor is produced.

Highly flammable • Poison! • Breathing the vapor, skin or eye contact, or swallowing the material can kill you • Firefighting gear (including SCBA) does not provide adequate protection. If exposure occurs, remove and isolate gear immediately and thoroughly decontaminate personnel • Containers may BLEVE when exposed to fire • Vapors may form explosive mixture with air • Vapors are heavier than air and will collect and stay in low areas • Vapors may travel long distances to ignition sources and flashback • Vapors in confined areas (e.g., tanks, sewers, buildings) may explode when exposed to fire • Polymerization hazard • Heat can induce polymerization with rapid release of energy; sealed containers may rupture explosively • May react with itself blocking relief valves; leading to tank explosions • Toxic products of combustion may include cyanide gas and nitrogen oxide • Do not put yourself in danger by entering a contaminated area to rescue a victim. *Warning:* Odor is not a reliable indicator of the presence of toxic amounts of acrylonitrile. The Odor threshold is 10-fold greater than the OSHA PEL.

Hazard Classification (based on NFPA-704 Rating System)
Health Hazards (Blue): 4; Flammability (Red): 3; Reactivity (Yellow): 2

EMERGENCY RESPONSE: See Appendix A (131)
Evacuation:
Public safety: Isolate spill area for at least 100 to 200 meters (330 to 660 feet) in all directions.
Spill: Increase, as necessary, the isolation distance shown above, in "Public safety."
Fire: Isolate for 800 meters (½ mile) in all directions, especially if tank, rail car, or tank truck is involved in fire.

EXPOSURE
Short-term effects: *SEEK MEDICAL ATTENTION.* **Vapor:** *POISONOUS IF INHALED.* Irritating to eyes. Move to fresh air. *IF BREATHING HAS STOPPED,* give artificial respiration; *avoid mouth-to-mouth resuscitation; use bag/mask apparatus.* IF breathing is difficult, administer oxygen. **Liquid:** *POISONOUS IF SWALLOWED.* Irritating to skin and eyes. Remove contaminated clothing and shoes. Flush affected areas with plenty of water. *IF IN EYES,* hold eyelids open and flush with plenty of water. *IF SWALLOWED* and victim is *CONSCIOUS AND ABLE TO SWALLOW,* have victim drink 4 to 8 ounces of water and have victim induce vomiting. *IF SWALLOWED* and victim is *UNCONSCIOUS OR HAVING CONVULSIONS,* do nothing except keep victim warm.
Note to physician or authorized medical personnel: (1) Jaundice and elevated serum levels of liver enzymes can occur 24 hours after exposure and may persist for several days (2) Consider the use of amyl nitrite perles if symptoms of cyanide poisoning develop. If symptoms indicate, initial treatment includes the cyanide antidote kit. In all cases, break an amyl nitrite perle in a gauze pad and hold lightly under victim's nose for 15 seconds, repeating 5 times at about 15-second intervals; if necessary (and if sodium nitrite infusions will be delayed), repeat procedure every 3 minutes with fresh perles until 3 or 4 have been used. Avoid breathing the vapor while administering it to the victim. Administer sodium nitrite IV, ASAP. The usual adult dose is 10 to 20 mL of a 3% solution infused over no less than 5 minutes; the average child dose is 0.15 to 0.20 mL/kg. Monitor blood pressure during administration, and slow the rate of infusion if hypotention develops. Next, infuse sodium thiosulfate IV. The usual adult dose is 50 mL of a 25% solution infused over 10 to 20 minutes; the average child dose is 1.65 mL/kg. Repeat with nitrite and thiosulfate as required.

HEALTH HAZARDS
Personal protective equipment (PPE): A-Level PPE. OSHA Table Z-1-A air contaminant. Chemical protective material(s) reported to have good to excellent resistance: Neoprene, polyurethane.
Recommendations for respirator selection: NIOSH
At any concentrations above the NIOSH REL, or where there is no REL, at any detectable concentration: SCBAF:PD,PP (any self-contained breathing apparatus that has a full facepiece and is operated in a pressure-demand or other positive-pressure mode); or SAF:PD,PP:ASCBA (any supplied-air respirator that has a full facepiece and is operated in a pressure-demand or other positive-pressure mode in combination with an auxiliary self-contained breathing apparatus operated in a pressure-demand or other positive pressure mode). *ESCAPE:* GMFOV [any air-purifying, full-facepiece respirator (gas mask) with a chin-style, front-or back-mounted organic vapor canister]; or SCBAE (any appropriate escape-type, self-contained breathing apparatus).
Exposure limits (TWA unless otherwise noted): ACGIH TLV 2 ppm (4.3 mg/m³); OSHA [1910.1045] 2 ppm; ceiling limit 10 ppm/15 minutes; NIOSH 1 ppm; ceiling 10 ppm/15 minutes; potential human carcinogen; reduce exposure to lowest feasible level; skin contact contributes significantly in overall exposure.
Toxicity by ingestion: Grade 3; LD_{50} 50 to 500 mg/kg (rat, guinea pig).
Long-term health effects: OSHA specifically regulated carcinogen; IARC probable carcinogen; NTP anticipated carcinogen.
Vapor (gas) irritant characteristics: Vapor is moderately irritating such that personnel will not usually tolerate moderate or high vapor concentrations.
Liquid or solid irritant characteristics: If spilled on clothing and allowed to remain, may cause smarting and reddening of the skin. Large amounts may be absorbed through the skin and cause poisoning.
Odor threshold: 1.6 ppm (Sense of smell fatigues rapidly). Odor can be detected above the PEL.
IDLH value: Suspected human carcinogen; 85 ppm.

FIRE DATA
Flash point: 31°F/1°C (oc); 32°F/0°C (cc).
Flammable limits in air: LEL: 3.05%; UEL: 17.0%.
Fire extinguishing agents not to be used: Water or foam may cause frothing.
Behavior in fire: May polymerize causing containers to rupture and explode.
Autoignition temperature: 896°F/480°C/753°K.
Electrical hazard: Class I, Group D. Due to low electric conductivity, this substance may generate electrostatic charges as a result of agitation and flow.

CHEMICAL REACTIVITY

Binary reactants: Forms explosive mixture with air. Strong oxidizers, acids, alkalis, bromine, amines. Attacks copper and copper alloys; aluminum in high concentrations. Penetrates leather, so contaminated leather shoes and gloves should be destroyed. Attacks aluminum in high concentrations. Reacts with strong oxidizers, acids and alkalis, peroxides (causes polymerization), bromine, amines. **Polymerization:** May occur spontaneously in absence of oxygen or on exposure to visible light, excessive heat; violently in the presence of alkali and peroxide. Pure is subject to self-polymerization with rapid pressure development. The commercial product is inhibited and may not be subject to this reaction; **Inhibitor of polymerization:** Methylhydroquinone (35-45 ppm).
Reactivity group: 15
Compatibility class: Substituted allyl

ENVIRONMENTAL DATA

Food chain concentration potential: Log P_{ow} = –0.88–0.25. Unlikely to accumulate.
Water pollution: DOT Appendix B, §172.101–marine pollutant. Harmful to aquatic life in very low concentrations. Fouling to shoreline. May be dangerous if it enters nearby water intakes; notify operators. Notify local health and wildlife officials. **Response to discharge:** Issue warning–poison. Highly flammable. Restrict access. Disperse and flush.

SHIPPING INFORMATION

Grades of purity: Technical: 98–100%; **Storage temperature:** Ambient; **Inert atmosphere:** None; **Venting:** Pressure-vacuum; **Stability during transport:** Stable.

NAS HAZARD CLASSIFICATION FOR BULK WATER TRANSPORTATION

FIRE: 3
HEALTH: Vapor irritant: 3; Liquid or solid irritant: 1; Poisons: 3
WATER POLLUTION: Human toxicity: 4; Aquatic toxicity: 3; Aesthetic effect: 2
REACTIVITY: Other chemicals: 3; Water: 0; Self-reaction: 3

PHYSICAL AND CHEMICAL PROPERTIES

Physical state @ 59°F/15°C and 1 atm: Liquid.
Molecular weight: 53.06
Boiling point @ 1 atm: 171°F/77.4°C/350.6°K.
Melting/Freezing point: –118°F/–83.6°C/189.6°K.
Critical temperature: 505°F/263°C/536°K.
Critical pressure: 660 psia = 45 atm = 4.6 MN/m².
Specific gravity (water = 1): 0.8075 @ 68°F/20°C (liquid).
Relative vapor density (air = 1): 1.8
Ratio of specific heats of vapor (gas): 1.151
Latent heat of vaporization: 265 Btu/lb = 147 cal/g = 6.16 x 10⁵ J/kg.
Heat of combustion: –14,300 Btu/lb = –7930 cal/g = 332 x 10⁵ J/kg.
Vapor pressure: (Reid) 3.5 psia; 83 mm.

ADIPIC ACID   REC. A:1000

SYNONYMS: ACIDO ADIPICO (Spanish); ACIFLOCTIN; ACINETTEN; ADILAC-TETTEN; ADIPINIC ACID; 1,4-BUTANEDICARBOXYLIC ACID; EEC No. 607-144-00-9; HEXANEDIOIC ACID; 1,6-HEXANEDIOIC ACID

IDENTIFICATION

CAS Number: 124-04-9
Formula: $C_6H_{10}O_4$; $HOOC(CH_2)_4COOH$
DOT ID Number: UN 3077; DOT Guide Number: 171
Proper Shipping Name: Environmentally hazardous substance, n.o.s.
Reportable Quantity (RQ): (CERCLA) 5000 lb/2270 kg

DESCRIPTION: White crystalline powder. Odorless. Sinks and mixes slowly with water.

Poison! • Combustible solid • The dust can cause severe eye irritation • Containers may BLEVE when exposed to fire • Concentrated dust in confined areas (e.g., tanks, sewers, buildings) may explode when exposed to fire • Toxic products of combustion may include nitrogen oxide

Hazard Classification (based on NFPA-704 Rating System)
Health Hazards (Blue): -; Flammability (Red): 1; Reactivity (Yellow): 0

EMERGENCY RESPONSE: See Appendix A (171)
Evacuation:
Public safety: Isolate the area of spill or leak for at least 10 to 25 meters (30 to 80 feet) in all directions.
Spill: Increase, in the downwind direction, as necessary, the distance shown under "Public Safety."
Fire: If any large container is involved in fire, isolate for at least 800 meters (½ mile) in all directions; also, consider initial evacuation for 800 meters (½ mile) in all directions.

EXPOSURE

Short-term effects: *SEEK MEDICAL ATTENTION*. **Dust:** Irritating to eyes, nose, and throat. *IF INHALED*, will, will cause coughing or difficult breathing. *IF IN EYES*, hold eyelids open and flush with plenty of water. *IF BREATHING HAS STOPPED*, give artificial respiration. IF breathing is difficult, administer oxygen. **Solid:** Irritating to skin and eyes. Harmful if swallowed. Remove contaminated clothing and shoes. Flush affected areas with plenty of water. *IF IN EYES*, hold eyelids open and flush with plenty of water. *IF SWALLOWED* and victim is *UNCONSCIOUS OR HAVING CONVULSIONS*, do nothing except keep victim warm.

HEALTH HAZARDS

Personal protective equipment (PPE): B-Level PPE. Chemical protective material(s) reported to have good to excellent resistance: butyl rubber, polycarbonate.
Exposure limits (TWA unless otherwise noted): ACGIH TLV 5 mg/m³.
Toxicity by ingestion: Grade 2; oral mouse LD_{50} = 1900 mg/kg.
Vapor (gas) irritant characteristics: Irritates eyes and respiratory tract.
Liquid or solid irritant characteristics: May cause eye irritation.

FIRE DATA

Flash point: Combustible solid. 300°F/149°C (oc); 385°F/196°C (cc).
Flammable limits in air: (dust) 10–15 mg/L.
Behavior in fire: Melts and may decompose to give volatile acidic vapors of valeric acid and other substances. Dust may form explosive mixture with air.
Autoignition temperature: 788°F/420°C/693°K.
Electrical hazard: Flow or agitation of substance may generate electrostatic charges due to low conductivity.

CHEMICAL REACTIVITY
Neutralizing agents for acids and caustics: Rinse with dilute sodium bicarbonate or soda ash solution.

ENVIRONMENTAL DATA
Food chain concentration potential: Log P_{ow} = 0.08. Unlikely to accumulate.
Water pollution: Dangerous to aquatic life in high concentrations. May be dangerous if it enters nearby water intakes; notify operators. Notify local health and wildlife officials. **Response to discharge:** Should be removed. Chemical and physical treatment.

SHIPPING INFORMATION
Grades of purity: Commercial, 99.8%; **Storage temperature:** Ambient; **Inert atmosphere:** None; **Venting:** Open; **Stability during transport:** Stable.

PHYSICAL AND CHEMICAL PROPERTIES
Physical state @ 59°F/15°C and 1 atm: Solid.
Molecular weight: 146.1
Boiling point @ 1 atm: Decomposes.
Melting/Freezing point: 304°F/151°C/424°K.
Specific gravity (water = 1): 1.36 @ 68°F/20°C (solid).
Relative vapor density (air = 1): 5.04
Heat of combustion: −8,242 Btu/lb = −4,579 cal/g = 191.6 x 10^5 J/kg.
Vapor pressure: Less than 0.0008 mm.

ADIPONITRILE REC. A:1050

SYNONYMS: ADIPIC ACID DINITRILE; ADIPIC ACID NITRILE; ADIPODINITRILE; ADIPONITRILO (Spanish); 1,4-DICYANOBUTANE; HEXANEDINITRILE; HEXANEDIOIC ACID, DINITRILE; TETRAMETHYLENE CYANIDE

IDENTIFICATION
CAS Number: 111-69-3
Formula: $C_6H_8N_2$; $CN(CH_2)_4CN$
DOT ID Number: UN 2205; DOT Guide Number: 153
Proper Shipping Name: Adiponitrile
Reportable Quantity (RQ): **(EHS)** 1 lb/0.454 kg

DESCRIPTION: Colorless to light yellow, oily liquid. Practically odorless. Floats on water surface; moderately soluble. Freezes at 36°F/2°C.

Poison! • Highly flammable • Exposure to heat produces cyanide gas • Breathing the vapor, skin or eye contact, or swallowing the material can kill you; produces cyanide in the body • Firefighting gear (including SCBA) does not provide adequate protection. If exposure occurs, remove and isolate gear immediately and thoroughly decontaminate personnel • Containers may BLEVE when exposed to fire • Vapors may form explosive mixture with air at 199°F/93°C • Vapors are heavier than air and will collect and stay in low areas • Vapors may travel long distances to ignition sources and flashback • Vapors in confined areas (e.g., tanks, sewers, buildings) may explode when exposed to fire • Toxic products of combustion may include nitrogen oxide and cyanide gas • Do not put yourself in danger by entering a contaminated area to rescue a victim.

Hazard Classification (based on NFPA-704 Rating System)
Health Hazards (Blue): 2; Flammability (Red): 2; Reactivity (Yellow): 1

EMERGENCY RESPONSE: See Appendix A (153)
Evacuation:
Public safety: Isolate the area of spill or leak for at least 25 to 50 meters (80 to 160 feet) in all directions.
Spill: Increase, in the downwind direction, as necessary, the distance shown under "Public Safety."
Fire: If tank, rail car, or tank truck is involved in fire, isolate for at least 800 meters (½ mile) in all directions; also, consider initial evacuation for 800 meters (½ mile) in all directions.

EXPOSURE
Short-term effects: *SEEK MEDICAL ATTENTION*. If artificial respiration is administered, *avoid mouth-to-mouth resuscitation; use bag/mask apparatus*. **Liquid or solid:** *HARMFUL IF ABSORBED THROUGH THE SKIN*. Irritating to skin and eyes. *IF SWALLOWED*, will cause nausea or vomiting. Remove contaminated clothing and shoes. Flush affected areas with plenty of water. *IF IN EYES*, hold eyelids open and flush with plenty of water. *IF SWALLOWED* and victim is *CONSCIOUS AND ABLE TO SWALLOW*, have victim drink 4 to 8 ounces of water.
Note to physician or authorized medical personnel: Consider the use of amyl nitrite perles if symptoms of cyanide poisoning develop. If symptoms indicate, initial treatment includes the cyanide antidote kit. In all cases, break an amyl nitrite perle in a gauze pad and hold lightly under victim's nose for 15 seconds, repeating 5 times at about 15-second intervals; if necessary (and if sodium nitrite infusions will be delayed), repeat procedure every 3 minutes with fresh perles until 3 or 4 have been used. Avoid breathing the vapor while administering it to the victim. Administer sodium nitrite IV, ASAP. The usual adult dose is 10 to 20 mL of a 3% solution infused over no less than 5 minutes; the average child dose is 0.15 to 0.20 mL/kg. Monitor blood pressure during administration, and slow the rate of infusion if hypotention develops. Next, infuse sodium thiosulfate IV. The usual adult dose is 50 mL of a 25% solution infused over 10 to 20 minutes; the average child dose is 1.65 mL/kg. Repeat with nitrite and thiosulfate as required.

HEALTH HAZARDS
Personal protective equipment (PPE): A-Level PPE. Chemical protective material(s) reported to offer minimal to poor protection. Teflon®.
Recommendations for respirator selection: NIOSH
40 ppm: SA (any supplied-air respirator). *100 ppm*: SA:CF (any supplied-air respirator operated in a continuous-flow mode). *200 ppm*: SCBAF (any self-contained breathing apparatus with a full facepiece); or SAF (any supplied-air respirator with a full facepiece). *250 ppm*: SAF:PD,PP (any supplied-air respirator that has a full facepiece and is operated in a pressure-demand or other *positive-pressure mode).EMERGENCY OR PLANNED ENTRY INTO UNKNOWN CONCENTRATIONS OR IDLH CONDITIONS*: SCBAF:PD,PP (any self-contained breathing apparatus that has a full facepiece and is operated in a pressure-demand or other positive-pressure mode); or SAF:PD,PP:ASCBA (any supplied-air respirator that has a full facepiece and is operated in a pressure-demand or other positive-pressure mode in combination with an auxiliary self-contained breathing apparatus operated in a pressure-demand or other positive pressure mode). *ESCAPE:* GMFOV [any air-purifying, full-facepiece respirator (gas mask) with a chin-style, front-or back-mounted organic vapor canister]; or SCBAE (any appropriate escape-type, self-contained breathing apparatus).
Exposure limits (TWA unless otherwise noted): ACGIH TLV 2 ppm (8.8 mg/m^3); NIOSH REL 4 ppm (18 mg/m^3).
Vapor (gas) irritant characteristics: If present in high concentrations, vapors cause a smarting of the eyes or respiratory

system; they may also cause more severe symptoms such as headache and convulsions, unconsciousness, and death.
Liquid or solid irritant characteristics: If spilled on clothing and allowed to remain, may cause smarting and reddening of the skin. If absorbed by skin may cause more severe symptoms such as headache and convulsions.

FIRE DATA
Flash point: 199°F/92°C (oc).
Flammable limits in air: LEL: 1.7%; UEL: 5.0%.
Autoignition temperature: 860°F/460°C/733°K.
Electrical hazard: Class I, Group D. Due to low electric conductivity, this substance may generate electrostatic charges as a result of agitation and flow.
Burning rate: 2.7 mm/min

CHEMICAL REACTIVITY
Binary reactants: Violent reaction with oxidizers (e.g. perchlorates, nitrates), strong acids. Decomposes above 194°F forming hydrogen cyanide.
Reactivity group: 37
Compatibility class: Nitriles

ENVIRONMENTAL DATA
Food chain concentration potential: Log P_{ow} = –0.35. Unlikely to accumulate.
Water pollution: DOT Appendix B, §172.101–marine pollutant. Dangerous to aquatic life in high concentrations. Fouling to shoreline. May be dangerous if it enters nearby water intakes; notify operators. Notify local health and pollution control officials.
Response to discharge: Issue warning–poison, water contaminant. Mechanical containment. Should be removed.

SHIPPING INFORMATION
Storage temperature: Ambient; **Inert atmosphere:** None; **Venting:** Open; **Stability during transport:** Stable.

NAS HAZARD CLASSIFICATION FOR BULK WATER TRANSPORTATION
FIRE: 1
HEALTH: Vapor irritant: 1; Liquid or solid irritant: 1; Poisons: 3
WATER POLLUTION: Human toxicity: 3; Aquatic toxicity: 2; Aesthetic effect: 3
REACTIVITY: Other chemicals: 2; Water: 0; Self-reaction: 0

PHYSICAL AND CHEMICAL PROPERTIES
Physical state @ 59°F/15°C and 1 atm: Liquid.
Molecular weight: 108.2
Boiling point @ 1 atm: 563°F/295°C/568°K.
Melting/Freezing point: 36°F/2°C/275°K.
Specific gravity (water = 1): 0.97 @ 77°F/25°C (liquid).
Latent heat of vaporization: (estimate).
Latent heat of vaporization: 240 Btu/lb = 134 cal/g = 5.59 x 10^5 J/kg.
Heat of combustion: –14,230 Btu/lb = –7910 cal/g = –331 x 10^5 J/kg.
Vapor pressure: (Reid) low; 0.002 mm.

ALDRIN   REC. A:1100

SYNONYMS: ALDREX; ALDREX-30; ALDRINA (Spanish); ALDRINE (French); ALDRITE; ALDROSOL; ALTOX; COMPOUND 118; DRINOX; HEXACHLOROHEXAHYDRO-*endo-exo*- DIMETH ANONAPHTHALENE; 1,2,3,4,10,10-HEXACHLORO-1,4,4A,5,8,8*A*-HEXAHYDRO-1,4,5,8-DIMETHANONAPHTHALENE; 1,2,3,4,10,10-HEXACHLORO-1,4,4A,5,8,8A -HEXAHYDRO-*exo*-1,4-*endo*-5,8-DIMETHANONAPHTHALENE; 1,2,3,4,10,10-HEXACHLORO-1,4,4A,5,8,8*A*-HEXAHYDRO-1,4-*endo-exo*-5,8-DIMETHANONAPHTHALENE; HHDN; OCTALENE; SEEDRIN; 1,2,3,4,10-10-HEXACHLORO-1,4,4A,5,8,8*A*-HEXAHYDRO-1,4,5,8-*endo,exo*-DIMETHANONAPHTHALENE; OCTALENE; RCRA No. P004; SEDRIN

IDENTIFICATION
CAS Number: 309-00-2
Formula: $C_{12}H_8Cl_6$
DOT ID Number: NA 2762 (liquid); NA 2761 (solid); DOT Guide Number: 131 (liquid); 151 (solid)
Proper Shipping Name: Aldrin, liquid; Aldrin, solid
Reportable Quantity (RQ): **(CERCLA)** 1 lb/0.454 kg

DESCRIPTION: Tan to dark brown crystalline solid or solution. Mild chemical odor. Solid sinks in water; solution floats on water surface; slightly soluble (0.003%).

Poison! • Breathing the dust, skin or eye contact, or swallowing the material can kill you • Firefighting gear (including SCBA) does not provide adequate protection. If exposure occurs, remove and isolate gear immediately and thoroughly decontaminate personnel • Toxic products of combustion may include hydrogen chloride and other chlorinated decomposition product • Do not put yourself in danger by entering a contaminated area to rescue a victim.

Hazard Classification (based on NFPA-704 Rating System)
Health Hazards (Blue): 4; Flammability (Red): 0; Reactivity (Yellow): 0

EMERGENCY RESPONSE: See Appendix A (131)
Evacuation:
Public safety: Isolate spill area for at least 100 to 200 meters (330 to 660 feet) in all directions.
Spill: Increase, as necessary, the isolation distance shown above, in "Public safety."
Fire: Isolate for 800 meters (½ mile) in all directions, especially if tank, rail car, or tank truck is involved in fire.

EXPOSURE
Short-term effects: *SEEK MEDICAL ATTENTION.* **Solid or solution:** *POISONOUS IF SWALLOWED OR IF SKIN IS EXPOSED.* Irritating to skin, eyes. Remove contaminated clothing and shoes. Flush affected areas with plenty of water. *IF IN EYES*, hold eyelids open and flush with plenty of water. *IF SWALLOWED* and victim is *CONSCIOUS AND ABLE TO SWALLOW*, have victim drink 4 to 8 ounces of water and have victim induce vomiting. *IF SWALLOWED* and victim is *UNCONSCIOUS OR HAVING CONVULSIONS*, do nothing except keep victim warm. *Note to physician or authorized medical personnel*: If symptoms indicate, administer anti-convulsant therapy (treatment may include barbiturates). Observe patient carefully because repeated treatment may be necessary.

HEALTH HAZARDS
Personal protective equipment (PPE): A-Level PPE. OSHA Table Z-1-A air contaminant. Chemical protective material(s) reported to have good to excellent resistance: butyl rubber, polycarbonate.
Recommendations for respirator selection: NIOSH *At any*

concentrations above the NIOSH REL, or where there is no REL, at any detectable concentration: SCBAF:PD,PP (any self-contained breathing apparatus that has a full facepiece and is operated in a pressure-demand or other positive-pressure mode); or SAF:PD,PP:ASCBA (any supplied-air respirator that has a full facepiece and is operated in a pressure-demand or other positive-pressure mode in combination with an auxiliary, self-contained breathing apparatus operated in a pressure-demand or other positive pressure mode). *ESCAPE:* GMFOVHiE [any air-purifying, full-facepiece respirator (gas mask) with a chin-style, front- or back-mounted organic vapor canister having a high-efficiency particulate filter]; or SCBAE (any appropriate escape-type, self-contained breathing apparatus).
Exposure limits (TWA unless otherwise noted): ACGIH TLV 0.25 mg/m^3; OSHA PEL 0.25 mg/m^3; NIOSH REL potential human carcinogen. Reduce to lowest detectable level; reduce exposure to lowest feasible level; skin contact contributes significantly in overall exposure.
Toxicity by ingestion: Grade 3; LD$_{50}$ to 500 mg/kg (rat).
Long-term health effects: Chronic exposure produces benign tumors in mice. IARC rating 3 limited animal evidence
Vapor (gas) irritant characteristics: Vapors cause irritation of the eyes or respiratory system if present in high concentrations. Effect may be temporary.
Liquid or solid irritant characteristics: If spilled on clothing and allowed to remain, may cause irritation of the skin.
IDLH value: Potential human carcinogen; 25 mg/m^3.

CHEMICAL REACTIVITY reported
Binary reactants: Concentrated mineral acids, active metals, acid catalysts, acid oxidizing agents, phenol.
Reactivity group: 36
Compatibility class: Halogenated hydrocarbons

ENVIRONMENTAL DATA
Food chain concentration potential: High for chlorinated hydrocarbons.
Water pollution: DOT Appendix B, §172.101- severe marine pollutant. Harmful to aquatic life in very low concentrations. May be dangerous if it enters nearby water intakes; notify operators. Notify local health and wildlife officials. **Response to discharge:** Issue warning–poison, water contaminant; liquid forms are flammable. Mechanical containment (of liquid form). Should be removed.

SHIPPING INFORMATION
Grades of purity: 20–95% aldrin, 5–80% inert ingredients. Several solutions in hydrocarbon solvents; **Stability during transport:** Stable.

NAS HAZARD CLASSIFICATION FOR BULK WATER TRANSPORTATION
FIRE: 0
HEALTH: Vapor irritant: 1; Liquid or solid irritant: 1; Poisons: 3
WATER POLLUTION: Human toxicity: 4; Aquatic toxicity: 4; Aesthetic effect: 2
REACTIVITY: Other chemicals: 2; Water: 0; Self-reaction: 0

PHYSICAL AND CHEMICAL PROPERTIES
Physical state @ 59°F/15°C and 1 atm: Solid.
Molecular weight: 364.93
Boiling point @ 1 atm: Decomposes.
Melting/Freezing point: 219°F/104°C/377°K.
Specific gravity (water = 1): 1.6 @ 68°F/20°C (solid).
Vapor pressure: 0.00008 mm.

ALKYLBENZENESULFONIC ACIDS REC. A:1150

SYNONYMS: DECYLBENZENESULFONIC ACID; UN; DO; TRI; TETRA; PENTA; DODECYLBENZENESULFONIC ACID; HEXABENZENESULFONIC ACID

IDENTIFICATION
CAS Number: (Varies by compound) 42615-29-2; 68411-30-3 (sodium salt, see also S:1400)
Formula: $CH_{2n+1}C_6H_4SO_3H$ (n = 10–16)

DESCRIPTION: Colorless to yellow liquid. Odorless. Mixes with water.

Combustible • Containers may BLEVE when exposed to fire •Vapors in confined areas (e.g., tanks, sewers, buildings) may explode when exposed to fire • Irritating to skin, eyes, and lungs • Toxic products of combustion may include sulfur oxides.

Hazard Classification (based on NFPA-704 Rating System)
Health Hazards (Blue): 0; Flammability (Red): 1; Reactivity (Yellow): 0

EMERGENCY RESPONSE: See Appendix A (171)
Evacuation:
Public safety: Isolate the area of spill or leak for at least 10 to 25 meters (30 to 80 feet) in all directions.
Spill: Increase, in the downwind direction, as necessary, the distance shown under "Public Safety."
Fire: If any large container is involved in fire, isolate for at least 800 meters (½ mile) in all directions; also, consider initial evacuation for 800 meters (½ mile) in all directions.

EXPOSURE
Short-term effects: *SEEK MEDICAL ATTENTION.* **Liquid:** Irritating to skin and eyes. *IF SWALLOWED*, will cause nausea. Remove contaminated clothing and shoes. Flush affected areas with plenty of water. *IF IN EYES*, hold eyelids open and flush with plenty of water. *IF SWALLOWED* and victim is *CONSCIOUS AND ABLE TO SWALLOW*, have victim drink 4 to 8 ounces of water. *IF SWALLOWED* and victim is *UNCONSCIOUS OR HAVING CONVULSIONS,* do nothing except keep victim warm.

Personal protective equipment (PPE): B-Level PPE. Goggles or face shield; rubber gloves.
Toxicity by ingestion: Grade 2; LD$_{50}$ = 0.5 to 5 g/kg (rat).
Vapor (gas) irritant characteristics: May cause irritation of the eyes and respiratory tract.
Liquid or solid irritant characteristics: May cause eye and skin irritation.

FIRE DATA
Flash point: 395°F/202°C (oc).
Fire extinguishing agents not to be used: Water may be ineffective.

CHEMICAL REACTIVITY
Binary reactants: May attack metals, forming flammable hydrogen gas. Dangerous if accumulated in confined space.
Neutralizing agents for acids and caustics: Flush with water, rinse with dilute sodium bicarbonate or soda ash solution.

ENVIRONMENTAL DATA
Food chain concentration potential: Negative; unlikely to accumulate.

Water pollution: Effect of low concentrations on aquatic life is unknown. May be dangerous if it enters nearby water intakes; notify operators. Notify local health and wildlife officials. **Response to discharge:** Issue warning–corrosive. Disperse and flush.

SHIPPING INFORMATION
Grades of purity: Commercial, 88–97%; **Storage temperature:** Ambient; **Inert atmosphere:** None; **Venting:** Open; **Stability during transport:** Stable.

PHYSICAL AND CHEMICAL PROPERTIES
Physical state @ 59°F/15°C and 1 atm: Liquid.
Molecular weight: 310-394
Boiling point @ 1 atm: Decomposes.
Specific gravity (water = 1): 1.0 to 1.4 @ 68°F/20°C (liquid).

ALLYL ALCOHOL   REC. A:1200

SYNONYMS: AA; ALCOOL ALLYLIQUE (French); ALLYLALKOHOL (German); ALLYLIC ALCOHOL; EEC No. 603-015-00-6; 3-HYDROXYPROPENE; ORVINYLCARBINOL; PROPENOL; 2-PROPENOL; 2-PROPEN-1-OL; PROPEN-1-OL-3; 1-PROPEN-3-OL; PROPENYL ALCOHOL; 2-PROPENYL ALCOHOL; VINYL CARBINOL; VINYL CARBINOL,2-PROPENOL; WEED DRENCH; RCRA No. P005

IDENTIFICATION
CAS Number: 107-18-6
Formula: C_3H_6O; $CH_2=CHCH_2OH$
DOT ID Number: UN 1098; DOT Guide Number: 131
Proper Shipping Name: Allyl alcohol
Reportable Quantity (RQ): **(CERCLA)** 100 lb/45.4 kg

DESCRIPTION: Colorless watery liquid. Sharp, mustard-like odor; causes tears. Floats on water surface; soluble; poisonous, flammable vapor is released.

Poison! • Highly flammable • Corrosive to the skin, eyes, and respiratory tract; contact with skin or eyes can cause burns and vision impairment; inhalation symptoms may be delayed • Firefighting gear (including SCBA) does not provide adequate protection. If exposure occurs, remove and isolate gear immediately and thoroughly decontaminate personnel • Containers may BLEVE when exposed to fire • Vapors may form explosive mixture with air • Vapors are heavier than air and will collect and stay in low areas • Vapors may travel long distances to ignition sources and flashback • Vapors in confined areas (e.g., tanks, sewers, buildings) may explode when exposed to fire • Toxic products of combustion may include carbon monoxide • Do not put yourself in danger by entering a contaminated area to rescue a victim.

Hazard Classification (based on NFPA-704 Rating System)
Health Hazards (Blue): 4; Flammability (Red): 3; Reactivity (Yellow): 1

EMERGENCY RESPONSE: See Appendix A (131)
Evacuation:
Public safety: Isolate spill area for at least 100 to 200 meters (330 to 660 feet) in all directions.
Spill: Small spill–First: Isolate in all directions 30 meters (100 feet); Then: Protect persons downwind. DAY: 0.2 km (0.1 mile); NIGHT: 0.2 km (0.1 mile). Large spill–First: Isolate in all directions 30 meters (100 feet); Then: Protect persons downwind, DAY: 0.3 km (0.2 mile); NIGHT: 0.6 km (0.4 mile).
Fire: Isolate for 800 meters (½ mile) in all directions, especially if tank, rail car, or tank truck is involved in fire.

EXPOSURE
Short-term effects: *SEEK MEDICAL ATTENTION.* **Vapor:** *POISONOUS IF INHALED OR IF ABSORBED THROUGH THE SKIN.* Irritating to eyes, nose, and throat. May cause lung edema; physical exertion will aggravate this condition. Move to fresh air. *IF IN EYES*, hold eyelids open and flush with plenty of water. *IF BREATHING HAS STOPPED*, give artificial respiration. IF breathing is difficult, administer oxygen. **Liquid:** *POISONOUS IF SWALLOWED OR IF ABSORBED THROUGH THE SKIN.* Will burn eyes. Remove contaminated clothing and shoes. Flush affected areas with plenty of water. *IF IN EYES*, hold eyelids open and flush with plenty of water. *IF SWALLOWED* and victim is *CONSCIOUS AND ABLE TO SWALLOW*, have victim drink 4 to 8 ounces of water.

Note to physician or authorized medical personnel: Medical observation is recommended for 24 to 48 hours after breathing overexposure, as pulmonary edema may be delayed. As first aid for pulmonary edema, consider administering a corticosteroid spray. Cigarette smoking may exacerbate pulmonary injury and should be discouraged for at least 72 hours following exposure.

HEALTH HAZARDS
Personal protective equipment (PPE): A-Level PPE. OSHA Table Z-1-A air contaminant. Chemical protective material(s) reported to have good to excellent resistance: butyl rubber, neoprene, Teflon®, Saranex®, chlorinated polyethylene, Viton®/Neoprene, PE, butyl rubber/neoprene, Viton®.
Recommendations for respirator selection: NIOSH/OSHA
20 ppm: SA:CF* (any supplied-air respirator operated in a continuous-flow mode); or PAPROV * [any powered, air-purifying respirator with organic vapor cartridge(s)]; or CCRFOV [any air-purifying, full-facepiece respirator (gas mask) with a chin-style, front-or back-mounted acid gas canister]; or GMFOV [any air-purifying, full-facepiece respirator (gas mask) with a chin-style, front- or back-mounted organic vapor canister]; or SCBAF (any self-contained breathing apparatus with a full facepiece); or SAF (any supplied-air respirator with a full facepiece); or SAF (any supplied-air respirator with a full facepiece). *EMERGENCY OR PLANNED ENTRY INTO UNKNOWN CONCENTRATIONS OR IDLH CONDITIONS:* SCBAF:PD,PP (any self-contained breathing apparatus that has a full facepiece and is operated in a pressure-demand or other positive-pressure mode); or SAF:PD,PP:ASCBA (any supplied-air respirator that has a full facepiece and is operated in a pressure-demand or other positive-pressure mode in combination with an auxiliary self-contained breathing apparatus operated in a pressure-demand or other positive-pressure mode). *ESCAPE:* GMFOV [any air-purifying, full-facepiece respirator (gas mask) with a chin-style, front- or back-mounted organic vapor canister]; or SCBAE (any appropriate escape-type, self-contained breathing apparatus). *Note:* Substance reported to cause eye irritation or damage; may require eye protection.
Exposure limits (TWA unless otherwise noted): ACGIH TLV 0.5 ppm; OSHA PEL 2 ppm (5 mg/m^3); NIOSH REL 2 ppm (5 mg/m^3); skin contact contributes significantly in overall exposure.
Short-term exposure limits (15-minute TWA): ACGIH STEL 4 ppm (9.5 mg/m^3); NIOSH STEL 4 ppm (10 mg/m^3); skin contact contributes significantly to overall exposure.
Toxicity by ingestion: Grade 3; LD_{50} = 50 to 500 mg/kg (mouse, rat).
Long-term health effects: May cause liver and kidney damage; possible nervous system damage.

Allyl bromide

Vapor (gas) irritant characteristics: Vapor is irritating such that personnel will not usually tolerate moderate or high vapor concentrations. High concentrations may cause lung edema; physical exertion will aggravate this condition.
Liquid or solid irritant characteristics: Causes smarting of the skin and first-degree burns on short exposure; may cause secondary burns on long exposure.
Odor threshold: 0.78 ppm.
IDLH value: 20 ppm.

FIRE DATA
Flash point: 90°F/32°C (oc); 70°F/21°C (cc).
Flammable limits in air: LEL: 2.5%; UEL 18%.
Fire extinguishing agents not to be used: Water may be ineffective, but spray can be used to cool containers.
Behavior in fire: Vapor heavier than air and may travel a considerable distance to a source of ignition and flashback.
Autoignition temperature: 713°F/378°C/651°K.
Electrical hazard: Class I, Group D. Due to low electric conductivity, this substance may generate electrostatic charges as a result of agitation and flow.
Burning rate: 2.7 mm/min

CHEMICAL REACTIVITY
Binary reactants: Mixes with air to form explosive mixture. Reacts with strong oxidizers, acids, carbon tetrachloride, peroxides. causes rubber to swell.
Polymerization: May be caused by high temperatures, oxidizers or peroxides.
Reactivity group: 15
Compatibility class: Substituted allyl

ENVIRONMENTAL DATA
Food chain concentration potential: Log P_{ow} = 0.18. Unlikely to accumulate.
Water pollution: Harmful to aquatic life in very low concentrations. May be dangerous if it enters nearby water intakes; notify operators. Notify local health and wildlife officials.
Response to discharge: Issue warning–high flammability, poison. Restrict access. Disperse and flush.

SHIPPING INFORMATION
Grades of purity: 98%; **Storage temperature:** Ambient; **Inert atmosphere:** None; **Venting:** Pressure-vacuum; **Stability during transport:** Stable at ordinary temperatures and pressures.

NAS HAZARD CLASSIFICATION FOR BULK WATER TRANSPORTATION
FIRE: 3
HEALTH: Vapor irritant: 3; Liquid or solid irritant: 2; Poisons: 3
WATER POLLUTION: Human toxicity: 2; Aquatic toxicity: 3; Aesthetic effect: 2
REACTIVITY: Other chemicals: 2; Water: 0; Self-reaction: 1

PHYSICAL AND CHEMICAL PROPERTIES
Physical state @ 59°F/15°C and 1 atm: Liquid.
Molecular weight: 58.08
Boiling point @ 1 atm: 206°F/96.9°C/370.1°K.
Melting/Freezing point: –200°F/–129°C/144°K.
Critical temperature: 521.4°F/271.9°C/545.1°K.
Critical pressure: 840 psia = 57 atm = 5.8 MN/m^2.
Specific gravity (water = 1): 0.852 @ 68°F/20°C (liquid).
Relative vapor density (air = 1): 2.0
Ratio of specific heats of vapor (gas): 1.12
Latent heat of vaporization: 295 Btu/lb = 164 cal/g = 6.87 x 10^5 J/kg.
Heat of combustion: –13,720 Btu/lb = –7620 cal/g = –319.0 x 10^5 J/kg.
Heat of solution: (estimate) Negligible.
Heat of polymerization: 100°C.
Vapor pressure: (Reid) 1.0 psia; 17 mm.

ALLYL BROMIDE   REC. A:1250

SYNONYMS: BROMALLYLENE; 3-BROMO-1-PROPENE; 3-BROMOPROPENE; 3-BROMOPROPYLENEENTIFICATION; BROMURO de ALILO (Spanish)

CAS Number: 106-95-6
Formula: C_3NO_5Br; $CH_2=CH \cdot CH_2Br$
DOT ID Number: UN 1099; DOT Guide Number: 131
Proper Shipping Name: Allyl bromide

DESCRIPTION: Colorless to light yellow liquid. Irritating odor. Sinks in water; slightly soluble. Flammable, irritating vapor is produced.

Poison! Corrosive to the skin, eyes, and respiratory tract; contact with skin or eyes can cause burns and vision impairment; inhalation symptoms may be delayed • Firefighting gear (including SCBA) does not provide adequate protection. If exposure occurs, remove and isolate gear immediately and thoroughly decontaminate personnel • Highly flammable • Containers may BLEVE when exposed to fire •Vapors may form explosive mixture with air •Vapors are heavier than air and will collect and stay in low areas • Vapors may travel long distances to ignition sources and flashback • Vapors in confined areas (e.g., tanks, sewers, buildings) may explode when exposed to fire • Severely irritating to skin, eyes, and respiratory tract; prolonged contact with the skin causes burns • Toxic products of combustion may include hydrogen bromide gas • Do not put yourself in danger by entering a contaminated area to rescue a victim.

Hazard Classification (based on NFPA-704 Rating System)
Health Hazards (Blue): 3; Flammability (Red): 3; Reactivity (Yellow): 1

EMERGENCY RESPONSE: See Appendix A (131)
Evacuation:
Public safety: Isolate spill area for at least 100 to 200 meters (330 to 660 feet) in all directions.
Spill: Increase, as necessary, the isolation distance shown above,in "Public safety."
Fire: Isolate for 800 meters (½ mile) in all directions, especially if tank, rail car, or tank truck is involved in fire.

EXPOSURE
Short-term effects: *SEEK MEDICAL ATTENTION*. **Vapor:** Irritating to eyes, nose, and throat. *IF INHALED*, will, will cause headache, dizziness, coughing or difficult breathing. Move victim to fresh air. If breathing is difficult, administer oxygen. **Liquid:** Irritating to skin and eyes. Remove contaminated clothing and shoes. Flush affected areas with plenty of water. *IF IN EYES*, hold eyelids open and flush with plenty of water. *IF SWALLOWED* and victim is *CONSCIOUS AND ABLE TO SWALLOW*, have victim drink 4 to 8 ounces of water.

HEALTH HAZARDS
Personal protective equipment (PPE): A-Level PPE.

Toxicity by ingestion: Grade 4; oral LD$_{50}$ = 30 mg/kg (guinea pig).
Vapor (gas) irritant characteristics: Vapors are moderately irritating such that personnel will not usually tolerate moderate or high vapor concentrations.
Liquid or solid irritant characteristics: Fairly severe skin irritant. May cause pain and second-degree burns after a few minutes of contact.
IDLH value: 3 ppm as bromine.

FIRE DATA
Flash point: 30°F/-1°C (cc).
Flammable limits in air: LEL: 4.4%; UEL: 7.3%.
Fire extinguishing agents not to be used: Water may be ineffective, but spray can be used to cool containers.
Autoignition temperature: 563°F/295°C/568°K.
Electrical hazard: Class I, Group D. Due to low electric conductivity, this substance may generate electrostatic charges as a result of agitation and flow.
Burning rate: 3.5 mm/min

CHEMICAL REACTIVITY
Binary reactants: Violent reaction with oxidizers, caustics, peroxides.
Polymerization: May be caused by oxidizers, peroxide or high temperatures.

ENVIRONMENTAL DATA
Food chain concentration potential: Negative; unlikely to accumulate.
Water pollution: DOT Appendix B, §172.101–marine pollutant. Effect of low concentrations on aquatic life is unknown. May be dangerous if it enters nearby water intakes; notify operators. Notify local health and wildlife officials. **Response to discharge:** Issue warning–high flammability. Restrict access. Should be removed. Chemical and physical treatment.

SHIPPING INFORMATION
Grades of purity: Commercial; **Storage temperature:** Ambient; **Venting:** Pressure-vacuum; **Stability during transport:** Stable.

NAS HAZARD CLASSIFICATION FOR BULK WATER TRANSPORTATION
FIRE: 3
HEALTH: Vapor irritant: 3; Liquid or solid irritant: 3; Poisons: 2
WATER POLLUTION: Human toxicity: 2; Aquatic toxicity: 2; Aesthetic effect: 2
REACTIVITY: Other chemicals: 2; Water: 0; Self-reaction: 1

PHYSICAL AND CHEMICAL PROPERTIES
Physical state @ 59°F/15°C and 1 atm: Liquid.
Molecular weight: 121
Boiling point @ 1 atm: 158°F/70°C/343°K.
Melting/Freezing point: −182°F/−119°C/154°K.
Specific gravity (water = 1): 1.4161 @ 68°F/20°C (liquid).
Liquid surface tension: 26.9 dynes/cm = 0.0269 N/m @ 68°F/20°C.
Liquid water interfacial tension: (estimate) 40 dynes/cm = 0.040 N/m @ 68°F/20°C.
Relative vapor density (air = 1): 4.2
Ratio of specific heats of vapor (gas): 1.1210
Latent heat of vaporization: (estimate) 110 Btu/lb = 59 cal/g = 2.5 x 10^5 J/kg.
Heat of combustion: (estimate) 6700 Btu/lb = 3700 cal/g = 150 x 10^5 J/kg.
Vapor pressure: About 111.0

ALLYL CHLORIDE   REC. A:1300

SYNONYMS: ALLYLCHLORID (German); ALLYL CHLORIDE; ALLYLE (CHLORURE d') (French); CHLORALLYLENE; 3-CHLOROPRENE; 1-CHLORO PROPENE-2; 1-CHLORO-2-PROPENE; 3-CHLOROPROPENE-1; 3-CHLOROPROPENE; 3-CHLORO-1-PROPENE; 3-CHLOROPROPYLENE; CLORURO de ALILO (Spanish); EEC No. 602-029-00-x; 1-CHLORO-2-PROPENE

IDENTIFICATION
CAS Number: 107-05-1
Formula: C_3H_5Cl; $CH_2=CHCH_2Cl$
DOT ID Number: UN 1100; DOT Guide Number: 131
Proper Shipping Name: Allyl chloride
Reportable Quantity (RQ): **(CERCLA)** 1000 lb/454 kg

DESCRIPTION: Colorless to yellow-brown or red liquid. Sharp, irritating odor. Floats on water surface; insoluble. Flammable, irritating vapor is produced.

Poison! • Highly flammable • Corrosive to the skin, eyes, and respiratory tract; contact with skin or eyes can cause burns and vision impairment; inhalation symptoms may be delayed • Firefighting gear (including SCBA) does not provide adequate protection. If exposure occurs, remove and isolate gear immediately and thoroughly decontaminate personnel • Containers may BLEVE when exposed to fire • Vapors may form explosive mixture with air • Vapors are heavier than air and will collect and stay in low areas • Vapors may travel long distances to ignition sources and flashback • Vapors in confined areas (e.g., tanks, sewers, buildings) may explode when exposed to fire • Toxic products of combustion may include hydrogen chloride and phosgene gas • Do not put yourself in danger by entering a contaminated area to rescue a victim.

Hazard Classification (based on NFPA-704 Rating System)
Health Hazards (Blue): 3; Flammability (Red): 3; Reactivity (Yellow): 1

EMERGENCY RESPONSE: See Appendix A (131)
Evacuation:
Public safety: Isolate spill area for at least 100 to 200 meters (330 to 660 feet) in all directions.
Spill: Increase, as necessary, the isolation distance shown above, in "Public safety."
Fire: Isolate for 800 meters (½ mile) in all directions, especially if tank, rail car, or tank truck is involved in fire.

EXPOSURE
Short-term effects: *SEEK MEDICAL ATTENTION.* **Vapor:** *POISONOUS IF INHALED OR IF SKIN IS EXPOSED.* Irritating to eyes, nose, and throat. May cause lung edema; physical exertion will aggravate this condition. Move to fresh air. *IF BREATHING HAS STOPPED,* give artificial respiration. IF breathing is difficult, administer oxygen. **Liquid:** *POISONOUS IF SWALLOWED OR IF SKIN IS EXPOSED.*
Will burn eyes. Remove contaminated clothing and shoes. Flush affected areas with plenty of water. *IF IN EYES,* hold eyelids open and flush with plenty of water. *IF SWALLOWED* and victim is *CONSCIOUS AND ABLE TO SWALLOW,* have victim drink 4 to 8 ounces of water and have victim induce vomiting. *IF SWALLOWED* and victim is *UNCONSCIOUS OR HAVING CONVULSIONS,* do nothing except keep victim warm.

Note to physician or authorized medical personnel: Medical observation is recommended for 24 to 48 hours after breathing overexposure, as pulmonary edema may be delayed. As first aid for pulmonary edema, consider administering a corticosteroid spray. Cigarette smoking may exacerbate pulmonary injury and should be discouraged for at least 72 hours following exposure.

HEALTH HAZARDS

Personal protective equipment (PPE): A-Level PPE. OSHA Table Z-1-A air contaminant. Chemical protective material(s) reported to offer minimal to poor protection: Chlorinated polyethylene, Teflon®, PV alcohol.
Recommendations for respirator selection: NIOSH/OSHA
25 ppm: SA:CF (any supplied-air respirator operated in a continuous-flow mode). *50 ppm:* SCBAF (any self-contained breathing apparatus with a full facepiece); or SAF (any supplied-air respirator with a full facepiece). *250 ppm:* SAF:PD,PP (any supplied-air respirator that has a full facepiece and is operated in a pressure-demand or other positive-pressure mode). *EMERGENCY OR PLANNED ENTRY INTO UNKNOWN CONCENTRATIONS OR IDLH CONDITIONS:* SCBAF:PD,PP (any self-contained breathing apparatus that has a full facepiece and is operated in a pressure-demand or other positive-pressure mode); or SAF:PD,PP:ASCBA (any supplied-air respirator that has a full facepiece and is operated in a pressure-demand or other positive-pressure mode in combination with an auxiliary, self-contained breathing apparatus operated in a pressure-demand or other positive-pressure mode. *ESCAPE:* GMFOV [any air-purifying, full-facepiece respirator (gas mask) with a chin-style, front-or back-mounted organic vapor canister]; or SCBAE (any appropriate escape-type, self-contained breathing apparatus). *Note:* Substance reported to cause eye irritation or damage; may require eye protection.
Exposure limits (TWA unless otherwise noted): ACGIH TLV 1 ppm (3 mg/m^3); NIOSH/OSHA 1 ppm (3 mg/m^3); OSHA 1 ppm (3 mg/m^3).
Short-term exposure limits (15-minute TWA): ACGIH STEL 2 ppm (6 mg/m^3); NIOSH STEL 2 ppm (6 mg/m^3); skin contact contributes significantly in overall exposure.
Toxicity by ingestion: Grade 2; LD$_{50}$ = 0.5 to 5 g/kg (rat).
Long-term health effects: Lung, liver and kidney damage in experimental animals. IARC rating 3
Vapor (gas) irritant characteristics: Vapor is irritating such that personnel will not usually tolerate moderate or high vapor concentrations. May cause lung edema; physical exertion will aggravate this condition.
Liquid or solid irritant characteristics: Severe eye and skin irritant. Causes smarting of the skin and first-degree burns on short exposure and may cause secondary burns on long exposure. Skin damage may not be apparent for some hours following contact; may cause pain.
Odor threshold: 0.47 ppm.
IDLH value: 250 ppm.

FIRE DATA
Flash point: –25°F/–32°K (cc).
Flammable limits in air: LEL: 2.9%; UEL: 11.1%.
Fire extinguishing agents not to be used: Water fog.
Autoignition temperature: 737°F/485°C/758°K.
Electrical hazard: Class I, Group D. Due to low electric conductivity, this substance may generate electrostatic charges as a result of agitation and flow.

CHEMICAL REACTIVITY

Reactivity with water: Slow decomposition forming hydrochloric acid.
Binary reactants: Violent reaction with strong oxidizers. Reacts with acids, amines, iron and aluminum chlorides, magnesium, zinc. Attacks steel and aluminum in presence of moisture.
Polymerization: May polymerize on exposure to light.
Reactivity group: 15
Compatibility class: Substituted allyl

ENVIRONMENTAL DATA
Water pollution: Harmful to aquatic life in very low concentrations. May be dangerous if it enters nearby water intakes; notify operators. Notify local health and wildlife officials.
Response to discharge: Issue warning–high flammability. Restrict access to spill site.

SHIPPING INFORMATION
Grades of purity: 97%; **Storage temperature:** Ambient; **Inert atmosphere:** None; **Venting:** Pressure-vacuum; **Stability during transport:** Stable.

NAS HAZARD CLASSIFICATION FOR BULK WATER TRANSPORTATION
FIRE: 3
HEALTH: Vapor irritant: 3; Liquid or solid irritant: 2; Poisons: 3
WATER POLLUTION: Human toxicity: 2; Aquatic toxicity: 1; Aesthetic effect: 2
REACTIVITY: Other chemicals: 2; Water: 0; Self-reaction: 1

PHYSICAL AND CHEMICAL PROPERTIES
Physical state @ 59°F/15°C and 1 atm: Liquid.
Molecular weight: 76.53
Boiling point @ 1 atm: 113°F/45°C/318°K.
Melting/Freezing point: –210.1°F/–134.5°C/138.7°K.
Critical temperature: 466°F/241°C/514°K.
Critical pressure: 690 psia = 47 atm = 4.8 MN/m^2.
Specific gravity (water = 1): 0.938 @ 68°F/20°C (liquid).
Liquid surface tension: 28.9 dynes/cm = 0.0289 N/m @ 15°C.
Relative vapor density (air = 1): 2.6
Ratio of specific heats of vapor (gas): 1.124
Heat of combustion: –9749 Btu/lb = –5416 cal/g = –226.8 x 10^5 J/kg.
Vapor pressure: (Reid) 10.3 psia; 295 mm.

ALLYL CHLOROFORMATE   REC. A:1350

SYNONYMS: ALLYL CHLOROCARBONATE; CARBONOCHLORIDIC ACID, 2-PROPENYL ESTER; CLOROFORMATO de ALILO (Spanish); 2-PROPENYL CHLORFORMATE

IDENTIFICATION
CAS Number: 2937-50-0
Formula: C$_4$H$_5$ClO$_2$; CH$_2$=CH·CH$_2$·O·COCl
DOT ID Number: UN 1722; DOT Guide Number: 155
Proper Shipping Name: Allyl chloroformate

DESCRIPTION: Colorless, watery liquid. Extremely irritating odor; causes tears. Sinks in water; reacts, forming allyl alcohol and chloroformic acid.

Highly flammable • Poison! •Corrosive •Firefighting gear (including SCBA) does not provide adequate protection. If exposure occurs, remove and isolate gear immediately and thoroughly decontaminate personnel • Containers may BLEVE when exposed to fire • Vapors may form explosive mixture with air • Vapors are heavier than air and will collect and stay in low areas • Vapors may travel long distances to ignition sources and flashback • Vapors in confined areas (e.g., tanks, sewers, buildings) may explode when exposed to fire • Severely Irritating to the skin, eyes, and respiratory tract • Toxic products of combustion may include phosgene gas • Do not put yourself in danger by entering a contaminated area to rescue a victim.

Hazard Classification (based on NFPA-704 Rating System)
Health Hazards (Blue): 3; Flammability (Red): 3; Reactivity (Yellow): 1

EMERGENCY RESPONSE: See Appendix A (155)
Evacuation:
Public safety: Isolate the area of spill or leak for at least 50 to 100 meters (160 to 330 feet) in all directions.
Spill: Small spill–First: Isolate in all directions 155 meters (500 feet). Then: Protect persons downwind, DAY: 1.3 km (0.8 mile); NIGHT: 2.7 km (1.7 miles). Large spill–First: Isolate in all directions 610 meters (2000 feet); Then: Protect persons downwind, DAY: 6.1 km (3.8 miles); NIGHT: 10.8 km (6.7 miles).
Fire: If tank, rail car, or tank truck is involved in fire, isolate for at least 800 meters (½ mile) in all directions; also, consider initial evacuation for 800 meters (½ mile) in all directions.

EXPOSURE
Short-term effects: *SEEK MEDICAL ATTENTION*. **Vapor:** Irritating to eyes, nose, and throat. *IF INHALED*, will, will cause difficult breathing. Move victim to fresh air. *IF BREATHING HAS STOPPED*, give artificial respiration; *avoid mouth-to-mouth resuscitation; use bag/mask apparatus*. If breathing is difficult, administer oxygen. **Liquid:** Irritating to skin and eyes. Harmful if swallowed. Remove contaminated clothing and shoes. Flush affected areas with plenty of water. *IF IN EYES*, hold eyelids open and flush with plenty of water. *IF SWALLOWED* and victim is *CONSCIOUS AND ABLE TO SWALLOW*, have victim drink 4 to 8 ounces of water.

HEALTH HAZARDS
Personal protective equipment (PPE): A-Level PPE. Chemical protective material(s) reported to have good to excellent resistance: butyl rubber.
Toxicity by ingestion: Grade 3; LD_{50} = 50 to 500 mg/kg.
Vapor (gas) irritant characteristics: Vapors are irritating such that personnel will not usually tolerate moderate or high vapor concentrations.
Liquid or solid irritant characteristics: Fairly severe skin irritant. May cause pain and second-degree burns after a few minutes of contact.
Odor threshold: 1.4 ppm.

FIRE DATA
Flash point: 92°F/33°C (oc); 88°F/31°C (cc).
Fire extinguishing agents not to be used: Water.
Electrical hazard: Class I, Group D.
Burning rate: 4.9 mm/min

CHEMICAL REACTIVITY
Reactivity with water: Reacts slowly, generating chloroformic acid and allyl alcohol.
Binary reactants: Reacts with peroxides, acids, oxidizers, caustics, amines, allyl alcohol. Corrodes metals

Neutralizing agents for acids and caustics: Flush with water, rinse with lime or sodium bicarbonate solution.
Polymerization: May be caused by high temperatures, peroxides, or oxidizers.

ENVIRONMENTAL DATA
Food chain concentration potential: Negative; unlikely to accumulate.
Water pollution: Effect of low concentrations on aquatic life is unknown. May be dangerous if it enters nearby water intakes; notify operators. Notify local health and wildlife officials. **Response to discharge:** Issue warning–corrosive. Restrict access. Disperse and flush.

SHIPPING INFORMATION
Grades of purity: Commercial, 97+%; **Storage temperature:** Keep cool; **Inert atmosphere:** None; **Venting:** Pressure-vacuum; **Stability during transport:** Stable.

NAS HAZARD CLASSIFICATION FOR BULK WATER TRANSPORTATION
FIRE: 3
HEALTH: Vapor irritant: 3; Liquid or solid irritant: 3; Poisons: 3
WATER POLLUTION: Human toxicity: 3; Aquatic toxicity: 3; Aesthetic effect: 3
REACTIVITY: Other chemicals: 3; Water: 0; Self-reaction: 1

PHYSICAL AND CHEMICAL PROPERTIES
Physical state @ 59°F/15°C and 1 atm: Liquid.
Molecular weight: 120.5
Boiling point @ 1 atm: 235°F/113°C/386°K.
Melting/Freezing point: –112°F/–80°C/193°K.
Specific gravity (water = 1): 1.139 @ 68°F/20°C (liquid).
Liquid surface tension: (estimate) 25 dynes/cm = 0.025 N/m @ 68°F/20°C.
Relative vapor density (air = 1): 4.15
Ratio of specific heats of vapor (gas): 1.0804
Latent heat of vaporization: (estimate) 100 Btu/lb = 56 cal/g = 2.3 x 10^5 J/kg.
Heat of combustion: (estimate) –7800 Btu/lb = –4300 cal/g = –180 x 10^5 J/kg.
Vapor pressure: 256 mm @140°F; 20 mm @ 77°F/25°C.

ALLYL TRICHLOROSILANE   REC. A:1400

SYNONYMS: ALILTRICLOROSILANO (Spanish); ALLYL TRICHOROSILANE, STABILIZED; ALLYLSILICONE TRICHLORIDE; ALLYLTRICHLOROSILANE; SILANE, TRICHLOROALLYL-; SILANE,TRICHLORO-2-PROPENYL-; ALLYLSILICONE TRICHLORIDE; TRICHLOROALLYLSILANE

IDENTIFICATION
CAS Number: 107-37-9
Formula: $C_3H_5Cl_3Si$; $CH_2=CH \cdot CH_2 \cdot SiCl_3$
DOT ID Number: UN 1724; DOT Guide Number: 155
Proper Shipping Name: Allyltrichlorosilane, stabilized

DESCRIPTION: Colorless liquid. Sharp, irritating odor like hydrochloric acid. Reacts violently with water forming hydrochloric acid and hydrogen chloride vapors.

Poison! • Breathing the vapor can kill you! Skin and eye contact can cause severe burns and blindness • Do not use water •

30 Aluminum chloride

Firefighting gear (including SCBA) does not provide adequate protection. If exposure occurs, remove and isolate gear immediately and thoroughly decontaminate personnel • Highly flammable • Containers may BLEVE when exposed to fire •Vapors may form explosive mixture with air •Vapors are heavier than air and will collect and stay in low areas • Vapors may travel long distances to ignition sources and flashback • Vapors in confined areas (e.g., tanks, sewers, buildings) may explode when exposed to fire • Severely irritating to skin, eyes, and respiratory tract; prolonged contact with the skin causes burns • Toxic products of combustion may include hydrogen chloride and phosgene gas • Do not put yourself in danger by entering a contaminated area to rescue a victim.

Hazard Classification (based on NFPA-704 Rating System)
Health Hazards (Blue): 3; Flammability (Red): 3; Reactivity (Yellow): 2; Special Notice (White): Water reactive

EMERGENCY RESPONSE: See Appendix A (155)
Evacuation:
Public safety: Isolate the area of spill or leak for at least 50 to 100 meters (160 to 330 feet) in all directions.
Spill: Increase, in the downwind direction, as necessary, the distance shown under "Public Safety."
Fire: If tank, rail car, or tank truck is involved in fire, isolate for at least 800 meters (½ mile) in all directions; also, consider initial evacuation for 800 meters (½ mile) in all directions.

EXPOSURE
Short-term effects: *SEEK MEDICAL ATTENTION.* **Vapor:** Irritating to eyes, nose, and throat. *IF INHALED*, will, will cause coughing or difficult breathing. Move victim to fresh air. *IF BREATHING HAS STOPPED*, give artificial respiration; *avoid mouth-to-mouth resuscitation; use bag/mask apparatus.* If breathing is difficult, administer oxygen. **Liquid:** Will burn skin and eyes. Harmful if swallowed. Remove contaminated clothing and shoes. Flush affected areas with plenty of water. *IF IN EYES*, hold eyelids open and flush with plenty of water. *IF SWALLOWED* and victim is *CONSCIOUS AND ABLE TO SWALLOW*, have victim drink 4 to 8 ounces of water. **Do NOT induce vomiting.**
Note to physician or authorized medical personnel: Medical observation is recommended for 24 to 48 hours after breathing overexposure, as pulmonary edema may be delayed. As first aid for pulmonary edema, consider administering a corticosteroid spray. Cigarette smoking may exacerbate pulmonary injury and should be discouraged for at least 72 hours following exposure.

HEALTH HAZARDS
Personal protective equipment (PPE): A-Level PPE. Acid-vapor-type respiratory protection; rubber gloves; chemical goggles; other equipment necessary to protect skin and eyes. Chemical protective material(s) reported to have good to excellent resistance: butyl rubber.
Toxicity by ingestion: Grade 3; LD_{50} = 50 to 500 mg/kg.
Vapor (gas) irritant characteristics: Vapors cause severe irritation of eyes and throat and can cause eye or lung injury. They cannot be tolerated even at low concentrations.
Liquid or solid irritant characteristics: Severe skin irritant. Causes second-and third-degree burns on short contact and is very injurious to the eyes.

FIRE DATA
Flash point: 100°F/38°C (oc); 95°F/35°C (cc).
Fire extinguishing agents not to be used: Water or foam.

Behavior in fire: Difficult to extinguish. Re-ignition may occur.
Burning rate: 2.2 mm/min

CHEMICAL REACTIVITY
Reactivity with water: Reacts vigorously, generating hydrogen chloride (hydrochloric acid).
Binary reactants: Corrodes metal; hydrochloric acid formed.
Neutralizing agents for acids and caustics: Flush with water, rinse with sodium bicarbonate.

ENVIRONMENTAL DATA
Food chain concentration potential: Unlikely to accumulate.
Water pollution: Effect of low concentrations on aquatic life is unknown. May be dangerous if it enters nearby water intakes; notify operators. Notify local health and wildlife officials. **Response to discharge:** Issue warning–corrosive. Restrict access. Disperse and flush.

SHIPPING INFORMATION
Grades of purity: Commercial; **Storage temperature:** Ambient; **Inert atmosphere:** None; **Venting:** Pressure-vacuum; **Stability during transport:** Stable.

NAS HAZARD CLASSIFICATION FOR BULK WATER TRANSPORTATION
FIRE: 3
HEALTH: Vapor irritant: 4; Liquid or solid irritant: 4; Poisons: 3
WATER POLLUTION: Human toxicity: 3; Aquatic toxicity: 3; Aesthetic effect: 2
REACTIVITY: Other chemicals: 3; Water: 4; Self-reaction: 4

PHYSICAL AND CHEMICAL PROPERTIES
Physical state @ 59°F/15°C and 1 atm: Liquid.
Molecular weight: 175.5
Boiling point @ 1 atm: 241°F/116°C/389°K.
Specific gravity (water = 1): 1.215 @ 68°F/20°C (liquid).
Liquid surface tension: (estimate) 20 dynes/cm = 0.020 N/m @ 68°F/20°C.
Relative vapor density (air = 1): 6.05
Ratio of specific heats of vapor (gas): 1.0863
Latent heat of vaporization: 97 Btu/lb = 54 cal/g = 2.3×10^5 J/kg.
Heat of combustion: (estimate) –5200 Btu/lb = –2900 cal/g = –120 $\times 10^5$ J/kg.

ALUMINUM CHLORIDE   **REC. A:1450**

SYNONYMS: ALUMINUMCHLORID (German); ALUMINUM TRICHLORIDE; CHLORURE d'ALUMINUM (French); CLORURO ALUMINICO ANHIDRO (Spanish); EEC No. 013-003-00-7; PEARSALL; TRICHLORO ALUMINUM

IDENTIFICATION
CAS Number: 7446-70-0
Formula: $AlCl_3$
DOT ID Number: UN 1726; DOT Guide Number: 137
Proper Shipping Name: Aluminum chloride, anhydrous

DESCRIPTION: Yellow-orange to grayish-white solid crystals or powder. Irritating odor like hydrochloric acid. Sinks in water; violent reaction producing heat, hydrogen chloride, and aluminum hydroxide.

Corrosive to the skin, eyes, and respiratory tract; contact with skin or eyes can cause burns and vision impairment; inhalation symptoms may be delayed • Firefighting gear (including SCBA) does not provide adequate protection. If exposure occurs, remove and isolate gear immediately and thoroughly decontaminate personnel • Toxic products of combustion may include hydrogen chloride.

Hazard Classification (based on NFPA-704 Rating System)
Health Hazards (Blue): 3; Flammability (Red): 0; Reactivity (Yellow): 2; Special Notice (White): Water reactive

EMERGENCY RESPONSE: See Appendix A (137)
Evacuation:
Public safety: Isolate the area of spill or leak for at least 50 to 100 meters (160 to 330 feet) in all directions.
Spill: Increase, in the downwind direction, as necessary, the distance shown under "PUBLIC SAFETY."
IF SPILLED IN WATER, Small spill–First: Isolate in all directions 30 meters (100 feet); Then: Protect persons downwind. DAY: 0.2 km (0.1 mile); NIGHT: 0.2 km (0.1 mile). Large spill–First: Isolate in all directions 60 meters (200 feet); Then: Protect persons downwind, DAY: 0.5 km (0.3 mile); NIGHT: 1.6 km (1.0 mile).
Fire: If any large container is involved in fire, isolate for at least 800 meters (½ mile) in all directions; also, consider initial evacuation for 800 meters (½ mile) in all directions.

EXPOSURE
Short-term effects: *SEEK MEDICAL ATTENTION*. **Dust:** Irritating to eyes, nose, and throat. Harmful if inhaled. Possible lung edema. Move to fresh air. *IF BREATHING HAS STOPPED*, give artificial respiration; *avoid mouth-to-mouth resuscitation; use bag/mask apparatus*. IF breathing is difficult, administer oxygen. **Solid:** Will burn skin and eyes. Harmful if swallowed. Remove contaminated clothing and shoes. Flush affected areas with plenty of water. *IF IN EYES*, hold eyelids open and flush with plenty of water. *IF SWALLOWED* and victim is *CONSCIOUS AND ABLE TO SWALLOW*, have victim drink 4 to 8 ounces of water. **Do NOT induce vomiting.**
Note to physician or authorized medical personnel: Medical observation is recommended for 24 to 48 hours after breathing overexposure, as pulmonary edema may be delayed. As first aid for pulmonary edema, consider administering a corticosteroid spray. Cigarette smoking may exacerbate pulmonary injury and should be discouraged for at least 72 hours following exposure.

HEALTH HAZARDS
Personal protective equipment (PPE): B-Level PPE. An acid-vapor canister mask should be carried in case of emergency. In certain applications, it may be advisable to wear this equipment on a routine basis. Chemical protective material(s) reported to have good to excellent resistance: butyl rubber, nitrile, neoprene, chlorinated polyethylene, polycarbonate, Viton®.
Recommendations for respirator selection:
EMERGENCY OR PLANNED ENTRY INTO UNKNOWN CONCENTRATIONS OR IDLH CONDITIONS: SCBAF:PD,PP (any self-contained breathing apparatus that has a full facepiece and is operated in a pressure-demand or other positive-pressure mode); or SAF:PD,PP:ASCBA (any supplied-air respirator that has a full facepiece and is operated in a pressure-demand or other positive-pressure mode in combination with an auxiliary self-contained breathing apparatus operated in a pressure-demand or other positive pressure mode). *ESCAPE:* GMFAG [any air-purifying, full-facepiece respirator (gas mask) with a chin-style, front- or back-mounted organic vapor cannister]; or SCBAE (any appropriate escape-type, self-contained breathing apparatus). **Note:* Reported to cause eye irritation or damage; may require eye protection.

Exposure limits (TWA unless otherwise noted): ACGIH TLV 2 mg/m^3; NIOSH REL 2 mg/m^3 as aluminum, soluble salt.
Vapor (gas) irritant characteristics: Vapor (of hydrogen chloride) is irritating such that personnel will not usually tolerate moderate or high vapor concentrations. Corrosive to the eyes, skin, and respiratory tract. Possible lung edema.
Liquid or solid irritant characteristics: Fairly severe skin irritant; may cause pain and second-degree burns after a few minutes of contact.
Odor threshold: 1–5 ppm; 0.256–10.1 ppm as hydrogen chloride.
IDLH value: 50 ppm as hydrogen chloride.

FIRE DATA
Fire extinguishing agents not to be used: Do not use water; possible violent reaction.

CHEMICAL REACTIVITY
Reactivity with water: Reacts with water, liberating hydrogen chloride gas and heat; becomes a strong acid.
Binary reactants: Reacts with organic materials and bases. If wet it attacks metals due to hydrochloric acid formation; flammable hydrogen is formed.
Neutralizing agents for acids and caustics: Hydrochloric acid formed by reaction with water can be flushed away with water. Rinse with sodium bicarbonate or lime solution.

ENVIRONMENTAL DATA
Water pollution: Harmful to aquatic life in very low concentrations. May be dangerous if it enters nearby water intakes; notify operators. Notify local health and wildlife officials.
Response to discharge: Issue warning–corrosive. Disperse and flush with care.

SHIPPING INFORMATION
Grades of purity: Pure: 99.7%; technical: 98.5%; **Stability during transport:** Stable if kept dry and protected from atmospheric moisture.

PHYSICAL AND CHEMICAL PROPERTIES
Physical state @ 59°F/15°C and 1 atm: Solid.
Molecular weight: 133.34
Boiling point @ 1 atm: Sublimes at 365°F/185°C/458°K.
Melting/Freezing point: 381°F/193.9°C/467.1°K @ 5.2 atm.
Specific gravity (water = 1): 2.44 @ 77°F/25°C (solid).
Heat of fusion: 63.6 cal/g.
Vapor pressure: 0.0031 mm.

ALUMINUM CHLORIDE SOLUTION   REC. A:1500

SYNONYMS: ALUMINUM TRICHLORIDE SOLUTION; TRICHLOROALUMINUM SOLUTION

IDENTIFICATION
CAS Number: 7446-70-0
Formula: AlCl$_3$; AlCl$_3$·H$_2$O
DOT ID Number: UN 2581; DOT Guide Number: 154
Proper Shipping Name: Aluminum chloride, solution

DESCRIPTION: Clear, colorless to amber liquid. Mild, pungent odor, like hydrochloric acid. Soluble in water.

Corrosive to the skin, eyes, and respiratory tract; contact with skin or eyes can cause burns and vision impairment; inhalation symptoms may be delayed • Firefighting gear (including SCBA)

does not provide adequate protection. If exposure occurs, remove and isolate gear immediately and thoroughly decontaminate personnel • Toxic products of combustion may include hydrogen chloride.

Hazard Classification (based on NFPA-704 Rating System)
Health Hazards (Blue): 1; Flammability (Red): 0; Reactivity (Yellow): 0

EMERGENCY RESPONSE: See Appendix A (154)
Evacuation:
Public safety: Isolate the area of spill or leak for at least 25 to 50 meters (80 to 160 feet) in all directions.
Spill: Increase, in the downwind direction, as necessary, the distance shown under "Public Safety."
Fire: If tank, rail car, or tank truck is involved in fire, isolate for at least 800 meters (½ mile) in all directions; also, consider initial evacuation for 800 meters (½ mile) in all directions.

EXPOSURE
Short-term effects: *SEEK MEDICAL ATTENTION.* **Vapor:** Corrosive. May burn eyes, and respiratory tract and skin. Move to fresh air. *IF BREATHING HAS STOPPED,* give artificial respiration; *avoid mouth-to-mouth resuscitation; use bag/mask apparatus.* IF breathing is difficult, administer oxygen. **Liquid:** Will burn skin and eyes. Harmful if swallowed. Remove contaminated clothing and shoes. Flush affected areas with plenty of water. *IF IN EYES,* hold eyelids open and flush with plenty of water. *IF SWALLOWED* and victim is *CONSCIOUS AND ABLE TO SWALLOW,* have victim drink 4 to 8 ounces of water.
Note to physician or authorized medical personnel: Medical observation is recommended for 24 to 48 hours after breathing overexposure, as pulmonary edema may be delayed. As first aid for pulmonary edema, consider administering a corticosteroid spray. Cigarette smoking may exacerbate pulmonary injury and should be discouraged for at least 72 hours following exposure.

HEALTH HAZARDS
Personal protective equipment (PPE): B-Level PPE. Wear impervious protective clothing and gloves to prevent skin contact. Wear splashproof chemical safety goggles and approved respirator. Chemical protective material(s) reported to have good to excellent resistance: butyl rubber, nitrile, neoprene, chlorinated polyethylene, polycarbonate, Viton®.
Toxicity by ingestion: Grade 3; oral LD_{50} = 770 mg/kg (mouse).
Exposure limits (TWA unless otherwise noted): ACGIH TLV 2 mg/m^3; NIOSH REL 2 mg/m^3 as aluminum, soluble salt.
Vapor (gas) irritant characteristics: Vapor is moderately irritating such that personnel will not usually tolerate moderate or high vapor concentrations.
Liquid or solid irritant characteristics: Fairly severe skin irritant; may cause pain and second-degree burns after a few minutes of contact.

CHEMICAL REACTIVITY reported.
Binary reactants: Corrosive to many metals, such as aluminum, steel, copper, and zinc. Reaction will produce flammable hydrogen gas.
Neutralizing agents for acids and caustics: Sodium carbonate (soda ash), lime (calcium hydroxide), or limestone (calcium carbonate).

ENVIRONMENTAL DATA
Water pollution: Effect of low concentrations on aquatic life is unknown. May be dangerous if it enters nearby water intakes; notify operators. Notify local health and wildlife officials. **Response to discharge:** Issue warning–corrosive. Chemical and physical treatment. Should be removed.

SHIPPING INFORMATION
Venting: None; **Stability during transport:** Stable.

PHYSICAL AND CHEMICAL PROPERTIES
Physical state @ 59°F/15°C and 1 atm: Liquid.
Molecular weight: 133.34
Boiling point @ 1 atm: 230°F/110°C/383°K.
Melting/Freezing point: –30°F/–34°C/239°K.
Specific gravity (water = 1): 1.28

ALUMINUM FLUORIDE　　　　　　　　REC. A:1550

SYNONYMS: FLUORURO ALUMINICO ANHIDRO (Spanish)

IDENTIFICATION
CAS Number: 7784-18-1
Formula: AlF_3; $AlF_3 \cdot 3H_2O$
DOT ID Number: UN 3077; DOT Guide Number: 171
Proper Shipping Name: Environmentally hazardous substance, solid, n.o.s.

DESCRIPTION: White solid powder or granules. Odorless. Sinks in water; slightly soluble.

Toxic • Avoid breathing dust in air; Irritating to the skin, eyes, and respiratory tract • Toxic products of combustion may include fluorides.

Hazard Classification (based on NFPA-704 Rating System)
Health Hazards (Blue): 3; Flammability (Red): 0; Reactivity (Yellow): 0

EMERGENCY RESPONSE: See Appendix A (171)
Evacuation:
Public safety: Isolate the area of spill or leak for at least 10 to 25 meters (30 to 80 feet) in all directions.
Spill: Increase, in the downwind direction, as necessary, the distance shown under "Public Safety."
Fire: If any large container is involved in fire, isolate for at least 800 meters (½ mile) in all directions; also, consider initial evacuation for 800 meters (½ mile) in all directions.

EXPOSURE
Short-term effects: Dust: If inhaled, irritating to nose and throat. Move to fresh air.
Note to physician or authorized medical personnel: For acute poisoning, oral administration of limewater, intravenous infusion of glucose, and intravenous injections of calcium gluconates.

HEALTH HAZARDS
Personal protective equipment (PPE): B-Level PPE.
Recommendations for respirator selection: NIOSH/OSHA as F. *12.5 mg/m^3:* DM (any dust and mist respirator). *25 mg/m^3:* DMXSQ* (any dust and mist respirator except single-use and quarter-mask respirators); or SA* (any supplied-air respirator). *62.5 mg/m^3:* SA:CF* [any supplied-air respirator operated in a continuous-flow mode)]; or PAPRDM*[+] *if not present as a fume* (any powered, air-purifying respirator with a dust and mist filter). *125 mg/m^3:* HiEF[+] (any air-purifying, full-facepiece respirator with a high-efficiency particulate filter); or SCBAF (any self-contained breathing apparatus with a full facepiece); or SAF (any supplied-air

respirator with a full facepiece). *250 mg/m³:* SA:PD,PP (any supplied-air respirator operated in a pressure-demand or other positive-pressure mode). *EMERGENCY OR PLANNED ENTRY INTO UNKNOWN CONCENTRATIONS OR IDLH CONDITIONS:* SCBAF:PD,PP (any self-contained breathing apparatus that has a full faceplate and is operated in a pressure-demand or other positive-pressure mode); or SAF:PD,PP:ASCBA (any supplied-air respirator that has a full facepiece and is operated in a pressure-demand or other positive-pressure mode in combination with an auxiliary, self-contained breathing apparatus operated in a pressure-demand or other positive-pressure mode). *ESCAPE:* HiEF⁺ (any air-purifying, full-facepiece respirator with a high-efficiency particulate filter); or SCBAE (any appropriate escape-type, self-contained breathing apparatus). *Notes:* *Substance reported to cause eye irritation or damage; may require eye protection. ⁺May need acid gas sorbent.
Toxicity by ingestion: LD_{LO} = 600 mg/kg (guinea pig).
Exposure limits (TWA unless otherwise noted): ACGIH TLV 2 mg/m³; NIOSH REL 2 mg/m³ as aluminum, soluble salt.
Long-term health effects: Skeletal fluorosis (bone abnormalities) in humans, working in aluminum plant for 12 years.
Vapor (gas) irritant characteristics: May cause irritation of the eyes and respiratory tract.
Liquid or solid irritant characteristics: May cause eye and skin irritation.
IDLH value: 250 mg/m³ as F.

CHEMICAL REACTIVITY
Reactivity with water: Forms sulfuric acid with water.
Binary reactants: Aqueous solution has a violent reaction with bases and many other materials. Dry material is weakly corrosive to carbon steel; aqueous solution attacks aluminum and other metals, forming hydrogen gas.

ENVIRONMENTAL DATA
Water pollution: Harmful to aquatic life in very low concentrations. May be dangerous if it enters nearby water intakes; notify operators. Notify local health and wildlife officials.
Response to discharge: Disperse and flush.

SHIPPING INFORMATION
Grades of purity: 90.7%; **Stability during transport:** Stable.

PHYSICAL AND CHEMICAL PROPERTIES
Physical state @ 59°F/15°C and 1 atm: Solid.
Boiling point @ 1 atm: 2799°F/1537°C.
Melting/Freezing point: 2356°F/1291°C.
Molecular weight: 83.98
Specific gravity (water = 1): 2.88 @ 77°F/25°C (solid).

ALUMINUM NITRATE　　　　　　　　　　　REC. A:1650

SYNONYMS: ALUMINUM(III) NITRATE; ALUMINUM SALT OF NITRIC ACID; ALUMINUM TRINITRATE; NITRAM; NITRATO ALUMINICO (Spanish); NITRIC ACID, ALUMINUM SALT; NITRIC ACID, ALUMINUM(3+); NORWAY SALTPETER; VARIOFORM I

IDENTIFICATION
CAS Number: 13473-90-0; 7784-27-2 (nonahydrate)
Formula: $N_3O_9 \cdot Al$; $Al(NO_3)_3 \cdot 9H_2O$; $N_3O_9 \cdot Al \cdot 9H_2O$ (nonahydrate)
DOT ID Number: UN 1438; DOT Guide Number: 140
Proper Shipping Name: Aluminum nitrate

DESCRIPTION: White solid. Odorless. Sinks and mixes slowly with water.

Strong oxidizer that may react spontaneously with low flash point organics or reducing agents. Heat forms oxygen; will increase the activity of an existing fire • May cause fire or explosion on contact with combustibles (wood, paper, oil, clothing, etc.). Toxic products of combustion may include nitrogen oxide.

Hazard Classification (based on NFPA-704 Rating System)
Health Hazards (Blue): 2; Flammability (Red): 0; Reactivity (Yellow): 1; Special Notice (White): Oxidizer

EMERGENCY RESPONSE: See Appendix A (140)
Evacuation:
Public safety: Isolate the area of spill or leak for at least 10 to 25 meters (30 to 80 feet) in all directions.
Spill: Consider initial downwind evacuation for at least 100 meters (330 feet).

Fire: If any large container is involved in fire, isolate for at least 800 meters (½ mile) in all directions; also, consider initial evacuation for 800 meters (½ mile) in all directions.

EXPOSURE
Short-term effects: *SEEK MEDICAL ATTENTION.* **Dust:** Irritating to eyes, nose, and throat. Harmful if inhaled. *IF IN EYES,* hold eyelids open and flush with plenty of water. *IF BREATHING HAS STOPPED,* give artificial respiration. IF breathing is difficult, administer oxygen. **Solid:** Irritating to skin and eyes. *IF SWALLOWED,* will cause nausea or vomiting. Remove contaminated clothing and shoes. Flush affected areas with plenty of water. *IF IN EYES,* hold eyelids open and flush with plenty of water. *IF SWALLOWED* and victim is *CONSCIOUS AND ABLE TO SWALLOW,* have victim drink 4 to 8 ounces of water. *IF SWALLOWED* and victim is *UNCONSCIOUS OR HAVING CONVULSIONS,* do nothing except keep victim warm.

HEALTH HAZARDS
Personal protective equipment (PPE): B-Level PPE. Goggles or face shield; dust respirator; rubber gloves.
Exposure limits (TWA unless otherwise noted): ACGIH TLV 2 mg/m³; NIOSH REL 2 mg/m³ as aluminum, soluble salt.
Toxicity by ingestion: Grade 3; oral rat LD_{50} = 264 mg/kg (nonahydrate).
Long-term health effects: Possible blood effects; eczema.
Vapor (gas) irritant characteristics: Irritates the eyes and respiratory tract. May cause unconsciousness.
Liquid or solid irritant characteristics: Irritation of the eyes and skin.

FIRE DATA
Behavior in fire: May increase the intensity of fire when in contact with combustible material.

CHEMICAL REACTIVITY
Reactivity with water: Dissolves and forms a solution of nitric acid.
Binary reactants: A strong oxidizer. Incompatible with combustibles, organics, reducing agents, acetonitrile. Aqueous solution is a strong acid; reacts with bases. Attacks metals in the presence of moisture.
Neutralizing agents for acids and caustics: Flush with water.

ENVIRONMENTAL DATA

Food chain concentration potential: Negative; unlikely to accumulate. reported.
Water pollution: Harmful to aquatic life in very low concentrations. May be dangerous if it enters nearby water intakes; notify operators. Notify local health and wildlife officials.
Response to discharge: Issue warning–water contaminant. Disperse and flush.

SHIPPING INFORMATION

Grades of purity: Reagent, 99+%; Technical; **Storage temperature:** Ambient; **Inert atmosphere:** None; **Venting:** Open; **Stability during transport:** Stable.

PHYSICAL AND CHEMICAL PROPERTIES

Physical state @ 59°F/15°C and 1 atm: Solid.
Molecular weight: 375.13
Boiling point @ 1 atm: 212°F/100°C/373°K (decomposes).
Melting/Freezing point: 163°F/73°C/346°K.
Specific gravity (water = 1): More than 1 @ 68°F/20°C(solid).
Heat of decomposition: 134°C.

ALUMINUM PHOSPHIDE   REC. A:1700

SYNONYMS: AIP; AL-PHOS; ALUMINUM PHOSPHITE; ALUMINUM MONOPHOSPHIDE; DELICIA; DELICIA GASTOXIN; DETIA; DETIA GAS EX-B; DETIA-EX-B; EEC No. 015-004-00-8; FOSFURO ALUMINICO (Spanish); PHOSPHURES D'ALUMIUM (French) PHOSTOXIN; QUICKPHOS; RCRA No. P006ENTIFICATION:

CAS Number: 20859-73-8
Formula: AlP
DOT ID Number: UN 1397; DOT Guide Number: 139
Proper Shipping Name: Aluminum phosphide
Reportable Quantity (RQ): **(CERCLA)** 100 lb/45.4 kg

DESCRIPTION: Dark gray or dark yellow crystalline solid. May have "fishy," phosphine odor, especially on contact with moist air or water. Sinks and reacts with water producing poisonous and spontaneously flammable phosphine gas and hydrogen chloride vapor.

Extremely flammable • Poison! • Dangerous when wet; do NOT use water based extinguishers • Breathing the vapors can kill • Skin and eye contact causes severe burns and blindness • Ignites spontaneously in moist air forming explosive phosphine gas • Firefighting gear (including SCBA) does not provide adequate protection. If exposure occurs, remove and isolate gear immediately and thoroughly decontaminate personnel • Vapors are heavier than air and will collect and stay in low areas • Vapors may travel long distances to source of ignition and flashback • Often shipped and stored dissolved in acetone • Exposure of cylinders to elevated temperatures, fire, and flame may cause cylinders to rupture or cause frangible disk to burst, releasing entire contents of cylinder • When combined with surface moisture this material is corrosive to most common metals • Toxic products of combustion may include oxides of phosphorus, phosphoric acid mist, and aluminum fumes. • May be reignited after fire is extinguished. Runoff may create fire or explosion hazard • Severely irritating to skin, eyes, and respiratory tract; prolonged contact with the skin causes burns • Do not put yourself in danger by entering a contaminated area to rescue a victim.

Hazard Classification (based on NFPA-704 Rating System)
Health Hazards (Blue): 4; Flammability (Red): 4; Reactivity (Yellow): 2; Special Notice (White): Water reactive

EMERGENCY RESPONSE: See Appendix A (139)
Evacuation:
Public safety: Isolate the area of spill or leak for at least 100 to 150 meters (330 to 490 feet) in all directions.
Spills: Increase, in the downwind direction, as necessary, the distance shown under "Public Safety."
IF SPILLED IN WATER: Small spill–First: Isolate in all directions 30 meters (100 feet); Then: Protect persons downwind, DAY: 0.2 km (0.1 mile); NIGHT: 0.8 km (0.5 mile). Large spill–First: Isolate in all directions 240 meters (800 feet); Then: Protect persons downwind, DAY: 2.4 km (1.5 mile); NIGHT: 6.4 km (4.0 miles).
Fire: If any large container is involved in fire, isolate for at least 800 meters (½ mile) in all directions; also, consider initial evacuation for 800 meters (½ mile) in all directions.

EXPOSURE

Short-term effects: *SEEK MEDICAL ATTENTION*. **Vapor:** *POISONOUS VAPOR PRODUCED IN REACTION WITH WATER MAY BE FATAL IF INHALED*. Severe pulmonary irritant and an acute systemic poison; may cause sudden or delayed death.
Note to physician or authorized medical personnel: Medical observation is recommended for 24 to 48 hours after breathing overexposure, as pulmonary edema may be delayed. As first aid for pulmonary edema, consider administering a corticosteroid spray. Cigarette smoking may exacerbate pulmonary injury and should be discouraged for at least 72 hours following exposure.
Dust: *POISONOUS; MAY BE FATAL IF INHALED*. Move to fresh air. *IF BREATHING HAS STOPPED,* give artificial respiration. IF breathing is difficult, administer oxygen. Effects may be delayed; keep under observation. **Solid:** *POISONOUS IF SWALLOWED OR SKIN EXPOSED. IF IN EYES OR ON SKIN*, flush with running water for at least 15 minutes; hold eyelids open periodically if appropriate. Remove and isolate contaminated clothing and shoes. *IF SWALLOWED* and victim is *UNCONSCIOUS OR HAVING CONVULSIONS,* do nothing except keep warm.

HEALTH HAZARDS

Personal protective equipment (PPE): A-Level PPE. Wear fully encapsulating protective clothing and approved positive-pressure, self-contained respirator.
Exposure limits (TWA unless otherwise noted): ACGIH TLV 2 mg/m^3; NIOSH REL 2 mg/m^3 as aluminum, soluble salt.
Toxicity by ingestion: Grade 4; LD$_{50}$ = 20 mg/kg (human).
Long-term health effects: Kidney and liver damage.
Vapor (gas) irritant characteristics: In the presence of moisture, it generates pyrophoric phosphine gas, a severe pulmonary irritant. High exposure may result in death from metabolic and nervous system failure.
Liquid or solid irritant characteristics: Irritates the eyes and respiratory tract.

FIRE DATA

Flash point: Noncombustible. See below, reactivity with water.
Fire extinguishing agents not to be used: Do not use water, foam or, CO$_2$.

CHEMICAL REACTIVITY

Reactivity with water: Reacts with water or steam to produce phosphine gas. Phosphine is highly toxic and spontaneously flammable; and its combustion products, such as oxides of phosphorus, are also highly toxic.

Binary reactants: Contact with moisture (including atmospheric moisture), water, steam, or alkali liberates spontaneously combustible phosphine gas. Contact with strong acids is violent and also yields phosphine gas. Violent reaction with chlorine, potassium nitrate. May react with the moisture in wood to produce spontaneously flammable phosphine gas.

ENVIRONMENTAL DATA
Water pollution: Effects of low concentration on aquatic life is unknown. May be dangerous if it enters nearby water intakes; notify operators. Notify local health and wildlife officials. **Response to discharge:** Issue warning–high flammability, poison, air and water contaminant. Restrict access to spill site. Evacuate area (large discharge only). Should be removed.

SHIPPING INFORMATION
Stability during transport: Stable if kept dry.

PHYSICAL AND CHEMICAL PROPERTIES
Physical state @ 59°F/15°C and 1 atm: Solid.
Molecular weight: 57.96
Melting/Freezing point; More than 1832°F/1000°C/1273°K.
Specific gravity (water = 1): 2.85 @ 77°F/25°C.

ALUMINUM SULFATE   REC. A:1750

SYNONYMS: ALUM; ALUMINUM ALUM; ALUMINUM TRISULFATE; ALUNOGENITE; CAKE ALUM; CAKE ALUMINUM; DIALUMINUM SULFATE; DIALUMINUM TRISULFATE; FILTER ALUM; PATENT ALUM; PATENT ALUMINUM; PERL ALUM; PAPER MAKER'S ALUM; PICKEL ALUM; SULFATO ALUMINICO (Spanish); SULFURIC ACID, ALUMINUM SALT

IDENTIFICATION
CAS Number: 10043-01-3
Formula: $O_{12}S_3 \cdot 2Al$; $Al_2(SO_4)_3 \cdot 18H_2O$
DOT ID Number: UN 3077; DOT Guide Number: 171
Proper Shipping Name: Environmentally hazardous substance, solid, n.o.s.
Reportable Quantity (RQ): **(CERCLA)** 5000 lb/2270 kg

DESCRIPTION: Gray-white solid. Odorless. Sinks and mixes slowly with water forming sulfuric acid.

Dust is extremely Irritating to the skin, eyes, and respiratory tract • Toxic products of combustion may include corrosive sulfur trioxide vapors.

Hazard Classification (based on NFPA-704 Rating System)
Health Hazards (Blue): 0; Flammability (Red): 0; Reactivity (Yellow): 0

EMERGENCY RESPONSE: See Appendix A (171)
Evacuation:
Public safety: Isolate the area of spill or leak for at least 10 to 25 meters (30 to 80 feet) in all directions.
Spill: Increase, in the downwind direction, as necessary, the distance shown under "Public Safety."
Fire: If any large container is involved in fire, isolate for at least 800 meters (½ mile) in all directions; also, consider initial evacuation for 800 meters (½ mile) in all directions.

EXPOSURE
Short-term effects: *SEEK MEDICAL ATTENTION.* **Dust:** Irritating to eyes, nose, and throat. *IF INHALED*, will, will cause difficult breathing. *IF IN EYES*, hold eyelids open and flush with plenty of water. *IF BREATHING HAS STOPPED*, give artificial respiration. IF breathing is difficult, administer oxygen. **Solid:** Irritating to skin and eyes. *IF SWALLOWED*, will cause nausea or vomiting. Remove contaminated clothing and shoes. Flush affected areas with plenty of water. *IF IN EYES*, hold eyelids open and flush with plenty of water. *IF SWALLOWED* and victim is *CONSCIOUS AND ABLE TO SWALLOW*, have victim drink 4 to 8 ounces of water. *IF SWALLOWED* and victim is *UNCONSCIOUS OR HAVING CONVULSIONS*, do nothing except keep victim warm.

HEALTH HAZARDS
Personal protective equipment (PPE): B-Level PPE. Dust respirator; goggles or face shield; rubber gloves.
Exposure limits (TWA unless otherwise noted): ACGIH TLV 2 mg/m³; NIOSH REL 2 mg/m³ as aluminum, soluble salt.
Toxicity by ingestion: Grade 2; oral mouse LD_{50} = 770 mg/kg.
Long-term health effects: Skin disorders (eczema).
Vapor (gas) irritant characteristics: Irritates the respiratory tract.
Liquid or solid irritant characteristics: Irritates the eyes.

FIRE DATA
Fire extinguishing agents not to be used: Water.

CHEMICAL REACTIVITY
Reactivity with water: Solution is a strong acid.
Binary reactants: Violent reaction with aluminum, magnesium. Reacts with bases. May corrode metals in presence of moisture.
Neutralizing agents for acids and caustics: Flush with water.

ENVIRONMENTAL DATA
Water pollution: Harmful to aquatic life in very low concentrations. May be dangerous if it enters water intake. Notify local health and wildlife officials. **Response to discharge:** Issue warning–water contaminant. Should be removed. Chemical and physical treatment.

SHIPPING INFORMATION
Grades of purity: Technical; **Storage temperature:** Ambient; **Inert atmosphere:** None; **Venting:** Open; **Stability during transport:** Stable.

PHYSICAL AND CHEMICAL PROPERTIES
Physical state @ 59°F/15°C and 1 atm: Solid.
Molecular weight: 342.0
Melting/Freezing point: Decomposes. 1409°F/765°C/1038°K.
Specific gravity (water = 1): 1.7 @ 68°F/20°C (solid).
Heat of decomposition: 765°C.
Heat of solution: –22.1 Btu/lb = –12.3 cal/g = 0.515 x 10^5 J/kg.

ALUMINUM SULFATE SOLUTION   REC. A:1800

SYNONYMS: ALUMINUM TRISULFATE SOLUTION; DIALUMINUM SULFATE SOLUTION; PAPER MAKER'S ALUM

IDENTIFICATION
CAS Number: 10043-01-3
Formula: $Al_2(SO_4)_3 \cdot 18H_2O$
DOT ID Number: UN 3082; DOT Guide Number: 171

Proper Shipping Name: Environmentally hazardous substance, liquid, n.o.s.
Reportable Quantity (RQ): 5000 lb/2270 kg

DESCRIPTION: Colorless liquid. Odorless.

Corrosive • Toxic products of combustion may include sulfur trioxide and sulfuric acid fumes.

Hazard Classification (based on NFPA-704 Rating System)
Health Hazards (Blue): 0; Flammability (Red): 0; Reactivity (Yellow): 0

EMERGENCY RESPONSE: See Appendix A (171)
Evacuation:
Public safety: Isolate the area of spill or leak for at least 10 to 25 meters (30 to 80 feet) in all directions.
Spill: Increase, in the downwind direction, as necessary, the distance shown under "Public Safety."
Fire: If any large container is involved in fire, isolate for at least 800 meters (½ mile) in all directions; also, consider initial evacuation for 800 meters (½ mile) in all directions.

EXPOSURE
Short-term effects: *SEEK MEDICAL ATTENTION*. **Vapor:** Irritating to eyes, nose, and throat. Move to fresh air. *IF BREATHING HAS STOPPED*, give artificial respiration. IF breathing is difficult, administer oxygen. **Liquid:** Irritating to skin and eyes. *IF SWALLOWED*, may cause nausea vomiting or diarrhea. Remove contaminated clothing and shoes. Flush affected areas with plenty of water. *IF IN EYES*, hold eyelids open and flush with plenty of water. *IF SWALLOWED and victim is CONSCIOUS AND ABLE TO SWALLOW*, have victim drink large quantity of water.
Induce vomiting. *IF SWALLOWED and victim is UNCONSCIOUS OR HAVING CONVULSIONS*, do nothing except keep victim warm.
Note to physician or authorized medical personnel: Medical observation is recommended for 24 to 48 hours after breathing overexposure, as pulmonary edema may be delayed. As first aid for pulmonary edema, consider administering a corticosteroid spray. Cigarette smoking may exacerbate pulmonary injury and should be discouraged for at least 72 hours following exposure.

HEALTH HAZARDS
Personal protective equipment (PPE): B-Level PPE. Wear impervious chemical protective clothing, including gloves to prevent skin contact with the liquid. Use splash-proof chemical safety goggles or face shield to prevent eye contact with liquid. Approved respirator.
Exposure limits (TWA unless otherwise noted): ACGIH TLV 2 mg/m^3; NIOSH REL 2 mg/m^3 as aluminum, soluble salt.
Toxicity by ingestion: Grade 2; oral rat LD$_{50}$ more than 5.0 g/kg.
Long-term health effects: May cause skin disorders (eczema).
Vapor (gas) irritant characteristics: May cause eye and respiratory tract irritation.

CHEMICAL REACTIVITY
Reactivity with water: Forms sulfuric acid with water.
Binary reactants: Aqueous solution has a violent reaction with bases and many other materials. Dry material is weakly corrosive to carbon steel; aqueous solution attacks aluminum, magnesium, and other metals, forming hydrogen gas.
Neutralizing agents for acids and caustics: Neutralize with lime.

Reactivity group: 43
Compatibility class: Miscellaneous water solutions.

ENVIRONMENTAL DATA
Water pollution: Effect of low concentrations on aquatic life is unknown. May be dangerous if it enters water intake. Notify local health and wildlife officials. **Response to discharge:** Should be removed. Chemical and physical treatment.

SHIPPING INFORMATION
Grades of purity: Technical grades of varying concentrations. **Storage temperature:** Ambient; **Inert atmosphere:** None; **Venting:** Open; **Stability during transport:** Stable.

PHYSICAL AND CHEMICAL PROPERTIES
Physical state @ 59°F/15°C and 1 atm: Liquid.
Molecular weight: 342.2
Boiling point @ 1 atm: 214°F/101°C/374°K.
Melting/Freezing point: 4°F/–16°C/257°K.
Specific gravity (water = 1): 1.29-1.34 @ 15°C (solid).
Relative vapor density (air = 1): 2.7

AMINOETHYLETHANOL AMINE   REC. A:1850

SYNONYMS: *N*-(2-AMINOETHYL) ETHANOLAMINE; 2-(2-AMINOETHOXY) ETHANOL; 2-[(2-AMINOETHYL)AMINO] ETHANOL; *N*-AMINOETHYLETHANOL AMINE; 1-(2-HYDROXYETHYLAMINO)-2-AMINOETHANE; HIDRIXIETILETILENIMIA (Spanish); 2-(2-HYDROXYETHYLAMINO)ETHYLAMINE; HYDROXYETHYLETHYLENEDIAMINE, *N*-β-; *N*-HYDROXETHYL-1,2-ETHANEDIAMINE; *N*-β-HYDROXYETHYLETHYLENEDIAMINE

IDENTIFICATION
CAS Number: 111-41-1
Formula: C$_4$H$_{12}$N$_2$O; HOCH$_2$CH$_2$NHCH$_2$CH$_2$NH$_2$
DOT ID Number: UN 3055; DOT Guide Number: 153
Proper Shipping Name: 2-(2-Aminoethoxy) ethanol

DESCRIPTION: Colorless liquid. Mild ammonia odor. Sinks and mixes with water.

Corrosive • Breathing the vapor can kill you; skin or eye contact causes severe burns, impaired vision, or blindness; effects of contact or inhalation may be delayed • Firefighting gear (including SCBA) does not provide adequate protection. If exposure occurs, remove and isolate gear immediately and thoroughly decontaminate personnel • Combustible • Heat or flame may cause explosion • Containers may BLEVE when exposed to fire • Vapors in confined areas (e.g., tanks, sewers, buildings) may explode when exposed to fire • Toxic products of combustion may include nitrogen oxides.

Hazard Classification (based on NFPA-704 Rating System)
Health Hazards (Blue): 2; Flammability (Red): 1; Reactivity (Yellow): 0

EMERGENCY RESPONSE: See Appendix A (153)
Evacuation:
Public safety: Isolate the area of spill or leak for at least 25 to 50 meters (80 to 160 feet) in all directions.
Spill: Increase, in the downwind direction, as necessary, the distance shown under "Public Safety."

Fire: If tank, rail car, or tank truck is involved in fire, isolate for at least 800 meters (½ mile) in all directions; also, consider initial evacuation for 800 meters (½ mile) in all directions.

EXPOSURE
Short-term effects: *SEEK MEDICAL ATTENTION.* **Fumes:** *IF BREATHING HAS STOPPED,* give artificial respiration; *avoid mouth-to-mouth resuscitation; use bag/mask apparatus.* May cause lung edema; physical exertion will aggravate this condition. **Liquid:** Will burn skin and eyes. Remove contaminated clothing and shoes. Flush affected areas with plenty of water. *IF IN EYES,* hold eyelids open and flush with plenty of water. *IF SWALLOWED* and victim is *CONSCIOUS AND ABLE TO SWALLOW,* have victim drink 4 to 8 ounces of water. **Do NOT induce vomiting.**
Note to physician or authorized medical personnel: Medical observation is recommended for 24 to 48 hours after breathing overexposure, as pulmonary edema may be delayed. As first aid for pulmonary edema, consider administering a corticosteroid spray. Cigarette smoking may exacerbate pulmonary injury and should be discouraged for at least 72 hours following exposure.

HEALTH HAZARDS
Personal protective equipment (PPE): B-Level PPE. Wear butyl rubber gloves, face shield and protective clothing.
Toxicity by ingestion: Grade 2; LD_{50} 0.5 to 5 g/kg.
Long-term health effects: Prolonged or repeated skin exposure may cause eczema or other skin problems.
Vapor (gas) irritant characteristics: Respiratory tract irritant. High concentrations of vapor may cause smarting of the eyes or respiratory system. May cause lung edema; physical exertion will aggravate this condition.
Liquid or solid irritant characteristics: Severe eye irritant. Skin irritant; may cause pain and second-degree burns after a few minutes of contact.

FIRE DATA
Flash point: 275°F/135°C (oc).
Flammable limits in air: LEL: 15%; UEL: 27%.
Fire extinguishing agents not to be used: Water or foam may cause frothing. Use water to cool exposed containers.
Autoignition temperature: 695°F/368°C/641°K.
Electrical hazard: Class I, Group C.

CHEMICAL REACTIVITY
Reactivity with water: Forms an organic base.
Binary reactants: Aqueous solution is an organic base. Violent reaction with oxidizers, cellulose nitrate (of high surface area). Incompatible with organic anhydrides, acrylates, alcohols, aldehydes, alkylene oxides, substituted allyls, cellulose nitrate, cresols, caprolactam solution, epichlorohydrin, ethylene dichloride, isocyanates, ketones, glycols, nitrates, phenols, vinyl acetate. Exothermic decomposition with maleic anhydride.
Neutralizing agents for acids and caustics: Dilute with water.
Reactivity group: 8
Compatibility class: Alkanolamines

ENVIRONMENTAL DATA
Food chain concentration potential: Log P_{ow} = –1.4. Unlikely to accumulate.
Water pollution: Effect of low concentrations on aquatic life is unknown. May be dangerous if it enters nearby water intakes; notify operators. Notify local health and pollution control officials.
Response to discharge: Disperse and flush.

SHIPPING INFORMATION
Storage temperature: Elevated; **Inert atmosphere:** None; **Venting:** Open; **Stability during transport:** Stable.

NAS HAZARD CLASSIFICATION FOR BULK WATER TRANSPORTATION
FIRE: 1
HEALTH: Vapor irritant: 1; Liquid or solid irritant: 3; Poisons: 1
WATER POLLUTION: Human toxicity: 1; Aquatic toxicity: 1; Aesthetic effect: 1
REACTIVITY: Other chemicals: 3; Water: 3; Self-reaction: 0

PHYSICAL AND CHEMICAL PROPERTIES
Physical state @ 59°F/15°C and 1 atm: Liquid.
Molecular weight: 104.15
Boiling point @ 1 atm: 469°F/243°C/516°K.
Melting/Freezing point: –108°F/–78°C/195°K.
Specific gravity (water = 1): 1.028 @ 77°F/25°C (liquid).
Relative vapor density (air = 1): 3.59
Ratio of specific heats of vapor (gas): (estimate) 1.053
Latent heat of vaporization: (estimate) 209 Btu/lb = 116 cal/g = 4.85×10^5 J/kg.
Heat of combustion: (estimate) –12,300 Btu/lb = –6860 cal/g = -287×10^5 J/kg.
Heat of solution: (estimate) –4 Btu/lb = –2 cal/g = -0.1×10^5 J/kg.
Vapor pressure: 0.001 mm (approximate).

N-AMINOETHYLPIPERAZINE   REC. A:1900

SYNONYMS: *N*-(2-AMINOETIL)PIPERAZINA (Spanish); *N*-(2-AMINOETHYL)PIPERAZINE; 1-(2-AMINOETHYL) PIPERAZINE; EEC No. 612-065-00-8; 1-PIPERAZINE ETHANAMINE

IDENTIFICATION
CAS Number: 140-31-8
Formula: $C_6H_{15}N_3$
DOT ID Number: UN 2815; DOT Guide Number: 153
Proper Shipping Name: *N*-Aminoethylpiperazine

DESCRIPTION: Light yellow liquid. Highly soluble in water, forming a strong base.

Corrosive • Breathing the vapor can kill you; skin or eye contact causes severe burns, impaired vision, or blindness; effects of contact or inhalation may be delayed • Firefighting gear (including SCBA) does not provide adequate protection. If exposure occurs, remove and isolate gear immediately and thoroughly decontaminate personnel • Combustible • Heat or flame may cause explosion • Containers may BLEVE when exposed to fire • Vapors in confined areas (e.g., tanks, sewers, buildings) may explode when exposed to fire • Toxic products of combustion may include nitrogen oxides.
Note: Combustible/poisonous gases may accumulate in tanks and hopper cars.

Hazard Classification (based on NFPA-704 Rating System)
Health Hazards (Blue): 2; Flammability (Red): 2; Reactivity (Yellow): 0

EMERGENCY RESPONSE: See Appendix A (153)
Evacuation:
Public safety: Isolate the area of spill or leak for at least 25 to 50 meters (80 to 160 feet) in all directions.
Spill: Increase, in the downwind direction, as necessary, the distance shown under "Public Safety."

Fire: If tank, rail car, or tank truck is involved in fire, isolate for at least 800 meters (½ mile) in all directions; also, consider initial evacuation for 800 meters (½ mile) in all directions.

EXPOSURE
Short-term effects: *SEEK MEDICAL ATTENTION.* **Vapor:** Irritating to eyes, nose, and throat. Move victim to fresh air. *IF BREATHING HAS STOPPED,* give artificial respiration; *avoid mouth-to-mouth resuscitation; use bag/mask apparatus.* IF breathing is difficult, administer oxygen. **Liquid:** Contact causes burns to skin and eyes. Remove and isolate contaminated clothing and shoes at the site. In case of contact with material, immediately flush skin or eyes with running water for at least 15 minutes. *IF SWALLOWED,* **Do NOT induce vomiting**. Keep victim quiet and maintain normal body temperature.

Note to physician or authorized medical personnel: Medical observation is recommended for 24 to 48 hours after breathing overexposure, as pulmonary edema may be delayed. As first aid for pulmonary edema, consider administering a corticosteroid spray. Cigarette smoking may exacerbate pulmonary injury and should be discouraged for at least 72 hours following exposure.

HEALTH HAZARDS
Personal protective equipment (PPE): B-Level PPE. Chemical protective material(s) reported to offer minimal to poor protection: butyl rubber.
Toxicity by ingestion: Grade 2: LD_{50} = 2.14 g/kg (rat).
Vapor (gas) irritant characteristics: Vapors cause severe irritation of eyes and throat and can cause lung injury. They cannot be tolerated even at low concentrations.

Liquid or solid irritant characteristics: Causes second- and third-degree burns on short contact, and is very injurious to the eyes.

FIRE DATA
Flash point: 199°F/93°C (oc).
Flammable limits in air: LEL: 1.6%; UEL 6.5%.
Autoignition temperature: More than 570°F/299°C/572°K.

CHEMICAL REACTIVITY
Binary reactants: Reacts with strong oxidizers, strong acids. Attacks metals including aluminum, cobalt, copper, and nickel.
Neutralizing agents for acids and caustics: Dry lime, soda ash
Reactivity group: 7
Compatibility class: Aliphatic amines.

ENVIRONMENTAL DATA
Water pollution: Effects of low concentrations on aquatic life are not known. May be dangerous if it enters nearby water intakes; notify operators. Notify local health and wildlife officials. Notify operators of local water intakes. **Response to discharge:** Issue warning–corrosive. Mechanical containment. Should be removed. Chemical and Physical treatment

SHIPPING INFORMATION
Grades of purity: 97%; **Stability during transport:** Stable.

NAS HAZARD CLASSIFICATION FOR BULK WATER TRANSPORTATION
FIRE: 1
HEALTH: Vapor irritant: 4; Liquid or solid irritant: 4; Poisons: 2
WATER POLLUTION: Human toxicity: 2; Aquatic toxicity: 3; Aesthetic effect: 2
REACTIVITY: Other chemicals: 1; Water: 0; Self-reaction: 0

PHYSICAL AND CHEMICAL PROPERTIES
Physical state @ 59°F/15°C and 1 atm: Liquid.
Molecular weight: 129.24
Boiling point @ 1 atm: 428°F/220°C/493°K.
Melting/Freezing point: –2°F/–19°C/254°K.
Specific gravity (water = 1): 0.9852 @ 68°F/20°C.
Relative vapor density (air = 1): 4.4
Vapor pressure: 0.075 mm.

2-AMINO-2-METHYL-1-PROPANOL REC. A:1925

SYNONYMS: 2-AMINODIMETHYLETHANOL β-AMINOISOBUTANOL; AMP; AMP-95; ISOBUTANOL AMINE; ISOBUTANOL-2-AMINE; 1-PROPANOL, 2-AMINO-2-METHYL-

IDENTIFICATION
CAS Number: 124-68-5
Formula: $C_4H_{11}NO$; $CH_3C(CH_3)(NH_2)CH_2OH$
DOT ID Number: NA 1993; DOT Guide Number: 128
Proper Shipping Name: Combustible liquid, n.o.s.

DESCRIPTION: Thick, colorless liquid or crystalline solid. Mixes with water.

Combustible • Containers may BLEVE when exposed to fire • Vapors may form explosive mixture with air • Vapors are heavier than air and will collect and stay in low areas • Vapors may travel long distances to ignition sources and flashback • Vapors in confined areas (e.g., tanks, sewers, buildings) may explode when exposed to fire • Irritating to the skin, eyes, and respiratory tract • Toxic products of combustion may include nitrogen oxide.

Hazard Classification (based on NFPA-704 Rating System)
Health Hazards (Blue): 2; Flammability (Red): 2; Reactivity (Yellow): 0

EMERGENCY RESPONSE: See Appendix A (128)
Evacuation:
Public safety: Isolate spill area for at least 25 to 50 meters (80 to 160 feet) in all directions.
Spill: Large spill–Consider initial downwind evacuation for at least 300 meters (1000 feet).
Fire: Isolate for 800 meters (½ mile) in all directions, especially if tank, rail car, or tank truck is involved in fire.

EXPOSURE
Short-term effects: *SEEK MEDICAL ATTENTION.* **Vapor:** Irritating to eyes, nose, and throat. Harmful if inhaled. Move to fresh air. *IF BREATHING HAS STOPPED,* give artificial respiration. IF breathing is difficult, administer oxygen. **Liquid:** Will burn skin and eyes. Harmful if swallowed. *IF IN EYES OR ON SKIN,* flush with plenty of water for at least 15 minutes. Remove and isolate contaminated clothing and shoes at the site.

HEALTH HAZARDS
Personal protective equipment (PPE): B-Level PPE. Self-contained breathing apparatus, rubber boots and heavy rubber gloves.
Toxicity by ingestion: Grade 2: LD_{LO} = 1 g/kg (rabbit).
Vapor (gas) irritant characteristics: Vapors cause severe irritation of eyes and throat and can cause eye and lung injury. They cannot be tolerated even at low concentrations.

Liquid or solid irritant characteristics: Severe skin irritant. Causes second- and third-degree burns on short contact and is very injurious to the eyes.

FIRE DATA
Flash point: 153°F/67°C (oc).

CHEMICAL REACTIVITY reported.
Binary reactants: Aqueous solution is a strong organic base. Incompatible with organic anhydrides, acrylates, alcohols, aldehydes, alkylene oxides, substituted allyls, cellulose nitrate, cresols, caprolactam solution, epichlorohydrin, ethylene dichloride, isocyanates, ketones, glycols, nitrates, phenols, vinyl acetate. Exothermic decomposition with maleic anhydride.
Reactivity group: 8
Compatibility class: Alkanolamines

ENVIRONMENTAL DATA
Water pollution: Effect of low concentrations on aquatic life is not known. May be dangerous if it enters nearby water intakes; notify operators. Notify local health and wildlife officials. **Response to discharge:** Restrict access. Should be removed. Chemical and physical treatment.

SHIPPING INFORMATION
Grades of purity: Different grades of purity up to 99+%; **Storage temperature:** Ambient; **Stability during transport:** Stable.

PHYSICAL AND CHEMICAL PROPERTIES
Physical state @ 59°F/15°C and 1 atm: Liquid.
Molecular weight: 89.14
Boiling point @ 1 atm: 329°F/165°C/438.2°K (99+% compound).
Melting/Freezing point: 88–90°F/31–32°C/304–305°K.
Specific gravity (water = 1): 0.935
Relative vapor density (air = 1): 3.0

4-AMINOPYRIDINE REC. A:1950

SYNONYMS: 4-AMINOPIRIDINA (Spanish); *p*-AMINOPYRIDINE; AVITROL; 4-PYRIDINAMINE; 4-PYRIDYLAMINE; RCRA No. P008

IDENTIFICATION
CAS Number: 504-24-5 (*p*-isomer); 504-29-0 (*o*-isomer); 462-08-8 (*m*-isomer)
Formula: $C_5H_6N_2$; $C_5NH_4NH_2$
DOT ID Number: UN 2671; DOT Guide Number: 153
Proper Shipping Name: Aminopyridines (*o*-; *m*-; *p*-)
Reportable Quantity (RQ): **(CERCLA)** 1000 lb/454 kg

DESCRIPTION: White solid (powder). Slight characteristic. Sinks and mixes with water; moderately soluble.

Poison! • Breathing the dust, skin or eye contact, or swallowing the material can kill you • Firefighting gear (including SCBA) does not provide adequate protection. If exposure occurs, remove and isolate gear immediately and thoroughly decontaminate personnel • Combustible • Do not put yourself in danger by entering a contaminated area to rescue a victim • Containers may explode in fire • Irritating to skin, eyes, and respiratory tract. Toxic products of combustion may include nitrogen oxides • Do not put yourself in danger by entering a contaminated area to rescue a victim.

Hazard Classification (based on NFPA-704 Rating System)
Health Hazards (Blue): 4; Flammability (Red): 1; Reactivity (Yellow): 0

EMERGENCY RESPONSE: See Appendix A (153)
Evacuation:
Public safety: Isolate the area of spill or leak for at least 25 to 50 meters (80 to 160 feet) in all directions.
Spill: Increase, in the downwind direction, as necessary, the distance shown under "Public Safety."
Fire: If tank, rail car, or tank truck is involved in fire, isolate for at least 800 meters (½ mile) in all directions; also, consider initial evacuation for 800 meters (½ mile) in all directions.

EXPOSURE
Short-term effects: *SEEK MEDICAL ATTENTION. POISONOUS. MAY BE FATAL IF INHALED OR ABSORBED THROUGH SKIN.*
Dust: Irritating to eyes, skin, and respiratory tract; may burn skin and eyes. Move victim to fresh air. *IF BREATHING HAS STOPPED,* give artificial respiration; *avoid mouth-to-mouth resuscitation; use bag/mask apparatus.* IF breathing is difficult, administer oxygen. **Solid:** Irritating to eyes and skin and may cause burns. *IF IN EYES OR ON SKIN*, flush immediately with running water for at least 15 minutes; hold eyelids open occasionally if appropriate. *Speed in removing material from skin is extremely important.* Remove and isolate contaminated clothing and shoes at the site. Keep victim quiet and maintain normal body temperature. Effects may be delayed; keep victim under observation. *IF SWALLOWED* and victim is *CONSCIOUS AND ABLE TO SWALLOW*, have victim induce vomiting by touching back of throat with a finger. *IF SWALLOWED* and victim is *UNCONSCIOUS OR HAVING CONVULSIONS*, do nothing except keep victim warm.

HEALTH HAZARDS
Personal protective equipment (PPE): A-Level PPE. OSHA Table Z-1-A air contaminant.
Recommendations for respirator selection: NIOSH/OSHA (2-aminopyridine).
5 ppm: SA* (any supplied-air respirator); or SCBAF (any self-contained breathing apparatus with a full facepiece). *EMERGENCY OR PLANNED ENTRY INTO UNKNOWN CONCENTRATIONS OR IDLH CONDITIONS:* SCBAF:PD,PP (any self-contained breathing apparatus that has a full facepiece and is operated in a pressure-demand or other positive-pressure mode); or SAF:PD,PP:ASCBA (any supplied-air respirator that has a full facepiece and is operated in a pressure-demand or other positive-pressure mode in combination with an auxiliary self-contained breathing apparatus operated in a pressure-demand or other positive pressure mode). *ESCAPE:* GMFOVHiE [any air-purifying, full-facepiece respirator (gas mask) with a chin-style, front-or back-mounted canister having a high efficiency particulate filter); or SCBAE (any appropriate escape-type, self-contained breathing apparatus). **Note:* Substance reported to cause eye irritation or damage; may require eye protection.
Exposure limits (TWA unless otherwise noted): ACGIH TLV 0.5 ppm (1.9 mg/m³).
Toxicity by ingestion: Grade 4; LD_{50} = 20 mg/kg (rat).
Vapor (gas) irritant characteristics: Irritates the respiratory tract.
Liquid or solid irritant characteristics: Irritates skin and eyes; may cause burns.

FIRE DATA
Flash point: 328°F/164°C (oc).

CHEMICAL REACTIVITY

Reactivity with water: Reacts with water, steam forming a strong base.

Binary reactants: Dust or powder may form explosive mixture with air. Violent reaction with strong oxidizers, strong acids. Aqueous solution incompatible with organic anhydrides, acrylates, alcohols, aldehydes, alkylene oxides, substituted allyls, cresols, caprolactam solution, epichlorohydrin, ethylene dichloride, glycols, isocyanates, ketones, maleic anhydride, nitrates, nitromethane, phenols, vinyl acetate.

ENVIRONMENTAL DATA

Water pollution: Harmful to aquatic life in very low concentrations. May be dangerous if it enters nearby water intakes; notify operators. Notify local health and wildlife officials.
Response to discharge: Issue warning–poison, water contaminant. Restrict access. Should be removed. Chemical and physical treatment.

SHIPPING INFORMATION

Grades of purity: 98%; **Storage temperature:** Ambient; **Stability during transport:** Stable.

PHYSICAL AND CHEMICAL PROPERTIES

Physical state @ 59°F/15°C and 1 atm: Solid.
Molecular weight: 94.12
Boiling point @ 1 atm: 524.3°F/273.5°C/546.7°K.
Melting/Freezing point: 311–316°F/155–158°C/428–431°K.
Specific gravity (water = 1): 1.2607 at 25.3°C (solid).

AMMONIA   REC. A:2000

SYNONYMS: AMMONIA, ANHYDROUS; AM-FOL; AMMONIAC (French); AMMONIALE (German); AMMONIUM HYDROXIDE; AMONIACO ANHIDRO (Spanish); ANHYDROUS AMMONIA; AQUA AMMONIA; EEC No. 007-001-00-5 (anhydrous); DAXAD-32S; EINECS 231-635-3; LIQUID AMMONIA; NITRO-SIL; R717; SPIRIT OF HARTSHORN

IDENTIFICATION

CAS Number: 7664-41-7 (anhydrous); 1336-21-6 (25% solution in water)
Formula: NH_3
DOT ID Number: UN 1005 (anhydrous); UN 3318 (> 50% ammonia); UN 2672 (>10% but <35% ammonia); UN 2073 (>35% but; DOT Guide Number: 125 <50% ammonia)
Proper Shipping Name: Ammonia solution, relative density less than 0.880 @ 15°C in water, with more than 50% ammonia (UN 3318); Ammonia solutions, relative density between 0.880 and 0.957 @ 15°C in water, with more than 10% but not more than 35% ammonia (UN 2672); Ammonia solutions, relative density less than 0.880 @ 15°C in water, with more than 35% but not more than 50% ammonia (UN 2073); Ammonia, anhydrous (UN 1005)
Reportable Quantity (RQ): **(CERCLA)** 100 lb/45.4 kg

DESCRIPTION: Colorless, compressed gas or cryogenic liquid. Extremely pungent odor. Floats and boils on water. Poisonous, visible vapor cloud is produced.

Poison! • Corrosive to the skin, eyes, and respiratory tract; contact with skin or eyes can cause burns and vision impairment; inhalation symptoms may be delayed • Firefighting gear (including SCBA) does not provide adequate protection. If exposure occurs, remove and isolate gear immediately and thoroughly decontaminate personnel • Do not add water to liquid ammonia; will increase evaporation • Always stay away from the ends of tanks • Flammable • Containers may BLEVE when exposed to fire • Gas is lighter than air • Gas may be visible or invisible, and is heavier than air and will collect and stay in low areas when cool • Gas in confined areas (e.g., tanks, sewers, buildings) may burn or explode when exposed to fire • Do not put yourself in danger by entering a contaminated area to rescue a victim • Although ammonia does not meet the DOT definition of a Flammable Gas (for labeling purposes), it should be treated as one (NIOSH).

Hazard Classification (based on NFPA-704 Rating System)
Health Hazards (Blue): 3; Flammability (Red): 1; Reactivity (Yellow): 0

EMERGENCY RESPONSE: See Appendix A (125)
Evacuation:
Public safety: See below.
Spill: Small spill–First: Isolate in all directions 30 meters (100 feet); Then: Protect persons downwind, DAY: 0.2 km (0.1 mile); NIGHT: 0.3 km (0.2 mile). Large spill–First: Isolate in all directions 95 meters (300 feet). Then: Protect persons downwind, DAY: 0.3 km (0.2 mile); NIGHT: 0.8 km (0.5 mile).
Fire: Isolate for 1600 meters (1 mile) in all directions, especially if tank, rail car, or tank truck is involved in fire.

EXPOSURE

Short-term effects: *SEEK MEDICAL ATTENTION.* **Vapor:** *POISONOUS IF INHALED.* Irritating to eyes, nose, and throat. May cause lung edema; physical exertion will aggravate this condition. Move to fresh air. *IF IN EYES*, hold eyelids open and flush with plenty of water. *IF BREATHING HAS STOPPED*, give artificial respiration; *avoid mouth-to-mouth resuscitation; use bag/mask apparatus.* IF breathing is difficult, administer oxygen. **Liquid:** Will burn skin and eyes. Harmful if swallowed. Will cause frostbite. Remove contaminated clothing and shoes. Flush affected areas with plenty of water. *DO NOT RUB AFFECTED AREAS. IF IN EYES*, hold eyelids open and flush with plenty of water. *IF SWALLOWED* and victim is *CONSCIOUS AND ABLE TO SWALLOW*, have victim drink 4 to 8 ounces of water.
Note to physician or authorized medical personnel: Medical observation is recommended for 24 to 48 hours after breathing overexposure, as upper airway obstruction or pulmonary edema may be delayed. As first aid for pulmonary edema, consider administering a corticosteroid spray. Cigarette smoking may exacerbate pulmonary injury and should be discouraged for at least 72 hours following exposure.

HEALTH HAZARDS

Personal protective equipment (PPE): A-Level PPE. OSHA Table Z-1-A air contaminant. Wear thermal protective clothing. Chemical protective material(s) reported to offer protection: butyl rubber, Nitrile, Neoprene, chlorinated polyethylene, PVC, polycarbonate, Teflon®, Viton®.
Recommendations for respirator selection: NIOSH
250 ppm: CCRS [any chemical cartridge respirator with cartridge(s) providing protection against the compound of concern)]; or SA (any supplied-air respirator). *300 ppm*: SA:CF (any supplied-air respirator operated in a continuous-flow mode); or PAPRS* [any powered, air-purifying respirator with cartridge(s) providing protection against the compound of concern]; or CCRFS [any chemical cartridge respirator with a full facepiece and cartridge(s) providing protection against the compound of concern]; or GMFS [any air-purifying, full-facepiece respirator (gas mask) with a chin-

style, front- or back-mounted canister providing protection against the compound of concern]; or SCBAF (any self-contained breathing apparatus with a full facepiece); or SAF (any supplied-air respirator with a full facepiece). *EMERGENCY OR PLANNED ENTRY INTO UNKNOWN CONCENTRATIONS OR IDLH CONDITIONS*: SCBAF:PD,PP (any self-contained breathing apparatus that has a full facepiece and is operated in a pressure-demand or other positive-pressure mode); or SAF:PD,PP:ASCBA (any supplied-air respirator that has a full facepiece and is operated in a pressure-demand or other positive-pressure mode in combination with an auxiliary self-contained breathing apparatus operated in a pressure-demand or other positive pressure mode). *ESCAPE:* GMFS [any air-purifying, full-facepiece respirator (gas mask) with a chin-style, front- or back-mounted canister providing protection against the compound of concern]; or SCBAE (any appropriate escape-type, self-contained breathing apparatus).
Note: Substance reported to cause eye irritation or damage; may require eye protection.
Exposure limits (TWA unless otherwise noted): ACGIH TLV 25 ppm (17 mg/m^3); NIOSH 25 ppm (18 mg/m^3).
Short-term exposure limits (15-minute TWA): ACGIH TLV 35 ppm (24 mg/m^3); OSHA STEL 50 ppm (35 mg/m^3); NIOSH STEL 35 ppm (27 mg/m^3).
Vapor (gas) irritant characteristics: Vapors cause severe eye or throat irritation and may cause eye or lung injury; vapors cannot be tolerated even at low concentrations. Lung edema may result.
Liquid or solid irritant characteristics: Causes smarting of the skin and first-degree burns on short exposure; may cause secondary burns on long exposure.
Odor threshold: 5.75 ppm.
IDLH value: 300 ppm.

FIRE DATA
Flash point: Gas. Indefinite below 32°F/0°C (sometimes difficult to ignite).
Flammable limits in air: LEL: 15%; UEL: 28%.

Fire extinguishing agents not to be used: Do not put water on liquid ammonia.
Behavior in fire: Sealed containers may rupture and explode.
Autoignition temperature: 1204°F/651°C/924°K.
Electrical hazard: Class I, Group D.
Burning rate: 1 mm/min
Stoichiometric air-to-fuel ratio: 6.050 (estimate).

CHEMICAL REACTIVITY
Reactivity with water: Dissolves with mild heat effect.
Binary reactants: Corrosive to copper, aluminum and galvanized surfaces; zinc. Reacts with strong oxidizers, halogens and acidic materials (violent reaction with strong acids). Reaction with halogens, salts of silver, mercury oxide may form shock-sensitive materials.
Neutralizing agents for acids and caustics: Dilute with water.
Reactivity group: 6
Compatibility class: Ammonia

ENVIRONMENTAL DATA
Food chain concentration potential: Log P_{ow} = –1.3. Unlikely to accumulate.
Water pollution: Harmful to aquatic life in very low concentrations. May be dangerous if it enters nearby water intakes; notify operators. Notify local health and wildlife officials.
Response to discharge: Issue warning–poison. Restrict access. Evacuate area. Disperse and flush.

SHIPPING INFORMATION
Grades of purity: Commercial, industrial, refrigeration, electronic, and metallurgical grades all have purity greater than 99.5%; **Storage temperature:** Ambient for pressurized ammonia; low temperature for ammonia at atmospheric pressure; **Inert atmosphere:** None; **Venting:** Safety relief 250 psi for ammonia under pressure. Pressure-vacuum for ammonia at atmospheric pressure; **Stability during transport:** Stable.

NAS HAZARD CLASSIFICATION FOR BULK WATER TRANSPORTATION
FIRE: 1
HEALTH: Vapor irritant: 4; Liquid or solid irritant: 2; Poisons: 2
WATER POLLUTION: Human toxicity: 2; Aquatic toxicity: 2; Aesthetic effect: 2
REACTIVITY: Other chemicals: 3; Water: 2; Self-reaction: 0

PHYSICAL AND CHEMICAL PROPERTIES
Physical state @ 59°F/15°C and 1 atm: Gas.
Molecular weight: 17.03
Boiling point @ 1 atm: –28°F/–33°C/240°K.
Melting/Freezing point: –108°F/–78°C/266°K.
Critical temperature: 271°F/133°C/406°K.
Critical pressure: 1636 psia = 111.3 atm = 11.27 MN/m^2.
Specific gravity (water = 1): 0.682 at –33.4°C (liquid).
Relative vapor density (air = 1): 0.612
Latent heat of vaporization: 589 Btu/lb = 327 cal/g = 13.7 x 10^5 J/kg.
Heat of combustion: –7992 Btu/lb –4440 cal/g = –185.9 x 10^5 J/kg.
Heat of solution: –232 Btu/lb = –129 cal/g = –5.40 x 10^5 J/kg.
Vapor pressure: (Reid) 211.9 psia; 400 mm @ 113°F/anhydrous).

AMMONIUM ACETATE   REC. A:2050

SYNONYMS: ACETATO AMONICO (Spanish); ACETIC ACID, AMMONIUM SALT

IDENTIFICATION
CAS Number: 631-61-8
Formula: $C_2H_4O_2 \cdot H_3N$; $NH_4C_2H_3O_2$
DOT ID Number: UN 3077; DOT Guide Number: 171
Proper Shipping Name: Environmentally hazardous substance, solid. n.o.s.
Reportable Quantity (RQ): **(CERCLA)** 5000 lb/2270 kg

DESCRIPTION: White solid. Weak ammonia odor. Sinks and mixes with water.

Toxic products of combustion may include oxides of nitrogen, ammonia, and acetic acid.

Hazard Classification (based on NFPA-704 Rating System)
Health Hazards (Blue): 1; Flammability (Red): 0; Reactivity (Yellow): 0

EMERGENCY RESPONSE: See Appendix A (171)
Evacuation:
Public safety: Isolate the area of spill or leak for at least 10 to 25 meters (30 to 80 feet) in all directions.
Spill: Increase, in the downwind direction, as necessary, the distance shown under "Public Safety."

42   Ammonium bicarbonate

Fire: If any large container is involved in fire, isolate for at least 800 meters (½ mile) in all directions; also, consider initial evacuation for 800 meters (½ mile) in all directions.

EXPOSURE
Short-term effects: *SEEK MEDICAL ATTENTION.* **Dust:** Irritating to eyes, nose, and throat. *IF INHALED*, will, will cause difficult breathing. *IF IN EYES*, hold eyelids open and flush with plenty of water. *IF BREATHING HAS STOPPED*, give artificial respiration. IF breathing is difficult, administer oxygen. **Solid:** Irritating to skin and eyes. *IF SWALLOWED*, will cause nausea. Remove contaminated clothing and shoes. Flush affected areas with plenty of water. *IF IN EYES*, hold eyelids open and flush with plenty of water. *IF SWALLOWED* and victim is *CONSCIOUS AND ABLE TO SWALLOW*, have victim drink 4 to 8 ounces of water. *IF SWALLOWED* and victim is *UNCONSCIOUS OR HAVING CONVULSIONS*, do nothing except keep victim warm.

HEALTH HAZARDS
Personal protective equipment (PPE): B-Level PPE. Dust mask; goggles or face shield; rubber gloves.
Vapor (gas) irritant characteristics: May cause respiratory tract irritation.
Liquid or solid irritant characteristics: May cause skin and eye irritation.

ENVIRONMENTAL DATA
Water pollution: Dangerous to aquatic life in high concentrations. May be dangerous if it enters nearby water intakes; notify operators. Notify local health and wildlife officials. **Response to discharge:** Disperse and flush.

SHIPPING INFORMATION
Grades of purity: Reagent; CP; Technical, 97+%; **Storage temperature:** Ambient; **Inert atmosphere:** None; **Venting:** Open; **Stability during transport:** Stable.

PHYSICAL AND CHEMICAL PROPERTIES
Physical state @ 59°F/15°C and 1 atm: Solid.
Molecular weight: 77.08
Boiling point @ 1 atm: Decomposes.
Specific gravity (water = 1): 1.17 @ 68°F/20°C (solid).
Heat of solution: -5.8 Btu/lb $= -3.2$ cal/g $= -0.13 \times 10^5$ J/kg.

AMMONIUM BENZOATE   REC. A:2100

SYNONYMS: BENZOATO AMONICO (Spanish); BENZOIC ACID, AMMONIUM SALT; VULNOC AB

IDENTIFICATION
CAS Number: 1863-63-4
Formula: $C_7H_5O_2 \cdot H_4N$; $C_6H_5COONH_4$
DOT ID Number: UN 3077; DOT Guide Number: 171
Proper Shipping Name: Environmentally hazardous, solid, n.o.s.
Reportable Quantity (RQ): **(CERCLA)** 5000 lb/2270 kg

DESCRIPTION: Colorless solid. Odorless. Sinks and mixes slowly with water; moderately soluble.

Combustible solid • Dust cloud may explode if ignited in an enclosed area • Containers may BLEVE when exposed to fire • Irritating to the skin, eyes, and respiratory tract • Toxic products of combustion may include nitrogen oxides and ammonia.

Hazard Classification (based on NFPA-704 Rating System)
Health Hazards (Blue): 2; Flammability (Red): 1; Reactivity (Yellow): 0

EMERGENCY RESPONSE: See Appendix A (171)
Evacuation:
Public safety: Isolate the area of spill or leak for at least 10 to 25 meters (30 to 80 feet) in all directions.
Spill: Increase, in the downwind direction, as necessary, the distance shown under "Public Safety."
Fire: If any large container is involved in fire, isolate for at least 800 meters (½ mile) in all directions; also, consider initial evacuation for 800 meters (½ mile) in all directions.

EXPOSURE
Short-term effects: *SEEK MEDICAL ATTENTION.* **Dust:** Irritating to eyes, nose, and throat. Harmful if inhaled. *IF IN EYES*, hold eyelids open and flush with plenty of water. *IF BREATHING HAS STOPPED*, give artificial respiration. IF breathing is difficult, administer oxygen. **Solid:** Irritating to skin and eyes. Remove contaminated clothing and shoes. Flush affected areas with plenty of water. *IF IN EYES*, hold eyelids open and flush with plenty of water. *IF SWALLOWED* and victim is *CONSCIOUS AND ABLE TO SWALLOW*, have victim drink 4 to 8 ounces of water. *IF SWALLOWED* and victim is *UNCONSCIOUS OR HAVING CONVULSIONS*, do nothing except keep victim warm.

HEALTH HAZARDS
Vapor (gas) irritant characteristics: May cause respiratory tract irritation.
Liquid or solid irritant characteristics: May cause eye and skin irritation.

CHEMICAL REACTIVITY
Reactivity with water: Basic solution is formed.
Binary reactants: Incompatible with acids, ferric salts, carbonates, or alkali hydroxides. Contact with acids, releases benzoic acid; contact with alkalis releases ammonia.
Water pollution: Effect of low concentrations on aquatic life is unknown. May be dangerous if it enters nearby water intakes; notify operators. Notify local health and wildlife officials. **Response to discharge:** Disperse and flush.

SHIPPING INFORMATION
Grades of purity: Technical; **Storage temperature:** Ambient; **Inert atmosphere:** None; **Venting:** Open; **Stability during transport:** Slowly releases ammonia gas on exposure to air, which may collect in closed container.

PHYSICAL AND CHEMICAL PROPERTIES
Physical state @ 59°F/15°C and 1 atm: Solid.
Molecular weight: 139.15
Boiling point @ 1 atm: 320°F/160°C/433°K (sublimes).
Melting/Freezing point: 388°F/198°C/471°K Decomposes.
Specific gravity (water = 1): 1.26 @ 77°F/25°C (solid).
Heat of solution: 34 Btu/lb $= 19$ cal/g $= 0.80 \times 10^5$ J/kg.

AMMONIUM BICARBONATE   REC. A:2150

SYNONYMS: ACID AMMONIUM CARBONATE; ACID AMMONIUM CARBONATE, MONOAMMONIUM SALT; AMMONIUM HYDROGEN CARBONATE; BICARBONATO AMONICO (Spanish); CARBONIC ACID, MONOAMMONIUM SALT

IDENTIFICATION
CAS Number: 1066-33-7
Formula: NH_4HCO_3; $HCO_3 \cdot H_4N$
DOT ID Number: UN 9084; DOT Guide Number: 154
Proper Shipping Name: Ammonium carbonate
Reportable Quantity (RQ): **(CERCLA)** 5000 lb/2270 kg

DESCRIPTION: Colorless crystalline solid or white powder. Strong ammonia odor. Sinks and mixes slowly with water.

Combustible solid • Decomposes above 91°F/34°C, forming ammonia gas which may collect in low areas. Irritating to the skin, eyes, and respiratory tract • Toxic products of combustion may include nitrogen oxides and ammonia gas.

Hazard Classification (based on NFPA-704 Rating System)
Health Hazards (Blue): 0; Flammability (Red): 0; Reactivity (Yellow): 0

EMERGENCY RESPONSE: See Appendix A (154)
Evacuation:
Public safety: Isolate the area of spill or leak for at least 25 to 50 meters (80 to 160 feet) in all directions.
Spill: Increase, in the downwind direction, as necessary, the distance shown under "Public Safety."
Fire: If tank, rail car, or tank truck is involved in fire, isolate for at least 800 meters (½ mile) in all directions; also, consider initial evacuation for 800 meters (½ mile) in all directions.

EXPOSURE
Short-term effects: *SEEK MEDICAL ATTENTION*. **Dust:** Irritating to eyes, nose, and throat. *IF INHALED, will*, will cause difficult breathing. lung edema may develop. *IF IN EYES*, hold eyelids open and flush with plenty of water.*IF BREATHING HAS STOPPED*, give artificial respiration; *avoid mouth-to-mouth resuscitation; use bag/mask apparatus*. IF breathing is difficult, administer oxygen. **Solid:** Irritating to skin and eyes. Harmful if swallowed. Remove contaminated clothing and shoes. Flush affected areas with plenty of water. *IF IN EYES*, hold eyelids open and flush with plenty of water. *IF SWALLOWED* and victim is *CONSCIOUS AND ABLE TO SWALLOW*, have victim drink 4 to 8 ounces of water. *IF SWALLOWED* and victim is *UNCONSCIOUS OR HAVING CONVULSIONS*, do nothing except keep victim warm.
Note to physician or authorized medical personnel: Medical observation is recommended for 24 to 48 hours after breathing overexposure, as pulmonary edema may be delayed. As first aid for pulmonary edema, consider administering a corticosteroid spray. Cigarette smoking may exacerbate pulmonary injury and should be discouraged for at least 72 hours following exposure.

HEALTH HAZARDS
Personal protective equipment (PPE): B-Level PPE. Work gloves; dust respirator; safety glasses or chemical safety goggles if dusty.
Vapor (gas) irritant characteristics: May cause respiratory tract irritation.
Liquid or solid irritant characteristics: May cause eye and skin irritation.

FIRE DATA
Behavior in fire: Containers may rupture and explode.

CHEMICAL REACTIVITY
Binary reactants: Reacts with alcohols, reducing agents and strong acids, sodium nitrite, magnesium. May attack copper, nickel, and zinc.
Water pollution: Affect of low concentrations on aquatic life is unknown. May be dangerous if it enters nearby water intakes; notify operators. Notify local health and wildlife officials. **Response to discharge:** Disperse and flush.

SHIPPING INFORMATION
Grades of purity: Food; Reagent; **Storage temperature:** Below 91°F/33°C; **Inert atmosphere:** None; **Venting:** Open. **Stability during transport:** Decomposes above 91°F/33°C with formation of ammonia gas, which may collect in closed containers.

PHYSICAL AND CHEMICAL PROPERTIES
Physical state @ 59°F/15°C and 1 atm: Solid.
Molecular weight: 79.07
Boiling point @ 1 atm: Decomposes.
Melting/Freezing point: 95°F/35°C/308°K (decomposes above 91°F/34°C).
Specific gravity (water = 1): 2.15 @ 68°F/20°C (solid).
Heat of decomposition: 91°F/34°C.
Heat of solution: 140 Btu/lb = 80 cal/g = 3.3×10^5 J/kg.

AMMONIUM BIFLUORIDE   REC. A:2200

SYNONYMS: AMMONIUM HYDROGEN FLUORIDE; AMMONIUM HYDROGENDIFLUORIDE; ACID AMMONIUM FLUORIDE; AMMONIUM ACID FLUORIDE; BIFLUORURO AMONICO (Spanish)

IDENTIFICATION
CAS Number: 1341-49-7
Formula: NH_4HF_2
DOT ID Number: UN 1727 (solid); UN 2817 (solution); DOT Guide Number: 154
Proper Shipping Name: Ammonium hydrogendifluoride, solid; Ammonium hydrogendifluoride, solution
Reportable Quantity (RQ): **(CERCLA)** 100 lb/45.4 kg

DESCRIPTION: White crystalline solid. Odorless. Sinks and mixes with water.

Toxic! • Corrosive • Contact with the molten material may cause severe burns of the skin or eyes • Effects of contact or inhalation may be delayed • Toxic products of combustion may include nitrogen oxides, fluorine, and ammonia.

Hazard Classification (based on NFPA-704 Rating System)
Health Hazards (Blue): 3; Flammability (Red): 0; Reactivity (Yellow): 0

EMERGENCY RESPONSE: See Appendix A (154)
Evacuation:
Public safety: Isolate the area of spill or leak for at least 25 to 50 meters (80 to 160 feet) in all directions.
Spill: Increase, in the downwind direction, as necessary, the distance shown under "Public Safety."
Fire: If tank, rail car, or tank truck is involved in fire, isolate for at least 800 meters (½ mile) in all directions; also, consider initial evacuation for 800 meters (½ mile) in all directions.

EXPOSURE
Short-term effects: *SEEK MEDICAL ATTENTION*. **Dust:** Irritating to eyes, nose, and throat. *IF INHALED*, will, will cause

44   Ammonium bisulfite

coughing or difficult breathing. *IF IN EYES*, hold eyelids open and flush with plenty of water. *IF BREATHING HAS STOPPED*, give artificial respiration; *avoid mouth-to-mouth resuscitation; use bag/mask apparatus*. IF breathing is difficult, administer oxygen. **Solid:** Will burn skin and eyes. *IF SWALLOWED*, will cause nausea or vomiting. Remove contaminated clothing and shoes. Flush affected areas with plenty of water. *IF IN EYES*, hold eyelids open and flush with plenty of water. *IF SWALLOWED* and victim is *CONSCIOUS AND ABLE TO SWALLOW*, have victim drink 4 to 8 ounces of water. *IF SWALLOWED* and victim is *UNCONSCIOUS OR HAVING CONVULSIONS*, do nothing except keep victim warm.

HEALTH HAZARDS
Personal protective equipment (PPE): A-Level PPE.
Recommendations for respirator selection: NIOSH/OSHA as F. *12.5 mg/m³:* DM (any dust and mist respirator). *25 mg/m³:* DMXSQ* (any dust and mist respirator except single-use and quarter-mask respirators); or SA* (any supplied-air respirator). *62.5 mg/m³:* SA:CF* [any supplied-air respirator operated in a continuous-flow mode)]; or PAPRDM*⁺ *if not present as a fume* (any powered, air-purifying respirator with a dust and mist filter). *125 mg/m³:* HiEF⁺ (any air-purifying, full-facepiece respirator with a high-efficiency particulate filter); or SCBAF (any self-contained breathing apparatus with a full facepiece); or SAF (any supplied-air respirator with a full facepiece). *250 mg/m³:* SA:PD,PP (any supplied-air respirator operated in a pressure-demand or other positive-pressure mode). *EMERGENCY OR PLANNED ENTRY INTO UNKNOWN CONCENTRATIONS OR IDLH CONDITIONS:* SCBAF:PD,PP (any self-contained breathing apparatus that has a full faceplate and is operated in a pressure-demand or other positive-pressure mode); or SAF:PD,PP:ASCBA (any supplied-air respirator that has a full facepiece and is operated in a pressure-demand or other positive-pressure mode in combination with an auxiliary, self-contained breathing apparatus operated in a pressure-demand or other positive-pressure mode). *ESCAPE:* HiEF⁺ (any air-purifying, full-facepiece respirator with a high-efficiency particulate filter); or SCBAE (any appropriate escape-type, self-contained breathing apparatus). *Notes:* *Substance reported to cause eye irritation or damage; may require eye protection. ⁺May need acid gas sorbent.

Exposure limits (TWA unless otherwise noted): ACGIH TLV, OSHA PEL, NIOSH REL: 2.5 mg/m³ as fluorides.
Toxicity by ingestion: Grade 3; LD_{50} = 50 mg/kg (guinea pig), 60 mg/kg (rat).
Vapor (gas) irritant characteristics: Irritation of the respiratory tract.
Liquid or solid irritant characteristics: Irritation of the eyes and skin.

FIRE DATA
Fire extinguishing agents not to be used: Do not apply water to adjacent fires.

CHEMICAL REACTIVITY
Reactivity with water: Dissolves and forms a weak solution of hydrofluoric acid.
Binary reactants: Incompatible with alkalis, acids. In presence of moisture will corrode glass, cement, and most metals. Flammable hydrogen gas may collect in enclosed spaces.
Neutralizing agents for acids and caustics: Flush with water, rinse with dilute solution of sodium bicarbonate or soda ash.
Water pollution: Effect of low concentrations on aquatic life is unknown. May be dangerous if it enters nearby water intakes; notify operators. Notify local health and wildlife officials.
Response to discharge: Issue warning–corrosive, water contaminant. Restrict access. Disperse and flush.

SHIPPING INFORMATION
Grades of purity: Pure, 99+%; Technical, 97–98.5%; **Storage temperature:** Ambient; **Inert atmosphere:** None; **Venting:** Open; **Stability during transport:** Stable.

PHYSICAL AND CHEMICAL PROPERTIES
Physical state @ 59°F/15°C and 1 atm: Solid.
Molecular weight: 57.04
Boiling point @ 1 atm: 463.1°F/239.5°C/512.7°K.
Melting/Freezing point: 258.0°F/125.6°C/398.8°K.
Specific gravity (water = 1): 1.5 @ 68°F/20°C (solid).
Heat of solution: 154 Btu/lb = 85.7 cal/g = 3.59×10^5 J/kg.

AMMONIUM BISULFITE   **REC. A:2250**

SYNONYMS: AMMONIUM HYDROGEN SULFITE; AMMONIUM HYDROSULFITE; AMMONIUM MONOSULFITE; AMMONIUM SULFITE, HYDROGEN; BISULFITO AMONICO (Spanish); MONOAMMONIUM SULFITE; SULFUROUS ACID, MONOAMMONIUM SALT

IDENTIFICATION
CAS Number: 10192-30-0
Formula: NH_4HSO_3
DOT ID Number: UN 2693; DOT Guide Number: 154
Proper Shipping Name: Ammonium bisulfite, solution
Reportable Quantity (RQ): **(CERCLA)** 5000 lb/2270 kg

DESCRIPTION: Colorless to yellow liquid. Sinks and mixes with water.

Toxic! • Corrosive to the skin, eyes, and respiratory tract; contact with skin or eyes can cause burns and vision impairment; inhalation symptoms may be delayed • Effects of contact or inhalation may be delayed • Toxic products of combustion may include sulfur oxides, nitrogen oxides, and ammonia.

Hazard Classification (based on NFPA-704 Rating System)
Health Hazards (Blue): 2; Flammability (Red): 0; Reactivity (Yellow): 0

EMERGENCY RESPONSE: See Appendix A (154)
Evacuation:
Public safety: Isolate the area of spill or leak for at least 25 to 50 meters (80 to 160 feet) in all directions.
Spill: Increase, in the downwind direction, as necessary, the distance shown under "Public Safety."
Fire: If tank, rail car, or tank truck is involved in fire, isolate for at least 800 meters (½ mile) in all directions; also, consider initial evacuation for 800 meters (½ mile) in all directions.

EXPOSURE
Short-term effects: *SEEK MEDICAL ATTENTION. IF INHALED, will*, lung edema may develop. *IF BREATHING HAS STOPPED*, give artificial respiration; *avoid mouth-to-mouth resuscitation; use bag/mask apparatus*. *IF IN THE EYES OR ON SKIN*, flush affected areas with plenty of water.
Note to physician or authorized medical personnel: Medical observation is recommended for 24 to 48 hours after breathing

overexposure, as pulmonary edema may be delayed. As first aid for pulmonary edema, consider administering a corticosteroid spray. Cigarette smoking may exacerbate pulmonary injury and should be discouraged for at least 72 hours following exposure.

HEALTH HAZARDS
Personal protective equipment (PPE): B-Level PPE. Rubber gloves, face shields or safety glasses, normal protective gear.
Vapor (gas) irritant characteristics: May cause difficult breathing and lung edema.
Liquid or solid irritant characteristics: May cause severe eye and skin irritation.

CHEMICAL REACTIVITY
Reactivity with water: Reacts with water forming corrosive acid solution and sulfur oxide fumes.
Binary reactants: React with acids forming corrosive acid solution and hydrogen sulfide fumes. Incompatible with lead diacetate, mercury(I) chloride.
Reactivity group: 43
Compatibility class: Miscellaneous water solutions.

ENVIRONMENTAL DATA
Water pollution: Effects of low concentrations on aquatic life are unknown. May be dangerous if it enters nearby water intakes; notify operators. Notify local health and wildlife officials.
Response to discharge: Disperse and flush.

PHYSICAL AND CHEMICAL PROPERTIES
Physical state @ 59°F/15°C and 1 atm: Normally available as an aqueous solution (usually 40%).
Molecular weight: 99.10
Boiling point @ 1 atm: Sublimes with decomposition @ 302°F/150°C/423°K.
Specific gravity (water = 1): 2.03 at room temperature (solid); 1.40 (72% aqueous solution).
Solubility: 620 g/100 ml @ 53°F/60°C; 71.8 g/100 ml @ 32°F/0°C in water
Heat of solution: Endothermic 43.6 Btu/lb = 24.2 cal/g = 1.01 x 10^5 J/kg (For dilute solution 1 mole/0.300 mole water, which is equal to 1.833 lb/volume lb of water).
Vapor pressure: 395 mm.

AMMONIUM BROMIDE   REC. A:2300

SYNONYMS: BROMURO AMONICO (Spanish); HYDROBROMIC ACID, MONO-AMMONIATE

IDENTIFICATION
CAS Number: 12124-97-0
Formula: NH_4Br

DESCRIPTION: White (becomes yellow on contact with air) solid crystals or granules. Odorless. Sinks and mixes with water; slightly soluble.

Irritates eyes, skin and respiratory tract and other health effects • Toxic products of combustion may include nitrogen oxides and bromine.

Hazard Classification (based on NFPA-704 Rating System)
Health Hazards (Blue): 1; Flammability (Red): 0; Reactivity (Yellow): 0

EMERGENCY RESPONSE: See Appendix A (171)
Evacuation:
Public safety: Isolate the area of spill or leak for at least 10 to 25 meters (30 to 80 feet) in all directions.
Spill: Increase, in the downwind direction, as necessary, the distance shown under "Public Safety."
Fire: If any large container is involved in fire, isolate for at least 800 meters (½ mile) in all directions; also, consider initial evacuation for 800 meters (½ mile) in all directions.

EXPOSURE
Short-term effects: *SEEK MEDICAL ATTENTION.* **Solid:** Dust irritating if breathed. Slightly irritating to skin and eyes. Harmful if swallowed. Move to fresh air. Flush affected areas with plenty of water. *IF SWALLOWED* and victim is *CONSCIOUS AND ABLE TO SWALLOW*, induce vomiting.

HEALTH HAZARDS
Personal protective equipment (PPE): A-Level PPE.
Toxicity by ingestion: Grade 2; LD_{50} = 0.5 to 5 g/kg.
Long-term health effects: Prolonged contact or inhalation of inorganic bromides causes acne-like bromoderma (bromide rash), especially of the hands and face. Other effects include emaciation, depression; and, in severe cases, psychosis and mental deterioration. Chronic bromide intoxication.
Liquid or solid irritant characteristics: Dust or liquid may cause eye, skin, and respiratory tract irritation.
Long-term health effects: Prolonged contact or inhalation of inorganic bromides causes acne-like bromoderma (bromide rash), especially of the hands and face. Other effects include emaciation, depression; and, in severe cases, psychosis and mental deterioration.

CHEMICAL REACTIVITY
Binary reactants: Incompatible with bromine trifluoride, potassium. May corrode metal at fire temperatures

ENVIRONMENTAL DATA
Water pollution: Effects of low concentrations on aquatic life are unknown. May be dangerous if it enters nearby water intakes; notify operators. Notify local health and wildlife officials. **Response to discharge:** Disperse and flush.

SHIPPING INFORMATION
Grades of purity: 99–99.5%; **Storage temperature:** Cool; **Stability during transport:** Stable.

PHYSICAL AND CHEMICAL PROPERTIES
Physical state @ 59°F/15°C and 1 atm: Solid.
Molecular weight: 97.95
Boiling point @ 1 atm: Sublimes 1007°F/542°C/815°K.
Melting/Freezing point: Sublimes without melting
Specific gravity (water = 1): 2.429 at room temperature
Heat of solution: Endothermic infinite dilution 76.0 Btu/lb = 42.2 cal/g = 1.77 x 10^5 J/kg.

AMMONIUM CARBAMATE   REC. A:2350

SYNONYMS: AMMONIUM AMINOFORMATE; ANHYDRIDE OF AMMONIUM CARBONATE; CARBAMATO AMONICO (Spanish); CARBAMIC ACID, MONOAMMONIUM SALT; CARBAMIC ACID, AMMONIUM SALT

46 Ammonium carbonate

IDENTIFICATION
CAS Number: 1111-78-0
Formula: NH_2COONH_4; $CH_3NO_2 \cdot H_3N$
DOT ID Number: UN 2757; DOT Guide Number: 151
Proper Shipping Name: Carbamate pesticides, solid, toxic, n.o.s.
Reportable Quantity (RQ): **(CERCLA)** 5000 lb/2270 kg

DESCRIPTION: Colorless, crystals or powder. Ammonia odor. Mixes with water; highly soluble.

Poison! (carbamate) • Combustible solid • Containers may BLEVE when exposed to fire • Irritating to eyes, mucous membranes of respiratory tract • Dust cloud may explode if ignited in an enclosed area • Decomposition products upon heating forming urea, nitrogen oxides, and ammonia gas.

Hazard Classification (based on NFPA-704 Rating System)
Health Hazards (Blue): 1; Flammability (Red): 1; Reactivity (Yellow): 1

EMERGENCY RESPONSE: See Appendix A (151)
Evacuation:
Public safety: Isolate the area of spill or leak for at least 25 to 50 meters (80 to 160 feet) in all directions.
Spill: Increase, in the downwind direction, as necessary, the distance shown under "Public Safety."
Fire: If tank, rail car, or tank truck is involved in fire, isolate for at least 800 meters (½ mile) in all directions; also, consider initial evacuation for 800 meters (½ mile) in all directions.

EXPOSURE
Short-term effects: *SEEK MEDICAL ATTENTION*. **Dust:** *IF BREATHING HAS STOPPED*, give artificial respiration; *avoid mouth-to-mouth resuscitation; use bag/mask apparatus*. Harmful if swallowed. Flush affected areas with plenty of water. *IF IN EYES*, hold eyelids open and flush with plenty of water. *IF SWALLOWED* and victim is *CONSCIOUS AND ABLE TO SWALLOW*, have victim drink 4 to 8 ounces of water.
Note to physician or authorized medical personnel. Administer atropine, 2 mg (1/30 gr) intramuscularly or intravenously as soon as any local or systemic signs or symptoms of an intoxication are noted; repeat the administration of atropine every 3 to 8 minutes until signs of atropinization (mydriasis, dry mouth, rapid pulse, hot and dry skin) occur; initiate treatment in children with 0.05 mg/kg of atropine; repeat at 5- to 10-minute intervals. Watch respiration, and remove bronchial secretions if they appear to be obstructing the airway; intubate if necessary.
Medical note: Due to the rapid regeneration of chlinesterase and the fact that 2-PAMCI may be contraindicated in the case of some carbamate poisonings, 2-PAMCI (Pralidoxime; Protopam) may not be needed.

HEALTH HAZARDS
Personal protective equipment (PPE): A-Level PPE. Self-contained breathing apparatus, rubber gloves, safety glasses, normal protective gear.
Recommendations for respirator selection: SCBAF:PD,PP (any self-contained breathing apparatus that has a full facepiece and is operated in a pressure-demand or other positive-pressure mode); or SAF:PD,PP:ASCBA (any supplied-air respirator that has a full facepiece and is operated in a pressure-demand or other positive-pressure mode in combination with an auxiliary, self-contained breathing apparatus operated in a pressure-demand or other positive pressure mode). *ESCAPE:* GMFOVHiE [any air-purifying, full-facepiece respirator (gas mask) with a chin-style, front- or back-mounted organic vapor canister having a high-efficiency particulate filter]; or SCBAE (any appropriate escape-type, self-contained breathing apparatus).
Exposure limits (TWA unless otherwise noted): ACGIH TLV 25 ppm (17 mg/m^3) as ammonia. Material decomposes in air releasing ammonia. Contains 44% ammonia.
Short-term exposure limits (15-minute TWA): ACGIH TLV 35 ppm (24 mg/m^3); NIOSH STEL 35 ppm (7 mg/m^3); OSHA STEL 50 ppm (35 mg/m^3) as ammonia
Long-term health effects: Carbamates are suspected carcinogens of lungs and hematopoietic organs.
Vapor (gas) irritant characteristics: Vapors are moderately irritating such that personnel will not usually tolerate moderate or high concentrations.
Liquid or solid irritant characteristics: May cause irritation of the eyes and respiratory tract.
Odor threshold: Less than 5 ppm as ammonia (detection); 46.8 ppm as ammonia (recognition) (CHRIS); 0.043–53 ppm as ammonia.

CHEMICAL REACTIVITY
Binary reactants: Strong oxidizers, strong acids.

ENVIRONMENTAL DATA
Water pollution: Harmful to aquatic life in very low concentrations. May be dangerous if it enters nearby water intakes; notify operators. Notify local health and wildlife officials. **Response to discharge:** Issue warning-EPA water contaminant. Disperse and flush.

SHIPPING INFORMATION
Storage temperature: Cool; **Stability during transport:** Unstable; decomposes in air, changing to ammonium carbonate. Volatilizes at 140°F/60°C.

PHYSICAL AND CHEMICAL PROPERTIES
Physical state @ 59°F/15°C and 1 atm: Solid.
Molecular weight: 78.07
Boiling point @ 1 atm: 140°F/60°C/333°K sublimes
Melting/Freezing point: 140°F/60°C/333°K sublimes
Heat of combustion: (estimate) @ 77°F/25°C-2612 Btu/lb = –1451 cal/g = 60.7 x 10^5 J/kg.
Heat of solution: (Endothermic) 86.9 Btu/lb = 48.3 cal/g = 2.02 x 10^5 J/kg.

AMMONIUM CARBONATE   REC. A:2400

SYNONYMS: AMMONIUMCARBONAT (German); CARBONATO AMONICO (Spanish); CARBONIC ACID, AMMONIUM SALT; CARBONIC ACID, DIAMMONIUM SALT; DIAMMONIUM CARBONATE; CRYSTAL AMMONIA; HARTSHORN; SAL VOLATILE

IDENTIFICATION
CAS Number: 506-87-6
Formula: $(NH_4)_2CO_3$; $(NH_4)HCO_3 \cdot (NH_4)CO_2 NH_2$
DOT ID Number: UN 9084; DOT Guide Number: 154
Proper Shipping Name: Ammonium carbonate
Reportable Quantity (RQ): **(CERCLA)** 5000 lb/2270 kg

DESCRIPTION: White solid. Strong ammonia odor. Sinks and mixes with water.

Irritating to the skin, eyes, and respiratory tract • Toxic products of combustion may include nitrogen oxides, ammonia gas.

Hazard Classification (based on NFPA-704 Rating System)
Health Hazards (Blue): 1; Flammability (Red): 0; Reactivity (Yellow): 0

EMERGENCY RESPONSE: See Appendix A (154)
Evacuation:
Public safety: Isolate the area of spill or leak for at least 25 to 50 meters (80 to 160 feet) in all directions.
Spill: Increase, in the downwind direction, as necessary, the distance shown under "Public Safety."
Fire: If tank, rail car, or tank truck is involved in fire, isolate for at least 800 meters (½ mile) in all directions; also, consider initial evacuation for 800 meters (½ mile) in all directions.
EXPOSURE
Short-term effects: *SEEK MEDICAL ATTENTION*. **Dust:** Irritating to eyes, nose, and throat. *IF INHALED*, will, will cause difficult breathing. *IF IN EYES*, hold eyelids open and flush with plenty of water. *IF BREATHING HAS STOPPED*, give artificial respiration; *avoid mouth-to-mouth resuscitation; use bag/mask apparatus*. IF breathing is difficult, administer oxygen. **Solid:** Irritating to skin and eyes. *IF SWALLOWED*, will cause nausea. Remove contaminated clothing and shoes. Flush affected areas with plenty of water. *IF IN EYES*, hold eyelids open and flush with plenty of water. *IF SWALLOWED* and victim is *CONSCIOUS AND ABLE TO SWALLOW*, have victim drink 4 to 8 ounces of water. *IF SWALLOWED* and victim is *UNCONSCIOUS OR HAVING CONVULSIONS*, do nothing except keep victim warm.

HEALTH HAZARDS
Personal protective equipment (PPE): B-Level PPE. Dust respirator; protection against ammonia vapors
Recommendations for respirator selection: See ammonia
Exposure limits (TWA unless otherwise noted): ACGIH TLV 25 ppm (17 mg/m^3) as ammonia gas; NIOSH 25 ppm (18 mg/m^3) as ammonia.
Short-term exposure limits (15-minute TWA): ACGIH STEL 35 ppm (24 mg/m^3); NIOSH STEL 35 ppm (27 mg/m^3); OSHA STEL 50 ppm (35 mg/m^3) as ammonia.
Vapor (gas) irritant characteristics: May cause irritation of the respiratory tract.
Liquid or solid irritant characteristics: May cause eye and skin irritation.
Odor threshold: Less than 5 ppm as ammonia gas.
IDLH value: 300 ppm as ammonia gas.

FIRE DATA
Behavior in fire: Decomposes, but reaction is not explosive.

CHEMICAL REACTIVITY
Binary reactants: Reacts with sodium nitrite, magnesium. Decomposes on exposure to air with loss of ammonia and CO_2, converting into ammonium bicarbonate.

ENVIRONMENTAL DATA
Water pollution: Effect of low concentrations on aquatic life is unknown. May be dangerous if it enters nearby water intakes; notify operators. Notify local health and wildlife officials.
Response to discharge: Disperse and flush

SHIPPING INFORMATION
Grades of purity: CP; NF; USP; Reagent; Technical; **Storage temperature:** Ambient; **Inert atmosphere:** None; **Venting:** Open; **Stability during transport:** Stable.

PHYSICAL AND CHEMICAL PROPERTIES
Physical state @ 59°F/15°C and 1 atm: Solid.
Molecular weight: 157.1
Boiling point @ 1 atm: Decomposes.
Melting/Freezing point: 136°F/58°C.
Specific gravity (water = 1): 1.5 @ 68°F/20°C (solid).

AMMONIUM CHLORIDE   REC. A:2450

SYNONYMS: AMCHLORIDE; AMMONERIC; AMCHLOR; AMMONIUM MURIATE; CLORURO AMONICO (Spanish); EEC No. 017-014-00-8; SAL AMMONIAC; SALMIAC

IDENTIFICATION
CAS Number: 12125-02-9
Formula: NH_4Cl
DOT ID Number: UN 9085; DOT Guide Number: 171
Proper Shipping Name: Ammonium chloride
Reportable Quantity (RQ): **(CERCLA)** 5000 lb/2270 kg

DESCRIPTION: White solid. Odorless. Sinks and mixes slowly with water.

Irritating to eyes, skin, and respiratory tract • Toxic products of combustion may include nitrogen oxides, hydrogen chloride, and ammonia gas.

Hazard Classification (based on NFPA-704 Rating System)
Health Hazards (Blue): 1; Flammability (Red): 0
REACTIVITY: 0

EMERGENCY RESPONSE: See Appendix A (171)
Evacuation:
Public safety: Isolate the area of spill or leak for at least 10 to 25 meters (30 to 80 feet) in all directions.
Spill: Increase, in the downwind direction, as necessary, the distance shown under "Public Safety."
Fire: If any large container is involved in fire, isolate for at least 800 meters (½ mile) in all directions; also, consider initial evacuation for 800 meters (½ mile) in all directions.

EXPOSURE
Short-term effects: *SEEK MEDICAL ATTENTION*. **Dust:** Irritating to eyes, nose, and throat. *IF INHALED*, will, will cause coughing or difficult breathing. *IF IN EYES*, hold eyelids open and flush with plenty of water. *IF BREATHING HAS STOPPED*, give artificial respiration. IF breathing is difficult, administer oxygen. **Solid:** Irritating to skin and eyes. *IF SWALLOWED*, will cause nausea. Remove contaminated clothing and shoes. Flush affected areas with plenty of water. *IF IN EYES*, hold eyelids open and flush with plenty of water. *IF SWALLOWED* and victim is *CONSCIOUS AND ABLE TO SWALLOW*, have victim drink 4 to 8 ounces of water. *IF SWALLOWED* and victim is *UNCONSCIOUS OR HAVING CONVULSIONS*, do nothing except keep victim warm.

HEALTH HAZARDS
Personal protective equipment (PPE): B-Level PPE. OSHA Table Z-1-A air contaminant. Gloves of any material; safety glasses or chemical safety goggles; dust mask or respirator as necessary.
Exposure limits (TWA unless otherwise noted): ACGIH TLV 10 mg/m^3; NIOSH REL 10 mg/m^3.

48 Ammonium chromate

Short-term exposure limits (15-minute TWA): ACGIH STEL 20 mg/m^3; NIOSH STEL 20 mg/m^3.
Toxicity by ingestion: Grade 2; oral rat LD$_{50}$ = 1650 mg/kg.
Vapor (gas) irritant characteristics: May irritate the respiratory tract.
Liquid or solid irritant characteristics: Skin, eye and respiratory system irritant

FIRE DATA
Behavior in fire: May volatilize and condense on cool surfaces.

CHEMICAL REACTIVITY
Reactivity with water: Contact can cause a violent reaction with evolution of heat and formation of hydrogen chloride.
Binary reactants: Can be self reactive; explosion may occur when closed containers are opened after long storage. Violent reaction with boron trifluoride, boron pentafluoride, bromine trifluoride, iodine heptafluoride, potassium chlorate. Mixture with hydrogen cyanide may form explosive nitrogen trichloride. Incompatible with alkalis, alkali carbonates, acids. Forms shock-sensitive compounds with salts of lead or silver. At fire temperature conditions fumes corrodes most metals including copper compounds.

ENVIRONMENTAL DATA
Water pollution: EPA water pollutant. Harmful to aquatic life in very low concentrations. May be dangerous if it enters nearby water intakes; notify operators. Notify local health and wildlife officials.
Response to discharge: Disperse and flush.

SHIPPING INFORMATION
Grades of purity: USP; Reagent; Technical, 99+%; **Storage temperature:** Ambient; **Inert atmosphere:** None; **Venting:** Open; **Stability during transport:** Stable.

PHYSICAL AND CHEMICAL PROPERTIES
Physical state @ 59°F/15°C and 1 atm: Solid.
Molecular weight: 53.5
Boiling point @ 1 atm: Sublimes
Melting/Freezing point: 630°F/332°C/605°K Sublimes
Specific gravity (water = 1): 1.53 @ 68°F/20°C (solid).
Heat of solution: 130 Btu/lb = 72 cal/g = 3.0 x 10^5 J/kg.
Vapor pressure: 1 mm @ 321°F/161°C.

AMMONIUM CHROMATE REC. A:2500

SYNONYMS: AMMONIUM CHROMATE(VI); CROMATO AMONICO (Spanish); DIAMMONIUM CHROMATE

IDENTIFICATION
CAS Number: 7788-98-9
Formula: $(NH_4)_2CrO_4$
DOT ID Number: UN 3077; DOT Guide Number: 171
Proper Shipping Name: Environmentally hazardous substances, solid, n.o.s.
Reportable Quantity (RQ): **(CERCLA)** 10 lb/4.54 kg

DESCRIPTION: Yellow crystalline material. Ammonia odor. Sinks and mixes with water forming an alkaline solution that releases free ammonia.

May explode when shocked or heated • Strong oxidizer which may react spontaneously with low flash point organics or reducing agents. Heat forms oxygen; will increase the activity of an existing fire • May cause fire or explosion on contact with combustibles (wood, paper, oil, clothing, etc.) • Toxic products of combustion may include nitrogen oxides and ammonia.

Hazard Classification (based on NFPA-704 Rating System)
Health Hazards (Blue): 2; Flammability (Red): 0; Reactivity (Yellow): 1; Special Notice (White): OXY

EMERGENCY RESPONSE: See Appendix A (171)
Evacuation:
Public safety: Isolate the area of spill or leak for at least 10 to 25 meters (30 to 80 feet) in all directions.
Spill: Increase, in the downwind direction, as necessary, the distance shown under "Public Safety."
Fire: If any large container is involved in fire, isolate for at least 800 meters (½ mile) in all directions; also, consider initial evacuation for 800 meters (½ mile) in all directions.

EXPOSURE
Short-term effects: *SEEK MEDICAL ATTENTION*. **Solid:** *POISONOUS IF SWALLOWED* OR INHALED. Irritating or corrosive to skin and mucous membranes.
Severely irritating to eyes. Remove contaminated clothing.
Flush affected areas with plenty of water. *IF IN EYES*, hold eyelids open and flush with plenty of water. *IF SWALLOWED* and victim is *CONSCIOUS AND ABLE TO SWALLOW*, have victim drink 4 to 8 ounces of water.

HEALTH HAZARDS
Personal protective equipment (PPE): B-Level PPE. Wear rubber gloves, industrial filter mask, face shield and safety glasses.
Recommendations for respirator selection: NIOSH
At any concentrations above the NIOSH REL, or where there is no REL, at any detectable concentration: SCBAF:PD,PP (any self-contained breathing apparatus that has a full facepiece and is operated in a pressure-demand or other positive-pressure mode); or SAF:PD,PP:ASCBA (any supplied-air respirator that has a full facepiece and is operated in a pressure-demand or other positive-pressure mode in combination with an auxiliary self-contained breathing apparatus operated in a pressure-demand or other positive pressure mode). *ESCAPE*: HiEF (any air-purifying, full-facepiece respirator with a high-efficiency particulate filter); or SCBAE (any appropriate escape-type, self-contained breathing apparatus).
Exposure limits (TWA unless otherwise noted): ACGIH TLV 0.05 mg/m^3 as water soluble Cr(VI) compounds; OSHA PEL 0.1 mg/m^3 as CrO$_3$; NIOSH 0.001 mg/m^3 as chromates; potential human carcinogen; reduce exposure to lowest feasible level.
Long-term health effects: Dust can cause lung cancer; a recognized carcinogen.

Vapor (gas) irritant characteristics: Causes severe irritation of the respiratory tract.
Liquid or solid irritant characteristics: Causes smarting of the skin and first-degree burns on short exposure; may cause second-degree burns on long exposure.
IDLH value: 15 mg/m^3 as chromium(VI).

FIRE DATA
Behavior in fire: A heat- and shock-sensitive explosive.

CHEMICAL REACTIVITY
Reactivity with water: Forms alkaline solution which evolves free ammonia; decomposes in hot water.
Reactivity with other materials: A powerful oxidizer. Contact with strong reducing agents such as hydrazine, alcohols, or ethers

can cause explosion. Violent reaction with combustible materials, finely divided metals, organic substances.
Neutralizing agents for acids and caustics: Dissolve in water. Cover with soda ash and mix. Neutralize with 6 M hydrochloric acid.

ENVIRONMENTAL DATA
Food chain concentration potential: High positive. Trout can accumulate hexavalent chromium at levels as low as 0.001 ppm. Half life in total human body 616 days.
Water pollution: Dangerous to aquatic life in high concentrations. May be dangerous if it enters nearby water intakes; notify operators. Notify local health and wildlife officials. **Response to discharge:** Issue warning-air and water pollutant. Chemical and physical treatment. Disperse and flush.

SHIPPING INFORMATION
Storage temperature: Ambient; **Stability during transport:** Stable-avoid shock, heat, and contact with reducing materials.

PHYSICAL AND CHEMICAL PROPERTIES
Physical state @ 59°F/15°C and 1 atm: Solid.
Molecular weight: 152.09
Boiling point @ 1 atm: 356°F/180°C/453.2°K (decomposes).
Melting/Freezing point: 356°F/180°C/453.2°K (decomposes).
Specific gravity (water = 1): 1.91 at 12°C.
Heat of solution: Endothermic (@ 77°F/25°C) 68.6 Btu/lb = 38.1 cal/g = 1.6×10^5 J/kg.

AMMONIUM CITRATE   REC. A:2550

SYNONYMS: AMMONIUM CITRATE, DIBASIC; CITRATO AMONICO DIBASICO (Spanish); CITRIC ACID, AMMONIUM SALT; CITRIC ACID, DIAMMONIUM SALT; DIAMMONIUM CITRATE; 1,2,3-PROPANE TRICARBOXYLIC ACID, 2-HYDROXY-, AMMONIUM SALT

IDENTIFICATION
CAS Number: 3012-65-5; 7632-50-0 (dibasic)
Formula: $(NH_4)_2HC_6H_5O_7$
DOT ID Number: UN 3077; DOT Guide Number: 171
Proper Shipping Name: Environmentally hazardous substances, solid, n.o.s.
Reportable Quantity (RQ): **(CERCLA)** 5000 lb/2270 kg

DESCRIPTION: White solid. Weak ammonia odor. Sinks and mixes with water.

Combustible solid • Dust cloud may explode if ignited in an enclosed area • Containers may BLEVE when exposed to fire • Irritating to the skin, eyes, and respiratory tract • Toxic products of combustion may include nitrogen oxides and ammonia gas.

Hazard Classification (based on NFPA-704 Rating System)
Health Hazards (Blue): 0; Flammability (Red): 1; Reactivity (Yellow): 0

EMERGENCY RESPONSE: See Appendix A (171)
Evacuation:
Public safety: Isolate the area of spill or leak for at least 10 to 25 meters (30 to 80 feet) in all directions.
Spill: Increase, in the downwind direction, as necessary, the distance shown under "Public Safety."
Fire: If any large container is involved in fire, isolate for at least 800 meters (½ mile) in all directions; also, consider initial evacuation for 800 meters (½ mile) in all directions.

EXPOSURE
Short-term effects: Dust: Irritating to eyes, nose, and throat. *IF INHALED*, will, will cause coughing or difficult breathing. *IF IN EYES*, hold eyelids open and flush with plenty of water. *IF BREATHING HAS STOPPED*, give artificial respiration. IF breathing is difficult, administer oxygen. **Solid:** Irritating to skin and eyes. *IF SWALLOWED*, will cause nausea. Remove contaminated clothing and shoes. Flush affected areas with plenty of water. *IF IN EYES*, hold eyelids open and flush with plenty of water. *IF SWALLOWED* and victim is *CONSCIOUS AND ABLE TO SWALLOW*, have victim drink 4 to 8 ounces of water and have victim induce vomiting. *IF SWALLOWED* and victim is *UNCONSCIOUS OR HAVING CONVULSIONS*, do nothing except keep victim warm.

HEALTH HAZARDS
Personal protective equipment (PPE): B-Level PPE. Guard against ammonia fumes. Butyl rubber is generally suitable for carbooxylic acid compounds.
Recommendations for respirator selection: NIOSH as ammonia *250 ppm*: CCRS [any chemical cartridge respirator with cartridge(s) providing protection against the compound of concern)]; or SA (any supplied-air respirator). *300 ppm*: SA:CF (any supplied-air respirator operated in a continuous-flow mode); or PAPRS* [any powered, air-purifying respirator with cartridge(s) providing protection against the compound of concern]; or CCRFS [any chemical cartridge respirator with a full facepiece and cartridge(s) providing protection against the compound of concern]; or GMFS [any air-purifying, full-facepiece respirator (gas mask) with a chin-style, front- or back-mounted canister providing protection against the compound of concern]; or SCBAF (any self-contained breathing apparatus with a full facepiece); or SAF (any supplied-air respirator with a full facepiece). *EMERGENCY OR PLANNED ENTRY INTO UNKNOWN CONCENTRATIONS OR IDLH CONDITIONS*: SCBAF:PD,PP (any self-contained breathing apparatus that has a full facepiece and is operated in a pressure-demand or other positive-pressure mode); or SAF:PD,PP:ASCBA (any supplied-air respirator that has a full facepiece and is operated in a pressure-demand or other positive-pressure mode in combination with an auxiliary self-contained breathing apparatus operated in a pressure-demand or other positive pressure mode). *ESCAPE:* GMFS [any air-purifying, full-facepiece respirator (gas mask) with a chin-style, front- or back-mounted canister providing protection against the compound of concern]; or SCBAE (any appropriate escape-type, self-contained breathing apparatus). *Note:* Substance reported to cause eye irritation or damage; may require eye protection.

Exposure limits (TWA unless otherwise noted): ACGIH TLV 25 ppm (17 mg/m³); NIOSH REL 5 ppm (18 mg/m³) as ammonia
Short-term exposure limits (15-minute TWA): ACGIH STEL 35 ppm (24 mg/m³); NIOSH STEL 35 ppm; OSHA STEL 50 ppm (35 mg/m³) as ammonia
Vapor (gas) irritant characteristics: May cause eye and respiratory tract irritation.
Liquid or solid irritant characteristics: May cause eye and respiratory tract irritation.
IDLH value: 300 ppm as ammonia

FIRE DATA
Flash point: Combustible solid.

ENVIRONMENTAL DATA
Water pollution: Effect of low concentrations on aquatic life is unknown. May be dangerous if it enters nearby water intakes; notify operators. Notify local health and wildlife officials.
Response to discharge: Disperse and flush.

SHIPPING INFORMATION
Grades of purity: Reagent; **Storage temperature:** Ambient; **Inert atmosphere:** None; **Venting:** Open; **Stability during transport:** Stable.

PHYSICAL AND CHEMICAL PROPERTIES
Physical state @ 59°F/15°C and 1 atm: Solid.
Molecular weight: 226
Boiling point @ 1 atm: Decomposes.
Specific gravity (water = 1): 1.48 @ 68°F/20°C (solid).

AMMONIUM DICHROMATE   REC. A:2600

SYNONYMS: AMMONIUM BICHROMATE; AMMONIUM DICHROMATE(VI); DICROMO AMONICO (Spanish); EEC No. 024-003-00-1

IDENTIFICATION
CAS Number: 7789-09-5
Formula: $Cr_2H_8N_2O_7$; $(NH_4)_2Cr_2O_7$
DOT ID Number: UN 1439; DOT Guide Number: 141
Proper Shipping Name: Ammonium dichromate
Reportable Quantity (RQ): **(CERCLA)** 10 lb/4.54 kg

DESCRIPTION: Yellow to orange-red crystals or powder. Odorless. Sinks and mixes with water.

Poison! • Explosion hazard! Closed containers readily rupture at decomposition temperature (decomposes below 338°F/170°C, self-sustaining at about 437°F/225°C; heat causes material to expand dramatically with evolution of heat and nitrogen oxides. Combustible • Containers may BLEVE when exposed to fire • Dust may form explosive mixture with air • Concentrated dust in confined areas (e.g., tanks, sewers, buildings) may explode when exposed to fire • Highly corrosive to skin and mucous membranes • Strong oxidizer which may react spontaneously with low flash point organics or reducing agents. Heat forms oxygen; will increase the activity of an existing fire • May cause fire or explosion on contact with combustibles (wood, paper, oil, clothing, etc.) • Toxic products of combustion may include nitrogen oxides and chromic oxide smoke (greenish color) • May cause lung cancer if inhaled as dust • Do not put yourself in danger by entering a contaminated area to rescue a victim.

Hazard Classification (based on NFPA-704 Rating System)
Health Hazards (Blue): 2; Flammability (Red): 1; Reactivity (Yellow): 1; Special Notice (White): OXY

EMERGENCY RESPONSE: See Appendix A (141)
Evacuation:
Public safety: Isolate the area of spill or leak for at least 10 to 25 meters (30 to 80 feet) in all directions.
Spill: Consider initial downwind evacuation for at least 100 meters (330 feet).
Fire: If any large container is involved in fire, isolate for at least 800 meters (½ mile) in all directions; also, consider initial evacuation for 800 meters (½ mile) in all directions.

EXPOSURE
Short-term effects: *SEEK MEDICAL ATTENTION*. **Dust:** Irritating to eyes, nose, and throat. *IF INHALED*, will, will cause coughing or difficult breathing. Move victim to fresh air. *IF IN EYES*, hold eyelids open and flush with plenty of water. IF breathing is difficult, administer oxygen. **Solid:** Irritating to skin and eyes. Harmful if swallowed. Remove contaminated clothing and shoes. Flush affected areas with plenty of water. *IF IN EYES*, hold eyelids open and flush with plenty of water. *IF SWALLOWED* and victim is *CONSCIOUS AND ABLE TO SWALLOW*, have victim drink 4 to 8 ounces of water and have victim induce vomiting. *IF SWALLOWED* and victim is *UNCONSCIOUS OR HAVING CONVULSIONS,* do nothing except keep victim warm. *Note to physician or authorized medical personnel*: Medical observation is recommended for 24 to 48 hours after breathing overexposure, as pulmonary edema may be delayed. As first aid for pulmonary edema, consider administering a corticosteroid spray. Cigarette smoking may exacerbate pulmonary injury and should be discouraged for at least 72 hours following exposure.

HEALTH HAZARDS
Personal protective equipment (PPE): B-Level PPE. Chemical protective material(s) reported to have good to excellent resistance: butyl rubber, polycarbonate.
Recommendations for respirator selection: NIOSH
At any concentrations above the NIOSH REL, or where there is no REL, at any detectable concentration: SCBAF:PD,PP (any self-contained breathing apparatus that has a full facepiece and is operated in a pressure-demand or other positive-pressure mode); or SAF:PD,PP:ASCBA (any supplied-air respirator that has a full facepiece and is operated in a pressure-demand or other positive-pressure mode in combination with an auxiliary self-contained breathing apparatus operated in a pressure-demand or other positive pressure mode). *ESCAPE*: HiEF (any air-purifying, full-facepiece respirator with a high-efficiency particulate filter); or SCBAE (any appropriate escape-type, self-contained breathing apparatus).
Exposure limits (TWA unless otherwise noted): ACGIH TLV 0.05 mg/m^3 as water soluble Cr(VI) compounds; OSHA PEL 0.1 mg/m^3 as CrO_3; NIOSH 0.001 mg/m^3 as chromates; potential human carcinogen; reduce exposure to lowest feasible level.
Long-term health effects: Has caused lung cancer in animals. May cause skin problems.
Vapor (gas) irritant characteristics: Severe irritation of the eyes and respiratory tract.
Liquid or solid irritant characteristics: Severe irritation of the eyes, skin, and respiratory tract.
IDLH value: 15 mg/m^3 as chromium(VI). Suspected human carcinogen.

FIRE DATA
Behavior in fire: Decomposes at about 356°F/180°C. Decomposition self-sustaining at about 1300°F/704°C with spectacular swelling and evolution of heat and nitrogen, leaving chromic oxide residue. Pressure of confined gases can burst closed containers explosively.
Autoignition temperature: 437°F/225°C/498°K.

CHEMICAL REACTIVITY
Reactivity with water: Forms a strong acid.
Binary reactants: A strong oxidizer. Reacts with acids, alcohol and reducing agents. Can ignite combustible material such as wood shavings. Keep solution away from bases and metals.

ENVIRONMENTAL DATA
Water pollution: Dangerous to aquatic life in high concentrations.

May be dangerous if it enters nearby water intakes; notify operators. Notify local health and wildlife officials. **Response to discharge:** Issue warning–high flammability, oxidizing material. Disperse and flush.

SHIPPING INFORMATION
Grades of purity: Analytical reagent 99.0%; technical (photolitho) grade; technical granular grade: 99.7%; C.P. granular grade: 99.8%; **Storage temperature:** Ambient; **Inert atmosphere:** None; **Venting:** None; **Stability during transport:** Stable.

PHYSICAL AND CHEMICAL PROPERTIES
Physical state @ 59°F/15°C and 1 atm: Solid.
Molecular weight: 252.06
Boiling point @ 1 atm: Decomposes.
Melting/Freezing point: 338°F/170°C/443°K (decomposes)
Specific gravity (water = 1): 2.15 @ 77°F/25°C (solid).
Heat of solution: 41 Btu/lb = 23 cal/g = 0.96×10^5 J/kg.

AMMONIUM FLUOBORATE   REC. A:2650

SYNONYMS: AMMONIUM BOROFLUORIDE; AMMONIUM TETRAFLUOBORATE

IDENTIFICATION
CAS Number: 13826-83-0
Formula: NH_4BF_4
DOT ID Number: UN 9088; DOT Guide Number: 154
Proper Shipping Name: Ammonium fluoborate
Reportable Quantity (RQ): **(CERCLA)** 5000 lb/2270 kg

DESCRIPTION: White to yellow solid crystals. Odorless. Sinks and mixes with water.

Poison! • Firefighting gear (including SCBA) does not provide adequate protection. If exposure occurs, remove and isolate gear immediately and thoroughly decontaminate personnel • Can cause irritation of eyes, respiratory passages; nosebleed, and nausea • Sublimes above 460°F/238°C yielding toxic fumes of nitrogen oxides, fluorine, and ammonia.

Hazard Classification (based on NFPA-704 Rating System)
Health Hazards (Blue): 4; Flammability (Red): 0; Reactivity (Yellow): 0

EMERGENCY RESPONSE: See Appendix A (154)
Evacuation:
Public safety: Isolate the area of spill or leak for at least 25 to 50 meters (80 to 160 feet) in all directions.
Spill: Increase, in the downwind direction, as necessary, the distance shown under "Public Safety."
Fire: If tank, rail car, or tank truck is involved in fire, isolate for at least 800 meters (½ mile) in all directions; also, consider initial evacuation for 800 meters (½ mile) in all directions.

EXPOSURE
Short-term effects: *SEEK MEDICAL ATTENTION*. **Vapor:** Irritating to eyes, nose, and throat. *IF INHALED*, will may cause nose bleeds and nausea. Move to fresh air. *IF BREATHING HAS STOPPED*, give artificial respiration; *avoid mouth-to-mouth resuscitation; use bag/mask apparatus*. **Dust:** Irritating to eyes, nose, and throat. *IF IN EYES*, hold eyelids open and flush with plenty of water. *IF SWALLOWED* and victim is *CONSCIOUS AND ABLE TO SWALLOW*, have victim drink water. Induce vomiting.

HEALTH HAZARDS
Personal protective equipment (PPE): B-Level PPE.
Recommendations for respirator selection: NIOSH/OSHA as F. *12.5 mg/m³:* DM (any dust and mist respirator). *25 mg/m³:* DMXSQ* (any dust and mist respirator except single-use and quarter-mask respirators); or SA* (any supplied-air respirator). *62.5 mg/m³:* SA:CF* [any supplied-air respirator operated in a continuous-flow mode)]; or PAPRDM*⁺ *if not present as a fume* (any powered, air-purifying respirator with a dust and mist filter). *125 mg/m³:* HiEF⁺ (any air-purifying, full-facepiece respirator with a high-efficiency particulate filter); or SCBAF (any self-contained breathing apparatus with a full facepiece); or SAF (any supplied-air respirator with a full facepiece). *250 mg/m³:* SA:PD,PP (any supplied-air respirator operated in a pressure-demand or other positive-pressure mode). *EMERGENCY OR PLANNED ENTRY INTO UNKNOWN CONCENTRATIONS OR IDLH CONDITIONS:* SCBAF:PD,PP (any self-contained breathing apparatus that has a full faceplate and is operated in a pressure-demand or other positive-pressure mode); or SAF:PD,PP:ASCBA (any supplied-air respirator that has a full facepiece and is operated in a pressure-demand or other positive-pressure mode in combination with an auxiliary, self-contained breathing apparatus operated in a pressure-demand or other positive-pressure mode). *ESCAPE:* HiEF⁺ (any air-purifying, full-facepiece respirator with a high-efficiency particulate filter); or SCBAE (any appropriate escape-type, self-contained breathing apparatus). *Notes:* *Substance reported to cause eye irritation or damage; may require eye protection. ⁺May need acid gas sorbent.
Exposure limits (TWA unless otherwise noted): ACGIH TLV, OSHA PEL, NIOSH REL: 2.5 mg/m³ as fluorides.
Liquid or solid irritant characteristics: Causes irritation of the eyes and respiratory system.

CHEMICAL REACTIVITY
Binary reactants: Reacts with acids, evolving hydrogen fluoride. reacts with alkalis, releasing ammonia. Corrosive to aluminum.

ENVIRONMENTAL DATA
Water pollution: Effects of low concentrations on aquatic life are unknown. May be dangerous if it enters nearby water intakes; notify operators. Notify local health and wildlife officials. **Response to discharge:** Should be removed. Disperse and flush.

SHIPPING INFORMATION
Grades of purity: 97.0% minimum; **Storage temperature:** Ambient (moderate); **Stability during transport:** Stable.

PHYSICAL AND CHEMICAL PROPERTIES
Physical state @ 59°F/15°C and 1 atm: Solid.
Molecular weight: 104.9
Boiling point @ 1 atm: Sublimes at 460°F/238°C/511°K.
Melting/Freezing point: 446°F/230°C/503°K.
Specific gravity (water = 1): 1.871 @ 15°C/1.85 at 17.5°C.
Latent heat of vaporization: Btu/lb = 144 = cal/g 80.2 = J/kg 3.36×10^5 (These data are for heat of sublimation).

AMMONIUM FLUORIDE   REC. A:2700

SYNONYMS: EEC No. 009-006-00-8; FLUOURO AMONICO (Spanish); NEUTRAL AMMONIUM FLUORIDE

Ammonium formate

IDENTIFICATION
CAS Number: 12125-01-8
Formula: NH_4F
DOT ID Number: UN 2505; DOT Guide Number: 154
Proper Shipping Name: Ammonium fluoride
Reportable Quantity (RQ): **(CERCLA)** 100 lb/45.4 kg

DESCRIPTION: White solid. Odorless. Sinks and mixes with water, form dilute solution of hydrofluoric acid.

Poison! • Breathing the dust, skin or eye contact, or swallowing the material can kill you • Firefighting gear (including SCBA) does not provide adequate protection. If exposure occurs, remove and isolate gear immediately and thoroughly decontaminate personnel • Toxic products of combustion may include ammonia, nitrogen oxides, and hydrogen fluoride • Do not put yourself in danger by entering a contaminated area to rescue a victim.

Hazard Classification (based on NFPA-704 Rating System)
Health Hazards (Blue): 3; Flammability (Red): 0; Reactivity (Yellow): 0

EMERGENCY RESPONSE: See Appendix A (154)
Evacuation:
Public safety: Isolate the area of spill or leak for at least 25 to 50 meters (80 to 160 feet) in all directions.
Spill: Increase, in the downwind direction, as necessary, the distance shown under "Public Safety."
Fire: If tank, rail car, or tank truck is involved in fire, isolate for at least 800 meters (½ mile) in all directions; also, consider initial evacuation for 800 meters (½ mile) in all directions.

EXPOSURE
Short-term effects: *SEEK MEDICAL ATTENTION*. **Dust or fume:** Irritating to eyes, nose, and throat. Possible lung edema. *IF INHALED*, will, will cause coughing or difficult breathing. *IF IN EYES*, hold eyelids open and flush with plenty of water. *IF BREATHING HAS STOPPED,* give artificial respiration; *avoid mouth-to-mouth resuscitation; use bag/mask apparatus.* IF breathing is difficult, administer oxygen. **Solid:** *POISONOUS IF SWALLOWED*. Will burn skin and eyes. Remove contaminated clothing and shoes. Flush affected areas with plenty of water. *IF IN EYES*, hold eyelids open and flush with plenty of water. *IF SWALLOWED* and victim is *CONSCIOUS AND ABLE TO SWALLOW*, have victim drink 4 to 8 ounces of water. *IF SWALLOWED* and victim is *UNCONSCIOUS OR HAVING CONVULSIONS,* do nothing except keep victim warm.
Note to physician or authorized medical personnel: Medical observation is recommended for 24 to 48 hours after breathing overexposure, as pulmonary edema may be delayed. As first aid for pulmonary edema, consider administering a corticosteroid spray. Cigarette smoking may exacerbate pulmonary injury and should be discouraged for at least 72 hours following exposure.

HEALTH HAZARDS
Personal protective equipment (PPE): A-Level PPE. Chemical protective material(s) reported to have good to excellent resistance: natural rubber, nitrile, Neoprene+natural rubber. Ammonium fluoride (30–70%): Viton®, neoprene, nitrile+PVC, nitrile, PVC.
Recommendations for respirator selection: NIOSH/OSHA as F. *12.5 mg/m³:* DM (any dust and mist respirator). *25 mg/m³:* DMXSQ* (any dust and mist respirator except single-use and quarter-mask respirators); or SA* (any supplied-air respirator). *62.5 mg/m³:* SA:CF* [any supplied-air respirator operated in a continuous-flow mode)]; or PAPRDM*+ *if not present as a fume* (any powered, air-purifying respirator with a dust and mist filter). *125 mg/m³:* HiEF+ (any air-purifying, full-facepiece respirator with a high-efficiency particulate filter); or SCBAF (any self-contained breathing apparatus with a full facepiece); or SAF (any supplied-air respirator with a full facepiece). *250 mg/m³:* SA:PD,PP (any supplied-air respirator operated in a pressure-demand or other positive-pressure mode). *EMERGENCY OR PLANNED ENTRY INTO UNKNOWN CONCENTRATIONS OR IDLH CONDITIONS:* SCBAF:PD,PP (any self-contained breathing apparatus that has a full faceplate and is operated in a pressure-demand or other positive-pressure mode); or SAF:PD,PP:ASCBA (any supplied-air respirator that has a full facepiece and is operated in a pressure-demand or other positive-pressure mode in combination with an auxiliary, self-contained breathing apparatus operated in a pressure-demand or other positive-pressure mode). *ESCAPE:* HiEF+ (any air-purifying, full-facepiece respirator with a high-efficiency particulate filter); or SCBAE (any appropriate escape-type, self-contained breathing apparatus). *Notes:* *Substance reported to cause eye irritation or damage; may require eye protection. +May need acid gas sorbent.
Exposure limits (TWA unless otherwise noted): ACGIH TLV, OSHA PEL, NIOSH REL: 2.5 mg/m³ as fluorides.
Long-term health effects: May cause bone disorders.
Vapor (gas) irritant characteristics: Corrosive to the eyes and respiratory tract. Possible lung edema.
Liquid or solid irritant characteristics: Corrosive.

FIRE DATA
Behavior in fire: May sublime when hot and condense on cool surfaces.

CHEMICAL REACTIVITY
Reactivity with water: Dissolves and forms dilute solution of hydrofluoric acid.
Binary reactants: Reacts with acids, alkalies. May corrode glass, cement, and most metals.

ENVIRONMENTAL DATA
Water pollution: Effect of low concentrations on aquatic life is unknown. May be dangerous if it enters nearby water intakes; notify operators. Notify local health and wildlife officials. **Response to discharge:** Issue warning–water contaminant. Disperse and flush.

SHIPPING INFORMATION
Grades of purity: Technical, 96.0%; Reagent; Electronic; Low sodium; **Storage temperature:** Ambient; **Inert atmosphere:** None; **Venting:** Open; **Stability during transport:** Stable.

PHYSICAL AND CHEMICAL PROPERTIES
Physical state @ 59°F/15°C and 1 atm: Solid.
Molecular weight: 37.04
Boiling point @ 1 atm: Decomposes.
Melting/Freezing point: Decomposes.
Specific gravity (water = 1): 1.32 @ 77°F/25°C (solid).
Heat of solution: 72 Btu/lb = 40 cal/g = 1.7×10^5 J/kg.

AMMONIUM FORMATE   REC. A:2750

SYNONYMS: FORMIATO AMONICO (Spanish); FORMIC ACID, AMMONIUM SALT

IDENTIFICATION
CAS Number: 540-69-2
Formula: $CH_2O_2 \cdot H_3N$; $HCOONH_4$

DESCRIPTION: White solid. Weak ammonia odor. Sinks and mixes slowly with water.

Combustible solid • Dust cloud may explode if ignited in an enclosed area • Containers may BLEVE when exposed to fire • Irritating to the skin, eyes, and respiratory tract • Toxic products of combustion may include irritating ammonia and toxic formic acid vapors.

Hazard Classification (based on NFPA-704 Rating System)
Health Hazards (Blue): 0; Flammability (Red): 1; Reactivity (Yellow): 0

EMERGENCY RESPONSE: See Appendix A (171)
Evacuation:
Public safety: Isolate the area of spill or leak for at least 10 to 25 meters (30 to 80 feet) in all directions.
Spill: Increase, in the downwind direction, as necessary, the distance shown under "Public Safety."
Fire: If any large container is involved in fire, isolate for at least 800 meters (½ mile) in all directions; also, consider initial evacuation for 800 meters (½ mile) in all directions.

EXPOSURE
Short-term effects: *SEEK MEDICAL ATTENTION.* **Dust:** Irritating to eyes, nose, and throat. *IF INHALED*, will, will cause coughing or difficult breathing. *IF IN EYES*, hold eyelids open and flush with plenty of water. *IF BREATHING HAS STOPPED*, give artificial respiration. IF breathing is difficult, administer oxygen.
Solid: Irritating to skin and eyes. *IF SWALLOWED*, will cause nausea. Remove contaminated clothing and shoes. Flush affected areas with plenty of water. *IF IN EYES*, hold eyelids open and flush with plenty of water. *IF SWALLOWED* and victim is *CONSCIOUS AND ABLE TO SWALLOW*, have victim drink 4 to 8 ounces of water. *IF SWALLOWED* and victim is *UNCONSCIOUS OR HAVING CONVULSIONS*, do nothing except keep victim warm.

Personal protective equipment (PPE): B-Level PPE. Dust mask; goggles or face shield; rubber gloves.
Toxicity by ingestion: Grade 2; oral LD_{50} = 2250 mg/kg (mouse).
Vapor (gas) irritant characteristics: May cause eye and respiratory tract irritation.
Liquid or solid irritant characteristics: Causes eye and skin irritation.

ENVIRONMENTAL DATA
Water pollution: Effect of low concentrations on aquatic life is unknown. May be dangerous if it enters nearby water intakes; notify operators. Notify local health and wildlife officials.
Response to discharge: Disperse and flush.

SHIPPING INFORMATION
Grades of purity: Analytical grade; organic chemical grade; **Storage temperature:** Ambient; **Inert atmosphere:** None; **Venting:** Open; **Stability during transport:** Stable.

PHYSICAL AND CHEMICAL PROPERTIES
Physical state @ 59°F/15°C and 1 atm: Solid.
Molecular weight: 63.06
Boiling point @ 1 atm: Decomposes.
Melting/Freezing point: 241°F/116°C/389°K.
Specific gravity (water = 1): 1.28 @ 77°F/25°C (solid).
Heat of solution: 84.7 Btu/lb −47.1 cal/g = −1.97 x 10^5 J/kg.

AMMONIUM GLUCONATE REC. A:2800

SYNONYMS: GLUCONATO AMONICO (Spanish)

IDENTIFICATION
Formula: $NH_4C_6H_{11}O_7$

White solid. Weak ammonia odor. Sinks and mixes with water.

Combustible solid • Dust cloud may explode if ignited in an enclosed area • Containers may BLEVE when exposed to fire • Irritating to the skin, eyes, and respiratory tract • Toxic products of combustion may include nitrogen oxides.

Hazard Classification (based on NFPA-704 Rating System)
Health Hazards (Blue): 0; Flammability (Red): 1; Reactivity (Yellow): 0

EMERGENCY RESPONSE: See Appendix A (171)
Evacuation:
Public safety: Isolate the area of spill or leak for at least 10 to 25 meters (30 to 80 feet) in all directions.
Spill: Increase, in the downwind direction, as necessary, the distance shown under "Public Safety."
Fire: If any large container is involved in fire, isolate for at least 800 meters (½ mile) in all directions; also, consider initial evacuation for 800 meters (½ mile) in all directions.

EXPOSURE
Short-term effects: *SEEK MEDICAL ATTENTION.* **Dust:** Irritating to eyes, nose, and throat. *IF INHALED*, will, will cause coughing or difficult breathing. *IF IN EYES*, hold eyelids open and flush with plenty of water. *IF BREATHING HAS STOPPED*, give artificial respiration. IF breathing is difficult, administer oxygen.
Solid: Irritating to skin and eyes. Harmful if swallowed. Remove contaminated clothing and shoes. Flush affected areas with plenty of water. *IF IN EYES*, hold eyelids open and flush with plenty of water. *IF SWALLOWED* and victim is *CONSCIOUS AND ABLE TO SWALLOW*, have victim drink 4 to 8 ounces of water. *IF SWALLOWED* and victim is *UNCONSCIOUS OR HAVING CONVULSIONS*, do nothing except keep victim warm.

HEALTH HAZARDS
Personal protective equipment (PPE): B-Level PPE. Respirator for nuisance dust
Vapor (gas) irritant characteristics: May cause eye and respiratory tract irritation.
Liquid or solid irritant characteristics: Causes eye and respiratory tract irritation.

FIRE DATA
Flash point: Combustible solid.

ENVIRONMENTAL DATA
Water pollution: Effect of low concentrations on aquatic life is unknown. May be dangerous if it enters nearby water intakes; notify operators. Notify local health and wildlife officials. **Response to discharge:** Disperse and flush.

SHIPPING INFORMATION
Grades of purity: Pure; **Storage temperature:** Ambient; **Inert atmosphere:** None; **Venting:** Open; **Stability during transport:** Stable.

PHYSICAL AND CHEMICAL PROPERTIES
Physical state @ 59°F/15°C and 1 atm: Solid.
Molecular weight: 213
Specific gravity (water = 1): More than 1 @ 68°F/20°C (solid).

AMMONIUM HYDROXIDE   REC. A:2850

SYNONYMS: AMMONIA WATER; AQUA AMMONIA; AQUEOUS AMMONIA; EEC No. 007-001-01-2; HIDROXIDO AMONICO (Spanish); HIDROXIDO AMONICO (Spanish) HOUSEHOLD AMMONIA

IDENTIFICATION
CAS Number: 1336-21-6
Formula: $H_4N \cdot HO$; $NH_4OH \cdot H_2O$
DOT ID Number: UN 2672 (10–35% ammonia in water); UN 2073 (35–50% ammonia in water); UN 1005 (>50% ammonia in water);
DOT Guide Number: 154 (10–35%); 125 (35–>50%)
Proper Shipping Name: Ammonia solutions, relative density between 0.880 and 0.957 at 15°C in water, with more than 10% but not more than 35% ammonia; Ammonia solutions, relative density less than 0.880 @ 15°C in water, with more than 35% but not more than 50% ammonia; Ammonia, anhydrous, liquefied or Ammonia solutions, relative density less than 0.880 @ 15°C in water, with more than 50% ammonia
Reportable Quantity (RQ): **(CERCLA)** 1000 lb/454 kg

DESCRIPTION: Colorless to milky-white liquid. Ammonia odor. Floats on water surface; soluble; reacts, producing mild heat; irritating vapor is produced.

Combustible • Containers may BLEVE when exposed to fire • Vapors are lighter than air and may hug the ground when cool • Vapors in confined areas (e.g., tanks, sewers, buildings) may explode when exposed to fire • Severely Irritating to the skin, eyes, and respiratory tract; skin and eye contact can cause severe burns and blindness • Firefighting gear (including SCBA) does not provide adequate protection. If exposure occurs, remove and isolate gear immediately and thoroughly decontaminate personnel • Vapors are lighter than air but will collect and stay in low areas • Containers may BLEVE when exposed to fire • Toxic products of combustion may include ammonia and nitrogen oxide gases • Do not put yourself in danger by entering a contaminated area to rescue a victim.

Hazard Classification (based on NFPA-704 Rating System)
Health Hazards (Blue): 2; Flammability (Red): 1; Reactivity (Yellow): 0

EMERGENCY RESPONSE: See Appendix A (154)
Evacuation:
Public safety: Isolate the area of spill or leak for at least 25 to 50 meters (80 to 160 feet) in all directions.
Spill: Increase, in the downwind direction, as necessary, the distance shown under "Public Safety."
Fire: If tank, rail car, or tank truck is involved in fire, isolate for at least 800 meters (½ mile) in all directions; also, consider initial evacuation for 800 meters (½ mile) in all directions.

EXPOSURE
Short-term effects: *SEEK MEDICAL ATTENTION*. **Vapor:** Irritating to skin, eyes, nose, and throat. *IF INHALED*, will, will cause nausea, vomiting, difficult breathing, or loss of consciousness. Move to fresh air. *IF IN EYES*, hold eyelids open and flush with plenty of water. *IF BREATHING HAS STOPPED*, give artificial respiration; *avoid mouth-to-mouth resuscitation; use bag/mask apparatus*. IF breathing is difficult, administer oxygen. **Liquid:** Will burn skin and eyes. Harmful is swallowed. Remove contaminated clothing and shoes. Flush affected areas with plenty of water. *IF IN EYES*, hold eyelids open and flush with plenty of water. *IF SWALLOWED* and victim is *CONSCIOUS AND ABLE TO SWALLOW*, have victim drink 4 to 8 ounces of water.

HEALTH HAZARDS
Personal protective equipment (PPE): B-Level PPE. Use of protective oil will reduce skin irritation from ammonia. Chemical protective material(s) reported to have good to excellent resistance: Styrene-butadiene, butyl rubber, neoprene, nitrile, chlorinated polyethylene.
Exposure limits (TWA unless otherwise noted): ACGIH TLV 25 ppm (17 mg/m^3); NIOSH 25 ppm as ammonia
Short-term exposure limits (15-minute TWA): ACGIH STEL 35 ppm (24 mg/m^3); NIOSH/OSHA STEL 50 ppm (35 mg/m^3) as ammonia
Toxicity by ingestion: Grade 3; oral rat, LD_{50} = 350 mg/kg.
Long-term health effects: Possible lung damage from prolonged or repeated exposure.
Vapor (gas) irritant characteristics: Vapors cause irritation such that personnel will find high concentrations intolerable.
Liquid or solid irritant characteristics: Causes smarting of the skin and first-degree burns on short exposure; may cause second-degree burns on long exposure.
Odor threshold: 50 ppm.
IDLH value: 300 ppm as ammonia.

FIRE DATA
Flash point: Noncombustible. Vapors will burn; difficult to ignite.

CHEMICAL REACTIVITY
Reactivity with water: Mild liberation of heat.
Binary reactants: Corrosive to copper, copper alloys, aluminum alloys, galvanized surfaces.
Neutralizing agents for acids and caustics: Dilute with water.
Reactivity group: 6
Compatibility class: Ammonia

ENVIRONMENTAL DATA
Water pollution: Harmful to aquatic life in very low concentrations. May be dangerous if it enters nearby water intakes; notify operators. Notify local health and wildlife officials.
Response to discharge: Issue warning-air contaminant. Restrict access. Disperse and flush.

SHIPPING INFORMATION
Grades of purity: Grade A: 29.4% NH_3; B: 25%; C: 15%. USP: 27 to 29%. CP: 28%; **Storage temperature:** Ambient; **Inert atmosphere:** None; **Venting:** Pressure-vacuum; **Stability during transport:** Stable.

PHYSICAL AND CHEMICAL PROPERTIES
Physical state @ 59°F/15°C and 1 atm: Liquid.
Specific gravity (water = 1): 0.89 @ 68°F/20°C (liquid).
Relative vapor density (air = 1): 0.6

AMMONIUM HYPOPHOSPHITE   REC. A:2900

SYNONYMS: HIPOFOSFITO AMONICO (Spanish); PHOSPHINIC ACID, AMMONIUM SALT

IDENTIFICATION
Formula: NH$_4$H$_2$PO$_2$

DESCRIPTION: White crystalline solid. Sinks and mixes with water.

Combustible solid • Dust cloud may explode if ignited in an enclosed area • Containers may BLEVE when exposed to fire • Irritating to the skin, eyes, and respiratory tract • Toxic products of combustion may include ammonia, phosphine, phosphorus oxides, and nitrogen oxides. Phosphine is a highly toxic gas which ignites spontaneously.

Hazard Classification (based on NFPA-704 Rating System)
Health Hazards (Blue): 1; Flammability (Red): 2; Reactivity (Yellow): 0

EMERGENCY RESPONSE: See Appendix A (171)
Evacuation:
Public safety: Isolate the area of spill or leak for at least 10 to 25 meters (30 to 80 feet) in all directions.
Spill: Increase, in the downwind direction, as necessary, the distance shown under "Public Safety."
Fire: If any large container is involved in fire, isolate for at least 800 meters (½ mile) in all directions; also, consider initial evacuation for 800 meters (½ mile) in all directions.

EXPOSURE
Short-term effects: May be irritating to the skin, eyes, and respiratory tract Decomposition product in fire is very dangerous.

HEALTH HAZARDS
Personal protective equipment (PPE): B-Level PPE.
Liquid or solid irritant characteristics: No appreciable hazard. Practically harmless to the skin.

ENVIRONMENTAL DATA
Water pollution: Effect of low concentrations on aquatic life is unknown. May be dangerous if it enters nearby water intakes; notify operators. Notify local health and wildlife officials.
Response to discharge: Disperse and flush.

SHIPPING INFORMATION
Grades of purity: 98% pure; **Storage temperature:** Cool; **Stability during transport:** Stable.

PHYSICAL AND CHEMICAL PROPERTIES
Physical state @ 59°F/15°C and 1 atm: Solid.
Molecular weight: 83.0271
Boiling point @ 1 atm: Decomposes. 464°F/240°C/513.2°K.
Melting/Freezing point: 392°F/200°C/473.2°K.
Specific gravity (water = 1): 1.634 at room temperature
Heat of solution: Endothermic at infinite dilution (77°F/25°C) 34.7 Btu/lb = 19.3 cal/g = 8.07 x 10^5 J/kg.

AMMONIUM IODIDE   REC. A:2950

SYNONYMS: YODURO AMONICO (Spanish)

IDENTIFICATION
CAS Number: 12027-06-4
Formula: NH$_4$I

DESCRIPTION: White solid. Odorless. Sinks and mixes with water.

Corrosive to the skin, eyes, and respiratory tract; contact with skin or eyes can cause burns and vision impairment; inhalation symptoms may be delayed • Toxic products of combustion may include hydrogen iodide, iodine, and oxides of nitrogen.

Hazard Classification (based on NFPA-704 Rating System)
Health Hazards (Blue): 1; Flammability (Red): 0; Reactivity (Yellow): 0

EMERGENCY RESPONSE: See Appendix A (171)
Evacuation:
Public safety: Isolate the area of spill or leak for at least 10 to 25 meters (30 to 80 feet) in all directions.
Spill: Increase, in the downwind direction, as necessary, the distance shown under "Public Safety."
Fire: If any large container is involved in fire, isolate for at least 800 meters (½ mile) in all directions; also, consider initial evacuation for 800 meters (½ mile) in all directions.

EXPOSURE
Short-term effects: *SEEK MEDICAL ATTENTION.* **Dust:** Irritating to eyes, nose, and throat. *IF INHALED*, will, will cause coughing or difficult breathing. *IF IN EYES*, hold eyelids open and flush with plenty of water. *IF BREATHING HAS STOPPED*, give artificial respiration. IF breathing is difficult, administer oxygen. **Solid:** Irritating to skin and eyes. Harmful if swallowed. Remove contaminated clothing and shoes. Flush affected areas with plenty of water. *IF IN EYES*, hold eyelids open and flush with plenty of water. *IF SWALLOWED* and victim is *CONSCIOUS AND ABLE TO SWALLOW*, have victim drink 4 to 8 ounces of water. *IF SWALLOWED* and victim is *UNCONSCIOUS OR HAVING CONVULSIONS*, do nothing except keep victim warm.

HEALTH HAZARDS
Personal protective equipment (PPE): B-Level PPE. Dust mask; goggles or face shield; rubber gloves.
Vapor (gas) irritant characteristics: May cause eye and respiratory tract irritation.
Liquid or solid irritant characteristics: May cause eye and skin irritation.

FIRE DATA
Behavior in fire: Compound may sublime in fire and condense on cold surfaces.

ENVIRONMENTAL DATA
Water pollution: Effect of low concentrations on aquatic life is unknown. May be dangerous if it enters nearby water intakes; notify operators. Notify local health and wildlife officials. **Response to discharge:** Issue warning–water contaminant. Disperse and flush.

SHIPPING INFORMATION
Grades of purity: NF; Reagent 99%; **Storage temperature:** Ambient; **Inert atmosphere:** None; **Venting:** Open; **Stability during transport:** Stable.

PHYSICAL AND CHEMICAL PROPERTIES
Physical state @ 59°F/15°C and 1 atm: Solid.
Molecular weight: 144.94
Specific gravity (water = 1): 2.56 @ 68°F/20°C (solid).
Heat of solution: 43 Btu/lb = 24 cal/g = 1.0 x 10^5 J/kg.

AMMONIUM LACTATE REC. A:3000

SYNONYMS: DL-LACTIC ACID, AMMONIUM SALT; AMMONIUM LACTATE SYRUP

IDENTIFICATION
Formula: $CH_3CH(OH)COONH_4$

DESCRIPTION: White solid or liquid. Odorless. Sinks and mixes with water.

Combustible • Containers may BLEVE when exposed to fire •Vapors in confined areas (e.g., tanks, sewers, buildings) may explode when exposed to fire • Irritating to the skin, eyes, and respiratory tract • Toxic products of combustion may include nitrogen oxides.

Hazard Classification (based on NFPA-704 Rating System)
Health Hazards (Blue): 0; Flammability (Red): 1; Reactivity (Yellow): 0

EMERGENCY RESPONSE: See Appendix A (171)
Evacuation:
Public safety: Isolate the area of spill or leak for at least 10 to 25 meters (30 to 80 feet) in all directions.
Spill: Increase, in the downwind direction, as necessary, the distance shown under "Public Safety."
Fire: If any large container is involved in fire, isolate for at least 800 meters (½ mile) in all directions; also, consider initial evacuation for 800 meters (½ mile) in all directions.

EXPOSURE
Short-term effects: *SEEK MEDICAL ATTENTION*. **Dust:** Irritating to eyes, nose, and throat. *IF INHALED*, will, will cause coughing or difficult breathing. *IF IN EYES*, hold eyelids open and flush with plenty of water. *IF BREATHING HAS STOPPED*, give artificial respiration. IF breathing is difficult, administer oxygen.
Liquid or solid: Irritating to skin and eyes. Harmful if swallowed. Remove contaminated clothing and shoes. Flush affected areas with plenty of water. *IF IN EYES*, hold eyelids open and flush with plenty of water. *IF SWALLOWED* and victim is *CONSCIOUS AND ABLE TO SWALLOW*, have victim drink 4 to 8 ounces of water. *IF SWALLOWED* and victim is *UNCONSCIOUS OR HAVING CONVULSIONS,* do nothing except keep victim warm.

HEALTH HAZARDS
Personal protective equipment (PPE): B-Level PPE. Dust mask; goggles or face shield; rubber gloves.

ENVIRONMENTAL DATA
Water pollution: Effect of low concentrations on aquatic life is unknown. May be dangerous if it enters nearby water intakes; notify operators. Notify local health and wildlife officials.
Response to discharge: Disperse and flush.

SHIPPING INFORMATION
Grades of purity: Pure; 60% syrup in water; **Storage temperature:** Ambient; **Inert atmosphere:** None; **Venting:** Open; **Stability during transport:** Stable.

PHYSICAL AND CHEMICAL PROPERTIES
Physical state @ 59°F/15°C and 1 atm: Solid.
Molecular weight: 107.11
Specific gravity (water = 1): 1.2 @ 15°C (solid).

AMMONIUM LAURYL SULFATE REC. A:3050

SYNONYMS: DODECYL SULFATE, AMMONIUM SALT; LAURYL AMMONIUM SULFATE; SULFURIC ACID, LAURYL ESTER, AMMONIUM SALT; SULFURIC ACID, MONODODECYL ESTER, AMMONIUM SALT

IDENTIFICATION
CAS Number: 2235-54-3; 32612-48-9
Formula: $C_{12}H_{26}O_4S \cdot H_3N$

DESCRIPTION: Light yellow liquid or solid. May float, or sink and mix with water.

Irritating to skin, eyes, and respiratory tract • Toxic products of combustion may include CO_2, carbon monoxide, and oxides of nitrogen and sulfur.

Hazard Classification (based on NFPA-704 Rating System)
Health Hazards (Blue): 2; Flammability (Red): 0; Reactivity (Yellow): 0

EMERGENCY RESPONSE: See Appendix A (171)
Evacuation:
Public safety: Isolate the area of spill or leak for at least 10 to 25 meters (30 to 80 feet) in all directions.
Spill: Increase, in the downwind direction, as necessary, the distance shown under "Public Safety."
Fire: If any large container is involved in fire, isolate for at least 800 meters (½ mile) in all directions; also, consider initial evacuation for 800 meters (½ mile) in all directions.

EXPOSURE
Short-term effects: *SEEK MEDICAL ATTENTION*. **Liquid:** Irritating to skin and eyes. Harmful if swallowed. Remove contaminated clothing and shoes. Flush affected areas with plenty of water. *IF IN EYES*, hold eyelids open and flush with plenty of water. *IF SWALLOWED* and victim is *CONSCIOUS AND ABLE TO SWALLOW*, have victim drink 4 to 8 ounces of water. *IF SWALLOWED* and victim is *UNCONSCIOUS OR HAVING CONVULSIONS,* do nothing except keep victim warm.

HEALTH HAZARDS
Personal protective equipment (PPE): B-Level PPE. Rubber gloves; goggles or face shield.

CHEMICAL REACTIVITY
Binary reactants: Violent reaction with aluminum, magnesium.

ENVIRONMENTAL DATA
Water pollution: Effect of low concentrations on aquatic life is unknown. May be dangerous if it enters nearby water intakes; notify operators. Notify local health and wildlife officials. **Response to discharge:** Disperse and flush.

SHIPPING INFORMATION
Grades of purity: 28-30% solution in water; **Storage temperature:** Ambient; **Inert atmosphere:** None; **Venting:** Open; **Stability during transport:** Stable.

PHYSICAL AND CHEMICAL PROPERTIES
Physical state @ 59°F/15°C and 1 atm: Liquid.
Molecular weight: 283 (solute only).
Specific gravity (water = 1): 1.03 @ 68°F/20°C (liquid).

AMMONIUM MOLYBDATE
REC. A:3100

SYNONYMS: AMMONIUM PARAMOLYBDATE; DIAMMONIUM MOLYBDATE; MOLIBDATO AMONICO (Spanish); MOLYBDIC ACID DIAMMONIUM SALT; MOLYBDIC ACID (58%).

IDENTIFICATION
CAS Number: 13106-76-8
Formula: $(NH_4)_6MO_7O_{24}\cdot 4H_2O$ (for CP and Reagent grades only).

DESCRIPTION: Colorless to greenish-yellow or white solid. Odorless. Sinks and mixes with water.

Hazard Classification (based on NFPA-704 Rating System)
Health Hazards (Blue): 1; Flammability (Red): 0; Reactivity (Yellow): 0

EMERGENCY RESPONSE: See Appendix A (171)
Evacuation:
Public safety: Isolate the area of spill or leak for at least 10 to 25 meters (30 to 80 feet) in all directions.
Spill: Increase, in the downwind direction, as necessary, the distance shown under "Public Safety."
Fire: If any large container is involved in fire, isolate for at least 800 meters (½ mile) in all directions; also, consider initial evacuation for 800 meters (½ mile) in all directions.

EXPOSURE
Short-term effects: *SEEK MEDICAL ATTENTION*. **Dust:** Irritating to eyes, nose, and throat. *IF INHALED*, will, will cause coughing, or difficult breathing. *IF IN EYES*, hold eyelids open and flush with plenty of water. *IF BREATHING HAS STOPPED*, give artificial respiration. IF breathing is difficult, administer oxygen. **Solid:** Irritating to skin and eyes. Harmful if swallowed. Remove contaminated clothing and shoes. Flush affected areas with plenty of water. *IF IN EYES*, hold eyelids open and flush with plenty of water. *IF SWALLOWED* and victim is *CONSCIOUS AND ABLE TO SWALLOW*, have victim drink 4 to 8 ounces of water. *IF SWALLOWED* and victim is *UNCONSCIOUS OR HAVING CONVULSIONS*, do nothing except keep victim warm.

HEALTH HAZARDS
Personal protective equipment (PPE): B-Level PPE. Dust mask; goggles or face shield; rubber gloves.
Recommendations for respirator selection: OSHA as molybdenum compounds.
25 mg/m^3: DM (any dust and mist respirator). 50 mg/m^3: DMXSQ (any dust and mist respirator except single-use and quarter mask respirators); or SA (any supplied-air respirator). 125 mg/m^3: SA:CF (any supplied-air respirator operated in a continuous-flow mode); or PAPRDM, *If not present as a fume* (any powered, air-purifying respirator with a dust and mist filter). 250 mg/m^3: HiEF (any air-purifying, full-facepiece respirator with a high-efficiency particulate filter); or SAT:CF (any supplied-air respirator that has a tight-fitting facepiece and is operated in a continuous-flow mode); or PAPRTHiE (any powered, air-purifying respirator with a tight-fitting facepiece and a high-efficiency particulate filter); or SCBAF (any self-contained breathing apparatus with a full facepiece); or SAF (any supplied-air respirator with a full facepiece). 1000 mg/m^3: SAF:PD,PP (any supplied-air respirator that has a full facepiece and is operated in a pressure-demand or other positive-pressure mode). *EMERGENCY OR PLANNED ENTRY INTO UNKNOWN CONCENTRATIONS OR IDLH CONDITIONS*: SCBAF:PD,PP (any self-contained breathing apparatus that has a full facepiece and is operated in a pressure-demand or other positive-pressure mode); or SAF:PD,PP:ASCBA (any supplied-air respirator that has a full facepiece and is operated in a pressure-demand or other positive-pressure mode in combination with an auxiliary self-contained breathing apparatus operated in a pressure-demand or other positive pressure mode). *ESCAPE:* HiEF (any air-purifying, full-facepiece respirator with a high-efficiency particulate filter); or SCBAE (any appropriate escape-type, self-contained breathing apparatus). *Note*: Substance reported to cause eye irritation or damage; may require eye protection.
Exposure limits (TWA unless otherwise noted): ACGIH TLV 5 mg/m^3 as molybdenum; OSHA PEL 5 mg/m^3 as molybdenum
Toxicity by ingestion: Grade 3; oral rat LD$_{50}$ = 333 mg/kg.
IDLH value: 1000 mg/m^3 as molybdenum.

ENVIRONMENTAL DATA
Water pollution: Effect of low concentrations on aquatic life is unknown. May be dangerous if it enters nearby water intakes; notify operators. Notify local health and wildlife officials. **Response to discharge:** Issue warning–water contaminant. Should be removed. Chemical and physical treatment.

SHIPPING INFORMATION
Grades of purity: Reagent; CP. A closely related substance is "molybdic acid, 85%"; **Storage temperature:** Ambient; **Inert atmosphere:** None; **Venting:** Open; **Stability during transport:** Stable.

PHYSICAL AND CHEMICAL PROPERTIES
Physical state @ 59°F/15°C and 1 atm: Solid.
Molecular weight: 123.6
Specific gravity (water = 1): 1.4 @ 68°F/20°C (solid).

AMMONIUM NITRATE
REC. A:3150

SYNONYMS: AMFO; AN/FO (NA 0331); NITRATO AMONICO (Spanish); NITRIC ACID, AMMONIUM SALT; AMMONIUM SALTPETER; NITRAM; NORWAY SALTPETER

IDENTIFICATION
CAS Number: 6484-52-2
Formula: NH_4NO_3
DOT ID Number: NA 2072; UN2071; UN2067; UN 2069; UN0222; UN 1942; UN2426; NA0331; DOT Guide Number: 140
Proper Shipping Name: Ammonium nitrate fertilizers; Ammonium nitrate fertilizers: (uniform nonsegregating mixtures of nitrogen/phosphate or nitrogen/potash types or complete fertilizers of nitrogen/phosphate/potash type, with not more than 70% ammonium nitrate and not more than 0.4% total added combustible material or with not more than 45% ammonium nitrate with unrestricted combustible material); Ammonium nitrate fertilizers [uniform nonsegregating mixtures of ammonium nitrate with added matter which is inorganic and chemically inert towards ammonium nitrate, with not less than 90% ammonium nitrate and not more than 0.2% combustible material (including organic material calculated as carbon), or with more than 70% but less than 90% ammonium nitrate and not more than 0.4% total combustible material]; Ammonium nitrate mixed fertilizers; Ammonium nitrate, [with more than 0.2% combustible substances, including any organic substance calculated as carbon, to the exclusion of any other added substance]; Ammonium nitrate, [with not more than 0.2% of combustible substances, including any organic substance calculated

58 Ammonium nitrate–phosphate mixture

as carbon, to the exclusion of any other added substance]; Ammonium nitrate, liquid [(hot concentrated solution)]; Ammonium nitrate-fuel oil mixture [containing only prilled ammonium nitrate and fuel oil]

DESCRIPTION: Colorless (pure) to gray or brown (fertilizer grades) solid pellets or flakes. Odorless. Sinks and mixes with water.
Ammonium nitrate/fuel oil mixture (NA 0331) is a slurry containing only prilled ammonium nitrate. Smells like fuel oil. No information on water solubility.

Reacts with organic contaminants forming shock-sensitive mixtures • Potentially explosive mixture; containers may detonate in fire • Heat forms oxygen; will increase the activity of an existing fire • May cause fire or explosion on contact with combustibles (wood, paper, oil, clothing, etc.) • May interfere with the body's ability to use oxygen. Irritating to the skin, eyes, and respiratory tract • Containers may BLEVE when exposed to fire • Toxic products of combustion may include nitrogen oxides.

Hazard Classification (based on NFPA-704 Rating System)
Health Hazards (Blue): 0; Flammability (Red): 0; Reactivity (Yellow): 3; Special Notice (White): OXY

Hazard Classification (based on NFPA-704 Rating System) (phosphate or sulfate mixture) Health Hazards (Blue): 1; Flammability (Red): 0; Reactivity (Yellow): 1; Special Notice (White): OXY

Hazard Classification (based on NFPA-704 Rating System) (urea solution) Health Hazards (Blue): 2; Flammability (Red): 0; Reactivity (Yellow): 1

Hazard Classification (based on NFPA-704 Rating System) (fuel oil mixture) Health Hazards (Blue): 2; Flammability (Red): 1; Reactivity (Yellow): 3; Special Notice (White): EXPLOSIVE

EMERGENCY RESPONSE: See Appendix A (140)
Evacuation:
Public safety: Isolate the area of spill or leak for at least 10 to 25 meters (30 to 80 feet) in all directions.
Spill: Consider initial downwind evacuation for at least 100 meters (330 feet).
Fire: If any large container is involved in fire, isolate for at least 800 meters (½ mile) in all directions; also, consider initial evacuation for 800 meters (½ mile) in all directions.

EXPOSURE
Short-term effects: *SEEK MEDICAL ATTENTION.* **Dust:** Irritating to eyes, nose, and throat. *IF INHALED*, will, may cause coughing or difficult breathing. Move to fresh air. *IF IN EYES*, hold eyelids open and flush with plenty of water. *IF BREATHING HAS STOPPED*, give artificial respiration. IF breathing is difficult, administer oxygen.

HEALTH HAZARDS
Personal protective equipment (PPE): B-Level PPE. Chemical protective material(s) reported to have good to excellent resistance: butyl rubber, polycarbonate.
Long-term health effects: Blood effects (abnormal hemoglobin) may result.
Liquid or solid irritant characteristics: May cause eye, skin and respiratory tract irritation.

FIRE DATA
Behavior in fire: Containers may rupture and explode in fires. Supports combustion of common organic fuels.

CHEMICAL REACTIVITY
Binary reactants: Strong reaction with strong oxidizers, acids powdered metals, reducing agents and organic materials.

ENVIRONMENTAL DATA
Water pollution: Effect of low concentrations on aquatic life is unknown. May be dangerous if it enters nearby water intakes; notify operators. Notify local health and wildlife officials. **Response to discharge:** Disperse and flush.

SHIPPING INFORMATION
Grades of purity: Pure grade; fertilizer grade (33.5% nitrogen); **Stability during transport:** If heated strongly, decomposes, giving off toxic gases and gases which support combustion. Undergoes detonation if heated under confinement.

PHYSICAL AND CHEMICAL PROPERTIES
Physical state @ 59°F/15°C and 1 atm: Solid.
Molecular weight: 80.05
Boiling point @ 1 atm: Decomposes 410°F/209.9°C/483°K.
Melting/Freezing point: 338°F/170°C/443°K.
Specific gravity (water = 1): 1.72 @ 68°F/20°C (solid).
Relative vapor density (air = 1): 2.8

AMMONIUM NITRATE-PHOSPHATE MIXTURE
REC. A:3200

IDENTIFICATION
CAS Number: 57608-40-9
Formula: NH_4NO_3-$(NH_4)_2HPO_4$-$NH_4H_2PO_4$-$CaHPO_4$-KCl-K_2SO_4
DOT ID Number: UN 2071; DOT Guide Number: 140
Proper Shipping Name: Ammonium nitrate fertilizers

DESCRIPTION: Grayish-white solid. Odorless. Sinks and mixes with water.

Heat forms oxygen; will increase the activity of an existing fire • May cause fire or explosion on contact with combustibles (wood, paper, oil, clothing, etc.) • Toxic products of combustion may include nitrogen oxides and phosphorus oxides.

Hazard Classification (based on NFPA-704 Rating System)
Health Hazards (Blue): 0; Flammability (Red): 0; Reactivity (Yellow): 3; Special Notice (White): OXY

EMERGENCY RESPONSE: See Appendix A (140)
Evacuation:
Public safety: Isolate the area of spill or leak for at least 10 to 25 meters (30 to 80 feet) in all directions.
Spill: Consider initial downwind evacuation for at least 100 meters (330 feet).
Fire: If any large container is involved in fire, isolate for at least 800 meters (½ mile) in all directions; also, consider initial evacuation for 800 meters (½ mile) in all directions.

EXPOSURE
Short-term effects: *SEEK MEDICAL ATTENTION.* **Dust:** Irritating to eyes, nose, and throat. *IF INHALED*, will, will cause coughing or difficult breathing. *IF IN EYES*, hold eyelids open and flush with plenty of water. *IF BREATHING HAS STOPPED*, give

artificial respiration. IF breathing is difficult, administer oxygen.
Solid: Irritating to skin and eyes. Harmful if swallowed. Remove contaminated clothing and shoes. Flush affected areas with plenty of water. *IF IN EYES*, hold eyelids open and flush with plenty of water. *IF SWALLOWED* and victim is *CONSCIOUS AND ABLE TO SWALLOW*, have victim drink 4 to 8 ounces of water. *IF SWALLOWED* and victim is *UNCONSCIOUS OR HAVING CONVULSIONS*, do nothing except keep victim warm.

HEALTH HAZARDS
Personal protective equipment (PPE): B-Level PPE. Chemical protective material(s) reported to have good to excellent resistance: butyl rubber, polycarbonate.
Liquid or solid irritant characteristics: Causes eye and respiratory tract irritation.

FIRE DATA
Fire extinguishing agents not to be used: Steam, inert gases, foam, dry chemical.
Behavior in fire: Containers may explode.

CHEMICAL REACTIVITY
Binary reactants: Corrodes metals.

ENVIRONMENTAL DATA
Water pollution: Effect of low concentrations on aquatic life is unknown. May be dangerous if it enters nearby water intakes; notify operators. Notify local health and wildlife officials.
Response to discharge: Issue warning–oxidizing material. Disperse and flush.

SHIPPING INFORMATION
Grades of purity: Several grades of varying composition, all of which contain less than 70% of ammonium nitrate, the only hazardous ingredient; **Storage temperature:** Ambient; **Inert atmosphere:** Ventilated (natural); **Venting:** Open; **Stability during transport:** Stable.

PHYSICAL AND CHEMICAL PROPERTIES
Physical state @ 59°F/15°C and 1 atm: Solid.
Specific gravity (water = 1): 1.8 @ 68°F/20°C (solid).

AMMONIUM NITRATE–SULFATE MIXTURE
REC. A:3250

IDENTIFICATION
CAS Number: 6484-52-2
Formula: $NH_4NO_3 \cdot H_3N$
DOT ID Number: UN 2069; UN 1942; UN 2426; DOT Guide Number: 140
Proper Shipping Name: Ammonium nitrate mixed fertilizers (UN 2069); Ammonium nitrate, with organic coating (UN 1942); Ammonium nitrate, liquid (hot concentrated solution) (UN 2426)

DESCRIPTION: Grayish-white solid. Odorless. Sinks and mixes with water.

Heat forms oxygen; will increase the activity of an existing fire • May cause fire or explosion on contact with combustibles (wood, paper, oil, clothing, etc.). Containers may BLEVE when exposed to fire • Toxic products of combustion may include sulfur and nitrogen oxides and ammonia gas.

Hazard Classification (based on NFPA-704 Rating System)
Health Hazards (Blue): 1; Flammability (Red): 0; Reactivity (Yellow): 1; Special Notice (White): OXY

EMERGENCY RESPONSE: See Appendix A (140)
Evacuation:
Public safety: Isolate the area of spill or leak for at least 10 to 25 meters (30 to 80 feet) in all directions.
Spill: Consider initial downwind evacuation for at least 100 meters (330 feet).
Fire: If any large container is involved in fire, isolate for at least 800 meters (½ mile) in all directions; also, consider initial evacuation for 800 meters (½ mile) in all directions.

EXPOSURE
Short-term effects: *SEEK MEDICAL ATTENTION.* **Dust:** Irritating to eyes, nose, and throat. *IF INHALED*, will, will cause coughing, or difficult breathing. *IF IN EYES*, hold eyelids open and flush with plenty of water. *IF BREATHING HAS STOPPED*, give artificial respiration. IF breathing is difficult, administer oxygen.
Solid: Irritating to skin and eyes. Harmful if swallowed. Remove contaminated clothing and shoes. Flush affected areas with plenty of water. *IF IN EYES*, hold eyelids open and flush with plenty of water. *IF SWALLOWED* and victim is *CONSCIOUS AND ABLE TO SWALLOW*, have victim drink 4 to 8 ounces of water. *IF SWALLOWED* and victim is *UNCONSCIOUS OR HAVING CONVULSIONS*, do nothing except keep victim warm.

HEALTH HAZARDS
Personal protective equipment (PPE): B-Level PPE. Chemical protective material(s) reported to have good to excellent resistance: butyl rubber, polycarbonate.
Toxicity by ingestion: Grade 3; oral rat LD_{50} = 58 mg/kg (ammonium sulfate).

FIRE DATA
Fire extinguishing agents not to be used: Steam, inert gases, foam, dry chemical.

CHEMICAL REACTIVITY
Binary reactants: Violent reaction with aluminum, magnesium. Reacts with oxidizers. Corrodes metals.

ENVIRONMENTAL DATA
Water pollution: Effect of low concentrations on aquatic life is unknown. Notify local health and wildlife officials. **Response to discharge:** Issue warning–oxidizing material. Disperse and flush.

SHIPPING INFORMATION
Grades of purity: Several grades of varying composition, all of which contain less than 70% of ammonium nitrate, the only hazardous ingredient; **Storage temperature:** Ambient; **Inert atmosphere:** Ventilated (natural); **Venting:** Open; **Stability during transport:** Stable.

PHYSICAL AND CHEMICAL PROPERTIES
Physical state @ 59°F/15°C and 1 atm: Solid.
Specific gravity (water = 1): 1.8 @ 68°F/20°C (solid).

AMMONIUM NITRATE-UREA SOLUTION   REC. A:3300

SYNONYMS: SOLAR NITROGEN SOLUTIONS; NITREX NITROGEN SOLUTIONS; AMMONIUM NITRATE-UREA SOLUTION (not containing ammonia)

IDENTIFICATION
CAS Number: 6484-52-2
Formula: $NH_4NO_3 \cdot NH_2CONH_2 \cdot H_2O$; $C_{17}H_{33}COONH_4 \cdot H_2O$

DESCRIPTION: Colorless liquid. Slight ammonia odor. Sinks and mixes with water.

Heat forms oxygen; will increase the activity of an existing fire • May cause fire or explosion on contact with combustibles (wood, paper, oil, clothing, etc.) • Containers may BLEVE when exposed to fire • Toxic products of combustion may include nitrogen oxides, carbon monoxide, and ammonia gas.

Hazard Classification (based on NFPA-704 Rating System)
Health Hazards (Blue): 2; Flammability (Red): 0; Reactivity (Yellow): 1

EMERGENCY RESPONSE: See Appendix A (171)
Evacuation:
Public safety: Isolate the area of spill or leak for at least 10 to 25 meters (30 to 80 feet) in all directions.
Spill: Increase, in the downwind direction, as necessary, the distance shown under "Public Safety."
Fire: If any large container is involved in fire, isolate for at least 800 meters (½ mile) in all directions; also, consider initial evacuation for 800 meters (½ mile) in all directions.

EXPOSURE
Short-term effects: *SEEK MEDICAL ATTENTION.* **Liquid:** Irritating to skin and eyes. Harmful if swallowed. Remove contaminated clothing and shoes. Flush affected areas with plenty of water. *IF IN EYES*, hold eyelids open and flush with plenty of water. *IF SWALLOWED* and victim is *CONSCIOUS AND ABLE TO SWALLOW*, have victim drink 4 to 8 ounces of water.

HEALTH HAZARDS
Personal protective equipment (PPE): B-Level PPE. Goggles or face shield; rubber gloves. May require eye protection.
Recommendations for respirator selection: NIOSH as ammonia *250 ppm*: CCRS [any chemical cartridge respirator with cartridge(s) providing protection against the compound of concern)]; or SA (any supplied-air respirator). *300 ppm*: SA:CF (any supplied-air respirator operated in a continuous-flow mode); or PAPRS* [any powered, air-purifying respirator with cartridge(s) providing protection against the compound of concern]; or CCRFS [any chemical cartridge respirator with a full facepiece and cartridge(s) providing protection against the compound of concern]; or GMFS [any air-purifying, full-facepiece respirator (gas mask) with a chin-style, front- or back-mounted canister providing protection against the compound of concern]; or SCBAF (any self-contained breathing apparatus with a full facepiece); or SAF (any supplied-air respirator with a full facepiece). *EMERGENCY OR PLANNED ENTRY INTO UNKNOWN CONCENTRATIONS OR IDLH CONDITIONS*: SCBAF:PD,PP (any self-contained breathing apparatus that has a full facepiece and is operated in a pressure-demand or other positive-pressure mode); or SAF:PD,PP:ASCBA (any supplied-air respirator that has a full facepiece and is operated in a pressure-demand or other positive-pressure mode in combination with an auxiliary self-contained breathing apparatus operated in a pressure-demand or other positive pressure mode). *ESCAPE*: GMFS [any air-purifying, full-facepiece respirator (gas mask) with a chin-style, front- or back-mounted canister providing protection against the compound of concern]; or SCBAE (any appropriate escape-type, self-contained breathing apparatus).

Note: Substance reported to cause eye irritation or damage; may require eye protection.
Exposure limits (TWA unless otherwise noted): ACGIH TLV 25 ppm (17 mg/m^3); NIOSH REL 5 ppm (18 mg/m^3) as ammonia
Short-term exposure limits (15-minute TWA): ACGIH TLV 35 ppm (24 mg/m^3); NIOSH STEL 35 ppm (18 mg/m^3); OSHA STEL 50 ppm (35 mg/m^3) as ammonia
Toxicity by ingestion: Grade 2; LD_{50} = 0.5 to 5 g/kg.
Vapor (gas) irritant characteristics: Vapors may be irritating to the eyes and throat of some individuals.
Liquid or solid irritant characteristics: If spilled on clothing and allowed to remain, may cause smarting and reddening of the skin.
IDLH value: 300 ppm as ammonia

FIRE DATA
Behavior in fire: Water of solution may evaporate, and remaining solids may then explode.

CHEMICAL REACTIVITY
Binary reactants: Sulfuric acid.
Reactivity group: 43
Compatibility class: Miscellaneous water solutions.

ENVIRONMENTAL DATA
Water pollution: Effect of low concentrations on aquatic life is unknown. May be dangerous if it enters nearby water intakes; notify operators. Notify local health and wildlife officials. **Response to discharge:** Disperse and flush.

SHIPPING INFORMATION
Grades of purity: Data are for nonpressure solutions containing 44.3% ammonium nitrate, 35.4% urea, and 20.3% water. Other grades contain 0–70% ammonium nitrate, 0–43% urea, 0–36.8% free ammonia, and water. Those containing more than 2% free ammonia are stored under pressure (0-30 psig at 104°F); for hazards of these, see Ammonium Hydroxide; **Storage temperature:** Ambient; **Inert atmosphere:** None; **Venting:** Open; if more than 2% free ammonia, then pressure-vacuum; **Stability during transport:** Stable.

NAS HAZARD CLASSIFICATION FOR BULK WATER TRANSPORTATION
FIRE: 0
HEALTH: Vapor irritant: 0; Liquid or solid irritant: 1; Poisons: 1
WATER POLLUTION: Human toxicity: 2; Aquatic toxicity: 2; Aesthetic effect: 0
REACTIVITY: Other chemicals: 2; Water: 0; Self-reaction: 0

PHYSICAL AND CHEMICAL PROPERTIES
Physical state @ 59°F/15°C and 1 atm: Liquid.
Boiling point @ 1 atm: More than 212°F/100°C/373°K.
Melting/Freezing point: 32°F/0°C/273°K.
Specific gravity (water = 1): 1.327 @ 68°F/20°C (liquid).
Heat of solution: 0.0 Btu/lb = 0.0 cal/g = 0.0 J/kg.
Vapor pressure: (Reid) Varies

AMMONIUM OLEATE   REC. A:3350

SYNONYMS: OLEATO AMONICO (Spanish); OLEIC ACID, AMMONIUM SALT; AMMONIA SOAP

IDENTIFICATION
CAS Number: 544-60-5

DESCRIPTION: Yellow-brown pasty solid. Weak ammonia odor. Sinks and mixes with water.

Combustible • Containers may BLEVE when exposed to fire • Dust or vapors in confined areas (e.g., tanks, sewers, buildings) may explode when exposed to fire • Irritating to the skin, eyes, and respiratory tract • Toxic products of combustion may include ammonia gas:

Hazard Classification (based on NFPA-704 Rating System)
Health Hazards (Blue): 0; Flammability (Red): 1; Reactivity (Yellow): 0

EMERGENCY RESPONSE: See Appendix A (171)
Evacuation:
Public safety: Isolate the area of spill or leak for at least 10 to 25 meters (30 to 80 feet) in all directions.
Spill: Increase, in the downwind direction, as necessary, the distance shown under "Public Safety."
Fire: If any large container is involved in fire, isolate for at least 800 meters (½ mile) in all directions; also, consider initial evacuation for 800 meters (½ mile) in all directions.

EXPOSURE
Short-term effects: *SEEK MEDICAL ATTENTION.* **Solid:** Irritating to skin and eyes. Remove contaminated clothing and shoes. Flush affected areas with plenty of water. *IF IN EYES,* hold eyelids open and flush with plenty of water. *IF SWALLOWED* and victim is *CONSCIOUS AND ABLE TO SWALLOW,* have victim drink 4 to 8 ounces of water. *IF SWALLOWED* and victim is *UNCONSCIOUS OR HAVING CONVULSIONS,* do nothing except keep victim warm.

HEALTH HAZARDS
Personal protective equipment (PPE): B-Level PPE. Goggles or face shield; rubber gloves.

CHEMICAL REACTIVITY
Binary reactants: Strong oxidizers.

ENVIRONMENTAL DATA
Water pollution: Effect of low concentrations on aquatic life is unknown. May be dangerous if it enters nearby water intakes; notify operators. Notify local health and wildlife officials.
Response to discharge: Disperse and flush.

SHIPPING INFORMATION
Grades of purity: 70% in water; Technical; **Storage temperature:** Ambient; **Inert atmosphere:** Ventilated (natural); **Venting:** Open; **Stability during transport:** Stable.

PHYSICAL AND CHEMICAL PROPERTIES
Physical state @ 59°F/15°C and 1 atm: Solid or Liquid.
Molecular weight: 299.5 (solute).
Specific gravity (water = 1): More than 1 @ 68°F/20°C (liquid or solid).

AMMONIUM OXALATE   REC. A:3400

SYNONYMS: EEC No. 607-007-00-3; OXALIC ACID, DIAMMONIUM SALT; DAIMMONIUM OXALATE; AMMONIUM OXALATE HYDRATE; AMMONIUM OXALATE MONOHYDRATE; OXALATO AMONICO (Spanish)

IDENTIFICATION
CAS Number: 6009-70-7; 5972-73-6; 14258-49-2
Formula: $(NH_4)_2C_2O_4 \cdot H_2O$
DOT ID Number: UN 2449; DOT Guide Number: 154
Proper Shipping Name: Oxalates, water soluble
Reportable Quantity (RQ): **(CERCLA)** 5000 lb/2270 kg

DESCRIPTION: White solid. Odorless. Sinks and mixes slowly with water.

Combustible • Containers may BLEVE when exposed to fire • Concentrated dust in confined areas (e.g., tanks, sewers, buildings) may explode when exposed to fire • Toxic products of combustion may include oxides of nitrogen.

Hazard Classification (based on NFPA-704 Rating System)
Health Hazards (Blue): 1; Flammability (Red): 1; Reactivity (Yellow): 0

EMERGENCY RESPONSE: See Appendix A (154)
Evacuation:
Public safety: Isolate the area of spill or leak for at least 25 to 50 meters (80 to 160 feet) in all directions.
Spill: Increase, in the downwind direction, as necessary, the distance shown under "Public Safety."
Fire: If tank, rail car, or tank truck is involved in fire, isolate for at least 800 meters (½ mile) in all directions; also, consider initial evacuation for 800 meters (½ mile) in all directions.

EXPOSURE
Short-term effects: *SEEK MEDICAL ATTENTION.* **Dust:** *POISONOUS IF INHALED OR IF SKIN IS EXPOSED. IF INHALED,* will, will cause coughing or difficult breathing. *IF IN EYES,* hold eyelids open and flush with plenty of water. *IF BREATHING HAS STOPPED,* give artificial respiration; *avoid mouth-to-mouth resuscitation; use bag/mask apparatus.* IF breathing is difficult, administer oxygen. **Solid:** *POISONOUS IF SWALLOWED OR IF SKIN IS EXPOSED.*
Will burn skin and eyes. Remove contaminated clothing and shoes. Flush affected areas with plenty of water. *IF IN EYES,* hold eyelids open and flush with plenty of water. *IF SWALLOWED* and victim is *CONSCIOUS AND ABLE TO SWALLOW,* have victim drink 4 to 8 ounces of water and have victim induce vomiting. *IF SWALLOWED* and victim is *UNCONSCIOUS OR HAVING CONVULSIONS,* do nothing except keep victim warm.
Note to physician or authorized medical personnel: Medical observation is recommended for 24 to 48 hours after breathing overexposure, as pulmonary edema may be delayed. As first aid for pulmonary edema, consider administering a corticosteroid spray. Cigarette smoking may exacerbate pulmonary injury and should be discouraged for at least 72 hours following exposure.

HEALTH HAZARDS
Personal protective equipment (PPE): B-Level PPE. Chemical protective material(s) reported to have good to excellent resistance: butyl rubber, polycarbonate.
Long-term health effects: Kidney damage.
Liquid or solid irritant characteristics: Corrosive to the eyes, skin, and respiratory tract.
IDLH value: 300 ppm as ammonia

CHEMICAL REACTIVITY
Binary reactants: Forms explosive substance with sodium hypochlorite. Corrosive to many metals.

ENVIRONMENTAL DATA
Water pollution: Effect of low concentrations on aquatic life is unknown. May be dangerous if it enters nearby water intakes; notify operators. Notify local health and wildlife officials. **Response to discharge:** Issue warning–poison, water contaminant. Restrict access. Should be removed. Chemical and physical treatment.

SHIPPING INFORMATION
Grades of purity: Pure, 98–100.5%; **Storage temperature:** Ambient; **Inert atmosphere:** None; **Venting:** Open; **Stability during transport:** Stable.

PHYSICAL AND CHEMICAL PROPERTIES
Physical state @ 59°F/15°C and 1 atm: Solid.
Molecular weight: 142.11
Boiling point @ 1 atm: Decomposes at 158°F/70°C/343°K.
Melting/Freezing point: Decomposes.
Specific gravity (water = 1): 1.50 at 18.5°C (solid).
Heat of solution: 101 Btu/lb = 56 cal/g = 2.3×10^5 J/kg.

AMMONIUM PENTABORATE   REC. A:3450

SYNONYMS: AMMONIUM PENTABORATE TETRAHYDRATE; AMMONIUM DECABORATE OCTAHYDRATE; PENTABORATO AMONICO (Spanish)

IDENTIFICATION
CAS Number: 12007-89-5
Formula: $NH_4B_5O_8 \cdot 4H_2O$; $(NH_4)_2B_{10}O_{16} \cdot 8H_2O$

DESCRIPTION: White solid. Odorless. Mixes slowly with water.

Irritates eyes, skin, and respiratory tract • Toxic products of combustion may include ammonia.

Hazard Classification (based on NFPA-704 Rating System)
Health Hazards (Blue): 2; Flammability (Red): 0; Reactivity (Yellow): 0

EMERGENCY RESPONSE: See Appendix A (171)
Evacuation:
Public safety: Isolate the area of spill or leak for at least 10 to 25 meters (30 to 80 feet) in all directions.
Spill: Increase, in the downwind direction, as necessary, the distance shown under "Public Safety."
Fire: If any large container is involved in fire, isolate for at least 800 meters (½ mile) in all directions; also, consider initial evacuation for 800 meters (½ mile) in all directions.

EXPOSURE
Short-term effects: *SEEK MEDICAL ATTENTION.* **Dust:** Irritating to eyes, nose, and throat. *IF INHALED*, will, will cause coughing or difficult breathing. *IF IN EYES*, hold eyelids open and flush with plenty of water. *IF BREATHING HAS STOPPED*, give artificial respiration. IF breathing is difficult, administer oxygen. **Solid:** Irritating to skin and eyes. *IF SWALLOWED*, will cause nausea and vomiting. Remove contaminated clothing and shoes. Flush affected areas with plenty of water. *IF IN EYES*, hold eyelids open and flush with plenty of water. *IF SWALLOWED* and victim is *CONSCIOUS AND ABLE TO SWALLOW*, have victim drink 4 to 8 ounces of water and have victim induce vomiting. *IF SWALLOWED* and victim is *UNCONSCIOUS OR HAVING CONVULSIONS*, do nothing except keep victim warm.

HEALTH HAZARDS
Personal protective equipment (PPE): B-Level PPE. Chemical protective material(s) reported to have good to excellent resistance: butyl rubber, polycarbonate.
Exposure limits (TWA unless otherwise noted): ACGIH TLV 10 mg/m³; NIOSH/OSHA 15 mg/m³ as boron oxide.
Liquid or solid irritant characteristics: May cause eye, skin and respiratory tract irritation.

CHEMICAL REACTIVITY
Reactivity with water: Slow reaction may form boric acid.

ENVIRONMENTAL DATA
Water pollution: Effect of low concentrations on aquatic life is unknown. May be dangerous if it enters nearby water intakes; notify operators. Notify local health and wildlife officials. **Response to discharge:** Disperse and flush.

SHIPPING INFORMATION
Grades of purity: Pure, 99%; Radio, 99.98%; Technical, 99.8%; **Storage temperature:** Ambient; **Inert atmosphere:** None; **Venting:** Open; **Stability during transport:** Stable.

PHYSICAL AND CHEMICAL PROPERTIES
Physical state @ 59°F/15°C and 1 atm: Solid.
Molecular weight: 272.20
Specific gravity (water = 1): 1.58 @ 15°C (solid).

AMMONIUM PERCHLORATE   REC. A:3500

SYNONYMS: PERCHLORIC ACID, AMMONIUM SALT; PERCLORATO AMONICO (Spanish)

IDENTIFICATION
CAS Number: 7790-98-9
Formula: NH_4ClO_4
DOT ID Number: UN 1442; DOT Guide Number: 143
Proper Shipping Name: Ammonium perchlorate

DESCRIPTION: White crystalline solid. Odorless. Sinks and mixes with water.

Severe explosion hazard; dangerous when exposed to heat (decomposition begins at 266°F/130°C; explodes at 716°F/380°C), shock, friction, or by spontaneous chemical reaction • Strong oxidizer which may react spontaneously with low flash point organics or reducing agents. Heat forms oxygen; will increase the activity of an existing fire • Containers may BLEVE when exposed to fire • May cause fire or explosion contact with combustibles (wood, paper, oil, clothing, etc.) • Toxic products of combustion may include nitrogen oxides, ammonia, and hydrogen chloride • Do not put yourself in danger by entering a contaminated area to rescue a victim.

Hazard Classification (based on NFPA-704 Rating System)
Health Hazards (Blue): 1; Flammability (Red): 0; Reactivity (Yellow): 4; Special Notice (White): OXY

EMERGENCY RESPONSE: See Appendix A (143)
Evacuation:
Public safety: Isolate the area of spill or leak for at least 50 to 100 meters (160 to 300 feet) in all directions.

Spill: Increase, in the downwind direction, as necessary, the distance shown under "Public Safety." *Note*: Do not let spill area dry until it has been determined that there is no perchlorates left in the area. Continue cooling after fire has been extinguished.
Fire: If any large container is involved in fire, isolate for at least 800 meters (½ mile) in all directions; also, consider initial evacuation for 800 meters (½ mile) in all directions.

EXPOSURE
Short-term effects: *SEEK MEDICAL ATTENTION.* **Solid:** Irritating to eyes and skin. Harmful if swallowed. remove contaminated clothing. Flush affected areas with plenty of water. *IF IN EYES*, hold eyelids open and flush with plenty of water. *IF SWALLOWED* and victim is *CONSCIOUS AND ABLE TO SWALLOW*, have victim drink 4 to 8 ounces of water.

HEALTH HAZARDS
Personal protective equipment (PPE): B-Level PPE. Chemical protective material(s) reported to have good to excellent resistance: butyl rubber, polycarbonate.
Toxicity by ingestion: Grade 2; oral rat LD_{50} = 3500 mg/kg.
Vapor (gas) irritant characteristics: Decomposition products are toxic and eye, skin and respiratory tract irritants.

FIRE DATA
Behavior in fire: Severe explosion hazard; decomposes at 266°F/130°C; explodes at 716°F/380°C.

CHEMICAL REACTIVITY
Binary reactants: Strong oxidizer. Becomes a sensitive high explosive when contaminated with reducing agents. Reacts, with combustible materials, strong acids, powdered metals; potentially explosive above 716°F/380°C. with dichromium trioxide, carbon, cadmium oxide, zinc oxide. If contaminated with carbonaceous materials, can become an explosive that is sensitive to shock and friction. Readily detonates or explodes.

ENVIRONMENTAL DATA
Water pollution: Effect of low concentrations on aquatic life is unknown. May be dangerous if it enters nearby water intakes; notify operators. Notify local health and wildlife officials.
Response to discharge: Issue warning–oxidizing material. Disperse and flush. Use dilute bisulfate solution.

PHYSICAL AND CHEMICAL PROPERTIES
Physical state @ 59°F/15°C and 1 atm: Solid.
Molecular weight: 117.49
Melting/Freezing point: Decomposes 464°F/240°C/513°K.
Specific gravity (water = 1): 1.95 @ 15°C (solid).

AMMONIUM PERSULFATE REC. A:3550

SYNONYMS: AMMONIUM PEROXYDISULFATE; PERSULFATE D'AMMONIUM (French); PEROXYDISULFANIC ACID, DIAMMONIUM SALT

IDENTIFICATION
CAS Number: 7727-54-0
Formula: $O_8S_2 \cdot 2H_4N$; $(NH_4)_2S_2O_8$
DOT ID Number: UN 1444; DOT Guide Number: 140
Proper Shipping Name: Ammonium persulfate

DESCRIPTION: Colorless to light straw solid. Mild, unpleasant, acrid odor. Sinks and mixes with water.

Corrosive to the skin, eyes, and respiratory tract; contact with skin or eyes can cause burns and vision impairment; inhalation symptoms may be delayed • A strong oxidizer; heat forms oxygen; will increase the activity of an existing fire • May cause fire or explosion on contact with combustibles (wood, paper, oil, clothing, etc.) • Decomposes above 248°F/120°C. Toxic products of combustion may include ammonia, sulfuric acid fumes, nitrogen oxides, and sulfur dioxide.

Hazard Classification (based on NFPA-704 Rating System)
Health Hazards (Blue): 1; Flammability (Red): 0; Reactivity (Yellow): 1; Special Notice (White): OXY

EMERGENCY RESPONSE: See Appendix A (140)
Evacuation:
Public safety: Isolate the area of spill or leak for at least 10 to 25 meters (30 to 80 feet) in all directions.
Spill: Consider initial downwind evacuation for at least 100 meters (330 feet).
Fire: If any large container is involved in fire, isolate for at least 800 meters (½ mile) in all directions; also, consider initial evacuation for 800 meters (½ mile) in all directions to a secure location.

EXPOSURE
Short-term effects: *SEEK MEDICAL ATTENTION.* **Dust:** Irritating to eyes, nose, and throat. Harmful if inhaled. *IF IN EYES*, hold eyelids open and flush with plenty of water. *IF BREATHING HAS STOPPED*, give artificial respiration. IF breathing is difficult, administer oxygen. Lung edema may develop. **Solid:** Irritating to skin and eyes. Harmful if swallowed. Remove contaminated clothing and shoes. Flush affected areas with plenty of water. *IF IN EYES*, hold eyelids open and flush with plenty of water. *IF SWALLOWED* and victim is *CONSCIOUS AND ABLE TO SWALLOW*, have victim drink 4 to 8 ounces of water. *IF SWALLOWED* and victim is *UNCONSCIOUS OR HAVING CONVULSIONS*, do nothing except keep victim warm.
Note to physician or authorized medical personnel: Medical observation is recommended for 24 to 48 hours after breathing overexposure, as pulmonary edema may be delayed. As first aid for pulmonary edema, consider administering a corticosteroid spray. Cigarette smoking may exacerbate pulmonary injury and should be discouraged for at least 72 hours following exposure.

HEALTH HAZARDS
Personal protective equipment (PPE): B-Level PPE. U.S. Bureau Mines approved toxic dust mask; chemical goggles; rubber gloves; neoprene-coated shoes. Chemical protective material(s) reported to have good to excellent resistance: butyl rubber, polycarbonate, chlorinated polyethylene, PVC, Neoprene, Viton®.
Toxicity by ingestion: Grade 2; oral rat LD_{50} = 820 mg/kg.
Long-term health effects: Skin disorders.
Vapor (gas) irritant characteristics: NA>
Liquid or solid irritant characteristics: May cause eye, skin and respiratory tract irritation
IDLH value: 300 ppm as ammonia

FIRE DATA
Behavior in fire: Decomposes with loss of oxygen that increases intensity of fire.

CHEMICAL REACTIVITY
Binary reactants: A strong oxidizer. Mixture with sodium peroxide produces a friction-, heat-, and water-sensitive explosive. Reacts with reducing agents, organic materials, fuels, oils, other

combustibles (grease, wood, and other materials); finely divided and powdered metals (especially aluminum or magnesium) with risk of fire and explosion.

ENVIRONMENTAL DATA
Water pollution: Dangerous to aquatic life in high concentrations. May be dangerous if it enters nearby water intakes; notify operators. Notify local health and wildlife officials. **Response to discharge:** Issue warning–oxidizing material. Disperse and flush.

SHIPPING INFORMATION
Grades of purity: Reagent; pure; **Storage temperature:** Ambient; **Inert atmosphere:** None; **Venting:** Open; **Stability during transport:** Stable.

PHYSICAL AND CHEMICAL PROPERTIES
Physical state @ 59°F/15°C and 1 atm: Solid.
Molecular weight: 228.20
Boiling point @ 1 atm: Decomposes at 248°F/120°C/393°K.
Melting/Freezing point: Decomposes.
Specific gravity (water = 1): 1.98 @ 68°F/20°C (solid).
Heat of decomposition: 120°C.
Heat of solution: 77 Btu/lb = 43 cal/g = 1.8×10^5 J/kg.

AMMONIUM PHOSPHATE REC. A:3600

SYNONYMS: AMMONIUM DIHYDROGEN PHOSPHATE; AMMONIUM BIPHOSPHATE; *sec*-AMMONIUM PHOSPHATE; AMMONIUM PHOSPHATE, DIBASIC; DIAMMONIUM HYDROGEN PHOSPHATE; DIAMMONIUM ORTHOPHOSPHATE; DIAMMONIUM PHOSPHATE; MONOAMMONIUM PHOSPHATE; MONOAMMONIUM ORTHOPHOSPHATE; FOSFATO AMONICO (Spanish)

IDENTIFICATION
CAS Number: 7783-28-0 (dibasic); 7772-76-1 (monobasic)
Formula: $NH_4H_2PO_4$; $(NH_4)_2HPO_4$

DESCRIPTION: White crystalline solid or powder. Diammonium-faint ammonia odor; monoammonium-faint acid odor. Sinks and mixes with water.

Highly irritating to eyes, skin, and respiratory tract • Toxic products of combustion may include phosphorus oxides and ammonia.

Hazard Classification (based on NFPA-704 Rating System)
Health Hazards (Blue): 0; Flammability (Red): 0; Reactivity (Yellow): 0

EMERGENCY RESPONSE: See Appendix A (171)
Evacuation:
Public safety: Isolate the area of spill or leak for at least 10 to 25 meters (30 to 80 feet) in all directions.
Spill: Increase, in the downwind direction, as necessary, the distance shown under "Public Safety."
Fire: If any large container is involved in fire, isolate for at least 800 meters (½ mile) in all directions; also, consider initial evacuation for 800 meters (½ mile) in all directions.

EXPOSURE
Short-term effects: *SEEK MEDICAL ATTENTION.* **Dust:** Irritating to eyes, nose, and throat. *IF INHALED*, will, will cause coughing or difficult breathing. *IF IN EYES*, hold eyelids open and flush with plenty of water. *IF BREATHING HAS STOPPED,* give artificial respiration. IF breathing is difficult, administer oxygen. **Solid:** Irritating to skin and eyes. Harmful if swallowed. Remove contaminated clothing and shoes. Flush affected areas with plenty of water. *IF IN EYES*, hold eyelids open and flush with plenty of water. *IF SWALLOWED* and victim is *CONSCIOUS AND ABLE TO SWALLOW*, have victim drink 4 to 8 ounces of water. *IF SWALLOWED* and victim is *UNCONSCIOUS OR HAVING CONVULSIONS,* do nothing except keep victim warm.

HEALTH HAZARDS
Personal protective equipment (PPE): B-Level PPE. Dust mask, protective gloves, and goggles. When diammonium phosphate is stored in closed area, self-contained breathing apparatus is required to protect against ammonia fumes.
Recommendations for respirator selection: NIOSH as ammonia *250 ppm*: CCRS [any chemical cartridge respirator with cartridge(s) providing protection against the compound of concern)]; or SA (any supplied-air respirator). *300 ppm*: SA:CF (any supplied-air respirator operated in a continuous-flow mode); or PAPRS* [any powered, air-purifying respirator with cartridge(s) providing protection against the compound of concern]; or CCRFS [any chemical cartridge respirator with a full facepiece and cartridge(s) providing protection against the compound of concern]; or GMFS [any air-purifying, full-facepiece respirator (gas mask) with a chin-style, front- or back-mounted canister providing protection against the compound of concern]; or SCBAF (any self-contained breathing apparatus with a full facepiece); or SAF (any supplied-air respirator with a full facepiece). *EMERGENCY OR PLANNED ENTRY INTO UNKNOWN CONCENTRATIONS OR IDLH CONDITIONS:* SCBAF:PD,PP (any self-contained breathing apparatus that has a full facepiece and is operated in a pressure-demand or other positive-pressure mode); or SAF:PD,PP:ASCBA (any supplied-air respirator that has a full facepiece and is operated in a pressure-demand or other positive-pressure mode in combination with an auxiliary self-contained breathing apparatus operated in a pressure-demand or other positive pressure mode). *ESCAPE:* GMFS [any air-purifying, full-facepiece respirator (gas mask) with a chin-style, front- or back-mounted canister providing protection against the compound of concern]; or SCBAE (any appropriate escape-type, self-contained breathing apparatus). *Note:* Substance reported to cause eye irritation or damage; may require eye protection.
Exposure limits (TWA unless otherwise noted): ACGIH TLV 25 ppm; NIOSH REL 5 ppm (18 mg/m^3) as ammonia
Short-term exposure limits (15-minute TWA): NIOSH STEL 35 ppm (18 mg/m^3); OSHA STEL 50 ppm (35 mg/m^3) as ammonia
Liquid or solid irritant characteristics: Irritation of the eyes, skin, and respiratory tract.
IDLH value: 300 ppm as ammonia

CHEMICAL REACTIVITY
Binary reactants: Strong oxidizers, strong bases. Reaction with air slowly releases ammonia.

ENVIRONMENTAL DATA
Water pollution: Effect of low concentrations on aquatic life is unknown. May be dangerous if it enters nearby water intakes; notify operators. Notify local health and wildlife officials. **Response to discharge:** Disperse and flush.

SHIPPING INFORMATION
Grades of purity: Reagent; Technical; **Storage temperature:** Ambient; **Inert atmosphere:** Ventilated (forced); **Venting:** Open; **Stability during transport:** Stable. *Note:* When diammonium phosphate is stored in closed area, self-contained breathing apparatus is required to protect against ammonia fumes.

PHYSICAL AND CHEMICAL PROPERTIES
Physical state @ 59°F/15°C and 1 atm: Solid.
Molecular weight: 115 (monoammonium); 132 (diammonium)
Boiling point @ 1 atm: Begins to decompose at 216°F/100°C/373°K.
Melting/Freezing point: Decomposes below melting point at 216°F, see above.
Specific gravity (water = 1): 1.8 @ 68°F/20°C (solid) (diammonium); 1.6 @ 68°F/20°C (solid) (monoammonium).
Heat of solution: 42 Btu/lb = 23 cal/g = 0.97×10^5 J/kg.

AMMONIUM PICRATE, WET   REC. A:3650

SYNONYMS: AMMONIUM CARBAZOATE; AMMONIUM PICRATE, DRY; AMMONIUM PICRATE, WET; AMMONIUM PICRATE (Yellow):; AMMONIUM PICRATE WETTED WITH LESS THAN 10% WATER; AMMONIUM PICRATE WETTED WITH MORE THAN 10% WATER; OBELINE PICRATE; PHENOL,2,4,6-TRINITRO-, AMMONIUM SALT; AMMONIUM PICRONITRATE; EXPLOSIVE D; PICRATO AMONICO (Spanish); PICTAROL; PICRIC ACID, AMMONIUM SALT; RCRA No. P009

IDENTIFICATION
CAS Number: 131-74-8
Formula: $C_6H_6N_4O_7$; $C_6H_2(NO_2)_3ONH_4$
DOT ID Number: UN 1310 (>10% water); UN 0004 (dry, or wetted with <10% water); DOT Guide Number: 113 (wetted material); 112 (dry material)
Proper Shipping Name: Ammonium picrate, wetted with not less than 10% water, by mass; Ammonium picrate, dry or wetted with less than 10% water, by mass (UN 0004)
Reportable Quantity (RQ): **(CERCLA)** 10 lb/4.54 kg

DESCRIPTION: Yellow solid (water wet crystals). Sinks and slowly mixes with water; forms an acid solution. Treat as an explosive; *DRIED OUT MATERIAL* may explode from heat, flame, friction.

Poisonous! • Keep material wet; dried out material (less than 10% water added) is a severe explosion hazard and will spontaneously explode if exposed to heat, flame, friction, or shock. *UNCONFINED* material may burn without detonation when ignited. *CONFINED* material will explode upon heating to its ignition temperature • Allergen • Containers may BLEVE when exposed to fire • Dust forms explosive mixture with air • Irritating to the skin, eyes, and respiratory tract • Toxic products of combustion may include oxides of nitrogen.

Hazard Classification (based on NFPA-704 Rating System)
Health Hazards (Blue): 3; Flammability (Red): 3; Reactivity (Yellow): 3; Special Notice (White): OXY

EMERGENCY RESPONSE (wetted material): See Glossary A (113).
Evacuation:
Public safety: Isolate spill area for at least 100 meters (330 feet) in all directions.
Spill: Consider initial downwind evacuation for at least 500 meters (⅓ mile).
Fire: Isolate for at least 800 meters (½ mile) in all directions, especially if tank, rail car, or tank truck is involved in fire.

EMERGENCY RESPONSE (dry material): See Glossary A (112)
Evacuation:
Public safety: Isolate spill area for at least 100 meters (330 feet) in all directions.
Spill: Consider initial downwind evacuation for at least 1600 meters (1 mile).
Fire: Isolate for at least 1600 meters (1 mile) in all directions, especially if tank, rail car, or tank truck is involved in fire.

EXPOSURE
Short-term effects: *SEEK MEDICAL ATTENTION*. **Dust:** Toxic via inhalation and percutaneous absorption. Irritating to eyes, skin and mucous membranes. Move victim to fresh air. *IF IN EYES OR ON SKIN*, flush with running water for at least 15 minutes; hold eyelids open if appropriate. *IF BREATHING HAS STOPPED*, give artificial respiration. IF breathing is difficult, administer oxygen. **Solid:** *POISONOUS IF SWALLOWED OR ABSORBED THROUGH SKIN*. Irritating to eyes and skin. *IF IN EYES OR ON SKIN*, flush with running water for at least 15 minutes. Hold eyelids open if appropriate. Use soap or mild detergent on skin. Remove and isolate contaminated clothing and shoes at the site. *IF SWALLOWED* and victim is *CONSCIOUS AND ABLE TO SWALLOW*, have victim drink water and induce vomiting. *IF SWALLOWED* and victim is *UNCONSCIOUS OR HAVING CONVULSIONS*, do nothing except keep victim quiet and maintain normal body temperature.

HEALTH HAZARDS
Personal protective equipment (PPE): B-Level PPE. Wear self-contained positive pressure breathing apparatus and full protective clothing.
Long-term health effects: An allergen. May cause liver and kidney damage.
Liquid or solid irritant characteristics: Irritating to skin and eyes.

FIRE DATA
Behavior in fire: *UNCONFINED* material burns without detonation when ignited. Confined material will explode upon heating to its ignition temperature.

CHEMICAL REACTIVITY
Binary reactants: Reacts with metals, concrete, and plaster to produce salts of picric acid that are much more sensitive to shock than ammonium picrate. Rate of reactivity with metal is increased by the presence of water.
Neutralizing agents for acids and caustics: Wet down with water and dike for later disposal. Ammonium picrate should be disposed of only by explosives experts.

ENVIRONMENTAL DATA
Water pollution: Harmful to aquatic life in very low concentrations. May be dangerous if it enters nearby intakes. Notify local health and wildlife officials. **Response to discharge:** Issue warning–poison, oxidizing material, and explosive. Restrict access. Evacuate area. Should be removed. Chemical and physical treatment.

SHIPPING INFORMATION
Grades of purity: 90%; 10% water (minimum); **Storage temperature:** Ambient; **Stability during transport:** Stable.

PHYSICAL AND CHEMICAL PROPERTIES
Physical state @ 59°F/15°C and 1 atm: Solid.
Molecular weight: 246.14

Boiling point @ 1 atm: Decomposes @ 793°F/423°C/696°K.
Specific gravity (water = 1): 1.719 at room temperature
Heat of combustion: –4,941 Btu/lb = –2,745 cal/g = –115x 10^5 J/kg.

AMMONIUM SILICOFLUORIDE   REC. A:3700

SYNONYMS: AMMONIUM FLUOROSILICATE; CRYPTOHALITE; FLUOSILICATE de AMMONIUM (French); SILICOFLURURO AMONICO (Spanish)

IDENTIFICATION
CAS Number: 1309-32-6; 16919-19-0
Formula: $(NH_4)_2SiF_6$; $F_6Si \cdot 2H_4N$
DOT ID Number: UN 2854; DOT Guide Number: 151
Proper Shipping Name: Ammonium fluorosilicate
Reportable Quantity (RQ): (CERCLA) 1000 lb/454 kg

DESCRIPTION: White solid. Odorless. Sinks and mixes slowly with water; forms corrosive hydrogen fluoride.

Poison! • Corrosive to the skin, eyes, and respiratory tract; contact with skin or eyes can cause burns and vision impairment; inhalation symptoms may be delayed • Firefighting gear (including SCBA) does not provide adequate protection. If exposure occurs, remove and isolate gear immediately and thoroughly decontaminate personnel • Toxic products of combustion may include ammonia, hydrogen fluoride, silicon tetrafluoride, and oxides of nitrogen.

Hazard Classification (based on NFPA-704 Rating System)
Health Hazards (Blue): 2; Flammability (Red): 0; Reactivity (Yellow): 0

EMERGENCY RESPONSE: See Appendix A (113)
Evacuation:
Public safety: Isolate the area of spill or leak for at least 25 to 50 meters (80 to 160 feet) in all directions.
Spill: Increase, in the downwind direction, as necessary, the distance shown under "Public Safety."
Fire: If tank, rail car, or tank truck is involved in fire, isolate for at least 800 meters (½ mile) in all directions; also, consider initial evacuation for 800 meters (½ mile) in all directions.

EXPOSURE
Short-term effects: *SEEK MEDICAL ATTENTION.* **Dust:** *POISONOUS IF INHALED OR IF SKIN IS EXPOSED. IF INHALED,* will, will cause coughing or difficult breathing. *IF BREATHING HAS STOPPED,* give artificial respiration; *avoid mouth-to-mouth resuscitation; use bag/mask apparatus. IF IN EYES,* hold eyelids open and flush with plenty of water. *IF BREATHING HAS STOPPED,* give artificial respiration. IF breathing is difficult, administer oxygen. **Solid:** *POISONOUS IF SWALLOWED OR IF SKIN IS EXPOSED.* Will burn skin and eyes. Harmful if swallowed. Remove contaminated clothing and shoes. Flush affected areas with plenty of water. *IF IN EYES,* hold eyelids open and flush with plenty of water. *IF SWALLOWED* and victim is *CONSCIOUS AND ABLE TO SWALLOW,* have victim drink 4 to 8 ounces of water and have victim induce vomiting. *IF SWALLOWED* and victim is *UNCONSCIOUS OR HAVING CONVULSIONS,* do nothing except keep victim warm.
Note to physician or authorized medical personnel: Medical observation is recommended for 24 to 48 hours after breathing overexposure, as pulmonary edema may be delayed. As first aid for pulmonary edema, consider administering a corticosteroid spray. Cigarette smoking may exacerbate pulmonary injury and should be discouraged for at least 72 hours following exposure.

HEALTH HAZARDS
Personal protective equipment (PPE): B-Level PPE. Chemical protective material(s) reported to have good to excellent resistance: butyl rubber, polycarbonate.
Recommendations for respirator selection: NIOSH/OSHA as F *12.5 mg/m^3:* DM (any dust and mist respirator). *25 mg/m^3:* DMXSQ* (any dust and mist respirator except single-use and quarter-mask respirators); or SA* (any supplied-air respirator). *62.5 mg/m^3:* SA:CF* [any supplied-air respirator operated in a continuous-flow mode)]; or PAPRDM*+ *if not present as a fume* (any powered, air-purifying respirator with a dust and mist filter). *125 mg/m^3:* HiEF+ (any air-purifying, full-facepiece respirator with a high-efficiency particulate filter); or SCBAF (any self-contained breathing apparatus with a full facepiece); or SAF (any supplied-air respirator with a full facepiece). *250 mg/m^3:* SA:PD,PP (any supplied-air respirator operated in a pressure-demand or other positive-pressure mode). *EMERGENCY OR PLANNED ENTRY INTO UNKNOWN CONCENTRATIONS OR IDLH CONDITIONS:* SCBAF:PD,PP (any self-contained breathing apparatus that has a full faceplate and is operated in a pressure-demand or other positive-pressure mode); or SAF:PD,PP:ASCBA (any supplied-air respirator that has a full facepiece and is operated in a pressure-demand or other positive-pressure mode in combination with an auxiliary, self-contained breathing apparatus operated in a pressure-demand or other positive-pressure mode). *ESCAPE:* HiEF+ (any air-purifying, full-facepiece respirator with a high-efficiency particulate filter); or SCBAE (any appropriate escape-type, self-contained breathing apparatus). *Notes:* *Substance reported to cause eye irritation or damage; may require eye protection. +May need acid gas sorbent.
Exposure limits (TWA unless otherwise noted): ACGIH TLV 2.5 mg/m^3 as fluorides; NIOSH/OSHA 2.5 mg/m^3 as inorganic fluorides.
Toxicity by ingestion: Grade 3; LD_{50} = 100 mg/kg (rat).
Long-term health effects: May cause bone disorders, pulmonary fibrosis.
Liquid or solid irritant characteristics: Irritates eyes, skin, and respiratory tract.
IDLH value: 250 mg/m^3 as F.

CHEMICAL REACTIVITY
Reactivity with water: Solution is acid (corrosive).
Binary reactants: Keep away from acids, acid fumes, and strong oxidizers.

ENVIRONMENTAL DATA
Water pollution: Effect of low concentrations on aquatic life is unknown. May be dangerous if it enters nearby water intakes; notify operators. Notify local health and wildlife officials. **Response to discharge:** Issue warning–water contaminant. Restrict access. Disperse and flush.

SHIPPING INFORMATION
Grades of purity: Pure, 99+%; Commercial, 98+%; **Storage temperature:** Ambient; **Inert atmosphere:** None; **Venting:** Open; **Stability during transport:** Stable.

PHYSICAL AND CHEMICAL PROPERTIES
Physical state @ 59°F/15°C and 1 atm: Solid.
Molecular weight: 178.14
Boiling point @ 1 atm: Decomposes.
Specific gravity (water = 1): 2.0 @ 68°F/20°C (solid).

Heat of solution: 85 Btu/lb = 47 cal/g = 2.0 x 10^5 J/kg.
Vapor pressure: Less than 0.075 mm.

AMMONIUM STEARATE   REC. A:3750

SYNONYMS: ESTEARATO AMONICO (Spanish); STEARIC ACID, AMMONIUM SALT

IDENTIFICATION
CAS Number: 1002-89-7
Formula: $C_{18}H_{35}COONH_4 \cdot H_2O$
DOT ID Number: UN 1325 (pure material only); DOT Guide Number: 133
Proper Shipping Name: Flammable solids, organic, n.o.s.

DESCRIPTION: White to tan, wax-like solid or paste. Weak ammonia odor. May float or sink in water; slightly soluble in cold water; soluble in very hot water.

Causes skin, eye, nose and lung irritation • Vapor may explode if ignited in an enclosed area • Containers may BLEVE when exposed to fire • Toxic products of combustion may include carbon monoxide, nitrogen oxides and ammonia gas.

Hazard Classification (based on NFPA-704 Rating System)
Health Hazards (Blue): 0; Flammability (Red): 1; Reactivity (Y)
Evacuation:
Public safety: Isolate spill area for at least 10 to 25 meters (30 to 80 feet) in all directions.
Spill: Consider initial downwind evacuation of at least 100 meters (330 feet).
Fire: Isolate for 800 meters (½ mile) in all directions, especially if tank, rail car, or tank truck is involved in fire.

EXPOSURE
Short-term effects: *SEEK MEDICAL ATTENTION.* **Liquid or solid:** Irritating to skin and eyes. *IF SWALLOWED*, will cause nausea. Remove contaminated clothing and shoes. Flush affected areas with plenty of water. *IF IN EYES*, hold eyelids open and flush with plenty of water. *IF SWALLOWED* and victim is *CONSCIOUS AND ABLE TO SWALLOW*, have victim drink 4 to 8 ounces of water. *IF SWALLOWED* and victim is *UNCONSCIOUS OR HAVING CONVULSIONS*, do nothing except keep victim warm.

HEALTH HAZARDS
Personal protective equipment (PPE): B-Level PPE. Chemical protective material(s) reported to have good to excellent resistance: butyl rubber, polycarbonate.

FIRE DATA
Flash point: More than 140°F/60°C/333°K (cc) (pure material only; solution not flammable).

CHEMICAL REACTIVITY
Binary reactants: Strong oxidizers.

ENVIRONMENTAL DATA
Water pollution: Effect of low concentrations on aquatic life is unknown. May be dangerous if it enters nearby water intakes; notify operators. Notify local health and wildlife officials.
Response to discharge: Disperse and flush.

SHIPPING INFORMATION
Grades of purity: 33% dispersion in water; technical paste;
Storage temperature: Ambient; **Inert atmosphere:** None; **Venting:** Open; **Stability during transport:** Stable.

PHYSICAL AND CHEMICAL PROPERTIES
Physical state @ 59°F/15°C and 1 atm: Solid or Liquid.
Molecular weight: 301.5
Boiling point @ 1 atm: Decomposes.
Specific gravity (water = 1): 1.01 @ 68°F/20°C (liquid).

AMMONIUM SULFAMATE   REC. A:3800

SYNONYMS: AMCIDE; AMICIDE; AMMAT or AMMATE HERBICIDE; AMMONIUM AMINOSULFONATE; AMMONIUM SULPHAMATE; AMMONIUM AMIDOSULPHATE; AMMONIUM SALZ DER AMIDOSULFONSAURE (German); AMMONIUM AMIDOSULFONATE; AMS; IKURIN; MONOAMMONIUM SULFAMATE; MONOAMMONIUM SALT OF SULFAMIC ACID; SULFAMATO AMONICO (Spanish); SULFAMINSAURE (German); SULFAMIC ACID, MONOAMMONIUM SALT

IDENTIFICATION
CAS Number: 7773-06-0
Formula: $NH_2SO_3NH_4$
DOT ID Number: UN 9089; DOT Guide Number: 171
Proper Shipping Name: Ammonium sulfamate
Reportable Quantity (RQ): **(CERCLA)** 5000 lb/2270 kg

DESCRIPTION: White salt or yellow to brownish-gray crystalline solid. Odorless. Sinks and mixes with water.

Explosion hazard when exposed to heat above 320°F/160°C or by spontaneous chemical reaction (hydrolysis) • Corrosive; dust can cause severe irritation • Strong oxidizer • Heat forms oxygen; will increase the activity of an existing fire • May cause fire or explosion contact with combustibles (wood, paper, oil, clothing, etc.). Toxic products of combustion may include oxides of sulfur and nitrogen, and ammonia gas.

Hazard Classification (based on NFPA-704 Rating System)
Health Hazards (Blue): 0 (FEMA); Flammability (Red): 0; Reactivity (Yellow): 0; Special Notice (White): Oxy

EMERGENCY RESPONSE: See Appendix A (133)
Evacuation:
Public safety: Isolate the area of spill or leak for at least 10 to 25 meters (30 to 80 feet) in all directions.
Spill: Increase, in the downwind direction, as necessary, the distance shown under "Public Safety."
Fire: If any large container is involved in fire, isolate for at least 800 meters (½ mile) in all directions; also, consider initial evacuation for 800 meters (½ mile) in all directions.

EXPOSURE
Short-term effects: *SEEK MEDICAL ATTENTION.* **Dust:** Irritating to eyes, nose, and throat. *IF INHALED*, will, will cause coughing or difficult breathing. *IF IN EYES*, hold eyelids open and flush with plenty of water. *IF BREATHING HAS STOPPED*, give artificial respiration. IF breathing is difficult, administer oxygen.
Solid: Irritating to skin and eyes. *IF SWALLOWED*, will cause nausea or vomiting. Remove contaminated clothing and shoes. Flush affected areas with plenty of water. *IF IN EYES*, hold eyelids open and flush with plenty of water. *IF SWALLOWED* and victim

68 Ammonium sulfate

is *CONSCIOUS AND ABLE TO SWALLOW*, have victim drink 4 to 8 ounces of water. *IF SWALLOWED* and victim is *UNCONSCIOUS OR HAVING CONVULSIONS*, do nothing except keep victim warm.

HEALTH HAZARDS
Personal protective equipment (PPE): B-Level PPE. OSHA Table Z-1-A air contaminant. Chemical protective material(s) reported to have good to excellent resistance: butyl rubber, polycarbonate.
Recommendations for respirator selection: NIOSH/OSHA
50 mg/m³: DM (any dust and mist respirator). *100 mg/m³:* DMXSQ (any dust and mist respirator except single-use and quarter mask respirators); or SA (any supplied-air respirator). *250 mg/m³:* SA:CF (any supplied-air respirator operated in a continuous-flow mode); or PAPRDM, *If not present as a fume* (any powered, air-purifying respirator with a dust and mist filter); *500 mg/m³:* SAT:CF (any supplied-air respirator that has a tight-fitting facepiece and is operated in a continuous-flow mode); or PAPRTHiE (any powered, air-purifying respirator with a tight-fitting facepiece and a high-efficiency particulate filter); or HiEF (any air-purifying, full-facepiece respirator with a high-efficiency particulate filter); or SCBAF (any self-contained breathing apparatus with a full facepiece); or SAF (any supplied-air respirator with a full facepiece). *1500 mg/m³:* SA:PD,PP (any supplied-air respirator operated in a pressure-demand or other positive-pressure mode). *EMERGENCY OR PLANNED ENTRY INTO UNKNOWN CONCENTRATIONS OR IDLH CONDITIONS:* SCBAF:PD,PP (any self-contained breathing apparatus that has a full facepiece and is operated in a pressure-demand or other positive-pressure mode); or SAF:PD,PP:ASCBA (any supplied-air respirator that has a full facepiece and is operated in a pressure-demand or other positive-pressure mode in combination with an auxiliary self-contained breathing apparatus operated in a pressure-demand or other positive pressure mode). *ESCAPE:* HiEF (any air-purifying, full-facepiece respirator with a high-efficiency particulate filter); or SCBAE (any appropriate escape-type, self-contained breathing apparatus).
*Note: Substance reported to cause eye irritation or damage; may require eye protection.
Exposure limits (TWA unless otherwise noted): ACGIH TLV 10 mg/m³; OSHA PEL 15 mg/m³ (total); 5 mg/m³ respirable fraction; NIOSH REL 10 mg/m³ (total); 5 mg/m³ respirable fraction
Toxicity by ingestion: Grade 2; oral rat LD_{50} = 1600 mg/kg.
Liquid or solid irritant characteristics: May cause irritation of the eyes, nose, and respiratory tract.
IDLH value: 1500 mg/m³.

CHEMICAL REACTIVITY
Reactivity with water: Elevated temperature causes a highly exothermic reaction.
Binary reactants: Acids or acid fumes form toxic sulfur oxides. Reacts with hot water.

ENVIRONMENTAL DATA
Water pollution: Dangerous to aquatic life in high concentrations. May be dangerous if it enters nearby water intakes; notify operators. Notify local health and wildlife officials. **Response to discharge:** Disperse and flush.

SHIPPING INFORMATION
Grades of purity: Reagent, 99.0%; Commercial, 80%; **Storage temperature:** Ambient; **Inert atmosphere:** None; **Venting:** Open; **Stability during transport:** Stable.

PHYSICAL AND CHEMICAL PROPERTIES

Physical state @ 59°F/15°C and 1 atm: Solid.
Molecular weight: 114.13
Boiling point @ 1 atm: Decomposes 320°F/160°C/433°K.
Melting/Freezing point: 268°F/131°C/404°K.
Specific gravity (water = 1): 1.77 @ 68°F/20°C (solid).
Vapor pressure: Low. 0 mm (approximate).

AMMONIUM SULFATE   REC. A:3850

SYNONYMS: AMMONIUM HYDROGEN SULFATE; AMMONIUM SULPHATE; DIAMMONIUM SULFATE; SULFATO AMONICO (Spanish); SULFURIC ACID, DIAMMONIUM SALT

IDENTIFICATION
CAS Number: 7783-20-2
Formula: $(NH_4)_2SO_4$
DOT ID Number: UN 2506; DOT Guide Number: 154
Proper Shipping Name: Ammonium hydrogen sulfate

DESCRIPTION: White solid. Odorless. Sinks and mixes with water; slightly acidic.

Corrosive to the skin, eyes, and respiratory tract • Decomposes above 455°F/235°C. Toxic products of combustion or decomposition include sulfur oxides; nitrogen oxides, and ammonia gas.

Hazard Classification (based on NFPA-704 Rating System)
Health Hazards (Blue): 0; Flammability (Red): 0; Reactivity (Yellow): 0

EMERGENCY RESPONSE: See Appendix A (154)
Evacuation:
Public safety: Isolate the area of spill or leak for at least 25 to 50 meters (80 to 160 feet) in all directions.
Spill: Increase, in the downwind direction, as necessary, the distance shown under "Public Safety."
Fire: If tank, rail car, or tank truck is involved in fire, isolate for at least 800 meters (½ mile) in all directions; also, consider initial evacuation for 800 meters (½ mile) in all directions.

EXPOSURE
Short-term effects: May cause eye, skin and respiratory tract irritation. *IF BREATHING HAS STOPPED*, give artificial respiration; *avoid mouth-to-mouth resuscitation; use bag/mask apparatus.*
Vapor or fume: High concentrations may cause difficult breathing and pulmonary edema.
Note to physician or authorized medical personnel: Medical observation is recommended for 24 to 48 hours after breathing overexposure, as pulmonary edema may be delayed. As first aid for pulmonary edema, consider administering a corticosteroid spray. Cigarette smoking may exacerbate pulmonary injury and should be discouraged for at least 72 hours following exposure.

HEALTH HAZARDS
Personal protective equipment (PPE): B-Level PPE. Approved respirator, eye and skin protection.
Toxicity by ingestion: Grade 3; LD_{50} = 58 mg/kg (rat).
Vapor (gas) irritant characteristics: Vapors may cause lung edema; physical exertion will aggravate this condition.
Liquid or solid irritant characteristics: May cause irritation of the eyes, skin, and respiratory tract.

CHEMICAL REACTIVITY
Reactivity with water: Aqueous solution is acidic.
Binary reactants: Violent reaction with aluminum and magnesium. Reacts with bases. When heated reacts, with nitrates, nitrites, and chlorates.
Reactivity group: 43
Compatibility class: Miscellaneous water solutions.

ENVIRONMENTAL DATA
Water pollution: Effect of low concentrations on aquatic life is unknown. May be dangerous if it enters nearby water intakes; notify operators. Notify local health and wildlife officials.
Response to discharge: Disperse and flush.

PHYSICAL AND CHEMICAL PROPERTIES
Physical state @ 59°F/15°C and 1 atm: Solid.
Molecular weight: 132.14
Melting/Freezing point: Decomposes. 455°F/235°C/508°K.
Specific gravity (water = 1): 1.78 @ 15°C (solid).

AMMONIUM SULFIDE REC. A:3900

SYNONYMS: AMMONIUM BISULFIDE; AMMONIUM HYDROGEN SULFIDE; AMMONIUM MONOSULFIDE; DIAMMONIUM SULFIDE; SULFURO AMONICO (Spanish)

IDENTIFICATION
CAS Number: 12135-76-1; 12124-99-1(bisulfide)
Formula: $(NH_4)_2S \cdot NH_4SH \cdot H_2O$; NH_4HS (bisulfide)
DOT ID Number: UN 2683; DOT Guide Number: 132
Proper Shipping Name: Ammonium sulfide, solution
Reportable Quantity (RQ): **(CERCLA)** 100 lb/45.4 kg

DESCRIPTION: Colorless to tan waxy solid, paste, or solution. Usually shipped and stored as a water solution (40–44%). Strong odor of rotten eggs and ammonia. Mixes with water; a large amount of irritating vapor is produced; reacts, forming hydrogen sulfide. Freezes at 0°F/–18°C.

Poison! • Corrosive • Highly flammable • Breathing the vapor can kill you • Severely irritating to skin, eyes, and respiratory tract; prolonged contact with the skin causes burns • Firefighting gear (including SCBA) does not provide adequate protection. If exposure occurs, remove and isolate gear immediately and thoroughly decontaminate personnel • Containers may BLEVE when exposed to fire •Vapors may form explosive mixture with air •Vapors are heavier than air and will collect and stay in low areas • Vapors may travel long distances to ignition sources and flashback • Vapors in confined areas (e.g., tanks, sewers, buildings) may explode when exposed to fire • At 100°F/38°C bubbles of hydrogen sulfide gas is released. If ignited, this will form corrosive sulfur dioxide gas. Toxic products of combustion also include nitrogen oxides, hydrogen sulfide, and ammonia • Do not put yourself in danger by entering a contaminated area to rescue a victim. *Note*: The bisulfide is pyrophoric.

Hazard Classification (based on NFPA-704 Rating System)
Health Hazards (Blue): 3; Flammability (Red): 3; Reactivity (Yellow): 0; Special Notice (White): Water reactive

EMERGENCY RESPONSE: See Appendix A (132)
Evacuation:
Public safety: Isolate spill area for at least 100 to 200 meters (330 to 660 feet) in all directions.
Spill: Increase, in the downwind direction, as necessary, the distance shown under "Public Safety."
Fire: Isolate for 800 meters (½ mile) in all directions, especially if tank, rail car, or tank truck is involved in fire.

EXPOSURE
Short-term effects: *SEEK MEDICAL ATTENTION.* **Vapor:** Irritating to eyes, nose, and throat. *IF INHALED*, will, will cause dizziness, headache, coughing, or difficult breathing. *IF IN EYES*, hold eyelids open and flush with plenty of water. *IF BREATHING HAS STOPPED*, give artificial respiration; *avoid mouth-to-mouth resuscitation; use bag/mask apparatus.* IF breathing is difficult, administer oxygen. Lung edema may develop. **Liquid:** Will burn skin and eyes. *IF SWALLOWED*, will cause nausea. Remove contaminated clothing and shoes. Flush affected areas with plenty of water. *IF IN EYES*, hold eyelids open and flush with plenty of water. *IF SWALLOWED* and victim is *CONSCIOUS AND ABLE TO SWALLOW*, have victim drink 4 to 8 ounces of water. *IF SWALLOWED* and victim is *UNCONSCIOUS OR HAVING CONVULSIONS*, do nothing except keep victim warm. **Do NOT induce vomiting.**

Note to physician or authorized medical personnel: Medical observation is recommended for 24 to 48 hours after breathing overexposure, as pulmonary edema may be delayed. As first aid for pulmonary edema, consider administering a corticosteroid spray. Cigarette smoking may exacerbate pulmonary injury and should be discouraged for at least 72 hours following exposure.

HEALTH HAZARDS
Personal protective equipment (PPE): B-Level PPE. Self-contained breathing apparatus; neoprene, rubber or plastic gloves; splash goggles; neoprene or rubber shoes; polycarbonate or butyl rubber chemical suit.
Recommendations for respirator selection: NIOSH/OSHA for hydrogen sulfide
100 ppm: PAPRS [any powered, air-purifying respirator with cartridge(s) providing protection against the compound of concern]; or GMFS [any air-purifying, full-facepiece respirator (gas mask) with a chin-style, front- or back-mounted canister providing protection against the compound of concern]; or SA (any supplied-air respirator); or SCBAF (any self-contained breathing apparatus with a full facepiece). *EMERGENCY OR PLANNED ENTRY INTO UNKNOWN CONCENTRATIONS OR IDLH CONDITIONS:* SCBAF:PD,PP (any self-contained breathing apparatus that has a full facepiece and is operated in a pressure-demand or other positive-pressure mode); or SAF:PD,PP:ASCBA (any supplied-air respirator that has a full facepiece and is operated in a pressure-demand or other positive-pressure mode in combination with an auxiliary self-contained breathing apparatus operated in a pressure-demand or other positive pressure mode). *ESCAPE:* GMFS [any air-purifying, full-facepiece respirator (gas mask) with a chin-style, front- or back-mounted canister providing protection against the compound of concern]; or SCBAE (any appropriate escape-type, self-contained breathing *apparatus). Note:* Substance reported to cause eye irritation or damage; may require eye protection.
Exposure limits (TWA unless otherwise noted): ACGIH TLV 10 ppm (14 mg/m^3); OSHA PEL ceiling 20 ppm; 50 ppm/10 minute maximum peak; NIOSH ceiling 10 ppm (15 mg/m^3)/10 minutes as hydrogen sulfide.
Short-term exposure limits (15-minute TWA): ACGIH STEL 15 ppm (21 mg/m^3).
Vapor (gas) irritant characteristics: May cause difficult breathing. Possible lung edema.
Liquid or solid irritant characteristics: May cause irritation of the eyes, skin, and respiratory tract.

Ammonium tartrate

FIRE DATA
Flash point: 72°F/22°C (cc).
Flammable limits in air: LEL: 4%; UEL: 46% (hydrogen sulfide).
Electrical hazard: Class I, Group D.

CHEMICAL REACTIVITY
Binary reactants: Reacts with acids to form hydrogen sulfide. Reacts with bases to form ammonia. Strong oxidizers may cause fire and explosion; strong nitric acid, metals. Severely corrodes aluminum, brass, bronze, copper, zinc, and related alloys.
Neutralizing agents for acids and caustics: Dilute with water. Do not attempt to neutralize with acid.

ENVIRONMENTAL DATA
Water pollution: Dangerous to aquatic life in high concentrations. May be dangerous if it enters nearby water intakes; notify operators. Notify local health and wildlife officials. **Response to discharge:** Issue warning-air contaminant, water contaminant. Restrict access. Should be removed. Chemical and physical treatment.

SHIPPING INFORMATION
Grades of purity: Technical, 45–50% in water; Reagent, 52–60% in water; **Storage temperature:** Ambient; **Inert atmosphere:** Ventilated (natural); **Venting:** Open (flame arrester); **Stability during transport:** Stable, but toxic hydrogen sulfide and ammonia gases may collect in enclosed spaces.

PHYSICAL AND CHEMICAL PROPERTIES
Physical state @ 59°F/15°C and 1 atm: Solid or pasty Liquid.
Molecular weight: 68.14 (solute).
Boiling point @ 1 atm: 104°F/40°C/313°K.
Melting/Freezing point: Decomposes; 244°F/118°C/391°K (bisulfide).
Specific gravity (water = 1): 0.99-1.01 @ 68°F/20°C (liquid).
Heat of solution: 95.0 Btu/lb = 52.8 cal/g = 2.21×10^5 J/kg.
Vapor pressure: 395 mm.

AMMONIUM SULFITE   REC. A:3950

SYNONYMS: SULFITO AMONICO (Spanish)

IDENTIFICATION
CAS Number: 10196-04-0
Formula: $(NH_4)_2SO_3$
DOT ID Number: UN 9090; DOT Guide Number: 171
Proper Shipping Name: Ammonium sulfite
Reportable Quantity (RQ): **(CERCLA)** 5000 lb/2270 kg

DESCRIPTION: White crystalline solid. Odorless. Initially sinks in water; soluble.

Irritating to skin, eyes, and respiratory tract • Toxic products of combustion may include nitrogen oxides, sulfur oxides, and ammonia gas.

Hazard Classification (based on NFPA-704 Rating System)
Health Hazards (Blue): 2; Flammability (Red): 0; Reactivity (Yellow): 0

EMERGENCY RESPONSE: See Appendix A (171)
Evacuation:
Public safety: Isolate the area of spill or leak for at least 10 to 25 meters (30 to 80 feet) in all directions.
Spill: Increase, in the downwind direction, as necessary, the distance shown under "Public Safety."
Fire: If any large container is involved in fire, isolate for at least 800 meters (½ mile) in all directions; also, consider initial evacuation for 800 meters (½ mile) in all directions.

EXPOSURE
Short-term effects: *SEEK MEDICAL ATTENTION*. **Dust:** Irritating to eyes, nose, and throat. *IF INHALED*, will, will cause coughing or difficult breathing. *IF IN EYES*, hold eyelids open and flush with plenty of water. *IF BREATHING HAS STOPPED*, give artificial respiration. IF breathing is difficult, administer oxygen. **Solid:** Irritating to skin and eyes. *IF SWALLOWED*, will cause nausea. Remove contaminated clothing and shoes. Flush affected areas with plenty of water. *IF IN EYES*, hold eyelids open and flush with plenty of water. *IF SWALLOWED* and victim is *CONSCIOUS AND ABLE TO SWALLOW*, have victim drink 4 to 8 ounces of water. *IF SWALLOWED* and victim is *UNCONSCIOUS OR HAVING CONVULSIONS*, do nothing except keep victim warm.

HEALTH HAZARDS
Personal protective equipment (PPE): B-Level PPE.
Liquid or solid irritant characteristics: Causes irritation of the eyes, nose, and respiratory tract.

CHEMICAL REACTIVITY
Reactivity with water: Solution is alkaline and evolves toxic hydrogen sulfide and/or ammonia, depending on pH.
Binary reactants: Reacts with acids to form hydrogen sulfide. Reacts with bases to form ammonia. Incompatible with lead diacetate, mercury(I) chloride. Attacks aluminum, brass, bronze, copper, tin, zinc, especially in the presence of moisture.
Neutralizing agents for acids and caustics: Dilute with water. Do not attempt to neutralize with acids.

ENVIRONMENTAL DATA
Water pollution: Dangerous to aquatic life in high concentrations. May be dangerous if it enters nearby water intakes; notify operators. Notify local health and wildlife officials. **Response to discharge:** Disperse and flush.

SHIPPING INFORMATION
Grades of purity: Anhydrous; Purified monohydrate; **Storage temperature:** Ambient; **Inert atmosphere:** None; **Venting:** Open; **Stability during transport:** Stable.

PHYSICAL AND CHEMICAL PROPERTIES
Physical state @ 59°F/15°C and 1 atm: Solid.
Molecular weight: 134.2
Specific gravity (water = 1): More than 1.1 @ 68°F/20°C (solid).
Heat of solution: 20.5 Btu/lb = 11.4 cal/g = 0.477×10^5 J/kg.

AMMONIUM TARTRATE   REC. A:4000

SYNONYMS: TARTRATO AMONICO (Spanish); 1-TARTARIC ACID, AMMONIUM SALT

IDENTIFICATION
CAS Number: 14307-43-8; 3164-29-2
Formula: $C_4H_{12}N_2O_6$
DOT ID Number: NA 9091; UN 3077; NA 9188; DOT Guide Number: 171

Proper Shipping Name: Ammonium tartrate (NA 9091); Environmentally hazardous substance, solid, n.o.s. (UN 3077); ORM-E, solid, n.o.s. (NA 9188)
Reportable Quantity (RQ): **(CERCLA)** 5000 lb/2270 kg

DESCRIPTION: White solid. Odorless. Sinks and mixes with water.

Combustible solid • Dust may explode if ignited in an enclosed area • Containers may BLEVE when exposed to fire • Toxic products of combustion may include carbon monoxide, nitrogen oxides and ammonia gas.

Hazard Classification (based on NFPA-704 Rating System)
Health Hazards (Blue): 1; Flammability (Red): 1; Reactivity (Yellow): 0

EMERGENCY RESPONSE: See Appendix A (171)
Evacuation:
Public safety: Isolate the area of spill or leak for at least 10 to 25 meters (30 to 80 feet) in all directions.
Spill: Increase, in the downwind direction, as necessary, the distance shown under "Public Safety."
Fire: If any large container is involved in fire, isolate for at least 800 meters (½ mile) in all directions; also, consider initial evacuation for 800 meters (½ mile) in all directions.

EXPOSURE
Short-term effects: *SEEK MEDICAL ATTENTION.* **Dust:** Irritating to eyes, nose, and throat. Harmful if inhaled. *IF IN EYES,* hold eyelids open and flush with plenty of water. *IF BREATHING HAS STOPPED,* give artificial respiration. IF breathing is difficult, administer oxygen. **Solid:** Irritating to skin and eyes. Harmful if swallowed. Remove contaminated clothing and shoes. Flush affected areas with plenty of water. *IF IN EYES,* hold eyelids open and flush with plenty of water. *IF SWALLOWED* and victim is *CONSCIOUS AND ABLE TO SWALLOW,* have victim drink 4 to 8 ounces of water. *IF SWALLOWED* and victim is *UNCONSCIOUS OR HAVING CONVULSIONS,* do nothing except keep victim warm.

HEALTH HAZARDS
Personal protective equipment (PPE): B-Level PPE. Approved respirator, goggles, or face shield; rubber gloves.

ENVIRONMENTAL DATA
Water pollution: Effect of low concentrations on aquatic life is unknown. May be dangerous if it enters nearby water intakes; notify operators. Notify local health and wildlife officials.
Response to discharge: Disperse and flush.

SHIPPING INFORMATION
Grades of purity: Analytical; Pure; **Storage temperature:** Ambient; **Inert atmosphere:** None; **Venting:** Open; **Stability during transport:** Stable.

PHYSICAL AND CHEMICAL PROPERTIES
Physical state @ 59°F/15°C and 1 atm: Solid.
Molecular weight: 184
Boiling point @ 1 atm: Decomposes.
Specific gravity (water = 1): 1.60 @ 77°F/25°C (solid).

AMMONIUM THIOCYANATE   **REC. A:4050**

SYNONYMS: AMMONIUM SULFOCYANATE; AMMONIUM RHODANATE; AMMONIUM RHODANIDE; AMMONIUM SULFOCYANIDE; THIOCYANIC ACID, AMMONIUM SALT; TIOCIANATO AMONICO (Spanish)

IDENTIFICATION
CAS Number: 1762-95-4
Formula: NH_4SCN
DOT ID Number: UN 3077 (solid); UN 3082 (liquid); NA 9188; DOT Guide Number: 171
Proper Shipping Name: Environmentally hazardous substance, solid or liquid, n.o.s.; ORM-E, solid, n.o.s. (NA 9188)
Reportable Quantity (RQ): **(CERCLA)** 5000 lb/2270 kg

DESCRIPTION: White solid or colorless solution in water. Odorless. Initially sinks in water; soluble.

Combustible solid • Solution not flammable • Irritating to skin, eyes, and respiratory tract • Toxic products of combustion may include hydrogen sulfide, nitrogen cyanide, oxide of nitrogen, and ammonia gas • Do not put yourself in danger by entering a contaminated area to rescue a victim.

Hazard Classification (based on NFPA-704 Rating System)
Solid: Health Hazards (Blue): 2; Flammability (Red): 1; Reactivity (Yellow): 0

EMERGENCY RESPONSE: See Appendix A (171)
Evacuation:
Public safety: Isolate the area of spill or leak for at least 10 to 25 meters (30 to 80 feet) in all directions.
Spill: Increase, in the downwind direction, as necessary, the distance shown under "Public Safety."
Fire: If any large container is involved in fire, isolate for at least 800 meters (½ mile) in all directions; also, consider initial evacuation for 800 meters (½ mile) in all directions.

EXPOSURE
Short-term effects: *SEEK MEDICAL ATTENTION.* **Vapor or dust:** Irritating to eyes, nose, and throat. *IF INHALED,* will, will cause coughing or difficult breathing. *IF IN EYES,* hold eyelids open and flush with plenty of water. *IF BREATHING HAS STOPPED,* give artificial respiration. IF breathing is difficult, administer oxygen. **Liquid or solid:** Irritating to skin and eyes. *IF SWALLOWED,* will cause nausea or vomiting. Remove contaminated clothing and shoes. Flush affected areas with plenty of water. *IF IN EYES,* hold eyelids open and flush with plenty of water. *IF SWALLOWED* and victim is *CONSCIOUS AND ABLE TO SWALLOW,* have victim drink 4 to 8 ounces of water. *IF SWALLOWED* and victim is *UNCONSCIOUS OR HAVING CONVULSIONS,* do nothing except keep victim warm.

HEALTH HAZARDS
Personal protective equipment (PPE): B-Level PPE.
Toxicity by ingestion: Grade 2; oral rat LD_{50} = 854 mg/kg.
Liquid or solid irritant characteristics: Causes irritation of the eyes, nose, and respiratory tract.

FIRE DATA
Flash point: Solid may be combustible; solution is not flammable.

ENVIRONMENTAL DATA
Water pollution: DOT Appendix B, §172.101–marine pollutant. Dangerous to aquatic life in high concentrations. May be dangerous if it enters nearby water intakes; notify operators. Notify local

health and wildlife officials. **Response to discharge:** Issue warning–water contaminant. Disperse and flush.

SHIPPING INFORMATION
Grades of purity: Reagent; Technical, 50–65% solution in water; **Storage temperature:** Ambient; **Inert atmosphere:** None; **Venting:** Open; **Stability during transport:** Stable.

PHYSICAL AND CHEMICAL PROPERTIES
Physical state @ 59°F/15°C and 1 atm: Solid or Liquid.
Molecular weight: 76.12
Boiling point @ 1 atm: (solution; solid decomposes) 239°F/115°C/388°K.
Melting/Freezing point: (solid) 320°F/160°C/433°K.
Specific gravity (water = 1): More than 1.1 @ 68°F/20°C (solid); 1.1-1.15 @ 68°F/20°C (solution).
Heat of solution: 133 Btu/lb = 74 cal/g = 3.1×10^5 J/kg.

AMMONIUM THIOSULFATE   REC. A:4100

SYNONYMS: AMMONIUM HYPOSULFITE; TIOSULFATO AMONICO (Spanish)

IDENTIFICATION
CAS Number: 7783-18-8
Formula: $(NH_4)_2S_2O_3$; $O_3S_2 \cdot 2H_4N$

DESCRIPTION: Colorless solid or water solution. Ammonia odor. Sinks and mixes with water.

Corrosive to the skin, eyes, and respiratory tract; contact with skin or eyes can cause burns and vision impairment; inhalation symptoms may be delayed • Toxic products of combustion may include hydrogen sulfide, oxides of nitrogen and sulfur, and ammonia gas.

Hazard Classification (based on NFPA-704 Rating System)
Health Hazards (Blue): 1; Flammability (Red): 0; Reactivity (Yellow): 0

EMERGENCY RESPONSE: See Appendix A (171)
Evacuation:
Public safety: Isolate the area of spill or leak for at least 10 to 25 meters (30 to 80 feet) in all directions.
Spill: Increase, in the downwind direction, as necessary, the distance shown under "Public Safety."
Fire: If any large container is involved in fire, isolate for at least 800 meters (½ mile) in all directions; also, consider initial evacuation for 800 meters (½ mile) in all directions.

EXPOSURE
Short-term effects: *SEEK MEDICAL ATTENTION.* **Dust:** Irritating to eyes, nose, and throat. *IF INHALED*, will, will cause difficult breathing. *IF IN EYES*, hold eyelids open and flush with plenty of water. *IF BREATHING HAS STOPPED*, give artificial respiration. IF breathing is difficult, administer oxygen.
Solution or solid: Irritating to skin and eyes. Harmful if swallowed. Remove contaminated clothing and shoes. Flush affected areas with plenty of water. *IF IN EYES*, hold eyelids open and flush with plenty of water. *IF SWALLOWED* and victim is *CONSCIOUS AND ABLE TO SWALLOW*, have victim drink 4 to 8 ounces of water. *IF SWALLOWED* and victim is *UNCONSCIOUS OR HAVING CONVULSIONS*, do nothing except keep victim warm.

HEALTH HAZARDS
Personal protective equipment (PPE): B-Level PPE.
Recommendations for respirator selection: NIOSH as ammonia *250 ppm*: CCRS [any chemical cartridge respirator with cartridge(s) providing protection against the compound of concern)]; or SA (any supplied-air respirator). *300 ppm*: SA:CF (any supplied-air respirator operated in a continuous-flow mode); or PAPRS* [any powered, air-purifying respirator with cartridge(s) providing protection against the compound of concern]; or CCRFS [any chemical cartridge respirator with a full facepiece and cartridge(s) providing protection against the compound of concern]; or GMFS [any air-purifying, full-facepiece respirator (gas mask) with a chin-style, front- or back-mounted canister providing protection against the compound of concern]; or SCBAF (any self-contained breathing apparatus with a full facepiece); or SAF (any supplied-air respirator with a full facepiece). *EMERGENCY OR PLANNED ENTRY INTO UNKNOWN CONCENTRATIONS OR IDLH CONDITIONS*: SCBAF:PD,PP (any self-contained breathing apparatus that has a full facepiece and is operated in a pressure-demand or other positive-pressure mode); or SAF:PD,PP:ASCBA (any supplied-air respirator that has a full facepiece and is operated in a pressure-demand or other positive-pressure mode in combination with an auxiliary self-contained breathing apparatus operated in a pressure-demand or other positive pressure mode). *ESCAPE*: GMFS [any air-purifying, full-facepiece respirator (gas mask) with a chin-style, front- or back-mounted canister providing protection against the compound of concern]; or SCBAE (any appropriate escape-type, self-contained breathing apparatus). *Note:* Substance reported to cause eye irritation or damage; may require eye protection.
Exposure limits (TWA unless otherwise noted): ACGIH TLV 25 ppm (17 mg/m^3); NIOSH 25 ppm (18 mg/m^3) as ammonia
Short-term exposure limits (15-minute TWA): ACGIH STEL 35 ppm (24 mg/m^3); NIOSH STEL 35 ppm (27 mg/m^3); OSHA STEL 50 ppm as ammonia
Liquid or solid irritant characteristics: Irritates the eyes, skin, and respiratory tract.
IDLH value: 300 ppm as ammonia

CHEMICAL REACTIVITY
Binary reactants: Violent reaction with aluminum, magnesium.
Reactivity group: 43
Compatibility class: Miscellaneous water solutions.

ENVIRONMENTAL DATA
Water pollution: Effect of low concentrations on aquatic life is unknown. May be dangerous if it enters nearby water intakes; notify operators. Notify local health and wildlife officials. **Response to discharge:** Issue warning–water contaminant. Disperse and flush.

SHIPPING INFORMATION
Grades of purity: Technical; Analytical; Technical, 60% solution in water; **Storage temperature:** Ambient; **Inert atmosphere:** Ventilated (natural); **Venting:** Open; **Stability during transport:** Stable, but toxic ammonia gas may collect in enclosed spaces.

PHYSICAL AND CHEMICAL PROPERTIES
Physical state @ 59°F/15°C and 1 atm: Solid or Liquid.
Molecular weight: 148.2
Boiling point @ 1 atm: Decomposes.
Specific gravity (water = 1): 2 @ 68°F/20°C (solid).

AMMONIUM THIOSULFATE SOLUTION   REC. A:4150

SYNONYMS: AMMONIUM HYPOSULFITE SOLUTION;

AMMONIUM HYPO SOLUTION; DIAMMONIUM THIOSULFATE; THIOSULFURIC ACID, DIAMMONIUM SALT

IDENTIFICATION
CAS Number: 7783-18-8
Formula: $(NH_4)_2S_2O_3$

DESCRIPTION: Colorless liquid. Ammonia odor. Soluble in water.

Toxic products of combustion may include hydrogen sulfide, oxides of nitrogen and sulfur, and ammonia gas.

EMERGENCY RESPONSE: See Appendix A (171)
Evacuation:
Public safety: Isolate the area of spill or leak for at least 10 to 25 meters (30 to 80 feet) in all directions.
Spill: Increase, in the downwind direction, as necessary, the distance shown under "Public Safety."
Fire: If any large container is involved in fire, isolate for at least 800 meters (½ mile) in all directions; also, consider initial evacuation for 800 meters (½ mile) in all directions.

EXPOSURE
Short-term effects: *SEEK MEDICAL ATTENTION*. **Solution:** Irritating to skin and eyes. Remove contaminated clothing and shoes. Flush affected areas with plenty of water. *IF IN EYES*, hold eyelids open and flush with plenty of water. *IF SWALLOWED* and victim is *CONSCIOUS AND ABLE TO SWALLOW*, have victim drink water and induce vomiting.

HEALTH HAZARDS
Personal protective equipment (PPE): B-Level PPE.
Vapor (gas) irritant characteristics: Vapors cause smarting of the eyes or respiratory system if present in high concentrations. The effect may be temporary.
Liquid or solid irritant characteristics: If spilled on clothing and allowed to remain, may cause smarting and reddening of skin.

CHEMICAL REACTIVITY
Binary reactants: Violent reaction with aluminum, magnesium. Reacts with strong oxidizers such as chlorates, nitrates, and nitrites to release toxic ammonia, hydrogen sulfide, and sulfur trioxide gases.
Reactivity group: 43
Compatibility class: Miscellaneous water solutions.

ENVIRONMENTAL DATA
Water pollution: Effect of low concentrations on aquatic life is unknown. May be dangerous if it enters nearby water intakes; notify operators. Notify local health and wildlife officials.
Response to discharge: Disperse and flush.

SHIPPING INFORMATION
Grades of purity: Technical, water solutions of varying concentrations; **Storage temperature:** Ambient; **Inert atmosphere:** Ventilated (natural); **Venting:** Open; **Stability during transport:** Stable, but toxic ammonia gas may collect in enclosed spaces.

PHYSICAL AND CHEMICAL PROPERTIES
Physical state @ 59°F/15°C and 1 atm: Liquid.
Molecular weight: 148.2
Boiling point @ 1 atm: 122°F/50°C/323°K.
Melting/Freezing point: 50°F/10°C/283°K.
Specific gravity (water = 1): 1.33 at 16°C.

AMYL ACETATE REC. A:4200

SYNONYMS: ACETATO de *n*-AMILO (Spanish); ACETATE d'AMYLE (French); ACETIC ACID, *n*-AMYL ESTER; AMYL ACETATE, mixed isomers; AMYL ACETIC ESTER; AMYL ACETIC ETHER; AMYAZETAT (German); BIRNENOEL; EEC No. 607-130-00-2; PEAR OIL; PENT ACETATE; PENTYL ACETATES; PENTYL ESTER OF ACETIC ACID; *n*-AMYL ACETATE; 1-PENTANOL ACETATE; PRIMARY AMYL ACETATE

IDENTIFICATION
CAS Number: 628-63-7
Formula: $C_7H_{14}O_2$; $CH_3COOC_5H_{11}$
DOT ID Number: UN 1104; DOT Guide Number: 129
Proper Shipping Name: Amyl acetates
Reportable Quantity (RQ): **(CERCLA)** 5000 lb/2270 kg

DESCRIPTION: Colorless to yellow watery liquid. Banana- or pear-like odor. Floats on the surface of water. Flammable, irritating vapor is produced.

Highly flammable • Containers may BLEVE when exposed to fire • Vapors may form explosive mixture with air • Vapors are heavier than air and will collect and stay in low areas • Vapors may travel long distances to ignition sources and flashback • Vapors in confined areas (e.g., tanks, sewers, buildings) may explode when exposed to fire • Irritating to the skin, eyes, and respiratory tract • Toxic products of combustion may include carbon monoxide.

Hazard Classification (based on NFPA-704 Rating System)
Health Hazards (Blue): 1; Flammability (Red): 3; Reactivity (Yellow): 0

EMERGENCY RESPONSE: See Appendix A (129)
Evacuation:
Public safety: Isolate spill area for at least 50 to 100 meters (160 to 330 feet) in all directions.
Spill: Large spill–Consider initial downwind evacuation for at least 300 meters (1000 feet).
Fire: Isolate for 800 meters (½ mile) in all directions, especially if tank, rail car, or tank truck is involved in fire.

EXPOSURE
Short-term effects: *SEEK MEDICAL ATTENTION*. **Vapor:** Irritating to eyes, nose, and throat. *IF INHALED*, will, will cause nausea, headache or dizziness. Move to fresh air. *IF BREATHING HAS STOPPED*, give artificial respiration. IF breathing is difficult, administer oxygen. **Liquid:** Irritating to skin and eyes. Remove contaminated clothing and shoes. Flush affected areas with plenty of water. *IF IN EYES*, hold eyelids open and flush with plenty of water. The use of alcoholic beverages may enhance the toxic effect.

HEALTH HAZARDS
Personal protective equipment (PPE): B-Level PPE. OSHA Table Z-1-A air contaminant. Chemical protective material(s) reported to have good to excellent resistance: butyl rubber, neoprene+natural rubber, PV alcohol.
Recommendations for respirator selection: NIOSH/OSHA
1000 ppm: CCRFOV* [any air-purifying, full-facepiece respirator (gas mask) with a chin-style, front- or back-mounted acid gas

canister]; or GMFOV [any air-purifying, full-facepiece respirator (gas mask) with a chin-style, front- or back-mounted organic vapor canister; PAPROV* [any powered, air-purifying respirator with organic vapor cartridge(s)]; or SA* (any supplied-air respirator); or SCBAF (any self-contained breathing apparatus with a full facepiece); or SAF (any supplied-air respirator with a full facepiece). *EMERGENCY OR PLANNED ENTRY INTO UNKNOWN CONCENTRATIONS OR IDLH CONDITIONS*: SCBAF:PD,PP (any self-contained breathing apparatus that has a full facepiece and is operated in a pressure-demand or other positive-pressure mode); or SAF:PD,PP:ASCBA (any supplied-air respirator that has a full facepiece and is operated in a pressure-demand or other positive-pressure mode in combination with an auxiliary self-contained breathing apparatus operated in a pressure-demand or other positive pressure mode). *ESCAPE:* GMFOV[any air-purifying, full-facepiece respirator (gas mask) with a chin-style, front- or back-mounted organic vapor canister]; or SCBAE (any appropriate escape-type, self-contained breathing apparatus).
Note: Substance reported to cause eye irritation or damage; may require eye protection.
Exposure limits (TWA unless otherwise noted): ACGIH TLV 50 ppm (266 mg/m^3); NIOSH/OSHA 100 ppm (525 mg/m^3).
Short-term exposure limits (15-minute TWA): ACGIH STEL 100 ppm (532 mg/m^3).
Toxicity by ingestion: Grade 1; LD$_{50}$ = 6.5 g/kg (rat).
Vapor (gas) irritant characteristics: Vapors cause smarting of the eyes or respiratory system if present in high concentration. The effect may be temporary.
Liquid or solid irritant characteristics: May cause irritation to the eyes, skin, and respiratory tract. If spilled on clothing and allowed to remain, may cause smarting and reddening of the skin.
Odor threshold: 0.067 ppm.
IDLH value: 1000 ppm.

FIRE DATA
Flash point: 77°F/25°C (cc).
Flammable limits in air: LEL: 1.1% @ 212°F/100°C; UEL: 7.5%.
Fire extinguishing agents not to be used: Water in straight hose stream will scatter and spread fire and should not be used
Autoignition temperature: 680°F/360°C/633°K.
Electrical hazard: Class I, Group D.
Burning rate: 4.1 mm/min

CHEMICAL REACTIVITY
Binary reactants: Heat may cause instability. Reacts with nitrates, strong oxidizers, alkalis, and acids. May attack certain plastics and coatings.
Reactivity group: 34
Compatibility class: Esters.

ENVIRONMENTAL DATA
Food chain concentration potential: Log P$_{ow}$ = 2.23. Unlikely to accumulate.
Water pollution: Effect of low concentrations on aquatic life is unknown. Fouling to shoreline. May be dangerous if it enters nearby water intakes; notify operators. Notify local health and pollution control officials. **Response to discharge:** Issue warning–high flammability. Mechanical containment. Chemical and physical treatment.

SHIPPING INFORMATION
Grades of purity: 85-96% (technical, commercial); **Storage temperature:** Ambient (cool); **Inert atmosphere:** None; **Stability during transport:** Stable.

PHYSICAL AND CHEMICAL PROPERTIES
Physical state @ 59°F/15°C and 1 atm: Liquid.
Molecular weight: 130.19
Boiling point @ 1 atm: 249°F/121°C/394°K.
Melting/Freezing point: –159.8°F/–72°C/201°K.
Specific gravity (water = 1): 0.876 @ 68°F/20°C (liquid).
Liquid surface tension: 12 dynes/cm = 0.012 N/m at 30°C.
Liquid water interfacial tension: (estimate) 50 dynes/cm = 0.05 N/m at 17°C.
Relative vapor density (air = 1): 4.5
Ratio of specific heats of vapor (gas): 1.1
Latent heat of vaporization: 140 Btu/lb = 75 cal/g = 3.1 x 10^5 J/kg.
Heat of combustion: –13,360 Btu/lb = –7423 cal/g = –310.8 x 10^5 J/kg.
Vapor pressure: (Reid) 0.1 psia; 4 mm.

sec-AMYL ACETATE   REC. A:4250

SYNONYMS: ACETATO de *sec*-AMILO (Spanish); ACETIC ACID, 2-PENTYL ESTER; 2-ACETOXYPENTANE; AMYL ACETATE; AMYLACETIC ESTER; BANANA OIL; EEC No. 607-130-00-2; ISOAMYL ETHANOATE; ISOPENTYL ACETATE; 3-METHYL-1-BUTANOL ACETATE; 1-METHYLBUTYL ACETATE; 1-METHYLBUTYL ETHANOATE; PEAR OIL; 2-PENTANOL ACETATE; 2-PENTYL ACETATE; 2-PENTYLACETATE

IDENTIFICATION
CAS Number: 626-38-0
Formula: C$_7$H$_{14}$O$_2$; CH$_3$COOCH(CH$_3$)CH$_2$C$_2$H$_3$
DOT ID Number: UN 1104; DOT Guide Number: 129
Proper Shipping Name: Amyl acetates
Reportable Quantity (RQ): **(CERCLA)** 5000 lb/2270 kg

DESCRIPTION: Colorless to yellow watery liquid. Banana odor. Floats on the surface of water. Flammable, irritating vapor is produced.

Highly flammable • Containers may BLEVE when exposed to fire • Vapors may form explosive mixture with air • Vapors are heavier than air and will collect and stay in low areas • Vapors may travel long distances to ignition sources and flashback • Vapors in confined areas (e.g., tanks, sewers, buildings) may explode when exposed to fire • Irritating to the skin, eyes, and respiratory tract • Toxic products of combustion may include carbon monoxide.

Hazard Classification (based on NFPA-704 Rating System)
Health Hazards (Blue): 1; Flammability (Red): 3; Reactivity (Yellow): 0

EMERGENCY RESPONSE: See Appendix A (129)
Evacuation:
Public safety: Isolate spill area for at least 50 to 100 meters (160 to 330 feet) in all directions.
Spill: Large spill–Consider initial downwind evacuation for at least 300 meters (1000 feet).
Fire: Isolate for 800 meters (½ mile) in all directions, especially if tank, rail car, or tank truck is involved in fire.

EXPOSURE
Short-term effects: *SEEK MEDICAL ATTENTION.* **Vapor:** Irritating to eyes, nose, and throat. *IF INHALED*, will, will cause nausea, headache or dizziness. Move to fresh air. *IF BREATHING*

HAS STOPPED, give artificial respiration. IF breathing is difficult, administer oxygen. **Liquid:** Irritating to skin and eyes. Remove contaminated clothing and shoes. Flush affected areas with plenty of water. *IF IN EYES,* hold eyelids open and flush with plenty of water.

HEALTH HAZARDS
Personal protective equipment (PPE): B-Level PPE. OSHA Table Z-1-A air contaminant. Chemical protective material(s) reported to have good to excellent resistance: butyl rubber, neoprene+natural rubber.

Recommendations for respirator selection: NIOSH/OSHA
1000 ppm: CCROV* [any chemical cartridge respirator with organic vapor cartridge(s)]; or GMFOV [any air-purifying, full-facepiece respirator (gas mask) with a chin-style, front- or back-mounted acid gas canister]; or PAPROV* [any powered, air-purifying respirator with organic vapor cartridge(s)]; or SA (any supplied-air respirator); or SCBAF (any self-contained breathing apparatus with a full facepiece). *EMERGENCY OR PLANNED ENTRY INTO UNKNOWN CONCENTRATIONS OR IDLH CONDITIONS:* SCBAF:PD,PP (any self-contained breathing apparatus that has a full facepiece and is operated in a pressure-demand or other positive-pressure mode); or SAF:PD,PP:ASCBA (any supplied-air respirator that has a full facepiece and is operated in a pressure-demand or other positive-pressure mode in combination with an auxiliary self-contained breathing apparatus operated in a pressure-demand or other positive pressure mode). *ESCAPE:* GMFOV[any air-purifying, full-facepiece respirator (gas mask) with a chin-style, front- or back-mounted organic vapor canister]; or SCBAE (any appropriate escape-type, self-contained breathing apparatus).
Note: Substance reported to cause eye irritation or damage; may require eye protection.
Exposure limits (TWA unless otherwise noted): ACGIH TLV 50 ppm (266 mg/m^3); NIOSH/OSHA 125 ppm (650 mg/m^3).
Short-term exposure limits (15-minute TWA): ACGIH STEL 100 ppm (532 mg/m^3).
Long-term health effects: May effect the nervous system; kidney and liver.
Vapor (gas) irritant characteristics: Vapors cause smarting of the eyes or respiratory system if present in high concentrations.
Liquid or solid irritant characteristics: If spilled on clothing and allowed to remain, may cause smarting and reddening of the skin.
Odor threshold: 0.08 ppm.
IDLH value: 1000 ppm.

FIRE DATA
Flash point: 89°F/32°C (cc).
Flammable limits in air: LEL: 1.1% @ 212°F/100°C; UEL: 7.5%.
Fire extinguishing agents not to be used: Water may be ineffective.
Behavior in fire: Vapor may explode if ignited in an enclosed space.
Autoignition temperature: 680–714°F/360–379°C/633–652°K.
Electrical hazard: Class I, Group D. Precautions should be taken to prevent the accumulation of static electricity.

CHEMICAL REACTIVITY
Binary reactants: Reacts with nitrates, strong oxidizers, strong alkalis, and strong acids. Will soften, and subsequently dissolve, a great many plastic materials, coatings, and rubber. Attacks asbestos.
Reactivity group: 34
Compatibility class: Esters.

ENVIRONMENTAL DATA
Water pollution: Harmful to aquatic life in very low concentrations. Fouling to shoreline. May be dangerous if it enters nearby water intakes; notify operators. Notify local health and pollution control officials. **Response to discharge:** Issue warning–high flammability. Mechanical containment. Chemical and physical treatment.

SHIPPING INFORMATION
Storage temperature: Ambient; **Stability during transport:** Stable.

PHYSICAL AND CHEMICAL PROPERTIES
Physical state @ 59°F/15°C and 1 atm: Liquid.
Molecular weight: 130.18
Boiling point @ 1 atm: 249°F/121°C/394°K.
Melting/Freezing point: –109°F/–79°C/195°K.
Critical temperature: 619°F/326°C/599°K.
Critical pressure: 411.6 psia = 28.0 atm = 2.83 MN/m^2.
Specific gravity (water = 1): 0.861-0.866 @ 68°F/20°C.
Liquid surface tension: 28.9 dynes/cm = 0.0289 N/m @ 68°F/20°C.
Liquid water interfacial tension: (estimate) 44.1 dynes/cm = 0.0441 N/m @ 68°F/20°C.
Relative vapor density (air = 1): 4.5
Ratio of specific heats of vapor (gas): (estimate) more than 1-1.1 @ 68°F/20°C (68°F).
Latent heat of vaporization: (estimate) 128.9 Btu/lb = 71.7 cal/g = 3.0 x 10^5 J/kg.
Heat of combustion: –14.402 Btu/lb = –8000 cal/g = –334.9 x 10^5 J/kg.
Vapor pressure: 7 mm.

---

tert-AMYL ACETATE   REC. A:4300

SYNONYMS: ACETATO de *terc*-AMILO (Spanish); ACETATO de AMILO TERCIARIO (Spanish); EEC No. 607-130-00-2; *tert*-PENTYL ACETATE

IDENTIFICATION
CAS Number: 625-16-1
Formula: $CH_3COOC(CH_3)_2C_2NO_5$
DOT ID Number: UN 1104; DOT Guide Number: 129
Proper Shipping Name: Amyl acetates
Reportable Quantity (RQ): **(CERCLA)** 5000 lb/2270 kg

DESCRIPTION: Colorless to yellow watery liquid. Fruity, banana-like odor. Floats on the surface of water; slightly soluble.

Highly flammable • Severely irritating to the skin, eyes, and respiratory tract • Containers may BLEVE when exposed to fire • Vapors may form explosive mixture with air • Vapors are heavier than air and will collect and stay in low areas • Vapors may travel long distances to ignition sources and flashback • Vapors in confined areas (e.g., tanks, sewers, buildings) may explode when exposed to fire • Poisonous if swallowed • Toxic products of combustion may include carbon monoxide.

Hazard Classification (based on NFPA-704 Rating System)
Health Hazards (Blue): 1; Flammability (Red): 3; Reactivity (Yellow): 0

EMERGENCY RESPONSE: See Appendix A (129)
Evacuation:
Public safety: Isolate spill area for at least 50 to 100 meters (160 to 330 feet) in all directions.
Spill: Large spill–Consider initial downwind evacuation for at least 300 meters (1000 feet).
Fire: Isolate for 800 meters (½ mile) in all directions, especially if tank, rail car, or tank truck is involved in fire.

EXPOSURE
Short-term effects: *SEEK MEDICAL ATTENTION.* **Vapor:** Irritating to eyes, nose, and throat. *IF INHALED*, will, will cause nausea, headache, or dizziness. Move to fresh air. *IF BREATHING HAS STOPPED,* give artificial respiration. IF breathing is difficult, administer oxygen. **Liquid:** Irritating to skin and eyes. Remove contaminated clothing and shoes. Flush affected areas with plenty of water. *IF IN EYES,* hold eyelids open and flush with plenty of water.

HEALTH HAZARDS
Personal protective equipment (PPE): B-Level PPE. Chemical protective material(s) reported to have good to excellent resistance: butyl rubber, neoprene+natural rubber.
Recommendations for respirator selection: NIOSH/OSHA as *sec*-amyl acetate
1000 ppm: CCROV [any chemical cartridge respirator with organic vapor cartridge(s)]; or GMFOV [any air-purifying, full-facepiece respirator (gas mask) with a chin-style, front- or back-mounted acid gas canister]; or PAPROV [any powered, air-purifying respirator with organic vapor cartridge(s)]; or SA (any supplied-air respirator); or SCBAF (any self-contained breathing apparatus with a full facepiece). *EMERGENCY OR PLANNED ENTRY INTO UNKNOWN CONCENTRATIONS OR IDLH CONDITIONS:* SCBAF:PD,PP (any self-contained breathing apparatus that has a full facepiece and is operated in a pressure-demand or other positive-pressure mode); or SAF:PD,PP:ASCBA (any supplied-air respirator that has a full facepiece and is operated in a pressure-demand or other positive-pressure mode in combination with an auxiliary self-contained breathing apparatus operated in a pressure-demand or other positive pressure mode). *ESCAPE:* GMFOV[any air-purifying, full-facepiece respirator (gas mask) with a chin-style, front- or back-mounted organic vapor canister]; or SCBAE (any appropriate escape-type, self-contained breathing apparatus). *Note:* Substance reported to cause eye irritation or damage; may require eye protection.
Exposure limits (TWA unless otherwise noted): ACGIH TLV 50 ppm (266 mg/m^3); NIOSH/OSHA 125 ppm (650 mg/m^3).
Short-term exposure limits (15-minute TWA): ACGIH STEL 100 ppm (532 mg/m^3).
Long-term health effects: May cause liver and kidney problems.
Vapor (gas) irritant characteristics: Vapors cause a smarting of the eyes or respiratory system if present in high concentrations. The effect may be temporary.
Liquid or solid irritant characteristics: If spilled on clothing and allowed to remain may cause smarting, reddening of the skin and severe irritation.
Odor threshold: 0.08 ppm (CHRIS).
IDLH value: 1000 ppm

FIRE DATA
Flash point: 77°F/25°C (cc).
Flammable limits in air: LEL: 1.1% @ 212°F/100°C; UEL: 7.5%.
Fire extinguishing agents not to be used: Water may be ineffective.

Behavior in fire: When exposed to flames can react vigorously with oxidizing material.
Autoignition temperature: (estimated) 715°F/379°C/652°K.
Electrical hazard: Class I, Group D.

CHEMICAL REACTIVITY
Reactivity group: 34
Compatibility class: Esters.

ENVIRONMENTAL DATA
Water pollution: Harmful to aquatic life in very low concentrations. Fouling to shoreline. May be dangerous if it enters nearby water intakes; notify operators. Notify local health and pollution control officials. **Response to discharge:** Issue warning–high flammability. Mechanical containment. Chemical and physical treatment.

SHIPPING INFORMATION
Storage temperature: Ambient; **Stability during transport:** Stable.

PHYSICAL AND CHEMICAL PROPERTIES
Physical state @ 59°F/15°C and 1 atm: Liquid.
Molecular weight: 130.18
Boiling point @ 1 atm: 256°F/125°C/398°K.
Melting/Freezing point: More than –148°F/–100°C/173°K.
Critical temperature: (estimate) 609°F/320.7°C/593.9°K.
Critical pressure: 395 psia = 26.9 atm = 2.73 MN/m^2.
Specific gravity (water = 1): 0.874 @ 66°F/19°C/292°K.
Liquid surface tension: (estimate) 29.2 dynes/cm = 0.0292 N/m @ 68°F/20°C.
Liquid water interfacial tension: 43.8 dynes/cm = 0.0438 N/m @ 68°F/20°C.
Relative vapor density (air = 1): 4.5
Ratio of specific heats of vapor (gas): (estimate) more than 1-1.1
Latent heat of vaporization: (estimate) 126.6 Btu/lb = 70.3 cal/g = 2.94 x 10^5 J/kg.
Heat of combustion: (estimate) –14.402 Btu/lb = –8000 cal/g = –334.9 x 10^5 J/kg.

---

n-AMYL ALCOHOL   REC. A:4350

SYNONYMS: ALCOHOL AMILICO (Spanish); ALCOHOL C-5; ALCOOL AMYLIQUE (French); AMYL ALCOHOL; 1-AMYL ALCOHOL; AMYL ALCOHOL, NORMAL; AMYLOL; *n*-BUTYL CARBINOL; EEC No. 603-006-00-7; PENTANOL; 1-PENTANOL; PENTANOL-1; *n*-PENTANOL; PENTAN-1-OL; PENTASOL; *prim-n*-AMYL ALCOHOL; PRIMARY AMYL ALCOHOL; PENTYL ALCOHOL; 1-PENTANOL

IDENTIFICATION
CAS Number: 71-41-0; can also apply to 6032-29-7 (*sec*-isomer)
Formula: $C_5H_{12}O$; $CH_3(CH_2)_3CH_2OH$
DOT ID Number: UN 1105; **DOT Guide Number:** 129
Proper Shipping Name: Amyl alcohols

DESCRIPTION: Colorless liquid. Sweet, pleasant odor, like alcohol; causes coughing. Floats on water surface; slightly soluble.

Highly flammable • Containers may BLEVE when exposed to fire •Vapors may form explosive mixture with air •Vapors are heavier than air and will collect and stay in low areas • Vapors may travel long distances to ignition sources and flashback • Vapors in confined areas (e.g., tanks, sewers, buildings) may explode when exposed to fire • Irritating to the skin, eyes, and respiratory tract • Toxic products of combustion may include carbon monoxide.

Hazard Classification (based on NFPA-704 Rating System)
Health Hazards (Blue): 1; Flammability (Red): 3; Reactivity (Yellow): 0

EMERGENCY RESPONSE: See Appendix A (129)
Evacuation:
Public safety: Isolate spill area for at least 50 to 100 meters (160 to 330 feet) in all directions.
Spill: Large spill–Consider initial downwind evacuation for at least 300 meters (1000 feet).
Fire: Isolate for 800 meters (½ mile) in all directions, especially if tank, rail car, or tank truck is involved in fire.

EXPOSURE
Short-term effects: *SEEK MEDICAL ATTENTION.* **Vapor:** Irritating to eyes, nose, and throat. *IF INHALED*, will, will cause coughing, nausea, headache, or difficult breathing. Move to fresh air. *IF BREATHING HAS STOPPED,* give artificial respiration. IF breathing is difficult, administer oxygen. **Liquid:** Irritating to eyes and skin. Harmful if swallowed.
Flush affected areas with plenty of water. *IF IN EYES*, hold eyelids open and flush with plenty of water. *IF SWALLOWED* and victim is *CONSCIOUS AND ABLE TO SWALLOW*, have victim drink 4 to 8 ounces of water. The use of alcoholic beverages may enhance the toxic effect.

HEALTH HAZARDS
Personal protective equipment (PPE): B-Level PPE. Chemical protective material(s) reported to have good to excellent resistance: PV alcohol, neoprene, neoprene+styrene-butadiene, polyurethane, styrene-butadiene, styrene-butadiene/neoprene, butyl rubber, neoprene+natural rubber, Viton®, Teflon®.
Short-term exposure limits (15-minute TWA): CHRIS 150 ppm for 30 minutes.
Toxicity by ingestion: Grade 2; LD_{50} = 0.5 to 5 g/kg.
Long-term health effects: May cause liver damage.
Vapor (gas) irritant characteristics: Vapors cause a smarting of the eyes or respiratory system if present in high concentrations. May cause unconsciousness.
Liquid or solid irritant characteristics: No appreciable hazard. Practically harmless to the skin.
Odor threshold: 0.12 ppm.

FIRE DATA
Flash point: 91°F/33°C (cc); 93°F/34°C (cc) (*sec*-isomer).
Flammable limits in air: LEL: 1.2%; UEL: 10.3%.
Fire extinguishing agents not to be used: Water may be ineffective.
Autoignition temperature: 572°F/300°C/573°K; 644°F/340°C/613°K (*sec*-isomer).
Electrical hazard: Class I, Group D.
Burning rate: 3.6 mm/min

CHEMICAL REACTIVITY
Binary reactants: Violent reaction with oxidizers, alkali metals.
Reactivity group: 20
Compatibility class: Alcohols, glycols

ENVIRONMENTAL DATA
Food chain concentration potential: Log P_{ow} = 1.30–1.4. Unlikely to accumulate.

Food chain concentration potWater pollution: Effect of low concentrations on aquatic life is unknown. Fouling to shoreline. May be dangerous if it enters nearby water intakes; notify operators. Notify local health and pollution control officials. **Response to discharge:** Disperse and flush.

SHIPPING INFORMATION
Grades of purity: 98%; 74% plus 25% 2-methyl-l-butanol; **Storage temperature:** Ambient; **Inert atmosphere:** None; **Venting:** Open (flame arrester); **Stability during transport:** Stable.

NAS HAZARD CLASSIFICATION FOR BULK WATER TRANSPORTATION
FIRE: 3
HEALTH: Vapor irritant: 1; Liquid or solid irritant: 0; Poisons: 2
WATER POLLUTION: Human toxicity: 2; Aquatic toxicity: 2; Aesthetic effect: 2
REACTIVITY: Other chemicals: 2; Water: 0; Self-reaction: 0

PHYSICAL AND CHEMICAL PROPERTIES
Physical state @ 59°F/15°C and 1 atm: Liquid.
Molecular weight: 88.15
Boiling point @ 1 atm: 280.2°F/137.9°C/411.1°K; 245°F/119°C/392°K (*sec*-isomer).
Melting/Freezing point: −110°F/−79°C/194°K; −58°F/−50°C/223.2°K (*sec*-isomer).
Critical temperature: 595°F/313°C/586°K.
Specific gravity (water = 1): 0.818 @ 15°C (liquid).
Liquid surface tension: 25.60 dynes/cm = 0.02560 N/m @ 68°F/20°C.
Liquid water interfacial tension: 5 dynes/cm = 0.005 N/m @ 68°F/20°C.
Relative vapor density (air = 1): 3.04
Ratio of specific heats of vapor (gas): 1.06
Latent heat of vaporization: 217.1 Btu/lb = 120.6 cal/g = 5.049 x 10^5 J/kg.
Heat of combustion: −16,200 Btu/lb = −9000 cal/g = −376.8 x 10^5 J/kg.
Vapor pressure: (Reid) 0.23 psia; 2.29 mm.

***n*-AMYL CHLORIDE** REC. A:4400

SYNONYMS: AMYL CHLORIDE; 1-CHLORPENTANE; CLORURO de *n*-AMILO (Spanish); *n*-BUTYLCARBINYL CHLORIDE; 1-PENTYL CHLORIDE; CHLORIDE OF AMYL

IDENTIFICATION
CAS Number: 543-59-9
Formula: $CH_3CH_2CH_2CH_2CH_2Cl$
DOT ID Number: UN 1107; DOT Guide Number: 129
Proper Shipping Name: Amyl chlorides

DESCRIPTION: Colorless to purple liquid. Pleasant, aromatic odor. Floats on water surface; insoluble. Flammable vapor is produced.

Highly flammable • Containers may BLEVE when exposed to fire • Vapors may form explosive mixture with air • Vapors are heavier than air and will collect and stay in low areas • Vapors may travel long distances to ignition sources and flashback • Vapors in confined areas (e.g., tanks, sewers, buildings) may explode when exposed to fire • Irritating to the skin, eyes, and respiratory tract •

Toxic products of combustion may include hydrogen chloride and phosgene gas.

Hazard Classification (based on NFPA-704 Rating System)
Health Hazards (Blue): 1; Flammability (Red): 3; Reactivity (Yellow): 0

EMERGENCY RESPONSE: See Appendix A (129)
Evacuation:
Public safety: Isolate spill area for at least 50 to 100 meters (160 to 330 feet) in all directions.
Spill: Large spill–Consider initial downwind evacuation for at least 300 meters (1000 feet).
Fire: Isolate for 800 meters (½ mile) in all directions, especially if tank, rail car, or tank truck is involved in fire.

EXPOSURE
Short-term effects: *SEEK MEDICAL ATTENTION*. **Vapor:** Irritating to eyes, nose, and throat. Move victim to fresh air.
Liquid: Irritating to skin and eyes. Remove contaminated clothing and shoes. Flush affected areas with plenty of water. *IF IN EYES*, hold eyelids open and flush with plenty of water. *IF SWALLOWED* and victim is *CONSCIOUS AND ABLE TO SWALLOW*, have victim drink 4 to 8 ounces of water.

HEALTH HAZARDS
Personal protective equipment (PPE): B-Level PPE. Protective goggles or face shield; rubber gloves.
Toxicity by ingestion: Grade 1; LD_{50} = 5 to 15 g/kg.
Vapor (gas) irritant characteristics: Vapors cause smarting of the eyes or respiratory system if present in high concentrations. The effect is temporary.
Liquid or solid irritant characteristics: If spilled on clothing and allowed to remain, may cause smarting and reddening of the skin.

FIRE DATA
Flash point: 55°F/13°C (oc); 34°F/1°C (cc).
Flammable limits in air: LEL: 1.4%; UEL: 8.6%.
Fire extinguishing agents not to be used: Water may be ineffective.
Autoignition temperature: 500°F/260°C/533°K.
Burning rate: 4.9 mm/min

ENVIRONMENTAL DATA
Water pollution: Effect of low concentrations on aquatic life is unknown. Fouling to shoreline. May be dangerous if it enters nearby water intakes; notify operators. Notify local health and wildlife officials. **Response to discharge:** Mechanical containment. Chemical and physical treatment.

SHIPPING INFORMATION
Grades of purity: Commercial; **Storage temperature:** Ambient; **Inert atmosphere:** None; **Venting:** Open (flame arrester); **Stability during transport:** Stable.

NAS HAZARD CLASSIFICATIONS FOR BULK WATER TRANSPORTATION
FIRE: 3
HEALTH: Vapor irritant: 1; Liquid or solid irritant: 1; Poisons: 2
WATER POLLUTION: Human toxicity: 1; Aquatic toxicity: 2; Aesthetic effect: 2
REACTIVITY: Other chemicals: 1; Water: 0; Self-reaction: 0

PHYSICAL AND CHEMICAL PROPERTIES
Physical state @ 59°F/15°C and 1 atm: Liquid.
Molecular weight: 106.6
Boiling point @ 1 atm: 226°F/108°C/381°K.
Melting/Freezing point: –146°F/–99°C/174°K.
Specific gravity (water = 1): 0.8834 @ 68°F/20°C (liquid).
Liquid surface tension: 24.9 dynes/cm = 0.0249 N/m @ 68°F/20°C.
Liquid water interfacial tension: (estimate) 35 dynes/cm = 0.035 N/m @ 68°F/20°C.
Relative vapor density (air = 1): 3.7
Ratio of specific heats of vapor (gas): 1.0650
Latent heat of vaporization: 132.1 Btu/lb = 73.40 cal/g = 3.073 x 10^5 J/kg.
Heat of combustion: –13,500 Btu/lb = –7500 cal/g = –314 x 10^5 J/kg.

n-AMYL MERCAPTAN   REC. A:4450

SYNONYMS: AMILMERCAPTANO (Spanish); AMYL HYDROSULFIDE; AMYL MERCAPTAN; AMYL SULFHYDRATE; AMYL THIOALCOHOL; 1-PENTANETHIOL; PENTYL MERCAPTAN

IDENTIFICATION
CAS Number: 110-66-7
Formula: $C_5H_{12}S$; $CH_3CH_2CH_2CH_2CH_2SH$
DOT ID Number: UN 1111; DOT Guide Number: 130
Proper Shipping Name: Amyl mercaptans

DESCRIPTION: Colorless to light yellow liquid. Strong, garlic odor. Floats on water surface; insoluble; reacts, forming flammable mercaptan vapors.

Highly flammable • Containers may BLEVE when exposed to fire • Vapors may form explosive mixture with air • Vapors are heavier than air and will collect and stay in low areas • Vapors may travel long distances to ignition sources and flashback • Vapors in confined areas (e.g., tanks, sewers, buildings) may explode when exposed to fire • Irritating to the skin, eyes, and respiratory tract • Toxic products of combustion may include sulfur dioxide gas.

Hazard Classification (based on NFPA-704 Rating System)
Health Hazards (Blue): 2; Flammability (Red): 3; Reactivity (Yellow): 0

EMERGENCY RESPONSE: See Appendix A (130)
Evacuation:
Public safety: Isolate spill area for at least 50 to 100 meters (160 to 330 feet) in all directions.
Spill: Large spill–Consider initial downwind evacuation for at least 300 meters (1000 feet).
Fire: Isolate for 800 meters (½ mile) in all directions, especially if tank, rail car, or tank truck is involved in fire.

EXPOSURE
Short-term effects: *SEEK MEDICAL ATTENTION*. **Vapor:** Move victim to fresh air. **Liquid:** Irritating to skin and eyes. Remove contaminated clothing and shoes. Flush affected areas with plenty of water. *IF IN EYES*, hold eyelids open and flush with plenty of water. *IF SWALLOWED* and victim is *CONSCIOUS AND ABLE TO SWALLOW*, have victim drink 4 to 8 ounces of water.

HEALTH HAZARDS
Personal protective equipment (PPE): B-Level PPE. Plastic

gloves; goggles. Natural rubber, neoprene, and nitrile rubber may offer some protection from alcohols.
Recommendations for respirator selection: NIOSH
5 ppm: CCRFOV [any chemical cartridge respirator with a full facepiece and organic vapor cartridges(s)]; or SA (any supplied-air respirator). *12.5 ppm:* SA:CF (any supplied-air respirator operated in a continuous-flow mode); or PAPROV [any powered, air-purifying respirator with organic vapor cartridge(s)]. *25 ppm:* CCRFOV [any chemical cartridge respirator with a full facepiece and organic vapor cartridges(s)]; or GMFOV [any air-purifying, full-facepiece respirator (gas mask) with a chin-style, front- or back-mounted acid gas canister]; or PAPRTOV [any powered, air-purifying respirator with a tight-fitting facepiece and organic vapor cartridges(s)]; SCBAF (any self-contained breathing apparatus with a full facepiece); or SAF (any supplied-air respirator with a full facepiece). *EMERGENCY OR PLANNED ENTRY INTO UNKNOWN CONCENTRATIONS OR IDLH CONDITIONS:* SCBAF:PD,PP (any self-contained breathing apparatus that has a full facepiece and is operated in a pressure-demand or other positive-pressure mode); or SAF:PD,PP:ASCBA (any supplied-air respirator that has a full facepiece and is operated in a pressure-demand or other positive-pressure mode in combination with an auxiliary self-contained breathing apparatus operated in a pressure-demand or other positive pressure mode). *ESCAPE:* GMFOV[any air-purifying, full-facepiece respirator (gas mask) with a chin-style, front- or back-mounted organic vapor cannister]; or SCBAE (any appropriate escape-type, self-contained breathing apparatus).
Exposure limits (TWA unless otherwise noted): NIOSH ceiling 0.5 ppm (2.1 mg/m^3)/15 minutes
Liquid or solid irritant characteristics: May cause eye and skin irritation.
Odor threshold: 0.3 mg/m^3.

FIRE DATA
Flash point: 65°F/18°C (oc).
Fire extinguishing agents not to be used: Water may be ineffective, but spray can be used to cool containers exposed to fire.
Behavior in fire: Floats on water surface; may travel to source of ignition and spread fire.
Electrical hazard: Class I, Group D.
Burning rate: 4.7 mm/min

CHEMICAL REACTIVITY
Binary reactants: Reacts with Oxidizers, reducing agents, alkali metals, calcium hypochlorite, concentrated nitric acid.

ENVIRONMENTAL DATA
Water pollution: DOT Appendix B, §172.101–marine pollutant. Effect of low concentrations on aquatic life is unknown. Fouling to shoreline. May be dangerous if it enters nearby water intakes; notify operators. Notify local health and wildlife officials.
Response to discharge: Issue warning–high flammability. Mechanical containment. Should be removed. Chemical and physical treatment.

SHIPPING INFORMATION
Grades of purity: 96.0+%; **Storage temperature:** Ambient; **Inert atmosphere:** None; **Venting:** Pressure-vacuum; **Stability during transport:** Stable.

PHYSICAL AND CHEMICAL PROPERTIES
Physical state @ 59°F/15°C and 1 atm: Liquid.
Molecular weight: 104.2
Boiling point @ 1 atm: 260°F/127°C/400°K.
Melting/Freezing point: –105°F/–76°C/197°K.
Critical temperature: 610°F/321°C/594°K.
Critical pressure: 508 psia = 34.5 atm = 3.50 MN/m^2.
Specific gravity (water = 1): 0.8392 @ 77°F/25°C (liquid).
Liquid surface tension: 26.8 dynes/cm = 0.0268 N/m @ 68°F/20°C.
Liquid water interfacial tension: (estimate) 35 dynes/cm = 0.035 N/m @ 68°F/20°C.
Relative vapor density (air = 1): 3.59
Ratio of specific heats of vapor (gas): 1.0622
Latent heat of vaporization: –171 Btu/lb = –94.9 cal/g = –3.97 x 10^5 J/kg.
Heat of combustion: –17,070 Btu/lb = –9,480 cal/g = –397 x 10^5 J/kg.
Vapor pressure: 14 mm @ 77°F/25°C.

n-AMYL METHYL KETONE   REC. A:4500

SYNONYMS: *n*-AMILMETILCETONA (Spanish); AMYL-METHYL-CETONE (French); AMYL METHYL KETONE; METHYL AMYL KETONE; METHYL-AMYL-CETONE (French); 2-HEPTANONE; 2-KETOHEPTANE; METIL *n*-AMIL CETONA (Spanish); METHYL PENTYL KETONE; PENTYL METHYL KETONE; METHYL *n*-AMYL KETONE

IDENTIFICATION
CAS Number: 110-43-0
Formula: $C_{13}H_{26}O_4$; $CH_3CH_2CH_2CH_2CH_2COCH_3$
DOT ID Number: UN 1110; DOT Guide Number: 127
Proper Shipping Name: *n*-Amyl methyl ketone

DESCRIPTION: Colorless or white liquid. Penetrating fruity odor. Floats and mixes slowly with water.

Highly flammable • Containers may BLEVE when exposed to fire • Vapors may form explosive mixture with air • Vapors are heavier than air and will collect and stay in low areas • Vapors may travel long distances to ignition sources and flashback • Vapors in confined areas (e.g., tanks, sewers, buildings) may explode when exposed to fire • Irritating to the skin, eyes, and respiratory tract • Toxic products of combustion may include carbon monoxide.

Hazard Classification (based on NFPA-704 Rating System)
Health Hazards (Blue): 1; Flammability (Red): 2; Reactivity (Yellow): 0

EMERGENCY RESPONSE: See Appendix A (127)
Evacuation:
Public safety: Isolate spill area for at least 25 to 50 meters (80 to 160 feet) in all directions.
Spill: Large spill–Consider initial downwind evacuation for at least 300 meters (1000 feet).
Fire: Isolate for 800 meters (½ mile) in all directions, especially if tank, rail car, or tank truck is involved in fire.

EXPOSURE
Short-term effects: *SEEK MEDICAL ATTENTION.* **Vapor:** Irritating to eyes, nose, and throat. *IF INHALED*, will, will cause dizziness, headache, or difficult breathing. *IF IN EYES*, hold eyelids open and flush with plenty of water. *IF BREATHING HAS STOPPED,* give artificial respiration. IF breathing is difficult, administer oxygen. **Liquid:** Irritating to skin and eyes. *IF SWALLOWED*, will cause nausea or vomiting. Remove contaminated clothing and shoes. Flush affected areas with plenty of water. *IF IN EYES*, hold eyelids open and flush with plenty of

water. *IF SWALLOWED* and victim is *CONSCIOUS AND ABLE TO SWALLOW*, have victim drink 4 to 8 ounces of water. *IF SWALLOWED* and victim is *UNCONSCIOUS OR HAVING CONVULSIONS*, do nothing except keep victim warm.

HEALTH HAZARDS
Personal protective equipment (PPE): B-Level PPE. OSHA Table Z-1-A air contaminant.
Recommendations for respirator selection: NIOSH/OSHA *800 ppm*: CCROV* [any chemical cartridge respirator with organic vapor cartridge(s)]; or PAPROV* [any powered, air-purifying respirator with organic vapor cartridge(s)]; or GMFOV [any air-purifying, full-facepiece respirator (gas mask) with a chin-style, front- or back-mounted acid gas canister]; or SA* (any supplied-air respirator); or SCBAF (any self-contained breathing apparatus with a full facepiece). *EMERGENCY OR PLANNED ENTRY INTO UNKNOWN CONCENTRATIONS OR IDLH CONDITIONS:* SCBAF:PD,PP (any self-contained breathing apparatus that has a full faceplate and is operated in a pressure-demand or other positive-pressure mode); or SAF:PD,PP:ASCBA (any supplied-air respirator that has a full facepiece and is operated in a pressure-demand or other positive-pressure mode in combination with an auxiliary self-contained breathing apparatus operated in a pressure-demand or other positive pressure mode). *ESCAPE:* GMFOV[any air-purifying, full-facepiece respirator (gas mask) with a chin-style, front- or back-mounted organic vapor canister or SCBAE (any appropriate escape-type, self-contained breathing apparatus). *Note:* Substance reported to cause eye irritation or damage; may require eye protection.
Exposure limits (TWA unless otherwise noted): ACGIH TLV 50 ppm (233 mg/m^3); NIOSH/OSHA 100 ppm (465 mg/m^3).
Toxicity by ingestion: Grade 2; oral LD$_{50}$ = 1670 mg/kg (rat).
Liquid or solid irritant characteristics: Causes irritation of the eyes and skin.
Odor threshold: 0.896 ppm.
IDLH value: 800 ppm.

FIRE DATA
Flash point: 117°F/47°C (oc); 102°F/39°C (cc).
Flammable limits in air: LEL: 1.11% @ 151°F/66°C; UEL: 7.9% @ 250°F/121°C.
Fire extinguishing agents not to be used: Water may be ineffective.
Autoignition temperature: 740°F/393°C/666°K.

CHEMICAL REACTIVITY
Binary reactants: Will attack some forms of plastic. Reacts with strong acids, alkalis, and oxidizers.

ENVIRONMENTAL DATA
Water pollution: Effect of low concentrations on aquatic life is unknown. Fouling to shoreline. May be dangerous if it enters nearby water intakes; notify operators. Notify local health and wildlife officials. **Response to discharge:** Issue warning–water contaminant. Mechanical containment. Should be removed. Chemical and physical treatment.

SHIPPING INFORMATION
Grades of purity: Technical; Pure; **Storage temperature:** Ambient; **Inert atmosphere:** Ventilated (natural); **Venting:** Open (flame arrester); **Stability during transport:** Stable.

PHYSICAL AND CHEMICAL PROPERTIES
Physical state @ 59°F/15°C and 1 atm: Liquid.
Molecular weight: 114.19
Boiling point @ 1 atm: 304.7°F/151.5°C/424.7°K.
Melting/Freezing point: –31°F/–35°C/238°K.
Specific gravity (water = 1): 0.8204 @ 15°C (liquid).
Liquid surface tension: 26.17 dynes/cm = 0.02617 N/m @ 77°F/25°C.
Liquid water interfacial tension: Melting/Freezing point: Dynes/cm = 0.0124 N/m @ 77°F/25°C.
Relative vapor density (air = 1): 3.94
Ratio of specific heats of vapor (gas): (estimate) 1.051
Latent heat of vaporization: 148.9 Btu/lb = 82.7 cal/g = 3.46 x 10^5 J/kg.
Vapor pressure: 3 mm.

n-AMYL NITRATE REC. A:4550

SYNONYMS: DIESEL IGNITION IMPROVER; MIXED PRIMARY AMYL NITRATES; NITRATE d'AMYLE (French); NITRATO de AMILO (Spanish); NITRIC ACID, PENTYL ESTER

IDENTIFICATION
CAS Number: 1002-16-0
Formula: C$_5$H$_{11}$ONO$_2$
DOT ID Number: UN 1112; DOT Guide Number: 140
Proper Shipping Name: Amyl nitrate

DESCRIPTION: Colorless to light straw liquid. Ether-like odor. May float or sink in water.

Highly flammable • Containers may BLEVE when exposed to fire • Vapors may form explosive mixture with air • Vapors are heavier than air and will collect and stay in low areas • Vapors may travel long distances to ignition sources and flashback • Vapors in confined areas (e.g., tanks, sewers, buildings) may explode when exposed to fire • Irritating to the skin, eyes, and respiratory tract; absorbed by the skin • A strong oxidizer. Heat forms oxygen; will increase the activity of an existing fire • May cause fire or explosion on contact with combustibles (wood, paper, oil, clothing, etc.). Toxic products of combustion may include nitrogen oxides.

Hazard Classification (based on NFPA-704 Rating System)
Health Hazards (Blue): 2; Flammability (Red): 2; Reactivity (Yellow): 0; Special Notice (White): OXY

EMERGENCY RESPONSE: See Appendix A (140)
Evacuation:
Public safety: Isolate the area of spill or leak for at least 10 to 25 meters (30 to 80 feet) in all directions.
Spill: Consider initial downwind evacuation for at least 100 meters (330 feet).
Fire: If any large container is involved in fire, isolate for at least 800 meters (½ mile) in all directions; also, consider initial evacuation for 800 meters (½ mile) in all directions.

EXPOSURE
Short-term effects: *SEEK MEDICAL ATTENTION.* **Vapor:** Irritating to eyes, nose, and throat. *IF INHALED*, will, will cause headache. *IF IN EYES*, hold eyelids open and flush with plenty of water. *IF BREATHING HAS STOPPED,* give artificial respiration. IF breathing is difficult, administer oxygen. Lung edema may develop. **Liquid:** Irritating to skin and eyes. *IF SWALLOWED*, will cause nausea or headache. Remove contaminated clothing and shoes. Flush affected areas with plenty of water. *IF IN EYES*, hold eyelids open and flush with plenty of water. *IF SWALLOWED* and victim is *CONSCIOUS AND ABLE TO SWALLOW*, have victim

drink 4 to 8 ounces of water and have victim induce vomiting. *IF SWALLOWED* and victim is *UNCONSCIOUS OR HAVING CONVULSIONS,* do nothing except keep victim warm.

Note to physician or authorized medical personnel: Medical observation is recommended for 24 to 48 hours after breathing overexposure, as pulmonary edema may be delayed. As first aid for pulmonary edema, consider administering a corticosteroid spray. Cigarette smoking may exacerbate pulmonary injury and should be discouraged for at least 72 hours following exposure.

HEALTH HAZARDS
Personal protective equipment (PPE): B-Level PPE. Respirator with canister for vapors at high concentrations.

FIRE DATA
Flash point: 120°F/49°C (oc).
Behavior in fire: Overheated material may detonate.
Electrical hazard: Class I, Group Unassigned

CHEMICAL REACTIVITY
Binary reactants: Reacts with reducing agents and acids. Form combustible mixture with wood or other combustibles. Attacks some plastics.

ENVIRONMENTAL DATA
Water pollution: Effect of low concentrations on aquatic life is unknown. Fouling to shoreline. May be dangerous if it enters nearby water intakes; notify operators. Notify local health and wildlife officials. Notify operators or nearby water intakes.
Response to discharge: Issue warning-air contaminant. Restrict access. Mechanical containment. Should be removed. Chemical and physical treatment.

SHIPPING INFORMATION
Grades of purity: Mixture containing *n*-amyl nitrate, 60%; *iso*-amyl nitrate, 5%, 2 methylbutyl nitrate, 35%; **Storage temperature:** Ambient; **Inert atmosphere:** None; **Venting:** Open; **Stability during transport:** Stable.

PHYSICAL AND CHEMICAL PROPERTIES
Physical state @ 59°F/15°C and 1 atm: Liquid.
Molecular weight: 133
Boiling point @ 1 atm: 292–314°F/144–156°C/417–429°K.
Melting/Freezing point: –190°F/–123°C/150°K.
Specific gravity (water = 1): 1.0 @ 68°F/20°C (liquid).
Relative vapor density (air = 1): 4.59

AMYL NITRITE   REC. A:4600

SYNONYMS: AMYL NITRITE; ISOPENTYL NITRITE; 3-METHYLBUTYL NITRITE; NITRITO de AMILO (Spanish); 1-NITROPENTANE; NITROUS ACID, PENTYL ESTER; PENTYL NITRITE

IDENTIFICATION
CAS Number: 463-04-7; 110-46-3 (*iso-*)
Formula: $(CH_3)_2CHCH_2CH_2ONO$; $C_5H_{11}NO_2$ (*iso-*)
DOT ID Number: UN 1113; DOT Guide Number: 129
Proper Shipping Name: Amyl nitrites

DESCRIPTION: Colorless to light yellow liquid. Pleasant, fruity odor. Floats on the surface of water. Poisonous gas (nitrogen oxide that is orange in color) is produced on contact with water.

Highly flammable • Containers may BLEVE when exposed to fire • Vapors may form explosive mixture with air • Vapors are heavier than air and will collect and stay in low areas • Vapors may travel long distances to ignition sources and flashback • Vapors in confined areas (e.g., tanks, sewers, buildings) may explode when exposed to fire • Irritating to the skin, eyes, and respiratory tract • Corrodes metals in the presence of moisture • Toxic products of combustion may include nitrogen oxides.

Hazard Classification (based on NFPA-704 Rating System)
Health Hazards (Blue): 1; Flammability (Red): 3; Reactivity (Yellow): 2

EMERGENCY RESPONSE: See Appendix A (129)
Evacuation:
Public safety: Isolate spill area for at least 50 to 100 meters (160 to 330 feet) in all directions.
Spill: Large spill–Consider initial downwind evacuation for at least 300 meters (1000 feet).
Fire: Isolate for 800 meters (½ mile) in all directions, especially if tank, rail car, or tank truck is involved in fire.

EXPOSURE
Short-term effects: *SEEK MEDICAL ATTENTION.* **Vapor:** *POISONOUS IF INHALED.*
Causes fall in blood pressure, headache, pulse throbbing, and weakness. Irritating to eyes, nose, and throat. Move to fresh air. *IF BREATHING HAS STOPPED,* give artificial respiration. IF breathing is difficult, administer oxygen. **Liquid:** Will burn skin and eyes. *IF SWALLOWED*, will cause dizziness, headache or loss of consciousness. Remove contaminated clothing and shoes. Flush affected areas with plenty of water. *IF IN EYES*, hold eyelids open and flush with plenty of water. *IF SWALLOWED* and victim is *CONSCIOUS AND ABLE TO SWALLOW*, have victim drink 4 to 8 ounces of water. **Do NOT induce vomiting**.

Note to physician or authorized medical personnel: Give milk and demulcents, induce emesis or perform gastric lavage: Give fluids: observe for methemoglobinemia. If needed, give methylene blue as a 1% solution intravenously, 1–2 mg/kg; an oral dose of 3–5 mg/kg. If severe, consider exchange transfusion with whole blood.

HEALTH HAZARDS
Personal protective equipment (PPE): B-Level PPE. Protective goggles or face shield; self-contained breathing apparatus; protective gloves and clothing.
Toxicity by ingestion: Grade 1; LD_{50} = 5 to 15 g/kg.
Long-term health effects: Methemoglobinemia may occur.
Vapor (gas) irritant characteristics: Mild irritant, but harmful if inhaled.
Liquid or solid irritant characteristics: Severe skin irritant. Causes second- and third-degree burns on short contact and is very injurious to the eyes.

FIRE DATA
Flash point: 0°F/–18°C (oc).
Fire extinguishing agents not to be used: Water.
Behavior in fire: Vapors may explode. Containers may rupture and explode.
Autoignition temperature: 410°F/210°C/483°K.
Electrical hazard: Class I, Group Unassigned.
Burning rate: 3.4 mm/min

CHEMICAL REACTIVITY
Reactivity with water: Decomposes on exposure to air, light, or water, evolving toxic oxides of nitrogen which are orange in color.

Binary reactants: A strong oxidizer. May be an explosion hazard when exposed to air and light. May corrode metals in the presence of moisture.

ENVIRONMENTAL DATA
Water pollution: Effect of low concentrations on aquatic life is unknown. Fouling to shoreline. May be dangerous if it enters nearby water intakes; notify operators. Notify local health and wildlife officials. **Response to discharge:** Issue warning–high flammability. Restrict access. Neutralize with lime, soda ash, crushed limestone. Disperse and flush.

SHIPPING INFORMATION
Grades of purity: Commercial; USP; **Storage temperature:** Ambient; **Inert atmosphere:** None; **Venting:** Pressure-vacuum. **Stability during transport:** Stable if kept sealed and not exposed to light.

NAS HAZARD CLASSIFICATIONS FOR BULK WATER TRANSPORTATION
FIRE: 3
HEALTH: Vapor irritant: 0; Liquid or solid irritant: 4; Poisons: 1
WATER POLLUTION: Human toxicity: 1; Aquatic toxicity: 1; Aesthetic effect: 2
REACTIVITY: Other chemicals: 3; Water: 0; Self-reaction: 4

PHYSICAL AND CHEMICAL PROPERTIES
Physical state @ 59°F/15°C and 1 atm: Liquid.
Molecular weight: 117.1
Boiling point @ 1 atm: 210°F/99°C/372°K.
Specific gravity (water = 1): 0.8758 @ 68°F/20°C (liquid).
Liquid surface tension: (estimate) 20 dynes/cm = 0.020 N/m @ 68°F/20°C.
Liquid water interfacial tension: (estimate) 40 dynes/cm = 0.040 N/m @ 68°F/20°C.
Relative vapor density (air = 1): 4
Ratio of specific heats of vapor (gas): 1.0709
Latent heat of vaporization: 212 Btu/lb = 118 cal/g = 4.94×10^5 J/kg.
Heat of combustion: –12,500 Btu/lb = –6930 cal/g = -290×10^5 J/kg.

n-AMYL TRICHLOROSILANE   REC. A:4700

SYNONYMS: AMILTRICLOROSILANO (Spanish); AMYL TRICHLOROSILANE; PENTYLSILICON TRICHLORIDE; PENTYLTRICHLOROSILANE; SILANE,TRICHLOROPENTYL-; SILANE, PENTYLTRICHLORO-; TRICHLOROAMYLSILANE; TRICHLOROPENTYLSILANE

IDENTIFICATION
CAS Number: 107-72-2
Formula: $C_5H_{11}Cl_3Si$; $CH_3CH_2CH_2CH_2CH_2SiCl_3$
DOT ID Number: UN 1728; DOT Guide Number: 155
Proper Shipping Name: Amyltrichlorosilane

DESCRIPTION: Colorless to yellow liquid. Sharp irritating odor; like hydrochloric acid. Reacts violently with water forming hydrochloric acid and hydrogen chloride vapor.

Poison! • Do not use water • Breathing the vapor can kill you; severely irritating to skin, eyes, and respiratory tract; contact with the skin and eyes causes burns and blindness • Firefighting gear (including SCBA) does not provide adequate protection. If exposure occurs, remove and isolate gear immediately and thoroughly decontaminate personnel • Combustible • Containers may BLEVE when exposed to fire •Vapors may form explosive mixture with air •Vapors are heavier than air and will collect and stay in low areas • Vapors may travel long distances to ignition sources and flashback • Vapors in confined areas (e.g., tanks, sewers, buildings) may explode when exposed to fire • Toxic products of combustion may include carbon monoxide, hydrogen chloride and phosgene • Do not put yourself in danger by entering a contaminated area to rescue a victim.

Hazard Classification (based on NFPA-704 Rating System)
Health Hazards (Blue): 3; Flammability (Red): 2; Reactivity (Yellow): 2; Special Notice (White): Water reactive

EMERGENCY RESPONSE: See Appendix A (155)
Evacuation:
Public safety: Isolate the area of spill or leak for at least 50 to 100 meters (160 to 330 feet) in all directions.
Spill: Increase, in the downwind direction, as necessary, the distance shown under "Public Safety."
Fire: If tank, rail car, or tank truck is involved in fire, isolate for at least 800 meters (½ mile) in all directions; also, consider initial evacuation for 800 meters (½ mile) in all directions.

EXPOSURE
Short-term effects: *SEEK MEDICAL ATTENTION.* **Vapor:** Irritating to eyes, nose, and throat. *IF INHALED*, will, will cause difficult breathing. Move victim to fresh air. *IF BREATHING HAS STOPPED*, give artificial respiration; *avoid mouth-to-mouth resuscitation; use bag/mask apparatus.* IF breathing is difficult, administer oxygen. **Liquid:** Will burn skin and eyes. Harmful if swallowed. Remove contaminated clothing and shoes. Flush affected areas with plenty of water. *IF IN EYES*, hold eyelids open and flush with plenty of water. *IF SWALLOWED* and victim is *CONSCIOUS AND ABLE TO SWALLOW*, have victim drink 4 to 8 ounces of water. **Do NOT induce vomiting.**

HEALTH HAZARDS
Personal protective equipment (PPE): B-Level PPE. Acid-vapor-type respiratory protection; rubber gloves.
Recommendations for respirator selection: *EMERGENCY OR PLANNED ENTRY INTO UNKNOWN CONCENTRATIONS OR IDLH CONDITIONS:* SCBAF:PD,PP (any self-contained breathing apparatus that has a full facepiece and is operated in a pressure-demand or other positive-pressure mode); or SAF:PD,PP:ASCBA (any supplied-air respirator that has a full facepiece and is operated in a pressure-demand or other positive-pressure mode in combination with an auxiliary self-contained breathing apparatus operated in a pressure-demand or other positive pressure mode). *ESCAPE:* GMFAG [any air-purifying, full-facepiece respirator (gas mask) with a chin-style, front- or back-mounted organic vapor cannister]; or SCBAE (any appropriate escape-type, self-contained breathing apparatus). *Note:* Substance reported to cause eye irritation or damage; may require eye protection.
Toxicity by ingestion: Grade 2; oral rat LD_{50} = 2340 mg/kg.
Vapor (gas) irritant characteristics: Vapors are irritating such that personnel will not usually tolerate moderate or high vapor concentrations.
Liquid or solid irritant characteristics: Severe skin irritant. Causes second-and third-degree burns on short contact and is very injurious to the eyes.

FIRE DATA
Flash point: 145°F/62°C (oc).

Fire extinguishing agents not to be used: Water, foam
Behavior in fire: Difficult to extinguish. Re-ignition may occur.
Burning rate: 2.5 mm/min

CHEMICAL REACTIVITY
Reactivity with water: Reacts vigorously to generate toxic hydrogen chloride gas (hydrochloric acid).
Binary reactants: Corrodes metal.
Neutralizing agents for acids and caustics: After flushing with water, rinse with sodium bicarbonate solution or lime water.

ENVIRONMENTAL DATA
Water pollution: Effect of low concentrations on aquatic life is unknown. May be dangerous if it enters nearby water intakes; notify operators. Notify local health and wildlife officials.
Response to discharge: Issue warning–corrosive. Restrict access. Disperse and flush.

SHIPPING INFORMATION
Grades of purity: Commercial; **Storage temperature:** Ambient; **Inert atmosphere:** None; **Venting:** Pressure-vacuum; **Stability during transport:** Stable.

NAS HAZARD CLASSIFICATIONS FOR BULK WATER TRANSPORTATION
FIRE: 2
HEALTH: Vapor irritant: 3; Liquid or solid irritant: 4; Poisons: 3
WATER POLLUTION: Human toxicity: 3; Aquatic toxicity: 3; Aesthetic effect: 2
REACTIVITY: Other chemicals: 3; Water: 4; Self-reaction: 1

PHYSICAL AND CHEMICAL PROPERTIES
Physical state @ 59°F/15°C and 1 atm: Liquid.
Molecular weight: 205.6
Boiling point @ 1 atm: 320°F/160°C/433°K.
Specific gravity (water = 1): 1.137 @ 77°F/25°C (liquid).
Liquid surface tension: (estimate) 20 dynes/cm = 0.020 N/m @ 68°F/20°C.
Relative vapor density (air = 1): 7.1
Latent heat of vaporization: (estimate) 86.8 Btu/lb = 48.2 cal/g = 2.02×10^5 J/kg.
Heat of combustion: (estimate) $-6,630$ Btu/lb = $-3,680$ cal/g = -154×10^5 J/kg.
Heat of solution: (estimate) 180 Btu/lb = 100 cal/g = 4.0×10^5 J/kg.

ANILINE REC. A:4750

SYNONYMS: AMINOBENZENE; AMINOPHEN; ANILINA (Spanish); ANILINE OIL; BENZENEAMINE; BLUE OIL; EEC No. 612-008-00-7; PHENYLAMINE; RCRA No. U012

IDENTIFICATION
CAS Number: 62-53-3
Formula: $C_6N_5NH_2$
DOT ID Number: UN 1547; DOT Guide Number: 153
Proper Shipping Name: Aniline
Reportable Quantity (RQ): **(CERCLA)** 5000 lb/2270 kg

DESCRIPTION: Colorless to yellow-brown oily liquid. Turns brown on exposure to light. Musty, fishy odor. Sinks slowly in water; slightly soluble.

Very flammable • Poison! • Inhalation of vapors, ingestion of liquid, or skin or eye contact with liquid can cause severe illness. Can interfere with the body's ability to use oxygen • May be able to polymerize •Containers may BLEVE when exposed to fire •Vapors may form explosive mixture with air •Vapors are heavier than air and will collect and stay in low areas • Vapors may travel long distances to ignition sources and flashback • Vapors in confined areas (e.g., tanks, sewers, buildings) may explode when exposed to fire • Decomposes above 363°F/184°C. Toxic products of combustion may include nitrogen oxides. *Warning:* Odor is not a reliable indicator of the presence of toxic amounts of aniline.

Hazard Classification (based on NFPA-704 Rating System)
Health Hazards (Blue): 3; Flammability (Red): 2; Reactivity (Yellow): 0

EMERGENCY RESPONSE: See Appendix A (153)
Evacuation:
Public safety: Isolate the area of spill or leak for at least 25 to 50 meters (80 to 160 feet) in all directions.
Spill: Increase, in the downwind direction, as necessary, the distance shown under "Public Safety."
Fire: If tank, rail car, or tank truck is involved in fire, isolate for at least 800 meters (½ mile) in all directions; also, consider initial evacuation for 800 meters (½ mile) in all directions.

EXPOSURE
Short-term effects: *SEEK MEDICAL ATTENTION.* **Liquid:** *POISONOUS IF SWALLOWED OR IF ABSORBED THROUGH THE SKIN.* Irritating to eyes. Remove contaminated clothing and shoes. Flush affected areas with plenty of water. *IF BREATHING HAS STOPPED*, give artificial respiration; *avoid mouth-to-mouth resuscitation; use bag/mask apparatus. IF IN EYES*, hold eyelids open and flush with plenty of water. *IF SWALLOWED* and victim is *CONSCIOUS AND ABLE TO SWALLOW*, have victim drink 4 to 8 ounces of water.
Note to physician or authorized medical personnel: Aniline converts the Fe^{+2} in hemoglobin to Fe^{+3}, causing methemoglobinemia and impaired oxygen transport. Production of methemoglobin may continue for up to 20 hours following exposure. When methemoglobin levels are 15 to 30%, the patient may become bluish in color. If ingested, Administer a slurry of activated charcoal if it has not been given previously. If activated charcoal is not available, induce vomiting by administering and emetic such as syrup of ipecac.

HEALTH HAZARDS
Personal protective equipment (PPE): A-Level PPE. OSHA Table Z-1-A air contaminant. Chemical protective material(s) reported to have good to excellent resistance: butyl rubber and polyvinyl alcohol. Also, chlorinated polyethylene, Teflon®, Saranex®, Silvershield®, and neoprene+natural rubber offers limited protection.
Recommendations for respirator selection: NIOSH
At any concentrations above the NIOSH REL, or where there is no REL, at any detectable concentration: SCBAF:PD,PP (any self-contained breathing apparatus that has a full facepiece and is operated in a pressure-demand or other positive-pressure mode); or SAF:PD,PP:ASCBA (any supplied-air respirator that has a full facepiece and is operated in a pressure-demand or other positive-pressure mode in combination with an auxiliary self-contained breathing apparatus operated in a pressure-demand or other positive pressure mode). *ESCAPE:* GMFOV [any air-purifying, full-facepiece respirator (gas mask) with a chin-style, front-or back-mounted organic vapor canister]; or SCBAE (any appropriate escape-type, self-contained breathing apparatus).

Anisoyl chloride

Exposure limits (TWA unless otherwise noted): ACGIH TLV 2 ppm (7.6 mg/m^3); OSHA 5 ppm (19 mg/m^3); skin contact contributes significantly in overall exposure; NIOSH potential human carcinogen; reduce exposure to lowest feasible level.
Toxicity by ingestion: Grade 3; LD_{50} = 50 to 500 mg/kg.
Long-term health effects: Suspected carcinogen. May cause anemia, and damage to the brain, heart, and kidneys.
Vapor (gas) irritant characteristics: Vapors cause smarting of the eyes or respiratory system if present in high concentrations.
Liquid or solid irritant characteristics: If spilled on clothing and allowed to remain, may cause smarting and reddening of the skin.
Odor threshold: 0.5 ppm.
IDLH value: Potential human carcinogen; 100 ppm.

FIRE DATA
Flash point: 168°F/76°C (oc), 158°F/70°C (cc).
Flammable limits in air: LEL: 1.3%; UEL: 11%.
Autoignition temperature: 1140°F/616°C/889°K.
Electrical hazard: Class I, Group D.
Burning rate: 3.0 mm/min

CHEMICAL REACTIVITY
Binary reactants: Violent reaction with strong oxidizers, strong acids, toluene diisocyanate, alkalis. Corrosive to copper and its alloys.
Neutralizing agents for acids and caustics: Flush with water and rinse with dilute acetic acid.
Reactivity group: 9
Compatibility class: Aromatic amines

ENVIRONMENTAL DATA
Food chain concentration potential: Log P_{ow} = 0.90. Unlikely to accumulate.
Water pollution: Dangerous to aquatic life in high concentrations. May be dangerous if it enters nearby water intakes; notify operators. Notify local health and wildlife officials. **Response to discharge:** Issue warning–poison, water contaminant. Restrict access. Should be removed. Chemical and physical treatment.

SHIPPING INFORMATION
Grades of purity: Commercial: 99.5%; **Storage temperature:** Ambient; **Inert atmosphere:** None; **Venting:** Pressure-vacuum; **Stability during transport:** Stable.

NAS HAZARD CLASSIFICATIONS FOR BULK WATER TRANSPORTATION
FIRE: 1
HEALTH: Vapor irritant: 1; Liquid or solid irritant: 1; Poisons: 3
WATER POLLUTION: Human toxicity: 3; Aquatic toxicity: 2; Aesthetic effect: 4
REACTIVITY: Other chemicals: 3; Water: 0; Self-reaction: 0

PHYSICAL AND CHEMICAL PROPERTIES
Physical state @ 59°F/15°C and 1 atm: Liquid.
Molecular weight: 93.13
Boiling point @ 1 atm: (decomposes) 363.6°F/184.2°C/457.4°K.
Melting/Freezing point: 21°F/–6°C/267°K.
Critical temperature: 798°F/426°C/699°K.
Critical pressure: 770 psia = 52.4 atm = 5.31 MN/m^2.
Specific gravity (water = 1): 1.022 @ 68°F/20°C (liquid).
Liquid surface tension: 45.5 dynes/cm = 0.0455 N/m @ 68°F/20°C.
Liquid water interfacial tension: 5.8 dynes/cm = 0.0058 N/m @ 68°F/20°C.
Relative vapor density (air = 1): 3.22
Ratio of specific heats of vapor (gas): 1.1
Latent heat of vaporization: 198 Btu/lb = 110 cal/g = 4.61 x 10^5 J/kg.
Heat of combustion: –14,980 Btu/lb = –8320 cal/g = –348.3 x 10^5 J/kg.
Vapor pressure: (Reid) 0.02 psia; 0.6 mm.

ANISOYL CHLORIDE   REC. A:4800

SYNONYMS: *p*-ANISOYL CHLORIDE; CLORURO de ANISOILO (Spanish); METHOXYBENZOYL CHLORIDE

IDENTIFICATION
CAS Number: 100-07-2
Formula: *p*-CH$_3$OC$_6$H$_5$COCl; C$_8$H$_7$ClO$_2$
DOT ID Number: UN 1729; DOT Guide Number: 156
Proper Shipping Name: Anisoyl chloride

DESCRIPTION: Clear to yellow needle-like crystals or amber liquid. Sharp, penetrating odor. Insoluble in water; reacts, forming hydrochloric acid and evolving hydrogen chloride fumes.

Combustible solid • Corrosive • May become unstable after prolonged storage; may be shock-sensitive or explode spontaneously at room temperature • Containers may BLEVE when exposed to fire • Dust or vapors in confined areas (e.g., tanks, sewers, buildings) may explode when exposed to fire • Irritating to the skin, eyes, and respiratory tract • Toxic products of combustion may include phosgene gas and hydrogen chloride.

Hazard Classification (based on NFPA-704 Rating System)
Health Hazards (Blue): 1; Flammability (Red): 1; Reactivity (Yellow): 1

EMERGENCY RESPONSE: See Appendix A (156)
Evacuation:
Public safety: Isolate the area of spill or leak for at least 50 to 100 meters (160 to 330 feet) in all directions.
Spill: Increase, in the downwind direction, as necessary, the distance shown under "Public Safety."
Fire: If tank, rail car, or tank truck is involved in fire, isolate for at least 800 meters (½ mile) in all directions; also, consider initial evacuation for 800 meters (½ mile) in all directions.

EXPOSURE
Short-term effects: *SEEK MEDICAL ATTENTION.* **Vapor:** Irritating to eyes, nose, and throat. Move victim to fresh air. *IF BREATHING HAS STOPPED,* give artificial respiration; *avoid mouth-to-mouth resuscitation; use bag/mask apparatus.* **Liquid:** Irritating to skin and eyes. Harmful if swallowed. Remove contaminated clothing and shoes. Flush affected areas with plenty of water. *IF IN EYES,* hold eyelids open and flush with plenty of water. *IF SWALLOWED* and victim is *CONSCIOUS AND ABLE TO SWALLOW*, have victim drink 4 to 8 ounces of water.

HEALTH HAZARDS
Personal protective equipment (PPE): A-Level PPE. Protective clothing: sealed chemical suit of butyl rubber.
Vapor (gas) irritant characteristics: Causes irritation of the respiratory tract.
Liquid or solid irritant characteristics: Causes severe irritation of the eyes, skin, and respiratory tract.

FIRE DATA
Flash point: Combustible solid.
Fire extinguishing agents not to be used: Water, foam
Behavior in fire: Containers may rupture and explode.

CHEMICAL REACTIVITY
Reactivity with water: Reacts slowly to generate hydrogen chloride (hydrochloric acid).
Binary reactants: Corrodes metal slowly. May explode at room temperature.
Neutralizing agents for acids and caustics: Flush with water, rinse with sodium bicarbonate or lime solution.

ENVIRONMENTAL DATA
Water pollution: Effect of low concentrations on aquatic life is unknown. May be dangerous if it enters nearby water intakes; notify operators. Notify local health and wildlife officials.
Response to discharge: Issue warning–corrosive. Restrict access. Disperse and flush

SHIPPING INFORMATION
Grades of purity: Commercial; **Storage temperature:** Ambient; **Inert atmosphere:** None; **Venting:** Pressure-vacuum.
Stability during transport: Stable, but may be a storage hazard.

PHYSICAL AND CHEMICAL PROPERTIES
Physical state @ 59°F/15°C and 1 atm: Liquid.
Molecular weight: 171.6
Boiling point @ 1 atm: 504°F/262°C/535°K.
Melting/Freezing point: 71°F/22°C/295°K.
Specific gravity (water = 1): 1.26 @ 68°F/20°C (liquid).
Liquid surface tension: (estimate) 25 dynes/cm = 0.025 N/m @ 68°F/20°C.
Heat of combustion: (estimate) –10,500 Btu/lb = –5830 cal/g = –244 x 10^5 J/kg.
Heat of solution: (estimate) 90 Btu/lb = 50 cal/g = 2.1 x 10^5 J/kg.

ANTHRACENE   REC. A:4850

SYNONYMS: ANTHRACEN (German); ANTHRACENO (Spanish); ANTHRACIN; GREEN OIL; PARANAPHTHALENE; TETRA OLIVE N2G

IDENTIFICATION
CAS Number: 120-12-7
Formula: $C_{13}H_{10}$
DOT ID Number: UN 3077; DOT Guide Number: 171
Proper Shipping Name: Environmentally hazardous substances, solid, n.o.s.
Reportable Quantity (RQ): **(CERCLA)** 5000 lb/2270 kg

DESCRIPTION: Colorless to yellow crystals with blue fluorescence. Weak aromatic odor. Sinks in water; insoluble.

Combustible solid • Containers may BLEVE when exposed to fire • Concentrated dust in confined areas (e.g., tanks, sewers, buildings) may explode when exposed to fire • Irritating to the skin, eyes, and respiratory tract • Toxic products of combustion may include carbon monoxide.

Hazard Classification (based on NFPA-704 Rating System)
Health Hazards (Blue): 0; Flammability (Red): 1; Reactivity (Yellow): 0

EMERGENCY RESPONSE: See Appendix A (171)
Evacuation:
Public safety: Isolate the area of spill or leak for at least 10 to 25 meters (30 to 80 feet) in all directions.
Spill: Increase, in the downwind direction, as necessary, the distance shown under "Public Safety."
Fire: If any large container is involved in fire, isolate for at least 800 meters (½ mile) in all directions; also, consider initial evacuation for 800 meters (½ mile) in all directions.

EXPOSURE
Short-term effects: *SEEK MEDICAL ATTENTION*. **Dust:** Irritating to eyes, nose, and throat. *IF INHALED*, will, will cause coughing or difficult breathing. *IF IN EYES*, hold eyelids open and flush with plenty of water. *IF BREATHING HAS STOPPED*, give artificial respiration. IF breathing is difficult, administer oxygen.
Solid: Irritating to skin and eyes. Harmful if swallowed. Remove contaminated clothing and shoes. Flush affected areas with plenty of water. *IF IN EYES*, hold eyelids open and flush with plenty of water. *IF SWALLOWED* and victim is *CONSCIOUS AND ABLE TO SWALLOW*, have victim drink 4 to 8 ounces of water. *IF SWALLOWED* and victim is *UNCONSCIOUS OR HAVING CONVULSIONS*, do nothing except keep victim warm.

HEALTH HAZARDS
Personal protective equipment (PPE): B-Level PPE. Sealed chemical suit of butyl rubber. Polycarbonate is also resistant.
Recommendations for respirator selection: *EMERGENCY OR PLANNED ENTRY INTO UNKNOWN CONCENTRATIONS OR IDLH CONDITIONS:* SCBAF:PD,PP (any self-contained breathing apparatus that has a full facepiece and is operated in a pressure-demand or other positive-pressure mode); or SAF:PD,PP:ASCBA (any supplied-air respirator that has a full facepiece and is operated in a pressure-demand or other positive-pressure mode in combination with an auxiliary self-contained breathing apparatus operated in a pressure-demand or other positive pressure mode). *ESCAPE:* GMFOV[any air-purifying, full-facepiece respirator (gas mask) with a chin-style, front- or back-mounted organic vapor cannister]; or SCBAE (any appropriate escape-type, self-contained breathing apparatus).
Long-term health effects: May cause skin disorders.
Liquid or solid irritant characteristics: Irritation of the eyes, skin, and respiratory tract.

FIRE DATA
Flash point: 250°F/121°C (cc).
Behavior in fire: Containers may rupture and explode.
Autoignition temperature: 1004°F/540°C/813°K.

CHEMICAL REACTIVITY
Binary reactants: Reacts violently with strong oxidizers; fluorine (explosive), chromic acid, hypochlorites.

ENVIRONMENTAL DATA
Food chain concentration potential: Log K_{ow} = 4.45–4.50. Values > 3.0 are likely to bioconcentrate in aquatic organisms and other living tissue, especially in fats.
Water pollution: Effect of low concentrations on aquatic life is unknown. May be dangerous if it enters nearby water intakes; notify operators. Notify local health and wildlife officials. **Response to discharge:** Should be removed. Chemical and physical treatment.

SHIPPING INFORMATION
Grades of purity: Various fluorescence grades; Scintillation grade;

Technical grade, 90–98%; commercial grade 90–95%; **Storage temperature:** Ambient; **Inert atmosphere:** None; **Venting:** Open; **Stability during transport:** Stable.

PHYSICAL AND CHEMICAL PROPERTIES
Physical state @ 59°F/15°C and 1 atm: Solid.
Molecular weight: 178.23
Boiling point @ 1 atm: 646.2°F/341.2°C/614.4°K.
Melting/Freezing point: 422°F/217°C/490°K.
Specific gravity (water = 1): 1.24 @ 68°F/20°C (solid).
Relative vapor density (air = 1): 6.15
Heat of combustion: $-17,100$ Btu/lb $= -9510$ cal/g $= -398 \times 10^5$ J/kg.
Heat of fusion: 38.70 cal/g.
Vapor pressure: 1 mmHg @ 293°F/145°C/418°K. (sublimes).

ANTIMONY PENTACHLORIDE REC. A:4900

SYNONYMS: ANTIMONIC CHLORIDE; ANTIMONPENTACHLORID (German); ANTIMONY(V) CHLORIDE, ANTIMONY PERCHLORIDE; EEC No. 051-022-00-3; BUTTER OF ANTIMONY; PENTACHLOROANTIMONY; PENTACLORURO de ANTIMONIO (Spanish); TENTACHLORURE D'ANTIMOINE (French); PERCHLORURE D'ANTIMOINE (French)

IDENTIFICATION
CAS Number: 7647-18-9
Formula: $SbCl_5$
DOT ID Number: UN 1730 (liquid); UN 1731 (solutions); DOT Guide Number: 157
Proper Shipping Name: Antimony pentachloride liquid; antimony pentachloride solutions
Reportable Quantity (RQ): **(CERCLA)** 1000 lb/454 kg

DESCRIPTION: Colorless to medium brown; yellow; red-brown liquid. Unpleasant, pungent odor. Sinks in water; forming hydrochloric acid with release of heat and hydrogen chloride fumes.

Corrosive • Breathing the vapor can kill you; skin or eye contact causes severe burns that can lead to blindness; effects of contact or inhalation may be delayed • Firefighting gear (including SCBA) does not provide adequate protection. If exposure occurs, remove and isolate gear immediately and thoroughly decontaminate personnel • Toxic products of combustion may include antimony and hydrogen chloride.

Hazard Classification (based on NFPA-704 Rating System)
Health Hazards (Blue): 3; Flammability (Red): 0; Reactivity (Yellow): 1

EMERGENCY RESPONSE: See Appendix A (157)
Evacuation:
Public safety: Isolate the area of spill or leak for at least 50 to 100 meters (160 to 330 feet) in all directions.
Spill: Increase, in the downwind direction, as necessary, the distance shown under "Public Safety."
Fire: If tank, rail car, or tank truck is involved in fire, isolate for at least 800 meters (½ mile) in all directions; also, consider initial evacuation for 800 meters (½ mile) in all directions.

EXPOSURE
Short-term effects: *SEEK MEDICAL ATTENTION*. **Vapor:** Irritating to eyes, nose, and throat. *IF INHALED*, will, will cause coughing or difficult breathing. Move victim to fresh air. *IF BREATHING HAS STOPPED*, give artificial respiration. IF breathing is difficult, administer oxygen. **Liquid:** Will burn skin and eyes. *IF SWALLOWED*, will cause nausea, vomiting, or loss of consciousness. Remove contaminated clothing and shoes. Flush affected areas with plenty of water. *IF IN EYES*, hold eyelids open and flush with plenty of water. *IF SWALLOWED* and victim is *CONSCIOUS AND ABLE TO SWALLOW*, have victim drink 4 to 8 ounces of water. **Do NOT induce vomiting**.

HEALTH HAZARDS
Personal protective equipment (PPE): A-Level PPE. Chemical protective material(s) reported to have good to excellent resistance: butyl rubber.
Recommendations for respirator selection: NIOSH/OSHA as antimony.
5 mg/m³: DMXSQ (*If not present as a fume*) (any dust and mist respirator except single-use and quarter mask respirators); or SA (any supplied-air respirator). *12.5 mg/m³*: SA:CF (any supplied-air respirator operated in a continuous-flow mode); PAPRDM (*If not present as a fume*) (any powered, air-purifying respirator with a dust and mist filter). *25 mg/m³*: HiEF (any air-purifying, full-facepiece respirator with a high-efficiency particulate filter); or SAT:CF (any supplied-air respirator that has a tight-fitting facepiece and is operated in a continuous-flow mode); or PAPRTHiE* (any powered, air-purifying respirator with a tight-fitting facepiece and a high-efficiency particulate filter); or SCBAF (any self-contained breathing apparatus with a full facepiece); or SAF (any supplied-air respirator with a full facepiece). *50 mg/m³*: SA:PD,PP* (any supplied-air respirator operated in a pressure-demand or other positive-pressure mode). *EMERGENCY OR PLANNED ENTRY INTO UNKNOWN CONCENTRATIONS OR IDLH CONDITIONS:* SCBAF:PD,PP (any self-contained breathing apparatus that has a full faceplate and is operated in a pressure-demand or other positive-pressure mode); or SAF:PD,PP:ASCBA (any supplied-air respirator that has a full facepiece and is operated in a pressure-demand or other positive-pressure mode in combination with an auxiliary self-contained breathing apparatus operated in a pressure-demand or other positive pressure mode). *ESCAPE:* HiEF (any air-purifying, full-facepiece respirator with a high-efficiency particulate filter); or SCBAE (any appropriate escape-type, self-contained breathing apparatus. *Note:* Substance reported to cause eye irritation or damage; may require eye protection.
Exposure limits (TWA unless otherwise noted): ACGIH TLV 0.5 mg/m³ as antimony; NIOSH/OSHA 0.5 mg/m³ as antimony.
Toxicity by ingestion: Grade 2; oral LD_{50} = 1115 mg/kg (rat), 900 mg/kg (guinea pig).
Long-term health effects: May cause kidney, heart and liver damage. Antimony poisoning may result.
Vapor (gas) irritant characteristics: Vapors are irritating such that personnel will not usually tolerate moderate or high vapor concentrations.
Liquid or solid irritant characteristics: Severe skin irritant; causes second-and third-degree burns on short contact and is very injurious to the eyes.
IDLH value: 50 mg/m³ as antimony.

FIRE DATA
Fire extinguishing agents not to be used: Keep away from water and water-based extinguishers, foam. Do not use water or foam on adjacent fires.

Behavior in fire: Irritating vapors of antimony and hydrogen chloride produced in fire.

CHEMICAL REACTIVITY
Reactivity with water: Reacts to form hydrogen chloride gas (hydrochloric acid).
Binary reactants: Reacts with bases and ammonia. Causes corrosion of many metals. Reacts with air forming heavier-than-air corrosive vapors.
Neutralizing agents for acids and caustics: Soda ash or soda ash-lime mixture.

ENVIRONMENTAL DATA
Water pollution: Effect of low concentrations on aquatic life is unknown. May be dangerous if it enters nearby water intakes; notify operators. Notify local health and wildlife officials.
Response to discharge: Issue warning–corrosive. Restrict access. Disperse and flush.

SHIPPING INFORMATION
Grades of purity: 99+%; **Storage temperature:** Ambient; **Inert atmosphere:** None; **Venting:** Pressure-vacuum; **Stability during transport:** Stable.

NAS HAZARD CLASSIFICATIONS FOR BULK WATER TRANSPORTATION
FIRE: 0
HEALTH: Vapor irritant: 3; Liquid or solid irritant: 4; Poisons: 3
WATER POLLUTION: Human toxicity:; Aquatic toxicity: 3; Aesthetic effect: 2
REACTIVITY: Other chemicals: 3; Water: 3; Self-reaction: 0

PHYSICAL AND CHEMICAL PROPERTIES
Physical state @ 59°F/15°C and 1 atm: Liquid.
Molecular weight: 299.05
Boiling point @ 1 atm: Decomposes.
Melting/Freezing point: 37°F/3°C/276°K.
Specific gravity (water = 1): 2.354 @ 68°F/20°C (liquid).
Liquid surface tension: (estimate) 15 dynes/cm = 0.015 N/m @ 68°F/20°C.
Latent heat of vaporization: 68.9 Btu/lb = 38.3 cal/g = 1.60×10^5 J/kg.
Heat of solution: –211.9 Btu/lb = –117.7 cal/g = -4.925×10^5 J/kg.
Heat of fusion: 8.0 cal/g.
Vapor pressure: 0.839 mm.

ANTIMONY PENTAFLUORIDE   REC. A:4950

SYNONYMS: ANTIMONY FLUORIDE; ANTIMONY(V) FLUORIDE; ANTIMONY(5+) FLUORIDE; ANTIMONY(V) PENTAFLUORIDE; ANTIMONY(5+) PENTAFLUORIDE; PENTAFLUORO ANTIMONY; PENTAFLUORURO de ANTIMONIO (Spanish)

IDENTIFICATION
CAS Number: 7783-70-2
Formula: SbF_5
DOT ID Number: UN 1732; DOT Guide Number: 157
Proper Shipping Name: Antimony pentafluoride
Reportable Quantity (RQ): **(EHS)** 1 lb/0.454 kg

DESCRIPTION: Colorless liquid. Sharp odor. Reacts violently with water forming hydrofluoric acid and poisonous hydrogen fluoride gas.

Corrosive • Breathing the vapor can kill you; skin or eye contact causes severe burns that can lead to blindness; effects of contact or inhalation may be delayed • Firefighting gear (including SCBA) does not provide adequate protection. If exposure occurs, remove and isolate gear immediately and thoroughly decontaminate personnel • Toxic products of combustion may include antimony and fluorine.

Hazard Classification (based on NFPA-704 Rating System)
Health Hazards (Blue): 4; Flammability (Red): 0; Reactivity (Yellow): 1

EMERGENCY RESPONSE: See Appendix A (157)
Evacuation:
Public safety: Isolate the area of spill or leak for at least 50 to 100 meters (160 to 330 feet) in all directions.
Spill: Increase, in the downwind direction, as necessary, the distance shown under "Public Safety."
Fire: If tank, rail car, or tank truck is involved in fire, isolate for at least 800 meters (½ mile) in all directions; also, consider initial evacuation for 800 meters (½ mile) in all directions.

EXPOSURE
Short-term effects: *SEEK MEDICAL ATTENTION*. **Vapor:** *POISONOUS IF INHALED*. Irritating to eyes, nose, and throat. Move victim to fresh air. *IF BREATHING HAS STOPPED,* give artificial respiration. IF breathing is difficult, administer oxygen. **Liquid:** *POISONOUS IF SWALLOWED*. Will burn skin and eyes. Remove contaminated clothing and shoes. Flush affected areas with plenty of water. *IF IN EYES*, hold eyelids open and flush with plenty of water. *IF SWALLOWED* and victim is *CONSCIOUS AND ABLE TO SWALLOW*, have victim drink 4 to 8 ounces of water. **Do NOT induce vomiting.**

HEALTH HAZARDS
Personal protective equipment (PPE): A-Level PPE. Sealed chemical suit of butyl rubber. Also polycarbonate is resistant.
Recommendations for respirator selection: NIOSH/OSHA as antimony.
5 mg/m³: DMXSQ (*If not present as a fume*) (any dust and mist respirator except single-use and quarter mask respirators); or SA (any supplied-air respirator).*12.5 mg/m³*: SA:CF (any supplied-air respirator operated in a continuous-flow mode); PAPRDM (*If not present as a fume)* (any powered, air-purifying respirator with a dust and mist filter). *25 mg/m³*: HiEF (any air-purifying, full-facepiece respirator with a high-efficiency particulate filter); or SAT:CF (any supplied-air respirator that has a tight-fitting facepiece and is operated in a continuous-flow mode); or PAPRTHiE* (any powered, air-purifying respirator with a tight-fitting facepiece and a high-efficiency particulate filter); or SCBAF (any self-contained breathing apparatus with a full facepiece); or SAF (any supplied-air respirator with a full facepiece). *50 mg/m³*: SA:PD,PP* (any supplied-air respirator operated in a pressure-demand or other positive-pressure mode). *EMERGENCY OR PLANNED ENTRY INTO UNKNOWN CONCENTRATIONS OR IDLH CONDITIONS:* SCBAF:PD,PP (any self-contained breathing apparatus that has a full faceplate and is operated in a pressure-demand or other positive-pressure mode); or SAF:PD,PP:ASCBA (any supplied-air respirator that has a full facepiece and is operated in a pressure-demand or other positive-pressure mode in combination with an auxiliary self-contained breathing apparatus

operated in a pressure-demand or other positive pressure *mode*). *ESCAPE:* HiEF (any air-purifying, full-facepiece respirator with a high-efficiency particulate filter); or SCBAE (any appropriate escape-type, self-contained breathing apparatus. *Note:* Substance reported to cause eye irritation or damage; may require eye protection.
Exposure limits (TWA unless otherwise noted): ACGIH TLV 0.5 mg/m^3; NIOSH/OSHA 0.5 mg/m^3 as antimony.
Long-term health effects: Antimony poisoning may result.
Vapor (gas) irritant characteristics: May cause eye, nose and respiratory tract irritation.
Liquid or solid irritant characteristics: May cause severe burns of the eyes and skin.
IDLH value: 50 mg/m^3 as antimony.

FIRE DATA
Fire extinguishing agents not to be used: Do not use water or foam on surrounding fires.
Behavior in fire: Gives off toxic hydrogen fluoride fumes when water is used to extinguish adjacent fire.

CHEMICAL REACTIVITY
Reactivity with water: Reacts vigorously forming toxic hydrogen fluoride (hydrofluoric acid).
Binary reactants: Reacts with organics and siliceous materials. When moisture is present, causes severe corrosion of glass and metals (forming explosive hydrogen gas). If confined and wet can cause explosion. May cause fire in contact with combustible material.
Neutralizing agents for acids and caustics: Flush with water, rinse with sodium bicarbonate or lime solution.

ENVIRONMENTAL DATA
Water pollution: Effect of low concentrations on aquatic life is unknown. May be dangerous if it enters nearby water intakes; notify operators. Notify local health and wildlife officials.
Response to discharge: Issue warning–corrosive, air contaminant. Restrict access. Disperse and flush.

SHIPPING INFORMATION
Grades of purity: Commercial; **Storage temperature:** Ambient; **Inert atmosphere:** None; **Venting:** Pressure-vacuum; **Stability during transport:** Stable.

PHYSICAL AND CHEMICAL PROPERTIES
Physical state @ 59°F/15°C and 1 atm: Liquid.
Molecular weight: 216.7
Boiling point @ 1 atm: 289°F/143°C/416°K.
Melting/Freezing point: 45°F/7°C/280°K.
Specific gravity (water = 1): 2.340 at 30°C (liquid).
Liquid surface tension: (estimate) 20 dynes/cm = 0.020 N/m @ 68°F/20°C.
Latent heat of vaporization: (estimate) 79 Btu/lb = 44 cal/g = 1.8 x 10^5 J/kg.
Vapor pressure: 10 mm.

ANTIMONY POTASSIUM TARTRATE   REC. A:5000

SYNONYMS: ANTIMONYL POTASSIUM TARTRATE; POTASSIUM ANTIMONYL-d-TARTRATE; EMETIQUE (French); POTASSIUM ANTIMONYL TARTRATE; TARTARIC ACID, ANTIMONY POTASSIUM SALT; TARTAR EMETIC; TARTARIZED ANTIMONY; TARTRATED ANTIMONY; TARTRATO de ANTIMONIO y POTASIO (Spanish); TASTOX

IDENTIFICATION
CAS Number: 28300-74-5
Formula: $C_8H_4O_{12}Sb_2 \cdot 3H_2O \cdot 2K$
DOT ID Number: UN 1551; DOT Guide Number: 151
Proper Shipping Name: Antimony potassium tartrate
Reportable Quantity (RQ): **(CERCLA)** 100 lb/45.4 kg

DESCRIPTION: White solid. Odorless. Sinks in water; slowly dissolved, forming a mildly acidic solution.

Poison! • Breathing the dust, skin or eye contact, or swallowing the material can kill you • Firefighting gear (including SCBA) does not provide adequate protection. If exposure occurs, remove and isolate gear immediately and thoroughly decontaminate personnel • Containers may BLEVE when exposed to fire • Concentrated dust in confined areas (e.g., tanks, sewers, buildings) may explode when exposed to fire • Toxic products of combustion may include carbon monoxide and fumes of antimony • Do not put yourself in danger by entering a contaminated area to rescue a victim.

Hazard Classification (based on NFPA-704 Rating System)
Health Hazards (Blue): 2; Flammability (Red): 0; Reactivity (Yellow): 0

EMERGENCY RESPONSE: See Appendix A (151)

Evacuation:
Public safety: Isolate the area of spill or leak for at least 25 to 50 meters (80 to 160 feet) in all directions.
Spill: Increase, in the downwind direction, as necessary, the distance shown under "Public Safety."
Fire: If tank, rail car, or tank truck is involved in fire, isolate for at least 800 meters (½ mile) in all directions; also, consider initial evacuation for 800 meters (½ mile) in all directions.

EXPOSURE
Short-term effects: *SEEK MEDICAL ATTENTION.* **Dust:** *POISONOUS IF INHALED OR IF SKIN IS EXPOSED. IF INHALED*, will, will cause dizziness, headache, coughing, or difficult breathing. *IF IN EYES*, hold eyelids open and flush with plenty of water. *IF BREATHING HAS STOPPED*, give artificial respiration; *avoid mouth-to-mouth resuscitation; use bag/mask apparatus*. IF breathing is difficult, administer oxygen. **Solid:** Will burn skin and eyes. *IF SWALLOWED*, will cause nausea, dizziness, loss of consciousness. Remove contaminated clothing and shoes. Flush affected areas with plenty of water. *IF IN EYES*, hold eyelids open and flush with plenty of water. *IF SWALLOWED* and victim is *CONSCIOUS AND ABLE TO SWALLOW*, have victim drink 4 to 8 ounces of water and have victim induce vomiting. *IF SWALLOWED* and victim is *UNCONSCIOUS OR HAVING CONVULSIONS*, do nothing except keep victim warm.

HEALTH HAZARDS
Personal protective equipment (PPE): A-Level PPE. Rubber gloves, protective clothing; safety goggles and face shield.
Recommendations for respirator selection: NIOSH/OSHA as antimony.
5 mg/m^3: DMXSQ (*If not present as a fume*) (any dust and mist respirator except single-use and quarter mask respirators); or SA (any supplied-air respirator). *12.5 mg/m^3*: SA:CF (any supplied-air respirator operated in a continuous-flow mode); PAPRDM (*If not present as a fume*) (any powered, air-purifying respirator with a dust and mist filter). *25 mg/m^3*: HiEF (any air-purifying, full-facepiece respirator with a high-efficiency particulate filter); or

SAT:CF (any supplied-air respirator that has a tight-fitting facepiece and is operated in a continuous-flow mode); or PAPRTHiE* (any powered, air-purifying respirator with a tight-fitting facepiece and a high-efficiency particulate filter); or SCBAF (any self-contained breathing apparatus with a full facepiece); or SAF (any supplied-air respirator with a full facepiece). *50 mg/m^3:* SA:PD,PP* (any supplied-air respirator operated in a pressure-demand or other positive-pressure mode). *EMERGENCY OR PLANNED ENTRY INTO UNKNOWN CONCENTRATIONS OR IDLH CONDITIONS:* SCBAF:PD,PP (any self-contained breathing apparatus that has a full faceplate and is operated in a pressure-demand or other positive-pressure mode); or SAF:PD,PP:ASCBA (any supplied-air respirator that has a full facepiece and is operated in a pressure-demand or other positive-pressure mode in combination with an auxiliary self-contained breathing apparatus operated in a pressure-demand or other positive pressure *mode). ESCAPE:* HiEF (any air-purifying, full-facepiece respirator with a high-efficiency particulate filter); or SCBAE (any appropriate escape-type, self-contained breathing apparatus. *Note:* Substance reported to cause eye irritation or damage; may require eye protection.
Exposure limits (TWA unless otherwise noted): ACGIH TLV 0.5 mg/m^3; NIOSH/OSHA PEL 0.5 mg/m^3 as antimony.
Toxicity by ingestion: Grade 3; oral rat LD$_{50}$ = 115 mg/kg.
Vapor (gas) irritant characteristics: Causes irritation of the respiratory tract.
Liquid or solid irritant characteristics: Causes irritation of the eyes, skin, and respiratory tract.
IDLH value: 50 mg/m^3 as antimony.

ENVIRONMENTAL DATA
Food chain concentration potential: High
Water pollution: Harmful to aquatic life in very low concentrations. May be dangerous if it enters nearby water intakes; notify operators. Notify local health and wildlife officials. Notify operators or nearby water intakes. **Response to discharge:** Issue warning–poison, water contaminant. Restrict access. Disperse and flush.

SHIPPING INFORMATION
Grades of purity: Pure, 99–103%; **Storage temperature:** Ambient; **Inert atmosphere:** None; **Venting:** Open; **Stability during transport:** Stable.

PHYSICAL AND CHEMICAL PROPERTIES
Physical state @ 59°F/15°C and 1 atm: Solid.
Molecular weight: 334
Specific gravity (water = 1): 2.60 @ 68°F/20°C (solid).

ANTIMONY TRIBROMIDE   REC. A:5050

SYNONYMS: ANTIMONOUS BROMIDE; EEC No. 051-003-00-9; STIBINE, TRIBROMO-; TRIBROMURO de ANTIMONIO (Spanish)

IDENTIFICATION
CAS Number: 7789-61-9
Formula: SbBr3
DOT ID Number: UN 1549; DOT Guide Number: 157
Proper Shipping Name: Antimony tribromide, solid; Antimony tribromide, Liquid
Reportable Quantity (RQ): **(CERCLA)** 1000 lb/454 kg

DESCRIPTION: Colorless to yellow crystalline solid. Sinks and mixes with water; decomposes forming antimony oxide and hydrogen bromide.

Corrosive to the skin, eyes, and respiratory tract; contact with skin or eyes can cause burns and vision impairment; inhalation symptoms may be delayed • Firefighting gear (including SCBA) does not provide adequate protection. If exposure occurs, remove and isolate gear immediately and thoroughly decontaminate personnel • Toxic products of combustion may include antimony and bromine.

Hazard Classification (based on NFPA-704 Rating System)
Health Hazards (Blue): 3; Flammability (Red): 0; Reactivity (Yellow): 1

EMERGENCY RESPONSE: See Appendix A (157)
Evacuation:
Public safety: Isolate the area of spill or leak for at least 50 to 100 meters (160 to 330 feet) in all directions.
Spill: Increase, in the downwind direction, as necessary, the distance shown under "Public Safety."
Fire: If tank, rail car, or tank truck is involved in fire, isolate for at least 800 meters (½ mile) in all directions; also, consider initial evacuation for 800 meters (½ mile) in all directions.

EXPOSURE
Short-term effects: *SEEK MEDICAL ATTENTION.* **Liquid or solid:** Irritating to skin, nose, and throat. Harmful, if swallowed. Move to fresh air. *IF SWALLOWED* and victim is *CONSCIOUS AND ABLE TO SWALLOW*, have victim drink 4 to 8 ounces of water. Flush affected areas with plenty of water.

HEALTH HAZARDS
Personal protective equipment (PPE): A-Level PPE. Gloves, dustproof clothing, goggles, and, where atmospheric exposure is high, respirators should be used.
Recommendations for respirator selection: NIOSH/OSHA as antimony.
5 mg/m^3: DMXSQ (*If not present as a fume*) (any dust and mist respirator except single-use and quarter mask respirators); or SA (any supplied-air respirator). *12.5 mg/m^3:* SA:CF (any supplied-air respirator operated in a continuous-flow mode); PAPRDM (*If not present as a fume*) (any powered, air-purifying respirator with a dust and mist filter). *25 mg/m^3:* HiEF (any air-purifying, full-facepiece respirator with a high-efficiency particulate filter); or SAT:CF (any supplied-air respirator that has a tight-fitting facepiece and is operated in a continuous-flow mode); or PAPRTHiE* (any powered, air-purifying respirator with a tight-fitting facepiece and a high-efficiency particulate filter); or SCBAF (any self-contained breathing apparatus with a full facepiece); or SAF (any supplied-air respirator with a full facepiece). *50 mg/m^3:* SA:PD,PP* (any supplied-air respirator operated in a pressure-demand or other positive-pressure mode). *EMERGENCY OR PLANNED ENTRY INTO UNKNOWN CONCENTRATIONS OR IDLH CONDITIONS:* SCBAF:PD,PP (any self-contained breathing apparatus that has a full faceplate and is operated in a pressure-demand or other positive-pressure mode); or SAF:PD,PP:ASCBA (any supplied-air respirator that has a full facepiece and is operated in a pressure-demand or other positive-pressure mode in combination with an auxiliary self-contained breathing apparatus operated in a pressure-demand or other positive pressure *mode). ESCAPE:* HiEF (any air-purifying, full-facepiece respirator with a high-efficiency particulate filter); or SCBAE (any appropriate escape-type, self-contained breathing apparatus. *Note:* Substance

Antimony trichloride

reported to cause eye irritation or damage; may require eye protection.
Exposure limits (TWA unless otherwise noted): ACGIH TLV 0.5 mg/m^3; NIOSH/OSHA PEL 0.5 mg/m^3 as antimony.
Long-term health effects: Dryness of throat, pain on swallowing. Occasional vomiting and persistent nausea, loss of appetite, weight loss, dermatitis, dizziness, diarrhea, and bloody stools. Causes pathologic changes in cardiac muscle of experimental animals. Possible heart, liver and kidney damage. Prolonged contact or inhalation of inorganic bromides causes acne-like bromoderma (bromide rash), especially of the hands and face. Other effects include emaciation, depression; and, in severe cases, psychosis and mental deterioration.
Vapor (gas) irritant characteristics: High concentrations may cause lung edema; physical exertion will aggravate this condition.
Liquid or solid irritant characteristics: Corrosive to the eyes, skin, and respiratory tract.
IDLH value: 50 mg/m^3 as antimony.

FIRE DATA
Flash point: Not combustible.
Fire extinguishing agents not to be used: Water reaction forms corrosive substances.

CHEMICAL REACTIVITY
Reactivity with water: Decomposes in water forming antimony oxide and hydrogen bromide
Binary reactants: Decomposes in air and alcohol.

ENVIRONMENTAL DATA
Food chain concentration potential: Antimony can be concentrated by a factor of 300 by marine life.
Water pollution: Harmful to aquatic life in very low concentrations. May be dangerous if it enters nearby water intakes; notify operators. Notify local health and pollution control officials.
Response to discharge: Issue warning–water contaminant. Disperse and flush.

SHIPPING INFORMATION
Stability during transport: Stable.

PHYSICAL AND CHEMICAL PROPERTIES
Physical state @ 59°F/15°C and 1 atm: Solid.
Molecular weight: 361.51
Boiling point @ 1 atm: 536°F/280°C/553.2°K.
Melting/Freezing point: 205.88°F/96.6°C/369.75°K.
Critical temperature: 1660.1°F/904.5°C/1177.65°K.
Critical pressure: 822.976 psia = 0.56 atm = 5.67 MN/m^2.
Specific gravity (water = 1): 4.148 at 23°C.
Heat of fusion: 9.7 cal/g.
Vapor pressure: Less than 0.075 mm.

ANTIMONY TRICHLORIDE   **REC. A:5100**

SYNONYMS: STIBINE, TRICHLORO-; ANTIMONIUS CHLORIDE; ANTIMONY(III) CHLORIDE; TRICHLORO STIBINE; ANTIMOINE (TRICHLORURE d') (French); ANTIMONY BUTTER; BUTTER OF ANTIMONY; CHLORID ANTIMONITY; STIBINE, TRICHLORO-; TRICLORURO de ANTIMONIO (Spanish); TRICHLOROSTIBINE; TRICHLORURE d' ANTIMOINE (French).

IDENTIFICATION
CAS Number: 10025-91-9
Formula: SbCl$_3$
DOT ID Number: UN 1733; DOT Guide Number: 157
Proper Shipping Name: Antimony trichloride, liquid; Antimony trichloride, solid; Antimony trichloride, solution
Reportable Quantity (RQ): **(CERCLA)** 1000 lb/454 kg

DESCRIPTION: White to pale yellow solid. Sharp, acrid odor. Sinks and mixes violently with water; forming hydrochloric acid.

Corrosive to the skin, eyes, and respiratory tract; contact with skin or eyes can cause burns and vision impairment; inhalation symptoms may be delayed • Firefighting gear (including SCBA) does not provide adequate protection. If exposure occurs, remove and isolate gear immediately and thoroughly decontaminate personnel • Toxic products of combustion may include toxic fumes of antimony oxide and hydrogen chloride • Do not put yourself in danger by entering a contaminated area to rescue a victim.

Hazard Classification (based on NFPA-704 Rating System)
Health Hazards (Blue): 3; Flammability (Red): 0; Reactivity (Yellow): 2; Special Notice (White): Water reactive

EMERGENCY RESPONSE: See Appendix A (157)
Evacuation:
Public safety: Isolate the area of spill or leak for at least 50 to 100 meters (160 to 330 feet) in all directions.
Spill: Increase, in the downwind direction, as necessary, the distance shown under "Public Safety."
Fire: If tank, rail car, or tank truck is involved in fire, isolate for at least 800 meters (½ mile) in all directions; also, consider initial evacuation for 800 meters (½ mile) in all directions.

EXPOSURE
Short-term effects: *SEEK MEDICAL ATTENTION.* **Dust:** *POISONOUS IF INHALED OR IF SKIN IS EXPOSED. IF INHALED,* will, will cause coughing, or difficult breathing. *IF IN EYES,* hold eyelids open and flush with plenty of water. *IF BREATHING HAS STOPPED,* give artificial respiration. IF breathing is difficult, administer oxygen. Lung edema may develop. **Solid:** Will burn skin and eyes. *IF SWALLOWED,* will cause nausea, vomiting, or loss of consciousness. Remove contaminated clothing and shoes. Flush affected areas with plenty of water. *IF IN EYES,* hold eyelids open and flush with plenty of water. *IF SWALLOWED and victim is CONSCIOUS AND ABLE TO SWALLOW,* have victim drink 4 to 8 ounces of water. *IF SWALLOWED and victim is UNCONSCIOUS OR HAVING CONVULSIONS,* do nothing except keep victim warm.
Note to physician or authorized medical personnel: Medical observation is recommended for 24 to 48 hours after breathing overexposure, as pulmonary edema may be delayed. As first aid for pulmonary edema, consider administering a corticosteroid spray. Cigarette smoking may exacerbate pulmonary injury and should be discouraged for at least 72 hours following exposure.

HEALTH HAZARDS
Personal protective equipment (PPE): A-Level PPE. *Note*: The respiratory system is the chief avenue of entrance of antimony and its compounds into the body. Chemical protective material(s) reported to have good to excellent resistance: butyl rubber, polycarbonate, polyethylene, fluoropolymers, PVC.
Recommendations for respirator selection: NIOSH/OSHA as antimony.
5 mg/m^3: DMXSQ (*If not present as a fume*) (any dust and mist respirator except single-use and quarter mask respirators); or SA (any supplied-air respirator).*12.5 mg/m^3*: SA:CF (any supplied-air

respirator operated in a continuous-flow mode); PAPRDM (*If not present as a fume*) (any powered, air-purifying respirator with a dust and mist filter). *25 mg/m³*: HiEF (any air-purifying, full-facepiece respirator with a high-efficiency particulate filter); or SAT:CF (any supplied-air respirator that has a tight-fitting facepiece and is operated in a continuous-flow mode); or PAPRTHiE* (any powered, air-purifying respirator with a tight-fitting facepiece and a high-efficiency particulate filter); or SCBAF (any self-contained breathing apparatus with a full facepiece); or SAF (any supplied-air respirator with a full facepiece). *50 mg/m³*: SA:PD,PP* (any supplied-air respirator operated in a pressure-demand or other positive-pressure mode). *EMERGENCY OR PLANNED ENTRY INTO UNKNOWN CONCENTRATIONS OR IDLH CONDITIONS*: SCBAF:PD,PP (any self-contained breathing apparatus that has a full faceplate and is operated in a pressure-demand or other positive-pressure mode); or SAF:PD,PP:ASCBA (any supplied-air respirator that has a full facepiece and is operated in a pressure-demand or other positive-pressure mode in combination with an auxiliary self-contained breathing apparatus operated in a pressure-demand or other positive pressure mode). *ESCAPE*: HiEF (any air-purifying, full-facepiece respirator with a high-efficiency particulate filter); or SCBAE (any appropriate escape-type, self-contained breathing apparatus. *Note:* Substance reported to cause eye irritation or damage; may require eye protection.
Exposure limits (TWA unless otherwise noted): ACGIH TLV 0.5 mg/m³; NIOSH/OSHA 0.5 mg/m³ as antimony.
Toxicity by ingestion: Grade 2; oral rat LD_{50} = 675 mg/kg.
Long-term health effects: Possible heart, liver and kidney damage.
Vapor (gas) irritant characteristics: May cause irritation of the respiratory tract and breathing difficulties. Possible lung edema.
Liquid or solid irritant characteristics: Corrosive to eyes, skin, and respiratory tract.
IDLH value: 50 mg/m³ as antimony.

FIRE DATA
Fire extinguishing agents not to be used: Do not use water on adjacent fires.

CHEMICAL REACTIVITY
Reactivity with water: Reacts vigorously to form a strong solution of hydrochloric acid and antimony oxychloride.
Binary reactants: Reacts with ammonia and bases. Corrodes most metals in presence of moisture. Reaction with air forms corrosive fumes which may collect in enclosed spaces.
Neutralizing agents for acids and caustics: Large amounts of water followed by sodium bicarbonate or soda ash solution.

ENVIRONMENTAL DATA
Food chain concentration potential: High
Water pollution: Harmful to aquatic life in very low concentrations. May be dangerous if it enters nearby water intakes; notify operators. Notify local health and wildlife officials.
Response to discharge: Issue warning–water contaminant, corrosive. Disperse and flush.

SHIPPING INFORMATION
Grades of purity: Commercial, 99+%; Analytical; Anhydrous;
Storage temperature: Ambient; **Inert atmosphere:** Padded;
Venting: Pressure-vacuum; **Stability during transport:** Stable.

PHYSICAL AND CHEMICAL PROPERTIES
Physical state @ 59°F/15°C and 1 atm: Solid.
Molecular weight: 228
Boiling point @ 1 atm: 433°F/223°C/496°K.
Melting/Freezing point: 163°F/73°C/346°K.
Specific gravity (water = 1): 3.14 @ 68°F/20°C (solid).
Heat of solution: –70 Btu/lb = –39 cal/g = –1.6 x 10⁵ J/kg.
Heat of fusion: 13.3 cal/g.
Vapor pressure: 0.119 mm.

ANTIMONY TRIFLUORIDE REC. A:5150

SYNONYMS: ANTIMONY(III) FLUORIDE; ANTIMOINE FLUORURE (French); ANTIMONOUS FLUORIDE; TRIFLUOROANTIMONY, STIBINE, TRIFLUORO-; TRIFLUORURO de ANTIMONIO (Spanish)

IDENTIFICATION
CAS Number: 7783-56-4
Formula: SbF_3
DOT ID Number: UN 1549; DOT Guide Number: 157
Proper Shipping Name: Antimony trifluoride, solid; Antimony trifluoride, solution
Reportable Quantity (RQ): **(CERCLA)** 1000 lb/454 kg

DESCRIPTION: White, solid crystals. Odorless. Sinks in water.

Corrosive • Breathing the vapor can kill you; skin or eye contact causes severe burns that can lead to blindness; effects of contact or inhalation may be delayed • Firefighting gear (including SCBA) does not provide adequate protection. If exposure occurs, remove and isolate gear immediately and thoroughly decontaminate personnel • Toxic products of combustion may include vapors of antimony and hydrogen fluoride.

Hazard Classification (based on NFPA-704 Rating System)
Health Hazards (Blue): 3; Flammability (Red): 0; Reactivity (Yellow): 0

EMERGENCY RESPONSE: See Appendix A (157)
Evacuation:
Public safety: Isolate the area of spill or leak for at least 50 to 100 meters (160 to 330 feet) in all directions.
Spill: Increase, in the downwind direction, as necessary, the distance shown under "Public Safety."
Fire: If tank, rail car, or tank truck is involved in fire, isolate for at least 800 meters (½ mile) in all directions; also, consider initial evacuation for 800 meters (½ mile) in all directions.

EXPOSURE
Short-term effects: *SEEK MEDICAL ATTENTION*. **Solid:** Irritating to skin and eyes. Harmful if swallowed. Remove contaminated clothing and shoes. Flush affected areas with plenty of water. *IF IN EYES*, hold eyelids open and flush with plenty of water. *IF SWALLOWED* and victim is *CONSCIOUS AND ABLE TO SWALLOW*, have victim drink 4 to 8 ounces of water.
Acute poisoning: irritation of the mouth, nose, stomach and intestines; vomiting, purging with bloody stools; slow pulse and low blood pressure; slow, shallow breathing; coma and convulsions sometimes followed by death from cardiac and respiratory exhaustion.

HEALTH HAZARDS
Personal protective equipment (PPE): A-Level PPE. Wear a sealed chemical suit of buryl rubber, polycarbonate.
Recommendations for respirator selection: NIOSH/OSHA as antimony.
5 mg/m³: DMXSQ (*If not present as a fume*) (any dust and mist

respirator except single-use and quarter mask respirators); or SA (any supplied-air respirator). *12.5 mg/m³*: SA:CF (any supplied-air respirator operated in a continuous-flow mode); PAPRDM (*If not present as a fume*) (any powered, air-purifying respirator with a dust and mist filter). *25 mg/m³*: HiEF (any air-purifying, full-facepiece respirator with a high-efficiency particulate filter); or SAT:CF (any supplied-air respirator that has a tight-fitting facepiece and is operated in a continuous-flow mode); or PAPRTHiE* (any powered, air-purifying respirator with a tight-fitting facepiece and a high-efficiency particulate filter); or SCBAF (any self-contained breathing apparatus with a full facepiece); or SAF (any supplied-air respirator with a full facepiece). *50 mg/m³*: SA:PD,PP* (any supplied-air respirator operated in a pressure-demand or other positive-pressure mode). *EMERGENCY OR PLANNED ENTRY INTO UNKNOWN CONCENTRATIONS OR IDLH CONDITIONS*: SCBAF:PD,PP (any self-contained breathing apparatus that has a full faceplate and is operated in a pressure-demand or other positive-pressure mode); or SAF:PD,PP:ASCBA (any supplied-air respirator that has a full facepiece and is operated in a pressure-demand or other positive-pressure mode in combination with an auxiliary self-contained breathing apparatus operated in a pressure-demand or other positive pressure mode). *ESCAPE*: HiEF (any air-purifying, full-facepiece respirator with a high-efficiency particulate filter); or SCBAE (any appropriate escape-type, self-contained breathing apparatus. *Note:* Substance reported to cause eye irritation or damage; may require eye protection.
Exposure limits (TWA unless otherwise noted): ACGIH TLV 0.5 mg/m³; NIOSH/OSHA PEL 0.5 mg/m³ as antimony.
Toxicity by ingestion: Grade 3; LD_{50} = 50 to 500 mg/kg (guinea pig).
Long-term health effects: Chronic poisoning: If any symptoms, however slight, are noticed, the affected individual should be removed from contact with the chemical and placed under the care of a physician who is versed in the treatment necessary. Symptoms include dryness of throat; pain on swallowing; occasional vomiting and persistent nausea; susceptibility to fainting; diarrhea, loss of appetite and weight; giddiness; dermatitis, either pustular or ulcerative; anemia
Vapor (gas) irritant characteristics: May cause irritation of the eyes and respiratory tract.
Liquid or solid irritant characteristics: Fairly severe skin irritant. May cause pain and second-degree burns after a few minutes of contact.
IDLH value: 50 mg/m³ as antimony.

CHEMICAL REACTIVITY
Binary reactants: Strong oxidizers.

ENVIRONMENTAL DATA
Water pollution: Harmful to aquatic life in very low concentrations. Fouling to shoreline. May be dangerous if it enters nearby water intakes; notify operators. Notify local health and wildlife officials. **Response to discharge:** Issue warning–water contaminant. Disperse and flush.

SHIPPING INFORMATION
Grades of purity: 98%; **Stability during transport:** Stable.

PHYSICAL AND CHEMICAL PROPERTIES
Physical state @ 59°F/15°C and 1 atm: Solid.
Boiling point @ 1 atm: 709°F/376°C.
Molecular weight: 178.75
Melting/Freezing point: 558°F/292°C/565°K.
Specific gravity (water = 1): 4.38 at 21°C (solid).

ANTIMONY TRIOXIDE REC. A:5200

SYNONYMS: ANTIMONIOUS OXIDE; ANTIMONY PEROXIDE; ANTIMONY SESQUIOXIDE; ANTIMONY, WHITE; DIANTIMONY TRIOXIDE; SENARMONTITE; VALENTINITE; EXITELITE; WEISSPIESSGLANZ; FLOWERS OF ANTIMONY; TRIOXIDO de ANTIMONIO (Spanish); WHITE ANTIMONY

IDENTIFICATION
CAS Number: 1309-64-4; 1327-33-9
Formula: Sb_2O_3
DOT ID Number: UN 9201; DOT Guide Number: 171
Proper Shipping Name: Antimony trioxide
Reportable Quantity (RQ): **(CERCLA)** 1000 lb/454 kg

DESCRIPTION: White solid crystalline solid or powder. Odorless. Insoluble in water.

Poison! • Breathing the dust, skin or eye contact, or swallowing the material can kill you • Firefighting gear (including SCBA) does not provide adequate protection. If exposure occurs, remove and isolate gear immediately and thoroughly decontaminate personnel • Do not put yourself in danger by entering a contaminated area to rescue a victim • Toxic products of combustion may include vapors of antimony.

Hazard Classification (based on NFPA-704 Rating System)
Health Hazards (Blue): 3; Flammability (Red): 0; Reactivity (Yellow): 0

EMERGENCY RESPONSE: See Appendix A (171)
Evacuation:
Public safety: Isolate the area of spill or leak for at least 10 to 25 meters (30 to 80 feet) in all directions.
Spill: Increase, in the downwind direction, as necessary, the distance shown under "Public Safety."
Fire: If any large container is involved in fire, isolate for at least 800 meters (½ mile) in all directions; also, consider initial evacuation for 800 meters (½ mile) in all directions.

EXPOSURE
Short-term effects: *SEEK MEDICAL ATTENTION*. **Dust:** *POISONOUS IF INHALED OR IF SKIN IS EXPOSED. IF INHALED*, will, will cause coughing, difficult breathing or loss of consciousness. *IF IN EYES*, hold eyelids open and flush with plenty of water. *IF BREATHING HAS STOPPED*, give artificial respiration. IF breathing is difficult, administer oxygen. **Solid:** POISONOUS IF SWALLOWED OR IF SKIN IS EXPOSED. *IF SWALLOWED*, will cause dizziness, nausea, vomiting, or loss of consciousness. Remove contaminated clothing and shoes. Flush affected areas with plenty of water. *IF IN EYES*, hold eyelids open and flush with plenty of water. *IF SWALLOWED* and victim is *CONSCIOUS AND ABLE TO SWALLOW*, have victim drink 4 to 8 ounces of water and have victim induce vomiting. *IF SWALLOWED* and victim is *UNCONSCIOUS OR HAVING CONVULSIONS,* do nothing except keep victim warm.

HEALTH HAZARDS
Personal protective equipment (PPE): A-Level PPE. Butyl rubber is generally suitable for peroxide compounds.
Recommendations for respirator selection: NIOSH/OSHA as antimony.
5 mg/m³: DMXSQ (*If not present as a fume*) (any dust and mist respirator except single-use and quarter mask respirators); or SA

(any supplied-air respirator). *12.5 mg/m³*: SA:CF (any supplied-air respirator operated in a continuous-flow mode); PAPRDM (*If not present as a fume*) (any powered, air-purifying respirator with a dust and mist filter). *25 mg/m³*: HiEF (any air-purifying, full-facepiece respirator with a high-efficiency particulate filter); or SAT:CF (any supplied-air respirator that has a tight-fitting facepiece and is operated in a continuous-flow mode); or PAPRTHiE* (any powered, air-purifying respirator with a tight-fitting facepiece and a high-efficiency particulate filter); or SCBAF (any self-contained breathing apparatus with a full facepiece); or SAF (any supplied-air respirator with a full facepiece). *50 mg/m³*: SA:PD,PP* (any supplied-air respirator operated in a pressure-demand or other positive-pressure mode). *EMERGENCY OR PLANNED ENTRY INTO UNKNOWN CONCENTRATIONS OR IDLH CONDITIONS*: SCBAF:PD,PP (any self-contained breathing apparatus that has a full faceplate and is operated in a pressure-demand or other positive-pressure mode); or SAF:PD,PP:ASCBA (any supplied-air respirator that has a full facepiece and is operated in a pressure-demand or other positive-pressure mode in combination with an auxiliary self-contained breathing apparatus operated in a pressure-demand or other positive pressure mode). *ESCAPE*: HiEF (any air-purifying, full-facepiece respirator with a high-efficiency particulate filter); or SCBAE (any appropriate escape-type, self-contained breathing apparatus. *Note*: Substance reported to cause eye irritation or damage; may require eye protection.

Exposure limits (TWA unless otherwise noted): ACGIH TLV 0.5 mg/m³; NIOSH/OSHA PEL 0.5 mg/m³ **as antimony.**
Toxicity by ingestion: Grade 0; oral rat LD_{50} = 20,000 mg/kg.
Long-term health effects: May cause lung and heart problems. Suspected human carcinogen. **Chronic poisoning:** If any symptoms, however slight, are noticed, the affected individual should be removed from contact with the chemical and placed under the care of a physician who is versed in the treatment necessary. Symptoms include dryness of throat; pain on swallowing; occasional vomiting and persistent nausea; susceptibility to fainting; diarrhea, loss of appetite and weight; giddiness; dermatitis, either pustular or ulcerative; anemia
Vapor (gas) irritant characteristics: A buildup of airborne particles can cause irritation of the eyes and respiratory tract.
Liquid or solid irritant characteristics: Causes eye, skin and respiratory tract irritation.
IDLH value: 50 mg/m³ as antimony.

CHEMICAL REACTIVITY
Binary reactants: Ignites and burns in heated air above 420°F/215°C. Violent reaction with bromine trifluoride. Reacts with chlorinated rubber, alcohols/glycols, organic and α-hydroxy acids (fruit acids), *o*-dihydric phenols; polyethylene glycol and other polyhydroxy compounds.

ENVIRONMENTAL DATA
Food chain concentration potential: High. Log K_{ow} unknown at this time.
Water pollution: Harmful to aquatic life in very low concentrations. May be dangerous if it enters nearby water intakes; notify operators. Notify local health and wildlife officials.
Response to discharge: Issue warning–poison. Restrict access. Should be removed. Chemical and physical treatment.

SHIPPING INFORMATION
Grades of purity: Reagent, 99.9+%; Technical grade; Pigment grade; Optical grade; **Storage temperature:** Ambient; **Inert atmosphere:** None; **Venting:** Open; **Stability during transport:** Stable.

PHYSICAL AND CHEMICAL PROPERTIES
Physical state @ 59°F/15°C and 1 atm: Solid.
Molecular weight: 291.50
Boiling point @ 1 atm: Sublimes. 2813°F/1545°C/1818°K.
Melting/Freezing point: 1211°F/655°C/928°K.
Specific gravity (water = 1): 5.2 @ 77°F/25°C (solid).
Heat of fusion: 46.3 cal/g.
Vapor pressure: 0.98 mm.

ARSENIC   REC. A:5250

SYNONYMS: ARSEN (German); ARSENICALS; ARSENIC-75; ARSENIC BLACK; ARSENIC, METALLIC; ARSENICO (Spanish); COLLOIDAL ARSENIC; GREY ARSENIC; METALLIC ARSENIC; RUBY ARSENIC; REALGAR; BUTTER OF ARSENIC

IDENTIFICATION
CAS Number: 7440-38-2
Formula: As
DOT ID Number: UN 1558; DOT Guide Number: 152
Proper Shipping Name: Arsenic
Reportable Quantity (RQ): **(CERCLA)** 1 lb/0.454 kg. Reporting is limited to those pieces of the metal having a diameter smaller than 100 micrometers (0.004 inches).

DESCRIPTION: Silver-gray crystalline solid. Sinks in water.

Poison! • Combustible solid • Breathing the dust, skin or eye contact, or swallowing the material can kill you • Firefighting gear (including SCBA) does not provide adequate protection. If exposure occurs, remove and isolate gear immediately and thoroughly decontaminate personnel • Powder or dust forms explosive mixture with air • Toxic products of combustion may include highly toxic arsenic trioxide and other forms of arsenic.

Hazard Classification (based on NFPA-704 Rating System)
Health Hazards (Blue): 3; Flammability (Red): 1; Reactivity (Yellow): 0

EMERGENCY RESPONSE: See Appendix A (152)
Evacuation:
Public safety: Isolate the area of spill or leak for at least 25 to 50 meters (80 to 160 feet) in all directions.
Spill: Increase, in the downwind direction, as necessary, the distance shown under "Public Safety."
Fire: If tank, rail car, or tank truck is involved in fire, isolate for at least 800 meters (½ mile) in all directions; also, consider initial evacuation for 800 meters (½ mile) in all directions.

EXPOSURE
Short-term effects: *SEEK MEDICAL ATTENTION.* **Dust:** *POISONOUS IF INHALED.* Move victim to fresh air. *IF IN EYES OR ON SKIN*, immediately flush with running water for at least 15 minutes; hold eyelids open if necessary. *IF BREATHING HAS STOPPED*, give artificial respiration; *avoid mouth-to-mouth resuscitation; use bag/mask apparatus.* IF breathing is difficult, administer oxygen. **Solid:** *POISONOUS IF SWALLOWED. IF IN EYES OR ON SKIN*, flush with running water for at least 15 minutes; hold eyelids open if necessary. *IF SWALLOWED* and victim is *CONSCIOUS* and has not vomited, induce vomiting with syrup of ipecac. *IF SWALLOWED* and victim is *UNCONSCIOUS OR HAVING CONVULSIONS*, do nothing except keep victim warm.

HEALTH HAZARDS

Personal protective equipment (PPE): A-Level PPE. Wear self-contained positive pressure breathing apparatus and full protective clothing.

Recommendations for respirator selection: NIOSH
At any concentrations above the NIOSH REL, or where there is no REL, at any detectable concentration: SCBAF:PD,PP (any self-contained breathing apparatus that has a full faceplate and is operated in a pressure-demand or other positive-pressure mode); or SAF:PD,PP:ASCBA (any supplied-air respirator that has a full facepiece and is operated in a pressure-demand or other positive-pressure mode in combination with an auxiliary self-contained breathing apparatus operated in a pressure-demand or other positive-pressure mode). *ESCAPE:* GMFAGHiE [any air-purifying, full-facepiece respirator (gas mask) with a chin-style, front-or back-mounted acid gas canister having a high-efficiency particulate filter]; or SCBAE (any appropriate escape-type, self-contained breathing apparatus).

Exposure limits (TWA unless otherwise noted): ACGIH TLV 0.01 mg/m^3; OSHA PEL [1910.1080] 0.010 mg/m^3; NIOSH REL ceiling 0.002 mg/m^3/15 min, as arsenic; potential human carcinogen; reduce exposure to lowest feasible level

Long-term health effects: NTP known carcinogen; IARC rating 1 Confirmed human carcinogen. Causes mutagenic, reproductive and tumorigenic effects along with damage to the gastrointestinal tract and degeneration of the liver and kidneys.

Vapor (gas) irritant characteristics: Fumes may cause pulmonary edema.

Liquid or solid irritant characteristics: Corrosive to the eyes, skin, and respiratory tract.

IDLH value: Potential human carcinogen; 5 mg/m^3.

FIRE DATA

Behavior in fire: Slight explosion hazard. Produces dense white fumes of highly toxic arsenic trioxide.

CHEMICAL REACTIVITY

Binary reactants: Arsenic gas, the most dangerous form of arsenic, is produced upon contact with an acid or acid fumes. Contact of dusts or powder with strong oxidizers can cause ignition or explosion.

ENVIRONMENTAL DATA

Food chain concentration potential: Bioaccumulated by fresh water and marine aquatic organisms.

Water pollution: The EPA drinking water standard of arsenic is 10 ppb (as As ion). May be dangerous if it enters nearby water intakes; notify operators. Notify local health and wildlife officials.

Response to discharge: Issue warning–poison. Restrict access. Should be removed. Chemical and physical treatment.

SHIPPING INFORMATION

Grades of purity: Crude, 90-95%; Refined, 99%; Semiconductor, 99.999%; **Storage temperature:** Ambient; ; **Stability during transport:** Stable.

PHYSICAL AND CHEMICAL PROPERTIES

Physical state @ 59°F/15°C and 1 atm: Solid.
Molecular weight: 74.9216
Boiling point @ 1 atm: Sublimes.1135°F/613°C/886°K.
Critical temperature: 1477°F/803°C/1076°K.
Critical pressure: 5027.4 psia = 342.0 atm = 34.6 MN/m^2.
Specific gravity (water = 1): 5.727 @ 77°F/25°C (solid).
Vapor pressure: 1 millimeter @ 701°F/372°C (sublimes).

ARSENIC ACID   REC. A:5300

SYNONYMS: ACIDO ARSENICO (Spanish); ARSENATE; ARSENIC PENTOXIDE; *o*-ARSENIC ACID; ORTHOARSENIC ACID; RCRA No. P010; ZOTOX

IDENTIFICATION

CAS Number: 1327-52-2; 7778-39-4
Formula: As$_2$O$_5$; H$_3$AsO$_4$·½H$_2$O
DOT ID Number: UN 1553 (liquid); UN 1554 (solid); DOT Guide Number: 154
Proper Shipping Name: Arsenic acid, liquid; Arsenic acid, solid
Reportable Quantity (RQ): (CERCLA) 1 lb/0.454 kg

DESCRIPTION: White or colorless crystalline solid or concentrated water solution. Odorless. Sinks and mixes with water.

Poison! • Combustible Breathing the vapors or dust, skin or eye contact, or swallowing the material can kill you • Firefighting gear (including SCBA) does not provide adequate protection. If exposure occurs, remove and isolate gear immediately and thoroughly decontaminate personnel • Toxic products of combustion may include fumes of arsenic.

Hazard Classification (based on NFPA-704 Rating System)
Health Hazards (Blue): 3; Flammability (Red): 0; Reactivity (Yellow): 0

EMERGENCY RESPONSE: See Appendix A (154)
Evacuation:
Public safety: Isolate the area of spill or leak for at least 25 to 50 meters (80 to 160 feet) in all directions.
Spill: Increase, in the downwind direction, as necessary, the distance shown under "Public Safety."
Fire: If tank, rail car, or tank truck is involved in fire, isolate for at least 800 meters (½ mile) in all directions; also, consider initial evacuation for 800 meters (½ mile) in all directions.

EXPOSURE

Short-term effects: *SEEK MEDICAL ATTENTION.* **Dust:** *POISONOUS IF INHALED.* Irritating to eyes, nose, and throat. Move victim to fresh air. *IF IN EYES*, hold eyelids open and flush with plenty of water. If artificial respiration is administered, *avoid mouth-to-mouth resuscitation; use bag/mask apparatus*. IF breathing is difficult, administer oxygen. **Solution or solid:** *POISONOUS IF SWALLOWED.* Irritating to skin and eyes. Remove contaminated clothing and shoes. Flush affected areas with plenty of water. *IF IN EYES*, hold eyelids open and flush with plenty of water. *IF SWALLOWED* and victim is *CONSCIOUS AND ABLE TO SWALLOW*, have victim drink 4 to 8 ounces of water and have victim induce vomiting. *IF SWALLOWED* and victim is *UNCONSCIOUS OR HAVING CONVULSIONS*, do nothing except keep victim warm.

HEALTH HAZARDS

Personal protective equipment (PPE): A-Level PPE. Calamine lotion and zinc oxide powder on hands and other skin areas. Chemical protective material(s) reported to have good to excellent resistance: butyl rubber, nitrile, neoprene, polycarbonate, chlorinated polyethylene, Viton®.

Recommendations for respirator selection: NIOSH
At any concentrations above the NIOSH REL, or where there is no REL, at any detectable concentration: SCBAF:PD,PP (any self-contained breathing apparatus that has a full faceplate and is

operated in a pressure-demand or other positive-pressure mode); or SAF:PD,PP:ASCBA (any supplied-air respirator that has a full facepiece and is operated in a pressure-demand or other positive-pressure mode in combination with an auxiliary self-contained breathing apparatus operated in a pressure-demand or other positive-pressure mode). *ESCAPE:* GMFAGHiE [any air-purifying, full-facepiece respirator (gas mask) with a chin-style, front-or back-mounted acid gas canister having a high-efficiency particulate filter]; or SCBAE (any appropriate escape-type, self-contained breathing apparatus).
Exposure limits (TWA unless otherwise noted): ACGIH TLV 0.01 mg/m^3; OSHA PEL [1910.1080] 0.010 mg/m^3; NIOSH REL ceiling 0.002 mg/m^3/15 min, as arsenic; potential human carcinogen; reduce exposure to lowest feasible level
Toxicity by ingestion: Grade 4; oral LD$_{50}$ = 48 mg/kg (young rats).
Long-term health effects: Arsenic compounds may be carcinogenic.
Vapor (gas) irritant characteristics: Vapors may be irritating to eyes and throat.
Liquid or solid irritant characteristics: If spilled on clothing and allowed to remain, may cause smarting and reddening of the skin.
IDLH value: Human carcinogen; 5 mg/m^3 as arsenic.

CHEMICAL REACTIVITY
Binary reactants: Corrodes metal to give off toxic arsine gas.
Neutralizing agents for acids and caustics: Flush with water, rinse with sodium bicarbonate or lime solution.

ENVIRONMENTAL DATA
Water pollution: DOT Appendix B, §172.101–marine pollutant. Effect of low concentrations on aquatic life is unknown. May be dangerous if it enters nearby water intakes; notify operators. Notify local health and wildlife officials. **Response to discharge:** Issue warning–poison, water contaminant. Restrict access. Should be removed. Chemical and physical treatment.

SHIPPING INFORMATION
Grades of purity: Commercial; **Storage temperature:** Ambient; **Inert atmosphere:** None; **Venting:** Pressure-vacuum; **Stability during transport:** Stable.

NAS HAZARD CLASSIFICATIONS FOR BULK WATER TRANSPORTATION
FIRE: 0
HEALTH: Vapor irritant: 0; Liquid or solid irritant: 1; Poisons: 4
WATER POLLUTION: Human toxicity: 4; Aquatic toxicity: 3; Aesthetic effect: 1
REACTIVITY: Other chemicals: 2; Water: 0; Self-reaction: 0

PHYSICAL AND CHEMICAL PROPERTIES
Physical state @ 59°F/15°C and 1 atm: Solid.
Molecular weight: 229.8
Specific gravity (water = 1): 2.2 @ 68°F/20°C (solid).
Heat of solution: 3.1 Btu/lb = 1.7 cal/g = 0.071 x 10^5 J/kg.

ARSENIC DISULFIDE   **REC. A:5350**

SYNONYMS: DISULFURO de ARSENICO (Spanish); REALGAR; RED ARSENIC GLASS; RED ARSENIC SULFIDE; RED ORPIMENT; RUBY ARSENIC

IDENTIFICATION
CAS Number: 1303-32-8
Formula: As$_2$S$_2$
DOT ID Number: UN 1557; DOT Guide Number: 152
Proper Shipping Name: Arsenic sulfides
Reportable Quantity (RQ): **(CERCLA)** 1 lb/0.454 kg

DESCRIPTION: Yellow, orange or red-brown powder; changes to red at @ 338°F/170°C. Odorless. Sinks in water; insoluble; reacts to emit toxic and flammable vapors.

Poison! • Combustible Breathing the vapors or dust, skin or eye contact, or swallowing the material can kill you • Firefighting gear (including SCBA) does not provide adequate protection. If exposure occurs, remove and isolate gear immediately and thoroughly decontaminate personnel • Toxic products of combustion may include highly toxic fumes of sulfur dioxide, hydrogen sulfide, and arsenic trioxide.

Hazard Classification (based on NFPA-704 Rating System)
Health Hazards (Blue): 3; Flammability (Red): 0; Reactivity (Yellow): 0

EMERGENCY RESPONSE: See Appendix A (152)
Evacuation:
Public safety: Isolate the area of spill or leak for at least 25 to 50 meters (80 to 160 feet) in all directions.
Spill: Increase, in the downwind direction, as necessary, the distance shown under "Public Safety."
Fire: If tank, rail car, or tank truck is involved in fire, isolate for at least 800 meters (½ mile) in all directions; also, consider initial evacuation for 800 meters (½ mile) in all directions.

EXPOSURE
Short-term effects: *SEEK MEDICAL ATTENTION.* **Dust:** *POISONOUS IF INHALED.* Harmful to skin. Move to fresh air. *IF BREATHING HAS STOPPED,* give artificial respiration; *avoid mouth-to-mouth resuscitation; use bag/mask apparatus.* IF breathing is difficult, administer oxygen. **Solid:** *POISONOUS IF SWALLOWED.* Will burn eyes and skin. Remove contaminated clothing and shoes. Flush affected areas with plenty of water. *IF IN EYES,* hold eyelids open and flush with plenty of water. *IF SWALLOWED* and victim is *CONSCIOUS AND ABLE TO SWALLOW,* have victim drink 4 to 8 ounces of water and have victim induce vomiting. *IF SWALLOWED* and victim is *UNCONSCIOUS OR HAVING CONVULSIONS,* do nothing except keep victim warm.

HEALTH HAZARDS
Personal protective equipment (PPE): A-Level PPE. OSHA Table Z-1-A air contaminant.
Recommendations for respirator selection: NIOSH
At any concentrations above the NIOSH REL, or where there is no REL, at any detectable concentration: SCBAF:PD,PP (any self-contained breathing apparatus that has a full faceplate and is operated in a pressure-demand or other positive-pressure mode); or SAF:PD,PP:ASCBA (any supplied-air respirator that has a full facepiece and is operated in a pressure-demand or other positive-pressure mode in combination with an auxiliary self-contained breathing apparatus operated in a pressure-demand or other positive-pressure mode). *ESCAPE:* GMFAGHiE [any air-purifying, full-facepiece respirator (gas mask) with a chin-style, front-or back-mounted acid gas canister having a high-efficiency particulate filter]; or SCBAE (any appropriate escape-type, self-contained breathing apparatus).
Exposure limits (TWA unless otherwise noted): ACGIH TLV 0.01 mg/m^3; OSHA PEL [1910.1080] 0.010 mg/m^3; NIOSH REL

ceiling 0.002 mg/m³/15 min, as arsenic; potential human carcinogen; reduce exposure to lowest feasible level
Toxicity by ingestion: Grade 4; LD_{50} less than 50 mg/kg.
Long-term health effects: NTP known carcinogen. Possible skin and lung cancer.
Vapor (gas) irritant characteristics: May cause eye and respiratory tract irritation.
Liquid or solid irritant characteristics: May cause irritation of the eyes, skin, and respiratory tract.
IDLH value: Potential human carcinogen; 5 mg/m³ as arsenic.

FIRE DATA
Behavior in fire: May ignite at very high temperatures.

CHEMICAL REACTIVITY
Reactivity with water: Reacts with water or steam to emitting toxic and flammable vapors.
Binary reactants: Highly toxic fumes of sulfur and arsenic emitted on contact with acid or acid fumes.
Reacts with hydrogen peroxide; sodium sulfide; nitric acid; sulfur; potassium nitrate. Reacts vigorously with oxidizing materials.

ENVIRONMENTAL DATA
Water pollution: Harmful to aquatic life in very low concentrations. May be dangerous if it enters nearby water intakes; notify operators. Notify local health and wildlife officials.
Response to discharge: Issue warning–poison, water contaminant. Restrict access. Should be removed. Chemical and physical treatment.

SHIPPING INFORMATION
Grades of purity: 99+%; Technical; **Storage temperature:** Ambient; **Inert atmosphere:** None; **Venting:** Open; **Stability during transport:** Stable.

PHYSICAL AND CHEMICAL PROPERTIES
Physical state @ 59°F/15°C and 1 atm: Solid.
Molecular weight: 214
Boiling point @ 1 atm: 1049°F/565°C/838°K.
Melting/Freezing point: 585°F/307°C/580°K.
Specific gravity (water = 1): 3.5 @ 68°F/20°C (solid).

ARSENIC PENTOXIDE   REC. A:5400

SYNONYMS: ANHYDRIDE ARSENIQUE (French); ARSENIC ANHYDRIDE; ARSENIC OXIDE; ARSENIC(V) OXIDE; ARSENIC ACID ANHYDRIDE; ARSENIC PENTAOXIDE; DIARSENIC PENTOXIDE; FOTOX; PENTOXIDO de ARSENICO (Spanish); RCRA No. P011

IDENTIFICATION
CAS Number: 1303-28-2
Formula: As_2O_5
DOT ID Number: UN 1559; DOT Guide Number: 151
Proper Shipping Name: Arsenic pentoxide
Reportable Quantity (RQ): **(CERCLA)** 1 lb/0.454 kg

DESCRIPTION: White amorphous solid. Odorless. Sinks and mixes with water; highly soluble.

Poison! • Combustible Breathing the vapors or dust, skin or eye contact, or swallowing the material can kill you • Firefighting gear (including SCBA) does not provide adequate protection. If exposure occurs, remove and isolate gear immediately and thoroughly decontaminate personnel • Toxic products of combustion may include highly toxic fumes of arsenic.

Hazard Classification (based on NFPA-704 Rating System)
Health Hazards (Blue): 3; Flammability (Red): 0; Reactivity (Yellow): 0

EMERGENCY RESPONSE: See Appendix A (151)
Evacuation:
Public safety: Isolate the area of spill or leak for at least 25 to 50 meters (80 to 160 feet) in all directions.
Spill: Increase, in the downwind direction, as necessary, the distance shown under "Public Safety."
Fire: If tank, rail car, or tank truck is involved in fire, isolate for at least 800 meters (½ mile) in all directions; also, consider initial evacuation for 800 meters (½ mile) in all directions.

EXPOSURE
Short-term effects: *SEEK MEDICAL ATTENTION.* **Dust:** *POISONOUS IF INHALED OR IF SKIN IS EXPOSED.* Move to fresh air. If artificial respiration is administered, *avoid mouth-to-mouth resuscitation; use bag/mask apparatus.* If breathing is difficult, administer oxygen. **Solid:** *POISONOUS IF SWALLOWED. IF SWALLOWED* and victim is *CONSCIOUS AND ABLE TO SWALLOW,* have victim drink 4 to 8 ounces of water, and have victim induce vomiting. *IF SWALLOWED* and victim is *UNCONSCIOUS OR HAVING CONVULSIONS,* do nothing except keep victim warm.

HEALTH HAZARDS
Personal protective equipment (PPE): A-Level PPE. OSHA Table Z-1-A air contaminant.
Recommendations for respirator selection: NIOSH *At any concentrations above the NIOSH REL, or where there is no REL, at any detectable concentration:* SCBAF:PD,PP (any self-contained breathing apparatus that has a full faceplate and is operated in a pressure-demand or other positive-pressure mode); or SAF:PD,PP:ASCBA (any supplied-air respirator that has a full facepiece and is operated in a pressure-demand or other positive-pressure mode in combination with an auxiliary self-contained breathing apparatus operated in a pressure-demand or other positive-pressure mode). *ESCAPE:* GMFAGHiE [any air-purifying, full-facepiece respirator (gas mask) with a chin-style, front-or back-mounted acid gas canister having a high-efficiency particulate filter]; or SCBAE (any appropriate escape-type, self-contained breathing apparatus).
Exposure limits (TWA unless otherwise noted): ACGIH TLV 0.01 mg/m³; OSHA PEL [1910.1080] 0.010 mg/m³; NIOSH REL ceiling 0.002 mg/m³/15 min, as arsenic; potential human carcinogen; reduce exposure to lowest feasible level
Toxicity by ingestion: Grade 4; LD_{50} below 50 mg/kg.
Long-term health effects: NTP known carcinogen. Exposure can be followed by tumor development. Cancers of the skin, lungs and ethmoids have been attributed to As exposure.
Vapor (gas) irritant characteristics: May cause irritation of eyes and respiratory tract.
Liquid or solid irritant characteristics: May cause burning sensation and tenderness in affected areas of the skin.
IDLH value: Potential human carcinogen; 5 mg/m³ as arsenic.

CHEMICAL REACTIVITY
Reactivity with water: Dissolves in water to give solutions of arsenic acid.
Binary reactants: Reacts with halogens, acids, zinc, and aluminum.

ENVIRONMENTAL DATA
Water pollution: Harmful to aquatic life in very low concentrations. May be dangerous if it enters nearby water intakes; notify operators. Notify local health and wildlife officials.
Response to discharge: Issue warning–poison, water contaminant. Restrict access. Should be removed. Chemical and physical treatment.

SHIPPING INFORMATION
Grades of purity: 56% Arsenic; **Stability during transport:** Stable.

PHYSICAL AND CHEMICAL PROPERTIES
Physical state @ 59°F/15°C and 1 atm: Solid.
Molecular weight: 229.82
Boiling point @ 1 atm: Decomposes.
Melting/Freezing point: 1472°F/800°C/1073°K.
Specific gravity (water = 1): 4.32 @ 77°F/25°C.

ARSENIC TRICHLORIDE   REC. A:5450

SYNONYMS: ARSENIC CHLORIDE; ARSENOUS CHLORIDE; ARSENOUS TRICHLORIDE; BUTTER OF ARSENIC; EEC No. 033-002-00-5; FUMING LIQUID ARSENIC; ARSENOUS CHLORIDE; CAUSTIC ARSENIC CHLORIDE; TRICHLOROARSINE; TRICLORURO de ARSENICO (Spanish)

IDENTIFICATION
CAS Number: 7784-34-1
Formula: $AsCl_3$
DOT ID Number: UN 1560; DOT Guide Number: 157
Proper Shipping Name: Arsenic trichloride
Reportable Quantity (RQ): **(CERCLA)** 1 lb/.0454 kg

DESCRIPTION: Colorless to yellow, oily liquid; fumes in air. Acrid odor. Sinks and decomposes in water forming hydrochloric acid, arsenic trioxide, arsenic acid, and heat; visible vapor cloud is produced.

Poison! • Combustible • Corrosive to the skin, eyes, and respiratory tract; contact with skin or eyes can cause burns and vision impairment; inhalation symptoms may be delayed • Firefighting gear (including SCBA) does not provide adequate protection. If exposure occurs, remove and isolate gear immediately and thoroughly decontaminate personnel • Reacts with most metals to produce explosive hydrogen gas • Toxic products of combustion may include arsenic, hydrogen chloride • Do not put yourself in danger by entering a contaminated area to rescue a victim.

Hazard Classification (based on NFPA-704 Rating System)
Health Hazards (Blue): 3; Flammability (Red): 0; Reactivity (Yellow): 0

EMERGENCY RESPONSE: See Appendix A (157)
Evacuation:
Public safety: Isolate the area of spill or leak for at least 50 to 100 meters (160 to 330 feet) in all directions.
Spill: Small spill–First: Isolate in all directions 30 meters (100 feet). Then: Protect persons downwind, DAY: 0.2 km (0.1 mile); NIGHT: 0.3 km (0.2 mile). Large spill–First: Isolate in all directions 60 meters (200 feet); Then: Protect persons downwind, DAY: 0.6 km (0.4 mile); NIGHT: 1.4 km (0.9 mile).
Fire: If tank, rail car, or tank truck is involved in fire, isolate for at least 800 meters (½ mile) in all directions; also, consider initial evacuation for 800 meters (½ mile) in all directions.

EXPOSURE
Short-term effects: *SEEK MEDICAL ATTENTION.* **Vapor:** *POISONOUS IF INHALED.* Move victim to fresh air. If artificial respiration is administered, *avoid mouth-to-mouth resuscitation; use bag/mask apparatus.* If breathing is difficult, administer oxygen. Lung edema may develop. **Liquid:** *POISONOUS IF SWALLOWED.* Irritating to skin and eyes. Remove contaminated clothing and shoes. Flush affected areas with plenty of water. *IF IN EYES,* hold eyelids open and flush with plenty of water. *IF SWALLOWED* and victim is *CONSCIOUS AND ABLE TO SWALLOW,* have victim drink 4 to 8 ounces of water and have victim induce vomiting. *IF SWALLOWED* and victim is *UNCONSCIOUS OR HAVING CONVULSIONS,* do nothing except keep victim warm.
Note to physician or authorized medical personnel: Medical observation is recommended for 24 to 48 hours after breathing overexposure, as pulmonary edema may be delayed. As first aid for pulmonary edema, consider administering a corticosteroid spray. Cigarette smoking may exacerbate pulmonary injury and should be discouraged for at least 72 hours following exposure.

HEALTH HAZARDS
Personal protective equipment (PPE): A-Level PPE. OSHA Table Z-1-A air contaminant. Chemical protective material(s) reported to have good to excellent resistance: PVC, chlorinated polyethylene, nitrile, Neoprene®, Viton®.
Recommendations for respirator selection: NIOSH *At any concentrations above the NIOSH REL, or where there is no REL, at any detectable concentration:* SCBAF:PD,PP (any self-contained breathing apparatus that has a full faceplate and is operated in a pressure-demand or other positive-pressure mode); or SAF:PD,PP:ASCBA (any supplied-air respirator that has a full facepiece and is operated in a pressure-demand or other positive-pressure mode in combination with an auxiliary self-contained breathing apparatus operated in a pressure-demand or other positive-pressure mode). *ESCAPE:* GMFAGHiE [any air-purifying, full-facepiece respirator (gas mask) with a chin-style, front-or back-mounted acid gas canister having a high-efficiency particulate filter]; or SCBAE (any appropriate escape-type, self-contained breathing apparatus).
Exposure limits (TWA unless otherwise noted): ACGIH TLV 0.01 mg/m³; OSHA PEL [1910.1080] 0.010 mg/m³; NIOSH REL ceiling 0.002 mg/m³/15 min, as arsenic; potential human carcinogen; reduce exposure to lowest feasible level
Toxicity by ingestion: Grade 3; oral rat LD_{50} = 138 mg/kg; fatal human dose 70-180 mg, depending on weight.
Long-term health effects: NTP known carcinogen. May cause damage to liver, kidney and skin.
Vapor (gas) irritant characteristics: May cause irritation to the eyes, respiratory system, difficult breathing. Possible lung edema.
Liquid or solid irritant characteristics: Corrosive to the eyes, skin, and respiratory tract.
IDLH value: Potential human carcinogen; 5 mg/m³ as arsenic.

FIRE DATA
Fire extinguishing agents not to be used: Avoid water on adjacent fires.

CHEMICAL REACTIVITY
Reactivity with water: Reacts to generate hydrogen chloride (hydrochloric acid), arsenic acid, and arsenic trioxide.

Binary reactants: Keep away from ammonia (especially anhydrous), acids, halogens, oxidizers. Reacts with light forming arsenic trioxide and hydrogen chlorine. Corrodes aluminum and zinc. Sunlight forms hydrogen chloride and arsenic trioxide.
Neutralizing agents for acids and caustics: Flush with water, rinse with sodium bicarbonate or lime solution.

ENVIRONMENTAL DATA
Water pollution: Effect of low concentrations on aquatic life is unknown. May be dangerous if it enters nearby water intakes; notify operators. Notify local health and wildlife officials.
Response to discharge: Issue warning–poison, water contaminant, corrosive. Restrict access. Disperse and flush.

SHIPPING INFORMATION
Grades of purity: Commercial; **Storage temperature:** Ambient; **Inert atmosphere:** None; **Venting:** Pressure-vacuum; **Stability during transport:** Stable.

PHYSICAL AND CHEMICAL PROPERTIES
Physical state @ 59°F/15°C and 1 atm: Liquid.
Molecular weight: 181.3
Boiling point @ 1 atm: 266.4°F/130.2°C/403.4°K.
Melting/Freezing point: 16.7°F/–8°C/265°K.
Specific gravity (water = 1): 2.156 @ 77°F/25°C (liquid).
Liquid surface tension: (estimate) 20 dynes/cm = 0.020 N/m @ 68°F/20°C.
Latent heat of vaporization: 88.31 Btu/lb = 49.06 cal/g = 2.054 x 10^5 J/kg.
Heat of solution: (estimate) –18 Btu/lb = –10 cal/g = –0.42 x 10^5 J/kg.
Heat of fusion: 13.3 cal/g.
Vapor pressure: 8.5 mm.

ARSENIC TRIOXIDE   REC. A:5500

SYNONYMS: ACIDE ARSENIEUX (French); ANHYDRIDE ARSENIEUX (French); ARSENIC BLANC (French); ARSENIC (III) OXIDE; ARSENIC SESQUIOXIDE; ARSENICUM ALBUM; ARSENIGEN SAURE (German); ARSENIOUS ACID; ARSENIOUS OXIDE; ARSENIOUS TRIOXIDE; ARSENITE; ARSENOLITE; ARSENOUS ACID; ARSENOUS ACID ANHYDRIDE; ARSENOUS ANHYDRIDE; ARSENOUS OXIDE; ARSENOUS OXIDE ANHYDRIDE; ARSENIC SESQUIOXIDE; ARSENTRIOXIDE; ARSODENT; EEC No. 033-003-00-0; CLAUDELITE; CLAUDETITE; CRUDE ARSENIC; DIARSENIC TRIOXIDE; POI2; TRIOXIDO de ARSENICO (Spanish); WHITE ARSENIC

IDENTIFICATION
CAS Number: 1327-53-3
Formula: As_2O_3
DOT ID Number: UN 1561; DOT Guide Number: 151
Proper Shipping Name: Arsenic trioxide
Reportable Quantity (RQ): **(CERCLA)** 1 lb/0.454 kg

DESCRIPTION: White crystalline solid crystals or powder. Odorless. Sinks and mixes slowly with water, forming hydrochloric acid.

Combustible • Corrosive to the skin, eyes, and respiratory tract; contact with skin or eyes can cause burns and vision impairment; inhalation symptoms may be delayed • Firefighting gear (including SCBA) does not provide adequate protection. If exposure occurs, remove and isolate gear immediately and thoroughly decontaminate personnel • Toxic products of combustion may include arsenic trioxide and arsine.

Hazard Classification (based on NFPA-704 Rating System)
Health Hazards (Blue): 3; Flammability (Red): 0; Reactivity (Yellow): 0

EMERGENCY RESPONSE: See Appendix A (151)
Evacuation:
Public safety: Isolate the area of spill or leak for at least 25 to 50 meters (80 to 160 feet) in all directions.
Spill: Increase, in the downwind direction, as necessary, the distance shown under "Public Safety."
Fire: If tank, rail car, or tank truck is involved in fire, isolate for at least 800 meters (½ mile) in all directions; also, consider initial evacuation for 800 meters (½ mile) in all directions.

EXPOSURE
Short-term effects: *SEEK MEDICAL ATTENTION.* **Dust:** *POISONOUS IF INHALED.* Irritating to eyes, nose, and throat. Move victim to fresh air. *IF IN EYES,* hold eyelids open and flush with plenty of water. If artificial respiration is administered, *avoid mouth-to-mouth resuscitation; use bag/mask apparatus.* If breathing is difficult, administer oxygen. **Solid:** *POISONOUS IF SWALLOWED.* Irritating to skin and eyes. Remove contaminated clothing and shoes. Flush affected areas with plenty of water. *IF IN EYES,* hold eyelids open and flush with plenty of water. *IF SWALLOWED* and victim is *CONSCIOUS AND ABLE TO SWALLOW,* have victim drink water, lime water, sweet milk, or raw eggs, followed by castor oil or any brisk cathartic or have victim induce vomiting. *IF SWALLOWED* and victim is *UNCONSCIOUS OR HAVING CONVULSIONS,* do nothing except keep victim warm.

HEALTH HAZARDS
Personal protective equipment (PPE): A-Level PPE. OSHA Table Z-1-A air contaminant. Bureau of Mines approved respirator; protective gloves, eye protection; full protective coveralls.
Recommendations for respirator selection: NIOSH
At any concentrations above the NIOSH REL, or where there is no REL, at any detectable concentration: SCBAF:PD,PP (any self-contained breathing apparatus that has a full faceplate and is operated in a pressure-demand or other positive-pressure mode); or SAF:PD,PP:ASCBA (any supplied-air respirator that has a full facepiece and is operated in a pressure-demand or other positive-pressure mode in combination with an auxiliary self-contained breathing apparatus operated in a pressure-demand or other positive-pressure mode). *ESCAPE:* GMFAGHiE [any air-purifying, full-facepiece respirator (gas mask) with a chin-style, front-or back-mounted acid gas canister having a high-efficiency particulate filter]; or SCBAE (any appropriate escape-type, self-contained breathing apparatus).
Exposure limits (TWA unless otherwise noted): ACGIH TLV 0.01 mg/m³; OSHA PEL [1910.1080] 0.010 mg/m³; NIOSH REL ceiling 0.002 mg/m³/15 min, as arsenic; potential human carcinogen; reduce exposure to lowest feasible level
Toxicity by ingestion: Grade 4; oral mouse LD_{50} = 45 mg/kg.
Long-term health effects: NTP known carcinogen. Possible liver, kidney and skin disorders, or cancer.
Liquid or solid irritant characteristics: Corrosive to the eyes, skin, and respiratory tract.
IDLH value: Potential human carcinogen; 5 mg/m³ as arsenic.

FIRE DATA
Fire extinguishing agents not to be used: Halogenated agents.
Behavior in fire: May volatilize and form toxic fumes of arsenic trioxide.

CHEMICAL REACTIVITY
Reactivity with water: Will form a basic solution.
Binary reactants: Reacts with chlorine trifluoride, fluorine, hydrogen fluoride, sodium hydroxide, acids, aluminum, halogens, and chemically active metals. Attacks metals in the presence of moisture.
Neutralizing agents for acids and caustics: Flush with water.

ENVIRONMENTAL DATA
Water pollution: Harmful to aquatic life in very low concentrations. May be dangerous if it enters nearby water intakes; notify operators. Notify local health and wildlife officials.
Response to discharge: Issue warning–poison, water contaminant. Restrict access. Should be removed. Chemical and physical treatment.

SHIPPING INFORMATION
Grades of purity: Refined: 99%; Crude: 95%; **Storage temperature:** Ambient; **Inert atmosphere:** None; **Venting:** Pressure-vacuum; **Stability during transport:** Stable.

PHYSICAL AND CHEMICAL PROPERTIES
Physical state @ 59°F/15°C and 1 atm: Solid.
Molecular weight: 197.8
Boiling point @ 1 atm: 855°F/457°C/730°K (sublimes).
Melting/Freezing point: 379°F/193°C/466°K (sublimes).
Specific gravity (water = 1): 3.7 @ 68°F/20°C (solid).
Heat of fusion: 22.2 cal/g.
Vapor pressure: 66 mm @ 312°C.

ARSENIC TRISULFIDE   REC. A:5550

SYNONYMS: ARSENIC SULFIDE; ARSENIC SESQUISULFIDE; ARSENIC YELLOW; DIARSENIC TRISULFIDE; KING'S GOLD; KING'S YELLOW; TRISULFURO de ARSENICO (Spanish); YELLOW ARSENIC SULFIDE

IDENTIFICATION
CAS Number: 1303-33-9
Formula: As_2S_3
DOT ID Number: NA 1557; DOT Guide Number: 152
Proper Shipping Name: Arsenic sulfide
Reportable Quantity (RQ): **(CERCLA)** 1 lb/0.454 kg

DESCRIPTION: Yellow-orange solid. Odorless. Sinks in water, forming hydrogen sulfide.

Poison! • Combustible Breathing the vapors or dust, skin or eye contact, or swallowing the material can kill you • Firefighting gear (including SCBA) does not provide adequate protection. If exposure occurs, remove and isolate gear immediately and thoroughly decontaminate personnel • Toxic products of combustion may include arsine, sulfur dioxide, hydrogen sulfide, and arsenic trioxide.

Hazard Classification (based on NFPA-704 Rating System)
Health Hazards (Blue): 3; Flammability (Red): 0; Reactivity (Yellow): 0

EMERGENCY RESPONSE: See Appendix A (152)
Evacuation:
Public safety: Isolate the area of spill or leak for at least 25 to 50 meters (80 to 160 feet) in all directions.
Spill: Increase, in the downwind direction, as necessary, the distance shown under "Public Safety."
Fire: If tank, rail car, or tank truck is involved in fire, isolate for at least 800 meters (½ mile) in all directions; also, consider initial evacuation for 800 meters (½ mile) in all directions.

EXPOSURE
Short-term effects: *SEEK MEDICAL ATTENTION*. **Dust:** *POISONOUS IF INHALED*. Harmful to skin. Move to fresh air. *IF BREATHING HAS STOPPED*, give artificial respiration; *avoid mouth-to-mouth resuscitation; use bag/mask apparatus*. IF breathing is difficult, administer oxygen. **Solid:** *POISONOUS IF SWALLOWED*. Will burn eyes and skin. Remove contaminated clothing and shoes. Flush affected areas with plenty of water. *IF IN EYES*, hold eyelids open and flush with plenty of water. *IF SWALLOWED* and victim is *CONSCIOUS AND ABLE TO SWALLOW*, have victim drink 4 to 8 ounces of water and have victim induce vomiting. *IF SWALLOWED* and victim is *UNCONSCIOUS OR HAVING CONVULSIONS*, do nothing except keep victim warm.

HEALTH HAZARDS
Personal protective equipment (PPE): A-Level PPE. OSHA Table Z-1-A air contaminant. Self-contained breathing apparatus; goggles; rubber gloves; clean protective clothing
Recommendations for respirator selection: NIOSH
At any concentrations above the NIOSH REL, or where there is no REL, at any detectable concentration: SCBAF:PD,PP (any self-contained breathing apparatus that has a full faceplate and is operated in a pressure-demand or other positive-pressure mode); or SAF:PD,PP:ASCBA (any supplied-air respirator that has a full facepiece and is operated in a pressure-demand or other positive-pressure mode in combination with an auxiliary self-contained breathing apparatus operated in a pressure-demand or other positive-pressure mode). *ESCAPE:* GMFAGHiE [any air-purifying, full-facepiece respirator (gas mask) with a chin-style, front-or back-mounted acid gas canister having a high-efficiency particulate filter]; or SCBAE (any appropriate escape-type, self-contained breathing apparatus).
Exposure limits (TWA unless otherwise noted): ACGIH TLV 0.01 mg/m^3; OSHA PEL [1910.1080] 0.010 mg/m^3; NIOSH REL ceiling 0.002 mg/m^3/15 min, as arsenic; potential human carcinogen; reduce exposure to lowest feasible level
Toxicity by ingestion: Grade 4; LD$_{50}$ less than 50 mg/kg.
Long-term health effects: NTP known carcinogen. Possible skin and lung cancer
IDLH value: Potential human carcinogen; 5 mg/m^3 as arsenic.

FIRE DATA
Fire extinguishing agents not to be used: Halogenated agents.
Behavior in fire: May ignite at very high temperatures.

CHEMICAL REACTIVITY
Reactivity with water: Forms hydrogen sulfide.
Binary reactants: Hydrogen peroxide, sodium sulfide, nitric acid, sulfur; potassium nitrate. Contact with acid fumes cause emission of fumes of sulfur and arsenic. Reacts vigorously on contact with oxidizing materials.

ENVIRONMENTAL DATA
Water pollution: Harmful to aquatic life in very low

concentrations. May be dangerous if it enters nearby water intakes; notify operators. Notify local health and wildlife officials. **Response to discharge:** Issue warning–poison, water contaminant. Restrict access. Should be removed. Chemical and physical treatment.

SHIPPING INFORMATION
Grades of purity: Technical; Pure, 99+%; Optical grade, 99.99+%; **Storage temperature:** Ambient; **Venting:** Open; **Stability during transport:** Stable.

PHYSICAL AND CHEMICAL PROPERTIES
Physical state @ 59°F/15°C and 1 atm: Solid.
Molecular weight: 246
Boiling point @ 1 atm: 1305°F/707°C/980°K.
Melting/Freezing point: 572°F/300°C/573°K.
Specific gravity (water = 1): 3.43 @ 68°F/20°C (solid).

ARSINE   REC. A:5575

SYNONYMS: AGENT SA; ARSENIC HYDRIDE; ARSENIC TRIHYDRIDE; ARSENIURETTED HYDROGEN; ARSENIOUS HYDRIDE; HYDROGEN ARSENIC; HYDROGEN ARSENIDE; SA

IDENTIFICATION
CAS Number: 7784-42-1
Formula: AsH_3
DOT ID Number: UN 2188; DOT Guide Number: 119
Proper Shipping Name: Arsine; SA
Reportable Quantity: **(EHS):** 1 lb/0.454 kg

DESCRIPTION: Colorless gas; thermally unstable (decomposes above 446°F/230°C with deposition of shiny, black arsenic); generally shipped in cylinders as a liquefied compressed gas. Mild, garlic-like or fishy odor. Slightly soluble in water. *Warning:* Odor is not an adequate indicator of arsine's presence and does not provide reliable warning of hazardous concentrations. Because this chemical is nonirritating and produces no immediate symptoms, persons exposed to hazardous levels may not be unaware of its presence.

Poison! (Hazard Zone A) • Extremely flammable • Breathing the vapor can kill you • Firefighting gear (including SCBA) provides NO protection. If exposure occurs, remove and isolate gear immediately and thoroughly decontaminate personnel • Exposure of cylinder(s) to fire, flame, or elevated temperatures may cause cylinders to rupture or cause frangible disc to burst, releasing entire contents of cylinder. Ruptured or venting cylinders may rocket through buildings or travel a considerable distance • Gas forms explosive mixture with air • Gas is heavier than air and will collect and stay in low areas • Gas may travel long distances to ignition sources and flashback • Gas or vapors in confined areas (e.g., tanks, sewers, buildings) may explode when exposed to fire • Toxic products of combustion include arsenic trioxide and other arsenic oxides • Do not put yourself in danger by entering a contaminated area to rescue a victim. Contact with the liquid (compressed gas) can cause frostbite injury to the skin or eyes. Exposure to 250 ppm for 30 minutes or 500 ppm for a few minutes can cause death.

Hazard Classification (based on NFPA-704 Rating System)
Health Hazards (Blue): 4; Flammability (Red): 4 ; Reactivity (Yellow): 2

EMERGENCY RESPONSE: See Appendix A (119)
Evacuation:
Public safety: See below.
Spill: Small spill–First: Isolate in all directions at least 60 meters (200 feet) in all directions; Then: Protect persons downwind, DAY: 0.5 km (0.3 mile); NIGHT: 2.1 km (1.3 miles). Large spill–First: Isolate in all directions 335 meters (1100 feet); Then: Protect persons downwind, DAY: 3.2 km (2.0 miles); NIGHT: 6.6 km (4.1 miles).
Fire: Isolate for 800 meters (½ mile) in all directions, especially if tank, rail car, or tank truck is involved in fire.

[Weaponized (Agent SA)] Small spill–First: Isolate in all directions at least 60 meters (200 feet) in all directions; Then: Protect persons downwind, DAY: 0.8 km (0.5 mile); NIGHT: 2.4 km (1.5 miles). Large spill–First: Isolate in all directions 400 meters (1300 feet); Then: Protect persons downwind, DAY: 4.0 km (2.5 miles); NIGHT: 8.0 km (5.0 miles).

EXPOSURE
Short-term effects: *CALL FOR MEDICAL AID.* Arsine is a highly toxic gas and may be fatal if inhaled in sufficient quantities. Although arsine is related to arsenic, it does not produce the usual signs and symptoms of arsenic poisoning. There are no antidotes for arsine poisoning. Inhalation is the major route of exposure. *IF BREATHING HAS STOPPED,* give artificial respiration; *avoid mouth-to-mouth resuscitation; use bag/mask apparatus.* Its primary toxic effect is due to hemolysis resulting in renal failure. Initially some patients may look relatively well. Common initial symptoms of exposure include malaise, headache, thirst, shivering, abdominal pain, and dyspnea. These symptoms usually occur within 30 to 60 minutes with heavy exposure, but can be delayed for 2 to 24 hours. Hemoglobinuria usually occurs within hours, jaundice within 1 or 2 days. Call your doctor or the Emergency Department if any unusual signs or symptoms develop within the next 24 to 72 hours, especially: unusual fatigue or weakness, shortness of breath; abnormal urine color (red or brown), stomach pain or tenderness, unusual skin color (yellow or bronze). *Skin Exposure* In case of frostbite injury, irrigate with lukewarm (108°F/42°C) water according to standard treatment. *Eye Exposure:* Red staining of the conjuctiva may be an early sign of arsine poisoning. In case of frostbite injury, do not remove clothes; ensure that thorough warming with lukewarm water or saline has been completed. Examine the eyes for corneal damage and treat appropriately. Immediately consult an ophthalmologist for patients who have corneal injuries.
Note to physician and medical personnel: **Do not administer arsenic chelating drugs.** Although BAL (British Anti-Lewisite, dimercaprol) and other chelating agents are acceptable for arsenic poisoning, they are not effective antidotes for arsine poisoning and are NOT recommended-if symptoms indicate, use exchange transfusion. All patients who have suspected arsine exposure should be carefully observed for 24 hours, including hourly urine output. Onset of hemolysis may be delayed for up to 24 hours, and acute renal failure may not become evident for as long as 72 hours after exposure. Also, medical observation is recommended for 24 to 48 hours after breathing overexposure, as pulmonary edema or bronchopneumonia may be delayed. As first aid for pulmonary edema, consider administering a corticosteroid spray. Cigarette smoking may exacerbate pulmonary injury and should be discouraged for at least 72 hours following exposure.

HEALTH HAZARDS
Personal protective equipment (PPE): A-Level PPE. Arsine is a highly toxic systemic poison. *Respiratory Protection*: Positive-

pressure, self-contained breathing apparatus (SCBA) is recommended in response situations that involve exposure to potentially unsafe levels of arsine. Full-facepiece respirators are recommended. *Skin Protection*: Although most authorities recommend special protective clothing; insulated gloves, *arsine gas is not absorbed through the skin* and does not cause skin irritation. However, contact with the liquid (compressed gas) can cause frostbite injury to the skin or eyes.

Recommendations for respirator selection: NIOSH
At any concentrations above the NIOSH REL, or where there is no REL, at any detectable concentration: SCBAF:PD,PP (any self-contained breathing apparatus that has a full facepiece and is operated in a pressure-demand or other positive-pressure mode); or SAF:PD,PP:ASCBA (any supplied-air respirator that has a full facepiece and is operated in a pressure-demand or other positive-pressure mode in combination with an auxiliary self-contained breathing apparatus operated in a pressure-demand or other positive pressure mode). *ESCAPE:* GMFS [any air-purifying, full-facepiece respirator (gas mask) with a chin-style, front- or back-mounted canister providing protection against the compound of concern]; or SCBAE (any appropriate escape-type, self-contained breathing apparatus).

Exposure limits (TWA unless otherwise noted): ACGIH TLV 0.05 ppm; OSHA PEL 0.05 ppm (0.2 mg/m^3); AIHA ERPG-2 (emergency response planning guideline) = 0.5 ppm (maximum airborne concentration below which it is believed that nearly all persons could be exposed for up to 1 hour without experiencing or developing irreversible or other serious health effects or symptoms that could impair their abilities to take protective action).

Short-term exposure limits (15-minute TWA): NIOSH [Ca] ceiling 0.002 mg/m^3.

Long-term health effects: Although arsine has not been classified for carcinogenic effects, arsenic compounds and metabolites have been classified as known human carcinogens by IARC and EPA. May cause lung and lymphatic cancer.

Odor threshold: 0.5 ppm. This is 10-fold greater than the OSHA Permissible Exposure Limit (PEL).

IDLH value: Potential human carcinogen; 3 ppm.

FIRE DATA
Flash point: Flammable gas.
Fire extinguishing agents not to be used: Halogens. Do not spray liquid with water.
Flammable limits in air: LEL: 5.1%; UEL: 78.0%.
Electrical hazard: Flow or agitation of substance may generate electrostatic charges due to low conductivity.

CHEMICAL REACTIVITY
Binary reactants: Halogens, oxidizers, strong acids. Moist arsine decomposes quickly on exposure to light.

ENVIRONMENTAL DATA
Water pollution: Water solubility: 20% @ 68°F/20°C. DOT Appendix B, §172.101–marine pollutant. Dangerous if it enters nearby water intakes; notify operators. Notify local health and wildlife officials. **Response to discharge:** Issue warning–poison, high flammability. Seek expert help for cleanup.

SHIPPING INFORMATION
Grades of purity: 99% pure (technical); may be mixed with other gases.; **Storage temperature:** Ambient, protect from heat and extreme shock; **Inert atmosphere:** None; **Stability during transport:** Stable.

PHYSICAL AND CHEMICAL PROPERTIES
Physical state @ 59°F/15°C and 1 atm:
Molecular weight: 77.95
Boiling point @ 1 atm: −80.4°F/−62.5°C/207.5°K.
Melting/Freezing point: −240°F/−116°C/157°K.
Relative vapor density (air = 1): 2.7
Vapor pressure: >760 mm; 11,000 mmHg @ 68°F/20°C.

ASPHALT   REC. A:5600

SYNONYMS: ASFALTO (Spanish); ASPHALT BITUMEN; ASPHALTIC BITUMEN; ASPHALT CEMENTS; ASPHALTUM; BITUMEN (European term); JUDEAN PITCH; MINERAL PITCH; PETROLEUM ASPHALT; PETROLEUM PITCH; ROAD ASPHALT; ROAD TAR; TARS Liquid.

IDENTIFICATION
CAS Number: 8052-42-4
Formula: Mixture of solid or semi-solid hydrocarbons
DOT ID Number: UN 1999; DOT Guide Number: 130
Proper Shipping Name: Asphalt, at or above its flashpoint

DESCRIPTION: Dark brown to black thick liquid (generally heated); rubbery solid is produced when cooled. Tar-like odor. May float or sink in water.

Combustible • Dust cloud may explode if ignited in an enclosed area • Containers may BLEVE when exposed to fire • Irritating to the skin, eyes, and respiratory tract • Toxic products of combustion may include carbon monoxide and carcinogenic fumes.

Hazard Classification (based on NFPA-704 Rating System)
Health Hazards (Blue): 0; Flammability (Red): 2; Reactivity (Yellow): 0

EMERGENCY RESPONSE: See Appendix A (130)
Evacuation:
Public safety: Isolate spill area for at least 50 to 100 meters (160 to 330 feet) in all directions.
Spill: Large spill–Consider initial downwind evacuation for at least 300 meters (1000 feet).
Fire: Isolate for 800 meters (½ mile) in all directions, especially if tank, rail car, or tank truck is involved in fire.

EXPOSURE
Short-term effects: Liquid: Will burn skin and eyes. Flush affected areas with plenty of water. Severe burns may result from contact with hot asphalt. If molten asphalt strikes the exposed skin, cool the skin immediately by quenching with cold water; do not try to scrub off adhering materials; *SEEK MEDICAL ATTENTION*. A burn should be covered with a sterile dressing, and the patient should be taken immediately to a hospital.

HEALTH HAZARDS
Personal protective equipment (PPE): B-Level PPE. Chemical protective material(s) reported to have good to excellent resistance: Viton/neoprene.

Recommendations for respirator selection: NIOSH
At any concentrations above the NIOSH REL, or where there is no REL, at any detectable concentration: SCBAF:PD,PP (any self-contained breathing apparatus that has a full facepiece and is operated in a pressure-demand or other positive-pressure mode); or SAF:PD,PP:ASCBA (any supplied-air respirator that has a full facepiece and is operated in a pressure-demand or other positive-

ASPHALT BLENDING STOCKS: ROOFERS FLUX
REC. A:5650

SYNONYMS: ASPHALT; ASPHALT (CUT BACK); ASPHALTUM; ASPHALTUM OIL; BITUMEN (European term); DUST-LAYING OIL; JUDEAN PITCH; LIQUID ASPHALTUM; FLUXING OIL; MINERAL PITCH; PETROLEUM PITCH; PETROLEUM TAILINGS; RESIDUAL OIL; ROAD ASPHALT; ROAD OIL; ROAD TAR

IDENTIFICATION
CAS Number: 8052-42-4
DOT ID Number: NA 1999; DOT Guide Number: 130
Proper Shipping Name: Asphalt, at or above its flash point

DESCRIPTION: Brown to black oily liquid (generally heated); rubbery solid when cooled. Tar odor. May float or sink in water.

Combustible • Dust cloud may explode if ignited in an enclosed area • Containers may BLEVE when exposed to fire • Irritating to the skin, eyes, and respiratory tract • Toxic products of combustion may include carbon monoxide and carcinogenic fumes.

Hazard Classification (based on NFPA-704 Rating System)
Health Hazards (Blue): 0; Flammability (Red): 1; Reactivity (Yellow): 0

EMERGENCY RESPONSE: See Appendix A (130)
Evacuation:
Public safety: Isolate spill area for at least 50 to 100 meters (160 to 330 feet) in all directions.
Spill: Large spill–Consider initial downwind evacuation for at least 300 meters (1000 feet).
Fire: Isolate for 800 meters (½ mile) in all directions, especially if tank, rail car, or tank truck is involved in fire.

EXPOSURE
Short-term effects: *SEEK MEDICAL ATTENTION*. **Liquid:** Will burn skin and eyes. Harmful if swallowed.
Flush affected areas with plenty of water. *IF IN EYES*, hold eyelids open and flush with plenty of water. *IF SWALLOWED* and victim is *CONSCIOUS AND ABLE TO SWALLOW*, have victim drink 4 to 8 ounces of water. **Do not induce vomiting;**, do NOT lavage. Administer 2-4 ounces of olive oil and 1-2 ounces of activated charcoal.

HEALTH HAZARDS
Personal protective equipment (PPE): B-Level PPE. Protective clothing; face and eye protection
Recommendations for respirator selection: NIOSH *At any concentrations above the NIOSH REL, or where there is no REL, at any detectable concentration:* SCBAF:PD,PP (any self-contained breathing apparatus that has a full facepiece and is operated in a pressure-demand or other positive-pressure mode); or SAF:PD,PP:ASCBA (any supplied-air respirator that has a full facepiece and is operated in a pressure-demand or other positive-pressure mode in combination with an auxiliary, self-contained breathing apparatus operated in a pressure-demand or other positive pressure mode). *ESCAPE:* GMFOVHiE [any air-purifying, full-facepiece respirator (gas mask) with a chin-style, front- or back-mounted organic vapor canister having a high-efficiency particulate filter]; or SCBAE (any appropriate escape-type, self-contained breathing apparatus).
Exposure limits (TWA unless otherwise noted): ACGIH TLV 0.5 mg/m^3; NIOSH (fumes) ceiling 5 mg/m^3/15 min; carcinogen
Toxicity by ingestion: Grade 1; LD$_{50}$ 5 to 15 g/kg.
Long-term health effects: Asphalt fumes, OSHA specifically regulated carcinogen.
Vapor (gas) irritant characteristics: Vapors cause a slight smarting of the eyes or respiratory system if present in high concentrations. The effect may be temporary.
Liquid or solid irritant characteristics: Causes smarting of the skin and first-degree burns on short exposure; may cause secondary burns on long exposure.
IDLH value: Human carcinogen. Has not been determined.

FIRE DATA
Flash point: 300-550°F/149-288°C (oc).
Fire extinguishing agents not to be used: Water or foam may cause frothing. Do not direct a solid stream of water into hot asphalt.
Autoignition temperature: 400-700°F/204-371°C/477-644°K.
Electrical hazard: Class I, Group D.

CHEMICAL REACTIVITY
Binary reactants: Nitric acid.
Reactivity group: 33
Compatibility class: Miscellaneous hydrocarbon mixtures

ENVIRONMENTAL DATA
Water pollution: Effect of low concentrations on aquatic life is unknown. Fouling TO SHORELINE. May be dangerous if it enters nearby water intakes; notify operators. Notify local health and pollution control officials. **Response to discharge:** Mechanical containment. Chemical and physical treatment.

SHIPPING INFORMATION
Grades of purity: Each of the following is available in several grades: Asphalt cement, rapid-curing liquid asphalt, medium-curing liquid asphalt, slow-curing liquid asphalt (road oil), emulsified asphalt, inverted asphaltic emulsion, oxidized (air-blown) asphalt; **Storage temperature:** Elevated; **Inert atmosphere:** None; **Venting:** Open (flame arrester); **Stability during transport:** Stable.

NAS HAZARD CLASSIFICATION FOR BULK WATER TRANSPORTATION
FIRE: 1
HEALTH: Vapor irritant: 1; Liquid or solid irritant: 2; Poisons: 1
WATER POLLUTION: Human toxicity: 0; Aquatic toxicity: 1; Aesthetic effect: 4
REACTIVITY: Other chemicals: 0; Water: 0; Self-reaction: 0

PHYSICAL AND CHEMICAL PROPERTIES
Physical state @ 59°F/15°C and 1 atm: Liquid.
Boiling point @ 1 atm: 694°F/371°C/644°K.
Specific gravity (water = 1): (estimate) 1.00 @ 68°F/20°C (liquid).
Liquid water interfacial tension: 70 dynes/cm = 0.07 N/m at 77°C.
Vapor pressure: (Reid) Varies.

Toxicity by ingestion: Grade 2; LD_{50} = 0.5 to 5 g/kg.
Long-term health effects: May be carcinogenic.
Vapor (gas) irritant characteristics: Vapors cause irritation of eyes or respiratory system if present in high concentrations. The effect may be temporary.
Liquid or solid irritant characteristics: Causes smarting of the skin and first-degree burns on short exposure; may cause secondary burns on long exposure.

FIRE DATA
Flash point: 300–550°F/149–288°C (cc)
Fire extinguishing agents not to be used: Water or foam may cause frothing.
Autoignition temperature: 400–700°F/204–371°C/477–644°K.

CHEMICAL REACTIVITY
Reactivity group: 33
Compatibility class: Miscellaneous hydrocarbon mixtures

ENVIRONMENTAL DATA
Water pollution: Effect of low concentrations on aquatic life is unknown. Fouling to shoreline. May be dangerous if it enters nearby water intakes; notify operators. Notify local health and pollution control officials. **Response to discharge:** Mechanical containment. Chemical and physical treatment.

SHIPPING INFORMATION
Storage temperature: Elevated; **Inert atmosphere:** None; **Venting:** Open (flame arrester); **Stability during transport:** Stable.

NAS HAZARD CLASSIFICATIONS FOR BULK WATER TRANSPORTATION
FIRE: 1
HEALTH: Vapor irritant: 1; Liquid or solid irritant: 2; Poisons: 1
WATER POLLUTION: Human toxicity: 0; Aquatic toxicity: 1; Aesthetic effect: 4
REACTIVITY: Other chemicals: 0; Water: 0; Self-reaction: 0

PHYSICAL AND CHEMICAL PROPERTIES
Physical state @ 59°F/15°C and 1 atm: Liquid.
Melting/Freezing point: 20-110°F/–7-43°C/266-316°K.
Specific gravity (water = 1): (estimate) 1.11 at 50°C (liquid).

ASPHALT BLENDING STOCKS: STRAIGHT RUN RESIDUE REC. A:5700

SYNONYMS: BITUMEN (European term); PETROLEUM RESIDUE; ROAD BINDER; SEAL-COATING MATERIAL; RESIDUAL ASPHALT; PETROLEUM PITCH; CARPETING MEDIUM

IDENTIFICATION
CAS Number: 8052-42-4
DOT ID Number: UN 1999; DOT Guide Number: 130
Proper Shipping Name: Tars, liquid, including road asphalt and oils, bitumen and cut backs.

DESCRIPTION: Brown to black oily liquid (generally heated); rubbery solid when cooled. Tar odor. May float or sink in water.

Combustible • Vapor may explode if ignited in an enclosed area • Containers may BLEVE when exposed to fire • Irritating to the skin, eyes, and respiratory tract • Toxic products of combustion may include carbon monoxide and carcinogenic fumes.

Hazard Classification (based on NFPA-704 Rating System)
Health Hazards (Blue): 0; Flammability (Red): 1; Reactivity (Yellow): 0

EMERGENCY RESPONSE: See Appendix A (130)
Evacuation:
Public safety: Isolate spill area for at least 50 to 100 meters (160 to 330 feet) in all directions.
Spill: Large spill–Consider initial downwind evacuation for at least 300 meters (1000 feet).
Fire: Isolate for 800 meters (½ mile) in all directions, especially if tank, rail car, or tank truck is involved in fire.

EXPOSURE
Short-term effects: *SEEK MEDICAL ATTENTION*. **Liquid:** Will burn skin and eyes.
Flush affected areas with plenty of water. *IF IN EYES*, hold eyelids open and flush with plenty of water. *IF SWALLOWED* and victim is *CONSCIOUS AND ABLE TO SWALLOW*, have victim drink 4 to 8 ounces of water. **Do NOT induce vomiting.**

HEALTH HAZARDS
Personal protective equipment (PPE): B-Level PPE. Protective clothing; eye and face protection
Recommendations for respirator selection: NIOSH *At any concentrations above the NIOSH REL, or where there is no REL, at any detectable concentration:* SCBAF:PD,PP (any self-contained breathing apparatus that has a full facepiece and is operated in a pressure-demand or other positive-pressure mode); or SAF:PD,PP:ASCBA (any supplied-air respirator that has a full facepiece and is operated in a pressure-demand or other positive-pressure mode in combination with an auxiliary, self-contained breathing apparatus operated in a pressure-demand or other positive pressure mode). *ESCAPE:* GMFOVHiE [any air-purifying, full-facepiece respirator (gas mask) with a chin-style, front- or back-mounted organic vapor canister having a high-efficiency particulate filter]; or SCBAE (any appropriate escape-type, self-contained breathing apparatus).
Exposure limits (TWA unless otherwise noted): ACGIH TLV 0.5 mg/m^3; NIOSH (fumes) ceiling 5 mg/m^3/15 min; carcinogen
Toxicity by ingestion: Grade 1; LD_{50} = 5 to 15 g/kg.
Vapor (gas) irritant characteristics: Vapors cause a slight smarting of the eyes or respiratory system if present in high concentrations. The effect may be temporary.
Liquid or solid irritant characteristics: Causes smarting of the skin and first-degree burns on short exposure; may cause secondary burns on long exposure.

FIRE DATA
Flash point: 400-600°F/204-316°C (oc).
Fire extinguishing agents not to be used: Water or foam may cause frothing.
Autoignition temperature: 450-700°F/232-371°C.

CHEMICAL REACTIVITY
Reactivity with water: May cause frothing.
Reactivity group: 33
Compatibility class: Miscellaneous hydrocarbon mixtures

ENVIRONMENTAL DATA
Water pollution: Effect of low concentrations on aquatic life is unknown. Fouling to shoreline. May be dangerous if it enters

nearby water intakes; notify operators. Notify local health and pollution control officials. **Response to discharge:** Mechanical containment. Chemical and physical treatment.

SHIPPING INFORMATION
Storage temperature: Elevated; **Inert atmosphere:** None; **Venting:** Open (flame arrester); **Stability during transport:** Stable.

NAS HAZARD CLASSIFICATION FOR BULK WATER TRANSPORTATION
FIRE: 1
HEALTH: Vapor irritant: 1; Liquid or solid irritant: 2; Poisons: 1
WATER POLLUTION: Human toxicity: 0; Aquatic toxicity: 1; Aesthetic effect: 4
REACTIVITY: Other chemicals: 0; Water: 0; Self-reaction: 0

PHYSICAL AND CHEMICAL PROPERTIES
Physical state @ 59°F/15°C and 1 atm: Solid.
Melting/Freezing point: 80–225°F/26–107°C/299–380°K.
Vapor pressure: (Reid) Varies

ATRAZINE   REC. A:5750

SYNONYMS: A 361; AATREX; AATREX 4L; AATREX 80W; 2-AETHYLAMINO-4-CHLOR-6-ISOPROPYLAMINO-1,3,5-TRIAZIN (German); AKTIKON; AKTIKON PK; AKTINIT A; AKTINIT PK; ARGEZIN; ATAZINAX; ATRANEX; ATRASINE; ATRATOL A; ATRAZINA (Spanish); ATRED; CANDEX; CEKUZINA-T; 2-CHLORO-4-ETHYLAMINEISOPROPYLAMINE-s-TRIAZINE; 1-CHLORO-3-ETHYLAMINO-5-ISOPROPYLAMINO-s-TRIAZINE; 1-CHLORO-3-ETHYLAMINO-5-ISOPROPYLAMINO-2,4,6-TRIAZINE; 2-CHLORO-4-ETHYLAMINO-6-ISOPROPYLAMINO-s-TRIAZINE; 2-CHLORO-4-ETHYLAMINO-6-ISOPROPYLAMINO-1,3,5-TRIAZINE; 6-CHLORO-N-ETHYL-N'-(1-METHYLETHYL)-1,3,5-TRIAZINE-2,4-DIAMINE; 2-CHLORO-4-(2-PROPYLAMINO)-6-ETHYLAMINO-s-TRIAZINE; CRISAZINE; CYAZIN; FENAMIN; FENAMINE; FENATROL; G30027; GEIGY 30,027; GESAPRIM; GESOPRIM; HUNGAZIN; HUNGAZIN PK; INAKOR; OLEOGESAPRIM; PENATROL; PRIMATOL; PRIMATOL A; PRIMAZE; RADAZIN; RADIZINE; STRAZINE; TRIAZINE A1294; s-TRIAZINE, ZEAZIN; 1,3,5-TRIAZINE-2,4-DIAMINE,6-CHLORO-N-ETHYL-N'-(1-METHYLETHYL)-; VECTAL; VECTAL SC; WEEDEX A; WONUK; ZEAZINE

IDENTIFICATION
CAS Number: 1912-24-9
Formula: $C_8H_{14}N_5Cl$
DOT ID Number: UN 2763; DOT Guide Number: 151
Proper Shipping Name: Triazine pesticide, solid, toxic ; Triazine pesticide, solid, poisonous

DESCRIPTION: White crystalline solid. Sinks in water.

Poison! (triazine pesticide) • Not combustible, but may be mixed with flammable liquids • Firefighting gear (including SCBA) does not provide adequate protection. If exposure occurs, remove and isolate gear immediately and thoroughly decontaminate personnel • Containers may BLEVE when exposed to fire • Toxic products of combustion may include nitrogen oxides and hydrogen chloride.

Hazard Classification (based on NFPA-704 Rating System)
Health Hazards (Blue): 1; Flammability (Red): 0; Reactivity (Yellow): 0

EMERGENCY RESPONSE: See Appendix A (151)
Evacuation:
Public safety: Isolate the area of spill or leak for at least 25 to 50 meters (80 to 160 feet) in all directions.
Spill: Increase, in the downwind direction, as necessary, the distance shown under "Public Safety."
Fire: If tank, rail car, or tank truck is involved in fire, isolate for at least 800 meters (½ mile) in all directions; also, consider initial evacuation for 800 meters (½ mile) in all directions.

EXPOSURE
Short-term effects: *SEEK MEDICAL ATTENTION*. **Dust:** *POISONOUS IF INHALED*. Irritating to eyes, nose, and throat. Move victim to fresh air. *IF IN EYES*, hold eyelids open and flush with plenty of water. *IF BREATHING HAS STOPPED*, give artificial respiration; *avoid mouth-to-mouth resuscitation; use bag/mask apparatus*. IF breathing is difficult, administer oxygen. **Solid:** *POISONOUS IF SWALLOWED*. Irritating to skin and eyes. Remove contaminated clothing and shoes. Flush affected areas with plenty of water. *IF IN EYES*, hold eyelids open and flush with plenty of water. *IF SWALLOWED* and victim is *CONSCIOUS AND ABLE TO SWALLOW*, have victim drink 4 to 8 ounces of water and have victim induce vomiting. *IF SWALLOWED* and victim is *UNCONSCIOUS OR HAVING CONVULSIONS*, do nothing except keep victim warm.

HEALTH HAZARDS
Personal protective equipment (PPE): A-Level PPE. OSHA Table Z-1-A air contaminant. Dust mask; goggles; rubber gloves.
Exposure limits (TWA unless otherwise noted): ACGIH TLV 5 mg/m³; NIOSH 5 mg/m³.
Toxicity by ingestion: Grade 2; oral rat LD_{50} = 3080 mg/kg.
Long-term health effects: Mutagen. Possible carcinogen.
Vapor (gas) irritant characteristics: May be irritating to eyes and throat.
Liquid or solid irritant characteristics: Causes smarting of the skin and first-degree burns on short exposure and may cause second-degree burns on long exposure.

CHEMICAL REACTIVITY
Binary reactants: Strong acids, strong bases

ENVIRONMENTAL DATA
Food chain concentration potential: Log P_{ow} = 2.64–2.74. Unlikely to accumulate.
Water pollution: Harmful to aquatic life in very low concentrations. May be dangerous if it enters nearby water intakes; notify operators. Notify local health and wildlife officials.
Response to discharge: Issue warning–water contaminant. Should be removed. Chemical and physical treatment.

SHIPPING INFORMATION
Grades of purity: Various grades, 70–80%. Mixtures with sodium chlorate and sodium metaborate; **Storage temperature:** Ambient; **Inert atmosphere:** None; **Venting:** Open; **Stability during transport:** Stable.

NAS HAZARD CLASSIFICATION FOR BULK WATER TRANSPORTATION
FIRE: 0
HEALTH: Vapor irritant: 0; Liquid or solid irritant: 2; Poisons: 0

WATER POLLUTION: Human toxicity: 2; Aquatic toxicity: 3; Aesthetic effect: 1
REACTIVITY: Other chemicals: 1; Water: 0; Self-reaction: 0

PHYSICAL AND CHEMICAL PROPERTIES
Physical state @ 59°F/15°C and 1 atm: Solid.
Molecular weight: 215.7
Boiling point @ 1 atm: Decomposes.
Melting/Freezing point: 340°F/171°C/444°K.
Specific gravity (water = 1): 1.19 @ 68°F/20°C (solid).
Heat of combustion: (estimate) -9500 Btu/lb = -5300 cal/g = -220×10^5 J/kg.
Vapor pressure: 0.0000003 mm.

AZINPHOS-METHYL REC. A:5800

SYNONYMS: AZINPHOS-METHYL; BAY 9027; BAYER 17147; CARFENE; COTNION METHYL; CRYSTHION 2L; CRYSTHYON; DBD; O,O-DIMETHYL-s-(4-oxo-1,2,3-BEZOTRIAZIN-3(4H)-YL METHYL)PHOSPHORODITHIOATE; GUSATHION; GUSATHION INSECTICIDE; METIL AZINFOS (Spanish); ENT23233; R 1582; GUSATHION M; GUTHION; GUTHION INSECTICIDE; METHYL AZINPHOS; METHYL GUTHION; METILTRIAZOTION

IDENTIFICATION
CAS Number: 86-50-0; some information may also apply to 2642-71-9 (azinphos-ethyl)
Formula: $C_{10}H_{12}N_3O_3PS_2$
DOT ID Number: UN 2783; DOT Guide Number: 152
Proper Shipping Name: Azinphos methyl
Reportable Quantity (RQ): **(CERCLA)** 1 lb/0.454 kg

DESCRIPTION: Brown solid. Sinks in water.

Poison! (organophosphate) • May be fatal if inhaled, swallowed, or absorbed through the skin. Effects of contact or inhalation may be delayed • Firefighting gear (including SCBA) does not provide adequate protection. If exposure occurs, remove and isolate gear immediately and thoroughly decontaminate personnel • Toxic products of combustion may include oxides of phosphorus, sulfur, and nitrogen • Do not put yourself in danger by entering a contaminated area to rescue a victim.

Hazard Classification (based on NFPA-704 Rating System)
Health Hazards (Blue): 3; Flammability (Red): 0; Reactivity (Yellow): 0

EMERGENCY RESPONSE: See Appendix A (152)
Evacuation:
Public safety: Isolate the area of spill or leak for at least 25 to 50 meters (80 to 160 feet) in all directions.
Spill: Increase, in the downwind direction, as necessary, the distance shown under "Public Safety."
Fire: If tank, rail car, or tank truck is involved in fire, isolate for at least 800 meters (½ mile) in all directions; also, consider initial evacuation for 800 meters (½ mile) in all directions.

EXPOSURE
Short-term effects: *SEEK MEDICAL ATTENTION*. **Dust:** *POISONOUS IF INHALED*. Causes sweating, constriction of pupils of eyes, asthmatic symptoms, cramps, weakness, convulsions, collapse. A cholinesterase inhibitor. *IF BREATHING HAS STOPPED*, give artificial respiration; *avoid mouth-to-mouth resuscitation; use bag/mask apparatus*. Move victim to fresh air. *IF IN EYES*, hold eyelids open and flush with plenty of water. IF breathing is difficult, administer oxygen. **Solid:** *POISONOUS IF SWALLOWED. IF SWALLOWED* and victim is *CONSCIOUS AND ABLE TO SWALLOW*, have victim drink 4 to 8 ounces of water and have victim induce vomiting. *IF SWALLOWED* and victim is *UNCONSCIOUS OR HAVING CONVULSIONS*, do nothing except keep victim warm. Absorbed by skin.

HEALTH HAZARDS
Personal protective equipment (PPE): A-Level PPE. OSHA Table Z-1-A air contaminant. Chemical protective material(s) reported to have good to excellent resistance: neoprene, nitrile. Also, PVC and natural rubber offers limited protection. Polycarbonate may be used where appropriate.
Recommendations for respirator selection: NIOSH/OSHA
$2~mg/m^3$: CCROVDMFu [any chemical cartridge respirator with organic vapor cartridge(s) in combination with a dust, mist, and fume filter]; or SA (any supplied-air respirator). $5~mg/m^3$: SA:CF (any supplied-air respirator operated in a continuous-flow mode); or PAPROVDMFu [any powered, air purifying respirator with organic vapor cartridge (s) in combination with a dust, mist, and fume filter]. $10~mg/m^3$: CCRFOVHiE (any chemical cartridge respirator with a full facepiece and organic vapor cartridge(s) in combination with a high-efficiency particulate filter]; or GMFOVHiE [any air-purifying, full-facepiece respirator (gas mask) with a chin-style, front- or back-mounted organic vapor canister having a high-efficiency particulate filter]; or PAPRTOVHiE [any powered, air-purifying respirator with a tight-fitting facepiece and organic vapor cartridge(s) in combination with a high-efficiency particulate filter]; or SAT:CF (any supplied-air respirator that has a tight-fitting facepiece and is operated in a continuous-flow mode); or SCBAF (any self-contained breathing apparatus with full facepiece); or SAF (any supplied-air respirator with a full facepiece). *EMERGENCY OR PLANNED ENTRY INTO UNKNOWN CONCENTRATIONS OR IDLH CONDITIONS*: SCBAF:PD,PP (any self-contained breathing apparatus that has a full facepiece and is operated in a pressure-demand or other positive-pressure mode); or SAF:PD,PP:ASCBA (any supplied-air respirator that has a full facepiece and is operated in a pressure-demand or other positive-pressure mode in combination with an auxiliary self-contained breathing apparatus operated in a pressure-demand or other positive pressure mode). *ESCAPE*: GMFOVHiE [any air-purifying, full-facepiece respirator (gas mask) with a chin-style, front-or back-mounted canister having a high efficiency particulate filter]; or SCBAE (any appropriate escape-type, self-contained breathing apparatus).
Exposure limits (TWA unless otherwise noted): ACGIH TLV 0.2 mg/m^3; NIOSH/OSHA 0.2 mg/m^3; skin contact contributes significantly in overall exposure.
Toxicity by ingestion: Grade 4; oral rat LD_{50} = 11-18.5 mg/kg.
Liquid or solid irritant characteristics: Causes eye irritation.
IDLH value: 10 mg/m^3.

CHEMICAL REACTIVITY
Binary reactants: Strong oxidizers, acids

ENVIRONMENTAL DATA
Water pollution: DOT Appendix B, §172.101- severe marine pollutant. Harmful to aquatic life in very low concentrations. May be dangerous if it enters nearby water intakes; notify operators. Notify local health and wildlife officials. **Response to discharge:** Issue warning–water contaminant. Should be removed. Chemical and physical treatment.

Barium carbonate

SHIPPING INFORMATION
Grades of purity: Technical; 50% wettable powder; water emulsions; **Storage temperature:** Ambient; **Inert atmosphere:** None; **Venting:** Open; **Stability during transport:** Stable.

PHYSICAL AND CHEMICAL PROPERTIES
Physical state @ 59°F/15°C and 1 atm: Solid.
Molecular weight: 317.3
Boiling point @ 1 atm: Decomposes.
Melting/Freezing point: 163°F/73°C/346°K.
Specific gravity (water = 1): 1.4 @ 68°F/20°C (solid).
Heat of combustion: (estimate) -8600 Btu/lb = -4800 cal/g = -200×10^5 J/kg.
Vapor pressure: 8×10^{-9} mm.

BARIUM CARBONATE   REC. B:0100

SYNONYMS: BARIUM CARBONATE; CARBONATO BARICO (Spanish); CARBONIC ACID, BARIUM SALT; C.I. PIGMENT WHITE 10; EEC No. 056-002-00-7

IDENTIFICATION
CAS Number: 513-77-9
Formula: $BaCO_3$
DOT ID Number: UN 1564; DOT Guide Number: 154
Proper Shipping Name: Barium compounds, n.o.s.

DESCRIPTION: White crystalline solid or powder. Odorless. Sinks in water; practically insoluble.

Irritates eyes, nose, and throat; skin contact causes dermatitis. Toxic products of combustion may include. carbon monoxide and oxides of barium.

Hazard Classification (based on NFPA-704 Rating System)
Health Hazards (Blue): 2; Flammability (Red): 0; Reactivity (Yellow): 0

EMERGENCY RESPONSE: See Appendix A (154)
Evacuation:
Public safety: Isolate the area of spill or leak for at least 25 to 50 meters (80 to 160 feet) in all directions.
Spill: Increase, in the downwind direction, as necessary, the distance shown under "Public Safety."
Fire: If tank, rail car, or tank truck is involved in fire, isolate for at least 800 meters (½ mile) in all directions; also, consider initial evacuation for 800 meters (½ mile) in all directions.

EXPOSURE
Short-term effects: *SEEK MEDICAL ATTENTION.* **Solid:** If swallowed, will cause nausea and vomiting. *IF BREATHING HAS STOPPED,* give artificial respiration; *avoid mouth-to-mouth resuscitation; use bag/mask apparatus. IF SWALLOWED* and victim is *CONSCIOUS AND ABLE TO SWALLOW,* have victim drink 4 to 8 ounces of water; Induce vomiting; give a 10% water solution of Epsom salt.
Medical Note: Alert doctor to possibility of barium poisoning, particularly if compound was swallowed. Have victim drink aqueous 10% solution of magnesium or sodium sulfate; for severe intoxication, calcium or a magnesium salt may have to be given I.V. with caution; treatment otherwise is supportive and symptomatic.

HEALTH HAZARDS
Personal protective equipment (PPE): B-Level PPE.
Recommendations for respirator selection: NIOSH/OSHA as barium.
5 mg/m^3: DMXSQ (any dust and mist respirator except single-use and quarter mask respirators); or SA (any supplied-air respirator). *12.5 mg/m^3*: SA:CF (any supplied-air respirator operated in a continuous-flow mode); or PAPRDM (any powered, air-purifying respirator with a dust and mist filter). *25 mg/m^3*: HiEF (any air-purifying, full-facepiece respirator with a high-efficiency particulate filter); or SAT:CF (any supplied-air respirator that has a tight-fitting facepiece and is operated in a continuous-flow mode); or PAPRTHiE (any powered, air-purifying respirator with a tight-fitting facepiece and a high-efficiency particulate filter); or SCBAF (any self-contained breathing apparatus with a full facepiece); or SAF (any supplied-air respirator with a full facepiece). *50 mg/m^3*: SAF:PD,PP (any supplied-air respirator that has a full facepiece and is operated in a pressure-demand or other positive-pressure mode). *EMERGENCY OR PLANNED ENTRY INTO UNKNOWN CONCENTRATIONS OR IDLH CONDITIONS*: SCBAF:PD,PP (any self-contained breathing apparatus that has a full facepiece and is operated in a pressure-demand or other positive-pressure mode); or SAF:PD,PP:ASCBA (any supplied-air respirator that has a full facepiece and is operated in a pressure-demand or other positive-pressure mode in combination with an auxiliary self-contained breathing apparatus operated in a pressure-demand or other positive pressure mode). *ESCAPE*: HiEF (any air-purifying, full-facepiece respirator with a high-efficiency particulate filter); or SCBAE (any appropriate escape-type, self-contained breathing apparatus).
Exposure limits (TWA unless otherwise noted): ACGIH TLV 0.5 mg/m^3; NIOSH/OSHA 0.5 mg/m^3 as soluble barium.
Toxicity by ingestion: Grade 2; LD_{50} = 0.5 to 5 g/kg (rabbit, rat, guinea pig).
Vapor (gas) irritant characteristics: Eye, skin and respiratory tract irritant.
Liquid or solid irritant characteristics: Eye, skin and respiratory tract irritant.
IDLH value: 50 mg/m^3 as barium.

CHEMICAL REACTIVITY
Binary reactants: Incompatible with strong acids, finely divided aluminum, bromine trifluoride, magnesium, sodium azide, silicon, oxidizers.

ENVIRONMENTAL DATA
Water pollution: Harmful to aquatic life in very low concentrations. Fouling to shoreline. May be dangerous if it enters nearby water intakes; notify operators. Notify local health and wildlife officials. **Response to discharge:** Should be removed. Chemical and physical treatment.

SHIPPING INFORMATION
Grades of purity: Reagent grade–99.0%; Ceramic grade–99.1%; Ceramic and chemical grade–99.3%; Glass grade–98.8%; Electronic ceramic grade–99.6%; **Stability during transport:** Stable.

PHYSICAL AND CHEMICAL PROPERTIES
Physical state @ 59°F/15°C and 1 atm: Solid.
Molecular weight: 197.35
Boiling point @ 1 atm: 2642°F/1450°C/1723°K (decomposes).
Melting/Freezing point: 3164°F/1740°C/2013°K.
Specific gravity (water = 1): 4.3 @ 68°F/20°C (solid).
Solubility in water: 0.025 lb/100 lb of water @ 68°F/20°C.
Vapor pressure: Very low. negligible.

BARIUM CHLORATE REC. B:0150

SYNONYMS: BARIUM CHLORATE MONOHYDRATE; CLORATO BARICO (Spanish); CHLORIC ACID, BARIUM SALT; EEC No. 017-003-00-8

IDENTIFICATION
CAS Number: 13477-00-4; 10294-38-9
Formula: $Ba(ClO_3)_2 \cdot H_2O$
DOT ID Number: UN 1445; DOT Guide Number: 141
Proper Shipping Name: Barium chlorate

DESCRIPTION: White solid. Odorless. Sinks and mixes with water.

Poison! • Heat may cause material to explode • A strong oxidizer; heat above 482°F/250°C releases oxygen, increasing the activity of an existing fire • May cause fire or explosion contact with combustibles (wood, paper, oil, clothing, etc.) • Toxic products of combustion may include hydrogen chloride and oxides of barium.

Hazard Classification (based on NFPA-704 Rating System)
Health Hazards (Blue): 2; Flammability (Red): 0; Reactivity (Yellow): 1; Special Notice (White): OXY

EMERGENCY RESPONSE: See Appendix A (141)
Evacuation:
Public safety: Isolate the area of spill or leak for at least 10 to 25 meters (30 to 80 feet) in all directions.
Spill: Consider initial downwind evacuation for at least 100 meters (330 feet). *Note* : Do not let spill area dry until it has been determined that there is no chlorates left in the area. Continue cooling after fire has been extinguished.
Fire: If any large container is involved in fire, isolate for at least 800 meters (½ mile) in all directions; also, consider initial evacuation for 800 meters (½ mile) in all directions.

EXPOSURE
Short-term effects: *SEEK MEDICAL ATTENTION*. **Dust:** *POISONOUS IF INHALED*. Irritating to eyes, nose, and throat. Move victim to fresh air. *IF IN EYES*, hold eyelids open and flush with plenty of water. IF breathing is difficult, administer oxygen. **Solid:** *POISONOUS IF SWALLOWED*. Irritating to skin and eyes. Remove contaminated clothing and shoes. Flush affected areas with plenty of water. *IF IN EYES*, hold eyelids open and flush with plenty of water. *IF SWALLOWED* and victim is *CONSCIOUS AND ABLE TO SWALLOW*, have victim drink 4 to 8 ounces of water; Induce vomiting; give a 10% water solution of Epsom salt. *IF SWALLOWED* and victim is *UNCONSCIOUS OR HAVING CONVULSIONS*, do nothing except keep victim warm.
Medical Note: Alert doctor to possibility of barium poisoning, particularly if compound was swallowed. Have victim drink aqueous 10% solution of magnesium or sodium sulfate; for severe intoxication, calcium or a magnesium salt may have to be given I.V. with caution; treatment otherwise is supportive and symptomatic.

HEALTH HAZARDS
Personal protective equipment (PPE): B-Level PPE. OSHA Table Z-1-A air contaminant. Goggles or face shield; dust respirator (U.S. Bureau of Mines or NIOSH/OSHA approved); rubberized shoes and gloves; coveralls or other suitable outer clothing.
Recommendations for respirator selection: NIOSH/OSHA as barium.
5 mg/m³: DMXSQ (any dust and mist respirator except single-use and quarter mask respirators); or SA (any supplied-air respirator). *12.5 mg/m³*: SA:CF (any supplied-air respirator operated in a continuous-flow mode); or PAPRDM (any powered, air-purifying respirator with a dust and mist filter). *25 mg/m³*: HiEF (any air-purifying, full-facepiece respirator with a high-efficiency particulate filter); or SAT:CF (any supplied-air respirator that has a tight-fitting facepiece and is operated in a continuous-flow mode); or PAPRTHiE (any powered, air-purifying respirator with a tight-fitting facepiece and a high-efficiency particulate filter); or SCBAF (any self-contained breathing apparatus with a full facepiece); or SAF (any supplied-air respirator with a full facepiece). *50 mg/m³*: SAF:PD,PP (any supplied-air respirator that has a full facepiece and is operated in a pressure-demand or other positive-pressure mode). *EMERGENCY OR PLANNED ENTRY INTO UNKNOWN CONCENTRATIONS OR IDLH CONDITIONS*: SCBAF:PD,PP (any self-contained breathing apparatus that has a full facepiece and is operated in a pressure-demand or other positive-pressure mode); or SAF:PD,PP:ASCBA (any supplied-air respirator that has a full facepiece and is operated in a pressure-demand or other positive-pressure mode in combination with an auxiliary self-contained breathing apparatus operated in a pressure-demand or other positive pressure mode). *ESCAPE:* HiEF (any air-purifying, full-facepiece respirator with a high-efficiency particulate filter); or SCBAE (any appropriate escape-type, self-contained breathing apparatus).
Exposure limits (TWA unless otherwise noted): ACGIH TLV 0.5 mg/m³; NIOSH/OSHA 0.5 mg/m³ as soluble barium.
Long-term health effects: Barium poisoning
Vapor (gas) irritant characteristics: Eye, skin and respiratory tract irritant.
Liquid or solid irritant characteristics: Eye, skin and respiratory tract irritant.
IDLH value: 50 mg/m³ as barium.

FIRE DATA
Behavior in fire: Risk of explosion when involved in a fire.

CHEMICAL REACTIVITY
Binary reactants: Strong reaction with strong oxidizers, sulfuric acid, ammonium compounds and finely divided metals. Can form explosive mixtures with combustible materials such as oil and wood; these can be ignited by friction or shock.

ENVIRONMENTAL DATA
Food chain concentration potential: Negative; unlikely to accumulate.
Water pollution: Effect of low concentrations on aquatic life is unknown. May be dangerous if it enters nearby water intakes; notify operators. Notify local health and wildlife officials. **Response to discharge:** Issue warning–oxidizing material, water contaminant. Should be removed. Chemical and physical treatment. Dispense and flush.

SHIPPING INFORMATION
Grades of purity: Technical; Reagent; **Storage temperature:** Ambient; **Inert atmosphere:** None; **Venting:** Open; **Stability during transport:** Stable.

PHYSICAL AND CHEMICAL PROPERTIES
Physical state @ 59°F/15°C and 1 atm: Solid.
Molecular weight: 332 (monohydrate).
Melting/Freezing point: 777°F/414°C/687°K.
Specific gravity (water = 1): 3.18 @ 68°F/20°C (solid).
Heat of solution: 36 Btu/lb = 20 cal/g = 0.84×10^5 J/kg.
Vapor pressure: Very low; negligible.

BARIUM CYANIDE   REC. B:0200

SYNONYMS: BARIUM CYANIDE, SOLID; BARIUM DICYANIDE; CIANURO BARICO (Spanish); RCRA No. P013

IDENTIFICATION
CAS Number: 542-62-1
Formula: $Ba(CN)_2$
DOT ID Number: UN 1565; DOT Guide Number: 157
Proper Shipping Name: Barium cyanide
Reportable Quantity (RQ): **(CERCLA)** 10 lb/4.54 kg

DESCRIPTION: White crystalline solid. Sinks and mixes with water; reacts, forming an acid solution.

Poison! • Breathing or swallowing the material can kill you; irritating to the eyes and skin.• Firefighting gear (including SCBA) does not provide adequate protection. If exposure occurs, remove and isolate gear immediately and thoroughly decontaminate personnel • Concentrated dust in confined areas (e.g., tanks, sewers, buildings) may explode when exposed to fire • Toxic products of combustion may include carbon monoxide, cyanides, and oxides of barium.

Hazard Classification (based on NFPA-704 Rating System)
Health Hazards (Blue): 3; Flammability (Red): 0; Reactivity (Yellow): 0

EMERGENCY RESPONSE: See Appendix A (157)
Evacuation:
Public safety: Isolate the area of spill or leak for at least 50 to 100 meters (160 to 330 feet) in all directions.
Spill: Increase, in the downwind direction, as necessary, the distance shown under "Public Safety."
Fire: If tank, rail car, or tank truck is involved in fire, isolate for at least 800 meters (½ mile) in all directions; also, consider initial evacuation for 800 meters (½ mile) in all directions.

EXPOSURE
Short-term effects: *CALL FOR MEDICAL AID* **Dust:** *POISONOUS IF INHALED OR IF SKIN IS EXPOSED*. Move to fresh air. *IF BREATHING HAS STOPPED,* give artificial respiration; *avoid mouth-to-mouth resuscitation; use bag/mask apparatus.* IF breathing is difficult, administer oxygen. **Solid:** *POISONOUS IF SWALLOWED OR IF SKIN IS EXPOSED.* Irritating to skin and eyes. Remove contaminated clothing and shoes. *IF IN EYES*, hold eyelids open and flush with plenty of water. *IF SWALLOWED* and victim is *CONSCIOUS AND ABLE TO SWALLOW*, have victim drink 4 to 8 ounces of water; Induce vomiting; give a 10% water solution of Epsom salt. *IF SWALLOWED* and victim is *UNCONSCIOUS OR HAVING CONVULSIONS,* do nothing except keep victim warm.
Note to physician or authorized medical personnel: Consider the use of amyl nitrite perles if symptoms of cyanide poisoning develop. If symptoms indicate, initial treatment includes the cyanide antidote kit. In all cases, break an amyl nitrite perle in a gauze pad and hold lightly under victim's nose for 15 seconds, repeating 5 times at about 15-second intervals; if necessary (and if sodium nitrite infusions will be delayed), repeat procedure every 3 minutes with fresh perles until 3 or 4 have been used. Avoid breathing the vapor while administering it to the victim. Administer sodium nitrite IV, ASAP. The usual adult dose is 10 to 20 mL of a 3% solution infused over no less than 5 minutes; the average child dose is 0.15 to 0.20 mL/kg. Monitor blood pressure during administration, and slow the rate of infusion if hypotention develops. Next, infuse sodium thiosulfate IV. The usual adult dose is 50 mL of a 25% solution infused over 10 to 20 minutes; the average child dose is 1.65 mL/kg. Repeat with nitrite and thiosulfate as required.
Medical Note: Alert doctor to possibility of barium poisoning, particularly if compound was swallowed. Have victim drink aqueous 10% solution of magnesium or sodium sulfate; for severe intoxication, calcium or a magnesium salt may have to be given I.V. with caution; treatment otherwise is supportive and symptomatic.

HEALTH HAZARDS
Personal protective equipment (PPE): A-Level PPE. OSHA Table Z-1-A air contaminant.
Recommendations for respirator selection: NIOSH/OSHA as barium.
5 mg/m³: DMXSQ (any dust and mist respirator except single-use and quarter mask respirators); or SA (any supplied-air respirator). *12.5 mg/m³*: SA:CF (any supplied-air respirator operated in a continuous-flow mode); or PAPRDM (any powered, air-purifying respirator with a dust and mist filter). *25 mg/m³*: HiEF (any air-purifying, full-facepiece respirator with a high-efficiency particulate filter); or SAT:CF (any supplied-air respirator that has a tight-fitting facepiece and is operated in a continuous-flow mode); or PAPRTHiE (any powered, air-purifying respirator with a tight-fitting facepiece and a high-efficiency particulate filter); or SCBAF (any self-contained breathing apparatus with a full facepiece); or SAF (any supplied-air respirator with a full facepiece). *50 mg/m³*: SAF:PD,PP (any supplied-air respirator that has a full facepiece and is operated in a pressure-demand or other positive-pressure mode). *EMERGENCY OR PLANNED ENTRY INTO UNKNOWN CONCENTRATIONS OR IDLH CONDITIONS:* SCBAF:PD,PP (any self-contained breathing apparatus that has a full facepiece and is operated in a pressure-demand or other positive-pressure mode); or SAF:PD,PP:ASCBA (any supplied-air respirator that has a full facepiece and is operated in a pressure-demand or other positive-pressure mode in combination with an auxiliary self-contained breathing apparatus operated in a pressure-demand or other positive pressure mode). *ESCAPE:* HiEF (any air-purifying, full-facepiece respirator with a high-efficiency particulate filter); or SCBAE (any appropriate escape-type, self-contained breathing apparatus).
Exposure limits (TWA unless otherwise noted): ACGIH 0.5 mg/m³; NIOSH/OSHA 0.5 mg/m³ as soluble barium.
Toxicity by ingestion: Grade 4 LD_{50} less than 50 mg/kg.
Long-term health effects: Exposure to small amounts of cyanide compounds over long periods of time is reported to cause loss of appetite, headache, weakness, nausea, dizziness, and irritation of upper respiratory tract and eyes.
Vapor (gas) irritant characteristics: Eye and respiratory tract irritant.
Liquid or solid irritant characteristics: Eye and skin irritant.
IDLH value: 50 mg/m³ as barium.

CHEMICAL REACTIVITY
Reactivity with water: Barium will precipitate as sulfate or carbonate. Cyanide will slowly convert to the less toxic cyanate. The cyanide ion is in equilibrium with HCN, a very weak acid.
Binary reactants: Cyanides in water are corrosive to metals.
Neutralizing agents for acids and caustics: Barium: sodium sulfate or CO_2 will precipitate insoluble salts. Cyanide: Ferric salts will precipitate. Lime will suppress hydrogen cyanide evolution and help convert to cyanates.

ENVIRONMENTAL DATA
Food chain concentration potential: Barium has been concentrated 150 times by goldfish. Half-life (humans)–65 days.

Water pollution: DOT Appendix B, §172.101–marine pollutant. Harmful to aquatic life in very low concentrations. May be dangerous if it enters nearby water intakes; notify operators. Notify local health and wildlife officials. **Response to discharge:** Issue warning–poison, air contaminant, water contaminant. Restrict access. Evacuate area. Should be removed. Chemical and physical treatment.

SHIPPING INFORMATION
Stability during transport: Stable under normal conditions.

PHYSICAL AND CHEMICAL PROPERTIES
Physical state @ 59°F/15°C and 1 atm: Solid.
Molecular weight: 189.40
Heat of combustion: -3604 Btu/lb = -1447 cal/g = -60.5×10^5 J/kg.
Heat of solution: Exothermic -26.6 Btu/lb = -14.8 cal/g = -6.19×10^5 J/kg.

BARIUM NITRATE   REC. B:0250

SYNONYMS: BARIUM DINITRATE; BARIUM NITRATE; BARIUM NITRATE; EEC No. 056-002-00-7; NITRATE de BARYUM (French); NITRATO BARICO (Spanish); NITRIC ACID, BARIUM SALT; NITROBARITE

IDENTIFICATION
CAS Number: 10022-31-8
Formula: $Ba(NO_3)_2$
DOT ID Number: UN 1446; DOT Guide Number: 141
Proper Shipping Name: Barium nitrate

DESCRIPTION: White solid. Odorless. Sinks and mixes with water.

Poison! • May explode in elevated temperatures or fire • A strong oxidizer; heat forms oxygen; will increase the activity of an existing fire • May cause fire or explosion contact with combustibles (wood, paper, oil, clothing, etc.) • Concentrated dust in confined areas (e.g., tanks, sewers, buildings) may explode when exposed to fire • Irritates eyes, skin, and respiratory tract • Toxic products of combustion may include oxides of barium and nitrogen • Do not put yourself in danger by entering a contaminated area to rescue a victim.

Hazard Classification (based on NFPA-704 Rating System)
Health Hazards (Blue): 2; Flammability (Red): 0; Reactivity (Yellow): 0; Special Notice (White): OXY

EMERGENCY RESPONSE: See Appendix A (141)
Evacuation:
Public safety: Isolate the area of spill or leak for at least 10 to 25 meters (30 to 80 feet) in all directions.
Spill: Consider initial downwind evacuation for at least 100 meters (330 feet).
Fire: If any large container is involved in fire, isolate for at least 800 meters (½ mile) in all directions; also, consider initial evacuation for 800 meters (½ mile) in all directions.

EXPOSURE
Short-term effects: *SEEK MEDICAL ATTENTION.* **Dust:** *POISONOUS IF INHALED.* Irritating to eyes, nose, and throat. Move victim to fresh air. *IF IN EYES,* hold eyelids open and flush with plenty of water. IF breathing is difficult, administer oxygen. **Solid:** *POISONOUS IF SWALLOWED.* Irritating to skin and eyes. Remove contaminated clothing and shoes. Flush affected areas with plenty of water. *IF IN EYES,* hold eyelids open and flush with plenty of water. *IF SWALLOWED* and victim is *CONSCIOUS AND ABLE TO SWALLOW,* have victim drink 4 to 8 ounces of water; Induce vomiting; give a 10% water solution of Epsom salt. *IF SWALLOWED* and victim is *UNCONSCIOUS OR HAVING CONVULSIONS,* do nothing except keep victim warm.
Medical Note: Alert doctor to possibility of barium poisoning, particularly if compound was swallowed. Have victim drink aqueous 10% solution of magnesium or sodium sulfate; for severe intoxication, calcium or a magnesium salt may have to be given I.V. with caution; treatment otherwise is supportive and symptomatic.

HEALTH HAZARDS
Personal protective equipment (PPE): B-Level PPE. OSHA Table Z-1-A air contaminant. Goggles or face shield; dust respirator; rubber gloves and shoes; suitable coveralls.
Recommendations for respirator selection: NIOSH/OSHA as barium.
5 mg/m³: DMXSQ (any dust and mist respirator except single-use and quarter mask respirators); or SA (any supplied-air respirator). *12.5 mg/m³*: SA:CF (any supplied-air respirator operated in a continuous-flow mode); or PAPRDM (any powered, air-purifying respirator with a dust and mist filter). *25 mg/m³*: HiEF (any air-purifying, full-facepiece respirator with a high-efficiency particulate filter); or SAT:CF (any supplied-air respirator that has a tight-fitting facepiece and is operated in a continuous-flow mode); or PAPRTHiE (any powered, air-purifying respirator with a tight-fitting facepiece and a high-efficiency particulate filter); or SCBAF (any self-contained breathing apparatus with a full facepiece); or SAF (any supplied-air respirator with a full facepiece). *50 mg/m³*: SAF:PD,PP (any supplied-air respirator that has a full facepiece and is operated in a pressure-demand or other positive-pressure mode). *EMERGENCY OR PLANNED ENTRY INTO UNKNOWN CONCENTRATIONS OR IDLH CONDITIONS:* SCBAF:PD,PP (any self-contained breathing apparatus that has a full facepiece and is operated in a pressure-demand or other positive-pressure mode); or SAF:PD,PP:ASCBA (any supplied-air respirator that has a full facepiece and is operated in a pressure-demand or other positive-pressure mode in combination with an auxiliary self-contained breathing apparatus operated in a pressure-demand or other positive pressure mode). *ESCAPE:* HiEF (any air-purifying, full-facepiece respirator with a high-efficiency particulate filter); or SCBAE (any appropriate escape-type, self-contained breathing apparatus).
Exposure limits (TWA unless otherwise noted): ACGIH 0.5 mg/m³; NIOSH/OSHA 0.5 mg/m³ as soluble barium.
Toxicity by ingestion: Grade 3; oral rat LD_{50} = 355 mg/kg.
Long-term health effects: Barium poisoning.
Vapor (gas) irritant characteristics: Eye and respiratory tract irritant.
Liquid or solid irritant characteristics: Eye and skin irritant. Contaminated clothing may be a fire hazard; flush with water.
IDLH value: 50 mg/m³ as barium.

FIRE DATA
Behavior in fire: Mixtures with combustible materials (including clothing) are readily ignited and may burn fiercely. Containers may explode.

CHEMICAL REACTIVITY
Binary reactants: Oxidizers, acids, aluminum-magnesium alloys; sulfur and finely divided metals may be shock-sensitive, causing a vigorous reaction. Contact with combustible material may cause fire.

Barium perchlorate

ENVIRONMENTAL DATA
Food chain concentration potential: Marine animals concentrate barium from seawater 7-100 times; marine plants 1000. Soybeans, and tomatoes accumulate soil barium 2-20 times (Prager).
Water pollution: Dangerous to aquatic life in high concentrations. May be dangerous if it enters nearby water intakes; notify operators. Notify local health and wildlife officials. **Response to discharge:** Issue warning–oxidizing material, water contaminant. Should be removed. Chemical and physical treatment. Disperse and flush.

SHIPPING INFORMATION
Grades of purity: Technical; Reagent; **Storage temperature:** Ambient; **Inert atmosphere:** None; **Venting:** Open; **Stability during transport:** Stable.

PHYSICAL AND CHEMICAL PROPERTIES
Physical state @ 59°F/15°C and 1 atm: Solid.
Molecular weight: 261.35
Boiling point @ 1 atm: Decomposes.
Melting/Freezing point: 1094°F/590°C/863°K.
Specific gravity (water = 1): 3.24 at 23°C (solid).
Heat of solution: 36 Btu/lb = 20 cal/g = 0.84×10^5 J/kg.
Heat of fusion: 22.6 cal/g (estimate).
Vapor pressure: Low

BARIUM PERCHLORATE   REC. B:0300

SYNONYMS: BARIUM PERCHLORATE TRIHYDRATE; EEC No. 017-007-00-X; PERCLORATO BARICO (Spanish); PERCHLORIC ACID, BARIUM SALT-3H$_2$O

IDENTIFICATION
CAS Number: 13465-95-7
Formula: Ba(ClO$_4$)$_2$·3H$_2$O
DOT ID Number: UN 1447; DOT Guide Number: 141
Proper Shipping Name: Barium perchlorate

DESCRIPTION: White solid. Odorless. Sinks and mixes with water.

Poison! • Heat may cause material to explode • A strong oxidizer; heat forms oxygen; will increase the activity of an existing fire • May cause fire or explosion contact with combustibles (wood, paper, oil, clothing, etc.) • Irritates eyes, skin, and respiratory tract • Toxic products of combustion may include hydrogen chloride.

Hazard Classification (based on NFPA-704 Rating System)
Health Hazards (Blue): 2; Flammability (Red): 0; Reactivity (Yellow): 0; Special Notice (White): OXY

EMERGENCY RESPONSE: See Appendix A (141)
Note: Do not let spill area dry until it has been determined that there is no perchlorates left in the area. Continue cooling after fire has been extinguished.
Evacuation:
Public safety: Isolate the area of spill or leak for at least 10 to 25 meters (30 to 80 feet) in all directions.
Spill: Consider initial downwind evacuation for at least 100 meters (330 feet).
Fire: If any large container is involved in fire, isolate for at least 800 meters (½ mile) in all directions; also, consider initial evacuation for 800 meters (½ mile).

EXPOSURE
Short-term effects: *SEEK MEDICAL ATTENTION*. **Dust:** *POISONOUS IF INHALED*. Irritating to eyes, nose, and throat. Move victim to fresh air. *IF IN EYES*, hold eyelids open and flush with plenty of water. IF breathing is difficult, administer oxygen. **Solid:** *POISONOUS IF SWALLOWED*. Irritating to skin and eyes. Remove contaminated clothing and shoes. Flush affected areas with plenty of water. *IF IN EYES*, hold eyelids open and flush with plenty of water. *IF SWALLOWED* and victim is *CONSCIOUS AND ABLE TO SWALLOW*, have victim drink 4 to 8 ounces of water; Induce vomiting; give a 10% water solution of Epsom salt. *IF SWALLOWED* and victim is *UNCONSCIOUS OR HAVING CONVULSIONS*, do nothing except keep victim warm.
Medical Note: Alert doctor to possibility of barium poisoning, particularly if compound was swallowed. Have victim drink aqueous 10% solution of magnesium or sodium sulfate; for severe intoxication, calcium or a magnesium salt may have to be given I.V. with caution; treatment otherwise is supportive and symptomatic.

HEALTH HAZARDS
Personal protective equipment (PPE): B-Level PPE. OSHA Table Z-1-A air contaminant. Goggles or face shield; dust respirator; rubber gloves and shoes; suitable coveralls.
Recommendations for respirator selection: NIOSH/OSHA as barium.
5 mg/m^3: DMXSQ (any dust and mist respirator except single-use and quarter mask respirators); or SA (any supplied-air respirator). *12.5 mg/m^3*: SA:CF (any supplied-air respirator operated in a continuous-flow mode); or PAPRDM (any powered, air-purifying respirator with a dust and mist filter). *25 mg/m^3*: HiEF (any air-purifying, full-facepiece respirator with a high-efficiency particulate filter); or SAT:CF (any supplied-air respirator that has a tight-fitting facepiece and is operated in a continuous-flow mode); or PAPRTHiE (any powered, air-purifying respirator with a tight-fitting facepiece and a high-efficiency particulate filter); or SCBAF (any self-contained breathing apparatus with a full facepiece); or SAF (any supplied-air respirator with a full facepiece). *50 mg/m^3*: SAF:PD,PP (any supplied-air respirator that has a full facepiece and is operated in a pressure-demand or other positive-pressure mode). *EMERGENCY OR PLANNED ENTRY INTO UNKNOWN CONCENTRATIONS OR IDLH CONDITIONS*: SCBAF:PD,PP (any self-contained breathing apparatus that has a full facepiece and is operated in a pressure-demand or other positive-pressure mode); or SAF:PD,PP:ASCBA (any supplied-air respirator that has a full facepiece and is operated in a pressure-demand or other positive-pressure mode in combination with an auxiliary self-contained breathing apparatus operated in a pressure-demand or other positive pressure mode). *ESCAPE*: HiEF (any air-purifying, full-facepiece respirator with a high-efficiency particulate filter); or SCBAE (any appropriate escape-type, self-contained breathing apparatus).
Exposure limits (TWA unless otherwise noted): ACGIH 0.5 mg/m^3; NIOSH/OSHA 0.5 mg/m^3 as soluble barium.
Toxicity by ingestion: Could be fatal.
Long-term health effects: Barium poisoning; possible skin disorders
Vapor (gas) irritant characteristics: Eye, skin and respiratory tract irritant.
Liquid or solid irritant characteristics: Eye and skin irritant.
IDLH value: 50 mg/m^3 as barium.

FIRE DATA
Behavior in fire: Containers may explode.

CHEMICAL REACTIVITY
Binary reactants: When mixed with combustible material, ammonia compounds or finely divided metals, can cause

explosions. Reaction with strong acids or combustible materials can cause explosions.

ENVIRONMENTAL DATA
Food chain concentration potential: Marine animals concentrate barium from seawater 7-100 times; marine plants 1000. Soybeans, and tomatoes accumulate soil barium 2-20 times (Prager).
Water pollution: Effect of low concentrations on aquatic life is unknown. May be dangerous if it enters nearby water intakes; notify operators. Notify local health and wildlife officials.
Response to discharge: Issue warning–oxidizing material, water contaminant. Should be removed. Chemical and physical treatment. Disperse and flush.

SHIPPING INFORMATION
Grades of purity: Technical; Reagent; **Storage temperature:** Ambient; **Inert atmosphere:** None; **Venting:** Open; **Stability during transport:** Stable.

PHYSICAL AND CHEMICAL PROPERTIES
Physical state @ 59°F/15°C and 1 atm: Solid.
Molecular weight: 390.35
Boiling point @ 1 atm: Decomposes.
Melting/Freezing point: Decomposes. 941°F/505°C/778°K (anhydrous); 778°F/400°C/673°K (trihydrate).
Specific gravity (water = 1): 3.2 @ 68°F/20°C (solid).
Heat of solution: 9 Btu/lb = 5 cal/g = 0.2×10^5 J/kg.

BARIUM PERMANGANATE   REC. B:0350

SYNONYMS: PERMANGANIC ACID, BARIUM SALT; BARIUM MANGANATE (VIII); PERMANGANATO BARICO (Spanish)

IDENTIFICATION
CAS Number: 7787-36-2
Formula: $Ba(MnO_4)_2$
DOT ID Number: UN 1448; DOT Guide Number: 141
Proper Shipping Name: Barium permanganate

DESCRIPTION: Dark purple to black solid. Odorless. Sinks and mixes with water.

Poison! • Heat forms oxygen; will increase the activity of an existing fire • May cause fire or explosion contact with combustibles (wood, paper, oil, clothing, etc.) • Concentrated dust in confined areas (e.g., tanks, sewers, buildings) may explode when exposed to fire • Irritates eyes, skin, and respiratory tract • Toxic products of combustion may include carbon monoxide, barium oxide, and manganese fume.

Hazard Classification (based on NFPA-704 Rating System)
Health Hazards (Blue): 2; Flammability (Red): 0; Reactivity (Yellow): 0; Special Notice (White): OXY

EMERGENCY RESPONSE: See Appendix A (141)
Evacuation:
Public safety: Isolate the area of spill or leak for at least 10 to 25 meters (30 to 80 feet) in all directions.
Spill: Consider initial downwind evacuation for at least 100 meters (330 feet).
Fire: If any large container is involved in fire, isolate for at least 800 meters (½ mile) in all directions; also, consider initial evacuation for 800 meters (½ mile) in all directions.

EXPOSURE
Short-term effects: *SEEK MEDICAL ATTENTION*. **Dust:** *POISONOUS IF INHALED*. Irritating to eyes, nose, and throat. Move victim to fresh air. *IF IN EYES*, hold eyelids open and flush with plenty of water. IF breathing is difficult, administer oxygen. **Solid:** *POISONOUS IF SWALLOWED*. Irritating to skin and eyes. Remove contaminated clothing and shoes. Flush affected areas with plenty of water. *IF IN EYES*, hold eyelids open and flush with plenty of water. *IF SWALLOWED* and victim is *CONSCIOUS AND ABLE TO SWALLOW*, have victim drink 4 to 8 ounces of water; Induce vomiting; give a 10% water solution of Epsom salt. *IF SWALLOWED* and victim is *UNCONSCIOUS OR HAVING CONVULSIONS*, do nothing except keep victim warm.
Medical Note: Alert doctor to possibility of barium poisoning, particularly if compound was swallowed. Have victim drink aqueous 10% solution of magnesium or sodium sulfate; for severe intoxication, calcium or a magnesium salt may have to be given I.V. with caution; treatment otherwise is supportive and symptomatic.
HEALTH HAZARDS
Personal protective equipment (PPE): B-Level PPE. OSHA Table Z-1-A air contaminant. Goggles or face shield; dust respirator; rubber gloves and shoes.
Recommendations for respirator selection: NIOSH/OSHA as barium.
5 mg/m³: DMXSQ (any dust and mist respirator except single-use and quarter mask respirators); or SA (any supplied-air respirator). *12.5 mg/m³*: SA:CF (any supplied-air respirator operated in a continuous-flow mode); or PAPRDM (any powered, air-purifying respirator with a dust and mist filter). *25 mg/m³*: HiEF (any air-purifying, full-facepiece respirator with a high-efficiency particulate filter); or SAT:CF (any supplied-air respirator that has a tight-fitting facepiece and is operated in a continuous-flow mode); or PAPRTHiE (any powered, air-purifying respirator with a tight-fitting facepiece and a high-efficiency particulate filter); or SCBAF (any self-contained breathing apparatus with a full facepiece); or SAF (any supplied-air respirator with a full facepiece). *50 mg/m³*: SAF:PD,PP (any supplied-air respirator that has a full facepiece and is operated in a pressure-demand or other positive-pressure mode). *EMERGENCY OR PLANNED ENTRY INTO UNKNOWN CONCENTRATIONS OR IDLH CONDITIONS*: SCBAF:PD,PP (any self-contained breathing apparatus that has a full facepiece and is operated in a pressure-demand or other positive-pressure mode); or SAF:PD,PP:ASCBA (any supplied-air respirator that has a full facepiece and is operated in a pressure-demand or other positive-pressure mode in combination with an auxiliary self-contained breathing apparatus operated in a pressure-demand or other positive pressure mode). *ESCAPE*: HiEF (any air-purifying, full-facepiece respirator with a high-efficiency particulate filter); or SCBAE (any appropriate escape-type, self-contained breathing apparatus).
Exposure limits (TWA unless otherwise noted): ACGIH 0.5 mg/m³; NIOSH/OSHA 0.5 mg/m³ as soluble barium.
Long-term health effects: Barium poisoning
Vapor (gas) irritant characteristics: Eye and respiratory tract irritant.
Liquid or solid irritant characteristics: Eye, skin and respiratory tract irritant.
IDLH value: 50 mg/m³ as barium.

FIRE DATA
Behavior in fire: Can increase the intensity of fire.

CHEMICAL REACTIVITY
Binary reactants: When mixed with combustible material, can ignite by friction or acids; may be spontaneously combustible.

112 Barium peroxide

ENVIRONMENTAL DATA
Food chain concentration potential: Marine animals concentrate barium from seawater 7-100 times; marine plants 1000. Soybeans, and tomatoes accumulate soil barium 2-20 times (Prager).
Water pollution: Harmful to aquatic life in very low concentrations. May be dangerous if it enters nearby water intakes; notify operators. Notify local health and wildlife officials.
Response to discharge: Issue warning–oxidizing material, water contaminant. Should be removed. Chemical and physical treatment. Disperse and flush.

SHIPPING INFORMATION
Grades of purity: Commercial; **Storage temperature:** Ambient; **Inert atmosphere:** None; **Venting:** Open; **Stability during transport:** Stable.

PHYSICAL AND CHEMICAL PROPERTIES
Physical state @ 59°F/15°C and 1 atm: Solid.
Molecular weight: 375
Boiling point @ 1 atm: Decomposes.
Specific gravity (water = 1): 3.77 @ 68°F/20°C (solid).

BARIUM PEROXIDE   **REC. B:0400**

SYNONYMS: BARIUM BINOXIDE, BARIUM DIOXIDE; BARIUMPEROXID (German); BARIUM SUPEROXIDE; DIOXYDE de BARYUM (French); EEC No. 056-001-00-1; PEROXYDE de BARYUM (French); PEROXIDO BARICO (Spanish)

IDENTIFICATION
CAS Number: 1304-29-6
Formula: BaO_2
DOT ID Number: UN 1449; DOT Guide Number: 141
Proper Shipping Name: Barium peroxide

DESCRIPTION: Light gray to tan solid. Odorless. Sinks in water; dissolves, forming hydrogen peroxide.

Poison! • Heat forms oxygen; will increase the activity of an existing fire • May cause fire or explosion contact with combustibles (wood, paper, oil, clothing, etc.) • Irritates eyes, skin, and respiratory tract. Toxic products of combustion may include oxides of barium.

Hazard Classification (based on NFPA-704 Rating System)
Health Hazards (Blue): 3; Flammability (Red): 0; Reactivity (Yellow): 1; Special Notice (White): OXY

EMERGENCY RESPONSE: See Appendix A (141)
Evacuation:
Public safety: Isolate the area of spill or leak for at least 10 to 25 meters (30 to 80 feet) in all directions.
Spill: Consider initial downwind evacuation for at least 100 meters (330 feet).
Fire: If any large container is involved in fire, isolate for at least 800 meters (½ mile) in all directions; also, consider initial evacuation for 800 meters (½ mile) in all directions.

EXPOSURE
Short-term effects: *SEEK MEDICAL ATTENTION.* **Dust:** *POISONOUS IF INHALED.* Irritating to eyes, nose, and throat. Move victim to fresh air. *IF IN EYES*, hold eyelids open and flush with plenty of water. IF breathing is difficult, administer oxygen.
Solid: *POISONOUS IF SWALLOWED.*
Will burn skin and eyes. Remove contaminated clothing and shoes. Flush affected areas with plenty of water. *IF IN EYES*, hold eyelids open and flush with plenty of water. *IF SWALLOWED* and victim is *CONSCIOUS AND ABLE TO SWALLOW*, have victim drink 4 to 8 ounces of water. **Do NOT induce vomiting.**
Medical Note: Alert doctor to possibility of barium poisoning, particularly if compound was swallowed. Have victim drink aqueous 10% solution of magnesium or sodium sulfate; for severe intoxication, calcium or a magnesium salt may have to be given I.V. with caution; treatment otherwise is supportive and symptomatic.

HEALTH HAZARDS
Personal protective equipment (PPE): B-Level PPE. Approved toxic dust respirator; liquid-proof PVC gloves; chemical safety goggles; full cover clothing. Butyl rubber is generally suitable for peroxide compounds.
Recommendations for respirator selection: NIOSH/OSHA as barium.
5 mg/m³: DMXSQ (any dust and mist respirator except single-use and quarter mask respirators); or SA (any supplied-air respirator). *12.5 mg/m³*: SA:CF (any supplied-air respirator operated in a continuous-flow mode); or PAPRDM (any powered, air-purifying respirator with a dust and mist filter). *25 mg/m³*: HiEF (any air-purifying, full-facepiece respirator with a high-efficiency particulate filter); or SAT:CF (any supplied-air respirator that has a tight-fitting facepiece and is operated in a continuous-flow mode); or PAPRTHiE (any powered, air-purifying respirator with a tight-fitting facepiece and a high-efficiency particulate filter); or SCBAF (any self-contained breathing apparatus with a full facepiece); or SAF (any supplied-air respirator with a full facepiece). *50 mg/m³*: SAF:PD,PP (any supplied-air respirator that has a full facepiece and is operated in a pressure-demand or other positive-pressure mode). *EMERGENCY OR PLANNED ENTRY INTO UNKNOWN CONCENTRATIONS OR IDLH CONDITIONS*: SCBAF:PD,PP (any self-contained breathing apparatus that has a full facepiece and is operated in a pressure-demand or other positive-pressure mode); or SAF:PD,PP:ASCBA (any supplied-air respirator that has a full facepiece and is operated in a pressure-demand or other positive-pressure mode in combination with an auxiliary self-contained breathing apparatus operated in a pressure-demand or other positive pressure mode). *ESCAPE:* HiEF (any air-purifying, full-facepiece respirator with a high-efficiency particulate filter); or SCBAE (any appropriate escape-type, self-contained breathing apparatus).
Exposure limits (TWA unless otherwise noted): ACGIH 0.5 mg/m³; NIOSH/OSHA 0.5 mg/m³ as soluble barium.
Toxicity by ingestion: May be fatal.
Long-term health effects: Barium poisoning; skin problems.
Vapor (gas) irritant characteristics: Eye and respiratory tract irritant. Saturated clothing may be a fire hazard.
Liquid or solid irritant characteristics: Eye, skin and respiratory tract irritant.
IDLH value: 50 mg/m³ as barium.

CHEMICAL REACTIVITY
Reactivity with water: Decomposes slowly. The reaction may not be hazardous.
Binary reactants: Corrodes metal slowly. If mixed with combustible material (including clothing) or finely divided metals, this material can ignite spontaneously or from shock or friction.

ENVIRONMENTAL DATA
Food chain concentration potential: Marine animals concentrate

barium from seawater 7-100 times; marine plants 1000. Soybeans, and tomatoes accumulate soil barium 2-20 times (Prager).
Water pollution: Effect of low concentrations on aquatic life is unknown. May be dangerous if it enters nearby water intakes; notify operators. Notify local health and wildlife officials.
Response to discharge: Issue warning–oxidizing material, water contaminant. Should be removed. Chemical and physical treatment.

SHIPPING INFORMATION
Grades of purity: Technical: 91-92.5% high-purity reagent; **Storage temperature:** Ambient; **Inert atmosphere:** None; **Venting:** Pressure-vacuum; **Stability during transport:** Stable.

PHYSICAL AND CHEMICAL PROPERTIES
Physical state @ 59°F/15°C and 1 atm: Solid.
Molecular weight: 169.4
Boiling point @ 1 atm: (Decomposes) 1472°F/800°C/1073°K.
Melting/Freezing point: 842°F/450°C/723°K.
Specific gravity (water = 1): 4.96 @ 68°F/20°C (solid).
Heat of decomposition: -194 Btu/lb = -108 cal/g = -4.52×10^5 J/kg.
Vapor pressure: Negligible.

BENZAL CHLORIDE   REC. B:0450

SYNONYMS: BENZYL DICHLORIDE; BENZYLENE CHLORIDE; BENZYLIDENE CHLORIDE; CHLOROBENZAL; CHLORURE de BENZYLIDENE (French); CLORURO de BENZAL (Spanish); (DICHLOROMETHYL)BENZENE; α,α-DICHLOROTOLUENE; EEC No. 602-058-00-8; TOLUENE, α,α-DICHLORO-; BENZENE, DICHLORO METHYL-; RCRA U017

IDENTIFICATION
CAS Number: 98-87-3
Formula: $C_7H_6Cl_2$; $C_6H_5CHCl_2$
DOT ID Number: UN 1886; DOT Guide Number: 156
Proper Shipping Name: Benzylidene chloride
Reportable Quantity (RQ): **(CERCLA)** 5000 lb/2270 kg

DESCRIPTION: Colorless to brown oily, liquid (crude). Pungent odor. Sinks in water; decomposes, forming hydrogen chloride fumes.

A tear gas • Corrosive to the skin, eyes, and respiratory tract; contact with skin or eyes can cause burns and vision impairment; inhalation symptoms may be delayed • Firefighting gear (including SCBA) does not provide adequate protection. If exposure occurs, remove and isolate gear immediately and thoroughly decontaminate personnel • Containers may BLEVE when exposed to fire • Toxic products of combustion may include hydrogen chloride and phosgene • Do not put yourself in danger by entering a contaminated area to rescue a victim.

Hazard Classification (based on NFPA-704 Rating System)
Health Hazards (Blue): 3; Flammability (Red): 2; Reactivity (Yellow): 1

EMERGENCY RESPONSE: See Appendix A (156)
Evacuation:
Public safety: Isolate the area of spill or leak for at least 50 to 100 meters (160 to 330 feet) in all directions.
Spill: Increase, in the downwind direction, as necessary, the distance shown under "Public Safety."
Fire: If tank, rail car, or tank truck is involved in fire, isolate for at least 800 meters (½ mile) in all directions; also, consider initial evacuation for 800 meters (½ mile) in all directions.

EXPOSURE
Short-term effects: *SEEK MEDICAL ATTENTION*. **Vapor:** May be fatal if inhaled or absorbed through skin. Lacrimator. Irritating to eyes and respiratory tract. Very high concentrations may cause central nervous system depression and lung edema. Effects may be delayed; keep under observation. Move victim to fresh air.*IF BREATHING HAS STOPPED*, give artificial respiration; *avoid mouth-to-mouth resuscitation; use bag/mask apparatus.* IF breathing is difficult, administer oxygen. **Liquid:** *MAY BE FATAL IF SWALLOWED OR ABSORBED THROUGH THE SKIN*. May burn skin and eyes. *IF IN EYES OR ON SKIN*, flush with running water for at least 15 minutes; hold eyelids open if appropriate. *Speed in removing material from skin is extremely important.* Remove and isolate contaminated clothing and shoes. Effects may be delayed; keep under observation. *IF SWALLOWED* and victim is *CONSCIOUS AND ABLE TO SWALLOW*, have victim drink several glasses of water. **Do NOT induce vomiting**.
Note to physician or authorized medical personnel: Medical observation is recommended for 24 to 48 hours after breathing overexposure, as pulmonary edema may be delayed. As first aid for pulmonary edema, consider administering a corticosteroid spray. Cigarette smoking may exacerbate pulmonary injury and should be discouraged for at least 72 hours following exposure.

HEALTH HAZARDS
Personal protective equipment (PPE): Wear positive pressure breathing apparatus and special protective clothing.
Toxicity by ingestion: Grade 2; LD_{50} = 3.249 g/kg (rat).
Long-term health effects: Possesses mutagenic and tumorigenic properties. Suspected animal carcinogen; indefinite human carcinogen. IARC substance, overall evaluation unassigned.
Vapor (gas) irritant characteristics: Highly irritating to the eyes and respiratory system.
Liquid or solid irritant characteristics: Strong irritant and lachrymator. May burn skin and eyes.

FIRE DATA
Flash point: 198°F/92°C (cc).
Behavior in fire: Supports combustion.
Autoignition temperature: 153°F/67°C/340°K.
Electrical hazard: Class I, Group D.

CHEMICAL REACTIVITY
Reactivity with water: Readily Hydrolyzes to benzaldehyde under neutral, acid or alkaline conditions.
Binary reactants: Reacts with common metals (except nickel and lead) to produce heat and resin formation (Friedel-Crafts self-condensation type products) along with toxic and corrosive hydrogen chloride. Heat build up causes the reaction to accelerate. Violent reaction with strong oxidizers. Contact with heat or acid forms hydrogen chloride and phosgene vapors.
Neutralizing agents for acids and caustics: Use sodium carbonate or lime to absorb residual spill material.
Polymerization: Can react with common metals (except nickel and lead) to produce resin formation (Friedel-Crafts self-condensation products) and hydrogen chloride; **Inhibitor of polymerization:** Propylene oxide

ENVIRONMENTAL DATA
Food chain concentration potential: Log P_{ow} = 3.22. Values > 3.0

are likely to bioconcentrate in aquatic organisms and other living tissue, especially in fats.
Water pollution: DOT Appendix B, §172.101–marine pollutant. Effects of low concentrations on aquatic life is unknown. May be dangerous if it enters nearby water intakes; notify operators. Notify local health and wildlife officials. **Response to discharge:** Issue warning–poison. Restrict access. Should be removed. Chemical and physical treatment.

SHIPPING INFORMATION
Grades of purity: 99%; 75–93% (crude); **Storage temperature:** Ambient; **Stability during transport:** Stable at atmospheric pressure and ambient temperature when kept free of reactive metals and moisture. Material (stabilized and unstabilized) should be consumed within 90 days. Exposure to moist air and/or heat reduces this period considerably to below 90 days.

PHYSICAL AND CHEMICAL PROPERTIES
Physical state @ 59°F/15°C and 1 atm: Liquid.
Molecular weight: 161.03
Boiling point @ 1 atm: 417°F/214°C/487°K.
Melting/Freezing point: 3°F/–16°C/257°K.
Specific gravity (water = 1): 1.2557 @ 14°C.
Liquid surface tension: 20.20 dynes/cm = 0.0202 N/m @ 203.5°C.
Relative vapor density (air = 1): 5.6 (calculated).
Latent heat of vaporization: 124 Btu/lb = 69 cal/g = 2.9×10^5 J/kg.
Vapor pressure: 0.3 mm.

BENZALDEHYDE   REC. B:0500

SYNONYMS: ALMOND ARTIFICIAL ESSENTIAL OIL; ARTIFICIAL ALMOND OIL; BENZALDEHIDO (Spanish); BENZENE CARBALDEHYDE; BENZENECARBONAL; BENZENE CARCABOXALDEHYDE; BENZENEMETHTAL; BENZOIC ALDEHYDE; EEC No. 605-012-00-5; OIL OF BITTER ALMOND; PHENYLMETHANAL

IDENTIFICATION
CAS Number: 100-52-7
Formula: C_7H_6O; C_6H_5CHO
DOT ID Number: UN 1989; DOT Guide Number: 129
Proper Shipping Name: Aldehydes, n.o.s.

DESCRIPTION: Colorless to pale yellow watery liquid. Bitter almond odor. May float or sink in water; insoluble.

Flammable • Highly irritating to the eyes and respiratory tract; inhalation symptoms of asthma may be delayed; this material has anesthetic properties (anyone with a history of asthma should avoid all contact with this chemical) • Containers may BLEVE when exposed to fire •Vapors may form explosive mixture with air • Vapors are heavier than air and will collect and stay in low areas • Vapors may travel long distances to ignition sources and flashback • Vapors in confined areas (e.g., tanks, sewers, buildings) may explode when exposed to fire • Toxic products of combustion may include carbon monoxide.

Hazard Classification (based on NFPA-704 Rating System)
Health Hazards (Blue): 2; Flammability (Red): 2; Reactivity (Yellow): 0

EMERGENCY RESPONSE: See Appendix A (129)
Evacuation:
Public safety: Isolate spill area for at least 50 to 100 meters (160 to 330 feet) in all directions.
Spill: Large spill–Consider initial downwind evacuation for at least 300 meters (1000 feet).
Fire: Isolate for 800 meters (½ mile) in all directions, especially if tank, rail car, or tank truck is involved in fire.

EXPOSURE
Short-term effects: *SEEK MEDICAL ATTENTION.* **Liquid:** Irritating to skin and eyes. Harmful if swallowed. Remove contaminated clothing and shoes. Flush affected areas with plenty of water. *IF IN EYES*, hold eyelids open and flush with plenty of water. *IF SWALLOWED* and victim is *CONSCIOUS AND ABLE TO SWALLOW*, have victim drink 4 to 8 ounces of water and have victim induce vomiting. *IF SWALLOWED* and victim is *UNCONSCIOUS OR HAVING CONVULSIONS*, do nothing except keep victim warm.
Note to physician or authorized medical personnel: Medical observation is recommended for several hours after breathing overexposure, as asthma symptoms may be delayed. Anyone with a history of asthma should NOT come in contact with this material.

HEALTH HAZARDS
Personal protective equipment (PPE): B-Level PPE. Full-face organic vapor respirator. Chemical protective material(s) reported to have good to excellent resistance: PV alcohol, Viton®. Also, butyl rubber and styrene-butadiene offers limited protection.
Exposure limits (TWA unless otherwise noted): 2 ppm (AIHA WEEL).
Toxicity by ingestion: Grade 2; LD_{50} = 0.5 to 5 g/kg.
Long-term health effects: Kidney and skin disorders.
Vapor (gas) irritant characteristics: Vapors cause severe irritation of eyes and throat and can cause eye and lung injury. They cannot be tolerated even at low concentrations.
Liquid or solid irritant characteristics: Eye and skin irritant. If spilled on clothing and allowed to remain, may cause smarting and reddening of the skin. Saturated clothing may be a fire hazard.
Odor threshold: 0.042 ppm.

FIRE DATA
Flash point: 165°F/74°C (oc), 145°F/63°C (cc).
Flammable limits in air: LEL: 1.1%; UEL: 3.7%.
Autoignition temperature: 377°F/192°C/465°K.
Burning rate: 3.8 mm/min

CHEMICAL REACTIVITY
Binary reactants: Reacts with strong oxidizers, combustible materials, aluminum, iron, bases. Possible self-ignition.

ENVIRONMENTAL DATA
Food chain concentration potential: Log P_{ow} = 1.48. Unlikely to accumulate.
Water pollution: DOT Appendix B, §172.101–marine pollutant. Effect of low concentrations on aquatic life is unknown. Fouling to shoreline. May be dangerous if it enters nearby water intakes; notify operators. Notify local health and pollution control officials. **Response to discharge:** Issue warning–water contaminant. Should be removed.

SHIPPING INFORMATION
Grades of purity: Technical grade, 98.0%; NF (FCC) grade, 98.0%; **Storage temperature:** Ambient; **Inert atmosphere:** Inerted; **Stability during transport:** Stable.

PHYSICAL AND CHEMICAL PROPERTIES
Physical state @ 59°F/15°C and 1 atm: Liquid.
Molecular weight: 106.12
Boiling point @ 1 atm: 354°F/179°C/452°K.
Melting/Freezing point: –14°F/–26°C/247°K.
Critical temperature: 666°F/352°C/625°K.
Critical pressure: 316 psia = 21.5 atm = 2.18 MN/m².
Specific gravity (water = 1): 1.046 @ 68°F/20°C (liquid).
Liquid surface tension: 40.0 dynes/cm = 0.040 N/m @ 68°F/20°C.
Liquid water interfacial tension: 15.5 dynes/cm = 0.0155 N/m @ 68°F/20°C.
Relative vapor density (air = 1): 3.7
Ratio of specific heats of vapor (gas): 1.1
Latent heat of vaporization: 156 Btu/lb = 86.5 cal/g = 3.62×10^5 J/kg.
Heat of combustion: –13,730 Btu/lb = –7630 cal/g = $–319.5 \times 10^5$ J/kg.
Vapor pressure: 1 mm.

BENZENE REC. B:0550

SYNONYMS: BENCENO (Spanish); BENZOL; BENZOLE; BENZELENE; CARBON OIL; CARBON NAPHTHA; COAL TAR NAPHTHA; CYCLOHEXATRIENE; EEC No. 601-020-00-8; PHENYL HYDRIDE; PHENE; COAL NAPHTHA; COAL NAPHTHA, PHENYL HYDRIDE; BENZELENE; MINERAL NAPHTHA; MOTOR BENZOL; PYROBENZOL; RCRA No. U109

IDENTIFICATION
CAS Number: 71-43-2
Formula: C_6H_6
DOT ID Number: UN 1114; DOT Guide Number: 128
Proper Shipping Name: Benzene
Reportable Quantity (RQ): **(CERCLA)** 10 lb/4.54 kg

DESCRIPTION: Colorless watery liquid. Gasoline-like odor. Floats on the surface of water. Flammable, irritating vapor is produced. *Note:* Benzene has a pleasant odor and narcotic effect and thus has poor warning properties.

Highly flammable • Poison! • Irritating to the skin, eyes, and respiratory tract; affects the central nervous system and affects the blood • Containers may BLEVE when exposed to fire • Vapors may form explosive mixture with air • Vapors are heavier than air and will collect and stay in low areas • Vapors may travel long distances to ignition sources and flashback • Vapors in confined areas (e.g., tanks, sewers, buildings) may explode when exposed to fire • Toxic products of combustion may include carbon monoxide.

Hazard Classification (based on NFPA-704 Rating System)
Health Hazards (Blue): 2; Flammability (Red): 3; Reactivity (Yellow): 0

EMERGENCY RESPONSE: See Appendix A (128)
Evacuation:
Public safety: Isolate spill area for at least 25 to 50 meters (80 to 160 feet) in all directions.
Spill: Large spill–Consider initial downwind evacuation for at least 300 meters (1000 feet).
Fire: Isolate for 800 meters (½ mile) in all directions, especially if tank, rail car, or tank truck is involved in fire.

EXPOSURE
Short-term effects: *SEEK MEDICAL ATTENTION.* **Vapor:** Irritating to eyes, nose, and throat. *IF INHALED*, will, will cause headache, difficult breathing, or loss of consciousness. Move to fresh air. *IF BREATHING HAS STOPPED*, give artificial respiration. IF breathing is difficult, administer oxygen. **Liquid:** Irritating to skin and eyes. Harmful if swallowed. Remove contaminated clothing and shoes. Flush affected areas with plenty of water. *IF IN EYES*, hold eyelids open and flush with plenty of water. *IF SWALLOWED* and victim is *CONSCIOUS AND ABLE TO SWALLOW*, have victim drink 4 to 8 ounces of water. Do NOT induce emesis. Administer a slurry of activated charcoal.

HEALTH HAZARDS
Personal protective equipment (PPE): OSHA Table Z-1-A air contaminant. OSHA Table Z-2 air contaminant. A-Level PPE. Chemical protective material(s) reported to offer minimal to poor protection: Butyl rubber/neoprene, PV acetate, Silvershield®, Viton®/neoprene.
Recommendations for respirator selection: NIOSH
At any concentrations above the NIOSH REL, or where there is no REL, at any detectable concentration: SCBAF:PD,PP (any self-contained breathing apparatus that has a full facepiece and is operated in a pressure-demand or other positive-pressure mode); or SAF:PD,PP:ASCBA (any supplied-air respirator that has a full facepiece and is operated in a pressure-demand or other positive-pressure mode in combination with an auxiliary self-contained breathing apparatus operated in a pressure-demand or other positive pressure mode). *ESCAPE:* GMFOV [any air-purifying, full-facepiece respirator (gas mask) with a chin-style, front-or back-mounted organic vapor canister]; or SCBAE (any appropriate escape-type, self-contained breathing apparatus).
Exposure limits (TWA unless otherwise noted): ACGIH TLV 0.5 ppm; suspected human carcinogen; OSHA PEL 1 ppm; NIOSH REL 0.1 ppm; potential human carcinogen.
Short-term exposure limits (15-minute TWA): ACGIH STEL 2.5 ppm; OSHA STEL 5 ppm; NIOSH STEL 1 ppm; potential human carcinogen
Toxicity by ingestion: Grade 3; LD_{50} = 50 to 500 mg/kg.
Long-term health effects: Blood effects; possible leukemia; liver and kidney damage. OSHA specifically regulated carcinogen. IARC carcinogen, rating 1. NTP known carcinogen.
Vapor (gas) irritant characteristics: If present in high concentrations, vapors may cause irritation of eyes or respiratory system. The effect may be temporary.
Liquid or solid irritant characteristics: If spilled on clothing and allowed to remain, may cause smarting and reddening of the skin.
Odor threshold: 4.68 ppm.
IDLH value: Potential human carcinogen; 500 ppm.

FIRE DATA
Flash point: 12°F/11°C (cc) Benzene is solid at 12°F/–11°C.
Flammable limits in air: LEL: 1.3%; UEL: 7.8%.
Fire extinguishing agents not to be used: Water may be ineffective
Behavior in fire: Containers may rupture and explode.
Autoignition temperature: 928°F/498°C/771°K.
Electrical hazard: Class I, Group D; Flow or agitation of substance may generate electrostatic charges due to low conductivity.
Burning rate: 6.0 mm/min

CHEMICAL REACTIVITY
Binary reactants: Reacts with strong oxidizers, many fluorides and perchlorates; nitric acid. Attacks plastics and rubber on prolonged exposure; first swells, then softens rubber.

116 Benzene hexachloride, gamma isomer

Reactivity group: 32
Compatibility class: Aromatic hydrocarbons

ENVIRONMENTAL DATA
Food chain concentration potential: Log P_{ow} = 1.9–2.13. Unlikely to accumulate.
Water pollution: DOT Appendix B, §172.101–marine pollutant. Harmful to aquatic life in very low concentrations. May be dangerous if it enters nearby water intakes; notify operators. Notify local health and wildlife officials. **Response to discharge:** Issue warning–high flammability. Restrict access.

SHIPPING INFORMATION
Grades of purity: Industrial pure: 99+%; Thiophene-free: 99+%; Nitration: 99+%; Industrial: 90%; 85+%; Reagent: 99+%; **Storage temperature:** Oper.; **Inert atmosphere:** None; **Venting:** Pressure-vacuum; **Stability during transport:** Stable.

NAS HAZARD CLASSIFICATION FOR BULK WATER TRANSPORTATION
FIRE: 3
HEALTH: Vapor irritant: 1; Liquid or solid irritant: 1; Poisons: 3
WATER POLLUTION: Human toxicity: 1; Aquatic toxicity: 3; Aesthetic effect: 2
REACTIVITY: Other chemicals: 1; Water: 0; Self-reaction: 0

PHYSICAL AND CHEMICAL PROPERTIES
Physical state @ 59°F/15°C and 1 atm: Liquid.
Molecular weight: 78.11
Boiling point @ 1 atm: 176°F/80.1°C/353.3°K.
Melting/Freezing point: 42.0°F/5.5°C/278.7°K.
Critical temperature: 552.0°F/288.9°C/562.1°K.
Critical pressure: 710 psia = 48.3 atm = 4.89 MN/m^2.
Specific gravity (water = 1): 0.879 @ 68°F/20°C (liquid).
Liquid surface tension: 28.9 dynes/cm = 0.0289 N/m @ 68°F/20°C.
Liquid water interfacial tension: 35.0 dynes/cm = 0.035 N/m @ 68°F/20°C.
Relative vapor density (air = 1): 2.7
Ratio of specific heats of vapor (gas): 1.061
Latent heat of vaporization: 169 Btu/lb = 94.1 cal/g = 3.94 x 10^5 J/kg.
Heat of combustion: –17,460 Btu/lb = –9698 cal/g = –406.0 x 10^5 J/kg.
Heat of fusion: 30.45 cal/g.
Vapor pressure: (Reid) 3.22 psia; 75 mm.

BENZENE HEXACHLORIDE, gamma isomer   REC. B:0600

SYNONYMS: AALINDAN; AFICIDE; AGROCIDE; AGROCIDE 2 or 6G or 7 or III or WP; AGRONEXIT; AGRISOL G-20; AMEISENATOD; AMEISENMITTEL MERCK; APARASIN; APHTIRIA; APLIDAL; ARBITEX; BBH; BEN-HEX; BENTOx 10; γ-BENZENE HEXACHLORIDE; BEXOL; BHC; γ-BHC; CELANEX; CHLORESENE; CODECHINE; DBH; DETMOL-EXTRAKT; DETOX 25; DEVORAN; DOL GRANULE; DRILL TOX-SPEZIAL AGLUKON; EEC No. 602-043-00-6; ENTOMOXAN; EXAGAMA; FORLIN; GALLOGAMA; GAMACID; GAMAPHEX; GAMENE; GAMMAHEXA; GAMMAHEXANE; GAMMALIN; GAMMALIN 20; GAMMATERR; GAMMEX; GAMMEXANE; GAMMOPAZ; GEXANE; HCCH; HCH; γ-HCH; HECLOTOX; HEXA; HEXACHLORAN; HEXACHLORANE; γ-HEXACHLORANE; γ-HEXACHLOROBENZENE; 1-A,2-α,3-β,4-A,5-A,6-β-HEXACHLOROCYCLOHEXANE; γ-HEXACHLOROCYCLOHEXANE; 1,2,3,4,5,6-HEXACHLOROCYCLOHEXANE, γ-isomer; HEXATOX; HEXAVERM; HEXICIDE; HEXYCLAN; HGI; HORTEX; INEXIT; ISOTOX; JACUTIN; KOKOTINE; KWELL; LENDINE; LENTOX; LIDENAL; LINDAFOR; LINDAGAM; LINDAGRAIN; LINDAGRANOX; LINDANE; γ-LINDANE; LINDANE; LINDAPOUDRE; LINDATOX; LINDOSEP; LINTOX; LOREXANE; MILBOL 49; MSZYCOL; NEO-SCABICIDOL; NEXEN FB; NEXIT; NEXIT-STARK; NEXOL-E; NICOCHLORAN; NOVIGAM; OMNITOX; PFLANZOL; QUELLADA; SANG GAMMA; SILVANOL; SPRITZ-RAPIDIN; SPRUEHPFLANZOL; STREUNEX; TAP 85; TRI-6; VITON; HEXACHLOROCYCLOHEXANE, GAMMA ISOMER; 1,2,3,4,5,6-HEXACHLOR-CYCLOHEXANE; γ-HEXACHLORO-CYCLOHEXANE; γ-BHC

IDENTIFICATION
CAS Number: 58-89-9 (γ-isomer); 608-73-1 (generic); 319-84-6 (α-isomer)
Formula: $C_6H_6Cl_6$
DOT ID Number: UN 2761; DOT Guide Number: 151
Proper Shipping Name: Lindane
Reportable Quantity (RQ): **(CERCLA)** 1 lb/0.454 kg

DESCRIPTION: Light to dark brown crystalline solid, powder, or solution. Usually dissolved in combustible liquid. Musty odor. Solid sinks in water; solution generally Floats on the surface of water.

Poison! (organochlorine) • Breathing the dust or swallowing the material can kill you; skin or eye contact causes irritation • Firefighting gear (including SCBA) does not provide adequate protection. If exposure occurs, remove and isolate gear immediately and thoroughly decontaminate personnel • Toxic products of combustion may include oxides of hydrogen chloride, hydrogen chloride, and phosgene.

Hazard Classification (based on NFPA-704 Rating System)
Health Hazards (Blue): 3; Flammability (Red): 0; Reactivity (Yellow): 0

EMERGENCY RESPONSE: See Appendix A (151)
Evacuation:
Public safety: Isolate the area of spill or leak for at least 25 to 50 meters (80 to 160 feet) in all directions.
Spill: Increase, in the downwind direction, as necessary, the distance shown under "Public Safety."
Fire: If tank, rail car, or tank truck is involved in fire, isolate for at least 800 meters (½ mile) in all directions; also, consider initial evacuation for 800 meters (½ mile) in all directions.

EXPOSURE
Short-term effects: *SEEK MEDICAL ATTENTION*. **Solid or solution:** *POISONOUS IF SWALLOWED*. Irritating to skin and eyes. Remove contaminated clothing and shoes. *IF BREATHING HAS STOPPED*, give artificial respiration; *avoid mouth-to-mouth resuscitation; use bag/mask apparatus*. Flush affected areas with plenty of water. *IF IN EYES*, hold eyelids open and flush with plenty of water. *IF SWALLOWED* and victim is *CONSCIOUS AND ABLE TO SWALLOW*, have victim drink 4 to 8 ounces of water. **Do NOT induce vomiting.** Administer a slurry of activated charcoal. *Note to physician or authorized medical personnel*: Gastric lavage and saline cathartics (not oil laxatives because they promote absorption). Sedatives: pentobarbital or phenobarbital in amounts

adequate to control convulsions. Calcium gluconate intravenously may be used in conjunction with sedatives to control convulsions. Rest and quiet. *Do NOT* use epinephrine because ventricular fibrillation may result.

HEALTH HAZARDS
Personal protective equipment (PPE): A-Level PPE. OSHA Table Z-1-A air contaminant (lindane).
Recommendations for respirator selection: NIOSH/OSHA
5 mg/m³: CCROVDMFu [any chemical cartridge respirator with organic vapor cartridge(s) in combination with a dust, mist, and fume filter]; or SA (any supplied-air respirator). *12.5 mg/m³*: SA:CF* (any supplied-air respirator operated in a continuous-flow mode); or PAPROVDMFu* [any powered, air purifying respirator with organic vapor cartridge (s) in combination with a dust, mist, and fume filter]. *25 mg/m³*: CCRFOVHiE [any chemical cartridge respirator with a full facepiece and organic vapor cartridge(s) in combination with a high-efficiency particulate filter]; or GMFOVHiE [any air-purifying, full-facepiece respirator (gas mask) with a chin-style, front- or back-mounted organic vapor canister having a high-efficiency particulate filter]; or PAPRTOVHiE* [any powered, air-purifying respirator with a tight-fitting facepiece and organic vapor cartridge (s) in combination with a high-efficiency particulate filter]; or SCBAF (any self-contained breathing apparatus with full facepiece); or SAF (any supplied-air respirator with a full facepiece). *50 mg/m³*: SAF:PD,PP (any supplied-air respirator that has a full facepiece and is operated in a pressure-demand or other positive-pressure mode). *EMERGENCY OR PLANNED ENTRY INTO UNKNOWN CONCENTRATIONS OR IDLH CONDITIONS:* SCBAF:PD,PP (any self-contained breathing apparatus that has a full facepiece and is operated in a pressure-demand or other positive-pressure mode); or SAF:PD,PP:ASCBA (any supplied-air respirator that has a full facepiece and is operated in a pressure-demand or other positive-pressure mode in combination with an auxiliary self-contained breathing apparatus operated in a pressure-demand or other positive pressure mode). *ESCAPE*: GMFOVHiE [any air-purifying, full-facepiece respirator (gas mask) with a chin-style, front-or back-mounted canister having a high efficiency particulate filter]; or SCBAE (any appropriate escape-type, self-ccntained breathing apparatus). *Note:* Substance reported to cause eye irritation or damage; may require eye protection.
Exposure limits (TWA unless otherwise noted): ACGIH PEL 0.5 mg/m³; OSHA PEL 0.5 mg/m³ as lindane; skin contact contributes significantly in overall exposure.
Toxicity by ingestion: Gamma isomer (Lindane): Grade 3; LD_{50} 50 to 500 mg/kg (rat); Technical mixture: Grade 2; LD_{50} 0.5 to 5 g/kg.
Long-term health effects: Confirmed animal carcinogen. IARC substance, overall evaluation unassigned. Mutagen to human lymphocytes. May effect the blood.
Vapor (gas) irritant characteristics: Eye, skin and respiratory tract irritant. Personnel will not usually tolerate moderate or high concentrations.
Liquid or solid irritant characteristics: Eye, skin and respiratory tract irritant. If spilled on clothing and allowed to remain, may cause smarting and reddening of the skin.
IDLH value: Suspected human carcinogen; 50 mg/m³.

CHEMICAL REACTIVITY
Binary reactants: Corrosive to metals.

ENVIRONMENTAL DATA
Food chain concentration potential: Log K_{ow} = variably reported at 3.6 to 3.72. Potential is high.

Water pollution: DOT Appendix B, §172.101- severe marine pollutant. Harmful to aquatic life in very low concentrations. Fouling to shoreline. May be dangerous if it enters nearby water intakes; notify operators. Notify local health and pollution control officials. **Response to discharge:** Issue warning–water contaminant, poison. Should be removed. Chemical and physical treatment.

SHIPPING INFORMATION
Grades of purity: Fortified grade: 40–45% gamma isomer; Lindane: pure gamma isomer; **Stability during transport:** Stable.

PHYSICAL AND CHEMICAL PROPERTIES
Physical state @ 59°F/15°C and 1 atm: Solid.
Molecular weight: 290.83
Boiling point @ 1 atm: Decomposes. 550°F/288°C/561°K.
Melting/Freezing point: 240°F/115°C/388°K.
Specific gravity (water = 1): 1.891 @ 66°F/19°C/292°K (solid).
Relative vapor density (air = 1): 1.1
Vapor pressure: 0.00001 mm.

BENZENE PHOSPHORUS DICHLORIDE   REC. B:0650

SYNONYMS: DICHLOROPHENYLPHOSPHINE; DICLORURO BENCENOFOSFOROSO (Spanish); PHENYL PHOSPHORUS DICHLORIDE; PHENYLPHOSPHONOUS DICHLORIDE; PHENYLPHOSPHINE DICHLORIDE; PHOSPHENYL CHLORIDE

IDENTIFICATION
CAS Number: 644-97-3
Formula: $C_6NO_5PCl_2$
DOT ID Number: UN 2798; DOT Guide Number: 137
Proper Shipping Name: Phenyl phosphorus dichloride

DESCRIPTION: Colorless liquid; fumes in air. Pungent, acrid odor. Sinks in water; reacts violently forming toxic hydrochloric acid.

Poison! • Do not use water • Breathing the vapor can kill you; skin or eye contact causes severe burns, impaired vision, or blindness • Firefighting gear (including SCBA) does not provide adequate protection. If exposure occurs, remove and isolate gear immediately and thoroughly decontaminate personnel • Containers may BLEVE when exposed to fire • Toxic products of combustion may include oxides of phosphorus and hydrogen chloride fumes • Do not put yourself in danger by entering a contaminated area to rescue a victim.

Hazard Classification (based on NFPA-704 Rating System)
Health Hazards (Blue): 3; Flammability (Red): 1; Reactivity (Yellow): 2; Special Notice (White): Water reactive

EMERGENCY RESPONSE: See Appendix A (137)
Evacuation:
Public safety: Isolate the area of spill or leak for at least 50 to 100 meters (160 to 330 feet) in all directions.
Spill: Increase, in the downwind direction, as necessary, the distance shown under "Public Safety."
Fire: If any large container is involved in fire, isolate for at least 800 meters (½ mile) in all directions; also, consider initial evacuation for 800 meters (½ mile) in all directions.

EXPOSURE
Short-term effects: *SEEK MEDICAL ATTENTION. GAS PRODUCED IN REACTION WITH WATER: POISONOUS IF INHALED.* Irritating to eyes, nose, and throat. May cause lung edema; physical exertion will aggravate this condition. Move to fresh air. *IF BREATHING HAS STOPPED,* give artificial respiration; *avoid mouth-to-mouth resuscitation; use bag/mask apparatus.* IF breathing is difficult, administer oxygen. **Liquid:** Irritating to skin and eyes. Harmful if swallowed. Remove contaminated clothing and shoes. Flush affected areas with plenty of water. *IF IN EYES,* hold eyelids open and flush with plenty of water. *IF SWALLOWED* and victim is *CONSCIOUS AND ABLE TO SWALLOW,* have victim drink 4 to 8 ounces of water.

Note to physician or authorized medical personnel: Medical observation is recommended for 24 to 48 hours after breathing overexposure, as pulmonary edema may be delayed. As first aid for pulmonary edema, consider administering a corticosteroid spray. Cigarette smoking may exacerbate pulmonary injury and should be discouraged for at least 72 hours following exposure.

HEALTH HAZARDS
Personal protective equipment (PPE): A-Level PPE. Self-contained breathing apparatus; acid-type canister mask; goggles and face shield; rubber gloves; protective clothing.
Vapor (gas) irritant characteristics: May cause severe eye and respiratory tract irritation. May cause lung edema; physical exertion will aggravate this condition.
Liquid or solid irritant characteristics: Severe eye and skin irritation and burns.

FIRE DATA
Flash point: 215°F/102°C (oc). Possibly lower due to presence of dissolved phosphorus.
Behavior in fire: Containers may rupture. Hot liquid is spontaneously flammable due to presence of dissolved phosphorus.
Autoignition temperature: 319°F/159°C/432°K.

CHEMICAL REACTIVITY
Reactivity with water: Reacts vigorously to form hydrogen chloride (hydrochloric acid).
Binary reactants: Corrodes metal, except 316 stainless steel, nickel, or Hastelloy.
Neutralizing agents for acids and caustics: Flush with water, rinse with sodium bicarbonate or lime solution.

ENVIRONMENTAL DATA
Food chain concentration potential: Negative; unlikely to accumulate.
Water pollution: Effect of low concentrations on aquatic life is unknown. May be dangerous if it enters nearby water intakes; notify operators. Notify local health and wildlife officials.
Response to discharge: Issue warning–corrosive. Restrict access. Disperse and flush.

SHIPPING INFORMATION
Grades of purity: Commercial; **Storage temperature:** Ambient; **Inert atmosphere:** None; **Venting:** Pressure-vacuum; **Stability during transport:** Stable.

PHYSICAL AND CHEMICAL PROPERTIES
Physical state @ 59°F/15°C and 1 atm: Liquid.
Molecular weight: 179.0
Boiling point @ 1 atm: 430°F/221°C/494°K.
Melting/Freezing point: –60°F/–51°C/222°K.
Specific gravity (water = 1): 1.140 @ 77°F/25°C (liquid).
Liquid surface tension: (estimate) 25 dynes/cm = 0.025 N/m @ 68°F/20°C.
Heat of combustion: (estimate) –8200 Btu/lb = –4500 cal/g = –190 x 10^5 J/kg.
Heat of solution: –72 Btu/lb = –40 cal/g = –1.7 x 10^5 J/kg.

BENZENE PHOSPHORUS THIODICHLORIDE
REC. B:0700

SYNONYMS: PHENYLPHOSPHONOTHIOIC DICHLORIDE; BENZENETHIOPHOSPHONYL CHLORIDE; PHENYLPHOSPHINE THIODICHLORIDE; TIODICLORURO BENCENOFOSFOROSO (Spanish)

IDENTIFICATION
CAS Number: 14684-25-4; 3497-00-5
Formula: $C_6NO_5PSCl_2$
DOT ID Number: UN 2799; DOT Guide Number: 137
Proper Shipping Name: Phenyl phosphorus thiodichloride

DESCRIPTION: Colorless to light yellow liquid. Unpleasant odor. Sinks and reacts in water; forming hydrogen chloride fumes.

Poison! • Heat forms oxygen; will increase the activity of an existing fire • May cause fire or explosion contact with combustibles (wood, paper, oil, clothing, etc.) • Containers may BLEVE when exposed to fire • Irritates eyes, skin, and respiratory tract • Toxic products of combustion may include oxides of phosphorus and sulfur, hydrogen chloride, and phosgene.

Hazard Classification (based on NFPA-704 Rating System)
Health Hazards (Blue): 3; Flammability (Red): 2; Reactivity (Yellow): 2; Special Notice (White): Water reactive

EMERGENCY RESPONSE: See Appendix A (137)
Evacuation:
Public safety: Isolate the area of spill or leak for at least 50 to 100 meters (160 to 330 feet) in all directions.
Spill: Increase, in the downwind direction, as necessary, the distance shown under "Public Safety."
Fire: If any large container is involved in fire, isolate for at least 800 meters (½ mile) in all directions; also, consider initial evacuation for 800 meters (½ mile) in all directions.

EXPOSURE
Short-term effects: *SEEK MEDICAL ATTENTION.* **Vapor produced in reaction with water:** *POISONOUS IF INHALED.* Irritating to eyes, nose, and throat. Move to fresh air. *IF BREATHING HAS STOPPED,* give artificial respiration; *avoid mouth-to-mouth resuscitation; use bag/mask apparatus.* IF breathing is difficult, administer oxygen. **Liquid:** Irritating to skin and eyes. Harmful if swallowed. Remove contaminated clothing and shoes. Flush affected areas with plenty of water. *IF IN EYES,* hold eyelids open and flush with plenty of water. *IF SWALLOWED* and victim is *CONSCIOUS AND ABLE TO SWALLOW,* have victim drink 4 to 8 ounces of water.

HEALTH HAZARDS
Personal protective equipment (PPE): B-Level PPE. Chemical protective material(s) reported to have good to excellent resistance: butyl rubber.
Vapor (gas) irritant characteristics: Eye and respiratory tract irritant.

Liquid or solid irritant characteristics: Severe eye and skin irritant.

FIRE DATA
Flash point: 252°F/122°C (oc).
Behavior in fire: Containers may rupture.
Autoignition temperature: 338°F/170°C/443°K.

CHEMICAL REACTIVITY
Reactivity with water: Forms hydrogen chloride fumes (hydrochloric acid). The reaction is slow unless water is hot.
Binary reactants: Corrodes metal slowly.
Neutralizing agents for acids and caustics: Flush with water, rinse with sodium bicarbonate or lime solution.

ENVIRONMENTAL DATA
Food chain concentration potential: Negative; unlikely to accumulate.
Water pollution: Effect of low concentrations on aquatic life is unknown. May be dangerous if it enters nearby water intakes; notify operators. Notify local health and wildlife officials.
Response to discharge: Issue warning–corrosive. Restrict access. Disperse and flush.

SHIPPING INFORMATION
Grades of purity: Commercial; **Storage temperature:** Ambient; **Inert atmosphere:** None; **Venting:** Pressure-vacuum; **Stability during transport:** Stable.

PHYSICAL AND CHEMICAL PROPERTIES
Physical state @ 59°F/15°C and 1 atm: Liquid.
Molecular weight: 211
Boiling point @ 1 atm: 518°F/270°C/543°K.
Melting/Freezing point: –11°F/–24°C/249°K.
Specific gravity (water = 1): 1.378 @ 68°F/20°C (liquid).
Liquid surface tension: (estimate) 25 dynes/cm = 0.025 N/m @ 68°F/20°C.
Heat of combustion: (estimate) –7700 Btu/lb = –4300 cal/g = –180 x 10^5 J/kg.
Heat of solution: (estimate) –9 Btu/lb = –5 cal/g = –0.2 x 10^5 J/kg.

BENZENE SULFONYL CHLORIDE   REC. B:0750

SYNONYMS: BENZENESULFONIC ACID CHLORIDE; BENZENESULFONYL CHLORIDE; BENZENE SULFONECHLORIDE; BENZENE SULFONE-CHLORIDE; BENZENOSULPHOCHLORIDE; CLORURO BENCENOFOSFOROSO (Spanish); BENZENE SULFOCHLORIDE; BSC-REFINED D; BENZENESULFONIC ACID CHLORIDE; RCRA No. U020

IDENTIFICATION
CAS Number: 98-09-9
Formula: $C_6NO_5SO_2Cl$
DOT ID Number: UN 2225; DOT Guide Number: 156
Proper Shipping Name: Benzene sulfonyl chloride
Reportable Quantity (RQ): **(CERCLA)** 100 lb/45.4 kg

DESCRIPTION: Colorless to light yellow liquid. Pungent odor. Sinks in water and rapidly hydrolyzes; insoluble. Solidifies at 32°F/0°C.

Poison! • May be fatal if inhaled, if swallowed, or if absorbed through the skin. Corrosive to the skin, eyes, and respiratory tract; contact with skin or eyes can cause burns and vision impairment; inhalation symptoms may be delayed • Firefighting gear (including SCBA) does not provide adequate protection. If exposure occurs, remove and isolate gear immediately and thoroughly decontaminate personnel • Containers may BLEVE when exposed to fire • Vapors are heavier than air and will collect and stay in low areas • Vapors in confined areas (e.g., tanks, sewers, buildings) may explode when exposed to fire • Toxic products of combustion may include oxides of sulfur, hydrogen chloride, and phosgene.

Hazard Classification (based on NFPA-704 Rating System)
Health Hazards (Blue): 3; Flammability (Red): 1; Reactivity (Yellow): 1

EMERGENCY RESPONSE: See Appendix A (156)
Evacuation:
Public safety: Isolate the area of spill or leak for at least 50 to 100 meters (160 to 330 feet) in all directions.
Spill: Increase, in the downwind direction, as necessary, the distance shown under "Public Safety."
Fire: If tank, rail car, or tank truck is involved in fire, isolate for at least 800 meters (½ mile) in all directions; also, consider initial evacuation for 800 meters (½ mile) in all directions.

EXPOSURE
Short-term effects: *SEEK MEDICAL ATTENTION*. **Vapor:** *MAY BE FATAL IF INHALED*. Irritating to eyes, skin and mucous membranes. Move victim to fresh air. *IF BREATHING HAS STOPPED*, give artificial respiration; *avoid mouth-to-mouth resuscitation; use bag/mask apparatus*. IF breathing is difficult, administer oxygen. **Liquid:** Will burn skin and eyes. *May BE FATAL IF SWALLOWED OR ABSORBED THROUGH SKIN. IF IN EYES OR ON SKIN*, flush with running water for at least 15 minutes; hold eyelids open if necessary. Wash skin with soap and water. Remove and isolate contaminated clothing and shoes at the site. Maintain normal body temperature and keep victim quiet. Hold victim for observation for delayed effects. *IF SWALLOWED* and victim is *UNCONSCIOUS OR HAVING CONVULSIONS*, do nothing except keep victim warm. Corrosive to the skin, eyes, and respiratory tract; contact with skin or eyes can cause burns and vision impairment; inhalation symptoms may be delayed

HEALTH HAZARDS
Personal protective equipment (PPE): B-Level PPE. Wear positive pressure breathing apparatus and special protective clothing.
Toxicity by ingestion: Grade 2; LD_{50} = 1.96 g/kg (rat).
Vapor (gas) irritant characteristics: Causes irritation of eyes, skin and mucous membranes.
Liquid or solid irritant characteristics: Severe eye and skin irritant. May cause corrosive burns to skin and eyes.

FIRE DATA
Flash point: More than 234°F/112°C (cc).
Behavior in fire: Cylinder may explode.

CHEMICAL REACTIVITY
Reactivity with water: Decomposes in hot water to produce corrosive and toxic hydrochloric acid and benzene sulfonic acid.
Binary reactants: Violent reaction with strong oxidizers, dimethyl sulfoxide, and methyl formamide. Corrosive to metals in the presence of moisture.
Neutralizing agents for acids and caustics: Sodium bicarbonate.
Reactivity group: Not compatible with Groups 5, 7, and 43.
Compatibility class: Special case

Benzenethiol

ENVIRONMENTAL DATA
Water pollution: Harmful to aquatic life in very low concentrations. May be dangerous if enters water intakes. Notify local health and wildlife officials. **Response to discharge:** Issue warning–corrosive, water contaminant. Restrict access. Should be removed. Chemical and physical treatment.

SHIPPING INFORMATION
Grades of purity: 96%; 99%; **Stability during transport:** Stable.

PHYSICAL AND CHEMICAL PROPERTIES
Physical state @ 59°F/15°C and 1 atm: Liquid.
Molecular weight: 176.62
Boiling point @ 1 atm: 485°F/252°C/525°K.
Melting/Freezing point: 58°F/15°C/288°K.
Specific gravity (water = 1): 1.3842 @ 15°C.
Relative vapor density (air = 1): 6.09 (estimate).
Vapor pressure: 1 mm.

BENZENETHIOL   REC. B:0800

SYNONYMS: FENILMERCAPTANO (Spanish); PHENYL MERCAPTAN; MERCAPTOBENZENE; PHENYLTHIOL; RCRA No. P014; THIOPHENOL

IDENTIFICATION
CAS Number: 108-98-5
Formula: C_6H_6S
DOT ID Number: UN 2337; DOT Guide Number: 131
Proper Shipping Name: Phenyl mercaptan
Reportable Quantity (RQ): **(CERCLA)** 100 lb/45.4 kg

DESCRIPTION: Colorless to pale yellow liquid. Nauseating odor; burnt rubber, stench; putrid. Sinks in water; reacts, forming flammable mercaptan vapors. Solid below 5°F/–15°C.

Flammable • Breathing the vapor, skin or eye contact, or swallowing the material may kill you • Firefighting gear (including SCBA) does not provide adequate protection. If exposure occurs, remove and isolate gear immediately and thoroughly decontaminate personnel • Containers may BLEVE when exposed to fire • Vapors may form explosive mixture with air • Vapors are heavier than air and will collect and stay in low areas • Vapors may travel long distances to ignition sources and flashback • Vapors in confined areas (e.g., tanks, sewers, buildings) may explode when exposed to fire • Irritating to the skin, eyes, and respiratory tract • Toxic products of combustion may include carbon monoxide, CO_2, sulfur oxides, and sulfides.

Hazard Classification (based on NFPA-704 Rating System)
Health Hazards (Blue): 1; Flammability (Red): 2; Reactivity (Yellow): 0

EMERGENCY RESPONSE: See Appendix A (131)
Evacuation:
Public safety: Isolate spill area for at least 100 to 200 meters (330 to 660 feet) in all directions.
Spill: Small spill–First: Isolate in all directions 30 meters (100 feet); Then: Protect persons downwind, DAY: 0.2 km (0.1 mile); NIGHT: 0.2 km (0.1 mile). Large spill–First: Isolate in all directions 30 meters (100 feet); Then: Protect persons downwind, DAY: 0.3 km (0.2 mile); NIGHT: 0.6 km (0.4 mile).
Fire: Isolate for 800 meters (½ mile) in all directions, especially if tank, rail car, or tank truck is involved in fire.

EXPOSURE
Short-term effects: *SEEK MEDICAL ATTENTION*. **Vapor:** POISONOUS; MAY BE FATAL IF INHALED OR ABSORBED THROUGH SKIN. Irritating to eyes, skin and mucous membranes. Over exposure may cause headache, dizziness, coughing, difficulty in breathing, nausea, and vomiting. Symptoms may be delayed. Move to fresh air. *IF BREATHING HAS STOPPED*, give artificial respiration. IF breathing is difficult, administer oxygen. **Liquid:** *POISONOUS. MAY BE FATAL IF SWALLOWED OR ABSORBED THROUGH SKIN.* May burn skin and eyes. *Speed in removing material from skin is extremely important. IF IN EYES OR ON SKIN*, immediately flush with running water for at least 15 minutes; hold eyelids open occasionally if appropriate. Remove and isolate contaminated clothing and shoes at the site. Effects may be delayed; keep victim under observation. *IF SWALLOWED* and victim is *CONSCIOUS AND ABLE TO SWALLOW*, have victim drink several
glasses of water and induce vomiting by touching back of throat. *IF SWALLOWED* and victim is *UNCONSCIOUS OR HAVING CONVULSIONS*, do nothing except keep victim warm.

HEALTH HAZARDS
Personal protective equipment (PPE): A-Level PPE. OSHA Table Z-1-A air contaminant. Wear special chemical protective clothing and approved respirator.
Recommendations for respirator selection: NIOSH
1 ppm: CCROV [any chemical cartridge respirator with organic vapor cartridge(s)]; or SA (any supplied-air respirator). *2.5 ppm:* SA:CF (any supplied-air respirator operated in a continuous-flow mode); or PAPROV [any powered, air-purifying respirator with organic vapor cartridge(s)]. *5 ppm:* CCRFOV [any air-purifying, full-facepiece respirator (gas mask) with a chin-style, front- or back-mounted acid gas canister]; or GMFOV [any air-purifying, full-facepiece respirator (gas mask) with a chin-style, front- or back-mounted organic vapor canister]; or PAPRTOV [any powered, air-purifying respirator with a tight-fitting facepiece and organic vapor cartridge(s)]; or SCBAF (any self-contained breathing apparatus with a full facepiece); or SAF (any supplied-air respirator with a full facepiece). *EMERGENCY OR PLANNED ENTRY INTO UNKNOWN CONCENTRATIONS OR IDLH CONDITIONS:* SCBAF:PD,PP (any self-contained breathing apparatus that has a full facepiece and is operated in a pressure-demand or other positive-pressure mode); or SAF:PD,PP:ASCBA (any supplied-air respirator that has a full facepiece and is operated in a pressure-demand or other positive-pressure mode in combination with an auxiliary self-contained breathing apparatus operated in a pressure-demand or other positive pressure mode). *ESCAPE:* GMFOV [any air-purifying, full-facepiece respirator (gas mask) with a chin-style, front- or back-mounted organic vapor canister]; or SCBAE (any appropriate escape-type, self-contained breathing apparatus).
Exposure limits (TWA unless otherwise noted): ACGIH TLV 0.5 ppm (2.3 mg/m^3); NIOSH 0.1 ppm (0.5 mg/m^3)/15 min.
Toxicity by ingestion: Grade 4; LD_{50} = 46 mg/kg (rat).
Long-term health effects: Lung, liver and kidney changes were found in mice after inhalation of high doses.
Vapor (gas) irritant characteristics: Nauseating. Vapors cause severe irritation of eyes and throat and can cause eye and lung injury.
Liquid or solid irritant characteristics: Severe eye and skin irritant.
Odor threshold: 0.061 mg/m^3.

FIRE DATA
Flash point: 132°F/56°C (cc).

CHEMICAL REACTIVITY
Binary reactants: Strong acids, and bases, calcium hypochlorite; alkali metal. Oxidizes on exposure to air. Mildly corrosive to carbon steel.
Neutralizing agents for acids and caustics: For small spills, soak up with absorbent; remove absorbent; treat spill area with sodium carbonate slurry. Remove it and treat area with sodium hypochlorite solution (commercial bleach); absorb liquid and remove.

ENVIRONMENTAL DATA
Food chain concentration potential: Log P_{ow} = 2.52. Unlikely to accumulate.
Water pollution: Effects of low concentration on aquatic life is unknown. May be dangerous if it enters nearby water intakes; notify operators. Notify local health and wildlife officials.
Response to discharge: Issue warning–poison. Restrict access. Should be removed. Chemical and physical treatment.

SHIPPING INFORMATION
Grades of purity: 97%; 99+%; **Storage temperature:** Ambient;
Stability during transport: In the absence of air, it is stable to 392°F/200°C; it oxidizes in air to yield diphenyl disulfide. Other oxidizing agents react similarly.

PHYSICAL AND CHEMICAL PROPERTIES
Physical state @ 59°F/15°C and 1 atm: Liquid.
Molecular weight: 110.18
Boiling point @ 1 atm: 334.4°F/168°C/441.2°K.
Melting/Freezing point: 5°F/–15°C/258°K.
Critical temperature: 778°F/415°C/688°K (estimate).
Critical pressure: 51.9 psia = 35.3 atm = 3.58 MN/m^2 (estimate).
Specific gravity (water = 1): 1.075 @ 68°F/20°C.
Liquid surface tension: 39.0 dynes/cm = 0.039 N/m @ 68°F/20°C.
Relative vapor density (air = 1): 3.8 (calculated).
Latent heat of vaporization: 156 Btu/lb = 86.5 cal/g = 3.6 x 10^5 J/kg.
Vapor pressure: 1 mm @ 65°F

BENZIDINE REC. B:0850

SYNONYMS: BENCIDINA (Spanish); *p,p*-BIANILINE; 4,4'-BIANILINE; (1,1'-BIFENYL)-4,4'-DIAMINE; (1,1'-BIPHENYL)-4,4'-DIAMINE; 4,4'-BIPHENYLDIAMINE; BIPHENYL, 4,4'-DIAMINO-; C.I. AZOIC DIAZO COMPONENT 112; *p,p'*-DIAMINOBIPHENYL; 4,4'-DIAMINOBIPHENYL; 4,4'-DIAMINO-1,1'-BIPHENYL; *p*-DIAMINODIPHENYL; 4,4'-DIAMINODIPHENYL; *p,p'*-DIANILINE; 4,4'-DIPHENYLENEDIAMINE; EEC No. 612-042-00-2; (1,1'-BIPHENYL)-4,4'DIAMINE; FAST CORINTH BASE B; C.I. AZOIC DIAZO; RCRA No. U021; COMPONENT 112

IDENTIFICATION
CAS Number: 92-87-5
Formula: $C_{12}H_{12}N_2$; $NH_2C_6H_4C_6H_4NH_2$
DOT ID Number: UN 1885; DOT Guide Number: 153
Proper Shipping Name: Benzidine
Reportable Quantity (RQ): **(CERCLA)** 1 lb/0.454 kg

DESCRIPTION: Grayish-yellow, white, reddish gray crystalline solid, powder, or leaflets; darkens on exposure to light. Sinks and very slowly mixes with water.

Poison! (aromatic amine) • Breathing the vapor, skin or eye contact, or swallowing the material can kill you • Firefighting gear (including SCBA) does not provide adequate protection. If exposure occurs, remove and isolate gear immediately and thoroughly decontaminate personnel • Containers may BLEVE when exposed to fire • Vapors are heavier than air and will collect and stay in low areas • Vapors in confined areas (e.g., tanks, sewers, buildings) may explode when exposed to fire • Toxic products of combustion may include nitrogen oxides • Do not put yourself in danger by entering a contaminated area to rescue a victim:

Hazard Classification (based on NFPA-704 Rating System)
Health Hazards (Blue): 2; Flammability (Red): 1; Reactivity (Yellow): 0

EMERGENCY RESPONSE: See Appendix A (153)
Evacuation:
Public safety: Isolate the area of spill or leak for at least 25 to 50 meters (80 to 160 feet) in all directions.
Spill: Increase, in the downwind direction, as necessary, the distance shown under "Public Safety."
Fire: If tank, rail car, or tank truck is involved in fire, isolate for at least 800 meters (½ mile) in all directions; also, consider initial evacuation for 800 meters (½ mile) in all directions.

EXPOSURE
Short-term effects: *SEEK MEDICAL ATTENTION.* **Dust:** *POISONOUS IF INHALED OR ABSORBED THROUGH THE SKIN.* May cause dermatitis, irritation or sensitization. *IF IN EYES OR ON SKIN,* flush with running water for at least 15 minutes; hold eyelids open if necessary. Wash skin with soap and water. *IF BREATHING HAS STOPPED,* give artificial respiration; *avoid mouth-to-mouth resuscitation; use bag/mask apparatus.* IF breathing is difficult, administer oxygen. May cause lung edema; physical exertion will aggravate this condition. **Solid:** *POISONOUS IF SWALLOWED OR ABSORBED THROUGH THE SKIN.* May cause contact dermatitis, irritation or sensitization. Ingestion may cause nausea and vomiting. *IF IN EYES OR ON SKIN,* flush with running water for at least 15 minutes; hold eyelids open if necessary. Wash skin with soap and water. *IF SWALLOWED* and victim is *UNCONSCIOUS OR HAVING CONVULSIONS,* do nothing except keep victim warm.
Note to physician or authorized medical personnel: Medical observation is recommended for 24 to 48 hours after breathing overexposure, as pulmonary edema may be delayed. As first aid for pulmonary edema, consider administering a corticosteroid spray. Cigarette smoking may exacerbate pulmonary injury and should be discouraged for at least 72 hours following exposure.

HEALTH HAZARDS
Personal protective equipment (PPE): A-Level PPE. OSHA Table Z-1-A air contaminant. Wear full protective clothing.
Recommendations for respirator selection: NIOSH
At any detectable concentration: SCBAF:PD,PP (any self-contained breathing apparatus that has a full facepiece and is operated in a pressure-demand or other positive-pressure mode); or SAF:PD,PP:ASCBA (any supplied-air respirator that has a full facepiece and is operated in a pressure-demand or other positive-pressure mode in combination with an auxiliary self-contained breathing apparatus operated in a pressure-demand or other positive pressure mode). *ESCAPE*: HiEF (any air-purifying, full-facepiece respirator with a high-efficiency particulate filter); or SCBAE (any appropriate escape-type, self-contained breathing apparatus).
Exposure limits (TWA unless otherwise noted): ACGIH TLV confirmed human carcinogen; NIOSH REL Potential human

carcinogen; reduce exposure to lowest feasible level. OSHA PEL [1910.1010] potential occupational carcinogen
Toxicity by ingestion: Grade 3; LD_{50} = 309 mg/kg (rat).
Long-term health effects: OSHA specifically regulated carcinogen. IARC carcinogen, rating 1. NTP known carcinogen. May cause mutagenic, tumorigenic and carcinogenic effects; liver and kidney damage; hemolysis and bone marrow depressions.
Liquid or solid irritant characteristics: Eye and skin irritant.

CHEMICAL REACTIVITY
Binary reactants: Oxidizers, strong acids.

ENVIRONMENTAL DATA
Food chain concentration potential: Log K_{ow} = 1.34. Unlikely to accumulate.
Water pollution: Harmful to aquatic life in very low concentrations. May be dangerous if it enters nearby water intakes; notify operators. Notify local health and wildlife officials.
Response to discharge: Issue warning–poison, water contaminant, air contaminant. Restrict access. Evacuate area. Should be removed. Chemical and physical treatment. Cover the spill with a 9:1 mixture of soda ash and sand.

SHIPPING INFORMATION
Storage temperature: Ambient; **Stability during transport:** Stable.

PHYSICAL AND CHEMICAL PROPERTIES
Physical state @ 59°F/15°C and 1 atm: Solid.
Molecular weight: 184.24
Boiling point @ 1 atm: 755°F/402°C/675°K.
Melting/Freezing point: 242°F/117°C/390°K.
Critical temperature: 1220°F/660°C/933°K (estimate).
Critical pressure: 479 psia = 32.6 atm = 3.30 MN/m² (estimate).
Specific gravity (water = 1): 1.25 @ 68°F
Relative vapor density (air = 1): 6.4 (estimate).
Vapor pressure: Very low.

BENZOIC ACID   REC. B:0900

SYNONYMS: ACIDO BENZOICO (Spanish); BENZENECARBOXYLIC ACID; CARBOXYLBENZENE; DRACYCLIC ACID

IDENTIFICATION
CAS Number: 65-85-0
Formula: C_6H_5COOH
DOT ID Number: UN 3082; DOT Guide Number: 171
Proper Shipping Name: Environmentally hazardous substance, solid, n.o.s.
Reportable Quantity (RQ): **(CERCLA)** 5000 lb/2270 kg

DESCRIPTION: White crystalline solid or powder. Faint pleasant, slight aromatic odor. Sinks in water.

Combustible solid • Containers may BLEVE when exposed to fire • Concentrated dust in confined areas (e.g., tanks, sewers, buildings) may explode when exposed to fire • Toxic products of combustion may include acrid carbon monoxide. Vapor from molten benzoic acid may form explosive mixture with air.

Hazard Classification (based on NFPA-704 Rating System)
Health Hazards (Blue): 2; Flammability (Red): 1; Reactivity (Yellow): 0

EMERGENCY RESPONSE: See Appendix A (171)
Evacuation:
Public safety: Isolate the area of spill or leak for at least 10 to 25 meters (30 to 80 feet) in all directions.
Spill: Increase, in the downwind direction, as necessary, the distance shown under "Public Safety."
Fire: If any large container is involved in fire, isolate for at least 800 meters (½ mile) in all directions; also, consider initial evacuation for 800 meters (½ mile) in all directions.

EXPOSURE
Short-term effects: *SEEK MEDICAL ATTENTION.* **Dust:** Irritating to nose and throat if inhaled. Move to fresh air. **Solid:** Irritating to skin and eyes.
Flush affected areas with plenty of water. *IF IN EYES*, hold eyelids open and flush with plenty of water.

HEALTH HAZARDS
Personal protective equipment (PPE): B-Level PPE. Butyl rubber is generally suitable for carbooxylic acid compounds.
Toxicity by ingestion: Grade 2; LD_{50} = 0.5 to 5 g/kg.
Vapor (gas) irritant characteristics: Eye and respiratory tract irritant.
Liquid or solid irritant characteristics: Eye and skin irritant. If spilled on clothing and allowed to remain, may cause smarting and reddening of the skin. Dust may irritate nose and eyes.

FIRE DATA
Flash point: 250°F/121°C (cc).
Fire extinguishing agents not to be used: None
Autoignition temperature: 1058°F/570°C/843°K.

ENVIRONMENTAL DATA
Food chain concentration potential: Log P_{ow} = 1.88. Unlikely to accumulate.
Water pollution: Harmful to aquatic life in very low concentrations. May be dangerous if it enters nearby water intakes; notify operators. Notify local health and wildlife officials. **Response to discharge:** Disperse and flush.

SHIPPING INFORMATION
Grades of purity: USP, FCC grade: 99.5–100.5%; **Stability during transport:** Stable.

PHYSICAL AND CHEMICAL PROPERTIES
Physical state @ 59°F/15°C and 1 atm: Solid.
Molecular weight: 122.12
Boiling point @ 1 atm: 481°F/249°C/522°K.
Melting/Freezing point: 252°F/122°C/396°K.
Critical temperature: 894°F/479°C/752°K.
Critical pressure: 660 psia = 45 atm = 4.6 MN/m².
Specific gravity (water = 1): 1.27 at 28°C (solid).
Relative vapor density (air = 1): 4.21
Heat of fusion: 33.89 cal/g.

BENZONITRILE   REC. B:0950

SYNONYMS: BENZONITRILO (Spanish); BENZENE, CYANO-; BENZENENITRILE; BENZOIC ACID NITRILE; CYANOBENZENE; EEC No. 608-012-00-3; PHENYL CYANIDE

IDENTIFICATION
CAS Number: 100-47-0
Formula: C_6H_5CN

Benzonitrile

DOT ID Number: UN 2224; DOT Guide Number: 152
Proper Shipping Name: Benzonitrile
Reportable Quantity (RQ): **(CERCLA)** 5000 lb/2270 kg

DESCRIPTION: Colorless liquid. Almond-like odor. May float or sink in water; slightly soluble.

Poison! • Breathing the vapor, skin or eye contact, or swallowing the material can kill you • Firefighting gear (including SCBA) does not provide adequate protection. If exposure occurs, remove and isolate gear immediately and thoroughly decontaminate personnel Containers may BLEVE when exposed to fire • Vapors are heavier than air and will collect and stay in low areas • Vapors in confined areas (e.g., tanks, sewers, buildings) may explode when exposed to fire • Toxic products of combustion may include hydrogen cyanide and nitrogen oxides • Do not put yourself in danger by entering a contaminated area to rescue a victim.

Hazard Classification (based on NFPA-704 Rating System)
Health Hazards (Blue): 4; Flammability (Red): 2; Reactivity (Yellow): 0

EMERGENCY RESPONSE: See Appendix A (152)
Evacuation:
Public safety: Isolate the area of spill or leak for at least 25 to 50 meters (80 to 160 feet) in all directions.
Spill: Increase, in the downwind direction, as necessary, the distance shown under "Public Safety."
Fire: If tank, rail car, or tank truck is involved in fire, isolate for at least 800 meters (½ mile) in all directions; also, consider initial evacuation for 800 meters (½ mile) in all directions.

EXPOSURE
Short-term effects: *SEEK MEDICAL ATTENTION.* **Vapor:** Irritating to eyes, nose, and throat. *IF INHALED*, will, will cause headache, difficult breathing or loss of consciousness. *IF IN EYES*, hold eyelids open and flush with plenty of water. *IF BREATHING HAS STOPPED*, give artificial respiration; *avoid mouth-to-mouth resuscitation; use bag/mask apparatus.* IF breathing is difficult, administer oxygen. **Liquid:** Irritating to skin and eyes. *IF SWALLOWED*, will cause headache, nausea, vomiting, or loss of consciousness. Remove contaminated clothing and shoes. Flush affected areas with plenty of water. *IF IN EYES*, hold eyelids open and flush with plenty of water. *IF SWALLOWED* and victim is *CONSCIOUS AND ABLE TO SWALLOW*, have victim drink 4 to 8 ounces of water. *IF SWALLOWED* and victim is *UNCONSCIOUS OR HAVING CONVULSIONS*, do nothing except keep victim warm.
Note to physician or authorized medical personnel: Consider the use of amyl nitrite perles if symptoms of cyanide poisoning develop. If symptoms indicate, initial treatment includes the cyanide antidote kit. In all cases, break an amyl nitrite perle in a gauze pad and hold lightly under victim's nose for 15 seconds, repeating 5 times at about 15-second intervals; if necessary (and if sodium nitrite infusions will be delayed), repeat procedure every 3 minutes with fresh perles until 3 or 4 have been used. Avoid breathing the vapor while administering it to the victim. Administer sodium nitrite IV, ASAP. The usual adult dose is 10 to 20 mL of a 3% solution infused over no less than 5 minutes; the average child dose is 0.15 to 0.20 mL/kg. Monitor blood pressure during administration, and slow the rate of infusion if hypotention develops. Next, infuse sodium thiosulfate IV. The usual adult dose is 50 mL of a 25% solution infused over 10 to 20 minutes; the average child dose is 1.65 mL/kg. Repeat with nitrite and thiosulfate as required.

Medical note: Exposed personnel should be checked periodically for chronic toxic effects.

HEALTH HAZARDS
Personal protective equipment (PPE): A-Level PPE. Chemical protective material(s) reported to have good to excellent resistance: butyl rubber, PV alcohol, and polycarbonate.
Recommendations for respirator selection: NIOSH as cyanides *25 mg/m³*: SA (any supplied-air respirator); or SCBAF (any self-contained breathing apparatus with full facepiece). *EMERGENCY OR PLANNED ENTRY INTO UNKNOWN CONCENTRATIONS OR IDLH CONDITIONS*: SCBAF:PD,PP (any self-contained breathing apparatus that has a full facepiece and is operated in a pressure-demand or other positive-pressure mode); or SAF:PD,PP:ASCBA (any supplied-air respirator that has a full facepiece and is operated in a pressure-demand or other positive-pressure mode in combination with an auxiliary self-contained breathing apparatus operated in a pressure-demand or other positive pressure mode). ESCAPE: GMFSHiE [any air-purifying, full-facepiece respirator (gas mask) with a chin-style, front- or back-mounted canister providing protection against the compound of concern and having a high efficiency particulate filter); or SCBAE (any appropriate escape-type, self-contained breathing apparatus).
Toxicity by ingestion: Grade 2; oral rat LD_{50} = 800 mg/kg.
IDLH value: 50 mg/m³ as cyanides.

FIRE DATA
Flash point: 167°F/75°C (cc).
Flammable limits in air: LEL: 1.4%; UEL: 7.2%.
Fire extinguishing agents not to be used: Water may be ineffective.
Autoignition temperature: 1022°F/550°C/823°K.
Burning rate: Difficult to burn

CHEMICAL REACTIVITY
Binary reactants: Reacts with strong acids. Attack some plastics.
Reactivity group: 37
Compatibility class: Nitriles

ENVIRONMENTAL DATA
Food chain concentration potential: Log P_{ow} = 1.57. Unlikely to accumulate.
Water pollution: DOT Appendix B, §172.101–marine pollutant. Effect of low concentrations on aquatic life is unknown. May be dangerous if it enters nearby water intakes; notify operators. Notify local health and wildlife officials. **Response to discharge:** Issue warning–water contaminant. Mechanical containment. Should be removed. Chemical and physical treatment.

SHIPPING INFORMATION
Grades of purity: Pure, 99+%; **Storage temperature:** Ambient; **Inert atmosphere:** Ventilated (natural); **Venting:** Open (flame arrester); **Stability during transport:** Stable.

PHYSICAL AND CHEMICAL PROPERTIES
Physical state @ 59°F/15°C and 1 atm: Liquid.
Molecular weight: 103.12
Boiling point @ 1 atm: 376°F/191°C/464°K.
Melting/Freezing point: 9°F/–13°C/260°K.
Critical temperature: 799°F/426°C/699°K.
Critical pressure: 611 psia = 41.6 atm = 4.22 MN/m²
Specific gravity (water = 1): 1.01 @ 77°F/25°C (liquid).
Liquid surface tension: 34.7 dynes/cm = 0.0347 N/m @ 77°F/25°C.
Relative vapor density (air = 1): 3.6

Ratio of specific heats of vapor (gas): 1.091
Latent heat of vaporization: 157.7 Btu/lb = 87.6 cal/g = 3.67 x 10^5 J/kg.
Heat of combustion: −15,100 Btu/lb = −8400 cal/g = −351 x 10^5 J/kg.

BENZOPHENONE   REC. B:1000

SYNONYMS: BENZOFENONA (Spanish); BENZOYL BENZENE; DIPHENYL KETONE; DIPHENYL METHANONE; α-OXODIPHENYLMETHANE; α-OXODITANE; PHENYL KETONE

IDENTIFICATION
CAS Number: 119-61-9
Formula: $C_6H_5COC_6NO_5$

DESCRIPTION: White solid. Flowery (rose-like) odor. May float or sink in water.

Combustible solid • Dust cloud may explode if ignited in an enclosed area • Containers may BLEVE when exposed to fire • Irritating to the skin, eyes, and respiratory tract • Toxic products of combustion may include carbon monoxide.

Hazard Classification (based on NFPA-704 Rating System)
Health Hazards (Blue): 1; Flammability (Red): 1; Reactivity (Yellow): 0

EMERGENCY RESPONSE: See Appendix A (171)

EXPOSURE
Short-term effects: *SEEK MEDICAL ATTENTION.* **Liquid or solid:** Irritating to skin and eyes. *IF SWALLOWED*, will cause nausea, or vomiting. Remove contaminated clothing and shoes. Flush affected areas with plenty of water. *IF IN EYES*, hold eyelids open and flush with plenty of water. *IF SWALLOWED* and victim is *CONSCIOUS AND ABLE TO SWALLOW*, have victim drink 4 to 8 ounces of water. *IF SWALLOWED* and victim is *UNCONSCIOUS OR HAVING CONVULSIONS,* do nothing except keep victim warm.

HEALTH HAZARDS
Personal protective equipment (PPE): B-Level PPE. Organic vapor respirator. Goggles or face shield; rubber gloves.
Exposure limits (TWA unless otherwise noted): 5 mg/m³ (AIHA WEEL)
Toxicity by ingestion: Grade 1; acute oral rat LD_{50} more than 10,000 mg/kg.
Liquid or solid irritant characteristics: Eye and skin irritant.

FIRE DATA
Fire extinguishing agents not to be used: Water may be ineffective.

CHEMICAL REACTIVITY
Binary reactants: Will attack some plastics.

ENVIRONMENTAL DATA
Food chain concentration potential: Log P_{ow} = 3.2. Values > 3.0 are likely to bioconcentrate in aquatic organisms and other living tissue, especially in fats.
Water pollution: Effect of low concentrations on aquatic life is unknown. Fouling to shoreline. May be dangerous if it enters nearby water intakes; notify operators. Notify local health and wildlife officials. **Response to discharge:** Issue warning–water contaminant. Mechanical containment. Should be removed. Chemical and physical treatment.

SHIPPING INFORMATION
Grades of purity: 99+%; **Storage temperature:** Ambient; **Inert atmosphere:** None; **Venting:** Open; **Stability during transport:** Stable.

PHYSICAL AND CHEMICAL PROPERTIES
Physical state @ 59°F/15°C and 1 atm: Solid.
Molecular weight: 182
Boiling point @ 1 atm: 582°F/306°C/579°K.
Melting/Freezing point: 118°F/48°C/321°K.
Specific gravity (water = 1): 1.085 at 50°C (liquid).
Liquid surface tension: 42 dynes/cm = 0.042 N/m at 50°C.
Latent heat of vaporization: 126.0 Btu/lb = 70.0 cal/g = 2.93 x 10^5 J/kg.
Heat of combustion: −15,400 Btu/lb = −8550 cal/g = −358 x 10^5 J/kg.
Heat of fusion: 25.53 cal/g.

p-BENZOQUINONE   REC. B:1050

SYNONYMS: BENZO-CHINON (German); *p*-BENZOQUINONA (Spanish); 1,4-BENZOQUINONE; BENZOQUINONE; CHINON (German); *p*-CHINON (German); CHINONE; CLYCLOHEXADEINEDIONE; 1,4-CYCLOHEXADIENEDIONE; 2,5-CYCLOHEXADIENE-1,4-DIONE; 1,4-DIOXYBENZENE; EEC No. 606-013-00-3; QUINONE; *p*-QUINONE; RCRA No. U197

IDENTIFICATION
CAS Number: 106-51-4
Formula: $C_6H_4O_2$; OC_6H_4O
DOT ID Number: UN 2587; DOT Guide Number: 153
Proper Shipping Name: Benzoquinone
Reportable Quantity (RQ): **(CERCLA)** 10 lb/4.54 kg

DESCRIPTION: Greenish-yellow yellow crystalline solid. Chlorine-like odor, acrid. Sinks and very slowly mixes with water; slight solubility.

Combustible solid •Poison! • Breathing the dust, skin or eye contact, or swallowing the material can kill you • Corrosive to the skin, eyes, and respiratory tract; contact with skin or eyes can cause burns and vision impairment; inhalation symptoms may be delayed • Heat or flame may cause explosion • Containers may explode in fire • Vapors are heavier than air and will collect and stay in low areas • Concentrated dust in confined areas (e.g., tanks, sewers, buildings) may explode when exposed to fire • Toxic products of combustion may include carbon monoxide.

Hazard Classification (based on NFPA-704 Rating System)
Health Hazards (Blue): 1; Flammability (Red): 2; Reactivity (Yellow): 1

EMERGENCY RESPONSE: See Appendix A (153)
Evacuation:
Public safety: Isolate the area of spill or leak for at least 25 to 50 meters (80 to 160 feet) in all directions.
Spill: Increase, in the downwind direction, as necessary, the distance shown under "Public Safety."

Fire: If tank, rail car, or tank truck is involved in fire, isolate for at least 800 meters (½ mile) in all directions; also, consider initial evacuation for 800 meters (½ mile) in all directions.

EXPOSURE
Short-term effects: *CALL FOR MEDICAL AID.* **Dust:** *POISONOUS; MAY BE FATAL IF INHALED.* Irritating to mucous membranes. Move victim to fresh air. *IF BREATHING HAS STOPPED,* give artificial respiration; *avoid mouth-to-mouth resuscitation; use bag/mask apparatus.* IF breathing is difficult, administer oxygen. May cause lung edema; physical exertion will aggravate this condition. **Solid:** *POISONOUS; MAY BE FATAL IF SWALLOWED OR ABSORBED THROUGH SKIN.* Can cause severe damage to the eyes, skin and mucous membranes. *IF IN EYES OR ON SKIN,* flush with running water for at least 15 minutes; hold eyelids open if necessary. Speed in removing material from skin is of extreme importance. Remove and isolate contaminated clothing and shoes at the site. Effects may be delayed; keep victim under observation. *IF SWALLOWED* and victim is *CONSCIOUS AND ABLE TO SWALLOW,* have victim drink large volumes of water and induce vomiting. *IF SWALLOWED* and victim is *UNCONSCIOUS OR HAVING CONVULSIONS,* do nothing except keep victim warm.
Note to physician or authorized medical personnel: Medical observation is recommended for 24 to 48 hours after breathing overexposure, as pulmonary edema may be delayed. As first aid for pulmonary edema, consider administering a corticosteroid spray. Cigarette smoking may exacerbate pulmonary injury and should be discouraged for at least 72 hours following exposure.

HEALTH HAZARDS
Personal protective equipment (PPE): A-Level PPE. OSHA Table Z-1-A air contaminant. Chemical protective material(s) reported to have good to excellent resistance: Saranex®
Recommendations for respirator selection: NIOSH/OSHA as quinone
10 mg/m³: SA:CF* (any supplied-air respirator operated in a continuous-flow mode). *20 mg/m³*: SCBAF (any self-contained breathing apparatus with a full facepiece); or SAF (any supplied-air respirator with a full facepiece). *100 mg/m³*: SAF:PD,PP (any supplied-air respirator that has a full facepiece and is operated in a pressure-demand or other positive-pressure mode). EMERGENCY OR PLANNED ENTRY INTO UNKNOWN CONCENTRATIONS OR IDLH CONDITIONS: SCBAF:PD,PP (any self-contained breathing apparatus that has a full facepiece and is operated in a pressure-demand or other positive-pressure mode); or SAF:PD,PP:ASCBA (any supplied-air respirator that has a full facepiece and is operated in a pressure-demand or other positive-pressure mode in combination with an auxiliary self-contained breathing apparatus operated in a pressure-demand or other positive pressure mode). ESCAPE: GMFOVHiE [any air-purifying, full-facepiece respirator (gas mask) with a chin-style, front-or back-mounted canister having a high efficiency particulate filter]; or SCBAE (any appropriate escape-type, self-contained breathing apparatus). *Note:* Substance causes eye irritation or damage; eye protection needed.
Exposure limits (TWA unless otherwise noted): ACGIH TLV 0.1 ppm (0.44 mg/m³); NIOSH/OSHA PEL 0.1 ppm (0.4 mg/m³).
Toxicity by ingestion: Grade 3; LD_{50} = 130 mg/kg (rat).
Long-term health effects: Eye damage; skin disorders. Causes mutagenic and tumorigenic effects. Indefinite animal carcinogen.
Vapor (gas) irritant characteristics: Vapors cause severe irritation of eyes and throat and can cause eye and lung injury.
Liquid or solid irritant characteristics: Severe skin irritant and very injurious to the eyes.

Odor threshold: 0.082–0.09 ppm.
IDLH value: 100 mg/m³.

FIRE DATA
Flash point: 100-200°F/38-93°C (cc). Actual flashpoint may depend on humidity level.
Behavior in fire: Cylinder may explode. In powder form, dust explosion may occur.
Autoignition temperature: 1040°F/560°C/833°K.
Electrical hazard: Flow or agitation of substance may generate electrostatic charges due to low conductivity.

CHEMICAL REACTIVITY
Binary reactants: Will attack some forms of plastics, rubber, and coatings. Reacts with bases and combustibles.

ENVIRONMENTAL DATA
Food chain concentration potential: Log P_{ow} = 0.2. Unlikely to accumulate.
Water pollution: Harmful to aquatic life in very low concentrations. May be dangerous if it enters nearby water intakes; notify operators. Notify local health and wildlife officials. **Response to discharge:** Issue warning–poison, water contaminant. Restrict access. Should be removed. Chemical and physical treatment.

SHIPPING INFORMATION
Grades of purity: 98%; **Storage temperature:** Ambient

PHYSICAL AND CHEMICAL PROPERTIES
Physical state @ 59°F/15°C and 1 atm: Solid.
Molecular weight: 108.10
Boiling point @ 1 atm: Sublimes
Melting/Freezing point: 240°F/116°C/ 389°K
Specific gravity (water = 1): 1.318 at 20KC
Relative vapor density (air = 1): 3.7
Heat of solution: –66.50 Btu/lb = –36.92 cal/g = –1.546X 10^5 J/kg.
Vapor pressure: 0.2 mm; 0.1mm @ 77°F/25°C.

BENZOYL CHLORIDE REC. B:1100

SYNONYMS: α-CHLOROBENZALDEHYDE; BENZALDEHYDE, α-CHLORO-; BENZENECARBONYL CHLORIDE; BENZOIC ACID, CHLORIDE; α-CHLOROBENZALDEHYDE; CLORURO de BENZOILO (Spanish); EEC No. 607-012-00-0

IDENTIFICATION
CAS Number: 98-88-4
Formula: C_6H_5COCl
DOT ID Number: UN 1736; DOT Guide Number: 137
Proper Shipping Name: Benzoyl chloride
Reportable Quantity (RQ): **(CERCLA)** 1000 lb/454 kg

DESCRIPTION: Colorless to slightly brown fuming liquid. Pungent odor. Sinks in water and reacts slowly producing benzoic and hydrochloric acid.

Combustible • A tear gas • Corrosive to the skin, eyes, and respiratory tract; contact with skin or eyes can cause burns and vision impairment; inhalation symptoms may be delayed; this material has anesthetic properties •Containers may BLEVE when exposed to fire • Vapors are heavier than air and will collect and stay in low areas • Vapors in confined areas (e.g., tanks, sewers, buildings) may explode when exposed to fire • Toxic products of combustion may include phosgene gas and hydrogen chloride.

Hazard Classification (based on NFPA-704 Rating System)
Health Hazards (Blue): 3; Flammability (Red): 2; Reactivity (Yellow): 2; Special Notice (White): Water reactive

EMERGENCY RESPONSE: See Appendix A (137)
Evacuation:
Public safety: Isolate the area of spill or leak for at least 50 to 100 meters (160 to 330 feet) in all directions.
Spill: Increase, in the downwind direction, as necessary, the distance shown under "Public Safety."
Fire: If any large container is involved in fire, isolate for at least 800 meters (½ mile) in all directions; also, consider initial evacuation for 800 meters (½ mile) in all directions.

EXPOSURE
Short-term effects: *SEEK MEDICAL ATTENTION*. **Vapor or fumes:** May cause lung edema; physical exertion will aggravate this condition. *IF BREATHING HAS STOPPED*, give artificial respiration; *avoid mouth-to-mouth resuscitation; use bag/mask apparatus*. **Liquid:** Will burn skin and eyes. Harmful if swallowed. Remove contaminated clothing and shoes. Flush affected areas with plenty of water. *IF IN EYES*, hold eyelids open and flush with plenty of water. *IF SWALLOWED* and victim is *CONSCIOUS AND ABLE TO SWALLOW*, have victim drink 4 to 8 ounces of water. **Do NOT induce vomiting.**
Note to physician or authorized medical personnel: Medical observation is recommended for 24 to 48 hours after breathing overexposure, as pulmonary edema may be delayed. As first aid for pulmonary edema, consider administering a corticosteroid spray. Cigarette smoking may exacerbate pulmonary injury and should be discouraged for at least 72 hours following exposure.

HEALTH HAZARDS
Personal protective equipment (PPE): B-Level PPE. Full protective clothing, including full-face respirator for acid gases and organic vapors (yellow GMC canister). Sealed chemical suits of butyl rubber, and PV alcohol offers limited protection.
Exposure limits (TWA unless otherwise noted): AIHA WEEL 1 ppm.
Long-term health effects: IARC rating 3; inadequate human evidence, inadequate animal evidence.
Vapor (gas) irritant characteristics: Vapors cause severe irritation of eyes and throat and can cause eye and lung injury. They cannot be tolerated even at low concentrations.
Liquid or solid irritant characteristics: Severe skin irritant. Causes second-and third-degree burns on short contact and is very injurious to the eyes.

FIRE DATA
Flash point: 162°F/72°C (oc).
Flammable limits in air: LEL: 2.5%; UEL: 27%.
Fire extinguishing agents not to be used: Water spray. Do not allow water to enter containers.
Behavior in fire: At fire temperatures the compound may react violently with water or steam.
Autoignition temperature: 1112°F/600°C/873°K.

CHEMICAL REACTIVITY
Reactivity with water: Slow reaction with water to produce hydrochloric acid fumes. Reaction much faster with steam.
Binary reactants: Reacts with alcohols, amines, oxidizers, dimethylsulfoxide. Slow corrosion of metals; forms hydrogen gas.

Neutralizing agents for acids and caustics: Soda ash and water: lime

ENVIRONMENTAL DATA
Food chain concentration potential: Negative; unlikely to accumulate.
Water pollution: Harmful to aquatic life in very low concentrations. May be dangerous if it enters nearby water intakes; notify operators. Notify local health and wildlife officials. **Response to discharge:** Issue warning–corrosive. Restrict access. Chemical and physical treatment.

SHIPPING INFORMATION
Grades of purity: 99+%; special grade; **Storage temperature:** Store in cool, dry area; **Venting:** Pressure-vacuum

NAS HAZARD CLASSIFICATION FOR BULK WATER TRANSPORTATION
FIRE: 1
HEALTH: Vapor irritant: 4; Liquid or solid irritant: 4; Poisons: 2
WATER POLLUTION: Human toxicity: 2; Aquatic toxicity: 2; Aesthetic effect: 2
REACTIVITY: Other chemicals: 3; Water: 4; Self-reaction: 0

PHYSICAL AND CHEMICAL PROPERTIES
Physical state @ 59°F/15°C and 1 atm: Liquid.
Molecular weight: 140.57
Boiling point @ 1 atm: 387°F/197.3°C/470.5°K.
Melting/Freezing point: 31°F/–1°C/273°K.
Specific gravity (water = 1): 1.211 @ 77°F/25°C (liquid).
Liquid surface tension: 36.3 dynes/cm = 0.0363 N/m @ 68°F/20°C.
Relative vapor density (air = 1): 4.9
Heat of combustion: –10,030 Btu/lb = –5570 cal/g = –233.2 x 10^5 J/kg.
Vapor pressure: 1 mm.

BENZYL ACETATE   **REC. B:1150**

SYNONYMS: ACETATO de BENCILO (Spanish); ACETIC ACID, PHENYLMETHYL ESTER; ACETIC ACID, BENZYL ESTER; ACETOMETHYLBENZENE; α-ACETOXYTOLUENE; BENZYL ETHANOATE; PHENYLMETHYL ACETATE

IDENTIFICATION
CAS Number: 140-11-4
Formula: $CH_3CO_2CH_2C_6NO_5$

DESCRIPTION: Colorless liquid. Pear-like odor. Slightly soluble in water.

Combustible • Containers may BLEVE when exposed to fire • Vapors in confined areas (e.g., tanks, sewers, buildings) may explode when exposed to fire • Irritating to the skin, eyes, and respiratory tract • Products of combustion may include CO.

Hazard Classification (based on NFPA-704 Rating System)
Health Hazards (Blue): 1; Flammability (Red): 1; Reactivity (Yellow): 0

EMERGENCY RESPONSE: See Appendix A (171)
Evacuation:
Public safety: Isolate the area of spill or leak for at least 10 to 25 meters (30 to 80 feet) in all directions.

Spill: Increase, in the downwind direction, as necessary, the distance shown under "Public Safety."
Fire: If any large container is involved in fire, isolate for at least 800 meters (½ mile) in all directions; also, consider initial evacuation for 800 meters (½ mile) in all directions.

EXPOSURE
Short-term effects: *SEEK MEDICAL ATTENTION.* **Vapor:** Irritating to eyes, nose, and throat. Move to fresh air. *IF BREATHING HAS STOPPED,* give artificial respiration. IF breathing is difficult, administer oxygen. **Liquid:** Irritating to skin and eyes. Remove contaminated clothing and shoes. Flush affected areas with plenty of water. *IF IN EYES,* hold eyelids open and flush with plenty of water.

HEALTH HAZARDS
Personal protective equipment (PPE): B-Level PPE. Self-contained breathing apparatus, rubber boots and heavy rubber gloves.
Exposure limits (TWA unless otherwise noted): ACGIH 10 ppm.
Toxicity by ingestion: LC50 = 245 ppm/8 hrs (cat); Grade 2: LD_{50} = 830 mg/kg (mouse).
Long-term health effects: IARC rating 3; no human data; limited animal evidence. May be carcinogenic.
Vapor (gas) irritant characteristics: Eye and respiratory tract irritant. Vapors cause irritation such that personnel will find high concentrations unpleasant. The effect may be temporary.
Liquid or solid irritant characteristics: Eye and skin irritant. If spilled on clothing and allowed to remain, may cause smarting and reddening of skin.

FIRE DATA
Flash point: 195°F/91°C (cc).
Autoignition temperature: 860°F/460°C/733°K.

CHEMICAL REACTIVITY
Binary reactants: Strong oxidizers.
Reactivity group: 34
Compatibility class: Esters.

ENVIRONMENTAL DATA
Food chain concentration potential: Log P_{ow} = 2.1. Unlikely to accumulate.
Water pollution: Effect of low concentrations on aquatic life is unknown. May be dangerous if it enters nearby water intakes; notify operators. Notify local health and pollution control agencies.
Response to discharge: Restrict access. Mechanical containment. Should be removed. Chemical and physical treatment.

SHIPPING INFORMATION
Grades of purity: 99+%; **Stability during transport:** Stable.

PHYSICAL AND CHEMICAL PROPERTIES
Physical state @ 59°F/15°C and 1 atm: Liquid.
Molecular weight: 150.18
Boiling point @ 1 atm: 417°F/214°C/487°K.
Melting/Freezing point: −60°F/−51°C/222°K.
Specific gravity (water = 1): 1.040
Relative vapor density (air = 1): 5.18
Vapor pressure: (Reid) Less than 0.01 psia; 1 mm.

BENZYL ALCOHOL **REC. B:1200**

SYNONYMS: ALCOHOL BENCILICO (Spanish); PHENYLCARBINOL; BENZENECARBINOL; EEC No. 603-057-00-5; *α*-HYDROXYTOLUENE; PHENYLCARBINOL; PHENYLMETHANOL; PHENYLMETHYL ALCOHOL

IDENTIFICATION
CAS Number: 100-51-6
Formula: C_7H_8O; $C_6H_5CH_2OH$
DOT ID Number: UN 3082; DOT Guide Number: 171
Proper Shipping Name: Environmentally hazardous, liquid, n.o.s.

DESCRIPTION: Colorless liquid. Mild, pleasant odor. May float or sink in water; moderately soluble.

Poison! • Containers may BLEVE when exposed to fire • Vapors in confined areas (e.g., tanks, sewers, buildings) may explode when exposed to fire • Irritates eyes, skin, and respiratory tract • Toxic products of combustion may include carbon monoxide.

Hazard Classification (based on NFPA-704 Rating System)
Health Hazards (Blue): 2; Flammability (Red): 1; Reactivity (Yellow): 0

EMERGENCY RESPONSE: See Appendix A (171)
Evacuation:
Public safety: Isolate the area of spill or leak for at least 10 to 25 meters (30 to 80 feet) in all directions.
Spill: Increase, in the downwind direction, as necessary, the distance shown under "Public Safety."
Fire: If any large container is involved in fire, isolate for at least 800 meters (½ mile) in all directions; also, consider initial evacuation for 800 meters (½ mile) in all directions.

EXPOSURE
Short-term effects: *SEEK MEDICAL ATTENTION.* **Vapor:** Irritating to eyes, nose, and throat. *IF INHALED,* will cause headache, coughing, or difficult breathing. *IF IN EYES,* hold eyelids open and flush with plenty of water. *IF BREATHING HAS STOPPED,* give artificial respiration. IF breathing is difficult, administer oxygen. **Liquid:** Irritating to skin and eyes. *IF SWALLOWED,* will cause nausea or vomiting. Remove contaminated clothing and shoes. Flush affected areas with plenty of water. *IF IN EYES,* hold eyelids and flush with plenty of water. *IF SWALLOWED* and victim is *CONSCIOUS AND ABLE TO SWALLOW,* have victim drink 4 to 8 ounces of water and have victim induce vomiting. *IF SWALLOWED* and victim is *UNCONSCIOUS OR HAVING CONVULSIONS,* do nothing except keep victim warm. The use of alcoholic beverages may enhance the toxic effect.

HEALTH HAZARDS
Personal protective equipment (PPE): B-Level PPE. Rubber gloves; chemical safety goggles.
Exposure limits (TWA unless otherwise noted): 10 ppm (AIHA WEEL)
Toxicity by ingestion: Grade 2; oral rat LD_{50} = 1230 mg/kg.
Vapor (gas) irritant characteristics: Eye, skin and respiratory tract irritant.
Liquid or solid irritant characteristics: Eye and skin irritant.
Odor threshold: 5.5 ppm.

FIRE DATA
Flash point: 220°F/104°C (oc); 213°F/101°C (cc).
Flammable limits in air: LEL: 1.3%; UEL: 13%.
Fire extinguishing agents not to be used: Water or foam may cause frothing.

128 Benzylamine

Autoignition temperature: 817°F/436°C/709°K.
Burning rate: 3.74 mm/min

CHEMICAL REACTIVITY
Binary reactants: Reacts with strong oxidizers. Will attack some plastics.

ENVIRONMENTAL DATA
Food chain concentration potential: Log P_{ow} = 1.10. Unlikely to accumulate.
Water pollution: Effect of low concentrations on aquatic life is unknown. May be dangerous if it enters nearby water intakes; notify operators. Notify local health and wildlife officials.
Response to discharge: Issue warning–water contaminant. Disperse and flush.

SHIPPING INFORMATION
Grades of purity: NF; Photographic; Technical; Textile; **Storage temperature:** Ambient; **Inert atmosphere:** None; **Venting:** Open; **Stability during transport:** Stable.

PHYSICAL AND CHEMICAL PROPERTIES
Physical state @ 59°F/15°C and 1 atm: Liquid.
Molecular weight: 108.13
Boiling point @ 1 atm: 401°F/205°C/478°K.
Melting/Freezing point: 5°F/–15°C/258°K.
Critical temperature: 757°F/403°C/676°K.
Critical pressure: 663 psia = 45.0 atm = 4.57 MN/m^2.
Specific gravity (water = 1): 1.050 at 15/15°C (liquid).
Liquid surface tension: 39.0 dynes/cm = 0.0390 N/m @ 68°F/20°C.
Relative vapor density (air = 1): 3.73
Ratio of specific heats of vapor (gas): 1.070
Latent heat of vaporization: 193 Btu/lb = 107 cal/g = 4.48 x 10^5 J/kg.
Heat of combustion: –14,850 Btu/lb = –8,260 cal/g = –345 x 10^5 J/kg.
Heat of fusion: 19.83 cal/g.
Vapor pressure: 0.01 mm.

BENZYLAMINE REC. B:1250

SYNONYMS: α-AMINOTOLUENE; BENCILAMINA (Spanish); PHENYLMETHYL AMINE

IDENTIFICATION
CAS Number: 100-46-9
Formula: C_7H_9N; $C_6H_5CH_2NH_2$
DOT ID Number: UN 2734; DOT Guide Number: 132
Proper Shipping Name: Amines, liquid, corrosive, n.o.s.

DESCRIPTION: Colorless to pale yellow liquid. Strong ammonia odor. Floats on the surface of water; soluble.

Combustible • Containers may BLEVE when exposed to fire • Vapors are heavier than air and will collect and stay in low areas • Vapors in confined areas (e.g., tanks, sewers, buildings) may explode when exposed to fire • Highly irritating to the eyes and respiratory tract; contact with burn skin and eyes • Toxic products of combustion may include carbon monoxide and nitrogen oxides.

Hazard Classification (based on NFPA-704 M Rating System)
Health Hazards (Blue): 2; Flammability (Red): 1; Reactivity (Yellow): 0

EMERGENCY RESPONSE: See Appendix A (132)
Evacuation:
Public safety: Isolate spill area for at least 50 to 100 meters (160 to 330 feet) in all directions.
Spill: Increase, as necessary, the isolation distance shown above, in "Public safety."
Fire: Isolate for 800 meters (½ mile) in all directions, especially if tank, rail car, or tank truck is involved in fire.

EXPOSURE
Short-term effects: *SEEK MEDICAL ATTENTION.* **Vapor:** Irritating to eyes, nose, and throat. *IF INHALED*, will, will cause dizziness, headache, or difficult breathing. *IF IN EYES*, hold eyelids open and flush with plenty of water. *IF BREATHING HAS STOPPED*, give artificial respiration. IF breathing is difficult, administer oxygen. **Liquid:** Will burn skin and eyes. Harmful if swallowed. Remove contaminated clothing and shoes. Flush affected areas with plenty of water. *IF IN EYES*, hold eyelids open and flush with plenty of water. *IF SWALLOWED* and victim is *CONSCIOUS AND ABLE TO SWALLOW*, have victim drink 4 to 8 ounces of water. *IF SWALLOWED* and victim is *UNCONSCIOUS OR HAVING CONVULSIONS*, do nothing except keep victim warm.

HEALTH HAZARDS
Personal protective equipment (PPE): B-Level PPE. Self-contained breathing apparatus; goggles or face shield; rubber gloves.
Recommendations for respirator selection: NIOSH as ammonia *250 ppm*: CCRS* [any chemical cartridge respirator with cartridge(s) providing protection against the compound of concern)]; or SA* (any supplied-air respirator). *300 ppm*: SA:CF (any supplied-air respirator operated in a continuous-flow mode); or PAPRS* [any powered, air-purifying respirator with cartridge(s) providing protection against the compound of concern]; or CCRFS [any chemical cartridge respirator with a full facepiece and cartridge(s) providing protection against the compound of concern]; or GMFS [any air-purifying, full-facepiece respirator (gas mask) with a chin-style, front- or back-mounted canister providing protection against the compound of concern]; or SCBAF (any self-contained breathing apparatus with a full facepiece); or SAF (any supplied-air respirator with a full facepiece). *EMERGENCY OR PLANNED ENTRY INTO UNKNOWN CONCENTRATIONS OR IDLH CONDITIONS:* SCBAF:PD,PP (any self-contained breathing apparatus that has a full facepiece and is operated in a pressure-demand or other positive-pressure mode); or SAF:PD,PP:ASCBA (any supplied-air respirator that has a full facepiece and is operated in a pressure-demand or other positive-pressure mode in combination with an auxiliary self-contained breathing apparatus operated in a pressure-demand or other positive pressure mode). *ESCAPE:* GMFS [any air-purifying, full-facepiece respirator (gas mask) with a chin-style, front- or back-mounted canister providing protection against the compound of concern]; or SCBAE (any appropriate escape-type, self-contained breathing apparatus). *Note:* Substance reported to cause eye irritation or damage; may require eye protection.
Vapor (gas) irritant characteristics: Eye and respiratory tract irritant.
Liquid or solid irritant characteristics: Eye and skin irritant.

FIRE DATA
Flash point: 168°F/76°C (oc).
Fire extinguishing agents not to be used: Water may be ineffective.
Burning rate: 4.13 mm/min

CHEMICAL REACTIVITY
Binary reactants: In presence of moisture may weakly corrode some metals. Liquid will attack some plastics.
Neutralizing agents for acids and caustics: Flush with water.

ENVIRONMENTAL DATA
Food chain concentration potential: Log K_{ow} = 1.09. Negative; unlikely to accumulate.
Water pollution: Harmful to aquatic life in very low concentrations. May be dangerous if it enters nearby water intakes; notify operators. Notify local health and wildlife officials.
Response to discharge: Issue warning-air contaminant, water contaminant. Restrict access. Disperse and flush.

SHIPPING INFORMATION
Grades of purity: Commercial, 98.5+%; **Storage temperature:** Ambient; **Inert atmosphere:** Ventilated (natural); **Venting:** Open (flame arrester); **Stability during transport:** Stable.

PHYSICAL AND CHEMICAL PROPERTIES
Physical state @ 59°F/15°C and 1 atm: Liquid.
Molecular weight: 107.16
Boiling point @ 1 atm: 364°F/185°C/458°K.
Melting/Freezing point: (approximate) –51°F/–46°C/227°K.
Specific gravity (water = 1): 0.98 @ 68°F/20°C (liquid).
Liquid surface tension: 39.5 dynes/cm = 0.0395 N/m @ 68°F/20°C.
Relative vapor density (air = 1): 3.70
Ratio of specific heats of vapor (gas): (estimate) 1.070
Dissociation constant: pK_a = 9.32
Latent heat of vaporization: 164 Btu/lb = 91 cal/g = 3.8 x 10^5 J/kg.
Heat of combustion: –16,260 Btu/lb = –9,040 cal/g = –378 x 10^5 J/kg.
Heat of solution: –43 Btu/lb = –24 cal/g = –1.0 x 10^5 J/kg.

BENZYL BROMIDE REC. B:1300

SYNONYMS: α-BROMOTOLUENE; ω-BROMOTOLUENE; p-(BROMOMETHYL)NITROBENZENE; BENZENE, (BROMOMETHYL)-; (BROMOMETHYL)BENZENE; BROMOPHENYLMETHANE; BROMURO de BENCILO (Spanish); EEC No. 602-057-00-2; TOLUENE, α-BROMO-

IDENTIFICATION
CAS Number: 100-39-0
Formula: $C_6H_5CH_2Br$
DOT ID Number: UN 1737; DOT Guide Number: 156
Proper Shipping Name: Benzyl bromide

DESCRIPTION: Colorless to yellow liquid. Sharp irritating odor; like tear gas. Sinks in water; insoluble; reacts slowly, forming bezyl alcohol and hydrogen bromide. Freezes at 25°F/–4°C.

Lachrymator • Combustible • Corrosive • Breathing the vapor can cause severe irritation; skin or eye contact causes severe burns, impaired vision, or blindness; effects of contact or inhalation may be delayed • Firefighting gear (including SCBA) does not provide adequate protection. If exposure occurs, remove and isolate gear immediately and thoroughly decontaminate personnel • Containers may BLEVE when exposed to fire • In the presence of metals may react violently with itself without warning • Toxic products of combustion may include hydrogen bromide, a powerful tear gas.

Hazard Classification (based on NFPA-704 Rating System)
Health Hazards (Blue): 2; Flammability (Red): 2; Reactivity (Yellow): 0

EMERGENCY RESPONSE: See Appendix A (156)
Evacuation:
Public safety: Isolate the area of spill or leak for at least 50 to 100 meters (160 to 330 feet) in all directions.
Spill: Increase, in the downwind direction, as necessary, the distance shown under "Public Safety."
Fire: If tank, rail car, or tank truck is involved in fire, isolate for at least 800 meters (½ mile) in all directions; also, consider initial evacuation for 800 meters (½ mile) in all directions.

EXPOSURE
Short-term effects: *SEEK MEDICAL ATTENTION.*
Vapors: May cause lung edema; physical exertion will aggravate this condition. Affects the nervous system. *IF BREATHING HAS STOPPED,* give artificial respiration; *avoid mouth-to-mouth resuscitation; use bag/mask apparatus.* **Liquid:** Irritating to skin and eyes. Harmful if swallowed. Remove contaminated clothing and shoes. Flush affected areas with plenty of water. *IF IN EYES,* hold eyelids open and flush with plenty of water. *IF SWALLOWED* and victim is *CONSCIOUS AND ABLE TO SWALLOW,* have victim drink 4 to 8 ounces of water.
Note to physician or authorized medical personnel: Medical observation is recommended for 24 to 48 hours after breathing overexposure, as pulmonary edema may be delayed. As first aid for pulmonary edema, consider administering a corticosteroid spray. Cigarette smoking may exacerbate pulmonary injury and should be discouraged for at least 72 hours following exposure.

HEALTH HAZARDS
Personal protective equipment (PPE): A-Level PPE. Self-contained breathing apparatus; goggles; rubber gloves; protective clothing. Chemical protective material(s) reported to have good to excellent resistance: butyl rubber, polycarbonate.
Recommendations for respirator selection: NIOSH/OSHA as hydrogen bromide
30 ppm: SA:CF* (any supplied-air respirator operated in a continuous-flow mode); or PAPRAG* [any powered, air-purifying respirator with acid gas cartridge(s)]; or GMFAG [any air-purifying, full-facepiece respirator (gas mask) with a chin-style, front- or back-mounted organic vapor cannister]; or SCBAF (any self-contained breathing apparatus with a full facepiece); or SAF (any supplied-air respirator with a full facepiece). *EMERGENCY OR PLANNED ENTRY INTO UNKNOWN CONCENTRATIONS OR IDLH CONDITIONS:* SCBAF:PD,PP (any self-contained breathing apparatus that has a full facepiece and is operated in a pressure-demand or other positive-pressure mode); or SAF:PD,PP:ASCBA (any supplied-air respirator that has a full facepiece and is operated in a pressure-demand or other positive-pressure mode in combination with an auxiliary self-contained breathing apparatus operated in a pressure-demand or other positive pressure mode). *ESCAPE:* GMFAG [any air-purifying, full-facepiece respirator (gas mask) with a chin-style, front- or back-mounted organic vapor cannister]; or SCBAE (any appropriate escape-type, self-contained breathing apparatus). *Note:* Substance causes eye irritation or damage; eye protection needed.
Exposure limits (TWA unless otherwise noted): ACGIH TLV ceiling 3 ppm (9.9 mg/m^3); OSHA PEL 3 ppm (10 mg/m^3); NIOSH REL ceiling 3 ppm (10 mg/m^3) as hydrogen bromide.
Long-term health effects: May cause liver and kidney damage.
Vapor (gas) irritant characteristics: Eye and respiratory tract edema.

130 Benzyl chloride

Liquid or solid irritant characteristics: Eye and skin irritant.
Odor threshold: 2 ppm as hydrogen bromide
IDLH value: 30 ppm as hydrogen bromide

FIRE DATA
Flash point: 174°F/79°C (cc).
Behavior in fire: Forms vapor that is a powerful tear gas.
Burning rate: 2.6 mm/min

CHEMICAL REACTIVITY
Reactivity with water: Reacts slowly to generate hydrogen bromide (hydrobromic acid).
Binary reactants: Decomposes rapidly in the presence of all common metals except nickel and lead, liberating heat and hydrogen bromide. Reacts with strong oxidizers and bases.
Neutralizing agents for acids and caustics: Rinse with sodium bicarbonate or lime solution.
Polymerization: Polymerizes with evolution of heat and hydrogen bromide when in contact with all common metals except nickel and lead.

ENVIRONMENTAL DATA
Food chain concentration potential: Log P_{ow} = 2.9. Values > 3.0 are likely to bioconcentrate in aquatic organisms and other living tissue, especially in fats.
Water pollution: Harmful to aquatic life in very low concentrations. May be dangerous if it enters nearby water intakes; notify operators. Notify local health and wildlife officials.
Response to discharge: Issue warning–corrosive. Restrict access. Should be removed. Chemical and physical treatment.

SHIPPING INFORMATION
Grades of purity: Commercial; **Storage temperature:** Ambient; **Inert atmosphere:** None; **Venting:** Pressure-vacuum; **Stability during transport:** Stable.

PHYSICAL AND CHEMICAL PROPERTIES
Physical state @ 59°F/15°C and 1 atm: Liquid.
Molecular weight: 171.0
Boiling point @ 1 atm: 388°F/198°C/471°K.
Melting/Freezing point: 25°F/–4°C/269°K.
Specific gravity (water = 1): 1.441 at 71°F/22°C (liquid).
Liquid surface tension: 32.3 dynes/cm = 0.0323 N/m @ 68°F/20°C.
Liquid water interfacial tension: (estimate) 35 dynes/cm = 0.035 N/m @ 68°F/20°C.
Relative vapor density (air = 1): 5.9
Latent heat of vaporization: 120 Btu/lb = 66.4 cal/g = 2.78 x 10^5 J/kg.
Heat of combustion: (estimate) –9000 Btu/lb = –5000 cal/g = –210 x 10^5 J/kg.
Heat of fusion: 20.86 cal/g.
Vapor pressure: 0.4 mm.

BENZYL CHLORIDE   REC. B:1350

SYNONYMS: α-CHLOROTOLUENE; BENZENE, (CHLOROMETHYL)-; BENZENE,CHLOROMETHYL-; α-CHLORTOLUOL (German); BENZYLE (CHLORURE de) (French); BENZYLCHLORID (German); CHLOROMETHYLBENZENE; CHLOROPHENYLMETHANE; CLORURO de BENCILO (Spanish); EEC No. 602-037-00-3; ω-CHLOROTOLUENE; CHLORURE de BENZYLE (French); RCRA No. P028; TOLYL CHLORIDE

IDENTIFICATION
CAS Number: 100-44-7
Formula: $C_6H_5CH_2Cl$
DOT ID Number: UN 1738; DOT Guide Number: 156
Proper Shipping Name: Benzyl chloride; Benzyl chloride unstabilized
Reportable Quantity (RQ): **(CERCLA)** 100 lb/45.4 kg

DESCRIPTION: Colorless to slightly yellow liquid. Sharp, irritating odor. Sinks in water; insoluble; reacts slowly forming hydrochloric acid.

Combustible • Corrosive • Breathing the vapor can cause severe irritation; skin or eye contact causes severe burns, impaired vision, or blindness; effects of contact or inhalation may be delayed • Firefighting gear (including SCBA) does not provide adequate protection. If exposure occurs, remove and isolate gear immediately and thoroughly decontaminate personnel • Containers may BLEVE when exposed to fire • May react with itself in the presence of metals, releasing heat and hydrogen chloride fumes • Toxic products of combustion may include hydrogen chloride and phosgene.

Hazard Classification (based on NFPA-704 Rating System)
Health Hazards (Blue): 3; Flammability (Red): 2; Reactivity (Yellow): 1

EMERGENCY RESPONSE: See Appendix A (156)
Evacuation:
Public safety: Isolate the area of spill or leak for at least 50 to 100 meters (160 to 330 feet) in all directions.
Spill: Increase, in the downwind direction, as necessary, the distance shown under "Public Safety."
Fire: If tank, rail car, or tank truck is involved in fire, isolate for at least 800 meters (½ mile) in all directions; also, consider initial evacuation for 800 meters (½ mile) in all directions.

EXPOSURE
Short-term effects: *SEEK MEDICAL ATTENTION.* Inhalation may cause lung edema; physical exertion will aggravate this condition. Affects the nervous system. *IF BREATHING HAS STOPPED*, give artificial respiration; *avoid mouth-to-mouth resuscitation; use bag/mask apparatus.* **Liquid:** Will burn skin and eyes. *IF SWALLOWED*, will cause nausea, and vomiting. Remove contaminated clothing and shoes. Flush affected areas with plenty of water. *IF IN EYES*, hold eyelids open and flush with plenty of water. *IF SWALLOWED* and victim is *CONSCIOUS AND ABLE TO SWALLOW*, have victim drink 4 to 8 ounces of water. **Do NOT induce vomiting.**
Note to physician or authorized medical personnel: Medical observation is recommended for 24 to 48 hours after breathing overexposure, as pulmonary edema may be delayed. As first aid for pulmonary edema, consider administering a corticosteroid spray. Cigarette smoking may exacerbate pulmonary injury and should be discouraged for at least 72 hours following exposure.

HEALTH HAZARDS
Personal protective equipment (PPE): A-Level PPE. OSHA Table Z-1-A air contaminant. Chemical protective material(s) reported to have good to excellent resistance: Teflon. Also, Viton®/neoprene offers limited protection.
Recommendations for respirator selection: NIOSH/OSHA
10 ppm: CCROVAG* [any chemical cartridge respirator with organic vapor and acid gas cartridge(s)]; or GMFOVAG [any air-purifying, full-facepiece respirator (gas mask) with a chin-style,

front- or back-mounted organic vapor or acid gas canister]; or PAPROVAG* [any powered, air-purifying respirator with organic vapor and acid gas cartridge(s)]; or SA* (any supplied-air respirator); or SCBAF (any self-contained breathing apparatus with a full facepiece).
EMERGENCY OR PLANNED ENTRY INTO UNKNOWN CONCENTRATIONS OR IDLH CONDITIONS: SCBAF:PD,PP (any self-contained breathing apparatus that has a full facepiece and is operated in a pressure-demand or other positive-pressure mode); or SAF:PD,PP:ASCBA (any supplied-air respirator that has a full facepiece and is operated in a pressure-demand or other positive-pressure mode in combination with an auxiliary self-contained breathing apparatus operated in a pressure-demand or other positive pressure mode). *ESCAPE:* GMFOVAG [any air-purifying, full-facepiece respirator (gas mask) with a chin-style, front- or back-mounted organic vapor or acid gas canister); or SCBAE (any appropriate escape-type, self-contained breathing apparatus).
Note: Substance causes eye irritation or damage; eye protection needed.
Exposure limits (TWA unless otherwise noted): ACGIH TLV 1 ppm (5.2 mg/m^3); OSHA 1 ppm (5 mg/m^3); NIOSH REL ceiling 1 ppm (5 mg/m^3)/15 minute
Toxicity by ingestion: Grade 2; oral rat LD$_{50}$ = 1231 mg/kg.
Long-term health effects: IARC overall evaluation unassigned; limited animal evidence. May cause liver and kidney damage.
Vapor (gas) irritant characteristics: Vapors cause severe irritation of eyes and throat and can cause eye and lung injury. They cannot be tolerated even at low concentrations.
Liquid or solid irritant characteristics: Severe skin irritant. Causes second-and third-degree burns on short contact and is very injurious to the eyes.
Odor threshold: 0.047 ppm.
IDLH value: 10 ppm.

FIRE DATA
Flash point: 165°F/74°C (oc); 153°F/67°C (cc).
Flammable limits in air: LEL: 1.1%; UEL: 14%.
Behavior in fire: Forms vapor that is a powerful tear gas.
Autoignition temperature: 1161°F/627°C/900°K.
Electrical hazard: Class I, Group D.
Burning rate: 4.2 mm/min

CHEMICAL REACTIVITY
Reactivity with water: Hydrolyzes to benzyl alcohol
Binary reactants: Reacts with oxidizers, acids, copper, aluminum, magnesium, iron, zinc, tin. Decomposes rapidly in the presence of most common metals (with the exception of nickel and lead), liberating heat and hydrogen chloride. Attacks some plastics.
Neutralizing agents for acids and caustics: Rinse with sodium bicarbonate or lime solution.
Polymerization: Polymerizes with evolution of heat and hydrogen chloride when in contact with all common metals except nickel and lead; **Inhibitor of polymerization:** Triethylamine, propylene oxide, or sodium carbonate.
Reactivity group: 36
Compatibility class: Halogenated hydrocarbons

ENVIRONMENTAL DATA
Food chain concentration potential: Log P$_{ow}$ = 2.3. Unlikely to accumulate.
Water pollution: Harmful to aquatic life in very low concentrations. May be dangerous if it enters nearby water intakes; notify operators. Notify local health and wildlife officials.
Response to discharge: Issue warning–corrosive. Restrict access. Should be removed. Chemical and physical treatment.

SHIPPING INFORMATION
Grades of purity: 98.5+%, either anhydrous or stabilized; **Storage temperature:** Ambient; **Inert atmosphere:** None; **Venting:** Pressure-vacuum; **Stability during transport:** Stable.

NAS HAZARD CLASSIFICATION FOR BULK WATER TRANSPORTATION
FIRE: 1
HEALTH: Vapor irritant: 4; Liquid or solid irritant: 4; Poisons: 2
WATER POLLUTION: Human toxicity: 2; Aquatic toxicity: 3; Aesthetic effect: 3
REACTIVITY: Other chemicals: 3; Water: 2; Self-reaction: 2

PHYSICAL AND CHEMICAL PROPERTIES
Physical state @ 59°F/15°C and 1 atm: Liquid.
Molecular weight: 126.6
Boiling point @ 1 atm: 354.9°F/179.4°C/452.6°K.
Melting/Freezing point: –38.6°F/–39.2°C/234.0°K.
Critical temperature: (estimate) 772°F/411°C/684°K.
Critical pressure: (estimate) 567 psia = 38.5 atm = 3.91 MN/m^2.
Specific gravity (water = 1): 1.10 @ 77°F/25°C (liquid).
Liquid surface tension: 37.5 dynes/cm = 0.0375 N/m @ 68°F/20°C.
Liquid water interfacial tension: (estimate) 30 dynes/cm = 0.030 N/m @ 68°F/20°C.
Relative vapor density (air = 1): 4.4
Ratio of specific heats of vapor (gas): 1.0689
Latent heat of vaporization: 130 Btu/lb = 70 cal/g = 2.9 x 10^5 J/kg.
Heat of combustion: –12,000 Btu/lb = –6700 cal/g = –280 x 10^5 J/kg.
Vapor pressure: (Reid) 0.07 psia; 11.8 mm.

BENZYL CHLOROFORMATE   REC. B:1400

SYNONYMS: BENZYLCARBONYL CHLORIDE; BENZYL CHLOROCARBONATE; BZCF; CARBOBENZOXY CHLORIDE; CHLOROFORMIC ACID, BENZYL ESTER; CHLOROFORMIC ACID, BENZYL ESTER; CLOROFORMIATO de BENCILO (Spanish); FORMIC ACID, CHLORO-, BENZYL ESTER

IDENTIFICATION
CAS Number: 501-53-1
Formula: C$_8$H$_7$ClO$_2$; C$_6$H$_5$CH$_2$OCOCl
DOT ID Number: UN 1739; DOT Guide Number: 137
Proper Shipping Name: Benzyl chloroformate

DESCRIPTION: Colorless to pale yellow, oily liquid. Sharp, irritating odor. Sinks in water; reacts with hot water or steam forming hydrochloric acid.

Combustible • Corrosive to the skin, eyes, and respiratory tract; contact with skin or eyes can cause burns and vision impairment; inhalation symptoms may be delayed • Containers may BLEVE when exposed to fire • Toxic products of combustion may include carbon monoxide, phosgene, hydrogen chloride, and benzyl chloride.

Hazard Classification (based on NFPA-704 Rating System)
Health Hazards (Blue): 2; Flammability (Red): 2; Reactivity (Yellow): 1

Benzyl dimethylamine

EMERGENCY RESPONSE: See Appendix A (137)
Evacuation:
Public safety: Isolate the area of spill or leak for at least 50 to 100 meters (160 to 330 feet) in all directions.
Spill: Increase, in the downwind direction, as necessary, the distance shown under "Public Safety."
Fire: If any large container is involved in fire, isolate for at least 800 meters (½ mile) in all directions; also, consider initial evacuation for 800 meters (½ mile) in all directions.

EXPOSURE
Short-term effects: *SEEK MEDICAL ATTENTION. IF BREATHING HAS STOPPED,* give artificial respiration; *avoid mouth-to-mouth resuscitation; use bag/mask apparatus*. Inhalation may cause lung edema to develop. **Liquid:** Irritating to skin and eyes. Harmful if swallowed. Remove contaminated clothing and shoes. Flush affected areas with plenty of water. *IF IN EYES*, hold eyelids open and flush with plenty of water. *IF SWALLOWED* and victim is *CONSCIOUS AND ABLE TO SWALLOW*, have victim drink 4 to 8 ounces of water.
Note to physician or authorized medical personnel: Medical observation is recommended for 24 to 48 hours after breathing overexposure, as pulmonary edema may be delayed. As first aid for pulmonary edema, consider administering a corticosteroid spray. Cigarette smoking may exacerbate pulmonary injury and should be discouraged for at least 72 hours following exposure.

HEALTH HAZARDS
Personal protective equipment (PPE): B-Level PPE. Self-contained breathing apparatus or acid-type canister mask, goggles or face shield, rubber gloves, protective clothing: Protective materials with good to excellent resistance: PVC, butyl rubber.
Toxicity by ingestion: Grade 3; LD_{50} = 50 to 500 mg/kg.
Vapor (gas) irritant characteristics: Vapors cause irritation such that personnel will find high concentrations unpleasant. The effect may be temporary.
Liquid or solid irritant characteristics: Causes smarting of the skin and first-degree burns on short exposure and may cause second-degree burns on long exposure.

FIRE DATA
Flash point: 227°F/108°C (cc); 176°F/80°C (oc) Vigorous decomposition occurs at these temperatures; thus these values are anomalous due to the effect of the decomposition products (benzyl chloride and CO_2).
Behavior in fire: Containers may explode.
Burning rate: 4.0 mm/min

CHEMICAL REACTIVITY
Reactivity with water: Forms hydrogen chloride (hydrochloric acid). The reaction is not very vigorous in cold water.
Binary reactants: Slow corrosion of metal.
Neutralizing agents for acids and caustics: Flush with water, rinse with sodium bicarbonate or lime solution.

ENVIRONMENTAL DATA
Food chain concentration potential: Negative; unlikely to accumulate.
Water pollution: DOT Appendix B, §172.101–marine pollutant. Effect of low concentrations on aquatic life is unknown. May be dangerous if it enters nearby water intakes; notify operators. Notify local health and wildlife officials. **Response to discharge:** Issue warning–corrosive. Restrict access. Should be removed. Chemical and physical treatment.

SHIPPING INFORMATION
Grades of purity: 97+%; **Storage temperature:** Ambient, in cool place; **Inert atmosphere:** None; **Venting:** Pressure-vacuum; **Stability during transport:** Stable.

NAS HAZARD CLASSIFICATION FOR BULK WATER TRANSPORTATION
FIRE: 1
HEALTH: Vapor irritant: 2; Liquid or solid irritant: 2; Poisons: 3
WATER POLLUTION: Human toxicity: 3; Aquatic toxicity: 3; Aesthetic effect: 3
REACTIVITY: Other chemicals: 2; Water: 3; Self-reaction: 3

PHYSICAL AND CHEMICAL PROPERTIES
Physical state @ 59°F/15°C and 1 atm: Liquid.
Molecular weight: 170.6
Boiling point @ 1 atm: (decomposes) 306°F/152°C/425°K.
Specific gravity (water = 1): 1.22 @ 68°F/20°C (liquid).
Liquid surface tension: (estimate) 25 dynes/cm = 0.025 N/m @ 68°F/20°C.
Latent heat of vaporization: (estimate) 90 Btu/lb = 50 cal/g = 2.1×10^5 J/kg.
Heat of combustion: (estimate) –10,000 Btu/lb = –5700 cal/g = -240×10^5 J/kg.

BENZYL DIMETHYLAMINE   REC. B:1450

SYNONYMS: *N*-BENCILDIMETILAMINA (Spanish); *N*-BENZYLDIMETHYLAMINE; α-(DIMETHYLAMINO)TOLUENE; *N,N*-DIMETHYL BENZYLAMINE; *N,N*-DIMETHYL BENZENE METHANAMINE; *N,N*-(DIMETHYL)-α-TOLUENEAMINE; CATALYST 9915

IDENTIFICATION
CAS Number: 103-83-3
Formula: $C_9H_{13}N$; $C_6H_5CH_2N(CH_3)_2$
DOT ID Number: UN 2619; DOT Guide Number: 132
Proper Shipping Name: Benzyldimethylamine

DESCRIPTION: Pale yellow to light brown liquid. Strong ammonia odor. Floats on water surface; soluble.

Combustible • Corrosive • Breathing the vapor can cause severe irritation; skin or eye contact causes severe burns, impaired vision, or blindness; effects of contact or inhalation may be delayed • Firefighting gear (including SCBA) does not provide adequate protection. If exposure occurs, remove and isolate gear immediately and thoroughly decontaminate personnel • Containers may BLEVE when exposed to fire • Toxic products of combustion may include carbon monoxide and nitrogen oxide.

Hazard Classification (based on NFPA-704 Rating System)
Health Hazards (Blue): 2; Flammability (Red): 2; Reactivity (Yellow): 0

EMERGENCY RESPONSE: See Appendix A (132)
Evacuation:
Public safety: Isolate spill area for at least 50 to 100 meters (160 to 330 feet) in all directions.
Spill: Increase, as necessary, the isolation distance shown above, in "Public safety."
Fire: Isolate for 800 meters (½ mile) in all directions, especially if tank, rail car, or tank truck is involved in fire.

EXPOSURE
Short-term effects: *SEEK MEDICAL ATTENTION*. Inhalation may cause lung edema; physical exertion will aggravate this condition. **Liquid:** Irritating to skin and eyes. Harmful if swallowed. Remove contaminated clothing and shoes. Flush affected areas with plenty of water. *IF IN EYES*, hold eyelids open and flush with plenty of water. *IF SWALLOWED* and victim is *CONSCIOUS AND ABLE TO SWALLOW*, have victim drink 4 to 8 ounces of water. *IF SWALLOWED* and victim is *UNCONSCIOUS OR HAVING CONVULSIONS*, do nothing except keep victim warm:

Note to physician or authorized medical personnel: Medical observation is recommended for 24 to 48 hours after breathing overexposure, as pulmonary edema may be delayed. As first aid for pulmonary edema, consider administering a corticosteroid spray. Cigarette smoking may exacerbate pulmonary injury and should be discouraged for at least 72 hours following exposure.

HEALTH HAZARDS
Personal protective equipment (PPE): B-Level PPE.
Toxicity by ingestion: Grade 3: LD_{50} = 265 mg/kg (rat).
Vapor (gas) irritant characteristics: Vapors cause severe irritation of eyes and throat and can cause eye and lung injury. They cannot be tolerated even at low concentrations.
Liquid or solid irritant characteristics: May cause eye and skin irritation.

FIRE DATA
Flash point: 170°F/77°C (oc).

CHEMICAL REACTIVITY
Binary reactants: May attack some forms of plastics

ENVIRONMENTAL DATA
Food chain concentration potential: Negative; unlikely to accumulate.
Water pollution: Effect of low concentrations on aquatic life is unknown. Fouling to shoreline. May be dangerous if it enters nearby water intakes; notify operators. Notify local health and wildlife officials. **Response to discharge:** Issue warning–poison, water contaminant. Restrict access. Mechanical containment. Absorbent (sand or vermiculite) should be removed. Chemical and physical treatment.

SHIPPING INFORMATION
Grades of purity: 99+%; **Storage temperature:** Store under nitrogen; **Inert atmosphere:** None; **Venting:** Open; **Stability during transport:** Stable.

NAS HAZARD CLASSIFICATION FOR BULK WATER TRANSPORTATION
FIRE: 2
HEALTH: Vapor irritant: 3; Liquid or solid irritant: 3; Poisons: 3
WATER POLLUTION: Human toxicity: 3; Aquatic toxicity:–; Aesthetic effect: 3
REACTIVITY: Other chemicals: 3; Water: 1; Self-reaction: 1

PHYSICAL AND CHEMICAL PROPERTIES
Physical state @ 59°F/15°C and 1 atm: Liquid.
Molecular weight: 135.21
Boiling point @ 1 atm: 405–420°F/207–216°C/480–489°K.
Melting/Freezing point: –103°F/–75°C/198.2°K.
Specific gravity (water = 1): 0.915 at 10°C (liquid).
Relative vapor density (air = 1): 4.66 (estimate).

BENZYL DIMETHYLOCTADECYL AMMONIUM CHLORIDE   REC. B:1500

SYNONYMS: BENZYLDIMETHYLSTEARYL-AMMONIUM CHLORIDE; *N,N*-DIMETHYL-*N*-OCTYLBENZENEMETHANAMINIUMCHLORIDE; STEARYLDIMETHYLBENZYL-AMMONIUM CHLORIDE; TALLOW BENZYL DIMETHYLAMMONIUM CHLORIDE

IDENTIFICATION
CAS Number: 959-55-7
Formula: $C_{17}H_{30}N \cdot Cl$; $(C_6H_5CH_2)(CH_3)_2(C_{18}H_{37})NCl$
DOT ID Number: UN 3082; DOT Guide Number: 171
Proper Shipping Name: Environmentally hazardous substance, liquid, n.o.s.

DESCRIPTION: White solid or thick liquid. Mild odor. Reacts slowly with water, forming hydrogen chloride:

Combustible • Containers may BLEVE when exposed to fire • Toxic products of combustion may include carbon monoxide, nitrogen oxide, ammonia, and hydrogen chloride.

EMERGENCY RESPONSE: See Appendix A (171)
Evacuation:
Public safety: Isolate the area of spill or leak for at least 10 to 25 meters (30 to 80 feet) in all directions.
Spill: Increase, in the downwind direction, as necessary, the distance shown under "Public Safety."
Fire: If any large container is involved in fire, isolate for at least 800 meters (½ mile) in all directions; also, consider initial evacuation for 800 meters (½ mile) in all directions.

EXPOSURE
Short-term effects: *SEEK MEDICAL ATTENTION*. **Liquid or solid:** Irritating to skin and eyes. *IF SWALLOWED*, will cause nausea or vomiting. Remove contaminated clothing and shoes. Flush affected areas with plenty of water. *IF IN EYES*, hold eyelids open and flush with plenty of water. *IF SWALLOWED* and victim is *CONSCIOUS AND ABLE TO SWALLOW*, have victim drink 4 to 8 ounces of water. *IF SWALLOWED* and victim is *UNCONSCIOUS OR HAVING CONVULSIONS*, do nothing except keep victim warm:**Do NOT induce vomiting.**

HEALTH HAZARDS
Personal protective equipment (PPE): B-Level PPE. Goggles or face shield; rubber gloves.
Toxicity by ingestion: Grade 2; oral rat LD_{50} = 4000 mg/kg.
Liquid or solid irritant characteristics: May cause eye and skin irritation.

CHEMICAL REACTIVITY
Binary reactants: Oxidizers, strong acids.

ENVIRONMENTAL DATA
Food chain concentration potential: Negative; unlikely to accumulate.
Water pollution: Effect of low concentrations on aquatic life is unknown. May be dangerous if it enters nearby water intakes; notify operators. Notify local health and wildlife officials. **Response to discharge:** Issue warning–water contaminant. Disperse and flush.

SHIPPING INFORMATION
Grades of purity: Pure, 95+%; 24–26% solution in water; **Storage**

temperature: Ambient; **Inert atmosphere:** None; **Venting:** Open; **Stability during transport:** Stable.

PHYSICAL AND CHEMICAL PROPERTIES
Physical state @ 59°F/15°C and 1 atm: Solid.
Molecular weight: 411
Boiling point @ 1 atm: Decomposes 248°F/120°C/393°K.
Specific gravity (water = 1): More than 1.1 @ 68°F/20°C (solid).

BENZYL TRIMETHYLAMMONIUM CHLORIDE
REC. B:1550

SYNONYMS: BTMAC; CLORURO de BENCILTRIMETILAMONIO (Spanish)

IDENTIFICATION
CAS Number: 56-93-9
Formula: $C_{10}H_{16}N \cdot Cl$; $C_6H_5CH_2N(CH_3)_3Cl \cdot H_2O$
DOT ID Number: UN 3082; DOT Guide Number: 171
Proper Shipping Name: Environmentally hazardous substance, liquid, n.o.s.

DESCRIPTION: Light yellow liquid. Mild almond odor. May float on water surface; soluble.

Combustible • Containers may BLEVE when exposed to fire • Toxic products of combustion may include carbon monoxide, nitrogen oxides, ammonia, hydrogen chloride, and phosgene.

Hazard Classification (based on NFPA-704 Rating System)
Health Hazards (Blue): 1; Flammability (Red): 1; Reactivity (Yellow): 0

EMERGENCY RESPONSE: See Appendix A (171)
Evacuation:
Public safety: Isolate the area of spill or leak for at least 10 to 25 meters (30 to 80 feet) in all directions.
Spill: Increase, in the downwind direction, as necessary, the distance shown under "Public Safety."
Fire: If any large container is involved in fire, isolate for at least 800 meters (½ mile) in all directions; also, consider initial evacuation for 800 meters (½ mile) in all directions.

EXPOSURE
Short-term effects: *SEEK MEDICAL ATTENTION.* **Liquid:** Irritating to skin and eyes. *IF SWALLOWED*, will cause nausea or vomiting. Remove contaminated clothing and shoes. Flush affected areas with plenty of water. *IF IN EYES*, hold eyelids open and flush with plenty of water. *IF SWALLOWED* and victim is *CONSCIOUS AND ABLE TO SWALLOW*, have victim drink 4 to 8 ounces of water. *IF SWALLOWED* and victim is *UNCONSCIOUS OR HAVING CONVULSIONS*, do nothing except keep victim warm.

HEALTH HAZARDS
Personal protective equipment (PPE): B-Level PPE. Goggles and rubber gloves.
Toxicity by ingestion: Poisonous. Grade 3; LD_{50} = 50 to 500 mg/kg.
Liquid or solid irritant characteristics: May cause eye and skin irritation.

FIRE DATA
Fire extinguishing agents not to be used: Water may be ineffective on fire.

CHEMICAL REACTIVITY
Binary reactants: Incompatible with oxidizers, strong acids, nitrates.

ENVIRONMENTAL DATA
Food chain concentration potential: Negative; unlikely to accumulate.
Water pollution: Effect of low concentrations on aquatic life is unknown. May be dangerous if it enters nearby water intakes; notify operators. Notify local health and wildlife officials. **Response to discharge:** Issue warning–water contaminant. Disperse and flush.

SHIPPING INFORMATION
Grades of purity: 50-60% solution in water; **Storage temperature:** Ambient; **Inert atmosphere:** None; **Venting:** Open; **Stability during transport:** Stable.

PHYSICAL AND CHEMICAL PROPERTIES
Physical state @ 59°F/15°C and 1 atm: Liquid.
Molecular weight: 172 (solute).
Boiling point @ 1 atm: (decomposes).
Specific gravity (water = 1): 1.07 @ 68°F/20°C (liquid).

BERYLLIUM
REC. B:1600

SYNONYMS: BERILIO (Spanish); BERYLLIUM DUST; BERYLLIUM, METAL POWDER; BERYLLIUM POWDER; GLUCINIUM; EEC No. 004-001-00-7; RCRA No. P015

IDENTIFICATION
CAS Number: 7440-41-7
Formula: Be
DOT ID Number: UN 1567; DOT Guide Number: 134
Proper Shipping Name: Beryllium, powder
Reportable Quantity (RQ): **(CERCLA)** 10 lb/4.54 kg. Reporting is limited to those pieces of the metal having a diameter smaller than 100 micrometers (0.004 inches).

DESCRIPTION: Gray or silvery solid. Odorless. Sinks in water; insoluble.

Poison! • Combustible solid; dust in air is a fire and explosion hazard • Breathing the dust, skin contact can cause chronic health problems. • Firefighting gear (including SCBA) does not provide adequate protection. If exposure occurs, remove and isolate gear immediately and thoroughly decontaminate personnel • Containers may BLEVE when exposed to fire • Concentrated dust in confined areas (e.g., tanks, sewers, buildings) may explode when exposed to fire • Toxic products of combustion may include beryllium. oxide fume.

Hazard Classification (based on NFPA-704 Rating System) (dust or powder) Health Hazards (Blue): 3; Flammability (Red): 1; Reactivity (Yellow): 0

EMERGENCY RESPONSE: See Appendix A (134)
Evacuation:
Public safety: Isolate spill area for at least 25 to 50 meters (30 to 160 feet) in all directions.
Spill: Consider initial downwind evacuation of at least 100 meters (330 feet).

Fire: Isolate for 800 meters (½ mile) in all directions, especially if tank, rail car, or tank truck is involved in fire.

EXPOSURE
Short-term effects: *SEEK MEDICAL ATTENTION.* **Dust:** *POISONOUS IF INHALED OR IF SKIN IS EXPOSED. IF INHALED,* will, will cause coughing or difficult breathing. *IF IN EYES,* hold eyelids open and flush with plenty of water. *IF BREATHING HAS STOPPED,* give artificial respiration; *avoid mouth-to-mouth resuscitation; use bag/mask apparatus.* IF breathing is difficult, administer oxygen. **Solid:** *POISONOUS IF SWALLOWED OR IF SKIN IS EXPOSED.* Remove contaminated clothing and shoes. Flush affected areas with plenty of water. *IF IN EYES,* hold eyelids open and flush with plenty of water. *IF SWALLOWED* and victim is *CONSCIOUS AND ABLE TO SWALLOW,* have victim drink 4 to 8 ounces of water and have victim induce vomiting. *IF SWALLOWED* and victim is *UNCONSCIOUS OR HAVING CONVULSIONS,* do nothing except keep victim warm.

HEALTH HAZARDS
Personal protective equipment (PPE): A-Level PPE. OSHA Table Z-1-A air contaminant. OSHA Table Z-2 air contaminant.
Recommendations for respirator selection: NIOSH
At any concentrations above the NIOSH REL, or where there is no REL, at any detectable concentration: SCBAF:PD,PP (any self-contained breathing apparatus that has a full facepiece and is operated in a pressure-demand or other positive-pressure mode); or SAF:PD,PP:ASCBA (any supplied-air respirator that has a full facepiece and is operated in a pressure-demand or other positive-pressure mode in combination with an auxiliary self-contained breathing apparatus operated in a pressure-demand or other positive pressure mode). *ESCAPE:* HiEF (any air-purifying, full-facepiece respirator with a high-efficiency particulate filter); or SCBAE (any appropriate escape-type, self-contained breathing apparatus).
Exposure limits (TWA unless otherwise noted): ACGIH TLV 0.002 mg/m^3; suspected human carcinogen; OSHA PEL 0.002 mg/m^3, ceiling 0.005 mg/m^3, 30-minute maximum peak 0.025 mg/m^3; NIOSH REL potential human carcinogen, not to exceed 0.0005 mg/m^3.
Short-term exposure limits (15-minute TWA): ACGIH STEL 0.01 mg/m^3.
Toxicity by ingestion: Grade 3; oral LD$_{50}$ = 100 mg/kg (mouse).
Long-term health effects: OSHA specifically regulated carcinogen. IARC probable carcinogen, rating 2A. NTP anticipated carcinogen. Berylliosis, a chronic disease of lungs, may occur from 3 months to 15 years after exposure. Skin disorders. Chronic systemic diseases of the liver, spleen, lymph nodes, bone, kidney, and other organs may also occur.
Liquid or solid irritant characteristics: Eye irritant.
IDLH value: 4 mg/m^3; potential human carcinogen.

FIRE DATA
Fire extinguishing agents not to be used: Water, foam, CO$_2$, halon chemicals.
Electrical hazard: Class II, Group E

CHEMICAL REACTIVITY
Binary reactants: May react with metals evolving flammable hydrogen gas. Reacts with acids, caustics, chlorinated hydrocarbons, oxidizers, molten lithium

ENVIRONMENTAL DATA
Water pollution: Effect of low concentrations on aquatic life is unknown. May be dangerous if it enters nearby water intakes; notify operators. Notify local health and wildlife officials. **Response to discharge:** Issue warning–poison, water contaminant. Restrict access. Should be removed. Chemical and physical treatment.

SHIPPING INFORMATION
Grades of purity: Grade AA, 99.96+%; Grade A, 99.87+%; Nuclear grade; **Storage temperature:** Ambient; **Inert atmosphere:** None; **Venting:** Open; **Stability during transport:** Stable.

PHYSICAL AND CHEMICAL PROPERTIES
Physical state @ 59°F/15°C and 1 atm: Solid.
Molecular weight: 9.01
Boiling point @ 1 atm: 5378°F/2970°C/3243°K.
Melting/Freezing point: 2332°F/1278°C/1551°K.
Specific gravity (water = 1): 1.85 @ 68°F/20°C (solid).
Heat of combustion: −28,000 Btu/lb −15,560 cal/g = −652 x 10^5 J/kg.
Heat of fusion: 260.0 cal/g.
Vapor pressure: 0 mm (approximate).

BERYLLIUM CHLORIDE REC. B:1650

SYNONYMS: BERYLLIUM DICHLORIDE; CLORURO de BERILO (Spanish)

IDENTIFICATION
CAS Number: 7787-47-5
Formula: BeCl$_2$
DOT ID Number: UN 1566; DOT Guide Number: 154
Proper Shipping Name: Beryllium chloride
Reportable Quantity (RQ): **(CERCLA)** 1 lb/0.454 kg

DESCRIPTION: White to green solid. Sharp odor. Sinks and mixes with water; reacts violently, evolving heat, and forming beryllium oxide and a solution of hydrochloric acid.

Poison! • Corrosive • Breathing the dust, skin or eye contact, or swallowing the material can kill you • Firefighting gear (including SCBA) does not provide adequate protection. If exposure occurs, remove and isolate gear immediately and thoroughly decontaminate personnel • Concentrated dust in confined areas (e.g., tanks, sewers, buildings) may explode when exposed to fire • Toxic products of combustion may include beryllium oxides and hydrogen chloride • Do not put yourself in danger by entering a contaminated area to rescue a victim.

Hazard Classification (based on NFPA-704 Rating System)
Health Hazards (Blue): 2; Flammability (Red): 0; Reactivity (Yellow): 2; Special Notice (White): Water reactive

EMERGENCY RESPONSE: See Appendix A (154)
Evacuation:
Public safety: Isolate the area of spill or leak for at least 25 to 50 meters (80 to 160 feet) in all directions.
Spill: Increase, in the downwind direction, as necessary, the distance shown under "Public Safety."
Fire: If tank, rail car, or tank truck is involved in fire, isolate for at least 800 meters (½ mile) in all directions; also, consider initial evacuation for 800 meters (½ mile) in all directions.

EXPOSURE
Short-term effects: *SEEK MEDICAL ATTENTION.* **Dust:** *POISONOUS IF INHALED OR IF SKIN IS EXPOSED. IF INHALED,* will, will cause coughing, difficult breathing, or loss of

consciousness. *IF IN EYES*, hold eyelids open and flush with plenty of water. *IF BREATHING HAS STOPPED*, give artificial respiration; *avoid mouth-to-mouth resuscitation; use bag/mask apparatus*. IF breathing is difficult, administer oxygen. **Solid:** *POISONOUS IF SWALLOWED OR IF SKIN IS EXPOSED*. Will burn skin and eyes. *IF SWALLOWED*, will cause nausea, coughing, or loss of consciousness. Remove contaminated clothing and shoes. Flush affected areas with plenty of water. *IF IN EYES*, hold eyelids open and flush with plenty of water. *IF SWALLOWED* and victim is *CONSCIOUS AND ABLE TO SWALLOW*, have victim drink 4 to 8 ounces of water and have victim induce vomiting. *IF SWALLOWED* and victim is *UNCONSCIOUS OR HAVING CONVULSIONS,* do nothing except keep victim warm.
Medical Note: cuts or puncture wounds in which beryllium may be embedded under the skin should be thoroughly cleansed immediately by a physician.

HEALTH HAZARDS
Personal protective equipment (PPE): OSHA Table Z-1-A air contaminant. OSHA Table Z-2 air contaminant. A-Level PPE.
Recommendations for respirator selection: NIOSH
At any concentrations above the NIOSH REL, or where there is no REL, at any detectable concentration: SCBAF:PD,PP (any self-contained breathing apparatus that has a full facepiece and is operated in a pressure-demand or other positive-pressure mode); or SAF:PD,PP:ASCBA (any supplied-air respirator that has a full facepiece and is operated in a pressure-demand or other positive-pressure mode in combination with an auxiliary self-contained breathing apparatus operated in a pressure-demand or other positive pressure mode). *ESCAPE:* HiEF (any air-purifying, full-facepiece respirator with a high-efficiency particulate filter); or SCBAE (any appropriate escape-type, self-contained breathing apparatus).
Exposure limits (TWA unless otherwise noted): ACGIH TLV 0.002 mg/m^3; suspected human carcinogen; OSHA PEL 0.002 mg/m^3, ceiling 0.005 mg/m^3, 30-minute maximum peak 0.025 mg/m^3; NIOSH REL potential human carcinogen, not to exceed 0.0005 mg/m^3.
Short-term exposure limits (15-minute TWA): ACGIH STEL 0.01 mg/m^3.
Toxicity by ingestion: Grade 3; oral rat LD$_{50}$ = 86 mg/kg.
Long-term health effects: OSHA specifically regulated carcinogen. IARC probable carcinogen, rating 2A. NTP anticipated carcinogen. Be produces a chronic systemic disease that primarily affects the lung but also can involve other organs such as lymph nodes, liver, bones, and kidney.
Liquid or solid irritant characteristics: Eye and skin irritant.
IDLH value: 4 mg/m^3 as beryllium; potential human carcinogen.

FIRE DATA
Fire extinguishing agents not to be used: Do not use water on adjacent fires.

CHEMICAL REACTIVITY
Reactivity with water: Reacts vigorously with evolution of heat. Forms beryllium oxide and hydrochloric acid solution. Corrodes most metals in presence of moisture. Flammable and explosive hydrogen gas may collect in enclosed spaces.
Neutralizing agents for acids and caustics: Flush with water, rinse with dilute solution of sodium bicarbonate or soda ash.

ENVIRONMENTAL DATA
Food chain concentration potential: Bioconcentration of 100-fold can occur under constant exposure. Not significant in spill conditions.
Water pollution: Harmful to aquatic life in very low concentrations. May be dangerous if it enters nearby water intakes; notify operators. Notify local health and wildlife officials. **Response to discharge:** Issue warning–poison, corrosive. Restrict access. Should be removed. Chemical and physical treatment.

SHIPPING INFORMATION
Grades of purity: Commercial, 99+%; **Storage temperature:** Ambient; **Inert atmosphere:** None; **Venting:** Open; **Stability during transport:** Stable.

PHYSICAL AND CHEMICAL PROPERTIES
Physical state @ 59°F/15°C and 1 atm: Solid.
Molecular weight: 79.9
Boiling point @ 1 atm: (sublimes) 968°F/520°C/793°K.
Melting/Freezing point: 824°F/440°C/713°K.
Specific gravity (water = 1): 1.90 @ 77°F/25°C (solid).
Heat of solution: –1000 Btu/lb = –557 cal/g = –23.3 x 10^5 J/kg.
Heat of fusion: 30 cal/g (estimate).

BERYLLIUM FLUORIDE   REC. B:1700

SYNONYMS: BERYLLIUM DIFLUORIDE; FLUORURO de BERILIO (Spanish)

IDENTIFICATION
CAS Number: 7787-49-7
Formula: BeF$_2$
DOT ID Number: UN 1566; DOT Guide Number: 154
Proper Shipping Name: Beryllium fluoride
Reportable Quantity (RQ): **(CERCLA)** 1 lb/0.454 kg

DESCRIPTION: White crystalline solid. Odorless. Sinks and mixes with water.

Poison! • Breathing the dust, skin or eye contact, or swallowing the material can kill you • Firefighting gear (including SCBA) does not provide adequate protection. If exposure occurs, remove and isolate gear immediately and thoroughly decontaminate personnel Toxic products of combustion may include fumes of unburned material and hydrogen fluoride • Do not put yourself in danger by entering a contaminated area to rescue a victim.

Hazard Classification (based on NFPA-704 Rating System)
Health Hazards (Blue): 2; Flammability (Red): 0; Reactivity (Yellow): 0

EMERGENCY RESPONSE: See Appendix A (154)
Evacuation:
Public safety: Isolate the area of spill or leak for at least 25 to 50 meters (80 to 160 feet) in all directions.
Spill: Increase, in the downwind direction, as necessary, the distance shown under "Public Safety."
Fire: If tank, rail car, or tank truck is involved in fire, isolate for at least 800 meters (½ mile) in all directions; also, consider initial evacuation for 800 meters (½ mile) in all directions.

EXPOSURE
Short-term effects: *SEEK MEDICAL ATTENTION*. **Dust:** *POISONOUS IF INHALED, IF SWALLOWED, OR IF SKIN IS EXPOSED*. Will burn eyes. Move to fresh air. *IF BREATHING HAS STOPPED*, give artificial respiration; *avoid mouth-to-mouth resuscitation; use bag/mask apparatus*. IF breathing is difficult, administer oxygen. **Solid:** *POISONOUS IF SWALLOWED OR IF SKIN IS EXPOSED*. Will burn eyes. Remove contaminated clothing

and shoes. Flush affected areas with plenty of water. *IF IN EYES*, hold eyelids open and flush with plenty of water. *IF SWALLOWED* and victim is *CONSCIOUS AND ABLE TO SWALLOW*, have victim drink 4 to 8 ounces of water and have victim induce vomiting. *IF SWALLOWED* and victim is *UNCONSCIOUS OR HAVING CONVULSIONS*, do nothing except keep victim warm.

HEALTH HAZARDS
Personal protective equipment (PPE): OSHA Table Z-1-A air contaminant. OSHA Table Z-2 air contaminant. A-Level PPE.
Recommendations for respirator selection: NIOSH
At any concentrations above the NIOSH REL, or where there is no REL, at any detectable concentration: SCBAF:PD,PP (any self-contained breathing apparatus that has a full facepiece and is operated in a pressure-demand or other positive-pressure mode); or SAF:PD,PP:ASCBA (any supplied-air respirator that has a full facepiece and is operated in a pressure-demand or other positive-pressure mode in combination with an auxiliary self-contained breathing apparatus operated in a pressure-demand or other positive pressure mode). *ESCAPE*: HiEF (any air-purifying, full-facepiece respirator with a high-efficiency particulate filter); or SCBAE (any appropriate escape-type, self-contained breathing apparatus).
Exposure limits (TWA unless otherwise noted): ACGIH TLV 0.002 mg/m^3; suspected human carcinogen; OSHA PEL 0.002 mg/m^3, ceiling 0.005 mg/m^3, 30-minute maximum peak 0.025 mg/m^3; NIOSH REL potential human carcinogen, not to exceed 0.0005 mg/m^3.
Short-term exposure limits (15-minute TWA): ACGIH STEL 0.01 mg/m^3.
Toxicity by ingestion: Grade 3; oral LD$_{50}$ = 100 mg/kg (mouse).
Long-term health effects: OSHA specifically regulated carcinogen. IARC Group 1, Human Sufficient Evidence. NTP carcinogen. Berylliosis of lungs may occur from 3 months to 15 years after exposure. Chronic systemic diseases of the liver, spleen, lymph nodes, bone, kidney, and other organs may also occur.
Vapor (gas) irritant characteristics: Eye and respiratory tract irritant.
Liquid or solid irritant characteristics: Eye and skin irritant.
IDLH value: 4 mg/m^3 as beryllium; potential human carcinogen.

CHEMICAL REACTIVITY
Binary reactants: Acids, caustics, chlorinated hydrocarbons, oxidizers, molten lithium.

ENVIRONMENTAL DATA
Food chain concentration potential: Bioconcentration of 100-fold under constant exposure only. Not significant under spill conditions.
Water pollution: Harmful to aquatic life in very low concentrations. May be dangerous if it enters nearby water intakes; notify operators. Notify local health and wildlife officials.
Response to discharge: Issue warning–poison, water contaminant. Restrict access. Should be removed. Chemical and physical treatment.

SHIPPING INFORMATION
Grades of purity: Purified, 99.99%; chemically pure, 99+%;
Storage temperature: Ambient; **Inert atmosphere:** None;
Venting: Open; **Stability during transport:** Stable.

PHYSICAL AND CHEMICAL PROPERTIES
Physical state @ 59°F/15°C and 1 atm: Solid.
Molecular weight: 47
Boiling point @ 1 atm: Sublimes 1472°F/800°C/1073°K.
Specific gravity (water = 1): 1.99 @ 68°F/20°C (solid).

Heat of solution: –180 Btu/lb = –98 cal/g = –4.1 x 10^5 J/kg.
Heat of fusion: 127.6 cal/g.

BERYLLIUM NITRATE   REC. B:1750

SYNONYMS: BERYLLIUM DINITRATE; BERYLLIUM NITRATE TRIHYDRATE; NITRATO de BERILO (Spanish); NITRIC ACID, BERYLLIUM SALT

IDENTIFICATION
CAS Number: 13597-99-4; 778-75-55
Formula: Be(NO$_3$)$_2$·3H$_2$O
DOT ID Number: UN 2464; DOT Guide Number: 141
Proper Shipping Name: Beryllium nitrate
Reportable Quantity (RQ): **(CERCLA)** 1 lb/0.454 kg

DESCRIPTION: White to pale yellow solid. Odorless. Sinks and mixes with water forming a weak acid solution.

Poison! • Breathing the dust, skin or eye contact, or swallowing the material can kill you • Firefighting gear (including SCBA) does not provide adequate protection. If exposure occurs, remove and isolate gear immediately and thoroughly decontaminate personnel • Heat forms oxygen; will increase the activity of an existing fire • May cause fire or explosion contact with combustibles (wood, paper, oil, clothing, etc.) • Toxic products of combustion may include oxides of nitrogen and beryllium • Do not put yourself in danger by entering a contaminated area to rescue a victim.

Hazard Classification (based on NFPA-704 Rating System)
Health Hazards (Blue): 2; Flammability (Red): 0; Reactivity (Yellow): 1; Special Notice (White): OXY

EMERGENCY RESPONSE: See Appendix A (141)
Evacuation:
Public safety: Isolate the area of spill or leak for at least 10 to 25 meters (30 to 80 feet) in all directions.
Spill: Consider initial downwind evacuation for at least 100 meters (330 feet).
Fire: If any large container is involved in fire, isolate for at least 800 meters (½ mile) in all directions; also, consider initial evacuation for 800 meters (½ mile) in all directions.

EXPOSURE
Short-term effects: *SEEK MEDICAL ATTENTION.* **Dust:** *POISONOUS IF INHALED OR IF SKIN IS EXPOSED. IF INHALED*, will, will cause coughing or difficult breathing. *IF IN EYES*, hold eyelids open and flush with plenty of water. *IF BREATHING HAS STOPPED*, give artificial respiration. IF breathing is difficult, administer oxygen. **Solid:** POISONOUS IF SWALLOWED OR IF SKIN IS EXPOSED. *IF SWALLOWED*, will cause nausea, vomiting, or loss of consciousness. Remove contaminated clothing and shoes. Flush affected areas with plenty of water. *IF IN EYES*, hold eyelids open and flush with plenty of water. *IF SWALLOWED* and victim is *CONSCIOUS AND ABLE TO SWALLOW*, have victim drink 4 to 8 ounces of water and have victim induce vomiting. *IF SWALLOWED* and victim is *UNCONSCIOUS OR HAVING CONVULSIONS*, do nothing except keep victim warm.

HEALTH HAZARDS
Personal protective equipment (PPE): OSHA Table Z-1-A air contaminant. OSHA Table Z-2 air contaminant. A-Level PPE.

Beryllium oxide

Chemical protective material(s) reported to have good to excellent resistance: butyl rubber, polycarbonate.
Recommendations for respirator selection: NIOSH
At any concentrations above the NIOSH REL, or where there is no REL, at any detectable concentration: SCBAF:PD,PP (any self-contained breathing apparatus that has a full facepiece and is operated in a pressure-demand or other positive-pressure mode); or SAF:PD,PP:ASCBA (any supplied-air respirator that has a full facepiece and is operated in a pressure-demand or other positive-pressure mode in combination with an auxiliary self-contained breathing apparatus operated in a pressure-demand or other positive pressure mode). *ESCAPE:* HiEF (any air-purifying, full-facepiece respirator with a high-efficiency particulate filter); or SCBAE (any appropriate escape-type, self-contained breathing apparatus).
Exposure limits (TWA unless otherwise noted): ACGIH TLV 0.002 mg/m^3; suspected human carcinogen; OSHA PEL 0.002 mg/m^3, ceiling 0.005 mg/m^3, 30-minute maximum peak 0.025 mg/m^3; NIOSH REL potential human carcinogen, not to exceed 0.0005 mg/m^3.
Short-term exposure limits (15-minute TWA): ACGIH STEL 0.01 mg/m^3.
Toxicity by ingestion: Poison.
Long-term health effects: OSHA specifically regulated carcinogen. IARC Group 1, Human Sufficient Evidence. NTP anticipated carcinogen. May cause chronic systemic disease of the lung as well as other organs such as liver, spleen, lymph nodes, bone, and kidney.
Liquid or solid irritant characteristics: Severe eye and skin irritant.
IDLH value: Potential human carcinogen; 4 mg/m^3 as beryllium.

FIRE DATA
Flash point: Not combustible, but may increase intensity of fire.

CHEMICAL REACTIVITY
Reactivity with water: Reacts to form weak solution of nitric acid; the reaction is not hazardous.
Binary reactants: Acids, caustics, chlorinated hydrocarbons, oxidizers, molten lithium. In presence of moisture will damage wood and corrode most metals.

ENVIRONMENTAL DATA
Food chain concentration potential: Bioconcentration of 100-fold can occur under *constant* exposure.
Water pollution: Harmful to aquatic life in very low concentrations. May be dangerous if it enters nearby water intakes; notify operators. Notify local health and wildlife officials.
Response to discharge: Issue warning–poison, water contaminant. Restrict access. Should be removed. Chemical and physical treatment.

SHIPPING INFORMATION
Grades of purity: Purified; **Storage temperature:** Ambient; **Inert atmosphere:** None; **Venting:** Open; **Stability during transport:** Stable.

PHYSICAL AND CHEMICAL PROPERTIES
Physical state @ 59°F/15°C and 1 atm: Solid.
Molecular weight: 205.1
Specific gravity (water = 1): 1.56 @ 68°F/20°C (solid).

BERYLLIUM OXIDE **REC. B:1800**

SYNONYMS: BERYLLIA; BROMELITE; BROMELLITE; BERYLLIUM MONOXIDE; EEC No. 004-002-00-2; OXIDO de BERILIO (Spanish); THERMALOX

IDENTIFICATION
CAS Number: 1304-56-9
Formula: BeO
DOT ID Number: UN 1566; DOT Guide Number: 154
Proper Shipping Name: Beryllium compounds, n.o.s.

DESCRIPTION: White solid. Odorless. Sinks in water.

Poison! • Breathing the dust, skin or eye contact, or swallowing the material can kill you • Firefighting gear (including SCBA) does not provide adequate protection. If exposure occurs, remove and isolate gear immediately and thoroughly decontaminate personnel • Toxic products of combustion may include toxic fumes of beryllium oxide.

Hazard Classification (based on NFPA-704 Rating System)
Health Hazards (Blue): 2; Flammability (Red): 0; Reactivity (Yellow): 0

EMERGENCY RESPONSE: See Appendix A (154)
Evacuation:
Public safety: Isolate the area of spill or leak for at least 25 to 50 meters (80 to 160 feet) in all directions.
Spill: Increase, in the downwind direction, as necessary, the distance shown under "Public Safety."
Fire: If tank, rail car, or tank truck is involved in fire, isolate for at least 800 meters (½ mile) in all directions; also, consider initial evacuation for 800 meters (½ mile) in all directions.

EXPOSURE
Short-term effects: *SEEK MEDICAL ATTENTION.* **Dust:** *POISONOUS IF INHALED. IF INHALED,* will, will cause coughing and difficult breathing. *IF IN EYES,* hold eyelids open and flush with plenty of water. *IF BREATHING HAS STOPPED,* give artificial respiration; *avoid mouth-to-mouth resuscitation; use bag/mask apparatus.* IF breathing is difficult, administer oxygen. May cause lung edema; physical exertion will aggravate this condition. **Solid:** *POISONOUS IF SWALLOWED.* Irritating to skin and eyes. Remove contaminated clothing and shoes. Flush affected areas with plenty of water. *IF IN EYES,* hold eyelids open and flush with plenty of water. *IF SWALLOWED* and victim is *CONSCIOUS AND ABLE TO SWALLOW,* have victim drink 4 to 8 ounces of water and have victim induce vomiting. *IF SWALLOWED* and victim is *UNCONSCIOUS OR HAVING CONVULSIONS,* do nothing except keep victim warm.
Note to physician or authorized medical personnel: Medical observation is recommended for 24 to 48 hours after breathing overexposure, as pulmonary edema may be delayed. As first aid for pulmonary edema, consider administering a corticosteroid spray. Cigarette smoking may exacerbate pulmonary injury and should be discouraged for at least 72 hours following exposure.

HEALTH HAZARDS
Personal protective equipment (PPE): OSHA Table Z-1-A air contaminant. OSHA Table Z-2 air contaminant. A-Level PPE. Chemical protective material(s) reported to have good to excellent resistance: butyl rubber, polycarbonate.
Recommendations for respirator selection: NIOSH
At any concentrations above the NIOSH REL, or where there is no REL, at any detectable concentration: SCBAF:PD,PP (any self-contained breathing apparatus that has a full facepiece and is operated in a pressure-demand or other positive-pressure mode); or

SAF:PD,PP:ASCBA (any supplied-air respirator that has a full facepiece and is operated in a pressure-demand or other positive-pressure mode in combination with an auxiliary self-contained breathing apparatus operated in a pressure-demand or other positive pressure mode). *ESCAPE*: HiEF (any air-purifying, full-facepiece respirator with a high-efficiency particulate filter); or SCBAE (any appropriate escape-type, self-contained breathing apparatus).
Exposure limits (TWA unless otherwise noted): ACGIH TLV 0.002 mg/m^3; suspected human carcinogen; OSHA PEL 0.002 mg/m^3, ceiling 0.005 mg/m^3, 30-minute maximum peak 0.025 mg/m^3; NIOSH REL potential human carcinogen, not to exceed 0.0005 mg/m^3.
Short-term exposure limits (15-minute TWA): ACGIH STEL 0.01 mg/m^3.
Toxicity by ingestion: Poison.
Long-term health effects: OSHA specifically regulated carcinogen. IARC Group 1, Human Sufficient Evidence. NTP anticipated carcinogen. Beryllium disease (berylliosis) may occur in lymph nodes, liver, spleen, kidney, etc. as well as lung. Skin disorders.
Vapor (gas) irritant characteristics: Eye and respiratory tract irritant.
Liquid or solid irritant characteristics: Severe eye and skin irritant.
IDLH value: Potential human carcinogen; 4 mg/m^3 as beryllium.

CHEMICAL REACTIVITY
Binary reactants: Acids, caustics, chlorinated hydrocarbons, oxidizers, molten lithium.

ENVIRONMENTAL DATA
Food chain concentration potential: Bioconcentration of 100-fold can occur under constant exposure. Not significant in spill conditions.
Water pollution: Effect of low concentrations on aquatic life is unknown. May be dangerous if it enters nearby water intakes; notify operators. Notify local health and wildlife officials.
Response to discharge: Issue warning–poison, water contaminant. Restrict access. Should be removed. Chemical and physical treatment.

SHIPPING INFORMATION
Grades of purity: Technical; Nuclear; **Storage temperature:** Ambient; **Inert atmosphere:** None; **Venting:** Open; **Stability during transport:** Stable.

PHYSICAL AND CHEMICAL PROPERTIES
Physical state @ 59°F/15°C and 1 atm: Solid.
Molecular weight: 25
Boiling point @ 1 atm: 7052°F/3900°C/4173°K.
Melting/Freezing point: 4568°C/2520°C/2793°K.
Specific gravity (water = 1): 3.0 @ 68°F/20°C (solid).
Heat of fusion: 679.7 cal/g.

BERYLLIUM SULFATE   REC. B:1850

SYNONYMS: BERYLLIUM SULFATE TETRAHYDRATE; SULFATO de BERILIO (Spanish)

IDENTIFICATION
CAS Number: 13510-49-1
Formula: BeSO$_4$H$_2$O
DOT ID Number: UN 1566; DOT Guide Number: 154
Proper Shipping Name: Beryllium compounds, n.o.s.

DESCRIPTION: White solid. Odorless. Sinks and mixes with water.

Poison! • Breathing the dust, skin or eye contact, or swallowing the material can kill you • Firefighting gear (including SCBA) does not provide adequate protection. If exposure occurs, remove and isolate gear immediately and thoroughly decontaminate personnel • Toxic products of combustion may include toxic fumes of beryllium oxide and sulfuric acid fumes.

Hazard Classification (based on NFPA-704 Rating System)
Health Hazards (Blue): 2; Flammability (Red): 0; Reactivity (Yellow): 0

EMERGENCY RESPONSE: See Appendix A (154)
Evacuation:
Public safety: Isolate the area of spill or leak for at least 25 to 50 meters (80 to 160 feet) in all directions.
Spill: Increase, in the downwind direction, as necessary, the distance shown under "Public Safety."
Fire: If tank, rail car, or tank truck is involved in fire, isolate for at least 800 meters (½ mile) in all directions; also, consider initial evacuation for 800 meters (½ mile) in all directions.

EXPOSURE
Short-term effects: *SEEK MEDICAL ATTENTION.* **Dust:** *POISONOUS IF INHALED OR IF SKIN IS EXPOSED. IF INHALED*, will, will cause coughing or difficult breathing. *IF IN EYES*, hold eyelids open and flush with plenty of water.*IF BREATHING HAS STOPPED,* give artificial respiration; *avoid mouth-to-mouth resuscitation; use bag/mask apparatus.* IF breathing is difficult, administer oxygen. **Solid:** Irritating to skin and eyes. Harmful if swallowed. Remove contaminated clothing and shoes. Flush affected areas with plenty of water. *IF IN EYES*, hold eyelids open and flush with plenty of water. *IF SWALLOWED* and victim is *CONSCIOUS AND ABLE TO SWALLOW*, have victim drink 4 to 8 ounces of water and have victim induce vomiting. *IF SWALLOWED* and victim is *UNCONSCIOUS OR HAVING CONVULSIONS,* do nothing except keep victim warm.

HEALTH HAZARDS
Personal protective equipment (PPE): OSHA Table Z-1-A air contaminant. OSHA Table Z-2 air contaminant. A-Level PPE. Chemical protective material(s) reported to have good to excellent resistance: butyl rubber, polycarbonate.
Recommendations for respirator selection: NIOSH
At any concentrations above the NIOSH REL, or where there is no REL, at any detectable concentration: SCBAF:PD,PP (any self-contained breathing apparatus that has a full facepiece and is operated in a pressure-demand or other positive-pressure mode); or SAF:PD,PP:ASCBA (any supplied-air respirator that has a full facepiece and is operated in a pressure-demand or other positive-pressure mode in combination with an auxiliary self-contained breathing apparatus operated in a pressure-demand or other positive pressure mode). *ESCAPE*: HiEF (any air-purifying, full-facepiece respirator with a high-efficiency particulate filter); or SCBAE (any appropriate escape-type, self-contained breathing apparatus).
Exposure limits (TWA unless otherwise noted): ACGIH TLV 0.002 mg/m^3; suspected human carcinogen; OSHA PEL 0.002 mg/m^3, ceiling 0.005 mg/m^3, 30-minute maximum peak 0.025 mg/m^3; NIOSH REL potential human carcinogen, not to exceed 0.0005 mg/m^3.
Short-term exposure limits (15-minute TWA): ACGIH STEL 0.01 mg/m^3.
Toxicity by ingestion: Grade 3; oral rat LD$_{50}$ = 82 mg/kg.

Long-term health effects: OSHA specifically regulated carcinogen. IARC Group 1, Human Sufficient Evidence. NTP anticipated carcinogen. Beryllium disease (berylliosis) may occur in the lymph nodes, liver, spleen, kidney, etc., as well as lung. Skin disorders.
Liquid or solid irritant characteristics: Severe eye and skin irritant.
IDLH value: Potential human carcinogen; 4 mg/m^3 as beryllium.

CHEMICAL REACTIVITY
Binary reactants: Violent reaction with aluminum, magnesium. Reacts with acids, caustics, chlorinated hydrocarbons, oxidizers, molten lithium.

ENVIRONMENTAL DATA
Food chain concentration potential: Bioconcentration of 100-fold can occur under constant exposure. Not significant in spill conditions.
Water pollution: Harmful to aquatic life in very low concentrations. May be dangerous if it enters nearby water intakes; notify operators. Notify local health and wildlife officials. **Response to discharge:** Issue warning–poison, water contaminant. Restrict access. Should be removed. Chemical and physical treatment.

SHIPPING INFORMATION
Grades of purity: High purity; Analytical grade; **Storage temperature:** Ambient; **Inert atmosphere:** None; **Venting:** Open; **Stability during transport:** Stable.

PHYSICAL AND CHEMICAL PROPERTIES
Physical state @ 59°F/15°C and 1 atm: Solid.
Molecular weight: 177.14
Specific gravity (water = 1): 1.71 at 11°C (solid).
Heat of solution: -11 Btu/lb $= -6$ cal/g $= -0.3 \times 10^5$ J/kg.

BISMUTH OXYCHLORIDE   REC. B:1900

SYNONYMS: BISMUTH CHLORIDE OXIDE; BASIC BISMUTH CHLORIDE; BISMUTHYL CHLORIDE; BISMUTH SUBCHLORIDE; CLORURO de BISMUTO (Spanish); PEARL WHITE

IDENTIFICATION
CAS Number: 7787-59-9
Formula: BiOCl

DESCRIPTION: White solid. Odorless. Sinks in water.

Irritating to skin, eyes, and respiratory tract • Toxic products of combustion may include hydrogen chloride and fumes of bismuth.

Hazard Classification (based on NFPA-704 Rating System)
Health Hazards (Blue): 0; Flammability (Red): 0; Reactivity (Yellow): 0

EMERGENCY RESPONSE: See Appendix A (171)
Evacuation:
Public safety: Isolate the area of spill or leak for at least 10 to 25 meters (30 to 80 feet) in all directions.
Spill: Increase, in the downwind direction, as necessary, the distance shown under "Public Safety."
Fire: If any large container is involved in fire, isolate for at least 800 meters (½ mile) in all directions; also, consider initial evacuation for 800 meters (½ mile) in all directions.

EXPOSURE
Short-term effects: *SEEK MEDICAL ATTENTION.* **Dust:** Irritating to eyes, nose, and throat. Harmful if inhaled. *IF IN EYES,* hold eyelids open and flush with plenty of water. *IF BREATHING HAS STOPPED,* give artificial respiration. IF breathing is difficult, administer oxygen. **Solid:** Irritating to skin and eyes. Harmful if swallowed. Remove contaminated clothing and shoes. Flush affected areas with plenty of water. *IF IN EYES,* hold eyelids open and flush with plenty of water. *IF SWALLOWED* and victim is *CONSCIOUS AND ABLE TO SWALLOW,* have victim drink 4 to 8 ounces of water. *IF SWALLOWED* and victim is *UNCONSCIOUS OR HAVING CONVULSIONS,* do nothing except keep victim warm.

HEALTH HAZARDS
Personal protective equipment (PPE): B-Level PPE. Goggles or face shield; protective gloves; dust mask.
Toxicity by ingestion: Grade 0; LD_{50} more than 21.5 g/kg (rat).
Liquid or solid irritant characteristics: Eye and skin irritant.

ENVIRONMENTAL DATA
Food chain concentration potential: Log K_{ow} = 3.32. Values > 3.0 are likely to bioconcentrate in aquatic organisms and other living tissue, especially in fats.
Water pollution: Effect of low concentrations on aquatic life is unknown. May be dangerous if it enters nearby water intakes; notify operators. Notify local health and wildlife officials. **Response to discharge:** Should be removed. Chemical and physical treatment.

SHIPPING INFORMATION
Grades of purity: Dry powder, 100%; aqueous concentrates; dispersions of solid in mineral oil or castor oil; **Storage temperature:** Ambient; **Inert atmosphere:** None; **Venting:** Open; **Stability during transport:** Stable.

PHYSICAL AND CHEMICAL PROPERTIES
Physical state @ 59°F/15°C and 1 atm: Solid.
Molecular weight: 260.4
Boiling point @ 1 atm: Decomposes.
Specific gravity (water = 1): 7.7 @ 68°F/20°C (solid).

BISPHENOL A   REC. B:1950

SYNONYMS: BISFENOL A (Spanish); DIFENYLOL PROPANE; *p,p'*-DIHYDROXYDIPHENYL-DIMETHYLMETHANE; 2,2-BIS(4-HYDROXYFENYL) PROPANE; 2,2-BIS(4-HYDROXYPHENYL) PROPANE; 4,4'-ISOPROPYLIDENDIPHENOL; 4,4-(1-METHYL ETHYLIDENE)BISPHENOL; UCAR BISPHENOL HP

IDENTIFICATION
CAS Number: 80-05-7
Formula: $C_{15}H_{16}O_2$; *p*-$HOC_6H_4C(CH_3)_2C_6H_4OH$

DESCRIPTION: White to cream solid flakes or powder. White to light brown. Weak medicine odor, like phenol. Sinks in water.

Poison! • Combustible solid • Containers may BLEVE when exposed to fire • Concentrated dust in confined areas (e.g., tanks, sewers, buildings) may explode when exposed to fire • Eye and respiratory tract irritant • Toxic products of combustion may include carbon monoxide.

Hazard Classification (based on NFPA-704 Rating System)
Health Hazards (Blue): 1; Flammability (Red): 1; Reactivity (Yellow): 0

EMERGENCY RESPONSE: See Appendix A (171)
Evacuation:
Public safety: Isolate the area of spill or leak for at least 10 to 25 meters (30 to 80 feet) in all directions.
Spill: Increase, in the downwind direction, as necessary, the distance shown under "Public Safety."
Fire: If any large container is involved in fire, isolate for at least 800 meters (½ mile) in all directions; also, consider initial evacuation for 800 meters (½ mile) in all directions.

EXPOSURE
Short-term effects: *SEEK MEDICAL ATTENTION*. **Dust:** Irritating to nose and throat if inhaled. Move to fresh air. **Solid:** Irritating to skin and eyes. Harmful if swallowed.
Flush affected areas with plenty of water. *IF IN EYES*, hold eyelids open and flush with plenty of water. *IF SWALLOWED* and victim is *CONSCIOUS AND ABLE TO SWALLOW*, have victim drink 4 to 8 ounces of water and have victim induce vomiting. *IF SWALLOWED* and victim is *UNCONSCIOUS OR HAVING CONVULSIONS*, do nothing except keep victim warm.

HEALTH HAZARDS
Personal protective equipment (PPE): B-Level PPE.
Toxicity by ingestion: Grade 2; LD_{50} = 0.5 to 5 g/kg (rat).
Long-term health effects: Lowered hemoglobin and erythrocyte (red blood cell) counts below normal in rats. Skin disorders.
Vapor (gas) irritant characteristics: Eye and respiratory tract irritant.
Liquid or solid irritant characteristics: Eye and skin irritant. If spilled on clothing and allowed to remain, may cause smarting and reddening of the skin.

FIRE DATA
Flash point: 415°F/213°C (oc).
Autoignition temperature: 1112°F/600°C/873°K.

CHEMICAL REACTIVITY
Binary reactants: Contact with strong oxidizers may cause fire and explosion.

ENVIRONMENTAL DATA
Food chain concentration potential: Log K_{ow} = 3.32. Values > 3.0 are likely to bioconcentrate in aquatic organisms and other living tissue, especially in fats.
Water pollution: Effect of low concentrations on aquatic life is unknown. May be dangerous if it enters nearby water intakes; notify operators. Notify local health and pollution control officials.
Response to discharge: Should be removed.

SHIPPING INFORMATION
Grades of purity: Commercial; high purity; **Stability during transport:** Stable.

PHYSICAL AND CHEMICAL PROPERTIES
Physical state @ 59°F/15°C and 1 atm: Solid.
Molecular weight: 228.28
Boiling point @ 1 atm: 428°F/220°C/493°K.
Melting/Freezing point: 315°F/157°C/430°K.
Specific gravity (water = 1): 1.195 @ 77°F/25°C (solid).
Vapor pressure: 0.7 mm.

BISPHENOL A DIGLYCIDYL ETHER   REC. B:2000

SYNONYMS: BISFENOL A DIGLICIDAL ETER (Spanish); BISPHENOL A EPICHLOROHYDRIN CONDENSATE; 2,2-BIS[4-(2,3-EPOXYPROPYLOXY)PHENYL]PROPANE; BIS(4-GLYCIDYLOXYPHENYL)DIMETHYAMETHANE; 2,2-BIS(p-GLYCIDYLOXYPHENYL)PROPANE; BIS(4-HYDROXYPHENYL)DIMETHYLMETHANE DIGLYCIDYL ETHER; 2,2-BIS(4-HYDROXYPHENYL)PROPANE,DIGLYCIDYL ETHER; D.E.R. 332; DIGLYCIDYL ETHER OF BISPHENOL A; EPI-REZ 508; EPI-REZ 510; EPON 828; EPOXIDE A; ERL–27774; 4,4'-ISOPROPYLIDENEDIPHENOL DIGLYCIDYL ETHER; 4,4'-ISOPROPYLIDENEDIPHENO EPICHLOROHYDRIN RESIN; 2,2'-(1-METHYLETHYLIDENE)BIS(4,1-PHENYLELEOXYMETHYLENE)BISOXIRANE

IDENTIFICATION
CAS Number: 1675-54-3
Formula: $C_{21}H_{24}O_4$
DOT ID Number: UN 2810; DOT Guide Number: 153
Proper Shipping Name: Poisonous liquid, n.o.s.

DESCRIPTION: Sticky, yellowish brown liquid (solvent curing agents of epoxy resins). Odorless. Sinks in water; insoluble.

Poison! • Combustible • Containers may BLEVE when exposed to fire • Vapor in confined areas (e.g., tanks, sewers, buildings) may explode when exposed to fire • Irritates the eyes, skin, and respiratory tract • Toxic products of combustion may include carbon monoxide.

Hazard Classification (based on NFPA-704 Rating System)
Health Hazards (Blue): 1; Flammability (Red): 2; Reactivity (Yellow): 0

EMERGENCY RESPONSE: See Appendix A (153)
Evacuation:
Public safety: Isolate the area of spill or leak for at least 25 to 50 meters (80 to 160 feet) in all directions.
Spill: Increase, in the downwind direction, as necessary, the distance shown under "Public Safety."
Fire: If tank, rail car, or tank truck is involved in fire, isolate for at least 800 meters (½ mile) in all directions; also, consider initial evacuation for 800 meters (½ mile) in all directions.

EXPOSURE
Short-term effects: *SEEK MEDICAL ATTENTION*. **Liquid:** Irritating to skin and eyes. Harmful if swallowed. Remove contaminated clothing and shoes. Flush affected areas with plenty of water. *IF BREATHING HAS STOPPED*, give artificial respiration; *avoid mouth-to-mouth resuscitation; use bag/mask apparatus*. *IF IN EYES*, hold eyelids open and flush with plenty of water. *IF SWALLOWED* and victim is *CONSCIOUS AND ABLE TO SWALLOW*, have victim drink 4 to 8 ounces of water. *IF SWALLOWED* and victim is *UNCONSCIOUS OR HAVING CONVULSIONS*, do nothing except keep victim warm.

HEALTH HAZARDS
Personal protective equipment (PPE): B-Level PPE. Chemical

protective material(s) reported to have good to excellent resistance: PV alcohol, Silvershield®.
Toxicity by ingestion: Grade 1; LD_{50} = 5 to 15 g/kg.
Long-term health effects: IARC unclassified, rating 3
Vapor (gas) irritant characteristics: Vapors are nonirritating to eyes and throat of most people.
Liquid or solid irritant characteristics: Eye and skin irritant. If spilled on clothing and allowed to remain, may cause smarting and reddening of skin.

FIRE DATA
Flash point: 175°F/79°C (oc).

CHEMICAL REACTIVITY
Binary reactants: May react with oxidizers.

ENVIRONMENTAL DATA
Food chain concentration potential: Log K_{ow} = 3.32. Values > 3.0 are likely to bioconcentrate in aquatic organisms and other living tissue, especially in fats.
Water pollution: Effect of low concentration on aquatic life is unknown. May be dangerous if it enters nearby water intakes; notify operators. Notify local health and wildlife officials.
Response to discharge: Should be removed. Chemical and physical treatment.

SHIPPING INFORMATION
Grades of purity: Commercial; **Storage temperature:** Ambient; **Inert atmosphere:** None; **Venting:** Open; **Stability during transport:** Stable.

NAS HAZARD CLASSIFICATION FOR BULK WATER TRANSPORTATION
FIRE: 1
HEALTH: Vapor irritant: 0; Liquid or solid irritant: 1; Poisons: 1
WATER POLLUTION: Human toxicity: 1; Aquatic toxicity: 0; Aesthetic effect: 0
REACTIVITY: Other chemicals: 2; Water: 0; Self-reaction: 1

PHYSICAL AND CHEMICAL PROPERTIES
Physical state @ 59°F/15°C and 1 atm: Liquid.
Molecular weight: 340 (decomposes).
Specific gravity (water = 1): 1.16 @ 68°F/20°C (liquid).
Heat of combustion: (estimate) –14,900 Btu/lb = –8300 cal/g = –350 x 10^5 J/kg.

BLISTER AGENTS   REC.B:2025

SYNONYMS: Sulfur Mustard Agents H and HD: BIS(β-CHLOROETHYL)SULFIDE; BIS(2-CHLOROETHYL)SULFIDE; 1-CHLORO-2-(-CHLOROETHYLTHIO)ETHANE; 2,2'-DICHLORODIETHYL SULFIDE; DI-2-CHLOROETHYL SULFIDE; 2,2'-DICHLOROETHYL SULFIDE; DICHLORO DIETHYL SULFIDE; ETHANE, 1,1'-THIOBIS(2-CHLORO-; GAS MOSTAZA (Spanish); IPRIT; KAMPSTOFF LOST; LOST; SENFGAS; MUSTARD GAS; MUSTARD HD; SCHWEFEL LOST; S-LOST; SULFIDE, BIS(2-CHLOROETHYL); SULFUR MUSTARD; SULFUR MUSTARD GAS; S-YPERITE; 1,1'-THIOBIS(2-CHLOROETHANE); YELLOW CROSS GAS; YELLOW CROSS LIQUID; YPERITE
Sulfur Mustard Agent HT: Mixture of BIS(2-CHLOROETHYL)SULFIDE and BIS[2-(2-CHLOROETHYLTHIO)-ETHYL]ETHER
Nitrogen Mustard Agent HN-1: BIS(2-CHLOROETHYL)ETHYLAMINE; 2-CHLORO-N-(2-CHLOROETHYL)-N-ETHYLETHANAMINE; 2,2'-DICHLOROTRIETHYLAMINE; ETHYLBIS(2-CHLOROETHYL)AMINE; ETHYL-S
Nitrogen Mustard Agent HN-2: BIS(β-CHLOROETHYL)METHYLAMINE; BIS(2-CHLOROETHYL)METHYLAMINE; N,N-BIS(2-CHLOROETHYL)METHYLAMINE; CARYOLYSIN; CHLORMETHINE; CLORAMIN; DICHLORAMINE; DICHLOREN (German); β,β-DICHLORODIETHYL-N-METHYLAMINE; DI(2-CHLOROETHYL)METHYLAMINE; 2,2'-DICHLORO-N-METHYLDIETHYLAMINE; EMBICHIN; MBA; MECHLORETHAMINE; N-METHYL-BIS-CHLORAETHYLAMIN (German); METHYLBIS(2-CHLOROETHYL)AMINE; N-METHYL-BIS(2-CHLOROETHYL)AMINE (MAK); N-METHYL-2,2'-DICHLORODIETHYLAMINE; METHYLDI(2-CHLOROETHYL)AMINE; N-METHYL-LOST; MUSTARGEN; MUSTINE; MUTAGEN; NITROGEN MUSTARD; N-LOST (German); TL 146
Nitrogen Mustard Agent HN-3: TRIS(2-CHLOROETHYL)AMINE; 2-CHLORO-N,N-BIS(2-CHLOROETHYL)ETHANAMINE; 2,2',2"-TRICHLOROTRIETHYLAMINE
Lewisite Agent L: ARSINE (2-CHLOROVINYL)DICHLORO-; ARSENOUS DICHLORIDE(2-CHLOROETHENYL)-; (2-CHLOROETHENYL)ARSONOUS DICHLORIDE; CHLOROVINYLARSINE DICHLORIDE; β-CHLOROVINYLBICHLOROARSINE; 2-CHLOROVINYLDICHLOROARSINE; (2-CHLOROVINYL)DICHLOROARSINE; DICHLORO(2-CHLOROVINYL)ARSINE; LEWISITE (ARSENIC COMPOUND); EA1034
Mustard-Lewisite Agent HL: SULFUR MUSTARD/LEWISITE

IDENTIFICATION
CAS Number: 505-60-2 (H or HD); 6392-89-8 (HT); 538-07-8 (HN-1); 51-75-2 (HN-2); 555-77-1(HN-3); 541-25-3 (L); (HL) not available
Formula: $C_4H_8Cl_2S$ (H or HD); $C_5H_{11}Cl_2N$ (HN-2); $C_2H_2AsCl_3$ (L)
DOT ID Number: UN2810 (H, HD, HL, HN-1, HN-2, HN-3, L) [weaponized]
Proper Shipping Name: H, HD, HL, HN-1, HN-2, HN-3, L (Lewisite), Mustard, Mustard-Lewisite
Reportable Quantity: **(EHS)** Lewisite 10 lb/4.54 kg; Mustard gas 500 lb/227 kg

DESCRIPTION: *Nitrogen mustards* are colorless when pure but are typically a yellow to brown oily substance. Odors are variable; Sweet, agreeable, slightly garlic- or mustard-like. HN-1 and HN-2 are very sparingly soluble in water; HN-3 is practically insoluble; releases corrosive vapors on contact with water or steam.
Sulfur mustards (H and HD) are colorless when pure but usually a pale yellow, dark brown or black oily liquid. The vapor is colorless. Agent HT is a clear yellowish liquid. Odors are slightly garlic or mustard-like. Although volatility is low, vapors can reach hazardous levels during warm weather. Practically insoluble in water (solubility in water 0.8 g/L at 68°F/20°C).
Lewisite (L) The pure material is an oily, colorless liquid; impure is amber to black. It remains a liquid at low temperatures and is persistent in colder climates. It has the odor of geraniums. Negligible solubility in water.
Mustard lewisite (HL) is a liquid mixture of distilled mustard (HD) and lewisite (L). Due to its low freezing point (developed to achieve a lower freezing point for ground dispersal and aerial spraying), the

a lower freezing point for ground dispersal and aerial spraying), the mixture remains a liquid in cold weather and at high altitudes. The mixture with the lowest freezing point consists of 63% lewisite/37% mustard. It has a garlic-like odor. Practically insoluble in water.

Note: These materials have been developed and used as a chemical warfare agent. Use M8 (Detection: Red) or M256-A1 Detector Kit (Detection limits: 3.0 mg/m^3 for mustards; 14 mg/m^3 for lewisite) if available. Notify U.S. Department of Defense: Army. *Warning:* Odor is not a reliable indicator of the presence of toxic amounts of mustard gas or lewisites.

Poison! • Nitrogen mustards are combustible • Breathing the vapor, skin or eye contact, or swallowing blister agents can kill you. Exposure (vapor or direct contact with the liquid) can cause permanent injury to the eyes, skin, mucous membranes at low concentrations • Firefighting gear (including SCBA) does not provide adequate protection. If exposure occurs, remove and isolate gear immediately and thoroughly decontaminate personnel • Containers may BLEVE when exposed to fire • Vapors are heavier than air and will collect and stay in low areas • Vapors in confined areas (e.g., tanks, sewers, buildings) may explode when exposed to fire • Do not put yourself in danger by entering a contaminated area to rescue a victim. • Sulfur mustards are stable at ambient temperatures but decompose at temperatures greater than 300°F/149°C. Toxic products of combustion may include carbon monoxide, sulfur oxides, hydrogen chloride. *Lewisite (L)* and *mustard/lewisite Mixture (HL)* emit toxic fumes of hydrogen chloride, sulfur oxides, and arsenic. Major hazard is direct contact and contact with vapors when heated to decomposition.

Hazard Classification (based on NFPA-704 Rating System)
Health Hazards (Blue): 4; Flammability (Red): 0-1; Reactivity (Yellow): 0-1

EMERGENCY RESPONSE: See Appendix A (153)
Evacuation:
Public safety: See below.
Spill: For H, HD, HT, HN-1, HN-2, HN-3, lewisite (L), mustard-lewisite (HL)
Small spills: First: Isolate in all directions 30 meters (100 feet); **Then:** Protect persons downwind, **DAY:** 0.2 km (0.1 mile); **NIGHT:** 0.2 km (0.1 mile).

EXPOSURE
Short-term effects: *CALL FOR MEDICAL AID.* Major hazard is direct contact and contact with vapors. There is NO antidote for nitrogen or sulfur mustard toxicity. For *sulfur mustards*, sodium thiosulfate given IV within minutes after exposure may prevent lethality. The antidote for lewisite is described below; it may be used to treat severe conditions but will not prevent lesions on the skin, eye, or airways. Persons who have been exposed to large amounts of lewisite and mustard lewisite will need immediate hospitalization. *Immediate decontamination reduces symptoms of exposure. Decontamination of all potentially exposed areas within 1 or 2 minutes following exposure is the only effective means of decreasing tissue damage.* Later decontamination is unlikely to improve the victim's condition but will protect other personnel from exposure. People whose skin or clothing is contaminated with nitrogen mustard or sulfur mustards can contaminate rescuers by direct contact or through off-gassing vapor. **Skin/Eye:** Exposure to *nitrogen mustard* vapor can cause injury to the eyes, skin, and mucous membranes at low concentrations. Direct contact with the liquid can cause skin and eye burns. *Nitrogen mustards* and *sulfur mustards* are absorbed by the skin causing erythema and blisters. Eye contact can cause conjunctivitis, incapacitating injury to the cornea and conjunctiva; blindness. **Inhalation:** may cause edema, ulceration, and respiratory impairment including necrosis of respiratory tract; damages the respiratory tract epithelium and may cause death. **Ingestion:** Causes nausea and vomiting. In cases of ingestion, **do not induce vomiting**. There is no evidence that administration of activated charcoal is beneficial. **Vapor:** Move victim to fresh air. *IF BREATHING HAS STOPPED,* give artificial respiration. If breathing is difficult, administer supplemental oxygen if cardiopulmonary compromise is suspected. Assist ventilation with a bag-valve-mask device equipped with a cannister or air filter if necessary. Flush the eyes immediately with water for about 5 to 10 minutes by tilting the head to the side, pulling eyelids apart with fingers, and pouring water slowly into eyes. Do not cover eyes with bandages. If exposure to vapor only is certain, remove outer clothing and wash exposed areas with soap and water or 0.5% solution of sodium hypochlorite. **Liquid:** Remove and double-bag contaminated clothing, personal belongings, and shoes and leave in Hot Zone. Wash affected areas with soap and water. If exposure to liquid agent is suspected, cut and remove all clothing and wash skin immediately with soap and water. If shower areas are available, showering with water alone will be adequate. However, in those cases where water is in short supply, and showers are not available, an alternative form of decontamination is to use 0.5% sodium hypochlorite solution or absorbent powders such as flour, talcum powder, or Fuller's earth. Produces delayed effects, including (1 to 12 hours) cough, edema of eyelids, erythema of skin, and severe pruritus. Mortality low, but permanent eye and lung damage may occur. For *nitrogen mustards*, potentially exposed individuals should be observed for 6 to 8 hours and, if signs or symptoms appear, be sent to the hospital.

Note to physician or authorized medical personnel: Pulmonary edema from nitrogen mustards is uncommon. Nevertheless, pulmonary edema may be delayed. Medical observation is recommended for 24 to 48 hours after inhalation overexposure. As first aid for pulmonary edema, a physician or authorized medical personnel may consider administering a corticosteroid spray. Cigarette smoking may exacerbate pulmonary injury and should be discouraged for at least 72 hours following exposure. Avoid drinking alcoholic beverages for at least 24 hours as alcohol may worsen injury to the stomach or have other effects.

Sulfur mustards: Although not an antidote, sodium thiosulfate given IV within minutes after exposure may prevent lethality.

Antidote/Lewisite: British Anti-Lewisite (BAL), also called dimercaprol, is a chelating agent shown to reduce systemic effects from Lewisite exposure. Due to toxic side effects, **BAL should be administered only to patients who have signs of shock or significant pulmonary injury.** Chelation therapy should be performed only by trained personnel. Consultation with the regional poison control center is recommended. The standard dosage regimen is 3 to 5 mg/kg IM every 4 hours for four doses. This regimen can be adjusted depending on the severity of the exposure and the symptoms. Contraindications to BAL include pre-existing renal disease, pregnancy (except in life threatening circumstances), and concurrent use of medicinal iron. Alkalization of the urine stabilizes the dimercaprol-metal complex and has been proposed to protect the kidneys during chelation therapy. If acute renal insufficiency develops, hemodialysis should be considered to remove the dimercaprol–arsenic complex. Side effects of BAL administered at 3 mg/kg are mostly pain at the injection site. At 5 mg/kg, the effects may include nausea; vomiting; headache; burning sensation of the lips, mouth, throat, and eyes; lacrimation; rhinorrhea; salivation; muscle aches; burning and tingling in the extremities; tooth pain; diaphoresis; chest pain; anxiety; and agitation.

HEALTH HAZARDS

Personal protective equipment (PPE): A-Level PPE and butyl rubber chemical protective gloves are recommended. Full skin protection and warfare-approved breathing equipment. If the proper equipment is not available, or if the rescuers have not been trained in its use, call for assistance from the U.S. Soldier and Biological Chemical Command, Edgewood Research Development and Engineering Center (from 0700-1630 EST call 410-671-4411, and from 1630-0700 EST call 410-278-5201; ask for the Staff Duty Officer).

Recommendations for respirator selection: Pressure-demand, self-contained breathing apparatus (SCBA) is recommended in response situations that involve exposure to any amount of nitrogen mustard.

Exposure limits (TWA unless otherwise noted): 0.003 mg/m^3 (HN-1 and sulfur mustards) as recommended by the Surgeon General's Working Group, U.S. Department of Health and Human Services. No standards exist for HN-2 or HN-3 nitrogen mustards.

Short-term exposure limits: The LCt_{50} (the product of concentration x time that is lethal to 50% of the exposed population by inhalation) is approximately 1500 mg-min/m^3 for HN-1 and HN-3, and 3000 mg-min/m^3 for HN-2. The median incapacitating dose for the eyes is 100 mg-min/m^3 for HN-2 and 200 mg-min/m^3 for HN-1 and HN-3. The LCt_{50} of sulfur mustards (the product of concentration x time that is lethal to 50% of the exposed population by inhalation) is approximately 1500 mg-min/m^3. When inhaled, these agents may cause systemic effects. The estimated Ct for airway injury is 100 to 200 mg-min/m^3. A Ct of 12 to 70 mg-min/m^3 produces eye lesions.

Toxicity by ingestion: Ingestion is an uncommon route for exposure but can lead to local effects such as esophageal or gastrointestinal burns and systemic absorption. The lethal dose of sulfur mustard is about 100 mg/kg or 1 to 1.5 teaspoons of liquid.

Short-term health effects: Both nitrogen and sulfur mustards are vesicants causing skin, eye, and respiratory tract injury. Although these agents cause cellular changes within several minutes of contact, the onset of pain and other clinical effects is delayed for hours (nitrogen mustards) or for 1 to 24 hours (sulfur mustards). *Nitrogen mustards*: The mechanisms of action are not clearly understood. They are highly reactive and combine rapidly with proteins, DNA, or other molecules. Therefore, within minutes following exposure intact mustard or its reactive metabolites are not found in tissue or biological fluids. *Sulfur mustards*: Direct contact with sulfur mustard liquid can cause skin and eye burns that develop an hour or more after exposure. A 10μg droplet is capable of producing blisters. Skin, eye, and air way exposure to vapor sulfur mustard and skin and eye exposure to liquid sulfur mustard may cause systemic toxicity. *Ingestion* Ingestion may cause local effects and systemic absorption. The onset of clinical symptoms and their time of onset depend on the severity of exposure. The death rate from exposure to sulfur mustard is low (2 to 3% during World War I). Death usually occurs between day 5 and 10 due to the pulmonary insufficiency complicated by infection due to immune system compromise. *Lewisite (L)* damages skin, eyes, and airways by direct contact. It inhibits many enzymes, in particular those with thiol groups, such as pyruvic oxidase, alcohol dehydrogenase, succinic oxidase, hexokinase, and succinic dehydrogenase. The exact mechanism by which lewisite damages cells is not known. *Mustard lewisite (HL)* shares the vesicant properties of *lewisite (L)* and the DNA alkylation and cross-linking properties of mustard. *Skin: lewisite (L)* liquid or vapor produces pain and skin irritation within seconds to minutes after contact. For liquid *lewisite (L)*, erythema occurs within 15 to 30 minutes after exposure and blisters start within several hours, developing fully by 12 to 18 hours. For the vapor, response times are a little longer. The lewisite blister starts as a small blister in the center of the erythematous area and expands to include the entire inflamed area. *mustard lewisite (HL)* also produces pain and irritation immediately, and erythema within 30 minutes. Blistering is delayed for hours and tends to cover the entire area of reddened skin. *Eyes: Lewisite (L)* vapor causes pain and blepharospasm on contact. Edema of the conjunctiva and eyelids follows, and the eyes may be swollen shut within an hour. With high doses, corneal damage and iritis may follow. Liquid lewisite causes severe eye damage on contact. *mustard lewisite (HL)* also causes ocular effects extremely rapidly. Lacrimation, photophobia, and inflammation of the conjunctiva and cornea may occur. *Respiratory:* Lewisite and *mustard lewisite (HL)* are extremely irritating to the respiratory tract mucosa. Burning nasal pain, epistaxis, sinus pain, laryngitis, cough and dyspnea may occur. Necrosis can cause pseudomembrane formation and local airway obstruction. Pulmonary edema may occur following exposure to high concentrations. *Gastrointestinal:* Ingestion or inhalation of lewisite may cause nausea and vomiting. Ingestion of *mustard lewisite (HL)* produces severe stomach pains, vomiting, and bloody stools after a 15-20 minute latency period. *Cardiovascular:* High-dose exposure to lewisite may cause "lewisite shock," a condition resulting from increased capillary permeability and subsequent intravascular fluid loss, hypovolemia, and organ congestion. *Hepatic* Hepatic necrosis may occur due to shock and hypoperfusion following exposure to high levels of lewisite. *Renal* Exposure to high levels of lewisite may cause decreased renal function secondary to hypotension. *Hematopoietic:* Systemic absorption of *mustard lewisite* may induce bone marrow suppression and an increased risk for fatal complicating infections.

Long-term health effects: IARC Summary: Overall evaluation: Group 1; Carcinogenicity in humans; sufficient evidence. Chemotherapeutic doses of HN-2 have been associated with menstrual irregularities, alopecia, hearing loss, tinnitus, jaundice, impaired spermatogenesis, generalized swelling, and hyperpigmentation. Systemic absorption of nitrogen mustard may induce bone marrow suppression and an increased risk for fatal complicating infections, hemorrhage, and anemia. Chronic respiratory and eye conditions may persist following exposure to large amounts of lewisite or *mustard lewisite (HL)*. Chronic exposure to lewisite may lead to arsenical poisoning (see Arsenic). Chronic exposure to *mustard lewisite (HL)* can cause immune sensitization and chronic lung impairment consisting of cough, shortness of breath, and chest pain. There is only anecdotal evidence for the potential carcinogenicity of lewisite. The data do not support classifying *lewisite (L)* as a suspected carcinogen. Repeated exposures to *mustard lewisite (HL)* over a long period of time may produce respiratory and skin cancer due to the mustard content. *mustard lewisite:* there are no specific data regarding the carcinogenicity. **Reproductive/developmental effects:** *lewisite (L)* inconclusive (because of limited human exposures). Animal studies show no clear evidence.

Vapor (gas) or Liquid or solid irritant characteristics: Nitrogen mustards: *Eyes:* Exposure to vapor or liquid may cause intense conjunctival and scleral inflammation, pain, swelling, lacrimation, photophobia, and corneal damage. High concentrations can cause burns and blindness. *Skin:* Direct skin exposure causes erythema and blistering. Generally, a rash will develop within several hours, followed by blistering within 6 to 12 hours. Prolonged contact, or short contact with large amounts, may result in second- and third-degree chemical burns. Ingestion may cause chemical burns of the gastrointestinal (GI) tract and hemorrhagic diarrhea. Nausea and vomiting may occur following ingestion, dermal, or inhalation exposure.

Odor threshold: 0.6 mg/m (nitrogen and sulfur mustards).

FIRE DATA
Flash point: *sulfur mustards*: 221°F/105°C (H and HD); 212°F/100°C (HT); *nitrogen mustards*: 221°F/105°C; *Lewisites*: L and HL do not burn easily.

Behavior in fire: HN-3 is the most stable of the *nitrogen mustards* but decomposes at temperatures above 493°F/256°C. *Sulfur mustards* are stable at ambient temperatures but decompose at temperatures above 300°F/149°C.

CHEMICAL REACTIVITY
Reaction with water: Hydrolyzes in aqueous solution to less toxic materials (half-life: 5 min @ 99°F/37°C).

Binary reactants: *Sulfur mustards:* Rapidly corrosive to brass and steel @149°F/65°C; they are destroyed by strong oxidizing agents. These agents hydrolyze to form hydrochloric acid (HCl) and thiodiglycol. *Nitrogen mustards:* HN-1 is corrosive to ferrous alloys above 149°F/68°C. HN-2 and HN-3 do not have any incompatible actions on metals or other materials. When heated *lewisite (L)* yields arsenic trichloride, tris-(2-chlorovinyl)arsine, and bis-(2-chlorovinyl)chloroarsine. *Mustard lewisite (HL)* is rapidly corrosive to brass at 149°F/65°C, and will corrode steel at a rate of 0.0001 inches of steel per month at 149°F/65°C. It will hydrolyze into hydrochloric acid, thiodiglycol, and nonvesicant arsenic compounds.

ENVIRONMENTAL DATA
Food chain concentration potential: Log K_{ow} = 1.4. Unlikely to accumulate.

Water pollution: *Nitrogen mustards and Sulfur mustards:* Effects of low concentrations on aquatic life is unknown. May smother benefic life. May be dangerous if it enters nearby water intakes; notify operators. If used as a weapon, utilize M272 Water Detection Kit (Detection limit 2.0 mg/L). Notify local health and wildlife officials. **Response to discharge:** Can be deactivated with bleaching powder, sodium hypochlorite. *Lewisite (L) and mustard lewisite (HL):* May be dangerous if it enters nearby water intakes; notify operators. Notify local health and wildlife officials. **Response to discharge:** Issue warning–poison. Restrict access. Should be removed. Chemical and physical treatment. The EPA drinking water standard of arsenic is 10 ppb (as As ion). For weapons testing, use an M272 Water Detection Kit. Detection limits nitrogen mustards, Sulfur mustards, *lewisite (L), and mustard lewisite (HL)*: 2.0 mg/L.

SHIPPING INFORMATION
Grades of purity: ICC: Class A poison; USCG: Poison A, poison gas label; IATA: Poison A, not acceptable passenger, or cargo. *Sulfur mustards* are vesicants and alkylating agents. H contains about 20–30% impurities (mostly sulfur); distilled mustard is known as HD and is nearly pure; HT is a mixture of 60% HD and 40% agent T (a closely related vesicant with a lower freezing point). *Sulfur mustards* evaporate slowly. **Storage temperature:** Ambient; **Inert atmosphere:** None; **Stability during transport:** Stable.

PHYSICAL AND CHEMICAL PROPERTIES
Physical state @ 59°F/15°C and 1 atm:

Molecular weight (daltons): *Sulfur mustards*: 159.08 (H); 159.08 (HD); 263.2 (T); *Nitrogen mustards*:170.08 (HN-1); 156.07 (HN-2); 204.54 (HN-3); *Lewisites*: 207.32 (L).

Boiling point @ 1 atm (760 mmHg): *Sulfur mustards*: 419°F/217.5°C (H and HD); >442°F/>228°C (HT); *Nitrogen mustards*: 381°F/194°C (HN-1); 167°F/75°C (HN-2); 493°F/256°C (HN-3) (decomposes); *Lewisites:* 374°F/190°C (L); Indefinite, but below 374°F/190°C (HL).

Melting/Freezing point: *Sulfur mustards*: 58.1°F (14.5°C) (H and HD); 32 to 34.3°F/0 to 1.3°C (HT) 29.2°F/–34°C ; *Nitrogen mustards*: (HN-1); –85 to –76°F/–65 to –60°C (HN-1); 25.3°F/–3.7°C (HN-1); *Lewisites*: 0.4°F/–18°C (L); 13°F/–25.4°C (purified mix), –43.6°F/–42°C (typical production batch) (HL)

Liquid density (water = 1): *Sulfur mustards*: 1.27 g/mL (H and HD); *Nitrogen mustards*: 1.09 g/mL @ 77°F/25°C (NH-1); 1.15 g/mL @ 68°F/20°C (HN-2); 1.24 g/mL @ 77°F/25°C (HN-3); 1.89 g/cm @ 77°F/25°C (L); 1.66 g/cm @ 68°F/20°C (HL).

Relative vapor density (air = 1): *Nitrogen mustards*: 5.9 (NH-1); 5.4 (NH-2); 7.1(NH-3); *Lewisites:* 7.1 (L); 6.5 (HL).

Vapor pressure (mmHg all @ 77°F/25°C): *Sulfur mustards*: 0.072 mmHg @ 68°/20°C (H); 0.11 mmHg @ 77°F/25°C (HD); *Nitrogen mustards*: 0.25 (HN-1); 0.427 (HN-2); 0.0109 (HN-3); *Lewisites:* 0.394 mmHg @ 68°F/20°C (L); 0.248 @ 68°F/20°C (HL)

Volatility: 4,480 mg/m³ 68°F/20°C [*lewisite (L)*].

BOILER COMPOUND   REC. B:2050

SYNONYMS: ALKAWAY LIQUID ALKALINE DERUSTER

IDENTIFICATION
DOT ID Number: UN 1760; DOT Guide Number: 154
Proper Shipping Name: Corrosive liquids, n.o.s.

DESCRIPTION: Colorless to brown liquid. Odorless or mild odor. Sinks and mixes with water.

Corrosive • See also sodium hydroxide

EMERGENCY RESPONSE: See Appendix A (154)
Evacuation:
Public safety: Isolate the area of spill or leak for at least 25 to 50 meters (80 to 160 feet) in all directions.
Spill: Increase, in the downwind direction, as necessary, the distance shown under "Public Safety."
Fire: If tank, rail car, or tank truck is involved in fire, isolate for at least 800 meters (½ mile) in all directions; also, consider initial evacuation for 800 meters (½ mile) in all directions.

EXPOSURE
Short-term effects: *SEEK MEDICAL ATTENTION*. **Liquid:** Will burn skin and eyes. Harmful if swallowed. Remove contaminated clothing and shoes. Flush affected areas with plenty of water. *IF BREATHING HAS STOPPED,* give artificial respiration; *avoid mouth-to-mouth resuscitation; use bag/mask apparatus. IF IN EYES*, hold eyelids open and flush with plenty of water. *IF SWALLOWED* and victim is *CONSCIOUS AND ABLE TO SWALLOW*, have victim drink 4 to 8 ounces of water. **Do NOT induce vomiting.**

HEALTH HAZARDS
Personal protective equipment (PPE): B-Level PPE. Goggles or face shield; rubber gloves; protective clothing.

Recommendations for respirator selection: NIOSH/OSHA as sodium hydroxide
10 mg/m³: SA:CF (any supplied-air respirator operated in a continuous-flow mode); or HiEF (any air-purifying, full-facepiece respirator with a high-efficiency particulate filter); PAPRDM (any powered, air-purifying respirator with a dust and mist filter); or SCBAF (any self-contained breathing apparatus with a full

facepiece); or SAF (any supplied-air respirator with a full facepiece). *EMERGENCY OR PLANNED ENTRY INTO UNKNOWN CONCENTRATIONS OR IDLH CONDITIONS:* SCBAF:PD,PP (any self-contained breathing apparatus that has a full facepiece and is operated in a pressure-demand or other positive-pressure mode); or SAF:PD,PP:ASCBA (any supplied-air respirator that has a full facepiece and is operated in a pressure-demand or other positive-pressure mode in combination with an auxiliary self-contained breathing apparatus operated in a pressure-demand or other positive pressure mode). *ESCAPE:* HiEF (any air-purifying, full-facepiece respirator with a high-efficiency particulate filter); or SCBAE (any appropriate escape-type, self-contained breathing apparatus). *Note*: Substance causes eye irritation or damage; eye protection needed.
Exposure limits (TWA unless otherwise noted): ACGIH TLV ceiling limit 2 mg/m^3 as sodium hydroxide; OSHA PEL 2 mg/m^3; NIOSH ceiling 2 mg/m^3 as sodium hydroxide
Vapor (gas) irritant characteristics: May cause eye and respiratory tract irritation.
Liquid or solid irritant characteristics: Severe eye and skin irritant.
IDLH value: 10 mg/m^3 as sodium hydroxide

FIRE DATA
Behavior in fire: Containers may rupture.

CHEMICAL REACTIVITY
Binary reactants: Attacks aluminum and zinc; the reaction may form flammable hydrogen gas.
Neutralizing agents for acids and caustics: Flush with water.

ENVIRONMENTAL DATA
Food chain concentration potential: Negative; unlikely to accumulate.
Water pollution: Effect of low concentrations on aquatic life is unknown. May be dangerous if it enters nearby water intakes; notify operators. Notify local health and wildlife officials.
Response to discharge: Issue warning–corrosive, water contaminant. Restrict access. Disperse and flush.

SHIPPING INFORMATION
Grades of purity: Various commercial grades, some of which contain chelating and complexing agents for metals; **Storage temperature:** Ambient, preferably 40–100°F; **Inert atmosphere:** None; **Venting:** Open; **Stability during transport:** Stable.

PHYSICAL AND CHEMICAL PROPERTIES
Physical state @ 59°F/15°C and 1 atm: Liquid.
Boiling point @ 1 atm: More than 220°F/104°C/377°K.
Specific gravity (water = 1): 1.48 @ 68°F/20°C (liquid).

BORIC ACID   REC. B:2100

SYNONYMS: ACIDO BORICO (Spanish); BORACIC ACID; BOROFAX; BORSAURE (German); ORTHOBORIC ACID; THREE ELEPHANT

IDENTIFICATION
CAS Number: 10043-35-3
Formula: H_3BO_3

DESCRIPTION: Colorless crystalline solid or white powder. Odorless. Sinks and mixes with water. Decomposes above 212°F/100°C forming boric anhydride and water.

Hazard Classification (based on NFPA-704 Rating System)
Health Hazards (Blue): 0; Flammability (Red): 0; Reactivity (Yellow): 0

EMERGENCY RESPONSE: See Appendix A (171)
Evacuation:
Public safety: Isolate the area of spill or leak for at least 10 to 25 meters (30 to 80 feet) in all directions.
Spill: Increase, in the downwind direction, as necessary, the distance shown under "Public Safety."
Fire: If any large container is involved in fire, isolate for at least 800 meters (½ mile) in all directions; also, consider initial evacuation for 800 meters (½ mile) in all directions.

EXPOSURE
Short-term effects: *SEEK MEDICAL ATTENTION.* Solid: Irritating to skin and eyes. *IF SWALLOWED*, will cause nausea or vomiting. Remove contaminated clothing and shoes. Flush affected areas with plenty of water. *IF IN EYES*, hold eyelids open and flush with plenty of water. *IF SWALLOWED* and victim is *CONSCIOUS AND ABLE TO SWALLOW*, have victim drink 4 to 8 ounces of water and have victim induce vomiting. *IF SWALLOWED* and victim is *UNCONSCIOUS OR HAVING CONVULSIONS*, do nothing except keep victim warm.

HEALTH HAZARDS
Personal protective equipment (PPE): B-Level PPE. Chemical protective material(s) reported to have good to excellent resistance: butyl rubber, nitrile, neoprene, Viton®
Exposure limits (TWA unless otherwise noted): ACGIH TLV 10 mg/m^3 as boron oxide; OSHA PEL 15 mg/m^3; NIOSH REL 10 mg/m^3 as boric oxide
Toxicity by ingestion: Grade 2; oral rat LD_{50} = 2660 mg/kg.
Long-term health effects: Causes skin problems (borism), brain and organ damage.
Vapor (gas) irritant characteristics: Possible eye irritation.
Liquid or solid irritant characteristics: Eye irritant.
IDLH value: 2000 mg/m^3 as boric oxide

CHEMICAL REACTIVITY
Binary reactants: Reacts with caustics.

ENVIRONMENTAL DATA
Food chain concentration potential: Negative; unlikely to accumulate.
Water pollution: Dangerous to aquatic life in high concentrations. May be dangerous if it enters nearby water intakes; notify operators. Notify local health and wildlife officials. **Response to discharge:** Issue warning–water contaminant. Should be removed. Chemical and physical treatment.

SHIPPING INFORMATION
Grades of purity: Radio, 99.98%; Technical, 99.9%; N.F., 99.5%; **Storage temperature:** Ambient; **Inert atmosphere:** None; **Venting:** Open; **Stability during transport:** Stable.

PHYSICAL AND CHEMICAL PROPERTIES
Physical state @ 59°F/15°C and 1 atm: Solid.
Molecular weight: 61.83
Boiling point @ 1 atm: (decomposes).
Melting/Freezing point: 338°F/170°C/443°K.
Specific gravity (water = 1): 1.51 at 14°C (solid).

Heat of decomposition: 212°F/100°C.
Heat of solution: –157 Btu/lb = –87 cal/g = –3.7 x 10^5 J/kg.
Vapor pressure: Less than 0.008 mm.

BORON TRIBROMIDE　　　　　　　　REC. B:2150

SYNONYMS: BORANE, TRIBROMO-; BORON BROMIDE; BORON TRIBROMIDE 6; TRIBROMURO de BORO (Spanish); TRONA

IDENTIFICATION
CAS Number: 10294-33-4
Formula: BBr3
DOT ID Number: UN 2692; DOT Guide Number: 157
Proper Shipping Name: Boron tribromide

DESCRIPTION: Colorless, fuming liquid. Sharp odor. Reacts violently with water; forming hydrobromic acid and fumes.

Poison! • Do not use water • Heat may cause explosion • Breathing the vapor can kill you; skin or eye contact causes severe burns, impaired vision, or blindness; effects of contact or inhalation may be delayed • Firefighting gear (including SCBA) does not provide adequate protection. If exposure occurs, remove and isolate gear immediately and thoroughly decontaminate personnel • Containers may BLEVE when exposed to fire • Toxic products of combustion may include hydrogen bromide and boron oxide • Do not put yourself in danger by entering a contaminated area to rescue a victim.

Hazard Classification (based on NFPA-704 Rating System)
Health Hazards (Blue): 3; Flammability (Red): 0; Reactivity (Yellow): 2; Special Notice (White): Water reactive

EMERGENCY RESPONSE: See Appendix A (157)
Evacuation:
Public safety: Isolate the area of spill or leak for at least 50 to 100 meters (160 to 330 feet) in all directions.
Spill: Small spill–First: Isolate in all directions 30 meters (100 feet) (land or water); Then: Protect persons downwind, DAY: 0.2 km (0.1 mile) (land or water); NIGHT: 0.3 km (0.2 mile) (land); 0.2 km (0.1 mile) (water). Large spill–First: Isolate in all directions 60 meters (200 feet) (land or water); Then: Protect persons downwind, DAY: 0.6 km (0.4 mile) (land); 0.5 km (0.3 mile) (water); NIGHT: 1.4 km (0.9 mile) (land); 1.6 km (1.0 mile) (water).
Fire: If tank, rail car, or tank truck is involved in fire, isolate for at least 800 meters (½ mile) in all directions; also, consider initial evacuation for 800 meters (½ mile) in all directions.

EXPOSURE
Short-term effects: *SEEK MEDICAL ATTENTION.* **Vapor:** Irritating to eyes, nose, and throat. *IF INHALED*, will, will cause coughing or difficult breathing. *IF IN EYES*, hold eyelids open and flush with plenty of water. *IF BREATHING HAS STOPPED*, give artificial respiration. IF breathing is difficult, administer oxygen. **Liquid:** Will burn skin and eyes. *IF SWALLOWED*, will cause nausea and vomiting. Remove contaminated clothing and shoes. Flush affected areas with plenty of water. *IF IN EYES*, hold eyelids open and flush with plenty of water. *IF SWALLOWED* and victim is *CONSCIOUS AND ABLE TO SWALLOW*, have victim drink 4 to 8 ounces of water. *IF SWALLOWED* and victim is *UNCONSCIOUS OR HAVING CONVULSIONS*, do nothing except keep victim warm.

HEALTH HAZARDS
Personal protective equipment (PPE): A-Level PPE. OSHA Table Z-1-A air contaminant.
Recommendations for respirator selection: NIOSH/OSHA as hydrogen bromide
30 ppm: SA:CF* (any supplied-air respirator operated in a continuous-flow mode); or PAPRAG* [any powered, air-purifying respirator with acid gas cartridge(s)]; or GMFAG [any air-purifying, full-facepiece respirator (gas mask) with a chin-style, front- or back-mounted organic vapor cannister]; or SCBAF (any self-contained breathing apparatus with a full facepiece); or SAF (any supplied-air respirator with a full facepiece). *EMERGENCY OR PLANNED ENTRY INTO UNKNOWN CONCENTRATIONS OR IDLH CONDITIONS:* SCBAF:PD,PP (any self-contained breathing apparatus that has a full facepiece and is operated in a pressure-demand or other positive-pressure mode); or SAF:PD,PP:ASCBA (any supplied-air respirator that has a full facepiece and is operated in a pressure-demand or other positive-pressure mode in combination with an auxiliary self-contained breathing apparatus operated in a pressure-demand or other positive pressure mode). *ESCAPE:* GMFAG [any air-purifying, full-facepiece respirator (gas mask) with a chin-style, front- or back-mounted organic vapor cannister]; or SCBAE (any appropriate escape-type, self-contained breathing apparatus). **Note:* Substance causes eye irritation or damage; eye protection needed.
Exposure limits (TWA unless otherwise noted): NIOSH ceiling 1 ppm (10 mg/m^3).
Short-term exposure limits (15-minute TWA): ACGIH ceiling 1 ppm.
Long-term health effects: Prolonged contact or inhalation of inorganic bromides causes acne-like bromoderma (bromide rash), especially of the hands and face. Other effects include emaciation, depression; and, in severe cases, psychosis and mental deterioration.
Vapor (gas), liquid, or solid irritant characteristics: Severe eye, skin and respiratory tract irritant.

FIRE DATA
Fire extinguishing agents not to be used: Do not use water or foam on adjacent fires.
Behavior in fire: Containers may rupture and explode.

CHEMICAL REACTIVITY
Reactivity with water: Forms boric acid and hydrobromic acid solution and fumes.
Binary reactants: Moisture, heat, potassium, sodium alcohols. Strongly attacks metals, rubber, and wood. Flammable hydrogen gas may collect in enclosed spaces.
Neutralizing agents for acids and caustics: Flush with water, rinse with dilute solution of sodium bicarbonate or soda ash.

ENVIRONMENTAL DATA
Food chain concentration potential: Negative; unlikely to accumulate.
Water pollution: Effect of low concentrations on aquatic life is unknown. May be dangerous if it enters nearby water intakes; notify operators. Notify local health and wildlife officials. **Response to discharge:** Issue warning–corrosive, air contaminant. Restrict access. Disperse and flush with care.

SHIPPING INFORMATION
Grades of purity: Epitaxial, 99.999+%; Pure, 99.99+%; Technical; **Storage temperature:** Ambient; **Inert atmosphere:** Padded; **Venting:** Pressure-vacuum; **Stability during transport:** Stable.

PHYSICAL AND CHEMICAL PROPERTIES
Physical state @ 59°F/15°C and 1 atm: Liquid.
Molecular weight: 250.5
Boiling point @ 1 atm: 196°F/91°C/364°K.
Melting/Freezing point: –51°F/–46°C/227°K.
Specific gravity (water = 1): 2.645 @ 68°F/20°C (liquid).
Liquid surface tension: 29.1 dynes/cm = 0.0291 N/m at 71°F/22°C.
Relative vapor density (air = 1): 8.64
Ratio of specific heats of vapor (gas): 1.140
Latent heat of vaporization: 52 Btu/lb = 29 cal/g = 1.2×10^5 J/kg.
Heat of fusion: 2.9 cal/g.
Vapor pressure: 40 mm 257°F/125°C.

BORON TRICHLORIDE   REC. B:2200

SYNONYMS: BORON CHLORIDE; EEC No. 005-002-00-5; TRICHLOROBORANE; TRICHLOROBORON; TRICLORURO de BORO (Spanish)

IDENTIFICATION
CAS Number: 10294-34-5
Formula: BCl_3
DOT ID Number: UN 1741; DOT Guide Number: 125
Proper Shipping Name: Boron trichloride
Reportable Quantity (RQ): **(EHS)** 1 lb/0.454 kg

DESCRIPTION: Colorless, compressed liquefied gas. Shipped in cylinders with special fittings. Sharp, choking acrid odor. Reacts violently with water forming hydrochloric acid and boric acid.

Poison! • Corrosive • Breathing the gas can kill you. Prolonged skin contact can cause severe burns, impaired vision, or blindness • Firefighting gear (including SCBA) does not provide adequate protection. If exposure occurs, remove and isolate gear immediately and thoroughly decontaminate personnel • Containers may BLEVE when exposed to fire • Concentrated dust in confined areas (e.g., tanks, sewers, buildings) may explode when exposed to fire • Attacks most metals in the presence of moisture • Toxic products of combustion may include hydrogen chloride and chlorine • Do not put yourself in danger by entering a contaminated area to rescue a victim.

Hazard Classification (based on NFPA-704 Rating System)
Health Hazards (Blue): 2; Flammability (Red): 0; Reactivity (Yellow): 0

EMERGENCY RESPONSE: See Appendix A (125)
Evacuation:
Public safety: See below.
Spill: Small spill–First: Isolate in all directions 30 meters (100 feet); Then: Protect persons downwind, DAY: 0.2 km (0.1 mile); NIGHT: 0.3 km (0.2 mile). Large spill–First: Isolate in all directions 60 meters (200 feet); Then: Protect persons downwind, DAY: 0.6 km (0.4 mile); NIGHT: 1.6 km (1.0 mile).
Fire: Isolate for 1600 meters (1 mile) in all directions, especially if tank, rail car, or tank truck is involved in fire.

EXPOSURE
Short-term effects: *SEEK MEDICAL ATTENTION.* **Vapor:** Irritating to eyes, nose, and throat. *IF INHALED*, will, will cause coughing or difficult breathing. Move victim to fresh air. *IF BREATHING HAS STOPPED*, give artificial respiration; avoid mouth-to-mouth resuscitation; use bag/mask apparatus. IF breathing is difficult, administer oxygen. Inhalation may cause lung edema may develop. **Liquid:** *POISONOUS IF SWALLOWED.* Will burn skin and eyes. Remove contaminated clothing and shoes. Flush affected areas with plenty of water. *IF IN EYES*, hold eyelids open and flush with plenty of water. *IF SWALLOWED* and victim is *CONSCIOUS AND ABLE TO SWALLOW*, have victim drink 4 to 8 ounces of water. **Do NOT induce vomiting.**
Note to physician or authorized medical personnel: Medical observation is recommended for 24 to 48 hours after breathing overexposure, as pulmonary edema may be delayed. As first aid for pulmonary edema, consider administering a corticosteroid spray. Cigarette smoking may exacerbate pulmonary injury and should be discouraged for at least 72 hours following exposure.

Evacuate area. Disperse and flush.

HEALTH HAZARDS
Personal protective equipment (PPE): A-Level PPE. Chemical protective material(s) reported to have good to excellent resistance: PVC offers limited protection.
Recommendations for respirator selection: NIOSH/OSHA as hydrogen chloride
50 ppm: CCRS* [any chemical cartridge respirator with cartridge(s) providing protection against the compound of concern)]; or GMFS [any air-purifying, full-facepiece respirator (gas mask) with a chin-style, front- or back-mounted canister providing protection against the compound of concern]; or PAPRS* [any powered, air-purifying respirator with cartridge(s) providing protection against the compound of concern]; or SA* (any supplied-air respirator); or SCBAF (any self-contained breathing apparatus with a full facepiece). *EMERGENCY OR PLANNED ENTRY INTO UNKNOWN CONCENTRATIONS OR IDLH CONDITIONS*: SCBAF:PD,PP (any self-contained breathing apparatus that has a full facepiece and is operated in a pressure-demand or other positive-pressure mode); or SAF:PD,PP:ASCBA (any supplied-air respirator that has a full facepiece and is operated in a pressure-demand or other positive-pressure mode in combination with an auxiliary self-contained breathing apparatus operated in a pressure-demand or other positive pressure mode). *ESCAPE*: GMFAG [any air-purifying, full-facepiece respirator (gas mask) with a chin-style, front- or back-mounted organic vapor cannister]; or SCBAE (any appropriate escape-type, self-contained breathing apparatus).
Note: Substance reported to cause eye irritation or damage; may require eye protection.
Exposure limits (TWA unless otherwise noted): ACGIH TLV ceiling 5 ppm (7.5 mg/m³); NIOSH/OSHA ceiling 5 ppm (7 mg/m³) as hydrogen chloride
Toxicity by ingestion: Grade 2; LD_{50} = 0.5 to 5 g/kg.
Long-term health effects: Liver and kidney damage. May cause brain damage.
Vapor (gas) irritant characteristics: Vapors cause severe irritation of eyes and throat and can cause eye or lung injury. They cannot be tolerated even at low concentrations.
Liquid or solid irritant characteristics: Severe skin irritant. Causes second- and third-degree burns on short contact and is very injurious to the eyes.
Odor threshold: Decomposes in moist air, releasing hydrochloric acid and decomposition products. Hydrochloric acid: 0.255–10.06 ppm.
IDLH value: 50 ppm as hydrogen chloride

FIRE DATA
Behavior in fire: Water applied to adjacent fires may cause toxic fumes of hydrogen chloride and boric acid.

CHEMICAL REACTIVITY
Reactivity with water: Reacts vigorously to liberate heat and forms hydrogen chloride fumes and boric acid.
Binary reactants: Vigorously attacks elastomers and packing materials. Viton, Tygon, Saran, or silastic elastomers and natural and synthetic rubbers are NOT recommended for service. Lead- and graphite-impregnated asbestos are to be avoided. In the presence of moisture, highly corrosive to most metals.
Neutralizing agents for acids and caustics: Flush with water, rinse with sodium bicarbonate or lime solution.

ENVIRONMENTAL DATA
Food chain concentration potential: Negative; unlikely to accumulate.
Water pollution: Effect of low concentrations on aquatic life is unknown. May be dangerous if it enters nearby water intakes; notify operators. Notify local health and wildlife officials.
Response to discharge: Issue warning–corrosive, air contaminant. Restrict access.

SHIPPING INFORMATION
Grades of purity: C.P. (99.9+%); **Storage temperature:** Ambient; **Inert atmosphere:** None; **Venting:** Pressure-vacuum; **Stability during transport:** Stable.

NAS HAZARD CLASSIFICATION FOR BULK WATER TRANSPORTATION
FIRE: 0
HEALTH: Vapor irritant: 4; Liquid or solid irritant: 4; Poisons: 3
WATER POLLUTION: Human toxicity: 2; Aquatic toxicity: 1; Aesthetic effect: 2
REACTIVITY: Other chemicals: 4; Water: 4; Self-reaction: 0

PHYSICAL AND CHEMICAL PROPERTIES
Physical state @ 59°F/15°C and 1 atm: Gas.
Molecular weight: 117.2
Boiling point @ 1 atm: 54.3°F/12.4°C/285.6°K.
Melting/Freezing point: −161°F/−107°C/166°K.
Critical temperature: 352°F/178°C/451°K.
Critical pressure: 566 psia = 38.5 atm = 3.90 MN/m^2.
Specific gravity (water = 1): 1.35 at 11°C (liquid).
Liquid surface tension: 16.7 dynes/cm = 0.0167 N/m @ 68°F/20°C.
Relative vapor density (air = 1): 4
Ratio of specific heats of vapor (gas): 1.1470
Latent heat of vaporization: 68.8 Btu/lb = 38.2 cal/g = 1.60 X 10^5 J/kg.
Heat of solution: −13,000 Btu/lb = −7200 cal/g = −300 x 10^5 J/kg.
Heat of fusion: 4.3 cal/g.
Vapor pressure: 985 mm.

BROMINE   REC. B:2250

SYNONYMS: BROMINE; BROM (German); BROME (French); BROMO (Spanish); EEC No. 035-001-00-5; MOLECULAR BROMINE

IDENTIFICATION
CAS Number: 7726-95-6
Formula: Br$_2$
DOT ID Number: UN 1744; DOT Guide Number: 154
Proper Shipping Name: Bromine; Bromine solutions
Reportable Quantity (RQ): **(EHS)** 1 lb/0.454 kg

DESCRIPTION: Fuming reddish-brown liquid. Sharp irritating odor. Sinks in water; soluble; reacts, forming hydrobromic acid and oxygen.

Poison! • Breathing the vapor can kill you. Corrosive to the skin and eyes; causes severe burns and blindness • Firefighting gear (including SCBA) does not provide adequate protection. If exposure occurs, remove and isolate gear immediately and thoroughly decontaminate personnel • Strong oxidizer which may react spontaneously with low flash point organics or reducing agents. Heat forms oxygen; will increase the activity of an existing fire • May cause fire or explosion contact with combustibles (wood, paper, oil, clothing, etc.) • Containers may BLEVE when exposed to fire • Toxic products of combustion may include bromine fumes • Do not put yourself in danger by entering a contaminated area to rescue a victim.

Hazard Classification (based on NFPA-704 Rating System)
Health Hazards (Blue): 3; Flammability (Red): 0; Reactivity (Yellow): 0; Special Notice (White): OXY

EMERGENCY RESPONSE: See Appendix A (154)
Evacuation:
Public safety: Isolate the area of spill or leak for at least 25 to 50 meters (80 to 160 feet) in all directions.
Spill: Small spill–First: Isolate in all directions 60 meters (200 feet); Then: Protect persons downwind, DAY: 0.3 km (0.2 mile); NIGHT: 1.1 km (0.7 mile). Large spill–First: Isolate in all directions 185 meters (600 feet); Then: Protect persons downwind, DAY: 1.6 km (1.0 mile); NIGHT: 4.0 km (2.5 miles).
Fire: If tank, rail car, or tank truck is involved in fire, isolate for at least 800 meters (½ mile) in all directions; also, consider initial evacuation for 800 meters (½ mile) in all directions.

EXPOSURE
Short-term effects: *SEEK MEDICAL ATTENTION.* **Vapor:** *Irritating to eyes, nose, and throat. IF INHALED*, will, will cause coughing, difficult breathing, or loss of consciousness. Move to fresh air. *IF BREATHING HAS STOPPED*, give artificial respiration, *avoid mouth-to-mouth resuscitation; use bag/mask apparatus.* IF breathing is difficult, administer oxygen. **Liquid:** Will burn skin and eyes. Harmful if swallowed. Remove contaminated clothing and shoes. Flush affected areas with plenty of water. *IF IN EYES*, hold eyelids open and flush with plenty of water. *IF SWALLOWED* and victim is *CONSCIOUS AND ABLE TO SWALLOW*, have victim drink 4 to 8 ounces of water.
Note to physician or authorized medical personnel: If there is obstruction to breathing establish airway by pulling tongue forward, inserting an airway tube, or doing a tracheostomy; begin artificial respiration; if difficulty in breathing is a result of pulmonary edema, treatment should be carried out with the patient in the sitting position. Administration of oxygen is most important
Medical observation is recommended for 24 to 48 hours after breathing overexposure, as pulmonary edema may be delayed. As first aid for pulmonary edema, consider administering a corticosteroid spray. Cigarette smoking may exacerbate pulmonary injury and should be discouraged for at least 72 hours following exposure.

HEALTH HAZARDS
Personal protective equipment (PPE): A-Level PPE. OSHA Table Z-1-A air contaminant. Chemical protective material(s) reported to have good to excellent resistance: Teflon® and Viton®. Neoprene offers minimal protection.

Bromine pentafluoride

Recommendations for respirator selection: NIOSH/OSHA
2.5 ppm: SA:CF£ (any supplied-air respirator operated in a continuous-flow mode); or PAPRS£* [any powered, air-purifying respirator with cartridge(s) providing protection against the compound of concern]. *3 ppm:* CCRFS [any chemical cartridge respirator with a full facepiece and cartridge(s) providing protection against the compound of concern]; or GMFS [any air-purifying, full-facepiece respirator (gas mask) with a chin-style, front- or back-mounted canister providing protection against the compound of concern]; or PAPRTS£* [any powered, air-purifying respirator with a tight-fitting facepiece and cartridge(s) providing protection against the compound of concern]; or SCBAF (any self-contained breathing apparatus with a full facepiece); or SAF (any supplied-air respirator with a full facepiece). *EMERGENCY OR PLANNED ENTRY INTO UNKNOWN CONCENTRATIONS OR IDLH CONDITIONS:* SCBAF:PD,PP (any self-contained breathing apparatus that has a full facepiece and is operated in a pressure-demand or other positive-pressure mode); or SAF:PD,PP:ASCBA (any supplied-air respirator that has a full facepiece and is operated in a pressure-demand or other positive-pressure mode in combination with an auxiliary, self-contained breathing apparatus operated in a pressure-demand or other positive-pressure mode). *ESCAPE:* GMFS* [any air-purifying, full-facepiece respirator (gas mask) with a chin-style, front- or back-mounted canister providing protection against the compound of concern]; or SCBAE (any appropriate escape-type, self-contained breathing apparatus). * *Note 1:* Substance causes eye irritation or damage; eye protection needed. £ *Note2:* Only nonoxidizable sorbents are allowed (not charcoal).
Exposure limits (TWA unless otherwise noted): ACGIH TLV 0.1 ppm (0.66 mg/m^3); OSHA PEL 0.1 ppm (0.7 mg/m^3); NIOSH REL 0.1 ppm (0.7 mg/m^3).
Short-term exposure limits (15-minute TWA): ACGIH STEL 0.2 ppm (1.3 mg/m^3); NIOSH STEL 0.3 ppm (2 mg/m^3).
Toxicity by ingestion: Human poison.
Long-term health effects: May cause skin problems.
Vapor (gas) irritant characteristics: Causes severe eye or throat irritations which can cause eye or lung injury; cannot be tolerated even at low concentrations.
Liquid or solid irritant characteristics: Severe skin irritant. Causes second- and third-degree burns on short contact; very injurious to the eyes.
Odor threshold: Less than 0.0099–0.46 ppm.
IDLH value: 3 ppm.

FIRE DATA
Behavior in fire: Containers may explode.

CHEMICAL REACTIVITY
Reactivity with water: Forms acidic hydrobromic acid solution and toxic fumes.
Binary reactants: Reacts violently with aluminum. May cause fire in contact with wood, cotton, straw. Iron, steel, stainless steel 316, and copper are corroded by bromine and are especially subject to attack by wet bromine. Of the plastics, only those which are highly fluorinated resist bromine attack. Titanium and Hastelloy C are chemical resistant.

ENVIRONMENTAL DATA
Food chain concentration potential: Negative; unlikely to accumulate.
Water pollution: Harmful to aquatic life in very low concentrations. May be dangerous if it enters nearby water intakes; notify operators. Notify local health and wildlife officials.

Response to discharge: Issue warning–poison, air contaminant. Restrict access. Evacuate area. Disperse and flush.

SHIPPING INFORMATION
Grades of purity: Commercial, technical; **Storage temperature:** Cool, but above 20°F/–7°C to prevent freezing; **Stability during transport:** Stable.

NAS HAZARD CLASSIFICATION FOR BULK WATER TRANSPORTATION
FIRE: 0
HEALTH: Vapor irritant: 4; Liquid or solid irritant: 4; Poisons: 4
WATER POLLUTION: Human toxicity: 4; Aquatic toxicity: 3; Aesthetic effect: 3
REACTIVITY: Other chemicals: 4; Water: 1; Self-reaction: 0

PHYSICAL AND CHEMICAL PROPERTIES
Physical state @ 59°F/15°C and 1 atm: Liquid.
Molecular weight: 159.81
Boiling point @ 1 atm: 138°F/58.8°C/332°K.
Melting/Freezing point: 19°F/–7.2°C/266°K.
Specific gravity (water = 1): 3.12 @ 68°F/20°C (liquid).
Liquid surface tension: 41 dynes/cm = 0.041 N/m @ 68°F/20°C.
Relative vapor density (air = 1): 5.5 @ 68°F/20°C.
Ratio of specific heats of vapor (gas): 1.3
Latent heat of vaporization: 80.6 Btu/lb = 44.8 cal/g = 1.88 x 10^5 J/kg.
Heat of fusion: 16.1 cal/g.
Vapor pressure: 175 mm.

BROMINE PENTAFLUORIDE   REC. B:2300

SYNONYMS: BROMINE FLUORIDE; PENTAFLUORURO de BROMO (Spanish)

IDENTIFICATION
CAS Number: 7789-30-2
Formula: BrF$_5$
DOT ID Number: UN 1745; DOT Guide Number: 144
Proper Shipping Name: Bromine pentafluoride

DESCRIPTION: Colorless liquefied gas. Highly irritating odor. Reacts explosively with water forming hydrofluoric acid and hydrogen fluoride gas.

Poison! • Do not use water • Breathing the gas can kill you. Skin or eye contact can cause severe burns and blindness • Firefighting gear (including SCBA) does not provide adequate protection. If exposure occurs, remove and isolate gear immediately and thoroughly decontaminate personnel • Strong oxidizer which may react spontaneously with low flash point organics or reducing agents. Heat forms oxygen; will increase the activity of an existing fire • May cause fire or explosion contact with combustibles (wood, paper, oil, clothing, etc.) • Containers may BLEVE when exposed to fire • Concentrated dust in confined areas (e.g., tanks, sewers, buildings) may explode when exposed to fire • Toxic products of combustion may include hydrogen fluoride and hydrogen bromide.

Hazard Classification (based on NFPA-704 Rating System)
Health Hazards (Blue): 4; Flammability (Red): 0; Reactivity (Yellow): 3; Special Notice (White): Water reactive, OXY

Bromine trifluoride

EMERGENCY RESPONSE: See Appendix A (144)
Evacuation:
Public safety: Isolate the area of spill or leak for at least 50 to 100 meters (160 to 300 feet) in all directions.
Spills: Small spill–First: Isolate in all directions 60 meters (200 feet) (land); 30 meters (100 feet) (water); Then: Protect persons downwind, DAY: 0.5 km (0.3 mile) (land); 0.2 km (0.1 mile) (water); NIGHT: 1.3 km (0.8 mile) (land); 0.8 km (0.5 mile) (water). Large spill–First: Isolate in all directions: 245 meters (800 feet) (land); 215 meters (700 feet) (water); Then: Protect persons downwind, DAY: 2.3 km (1.4 miles) (land); 1.9 km (1.2 miles) (water); NIGHT: 5.0 km (3.1 miles) (land); 4.2 km (2.6 miles) (water).
Fire: If any large container is involved in fire, isolate for at least 800 meters (½ mile) in all directions; also, consider initial evacuation for 800 meters (½ mile) in all directions.

EXPOSURE

Short-term effects: *GET MEDICAL ATTENTION IMMEDIATELY FOR ANY EXPOSURE TO THIS CHEMICAL, EVEN IF NO ADVERSE EFFECTS ARE EVIDENT.* **Vapor:** Irritating to eyes, nose, and throat. *IF INHALED*, will, will cause coughing or difficult breathing. *IF IN EYES*, hold eyelids open and flush with plenty of water. *IF BREATHING HAS STOPPED,* give artificial respiration; *avoid mouth-to-mouth resuscitation; use bag/mask apparatus.* IF breathing is difficult, administer oxygen. **Liquid:** Will burn skin and eyes. Harmful if swallowed. Remove contaminated clothing and shoes. Flush affected areas with plenty of water. *IF IN EYES*, hold eyelids open and flush with plenty of water. *IF SWALLOWED* and victim is *CONSCIOUS AND ABLE TO SWALLOW,* have victim drink 4 to 8 ounces of water. *IF SWALLOWED* and victim is *UNCONSCIOUS OR HAVING CONVULSIONS,* do nothing except keep victim warm.
Medical observation is recommended for 24 to 48 hours after breathing overexposure, as pulmonary edema may be delayed. As first aid for pulmonary edema, consider administering a corticosteroid spray. Cigarette smoking may exacerbate pulmonary injury and should be discouraged for at least 72 hours following exposure.

HEALTH HAZARDS

Personal protective equipment (PPE): A-Level PPE. OSHA Table Z-1-A air contaminant. Self-contained breathing apparatus, acid suit, and gloves.
Recommendations for respirator selection: NIOSH/OSHA as F. *12.5 mg/m³:* DM (any dust and mist respirator). *25 mg/m³:* DMXSQ* (any dust and mist respirator except single-use and quarter-mask respirators); or SA* (any supplied-air respirator). *62.5 mg/m³:* SA:CF* [any supplied-air respirator operated in a continuous-flow mode)]; or PAPRDM*⁺ *if not present as a fume* (any powered, air-purifying respirator with a dust and mist filter). *125 mg/m³:* HiEF + (any air-purifying, full-facepiece respirator with a high-efficiency particulate filter); or SCBAF (any self-contained breathing apparatus with a full facepiece); or SAF (any supplied-air respirator with a full facepiece). *250 mg/m³:* SA:PD,PP (any supplied-air respirator operated in a pressure-demand or other positive-pressure mode). *EMERGENCY OR PLANNED ENTRY INTO UNKNOWN CONCENTRATIONS OR IDLH CONDITIONS:* SCBAF:PD,PP (any self-contained breathing apparatus that has a full faceplate and is operated in a pressure-demand or other positive-pressure mode); or SAF:PD,PP:ASCBA (any supplied-air respirator that has a full facepiece and is operated in a pressure-demand or other positive-pressure mode in combination with an auxiliary, self-contained breathing apparatus operated in a pressure-demand or other positive-pressure mode). *ESCAPE:* HiEF+ (any air-purifying, full-facepiece respirator with a high-efficiency particulate filter); or SCBAE (any appropriate escape-type, self-contained breathing apparatus). *Notes:* *Substance reported to cause eye irritation or damage; may require eye protection. ⁺May need acid gas sorbent.
Exposure limits (TWA unless otherwise noted): ACGIH TLV 0.1 ppm (0.72 mg/m³); NIOSH REL 0.1 ppm (0.7 mg/m³).
Toxicity by ingestion: Boron compounds are toxic to humans.
Long-term health effects: May effect the central nervous system.
Vapor (gas) irritant characteristics: Severe eye and respiratory tract irritant.
Liquid or solid irritant characteristics: Severe eye and skin irritant.

FIRE DATA

Fire extinguishing agents not to be used: Water used directly on material will cause a violent reaction.
Behavior in fire: Containers may burst when exposed to heat of fire.

CHEMICAL REACTIVITY

Reactivity with water: Reacts violently with water, evolving hydrogen fluoride, an extremely irritating and corrosive gas.
Binary reactants: Reacts with acids, alkalies, halogens, arsenic, selenium, salts, sulfur, glass, organic materials. A strong oxidizer that may react violently with many metals, common materials such as cotton and straw, and materials of construction such as wood, glass, some plastics.
Neutralizing agents for acids and caustics: Flush with water.

ENVIRONMENTAL DATA

Food chain concentration potential: Negative; unlikely to accumulate.
Water pollution: Effect of low concentrations on aquatic life is unknown. May be dangerous if it enters nearby water intakes; notify operators. Notify local health and wildlife officials. **Response to discharge:** Issue warning–corrosive, air contaminant. Restrict access. Evacuate area. Disperse and flush.

SHIPPING INFORMATION

Grades of purity: Technical, 98.0+%; Pure, 99.9%; **Storage temperature:** Ambient; **Inert atmosphere:** Padded; **Venting:** Safety relief; **Stability during transport:** Stable.

PHYSICAL AND CHEMICAL PROPERTIES

Physical state @ 59°F/15°C and 1 atm: Liquid.
Molecular weight: 174.9
Boiling point @ 1 atm: 106°F/41°C/314°K.
Melting/Freezing point: –76°F/–60°C/213°K.
Critical temperature: 387°F/197°C/470°K.
Specific gravity (water = 1): 2.48 @ 68°F/20°C (liquid).
Relative vapor density (air = 1): 6.03
Ratio of specific heats of vapor (gas): 1.089 @ 77°F/25°C.
Latent heat of vaporization: 76.8 Btu/lb = 42.7 cal/g = 1.79×10^5 J/kg.
Heat of fusion: 7.07 cal/g.
Vapor pressure: 328 mm.

BROMINE TRIFLUORIDE　　　　　　　　　　**REC. B:2350**

SYNONYMS: BORON FLUORIDE; BROMINE FLUORIDE; TRIFLUOROBORANE; TRIFLUORURO de BROMO (Spanish)

Bromine trifluoride

IDENTIFICATION
CAS Number: 7787-71-5
Formula: BrF_3
DOT ID Number: UN 1746; DOT Guide Number: 144
Proper Shipping Name: Bromine trifluoride

DESCRIPTION: Colorless to gray-yellow liquid. Extremely irritating odor. Reacts violently with water, even at low temperatures, forming hydrogen fluoride vapors.

Poison! • Corrosive • Breathing the vapor, skin or eye contact, or swallowing the material can kill you • Firefighting gear (including SCBA) does not provide adequate protection. If exposure occurs, remove and isolate gear immediately and thoroughly decontaminate personnel • *Exposure of cylinders to fire and flame or elevated temperatures may cause cylinders to rupture or cause frangible disc to burst, releasing entire contents of cylinder. Ruptures or venting cylinders may rocket through buildings and/or travel a considerable distance* • Strong oxidizer which may react spontaneously with low flash point organics or reducing agents. Heat forms oxygen; will increase the activity of an existing fire • May cause fire or explosion contact with combustibles (wood, paper, oil, clothing, etc.) • Vapors are heavier than air and will collect and stay in low areas • Vapors in confined areas (e.g., tanks, sewers, buildings) may explode when exposed to fire • Toxic products of combustion may include hydrogen bromide and hydrogen fluoride fumes • *DO NOT ATTEMPT RESCUE* • Freezes at 48°F/8.8°C.

Hazard Classification (based on NFPA-704 Rating System)
Health Hazards (Blue): 4; Flammability (Red): 0; Reactivity (Yellow): 3; Special Notice (White): OXY, Water reactive

EMERGENCY RESPONSE: See Appendix A (144)
Evacuation:
Public safety: Isolate the area of spill or leak for at least 50 to 100 meters (160 to 300 feet) in all directions.
Spills: Small spill–First: Isolate in all directions 30 meters (100 feet) (land or water); Then: Protect persons downwind, DAY: 0.2 km (0.1 mile) (land or water); NIGHT: 0.3 km (0.2 mile) (land); 0.6 km (0.4 mile) (water). Large spill–First: Isolate in all directions 60 meters (200 feet) (land); 185 meters (600 feet) (water); Then: Protect persons downwind, DAY: 0.3 km (0.2 mile) (land); 2.1 km (1.3 miles) (water); NIGHT: 0.8 km (0.5 mile) (land); 5.5 km (3.4 miles) (water).
Fire: If any large container is involved in fire, isolate for at least 800 meters (½ mile) in all directions; also, consider initial evacuation for 800 meters (½ mile) in all directions.

EXPOSURE
Short-term effects: *SEEK MEDICAL ATTENTION.* **Vapor:** *POISONOUS IF INHALED.* Irritating to eyes, nose, and throat. Move victim to fresh air. *IF IN EYES*, hold eyelids open and flush with plenty of water. *IF BREATHING HAS STOPPED*, give artificial respiration; *avoid mouth-to-mouth resuscitation; use bag/mask apparatus.* IF breathing is difficult, administer oxygen. Inhalation may cause lung edema to develop. **Liquid:** *POISONOUS IF SWALLOWED.* Will burn skin and eyes. Remove contaminated clothing and shoes. Flush affected areas with plenty of water. *IF IN EYES*, hold eyelids open and flush with plenty of water. *IF SWALLOWED* and victim is *CONSCIOUS AND ABLE TO SWALLOW*, have victim drink 4 to 8 ounces of water. **Do NOT induce vomiting.**
Note to physician or authorized medical personnel: Medical observation is recommended for 24 to 48 hours after breathing overexposure, as pulmonary edema may be delayed. As first aid for pulmonary edema, consider administering a corticosteroid spray. Cigarette smoking may exacerbate pulmonary injury and should be discouraged for at least 72 hours following exposure.

HEALTH HAZARDS
Personal protective equipment (PPE): A-Level PPE.
Recommendations for respirator selection: NIOSH/OSHA as F. *12.5 mg/m³:* DM (any dust and mist respirator). *25 mg/m³:* DMXSQ* (any dust and mist respirator except single-use and quarter-mask respirators); or SA* (any supplied-air respirator). *62.5 mg/m³:* SA:CF* [any supplied-air respirator operated in a continuous-flow mode)]; or PAPRDM*+ *if not present as a fume* (any powered, air-purifying respirator with a dust and mist filter). *125 mg/m³:* HiEF+ (any air-purifying, full-facepiece respirator with a high-efficiency particulate filter); or SCBAF (any self-contained breathing apparatus with a full facepiece); or SAF (any supplied-air respirator with a full facepiece). *250 mg/m³:* SA:PD,PP (any supplied-air respirator operated in a pressure-demand or other positive-pressure mode). *EMERGENCY OR PLANNED ENTRY INTO UNKNOWN CONCENTRATIONS OR IDLH CONDITIONS:* SCBAF:PD,PP (any self-contained breathing apparatus that has a full faceplate and is operated in a pressure-demand or other positive-pressure mode); or SAF:PD,PP:ASCBA (any supplied-air respirator that has a full facepiece and is operated in a pressure-demand or other positive-pressure mode in combination with an auxiliary, self-contained breathing apparatus operated in a pressure-demand or other positive-pressure mode). *ESCAPE:* HiEF+ (any air-purifying, full-facepiece respirator with a high-efficiency particulate filter); or SCBAE (any appropriate escape-type, self-contained breathing apparatus). *Notes:* *Substance reported to cause eye irritation or damage; may require eye protection. +May need acid gas sorbent.
Exposure limits (TWA unless otherwise noted): ACGIH TLV 2.5 mg/m³ as fluorides; NIOSH/OSHA 2.5 mg/m³ as inorganic fluorides.
Vapor (gas) irritant characteristics: Severe respiratory tract irritant.
Liquid or solid irritant characteristics: Severe eye and skin irritant.
IDLH value: 250 mg/m³ as F.

FIRE DATA
Fire extinguishing agents not to be used: Water, foam

CHEMICAL REACTIVITY
Reactivity with water: Reacts violently to generate toxic hydrogen fluoride gas (hydrofluoric acid).
Binary reactants: Reacts with organic matter, acids, halogens, salts, alkalies, metal oxides. Will cause severe corrosion of common metals and glass. May cause fire in contact with organic materials such as wood, cotton, or straw.
Neutralizing agents for acids and caustics: Flush with water, rinse with sodium bicarbonate or lime solution.

ENVIRONMENTAL DATA
Food chain concentration potential: Negative; unlikely to accumulate.
Water pollution: Effect of low concentrations on aquatic life is unknown. May be dangerous if it enters nearby water intakes; notify operators. Notify local health and wildlife officials. **Response to discharge:** Issue warning–corrosive, air contaminant. Restrict access. Evacuate area. Disperse and flush.

SHIPPING INFORMATION
Grades of purity: 98+%; **Storage temperature:** Ambient; **Inert atmosphere:** None; **Venting:** Pressure-vacuum; **Stability during transport:** Stable.

PHYSICAL AND CHEMICAL PROPERTIES
Physical state @ 59°F/15°C and 1 atm: Liquid.
Molecular weight: 136.9
Boiling point @ 1 atm: 275°F/135°C/408°K.
Melting/Freezing point: 47.8°F/8.8°C/282.0°K.
Critical temperature: (estimate) 621°F/327°C/600°K.
Specific gravity (water = 1): 2.81 @ 68°F/20°C (liquid).
Liquid surface tension: 36.3 dynes/cm = 0.0363 N/m @ 68°F/20°C.
Relative vapor density (air = 1): 4.7
Ratio of specific heats of vapor (gas): 1.1428
Latent heat of vaporization: 130 Btu/lb = 74 cal/g = 3.1×10^5 J/kg.

BROMOACETONE   REC. B:2400

SYNONYMS: ACETYLMETHYL BROMIDE; ACETONYL BROMIDE; BROMOACETONA (Spanish); BPOMOPROPANE; 1-BROMO-2-PROPANONE; BROMO-2-PROPANONE; BROMOMETHYL METHYL KETONE; MONOBROMOACETONE; RCRA No. P017

IDENTIFICATION
CAS Number: 598-31-2
Formula: C_3NO_5BrO
DOT ID Number: UN 1569; **DOT Guide Number:** 131
Proper Shipping Name: Bromoacetone
Reportable Quantity (RQ): (CERCLA) 1000 lb/454 kg

DESCRIPTION: Colorless liquid, turns violet on exposure to air. Pungent odor. Sinks in water; practically insoluble.

A tear gas • Corrosive to the skin, eyes, and respiratory tract; contact with skin or eyes can cause burns and vision impairment; inhalation symptoms may be delayed • Firefighting gear (including SCBA) does not provide adequate protection. If exposure occurs, remove and isolate gear immediately and thoroughly decontaminate personnel • Containers may BLEVE when exposed to fire • Decomposes in temperatures above 276°F/136°C • Toxic products of combustion may include hydrogen bromide.

Hazard Classification (based on NFPA-704 Rating System)
Health Hazards (Blue): 2; Flammability (Red): 2; Reactivity (Yellow): 0

EMERGENCY RESPONSE: See Appendix A (131)
Evacuation:
Public safety: Isolate spill area for at least 100 to 200 meters (330 to 660 feet) in all directions.
Spill: Small spill–First: Isolate in all directions 30 meters (100 feet); Then: Protect persons downwind, DAY: 0.2 km (0.1 mile); NIGHT: 0.3 km (0.2 mile). Large spill–First: Isolate in all directions 95 meters (300 feet). Then: Protect persons downwind, DAY: 0.8 km (0.5 mile); NIGHT: 1.9 km (1.2 miles).
Fire: Isolate for 800 meters (½ mile) in all directions, especially if tank, rail car, or tank truck is involved in fire.

EXPOSURE
Short-term effects: *SEEK MEDICAL ATTENTION.* **Vapor:** Extremely irritating to the eyes, nose, throat, and upper respiratory system. *MAY BE HARMFUL IF INHALED OR ABSORBED THROUGH THE SKIN.* Move victim to fresh air. *IF BREATHING HAS STOPPED,* give artificial respiration. IF breathing is difficult, administer oxygen. **Liquid:** Corrosive to eyes, skin, and upper respiratory tract. *HARMFUL IF SWALLOWED OR ABSORBED THROUGH THE SKIN. IF IN EYES,* hold eyelids open and flush with water for at least 15 minutes. Remove contaminated clothing and shoes, flush affected areas with plenty of water for at least 15 minutes. *IF SWALLOWED*: Do nothing except keep victim warm. **Do NOT induce vomiting.**
Note to physician or authorized medical personnel: Medical observation is recommended for 24 to 48 hours after breathing overexposure, as pulmonary edema may be delayed. As first aid for pulmonary edema, consider administering a corticosteroid spray. Cigarette smoking may exacerbate pulmonary injury and should be discouraged for at least 72 hours following exposure.

HEALTH HAZARDS
Personal protective equipment (PPE): A-Level PPE. Self-contained breathing apparatus, chemical-resistant gloves, rubber boots, full protective clothing.
Vapor (gas) irritant characteristics: Vapors cause severe irritation of eyes and throat and can cause eye and lung injury. They cannot be tolerated even at low concentrations.
Liquid or solid irritant characteristics: Severe skin irritant. Causes second- and third-degree burns on short contact and is very injurious to the eyes.

FIRE DATA
Flash point: 113°F/45°C.
Fire extinguishing agents not to be used: Water may be ineffective.

CHEMICAL REACTIVITY
Binary reactants: Violent reaction with strong oxidizers.

ENVIRONMENTAL DATA
Water pollution: DOT Appendix B, §172.101–marine pollutant. Toxic to aquatic life in low concentrations. May be dangerous if it enters nearby water intakes; notify operators. Notify local health and wildlife officials. Notify operators of local water intakes.
Response to discharge: Evacuate area. Issue warning–high flammability, poison, corrosive. Should be removed. Chemical and physical treatment.

SHIPPING INFORMATION
Inert atmosphere: None; **Venting:** None; **Stability during transport:** Stable.

NAS HAZARD CLASSIFICATION FOR BULK WATER TRANSPORTATION
FIRE: 2
HEALTH: Vapor irritant: 4; Liquid or solid irritant: 4; Poisons:–
WATER POLLUTION: Human toxicity:–; Aquatic toxicity: 3; Aesthetic effect: 3
REACTIVITY: Other chemicals: 1; Water: 0; Self-reaction: 0

PHYSICAL AND CHEMICAL PROPERTIES
Physical state @ 59°F/15°C and 1 atm: Liquid.
Molecular weight: 136.98
Boiling point @ 1 atm: 277°F/136°C/409°K.
Melting/Freezing point: –33.7°F/–36.5°C/236.7°K.
Specific gravity (water = 1): 1.634 at 23°C.

Relative vapor density (air = 1): 4.72
Vapor pressure: 9 mm.

BROMOACETYL BROMIDE REC. B:2450

SYNONYMS: BROMOETHANOYL BROMIDE

IDENTIFICATION
CAS Number: 598-21-0
Formula: $BrCH_2CO_2Br$
DOT ID Number: UN 2513; DOT Guide Number: 156
Proper Shipping Name: Bromoacetyl bromide

DESCRIPTION: Colorless to pale yellow liquid. Sharp, extremely irritating odor. Sinks in water; reacts violently forming hydrobromic acid. Irritating vapor is produced.

Avoid the use of water • Corrosive to the skin, eyes or respiratory tract; inhalation may cause lung edema; effects of inhalation may be delayed • Firefighting gear (including SCBA) does not provide adequate protection. If exposure occurs, remove and isolate gear immediately and thoroughly decontaminate personnel • Toxic products of combustion may include hydrogen bromide and bromophosgene vapors. Highly irritating (tear gas) vapors released when heated. Do not put yourself in danger by entering a contaminated area to rescue a victim.

Hazard Classification (based on NFPA-704 Rating System)
Health Hazards (Blue): 3; Flammability (Red): 0; Reactivity (Yellow): 0; Special Notice (White): Water reactive

EMERGENCY RESPONSE: See Appendix A (156)
Evacuation:
Public safety: Isolate the area of spill or leak for at least 50 to 100 meters (160 to 330 feet) in all directions.
Spill: Increase, in the downwind direction, as necessary, the distance shown under "Public Safety."
Fire: If tank, rail car, or tank truck is involved in fire, isolate for at least 800 meters (½ mile) in all directions; also, consider initial evacuation for 800 meters (½ mile) in all directions.

EXPOSURE
Short-term effects: *SEEK MEDICAL ATTENTION.* **Vapor:** Irritating to eyes, nose, and throat. Move victim to fresh air. *IF BREATHING HAS STOPPED,* give artificial respiration; *avoid mouth-to-mouth resuscitation; use bag/mask apparatus.* If breathing is difficult, administer oxygen. **Liquid:** Will burn skin and eyes. Harmful if swallowed. Remove contaminated clothing and shoes. Flush affected areas with plenty of water. *IF IN EYES,* hold eyelids open and flush with plenty of water. *IF SWALLOWED* and victim is *CONSCIOUS AND ABLE TO SWALLOW,* have victim drink 4 to 8 ounces of water. **Do NOT induce vomiting.** *Note to physician or authorized medical personnel:* Medical observation is recommended for 24 to 48 hours after breathing overexposure, as pulmonary edema may be delayed. As first aid for pulmonary edema, consider administering a corticosteroid spray. Cigarette smoking may exacerbate pulmonary injury and should be discouraged for at least 72 hours following exposure.

HEALTH HAZARDS
Personal protective equipment (PPE): A-Level PPE. Acid-type canister mask; self-contained breathing apparatus (full face); rubber gloves and full protective clothing.
Vapor (gas) irritant characteristics: Vapors are moderately irritating, such that personnel will not usually tolerate moderate or high vapor concentrations.
Liquid or solid irritant characteristics: Severe skin irritant. Causes second- and third-degree burns on short contact and is very injurious to the eyes.

FIRE DATA
Fire extinguishing agents not to be used: Do not use water on adjacent fires.

CHEMICAL REACTIVITY
Reactivity with water: Reacts vigorously to generate hydrogen bromide (hydrobromic acid).
Binary reactants: Will react with surface moisture to generate hydrogen bromide, which is corrosive to metals.
Neutralizing agents for acids and caustics: Flush with water, rinse with sodium bicarbonate or lime solution.

ENVIRONMENTAL DATA
Food chain concentration potential: Negative; unlikely to accumulate.
Water pollution: Effect of low concentrations on aquatic life is unknown. May be dangerous if it enters nearby water intakes; notify operators. Notify local health and wildlife officials. **Response to discharge:** Issue warning-air contaminant, corrosive. Restrict access. Disperse and flush.

SHIPPING INFORMATION
Grades of purity: Commercial; **Storage temperature:** Ambient; **Inert atmosphere:** None; **Venting:** Pressure-vacuum; **Stability during transport:** Stable.

NAS HAZARD CLASSIFICATION FOR BULK WATER TRANSPORTATION
FIRE: 0
HEALTH: Vapor irritant: 3; Liquid or solid irritant: 4; Poisons: 3
WATER POLLUTION: Human toxicity: 2; Aquatic toxicity: 2; Aesthetic effect: 2
REACTIVITY: Other chemicals: 2; Water: 3; Self-reaction: 0

PHYSICAL AND CHEMICAL PROPERTIES
Physical state @ 59°F/15°C and 1 atm: Liquid.
Molecular weight: 201.85
Boiling point @ 1 atm: 298°F/148°C/421°K.
Specific gravity (water = 1): 2.317 @ 68°F/20°C (liquid).

BROMOBENZENE REC. B:2500

SYNONYMS: BROMOBENCENO (Spanish); BROMOBENZOL; EEC No. 602-060-00-9; MONOBROMOBENZENE; PHENYL BROMIDE

IDENTIFICATION
CAS Number: 108-86-1
DOT ID Number: UN 2514; DOT Guide Number: 129
Proper Shipping Name: Bromobenzene

DESCRIPTION: Colorless liquid. Pleasant aromatic odor. Sinks in water; insoluble

Combustible • Containers may BLEVE when exposed to fire •Vapors may form explosive mixture with air • Vapors are heavier than air and will collect and stay in low areas • Vapors may travel long distances to ignition sources and flashback • Vapors in confined areas (e.g., tanks, sewers, buildings) may explode when exposed to fire • Irritating to the skin, eyes, and respiratory tract • Toxic products of combustion may include hydrogen bromide gas.

Hazard Classification (based on NFPA-704 Rating System)
Health Hazards (Blue): 2; Flammability (Red): 2; Reactivity (Yellow): 0

EMERGENCY RESPONSE: See Appendix A (129)
Evacuation:
Public safety: Isolate spill area for at least 50 to 100 meters (160 to 330 feet) in all directions.
Spill: Large spill–Consider initial downwind evacuation for at least 300 meters (1000 feet).
Fire: Isolate for 800 meters (½ mile) in all directions, especially if tank, rail car, or tank truck is involved in fire.

EXPOSURE
Short-term effects: *SEEK MEDICAL ATTENTION.* **Liquid:** Irritating to skin and eyes. Affects the nervous system. *HARMFUL IF ABSORBED THROUGH THE SKIN.* Harmful if swallowed; droplets in lungs may cause pneumonia. Remove contaminated clothing and shoes. Flush affected areas with plenty of water. *IF IN EYES,* hold eyelids open and flush with plenty of water. *IF SWALLOWED* and victim is *CONSCIOUS AND ABLE TO SWALLOW,* have victim drink 4 to 8 ounces of water. Chemical pneumonitis may develop.

HEALTH HAZARDS
Personal protective equipment (PPE): B-Level PPE. Protective materials with good to excellent resistance: PV alcohol, Viton®.
Long-term health effects: Possible organ damage; liver.
Vapor (gas) irritant characteristics: Eye and respiratory tract irritant.
Liquid or solid irritant characteristics: Eye and skin irritant.

FIRE DATA
Flash point: 124°F/51°C (cc).
Autoignition temperature: 1049°F/565°C/838°K.
Electrical hazard: Flow or agitation of substance may generate electrostatic charges due to low conductivity.
Burning rate: 3.8 mm/min

CHEMICAL REACTIVITY
Binary reactants: Reacts with strong oxidants (possibly violent); alkali metals.

ENVIRONMENTAL DATA
Food chain concentration potential: Log P_{ow} = 3.0. Values > 3.0 are likely to bioconcentrate in aquatic organisms and other living tissue, especially in fats.
Water pollution: DOT Appendix B, §172.101–marine pollutant. Effect of low concentrations on aquatic life is unknown. May be dangerous if it enters nearby water intakes; notify operators. Notify local health and wildlife officials. **Response to discharge:** Issue warning–water contaminant. Should be removed. Chemical and physical treatment.

SHIPPING INFORMATION
Grades of purity: Commercial; **Storage temperature:** Ambient; **Inert atmosphere:** None; **Venting:** Open (flame arrester); **Stability during transport:** Stable.

PHYSICAL AND CHEMICAL PROPERTIES
Physical state @ 59°F/15°C and 1 atm: Liquid.
Molecular weight: 157
Boiling point @ 1 atm: 313°F/156°C/429°K.
Melting/Freezing point: –23.1°F/–30.6°C/242.6°K.
Critical temperature: 747°F/397°C/670°K.
Critical pressure: 655 psia = 44.6 atm = 4.52 MN/m².
Specific gravity (water = 1): 1.49 @ 77°F/25°C (liquid).
Liquid surface tension: 36 dynes/cm = 0.036 N/m @ 68°F/20°C.
Liquid water interfacial tension: (estimate) 30 dynes/cm = 0.030 N/m @ 68°F/20°C.
Relative vapor density (air = 1): 5.4
Ratio of specific heats of vapor (gas): 1.0931
Latent heat of vaporization: 104 Btu/lb = 58 cal/g = 2.4 x 10^5 J/kg.
Heat of combustion: –8,510 Btu/lb = –4,730 cal/g = –198 x 10^5 J/kg.
Heat of fusion: 16.17 cal/g.
Vapor pressure: 3 mm.

1-BROMOBUTANE   REC. B:2550

SYNONYMS: 1-BROMOBUTANO (Spanish); *n*-BROMOBUTANE; BUTYL BROMIDE; *n*-BUTYL BROMIDE

IDENTIFICATION
CAS Number: 109-65-9
Formula: C_4H_9Br; 1-C_4H_9Br
DOT ID Number: UN 1126; DOT Guide Number: 129
Proper Shipping Name: *n*-Butyl bromide

DESCRIPTION: Colorless to pale yellow liquid. Odorless. Sinks in water; insoluble.

Highly flammable • Containers may BLEVE when exposed to fire • Vapors may form explosive mixture with air • Vapors are heavier than air and will collect and stay in low areas • Vapors may travel long distances to ignition sources and flashback • Vapors in confined areas (e.g., tanks, sewers, buildings) may explode when exposed to fire • Irritating to the skin, eyes, and respiratory tract • Toxic products of combustion may include toxic hydrogen bromide vapors.

Hazard Classification (based on NFPA-704 Rating System)
Health Hazards (Blue): 2; Flammability (Red): 3; Reactivity (Yellow): 0

EMERGENCY RESPONSE: See Appendix A (129)
Evacuation:
Public safety: Isolate spill area for at least 50 to 100 meters (160 to 330 feet) in all directions.
Spill: Large spill–Consider initial downwind evacuation for at least 300 meters (1000 feet).
Fire: Isolate for 800 meters (½ mile) in all directions, especially if tank, rail car, or tank truck is involved in fire.

EXPOSURE
Short-term effects: *SEEK MEDICAL ATTENTION.* **Vapor:** May be harmful if inhaled or absorbed through skin. Irritating to the eyes, upper respiratory tract, nose and throat. Move victim to fresh air. *IF BREATHING HAS STOPPED,* give artificial respiration. IF breathing is difficult, administer oxygen. **Liquid:** Irritating to eyes and skin. *MAY BE HARMFUL IF SWALLOWED OR ABSORBED THROUGH THE SKIN.* Remove contaminated clothing and shoes. Flush affected areas with water. Wash with soap and water. *IF IN EYES:* immediately flush with water for at least 15 minutes. *IF*

2-Bromobutane

SWALLOWED and victim is CONSCIOUS: Have victim drink water and induce vomiting. IF SWALLOWED and victim is UNCONSCIOUS OR HAVING CONVULSIONS: Do nothing except keep victim warm.

HEALTH HAZARDS
Personal protective equipment (PPE): B-Level PPE. Approved respirator, chemical safety goggles, rubber gloves. Chemical protective material(s) reported to offer minimal protection: nitrile, Viton®.
Toxicity by ingestion: Grade 2: LD_{50} = 4.45 g/kg (rat, ipr).
Vapor (gas) irritant characteristics: Eye and respiratory tract irritant. Vapors cause smarting of the eyes or respiratory system if present in high concentrations. The effect is temporary.
Liquid or solid irritant characteristics: Eye and skin irritant. If spilled on clothing and allowed to remain, may cause smarting and reddening of skin.

FIRE DATA
Flash point: 65°F/18°C (cc).
Flammable limits in air: LEL: 2.5%; UEL: 6.6% (both @ 212°F/100°C).
Fire extinguishing agents not to be used: Water may be ineffective against fire.
Autoignition temperature: 509°F/265°C/538°K.

CHEMICAL REACTIVITY
Binary reactants: Oxidizers, strong acids.

ENVIRONMENTAL DATA
Water pollution: Effects of low concentrations on aquatic life are not known. May be dangerous if it enters nearby water intakes; notify operators. Notify local health and wildlife officials.
Response to discharge: Issue warning–high flammability. Should be removed. Chemical and physical treatment.

SHIPPING INFORMATION
Grades of purity: 99%; **Venting:** None; **Stability during transport:** Stable.

NAS HAZARD CLASSIFICATION FOR BULK WATER TRANSPORTATION
FIRE: 3
HEALTH: Vapor irritant: 1; Liquid or solid irritant: 1; Poisons: 2
WATER POLLUTION: Human toxicity: 2; Aquatic toxicity:–; Aesthetic effect: 1
REACTIVITY: Other chemicals: 1; Water: 0; Self-reaction: 0

PHYSICAL AND CHEMICAL PROPERTIES
Physical state @ 59°F/15°C and 1 atm: Liquid.
Molecular weight: 137.04
Boiling point @ 1 atm: 214°F/101.4°C/375°K.
Melting/Freezing point: –170°F/–112.4°C/161°K.
Specific gravity (water = 1): 1.276 @ 68°F/20°C.
Liquid surface tension: 26.5 dyne/cm = 0.026 N/m @ 68°F/20°C.
Relative vapor density (air = 1): 4.72
Vapor pressure: (Reid) 1.5 psia.

2-BROMOBUTANE REC. B:2600

SYNONYMS: 2-BROMOBUTANO (Spanish); *sec*-BUTYL BROMIDE; METHYL ETHYL BROMO-METHANE

IDENTIFICATION
CAS Number: 78-76-2
Formula: C_4H_9Br; 2-C_4H_9Br
DOT ID Number: UN 2339; DOT Guide Number: 130
Proper Shipping Name: 2-Bromobutane

DESCRIPTION: Colorless liquid. Sinks in water.

Hazard Classification (based on NFPA-704 Rating System)
Health Hazards (Blue): 2; Flammability (Red): 3; Reactivity (Yellow): 0

Highly flammable • Containers may BLEVE when exposed to fire • Vapors may form explosive mixture with air • Vapors are heavier than air and will collect and stay in low areas • Vapors may travel long distances to ignition sources and flashback • Vapors in confined areas (e.g., tanks, sewers, buildings) may explode when exposed to fire • Irritating to the skin, eyes, and respiratory tract • Toxic products of combustion may include hydrogen bromide.

EMERGENCY RESPONSE: See Appendix A (130)
Evacuation:
Public safety: Isolate spill area for at least 50 to 100 meters (160 to 330 feet) in all directions.
Spill: Large spill–Consider initial downwind evacuation for at least 300 meters (1000 feet).
Fire: Isolate for 800 meters (½ mile) in all directions, especially if tank, rail car, or tank truck is involved in fire.

EXPOSURE
Short-term effects: *SEEK MEDICAL ATTENTION*. **Vapor:** *HARMFUL IF INHALED OR ABSORBED THROUGH THE SKIN*. Irritating to the eyes, nose, throat, and upper respiratory tract. Move victim to fresh air. *IF BREATHING HAS STOPPED*, give artificial respiration. IF breathing is difficult, administer oxygen. **Liquid:** *MAY BE HARMFUL IF SWALLOWED OR ABSORBED THROUGH THE SKIN*. Irritating to the eyes and skin. Remove contaminated clothing and shoes. *IF IN EYES*: immediately flush with running water for at least 15 minutes. *IF SWALLOWED* and victim is CONSCIOUS: Have victim drink water and induce vomiting. *IF SWALLOWED* and victim is UNCONSCIOUS OR HAVING CONVULSIONS: Do nothing except keep victim warm.

HEALTH HAZARDS
Personal protective equipment (PPE): B-Level PPE. Approved respirator, chemical-resistant gloves, safety goggles, other protective clothing. Chemical protective material(s) reported to offer minimal protection: nitrile, Viton®.
Toxicity by ingestion: Grade 2: TDLo = 3.0 g/kg (ipr, mouse).
Long-term health effects: Suspected carcinogen.
Vapor (gas) irritant characteristics: Eye and respiratory tract irritant. Vapors cause smarting of the eyes or respiratory system if present in high concentrations. The effect is temporary.
Liquid or solid irritant characteristics: Eye and skin irritant. If spilled on clothing and allowed to remain, may cause smarting and reddening of skin.

FIRE DATA
Flash point: 70°F/21°C (cc).
Fire extinguishing agents not to be used: Water may be ineffective against fire.

CHEMICAL REACTIVITY
Binary reactants: Violent reaction with strong acids, strong oxidizers.

ENVIRONMENTAL DATA
Water pollution: Effect of low concentrations on aquatic life are not known. May be dangerous if it enters nearby water intakes; notify operators. Notify local health and wildlife officials. Notify operators of local water intakes. **Response to discharge:** Issue warning–high flammability. Should be removed. Chemical and physical treatment.

SHIPPING INFORMATION
Grades of purity: 98%; **Inert atmosphere:** None; **Venting:** None; **Stability during transport:** Stable.

NAS HAZARD CLASSIFICATION FOR BULK WATER TRANSPORTATION
FIRE: 4
HEALTH: Vapor irritant: 1; Liquid or solid irritant: 1; Poisons: 2
WATER POLLUTION: Human toxicity: 2; Aquatic toxicity:; Aesthetic effect: 1
REACTIVITY: Other chemicals: 1; Water: 0; Self-reaction: 0

PHYSICAL AND CHEMICAL PROPERTIES
Physical state @ 59°F/15°C and 1 atm: Liquid.
Molecular weight: 137.04
Boiling point @ 1 atm: 196.5°F/91.4°C/365°K.
Melting/Freezing point: −169.4°F/−111.9°C/161.3°K.
Specific gravity (water = 1): 1.258 @ 68°F/20°C.
Liquid surface tension: 25.3 dyne/cm = 0.025 N/m @ 68°F/20°C.
Heat of fusion: 21.62 Btu/lb = 12.01 cal/g = 0.5 x 10^5 J/kg.

BROMOFORM   REC. B:2650

SYNONYMS: BROMOFORME (French); METHENYL TRIBROMIDE; METHANE, TRIBROMO-; METHYL TRIBROMIDE; TRIBROMMETHAN (German); FORMYL TRIBROMIDE; METHENYL TRIBROMIDE; METHYLENE TRIBROMIDE; TRIBROMOMETHANE; RCRA No. U225

IDENTIFICATION
CAS Number: 75-25-2
Formula: $CHBr_3$
DOT ID Number: UN 2515; DOT Guide Number: 159
Proper Shipping Name: Bromoform
Reportable Quantity (RQ): **(CERCLA)** 100 lb/45.4 kg

DESCRIPTION: Colorless to yellow liquid. Sweetish, chloroform-like odor. Sinks and very slowly mixes with water.

Irritating to the skin, eyes, and respiratory tract • May be a polymerization hazard • Liquid will attack some forms of plastics, rubber, and coatings • Toxic products of combustion may include hydrogen bromide and bromine.

Hazard Classification (based on NFPA-704 Rating System)
Health Hazards (Blue): 1; Flammability (Red): 0; Reactivity (Yellow): 0

EMERGENCY RESPONSE: See Appendix A (159)
Evacuation:
Public safety: Isolate the area of spill or leak for at least 25 to 50 meters (80- 160 feet) in all directions.
Spill: Consider initial downwind evacuation for at least 100 meters (330 feet); increase, in the downwind direction, as necessary.
Fire: If tank, rail car, or tank truck is involved in fire, isolate for at least 800 meters (½ mile) in all directions; also, consider initial evacuation for 800 meters (½ mile) in all directions.

EXPOSURE
Short-term effects: *SEEK MEDICAL ATTENTION*. **Vapor:** Harmful if inhaled; narcotic effects. Lachrymator; irritating to eyes, skin, and respiratory tract. Move victim to fresh air. *IF BREATHING HAS STOPPED*, give artificial respiration; *avoid mouth-to-mouth resuscitation; use bag/mask apparatus*. IF breathing is difficult, administer oxygen. **Liquid:** *HARMFUL IF SWALLOWED OR ABSORBED THROUGH THE SKIN*. Harmful if swallowed, contacts skin or eyes, or is absorbed through skin. *IF IN EYES OR ON SKIN*, flush with running water for at least 15 minutes; hold eyelids open if necessary. Remove and isolate contaminated clothing and shoes at the site. *IF SWALLOWED* and victim is *CONSCIOUS AND ABLE TO SWALLOW*, have victim take syrup of ipecac to induce vomiting. *IF SWALLOWED* and victim is *UNCONSCIOUS OR HAVING CONVULSIONS*, do nothing except keep victim warm.

HEALTH HAZARDS
Personal protective equipment (PPE): B-Level PPE. OSHA Table Z-1-A air contaminant. Full face organic vapor respirator. Chemical protective material(s) reported to offer minimal protection: Viton®.
Recommendations for respirator selection: NIOSH/OSHA
12.5 ppm: SA:CF£ (any supplied-air respirator operated in a continuous-flow mode); or PAPROV £[any powered, air-purifying respirator with organic vapor cartridge(s)]. *25 ppm:* CCRFOV [any chemical cartridge respirator with a full facepiece and organic vapor cartridge(s)]; or GMFOV [any air-purifying, full-facepiece respirator (gas mask) with a chin-style, front- or back-mounted acid gas canister]; or PAPRTOV£[any powered, air-purifying respirator with a tight-fitting facepiece and organic vapor cartridge(s)]; or SCBAF (any self-contained breathing apparatus with a full facepiece); or SAF (any supplied-air respirator with a full facepiece). *850 ppm:* SAF:PD,PP (any supplied-air respirator that has a full facepiece and is operated in a pressure-demand or other positive-pressure mode). *EMERGENCY OR PLANNED ENTRY INTO UNKNOWN CONCENTRATIONS OR IDLH CONDITIONS:* SCBAF:PD,PP (any self-contained breathing apparatus that has a full facepiece and is operated in a pressure-demand or other positive-pressure mode); or SAF:PD,PP:ASCBA (any supplied-air respirator that has a full facepiece and is operated in a pressure-demand or other positive-pressure mode in combination with an auxiliary self-contained breathing apparatus operated in a pressure-demand or other positive pressure mode). *ESCAPE:* GMFOV [any air-purifying, full-facepiece respirator (gas mask) with a chin-style, front- or back-mounted organic vapor canister]; or SCBAE (any appropriate escape-type, self-contained breathing apparatus).
Note: Substance causes eye irritation or damage; eye protection needed.
Exposure limits (TWA unless otherwise noted): ACGIH TLV 0.5 ppm (5.2 mg/m³); NIOSH/OSHA 0.5 ppm (5 mg/m³); skin contact contributes significantly in overall exposure.
Toxicity by ingestion: Grade 2; LD_{50} = 1.147 g/kg (rat).
Long-term health effects: Causes mutagenic and tumorigenic effects. May cause liver damage and depression of the central nervous system.
Vapor (gas) irritant characteristics: Irritating to eyes, skin, pharynx, larynx, and respiratory tract; lachrymator.
Liquid or solid irritant characteristics: Irritating to eyes and skin.
Odor threshold: 0.9–6.2 mg/m³.
IDLH value: 850 ppm.

2-Bromopentane

CHEMICAL REACTIVITY
Binary reactants: Reacts with some forms of plastics, rubber, and coatings. Reacts with lithium, sodium, potassium, calcium, powdered aluminum, zinc and magnesium, strong caustics, acetone. Gradually decomposes, acquiring yellow color; air and light accelerate decomposition.
Polymerization: Protect from light and oxygen. Stabilized with 75 to 125 ppm diphenylamine or 1–3% ethanol.

ENVIRONMENTAL DATA
Food chain concentration potential: Log P_{ow} = 2.4. Unlikely to accumulate.
Water pollution: DOT Appendix B, §172.101–marine pollutant. Harmful to aquatic life in very low concentrations. May be dangerous if it enters nearby water intakes; notify operators. Notify local health and wildlife officials. **Response to discharge:** Issue warning–poison, water contaminant. Should be removed. Chemical and physical treatment.

SHIPPING INFORMATION
Grades of purity: 96%; 99%; **Storage temperature:** Ambient; **Stability during transport:** Stable when stabilized.

PHYSICAL AND CHEMICAL PROPERTIES
Physical state @ 59°F/15°C and 1 atm: Liquid.
Molecular weight: 252.75
Boiling point @ 1 atm: 302.9°F/150.5°C/423.7°K.
Melting/Freezing point: 47.3°F/8.5°C/281.7°K.
Critical temperature: 796°F/425°C/698°K (estimate).
Specific gravity (water = 1): 2.8912 @ 68°F/20°C.
Liquid surface tension: 41.53 dynes/cm = 0.04153 N/m @ 68°F/20°C.
Relative vapor density (air = 1): 8.7
Latent heat of vaporization: 68.9 Btu/lb = 38.3 cal/g = 1.60X105 KJ/kg.
Vapor pressure: 5 mm.

2-BROMOPENTANE   REC. B:2700

SYNONYMS: AMYL BROMIDE; 2-BROMOPENTANO (Spanish); 2-PENTYLBROMIDE

IDENTIFICATION
CAS Number: 107-81-3
Formula: $C_5H_{11}Br$; $CH_3CHBr(C_2H_4)CH_3$
DOT ID Number: UN 2343; DOT Guide Number: 128
Proper Shipping Name: 2-Bromopentane

DESCRIPTION: Colorless liquid. Sinks in water.

Highly flammable • Narcotic effects in high concentrations • Containers may BLEVE when exposed to fire •Vapors may form explosive mixture with air • Vapors are heavier than air and will collect and stay in low areas • Vapors may travel long distances to ignition sources and flashback • Vapors in confined areas (e.g., tanks, sewers, buildings) may explode when exposed to fire • Irritating to the skin, eyes, and respiratory tract • Toxic products of combustion may include bromine and phosgene.

Hazard Classification (based on NFPA-704 Rating System)
Health Hazards (Blue): 1; Flammability (Red): 3; Reactivity (Yellow): 0

EMERGENCY RESPONSE: See Appendix A (128)
Evacuation:
Public safety: Isolate spill area for at least 25 to 50 meters (80 to 160 feet) in all directions.
Spill: Large spill–Consider initial downwind evacuation for at least 300 meters (1000 feet).
Fire: Isolate for 800 meters (½ mile) in all directions, especially if tank, rail car, or tank truck is involved in fire.

EXPOSURE
Short-term effects: *SEEK MEDICAL ATTENTION*. **Vapor:** May be harmful if inhaled or absorbed through the skin. Irritating to the eyes, nose, and throat. Move victim to fresh air. *IF BREATHING HAS STOPPED*, give artificial respiration. IF breathing is difficult, administer oxygen. **Liquid:** *MAY BE HARMFUL IF SWALLOWED OR ABSORBED THROUGH THE SKIN*. Irritating to the eyes and skin. *IF IN EYES*, flush with plenty of water for at least 15 minutes. Remove contaminated clothing and shoes. Flush affected areas with plenty of water. *IF SWALLOWED*, do nothing except keep victim warm.

HEALTH HAZARDS
Personal protective equipment (PPE): B-Level PPE. Approved respirator, safety goggles, chemical-resistant gloves, other protective clothing.
Toxicity by ingestion: Grade 3: LD_{50} = 150 mg/kg (mouse).
Long-term health effects: May cause liver damage.
Vapor (gas) irritant characteristics: Eye and respiratory tract irritant. Vapors cause smarting of the eyes or respiratory system if present in high concentrations. The effect is temporary.
Liquid or solid irritant characteristics: Eye and skin irritant. If spilled on clothing and allowed to remain, may cause smarting and reddening of skin.

FIRE DATA
Flash point: 90°F/32°C (cc).
Fire extinguishing agents not to be used: Water may be ineffective against fire.

CHEMICAL REACTIVITY
Binary reactants: Oxidizers, strong acids.

ENVIRONMENTAL DATA
Water pollution: Effects of low concentrations on aquatic life are not known. May be dangerous if it enters nearby water intakes; notify operators. Notify local health and wildlife officials. **Response to discharge:** Issue warning–high flammability-Should be removed. Chemical and physical treatment.

SHIPPING INFORMATION
Grades of purity: 95%; **Inert atmosphere:** None; **Venting:** None; **Stability during transport:** Stable.

NAS HAZARD CLASSIFICATIONS FOR BULK WATER TRANSPORTATION
FIRE: 3
HEALTH: Vapor irritant: 1; Liquid or solid irritant: 1; Poisons: 3
WATER POLLUTION: Human toxicity: 3; Aquatic toxicity:–; Aesthetic effect: 1
REACTIVITY: Other chemicals: 1; Water: 0; Self-reaction: 0

PHYSICAL AND CHEMICAL PROPERTIES
Physical state @ 59°F/15°C and 1 atm: Liquid.
Molecular weight: 151.07

Boiling point @ 1 atm: 128°F/53°C/326°K.
Melting/Freezing point: −140°F/−95.5°C/177.7°K.
Specific gravity (water = 1): 1.208 @ 68°F/20°C.

1-BROMOPROPANE REC. B:2750

SYNONYMS: 1-BROMOPROPANO (Spanish); *n*-PROPYLBROMIDE; PROPYLBROMIDE

IDENTIFICATION
CAS Number: 106-94-5
Formula: C_3H_7Br
DOT ID Number: UN 2344; DOT Guide Number: 132
Proper Shipping Name: 2-Bromopropane

DESCRIPTION: Colorless liquid. Sinks in water; slowly forms propanol.

Extremely flammable • Containers may BLEVE when exposed to fire • Vapors may form explosive mixture with air • Vapors are heavier than air and will collect and stay in low areas • Vapors may travel long distances to ignition sources and flashback • Vapors in confined areas (e.g., tanks, sewers, buildings) may explode when exposed to fire • Irritating to the skin, eyes, and respiratory tract • Toxic products of combustion may include hydrogen bromide gas.

Hazard Classification (based on NFPA-704 Rating System)
Health Hazards (Blue): 2; Flammability (Red): 3; Reactivity (Yellow): 0

EMERGENCY RESPONSE: See Appendix A (132)
Evacuation:
Public safety: Isolate spill area for at least 50 to 100 meters (160 to 330 feet) in all directions.
Spill: Increase, as necessary, the isolation distance shown above, in "Public safety."
Fire: Isolate for 800 meters (½ mile) in all directions, especially if tank, rail car, or tank truck is involved in fire.

EXPOSURE
Short-term effects: *SEEK MEDICAL ATTENTION.* **Vapor:** Irritating to the eyes, nose, and throat. May be harmful if inhaled or absorbed through the skin. Move victim to fresh air. *IF BREATHING HAS STOPPED,* give artificial respiration. IF breathing is difficult, administer oxygen. **Liquid:** Irritating to the skin and eyes. *MAY BE HARMFUL IF SWALLOWED OR ABSORBED THROUGH THE SKIN. IF IN EYES*, flush with plenty of water for at least 15 minutes. Remove contaminated clothing and shoes. Flush affected areas with plenty of water. *IF SWALLOWED*, do nothing except keep victim warm.

HEALTH HAZARDS
Personal protective equipment (PPE): B-Level PPE. Approved respirator, rubber gloves, chemical safety goggles, other protective clothing.
Toxicity by ingestion: Grade 2: LD_{50} = 2.95 g/kg (rat).
Vapor (gas) irritant characteristics: Eye and respiratory tract irritant. Vapors cause smarting of the eyes or respiratory system if present in high concentrations. The effect is temporary.
Liquid or solid irritant characteristics: Eye and skin irritant. If spilled on clothing and allowed to remain, may cause smarting and reddening of skin.

FIRE DATA
Flash point: 78°F/26°C (cc).
Fire extinguishing agents not to be used: Water may be ineffective against fire.
Behavior in fire: 914°F/490°C/763°K.

CHEMICAL REACTIVITY
Binary reactants: Strong oxidizers.
Reactivity group: 36
Compatibility class: Halogenated hydrocarbons (aliphatic).

ENVIRONMENTAL DATA
Water pollution: Effects of low concentrations on aquatic life are not known. May be dangerous if it enters nearby water intakes; notify operators. Notify local health and wildlife officials. **Response to discharge:** Evacuate area. Issue warning–high flammability. Should be removed. Chemical and physical treatment.

SHIPPING INFORMATION
Grades of purity: 99%; **Inert atmosphere:** None; **Venting:** None; **Stability during transport:** Stable.

NAS HAZARD CLASSIFICATION FOR BULK WATER TRANSPORTATION
FIRE: 4
HEALTH: Vapor irritant: 1; Liquid or solid irritant: 1; Poisons: 2
WATER POLLUTION: Human toxicity: 2; Aquatic toxicity:–; Aesthetic effect: 1
REACTIVITY: Other chemicals: 2; Water: 0; Self-reaction: 0

PHYSICAL AND CHEMICAL PROPERTIES
Physical state @ 59°F/15°C and 1 atm: Liquid.
Molecular weight: 123.01
Boiling point @ 1 atm: 160°F/70.9°C/344°K.
Melting/Freezing point: −166°F/−110°C/163°K.
Specific gravity (water = 1): 1.3537 @ 68°F/20°C.
Liquid surface tension: 25.9 dyne/cm = 0.026 N/m @ 68°F/20°C.
Relative vapor density (air = 1): 4.34
Heat of combustion: −7276 Btu/lb = −4042 cal/g = 169 x 10^5 J/kg.
Vapor pressure: (Reid) 5.3 psia.

BRUCINE REC. B:2800

SYNONYMS: BRUCINA (Spanish); (-)BRUCINE; (-)BRUCINE DIHYDRATE; BRUCINE HYDRATE; BRUCINE QUARTERNARY HYDRATE; DIMETHOXY STRYCHNINE; 2,3-DIMETHOXYSTRICHNIDIN-10-ONE; 2,3-DIMETHOXYSTRYCHNINE; 10,11-DIMETHOXYSTRYCHNINE; 10,11-DIMETHYLSTRYCHNINE; EEC No. 614-006-00-1; RCRA No. P018

IDENTIFICATION
CAS Number: 357-57-3
Formula: $C_{23}H_{26}N_2O_4$; $C_{23}H_{26}N_2O_4 \cdot 2H_2O$
DOT ID Number: UN 1570; DOT Guide Number: 152
Proper Shipping Name: Brucine
Reportable Quantity (RQ): **(CERCLA)** 100 lb/45.4 kg

DESCRIPTION: White solid. Odorless. Sinks in water; practically insoluble.

Combustible solid • Containers may BLEVE when exposed to fire • Dust cloud may explode if ignited in an enclosed area • Irritating to the skin, eyes, and respiratory tract; affects the central nervous to the skin, eyes, and respiratory tract; affects the central nervous system • Toxic products of combustion may include nitrogen oxides.

Hazard Classification (based on NFPA-704 Rating System)
Health Hazards (Blue): 2; Flammability (Red): 1; Reactivity (Yellow): 0

EMERGENCY RESPONSE: See Appendix A (152)
Evacuation:
Public safety: Isolate the area of spill or leak for at least 25 to 50 meters (80 to 160 feet) in all directions.
Spill: Increase, in the downwind direction, as necessary, the distance shown under "Public Safety."
Fire: If tank, rail car, or tank truck is involved in fire, isolate for at least 800 meters (½ mile) in all directions; also, consider initial evacuation for 800 meters (½ mile) in all directions.

EXPOSURE
Short-term effects: *SEEK MEDICAL ATTENTION.* **Dust:** *POISONOUS IF INHALED.* Irritating to eyes, nose, and throat. *IF IN EYES,* hold eyelids open and flush with plenty of water. *IF BREATHING HAS STOPPED,* give artificial respiration; *avoid mouth-to-mouth resuscitation; use bag/mask apparatus.* IF breathing is difficult, administer oxygen. **Solid:** POISONOUS IF SWALLOWED OR IF SKIN IS EXPOSED. *IF SWALLOWED,* will cause nausea and vomiting. Remove contaminated clothing and shoes. Flush affected areas with plenty of water. *IF IN EYES,* hold eyelids open and flush with plenty of water. *IF SWALLOWED* and victim is *CONSCIOUS AND ABLE TO SWALLOW,* have victim drink 4 to 8 ounces of water and have victim induce vomiting. *IF SWALLOWED* and victim is *UNCONSCIOUS OR HAVING CONVULSIONS,* do nothing except keep victim warm.

HEALTH HAZARDS
Personal protective equipment (PPE): B-Level PPE. Dust mask; goggles or face shield; rubber gloves.
Exposure limits (TWA unless otherwise noted): NIOSH/OSHA 0.15 mg/m^3 as strychnine
Toxicity by ingestion: Grade 4; oral rat LD$_{50}$ = 1 mg/kg.
Long-term health effects: A strychnine-like poison.
Vapor (gas) irritant characteristics: Eye and respiratory tract irritant.
Liquid or solid irritant characteristics: Eye and respiratory tract irritant.
IDLH value: 3 mg/m^3 as strychnine

FIRE DATA
Flash point: Combustible solid.
Electrical hazard: Flow or agitation of substance may generate electrostatic charges due to low conductivity.

CHEMICAL REACTIVITY
Binary reactants: Strong oxidizers.

ENVIRONMENTAL DATA
Food chain concentration potential: Log P$_{ow}$ = 0.4–0.98. Unlikely to accumulate.
Water pollution: Effect of low concentrations on aquatic life is unknown. May be dangerous if it enters nearby water intakes; notify operators. Notify local health and wildlife officials.
Response to discharge: Issue warning–poison, water contaminant. Restrict access. Should be removed. Chemical and physical treatment.

SHIPPING INFORMATION
Grades of purity: Pure; **Storage temperature:** Ambient; **Inert atmosphere:** None; **Venting:** Open; **Stability during transport:** Stable.

PHYSICAL AND CHEMICAL PROPERTIES
Physical state @ 59°F/15°C and 1 atm: Solid.
Molecular weight: 394.4
Boiling point @ 1 atm: Decomposes.
Melting/Freezing point: 352°F/178°C/451°K.
Specific gravity (water = 1): More than 1 @ 68°F/20°C (solid).
Heat of combustion: –13,400 Btu/lb = –7440 cal/g = –311 x 10^5 J/kg.

BUTADIENE   REC. B:2850

SYNONYMS: BIETHYLENE; BIVINYL; 1,2-BUTADIENE; 1,3-BUTADIENE; BUTA-1,3-DIENE; α-γ-BUTADIENE; 1,3-BUTADIENO (Spanish); DIVINYL; EEC No. 601-013-00-X; ERYTHRENE; PYRROLYLENE; VINYLETHYLENE

IDENTIFICATION
CAS Number: 106-99-0 (1,3-isomer); 590-19-2 (1,2-isomer)
Formula: C$_4$H$_6$; CH$_2$=CHCH=CH$_2$
DOT ID Number: UN 1010; DOT Guide Number: 116P
Proper Shipping Name: Butadienes, inhibited
Reportable Quantity (RQ): **(CERCLA)** 1 lb/0.454 kg

DESCRIPTION: Colorless, liquefied compressed gas. Often shipped and stored dissolved in acetone. Gasoline-like odor. Floats and boils on water; flammable visible vapor cloud is produced.

Extremely flammable • Polymerization hazard • Heat can induce polymerization with rapid release of energy; sealed containers may rupture explosively • Vapors may cause dizziness or asphyxiation without warning • Contact with liquid may cause frostbite • May react with itself blocking relief valves; leading to tank explosions • Forms explosive mixture with air • Reacts explosively with many materials • Gas is heavier than air, and spread along ground • Gas may travel to source of ignition and flashback • Exposure of cylinders to elevated temperatures, fire, and flame may cause cylinders to rupture or cause frangible disk to burst, releasing entire contents of cylinder • Ruptured or venting cylinders may rocket through buildings and/or travel a considerable distance.

Hazard Classification (based on NFPA-704 Rating System)
Health Hazards (Blue): 2; Flammability (Red): 4; Reactivity (Yellow): 2

EMERGENCY RESPONSE: See Appendix A (116)
Evacuation:
Public safety: Isolate spill area for at least 100 meters (330 feet) in all directions.
Spill: Consider initial downwind evacuation for at least 800 meters (½ mile).
Fire: Isolate for 1600 meters (1 mile) in all directions, especially if tank, rail car, or tank truck is involved in fire.

EXPOSURE
Short-term effects: *SEEK MEDICAL ATTENTION.* **Vapor:** Irritating to eyes, nose, and throat. Move to fresh air. **Liquid:** Irritating to skin and eyes. Remove contaminated clothing and shoes. Flush affected areas with plenty of water. *IF IN EYES,* hold eyelids open and flush with plenty of water.

HEALTH HAZARDS
Personal protective equipment (PPE): B-Level PPE. OSHA Table Z-1-A air contaminant. Chemical-type safety goggles; rescue harness and life line for those entering a tank or enclosed storage space; hose mask with hose inlet in a vapor-free atmosphere; self-contained breathing apparatus; rubber suit. Wear thermal protective clothing. Chemical protective material(s) reported to have minimal to poor protection: butyl rubber, Viton®.
Recommendations for respirator selection: NIOSH:
At any detectable concentration: SCBAF:PD,PP (any MSHA/NIOSH approved self-contained breathing apparatus that has a full facepiece and is operated in a pressure-demand or other positive-pressure mode); or SAF:PD,PP:ASCBA (any supplied-air respirator that has a full facepiece and is operated in a pressure-demand or other positive-pressure mode in combination with an auxiliary, self-contained breathing apparatus operated in a pressure-demand or other positive pressure mode).
ESCAPE: GMFS [any air-purifying, full-facepiece respirator (gas mask) with a chin-style, front- or back-mounted canister providing protection against the compound of concern]; or SCBAE (any appropriate escape-type, self-contained breathing apparatus).
Exposure limits (TWA unless otherwise noted): ACGIH TLV 2 ppm (4.4 mg/m^3); suspected human carcinogen; OSHA PEL 1000 ppm (2200 mg/m^3); NIOSH REL potential human carcinogen; reduce exposure to lowest possible concentrations.
Short-term exposure limits (15-minute TWA): OSHA 5 ppm.
Long-term health effects: IARC possible carcinogen, rating 2B; NTP anticipated carcinogen.
Vapor (gas) irritant characteristics: Eye, skin and respiratory tract irritant. Vapors cause smarting of the eyes or respiratory system if present in high concentrations. The effect may be temporary.
Liquid or solid irritant characteristics: Eye and skin irritant. If spilled on clothing and allowed to remain, may cause smarting and reddening of the skin because of frostbite.
Odor threshold: 0.45 ppm.
IDLH value: Potential human carcinogen; 2000 ppm [10% LEL].

FIRE DATA
Flash point: <0°C (1,2-isomer); –105°F/–76°C (1,3-isomer).
Flammable limits in air: LEL: 2.0%; UEL: 11.5%.
(NFPA). Use flooding quantities of water as spray or fog.
Behavior in fire: Vapors heavier than air and may travel a considerable distance to a source of ignition and flashback.
Containers may rupture and explode in a fire from polymerization.
Autoignition temperature: 788°F/420°C/693°K.
Electrical hazard: Class 1, Group B. Flow or agitation of substance may generate electrostatic charges due to low conductivity.
Burning rate: 8.0 mm/min
Stoichiometric air-to-fuel ratio: 13.96 (estimate).

CHEMICAL REACTIVITY
Binary reactants: Reacts with oxidizers, phenol, chlorine dioxide, crotonaldehyde. Unsafe in contact with acetylide-forming materials such as monel, copper and copper alloys.
Polymerization: Stable when inhibitors present; **Inhibitor of polymerization:** *tert*-Butylcatechol (0.01-0.02%).
Reactivity group: 30
Compatibility class: Olefins

ENVIRONMENTAL DATA
Food chain concentration potential: Log P_{ow} = 2.0. Unlikely to accumulate.
Water pollution: Not harmful to aquatic life. May be dangerous if it enters nearby water intakes; notify operators. **Response to discharge:** Issue warning–high flammability. Evacuate area.

SHIPPING INFORMATION
Grades of purity: Research grade: 99.86 mole%; 99.5 mole%; Rubber grade: 99.0 mole%; Commercial: 98%; **Storage temperature:** Ambient; **Inert atmosphere:** None; **Venting:** Safety relief; **Stability during transport:** Stable when inhibited.

NAS HAZARD CLASSIFICATION FOR BULK WATER TRANSPORTATION
FIRE: 4
HEALTH: Vapor irritant: 1; Liquid or solid irritant: 1; Poisons: 1
WATER POLLUTION: Human toxicity: 0; Aquatic toxicity: 1; Aesthetic effect: 1
REACTIVITY: Other chemicals: 2; Water: 0; Self-reaction: 3

PHYSICAL AND CHEMICAL PROPERTIES
Physical state @ 59°F/15°C and 1 atm: Gas.
Molecular weight: 54.1
Boiling point @ 1 atm: 24.1°F/–4.4°C/268.8°K.
Melting/Freezing point: –164°F/–108.9°C/164.3°K.
Critical temperature: 306°F/152°C/425°K.
Critical pressure: 628 psia = 42.7 atm = 4.32 MN/m^2.
Specific gravity (water = 1): 0.621 @ 68°F/20°C (liquid AT 21°F).
Liquid surface tension: 13.4 dynes/cm = 0.0134 N/m at 71°F/22°C.
Liquid water interfacial tension: (estimate) 67 dynes/cm = 0.067 N/m at 71°F/22°C.
Relative vapor density (air = 1): 1.88 @ 68°F/20°C.
Ratio of specific heats of vapor (gas): 1.1
Latent heat of vaporization: 180 Btu/lb = 100 cal/g = 4.19 x 10^5 J/kg.
Heat of combustion: –19,008 Btu/lb = –10,560 cal/g = –442.13 x 10^5 J/kg.
Heat of polymerization: –549 Btu/lb = –305 cal/g = **-Liquid surface tension:** x 10^5 J/kg.
Heat of fusion: 35.28 cal/g.
Vapor pressure: (Reid) 61 psia; 75 psia @ 115°F/46°C; 1860 mm.

BUTANE   REC. B:2900

SYNONYMS:; BUTANO (Spanish); BUTYL HYDRIDE; DIETHYL; DIETHYL, EEC No. 601-004-00-0; LIQUEFIED PETROLEUM GAS; METHYLETHYLMETHANE

IDENTIFICATION
CAS Number: 106-97-8
Formula: *n*-C$_4$H$_{10}$; CH$_3$CH$_2$CH$_2$CH$_3$
DOT ID Number: UN 1011; UN 1075; DOT Guide Number: 115
Proper Shipping Name: Butane; Butane mixtures

DESCRIPTION: Colorless, liquefied compressed gas. Gasoline-like odor. Floats and boils on water; flammable, visible vapor cloud is formed.

Extremely flammable • Forms explosive mixture with air • Reacts explosively with many materials • Gas from liquefied gas are initially heavier than air and spread along ground • Gas may travel to source of ignition and flashback • Exposure of cylinders to elevated temperatures, fire, and flame may cause cylinders to rupture or cause frangible disk to burst, releasing entire contents of cylinder • Ruptured or venting cylinders may rocket through buildings and/or travel a considerable distance • Vapors may cause dizziness or asphyxiation without warning • Contact with liquid may cause frostbite.

Hazard Classification (based on NFPA-704 Rating System)
Health Hazards (Blue): 1; Flammability (Red): 4; Reactivity (Yellow): 0

EMERGENCY RESPONSE: See Appendix A (115)
Evacuation:
Public safety: Isolate spill area for at least 50 to 100 meters (160 to 330 feet) in all directions.
Spill: Consider initial downwind evacuation for at least 800 meters (½ mile).
Fire: Isolate for 1600 meters (1 mile) in all directions, especially if tank, rail car, or tank truck is involved in fire.

EXPOSURE
Short-term effects: *SEEK MEDICAL ATTENTION*. **Vapor:** If inhaled, will cause dizziness or difficult breathing. Not irritating to eyes, nose or throat. Move to fresh air. *IF BREATHING HAS STOPPED,* give artificial respiration. IF breathing is difficult, administer oxygen. **Liquid:** Will cause frostbite. Flush affected areas with plenty of water. *DO NOT RUB AFFECTED AREAS.*
Inhalation: Guard against self-injury if stuporous, confused, or anesthetized. Apply artificial respiration if not breathing. Avoid administration of epinephrine or other sympathomimetic amines. Prevent aspirations of vomitus by proper positioning of the head. Give symptomatic and supportive treatment.

HEALTH HAZARDS
Personal protective equipment (PPE): B-Level PPE. OSHA Table Z-1-A air contaminant. Self-contained breathing apparatus and safety goggles. Wear thermal protective clothing. Chemical protective material(s) reported to offer protection: neoprene, polyurethane. Also, nitrile, neoprene+Styrene-butadiene, and Viton® offers minimal to poor protection.
Exposure limits (TWA unless otherwise noted): ACGIH TLV 800 ppm (1900 mg/m^3); NIOSH 800 ppm (1900 mg/m^3).
Liquid or solid irritant characteristics: Except for frostbite, otherwise practically harmless to the skin.
Odor threshold: 1200–5000 ppm.

FIRE DATA
Flash point: –76°F/–60°C.
Flammable limits in air: LEL: 1.9%; UEL: 8.4%.
Autoignition temperature: 550°F/287°C/560°K.
Electrical hazard: Class 1, Group D. Flow or agitation of substance may generate electrostatic charges due to low conductivity.
Burning rate: 7.9 mm/min
Adiabatic flame temperature: 2435°F/1335°C (estimate).
Stoichiometric air-to-fuel ratio: 15.35 (estimate).

CHEMICAL REACTIVITY
Binary reactants: Strong oxidizers, chlorine, fluorine; nickel carbonyl + oxygen.
Reactivity group: 31
Compatibility class: Paraffins

ENVIRONMENTAL DATA
Food chain concentration potential: Negative; unlikely to accumulate.
Water pollution: Not harmful to aquatic life. May be dangerous if it enters nearby water intakes; notify operators. **Response to discharge:** Issue warning–high flammability. Restrict access. Evacuate area.

SHIPPING INFORMATION
Grades of purity: Research: 99.95%; Pure: 99.4%; Technical: 97.6%; **Storage temperature:** Ambient; **Inert atmosphere:** None; **Venting:** Safety relief; **Stability during transport:** Stable.

NAS HAZARD CLASSIFICATION FOR BULK WATER TRANSPORTATION
FIRE: 4
HEALTH: Vapor irritant: 0; Liquid or solid irritant: 0; Poisons: 0
WATER POLLUTION: Human toxicity: 0; Aquatic toxicity: 0; Aesthetic effect: 0
REACTIVITY: Other chemicals: 0; Water: 0; Self-reaction: 0

PHYSICAL AND CHEMICAL PROPERTIES
Physical state @ 59°F/15°C and 1 atm: Gas.
Molecular weight: 58.12
Boiling point @ 1 atm: 31.1°F/–0.48°C/272.72°K.
Melting/Freezing point: –216°F/–138°C/135°K.
Critical temperature: 306°F/152°C/425°K.
Critical pressure: 550.8 psia = 37.47 atm = 3.796 MN/m^2.
Specific gravity (water = 1): 0.60 at 0°C (liquid).
Liquid surface tension: 14.7 dynes/cm = 0.0147 N/m at 0°C.
Liquid water interfacial tension: (estimate) 65 dynes/cm = 0.065 N/m at 71°F/22°C.
Relative vapor density (air = 1): 2.0
Ratio of specific heats of vapor (gas): 1.092
Latent heat of vaporization: 170 Btu/lb = 92 cal/g = 3.9 x 10^5 J/kg.
Heat of combustion: –19,512 Btu/lb = –10,840 cal/g = –453.85 x 10^5 J/kg.
Heat of fusion: 19.18 cal/g.
Vapor pressure: (Reid) 52.4 psia; 2 mm.

1,4-BUTANEDIOL   REC. B:2950

SYNONYMS: BUTANE-1,4-DIOL; 1,4-BUTYLENEGLYCOL; 1,4-DIHYDROXYBUTANE; TETRAMETHYLENE GLYCOL; 1,4-TETRAMETHYLENE GLYCOL

IDENTIFICATION
CAS Number: 110-63-4 (1,4-isomer); 584-03-2 (1,2-isomer); 107-88-0 (1,3-isomer); 513-85-9 (2,3-isomer)
Formula: $C_4H_{10}O_2$; $HOCH_2(CH_2)_2CH_2OH$

DESCRIPTION: Colorless, thick liquid. Odorless. Sinks and mixes with water; highly soluble.

Combustible • Containers may BLEVE when exposed to fire • Vapors may form explosive mixture with air • Vapors are heavier than air and will collect and stay in low areas • Vapors may travel long distances to ignition sources and flashback • Vapors in confined areas (e.g., tanks, sewers, buildings) may explode when exposed to fire • Irritating to the skin, eyes, and respiratory tract • Toxic products of combustion may include carbon monoxide.

Hazard Classification (based on NFPA-704 Rating System)
Health Hazards (Blue): 1; Flammability (Red): 1; Reactivity (Yellow): 0

EMERGENCY RESPONSE: See Appendix A (171)
Evacuation:

Public safety: Isolate spill area for at least 25 to 50 meters (80 to 160 feet) in all directions.
Spill: Large spill–Consider initial downwind evacuation for at least 300 meters (1000 feet).
Fire: Isolate for 800 meters (½ mile) in all directions, especially if tank, rail car, or tank truck is involved in fire.

EXPOSURE
Short-term effects: *SEEK MEDICAL ATTENTION.* **Liquid or solid:** Irritating to skin or eyes. Harmful if swallowed.
Flush affected areas with plenty of water. *IF SWALLOWED* and victim is *CONSCIOUS AND ABLE TO SWALLOW*, have victim drink wate or milk.

HEALTH HAZARDS
Personal protective equipment (PPE): B-Level PPE.
Toxicity by ingestion: Grade 2; LD_{50} = 0.5 to 5 g/kg (rat).
Long-term health effects: Kidney damage.
Liquid or solid irritant characteristics: Eye irritant.

FIRE DATA
Flash point: 250°F/121°C (oc); 194°F/90°C (1,2-isomer); 229°F/109°C (1,3-isomer); 185°F/85°C (2,3-isomer)
Fire extinguishing agents not to be used: Water or foam may cause frothing
Autoignition temperature: 756°F/402°C; 707°F/375°C (1,3-isomer).

CHEMICAL REACTIVITY
Binary reactants: Reacts with strong oxidizers.

ENVIRONMENTAL DATA
Food chain concentration potential: Log P_{ow} = –0.92. Unlikely to accumulate.
Water pollution: Effect of low concentrations on aquatic life is unknown. May be dangerous if it enters nearby water intakes; notify operators. Notify local health and pollution control officials.
Response to discharge: Disperse and flush.

SHIPPING INFORMATION
Grades of purity: Regular grade: 99%; Anhydrous grade: 99.3%; **Storage temperature:** 75°–100°F; **Stability during transport:** Stable.

PHYSICAL AND CHEMICAL PROPERTIES
Physical state @ 59°F/15°C and 1 atm: Liquid.
Molecular weight: 90.12
Boiling point @ 1 atm: 442°F/228°C/501°K.
Melting/Freezing point: 68.2°F/20.1°C/293.3°K.
Critical temperature: 716°F/380°C/653°K.
Critical pressure: 720 psia = 49 atm = 5.0 MN/m^2.
Specific gravity (water = 1): 1.017 @ 68°F/20°C (liquid).
Relative vapor density (air = 1): 3.1
Heat of combustion: (estimate) –11,900 Btu/lb = –6630 cal/g = –277 x 10^5 J/kg.
Vapor pressure: Less than 0.08 mm.

1,4-BUTENEDIOL **REC. B:3000**

SYNONYMS: 2-BUTENE-1,4-DIOL; 1,4-DIHYDROXY-2-BUTENE; *cis*-2-BUTENE-1,4-DIOL

IDENTIFICATION

CAS Number: 110-64-5
Formula: $C_4H_8O_2$; $HOCH_2CH=CHCH_2OH$

DESCRIPTION: Light yellow, thick liquid. Odorless. Sinks and mixes with water.

Combustible • Containers may BLEVE when exposed to fire • Vapors are heavier than air and will collect and stay in low areas • Vapors in confined areas (e.g., tanks, sewers, buildings) may explode when exposed to fire • Irritating to the skin, eyes, and respiratory tract • Toxic products of combustion may include carbon monoxide.

Hazard Classification (based on NFPA-704 Rating System)
Health Hazards (Blue): 1; Flammability (Red): 1; Reactivity (Yellow): 0

EMERGENCY RESPONSE: See Appendix A (171)
Evacuation:
Public safety: Isolate the area of spill or leak for at least 10 to 25 meters (30 to 80 feet) in all directions.
Spill: Increase, in the downwind direction, as necessary, the distance shown under "Public Safety."
Fire: If any large container is involved in fire, isolate for at least 800 meters (½ mile) in all directions; also, consider initial evacuation for 800 meters (½ mile) in all directions.

EXPOSURE
Short-term effects: *SEEK MEDICAL ATTENTION.* **Liquid:** Irritating to skin and eyes. Harmful if swallowed. Remove contaminated clothing and shoes. Flush affected areas with plenty of water. *IF IN EYES*, hold eyelids open and flush with plenty of water. *IF SWALLOWED* and victim is *CONSCIOUS AND ABLE TO SWALLOW*, have victim drink 4 to 8 ounces of water.

HEALTH HAZARDS
Personal protective equipment (PPE): B-Level PPE. Eye protection. Chemical protective material(s) reported to offer minimal to poor protection: Viton®/neoprene, butyl rubber,/neoprene.
Toxicity by ingestion: Grade 2; LD_{50} = 0.5 to 5 g/kg.
Liquid or solid irritant characteristics: If spilled on clothing and allowed to remain, may cause smarting and reddening of the skin.

FIRE DATA
Flash point: 250°F/121°C (oc).
Fire extinguishing agents not to be used: Foam or water may cause frothing

CHEMICAL REACTIVITY
Binary reactants: Strong oxidizers.
Reactivity group: 20
Compatibility class: Alcohols, glycols

ENVIRONMENTAL DATA
Food chain concentration potential: Log K_{ow} = –1.3 (estimate). Unlikely to accumulate.
Water pollution: Effect of low concentrations on aquatic life is unknown. May be dangerous if it enters nearby water intakes; notify operators. Notify local health and pollution control officials.
Response to discharge: Disperse and flush.

SHIPPING INFORMATION
Grades of purity: 95%; **Storage temperature:** Above 45°F; **Inert atmosphere:** Inerted; **Stability during transport:** Stable.

PHYSICAL AND CHEMICAL PROPERTIES
Physical state @ 59°F/15°C and 1 atm: Liquid.
Molecular weight: 88.11
Boiling point @ 1 atm: 442°F/228°C/501°K.
Melting/Freezing point: 45°F/7°C/280°K.
Specific gravity (water = 1): 1.07 @ 77°F/25°C (liquid).
Relative vapor density (air = 1): 3.1
Heat of combustion: (estimate) –10,8000 Btu/lb = –5980 cal/g = –250 x 10^5 J/kg.
Heat of solution: (estimate) 9 Btu/lb = 5 cal/g = 0.2 x 10^5 J/kg.

2-BUTANONE PEROXIDE   REC. B:3050

SYNONYMS: 2-BUTANONEPEROXIDE; BUTANOX® M50; CHALOXYD; HI-POINT 90; LUPERSOL; MEKP; MEK PEROXIDE; METHYL ETHYL KETONE HYDROPEROXIDE; METHYL ETHYL KETONE PEROXIDE; QUICKSET EXTRA; SPRAYSET MEKP; THERMACURE; ETHYL METHYL KETONE PEROXIDE; KETONOX; M105; LPT; MEKP-HA 1; MEKP-LA 1; PEROXIDO de METIL ETIL CETONA (Spanish); RCRA No. U160

IDENTIFICATION
CAS Number: 1338-23-4
Formula: $[(CH_3)_2CHCHOH-O]_2$
DOT ID Number: UN 2550; DOT Guide Number: 147
Proper Shipping Name: Methyl ethyl ketone peroxide
Reportable Quantity (RQ): **(CERCLA)** 10 lb/4.54 kg

DESCRIPTION: Colorless liquid. May be diluted with 40% dimethylphthalate, cyclohexane peroxide, or diallyl phthalate to reduce sensitivity to shock. Acetone-like odor. Not soluble in water.

Heat, shock, or contamination may cause explosion • Corrosive to the skin, eyes or respiratory tract; eye contact may cause blindness; inhalation can cause lung edema; effects of inhalation may be delayed • Firefighting gear (including SCBA) does not provide adequate protection. If exposure occurs, remove and isolate gear immediately and thoroughly decontaminate personnel • Strong oxidizer which may react spontaneously with low flash point organics or reducing agents. Heat forms oxygen; will increase the activity of an existing fire • May cause fire or explosion contact with combustibles (wood, paper, oil, clothing, etc.) • Containers may BLEVE when exposed to fire • Toxic products of combustion may include carbon monoxide.

Hazard Classification (based on NFPA-704 Rating System)
Health Hazards (Blue): 1; Flammability (Red): 2; Reactivity (Yellow): 3

EMERGENCY RESPONSE: See Appendix A (147)
Evacuation:
Public safety: Isolate the area of spill or leak for at least 25 to 50 meters (80 to 160 feet) in all directions.
Spill: Consider initial evacuation for at least 250 meters (800 feet).
Fire: If tank, rail car, or tank truck is involved in fire, isolate for at least 800 meters (½ mile) in all directions; also, consider initial evacuation for 800 meters (½ mile) in all directions.

EXPOSURE
Short-term effects: *SEEK MEDICAL ATTENTION.* **Vapor:** Vapor extremely irritating. *CONTACT OF VAPOR WITH EYES MAY CAUSE BLINDNESS.* Move victim to fresh air. *IF BREATHING HAS STOPPED,* give artificial respiration. IF breathing is difficult, administer oxygen. **Liquid:** Harmful if swallowed or absorbed through skin. *IF IN EYES,* hold eyelids open, flush with running water for at least 15 minutes. Flush affected areas with plenty of water. Wash skin with soap and water. Remove and isolate contaminated clothing and shoes at the site. *IF SWALLOWED,* keep victim quiet and maintain normal body temperature.

Note to physician or authorized medical personnel: Medical observation is recommended for 24 to 48 hours after breathing overexposure, as pulmonary edema may be delayed. As first aid for pulmonary edema, consider administering a corticosteroid spray. Cigarette smoking may exacerbate pulmonary injury and should be discouraged for at least 72 hours following exposure.

HEALTH HAZARDS
Personal protective equipment (PPE): A-Level PPE. Chemical protective material(s) reported to offer protection: Butyl rubber, neoprene, Viton®.
Exposure limits (TWA unless otherwise noted): See below for ceiling limits.
Short-term exposure limits (15-minute TWA): ACGIH STEL ceiling 0.2 ppm (1.5 mg/m^3); NIOSH ceiling 0.2 ppm (1.5 mg/m^3).
Toxicity by ingestion: Grade 3: LD_{50} = 470 mg/kg (mouse).
Long-term health effects: Possible liver and kidney damage.
Vapor (gas) irritant characteristics: Vapors cause severe irritation of eyes and throat and can cause eye and lung injury. They cannot be tolerated even at low concentrations.
Liquid or solid irritant characteristics: Severe skin irritant. Causes second- and third-degree burns on short contact and is very injurious to the eyes.

FIRE DATA
Flash point: 125-200°F/52-93°C (oc); 180°F/82°C (cc) (60% MEKP).
Behavior in fire: Explosive.
Electrical hazard: Class I, Group D. Flow or agitation of substance may generate electrostatic charges due to low conductivity.

CHEMICAL REACTIVITY
Binary reactants: Contact with other materials may cause explosion. Organic materials, heat, flame, sunlight, salts, strong acids and bases, heavy metal oxides, trace contaminants. Saturated clothing can become a fire hazard; flush with water.
Neutralizing agents for acids and caustics: Dry lime, soda ash.

ENVIRONMENTAL DATA
Food chain concentration potential: Log K_{ow} = 0.91 (estimate). Unlikely to accumulate.
Water pollution: Effects of low concentrations on aquatic life are not known. May be harmful if it enters water intakes. Notify local health and wildlife officials. **Response to discharge:** Evacuate area. Issue warning–corrosive. Mechanical containment. Should be removed. Chemical and physical treatment.

SHIPPING INFORMATION
Grades of purity: 50% in dimethyl phthalate; 60% MEKP;
Storage temperature: Refrigerate; **Venting:** Vent periodically

NAS HAZARD CLASSIFICATION FOR BULK WATER TRANSPORTATION
FIRE: 1
HEALTH: Vapor irritant: 4; Liquid or solid irritant: 4; Poisons: 3

WATER POLLUTION: Human toxicity: 3; Aquatic toxicity: 4; Aesthetic effect: 2
REACTIVITY: Other chemicals: 4; Water: 0; Self-reaction: 4

PHYSICAL AND CHEMICAL PROPERTIES
Physical state @ 59°F/15°C and 1 atm: Liquid.
Molecular weight: 176.22
Boiling point @ 1 atm: Decomposes. 244°F/118°C/391°K.
Specific gravity (water = 1): 1.12 @ 68°F/20°C.
Relative vapor density (air = 1): 6.08 (estimate).

BUTYL ACETATE REC. B:3100

SYNONYMS: ACETATE de BUTYLE (French); ACETATO de BUTILO (Spanish); ACETIC ACID, *n*-BUTYL ESTER; *n*-BUTYL ACETATE; 1-BUTYL ACETATE; BUTYLE (ACETATE de) (French); BUTYL ETHANOATE; *n*-BUTYL ESTER OF ACETIC ACID; EEC No. 607-025-00-1

IDENTIFICATION
CAS Number: 123-86-4
Formula: $C_6H_{12}O_2$; $CH_3COO(CH_2)_3CH_3$
DOT ID Number: UN 1123; DOT Guide Number: 129
Proper Shipping Name: Butyl acetates
Reportable Quantity (RQ): **(CERCLA)** 5000 lb/2270 kg

DESCRIPTION: Colorless watery liquid. Pleasant fruity odor (in low concentrations), disagreeable (in higher concentrations). Floats on water surface; slowly forms *n*-butanol and acetic acid.

Highly flammable • Containers may BLEVE when exposed to fire • Vapors may form explosive mixture with air • Vapors are heavier than air and will collect and stay in low areas • Vapors may travel long distances to ignition sources and flashback • Vapors in confined areas (e.g., tanks, sewers, buildings) may explode when exposed to fire • Irritating to the skin, eyes, and respiratory tract. Vapor has a narcotic effect at high concentrations • Toxic products of combustion may include carbon monoxide.

Hazard Classification (based on NFPA-704 Rating System)
Health Hazards (Blue): 1; Flammability (Red): 3; Reactivity (Yellow): 0

EMERGENCY RESPONSE: See Appendix A (129)
Evacuation:
Public safety: Isolate spill area for at least 50 to 100 meters (160 to 330 feet) in all directions.
Spill: Large spill–Consider initial downwind evacuation for at least 300 meters (1000 feet).
Fire: Isolate for 800 meters (½ mile) in all directions, especially if tank, rail car, or tank truck is involved in fire.

EXPOSURE
Short-term effects: *SEEK MEDICAL ATTENTION.* **Vapor:** Irritating to eyes, nose, and throat. *IF INHALED*, will, will cause nausea, dizziness, headache or difficult breathing. Move to fresh air. *IF BREATHING HAS STOPPED,* give artificial respiration. IF breathing is difficult, administer oxygen. **Liquid:** Irritating to skin and eyes. Harmful if swallowed. Remove contaminated clothing and shoes. Flush affected areas with plenty of water. *IF IN EYES,* hold eyelids open and flush with plenty of water. *IF SWALLOWED* and victim is *CONSCIOUS AND ABLE TO SWALLOW,* have victim drink 4 to 8 ounces of water, and have victim induce vomiting. *IF SWALLOWED* and victim is *UNCONSCIOUS OR HAVING CONVULSIONS,* do nothing except keep victim warm.

HEALTH HAZARDS
Personal protective equipment (PPE): B-Level PPE. OSHA Table Z-1-A air contaminant. Chemical protective material(s) reported to offer protection: PV alcohol, Teflon®, Silvershield®. Also, chlorinated polyethylene and butyl rubber offers limited protection.
Recommendations for respirator selection: NIOSH/OSHA
1500 ppm: CCROV* [any chemical cartridge respirator with organic vapor cartridge(s)]; or SA* (any supplied-air respirator). *1700 ppm:* SA:CF* (any supplied-air respirator operated in a continuous-flow mode); or PAPROV* [any powered, air-purifying respirator with organic vapor cartridge(s)]; or CCRFOV [any air-purifying, full-facepiece respirator (gas mask) with a chin-style, front- or back-mounted acid gas canister]; or GMFOV [any air-purifying, full-facepiece respirator (gas mask) with a chin-style, front- or back-mounted acid gas canister]; or SCBAF (any self-contained breathing apparatus with a full facepiece); or SAF (any supplied-air respirator with a full facepiece). *EMERGENCY OR PLANNED ENTRY INTO UNKNOWN CONCENTRATIONS OR IDLH CONDITIONS:* SCBAF:PD,PP (any self-contained breathing apparatus that has a full facepiece and is operated in a pressure-demand or other positive-pressure mode); or SAF:PD,PP:ASCBA (any supplied-air respirator that has a full facepiece and is operated in a pressure-demand or other positive-pressure mode in combination with an auxiliary self-contained breathing apparatus operated in a pressure-demand or other positive pressure mode). *ESCAPE:* GMFOV [any air-purifying, full-facepiece respirator (gas mask) with a chin-style, front- or back-mounted organic vapor canister]; or SCBAE (any appropriate escape-type, self-contained breathing apparatus). *Note:* Substance reported to cause eye irritation or damage; may require eye protection.
Exposure limits (TWA unless otherwise noted): ACGIH TLV 150 ppm (713 mg/m^3); NIOSH/OSHA 150 ppm (710 mg/m^3).
Short-term exposure limits (15-minute TWA): ACGIH STEL 200 ppm (950 mg/m^3); NIOSH STEL 200 ppm (950 mg/m^3).
Toxicity by ingestion: Grade 2; LD_{50} = 0.5 to 5 g/kg.
Long-term health effects: May be a chronic poison.
Vapor (gas) irritant characteristics: Eye and respiratory tract irritant. Vapors cause smarting of the eyes or respiratory system if present in high concentrations. The effect is temporary.
Liquid or solid irritant characteristics: If spilled on clothing and allowed to remain, may cause smarting and reddening of the skin.
Odor threshold: 0.06–7.0 ppm.
IDLH value: 1700 ppm [10% LEL].

FIRE DATA
Flash point: 99°F/37°C (oc); 72°F/71°F/22°C (cc).
Flammable limits in air: LEL: 1.7%; UEL: 15%.
Fire extinguishing agents not to be used: Water in straight hose stream will scatter and spread fire and should not be used.
Autoignition temperature: 797°F/425°C/698°K.
Electrical hazard: Class I, Group D. Flow or agitation of substance may generate electrostatic charges due to low conductivity.
Burning rate: 4.4 mm/min

CHEMICAL REACTIVITY
Reactivity with water: Air moisture can cause decomposition and production of irritating and toxic fumes.
Binary reactants: Nitrates, strong oxidizers, alkalis, and acids. Attacks some plastics.
Reactivity group: 34
Compatibility class: Esters.

ENVIRONMENTAL DATA
Food chain concentration potential: Log K_{ow} = 1.85. Unlikely to accumulate.
Water pollution: Effect of low concentrations on aquatic life is unknown. Fouling to shoreline. May be dangerous if it enters nearby water intakes; notify operators. Notify local health and pollution control officials. **Response to discharge:** Issue warning-flammability. Mechanical containment.

SHIPPING INFORMATION
Grades of purity: Urethane: 99.5%; pure: 98%; commercial: 90-92%; **Storage temperature:** Ambient; **Inert atmosphere:** None; **Venting:** Open (flame arrester); **Stability during transport:** Stable.

NAS HAZARD CLASSIFICATION FOR BULK WATER TRANSPORTATION
FIRE: 3
HEALTH: Vapor irritant: 1; Liquid or solid irritant: 1; Poisons: 2
WATER POLLUTION: Human toxicity: 0; Aquatic toxicity: 2; Aesthetic effect: 2
REACTIVITY: Other chemicals: 1; Water: 0; Self-reaction: 0

PHYSICAL AND CHEMICAL PROPERTIES
Physical state @ 59°F/15°C and 1 atm: Liquid.
Molecular weight: 116.16
Boiling point @ 1 atm: 259°F/126°C/399°K.
Melting/Freezing point: −107°F/−77°C/196°K.
Critical temperature: 582.6°F/305.9°C/579.1°K.
Critical pressure: 455 psia = 31 atm = 3.1 MN/m².
Specific gravity (water = 1): 0.875 @ 68°F/20°C (liquid).
Liquid surface tension: 14.5 dynes/cm = 0.0145 N/m @ 77°F/25°C.
Liquid water interfacial tension: (estimate) 57 dynes/cm = 0.057 N/m at 71°F/22°C.
Relative vapor density (air = 1): 4.1
Ratio of specific heats of vapor (gas): 1.058
Latent heat of vaporization: 133 Btu/lb = 73.9 cal/g = 3.09 x 10^5 J/kg.
Heat of combustion: −13,130 Btu/lb = −7294 cal/g = −305.4 x 10^5 J/kg.
Vapor pressure: (Reid) 0.5 psia; 10 mm.

sec-BUTYL ACETATE　　　　　　　　　　　REC. B:3150

SYNONYMS: ACETATE de BUTYLE SECONDAIRE (French); ACETATO de *sec*-BUTILO (Spanish); ACETIC *sec*-BUTYLESTER; ACETIC ACID, 2-BUTOXY ESTER; ACETIC ACID, 1-METHYLPROPYL ESTER; 2-BUTYL ACETATE; *sec*-BUTYL ALCOHOL ACETATE; EEC No. 607-026-00-7; 1-METHYL PROPYL ACETATE; ACETIC ACID, *sec*-BUTYL ESTER

IDENTIFICATION
CAS Number: 105-46-4
Formula: $C_6H_{12}O_2$; $CH_3COOCH(CH_3)CH_2CH_3$
DOT ID Number: UN 1123; DOT Guide Number: 129
Proper Shipping Name: Butyl acetates
Reportable Quantity (RQ): **(CERCLA)** 5000 lb/2270 kg

DESCRIPTION: Colorless watery liquid. Pleasant, fruity odor. Floats on the surface of water; slightly soluble; forms *sec*-butanol and acetic acid.

Highly flammable • Containers may BLEVE when exposed to fire • Vapors may form explosive mixture with air • Vapors are heavier than air and will collect and stay in low areas • Vapors may travel long distances to ignition sources and flashback • Vapors in confined areas (e.g., tanks, sewers, buildings) may explode when exposed to fire • Irritating to the skin, eyes, and respiratory tract; narcotic effects at high concentrations • Toxic products of combustion may include carbon monoxide.

Hazard Classification (based on NFPA-704 Rating System)
Health Hazards (Blue): 1; Flammability (Red): 3; Reactivity (Yellow): 0

EMERGENCY RESPONSE: See Appendix A (129)
Evacuation:
Public safety: Isolate spill area for at least 50 to 100 meters (160 to 330 feet) in all directions.
Spill: Large spill–Consider initial downwind evacuation for at least 300 meters (1000 feet).
Fire: Isolate for 800 meters (½ mile) in all directions, especially if tank, rail car, or tank truck is involved in fire.

EXPOSURE
Short-term effects: *SEEK MEDICAL ATTENTION.* **Vapor:** Irritating to eyes, nose, and throat. *IF INHALED*, will, will cause nausea, headache or difficult breathing. Move to fresh air. *IF BREATHING HAS STOPPED,* give artificial respiration. IF breathing is difficult, administer oxygen. **Liquid:** Irritating to skin and eyes. Remove contaminated clothing and shoes. Flush affected areas with plenty of water. *IF IN EYES*, hold eyelids open and flush with plenty of water.

HEALTH HAZARDS
Personal protective equipment (PPE): B-Level PPE. OSHA Table Z-1-A air contaminant. Chemical protective material(s) reported to offer protection (n-butyl acetate): PV alcohol, Teflon®, Silvershield®. Also, chlorinated polyethylene and butyl rubber offers limited protection The following materials are NOT recommended for service (n-butyl acetate): natural rubber, PVC, neoprene, nitrile, nitrile+PVC, Viton®.
Recommendations for respirator selection: NIOSH/OSHA
1700 ppm: SA:CF* (any supplied-air respirator operated in a continuous-flow mode); or PAPROV* [any powered, air-purifying respirator with organic vapor cartridge(s)]; or CCRFOV [any air-purifying, full-facepiece respirator (gas mask) with a chin-style, front- or back-mounted acid gas canister]; GMFOV [any air-purifying, full-facepiece respirator (gas mask) with a chin-style, front- or back-mounted acid gas canister]; or SCBAF (any self-contained breathing apparatus with a full facepiece); or SAF (any supplied-air respirator with a full facepiece). *EMERGENCY OR PLANNED ENTRY INTO UNKNOWN CONCENTRATIONS OR IDLH CONDITIONS:* SCBAF:PD,PP (any self-contained breathing apparatus that has a full facepiece and is operated in a pressure-demand or other positive-pressure mode); or SAF:PD,PP:ASCBA (any supplied-air respirator that has a full facepiece and is operated in a pressure-demand or other positive-pressure mode in combination with an auxiliary self-contained breathing apparatus operated in a pressure-demand or other positive pressure mode). *ESCAPE:* GMFOV [any air-purifying, full-facepiece respirator (gas mask) with a chin-style, front- or back-mounted organic vapor canister]; or SCBAE (any appropriate escape-type, self-contained breathing apparatus). *Note:* Substance causes eye irritation or damage; eye protection needed.
Exposure limits (TWA unless otherwise noted): ACGIH TLV 200 ppm (950 mg/m³); NIOSH/OSHA 200 ppm (950 mg/m³).

Vapor (gas) irritant characteristics: Eye and respiratory tract irritant. Vapors cause smarting of the eyes or respiratory system if present in high concentrations. The effect may be temporary.
Liquid or solid irritant characteristics: If spilled on clothing and allowed to remain, may cause smarting and reddening of the skin.
Odor threshold: Approximately 5 ppm.
IDLH value: 1700 ppm [10% LEL].

FIRE DATA
Flash point: 88°F/31°C (oc); 62°F/17°C (cc).
Flammable limits in air: LEL: 1.7%; UEL: 9.8%.
Fire extinguishing agents not to be used: Water may be ineffective
Electrical hazard: Class I, Group D.
Burning rate: 4.4 mm/min (approximate).

CHEMICAL REACTIVITY
Reactivity with water: Slow decomposition forming acetic acid and *sec*-butyl alcohol.
Binary reactants: Nitrates, strong oxidizers, alkalis, and acids. Softens and dissolves rubber many and plastics.
Reactivity group: 34
Compatibility class: Esters.

ENVIRONMENTAL DATA
Food chain concentration potential: Log P_{ow} = 1.51. Unlikely to accumulate.
Water pollution: Effect of low concentrations on aquatic life is unknown. Fouling to shoreline. May be dangerous if it enters nearby water intakes; notify operators. Notify local health and pollution control officials. **Response to discharge:** Issue warning–high flammability. Mechanical containment.

SHIPPING INFORMATION
Grades of purity: Technical and Pure; **Storage temperature:** Ambient; **Inert atmosphere:** None; **Venting:** Open (flame arrester) or pressure-vacuum; **Stability during transport:** Stable.

NAS HAZARD CLASSIFICATION FOR BULK WATER TRANSPORTATION
FIRE: 3
HEALTH: Vapor irritant: 1; Liquid or solid irritant: 1; Poisons: 2
WATER POLLUTION: Human toxicity: 1; Aquatic toxicity: 2; Aesthetic effect: 2
REACTIVITY: Other chemicals: 1; Water: 0; Self-reaction: 0

PHYSICAL AND CHEMICAL PROPERTIES
Physical state @ 59°F/15°C and 1 atm: Liquid.
Molecular weight: 116.16
Boiling point @ 1 atm: 234°F/112°C/385°K.
Melting/Freezing point: –100°F/–73.5°C/199.7°K.
Critical temperature: 550°F/288°C/561°K.
Critical pressure: 469 psia = 32 atm = 3.2 MN/m^2.
Specific gravity (water = 1): 0.872 @ 68°F/20°C (liquid).
Liquid surface tension: 23.3 dynes/cm = 0.0233 N/m at 21°C.
Liquid water interfacial tension: (estimate) 58 dynes/cm = 0.058 N/m at 17°C.
Relative vapor density (air = 1): 4.0
Ratio of specific heats of vapor (gas): 1.061
Latent heat of vaporization: (estimate) 130 Btu/lb = 74 cal/g = 3.1 x 10^5 J/kg.
Heat of combustion: (estimate) –13,100 Btu/lb = –7300 cal/g = –305 x 10^5 J/kg.
Vapor pressure: (Reid) 1.0 psia; 20 mm.

tert-BUTYL ACETATE REC. B:3200

SYNONYMS: ACETIC ACID-*tert*-BUTYL ESTER; ACETIC ACID, 1,1-DIMETHYLETHYL ESTER; ACETATO de *terc*-BUTILO (Spanish); TEXACO LEAD APPRECIATOR; TLA; *tert*-BUTYL ESTER OF ACETIC ACID

IDENTIFICATION
CAS Number: 540-88-5
Formula: $C_6H_{12}O_2$; $CH_3COOC(CH_3)_3C_6H_{12}O_2$
DOT ID Number: 1123; DOT Guide Number: 129
Proper Shipping Name: Butyl acetates
Reportable Quantity (RQ): **(CERCLA)** 5000 lb/2270 kg

DESCRIPTION: Colorless liquid. Strong, fruity odor. Floats on water surface; insoluble. Flammable, irritating vapor is produced.

Highly flammable • Containers may BLEVE when exposed to fire • Vapors may form explosive mixture with air • Vapors are heavier than air and will collect and stay in low areas • Vapors may travel long distances to ignition sources and flashback • Vapors in confined areas (e.g., tanks, sewers, buildings) may explode when exposed to fire • Irritating to the skin, eyes, and respiratory tract; narcotic effects at high concentrations • Toxic products of combustion may include carbon monoxide.

Hazard Classification (based on NFPA-704 Rating System)
Health Hazards (Blue): 0; Flammability (Red): 3; Reactivity (Yellow): 0

EMERGENCY RESPONSE: See Appendix A (129)
Evacuation:
Public safety: Isolate spill area for at least 50 to 100 meters (160 to 330 feet) in all directions.
Spill: Large spill–Consider initial downwind evacuation for at least 300 meters (1000 feet).
Fire: Isolate for 800 meters (½ mile) in all directions, especially if tank, rail car, or tank truck is involved in fire.

EXPOSURE
Short-term effects: *SEEK MEDICAL ATTENTION*. **Vapor:** Irritating to eyes, nose, and throat. *IF INHALED*, will, will cause nausea, headache, or difficult breathing. Move to fresh air. *IF BREATHING HAS STOPPED*, give artificial respiration. IF breathing is difficult, administer oxygen. **Liquid:** Irritating to eyes. Harmful if swallowed.
Flush affected areas with plenty of water. *IF IN EYES*, hold eyelids open and flush with plenty of water.

HEALTH HAZARDS
Personal protective equipment (PPE): B-Level PPE. OSHA Table Z-1-A air contaminant. Chemical protective material(s) reported to offer protection (n-butyl acetate): PV alcohol, Teflon®, Silvershield®. Also, chlorinated polyethylene and butyl rubber offers limited protection The following materials are NOT recommended for service (n-butyl acetate): natural rubber, PVC, neoprene, nitrile, nitrile+PVC, Viton®.
Recommendations for respirator selection: NIOSH/OSHA
1500 ppm: SA:CF* (any supplied-air respirator operated in a continuous-flow mode); or PAPROV* [any powered, air-purifying respirator with organic vapor cartridge(s)]; or CCRFOV [any chemical cartridge respirator with a full facepiece and organic vapor cartridge(s)]; or GMFOV [any air-purifying, full-facepiece respirator (gas mask) with a chin-style, front- or back-mounted acid gas canister]; or SCBAF (any self-contained breathing apparatus

n-Butyl acrylate

with a full facepiece); or SAF (any supplied-air respirator with a full facepiece). *EMERGENCY OR PLANNED ENTRY INTO UNKNOWN CONCENTRATIONS OR IDLH CONDITIONS:* SCBAF:PD,PP (any self-contained breathing apparatus that has a full facepiece and is operated in a pressure-demand or other positive-pressure mode); or SAF:PD,PP:ASCBA (any supplied-air respirator that has a full facepiece and is operated in a pressure-demand or other positive-pressure mode in combination with an auxiliary self-contained breathing apparatus operated in a pressure-demand or other positive pressure mode). *ESCAPE:* GMFOV [any air-purifying, full-facepiece respirator (gas mask) with a chin-style, front- or back-mounted organic vapor canister or SCBAE (any appropriate escape-type, self-contained breathing apparatus). *Note:* Substance causes eye irritation or damage; eye protection needed.
Exposure limits (TWA unless otherwise noted): ACGIH TLV 200 ppm (950 mg/m^3); NIOSH/OSHA 200 ppm (950 mg/m^3).
Toxicity by ingestion: Poison.
Long-term health effects: Minor allergen and irritant.
Vapor (gas) irritant characteristics: Vapors cause smarting of eyes or respiratory system, if present in high concentrations the effect is temporary.
Liquid or solid irritant characteristics: Eye irritant.
Odor threshold: 4–47 ppm; detection in water 4 ppb.
IDLH value: 1500 ppm.

FIRE DATA
Flash point: 71°F/22°C.
Fire extinguishing agents not to be used: Water may be ineffective.
Behavior in fire: Containers may rupture and explode.
Electrical hazard: Class I. Group D.

CHEMICAL REACTIVITY
Binary reactants: Nitrates, strong oxidizers, alkalis, and acids.

ENVIRONMENTAL DATA
Food chain concentration potential: Log P_{ow} = 1.4. Unlikely to accumulate.
Water pollution: Effects of low concentrations on aquatic life are unknown. Fouling to shoreline. May be dangerous if it enters nearby water intakes; notify operators. Notify local health and pollution control authorities. **Response to discharge:** Issue warning–high flammability. Mechanical containment.

PHYSICAL AND CHEMICAL PROPERTIES
Physical state @ 59°F/15°C and 1 atm: Liquid.
Molecular weight: 116.16
Boiling point @ 1 atm: 208°F/97.8°C/370.8°K.
Specific gravity (water = 1): 0.8665 @ 68°F/20°C; 0.8593 @ 77°F/25°C.

n-BUTYL ACRYLATE   REC. B:3250

SYNONYMS: ACRILATO de *n*-BUTILO (Spanish); ACRYLIC ACID, BUTYL ESTER; ACRYLIC ACID, *n*-BUTYL ESTER; BUTYL ACRYLATE; BUTYL 2-PROPENOATE; 2-PROPENOIC ACID, BUTYL ESTER; *n*-BUTYL 2-PROPENOATE; EEC No. 607-062-00-3

IDENTIFICATION
CAS Number: 141-32-2
Formula: $C_7H_{12}O_2$; CH_2=CHCOO$(CH_2)_3CH_3$
DOT ID Number: UN 2348; DOT Guide Number: 129P
Proper Shipping Name: Butylacrylate

DESCRIPTION: Colorless watery liquid. Sharp, fragrant, acrylic odor. Floats on the surface of water.

Flammable • Corrosive to the skin, eyes, and respiratory tract; contact with skin or eyes can cause burns and vision impairment; inhalation symptoms may be delayed (lung edema; skin and eye sensitization) • Containers may BLEVE when exposed to fire • Vapors may form explosive mixture with air • Vapors are heavier than air and will collect and stay in low areas • Vapors may travel long distances to ignition sources and flashback • Vapors in confined areas (e.g., tanks, sewers, buildings) may explode when exposed to fire • Polymerization hazard • Heat can induce polymerization with rapid release of energy; sealed containers may rupture explosively • May react with itself blocking relief valves; leading to tank explosions • Toxic products of combustion may include carbon monoxide.

Hazard Classification (based on NFPA-704 Rating System)
Health Hazards (Blue): 2; Flammability (Red): 2; Reactivity (Yellow): 2

EMERGENCY RESPONSE: See Appendix A (129)
Evacuation:
Public safety: Isolate spill area for at least 50 to 100 meters (160 to 330 feet) in all directions.
Spill: Large spill–Consider initial downwind evacuation for at least 300 meters (1000 feet).
Fire: Isolate for 800 meters (½ mile) in all directions, especially if tank, rail car, or tank truck is involved in fire.

EXPOSURE
Short-term effects: *SEEK MEDICAL ATTENTION.*
Vapors/fumes: May cause lung edema; physical exertion will aggravate this condition. **Liquid:** Extremely irritating to skin and eyes. Harmful if swallowed. Remove contaminated clothing and shoes. Flush affected areas with plenty of water. *IF IN EYES*, hold eyelids open and flush with plenty of water. *IF SWALLOWED* and victim is *CONSCIOUS* have victim drink 4 to 8 ounces of water. *IF SWALLOWED* and victim is *UNCONSCIOUS OR HAVING CONVULSIONS,* do nothing except keep victim warm.
Note to physician or authorized medical personnel: Medical observation is recommended for 24 to 48 hours after breathing overexposure, as pulmonary edema may be delayed. As first aid for pulmonary edema, consider administering a corticosteroid spray. Cigarette smoking may exacerbate pulmonary injury and should be discouraged for at least 72 hours following exposure.

HEALTH HAZARDS
Personal protective equipment (PPE): B-Level PPE. OSHA Table Z-1-A air contaminant. Self-contained breathing apparatus, rubber gloves, acid goggles. Chemical protective material(s) reported to offer minimal to poor protection: Teflon®.
Exposure limits (TWA unless otherwise noted): ACGIH TLV 2 ppm; NIOSH 10 ppm (55 mg/m^3).
Toxicity by ingestion: Grade 2; LD_{50} = 0.5 to 5 g/kg (rat).
Long-term health effects: IARC rating 3; no human data; animal evidence inadequate. Repeated contact causes skin disorders, sensitization.
Vapor (gas) irritant characteristics: Eye and respiratory tract irritant. Vapors cause smarting of the eyes or respiratory system if present in high concentrations. The effect is temporary.
Liquid or solid irritant characteristics: Severe eye and skin irritant. If spilled on clothing and allowed to remain, may cause smarting and reddening of the skin.
Odor threshold: 0.001–0.10 ppm.

FIRE DATA
Flash point: 118°F/48°C (oc); 103°F/39°C (cc).
Flammable limits in air: LEL: 1.5%; UEL: 9.9%.
Fire extinguishing agents not to be used: Water may be ineffective or cause foaming.
Autoignition temperature: 559°F/292°C/565°K.
Electrical hazard: Class I, Group D.
Burning rate: 4.7 mm/min

CHEMICAL REACTIVITY
Binary reactants: Strong oxidants (may be violent), strong acids, aliphatic amines, alkanolamines.
Polymerization: Will polymerize on application of heat or exposure to light; uncontrolled bulk polymerization can be explosive; **Inhibitor of polymerization:** Methyl ether of hydroquinone: 15–100 ppm. Store in contact with air.
Reactivity group: 14
Compatibility class: Acrylate

ENVIRONMENTAL DATA
Food chain concentration potential: Negative; unlikely to accumulate.
Food chain concentration potential: Log P_{ow} = 2.4. Unlikely to accumulate.
Water pollution: Effect of low concentrations on aquatic life is unknown. Fouling to shoreline. May be dangerous if it enters nearby water intakes; notify operators. Notify local health and pollution control officials. **Response to discharge:** Mechanical containment. Chemical and physical treatment.

SHIPPING INFORMATION
Grades of purity: 99+%; **Storage temperature:** Ambient; **Inert atmosphere:** None; **Venting:** Pressure-vacuum; **Stability during transport:** Stable.

NAS HAZARD CLASSIFICATION FOR BULK WATER TRANSPORTATION
FIRE: 2
HEALTH: Vapor irritant: 1; Liquid or solid irritant: 1; Poisons: 1
WATER POLLUTION: Human toxicity: 1; Aquatic toxicity: 2; Aesthetic effect: 2
REACTIVITY: Other chemicals: 2; Water: 0; Self-reaction: 3

PHYSICAL AND CHEMICAL PROPERTIES
Physical state @ 59°F/15°C and 1 atm: Liquid.
Molecular weight: 128.17
Boiling point @ 1 atm: 260°F/127°C/400°K.
Melting/Freezing point: −83°F/−64°C/209°K.
Critical temperature: 621°F/327°C/600°K.
Critical pressure: 426 psia = 29 atm = 2.9 MN/m^2.
Specific gravity (water = 1): 0.899 @ 68°F/20°C (liquid).
Liquid surface tension: (estimate) 20 dynes/cm = 0.020 N/m at 27°C.
Liquid water interfacial tension: (estimate) 60 dynes/cm = 0.060 N/m at 27°C.
Relative vapor density (air = 1): 4.4
Ratio of specific heats of vapor (gas): 1.080
Latent heat of vaporization: 120 Btu/lb = 66.4 cal/g = 2.78 x 10^5 J/kg.
Heat of combustion: −13,860 Btu/lb = −7700 cal/g = −322.4 x 10^5 J/kg.
Heat of polymerization: −25.9 Btu/lb = −144 cal/g = −6.03 x 10^5 J/kg.
Vapor pressure: (Reid) 0.2 psia; 4 mm.

n-BUTYL ALCOHOL   REC. B:3350

SYNONYMS: ALCOHOL n-BUTILICO (Spanish); ALCOHOL C-4; ALCOOL BUTYLIQUE (French); BUTANOL; 1-BUTANOL; n-BUTANOL; BUTYL ALCOHOL; PROPYL CARBINOL; BUTAN-1-OL; BUTYL HYDROXIDE; CCS 203; EEC No. 603-004-00-6; 1-HYDROXYBUTANE; METHYLOLPROPANE; NBA; PROPYL CARBINOL; n-PROPYL CARBINOL; PROPYLMETHANOL; RCRA No. U031

IDENTIFICATION
CAS Number: 71-36-3
Formula: $C_4H_{10}O$; $CH_3(CH_2)_2CH_2OH$
DOT ID Number: UN 1120; DOT Guide Number: 129
Proper Shipping Name: Butanols
Reportable Quantity (RQ): **(CERCLA)** 5000 lb/2270 kg

DESCRIPTION: Colorless, watery liquid. Alcohol odor. Floats and mixes slowly with water. Flammable, irritating vapor is produced.

Highly flammable • Containers may BLEVE when exposed to fire • Vapors may form explosive mixture with air • Vapors are heavier than air and will collect and stay in low areas • Vapors may travel long distances to ignition sources and flashback • Vapors in confined areas (e.g., tanks, sewers, buildings) may explode when exposed to fire • Irritating to the skin, eyes, and respiratory tract • Toxic products of combustion may include carbon monoxide.

Hazard Classification (based on NFPA-704 Rating System)
Health Hazards (Blue): 1; Flammability (Red): 3; Reactivity (Yellow): 0

EMERGENCY RESPONSE: See Appendix A (129)
Evacuation:
Public safety: Isolate spill area for at least 50 to 100 meters (160 to 330 feet) in all directions.
Spill: Large spill–Consider initial downwind evacuation for at least 300 meters (1000 feet).
Fire: Isolate for 800 meters (½ mile) in all directions, especially if tank, rail car, or tank truck is involved in fire.

EXPOSURE
Short-term effects: *SEEK MEDICAL ATTENTION*. **Vapor:** Irritating to eyes, nose, and throat. *IF INHALED*, will, will cause nausea, headache, dizziness. Move to fresh air. *IF BREATHING HAS STOPPED*, give artificial respiration. IF breathing is difficult, administer oxygen. **Liquid:** *HARMFUL IF SWALLOWED OR ABSORBED THROUGH THE SKIN*. Irritating to skin and eyes. Harmful if swallowed. Remove contaminated clothing and shoes. Flush affected areas with plenty of water. *IF IN EYES*, hold eyelids open and flush with plenty of water. *IF SWALLOWED* and victim is *CONSCIOUS AND ABLE TO SWALLOW*, have victim drink 4 to 8 ounces of water.

HEALTH HAZARDS
Personal protective equipment (PPE): B-Level PPE. OSHA Table Z-1-A air contaminant. Full face organic vapor mask. Chemical protective material(s) reported to offer protection: neoprene, nirtile, polyethylene, butyl rubber, Teflon®. Also, polyurethane, chlorinated polyethylene, PV alcohol, PVC, and Viton offers limited protection
Recommendations for respirator selection: NIOSH/OSHA
1250 ppm: SA:CF* [any supplied-air respirator operated in a continuous-flow mode); or PAPROV* (any powered, air-purifying

respirator with organic vapor cartridge(s)]. *1400 ppm*: CCRFOV [any chemical cartridge respirator with a full facepiece and organic vapor cartridge(s)]; or GMFOV [any air-purifying, full-facepiece respirator (gas mask) with a chin-style, front- or back-mounted acid gas canister]; or PAPRTOV* [any powered, air-purifying respirator with a tight-fitting facepiece and organic vapor cartridge(s)]; SCBAF (any self-contained breathing apparatus with a full facepiece); or SAF (any supplied-air respirator with a full facepiece). *EMERGENCY OR PLANNED ENTRY INTO UNKNOWN CONCENTRATIONS OR IDLH CONDITIONS:* SCBAF:PD,PP (any self-contained breathing apparatus that has a full facepiece and is operated in a pressure-demand or other positive-pressure mode); or SAF:PD,PP:ASCBA (any supplied-air respirator that has a full facepiece and is operated in a pressure-demand or other positive-pressure mode in combination with an auxiliary self-contained breathing apparatus operated in a pressure-demand or other positive pressure mode). *ESCAPE:* GMFOV [any air-purifying, full-facepiece respirator (gas mask) with a chin-style, front- or back-mounted organic vapor canister]; or SCBAE (any appropriate escape-type, self-contained breathing apparatus). *Note:* Substance causes eye irritation or damage; eye protection needed.
Exposure limits (TWA unless otherwise noted): OSHA PEL 100 ppm (300 mg/m^3); NIOSH REL ceiling 50 ppm (150 mg/m^3); skin contact contributes significantly in overall exposure.
Short-term exposure limits (15-minute TWA): ACGIH TLV ceiling 50 ppm (152 mg/m^3)*; skin contact contributes significantly in overall exposure. *25 ppm proposed.
Toxicity by ingestion: Grade 2; LD$_{50}$ = 0.5 to 5 g/kg (rat).
Vapor (gas) irritant characteristics: Vapors cause smarting of the eyes or respiratory system if present in high concentrations. The effect may be temporary.
Liquid or solid irritant characteristics: Minimum hazard. If spilled on clothing and allowed to remain, may cause smarting and reddening of the skin.
Odor threshold: 0.12–11 ppm.
IDLH value: 1400 ppm [10% LEL].

FIRE DATA
Flash point: 84°F/29°C (cc); 98°F/37°C (oc).
Flammable limits in air: LEL: 1.4%; UEL 11.2%.
Autoignition temperature: 650°F/343°C/616°K.
Electrical hazard: Class I, Group D. Flow or agitation of substance may generate electrostatic charges due to low conductivity.
Burning rate: 3.2 mm/min

CHEMICAL REACTIVITY
Binary reactants: Strong oxidizers, strong mineral acids, alkali metals, halogens. Attacks many plastics. Water free *n*-butanol reacts with aluminum at temperatures above 120°F/88°C.
Reactivity group: 20
Compatibility class: Alcohols, glycols

ENVIRONMENTAL DATA
Food chain concentration potential: Negative.
Water pollution: Dangerous to aquatic life in high concentrations. May be dangerous if it enters nearby water intakes; notify operators. Notify local health and pollution control officials.
Response to discharge: Restrict access. Disperse and flush.

SHIPPING INFORMATION
Grades of purity: 99+%; **Storage temperature:** Ambient; **Inert atmosphere:** None; **Venting:** Open (flame arrester); **Stability during transport:** Stable.

NAS HAZARD CLASSIFICATION FOR BULK WATER TRANSPORTATION
FIRE: 3
HEALTH: Vapor irritant: 1; Liquid or solid irritant: 1; Poisons: 2
WATER POLLUTION: Human toxicity: 2; Aquatic toxicity: 2; Aesthetic effect: 2
REACTIVITY: Other chemicals: 2; Water: 0; Self-reaction: 0

PHYSICAL AND CHEMICAL PROPERTIES
Physical state @ 59°F/15°C and 1 atm: Liquid.
Molecular weight: 74.12
Boiling point @ 1 atm: 243.9°F/117.7°C/390.9°K.
Melting/Freezing point: –129°F/–89.3°C/183.9°K.
Critical temperature: 553.6°F/289.8°C/563.0°K.
Critical pressure: 640.2 psia = 43.55 atm = 4.412 MN/m^2.
Specific gravity (water = 1): 0.810 @ 68°F/20°C (liquid).
Liquid surface tension: 24.6 dynes/cm = 0.0246 N/m @ 68°F/20°C.
Liquid water interfacial tension: (estimate) 56 dynes/cm = 0.056 N/m at 27°C.
Relative vapor density (air = 1): 2.6
Ratio of specific heats of vapor (gas): 1.083
Latent heat of vaporization: 256 Btu/lb = 142 cal/g = 5.95 x 10^5 J/kg.
Heat of combustion: –14,230 Btu/lb = –7906 cal/g = –331.0 x 10^5 J/kg.
Heat of fusion: 29.93 cal/g.
Vapor pressure: (Reid) 0.3 psia; 5 mm.

---

sec-BUTYL ALCOHOL REC. B:3400

SYNONYMS: ALCOHOL *sec*-BUTILICO (Spanish); ALCOHOL C-4; ALCOOL BUTYLIQUE SECONDAIRE (French): *sec*-BUTANOL; *dl*-2-BUTANOL; BUTAN-2-OL; BUTANOL-2; 2-BUTANOL; BUTANOL SECONDAIRE (French); 2-BUTYL ALCOHOL; BUTYLENE HYDRATE; CCS 301; ETHYL METHYL CARBINOL; 2-HYDROXYBUTANE; METHYLETHYLCARBINOL; S.B.A; 1-METHYPROPYL ALCOHOL; METHYL ETHYL CARBINOL; METHYLETHYLCARBINOL; SBA0

IDENTIFICATION
CAS Number: 78-92-2
Formula: $C_4H_{10}O$; $CH_3CH_2CH(OH)CH_3$
DOT ID Number: UN 1120; DOT Guide Number: 129
Proper Shipping Name: Butanols

DESCRIPTION: Colorless, watery liquid. Strong, pleasant, alcohol odor. Floats and mixes slowly with water.

Highly flammable • Containers may BLEVE when exposed to fire • Vapors may form explosive mixture with air • Vapors are heavier than air and will collect and stay in low areas • Vapors may travel long distances to ignition sources and flashback • Vapors in confined areas (e.g., tanks, sewers, buildings) may explode when exposed to fire • Irritating to the skin, eyes, and respiratory tract • Toxic products of combustion may include carbon monoxide.

Hazard Classification (based on NFPA-704 Rating System)
Health Hazards (Blue): 1; Flammability (Red): 3; Reactivity (Yellow): 0

EMERGENCY RESPONSE: See Appendix A (129)
Evacuation:
Public safety: Isolate spill area for at least 50 to 100 meters (160 to 330 feet) in all directions.
Spill: Large spill–Consider initial downwind evacuation for at least 300 meters (1000 feet).
Fire: Isolate for 800 meters (½ mile) in all directions, especially if tank, rail car, or tank truck is involved in fire.

EXPOSURE
Short-term effects: *SEEK MEDICAL ATTENTION.* **Vapor:** Irritating to eyes, nose, and throat. *IF INHALED*, will, will cause headache, dizziness or difficult breathing. Move to fresh air. *IF BREATHING HAS STOPPED*, give artificial respiration. IF breathing is difficult, administer oxygen. **Liquid:** Irritating to eyes. *IF IN EYES*, hold eyelids open and flush with plenty of water.

HEALTH HAZARDS
Personal protective equipment (PPE): B-Level PPE. OSHA Table Z-1-A air contaminant. Chemical protective material(s) reported to offer protection: butyl rubber. Also, natural rubber, neoprene, and nitrile rubber may offer some protection from alcohols. See also *n*-butyl alcohol.
Recommendations for respirator selection: NIOSH/OSHA *1000 ppm*: CCROV [any chemical cartridge respirator with organic vapor cartridge(s)]; SA* (any supplied-air respirator). *2000 ppm*: SA:CF* (any supplied-air respirator operated in a continuous-flow mode); or PAPROV* [any powered, air-purifying respirator with organic vapor cartridge(s)]; or CCRFOV [any chemical cartridge respirator with a full facepiece and organic vapor cartridge(s)]; or GMFOV [any air-purifying, full-facepiece respirator (gas mask) with a chin-style, front- or back-mounted acid gas canister]; or SCBAF (any self-contained breathing apparatus with a full facepiece); or SAF (any supplied-air respirator with a full facepiece). *EMERGENCY OR PLANNED ENTRY INTO UNKNOWN CONCENTRATIONS OR IDLH CONDITIONS*: SCBAF:PD,PP (any self-contained breathing apparatus that has a full facepiece and is operated in a pressure-demand or other positive-pressure mode); or SAF:PD,PP:ASCBA (any supplied-air respirator that has a full facepiece and is operated in a pressure-demand or other positive-pressure mode in combination with an auxiliary self-contained breathing apparatus operated in a pressure-demand or other positive pressure mode). *ESCAPE*: GMFOV [any air-purifying, full-facepiece respirator (gas mask) with a chin-style, front- or back-mounted organic vapor canister]; or SCBAE (any appropriate escape-type, self-contained breathing apparatus). *Note:* Substance reported to cause eye irritation or damage; may require eye protection.
Exposure limits (TWA unless otherwise noted): ACGIH TLV 100 ppm (303 mg/m^3); OSHA PEL 150 ppm (450 mg/m3); NIOSH REL 100 ppm (305 mg/m^3).
Short-term exposure limits (15-minute TWA): NIOSH STEL 150 ppm (455 mg/m^3).
Toxicity by ingestion: Grade 1; LD$_{50}$ = 5 to 15 g/kg (rat-single oral dose).
Long-term health effects: Possible skin problems.
Vapor (gas) irritant characteristics: Vapors cause smarting of the eyes or respiratory system if present in high concentrations. The effect may be temporary.
Liquid or solid irritant characteristics: Severe eye irritant.
Odor threshold: 0.1–14 ppm.
IDLH value: 2000 ppm.

FIRE DATA
Flash point: 74°F/24°C (cc).
Flammable limits in air: LEL: 1.7%; UEL: 9.8% (both @ 212°F/100°C).
Fire extinguishing agents not to be used: Water maybe ineffective on fire.
Autoignition temperature: 761°F/405°C/678°K.
Electrical hazard: Class I, Group D.
Burning rate: 3.1 mm/min

CHEMICAL REACTIVITY
Binary reactants: Strong oxidizers, organic peroxides, perchloric acid, permonosulfuric acid.
Reactivity group: 20
Compatibility class: Alcohols, glycols

ENVIRONMENTAL DATA
Food chain concentration potential: Log P$_{ow}$ = 0.6–8.1. Unlikely to accumulate.
Water pollution: Effect of low concentrations on aquatic life is unknown. May be dangerous if it enters nearby water intakes; notify operators. Notify local health and pollution control officials.
Response to discharge: Restrict access. Disperse and flush.

SHIPPING INFORMATION
Grades of purity: 99+%; **Storage temperature:** Ambient; **Inert atmosphere:** None; **Venting:** Open (flame arrester) or pressure-vacuum; **Stability during transport:** Stable.

NAS HAZARD CLASSIFICATION FOR BULK WATER TRANSPORTATION
FIRE: 3
HEALTH: Vapor irritant: 1; Liquid or solid irritant: 0; Poisons: 1
WATER POLLUTION: Human toxicity: 1; Aquatic toxicity: 1; Aesthetic effect: 2
REACTIVITY: Other chemicals: 2; Water: 0; Self-reaction: 0

PHYSICAL AND CHEMICAL PROPERTIES
Physical state @ 59°F/15°C and 1 atm: Liquid.
Molecular weight: 74.12
Boiling point @ 1 atm: 201°F/94°C/367°K.
Melting/Freezing point: −174.5°F/−114.7°C/158.5°K.
Critical temperature: 505.0°F/262.8°C/536.0°K.
Critical pressure: 608.4 psia = 41.39 atm = 4.193 MN/m^2.
Specific gravity (water = 1): 0.807 @ 68°F/20°C (liquid).
Liquid surface tension: 23.0 dynes/cm = 0.023 N/m @ 68°F/20°C.
Ratio of specific heats of vapor (gas): 1.080
Latent heat of vaporization: 243 Btu/lb = 135 cal/g = 5.65 x 10^5 J/kg.
Heat of combustion: −15,500 Btu/lb = −8600 cal/g = −360 x 10^5 J/kg.
Vapor pressure: (Reid) 0.2 psia; 12 mm.

---

tert-BUTYL ALCOHOL   REC. B:3450

SYNONYMS: ALCOHOL *terc*-BUTILICO (Spanish); ALCOHOL C-4; ALCOOL BUTYLIQUE TERTIAIRE (French); *tert*-BUTANOL; BUTANOL TERTIAIRE (French); *tert*-BUTYL HYDROXIDE; 1,1-DIMETHYLETHANOL; EEC No. 603-005-00-1; METHANOL, TRIMETHYL-; 2-METHYL-2-PROPANOL; 2-PROPANOL, 2-METHYL-; TBA; TRIMETHYL CARBINOL

IDENTIFICATION
CAS Number: 75-65-0
Formula: C$_4$H$_{10}$O; (CH$_3$)$_3$COH

tert-Butyl alcohol

DOT ID Number: UN 1120; DOT Guide Number: 129
Proper Shipping Name: Butanols

DESCRIPTION: Colorless, oily liquid. Sharp camphor-like odor. Floats on the surface of water; soluble. Flammable, irritating vapor is produced.

Highly flammable • Containers may BLEVE when exposed to fire • Vapors may form explosive mixture with air • Vapors are heavier than air and will collect and stay in low areas • Vapors may travel long distances to ignition sources and flashback • Vapors in confined areas (e.g., tanks, sewers, buildings) may explode when exposed to fire • Irritating to the skin, eyes, and respiratory tract • Toxic products of combustion may include carbon monoxide.

Hazard Classification (based on NFPA-704 Rating System)
Health Hazards (Blue): 1; Flammability (Red): 3; Reactivity (Yellow): 0

EMERGENCY RESPONSE: See Appendix A (129)
Evacuation:
Public safety: Isolate spill area for at least 50 to 100 meters (160 to 330 feet) in all directions.
Spill: Large spill–Consider initial downwind evacuation for at least 300 meters (1000 feet).
Fire: Isolate for 800 meters (½ mile) in all directions, especially if tank, rail car, or tank truck is involved in fire.

EXPOSURE
Short-term effects: *SEEK MEDICAL ATTENTION.* **Vapor:** Irritating to eyes, nose, throat. *IF INHALED*, will, will cause dizziness, difficult breathing. Move to fresh air. *IF BREATHING HAS STOPPED*, give artificial respiration. IF breathing is difficult, administer oxygen. **Liquid:** Irritating to skin and eyes. Harmful if swallowed. Remove contaminated clothing and shoes. Flush affected areas with plenty of water. *IF IN EYES*, hold eyelids open and flush with plenty of water. *IF SWALLOWED* and victim is *CONSCIOUS AND ABLE TO SWALLOW*, have victim drink 4 to 8 ounces of water.

HEALTH HAZARDS
Personal protective equipment (PPE): B-Level PPE. OSHA Table Z-1-A air contaminant. Chemical protective material(s) reported to offer protection: butyl rubber. See also *n*-butyl alcohol.
Recommendations for respirator selection: NIOSH/OSHA
1600 ppm: SA:CF (any supplied-air respirator operated in a continuous-flow mode); or PAPROV [any powered, air-purifying respirator with organic vapor cartridge(s)]; or CCRFOV [any chemical cartridge respirator with a full facepiece and organic vapor cartridge(s)]; or GMFOV[any air-purifying, full-facepiece respirator (gas mask) with a chin-style, front- or back-mounted acid gas canister]; or SCBAF (any self-contained breathing apparatus with a full facepiece); or SAF (any supplied-air respirator with a full facepiece). *EMERGENCY OR PLANNED ENTRY INTO UNKNOWN CONCENTRATIONS OR IDLH CONDITIONS:* SCBAF:PD,PP (any self-contained breathing apparatus that has a full facepiece and is operated in a pressure-demand or other positive-pressure mode); or SAF:PD,PP:ASCBA (any supplied-air respirator that has a full facepiece and is operated in a pressure-demand or other positive-pressure mode in combination with an auxiliary self-contained breathing apparatus operated in a pressure-demand or other positive pressure mode). *ESCAPE:* GMFOV[any air-purifying, full-facepiece respirator (gas mask) with a chin-style, front- or back-mounted organic vapor canister]; or SCBAE (any appropriate escape-type, self-contained breathing apparatus). *Note:* Substance causes eye irritation or damage; eye protection needed.
Exposure limits (TWA unless otherwise noted): ACGIH TLV 100 ppm; NIOSH/OSHA 100 ppm (300 mg/m^3).
Short-term exposure limits (15-minute TWA): NIOSH STEL 150 ppm (450 mg/m^3).
Toxicity by ingestion: Grade 2; LD_{50} = 0.5 to 5.0 g/kg (rat).
Vapor (gas) irritant characteristics: Eye and respiratory tract irritant. Vapors cause smarting of the eyes or respiratory system if present in high concentrations. The effect may be temporary.
Liquid or solid irritant characteristics: Eye and skin irritant. Practically harmless to the skin.
Odor threshold: 21.5 ppm.
IDLH value: 1600 ppm.

FIRE DATA
Flash point: 61°F/16°C (oc); 52°F/11°C (cc).
Flammable limits in air: LEL: 2.35%; UEL: 8.00%.
Fire extinguishing agents not to be used: Water may be ineffective on fire
Autoignition temperature: 896°F/480°C/753°K.
Electrical hazard: Class I, Group D.
Burning rate: 3.4 mm/min

CHEMICAL REACTIVITY
Binary reactants: Forms explosive mixture with air. Reacts with strong mineral acids, alkali metals, strong hydrochloric acid, oxidizers. Attacks plastics.
Reactivity group: 20
Compatibility class: Alcohols, glycols

ENVIRONMENTAL DATA
Food chain concentration potential: Log P_{ow} = 0.35–0.89. Unlikely to accumulate.
Water pollution: Effect of low concentrations on aquatic life is unknown. May be dangerous if it enters nearby water intakes; notify operators. Notify local health and pollution control officials.
Response to discharge: Issue warning–high flammability. Restrict access. Disperse and flush.

SHIPPING INFORMATION
Grades of purity: 99+%; **Storage temperature:** Ambient; **Inert atmosphere:** None; **Venting:** Open (flame arrester) or pressure-vacuum; **Stability during transport:** Stable.

NAS HAZARD CLASSIFICATION FOR BULK WATER TRANSPORTATION
FIRE: 3
HEALTH: Vapor irritant: 1; Liquid or solid irritant: 0; Poisons: 1
WATER POLLUTION: Human toxicity: 2; Aquatic toxicity: 1; Aesthetic effect: 2
REACTIVITY: Other chemicals: 2; Water: 0; Self-reaction: 0

PHYSICAL AND CHEMICAL PROPERTIES
Physical state @ 59°F/15°C and 1 atm: Liquid.
Molecular weight: 74.12
Boiling point @ 1 atm: 181°F/82.6°C/355.8°K.
Melting/Freezing point: 78.3°F/25.7°C/298.9°K.
Critical temperature: 451°F/233°C/506°K.
Critical pressure: 576 psia = 39.2 atm = 3.97 MN/m^2.
Specific gravity (water = 1): 0.78 at 26°C (liquid); 0.79 at 68°F/20°F (solid).
Liquid surface tension: 20.7 dynes/cm = 0.0207 N/m at 77°F/25°C.
Relative vapor density (air = 1): 2.6

Ratio of specific heats of vapor (gas): 1.080
Latent heat of vaporization: 234 Btu/lb = 130 cal/g = 5.44 x 10^5 J/kg.
Heat of combustion: –14,000 Btu/lb = –7780 cal/g = –325.7 x 10^5 J/kg.
Heat of fusion: 21.88 cal/g.
Vapor pressure: (Reid) 1.8 psia; 3 mm.

n-BUTYLAMINE   REC. B:3500

SYNONYMS: 1-AMINOBUTANE; 1-BUTANAMINE; *n*-BUTILAMINA (Spanish); *n*-BUTYLAMIN (German); BUTYLAMINE; EEC No. 612-005-00-0; MONOBUTYLAMINE; MONO-*n*-BUTYLAMINE; NORVALAMINE

IDENTIFICATION
CAS Number: 109-73-9
Formula: $C_4H_{11}N$; $CH_3(CH_2)_3NH_2$
DOT ID Number: UN 1125; DOT Guide Number: 132
Proper Shipping Name: *n*-Butylamine
Reportable Quantity (RQ): 1000 lb/454 kg

DESCRIPTION: Colorless liquid. Fishy, ammonia-like odor. Mixes with water.

Extremely flammable • Corrosive to the skin, eyes, and respiratory tract; contact with skin or eyes can cause burns and vision impairment; inhalation symptoms may be delayed • Containers may BLEVE when exposed to fire •Vapors may form explosive mixture with air • Vapors are heavier than air and will collect and stay in low areas • Vapors may travel long distances to ignition sources and flashback • Vapors in confined areas (e.g., tanks, sewers, buildings) may explode when exposed to fire • Toxic products of combustion may include oxides of nitrogen, amine, CO_2, and other gases.

Hazard Classification (based on NFPA-704 Rating System)
Health Hazards (Blue): 3; Flammability (Red): 3; Reactivity (Yellow): 0

EMERGENCY RESPONSE: See Appendix A (132)
Evacuation:
Public safety: Isolate spill area for at least 50 to 100 meters (160 to 330 feet) in all directions.
Spill: Increase, as necessary, the isolation distance shown above,in "Public safety."
Fire: Isolate for 800 meters (½ mile) in all directions, especially if tank, rail car, or tank truck is involved in fire.

EXPOSURE
Short-term effects: *SEEK MEDICAL ATTENTION.* **Vapor:** *POISONOUS IF INHALED OR IF SKIN IS EXPOSED. IF INHALED*, will, will cause dizziness, headache, coughing, or difficult breathing. *IF IN EYES*, hold eyelids open and flush with plenty of water. *IF BREATHING HAS STOPPED,* give artificial respiration. IF breathing is difficult, administer oxygen. **Liquid:***HARMFUL IF SWALLOWED OR ABSORBED THROUGH THE SKIN.* Will burn skin and eyes. *IF SWALLOWED*, will cause nausea and vomiting. Remove contaminated clothing and shoes. Flush affected areas with plenty of water. *IF IN EYES*, hold eyelids open and flush with plenty of water. *IF SWALLOWED* and victim is *CONSCIOUS AND ABLE TO SWALLOW*, have victim drink 4 to 8 ounces of water. *IF SWALLOWED* and victim is *UNCONSCIOUS OR HAVING CONVULSIONS* do nothing except keep victim warm.
Note to physician or authorized medical personnel: Medical observation is recommended for 24 to 48 hours after breathing overexposure, as pulmonary edema may be delayed. As first aid for pulmonary edema, consider administering a corticosteroid spray. Cigarette smoking may exacerbate pulmonary injury and should be discouraged for at least 72 hours following exposure.

HEALTH HAZARDS
Personal protective equipment (PPE): A-Level PPE. OSHA Table Z-1-A air contaminant. Ammonia-methylamine or organic vapor respirator. Chemical protective material(s) reported to offer protection: Teflon®.
Recommendations for respirator selection: NIOSH/OSHA
50 ppm: CCRS* [any chemical cartridge respirator with cartridge(s) providing protection against the compound of concern]; or SA (any supplied-air respirator).*125 ppm*: SA:CF* (any supplied-air respirator operated in a continuous-flow mode); or PAPRS* [any powered, air-purifying respirator with cartridge(s) providing protection against the compound of concern]. *250 ppm*: CCRFS [any chemical cartridge respirator with a full facepiece and cartridge(s) providing protection against the compound of concern]; or GMFS [any air-purifying, full-facepiece respirator (gas mask) with a chin-style, front- or back-mounted canister providing protection against the compound of concern]; or PAPRTS* [any powered, air-purifying respirator with a tight-fitting facepiece and cartridge(s) providing protection against the compound of concern]; or SCBAF (any self-contained breathing apparatus with a full facepiece); or SAF (any supplied-air respirator with a full facepiece). *300 ppm*: SAF:PD,PP (any supplied-air respirator that has a full facepiece and is operated in a pressure-demand or other positive-pressure mode). *EMERGENCY OR PLANNED ENTRY INTO UNKNOWN CONCENTRATIONS OR IDLH CONDITIONS*: SCBAF:PD,PP (any self-contained breathing apparatus that has a full facepiece and is operated in a pressure-demand or other positive-pressure mode); or SAF:PD,PP:ASCBA (any supplied-air respirator that has a full facepiece and is operated in a pressure-demand or other positive-pressure mode in combination with an auxiliary self-contained breathing apparatus operated in a pressure-demand or other positive pressure mode). *ESCAPE*: GMFS [any air-purifying, full-facepiece respirator (gas mask) with a chin-style, front- or back-mounted canister providing protection against the compound of concern]; or SCBAE (any appropriate escape-type, self-contained breathing apparatus). *Note*: Substance reported to cause eye irritation or damage; may require eye protection.
Exposure limits (TWA unless otherwise noted): NIOSH/OSHA 5 ppm (15 mg/m^3); skin contact contributes significantly in overall exposure.
Short-term exposure limits (15-minute TWA): ACGIH TLV ceiling 5 ppm (15 mg/m^3) (skin)
Toxicity by ingestion: Grade 3; oral LD_{50} = 500 mg/kg (rat).
Vapor (gas) irritant characteristics: Severe eye and respiratory tract irritant.
Liquid or solid irritant characteristics: Severe eye and skin irritant.
Odor threshold: 0.08 ppm.
IDLH value: 300 ppm.

FIRE DATA
Flash point: 30°F/–1°C (oc); 10°F/–12°C (cc).
Flammable limits in air: LEL: 1.7%; UEL: 9.8%.
Fire extinguishing agents not to be used: Water may be ineffective.
Behavior in fire: Containers may rupture and explode.
Autoignition temperature: 594°F/312°C/585°K.

Electrical hazard: Class I, Group D.
Burning rate: 5.79 mm/min

CHEMICAL REACTIVITY
Binary reactants: Reacts with oxidizers, strong acids, halogenated compounds. Corrodes light metals, especially copper, aluminum and their alloys, in presence of moisture.
Neutralizing agents for acids and caustics: Flush with water.
Reactivity group: 7
Compatibility class: Aliphatic amines

ENVIRONMENTAL DATA
Food chain concentration potential: Log P_{ow} = 0.8–0.97. Unlikely to accumulate.
Water pollution: Harmful to aquatic life in very low concentrations. May be dangerous if it enters nearby water intakes; notify operators. Notify local health and wildlife officials.
Response to discharge: Issue warning–high flammability, air contaminant, water contaminant. Restrict access. Disperse and flush.

SHIPPING INFORMATION
Grades of purity: Pure, 100%; **Storage temperature:** Ambient; **Inert atmosphere:** None; **Venting:** Open (flame arrester); **Stability during transport:** Stable.

PHYSICAL AND CHEMICAL PROPERTIES
Physical state @ 59°F/15°C and 1 atm: Liquid.
Molecular weight: 73.14
Boiling point @ 1 atm: 171.3°F/77.4°C/350.6°K.
Melting/Freezing point: –58°F/–50°C/223°K.
Critical temperature: 484°F/251°C/524°K.
Critical pressure: 603 psia = 41 atm = 4.16 MN/m^2.
Specific gravity (water = 1): 0.741 @ 68°F/20°C (liquid).
Liquid surface tension: 53.11 dynes/cm = 0.05311 N/m @ 68°F/20°C.
Relative vapor density (air = 1): 2.5
Ratio of specific heats of vapor (gas): (estimate) 1.071
Latent heat of vaporization: 180 Btu/lb = 100 cal/g = 4.2 x 10^5 J/kg.
Heat of combustion: –17,595 Btu/lb = –9,775 cal/g = –409.0 x 10^5 J/kg.
Heat of solution: –137 Btu/lb = –76.2 cal/g = –3.19 x 10^5 J/kg.
Vapor pressure: (Reid) 1.39 psia; 82 mm.

---

sec-BUTYLAMINE   REC. B:3550

SYNONYMS: 2-AB; 2-AMINOBUTANE; 2-BUTANAMINE; *sec*-BUTILAMINA (Spanish); 1-METHYLPROPYLAMINE; TUTANE

IDENTIFICATION
CAS Number: 13952-84-6; may also apply to 513-49-5
Formula: C$_4$H$_{11}$N: CH$_3$CH$_2$CH(CH$_3$)NH$_2$
DOT ID Number: UN 2733; UN 2734; DOT Guide Number: 132
Proper Shipping Name: Alkylamines, n.o.s.
Reportable Quantity (RQ): **(CERCLA)** 1000 lb/454 kg

DESCRIPTION: White liquid. Ammonia-like odor. Mixes with water.

Extremely flammable • Corrosive to the skin, eyes, and respiratory tract; contact with skin or eyes can cause burns and vision impairment; inhalation symptoms may be delayed • Containers may BLEVE when exposed to fire •Vapors may form explosive mixture with air • Vapors are heavier than air and will collect and stay in low areas • Vapors may travel long distances to ignition sources and flashback • Vapors in confined areas (e.g., tanks, sewers, buildings) may explode when exposed to fire • Toxic products of combustion may include nitrogen oxides.

Hazard Classification (based on NFPA-704 Rating System)
Health Hazards (Blue): 3; Flammability (Red): 3; Reactivity (Yellow): 0

EMERGENCY RESPONSE: See Appendix A (132)
Evacuation:
Public safety: Isolate spill area for at least 50 to 100 meters (160 to 330 feet) in all directions.
Spill: Increase, as necessary, the isolation distance shown above, in "Public safety."
Fire: Isolate for 800 meters (½ mile) in all directions, especially if tank, rail car, or tank truck is involved in fire.

EXPOSURE
Short-term effects: *SEEK MEDICAL ATTENTION*. **Vapor:** Irritating to eyes, nose, and throat. *IF INHALED*, will, will cause coughing or difficult breathing. *IF IN EYES*, hold eyelids open and flush with plenty of water. *IF BREATHING HAS STOPPED*, give artificial respiration. IF breathing is difficult, administer oxygen. **Liquid:** Will burn skin and eyes. *IF SWALLOWED*, will cause nausea and vomiting
Remove contaminated clothing and shoes. Flush affected areas with plenty of water. *IF IN EYES*, hold eyelids open and flush with plenty of water. *IF SWALLOWED* and victim is *CONSCIOUS AND ABLE TO SWALLOW*, have victim drink 4 to 8 ounces of water and have victim induce vomiting. *IF SWALLOWED* and victim is *UNCONSCIOUS OR HAVING CONVULSIONS*, do nothing except keep victim warm.
Note to physician or authorized medical personnel: Medical observation is recommended for 24 to 48 hours after breathing overexposure, as pulmonary edema may be delayed. As first aid for pulmonary edema, consider administering a corticosteroid spray. Cigarette smoking may exacerbate pulmonary injury and should be discouraged for at least 72 hours following exposure.

HEALTH HAZARDS
Personal protective equipment (PPE): A-Level PPE. Chemical safety goggles; rubber gloves and apron; nonsparking shoes.
Exposure limits (TWA unless otherwise noted): ACGIH TLV ceiling 5 ppm (15 mg/m^3) (skin) as *n*-butylamine; NIOSH/OSHA ceiling 5 ppm (15 mg/m^3); skin contact contributes significantly in overall exposure as *n*-butylamine.
Toxicity by ingestion: Grade 3; oral LD$_{50}$ = 380 mg/kg (rat).
Vapor (gas) irritant characteristics: Eye and respiratory tract irritant. Vapors are irritating such that personnel will not usually tolerate moderate or high concentrations.
Liquid or solid irritant characteristics: Severe eye and skin irritant. Causes second-and-third-degree burns on short contact and is very injurious to the eyes.
Odor threshold: 0.08 ppm.
IDLH value: 2000 ppm.

FIRE DATA
Flash point: 16°F/–9°C (cc).
Fire extinguishing agents not to be used: Water may be ineffective.
Behavior in fire: Containers may rupture and explode.
Autoignition temperature: 712°F/378°C/651°K.

Electrical hazard: Class I, Group D.
Burning rate: 6.18 mm/min

CHEMICAL REACTIVITY
Binary reactants: Reacts with strong acids, alcohol, glycol, phenols, cresols, and many other chemicals and compounds. May corrode some metals in presence of water.
Neutralizing agents for acids and caustics: Flush with water.
Reactivity group: 7
Compatibility class: Aliphatic amines

ENVIRONMENTAL DATA
Food chain concentration potential: Negative; unlikely to accumulate.
Water pollution: Effect of low concentrations on aquatic life is unknown. May be dangerous if it enters nearby water intakes; notify operators. Notify local health and wildlife officials.
Response to discharge: Issue warning–high flammability, air contaminant, water contaminant. Restrict access. Disperse and flush.

SHIPPING INFORMATION
Grades of purity: Pure; **Storage temperature:** Ambient; **Inert atmosphere:** None; **Venting:** Open (flame arrester); **Stability during transport:** Stable.

NAS HAZARD CLASSIFICATION FOR BULK WATER TRANSPORTATION
FIRE: 3
HEALTH: Vapor irritant: 3; Liquid or solid irritant: 4; Poisons: 4
WATER POLLUTION: Human toxicity: 2; Aquatic toxicity: 2; Aesthetic effect: 1
REACTIVITY: Other chemicals: 3; Water: 0; Self-reaction: 0

PHYSICAL AND CHEMICAL PROPERTIES
Physical state @ 59°F/15°C and 1 atm: Liquid.
Molecular weight: 73.1
Boiling point @ 1 atm: 145°F/63°C/336°K.
Melting/Freezing point: −155°F/−104°C/169°K.
Specific gravity (water = 1): 0.721 @ 68°F/20°C (liquid).
Liquid surface tension: 22.42 dynes/cm = 0.02242 N/m @ 68°F/20°C.
Relative vapor density (air = 1): 2.52
Ratio of specific heats of vapor (gas): (estimate) 1.073 @ 68°F/20°C.
Latent heat of vaporization: 178.09 Btu/lb = 98.94 cal/g.
Latent heat of vaporization: 4.160×10^5 J/kg.
Heat of combustion: −17,600 Btu/lb = −9780 cal/g = 409×10^5 J/kg.
Heat of solution: −170 Btu/lb = −93 cal/g = -3.9×10^5 J/kg.
Vapor pressure: (Reid) 6.1 psia.

tert-BUTYLAMINE   REC. B:3600

SYNONYMS: 2-AMINOISOBUTANE; 2-AMINO-2-METHYLPROPANE; BUTYLAMINE, TERTIARY; 1,1-*terc*-BUTILAMINA (Spanish); DIMETHYLETHYLAMINE; 2-METHYL-2-PROPANAMINE; TRIMETHYLAMINOMETHANE; TRIMETHYLCARBINYLAMINE; TBA

IDENTIFICATION
CAS Number: 75-64-9
Formula: $C_4H_{11}N$; $(CH_3)_3CNH_2$

DOT ID Number: UN 2733; UN 2734; DOT Guide Number: 132
Proper Shipping Name: Alkylamines, n.o.s.; Amines, liquid, corrosive, flammable, n.o.s.
Reportable Quantity (RQ): **(CERCLA)** 1000 lb/454 kg

DESCRIPTION: Colorless liquid. Ammonia-like odor. Floats on the surface of water; soluble. Flammable, irritating vapor is produced.

Extremely flammable • Corrosive to the skin, eyes, and respiratory tract; contact with skin or eyes can cause burns and vision impairment; inhalation symptoms may be delayed • Containers may BLEVE when exposed to fire • Vapors may form explosive mixture with air • Vapors are heavier than air and will collect and stay in low areas • Vapors may travel long distances to ignition sources and flashback • Vapors in confined areas (e.g., tanks, sewers, buildings) may explode when exposed to fire.

Hazard Classification (based on NFPA-704 Rating System)
Health Hazards (Blue): 2; Flammability (Red): 4; Reactivity (Yellow): 0

EMERGENCY RESPONSE: See Appendix A (132)
Evacuation:
Public safety: Isolate spill area for at least 50 to 100 meters (160 to 330 feet) in all directions.
Spill: Increase, as necessary, the isolation distance shown above, in "Public safety."
Fire: Isolate for 800 meters (½ mile) in all directions, especially if tank, rail car, or tank truck is involved in fire.

EXPOSURE
Short-term effects: *SEEK MEDICAL ATTENTION.* **Vapor:** Irritating to eyes, nose, and throat. *IF INHALED*, will, will cause difficult breathing. Move to fresh air. *IF BREATHING HAS STOPPED,* give artificial respiration. IF breathing is difficult, administer oxygen. **Liquid:** Irritating to skin and eyes. Harmful if swallowed. Remove contaminated clothing and shoes. Flush affected areas with plenty of water. *IF IN EYES*, hold eyelids open and flush with plenty of water. *IF SWALLOWED* and victim is *CONSCIOUS AND ABLE TO SWALLOW*, have victim drink 4 to 8 ounces of water.
Note to physician or authorized medical personnel: Medical observation is recommended for 24 to 48 hours after breathing overexposure, as pulmonary edema may be delayed. As first aid for pulmonary edema, consider administering a corticosteroid spray. Cigarette smoking may exacerbate pulmonary injury and should be discouraged for at least 72 hours following exposure.

Evacuate area. Disperse and flush.

HEALTH HAZARDS
Personal protective equipment (PPE): A-Level PPE. Chemical protective material(s) reported to offer protection: butyl rubber.
Exposure limits (TWA unless otherwise noted): ACGIH TLV ceiling 5 ppm (15 mg/m^3) as *n*-butylamine; NIOSH/OSHA ceiling 5 ppm (15 mg/m^3) as *n*-butylamine.
Toxicity by ingestion: Grade 3; oral LD$_{50}$ = 180 mg/kg (rat).
Long-term health effects: Kidney damage; skin disorders.
Vapor (gas) irritant characteristics: Severe eye, skin and respiratory tract irritant.
Liquid or solid irritant characteristics: Severe eye, skin and respiratory tract irritant.
Odor threshold: 0.08 ppm.
IDLH value: 300 ppm.

176 Butyl benzyl phthalate

FIRE DATA
Flash point: 16°F/9°C (cc).
Flammable limits in air: LEL: 1.7%; UEL: 8.9% (both @ 212°F/100°C.
Fire extinguishing agents not to be used: Water may be ineffective.
Autoignition temperature: 716°F/380°C/653°K.
Electrical hazard: Class I, Group D.
Burning rate: 7 mm/min

CHEMICAL REACTIVITY
Binary reactants: Strong oxidizers, strong acids, alcohols, nitriles, phenols, glycol ethers. Liquid will attack some forms of plastics and may corrode some metals in presence of water.
Neutralizing agents for acids and caustics: Flush with water.
Reactivity group: 7
Compatibility class: Aliphatic amines

ENVIRONMENTAL DATA
Food chain concentration potential: Log P_{ow} = 0.4. Unlikely to accumulate.
Water pollution: Effect of low concentrations on aquatic life is unknown. May be dangerous if it enters nearby water intakes; notify operators. Notify local health and wildlife officials.
Response to discharge: Issue warning–high flammability, air contaminant. Restrict access.

SHIPPING INFORMATION
Grades of purity: 99+%; **Storage temperature:** Ambient; **Inert atmosphere:** None; **Venting:** Open; **Stability during transport:** Stable.

PHYSICAL AND CHEMICAL PROPERTIES
Physical state @ 59°F/15°C and 1 atm: Liquid.
Molecular weight: 73.14
Boiling point @ 1 atm: 113°F/45°C/318°K.
Specific gravity (water = 1): 0.696 @ 68°F/20°C (liquid).
Liquid surface tension: 19 dynes/cm = 0.019 N/m @ 68°F/20°C.
Relative vapor density (air = 1): 2.5
Latent heat of vaporization: 167.0 Btu/lb = 92.8 cal/g = 3.88 x 10^5 J/kg.
Heat of combustion: –17,600 Btu/lb = –9790 cal/g = –410 x 10^5 J/kg.
Heat of solution: –170 Btu/lb = –96 cal/g = –4.0 x 10^5 J/kg.
Vapor pressure: (Reid) 11 psia; 290 mm.

BUTYL BENZYL PHTHALATE REC. B:3650

SYNONYMS: BBP; BENZYL *n*-BUTYL PHTHALATE; FTALATO de BUTILBENCILO (Spanish); PHTHALIC ACID, BENZYL BUTYL ETHER; SATICIZER 160; SICOL 160

IDENTIFICATION
CAS Number: 85-68-7
Formula: $C_{19}H_{21}O_4$
DOT ID Number: UN 3082; DOT Guide Number: 171
Proper Shipping Name: Environmentally hazardous substances, n.o.s.
Reportable Quantity (RQ): **(CERCLA)** 100 lb/45.4 kg

DESCRIPTION: Colorless liquid. Slight odor. Sinks in water; soluble.

Combustible • Containers may BLEVE when exposed to fire • Vapors in confined areas (e.g., tanks, sewers, buildings) may explode when exposed to fire • Irritating to the skin, eyes, and respiratory tract • Toxic products of combustion may include carbon monoxide.

Hazard Classification (based on NFPA-704 Rating System)
Health Hazards (Blue): 1; Flammability (Red): 1; Reactivity (Yellow): 0

EMERGENCY RESPONSE: See Appendix A (171)
Evacuation:
Public safety: Isolate the area of spill or leak for at least 10 to 25 meters (30 to 80 feet) in all directions.
Spill: Increase, in the downwind direction, as necessary, the distance shown under "Public Safety."
Fire: If any large container is involved in fire, isolate for at least 800 meters (½ mile) in all directions; also, consider initial evacuation for 800 meters (½ mile) in all directions.

EXPOSURE
Short-term effects: *SEEK MEDICAL ATTENTION*. **Liquid:** Irritating to skin and eyes. Remove contaminated clothing and shoes. Flush affected areas with plenty of water.

HEALTH HAZARDS
Personal protective equipment (PPE): B-Level PPE.
Toxicity by ingestion: Grade 1; oral rat LD_{50} = 13,500 mg/kg.
Vapor (gas) irritant characteristics: Vapors are nonirritating to eyes and throat of most people.
Liquid or solid irritant characteristics: No appreciable hazard; practically harmless to skin.

FIRE DATA
Flash point: 390°F/199°C (oc).
Fire extinguishing agents not to be used: Water or foam may cause frothing.
Electrical hazard: Class I, Group D.

CHEMICAL REACTIVITY
Binary reactants: Destructive to rubber and paint.

ENVIRONMENTAL DATA
Food chain concentration potential: Log K_{ow} = 4.77. Values of > 3.0 are likely to accumulate.
Water pollution: DOT Appendix B, §172.101–marine pollutant. Effect of low concentrations on aquatic life is unknown. May be dangerous if it enters nearby water intakes; notify operators. Notify local health and wildlife officials. **Response to discharge:** Should be removed. Chemical and physical treatment. Confine and absorb on suitable material such as sawdust, clay, or Filtercel®. May be incinerated.

SHIPPING INFORMATION
Grades of purity: Commercial; **Storage temperature:** Ambient; **Inert atmosphere:** None; **Venting:** Open; **Stability during transport:** Stable.

NAS HAZARD CLASSIFICATION FOR BULK WATER TRANSPORTATION
FIRE: 1
HEALTH: Vapor irritant: 0; Liquid or solid irritant: 0; Poisons: 0
WATER POLLUTION: Human toxicity: 0; Aquatic toxicity: 1; Aesthetic effect: 3
REACTIVITY: Other chemicals: 1; Water: 0; Self-reaction: 0

PHYSICAL AND CHEMICAL PROPERTIES
Physical state @ 59°F/15°C and 1 atm: Liquid.
Molecular weight: 313
Boiling point @ 1 atm: 698°F/370°C/643°K.
Specific gravity (water = 1): 1.12 @ 68°F/20°C (liquid).
Heat of combustion: $-14{,}550$ Btu/lb $= -8{,}090$ cal/g $= -338 \times 10^5$ J/kg.
Vapor pressure: (Reid) Low

BUTYL BUTYRATE   REC. B:3700

SYNONYMS: BUTANOIC ACID, BUTYL ESTER; BUTIRATO de *n*-BUTILO (Spanish); *n*-BUTYL *n*-BUTANOATE; *n*-BUTYL *n*-BUTYRATE; BUTYL BUTANOATE; BUTYRIC ACID, BUTYL ESTER; EEC No. 607-031-00-4

IDENTIFICATION
CAS Number: 109-21-7
Formula: $C_8H_{16}O_2$; $C_3H_7CO_2C_4H_9$
DOT ID Number: UN 1993; DOT Guide Number: 128
Proper Shipping Name: Combustible liquid, n.o.s.

DESCRIPTION: Colorless liquid. Ester or pineapple odor. Floats on the surface of water.

Flammable • Containers may BLEVE when exposed to fire • Vapors may form explosive mixture with air • Vapors are heavier than air and will collect and stay in low areas • Vapors may travel long distances to ignition sources and flashback • Vapors in confined areas (e.g., tanks, sewers, buildings) may explode when exposed to fire • Irritating to the skin, eyes, and respiratory tract • Toxic products of combustion may include carbon monoxide.

Hazard Classification (based on NFPA-704 Rating System)
Health Hazards (Blue): 2; Flammability (Red): 2; Reactivity (Yellow): 0

EMERGENCY RESPONSE: See Appendix A (128)
Evacuation:
Public safety: Isolate spill area for at least 25 to 50 meters (80 to 160 feet) in all directions.
Spill: Large spill–Consider initial downwind evacuation for at least 300 meters (1000 feet).
Fire: Isolate for 800 meters (½ mile) in all directions, especially if tank, rail car, or tank truck is involved in fire.

EXPOSURE
Short-term effects: *SEEK MEDICAL ATTENTION.* **Vapor:** Irritating to eyes, nose, and throat. *IF INHALED*, will, will cause headache or dizziness. *IF IN EYES*, hold eyelids open and flush with plenty of water. *IF BREATHING HAS STOPPED*, give artificial respiration. IF breathing is difficult, administer oxygen. **Liquid:** Irritating to skin and eyes. *IF SWALLOWED*, will cause nausea, vomiting, dizziness, or headache. Remove contaminated clothing and shoes. Flush affected areas with plenty of water. *IF IN EYES*, hold eyelids open and flush with plenty of water. *IF SWALLOWED* and victim is *CONSCIOUS AND ABLE TO SWALLOW*, have victim drink 4 to 8 ounces of water and induce vomiting. *IF SWALLOWED* and victim is UNCONSCIOUS OR HAVING CONVULSIONS: Do nothing except keep victim warm.

HEALTH HAZARDS
Personal protective equipment (PPE): B-Level PPE. All-purpose canister mask or chemical cartridge respirator; glass or face shield; rubber gloves.
Toxicity by ingestion: Grade 1: LD_{50} = 9.5 g/kg (rabbit).
Vapor (gas) irritant characteristics: Eye and respiratory tract irritant.
Liquid or solid irritant characteristics: Eye irritant.

FIRE DATA
Flash point: 128°F/53°C (cc).
Fire extinguishing agents not to be used: Water may be ineffective

CHEMICAL REACTIVITY
Binary reactants: Violent reaction with strong oxidizers. May attack some forms of plastics.
Reactivity group: 34
Compatibility class: Esters.

ENVIRONMENTAL DATA
Water pollution: DOT Appendix B, §172.101–marine pollutant. Effect of low concentrations on aquatic life is unknown. Fouling to shoreline. May be dangerous if it enters nearby water intakes; notify operators. Notify local health and wildlife officials. Notify operators of local water intakes. **Response to discharge:** Issue warning–high flammability. Restrict access. Mechanical containment. Should be removed. Chemical and physical treatment.

SHIPPING INFORMATION
Grades of purity: 98%; **Storage temperature:** Ambient; **Stability during transport:** Stable.

NAS HAZARD CLASSIFICATION FOR BULK WATER TRANSPORTATION
FIRE: 2
HEALTH: Vapor irritant: 2; Liquid or solid irritant: 1; Poisons: 1
WATER POLLUTION: Human toxicity: 1; Aquatic toxicity:–; Aesthetic effect: 2
REACTIVITY: Other chemicals: 2; Water: 1; Self-reaction: 1

PHYSICAL AND CHEMICAL PROPERTIES
Physical state @ 59°F/15°C and 1 atm: Liquid.
Molecular weight: 144.21
Boiling point @ 1 atm: 332°F/166.6°C/440°K.
Melting/Freezing point: $-133°F/-91.5°C/182°K$.
Specific gravity (water = 1): 0.872 @ 20°C.
Relative vapor density (air = 1): 4.97
Latent heat of vaporization: 195.9 Btu/lb = 108.8 cal/g = 4.56×10^5 J/kg.
Vapor pressure: (Reid) 0.2 psia; 1.4 mm.

BUTYL CHLORIDE   REC. B:3750

SYNONYMS: BUTANE, 1-CHLORO-; *n*-BUTYL CHLORIDE; 1-CHLOROBUTANE; CHLORURE de BUTYLE (French); CLORURO de *sec*-BUTILO (Spanish); *n*-PROPYLCARBINYL CHLORIDE

IDENTIFICATION
CAS Number: 109-69-3
Formula: C_4H_9Cl
DOT ID Number: UN 1127; DOT Guide Number: 130
Proper Shipping Name: Butyl chloride

DESCRIPTION: Colorless liquid. Unpleasant, chlorine odor. Floats on water surface; insoluble; reacts slowly forming hydrochloric acid.

Highly flammable • Containers may BLEVE when exposed to fire •Vapors may form explosive mixture with air • Vapors are heavier than air and will collect and stay in low areas • Vapors may travel long distances to ignition sources and flashback • Vapors in confined areas (e.g., tanks, sewers, buildings) may explode when exposed to fire • Irritating to the skin, eyes, and respiratory tract • Toxic products of combustion may include hydrogen chloride and phosgene gas.

Hazard Classification (based on NFPA-704 Rating System)
Health Hazards (Blue): 2; Flammability (Red): 3; Reactivity (Yellow): 0

EMERGENCY RESPONSE: See Appendix A (130)
Evacuation:
Public safety: Isolate spill area for at least 50 to 100 meters (160 to 330 feet) in all directions.
Spill: Large spill–Consider initial downwind evacuation for at least 300 meters (1000 feet).
Fire: Isolate for 800 meters (½ mile) in all directions, especially if tank, rail car, or tank truck is involved in fire.

EXPOSURE
Short-term effects: *SEEK MEDICAL ATTENTION.* **Vapor:** May be harmful if inhaled or absorbed through the skin. Move victim to fresh air. *IF BREATHING HAS STOPPED,* give artificial respiration. IF breathing is difficult, administer oxygen. **Liquid:** *MAY BE HARMFUL IF SWALLOWED OR ABSORBED THROUGH THE SKIN. IF IN EYES,* immediately flush with plenty of water for 15 minutes. *IF SWALLOWED* and victim is *CONSCIOUS AND ABLE TO SWALLOW,* have victim drink 4 to 8 ounces of water and induce vomiting. *IF SWALLOWED* and victim is *UNCONSCIOUS OR HAVING CONVULSIONS,* do nothing except keep victim warm.

HEALTH HAZARDS
Personal protective equipment (PPE): B-Level PPE. Chemical protective material(s) reported to offer protection: PV alcohol. Also, butyl rubber/neoprene and Viton® offers limited protection
Toxicity by ingestion: Grade 2: LD_{50} = 2.67 g/kg (rat).
Vapor (gas) irritant characteristics: Eye and respiratory tract irritant. Vapors cause smarting of the eyes or respiratory system if present in high concentrations. The effect is temporary.
Liquid or solid irritant characteristics: Eye and skin irritant. If spilled on clothing and allowed to remain, may cause smarting and reddening of skin.

FIRE DATA
Flash point: 15°F/–9°C (cc).
Flammable limits in air: LEL: 1.8%; UEL: 10.1%.
Fire extinguishing agents not to be used: Water may be ineffective against fire.
Autoignition temperature: 464°F/240°C/513°K.

CHEMICAL REACTIVITY
Reactivity with water: Slow decomposition produces hydrochloric acid.
Binary reactants: Reacts with oxidizers, alkali metals; finely divided metals may cause fire or explosion. Attacks aluminum. in presence of moisture. Reacts with many forms of plastic.

ENVIRONMENTAL DATA
Food chain concentration potential: Log P_{ow} = 2.4. Unlikely to accumulate.
Water pollution: Effect of low concentrations on aquatic life are not known. May be dangerous if it enters nearby water intakes; notify operators. Notify local health and wildlife officials. **Response to discharge:** Issue warning–high flammability. Mechanical containment. Should be removed. Chemical and physical treatment.

SHIPPING INFORMATION
Grades of purity: 99.5+%; **Inert atmosphere:** None; **Venting:** None; **Stability during transport:** Stable.

NAS HAZARD CLASSIFICATION FOR BULK WATER TRANSPORTATION
FIRE: 3
HEALTH: Vapor irritant: 1; Liquid or solid irritant: 1; Poisons: 2
WATER POLLUTION: Human toxicity: 2; Aquatic toxicity:–; Aesthetic effect: 1
REACTIVITY: Other chemicals: 1; Water: 0; Self-reaction: 0

PHYSICAL AND CHEMICAL PROPERTIES
Physical state @ 59°F/15°C and 1 atm: Liquid.
Molecular weight: 92.58
Boiling point @ 1 atm: 173°F/78.4°C/352°K.
Melting/Freezing point: –190°F/–123°C/150°K.
Specific gravity (water = 1): 0.9
Relative vapor density (air = 1): 3.20
Vapor pressure: (Reid) 3.6 psia; 80 mm.

n-BUTYL CHLOROFORMATE REC. B:3800

SYNONYMS: CHLOROFORMIC ACID, *n*-BUTYL ESTER; *n*-BUTYL CHLOROCARBONATE; CHLOROCARBONIC ACID, *n*-BUTYL ESTER; CARBONOCHLORIDIC ACID, BUTYL ESTER

IDENTIFICATION
CAS Number: 592-34-7
Formula: $C_5H_9ClO_2$
DOT ID Number: UN 2743; DOT Guide Number: 155
Proper Shipping Name: *n*-Butyl chloroformate

DESCRIPTION: Colorless to light yellow liquid. Acrid, unpleasant odor. Sinks and reacts in water. Flammable, irritating vapor is produced.

Poison! • Breathing the vapor can kill you • Firefighting gear (including SCBA) does not provide adequate protection. If exposure occurs, remove and isolate gear immediately and thoroughly decontaminate personnel • Combustible • Containers may BLEVE when exposed to fire •Vapors may form explosive mixture with air •Vapors are heavier than air and will collect and stay in low areas • Vapors may travel long distances to ignition sources and flashback • Vapors in confined areas (e.g., tanks, sewers, buildings) may explode when exposed to fire • Severely irritating to skin, eyes, and respiratory tract; prolonged contact with the skin causes burns • Toxic products of combustion may include hydrogen chloride and phosgene • Do not use water; reacts to form hydrogen chloride vapor • Do not put yourself in danger by entering a contaminated area to rescue a victim.

Hazard Classification (based on NFPA-704 Rating System)
Health Hazards (Blue): 2; Flammability (Red): 3; Reactivity (Yellow): 1

EMERGENCY RESPONSE: See Appendix A (155)
Evacuation:
Public safety: Isolate the area of spill or leak for at least 50 to 100 meters (160 to 330 feet) in all directions.
Spill: Small spill–First: Isolate in all directions 30 meters (100 feet); Then: Protect persons downwind, DAY: 0.2 km (0.1 mile); NIGHT: 0.2 km (0.1 mile). Large spill–First: Isolate in all directions 30 meters (100 feet); Then: Protect persons downwind, DAY: 0.3 km (0.2 mile); NIGHT: 0.5 km (0.3 mile).
Fire: If tank, rail car, or tank truck is involved in fire, isolate for at least 800 meters (½ mile) in all directions; also, consider initial evacuation for 800 meters (½ mile) in all directions.

EXPOSURE
Short-term effects: *SEEK MEDICAL ATTENTION.* **Vapor:** Irritating to eyes, nose, and throat. *IF INHALED*, will, will cause difficult breathing. Move victim to fresh air. *IF BREATHING HAS STOPPED,* give artificial respiration; *avoid mouth-to-mouth resuscitation; use bag/mask apparatus.* IF breathing is difficult, administer oxygen. **Liquid:** Will burn skin and eyes. Remove contaminated clothing and shoes. Flush affected areas with plenty of water. *IF IN EYES*, hold eyelids open and flush with plenty of water. *POISONOUS IF SWALLOWED.* **Do NOT induce vomiting but immediately administer slurry of activated charcoal.**

HEALTH HAZARDS
Personal protective equipment (PPE): A-Level PPE. Acid- or organic-canister mask or self-contained breathing apparatus; goggles or face shield; plastic gloves.
Vapor (gas) irritant characteristics: Eye and respiratory tract irritant.
Liquid or solid irritant characteristics: Severe skin and eye irritant.

FIRE DATA
Flash point: 77°F/25°C (cc).
Behavior in fire: distance to a source of ignition and flashback.

CHEMICAL REACTIVITY
Reactivity with water: Reacts slowly, evolving hydrogen chloride (hydrochloric acid). Reaction can be hazardous if water is hot.
Neutralizing agents for acids and caustics: Flood with water, rinse with sodium bicarbonate or lime solution.
Reactivity group: 34
Compatibility class: Esters.

ENVIRONMENTAL DATA
Food chain concentration potential: Negative; unlikely to accumulate.
Water pollution: Effect of low concentrations on aquatic life is unknown. May be dangerous if it enters nearby water intakes; notify operators. Notify local health and wildlife officials.
Response to discharge: Issue warning–corrosive, poison, high flammability. Restrict access. Disperse and flush.

SHIPPING INFORMATION
Grades of purity: 97+%; **Storage temperature:** Ambient; **Inert atmosphere:** None; **Venting:** Pressure-vacuum; **Stability during transport:** Stable.

NAS HAZARD CLASSIFICATION FOR BULK WATER TRANSPORTATION
FIRE: 3
HEALTH: Vapor irritant: 3; Liquid or solid irritant: 4; Poisons:–
WATER POLLUTION: Human toxicity:–; Aquatic toxicity:–; Aesthetic effect: 1
REACTIVITY: Other chemicals: 1; Water: 0; Self-reaction: 0

PHYSICAL AND CHEMICAL PROPERTIES
Physical state @ 59°F/15°C and 1 atm: Liquid.
Molecular weight: 136.58
Boiling point @ 1 atm: 280.4°F/138°C/411.2°K.
Specific gravity (water = 1): 1.0513 @ 68°F/20°C (liquid).

BUTYLENE REC. B:3850

SYNONYMS: BUTILENO (Spanish); *α*-BUTYLENE; BUTENE; 1-BUTENE; *n*-BUTENE; 1-BUTYLENE; ETHYLETHYLENE

IDENTIFICATION
CAS Number: 25167-67-3
Formula: $CH_3CH_2CH=CH_2$
DOT ID Number: UN 1012; DOT Guide Number: 115
Proper Shipping Name: Butylene

DESCRIPTION: Colorless, liquefied compressed gas, colorless. Gasoline-like odor. Floats and boils on water; insoluble; flammable, visible vapor cloud is produced.

Extremely flammable • Forms explosive mixture with air • Reacts explosively with many materials • Gas from liquefied gas are initially heavier than air and spread along ground • Gas may travel to source of ignition and flashback • Gas in confined areas (e.g., tanks, sewers, buildings) may explode when exposed to fire • Ruptured or venting cylinders may rocket through buildings and/or travel a considerable distance • Irritating to the skin, eyes, and respiratory tract • Vapors may cause dizziness or asphyxiation without warning • Contact with liquid may cause frostbite.

Hazard Classification (based on NFPA-704 Rating System)
Health Hazards (Blue): 1; Flammability (Red): 4; Reactivity (Yellow): 0

EMERGENCY RESPONSE: See Appendix A (115)
Evacuation:
Public safety: Isolate spill area for at least 50 to 100 meters (160 to 330 feet) in all directions.
Spill: Consider initial downwind evacuation for at least 800 meters (½ mile).
Fire: Isolate for 1600 meters (1 mile) in all directions, especially if tank, rail car, or tank truck is involved in fire.

EXPOSURE
Short-term effects: *SEEK MEDICAL ATTENTION.* **Vapor:** If inhaled, will cause dizziness and difficult breathing. Move to fresh air. *IF BREATHING HAS STOPPED,* give artificial respiration. IF breathing is difficult, administer oxygen. **Liquid:** Will cause frostbite. Flush affected areas with plenty of water. *DO NOT RUB AFFECTED AREAS.*

HEALTH HAZARDS
Personal protective equipment (PPE): B-Level PPE. Chemical goggles, gloves, self-contained breathing apparatus or organic canister. Wear thermal protective clothing. Chemical protective

material(s) reported to offer minimal to poor protection: neoprene, nitrile, PVC, styrene-butadiene
Vapor (gas) irritant characteristics: Vapors are nonirritating to the eyes and throat.
Liquid or solid irritant characteristics: No appreciable hazard. Practically harmless to the skin because it is very volatile and evaporates quickly; however, contact with liquid may cause frostbite.

FIRE DATA
Flash point: Gas; 110°F/43°C (cc).
Flammable limits in air: LEL: 1.6%; UEL: 10.0%.
Autoignition temperature: 725°F/385°C/658°K.
Electrical hazard: Class I, Group D.
Burning rate: 8.8 mm/min
Adiabatic flame temperature: 2493°F/1367°C (estimate).
Stoichiometric air-to-fuel ratio: 14.68 (estimate).

CHEMICAL REACTIVITY
Binary reactants: Reacts with oxidizers, strong acids.
Reactivity group: 30
Compatibility class: Olefins

ENVIRONMENTAL DATA
Food chain concentration potential: Negative; unlikely to accumulate.
Water pollution: Not harmful to aquatic life. May be dangerous if it enters nearby water intakes; notify operators. **Response to discharge:** Issue warning–high flammability. Restrict access. Evacuate area.

SHIPPING INFORMATION
Storage temperature: Ambient; **Inert atmosphere:** None; **Venting:** Safety relief; **Stability during transport:** Stable.

PHYSICAL AND CHEMICAL PROPERTIES
Physical state @ 59°F/15°C and 1 atm: Gas.
Molecular weight: 56.10
Boiling point @ 1 atm: 20.7°F/–6.3°C/266.9°K.
Melting/Freezing point: –302°F/–186°C/87°K.
Critical temperature: 295.5°F/146.4°C/419.6°K.
Critical pressure: 584 psia = 39.7 atm = 4.02 MN/m^2.
Specific gravity (water = 1): 0.595 @ 68°F/20°C (liquid).
Liquid surface tension: 12.5 dynes/cm = 0.0125 N/m @ 68°F/20°C.
Liquid water interfacial tension: (estimate) 68 dynes/cm = 0.068 N/m at 0°C.
Relative vapor density (air = 1): 1.9
Ratio of specific heats of vapor (gas): 1.104
Latent heat of vaporization: 168 Btu/lb = 93.4 cal/g = 3.91 x 10^5 J/kg.
Heat of combustion: –19,487 Btu/lb = –10,286 cal/g = –453.26 x 10^5 J/kg.
Vapor pressure: (Reid) 62.5 psia.

1,3-BUTYLENE GLYCOL   **REC. B:3900**

SYNONYMS: BUTANE-1,3-DIOL; 1,3-BUTANEDIOL; 1,3-BUTILENGLICOL (Spanish); β-BUTYLENE GLYCOL; 1,3-DIHYDROXYBUTANE; METHYLTRIMETHYLENE GLYCOL

IDENTIFICATION
CAS Number: 107-88-0
Formula: C$_4$H$_{10}$O$_2$

DESCRIPTION: Colorless liquid or solid (depending upon temperature). Odorless.

Combustible • Containers may BLEVE when exposed to fire • Vapors in confined areas (e.g., tanks, sewers, buildings) may explode when exposed to fire • Irritating to the skin, eyes, and respiratory tract • Thermally unstable; may form flammable tetrahydrofuran above 300°F/149°C. Toxic products of combustion may also include carbon monoxide.

Hazard Classification (based on NFPA-704 Rating System)
Health Hazards (Blue): 0; Flammability (Red): 2; Reactivity (Yellow): -

EMERGENCY RESPONSE: See Appendix A (171)
Evacuation:
Public safety: Isolate the area of spill or leak for at least 10 to 25 meters (30 to 80 feet) in all directions.
Spill: Increase, in the downwind direction, as necessary, the distance shown under "Public Safety."
Fire: If any large container is involved in fire, isolate for at least 800 meters (½ mile) in all directions; also, consider initial evacuation for 800 meters (½ mile) in all directions.

EXPOSURE
Short-term effects: SEEK MEDICAL ATTENTION. **Liquid or solid:** Irritating to skin or eyes. Harmful if swallowed.
Flush affected areas with plenty of water. IF SWALLOWED and victim is CONSCIOUS AND ABLE TO SWALLOW, have victim drink 4 to 8 ounces of water.

HEALTH HAZARDS
Personal protective equipment (PPE): B-Level PPE. Eye protection and gloves. Chemical protective material(s) reported to offer minimal to poor protection: Viton®/neoprene, butyl rubber,/neoprene.
Toxicity by ingestion: Grade 1; LD$_{50}$ = 23 g/kg (rat).
Vapor (gas) irritant characteristics: Vapors are nonirritating to eyes and throat.
Liquid or solid irritant characteristics: Eye irritant. Practically harmless to the skin.

FIRE DATA
Flash point: 250°F/121°C (cc).
Fire extinguishing agents not to be used: Water or foam may cause frothing.
Behavior in fire: Unstable with heat; may form flammable tetrahydrofuran at 300°F/149°C.
Autoignition temperature: 707°F/375°C/980°K.
Electrical hazard: Class I, Group D.

CHEMICAL REACTIVITY
Binary reactants: Incompatible with strong acids, strong alkalies, aliphatic amines, isocyanates, oxidizers.
Reactivity group: 20
Compatibility class: Alcohols, glycols

ENVIRONMENTAL DATA
Food chain concentration potential: Log P$_{ow}$ = –1.1. Unlikely to accumulate.
Water pollution: Effect of low concentrations on aquatic life is unknown. May be dangerous if it enters nearby water intakes; notify operators. Notify local health and pollution control officials.
Response to discharge: Should be removed.

SHIPPING INFORMATION
Grades of purity: Regular and anhydrous grades; technical;
Storage temperature: Ambient; **Inert atmosphere:** None.

NAS HAZARD CLASSIFICATION FOR BULK WATER TRANSPORTATION
FIRE: 1
HEALTH: Vapor irritant: 0; Liquid or solid irritant: 0; Poisons: 0
WATER POLLUTION: Human toxicity: 0; Aquatic toxicity: 2; Aesthetic effect: 1
REACTIVITY: Other chemicals: 2; Water: 0; Self-reaction: 0

PHYSICAL AND CHEMICAL PROPERTIES
Physical state @ 59°F/15°C and 1 atm: Solid or Liquid.
Molecular weight: 90.12
Boiling point @ 1 atm: 378°F/192°C/465°K.
Melting/Freezing point: −58°F/−50°C/223°K.
Specific gravity (water = 1): 1.017 @ 68°F/20°C (liquid).
Relative vapor density (air = 1): 3.1
Vapor pressure: 0.6 mm.

1,4-BUTYLENE GLYCOL   REC. B:3950

SYNONYMS: BUTANE-1,4-DIOL; 1,4-BUTANEDIOL; β-BUTYLENE GLYCOL; 1,4-BUTILENGLICOL (Spanish); 1,4-DIHYDROXYBUTANE; 1,4 TETRAMETHYLENE GLYCOL

IDENTIFICATION
CAS Number: 110-63-4
Formula: $C_4H_{10}O_2$; $HOCH_2(CH_2)_2CH_2OH$

DESCRIPTION: Colorless liquid or solid (depending upon temperature). Odorless.

Combustible • Containers may BLEVE when exposed to fire • Vapors may form explosive mixture with air • Vapors are heavier than air and will collect and stay in low areas • Vapors may travel long distances to ignition sources and flashback • Vapors in confined areas (e.g., tanks, sewers, buildings) may explode when exposed to fire • Irritating to the skin, eyes, and respiratory tract • Thermally unstable; may form flammable tetrahydrofuran above 300°F/149°C. Toxic products of combustion may also include carbon monoxide.

Hazard Classification (based on NFPA-704 Rating System)
Health Hazards (Blue): 1; Flammability (Red): 1; Reactivity (Yellow): 0

EMERGENCY RESPONSE: See Appendix A (171)
Evacuation:
Public safety: Isolate the area of spill or leak for at least 10 to 25 meters (30 to 80 feet) in all directions.
Spill: Increase, in the downwind direction, as necessary, the distance shown under "Public Safety."
Fire: If any large container is involved in fire, isolate for at least 800 meters (½ mile) in all directions; also, consider initial evacuation for 800 meters (½ mile) in all directions.

EXPOSURE
Short-term effects: *SEEK MEDICAL ATTENTION.* **Liquid or solid:** Irritating to skin or eyes. Harmful if swallowed.
Flush affected areas with plenty of water. *IF SWALLOWED* and victim is *CONSCIOUS AND ABLE TO SWALLOW*, have victim drink 4 to 8 ounces of water.

HEALTH HAZARDS
Personal protective equipment (PPE): B-Level PPE. Eye protection and gloves. Chemical protective material(s) reported to offer minimal to poor protection: Viton®/neoprene, butyl rubber,/neoprene.
Toxicity by ingestion: Grade 2; LD_{50} = 0.5 to 5 g/kg (rat).
Long-term health effects: Kidney damage.
Vapor (gas) irritant characteristics: Eye irritant.
Liquid or solid irritant characteristics: Eye irritant. Practically harmless to the skin.

FIRE DATA
Flash point: 185-311°F/85-155°C (oc).
Fire extinguishing agents not to be used: Water or foam may cause frothing.
Autoignition temperature: 671°F/355°C/628°K.
Electrical hazard: Class I, Group D.

CHEMICAL REACTIVITY
Binary reactants: Incompatible with strong acids, strong caustics, aliphatic amines, isocyanates and strong oxidizers.
Reactivity group: 20
Compatibility class: Alcohols, glycols

ENVIRONMENTAL DATA
Food chain concentration potential: Log P_{ow} = −0.03 to −1.3. Unlikely to accumulate.
Water pollution: Effect of low concentrations on aquatic life is unknown. May be dangerous if it enters nearby water intakes; notify operators. Notify local health and pollution control officials.
Response to discharge: Should be removed.

SHIPPING INFORMATION
Grades of purity: Regular and anhydrous grades; technical;
Storage temperature: Ambient; **Inert atmosphere:** None.

NAS HAZARD CLASSIFICATION FOR BULK WATER TRANSPORTATION
FIRE: 1
HEALTH: Vapor irritant: 0; Liquid or solid irritant: 0; Poisons: 0
WATER POLLUTION: Human toxicity: 0; Aquatic toxicity: 2; Aesthetic effect: 1
REACTIVITY: Other chemicals: 2; Water: 0; Self-reaction: 0

PHYSICAL AND CHEMICAL PROPERTIES
Physical state @ 59°F/15°C and 1 atm: Solid or liquid.
Molecular weight: 90.12
Boiling point @ 1 atm: 442°F/228°C/501°K.
Melting/Freezing point: 66-68°F/19-20°C/292-293°K **Critical temperature:** 716°F/380°C/653°K.
Critical pressure: 720 psia = 49 atm = 5.0 MN/m^2.
Specific gravity (water = 1): 1.017 @ 68°F/20°C (liquid).
Relative vapor density (air = 1): 3.1
Heat of combustion: (estimate) −11,900 Btu/lb = −6630 cal/g = 277 x 10^5 J/kg.
Vapor pressure: Less than 0.08 mm.

2,3-BUTYLENE GLYCOL   REC. B:4000

SYNONYMS: 2,3-BUTANEDIOL; BUTYLENE GLYCOL (pseudo); 2,3-BUTILENGLICOL (Spanish); 2,3-DIHYDROXY BUTANE; DIMETHYLENE GLYCOL

1,2-Butylene oxide

IDENTIFICATION
CAS Number: 513-85-9
Formula: $C_4H_{10}O_2$

DESCRIPTION: Colorless liquid or solid (depending on temperature). Odorless.

Combustible • Containers may BLEVE when exposed to fire • Vapors may form explosive mixture with air • Vapors are heavier than air and will collect and stay in low areas • Vapors may travel long distances to ignition sources and flashback • Vapors in confined areas (e.g., tanks, sewers, buildings) may explode when exposed to fire • Irritating to the skin, eyes, and respiratory tract • Thermally unstable; may form flammable tetrahydrofuran above 300°F/149°C. Toxic products of combustion may also include carbon monoxide.

Hazard Classification (based on NFPA-704 Rating System)
Health Hazards (Blue): 0; Flammability (Red): 2; Reactivity (Yellow): -

EMERGENCY RESPONSE: See Appendix A (171)
Evacuation:
Public safety: Isolate the area of spill or leak for at least 10 to 25 meters (30 to 80 feet) in all directions.
Spill: Increase, in the downwind direction, as necessary, the distance shown under "Public Safety."
Fire: If any large container is involved in fire, isolate for at least 800 meters (½ mile) in all directions; also, consider initial evacuation for 800 meters (½ mile) in all directions.

EXPOSURE
Short-term effects: *SEEK MEDICAL ATTENTION.* **Liquid or solid:** Irritating to skin or eyes. Harmful if swallowed.
Flush affected areas with plenty of water. *IF SWALLOWED* and victim is *CONSCIOUS AND ABLE TO SWALLOW*, have victim drink 4 to 8 ounces of water.

HEALTH HAZARDS
Personal protective equipment (PPE): B-Level PPE. Chemical protective material(s) reported to offer minimal to poor protection: Viton®/neoprene, butyl rubber,/neoprene.
Toxicity by ingestion: Grade 2; LD_{50} = 5.462 g/kg (mouse).
Long-term health effects: Kidney damage.
Vapor (gas) irritant characteristics: Vapors are nonirritating to eyes and throat.
Liquid or solid irritant characteristics: Eye irritant. Practically harmless to the skin.

FIRE DATA
Flash point: 185-311°F/85-155°C (oc).
Fire extinguishing agents not to be used: Water or foam may cause frothing.
Electrical hazard: Class I, Group D.

CHEMICAL REACTIVITY
Binary reactants: Incompatible with strong acids, strong caustics, aliphatic amines, isocyanates and strong oxidizers.
Reactivity group: 20
Compatibility class: Alcohols, glycols

ENVIRONMENTAL DATA
Food chain concentration potential: Negative; unlikely to accumulate.
Water pollution: Effect of low concentrations on aquatic life is unknown. May be dangerous if it enters nearby water intakes; notify operators. Notify local health and pollution control officials.
Response to discharge: Should be removed.

SHIPPING INFORMATION
Grades of purity: Regular and anhydrous grades; technical;
Storage temperature: Ambient; **Inert atmosphere:** None.

NAS HAZARD CLASSIFICATION FOR BULK WATER TRANSPORTATION
FIRE: 1
HEALTH: Vapor irritant: 0; Liquid or solid irritant: 0; Poisons: 0
WATER POLLUTION: Human toxicity: 0; Aquatic toxicity: 2; Aesthetic effect: 1
REACTIVITY: Other chemicals: 2; Water: 0; Self-reaction: 0

PHYSICAL AND CHEMICAL PROPERTIES
Physical state @ 59°F/15°C and 1 atm: Solid or Liquid.
Molecular weight: 90.12
Boiling point @ 1 atm: 356°F/180°C/453°K.
Melting/Freezing point: 66°F/19°C/292°K.
Specific gravity (water = 1): 1.017 @ 68°F/20°C (liquid).
Relative vapor density (air = 1): 3.1
Vapor pressure: 0.2 mm.

1,2-BUTYLENE OXIDE   REC. B:4050

SYNONYMS: α-BUTYLENE OXIDE; 1-BUTENE OXIDE; 1-BUTYLENE OXIDE; 1,2-BUTYLENE OXIDE, stabilized; 1,2-EPOXYBUTANE; OXIDO de 1,2-BUTILENO (Spanish)

IDENTIFICATION
CAS Number: 106-88-7; 3266-23-7 (2-); 1758-33-4 (*cis*-2,3-)
Formula: C_4H_8O; $C_2H_5CHCH_2O$
DOT ID Number: UN 3022; DOT Guide Number: 127P
Proper Shipping Name: 1,2-Butylene oxide, stabilized
Reportable Quantity (RQ): **(CERCLA)** 1 lb/0.454 kg

DESCRIPTION: Colorless liquid. Sharp, disagreeable odor. Floats on the surface of water; moderately soluble.

Highly flammable • Thermodynamically unstable; decomposition is rapid • Corrosive to the skin, eyes, and respiratory tract; contact with skin or eyes can cause burns and vision impairment; inhalation symptoms may be delayed • Polymerization hazard • Heat can induce polymerization with rapid release of energy; sealed containers may rupture explosively • May react with itself blocking relief valves; leading to tank explosions • Containers may BLEVE when exposed to fire • Vapors may form explosive mixture with air • Vapors are heavier than air and will collect and stay in low areas • Vapors may travel long distances to ignition sources and flashback • Vapors in confined areas (e.g., tanks, sewers, buildings) may explode when exposed to fire • Forms unstable and explosive peroxides on contact with heat and light • Toxic products of combustion may include carbon monoxide.

Hazard Classification (based on NFPA-704 Rating System)
Health Hazards (Blue): 2; Flammability (Red): 3; Reactivity (Yellow): 2

EMERGENCY RESPONSE: See Appendix A (127)
Evacuation:
Public safety: Isolate spill area for at least 25 to 50 meters (80 to 160 feet) in all directions.

Spill: Large spill–Consider initial downwind evacuation for at least 300 meters (1000 feet).
Fire: Isolate for 800 meters (½ mile) in all directions, especially if tank, rail car, or tank truck is involved in fire.

EXPOSURE
Short-term effects: *SEEK MEDICAL ATTENTION.* **Vapor:** Irritating to eyes, nose, and throat. *IF INHALED*, will, will cause coughing or difficult breathing. *IF IN EYES*, hold eyelids open and flush with plenty of water. *IF BREATHING HAS STOPPED,* give artificial respiration. IF breathing is difficult, administer oxygen. **Liquid:** Will burn skin and eyes. *IF SWALLOWED*, will cause nausea and vomiting. Remove contaminated clothing and shoes. Flush affected areas with plenty of water. *IF IN EYES*, hold eyelids open and flush with plenty of water. *IF SWALLOWED* and victim is *CONSCIOUS AND ABLE TO SWALLOW*, have victim drink 4 to 8 ounces of water and have victim induce vomiting. *IF SWALLOWED* and victim is *UNCONSCIOUS OR HAVING CONVULSIONS,* do nothing except keep victim warm.

HEALTH HAZARDS
Personal protective equipment (PPE): A-Level PPE. Clean protective clothing; rubber gloves; chemical worker's goggles; self-contained breathing apparatus). Wear thermal protective clothing. Chemical protective material(s) reported to offer minimal protection: butyl rubber. Other chemical resistant materials: polycarbonate.
clothing and wash before reuse.
Toxicity by ingestion: Grade 2; oral LD_{50} = 1410 mg/kg (rat).
Long-term health effects: IARC unclassified, rating 3; limited animal evidence.
Vapor (gas) irritant characteristics: Eye and respiratory tract irritant.
Liquid or solid irritant characteristics: Eye and skin irritant. Possible frostbite.

FIRE DATA
Flash point: Less than –20°F/–29°C; (oc); –7°F/–22°C (cc).
Flammable limits in air: LEL: 1.5%; UEL: 18.3%.
Fire extinguishing agents not to be used: Water may be ineffective.
Behavior in fire: Thermodynamically unstable; decomposition is rapid. Containers may explode in fire.
Autoignition temperature: 822°F/439°C/712°K.
Electrical hazard: Class I, Group D.

CHEMICAL REACTIVITY
Binary reactants: Reacts violently with oxidizers, hydroxides, acids, metal chlorides, and other metal catalysts.
Polymerization: May occur when in contact with acids, bases and certain salts.

ENVIRONMENTAL DATA
Food chain concentration potential: Log P_{ow} = –0.02. Unlikely to accumulate.
Water pollution: Effect of low concentrations on aquatic life is unknown. May be dangerous if it enters nearby water intakes; notify operators. Notify local health and wildlife officials.
Response to discharge: Issue warning–high flammability, air contaminant. Restrict access. Disperse and flush.

SHIPPING INFORMATION
Grades of purity: Technical, 99%; **Storage temperature:** Ambient; **Inert atmosphere:** Packages under nitrogen gas; **Venting:** Pressure-vacuum; **Stability during transport:** Stable.

PHYSICAL AND CHEMICAL PROPERTIES
Physical state @ 59°F/15°C and 1 atm: Liquid.
Molecular weight: 72
Boiling point @ 1 atm: 145°F/63°C/336°K.
Melting/Freezing point: Less than –58°F/–50°C/223°K.
Specific gravity (water = 1): 0.826 @ 77°F/25°C (liquid).
Relative vapor density (air = 1): 2.49
Latent heat of vaporization: (estimate) 180 Btu/lb = 100 cal/g = 4.2×10^5 J/kg.
Heat of combustion: –15,200 Btu/lb = –8,470 cal/g = -354×10^5 J/kg.
Vapor pressure: (Reid) 5.8 psia.

n-BUTYL FORMATE REC. B:4100

SYNONYMS: BUTYL FORMATE; BUTYL METHANOATE; EEC No. 607-017-00-8; FORMIATO de BUTILO (Spanish); FORMIC ACID, BUTYL ESTER

IDENTIFICATION
CAS Number: 592-84-7
Formula: $C_5H_{10}O_2$; $HCOO(CH_2)_3CH_3$
DOT ID Number: 1128; DOT Guide Number: 129
Proper Shipping Name: Butyl formate

DESCRIPTION: Colorless liquid. Floats on the surface of water.

Highly flammable • Corrosive to the skin, eyes, and respiratory tract; contact with skin or eyes can cause burns and vision impairment; inhalation symptoms may be delayed • Containers may BLEVE when exposed to fire •Vapors may form explosive mixture with air • Vapors are heavier than air and will collect and stay in low areas • Vapors may travel long distances to ignition sources and flashback • Vapors in confined areas (e.g., tanks, sewers, buildings) may explode when exposed to fire •Combustion products include carbon monoxide.

Hazard Classification (based on NFPA-704 Rating System)
Health Hazards (Blue): 2; Flammability (Red): 3; Reactivity (Yellow): 0

EMERGENCY RESPONSE: See Appendix A (129)
Evacuation:
Public safety: Isolate spill area for at least 50 to 100 meters (160 to 330 feet) in all directions.
Spill: Large spill–Consider initial downwind evacuation for at least 300 meters (1000 feet).
Fire: Isolate for 800 meters (½ mile) in all directions, especially if tank, rail car, or tank truck is involved in fire.

EXPOSURE
Short-term effects: *SEEK MEDICAL ATTENTION.* **Vapor:** Move victim to fresh air. *IF BREATHING HAS STOPPED,* give artificial respiration. IF breathing is difficult, administer oxygen. May cause lung edema; physical exertion will aggravate this condition. **Liquid:** Irritating to skin and eyes. Overexposures may have a narcotic effect. Remove contaminated clothing and shoes. Wash affected areas with soap and water. *IF IN EYES*, hold eyelids open and flush with plenty of water.
Note to physician or authorized medical personnel: Medical observation is recommended for 24 to 48 hours after breathing overexposure, as pulmonary edema may be delayed. As first aid for pulmonary edema, consider administering a corticosteroid spray.

Cigarette smoking may exacerbate pulmonary injury and should be discouraged for at least 72 hours following exposure.

Personal protective equipment (PPE): Full impervious protective clothing, including boots and gloves. Where splashing is possible wear full face shield or chemical safety goggles. Use approved respirator to protect against vapors.
Toxicity by ingestion: Grade 2; oral rabbit LD_{50} = 2.656 g/kg.
Vapor (gas) irritant characteristics: Eye and respiratory tract irritant. Vapors cause smarting of the eyes or respiratory system if present in high concentrations. The effect may be temporary.
Liquid or solid irritant characteristics: Severe eye irritant. Skin irritant. If spilled on clothing and allowed to remain, may cause smarting and reddening of the skin.

FIRE DATA
Flash point: 64°F/18°C (cc).
Flammable limits in air: LEL: 1.7%; UEL: 8.2%.
Fire extinguishing agents not to be used: Water.
Autoignition temperature: 612°F/322°C/595°K.
Electrical hazard: Flow or agitation of substance may generate electrostatic charges due to low conductivity.
Stoichiometric air-to-fuel ratio: 31.0

CHEMICAL REACTIVITY
Binary reactants: Explosive mixture in air. Violent reaction with strong oxidizers. Reacts with strong acids.
Reactivity group: 34
Compatibility class: Esters.

ENVIRONMENTAL DATA
Water pollution: Effect of low concentrations on aquatic life is unknown. May be dangerous if it enters nearby water intakes; notify operators. Notify local health and wildlife officials.
Response to discharge: Issue warning–high flammability.

SHIPPING INFORMATION
Grades of purity: 97%; technical. **Storage temperature:** Ambient. **Inert atmosphere:** None.

PHYSICAL AND CHEMICAL PROPERTIES
Physical state @ 59°F/15°C and 1 atm: Liquid.
Molecular weight: 102.15
Boiling point @ 1 atm: 224.6°F/107°C/380°K.
Melting/Freezing point: –130°F/–90°C/183°K.
Specific gravity (water = 1): 0.885-0.9108
Relative vapor density (air = 1): 3.5
Vapor pressure: 20 mm.

n-BUTYL GLYCIDYL ETHER   REC. B:4150

SYNONYMS: *n*-BGE; BGE; EEC No. 603-039-00-7; 2,3-EPOXYPROPYL BUTYL ETHER; 1,2-EPOXY-3-BUTOXY PROPANE; 2,3-EPOXYPROPYL BUTYL ETHER; ETER de *n*-BUTILGLICIDIL (Spanish); 1-BUTOXY-2,3-EPOXYPROPANE; GYLCIDY BUTYL ETHER

IDENTIFICATION
CAS Number: 2426-08-6
Formula: $C_7H_{14}O_2$; $CH_3(CH_2)_3OCH_2CHOCH_2$
DOT ID Number: 1993; DOT Guide Number: 128
Proper Shipping Name: Combustible liquid, n.o.s.

DESCRIPTION: Colorless to pale yellow liquid. Strong, slightly unpleasant odor. Soluble in water.

Highly flammable • Containers may BLEVE when exposed to fire • Vapors may form explosive mixture with air • Vapors are heavier than air and will collect and stay in low areas • Vapors may travel long distances to ignition sources and flashback • Vapors in confined areas (e.g., tanks, sewers, buildings) may explode when exposed to fire • Irritating to the skin, eyes, and respiratory tract • Toxic products of combustion may include carbon monoxide.

Hazard Classification (based on NFPA-704 Rating System)
Health Hazards (Blue): 0; Flammability (Red): 2; Reactivity (Yellow): 0

EMERGENCY RESPONSE: See Appendix A (128)
Evacuation:
Public safety: Isolate spill area for at least 25 to 50 meters (80 to 160 feet) in all directions.
Spill: Large spill–Consider initial downwind evacuation for at least 300 meters (1000 feet).
Fire: Isolate for 800 meters (½ mile) in all directions, especially if tank, rail car, or tank truck is involved in fire.

EXPOSURE
Short-term effects: *SEEK MEDICAL ATTENTION*. **Liquid:** Move victim to fresh air. Remove contaminated clothing and shoes. Wash affected areas with plenty of soap and water. *IF IN EYES*, hold eyelids open and flush with plenty of water. *IF SWALLOWED* and victim is *CONSCIOUS AND ABLE TO SWALLOW*, induce vomiting.

HEALTH HAZARDS
Personal protective equipment (PPE): OSHA Table Z-1-A air contaminant. Chemical protective clothing, gloves, face shields, and approved respirator.
Recommendations for respirator selection: NIOSH
56 ppm: CCROV [any chemical cartridge respirator with organic vapor cartridge(s)]; or SA (any supplied-air respirator). *140 ppm:* SA:CF (any supplied-air respirator operated in a continuous-flow mode); or PAPROV [any powered, air-purifying respirator with organic vapor cartridge(s)]. *250 ppm:* CCRFOV [any chemical cartridge respirator with a full facepiece and organic vapor cartridge(s)]; or GMFOV [any air-purifying, full-facepiece respirator (gas mask) with a chin-style, front- or back-mounted acid gas canister]; or PAPRTOV [any powered, air-purifying respirator with a tight-fitting facepiece and organic vapor cartridge(s)]; or SCBAF (any self-contained breathing apparatus with a full facepiece); or SAF (any supplied-air respirator with a full facepiece). *EMERGENCY OR PLANNED ENTRY INTO UNKNOWN CONCENTRATIONS OR IDLH CONDITIONS:* SCBAF:PD,PP (any self-contained breathing apparatus that has a full facepiece and is operated in a pressure-demand or other positive-pressure mode); or SAF:PD,PP:ASCBA (any supplied-air respirator that has a full facepiece and is operated in a pressure-demand or other positive-pressure mode in combination with an auxiliary self-contained breathing apparatus operated in a pressure-demand or other positive pressure mode). *ESCAPE:* GMFOV[any air-purifying, full-facepiece respirator (gas mask) with a chin-style, front- or back-mounted organic vapor canister or SCBAE (any appropriate escape-type, self-contained breathing apparatus). *Note:* Substance reported to cause eye irritation or damage; may require eye protection.
Exposure limits (TWA unless otherwise noted): ACGIH TLV 25

ppm (133 mg/m^3); OSHA PEL 50 ppm (270 mg/m^3); NIOSH ceiling 5.6 ppm (30 mg/m^3)/15 min.
Toxicity by ingestion: Grade 2; LD$_{50}$ = 2.05 g/kg(rat).
Long-term health effects: Mutagenic in bacterial test systems, and DNA damage was induced in vitro in human white blood cells. Skin and lung disorders.
Vapor (gas) irritant characteristics: Eye and respiratory tract irritant. Vapors cause irritation such that personnel will find high concentrations unpleasant.
Liquid or solid irritant characteristics: Eye and skin irritant. Causes smarting of the skin and first-degree burns on short exposure; may cause second-degree burns on long exposure.
IDLH value: 250 ppm.

FIRE DATA
Flash point: 130°F/55°C (cc).
Fire extinguishing agents not to be used: Solid stream of water may cause frothing.
Behavior in fire: Containers may burst.
Electrical hazard: May cause some plastics, coatings, and rubbers (insulators) to deteriorate.
Stoichiometric air-to-fuel ratio: 45.2

CHEMICAL REACTIVITY
Binary reactants: Contact with strong oxidizers may cause fires and explosions. May form explosive peroxides upon contact with air. Contact with strong caustics may cause polymerization with the release of heat, which may cause the container to burst.
Polymerization: Contact with strong caustics or heat may cause polymerization with the liberation of heat.

ENVIRONMENTAL DATA
Water pollution: Effects of low concentrations on aquatic life is unknown. May be dangerous if it enters nearby water intakes; notify operators. Notify local health and wildlife officials.
Response to discharge: Mechanical containment. Should be removed. Chemical and physical treatment.

SHIPPING INFORMATION
Grades of purity: 97–99%; **Storage temperature:** Less than 75°F/24°C; **Inert atmosphere:** Nitrogen blanket; **Stability during transport:** Stable.

PHYSICAL AND CHEMICAL PROPERTIES
Physical state @ 59°F/15°C and 1 atm: Liquid.
Molecular weight: 130.21
Boiling point @ 1 atm: 327°F/164°C/437°K.
Specific gravity (water = 1): 0.91
Vapor pressure: 3 mm @ 77°F/25°C.

tert-BUTYL HYDROPEROXIDE   REC. B:4200

SYNONYMS: CADOX TBH; 1,1-DIMETHYLETHYL HYDROPEROXIDE; HIDROPEROXIDO de *terc*-BUTILO (Spanish); HYDROPEROXY-2-METHYL PROPANE; TBHP-70

IDENTIFICATION
CAS Number: 75-91-2
Formula: $C_4H_{10}O_2$; $(CH_3)_3C \cdot O \cdot OH$
DOT ID Number: UN 2093; UN 2094; DOT Guide Number: 147
Proper Shipping Name: *tert*-Butyl hydroperoxide

DESCRIPTION: Colorless to pale yellow liquid. Odorless. Floats and mixes slowly with water.

Extremely flammable • Heat, shock, or contamination may cause explosion • Strong oxidizer which may react spontaneously with low flash point organics or reducing agents. Heat forms oxygen; will increase the activity of an existing fire • May cause fire or explosion on contact with combustibles (wood, paper, oil, clothing, etc.) • Containers may BLEVE when exposed to fire •Vapors may form explosive mixture with air • Vapors are heavier than air and will collect and stay in low areas • Vapors may travel long distances to ignition sources and flashback • Vapors in confined areas (e.g., tanks, sewers, buildings) may explode when exposed to fire • Irritating to the skin, eyes, and respiratory tract • Toxic products of combustion may include carbon monoxide.

Hazard Classification (based on NFPA-704 Rating System)
Health Hazards (Blue): 1; Flammability (Red): 4; Reactivity (Yellow): 4; Special Notice (White): OXY

EMERGENCY RESPONSE: See Appendix A. If material or contaminated runoff enters waterways, notify downst)
Evacuation:
Public safety: Isolate the area of spill or leak for at least 25 to 50 meters (80 to 160 feet) in all directions.
Spill: Consider initial evacuation for at least 250 meters (800 feet).
Fire: If tank, rail car, or tank truck is involved in fire, isolate for at least 800 meters (½ mile) in all directions; also, consider initial evacuation for 800 meters (½ mile) in all directions.

EXPOSURE
Short-term effects: *SEEK MEDICAL ATTENTION*. **Vapor:** Irritating to eyes, nose, and throat. Move to fresh air. *IF BREATHING HAS STOPPED,* give artificial respiration. IF breathing is difficult, administer oxygen. **Liquid:** Irritating to skin and eyes. Harmful if swallowed. Remove contaminated clothing and shoes. Flush affected areas with plenty of water. *IF IN EYES,* hold eyelids open and flush with plenty of water. *IF SWALLOWED* and victim is *CONSCIOUS AND ABLE TO SWALLOW*, have victim drink 4 to 8 ounces of water, and have victim induce vomiting. *IF SWALLOWED* and victim is *UNCONSCIOUS OR HAVING CONVULSIONS,* do nothing except keep victim warm.

HEALTH HAZARDS
Personal protective equipment (PPE): Goggles, well-fitting gloves, barrier creams. **Butyl rubber is generally suitable for** peroxide compounds.
Toxicity by ingestion: Grade 3; LD$_{50}$ = 50 to 500 mg/kg.
Vapor (gas) irritant characteristics: Vapors cause smarting of the eyes or respiratory system if present in high concentrations. The effect may be temporary.
Liquid or solid irritant characteristics: Powerful irritant of skin and eyes.

FIRE DATA
Flash point: 100°F/37°C (oc).
Behavior in fire: May explode in fire

CHEMICAL REACTIVITY
Binary reactants: Reacts vigorously with easily oxidized materials, including wood, cloth, oils, and some metals.

ENVIRONMENTAL DATA
Food chain concentration potential: Negative; unlikely to accumulate.
Water pollution: Effect of low concentrations on aquatic life is unknown. May be dangerous if it enters nearby water intakes; notify operators. Notify local health and pollution control officials.

Response to discharge: Issue warning–high flammability. Restrict access. Mechanical containment. Chemical and physical treatment.

SHIPPING INFORMATION
Grades of purity: 70–90%; **Storage temperature:** 65–85°F/18–32°C; **Stability during transport:** Shock and heat-sensitive; self-accelerating decomposition at 200°F/93°C.

NAS HAZARD CLASSIFICATION FOR BULK WATER TRANSPORTATION
FIRE: 2
HEALTH: Vapor irritant: 1; Liquid or solid irritant: 1; Poisons: 2
WATER POLLUTION: Human toxicity: 3; Aquatic toxicity: 1; Aesthetic effect: 4
REACTIVITY0: Other chemicals: 0; Water: 4; Self-reaction: 0

PHYSICAL AND CHEMICAL PROPERTIES
Physical state @ 59°F/15°C and 1 atm: Liquid.
Molecular weight: 90.12
Boiling point @ 1 atm: Decomposes.
Melting/Freezing point: −31°F/−35°C/238°K.
Specific gravity (water = 1): 0.880 @ 77°F/25°C (liquid).
Heat of combustion: (estimate) −13,000 Btu/lb = −7200 cal/g = −300 x 10^5 J/kg.
Heat of decomposition: −675 Btu/lb = −375 cal/g = −15.7 x 10^5 J/kg.

BUTYL LACTATE   REC. B:4350

SYNONYMS: BUTYL-α-HYDROXYPROPIONATE; n-BUTYL LACTATE; LACTATO de n-BUTILO (Spanish); LACTIC ACID, BUTYL ESTER

IDENTIFICATION
CAS Number: 138-22-7
Formula: $C_7H_{14}O_3$; $CH_3CH(OH)COO(CH_2)_3CH_3$
DOT ID Number: UN 1993; DOT Guide Number: 128
Proper Shipping Name: Combustible liquid, n.o.s.

DESCRIPTION: Clear, colorless to white liquid. Mild, transient odor. Slightly soluble in water.

Combustible • Containers may BLEVE when exposed to fire • Vapors may form explosive mixture with air • Vapors are heavier than air and will collect and stay in low areas • Vapors may travel long distances to ignition sources and flashback • Vapors in confined areas (e.g., tanks, sewers, buildings) may explode when exposed to fire • Irritating to the skin, eyes, and respiratory tract • Toxic products of combustion may include carbon monoxide.

Hazard Classification (based on NFPA-704 Rating System)
Health Hazards (Blue): 1; Flammability (Red): 2; Reactivity (Yellow): 0

EMERGENCY RESPONSE: See Appendix A (128)
Evacuation:
Public safety: Isolate spill area for at least 25 to 50 meters (80 to 160 feet) in all directions.
Spill: Large spill–Consider initial downwind evacuation for at least 300 meters (1000 feet).
Fire: Isolate for 800 meters (½ mile) in all directions, especially if tank, rail car, or tank truck is involved in fire.

EXPOSURE

Short-term effects: *SEEK MEDICAL ATTENTION*. **Liquid:** Move victim to fresh air. Remove contaminated clothing and shoes. Flush affected areas with plenty of soap and water. *IF IN EYES*, hold eyelids open and flush with plenty of water.

HEALTH HAZARDS
Personal protective equipment (PPE): OSHA Table Z-1-A air contaminant. Wear impervious rubber gloves and safety glasses. Respiratory protection not required at ambient temperatures, based on volatility.
Exposure limits (TWA unless otherwise noted): ACGIH TLV 5 ppm (30 mg/m^3); NIOSH REL 5 ppm (25 mg/m^3).
Vapor (gas) irritant characteristics: Eye irritant. Vapors cause irritation such that personnel will find high concentrations unpleasant. The effect may be temporary.
Liquid or solid irritant characteristics: Eye and skin irritant. If spilled on clothing and allowed to remain, may cause smarting and reddening of the skin.

FIRE DATA
Flash point: 160°F/71°C (oc).
Fire extinguishing agents not to be used: Water.
Autoignition temperature: 720°F/382°C/655°K.

CHEMICAL REACTIVITY
Binary reactants: Avoid contact with strong oxidizing agents and strong bases.

ENVIRONMENTAL DATA
Water pollution: Effects of low concentrations on aquatic life is unknown. May be dangerous if it enters nearby water intakes; notify operators. Notify local health and wildlife officials. **Response to discharge:** Restrict access. Mechanical containment. Should be removed. Chemical and physical treatment.

SHIPPING INFORMATION
Grades of purity: Technical grades; **Storage temperature:** Ambient; **Inert atmosphere:** None.

PHYSICAL AND CHEMICAL PROPERTIES
Physical state @ 59°F/15°C and 1 atm: Liquid.
Molecular weight: 146.19
Boiling point @ 1 atm: 320°F/160°C/433°K.
Melting/Freezing point: −45°F/−43°C/230°K.
Specific gravity (water = 1): 0.98
Relative vapor density (air = 1): 5.04
Vapor pressure: 0.4 mm.

n-BUTYL MERCAPTAN   REC. B:4400

SYNONYMS: BUTANETHIOL; BUTANE-THIOL; n-BUTANETHIOL; n-BUTILMERCAPTANO (Spanish); n-BUTYL THIOALCOHOL; 1-MERCAPTOBUTANE; 1-BUTANETHIOL; THIOBUTYL ALCOHOL

IDENTIFICATION
CAS Number: 109-79-5
Formula: $C_4H_{10}S$; $CH_3CH_2CH_2CH_2SH$
DOT ID Number: UN 2347; DOT Guide Number: 128
Proper Shipping Name: Butyl mercaptans

DESCRIPTION: Colorless to yellow liquid. Obnoxious garlic-, rotten cabbage-like odor. Floats on water surface; poisonous and flammable vapor is produced.

n-Butyl mercaptan

Highly flammable • Containers may BLEVE when exposed to fire • Vapors may form explosive mixture with air • Vapors are heavier than air and will collect and stay in low areas • Vapors may travel long distances to ignition sources and flashback • Vapors in confined areas (e.g., tanks, sewers, buildings) may explode when exposed to fire • Irritating to the skin, eyes, and respiratory tract • Toxic products of combustion may include carbon monoxide and sulfur oxides.

Hazard Classification (based on NFPA-704 Rating System)
Health Hazards (Blue): 2; Flammability (Red): 3; Reactivity (Yellow): 0

EMERGENCY RESPONSE: See Appendix A (128)
Evacuation:
Public safety: Isolate spill area for at least 25 to 50 meters (80 to 160 feet) in all directions.
Spill: Large spill–Consider initial downwind evacuation for at least 300 meters (1000 feet).
Fire: Isolate for 800 meters (½ mile) in all directions, especially if tank, rail car, or tank truck is involved in fire.

EXPOSURE
Short-term effects: *SEEK MEDICAL ATTENTION.* **Vapor:** POISONOUS IF INHALED. Irritating to eyes. Move victim to fresh air. *IF BREATHING HAS STOPPED,* give artificial respiration. IF breathing is difficult, administer oxygen. **Liquid:** Irritating to skin and eyes. Harmful if swallowed. Remove contaminated clothing and shoes. Flush affected areas with plenty of water. *IF IN EYES,* hold eyelids open and flush with plenty of water. *IF SWALLOWED* and victim is *CONSCIOUS AND ABLE TO SWALLOW,* have victim drink 4 to 8 ounces of water and have victim induce vomiting. *IF SWALLOWED* and victim is *UNCONSCIOUS OR HAVING CONVULSIONS,* do nothing except keep victim warm.
Note to physician or authorized medical personnel: Medical observation is recommended for 24 to 48 hours after breathing overexposure, as pulmonary edema may be delayed. As first aid for pulmonary edema, consider administering a corticosteroid spray. Cigarette smoking may exacerbate pulmonary injury and should be discouraged for at least 72 hours following exposure.

HEALTH HAZARDS
Personal protective equipment (PPE): OSHA Table Z-1-A air contaminant. Plastic gloves, goggles; self-contained breathing apparatus. Natural rubber, neoprene, and nitrile rubber may offer some protection from alcohols.
Recommendations for respirator selection: NIOSH/OSHA
5 ppm: CCRFOV [any chemical cartridge respirator with a full facepiece and organic vapor cartridges(s)]; or SA (any supplied-air respirator). *12.5 ppm:* SA:CF (any supplied-air respirator operated in a continuous-flow mode); or PAPROV [any powered, air-purifying respirator with organic vapor cartridge(s)]. *25 ppm:* CCRFOV[any chemical cartridge respirator with a full facepiece and organic vapor cartridge(s)]; or GMFOV [any air-purifying, full-facepiece respirator (gas mask) with a chin-style, front- or back-mounted acid gas canister]; or PAPRTOV [any powered, air-purifying respirator with a tight-fitting facepiece and organic vapor cartridge(s)]; or SCBAF (any self-contained breathing apparatus with a full facepiece); or SAF (any supplied-air respirator with a full facepiece). *500 ppm:* SA:PD,PP (any supplied-air respirator operated in a pressure-demand or other positive-pressure mode). *EMERGENCY OR PLANNED ENTRY INTO UNKNOWN CONCENTRATIONS OR IDLH CONDITIONS:* SCBAF:PD,PP (any self-contained breathing apparatus that has a full facepiece and is operated in a pressure-demand or other positive-pressure mode); or SAF:PD,PP:ASCBA (any supplied-air respirator that has a full facepiece and is operated in a pressure-demand or other positive-pressure mode in combination with an auxiliary self-contained breathing apparatus operated in a pressure-demand or other positive pressure mode). *ESCAPE:* GMFOV[any air-purifying, full-facepiece respirator (gas mask) with a chin-style, front- or back-mounted organic vapor canister or SCBAE (any appropriate escape-type, self-contained breathing apparatus). *Note:* Substance reported to cause eye irritation or damage; may require eye protection.
Exposure limits (TWA unless otherwise noted): ACGIH TLV 0.5 ppm (1.8 mg/m^3); OSHA PEL 10 ppm (35 mg/m^3); NIOSH ceiling 0.5 ppm (1.8 mg/m^3)/15 min.
Toxicity by ingestion: Grade 2; oral LD_{50} = 1500 mg/kg (rat).
Liquid or solid irritant characteristics: Eye and skin irritant
Odor threshold: 0.00073–0.001 ppm.
IDLH value: 500 ppm.

FIRE DATA
Flash point: 55°F/13°C (oc); 35°F/2°C (cc).
Fire extinguishing agents not to be used: Water.
Burning rate: 7.4 mm/min

CHEMICAL REACTIVITY
Binary reactants: Violent reaction with strong oxidizers, strong acids, alkalis, alkali metals. Incompatible with aliphatic amines, ethylene oxide, isocyanates. Attacks some forms of coatings and plastics.

ENVIRONMENTAL DATA
Food chain concentration potential: Negative; unlikely to accumulate.
Water pollution: DOT Appendix B, §172.101–marine pollutant. Harmful to aquatic life in very low concentrations. Fouling to shoreline. May be dangerous if it enters nearby water intakes; notify operators. Notify local health and wildlife officials. **Response to discharge:** Issue warning–high flammability. Restrict access. Mechanical containment. Should be removed. Chemical and physical treatment.

SHIPPING INFORMATION
Grades of purity: 98+%; **Storage temperature:** Ambient; **Inert atmosphere:** None; **Venting:** Stable

PHYSICAL AND CHEMICAL PROPERTIES
Physical state @ 59°F/15°C and 1 atm: Liquid.
Molecular weight: 90.2
Boiling point @ 1 atm: 208°F/98°C/371°K.
Melting/Freezing point: –176.2°F/–115.7°C/157.5°K.
Critical temperature: 554°F/290°C/563°K.
Critical pressure: 572 psia = 38.9 atm = 3.94 MN/m^2.
Specific gravity (water = 1): 0.83 @ 68°F/20°C (liquid).
Liquid surface tension: 26.1 dynes/cm = 0.0261 N/m @ 68°F/20°C.
Liquid water interfacial tension: 30 dynes/cm = 0.030 N/m @ 68°F/20°C.
Relative vapor density (air = 1): 3.1
Ratio of specific heats of vapor (gas): 1.0770 at 16°C.
Latent heat of vaporization: 154.0 Btu/lb = 85.58 cal/g = 3.583 x 10^5 J/kg.
Heat of combustion: –16,601 Btu/lb = –9,223 cal/g = –386 x 10^5 J/kg.
Vapor pressure: 35 mm.

BUTYL METHACRYLATE   REC. B:4450

SYNONYMS: *n*-BUTYL METHACRYLATE; BUTYL 2-METHACRYLATE; *n*-BUTYL α-METHYLACRYLATE; BUTYL 2-METHYL-2-PROPENOATE; *n*-BUTYL METHACRYLATE; 2-PROPENIC ACID, 2-METHYL-, BUTYL ESTER; 2-METHYL-BUTYLACRYLATE; BUTYL-2-METHYL-2-PROPENOATE; EEC No. 607-033-00-5; METACRILATO de *n*-BUTILO (Spanish); METHACRYLATE de BUTYLE (French); METHACRYLIC ACID, BUTYL ESTER; METHACRYLSAEURE BUTYL ESTER (German)

IDENTIFICATION
CAS Number: 97-88-1
Formula: $C_8H_{14}O_2$; $CH_2 = C(CH_3)COOCH_2CH_2CH_2CH_3$
DOT ID Number: UN 2227; DOT Guide Number: 129P
Proper Shipping Name: *n*-Butyl methacrylate, inhibited

DESCRIPTION: Colorless liquid. Mild acrylate or ester odor. Floats on the surface of water; insoluble.

Flammable • Containers may BLEVE when exposed to fire • Vapors may form explosive mixture with air • Vapors are heavier than air and will collect and stay in low areas • Vapors may travel long distances to ignition sources and flashback • Vapors in confined areas (e.g., tanks, sewers, buildings) may explode when exposed to fire • Polymerization hazard • Heat can induce polymerization with rapid release of energy; sealed containers may rupture explosively • May react with itself blocking relief valves; leading to tank explosions • Irritating to the skin, eyes, and respiratory tract; inhalation symptoms may be delayed • Toxic products of combustion may include carbon monoxide.

Hazard Classification (based on NFPA-704 Rating System)
Health Hazards (Blue): 2; Flammability (Red): 2; Reactivity (Yellow): 0

EMERGENCY RESPONSE: See Appendix A (129)
Evacuation:
Public safety: Isolate spill area for at least 50 to 100 meters (160 to 330 feet) in all directions.
Spill: Large spill–Consider initial downwind evacuation for at least 300 meters (1000 feet).
Fire: Isolate for 800 meters (½ mile) in all directions, especially if tank, rail car, or tank truck is involved in fire.

EXPOSURE
Short-term effects: *SEEK MEDICAL ATTENTION.* **Vapor:** May cause lung edema; physical exertion will aggravate this condition. **Liquid:** Irritating to skin and eyes. Harmful if swallowed. Remove contaminated clothing and shoes. Flush affected areas with plenty of water. *IF IN EYES*, hold eyelids open and flush with plenty of water. *IF SWALLOWED* and victim is *CONSCIOUS AND ABLE TO SWALLOW*, have victim drink 4 to 8 ounces of water and have victim induce vomiting. *IF SWALLOWED* and victim is *UNCONSCIOUS OR HAVING CONVULSIONS*, do nothing except keep victim warm.
Note to physician or authorized medical personnel: Medical observation is recommended for 24 to 48 hours after breathing overexposure, as pulmonary edema may be delayed. As first aid for pulmonary edema, consider administering a corticosteroid spray. Cigarette smoking may exacerbate pulmonary injury and should be discouraged for at least 72 hours following exposure.

HEALTH HAZARDS
Personal protective equipment (PPE): Self-contained respirator; impervious gloves; chemical splash goggles.
Toxicity by ingestion: Grade 0; LD_{50} = > 15 g/kg.
Long-term health effects: Birth defects in rats (gross and skeletal abnormalities).
Vapor (gas) irritant characteristics: Respiratory tract irritant.
Liquid or solid irritant characteristics: Eye and skin irritant.

FIRE DATA
Flash point: 126°F/52°C (oc).
Flammable limits in air: LEL: 2%: UEL: 8% (estimate).
Fire extinguishing agents not to be used: Water may be ineffective.
Behavior in fire: Containers may explode.
Autoignition temperature: 562°F/294°C/567°K.
Electrical hazard: Class I, Group D. Flow or agitation of substance may generate electrostatic charges due to low conductivity.
Burning rate: 4.8 mm/min

CHEMICAL REACTIVITY
Binary reactants: Oxidizers may cause polymerization.
Polymerization: May occur when heated, exposed to light or on contact with oxidizers; **Inhibitor of polymerization:** 9–15 ppm monomethyl ether of hydroquinone; 90–120 ppm hydroquinone
Reactivity group: 14
Compatibility class: Acrylate

ENVIRONMENTAL DATA
Food chain concentration potential: Log P_{ow} = 2.9. Values > 3.0 are likely to bioconcentrate in aquatic organisms and other living tissue, especially in fats.
Water pollution: Effect of low concentrations on aquatic life is unknown. Fouling to shoreline. May be dangerous if it enters nearby water intakes; notify operators. Notify local health and wildlife officials. **Response to discharge:** Mechanical containment. Should be removed. Chemical and physical treatment.

SHIPPING INFORMATION
Grades of purity: 98.5+%; **Storage temperature:** Ambient; **Inert atmosphere:** None; **Venting:** Pressure-vacuum; **Stability during transport:** Stable.

PHYSICAL AND CHEMICAL PROPERTIES
Physical state @ 59°F/15°C and 1 atm: Liquid.
Molecular weight: 142.2
Boiling point @ 1 atm: 325°F/163°C/436°K.
Melting/Freezing point: Less than 32°F/0°C/273°F
Specific gravity (water = 1): 0.8975 @ 68°F/20°C (liquid).
Liquid surface tension: (estimate) 30 dynes/cm = 0.030 N/m @ 68°F/20°C.
Liquid water interfacial tension: (estimate) 35 dynes/cm = 0.035 N/m @ 68°F/20°C.
Relative vapor density (air = 1): 4.9
Heat of combustion: (estimate) –14,800 Btu/lb = –8230 cal/g = –344 x 10^5 J/kg.
Heat of polymerization: –180 Btu/lb = –100 cal/g = –4.2 x 10^5 J/kg.
Vapor pressure: (Reid) Low; 3.5 mm.

n-BUTYL PROPIONATE   REC. B:4550

SYNONYMS: BUTYL PROPANOATE; EEC No. 607-029-00-3; PROPANOIC ACID BUTYL ESTER; PROPIONATO de *n*-

BUTILO (Spanish); PROPIONIC ACID BUTYL ESTER; PROPANOIC ACID BUTYL ESTER

IDENTIFICATION
CAS Number: 590-01-2
Formula: $C_7H_{14}O_2$; $CH_3CH_2COO(CH_2)_3CH_3$
DOT ID Number: UN 1914; DOT Guide Number: 130
Proper Shipping Name: Butylpropionate

DESCRIPTION: Colorless to straw yellow liquid.

Highly flammable • Containers may BLEVE when exposed to fire • Vapors may form explosive mixture with air • Vapors are heavier than air and will collect and stay in low areas • Vapors may travel long distances to ignition sources and flashback • Vapors in confined areas (e.g., tanks, sewers, buildings) may explode when exposed to fire • Irritating to the skin, eyes, and respiratory tract • Toxic products of combustion may include carbon monoxide.

Hazard Classification (based on NFPA-704 Rating System)
Health Hazards (Blue): 2; Flammability (Red): 3; Reactivity (Yellow): 0

EMERGENCY RESPONSE: See Appendix A (130)
Evacuation:
Public safety: Isolate spill area for at least 50 to 100 meters (160 to 330 feet) in all directions.
Spill: Large spill–Consider initial downwind evacuation for at least 300 meters (1000 feet).
Fire: Isolate for 800 meters (½ mile) in all directions, especially if tank, rail car, or tank truck is involved in fire.

EXPOSURE
Short-term effects: *SEEK MEDICAL ATTENTION*. **Vapor:** Move victim to fresh air. If breathing is difficult, administer oxygen. **Liquid:** Irritating to skin and eyes. Remove contaminated clothing and shoes. Flush skin with water. *IF IN EYES*, hold eyelids open and flush with plenty of water. Can be absorbed by the skin.

Personal protective equipment (PPE): Wear full impervious protective clothing and approved respirator. Where splashing is possible wear full face shield or chemical safety goggles. Use approved respirator to protect against vapors.
Vapor (gas) irritant characteristics: Eye and respiratory tract irritant.
Liquid or solid irritant characteristics: Eye and skin irritant.

FIRE DATA
Flash point: 90°F/32°C (cc).
Fire extinguishing agents not to be used: Water.
Autoignition temperature: 799°F/426°C/699°K.

CHEMICAL REACTIVITY
Binary reactants: Oxidizers, strong acids.

ENVIRONMENTAL DATA
Water pollution: Effect of low concentrations on aquatic life is unknown. May be dangerous if it enters nearby water intakes; notify operators. Notify local health and wildlife officials.
Response to discharge: Should be removed.

SHIPPING INFORMATION
Grades of purity: 99%; varying concentrations available; **Storage temperature:** Ambient; **Inert atmosphere:** None; **Stability during transport:** Stable.

PHYSICAL AND CHEMICAL PROPERTIES
Physical state @ 59°F/15°C and 1 atm: Liquid.
Molecular weight: 130.19
Boiling point @ 1 atm: 296.2°F/146.8°C/419.8°K.
Melting/Freezing point: –128KF = –89°C/184°K.
Specific gravity (water = 1): 0.8754
Relative vapor density (air = 1): 4.5
Vapor pressure: 3 mm.

p-tert-BUTYLPHENOL   REC. B:4600

SYNONYMS: *p-terc*-BUTILFENOL (Spanish); 4-*tert*-BUTYLPHENOL; 4-(1,1-DEMETHYLETHYL)PHENOL; BUTYLPHEN; 1-HYDROXY-4-*tert*-BUTYLBENZENE; UCAR BUTYLPHENOL 4-T

IDENTIFICATION
CAS Number: 98-54-4
Formula: $C_{10}H_{14}O$; $1,4-(CH_3)_3CC_6H_4OH$
DOT ID Number: UN 2229; DOT Guide Number: 153
Proper Shipping Name: Butylphenols, solid

DESCRIPTION: White solid. Disinfectant-like odor. May float or sink in water.

Combustible solid • Dust cloud may explode if ignited in an enclosed area • Containers may BLEVE when exposed to fire • Irritating to the skin, eyes, and respiratory tract • Toxic products of combustion may include smoke and irritating vapors of unburned chemical.

Hazard Classification (based on NFPA-704 Rating System)
Health Hazards (Blue): 1; Flammability (Red): 1; Reactivity (Yellow): 0

EMERGENCY RESPONSE: See Appendix A (153)
Evacuation:
Public safety: Isolate the area of spill or leak for at least 25 to 50 meters (80 to 160 feet) in all directions.
Spill: Increase, in the downwind direction, as necessary, the distance shown under "Public Safety."
Fire: If tank, rail car, or tank truck is involved in fire, isolate for at least 800 meters (½ mile) in all directions; also, consider initial evacuation for 800 meters (½ mile) in all directions.

EXPOSURE
Short-term effects: *SEEK MEDICAL ATTENTION*. **Dust:** Irritating to eyes, nose, and throat. *IF INHALED*, will, will cause difficult breathing. *IF IN EYES*, hold eyelids open and flush with plenty of water. *IF BREATHING HAS STOPPED,* give artificial respiration; *avoid mouth-to-mouth resuscitation; use bag/mask apparatus.* IF breathing is difficult, administer oxygen. **Solid:** Will burn skin and eyes. *IF SWALLOWED*, will cause nausea and vomiting. Remove contaminated clothing and shoes. Flush affected areas with plenty of water. *IF IN EYES*, hold eyelids open and flush with plenty of water. *IF SWALLOWED* and victim is *CONSCIOUS AND ABLE TO SWALLOW*, have victim drink 4 to 8 ounces of water and have victim induce vomiting. *IF SWALLOWED* and victim is *UNCONSCIOUS OR HAVING CONVULSIONS,* do nothing except keep victim warm.

HEALTH HAZARDS
Personal protective equipment (PPE): Chemical workers' goggles; clean, body-protecting clothing.

Exposure limits (TWA unless otherwise noted): MAK (DFG) 0.5 mg/m³.
Toxicity by ingestion: Grade 2; oral LD_{50} = 3250 mg/kg (rat).
Vapor (gas) irritant characteristics: Respiratory tract irritant.
Liquid or solid irritant characteristics: Corrosive to eyes and wet skin.

FIRE DATA
Flash point: 235°F/113°C (cc)(liquid).
Fire extinguishing agents not to be used: Water may be ineffective

CHEMICAL REACTIVITY
Binary reactants: Reacts with oxidizers, acids, caustics, amines, amides.

ENVIRONMENTAL DATA
Food chain concentration potential: Negative; unlikely to accumulate.
Water pollution: DOT Appendix B, §172.101–marine pollutant. Effect of low concentrations on aquatic life is unknown. May be dangerous if it enters nearby water intakes; notify operators. Notify local health and wildlife officials. **Response to discharge:** Mechanical containment. Should be removed. Chemical and physical treatment.

SHIPPING INFORMATION
Grades of purity: Technical, 98.5%; **Storage temperature:** Ambient; **Inert atmosphere:** None; **Venting:** Open; **Stability during transport:** Stable.

PHYSICAL AND CHEMICAL PROPERTIES
Physical state @ 59°F/15°C and 1 atm: Solid.
Molecular weight: 150
Boiling point @ 1 atm: 463°F/240°C/513°C.
Melting/Freezing point: 210°F/99°C/372°K.
Specific gravity (water = 1): 1037 @ 77°F/25°C (solid).
Heat of combustion: (estimate) –16,900 Btu/lb = –9410 cal/g = –394 x 10^5 J/kg.

BUTYL TOLUENE   **REC. B:4650**

SYNONYMS: *p*-METHYL-*tert*-BUTYLBENZENE; 1-METHYL-4-*tert*-BUTYLBENZENE; *p*-*tert*-BUTYLTOLUENE; 4-*tert*-BUTYLTOLUENE; 1-(1,1-DIMETYLETHYL)-4-METHYLBENZENE; TOLUENE, p-*tert*-BUTYL; TBT

IDENTIFICATION
CAS Number: 98-51-1
Formula: $C_{11}H_{16}$; $(CH_3)C_6H_4(C_4H_9)$
DOT ID Number: UN 2667; DOT Guide Number: 131
Proper Shipping Name: Butyltoluenes

DESCRIPTION: Colorless liquid. Aromatic gasoline-like odor.

Poison! • Breathing the vapor can kill you • Firefighting gear (including SCBA) does not provide adequate protection. If exposure occurs, remove and isolate gear immediately and thoroughly decontaminate personnel • Flammable • Containers may BLEVE when exposed to fire • Vapors may form explosive mixture with air • Vapors are heavier than air and will collect and stay in low areas • Vapors may travel long distances to ignition sources and flashback • Vapors in confined areas (e.g., tanks, sewers, buildings) may explode when exposed to fire • Irritating to the skin, eyes, and respiratory tract • Toxic products of combustion may include carbon monoxide.

Hazard Classification (based on NFPA-704 Rating System)
Health Hazards (Blue): 2; Flammability (Red): 2; Reactivity (Yellow): 0

EMERGENCY RESPONSE: See Appendix A (131)
Evacuation:
Public safety: Isolate spill area for at least 100 to 200 meters (330 to 660 feet) in all directions.
Spill: Increase, as necessary, the isolation distance shown above, in "Public safety."
Fire: Isolate for 800 meters (½ mile) in all directions, especially if tank, rail car, or tank truck is involved in fire.

EXPOSURE
Short-term effects: *SEEK MEDICAL ATTENTION.* **Vapor:** Irritating to skin, eyes, and respiratory tract. Move to fresh air. *IF BREATHING HAS STOPPED,* give artificial respiration. IF breathing is difficult, administer oxygen. *IF IN EYES*, hold eyelids open and flush with plenty of water. **Liquid:** Irritating to skin, eyes, and respiratory tract. Remove contaminated clothing and shoes. Flush affected areas with plenty of water. *IF IN EYES*, hold eyelids open and flush with plenty of water. *IF SWALLOWED,***Do NOT induce vomiting**.

HEALTH HAZARDS
Personal protective equipment (PPE): A-Level PPE. OSHA Table Z-1-A air contaminant. Chemical safety goggles, compatible chemical resistant gloves, approved respirator.
Recommendations for respirator selection: NIOSH/OSHA
100 ppm: SA:CF* (any supplied-air respirator operated in a continuous-flow mode); PAPROV* [any powered, air-purifying respirator with organic vapor cartridge(s)]; or CCRFOV [any chemical cartridge respirator with a full facepiece and organic vapor cartridge(s)]; GMFOV [any air-purifying, full-facepiece respirator (gas mask) with a chin-style, front- or back-mounted acid gas canister]; or SCBAF (any self-contained breathing apparatus with a full facepiece); or SAF (any supplied-air respirator with a full *facepiece). EMERGENCY OR PLANNED ENTRY INTO UNKNOWN CONCENTRATIONS OR IDLH CONDITIONS:* SCBAF:PD,PP (any self-contained breathing apparatus that has a full facepiece and is operated in a pressure-demand or other positive-pressure mode); or SAF:PD,PP:ASCBA (any supplied-air respirator that has a full facepiece and is operated in a pressure-demand or other positive-pressure mode in combination with an auxiliary self-contained breathing apparatus operated in a pressure-demand or other positive pressure mode). *ESCAPE:* GMFOV [any air-purifying, full-facepiece respirator (gas mask) with a chin-style, front- or back-mounted organic vapor canister]; or SCBAE (any appropriate escape-type, self-contained breathing apparatus). *Note:* Substance causes eye irritation or damage; eye protection needed.
Exposure limits (TWA unless otherwise noted): ACGIH TLV 1 ppm (6.1 mg/m³); NIOSH/OSHA PEL 10 ppm (60 mg/m³).
Short-term exposure limits (15-minute TWA): NIOSH STEL 20 ppm (120 mg/m³).
Toxicity by ingestion: Grade 2: LD_{50} = 0.9g/kg (mouse).
Long-term health effects: Organ damage; liver, kidney.
Vapor (gas) irritant characteristics: Eye, and respiratory tract and mucous membrane irritant. Vapors cause smarting of the eyes or respiratory system if present in high concentrations.
Liquid or solid irritant characteristics: Eye and skin irritant. If spilled on clothing and allowed to remain, may cause smarting and reddening of skin.

Odor threshold: 5 ppm.
IDLH value: 100 ppm.

FIRE DATA
Flash point: 155°F/68°C (cc).

CHEMICAL REACTIVITY
Binary reactants: Oxidizers, nitric acid
Reactivity group: 32
Compatibility class: Aromatic hydrocarbons

ENVIRONMENTAL DATA
Water pollution: DOT Appendix B, §172.101–marine pollutant. Effect of low concentrations on aquatic life is unknown. Fouling to shoreline. May be dangerous if it enters nearby water intakes; notify operators. Notify local health and wildlife officials.
Response to discharge: Restrict access. Mechanical containment. Should be removed. Chemical and physical treatment.

SHIPPING INFORMATION
Stability during transport: Stable.

PHYSICAL AND CHEMICAL PROPERTIES
Physical state @ 59°F/15°C and 1 atm: Liquid.
Molecular weight: 148.25
Boiling point @ 1 atm: 372–378°F/189–192°C/462–465°K.
Melting/Freezing point: –62°F/–52°C/221°K.
Specific gravity (water = 1): 0.853
Relative vapor density (air = 1): 5.11
Vapor pressure: 0.7 mm @ 77°F/25°C.

BUTYLTRICHLOROSILANE   REC. B:4700

SYNONYMS: BUTILTRICLOROSILANO (Spanish); *n*-BUTYLTRICHLOROSILANE

IDENTIFICATION
CAS Number: 7521-80-4
Formula: $CH_3CH_2CH_2CH_2SiCl_3$; $C_4H_9Cl_3Si$
DOT ID Number: UN 1747; DOT Guide Number: 155
Proper Shipping Name: Butyltrichlorosilane

DESCRIPTION: Colorless liquid. Pungent odor; like hydrochloric acid. Soluble in water; reacts violently forming hydrochloric acid and hydrogen chloride fumes.

Poison! • Very flammable • Do not use water • Breathing the vapor can kill you • Firefighting gear (including SCBA) does not provide adequate protection. If exposure occurs, remove and isolate gear immediately and thoroughly decontaminate personnel • Containers may BLEVE when exposed to fire • Vapors may form explosive mixture with air • Vapors are heavier than air and will collect and stay in low areas • Vapors may travel long distances to ignition sources and flashback • Vapors in confined areas (e.g., tanks, sewers, buildings) may explode when exposed to fire • Severely irritating to skin, eyes, and respiratory tract; prolonged contact with the skin or eyes can cause burns and blindness • Toxic products of combustion may include hydrogen chloride and phosgene • Do not put yourself in danger by entering a contaminated area to rescue a victim.

Hazard Classification (based on NFPA-704 Rating System)
Health Hazards (Blue): 2; Flammability (Red): 2; Reactivity (Yellow): 0; Special Notice (White): Water reactive

EMERGENCY RESPONSE: See Appendix A (155)
Evacuation:
Public safety: Isolate the area of spill or leak for at least 50 to 100 meters (160 to 330 feet) in all directions.
Spill: Increase, in the downwind direction, as necessary, the distance shown under "Public Safety."
Fire: If tank, rail car, or tank truck is involved in fire, isolate for at least 800 meters (½ mile) in all directions; also, consider initial evacuation for 800 meters (½ mile) in all directions.

EXPOSURE
Short-term effects: *SEEK MEDICAL ATTENTION.* **Vapor:** Irritating to eyes, nose, and throat. Harmful if inhaled. Move victim to fresh air. *IF BREATHING HAS STOPPED,* give artificial respiration; *avoid mouth-to-mouth resuscitation; use bag/mask apparatus.* If breathing is difficult, administer oxygen. **Liquid:** Will burn skin and eyes. Harmful if swallowed. Remove contaminated clothing and shoes. Flush affected areas with plenty of water. *IF IN EYES,* hold eyelids open and flush with plenty of water. *IF SWALLOWED* and victim is *CONSCIOUS AND ABLE TO SWALLOW,* have victim drink 4 to 8 ounces of water. **Do NOT induce vomiting.**

HEALTH HAZARDS
Personal protective equipment (PPE): Acid-vapor-type respiratory protection; rubber gloves, chemical worker's goggles, and other protective equipment as necessary to protect skin and eyes.
Recommendations for respirator selection: NIOSH/OSHA as hydrogen chloride
50 ppm: CCRS* [any chemical cartridge respirator with cartridge(s) providing protection against the compound of concern)]; or GMFS [any air-purifying, full-facepiece respirator (gas mask) with a chin-style, front- or back-mounted canister providing protection against the compound of concern]; or PAPRS* [any powered, air-purifying respirator with cartridge(s) providing protection against the compound of concern]; or SA* (any supplied-air respirator); or SCBAF (any self-contained breathing apparatus with a full facepiece). *EMERGENCY OR PLANNED ENTRY INTO UNKNOWN CONCENTRATIONS OR IDLH CONDITIONS*: SCBAF:PD,PP (any self-contained breathing apparatus that has a full facepiece and is operated in a pressure-demand or other positive-pressure mode); or SAF:PD,PP:ASCBA (any supplied-air respirator that has a full facepiece and is operated in a pressure-demand or other positive-pressure mode in combination with an auxiliary self-contained breathing apparatus operated in a pressure-demand or other positive pressure mode). *ESCAPE:* GMFAG [any air-purifying, full-facepiece respirator (gas mask) with a chin-style, front- or back-mounted organic vapor cannister]; or SCBAE (any appropriate escape-type, self-contained breathing apparatus).
**Note:* Substance reported to cause eye irritation or damage; may require eye protection.
Exposure limits (TWA unless otherwise noted): ACGIH TLV 5 ppm; NIOSH/OSHA ceiling 5 PPM (7 mg/m^3) as hydrogen chloride
Vapor (gas) irritant characteristics: Respiratory tract irritant.
Liquid or solid irritant characteristics: Eye and skin irritant.
Odor threshold: 0.255-10.6 as hydrogen chloride
IDLH value: 50 ppm as hydrogen chloride

FIRE DATA
Flash point: 130°F/55°C (oc); 126°F/52°C (cc).
Fire extinguishing agents not to be used: Water, foam
Behavior in fire: Difficult to extinguish. Re-ignition may occur.
Burning rate: 2.2 mm/min

192 Butyraldehyde

CHEMICAL REACTIVITY
Reactivity with water: Reacts vigorously to generate hydrogen chloride (hydrochloric acid).
Binary reactants: Will react with common metals to evolve hydrogen chloride and cause severe corrosion.
Neutralizing agents for acids and caustics: Flush with water, rinse with sodium bicarbonate or lime solution.

ENVIRONMENTAL DATA
Food chain concentration potential: Negative; unlikely to accumulate.
Water pollution: Effect of low concentrations on aquatic life is unknown. May be dangerous if it enters nearby water intakes; notify operators. Notify local health and wildlife officials.
Response to discharge: Issue warning–corrosive. Restrict access. Disperse and flush.

SHIPPING INFORMATION
Grades of purity: 95+%; **Storage temperature:** Ambient; **Inert atmosphere:** None; **Venting:** Pressure-vacuum; **Stability during transport:** Stable.

PHYSICAL AND CHEMICAL PROPERTIES
Physical state @ 59°F/15°C and 1 atm: Liquid.
Molecular weight: 191.5
Boiling point @ 1 atm: 288°F/142°C/415°K.
Specific gravity (water = 1): 1.16 @ 68°F/20°C (liquid).
Liquid surface tension: (estimate) 25 dynes/cm = 0.025 N/m @ 68°F/20°C.
Relative vapor density (air = 1): 6.5
Latent heat of vaporization: (estimate) 81 Btu/lb = 45 cal/g = 1.9 x 10^5 J/kg.
Heat of combustion: (estimate) –4300 Btu/lb = –2400 cal/g = –100 x 10^5 J/kg.

1,4-BUTYNEDIOL　　　　　　　　　　　　　　　　**REC. B:4750**

SYNONYMS: 1,4-BUTINODIOL (Spanish); 2-BUTYNE-1,4-DIOL; 1,4-DIHYDROXY-2-BUTYNE

IDENTIFICATION
CAS Number: 110-65-6
Formula: $C_4H_6O_2$
DOT ID Number: UN 2716; DOT Guide Number: 153
Proper Shipping Name: 1,4-Butynediol

DESCRIPTION: White to pale yellow crystals crystalline solid, or brownish-yellow watery solution (usually 35%). Sinks and mixes with water.

Combustible • Containers may explode when exposed to fire • Vapors are heavier than air and will collect and stay in low areas • Dust or vapors in confined areas (e.g., tanks, sewers, buildings) may explode when exposed to fire • Severely Irritating to the skin, eyes, and respiratory tract • Toxic products of combustion may include carbon monoxide and irritating vapors of unburned chemical.

Hazard Classification (based on NFPA-704 Rating System)
Health Hazards (Blue): 0; Flammability (Red): 1; Reactivity (Yellow): 0

EMERGENCY RESPONSE: See Appendix A (153)
Evacuation:
Public safety: Isolate the area of spill or leak for at least 25 to 50 meters (80 to 160 feet) in all directions.
Spill: Increase, in the downwind direction, as necessary, the distance shown under "Public Safety."
Fire: If tank, rail car, or tank truck is involved in fire, isolate for at least 800 meters (½ mile) in all directions; also, consider initial evacuation for 800 meters (½ mile) in all directions.

EXPOSURE
Short-term effects: *SEEK MEDICAL ATTENTION.* **Liquid or solid:** Irritating to skin and eyes. Harmful if swallowed. Remove contaminated clothing and shoes. Flush affected areas with plenty of water. *IF BREATHING HAS STOPPED*, give artificial respiration; *avoid mouth-to-mouth resuscitation; use bag/mask apparatus. IF IN EYES*, hold eyelids open and flush with plenty of water. *IF SWALLOWED* and victim is *CONSCIOUS AND ABLE TO SWALLOW*, have victim drink 4 to 8 ounces of water.

HEALTH HAZARDS
Personal protective equipment (PPE): Neoprene rubber gloves and safety goggles or face shield.
Eye contact: immediately wash with water for at least 15 minutes and *GET MEDICAL ATTENTION.*
Toxicity by ingestion: Grade 3; LD_{50} = 50 to 500 mg/kg.
Liquid or solid irritant characteristics: If spilled on clothing and allowed to remain. may cause smarting and reddening of the skin.

FIRE DATA
Flash point: 263°F/128°C (oc) (pure butynediol).

ENVIRONMENTAL DATA
Food chain concentration potential: Negative; unlikely to accumulate.
Water pollution: Effect of low concentrations on aquatic life is unknown. May be dangerous if it enters nearby water intakes; notify operators. Notify local health and pollution control officials.
Response to discharge: Disperse and flush.

SHIPPING INFORMATION
Grades of purity: Tech. Flake: 96%, 35% solution; **Stability during transport:** Stable.

PHYSICAL AND CHEMICAL PROPERTIES
Physical state @ 59°F/15°C and 1 atm: Solid.
Molecular weight: 36.09
Boiling point @ 1 atm: 460°F/238°C/511°K.
Specific gravity (water = 1): 1.07 @ 68°F/20°C (solid).
Heat of combustion: –11,020 Btu/lb = –6120 cal/g = –256.2 x 10^5 J/kg.

BUTYRALDEHYDE　　　　　　　　　　　　　　　**REC. B:4800**

SYNONYMS: ALDEHYDE BUTYRIQUE (French); BUTAL; n-BUTALDEHYDE; BUTANALDEHYDE; n-BUTIRALDEHIDO (Spanish); n-BUTYRALDEHYDE; BUTALDEHYDE; BUTALYDE; BUTANAL; BUTYRAL; BUTYRAL BUTYRIC ALDEHYDE; BUTYRALDEHYD (German); BUTYL ALDEHYDE; n-BUTYL ALDEHYDE; BUTYRIC ACID; EEC No. 605-006-00-2

IDENTIFICATION
CAS Number: 123-72-8
Formula: C_4H_8O; $CH_3CH_2CH_2CHO$
DOT ID Number: UN 1129; DOT Guide Number: 129
Proper Shipping Name: Butyraldehyde

DESCRIPTION: Colorless, watery liquid. Pungent, intense, suffocating odor. Floats and mixes slowly with water. Flammable, irritating vapor is produced.

Highly flammable • Corrosive to the eyes and respiratory tract; contact with eyes can cause burns and vision impairment; inhalation symptoms may be delayed; this material has anesthetic properties • Containers may BLEVE when exposed to fire •Forms explosive peroxides upon exposure to air and excessive heat • Vapors may form explosive mixture with air • Vapors are heavier than air and will collect and stay in low areas • Vapors may travel long distances to ignition sources and flashback • Vapors in confined areas (e.g., tanks, sewers, buildings) may explode when exposed to fire • Toxic products of combustion may include carbon monoxide.

Hazard Classification (based on NFPA-704 Rating System)
Health Hazards (Blue): 3; Flammability (Red): 3; Reactivity (Yellow): 0

EMERGENCY RESPONSE: See Appendix A (129)
Evacuation:
Public safety: Isolate spill area for at least 50 to 100 meters (160 to 330 feet) in all directions.
Spill: Large spill–Consider initial downwind evacuation for at least 300 meters (1000 feet).
Fire: Isolate for 800 meters (½ mile) in all directions, especially if tank, rail car, or tank truck is involved in fire.

EXPOSURE
Short-term effects: *SEEK MEDICAL ATTENTION.* **Vapor:** Irritating to eyes, nose, and throat. *IF INHALED,* will, will cause nausea, vomiting, headache or loss of consciousness. Move to fresh air. *IF BREATHING HAS STOPPED,* give artificial respiration. IF breathing is difficult, administer oxygen. May cause lung edema; physical exertion will aggravate this condition. **Liquid:** Irritating to skin. Will burn eyes. Harmful if swallowed. Remove contaminated clothing and shoes. Flush affected areas with plenty of water. *IF IN EYES,* hold eyelids open and flush with plenty of water. *IF SWALLOWED* and victim is *CONSCIOUS AND ABLE TO SWALLOW,* have victim drink 4 to 8 ounces of water.
Note to physician or authorized medical personnel: Medical observation is recommended for 24 to 48 hours after breathing overexposure, as pulmonary edema may be delayed. As first aid for pulmonary edema, consider administering a corticosteroid spray. Cigarette smoking may exacerbate pulmonary injury and should be discouraged for at least 72 hours following exposure.

HEALTH HAZARDS
Personal protective equipment (PPE): B-Level PPE. Protective goggles, gloves, and organic canister gas mask. Chemical protective material(s) reported to offer protection: butyl rubber, Teflon®.
Exposure limits (TWA unless otherwise noted): 25 ppm (AIHA WEEL).
Toxicity by ingestion: Grade 1; LD_{50} = 5 to 15 g/kg (rat).
Vapor (gas) irritant characteristics: Vapors cause moderate irritation such that personnel will find high concentrations unpleasant. The effect is temporary.
Liquid or solid irritant characteristics: Minimum hazard. If spilled on clothing and allowed to remain, may cause smarting and reddening of the skin.
Odor threshold: 0.0046 ppm.
IDLH value: 19,000 ppm.

FIRE DATA
Flash point: 10°F/–12°C (cc).
Flammable limits in air: LEL: 2.5%; UEL: 12.5%.
Behavior in fire: Fires are difficult to control due to ease of reignition.
Autoignition temperature: 446°F/230°C/503°K.
Electrical hazard: Class I, Group C.
Burning rate: 4.4 mm/min

CHEMICAL REACTIVITY
Binary reactants: Possible self-reaction in air; undergoes rapid oxidation to butyric acid in air. Violent reaction with strong oxidizers. May corrode steel due to corrosive action of butyric acid.
Polymerization: May occur in presence of heat, acids, or alkalis
Reactivity group: 19
Compatibility class: Aldehyde

ENVIRONMENTAL DATA
Food chain concentration potential: Log P_{ow} = 0.88–1.2. Unlikely to accumulate.
Water pollution: DOT Appendix B, §172.101–marine pollutant. Effect of low concentrations on aquatic life is unknown. May be dangerous if it enters nearby water intakes; notify operators. Notify local health and pollution control officials. **Response to discharge:** Issue warning–high ly flammable. Restrict access. Disperse and flush.

SHIPPING INFORMATION
Grades of purity: Water saturated: 97%; dry: 99.5%; **Storage temperature:** Ambient; **Inert atmosphere:** None; **Venting:** Pressure-vacuum; **Stability during transport:** Stable.

NAS HAZARD CLASSIFICATION FOR BULK WATER TRANSPORTATION
FIRE: 3
HEALTH: Vapor irritant: 2; Liquid or solid irritant: 1; Poisons: 2
WATER POLLUTION: Human toxicity: 1; Aquatic toxicity: 3; Aesthetic effect: 3
REACTIVITY: Other chemicals: 2; Water: 0; Self-reaction: 1

PHYSICAL AND CHEMICAL PROPERTIES
Physical state @ 59°F/15°C and 1 atm: Liquid.
Molecular weight: 72.11
Boiling point @ 1 atm: 167°F/74.8°C/348.0°K.
Melting/Freezing point: –146°F/–99°C/174°K.
Critical temperature: 484°F/251°C/524°K.
Critical pressure: 590 psia = 40 atm = 4.1 MN/m².
Specific gravity (water = 1): 0.803 @ 68°F/20°C (liquid).
Liquid surface tension: 24.6 dynes/cm = 0.0246 N/m @ 68°F/20°C.
Liquid water interfacial tension: 5.7 dynes/cm = 0.0057 N/m at 22.3°C.
Relative vapor density (air = 1): 2.5
Ratio of specific heats of vapor (gas): 1.089
Latent heat of vaporization: 184 Btu/lb = 102 cal/g = 4.27×10^5 J/kg.
Heat of combustion: –15,210 Btu/lb = –8450 cal/g = -353.8×10^5 J/kg.
Vapor pressure: (Reid) 4.8 psia; 94 mm.

iso-BUTYRALDEHYDE   REC. B:4850

SYNONYMS: *iso*-BUTIRALDEHIDO (Spanish); *iso*-BUTYL ALDEHYDE; ISOBUTANAL; ISOBUTYRALDEHYDE;

iso-Butyraldehyde

ISOBUTYL ALDEHYDE; ISOBUTYRIC ALDEHYDE; 2-METHYLPROPANAL; VALINE ALDEHYDE

IDENTIFICATION
CAS Number: 78-84-2
Formula: C_4H_8O; $(CH_3)_2CHCHO$
DOT ID Number: UN 2045; DOT Guide Number: 129
Proper Shipping Name: Isobutyraldehyde

DESCRIPTION: Colorless watery liquid. Pleasant gasoline-like odor. Floats and mixes slowly with water.

Highly flammable • Corrosive to the eyes and respiratory tract; contact with eyes can cause burns and vision impairment; inhalation symptoms may be delayed; this material has anesthetic properties • Containers may BLEVE when exposed to fire •Forms explosive peroxides upon exposure to air and excessive heat • Vapors may form explosive mixture with air • Vapors are heavier than air and will collect and stay in low areas • Vapors may travel long distances to ignition sources and flashback • Vapors in confined areas (e.g., tanks, sewers, buildings) may explode when exposed to fire • Toxic products of combustion may include carbon monoxide.

Hazard Classification (based on NFPA-704 Rating System)
Health Hazards (Blue): 2; Flammability (Red): 3; Reactivity (Yellow): 0

EMERGENCY RESPONSE: See Appendix A (129)
Evacuation:
Public safety: Isolate spill area for at least 50 to 100 meters (160 to 330 feet) in all directions.
Spill: Large spill–Consider initial downwind evacuation for at least 300 meters (1000 feet).
Fire: Isolate for 800 meters (½ mile) in all directions, especially if tank, rail car, or tank truck is involved in fire.

EXPOSURE
Short-term effects: *SEEK MEDICAL ATTENTION*. **Vapor:** Irritating to eyes, nose, and throat. *IF INHALED*, will, will cause nausea, vomiting, headache or loss of consciousness. Move to fresh air. *IF BREATHING HAS STOPPED*, give artificial respiration. IF breathing is difficult, administer oxygen. May cause lung edema; physical exertion will aggravate this condition. **Liquid:** Irritating to skin. Will burn eyes. Harmful if swallowed. Remove contaminated clothing and shoes. Flush affected areas with plenty of water. *IF IN EYES*, hold eyelids open and flush with plenty of water. *IF SWALLOWED* and victim is *CONSCIOUS AND ABLE TO SWALLOW*, have victim drink 4 to 8 ounces of water.
Note to physician or authorized medical personnel: Medical observation is recommended for 24 to 48 hours after breathing overexposure, as pulmonary edema may be delayed. As first aid for pulmonary edema, consider administering a corticosteroid spray. Cigarette smoking may exacerbate pulmonary injury and should be discouraged for at least 72 hours following exposure.

HEALTH HAZARDS
Personal protective equipment (PPE): Appropriate protective clothing, including rubber gloves, rubber shoes and protective eye wear. Chemical protective material(s) reported to offer protection: natural rubber, PVC, Teflon. Also, butyl rubber is generally suitable for aldehydes.
Toxicity by ingestion: Grade 2; LD_{50} = 0.5 to 5 mg/kg.
Vapor (gas) irritant characteristics: Vapors cause irritation such that personnel will find high concentrations unpleasant. The effect is temporary.
Liquid or solid irritant characteristics: Eye irritant. If spilled on clothing and allowed to remain may cause smarting and reddening of the skin.
Odor threshold: 0.047 ppm.

FIRE DATA
Flash point: –40°F/–40°C (cc).
Flammable limits in air: LEL: 1.6%; UEL: 10.5%.
Behavior in fire: Fires are difficult to control due to ease of reignition.
Autoignition temperature: 435°F/224°C/497°K.
Electrical hazard: Class I, Group C.
Burning rate: 4.8 mm/min

CHEMICAL REACTIVITY
Binary reactants: Strong acids, aliphatic amines. Undergoes rapid oxidation to butyric acid; may corrode mild steel due to corrosive action of butyric acid.
Reactivity group: 19
Compatibility class: Aldehyde

ENVIRONMENTAL DATA
Food chain concentration potential: Negative; unlikely to accumulate.
Water pollution: DOT Appendix B, §172.101–marine pollutant. Effect of low concentrations on aquatic life is unknown. May be dangerous if it enters nearby water intakes; notify operators. Notify local health and pollution control officials. **Response to discharge:** Issue warning–high flammability. Restrict access. Disperse and flush.

SHIPPING INFORMATION
Grades of purity: Dry grade: 98.0%; Wet grade: 96.0%; Commercial: 97%; **Storage temperature:** Ambient; **Inert atmosphere:** None; **Venting:** Pressure-vacuum; **Stability during transport:** Stable.

NAS HAZARD CLASSIFICATION FOR BULK WATER TRANSPORTATION
FIRE: 3
HEALTH: Vapor irritant: 2; Liquid or solid irritant: 1; Poisons: 2
WATER POLLUTION: Human toxicity: 2; Aquatic toxicity: 2; Aesthetic effect: 3
REACTIVITY: Other chemicals: 2; Water: 0; Self-reaction: 1

PHYSICAL AND CHEMICAL PROPERTIES
Physical state @ 59°F/15°C and 1 atm: Liquid.
Molecular weight: 72.11
Boiling point @ 1 atm: 147°F/64.1°C/337.3°K.
Melting/Freezing point: –112°F/–80°C/193°K.
Critical temperature: 464°F/240°C/513°K.
Critical pressure: 600 psia = 41 atm = 4.2 MN/m^2.
Specific gravity (water = 1): 0.791 @ 68°F/20°C (liquid).
Liquid surface tension: 22.0 dynes/cm = 0.0220 N/m at 24°C.
Liquid water interfacial tension: 7.2 dynes/cm = 0.0072 N/m at 22.7°C.
Relative vapor density (air = 1): 2.5
Ratio of specific heats of vapor (gas): 1.093
Latent heat of vaporization: 180 Btu/lb = 98 cal/g = 4.1 x 10^5 J/kg.
Heat of combustion: –13,850 Btu/lb = –7693 cal/g = –322.1 x 10^5 J/kg.
Vapor pressure: (Reid) 5.0 psia.

BUTYRIC ACID REC. B:4900

SYNONYMS: ACIDO BUTRICO (Spanish); BUTANIC ACID; BUTANOIC ACID; BUTTERSAEURE (German); *n*-BUTYRIC ACID; EEC No. 607-135-00-X; ETHYLACETIC ACID; 1-PROPANECARBOXYIC ACID; PROPYLFORMIC ACID

IDENTIFICATION
CAS Number: 107-92-6
Formula: $C_4H_8O_2$; $CH_3CH_2CH_2COOH$
DOT ID Number: UN 2820; DOT Guide Number: 153
Proper Shipping Name: Butyric acid
Reportable Quantity (RQ): 5000 lb/2270 kg

DESCRIPTION: Clear, colorless liquid. Rancid butter odor. Floats on the surface of water; soluble.

Flammable • Containers may BLEVE when exposed to fire • Vapors may form explosive mixture with air • Vapors are heavier than air and will collect and stay in low areas • Vapors may travel long distances to ignition sources and flashback • Vapors in confined areas (e.g., tanks, sewers, buildings) may explode when exposed to fire • Irritating to the skin, eyes, and respiratory tract • Toxic products of combustion may include carbon monoxide.

Hazard Classification (based on NFPA-704 Rating System)
Health Hazards (Blue): 3; Flammability (Red): 2; Reactivity (Yellow): 0

EMERGENCY RESPONSE: See Appendix A (153)
Evacuation:
Public safety: Isolate the area of spill or leak for at least 25 to 50 meters (80 to 160 feet) in all directions.
Spill: Increase, in the downwind direction, as necessary, the distance shown under "Public Safety."
Fire: If tank, rail car, or tank truck is involved in fire, isolate for at least 800 meters (½ mile) in all directions; also, consider initial evacuation for 800 meters (½ mile) in all directions.

EXPOSURE
Short-term effects: *SEEK MEDICAL ATTENTION*. **Vapor:** Irritating to eyes, nose, and throat. *IF INHALED*, will, will cause coughing or difficult breathing. *IF IN EYES*, hold eyelids open and flush with plenty of water. *IF BREATHING HAS STOPPED*, give artificial respiration; *avoid mouth-to-mouth resuscitation; use bag/mask apparatus*. IF breathing is difficult, administer oxygen.
Liquid: Will burn skin and eyes. *IF SWALLOWED*, will cause nausea and vomiting. Remove contaminated clothing and shoes. Flush affected areas with plenty of water. *IF IN EYES*, hold eyelids open and flush with plenty of water. *IF SWALLOWED* and victim is *CONSCIOUS AND ABLE TO SWALLOW*, have victim drink 4 to 8 ounces of water and have victim induce vomiting. *IF SWALLOWED* and victim is *UNCONSCIOUS OR HAVING CONVULSIONS*, do nothing except keep victim warm.

HEALTH HAZARDS
Personal protective equipment (PPE): B-Level PPE. Self-contained breathing apparatus; rubber gloves; vapor-proof plastic goggles; impervious apron and boots. Chemical protective material(s) reported to offer protection: butyl rubber, Viton. Also, chlorinated polyethylene and neoprene offers limited protection.
Toxicity by ingestion: Grade 2; oral LD_{50} = 2,940 mg/kg (rat).
Vapor (gas) irritant characteristics: Eye and respiratory tract irritant. Vapors cause irritation such that personnel will find high concentrations unpleasant. The effect is temporary.
Liquid or solid irritant characteristics: Severe skin irritant. May cause pain and second-degree burns after a few minutes of contact.
Odor threshold: 0.001 ppm.

FIRE DATA
Flash point: 166°F/75°C (oc); 161°F/72°C (cc).
Flammable limits in air: LEL: 2.0%; UEL: 10.0%.
Fire extinguishing agents not to be used: Water may be ineffective.
Autoignition temperature: 830°F/443°C/716°K.
Electrical hazard: Class I, Group D.
Burning rate: 2.7 mm/min

CHEMICAL REACTIVITY
Binary reactants: Reacts with strong oxidizers and bases.
Neutralizing agents for acids and caustics: Flush with water.

ENVIRONMENTAL DATA
Food chain concentration potential: Log P_{ow} = 0.8. Seafood may be tainted following a spill, but it is unlikely for this chemical to concentrate in food chain.
Water pollution: Dangerous to aquatic life in high concentrations. May be dangerous if it enters nearby water intakes; notify operators. Notify local health and wildlife officials. **Response to discharge:** Issue warning–water contaminant. Disperse and flush.

SHIPPING INFORMATION
Grades of purity: Commercial, 99.5+%; **Storage temperature:** Ambient; **Inert atmosphere:** None; **Venting:** Open; **Stability during transport:** Stable.

NAS HAZARD CLASSIFICATION FOR BULK WATER TRANSPORTATION
FIRE: 1
HEALTH: Vapor irritant: 2; Liquid or solid irritant: 3; Poisons: 0
WATER POLLUTION: Human toxicity: 1; Aquatic toxicity: 2; Aesthetic effect: 3
REACTIVITY: Other chemicals: 2; Water: 0; Self-reaction: 0

PHYSICAL AND CHEMICAL PROPERTIES
Physical state @ 59°F/15°C and 1 atm: Liquid.
Molecular weight: 88.1
Boiling point @ 1 atm: 327°F/164°C/437°K.
Melting/Freezing point: 23°F/–5°C/268°K.
Critical temperature: 671°F/355°C/628°K.
Critical pressure: 764 psia = 52 atm = 5.3 MN/m^2.
Specific gravity (water = 1): 0.958 @ 68°F/20°C.
Liquid surface tension: 26.74 dynes/cm = 0.02674 N/m @ 68°F/20°C.
Relative vapor density (air = 1): 3.0
Ratio of specific heats of vapor (gas): 1.079 @ 68°F/20°C.
Latent heat of vaporization: 167 Btu/lb = 92.7 cal/g = 3.88 x 10^5 J/kg.
Heat of combustion: –10,620 Btu/lb = –5900 cal/g = –247 x 10^5 J/kg.
Heat of solution: –82 Btu/lb = –45 cal/g = –1.9 x 10^5 J/kg.
Heat of fusion: 30.04 cal/g.
Vapor pressure: 0.43 mm.

BUTYRONITRILE REC. B:4950

SYNONYMS: BUTIRONITRILO (Spanish); BUTANENITRILE; PROPYL CYANIDE; BUTYRIC ACID NITRILE; CYANOPROPANE; PROPYL CYANIDE; EEC No. 608-005-00-5

Butyronitrile

IDENTIFICATION
CAS Number: 109-74-0
Formula: C_3H_7CN; $CH_3CH_2CH_2CN$
DOT ID Number: UN 2411; DOT Guide Number: 131
Proper Shipping Name: Butyronitrile

DESCRIPTION: Colorless liquid. Sharp suffocating odor. Soluble in water; hot surfaces or hot water forms hydrogen cyanide.

Poison! • Breathing the vapor, skin or eye contact, or swallowing the material can kill you; converted to cyanide in the body • Highly flammable • Firefighting gear (including SCBA) does not provide adequate protection. If exposure occurs, remove and isolate gear immediately and thoroughly decontaminate personnel • Containers may BLEVE when exposed to fire •Vapors may form explosive mixture with air • Vapors are heavier than air and will collect and stay in low areas • Vapors may travel long distances to ignition sources and flashback • Vapors in confined areas (e.g., tanks, sewers, buildings) may explode when exposed to fire • Irritating to the skin, eyes, and respiratory tract • Toxic products of combustion may include cyanide, nitrogen oxides, and carbon monoxide.

Hazard Classification (based on NFPA-704 Rating System)
Health Hazards (Blue): 3; Flammability (Red): 3; Reactivity (Yellow): 0

EMERGENCY RESPONSE: See Appendix A (131)
Evacuation:
Public safety: Isolate spill area for at least 100 to 200 meters (330 to 660 feet) in all directions.
Spill: Increase, as necessary, the isolation distance shown above, in "Public safety."
Fire: Isolate for 800 meters (½ mile) in all directions, especially if tank, rail car, or tank truck is involved in fire.

EXPOSURE
Short-term effects: *SEEK MEDICAL ATTENTION.* **Vapor:** Irritating to eyes, nose, throat and skin. *IF INHALED*, will, remove to fresh air. If not breathing, give artificial respiration; *avoid mouth-to-mouth resuscitation; use bag/mask apparatus.* IF breathing difficult, administer oxygen. **Liquid:** Irritating to the skin. Remove contaminated clothing and shoes. Flush affected areas with plenty of water. *IF IN EYES*, hold eyelids open and flush with plenty of water. *IF SWALLOWED* and victim is *CONSCIOUS AND ABLE TO SWALLOW*, have victim drink 4 to 8 ounces of water and induce vomiting. *IF SWALLOWED* and victim is UNCONSCIOUS: Do nothing except keep victim warm. *Note to physician or authorized medical personnel:* Consider the use of amyl nitrite perles if symptoms of cyanide poisoning develop. If symptoms indicate, initial treatment includes the cyanide antidote kit. In all cases, break an amyl nitrite perle in a gauze pad and hold lightly under victim's nose for 15 seconds, repeating 5 times at about 15-second intervals; if necessary (and if sodium nitrite infusions will be delayed), repeat procedure every 3 minutes with fresh perles until 3 or 4 have been used. Avoid breathing the vapor while administering it to the victim. Administer sodium nitrite IV, ASAP. The usual adult dose is 10 to 20 mL of a 3% solution infused over no less than 5 minutes; the average child dose is 0.15 to 0.20 mL/kg. Monitor blood pressure during administration, and slow the rate of infusion if hypotention develops. Next, infuse sodium thiosulfate IV. The usual adult dose is 50 mL of a 25% solution infused over 10 to 20 minutes; the average child dose is 1.65 mL/kg. Repeat with nitrite and thiosulfate as required.

HEALTH HAZARDS
Personal protective equipment (PPE): A-Level PPE. Goggles, rubber gloves, self-contained respirator, protective clothing.
Recommendations for respirator selection: NIOSH
80 ppm: CCROV [any chemical cartridge respirator with organic vapor cartridge(s)]; or SA (any supplied-air respirator). *200 ppm*: SA:CF (any supplied-air respirator operated in a continuous-flow mode); or PAPROV [any powered, air-purifying respirator with organic vapor cartridge(s)]. *400 ppm*: CCRFOV [any chemical cartridge respirator with a full facepiece and organic vapor cartridge(s)]; or GMFOV [any air-purifying, full-facepiece respirator (gas mask) with a chin-style, front- or back-mounted acid gas canister]; or PAPRTOV [any powered, air-purifying respirator with a tight-fitting facepiece and organic vapor cartridge(s)]; or SCBAF (any self-contained breathing apparatus with a full facepiece); or SAF (any supplied-air respirator with a full facepiece). *1000 ppm:* SAF:PD,PP (any supplied-air respirator that has a full facepiece and is operated in a pressure-demand or other positive-pressure mode). *EMERGENCY OR PLANNED ENTRY INTO UNKNOWN CONCENTRATIONS OR IDLH CONDITIONS:* SCBAF:PD,PP (any self-contained breathing apparatus that has a full facepiece and is operated in a pressure-demand or other positive-pressure mode); or SAF:PD,PP:ASCBA (any supplied-air respirator that has a full facepiece and is operated in a pressure-demand or other positive-pressure mode in combination with an auxiliary self-contained breathing apparatus operated in a pressure-demand or other positive pressure mode). *ESCAPE:* GMFOV[any air-purifying, full-facepiece respirator (gas mask) with a chin-style, front- or back-mounted organic vapor canister]; or SCBAE (any appropriate escape-type, self-contained breathing apparatus).
Exposure limits (TWA unless otherwise noted): NIOSH REL 8 ppm ($22\ mg/m^3$).
Toxicity by ingestion: Grade 4: LD_{50} = 28 mg/kg (mouse).
Vapor (gas) irritant characteristics: Vapors cause severe irritation of eyes and throat and can cause eye and lung injury. They cannot be tolerated even at low concentrations.
Liquid or solid irritant characteristics: If spilled on clothing and allowed to remain, may cause smarting and reddening of skin.

FIRE DATA
Flash point: 79°F/26°C (oc).
Behavior in fire: Containers may rupture and explode.
Autoignition temperature: 935°F/501°C/774°K.

CHEMICAL REACTIVITY
Binary reactants: Contact with hot surfaces or steam forms hydrogen cyanide gas. Exothermic reaction with strong acids or strong oxidizers (forms toxic hydrogen cyanide gas). May accumulate static electrical charges; may cause ignition of its own vapors.
Reactivity group: 37
Compatibility class: Nitriles

ENVIRONMENTAL DATA
Water pollution: DOT Appendix B, §172.101–marine pollutant. Harmful to aquatic life in low concentrations. May be dangerous if it enters nearby water intakes; notify operators. Fouling to shoreline. Notify local health and wildlife officials. Notify operators of local water intakes. **Response to discharge:** Issue warning–poison. Restrict access. Evacuate area. Mechanical containment. Should be removed. Chemical and physical treatment.

SHIPPING INFORMATION
Grades of purity: 99 +%; **Stability during transport:** Stable.

NAS HAZARD CLASSIFICATION FOR BULK WATER TRANSPORTATION
FIRE: 3
HEALTH: Vapor irritant: 4; Liquid or solid irritant: 2; Poisons: 4
WATER POLLUTION: Human toxicity: 4; Aquatic toxicity:3; Aesthetic effect: 1
REACTIVITY: Other chemicals: 2; Water: 0; Self-reaction: 0

PHYSICAL AND CHEMICAL PROPERTIES
Physical state @ 59°F/15°C and 1 atm: Liquid.
Molecular weight: 69.10
Boiling point @ 1 atm: 244°F/118°C/391°K.
Melting/Freezing point: −171°F/−113°C/160°K.
Critical temperature: 588.2°F/309°C/582.2°K.
Critical pressure: 549.8 psia = 37.4 atm = 3.9 MN/m^2.
Specific gravity (water = 1): 0.7936 @ 68°F/20°C.
Relative vapor density (air = 1): 2.4
Latent heat of vaporization: 206.8 Btu/lb = 114.9 cal/g = 4.8 x 10^5 J/kg.
Heat of combustion: −15,975 Btu/lb = −8875 cal/g = −372 x 10^5 J/kg.
Vapor pressure: (Reid) 0.4 psia; 14 mm.

BUTYRYL CHLORIDE   **REC. B:5000**

SYNONYMS: BUTANOYL CHLORIDE; BUTYROYL CHLORIDE; *n*-BUTYRYL CHLORIDE; CLORURO de BUTIRILO (Spanish); EEC No. 607-136-00-5

IDENTIFICATION
CAS Number: 141-75-3
Formula: C_4H_7ClO
DOT ID Number: UN 2353; DOT Guide Number: 132
Proper Shipping Name: Butyryl chloride

DESCRIPTION: Colorless liquid. Sharp, irritating odor. Sinks in water; slowly dissolves; decomposes forming hydrochloric acid and hydrogen chloride gas.

Corrosive to the skin, eyes, and respiratory tract; contact with skin or eyes can cause burns and vision impairment; inhalation symptoms may be delayed • Firefighting gear (including SCBA) does not provide adequate protection. If exposure occurs, remove and isolate gear immediately and thoroughly decontaminate personnel • Vapors are heavier than air and will collect and stay in low areas • Vapors in confined areas (e.g., tanks, sewers, buildings) may explode when exposed to fire • Containers may BLEVE when exposed to fire • Toxic products of combustion may include phosgene and hydrogen chloride • Do not put yourself in danger by entering a contaminated area to rescue a victim.

Hazard Classification (based on NFPA-704 Rating System)
Health Hazards (Blue): 2; Flammability (Red): 3; Reactivity (Yellow): 0

EMERGENCY RESPONSE: See Appendix A (132)
Evacuation:
Public safety: Isolate spill area for at least 50 to 100 meters (160 to 330 feet) in all directions.
Spill: Increase, as necessary, the isolation distance shown above, in "Public safety."
Fire: Isolate for 800 meters (½ mile) in all directions, especially if tank, rail car, or tank truck is involved in fire.

EXPOSURE
Short-term effects: *SEEK MEDICAL ATTENTION.* **Vapor:** Highly irritating to skin, eyes, and mucous membranes. Harmful if inhaled or absorbed through the skin. Move victim to fresh air. *IF BREATHING HAS STOPPED,* give artificial respiration. IF breathing is difficult, administer oxygen. May cause lung edema; physical exertion will aggravate this condition. **Liquid:** Harmful if swallowed or absorbed through the skin. Corrosive to the skin and eyes. *IF IN EYES,* hold eyelids open, flush with water for at least 15 minutes. Remove contaminated clothing and shoes, flush affected areas with water. *IF SWALLOWED,* do nothing except keep victim warm. **Do not induce vomiting.**
Note to physician or authorized medical personnel: Medical observation is recommended for 24 to 48 hours after breathing overexposure, as pulmonary edema may be delayed. As first aid for pulmonary edema, consider administering a corticosteroid spray. Cigarette smoking may exacerbate pulmonary injury and should be discouraged for at least 72 hours following exposure.

HEALTH HAZARDS
Personal protective equipment (PPE): B-Level PPE. Approved respirator, chemical-resistant gloves, safety goggles, other protective clothing.
Vapor (gas) irritant characteristics: Vapors cause severe irritation of the eyes and throat and can cause eye and lung injury. They cannot be tolerated even at low concentrations.
Liquid or solid irritant characteristics: Severe skin and eye irritant. Causes second- and third-degree burns on short contact and is very injurious to the eyes.

FIRE DATA
Flash point: 71°F/22°C (cc).
Fire extinguishing agents not to be used: Do not use water or foam to extinguish fire.

CHEMICAL REACTIVITY
Reactivity with water: Decomposes to produce hydrogen chloride gas, which may react with metals to produce explosive hydrogen gas.
Binary reactants: Violent reaction with oxidizers. Corrosive to metals. Forms hydrogen chloride in air.
Neutralizing agents for acids and caustics: Sodium carbonate, slaked lime, soda ash.

ENVIRONMENTAL DATA
Water pollution: Effects of low concentrations on aquatic life are not known. May be dangerous if it enters nearby water intakes; notify operators. Notify local health and wildlife officials. **Response to discharge:** Evacuate area. Issue warning–high flammability, corrosive. Should be removed. Chemical and physical treatment.

SHIPPING INFORMATION
Grades of purity: 98%; **Inert atmosphere:** None; **Venting:** None; **Stability during transport:** Stable.

NAS HAZARD CLASSIFICATION FOR BULK WATER TRANSPORTATION
FIRE: 3
HEALTH: Vapor irritant: 4; Liquid or solid irritant: 4; Poisons:–
WATER POLLUTION: Human toxicity:–; Aquatic toxicity:–; Aesthetic effect: 1
REACTIVITY: Other chemicals: 2; Water: 4; Self-reaction: 0

PHYSICAL AND CHEMICAL PROPERTIES
Physical state @ 59°F/15°C and 1 atm: Liquid.

Molecular weight: 106.51
Boiling point @ 1 atm: 216°F/102°C/375°K.
Melting/Freezing point: –128°F/–89°C/184°K.
Specific gravity (water = 1): 1.0277 @ 68°F/20°C.
Relative vapor density (air = 1): 3.67
Vapor pressure: 30 mm.

CACODYLIC ACID   REC. C:0100

SYNONYMS: ACIDO CACODILICO (Spanish); HYDROXYDIMETHYLARSINE OXIDE; DIMETHYLARSINIC ACID; ANSAR; SILVISAR 510

IDENTIFICATION
CAS Number: 75-60-5
Formula: $(CH_3)_2AsOOH$
DOT ID Number: UN 1572; DOT Guide Number: 151
Proper Shipping Name: Cacodylic acid
Reportable Quantity (RQ): 1 lb/0.454 kg

DESCRIPTION: Colorless solid or dyed blue water solution. Sinks and mixes with water forming a corrosive solution.

Poison! • Breathing the dust or swallowing the material can kill you; skin or eye contact may cause severe irritation and burns • Firefighting gear (including SCBA) does not provide adequate protection. If exposure occurs, remove and isolate gear immediately and thoroughly decontaminate personnel • Toxic products of combustion may include carbon monoxide and arsenic fumes.

Hazard Classification (based on NFPA-704 Rating System)
Health Hazards (Blue): 1; Flammability (Red): 0; Reactivity (Yellow): 0

EMERGENCY RESPONSE: See Appendix A (151)
Evacuation:
Public safety: Isolate the area of spill or leak for at least 25 to 50 meters (80 to 160 feet) in all directions.
Spill: Increase, in the downwind direction, as necessary, the distance shown under "Public Safety."
Fire: If tank, rail car, or tank truck is involved in fire, isolate for at least 800 meters (½ mile) in all directions; also, consider initial evacuation for 800 meters (½ mile) in all directions.

EXPOSURE
Short-term effects: *SEEK MEDICAL ATTENTION.* **Dust:** *POISONOUS IF INHALED OR IF SKIN IS EXPOSED.* Move victim to fresh air. *IF IN EYES*, hold eyelids open and flush with plenty of water. If artificial respiration is administered, *avoid mouth-to-mouth resuscitation; use bag/mask apparatus.* **Solid:** *POISONOUS IF SWALLOWED OR IF SKIN IS EXPOSED.* Remove contaminated clothing and shoes. Flush affected areas with plenty of water. *IF IN EYES*, hold eyelids open and flush with plenty of water. *IF SWALLOWED* and victim is *CONSCIOUS AND ABLE TO SWALLOW*, have victim drink 4 to 8 ounces of water and have victim induce vomiting. *IF SWALLOWED* and victim is *UNCONSCIOUS OR HAVING CONVULSIONS*, do nothing except keep victim warm.

HEALTH HAZARDS
Personal protective equipment (PPE): A-Level PPE. Dust respirator; goggles; protective clothing.

Recommendations for respirator selection: NIOSH (as inorganic arsenic; *FOR REFERENCE ONLY*).
At any concentrations above the NIOSH REL, or where there is no REL, at any detectable concentration: SCBAF:PD,PP (any self-contained breathing apparatus that has a full faceplate and is operated in a pressure-demand or other positive-pressure mode); or SAF:PD,PP:ASCBA (any supplied-air respirator that has a full facepiece and is operated in a pressure-demand or other positive-pressure mode in combination with an auxiliary self-contained breathing apparatus operated in a pressure-demand or other positive-pressure mode). *ESCAPE:* GMFAGHiE [any air-purifying, full-facepiece respirator (gas mask) with a chin-style, front-or back-mounted acid gas canister having a high-efficiency particulate filter]; or SCBAE (any appropriate escape-type, self-contained breathing apparatus).
Exposure limits (TWA unless otherwise noted): OSHA PEL 0.5 mg/m³ as organic arsenic.
Toxicity by ingestion: Grade 2; oral rat LD_{50} = 700 mg/kg.
Long-term health effects: Possible arsenic poisoning
IDLH value: 5 mg/m³ as inorganic arsenic; known human carcinogen. Not determined for organic arsenic.

FIRE DATA
Behavior in fire: May form toxic oxides of arsenic when heated.

CHEMICAL REACTIVITY
zzzxxx

ENVIRONMENTAL DATA
Water pollution: Effect of low concentrations on aquatic life is unknown. May be dangerous if it enters nearby water intakes; notify operators. Notify local health and wildlife officials. **Response to discharge:** Issue warning–poison, water contaminant. Should be removed. Chemical and physical treatment.

SHIPPING INFORMATION
Grades of purity: Commercial: 50% solution in water, dyed blue; **Storage temperature:** Ambient; **Inert atmosphere:** None; **Venting:** Open (flame arrester); **Stability during transport:** Stable.

PHYSICAL AND CHEMICAL PROPERTIES
Physical state @ 59°F/15°C and 1 atm: Solid.
Molecular weight: 138
Boiling point @ 1 atm: More than 392°F/200°C/473°K.
Specific gravity (water = 1): More than 1.1 (estimate) @ 68°F/20°C (solid).
Heat of combustion: (estimate) –6000 Btu/lb = –3300 cal/g = –140 x 10⁵ J/kg.
Heat of solution: (estimate) –54 Btu/lb = –30 cal/g = –1.3 x 10⁵ J/kg.

CADMIUM ACETATE   REC. C:0150

SYNONYMS: ACETO CADMIO (Spanish); ACETIC ACID, CADMIUM SALT; BIS(ACETOXY)CADMIUM; CADMIUM (II) ACETATE; CADMIUM ACETATE DIHYDRATE; CADMIUM DIACETATE; EEC No. 048-001-00-5

IDENTIFICATION
CAS Number: 543-90-8
Formula: $Cd(C_2H_3O_2)_2 \cdot 2H_2O$
DOT ID Number: UN 2570; DOT Guide Number: 154

Proper Shipping Name: Cadmium compounds
Reportable Quantity (RQ): **(CERCLA)** 10 lb/4.45 kg

DESCRIPTION: Colorless crystalline solid. Odorless. Sinks and mixes with water.

Poison! • Breathing the dust or swallowing the material can kill you; irritating to the skin, eyes, and respiratory tract; inhalation symptoms may be delayed • Firefighting gear (including SCBA) does not provide adequate protection. If exposure occurs, remove and isolate gear immediately and thoroughly decontaminate personnel • Concentrated dust in confined areas (e.g., tanks, sewers, buildings) may explode when exposed to fire • Toxic products of combustion may include carbon monoxide and cadmium oxide • Do not put yourself in danger by entering a contaminated area to rescue a victim.

Hazard Classification (based on NFPA-704 Rating System)
Health Hazards (Blue): 2; Flammability (Red): 0; Reactivity (Yellow): 0

EMERGENCY RESPONSE: See Appendix A (154)
Evacuation:
Public safety: Isolate the area of spill or leak for at least 25 to 50 meters (80 to 160 feet) in all directions.
Spill: Increase, in the downwind direction, as necessary, the distance shown under "Public Safety."
Fire: If tank, rail car, or tank truck is involved in fire, isolate for at least 800 meters (½ mile) in all directions; also, consider initial evacuation for 800 meters (½ mile) in all directions.

EXPOSURE
Short-term effects: *SEEK MEDICAL ATTENTION.* **Dust:** *POISONOUS IF INHALED. IF INHALED,* will, will cause coughing or difficult breathing. May cause lung edema; physical exertion will aggravate this condition. *IF IN EYES,* hold eyelids open and flush with plenty of water. *IF BREATHING HAS STOPPED,* give artificial respiration; *avoid mouth-to-mouth resuscitation; use bag/mask apparatus.* IF breathing is difficult, administer oxygen. **Solid:** *POISONOUS IF SWALLOWED. IF SWALLOWED* and victim is *CONSCIOUS AND ABLE TO SWALLOW,* have victim drink 4 to 8 ounces of water and have victim induce vomiting. *IF SWALLOWED* and victim is *UNCONSCIOUS OR HAVING CONVULSIONS,* do nothing except keep victim warm.
Note to physician or authorized medical personnel: Medical observation is recommended for 24 to 48 hours after breathing overexposure, as pulmonary edema may be delayed. As first aid for pulmonary edema, consider administering a corticosteroid spray. Cigarette smoking may exacerbate pulmonary injury and should be discouraged for at least 72 hours following exposure.

HEALTH HAZARDS
Personal protective equipment (PPE): A-Level PPE. OSHA Table Z-1-A air contaminant.
Recommendations for respirator selection: NIOSH
At any concentrations above the NIOSH REL, or where there is no REL, at any detectable concentration: SCBAF:PD,PP (any self-contained breathing apparatus that has a full facepiece and is operated in a pressure-demand or other positive-pressure mode); or SAF:PD,PP:ASCBA (any supplied-air respirator that has a full facepiece and is operated in a pressure-demand or other positive-pressure mode in combination with an auxiliary self-contained breathing apparatus operated in a pressure-demand or other positive pressure mode). *ESCAPE:* HiEF (any air-purifying, full-facepiece respirator with a high-efficiency particulate filter); or SCBAE (any appropriate escape-type, self-contained breathing apparatus).
Exposure limits (TWA unless otherwise noted): ACGIH TLV 0.01 mg/m^3; 0.002 mg/m^3 (respirable dust); OSHA PEL (fume) 0.005 mg/m^3; NIOSH potential human carcinogen; reduce to lowest feasible concentration.
Toxicity by ingestion: Grade 4; LD$_{50}$ less than 50 mg/kg.
Long-term health effects: Liver, lung, blood and kidney damage has followed respiratory exposures to cadmium. Cadmium compounds are IARC probable carcinogens, rating 2A; NTP anticipated carcinogens.
Long-term health effects: Delayed liver, kidney, and lung damage has followed respiratory exposures to cadmium salts in industry.
Vapor (gas) irritant characteristics: Eye and respiratory tract irritant.
Liquid or solid irritant characteristics: Eye and skin irritant.
IDLH value: Suspected human carcinogen; 9 mg/m^3, as cadmium.

CHEMICAL REACTIVITY
Binary reactants: Cadmium compounds may react with strong oxidizers; elemental sulfur, selenium, and tellurium.

ENVIRONMENTAL DATA
Food chain concentration potential: Concentrated by shellfish
Water pollution: DOT Appendix B, §172.101–marine pollutant. Effect of low concentrations on aquatic life is unknown. May be dangerous if it enters nearby water intakes; notify operators. Notify local health and wildlife officials. **Response to discharge:** Issue warning–water contaminant. Disperse and flush.

SHIPPING INFORMATION
Grades of purity: Pure, 98%; Reagent; **Storage temperature:** Ambient; **Inert atmosphere:** None; **Venting:** Open; **Stability during transport:** Stable.

PHYSICAL AND CHEMICAL PROPERTIES
Physical state @ 59°F/15°C and 1 atm: Solid.
Molecular weight: 266.52
Boiling point @ 1 atm: (decomposes).
Melting/Freezing point: 493°F/256°C/529°K (anhydrous).
Specific gravity (water = 1): 2.34 @ 68°F/20°C (solid).

CADMIUM BROMIDE   REC. C:0200

SYNONYMS: BROMURO de CADMIO (Spanish); CADMIUM BROMIDE TETRAHYDRATE

IDENTIFICATION
CAS Number: 7789-42-6
Formula: CdBr$_2$·4H$_2$O; Br$_2$Cd
DOT ID Number: UN 2570; DOT Guide Number: 154
Proper Shipping Name: Cadmium compounds
Reportable Quantity (RQ): **(CERCLA)** 10 lb/4.54 kg

DESCRIPTION: White solid. Odorless. Mixes with water.

Poison! • Breathing the dust or swallowing the material can kill you • Firefighting gear (including SCBA) does not provide adequate protection. If exposure occurs, remove and isolate gear immediately and thoroughly decontaminate personnel • Concentrated dust in confined areas (e.g., tanks, sewers, buildings) may explode when exposed to fire • Toxic products of combustion may include bromine and cadmium oxide • Do not put yourself in danger by entering a contaminated area to rescue a victim.

Hazard Classification (based on NFPA-704 Rating System)
Health Hazards (Blue): 2; Flammability (Red): 0; Reactivity (Yellow): 0

EMERGENCY RESPONSE: See Appendix A (154)
Evacuation:
Public safety: Isolate the area of spill or leak for at least 25 to 50 meters (80 to 160 feet) in all directions.
Spill: Increase, in the downwind direction, as necessary, the distance shown under "Public Safety."
Fire: If tank, rail car, or tank truck is involved in fire, isolate for at least 800 meters (½ mile) in all directions; also, consider initial evacuation for 800 meters (½ mile) in all directions.

EXPOSURE
Short-term effects: *SEEK MEDICAL ATTENTION.* **Dust:** *POISONOUS IF INHALED. IF INHALED,* will, will cause coughing or difficult breathing. *IF IN EYES,* hold eyelids open and flush with plenty of water. *IF BREATHING HAS STOPPED,* give artificial respiration; *avoid mouth-to-mouth resuscitation; use bag/mask apparatus.* IF breathing is difficult, administer oxygen. **Solid:** *POISONOUS IF SWALLOWED.* Irritating to skin and eyes. *IF SWALLOWED,* will cause nausea, vomiting, and loss of consciousness. Remove contaminated clothing and shoes. Flush affected areas with plenty of water. *IF IN EYES,* hold eyelids open and flush with plenty of water. *IF SWALLOWED* and victim is *CONSCIOUS AND ABLE TO SWALLOW,* have victim drink 4 to 8 ounces of water and have victim induce vomiting. *IF SWALLOWED* and victim is *UNCONSCIOUS OR HAVING CONVULSIONS,* do nothing except keep victim warm.

HEALTH HAZARDS
Personal protective equipment (PPE): Cadmium dust is a A-Level PPE. OSHA Table Z-1-A air contaminant.
Recommendations for respirator selection: NIOSH
At any concentrations above the NIOSH REL, or where there is no REL, at any detectable concentration: SCBAF:PD,PP (any self-contained breathing apparatus that has a full facepiece and is operated in a pressure-demand or other positive-pressure mode); or SAF:PD,PP:ASCBA (any supplied-air respirator that has a full facepiece and is operated in a pressure-demand or other positive-pressure mode in combination with an auxiliary self-contained breathing apparatus operated in a pressure-demand or other positive pressure mode). *ESCAPE:* HiEF (any air-purifying, full-facepiece respirator with a high-efficiency particulate filter); or SCBAE (any appropriate escape-type, self-contained breathing apparatus).
Exposure limits (TWA unless otherwise noted): ACGIH TLV 0.01 mg/m^3; 0.002 mg/m^3 (respirable dust); OSHA PEL (fume) 0.005 mg/m^3; NIOSH potential human carcinogen; reduce to lowest feasible concentration.
Toxicity by ingestion: Grade 4; LD_{50} less than 50 mg/kg.
Long-term health effects: Delayed liver, kidney, and lung damage has followed respiratory exposure to cadmium salts in industry. Cadmium compounds are IARC probable carcinogens, rating 2A; NTP anticipated carcinogen. Prolonged contact or inhalation of inorganic bromides causes acne-like bromoderma (bromide rash), especially of the hands and face. Other effects include emaciation, depression; and, in severe cases, psychosis and mental deterioration.
Vapor (gas) irritant characteristics: Eye and respiratory tract irritant.
Liquid or solid irritant characteristics: Eye and skin irritant.
IDLH value: Suspected human carcinogen; 9 mg/m^3, as cadmium.

CHEMICAL REACTIVITY
Binary reactants: Cadmium compounds may react with strong oxidizers; elemental sulfur, selenium and tellurium.

ENVIRONMENTAL DATA
Food chain concentration potential: Concentrated by shellfish.
Water pollution: DOT Appendix B, §172.101–marine pollutant. Effect of low concentrations on aquatic life is unknown. May be dangerous if it enters nearby water intakes; notify operators. Notify local health and wildlife officials. **Response to discharge:** Issue warning–water contaminant. Disperse and flush.

SHIPPING INFORMATION
Grades of purity: Commercial, 99.5%; Anhydrous, 99+%; **Storage temperature:** Ambient; **Inert atmosphere:** None; **Venting:** Open; **Stability during transport:** Stable.

PHYSICAL AND CHEMICAL PROPERTIES
Physical state @ 59°F/15°C and 1 atm: Solid.
Molecular weight: 344.27
Boiling point @ 1 atm: (decomposes).
Specific gravity (water = 1): More than 1.1 @ 68°F/20°C (solid).
Heat of solution: -2.3 Btu/lb $= -1.3$ cal/g $= -0.054 \times 10^5$ J/kg.
Heat of fusion: 18.4 cal/g.

CADMIUM CHLORIDE   **REC. C:0250**

SYNONYMS: CADDY; CADMIUM DICHLORIDE; CLORURO de CADMIO (Spanish); KADMIUM CHLORID (German); VI-CAD

IDENTIFICATION
CAS Number: 10108-64-2
Formula: $CdCl_2$
DOT ID Number: UN 2570; DOT Guide Number: 154
Proper Shipping Name: Cadmium compounds
Reportable Quantity (RQ): **(CERCLA)** 10 lb/4.54 kg

DESCRIPTION: White crystalline solid. Odorless. Sinks and mixes with water.

Poison! • Breathing the dust or swallowing the material can kill you • Firefighting gear (including SCBA) does not provide adequate protection. If exposure occurs, remove and isolate gear immediately and thoroughly decontaminate personnel • Concentrated dust in confined areas (e.g., tanks, sewers, buildings) may explode when exposed to fire • Toxic products of combustion may include hydrogen chloride and cadmium oxide • Do not put yourself in danger by entering a contaminated area to rescue a victim.

Hazard Classification (based on NFPA-704 Rating System)
Health Hazards (Blue): 2; Flammability (Red): 0; Reactivity (Yellow): 0

EMERGENCY RESPONSE: See Appendix A (154)
Evacuation:
Public safety: Isolate the area of spill or leak for at least 25 to 50 meters (80 to 160 feet) in all directions.
Spill: Increase, in the downwind direction, as necessary, the distance shown under "Public Safety."

Fire: If tank, rail car, or tank truck is involved in fire, isolate for at least 800 meters (½ mile) in all directions; also, consider initial evacuation for 800 meters (½ mile) in all directions.

EXPOSURE
Short-term effects: *SEEK MEDICAL ATTENTION.* **Solid:** Harmful if swallowed. *IF BREATHING HAS STOPPED,* give artificial respiration; *avoid mouth-to-mouth resuscitation; use bag/mask apparatus. IF SWALLOWED* and victim is *CONSCIOUS AND ABLE TO SWALLOW,* have victim drink 4 to 8 ounces of water, have victim induce vomiting. *IF SWALLOWED* and victim is *UNCONSCIOUS OR HAVING CONVULSIONS,* do nothing except keep victim warm.

HEALTH HAZARDS
Personal protective equipment (PPE): Cadmium dust is a A-Level PPE. OSHA Table Z-1-A air contaminant.
Recommendations for respirator selection: NIOSH
At any concentrations above the NIOSH REL, or where there is no REL, at any detectable concentration: SCBAF:PD,PP (any self-contained breathing apparatus that has a full facepiece and is operated in a pressure-demand or other positive-pressure mode); or SAF:PD,PP:ASCBA (any supplied-air respirator that has a full facepiece and is operated in a pressure-demand or other positive-pressure mode in combination with an auxiliary self-contained breathing apparatus operated in a pressure-demand or other positive pressure mode). *ESCAPE:* HiEF (any air-purifying, full-facepiece respirator with a high-efficiency particulate filter); or SCBAE (any appropriate escape-type, self-contained breathing apparatus).
Note to physician or authorized medical personnel: Physician may consider using atropine, opiates, and fluid therapy. CaNa$_2$- EDTA [(ethylenedinitrilo)tetraacetic acid] has been effective in acutely poisoned animals and in a few humans. BAL has been found sufficiently effective in animal experiments to justify its use in human intoxication. Since the BAL-cadmium complex has a nephrotoxic action, the physician will have to decide whether or not to use this drug.
Exposure limits (TWA unless otherwise noted): ACGIH TLV 0.01 mg/m^3; 0.002 mg/m^3 (respirable dust); OSHA PEL (fume) 0.005 mg/m^3; NIOSH potential human carcinogen; reduce to lowest feasible concentration.
Toxicity by ingestion: Grade 4; LD$_{50}$ below 50 mg/kg.
Long-term health effects: NTP anticipated carcinogen
Vapor (gas) irritant characteristics: May cause eye and respiratory tract irritation.
Liquid or solid irritant characteristics: Eye and skin irritant. Causes smarting of the skin and first-degree burns on short exposure; may cause second-degree burns on long exposure.
IDLH value: Suspected human carcinogen; 9 mg/m^3, as cadmium.

CHEMICAL REACTIVITY
Binary reactants: Violent reaction with bromine trifluoride or potassium. Contact with acids or acid fumes forms toxic fumes of chlories. Incompatible with strong oxidizers, elemental sulfur, selenium, tellurium. Attacks some steels causing stress corrosion and pitting.

ENVIRONMENTAL DATA
Water pollution: DOT Appendix B, §172.101–marine pollutant. Harmful to aquatic life in very low concentrations. May be dangerous if it enters nearby water intakes; notify operators. Notify local health and wildlife officials. **Response to discharge:** Issue warning–water contaminant.

SHIPPING INFORMATION
Inert atmosphere: None; **Venting:** Open; **Stability during transport:** Stable.

PHYSICAL AND CHEMICAL PROPERTIES
Physical state @ 59°F/15°C and 1 atm: Solid.
Molecular weight: 228.35
Specific gravity (water = 1): 4.05 @ 77°F/25°C (solid).
Heat of fusion: 28.8 cal/g.

CADMIUM FLUOROBORATE REC. C:0300

SYNONYMS: CADMIUM FLUOBORATE; FLUOBORATO de CADMIO (Spanish)

IDENTIFICATION
CAS Number: 14486-19-2
Formula: Cd(BF$_4$)$_2$·H$_2$O
DOT ID Number: UN 2570; DOT Guide Number: 154
Proper Shipping Name: Cadmium compounds

DESCRIPTION: Colorless liquid. Odorless. Sinks and mixes with water.

Poison! • Breathing the dust or swallowing the material can kill you • Firefighting gear (including SCBA) does not provide adequate protection. If exposure occurs, remove and isolate gear immediately and thoroughly decontaminate personnel • Concentrated dust in confined areas (e.g., tanks, sewers, buildings) may explode when exposed to fire • Toxic products of combustion may include hydrogen fluoride and cadmium oxide • Do not put yourself in danger by entering a contaminated area to rescue a victim.

Hazard Classification (based on NFPA-704 Rating System)
Health Hazards (Blue): 2; Flammability (Red): 0; Reactivity (Yellow): 0

EMERGENCY RESPONSE: See Appendix A (154)
Evacuation:
Public safety: Isolate the area of spill or leak for at least 25 to 50 meters (80 to 160 feet) in all directions.
Spill: Increase, in the downwind direction, as necessary, the distance shown under "Public Safety."
Fire: If tank, rail car, or tank truck is involved in fire, isolate for at least 800 meters (½ mile) in all directions; also, consider initial evacuation for 800 meters (½ mile) in all directions.

EXPOSURE
Short-term effects: *SEEK MEDICAL ATTENTION.* **Vapor:** *POISONOUS IF INHALED. IF INHALED,* will, will cause coughing or difficult breathing. *IF IN EYES,* hold eyelids open and flush with plenty of water. *IF BREATHING HAS STOPPED,* give artificial respiration; *avoid mouth-to-mouth resuscitation; use bag/mask apparatus.* IF breathing is difficult, administer oxygen. **Liquid:** Irritating to skin and eyes. Harmful if swallowed. Remove contaminated clothing and shoes. Flush affected areas with plenty of water. *IF IN EYES,* hold eyelids open and flush with plenty of water. *IF SWALLOWED* and victim is *CONSCIOUS AND ABLE TO SWALLOW,* have victim drink 4 to 8 ounces of water and have victim induce vomiting. *IF SWALLOWED* and victim is *UNCONSCIOUS OR HAVING CONVULSIONS,* do nothing except keep victim warm.

202 Cadmium(II) nitrate, tetrahydrate

HEALTH HAZARDS
Personal protective equipment (PPE): A-Level PPE.
Recommendations for respirator selection: NIOSH
At any concentrations above the NIOSH REL, or where there is no REL, at any detectable concentration: SCBAF:PD,PP (any self-contained breathing apparatus that has a full facepiece and is operated in a pressure-demand or other positive-pressure mode); or SAF:PD,PP:ASCBA (any supplied-air respirator that has a full facepiece and is operated in a pressure-demand or other positive-pressure mode in combination with an auxiliary self-contained breathing apparatus operated in a pressure-demand or other positive pressure mode). *ESCAPE:* HiEF (any air-purifying, full-facepiece respirator with a high-efficiency particulate filter); or SCBAE (any appropriate escape-type, self-contained breathing apparatus).
Exposure limits (TWA unless otherwise noted): ACGIH TLV 0.01 mg/m^3; 0.002 mg/m^3 (respirable dust); OSHA PEL (fume) 0.005 mg/m^3; NIOSH potential human carcinogen; reduce to lowest feasible concentration.
Toxicity by ingestion: Grade 3; LD$_{50}$ = 250 mg/kg (rat).
Long-term health effects: Delayed liver, kidney, and lung damage has followed respiratory exposure to cadmium salts in industry. Cadmium compounds are IARC probable carcinogens, rating 2A; NTP anticipated carcinogen.
Vapor (gas) irritant characteristics: Eye and respiratory tract irritant.
Liquid or solid irritant characteristics: Eye and skin irritant.
IDLH value: Suspected human carcinogen; 9 mg/m^3, as cadmium.

CHEMICAL REACTIVITY
Binary reactants: Cadmium compounds may react with strong oxidizers; elemental sulfur, selenium and tellurium.

ENVIRONMENTAL DATA
Food chain concentration potential: Concentrated by shellfish
Water pollution: DOT Appendix B, §172.101–marine pollutant. Effect of low concentrations on aquatic life is unknown. May be dangerous if it enters nearby water intakes; notify operators. Notify local health and wildlife officials. **Response to discharge:** Issue warning–water contaminant. Disperse and flush.

SHIPPING INFORMATION
Grades of purity: Commercial, 50% solution in water; **Storage temperature:** Ambient; **Inert atmosphere:** None; **Venting:** Open; **Stability during transport:** Stable.

PHYSICAL AND CHEMICAL PROPERTIES
Physical state @ 59°F/15°C and 1 atm: Liquid.
Molecular weight: 286 (solute).
Specific gravity (water = 1): 1.60 @ 68°F/20°C (liquid).

CADMIUM(II) NITRATE, tetrahydrate REC. C:0350

SYNONYMS: NITRATO de CADMIO(II) (Spanish); NITRIC ACID, CADMIUM SALT, TETRAHYDRATE

IDENTIFICATION
CAS Number: 10022-68-1
Formula: Cd(NO$_3$)$_2$·4H$_2$O
DOT ID Number: UN 2570; DOT Guide Number: 154
Proper Shipping Name: Cadmium compounds

DESCRIPTION: White solid. Odorless. Sinks in water.

Poison! • Breathing the dust or swallowing the material can kill you • Firefighting gear (including SCBA) does not provide adequate protection. If exposure occurs, remove and isolate gear immediately and thoroughly decontaminate personnel • Heat forms oxygen; will increase the activity of an existing fire • May cause fire or explosion contact with combustibles (wood, paper, oil, clothing, etc.) • Concentrated dust in confined areas (e.g., tanks, sewers, buildings) may explode when exposed to fire • Toxic products of combustion may include nitrogen and cadmium oxides • Do not put yourself in danger by entering a contaminated area to rescue a victim.

Hazard Classification (based on NFPA-704 Rating System)
Health Hazards (Blue): 2; Flammability (Red): 0; Reactivity (Yellow): 0

EMERGENCY RESPONSE: See Appendix A (154)
Evacuation:
Public safety: Isolate the area of spill or leak for at least 25 to 50 meters (80 to 160 feet) in all directions.
Spill: Increase, in the downwind direction, as necessary, the distance shown under "Public Safety."
Fire: If tank, rail car, or tank truck is involved in fire, isolate for at least 800 meters (½ mile) in all directions; also, consider initial evacuation for 800 meters (½ mile) in all directions.

EXPOSURE
Short-term effects: *SEEK MEDICAL ATTENTION.* **Dust:** *POISONOUS IF INHALED. IF INHALED,* will, will cause headache, coughing, or difficult breathing. *IF IN EYES,* hold eyelids open and flush with plenty of water. *IF BREATHING HAS STOPPED,* give artificial respiration; *avoid mouth-to-mouth resuscitation; use bag/mask apparatus.* IF breathing is difficult, administer oxygen. **Solid:** *POISONOUS IF SWALLOWED.* Irritating to skin and eyes. *IF SWALLOWED,* will cause nausea and vomiting. Remove contaminated clothing and shoes. Flush affected areas with plenty of water. *IF IN EYES,* hold eyelids open and flush with plenty of water. *IF SWALLOWED* and victim is *CONSCIOUS AND ABLE TO SWALLOW,* have victim drink 4 to 8 ounces of water and have victim induce vomiting. IF SWALLOWED and victim is *UNCONSCIOUS OR HAVING CONVULSIONS,* do nothing except keep victim warm.

HEALTH HAZARDS
Personal protective equipment (PPE): Cadmium dust is a A-Level PPE. OSHA Table Z-1-A air contaminant.
Recommendations for respirator selection: NIOSH
At any concentrations above the NIOSH REL, or where there is no REL, at any detectable concentration: SCBAF:PD,PP (any self-contained breathing apparatus that has a full facepiece and is operated in a pressure-demand or other positive-pressure mode); or SAF:PD,PP:ASCBA (any supplied-air respirator that has a full facepiece and is operated in a pressure-demand or other positive-pressure mode in combination with an auxiliary self-contained breathing apparatus operated in a pressure-demand or other positive pressure mode). *ESCAPE:* HiEF (any air-purifying, full-facepiece respirator with a high-efficiency particulate filter); or SCBAE (any appropriate escape-type, self-contained breathing apparatus).
Exposure limits (TWA unless otherwise noted): ACGIH TLV 0.01 mg/m^3; 0.002 mg/m^3 (respirable dust); OSHA PEL (fume) 0.005 mg/m^3; NIOSH potential human carcinogen; reduce to lowest feasible concentration.
Toxicity by ingestion: Grade 3; oral mouse LD$_{50}$ = 100 mg/kg.
Long-term health effects: Delayed liver, lung, and kidney damage has followed respiratory exposures to cadmium salts in industry.

Cadmium compounds are IARC probable carcinogens, rating 2A; NTP anticipated carcinogen.
Vapor (gas) irritant characteristics: Eye and respiratory tract irritant.
Liquid or solid irritant characteristics: Eye and skin irritant.
IDLH value: Suspected human carcinogen; 9 mg/m^3, as cadmium.

CHEMICAL REACTIVITY
Binary reactants: violent reaction with hydrazinium nitrate, hydrogen peroxide, hydrogen sulfide, hydrogen trisulfide, lithium. Cadmium compounds may react with strong oxidizers; elemental sulfur, selenium and tellurium.

ENVIRONMENTAL DATA
Food chain concentration potential: Shellfish concentrate 900–1600 times
Water pollution: DOT Appendix B, §172.101–marine pollutant. Harmful to aquatic life in very low concentrations. May be dangerous if it enters nearby water intakes; notify operators. Notify local health and wildlife officials. **Response to discharge:** Issue warning–water contaminant. Disperse and flush.

SHIPPING INFORMATION
Grades of purity: Technical; **Storage temperature:** Ambient; **Inert atmosphere:** None; **Venting:** Open; **Stability during transport:** Stable.

PHYSICAL AND CHEMICAL PROPERTIES
Physical state @ 59°F/15°C and 1 atm: Solid.
Molecular weight: 308.47
Boiling point @ 1 atm: Decomposes.
Melting/Freezing point: 138°F/59°C/332°K.
Specific gravity (water = 1): 2.45 @ 68°F/20°C (solid).
Heat of solution: 29.7 Btu/lb = 16.5 cal/g = 0.691 x 10^5 J/kg.

CADMIUM OXIDE REC. C:0400

SYNONYMS: CADMIUM FUME; CADMIUM MONOXIDE; CADMIUM OXIDE FUME; EEC No. 048-002-00-0

IDENTIFICATION
CAS Number: 1306-19-0
Formula: CdO
DOT ID Number: UN 2570; DOT Guide Number: 154
Proper Shipping Name: Cadmium compounds
Reportable Quantity (RQ): **(EHS)** 1 lb/0.454 kg

DESCRIPTION: Brown-red to yellow-brown powder. Odorless. Sinks in water.

Poison! • Breathing the dust or swallowing the material can kill you; irritating to the skin and eyes; inhalation symptoms may be delayed • Firefighting gear (including SCBA) does not provide adequate protection. If exposure occurs, remove and isolate gear immediately and thoroughly decontaminate personnel • Concentrated dust in confined areas (e.g., tanks, sewers, buildings) may explode when exposed to fire • Toxic products of combustion may include cadmium oxide vapors • Do not put yourself in danger by entering a contaminated area to rescue a victim.

Hazard Classification (based on NFPA-704 Rating System)
Health Hazards (Blue): 2; Flammability (Red): 0; Reactivity (Yellow): 0

EMERGENCY RESPONSE: See Appendix A (154)
Evacuation:
Public safety: Isolate the area of spill or leak for at least 25 to 50 meters (80 to 160 feet) in all directions.
Spill: Increase, in the downwind direction, as necessary, the distance shown under "Public Safety."
Fire: If tank, rail car, or tank truck is involved in fire, isolate for at least 800 meters (½ mile) in all directions; also, consider initial evacuation for 800 meters (½ mile) in all directions.

EXPOSURE
Short-term effects: *SEEK MEDICAL ATTENTION.* **Dust:** *POISONOUS IF INHALED. IF INHALED*, will, will cause coughing. *May cause lung edema; physical exertion will aggravate this condition. IF IN EYES*, hold eyelids open and flush with plenty of water. *IF BREATHING HAS STOPPED*, give artificial respiration; *avoid mouth-to-mouth resuscitation; use bag/mask apparatus*. IF breathing is difficult, administer oxygen. **Solid:** Irritating to skin and eyes. *IF SWALLOWED*, will cause nausea and vomiting. Remove contaminated clothing and shoes. Flush affected areas with plenty of water. *IF IN EYES*, hold eyelids open and flush with plenty of water. *IF SWALLOWED* and victim is *CONSCIOUS AND ABLE TO SWALLOW*, have victim drink 4 to 8 ounces of water and have victim induce vomiting. *IF SWALLOWED* and victim is *UNCONSCIOUS OR HAVING CONVULSIONS*, do nothing except keep victim warm.
Note to physician or authorized medical personnel: Medical observation is recommended for 24 to 48 hours after breathing overexposure, as pulmonary edema may be delayed. As first aid for pulmonary edema, consider administering a corticosteroid spray. Cigarette smoking may exacerbate pulmonary injury and should be discouraged for at least 72 hours following exposure.

HEALTH HAZARDS
Personal protective equipment (PPE): A-Level PPE. Cadmium dust is a OSHA Table Z-1-A air contaminant. Chemical protective material(s) reported to offer protection (solid): neoprene, nitrile.
Recommendations for respirator selection: NIOSH
At any concentrations above the NIOSH REL, or where there is no REL, at any detectable concentration: SCBAF:PD,PP (any self-contained breathing apparatus that has a full facepiece and is operated in a pressure-demand or other positive-pressure mode); or SAF:PD,PP:ASCBA (any supplied-air respirator that has a full facepiece and is operated in a pressure-demand or other positive-pressure mode in combination with an auxiliary self-contained breathing apparatus operated in a pressure-demand or other positive pressure mode). *ESCAPE:* HiEF (any air-purifying, full-facepiece respirator with a high-efficiency particulate filter); or SCBAE (any appropriate escape-type, self-contained breathing apparatus).
Exposure limits (TWA unless otherwise noted): ACGIH TLV 0.01 mg/m^3; 0.002 mg/m^3 (respirable dust); OSHA PEL (fume) 0.005 mg/m^3; NIOSH potential human carcinogen; reduce to lowest feasible concentration.
Toxicity by ingestion: Grade 3; oral rat LD$_{50}$ = 72 mg/kg.
Long-term health effects: Delayed liver, lung, blood and kidney damage has followed respiratory exposures to cadmium salts in industry. Olfactory sense may be impaired. Cadmium compounds are IARC probable carcinogens, rating 2A; NTP anticipated carcinogen.
Vapor (gas) irritant characteristics: Eye and respiratory tract irritant.
Liquid or solid irritant characteristics: Eye, skin and respiratory tract irritant.
IDLH value: Suspected human carcinogen; 9 mg/m^3, as cadmium.

Cadmium sulfate

CHEMICAL REACTIVITY
Binary reactants: Cadmium compounds may react with strong oxidizers; elemental sulfur, selenium and tellurium.

ENVIRONMENTAL DATA
Food chain concentration potential: Concentrated by shellfish
Water pollution: DOT Appendix B, §172.101–marine pollutant. Effect of low concentrations on aquatic life is unknown. May be dangerous if it enters nearby water intakes; notify operators. Notify local health and wildlife officials. **Response to discharge:** Should be removed. Chemical and physical treatment.

SHIPPING INFORMATION
Grades of purity: Reagent; technical; **Storage temperature:** Ambient; **Inert atmosphere:** None; **Venting:** Open; **Stability during transport:** Stable.

PHYSICAL AND CHEMICAL PROPERTIES
Physical state @ 59°F/15°C and 1 atm: Solid.
Molecular weight: 128.4
Boiling point @ 1 atm: 2838°F/1559°C/1832°K (decomposes).
Melting/Freezing point: More than 2260°F/1230°C/1503°K.
Specific gravity (water = 1): 6.95 @ 68°F/20°C (solid).
Vapor pressure: 0.000001 mm @ 130°F

CADMIUM SULFATE　　　　　　　　REC. C:0450

SYNONYMS: CADMIUM SULFATE; CADMIUM SULPHATE; SULPHURIC ACID, CADMIUM SALT; SULFATO de CADMIO (Spanish)

IDENTIFICATION
CAS Number: 10124-36-4
Formula: $CdSO_4$
DOT ID Number: UN 2570; DOT Guide Number: 154
PROPER 45IPPING NAME: Cadmium compounds

DESCRIPTION: White solid. Odorless. Sinks and mixes slowly with water.

Poison! • Breathing the dust or swallowing the material can kill you • Firefighting gear (including SCBA) does not provide adequate protection. If exposure occurs, remove and isolate gear immediately and thoroughly decontaminate personnel • Concentrated dust in confined areas (e.g., tanks, sewers, buildings) may explode when exposed to fire • Toxic products of combustion may include cadmium, hydrogen sulfide, and sulfur oxides • Do not put yourself in danger by entering a contaminated area to rescue a victim.

Hazard Classification (based on NFPA-704 Rating System)
Health Hazards (Blue): 2; Flammability (Red): 0; Reactivity (Yellow): 0

EMERGENCY RESPONSE: See Appendix A (154)
Evacuation:
Public safety: Isolate the area of spill or leak for at least 25 to 50 meters (80 to 160 feet) in all directions.
Spill: Increase, in the downwind direction, as necessary, the distance shown under "Public Safety."
Fire: If tank, rail car, or tank truck is involved in fire, isolate for at least 800 meters (½ mile) in all directions; also, consider initial evacuation for 800 meters (½ mile) in all directions.

EXPOSURE
Short-term effects: *SEEK MEDICAL ATTENTION.* **Dust:** *POISONOUS IF INHALED. IF INHALED*, will, will cause headache, coughing, or difficult breathing. *IF IN EYES*, hold eyelids open and flush with plenty of water. *IF BREATHING HAS STOPPED*, give artificial respiration; *avoid mouth-to-mouth resuscitation; use bag/mask apparatus.* IF breathing is difficult, administer oxygen. **Solid:** *POISONOUS IF SWALLOWED.* Irritating to skin and eyes. *IF SWALLOWED*, will cause nausea and vomiting. Remove contaminated clothing and shoes. Flush affected areas with plenty of water. *IF IN EYES*, hold eyelids open and flush with plenty of water. *IF SWALLOWED* and victim is *CONSCIOUS AND ABLE TO SWALLOW*, have victim drink 4 to 8 ounces of water and have victim induce vomiting. *IF SWALLOWED* and victim is *UNCONSCIOUS OR HAVING CONVULSIONS*, do nothing except keep victim warm.

HEALTH HAZARDS
Personal protective equipment (PPE): Cadmium dust is a A-Level PPE. OSHA Table Z-1-A air contaminant.
Recommendations for respirator selection: NIOSH
At any concentrations above the NIOSH REL, or where there is no REL, at any detectable concentration: SCBAF:PD,PP (any self-contained breathing apparatus that has a full facepiece and is operated in a pressure-demand or other positive-pressure mode); or SAF:PD,PP:ASCBA (any supplied-air respirator that has a full facepiece and is operated in a pressure-demand or other positive-pressure mode in combination with an auxiliary self-contained breathing apparatus operated in a pressure-demand or other positive pressure mode). *ESCAPE:* HiEF (any air-purifying, full-facepiece respirator with a high-efficiency particulate filter); or SCBAE (any appropriate escape-type, self-contained breathing apparatus).
Exposure limits (TWA unless otherwise noted): ACGIH TLV 0.01 mg/m^3; 0.002 mg/m^3 (respirable dust); OSHA PEL (fume) 0.005 mg/m^3; NIOSH potential human carcinogen; reduce to lowest feasible concentration.
Toxicity by ingestion: Grade 3; oral mouse LD_{50} = 88 mg/kg.
Long-term health effects: Delayed liver, kidney, and lung damage has followed respiratory exposures to cadmium salts in industry. Cadmium compounds are IARC probable carcinogens, rating 2A; NTP anticipated carcinogen.
Vapor (gas) irritant characteristics: Eye and respiratory tract irritant.
Liquid or solid irritant characteristics: Eye, skin and respiratory tract irritant.
IDLH value: Suspected human carcinogen; 9 mg/m^3, as cadmium.

CHEMICAL REACTIVITY
Binary reactants: Violent reaction with aluminum, magnesium. Cadmium compounds may react with strong oxidizers; elemental sulfur, selenium and tellurium.

ENVIRONMENTAL DATA
Food chain concentration potential: Shellfish concentrate cadmium 900-1600 times
Water pollution: DOT Appendix B, §172.101–marine pollutant. Dangerous to aquatic life in high concentrations. May be dangerous if it enters nearby water intakes; notify operators. Notify local health and wildlife officials. **Response to discharge:** Issue warning–water contaminant. Disperse and flush.

SHIPPING INFORMATION
Grades of purity: Technical; 8/3 Hydrate grade; Reagent; **Storage temperature:** Ambient; **Inert atmosphere:** None; **Venting:** Open; **Stability during transport:** Stable.

PHYSICAL AND CHEMICAL PROPERTIES
Physical state @ 59°F/15°C and 1 atm: Solid.
Molecular weight: 208.46
Boiling point @ 1 atm: Decomposes.
Melting/Freezing point: 1832°F/1000°C/1273°K.
Specific gravity (water = 1): 4.7 @ 68°F/20°C (solid).
Heat of solution: –92 Btu/lb = –51.3 cal/g = –2.15 x 10^5 J/kg.
Heat of fusion: 22.9 cal/g.

CALCIUM   REC. C:0500

SYNONYMS: CALCICAT; CALCIO (Spanish); CALCIUM METAL, CRYSTALINE; EEC No. 020-001-00-X

IDENTIFICATION
CAS Number: 7440-70-2
Formula: Ca
DOT ID Number: UN 1401; DOT Guide Number: 138
Proper Shipping Name: Calcium

DESCRIPTION: Silvery-gray powder; turns to grayish-white on exposure to air. Odorless. Sinks in water; reacts violently with water generating flammable hydrogen gas.

Combustible solid; reacts with air forming flammable hydrogen gas • Dust cloud may explode if ignited in an enclosed area • Containers may BLEVE when exposed to fire • Irritating to the skin, eyes, and respiratory tract • Toxic products of combustion may include smoke and irritating vapors of unburned chemical.

Hazard Classification (based on NFPA-704 Rating System)
Health Hazards (Blue): 3; Flammability (Red): 1; Reactivity (Yellow): 2; Special Notice (White): Water reactive

EMERGENCY RESPONSE: See Appendix A (138)
Evacuation:
Public safety: Isolate the area of spill or leak for at least 50 to 100 meters (160 to 330 feet) in all directions.
Spill: Consider initial downwind evacuation for at least 250 meters (800 feet).
Fire: If any large container is involved in fire, isolate for at least 800 meters (½ mile) in all directions; also, consider initial evacuation for 800 meters (½ mile) in all directions.

EXPOSURE
Short-term effects: *SEEK MEDICAL ATTENTION*. **Solid:** Will burn skin and eyes. Remove contaminated clothing and shoes. Flush affected areas with plenty of water. *IF IN EYES*, hold eyelids open and flush with plenty of water.

HEALTH HAZARDS
Personal protective equipment (PPE): B-Level PPE. Safety goggles and rubber gloves.
Liquid or solid irritant characteristics: Severe eye irritant. Skin irritant.

FIRE DATA
Fire extinguishing agents not to be used: Water, halogenated hydrocarbons, halons, dry chemical, CO_2, foam
Behavior in fire: Burns violently, especially if finely divided.
Autoignition temperature: 1436–1490°F/780–810°C.

CHEMICAL REACTIVITY
Reactivity with water: Reacts to form flammable hydrogen gas, which may ignite. Warm water reaction may be violent.
Binary reactants: Reacts with oxidizers, acids, alcohols, halogenated hydrocarbons, carbonates, and other substances. Reacts with moist air water vapor to form hydrogen. Shock-sensitive compounds formed with oxides of iron and copper.
Neutralizing agents for acids and caustics: Flush with water.

ENVIRONMENTAL DATA
Food chain concentration potential: Negative; unlikely to accumulate.
Water pollution: Dangerous to aquatic life in high concentrations. May be dangerous if it enters nearby water intakes; notify operators. Notify local health and wildlife officials. **Response to discharge:** Issue warning–high flammability. Absorb in bone-dry anhydrous soda.

SHIPPING INFORMATION
Grades of purity: Commercial, 99.5%; redistilled 99.9%; **Storage temperature:** Ambient; **Inert atmosphere:** None; **Venting:** Airtight, sealed containers must be in a ventilated area; **Stability during transport:** Stable.

PHYSICAL AND CHEMICAL PROPERTIES
Physical state @ 59°F/15°C and 1 atm: Solid.
Molecular weight: 40.1
Boiling point @ 1 atm: 4755°F/1440°C/1713°K.
Melting/Freezing point: 1562°F/850°C/1,123°K.
Specific gravity (water = 1): 1.55 @ 68°F/20°C (solid).
Heat of combustion: –6790 Btu/lb = –3,770 cal/g = –158 x 10^5 J/kg.
Heat of fusion: 55.7 cal/g.
Vapor pressure: 10 mm @ 1795°F/979°C.

CALCIUM ARSENATE   REC. C:0550

SYNONYMS: ARSENIATO CALCICO (Spanish); ARSENIC ACID, CALCIUM SALT; ARSENATE DE CALCIUM (French); CALCIUM ARSENATE; CALCIUM ORTHOARSENATE; CALCIUM SALT of ARSENIC ACID; CUCUMBER DUST; KALZIUMARSENIAT (German); TRICALCIUMARSENAT (German); TRICALCIUM ARSENATE; TRICALCIUM *o*-ARSENATE; TRICALCIUM ORTHO-ARSENATE

IDENTIFICATION
CAS Number: 7778-44-1; 10103-62-5
Formula: $As_2O_8 \cdot 3Ca$
DOT ID Number: UN 1573; DOT Guide Number: 151
Proper Shipping Name: Calcium arsenate
Reportable Quantity (RQ): **(CERCLA)** 1 lb/0.454 kg

DESCRIPTION: White solid. Odorless. Sinks in water.

Poison! • Breathing the dust, skin or eye contact, or swallowing the material can kill you • Firefighting gear (including SCBA) does not provide adequate protection. If exposure occurs, remove and isolate gear immediately and thoroughly decontaminate personnel • Concentrated dust in confined areas (e.g., tanks, sewers, buildings) may explode when exposed to fire • Toxic products of combustion may include arsenic • Do not put yourself in danger by entering a contaminated area to rescue a victim.

Hazard Classification (based on NFPA-704 Rating System)
Health Hazards (Blue): 4; Flammability (Red): 0; Reactivity (Yellow): 0

EMERGENCY RESPONSE: See Appendix A (151)
Evacuation:
Public safety: Isolate the area of spill or leak for at least 25 to 50 meters (80 to 160 feet) in all directions.
Spill: Increase, in the downwind direction, as necessary, the distance shown under "Public Safety."
Fire: If tank, rail car, or tank truck is involved in fire, isolate for at least 800 meters (½ mile) in all directions; also, consider initial evacuation for 800 meters (½ mile) in all directions.

EXPOSURE
Short-term effects: *SEEK MEDICAL ATTENTION.* **Dust:** *POISONOUS IF INHALED. IF INHALED*, will, will cause coughing or difficult breathing. *IF IN EYES*, hold eyelids open and flush with plenty of water. If artificial respiration is administered, *avoid mouth-to-mouth resuscitation; use bag/mask apparatus*. IF breathing is difficult, administer oxygen. **Solid:** POISONOUS IF SWALLOWED. *IF SWALLOWED*, will cause nausea and vomiting. Remove contaminated clothing and shoes. Flush affected areas with plenty of water. *IF IN EYES*, hold eyelids open and flush with plenty of water. *IF SWALLOWED* and victim is *CONSCIOUS AND ABLE TO SWALLOW*, have victim drink 4 to 8 ounces of water and have victim induce vomiting. *IF SWALLOWED* and victim is *UNCONSCIOUS OR HAVING CONVULSIONS*, do nothing except keep victim warm.

HEALTH HAZARDS
Personal protective equipment (PPE): Arsenic, inorganic compounds are OSHA Table Z-1-A air contaminants. A-Level PPE.
Recommendations for respirator selection: NIOSH
At any concentrations above the NIOSH REL, or where there is no REL, at any detectable concentration: SCBAF:PD,PP (any self-contained breathing apparatus that has a full faceplate and is operated in a pressure-demand or other positive-pressure mode); or SAF:PD,PP:ASCBA (any supplied-air respirator that has a full facepiece and is operated in a pressure-demand or other positive-pressure mode in combination with an auxiliary self-contained breathing apparatus operated in a pressure-demand or other positive-pressure mode). *ESCAPE:* GMFAGHiE [any air-purifying, full-facepiece respirator (gas mask) with a chin-style, front-or back-mounted acid gas canister having a high-efficiency particulate filter]; or SCBAE (any appropriate escape-type, self-contained breathing apparatus).
Exposure limits (TWA unless otherwise noted): ACGIH TLV 0.01 mg/m^3; OSHA PEL [1910.1080] 0.010 mg/m^3; NIOSH REL ceiling 0.002 mg/m^3/15 min, as arsenic; potential human carcinogen; reduce exposure to lowest feasible level
Short-term exposure limits (15-minute TWA): 1.5 mg/m^3 as arsenic.
Toxicity by ingestion: Grade 4; oral rat LD$_{50}$ = 20 mg/kg.
Long-term health effects: Arsenic is an NTP known carcinogen. Arsenic compounds may cause skin and lung cancer.
Vapor (gas) irritant characteristics: Eye and respiratory tract irritant.
Liquid or solid irritant characteristics: Eye irritant.
IDLH value: 5 mg/m^3 as inorganic arsenic; known human carcinogen.

ENVIRONMENTAL DATA
Food chain concentration potential: Possible bioaccumulation problem
Water pollution: DOT Appendix B, §172.101–marine pollutant. Harmful to aquatic life in very low concentrations. May be dangerous if it enters nearby water intakes; notify operators. Notify local health and wildlife officials. **Response to discharge:** Issue warning–poison, water contaminant. Restrict access. Should be removed. Chemical and physical treatment.

SHIPPING INFORMATION
Grades of purity: 70%, containing calcium carbonate and calcium hydroxide (limestone and slaked lime); **Storage temperature:** Ambient; **Inert atmosphere:** None; **Venting:** Open; **Stability during transport:** Stable.

PHYSICAL AND CHEMICAL PROPERTIES
Physical state @ 59°F/15°C and 1 atm: Solid.
Molecular weight: 398
Boiling point @ 1 atm: Decomposes.
Specific gravity (water = 1): 3.62 @ 68°F/20°C (solid).
Vapor pressure: 0 mm (approximate).

CALCIUM ARSENITE   REC. C:0600

SYNONYMS: ARSENITO CALCICO (Spanish); ARSENOUS ACID, CALCIUM SALT

IDENTIFICATION
CAS Number: 52740-16-6; 15194-98-6
Formula: Variable composition CaAsO$_3$H; AsO$_4$·Ca; As$_2$O$_6$·3Ca
DOT ID Number: UN 1574; DOT Guide Number: 151
Proper Shipping Name: Calcium arsenite, solid
Reportable Quantity (RQ): **(CERCLA)** 1 lb/0.454 kg

DESCRIPTION: White granular powder. Odorless.

Poison! • Breathing the dust, skin or eye contact, or swallowing the material can kill you • Firefighting gear (including SCBA) does not provide adequate protection. If exposure occurs, remove and isolate gear immediately and thoroughly decontaminate personnel • Concentrated dust in confined areas (e.g., tanks, sewers, buildings) may explode when exposed to fire • Toxic products of combustion may include arsenic • Do not put yourself in danger by entering a contaminated area to rescue a victim.

Hazard Classification (based on NFPA-704 Rating System)
Health Hazards (Blue): 4; Flammability (Red): 0; Reactivity (Yellow): 0

EMERGENCY RESPONSE: See Appendix A (151)
Evacuation:
Public safety: Isolate the area of spill or leak for at least 25 to 50 meters (80 to 160 feet) in all directions.
Spill: Increase, in the downwind direction, as necessary, the distance shown under "Public Safety."
Fire: If tank, rail car, or tank truck is involved in fire, isolate for at least 800 meters (½ mile) in all directions; also, consider initial evacuation for 800 meters (½ mile) in all directions.

EXPOSURE
Short-term effects: *SEEK MEDICAL ATTENTION.* If artificial respiration is administered, *avoid mouth-to-mouth resuscitation; use bag/mask apparatus*. **Solid:** *POISONOUS IF SWALLOWED.* Irritating to skin, eyes, and nose. Move to fresh air. Remove contaminated clothing and shoes. Flush affected areas with plenty of water. *IF SWALLOWED* and victim is *CONSCIOUS AND ABLE TO SWALLOW*, have victim drink 4 to 8 ounces of water and have victim induce vomiting.

HEALTH HAZARDS
Personal protective equipment (PPE): A-Level PPE. OSHA Table Z-1-A air contaminant. Hand and arm protection, protective clothing, waterproof boots, respiratory protective equipment, and eye protection.
Recommendations for respirator selection: NIOSH
At any concentrations above the NIOSH REL, or where there is no REL, at any detectable concentration: SCBAF:PD,PP (any self-contained breathing apparatus that has a full faceplate and is operated in a pressure-demand or other positive-pressure mode); or SAF:PD,PP:ASCBA (any supplied-air respirator that has a full facepiece and is operated in a pressure-demand or other positive-pressure mode in combination with an auxiliary self-contained breathing apparatus operated in a pressure-demand or other positive-pressure mode). *ESCAPE:* GMFAGHiE [any air-purifying, full-facepiece respirator (gas mask) with a chin-style, front-or back-mounted acid gas canister having a high-efficiency particulate filter]; or SCBAE (any appropriate escape-type, self-contained breathing apparatus).
Exposure limits (TWA unless otherwise noted): ACGIH TLV 0.01 mg/m^3; OSHA PEL [1910.1080] 0.010 mg/m^3; NIOSH REL ceiling 0.002 mg/m^3/15 min, as arsenic; potential human carcinogen; reduce exposure to lowest feasible level
Long-term health effects: Available studies point consistently to a causal relationship between skin cancer and heavy exposure to inorganic arsenic. An increased frequency of deaths from lung cancer has been found in occupational groups exposed to high levels of inorganic arsenic compounds. Arsenic is an NTP known carcinogen.
Vapor (gas) irritant characteristics: Eye and respiratory tract irritant.
Liquid or solid irritant characteristics: Eye and skin irritant.
IDLH value: 5 mg/m^3 as inorganic arsenic; known human carcinogen.

ENVIRONMENTAL DATA
Water pollution: DOT Appendix B, §172.101–marine pollutant. Harmful to aquatic life in very low concentrations. Notify local health and wildlife officials. **Response to discharge:** Issue warning–poison. Restrict access. Should be removed. Chemical and physical treatment.

SHIPPING INFORMATION
Stability during transport: Atmospheric CO_2 may cause decomposition.

PHYSICAL AND CHEMICAL PROPERTIES
Physical state @ 59°F/15°C and 1 atm: Solid.
Molecular weight: 164.0

CALCIUM CARBIDE   REC. C:0650

SYNONYMS: ACETYLENOGEN; ACTYLENOGEN; CALCIUM ACETYLIDE; CALCIUM DICARBIDE; CARBIDE; CARBURO CALCICO (Spanish); EEC No. 006-004-00-9

IDENTIFICATION
CAS Number: 75-20-7
Formula: CaC_2
DOT ID Number: UN 1402; DOT Guide Number: 138
Proper Shipping Name: Calcium carbide
Reportable Quantity (RQ): **(CERCLA)** 10 lb/4.54 kg

DESCRIPTION: Gray to bluish black powder or granules. Garlic odor. Sinks in water; bubbles appear on surface as flammable acetylene gas and corrosive calcium hydroxide (lime) is produced.

Reacts with moisture, including moisture in the air, forming explosive acetylene gas and corrosive lime • Do not use water, foam, or water-based extinguishers • Corrosive to the skin, eyes, and respiratory tract; contact with skin or eyes can cause burns and vision impairment; inhalation symptoms may be delayed • Firefighting gear (including SCBA) does not provide adequate protection. If exposure occurs, remove and isolate gear immediately and thoroughly decontaminate personnel • Containers may BLEVE when exposed to fire • Vapors may form explosive mixture with air • Vapors are heavier than air and will collect and stay in low areas • Vapors may travel long distances to ignition sources and flashback • Vapors in confined areas (e.g., tanks, sewers, buildings) may explode when exposed to fire • Toxic products of combustion. include carbon monoxide.

Hazard Classification (based on NFPA-704 Rating System)
Health Hazards (Blue): 3; Flammability (Red): 3; Reactivity (Yellow): 2; Special Notice (White): Water reactive

EMERGENCY RESPONSE: See Appendix A (138)
Evacuation:
Public safety: Isolate the area of spill or leak for at least 50 to 100 meters (160 to 330 feet) in all directions.
Spill: Consider initial downwind evacuation for at least 250 meters (800 feet).
Fire: If any large container is involved in fire, isolate for at least 800 meters (½ mile) in all directions; also, consider initial evacuation for 800 meters (½ mile) in all directions.

EXPOSURE
Short-term effects: *SEEK MEDICAL ATTENTION.* **Solid:** Irritating to skin and eyes. Remove contaminated clothing and shoes. Flush affected areas with plenty of water. *IF IN EYES*, hold eyelids open and flush with plenty of water.
Note to physician or authorized medical personnel: Medical observation is recommended for 24 to 48 hours after breathing overexposure, as pulmonary edema may be delayed. As first aid for pulmonary edema, consider administering a corticosteroid spray. Cigarette smoking may exacerbate pulmonary injury and should be discouraged for at least 72 hours following exposure.

HEALTH HAZARDS
Personal protective equipment (PPE): B-Level PPE. Chemical safety goggles and (for those exposed to unusually dusty operations) a positive pressure self contained breathing apparatus or respirator such as those approved by the U.S. Bureau of Mines for "nuisance dusts."
Vapor (gas) irritant characteristics: May cause eye and respiratory tract irritation.
Liquid or solid irritant characteristics: Severe eye irritant. Skin irritant. If spilled on clothing and allowed to remain, may cause smarting and reddening of the skin.

FIRE DATA if kept absolutely dry.
Fire extinguishing agents not to be used: Water, vaporizing liquid, foam, or CO_2

CHEMICAL REACTIVITY
Reactivity with water: Reacts vigorously with water to form the highly flammable acetylene gas, which may ignite spontaneously.

Binary reactants: Reacts with copper and brass to form explosive compound. Incompatible with hydrogen chloride (gas), selenium. magnesium, silver nitrate, sodium peroxide. Flammable gas formed in moist air.

ENVIRONMENTAL DATA
Food chain concentration potential: Negative; unlikely to accumulate.
Water pollution: Harmful to aquatic life in very low concentrations. May be dangerous if it enters nearby water intakes; notify operators. Notify local health and wildlife officials. **Response to discharge:** Issue warning–poison, water contaminant. Restrict access. Should be removed. Chemical and physical treatment.

SHIPPING INFORMATION
Stability during transport: Stable in absence of moisture.

PHYSICAL AND CHEMICAL PROPERTIES
Physical state @ 59°F/15°C and 1 atm: Solid.
Molecular weight: 64.10
Boiling point @ 1 atm: 4172°F/2300°C/2573°K.
Melting/Freezing point: More than 810°F/450°C/723°K.
Specific gravity (water = 1): 2.22 at 18°C (solid).

CALCIUM CHLORATE   REC. C:0700

SYNONYMS: CHLORATE de CALCIUM (French); CHLORIC ACID, CALCIUM SALT; CLORATO CALCICO (Spanish)

IDENTIFICATION
CAS Number: 10137-74-3
Formula: $Ca(ClO_3)_2$
DOT ID Number: UN 1452; DOT Guide Number: 140
Proper Shipping Name: Calcium chlorate

DESCRIPTION: White to light yellow hygroscopic crystals. Odorless. Sinks and mixes with water.

Poison! • Breathing the dust, or swallowing the material can kill you; irritating to the eyes and skin • May interfere with the body's ability to use oxygen • Firefighting gear (including SCBA) does not provide adequate protection. If exposure occurs, remove and isolate gear immediately and thoroughly decontaminate personnel • Heat may cause containers to explode • Strong oxidizer which may react spontaneously with low flash point organics or reducing agents. Heat forms oxygen; will increase the activity of an existing fire • May cause fire or explosion contact with combustibles (wood, paper, oil, clothing, etc.) • Toxic products of combustion may include hydrogen chloride.

Hazard Classification (based on NFPA-704 Rating System)
Health Hazards (Blue): 1; Flammability (Red): 0; Reactivity (Yellow): 2; Special Notice (White): OXY

EMERGENCY RESPONSE: See Appendix A (140)
Note: Do not let spill area dry until it has been determined that there is no chlorates left in the area. Continue cooling after fire has been extinguished.
Evacuation:
Public safety: Isolate the area of spill or leak for at least 10 to 25 meters (30 to 80 feet) in all directions.
Spill: Consider initial downwind evacuation for at least 100 meters (330 feet).
Fire: If any large container is involved in fire, isolate for at least 800 meters (½ mile) in all directions; also, consider initial evacuation for 800 meters (½ mile) in all directions.

EXPOSURE
Short-term effects: *SEEK MEDICAL ATTENTION*. **Dust:** Irritating to eyes, nose, and throat. Move victim to fresh air. *IF IN EYES*, hold eyelids open and flush with plenty of water. **Solid:** Irritating to skin and eyes. Harmful if swallowed. Remove contaminated clothing and shoes. Flush affected areas with plenty of water. *IF IN EYES*, hold eyelids open and flush with plenty of water. *IF SWALLOWED* and victim is *CONSCIOUS AND ABLE TO SWALLOW*, have victim drink 4 to 8 ounces of water and have victim induce vomiting. *IF SWALLOWED* and victim is *UNCONSCIOUS OR HAVING CONVULSIONS*, do nothing except keep victim warm.

HEALTH HAZARDS
Personal protective equipment (PPE): A-Level PPE. Goggles or face shield; dust respirator; coveralls or other protective clothing.
Toxicity by ingestion: Grade 2; oral LD_{50} = 4500 mg/kg (rat).
Long-term health effects: Heart, liver, blood and kidney injury may occur.
Vapor (gas) irritant characteristics: Eye and respiratory tract irritant.
Liquid or solid irritant characteristics: Eye and skin irritant.

CHEMICAL REACTIVITY
Binary reactants: A strong oxidizer. Reacts violently with combustible materials; finely divided metals such as aluminum and copper; salts of ammonia, acids, sulfur, cyanides, manganese dioxide, sulfur dioxide, organic acids, metal sulfides, organic materials. The mixture may ignite from friction.

ENVIRONMENTAL DATA
Food chain concentration potential: Negative; unlikely to accumulate.
Water pollution: Effect of low concentrations on aquatic life is unknown. May be dangerous if it enters nearby water intakes; notify operators. Notify local health and wildlife officials. **Response to discharge:** Issue warning–oxidizing material, water contaminant. Should be removed. Chemical and physical treatment. Disperse and flush.

SHIPPING INFORMATION
Grades of purity: Commercial. May be shipped as dihydrate;
Storage temperature: Ambient; **Inert atmosphere:** None;
Venting: Open; **Stability during transport:** Stable.

PHYSICAL AND CHEMICAL PROPERTIES
Physical state @ 59°F/15°C and 1 atm: Solid.
Molecular weight: 243.0
Boiling point @ 1 atm: Decomposes.
Melting/Freezing point: 212°F/100°C/373°K.
Specific gravity (water = 1): 2.710 at 0°C (solid).
Heat of solution: (estimate) –54 Btu/lb = –30 cal/g = –1.3 x 10^5 J/kg.

CALCIUM CHLORIDE   REC. C:0750

SYNONYMS: CALCIUM CHLORIDE HYDRATES; CAL PLUS; CALTAC; CLORURO CALCICO (Spanish); DOWFLAKE; EEC No. 017-013-00-2; LIQUIDOW; PELADOW; SNOMELT; SUPERFLAKE, ANHYDROUS

IDENTIFICATION
CAS Number: 10043-52-4
Formula: $CaCl_2$; $CaCl_2 \cdot xH_2O$ where x = 0 to 6

DESCRIPTION: White to off-white crystalline solid or water solution. Odorless. Sinks and mixes with water; violent reaction and heat.

Irritating to the skin, eyes, and respiratory tract • Toxic products of combustion may include hydrogen chloride.

Hazard Classification (based on NFPA-704 Rating System)
Health Hazards (Blue): 1; Flammability (Red): 0; Reactivity (Yellow): 0

EMERGENCY RESPONSE: See Appendix A (140)
Evacuation:
Public safety: Isolate the area of spill or leak for at least 10 to 25 meters (30 to 80 feet) in all directions.
Spill: Increase, in the downwind direction, as necessary, the distance shown under "Public Safety."
Fire: If any large container is involved in fire, isolate for at least 800 meters (½ mile) in all directions; also, consider initial evacuation for 800 meters (½ mile) in all directions.

EXPOSURE
Short-term effects: *SEEK MEDICAL ATTENTION.* **Solution or solid:** Will burn skin and eyes. *IF SWALLOWED,* will cause nausea and vomiting. Remove contaminated clothing and shoes. Flush affected areas with plenty of water. IF IN EYES, hold eyelids open and flush with plenty of water. IF SWALLOWED and victim is *CONSCIOUS AND ABLE TO SWALLOW*, have victim drink 4 to 8 ounces of water. IF SWALLOWED and victim is *UNCONSCIOUS OR HAVING CONVULSIONS,* do nothing except keep victim warm.

HEALTH HAZARDS
Personal protective equipment (PPE): B-Level PPE. Close-fitting safety goggles or face shield, dust-type respirator, protective rubber gloves.
Toxicity by ingestion: Grade 2; oral LD_{50} = 1000 mg/kg (rat).
Vapor (gas) irritant characteristics: Eye and respiratory tract irritant.
Liquid or solid irritant characteristics: Eye and skin irritant.

CHEMICAL REACTIVITY
Reactivity with water: Anhydrous grade dissolves with evolution of some heat.
Binary reactants: Metals will slowly corrode in aqueous solutions. attacks metals and construction materials.
Reactivity group: 43
Compatibility class: Miscellaneous water solutions.

ENVIRONMENTAL DATA
Food chain concentration potential: Negative; unlikely to accumulate.
Water pollution: Dangerous to aquatic life in high concentrations. May be dangerous if it enters nearby water intakes; notify operators. Notify local health and wildlife officials. **Response to discharge:** Disperse and flush.

SHIPPING INFORMATION
Grades of purity: Anhydrous 90–97%; water solutions containing 51–86%; **Storage temperature:** Ambient; **Inert atmosphere:** None; **Venting:** Open; **Stability during transport:** Stable.

PHYSICAL AND CHEMICAL PROPERTIES
Physical state @ 59°F/15°C and 1 atm: Solid.
Molecular weight: 110.99 (solute).
Boiling point @ 1 atm: 2910°F/1600°C/1873°K.
Melting/Freezing point: 1390°F/770°C/1043°K.
Specific gravity (water = 1): 2.15 @ 68°F/20°C (solid).
Heat of solution: –292 Btu/lb = –162 cal/g = –6.79 x 10^5 J/kg.
Heat of fusion: 55 cal/g.

CALCIUM CHROMATE   REC. C:0800

SYNONYMS: CALCIUM CHROMATE(VI); CALCIUM CHROME YELLOW; CALCIUM CHROMATE DIHYDRATE; CALCIUM CHROMIUM OXIDE; CALCIUM MONOCHROMATE; C.I.PIGMENT YELLOW 33; GELBIN YELLOW ULTRAMARINE; CHROMIC ACID, CALCIUM SALT; CROMATO CALCICO (Spanish); STEINBUHL YELLOW; RCRA No. U032

IDENTIFICATION
CAS Number: 13765-19-0
Formula: $CrO_4 \cdot Ca$; $CaCrO_4 \cdot 2H_2O$
DOT ID Number: UN 9096; DOT Guide Number: 171
Proper Shipping Name: Calcium chromate
Reportable Quantity (RQ): **(CERCLA)** 10 lb/4.54 kg

DESCRIPTION: Yellow crystalline solid. Odorless. Sinks and mixes slowly with water.

Strong oxidizer which may react spontaneously with low flash point organics or reducing agents. Heat forms oxygen; will increase the activity of an existing fire • May cause fire or explosion on contact with combustibles (wood, paper, oil, clothing, etc.) • Toxic products of combustion may include chromium fume.

Hazard Classification (based on NFPA-704 Rating System)
Health Hazards (Blue): 2; Flammability (Red): 0; Reactivity (Yellow): 0; Special Notice (White): OXY

EMERGENCY RESPONSE: See Appendix A (171)
Evacuation:
Public safety: Isolate the area of spill or leak for at least 10 to 25 meters (30 to 80 feet) in all directions.
Spill: Increase, in the downwind direction, as necessary, the distance shown under "Public Safety."
Fire: If any large container is involved in fire, isolate for at least 800 meters (½ mile) in all directions; also, consider initial evacuation for 800 meters (½ mile) in all directions.

EXPOSURE
Short-term effects: *SEEK MEDICAL ATTENTION.* **Dust:** Irritating to eyes, nose, and throat. *IF INHALED*, will, will cause coughing or difficult breathing. *IF IN EYES*, hold eyelids open and flush with plenty of water. *IF BREATHING HAS STOPPED,* give artificial respiration. IF breathing is difficult, administer oxygen. **Solid:** Will burn skin and eyes. Harmful if swallowed. Remove contaminated clothing and shoes. Flush affected areas with plenty of water. *IF IN EYES*, hold eyelids open and flush with plenty of water. *IF SWALLOWED* and victim is *CONSCIOUS AND ABLE TO SWALLOW*, have victim drink 4 to 8 ounces of water and have victim induce vomiting. *IF SWALLOWED* and victim is *UNCONSCIOUS OR HAVING CONVULSIONS,* do nothing except keep victim warm.

HEALTH HAZARDS
Personal protective equipment (PPE): B-Level PPE. Dust mask; goggles or face shield; protective gloves.
Recommendations for respirator selection: NIOSH
At any concentrations above the NIOSH REL, or where there is no REL, at any detectable concentration: SCBAF:PD,PP (any self-contained breathing apparatus that has a full facepiece and is operated in a pressure-demand or other positive-pressure mode); or SAF:PD,PP:ASCBA (any supplied-air respirator that has a full facepiece and is operated in a pressure-demand or other positive-pressure mode in combination with an auxiliary self-contained breathing apparatus operated in a pressure-demand or other positive pressure mode). *ESCAPE:* HiEF (any air-purifying, full-facepiece respirator with a high-efficiency particulate filter); or SCBAE (any appropriate escape-type, self-contained breathing apparatus).
Exposure limits (TWA unless otherwise noted): ACGIH TLV 0.001 mg/m^3; OSHA PEL 0.1 mg/m^3 as CrO$_3$; NIOSH 0.001 mg/m^3 as chromates; potential human carcinogen; reduce exposure to lowest feasible level.
Toxicity by ingestion: Grade 3; LD$_{50}$ = 50 to 500 mg/kg.
Long-term health effects: NTP anticipated carcinogen. Lung cancer may develop.
Vapor (gas) irritant characteristics: Eye and respiratory tract irritant.
Liquid or solid irritant characteristics: Eye and skin irritant.
IDLH value: 15 mg/m^3 as chromium(VI).

FIRE DATA
Behavior in fire: The hydrated salt loses water when hot and changes color, but no increase in hazard occurs.

ENVIRONMENTAL DATA
Food chain concentration potential: Bioconcentration up to 2000-fold possible under constant exposure. May not be significant under spill conditions.
Water pollution: Effect of low concentrations on aquatic life is unknown. May be dangerous if it enters nearby water intakes; notify operators. Notify local health and wildlife officials.
Response to discharge: Issue warning–water contaminant. Disperse and flush.

SHIPPING INFORMATION
Grades of purity: Technical; **Storage temperature:** Ambient; **Inert atmosphere:** None; **Venting:** Open; **Stability during transport:** Stable.

PHYSICAL AND CHEMICAL PROPERTIES
Physical state @ 59°F/15°C and 1 atm: Solid.
Molecular weight: 192.1
Specific gravity (water = 1): More than 1 @ 68°F/20°C (solid).
Heat of solution: –73 Btu/lb = –41 cal/g = –1.7 x 10^5 J/kg.

CALCIUM CYANIDE REC. C:0850

SYNONYMS: CALCIUM CYANIDE MIXTURE; CALCIUM CYANIDE MIXTURE, SOLID; CIANURO CALCICO (Spanish); CYANIDE OF CALCIUM; CYANOGAS G-FUMIGANT; CYANOGAS A-DUST; RCRA No. P021

IDENTIFICATION
CAS Number: 592-01-8
Formula: Ca(CN)$_2$ plus inert ingredients
DOT ID Number: UN 1575; DOT Guide Number: 157
Proper Shipping Name: Calcium cyanide
Reportable Quantity (RQ): **(CERCLA)** 10 lb/4.45 kg

DESCRIPTION: White to gray or black solid. Compound reacts with moisture in air to form hydrogen cyanide gas, which has a characteristic almond-like odor. Sinks and mixes with water.

Poison! • Breathing the vapor, skin or eye contact, or swallowing the material can kill you • Firefighting gear (including SCBA) does not provide adequate protection. If exposure occurs, remove and isolate gear immediately and thoroughly decontaminate personnel • Toxic products of combustion may include hydrogen cyanide.

Hazard Classification (based on NFPA-704 Rating System)
Health Hazards (Blue): 3; Flammability (Red): 0; Reactivity (Yellow): 1

EMERGENCY RESPONSE: See Appendix A (157)
Evacuation:
Public safety: Isolate the area of spill or leak for at least 50 to 100 meters (160 to 330 feet) in all directions.
Spill: Increase, in the downwind direction, as necessary, the distance shown under "Public Safety."
Fire: If tank, rail car, or tank truck is involved in fire, isolate for at least 800 meters (½ mile) in all directions; also, consider initial evacuation for 800 meters (½ mile) in all directions.

EXPOSURE
Short-term effects: *SEEK MEDICAL ATTENTION.* **Dust:** POISONOUS IF INHALED. Irritating to eyes, nose, and throat. Move victim to fresh air. *IF IN EYES*, hold eyelids open and flush with plenty of water. *IF BREATHING HAS STOPPED*, give artificial respiration; *avoid mouth-to-mouth resuscitation; use bag/mask apparatus.* IF breathing is difficult, administer oxygen. **Solid:** *POISONOUS IF SWALLOWED.* Irritating to skin and eyes. Remove contaminated clothing and shoes. Flush affected areas with plenty of water. *IF IN EYES*, hold eyelids open and flush with plenty of water. *IF SWALLOWED* and victim is *CONSCIOUS AND ABLE TO SWALLOW*, have victim drink 4 to 8 ounces of water and have victim induce vomiting. *IF SWALLOWED* and victim is *UNCONSCIOUS OR HAVING CONVULSIONS*, do nothing except keep victim warm.
Note to physician or authorized medical personnel: Consider the use of amyl nitrite perles if symptoms of cyanide poisoning develop. If symptoms indicate, initial treatment includes the cyanide antidote kit. In all cases, break an amyl nitrite perle in a gauze pad and hold lightly under victim's nose for 15 seconds, repeating 5 times at about 15-second intervals; if necessary (and if sodium nitrite infusions will be delayed), repeat procedure every 3 minutes with fresh perles until 3 or 4 have been used. Avoid breathing the vapor while administering it to the victim. Administer sodium nitrite IV, ASAP. The usual adult dose is 10 to 20 mL of a 3% solution infused over no less than 5 minutes; the average child dose is 0.15 to 0.20 mL/kg. Monitor blood pressure during administration, and slow the rate of infusion if hypotention develops. Next, infuse sodium thiosulfate IV. The usual adult dose is 50 mL of a 25% solution infused over 10 to 20 minutes; the average child dose is 1.65 mL/kg. Repeat with nitrite and thiosulfate as required.

HEALTH HAZARDS
Personal protective equipment (PPE): A-Level PPE.
Recommendations for respirator selection: NIOSH as cyanides *25 mg/m^3:* SA (any supplied-air respirator); or SCBAF (any self-contained breathing apparatus with full facepiece). *EMERGENCY*

OR PLANNED ENTRY INTO UNKNOWN CONCENTRATIONS OR IDLH CONDITIONS: SCBAF:PD,PP (any self-contained breathing apparatus that has a full facepiece and is operated in a pressure-demand or other positive-pressure mode); or SAF:PD,PP:ASCBA (any supplied-air respirator that has a full facepiece and is operated in a pressure-demand or other positive-pressure mode in combination with an auxiliary self-contained breathing apparatus operated in a pressure-demand or other positive pressure mode). *ESCAPE:* GMFSHiE [any air-purifying, full-facepiece respirator (gas mask) with a chin-style, front- or back-mounted canister providing protection against the compound of concern and having a high efficiency particulate filter); or SCBAE (any appropriate escape-type, self-contained breathing apparatus).

Toxicity by ingestion: Grade 4; oral LD_{50} = 39 mg/kg (rat).
IDLH value: 50 mg/m^3 as cyanide

FIRE DATA
Fire extinguishing agents not to be used: Do not use water or CO_2 on adjacent fires.

CHEMICAL REACTIVITY
Reactivity with water: Releases very poisonous hydrogen cyanide gas slowly on contact with water. Release is rapid if acid is also present.
Binary reactants: Cyanides may react with acids, sluorine, magnesium, nitrates, nitrites, nitric acid.

ENVIRONMENTAL DATA
Food chain concentration potential: Negative; unlikely to accumulate.
Water pollution: DOT Appendix B, §172.101–marine pollutant. Harmful to aquatic life in very low concentrations. May be dangerous if it enters nearby water intakes; notify operators. Notify local health and wildlife officials. **Response to discharge:** Issue warning–poison, air contaminant, water contaminant. Restrict access. Evacuate area. Should be removed. Chemical and physical treatment.

SHIPPING INFORMATION
Grades of purity: 42% with 58% inert ingredients. May contain up to 3% calcium carbide, which releases flammable acetylene gas when wet; **Storage temperature:** Ambient; **Inert atmosphere:** None; **Venting:** Well-sealed containers in ventilated area; ,
Stability during transport: Stable if kept dry.

PHYSICAL AND CHEMICAL PROPERTIES
Physical state @ 59°F/15°C and 1 atm: Solid.
Molecular weight: 92.10
Boiling point @ 1 atm: Decomposes.
Melting/Freezing point: 662°F/350°C/623°K.
Specific gravity (water = 1): 1.853 @ 68°F/20°C (solid).
Heat of solution: –264 Btu/lb = –147 cal/g = –6.14 x 10^5 J/kg.

CALCIUM FLUORIDE   REC. C:0900

SYNONYMS: CALCIUM DIFLUORIDE; FLUORURO CALCICO (Spanish); EINECS No. 232-188-7; FLUOSPAR; FLUORSPAR; MET-SPAR

IDENTIFICATION
CAS Number: 7789-75-5
Formula: CaF_2

DESCRIPTION: Gray powder or granules. Odorless. Sinks in water; practically insoluble.

Poison! • Breathing the dust, skin or eye contact, or swallowing the material can kill you • Firefighting gear (including SCBA) does not provide adequate protection. If exposure occurs, remove and isolate gear immediately and thoroughly decontaminate personnel • Containers may BLEVE when exposed to fire • Concentrated dust in confined areas (e.g., tanks, sewers, buildings) may explode when exposed to fire • Toxic products of combustion may include fluorides.

Hazard Classification (based on NFPA-704 Rating System)
Health Hazards (Blue): 2; Flammability (Red): 0; Reactivity (Yellow): 0

EMERGENCY RESPONSE: See Appendix A (171)
Evacuation:
Public safety: Isolate the area of spill or leak for at least 10 to 25 meters (30 to 80 feet) in all directions.
Spill: Increase, in the downwind direction, as necessary, the distance shown under "Public Safety."
Fire: If any large container is involved in fire, isolate for at least 800 meters (½ mile) in all directions; also, consider initial evacuation for 800 meters (½ mile) in all directions.

EXPOSURE
Short-term effects: *SEEK MEDICAL ATTENTION*. **Solid:** Harmful if swallowed. *IF SWALLOWED* and victim is *CONSCIOUS AND ABLE TO SWALLOW*, have victim drink 4 to 8 ounces of water.

HEALTH HAZARDS
Personal protective equipment (PPE): A-Level PPE.
Recommendations for respirator selection: NIOSH/OSHA as F. *12.5 mg/m^3:* DM (any dust and mist respirator). *25 mg/m^3:* DMXSQ* (any dust and mist respirator except single-use and quarter-mask respirators); or SA* (any supplied-air respirator). *62.5 mg/m^3:* SA:CF* [any supplied-air respirator operated in a continuous-flow mode)]; or PAPRDM*$^+$ *if not present as a fume* (any powered, air-purifying respirator with a dust and mist filter). *125 mg/m^3:* HiEF + (any air-purifying, full-facepiece respirator with a high-efficiency particulate filter); or SCBAF (any self-contained breathing apparatus with a full facepiece); or SAF (any supplied-air respirator with a full facepiece). *250 mg/m^3:* SA:PD,PP (any supplied-air respirator operated in a pressure-demand or other positive-pressure mode). *EMERGENCY OR PLANNED ENTRY INTO UNKNOWN CONCENTRATIONS OR IDLH CONDITIONS:* SCBAF:PD,PP (any self-contained breathing apparatus that has a full faceplate and is operated in a pressure-demand or other positive-pressure mode); or SAF:PD,PP:ASCBA (any supplied-air respirator that has a full facepiece and is operated in a pressure-demand or other positive-pressure mode in combination with an auxiliary, self-contained breathing apparatus operated in a pressure-demand or other positive-pressure mode). *ESCAPE:* HiEF+ (any air-purifying, full-facepiece respirator with a high-efficiency particulate filter); or SCBAE (any appropriate escape-type, self-contained breathing apparatus). *Notes:* *Substance reported to cause eye irritation or damage; may require eye protection. $^+$May need acid gas sorbent.
Exposure limits (TWA unless otherwise noted): ACGIH TLV 2.5 mg/m^3 as fluorides; NIOSH/OSHA 2.5 mg/m^3 as inorganic fluorides.
Toxicity by ingestion: Grade 2; LD_{50} = 0.5 to 5 g/kg.

Calcium hypochlorite

Vapor (gas) irritant characteristics: Eye and respiratory tract irritation.
Liquid or solid irritant characteristics: May cause eye and skin irritation.
IDLH value: 250 mg/m^3 as F.

ENVIRONMENTAL DATA
/tinca vulgaris/lethal/fresh **water.**
Water pollution: Dangerous to aquatic life in high concentrations. May be dangerous if it enters nearby water intakes; notify operators. Notify local health and pollution control officials.
Response to discharge: Disperse and flush.

SHIPPING INFORMATION
Grades of purity: Acid grade: 97.4%; Ceramic grade: 91.5%; Fine powder (dry or damp cake); Gravel fluorspar; Pellet; **Stability during transport:** Stable.

PHYSICAL AND CHEMICAL PROPERTIES
Physical state @ 59°F/15°C and 1 atm: Solid.
Molecular weight: 78.08.
Specific gravity (water = 1): 3.18 @ 68°F/20°C (solid).
Heat of fusion: 52.5 cal/g.

CALCIUM HYDROXIDE   REC. C:0950

SYNONYMS: BELL MINE; CALCIUM HYDRATE; CARBOXIDE; HIDROXIDO CALCICO (Spanish); HYDRATED KEMIKAL; LIME WATER; SLAKED LIME

IDENTIFICATION
CAS Number: 1305-62-0
–Formula: Ca(OH)$_2$

DESCRIPTION: White powder or granules. Odorless. Sinks in water; slightly soluble. Readily absorbs CO$_2$ from the air forming calcium carbonate.

Corrosive to the skin, eyes or respiratory tract; inhalation can cause lung edema; effects of inhalation may be delayed • Firefighting gear (including SCBA) does not provide adequate protection. If exposure occurs, remove and isolate gear immediately and thoroughly decontaminate personnel • Toxic products of combustion may include calcium oxide.

Hazard Classification (based on NFPA-704 Rating System)
Health Hazards (Blue): 1; Flammability (Red): 0; Reactivity (Yellow): 0

EMERGENCY RESPONSE: See Appendix A (171)
Evacuation:
Public safety: Isolate the area of spill or leak for at least 10 to 25 meters (30 to 80 feet) in all directions.
Spill: Increase, in the downwind direction, as necessary, the distance shown under "Public Safety."
Fire: If any large container is involved in fire, isolate for at least 800 meters (½ mile) in all directions; also, consider initial evacuation for 800 meters (½ mile) in all directions.

EXPOSURE
Short-term effects: *SEEK MEDICAL ATTENTION.* **Dust:** Irritating to nose and throat if inhaled. Move to fresh air. **Solid:** Will burn skin and eyes. Harmful if swallowed. Remove contaminated clothing and shoes. Flush affected areas with plenty of water. *IF IN EYES*, hold eyelids open and flush with plenty of water. *IF SWALLOWED* and victim is *CONSCIOUS AND ABLE TO SWALLOW*, have victim drink 4 to 8 ounces of water. **Do NOT induce vomiting.**

HEALTH HAZARDS
Personal protective equipment (PPE): B-Level PPE. OSHA Table Z-1-A air contaminant. Chemical protective material(s) reported to offer protection: natural rubber, neoprene, nitrile.
Exposure limits (TWA unless otherwise noted): ACGIH TLV 5 mg/m^3; OSHA PEL 15 mg/m^3 (total) 5 mg/m^3 (respirable); NIOSH 5 mg/m^3.
Toxicity by ingestion: Grade 1; LD$_{50}$ = 5 to 15 g/kg (rat).
Vapor (gas) irritant characteristics: Eye and respiratory tract irritant.
Liquid or solid irritant characteristics: Eye and skin irritant.

CHEMICAL REACTIVITY
Binary reactants: Maleic anhydride (may cause explosive decomposition), phosphorus, nitromethane, nitroparrafins, nitropropane. Forms explosive products with nitroethane and water. Attacks some metals.

ENVIRONMENTAL DATA
Food chain concentration potential: Negative; unlikely to accumulate.
Water pollution: Harmful to aquatic life in very low concentrations. May be dangerous if it enters nearby water intakes; notify operators. Notify local health and wildlife officials. **Response to discharge:** Disperse and flush.

SHIPPING INFORMATION
Grades of purity: Agricultural: 65–71%; industrial: 70–73%; chemical: 71–73%; **Stability during transport:** Stable.

PHYSICAL AND CHEMICAL PROPERTIES
Physical state @ 59°F/15°C and 1 atm: Solid.
Molecular weight: 74.09.
Boiling point @ 1 atm: Decomposes.
Specific gravity (water = 1): 2.24 @ 68°F/20°C (solid).
Vapor pressure: 0 mm (approximate).

CALCIUM HYPOCHLORITE   REC. C:1000

SYNONYMS: B-K POWDER; BLEACHING POWDER; CALCIUM CHLOROHYDROCHLORITE; CAL HYPO; CALCIUM HYPOCHLORIDE; CALCIUM OXYCHLORIDE; CAPORIT; CCH; CHLORIDE of LIME; CHLORIDE OF LIME; CHLORINATED LIME; CHLORINATED LIME; EEC No. 017-012-00-7; HIPOCLORITO CALCICO (Spanish); HTH; HYCHLOR; HYPOCHLOROUS ACID, CALCIUM; LIME CHLORIDE; LO-BAX; LOSANTIN; PERCHLORON; PITTCIDE; PITTCHLOR; SENTRY; HTH DRY CHLORINE; NEUTRAL ANHYDROUS CALCIUM HYPOCHLORITE

IDENTIFICATION
CAS Number: 7778-54-3
Formula: Ca(OCl)$_2$
DOT ID Number: UN 2208; UN 1748 (mixtures, dry); UN 2880 (hydrated); DOT Guide Number: 140
Proper Shipping Name: Calcium hypochlorite, hydrated or Calcium hypochlorite mixtures dry, with not less than 5.5% but more than 10% water (2880); Calcium hypochlorite, dry; Calcium hypochlorite mixtures dry, with more than 39% available chlorine

(8.8% available oxygen) (1748); Calcium hypochlorite, dry or with more than 10% but not more than 39% available chlorine (2208) Reportable Quantity (RQ): 10 lb/4.45 kg

DESCRIPTION: White granules or powder. Odor like household bleaching powder. Sinks in water; soluble; forms hydrochloric acid and corrosive chlorine gas.

Serious inhalation hazard if spilled in water. Corrosive to the skin, eyes, and respiratory tract; contact with skin or eyes can cause burns and vision impairment; inhalation symptoms may be delayed • Strong oxidizer which may react spontaneously with low flash point organics or reducing agents. Heat forms oxygen; will increase the activity of an existing fire • May cause fire or explosion on contact with combustibles (wood, paper, oil, clothing, etc.). If a fire starts, it may be difficult to extinguish • Toxic products of combustion may include hydrogen chloride fumes; may explode.

Hazard Classification (based on NFPA-704 Rating System)
Health Hazards (Blue): 3; Flammability (Red): 0; Reactivity (Yellow): 1; Special Notice (White): OXY

EMERGENCY RESPONSE: See Appendix A (140)
Evacuation:
Public safety: Isolate the area of spill or leak for at least 10 to 25 meters (30 to 80 feet) in all directions.
Spill: Consider initial downwind evacuation for at least 100 meters (330 feet).
IF SPILLED IN WATER, dangerous up to 10 km (6.0 miles) downwind.
Fire: If any large container is involved in fire, isolate for at least 800 meters (½ mile) in all directions; also, consider initial evacuation for 800 meters (½ mile) in all directions.

EXPOSURE
Short-term effects: *SEEK MEDICAL ATTENTION.* **Fumes:** May cause lung edema; physical exertion will aggravate this condition. **Solid:** Irritating to skin and eyes. *IF SWALLOWED,* will cause nausea, vomiting, or loss of consciousness. Remove contaminated clothing and shoes. Flush affected areas with plenty of water. *IF IN EYES,* hold eyelids open and flush with plenty of water. *IF SWALLOWED* and victim is *CONSCIOUS AND ABLE TO SWALLOW,* have victim drink 4 to 8 ounces of water. *IF SWALLOWED* and victim is *UNCONSCIOUS OR HAVING CONVULSIONS,* do nothing except keep victim warm. **Do NOT induce vomiting.**
Note to physician or authorized medical personnel: Medical observation is recommended for 24 to 48 hours after breathing overexposure, as pulmonary edema may be delayed. As first aid for pulmonary edema, consider administering a corticosteroid spray. Cigarette smoking may exacerbate pulmonary injury and should be discouraged for at least 72 hours following exposure.

HEALTH HAZARDS
Personal protective equipment (PPE): B-Level PPE. Face mask or protective goggles, approved, self-contained breathing apparatus.
Toxicity by ingestion: Grade 0; LD_{50} = > 15 g/kg.
Long-term health effects: May cause dermatitis; chronic lung irritation.
Vapor (gas) irritant characteristics: Eye and respiratory tract irritant.
Liquid or solid irritant characteristics: Irritates eyes, skin, and mucous membranes.

FIRE DATA
Fire extinguishing agents not to be used: Do not use agents containing mono-ammonium phosphate for other ammonium compounds.

CHEMICAL REACTIVITY
Reactivity with water: Reaction releases chlorine gas.
Binary reactants: May cause fire in contact with wood, cloth, straw, or other combustible materials. Corrosive to most metals. Forms explosive mixture with ammonia and amines. Reacts with acids, organic sulfides, sulfur, nitrogen containing compounds, phenol, diethylene glycol, glycerol, thiols. Slowly decomposes in air.
Neutralizing agents for acids and caustics: Dilute with water.

ENVIRONMENTAL DATA
Water pollution: Harmful to aquatic life in very low concentrations. May be dangerous if it enters nearby water intakes; notify operators. Notify local health and wildlife officials. **Response to discharge:** Issue warning–corrosive. Disperse and flush.

SHIPPING INFORMATION
Grades of purity: 70% (self-propagating); 65% (nonpropagating); **Stability during transport:** Stable if kept cool and out of direct sunlight. The 70% grade may decompose violently if exposed to heat or direct sunlight. Gives off chlorine and chlorine monoxide above 350°F/177°C.

PHYSICAL AND CHEMICAL PROPERTIES
Physical state @ 59°F/15°C and 1 atm: Oxidizer
Molecular weight: 174.98
Melting/Freezing point: Decomposes 350°F/177°C/450°K.
Specific gravity (water = 1): 2.35 @ 68°F/20°C (solid).

CALCIUM NITRATE   REC. C:1050

SYNONYMS: CALCIUM NITRATE; CALCIUM(II) NITRATE; NITRIC ACID, CALCIUM SALT; LIME SALTPETER; NORWEGIAN SALTPETER; NITRATO CALCICO (Spanish); NITROCALCITE; CALCIUM NITRATE TETRAHYDRATE

IDENTIFICATION
CAS Number: 10124-37-5
Formula: $Ca(NO_3)_2 \cdot 4H_2O$
DOT ID Number: UN 1454; DOT Guide Number: 140
Proper Shipping Name: Calcium nitrate

DESCRIPTION: Colorless crystalline solid. Odorless. Sinks in water; highly soluble.

Dust is corrosive to skin, eyes, and respiratory tract; inhalation effects may be delayed • Potential explosion hazard. Reacts with organic contaminants forming shock-sensitive mixtures • Strong oxidizer which may react spontaneously with low flash point organics or reducing agents. Heat forms oxygen; will increase the activity of an existing fire • May cause fire or explosion on contact with combustibles (wood, paper, oil, clothing, etc.) •Toxic products of combustion may include nitrogen oxides.

Hazard Classification (based on NFPA-704 Rating System)
Health Hazards (Blue): 2; Flammability (Red): 0; Reactivity (Yellow): 1; Special Notice (White): OXY

214 Calcium oxide

EMERGENCY RESPONSE: See Appendix A.
Evacuation:
Public safety: Isolate the area of spill or leak for at least 10 to 25 meters (30 to 80 feet) in all directions.
Spill: Consider initial downwind evacuation for at least 100 meters (330 feet).
Fire: If any large container is involved in fire, isolate for at least 800 meters (½ mile) in all directions; also, consider initial evacuation for 800 meters (½ mile) in all directions.

EXPOSURE
Short-term effects: *SEEK MEDICAL ATTENTION*. **Dust:** Irritating to eyes, nose, and throat. Move victim to fresh air. *IF IN EYES*, hold eyelids open and flush with plenty of water. **Solid:** Irritating to skin and eyes. Poisonous if swallowed. Remove contaminated clothing and shoes. Flush affected areas with plenty of water. *IF IN EYES*, hold eyelids open and flush with plenty of water. *IF SWALLOWED* and victim is *CONSCIOUS AND ABLE TO SWALLOW*, have victim drink 4 to 8 ounces of water.
Note to physician or authorized medical personnel: Medical observation is recommended for 24 to 48 hours after breathing overexposure, as pulmonary edema may be delayed. As first aid for pulmonary edema, consider administering a corticosteroid spray. Cigarette smoking may exacerbate pulmonary injury and should be discouraged for at least 72 hours following exposure.

HEALTH HAZARDS
Personal protective equipment (PPE): B-Level PPE. Face shield, dust respirator and rubber gloves.
Long-term health effects: Ingestion may cause blood disorders (formation of methemoglobin).
Vapor (gas) irritant characteristics: Eye and respiratory tract irritant.
Liquid or solid irritant characteristics: Eye and skin irritant.

FIRE DATA
Flash point: Non combustible, but will support combustion by liberation of oxygen.

CHEMICAL REACTIVITY
Binary reactants: Contact with organic materials may form shock-sensitive explosive material. Contact with combustible material may cause fire. Corrosive to many substances including aluminum, cyanides, esters, phosphorus, sodium hypophosphate, thyocyanates.

ENVIRONMENTAL DATA
Food chain concentration potential: Negative; unlikely to accumulate.
Water pollution: Dangerous to aquatic life in high concentrations. May be dangerous if it enters nearby water intakes; notify operators. Notify local health and wildlife officials. **Response to discharge:** Issue warning–oxidizing material. Disperse and flush.

SHIPPING INFORMATION
Grades of purity: Analytical reagent (99.0+%); purified; technical; **Storage temperature:** Ambient; **Inert atmosphere:** None; **Venting:** Ambient; **Stability during transport:** Stable.

PHYSICAL AND CHEMICAL PROPERTIES
Physical state @ 59°F/15°C and 1 atm: Solid.
Molecular weight: 164
Boiling point @ 1 atm: Decomposes.
Melting/Freezing point: 1042°F/561°C/834°K.
Specific gravity (water = 1): 2.50 at 18°C (solid).

Heat of solution: (estimate) –90 Btu/lb = –50 cal/g = –2.1 x 10^5 J/kg.
Heat of fusion: 31.2 cal/g.

CALCIUM OXIDE REC. C:1100

SYNONYMS: BURNED LIME; BURNT LIME; CALCIA; CALX; LIME; LIME, BURNED; LIME, BURNT; LIME, UNSLAKED; OXIDO CALCICO (Spanish); OXYDE de CALCIUM (French); PEBBLE LIME; QUICKLIME; UNSLAKED LIME

IDENTIFICATION
CAS Number: 1305-78-8
Formula: CaO
DOT ID Number: UN 1910; DOT Guide Number: 157
Proper Shipping Name: Calcium oxide

DESCRIPTION: White to gray lumps or granular powder. Odorless. Sinks and reacts violently with water. Reacts violently with water (appears to boil) generating heat; forming caustic calcium hydroxide; powder may cause explosion.

Corrosive to the skin, eyes or respiratory tract; inhalation can cause lung edema; effects of inhalation may be delayed • Firefighting gear (including SCBA) does not provide adequate protection. If exposure occurs, remove and isolate gear immediately and thoroughly decontaminate personnel • Strong oxidizer which may react spontaneously with low flash point organics or reducing agents. Heat forms oxygen; will increase the activity of an existing fire • May cause fire or explosion contact with combustibles (wood, paper, oil, clothing, etc.) • Combustion products include corrosive gases.

Hazard Classification (based on NFPA-704 Rating System)
Health Hazards (Blue): 1; Flammability (Red): 0; Reactivity (Yellow): 0

EMERGENCY RESPONSE: See Appendix A (140)
Evacuation:
Public safety: Isolate the area of spill or leak for at least 50 to 100 meters (160 to 330 feet) in all directions.
Spill: Increase, in the downwind direction, as necessary, the distance shown under "Public Safety."
Fire: If tank, rail car, or tank truck is involved in fire, isolate for at least 800 meters (½ mile) in all directions; also, consider initial evacuation for 800 meters (½ mile) in all directions.

EXPOSURE
Short-term effects: *SEEK MEDICAL ATTENTION*. **Dust:** Irritating to nose and throat. May cause lung edema; physical exertion will aggravate this condition. Move to fresh air. **Solid:** Will burn skin and eyes. Harmful if swallowed. Remove contaminated clothing and shoes. Flush affected areas with plenty of water. *IF IN EYES*, hold eyelids open and flush with plenty of water. *IF SWALLOWED* and victim is *CONSCIOUS AND ABLE TO SWALLOW*, have victim drink 4 to 8 ounces of water. **Do NOT induce vomiting.**
Note to physician or authorized medical personnel: Medical observation is recommended for 24 to 48 hours after breathing overexposure, as pulmonary edema may be delayed. As first aid for pulmonary edema, consider administering a corticosteroid spray. Cigarette smoking may exacerbate pulmonary injury and should be discouraged for at least 72 hours following exposure.

HEALTH HAZARDS

Personal protective equipment (PPE): OSHA Table Z-1-A air contaminant. B-Level PPE. Protective gloves, close fitting safety goggles, and type of respirator prescribed for fine dust.
Recommendations for respirator selection: NIOSH
10 mg/m^3: DM (any dust and mist respirator). *20 mg/m^3*: DMXSQ (any dust and mist respirator except single-use and quarter mask respirators); or SA (any supplied-air respirator). *25 mg/m^3*: SA:CF (any supplied-air respirator operated in a continuous-flow mode); or PAPRHiE (any powered, air-purifying respirator with a high-efficiency particulate filter); or HiEF (any air-purifying, full-facepiece respirator with a high-efficiency particulate filter); or SCBAF (any self-contained breathing apparatus with a full facepiece); or SAF (any supplied-air respirator with a full facepiece). *EMERGENCY OR PLANNED ENTRY INTO UNKNOWN CONCENTRATIONS OR IDLH CONDITIONS:* SCBAF:PD,PP (any self-contained breathing apparatus that has a full facepiece and is operated in a pressure-demand or other positive-pressure mode); or SAF:PD,PP:ASCBA (any supplied-air respirator that has a full facepiece and is operated in a pressure-demand or other positive-pressure mode in combination with an auxiliary self-contained breathing apparatus operated in a pressure-demand or other positive pressure mode). *ESCAPE*: HiEF (any air-purifying, full-facepiece respirator with a high-efficiency particulate filter); or SCBAE (any appropriate escape-type, self-contained breathing apparatus).
Exposure limits (TWA unless otherwise noted): ACGIH TLV 2 mg/m^3; OSHA 5 mg/m^3; NIOSH 2 mg/m^3.
Vapor (gas) irritant characteristics: Eye and respiratory tract.
Liquid or solid irritant characteristics: Eye and skin irritant. Causes smarting of the skin and first-degree burns on short exposure and may cause secondary burns on long exposure.
IDLH value: 25 mg/m^3.

FIRE DATA

Fire extinguishing agents not to be used: Do not use halogenated chemical or CO_2.

CHEMICAL REACTIVITY

Reactivity with water: Water liberates heat; heat may cause ignition of combustibles. Material swells during reaction; forms highly irritating calcium hydroxide.
Binary reactants: Reacts with acids, fluorine, phosphorus oxide, ethanol, acids, light.

ENVIRONMENTAL DATA

Water pollution: Harmful to aquatic life in very low concentrations. May be dangerous if it enters nearby water intakes; notify operators. Notify local health and wildlife officials.
Response to discharge: Issue warning–corrosive. Restrict access. Chemical and physical treatment.

SHIPPING INFORMATION

Grades of purity: 96–97%; **Stability during transport:** Stable.

PHYSICAL AND CHEMICAL PROPERTIES

Physical state @ 59°F/15°C and 1 atm: Solid.
Molecular weight: 56.08
Boiling point @ 1 atm: 5162°F/2850° = 3123°K.
Melting/Freezing point: 4737°F/2614°C/2887°K.
Specific gravity (water = 1): 3.34 @ 68°F/20°C (solid).
Heat of fusion: 218.1 cal/g.
Vapor pressure: 0 mm (approximate).

CALCIUM PEROXIDE REC. C:1150

SYNONYMS: CALCIUM DIOXIDE; CALCIUM SUPEROXIDE; PEROXIDO CALCICO (Spanish)

IDENTIFICATION

CAS Number: 1305-79-9
Formula: CaO_2
DOT ID Number: UN 1457; DOT Guide Number: 140
Proper Shipping Name: Calcium peroxide

DESCRIPTION: Yellow to white powder. Odorless. Sinks in water.

Strong oxidizer which may react spontaneously with low flash point organics or reducing agents. Heat forms oxygen; will increase the activity of an existing fire • May cause fire or explosion on contact with combustibles (wood, paper, oil, clothing, etc.).

Hazard Classification (based on NFPA-704 Rating System)
Health Hazards (Blue): 1; Flammability (Red): 0; Reactivity (Yellow): 1; Special Notice (White): OXY

EMERGENCY RESPONSE: See Appendix A (140)
Evacuation:
Public safety: Isolate the area of spill or leak for at least 10 to 25 meters (30 to 80 feet) in all directions.
Spill: Consider initial downwind evacuation for at least 100 meters (330 feet).
Fire: If any large container is involved in fire, isolate for at least 800 meters (½ mile) in all directions; also, consider initial evacuation for 800 meters (½ mile) in all directions.

EXPOSURE

Short-term effects: *SEEK MEDICAL ATTENTION*. **Dust:** Irritating to eyes, nose, and throat. Move victim to fresh air. *IF IN EYES*, hold eyelids open and flush with plenty of water. **Solid:** Irritating to skin and eyes. Harmful if swallowed. Remove contaminated clothing and shoes. Flush affected areas with plenty of water. *IF IN EYES*, hold eyelids open and flush with plenty of water. *IF SWALLOWED* and victim is *CONSCIOUS AND ABLE TO SWALLOW*, have victim drink 4 to 8 ounces of water.

HEALTH HAZARDS

Personal protective equipment (PPE): A-Level PPE. Approved respirator toxic dust respirator; general-purpose gloves; chemical safety goggles; full cover clothing. Butyl rubber is generally suitable for peroxide compounds.
Vapor (gas) irritant characteristics: Eye and respiratory tract irritant.
Liquid or solid irritant characteristics: Eye and skin irritant.

FIRE DATA, but may cause fire on contact with combustible material.
Behavior in fire: Can increase severity of fire. Containers may explode.

CHEMICAL REACTIVITY

Reactivity with water: Reacts very slowly with water at room temperature to form limewater and oxygen gas.
Binary reactants: Cellulose; hydrogen peroxide. Heavy metals and dirt can accelerate decomposition to lime and oxygen. Reacts with reducing agents, organic material, thiocyanates.
Neutralizing agents for acids and caustics: Flush with water.

ENVIRONMENTAL DATA
Food chain concentration potential: Negative; unlikely to accumulate.
Water pollution: Effect of low concentrations on aquatic life is unknown. May be dangerous if it enters nearby water intakes; notify operators. Notify local health and wildlife officials.
Response to discharge: Issue warning–oxidizing material. Disperse and flush.

SHIPPING INFORMATION
Grades of purity: Commercial: 60+%; **Storage temperature:** Stable; **Inert atmosphere:** None; **Venting:** Pressure-vacuum; **Stability during transport:** Stable.

PHYSICAL AND CHEMICAL PROPERTIES
Physical state @ 59°F/15°C and 1 atm: Solid.
Molecular weight: 72.1
Boiling point @ 1 atm: Decomposes.
Specific gravity (water = 1): 2.92 @ 77°F/25°C (solid).
Heat of decomposition: -135 Btu/lb = -75 cal/g = -3.1×10^5 J/kg.

CALCIUM PHOSPHATE   REC. C:1200

SYNONYMS: FOSFATO CALCICO (Spanish); MCP; MONOCALCIUM PHOSPHATE MONOHYDRATE; CALCIUM BIPHOSPHATE; DCP; DICALCIUM PHOSPHATE; CALCIUM PYROPHOSPHATE; CALCIUM SUPERPHOSPHATE

IDENTIFICATION
CAS Number: 10103-46-5; 7757-93-9 (dibasic); 7758-23-8 (monobasic)
Formula: $CaH_4(PO_4)_2 \cdot H_2O$; $Ca_2P_2O_7$

DESCRIPTION: White solid. Odorless. Sinks and mixes with water.

Hazard Classification (based on NFPA-704 Rating System)
Health Hazards (Blue): 0; Flammability (Red): 0; Reactivity (Yellow): 0

EMERGENCY RESPONSE: See Appendix A (140)
Evacuation:
Public safety: Isolate the area of spill or leak for at least 10 to 25 meters (30 to 80 feet) in all directions.
Spill: Increase, in the downwind direction, as necessary, the distance shown under "Public Safety."
Fire: If any large container is involved in fire, isolate for at least 800 meters (½ mile) in all directions; also, consider initial evacuation for 800 meters (½ mile) in all directions.

EXPOSURE
Short-term effects: *SEEK MEDICAL ATTENTION.* **Dust:** Irritating to eyes, nose, and throat. *IF INHALED*, will, will cause coughing or difficult breathing. *IF IN EYES*, hold eyelids open and flush with plenty of water. *IF BREATHING HAS STOPPED*, give artificial respiration. IF breathing is difficult, administer oxygen.
Solid: Irritating to skin and eyes. *IF SWALLOWED*, will cause nausea and vomiting. Remove contaminated clothing and shoes. Flush affected areas with plenty of water. *IF IN EYES*, hold eyelids open and flush with plenty of water. *IF SWALLOWED* and victim is *CONSCIOUS AND ABLE TO SWALLOW*, have victim drink 4 to 8 ounces of water and have victim induce vomiting. *IF SWALLOWED* and victim is *UNCONSCIOUS OR HAVING CONVULSIONS,* do nothing except keep victim warm.

HEALTH HAZARDS
Personal protective equipment (PPE): CA-Level PPE. Dust mask, goggles, and gloves.
Toxicity by ingestion: Grade 0; LD_{50} = > 15 g/kg.
Vapor (gas) irritant characteristics: Eye and respiratory tract irritant.
Liquid or solid irritant characteristics: Eye and skin irritant.

CHEMICAL REACTIVITY
Binary reactants: Some calcium phosphates form acid solutions in water. These may attack metals with formation of flammable hydrogen gas, which may collect in enclosed spaces.
Neutralizing agents for acids and caustics: Flush with water.

ENVIRONMENTAL DATA
Food chain concentration potential: Negative; unlikely to accumulate.
Water pollution: Effect of low concentrations on aquatic life is unknown. May be dangerous if it enters nearby water intakes; notify operators. Notify local health and wildlife officials. **Response to discharge:** Disperse and flush.

SHIPPING INFORMATION
Grades of purity: NF; USP; Dentrifrice; Reagent; **Storage temperature:** Ambient; **Inert atmosphere:** None; **Venting:** Open; **Stability during transport:** Stable.

PHYSICAL AND CHEMICAL PROPERTIES
Physical state @ 59°F/15°C and 1 atm: Solid.
Molecular weight: Monocalcium phosphate: 252.16; Dicalcium phosphate: 136.06; Calcium pyrophosphate: 254
Specific gravity (water = 1): 2-3 @ 68°F/20°C (solid).

CALCIUM PHOSPHIDE   REC. C:1250

SYNONYMS: EEC No. 015-003-00-2; FOSFURO CALCICO (Spanish); PHOTOPHOR; TRICALCIUM DIPHOSPHIDE

IDENTIFICATION
CAS Number: 1305-99-3
Formula: Ca_3P_2
DOT ID Number: UN 1360; DOT Guide Number: 139
Proper Shipping Name: Calcium phosphide

DESCRIPTION: Red-brown crystalline solid or gray lumps. Musty odor, like acetylene. Reacts violently with water forming toxic and flammable phosphine and phosphine dimer; risk of fire and explosions.

Extremely flammable • Poison • Breathing the vapors can kill • skin and eye contact causes severe burns and blindness • Ignites spontaneously in moist air (forms phosphine) • Do not use water or water-based extinguishers • Firefighting gear (including SCBA) does not provide adequate protection. If exposure occurs, remove and isolate gear immediately and thoroughly decontaminate personnel • Vapors are heavier than air and will collect and stay in low areas • Vapors may travel long distances to source of ignition and flashback • Often shipped and stored dissolved in acetone • Exposure of cylinders to elevated temperatures, fire, and flame may cause cylinders to rupture or cause frangible disk to burst, releasing entire contents of cylinder • Ruptured or venting cylinders may rocket through buildings and/or travel a considerable distance • When combined with surface moisture this material is corrosive to most common metals • Dense smoke and toxic phosphine fumes may be formed in fires. Phosphine fumes may ignite. Toxic products of combustion may also include hydrogen chloride and phosphorus oxides • Severely irritating to skin, eyes, and respiratory tract; prolonged contact with the skin causes burns • Do not put yourself in danger by entering a contaminated area to rescue a victim.

Hazard Classification (based on NFPA-704 Rating System)
Health Hazards (Blue): 4; Flammability (Red): 0; Reactivity (Yellow): 2; Special Notice (White): Water reactive

EMERGENCY RESPONSE: See Appendix A (139)
Evacuation:
Public safety: Isolate the area of spill or leak for at least 100 to 150 meters (330 to 490 feet) in all directions.
Spills: *IF SPILLED IN WATER*: Small spill–First: Isolate in all directions 30 meters (100 feet); Then: Protect persons downwind, DAY: 0.2 km (0.1 mile); NIGHT: 0.8 km (0.5 mile). Large spill–First: Isolate in all directions 215 meters (700 feet); Then: Protect persons downwind, DAY: 2.1 km (1.3 miles); NIGHT: 5.3 km (3.3 mile).
Fire: If any large container is involved in fire, isolate for at least 800 meters (½ mile) in all directions; also, consider initial evacuation for 800 meters (½ mile) in all directions.

EXPOSURE
Short-term effects: *SEEK MEDICAL ATTENTION. VAPOR PRODUCED IN REACTION WITH WATER: POISONOUS IF INHALED.* Irritating to eyes, nose, and throat. Move to fresh air. *IF BREATHING HAS STOPPED,* give artificial respiration. IF breathing is difficult, administer oxygen. **Dust:** Irritating to eyes. Move victim to fresh air. *IF IN EYES*, hold eyelids open and flush with plenty of water. **Solid:** Irritating to skin and eyes. Harmful if swallowed. Remove contaminated clothing and shoes. Flush affected areas with plenty of water. *IF IN EYES*, hold eyelids open and flush with plenty of water. *IF SWALLOWED* and victim is *CONSCIOUS AND ABLE TO SWALLOW*, have victim drink 4 to 8 ounces of water.

HEALTH HAZARDS
Personal protective equipment (PPE): A-Level PPE. Dust respirator; protective gloves and clothing; face shield or eye protection in combination with BREATHING protection.
Toxicity by ingestion: Poisonous
Long-term health effects: Liver, heart and kidney injury may occur. The substance effects the body's metabolism. Serious exposure may result in death.
Vapor (gas) irritant characteristics: Eye and respiratory tract irritant.
Liquid or solid irritant characteristics: Eye and skin irritant.
Odor threshold: 1–100 mg/m^3.

FIRE DATA but may ignite spontaneously if wet.
Fire extinguishing agents not to be used: Water, foam, water-based extinguishing agents.

CHEMICAL REACTIVITY
Reactivity with water: Reacts vigorously, generating phosphine (a poisonous, spontaneously flammable gas).
Binary reactants: Reacts with oxidizers, chlorine, sulfur. Can react with surface moisture to evolve phosphine, which is toxic and spontaneously flammable; with oxygen at elevated temperatures.

ENVIRONMENTAL DATA
Food chain concentration potential: Negative; unlikely to accumulate.
Water pollution: Effect of low concentrations on aquatic life is unknown. May be dangerous if it enters nearby water intakes; notify operators. Notify local health and wildlife officials. **Response to discharge:** Issue warning–high flammability, poison, air contaminant, water contaminant. Restrict access. Evacuate area (large discharge only). Should be removed.

SHIPPING INFORMATION
Grades of purity: Commercial; **Storage temperature:** Ambient; **Inert atmosphere:** None; **Venting:** Sealed containers must be in well-ventilated area; **Stability during transport:** Stable if dry.

PHYSICAL AND CHEMICAL PROPERTIES
Physical state @ 59°F/15°C and 1 atm: Solid.
Molecular weight: 182.2
Boiling point @ 1 atm: Decomposes.
Melting/Freezing point: (approximate) 2910°F/1600°C/1870°K.
Specific gravity (water = 1): 2.51 @ 68°F/20°C (solid).

CALCIUM RESINATE   **REC. C:1300**

SYNONYMS: CALCIUM ABIETATE; CALCIUM ROSIN; CALCIUM ABIETATE; CALCIUM LIMED WOOD ROSIN; LIMED ROSIN; METALLIC RESINATE; RESINATO CALCICO (Spanish)

IDENTIFICATION
CAS Number: 9007-13-0
Formula: Ca(C$_{44}$H$_6$2O$_4$)$_2$
DOT ID Number: UN 1313; DOT Guide Number: 133
Proper Shipping Name: Calcium resinate

DESCRIPTION: Yellow to very dark brown solid. Odorless. Sinks in water.

Flammable • Containers may BLEVE when exposed to fire • Vapors may form explosive mixture with air • Vapors are heavier than air and will collect and stay in low areas • Vapors may travel long distances to ignition sources and flashback • Vapors in confined areas (e.g., tanks, sewers, buildings) may explode when exposed to fire • Irritating to the skin, eyes, and respiratory tract • Toxic products of combustion may include carbon monoxide.

Hazard Classification (based on NFPA-704 Rating System)
Health Hazards (Blue): 0; Flammability (Red): 2; Reactivity (Yellow): 1

EMERGENCY RESPONSE: See Appendix A (133)
Evacuation:
Public safety: Isolate spill area for at least 10 to 25 meters (30 to 80 feet) in all directions.
Spill: Consider initial downwind evacuation of at least 100 meters (330 feet).
Fire: Isolate for 800 meters (½ mile) in all directions, especially if tank, rail car, or tank truck is involved in fire.

EXPOSURE
Short-term effects: *SEEK MEDICAL ATTENTION.* **Dust:** Irritating to eyes, nose, and throat. *IF INHALED*, will, will cause coughing or difficult breathing. *IF IN EYES*, hold eyelids open and flush with plenty of water. *IF BREATHING HAS STOPPED*, give

artificial respiration. IF breathing is difficult, administer oxygen.
Solid: Harmful if swallowed. Remove contaminated clothing and shoes. Flush affected areas with plenty of water. *IF IN EYES*, hold eyelids open and flush with plenty of water. *IF SWALLOWED* and victim is *CONSCIOUS AND ABLE TO SWALLOW*, have victim drink 4 to 8 ounces of water and have victim induce vomiting. *IF SWALLOWED* and victim is *UNCONSCIOUS OR HAVING CONVULSIONS,* do nothing except keep victim warm.

HEALTH HAZARDS
Personal protective equipment (PPE): B-Level PPE. Dust mask; goggles or face shield; gloves.
Vapor (gas) irritant characteristics: Eye and respiratory tract irritation.
Liquid or solid irritant characteristics: Eye and skin irritant.

FIRE DATA
Flash point: Combustible solid.
Fire extinguishing agents not to be used: Water.
Autoignition temperature: 480°F/249°C/522°K (may ignite spontaneously).

CHEMICAL REACTIVITY
Binary reactants: Oxidizing materials.

ENVIRONMENTAL DATA
Food chain concentration potential: Negative; unlikely to accumulate.
Water pollution: Effect of low concentrations on aquatic life is unknown. May be dangerous if it enters nearby water intakes; notify operators. Notify local health and wildlife officials.
Response to discharge: Should be removed. Chemical and physical treatment.

SHIPPING INFORMATION
Grades of purity: Commercial; **Storage temperature:** Ambient; **Inert atmosphere:** None; **Venting:** Open; **Stability during transport:** Stable.

PHYSICAL AND CHEMICAL PROPERTIES
Physical state @ 59°F/15°C and 1 atm: Solid.
Molecular weight: 643 (approximate).
Boiling point @ 1 atm: More than 600°F/316°C/589°K.
Specific gravity (water = 1): 1.13 @ 77°F/25°C (solid).

CAMPHENE   REC. C:1350

SYNONYMS: CANFENO (Spanish); 2,2-DIMETHYL-3-METHYLENE-; BICYCLO-(2.2.1)HEPTANE; 2-2-DIMETHYL-3-METHYLENE NORBORANE; 3,3-DIMETHYLENENORCAMPHENE; 3,3-DIMETHYL-2-METHYLENE NORCAMPHANE

IDENTIFICATION
CAS Number: 79-92-5
Formula: $C_{10}H_{16}$
DOT ID Number: UN 9011; DOT Guide Number: 133
Proper Shipping Name: Camphene

DESCRIPTION: White solid. Camphor-like odor. Floats on the surface of water.

Flammable • Containers may BLEVE when exposed to fire • Vapors may form explosive mixture with air • Vapors are heavier than air and will collect and stay in low areas • Vapors may travel long distances to ignition sources and flashback • Vapors in confined areas (e.g., tanks, sewers, buildings) may explode when exposed to fire • Irritating to the skin, eyes, and respiratory tract • Toxic products of combustion may include carbon monoxide.

Hazard Classification (based on NFPA-704 Rating System)
Health Hazards (Blue): 2; Flammability (Red): 2; Reactivity (Yellow): 0

EMERGENCY RESPONSE: See Appendix A (133)
Evacuation:
Public safety: Isolate spill area for at least 10 to 25 meters (30 to 80 feet) in all directions.
Spill: Consider initial downwind evacuation of at least 100 meters (330 feet).
Fire: Isolate for 800 meters (½ mile) in all directions, especially if tank, rail car, or tank truck is involved in fire.

EXPOSURE
Short-term effects: *SEEK MEDICAL ATTENTION.* **Dust:** Irritating to eyes, nose, and throat. *IF INHALED*, will, will cause headache or difficult breathing. *IF IN EYES*, hold eyelids open and flush with plenty of water. *IF BREATHING HAS STOPPED,* give artificial respiration. IF breathing is difficult, administer oxygen.
Solid: Irritating to skin and eyes. Harmful if swallowed. Remove contaminated clothing and shoes. Flush affected areas with plenty of water. *IF IN EYES*, hold eyelids open and flush with plenty of water. *IF SWALLOWED* and victim is *CONSCIOUS AND ABLE TO SWALLOW*, have victim drink 4 to 8 ounces of water. *IF SWALLOWED* and victim is *UNCONSCIOUS OR HAVING CONVULSIONS,* do nothing except keep victim warm.

HEALTH HAZARDS
Personal protective equipment (PPE): B-Level PPE. Gloves and face shield.
Long-term health effects: Mutation data reported.
Vapor (gas) irritant characteristics: Eye and respiratory tract irritant.
Liquid or solid irritant characteristics: Eye and skin irritant.

FIRE DATA
Flash point: 108°F/42°C (oc); 92°F/33°C (cc) (typical).
Fire extinguishing agents not to be used: Water.
Behavior in fire: Vapor may react with oxidizing materials.

CHEMICAL REACTIVITY
Binary reactants: Oxidizing materials

ENVIRONMENTAL DATA
Food chain concentration potential: Negative; unlikely to accumulate.
Water pollution: Effect of low concentrations on aquatic life is unknown. May be dangerous if it enters nearby water intakes; notify operators. Notify local health and wildlife officials. **Response to discharge:** Mechanical containment. Should be removed. Chemical and physical treatment.

SHIPPING INFORMATION
Grades of purity: Commercial, 75+%; **Storage temperature:** Ambient; **Inert atmosphere:** None; **Venting:** Open; **Stability during transport:** Stable.

PHYSICAL AND CHEMICAL PROPERTIES
Physical state @ 59°F/15°C and 1 atm: Solid.

Molecular weight: 136
Boiling point @ 1 atm: 310°F/154°C/427°K.
Melting/Freezing point: 122°F/50°C/323°K.
Specific gravity (water = 1): 0.87 @ 15°C (solid).
Heat of combustion: $-19,400$ Btu/lb $= -10,800$ cal/g $= -452 \times 10^5$ J/kg.

CAMPHOR OIL   REC. C:1400

SYNONYMS: ALCANFOR (Spanish); BICYCLO 2.2.1 HEPTAN-2-ONE,1,7,7-TRIMETHYL-; BORNANE, 2-oxo-; 2-BORNANONE; 2-CAMPHANONE; 2-CAMPHORONE; CAMPHOR, NATURAL; FORMOSA CAMPHOR; GUM CAMPHOR; HUILE de CAMPHRE (French); JAPAN CAMPHOR; KAMPFER (German); 2-KETO-1,7,7-TRIMETHYLNORCAMPHANE; LAUREL CAMPHOR; LIQUID CAMPHOR; LIQUID IMPURE CAMPHOR; LIQUID GUM CAMPHOR; MATRICARIA CAMPHOR; NORCAMPHOR, SYNTHETIC CAMPHOR; 1,7,7-TRIMETHYL-; 1,7,7-TRIMETHYLBICYCLO(2.2.1)-2-HEPTANONE; 1,7,7-TRIMETHYLBICYCLO(2,2,1)HEPTANONE-2; 1,7,7-TRIMETHYLNORCAMPHOR; WHITE CAMPHOR OIL; WHITE OIL OF CAMPHOR

IDENTIFICATION
CAS Number: 8008-51-3; 76-22-2
Formula: $C_{10}H_{16}O$
DOT ID Number: UN 1130; DOT Guide Number: 128
Proper Shipping Name: Camphor oil

DESCRIPTION: Colorless or brown or blue oily liquid. Penetrating camphor odor. Usually Floats on the surface of water; practically insoluble.

Flammable • Containers may BLEVE when exposed to fire •Vapors may form explosive mixture with air • Vapors are heavier than air and will collect and stay in low areas • Vapors may travel long distances to ignition sources and flashback • Vapors in confined areas (e.g., tanks, sewers, buildings) may explode when exposed to fire • Irritating to the skin, eyes, and respiratory tract • Toxic products of combustion may include carbon monoxide and soot.

Hazard Classification (based on NFPA-704 Rating System)
Health Hazards (Blue): 2; Flammability (Red): 2; Reactivity (Yellow): 0

EMERGENCY RESPONSE: See Appendix A (128)
Evacuation:
Public safety: Isolate spill area for at least 25 to 50 meters (80 to 160 feet) in all directions.
Spill: Large spill–Consider initial downwind evacuation for at least 300 meters (1000 feet).
Fire: Isolate for 800 meters (½ mile) in all directions, especially if tank, rail car, or tank truck is involved in fire.

EXPOSURE
Short-term effects: *SEEK MEDICAL ATTENTION.* **Vapor:** Not irritating to eyes, nose or throat. Move to fresh air. **Liquid:** Irritating to skin and eyes. *IF SWALLOWED*, will cause nausea, vomiting, or loss of consciousness. Remove contaminated clothing and shoes. Flush affected areas with plenty of water. *IF IN EYES*, hold eyelids open and flush with plenty of water. *IF SWALLOWED* and victim is *CONSCIOUS AND ABLE TO SWALLOW*, have victim drink 4 to 8 ounces of water.

HEALTH HAZARDS
Personal protective equipment (PPE): Synthetic camphor is an OSHA Table Z-1-A air contaminant. B-Level PPE (camphor oil); . A-Level PPE (camphor).
Recommendations for respirator selection: NIOSH/OSHA *50 mg/m³:* SA:CF (any supplied-air respirator operated in a continuous-flow mode); PAPROVDM [any powered, air-purifying respirator with organic vapor cartridge(s) in combination with a dust and mist filter]. *100 mg/m³:* CCRFOVHiE [any chemical cartridge respirator with a full facepiece and organic vapor cartridge(s) in combination with a high efficiency partiulate filter); or GMFOVHiE [any air-purifying, full-facepiece respirator (gas mask) with a chin-style, front- or back-mounted organic vapor canister having a high-efficiency particulate filter]; or PAPRTOVHiE [any powered, air-purifying respirator with a tight-fitting facepiece and organic vapor cartridge(s) in combonation with a high-efficiency particulate filter]; or SCBAF (any self-contained breathing apparatus with a full facepiece); or SAF (any supplied-air respirator with a full facepiece); *200 mg/m³:* SAF:PD,PP (any supplied-air respirator that has a full facepiece and is operated in a pressure-demand or other positive-pressure mode). *EMERGENCY OR PLANNED ENTRY INTO UNKNOWN CONCENTRATIONS OR IDLH CONDITIONS:* SCBAF:PD,PP (any self-contained breathing apparatus that has a full facepiece and is operated in a pressure-demand or other positive-pressure mode); or SAF:PD,PP:ASCBA (any supplied-air respirator that has a full facepiece and is operated in a pressure-demand or other positive-pressure mode in combination with an auxiliary self-contained breathing apparatus operated in a pressure-demand or other positive pressure mode). *ESCAPE:* GMFOVHiE [any air-purifying, full-facepiece respirator (gas mask) with a chin-style, front- or back-mounted organic vapor canister having a high-efficiency particulate filter]; or SCBAE (any appropriate escape-type, self-contained breathing apparatus). *Note*: Substance causes eye irritation or damage; eye protection needed.
Exposure limits (TWA unless otherwise noted): ACGIH TLV 2 ppm (12 mg/m³) (synthetic); NIOSH/OSHA 2 mg/m³ (synthetic).
Short-term exposure limits (15-minute TWA): ACGIH STEL 4 ppm (19 mg/m³) (synthetic).
Toxicity by ingestion: Poisonous. May cause tremors, convulsions and respiratory problems.
Vapor (gas) irritant characteristics: Eye and respiratory tract irritant.
Liquid or solid irritant characteristics: Eye and skin irritant. If spilled on clothing and allowed to remain, may cause smarting and reddening of the skin.
Odor threshold: 0.079 ppm.
IDLH value: 200 mg/m³ as camphor (synthetic).

FIRE DATA
Flash point: 117°F/47°C (cc); (synthetic camphor) 150°F/66°C (cc).
Flammable limits in air: LEL: 0.6%; UEL: 4.5%; synthetic camphor LEL: 0.6%; UEL: 3.5%.
Behavior in fire: The solid often evaporates without first melting.
Autoignition temperature: 870°F/466°C/739°K.
Electrical hazard: Flow or agitation of substance may generate electrostatic charges.

CHEMICAL REACTIVITY
Binary reactants: Oxydizers; especially chromic anhydride and potassium permanganate

Reactivity group: 18
Compatibility class: Ketones

ENVIRONMENTAL DATA
Food chain concentration potential: Negative; unlikely to accumulate.
Water pollution: DOT Appendix B, §172.101–marine pollutant. Effect of low concentrations on aquatic life is unknown. Fouling to shoreline. May be dangerous if it enters nearby water intakes; notify operators. Notify local health and pollution control officials.
Response to discharge: Issue warning–water contaminant. Mechanical containment. Should be removed.

SHIPPING INFORMATION
Grades of purity: Each lot of camphor oil has a unique composition, which varies with the season of the year and the country of origin. At least a dozen grades are known; all have Chinese names. Most camphor sold in the United States is synthetic and is quite pure; **Storage temperature:** Ambient; **Inert atmosphere:** None; **Venting:** Open; **Stability during transport:** Stable.

NAS HAZARD CLASSIFICATION FOR BULK WATER TRANSPORTATION
FIRE: 2
HEALTH: Vapor irritant: 0; Liquid or solid irritant: 1; Poisons: 1
WATER POLLUTION: Human toxicity: 3; Aquatic toxicity: 1; Aesthetic effect: 2
REACTIVITY: Other chemicals: 2; Water: 0; Self-reaction: 0

PHYSICAL AND CHEMICAL PROPERTIES
Physical state @ 59°F/15°C and 1 atm: Liquid.
Molecular weight: 152.3
Boiling point @ 1 atm: 347–392°F/175–200°C/448–473°K.
Melting/Freezing point: 349°F/176°C/449°K; 334–339°F/168–171°C/441–444°K (synthetic grade).
Specific gravity (water = 1): 0.923 @ 77°F/25°C (liquid).
Relative vapor density (air = 1): 5.3
Vapor pressure: 0.2 mm.

CAPROLACTAM   REC. C:1450

SYNONYMS: AMINOCAPROICLACTAM; CAPROLACTAMA (Spanish); 2-KETOHEXAMETHYLENIMINE; 2-OXOHEXAMETHYLENIMINE; HEXAHYDRO-2H-AZEPINE-2-ONE

IDENTIFICATION
CAS Number: 105-60-2
Formula: $C_6H_{11}NO$; $HNCH_2(CH_2)_4CO$
DOT ID Number: UN 3082; DOT Guide Number: 171
Proper Shipping Name: Environmentally hazardous substances, liquid, n.o.s.
Reportable Quantity (RQ): **(CERCLA)** 1 lb/0.454 kg

DESCRIPTION: White crystalline crystalline solid or flakes. Usually stored and transported in a molten state. Mild, unpleasant odor. Sinks and mixes with water.

Combustible solid • Containers may BLEVE when exposed to fire • Irritating to the skin, eyes, and respiratory tract • Thermally unstable above 400°F/205°C (decomposes). Toxic products of combustion may include nitrogen oxides and anhydrous ammonia.
Hazard Classification (based on NFPA-704 Rating System) Health Hazards (Blue): 1; Flammability (Red): 1; Reactivity (Yellow): 0

EMERGENCY RESPONSE: See Appendix A (171)
Evacuation:
Public safety: Isolate the area of spill or leak for at least 10 to 25 meters (30 to 80 feet) in all directions.
Spill: Increase, in the downwind direction, as necessary, the distance shown under "Public Safety."
Fire: If any large container is involved in fire, isolate for at least 800 meters (½ mile) in all directions; also, consider initial evacuation for 800 meters (½ mile) in all directions.

EXPOSURE
Short-term effects: *SEEK MEDICAL ATTENTION.* **Liquid:** Irritating to skin and eyes. Harmful if swallowed. Remove contaminated clothing and shoes. Flush affected areas with plenty of water. *IF IN EYES*, hold eyelids open and flush with plenty of water. *IF SWALLOWED* and victim is *CONSCIOUS AND ABLE TO SWALLOW*, have victim drink 4 to 8 ounces of water.

HEALTH HAZARDS
Personal protective equipment (PPE): OSHA Table Z-1-A air contaminant. B-Level PPE.
Exposure limits (TWA unless otherwise noted): ACGIH TLV 1 mg/m^3 (particulate); 5 ppm (23 mg/m^3) (vapor); NIOSH REL 1 mg/m^3 (dust); 0.22 ppm (1 mg/m^3) (vapor). Significant vapor concentrations would be expected only at elevated temperatures.
Short-term exposure limits (15-minute TWA): ACGIH STEL 3 mg/m^3(particulate); 10 ppm (46 mg/m^3) (vapor); NIOSH STEL 3 mg/m^3 (dust); 0.66 ppm (3 mg/m^3).
Toxicity by ingestion: Moderately toxic. Grade 2; oral rat LD$_{50}$ = 2140 mg/kg.
Long-term health effects: IARC probably noncarcinogenic substance. Mutation data reported.
Vapor (gas) irritant characteristics: Eye and respiratory tract irritant.
Liquid or solid irritant characteristics: Skin and eye irritation.
Odor threshold: 3 mg/m^3.

FIRE DATA
Flash point: 257°F/125°C (oc); 282°F/139°C (cc).
Flammable limits in air: LEL: 1.4%; UEL: 8.0%.
Autoignition temperature: 707°F/375°C/648°K.
Burning rate: 2.4 mm/min

CHEMICAL REACTIVITY
Binary reactants: Contact with strong oxidizers may cause fire and explosion.
Reactivity group: 22
Compatibility class: Caprolactam solution

ENVIRONMENTAL DATA
Food chain concentration potential: Log P$_{ow}$ = –0.19. Unlikely to accumulate.
Water pollution: Effect of low concentrations on aquatic life is unknown. May be dangerous if it enters nearby water intakes; notify operators. Notify local health and wildlife officials. **Response to discharge:** Issue warning–water contaminant. Disperse and flush.

SHIPPING INFORMATION
Grades of purity: 99%; **Storage temperature:** 167°F/75°C; **Inert atmosphere:** Nitrogen cushion; **Venting:** Pressure-vacuum; **Stability during transport:** Stable.

NAS HAZARD CLASSIFICATION FOR BULK WATER TRANSPORTATION
FIRE: 1
HEALTH: Vapor irritant: 0; Liquid or solid irritant: 0; Poisons: 4
WATER POLLUTION: Human toxicity: 2; Aquatic toxicity:–; Aesthetic effect: 0
REACTIVITY: Other chemicals: 2; Water: 0; Self-reaction: 1

PHYSICAL AND CHEMICAL PROPERTIES
Physical state @ 59°F/15°C and 1 atm: Solid.
Molecular weight: 113.2
Boiling point @ 1 atm: 515°F/268°C/541°K.
Melting/Freezing point: 154°F/68°C/341°K.
Critical temperature: 944°F/507°C/780°K.
Critical pressure: 660 psia = 45 atm = 4.6 MN/m^2.
Specific gravity (water = 1): 1.02 at 77°C (liquid).
Liquid surface tension: (estimate) 20 dynes/cm = 0.020 N/m at 77°C.
Liquid water interfacial tension: (estimate) 40 dynes/cm = 0.040 N/m at 77°C.
Relative vapor density (air = 1): 3.9
Latent heat of vaporization: 209 Btu/lb = 116 cal/g = 4.85 x 10^5 J/kg.
Heat of combustion: –13,700 Btu/lb = –7640 cal/g = –320 x 10^5 J/kg.
Heat of polymerization: –324 Btu/lb = –180 cal/g = –7.5 x 10^5 J/kg.
Vapor pressure: (Reid) 0.45 psia; 0.00000008 mm.

CAPTAN   REC. C:1475

SYNONYMS: AMERCIDE; CAPTAF; CAPTAF 85W; CAPTANCAPTENEET 26,538; CAPTANE; CAPTEX; ESSOFUNGICIDE 406; FLIT 406; FUNGUS BAN TYPE II; GLYODEX 3722; LE CAPTANE (French); MERPAN; NCI-0077; ORTHOCIDE; ORTHOCIDE 7.5; ORTHOCIDE 50; ORTHOCIDE 406; SR406; STAUFFER CAPTAN; *N*-TRICHLOROMETHYLMERCAPTO-4-CYCLOHEXENE-1,2-DICARBOXIMIDE; *N*-(TRICHLOROMETHYLMERCAPTO)-δ(SUP 4)-TETRAHYDROPHTHALIMIDE; *N*-TRICHLOROMETHYLTHIOCYCLOHEX-4-ENE-1,2-DICARBOXIMIDE; *N*-TRICHLOROMETHYLTHIO-*cis*-δ(sup4)-CYCLOHEXENE-1,2-DICARBOXIMIDE; *N*-[(TRICHLOROMETHYL)THIO]-4-CYCLOHEXENE-1,2-DICARBOXIMIDE; *N*-[(TRICHLOROMETHYL)THIO]TETRAHYDROPHTHALIMIDE; *N*-TRICHLOROMETHYLTHIO-3*A*,4,7,7A-TETRAHYDROPHTHALIMIDE; VANCIDE; VANCIDE 89; VANGARD K; VONDCAPTAN

IDENTIFICATION
CAS Number: 133-06-2
Formula: $C_9H_8Cl_3NO_2S$
DOT ID Number: UN 3077; DOT Guide Number: 171
Proper Shipping Name: Environmentally hazardous substances, solid, n.o.s.
Reportable Quantity (RQ): **(CERCLA)** 10 lb/4.54 kg

DESCRIPTION: White to buff-brown crystalline solid. Slight odor. Sinks in water; practically insoluble. Reacts with water forming hydrogen chloride.

Poison! (thiophthalimide fungicide) • Combustible solid • Breathing the dust, skin or eye contact can cause irritation; swallowing the material can kill you • Firefighting gear (including SCBA) does not provide adequate protection. If exposure occurs, remove and isolate gear immediately and thoroughly decontaminate personnel • Containers may BLEVE when exposed to fire • Concentrated dust in confined areas (e.g., tanks, sewers, buildings) may explode when exposed to fire • Toxic products of combustion may include sulfur dioxide, nitrogen oxides and hydrogen chloride.

Hazard Classification (based on NFPA-704 Rating System)
Health Hazards (Blue): 3; Flammability (Red): 2; Reactivity (Yellow): 0

EMERGENCY RESPONSE: See Appendix A (171)
Evacuation:
Public safety: Isolate the area of spill or leak for at least 10 to 25 meters (30 to 80 feet) in all directions.
Spill: Increase, in the downwind direction, as necessary, the distance shown under "Public Safety."
Fire: If any large container is involved in fire, isolate for at least 800 meters (½ mile) in all directions; also, consider initial evacuation for 800 meters (½ mile) in all directions.

EXPOSURE
Short-term effects: *SEEK MEDICAL ATTENTION*. **Dust:** *POISONOUS IF INHALED OR IF SKIN IS EXPOSED*. Move victim to fresh air. *IF IN EYES*, hold eyelids open and flush with plenty of water. *IF BREATHING HAS STOPPED*, give artificial respiration. IF breathing is difficult, administer oxygen. **Solid:** *POISONOUS IF SWALLOWED OR IF SKIN IS EXPOSED*. Remove contaminated clothing and shoes. Flush affected areas with plenty of water. *IF IN EYES*, hold eyelids open and flush with plenty of water. *IF SWALLOWED* and victim is *CONSCIOUS AND ABLE TO SWALLOW*, have victim drink 4 to 8 ounces of water and have victim induce vomiting. *IF SWALLOWED* and victim is *UNCONSCIOUS OR HAVING CONVULSIONS*, do nothing except keep victim warm.

HEALTH HAZARDS
Personal protective equipment (PPE): OSHA Table Z-1-A air contaminant. B-Level PPE. Dust mask, rubber gloves, and goggles.
Recommendations for respirator selection: NIOSH
At any concentrations above the NIOSH REL, or where there is no REL, at any detectable concentration: SCBAF:PD,PP (any self-contained breathing apparatus that has a full facepiece and is operated in a pressure-demand or other positive-pressure mode); or SAF:PD,PP:ASCBA (any supplied-air respirator that has a full facepiece and is operated in a pressure-demand or other positive-pressure mode in combination with an auxiliary self-contained breathing apparatus operated in a pressure-demand or other positive pressure mode). *ESCAPE:* GMFOV [any air-purifying, full-facepiece respirator (gas mask) with a chin-style, front-or back-mounted organic vapor canister]; or SCBAE (any appropriate escape-type, self-contained breathing apparatus).
Exposure limits (TWA unless otherwise noted): ACGIH TLV 5 mg/m^3; NIOSH REL 5 mg/m^3; potential human carcinogen; reduce exposure to lowest feasible levels.
Toxicity by ingestion: Toxic oral rat LD$_{50}$ = 15g/kg.
Long-term health effects: Mutation data reported. No human carcinogenic data.
Liquid or solid irritant characteristics: Eye irritant.

CHEMICAL REACTIVITY
Binary reactants: Incompatible with tetraethyl pyrophosphate, parathion, strong alkaline materials (e.g., hydrated lime). Corrosive to metals.

Carbofuran

ENVIRONMENTAL DATA
Food chain concentration potential: Log K_{ow} = 2.4. Unlikely to accumulate.
Water pollution: Harmful to aquatic life in very low concentrations. May be dangerous if it enters nearby water intakes; notify operators. Notify local health and wildlife officials.
Response to discharge: Issue warning–poison, water contaminant. Restrict access. Should be removed. Chemical and physical treatment. Clean water act.

SHIPPING INFORMATION
Grades of purity: Technical: 90-97%; also available as dusts, wettable powders, and aqueous suspension; **Storage temperature:** Ambient; **Inert atmosphere:** None; **Venting:** Open (flame arrester); **Stability during transport:** Stable.

PHYSICAL AND CHEMICAL PROPERTIES
Physical state @ 59°F/15°C and 1 atm: Solid.
Molecular weight: 300.6
Boiling point @ 1 atm: Decomposes.
Melting/Freezing point: 338°F/170°C/443°K.
Specific gravity (water = 1): 1.74 @ 68°F/20°C (solid).
Heat of combustion: (estimate) –7100 Btu/lb = –3,940 cal/g = –165 x 10^5 J/kg.
Vapor pressure: 0 mm (approximate).

CARBOFURAN   REC. C:1500

SYNONYMS: BAY 70143; CARBAMIC ACID, METHYL-, 2,2-DIMETHYL-2,3-DIHYDROBENZOFURAN-7-YL ESTER; CARBOFURANO (Spanish); CARBOSIP 5G; CURATERR; D 1221; NEX; CRISFURAN; 2,3-DIHYDRO-2,2-DIMETHYLBENZOFURANYL-7-N-METHYLCARBAMATE; 2,3-DIHYDRO-2,2-DIMETHYL-7-BENZOFURANYL METHYLCARBAMATE; 2,2-DIMETHYL-2,2-DIHYDROBENZOFURANYL-7N-METHYLCARBAMATE; FMC 10242; FURADAN; FURODAN; METHYLCARBAMATE; NIA 10242; NIAGARA 10242; PILLARFURAN; YALTOX

IDENTIFICATION
CAS Number: 1563-66-2
Formula: $C_{12}H_{15}NO_3$
DOT ID Number: UN 2757; DOT Guide Number: 151
Proper Shipping Name: Carbofuran
Reportable Quantity (RQ): **(CERCLA)** 10 lb/4.54 kg

DESCRIPTION: White crystalline solid. Odorless. Mixes and sinks in water.

Poison! (carbamate) • May be fatal if inhaled, swallowed, or absorbed through the skin. Effects of contact or inhalation may be delayed • Effects of contact or inhalation may be delayed • Firefighting gear (including SCBA) does not provide adequate protection. If exposure occurs, remove and isolate gear immediately and thoroughly decontaminate personnel • Containers may BLEVE when exposed to fire • Toxic products of combustion may include nitrogen oxides and phosphorus.

Hazard Classification (based on NFPA-704 Rating System)
Health Hazards (Blue): 4; Flammability (Red): 1; Reactivity (Yellow): 0

EMERGENCY RESPONSE: See Appendix A (151)
Evacuation:

Public safety: Isolate the area of spill or leak for at least 25 to 50 meters (80 to 160 feet) in all directions.
Spill: Increase, in the downwind direction, as necessary, the distance shown under "Public Safety."
Fire: If tank, rail car, or tank truck is involved in fire, isolate for at least 800 meters (½ mile) in all directions; also, consider initial evacuation for 800 meters (½ mile) in all directions.

EXPOSURE
Short-term effects: *SEEK MEDICAL ATTENTION*. **Dust or Solid:** *POISONOUS IF INHALED OR SWALLOWED*. Move to fresh air. *IF BREATHING HAS STOPPED,* give artificial respiration; *avoid mouth-to-mouth resuscitation; use bag/mask apparatus.* IF breathing is difficult, administer oxygen. Remove contaminated clothing and shoes. Flush affected areas with plenty of water. *IF IN EYES,* hold eyelids open and flush with plenty of water. *IF SWALLOWED* and victim is *CONSCIOUS AND ABLE TO SWALLOW,* have victim drink 4 to 8 ounces of water or induce vomiting by giving a tablespoon of salt in a glass of warm water. Repeat until vomitus is clear. Gastric lavage or syrup of ipecac may be warranted if vomiting is not prompt and profuse. *IF ON SKIN,* wash with soap and water followed by alcohol washing and a final soap washing.
Note to physician or authorized medical personnel: Physician may administer one (1) drop of homatropine into conjunctival sac may relieve miosis and loss of accomodation.
Note to physician or authorized medical personnel. Administer atropine, 2 mg (1/30 gr) intramuscularly or intravenously as soon as any local or systemic signs or symptoms of an intoxication are noted; repeat the administration of atropine every 3 to 8 minutes until signs of atropinization (mydriasis, dry mouth, rapid pulse, hot and dry skin) occur; initiate treatment in children with 0.05 mg/kg of atropine; repeat at 5- to 10-minute intervals. Watch respiration, and remove bronchial secretions if they appear to be obstructing the airway; intubate if necessary.
Medical note: Due to the rapid regeneration of chlolinesterase and the fact that 2-PAMCI may be contraindicated in the case of some carbamate poisonings, 2-PAMCI (Pralidoxime; Protopam) may not be needed.

HEALTH HAZARDS
Personal protective equipment (PPE): OSHA Table Z-1-A air contaminant. A-Level PPE.
Exposure limits (TWA unless otherwise noted): ACGIH TLV 0.1 mg/m³; NIOSH 0.1 mg/m³; skin contact contributes significantly in overall exposure.
Toxicity by ingestion: Poison. Grade 4. LD_{50} = less than 50 mg/kg.
Long-term health effects: Prenatal exposure initiated persistent postnatal endocrine dysfunction in mice. Mutation data reported.

FIRE DATA
Fire extinguishing agents not to be used: Water streams applied to adjacent fires will spread contamination of pesticide over wide area.
Behavior in fire: Very toxic dust and irritating vapors produced at fire temperatures.

CHEMICAL REACTIVITY
Binary reactants: Alkaline substances, acid, strong oxidizers.

ENVIRONMENTAL DATA
Food chain concentration potential: Log P_{ow} = 2.32. Unlikely to accumulate.
Water pollution: Harmful to aquatic life at very low concentrations. Fouling to shoreline. May be dangerous if it enters

nearby water intakes; notify operators. Notify local health and wildlife officials. **Response to discharge:** Issue warning–poison, water contaminant. Restrict access. Should be removed. Physical and chemical treatment.

SHIPPING INFORMATION
Grades of purity: Technical 98%; 10% granular formulation; 50%, 75%, or 80% wettable powder; 4 lb/gal flowable paste; **Storage temperature:** Keep away from heat and water; **Stability during transport:** Stable.

PHYSICAL AND CHEMICAL PROPERTIES
Physical state @ 59°F/15°C and 1 atm: Solid.
Molecular weight: 221.3
Melting/Freezing point: 302–307.4°F/150–153°C/423.2–426.2°K.
Specific gravity (water = 1): 1.180 @ 68°F/20°C.
Relative vapor density (air = 1): (estimate) 7.9.
Ratio of specific heats of vapor (gas): (estimate) more than 1
Vapor pressure: 0.000003 mm.

CARBARYL (SEVIN)   REC. C:1550

SYNONYMS: CARBARILO (Spanish); CARBATOX; CARBATOX-60; CARBATOX 75; CARPOLIN; CRAG SEVIN; DENAPON; DICARBAM; EEC No. 006-011-00-7; EXPERIMENTAL INSECTICIDE 7744; GAMONIL; GERMAIN'S; HEXAVIN; KARBASPRAY; KARBATOX; KARBOSEP; N-METHYLCARBAMATE de 1-NAPHTYLE (French); METHYLCARBAMATE 1-NAPHTHALENOL; METHYLCARBAMATE 1-NAPHTHALENOL, METHYCARBAMATE; 1-NAPHTHOL; METHYLCARBAMIC ACID, 1-NAPHTHYL ESTER; CAPROLIN; N-METHYL-1-NAPHTHYL-CARBAMAT (German); N-METHYL-α-NAPHTHYLCARBAMATE; N-METHYL-1-NAPHTHYL CARBAMATE; N-METHYL-α-NAPHTHYLURETHAN; CARBARYL; NAC; 1-NAPHTHOL N-METHYLCARBAMATE; α-NAPHTHYL-N-METHYLCARBAMATE; 1-NAPHTHYLMETHYLCARBAMATE; 1-NAPHTHYL-N-METHYLCARBAMATE; OMS-29; PANAM; RAVYON; SEPTENE; SEVIMOL; SEVIN® 50W; SOK; TERCYL; TRICARNAM; COMPOUND 7744; UC 7744; UNION CARBIDE 7,744

IDENTIFICATION
CAS Number: 63-25-2
Formula: $C_{12}H_{11}NO_2$
DOT ID Number: UN 2757; DOT Guide Number: 151
Proper Shipping Name: Carbaryl
Reportable Quantity (RQ): **(CERCLA)** 100 lb/45.4 kg

DESCRIPTION: White liquid or off-white solid. May be shipped and stored as a paste or a suspension in water. Weak odor. Solid sinks in water; insoluble.

Poison! (carbamate) • Prolonged skin contact, swallowing the material or breathing the dust can kill you • Effects of contact or inhalation may be delayed • Firefighting gear (including SCBA) does not provide adequate protection. If exposure occurs, remove and isolate gear immediately and thoroughly decontaminate personnel • Toxic products of combustion may include methylamine, carbon monoxide, and nitrogen oxides • Do not put yourself in danger by entering a contaminated area to rescue a victim.

Hazard Classification (based on NFPA-704 Rating System)
Health Hazards (Blue): 2; Flammability (Red): 0; Reactivity (Yellow): 0

EMERGENCY RESPONSE: See Appendix A (151)
Evacuation:
Public safety: Isolate the area of spill or leak for at least 25 to 50 meters (80 to 160 feet) in all directions.
Spill: Increase, in the downwind direction, as necessary, the distance shown under "Public Safety."
Fire: If tank, rail car, or tank truck is involved in fire, isolate for at least 800 meters (½ mile) in all directions; also, consider initial evacuation for 800 meters (½ mile) in all directions.

EXPOSURE
Short-term effects: *SEEK MEDICAL ATTENTION.* **Solid or solution: Ingestion and Absorption:** Eye pupils are small; blurred vision; runny nose; cough; shortness of breath; pain; diarrhea, nausea and vomiting; increased blood pressure, hypermotility, hallucinations; loss of consciousness; convulsions; breathing stops; death.
Treatment of exposure: The contaminated victim poses a health risk to the responder. Decontaminate the victim from a safe distance with a stream of water; have victim remove clothing if possible; provide Basic Life Support/CPR as needed. Further decontaminate the victim as follows: **Inhalation:** Remove the victim to fresh air and give oxygen if available. *IF BREATHING HAS STOPPED,* give artificial respiration; *avoid mouth-to-mouth resuscitation; use bag/mask apparatus.* **Skin:** Remove and isolate contaminated clothing (including shoes) and leave them in the Hot Zone; wash skin with soap and water and large volumes of water for at least 15 minutes. *Skin can also be decontaminated with diluted hypochlorite solution, U.S. Army M291 kit, and M258(A1) skin decontamination kit.* **Eye:** Rinse with large volumes of water or saline for at least 15 minutes. **Swallowed:** Do NOT make victim vomit.
Note to physician and medical personnel: If symptoms indicate, initial treatment for an adult includes atropine, 2 mg (1/30 gr) intramuscularly or intravenously as soon as any local or systemic signs or symptoms of an intoxication are noted; repeat the administration of atropine every 3 to 8 minutes until signs of atropinization (mydriasis, dry mouth, rapid pulse, hot and dry skin) occur; initiate treatment in children with 0.05 mg/kg of atropine; repeat at 5- to 10-minute intervals. Watch respiration, and remove bronchial secretions if they appear to be obstructing the airway; intubate if necessary. Also Diazepam, an anticonvulsant, might be needed. For adults, the dose is 5 to 10 mg (slow IV), repeated every 12 to 15 minutes up to three (3) doses maximum. For children, the dose is 0.2 to 0.5 mg/kg. *Do NOT use 2-PAMCl in the case of carbaryl (Sevin) poisoning.*
Note to physician or authorized medical personnel: If inhaled, medical observation is recommended for 24 to 48 hours after breathing overexposure, as pulmonary edema may be delayed. As first aid for pulmonary edema, consider administering a

HEALTH HAZARDS
Personal protective equipment (PPE): OSHA Table Z-1-A air contaminant. Chemical protective material(s) reported to offer protection: natural rubber, neoprene, nitrile, PVC. Also, Polyvinyl alcohol offers limited protection. If the proper equipment is not available, or if the rescuers have not been trained in its use, call for assistance from the U.S. Soldier and Biological Chemical Command, Edgewood Research Development and Engineering Center (from 0700-1630 EST call 410-671-4411, and from 1630-0700 EST call 410-278-5201; ask for the Staff Duty Officer).

Carbon dioxide

Recommendations for respirator selection: NIOSH/OSHA
50 mg/m³: SA* (any supplied-air respirator). *100 mg/m³*: SA:CF* (any supplied-air respirator operated in a continuous-flow mode); or SCBAF (any self-contained breathing apparatus with a full facepiece); or SAF (any supplied-air respirator with a full facepiece). *EMERGENCY OR PLANNED ENTRY INTO UNKNOWN CONCENTRATIONS OR IDLH CONDITIONS:* SCBAF:PD,PP (any self-contained breathing apparatus that has a full facepiece and is operated in a pressure-demand or other positive-pressure mode); or SAF:PD,PP:ASCBA (any supplied-air respirator that has a full facepiece and is operated in a pressure-demand or other positive-pressure mode in combination with an auxiliary self-contained breathing apparatus operated in a pressure-demand or other positive pressure mode. *ESCAPE:* GMFOVHiE [any air-purifying, full-facepiece respirator (gas mask) with a chin-style, front- or back-mounted organic vapor canister having a high-efficiency particulate filter]; or SCBAE (any appropriate escape-type, self-contained breathing apparatus). *Note*: Substance reported to cause eye irritation or damage; may require eye protection.
Exposure limits (TWA unless otherwise noted): ACGIH TLV 5 mg/m³; NIOSH/OSHA 5 mg/m³.
Toxicity by ingestion: Poison. Grade 2; LD_{50} = 0.5 to 5 g/kg (rat LD_{50} 0.51 g/kg).
Long-term health effects: Liver damage to rats at high dose by mouth. Mutation data reported.
Vapor (gas) irritant characteristics: Eye and respiratory tract irritant. High levels are dangerous.
Liquid or solid irritant characteristics: Eye and skin irritant.
IDLH value: 100 mg/m³.

FIRE DATA
Flash point: Combustible solid.
Fire extinguishing agents not to be used: Water streams applied to adjacent fires will spread contamination of pesticide over wide area.
Behavior in fire: Containers may explode.

CHEMICAL REACTIVITY
Binary reactants: Incompatible with strong oxidizers, alkalis, strong alkaline pesticides. Decomposes below boiling point.

ENVIRONMENTAL DATA
Food chain concentration potential: Log K_{ow} = 2.36. Unlikely to accumulate.
Water pollution: DOT Appendix B, §172.101–marine pollutant. Harmful to aquatic life in very low concentrations. Fouling to shoreline. May be dangerous if it enters nearby water intakes; notify operators. Notify local health and wildlife officials.
Response to discharge: Issue warning–water contaminant. Should be removed.

SHIPPING INFORMATION
Grades of purity: Sevin 50 W wettable powder; Sevin sprayable 80% powder; Sevin 4 oil; **Stability during transport:** Stable.

PHYSICAL AND CHEMICAL PROPERTIES
Physical state @ 59°F/15°C and 1 atm: Liquid or solid.
Molecular weight: 202.2
Boiling point @ 1 atm: 200°F/93°C/366°K (decomposes).
Melting/Freezing point: 288°F/142°C/415°K.
Specific gravity (water = 1): 1.23 @ 68°F/20°C (solid).
Surface tension: 37.1 dynes/cm @ 77°F/25°C.
Relative vapor density (air = 1): 0.9
Vapor pressure: Less than 0.00004 mm @ 77°F/25°C.

CARBON DIOXIDE (CO_2)   REC. C:1650

SYNONYMS: ACID GAS; ANHYDRIDE CARBONIQUE (French); CARBONIC ACID GAS; CARBONIC ANHYDRIDE; DIOXIDO de CARBONO (Spanish); DRY ICE (solid); KOHLENDIOXYD (German); KOHLENSAURE (German)

IDENTIFICATION
CAS Number: 124-38-9
Formula: CO_2
DOT ID Number: UN 1013 (gas, compressed); UN 1845 (solid, dry ice); UN 2187 (refrigerated liquid); DOT Guide Number: 120
Proper Shipping Name: Carbon dioxide; Carbon dioxide, compressed; Carbon dioxide, refrigerated liquid; Carbon dioxide, solid or dry ice

DESCRIPTION: Colorless gas, white solid (dry ice) or cryogenic liquid. Odorless. Solid sinks and boils in water; liquid floats on the surface; insoluble; visible vapor cloud is produced.

Contact with liquid may cause frostbite • Replaces oxygen in enclosed areas; vapors may cause dizziness or asphyxiation without warning • Gas is heavier than air and spread along ground • Containers may BLEVE when exposed to fire • *Warning*: Odor is not a reliable indicator of the presence of toxic amounts of carbon dioxide gas.

Hazard Classification (based on NFPA-704 Rating System)
Health Hazards (Blue): 1; Flammability (Red): 0; Reactivity (Yellow): 0

EMERGENCY RESPONSE: See Appendix A (120)
Evacuation:
Public safety: Isolate spill area for at least 100 meters (330 feet) in all directions.
Spill: Consider initial downwind evacuation for at least 25 meters (80 feet).
Fire: Isolate for 800 meters (½ mile) in all directions, especially if tank, rail car, or tank truck is involved in fire.

EXPOSURE
Short-term effects: *SEEK MEDICAL ATTENTION.* **Vapor:** IF inhaled, will cause dizziness, or difficult breathing. More than 12% CO_2 can cause unconsciousness and death. Move victim to fresh air. If breathing is difficult, administer oxygen. **Liquid or solid:** Will cause frostbite. Flush affected areas with plenty of water. *DO NOT RUB AFFECTED AREAS.*

HEALTH HAZARDS
Personal protective equipment (PPE): A-Level PPE. OSHA Table Z-1-A air contaminant. Wear approved self-contained breathing apparatus in excessively high CO_2 concentration areas. Wear thermal protective clothing.
Recommendations for respirator selection: NIOSH/OSHA
40,000 ppm: SA (any supplied-air respirator); or SCBAF (any self-contained breathing apparatus with a full facepiece). *EMERGENCY OR PLANNED ENTRY INTO UNKNOWN CONCENTRATIONS OR IDLH CONDITIONS:* SCBAF:PD,PP (any self-contained breathing apparatus that has a full facepiece and is operated in a pressure-demand or other positive-pressure mode); or SAF:PD,PP:ASCBA (any supplied-air respirator that has a full facepiece and is operated in a pressure-demand or other positive-pressure mode in combination with an auxiliary self-contained breathing apparatus operated in a pressure-demand or other positive

pressure mode). *ESCAPE*: SCBAE (any appropriate escape-type, self-contained breathing apparatus).
Exposure limits (TWA unless otherwise noted): ACGIH TLV 5000 ppm (9000 mg/m^3); NIOSH 5000 ppm (9000 mg/m^3); NIOSH/OSHA 5000 ppm (9000 mg/m^3).
Short-term exposure limits (15-minute TWA): ACGIH STEL 30,000 ppm (54,000 mg/m^3): NIOSH STEL 30,000 ppm (54,000 mg/m^3); OSHA STEL 5000 ppm (9000 mg/m^3).
Odor threshold: 74,000 ppm.
IDLH value: 40,000 ppm.

FIRE DATA
Behavior in fire: Containers may explode when heated.
Electrical hazard: Release of gas from container can cause static electricity to build up and cause ignition of nearby combustible materials.

CHEMICAL REACTIVITY
Reactivity with water: Forms carbonic acid
Binary reactants: Dusts of various metals such as magnesium, zirconium, titanium, aluminum, chromium and manganese are ignitable and explosive when suspended in CO_2. violent reaction with ammonia and amines.

ENVIRONMENTAL DATA
Food chain concentration potential: Negative; unlikely to accumulate.
Water pollution: Not harmful to aquatic life. **Response to discharge:** Restrict access. Disperse and flush.

SHIPPING INFORMATION
Grades of purity: Research: 99.995+%; Instrument: 99.99+%; Bone Dry: 99.95+%; Commercial: 99.5+%; **Storage temperature:** Ambient; **Inert atmosphere:** None; **Venting:** Liquid-safety relief; solid-open; **Stability during transport:** Stable.

PHYSICAL AND CHEMICAL PROPERTIES
Physical state @ 59°F/15°C and 1 atm: Gas.
Molecular weight: 44.0
Boiling point @ 1 atm: –174°F/–114°C/158°K (sublimes).
Melting/Freezing point: –109°F/–78°C/195°K.
Critical temperature: 88°F/31°C/304°K.
Critical pressure: 1.070 psia = 72.9 atm = 7.40 MN/m^2.
Specific gravity (water = 1): 1.56 at –79°C (solid).
Relative vapor density (air = 1): 1.53
Ratio of specific heats of vapor (gas): 1.0474
Latent heat of vaporization: 150 Btu/lb = 83 cal/g = 3.5 x 10^5 J/kg.
Heat of fusion: 43.2 cal/g.
Vapor pressure: 43–44 mm.

CARBON DISULFIDE   REC. C:1700

SYNONYMS: CARBON BISULFIDE; CARBON BISULPHIDE; CARBON DISULPHIDE; CARBONE (SUFURE DE) (French); CARBON SULFIDE; DITHIOCARBONIC ANHYDRIDE; DISULFURO de CARBONO (Spanish); EEC No. 006-003-00-3; KOHLENDISULFID (SCHWEFELKOHLENSTOFF) (German); NCI-4591; SCHWEFELKOHLENSTOFF (German); SULPHOCARBONIC ANHYDRIDE; WEEVILTOX; RCRA No. P022

IDENTIFICATION
CAS Number: 75-15-0
Formula: CS_2
DOT ID Number: UN 1131; DOT Guide Number: 131
Proper Shipping Name: Carbon disulfide
Reportable Quantity (RQ): **(CERCLA)** 100 lb/45.4 kg

DESCRIPTION: Colorless to faint yellow liquid. Sweet, ether-like odor. Reagent grades smell like rotten egg or decaying cabbage. Sinks in water; practically insoluble.

Shock or contact with hot surfaces can cause explosive decomposition • Inhalation of the vapors, absorption through the skin, or swallowing the liquid can cause severe illness • Very irritating to skin, eyes, and respiratory tract; prolonged contact with the skin will cause burns • Extremely flammable • Firefighting gear (including SCBA) does not provide adequate protection. If exposure occurs, remove and isolate gear immediately and thoroughly decontaminate personnel • Containers may BLEVE when exposed to fire • Vapors can form explosive mixture with air • Vapors are heavier than air and will collect and stay in low areas • Vapors may travel long distances to ignition sources and flashback • Vapors in confined areas (e.g., tanks, sewers, buildings) may explode when exposed to fire • Irritating to the skin, eyes, and respiratory tract • Toxic products of combustion may include carbon monoxide and sulfur oxides • Corrosive to plastics and rubber.

Hazard Classification (based on NFPA-704 Rating System)
Health Hazards (Blue): 3; Flammability (Red): 4; Reactivity (Yellow): 0

EMERGENCY RESPONSE: See Appendix A (131)
Evacuation:
Public safety: Isolate spill area for at least 100 to 200 meters (330 to 660 feet) in all directions.
Spill: Increase, as necessary, the isolation distance shown above, in "Public safety."
Fire: Isolate for 800 meters (½ mile) in all directions, especially if tank, rail car, or tank truck is involved in fire.

EXPOSURE
Short-term effects: *SEEK MEDICAL ATTENTION. ABSORBED THROUGH THE SKIN.* The use of alcoholic beverages may enhance toxic effects. **Vapor:** Irritating to eyes, nose, and throat. *IF INHALED*, will, will cause nausea, vomiting, difficult breathing, or loss of consciousness. Move to fresh air. *IF BREATHING HAS STOPPED*, give artificial respiration. IF breathing is difficult, administer oxygen. **Liquid:** Will burn skin and eyes. Harmful if swallowed. Remove contaminated clothing and shoes. Flush affected areas with plenty of water. *IF IN EYES*, hold eyelids open and flush with plenty of water. *IF SWALLOWED* and victim is *CONSCIOUS AND ABLE TO SWALLOW*, have victim drink 4 to 8 ounces of water and have victim induce vomiting. *IF SWALLOWED* and victim is *UNCONSCIOUS OR HAVING CONVULSIONS*, do nothing except keep victim warm.

HEALTH HAZARDS
Personal protective equipment (PPE): A-Level PPE. OSHA Table Z-1-A air contaminant. On OSHA Table Z-2 list. Organic vapor respirator. Chemical protective material(s) reported to offer protection: PV alcohol, Viton®. Also, Teflon® offers limited protection.
Recommendations for respirator selection: NIOSH
10 ppm: CCROV [any chemical cartridge respirator with organic vapor cartridge(s)]; or SA (any supplied-air respirator). *25 ppm:* SA:CF (any supplied-air respirator operated in a continuous-flow

mode); or PAPROV [any powered, air-purifying respirator with organic vapor cartridge(s)]. *50 ppm:* CCRFOV [any chemical cartridge respirator with a full facepiece and organic vapor cartridges(s)]; or GMFOV [any air-purifying, full-facepiece respirator (gas mask) with a chin-style, front- or back-mounted acid gas canister]; or PAPRTOV [any powered, air-purifying respirator with a tight-fitting facepiece and organic vapor cartridges(s)]; or SCBAF (any self-contained breathing apparatus with a full facepiece); or SAF (any supplied-air respirator with a full facepiece). *500 ppm:* SA:PD,PP (any supplied-air respirator operated in a pressure-demand or other positive-pressure mode). *EMERGENCY OR PLANNED ENTRY INTO UNKNOWN CONCENTRATIONS OR IDLH CONDITIONS:* SCBAF:PD,PP (any self-contained breathing apparatus that has a full facepiece and is operated in a pressure-demand or other positive-pressure mode); or SAF:PD,PP:ASCBA (any supplied-air respirator that has a full facepiece and is operated in a pressure-demand or other positive-pressure mode in combination with an auxiliary self-contained breathing apparatus operated in a pressure-demand or other positive pressure mode). *ESCAPE:* GMFOV [any air-purifying, full-facepiece respirator (gas mask) with a chin-style, front- or back-mounted acid gas canister]; or SCBAE (any appropriate escape-type, self-contained breathing apparatus).
Exposure limits (TWA unless otherwise noted): ACGIH TLV 10 ppm (31 mg/m^3); OSHA PEL 20 ppm (60 mg/m^3); ceiling 30 ppm; NIOSH REL 1 ppm (3 mg/m^3); skin contact contributes significantly in overall exposure.
Short-term exposure limits (15-minute TWA): NIOSH STEL 10 ppm (30 mg/m^3); skin contact contributes significantly in overall exposure. OSHA 30-minute maximum peak 100 ppm.
Toxicity by ingestion: Poison. Grade 2; rat LD$_{50}$ = 0.1-0.99 g/kg The use of alcoholic beverages enhances the toxic effect.
Long-term health effects: Nonspecific liver cell damage in rats; higher incidence of upper respiratory disease in humans. Chronic poisoning may result in permanent central nervous system damage.
Vapor (gas) irritant characteristics: Eye and respiratory tract irritant. Vapors cause irritation such that personnel will find high concentrations unpleasant.
Liquid or solid irritant characteristics: Eye and skin irritant. Causes smarting of the skin and first-degree burns on short exposure and may cause secondary burns on long exposure.
Odor threshold: 0.21 ppm.
IDLH value: 500 ppm.

FIRE DATA
Flash point: −22°F/−30°C (cc).
Flammable limits in air: LEL: 1.3%; UEL: 50%.
Fire extinguishing agents not to be used: Water may be ineffective on fire.
Autoignition temperature: 203°F/95°C/368°K.
Electrical hazard: Cannot be classified in conventional groups (i.e., A, B, C, etc.). Due to low electric conductivity, this substance may generate electrostatic charges as a result of agitation and flow. Contact of the liquid or vapor with a steam line or the surface of a lighted electric light bulb could result in ignition. No electrical equipment allowed.
Burning rate: 2.7 mm/min

CHEMICAL REACTIVITY
Binary reactants: Strong oxidizers; chemically active metals such as sodium, potassium and zinc; azides; rust; halogens; amines. Vapors may be ignited by heat or contact with ordinary light bulb or other heated objects. May be shock-sensitive. Softens rubber and many plastic materials. May be corrosive to metals of construction due to impurities.

Reactivity group: 38
Compatibility class: Carbon disulfide

ENVIRONMENTAL DATA
Food chain concentration potential: Log P_{ow} = 2.1 (estimate). Unlikely to accumulate.
Water pollution: Harmful to aquatic life in very low concentrations. May be dangerous if it enters nearby water intakes; notify operators. Notify local health and wildlife officials. **Response to discharge:** Issue warning–high flammability. Restrict access. Evacuate area.

SHIPPING INFORMATION
Grades of purity: Commercial; technical; USP; **Storage temperature:** Ambient; **Inert atmosphere:** Inerted; **Venting:** Pressure-vacuum; **Stability during transport:** Stable.

NAS HAZARD CLASSIFICATION FOR BULK WATER TRANSPORTATION
FIRE: 4
HEALTH: Vapor irritant: 2; Liquid or solid irritant: 2; Poisons: 3
WATER POLLUTION: Human toxicity: 1; Aquatic toxicity: 2; Aesthetic effect: 3
REACTIVITY: Other chemicals: 2; Water: 0; Self-reaction: 0

PHYSICAL AND CHEMICAL PROPERTIES
Physical state @ 59°F/15°C and 1 atm: Liquid.
Molecular weight: 76.14
Boiling point @ 1 atm: 115°F/46.3°C/319.5°K.
Melting/Freezing point: −169°F/−112°C/162°K.
Critical temperature: 523°F/273°C/546°K.
Critical pressure: 1100 psia = 76 atm = 7.7 MN/m^2.
Specific gravity (water = 1): 1.26 @ 68°F/20°C (liquid).
Liquid surface tension: 32 dynes/cm = 0.032 N/m @ 68°F/20°C.
Liquid water interfacial tension: 48.4 dynes/cm = 0.0484 N/m @ 68°F/20°C.
Relative vapor density (air = 1): 2.6
Ratio of specific heats of vapor (gas): 1.292
Latent heat of vaporization: 153 Btu/lb = 85 cal/g = 3.559 x 10^5 J/kg.
Heat of combustion: −5814 Btu/lb = −3230 cal/g = −135.2 x 10^5 J/kg.
Heat of fusion: 13.80 cal/g.
Vapor pressure: (Reid) 10.3 psia; 297 mm.

CARBON MONOXIDE REC. C:1750

SYNONYMS: CARBONE (OXYDE de) (French); CARBONIC OXIDE; CARBON OXIDE; CO; EEC No. 006-001-00-2; EXHAUST GAS; FLUE GAS; KOHLENMONOXID (German); MONOXIDO de CARBONO (Spanish); OXYDE de CARBONE (French); MONOXIDE

IDENTIFICATION
CAS Number: 630-08-0
Formula: CO
DOT ID Number: UN 1016; UN 9202 (cryogenic liquid); DOT Guide Number: 119 (gas); 168 (cryogenic liquid)
Proper Shipping Name: Carbon monoxide; Carbon monoxide, refrigerated Liquid.

DESCRIPTION: Colorless compressed gas or cryogenic liquid. Colorless. Odorless. Liquid floats and boils on water; poisonous, flammable visible vapor cloud is produced.

Carbon monoxide

Poison! Breathing the gas can kill you. A chemical asphyxiant, CO_2 poisons the hemoglobin and prevents blood from delivering oxygen to vital organs • Extremely flammable; flame is practically invisible • Gas is much lighter than air, and will disperse slowly unless confined. Under some conditions the gas is heavier than air and will collect and stay in low areas • Contact with liquid may cause frostbite • Do not put yourself in danger by entering a contaminated area to rescue a victim.

Warning: Odor is not an indicator of the presence of toxic amounts of carbon monoxide gas. Inhalation of air containing the following amounts can be fatal: 0.4% (1 hour); 1% (1 minute).

Hazard Classification (based on NFPA-704 Rating System)
Health Hazards (Blue): 3; Flammability (Red): 4; Reactivity (Yellow): 0

EMERGENCY RESPONSE, gas: See Appendix A (119)
Evacuation:
Public safety: See below.
Spill: Small spill–First: Isolate in all directions 30 meters (100 feet); Then: Protect persons downwind, DAY: 0.2 km (0.1 mile); NIGHT: 0.2 km (0.1 mile). Large spill–First: Isolate in all directions 125 meters (400 feet); Then: Protect persons downwind, DAY: 0.6 km (0.4 mile); NIGHT: 1.8 km (1.1 miles).
Fire: Isolate for 1600 meters (1 mile) in all directions, especially if tank, rail car, or tank truck is involved in fire.

EMERGENCY RESPONSE, cryogenic liquid: See Appendix A (168)
Evacuation:
Public safety: Isolate the area of spill or leak for at least 100 to 200 meters (330 to 660 feet) in all directions.
Spill: Small spill–First: Isolate in all directions 30 meters (100 feet); Then: Protect persons downwind, DAY: 0.2 km (0.1 mile); NIGHT: 0.2 km (0.1 mile). Large spill–First: Isolate in all directions 125 meters (400 feet); Then: Protect persons downwind, DAY: 0.6 km (0.4 mile); NIGHT: 1.8 km (1.1 miles).
Fire: If tank, rail car, or tank truck is involved in fire, isolate for at least 800 meters (½ mile) in all directions; also, consider initial evacuation for 800 meters (½ mile) in all directions.

EXPOSURE
Short-term effects: *SEEK MEDICAL ATTENTION.* **Vapor:** *POISONOUS IF INHALED.* Move victim to fresh air. *IF BREATHING HAS STOPPED,* give artificial respiration; *avoid mouth-to-mouth resuscitation; use bag/mask apparatus.* If breathing is difficult, administer oxygen. **Liquid:** Will cause frostbite. Flush affected areas with plenty of water. *DO NOT RUB AFFECTED AREAS.*

HEALTH HAZARDS
Personal protective equipment (PPE): OSHA Table Z-1-A air contaminant. A-Level PPE. Self-contained breathing apparatus; safety glasses and safety shoes. Wear thermal protective clothing.
Recommendations for respirator selection: NIOSH/OSHA *350 ppm:* SA (any supplied-air respirator). *875 ppm:* SA:CF (any supplied-air respirator operated in a continuous-flow mode).*1200 ppm:* GMFS** [any air-purifying, full-facepiece respirator (gas mask) with a chin-style, front- or back-mounted canister providing protection against the compound of concern]; or SCBAF (any self-contained breathing apparatus with a full facepiece); or SAF (any supplied-air respirator with a full facepiece). *EMERGENCY OR PLANNED ENTRY INTO UNKNOWN CONCENTRATIONS OR IDLH CONDITIONS:* SCBAF:PD,PP (any self-contained breathing apparatus that has a full facepiece and is operated in a pressure-demand or other positive-pressure mode); or SAF:PD,PP:ASCBA (any supplied-air respirator that has a full facepiece and is operated in a pressure-demand or other positive-pressure mode in combination with an auxiliary self-contained breathing apparatus operated in a pressure-demand or other positive pressure mode). *ESCAPE:* GMFS** [any air-purifying, full-facepiece respirator (gas mask) with a chin-style, front- or back-mounted canister providing protection against the compound of concern]; or SCBAE (any appropriate escape-type, self-contained breathing apparatus). **Note*: End of service life indicator (ESLI) required.
Exposure limits (TWA unless otherwise noted): ACGIH TLV 25 (29 mg/m^3); OSHA PEL 50 ppm (55 mg/m^3); NIOSH 35 ppm (40 mg/m^3); ceiling 200 ppm (229 mg/m^3).
Short-term exposure limits (15-minute TWA): 400 ppm, 15 minutes
Long-term health effects: Toxicity from overexposure persists for many days. May cause brain damage. Possible blood disorders.
Liquid or solid irritant characteristics: Rapid evaporation of liquid causes frostbite.
IDLH value: 1200 ppm.

FIRE DATA
Flash point: Flammable gas.
Flammable limits in air: LEL: 12%; UEL: 74%.
self-contained breathing apparatus) with dry chemicals or CO_2. Use water spray to keep containers cool.
Fire extinguishing agents not to be used: Do not use halogenated compounds.
Behavior in fire: Flame has very little color. Containers may explode in fire.
Autoignition temperature: 1128°F/609°C/882°K.
Electrical hazard: Class I, Group C. Unless protected, electrical equipment and lighting may cause explosion.
Adiabatic flame temperature: 2701°F/1483°C (estimate).
Stoichiometric air-to-fuel ratio: 2.451 (estimate).

CHEMICAL REACTIVITY
Binary reactants: Strong oxidizers, bromine trifluoride, chlorine trifluoride, lithium, halogenated compounds.

ENVIRONMENTAL DATA
Food chain concentration potential: Negative; unlikely to accumulate.
Water pollution: Harmful to aquatic life in very low concentrations. May be dangerous if it enters nearby water intakes; notify operators. Notify local health and wildlife officials. **Response to discharge:** Issue warning–poison, high flammability. Restrict access. Evacuate area.

SHIPPING INFORMATION

Grades of purity: Liquid: 98.6+%; **Gas:** Research High Purity; CP (99.5%); Technical (99.0+%); Commercial (97.5+%); **Storage temperature:** Ambient (for gas); –312.7°F/for liquid); **Inert atmosphere:** None; **Venting:** Safety relief; **Stability during transport:** Stable.

PHYSICAL AND CHEMICAL PROPERTIES
Physical state @ 59°F/15°C and 1 atm: Gas.
Molecular weight: 28.0
Boiling point @ 1 atm: –313°F/–192°C/82°K.
Melting/Freezing point: –337°F/–205°C/478°K.
Critical temperature: –220°F/–140°C/133°K.
Critical pressure: 507.5 psia = 34.51 atm = 3.502 MN/m^2.

Specific gravity (water = 1): 0.791 at –191.5°C (liquid).
Liquid surface tension: 9.8 dynes/cm = 0.098 N/m at –193°C.
Relative vapor density (air = 1): 0.1
Ratio of specific heats of vapor (gas): 1.3962
Latent heat of vaporization: 92.8 Btu/lb = 51.6 cal/g = 2.16 x 10^5 J/kg.
Heat of combustion: –4,343 Btu/lb = –2,412 cal/g = –101 x 10^5 J/kg.
Heat of fusion: 7.13 cal/g.
Vapor pressure: More than 750 mm.

CARBON OXYFLUORIDE   REC. C:1800

SYNONYMS: CARBONYL FLUORIDE; CARBON DIFLUORIDE OXIDE; CARBON FLUORIDE OXIDE; CARBONYL DIFLUORIDE; FLUOPHOSGENE; CARBONIC DIFLUORIDE; FLUORURO de CARBONILO (Spanish); FLUOROFORMYL FLUORIDE; FLUOROPHOSGENE; RCRA U033

IDENTIFICATION
CAS Number: 353-50-4
Formula: CF_2O
DOT ID Number: UN 2417; DOT Guide Number: 125
Proper Shipping Name: Carbonyl fluoride
Reportable Quantity (RQ): **(CERCLA)** 1000 lb/454 kg

DESCRIPTION: Colorless liquefied compressed gas or light yellow liquid. Sharp. pungent odor.

Poison! • Breathing the vapor, skin or eye contact, or swallowing the material can kill you • Firefighting gear (including SCBA) does not provide adequate protection. If exposure occurs, remove and isolate gear immediately and thoroughly decontaminate personnel • Containers may BLEVE when exposed to fire • Toxic products of combustion may include hydrogen fluoride • Do not put yourself in danger by entering a contaminated area to rescue a victim • Contact with liquid may cause severe injury or frostbite.

Hazard Classification (based on NFPA-704 Rating System)
Health Hazards (Blue): 4; Flammability (Red): 1; Reactivity (Yellow): 0

EMERGENCY RESPONSE: See Appendix A (125)
Evacuation:
Public safety: Isolate spill area for at least 100 to 200 meters (330 to 660 feet) in all directions.
Spill: Small spill–First: Isolate in all directions 30 meters (100 feet); Then: Protect persons downwind, DAY: 0.2 km (0.1 mile); NIGHT: 1.1 km (0.7 mile). Large spill–First: Isolate in all directions 125 meters (400 feet); Then: Protect persons downwind, DAY: 1.0 km (0.6 mile); NIGHT: 3.1 km (1.9 miles).
Fire: Isolate for 1600 meters (1 mile) in all directions, especially if tank, rail car, or tank truck is involved in fire.

EXPOSURE
Short-term effects: *SEEK MEDICAL ATTENTION.* **Vapor:** *POISONOUS IF INHALED.* Irritating to eyes, nose, and throat. Effects may be delayed. Move to fresh air. *IF BREATHING HAS STOPPED,* give artificial respiration; *avoid mouth-to-mouth resuscitation; use bag/mask apparatus.* IF breathing is difficult, administer oxygen. Maintain absolute rest until medical aid arrives.

HEALTH HAZARDS
Personal protective equipment (PPE): A-Level PPE. OSHA Table Z-1-A air contaminant.
Exposure limits (TWA unless otherwise noted): ACGIH TLV 2 ppm (5.4 mg/m^3); NIOSH REL 2 ppm (5 mg/m^3).
Short-term exposure limits (15-minute TWA): NIOSH STEL 5 ppm (13 mg/m^3).
Toxicity by ingestion: Poison.
Long-term health effects: May cause bone damage.
Vapor (gas) irritant characteristics: Vapors cause severe irritation of eyes and throat and can cause eye and lung injury. They cannot be tolerated even at low concentrations.
Liquid or solid irritant characteristics: Severe irritant to all tissues.

CHEMICAL REACTIVITY
Reactivity with water: Reacts to form hydrogen fluoride & CO_2
Binary reactants: Heat, moisture, hexafluoroisopropylideneaminolithium.
Neutralizing agents for acids and caustics: Can be absorbed in caustic soda solution. One ton of Carbon oxyfluoride requires 2,480 lb of caustic soda dissolved in 1000 gals. of water.

ENVIRONMENTAL DATA
Food chain concentration potential: Negative; unlikely to accumulate.
Water pollution: Effects of low concentrations on aquatic life is unknown. May be dangerous if it enters nearby water intakes; notify operators. Notify local health and wildlife officials. Notify operators of local water intakes. **Response to discharge:** Issue warning–poison. Restrict access. Evacuate area.

SHIPPING INFORMATION
Grades of purity: Commercial, 100%; **Storage temperature:** Ambient; **Inert atmosphere:** None; **Venting:** Safety relief; **Stability during transport:** Stable.

NAS HAZARD CLASSIFICATION FOR BULK WATER TRANSPORTATION
FIRE: 0
HEALTH: Vapor irritant: 4; Liquid or solid irritant: 2; Poisons: 4
WATER POLLUTION: Human toxicity: 4; Aquatic toxicity: 3; Aesthetic effect: 0
REACTIVITY: Other chemicals: 3; Water: 1; Self-reaction: 1

PHYSICAL AND CHEMICAL PROPERTIES
Physical state @ 59°F/15°C and 1 atm: Gas.
Molecular weight: 66.01
Boiling point @ 1 atm: –117°F/–83°C/190°K.
Melting/Freezing point: –173°F/–114°C/159°K.
Specific gravity (water = 1): 1.139 at –114°C.
Relative vapor density (air = 1): 2.28
Vapor pressure: 55.4 atm.

CARBON TETRACHLORIDE   REC. C:1850

SYNONYMS: BENZINOFORM; CARBONA; CARBON CHLORIDE; CARBON TET; EEC No. 602-008-00-5; FASCIOLIN; FLUKOIDS; FREON 10; HALON 104; METHANE TETRACHLORIDE; METHANE, TETRACHLORO-; NECATORINA; NECATORINE; PERCHLOROMETHANE; R 10; STCC; TETRACHLOORMETAN; TETRACHLORKOHLENSTOFF, TETRA (German); TETRACHLORMETHAN (German); TETRACHLOROCARBON;

TETRACHLOROMETHANE; TETRACHLORURE de CARBONE (French); TETRACLORURO de CARBONO (Spanish); TETRAFINOL; TETRAFORM; TETRASOL; UNIVERM; VERMOESTRICID

IDENTIFICATION
CAS Number: 56-23-5
Formula: CCl_4
DOT ID Number: UN 1846; DOT Guide Number: 151
Proper Shipping Name: Carbon tetrachloride
Reportable Quantity (RQ): **(CERCLA)** 10 lb/4.54 kg

DESCRIPTION: Colorless watery liquid. Sweet odor; resembles ether or chloroform. Sinks in water; insoluble; poisonous vapors are produced.

Containers may BLEVE when exposed to fire • Vapors are heavier than air and will collect and stay in low areas • Irritating to the skin, eyes, and respiratory tract • Toxic products of combustion may include hydrogen chloride and phosgene.

Hazard Classification (based on NFPA-704 Rating System)
Health Hazards (Blue): 3; Flammability (Red): 0; Reactivity (Yellow): 0

EMERGENCY RESPONSE: See Appendix A (151)
Evacuation:
Public safety: Isolate the area of spill or leak for at least 25 to 50 meters (80 to 160 feet) in all directions.
Spill: Increase, in the downwind direction, as necessary, the distance shown under "Public Safety."
Fire: If tank, rail car, or tank truck is involved in fire, isolate for at least 800 meters (½ mile) in all directions; also, consider initial evacuation for 800 meters (½ mile) in all directions.

EXPOSURE
Short-term effects: *SEEK MEDICAL ATTENTION. ABSORBED THROUGH THE SKIN.* **Vapor:** *POISONOUS IF INHALED.* Irritating to eyes. Move to fresh air. *IF BREATHING HAS STOPPED*, give artificial respiration; *avoid mouth-to-mouth resuscitation; use bag/mask apparatus.* IF breathing is difficult, administer oxygen. **Liquid:** *POISONOUS IF SWALLOWED.* Irritating to skin and eyes. Remove contaminated clothing and shoes. Flush affected areas with plenty of water. *IF IN EYES*, hold eyelids open and flush with plenty of water. *IF SWALLOWED* and victim is *CONSCIOUS AND ABLE TO SWALLOW*, have victim drink 4 to 8 ounces of water and have victim induce vomiting. *IF SWALLOWED* and victim is *UNCONSCIOUS OR HAVING CONVULSIONS*, do nothing except keep victim warm. The use of alcoholic beverages may enhance the toxic effect.

HEALTH HAZARDS
Personal protective equipment (PPE): A-Level PPE. OSHA Table Z-1-A air contaminant. OSHA Table Z-2 air contaminant. Full face organic vapor respirator. Chemical protective material(s) reported to offer protection: PV alcohol®, Viton®, Teflon®. Also, nitrile rubber is generally suitable for freons.
Recommendations for respirator selection: NIOSH
At any concentrations above the NIOSH REL, or where there is no REL, at any detectable concentration: SCBAF:PD,PP (any self-contained breathing apparatus that has a full facepiece and is operated in a pressure-demand or other positive-pressure mode); or SAF:PD,PP:ASCBA (any supplied-air respirator that has a full facepiece and is operated in a pressure-demand or other positive-pressure mode in combination with an auxiliary self-contained breathing apparatus operated in a pressure-demand or other positive pressure mode). *ESCAPE:* GMFOV [any air-purifying, full-facepiece respirator (gas mask) with a chin-style, front-or back-mounted organic vapor canister]; or SCBAE (any appropriate escape-type, self-contained breathing apparatus).
Exposure limits (TWA unless otherwise noted): ACGIH TLV 5 ppm (31 mg/m^3) animal carcinogen; OSHA PEL 10 ppm, ceiling 25 ppm; 200 ppm (5 minute peak in any 4 hours); NIOSH REL potential human carcinogen.
Short-term exposure limits (15-minute TWA): ACGIH STEL 10 ppm (63 mg/m^3) animal carcinogen; NIOSH STEL 2 ppm (12.6 mg/m^3) [60 minutes] suspected human carcinogen.
Toxicity by ingestion: Grade 2; LD_{50} = 0.5 to 5 g/kg (rat).
Long-term health effects: Causes severe liver damage and death if ingested. IARC possible carcinogen, rating 2B; OSHA specifically regulated carcinogen; NTP anticipated carcinogen.
Vapor (gas) irritant characteristics: Vapors cause irritation such that personnel will find high concentrations unpleasant. The effect is temporary.
Liquid or solid irritant characteristics: Eye and skin irritant. If spilled on clothing and allowed to remain, may cause smarting and reddening of the skin.
Odor threshold: Greater than 10 ppm. Literature reporting 50 ppm notes that Odor threshold is not considered adequate warning of potentially dangerous vapor concentrations.
IDLH value: Potential human carcinogen; 200 ppm.

FIRE DATA
Electrical hazard: Due to low electric conductivity, this substance may generate electrostatic charges as a result of agitation and flow.

CHEMICAL REACTIVITY
Reactivity with water: Solution is corrosive.
Binary reactants: Reacts explosively with chemically active alkali metals such as sodium, potassium and magnesium; fluorine; aluminum. Corrosive to rubber, plastic materials, most iron-, aluminum-, and copper-based alloys.
Reactivity group: 36
Compatibility class: Halogenated hydrocarbons

ENVIRONMENTAL DATA
Food chain concentration potential: Log P_{ow} = 2.6. Unlikely to accumulate.
Water pollution: DOT Appendix B, §172.101–marine pollutant. Effect of low concentrations on aquatic life is unknown. May be dangerous if it enters nearby water intakes; notify operators. Notify local health and pollution control officials. **Response to discharge:** Issue warning–poison. Restrict access. Should be removed.

SHIPPING INFORMATION
Grades of purity: Commercial; technical; USP; **Storage temperature:** Ambient, cool; **Inert atmosphere:** None; **Venting:** Pressure-vacuum; **Stability during transport:** Stable.

NAS HAZARD CLASSIFICATION FOR BULK WATER TRANSPORTATION
FIRE: 0
HEALTH: Vapor irritant: 2; Liquid or solid irritant: 1; Poisons: 4
WATER POLLUTION: Human toxicity: 2; Aquatic toxicity: 2; Aesthetic effect: 2
REACTIVITY: Other chemicals: 1; Water: 0; Self-reaction: 0

PHYSICAL AND CHEMICAL PROPERTIES
Physical state @ 59°F/15°C and 1 atm: Liquid.
Molecular weight: 153.83

Boiling point @ 1 atm: 170°F/77°C/350°K.
Melting/Freezing point: –9°F/–23°C/250°K.
Critical temperature: 541°F/283°C/556°K.
Critical pressure: 660 psia = 45 atm = 4.6 MN/m^2.
Specific gravity (water = 1): 1.59 @ 68°F/20°C (liquid).
Liquid surface tension: 27.0 dynes/cm = 0.027 N/m @ 68°F/20°C.
Liquid water interfacial tension: 45.0 dynes/cm = 0.045 N/m @ 68°F/20°C.
Relative vapor density (air = 1): 5.3
Ratio of specific heats of vapor (gas): 1.111
Latent heat of vaporization: 84.2 Btu/lb = 46.8 cal/g = 1.959 x 10^5 J/kg.
Heat of fusion: 5.09 cal/g.
Vapor pressure: (Reid) 3.8 psia; 91 mm.

CARENE   REC. C:1900

SYNONYMS: 3-CARENE; ISODIPRENE; δ-3-CARENO (Spanish); 4,7,7-TRIMETHYL-3-NORCARENE; 3,7,7-TRIMETHYLBICYCLO (0,1,4) HEPT-3-ENE

IDENTIFICATION
CAS Number: 13466-78-9; 498-15-7
Formula: $C_{10}H_{16}$
DOT ID Number: UN 1993; DOT Guide Number: 128
Proper Shipping Name: Combustible liquid, n.o.s.

DESCRIPTION: Colorless liquid. Sweet, turpentine-like odor. Floats on water surface; insoluble.

Combustible • Containers may BLEVE when exposed to fire • Vapors may form explosive mixture with air • Vapors are heavier than air and will collect and stay in low areas • Vapors may travel long distances to ignition sources and flashback • Vapors in confined areas (e.g., tanks, sewers, buildings) may explode when exposed to fire • Irritating to the skin, eyes, and respiratory tract • Toxic products of combustion may include carbon monoxide.

Hazard Classification (based on NFPA-704 Rating System)
Health Hazards (Blue): 0; Flammability (Red): 2; Reactivity (Yellow): 0

EMERGENCY RESPONSE: See Appendix A (128)
Evacuation:
Public safety: Isolate spill area for at least 25 to 50 meters (80 to 160 feet) in all directions.
Spill: Large spill–Consider initial downwind evacuation for at least 300 meters (1000 feet).
Fire: Isolate for 800 meters (½ mile) in all directions, especially if tank, rail car, or tank truck is involved in fire.

EXPOSURE
Short-term effects: *SEEK MEDICAL ATTENTION*. **Vapor:** Irritating to eyes, nose, and throat. *IF INHALED*, will, will cause headache, coughing or difficult breathing. *IF IN EYES*, hold eyelids open and flush with plenty of water. *IF BREATHING HAS STOPPED*, give artificial respiration. IF breathing is difficult, administer oxygen. **Liquid:** Irritating to skin and eyes. *IF SWALLOWED*, will cause nausea and vomiting. Remove contaminated clothing and shoes. Flush affected areas with plenty of water. *IF IN EYES*, hold eyelids open and flush with plenty of water. *IF SWALLOWED* and victim is *CONSCIOUS AND ABLE TO SWALLOW*, have victim drink 4 to 8 ounces of water and have victim induce vomiting. *IF SWALLOWED* and victim is *UNCONSCIOUS OR HAVING CONVULSIONS*, do nothing except keep victim warm.

HEALTH HAZARDS
Personal protective equipment (PPE): B-Level PPE. Organic canister); or air–supplied mask; goggles or face shield; rubber gloves.
Toxicity by ingestion: Grade 2; oral rat LD_{50} = 4.8 g/kg.
Liquid or solid irritant characteristics: Eye and skin irritant.

FIRE DATA
Fire extinguishing agents not to be used: Water may be ineffective on fire.

CHEMICAL REACTIVITY
Binary reactants: Will attack some forms of plastics

ENVIRONMENTAL DATA
Food chain concentration potential: Negative; unlikely to accumulate.
Water pollution: Effect of low concentrations on aquatic life is unknown. Fouling to shoreline. May be dangerous if it enters nearby water intakes; notify operators. Notify local health and wildlife officials. **Response to discharge:** Issue warning–high flammability. Mechanical containment. Should be removed. Chemical and physical treatment.

SHIPPING INFORMATION
Grades of purity: Commercial; **Storage temperature:** Ambient; **Inert atmosphere:** None; **Venting:** Open; **Stability during transport:** Stable.

PHYSICAL AND CHEMICAL PROPERTIES
Physical state @ 59°F/15°C and 1 atm: Liquid.
Molecular weight: 136
Boiling point @ 1 atm: 338°F/170°C/443°K.
Specific gravity (water = 1): 0.860 @ 68°F/20°C (liquid).
Heat of combustion: (estimate) –19,370 Btu/lb = 10,760 cal/g = –450 x 10^5 J/kg.

CATECHOL   REC. C:1950

SYNONYMS: BENZENE, o-IHYDROXY-; o-BENZENEDIOL; 1,2-BENZENEDIOL; CATACOL (Spanish); ATION BASE 26; o-IHYDROXYBENZENE; 1,2-DIHYDROXYBENZENE; o-DIOXYBENZENE; o-DIPHENOL; DURAFUR DEVELOPER C; FOURAMINE PCH; FOURRINE 68; o-HYDROQUINONE; o-HYDROXYPHENOL; 2-HYDROXYPHENOL; OXYPHENIC ACID; PELAGOL GREY C; o-PHENYLENEDIOL; PYROCATECHIN; PYROCATECHINE; PYROCATECHINIC ACID; PYROCATECHOL; PYROCATECHUIC ACID

IDENTIFICATION
CAS Number: 120-80-9
Formula: $C_6H_6O_2$; 1,2-HOC_6H_4OH
DOT ID Number: UN 3077; DOT Guide Number: 171
Proper Shipping Name: Environmentally hazardous substances, solid, n.o.s.
Reportable Quantity (RQ): **(CERCLA)** 1 lb/0.454 kg

DESCRIPTION: Colorless crystalline solid; turning brown in air and light. Faint odor. Sinks and mixes with water; soluble.

Combustible, highly volatile solid • Corrosive to the skin, eyes, and respiratory tract; inhalation symptoms may be delayed • Dust cloud may explode if ignited in an enclosed area • Containers may BLEVE when exposed to fire • Toxic products of combustion may include carbon monoxide.

Hazard Classification (based on NFPA-704 Rating System)
Health Hazards (Blue): 2; Flammability (Red): 1; Reactivity (Yellow): 0

EMERGENCY RESPONSE: See Appendix A (171)
Evacuation:
Public safety: Isolate the area of spill or leak for at least 10 to 25 meters (30 to 80 feet) in all directions.
Spill: Increase, in the downwind direction, as necessary, the distance shown under "Public Safety."
Fire: If any large container is involved in fire, isolate for at least 800 meters (½ mile) in all directions; also, consider initial evacuation for 800 meters (½ mile) in all directions.

EXPOSURE
Short-term effects: *SEEK MEDICAL ATTENTION. ABSORBED THROUGH THE SKIN.* **Dust:** Irritating to eyes, nose, and throat. *IF INHALED,* will, will cause coughing or difficult breathing. *IF IN EYES,* hold eyelids open and flush with plenty of water. *IF BREATHING HAS STOPPED,* give artificial respiration. IF breathing is difficult, administer oxygen. **Solid:** Will burn skin and eyes. Harmful if swallowed. Remove contaminated clothing and shoes. Flush affected areas with plenty of water. *IF IN EYES,* hold eyelids open and flush with plenty of water. *IF SWALLOWED* and victim is *CONSCIOUS AND ABLE TO SWALLOW,* have victim drink 4 to 8 ounces of water and have victim induce vomiting. *IF SWALLOWED* and victim is *UNCONSCIOUS OR HAVING CONVULSIONS,* do nothing except keep victim warm.
Note to physician or authorized medical personnel: Medical observation is recommended for 24 to 48 hours after breathing overexposure, as pulmonary edema may be delayed. As first aid for pulmonary edema, consider administering a corticosteroid spray. Cigarette smoking may exacerbate pulmonary injury and should be discouraged for at least 72 hours following exposure.

HEALTH HAZARDS
Personal protective equipment (PPE): B-Level PPE. OSHA Table Z-1-A air contaminant. Organic vapor respirator; rubber gloves, apron, and boots; face shield.
Exposure limits (TWA unless otherwise noted): ACGIH TLV 5 ppm (23 mg/m^3); NIOSH REL 5 ppm (20 mg/m^3); skin contact contributes significantly in overall exposure.
Toxicity by ingestion: Poison. Grade 2; LD_{50} = 260 mg/kg (rat).
Long-term health effects: Causes tumors in mice; kidney damage, dermatitis; an allergen.
Vapor (gas) irritant characteristics: Mists can cause eye, skin and respiratory tract irritation.
Liquid or solid irritant characteristics: Extremely corrosive to eyes and skin.

FIRE DATA
Flash point: (liquid) 278°F/137°C (oc); 261°F/127°C (cc).
Fire extinguishing agents not to be used: Water and foam may be ineffective.

CHEMICAL REACTIVITY
Binary reactants: Vigorous reaction with oxidizing materials and nitric acid.

ENVIRONMENTAL DATA
Food chain concentration potential: Log P_{ow} = 0.89. Unlikely to accumulate.
Water pollution: Effect of low concentrations on aquatic life is unknown. May be dangerous if it enters nearby water intakes; notify operators. Notify local health and wildlife officials. **Response to discharge:** Issue warning–water contaminant. Disperse and flush.

SHIPPING INFORMATION
Grades of purity: CP-high purity, 99.3+%; XP-extremely high purity, 99.8+%; **Storage temperature:** Ambient; **Inert atmosphere:** None; **Venting:** Open; **Stability during transport:** Stable.

PHYSICAL AND CHEMICAL PROPERTIES
Physical state @ 59°F/15°C and 1 atm: Solid.
Molecular weight: 110.11
Boiling point @ 1 atm: 474°F/246°C/419°K.
Melting/Freezing point: 220°F/104°C/378°K.
Specific gravity (water = 1): 1.344 @ 68°F/20°C (solid).
Heat of combustion: –11200 Btu/lb = –6,220 cal/g = –260 x 10^5 J/kg.
Heat of fusion: 49.40 cal/g.
Vapor pressure: 10 mm @ 118°F

CHARCOAL   REC. C:2000

SYNONYMS: ACTIVATED CARBON; ACTIVATED CHARCOAL; ANIMAL CARBON; CARBON, ACTIVATED; CARBON ACTIVO (Spanish); RBORAFFIN; CARBORAFINE; CHARCOAL, ACTIVATED; CHARCOAL, SHELL; MADERA CARBON (Spanish); MINERAL CARBON; PINON WOOD CHARCOAL; SHELL CHARCOAL; WOOD CHARCOAL; VEGETABLE CARBON; NUCHAR 722

IDENTIFICATION
CAS Number: 64365-11-3 (activated); 16291-96-6 (briquettes, or screenings from wood of Pinon tree, or screenings other than pinon wood); 7440-44-0 (purified)
Formula: C
DOT ID Number: UN 1361 (Charcoal, Carbon); UN 1362 (Carbon, activated); DOT Guide Number: 133
Proper Shipping Name: Carbon, animal or vegetable origin; Charcoal briquettes, shell, screening, wood, etc; Carbon, activated.

DESCRIPTION: Black powder, lumps, sticks, or grains. Odorless. May float or sink in water.

Flammable solid • Dust cloud may explode if ignited in an enclosed area • Containers may BLEVE when exposed to fire • Irritating to the skin, eyes, and respiratory tract • Toxic products of combustion may include smoke and irritating vapors of unburned material.

Hazard Classification (based on NFPA-704 Rating System)
Health Hazards (Blue): 0; Flammability (Red): 1; Reactivity (Yellow): 0

EMERGENCY RESPONSE: See Appendix A (133)
Evacuation:
Public safety: Isolate spill area for at least 10 to 25 meters (30 to 80 feet) in all directions.
Spill: Consider initial downwind evacuation of at least 100 meters (330 feet).

Fire: Isolate for 800 meters (½ mile) in all directions, especially if tank, rail car, or tank truck is involved in fire.

EXPOSURE
Short-term effects: Dust: Irritating to eyes, nose, and throat. Move victim to fresh air. *IF IN EYES*, hold eyelids open and flush with plenty of water.

HEALTH HAZARDS
Personal protective equipment (PPE): B-Level PPE. Respirator for dust
Toxicity by ingestion: Nontoxic (actually used in therapy of poisoning cases).

FIRE DATA
Flash point: Flammable solid; may ignite spontaneously in air.
Autoignition temperature: 600–750°F/316–390°C/589–663°K.
Electrical hazard: Class I, Group F.

CHEMICAL REACTIVITY
Binary reactants: Calcium hypochlorite, oxidizing materials, unsaturated oils.

ENVIRONMENTAL DATA
Water pollution: Effect of low concentrations on aquatic life is unknown. May be dangerous if it enters nearby water intakes; notify operators. Notify local health and wildlife officials.
Response to discharge: Issue warning–high flammability. Disperse and flush.

SHIPPING INFORMATION
Grades of purity: Various grades; those containing appreciable volatile material are more likely to catch fire. All shipments must be exposed to air and so certified; **Storage temperature:** Ambient; **Inert atmosphere:** None; **Venting:** Open (flame arrester); **Stability during transport:** Stable.

PHYSICAL AND CHEMICAL PROPERTIES
Physical state @ 59°F/15°C and 1 atm: Solid.
Molecular weight: 12
Boiling point @ 1 atm: Very high.
Specific gravity (water = 1): 2 @ 68°F/20°C (solid).
Heat of combustion: 14,100 Btu/lb = 7830 cal/g = 328 x 10^5 J/kg.

CHLORDANE   REC. C:2050

SYNONYMS: ASPON-CHLORDANE; BELT; CD 68; CHLORDAN; γ-CHLORDAN; CHLORINDAN; CHLOR KIL; CHLORODANE; CLORDAN (Spanish); CORODANE; CORTILAN-NEU; DICHLOROCHLORDENE; DOWCHLOR; HCS 3260; KYPCHLOR; M 140; M 410; 4,7-METHANO-1H-INDENE,1,2,4,5,6,7,8,8-OCTACHLORO-2,3,3A,4,7,7A-HEXAHYDRO-; NIRAN; OCTACHLOR; OCTACHLORODIHYDRODICYCLOPENTADIENE; 1,2,4,5,6,7,8,8-OCTACHLORO-2,3,3A,4,7,7A-HEXAHYDRO-4,7-METHANOINDENE; 1,2,4,5,6,7,8,8-OCTACHLORO-2,3,3A,4,7,7A-HEXAHYDRO-4,7-METHANO-1H-INDENE; 1,2,4,5,6,7,8,8-OCTACHLORO-3A,4,7,7A-HEXAHYDRO-4,7-METHYLENE INDANE; OCTACHLORO-4,7-METHANOHYDROINDANE; OCTACHLORO-4,7-METHANOTETRAHYDROINDANE; 1,2,4,5,6,7,8,8-OCTACHLORO-4,7-METHANO-3A,4,7,7A-TETRAHYDROINDANE; 1,2,4,5,6,7,8,8-OCTACHLORO-3A,4,7,7A-TETRAHYDRO-4,7-METHANOINDAN; 1,2,4,5,6,7,8,8-OCTACHLORO-3A,4,7,7A-TETRAHYDRO-4,7-METHANOINDANE; 1,2,4,5,6,7,10,10-OCTACHLORO-4,7,8,9-TETRAHYDRO-4,7-METHYLENEINDANE; 1,2,4,5,6,7,8,8-OCTACHLOR-3A,4,7,7A-TETRAHYDRO-4,7-*endo*-METHANO-INDAN (German); OCTA-KLOR; OKTATERR; ORTHO-KLOR; SD 5532; SHELL SD-5532; SYNKLOR; TAT CHLOR 4; TOPICHLOR 20; TOPICLOR; TOPICLOR 20; TOXICHLOR; VELSICOL 1068; 1,2,4,5,6,7,8,8-OCTACHLORO-2,3,3A,4,7,7A-HEXAHYDRO-4,7-METHANOINDENE; RCRA No. U036

IDENTIFICATION
CAS Number: 57-74-9; may also apply to 5103-71-9 [α- (*cis*-) isomer]; 5566-34-7 [γ- (*trans*-) isomer]; 5103-74-2 [β- (*trans*-) isomer]
Formula: $C_{10}H_6Cl_{18}$
DOT ID Number: UN 2762; DOT Guide Number: 131
Proper Shipping Name: Organochlorine pesticide, liquid, flammable, poisonous
Reportable Quantity (RQ): **(CERCLA)** 1 lb/0.454 kg

DESCRIPTION: Thick amber-to-brown liquid or a white powder. At room temperature it is almost odorless or may have a slight chlorine-like odor. Sinks in water; insoluble. Banned in the United States since 1988. However, old containers are still found in warehouses, landfills, garages, etc.

Poison! (organochlorine) • Breathing the vapor, skin or eye contact, or swallowing the material can kill you • Firefighting gear (including SCBA) does not provide adequate protection. If exposure occurs, remove and isolate gear immediately and thoroughly decontaminate personnel • Containers may BLEVE when exposed to fire • Vapors are heavier than air and will collect and stay in low areas • Vapors in confined areas (e.g., tanks, sewers, buildings) may explode when exposed to fire • Toxic products of combustion may include chlorine • Do not put yourself in danger by entering a contaminated area to rescue a victim.

Hazard Classification (based on NFPA-704 Rating System)
Health Hazards (Blue): 2; Flammability (Red): 3; Reactivity (Yellow): 0

EMERGENCY RESPONSE: See Appendix A (131)
Evacuation:
Public safety: Isolate spill area for at least 100 to 200 meters (330 to 660 feet) in all directions.
Spill: Increase, as necessary, the isolation distance shown above, in "Public safety."
Fire: Isolate for 800 meters (½ mile) in all directions, especially if tank, rail car, or tank truck is involved in fire.

EXPOSURE
Short-term effects: *SEEK MEDICAL ATTENTION*. There is no specific antidote for chlordane poisoning. Treatment consists of management of seizures and other measures to support respiratory and cardiovascular function. **Liquid or solution:** *POISONOUS IF SWALLOWED OR IF SKIN IS EXPOSED*. Irritating to skin and eyes. Remove contaminated clothing and shoes. Flush affected areas with plenty of water. *DO NOT RUB AFFECTED AREAS. IF IN EYES*, hold eyelids open and flush with plenty of water. *IF SWALLOWED* and victim is *CONSCIOUS AND ABLE TO SWALLOW*, have victim drink 4 to 8 ounces of water. **Do not induce vomiting** because the patient is at risk of CNS depression or seizures, which may lead to pulmonary aspiration during vomiting. If the exposure is recent (within 1–2 hours) and the patient is conscious and able to swallow, administer a slurry of

activated charcoal (at 1 gm/kg, usual adult dose 60–90 g, child dose 25–50 g) if it has not been given previously. A soda can and straw may be of assistance when offering charcoal to a child (The efficacy of activated charcoal for chlordane poisoning is uncertain.) *IF SWALLOWED* and victim is *UNCONSCIOUS OR HAVING CONVULSIONS,* do nothing except keep victim warm.

Note to physician or authorized medical personnel: Acute inhalation of chlordane is unlikely because of the material's low vapor pressure at ordinary temperatures. Toxic effects can occur after acute inhalation of a spray or mist containing chlordane or from inhalation of the solvents used to dissolve chlordane.

HEALTH HAZARDS

Personal protective equipment (PPE): A-Level PPE. OSHA Table Z-1-A air contaminant. Chemical-protective clothing is recommended because skin irritation and dermal absorption may occur and may contribute to systemic toxicity. Chlordane is absorbed through the skin and can attack rubber and several kinds of plastics. For chlordane hazards, NIOSH recommends suits made of CPF3™ (Kappler Co), and Trellchem HPS™ (Trelleborg Co) and gloves or boots made of Teflon™ (DuPont Co). NTP recommends Tyvek-type clothing or sleeves and gloves made of Viton®, nitrile, PV alcohol, or neoprene.

Recommendations for respirator selection: NIOSH
At any concentrations above the NIOSH REL, or where there is no REL, at any detectable concentration: SCBAF:PD,PP (any self-contained breathing apparatus that has a full facepiece and is operated in a pressure-demand or other positive-pressure mode); or SAF:PD,PP:ASCBA (any supplied-air respirator that has a full facepiece and is operated in a pressure-demand or other positive-pressure mode in combination with an auxiliary, self-contained breathing apparatus operated in a pressure-demand or other positive pressure mode). *ESCAPE:* GMFOVHiE [any air-purifying, full-facepiece respirator (gas mask) with a chin-style, front- or back-mounted organic vapor canister having a high-efficiency particulate filter]; or SCBAE (any appropriate escape-type, self-contained breathing apparatus).

Exposure limits (TWA unless otherwise noted): ACGIH TLV 0.5 mg/m^3; NIOSH/OSHA 0.5 mg/m^3; OSHA PEL 0.5 mg/m^3; NIOSH 0.5 mg/m^3; potential human carcinogen; skin contact contributes significantly in overall exposure

Toxicity by ingestion: Poison. Grade 3; oral LD$_{50}$ = 283 mg/kg (rat).

Long-term health effects: Possible liver damage; loss of appetite and weight. Mutation data.

Liquid or solid irritant characteristics: Eye and skin irritant.

IDLH value: Potential human carcinogen; 100 mg/m^3.

FIRE DATA

Flash point: (solution) 225°F/107°C (oc); 132°F/56°C (cc)

Flammable limits in air: (kerosene solution) LEL: 0.7%; UEL: 5%.

Fire extinguishing agents not to be used: Water may be ineffective on solution fire.

Special hazards of combustion products: Water streams applied to adjacent fires will spread contamination of environmentally hazardous substance over wide area.

Autoignition temperature: 410°F/210°C/483°K (kerosene solvent).

CHEMICAL REACTIVITY

Binary reactants: Strong oxidizers; alkaline reagents.

ENVIRONMENTAL DATA

Food chain concentration potential: Log K$_{ow}$ = 2.78. Values above 3.0 are very likely to accumulate in living tissues and especially in fats. Unlikely to accumulate.

Water pollution: DOT Appendix B, §172.101–marine pollutant. Harmful to aquatic life in very low concentrations. May be dangerous if it enters nearby water intakes; notify operators. Notify local health and wildlife officials. **Response to discharge:** Issue warning–poison. Restrict access. Should be removed. Chemical and physical treatment.

SHIPPING INFORMATION

Grades of purity: Technical. A variety of dusts, powders, and solutions in kerosene containing 2–80% chlordane are shipped; **Storage temperature:** Ambient; **Inert atmosphere:** None; **Venting:** Open (flame arrester); **Stability during transport:** Stable if kept cool.

PHYSICAL AND CHEMICAL PROPERTIES (undiluted, technical-grade chlordane).

Physical state @ 59°F/15°C and 1 atm: Liquid.
Molecular weight: 409.8
Boiling point @ 1 atm: Decomposes (347°F/175°C at 2 mmHg)
Melting/Freezing point: 217-228°F/103-109°K = 376-383°K.
Specific gravity (water = 1): 1.6 @ 77°F/25°C (liquid).
Liquid surface tension: (estimate) 25 dynes/cm = 0.025 N/m @ 68°F/20°C.
Liquid water interfacial tension: (estimate) 50 dynes/cm = 0.05 N/m @ 68°F/20°C.
Heat of combustion: (estimate) –4000 Btu/lb = –2200 cal/g = –93 x 10^5 J/kg.
Vapor pressure: 0.00001 mm.

CHLORINE   REC. C:2100

SYNONYMS: BERTHOLITE; CHLOR (German); CHLORE (French); CHLORINE MOLECULAR; CLORO (Spanish); DICHLORINE; EEC No. 017-001-00-7; MOLECULAR CHLORINE

IDENTIFICATION
CAS Number: 7782-50-5
Formula: Cl$_2$
DOT ID Number: UN 1017; DOT Guide Number: 124
Proper Shipping Name: Chlorine
Reportable Quantity (RQ): **(CERCLA)** 10 lb/4.54 kg

DESCRIPTION: Greenish-yellow gas. Shipped and stored as a liquefied compressed gas. Irritating, pungent, choking odor, like bleach. Sinks and boils in water; slightly soluble; reacts, forming toxic hypochlorous acid.

Poison! • Breathing the gas can kill you. Corrosive to skin, eyes, and respiratory tract; skin and eye contact causes severe burns and blindness; inhalation symptoms may be delayed • Firefighting gear (including SCBA) provides NO protection. If exposure occurs, remove and isolate gear immediately and thoroughly decontaminate personnel • Strong oxidizer which may react spontaneously with low flash point organics or reducing agents. Heat forms oxygen; will increase the activity of an existing fire • May cause fire or explosion contact with combustibles (wood, paper, oil, clothing, etc.)• Containers may explode when exposed to fire • Contact with liquid may cause frostbite • Do not put yourself in danger by entering a contaminated area to rescue a victim • Toxic products of combustion may include unburned vapors of chlorine.

Chlorine

Hazard Classification (based on NFPA-704 Rating System)
Health Hazards (Blue): 4; Flammability (Red): 0; Reactivity (Yellow): 0; Special Notice (White): OXY

EMERGENCY RESPONSE: See Appendix A (124)
Evacuation:
Public safety: Isolate spill area for at least 100 to 200 meters (330 to 660 feet) in all directions.
Spill: Small spill–First: Isolate in all directions 30 meters (100 feet); Then: Protect persons downwind, DAY: 0.3 km (0.2 mile); NIGHT: 1.1 km (0.7 mile). Large spill–First: Isolate in all directions 275 meters (900 feet); Then: Protect persons downwind, DAY: 2.7 km (1.7 miles); NIGHT: 6.8 km (4.2 miles).
Fire: Isolate for 800 meters (½ mile) in all directions, especially if tank, rail car, or tank truck is involved in fire.

EXPOSURE
Short-term effects: *SEEK MEDICAL ATTENTION*. **Vapor:** *POISONOUS IF INHALED*. Will burn eyes. Move to fresh air. *IF BREATHING HAS STOPPED*, give artificial respiration; *avoid mouth-to-mouth resuscitation; use bag/mask apparatus*. IF breathing is difficult, administer oxygen. *IF IN EYES*, hold eyelids open and flush with plenty of water. **Liquid:** Will burn skin and eyes. Will cause frostbite. Flush affected areas with plenty of water. *IF IN EYES*, hold eyelids open and flush with plenty of water. *DO NOT RUB AFFECTED AREAS*.
Note to physician or authorized medical personnel: To relieve irritation of the respiratory tract consider a spray of 10% sodium thiosulfate. Keep the victim lying down. The slightest exertion, including walking, may result in cardiac arrest. Medical observation is recommended for 24 to 48 hours after breathing overexposure, as pulmonary edema or bronchopneumonia may be delayed. As first aid for pulmonary edema, consider administering a corticosteroid spray. Cigarette smoking may exacerbate pulmonary injury and should be discouraged for at least 72 hours following exposure.

HEALTH HAZARDS
Personal protective equipment (PPE): A-Level PPE. OSHA Table Z-1-A air contaminant. Quick-opening safety shower and eye fountain; respiratory equipment approved for chlorine service. Wear safety goggles at all times when in vicinity of liquid chlorine. Wear thermal protective clothing. Chemical protective material(s) reported to offer minimal to poor protection: butyl rubber, Saranex®, Viton®, butyl rubber/neoprene, neoprene, Teflon®.
Recommendations for respirator selection: NIOSH/OSHA
5 ppm: CCRS [any chemical cartridge respirator with cartridge(s) providing protection against the compound of concern)]; or SA* (any supplied-air respirator). *10 ppm:* SA:CF (any supplied-air respirator operated in a continuous-flow mode); or PAPRS* [any powered, air-purifying respirator with cartridge(s) providing protection against the compound of concern]; or CCRFS [any chemical cartridge respirator with a full facepiece and cartridge(s) providing protection against the compound of concern]; or GMFS [any air-purifying, full-facepiece respirator (gas mask) with a chin-style, front- or back-mounted canister providing protection against the compound of concern]; or SCBAF (any self-contained breathing apparatus with a full facepiece); or SAF (any supplied-air respirator with a full facepiece). *EMERGENCY OR PLANNED ENTRY INTO UNKNOWN CONCENTRATIONS OR IDLH CONDITIONS:* SCBAF:PD,PP (any self-contained breathing apparatus that has a full facepiece and is operated in a pressure-demand or other positive-pressure mode); or SAF:PD,PP:ASCBA (any supplied-air respirator that has a full facepiece and is operated in a pressure-demand or other positive-pressure mode in combination with an auxiliary self-contained breathing apparatus operated in a pressure-demand or other positive pressure mode). *ESCAPE:* GMFS [any air-purifying, full-facepiece respirator (gas mask) with a chin-style, front- or back-mounted canister providing protection against the compound of concern]; or SCBAE (any appropriate escape-type, self-contained breathing apparatus). **Note*: Substance reported to cause eye irritation or damage; may require eye protection.
Exposure limits (TWA unless otherwise noted): ACGIH TLV 0.5 ppm (1.5 mg/m^3); OSHA PEL ceiling 1 ppm (0.3 mg/m^3); NIOSH ceiling 0.5 ppm (1.45 mg/m^3)/15 minutes
Short-term exposure limits (15-minute TWA): ACGIH STEL 1 ppm (2.9 mg/m^3).
ingestion unlikely (chlorine is a gas above –34.5°C).
Long-term health effects: May cause lung edema; physical exertion will aggravate this condition.
Vapor (gas) irritant characteristics: Vapors cause severe irritation of eyes and throat and can cause eye and lung injury. They cannot be tolerated even at low concentrations.
Liquid or solid irritant characteristics: Causes smarting of the skin and first-degree burns on short exposure; may cause secondary burns on long exposure.
Odor threshold: 0.08–30 ppm.
IDLH value: 10 ppm.

FIRE DATA
Behavior in fire: Most combustibles will burn in chlorine, although gas is not flammable. Containers may vent rapidly or rupture and explode.

CHEMICAL REACTIVITY
Reactivity with water: Forms a corrosive acidic solution.
Binary reactants: Reacts explosively or forms explosive compounds with many common substances such as acetylene, ether, turpentine, ammonia, fuel gas, hydrogen and finely divided metals. Reacts vigorously with most metals at high temperature, especially finely divided metals. Attacks copper and copper alloys; synthetic rubber.
Reactivity group: Due to very high reactivity or unusual conditions of carriage or potential compatibility problems this substance is not assigned a specific group in the compatibility chart.

ENVIRONMENTAL DATA
Food chain concentration potential: Negative; unlikely to accumulate.
Water pollution: DOT Appendix B, §172.101–marine pollutant. Harmful to aquatic life in very low concentrations. May be dangerous if it enters nearby water intakes; notify operators. Notify local health and wildlife officials. **Response to discharge:** Issue warning–poison. Restrict access. Evacuate area.

SHIPPING INFORMATION
Grades of purity: Research purity; ultrahigh purity; high purity; **Storage temperature:** Ambient; **Inert atmosphere:** None; **Venting:** Safety relief (300 psi); **Stability during transport:** Stable.

NAS HAZARD CLASSIFICATION FOR BULK WATER TRANSPORTATION
FIRE: 0
HEALTH: Vapor irritant: 4; Liquid or solid irritant: 2; Poisons: 4
WATER POLLUTION: Human toxicity: 2; Aquatic toxicity: 3; Aesthetic effect: 2
REACTIVITY: Other chemicals: 4; Water: 1; Self-reaction: 0

PHYSICAL AND CHEMICAL PROPERTIES
Physical state @ 59°F/15°C and 1 atm: Gas.
Molecular weight: 70.91
Boiling point @ 1 atm: –29°F/–34°C/239°K.
Melting/Freezing point: –150°F/–101°C/172°K.
Critical temperature: 291°F/144°C/417°K.
Critical pressure: 1118 psia = 76.05 atm = 7.704 MN/m^2.
Specific gravity (water = 1): 1.424 @ 15°C (liquid).
Liquid surface tension: 26.55 dynes/cm at –35.3°C.
Relative vapor density (air = 1): 2.47
Ratio of specific heats of vapor (gas): 1.325
Latent heat of vaporization: 124 Btu/lb = 68.7 cal/g = 2.87 x 10^5 J/kg.
Heat of fusion: 22.8 cal/g.
Vapor pressure: (Reid) 155 psia; 5170 mm.

CHLORINE TRIFLUORIDE   REC. C:2150

SYNONYMS: CTF; CHLORINE FLUORIDE; CHLOROTRIFLUORIDE; TRIFLUORURE de CHLORE (French); TRIFLUORURO de CLORO (Spanish); TRIFLUOROCHLORINE

IDENTIFICATION
CAS Number: 7790-91-2
Formula: ClF$_3$
DOT ID Number: UN 1749; DOT Guide Number: 124
Proper Shipping Name: Chlorine trifluoride

DESCRIPTION: Greenish-yellow fuming liquid below 53°F/12°C. Colorless liquefied compressed gas above 53°F/12°C. Strong sweetish odor. Reacts violently with water producing toxic chlorine gas and hydrofluoric acid.

Poison! • Breathing the vapor can kill you; skin and eye contact causes severe burns and blindness. Firefighting gear (including SCBA) does not provide adequate protection. If exposure occurs, remove and isolate gear immediately and thoroughly decontaminate personnel • Noncombustible but may cause combustible to ignite or explode; contact with organic material may result in SPONTANEOUS ignition • Gas explodes on contact with water or combustible materials • Containers may BLEVE when exposed to fire • Vapors are heavier than air and will collect and stay in low areas • Toxic products of combustion may include chlorine and fluorine • Do not attempt rescue • Corrosive to metals and rubber; reacts vigorously with glass, sand, and concrete.

Hazard Classification (based on NFPA-704 Rating System)
Health Hazards (Blue): 4; Flammability (Red): 0; Reactivity (Yellow): 3; Special Notice (White): OXY and Water reactive

EMERGENCY RESPONSE: See Appendix A (124)
Evacuation:
Public safety: Isolate spill area for at least 100 to 200 meters (330 to 660 feet) in all directions.
Spill: Small spill–First: Isolate in all directions 60 meters (200 feet); Then: Protect persons downwind, DAY: 0.5 km (0.3 mile); NIGHT: 1.6 km (1.0 mile). Large spill–First: Isolate in all directions 335 meters (1100 feet); Then: Protect persons downwind, DAY: 3.4 km (2.1 miles); NIGHT: 7.7 km (4.8 mile).
Fire: Isolate for 800 meters (½ mile) in all directions, especially if tank, rail car, or tank truck is involved in fire.

EXPOSURE
Short-term effects: *SEEK MEDICAL ATTENTION.* **Vapor:** *POISONOUS IF INHALED.* Irritating to skin, eyes, nose, and throat. Move victim to fresh air. *IF BREATHING HAS STOPPED,* give artificial respiration. IF breathing is difficult, administer oxygen. **Liquid:** *POISONOUS IF SWALLOWED.* Will burn skin and eyes. Remove contaminated clothing and shoes. Flush affected areas with plenty of water. *DO NOT RUB AFFECTED AREAS. IF IN EYES*, hold eyelids open and flush with plenty of water. *IF SWALLOWED* and victim is *CONSCIOUS AND ABLE TO SWALLOW*, have victim drink 4 to 8 ounces of water. **Do NOT induce vomiting.**

HEALTH HAZARDS
Personal protective equipment (PPE): A-Level PPE. OSHA Table Z-1-A air contaminant. Neoprene gloves and protective clothing made of glass fiber and Teflon, including full hood; self-contained breathing apparatus with full face mask.
Recommendations for respirator selection: NIOSH/OSHA
2.5 ppm: SA:CF* (any supplied-air respirator operated in a continuous-flow mode). *5 ppm*: SCBAF (any self-contained breathing apparatus with a full facepiece); or SAF (any supplied-air respirator with a full facepiece). *20 ppm*: SAF:PD,PP (any supplied-air respirator that has a full facepiece and is operated in a pressure-demand or other positive-pressure mode). *EMERGENCY OR PLANNED ENTRY INTO UNKNOWN CONCENTRATIONS OR IDLH CONDITIONS*: SCBAF:PD,PP (any self-contained breathing apparatus that has a full facepiece and is operated in a pressure-demand or other positive-pressure mode); or SAF:PD,PP:ASCBA (any supplied-air respirator that has a full facepiece and is operated in a pressure-demand or other positive-pressure mode in combination with an auxiliary self-contained breathing apparatus operated in a pressure-demand or other positive pressure mode). *ESCAPE:* GMFS [any air-purifying, full-facepiece respirator (gas mask) with a chin-style, front- or back-mounted canister providing protection against the compound of concern]; or SCBAE (any appropriate escape-type, self-contained breathing apparatus). *Note: Substance causes eye irritation or damage; eye protection needed.
Short-term exposure limits (15-minute TWA): ACGIH STEL ceiling 0.1 ppm (0.38 mg/m^3); OSHA/NIOSH ceiling 0.1 ppm (0.4 mg/m^3).
Toxicity by ingestion: Grade 4; LD$_{50}$ less than 50 mg/kg.
Vapor (gas) irritant characteristics: Vapors cause severe irritation of eyes and throat and can cause eye or lung injury. They cannot be tolerated even at low concentrations.
Liquid or solid irritant characteristics: Severe skin irritant, causes second-and third-degree burns on short contact and is very injurious to the eyes.
IDLH value: 20 ppm.

FIRE DATA
Fire extinguishing agents not to be used: Reaction with water may be violent.
Behavior in fire: Containers may explode.

CHEMICAL REACTIVITY
Reactivity with water: Reacts explosively with water, evolving hydrogen fluoride and chlorine gases.
Binary reactants: Reacts violently with oxidizers, water, acids, sand, metal oxides, organic matter, chlorofluorocarbons, silicon-containing materials. Causes ignition of all combustible materials and even sand, asbestos, glass or concrete. Very similar to fluorine gas.

ENVIRONMENTAL DATA
Food chain concentration potential: Negative; unlikely to accumulate.
Water pollution: Effects of low concentrations on aquatic life is unknown. May be dangerous if it enters nearby water intakes; notify operators. Notify local health and wildlife officials.
Response to discharge: Issue warning–poison, air contaminant, water contaminant, corrosive, oxidizing material. Restrict access. Evacuate area.

SHIPPING INFORMATION
Grades of purity: 99+%; **Storage temperature:** Ambient; **Inert atmosphere:** None; **Venting:** Safety relief; **Stability during transport:** Stable.

NAS HAZARD CLASSIFICATION FOR BULK WATER TRANSPORTATION
FIRE: 0
HEALTH: Vapor irritant: 4; Liquid or solid irritant: 4; Poisons: 4
WATER POLLUTION: Human toxicity: 4; Aquatic toxicity: 2; Aesthetic effect: 3
REACTIVITY: Other chemicals: 4; Water: 4; Self-reaction: 0

PHYSICAL AND CHEMICAL PROPERTIES
Physical state @ 59°F/15°C and 1 atm: Gas.
Molecular weight: 92.5
Boiling point @ 1 atm: 53°F/12°C/285°K.
Melting/Freezing point: –105°F/–76°C/197°K.
Critical temperature: 307°F/153°C/426°K.
Critical pressure: 837 psia = 56.9 atm = 5.77 MN/m^2.
Specific gravity (water = 1): 1.77 at 53°F/liquid).
Liquid surface tension: 26.6 dynes/cm = 0.0266 N/m at 0°C.
Relative vapor density (air = 1): 3.21
Ratio of specific heats of vapor (gas): 1.2832
Latent heat of vaporization: 128 Btu/lb = 71.2 cal/g = 2.98 x 10^5 J/kg.
Vapor pressure: 1.4 atm.

CHLOROACETALDEHYDE   REC. C:2200

SYNONYMS: ACETALDEHYDE, CHLORO-; CHLOROACETALDEHYDE (40% solution); 2-CHLOROACETALDEHYDE; CHLOROACETALDEHYDE MONOMER; CHLOROETHANAL; 2-CHLOROETHANAL; 2-CHLORO-1-ETHANAL; CLOROACETALDEHIDO (Spanish); MONOCHLOROACETALDEHYDE; RCRA No. P023

IDENTIFICATION
CAS Number: 107-20-0
Formula: C_2H_3ClO; $ClCH_2$
DOT ID Number: UN 2232; DOT Guide Number: 153
Proper Shipping Name: Chloroacetaldehyde
Reportable Quantity (RQ): **(CERCLA)** 1000lb/454 kg

DESCRIPTION: Clear, colorless liquid. Pungent, irritating odor. Sinks and mixes with water. Large amounts of irritating vapor produced.

Highly flammable • Severely irritating to skin, eyes, and respiratory tract; skin and eye contact can cause severe burns and blindness • Firefighting gear (including SCBA) does not provide adequate protection. If exposure occurs, remove and isolate gear immediately and thoroughly decontaminate personnel • Containers may BLEVE when exposed to fire • Vapors may form explosive mixture with air • Vapors are heavier than air and will collect and stay in low areas • Vapors may travel long distances to ignition sources and flashback • Vapors in confined areas (e.g., tanks, sewers, buildings) may explode when exposed to fire • Highly irritating to the eyes and respiratory tract; this material has anesthetic properties • Toxic products of combustion may include hydrogen chloride.

Hazard Classification (based on NFPA-704 Rating System)
Health Hazards (Blue): 3; Flammability (Red): 2; Reactivity (Yellow): 1

EMERGENCY RESPONSE: See Appendix A (153)
Evacuation:
Public safety: Isolate the area of spill or leak for at least 25 to 50 meters (80 to 160 feet) in all directions.
Spill: Small spill–First: Isolate in all directions 30 meters (100 feet); Then: Protect persons downwind, DAY: 0.2 km (0.1 mile); NIGHT: 0.5 km (0.3 mile). Large spill–First: Isolate in all directions 60 meters (200 feet); Then: Protect persons downwind, DAY: 0.6 km (0.4 mile); NIGHT: 1.6 km (1.0 mile).
Fire: If tank, rail car, or tank truck is involved in fire, isolate for at least 800 meters (½ mile) in all directions; also, consider initial evacuation for 800 meters (½ mile) in all directions.

EXPOSURE
Short-term effects: *SEEK MEDICAL ATTENTION.* **Vapor:** *POISONOUS IF INHALED OR ABSORBED THROUGH THE SKIN.* Irritating to eyes, nose, throat, lungs, and skin. Move to fresh air. *IF BREATHING HAS STOPPED,* give artificial respiration; *avoid mouth-to-mouth resuscitation; use bag/mask apparatus.* IF breathing is difficult, administer oxygen. **Liquid:** *POISONOUS IF SWALLOWED OR ABSORBED THROUGH SKIN.* Contact may cause burns to skin and eyes. *IF IN EYES OR ON SKIN,* immediately flush with running water for at least 15 minutes; lift eyelids occasionally if appropriate. Speed in removing material from skin is of extreme importance. *IF SWALLOWED* and victim is *CONSCIOUS AND ABLE TO SWALLOW,* get victim to induce vomiting by touching back of throat or taking syrup of Ipecac. *IF SWALLOWED* and victim is *UNCONSCIOUS OR HAVING CONVULSIONS,* do nothing except keep victim warm.

HEALTH HAZARDS
Personal protective equipment (PPE): A-Level PPE. OSHA Table Z-1-A air contaminant. Full face organic vapor respirator.
Recommendations for respirator selection: NIOSH/OSHA
10 ppm: CCROV* [any chemical cartridge respirator with organic vapor cartridge(s)]; or SA* (any supplied-air respirator). *25 ppm:* SA:CF (any supplied-air respirator operated in a continuous-flow mode); or PAPROV [any powered, air-purifying respirator with organic vapor cartridge(s)]. *45 ppm:* CCRFOV [any chemical cartridge respirator with a full facepiece and organic vapor cartridge(s)]; or GMFOV [any air-purifying, full-facepiece respirator (gas mask) with a chin-style, front- or back-mounted acid gas canister]; or PAPRTOV[(any powered, air-purifying respirator with a tight-fitting facepiece and organic vapor cartridge(s)]; or SCBAF (any self-contained breathing apparatus with a full facepiece); or SAF (any supplied-air respirator with a full facepiece). *EMERGENCY OR PLANNED ENTRY INTO UNKNOWN CONCENTRATIONS OR IDLH CONDITIONS:* SCBAF:PD,PP (any self-contained breathing apparatus that has a full facepiece and is operated in a pressure-demand or other positive-pressure mode); or SAF:PD,PP:ASCBA (any supplied-air respirator that has a full facepiece and is operated in a pressure-demand or other positive-pressure mode in combination with an

auxiliary self-contained breathing apparatus operated in a pressure-demand or other positive pressure mode). *ESCAPE*: GMFOV[any air-purifying, full-facepiece respirator (gas mask) with a chin-style, front- or back-mounted organic vapor cannister]; or SCBAE (any appropriate escape-type, self-contained breathing apparatus).
*Note: Substance reported to cause eye irritation or damage; may require eye protection.
Exposure limits (TWA unless otherwise noted): ACGIH TLV ceiling 1 ppm (3.2 mg/m^3); NIOSH/OSHA ceiling 1 ppm (3 mg/m^3).
Toxicity by ingestion: Grade 4; LD_{50} = 23 mg/kg (rat).
Long-term health effects: Showed mutagenic properties in the Ames test and in Chinese hamsters and rats.
Short-term exposure limits (15-minute TWA): ACGIH ceiling 1 ppm; NIOSH/OSHA ceiling 1 ppm (3 mg/m^3).
Vapor (gas) irritant characteristics: Vapors cause severe irritation of eyes and throat and can cause eye and lung injury. They cannot be tolerated even at low concentrations.
Liquid or solid irritant characteristics: Severe skin irritant. Causes second- and third-degree burns on short contact and is very injurious to the eyes.
Odor threshold: Less than 1 ppm.
IDLH value: 45 ppm.

FIRE DATA
Flash point: 190°F/88°C (cc) (pure substance)
Autoignition temperature: 190°F/88°C

CHEMICAL REACTIVITY
Reactivity with water: Reacts with water to form a hydrate; some heat is liberated. Reaction occurs when the concentration in water exceeds 50%.
Binary reactants: Contact with acids and oxidizing materials may cause fires or explosions.

ENVIRONMENTAL DATA
Food chain concentration potential: Log K_{ow} = 0.37. Unlikely to accumulate.
Water pollution: Effects of low concentrations on aquatic life is unknown. May be dangerous if it enters nearby water intakes; notify operators. Notify local health and wildlife officials.
Response to discharge: Issue warning–poison, water contaminant. Restrict access. Should be removed. Chemical and physical treatment.

SHIPPING INFORMATION
Grades of purity: 40% Aqueous solution; **Stability during transport:** Stable.

PHYSICAL AND CHEMICAL PROPERTIES
Physical state @ 59°F/15°C and 1 atm: Liquid.
Molecular weight: 78.5
Boiling point @ 1 atm: 185°F/85°C/358°K.
Melting/Freezing point: –3°F/–19°C/254°K (40% solution)
Specific gravity (water = 1): 1.19 @ 77°F/25°C (40% solution).
Relative vapor density (air = 1): 2.7
Vapor pressure: 100 mm.

CHLOROACETOPHENONE REC. C:2250

SYNONYMS: α-CHLOROACETOPHENONE; 2-CHLOROACETOPHENONE; ω-CHLOROACETOPHENONE; CHLOROMETHYL PHENYL KETONE; α-CLOROACETOFENONA (Spanish); CN; MACE®; PHENACYL CHLORIDE; PHENYL CHLOROMETHYL KETONE; TEAR GAS

IDENTIFICATION
CAS Number: 532-27-4
Formula: C_8H_7ClO; $C_6H_5COCH_2Cl$
DOT ID Number: UN 1697; DOT Guide Number: 153
Proper Shipping Name: Chloroacetophenone; Chloroacetophenone, solid; Chloroacetophenone, Liquid
Reportable Quantity (RQ): **(CERCLA)** 1 lb/0.454 kg

DESCRIPTION: Colorless, white or gray crystalline solid. Flower-like (locust blossom) smell at low concentrations; irritating smell at high concentrations. Sinks in water; insoluble; reacts slowly, generating hydrogen chloride. A tear gas, used for riot control.

Combustible solid • Severely irritating to the eyes; also irritating to the skin, and respiratory tract; prolonged contact causes burns • Vapors are heavier than air and will collect and stay in low areas • Containers may explode when exposed to fire • Irritating to the skin, eyes, and respiratory tract • Toxic products of combustion may include hydrogen chloride.

Hazard Classification (based on NFPA-704 Rating System)
Health Hazards (Blue): 2; Flammability (Red): 1; Reactivity (Yellow): 0

EMERGENCY RESPONSE: See Appendix A (153)
Evacuation:
Public safety: Isolate the area of spill or leak for at least 25 to 50 meters (80 to 160 feet) in all directions.
Spill: Increase, in the downwind direction, as necessary, the distance shown under "Public Safety." For CN [weaponized]: Small spill–First: Isolate in all directions 30 meters (100 feet); Then: Protect persons downwind, DAY: 0.2 km (0.1 mile); NIGHT: 0.5 km (0.3 mile). Large spill–First: Isolate in all directions 125 meters (400 feet); Then: Protect persons downwind, DAY: 1.1 km (0.7 mile); NIGHT: 3.2 km (2.0 miles).
Fire: If tank, rail car, or tank truck is involved in fire, isolate for at least 800 meters (½ mile) in all directions; also, consider initial evacuation for 800 meters (½ mile) in all directions.

EXPOSURE
Short-term effects: *CALL FOR MEDICAL AID* Dust: *POISONOUS IF INHALED OR IF SKIN IS EXPOSED*. Irritating to eyes, nose, and throat. Move victim to fresh air. *IF BREATHING HAS STOPPED*, give artificial respiration; *avoid mouth-to-mouth resuscitation; use bag/mask apparatus. IF IN EYES*, hold eyelids open and flush with plenty of water or a 2% solution of sodium bicarbonate or boric acid solution. Solid: *POISONOUS IF SWALLOWED*. Irritating to skin and eyes. Remove contaminated clothing and shoes. Flush affected areas with plenty of water. *IF IN EYES*, hold eyelids open and flush with plenty of water. *IF SWALLOWED* and victim is *CONSCIOUS AND ABLE TO SWALLOW*, have victim drink 4 to 8 ounces of water and have victim induce vomiting. *IF SWALLOWED* and victim is *UNCONSCIOUS OR HAVING CONVULSIONS*, do nothing except keep victim warm.

HEALTH HAZARDS
Personal protective equipment (PPE): B-Level PPE. OSHA Table Z-1-A air contaminant.
Recommendations for respirator selection: NIOSH/OSHA
3 mg/m^3: CCROVDM [any chemical cartridge respirator with

organic vapor cartridge(s) in combination with a dust and mist filter]; SA (any supplied-air respirator). *7.5 mg/m³*: SA:CF * (any supplied-air respirator operated in a continuous-flow mode); or PAPROVDM * [any powered, air-purifying respirator with organic vapor cartridge(s) in combination with a dust and mist filter]. *15 mg/m³*: CCRFOVHiE [any chemical cartridge respirator with a full facepiece and organic vapor cartridge(s) in combination with a high efficiency particulate filter]; or GMFSHiE[[any air-purifying, full-facepiece respirator (gas mask) with a chin-style, front- or back-mounted canister providing protection against the compound of concern and having a high-efficiency particulate filter]; or SCBAF (any self-contained breathing apparatus with a full facepiece); or SAF (any supplied-air respirator with a full facepiece). *EMERGENCY OR PLANNED ENTRY INTO UNKNOWN CONCENTRATIONS OR IDLH CONDITIONS:* SCBAF:PD,PP (any self-contained breathing apparatus that has a full facepiece and is operated in a pressure-demand or other positive-pressure mode); or SAF:PD,PP:ASCBA (any supplied-air respirator that has a full facepiece and is operated in a pressure-demand or other positive-pressure mode in combination with an auxiliary self-contained breathing apparatus operated in a pressure-demand or other positive pressure mode). *ESCAPE:* GMFSHiE [any air-purifying, full-facepiece respirator (gas mask) with a chin-style, front- or back-mounted canister providing protection against the compound of concern and having a high-efficiency particulate filter]; or SCBAE (any appropriate escape-type, self-contained breathing apparatus). **Note:* Substance causes eye irritation or damage; eye protection needed.
Exposure limits (TWA unless otherwise noted): ACGIH TLV 0.05 ppm (0.32 mg/m³); NIOSH/OSHA 0.3 (0.05 ppm).
Toxicity by ingestion: Grade 3; oral LD_{50} = 52 mg/kg (rat).
Long-term health effects: Fatty infiltration of liver
Vapor (gas) irritant characteristics: Eye and respiratory tract irritant.
Liquid or solid irritant characteristics: Severe eye irritant. Skin irritant.
Odor threshold: 0.03–0.1 ppm.
IDLH value: 15 mg/m³.

FIRE DATA
Flash point: Combustible solid; 244°F/118°C (cc) (solution only).

CHEMICAL REACTIVITY
Reactivity with water: Reacts slowly, generating hydrogen chloride.
Binary reactants: Reacts slowly with metals, causing mild corrosion. Reacts with strong oxidizers.

ENVIRONMENTAL DATA
Food chain concentration potential: Negative; unlikely to accumulate.
Food chain concentration potential: Log P_{ow} = 2.1. Unlikely to accumulate.
Water pollution: Effect of low concentrations on aquatic life is unknown. May be dangerous if it enters nearby water intakes; notify operators. Notify local health and wildlife officials.
Response to discharge: Issue warning-air contaminant. Restrict access. Disperse and flush.

SHIPPING INFORMATION
Grades of purity: Sometimes shipped as a solution in an organic solvent; **Storage temperature:** Ambient; **Inert atmosphere:** None; **Venting:** Pressure-vacuum; **Stability during transport:** Stable.

PHYSICAL AND CHEMICAL PROPERTIES
Physical state @ 59°F/15°C and 1 atm: Solid.
Molecular weight: 154.6
Boiling point @ 1 atm: 477°F/247°C/520°K.
Melting/Freezing point: 68-138°F/20-59°C/293-332°K.
Specific gravity (water = 1): 1.32 @ 15°C (solid).
Relative vapor density (air = 1): 5.3
Heat of combustion: (estimate) –9,340 Btu/lb = –5,190 cal/g = –217 x 10^5 J/kg.
Vapor pressure: 0.005 mm.

CHLOROACETYL CHLORIDE REC. C:2300

SYNONYMS: ACETYL CHLORIDE, CHLORO-; CHLORACETYL CHLORIDE; CHLOROACETIC ACID CHLORIDE; CHLOROACETIC CHLORIDE; CHLORURE de CHLORACETYLE (French); CLORURO de CLOROACETILO (Spanish); EEC No. 607-080-00-1; MONOCHLOROACETYL CHLORIDE

IDENTIFICATION
CAS Number: 79-04-9
Formula: $C_2H_2Cl_2O$; $ClCH_2COCl$
DOT ID Number: UN 1752; DOT Guide Number: 156
Proper Shipping Name: Chloroacetyl chloride

DESCRIPTION: Colorless to pale yellow liquid. Sharp, extremely irritating odor. Reacts violently with water; decomposes forming chloroacetic acid and hydrogen chloride gas.

Corrosive to skin, eyes, and respiratory tract; skin or eye contact causes severe burns, impaired vision, or blindness; effects of contact or inhalation may be delayed • Do not use water or water-based extinguishers • Firefighting gear (including SCBA) does not provide adequate protection. If exposure occurs, remove and isolate gear immediately and thoroughly decontaminate personnel • Vapors are heavier than air and will collect and stay in low areas • Toxic products of combustion may include phosgene and hydrogen chloride • Corrosive to metals.

Hazard Classification (based on NFPA-704 Rating System)
Health Hazards (Blue): 3; Flammability (Red): 0; Reactivity (Yellow): 1

EMERGENCY RESPONSE: See Appendix A (156)
Evacuation:
Public safety: Isolate the area of spill or leak for at least 50 to 100 meters (160 to 330 feet) in all directions.
Spill: Small spill–First: Isolate in all directions 30 meters (100 feet) (land or water); Then: Protect persons downwind, DAY: 0.2 km (0.1 mile) (land or water); NIGHT: 0.5 km (0.3 mile) (land); 0.2 km (0.1 mile) (water). Large spill–First: Isolate in all directions 95 meters (300 feet) (land); 60 meters (200 feet) (water); Then: Protect persons downwind, DAY: 0.8 km (0.5 mile) (land); 0.3 km (0.2 mile) (water); NIGHT: 1.6 km (1.0 mile) (land); 1.3 km (0.8 mile) (water).
Fire: If tank, rail car, or tank truck is involved in fire, isolate for at least 800 meters (½ mile) in all directions; also, consider initial evacuation for 800 meters (½ mile) in all directions.

EXPOSURE
Short-term effects: *SEEK MEDICAL ATTENTION.* **Vapor:** Irritating to eyes, nose, and throat. Move victim to fresh air. *IF BREATHING HAS STOPPED,* give artificial respiration; *avoid*

mouth-to-mouth resuscitation; use bag/mask apparatus. If breathing is difficult, administer oxygen. **Liquid:** *HARMFUL IF SWALLOWED OR ABSORBED THROUGH THE SKIN*. Will burn skin and eyes. Harmful if swallowed. Remove contaminated clothing and shoes. Flush affected areas with plenty of water. *IF IN EYES*, hold eyelids open and flush with plenty of water. *IF SWALLOWED* and victim is *CONSCIOUS AND ABLE TO SWALLOW*, have victim drink 4 to 8 ounces of water. **Do NOT induce vomiting.**

Note to physician or authorized medical personnel: Medical observation is recommended for 24 to 48 hours after breathing overexposure, as pulmonary edema may be delayed. As first aid for pulmonary edema, consider administering a corticosteroid spray. Cigarette smoking may exacerbate pulmonary injury and should be discouraged for at least 72 hours following exposure.

HEALTH HAZARDS
Personal protective equipment (PPE): B-Level PPE. Full face organic vapor/acid-type canister mask; self-contained breathing apparatus.
Exposure limits (TWA unless otherwise noted): ACGIH TLV 0.05 ppm (0.23 mg/m^3); NIOSH REL 0.05 ppm (0.2 mg/m^3).
Short-term exposure limits (15-minute TWA): ACGIH STEL 0.15 ppm (0.69 mg/m^3) [Skin].
Toxicity by ingestion: Grade 2; LD$_{50}$ = 0.5 to 5 g/kg.
Vapor (gas) irritant characteristics: Corrosive to eyes, skin, and respiratory tract. Vapors are irritating, such that personnel will not usually tolerate moderate or high vapor concentrations.
Liquid or solid irritant characteristics: Severe skin irritant. Causes second-and third-degree burns on short contact and is very injurious to the eyes.

FIRE DATA
Fire extinguishing agents not to be used: Reacts violently with water or foam. Do not use water or foam on adjacent fires.

CHEMICAL REACTIVITY
Reactivity with water: Reacts vigorously and decomposes to form chloroacetic acid and hydrogen chloride gas.
Binary reactants: Violent reaction with combustible materials, alcohols, bases, amines, alkali metals, sodium amide, finely divided metals. Reacts with surface moisture to generate hydrogen chloride, which is corrosive to metals.
Neutralizing agents for acids and caustics: Flush with water, rinse with sodium bicarbonate or lime solution.

ENVIRONMENTAL DATA
Food chain concentration potential: Negative; unlikely to accumulate.
Water pollution: Effect of low concentrations on aquatic life is unknown. May be dangerous if it enters nearby water intakes; notify operators. Notify local health and wildlife officials.
Response to discharge: Issue warning-air contaminant, corrosive. Restrict access. Disperse and flush.

SHIPPING INFORMATION
Grades of purity: Commercial; **Storage temperature:** Ambient; **Inert atmosphere:** None; **Venting:** Pressure-vacuum; **Stability during transport:** Stable.

NAS HAZARD CLASSIFICATION FOR BULK WATER TRANSPORTATION
FIRE: 0
HEALTH: Vapor irritant: 3; Liquid or solid irritant: 4; Poisons: 3
WATER POLLUTION: Human toxicity: 2; Aquatic toxicity: 2; Aesthetic effect: 2
REACTIVITY: Other chemicals: 2; Water: 3; Self-reaction: 0

PHYSICAL AND CHEMICAL PROPERTIES
Physical state @ 59°F/15°C and 1 atm: Liquid.
Molecular weight: 112.9
Boiling point @ 1 atm: 225°F/107°C/380°K.
Melting/Freezing point: –8°F/–22°C/251°K.

Critical pressure: 1.423 @ 68°F/20°C (liquid).
Specific gravity (water = 1): 1.42 @ 68°F/20°C (liquid).
Liquid surface tension: (estimate) 25 dynes/cm = 0.025 N/m @ 68°F/20°C.
Relative vapor density (air = 1): 3.9
Ratio of specific heats of vapor (gas): 1.1191
Latent heat of vaporization: 166 Btu/lb = 92.0 cal/g = 3.85 x 10^5 J/kg.
Heat of combustion: (estimate) –4000 Btu/lb = –2000 cal/g = –90 x 10^5 J/kg.
Heat of solution: (estimate) –54 Btu/lb = –30 cal/g = –1.3 x 10^5 J/kg.
Vapor pressure: 19 mm.

p-CHLOROANILINE   REC. C:2350

SYNONYMS: 1-AMINO-4-CHLOROBENZENE; *p*-CHLOROAMINOBENZENE; 4-CHLORO-1-AMINOBENZENE; 4-CHLOROBENZENEAMINE; 4-CHLOROPHENYLAMINE; 4-CHLOROANILINE; *p*-CLOROANILINA (Spanish); EEC No. 612-010-00-8

IDENTIFICATION
CAS Number: 106-47-8; can also apply to 95-51-2 (*o*-isomer) and 108-42-9 (*m*-isomer)
Formula: C$_6$H$_6$ClN; 4-ClC$_6$H$_4$NH$_2$
DOT ID Number: UN 2018; DOT Guide Number: 152
Proper Shipping Name: Chloroanilines, solid
Reportable Quantity (RQ): **(CERCLA)** 1000 lb/454 kg

DESCRIPTION: Colorless to yellowish-white crystalline solid; darkens to light brown in air. Mild, sweet, amine-like odor. Sinks in water; all isomers are practically insoluble in water. The *o*- and *m*-isomers are yellow liquids; turning brown on contact with air.

Combustible solid • Poison! • Containers may BLEVE when exposed to fire • Vapors are heavier than air and will collect and stay in low areas • Irritates eyes, skin, and respiratory tract • Toxic products of combustion may include hydrogen chloride and oxides of nitrogen.

Hazard Classification (based on NFPA-704 Rating System)
Health Hazards (Blue): 2; Flammability (Red): 1; Reactivity (Yellow): 0

EMERGENCY RESPONSE: See Appendix A (152)
Evacuation:
Public safety: Isolate the area of spill or leak for at least 25 to 50 meters (80 to 160 feet) in all directions.
Spill: Increase, in the downwind direction, as necessary, the distance shown under "Public Safety."
Fire: If tank, rail car, or tank truck is involved in fire, isolate for at least 800 meters (½ mile) in all directions; also, consider initial evacuation for 800 meters (½ mile) in all directions.

EXPOSURE
Short-term effects:
Solid and dust: *POISONOUS IF SWALLOWED OR IF SKIN IS EXPOSED.* Remove contaminated clothing and shoes. Flush affected areas with plenty of water. *IF BREATHING HAS STOPPED,* give artificial respiration; *avoid mouth-to-mouth resuscitation; use bag/mask apparatus.* IF IN EYES, hold eyelids open and flush with plenty of water. *IF SWALLOWED* and victim is *CONSCIOUS AND ABLE TO SWALLOW,* have victim drink 4 to 8 ounces of water and have victim induce vomiting. *IF SWALLOWED* and victim is *UNCONSCIOUS OR HAVING CONVULSIONS,* do nothing except keep victim warm.

HEALTH HAZARDS
Personal protective equipment (PPE): B-Level PPE. Rubber gloves; chemical goggles; protective clothing; dust respirator.
Toxicity by ingestion: Grade 3; oral LD_{50} = 300 mg/kg (rat).
Long-term health effects: Possible liver and kidney damage.
Vapor (gas) irritant characteristics: Eye, skin and respiratory tract irritant.
Liquid or solid irritant characteristics: Eye and skin irritant.

FIRE DATA
Flash point: (Combustible solid), more than 220°F/104°C (oc). 227°F/108°C (*o*-isomer).

CHEMICAL REACTIVITY
Binary reactants: Reacts with nitric acid, peroxides, oxidizers, acetic anhydride, chlorosulfonic acid, oleum, ozone.

ENVIRONMENTAL DATA
Food chain concentration potential: Log P_{ow} = 1.89. Unlikely to accumulate.
Water pollution: Effect of low concentrations on aquatic life is unknown. May be dangerous if it enters nearby water intakes; notify operators. Notify local health and wildlife officials.
Response to discharge: Issue warning–poison. Restrict access. Should be removed. Chemical and physical treatment.

SHIPPING INFORMATION
Grades of purity: 99.0%, Technical; **Storage temperature:** Ambient; Store containers in a well-ventilated area; **Inert atmosphere:** None; **Venting:** Open (flame arrester); **Stability during transport:** Stable.

PHYSICAL AND CHEMICAL PROPERTIES
Physical state @ 59°F/15°C and 1 atm: Solid.
Molecular weight: 127.6
Boiling point @ 1 atm: 446°F/230°C/503°K.
Melting/Freezing point: 158°F/70°C/343°K.
Specific gravity (water = 1): 1.43 @ 66°F/19°C/292°K (solid).
Relative vapor density (air = 1): 4.4
Heat of combustion: (estimate) –11,000 Btu/lb = –6000 cal/g = –250 x 10^5 J/kg.
Vapor pressure: 0.02 mm.

CHLOROBENZENE　　　　　　　　　　**REC. C:2400**

SYNONYMS: CHLOROBENZOL; CLOROBENCENO (Spanish); BENZENE, CHLORO-; BENZENE CHLORIDE; CHLORBENZEN; CHLOROBENZOL; EEC No. 602-033-00-1; MCB; MONOCHLORBENZENE; MONOCHLORBENZOL (German); PHENYL CHLORIDE; RCRA No. U037

IDENTIFICATION
CAS Number: 108-90-7
Formula: C_6H_5Cl
DOT ID Number: UN 1134; DOT Guide Number: 128
Proper Shipping Name: Chlorobenzene
Reportable Quantity (RQ): **(CERCLA)** 100 lb/45.4 kg

DESCRIPTION: Colorless, watery liquid. Sweet, almond odor. Sinks in water; insoluble. Flammable vapor is produced.

Highly flammable • Containers may BLEVE when exposed to fire • Vapors may form explosive mixture with air • Vapors are heavier than air and will collect and stay in low areas • Vapors may travel long distances to ignition sources and flashback • Vapors in confined areas (e.g., tanks, sewers, buildings) may explode when exposed to fire • Irritating to the skin, eyes, and respiratory tract • Toxic products of combustion may include carbon monoxide; hydrogen chloride and phosgene.

Hazard Classification (based on NFPA-704 Rating System)
Health Hazards (Blue): 2; Flammability (Red): 3; Reactivity (Yellow): 0

EMERGENCY RESPONSE: See Appendix A (128)
Evacuation:
Public safety: Isolate spill area for at least 25 to 50 meters (80 to 160 feet) in all directions.
Spill: Large spill–Consider initial downwind evacuation for at least 300 meters (1000 feet).
Fire: Isolate for 800 meters (½ mile) in all directions, especially if tank, rail car, or tank truck is involved in fire.

EXPOSURE
Short-term effects: *SEEK MEDICAL ATTENTION.* **Vapor:** If inhaled, will cause coughing or dizziness.
Not irritating to eyes, nose, and throat. Move to fresh air. *IF BREATHING HAS STOPPED,* give artificial respiration. IF breathing is difficult, administer oxygen. **Liquid:** Irritating to skin and eyes. Harmful if swallowed. Remove contaminated clothing and shoes. Flush affected areas with plenty of water. *IF IN EYES,* hold eyelids open and flush with plenty of water. *IF SWALLOWED* and victim is *CONSCIOUS AND ABLE TO SWALLOW,* have victim drink 4 to 8 ounces of water.

HEALTH HAZARDS
Personal protective equipment (PPE): B-Level PPE. OSHA Table Z-1-A air contaminant. Chemical protective material(s) reported to have good to excellent resistance: Teflon®, Viton®.
Recommendations for respirator selection: OSHA
1000 ppm: SA:CF* (any supplied-air respirator operated in a continuous-flow mode); or PAPROV* [any powered, air-purifying respirator with organic vapor cartridge(s)]; or CCRFOV [any air-purifying, full-facepiece respirator (gas mask) with a chin-style, front- or back-mounted acid gas canister]; or GMFOV [any air-purifying, full-facepiece respirator (gas mask) with a chin-style, front- or back-mounted organic vapor canister]; or SCBAF (any self-contained breathing apparatus with a full facepiece); or SAF (any supplied-air respirator with a full facepiece). *EMERGENCY OR PLANNED ENTRY INTO UNKNOWN CONCENTRATIONS OR IDLH CONDITIONS:* SCBAF:PD,PP (any self-contained breathing apparatus that has a full facepiece and is operated in a pressure-demand or other positive-pressure mode); or SAF:PD,PP:ASCBA (any supplied-air respirator that has a full facepiece and is operated in a pressure-demand or other positive-pressure mode in combination with an auxiliary self-contained

breathing apparatus operated in a pressure-demand or other positive pressure mode). *ESCAPE:* GMFOV[any air-purifying, full-facepiece respirator (gas mask) with a chin-style, front- or back-mounted organic vapor canister]; or SCBAE (any appropriate escape-type, self-contained breathing apparatus). *Note*: Substance causes eye irritation or damage; eye protection needed.
Exposure limits (TWA unless otherwise noted): ACGIH TLV 10 ppm (46 mg/m^3); OSHA PEL 75 ppm (350 mg/m^3).
Toxicity by ingestion: Grade 2; LD_{50} = 0.5 to 5 g/kg (rat, rabbit).
Long-term health effects: Liver and kidney damage. Possible dermatitis.
Vapor (gas) irritant characteristics: Eye and respiratory tract irritant.
Liquid or solid irritant characteristics: Eye and skin irritant. If spilled on clothing and allowed to remain, may cause smarting and reddening of the skin.
Odor threshold: 1.3 ppm.
IDLH value: 1000 ppm.

FIRE DATA
Flash point: 97°F/oc); 82°F/29°C (cc).
Flammable limits in air: LEL: 1.3%; UEL: 9.6%.
Fire extinguishing agents not to be used: Water may not be effective.
Behavior in fire: Containers may explode.
Autoignition temperature: 1184°F/640°C/913°K.
Electrical hazard: Class I, Group D. Due to low electric conductivity, this substance may generate electrostatic charges as a result of agitation and flow.
Burning rate: (estimate) 4.6 mm/min

CHEMICAL REACTIVITY
Binary reactants: Strong oxidizers or alkali metals may cause a violent reaction. Attacks rubber.
Reactivity group: 36
Compatibility class: Halogenated hydrocarbons

ENVIRONMENTAL DATA
Food chain concentration potential: Log P_{ow} = −0.1 to 0.22. Unlikely to accumulate.
Water pollution: Harmful to aquatic life in very low concentrations. May be dangerous if it enters nearby water intakes; notify operators. Notify local health and wildlife officials.
Response to discharge: Should be removed. Chemical and physical treatment.

SHIPPING INFORMATION
Grades of purity: 99.5%; technical; **Storage temperature:** Ambient; **Inert atmosphere:** None; **Venting:** Pressure-vacuum; **Stability during transport:** Stable.

NAS HAZARD CLASSIFICATION FOR BULK WATER TRANSPORTATION
FIRE: 3
HEALTH: Vapor irritant: 0; Liquid or solid irritant: 1; Poisons: 2
WATER POLLUTION: Human toxicity: 1; Aquatic toxicity: 3; Aesthetic effect: 2
REACTIVITY: Other chemicals: 1; Water: 0; Self-reaction: 0

PHYSICAL AND CHEMICAL PROPERTIES
Physical state @ 59°F/15°C and 1 atm: Liquid.
Molecular weight: 112.56
Boiling point @ 1 atm: 270°F/132°C/405°K.
Melting/Freezing point: −50°F/−46°C/228°K.
Critical temperature: 678°F/359°C/632°K.
Critical pressure: 656 psia = 44.6 atm = 4.52 MN/m^2.
Specific gravity (water = 1): 1.11 @ 68°F/20°C (liquid).
Liquid surface tension: 33 dynes/cm = 0.033 N/m @ 77°F/25°C.
Liquid water interfacial tension: 37.41 dynes/cm = 0.03741 N/m @ 68°F/20°C.
Relative vapor density (air = 1): 4.0
Ratio of specific heats of vapor (gas): 1.094
Latent heat of vaporization: 135 Btu/lb = 75 cal/g = 3.140 x 10^5 J/kg.
Heat of combustion: (estimate) 12,000 Btu/lb = 6700 cal/g = 280 x 10^5 J/kg.
Heat of fusion: 20.40 cal/g.
Vapor pressure: (Reid) 0.5 psia; 9mm.

4-CHLOROBUTYRONITRILE   REC. C:2450

SYNONYMS: BUTANENITRILE, 4-CHLORO-; BUTYRONITRILE, 4-CHLORO-; γ-CHLOROBUTYRONITRILE

IDENTIFICATION
CAS Number: 628-20-6
Formula: C_4H_6ClN; $CH_2ClCH_2CH_2CN$+$CH_2BrCH_2CH_2CN$
DOT ID Number: UN 1993; DOT Guide Number: 128
Proper Shipping Name: Combustible liquid, n.o.s.

DESCRIPTION: White to light yellow liquid. Sinks in water.

Combustible • Containers may BLEVE when exposed to fire • Vapors may form explosive mixture with air • Vapors are heavier than air and will collect and stay in low areas • Vapors may travel long distances to ignition sources and flashback • Vapors in confined areas (e.g., tanks, sewers, buildings) may explode when exposed to fire • Irritating to the skin, eyes, and respiratory tract • Toxic products of combustion may include nitrogen oxides and hydrogen chloride.

Hazard Classification (based on NFPA-704 Rating System)
Health Hazards (Blue): 3; Flammability (Red): 2; Reactivity (Yellow): 0

EMERGENCY RESPONSE: See Appendix A (128)
Evacuation:
Public safety: Isolate spill area for at least 25 to 50 meters (80 to 160 feet) in all directions.
Spill: Large spill–Consider initial downwind evacuation for at least 300 meters (1000 feet).
Fire: Isolate for 800 meters (½ mile) in all directions, especially if tank, rail car, or tank truck is involved in fire.

EXPOSURE
Short-term effects: *SEEK MEDICAL ATTENTION.* **Vapor:** Irritating to eyes, nose, and throat. *IF INHALED*, will, will cause coughing or difficult breathing. *IF IN EYES*, hold eyelids open and flush with plenty of water. *IF BREATHING HAS STOPPED,* give artificial respiration; *avoid mouth-to-mouth resuscitation; use bag/mask apparatus*. IF breathing is difficult, administer oxygen.
Liquid: Irritating to skin and eyes. *IF SWALLOWED*, will cause nausea and vomiting. Remove contaminated clothing and shoes. Flush affected areas with plenty of water. *IF IN EYES*, hold eyelids open and flush with plenty of water. *IF SWALLOWED* and victim is *CONSCIOUS AND ABLE TO SWALLOW*, have victim drink 4 to 8 ounces of water and have victim induce vomiting. *IF SWALLOWED* and victim is *UNCONSCIOUS OR HAVING CONVULSIONS,* do nothing except keep victim warm.

Note to physician or authorized medical personnel: Consider the use of amyl nitrite perles if symptoms of cyanide poisoning develop. If symptoms indicate, initial treatment includes the cyanide antidote kit. In all cases, break an amyl nitrite perle in a gauze pad and hold lightly under victim's nose for 15 seconds, repeating 5 times at about 15-second intervals; if necessary (and if sodium nitrite infusions will be delayed), repeat procedure every 3 minutes with fresh perles until 3 or 4 have been used. Avoid breathing the vapor while administering it to the victim. Administer sodium nitrite IV, ASAP. The usual adult dose is 10 to 20 mL of a 3% solution infused over no less than 5 minutes; the average child dose is 0.15 to 0.20 mL/kg. Monitor blood pressure during administration, and slow the rate of infusion if hypotention develops. Next, infuse sodium thiosulfate IV. The usual adult dose is 50 mL of a 25% solution infused over 10 to 20 minutes; the average child dose is 1.65 mL/kg. Repeat with nitrite and thiosulfate as required.

HEALTH HAZARDS
Personal protective equipment (PPE): B-Level PPE.
Toxicity by ingestion: Grade 3; LD_{50} = 50–400 mg/kg (rat).
Vapor (gas) irritant characteristics: Eye and respiratory tract irritant.
Liquid or solid irritant characteristics: Eye irritant. Possible skin irritant.

CHEMICAL REACTIVITY
Binary reactants: May attack some forms of plastics.
Reactivity group: 37
Compatibility class: Nitriles

ENVIRONMENTAL DATA
Food chain concentration potential: Negative; unlikely to accumulate.
Water pollution: Effect of low concentrations on aquatic life is unknown. May be dangerous if it enters nearby water intakes; notify operators. Notify local health and wildlife officials.
Response to discharge: Issue warning–water contaminant. Should be removed. Chemical and physical treatment.

SHIPPING INFORMATION
Grades of purity: 50%, + 40% 4-bromobutyronitrile + 8% glutaronitrile. Major components have same hazard ratings;
Storage temperature: Ambient; **Inert atmosphere:** None; **Venting:** Open; **Stability during transport:** Stable.

PHYSICAL AND CHEMICAL PROPERTIES
Physical state @ 59°F/15°C and 1 atm: Liquid.
Molecular weight: 103.55
Boiling point @ 1 atm: 374°F/190°C/463°K.
Specific gravity (water = 1): 1.22 @ 68°F/20°C (liquid).
Relative vapor density (air = 1): 3.57
Ratio of specific heats of vapor (gas): (estimate) 1.080 @ 68°F/20°C.
Latent heat of vaporization: (estimate) 185 Btu/lb = 103 cal/g = 4.31×10^5 J/kg.

CHLOROFORM REC. C:2500

SYNONYMS: CHLOROFORME (French); CLOROFORMO (Spanish); EEC No. 602-006-00-4; FORMYL TRICHLORIDE; FREON 20; METHANE TRICHLORIDE; METHANE, TRICHLORO-; METHENYL TRICHLORIDE; METHYL TRICHLORIDE; R 20; R 20 (REFRIGERANT); TCM; TRICHLOROFORM; TRICHLOROMETHANE; RCRA No. U044

IDENTIFICATION
CAS Number: 67-66-3
Formula: $CHCl_3$
DOT ID Number: UN 1888; DOT Guide Number: 151
Proper Shipping Name: Chloroform
Reportable Quantity (RQ): **(CERCLA)** 10 lb/4.54 kg

DESCRIPTION: Clear, colorless liquid. Sweet odor, like ether. Sinks in water; insoluble. Large amounts of irritating vapor is produced.

Vapors are heavier than air and will collect and stay in low areas • Containers may BLEVE when exposed to fire • Toxic products of combustion may include phosgene and hydrogen chloride • Do not put yourself in danger by entering a contaminated area to rescue a victim.

Hazard Classification (based on NFPA-704 Rating System)
Health Hazards (Blue): 2; Flammability (Red): 0; Reactivity (Yellow): 0

EMERGENCY RESPONSE: See Appendix A (151)
Evacuation:
Public safety: Isolate the area of spill or leak for at least 25 to 50 meters (80 to 160 feet) in all directions.
Spill: Increase, in the downwind direction, as necessary, the distance shown under "Public Safety."
Fire: If tank, rail car, or tank truck is involved in fire, isolate for at least 800 meters (½ mile) in all directions; also, consider initial evacuation for 800 meters (½ mile) in all directions.

EXPOSURE
Short-term effects: *SEEK MEDICAL ATTENTION.* **Vapor:** Irritating to eyes, nose, and throat. *IF INHALED*, will, will cause headache, nausea, dizziness, or loss of consciousness. Move to fresh air. *IF BREATHING HAS STOPPED*, give artificial respiration; *avoid mouth-to-mouth resuscitation; use bag/mask apparatus*. IF breathing is difficult, administer oxygen. **Liquid:** Irritating to skin and eyes. Harmful if swallowed. Remove contaminated clothing. Flush affected areas with plenty of water. *IF IN EYES*, hold eyelids open and flush with plenty of water. *IF SWALLOWED* and victim is *CONSCIOUS AND ABLE TO SWALLOW*, have victim drink 4 to 8 ounces of water and have victim induce vomiting. *IF SWALLOWED* and victim is *UNCONSCIOUS AND HAVING CONVULSIONS*, do nothing except keep victim warm.

HEALTH HAZARDS
Personal protective equipment (PPE): A-Level PPE. OSHA Table Z-1-A air contaminant. Chemical protective material(s) reported to have good to excellent resistance: PV alcohol, Teflon®. Also, Viton®, Viton®/clorobutyl, Viton®/neoprene, nitrile, and polyurethane offers limited protection.
Recommendations for respirator selection: NIOSH
At any concentrations above the NIOSH REL, or where there is no REL, at any detectable concentration: SCBAF:PD,PP (any self-contained breathing apparatus that has a full facepiece and is operated in a pressure-demand or other positive-pressure mode); or SAF:PD,PP:ASCBA (any supplied-air respirator that has a full facepiece and is operated in a pressure-demand or other positive-pressure mode in combination with an auxiliary self-contained breathing apparatus operated in a pressure-demand or other positive pressure mode). *ESCAPE:* GMFOV [any air-purifying, full-

facepiece respirator (gas mask) with a chin-style, front-or back-mounted organic vapor canister]; or SCBAE (any appropriate escape-type, self-contained breathing apparatus).
Exposure limits (TWA unless otherwise noted): ACGIH TLV 10 ppm (49 mg/m^3) suspected human carcinogen; OSHA PEL ceiling 50 ppm (240 mg/m^3); NIOSH REL potential human carcinogen; reduce exposure to lowest feasible level.
Short-term exposure limits (15-minute TWA): NIOSH STEL 2 ppm (9.78 mg/m^3) [60 minutes]; potential human carcinogen
Toxicity by ingestion: Grade 2; LD_{50} = 0.5 to 5 g/kg.
Long-term health effects: OSHA specifically regulated carcinogen. IARC possible carcinogen, rating 2B; NTP anticipated carcinogen. Nervous system, kidney and liver damage may occur.
Vapor (gas) irritant characteristics: Eye and respiratory tract irritation. Vapors cause irritation such that personnel will find high concentrations unpleasant. The effect is temporary.
Liquid or solid irritant characteristics: eye and skin irritant. If spilled on clothing and allowed to remain, may cause smarting and reddening of the skin.
Odor threshold: 206 ppm.
IDLH value: Potential human carcinogen; 500 ppm.

FIRE DATA
Electrical hazard: Class I, Group D. Due to low electric conductivity, this substance may generate electrostatic charges as a result of agitation and flow.

CHEMICAL REACTIVITY
Binary reactants: Strong caustics; aluminum, magnesium, potassium and chemically active metals; aluminum chloride, ethylene, strong oxidizers, alcohols. Attacks many forms of plastic.
Reactivity group: 36
Compatibility class: Halogenated hydrocarbons

ENVIRONMENTAL DATA
Food chain concentration potential: Log K_{ow} = variably listed at 1.98–2.59. Unlikely to accumulate.
Water pollution: Effect of low concentrations on aquatic life is unknown. May be dangerous if it enters nearby water intakes; notify operators. Notify local health and pollution control officials.
Response to discharge: Issue warning-air contaminant. Restrict access. Should be removed.

SHIPPING INFORMATION
Grades of purity: Technical, USP; **Storage temperature:** Ambient; **Inert atmosphere:** None; **Venting:** Open; **Stability during transport:** Stable.

NAS HAZARD CLASSIFICATION FOR BULK WATER TRANSPORTATION
FIRE: 1
HEALTH: Vapor irritant: 2; Liquid or solid irritant: 1; Poisons: 2
WATER POLLUTION: Human toxicity: 1; Aquatic toxicity: 2; Aesthetic effect: 2
REACTIVITY: Other chemicals: 1; Water: 0; Self-reaction: 0

PHYSICAL AND CHEMICAL PROPERTIES
Physical state @ 59°F/15°C and 1 atm: Liquid.
Molecular weight: 119.39
Boiling point @ 1 atm: 142°F/61°C/334°K.
Melting/Freezing point: –82°F/–64°C/210°K.
Critical temperature: 506°F/263°C/536°K.
Critical pressure: 790 psia = 54 atm = 5.5 MN/m^2.
Specific gravity (water = 1): 1.48 @ 68°F/20°C (liquid).
Liquid surface tension: 27.1 dynes/cm = 0.0271 N/m @ 68°F/20°C.
Liquid water interfacial tension: 32.8 dynes/cm = 0.0328 N/m @ 68°F/20°C.
Relative vapor density (air = 1): 4.1
Ratio of specific heats of vapor (gas): 1.146
Latent heat of vaporization: 106.7 Btu/lb = 59.3 cal/g = 2.483 x 10^5 J/kg.
Heat of fusion: 17.62 cal/g.
Vapor pressure: (Reid) 6.39 psia; 160 mm.

CHLOROHYDRINS   REC. C:2550

SYNONYMS: 1-CHLORO-2,3-EPOXYPROPANE; 3-CHLORO-1,2-EPOXYPROPANE; CHLOROMETHYLOXIRANE; 2-CHLOROPROPYLENE OXIDE; γ-CHLOROPROPYLENE OXIDE; CLOROHYDRINA (Spanish); CRUDE EPICHLOROHYDRIN; EEC No. 603-026-00-6; OXIRANE,(CHLOROMETHYL)-; PROPANE, 1-CHLORO-2,3-EPOXY; RCRA No. U041

IDENTIFICATION
CAS Number: 106-89-8
Formula: C_3H_5ClO; $O \cdot CH_2CH \cdot CH_2Cl$
DOT ID Number: UN 2023; DOT Guide Number: 131P
Proper Shipping Name: Epichlorohydrin
Reportable Quantity (RQ): **(CERCLA)** 100 lb/45.4 kg

DESCRIPTION: Clear, colorless watery liquid. Irritating, chloroform-like odor. Sinks and mixes with water; poisonous vapor is produced.

Flammable • Firefighting gear (including SCBA) does not provide adequate protection. If exposure occurs, remove and isolate gear immediately and thoroughly decontaminate personnel • Extremely irritating to skin, eyes, and respiratory tract; skin and eye contact causes severe burns and blindness • Polymerization hazard • Heat can induce polymerization with rapid release of energy; sealed containers may rupture explosively • May react with itself blocking relief valves; leading to tank explosions • Containers may BLEVE when exposed to fire • Vapors may form explosive mixture with air • Vapors are heavier than air and will collect and stay in low areas • Vapors may travel long distances to ignition sources and flashback • Vapors in confined areas (e.g., tanks, sewers, buildings) may explode when exposed to fire • Irritating to the skin, eyes, and respiratory tract • Toxic products of combustion may include hydrogen chloride.

Hazard Classification (based on NFPA-704 Rating System)
Health Hazards (Blue): 3; Flammability (Red): 2; Reactivity (Yellow): 2

EMERGENCY RESPONSE: See Appendix A (131)
Evacuation:
Public safety: Isolate spill area for at least 100 to 200 meters (330 to 660 feet) in all directions.
Spill: Increase, as necessary, the isolation distance shown above, in "Public safety."
Fire: Isolate for 800 meters (½ mile) in all directions, especially if tank, rail car, or tank truck is involved in fire.

EXPOSURE
Short-term effects: *SEEK MEDICAL ATTENTION.* **Vapor:** *POISONOUS IF INHALED.* Irritating to eyes, nose, and throat.

244   Chloromethyl methyl ether

Move to fresh air. *IF BREATHING HAS STOPPED,* give artificial respiration. IF breathing is difficult, administer oxygen. **Liquid:** *POISONOUS IF SWALLOWED.* Irritating to skin and eyes. Remove contaminated clothing and shoes. Flush affected areas with plenty of water. *IF IN EYES,* hold eyelids open and flush with plenty of water. *IF SWALLOWED* and victim is *CONSCIOUS AND ABLE TO SWALLOW,* have victim drink 4 to 8 ounces of water and have victim induce vomiting. *IF SWALLOWED* and victim is *UNCONSCIOUS OR HAVING CONVULSIONS,* do nothing except keep victim warm.

HEALTH HAZARDS
Personal protective equipment (PPE): A-Level PPE. OSHA Table Z-1-A air contaminant.
Recommendations for respirator selection: NIOSH
At any concentrations above the NIOSH REL, or where there is no REL, at any detectable concentration: SCBAF:PD,PP (any self-contained breathing apparatus that has a full facepiece and is operated in a pressure-demand or other positive-pressure mode); or SAF:PD,PP:ASCBA (any supplied-air respirator that has a full facepiece and is operated in a pressure-demand or other positive-pressure mode in combination with an auxiliary self-contained breathing apparatus operated in a pressure-demand or other positive pressure mode). *ESCAPE:* GMFOVAG [any air-purifying, full-facepiece respirator (gas mask) with a chin-style, front- or back-mounted organic vapor or acid gas canister); or SCBAE (any appropriate escape-type, self-contained breathing apparatus).
Exposure limits (TWA unless otherwise noted): ACGIH TLV 0.1 ppm (3.8 mg/m^3); OSHA PEL 5 ppm (19 mg/m^3); skin contact contributes significantly in overall exposure; NIOSH potential human carcinogen; reduce exposure to lowest feasible level.
Toxicity by ingestion: Grade 3; LD$_{50}$ = 50 to 500 mg/kg.
Long-term health effects: IARC probable carcinogen, rating 2A; NTP anticipated carcinogen. Liver, kidney, skin and glandular disorders. May cause temporary sterility.
Vapor (gas) irritant characteristics: Severe eye and respiratory tract irritant. Vapors cause irritation such that personnel will find high concentrations unpleasant.
Liquid or solid irritant characteristics: Severe eye irritant. Skin irritant. If spilled on clothing and allowed to remain, may cause smarting and reddening of the skin.
Odor threshold: 10 ppm.
IDLH value: 75 ppm.

FIRE DATA
Flash point: 93°F/34°C (oc); 99°F/37°C (cc).
Flammable limits in air: LEL: 3.8%; UEL: 21%.
Fire extinguishing agents not to be used: Avoid use of dry chemical if fire occurs in container with confined vent.
Behavior in fire: Containers may rupture and explode.
Autoignition temperature: 804°F/429°C/702°K.
Electrical hazard: Class I, Group C.
Burning rate: 2.6 mm/min

CHEMICAL REACTIVITY
Reactivity with water: Mild reaction; may not be hazardous.
Binary reactants: Strong oxidizers, strong acids, aniline, isopropylamine, potassium-*tert*-butoxide, certain salts, caustics, zinc, powdered metals, aluminum. Forms explosive mixture with trichloroethylene. Dissolves most paints, causes rubber to swell. The wet product will pit carbon steel.
Polymerization: Can polymerize in presence of strong acids, bases, metallic halides, elevated temperatures.
Reactivity group: 17
Compatibility class: Epichlorohydrins

ENVIRONMENTAL DATA
Food chain concentration potential: Negative; unlikely to accumulate.
Water pollution: DOT Appendix B, §172.101–marine pollutant. Effect of low concentrations on aquatic life is unknown. May be dangerous if it enters nearby water intakes; notify operators. Notify local health and pollution control officials. **Response to discharge:** Issue warning–water contaminant. Restrict access. Disperse and flush.

SHIPPING INFORMATION
Grades of purity: 90% epichlorohydrin. The balance is water (2.5%), 1,2,3-trichloropropane (5%), glycerol (1.8%), isopropyl chloride (0.5%), *n*-propylchloride (0.8%) and others (1.0%); **Storage temperature:** Ambient; **Inert atmosphere:** None; **Venting:** Pressure-vacuum; **Stability during transport:** Stable.

NAS HAZARD CLASSIFICATION FOR BULK WATER TRANSPORTATION
FIRE: 3
HEALTH: Vapor irritant: 3; Liquid or solid irritant: 3; Poisons: 4
WATER POLLUTION: Human toxicity: 3; Aquatic toxicity: 3; Aesthetic effect: 2
REACTIVITY: Other chemicals: 3; Water: 1; Self-reaction: 2

PHYSICAL AND CHEMICAL PROPERTIES
Physical state @ 59°F/15°C and 1 atm: Liquid.
Molecular weight: 92.5
Boiling point @ 1 atm: 240°F/116°C/389°K.
Melting/Freezing point: –118°F/–48°C/225°K.
Specific gravity (water = 1): 1.18 at 68°F/20°C (liquid).
Relative vapor density (air = 1): 3.2
Latent heat of vaporization: (estimate) 142 Btu/lb = 78.8 cal/g = 3.30 x 10^5 J/kg.
Heat of combustion: (approximate) –8100 Btu/lb = –4500 cal/g = –190 x 10^5 J/kg.
Vapor pressure: (Reid) 0.3 psia; 13 mm.

CHLOROMETHYL METHYL ETHER   REC. C:2600

SYNONYMS: CLOROMETILMETIL ETER (Spanish); CMME; DIMETHYLCHLOROETHER; ETHER, DIMETHYL CHLORO; EEC No. 603-075-00-3; ETHER METHYLIQUE MONOCHLORE (French); METHYLCHLOROMETHYL ETHER; METHYLCHLOROMETHYL ETHER, ANHYDROUS; METHANE, CHLOROMETHOXY-; METHOXYMETHYL CHLORIDE; MONOCHLOROMETHYL ETHER; MONOCHLORODIMETHYL ETHER; RCRA No. U046

IDENTIFICATION
CAS Number: 107-30-2
Formula: ClCH$_2$OCH$_3$
DOT ID Number: UN 1239; DOT Guide Number: 131
Proper Shipping Name: Methylchloromethyl ether
Reportable Quantity (RQ): **(CERCLA)** 1 lb/0.454 kg

DESCRIPTION: Colorless liquid. Irritating odor. May float or sink in water; reacts slowly forming formaldehyde and hydrochloric acid; heated water will speed up this reaction.

Flammable • Corrosive to the skin, eyes, and respiratory tract; contact with skin or eyes can cause burns and vision impairment; inhalation symptoms may be delayed. May contain up to 7% bis(chloromethyl)ether, a confirmed human carcinogen • Unstable peroxides may accumulate after prolonged storage in presence of air; peroxides may be detonated by heating, impact, or friction • Containers may BLEVE when exposed to fire •Vapors may form explosive mixture with air • Vapors are heavier than air and will collect and stay in low areas • Vapors may travel long distances to ignition sources and flashback • Vapors in confined areas (e.g., tanks, sewers, buildings) may explode when exposed to fire • Very Irritating to the skin, eyes, and respiratory tract • Toxic products of combustion may include carbon monoxide and hydrogen chloride.

Hazard Classification (based on NFPA-704 Rating System)
Health Hazards (Blue): 2; Flammability (Red): 3; Reactivity (Yellow): 1

EMERGENCY RESPONSE: See Appendix A (131)
Evacuation:
Public safety: Isolate spill area for at least 100 to 200 meters (330 to 660 feet) in all directions.
Spill: Small spill–First: Isolate in all directions 30 meters (100 feet); Then: Protect persons downwind, DAY: 0.2 km (0.1 mile); NIGHT: 0.6 km (0.4 mile). Large spill–First: Isolate in all directions 125 meters (400 feet); Then: Protect persons downwind, DAY: 1.1 km (0.7 mile); NIGHT: 2.7 km (1.7 miles).
Fire: Isolate for 800 meters (½ mile) in all directions, especially if tank, rail car, or tank truck is involved in fire.

EXPOSURE
Short-term effects: *SEEK MEDICAL ATTENTION.* **Vapor:** Irritating to eyes, nose, and throat. *IF INHALED*, will, will cause difficult breathing. Move victim to fresh air. If breathing is difficult, administer oxygen. **Liquid:** Will burn skin and eyes. Harmful if swallowed. Remove contaminated clothing and shoes.Flush affected areas with plenty of water. IF IN EYES, hold eyelids open and flush with plenty of water. IF SWALLOWED and victim is *CONSCIOUS AND ABLE TO SWALLOW*, have victim drink 4 to 8 ounces of water. **Do NOT induce vomiting.**

HEALTH HAZARDS
Personal protective equipment (PPE): A-Level PPE. OSHA Table Z-1-A air contaminant.
Recommendations for respirator selection: NIOSH
At any concentrations above the NIOSH REL, or where there is no REL, at any detectable concentration: SCBAF:PD,PP (any self-contained breathing apparatus that has a full facepiece and is operated in a pressure-demand or other positive-pressure mode); or SAF:PD,PP:ASCBA (any supplied-air respirator that has a full facepiece and is operated in a pressure-demand or other positive-pressure mode in combination with an auxiliary self-contained breathing apparatus operated in a pressure-demand or other positive pressure mode). *ESCAPE:* GMFOV [any air-purifying, full-facepiece respirator (gas mask) with a chin-style, front-or back-mounted organic vapor canister]; or SCBAE (any appropriate escape-type, self-contained breathing apparatus).
Exposure limits (TWA unless otherwise noted): ACGIH TLV suspected human carcinogen; OSHA suspected carcinogen; NIOSH see [29 CFR 1910.1006]; potential human carcinogen; reduce exposure to lowest feasible level.
Toxicity by ingestion: Grade 2; oral LD_{50} = 817 mg/kg (rat).
Long-term health effects: Considered to be lung cancer-producing. NTP known carcinogen; IARC carcinogen, rating 1. May cause liver, kidney and/or nervous system injury.
Vapor (gas) irritant characteristics: Eye and respiratory tract irritant. Vapors are irritating such that personnel will not usually tolerate moderate or high concentrations.
Liquid or solid irritant characteristics: Severe eye irritant. Fairly severe skin irritant. May cause pain and second-degree burns after a few minutes of contact.
IDLH value: Potential human carcinogen.

FIRE DATA
Flash point: 32°F/0°C (oc).
Fire extinguishing agents not to be used: Water may be ineffective.
Behavior in fire: Unburned material may form powerful tear gas. When wet, also forms irritating formaldehyde gas.
Burning rate: 3.0 mm/min

CHEMICAL REACTIVITY
Reactivity with water: Reacts to evolve formaldehyde and hydrogen chloride. The reaction is not violent.
Binary reactants: Will react with surface moisture to evolve hydrogen chloride, which is corrosive to metal.
Neutralizing agents for acids and caustics: Flood with water. Rinse with sodium bicarbonate or lime solution.

ENVIRONMENTAL DATA
Food chain concentration potential: Negative; unlikely to accumulate.
Water pollution: Effect of low concentrations on aquatic life is unknown. Fouling to shoreline. May be dangerous if it enters nearby water intakes; notify operators. Notify local health and wildlife officials. **Response to discharge:** Issue warning–poison, high flammability, air contaminant, corrosive. Restrict access.

SHIPPING INFORMATION
Grades of purity: Commercial; **Storage temperature:** Ambient; **Inert atmosphere:** None; **Venting:** Pressure-vacuum; **Stability during transport:** Stable.

NAS HAZARD CLASSIFICATION FOR BULK WATER TRANSPORTATION
FIRE: 3
HEALTH: Vapor irritant: 3; Liquid or solid irritant: 3; Poisons: 3
WATER POLLUTION: Human toxicity: 2; Aquatic toxicity: 2; Aesthetic effect: 2
REACTIVITY: Other chemicals: 3; Water: 3; Self-reaction: 0

PHYSICAL AND CHEMICAL PROPERTIES
Physical state @ 59°F/15°C and 1 atm: Liquid.
Molecular weight: 80.5
Boiling point @ 1 atm: 140°F/60°C/333°K.
Melting/Freezing point: –154°F/–104°C/170°K.
Specific gravity (water = 1): 1.07 @ 77°F/25°C (liquid).
Liquid surface tension: (estimate) 30 dynes/cm = 0.030 N/m @ 68°F/20°C.
Relative vapor density (air = 1): 2.8
Ratio of specific heats of vapor (gas): 1.1195
Latent heat of vaporization: (estimate) 154 Btu/lb = 85.6 cal/g = 3.58×10^5 J/kg.
Heat of combustion: (estimate) –7300 Btu/lb = –4100 cal/g = –170 $\times 10^5$ J/kg.
Vapor pressure: 192 mm @ 70°F

o-CHLORONITROBENZENE   REC. C:2650

SYNONYMS: CHLORO-*o*-NITROBENZENE; 1-CHLORO-2-NITROBENZENE; 1-CHLORO-2(3)-NITROBENZENE; CHLORONITROBENZENES; CHLORONITROBENZENES(-o);

2-CHLORONITROBENZENE; 2-CHLORO-1-NITROBENZENE; *o*-CLORONITROBENCENO (Spanish); *o*-NITROCHLOROBENZENE; ONCB

IDENTIFICATION
CAS Number: 88-73-3 (*o*-isomer); may also apply to 121-73-3 (*m*-isomer); 100-00-5 (*p*-isomer)
Formula: $C_6H_4ClNO_2$
DOT ID Number: UN 1578; DOT Guide Number: 152
Proper Shipping Name: Chloronitrobenzene, ortho, Liquid.

DESCRIPTION: Yellow crystalline solid or powder. Aromatic odor. Sinks in water; insoluble.

Combustible solid • Severely irritating to skin, eyes, and respiratory tract; skin and eye contact causes severe burns and blindness; may form methemoglobin • Firefighting gear (including SCBA) does not provide adequate protection. If exposure occurs, remove and isolate gear immediately and thoroughly decontaminate personnel • Dust cloud may explode if ignited in an enclosed area • Containers may explode when exposed to fire • Toxic products of combustion may include hydrogen chloride, nitrogen oxides, and phosgene.

Hazard Classification (based on NFPA-704 Rating System)
Health Hazards (Blue): 3; Flammability (Red): 1; Reactivity (Yellow): 0

EMERGENCY RESPONSE: See Appendix A (152)
Evacuation:
Public safety: Isolate the area of spill or leak for at least 25 to 50 meters (80 to 160 feet) in all directions.
Spill: Increase, in the downwind direction, as necessary, the distance shown under "Public Safety."
Fire: If tank, rail car, or tank truck is involved in fire, isolate for at least 800 meters (½ mile) in all directions; also, consider initial evacuation for 800 meters (½ mile) in all directions.

EXPOSURE
Short-term effects: *SEEK MEDICAL ATTENTION.* **Dust:** Irritating to eyes, nose, and throat. *IF INHALED*, will can cause headache, languor, cyanosis, shallow respiration, and coma. Move to fresh air. *IF BREATHING HAS STOPPED*, give artificial respiration; *avoid mouth-to-mouth resuscitation; use bag/mask apparatus.* IF breathing is difficult, administer oxygen. **Solid:** Irritating to skin and eyes. *POISONOUS IF SWALLOWED OR SKIN IS EXPOSED.* Remove contaminated clothing and shoes. Flush affected area with plenty of water. *IF IN EYES*, hold eyelids open and flush with plenty of water. *IF SWALLOWED* and victim is *CONSCIOUS AND ABLE TO SWALLOW*, have victim drink 4 to 8 ounces of water and have victim induce vomiting.

HEALTH HAZARDS
Personal protective equipment (PPE): A-Level PPE. Butyl rubber gloves.
Toxicity by ingestion: Poison. Grade 3; LD_{50} = 50 to 500 mg/kg.
Long-term health effects: Cumulative poison. Weight loss, anemia, weakness and irritability. May effect the blood. Liver, kidney and skin disorders may occur.
Vapor (gas) irritant characteristics: Eye and respiratory tract irritant.
Liquid or solid irritant characteristics: Severe eye irritant. Skin irritant.
IDLH value: 1000 ppm.

FIRE DATA
Flash point: (para-, ortho-) 261°F/128°C (cc).
Flammable limits in air: LEL: 1.38%; UEL: 8.8%.
Fire extinguishing agents not to be used: Water may cause frothing.
Autoignition temperature: 491°F/255°C/528°K (approximate). 500°F/260°C/533°k (meta-).

CHEMICAL REACTIVITY
Binary reactants: Reacts with alkalis, ammonia, reducing agents.
Reactivity group: 42
Compatibility class: Nitrocompounds

ENVIRONMENTAL DATA
Food chain concentration potential: Log P_{ow} = 2.19. Unlikely to accumulate.
Water pollution: Dangerous to aquatic life in high concentrations. May be dangerous if it enters nearby water intakes; notify operators. Notify local health and wildlife officials. **Response to discharge:** Issue warning–poison, water contaminant. Restrict access. Should be removed. Chemical and physical treatment.

SHIPPING INFORMATION
Storage temperature: Cool; **Stability during transport:** Stable.

PHYSICAL AND CHEMICAL PROPERTIES
Physical state @ 59°F/15°C and 1 atm: Solid.
Molecular weight: 157.56
Boiling point @ 1 atm: 475°F/246°C/519°K.
Melting/Freezing point: 91°F/33°C/364°K.
Specific gravity (water = 1): 1.368 at 71°F/22°C (solid).
Liquid surface tension: 43.63 dynes/cm = 0.04363 N/m at 35°C.
Relative vapor density (air = 1): 5.4
Vapor pressure: Less than 0.08 mm.

---

o-CHLOROPHENOL   REC. C:2700

SYNONYMS: 1-CHLORO-2-HYDROXYBENZENE; 2-CHLOROPHENOL; 2-CHLORO-1-HYDROXYBENZENE; *o*-CHLRPHENOL (German); EEC No. 604-008-00-0; 2-HYDROXYCHLOROBENZENE; PHENOL, *o*-CHLORO; *o*-MONOCHLOROPHENOL; PHENOL, 2-CHLORO-; PHENOL,2-CHLORO-; PHENOL, *o*-CHLORO-; RCRA No. U048

IDENTIFICATION
CAS Number: 95-57-8
Formula: C_6H_5ClO
DOT ID Number: UN 2021; DOT Guide Number: 153
Proper Shipping Name: Chlorophenols, Liquid
Reportable Quantity (RQ): **(CERCLA)** 100 lb/45.4 kg

DESCRIPTION: Colorless or amber liquid. Unpleasant, penetrating odor. Sinks and slowly mixes.

Combustible • Poison! • Corrosive • Breathing the vapor can kill you; skin or eye contact causes severe burns, impaired vision, or blindness; effects of contact or inhalation may be delayed • Firefighting gear (including SCBA) does not provide adequate protection. If exposure occurs, remove and isolate gear immediately and thoroughly decontaminate personnel • Containers may BLEVE when exposed to fire • Toxic products of combustion may include hydrogen chloride

Hazard Classification (based on NFPA-704 Rating System)
Health Hazards (Blue): 3; Flammability (Red): 2; Reactivity (Yellow): 0

EMERGENCY RESPONSE: See Appendix A (153)
Evacuation:
Public safety: Isolate the area of spill or leak for at least 25 to 50 meters (80 to 160 feet) in all directions.
Spill: Increase, in the downwind direction, as necessary, the distance shown under "Public Safety."
Fire: If tank, rail car, or tank truck is involved in fire, isolate for at least 800 meters (½ mile) in all directions; also, consider initial evacuation for 800 meters (½ mile) in all directions.
EXPOSURE
Short-term effects: *SEEK MEDICAL ATTENTION.* **Vapor:** *POISONOUS. MAY BE FATAL IF INHALED OR ABSORBED THROUGH SKIN.* Inhalation can cause liver and kidney damage. Irritating to skin and eyes. Move victim to fresh air. *IF BREATHING HAS STOPPED,* give artificial respiration; *avoid mouth-to-mouth resuscitation; use bag/mask apparatus.* IF breathing is difficult, administer oxygen. **Liquid:** *POISONOUS. MAY BE FATAL IF SWALLOWED OR ABSORBED THROUGH SKIN.* Can cause severe skin and eye irritation; may cause burns. *IF IN EYES OR ON SKIN,* flush contaminated area with running water for at least 15 minutes; hold upper and lower eyelids open occasionally if appropriate. *SPEED IN REMOVING MATERIAL FROM SKIN IS EXTREMELY IMPORTANT.* Remove and isolate contaminated clothing and shoes. *IF SWALLOWED* and victim is *UNCONSCIOUS OR HAVING CONVULSIONS,* do nothing except keep victim warm.

HEALTH HAZARDS
Personal protective equipment (PPE): B-Level PPE. Chemical protective material(s) reported to have good to excellent resistance: butyl rubber, polycarbonate.
Toxicity by ingestion: Poison. Grade 2; LD_{50} = 670 mg/kg (mouse; rat).
Long-term health effects: It produced tumorigenic effects and reproductive effects. Rat toxicity studies showed marked injury to the kidneys, fatty infiltration of the liver, and hemorrhages in the intestines.
Inhalation can cause liver and kidney damage.
Vapor (gas) irritant characteristics: Eye and respiratory tract irritant. The vapors are irritating and toxic.
Liquid or solid irritant characteristics: Severe eye irritant. Strong irritant to tissue.
Odor threshold: 0.019 mg/m^3.

FIRE DATA
Flash point: 147°F/64°C (cc).
Electrical hazard: Class I, Group D.

CHEMICAL REACTIVITY
Binary reactants: Oxidizers, copper, copper alloys, aluminum, organic acids.
Neutralizing agents for acids and caustics: Sodium bicarbonate

ENVIRONMENTAL DATA
Food chain concentration potential: Log P_{ow} = 2.12–2.2. Unlikely to accumulate.
Water pollution: Harmful to aquatic life in very low concentrations. May be dangerous if it enters nearby water intakes; notify operators. Notify local health and wildlife officials.
Response to discharge: Issue warning–poison, water contaminant. Restrict access. Remove. Chemical and physical treatment.

SHIPPING INFORMATION
Stability during transport: Stable.

PHYSICAL AND CHEMICAL PROPERTIES
Physical state @ 59°F/15°C and 1 atm: Liquid.
Molecular weight: 128.56
Boiling point @ 1 atm: 346°F/175°C/448°K.
Melting/Freezing point: 49°F/9°C/283°K.
Specific gravity (water = 1): 1.25 @ 77°F/25°C.
Liquid surface tension: 40.3 dynes/cm = 0.040 N/m @ 68°F/20°C.
Relative vapor density (air = 1): 4.5
Latent heat of vaporization: 144.8 Btu/lb = 80.4 cal/g = 3.4 x 10^6 J/kg.
Vapor pressure: 1 mm.

p-CHLOROPHENOL   REC. C:2750

SYNONYMS: 4-CHLORO-1-HYDROXYBENZENE; CHLOROPHENATE; 4-CHLOROPHENOL; *p*- CLOROFENOL (Spanish); EEC No. 604-008-00-0; 4-HYDROXYCHLOROBENZENE; PARACHLOROPHENOL

IDENTIFICATION
CAS Number: 106-48-9
Formula: C_6H_5ClO; 1,4-ClC_6H_4OH
DOT ID Number: UN 2020; DOT Guide Number: 153
Proper Shipping Name: Chlorophenols, solid.

DESCRIPTION: White or pale yellow crystalline solid. Medicinal odor. Sinks in water.

Combustible solid • Poison! • Absorption through the skin or swallowing the material can kill you; irritates the skin, eyes, and respiratory tract • Firefighting gear (including SCBA) does not provide adequate protection. If exposure occurs, remove and isolate gear immediately and thoroughly decontaminate personnel • Containers may BLEVE when exposed to fire • Toxic products of combustion may include hydrogen chloride.

Hazard Classification (based on NFPA-704 Rating System)
Health Hazards (Blue): 3; Flammability (Red): 2; Reactivity (Yellow): 0

EMERGENCY RESPONSE: See Appendix A (153)
Evacuation:
Public safety: Isolate the area of spill or leak for at least 25 to 50 meters (80 to 160 feet) in all directions.
Spill: Increase, in the downwind direction, as necessary, the distance shown under "Public Safety."
Fire: If tank, rail car, or tank truck is involved in fire, isolate for at least 800 meters (½ mile) in all directions; also, consider initial evacuation for 800 meters (½ mile) in all directions.

EXPOSURE
Short-term effects: *SEEK MEDICAL ATTENTION.* **Dust:** Absorbed through the skin. Irritating to eyes, nose, and throat. *IF INHALED,* will, will cause headache or dizziness. *IF IN EYES,* hold eyelids open and flush with plenty of water. *IF BREATHING HAS STOPPED,* give artificial respiration; *avoid mouth-to-mouth resuscitation; use bag/mask apparatus.* IF breathing is difficult, administer oxygen. **Solid:** Will burn skin and eyes. *IF SWALLOWED,* will cause nausea and vomiting. Remove contaminated clothing and shoes. Flush affected areas with plenty of water. *IF IN EYES,* hold eyelids open and flush with plenty of

water. *IF SWALLOWED* and victim is *CONSCIOUS AND ABLE TO SWALLOW*, have victim drink 4 to 8 ounces of water. *IF SWALLOWED* and victim is *UNCONSCIOUS OR HAVING CONVULSIONS*, do nothing except keep victim warm. **Do NOT induce vomiting.**

HEALTH HAZARDS
Personal protective equipment (PPE): A-Level PPE. Avoid all skin contact. Chemical protective material(s) reported to have good to excellent resistance: butyl rubber, polycarbonate.
Toxicity by ingestion: Grade 3; oral LD_{50} = 500 mg/kg (rat).
Long-term health effects: Mutation data. Possible liver and kidney damage; chloracne.
Vapor (gas) irritant characteristics: Eye and respiratory tract irritant.
Liquid or solid irritant characteristics: Severe eye and skin irritant.
Odor threshold: 30 ppm.

FIRE DATA
Flash point: 250°F/121°C.
Electrical hazard: Class I, Group D.

CHEMICAL REACTIVITY
Binary reactants: Reacts with oxidizers, copper and copper alloys and aluminum.

ENVIRONMENTAL DATA
Food chain concentration potential: Log P_{ow} = 2.39. Unlikely to accumulate.
Water pollution: Harmful to aquatic life in very low concentrations. May be dangerous if it enters nearby water intakes; notify operators. Notify local health and wildlife officials.
Response to discharge: Issue warning–water contaminant. Should be removed. Chemical and physical treatment.

SHIPPING INFORMATION
Grades of purity: Pure, 99%; **Storage temperature:** Ambient; **Inert atmosphere:** None; **Venting:** Open; **Stability during transport:** Stable.

PHYSICAL AND CHEMICAL PROPERTIES
Physical state @ 59°F/15°C and 1 atm: Solid.
Molecular weight: 128.6
Boiling point @ 1 atm: 428°F/220°C/493°K.
Melting/Freezing point: 109°F/43°C/316°K.
Specific gravity (water = 1): 1.31 @ 68°F/20°C (solid).
Relative vapor density (air = 1): 4.5
Latent heat of vaporization: 160 Btu/lb = 89 cal/g = 3.7×10^5 J/kg.
Heat of combustion: –9,330 Btu/lb = –5,180 cal/g = -217×10^5 J/kg.
Vapor pressure: 0.1 mm.

CHLOROPICRIN   **REC. C:2800**

SYNONYMS: ACQUINITE; CHLOR-O-PIC; CHLOROPICRINE (French); CHLORPIKRIN (German); CHLRPICRINA (Spanish); EEC No. 610-001-00-3; LARVACIDE; METHANE, TRICHLORONITRO-; MYCROLYSIN; NITROTRICHLOROMETHANE; NITROCHLOROFORM; PICFUME; PIC-CHLOR; PICRIDE; PROFUME A; PS; TRICHLORONITROMETHANE; TRICHLOR

IDENTIFICATION
CAS Number: 76-06-2
Formula: Cl_3CNO_2
DOT ID Number: UN 1580; DOT Guide Number: 154
Proper Shipping Name: Chloropicrin

DESCRIPTION: Colorless to pale yellow, oily liquid. Odor like licorice. Intensely irritating odor which causes a pronounced secretion of tears and lung irritation. Sinks in water; insoluble. Has been used as a lung irritant in warfare.

Poison! • Corrosive • Rapid heating can cause explosion; decomposes explosively when heated above 234°F/112°C • Breathing the vapor can kill you; skin or eye contact causes severe burns, impaired vision, or blindness; affects the blood • Firefighting gear (including SCBA) does not provide adequate protection. If exposure occurs, remove and isolate gear immediately and thoroughly decontaminate personnel • Containers may BLEVE when exposed to fire • Compound forms a powerful tear gas when heated. Toxic products of combustion may include phosgene, hydrogen chloride, and nitrogen oxides • Do NOT attempt rescue.

Hazard Classification (based on NFPA-704 Rating System)
Health Hazards (Blue): 4; Flammability (Red): 0; Reactivity (Yellow): 3

EMERGENCY RESPONSE: See Appendix A (154)
Evacuation:
Public safety: Isolate the area of spill or leak for at least 25 to 50 meters (80 to 160 feet) in all directions.
Spill: Small spill–First: Isolate in all directions 60 meters (200 feet); Then: Protect persons downwind, DAY: 0.5 km (0.3 mile); NIGHT: 1.3 km (0.8 mile). Large spill–First: Isolate in all directions 185 meters (600 feet); Then: Protect persons downwind, DAY: 1.8 km (1.1 miles); NIGHT: 4.0 km (2.5 miles).
Fire: If tank, rail car, or tank truck is involved in fire, isolate for at least 800 meters (½ mile) in all directions; also, consider initial evacuation for 800 meters (½ mile) in all directions.

EXPOSURE
Short-term effects: *SEEK MEDICAL ATTENTION*. **Vapor:** *POISONOUS IF INHALED*. Extremely irritating to eyes, nose, and lungs. Move victim to fresh air. *IF BREATHING HAS STOPPED, avoid mouth-to-mouth resuscitation; use bag/mask apparatus.* Artificial respiration may cause lung damage. IF breathing is difficult, administer oxygen. May cause lung edema; physical exertion will aggravate this condition. **Liquid:** *POISONOUS IF SWALLOWED OR IF SKIN IS EXPOSED*. Will burn skin and eyes. Remove contaminated clothing and shoes. Flush affected areas with plenty of water. *IF IN EYES*, hold eyelids open and flush with plenty of water. *IF SWALLOWED* and victim is *CONSCIOUS AND ABLE TO SWALLOW*, have victim drink 4 to 8 ounces of water. **Do NOT induce vomiting.**
Note to physician or authorized medical personnel: To relieve irritation of the respiratory tract consider a spray of 10% sodium thiosulfate. Keep the victim lying down. The slightest exertion, including walking, may result in cardiac arrest. Medical observation is recommended for 24 to 48 hours after breathing overexposure, as pulmonary edema or bronchopneumonia may be delayed. As first aid for pulmonary edema, consider administering a corticosteroid spray. Cigarette smoking may exacerbate pulmonary injury and should be discouraged for at least 72 hours following exposure.

HEALTH HAZARDS
Personal protective equipment (PPE): A-Level PPE. OSHA

Table Z-1-A air contaminant. Chemical protective material(s) reported to have good to excellent resistance: Teflon®.
Recommendations for respirator selection: NIOSH/OSHA
2 ppm: SA:CF* (any supplied-air respirator operated in a continuous-flow mode); or PAPROV* [any powered, air-purifying respirator with organic vapor cartridge(s)]; or CCRFOV [any chemical cartridge respirator with a full facepiece and organic vapor cartridge(s)]; or GMFOV [any air-purifying, full-facepiece respirator (gas mask) with a chin-style, front- or back-mounted acid gas canister]; or SCBAF (any self-contained breathing apparatus with a full facepiece); or SAF (any supplied-air respirator with a full facepiece). *EMERGENCY OR PLANNED ENTRY INTO UNKNOWN CONCENTRATIONS OR IDLH CONDITIONS:* SCBAF:PD,PP (any self-contained breathing apparatus that has a full facepiece and is operated in a pressure-demand or other positive-pressure mode); or SAF:PD,PP:ASCBA (any supplied-air respirator that has a full facepiece and is operated in a pressure-demand or other positive-pressure mode in combination with an auxiliary self-contained breathing apparatus operated in a pressure-demand or other positive pressure mode). ESCAPE: GMFOV[any air-purifying, full-facepiece respirator (gas mask) with a chin-style, front- or back-mounted organic vapor cannister]; or SCBAE (any appropriate escape-type, self-contained breathing apparatus). *Note:* Substance causes eye irritation or damage; eye protection needed.
Exposure limits (TWA unless otherwise noted): ACGIH TLV 0.1 ppm (6.7 mg/m^3); NIOSH/OSHA 0.1 ppm (0.7 mg/m^3).
Toxicity by ingestion: Poison. Grade 3; oral LD$_{50}$ = 250 mg/kg (rat).
Long-term health effects: Possible blood damage and heart muscle injury. Mutation data reported.
Vapor (gas) irritant characteristics: Vapors cause severe irritation of eyes and throat and can cause eye or lung injury. Cannot be tolerated even at low concentrations.
Liquid or solid irritant characteristics: Severe eye and skin irritant. Causes second-and third-degree burns on short contact and is very injurious to the eyes.
Odor threshold: 0.8 ppm. *Note:* This value is above published exposure limits.
IDLH value: 2 ppm.

FIRE DATA
Fire extinguishing agents not to be used: Water may not be effective.
Behavior in fire: Heated material may detonate under fire conditions. Closed containers may rupture and explode.

CHEMICAL REACTIVITY
Binary reactants: A strong oxidizer. May be self-reactive; can decompose rapidly and explode in rapidly elevating temperatures and when subjected to shock. Reacts with combustibles, propargyl bromide, alkaline earth metals, alkali metals, finely divided metals, and reducing agents.

ENVIRONMENTAL DATA
Food chain concentration potential: Log P$_{ow}$ = 2.38–2.4. Unlikely to accumulate.
Water pollution: DOT Appendix B, §172.101–marine pollutant. Effect of low concentrations on aquatic life is unknown. May be dangerous if it enters nearby water intakes; notify operators. Notify local health and wildlife officials. **Response to discharge:** Issue warning–poison, water contaminant, air contaminant. Restrict access; evacuate area. Should be removed. Chemical and physical treatment.

SHIPPING INFORMATION
Grades of purity: 99%; **Storage temperature:** Ambient; **Inert atmosphere:** None; **Venting:** Pressure-vacuum
Stability during transport: Stable. Shock or friction may cause decomposition.

NAS HAZARD CLASSIFICATION FOR BULK WATER TRANSPORTATION
FIRE: 0
HEALTH: Vapor irritant: 4; Liquid or solid irritant: 4; Poisons: 4
WATER POLLUTION: Human toxicity: 4; Aquatic toxicity: 3; Aesthetic effect: 4
REACTIVITY: Other chemicals: 1; Water: 0; Self-reaction: 0

PHYSICAL AND CHEMICAL PROPERTIES
Physical state @ 59°F/15°C and 1 atm: Liquid.
Molecular weight: 164.4
Boiling point @ 1 atm: 234°F/112°C/385°K.
Melting/Freezing point: –147°F/–64°C.
Specific gravity (water = 1): 1.64 @ 77°F/25°C (liquid).
Liquid surface tension: 32.3 dynes/cm = 0.0323 N/m @ 68°F/20°C.
Liquid water interfacial tension: (estimate) 30 dynes/cm = 0.03 N/m @ 68°F/20°C.
Relative vapor density (air = 1): 5.7
Ratio of specific heats of vapor (gas): 1.0991
Latent heat of vaporization: 103 Btu/lb = 57.3 cal/g = 2.4 x 10^5 J/kg.
Heat of fusion: 48.16 cal/g.
Vapor pressure: 18 mm.

CHLOROPRENE   REC. C:2850

SYNONYMS: CHLOROBUTADIENE; 2-CHLOROBUTADIENE; 2-CHLORO-1,3-BUTADIENE; 2-CHLORO-1,3-DIENE; β-CHLOROPRENE; β-CHLOROPRENE; β-CLOROPRENO (Spanish); EEC No. 602-036-00-8; NEOPRENE

IDENTIFICATION
CAS Number: 126-99-8
Formula: C$_4$H$_5$Cl; CH$_2$CHCCICH$_2$
DOT ID Number: UN 1991; DOT Guide Number: 131P
Proper Shipping Name: Chloroprene, inhibited
Reportable Quantity (RQ): **(CERCLA)** 1 lb/0.454 kg

DESCRIPTION: Colorless liquid. Pungent, ether-like. Floats and mixes slowly with water; slightly soluble. Flammable, irritating vapor is produced.

Highly flammable • Polymerization hazard • Heat can induce polymerization with rapid release of energy; sealed containers may rupture explosively • May react with itself blocking relief valves; leading to tank explosions • Containers may BLEVE when exposed to fire •Vapors may form explosive mixture with air • Vapors are heavier than air and will collect and stay in low areas • Vapors may travel long distances to ignition sources and flashback • Vapors in confined areas (e.g., tanks, sewers, buildings) may explode when exposed to fire • Irritating to the skin, eyes, and respiratory tract • Toxic products of combustion may include phosgene and hydrogen chloride.

Hazard Classification (based on NFPA-704 Rating System)
Health Hazards (Blue): 2; Flammability (Red): 3; Reactivity (Yellow): 0

EMERGENCY RESPONSE: See Appendix A (131)
Evacuation:
Public safety: Isolate spill area for at least 100 to 200 meters (330 to 660 feet) in all directions.
Spill: Increase, as necessary, the isolation distance shown above, in "Public safety."
Fire: Isolate for 800 meters (½ mile) in all directions, especially if tank, rail car, or tank truck is involved in fire.

EXPOSURE
Short-term effects: *SEEK MEDICAL ATTENTION. ABSORBED THROUGH THE SKIN.* **Vapor:** Irritating to eyes, nose, and throat. *IF INHALED*, will, will cause difficult breathing and asphyxia. Move to fresh air. *IF BREATHING HAS STOPPED*, give artificial respiration. IF breathing is difficult, administer oxygen. **Liquid:** Irritating to skin and eyes. Harmful if swallowed. Remove contaminated clothing and shoes. Flush affected areas with plenty of water. *IF IN EYES*, hold eyelids open and flush with plenty of water. *IF SWALLOWED* and victim is *CONSCIOUS AND ABLE TO SWALLOW*, have victim drink 4 to 8 ounces of water.

HEALTH HAZARDS
Personal protective equipment (PPE): A-Level PPE. Full face organic vapor respirator. Chemical protective material(s) reported to have good to excellent resistance: PV alcohol, Viton®.
Recommendations for respirator selection: NIOSH
At any concentrations above the NIOSH REL, or where there is no REL, at any detectable concentration: SCBAF:PD,PP (any self-contained breathing apparatus that has a full facepiece and is operated in a pressure-demand or other positive-pressure mode); or SAF:PD,PP:ASCBA (any supplied-air respirator that has a full facepiece and is operated in a pressure-demand or other positive-pressure mode in combination with an auxiliary self-contained breathing apparatus operated in a pressure-demand or other positive pressure mode). *ESCAPE:* GMFOV [any air-purifying, full-facepiece respirator (gas mask) with a chin-style, front-or back-mounted organic vapor canister]; or SCBAE (any appropriate escape-type, self-contained breathing apparatus).
Exposure limits (TWA unless otherwise noted): ACGIH TLV 10 ppm (36 mg/m^3); OSHA PEL 25 ppm (90 mg/m^3); NIOSH ceiling 1 ppm (3.6 mg/m^3)/15 min; potential human carcinogen; reduce exposure to lowest feasible level; skin contact contributes significantly in overall exposure; suspected carcinogen.
Toxicity by ingestion: Grade 3; LD$_{50}$ = 50 to 500 mg/kg.
Long-term health effects: May cause kidney and liver damage, dermatitis, conjunctivitis, corneal necrosis, anemia, loss of hair (temporary), nervousness, and irritability. CNS depression and significant injury to lungs, liver, and kidneys. Suspected carcinogen, and mutagen. In animal experiments has caused degenerative changes of reproductive organs with the males being more susceptible.
Vapor (gas) irritant characteristics: Eye and respiratory tract irritant. Vapors are irritating such that personnel will not usually tolerate moderate or high concentrations.
Liquid or solid irritant characteristics: Eye and skin irritant.
Odor threshold: 0.1–138 ppm
IDLH value: 300 ppm; potential human carcinogen

FIRE DATA
Flash point: –4°F/–20°C (cc).
Flammable limits in air: LEL: 2.6%; UEL: 20%.
Fire extinguishing agents not to be used: Water may be ineffective.
Behavior in fire: Possible polymerization and container explosion.

Electrical hazard: Due to low electric conductivity, this substance may generate electrostatic charges as a result of agitation and flow.

CHEMICAL REACTIVITY
Binary reactants: Peroxides and other oxidizers, finely divided metals. Attacks some plastic materials.
Polymerization: Polymerizes readily under the influence of light and catalysts; at room temperature, unless inhibited with antioxidants; **Inhibitor of polymerization:** Antioxidants. May still polymerize at high temperatures.

ENVIRONMENTAL DATA
Food chain concentration potential: Log P$_{ow}$ = 2.09. Unlikely to accumulate.
Water pollution: Harmful to aquatic life in very low concentrations. May be dangerous if it enters nearby water intakes; notify operators. Notify local health and wildlife officials. **Response to discharge:** Issue warning–high flammability. Evacuate area. Chemical and physical treatment.

SHIPPING INFORMATION
Storage temperature: Cool

PHYSICAL AND CHEMICAL PROPERTIES
Physical state @ 59°F/15°C and 1 atm: Liquid.
Molecular weight: 88.54
Boiling point @ 1 atm: 139°F/59°C/333°K.
Melting/Freezing point: –266°F/–130°C/143°K.
Specific gravity (water = 1): 0.9583 @ 68°F/20°C.
Relative vapor density (air = 1): 3.0
Latent heat of vaporization: Estimated at 164 Btu/lb = 91.2 cal/g = 3.8 x 10^5 J/kg.
Heat of polymerization: per mode of monomer at 61.3°C (142.34°F) 3.30 Btu/lb = 183.6 cal/g = 7.68 x 10^5 J/kg.
Vapor pressure: 190 mm.

2-CHLOROPROPIONIC ACID   **REC. C:2900**

SYNONYMS: α-CHLOROPROPIONIC ACID; PROPANOIC ACID, 2-CHLORO-; EEC No. 607-139-00-1

IDENTIFICATION
CAS Number: 598-78-7
Formula: C$_3$H$_5$ClO$_2$; CH$_3$CHClCOOH
DOT ID Number: UN 2511; DOT Guide Number: 153
Proper Shipping Name: 2-Chloropropionic acid

DESCRIPTION: Pale yellow liquid. Slight pungent odor. Sinks and mixes with water.

Combustible • Corrosive to the skin, eyes, and respiratory tract; skin or eye contact causes severe burns, impaired vision, or blindness; effects of contact or inhalation may be delayed • Containers may BLEVE when exposed to fire • Vapors in confined areas (e.g., tanks, sewers, buildings) may explode when exposed to fire • Extremely irritating to skin, eyes, and respiratory tract; skin and eye contact causes severe burns and blindness • Toxic products of combustion may include hydrogen chloride • Corrosive to many metals.

Hazard Classification (based on NFPA-704 Rating System)
Health Hazards (Blue): 2; Flammability (Red): 1; Reactivity (Yellow): 0

EMERGENCY RESPONSE: See Appendix A (153)
Evacuation:
Public safety: Isolate the area of spill or leak for at least 25 to 50 meters (80 to 160 feet) in all directions.
Spill: Increase, in the downwind direction, as necessary, the distance shown under "Public Safety."
Fire: If tank, rail car, or tank truck is involved in fire, isolate for at least 800 meters (½ mile) in all directions; also, consider initial evacuation for 800 meters (½ mile) in all directions.

EXPOSURE
Short-term effects: *SEEK MEDICAL ATTENTION. HARMFUL IF ABSORBED THROUGH THE SKIN.* **Vapor:** Harmful if inhaled. May cause lung and eye injury. *IF BREATHING HAS STOPPED,* give artificial respiration; *avoid mouth-to-mouth resuscitation; use bag/mask apparatus.* If breathing is difficult, administer oxygen. May cause lung edema; physical exertion will aggravate this condition. **Liquid:** Will burn skin and eyes. Harmful if absorbed through skin. Harmful if swallowed. Remove contaminated clothing and shoes at the site. Flush with running water for at least 15 minutes; hold eyelids open if necessary. *IF IN EYES OR ON SKIN,* wash skin with soap and water. *IF SWALLOWED* and victim is *CONSCIOUS AND ABLE TO SWALLOW,* have victim drink 4 to 8 ounces of water. **Do NOT induce vomiting.** *IF SWALLOWED* and victim is *UNCONSCIOUS OR HAVING CONVULSIONS,* do nothing except keep victim warm.
Note to physician or authorized medical personnel: Medical observation is recommended for 24 to 48 hours after breathing overexposure, as pulmonary edema may be delayed. As first aid for pulmonary edema, consider administering a corticosteroid spray. Cigarette smoking may exacerbate pulmonary injury and should be discouraged for at least 72 hours following exposure.

HEALTH HAZARDS
Personal protective equipment (PPE): B-Level PPE. Organic vapor/acid gas respirator and full protective clothing.
Exposure limits (TWA unless otherwise noted): ACGIH TLV 0.1 ppm.
Toxicity by ingestion: Grade 2; LD_{50} = 500 mg/kg (rat).
Long-term health effects: Lung damage.
Vapor (gas) irritant characteristics: Vapors cause severe irritation of eyes and throat and can cause eye and lung injury. They cannot be tolerated even at low concentrations.
Liquid or solid irritant characteristics: Corrosive eye irritant. Causes smarting of the skin and first-degree burns on short exposures; may cause second-degree burns on long exposure.

FIRE DATA
Flash point: 225°F/107°C (cc).
Autoignition temperature: 932°F/500°C/773°K.
Electrical hazard: Not applicable.

CHEMICAL REACTIVITY
Reactivity with water: An acid in solution.
Binary reactants: Only aluminum, stainless steel, steel covered with a protective lining, or coating may contact the liquid or vapor. Reacts with alkaline material and strong oxidizers.
Reactivity group: 4
Compatibility class: Organic acids

ENVIRONMENTAL DATA
Food chain concentration potential: Log P_{ow} = 0.9. Unlikely to accumulate.
Water pollution: Effect of low concentrations on aquatic life is unknown. May be dangerous if it enters nearby water intakes; notify operators. Notify local health and wildlife officials. **Response to discharge:** Issue warning–corrosive, water contaminant. Restrict access. Should be removed. Chemical and physical treatment.

SHIPPING INFORMATION
Grades of purity: 99%; **Storage temperature:** Ambient; **Inert atmosphere:** No; **Venting:** Open; **Stability during transport:** Stable.

PHYSICAL AND CHEMICAL PROPERTIES
Physical state @ 59°F/15°C and 1 atm: Liquid.
Molecular weight: 108.53
Boiling point @ 1 atm: 366°F/186°C/459°K.
Melting/Freezing point: −54°F/−12°C/261°K.
Critical temperature: 750°F/399°C/672°K (estimate).
Specific gravity (water = 1): 1.2585 @ 68°F/20°C.
Relative vapor density (air = 1): 3.7
Latent heat of vaporization: 175 Btu/lb = 97 cal/g = 4.06X 10^5 J/kg.
Vapor pressure: 0.25 mm.

3-CHLOROPROPIONIC ACID REC. C:2950

SYNONYMS: ACIDO 3-CLOROPROPIONICO (Spanish); 3-CHLOROPROPANOIC ACID; β-CHLOROPROPIONIC ACID; EEC No. 607-139-00-1; β-MONOCHLOROPROPIONIC ACID; PROPIONIC ACID, 3-CHLORO-

IDENTIFICATION
CAS Number: 107-94-8
Formula: $C_3H_5ClO_2$; $ClCH_2CH_2COOH$
DOT ID Number: UN 2511; DOT Guide Number: 153
Proper Shipping Name: Chloropropionic acid

DESCRIPTION: Off-white to tan crystalline solid. Sharp odor. Sinks and mixes with water.

Combustible • Containers may BLEVE when exposed to fire • Vapors in confined areas (e.g., tanks, sewers, buildings) may explode when exposed to fire • Extremely irritating to skin, eyes, and respiratory tract; skin and eye contact causes severe burns and blindness • Toxic products of combustion may include hydrogen chloride • Corrosive to many metals.

Hazard Classification (based on NFPA-704 Rating System)
Health Hazards (Blue): 2; Flammability (Red): 1; Reactivity (Yellow): 0

EMERGENCY RESPONSE: See Appendix A (153)
Evacuation:
Public safety: Isolate the area of spill or leak for at least 25 to 50 meters (80 to 160 feet) in all directions.
Spill: Increase, in the downwind direction, as necessary, the distance shown under "Public Safety."
Fire: If tank, rail car, or tank truck is involved in fire, isolate for at least 800 meters (½ mile) in all directions; also, consider initial evacuation for 800 meters (½ mile) in all directions.

EXPOSURE
Short-term effects: *SEEK MEDICAL ATTENTION. IF BREATHING HAS STOPPED,* give artificial respiration; *avoid mouth-to-mouth resuscitation; use bag/mask apparatus.* **Solid:** Irritating to skin and eyes. Harmful if swallowed. *IF IN EYES OR*

ON SKIN, flush with running water for at least 15 minutes; hold eyelids open if necessary. *IF SWALLOWED* and victim is *UNCONSCIOUS OR HAVING CONVULSIONS,* do nothing except keep victim warm.

HEALTH HAZARDS
Personal protective equipment (PPE): B-Level PPE. Wear self-contained positive pressure breathing apparatus and full protective clothing.
Toxicity by ingestion: Grade 2; LD_{50} = more than 2.0 g/kg (mouse).
Long-term health effects: Tumorigenic toward the lungs of mice. Possible mutagen.
Vapor (gas) irritant characteristics: Eye and respiratory tract irritant.
Liquid or solid irritant characteristics: Severe eye and skin irritant. Causes smarting of the skin and first-degree burns on short exposure; may cause second-degree burns on long exposures.

FIRE DATA
Flash point: More than 230°F/110°C (cc).

CHEMICAL REACTIVITY
Binary reactants: Only aluminum, stainless steel or steel covered with a protective lining or coating may contact the liquid or vapor.
Reactivity group: 4
Compatibility class: Organic acids

ENVIRONMENTAL DATA
Water pollution: Effects of low concentrations on aquatic life is unknown. May be dangerous if it enters nearby water intakes; notify operators. Notify local health and wildlife officials.
Response to discharge: Issue warning–corrosive, water contaminant. Restrict access. Should be removed. Chemical and physical treatment.

SHIPPING INFORMATION
Grades of purity: 99%; **Storage temperature:** Ambient ; **Venting:** Open; **Stability during transport:** Stable.

PHYSICAL AND CHEMICAL PROPERTIES
Physical state @ 59°F/15°C and 1 atm: Solid.
Molecular weight: 108.53
Boiling point @ 1 atm: 392°F/200°C/473.2°K.
Melting/Freezing point: 106°F/41°C/314°K.
Critical temperature: 787°F/420°C/693°K (estimate).
Specific gravity (water = 1): 1.26

CHLOROSULFONIC ACID   REC. C:3000

SYNONYMS: ACIDO CLOROSULFONICO (Spanish); CHLOROSULFURIC ACID; EEC No. 016-017-00-1; MONOCHLOROSULFURIC ACID; SULFONIC ACID, MONOCHLORIDE; SULFURIC CHLOROHYDRIN

IDENTIFICATION
CAS Number: 7790-94-5
Formula: $ClSO_3H$
DOT ID Number: UN 1754; DOT Guide Number: 137
Proper Shipping Name: Chlorosulfonic acid (with or without sulfur trioxide)
Reportable Quantity (RQ): **(CERCLA)** 1000 lb/454 kg

DESCRIPTION: Colorless to pale yellow liquid. Sharp, choking odor. Reacts violently with water; appears to explode, forming toxic hydrochloric acid and sulfuric acids.

Poison! • Corrosive • Breathing the vapor can kill you; skin or eye contact causes severe burns, impaired vision, or blindness; symptoms from inhalation may be delayed • Do not use water • Firefighting gear (including SCBA) provides NO protection. If exposure occurs, remove and isolate gear immediately and thoroughly decontaminate personnel • Extremely strong oxidizer. Heat forms oxygen; will increase the activity of an existing fire • Vapors are heavier than air and will collect and stay in low areas • May cause fire or explosion contact with combustibles (wood, paper, oil, clothing, etc.) • Containers may BLEVE when exposed to fire • Toxic products of combustion may include sulfur oxides and hydrogen chloride • Do NOT attempt rescue.

Hazard Classification (based on NFPA-704 Rating System)
Health Hazards (Blue): 4; Flammability (Red): 0; Reactivity (Yellow): 2; Special Notice (White): Water reactive

EMERGENCY RESPONSE: See Appendix A (137)
Evacuation:
Public safety: See below.
Spill: Small spill–First: Isolate in all directions 30 meters (100 feet) (land or water); Then: Protect persons downwind, DAY: 0.2 km (0.1 mile) (land or water); NIGHT: 0.2 km (0.1 mile) (land or water). Large spill–First: Isolate in all directions 30 meters (100 feet) (land); 60 meters (200 feet) (water); Then: Protect persons downwind, DAY: 0.2 km (0.1 mile) (land); 0.5 km (0.3 mile) (water); NIGHT: 0.5 km (0.3 mile) (land); 1.4 km (0.9 mile) (water).
Fire: If any large container is involved in fire, isolate for at least 800 meters (½ mile) in all directions; also, consider initial evacuation for 800 meters (½ mile) in all directions.

EXPOSURE
Short-term effects: *SEEK MEDICAL ATTENTION.* **Vapor:** Irritating to eyes, nose, and throat. Harmful if inhaled. Move to fresh air. *IF BREATHING HAS STOPPED,* give artificial respiration; *avoid mouth-to-mouth resuscitation; use bag/mask apparatus.* IF breathing is difficult, administer oxygen. May cause lung edema; physical exertion will aggravate this condition.
Liquid: Will burn skin and eyes. Harmful if swallowed. Remove contaminated clothing and shoes. Flush affected areas with plenty of water. *IF IN EYES,* hold eyelids open and flush with plenty of water. *IF SWALLOWED* and victim is *CONSCIOUS AND ABLE TO SWALLOW,* have victim drink 4 to 8 ounces of water. **Do NOT induce vomiting.**
Note to physician or authorized medical personnel: Medical observation is recommended for 24 to 48 hours after breathing overexposure, as pulmonary edema may be delayed. As first aid for pulmonary edema, consider administering a corticosteroid spray. Cigarette smoking may exacerbate pulmonary injury and should be discouraged for at least 72 hours following exposure.

HEALTH HAZARDS
Personal protective equipment (PPE): A-Level PPE. Full face acid gas respirator. Chemical protective material(s) reported to have good to excellent resistance: Teflon®, Saranex®, and PE (protection may be limited).
Exposure limits (TWA unless otherwise noted): 0.3 ppm (AIHA WEEL).
Toxicity by ingestion: Strong corrosive.

Vapor (gas) irritant characteristics: Severe eye and throat irritant. Can cause eye or lung injury and cannot be tolerated even at low concentrations.
Liquid or solid irritant characteristics: Severe eye and skin irritant. Causes second-and third-degree burns on short contact; very injurious to the eyes.
Odor threshold: 1-5 ppm.

FIRE DATA
Fire extinguishing agents not to be used: Water. Highly acidic runoff will be formed.
Behavior in fire: Although nonflammable, it may ignite other combustibles. Contact with water and metal produces explosive hydrogen gas.
Electrical hazard: Class I, Group B. Based upon possible hydrogen gas generation should a leak or spill occur.

CHEMICAL REACTIVITY
Reactivity with water: Reacts violently with water, forming hydrochloric acid (vapor) and sulfuric acid.
Binary reactants: Reacts with organic materials, acids, acrolein, alcohols, ammonia, anhydrides, esters, ketones, hydrogen peroxide, finely divided metals or powders, nitriles. Hydrogen, a highly flammable and explosive gas, is generated by the action of the acid on most metals. Contact with combustible materials may cause ignition.
Neutralizing agents for acids and caustics: Although the acid reacts violently with water, flooding (from a distance) must be carried out before neutralizing with lime water or sodium bicarbonate solution.
Reactivity group: Unsafe combinations include: sulfuric acid. caustics, ammonia, aliphatic amines, alkanolamines; aromatic amines,
Compatibility class: Special case amides, organic anhydrides, isocyanates, vinyl acetate, alkylene oxides, epichlorohydrin

ENVIRONMENTAL DATA
Food chain concentration potential: Negative; unlikely to accumulate.
Water pollution: Dangerous to aquatic life in high concentrations. May be dangerous if it enters nearby water intakes; notify operators. Notify local health and pollution control officials.
Response to discharge: Issue warning-air contaminant, corrosive. Restrict access. Chemical and physical treatment.

SHIPPING INFORMATION
Grades of purity: Technical; **Storage temperature:** Ambient; **Inert atmosphere:** None; **Venting:** Pressure-vacuum; **Stability during transport:** Stable.

NAS HAZARD CLASSIFICATION FOR BULK WATER TRANSPORTATION
FIRE: 0
HEALTH: Vapor irritant: 4; Liquid or solid irritant: 4; Poisons: 4
WATER POLLUTION: Human toxicity: 2; Aquatic toxicity: 3; Aesthetic effect: 2
REACTIVITY: Other chemicals: 4; Water: 4; Self-reaction: 0

PHYSICAL AND CHEMICAL PROPERTIES
Physical state @ 59°F/15°C and 1 atm: Liquid.
Molecular weight: 116.53
Boiling point @ 1 atm: 311°F/155°C/428°K.
Melting/Freezing point: −112°F/−80°C/193°K.
Specific gravity (water = 1): 1.75 @ 68°F/20°C (liquid).
Relative vapor density (air = 1): 4.0

Latent heat of vaporization: (estimate) 198 Btu/lb = 110 cal/g = 4.6×10^5 J/kg.
Vapor pressure: (Reid) 0.03 psia; 0.4 mm.

m-CHLOROTOLUENE   REC. C:3050

SYNONYMS: 3-CHLOROTOLUENE; m-TOLYL CHLORIDE; 1-CHLORO-3-METHYLBENZENE; 3-CHLORO-1-METHYLBENZENE; m-CLOROTOLUENO (Spanish)

IDENTIFICATION
CAS Number: 108-41-8
Formula: C_7H_7Cl; $ClC_6H_4CH_3$
DOT ID Number: UN 2238; DOT Guide Number: 128
Proper Shipping Name: Chlorotoluenes

DESCRIPTION: Colorless liquid. Aromatic odor. Sinks in water; insoluble.

Highly flammable • Corrosive to the skin, eyes, and respiratory tract; contact with skin or eyes can cause burns and vision impairment; inhalation symptoms may be delayed • Containers may BLEVE when exposed to fire •Vapors may form explosive mixture with air • Vapors are heavier than air and will collect and stay in low areas • Vapors may travel long distances to ignition sources and flashback • Vapors in confined areas (e.g., tanks, sewers, buildings) may explode when exposed to fire • Toxic products of combustion may include hydrogen chloride.

Hazard Classification (based on NFPA-704 Rating System)
Health Hazards (Blue): 2; Flammability (Red): 2; Reactivity (Yellow): 0

EMERGENCY RESPONSE: See Appendix A (128)
Evacuation:
Public safety: Isolate spill area for at least 25 to 50 meters (80 to 160 feet) in all directions.
Spill: Large spill–Consider initial downwind evacuation for at least 300 meters (1000 feet).
Fire: Isolate for 800 meters (½ mile) in all directions, especially if tank, rail car, or tank truck is involved in fire.

EXPOSURE
Short-term effects: *SEEK MEDICAL ATTENTION*. **Vapor:** May be harmful if inhaled or absorbed through the skin. Irritating to eyes, skin, nose and throat. Move to fresh air. *IF BREATHING HAS STOPPED*, give artificial respiration. IF breathing is difficult, administer oxygen.May cause lung edema; physical exertion will aggravate this condition. **Liquid:** Harmful if swallowed or absorbed through skin. Irritating to skin and eyes. *IF IN EYES OR ON SKIN*, flush with running water for at least 15 minutes; hold eyelids open if necessary. Remove and isolate contaminated clothing and shoes at the site. *IF SWALLOWED* and victim is *CONSCIOUS AND ABLE TO SWALLOW*, have victim drink 4 to 8 ounces of water. **Do NOT induce vomiting.** *IF SWALLOWED* and victim is *UNCONSCIOUS OR HAVING CONVULSIONS*, do nothing except keep victim warm.
Note to physician or authorized medical personnel: Medical observation is recommended for 24 to 48 hours after breathing overexposure, as pulmonary edema may be delayed. As first aid for pulmonary edema, consider administering a corticosteroid spray. Cigarette smoking may exacerbate pulmonary injury and should be discouraged for at least 72 hours following exposure.

254 o-Chlorotoluene

HEALTH HAZARDS
Personal protective equipment (PPE): B-Level PPE. Wear self-contained positive pressure breathing apparatus and full protective clothing.
Exposure limits (TWA unless otherwise noted): ACGIH TLV 50 ppm (59 mg/m^3); NIOSH 50 ppm (250 mg/m^3) as o-chlorotoluene 95-49-8.
Short-term exposure limits (15-minute TWA): NIOSH STEL 75 ppm as o-chlorotoluene.
Long-term health effects: Prolonged and repeated vapor exposure may result in systemic toxic effects.
Vapor (gas) irritant characteristics: Eye and respiratory tract irritant. Vapors are irritating such that personnel will not usually tolerate moderate or high concentrations.
Liquid or solid irritant characteristics: Eye and skin irritant. If spilled on clothing and allowed to remain, may cause smarting and reddening of skin.

FIRE DATA
Flash point: 123°F/51°C (cc).
Behavior in fire: May produce toxic and irritating chloride fumes.

CHEMICAL REACTIVITY
Reactivity group: 36
Compatibility class: Halogenated hydrocarbons

ENVIRONMENTAL DATA
Water pollution: DOT Appendix B, §172.101–marine pollutant. Effects of low concentrations on aquatic life is unknown. May be dangerous if it enters nearby water intakes; notify operators. Notify local health and wildlife officials. **Response to discharge:** Issue warning–high flammability, water contaminant. Restrict access. Should be removed. Chemical and physical treatment.

SHIPPING INFORMATION
Storage temperature: Ambient; **Stability during transport:** Stable.

PHYSICAL AND CHEMICAL PROPERTIES
Physical state @ 59°F/15°C and 1 atm: Liquid.
Molecular weight: 126.59
Boiling point @ 1 atm: 324°F/162°C/435°K.
Melting/Freezing point: –54°F/–48°C/226°K.
Critical temperature: 730°F/388°C/161°K (estimate).
Critical pressure: 567 psia = 38.6 atm = 3.91 MN/m^2 (estimate).
Specific gravity (water = 1): 1.0722 @ 68°F/20°C.
Relative vapor density (air = 1): 4.4
Latent heat of vaporization: 143 Btu/lb = 79.6 cal/g = 3.3 x 10^5 J/kg.

o-CHLOROTOLUENE   REC. C:3100

SYNONYMS: BENZENE, 1-CHLORO-2-METHYL-; 2-CHLORO-1-METHYLBENZENE; 2-CHLOROTOLUENE; o-CLOROTOLUENO (Spanish); 1-METHYL-2-CHLOROBENZENE; 2-METHYLCHLOROBENZENE; TOLUENE, o-CHLORO-; o-TOLYLCHLORIDE; BENZENE, 1-CHLORO-2-METHYL; o-TOLYL CHLORIDE

IDENTIFICATION
CAS Number: 95-49-8
Formula: C$_7$H$_7$Cl; CH$_3$C$_6$H$_4$Cl
DOT ID Number: UN 2238; DOT Guide Number: 128
Proper Shipping Name: Chlorotoluenes

DESCRIPTION: Colorless liquid. Aromatic odor. Sinks in water; insoluble.

Highly flammable •Corrosive to the skin, eyes, and respiratory tract; contact with skin or eyes can cause burns and vision impairment; inhalation symptoms may be delayed • Containers may BLEVE when exposed to fire •Vapors may form explosive mixture with air • Vapors are heavier than air and will collect and stay in low areas • Vapors may travel long distances to ignition sources and flashback • Vapors in confined areas (e.g., tanks, sewers, buildings) may explode when exposed to fire • Toxic products of combustion may include carbon monoxide and hydrogen chloride.

Hazard Classification (based on NFPA-704 Rating System)
Health Hazards (Blue): 2; Flammability (Red): 2; Reactivity (Yellow): 0

EMERGENCY RESPONSE: See Appendix A (128)
Evacuation:
Public safety: Isolate spill area for at least 25 to 50 meters (80 to 160 feet) in all directions.
Spill: Large spill–Consider initial downwind evacuation for at least 300 meters (1000 feet).
Fire: Isolate for 800 meters (½ mile) in all directions, especially if tank, rail car, or tank truck is involved in fire.

EXPOSURE
Short-term effects: *SEEK MEDICAL ATTENTION*. **Vapor:** May be harmful if inhaled or absorbed through the skin. Irritating to eyes, skin, nose and throat. Move to fresh air. *IF BREATHING HAS STOPPED*, give artificial respiration. IF breathing is difficult, administer oxygen. May cause lung edema; physical exertion will aggravate this condition. **Liquid:** Irritating to skin and eyes. Harmful if swallowed or absorbed through the skin. *IF IN EYES OR ON SKIN*, flush with running water for at least 15 minutes; hold eyelids open if necessary. *IF SWALLOWED* and victim is *CONSCIOUS AND ABLE TO SWALLOW*, have victim drink 4 to 8 ounces of water. **Do NOT induce vomiting.** *IF SWALLOWED* and victim is *UNCONSCIOUS OR HAVING CONVULSIONS*, do nothing except keep victim warm. Remove and isolate contaminated clothing and shoes at the site.
Note to physician or authorized medical personnel: Medical observation is recommended for 24 to 48 hours after breathing overexposure, as pulmonary edema may be delayed. As first aid for pulmonary edema, consider administering a corticosteroid spray. Cigarette smoking may exacerbate pulmonary injury and should be discouraged for at least 72 hours following exposure.

HEALTH HAZARDS
Personal protective equipment (PPE): B-Level PPE. OSHA Table Z-1-A air contaminant. Wear self-contained positive pressure breathing apparatus and full protective clothing. Chemical protective material(s) reported to have good to excellent resistance: Viton®.
Exposure limits (TWA unless otherwise noted): ACGIH TLV 50 ppm (259 mg/m^3); NIOSH 50 ppm (250 mg/m^3).
Short-term exposure limits (15-minute TWA): NIOSH STEL 75 ppm (375 mg/m^3).
Toxicity by ingestion: Grade 2; LD$_{50}$ = 0.5 to 5 g/kg.
Long-term health effects: Prolonged and repeated vapor exposure may produce systemic effects.
Vapor (gas) irritant characteristics: Eye and respiratory tract irritant. Vapors are irritating such that personnel will not usually tolerate moderate or high concentrations.

Liquid or solid irritant characteristics: Eye and skin irritant. If spilled on clothing and allowed to remain, may cause smarting and reddening of skin.
Odor threshold: 0.32 ppm.

FIRE DATA
Flash point: 126°F/52°C (oc); 96°F/36°C (cc).
Behavior in fire: Container may explode in heat of fire.

CHEMICAL REACTIVITY
Reactivity with water: Reacts with water.
Binary reactants: Reacts with acids, alkalis, oxidizers, reducing materials, water.
Reactivity group: 36
Compatibility class: Halogenated hydrocarbons

ENVIRONMENTAL DATA
Food chain concentration potential: Log P_{ow} = 3.4. Values > 3.0 are likely to bioconcentrate in aquatic organisms and other living tissue, especially in fats.
Water pollution: DOT Appendix B, §172.101–marine pollutant. Effects of low concentration on aquatic life is unknown. May be dangerous if it enters nearby water intakes; notify operators. Notify local health and wildlife officials. **Response to discharge:** Issue warning–high flammability., water contaminant. Restrict access. Should be removed. Chemical and physical treatment.

SHIPPING INFORMATION
Grades of purity: 98%; **Storage temperature:** Ambient; **Stability during transport:** Stable.

PHYSICAL AND CHEMICAL PROPERTIES
Physical state @ 59°F/15°C and 1 atm: Liquid.
Molecular weight: 126.59
Boiling point @ 1 atm: 319°F/159°C/432°K.
Melting/Freezing point: –31°F/–35°C/238°K.
Critical temperature: 719°F/382°C/655°K (estimate).
Critical pressure: 567 psia = 38.6 atm = 3.91 MN/m² (estimate).
Specific gravity (water = 1): 1.0825 @ 68°F/20°C.
Liquid surface tension: 33440 dynes/cm = 0.03344 N/m @ 68°F/20°C.
Relative vapor density (air = 1): 4.4
Latent heat of vaporization: 146 Btu/lb = 81.2 cal/g = 3.4 x 10⁵ J/kg.
Vapor pressure: 4 mm @ 77°F/25°C.

p-CHLOROTOLUENE   REC. C:3150

SYNONYMS: *p*-TOLYL CHLORIDE; 4-CHLORO-1-METHYLBENZENE; 4-CHLOROTOLUENE; 1-CHLORO-4-METHYLBENZENE; *p*-CLOROTOLUENO (Spanish)

IDENTIFICATION
CAS Number: 106-43-4
Formula: C₇H₇Cl; CH₃C₆H₄Cl
DOT ID Number: UN 2238; DOT Guide Number: 128
Proper Shipping Name: Chlorotoluenes

DESCRIPTION: Colorless liquid. Aromatic odor. Sinks slowly in water; insoluble.

Highly flammable • Corrosive to the skin, eyes, and respiratory tract; contact with skin or eyes can cause burns and vision impairment; inhalation symptoms may be delayed • Containers may BLEVE when exposed to fire •Vapors may form explosive mixture with air • Vapors are heavier than air and will collect and stay in low areas • Vapors may travel long distances to ignition sources and flashback • Vapors in confined areas (e.g., tanks, sewers, buildings) may explode when exposed to fire • Toxic products of combustion may include hydrogen chloride.

Hazard Classification (based on NFPA-704 Rating System)
Health Hazards (Blue): 2; Flammability (Red): 2; Reactivity (Yellow): 0

EMERGENCY RESPONSE: See Appendix A (128)
Evacuation:
Public safety: Isolate spill area for at least 25 to 50 meters (80 to 160 feet) in all directions.
Spill: Large spill–Consider initial downwind evacuation for at least 300 meters (1000 feet).
Fire: Isolate for 800 meters (½ mile) in all directions, especially if tank, rail car, or tank truck is involved in fire.

EXPOSURE
Short-term effects: *SEEK MEDICAL ATTENTION.* **Vapor:** May cause lung edema; physical exertion will aggravate this condition. **Liquid:** Irritating to skin and eyes. Harmful if swallowed. Remove contaminated clothing and shoes. Flush affected areas with plenty of water. *IF IN EYES*, hold eyelids open and flush with plenty of water. *IF SWALLOWED* and victim is *CONSCIOUS AND ABLE TO SWALLOW*, have victim drink 4 to 8 ounces of water and induce vomiting.
Note to physician or authorized medical personnel: Medical observation is recommended for 24 to 48 hours after breathing overexposure, as pulmonary edema may be delayed. As first aid for pulmonary edema, consider administering a corticosteroid spray. Cigarette smoking may exacerbate pulmonary injury and should be discouraged for at least 72 hours following exposure.

HEALTH HAZARDS
Personal protective equipment (PPE): B-Level PPE. Respirator with proper filter, goggles. Chemical protective material(s) reported to have good to excellent resistance: Viton®.
Exposure limits (TWA unless otherwise noted): ACGIH TLV 50 ppm (59 mg/m³); NIOSH 50 ppm (250 mg/m³) as *o*-chlorotoluene 95-49-8.
Short-term exposure limits (15-minute TWA): NIOSH STEL 75 ppm as *o*-chlorotoluene.
Vapor (gas) irritant characteristics: Eye and respiratory tract irritant.
Liquid or solid irritant characteristics: Eye and skin irritant.

FIRE DATA
Flash point: 120°F/49°C (oc).

CHEMICAL REACTIVITY
Binary reactants: Acids, alkalis, oxidizers, reducing materials.

ENVIRONMENTAL DATA
Food chain concentration potential: Log P_{ow} = 3.4. Values > 3.0 are likely to bioconcentrate in aquatic organisms and other living tissue, especially in fats.
Water pollution: DOT Appendix B, §172.101–marine pollutant. Harmful to aquatic life in very low concentrations. May be dangerous if it enters nearby water intakes; notify operators. Notify local health and wildlife officials. **Response to discharge:** Restrict access. Chemical and physical treatment. Disperse and flush.

4-CHLORO-*o*-TOLUIDINE REC. C:3200

SYNONYMS: 2-AMINO-4-CHLOROTOLUENE; 1-AMINO-3-CHLORO-6-METHYLBENZENE; 2-AMINO-5-CHLOROTOLUENE; BENZENAMINE, 5-CHLORO-2-METHYL-; CHLOROTOLUIDINE; 4-CHLORO-2-AMINOTOLUENE; 3-CHLORO-6-METHYLANILINE; 5-CHLORO-*o*-TOLUIDINE; 5-CHLORO-2-AMINOTOLUENE; 4-CHLORO-2-METHYLANILINE; CLORO-*o*-TOLUIDINA (Spanish); FAST RED TR BASE; RED TR BASE

IDENTIFICATION
CAS Number: 95-79-4
Formula: C_7H_8ClN; $Cl-CC_6H_3(CH_3)-NH_2$
DOT ID Number: UN 2239; DOT Guide Number: 153
Proper Shipping Name: Chlorotoluidines, solid.

DESCRIPTION: Gray to white solid. Weak fishy odor. Sinks in water.

Combustible solid • Corrosive • Poison! • Breathing the dust or swallowing the material can kill you; skin and eye contact may cause illness • Firefighting gear (including SCBA) does not provide adequate protection. If exposure occurs, remove and isolate gear immediately and thoroughly decontaminate personnel • Dust cloud may explode if ignited in an enclosed area • Containers may BLEVE when exposed to fire • Toxic products of combustion may include hydrogen chloride and nitrogen oxides.

Hazard Classification (based on NFPA-704 Rating System)
Health Hazards (Blue): 1; Flammability (Red): 1; Reactivity (Yellow): 0

EMERGENCY RESPONSE: See Appendix A (153)
Evacuation:
Public safety: Isolate the area of spill or leak for at least 25 to 50 meters (80 to 160 feet) in all directions.
Spill: Increase, in the downwind direction, as necessary, the distance shown under "Public Safety."
Fire: If tank, rail car, or tank truck is involved in fire, isolate for at least 800 meters (½ mile) in all directions; also, consider initial evacuation for 800 meters (½ mile) in all directions.

EXPOSURE
Short-term effects: *SEEK MEDICAL ATTENTION.* **Dust:** *POISONOUS IF INHALED.* Move victim to fresh air. *IF IN EYES,* hold eyelids open and flush with plenty of water. *IF BREATHING HAS STOPPED,* give artificial respiration; *avoid mouth-to-mouth resuscitation; use bag/mask apparatus.* IF breathing is difficult, administer oxygen. **Solid:** *POISONOUS IF SWALLOWED.* Remove contaminated clothing and shoes. Flush affected areas with plenty of water. *IF IN EYES,* hold eyelids open and flush with plenty of water. *IF SWALLOWED* and victim is *CONSCIOUS AND ABLE TO SWALLOW,* have victim drink 4 to 8 ounces of water and have victim induce vomiting. *IF SWALLOWED* and victim is *UNCONSCIOUS OR HAVING CONVULSIONS,* do nothing except keep victim warm.

HEALTH HAZARDS
Personal protective equipment (PPE): A-Level PPE. For a large spill A-Level might be considered.
Toxicity by ingestion: Grade 3; oral LD_{50} = 464 mg/kg (rat).
Long-term health effects: Suspected human carcinogen.
Liquid or solid irritant characteristics: Eye irritant.

FIRE DATA
Flash point: Combustible solid.

ENVIRONMENTAL DATA
Water pollution: DOT Appendix B, §172.101–marine pollutant. Effects of low concentrations on aquatic life is unknown. May be dangerous if it enters nearby water intakes; notify operators. Notify local health and wildlife officials. **Response to discharge:** Issue warning–poison, water contaminant. Restrict access. Should be removed. Chemical and physical treatment.

SHIPPING INFORMATION
Grades of purity: 99%; **Storage temperature:** Ambient; **Inert atmosphere:** None; **Venting:** Pressure-vacuum; **Stability during transport:** Stable.

PHYSICAL AND CHEMICAL PROPERTIES
Physical state @ 59°F/15°C and 1 atm: Liquid.
Molecular weight: 126.6
Boiling point @ 1 atm: 324°F/162°C/435.2°K.
Melting/Freezing point: 46°F/8°C/281°K.
Specific gravity (water = 1): 1.0697 @ 68°F/20°C.
Liquid surface tension: 32.24 dynes/cm = 0.03224 N/m @ 77°F/25°C.
Relative vapor density (air = 1): 4.36 (estimated).
Latent heat of vaporization: At boiling point 136.8 Btu/lb = 76 cal/g = 3.18×10^5 J/kg.

CHROMIC ACETATE REC. C:3250

SYNONYMS: ACETATO CROMICO (Spanish); ACETIC ACID, CHROMIUM(3+) SALT; CHROMIC ACETATE(III); CHROMIUM ACETATE; CHROMIUM(III) ACETATE; CHROMIUM TRIACETATE; ACETIC ACID, CHROMIUM SALT; CHROMIC(III) ACETATE

IDENTIFICATION
CAS Number: 1066-30-4
Formula: $Cr(C_2H_3O_2)_3$; $Cr(C_2H_3O_2)_3 \cdot H_2O$; $CrOH(C_2H_3O_2)_2$ (basic acetate)
DOT ID Number: UN 9101; DOT Guide Number: 171
Proper Shipping Name: Chromic acetate
Reportable Quantity (RQ): **(CERCLA)** 1000 lb/454 kg

DESCRIPTION: Dark green to violet powder or blue-green paste (aqueous solution). Acetic acid odor. Sinks and mixes with water.

Poison! • Corrosive to the skin, eyes, and respiratory tract; contact with skin or eyes can cause burns and vision impairment; inhalation symptoms may be delayed • Toxic products of combustion may include carbon monoxide and chromium fume.

Hazard Classification (based on NFPA-704 Rating System)
Health Hazards (Blue): 1; Flammability (Red): 0; Reactivity (Yellow): 0

PHYSICAL AND CHEMICAL PROPERTIES
Physical state @ 59°F/15°C and 1 atm: Solid.
Molecular weight: 141.6
Boiling point @ 1 atm: 466°F/241°C/514°K.
Melting/Freezing point: 77°F/25°C/298°K.
Specific gravity (water = 1): (estimate) more than 1.1 @ 68°F/20°C (solid).

EMERGENCY RESPONSE: See Appendix A (171)
Evacuation:
Public safety: Isolate the area of spill or leak for at least 10 to 25 meters (30 to 80 feet) in all directions.
Spill: Increase, in the downwind direction, as necessary, the distance shown under "Public Safety."
Fire: If any large container is involved in fire, isolate for at least 800 meters (½ mile) in all directions; also, consider initial evacuation for 800 meters (½ mile) in all directions.

EXPOSURE
Short-term effects: *SEEK MEDICAL ATTENTION.* **Dust:** Harmful if inhaled. Move to fresh air. *IF BREATHING HAS STOPPED,* give artificial respiration. May cause lung edema; physical exertion will aggravate this condition. **Liquid or solid:** Irritating to skin and eyes. Harmful if swallowed. Remove contaminated clothing and shoes. Flush affected area with plenty of water. *IF IN EYES*, hold eyelids open and flush with plenty of water. *IF SWALLOWED* and victim is *CONSCIOUS AND ABLE TO SWALLOW*, have victim drink 4 to 8 ounces of water.
Note to physician or authorized medical personnel: Medical observation is recommended for 24 to 48 hours after breathing overexposure, as pulmonary edema may be delayed. As first aid for pulmonary edema, consider administering a corticosteroid spray. Cigarette smoking may exacerbate pulmonary injury and should be discouraged for at least 72 hours following exposure.

HEALTH HAZARDS
Personal protective equipment (PPE): B-Level PPE. Wear protective clothing and approved respirator.
Recommendations for respirator selection:
2.5 mg/m³: DM (any dust and mist respirator). *5 mg/m³*: DMXSQ (any dust and mist respirator except single-use and quarter mask respirators); or SA (any supplied-air respirator); *12.5 mg/m³*: SA:CF (any supplied-air respirator operated in a continuous-flow mode); or PAPRDM (any powered, air-purifying respirator with a dust and mist filter). *25 mg/m³*: HiEF (any air-purifying, full-facepiece respirator with a high-efficiency particulate filter); or PAPRTHiE (any powered, air-purifying respirator with a tight-fitting facepiece and a high-efficiency particulate filter); or SCBAF (any self-contained breathing apparatus with a full facepiece); or SAF (any supplied-air respirator with a full facepiece). *EMERGENCY OR PLANNED ENTRY INTO UNKNOWN CONCENTRATIONS OR IDLH CONDITIONS*: SCBAF:PD,PP (any self-contained breathing apparatus that has a full facepiece and is operated in a pressure-demand or other positive-pressure mode); or SAF:PD,PP:ASCBA (Any supplied-air respirator that has a full facepiece and is operated in a pressure-demand or other positive-pressure mode in combination with an auxiliary self-contained breathing apparatus operated in a pressure-demand or other positive pressure mode). *ESCAPE:* HiEF (Any air-purifying, full-facepiece respirator with a high-efficiency particulate filter); or SCBAE (Any appropriate escape-type, self-contained breathing apparatus). *Note:* * Substance reported to cause eye irritation or damage; may require eye protection.
Exposure limits (TWA unless otherwise noted): OSHA PEL ceiling 0.1 mg/m³ as chromium trioxide; NIOSH 0.001 mg/m³ as chromates; potential human carcinogen; reduce exposure to lowest feasible level
Long-term health effects: Confirmed carcinogen. Mutation data.
Vapor (gas) irritant characteristics: Eye and respiratory tract irritant.
Liquid or solid irritant characteristics: Eye and skin irritant.
IDLH value: 15 mg/m³ as chromium.

CHEMICAL REACTIVITY
Binary reactants: Strong oxidizers.

ENVIRONMENTAL DATA
Food chain concentration potential: High

ENVIRONMENTAL DATA
Water pollution: Harmful to aquatic life in very low concentrations. May be dangerous if it enters nearby water intakes; notify operators. Notify local health and wildlife officials. **Response to discharge:** Issue warning–water contaminant. Chemical and physical treatment. Disperse and flush.

SHIPPING INFORMATION
Storage temperature: Ambient; **Stability during transport:** Stable.

PHYSICAL AND CHEMICAL PROPERTIES
Physical state @ 59°F/15°C and 1 atm: Solid.
Molecular weight: 229.14 (anhydrous); 247.16 (hydrate).
Boiling point @ 1 atm: 212°F/100°C/373°K For aqueous solution
Specific gravity (water = 1): 1.30

CHROMIC ANHYDRIDE   REC. C:3300

SYNONYMS: ANHIDRIDO CROMICO (Spanish); ANHYDRIDE CHROMIQUE (French); CHROMIC ACID; CHROMIC(VI) ACID; CHROMIC TRIOXIDE; CHROMIC OXIDE; CHROMIUM OXIDE; CHROMIUM(VI) OXIDE; CHROMIUM(VI) OXIDE; CHROMIUM TRIOXIDE; CHROMIUM TRIOXIDE, ANHYDROUS; CHROMSAUREANHYDRID (German); EEC No. 024-001-00-0; MONOCHROMIUM OXIDE; MONOCHROMIUM TRIOXIDE; PURATRONIC CHROMIUM TRIOXIDE

IDENTIFICATION
CAS Number: 1333-82-0
Formula: CrO_3
DOT ID Number: UN 1463; DOT Guide Number: 141
Proper Shipping Name: Chromium trioxide, anhydrous
Reportable Quantity (RQ): **(CERCLA)** 10 lb/4.54 kg

DESCRIPTION: Solid flakes or powder. Dark red. Odorless. Sinks and mixes with water.

Poison! • Breathing the dust or swallowing the material can kill you; skin or eye contact causes burns • Firefighting gear (including SCBA) does not provide adequate protection. If exposure occurs, remove and isolate gear immediately and thoroughly decontaminate personnel • A strong oxidizer; heat forms oxygen; will increase the activity of an existing fire • Can cause fire or explosion contact with combustibles (wood, paper, oil, clothing, etc.) and many other materials • Containers may BLEVE when exposed to fire • Toxic products of combustion may include carbon monoxide and chromium fume.

Hazard Classification (based on NFPA-704 Rating System)
Health Hazards (Blue): 1; Flammability (Red): 0; Reactivity (Yellow): 2; Special Notice (White): OXY

EMERGENCY RESPONSE: See Appendix A (141)
Evacuation:
Public safety: Isolate the area of spill or leak for at least 10 to 25 meters (30 to 80 feet) in all directions.

Chromic sulfate

Spill: Consider initial downwind evacuation for at least 100 meters (330 feet).
Fire: If any large container is involved in fire, isolate for at least 800 meters (½ mile) in all directions; also, consider initial evacuation for 800 meters (½ mile) in all directions.

EXPOSURE
Short-term effects: *SEEK MEDICAL ATTENTION.* **Solid:** Will burn skin and eyes.
Harmful if swallowed. Remove contaminated clothing and shoes. Flush affected areas with plenty of water. *IF IN EYES*, hold eyelids open and flush with plenty of water. *IF SWALLOWED* and victim is *CONSCIOUS AND ABLE TO SWALLOW*, have victim drink 4 to 8 ounces of water. **Do NOT induce vomiting.**

HEALTH HAZARDS
Personal protective equipment (PPE): A-Level PPE. Chemical protective material(s) reported to have good to excellent resistance: butyl rubber, PV alcohol, PVC, nitrile+PVC. Also, nitrile and PE offers limited protection For < 30% solutions neoprene and natural rubber compounds are recommended. For up to 70% solutions nitrile+ PVC offers limited protection
Recommendations for respirator selection: NIOSH
At any concentrations above the NIOSH REL, or where there is no REL, at any detectable concentration: SCBAF:PD,PP (any self-contained breathing apparatus that has a full facepiece and is operated in a pressure-demand or other positive-pressure mode); or SAF:PD,PP:ASCBA (any supplied-air respirator that has a full facepiece and is operated in a pressure-demand or other positive-pressure mode in combination with an auxiliary self-contained breathing apparatus operated in a pressure-demand or other positive pressure mode). *ESCAPE:* HiEF (any air-purifying, full-facepiece respirator with a high-efficiency particulate filter); or SCBAE (any appropriate escape-type, self-contained breathing apparatus).
Exposure limits (TWA unless otherwise noted): ACGIH TLV 0.05 mg/m^3 as water soluble chromium(VI) compounds; OSHA PEL ceiling 0.1 mg/m^3 as chromium trioxide; NIOSH 0.001 mg/m^3 as chromates; potential human carcinogen; reduce exposure to lowest feasible level
Toxicity by ingestion: Grade 3; LD$_{50}$ = 50 to 500 mg/kg.
Long-term health effects: NTP anticipated carcinogen. Possible lung cancer. Skin disorders.
Vapor (gas) irritant characteristics: Severe eye and respiratory tract irritant.
Liquid or solid irritant characteristics: Severe skin and eye irritant. Causes second- and third-degree burns on short contact; very injurious to the eyes.
IDLH value: 15 mg/m^3 as chromium.

FIRE DATA
Behavior in fire: Containers may explode
Autoignition temperature: May ignite organic materials on contact.

CHEMICAL REACTIVITY
Reactivity with water: Aqueous solution is a strong acid.
Binary reactants: Reacts with bases. May react with organic materials rapidly enough to generate sufficient heat to cause ignition. Prolonged contact, particularly on wood floors, may produce a fire hazard. Corrosive to metals, especially in the presence of water or water vapor.
Neutralizing agents for acids and caustics: Flood with water, rinse with sodium bicarbonate solution.

ENVIRONMENTAL DATA
Food chain concentration potential: Unlikely to accumulate.
Water pollution: Harmful to aquatic life in very low concentrations. May be dangerous if it enters nearby water intakes; notify operators. Notify local health and wildlife officials. **Response to discharge:** Issue warning–water contaminant. Disperse and flush.

SHIPPING INFORMATION
Grades of purity: Technical; technical flake: 99.75%; **Stability during transport:** Stable.

PHYSICAL AND CHEMICAL PROPERTIES
Physical state @ 59°F/15°C and 1 atm: Solid.
Molecular weight: 100.01
Boiling point @ 1 atm: 388°F/198°C/471°K (decomposes).
Melting/Freezing point: 387°F/195°C/468°K.
Specific gravity (water = 1): 2.70 @ 68°F/20°C (solid).
Heat of fusion: 37.7 cal/g.
Vapor pressure: (Reid) Very low.

CHROMIC SULFATE REC. C:3350

SYNONYMS: CHROMIC SULPHATE; CHROMIUM(III) SULFATE; CHROMIUM SULFATE; CHROMIUM SULPHATE; C.I.77305; DICHROMIUM SULFATE; DICHROMIUM SULPHATE; DICHROMIUM TRISULFATE; DICHROMIUM TRISULPHATE; SULFATO CHROMICO (Spanish); SULFURIC ACID, CHROMIUM(3+) SALT

IDENTIFICATION
CAS Number: 10101-53-8
Formula: Cr$_2$(SO$_4$)$_3$; Cr$_2$(SO$_4$)$_3$·10H$_2$O (technical)
DOT ID Number: UN 9100; DOT Guide Number: 171
Proper Shipping Name: Chromic sulfate
Reportable Quantity (RQ): **(CERCLA)** 1000 lb/454 kg

DESCRIPTION: Peach (anhydrous), violet (hydrated), or dark green (technical) solid. Odorless. Sinks and mixes slowly with water; practically insoluble.

Poison! • Corrosive • Breathing the dust or swallowing the material can kill you • Has a corrosive action on the skin and mucous membranes, lungs. Effects of inhalation may be delayed • Firefighting gear (including SCBA) does not provide adequate protection. If exposure occurs, remove and isolate gear immediately and thoroughly decontaminate personnel • Toxic products of combustion may include sulfur oxide and chromium fume.

Hazard Classification (based on NFPA-704 Rating System)
Health Hazards (Blue): 1; Flammability (Red): 0; Reactivity (Yellow): 0

EMERGENCY RESPONSE: See Appendix A (171)
Evacuation:
Public safety: Isolate the area of spill or leak for at least 10 to 25 meters (30 to 80 feet) in all directions.
Spill: Increase, in the downwind direction, as necessary, the distance shown under "Public Safety."
Fire: If any large container is involved in fire, isolate for at least 800 meters (½ mile) in all directions; also, consider initial evacuation for 800 meters (½ mile) in all directions.

EXPOSURE

Short-term effects: *SEEK MEDICAL ATTENTION.* **Dust:** Harmful if inhaled. Move to fresh air. *IF BREATHING HAS STOPPED,* give artificial respiration. Lung edema may develop. **Solid:** Irritating to skin and eyes. Harmful if swallowed. Remove contaminated clothing and shoes. Flush affected areas with plenty of water. *IF IN EYES* hold eyelids open and flush with plenty of water. *IF SWALLOWED* and victim is *CONSCIOUS AND ABLE TO SWALLOW,* have victim drink 4 to 8 ounces of water.

Note to physician or authorized medical personnel: Medical observation is recommended for 24 to 48 hours after breathing overexposure, as pulmonary edema may be delayed. As first aid for pulmonary edema, consider administering a corticosteroid spray. Cigarette smoking may exacerbate pulmonary injury and should be discouraged for at least 72 hours following exposure.

HEALTH HAZARDS

Personal protective equipment (PPE): B-Level PPE. Rubber gloves, safety glasses, laboratory coat, dust mask.
Recommendations for respirator selection: NIOSH/OSHA *2.5 mg/m³:* DM (any dust and mist respirator). *5 mg/m³:* DMXSQ (any dust and mist respirator except single-use and quarter mask respirators); or SA (any supplied-air respirator); *12.5 mg/m³:* SA:CF (any supplied-air respirator operated in a continuous-flow mode); or PAPRDM (any powered, air-purifying respirator with a dust and mist filter). *25 mg/m³:* HiEF (any air-purifying, full-facepiece respirator with a high-efficiency particulate filter); or PAPRTHiE (any powered, air-purifying respirator with a tight-fitting facepiece and a high-efficiency particulate filter); or SCBAF (any self-contained breathing apparatus with a full facepiece); or SAF (any supplied-air respirator with a full facepiece). *EMERGENCY OR PLANNED ENTRY INTO UNKNOWN CONCENTRATIONS OR IDLH CONDITIONS:* SCBAF:PD,PP (any self-contained breathing apparatus that has a full facepiece and is operated in a pressure-demand or other positive-pressure mode); or SAF:PD,PP:ASCBA (Any supplied-air respirator that has a full facepiece and is operated in a pressure-demand or other positive-pressure mode in combination with an auxiliary self-contained breathing apparatus operated in a pressure-demand or other positive pressure mode). *ESCAPE:* HiEF (Any air-purifying, full-facepiece respirator with a high-efficiency particulate filter); or SCBAE (Any appropriate escape-type, self-contained breathing apparatus). *Note:* * Substance reported to cause eye irritation or damage; may require eye protection.
Exposure limits (TWA unless otherwise noted): ACGIH TLV 0.05 mg/m³; NIOSH/OSHA 0.5 mg/m³ as chromium(III).
Toxicity by ingestion: Grade 2; LD_{50} = 0.5 to 5 mg/kg.
Long-term health effects: A potential human carcinogen.
Vapor (gas) irritant characteristics: Severe eye and respiratory tract irritant.
Liquid or solid irritant characteristics: Severe eye and skin irritant.
IDLH value: 25 mg/m³ as chromium(III).

FIRE DATA

Fire extinguishing agents not to be used: Water streams applied to adjacent fires will spread contamination of environmentally hazardous substance over wide area.

CHEMICAL REACTIVITY

Binary reactants: Violent reaction with aluminum, magnesium.
Neutralizing agents for acids and caustics: Add water slowly, stir in slight excess of soda ash. Let stand 24 hours. Neutralize with 6m hydrochloric acid. Flush with large excess of water.

ENVIRONMENTAL DATA

Water pollution: Harmful to aquatic life in very low concentrations. May be dangerous if it enters nearby water intakes; notify operators. Notify local health and wildlife officials. **Response to discharge:** Issue warning–water contaminant. Chemical and physical treatment. Disperse and flush.

SHIPPING INFORMATION

Storage temperature: Store in cool, dry place; **Stability during transport:** Stable.

PHYSICAL AND CHEMICAL PROPERTIES

Physical state @ 59°F/15°C and 1 atm: Solid.
Molecular weight: 392.20
Boiling point @ 1 atm: Loses water of hydration @ 100°C $Cr_2(SO_4)_3 \cdot 18$ loses 12; $Cr_2(SO_4)_3 \cdot 15$ loses 10
Melting/Freezing point: 212°F/100°C/373°K.
Specific gravity (water = 1): 3.012 at room temperature for anhydrous salt; Hydrated: 1.867 at 17°C for 15 H_2O; 1.7 @ 71°F/22°C for 18 H_2O

CHROMOUS CHLORIDE   REC. C:3400

SYNONYMS: CHROMIUM DICHLORIDE; DICLORURO CROMOSO (Spanish)

IDENTIFICATION

CAS Number: 10049-05-5
Formula: $CrCl_2$
DOT ID Number: UN 9102; DOT Guide Number: 171
Proper Shipping Name: Chromous chloride
Reportable Quantity (RQ): **(CERCLA)** 1000 lb/454 kg

DESCRIPTION: Bright blue crystalline solid. Sinks and mixes with water; forms an acid solution.

Poison! • Corrosive • Breathing the dust or swallowing the material can kill you • Has a corrosive action on the skin and mucous membranes, lungs; effects of inhalation may be delayed • Firefighting gear (including SCBA) does not provide adequate protection. If exposure occurs, remove and isolate gear immediately and thoroughly decontaminate personnel • Containers may explode when exposed to fire (probably caused by pressure of hydrogen by reduction of water) • Toxic products of combustion may include hydrogen chloride and chromium fume.

Hazard Classification (based on NFPA-704 Rating System)
Health Hazards (Blue): 1; Flammability (Red): 0; Reactivity (Yellow): 0

EMERGENCY RESPONSE: See Appendix A (171)
Evacuation:
Public safety: Isolate the area of spill or leak for at least 10 to 25 meters (30 to 80 feet) in all directions.
Spill: Increase, in the downwind direction, as necessary, the distance shown under "Public Safety."
Fire: If any large container is involved in fire, isolate for at least 800 meters (½ mile) in all directions; also, consider initial evacuation for 800 meters (½ mile) in all directions.

EXPOSURE

Short-term effects: *SEEK MEDICAL ATTENTION.* **Dust:** Harmful if inhaled. Move to fresh air. *IF BREATHING HAS STOPPED,* give artificial respiration. Lung edema may develop. **Solid:** Irritating to

Chromyl chloride

skin and eyes. Harmful if swallowed. Flush affected areas with plenty of water. *IF IN EYES*, hold eyelids open and flush with plenty of water. *IF SWALLOWED* and victim is *CONSCIOUS AND ABLE TO SWALLOW*, have victim drink 4 to 8 ounces of water.

Note to physician or authorized medical personnel: Medical observation is recommended for 24 to 48 hours after breathing overexposure, as pulmonary edema may be delayed. As first aid for pulmonary edema, consider administering a corticosteroid spray. Cigarette smoking may exacerbate pulmonary injury and should be discouraged for at least 72 hours following exposure.

HEALTH HAZARDS
Personal protective equipment (PPE): A-Level PPE.
Recommendations for respirator selection: NIOSH/OSHA
2.5 mg/m^3: DM* (any dust and mist respirator). *5 mg/m^3*: DMXSQ (any dust and mist respirator except single-use and quarter mask respirators); or SA* (any supplied-air respirator); *12.5 mg/m^3*: SA:CF (any supplied-air respirator operated in a continuous-flow mode); or PAPRDM* (any powered, air-purifying respirator with a dust and mist filter). *25 mg/m^3*: HiEF (any air-purifying, full-facepiece respirator with a high-efficiency particulate filter); or PAPRTHiE (any powered, air-purifying respirator with a tight-fitting facepiece and a high-efficiency particulate filter); or SCBAF (any self-contained breathing apparatus with a full facepiece); or SAF (any supplied-air respirator with a full facepiece). *250 mg/m^3*: (any supplied-air respirator that has a full facepiece and is operated in a pressure-demand or other positive-pressure mode). *EMERGENCY OR PLANNED ENTRY INTO UNKNOWN CONCENTRATIONS OR IDLH CONDITIONS*: SCBAF:PD,PP (any self-contained breathing apparatus that has a full facepiece and is operated in a pressure-demand or other positive-pressure mode); or SAF:PD,PP:ASCBA (Any supplied-air respirator that has a full facepiece and is operated in a pressure-demand or other positive-pressure mode in combination with an auxiliary self-contained breathing apparatus operated in a pressure-demand or other positive pressure mode). *ESCAPE*: HiEF (Any air-purifying, full-facepiece respirator with a high-efficiency particulate filter); or SCBAE (Any appropriate escape-type, self-contained breathing apparatus). *Note:* Substance reported to cause eye irritation or damage; may require eye protection.
Exposure limits (TWA unless otherwise noted): ACGIH TLV 0.5 mg/m^3; NIOSH/OSHA 0.5 mg/m^3 as chromium(II).
Toxicity by ingestion: Grade 2; LD$_{50}$ = 0.5 to 5 g/kg.
Long-term health effects: Possible carcinogen and mutagen.
Vapor (gas) irritant characteristics: Eye and respiratory tract irritation.
Liquid or solid irritant characteristics: Eye and skin irritant.
IDLH value: 250 mg/m^3 as chromium(II).

FIRE DATA
Fire extinguishing agents not to be used: Water streams applied to adjacent fires will spread contamination of environmentally hazardous substance over wide area.

CHEMICAL REACTIVITY
Reactivity with water: On standing in solution it is oxidized by water with liberation of hydrogen. Powerful reducing agent; violent reaction with oxidizers. Keep well closed.

Neutralizing agents for acids and caustics: Mix with equal volume of soda ash and add water. Add calcium hypochlorite. Add more water and let stand for two hours. Neutralize oxidized solution (Check with litmus and neutralize with 6M HCl or 6M NaOH.) Flush with large excess of water.

ENVIRONMENTAL DATA
Water pollution: Harmful to aquatic life in very low concentrations. May be dangerous if it enters nearby water intakes; notify operators. Notify local health and wildlife officials. **Response to discharge:** Issue warning–water contaminant. Chemical and physical treatment. Disperse and flush.

SHIPPING INFORMATION
Venting: Vented storage recommended; **Stability during transport:** Very hygroscopic; stable in dry air but oxidizes rapidly if moist. Keep container well closed.

PHYSICAL AND CHEMICAL PROPERTIES
Physical state @ 59°F/15°C and 1 atm: Solid.
Molecular weight: 122.92
Melting/Freezing point: 1515°F/824°C/1097°K.
Specific gravity (water = 1): 2.751 at 14°C; 2.878 @ 77°F/25°C.
Heat of solution: For anhydrous CrCl$_2$ –273 Btu/lb = –151.6 cal/g = –6.34 x 10^5 J/kg.
Heat of fusion: 65.9 cal/g.

CHROMYL CHLORIDE REC. C:3450

SYNONYMS: CHLOROCHROMIC ANHYDRIDE; CHROMOCHROMIC ANHYDRIDE; CHROMIC OXYCHLORIDE; CHROMIUM DICHLORIDE DIOXIDE; CHROMIUM CHLORIDE OXIDE; CHROMIUM DIOXIDE DICHLORIDE; CHROMIUM(VI) DIOXYCHLORIDE; CHROMIUM OXYCHLORIDE; CICHLORODIOXO CHROMIUM; CLORURO de CROMILO (Spanish); DICHLORO-DIOXOCHROMIUM; DICHLORODIOXOCHROMIUM; DIOXODICHLOROCHROMIUM

IDENTIFICATION
CAS Number: 14977-61-8; 7791-14-2
Formula: CrO$_2$Cl$_2$
DOT ID Number: UN 1758; DOT Guide Number: 137
Proper Shipping Name: Chromium oxychloride

DESCRIPTION: Dark red liquid that fumes in moist air. Acrid, musty, burning odor. Reacts violently with water forming chromic acid, hydrochloric acid (hydrogen chloride vapors), chromic chloride, and chlorine gas. Irritating visible vapor cloud is produced.

Poison! • Corrosive • Breathing the dust or swallowing the material can kill you; skin or eye contact causes inflammation, blistering, burns, and blindness • Firefighting gear (including SCBA) does not provide adequate protection. If exposure occurs, remove and isolate gear immediately and thoroughly decontaminate personnel • A strong oxidizer; heat forms oxygen; will increase the activity of an existing fire • Can cause fire or explosion contact with combustibles (wood, paper, oil, clothing, etc.) and many other materials • Containers may explode when exposed to fire • Toxic products of combustion may include hydrogen chloride and chromium fume • Severely corrodes common metals.

Hazard Classification (based on NFPA-704 Rating System)
Health Hazards (Blue): 1; Flammability (Red): 0; Reactivity (Yellow): 2; Special Notice (White): OXY, Water reactive

EMERGENCY RESPONSE: See Appendix A (137)
Evacuation:
Public safety: Isolate the area of spill or leak for at least 50 to 100

meters (160 to 330 feet) in all directions. For water contact see below.

Spill: FOR SPILL IN WATER. Small spill–First: Isolate in all directions 30 meters (100 feet) (water); Then: Protect persons downwind, DAY: 0.2 km (0.1 mile) (water); NIGHT: 0.2 km (0.1 mile) (water). Large spill–First: Isolate in all directions 60 meters (200 feet) (water); Then: Protect persons downwind, DAY: 0.3 km (0.2 mile) (water); NIGHT: 1.3 km (1.3 miles) (water).

Fire: Isolate for 800 meters (½ mile) in all directions, especially if tank, rail car, or tank truck is involved in fire.

EXPOSURE

Short-term effects: *SEEK MEDICAL ATTENTION.* **Vapor:** Irritating to eyes, nose, and throat. *IF INHALED,* will, will cause difficult breathing. Move victim to fresh air. *IF BREATHING HAS STOPPED,* give artificial respiration; *avoid mouth-to-mouth resuscitation; use bag/mask apparatus.* If breathing is difficult, administer oxygen. **Liquid:** *POISONOUS IF SWALLOWED.* Will burn skin and eyes. Remove contaminated clothing and shoes. Flush affected areas with plenty of water. IF IN EYES, hold eyelids open and flush with plenty of water. IF SWALLOWED and victim is *CONSCIOUS AND ABLE TO SWALLOW,* have victim drink 4 to 8 ounces of water. **Do NOT induce vomiting.**

HEALTH HAZARDS

Personal protective equipment (PPE): A-Level PPE. Self-contained breathing apparatus (full face); rubber gloves; protective clothing.

Recommendations for respirator selection: NIOSH
At any concentrations above the NIOSH REL, or where there is no REL, at any detectable concentration: SCBAF:PD,PP (any self-contained breathing apparatus that has a full facepiece and is operated in a pressure-demand or other positive-pressure mode); or SAF:PD,PP:ASCBA (any supplied-air respirator that has a full facepiece and is operated in a pressure-demand or other positive-pressure mode in combination with an auxiliary self-contained breathing apparatus operated in a pressure-demand or other positive pressure mode). *ESCAPE:* GMFOV [any air-purifying, full-facepiece respirator (gas mask) with a chin-style, front-or back-mounted organic vapor canister]; or SCBAE (any appropriate escape-type, self-contained breathing apparatus).

Exposure limits (TWA unless otherwise noted): ACGIH TLV 0.025 ppm (0.16 mg/m^3); NIOSH 0.001 mg/m^3 as chromium(VI); potential human carcinogen; reduce exposure to lowest feasible level.

Toxicity by ingestion: Grade 4; LD$_{50}$ less than 50 mg/kg.

Long-term health effects: OSHA specifically regulated carcinogen. May cause lung cancer. Kidney and liver damage.

Vapor (gas) irritant characteristics: Vapors cause severe irritation of eyes and throat and can cause eye and lung injury. They cannot be tolerated even at low concentrations.

Liquid or solid irritant characteristics: Severe eye and skin irritant. Causes second- and third-degree burns on short contact and is very injurious to the eyes.

FIRE DATA

Fire extinguishing agents not to be used: Do not use water on adjacent fires unless fully protected against toxic fumes.

Behavior in fire: Vapors are very irritating to eyes and mucous membranes. May increase severity of fire.

CHEMICAL REACTIVITY

Reactivity with water: Reacts violently to form hydrogen chloride (hydrochloric acid), chlorine gases, chromic chloride, and chromic acid.

Binary reactants: Combustible substances, halides, phosphorus, turpentine. Will cause severe corrosion of common metals.

Neutralizing agents for acids and caustics: Flood with water. Rinse with sodium bicarbonate or lime solution.

ENVIRONMENTAL DATA

Food chain concentration potential: Negative; unlikely to accumulate.

Water pollution: Effect of low concentrations on aquatic life is unknown. May be dangerous if it enters nearby water intakes; notify operators. Notify local health and wildlife officials. **Response to discharge:** Issue warning–poison, corrosive, air contaminant, water contaminant, oxidizing substance. Restrict access. Evacuate area. Disperse and flush with care.

SHIPPING INFORMATION

Grades of purity: 99.5+%; **Storage temperature:** Ambient; **Inert atmosphere:** None; **Venting:** Pressure-vacuum; **Stability during transport:** Stable.

NAS HAZARD CLASSIFICATION FOR BULK WATER TRANSPORTATION

FIRE: 0
HEALTH: Vapor irritant: 4; Liquid or solid irritant: 4; Poisons: 4
WATER POLLUTION: Human toxicity: 4; Aquatic toxicity: 4; Aesthetic effect: 2
REACTIVITY: Other chemicals: 4; Water: 4; Self-reaction: 0

PHYSICAL AND CHEMICAL PROPERTIES

Physical state @ 59°F/15°C and 1 atm: Liquid.
Molecular weight: 154.9
Boiling point @ 1 atm: 241°F/116°C/389°K.
Melting/Freezing point: –142°F/–97°C/177°K.
Specific gravity (water = 1): 1.91 @ 68°F/20°C (liquid).
Liquid surface tension: 36.61 dynes/cm = 0.03661 N/m @ 66°F/19°C/292°K.
Relative vapor density (air = 1): 5.3
Ratio of specific heats of vapor (gas): 1.2832
Latent heat of vaporization: 113 Btu/lb = 62.6 cal/g = 2.62 x 10^5 J/kg.
Heat of solution: –279 Btu/lb = –155 cal/g = –6.48 x 10^5 J/kg.
Vapor pressure: 20 mm.

CITRIC ACID	REC. C:3500

SYNONYMS: ACIDO CITRICO (Spanish); 2-HYDROXY-1, 2, 3-PROPANE-TRICARBOXYLIC ACID; β-HYDROXYTRICARBALLYLIC ACID; β-HYDROXYTRICARBOXYLIC ACID; 2-HYDROXY-1,2,3-PROPANETRICARBOXYLIC ACID

IDENTIFICATION

CAS Number: 77-92-9
Formula: C$_6$H$_8$O$_7$; HOC(CH$_2$CO$_2$H)$_2$CO$_2$H

DESCRIPTION: Colorless solid. Odorless. Sinks and mixes in water forming an acid solution.

Combustible solid • Dust cloud may explode if ignited in an enclosed area • Containers may BLEVE when exposed to fire • Irritating to the skin, eyes, and respiratory tract • Toxic products of combustion may include carbon monoxide.

Cobalt acetate

Hazard Classification (based on NFPA-704 Rating System)
Health Hazards (Blue): 0; Flammability (Red): 1; Reactivity (Yellow): 0

EMERGENCY RESPONSE: See Appendix A (171)
Evacuation:
Public safety: Isolate the area of spill or leak for at least 10 to 25 meters (30 to 80 feet) in all directions.
Spill: Increase, in the downwind direction, as necessary, the distance shown under "Public Safety."
Fire: If any large container is involved in fire, isolate for at least 800 meters (½ mile) in all directions; also, consider initial evacuation for 800 meters (½ mile) in all directions.

EXPOSURE
Short-term effects: *SEEK MEDICAL ATTENTION.* **Dust:** Irritating to eyes, nose, and throat. *IF INHALED*, will, will cause coughing or difficult breathing. IF IN EYES, hold eyelids open and flush with plenty of water. *IF BREATHING HAS STOPPED*, give artificial respiration. If breathing is difficult, administer oxygen. **Solid:** Irritating to skin and eyes. Harmful if swallowed. Remove contaminated clothing and shoes. Flush affected areas with plenty of water. IF IN EYES, hold eyelids open and flush with plenty of water. IF SWALLOWED and victim is *CONSCIOUS AND ABLE TO SWALLOW*, have victim drink 4 to 8 ounces of water. IF SWALLOWED and victim is *UNCONSCIOUS OR HAVING CONVULSIONS*, do nothing except keep victim warm.

HEALTH HAZARDS
Personal protective equipment (PPE): B-Level PPE. Chemical protective material(s) reported to have good to excellent resistance: butyl rubber, natural rubber, neoprene, nitrile, nitrile+PVC, PVC, polyurethane, styrene-butdiene rubber. Also, Viton® offers limited protection.
Toxicity by ingestion: Grade 1; oral LD_{50} = 11.7 g/kg (rat).
Vapor (gas) irritant characteristics: Eye and respiratory tract irritant.
Liquid or solid irritant characteristics: Eye irritant.

FIRE DATA
Flash point: Combustible solid.
Flammable limits in air: LEL: 0.28; UEL: 2.29 kg/m3 (dust).
Behavior in fire: Melts and decomposes. The reaction may not be hazardous.
Autoignition temperature: 1850°F/1010°C/1283°K (powder).

CHEMICAL REACTIVITY
Reactivity with water: Aqueous solution is acidic.
Binary reactants: Reacts with bases and strong oxidizers. Will corrode copper, zinc, aluminum and their alloys. May have an explosive reaction with metal nitrates.

ENVIRONMENTAL DATA
Food chain concentration potential: Log P_{ow} = –1.68. Negative; unlikely to accumulate.
Water pollution: Dangerous to aquatic life in high concentrations. May be dangerous if it enters nearby water intakes; notify operators. Notify local health and wildlife officials. **Response to discharge:** Disperse and flush.

SHIPPING INFORMATION
Grades of purity: USP; Reagent; Monohydrate grade; **Storage temperature:** Ambient; **Inert atmosphere:** None; **Venting:** Open; **Stability during transport:** Stable.

PHYSICAL AND CHEMICAL PROPERTIES
Physical state @ 59°F/15°C and 1 atm: Solid.
Molecular weight: 192.1
Boiling point @ 1 atm: Decomposes below boiling point.
Melting/Freezing point: 307°F/153°C/426°K.
Specific gravity (water = 1): 1.54 @ 68°F/20°C (solid).
Heat of combustion: –4000 Btu/lb = –2220 cal/g = –93.0 J/kg.
Vapor pressure: 0.08 mm.

COBALT ACETATE   REC. C:3550

SYNONYMS: ACETATO de COBALTO (Spanish); BIS(ACETO)COBALT; ACETIC ACID, COBALT(2+) SALT; COBALT ACETATE TETRAHYDRATE; COBALT DIACETATE TETRAHYDRATE; COBALTOUS ACETATE TETRAHYDRATE; COBALT(II) ACETATE; COBALT(2+) ACETATE

IDENTIFICATION
CAS Number: 6147-53-1 (tetrahydrate); 71-48-7 (diacetate); 917-69-1 (triacetate)
Formula: $C_4H_6O_4 \cdot Co$ (diacetate); $C_6H_{12}O_6 \cdot Co$ (triacetate); $Co(C_2H_3O_2)_2 \cdot 4H_2O$
DOT ID Number: UN 3077; DOT Guide Number: 171
Proper Shipping Name: Environmentally hazardous substances, solid, n.o.s.
Reportable Quantity (RQ): **(CERCLA)** 1 lb/0.454 kg

DESCRIPTION: Pink or red crystalline solid. Acetic acid or vinegar-like odor. Sinks and mixes with water.

Irritating to the skin, eyes, and respiratory tract • Toxic products of combustion may include carbon monoxide and cobalt fume.

Hazard Classification (based on NFPA-704 Rating System)
Health Hazards (Blue): 1; Flammability (Red): 0; Reactivity (Yellow): 0

EMERGENCY RESPONSE: See Appendix A (171)
Evacuation:
Public safety: Isolate the area of spill or leak for at least 10 to 25 meters (30 to 80 feet) in all directions.
Spill: Increase, in the downwind direction, as necessary, the distance shown under "Public Safety."
Fire: If any large container is involved in fire, isolate for at least 800 meters (½ mile) in all directions; also, consider initial evacuation for 800 meters (½ mile) in all directions.

EXPOSURE
Short-term effects: *SEEK MEDICAL ATTENTION.* **Dust:** Irritating to eyes, nose, and throat. *IF INHALED*, will, will cause coughing or difficult breathing. *IF IN EYES*, hold eyelids open and flush with plenty of water. *IF BREATHING HAS STOPPED*, give artificial respiration. IF breathing is difficult, administer oxygen. **Solid:** Irritating to skin and eyes. *IF SWALLOWED*, will cause nausea and vomiting. Remove contaminated clothing and shoes. Flush affected areas with plenty of water. *IF IN EYES*, hold eyelids open and flush with plenty of water. *IF SWALLOWED* and victim is *CONSCIOUS AND ABLE TO SWALLOW*, have victim drink 4 to 8 ounces of water and have victim induce vomiting. *IF SWALLOWED* and victim is *UNCONSCIOUS OR HAVING CONVULSIONS*, do nothing except keep victim warm.

HEALTH HAZARDS

Personal protective equipment (PPE): B-Level PPE. Dust respirator; rubber gloves; goggles or face shield; protective clothing.

Recommendations for respirator selection: NIOSH/OSHA as metal dust and fume.

0.25 mg/m^3: DM* *If not present as a fume* (any dust and mist respirator). 0.5 mg/m^3: DMXSQ* *If not present as a fume* (any dust and mist respirator except single-use and quarter mask respirators); or DMFu (any dust, mist and fume respirator); or SA* (any supplied-air respirator). 1.25 mg/m^3: SA:CF* (any supplied-air respirator operated in a continuous-flow mode); or PAPRDM* *If not present as a fume* (any powered, air-purifying respirator with a dust and mist filter); or PAPRDMFu* (any powered, air-purifying respirator with a dust, mist, and fume filter). 2.5 mg/m^3: HiEF (any air-purifying, full-facepiece respirator with a high-efficiency particulate filter); or SCBAF (any self-contained breathing apparatus with a full facepiece); or SAF (any supplied-air respirator with a full facepiece). 20 mg/m^3: SAF:PD,PP (any supplied-air respirator that has a full facepiece and is operated in a pressure-demand or other positive-pressure mode). *EMERGENCY OR PLANNED ENTRY INTO UNKNOWN CONCENTRATIONS OR IDLH CONDITIONS:* SCBAF:PD,PP (any self-contained breathing apparatus that has a full facepiece and is operated in a pressure-demand or other positive-pressure mode); or SAF:PD,PP:ASCBA (any supplied-air respirator that has a full facepiece and is operated in a pressure-demand or other positive-pressure mode in combination with an auxiliary self-contained breathing apparatus operated in a pressure-demand or other positive pressure mode). *ESCAPE:* HiEF (any air-purifying, full-facepiece respirator with a high-efficiency particulate filter); or SCBAE (any appropriate escape-type, self-contained breathing apparatus). *Note:* Substance reported to cause eye irritation or damage; may require eye protection.

Toxicity by ingestion: Grade 3; LD$_{50}$ = 50 to 500 mg/kg.

Long-term health effects: Cobalt is a confirmed carcinogen. Skin (eczema), blood, kidney, lung, and liver disorders.

Vapor (gas) irritant characteristics: Eye and respiratory tract irritant.

Liquid or solid irritant characteristics: Eye and skin irritant.

IDLH value: 20 mg/m^3 as cobalt.

CHEMICAL REACTIVITY

Binary reactants: Cobalt compounds may react with acetylene, oxidizers.

ENVIRONMENTAL DATA

Food chain concentration potential: Bioconcentration of 200–1000-fold only under constant exposure. Not significant in spill conditions.

Water pollution: Effects of low concentrations on aquatic life is unknown. May be dangerous if it enters nearby water intakes; notify operators. Notify local health and wildlife officials.

Response to discharge: Issue warning–water contaminant. Disperse and flush.

SHIPPING INFORMATION

Grades of purity: Technical; Reagent; **Storage temperature:** Ambient; **Inert atmosphere:** None; **Venting:** Open; **Stability during transport:** Stable.

PHYSICAL AND CHEMICAL PROPERTIES

Physical state @ 59°F/15°C and 1 atm: Solid.
Molecular weight: 249.1
Boiling point @ 1 atm: Decomposes.
Melting/Freezing point: 284°F/140°C/413°K.
Specific gravity (water = 1): 1.71 @ 68°F/20°C (solid).

COBALT BROMIDE REC. C:3600

SYNONYMS: BROMURO de COBALTO (Spanish); COBALT DIBROMIDE; COBALT(II) BROMIDE; COBALTOUS BROMIDE

IDENTIFICATION

CAS Number: 7789-43-7
Formula: CoBr$_2$
DOT ID Number: UN 3077; DOT Guide Number: 171
Proper Shipping Name: Environmentally hazardous substances, solid, n.o.s.

DESCRIPTION: Red-violet crystalline solid. Anhydrous form is green. Slight odor. Sinks and mixes with water.

Irritating to the skin, eyes, and respiratory tract • Toxic products of combustion may include bromide fumes.

Hazard Classification (based on NFPA-704 Rating System)
Health Hazards (Blue): 2; Flammability (Red): 0; Reactivity (Yellow): 0

EMERGENCY RESPONSE: See Appendix A (171)
Evacuation:
Public safety: Isolate the area of spill or leak for at least 10 to 25 meters (30 to 80 feet) in all directions.
Spill: Increase, in the downwind direction, as necessary, the distance shown under "Public Safety."
Fire: If any large container is involved in fire, isolate for at least 800 meters (½ mile) in all directions; also, consider initial evacuation for 800 meters (½ mile) in all directions.

EXPOSURE

Short-term effects: *SEEK MEDICAL ATTENTION.* **Solid:** Irritating to skin and eyes. Harmful if swallowed.
Flush affected area with plenty of water. *IF IN EYES*, hold eyelids open and flush with plenty of water. *IF SWALLOWED* and victim is *CONSCIOUS AND ABLE TO SWALLOW*, have victim drink 4 to 8 ounces of water and induce vomiting.

HEALTH HAZARDS

Personal protective equipment (PPE): B-Level PPE. Prevent contact, use rubber gloves, protective clothing, barrier creams, chemical dust mask, and safety goggles.

Recommendations for respirator selection: NIOSH/OSHA as metal dust and fume.

0.25 mg/m^3: DM* *If not present as a fume* (any dust and mist respirator). 0.5 mg/m^3: DMXSQ* *If not present as a fume* (any dust and mist respirator except single-use and quarter mask respirators); or DMFu (any dust, mist and fume respirator); or SA* (any supplied-air respirator). 1.25 mg/m^3: SA:CF* (any supplied-air respirator operated in a continuous-flow mode); or PAPRDM* *If not present as a fume* (any powered, air-purifying respirator with a dust and mist filter); or PAPRDMFu* (any powered, air-purifying respirator with a dust, mist, and fume filter). 2.5 mg/m^3: HiEF (any air-purifying, full-facepiece respirator with a high-efficiency particulate filter); or SCBAF (any self-contained breathing apparatus with a full facepiece); or SAF (any supplied-air respirator with a full facepiece). 20 mg/m^3: SAF:PD,PP (any supplied-air respirator that has a full facepiece and is operated in a pressure-

Cobalt chloride

demand or other positive-pressure mode). *EMERGENCY OR PLANNED ENTRY INTO UNKNOWN CONCENTRATIONS OR IDLH CONDITIONS:* SCBAF:PD,PP (any self-contained breathing apparatus that has a full facepiece and is operated in a pressure-demand or other positive-pressure mode); or SAF:PD,PP:ASCBA (any supplied-air respirator that has a full facepiece and is operated in a pressure-demand or other positive-pressure mode in combination with an auxiliary self-contained breathing apparatus operated in a pressure-demand or other positive pressure mode). *ESCAPE:* HiEF (any air-purifying, full-facepiece respirator with a high-efficiency particulate filter); or SCBAE (any appropriate escape-type, self-contained breathing apparatus). *Note:* Substance reported to cause eye irritation or damage; may require eye protection.

Exposure limits (TWA unless otherwise noted): ACGIH TLV 0.02 mg/m^3 confirmed animal carcinogen; OSHA PEL 0.1 mg/m^3; NIOSH REL 0.05 mg/m^3 as inorganic cobalt

Toxicity by ingestion: Grade 2; LD_{50} = 0.5 to 5 g/kg.

Long-term health effects: Prolonged contact or inhalation of inorganic bromides causes acne-like bromoderma (bromide rash), especially of the hands and face. Other effects include emaciation, depression; and, in severe cases, psychosis and mental deterioration. Cobalt is a confirmed carcinogen.

Vapor (gas) irritant characteristics: Eye irritant.

Liquid or solid irritant characteristics: Eye and skin irritant.

IDLH value: 20 mg/m^3 as cobalt.

CHEMICAL REACTIVITY

Binary reactants: Reacts with acetylene, oxidizers.

ENVIRONMENTAL DATA

Food chain concentration potWater pollution: Harmful to aquatic life in very low concentrations. May be dangerous if it enters nearby water intakes; notify operators. Notify local health and wildlife officials. **Response to discharge:** Issue warning–water contaminant. Disperse and flush.

SHIPPING INFORMATION

Storage temperature: Cool; **Stability during transport:** Stable.

PHYSICAL AND CHEMICAL PROPERTIES

Physical state @ 59°F/15°C and 1 atm: Solid.
Molecular weight: 218.77
Boiling point @ 1 atm: Loses 4 H_2O at 100°C and all H_2O at 130°C.
Melting/Freezing point: Anhydride: 1252.4°F/678°C/951.2°K. Hexahydrate: 118°F/48°C/321°K.
Specific gravity (water = 1): 4.909 @ 77°F/25°C; 2.46 at room temperature (hexahydrate).
Heat of solution: For anhydrous $CoBr_2$ (exothermic) –151 Btu/lb = –84.1 cal/g = –3.5 x 10^5 J/kg.

COBALT CHLORIDE   **REC. C:3650**

SYNONYMS: CLORURO COBALTOSO (Spanish); COBALTOUS CHLORIDE; COBALTOUS CHLORIDE HEXAHYDRATE; COBALTOUS CHLORIDE DIHYDRATE; COBALT(II) CHLORIDE; COBALT(2+) CHLORIDE; COBALT MURIATE; KOBALT CHLORID (German)

IDENTIFICATION

CAS Number: 7646-79-9; 7791-13-1 (hexahydrate)
Formula: $CoCl_2$; $Cl_2Co·6H_2O$ (hexahydrate)
DOT ID Number: UN 3077; **DOT Guide Number:** 171
Proper Shipping Name: Environmentally hazardous substances, solid, n.o.s.
Reportable Quantity (RQ): **(CERCLA)** 1 lb/0.454 kg

DESCRIPTION: Pink to dark red crystals or blue powder. Slight sharp odor. Sinks and mixes with water forming an acid solution.

Poison! • Breathing the dust, skin or eye contact, or swallowing the material may cause illness and possible death • Firefighting gear (including SCBA) does not provide adequate protection. If exposure occurs, remove and isolate gear immediately and thoroughly decontaminate personnel • Irritating to the skin, eyes, and respiratory tract • Toxic products of combustion may include cobalt oxide and hydrogen chloride fumes.

Hazard Classification (based on NFPA-704 Rating System)
Health Hazards (Blue): 3; Flammability (Red): 0; Reactivity (Yellow): 0

EMERGENCY RESPONSE: See Appendix A (171)
Evacuation:
Public safety: Isolate the area of spill or leak for at least 10 to 25 meters (30 to 80 feet) in all directions.
Spill: Increase, in the downwind direction, as necessary, the distance shown under "Public Safety."
Fire: If any large container is involved in fire, isolate for at least 800 meters (½ mile) in all directions; also, consider initial evacuation for 800 meters (½ mile) in all directions.

EXPOSURE

Short-term effects: *SEEK MEDICAL ATTENTION.* **Dust:** Irritating to eyes, nose, and throat. *IF INHALED*, will, will cause coughing or difficult breathing. *IF IN EYES*, hold eyelids open and flush with plenty of water. *IF BREATHING HAS STOPPED*, give artificial respiration. IF breathing is difficult, administer oxygen. **Solid:** Irritating to skin and eyes. *IF SWALLOWED*, will cause nausea and vomiting. Remove contaminated clothing and shoes. Flush affected areas with plenty of water. *IF IN EYES*, hold eyelids open and flush with plenty of water. *IF SWALLOWED* and victim is *CONSCIOUS AND ABLE TO SWALLOW*, have victim drink 4 to 8 ounces of water and have victim induce vomiting. *IF SWALLOWED* and victim is *UNCONSCIOUS OR HAVING CONVULSIONS,* do nothing except keep victim warm.

HEALTH HAZARDS

Personal protective equipment (PPE): A-Level PPE.
Recommendations for respirator selection: NIOSH/OSHA as metal dust and fume.
0.25 mg/m^3: DM* *If not present as a fume* (any dust and mist respirator). 0.5 mg/m^3: DMXSQ* *If not present as a fume* (any dust and mist respirator except single-use and quarter mask respirators); or DMFu (any dust, mist and fume respirator); or SA* (any supplied-air respirator). 1.25 mg/m^3: SA:CF* (any supplied-air respirator operated in a continuous-flow mode); or PAPRDM* *If not present as a fume* (any powered, air-purifying respirator with a dust and mist filter); or PAPRDMFu* (any powered, air-purifying respirator with a dust, mist, and fume filter). 2.5 mg/m^3: HiEF (any air-purifying, full-facepiece respirator with a high-efficiency particulate filter); or SCBAF (any self-contained breathing apparatus with a full facepiece); or SAF (any supplied-air respirator with a full facepiece). 20 mg/m^3: SAF:PD,PP (any supplied-air respirator that has a full facepiece and is operated in a pressure-demand or other positive-pressure mode). *EMERGENCY OR PLANNED ENTRY INTO UNKNOWN CONCENTRATIONS OR IDLH CONDITIONS:* SCBAF:PD,PP (any self-contained breathing

apparatus that has a full facepiece and is operated in a pressure-demand or other positive-pressure mode); or SAF:PD,PP:ASCBA (any supplied-air respirator that has a full facepiece and is operated in a pressure-demand or other positive-pressure mode in combination with an auxiliary self-contained breathing apparatus operated in a pressure-demand or other positive pressure mode). *ESCAPE:* HiEF (any air-purifying, full-facepiece respirator with a high-efficiency particulate filter); or SCBAE (any appropriate escape-type, self-contained breathing apparatus). *Note:* Substance reported to cause eye irritation or damage; may require eye protection.
Exposure limits (TWA unless otherwise noted): ACGIH TLV 0.02 mg/m^3 confirmed animal carcinogen; OSHA PEL 0.1 mg/m^3; NIOSH REL 0.05 mg/m^3 as inorganic cobalt
Toxicity by ingestion: Grade 3; LD$_{50}$ = 50 to 500 mg/kg.
Long-term health effects: May cause thyroid gland problems; goiter; anorexia and weight loss. Heart disorders and kidney damage may occur. Mutation data reported.
Vapor (gas) irritant characteristics: Respiratory tract irritation and possible permanent damage. Eye irritant.
Liquid or solid irritant characteristics: Eye and skin irritant.
IDLH value: 20 mg/m^3 as cobalt.

CHEMICAL REACTIVITY
Binary reactants: Cobalt compounds react with oxidizers, acetylene, metals.

ENVIRONMENTAL DATA
Food chain concentration potential: Bioconcentration of 200–1000-fold only under constant exposure. Not significant in spill conditions.
Water pollution: Dangerous to aquatic life in high concentrations. May be dangerous if it enters nearby water intakes; notify operators. Notify local health and wildlife officials. **Response to discharge:** Issue warning–water contaminant. Disperse and flush.

SHIPPING INFORMATION
Grades of purity: Anhydrous, 100.1%. May also be shipped as dihydrate and hexahydrate; **Storage temperature:** Ambient; **Inert atmosphere:** None; **Venting:** Open; **Stability during transport:** Stable. Keep hexahydrate away from moisture.

PHYSICAL AND CHEMICAL PROPERTIES
Physical state @ 59°F/15°C and 1 atm: Solid.
Molecular weight: 237.9
Boiling point @ 1 atm: 752°F/400°C/673°K (decomposes).
Melting/Freezing point: 187°F/86°C/359°K.
Specific gravity (water = 1): 1.924 @ 68°F/20°C (solid).
Heat of decomposition: 266°F/130°C.
Heat of solution: 22 Btu/lb = 12 cal/g = 0.50 x 10^5 J/kg.
Heat of fusion: 56.9 cal/g.

COBALT FLUORIDE REC. C:3700

SYNONYMS: COBALT DIFLUORIDE; COBALT(II) FLUORIDE; COBALTOUS FLUORIDE; FLUORURO COBALTICO (Spanish)

IDENTIFICATION
CAS Number: 10026-17-2
Formula: CoF$_2$
DOT ID Number: UN 3077; DOT Guide Number: 171
Proper Shipping Name: Environmentally hazardous substances, solid, n.o.s.
Reportable Quantity (RQ): **(CERCLA)** 1 lb/0.454 kg

DESCRIPTION: Violet to rose-red crystalline solid. Sinks and mixes slowly with water; slightly soluble.

Poison! • Breathing the dust, skin or eye contact, or swallowing the material may cause illness • Firefighting gear (including SCBA) does not provide adequate protection. If exposure occurs, remove and isolate gear immediately and thoroughly decontaminate personnel • Irritating to the skin, eyes, and respiratory tract • Toxic products of combustion may include cobalt oxide and fluoride fumes.

Hazard Classification (based on NFPA-704 Rating System)
Health Hazards (Blue): 2; Flammability (Red): 0; Reactivity (Yellow): 0

EMERGENCY RESPONSE: See Appendix A (171)
Evacuation:
Public safety: Isolate the area of spill or leak for at least 10 to 25 meters (30 to 80 feet) in all directions.
Spill: Increase, in the downwind direction, as necessary, the distance shown under "Public Safety."
Fire: If any large container is involved in fire, isolate for at least 800 meters (½ mile) in all directions; also, consider initial evacuation for 800 meters (½ mile) in all directions.

EXPOSURE
Short-term effects: *SEEK MEDICAL ATTENTION.* **Solid:** Irritating to skin and eyes. Harmful if swallowed.
Flush affected areas with plenty of water. *IF IN EYES*, hold eyelids open and flush with plenty of water. *IF SWALLOWED* and victim is *CONSCIOUS AND ABLE TO SWALLOW*, have victim drink 4 to 8 ounces of water and have victim induce vomiting.

HEALTH HAZARDS
Personal protective equipment (PPE): B-Level PPE. Safety glasses, polyvinyl chloride gloves, approved respirator.
Recommendations for respirator selection: NIOSH/OSHA
0.25 mg/m^3: DM* *If not present as a fume* (any dust and mist respirator). *0.5 mg/m^3*: DMXSQ* *If not present as a fume* (any dust and mist respirator except single-use and quarter mask respirators); or DMFu (any dust, mist and fume respirator); or SA* (any supplied-air respirator). *1.25 mg/m^3*: SA:CF* (any supplied-air respirator operated in a continuous-flow mode); or PAPRDM* *If not present as a fume* (any powered, air-purifying respirator with a dust and mist filter); or PAPRDMFu* (any powered, air-purifying respirator with a dust, mist, and fume filter). *2.5 mg/m^3*: HiEF (any air-purifying, full-facepiece respirator with a high-efficiency particulate filter); or SCBAF (any self-contained breathing apparatus with a full facepiece); or SAF (any supplied-air respirator with a full facepiece). 20 mg/m^3: SAF:PD,PP (any supplied-air respirator that has a full facepiece and is operated in a pressure-demand or other positive-pressure mode). *EMERGENCY OR PLANNED ENTRY INTO UNKNOWN CONCENTRATIONS OR IDLH CONDITIONS:* SCBAF:PD,PP (any self-contained breathing apparatus that has a full facepiece and is operated in a pressure-demand or other positive-pressure mode); or SAF:PD,PP:ASCBA (any supplied-air respirator that has a full facepiece and is operated in a pressure-demand or other positive-pressure mode in combination with an auxiliary self-contained breathing apparatus operated in a pressure-demand or other positive pressure mode). *ESCAPE:* HiEF (any air-purifying, full-facepiece respirator with a

high-efficiency particulate filter); or SCBAE (any appropriate escape-type, self-contained breathing apparatus). *Note:* Substance reported to cause eye irritation or damage; may require eye protection.
Exposure limits (TWA unless otherwise noted): ACGIH TLV 0.02 mg/m^3 confirmed animal carcinogen; OSHA PEL 0.1 mg/m^3; NIOSH REL 0.05 mg/m^3 as inorganic cobalt.
Toxicity by ingestion: Grade 3; LD$_{50}$ = 50 to 500 mg/kg.
Long-term health effects: Loss of weight, anorexia, anemia, wasting and cachexia, and dental effects are common in chronic fluoride poisoning.
Vapor (gas) irritant characteristics: May cause respiratory tract irritation and lung damage.
Liquid or solid irritant characteristics: Eye and skin damage.
IDLH value: 20 mg/m^3 as cobalt.

CHEMICAL REACTIVITY
Binary reactants: Reacts with oxidizers, acetylene.

ENVIRONMENTAL DATA
Food chain concentration potential: Organisms can concentrate cobalt 1070 to 1500 times.
Water pollution: Harmful to aquatic life in very low concentrations. May be harmful if it enters water intakes. Notify local health and wildlife officials. **Response to discharge:** Issue warning–water contaminant. Should be removed.

SHIPPING INFORMATION
Grades of purity: 98%; **Stability during transport:** Stable.

PHYSICAL AND CHEMICAL PROPERTIES
Physical state @ 59°F/15°C and 1 atm: Solid.
Molecular weight: 96.94
Boiling point @ 1 atm: Volatilizes at about 1400°C.
Melting/Freezing point: 2192°F/1200°C/1473.2°K.
Specific gravity (water = 1): 4.46 @ 77°F/25°C.
Heat of solution: For anhydrous CoF$_2$ (exothermic) –271 Btu/lb = 150.6 cal/g = –6.3 x 10^5 J/kg.
Heat of fusion: 92.9 cal/g.

COBALT FORMATE   **REC. C:3750**

SYNONYMS: COBALT DIFORMATE; COBALTOUS FORMATE; FORMIATO COBALTOSO (Spanish)

IDENTIFICATION
CAS Number: 544-18-3
Formula: Co(HCOO)$_2$·2H$_2$O
DOT ID Number: UN 9104; DOT Guide Number: 171
Proper Shipping Name: Cobalt formate
Reportable Quantity (RQ): **(CERCLA)** 1000 lb/454 kg

DESCRIPTION: Red solid (dihydrate). Sinks and mixes with water.

Poison! • Breathing the dust, skin or eye contact, or swallowing the material may cause illness • Firefighting gear (including SCBA) does not provide adequate protection. If exposure occurs, remove and isolate gear immediately and thoroughly decontaminate personnel • Irritating to the skin, eyes, and respiratory tract • Toxic products of combustion may include cobalt oxide and carbon monoxide.

Hazard Classification (based on NFPA-704 Rating System)
Health Hazards (Blue): 2; Flammability (Red): 0; Reactivity (Yellow): 0

EMERGENCY RESPONSE: See Appendix A (171)
Evacuation:
Public safety: Isolate the area of spill or leak for at least 10 to 25 meters (30 to 80 feet) in all directions.
Spill: Increase, in the downwind direction, as necessary, the distance shown under "Public Safety."
Fire: If any large container is involved in fire, isolate for at least 800 meters (½ mile) in all directions; also, consider initial evacuation for 800 meters (½ mile) in all directions.

EXPOSURE
Short-term effects: *SEEK MEDICAL ATTENTION.* **Dust:** Irritating to skin and eyes. Harmful if inhaled. Move to fresh air. *IF BREATHING HAS STOPPED,* give artificial respiration. **Solid:** Will burn skin and eyes. Harmful if swallowed. Remove contaminated clothing and shoes. Flush affected area with plenty of water. *IF IN EYES,* hold eyelids open and flush with plenty of water. *IF SWALLOWED* and victim is *CONSCIOUS AND ABLE TO SWALLOW,* have victim drink 4 to 8 ounces of water.

HEALTH HAZARDS
Personal protective equipment (PPE): B-Level PPE.
Recommendations for respirator selection: NIOSH/OSHA as metal dust and fume.
0.25 mg/m^3: DM* *If not present as a fume* (any dust and mist respirator). 0.5 mg/m^3: DMXSQ* *If not present as a fume* (any dust and mist respirator except single-use and quarter mask respirators); or DMFu (any dust, mist and fume respirator); or SA* (any supplied-air respirator). 1.25 mg/m^3: SA:CF* (any supplied-air respirator operated in a continuous-flow mode); or PAPRDM* *If not present as a fume* (any powered, air-purifying respirator with a dust and mist filter); or PAPRDMFu* (any powered, air-purifying respirator with a dust, mist, and fume filter). 2.5 mg/m^3: HiEF (any air-purifying, full-facepiece respirator with a high-efficiency particulate filter); or SCBAF (any self-contained breathing apparatus with a full facepiece); or SAF (any supplied-air respirator with a full facepiece). 20 mg/m^3: SAF:PD,PP (any supplied-air respirator that has a full facepiece and is operated in a pressure-demand or other positive-pressure mode). *EMERGENCY OR PLANNED ENTRY INTO UNKNOWN CONCENTRATIONS OR IDLH CONDITIONS:* SCBAF:PD,PP (any self-contained breathing apparatus that has a full facepiece and is operated in a pressure-demand or other positive-pressure mode); or SAF:PD,PP:ASCBA (any supplied-air respirator that has a full facepiece and is operated in a pressure-demand or other positive-pressure mode in combination with an auxiliary self-contained breathing apparatus operated in a pressure-demand or other positive pressure mode). *ESCAPE:* HiEF (any air-purifying, full-facepiece respirator with a high-efficiency particulate filter); or SCBAE (any appropriate escape-type, self-contained breathing apparatus). *Note:* Substance reported to cause eye irritation or damage; may require eye protection.
Long-term health effects: May cause dermatitis and hypersensitivity of skin.
Liquid or solid irritant characteristics: Severe eye irritant. Causes smarting of the skin and first-degree burns on short exposure; may cause second-degree burns on long exposure.
IDLH value: 20 mg/m^3 as cobalt.

FIRE DATA
Behavior in fire: Decomposes at 347°F/175°C/448°C.

CHEMICAL REACTIVITY
Binary reactants: Cobalt reacts with strong oxidizers, ammonium nitrate, acetylene.

ENVIRONMENTAL DATA
Food chain concentration potential: Microorganisms can concentrate cobalt in water up to 1070 to 1500 times.
Water pollution: Harmful to aquatic life in very low concentrations. May be dangerous if it enters nearby water intakes; notify operators. Notify local health and wildlife officials.
Response to discharge: Issue warning–water contaminant. Disperse and flush.

SHIPPING INFORMATION
Storage temperature: Ambient; **Stability during transport:** Stable.

PHYSICAL AND CHEMICAL PROPERTIES
Physical state @ 59°F/15°C and 1 atm: Solid.
Molecular weight: 148.98 anhydrous; 185.00 dihydrate
Boiling point @ 1 atm: Becomes anhydrous at 284°F/140°C/413°K; Decomposes at @ 1 atm: 347°F/175°C/448°K.
Specific gravity (water = 1): 2.13 at 71°F/22°C dihydrate

COBALT NITRATE   REC. C:3800

SYNONYMS: COBALTOUS NITRATE; COBALT(II) NITRATE; COBALT(2+) NITRATE; COBALTOUS NITRATE, HEXAHYDRATE; NITRATO COBALTOSO (Spanish); NITRIC ACID, COBALT(2+) SALT

IDENTIFICATION
CAS Number: 10141-05-6; 10026-22-9 (hexahydrate)
Formula: CoN_2O_6; $Co(NO_3)_2 \cdot 6H_2O$
DOT ID Number: UN 1477; DOT Guide Number: 140
Proper Shipping Name: Nitrates, inorganic, n.o.s.

DESCRIPTION: Pink powder or red crystalline solid. Odorless. Sinks and mixes with water; forms acid solution.

Poison! • Breathing the dust, skin or eye contact, or swallowing the material may cause illness • Firefighting gear (including SCBA) does not provide adequate protection. If exposure occurs, remove and isolate gear immediately and thoroughly decontaminate personnel • Strong oxidizer which may react spontaneously with low flash point organics or reducing agents. Heat forms oxygen; will increase the activity of an existing fire • May cause fire or explosion on contact with combustibles (wood, paper, oil, clothing, etc.) • Irritating to the skin, eyes, and respiratory tract • Toxic products of combustion may include cobalt oxide and nitrogen oxides.

Hazard Classification (based on NFPA-704 Rating System)
Health Hazards (Blue): 2; Flammability (Red): 0; Reactivity (Yellow): 1; Special Notice (White): OXY

EMERGENCY RESPONSE: See Appendix A (140)
Evacuation:
Public safety: Isolate the area of spill or leak for at least 10 to 25 meters (30 to 80 feet) in all directions.
Spill: Consider initial downwind evacuation for at least 100 meters (330 feet).
Fire: If any large container is involved in fire, isolate for at least 800 meters (½ mile) in all directions; also, consider initial evacuation for 800 meters (½ mile) in all directions.

EXPOSURE
Short-term effects: *SEEK MEDICAL ATTENTION.* **Dust:** Irritating to eyes, nose, and throat. *IF INHALED*, will, will cause coughing or difficult breathing. If in eyes, hold eyelids open and flush with plenty of water. *IF BREATHING HAS STOPPED*, give artificial respiration. If breathing is difficult, administer oxygen. **Solid:** Irritating to skin and eyes. *IF SWALLOWED*, will cause nausea and vomiting. Remove contaminated clothing and shoes. Flush affected areas with plenty of water. *IF IN EYES*, hold eyelids open and flush with plenty of water. *IF SWALLOWED* and victim is *CONSCIOUS AND ABLE TO SWALLOW*, have victim drink 4 to 8 ounces of water and have victim induce vomiting. *IF SWALLOWED* and victim is *UNCONSCIOUS OR HAVING CONVULSIONS,* do nothing except keep victim warm.

HEALTH HAZARDS
Personal protective equipment (PPE): A-Level PPE.
Recommendations for respirator selection: NIOSH/OSHA as metal dust and fume.
0.25 mg/m³: DM* *If not present as a fume* (any dust and mist respirator). 0.5 mg/m³: DMXSQ* *If not present as a fume* (any dust and mist respirator except single-use and quarter mask respirators); or DMFu (any dust, mist and fume respirator); or SA* (any supplied-air respirator). 1.25 mg/m³: SA:CF* (any supplied-air respirator operated in a continuous-flow mode); or PAPRDM* *If not present as a fume* (any powered, air-purifying respirator with a dust and mist filter); or PAPRDMFu* (any powered, air-purifying respirator with a dust, mist, and fume filter). 2.5 mg/m³: HiEF (any air-purifying, full-facepiece respirator with a high-efficiency particulate filter); or SCBAF (any self-contained breathing apparatus with a full facepiece); or SAF (any supplied-air respirator with a full facepiece). 20 mg/m³: SAF:PD,PP (any supplied-air respirator that has a full facepiece and is operated in a pressure-demand or other positive-pressure mode). *EMERGENCY OR PLANNED ENTRY INTO UNKNOWN CONCENTRATIONS OR IDLH CONDITIONS:* SCBAF:PD,PP (any self-contained breathing apparatus that has a full facepiece and is operated in a pressure-demand or other positive-pressure mode); or SAF:PD,PP:ASCBA (any supplied-air respirator that has a full facepiece and is operated in a pressure-demand or other positive-pressure mode in combination with an auxiliary self-contained breathing apparatus operated in a pressure-demand or other positive pressure mode). *ESCAPE:* HiEF (any air-purifying, full-facepiece respirator with a high-efficiency particulate filter); or SCBAE (any appropriate escape-type, self-contained breathing apparatus). *Note:* Substance reported to cause eye irritation or damage; may require eye protection.
Exposure limits (TWA unless otherwise noted): ACGIH TLV 0.02 mg/m³ confirmed animal carcinogen; OSHA PEL 0.1 mg/m³; NIOSH REL 0.05 mg/m³ as inorganic cobalt
Toxicity by ingestion: Grade 3; LD_{50} about 400 mg/kg (rabbit).
Long-term health effects: Causes malignant tumors in rabbits. Kidney, blood and glandular (thyroid) damage.
Vapor (gas) irritant characteristics: Respiratory irritant; possible permanent damage.
Liquid or solid irritant characteristics: Eye and skin irritant.
IDLH value: 20 mg/m³.

CHEMICAL REACTIVITY
Binary reactants: A strong oxidizer; contact with wood and paper may cause fire; contact with carbon may cause explosion.

ENVIRONMENTAL DATA
Food chain concentration potential: Bioconcentration of 200–1000-fold only under constant exposure. Not significant in spill conditions.
Water pollution: Harmful to aquatic life in very low concentrations. May be dangerous if it enters nearby water intakes; notify operators. Notify local health and wildlife officials.
Response to discharge: Issue warning–water contaminant. Disperse and flush.

SHIPPING INFORMATION
Grades of purity: Technical hexahydrate. May also be shipped as anhydrous solid; **Storage temperature:** Ambient; **Inert atmosphere:** None; **Venting:** Open; **Stability during transport:** Stable.

PHYSICAL AND CHEMICAL PROPERTIES
Physical state @ 59°F/15°C and 1 atm: Solid.
Molecular weight: 291.04
Boiling point @ 1 atm: 167°F/75°C/348°K (decomposes).
Melting/Freezing point: 131°F/55°C/328°K.
Specific gravity (water = 1): 1.54 @ 68°F/20°C (solid).
Heat of solution: 31 Btu/lb = 17 cal/g = 0.71×10^5 J/kg.

COBALT SULFAMATE   REC. C: 3850

SYNONYMS: COBALT AMINO SULFONATE; COBALTOUS SULFAMATE; SULFAMIC ACID, COBALT SALT

IDENTIFICATION
CAS Number: 16107-41-3
Formula: $CoS_2O_6N_2H_6$
DOT ID Number: UN 9105; DOT Guide Number: 171
Proper Shipping Name: Cobaltous Sulfamate

DESCRIPTION: Red solid. Odorless. Sinks and mixes with water.

Poison! • Breathing the dust, skin or eye contact, or swallowing the material may cause illness • Firefighting gear (including SCBA) does not provide adequate protection. If exposure occurs, remove and isolate gear immediately and thoroughly decontaminate personnel • Irritating to the skin, eyes, and respiratory tract • Toxic products of combustion may include cobalt oxide nitrogen oxide, and sulfur oxide.

Hazard Classification (based on NFPA-704 Rating System)
Health Hazards (Blue): 2; Flammability (Red): 0; Reactivity (Yellow): 0

EMERGENCY RESPONSE: See Appendix A (171)
Evacuation:
Public safety: Isolate the area of spill or leak for at least 10 to 25 meters (30 to 80 feet) in all directions.
Spill: Increase, in the downwind direction, as necessary, the distance shown under "Public Safety."
Fire: If any large container is involved in fire, isolate for at least 800 meters (½ mile) in all directions; also, consider initial evacuation for 800 meters (½ mile) in all directions.

EXPOSURE
Short-term effects: *SEEK MEDICAL ATTENTION.* **Dust:** Irritating to eyes, nose, and throat. *IF INHALED*, will, will cause coughing or difficult breathing. *IF IN EYES*, hold eyelids open and flush with plenty of water. *IF BREATHING HAS STOPPED*, give artificial respiration. IF breathing is difficult, administer oxygen. **Solid:** Irritating to skin and eyes. *IF SWALLOWED*, will cause nausea and vomiting. Remove contaminated clothing and shoes. Flush affected areas with plenty of water. *IF IN EYES*, hold eyelids open and flush with plenty of water. *IF SWALLOWED* and victim is *CONSCIOUS AND ABLE TO SWALLOW*, have victim drink 4 to 8 ounces of water and induce vomiting. *IF SWALLOWED* and victim is *UNCONSCIOUS OR HAVING CONVULSIONS*, do nothing except keep victim warm.

HEALTH HAZARDS
Personal protective equipment (PPE): A-Level PPE. Approved respirator, rubber gloves, safety goggles, protective clothing.
Recommendations for respirator selection: NIOSH/OSHA as metal dust and fume.
0.25 mg/m³: DM* *If not present as a fume* (any dust and mist respirator). 0.5 mg/m³: DMXSQ* *If not present as a fume* (any dust and mist respirator except single-use and quarter mask respirators); or DMFu (any dust, mist and fume respirator); or SA* (any supplied-air respirator). 1.25 mg/m³: SA:CF* (any supplied-air respirator operated in a continuous-flow mode); or PAPRDM* *If not present as a fume* (any powered, air-purifying respirator with a dust and mist filter); or PAPRDMFu* (any powered, air-purifying respirator with a dust, mist, and fume filter). 2.5 mg/m³: HiEF (any air-purifying, full-facepiece respirator with a high-efficiency particulate filter); or SCBAF (any self-contained breathing apparatus with a full facepiece); or SAF (any supplied-air respirator with a full facepiece). 20 mg/m³: SAF:PD,PP (any supplied-air respirator that has a full facepiece and is operated in a pressure-demand or other positive-pressure mode). *EMERGENCY OR PLANNED ENTRY INTO UNKNOWN CONCENTRATIONS OR IDLH CONDITIONS:* SCBAF:PD,PP (any self-contained breathing apparatus that has a full facepiece and is operated in a pressure-demand or other positive-pressure mode); or SAF:PD,PP:ASCBA (any supplied-air respirator that has a full facepiece and is operated in a pressure-demand or other positive-pressure mode in combination with an auxiliary self-contained breathing apparatus operated in a pressure-demand or other positive pressure mode). *ESCAPE:* HiEF (any air-purifying, full-facepiece respirator with a high-efficiency particulate filter); or SCBAE (any appropriate escape-type, self-contained breathing apparatus). *Note:* Substance reported to cause eye irritation or damage; may require eye protection.
Exposure limits (TWA unless otherwise noted): ACGIH TLV 0.02 mg/m³ confirmed animal carcinogen; OSHA PEL 0.1 mg/m³; NIOSH 0.05 mg/m³ as inorganic cobalt
Toxicity by ingestion: Grade 3: LD_{50} about 400 mg/kg (rabbit), as cobalt.
Long-term health effects: Causes malignant tumors in rabbits; lung damage.
Vapor (gas) irritant characteristics: Respiratory tract irritant; possible permanent damage.
Liquid or solid irritant characteristics: Eye and skin irritant.
IDLH value: 20 mg/m³ as cobalt.

CHEMICAL REACTIVITY
Binary reactants: Acids or acid fumes form toxic sulfur oxides. Contact with wood or paper may cause fire. Reacts with reducing agents.

ENVIRONMENTAL DATA
Food chain concentration potential: Bioconcentration of 200–1000-fold only under constant exposure. Not significant under spill conditions.

Water pollution: Harmful to aquatic life in very low concentrations. May be dangerous if it enters nearby water intakes; notify operators. Notify local health and wildlife officials. Notify operators of local water intakes. **Response to discharge:** Issue warning–water contaminant. Disperse and flush.

SHIPPING INFORMATION
Grades of purity: Technical trihydrate. May be shipped as anhydrous solid; **Storage temperature:** Ambient; **Inert atmosphere:** None; **Venting:** Open; **Stability during transport:** Stable.

NAS HAZARD CLASSIFICATION FOR BULK WATER TRANSPORTATION
FIRE: 0
HEALTH: Vapor irritant: 4; Liquid or solid irritant: 1; Poisons: 4
WATER POLLUTION: Human toxicity: 3; Aquatic toxicity: 3; Aesthetic effect: 1
REACTIVITY: Other chemicals: 1; Water: 0; Self-reaction: 0

PHYSICAL AND CHEMICAL PROPERTIES
Physical state @ 59°F/15°C and 1 atm: Solid.
Molecular weight: 253.13

COBALT SULFATE   REC. C:3900

SYNONYMS: COBALTOUS SULFATE HEPTAHYDRATE; COBALT(II) SULFATE; COBALT(2+) SULFATE; BIEBERITE; SULFATO de COBALTO (Spanish); SULFURIC ACID, COBALT(II) SALT; SULFURIC ACID, COBALT(2+) SALT

IDENTIFICATION
CAS Number: 10124-43-3
Formula: $CoSO_4 \cdot 7H_2O$
DOT ID Number: UN 3044; DOT Guide Number: 171
Proper Shipping Name: Environmentally hazardous substances, solid, n.o.s.
Reportable Quantity (RQ): **(CERCLA)** 1 lb/0.454 kg

DESCRIPTION: Dark blue crystalline solid. Odorless. Sinks and mixes with water.

Poison! • Breathing the dust, skin or eye contact, or swallowing the material may cause illness • Firefighting gear (including SCBA) does not provide adequate protection. If exposure occurs, remove and isolate gear immediately and thoroughly decontaminate personnel • Irritating to the skin, eyes, and respiratory tract • Toxic products of combustion may include cobalt oxide and sulfur oxide.

Hazard Classification (based on NFPA-704 Rating System)
Health Hazards (Blue): 2; Flammability (Red): 0; Reactivity (Yellow): 0

EMERGENCY RESPONSE: See Appendix A (171)
Evacuation:
Public safety: Isolate the area of spill or leak for at least 10 to 25 meters (30 to 80 feet) in all directions.
Spill: Increase, in the downwind direction, as necessary, the distance shown under "Public Safety."
Fire: If any large container is involved in fire, isolate for at least 800 meters (½ mile) in all directions; also, consider initial evacuation for 800 meters (½ mile) in all directions.

EXPOSURE
Short-term effects: *SEEK MEDICAL ATTENTION*. **Dust:** Irritating to eyes, nose, and throat. *IF INHALED*, will, will cause coughing or difficult breathing. *IF IN EYES*, hold eyelids open and flush with plenty of water. *IF BREATHING HAS STOPPED*, give artificial respiration. IF breathing is difficult, administer oxygen. **Solid:** Irritating to skin and eyes. *IF SWALLOWED*, will cause nausea and vomiting. Remove contaminated clothing and shoes. Flush affected areas with plenty of water. *IF IN EYES*, hold eyelids open and flush with plenty of water. *IF SWALLOWED* and victim is *CONSCIOUS AND ABLE TO SWALLOW*, have victim drink 4 to 8 ounces of water and have victim induce vomiting. *IF SWALLOWED* and victim is *UNCONSCIOUS OR HAVING CONVULSIONS*, do nothing except keep victim warm.

HEALTH HAZARDS
Personal protective equipment (PPE): B-Level PPE. Bureau Mines approved respirator; goggles; protective gloves.
Recommendations for respirator selection: NIOSH/OSHA as metal dust and fume.
0.25 mg/m^3: DM* *If not present as a fume* (any dust and mist respirator). 0.5 mg/m^3: DMXSQ* *If not present as a fume* (any dust and mist respirator except single-use and quarter mask respirators); or DMFu (any dust, mist and fume respirator); or SA* (any supplied-air respirator). 1.25 mg/m^3: SA:CF* (any supplied-air respirator operated in a continuous-flow mode); or PAPRDM* *If not present as a fume* (any powered, air-purifying respirator with a dust and mist filter); or PAPRDMFu* (any powered, air-purifying respirator with a dust, mist, and fume filter). 2.5 mg/m^3: HiEF (any air-purifying, full-facepiece respirator with a high-efficiency particulate filter); or SCBAF (any self-contained breathing apparatus with a full facepiece); or SAF (any supplied-air respirator with a full facepiece). 20 mg/m^3: SAF:PD,PP (any supplied-air respirator that has a full facepiece and is operated in a pressure-demand or other positive-pressure mode). *EMERGENCY OR PLANNED ENTRY INTO UNKNOWN CONCENTRATIONS OR IDLH CONDITIONS:* SCBAF:PD,PP (any self-contained breathing apparatus that has a full facepiece and is operated in a pressure-demand or other positive-pressure mode); or SAF:PD,PP:ASCBA (any supplied-air respirator that has a full facepiece and is operated in a pressure-demand or other positive-pressure mode in combination with an auxiliary self-contained breathing apparatus operated in a pressure-demand or other positive pressure mode). *ESCAPE:* HiEF (any air-purifying, full-facepiece respirator with a high-efficiency particulate filter); or SCBAE (any appropriate escape-type, self-contained breathing apparatus). *Note:* Substance reported to cause eye irritation or damage; may require eye protection.
Exposure limits (TWA unless otherwise noted): ACGIH TLV 0.02 mg/m^3 confirmed animal carcinogen; OSHA PEL 0.1 mg/m^3; NIOSH 0.05 mg/m^3 as inorganic cobalt
Toxicity by ingestion: Grade 3; LD$_{50}$ = 50 to 500 mg/kg.
Long-term health effects: Blood, lung, kidney or thyroid damage may occur.
Vapor (gas) irritant characteristics: Respiratory tract irritation; possible permanent damage.
Liquid or solid irritant characteristics: Eye and skin irritant.
IDLH value: 20 mg/m^3.

CHEMICAL REACTIVITY
Binary reactants: Violent reaction with aluminum, magnesium. Reacts with oxidizers, acetylene.

ENVIRONMENTAL DATA
Food chain concentration potential: Bioconcentration of 200–1000-fold only under constant exposure.

Water pollution: Harmful to aquatic life in very low concentrations. May be dangerous if it enters nearby water intakes; notify operators. Notify local health and wildlife officials. **Response to discharge:** Issue warning–water contaminant. Disperse and flush.

SHIPPING INFORMATION
Grades of purity: Technical; reagent; may also be shipped as a monohydrate or hexahydrate; **Storage temperature:** Ambient; **Inert atmosphere:** None; **Venting:** Open; **Stability during transport:** Stable.

PHYSICAL AND CHEMICAL PROPERTIES
Physical state @ 59°F/15°C and 1 atm: Solid.
Molecular weight: 281.1
Boiling point @ 1 atm: (decomposes).
Melting/Freezing point: 1355°F/735°C/1008°K (decomposes).
Specific gravity (water = 1): 1.948 @ 68°F/20°C (solid).
Heat of solution: 23 Btu/lb = 13 cal/g = 0.54×10^5 J/kg.

COLLODION   REC. C:3950

SYNONYMS: CELLOIDIN; CELLULOSE NITRATE SOLUTION; COLLODION COTTON; COLODION (Spanish); GUN COTTON; NITROCELLULOSE; NITROCELLULOSE GUM; NITROCELLULOSE SOLUTION; PYROXYLIN SOLUTION; BOX TOE GUM; XYLODIDIN

IDENTIFICATION
CAS Number: 9004-70-0
Formula: $C_{12}H_{16}(ONO_2)_4O_6$
DOT ID Number: UN 2059; DOT Guide Number: 127
Proper Shipping Name: Collodion

DESCRIPTION: Clear, colorless thick, colorless liquid. Ether- or alcohol-like odor. Odor depends on solvent used. Floats on the surface of water. Flammable, irritating vapor is produced.

Extremely flammable • Explosion hazard; if solvent evaporates the dry material (nitrocellulose) becomes shock-sensitive; may burn explosively (deflagrate) • Containers may BLEVE when exposed to fire • Vapors may form explosive mixture with air • Vapors are heavier than air and will collect and stay in low areas • Vapors may travel long distances to ignition sources and flashback • Vapors in confined areas (e.g., tanks, sewers, buildings) may explode when exposed to fire • Irritating to the skin, eyes, and respiratory tract • Toxic products of combustion may include hydrogen cyanide, nitrogen oxide, and carbon monoxide.

Hazard Classification (based on NFPA-704 Rating System)
Health Hazards (Blue): 1; Flammability (Red): 4; Reactivity (Yellow): 0

EMERGENCY RESPONSE: See Appendix A (127)
Evacuation:
Public safety: Isolate spill area for at least 25 to 50 meters (80 to 160 feet) in all directions.
Spill: Large spill–Consider initial downwind evacuation for at least 300 meters (1000 feet).
Fire: Isolate for 800 meters (½ mile) in all directions, especially if tank, rail car, or tank truck is involved in fire.

EXPOSURE
Short-term effects: *SEEK MEDICAL ATTENTION.* **Vapor:** Irritating to eyes, nose, and throat. *IF INHALED*, will, will cause dizziness, difficult breathing, or loss of consciousness. Move victim to fresh air. *IF BREATHING HAS STOPPED*, give artificial respiration; *avoid mouth-to-mouth resuscitation; use bag/mask apparatus*. IF breathing is difficult, administer oxygen.

HEALTH HAZARDS
Personal protective equipment (PPE): B-Level PPE. Self-contained breathing apparatus; rubber gloves; goggles or face shield.
Exposure limits (TWA unless otherwise noted): ACGIH TLV 10 mg/m^3 as cellulose; OSHA PEL 15 mg/m^3 (total); 5 mg/m^3 (respirable); NIOSH REL OSHA PEL 10 mg/m^3 (total); 5 mg/m^3 (respirable) as cellulose
Toxicity by ingestion: Grade 0; LD_{50} = > 15 g/kg.
Long-term health effects: May effect the central nervous system.
Vapor (gas) irritant characteristics: Vapors cause smarting of the eyes or respiratory system if present in high concentrations. The effect is temporary.
Liquid or solid irritant characteristics: May cause eye and skin irritation. Practically harmless to the skin.
IDLH value: 19,000 ppm.

FIRE DATA
Flash point: −64°F/−53°C (cc).
Flammable limits in air: LEL: 1.7%; UEL: 48%.
Fire extinguishing agents not to be used: Water may be ineffective.
Autoignition temperature: 338°F/206°C/479°K.
Electrical hazard: Class I, Group C.

CHEMICAL REACTIVITY
Binary reactants: Reacts violently with strong oxidizers; may cause fire and explosion hazards.

ENVIRONMENTAL DATA
Food chain concentration potential: Negative; unlikely to accumulate.
Water pollution: Effect of low concentrations on aquatic life is unknown. Fouling to shoreline. May be dangerous if it enters nearby water intakes; notify operators. Notify local health and wildlife officials. **Response to discharge:** Issue warning–high flammability. Restrict access. Mechanical containment. Should be removed. Chemical and physical treatment.

SHIPPING INFORMATION
Grades of purity: USP. All grades contain less than 60% nitrocellulose by weight; **Storage temperature:** Ambient; **Inert atmosphere:** None; **Venting:** Pressure-vacuum; **Stability during transport:** Stable.

NAS HAZARD CLASSIFICATION FOR BULK WATER TRANSPORTATION
FIRE: 4
HEALTH: Vapor irritant: 1; Liquid or solid irritant: 0; Poisons: 2
WATER POLLUTION: Human toxicity: 0; Aquatic toxicity: 1; Aesthetic effect: 1
REACTIVITY: Other chemicals: 1; Water: 0; Self-reaction: 0

PHYSICAL AND CHEMICAL PROPERTIES (based on ether).
Physical state @ 59°F/15°C and 1 atm: Liquid.
Boiling point @ 1 atm: 93°F/34°C/307°K (ether solvent).
Melting/Freezing point: −240°F/−116°C/157°K (approximate).
Specific gravity (water = 1): 0.772 @ 77°F/25°C (liquid).

Relative vapor density (air = 1): 2.6
Vapor pressure: 450 mm.

COPPER ACETATE REC. C:4000

SYNONYMS: ACETATE de CUIVRE (French); ACETIC ACID, COPPER (2+); ACETIC ACID, CUPRIC SALT; ACETATO de COBRE (Spanish); COPPER(2+) ACETATE; COPPER(II) ACETATE; COPPER DIACETATE; COPPER(2+) DIACETATE; CRYSTALLIZED VERDIGRIS; CRYSTALS OF VENUS; CUPRIC ACETATE; CUPRIC DIACETATE; NEUTRAL VERDIGRIS; CUPRIC ACETATE MONOHYDRATE

IDENTIFICATION
CAS Number: 142-71-2; 6046-93-1 (monohydrate)
Formula: $Cu(C_2H_3O_2)_2H_2O$
DOT ID Number: UN 9106; DOT Guide Number: 171
Proper Shipping Name: Cupric acetate
Reportable Quantity (RQ): **(CERCLA)** 100 lb/45.4 kg

DESCRIPTION: Bluish-green powder and crystalline solid. Odorless. Mixes with water.

The dust or solution is Irritating to the skin, eyes, and respiratory tract • Toxic products of combustion may include acetic acid, carbon monoxide, and copper oxide.

Hazard Classification (based on NFPA-704 Rating System)
Health Hazards (Blue): 1; Flammability (Red): 0; Reactivity (Yellow): 0

EMERGENCY RESPONSE: See Appendix A (171)
Evacuation:
Public safety: Isolate the area of spill or leak for at least 10 to 25 meters (30 to 80 feet) in all directions.
Spill: Increase, in the downwind direction, as necessary, the distance shown under "Public Safety."
Fire: If any large container is involved in fire, isolate for at least 800 meters (½ mile) in all directions; also, consider initial evacuation for 800 meters (½ mile) in all directions.

EXPOSURE
Short-term effects: *SEEK MEDICAL ATTENTION.* **Dust:** Irritating to eyes, nose, and throat. *IF INHALED,* will, will cause coughing or difficult breathing. *IF IN EYES,* hold eyelids open and flush with plenty of water. *IF BREATHING HAS STOPPED,* give artificial respiration. IF breathing is difficult, administer oxygen. **Solid:** Will burn eyes. Irritating to eyes. *IF SWALLOWED,* will cause nausea, vomiting, or loss of consciousness. Remove contaminated clothing and shoes. Flush affected areas with plenty of water. *IF IN EYES,* hold eyelids open and flush with plenty of water. *IF SWALLOWED* and victim is *CONSCIOUS AND ABLE TO SWALLOW,* have victim drink 4 to 8 ounces of water and have victim induce vomiting. *IF SWALLOWED* and victim is *UNCONSCIOUS OR HAVING CONVULSIONS,* do nothing except keep victim warm.

HEALTH HAZARDS
Personal protective equipment (PPE): B-Level PPE. OSHA Table Z-1-A air contaminant. Dust mask; goggles or face shield; protective gloves.
Recommendations for respirator selection: NIOSH/OSHA as copper fume

NIOSH/OSHA: *Up to 1 mg/m^3:* DMFu (any dust, mist, and fume respirator; or SA (any supplied-air respirator). *2.5 mg/m^3:* SA:CF (any supplied-air respirator operated in a continuous-flow mode); or PAPRDMFu (any powered, air-purifying respirator with a dust, mist, and fume filter). *5 mg/m^3:* HiEF (any air-purifying, full-facepiece respirator with a high-efficiency particulate filter); or SAT:CF (any supplied-air respirator that has a tight-fitting facepiece and is operated in a continuous-flow mode); or PAPRTHiE (any powered, air-purifying respirator with a tight-fitting facepiece and a high-efficiency particulate filter); or SCBAF (any self-contained breathing apparatus with a full facepiece); or SAF (any supplied-air respirator with a full facepiece). *100 mg/m^3:* SAF:PD,PP (any supplied-air respirator that has a full facepiece and is operated in a pressure-demand or other positive-pressure mode). *EMERGENCY OR PLANNED ENTRY INTO UNKNOWN CONCENTRATIONS OR IDLH CONDITIONS:* SCBAF:PD,PP (any self-contained breathing apparatus that has a full facepiece and is operated in a pressure-demand or other positive-pressure mode); or SAF:PD,PP:ASCBA (Any supplied-air respirator that has a full facepiece and is operated in a pressure-demand or other positive-pressure mode in combination with an auxiliary self-contained breathing apparatus operated in a pressure-demand or other positive pressure mode). *ESCAPE:* HiEF (Any air-purifying, full-facepiece respirator with a high-efficiency particulate filter); or SCBAE (Any appropriate escape-type, self-contained breathing apparatus).
Exposure limits (TWA unless otherwise noted): As copper: ACGIH TLV 0.2 mg/m^3 (fume); 1 mg/m^3 (dusts and mists); NIOSH/OSHA 0.1 mg/m^3 (fume); 1 mg/m^3 as (dusts and mists).
Toxicity by ingestion: Grade 2; LD_{50} = 0.5 to 5 g/kg (rat).
Long-term health effects: Kidney, skin, blood and liver disorders.
Vapor (gas) irritant characteristics: Dust is and eye and respiratory tract irritant.
Liquid or solid irritant characteristics: Eye and skin irritant.
IDLH value: 100 mg/m^3 as copper.

CHEMICAL REACTIVITY
Binary reactants: Copper reacts with acetylides, ammonium nitrate, azides, bromates, chlorates, chlorine, fluorine, hydrazine, peroxides, hydrogen sulfide, ethylene oxide, mercurous chloride, strong acids.

ENVIRONMENTAL DATA
Food chain concentration potential: Copper known to be accumulated by shellfish. Copper is concentrated by plankton by factors of 1000 to 5000 or more.
Water pollution: Effect of low concentrations on aquatic life is unknown. May be dangerous if it enters nearby water intakes; notify operators. Notify local health and wildlife officials. **Response to discharge:** Issue warning–water contaminant. Disperse and flush.

SHIPPING INFORMATION
Grades of purity: Technical, 95-99%; Reagent, 99%; **Storage temperature:** Ambient; **Inert atmosphere:** None; **Venting:** Open; **Stability during transport:** Stable.

PHYSICAL AND CHEMICAL PROPERTIES
Physical state @ 59°F/15°C and 1 atm: Solid.
Molecular weight: 199.65
Boiling point @ 1 atm: 465°F/241°C (decomposes).
Melting/Freezing point: 239°F/115°C/388°K.
Specific gravity (water = 1): 1.9 @ 68°F/20°C (solid).
Heat of decomposition: 465°F/241°C.
Vapor pressure: Negligible.

COPPER ACETOARSENITE REC. C:4050

SYNONYMS: ACETOARSENITE de CUIVRE (French); ACETOARSENITO de COBRE (Spanish); CUPRIC ACETOARSENITE; EMERALD GREEN; IMPERIAL GREEN; KING'S GREEN; MEADOW GREEN; MITIS GREEN; MOSS GREEN; PARIS GREEN; PARROT GREEN; SCHWEINFURTH GREEN; VIENNA GREEN

IDENTIFICATION
CAS Number: 12002-03-8; 1299-88-3
Formula: $3Cu(AsO_2)_2 \cdot Cu(C_2H_3O_2)_2$
DOT ID Number: UN 1585; DOT Guide Number: 151
Proper Shipping Name: Copper acetoarsenite
Reportable Quantity (RQ): **(CERCLA)** 1 lb/0.454 kg

DESCRIPTION: Emerald green powder. Odorless. Sinks and mixes slowly with water.

Poison! • Breathing the dust, skin or eye contact, or swallowing the material can kill you • Firefighting gear (including SCBA) does not provide adequate protection. If exposure occurs, remove and isolate gear immediately and thoroughly decontaminate personnel • Irritating to the skin, eyes, and respiratory tract • Toxic products of combustion may include oxides of arsenic and copper.

Hazard Classification (based on NFPA-704 Rating System)
Health Hazards (Blue): 3; Flammability (Red): 0; Reactivity (Yellow): 0

EMERGENCY RESPONSE: See Appendix A (151)
Evacuation:
Public safety: Isolate the area of spill or leak for at least 25 to 50 meters (80 to 160 feet) in all directions.
Spill: Increase, in the downwind direction, as necessary, the distance shown under "Public Safety."
Fire: If tank, rail car, or tank truck is involved in fire, isolate for at least 800 meters (½ mile) in all directions; also, consider initial evacuation for 800 meters (½ mile) in all directions.

EXPOSURE
Short-term effects: *SEEK MEDICAL ATTENTION.* **Dust:** *POISONOUS IF INHALED.* Irritating to eyes, nose, and throat. Move victim to fresh air. *IF IN EYES*, hold eyelids open and flush. with plenty of water. If artificial respiration is administered, *avoid mouth-to-mouth resuscitation; use bag/mask apparatus.* IF breathing is difficult, administer oxygen. **Solid:** *POISONOUS IF SWALLOWED.* Irritating to skin and eyes. Remove contaminated clothing and shoes. Flush affected areas with plenty of water. *IF IN EYES*, hold eyelids open and flush. with plenty of water. *IF SWALLOWED* and victim is *CONSCIOUS AND ABLE TO SWALLOW*, have victim drink 4 to 8 ounces of water and have victim induce vomiting. *IF SWALLOWED* and victim is *UNCONSCIOUS OR HAVING CONVULSIONS*, do nothing except keep victim warm.

HEALTH HAZARDS
Personal protective equipment (PPE): A-Level PPE. OSHA Table Z-1-A air contaminant.
Recommendations for respirator selection: NIOSH as inorganic arsenic (for reference only)
At any concentrations above the NIOSH REL, or where there is no REL, at any detectable concentration: SCBAF:PD,PP (any self-contained breathing apparatus that has a full faceplate and is operated in a pressure-demand or other positive-pressure mode); or SAF:PD,PP:ASCBA (any supplied-air respirator that has a full facepiece and is operated in a pressure-demand or other positive-pressure mode in combination with an auxiliary self-contained breathing apparatus operated in a pressure-demand or other positive-pressure mode). *ESCAPE:* GMFAGHiE [any air-purifying, full-facepiece respirator (gas mask) with a chin-style, front-or back-mounted acid gas canister having a high-efficiency particulate filter]; or SCBAE (any appropriate escape-type, self-contained breathing apparatus).

Exposure limits (TWA unless otherwise noted): OSHA PEL 0.5 mg/m^3 as organic arsenic. As copper: ACGIH TLV 0.2 mg/m^3 (fume); 1 mg/m^3 (dusts and mists); NIOSH/OSHA 0.1 mg/m^3 (fume); 1 mg/m^3 as (dusts and mists).
Toxicity by ingestion: Grade 4; oral LD$_{50}$ = 22 mg/kg (rat).
Long-term health effects: Arsenic poisoning. Arsenic is a known NTP carcinogen.
Liquid or solid irritant characteristics: Eye irritant.
IDLH value: 5 mg/m^3 as inorganic arsenic; known human carcinogen. Level for organic arsenic has not been determined.

CHEMICAL REACTIVITY
Binary reactants: Reacts with chlorine, fluorine, peroxides.

ENVIRONMENTAL DATA
Water pollution: DOT Appendix B, §172.101–marine pollutant. Effects of low concentrations on aquatic life is unknown. May be dangerous if it enters nearby water intakes; notify operators. Notify local health and wildlife officials. **Response to discharge:** Issue warning–poison, water contaminant. Restrict access. Should be removed. Chemical and physical treatment.

SHIPPING INFORMATION
Grades of purity: Commercial; **Storage temperature:** Ambient; **Inert atmosphere:** None; **Venting:** Open; **Stability during transport:** Stable.

PHYSICAL AND CHEMICAL PROPERTIES
Physical state @ 59°F/15°C and 1 atm: Solid.
Molecular weight: 1014
Boiling point @ 1 atm: Decomposes.
Specific gravity (water = 1): (estimate) more than 1.1 @ 68°F/20°C (solid).
Vapor pressure: Negligible.

COPPER ARSENITE REC. C:4100

SYNONYMS: ARSENITO de COBRE (Spanish); ARSONIC ACID, COPPER(2+) SALT; CUPRIC ARSENITE; CUPRIC GREEN; COPPER ORTHOARSENITE; SCHEELE'S GREEN; SCHEELE'S MINERAL; SWEDISH GREEN

IDENTIFICATION
CAS Number: 10290-12-7; 1302-97-2
Formula: $CuHAsO_3$
DOT ID Number: UN 1586; DOT Guide Number: 151
Proper Shipping Name: Copper arsenite

DESCRIPTION: Green or yellowish-green solid. Odorless. Sinks in water.

Poison! • Breathing the dust, skin or eye contact, or swallowing the material can kill you • Firefighting gear (including SCBA) does not provide adequate protection. If exposure occurs, remove and isolate gear immediately and thoroughly decontaminate personnel •

Irritating to the skin, eyes, and respiratory tract • Toxic products of combustion may include arsenic and copper oxide.

Hazard Classification (based on NFPA-704 Rating System)
Health Hazards (Blue): 4; Flammability (Red): 0; Reactivity (Yellow): 0

EMERGENCY RESPONSE: See Appendix A (151)
Evacuation:
Public safety: Isolate the area of spill or leak for at least 25 to 50 meters (80 to 160 feet) in all directions.
Spill: Increase, in the downwind direction, as necessary, the distance shown under "Public Safety."
Fire: If tank, rail car, or tank truck is involved in fire, isolate for at least 800 meters (½ mile) in all directions; also, consider initial evacuation for 800 meters (½ mile) in all directions.

EXPOSURE
Short-term effects: *SEEK MEDICAL ATTENTION.* **Dust:** *POISONOUS IF INHALED.* Irritating to eyes, nose, and throat. Move victim to fresh air. *IF IN EYES*, hold eyelids open and flush with plenty of water. If artificial respiration is administered, *avoid mouth-to-mouth resuscitation; use bag/mask apparatus.* IF breathing is difficult, administer oxygen. **Solid:** *POISONOUS IF SWALLOWED.* Irritating to skin and eyes. Remove contaminated clothing and shoes. Flush affected areas with plenty of water. *IF IN EYES*, hold eyelids open and flush with plenty of water. *IF SWALLOWED* and victim is *CONSCIOUS AND ABLE TO SWALLOW*, have victim drink 4 to 8 ounces of water and have victim induce vomiting. *IF SWALLOWED* and victim is *UNCONSCIOUS OR HAVING CONVULSIONS*, do nothing except keep victim warm.

HEALTH HAZARDS
Personal protective equipment (PPE): A-Level PPE. OSHA Table Z-1-A air contaminant.
Recommendations for respirator selection: NIOSH
At any concentrations above the NIOSH REL, or where there is no REL, at any detectable concentration: SCBAF:PD,PP (any self-contained breathing apparatus that has a full faceplate and is operated in a pressure-demand or other positive-pressure mode); or SAF:PD,PP:ASCBA (any supplied-air respirator that has a full facepiece and is operated in a pressure-demand or other positive-pressure mode in combination with an auxiliary self-contained breathing apparatus operated in a pressure-demand or other positive-pressure mode). *ESCAPE:* GMFAGHiE [any air-purifying, full-facepiece respirator (gas mask) with a chin-style, front-or back-mounted acid gas canister having a high-efficiency particulate filter]; or SCBAE (any appropriate escape-type, self-contained breathing apparatus).
Exposure limits (TWA unless otherwise noted): ACGIH TLV 0.01 mg/m^3; OSHA PEL [1910.1080] 0.010 mg/m^3; NIOSH REL ceiling 0.002 mg/m^3/15 min, as arsenic; potential human carcinogen; reduce exposure to lowest feasible level. As copper: ACGIH TLV 0.2 mg/m^3 (fume); 1 mg/m^3 (dusts and mists); NIOSH/OSHA 0.1 mg/m^3 (fume); 1 mg/m^3 as (dusts and mists).
Toxicity by ingestion: Grade 3; LD$_{50}$ = 50 to 500 mg/kg.
Long-term health effects: Arsenic poisoning. Arsenic is a known NTP carcinogen.
Liquid or solid irritant characteristics: Eye irritant.
IDLH value: 5 mg/m^3 as arsenic.

CHEMICAL REACTIVITY
Binary reactants: Reacts with chlorine, fluorine, peroxides.

ENVIRONMENTAL DATA
Water pollution: DOT Appendix B, §172.101–marine pollutant. Effect of low concentrations on aquatic life is unknown. May be dangerous if it enters nearby water intakes; notify operators. Notify local health and wildlife officials. **Response to discharge:** Issue warning–poison, water contaminant. Restrict access. Should be removed. Chemical and physical treatment.

SHIPPING INFORMATION
Grades of purity: Commercial; **Storage temperature:** Ambient; **Inert atmosphere:** None; **Venting:** Open; **Stability during transport:** Stable.

PHYSICAL AND CHEMICAL PROPERTIES
Physical state @ 59°F/15°C and 1 atm: Solid.
Molecular weight: 277.4
Boiling point @ 1 atm: Decomposes.
Specific gravity (water = 1): (estimate) more than 1.1 @ 68°F/20°C (solid).

COPPER BROMIDE   REC. C:4150

SYNONYMS: BROMURO de COBRE (Spanish); CUPRIC BROMIDE, ANHYDROUS

IDENTIFICATION
CAS Number: 7789-45-9
Formula: CuBr$_2$

DESCRIPTION: Black solid. Odorless. Sinks and mixes with water.

Irritating to the skin, eyes, and respiratory tract • Toxic products of combustion may include hydrogen bromide and copper metal fume.

Hazard Classification (based on NFPA-704 Rating System)
Health Hazards (Blue): 2; Flammability (Red): 0; Reactivity (Yellow): 0

EMERGENCY RESPONSE: See Appendix A (171)
Evacuation:
Public safety: Isolate the area of spill or leak for at least 10 to 25 meters (30 to 80 feet) in all directions.
Spill: Increase, in the downwind direction, as necessary, the distance shown under "Public Safety."
Fire: If any large container is involved in fire, isolate for at least 800 meters (½ mile) in all directions; also, consider initial evacuation for 800 meters (½ mile) in all directions.

EXPOSURE
Short-term effects: *SEEK MEDICAL ATTENTION.* **Dust:** Irritating to eyes, nose, and throat. *IF INHALED*, will, will cause coughing or difficult breathing. *IF IN EYES*, hold eyelids open and flush with plenty of water. *IF BREATHING HAS STOPPED*, give artificial respiration. IF breathing is difficult, administer oxygen. **Solid:** Will burn eyes. Irritating to skin. Remove contaminated clothing and shoes. Flush affected areas with plenty of water. *IF IN EYES*, hold eyelids open and flush with plenty of water. *IF SWALLOWED* and victim is *CONSCIOUS AND ABLE TO SWALLOW*, have victim drink 4 to 8 ounces of water and have victim induce vomiting. *IF SWALLOWED* and victim is *UNCONSCIOUS OR HAVING CONVULSIONS*, do nothing except keep victim warm.

HEALTH HAZARDS

Personal protective equipment (PPE): B-Level PPE. OSHA Table Z-1-A air contaminant. Dust mask; goggles or face shield; protective gloves.

Recommendations for respirator selection: NIOSH/OSHA as copper fume

NIOSH/OSHA: *Up to 1 mg/m³:* DMFu (any dust, mist, and fume respirator; or SA (any supplied-air respirator). *2.5 mg/m³:* SA:CF (any supplied-air respirator operated in a continuous-flow mode); or PAPRDMFu (any powered, air-purifying respirator with a dust, mist, and fume filter). *5 mg/m³:* HiEF (any air-purifying, full-facepiece respirator with a high-efficiency particulate filter); or SAT:CF (any supplied-air respirator that has a tight-fitting facepiece and is operated in a continuous-flow mode); or PAPRTHiE (any powered, air-purifying respirator with a tight-fitting facepiece and a high-efficiency particulate filter); or SCBAF (any self-contained breathing apparatus with a full facepiece); or SAF (any supplied-air respirator with a full facepiece). *100 mg/m³:* SAF:PD,PP (any supplied-air respirator that has a full facepiece and is operated in a pressure-demand or other positive-pressure mode). *EMERGENCY OR PLANNED ENTRY INTO UNKNOWN CONCENTRATIONS OR IDLH CONDITIONS:* SCBAF:PD,PP (any self-contained breathing apparatus that has a full facepiece and is operated in a pressure-demand or other positive-pressure mode); or SAF:PD,PP:ASCBA (Any supplied-air respirator that has a full facepiece and is operated in a pressure-demand or other positive-pressure mode in combination with an auxiliary self-contained breathing apparatus operated in a pressure-demand or other positive pressure mode). *ESCAPE:* HiEF (Any air-purifying, full-facepiece respirator with a high-efficiency particulate filter); or SCBAE (Any appropriate escape-type, self-contained breathing apparatus).

Exposure limits (TWA unless otherwise noted): As copper: ACGIH TLV 0.2 mg/m³ (fume); 1 mg/m³ (dusts and mists); NIOSH/OSHA 0.1 mg/m³ (fume); 1 mg/m³ as (dusts and mists).

Long-term health effects: Prolonged contact or inhalation of inorganic bromides causes acne-like bromoderma (bromide rash), especially of the hands and face. Other effects include emaciation, depression; and, in severe cases, psychosis and mental deterioration.

Toxicity by ingestion: Grade 3; LD_{50} = 50 to 500 mg/kg.

Liquid or solid irritant characteristics: Eye, skin and respiratory tract irritant.

IDLH value: 100 mg/m³ as copper.

CHEMICAL REACTIVITY

Binary reactants: Reacts with chlorine, fluorine, peroxides.

ENVIRONMENTAL DATA

Food chain concentration potential: Copper known to be accumulated by shellfish. Copper is concentrated by plankton by factors of 1000 to 5000 or more.

Water pollution: Effect of low concentrations on aquatic life is unknown. May be dangerous if it enters nearby water intakes; notify operators. Notify local health and wildlife officials.

Response to discharge: Issue warning–water contaminant. Disperse and flush.

SHIPPING INFORMATION

Grades of purity: Pure, 99%, Reagent; **Storage temperature:** Ambient; **Inert atmosphere:** None; **Venting:** Open; **Stability during transport:** Stable.

PHYSICAL AND CHEMICAL PROPERTIES

Physical state @ 59°F/15°C and 1 atm: Solid.

Molecular weight: 223.35
Boiling point @ 1 atm: (decomposes).
Melting/Freezing point: 928°F/498°C/771°K.
Specific gravity (water = 1): 4.77 @ 68°F/20°C (solid).
Heat of solution: –70.9 Btu/lb = –39.4 cal/g = –1.65 x 10^5 J/kg.

COPPER BROMIDE(OUS)   REC. C:4200

SYNONYMS: COPPER MONOBROMIDE

IDENTIFICATION
CAS Number: 7787-70-4
Formula: CuBr

DESCRIPTION: White powder or crystals; turns green to dark blue on exposure to sunlight. Sinks and mixes slowly with water.

Irritating to the skin, eyes, and respiratory tract • Toxic products of combustion may include bromine and copper oxide.

Hazard Classification (based on NFPA-704 Rating System)
Health Hazards (Blue): 2; Flammability (Red): 0; Reactivity (Yellow): 0

EMERGENCY RESPONSE: See Appendix A (171)
Evacuation:
Public safety: Isolate the area of spill or leak for at least 10 to 25 meters (30 to 80 feet) in all directions.
Spill: Increase, in the downwind direction, as necessary, the distance shown under "Public Safety."
Fire: If any large container is involved in fire, isolate for at least 800 meters (½ mile) in all directions; also, consider initial evacuation for 800 meters (½ mile) in all directions.

EXPOSURE
Short-term effects: *SEEK MEDICAL ATTENTION.* **Solid:** Irritating to skin and eyes. *IF SWALLOWED*, will cause pain, nausea, and vomiting.
Flush affected areas with plenty of water. *IF IN EYES*, hold eyelids open and flush with plenty of water. *IF SWALLOWED* and victim is *CONSCIOUS AND ABLE TO SWALLOW*, have victim drink 4 to 8 ounces of water and induce vomiting.

HEALTH HAZARDS
Personal protective equipment (PPE): B-Level PPE. OSHA Table Z-1-A air contaminant. Rubber gloves, safety glasses, and laboratory coat.
Recommendations for respirator selection: NIOSH/OSHA as copper fume
NIOSH/OSHA: *Up to 1 mg/m³:* DMFu (any dust, mist, and fume respirator; or SA (any supplied-air respirator). *2.5 mg/m³:* SA:CF (any supplied-air respirator operated in a continuous-flow mode); or PAPRDMFu (any powered, air-purifying respirator with a dust, mist, and fume filter). *5 mg/m³:* HiEF (any air-purifying, full-facepiece respirator with a high-efficiency particulate filter); or SAT:CF (any supplied-air respirator that has a tight-fitting facepiece and is operated in a continuous-flow mode); or PAPRTHiE (any powered, air-purifying respirator with a tight-fitting facepiece and a high-efficiency particulate filter); or SCBAF (any self-contained breathing apparatus with a full facepiece); or SAF (any supplied-air respirator with a full facepiece). *100 mg/m³:* SAF:PD,PP (any supplied-air respirator that has a full facepiece and is operated in a pressure-demand or other positive-pressure mode). *EMERGENCY OR PLANNED ENTRY INTO UNKNOWN*

CONCENTRATIONS OR IDLH CONDITIONS: SCBAF:PD,PP (any self-contained breathing apparatus that has a full facepiece and is operated in a pressure-demand or other positive-pressure mode); or SAF:PD,PP:ASCBA (Any supplied-air respirator that has a full facepiece and is operated in a pressure-demand or other positive-pressure mode in combination with an auxiliary self-contained breathing apparatus operated in a pressure-demand or other positive pressure mode). *ESCAPE:* HiEF (Any air-purifying, full-facepiece respirator with a high-efficiency particulate filter); or SCBAE (Any appropriate escape-type, self-contained breathing apparatus).

Exposure limits (TWA unless otherwise noted): As copper: ACGIH TLV 0.2 mg/m^3 (fume); 1 mg/m^3 (dusts and mists); NIOSH/OSHA 0.1 mg/m^3 (fume); 1 mg/m^3 as (dusts and mists).

Toxicity by ingestion: Grade 3; LD_{50} = 50 to 500 mg/kg.

Long-term health effects: Prolonged contact or inhalation of inorganic bromides causes acne-like bromoderma (bromide rash), especially of the hands and face. Other effects include emaciation, depression; and, in severe cases, psychosis and mental deterioration.

Vapor (gas) irritant characteristics: Eye and respiratory tract irritant.

Liquid or solid irritant characteristics: Eye and skin irritant.

IDLH value: 100 mg/m^3 as copper.

CHEMICAL REACTIVITY
Binary reactants: Reacts with chlorine, fluorine, peroxides.

ENVIRONMENTAL DATA
Food chain concentration potential: Copper known to be accumulated by shellfish. Copper is concentrated by plankton by factors of 1000 to 5000 or more.

Water pollution: Harmful to aquatic life in very low concentrations. May be dangerous if it enters nearby water intakes; notify operators. Notify local health and wildlife officials.

Response to discharge: Issue warning–water contaminant. Disperse and flush.

SHIPPING INFORMATION
Stability during transport: Keep tightly closed in a dark place.

PHYSICAL AND CHEMICAL PROPERTIES
Physical state @ 59°F/15°C and 1 atm: Solid.
Molecular weight: 143.46 286.91 for dimer (Cu_2Br_2).
Boiling point @ 1 atm: 2453°F/1345°C/1618.2°K.
Melting/Freezing point: 939.2°F/504°C/777.2°K.
Specific gravity (water = 1): 4.72 @ 77°F/25°C.
Heat of fusion: 16.03 cal/g.

COPPER CHLORIDE   REC. C:4250

SYNONYMS: CLORURO de COBRE (Spanish); COPPER(2+) CHLORIDE; COPPER(II) CHLORIDE; CUPRIC CHLORIDE; CUPRIC CHLORIDE DIHYDRATE

IDENTIFICATION
CAS Number: 1344-67-8; 7447-39-4 (II); 7758-89-6 (I)
Formula: ClCu (I); $CuCl_2 \cdot 2H_2O$; Cl_2Cu (II)
DOT ID Number: UN 2802; DOT Guide Number: 154
Proper Shipping Name: Copper chloride
Reportable Quantity (RQ): 10 lb/4.54 kg

DESCRIPTION: White crystalline solid or powder. Turns blue-green on exposure to moist air. Bichloride turns yellow-brown. Odorless. Sinks and mixes with water; reacts, forming hydrochloric acid solution.

Corrosive • Breathing the dust can cause irritation; skin contact causes severe irritation; corrosive to the eyes causing burns and impaired vision; effects of inhalation may be delayed • Firefighting gear (including SCBA) does not provide adequate protection. If exposure occurs, remove and isolate gear immediately and thoroughly decontaminate personnel • Toxic products of combustion may include highly toxic hydrogen chloride and copper oxide.

Hazard Classification (based on NFPA-704 Rating System)
Health Hazards (Blue): 2; Flammability (Red): 0; Reactivity (Yellow): 0

EMERGENCY RESPONSE: See Appendix A (154)
Evacuation:
Public safety: Isolate the area of spill or leak for at least 25 to 50 meters (80 to 160 feet) in all directions.
Spill: Increase, in the downwind direction, as necessary, the distance shown under "Public Safety."
Fire: If tank, rail car, or tank truck is involved in fire, isolate for at least 800 meters (½ mile) in all directions; also, consider initial evacuation for 800 meters (½ mile) in all directions.

EXPOSURE
Short-term effects: *SEEK MEDICAL ATTENTION.* **Dust:** Irritating to eyes, nose, and throat. *IF INHALED*, will, will cause coughing or difficult breathing. If in eyes, hold eyelids open and flush with plenty of water. *IF BREATHING HAS STOPPED,* give artificial respiration; *avoid mouth-to-mouth resuscitation; use bag/mask apparatus.* If breathing is difficult, administer oxygen. **Solid:** Will burn eyes. Irritating to eyes. *IF SWALLOWED*, will cause nausea and vomiting. Remove contaminated clothing and shoes. Flush affected areas with plenty of water. *IF IN EYES*, hold eyelids open and flush with plenty of water. *IF SWALLOWED* and victim is *CONSCIOUS AND ABLE TO SWALLOW*, have victim drink 4 to 8 ounces of water and have victim induce vomiting. *IF SWALLOWED* and victim is *UNCONSCIOUS OR HAVING CONVULSIONS,* do nothing except keep victim warm.

HEALTH HAZARDS
Personal protective equipment (PPE): A-Level PPE. OSHA Table Z-1-A air contaminant. Chemical protective material(s) reported to offer minimal to poor protection: natural rubber, nitrile, neoprene, PVC. Also, butyl rubber and nitrile+PVC offers limited protection.

Recommendations for respirator selection: NIOSH/OSHA as copper fume
Up to 1 mg/m^3: DMFu (any dust, mist, and fume respirator; or SA (any supplied-air respirator). *2.5 mg/m^3:* SA:CF (any supplied-air respirator operated in a continuous-flow mode); or PAPRDMFu (any powered, air-purifying respirator with a dust, mist, and fume filter). *5 mg/m^3:* HiEF (any air-purifying, full-facepiece respirator with a high-efficiency particulate filter); or SAT:CF (any supplied-air respirator that has a tight-fitting facepiece and is operated in a continuous-flow mode); or PAPRTHiE (any powered, air-purifying respirator with a tight-fitting facepiece and a high-efficiency particulate filter); or SCBAF (any self-contained breathing apparatus with a full facepiece); or SAF (any supplied-air respirator with a full facepiece). *100 mg/m^3:* SAF:PD,PP (any supplied-air respirator that has a full facepiece and is operated in a pressure-demand or other positive-pressure mode). *EMERGENCY OR PLANNED ENTRY INTO UNKNOWN CONCENTRATIONS OR*

Copper cyanide

IDLH CONDITIONS: SCBAF:PD,PP (any self-contained breathing apparatus that has a full facepiece and is operated in a pressure-demand or other positive-pressure mode); or SAF:PD,PP:ASCBA (Any supplied-air respirator that has a full facepiece and is operated in a pressure-demand or other positive-pressure mode in combination with an auxiliary self-contained breathing apparatus operated in a pressure-demand or other positive pressure mode). *ESCAPE:* HiEF (Any air-purifying, full-facepiece respirator with a high-efficiency particulate filter); or SCBAE (Any appropriate escape-type, self-contained breathing apparatus).

Exposure limits (TWA unless otherwise noted): As copper: ACGIH TLV 0.2 mg/m^3 (fume); 1 mg/m^3 (dusts and mists); NIOSH/OSHA 0.1 mg/m^3 (fume); 1 mg/m^3 as (dusts and mists).

Toxicity by ingestion: Grade 3; LD$_{50}$ = 50 to 500 mg/kg.

Long-term health effects: Causes liver damage in rabbits. Can cause skin, blood, kidney, and liver disorders

Vapor (gas) irritant characteristics: Eye and respiratory tract irritant.

Liquid or solid irritant characteristics: Eye and skin irritant.

IDLH value: 100 mg/m^3 as copper.

CHEMICAL REACTIVITY

Binary reactants: In presence of moisture may corrode metals. Shock-sensitive mixture formed with potassium; sodium.

Neutralizing agents for acids and caustics: Flush with water, rinse with dilute solution of sodium bicarbonate or soda ash.

ENVIRONMENTAL DATA

Food chain concentration potential: Copper known to be accumulated by shellfish. Copper is concentrated by plankton by factors of 1000 to 5000 or more.

Water pollution: DOT Appendix B, §172.101- solution is a severe marine pollutant. Harmful to aquatic life in very low concentrations. May be dangerous if it enters nearby water intakes; notify operators. Notify local health and wildlife officials.

Response to discharge: Issue warning–water contaminant. Disperse and flush.

SHIPPING INFORMATION

Grades of purity: Reagent; C.P; Technical; **Storage temperature:** Ambient; **Inert atmosphere:** None; **Venting:** Open; **Stability during transport:** Stable.

PHYSICAL AND CHEMICAL PROPERTIES

Physical state @ 59°F/15°C and 1 atm: Solid.
Molecular weight: 170.48 (dihydrate).
Boiling point @ 1 atm: 1823°F/995°C/1268°K (decomposes).
Melting/Freezing point: 932°F/500°C/773°K.
Specific gravity (water = 1): 2.54 @ 68°F/20°C (solid).
Heat of decomposition: More than 572°F/300°C.
Heat of fusion: 24.7 cal/g.

COPPER CYANIDE REC. C:4300

SYNONYMS: CIANURO de COBRE (Spanish); CUPRICIN; CUPROUS CYANIDE; RCRA No. P029

IDENTIFICATION

CAS Number: 544-92-3; 14763-77-0
Formula: CuCN; C$_2$CuN$_2$ (II)
DOT ID Number: UN 1587; DOT Guide Number: 151
Proper Shipping Name: Copper cyanide
Reportable Quantity (RQ): **(CERCLA)** 10 lb/4.54 kg

DESCRIPTION: White to cream-colored powder or dark green crystalline solid. Sinks in water; practically insoluble; slowly forms hydrogen cyanide.

Poison! • Breathing the vapor or dust, skin or eye contact, or swallowing the material can kill you • Firefighting gear (including SCBA) does not provide adequate protection. If exposure occurs, remove and isolate gear immediately and thoroughly decontaminate personnel • Toxic products of combustion may include hydrogen cyanide gas, nitrogen, and copper oxides.

Hazard Classification (based on NFPA-704 Rating System)
Health Hazards (Blue): 4; Flammability (Red): 0; Reactivity (Yellow): 0

EMERGENCY RESPONSE: See Appendix A (151)
Evacuation:
Public safety: Isolate the area of spill or leak for at least 25 to 50 meters (80 to 160 feet) in all directions.
Spill: Increase, in the downwind direction, as necessary, the distance shown under "Public Safety."
Fire: If tank, rail car, or tank truck is involved in fire, isolate for at least 800 meters (½ mile) in all directions; also, consider initial evacuation for 800 meters (½ mile) in all directions.

EXPOSURE

Short-term effects: *SEEK MEDICAL ATTENTION.* **Dust:** Irritating to eyes, nose, and throat. *IF INHALED*, will, will cause dizziness or loss of consciousness. *IF IN EYES*, hold eyelids open and flush with plenty of water. *IF BREATHING HAS STOPPED*, give artificial respiration; *avoid mouth-to-mouth resuscitation; use bag/mask apparatus*. IF breathing is difficult, administer oxygen. **Solid:** *POISONOUS IF SWALLOWED*. Irritating to skin and eyes. *IF SWALLOWED*, will cause dizziness and loss of consciousness. Remove contaminated clothing and shoes. Flush affected areas with plenty of water. *IF IN EYES*, hold eyelids open and flush with plenty of water. *IF SWALLOWED* and victim is *CONSCIOUS AND ABLE TO SWALLOW*, have victim drink 4 to 8 ounces of water and have victim induce vomiting. *IF SWALLOWED* and victim is *UNCONSCIOUS OR HAVING CONVULSIONS,* do nothing except keep victim warm.

Note to physician or authorized medical personnel: Consider the use of amyl nitrite perles if symptoms of cyanide poisoning develop. If symptoms indicate, initial treatment includes the cyanide antidote kit. In all cases, break an amyl nitrite perle in a gauze pad and hold lightly under victim's nose for 15 seconds, repeating 5 times at about 15-second intervals; if necessary (and if sodium nitrite infusions will be delayed), repeat procedure every 3 minutes with fresh perles until 3 or 4 have been used. Avoid breathing the vapor while administering it to the victim. Administer sodium nitrite IV, ASAP. The usual adult dose is 10 to 20 mL of a 3% solution infused over no less than 5 minutes; the average child dose is 0.15 to 0.20 mL/kg. Monitor blood pressure during administration, and slow the rate of infusion if hypotention develops. Next, infuse sodium thiosulfate IV. The usual adult dose is 50 mL of a 25% solution infused over 10 to 20 minutes; the average child dose is 1.65 mL/kg. Repeat with nitrite and thiosulfate as required.

HEALTH HAZARDS

Personal protective equipment (PPE): A-Level PPE.
Recommendations for respirator selection: NIOSH as cyanides *25 mg/m^3:* SA (any supplied-air respirator); or SCBAF (any self-contained breathing apparatus with full facepiece). *EMERGENCY OR PLANNED ENTRY INTO UNKNOWN CONCENTRATIONS*

OR IDLH CONDITIONS: SCBAF:PD,PP (any self-contained breathing apparatus that has a full facepiece and is operated in a pressure-demand or other positive-pressure mode); or SAF:PD,PP:ASCBA (any supplied-air respirator that has a full facepiece and is operated in a pressure-demand or other positive-pressure mode in combination with an auxiliary self-contained breathing apparatus operated in a pressure-demand or other positive pressure mode). *ESCAPE:* GMFSHiE [any air-purifying, full-facepiece respirator (gas mask) with a chin-style, front- or back-mounted canister providing protection against the compound of concern and having a high efficiency particulate filter); or SCBAE (any appropriate escape-type, self-contained breathing apparatus).

Exposure limits (TWA unless otherwise noted): As copper: ACGIH TLV 0.2 mg/m^3 (fume); 1 mg/m^3 (dusts and mists); NIOSH/OSHA 0.1 mg/m^3 (fume); 1 mg/m^3 as (dusts and mists).
Toxicity by ingestion: Grade 4; LD_{50} less than 50 mg/kg.
Liquid or solid irritant characteristics: Eye irritant.
IDLH value: 50 mg/m^3 as cyanide

CHEMICAL REACTIVITY
Reactivity with water: May form hydrogen cyanide.
Binary reactants: Reacts with chlorine, fluorine, peroxides.

ENVIRONMENTAL DATA
Food chain concentration potential: Copper known to be accumulated by shellfish. Insoluble cyanide compounds may bioaccumulate in aquatic organisms. Copper is concentrated by plankton by factors of 1000 to 5000 or more.
Water pollution: DOT Appendix B, §172.101- severe marine pollutant. Effect of low concentrations on aquatic life is unknown. May be dangerous if it enters nearby water intakes; notify operators. Notify local health and wildlife officials. **Response to discharge:** Issue warning–poison, water contaminant. Restrict access. Should be removed. Chemical and physical treatment.

SHIPPING INFORMATION
Grades of purity: Technical; C.P; **Storage temperature:** Ambient; In presence of moisture, toxic hydrogen cyanide gas may collect in enclosed spaces; **Inert atmosphere:** None; **Venting:** Closed container; **Stability during transport:** Stable.

PHYSICAL AND CHEMICAL PROPERTIES
Physical state @ 59°F/15°C and 1 atm: Solid.
Molecular weight: 89.56
Boiling point @ 1 atm: (decomposes).
Specific gravity (water = 1): 2.92 @ 68°F/20°C (solid).
Heat of fusion: 30.1 cal/g.

COPPER FLUOROBORATE   REC. C:4350

SYNONYMS: CUPRIC FLUOBORATE SOLUTION; COPPER BOROFLUORIDE SOLUTION; COPPER(II) FLUOBORATE SOLUTION

IDENTIFICATION
CAS Number: 14735-84-3
Formula: $Cu(BF_4)_2 \cdot H_2O$

DESCRIPTION: Clear, dark blue liquid. Odorless. Sinks and mixes with water.

Poison! • Breathing the vapor can cause severe illness; skin or eye contact causes severe burns, impaired vision, or blindness • Firefighting gear (including SCBA) does not provide adequate protection. If exposure occurs, remove and isolate gear immediately and thoroughly decontaminate personnel • Containers may BLEVE when exposed to fire • Toxic products of combustion may include copper oxides, boron, and fluoride.

Hazard Classification (based on NFPA-704 Rating System)
Health Hazards (Blue): 3; Flammability (Red): 0; Reactivity (Yellow): 0

EMERGENCY RESPONSE: See Appendix A (171)
Evacuation:
Public safety: Isolate the area of spill or leak for at least 10 to 25 meters (30 to 80 feet) in all directions.
Spill: Increase, in the downwind direction, as necessary, the distance shown under "Public Safety."
Fire: If any large container is involved in fire, isolate for at least 800 meters (½ mile) in all directions; also, consider initial evacuation for 800 meters (½ mile) in all directions.

EXPOSURE
Short-term effects: *SEEK MEDICAL ATTENTION.* **Vapor:** Irritating to eyes, nose, and throat. *IF INHALED*, will, will cause coughing or difficult breathing. *IF IN EYES*, hold eyelids open and flush with plenty of water. *IF BREATHING HAS STOPPED*, give artificial respiration. IF breathing is difficult, administer oxygen. **Liquid:** Irritating to skin and eyes. *IF SWALLOWED*, will cause nausea and vomiting. Remove contaminated clothing and shoes. Flush affected areas with plenty of water. *IF IN EYES*, hold eyelids open and flush with plenty of water. *IF SWALLOWED* and victim is *CONSCIOUS AND ABLE TO SWALLOW*, have victim drink 4 to 8 ounces of water and have victim induce vomiting. *IF SWALLOWED* and victim is *UNCONSCIOUS OR HAVING CONVULSIONS,* do nothing except keep victim warm.

HEALTH HAZARDS
Personal protective equipment (PPE): A-Level PPE. OSHA Table Z-1-A air contaminant. Goggles or face shield; rubber apron and gloves.
Recommendations for respirator selection: NIOSH/OSHA as copper fume
NIOSH/OSHA: *Up to 1 mg/m^3:* DMFu (any dust, mist, and fume respirator; or SA (any supplied-air respirator). *2.5 mg/m^3:* SA:CF (any supplied-air respirator operated in a continuous-flow mode); or PAPRDMFu (any powered, air-purifying respirator with a dust, mist, and fume filter). *5 mg/m^3:* HiEF (any air-purifying, full-facepiece respirator with a high-efficiency particulate filter); or SAT:CF (any supplied-air respirator that has a tight-fitting facepiece and is operated in a continuous-flow mode); or PAPRTHiE (any powered, air-purifying respirator with a tight-fitting facepiece and a high-efficiency particulate filter); or SCBAF (any self-contained breathing apparatus with a full facepiece); or SAF (any supplied-air respirator with a full facepiece). *100 mg/m^3:* SAF:PD,PP (any supplied-air respirator that has a full facepiece and is operated in a pressure-demand or other positive-pressure mode). *EMERGENCY OR PLANNED ENTRY INTO UNKNOWN CONCENTRATIONS OR IDLH CONDITIONS:* SCBAF:PD,PP (any self-contained breathing apparatus that has a full facepiece and is operated in a pressure-demand or other positive-pressure mode); or SAF:PD,PP:ASCBA (Any supplied-air respirator that has a full facepiece and is operated in a pressure-demand or other positive-pressure mode in combination with an auxiliary self-contained breathing apparatus operated in a pressure-demand or other positive pressure mode). *ESCAPE:* HiEF (Any air-purifying, full-facepiece

respirator with a high-efficiency particulate filter); or SCBAE (Any appropriate escape-type, self-contained breathing apparatus).
Exposure limits (TWA unless otherwise noted): As copper: ACGIH TLV 0.2 mg/m^3 (fume); 1 mg/m^3 (dusts and mists); NIOSH/OSHA 0.1 mg/m^3 (fume); 1 mg/m^3 as (dusts and mists).
Toxicity by ingestion: Grade 3; LD$_{50}$ = 50 to 500 mg/kg.
Vapor (gas) irritant characteristics: Eye and respiratory tract irritant.
Liquid or solid irritant characteristics: Severe eye and skin irritant.
IDLH value: 100 mg/m^3 as copper.

CHEMICAL REACTIVITY
Binary reactants: Reacts with chlorine, fluorine, peroxides. May corrode some metals.
Neutralizing agents for acids and caustics: Flush with water, rinse with dilute solution of sodium bicarbonate or soda ash.

ENVIRONMENTAL DATA
Food chain concentration potential: Copper known to be accumulated by shellfish. Copper is concentrated by plankton by factors of 1000 to 5000 or more.
Water pollution: Effect of low concentrations on aquatic life is unknown. May be dangerous if it enters nearby water intakes; notify operators. Notify local health and wildlife officials.
Response to discharge: Issue warning–water contaminant. Disperse and flush.

SHIPPING INFORMATION
Grades of purity: Technical; C.P. 45–50% solutions in water; **Storage temperature:** Ambient; **Inert atmosphere:** None; **Venting:** Open; **Stability during transport:** Stable.

PHYSICAL AND CHEMICAL PROPERTIES
Physical state @ 59°F/15°C and 1 atm: Liquid.
Molecular weight: 237.16 (solute only).
Boiling point @ 1 atm: (approximate) 212°F/100°C/373°K.
Specific gravity (water = 1): 1.54 @ 68°F/20°C (liquid).

COPPER FORMATE REC. C:4400

SYNONYMS: FORMIATO de COBRE (Spanish); TUBERCUPROSE; CUPRIC DIFORMATE

IDENTIFICATION
CAS Number: 544-19-4
Formula: Cu(HCOO)$_2$

DESCRIPTION: Powder blue, turquoise, royal blue crystalline powder. Sinks and mixes with water; may form formic acid.

Irritating to the skin, eyes, and respiratory tract • Toxic products of combustion may include carbon monoxide and copper oxide.

Hazard Classification (based on NFPA-704 Rating System)
Health Hazards (Blue): 2; Flammability (Red): 0; Reactivity (Yellow): 0

EMERGENCY RESPONSE: See Appendix A (171)
Evacuation:
Public safety: Isolate the area of spill or leak for at least 10 to 25 meters (30 to 80 feet) in all directions.
Spill: Increase, in the downwind direction, as necessary, the distance shown under "Public Safety."
Fire: If any large container is involved in fire, isolate for at least 800 meters (½ mile) in all directions; also, consider initial evacuation for 800 meters (½ mile) in all directions.

EXPOSURE
Short-term effects: *SEEK MEDICAL ATTENTION*. **Dust or Solid:** Irritating to skin, eyes, and nose. Harmful if swallowed. Move to fresh air. *IF IN EYES*, hold eyelids open and flush with plenty of water. Flush affected areas with plenty of water. *IF SWALLOWED* and victim is *CONSCIOUS AND ABLE TO SWALLOW*, have victim drink 4 to 8 ounces of water.

HEALTH HAZARDS
Personal protective equipment (PPE): B-Level PPE. OSHA Table Z-1-A air contaminant.
Recommendations for respirator selection: NIOSH/OSHA as copper fume
NIOSH/OSHA: *Up to 1 mg/m^3:* DMFu (any dust, mist, and fume respirator; or SA (any supplied-air respirator). *2.5 mg/m^3:* SA:CF (any supplied-air respirator operated in a continuous-flow mode); or PAPRDMFu (any powered, air-purifying respirator with a dust, mist, and fume filter). *5 mg/m^3:* HiEF (any air-purifying, full-facepiece respirator with a high-efficiency particulate filter); or SAT:CF (any supplied-air respirator that has a tight-fitting facepiece and is operated in a continuous-flow mode); or PAPRTHiE (any powered, air-purifying respirator with a tight-fitting facepiece and a high-efficiency particulate filter); or SCBAF (any self-contained breathing apparatus with a full facepiece); or SAF (any supplied-air respirator with a full facepiece). *100 mg/m^3:* SAF:PD,PP (any supplied-air respirator that has a full facepiece and is operated in a pressure-demand or other positive-pressure mode). *EMERGENCY OR PLANNED ENTRY INTO UNKNOWN CONCENTRATIONS OR IDLH CONDITIONS:* SCBAF:PD,PP (any self-contained breathing apparatus that has a full facepiece and is operated in a pressure-demand or other positive-pressure mode); or SAF:PD,PP:ASCBA (Any supplied-air respirator that has a full facepiece and is operated in a pressure-demand or other positive-pressure mode in combination with an auxiliary self-contained breathing apparatus operated in a pressure-demand or other positive pressure mode). *ESCAPE:* HiEF (Any air-purifying, full-facepiece respirator with a high-efficiency particulate filter); or SCBAE (Any appropriate escape-type, self-contained breathing apparatus).
Exposure limits (TWA unless otherwise noted): As copper: ACGIH TLV 0.2 mg/m^3 (fume); 1 mg/m^3 (dusts and mists); NIOSH/OSHA 0.1 mg/m^3 (fume); 1 mg/m^3 as (dusts and mists).
Liquid or solid irritant characteristics: Eye and skin irritant.
IDLH value: 100 mg/m^3 as copper.

CHEMICAL REACTIVITY
Binary reactants: Reacts with chlorine, fluorine, peroxides.

ENVIRONMENTAL DATA
Water pollution: Effects of low concentrations on aquatic life are unknown. May be dangerous if it enters nearby water intakes; notify operators. Notify local health and wildlife officials. **Response to discharge:** Issue warning–water contaminant. Disperse and flush.

SHIPPING INFORMATION
Storage temperature: Ambient; **Stability during transport:** Stable.

PHYSICAL AND CHEMICAL PROPERTIES
Physical state @ 59°F/15°C and 1 atm: Solid.
Molecular weight: 153.58
Specific gravity (water = 1): 1.831 at 15–20°C.

COPPER GLYCINATE　　　　　　　REC. C:4450

SYNONYMS: BIS(GLYCINATO) COPPER; CUPRIC AMINOACETATE; GLICINATO de COBRE (Spanish); GLYCINE COPPER COMPLEX; GLYCOCOLL-COPPER

IDENTIFICATION
Formula: $Cu(H_2NCH_2COO)_2$ (anhydrous); $Cu(H_2NCH_2COO)_2 \cdot 2H_2O$ (dihydrate); $Cu(H_2NCH_2COO)_2 \cdot H_2O$ (hydrate)

DESCRIPTION: Deep blue (hydrate); Light blue (dihydrate) solid. Blue. Mixes in water.

Combustible solid • Dust cloud may explode if ignited in an enclosed area • Containers may BLEVE when exposed to fire • Irritating to the skin, eyes, and respiratory tract • Toxic products of combustion may include nitrogen oxides and copper oxide.

Hazard Classification (based on NFPA-704 Rating System)
Health Hazards (Blue): 1; Flammability (Red): 1; Reactivity (Yellow): 0

EMERGENCY RESPONSE: See Appendix A (171)
Evacuation:
Public safety: Isolate the area of spill or leak for at least 10 to 25 meters (30 to 80 feet) in all directions.
Spill: Increase, in the downwind direction, as necessary, the distance shown under "Public Safety."
Fire: If any large container is involved in fire, isolate for at least 800 meters (½ mile) in all directions; also, consider initial evacuation for 800 meters (½ mile) in all directions.

EXPOSURE
Short-term effects: *SEEK MEDICAL ATTENTION*. **Solid:** Harmful if swallowed. Irritating to skin and eyes.
Flush affected areas with plenty of water. *IF IN EYES*, hold eyelids open and flush with plenty of water. *IF SWALLOWED* and victim is *CONSCIOUS AND ABLE TO SWALLOW*, have victim drink 4 to 8 ounces of water and induce vomiting.

HEALTH HAZARDS
Personal protective equipment (PPE): B-Level PPE. OSHA Table Z-1-A air contaminant.
Recommendations for respirator selection: NIOSH/OSHA as copper fume
NIOSH/OSHA: *Up to 1 mg/m³:* DMFu (any dust, mist, and fume respirator; or SA (any supplied-air respirator). *2.5 mg/m³:* SA:CF (any supplied-air respirator operated in a continuous-flow mode); or PAPRDMFu (any powered, air-purifying respirator with a dust, mist, and fume filter). *5 mg/m³:* HiEF (any air-purifying, full-facepiece respirator with a high-efficiency particulate filter); or SAT:CF (any supplied-air respirator that has a tight-fitting facepiece and is operated in a continuous-flow mode); or PAPRTHiE (any powered, air-purifying respirator with a tight-fitting facepiece and a high-efficiency particulate filter); or SCBAF (any self-contained breathing apparatus with a full facepiece); or SAF (any supplied-air respirator with a full facepiece). *100 mg/m³:* SAF:PD,PP (any supplied-air respirator that has a full facepiece and is operated in a pressure-demand or other positive-pressure mode). *EMERGENCY OR PLANNED ENTRY INTO UNKNOWN CONCENTRATIONS OR IDLH CONDITIONS:* SCBAF:PD,PP (any self-contained breathing apparatus that has a full facepiece and is operated in a pressure-demand or other positive-pressure mode); or SAF:PD,PP:ASCBA (Any supplied-air respirator that has a full facepiece and is operated in a pressure-demand or other positive-pressure mode in combination with an auxiliary self-contained breathing apparatus operated in a pressure-demand or other positive pressure mode). *ESCAPE:* HiEF (Any air-purifying, full-facepiece respirator with a high-efficiency particulate filter); or SCBAE (Any appropriate escape-type, self-contained breathing apparatus).
Exposure limits (TWA unless otherwise noted): As copper: ACGIH TLV 0.2 mg/m³ (fume); 1 mg/m³ (dusts and mists); NIOSH/OSHA 0.1 mg/m³ (fume); 1 mg/m³ as (dusts and mists).
Long-term health effects: Copper poisoning in animals leads to injury of the liver, kidneys, and spleen.
Liquid or solid irritant characteristics: Eye and skin irritant.
IDLH value: 100 mg/m³.

CHEMICAL REACTIVITY
Binary reactants: Copper compounds react with chlorine, fluorine, peroxides.

ENVIRONMENTAL DATA
Water pollution: Effects of low concentrations on aquatic life are unknown. May be dangerous if it enters nearby water intakes; notify operators. Notify local health and pollution control officials.
Response to discharge: Disperse and flush.

PHYSICAL AND CHEMICAL PROPERTIES
Physical state @ 59°F/15°C and 1 atm: Solid.
Molecular weight: 211.66; 229.67 monohydrate; 247.69 dihydrate.
Boiling point @ 1 atm: Decomposes with gas evolution.
Melting/Freezing point: Monohydrate·H₂O at 253°F/123°C/396°K; chars at 415°F/213°C/486°K; decomposes at 442°F/228°C/501°K; Dihydrate·H₂O at 217°F/103°C/376°K; 2H₂O at 284°F/140°C/413°K; decomposes at 437°F/225°C/498°K.

COPPER IODIDE　　　　　　　REC. C:4500

SYNONYMS: CUPROUS IODIDE; MARSHITE; YODURO de COBRE (Spanish)

IDENTIFICATION
CAS Number: 7681-65-4
Formula: CuI

DESCRIPTION: Tan or off-white solid. Odorless. Sinks in water; insoluble.

Toxic • Irritating to the skin, eyes, and respiratory tract • Toxic products of combustion may include hydrogen iodide or iodine vapors, and copper fume.

Hazard Classification (based on NFPA-704 Rating System)
Health Hazards (Blue): 2; Flammability (Red): 0; Reactivity (Yellow): 0

EMERGENCY RESPONSE: See Appendix A (171)
Evacuation:
Public safety: Isolate the area of spill or leak for at least 10 to 25 meters (30 to 80 feet) in all directions.
Spill: Increase, in the downwind direction, as necessary, the distance shown under "Public Safety."
Fire: If any large container is involved in fire, isolate for at least 800 meters (½ mile) in all directions; also, consider initial evacuation for 800 meters (½ mile) in all directions.

EXPOSURE
Short-term effects: *SEEK MEDICAL ATTENTION.* **Dust:** Irritating to eyes, nose, and throat. *IF INHALED,* will, will cause coughing or difficult breathing. *IF IN EYES,* hold eyelids open and flush with plenty of water. *IF BREATHING HAS STOPPED,* give artificial respiration. IF breathing is difficult, administer oxygen. **Solid:** *POISONOUS IF SWALLOWED.* Irritating to skin and eyes. *IF SWALLOWED,* will cause nausea, vomiting, or loss of consciousness. Remove contaminated clothing and shoes. Flush affected areas with plenty of water. *IF IN EYES,* hold eyelids open and flush with plenty of water. *IF SWALLOWED* and victim is *CONSCIOUS AND ABLE TO SWALLOW,* have victim drink 4 to 8 ounces of water and have victim induce vomiting. *IF SWALLOWED* and victim is *UNCONSCIOUS OR HAVING CONVULSIONS,* do nothing except keep victim warm.

HEALTH HAZARDS
Personal protective equipment (PPE): B-Level PPE. OSHA Table Z-1-A air contaminant. Dust mask; goggles or face shield; protective gloves.
Recommendations for respirator selection: NIOSH/OSHA
Up to 1 mg/m³: DMFu (any dust, mist, and fume respirator; or SA (any supplied-air respirator). *2.5 mg/m³:* SA:CF (any supplied-air respirator operated in a continuous-flow mode); or PAPRDMFu (any powered, air-purifying respirator with a dust, mist, and fume filter). *5 mg/m³:* HiEF (any air-purifying, full-facepiece respirator with a high-efficiency particulate filter); or SAT:CF (any supplied-air respirator that has a tight-fitting facepiece and is operated in a continuous-flow mode); or PAPRTHiE (any powered, air-purifying respirator with a tight-fitting facepiece and a high-efficiency particulate filter); or SCBAF (any self-contained breathing apparatus with a full facepiece); or SAF (any supplied-air respirator with a full facepiece). *100 mg/m³:* SAF:PD,PP (any supplied-air respirator that has a full facepiece and is operated in a pressure-demand or other positive-pressure mode). *EMERGENCY OR PLANNED ENTRY INTO UNKNOWN CONCENTRATIONS OR IDLH CONDITIONS:* SCBAF:PD,PP (any self-contained breathing apparatus that has a full facepiece and is operated in a pressure-demand or other positive-pressure mode); or SAF:PD,PP:ASCBA (Any supplied-air respirator that has a full facepiece and is operated in a pressure-demand or other positive-pressure mode in combination with an auxiliary self-contained breathing apparatus operated in a pressure-demand or other positive pressure mode). *ESCAPE:* HiEF (Any air-purifying, full-facepiece respirator with a high-efficiency particulate filter); or SCBAE (Any appropriate escape-type, self-contained breathing apparatus).
Exposure limits (TWA unless otherwise noted): As copper: ACGIH TLV 0.2 mg/m³ (fume); 1 mg/m³ (dusts and mists); NIOSH/OSHA 0.1 mg/m³ (fume); 1 mg/m³ as (dusts and mists).
Toxicity by ingestion: Grade 3; LD_{50} = 50 to 500 mg/kg.
Vapor (gas) irritant characteristics: Eye and respiratory tract irritant.
Liquid or solid irritant characteristics: Eye and skin irritant.
IDLH value: 100 mg/m³ as copper.

ENVIRONMENTAL DATA
Food chain concentration potential: Copper known to be accumulated by shellfish. Copper is concentrated by plankton by factors of 1000 to 5000 or more.
Water pollution: Effect of low concentrations on aquatic life is unknown. May be dangerous if it enters nearby water intakes; notify operators. Notify local health and wildlife officials.
Response to discharge: Should be removed. Chemical and physical treatment.

SHIPPING INFORMATION
Grades of purity: Anhydrous, 99+%; **Storage temperature:** Ambient; **Inert atmosphere:** None; **Venting:** Open; **Stability during transport:** Stable.

PHYSICAL AND CHEMICAL PROPERTIES
Physical state @ 59°F/15°C and 1 atm: Solid.
Molecular weight: 190.4
Boiling point @ 1 atm: 2354°F = 1290°C/1,563°K.
Melting/Freezing point: 1121°F/605°C/878°K.
Specific gravity (water = 1): 5.62 @ 68°F/20°C (solid).
Heat of fusion: 13.6 cal/g.

COPPER LACTATE REC. C:4550

SYNONYMS: LACTATO de COBRE (Spanish); CUPRIC LACTATE

IDENTIFICATION
Formula: $Cu(C_3H_5O_3)_2 \cdot 2H_2O$

DESCRIPTION: Dark blue, green to blue crystalline solid or powder. Sinks and mixes with water.

Combustible solid • Dust cloud may explode if ignited in an enclosed area • Containers may BLEVE when exposed to fire • Irritating to the skin, eyes, and respiratory tract • Toxic products of combustion may include carbon monoxide and copper fume.

Hazard Classification (based on NFPA-704 Rating System)
Health Hazards (Blue): 1; Flammability (Red): 1; Reactivity (Yellow): 0

EMERGENCY RESPONSE: See Appendix A (171)
Evacuation:
Public safety: Isolate the area of spill or leak for at least 10 to 25 meters (30 to 80 feet) in all directions.
Spill: Increase, in the downwind direction, as necessary, the distance shown under "Public Safety."
Fire: If any large container is involved in fire, isolate for at least 800 meters (½ mile) in all directions; also, consider initial evacuation for 800 meters (½ mile) in all directions.

EXPOSURE
Short-term effects: *SEEK MEDICAL ATTENTION.* **Solid:** Harmful if swallowed. Irritating to skin and eyes. Flush affected areas with plenty of water. *IF IN EYES,* hold eyelids open and flush with plenty of water. *IF SWALLOWED* and victim is *CONSCIOUS AND ABLE TO SWALLOW,* have victim drink 4 to 8 ounces of water and induce vomiting.

HEALTH HAZARDS
Personal protective equipment (PPE): B-Level PPE. OSHA Table Z-1-A air contaminant.
Recommendations for respirator selection: NIOSH/OSHA as copper fume
NIOSH/OSHA: *Up to 1 mg/m³:* DMFu (any dust, mist, and fume respirator; or SA (any supplied-air respirator). *2.5 mg/m³:* SA:CF (any supplied-air respirator operated in a continuous-flow mode); or PAPRDMFu (any powered, air-purifying respirator with a dust, mist, and fume filter). *5 mg/m³:* HiEF (any air-purifying, full-facepiece respirator with a high-efficiency particulate filter); or SAT:CF (any supplied-air respirator that has a tight-fitting facepiece and is operated in a continuous-flow mode); or

PAPRTHiE (any powered, air-purifying respirator with a tight-fitting facepiece and a high-efficiency particulate filter); or SCBAF (any self-contained breathing apparatus with a full facepiece); or SAF (any supplied-air respirator with a full facepiece). *100 mg/m³:* SAF:PD,PP (any supplied-air respirator that has a full facepiece and is operated in a pressure-demand or other positive-pressure mode). *EMERGENCY OR PLANNED ENTRY INTO UNKNOWN CONCENTRATIONS OR IDLH CONDITIONS:* SCBAF:PD,PP (any self-contained breathing apparatus that has a full facepiece and is operated in a pressure-demand or other positive-pressure mode); or SAF:PD,PP:ASCBA (Any supplied-air respirator that has a full facepiece and is operated in a pressure-demand or other positive-pressure mode in combination with an auxiliary self-contained breathing apparatus operated in a pressure-demand or other positive pressure mode). *ESCAPE:* HiEF (Any air-purifying, full-facepiece respirator with a high-efficiency particulate filter); or SCBAE (Any appropriate escape-type, self-contained breathing apparatus).
Exposure limits (TWA unless otherwise noted): As copper: ACGIH TLV 0.2 mg/m³ (fume); 1 mg/m³ (dusts and mists); NIOSH/OSHA 0.1 mg/m³ (fume); 1 mg/m³ as (dusts and mists).
Long-term health effects: Copper poisoning in animals leads to injury of the liver, kidneys, and spleen.
Liquid or solid irritant characteristics: Eye and skin irritant.
IDLH value: 100 mg/m³ as copper.

CHEMICAL REACTIVITY
Binary reactants: Copper compounds react with chlorine, fluorine and peroxides.

ENVIRONMENTAL DATA
Water pollution: Effects of low concentrations on aquatic life are unknown. May be dangerous if it enters nearby water intakes; notify operators. Notify local health and wildlife officials.
Response to discharge: Issue warning–water contaminant. Disperse and flush.

PHYSICAL AND CHEMICAL PROPERTIES
Physical state @ 59°F/15°C and 1 atm: Solid.
Molecular weight: 277.71

COPPER NAPHTHENATE   REC. C:4600

SYNONYMS: CUPRINOL; NAFTENATO de COBRE (Spanish); NAPHTHENIC ACID, COPPER SALT; PAINT DRIER; WITTOX-C

IDENTIFICATION
CAS Number: 1338-02-9
Formula: An organic mixture
DOT ID Number: UN 1993; DOT Guide Number: 128
Proper Shipping Name: Flammable liquids, n.o.s.

DESCRIPTION: Dark green liquid or blue-green solid. Gasoline-like odor. May float or sink in water; insoluble.

Poison! (fungicide) • Flammable • Containers may BLEVE when exposed to fire •Vapors may form explosive mixture with air • Vapors are heavier than air and will collect and stay in low areas • Vapors may travel long distances to ignition sources and flashback • Vapors in confined areas (e.g., tanks, sewers, buildings) may explode when exposed to fire • Irritating to the skin, eyes, and respiratory tract • Toxic products of combustion may include copper fume and carbon monoxide.

Hazard Classification (based on NFPA-704 Rating System)
Health Hazards (Blue): 0; Flammability (Red): 2; Reactivity (Yellow): 0

EMERGENCY RESPONSE: See Appendix A (128)
Evacuation:
Public safety: Isolate spill area for at least 25 to 50 meters (80 to 160 feet) in all directions.
Spill: Large spill–Consider initial downwind evacuation for at least 300 meters (1000 feet).
Fire: Isolate for 800 meters (½ mile) in all directions, especially if tank, rail car, or tank truck is involved in fire.

EXPOSURE
Short-term effects: *SEEK MEDICAL ATTENTION.* **Liquid:** Irritating to skin and eyes. Harmful if swallowed. Remove contaminated clothing and shoes. Flush affected areas with plenty of water. *IF IN EYES,* hold eyelids open and flush with plenty of water. *IF SWALLOWED* and victim is *CONSCIOUS AND ABLE TO SWALLOW,* have victim drink 4 to 8 ounces of water. **Do NOT induce vomiting.**

HEALTH HAZARDS
Personal protective equipment (PPE): B-Level PPE. OSHA Table Z-1-A air contaminant.
Recommendations for respirator selection: NIOSH/OSHA as copper fume
NIOSH/OSHA: *Up to 1 mg/m³:* DMFu (any dust, mist, and fume respirator; or SA (any supplied-air respirator). *2.5 mg/m³:* SA:CF (any supplied-air respirator operated in a continuous-flow mode); or PAPRDMFu (any powered, air-purifying respirator with a dust, mist, and fume filter). *5 mg/m³:* HiEF (any air-purifying, full-facepiece respirator with a high-efficiency particulate filter); or SAT:CF (any supplied-air respirator that has a tight-fitting facepiece and is operated in a continuous-flow mode); or PAPRTHiE (any powered, air-purifying respirator with a tight-fitting facepiece and a high-efficiency particulate filter); or SCBAF (any self-contained breathing apparatus with a full facepiece); or SAF (any supplied-air respirator with a full facepiece). *100 mg/m³:* SAF:PD,PP (any supplied-air respirator that has a full facepiece and is operated in a pressure-demand or other positive-pressure mode). *EMERGENCY OR PLANNED ENTRY INTO UNKNOWN CONCENTRATIONS OR IDLH CONDITIONS:* SCBAF:PD,PP (any self-contained breathing apparatus that has a full facepiece and is operated in a pressure-demand or other positive-pressure mode); or SAF:PD,PP:ASCBA (Any supplied-air respirator that has a full facepiece and is operated in a pressure-demand or other positive-pressure mode in combination with an auxiliary self-contained breathing apparatus operated in a pressure-demand or other positive pressure mode). *ESCAPE:* HiEF (Any air-purifying, full-facepiece respirator with a high-efficiency particulate filter); or SCBAE (Any appropriate escape-type, self-contained breathing apparatus).
Exposure limits (TWA unless otherwise noted): As copper: ACGIH TLV 0.2 mg/m³ (fume); 1 mg/m³ (dusts and mists); NIOSH/OSHA 0.1 mg/m³ (fume); 1 mg/m³ as (dusts and mists).
Toxicity by ingestion: Grade 1; oral rat LD$_{50}$ = 4-6 g/kg.
Vapor (gas) irritant characteristics: Eye and respiratory tract irritant.
Liquid or solid irritant characteristics: Eye and skin irritant. If spilled on clothing and allowed to remain, may cause smarting and reddening of skin.
IDLH value: 100 mg/m³ as copper.

FIRE DATA
Flash point: 100°F/38°C (cc) (typical).

Flammable limits in air: LEL: 0.8%; UEL: 5.0% (mineral spirits).
Fire extinguishing agents not to be used: Water may be ineffective.
Autoignition temperature: 540°F/282°C/555°K as mineral spirits.
Burning rate: 4 mm/min

CHEMICAL REACTIVITY
Binary reactants: Violent reaction with strong oxidizers, acids.

ENVIRONMENTAL DATA
Water pollution: Harmful to aquatic life in very low concentrations. Fouling to shoreline. May be dangerous if it enters nearby water intakes; notify operators. Notify local health and wildlife officials. **Response to discharge:** Issue warning–water contaminant. Mechanical containment. Should be removed. Chemical and physical treatment.

SHIPPING INFORMATION
Grades of purity: 8% in mineral spirits or mineral oil. 5% in mineral spirits. May float instead of sink in water; **Storage temperature:** Ambient; **Inert atmosphere:** None; **Venting:** Open (flame arrester); **Stability during transport:** Stable.

PHYSICAL AND CHEMICAL PROPERTIES
Physical state @ 59°F/15°C and 1 atm: Liquid.
Molecular weight: Mixture
Boiling point @ 1 atm: 310-395°F = 154-202°C/427-475°K.
Specific gravity (water = 1): 0.93-1.05 @ 77°F/25°C (liquid).
Liquid surface tension: 20 dynes/cm = 0.020 N/m @ 68°F/20°C.
Liquid water interfacial tension: 45 dynes/cm = 0.045 N/m @ 68°F/20°C.
Heat of combustion: (estimate) –17,600 Btu/lb = –9800 cal/g = –410 x 10^5 J/kg.

COPPER NITRATE REC. C:4650

SYNONYMS: COPPER DINITRATE; COPPER (2+) NITRATE; COPPER(II) NITRATE; NITRIC ACID, COPPER (II) SALT; NITRIC ACID, COPPER (2+) SALT; CUPRIC DINITRATE; CUPRIC NITRATE; CUPRIC NITRATE TRIHYDRATE; NITRATO de COBRE (Spanish)

IDENTIFICATION
CAS Number: 3251-23-8
Formula: CuN_2O_6; $Cu(NO_3)_2 \cdot 3H_2O$
DOT ID Number: UN 1477; DOT Guide Number: 140
Proper Shipping Name: Nitrates, inorganic, n.o.s.

DESCRIPTION: Blue solid. Odorless. Sinks and mixes with water.

Severely irritates the eyes, skin, and respiratory tract • Strong oxidizer which may react spontaneously with low flash point organics or reducing agents • Heat forms oxygen; will increase the activity of an existing fire • May cause fire or explosion on contact with combustibles (wood, paper, oil, clothing, etc.) • Toxic products of combustion may include nitrogen oxides and copper oxides.

Hazard Classification (based on NFPA-704 Rating System)
Health Hazards (Blue): 1; Flammability (Red): 0; Reactivity (Yellow): 1; Special Notice (White): OXY

EMERGENCY RESPONSE: See Appendix A (140)
Evacuation:
Public safety: Isolate the area of spill or leak for at least 10 to 25 meters (30 to 80 feet) in all directions.
Spill: Consider initial downwind evacuation for at least 100 meters (330 feet).
Fire: If any large container is involved in fire, isolate for at least 800 meters (½ mile) in all directions; also, consider initial evacuation for 800 meters (½ mile) in all directions.

EXPOSURE
Short-term effects: *SEEK MEDICAL ATTENTION.* **Dust:** Irritating to eyes, nose, and throat. *IF INHALED*, will, will cause coughing or difficult breathing. *IF IN EYES*, hold eyelids open and flush with plenty of water. *IF BREATHING HAS STOPPED*, give artificial respiration. IF breathing is difficult, administer oxygen. **Solid:** Will burn eyes. *IF SWALLOWED*, will cause nausea, vomiting, or loss of c consciousness. Remove contaminated clothing and shoes. Flush affected areas with plenty of water. *IF IN EYES*, hold eyelids open and flush with plenty of water. *IF SWALLOWED* and victim is *CONSCIOUS AND ABLE TO SWALLOW*, have victim drink 4 to 8 ounces of water and have victim induce vomiting. *IF SWALLOWED* and victim is *UNCONSCIOUS OR HAVING CONVULSIONS*, do nothing except keep victim warm.

HEALTH HAZARDS
Personal protective equipment (PPE): B-Level PPE. OSHA Table Z-1-A air contaminant.
Recommendations for respirator selection: NIOSH/OSHA as copper fume
NIOSH/OSHA: *Up to 1 mg/m^3:* DMFu (any dust, mist, and fume respirator; or SA (any supplied-air respirator). *2.5 mg/m^3:* SA:CF (any supplied-air respirator operated in a continuous-flow mode); or PAPRDMFu (any powered, air-purifying respirator with a dust, mist, and fume filter). *5 mg/m^3:* HiEF (any air-purifying, full-facepiece respirator with a high-efficiency particulate filter); or SAT:CF (any supplied-air respirator that has a tight-fitting facepiece and is operated in a continuous-flow mode); or PAPRTHiE (any powered, air-purifying respirator with a tight-fitting facepiece and a high-efficiency particulate filter); or SCBAF (any self-contained breathing apparatus with a full facepiece); or SAF (any supplied-air respirator with a full facepiece). *100 mg/m^3:* SAF:PD,PP (any supplied-air respirator that has a full facepiece and is operated in a pressure-demand or other positive-pressure mode). *EMERGENCY OR PLANNED ENTRY INTO UNKNOWN CONCENTRATIONS OR IDLH CONDITIONS:* SCBAF:PD,PP (any self-contained breathing apparatus that has a full facepiece and is operated in a pressure-demand or other positive-pressure mode); or SAF:PD,PP:ASCBA (Any supplied-air respirator that has a full facepiece and is operated in a pressure-demand or other positive-pressure mode in combination with an auxiliary self-contained breathing apparatus operated in a pressure-demand or other positive pressure mode). *ESCAPE:* HiEF (Any air-purifying, full-facepiece respirator with a high-efficiency particulate filter); or SCBAE (Any appropriate escape-type, self-contained breathing apparatus).
Exposure limits (TWA unless otherwise noted): As copper: ACGIH TLV 0.2 mg/m^3 (fume); 1 mg/m^3 (dusts and mists); NIOSH/OSHA 0.1 mg/m^3 (fume); 1 mg/m^3 as (dusts and mists).
Toxicity by ingestion: Grade 2; LD_{50} = 0.5–5 g.kg. Forms methemoglobin; possible blood effects.
Vapor (gas) irritant characteristics: Eye and respiratory tract irritant.
Liquid or solid irritant characteristics: Eye and skin irritant.
IDLH value: 100 mg/m^3 as copper.

CHEMICAL REACTIVITY
Reactivity with water: Forms acid solution.
Binary reactants: Mixtures with wood, paper, cloth and other combustibles may catch fire. Reacts with bases, acetic anhydride, ether. Attacks metals creating heat.

ENVIRONMENTAL DATA
Food chain concentration potential: Copper known to be accumulated by shellfish. Copper is concentrated by plankton by factors of 1000 to 5000 or more. Nitrates persist for prolonged periods in natural waters.
Water pollution: Harmful to aquatic life in very low concentrations. May be dangerous if it enters nearby water intakes; notify operators. Notify local health and wildlife officials.
Response to discharge: Issue warning–water contaminant. Disperse and flush.

SHIPPING INFORMATION
Grades of purity: Pure, 100%; Technical; Reagent. May also be shipped as anhydrous grade; **Storage temperature:** Ambient; **Inert atmosphere:** None; **Venting:** Open; **Stability during transport:** Stable.

PHYSICAL AND CHEMICAL PROPERTIES
Physical state @ 59°F/15°C and 1 atm: Solid.
Molecular weight: 241.60
Boiling point @ 1 atm: 330°F/167°C/440°K (decomposes).
Melting/Freezing point: 238°F/115°C/388°K.
Specific gravity (water = 1): 2.32 @ 68°F/20°C (solid).
Heat of decomposition: More than 330°F/167°C/440°K.

COPPER OXALATE   REC. C:4700

SYNONYMS: COPPER(I) OXALATE; CUPRIC OXALATE, CUPRIC OXALATE HEMIHYDRATE; ETHANEDIOIC ACID, COPPER(II) SALT (1:1); OXALIC ACID, COPPER(II) SALT (1:1); OXALATO de COBRE (Spanish)

IDENTIFICATION
CAS Number: 53421-36-6; 5893-66-3; 814-91-5 (II)
Formula: CuC_2O_4; $CuC_2O_4 \cdot \tfrac{1}{2}H_2O$; $C_2H_2O_4 \cdot Cu$ (II); $C_2Cu_2O_4$ (I)
DOT ID Number: UN 2775; DOT Guide Number: 151
Proper Shipping Name: Copper based pesticide, solid, poisonous

DESCRIPTION: Bluish-green powder. Odorless. Sinks in water; insoluble.

Corrosive to skin, eyes, and respiratory tract; inhalation symptoms may be delayed • Containers may BLEVE when exposed to fire • Dust may explode when exposed to fire • Toxic products of combustion may include copper oxide and carbon monoxide.

Hazard Classification (based on NFPA-704 Rating System)
Health Hazards (Blue): 2; Flammability (Red): 0; Reactivity (Yellow): 0

EMERGENCY RESPONSE: See Appendix A (151)
Evacuation:
Public safety: Isolate the area of spill or leak for at least 25 to 50 meters (80 to 160 feet) in all directions.
Spill: Increase, in the downwind direction, as necessary, the distance shown under "Public Safety."
Fire: If tank, rail car, or tank truck is involved in fire, isolate for at least 800 meters (½ mile) in all directions; also, consider initial evacuation for 800 meters (½ mile) in all directions.

EXPOSURE
Short-term effects: *SEEK MEDICAL ATTENTION.* **Dust:** Irritating to eyes, nose, and throat. *IF INHALED*, will, will cause coughing or difficult breathing. *IF IN EYES*, hold eyelids open and flush with plenty of water. *IF BREATHING HAS STOPPED,* give artificial respiration; *avoid mouth-to-mouth resuscitation; use bag/mask apparatus.* IF breathing is difficult, administer oxygen. May cause lung edema or glottis edema; physical exertion will aggravate this condition. **Solid:** Irritating to skin and eyes. *IF SWALLOWED*, will cause nausea, vomiting, or loss of consciousness. Remove contaminated clothing and shoes. Flush affected areas with plenty of water. *IF IN EYES*, hold eyelids open and flush with plenty of water. *IF SWALLOWED* and victim is *CONSCIOUS AND ABLE TO SWALLOW*, have victim drink 4 to 8 ounces of water and have victim induce vomiting. *IF SWALLOWED* and victim is *UNCONSCIOUS OR HAVING CONVULSIONS,* do nothing except keep victim warm.
Note to physician or authorized medical personnel: Medical observation is recommended for 24 to 48 hours after breathing overexposure, as pulmonary edema may be delayed. As first aid for pulmonary edema, consider administering a corticosteroid spray. Cigarette smoking may exacerbate pulmonary injury and should be discouraged for at least 72 hours following exposure. Also, watch out for edema of the glottis and delayed constriction of esophagus.

HEALTH HAZARDS
Personal protective equipment (PPE): B-Level PPE. OSHA Table Z-1-A air contaminant.
Recommendations for respirator selection: NIOSH/OSHA as copper fume
Up to 1 mg/m^3: DMFu (any dust, mist, and fume respirator; or SA (any supplied-air respirator). *2.5 mg/m^3:* SA:CF (any supplied-air respirator operated in a continuous-flow mode); or PAPRDMFu (any powered, air-purifying respirator with a dust, mist, and fume filter). *5 mg/m^3:* HiEF (any air-purifying, full-facepiece respirator with a high-efficiency particulate filter); or SAT:CF (any supplied-air respirator that has a tight-fitting facepiece and is operated in a continuous-flow mode); or PAPRTHiE (any powered, air-purifying respirator with a tight-fitting facepiece and a high-efficiency particulate filter); or SCBAF (any self-contained breathing apparatus with a full facepiece); or SAF (any supplied-air respirator with a full facepiece). *100 mg/m^3:* SAF:PD,PP (any supplied-air respirator that has a full facepiece and is operated in a pressure-demand or other positive-pressure mode). *EMERGENCY OR PLANNED ENTRY INTO UNKNOWN CONCENTRATIONS OR IDLH CONDITIONS:* SCBAF:PD,PP (any self-contained breathing apparatus that has a full facepiece and is operated in a pressure-demand or other positive-pressure mode); or SAF:PD,PP:ASCBA (Any supplied-air respirator that has a full facepiece and is operated in a pressure-demand or other positive-pressure mode in combination with an auxiliary self-contained breathing apparatus operated in a pressure-demand or other positive pressure mode). *ESCAPE:* HiEF (Any air-purifying, full-facepiece respirator with a high-efficiency particulate filter); or SCBAE (Any appropriate escape-type, self-contained breathing apparatus).
Exposure limits (TWA unless otherwise noted): As copper: ACGIH TLV 0.2 mg/m^3 (fume); 1 mg/m^3 (dusts and mists); NIOSH/OSHA 0.1 mg/m^3 (fume); 1 mg/m^3 as (dusts and mists).
Long-term health effects: Oxalates may cause kidney damage.
Vapor (gas) irritant characteristics: Eye and respiratory tract irritant.

Liquid or solid irritant characteristics: Eye irritant.
IDLH value: 100 mg/m^3 as copper.

CHEMICAL REACTIVITY
Binary reactants: Copper compounds react with chlorine, fluorine, peroxides.

ENVIRONMENTAL DATA
Food chain concentration potential: Copper known to be accumulated by shellfish. Copper is concentrated by plankton by factors of 1000 to 5000, or more.
Water pollution: Effect of low concentrations on aquatic life is unknown. May be dangerous if it enters nearby water intakes; notify operators. Notify local health and wildlife officials.
Response to discharge: Should be removed. Chemical and physical treatment.

SHIPPING INFORMATION
Grades of purity: 98+%; **Storage temperature:** Ambient; **Inert atmosphere:** None; **Venting:** Open; **Stability during transport:** Stable.

PHYSICAL AND CHEMICAL PROPERTIES
Physical state @ 59°F/15°C and 1 atm: Solid.
Molecular weight: 160.6
Specific gravity (water = 1): More than 1 @ 68°F/20°C (solid).

COPPER SUBACETATE   REC. C:4750

SYNONYMS: CUPRIC ACETATE, BASIC; BASIC COPPER ACETATE; COMMON VERDIGRIS; BLUE VERDIGRIS; GREEN VERDIGRIS; FRENCH VERDIGRIS; SUBACETATO de COBRE (Spanish)

IDENTIFICATION
CAS Number: See copper acetate C:4000
Formula: $Cu(C_2H_3O_2)_2 \cdot CuO \cdot 6H_2O$

DESCRIPTION: Blue to green crystalline solid or powder. Sinks and mixes slowly with water.

Hazard Classification (based on NFPA-704 Rating System)
Health Hazards (Blue): 1; Flammability (Red): 0; Reactivity (Yellow): 0

EMERGENCY RESPONSE: See Appendix A (171)
Evacuation:
Public safety: Isolate the area of spill or leak for at least 10 to 25 meters (30 to 80 feet) in all directions.
Spill: Increase, in the downwind direction, as necessary, the distance shown under "Public Safety."
Fire: If any large container is involved in fire, isolate for at least 800 meters (½ mile) in all directions; also, consider initial evacuation for 800 meters (½ mile) in all directions.

EXPOSURE
Short-term effects: *SEEK MEDICAL ATTENTION.* **Solid:** Harmful if swallowed. Irritating to skin and eyes. Flush affected areas with plenty of water. *IF IN EYES,* hold eyelids open and flush with plenty of water. *IF SWALLOWED* and victim is *CONSCIOUS AND ABLE TO SWALLOW,* have victim drink 4 to 8 ounces of water and induce vomiting.

HEALTH HAZARDS
Personal protective equipment (PPE): B-Level PPE. OSHA Table Z-1-A air contaminant.
Recommendations for respirator selection: NIOSH/OSHA as copper fume
Up to 1 mg/m^3: DMFu (any dust, mist, and fume respirator; or SA (any supplied-air respirator). *2.5 mg/m^3:* SA:CF (any supplied-air respirator operated in a continuous-flow mode); or PAPRDMFu (any powered, air-purifying respirator with a dust, mist, and fume filter). *5 mg/m^3:* HiEF (any air-purifying, full-facepiece respirator with a high-efficiency particulate filter); or SAT:CF (any supplied-air respirator that has a tight-fitting facepiece and is operated in a continuous-flow mode); or PAPRTHiE (any powered, air-purifying respirator with a tight-fitting facepiece and a high-efficiency particulate filter); or SCBAF (any self-contained breathing apparatus with a full facepiece); or SAF (any supplied-air respirator with a full facepiece). *100 mg/m^3:* SAF:PD,PP (any supplied-air respirator that has a full facepiece and is operated in a pressure-demand or other positive-pressure mode). *EMERGENCY OR PLANNED ENTRY INTO UNKNOWN CONCENTRATIONS OR IDLH CONDITIONS:* SCBAF:PD,PP (any self-contained breathing apparatus that has a full facepiece and is operated in a pressure-demand or other positive-pressure mode); or SAF:PD,PP:ASCBA (Any supplied-air respirator that has a full facepiece and is operated in a pressure-demand or other positive-pressure mode in combination with an auxiliary self-contained breathing apparatus operated in a pressure-demand or other positive pressure mode). *ESCAPE:* HiEF (Any air-purifying, full-facepiece respirator with a high-efficiency particulate filter); or SCBAE (Any appropriate escape-type, self-contained breathing apparatus).
Exposure limits (TWA unless otherwise noted): As copper: ACGIH TLV 0.2 mg/m^3 (fume); 1 mg/m^3 (dusts and mists); NIOSH/OSHA 0.1 mg/m^3 (fume); 1 mg/m^3 as (dusts and mists).
Long-term health effects: Copper poisoning in animals leads to injury of liver, kidneys, and spleen.
Liquid or solid irritant characteristics: Eye and skin irritant.
IDLH value: 100 mg/m^3 as copper.

ENVIRONMENTAL DATA
Water pollution: Effects of low concentrations on aquatic life are unknown. May be dangerous if it enters nearby water intakes; notify operators. Notify local health and wildlife officials. **Response to discharge:** Issue warning–water contaminant. Disperse and flush.

PHYSICAL AND CHEMICAL PROPERTIES
Physical state @ 59°F/15°C and 1 atm: Solid.
Molecular weight: 369.26

COPPER SULFATE   REC. C:4800

SYNONYMS: BCS COPPER FUNGICIDE; BLUE COPPER; BLUE STONE; BLUE VITRIOL; COPPER MONOSULFATE; COPPER(2+) SULFATE; COPPER(II) SULFATE; COPPER SULFATE PENTAHYDRATE; CP BASIC SULFATE; CUPRIC SULFATE ANHYDROUS; CUPRIC SULPHATE; KUPPERSULFAT (German); ROMAN VITRIOL; SULFATE de CUIVRE (French); SULFATO de COBRE (Spanish); SULFURIC ACID, COPPER(2+) SALT; TNCS 53; TRIANGLE; SULFATE OF COPPER

IDENTIFICATION
CAS Number: 7758-98-7; 7758-99-8 (pentahydrate)
Formula: $CuSO_4 \cdot 5H_2O$; $O_4S \cdot Cu \cdot 5H_2O$ (pentahydrate)
DOT ID Number: NA 9109; DOT Guide Number: 171

Proper Shipping Name: Cupric Sulfate
Reportable Quantity (RQ): **(CERCLA)** 10 lb/4.54 kg

DESCRIPTION: White to blue crystalline granules or powder. Odorless. Sinks in water; soluble; forms acid solution.

Corrosive to the eyes Irritates the skin and respiratory tract; contact with the eyes can cause burns and vision impairment • Toxic products of combustion may include sulfur oxide and copper oxide.

Hazard Classification (based on NFPA-704 Rating System)
Health Hazards (Blue): 2; Flammability (Red): 0; Reactivity (Yellow): 0

EMERGENCY RESPONSE: See Appendix A (171)
Evacuation:
Public safety: Isolate the area of spill or leak for at least 10 to 25 meters (30 to 80 feet) in all directions.
Spill: Increase, in the downwind direction, as necessary, the distance shown under "Public Safety."
Fire: If any large container is involved in fire, isolate for at least 800 meters (½ mile) in all directions; also, consider initial evacuation for 800 meters (½ mile) in all directions.

EXPOSURE
Short-term effects: *SEEK MEDICAL ATTENTION.* **Solid:** If swallowed, will cause nausea, vomiting, or loss of consciousness. *IF SWALLOWED* and victim is *CONSCIOUS AND ABLE TO SWALLOW*, have victim drink 4 to 8 ounces of water and have victim induce vomiting. *IF SWALLOWED* and victim is *UNCONSCIOUS OR HAVING CONVULSIONS*, do nothing except keep victim warm.

HEALTH HAZARDS
Personal protective equipment (PPE): A-Level PPE. OSHA Table Z-1-A air contaminant. Chemical protective material(s) reported to offer minimal to poor protection: neoprene, PVC. Also, butyl rubber, nitrile, and nitrile+PVC offers limited protection
Recommendations for respirator selection: NIOSH/OSHA as copper fume
NIOSH/OSHA: *Up to 1 mg/m^3:* DMFu (any dust, mist, and fume respirator; or SA (any supplied-air respirator). *2.5 mg/m^3:* SA:CF (any supplied-air respirator operated in a continuous-flow mode); or PAPRDMFu (any powered, air-purifying respirator with a dust, mist, and fume filter). *5 mg/m^3:* HiEF (any air-purifying, full-facepiece respirator with a high-efficiency particulate filter); or SAT:CF (any supplied-air respirator that has a tight-fitting facepiece and is operated in a continuous-flow mode); or PAPRTHiE (any powered, air-purifying respirator with a tight-fitting facepiece and a high-efficiency particulate filter); or SCBAF (any self-contained breathing apparatus with a full facepiece); or SAF (any supplied-air respirator with a full facepiece). *100 mg/m^3:* SAF:PD,PP (any supplied-air respirator that has a full facepiece and is operated in a pressure-demand or other positive-pressure mode). *EMERGENCY OR PLANNED ENTRY INTO UNKNOWN CONCENTRATIONS OR IDLH CONDITIONS:* SCBAF:PD,PP (any self-contained breathing apparatus that has a full facepiece and is operated in a pressure-demand or other positive-pressure mode); or SAF:PD,PP:ASCBA (Any supplied-air respirator that has a full facepiece and is operated in a pressure-demand or other positive-pressure mode in combination with an auxiliary self-contained breathing apparatus operated in a pressure-demand or other positive pressure mode). *ESCAPE:* HiEF (Any air-purifying, full-facepiece respirator with a high-efficiency particulate filter); or SCBAE (Any appropriate escape-type, self-contained breathing apparatus).
Exposure limits (TWA unless otherwise noted): As copper: ACGIH TLV 0.2 mg/m^3 (fume); 1 mg/m^3 (dusts and mists); NIOSH/OSHA 0.1 mg/m^3 (fume); 1 mg/m^3 as (dusts and mists).
Toxicity by ingestion: Grade 3; LD$_{50}$ = 50 to 500 mg/kg (rat).
Long-term health effects: Causes liver, kidney and testicular damage in rats. Can cause skin and blood disorders; kidney and liver damage. An allergen.
Vapor (gas) irritant characteristics: Eye and respiratory tract irritant.
Liquid or solid irritant characteristics: Severe eye irritant. Skin irritant. Causes smarting of the skin and first-degree burns on short exposure; may cause second-degree burns on long exposure.
IDLH value: 100 mg/m^3 as copper.

CHEMICAL REACTIVITY
Reactivity with water: Solution is acidic.
Binary reactants: Violent reaction with aluminum, magnesium. Reacts with strong bases, hydroxylamine.

ENVIRONMENTAL DATA
Water pollution: DOT Appendix B, §172.101- anhydrous, hydrates are severe marine pollutants. Harmful to aquatic life in very low concentrations. May be dangerous if it enters nearby water intakes; notify operators. Notify local health and wildlife officials.
Response to discharge: Issue warning–water contaminant. Disperse and flush.

SHIPPING INFORMATION
Storage temperature: Ambient; **Inert atmosphere:** None; **Stability during transport:** Stable.

PHYSICAL AND CHEMICAL PROPERTIES
Physical state @ 59°F/15°C and 1 atm: Solid.
Molecular weight: 249.7
Specific gravity (water = 1): 2.29 @ 15°C (solid).

COPPER SULFATE, AMMONIATED   **REC. C:4850**

SYNONYMS: AMMONIO-CUPRIC SULFATE; AMMONIUM CUPRIC SULFATE; COPPER AMINOSULFATE; COPPER AMMONIUM SULFATE; CUPRIC AMINE SULFATE; CUPRAMMONIUM SULFATE; SULFATO de COBRE AMONIACAL (Spanish); TETRAAMINE COPPER SULFATE

IDENTIFICATION
CAS Number: 10380-29-7
Formula: $CuSO_4 \cdot 4NH_3 \cdot H_2O$; $Cu(NH_3)_4SO_4 \cdot H_2O$
DOT ID Number: UN 9110; DOT Guide Number: 171
Proper Shipping Name: Cupric sulfate, ammoniated
Reportable Quantity (RQ): **(CERCLA)** 10 lb/4.54 kg

DESCRIPTION: Dark blue crystalline powder. Ammonia odor. Sinks in water; soluble.

Corrosive to the eyes Irritates the skin and respiratory tract; contact with the eyes can cause burns and vision impairment • Toxic products of combustion may include ammonia gas, sulfur oxide, and copper oxide.

Hazard Classification (based on NFPA-704 Rating System)
Health Hazards (Blue): 1; Flammability (Red): 0; Reactivity (Yellow): 0

286 Copper tartrate

EMERGENCY RESPONSE: See Appendix A (171)
Evacuation:
Public safety: Isolate the area of spill or leak for at least 10 to 25 meters (30 to 80 feet) in all directions.
Spill: Increase, in the downwind direction, as necessary, the distance shown under "Public Safety."
Fire: If any large container is involved in fire, isolate for at least 800 meters (½ mile) in all directions; also, consider initial evacuation for 800 meters (½ mile) in all directions.

EXPOSURE
Short-term effects: *SEEK MEDICAL ATTENTION.* **Solid:** Harmful if swallowed. Irritating to skin and eyes.
Flush affected areas with plenty of water. *IF IN EYES*, hold eyelids open and flush with plenty of water. *IF SWALLOWED* and victim is *CONSCIOUS AND ABLE TO SWALLOW*, have victim drink 4 to 8 ounces of water and induce vomiting.

HEALTH HAZARDS
Personal protective equipment (PPE): A-Level PPE. OSHA Table Z-1-A air contaminant.
Recommendations for respirator selection: NIOSH/OSHA
Up to 1 mg/m³: DMFu (any dust, mist, and fume respirator; or SA (any supplied-air respirator). *2.5 mg/m³:* SA:CF (any supplied-air respirator operated in a continuous-flow mode); or PAPRDMFu (any powered, air-purifying respirator with a dust, mist, and fume filter). *5 mg/m³:* HiEF (any air-purifying, full-facepiece respirator with a high-efficiency particulate filter); or SAT:CF (any supplied-air respirator that has a tight-fitting facepiece and is operated in a continuous-flow mode); or PAPRTHiE (any powered, air-purifying respirator with a tight-fitting facepiece and a high-efficiency particulate filter); or SCBAF (any self-contained breathing apparatus with a full facepiece); or SAF (any supplied-air respirator with a full facepiece). *100 mg/m³:* SAF:PD,PP (any supplied-air respirator that has a full facepiece and is operated in a pressure-demand or other positive-pressure mode). *EMERGENCY OR PLANNED ENTRY INTO UNKNOWN CONCENTRATIONS OR IDLH CONDITIONS:* SCBAF:PD,PP (any self-contained breathing apparatus that has a full facepiece and is operated in a pressure-demand or other positive-pressure mode); or SAF:PD,PP:ASCBA (Any supplied-air respirator that has a full facepiece and is operated in a pressure-demand or other positive-pressure mode in combination with an auxiliary self-contained breathing apparatus operated in a pressure-demand or other positive pressure mode). *ESCAPE:* HiEF (Any air-purifying, full-facepiece respirator with a high-efficiency particulate filter); or SCBAE (Any appropriate escape-type, self-contained breathing apparatus).
Exposure limits (TWA unless otherwise noted): As copper: ACGIH TLV 0.2 mg/m³ (fume); 1 mg/m³ (dusts and mists); NIOSH/OSHA 0.1 mg/m³ (fume); 1 mg/m³ as (dusts and mists).
Long-term health effects: Copper poisoning in animals leads to injury of liver, kidneys, and spleen.
Vapor (gas) irritant characteristics: Eye and respiratory tract irritant.
Liquid or solid irritant characteristics: may cause eye and skin irritation.
IDLH value: 100 mg/m³ as copper.

CHEMICAL REACTIVITY
Reactivity with water: May release ammonia while present in water.
Binary reactants: Violent reaction with aluminum, magnesium, hydroxylamine. Copper compounds react with chlorine, fluorine, peroxides.

ENVIRONMENTAL DATA
Water pollution: Effects of low concentrations on aquatic life are unknown. May be dangerous if it enters nearby water intakes; notify operators. Notify local health and wildlife officials. **Response to discharge:** Issue warning–water contaminant. Disperse and flush.

PHYSICAL AND CHEMICAL PROPERTIES
Physical state @ 59°F/15°C and 1 atm: Solid.
Molecular weight: 227.73 (anhydrous); 245.8 (monohydrate).
Boiling point @ 1 atm: Hydrate loses water and 2NH$_3$ at 248°F/120°C and remaining 2NH$_3$ @ 320°F/160°C.
Melting/Freezing point: 230°F/110°C/383°K (decomposes).
Specific gravity (water = 1): 1.81 @ 68°F/20°C; 1.79 @ 77°F/25°C.

COPPER TARTRATE REC. C:4900

SYNONYMS: CUPRIC TARTRATE; TARTARIC ACID, COPPER SALT

IDENTIFICATION
CAS Number: 815-82-7
Formula: CuC$_4$H$_4$O$_6$
DOT ID Number: UN 9111; DOT Guide Number: 171
Proper Shipping Name: Cupric tartrate
Reportable Quantity (RQ): **(CERCLA)** 100 lb/45.4 kg

DESCRIPTION: Green to blue powder. Odorless. Sinks and slowly mixes with water.

Irritating to the skin, eyes, and respiratory tract • Toxic products of combustion may include copper oxide and carbon monoxide.

Hazard Classification (based on NFPA-704 Rating System)
Health Hazards (Blue): 1; Flammability (Red): 0; Reactivity (Yellow): 0

EMERGENCY RESPONSE: See Appendix A (171)
Evacuation:
Public safety: Isolate the area of spill or leak for at least 10 to 25 meters (30 to 80 feet) in all directions.
Spill: Increase, in the downwind direction, as necessary, the distance shown under "Public Safety."
Fire: If any large container is involved in fire, isolate for at least 800 meters (½ mile) in all directions; also, consider initial evacuation for 800 meters (½ mile) in all directions.

EXPOSURE
Short-term effects: *SEEK MEDICAL ATTENTION.* **Solid:** Harmful if swallowed. Irritating to skin and eyes.
Flush affected areas with plenty of water. *IF IN EYES*, hold eyelids open and flush with plenty of water. *IF SWALLOWED* and victim is *CONSCIOUS AND ABLE TO SWALLOW*, have victim drink 4 to 8 ounces of water and induce vomiting.

HEALTH HAZARDS
Personal protective equipment (PPE): A-Level PPE. OSHA Table Z-1-A air contaminant.
Recommendations for respirator selection: NIOSH/OSHA as copper fume
NIOSH/OSHA: *Up to 1 mg/m³:* DMFu (any dust, mist, and fume respirator; or SA (any supplied-air respirator). *2.5 mg/m³:* SA:CF (any supplied-air respirator operated in a continuous-flow mode);

or PAPRDMFu (any powered, air-purifying respirator with a dust, mist, and fume filter). *5 mg/m³:* HiEF (any air-purifying, full-facepiece respirator with a high-efficiency particulate filter); or SAT:CF (any supplied-air respirator that has a tight-fitting facepiece and is operated in a continuous-flow mode); or PAPRTHiE (any powered, air-purifying respirator with a tight-fitting facepiece and a high-efficiency particulate filter); or SCBAF (any self-contained breathing apparatus with a full facepiece); or SAF (any supplied-air respirator with a full facepiece). *100 mg/m³:* SAF:PD,PP (any supplied-air respirator that has a full facepiece and is operated in a pressure-demand or other positive-pressure mode). *EMERGENCY OR PLANNED ENTRY INTO UNKNOWN CONCENTRATIONS OR IDLH CONDITIONS:* SCBAF:PD,PP (any self-contained breathing apparatus that has a full facepiece and is operated in a pressure-demand or other positive-pressure mode); or SAF:PD,PP:ASCBA (Any supplied-air respirator that has a full facepiece and is operated in a pressure-demand or other positive-pressure mode in combination with an auxiliary self-contained breathing apparatus operated in a pressure-demand or other positive pressure mode). *ESCAPE:* HiEF (Any air-purifying, full-facepiece respirator with a high-efficiency particulate filter); or SCBAE (Any appropriate escape-type, self-contained breathing apparatus).

Exposure limits (TWA unless otherwise noted): As copper: ACGIH TLV 0.2 mg/m³ (fume); 1 mg/m³ (dusts and mists); NIOSH/OSHA 0.1 mg/m³ (fume); 1 mg/m³ as (dusts and mists).
Long-term health effects: Copper poisoning in animals leads to injury of liver, kidneys, and spleen.
Vapor (gas) irritant characteristics: Eye irritant.
Liquid or solid irritant characteristics: Eye and skin irritant.
IDLH value: 100 mg/m³ as copper.

CHEMICAL REACTIVITY
Binary reactants: Copper compounds react with chlorine, fluorine and peroxides.

ENVIRONMENTAL DATA
Water pollution: Effects of low concentrations on aquatic life are unknown. May be dangerous if it enters nearby water intakes; notify operators. Notify local health and wildlife officials.
Response to discharge: Issue warning–water contaminant. Disperse and flush.

PHYSICAL AND CHEMICAL PROPERTIES
Physical state @ 59°F/15°C and 1 atm: Solid.
Molecular weight: 211.61; 265.66 trihydrate decomposes on heating
Boiling point @ 1 atm: Trihydrate
Specific gravity (water = 1): More than 1.

COUMAPHOS   REC. C:5000

SYNONYMS: AGRIDIP; ASUNTHOL; ASUNTOL; AZUNTHOL; BAYER 21/199; BAYMIX; BAYMIX 50; 3-CHLORO-7-HYDROXY-4-METHYL-COUMARIN *O,O*-DIETHYL PHOSPHOROTHIOATE; 3-CHLORO-7-HYDROXY-4-METHYL-COUMARIN *O*-ESTER WITH *O,O*-DIETHYLPHOSPHOROTHIOATE; 3-CHLORO-4-METHYL-7-COUMARINYLDIETHYL PHOSPHOROTHIOATE; *O*-3-CHLORO-4-METHYL-7-COUMARINYL*O,O*-DIETHYL PHOSPHOROTHIOATE; 3-CHLORO-4-METHYL-7-HYDROXYCOUMARINDIETHYL THIOPHOSPHORIC ACID ESTER; 3-CHLORO-4-METHYLUMBELLIFERONE*O*-ESTER WITH *O,O*-DIETHYL PHOSPHOROTHIOATE; CO-RAL; COUMAFOS; CUMAFOS (Spanish); *O,O*-DIAETHYL-*O*-(3-CHLOR-4-METHYL-CUMARIN-7-YL)-MONOTHIOPHOSPHAT (German); *O,O*-DIETHYLO-(3-CHLORO-4-METHYL-7-COUMARINYL)PHOSPHOROTHIOATE; *O,O*-DIETHYLO-(3-CHLORO-4-METHYLCOUMARINYL-7)THIOPHOSPHATE; *O,O*-DIETHYLO-(3-CHLORO-4-METHYL-2-*oxo*-2H-BENZOPYRAN-7-YL)PHOSPHOROTHIOATE; *O,O*-DIETHYL 3-CHLORO-4-METHYL-7-UMBELLIFERONE THIOPHOSPHATE; *O,O*-DIETHYLO-(3-CHLORO-4-METHYLUMBELLIFERYL)PHOSPHOROTHIOATE; DIETHYL 3-CHLORO-4-METHYLUMBELLIFERYL THIONOPHOSPHATE; DIETHYLTHIOPHOSPHORIC ACID ESTER OF 3-CHLORO-4-METHYL-7-HYDROXYCOUMARIN; MELDANE; MELDONE; MUSCATOX; PHOSPHOROTHIOIC ACID, *O,O*-DIETHYL ESTER, *O*-ESTER WITH 3-CHLORO-7-HYDROXY-4-METHYLCOUMARIN; RESITOX; SUNTOL; THIOPHOSPHATE de *O,O*-DIETHYLE ET de *O*-(3-CHLORO-4-METHYL-7-COUMARINYLE) (French); UMBETHION; 3-CHLORO-7-HYDROXY-4-METHYL-COUMARIN-*O,O*-DIETHYLPHOSPHOROTHIONATE; CO-RAL; *O,O*-DIETHYL; *O*-(3-CHLORO-4-METHYL-2-*oxo*-(2H)-1-BENZOPYRAN-7-YL) PHOSPHOROTHIOATE

IDENTIFICATION
CAS Number: 56-72-4
Formula: $C_{13}H_{16}ClO_5PS$
DOT ID Number: UN 2783; DOT Guide Number: 152
Proper Shipping Name: Coumaphos
Reportable Quantity (RQ): **(CERCLA)** 10 lb/4.54 kg

DESCRIPTION: Tan crystalline solid. Weak sulfurous odor. Solid sinks in water.

Poison! (organophosphate) • Combustible • Breathing the vapor, skin or eye contact, or swallowing the material can kill you • Firefighting gear (including SCBA) does not provide adequate protection. If exposure occurs, remove and isolate gear immediately and thoroughly decontaminate personnel • Containers may BLEVE when exposed to fire •Dust cloud may explode if ignited in an enclosed area • Toxic products of combustion may include hydrogen chloride, carbon monoxide, and oxides of sulfur and phosphorus • Do not put yourself in danger by entering a contaminated area to rescue a victim.

Hazard Classification (based on NFPA-704 Rating System)
Health Hazards (Blue): 3; Flammability (Red): 1; Reactivity (Yellow): 0

EMERGENCY RESPONSE: See Appendix A (152)
Evacuation:
Public safety: Isolate the area of spill or leak for at least 25 to 50 meters (80 to 160 feet) in all directions.
Spill: Increase, in the downwind direction, as necessary, the distance shown under "Public Safety."
Fire: If tank, rail car, or tank truck is involved in fire, isolate for at least 800 meters (½ mile) in all directions; also, consider initial evacuation for 800 meters (½ mile) in all directions.

EXPOSURE
Short-term effects: *SEEK MEDICAL ATTENTION.* **Solid or dust:** *POISONOUS IF SWALLOWED OR IF SKIN IS EXPOSED.* Eye pupils are small; blurred vision; runny nose; cough; shortness of breath; pain; diarrhea, nausea and vomiting; increased blood pressure, hypermotility, hallucinations; loss of consciousness;

convulsions; breathing stops; death. *IF BREATHING HAS STOPPED*, give artificial respiration; *avoid mouth-to-mouth resuscitation; use bag/mask apparatus*. **Skin:** *Remove and double-bag contaminated clothing and shoes and leave in Hot Zone for later incineration by hazardous materials experts.* Flush affected areas with plenty of water. *Skin can also be decontaminated with diluted hypochlorite solution, U.S. Army M291 kit, and M258(A1) skin decontamination kit.* **Eyes:** hold eyelids open and flush with plenty of water. *IF SWALLOWED* and victim is *CONSCIOUS AND ABLE TO SWALLOW*, have victim drink 4 to 8 ounces of water. **Do NOT induce vomiting but immediately administer slurry of activated charcoal**. *IF SWALLOWED* and victim is *UNCONSCIOUS OR HAVING CONVULSIONS*, do nothing except keep victim warm.

Note to physician or authorized medical personnel. Administer atropine, 2 mg (1/30 gr) intramuscularly or intravenously as soon as any local or systemic signs or symptoms of an intoxication are noted; repeat the administration of atropine every 3 to 8 minutes until signs of atropinization (mydriasis, dry mouth, rapid pulse, hot and dry skin) occur; initiate treatment in children with 0.05 mg/kg of atropine; repeat at 5- to 10-minute intervals. Watch respiration, and remove bronchial secretions if they appear to be obstructing the airway; intubate if necessary. Pralidoxime must be administered within minutes to a few hours following exposure (depending on the specific agent) to be effective. Give 2-PAMCI (Pralidoxime; Protopam), 2.5 g in 100 mL of sterile water or in 5% dextrose and water, intravenously, slowly, in 15–30 minutes; if sufficient fluid is not available, give 1 g of 2-PAMCI in 3 mL of distilled water by deep intramuscular injection; repeat this every half hour if respiration weakens or if muscle fasciculation or convulsions recur. Also Diazepam, an anticonvulsant, might be considered.

Medical observation is recommended for 24 to 48 hours after breathing overexposure, as pulmonary edema may be delayed. As first aid for pulmonary edema, consider administering a corticosteroid spray. Cigarette smoking may exacerbate pulmonary injury and should be discouraged for at least 72 hours following exposure.

HEALTH HAZARDS
Personal protective equipment (PPE): A-Level PPE. Chemical-protective clothing and butyl rubber gloves are recommended when skin contact is possible because liquid is rapidly absorbed through the skin and may cause systemic toxicity.
Recommendations for respirator selection: *EMERGENCY OR PLANNED ENTRY INTO UNKNOWN CONCENTRATIONS OR IDLH CONDITIONS:* SCBAF:PD,PP (any self-contained breathing apparatus that has a full facepiece and is operated in a pressure-demand or other positive-pressure mode); or SAF:PD,PP:ASCBA (any supplied-air respirator that has a full facepiece and is operated in a pressure-demand or other positive-pressure mode in combination with an auxiliary self-contained breathing apparatus operated in a pressure-demand or other positive pressure mode). *ESCAPE:* GMFOVHiE [any air-purifying, full-facepiece respirator (gas mask) with a chin-style, front-or back-mounted canister having a high efficiency particulate filter]; or SCBAE (any appropriate escape-type, self-contained breathing apparatus).
Toxicity by ingestion: Grade 4; oral LD_{50} = 16 mg/kg (rat).
Liquid or solid irritant characteristics: Eye and skin irritant.
Odor threshold: 0.02 ppm.

CHEMICAL REACTIVITY
Binary reactants: Reacts with oxidizers.

ENVIRONMENTAL DATA
Food chain concentration potential: Log K_{ow} = 4.13. Values above 3.0 are most likely to accumulate.
Water pollution: DOT Appendix B, §172.101- severe marine pollutant. Harmful to aquatic life in very low concentrations. Fouling to shoreline. May be dangerous if it enters nearby water intakes; notify operators. Notify local health and wildlife officials. For weapons testing use an M272 Water Detection Kit. **Response to discharge:** Issue warning–water contaminant. Restrict access. Should be removed. Chemical and physical treatment.

SHIPPING INFORMATION
Grades of purity: 5–50% active ingredient, the balance is inert solids; **Storage temperature:** Ambient; **Inert atmosphere:** None; **Venting:** Open; **Stability during transport:** Stable.

PHYSICAL AND CHEMICAL PROPERTIES
Melting/Freezing point: 199°F/93°C/366°K.
Specific gravity (water = 1): 1.474 @ 68°F/20°C (solid).
Physical state @ 59°F/15°C and 1 atm: Solid.
Molecular weight: 362.5

CREOSOTE, COAL TAR   REC. C:5050

SYNONYMS: AWPA 1; BRICK OIL; COAL TAR OIL; COAL TAR CREOSOTE; CREOSOTA de ALQUITRAN de HULLA (Spanish); CREOSOTE FROM COAL TAR; CREOSOTE OIL; CREOSOTE P1; CREOSOTUM; CRESYLIC CREOSOTE; DEAD OIL; HEAVY OIL; LIQUID PITCH OIL; NAPHTHALENE OIL; PRESERV-O-SOTE; RCRA No. U051; TAR OIL; WASH OIL (coal tar pitch volatiles)

IDENTIFICATION
CAS Number: 8001-58-9; 8007-45-2; 65996-93-2 (coal tar pitches and volatiles)
Formula: Mixture
DOT ID Number: UN 1993; DOT Guide Number: 128
Proper Shipping Name: Coal tar distillate, flammable
Reportable Quantity (RQ): **(CERCLA)** 1 lb/0.454 kg

DESCRIPTION: Liquid, yellow to black. Tarry odor, like creosote. May float or sink in water.

Combustible • Containers may BLEVE when exposed to fire •Vapors may form explosive mixture with air • Vapors are heavier than air and will collect and stay in low areas • Vapors may travel long distances to ignition sources and flashback • Vapors in confined areas (e.g., tanks, sewers, buildings) may explode when exposed to fire • Irritating to the skin, eyes, and respiratory tract • Combustion products include heavy, irritating back smoke and carbon monoxide.

Hazard Classification (based on NFPA-704 Rating System)
Health Hazards (Blue): 2; Flammability (Red): 2; Reactivity (Yellow): 0

EMERGENCY RESPONSE: See Appendix A (128)
Evacuation:
Public safety: Isolate spill area for at least 25 to 50 meters (80 to 160 feet) in all directions.
Spill: Large spill–Consider initial downwind evacuation for at least 300 meters (1000 feet).
Fire: Isolate for 800 meters (½ mile) in all directions, especially if tank, rail car, or tank truck is involved in fire.

EXPOSURE
Short-term effects: *SEEK MEDICAL ATTENTION.* **Liquid:** Irritating to skin and eyes. Harmful if swallowed. Remove contaminated clothing and shoes. Flush affected areas with plenty of water. *IF IN EYES*, hold eyelids open and flush with plenty of water. *IF SWALLOWED* and victim is *CONSCIOUS AND ABLE TO SWALLOW*, have victim drink 4 to 8 ounces of water and have victim induce vomiting. *IF SWALLOWED* and victim is *UNCONSCIOUS OR HAVING CONVULSIONS*, do nothing except keep victim warm.

HEALTH HAZARDS
Personal protective equipment (PPE): All-service canister mask; gloves; chemical safety goggles and/or face shield; overalls or apron; barrier creams. Chemical protective material(s) reported to have good to excellent resistance: butyl rubber, nitrile, neoprene, Teflon®, Viton®. Also, PVC, styrene-butadiene, butyl/neoprene, nitrile+PVC, Viton®./neoprene, and neoprene offer limited protection

Recommendations for respirator selection: NIOSH *At any concentrations above the NIOSH REL, or where there is no REL, at any detectable concentration:* SCBAF:PD,PP (any self-contained breathing apparatus that has a full facepiece and is operated in a pressure-demand or other positive-pressure mode); or SAF:PD,PP:ASCBA (any supplied-air respirator that has a full facepiece and is operated in a pressure-demand or other positive-pressure mode in combination with an auxiliary, self-contained breathing apparatus operated in a pressure-demand or other positive pressure mode). *ESCAPE:* GMFOVHiE [any air-purifying, full-facepiece respirator (gas mask) with a chin-style, front- or back-mounted organic vapor canister having a high-efficiency particulate filter]; or SCBAE (any appropriate escape-type, self-contained breathing apparatus).

Exposure limits (TWA unless otherwise noted): ACGIH TLV 0.2 mg/m^3 as benzene soluble aerosol (confirmed human carcinogen); OSHA PEL [1910.1002] 0.2 mg/m^3 (benzene-soluble fraction); NIOSH 0.1 mg/m^3 (cyclohexane-extractable fraction); carcinogen.

Toxicity by ingestion: Grade 2; LD$_{50}$ = 0.5 to 5 g/kg.

Long-term health effects: IARC carcinogen; rating 1; Repeated exposures may cause cancer of skin.

Vapor (gas) irritant characteristics: Eye and respiratory tract irritant. Vapors cause moderate irritation such that personnel will find high concentrations unpleasant. The effect is temporary.

Liquid or solid irritant characteristics: Fairly severe eye and skin irritant. May cause pain and second-degree burns after a few minutes of contact.

IDLH value: 80 mg/m^3 as coal tar pitch volatiles. Potential human carcinogen.

FIRE DATA
Flash point: 165°F/74°C (cc).
Fire extinguishing agents not to be used: Water may be ineffective.
Autoignition temperature: 637°F/336°C/609°K.
Electrical hazard: Class I, Group D.

CHEMICAL REACTIVITY
Binary reactants: Oxidizers, acids.
Reactivity group: 21
Compatibility class: Phenols, cresols

ENVIRONMENTAL DATA
Food chain concentration potential: Negative; unlikely to accumulate.

Water pollution: DOT Appendix B, §172.101–marine pollutant. Effect of low concentrations on aquatic life is unknown. Fouling to shoreline. May be dangerous if it enters nearby water intakes; notify operators. Notify local health and wildlife officials. **Response to discharge:** Issue warning–water contaminant. Mechanical containment. Should be removed. Chemical and physical treatment.

SHIPPING INFORMATION
Grades of purity: Whole creosote or various fractions, depending on boiling point. All have similar properties; **Storage temperature:** Ambient; **Inert atmosphere:** None; **Venting:** Open (flame arrester); **Stability during transport:** Stable.

NAS HAZARD CLASSIFICATION FOR BULK WATER TRANSPORTATION
FIRE: 1
HEALTH: Vapor irritant: 2; Liquid or solid irritant: 3; Poisons: 2
WATER POLLUTION: Human toxicity: 2; Aquatic toxicity: 3; Aesthetic effect: 4
REACTIVITY: Other chemicals: 1; Water: 0; Self-reaction: 0

PHYSICAL AND CHEMICAL PROPERTIES
Physical state @ 59°F/15°C and 1 atm: Liquid.
Molecular weight: Mixture
Boiling point @ 1 atm: More than 356°F/180°C/353°K.
Specific gravity (water = 1): 1.05-1.09 @ 15°C (liquid).
Liquid surface tension: (estimate) 15 dynes/cm = 0.015 N/m @ 68°F/20°C.
Liquid water interfacial tension: (estimate) 20 dynes/cm = 0.020 N/m @ 68°F/20°C.
Heat of combustion: (estimate) –12,500 Btu/lb = –6900 cal/g = –290 x 10^5 J/kg.
Vapor pressure: (Reid) Low

CRESOLS   REC. C:5100

SYNONYMS: ACEDE CRESYLIQUE (French); BACILLOL; CRESYLIC ACID; HYDROXYTOLUOLE (German); ISOMERIC MIXTURE OF CRESOLS; KRESOLE (German); PHENOL, METHYL-; METHYLPHENOL; 2-METHYL PHENOL; 3-METHYL PHENOL; 4-METHYL PHENOL; TEKRESOL; AR-TOLUENOL; TRICRESOL; CRESYLIC ACIDS; HYDROXYTOLUENES; METHYLPHENOLS; OXYTOLUENES; TAR ACIDS; RCRA No. U0520
m-cresol: 3-CRESOL; *m*-CRESYLIC ACID; *m*-METHYLPHENOL; 3-HYDROXYTOLUENE
o-cresol: EEC No. 604-004-00-9; *o*-HYDROXYTOLUENE; 2-METHYLPHENOL; *o*-TOLUOL; 2-CRESOL
p-cresol: *p*-METHYLPHENOL; 4-HYDROXYTOLUENE; *p*-TOLUOL; *p*-METHYLHYDROXYBENZENE

CAS Number: 1319-77-3 (isomeric mixtures); 108-39-4 (*m*-isomer); 95-48-7 (*o*-isomer); 106-44-5 (*p*-isomer)
Formula: C$_7$H$_8$O; CH$_3$C$_6$H$_4$OH
DOT ID Number: UN 2076; DOT Guide Number: 153
Proper Shipping Name: Cresols
Reportable Quantity (RQ): **(CERCLA)** 1000 lb/454 kg

DESCRIPTION: Colorless or dark yellow liquid (*m*-isomers), or crystalline solid (*o*- and *p*-isomers). Sweet tarry, phenolic odor. Sinks in water; soluble.

Poison! • Flammable • Corrosive to the skin, eyes, and respiratory tract; contact with skin or eyes can cause burns and vision impairment (especially in direct sunlight); inhalation symptoms may be delayed • Firefighting gear (including SCBA) does not provide adequate protection. If exposure occurs, remove and isolate gear immediately and thoroughly decontaminate personnel • Containers may BLEVE when exposed to fire • Vapors and dust (from p-isomer) are heavier than air and will collect and stay in low areas • Vapors or dust (from p-isomer) in confined areas (e.g., tanks, sewers, buildings) may explode when exposed to fire • Toxic products of combustion may include carbon monoxide • Do not put yourself in danger by entering a contaminated area to rescue a victim.

Hazard Classification (based on NFPA-704 Rating System)
Health Hazards (Blue): 3; Flammability (Red): 2; Reactivity (Yellow): 0

EMERGENCY RESPONSE: See Appendix A (153)
Evacuation:
Public safety: Isolate the area of spill or leak for at least 25 to 50 meters (80 to 160 feet) in all directions.
Spill: Increase, in the downwind direction, as necessary, the distance shown under "Public Safety."
Fire: If tank, rail car, or tank truck is involved in fire, isolate for at least 800 meters (½ mile) in all directions; also, consider initial evacuation for 800 meters (½ mile) in all directions.

EXPOSURE
Short-term effects: *SEEK MEDICAL ATTENTION. ABSORBED THROUGH THE SKIN. IF BREATHING HAS STOPPED,* give artificial respiration; *avoid mouth-to-mouth resuscitation; use bag/mask apparatus.* **Liquid or dust** (*p*-isomer): Will burn skin and eyes. Harmful if absorbed through the skin or swallowed. Remove contaminated clothing and shoes. Flush affected areas with plenty of water. *IF IN EYES*, hold eyelids open and flush with plenty of water. *IF SWALLOWED* and victim is *CONSCIOUS AND ABLE TO SWALLOW,* have victim drink 4 to 8 ounces of water. **Do NOT induce vomiting.** Inhalation may cause lung edema; physical exertion will aggravate this condition.
Note to physician or authorized medical personnel: Medical observation is recommended for 24 to 48 hours after breathing overexposure, as pulmonary edema may be delayed. As first aid for pulmonary edema, consider administering a corticosteroid spray. Cigarette smoking may exacerbate pulmonary injury and should be discouraged for at least 72 hours following exposure.

HEALTH HAZARDS
Personal protective equipment (PPE): A-Level PPE. OSHA Table Z-1-A air contaminant. Organic vapor respirator. Sealed chemical protective materials recommended (*m*-isomer): neoprene, Teflon®. Also, nitrile, PE, neoprene/natural rubber, and Saranex® offers limited protection. Sealed chemical protective materials recommended (isomeric mixtures): butyl rubber, neoprene, Saranex®, Viton®.
Recommendations for respirator selection: NIOSH
23 ppm: CCROVDM [any chemical cartridge respirator with organic vapor cartridge(s) in combination with a dust and mist filter]; SA (any supplied-air respirator). *57.5 ppm*: SA:CF (any supplied-air respirator operated in a continuous-flow mode); or PAPROVDM [any powered, air-purifying respirator with organic vapor cartridge(s) in combination with a dust and mist filter]. *115 ppm*: CCRFOVHiE [any chemical cartridge respirator with a full facepiece and organic vapor cartridge(s) in combination with a high efficiency particulate filter]; or GMFOVHiE [any air-purifying, full-facepiece respirator (gas mask) with a chin-style, front- or back-mounted organic vapor canister having a high-efficiency particulate filter]; or PAPRTHiE* (any powered, air-purifying respirator with a tight-fitting facepiece and a high-efficiency particulate filter); or SAT:CF* (any supplied-air respirator that has a tight-fitting facepiece and is operated in a continuous-flow mode); or SCBAF (any self-contained breathing apparatus with a full facepiece); or SAF (any supplied-air respirator with a full facepiece). *250 ppm*: SAF:PD,PP (any supplied-air respirator that has a full facepiece and is operated in a pressure-demand or other positive-pressure mode). *EMERGENCY OR PLANNED ENTRY INTO UNKNOWN CONCENTRATIONS OR IDLH CONDITIONS:* SCBAF:PD,PP (any self-contained breathing apparatus that has a full facepiece and is operated in a pressure-demand or other positive-pressure mode); or SAF:PD,PP:ASCBA (any supplied-air respirator that has a full facepiece and is operated in a pressure-demand or other positive-pressure mode in combination with an auxiliary self-contained breathing apparatus operated in a pressure-demand or other positive pressure mode). *ESCAPE:* GMFOVHiE [any air-purifying, full-facepiece respirator (gas mask) with a chin-style, front- or back-mounted organic vapor canister having a high-efficiency particulate filter]; or SCBAE (any appropriate escape-type, self-contained breathing apparatus). *Note:* Substance reported to cause eye irritation or damage; may require eye protection.
Exposure limits (TWA unless otherwise noted): ACGIH TLV 5 ppm (22 mg/m^3); OSHA PEL 5 PPM (22 mg/m^3); NIOSH 2.3 ppm (10 mg/m^3); skin contact contributes to overall exposure.
Toxicity by ingestion: Grade 2; LD_{50} = 0.5 to 5 g/kg (rat, rabbit); Grade 3; LD_{50} = 50 to 500 mg/kg (*o*- and *p*-isomer).
Long-term health effects: May produce neoplasms or act as tumor promoters. Central nervous system damage and chronic gastritis, possible liver and kidney damage, and lesions of heart and brain. Dermatitis may result. Suspected carcinogens.
Vapor (gas) irritant characteristics: Eye and respiratory tract irritant. Vapors cause irritation such that personnel will find high concentrations unpleasant.
Liquid or solid irritant characteristics: Severe eye and skin irritant; may cause pain and second-degree burns after a few minutes of contact.
Odor threshold: 0.00005-0.008 ppm; 0.65 ppm detection in water; 0.26 ppm recognition in air (*o*-isomer); 0.2 ppm recognition in air; 0.46 ppb detection in air (*p*-isomer).
IDLH value: 250 ppm.

FIRE DATA
Flash point: 178°F/81°C (cc) (*o*-isomer); 187°F/558°C (cc) (*m*- and *p*-isomer).
Flammable limits in air: LEL: 1.06% (302°F/150°C); UEL: 1.1% (*m*- and *p*-isomer); LEL: 1.4% (300°F/147°C); UEL: ? (*o*-isomer).
Autoignition temperature: 1100°F/599°C/872°K (*o*-isomer); 1038°F/558°C/831°K (*m*- or *p*-isomer).
Electrical hazard: Class I, Group D.

CHEMICAL REACTIVITY
Binary reactants: Strong oxidizers, strong acids, acetaldehyde, alkalis, aliphatic amines, amides, chlorosulfonic acid, fuming sulfuric acid (oleum). Liquid attacks most plastics, rubber, and coatings.
Reactivity group: 21
Compatibility class: Phenols, Creosols

ENVIRONMENTAL DATA
Food chain concentration potential: Log P_{ow} = 1.94–1.97. Unlikely to accumulate.
Water pollution: DOT Appendix B, §172.101–marine pollutant. Harmful to aquatic life in very low concentrations. May be

dangerous if it enters nearby water intakes; notify operators. Notify local health and wildlife officials. **Response to discharge:** Issue warning–water contaminant, poison. Restrict access. Should be removed. Chemical and physical treatment.

SHIPPING INFORMATION
Grades of purity: USP Liquid (mixed isomers) phenol-cresol mixtures; *o*–cresol : 80–98% containing phenol; *m*-cresol: 60 to 98% containing other cresols and xylenols; *p*-cresol: 92 to 98% containing *m*-cresol; meta-para-cresol containing *o*-cresol and xylenols; "Resin" cresols containing phenols and xylenols; cresylic acids containing xylenols, cresols, and phenols; 80-98% containing 2-20% phenol. *o*-cresol: 99.2% with 0.2% phenol and 0.6% *m*- and *p*-isomers; **Storage temperature:** Ambient; **Inert atmosphere:** None; **Venting:** Open; **Stability during transport:** Stable.

NAS HAZARD CLASSIFICATION FOR BULK WATER TRANSPORTATION
FIRE: 1
HEALTH: Vapor irritant: 2; Liquid or solid irritant: 3; Poisons: 2
WATER POLLUTION: Human toxicity: 1; Aquatic toxicity: 3; Aesthetic effect: 4
REACTIVITY: Other chemicals: 2; Water: 0; Self-reaction: 0

PHYSICAL AND CHEMICAL PROPERTIES
Physical state @ 59°F/15°C and 1 atm: Liquid (*m*-isomer); Solid (*o*- and *p*-isomers).
Molecular weight: 108.134
Boiling point @ 1 atm: >350°F/>177°C/>450°K (*m*-isomer); 376°F/191°C/464.2°K (*o*-isomer); 395.46°F/201.92°C/475°K.
Melting/Freezing point: 54°F/13°C/286°K (*m*-isomer); 88°F/31°C/304.2°K (*o*-isomer); 94.6°F/34.78°C/307.93°K (*p*-isomer).
Critical temperature: 795.9°F/424.4°C/697.6°K (*o*-isomer); 808.5°F/431.4°C/704.6°K (*p*-isomer).
Critical pressure: 726.0 psia = 49.4 atm = 5.00 MN/m² (*o*-isomer); 746.7 psia = 50.8 atm = 5.15 MN/m² (*p*-isomer).
Specific gravity (water = 1): 1.0336 @ 68°F/20°C (*m*-isomer); 1.05 @ 68°F/20°C (*o*-isomer); 1.04 @ 68°F/20°C (*p*-isomer).
Liquid surface tension: 41.7 dynes/cm; 0.0417 N/m @ 68°F/20°C; 37 dynes/cm = 0.037 N/m @ 68°F/20°C (*m*-isomer); 40.3 dynes/cm = 0.0403 N/m @ 68°F/20°C (*o*-isomer); 41.8 dynes/cm = 0.041 N/m at 40°C (*p*-isomer).
Ratio of specific heats of vapor (gas): 1.073
Liquid water interfacial tension: 31.3 dynes/cm = 0.0313 N/m @ 68°F/20°C (*m*-isomer); 32.7 dynes/cm = 0.0327 N/m @ 68°F/20°C (*o*-isomer); 31.2 dynes/cm = 0.0312 N/m at 40°C (*p*-isomer).
Relative vapor density (air = 1): 3.72
Ratio of specific heats of vapor (gas): 1.073 (mixed isomers); >1- 1.05 (estimate).
Latent heat of vaporization: 181.1 Btu/lb = 100.6 cal/g = 4.2 x 10^5 J/kg; (estimate, mixed isomers) 200 Btu/lb = 110 cal/g = 4.6 x 10^5 J/kg (*m*-isomer); 178.4 Btu/lb = 99.12 cal/g = 4.15 x 10^5 J/kg (*o*-isomer); 188.7 Btu/lb = 104.85 cal/g = 4.39 x 10^5 J/kg (*p*-isomer).
Heat of combustion: –14,720 to –14,740 Btu/lb = –8180 to –8190 cal/g = –342.5 to –342.9 x 10^5 J/kg (mixed isomers); –14036 Btu/lb = –7798 cal/g = –326 x 10^5 J/kg (*m*-isomer); –13994 Btu/lb = –7774 cal/g = –325 x 10^5 J/kg (*o*-isomer); –14014 Btu/lb = –7786 cal/g = –326 x 10^5 J/kg (*p*-isomer).
Heat of fusion: 26.28 cal/g (*p*-isomer).
Vapor pressure: (Reid) 0.03 psia (mixed isomers); 0.2 mm; 0.14 mm @ 77°F/25°C (*m*-isomer); 0.29 mm @ 77°F/25°C (*o*-isomer); 0.11 mm @ 77°F/25°C (*p*-isomer).

CRESYLATE SPENT CAUSTIC SOLUTION   REC. C:5300

SYNONYMS: RCRA No. D026

IDENTIFICATION
Formula: Mixture
Reportable Quantity (RQ): 1000 lb/454 kg as cresol

DESCRIPTION: Liquid. Mixes with water.

Noncombustible, but if heated to dryness, the resulting solids may ignite spontaneously in air, yielding toxic and corrosive fumes containing sodium monoxide. Spilled material may ignite in air if allowed to dry (after the water evaporates) • Corrosive to the skin, eyes, and respiratory tract; contact with skin or eyes can cause burns and vision impairment; inhalation symptoms may be delayed • Firefighting gear (including SCBA) does not provide adequate protection. If exposure occurs, remove and isolate gear immediately and thoroughly decontaminate personnel • Containers may BLEVE when exposed to fire • Vapors are heavier than air and will collect and stay in low areas • Vapors in confined areas (e.g., tanks, sewers, buildings) may explode when exposed to fire • Toxic products of combustion may include sodium monoxide and nitrogen oxides.

Hazard Classification (based on NFPA-704 Rating System)
Health Hazards (Blue): 3; Flammability (Red): 0; Reactivity (Yellow): 0

EMERGENCY RESPONSE: See Appendix A (171)
Evacuation:
Public safety: Isolate the area of spill or leak for at least 10 to 25 meters (30 to 80 feet) in all directions.
Spill: Increase, in the downwind direction, as necessary, the distance shown under "Public Safety." *Note:* Spilled material may ignite in air if allowed to dry.
Fire: If any large container is involved in fire, isolate for at least 800 meters (½ mile) in all directions; also, consider initial evacuation for 800 meters (½ mile) in all directions.

EXPOSURE
Short-term effects: *SEEK MEDICAL ATTENTION.* **Liquid:** Contact causes burns to skin and eyes. Poisonous if swallowed. Remove and isolate contaminated clothing and shoes at the site. *IF IN EYES OR ON SKIN,* flush with running water for at least 15 minutes; old eyelids open periodically if appropriate. *IF SWALLOWED* and victim is *CONSCIOUS AND ABLE TO SWALLOW,* have victim drink 4 to 8 ounces of water. **Do NOT induce vomiting.** *IF SWALLOWED* and victim is *UNCONSCIOUS OR HAVING CONVULSIONS,* do nothing except keep victim warm.
Note to physician or authorized medical personnel: Medical observation is recommended for 24 to 48 hours after breathing overexposure, as pulmonary edema may be delayed. As first aid for pulmonary edema, consider administering a corticosteroid spray. Cigarette smoking may exacerbate pulmonary injury and should be discouraged for at least 72 hours following exposure.

HEALTH HAZARDS
Personal protective equipment (PPE): A-Level PPE. Wear protective clothing and approved respirator.
Vapor (gas) irritant characteristics: Eye and respiratory tract irritation.
Liquid or solid irritant characteristics: Severe eye, skin and respiratory tract irritant.

FIRE DATA
Flash point: Dry salt ignites spontaneously in air.
Fire extinguishing agents not to be used: Keep away from dry chemicals containing ammonium salts or urea.
Behavior in fire: Noncombustible; however, if heated to dryness, resulting solids may ignite spontaneously in air to yield toxic and corrosive fumes containing sodium monoxide. Spilled material may ignite in air if allowed to dry (after the water evaporates).

CHEMICAL REACTIVITY
Binary reactants: Contact not permitted with copper, copper alloys, zinc or aluminum. These compounds may react with caustic solutions to generate toxic ammonia gas.
Reactivity group: 5
Compatibility class: Caustics

ENVIRONMENTAL DATA
Water pollution: Effects on low concentrations on aquatic life is unknown. May be dangerous if it enters nearby water intakes; notify operators. Notify local health and wildlife officials.
Response to discharge: Issue warning–water contaminant, corrosive. Restrict access. Should be removed. Chemical and physical treatment.

SHIPPING INFORMATION
Storage temperature: Ambient; **Inert atmosphere:** None; **Venting:** Open; **Stability during transport:** Stable.

PHYSICAL AND CHEMICAL PROPERTIES
Physical state @ 59°F/15°C and 1 atm: Liquid.

CRESYL GLYCIDYL ETHER   REC. C: 5350

SYNONYMS: CRESOL, EPOXYPROPYL ETHER; CRESYL GLICIDE ETHER; GLYCIDYL METHYLPHENYL ETHER; TOLYL EPOXYPROPYL ETHER; TOLYL GLYCIDYL ETHER

IDENTIFICATION
CAS Number: 26447-14-3
Formula: $C_{10}H_{12}O_2$; $CH_3C_6H_4$—O—CH_2—CH—CH_2—O

DESCRIPTION: White liquid. Sinks and mixes with water.

Flammable • Poisonous if swallowed or inhaled; toxic by skin absorption; irritating in low concentrations • may form explosive peroxides upon exposure to air • Containers may BLEVE when exposed to fire • Vapors may form explosive mixture with air • Vapors are heavier than air and will collect and stay in low areas • Containers may BLEVE when exposed to fire • Vapors in confined areas (e.g., tanks, sewers, buildings) may explode when exposed to fire • • Toxic products of combustion may include carbon monoxide.

Hazard Classification (based on NFPA-704 Rating System)
Health Hazards (Blue): 1; Flammability (Red): 2; Reactivity (Yellow): 0

EMERGENCY RESPONSE: See Appendix A (171)
Evacuation:
Public safety: Isolate the area of spill or leak for at least 10 to 25 meters (30 to 80 feet) in all directions.
Spill: Increase, in the downwind direction, as necessary, the distance shown under "Public Safety."
Fire: If any large container is involved in fire, isolate for at least 800 meters (½ mile) in all directions; also, consider initial evacuation for 800 meters (½ mile) in all directions.

EXPOSURE
Short-term effects: *SEEK MEDICAL ATTENTION*. **Dust:** Irritating to eyes, nose, and throat. Harmful if inhaled. *IF IN EYES*, hold eyelids open and flush with plenty of water. *IF BREATHING HAS STOPPED*, give artificial respiration. IF breathing is difficult, administer oxygen. **Solid:** Irritating to skin and eyes. Harmful if swallowed. Remove contaminated clothing and shoes. Flush affected areas with plenty of water. *IF IN EYES*, hold eyelids open and flush with plenty of water. *IF SWALLOWED* and victim is *CONSCIOUS AND ABLE TO SWALLOW*, have victim drink 4 to 8 ounces of water. *IF SWALLOWED* and victim is *UNCONSCIOUS OR HAVING CONVULSIONS*, do nothing except keep victim warm.

HEALTH HAZARDS
Personal protective equipment (PPE): B-Level PPE. Organic canister mask or air pack; rubber gloves; goggles or face shield; body-covering clothing.
Long-term health effects: Skin allergy.
Liquid or solid irritant characteristics: Eye and skin irritant.

FIRE DATA
Fire extinguishing agents not to be used: Water may be ineffective.

CHEMICAL REACTIVITY
Binary reactants: May attack some forms of plastics. Cresol compounds may react with oxidizers, alkalis, sodium azide, acetylene.

ENVIRONMENTAL DATA
Food chain concentration potential: Negative; unlikely to accumulate.
Water pollution: Effect of low concentrations on aquatic life is unknown. May be dangerous if it enters nearby water intakes; notify operators. Notify local health and wildlife officials. **Response to discharge:** Issue warning–water contaminant. Should be removed. Chemical and physical treatment.

SHIPPING INFORMATION
Grades of purity: Technical, 100%; **Storage temperature:** Ambient; **Inert atmosphere:** None; **Venting:** Open; **Stability during transport:** Stable.

PHYSICAL AND CHEMICAL PROPERTIES
Physical state @ 59°F/15°C and 1 atm: Liquid.
Molecular weight: 164
Boiling point @ 1 atm: (approximate) 498°F/259°C/532°K.
Flash point: 200°F/93°C (oc).
Specific gravity (water = 1): 1.09 @ 68°F/20°C (liquid).
Heat of combustion: (estimate) –16,500 Btu/lb = –9190 cal/g = –384 x 10^5 J/kg.

CROTONALDEHYDE   REC. C:5400

SYNONYMS: 2-BUTENAL; *trans*-2-BUTENAL; CROTENALDEHYDE; CROTONALDEHIDO (Spanish); CROTONIC ALDEHYDE; EEC No. 605-009-00-9; β-METHYL ACROLEIN; PROPYLENE ALDEHYDE; RCRA No. U053

Crotonaldehyde

IDENTIFICATION
CAS Number: 4170-30-3; 123-73-9
Formula: C_4H_6O; $CH_3CH=CHCHO$
DOT ID Number: UN 1143; DOT Guide Number: 131P
Proper Shipping Name: Crotonaldehyde, stabilized
Reportable Quantity (RQ): **(CERCLA)** 100 lb/45.4 kg

DESCRIPTION: Clear watery liquid. Turns yellow on contact with air. Pungent, suffocating, tar-like odor. Floats and mixes slowly with water.

Poison! • Highly flammable • Strong lacrimator • Corrosive • Breathing the vapor can kill you; corrosive to skin, nose, eyes, and lungs; skin and eye contact can cause severe burns, impaired vision, and blindness; inhalation symptoms may be delayed; this material has anesthetic properties; *vapor is extremely damaging to the eyes* • Polymerization hazard • Forms explosive peroxides upon exposure to air • Heat or exposure to alkali can induce polymerization with rapid release of energy; sealed containers may rupture explosively • May react with itself blocking relief valves; leading to tank explosions • Containers may BLEVE when exposed to fire • Vapors may form explosive mixture with air • Vapors are heavier than air and will collect and stay in low areas • Vapors may travel long distances to ignition sources and flashback • Vapors in confined areas (e.g., tanks, sewers, buildings) may explode when exposed to fire • Toxic products of combustion may include carbon monoxide • Do not put yourself in danger by entering a contaminated area to rescue a victim.

Hazard Classification (based on NFPA-704 Rating System)
Health Hazards (Blue): 4; Flammability (Red): 3; Reactivity (Yellow): 2

EMERGENCY RESPONSE: See Appendix A (131)
Evacuation:
Public safety: Isolate spill area for at least 100 to 200 meters (330 to 660 feet) in all directions.
Spill: Small spill–First: Isolate in all directions 30 meters (100 feet); Then: Protect persons downwind, DAY: 0.2 km (0.1 mile); NIGHT: 0.2 km (0.1 mile). Large spill–First: Isolate in all directions 30 meters (100 feet); Then: Protect persons downwind, DAY: 0.3 km (0.2 mile); NIGHT: 0.8 km (0.5 mile).
Fire: Isolate for 800 meters (½ mile) in all directions, especially if tank, rail car, or tank truck is involved in fire.

EXPOSURE
Short-term effects: *SEEK MEDICAL ATTENTION.* **Vapor:** Irritating to eyes, nose, and throat. *IF INHALED*, will, will cause coughing, nausea, vomiting, or loss of consciousness. Move to fresh air. *IF BREATHING HAS STOPPED*, give artificial respiration. IF breathing is difficult, administer oxygen. May cause lung edema; physical exertion will aggravate this condition. **Liquid:** Will burn skin and eyes. Harmful if swallowed. Remove contaminated clothing and shoes. Flush affected areas with plenty of water. *IF IN EYES*, hold eyelids open and flush with plenty of water. *IF SWALLOWED* and victim is *CONSCIOUS AND ABLE TO SWALLOW*, have victim drink 4 to 8 ounces of water. **Do NOT induce vomiting.**
Note to physician or authorized medical personnel: Medical observation is recommended for 24 to 48 hours after breathing overexposure, as pulmonary edema may be delayed. As first aid for pulmonary edema, consider administering a corticosteroid spray. Cigarette smoking may exacerbate pulmonary injury and should be discouraged for at least 72 hours following exposure.

HEALTH HAZARDS
Personal protective equipment (PPE): A Level PPE. OSHA Table Z-1-A air contaminant. Chemical protective material(s) reported to have good to excellent resistance: butyl rubber, Teflon®.
Recommendations for respirator selection: NIOSH/OSHA
20 ppm: CCRFOV* [any air-purifying, full-facepiece respirator (gas mask) with a chin-style, front- or back-mounted acid gas canister]; or SA* (any supplied-air respirator). *50 ppm*: SA:CF* (any supplied-air respirator operated in a continuous-flow mode); or PAPROV* [any powered, air-purifying respirator with organic vapor cartridge(s)]; or CCRFOV [any air-purifying, full-facepiece respirator (gas mask) with a chin-style, front- or back-mounted acid gas canister]; or GMFOV [any air-purifying, full-facepiece respirator (gas mask) with a chin-style, front- or back-mounted organic vapor canister]; or SCBAF (any self-contained breathing apparatus with a full facepiece); or SAF (any supplied-air respirator with a full facepiece). *EMERGENCY OR PLANNED ENTRY INTO UNKNOWN CONCENTRATIONS OR IDLH CONDITIONS*: SCBAF:PD,PP (any self-contained breathing apparatus that has a full facepiece and is operated in a pressure-demand or other positive-pressure mode); or SAF:PD,PP:ASCBA (any supplied-air respirator that has a full facepiece and is operated in a pressure-demand or other positive-pressure mode in combination with an auxiliary self-contained breathing apparatus operated in a pressure-demand or other positive pressure mode). *ESCAPE:* GMFOV [any air-purifying, full-facepiece respirator (gas mask) with a chin-style, front- or back-mounted organic vapor canister]; or SCBAE (any appropriate escape-type, self-contained breathing apparatus). *Note*: Substance reported to cause eye irritation or damage; may require eye protection.
Exposure limits (TWA unless otherwise noted): OSHA PEL 2 ppm (6 mg/m^3); NIOSH REL 2 ppm (6 mg/m^3).
Short-term exposure limits (15-minute TWA): ACGIH STEL 0.3 ppm.
Toxicity by ingestion: Grade 3; LD_{50} = 50 to 500 mg/kg.
Long-term health effects: Suspected carcinogen.
Vapor (gas) irritant characteristics: Eye and respiratory tract irritant. Vapor is irritating such that personnel will not usually tolerate moderate or high vapor concentrations.
Liquid or solid irritant characteristics: Severe eye irritant. Fairly severe skin irritant; may cause pain and second-degree burns after a few minutes of contact.
Odor threshold: 0.06–0.20 ppm.
IDLH value: 50 ppm.

FIRE DATA
Flash point: 59°F/15°C (oc); 45°F/7°C (cc).
Flammable limits in air: LEL: 2.1%; UEL: 15.5%.
Fire extinguishing agents not to be used: Water may be ineffective on fire.
Behavior in fire: Heat may cause polymerization; containers may rupture and explode.
Autoignition temperature: 450°F/232°C/505°K.
Electrical hazard: Class I, Group C.
Burning rate: 3.3 mm/min

CHEMICAL REACTIVITY
Binary reactants: Reacts with caustics, ammonia, strong oxidizers, some plastics. May form explosive peroxides.
Polymerization: May polymerize in high heat such as fire conditions, or condense with evolution of heat in presence of alkalies, amines, or acids.
Reactivity group: 19
Compatibility class: Aldehyde

ENVIRONMENTAL DATA
Food chain concentration potential: Log P_{ow} = 0.6. Unlikely to accumulate.
Water pollution: DOT Appendix B, §172.101–marine pollutant. DOT Appendix B, §172.101–marine pollutant. Effects of low concentrations on aquatic life is unknown. Fouling to shoreline. May be dangerous if it enters nearby water intakes; notify operators. Notify local health and pollution control officials.
Response to discharge: Issue warning–high flammability, water contaminant. Restrict access. Disperse and flush.

SHIPPING INFORMATION
Grades of purity: 98.0%; **Storage temperature:** Ambient; **Inert atmosphere:** None; **Venting:** Pressure-vacuum; **Stability during transport:** Stable if kept cool. Avoid heat, light and contact with polymerization initiators.

NAS HAZARD CLASSIFICATION FOR BULK WATER TRANSPORTATION
FIRE: 3
HEALTH: Vapor irritant: 3; Liquid or solid irritant: 3; Poisons: 3
WATER POLLUTION: Human toxicity: 3; Aquatic toxicity: 3; Aesthetic effect: 3
REACTIVITY: Other chemicals: 2; Water: 0; Self-reaction: 1

PHYSICAL AND CHEMICAL PROPERTIES
Physical state @ 59°F/15°C and 1 atm: Liquid.
Molecular weight: 70.09
Boiling point @ 1 atm: 216°F/102°C/375°K.
Melting/Freezing point: –100°F/–75°C/198°K.
Critical temperature: 563°F/295°C/568°K.
Critical pressure: 630 psia = 43 atm = 4.4 MN/m^2.
Specific gravity (water = 1): 0.87 @ 68°F/20°C (liquid).
Relative vapor density (air = 1): 2.4
Ratio of specific heats of vapor (gas): 1.104
Latent heat of vaporization: 200 Btu/lb = 111 cal/g = 4.65 x 10^5 J/kg.
Heat of combustion: –14,000 Btu/lb = –7760 cal/g = –325 x 10^5 J/kg.
Vapor pressure: (Reid) 1.5 psia; 19 mm.

CUMENE   REC. C:5450

SYNONYMS: BENZENE, (1-METHYLETHYL-)-; CUMENO (Spanish); CUMOL; EEC No. 601-024-00-X; ISOPROPYL BENZENE; ISOPROPYL BENZENE; ISOPROPYLBENZENE; ISOPROPYLBENZOL; 2-PHENYLPROPANE; (1-METHYLETHYL)BENZENE; RCRA No. U055

IDENTIFICATION
CAS Number: 98-82-8
Formula: C_9H_{12}; $C_6H_5CH(CH_3)_2$
DOT ID Number: UN 1918; DOT Guide Number: 131
Proper Shipping Name: Isopropylbenzene
Reportable Quantity (RQ): **(CERCLA)** 5000 lb/2270 kg

DESCRIPTION: Colorless, watery liquid. Sharp, penetrating, gasoline-like odor. Floats on water surface; insoluble. Forms spontaneously explosive cumene hydroperoxide on long contact with air.

Highly flammable • Containers may BLEVE when exposed to fire • Vapors may form explosive mixture with air • Vapors are heavier than air and will collect and stay in low areas • Vapors may travel long distances to ignition sources and flashback • Containers may BLEVE when exposed to fire • Vapors in confined areas (e.g., tanks, sewers, buildings) may explode when exposed to fire • Toxic if swallowed, inhaled, or by skin contact; Irritating to the skin, eyes, and respiratory tract • Toxic products of combustion may include carbon monoxide.

Hazard Classification (based on NFPA-704 Rating System)
Health Hazards (Blue): 2; Flammability (Red): 3; Reactivity (Yellow): 1

EMERGENCY RESPONSE: See Appendix A (131)
Evacuation:
Public safety: Isolate the area of spill or leak for at least 10 to 25 meters (30 to 80 feet) in all directions.
Spill: Increase, in the downwind direction, as necessary, the distance shown under "Public Safety."
Fire: If any large container is involved in fire, isolate for at least 800 meters (½ mile) in all directions; also, consider initial evacuation for 800 meters (½ mile) in all directions.

EXPOSURE
Short-term effects: *SEEK MEDICAL ATTENTION*. **Liquid:** Irritating to skin and eyes. Harmful if swallowed. Remove contaminated clothing and shoes. Flush affected areas with plenty of water. *IF IN EYES*, hold eyelids open and flush with plenty of water. *IF SWALLOWED* and victim is *CONSCIOUS AND ABLE TO SWALLOW*, have victim drink 4 to 8 ounces of water.

HEALTH HAZARDS
Personal protective equipment (PPE): A-Level PPE. OSHA Table Z-1-A air contaminant. Chemical protective material(s) reported to have good to excellent resistance: Chlorinated polyethylene, Viton®. Also, neoprene and Viton®/neoprene materials offers limited protection
Recommendations for respirator selection: NIOSH/OSHA
500 ppm: CCROV* [any chemical cartridge respirator with organic vapor cartridge(s)]; or SA* (any supplied-air respirator). *900 ppm*: SA:CF* (any supplied-air respirator operated in a continuous-flow mode); or PAPROV* [any powered, air-purifying respirator with organic vapor cartridge(s)]; or CCRFOV [any chemical cartridge respirator with a full facepiece and organic vapor cartridge(s)]; or GMFOV [any air-purifying, full-facepiece respirator (gas mask) with a chin-style, front- or back-mounted acid gas canister]; or SCBAF (any self-contained breathing apparatus with a full facepiece); or SAF (any supplied-air respirator with a full facepiece). *EMERGENCY OR PLANNED ENTRY INTO UNKNOWN CONCENTRATIONS OR IDLH CONDITIONS*: SCBAF:PD,PP (any self-contained breathing apparatus that has a full facepiece and is operated in a pressure-demand or other positive-pressure mode); or SAF:PD,PP:ASCBA (any supplied-air respirator that has a full facepiece and is operated in a pressure-demand or other positive-pressure mode in combination with an auxiliary self-contained breathing apparatus operated in a pressure-demand or other positive pressure mode). *ESCAPE:* GMFOV [any air-purifying, full-facepiece respirator (gas mask) with a chin-style, front- or back-mounted organic vapor cannister]; or SCBAE (any appropriate escape-type, self-contained breathing apparatus). *Note:* Substance reported to cause eye irritation or damage; may require eye protection.
Exposure limits (TWA unless otherwise noted): ACGIH TLV 50 ppm (246 mg/m^3); NIOSH/OSHA 50 ppm (245 mg/m^3); skin contact contributes significantly in overall exposure.
Toxicity by ingestion: Grade 3; LD_{50} = 50 to 500 mg/kg.

Vapor (gas) irritant characteristics: Vapors cause smarting of the eyes or respiratory system if present in high concentrations. The effect is temporary.
Liquid or solid irritant characteristics: Eye and skin irritant. If spilled on clothing and allowed to remain, may cause smarting and reddening of the skin.
Odor threshold: 0.008–0.13 ppm.
IDLH value: 900 ppm [10% LEL].

FIRE DATA
Flash point: 99°F/37°C (cc)
Flammable limits in air: LEL: 0.9%; UEL: 6.5%.
Autoignition temperature: 797°F/423°C/696°K.
Electrical hazard: Class I, Group D. Due to low electric conductivity, this substance may generate electrostatic charges as a result of agitation and flow.
Burning rate: 5.0 mm/min

CHEMICAL REACTIVITY
Binary reactants: Oxidizers, nitric acid, sulfuric, acid. Cumene hydroperoxide is formed upon long exposure to air. Attacks rubber.
Reactivity group: 32
Compatibility class: Aromatic hydrocarbons

ENVIRONMENTAL DATA
Food chain concentration potential: Log P_{ow} = 3.7. Values > 3.0 are likely to bioconcentrate in aquatic organisms and other living tissue, especially in fats.
Water pollution: DOT Appendix B, §172.101–marine pollutant. Effect of low concentrations on aquatic life is unknown. Fouling to shoreline. May be dangerous if it enters nearby water intakes; notify operators. Notify local health and pollution control officials.
Response to discharge: Mechanical containment. Should be removed. Chemical and physical treatment.

SHIPPING INFORMATION
Grades of purity: Research grade; pure grade; technical grade; **Storage temperature:** Ambient; **Inert atmosphere:** None; **Venting:** Open (flame arrester); **Stability during transport:** Stable.

NAS HAZARD CLASSIFICATION FOR BULK WATER TRANSPORTATION
FIRE: 2
HEALTH: Vapor irritant: 1; Liquid or solid irritant: 1; Poisons: 1
WATER POLLUTION: Human toxicity: 1; Aquatic toxicity: 3; Aesthetic effect: 2
REACTIVITY: Other chemicals: 1; Water: 0; Self-reaction: 0

PHYSICAL AND CHEMICAL PROPERTIES
Physical state @ 59°F/15°C and 1 atm: Liquid.
Molecular weight: 120.19
Boiling point @ 1 atm: 306.3°F/152.4°C/425.6°K.
Melting/Freezing point: –140.9°F/–96.1°C/177.1°K.
Critical temperature: 676.2°F/357.9°C/631.1°K.
Critical pressure: 465.5 psia = 31.67 atm = 3.208 MN/m².
Specific gravity (water = 1): 0.866 @ 15°C (liquid).
Liquid surface tension: 28.2 dynes/cm = 0.0282 N/m @ 68°F/20°C.
Liquid water interfacial tension: 54.6 dynes/cm = 0.0546 N/m at 22.7°
Relative vapor density (air = 1): 4.1
Ratio of specific heats of vapor (gas): 1.059
Latent heat of vaporization: 134 Btu/lb = 74.6 cal/g = 3.12 x 10^5 J/kg.
Heat of combustion: –17,710 Btu/lb = –9840 cal/g = –412.0 x 10^5 J/kg.
Vapor pressure: (Reid) 0.5 psia; 8 mm.

CUMENE HYDROPEROXIDE   REC. C:5500

SYNONYMS: α, α-DIMETHYLBENZENE HYDROPEROXIDE; α, α-DIMETHYLBENZYL HYDROPEROXIDE; CHP; α-CUMENYL HYDROPEROXIDE; DIMETHYLBENZYL HYDROPEROXIDE; EEC No. 617-002-00-8; HIDROPEROXIDO de CUMENO (Spanish); HYDROPEROXIDE,1-METHYL-1-PHENYLETHYL-; ISOPROPYLBENZENE HYDROPEROXIDE

IDENTIFICATION
CAS Number: 80-15-9
Formula: $C_9H_{12}O_2$; $C_6H_5C(OOH)(CH_3)_2 \cdot C_6H_5CH(CH_3)_2$ (mixture)
DOT ID Number: UN 2116; DOT Guide Number: 147
Proper Shipping Name: Cumene hydroperoxide
Reportable Quantity (RQ): **(CERCLA)** 10 lb/4.54 kg

DESCRIPTION: Colorless to light yellow liquid. Sharp, irritating odor. Sinks in water; slightly soluble.

Highly flammable • Heat, shock, or contamination may cause explosion • Strong oxidizer which may react spontaneously with low flash point organics or reducing agents. Heat forms oxygen; will increase the activity of an existing fire • May cause fire or explosion contact with combustibles (wood, paper, oil, clothing, etc.) • Containers may BLEVE when exposed to fire • Vapors may form explosive mixture with air • Vapors are heavier than air and will collect and stay in low areas • Vapors may travel long distances to ignition sources and flashback • Containers may BLEVE when exposed to fire • Vapors in confined areas (e.g., tanks, sewers, buildings) may explode when exposed to fire • Irritating to the skin, eyes, and respiratory tract • Toxic products of combustion may include carbon monoxide and phenol • Do NOT attempt rescue.

Hazard Classification (based on NFPA-704 Rating System)
Health Hazards (Blue): 1; Flammability (Red): 2; Reactivity (Yellow): 4; Special Notice (White): OXY

EMERGENCY RESPONSE: See Appendix A (147)
Evacuation:
Public safety: Isolate the area of spill or leak for at least 25 to 50 meters (80 to 160 feet) in all directions.
Spill: Consider initial evacuation for at least 250 meters (800 feet).
Fire: If tank, rail car, or tank truck is involved in fire, isolate for at least 800 meters (½ mile) in all directions; also, consider initial evacuation for 800 meters (½ mile) in all directions.

EXPOSURE
Short-term effects: *SEEK MEDICAL ATTENTION. ABSORBED THROUGH THE SKIN.* **Vapor:** Irritating to eyes, nose, and throat. *IF INHALED*, will, will cause headache or coughing. Move victim to fresh air. If breathing is difficult, administer oxygen. **Liquid:** Irritating to skin and eyes. Harmful if swallowed. Remove contaminated clothing and shoes. Flush affected areas with plenty of water. *IF IN EYES*, hold eyelids open and flush. with plenty of water. *IF SWALLOWED* and victim is *CONSCIOUS AND ABLE TO SWALLOW*, have victim drink 4 to 8 ounces of water and have victim induce vomiting. *IF SWALLOWED* and victim is

UNCONSCIOUS OR HAVING CONVULSIONS, do nothing except keep victim warm.

HEALTH HAZARDS
Personal protective equipment (PPE): B-Level PPE. Full face organic vapor respirator. Chemical protective material(s) reported to have good to excellent resistance: Teflon®. Also, butyl rubber is generally suitable for peroxide compounds.
Exposure limits (TWA unless otherwise noted): ACGIH TLV 50 ppm (246 mg/m^3); NIOSH/OSHA PEL 50 ppm (245 mg/m^3) as cumene; skin contact contributes significantly in overall exposure.
Toxicity by ingestion: Grade 3; oral LD_{50} = 382 mg/kg (rat).
Vapor (gas) irritant characteristics: Eye and respiratory tract irritant.
Liquid or solid irritant characteristics: Eye and skin irritant.

FIRE DATA
Flash point: 147°F/64°C (oc); 175°F/79°C (cc)
Fire extinguishing agents not to be used: Water may be ineffective.
Behavior in fire: May decompose violently when heated. Burning rate becomes more rapids as fire burns. Containers may explode.
Autoignition temperature: decomposes violently at temperature above 300°F/149°C.
Electrical hazard: Flow or agitation of substance may generate electrostatic charges due to low conductivity.

CHEMICAL REACTIVITY
Binary reactants: A strong oxidizer. Reacts violently with reducing agents, acids, combustible materials, metallic salts of cobalt, organic materials, copper, lead. Explosive decomposition may occur above 120°F/50°C. Corrodes or reacts with materials containing metals.

ENVIRONMENTAL DATA
Food chain concentration potential: Negative; unlikely to accumulate.
Water pollution: Effects of low concentrations on aquatic life is unknown. Fouling to shoreline. May be dangerous if it enters nearby water intakes; notify operators. Notify local health and wildlife officials. **Response to discharge:** Issue warning–oxidizing material, water contaminant. Mechanical containment. Should be removed. Chemical and physical treatment.

SHIPPING INFORMATION
Grades of purity: 77-85%, the balance is cumene hydrocarbon; **Storage temperature:** Below 125°F/52°C; **Inert atmosphere:** None; **Venting:** Containers must be stored in well-ventilated area; **Stability during transport:** Stable if kept below 125°F/52°C and out of direct sunlight.

PHYSICAL AND CHEMICAL PROPERTIES
Physical state @ 59°F/15°C and 1 atm: Liquid.
Molecular weight: Mixture
Boiling point @ 1 atm: (decomposes; EXPLODES ON HEATING); 302°F/150°C/423°K.
Melting/Freezing point: 16°F/–9°C/264°K.
Specific gravity (water = 1): 1.03 @ 77°F/25°C (liquid).
Liquid surface tension: (estimate) 25 dynes/cm = 0.025 N/m @ 68°F/20°C.
Liquid water interfacial tension: (estimate) 30 dynes/cm = 0.030 N/m @ 68°F/20°C.
Relative vapor density (air = 1): 5.3
Heat of combustion: (estimate) –13,300 Btu/lb = –7400 cal/g = –310 x 10^5 J/kg.
Heat of decomposition: –855 Btu/lb = –475 cal/g = –19.9 x 10^5 J/kg.

CUPRIETHYLENEDIAMINE SOLUTION REC. C:5550

SYNONYMS: CUPRIETILENDIAMINA (Spanish); CUPRIETHYLENE DIAMINE HYDROXIDE SOLUTION; 1,2-DIAMINOETHANE COPPER COMPLEX

IDENTIFICATION
CAS Number: 13426-91-0
Formula: $C_2H_{10}N_2 \cdot xCu$; $Cu(OH)_2$ –$NH_2CH_2CH_2NH_2$ –H_2O
DOT ID Number: UN 1761; DOT Guide Number: 154
Proper Shipping Name: Cupriethylenediamine solution

DESCRIPTION: Blue to dark purple liquid (may contain red or blue sediment). Fishy odor; like ammonia. Sinks and mixes with water. Irritating vapor is produced.

Combustible • Moderately toxic (fungicide) • Corrosive • Highly irritating to the skin, eyes, and respiratory tract • Containers may BLEVE when exposed to fire • Vapors in confined areas (e.g., tanks, sewers, buildings) may explode when exposed to fire • Toxic products of combustion may include nitrogen oxides and copper oxide.

Hazard Classification (based on NFPA-704 Rating System)
Health Hazards (Blue): 1; Flammability (Red): 1; Reactivity (Yellow): 0

EMERGENCY RESPONSE: See Appendix A (154)
Evacuation:
Public safety: Isolate the area of spill or leak for at least 25 to 50 meters (80 to 160 feet) in all directions.
Spill: Increase, in the downwind direction, as necessary, the distance shown under "Public Safety."
Fire: If tank, rail car, or tank truck is involved in fire, isolate for at least 800 meters (½ mile) in all directions; also, consider initial evacuation for 800 meters (½ mile) in all directions.

EXPOSURE
Short-term effects: *SEEK MEDICAL ATTENTION.* **Vapor:** Irritating to eyes, nose, and throat. *IF INHALED*, will, will cause difficult breathing. Move victim to fresh air. *IF BREATHING HAS STOPPED,* give artificial respiration; *avoid mouth-to-mouth resuscitation; use bag/mask apparatus.* If breathing is difficult, administer oxygen. **Liquid:** *POISONOUS IF SWALLOWED OR IF SKIN IS EXPOSED.* Irritating to skin and eyes. Remove contaminated clothing and shoes. Flush affected areas with plenty of water. *IF IN EYES*, hold eyelids open and flush with plenty of water. *IF SWALLOWED* and victim is *CONSCIOUS AND ABLE TO SWALLOW,* have victim drink 4 to 8 ounces of water and have victim induce vomiting. *IF SWALLOWED* and victim is *UNCONSCIOUS OR HAVING CONVULSIONS,* do nothing except keep victim warm.

HEALTH HAZARDS
Personal protective equipment (PPE): A-Level PPE. Goggles or face shield; organic canister mask; rubber gloves; protective clothing.
Exposure limits (TWA unless otherwise noted): As copper: ACGIH TLV 0.2 mg/m^3 (fume); NIOSH/OSHA 0.1 mg/m^3 (fume).
Long-term health effects: Copper poisoning in animals leads to injury of liver, kidneys, and spleen.

Vapor (gas) irritant characteristics: Eye and respiratory tract irritant.
Liquid or solid irritant characteristics: Severe eye irritant. Skin irritant.
IDLH value: 100 mg/m^3 as copper.

FIRE DATA
Behavior in fire: Irritating vapors of ethylenediamine may be produced when heated.

CHEMICAL REACTIVITY
Binary reactants: Dissolves cotton, wood, and other cellulosic materials. Corrosive to copper, aluminum, zinc, and tin.
Neutralizing agents for acids and caustics: Flush with water.

ENVIRONMENTAL DATA
Food chain concentration potential: Negative; unlikely to accumulate.
Water pollution: Effect of low concentrations on aquatic life is unknown. May be dangerous if it enters nearby water intakes; notify operators. Notify local health and wildlife officials.
Response to discharge: Issue warning–corrosive, water contaminant. Restrict access. Disperse and flush.

SHIPPING INFORMATION
Grades of purity: Commercial; **Storage temperature:** Ambient; **Inert atmosphere:** Nitrogen; **Venting:** Pressure-vacuum; **Stability during transport:** Stable.

PHYSICAL AND CHEMICAL PROPERTIES
Physical state @ 59°F/15°C and 1 atm: Liquid.
Molecular weight: Mixture
Boiling point @ 1 atm: (approximate) 212°F/100°C/373°K.
Specific gravity (water = 1): (estimate) more than 1.1 @ 68°F/20°C (liquid).

CYANOACETIC ACID   **REC. C:5600**

SYNONYMS: ACIDO CIANOACETICO (Spanish); CAA; CYANACETIC ACID; MALONIC MONONITRILE; RCRA No. P030

IDENTIFICATION
CAS Number: 372-09-8
Formula: CNCH$_2$COOH
DOT ID Number: UN 3276; DOT Guide Number: 151
Proper Shipping Name: Nitriles, poisonous, n.o.s.
Reportable Quantity (RQ): **(CERCLA)** 10 lb/4.54 kg

DESCRIPTION: White crystalline solid or yellowish-brown solution. Unpleasant odor. Sinks and mixes with water forming an acid solution.

Poison! • Combustible • Corrosive (liquid) • Breathing the vapor, skin or eye contact, or swallowing the material can cause illness and may kill you • Firefighting gear (including SCBA) does not provide adequate protection. If exposure occurs, remove and isolate gear immediately and thoroughly decontaminate personnel • Containers may BLEVE when exposed to fire • Vapors or dusts are heavier than air and will collect and stay in low areas • Vapors in confined areas (e.g., tanks, sewers, buildings) may explode when exposed to fire • Decomposes at more than 300°F/148°C forming toxic oxides of nitrogen, and toxic and flammable acetonitrile vapors. Toxic products of combustion may also include cyanide.

Hazard Classification (based on NFPA-704 Rating System)
Health Hazards (Blue): 3; Flammability (Red): 1; Reactivity (Yellow): 0

EMERGENCY RESPONSE: See Appendix A (151)
Evacuation:
Public safety: Isolate the area of spill or leak for at least 25 to 50 meters (80 to 160 feet) in all directions.
Spill: Increase, in the downwind direction, as necessary, the distance shown under "Public Safety."
Fire: If tank, rail car, or tank truck is involved in fire, isolate for at least 800 meters (½ mile) in all directions; also, consider initial evacuation for 800 meters (½ mile) in all directions.

EXPOSURE
Short-term effects: *SEEK MEDICAL ATTENTION.* **Vapor:** Irritating to eyes, nose, and throat. Harmful if inhaled. *IF IN EYES,* hold eyelids open and flush with plenty of water. *IF BREATHING HAS STOPPED,* give artificial respiration; *avoid mouth-to-mouth resuscitation; use bag/mask apparatus.* IF breathing is difficult, administer oxygen. May cause lung edema; physical exertion will aggravate this condition. **Liquid:** Irritating to skin and eyes. Harmful if swallowed. Remove contaminated clothing and shoes. Flush affected areas with plenty of water. *IF IN EYES,* hold eyelids open and flush with plenty of water. *IF SWALLOWED* and victim is *CONSCIOUS AND ABLE TO SWALLOW,* have victim drink 4 to 8 ounces of water. *IF SWALLOWED* and victim is *UNCONSCIOUS OR HAVING CONVULSIONS,* do nothing except keep victim warm.
Note to physician or authorized medical personnel: Consider the use of amyl nitrite perles if symptoms of cyanide poisoning develop. If symptoms indicate, initial treatment includes the cyanide antidote kit. In all cases, break an amyl nitrite perle in a gauze pad and hold lightly under victim's nose for 15 seconds, repeating 5 times at about 15-second intervals; if necessary (and if sodium nitrite infusions will be delayed), repeat procedure every 3 minutes with fresh perles until 3 or 4 have been used. Avoid breathing the vapor while administering it to the victim. Administer sodium nitrite IV, ASAP. The usual adult dose is 10 to 20 mL of a 3% solution infused over no less than 5 minutes; the average child dose is 0.15 to 0.20 mL/kg. Monitor blood pressure during administration, and slow the rate of infusion if hypotention develops. Next, infuse sodium thiosulfate IV. The usual adult dose is 50 mL of a 25% solution infused over 10 to 20 minutes; the average child dose is 1.65 mL/kg. Repeat with nitrite and thiosulfate as required.

HEALTH HAZARDS
Personal protective equipment (PPE): Level-A PPE.
Recommendations for respirator selection: NIOSH as cyanides *25 mg/m^3:* SA (any supplied-air respirator); or SCBAF (any self-contained breathing apparatus with full facepiece). *EMERGENCY OR PLANNED ENTRY INTO UNKNOWN CONCENTRATIONS OR IDLH CONDITIONS:* SCBAF:PD,PP (any self-contained breathing apparatus that has a full facepiece and is operated in a pressure-demand or other positive-pressure mode); or SAF:PD,PP:ASCBA (any supplied-air respirator that has a full facepiece and is operated in a pressure-demand or other positive-pressure mode in combination with an auxiliary self-contained breathing apparatus operated in a pressure-demand or other positive pressure mode). *ESCAPE:* GMFSHiE [any air-purifying, full-facepiece respirator (gas mask) with a chin-style, front- or back-

mounted canister providing protection against the compound of concern and having a high efficiency particulate filter); or SCBAE (any appropriate escape-type, self-contained breathing apparatus).
Exposure limits (TWA unless otherwise noted): ACGIH TLV 5 mg/m^3; OSHA PEL 5 mg/m^3; NIOSH ceiling 5 mg/m^3 (4.7 ppm) [10 minutes] as cyanides; skin contact contributes significantly in overall exposure
Liquid or solid irritant characteristics: Eye irritant. may cause skin irritation.
IDLH value: 25 mg/m^3 as cyanide

FIRE DATA
Flash point: 226°F/107°C (cc).
Special hazards of combustion products: Decomposes at more than 300°F/149°C, forming toxic oxides of nitrogen and forming toxic and flammable acetonitrile vapors.

CHEMICAL REACTIVITY
Binary reactants: Violent reaction with furfuryl alcohol. Reacts with oxidizers, acids, caustics, reducing agents.

ENVIRONMENTAL DATA
Food chain concentration potential: Negative; unlikely to accumulate.
Water pollution: Harmful to aquatic life in very low concentrations. May be dangerous if it enters nearby water intakes; notify operators. Notify local health and wildlife officials.
Response to discharge: Issue warning–water contaminant. Disperse and flush.

SHIPPING INFORMATION
Grades of purity: 98+%; **Storage temperature:** Ambient; **Inert atmosphere:** None; **Venting:** Open; **Stability during transport:** Stable.

PHYSICAL AND CHEMICAL PROPERTIES
Physical state @ 59°F/15°C and 1 atm: Solid.
Molecular weight: 85.06
Boiling point @ 1 atm: 226°F/108°C/381°K.
Melting/Freezing point: 151°F/66°C/339°K.
Specific gravity (water = 1): More than 1.1 @ 68°F/20°C (solid).
Heat of combustion: -6300 Btu/lb = -3500 cal/g = -146×10^5 J/kg.

CYANOGEN   REC. C:5650

SYNONYMS: CARBON NITRIDE; CIANOGENO (Spanish); CYANOGENE (French); DICYAN; DICYANOGEN; ETHANEDINITRILE; MONOCYANOGEN; NITRILOACETONITRILE; OXALIC ACID DINITRILE; OXALONITRILE; OXALYL CYANIDE; PRUSSITE; ETHANE DINITRILE; DICYAN; RCRA No. P031

IDENTIFICATION
CAS Number: 460-19-5
Formula: $(CN)_2$
DOT ID Number: UN 1026; DOT Guide Number: 119
Proper Shipping Name: Cyanogen; Cyanogen, liquefied
Reportable Quantity (RQ): **(CERCLA)** 100 lb/45.4 kg

DESCRIPTION: Colorless gas. Shipped and stored as a liquefied compressed gas. Almond-like odor; may not be sufficiently strong to provide an adequate warning. Floats and boils on water; poisonous, flammable visible vapor cloud is produced. Poison! • Extremely flammable • May be fatal if inhaled or absorbed through the skin; converted to cyanide in the body • Firefighting gear (including SCBA) does not provide adequate protection. If exposure occurs, remove and isolate gear immediately and thoroughly decontaminate personnel • Containers may BLEVE or explode when exposed to fire • Gas is heavier than air and will collect and stay in low areas • Gas may travel to source of ignition and flashback • Gas in confined areas (e.g., tanks, sewers, buildings) may explode when exposed to fire • Contact with liquid may cause frostbite • Toxic products of combustion may include cyanide and nitrogen oxides • Do NOT attempt rescue. *Warning:* Odor is not a reliable indicator of the presence of toxic amounts of cyanogen gas.

Hazard Classification (based on NFPA-704 Rating System)
Health Hazards (Blue): 4; Flammability (Red): 4; Reactivity (Yellow): 2

EMERGENCY RESPONSE: See Appendix A (119)
Evacuation:
Public safety: Isolate spill area for at least 100 to 200 meters (330 to 660 feet) in all directions.
Spill: Small spill–First: Isolate in all directions 30 meters (100 feet); Then: Protect persons downwind, DAY: 0.3 km (0.2 mile); NIGHT: 1.1 km (0.7 mile). Large spill–First: Isolate in all directions 305 meters (1000 feet); Then: Protect persons downwind, DAY: 3.1 km (1.9 miles); NIGHT: 7.7 km (4.8 mile).
Fire: Isolate for 1600 meters (1 mile) in all directions, especially if tank, rail car, or tank truck is involved in fire.

EXPOSURE
Short-term effects: *SEEK MEDICAL ATTENTION.* **Vapor:** *POISONOUS IF INHALED.* Irritating to eyes. Move victim to fresh air. *IF IN EYES,* hold eyelids open and flush with plenty of water. *IF BREATHING HAS STOPPED,* give artificial respiration; *avoid mouth-to-mouth resuscitation; use bag/mask apparatus.* **Liquid:** *POISONOUS IF SWALLOWED.* Will cause frostbite. Remove contaminated clothing and shoes. Flush affected areas with plenty of water. *DO NOT RUB AFFECTED AREAS. IF IN EYES,* hold eyelids open and flush with plenty of water. *IF SWALLOWED* and victim is *CONSCIOUS AND ABLE TO SWALLOW*, have victim drink 4 to 8 ounces of water and have victim induce vomiting. *IF SWALLOWED* and victim is *UNCONSCIOUS OR HAVING CONVULSIONS,* do nothing except keep victim warm.
Note to physician or authorized medical personnel: Consider the use of amyl nitrite perles if symptoms of cyanide poisoning develop. If symptoms indicate, initial treatment includes the cyanide antidote kit. In all cases, break an amyl nitrite perle in a gauze pad and hold lightly under victim's nose for 15 seconds, repeating 5 times at about 15-second intervals; if necessary (and if sodium nitrite infusions will be delayed), repeat procedure every 3 minutes with fresh perles until 3 or 4 have been used. Avoid breathing the vapor while administering it to the victim. Administer sodium nitrite IV, ASAP. The usual adult dose is 10 to 20 mL of a 3% solution infused over no less than 5 minutes; the average child dose is 0.15 to 0.20 mL/kg. Monitor blood pressure during administration, and slow the rate of infusion if hypotention develops. Next, infuse sodium thiosulfate IV. The usual adult dose is 50 mL of a 25% solution infused over 10 to 20 minutes; the average child dose is 1.65 mL/kg. Repeat with nitrite and thiosulfate as required.

HEALTH HAZARDS
Personal protective equipment (PPE): OSHA Table Z-1-A air contaminant. Self-contained breathing apparatus; rubber gloves;

rubber protective clothing; rubber-soled shoes. Wear thermal protective clothing.
Exposure limits (TWA unless otherwise noted): ACGIH TLV 10 ppm (21 mg/m^3); NIOSH REL 10 ppm (20 mg/m^3).
Long-term health effects: may effect the sense of smell.
Liquid or solid irritant characteristics: Eye irritant.

FIRE DATA
Flash point: Flammable gas.
Flammable limits in air: LEL: 6.6%; UEL: 32%.
Special hazards of combustion products: Unburned vapors of hydrogen cyanide and oxides of nitrogen are highly toxic.
Behavior in fire: Containers may explode in fire, releasing highly toxic gas.
Electrical hazard: Class I, Group B
Stoichiometric air-to-fuel ratio: 5.280 (estimate).

CHEMICAL REACTIVITY
Reactivity with water: Hydrolyzed to form hydrogen cyanide, ammonia or oxalic acid.
Binary reactants: Reacts with acids, oxidizers (explosively). Attacks some metals in presence of moisture.

ENVIRONMENTAL DATA
Food chain concentration potential: Negative; unlikely to accumulate.
Water pollution: DOT Appendix B, §172.101–marine pollutant. Effects of low concentrations on aquatic life is unknown. May be dangerous if it enters nearby water intakes; notify operators. Notify local health and wildlife officials. **Response to discharge:** Issue warning–poison, high flammability, air contaminant, water contaminant. Restrict access. Evacuate area.

SHIPPING INFORMATION
Grades of purity: 98.5%; **Storage temperature:** Cool ambient; **Inert atmosphere:** None; **Venting:** Store containers in well-ventilated area; **Stability during transport:** Stable.

PHYSICAL AND CHEMICAL PROPERTIES
Physical state @ 59°F/15°C and 1 atm: Gas.
Molecular weight: 52.0
Boiling point @ 1 atm: –6.1°F/–21.1°C/252.1°K.
Melting/Freezing point: –18.2°F/–27.9°C/245.3°K.
Critical temperature: 259.9°F/126.6°C/399.8°K.
Critical pressure: 857 psia = 58.2 atm = 5.91 MN/m^2.
Specific gravity (water = 1): 0.954 at –6°F/–21°C (liquid).
Liquid surface tension: 22 dynes/cm = 0.022 N/m at –21°C.
Relative vapor density (air = 1): 1.82
Ratio of specific heats of vapor (gas): 1.205 @ 77°F/25°C.
Latent heat of vaporization: 200 Btu/lb = 111 cal/g = 4.65 x 10^5 J/kg.
Heat of combustion: –9,059 Btu/lb = –5,033 cal/g = –210.6 x 10^5 J/kg.
Heat of solution: 2520 Btu/lb = 1400 cal/g = 58.5 x 10^5 J/kg.
Vapor pressure: 5.1 atm @ 70°F

CYANOGEN BROMIDE REC. C:5700

SYNONYMS: BROMINE CYANIDE; BROMOCYAN; BROMOCYANOGEN; BROMURE de CYANOGEN (French); BROMURO de CIANOGENO (Spanish); CAMPILIT; CYANOBROMIDE; CYANOGEN MONOBROMIDE; TL822; RCRA No. U246

IDENTIFICATION
CAS Number: 506-68-3
Formula: BrCN
DOT ID Number: UN 1889; DOT Guide Number: 157
Proper Shipping Name: Cyanogen bromide
Reportable Quantity (RQ): **(CERCLA)** 1000 lb/454 kg

DESCRIPTION: Colorless crystalline solid. Penetrating odor. Sinks and mixes with water; decomposes slowly producing toxic hydrogen cyanide, bromine gas, and hydrogen bromide. Solid produces a large amounts of vapor.

Poison! • Corrosive • Thermally unstable; impure material tends to explode (Merck) • Breathing the vapors or dust, skin or eye contact, or swallowing the material can kill you; converted to cyanide in the body • Firefighting gear (including SCBA) provide NO protection. If exposure occurs, remove and isolate gear immediately and thoroughly decontaminate personnel • Containers may explode when exposed to fire • Concentrated dust in confined areas (e.g., tanks, sewers, buildings) may explode when exposed to fire • Toxic products of combustion may include cyanide, nitrogen oxide, hydrogen bromide • Do NOT attempt rescue.

Hazard Classification (based on NFPA-704 Rating System)
Health Hazards (Blue): 4; Flammability (Red): 0; Reactivity (Yellow): 1

EMERGENCY RESPONSE: See Appendix A (157)
Evacuation:
Public safety: Isolate the area of spill or leak for at least 50 to 100 meters (160 to 330 feet) in all directions.
Spill: Increase, in the downwind direction, as necessary, the distance shown under "Public Safety."
Fire: If tank, rail car, or tank truck is involved in fire, isolate for at least 800 meters (½ mile) in all directions; also, consider initial evacuation for 800 meters (½ mile) in all directions.

EXPOSURE
Short-term effects: *SEEK MEDICAL ATTENTION.* **Dust:** *POISONOUS IF INHALED OR IF SKIN IS EXPOSED.* Irritating to eyes. Move to fresh air. *IF BREATHING HAS STOPPED,* give artificial respiration; *avoid mouth-to-mouth resuscitation; use bag/mask apparatus.* IF breathing is difficult, administer oxygen. **Solids:** *POISONOUS IF SWALLOWED.* Will burn skin and eyes. Remove contaminated clothing and shoes. Flush affected areas with plenty of water. *IF IN EYES,* hold eyelids open and flush with plenty of water. *IF SWALLOWED* and victim is *CONSCIOUS AND ABLE TO SWALLOW,* have victim drink 4 to 8 ounces of water. **Do NOT induce vomiting.**
Note to physician or authorized medical personnel: Medical observation is recommended for 24 to 48 hours after breathing overexposure, as pulmonary edema may be delayed. As first aid for pulmonary edema, consider administering a corticosteroid spray.
Note to physician or authorized medical personnel: Consider the use of amyl nitrite perles if symptoms of cyanide poisoning develop. If symptoms indicate, initial treatment includes the cyanide antidote kit. In all cases, break an amyl nitrite perle in a gauze pad and hold lightly under victim's nose for 15 seconds, repeating 5 times at about 15-second intervals; if necessary (and if sodium nitrite infusions will be delayed), repeat procedure every 3 minutes with fresh perles until 3 or 4 have been used. Avoid breathing the vapor while administering it to the victim. Administer sodium nitrite IV, ASAP. The usual adult dose is 10 to 20 mL of a 3% solution infused over no less than 5 minutes; the average child dose is 0.15 to 0.20 mL/kg. Monitor blood pressure during

administration, and slow the rate of infusion if hypotention develops. Next, infuse sodium thiosulfate IV. The usual adult dose is 50 mL of a 25% solution infused over 10 to 20 minutes; the average child dose is 1.65 mL/kg. Repeat with nitrite and thiosulfate as required.

HEALTH HAZARDS
Personal protective equipment (PPE): A-Level PPE. Wear chemical protective suit with self-contained breathing apparatus.
Recommendations for respirator selection: NIOSH as cyanides 25 mg/m^3: SA (any supplied-air respirator); or SCBAF (any self-contained breathing apparatus with full facepiece). *EMERGENCY OR PLANNED ENTRY INTO UNKNOWN CONCENTRATIONS OR IDLH CONDITIONS:* SCBAF:PD,PP (any self-contained breathing apparatus that has a full facepiece and is operated in a pressure-demand or other positive-pressure mode); or SAF:PD,PP:ASCBA (any supplied-air respirator that has a full facepiece and is operated in a pressure-demand or other positive-pressure mode in combination with an auxiliary self-contained breathing apparatus operated in a pressure-demand or other positive pressure mode). *ESCAPE:* GMFSHiE [any air-purifying, full-facepiece respirator (gas mask) with a chin-style, front- or back-mounted canister providing protection against the compound of concern and having a high efficiency particulate filter); or SCBAE (any appropriate escape-type, self-contained breathing apparatus).
Exposure limits (TWA unless otherwise noted): ACGIH TLV 5 mg/m^3; OSHA PEL 5 mg/m^3; NIOSH ceiling 5 mg/m^3 (4.7 ppm) [10 minutes] as cyanides; skin contact contributes significantly in overall exposure
Long-term health effects: Workers exposed to solutions may develop dermatitis.
Vapor (gas) irritant characteristics: Vapors cause severe irritation of eyes and throat and can cause eye and lung injury. They cannot be tolerated even at low concentrations.
Liquid or solid irritant characteristics: Severe eye and skin irritant. Causes second-and third-degree burns on short contact; very injurious to the eyes.

FIRE DATA, but impure material is thermally unstable and may explode.

CHEMICAL REACTIVITY
Reactivity with water: Slow hydrolysis, releasing toxic hydrogen cyanide, bromine gas, and hydrogen bromide.
Binary reactants: Reacts with acids, releasing hydrogen cyanide.
Neutralizing agents for acids and caustics: Strong bleaching powder solution; let stand 24 hours; remove.

ENVIRONMENTAL DATA
Food chain concentration potential: Negative; unlikely to accumulate.
Water pollution: DOT Appendix B, §172.101–marine pollutant. Harmful to aquatic life in very low concentrations. May be dangerous if it enters nearby water intakes; notify operators. Notify local health and wildlife officials. **Response to discharge:** Issue warning–poison. Restrict access. Disperse and flush.

SHIPPING INFORMATION
Stability during transport: Stable.

NAS HAZARD CLASSIFICATION FOR BULK WATER TRANSPORTATION
FIRE: 0
HEALTH: Vapor irritant: 4; Liquid or solid irritant: 4; Poisons: 4
WATER POLLUTION: Human toxicity: 4; Aquatic toxicity: 3; Aesthetic effect: 3
REACTIVITY: Other chemicals: 3; Water: 1; Self-reaction: 1

PHYSICAL AND CHEMICAL PROPERTIES
Physical state @ 59°F/15°C and 1 atm: Solid.
Molecular weight: 105.93
Boiling point @ 1 atm: 144°F/62°C/335°K.
Melting/Freezing point: 120–124°F/49–51°C/322–324°K.
Specific gravity (water = 1): 2.015 @ 68°F/20°C (solid).
Relative vapor density (air = 1): 3.6
Vapor pressure: 98 mm.

CYANOGEN CHLORIDE REC. C:5750

SYNONYMS: CHLORCYAN; CHLORINE CYANIDE; CHLOROCYAN; CHLOROCYANIDE; CHLOROCYANOGEN; CHLORURE de CYANOGENE (French); CICN; CK; CLORURO de CIANOGENO (Spanish); CNCI; RCRA No. P033

IDENTIFICATION
CAS Number: 506-77-4
Formula: CNCl
DOT ID Number: UN 1589; DOT Guide Number: 125
Proper Shipping Name: Cyanogen chloride, inhibited; CK
Reportable Quantity (RQ): **(CERCLA)** 10 lb/4.54 kg

DESCRIPTION: Colorless gas. Shipped and stored as compressed gas; becomes liquid below 56°F/13°C and a solid below 20°F/–7°C. Sharp, pungent odor; choking, lachrymatory. Slightly soluble in water; reacts slowly forming poisonous vapor cloud of hydrogen cyanide.
Note: Has been used as a chemical warfare agent. If used as a weapon, notify U.S. Department of Defense: Army. Damage and/or death may occur before chemical detection can take place. Use M256-A1 Detector Kit (Detection level: 10 mg/m^3) if available.

Poison! • Corrosive • Breathing the vapor, skin or eye contact, or swallowing the material can kill you; converted to cyanide in the body • Firefighting gear (including SCBA) provides NO protection. If exposure occurs, remove and isolate gear immediately and thoroughly decontaminate personnel • Polymerization hazard • Containers may BLEVE when exposed to fire • Contact with liquid may cause frostbite • Vapors are heavier than air and will collect and stay in low areas including basements • Toxic products of combustion may include chlorine gas, cyanide, and nitrogen oxides • Do NOT attempt rescue.

Hazard Classification (based on NFPA-704 Rating System)
Health Hazards (Blue): 4; Flammability (Red): 0; Reactivity (Yellow): 2

EMERGENCY RESPONSE: See Appendix A (125)
Evacuation:
Public safety: Isolate spill area for at least 100 to 200 meters (330 to 660 feet) in all directions.
Spill: Small spill–First: Isolate in all directions 60 meters (200 feet) (weaponized or spill); Then: Protect persons downwind, DAY: 0.6 km (0.4 mile) (weaponized); 0.8 km (0.5 mile) (spill); NIGHT: 1.8 km (1.1 miles) (weaponized); 0.2 km (0.1 mile) (spill). Large spill–First: Isolate in all directions 400 meters (1300 feet); (weaponized); 275 meters (900 feet) (spill); Then: Protect persons downwind, DAY: 4.0 km (2.5 miles) (weaponized); 2.7 km (1.7

miles) (spill); NIGHT: 8.0 km (5.0 miles) (weaponized); 6.8 km 6.8 km (4.2 miles) (spill).
Fire: Isolate for 1600 meters (1 mile) in all directions, especially if tank, rail car, or tank truck is involved in fire.

EXPOSURE
Short-term effects: *SEEK MEDICAL ATTENTION.* **Vapor:** *POISONOUS IF INHALED OR IF SKIN IS EXPOSED.* Irritating to eyes. Move to fresh air. *IF BREATHING HAS STOPPED,* give artificial respiration; *avoid mouth-to-mouth resuscitation; use bag/mask apparatus.* If breathing is difficult, administer oxygen. **Liquid:** *POISONOUS IF SWALLOWED.* Will burn skin and eyes. Remove contaminated clothing and shoes. Flush affected areas with plenty of water. *IF IN EYES,* hold eyelids open and flush with plenty of water. *IF SWALLOWED* and victim is *CONSCIOUS AND ABLE TO SWALLOW,* have victim drink 4 to 8 ounces of water. **Do NOT induce vomiting**.
Note to physician or authorized medical personnel: Consider the use of amyl nitrite perles if symptoms of cyanide poisoning develop. If symptoms indicate, initial treatment includes the cyanide antidote kit. In all cases, break an amyl nitrite perle in a gauze pad and hold lightly under victim's nose for 15 seconds, repeating 5 times at about 15-second intervals; if necessary (and if sodium nitrite infusions will be delayed), repeat procedure every 3 minutes with fresh perles until 3 or 4 have been used. Avoid breathing the vapor while administering it to the victim. Administer sodium nitrite IV, ASAP. The usual adult dose is 10 to 20 mL of a 3% solution infused over no less than 5 minutes; the average child dose is 0.15 to 0.20 mL/kg. Monitor blood pressure during administration, and slow the rate of infusion if hypotention develops. Next, infuse sodium thiosulfate IV. The usual adult dose is 50 mL of a 25% solution infused over 10 to 20 minutes; the average child dose is 1.65 mL/kg. Repeat with nitrite and thiosulfate as required.
Note to physician or authorized medical personnel: Medical observation is recommended for 24 to 48 hours after breathing overexposure, as pulmonary edema may be delayed. As first aid for pulmonary edema, consider administering a corticosteroid spray. Cigarette smoking may exacerbate pulmonary injury and should be discouraged for at least 72 hours following exposure.

HEALTH HAZARDS
Personal protective equipment (PPE): A-Level PPE. OSHA Table Z-1-A air contaminant. Wear thermal protective clothing. If the proper equipment is not available, or if the rescuers have not been trained in its use, call for assistance from the U.S. Soldier and Biological Chemical Command, Edgewood Research Development and Engineering Center (from 0700-1630 EST call 410-671-4411, and from 1630-0700 EST call 410-278-5201; ask for the Staff Duty Officer).
Exposure limits (TWA unless otherwise noted): NIOSH REL ceiling 0.3 mg/m^3 (0.6 ppm).
Short-term exposure limits (15-minute TWA): ACGIH STEL ceiling 0.3 ppm (0.75 mg/m^3).
Long-term health effects: May cause dermatitis, loss of appetite, headache, upper respiratory irritation in humans.
Vapor (gas) irritant characteristics: Vapors cause severe irritation of eyes and throat and can cause eye and lung injury. They cannot be tolerated even at low concentrations.
Liquid or solid irritant characteristics: Severe eye and skin irritant. Causes second-and-third-degree burns on short contact; very injurious to the eyes.
Odor threshold: 1 ppm.

FIRE DATA
Behavior in fire: May explode.

CHEMICAL REACTIVITY
Binary reactants: Violent reaction with alcohols, acids, amines, strong alkalis, olefins, strong oxidizers. Alkaline conditions will convert this chemical to cyanide. Corrodes brass, copper, bronze.
Polymerization: Violent polymerization can be caused by chlorine or moisture. In crude form, trimerizes violently if catalyzed by traces of hydrogen chloride or ammonium chloride.

ENVIRONMENTAL DATA
Food chain concentration potential: Log P_{ow} = 0.72. Unlikely to accumulate.
Water pollution: DOT Appendix B, §172.101–marine pollutant. If used as a weapon, use an M272 Water Detection Kit: Detection limits: 20 mg/L. Harmful to aquatic life in very low concentrations. May be dangerous if it enters nearby water intakes; notify operators. Notify local health and wildlife officials. **Response to discharge:** Issue warning–poison. Restrict access. Evacuate area.

SHIPPING INFORMATION
Storage temperature: Ambient; prolonged storage may cause the formation of polymers. **Stability during transport:** Stable.

PHYSICAL AND CHEMICAL PROPERTIES
Molecular weight: 61.48
Boiling point @ 1 atm: 56°F/13°C/286°K.
Melting/Freezing point: 20°F/–7°C/266°K.
Specific gravity (water = 1): 1.222 at 0°C (liquid).
Liquid surface tension: 24.6 dynes/cm = 0.0246 N/m at 10°C.
Relative vapor density (air = 1): 2.1
Ratio of specific heats of vapor (gas): 1.229
Latent heat of vaporization: 191.3 Btu/lb = 106.3 cal/g = 4.451 x 10^5 J/kg.
Vapor pressure: 1010 mm.

CYCLOHEPTANE   **REC. C:5800**

SYNONYMS: CICLOHEPTANO (Spanish); HEPTAMETHYLENE; SUBERANE

IDENTIFICATION
CAS Number: 291-64-5
Formula: C$_7$H$_{14}$
DOT ID Number: UN 2241; DOT Guide Number: 128
Proper Shipping Name: Cycloheptane

DESCRIPTION: Colorless, oily liquid. Hydrocarbon-like odor. Floats on the surface of water; insoluble. Freezes at 10°F/–12°C.

Highly flammable • Narcotic effects when inhaled • Containers may BLEVE when exposed to fire •Vapors may form explosive mixture with air • Vapors are heavier than air and will collect and stay in low areas • Vapors may travel long distances to ignition sources and flashback • Vapors in confined areas (e.g., tanks, sewers, buildings) may explode when exposed to fire • Irritating to the skin, eyes, and respiratory tract • Toxic products of combustion may include carbon monoxide.

Hazard Classification (based on NFPA-704 Rating System)
Health Hazards (Blue): 0; Flammability (Red): 3; Reactivity (Yellow): 0

EMERGENCY RESPONSE: See Appendix A (128)
Evacuation:
Public safety: Isolate spill area for at least 25 to 50 meters (80 to 160 feet) in all directions.
Spill: Large spill–Consider initial downwind evacuation for at least 300 meters (1000 feet).
Fire: Isolate for 800 meters (½ mile) in all directions, especially if tank, rail car, or tank truck is involved in fire.

EXPOSURE
Short-term effects: *SEEK MEDICAL ATTENTION.* The use of alcoholic beverages may enhance toxic effects. **Vapor:** Irritating to eyes, nose, and throat. *IF INHALED,* will, will cause dizziness, nausea, vomiting, or loss of consciousness. Move to fresh air. *IF BREATHING HAS STOPPED,* give artificial respiration. IF breathing is difficult, administer oxygen. **Liquid:** Irritating to skin and eyes. Harmful if swallowed. Remove contaminated clothing and shoes. Flush affected areas with plenty of water. *IF IN EYES,* hold eyelids open and flush with plenty of water.

HEALTH HAZARDS
Personal protective equipment (PPE): B-Level PPE. Self-contained breathing apparatus, rubber boots and heavy rubber gloves.
Recommendations for respirator selection: *EMERGENCY OR PLANNED ENTRY INTO UNKNOWN CONCENTRATIONS OR IDLH CONDITIONS:* SCBAF:PD,PP (any self-contained breathing apparatus that has a full facepiece and is operated in a pressure-demand or other positive-pressure mode); or SAF:PD,PP:ASCBA (any supplied-air respirator that has a full facepiece and is operated in a pressure-demand or other positive-pressure mode in combination with an auxiliary self-contained breathing apparatus operated in a pressure-demand or other positive pressure mode). *ESCAPE:* GMFOV[any air-purifying, full-facepiece respirator (gas mask) with a chin-style, front- or back-mounted organic vapor cannister]; or SCBAE (any appropriate escape-type, self-contained breathing apparatus).
Note: Substance causes eye irritation or damage; eye protection needed.
Vapor (gas) irritant characteristics: Vapors cause smarting of the eyes or respiratory system if present in high concentrations. The effect is temporary.
Liquid or solid irritant characteristics: May cause eye and skin irritation. If spilled on clothing and allowed to remain, may cause smarting and reddening of skin.

FIRE DATA
Flash point: 43°F/6°C.
Flammable limits in air: LEL: 1.1%; UEL: 6.7%.
Fire extinguishing agents not to be used: Water may be ineffective.
Special hazards of combustion products: Vapor may travel considerable distance to a source of ignition and flashback. Container explosion may occur under fire conditions. Forms explosive mixtures in air.

CHEMICAL REACTIVITY
Reactivity group: 31
Compatibility class: Paraffins

ENVIRONMENTAL DATA
Water pollution: The effect of low concentration on aquatic life is not known. Fouling to shoreline. May be dangerous if it enters nearby water intakes; notify operators. Notify local health and pollution control officials. **Response to discharge:** Issue warning-flammable. Restrict access. Evacuate area. Mechanical containment. Should be removed. Chemical and physical treatment.

SHIPPING INFORMATION
Grades of purity: 98%; **Storage temperature:** Ambient; **Stability during transport:** Stable.

PHYSICAL AND CHEMICAL PROPERTIES
Physical state @ 59°F/15°C and 1 atm: Liquid.
Molecular weight: 98.19
Boiling point @ 1 atm: 245°F/119°C/392°K.
Melting/Freezing point: 10°F/–12°C/261°K.
Specific gravity (water = 1): 0.811
Relative vapor density (air = 1): 3.39
Vapor pressure: (Reid) 0.851 psia.

CYCLOHEXANE   REC. C:5850

SYNONYMS: BENZENE HEXAHYDRIDE; BENZENE, HEXAHYDRO-; CICLOHEXANO (Spanish); CYCLOHEXAN (German); EEC No. 601-017-00-1; HEXAHYDROBENZENE, HEXAMETHYLENE; HEXAHYDROBENZENE; HEXAMETHYLENE; HEXANAPHTHENE; RCRA No. U056

IDENTIFICATION
CAS Number: 110-82-7
Formula: C_6H_{12}
DOT ID Number: UN 1145; DOT Guide Number: 128
Proper Shipping Name: Cyclohexane
Reportable Quantity (RQ): **(CERCLA)** 1000 lb/454 kg

DESCRIPTION: Colorless, watery liquid. Gasoline- or benzene-like odor. Floats on water surface; insoluble. Flammable irritating vapor is produced. Freezes at 44°F/7°C.

Highly flammable • Containers may BLEVE when exposed to fire •Vapors may form explosive mixture with air • Vapors are heavier than air and will collect and stay in low areas • Vapors may travel long distances to ignition sources and flashback • Vapors in confined areas (e.g., tanks, sewers, buildings) may explode when exposed to fire • Irritating to the skin, eyes, and respiratory tract; this material has anesthetic qualities • Toxic products of combustion may include carbon monoxide.

Hazard Classification (based on NFPA-704 Rating System)
Health Hazards (Blue): 1; Flammability (Red): 3; Reactivity (Yellow): 0

EMERGENCY RESPONSE: See Appendix A (128)
Evacuation:
Public safety: Isolate spill area for at least 25 to 50 meters (80 to 160 feet) in all directions.
Spill: Large spill–Consider initial downwind evacuation for at least 300 meters (1000 feet).
Fire: Isolate for 800 meters (½ mile) in all directions, especially if tank, rail car, or tank truck is involved in fire.

EXPOSURE
Short-term effects: *SEEK MEDICAL ATTENTION.* **Vapor:** Irritating to eyes, nose, and throat. *IF INHALED,* will, will cause dizziness, nausea, vomiting, or loss of consciousness. Move to fresh air. *IF BREATHING HAS STOPPED,* give artificial respiration. IF breathing is difficult, administer oxygen. **Liquid:** Irritating to skin and eyes. Harmful if swallowed. Remove contaminated clothing

and shoes. Flush affected areas with plenty of water. *IF IN EYES*, hold eyelids open and flush with plenty of water. *IF SWALLOWED* and victim is *CONSCIOUS AND ABLE TO SWALLOW*, have victim drink 4 to 8 ounces of water.

HEALTH HAZARDS
Personal protective equipment (PPE): B-Level PPE. OSHA Table Z-1-A air contaminant. Chemical protective material(s) reported to have good to excellent resistance: nitrile, Teflon®, Viton®. Also, Chlorinated polyethylene and Viton®/neoprene materials offers limited protection
Recommendations for respirator selection: NIOSH/OSHA *1300 ppm:* SA:CF* (any supplied-air respirator operated in a continuous-flow mode); or PAPROV* [any powered, air-purifying respirator with organic vapor cartridge(s)]; or CCRFOV [any chemical cartridge respirator with a full facepiece and organic vapor cartridge(s)]; or GMFOV [any air-purifying, full-facepiece respirator (gas mask) with a chin-style, front- or back-mounted acid gas canister]; or SCBAF (any self-contained breathing apparatus with a full facepiece; or SAF (any supplied-air respirator with a full facepiece). *EMERGENCY OR PLANNED ENTRY INTO UNKNOWN CONCENTRATIONS OR IDLH CONDITIONS:* SCBAF:PD,PP (any self-contained breathing apparatus that has a full facepiece and is operated in a pressure-demand or other positive-pressure mode); or SAF:PD,PP:ASCBA (any supplied-air respirator that has a full facepiece and is operated in a pressure-demand or other positive-pressure mode in combination with an auxiliary self-contained breathing apparatus operated in a pressure-demand or other positive pressure mode). *ESCAPE:* GMFOV[any air-purifying, full-facepiece respirator (gas mask) with a chin-style, front- or back-mounted organic vapor cannister]; or SCBAE (any appropriate escape-type, self-contained breathing apparatus).
**Note:* Substance causes eye irritation or damage; eye protection needed.
Exposure limits (TWA unless otherwise noted): ACGIH TLV 300 ppm (1030 mg/m^3); NIOSH/OSHA 300 ppm (1050 mg/m^3).
Toxicity by ingestion: Grade 2; LD$_{50}$ = 0.5 to 5 g/kg.
Vapor (gas) irritant characteristics: Vapors cause smarting of the eyes or respiratory system if present in high concentrations. The effect is temporary.
Liquid or solid irritant characteristics: If spilled on clothing and allowed to remain, may cause smarting and reddening of the skin.
Odor threshold: 780 ppm.
IDLH value: 1300 ppm.

FIRE DATA
Flash point: –4°F/–20°C (cc).
Flammable limits in air: LEL: 1.33%; UEL: 8.35%.
Fire extinguishing agents not to be used: Water may be ineffective on fire.
Autoignition temperature: 473°F/245°C/518°K.
Electrical hazard: Class I, Group D. Due to low electric conductivity, this substance may generate electrostatic charges as a result of agitation and flow.
Burning rate: 6.9 mm/min

CHEMICAL REACTIVITY
Binary reactants: Reacts with oxidizers.
Reactivity group: 31
Compatibility class: Paraffins

ENVIRONMENTAL DATA
Food chain concentration potential: Negative; unlikely to accumulate.
Water pollution: Dangerous to aquatic life in high concentrations.
Fouling to shoreline. May be dangerous if it enters nearby water intakes; notify operators. Notify local health and pollution control officials. **Response to discharge:** Issue warning–high flammability. Evacuate area. Disperse and flush.

SHIPPING INFORMATION
Grades of purity: Research grades: 99.5%, 98.0%; commercial: 85–98%; **Storage temperature:** Ambient; **Inert atmosphere:** None; **Venting:** Open (flame arrester) or pressure-vacuum; **Stability during transport:** Stable.

NAS HAZARD CLASSIFICATION FOR BULK WATER TRANSPORTATION
FIRE: 3
HEALTH: Vapor irritant: 1; Liquid or solid irritant: 1; Poisons: 2
WATER POLLUTION: Human toxicity: 1; Aquatic toxicity: 2; Aesthetic effect: 2
REACTIVITY: Other chemicals: 0; Water: 0; Self-reaction: 0

PHYSICAL AND CHEMICAL PROPERTIES
Physical state @ 59°F/15°C and 1 atm: Liquid.
Molecular weight: 84.16
Boiling point @ 1 atm: 177°F/81°C/354°K.
Melting/Freezing point: 44°F/7°C/280°K.
Critical temperature: 537°F/281°C/554°K.
Critical pressure: 591 psia = 40.2 atm = 4.07 MN/m^2.
Specific gravity (water = 1): 0.779 @ 68°F/20°C (liquid).
Liquid surface tension: 24.6 dynes/cm = 0.0246 N/m @ 68°F/20°C.
Liquid water interfacial tension: 50 dynes/cm = 0.050 N/m @ 77°F/25°C.
Relative vapor density (air = 1): 2.9
Ratio of specific heats of vapor (gas): 1.087
Latent heat of vaporization: 150 Btu/lb = 85 cal/g = 3.6 x 10^5 J/kg.
Heat of combustion: –18,684 Btu/lb = –10,380 cal/g = –434.59 x 10^5 J/kg.
Heat of fusion: 7.47 cal/g.
Vapor pressure: (Reid) 3.3 psia; 78 mm.

CYCLOHEXANOL   REC. C:5900

SYNONYMS: ADRONAL; ANOL; CICLOHEXANOL (Spanish); 1-CYCLOHEXANOL; CYCLOHEXYL ALCOHOL; CYCLOHEXENE; EEC No. 603-009-00-3; HEXAHYDROPHENOL; HEXALIN; HYDRALIN; HYDROPHENOL; HYDROXYCYCLOHEXANE; NAXOL; PHENOL, HEXAHYDRO-

IDENTIFICATION
CAS Number: 108-93-0
Formula: $C_6H_{12}O$; $(CH_2)_5CHOH$
DOT ID Number: NA 1993; DOT Guide Number: 128
Proper Shipping Name: Combustible liquid, n.o.s.

DESCRIPTION: Sticky solid or colorless to pale yellow oily liquid above 75°F/24°C. Alcohol odor; like camphor. Floats and mixes slowly with water.

Combustible • Containers may BLEVE when exposed to fire • Vapors may form explosive mixture with air • Vapors are heavier than air and will collect and stay in low areas • Vapors may travel long distances to ignition sources and flashback • Vapors in confined areas (e.g., tanks, sewers, buildings) may explode when exposed to fire • Irritating to the skin, eyes, and respiratory tract • Toxic products of combustion may include carbon monoxide.

Hazard Classification (based on NFPA-704 Rating System)
Health Hazards (Blue): 1; Flammability (Red): 2; Reactivity (Yellow): 0

EMERGENCY RESPONSE: See Appendix A (128)
Evacuation:
Public safety: Isolate spill area for at least 25 to 50 meters (80 to 160 feet) in all directions.
Spill: Large spill–Consider initial downwind evacuation for at least 300 meters (1000 feet).
Fire: Isolate for 800 meters (½ mile) in all directions, especially if tank, rail car, or tank truck is involved in fire.

EXPOSURE
Short-term effects: *SEEK MEDICAL ATTENTION.* **Liquid or solid:** *ABSORBED THROUGH THE SKIN.* Will burn skin and eyes. Harmful if swallowed or absorbed through the skin. Remove contaminated clothing and shoes. Flush affected areas with plenty of water. *IF IN EYES*, hold eyelids open and flush with plenty of water. *IF SWALLOWED* and victim is *CONSCIOUS AND ABLE TO SWALLOW*, have victim drink 4 to 8 ounces of water.

HEALTH HAZARDS
Personal protective equipment (PPE): B-Level PPE. OSHA Table Z-1-A air contaminant. Organic vapor respirator. Chemical protective material(s) reported to have good to excellent resistance: nitrile, PV alcohol, Viton®, PE, PVC, Silvershield®. Also, chlorinated polyethylene, Viton/neoprene, butryl/neoprene offers limited protection
Recommendations for respirator selection: NIOSH/OSHA
400 ppm: CCROV* [any chemical cartridge respirator with organic vapor cartridge(s)]; or PAPROV* [any powered, air-purifying respirator with organic vapor cartridge(s)]; or GMFOV[any air-purifying, full-facepiece respirator (gas mask) with a chin-style, front- or back-mounted organic vapor cannister]; or SA* (any supplied-air respirator); or SCBAF (any self-contained breathing apparatus with a full facepiece). *EMERGENCY OR PLANNED ENTRY INTO UNKNOWN CONCENTRATIONS OR IDLH CONDITIONS:* SCBAF:PD,PP (any self-contained breathing apparatus that has a full facepiece and is operated in a pressure-demand or other positive-pressure mode); or SAF:PD,PP:ASCBA (any supplied-air respirator that has a full facepiece and is operated in a pressure-demand or other positive-pressure mode in combination with an auxiliary self-contained breathing apparatus operated in a pressure-demand or other positive pressure mode). *ESCAPE:* GMFOV [any air-purifying, full-facepiece respirator (gas mask) with a chin-style, front- or back-mounted organic vapor canister]; or SCBAE (any appropriate escape-type, self-contained breathing apparatus).
Note: Substance reported to cause eye irritation or damage; may require eye protection.
Exposure limits (TWA unless otherwise noted): ACGIH TLV 50 ppm (206 mg/m^3); NIOSH/OSHA 50 ppm (200 mg/m^3; NIOSH REL 50 ppm (200 mg/m^3); skin contact contributes significantly in overall exposure.
Toxicity by ingestion: Grade 2; LD_{50} = 0.5 to 5 g/kg.
Long-term health effects: Liver, lung and kidney damage.
Vapor (gas) irritant characteristics: Eye and respiratory tract irritant. Vapors cause smarting of the eyes or respiratory system if present in high concentrations. The effect is temporary.
Liquid or solid irritant characteristics: Eye and skin irritant.

Causes smarting of the skin and first-degree burns on short exposure and may cause secondary burns on long exposure.
Odor threshold: 0.16 ppm.
IDLH value: 400 ppm.

FIRE DATA
Flash point: 160°F/71°C (oc); 154°F/68°C (cc)
Flammable limits in air: LEL: 1.21% (estimate).
Autoignition temperature: 572°F/300°C/573°K.
Electrical hazard: Class I, Group D.
Burning rate: 3.9 mm/min

CHEMICAL REACTIVITY
Binary reactants: Reacts with strong oxidizers (such as nitric acid, hydrogen peroxide).
Reactivity group: 20
Compatibility class: Alcohols, glycols

ENVIRONMENTAL DATA
Food chain concentration potential: Log P_{ow} = 1.23. Unlikely to accumulate.
Water pollution: Effect of low concentrations on aquatic life is unknown. May be dangerous if it enters nearby water intakes; notify operators. Notify local health and pollution control officials.
Response to discharge: Disperse and flush.

SHIPPING INFORMATION
Grades of purity: Technical; pure; **Stability during transport:** Stable.

NAS HAZARD CLASSIFICATION FOR BULK WATER TRANSPORTATION
FIRE: 1
HEALTH: Vapor irritant: 1; Liquid or solid irritant: 2; Poisons: 1
WATER POLLUTION: Human toxicity: 1; Aquatic toxicity: 3; Aesthetic effect: 2
REACTIVITY: Other chemicals: 2; Water: 0; Self-reaction: 0

PHYSICAL AND CHEMICAL PROPERTIES
Physical state @ 59°F/15°C and 1 atm: Solid.
Molecular weight: 100.16
Boiling point @ 1 atm: 322°F/161°C/434°K.
Melting/Freezing point: 75°F/24°C/297°K.
Critical temperature: 666°F/352°C/625°K.
Critical pressure: 540 psia = 37 atm = 3.7 MN/m^2.
Specific gravity (water = 1): 0.947 @ 68°F/20°C (liquid).
Liquid surface tension: 34.2 dynes/cm = 0.0342 N/m at 16.2°C.
Liquid water interfacial tension: 3.9 dynes/cm = 0.0039 N/m @ 77°F/25°C.
Relative vapor density (air = 1): 3.4
Ratio of specific heats of vapor (gas): 1.071
Latent heat of vaporization: 196 Btu/lb = 109 cal/g = 4.56 x 10^5 J/kg.
Heat of combustion: –16,000 Btu/lb = –8910 cal/g = –373 x 10^5 J/kg.
Heat of fusion: 4.19 cal/g.
Vapor pressure: (Reid) 0.1 psia; 1 mm.

CYCLOHEXANONE REC. C:5950

SYNONYMS: ANOL; ANONE; CICLOHEXANONA (Spanish); CYCLOHEXENE; CYCLOHEXYL ALCOHOL; CYCLOHEXYL KETONE; EEC No. 606-010-00-7; HEXAHYDROPHENOL; HYDROOXYCYCLOHEXANE; HEXALIN; HEXANON;

HYDRALIN; HYTROL O; KETOHEXAMETHYLENE; NADONE; PIMELIC KETONE; RCRA No. U057; SEXTONE

IDENTIFICATION
CAS Number: 108-94-1
Formula: $C_6H_{10}O$; $(CH_2)_5CO$
DOT ID Number: UN 1915; DOT Guide Number: 127
Proper Shipping Name: Cyclohexanone
Reportable Quantity (RQ): **(CERCLA)** 5000 lb/2270 kg

DESCRIPTION: Colorless to pale yellow watery liquid. Sweet, peppermint odor or acetone-like odor. Floats and mixes slowly with water.

Flammable • May be able to form unstable peroxides after prolonged storage in air; peroxides may be detonated by heating, impact, or friction. • Containers may BLEVE when exposed to fire • Vapors may form explosive mixture with air • Vapors are heavier than air and will collect and stay in low areas • Vapors may travel long distances to ignition sources and flashback • Vapors in confined areas (e.g., tanks, sewers, buildings) may explode when exposed to fire • Irritating to the skin, eyes, and respiratory tract • Toxic products of combustion may include carbon monoxide.

Hazard Classification (based on NFPA-704 Rating System)
Health Hazards (Blue): 1; Flammability (Red): 2; Reactivity (Yellow): 0

EMERGENCY RESPONSE: See Appendix A (127)
Evacuation:
Public safety: Isolate spill area for at least 25 to 50 meters (80 to 160 feet) in all directions.
Spill: Large spill–Consider initial downwind evacuation for at least 300 meters (1000 feet).
Fire: Isolate for 800 meters (½ mile) in all directions, especially if tank, rail car, or tank truck is involved in fire.

EXPOSURE
Short-term effects: *SEEK MEDICAL ATTENTION.* **Liquid:** *HARMFUL IF SWALLOWED OR ABSORBED THROUGH THE SKIN.* Will burn skin and eyes. Harmful if swallowed. Remove contaminated clothing and shoes. Flush affected areas with plenty of water. *IF IN EYES*, hold eyelids open and flush with plenty of water. *IF SWALLOWED* and victim is *CONSCIOUS AND ABLE TO SWALLOW*, have victim drink 4 to 8 ounces of water. The use of alcoholic beverages may enhance toxic effects.

HEALTH HAZARDS
Personal protective equipment (PPE): B-Level PPE. OSHA Table Z-1-A air contaminant. Organic vapor respirator. Chemical protective material(s) reported to have good to excellent resistance: butyl rubber, PV alcohol, Teflon®, Silvershield®. Also, natural rubber, neoprene, and nitrile rubber may offer some protection from alcohols.
Recommendations for respirator selection: NIOSH
625 ppm: SA:CF* (any supplied-air respirator operated in a continuous-flow mode); or PAPROV* [any powered, air-purifying respirator with organic vapor cartridge(s)]. *700 ppm*: CCRFOV [any chemical cartridge respirator with a full facepiece and organic vapor cartridge(s)]; or GMFOV[any air-purifying, full-facepiece respirator (gas mask) with a chin-style, front- or back-mounted organic vapor cannister]; or PAPRTOV* [any powered, air-purifying respirator with a tight-fitting facepiece and organic vapor cartridge(s)]; or SCBAF (any self-contained breathing apparatus with a full facepiece); or SAF (any supplied-air respirator with a full facepiece). *EMERGENCY OR PLANNED ENTRY INTO UNKNOWN CONCENTRATIONS OR IDLH CONDITIONS:* SCBAF:PD,PP (any self-contained breathing apparatus that has a full facepiece and is operated in a pressure-demand or other positive-pressure mode); or SAF:PD,PP:ASCBA (any supplied-air respirator that has a full facepiece and is operated in a pressure-demand or other positive-pressure mode in combination with an auxiliary self-contained breathing apparatus operated in a pressure-demand or other positive pressure mode). *ESCAPE:* GMFOV [any air-purifying, full-facepiece respirator (gas mask) with a chin-style, front- or back-mounted organic vapor canister]; or SCBAE (any appropriate escape-type, self-contained breathing apparatus). **Note*: Substance causes eye irritation or damage; eye protection needed.
Exposure limits (TWA unless otherwise noted): ACGIH TLV 25 ppm (100 mg/m³); OSHA PEL 50 ppm (200 mg/m³); NIOSH REL 25 ppm (100 mg/m³); skin contact contributes significantly in overall exposure.
Toxicity by ingestion: Grade 2; LD_{50} = 0.5 to 5 g/kg.
Long-term health effects: IARC rating 3 human carcinogenicity not classifiable. May affect the nervous system.
Vapor (gas) irritant characteristics: Eye and respiratory tract irritant. Vapor is moderately irritating such that personnel will not usually tolerate moderate or high vapor concentrations.
Liquid or solid irritant characteristics: Eye and skin irritant. Causes smarting of the skin and first-degree burns on short exposure and may cause secondary burns on long exposure.
Odor threshold: 0.12 ppm.
IDLH value: 700 ppm.

FIRE DATA
Flash point: 129°F/54°C (oc); 111°F/44°C (cc).
Flammable limits in air: LEL: 1.1% @ 212°F; UEL: 9.4%.
Autoignition temperature: 788°F/420°C/693°K.
Electrical hazard: Due to low electric conductivity, this substance may generate electrostatic charges as a result of agitation and flow.
Burning rate: 4.2 mm/min

CHEMICAL REACTIVITY
Binary reactants: Oxidizers; nitric acid. Attacks copper and its alloys; some plastics.
Reactivity group: 18
Compatibility class: Ketone
Polymerization: Unstable peroxides may accumulate after prolonged storage in presence of air.

ENVIRONMENTAL DATA
Food chain concentration potential: Log P_{ow} = 0.83. Unlikely to accumulate.
Water pollution: Effect of low concentrations on aquatic life is unknown. May be dangerous if it enters nearby water intakes; notify operators. Notify local health and pollution control officials.
Response to discharge: Disperse and flush.

SHIPPING INFORMATION
Grades of purity: Technical: 99.87%; **Stability during transport:** Stable.

NAS HAZARD CLASSIFICATION FOR BULK WATER TRANSPORTATION
FIRE: 2
HEALTH: Vapor irritant: 3; Liquid or solid irritant: 2; Poisons: 1
WATER POLLUTION: Human toxicity: 1; Aquatic toxicity: 3; Aesthetic effect: 2
REACTIVITY: Other chemicals: 2; Water: 0; Self-reaction: 0

306   Cyclohexenyl trichlorosilane

PHYSICAL AND CHEMICAL PROPERTIES
Physical state @ 59°F/15°C and 1 atm: Liquid.
Molecular weight: 98.15
Boiling point @ 1 atm: 312.4°F/155.8°C/429.0°K.
Melting/Freezing point: –49°F/–45°C/228°K.
Critical temperature: 673°F/356°C/629°K.
Critical pressure: 560 psia = 38 atm = 3.8 MN/m^2.
Specific gravity (water = 1): 0.945 @ 68°F/20°C (liquid).
Liquid surface tension: 34 dynes/cm = 0.034 N/m @ 68°F/20°C.
Liquid water interfacial tension: 90 dynes/cm = 0.090 N/m at 22.7°C.
Relative vapor density (air = 1): 3.5
Ratio of specific heats of vapor (gas): 1.084
Latent heat of vaporization: 160 Btu/lb = 91 cal/g = 3.8 x 10^5 J/kg.
Heat of combustion: –15,430 Btu/lb = –8570 cal/g = –358.8 x 10^5 J/kg.
Vapor pressure: (Reid) 0.8 psia; 5 mm.

CYCLOHEXANONE PEROXIDE   **REC. C:6000**

SYNONYMS: CADOX HDP; DICYCLOHEXANONE DIPEROXIDE; 1-HYDROPEROXYCYCLOHEXYL 1-HYDROXYCYCLOHEXYL PEROXIDE; LUPERCO; JDB-50-T; PEROXIDO de CICLOHEXANONA (Spanish)

IDENTIFICATION
CAS Number: 78-18-2
Formula: $C_{12}H_{22}O_5$; $C_6H_{10}(OOH)\cdot OO\cdot C_6H_{10}OH$ in dibutyl phthalate
DOT ID Number: UN 2119; DOT Guide Number: 147
Proper Shipping Name: Cyclohexanone peroxide, not more than 90%, with not less than 10% water.

DESCRIPTION: Colorless or white, thick liquid or paste. Odorless. Sinks in water.

Flammable • Heat, shock, or contamination may cause explosion • Containers may BLEVE when exposed to fire •Vapors may form explosive mixture with air • Vapors are heavier than air and will collect and stay in low areas • Vapors may travel long distances to ignition sources and flashback • Containers may BLEVE when exposed to fire • Vapors in confined areas (e.g., tanks, sewers, buildings) may explode when exposed to fire • Irritating to the skin, eyes, and respiratory tract • Toxic products of combustion may include carbon monoxide.

Hazard Classification (based on NFPA-704 Rating System)
Health Hazards (Blue): 1; Flammability (Red): 2; Reactivity (Yellow): 2; Special Notice: OXY

EMERGENCY RESPONSE: See Appendix A (147)
Evacuation:
Public safety: Isolate the area of spill or leak for at least 25 to 50 meters (80 to 160 feet) in all directions.
Spill: Consider initial evacuation for at least 250 meters (800 feet).
Fire: If tank, rail car, or tank truck is involved in fire, isolate for at least 800 meters (½ mile) in all directions; also, consider initial evacuation for 800 meters (½ mile) in all directions.

EXPOSURE
Short-term effects: *SEEK MEDICAL ATTENTION.* **Liquid:** Irritating to skin and eyes. Harmful if swallowed. Remove contaminated clothing and shoes. Flush affected areas with plenty of water. *IF IN EYES*, hold eyelids open and flush with plenty of water. *IF SWALLOWED* and victim is *CONSCIOUS AND ABLE TO SWALLOW*, have victim drink 4 to 8 ounces of water.

HEALTH HAZARDS
Personal protective equipment (PPE): B-Level PPE. Goggles or face shield; rubber gloves; protective clothing. Butyl rubber is generally suitable for peroxide compounds.
Toxicity by ingestion: Grade 2; LD$_{50}$ = 0.5 to 5 g/kg.
Liquid or solid irritant characteristics: Eye and skin irritant.

FIRE DATA
Flash point: 315°F/157°C (cc) (dibutyl phthalate).
Behavior in fire: May explode.
Autoignition temperature: 757°F/403°C/676°K (dibutyl phthalate).

CHEMICAL REACTIVITY
Binary reactants: Reacts with reducing agents, caustics, ammonia, combustible materials.

ENVIRONMENTAL DATA
Food chain concentration potential: Negative; unlikely to accumulate.
Water pollution: Effect of low concentrations on aquatic life is unknown. Fouling to shoreline. May be dangerous if it enters nearby water intakes; notify operators. Notify local health and wildlife officials. **Response to discharge:** Issue warning–oxidizing material, water contaminant. Mechanical containment. Should be removed. Chemical and physical treatment.

SHIPPING INFORMATION
Grades of purity: 45% paste with dibutyl phthalate; **Storage temperature:** Cool ambient; **Inert atmosphere:** None; **Venting:** Open (flame arrester); **Stability during transport:** Stable.

PHYSICAL AND CHEMICAL PROPERTIES
Physical state @ 59°F/15°C and 1 atm: Solid or Liquid.
Boiling point @ 1 atm: (decomposes).
Specific gravity (water = 1): 1.05 @ 68°F/20°C (liquid).
Liquid surface tension: (estimate) 30 dynes/cm = 0.030 N/m @ 68°F/20°C.
Liquid water interfacial tension: (estimate) 35 dynes/cm = 0.035 N/m @ 68°F/20°C.
Heat of combustion: (estimate) –14,000 Btu/lb = –7900 cal/g = –330 x 10^5 J/kg.

CYCLOHEXENYL TRICHLOROSILANE   **REC. C:6050**

SYNONYMS: CICLOHEXENILTRICLOROSILANO (Spanish); CYCLOHEXENYLTRICHLOROSILANE

IDENTIFICATION
CAS Number: 10137-69-6
Formula: $C_6H_9SiCl_3$
DOT ID Number: UN 1762; DOT Guide Number: 156
Proper Shipping Name: Cyclohexenyltrichlorosilane

DESCRIPTION: Colorless liquid fumes in moist air emitting hydrogen chloride. Sharp, irritating odor; like hydrochloric acid. Reacts with water releasing hydrogen chloride vapor.

Poison! • Combustible • Corrosive • Do not use water • Breathing the vapor can kill you • Firefighting gear (including SCBA) does not provide adequate protection. If exposure occurs, remove and isolate gear immediately and thoroughly decontaminate personnel • Containers may BLEVE when exposed to fire •Vapors may form explosive mixture with air •Vapors are heavier than air and will collect and stay in low areas • Vapors in confined areas (e.g., tanks, sewers, buildings) may explode when exposed to fire • Severely irritating to skin, eyes, and respiratory tract; prolonged contact with the skin causes burns • Toxic products of combustion may include hydrogen chloride and phosgene • Do not put yourself in danger by entering a contaminated area to rescue a victim.

Hazard Classification (based on NFPA-704 Rating System)
Health Hazards (Blue): 2; Flammability (Red): 2; Reactivity (Yellow): 1; Special Notice (White): Water reactive

EMERGENCY RESPONSE: See Appendix A (156)
Evacuation:
Public safety: Isolate the area of spill or leak for at least 50 to 100 meters (160 to 330 feet) in all directions.
Spill: Increase, in the downwind direction, as necessary, the distance shown under "Public Safety."
Fire: If tank, rail car, or tank truck is involved in fire, isolate for at least 800 meters (½ mile) in all directions; also, consider initial evacuation for 800 meters (½ mile) in all directions.

EXPOSURE
Short-term effects: *SEEK MEDICAL ATTENTION.* **Gas produced in reaction with water:** *POISONOUS IF INHALED.* Irritating to eyes, nose, and throat. Move victim to fresh air. *IF BREATHING HAS STOPPED,* give artificial respiration; *avoid mouth-to-mouth resuscitation; use bag/mask apparatus.* IF breathing is difficult, administer oxygen. **Liquid:** Will burn skin and eyes. Harmful if swallowed. Remove contaminated clothing and shoes. Flush affected areas with plenty of water. *IF IN EYES,* hold eyelids open and flush with plenty of water. *IF SWALLOWED* and victim is *CONSCIOUS AND ABLE TO SWALLOW,* have victim drink 4 to 8 ounces of water. **Do NOT induce vomiting.**

HEALTH HAZARDS
Personal protective equipment (PPE): B-Level PPE.
Recommendations for respirator selection: *EMERGENCY OR PLANNED ENTRY INTO UNKNOWN CONCENTRATIONS OR IDLH CONDITIONS:* SCBAF:PD,PP (any self-contained breathing apparatus that has a full facepiece and is operated in a pressure-demand or other positive-pressure mode); or SAF:PD,PP:ASCBA (any supplied-air respirator that has a full facepiece and is operated in a pressure-demand or other positive-pressure mode in combination with an auxiliary self-contained breathing apparatus operated in a pressure-demand or other positive pressure mode). *ESCAPE:* GMFAG [any air-purifying, full-facepiece respirator (gas mask) with a chin-style, front- or back-mounted organic vapor cannister]; or SCBAE (any appropriate escape-type, self-contained breathing apparatus). **Note:* Substance reported to cause eye irritation or damage; may require eye protection.
Exposure limits (TWA unless otherwise noted): ACGIH TLV ceiling 5 ppm; NIOSH/OSHA ceiling 5 ppm (7 mg/m^3) as hydrogen chloride
Toxicity by ingestion: Grade 2; oral LD$_{50}$ = 2,830 mg/kg (rat).
Vapor (gas) irritant characteristics: Eye and respiratory tract irritant.
Liquid or solid irritant characteristics: Severe eye and skin irritant; causes burns.
IDLH value: 50 ppm as hydrogen chloride.

FIRE DATA
Flash point: 200°F/93°C (oc).
Fire extinguishing agents not to be used: Water or foam
Behavior in fire: Difficult to extinguish. Re-ignition may occur. Water applied to adjacent fires will produce hydrogen chloride upon contact with this material.

CHEMICAL REACTIVITY
Reactivity with water: Reacts to generate hydrogen chloride (hydrochloric acid).
Binary reactants: Corrodes metals by reacting with surface moisture and generating hydrogen chloride. Reacts with strong bases.
Neutralizing agents for acids and caustics: Flush with water, rinse with sodium bicarbonate or lime solution.

ENVIRONMENTAL DATA
Food chain concentration potential: Negative; unlikely to accumulate.
Water pollution: Effect of low concentrations on aquatic life is unknown. May be dangerous if it enters nearby water intakes; notify operators. Notify local health and wildlife officials. **Response to discharge:** Issue warning–corrosive, water contaminant. Restrict access. Disperse and flush with care.

SHIPPING INFORMATION
Grades of purity: Commercial; **Storage temperature:** Ambient; **Inert atmosphere:** None; **Stability during transport:** Stable.

PHYSICAL AND CHEMICAL PROPERTIES
Physical state @ 59°F/15°C and 1 atm: Liquid.
Molecular weight: 215.6
Boiling point @ 1 atm: 396°F/202°C/475°K.
Melting/Freezing point: (estimate) less than 77°F/25°C/248°K.
Specific gravity (water = 1): 1.23 @ 68°F/20°C (liquid).
Liquid surface tension: (estimate) 30 dynes/cm = 0.030 N/m @ 68°F/20°C.
Heat of combustion: (estimate) –78 Btu/lb = –43 cal/g = –1.8 x 10^5 J/kg.

CYCLOHEXYLAMINE   REC. C:6100

SYNONYMS: AMINOCYCLOHEXANE; AMINOHEXAHYDROBENZENE; ANILINE, HEXAHYDRO-; CHA; CYCLOHEXANAMINE; HEXAHYDROANILINE; HEXAHYDROBENZENAMINE; EEC No. 612-050-00-6

IDENTIFICATION
CAS Number: 108-91-8
Formula: (CH$_2$)$_5$CHNH$_2$
DOT ID Number: UN 2357; DOT Guide Number: 132
Proper Shipping Name: Cyclohexylamine
Reportable Quantity (RQ): **(EHS)** 1 lb/0.454 kg

DESCRIPTION: Colorless or yellow liquid. Strong, fishy odor. Floats on water surface; soluble.

Highly flammable • Extremely irritating to skin, eyes, and respiratory tract; prolonged contact can cause burns • Firefighting gear (including SCBA) does not provide adequate protection. If exposure occurs, remove and isolate gear immediately and thoroughly decontaminate personnel • Containers may BLEVE when exposed to fire •Vapors may form explosive mixture with air • Vapors are heavier than air and will collect and stay in low areas

- Vapors may travel long distances to ignition sources and flashback • Containers may BLEVE when exposed to fire • Vapors in confined areas (e.g., tanks, sewers, buildings) may explode when exposed to fire • Toxic products of combustion may include nitrogen oxide and carbon monoxide.

Hazard Classification (based on NFPA-704 Rating System)
Health Hazards (Blue): 3; Flammability (Red): 3; Reactivity (Yellow): 0

EMERGENCY RESPONSE: See Appendix A (132)
Evacuation:
Public safety: Isolate spill area for at least 50 to 100 meters (160 to 330 feet) in all directions.
Spill: Increase, as necessary, the isolation distance shown above, in "Public safety."
Fire: Isolate for 800 meters (½ mile) in all directions, especially if tank, rail car, or tank truck is involved in fire.

EXPOSURE
Short-term effects: *SEEK MEDICAL ATTENTION*. **Liquid:** Will burn skin and eyes. Harmful if swallowed. Remove contaminated clothing and shoes. Flush affected areas with plenty of water. *IF IN EYES*, hold eyelids open and flush with plenty of water. *IF SWALLOWED* and victim is *CONSCIOUS AND ABLE TO SWALLOW*, have victim drink 4 to 8 ounces of water. **Do NOT induce vomiting.**

HEALTH HAZARDS
Personal protective equipment (PPE): B-Level PPE. OSHA Table Z-1-A air contaminant. Rubber gloves, chemical goggles, approved Bureau of Mines or NIOSH respirator for organic vapors.
Exposure limits (TWA unless otherwise noted): ACGIH TLV 10 ppm (41 mg/m^3); NIOSH REL 10 ppm (40 mg/m^3).
Toxicity by ingestion: Grade 3; LD$_{50}$ = 50 to 500 mg/kg.
Long-term health effects: Produced cancer of the bladder in the rat.
Vapor (gas) irritant characteristics: Eye and respiratory tract irritant. Vapor is irritating such that personnel will not usually tolerate moderate or high vapor concentration.
Liquid or solid irritant characteristics: Severe eye and skin irritant. Causes second- and third-degree burns on short contact; very injurious to the eyes.
Odor threshold: 2.6 ppm.

FIRE DATA
Flash point: 88°F/31°C (cc).
Flammable limits in air: LEL: 1.5%; UEL: 9.4%.
Fire extinguishing agents not to be used: Water may be ineffective and may create an environmental hazard.
Autoignition temperature: 560°F/293°C/566°K.
Electrical hazard: Class I, Group D.
Burning rate: 5.0 mm/min

CHEMICAL REACTIVITY
Binary reactants: Reacts with oxidizers, organic compounds, acid anhydrides, acid chlorides, strong acids, halogenated compounds, lead. Corrosive to copper, aluminum, zinc, their alloys, and galvanized steel.
Neutralizing agents for acids and caustics: Flush with water.
Reactivity group: 7
Compatibility class: Aliphatic amines

ENVIRONMENTAL DATA
Food chain concentration potential: Log P$_{ow}$ = 1.5.

Water pollution: Effect of low concentrations on aquatic life is unknown. May be dangerous if it enters nearby water intakes; notify operators. Notify local health and pollution control officials.
Response to discharge: Disperse and flush.

SHIPPING INFORMATION
Grades of purity: Commercial; **Storage temperature:** Ambient; **Inert atmosphere:** None; **Venting:** None; **Stability during transport:** Stable

PHYSICAL AND CHEMICAL PROPERTIES
Physical state @ 59°F/15°C and 1 atm: Liquid.
Molecular weight: 99.18
Boiling point @ 1 atm: 274.1°F/134.5°C/407.7°K.
Melting/Freezing point: 0.1°F/–17.7°C/255.5°K.
Critical temperature: 648°F/342°C/615°K.
Specific gravity (water = 1): 0.865 @ 68°F/20°C (liquid).
Relative vapor density (air = 1): 3.4
Latent heat of vaporization: 158 Btu/lb = 87.6 cal/g = 3.67 x 10^5 J/kg.
Heat of combustion: (estimate) –18,000 Btu/lb = –10,000 cal/g = –420 x 10^5 J/kg.
Heat of solution: (estimate) –4 Btu/lb = –2 cal/g = –0.1 x 10^5 J/kg.
Vapor pressure: 11 mm.

CYCLOHEXYL ACETATE   REC. C:6150

SYNONYMS: ACETATO de CICLOHEXANILO (Spanish); CYCLOHEXANYL ACETATE; ACETIC ACID, CYCLOHEXYL ESTER

IDENTIFICATION
CAS Number: 622-45-7
Formula: C$_8$H$_{14}$O$_2$; CH$_3$CO$_2$C$_6$H$_{11}$
DOT ID Number: UN 2243; DOT Guide Number: 128
Proper Shipping Name: Cyclohexyl acetate

DESCRIPTION: Colorless liquid.

Flammable • Containers may BLEVE when exposed to fire • Vapors may form explosive mixture with air • Vapors are heavier than air and will collect and stay in low areas • Vapors may travel long distances to ignition sources and flashback • Vapors in confined areas (e.g., tanks, sewers, buildings) may explode when exposed to fire • Irritating to the skin, eyes, and respiratory tract • Toxic products of combustion may include carbon monoxide.

Hazard Classification (based on NFPA-704 Rating System)
Health Hazards (Blue): 1; Flammability (Red): 2; Reactivity (Yellow): 0

EMERGENCY RESPONSE: See Appendix A (128)
Evacuation:
Public safety: Isolate spill area for at least 25 to 50 meters (80 to 160 feet) in all directions.
Spill: Large spill–Consider initial downwind evacuation for at least 300 meters (1000 feet).
Fire: Isolate for 800 meters (½ mile) in all directions, especially if tank, rail car, or tank truck is involved in fire.

EXPOSURE
Short-term effects: *SEEK MEDICAL ATTENTION*. **Vapor:** May cause irritation. Move to fresh air. If victim is not breathing, give artificial respiration. IF breathing is difficult, administer oxygen.

Liquid: Irritating to skin and eyes. Remove contaminated clothing and shoes. Flush affected areas with plenty of water. *IF IN EYES*, hold eyelids open and flush with plenty of water.

HEALTH HAZARDS
Personal protective equipment (PPE): Self-contained breathing apparatus, rubber boots, and heavy rubber gloves.
Toxicity by ingestion: Grade 1: LD_{50} = 6.73 g/kg (rat).
Vapor (gas) irritant characteristics: Vapors cause smarting of the eyes or respiratory system if present in high concentrations. The effect is temporary.
Liquid or solid irritant characteristics: Eye and skin irritant. If spilled on clothing and allowed to remain, may cause smarting and reddening of skin.

FIRE DATA
Flash point: 136°F/58°C (cc).

CHEMICAL REACTIVITY
Reactivity group: 34
Compatibility class: Esters.

ENVIRONMENTAL DATA
Water pollution: Effect of low concentrations on aquatic life is unknown. Fouling to shoreline. May be dangerous if it enters nearby water intakes; notify operators. Notify local health and wildlife officials. **Response to discharge:** Restrict access. Evacuate area. Should be removed. Mechanical containment. Chemical and physical treatment.

SHIPPING INFORMATION
Storage temperature: Ambient.

PHYSICAL AND CHEMICAL PROPERTIES
Physical state @ 59°F/15°C and 1 atm: Liquid.
Molecular weight: 142.2
Boiling point @ 1 atm: 341.6–343.4°F/172–173°C/445.2–446.2°K.
Specific gravity (water = 1): 0.966
Relative vapor density (air = 1): 4.9

CYCLOPENTANE   REC. C:6200

SYNONYMS: CICLOPENTANO (Spanish); EEC No. 601-030-00-2; PENTAMETHYLENE

IDENTIFICATION
CAS Number: 287-92-3
Formula: C_5H_{10}
DOT ID Number: UN 1146; DOT Guide Number: 128
Proper Shipping Name: Cyclopentane

DESCRIPTION: Colorless watery liquid. Mild, sweet odor; like gasoline. Floats on water surface; soluble. Flammable, irritating vapor is produced.

Highly flammable • Containers may BLEVE when exposed to fire • Vapors may form explosive mixture with air • Vapors are heavier than air and will collect and stay in low areas • Vapors may travel long distances to ignition sources and flashback • Vapors in confined areas (e.g., tanks, sewers, buildings) may explode when exposed to fire • Irritating to the skin, eyes, and respiratory tract • Toxic products of combustion may include carbon monoxide.

Hazard Classification (based on NFPA-704 Rating System)
Health Hazards (Blue): 1; Flammability (Red): 3; Reactivity (Yellow): 0

EMERGENCY RESPONSE: See Appendix A (128)
Evacuation:
Public safety: Isolate spill area for at least 25 to 50 meters (80 to 160 feet) in all directions.
Spill: Large spill–Consider initial downwind evacuation for at least 300 meters (1000 feet).
Fire: Isolate for 800 meters (½ mile) in all directions, especially if tank, rail car, or tank truck is involved in fire.

EXPOSURE
Short-term effects: *SEEK MEDICAL ATTENTION*. **Vapor:** Irritating to eyes, nose, and throat. *IF INHALED*, will, will cause dizziness, nausea, vomiting, difficult breathing, or loss of consciousness. Move victim to fresh air. *IF BREATHING HAS STOPPED*, give artificial respiration. IF breathing is difficult, administer oxygen. **Liquid:** Irritating to skin and eyes. Harmful if swallowed. Remove contaminated clothing and shoes. Flush affected areas with plenty of water. *IF IN EYES*, hold eyelids open and flush with plenty of water. *IF SWALLOWED* and victim is *CONSCIOUS AND ABLE TO SWALLOW*, have victim drink 4 to 8 ounces of water. **Do NOT induce vomiting**.

HEALTH HAZARDS
Personal protective equipment (PPE): B-Level PPE. OSHA Table Z-1-A air contaminant. Hydrocarbon canister, supplied-air, or hose mask; rubber or plastic gloves; chemical goggles or face shield.
Recommendations for respirator selection: *EMERGENCY OR PLANNED ENTRY INTO UNKNOWN CONCENTRATIONS OR IDLH CONDITIONS:* SCBAF:PD,PP (any self-contained breathing apparatus that has a full facepiece and is operated in a pressure-demand or other positive-pressure mode); or SAF:PD,PP:ASCBA (any supplied-air respirator that has a full facepiece and is operated in a pressure-demand or other positive-pressure mode in combination with an auxiliary self-contained breathing apparatus operated in a pressure-demand or other positive pressure mode). *ESCAPE:* GMFOV[any air-purifying, full-facepiece respirator (gas mask) with a chin-style, front- or back-mounted organic vapor cannister]; or SCBAE (any appropriate escape-type, self-contained breathing apparatus). *Note*: Substance causes eye irritation or damage; eye protection needed.
Exposure limits (TWA unless otherwise noted): ACGIH TLV 600 ppm (1750 mg/m^3); NIOSH REL 600 ppm (1720 mg/m^3).
Toxicity by ingestion: Grade 2; LD_{50} = 0.5 to 5 g/kg.
Vapor (gas) irritant characteristics: Vapors cause smarting of the eyes or respiratory system if present in high concentrations.
Liquid or solid irritant characteristics: Eye and skin irritant. If spilled on clothing and allowed to remain, may cause smarting and reddening of the skin.

FIRE DATA
Flash point: –35°F/–37°C (cc).
Flammable limits in air: LEL: 1.1%; UEL: 8.7%.
Fire extinguishing agents not to be used: Water may be ineffective.
Behavior in fire: Containers may explode.
Autoignition temperature: 682°F/361°C/634°K.
Electrical hazard: Due to low electric conductivity, this substance may generate electrostatic charges as a result of agitation and flow.
Burning rate: 7.9 mm/min

CHEMICAL REACTIVITY
Binary reactants: Strong oxidizers; ultraviolet light.

ENVIRONMENTAL DATA
Food chain concentration potential: Negative; unlikely to accumulate.
Water pollution: Effect of low concentrations on aquatic life is unknown. Fouling to shoreline. May be dangerous if it enters nearby water intakes; notify operators. Notify local health and wildlife officials. **Response to discharge:** Issue warning–high flammability. Evacuate area. Disperse and flush.

SHIPPING INFORMATION
Grades of purity: Commercial; 60% (remainder consists of hydrocarbons of similar boiling point); Research: 99+%; **Storage temperature:** Ambient; **Inert atmosphere:** None; **Venting:** Pressure-vacuum; **Stability during transport:** Stable.

NAS HAZARD CLASSIFICATION FOR BULK WATER TRANSPORTATION
FIRE: 3
HEALTH: Vapor irritant: 1; Liquid or solid irritant: 1; Poisons: 1
WATER POLLUTION: Human toxicity: 2; Aquatic toxicity: 1; Aesthetic effect: 0
REACTIVITY: Other chemicals: 2; Water: 0; Self-reaction: 0

PHYSICAL AND CHEMICAL PROPERTIES
Physical state @ 59°F/15°C and 1 atm: Liquid.
Molecular weight: 70.2
Boiling point @ 1 atm: 121°F/49°C/323°K.
Melting/Freezing point: –137°F/–94°C/–179°K.
Critical temperature: 462°F/239°C/512°K.
Critical pressure: 654 psia = 44.4 atm = 4.51 MN/m^2.
Specific gravity (water = 1): 0.74 @ 68°F/20°C (liquid).
Liquid surface tension: 23 dynes/cm = 0.023 N/m @ 68°F/20°C.
Liquid water interfacial tension: (estimate) 28 dynes/cm = 0.028 N/m @ 68°F/20°C.
Relative vapor density (air = 1): 2.4
Ratio of specific heats of vapor (gas): 1.1217
Latent heat of vaporization: 170 Btu/lb = 94 cal/g = 3.9 x 10^5 J/kg.
Heat of combustion: –19,900 Btu/lb = –11,110 cal/g = –465 x 10^5 J/kg.
Heat of fusion: 2.07 cal/g.
Vapor pressure: 400 mm @ 88°F

CYCLOPENTENE   REC. C:6250

SYNONYMS: CICLOPENTENO (Spanish); RTECS No. GY5950000

IDENTIFICATION
CAS Number: 142-29-0
Formula: C_5H_8
DOT ID Number: UN 2246; DOT Guide Number: 128
Proper Shipping Name: Cyclopentene

DESCRIPTION: Colorless liquid. Gasoline-like odor. Floats on the surface of water; insoluble. Produces large amounts of vapor.

Highly flammable • May be able to form unstable peroxides; peroxides may be detonated by heating, impact, or friction. • Containers may BLEVE when exposed to fire •Vapors may form explosive mixture with air • Vapors are heavier than air and will collect and stay in low areas • Vapors may travel long distances to ignition sources and flashback • Vapors in confined areas (e.g., tanks, sewers, buildings) may explode when exposed to fire • Irritating to the skin, eyes, and respiratory tract • Boils at 112°F/44°C • Toxic products of combustion may include carbon monoxide.

Hazard Classification (based on NFPA-704 Rating System)
Health Hazards (Blue): 1; Flammability (Red): 3; Reactivity (Yellow): 1

EMERGENCY RESPONSE: See Appendix A (128)
Evacuation:
Public safety: Isolate spill area for at least 25 to 50 meters (80 to 160 feet) in all directions.
Spill: Large spill–Consider initial downwind evacuation for at least 300 meters (1000 feet).
Fire: Isolate for 800 meters (½ mile) in all directions, especially if tank, rail car, or tank truck is involved in fire.

EXPOSURE
Short-term effects: *SEEK MEDICAL ATTENTION.* **Vapor:** If inhaled, will cause dizziness, difficult breathing, or loss of consciousness. Move to fresh air. *IF BREATHING HAS STOPPED*, give artificial respiration. IF breathing is difficult, administer oxygen. **Liquid:** Irritating to skin and eyes. Harmful if swallowed. Remove contaminated clothing and shoes. Flush affected areas with plenty of water. *IF IN EYES*, hold eyelids open and flush with plenty of water. *IF SWALLOWED*, may cause chemical pneumonitis.

HEALTH HAZARDS
Personal protective equipment (PPE): B-Level PPE.
Recommendations for respirator selection: *EMERGENCY OR PLANNED ENTRY INTO UNKNOWN CONCENTRATIONS OR IDLH CONDITIONS:* SCBAF:PD,PP (any self-contained breathing apparatus that has a full facepiece and is operated in a pressure-demand or other positive-pressure mode); or SAF:PD,PP:ASCBA (any supplied-air respirator that has a full facepiece and is operated in a pressure-demand or other positive-pressure mode in combination with an auxiliary self-contained breathing apparatus operated in a pressure-demand or other positive pressure mode). *ESCAPE:* GMFOV[any air-purifying, full-facepiece respirator (gas mask) with a chin-style, front- or back-mounted organic vapor cannister]; or SCBAE (any appropriate escape-type, self-contained breathing apparatus). *Note*: Substance causes eye irritation or damage; eye protection needed.
Toxicity by ingestion: Grade 2: LD_{50} = 1.656 g/kg (rat).
Vapor (gas) irritant characteristics: Eye and respiratory tract irritant. Vapors cause smarting of the eyes or respiratory system if present in high concentrations. The effect is temporary.
Liquid or solid irritant characteristics: Eye and skin irritant. If spilled on clothing and allowed to remain, may cause smarting and reddening of skin.

FIRE DATA
Flash point: –20°F/–29°C (cc).
Fire extinguishing agents not to be used: Water may be ineffective.
Autoignition temperature: 743°F/395°C/668°K.
Electrical hazard: Due to low electric conductivity, this substance may generate electrostatic charges as a result of agitation and flow.

CHEMICAL REACTIVITY
Binary reactants: Strong oxidizers and many other compounds.

Polymerization: May form unstable peroxides; able to polymerize.
Reactivity group: 30
Compatibility class: Olefins

ENVIRONMENTAL DATA
Water pollution: Effect of low concentrations on aquatic life is unknown. Fouling to shoreline. May be dangerous if it enters nearby water intakes; notify operators. Notify local health and wildlife officials. **Response to discharge:** Issue warning–high flammability. Restrict access-Flammable liquid. Evacuate areas. Mechanical containment. Should be removed. Chemical and physical treatment.

SHIPPING INFORMATION
Grades of purity: 99%; **Storage temperature:** Refrigerate; **Stability during transport:** Stable.

PHYSICAL AND CHEMICAL PROPERTIES
Physical state @ 59°F/15°C and 1 atm: Liquid.
Molecular weight: 68.12
Boiling point @ 1 atm: 112°F/44°C/317.2°K.
Melting/Freezing point: –211°F/–135°C/138.2°K.
Specific gravity (water = 1): 0.774
Relative vapor density (air = 1): 2.35
Vapor pressure: 320 mm.

CYCLOPROPANE   REC. C:6300

SYNONYMS: CICLOPROPANO (Spanish); CYCLOPROPANE, Liquefied; EEC No. 601-016-00-6; TRIMETHYLENE

IDENTIFICATION
CAS Number: 75-19-4
Formula: C_3H_6
DOT ID Number: UN 1027; DOT Guide Number: 115
Proper Shipping Name: Cyclopropane, liquefied

DESCRIPTION: Colorless compressed gas. May be stored and shipped as a cryogenic liquid. Mild sweet odor; like light petroleum solvent. Floats and boils on water; moderately soluble; flammable visible vapor cloud is produced.

Extremely flammable • Forms explosive mixture with air • Reacts explosively with many materials • Gas from liquefied gas are initially heavier than air and spread along ground • Gas may travel to source of ignition and flashback • Exposure of cylinders to elevated temperatures, fire, and flame may cause cylinders to rupture or cause frangible disk to burst, releasing entire contents of cylinder; that may rocket through buildings and/or travel a considerable distance •Irritating to the skin, eyes, and respiratory tract • Vapors may cause dizziness or asphyxiation without warning • Contact with liquid may cause frostbite.

Hazard Classification (based on NFPA-704 Rating System)
Health Hazards (Blue): 1; Flammability (Red): 4; Reactivity (Yellow): 0

EMERGENCY RESPONSE: See Appendix A (115)
Evacuation:
Public safety: Isolate spill area for at least 50 to 100 meters (160 to 330 feet) in all directions.
Spill: Consider initial downwind evacuation for at least 800 meters (½ mile).
Fire: Isolate for 1600 meters (1 mile) in all directions, especially if tank, rail car, or tank truck is involved in fire.

EXPOSURE
Short-term effects: *SEEK MEDICAL ATTENTION.* **Vapor:** *IF inhaled, will cause difficult breathing. Move victim to fresh air. IF BREATHING HAS STOPPED,* give artificial respiration. IF breathing is difficult, administer oxygen. **Liquid:** Will cause frostbite. Flush affected areas with plenty of water. *DO NOT RUB AFFECTED AREAS.*

HEALTH HAZARDS
Personal protective equipment (PPE): B-Level PPE. Self-contained breathing apparatus for high concentrations of vapor; safety goggles or face shield. Wear thermal protective clothing.
Recommendations for respirator selection: *EMERGENCY OR PLANNED ENTRY INTO UNKNOWN CONCENTRATIONS OR IDLH CONDITIONS:* SCBAF:PD,PP (any self-contained breathing apparatus that has a full facepiece and is operated in a pressure-demand or other positive-pressure mode); or SAF:PD,PP:ASCBA (any supplied-air respirator that has a full facepiece and is operated in a pressure-demand or other positive-pressure mode in combination with an auxiliary self-contained breathing apparatus operated in a pressure-demand or other positive pressure mode). *ESCAPE:* GMFOV[any air-purifying, full-facepiece respirator (gas mask) with a chin-style, front- or back-mounted organic vapor cannister]; or SCBAE (any appropriate escape-type, self-contained breathing apparatus).
Liquid or solid irritant characteristics: Rapid evaporation of liquid may cause frostbite.

FIRE DATA
Flash point: Flammable gas.
Flammable limits in air: LEL: 2.4%; UEL: 10.3%.
Behavior in fire: Containers may explode.
Autoignition temperature: 932°F/500°C.
Electrical hazard: Class I, Group C. Due to low electric conductivity, may generate electrostatic charges from agitation and flow.
Stoichiometric air-to-fuel ratio: 14.67 (estimate).

CHEMICAL REACTIVITY
Binary reactants: Oxidizers.

ENVIRONMENTAL DATA
Food chain concentration potential: Negative; unlikely to accumulate.
Water pollution: Not harmful to aquatic life. **Response to discharge:** Issue warning–high flammability. Restrict access. Evacuate area.

SHIPPING INFORMATION
Grades of purity: 99.5+%; USP; **Storage temperature:** Ambient; **Inert atmosphere:** None; **Venting:** Safety relief; **Stability during transport:** Stable.

PHYSICAL AND CHEMICAL PROPERTIES
Physical state @ 59°F/15°C and 1 atm: Gas.
Molecular weight: 42.1
Boiling point @ 1 atm: –27°F/–33°C/240°K.
Melting/Freezing point: –197°F/–127°C/146°K.
Critical temperature: 257°F/125°C/398°K.
Critical pressure: 798 psia = 54.2 atm = 5.50 MN/m^2.
Specific gravity (water = 1): 0.676 at –33°C (liquid).

Liquid surface tension: 22 dynes/cm = 0.022 N/m at –40°F/–40°C.
Relative vapor density (air = 1): 1.48
Ratio of specific heats of vapor (gas): 1.1790
Latent heat of vaporization: 203 Btu/lb = 113 cal/g = 4.73 x 10^5 J/kg.
Heat of combustion: 21,247 Btu/lb = –11,804 cal/g = 493.88 x 10^5 J/kg.
Heat of fusion: 30.92 cal/g.
Vapor pressure: 4850 mm.

p-CYMENE   REC. C:6350

SYNONYMS: CIMENO (Spanish); CYMENE; CYMOL; DOLCYMENE; 4-ISOPROPYL-1-METHYL BENZENE; ISOPROPYLTOLUOL; *p*-ISOPROPYLTOLUENE; 4-ISOPROPYL TOLUENE; METHYL PROPYL BENZENE; 1-METHYL-4-ISOPROPYLBENZENE; PARA-CYMENE; PARA-CYMOL

IDENTIFICATION
CAS Number: 99-87-6 (*p*-isomer); 535-77-3 (*m*-isomer); 527-84-4 (*o*-isomer)
Formula: $C_{10}H_{14}$; *p*-$CH_3C_6H_4CH(CH_3)_2$
DOT ID Number: UN 2046; DOT Guide Number: 130
Proper Shipping Name: Cymenes

DESCRIPTION: Colorless to pale yellow liquid. Mild pleasant odor; like solvent. Floats on the surface of water.

Highly flammable • Containers may BLEVE when exposed to fire • Vapors may form explosive mixture with air • Vapors are heavier than air and will collect and stay in low areas • Vapors may travel long distances to ignition sources and flashback • Vapors in confined areas (e.g., tanks, sewers, buildings) may explode when exposed to fire • Irritating to the skin, eyes, and respiratory tract • Toxic products of combustion may include carbon monoxide and CO_2.

Hazard Classification (based on NFPA-704 Rating System)
Health Hazards (Blue): 2; Flammability (Red): 2; Reactivity (Yellow): 0

EMERGENCY RESPONSE: See Appendix A (130)
Evacuation:
Public safety: Isolate spill area for at least 50 to 100 meters (160 to 330 feet) in all directions.
Spill: Large spill–Consider initial downwind evacuation for at least 300 meters (1000 feet).
Fire: Isolate for 800 meters (½ mile) in all directions, especially if tank, rail car, or tank truck is involved in fire.

EXPOSURE
Short-term effects: *SEEK MEDICAL ATTENTION.* **Liquid:** Irritating to skin and eyes. Remove contaminated clothing and shoes. Flush affected areas with plenty of water. *IF IN EYES*, hold eyelids open and flush with plenty of water. *IF SWALLOWED* and victim is *CONSCIOUS AND ABLE TO SWALLOW*, do NOT induce vomiting; *SEEK MEDICAL ATTENTION.* Chemical pneumonitis may develop.

HEALTH HAZARDS
Personal protective equipment (PPE): Self-contained or air-line breathing apparatus; solvent-resistant gloves; chemical splash goggles. Chemical protective material(s) reported to offer minimal to poor protection: Viton®/neoprene.
Toxicity by ingestion: Grade 2; oral rat LD_{50} = 4750 mg/kg Oral human TDLO = 86 mg/kg (affects central nervous system).
Long-term health effects: Possible central nervous system damage.
Vapor (gas) irritant characteristics: Eye and respiratory tract irritant.
Liquid or solid irritant characteristics: Eye and skin irritant. If spilled on clothing and allowed to remain, may cause smarting and reddening of the skin.

FIRE DATA
Flash point: 140°F/60°C (oc); 117°F/47°C (cc).
Flammable limits in air: LEL: 0.7% @ 212°F/100°C; UEL: 5.6%.
Fire extinguishing agents not to be used: Water may be ineffective.
Autoignition temperature: 817°F/436°C/709°K.
Electrical hazard: Class I, Group D. Due to low electric conductivity, this substance may generate electrostatic charges as a result of agitation and flow.
Burning rate: 6.1 mm/min

CHEMICAL REACTIVITY
Binary reactants: Reacts with strong oxidizers. Will cause rubber to swell and soften.
Reactivity group: 32
Compatibility class: Aromatic hydrocarbons

ENVIRONMENTAL DATA
Water pollution: DOT Appendix B, §172.101–marine pollutant. Effect of low concentrations on aquatic life is unknown. Fouling to shoreline. May be dangerous if it enters nearby water intakes; notify operators. Notify local health and wildlife officials. **Response to discharge:** Issue warning–water contaminant. Mechanical containment. Should be removed. Chemical and physical treatment.

SHIPPING INFORMATION
Grades of purity: 95+%; **Storage temperature:** Ambient; **Inert atmosphere:** None; **Venting:** Open (flame arrester); **Stability during transport:** Stable.

NAS HAZARD CLASSIFICATION FOR BULK WATER TRANSPORTATION
FIRE: 2
HEALTH: Vapor irritant: 0; Liquid or solid irritant: 1; Poisons: 1
WATER POLLUTION: Human toxicity: 1; Aquatic toxicity: 1; Aesthetic effect: 2
REACTIVITY: Other chemicals: 1; Water: 0; Self-reaction: 0

PHYSICAL AND CHEMICAL PROPERTIES
Physical state @ 59°F/15°C and 1 atm: Liquid.
Molecular weight: 134.2
Boiling point @ 1 atm: 351°F/177°C/450°K.
Melting/Freezing point: –90°F/–68°C/205°K.
Specific gravity (water = 1): 0.857 @ 68°F/20°C (liquid).
Liquid surface tension: 28.09 dynes/cm = 0.02809 N/m @ 68°F/20°C.
Liquid water interfacial tension: 36.41 dynes/cm = 0.03641 N/m @ 68°F/20°C.
Relative vapor density (air = 1): 4.6
Latent heat of vaporization: 122 Btu/lb = 67.8 cal/g = 2.84 x 10^5 J/kg.
Heat of combustion: –18,800 Btu/lb = –10,400 cal/g = –437 x 10^5 J/kg.

Heat of fusion: 17.10 cal/g.
Vapor pressure: (Reid) Low; 1.5 mm.

2,4-D   REC. D:0100

SYNONYMS: 2,4-D ACID; ACIDE 2,4-DICHLORO PHENOXYACETIQUE (French); ACIDO 2,4-DICLOROFENOXIACETICO (Spanish); AGROTECT; AMIDOX; AMOXONE; AQUA-KLEEN; BH 2,4-D; BRUSH-RHAP; BUSH KILLER; β-SELEKTONON; CHIPCO TURF HERBICIDE "D"; CHLOROXONE; CROP RIDER; CROTILIN; D 50; 2,4-D, SALTS AND ESTERS; DACAMINE; DEBROUSSAILLANT 600; DECAMINE; DED-WEED; DED-WEED LV-69; DESORMONE; DICHLOROPHENOXYACETIC ACID; 2,4-DICHLORPHENOXYACETIC ACID; 2,4-DICHLOROPHENOXYACETIC ACID, SALTS AND ESTERS; (2,4-DICHLOR-PHENOXY)-ESSIGSAEURE (German); DICOPUR; DICOTOX; DINOXOL; DMA-4; DORMONE; EEC No. 607-039-00-8; EMULSAMINE BK; EMULSAMINE E-3; ENVERT 171; ENVERT DT; ESTERON 44; ESTERON 99; ESTERON 76 BE; ESTERONE FOUR; ESTONE; FERNESTA; FERNIMINE; FERNOXONE; FERXONE; FOREDEX 75; FORMULA 40; HEDONAL; HERBIDAL; IPANER; KROTILINE; LAWN-KEEP; MACRONDRAY; MIRACLE; MONOSAN; MOXONE; NETAGRONE; NETAGRONE 600; PENNAMINE; PENNAMINE D; PHENOX; PIELIK; PLANOTOX; PLANTGARD; RHODIA; SALVO; SPRITZ-HORMIN/2,4-D; SPRITZ-HORMIT/2,4-D; SUPER D WEEDONE; SUPERORMONE CONCENTRE; TRANSAMINE; TRIBUTON; U 46; U-5043; U 46DP; VERGEMASTER; VERTON; VERTON D; VERTON 2D; VERTRON 2D; VIDON 638; VISKO; VISKO-RHAP; VISKO-RHAP LOW DRIFT HERBICIDES; VISKO-RHAP LOW VOLATILE 4L; WEED-AG-BAR; WEEDAR; WEEDAR-64; WEED-B-GON; WEEDEZ WONDER BAR; WEEDONE; WEEDONE LV4; WEED-RHAP; WEED TOX; WEEDTROL; RCRA No. U240

IDENTIFICATION
CAS Number: 94-75-7
Formula: $C_8H_6Cl_2O_3$; 2, 4-$Cl_2C_6H_3OCH_2COOH$
DOT ID Number: UN 2765; DOT Guide Number: 152
Proper Shipping Name: 2,4-Dichlorphenoxyacetic acid
Reportable Quantity (RQ): **(CERCLA)** 100 lb/45.4 kg

DESCRIPTION: White to tan solid. Noncombustible but may be dissolved in a flammable liquid. Odorless. Sinks in water; insoluble.

Poison! (chlorophenoxy pesticide) • Combustible • Breathing or swallowing the dust can cause illness • Containers may BLEVE when exposed to fire • Concentrated dust in confined areas (e.g., tanks, sewers, buildings) may explode when exposed to fire • Toxic products of combustion may include carbon monoxide and hydrogen chloride.

Hazard Classification (based on NFPA-704 Rating System)
Health Hazards (Blue): 2; Flammability (Red): 1; Reactivity (Yellow): 0

EMERGENCY RESPONSE: See Appendix A (152)
Evacuation:
Public safety: Isolate the area of spill or leak for at least 25 to 50 meters (80 to 160 feet) in all directions.

Spill: Increase, in the downwind direction, as necessary, the distance shown under "Public Safety."
Fire: If tank, rail car, or tank truck is involved in fire, isolate for at least 800 meters (½ mile) in all directions; also, consider initial evacuation for 800 meters (½ mile) in all directions.

EXPOSURE
Short-term effects: *SEEK MEDICAL ATTENTION. IF BREATHING HAS STOPPED,* give artificial respiration; *avoid mouth-to-mouth resuscitation; use bag/mask apparatus.* **Solid:** *POISONOUS IF SWALLOWED. IF SWALLOWED* and victim is *CONSCIOUS AND ABLE TO SWALLOW,* have victim drink 4 to 8 ounces of water and have victim induce vomiting. *IF SWALLOWED* and victim is *UNCONSCIOUS OR HAVING CONVULSIONS,* do nothing except keep victim warm.

HEALTH HAZARDS
Personal protective equipment (PPE): A-Level PPE. OSHA Table Z-1-A air contaminant. Chemical protective material(s) reported to have good to excellent resistance: natural rubber, neoprene, nitrile, PV alcohol, PVC.
Recommendations for respirator selection: NIOSH/OSHA
100 mg/m³: CCROVDMFu [any chemical cartridge respirator with organic vapor cartridge(s) in combination with a dust, mist, and fume filter]; or GMFOVHiE [any air-purifying, full-facepiece respirator (gas mask) with a chin-style, front- or back-mounted organic vapor canister having a high-efficiency particulate filter]; or PAPROVDMFu [any powered, air-purifying respirator with organic vapor cartridge(s) in combination with a dust, mist, and fume filter]; or SA (any supplied-air respirator); or SCBAF (any self-contained breathing apparatus with a full facepiece); or SAF (any supplied-air respirator with a full facepiece); or SAT:CF (any supplied-air respirator that has a tight-fitting facepiece and is operated in a continuous-flow mode). *EMERGENCY OR PLANNED ENTRY INTO UNKNOWN CONCENTRATIONS OR IDLH CONDITIONS:* SCBAF:PD,PP (any self-contained breathing apparatus that has a full facepiece and is operated in a pressure-demand or other positive-pressure mode); or SAF:PD,PP:ASCBA (any supplied-air respirator that has a full facepiece and is operated in a pressure-demand or other positive-pressure mode in combination with an auxiliary self-contained breathing apparatus operated in a pressure-demand or other positive pressure mode). *ESCAPE:* GMFOVHiE [any air-purifying, full-facepiece respirator (gas mask) with a chin-style, front- or back-mounted organic vapor canister having a high-efficiency particulate filter]; or SCBAE (any appropriate escape-type, self-contained breathing apparatus).
Exposure limits (TWA unless otherwise noted): ACGIH TLV 10 mg/m³; NIOSH/OSHA 10 mg/m³.
Toxicity by ingestion: Grade 3; oral rat LD_{50} = 375 mg/kg (rat), 80 mg/kg (human).
Long-term health effects: Causes birth defects in some laboratory animals. Skin disorders. IARC overall evaluation unassigned; animal evidence inadequate. Organ damage including liver and kidney.
Vapor (gas) irritant characteristics: Eye and respiratory tract irritant.
Liquid or solid irritant characteristics: Eye and skin irritant.
IDLH value: 100 mg/m³.

FIRE DATA
Behavior in fire: Water streams applied to adjacent fires will spread contamination of pesticide over wide area.

CHEMICAL REACTIVITY
Binary reactants: Strong oxidizers.

Neutralizing agents for acids and caustics: Flush with water, rinse with sodium bicarbonate or lime solution.
Reactivity group: 43
Compatibility class: Miscellaneous water solutions.

ENVIRONMENTAL DATA

Food chain concentration potential: Log P_{ow} = 2.8. Values > 3.0 are likely to bioconcentrate in aquatic organisms and other living tissue, especially in fats.
Water pollution: DOT Appendix B, §172.101–marine pollutant. Dangerous to aquatic life in high concentrations. May be dangerous if it enters nearby water intakes; notify operators. Notify local health and wildlife officials. **Response to discharge:** Issue warning–poison, water contaminant Restrict access. Should be removed. Chemical and physical treatment.

SHIPPING INFORMATION

Grades of purity: 98+%; **Storage temperature:** Ambient; **Inert atmosphere:** None; **Venting:** Open; **Stability during transport:** Stable.

PHYSICAL AND CHEMICAL PROPERTIES

Physical state @ 59°F/15°C and 1 atm: Solid.
Molecular weight: 221.0
Boiling point @ 1 atm: 350°F/179°C/452°K (decomposes).
Melting/Freezing point: 286°F/141°C/314°K.
Specific gravity (water = 1): 1.563 @ 68°F/20°C (solid).
Relative vapor density (air = 1): 7.6
Heat of combustion: estimate –7700 Btu/lb = –4300 cal/g = –180 x 10^5 J/kg.
Vapor pressure: 0.4 mm @ 320°F/–1.1°C.

2,4-D (ESTERS)   **REC. D:0150**

SYNONYMS: (2,4-DICHLOROPHENOXY)ACETIC ACID, ISOPROPYL ESTER; BUTYL 2,4-D ISOPROPYL ESTER; ESTERON 44; WEEDONE 128; 2,4-DICHLOROPHENOXYACETATE; ISOPROPYL 2,4-DICHLOROPHENOXY ACETATE; 2,4-DICHLOROPHENOXYACETIC ACID, BUTOXYETHYL ESTER

IDENTIFICATION

CAS Number: 94-11-1
Formula: 2,4-$Cl_2C_6H_3OCH_2COOR$, where R = C_4H_9, C_3H_7, or $CH_2CH_2OC_4H_9$
DOT ID Number: UN 3000; DOT Guide Number: 152
Proper Shipping Name: Phenoxy pesticides, liquid, toxic
Reportable Quantity (RQ): **(CERCLA)** 100 lb/45.4 kg

DESCRIPTION: White to yellowish brown liquid. Fuel oil-like odor. Sinks in water; insoluble.

Poison! (phenoxy pesticide) • Combustible • Breathing or swallowing the dust can cause illness • Containers may BLEVE when exposed to fire • Concentrated dust in confined areas (e.g., tanks, sewers, buildings) may explode when exposed to fire • Toxic products of combustion may include hydrogen chloride.

Hazard Classification (based on NFPA-704 M Rating System)
Health Hazards (Blue): 2; Flammability (Red): 1; Reactivity (Yellow): 0

EMERGENCY RESPONSE: See Appendix A (152)
Evacuation:
Public safety: Isolate the area of spill or leak for at least 25 to 50 meters (80 to 160 feet) in all directions.
Spill: Increase, in the downwind direction, as necessary, the distance shown under "Public Safety."
Fire: If tank, rail car, or tank truck is involved in fire, isolate for at least 800 meters (½ mile) in all directions; also, consider initial evacuation for 800 meters (½ mile) in all directions.

EXPOSURE

Short-term effects: *SEEK MEDICAL ATTENTION. IF BREATHING HAS STOPPED*, give artificial respiration; *avoid mouth-to-mouth resuscitation; use bag/mask apparatus*. **Solid:** *POISONOUS IF SWALLOWED. IF SWALLOWED* and victim is *CONSCIOUS AND ABLE TO SWALLOW*, have victim drink 4 to 8 ounces of water and have victim induce vomiting. *IF SWALLOWED* and victim is *UNCONSCIOUS OR HAVING CONVULSIONS*, do nothing except keep victim warm.

HEALTH HAZARDS

Personal protective equipment (PPE): B-Level PPE. Face shield or goggles, rubber gloves. Wear sealed chemical clothes. Chemical protective material(s) reported to have good to excellent resistance: natural rubber, neoprene, nitrile, PVC.
Recommendations for respirator selection: NIOSH as 2,4-D. 100 mg/m^3: CCROVDMFu [any chemical cartridge respirator with organic vapor cartridge(s) in combination with a dust, mist, and fume filter]; or GMFOVHiE [any air-purifying, full-facepiece respirator (gas mask) with a chin-style, front- or back-mounted organic vapor canister having a high-efficiency particulate filter]; or PAPROVDMFu [any powered, air-purifying respirator with organic vapor cartridge(s) in combination with a dust, mist, and fume filter]; or SA (any supplied-air respirator); or SCBAF (any self-contained breathing apparatus with a full facepiece); or SAF (any supplied-air respirator with a full facepiece); or SAT:CF (any supplied-air respirator that has a tight-fitting facepiece and is operated in a continuous-flow mode). *EMERGENCY OR PLANNED ENTRY INTO UNKNOWN CONCENTRATIONS OR IDLH CONDITIONS:* SCBAF:PD,PP (any self-contained breathing apparatus that has a full facepiece and is operated in a pressure-demand or other positive-pressure mode); or SAF:PD,PP:ASCBA (any supplied-air respirator that has a full facepiece and is operated in a pressure-demand or other positive-pressure mode in combination with an auxiliary self-contained breathing apparatus operated in a pressure-demand or other positive pressure mode). *ESCAPE:* GMFOVHiE [any air-purifying, full-facepiece respirator (gas mask) with a chin-style, front- or back-mounted organic vapor canister having a high-efficiency particulate filter]; or SCBAE (any appropriate escape-type, self-contained breathing apparatus).
Exposure limits (TWA unless otherwise noted): ACGIH TLV 10 mg/m^3; NIOSH/OSHA 10 mg/m^3 as 2,4-D.
Toxicity by ingestion: Grade 2 or 3; LD_{50} = 320–617 mg/kg.
Liquid or solid irritant characteristics: Eye irritant.
IDLH value: 100 mg/m^3 as 2,4-D.

FIRE DATA

Flash point: More than 175°F/79°C (oc).
Fire extinguishing agents not to be used: Water may be ineffective.

CHEMICAL REACTIVITY

Binary reactants: Oxidizers. May attack some forms of plastics and coatings.

ENVIRONMENTAL DATA
Food chain concentration potential: Negative; unlikely to accumulate.
Water pollution: Dangerous to aquatic life in high concentrations. May be dangerous if it enters nearby water intakes; notify operators. Notify local health and wildlife officials. **Response to discharge:** Issue warning–water contaminant. Should be removed. Chemical and physical treatment.

SHIPPING INFORMATION
Grades of purity: Technical, 99%; 64% in petroleum oil; **Storage temperature:** Ambient; **Inert atmosphere:** None; **Venting:** Open; **Stability during transport:** Stable.

PHYSICAL AND CHEMICAL PROPERTIES
Physical state @ 59°F/15°C and 1 atm: Liquid.
Molecular weight: 234-291
Boiling point @ 1 atm: Very high.
Specific gravity (water = 1): 1.088-1.237 @ 68°F/20°C (liquid).

DALAPON   REC. D:0200

SYNONYMS: ACIDO 2,2-DICLOROPROPIONICO (Spanish); ALATEX; BASFAPON; BASFAPON B; BASFAPON/BASFAPON N; BH DALAPON; BASINEX; BH DALAPON; CRISAPON; DALAPON 85; DED-WEED; DEVIPON; α-DICHLOROPROPIONIC ACID; α,α-DICHLOROPROPIONIC ACID; 2,2-DICHLOROPROPIONIC ACID; DOWPON; DOWPON M; GRAMEVIN; KENAPON; LIROPON; PROPANOIC ACID, 2,2-DICHLORO-; PROPROP; RADAPON; REVENGE; S95; UNIPON; 2,2-DICHLOROPROPANOIC ACID

IDENTIFICATION
CAS Number: 75-99-0
Formula: CH_3CCl_2COOH
DOT ID Number: UN 1760; DOT Guide Number: 154
Proper Shipping Name: 2,2-Dichloropropionic acid
Reportable Quantity (RQ): **(CERCLA)** 5000 lb/2270 kg

DESCRIPTION: Colorless liquid. Acrid odor. Sinks and mixes with water forming hydrochloric and pyruvic acid solution.

Corrosive • Poison! (aliphatic acid herbicide) Combustible • Breathing or swallowing the vapor or solution can cause illness; vapor is irritating to eyes, skin, and respiratory tract; contact with liquid will burn skin and eyes • Firefighting gear (including SCBA) does not provide adequate protection. If exposure occurs, remove and isolate gear immediately and thoroughly decontaminate personnel • Heat or flame may cause explosion • Containers may BLEVE when exposed to fire • Vapors in confined areas (e.g., tanks, sewers, buildings) may explode when exposed to fire • Toxic products of combustion may include hydrogen chloride.

Hazard Classification (based on NFPA-704 Rating System)
Health Hazards (Blue): 1; Flammability (Red): 1; Reactivity (Yellow): 1

EMERGENCY RESPONSE: See Appendix A (154)
Evacuation:
Public safety: Isolate the area of spill or leak for at least 25 to 50 meters (80 to 160 feet) in all directions.
Spill: Increase, in the downwind direction, as necessary, the distance shown under "Public Safety."
Fire: If tank, rail car, or tank truck is involved in fire, isolate for at least 800 meters (½ mile) in all directions; also, consider initial evacuation for 800 meters (½ mile) in all directions.

EXPOSURE
Short-term effects: *SEEK MEDICAL ATTENTION*. **Vapor:** Irritating to eyes, nose, and throat. Move to fresh air. *IF IN EYES*, hold eyelids open and flush with plenty of water. *IF BREATHING HAS STOPPED,* give artificial respiration; *avoid mouth-to-mouth resuscitation; use bag/mask apparatus*. If breathing is difficult, administer oxygen. **Liquid:** Will burn skin and eyes. Harmful if swallowed. Remove contaminated clothing and shoes. Flush affected areas with plenty of water. *IF IN EYES*, hold eyelids open and flush with plenty of water. *IF SWALLOWED* and victim is *CONSCIOUS AND ABLE TO SWALLOW*, have victim drink 4 to 8 ounces of water. **Do NOT induce vomiting**.

HEALTH HAZARDS
Personal protective equipment (PPE): B-Level PPE. OSHA Table Z-1-A air contaminant.
Exposure limits (TWA unless otherwise noted): ACGIH TLV 1 ppm (5.8 mg/m^3); NIOSH REL 1 ppm (6 gm/m^3).
Toxicity by ingestion: Grade 2; oral LD_{50} = 3.65 g/kg (mouse), 7.57 g/kg (rat).
Vapor (gas) irritant characteristics: Severe eye and respiratory tract irritant.
Liquid or solid irritant characteristics: Severe eye and skin irritant.
Odor threshold: 2500 mg/m^3.

FIRE DATA
Fire extinguishing agents not to be used: Water may be ineffective.
Behavior in fire: Volatilizes with steam

CHEMICAL REACTIVITY
Reactivity with water: Reacts slowly to form hydrochloric and pyruvic acids. The reaction may not be hazardous.
Binary reactants: Very corrosive to aluminum and copper alloys. Flammable and explosive hydrogen gas may form in enclosed spaces.
Neutralizing agents for acids and caustics: Flush with water; rinse with dilute sodium bicarbonate or soda ash solution.

ENVIRONMENTAL DATA
Food chain concentration potential: Log K_{ow} = 0.8. Unlikely to accumulate.
Water pollution: Harmful to aquatic life in very low concentrations. May be dangerous if it enters nearby water intakes; notify operators. Notify local health and wildlife officials. **Response to discharge:** Issue warning–corrosive. Restrict access. Disperse and flush.

SHIPPING INFORMATION
Grades of purity: Technical grade, 90%; solid formulations of sodium and magnesium salts are sometimes referred to as Dalapon and are much less corrosive; **Storage temperature:** 70-90°F; **Inert atmosphere:** None; **Venting:** Open; **Stability during transport:** Stable.

PHYSICAL AND CHEMICAL PROPERTIES
Physical state @ 59°F/15°C and 1 atm: Liquid.
Molecular weight: 143

Boiling point @ 1 atm: 374°F/190°C/463°K.
Melting/Freezing point: 46°F/8°C/281°K.
Specific gravity (water = 1): 1.39 at 23°C (liquid).
Relative vapor density (air = 1): 4.9

DDD REC. D:0250

SYNONYMS: 1,1-BIS(p-CHLOROPHENYL)-2,2-DICHLOROETHANE; p,p'-DDD; 1,1-DICHLORO-2,2-BIS(p-CHLORO-PHENYL) ETHANE; DICHLORODIPHENYLDICHLORO ETHANE; 1,1-DICLORO-2,2-BIS(p-ETILFENIL)ETANO (Spanish); DILENE; ME-1700; RHOTHANE; p,p'-TDE; TDE; TETRACHLORODIPHENYLETHANE

IDENTIFICATION
CAS Number: 72-54-8
Formula: $C_{13}H_{10}Cl_4$; $(4\text{-ClC}_6H_4)_2CH\cdot CHCl_2$
DOT ID Number: UN 2761; DOT Guide Number: 151
Proper Shipping Name: Organochlorine pesticides, solid, toxic
Reportable Quantity (RQ): **(CERCLA)** 1lb/0.454 kg

DESCRIPTION: White solid. Sinks in water; insoluble.

Poison! (organochlorine) • Breathing the dust, skin or eye contact, or swallowing the material can kill you • Firefighting gear (including SCBA) does not provide adequate protection. If exposure occurs, remove and isolate gear immediately and thoroughly decontaminate personnel • Containers may BLEVE when exposed to fire • Concentrated dust in confined areas (e.g., tanks, sewers, buildings) may explode when exposed to fire • Toxic products of combustion may include hydrogen chloride • Do not put yourself in danger by entering a contaminated area to rescue a victim.

Hazard Classification (based on NFPA-704 Rating System)
Health Hazards (Blue): 2; Flammability (Red): 1; Reactivity (Yellow): 0

EMERGENCY RESPONSE: See Appendix A (151)
Evacuation:
Public safety: Isolate the area of spill or leak for at least 25 to 50 meters (80 to 160 feet) in all directions.
Spill: Increase, in the downwind direction, as necessary, the distance shown under "Public Safety."
Fire: If tank, rail car, or tank truck is involved in fire, isolate for at least 800 meters (½ mile) in all directions; also, consider initial evacuation for 800 meters (½ mile) in all directions.

EXPOSURE
Short-term effects: *SEEK MEDICAL ATTENTION.* **Dust:** Irritating to eyes, nose, and throat. Harmful if inhaled. *IF IN EYES*, hold eyelids open and flush with plenty of water. *IF BREATHING HAS STOPPED*, give artificial respiration; *avoid mouth-to-mouth resuscitation; use bag/mask apparatus*. If breathing is difficult, administer oxygen. **Solid:** Irritating to skin and eyes. Harmful if swallowed. Remove contaminated clothing and shoes. Flush affected areas with plenty of water. *IF IN EYES*, hold eyelids open and flush with plenty of water. *IF SWALLOWED* and victim is *CONSCIOUS AND ABLE TO SWALLOW*, have victim drink 4 to 8 ounces of water. **Do NOT induce vomiting.** Administer a slurry of activated charcoal. *IF SWALLOWED* and victim is *UNCONSCIOUS OR HAVING CONVULSIONS*, do nothing except keep victim warm.

HEALTH HAZARDS
Personal protective equipment (PPE): A-Level PPE.
Recommendations for respirator selection:
AT ANY CONCENTRATIONS ABOVE THE NIOSH REL, OR WHERE THERE IS NO REL, AT ANY DETECTABLE CONCENTRATION: SCBAF:PD,PP (any self-contained breathing apparatus that has a full facepiece and is operated in a pressure-demand or other positive-pressure mode); or SAF:PD,PP:ASCBA (any supplied-air respirator that has a full facepiece and is operated in a pressure-demand or other positive-pressure mode in combination with an auxiliary, self-contained breathing apparatus operated in a pressure-demand or other positive pressure mode). *ESCAPE:* GMFOVHiE [any air-purifying, full-facepiece respirator (gas mask) with a chin-style, front- or back-mounted organic vapor canister having a high-efficiency particulate filter]; or SCBAE (any appropriate escape-type, self-contained breathing apparatus).
Toxicity by ingestion: Grade 2; oral LD_{50} = 1.2 g/kg (mouse), 3.4 g/kg (rat).
Long-term health effects: IARC animal carcinogen.
Liquid or solid irritant characteristics: Eye irritant.

FIRE DATA
Behavior in fire: Water streams applied to adjacent fires will spread contamination of pesticide over wide area.

CHEMICAL REACTIVITY
Binary reactants: Oxidizers.

ENVIRONMENTAL DATA
Food chain concentration potential: High
Water pollution: Harmful to aquatic life in very low concentrations. May be dangerous if it enters nearby water intakes; notify operators. Notify local health and wildlife officials. **Response to discharge:** Issue warning–water contamination. Should be removed. Chemical and physical treatment.

SHIPPING INFORMATION
Grades of purity: Technical; **Storage temperature:** Ambient; **Inert atmosphere:** None; **Venting:** Open; **Stability during transport:** Stable.

PHYSICAL AND CHEMICAL PROPERTIES
Physical state @ 59°F/15°C and 1 atm: Solid.
Molecular weight: 320
Boiling point @ 1 atm: (decomposes).
Melting/Freezing point: 234°F/112°C/385°K.
Specific gravity (water = 1): 1.476 @ 68°F/20°C (solid).
Relative vapor density (air = 1): 11.1

DDT REC. D:0300

SYNONYMS: AGRITAN; ANOFEX; ARKOTINE; AZOTOX; BENZENE,1,1'-(2,2,2-TRICHLOROETHYLIDENE)BIS(4-CHLORO-; α,α-BIS(p-CHLOROPHENYL)-β,β,β-TRICHLORETHANE; 1,1-BIS-(p-CHLOROPHENYL)-2,2,2-TRICHLOROETHANE; 2,2-BIS(p-CHLOROPHENYL)-1,1-TRICHLOROETHANE; DICHLORODIPHENYL TRICHLOROETHANE-2,2-BIS(p-CHLOROPHENYL)-1,1,1-TRICHLOROETHANE; BOSAN SUPRA; BOVIDERMOL; CHLOROPHENOTHAN; CHLOROPHENOTHANE; CHLOROPHENOTOXUM; CITOX; CLOFENOTANE; p,p'-DDT; 4,4' DDT; DEDELO; DEOVAL; DETOX; DETOXAN; DIBOVAN; DICHLORODIPHENYLTRICHLOROETHANE; p,p'-DICHLORODIPHENYLTRICHLOROETHANE; 4,4'-

DICHLORODIPHENYLTRICHLOROETHANE; DICLORODIFENILTRICLOROETANO (Spanish); DICOPHANE; DIDIGAM; DIDIMAC; DIPHENYLTRICHLOROETHANE; DODAT; DYKOL; ESTONATE; GENITOX; GESAFID; GESAPON; GESAREX; GESAROL; GUESAROL; GYRON; HAVERO-EXTRA; HILDIT; IVORAN; IXODEX; KOPSOL; MICRO DDT 75; MUTOXIN; NEOCID; PARACHLOROCIDUM; PEB1; PENTACHLORIN; PENTECH; RUKSEAM; SANTOBANE; 1,1,1-TRICHLOR-2,2-BIS(4-CHLOR-PHENYL)-AETHAN (German); TRICHLOROBIS(4-CHLOROPHENYL)ETHANE; 1,1,1-TRICHLORO-2,2-BIS(*p*-CHLOROPHENYL)ETHANE; 1,1,1-TRICHLORO-2,2-DI(4-CHLOROPHENYL)-ETHANE; ZEIDANE; ZERDANE; RCRA No. U061

IDENTIFICATION
CAS Number: 50-29-3
Formula: $C_{13}H_9Cl_5$; $(p\text{-}ClC_6H_4)_2CHCCl_3$
DOT ID Number: UN 2761; DOT Guide Number: 151
Proper Shipping Name: Dichlorodiphenyltrichloroethane (DDT); DDT
Reportable Quantity (RQ): **(CERCLA)** 1 lb/0.454 kg

DESCRIPTION: Colorless solid. Odorless. Sinks in water; insoluble. All uses have been banned since 1972 by the US EPA.

Poison! (organochlorine) • Breathing the dust, skin or eye contact, or swallowing the material can cause illness • Firefighting gear (including SCBA) does not provide adequate protection. If exposure occurs, remove and isolate gear immediately and thoroughly decontaminate personnel • Containers may BLEVE when exposed to fire • Concentrated dust in confined areas (e.g., tanks, sewers, buildings) may explode when exposed to fire • Toxic products of combustion may include hydrogen chloride.

Hazard Classification (based on NFPA-704 Rating System)
Health Hazards (Blue): 2; Flammability (Red): 1; Reactivity (Yellow): 0

EMERGENCY RESPONSE: See Appendix A (151)
Evacuation:
Public safety: Isolate the area of spill or leak for at least 25 to 50 meters (80 to 160 feet) in all directions.
Spill: Increase, in the downwind direction, as necessary, the distance shown under "Public Safety."
Fire: If tank, rail car, or tank truck is involved in fire, isolate for at least 800 meters (½ mile) in all directions; also, consider initial evacuation for 800 meters (½ mile) in all directions.

EXPOSURE
Short-term effects: *SEEK MEDICAL ATTENTION.* **Solids:** Irritating to skin and eyes. *IF SWALLOWED*, will cause nausea, vomiting, headache, or loss of consciousness. Remove contaminated clothing and shoes. Flush affected areas with plenty of water. *IF IN EYES*, hold eyelids open and flush with plenty of water. *IF BREATHING HAS STOPPED*, give artificial respiration; *avoid mouth-to-mouth resuscitation; use bag/mask apparatus.* **Do NOT induce vomiting.** Administer a slurry of activated charcoal. *IF SWALLOWED* and victim is *CONSCIOUS AND ABLE TO SWALLOW*, have victim drink 4 to 8 ounces of water.

HEALTH HAZARDS
Personal protective equipment (PPE): B-Level PPE. OSHA Table Z-1-A air contaminant.
Recommendations for respirator selection: NIOSH *At any concentrations above the NIOSH REL, or where there is no REL, at any detectable concentration:* SCBAF:PD,PP (any self-contained breathing apparatus that has a full facepiece and is operated in a pressure-demand or other positive-pressure mode); or SAF:PD,PP:ASCBA (any supplied-air respirator that has a full facepiece and is operated in a pressure-demand or other positive-pressure mode in combination with an auxiliary, self-contained breathing apparatus operated in a pressure-demand or other positive pressure mode). *ESCAPE:* GMFOVHiE [any air-purifying, full-facepiece respirator (gas mask) with a chin-style, front- or back-mounted organic vapor canister having a high-efficiency particulate filter]; or SCBAE (any appropriate escape-type, self-contained breathing apparatus).
Exposure limits (TWA unless otherwise noted): ACGIH TLV 1 mg/m³; OSHA 1 mg/m³; NIOSH potential human carcinogen 0.5 mg/m³; skin contact contributes significantly in overall exposure.
Toxicity by ingestion: Grade 3; LD_{50} = 50 to 500 mg/kg (rat).
Long-term health effects: Disrupts hormonal function. IARC possible carcinogen, rating 2B, human evidence inadequate; animal evidence sufficient. NTP anticipated carcinogen.
Liquid or solid irritant characteristics: Eye and skin irritant. If spilled on clothing and allowed to remain, may cause smarting and reddening of the skin.
IDLH value: Potential human carcinogen; 500 mg/m³.

FIRE DATA
Flash point: 162–171°F/cc).
Behavior in fire: Water streams applied to adjacent fires will spread contamination of pesticide over wide area.

CHEMICAL REACTIVITY
Binary reactants: Strong oxidizers, alkalis

ENVIRONMENTAL DATA
Food chain concentration potential: High. Breaks down, forming DDE which can persist in animal tissue for decades.
Water pollution: DOT Appendix B, §172.101- severe marine pollutant. Harmful to aquatic life in very low concentrations. May be dangerous if it enters nearby water intakes; notify operators. Notify local health and wildlife officials. **Response to discharge:** Issue warning–water contaminant. Should be removed.

SHIPPING INFORMATION
Grades of purity: Technical; **Stability during transport:** Stable.

PHYSICAL AND CHEMICAL PROPERTIES
Physical state @ 59°F/15°C and 1 atm: Solid.
Molecular weight: 354.5
Boiling point @ 1 atm: 230°F/110°C/383°K (decomposes).
Melting/Freezing point: 226°F/108°C/381°K.
Specific gravity (water = 1): 0.99 @ 68°F/20°C (solid).
Vapor pressure: 0.0000002 mm.

DECABORANE   REC. D:0350

SYNONYMS: BORON HYDRIDE; DECABORANO (Spanish); DECABORANE(14); DECARBORON TETRADECAHYDRIDE

IDENTIFICATION
CAS Number: 17702-41-9
Formula: $B_{10}H_{14}$
DOT ID Number: UN 1868; DOT Guide Number: 134
Proper Shipping Name: Decaborane
Reportable Quantity (RQ): **(EHS)** 1 lb/0.454 kg

Decaborane

DESCRIPTION: White solid. Sharp odor. Floats on the surface of water.

Poison! • Breathing the vapor, skin or eye contact, or swallowing the material can kill you • Firefighting gear (including SCBA) does not provide adequate protection. If exposure occurs, remove and isolate gear immediately and thoroughly decontaminate personnel • Containers may BLEVE when exposed to fire • Toxic products of combustion may include boron oxide • Do not put yourself in danger by entering a contaminated area to rescue a victim.

Hazard Classification (based on NFPA-704 Rating System)
Health Hazards (Blue): 3; Flammability (Red): 2; Reactivity (Yellow): 1

EMERGENCY RESPONSE: See Appendix A (134)
Evacuation:
Public safety: Isolate spill area for at least 25 to 50 meters (30 to 160 feet) in all directions.
Spill: Consider initial downwind evacuation of at least 100 meters (330 feet).
Fire: Isolate for 800 meters (½ mile) in all directions, especially if tank, rail car, or tank truck is involved in fire.
EXPOSURE
Short-term effects: *SEEK MEDICAL ATTENTION.* **Dust:** *POISONOUS IF INHALED OR IF SKIN IS EXPOSED.* Move victim to fresh air. *IF IN EYES*, hold eyelids open and flush with plenty of water. If breathing is difficult, administer oxygen. *IF BREATHING HAS STOPPED*, give artificial respiration; *avoid mouth-to-mouth resuscitation; use bag/mask apparatus.* **Solid:** *POISONOUS IF SWALLOWED OR IF SKIN IS EXPOSED.* Remove contaminated clothing and shoes. Flush affected areas with plenty of water. *IF IN EYES*, hold eyelids open and flush with plenty of water. *IF SWALLOWED* and victim is *CONSCIOUS AND ABLE TO SWALLOW*, have victim drink 4 to 8 ounces of water and have victim induce vomiting. *IF SWALLOWED* and victim is *UNCONSCIOUS OR HAVING CONVULSIONS*, do nothing except keep victim warm.
Note to physician or authorized medical personnel: Medical observation is recommended for 24 to 48 hours after breathing overexposure, as pulmonary edema may be delayed. As first aid for pulmonary edema, consider administering a corticosteroid spray. Cigarette smoking may exacerbate pulmonary injury and should be discouraged for at least 72 hours following exposure.
Note to physician or authorized medical personnel: Treat symptomatically; administration of methocarbamol or other muscle relaxant may be helpful immediately following exposure and in the absence of symptoms.

HEALTH HAZARDS
Personal protective equipment (PPE): A-Level PPE. OSHA Table Z-1-A air contaminant.
Recommendations for respirator selection: NIOSH/OSHA
3 mg/m³: SA (any supplied-air respirator). *7.5 mg/m³:* SA:CF (any supplied-air respirator operated in a continuous-flow mode). *15 mg/m³:* SAT:CF (any supplied-air respirator that has a tight-fitting facepiece and is operated in a continuous-flow mode); or SCBAF (any self-contained breathing apparatus with a full facepiece); or SAF (any supplied-air respirator with a full facepiece). *EMERGENCY OR PLANNED ENTRY INTO UNKNOWN CONCENTRATIONS OR IDLH CONDITIONS:* SCBAF:PD,PP (any self-contained breathing apparatus that has a full facepiece and is operated in a pressure-demand or other positive-pressure mode); or SAF:PD,PP:ASCBA (any supplied-air respirator that has a full facepiece and is operated in a pressure-demand or other positive-pressure mode in combination with an auxiliary self-contained breathing apparatus operated in a pressure-demand or other positive pressure mode). *ESCAPE:* GMFOVHiE [any air-purifying, full-facepiece respirator (gas mask) with a chin-style, front- or back-mounted organic vapor canister having a high-efficiency particulate filter]; or SCBAE (any appropriate escape-type, self-contained breathing apparatus).
Exposure limits (TWA unless otherwise noted): ACGIH TLV 0.05 ppm (0.25 mg/m³); NIOSH/OSHA 0.3 mg/m³ (0.05 ppm); skin contact contributes significantly in overall exposure.
Short-term exposure limits (15-minute TWA): ACGIH STEL 0.15 (0.75 mg/m³); NIOSH/OSHA STEL 0.9 mg/m³ (0.15 ppm); skin contact contributes significantly in overall exposure.
Toxicity by ingestion: Grade 4; oral LD_{50} = 40 mg/kg (mouse).
Liquid or solid irritant characteristics: Eye and skin irritant.
Odor threshold: 0.05 ppm.
IDLH value: 15 mg/m³.

FIRE DATA
Flash point: Combustible solid. 176°F/80°C (cc).
Fire extinguishing agents not to be used: Halogenated extinguishing agents
Behavior in fire: May explode when hot. Burns with a green-colored flame.
Autoignition temperature: 300°F/149°C/422°K.

CHEMICAL REACTIVITY
Reactivity with water: Reacts slowly to form flammable hydrogen gas, which can accumulate in closed area.
Binary reactants: May react with metals evolving flammable hydrogen gas. May ignite spontaneously on exposure to air. Reacts with oxidizers, halogenated compounds (especially carbon tetrachloride). Corrosive to natural rubber, some synthetic rubbers, some greases, and some lubricants.
Neutralizing agents for acids and caustics: Flush with 3% aqueous ammonia solution, then with water. Methyl alcohol may also be used.

ENVIRONMENTAL DATA
Food chain concentration potential: Negative; unlikely to accumulate.
Water pollution: Effect of low concentrations on aquatic life is unknown. Fouling to shoreline. May be dangerous if it enters nearby water intakes; notify operators. Notify local health and wildlife officials. **Response to discharge:** Issue warning–high flammability, water contaminant, poison. Restrict access. Mechanical containment. Should be removed. Chemical and physical treatment.

SHIPPING INFORMATION
Grades of purity: Technical: 95+%; High purity: 99+%; **Storage temperature:** Ambient; **Inert atmosphere:** None; **Venting:** Pressure-vacuum; **Stability during transport:** Stable.

PHYSICAL AND CHEMICAL PROPERTIES
Physical state @ 59°F/15°C and 1 atm: Solid.
Molecular weight: 122.2
Boiling point @ 1 atm: 415°F/213°C/486°K.
Melting/Freezing point: 210°F/99°C/372°K.
Specific gravity (water = 1): 0.94 @ 77°F/25°C (solid).
Heat of combustion: $-28,699$ Btu/lb = $-15,944$ cal/g = -667.10×10^5 J/kg.
Heat of decomposition: -279 Btu/lb = -155 cal/g = -6.49×10^5 J/kg.
Vapor pressure: 0.2 mm.

DECAHYDRONAPHTHALENE (cis/trans)   REC. D:0400

SYNONYMS: BICYCLO[4.4.0]DECANE; DEC; DECAHIDRONAFTALENO (Spanish); DECALIN (*cis/trans*); DECALIN SOLVENT; DE KALIN; NAPTHALANE (do not confuse with naphthalene); NAPHTHANE; PERHYDRONAPTHALENE

IDENTIFICATION
CAS Number: 91-17-8
Formula: $C_{10}H_{18}$
DOT ID Number: UN 1147; DOT Guide Number: 128
Proper Shipping Name: Decahydronaphthalene

DESCRIPTION: Colorless liquid. Turpentine-like odor. Floats on the surface of water; insoluble.

Highly flammable • Containers may BLEVE when exposed to fire •Vapors may form explosive mixture with air • Vapors are heavier than air and will collect and stay in low areas • Vapors may travel long distances to ignition sources and flashback • Vapors in confined areas (e.g., tanks, sewers, buildings) may explode when exposed to fire • Irritating to the skin, eyes, and respiratory tract • Toxic products of combustion may include carbon monoxide.

Hazard Classification (based on NFPA-704 Rating System)
Health Hazards (Blue): 2; Flammability (Red): 2; Reactivity (Yellow): 0

EMERGENCY RESPONSE: See Appendix A (128)
Evacuation:
Public safety: Isolate spill area for at least 25 to 50 meters (80 to 160 feet) in all directions.
Spill: Large spill–Consider initial downwind evacuation for at least 300 meters (1000 feet).
Fire: Isolate for 800 meters (½ mile) in all directions, especially if tank, rail car, or tank truck is involved in fire.
EXPOSURE
Short-term effects: *SEEK MEDICAL ATTENTION.* **Liquid:** Irritating to skin and eyes. *IF SWALLOWED*, will cause headache, nausea, or vomiting. Drops in the lungs may cause pneumonia. Remove contaminated clothing and shoes. Flush affected areas with plenty of water. *IF IN EYES*, hold eyelids open and flush with plenty of water. *IF SWALLOWED* and victim is *CONSCIOUS AND ABLE TO SWALLOW*,**Do NOT induce vomiting.** Chemical pneumonitis may develop. The consumption of alcoholic beverages enhances the toxic effect.

HEALTH HAZARDS
Personal protective equipment (PPE): Air mask or self-contained breathing apparatus if in enclosed tank; rubber gloves or protective cream; goggles or face shield.
Toxicity by ingestion: Grade 2; oral LD_{50} = 4,170 mg/kg (rat).
Long-term health effects: May effect the central nervous system.
Vapor (gas) irritant characteristics: Eye and respiratory tract irritant.
Liquid or solid irritant characteristics: Eye and skin irritant.

FIRE DATA
Flash point: 134°F/57°C (oc); 122°F/50°C (*cis-*); 142°F/61°C (*trans-*)
Flammable limits in air: LEL: 0.7%; UEL: 5.4% both @ 212°F/100°C (*trans-*); 0.7%; UEL: 4.9% (*cis-*)
Autoignition temperature: 482°F/250°C/523°K.
Electrical hazard: Class I, Group D Substance may generate electrostatic charges due to low conductivity.
Burning rate: 5.9 mm/min

CHEMICAL REACTIVITY
Binary reactants: Reacts with strong oxidizers.

ENVIRONMENTAL DATA
Water pollution: Effect of low concentrations on aquatic life is unknown. Fouling to shoreline. May be dangerous if it enters nearby water intakes; notify operators. Notify local health and wildlife officials. **Response to discharge:** Mechanical containment. Should be removed. Chemical and physical treatment.

SHIPPING INFORMATION
Grades of purity: Technical: mixture of *cis*-(35%) and *trans*-(65%) isomers. Properties of all such mixtures are very similar. Spectro grade; **Storage temperature:** Ambient; **Inert atmosphere:** None; **Venting:** Open (flame arrester); **Stability during transport:** Stable.

PHYSICAL AND CHEMICAL PROPERTIES
Physical state @ 59°F/15°C and 1 atm: Liquid.
Molecular weight: 138.2
Boiling point @ 1 atm: 383°F/195°C/468°K (*cis-*)
Melting/Freezing point: –44°F/–42°C/231°K (*cis-*); –22°F/–30°C/243°K (*trans-*)
Specific gravity (water = 1): 0.89 @ 68°F/20°C (liquid).
Liquid surface tension: 30 dynes/cm = 0.030 N/m @ 68°F/20°C.
Liquid water interfacial tension: 51.5 dynes/cm = 0.0515 N/m @ 68°F/20°C.
Relative vapor density (air = 1): 4.8
Latent heat of vaporization: 130 Btu/lb = 71 cal/g = 3.0×10^5 J/kg.
Heat of combustion: –19,200 Btu/lb = –10,700 cal/g = $–447 \times 10^5$ J/kg.
Vapor pressure: 1.3 mbar @ 71°F/22°C.

DECALDEHYDE   REC. D:0450

SYNONYMS: ALDEHYDE C-10; CAPRIC ALDEHYDE; CAPRALDEHYDE; DECALDEHIDO (Spanish); *n*-DECYL ALDEHYDE; 1-DECANAL; DECANAL (mixed isomers); DECANAL; 1-DECYL ALDEHYDE

IDENTIFICATION
CAS Number: 112-31-2
Formula: $CH_3(CH_2)_8CHO$
DOT ID Number: NA 1993; DOT Guide Number: 128
Proper Shipping Name: Combustible liquid, n.o.s.

DESCRIPTION: Colorless to light yellow liquid. Pleasant odor. Floats on the surface of water.

Combustible • Containers may BLEVE when exposed to fire •Vapors may form explosive mixture with air • Vapors are heavier than air and will collect and stay in low areas • Vapors may travel long distances to ignition sources and flashback • Vapors in confined areas (e.g., tanks, sewers, buildings) may explode when exposed to fire • Highly irritating to the eyes and respiratory tract; this material has anesthetic properties • Toxic products of combustion may include carbon monoxide.

Hazard Classification (based on NFPA-704 Rating System)
Health Hazards (Blue): 0; Flammability (Red): 2; Reactivity (Yellow): 0

EMERGENCY RESPONSE: See Appendix A (128)
Evacuation:
Public safety: Isolate spill area for at least 25 to 50 meters (80 to 160 feet) in all directions.
Spill: Large spill–Consider initial downwind evacuation for at least 300 meters (1000 feet).
Fire: Isolate for 800 meters (½ mile) in all directions, especially if tank, rail car, or tank truck is involved in fire.

EXPOSURE
Short-term effects: *SEEK MEDICAL ATTENTION.* **Liquid:** Irritating to skin and eyes. Remove contaminated clothing and shoes. Flush affected areas with plenty of water. *IF IN EYES*, hold eyelids open and flush with plenty of water.

HEALTH HAZARDS
Personal protective equipment (PPE): Protective clothing and chemical goggles. Chemical protective material(s) reported to offer minimal to poor protection: butyl rubber, neoprene, Viton®, Viton®/neoprene, butyl rubber/neoprene, Saranex®. Also, PVC offers limited protection.
Toxicity by ingestion: Grade 0; LD_{50} more than 33.3 g/kg (rat).
Liquid or solid irritant characteristics: Eye and skin irritant.
Odor threshold: 0.168 ppm.

FIRE DATA
Flash point: 185°F/85°C (oc).
Electrical hazard: Class I, Group C.

CHEMICAL REACTIVITY
Binary reactants: Reacts with oxidizers. Incompatible with galvanized iron, acids, caustics, amines.
Reactivity group: 19
Compatibility class: Aldehydes

ENVIRONMENTAL DATA
Food chain concentration potential: Negative; unlikely to accumulate.
Water pollution: DOT Appendix B, §172.101–marine pollutant. Effect of low concentrations on aquatic life is unknown. Fouling to shoreline. May be dangerous if it enters nearby water intakes; notify operators. Notify local health and pollution control officials.
Response to discharge: Mechanical containment. Should be removed. Chemical and physical treatment.

SHIPPING INFORMATION
Storage temperature: Ambient; **Inert atmosphere:** None; **Venting:** Open (flame arrester); **Stability during transport:** Stable.

PHYSICAL AND CHEMICAL PROPERTIES
Physical state @ 59°F/15°C and 1 atm: Liquid.
Molecular weight: 145.3
Boiling point @ 1 atm: 404-410°F/207-210°C/480-483°K.
Melting/Freezing point: 64°F/18°C/291°K.
Specific gravity (water = 1): 0.830 @ 15°C (liquid).
Liquid surface tension: 28.0 dynes/cm = 0.0280 N/m at 24°C.
Liquid water interfacial tension: 16.9 dynes/cm = 0.0169 N/m at 22.7°C.
Ratio of specific heats of vapor (gas): 1.036

Heat of combustion: estimate $-18,000$ Btu/lb = $-10,000$ cal/g = -424×10^5 J/kg.

DECANE REC. D:0500

SYNONYMS: *n*-DECANE; *n*-DECANO (Spanish)

IDENTIFICATION
CAS Number: 124-18-5
Formula: $C_{10}H_{22}$; $CH_3(CH_2)_8CH_3$
DOT ID Number: UN 2247; DOT Guide Number: 128
Proper Shipping Name: *n*-Decane

DESCRIPTION: Colorless liquid. Floats on the surface of water.

Highly flammable • Containers may BLEVE when exposed to fire • Vapors may form explosive mixture with air • Vapors are heavier than air and will collect and stay in low areas • Vapors may travel long distances to ignition sources and flashback • Vapors in confined areas (e.g., tanks, sewers, buildings) may explode when exposed to fire • Irritating to the skin, eyes, and respiratory tract • Toxic products of combustion may include carbon monoxide.

Hazard Classification (based on NFPA-704 Rating System)
Health Hazards (Blue): 0; Flammability (Red): 2; Reactivity (Yellow): 0

EMERGENCY RESPONSE: See Appendix A (128)
Evacuation:
Public safety: Isolate spill area for at least 25 to 50 meters (80 to 160 feet) in all directions.
Spill: Large spill–Consider initial downwind evacuation for at least 300 meters (1000 feet).
Fire: Isolate for 800 meters (½ mile) in all directions, especially if tank, rail car, or tank truck is involved in fire.

EXPOSURE
Short-term effects: *SEEK MEDICAL ATTENTION.* **Vapor:** Move victim to fresh air. Remove contaminated clothing and shoes. Wash affected areas with plenty of soap and water. *IF SWALLOWED*, **Do NOT induce vomiting.** If breathing is difficult, administer oxygen.
Personal protective equipment (PPE): Approved self-contained respirator, safety glasses, impervious gloves, and chemical protective clothing. Chemical protective material(s) reported to offer minimal to poor protection: Viton®/neoprene.

HEALTH HAZARDS
Personal protective equipment (PPE): Chemical protective material(s) reported to offer minimal to poor protection: Viton®/neoprene.
Long-term health effects: Continued skin contact may cause dermatitis and hair loss. Enhances carcinogenicity of known carcinogens and has demonstrated tumor promoting activity when tested dermally in mice.
Vapor (gas) irritant characteristics: Vapors cause moderate irritation such that personnel will find high concentrations unpleasant. The effect is temporary.
Liquid or solid irritant characteristics: Eye irritant. If spilled on clothing and allowed to remain, may cause smarting and reddening of the skin.

FIRE DATA
Flash point: 115°F/46°C (cc).
Flammable limits in air: LEL: 0.82%; UEL: 5.4%.

Fire extinguishing agents not to be used: Do not spray water directly on fire-it will scatter and spread the fire.
Autoignition temperature: 410°F/210°C/483°K.

ENVIRONMENTAL DATA
Water pollution: Effects of low concentrations on aquatic life is unknown. May be dangerous if it enters nearby water intakes; notify operators. Notify local health and wildlife officials.
Response to discharge: Restrict access. Mechanical containment. Should be removed. Chemical and physical treatment.

SHIPPING INFORMATION
Grades of purity: Pure or technical grades; **Storage temperature:** Ambient; **Inert atmosphere:** None; **Stability during transport:** Stable.

PHYSICAL AND CHEMICAL PROPERTIES
Physical state @ 59°F/15°C and 1 atm: Liquid.
Molecular weight: 142.29
Boiling point @ 1 atm: 345°F/174°C/447°K.
Melting/Freezing point: –22°F/–30°C/243°K.
Specific gravity (water = 1): 0.73 @ 60°F

DECANOIC ACID   **REC. D:0550**

SYNONYMS: ACIDO DECANOICO (Spanish); CAPRIC ACID; *n*-CAPRIC ACID; CAPRINIC ACID; CAPRYNIC ACID; *n*-DECANOIC ACID; *n*-DECOIC ACID; *n*-DECYLIC ACID; HEXACID 1095; NEO-FAT 10; 1-NONANECARBOXYLIC ACID

IDENTIFICATION
CAS Number: 334-48-5
Formula: $C_{10}H_{20}O_2$; $CH_3(CH_2)_8CO_2H$

DESCRIPTION: White crystalline solid. Rancid odor. Insoluble in water.

Combustible solid • Corrosive to the skin, eyes, and respiratory tract; contact with skin or eyes can cause burns and vision impairment; inhalation symptoms may be delayed • Dust cloud may explode if ignited in an enclosed area • Containers may BLEVE when exposed to fire • Toxic products of combustion may include carbon monoxide.

Hazard Classification (based on NFPA-704 Rating System)
Health Hazards (Blue): 0; Flammability (Red): 1; Reactivity (Yellow): 0

EMERGENCY RESPONSE: See Appendix A (171)
Evacuation:
Public safety: Isolate the area of spill or leak for at least 10 to 25 meters (30 to 80 feet) in all directions.
Spill: Increase, in the downwind direction, as necessary, the distance shown under "Public Safety."
Fire: If any large container is involved in fire, isolate for at least 800 meters (½ mile) in all directions; also, consider initial evacuation for 800 meters (½ mile) in all directions.

EXPOSURE
Short-term effects: *SEEK MEDICAL ATTENTION.* **Vapor:** Irritating to eyes, nose, and throat. *IF INHALED*, will, will cause coughing or difficult breathing. *IF IN EYES*, hold eyelids open and flush with plenty of water. *IF BREATHING HAS STOPPED*, give artificial respiration. If breathing is difficult, administer oxygen. May cause lung edema; physical exertion will aggravate this condition. **Liquid or solid:** Will burn skin and eyes. *IF SWALLOWED*, will cause nausea and vomiting. Remove contaminated clothing and shoes. Flush affected areas with plenty of water. *IF IN EYES*, hold eyelids open and flush with plenty of water.
Note to physician or authorized medical personnel: Medical observation is recommended for 24 to 48 hours after breathing overexposure, as pulmonary edema may be delayed. As first aid for pulmonary edema, consider administering a corticosteroid spray. Cigarette smoking may exacerbate pulmonary injury and should be discouraged for at least 72 hours following exposure.

HEALTH HAZARDS
Personal protective equipment (PPE): Respirator, chemical safety goggles, rubber boots, and heavy rubber gloves. Butyl rubber is generally suitable for carbooxylic acid compounds.
Toxicity by ingestion: Grade 3: LD_{50} = 129 mg/kg mouse, intravenous.
Vapor (gas) irritant characteristics: Vapors cause severe irritation of eyes and throat and can cause eye and lung injury. They cannot be tolerated even at low concentrations.
Liquid or solid irritant characteristics: Severe eye and skin irritant. Causes second- and third-degree burns on short contact and is very injurious to the eyes.

FIRE DATA
Flash point: More than 230°F/110°C (cc).
Fire extinguishing agents not to be used: Water may not be effective.

CHEMICAL REACTIVITY
Binary reactants: Corrosive solution, attacks most common metals.
Neutralizing agents for acids and caustics: Sodium bicarbonate solution.
Reactivity group: 4
Compatibility class: Organic acids

ENVIRONMENTAL DATA
Water pollution: May be dangerous to aquatic life in high concentrations. May be dangerous if it enters nearby water intakes; notify operators. Notify local health and wildlife officials. **Response to discharge:** Restrict access. Mechanical containment. Should be removed.

SHIPPING INFORMATION
Grades of purity: 99+%; **Storage temperature:** Ambient; **Stability during transport:** Stable.

PHYSICAL AND CHEMICAL PROPERTIES
Physical state @ 59°F/15°C and 1 atm: Solid.
Molecular weight: 172.27
Boiling point @ 1 atm: 514–518°F/268–270°C/541–543°K.
Melting/Freezing point: 88–90°F/31–32°C/304–305°K.
Specific gravity (water = 1): 0.893
Relative vapor density (air = 1): 5.94

1-DECENE   **REC. D:0600**

SYNONYMS: DECENE; *α*-DECENE; *α*-DECENO (Spanish)

322 *n*-Decyl acrylate

IDENTIFICATION
CAS Number: 872-05-9
Formula: C_9H_{20}; $CH_2=CH(CH_2)_7CH_3$
DOT ID Number: UN 1993; DOT Guide Number: 128
Proper Shipping Name: Flammable liquid, n.o.s.

DESCRIPTION: Colorless watery liquid. Pleasant odor. Floats on the surface of water.

Flammable • Containers may BLEVE when exposed to fire • Vapors may form explosive mixture with air • Vapors are heavier than air and will collect and stay in low areas • Vapors may travel long distances to ignition sources and flashback • Vapors in confined areas (e.g., tanks, sewers, buildings) may explode when exposed to fire • Irritating to the skin, eyes, and respiratory tract • Toxic products of combustion may include carbon monoxide.

Hazard Classification (based on NFPA-704 Rating System)
Health Hazards (Blue): 0; Flammability (Red): 2; Reactivity (Yellow): 0

EMERGENCY RESPONSE: See Appendix A (128)
Evacuation:
Public safety: Isolate spill area for at least 25 to 50 meters (80 to 160 feet) in all directions.
Spill: Large spill–Consider initial downwind evacuation for at least 300 meters (1000 feet).
Fire: Isolate for 800 meters (½ mile) in all directions, especially if tank, rail car, or tank truck is involved in fire.

EXPOSURE
Short-term effects: *SEEK MEDICAL ATTENTION.* **Liquid:** Irritating to skin and eyes. Remove contaminated clothing and shoes. Flush affected areas with plenty of water. *IF IN EYES*, hold eyelids open and flush with plenty of water.

HEALTH HAZARDS
Personal protective equipment (PPE): Organic canister); or air–supplied mask, goggles or face shield.
Vapor (gas) irritant characteristics: Eye and respiratory tract irritant. Slight smarting of eyes and respiratory system at high concentrations. The effect is temporary.
Liquid or solid irritant characteristics: If spilled on clothing and allowed to remain, may cause smarting and reddening of the skin.

FIRE DATA
Flash point: 128°F/53°C (oc).
Autoignition temperature: 455°F/235°C/508°K.
Burning rate: 6.0 mm/min

CHEMICAL REACTIVITY
Binary reactants: Oxidizers.

ENVIRONMENTAL DATA
Food chain concentration potential: Negative; unlikely to accumulate.
Water pollution: Effect of low concentrations on aquatic life is unknown. Fouling to shoreline. May be dangerous if it enters nearby water intakes; notify operators. Notify local health and pollution control officials. **Response to discharge:** Mechanical containment. Should be removed. Chemical and physical treatment.

SHIPPING INFORMATION
Grades of purity: Technical: 95–99%; **Storage temperature:** Ambient; **Inert atmosphere:** None; **Venting:** Open (flame arrester); **Stability during transport:** Stable.

PHYSICAL AND CHEMICAL PROPERTIES
Physical state @ 59°F/15°C and 1 atm: Liquid.
Molecular weight: 140.2
Boiling point @ 1 atm: 339.1°F/170.6°C/443.8°K.
Melting/Freezing point: −87.3°F/−66.3°C/206.9°K.
Specific gravity (water = 1): 0.741 @ 68°F/20°C (liquid).
Liquid surface tension: 24 dynes/cm = 0.024 N/m @ 68°F/20°C.
Liquid water interfacial tension: 28 dynes/cm = 0.028 N/m at 22.7°C.
Relative vapor density (air = 1): 4.8
Ratio of specific heats of vapor (gas): 1.039
Latent heat of vaporization: 119 Btu/lb = 65.9 cal/g = 2.76×10^5 J/kg.
Heat of combustion: −19,107 Btu/lb = −10,615 cal/g = -444.43×10^5 J/kg.

n-DECYL ACRYLATE   REC. D:0650

SYNONYMS: ACRYLIC ACID, DECYL ESTER; ACRYLIC ACID, *n*-DECYL ESTER; DECIL ACRILATO (Spanish); 2-PROPENOIC ACID, DECYL ESTER

IDENTIFICATION
CAS Number: 2156-96-9
Formula: $C_{13}H_{24}O_2$; $CH_3(CH_2)_9OCOCH=CH_2$

DESCRIPTION: Liquid. Floats on the surface of water.

Combustible • Firefighting gear (including SCBA) does not provide adequate protection. If exposure occurs, remove and isolate gear immediately and thoroughly decontaminate personnel • Containers may BLEVE when exposed to fire • Vapors in confined areas (e.g., tanks, sewers, buildings) may explode when exposed to fire • Irritating to the skin, eyes, and respiratory tract • Toxic products of combustion may include carbon monoxide and toxic acrylic acid, one of the most serious eye injury chemicals and a severe skin irritant.

Hazard Classification (based on NFPA-704 Rating System)
Health Hazards (Blue): 2; Flammability (Red): 1; Reactivity (Yellow): 0

EMERGENCY RESPONSE: See Appendix A (171)
Evacuation:
Public safety: Isolate the area of spill or leak for at least 10 to 25 meters (30 to 80 feet) in all directions.
Spill: Increase, in the downwind direction, as necessary, the distance shown under "Public Safety."
Fire: If any large container is involved in fire, isolate for at least 800 meters (½ mile) in all directions; also, consider initial evacuation for 800 meters (½ mile) in all directions.

EXPOSURE
Short-term effects: *SEEK MEDICAL ATTENTION.* **Liquid:** Irritating to skin and eyes. Harmful if swallowed. *IF IN EYES OR ON SKIN:* Flush with running water for at least 15 minutes; hold eyelids open if necessary. Wash skin with soap and water. Remove and isolate contaminated clothing and shoes at the site. *IF SWALLOWED* and victim is *CONSCIOUS AND ABLE TO SWALLOW*, have victim drink several glasses of water and induce

vomiting. *IF SWALLOWED* and victim is *UNCONSCIOUS OR HAVING CONVULSIONS,* do nothing except keep victim warm.

HEALTH HAZARDS
Personal protective equipment (PPE): Wear self-contained positive pressure breathing apparatus and full protective clothing.
Skin or Ingestion: Prolonged contact may cause severe damage to tissues. May be fatal if swallowed or absorbed through skin.
Toxicity by ingestion: Grade 1; LD_{50} = 6.46 g/kg (rat).
Liquid or solid irritant characteristics: Severe skin irritant.

FIRE DATA
Flash point: 441°F/227°C (oc).
Electrical hazard: Class I, Group D.
Stoichiometric air-to-fuel ratio: 12 estimate.

CHEMICAL REACTIVITY
Binary reactants: Incompatible with copper, copper alloys, zinc, galvanized steel, alloys having more than 10% zinc by weight, strong oxidizing agents and polymerization initiators. Will swell and soften certain rubbers, and soften and remove certain paints.
Neutralizing agents for acids and caustics: Not applicable
Polymerization: Will polymerize unless inhibited. Inhibited material may polymerize if heated, if cooled so that the inhibitor crystallizes, if stored in an oxygen-free atmosphere, or if stored in contact with copper and copper alloys, zinc and zinc alloys with more than 10% zinc, and galvanized steel. Strong oxidizers and other contaminants may also initiate this reaction.
Reactivity group: 14
Compatibility class: Acrylates

ENVIRONMENTAL DATA
Water pollution: DOT Appendix B, §172.101–marine pollutant. Effects of low concentrations on aquatic life is unknown. May be dangerous if it enters nearby water intakes; notify operators. Fouling to shoreline. Notify local health and wildlife officials.
Response to discharge: Issue warning–water contaminant. Restrict access. Mechanical containment. Should be removed. Chemical and physical treatment.

SHIPPING INFORMATION
Storage temperature: Ambient; **Inert atmosphere:** None; **Venting:** Open; **Stability during transport:** Stable. Avoid heat, light and contact with polymerization initiators.

PHYSICAL AND CHEMICAL PROPERTIES
Physical state @ 59°F/15°C and 1 atm: Liquid.
Molecular weight: 212.37
Boiling point @ 1 atm: 316°F/158°C/431°K.
Melting/Freezing point: Less than 32°F/0°C/273°K.
Specific gravity (water = 1): 0.8781 @ 68°F/20°C.
Relative vapor density (air = 1): 4.8
Vapor pressure: (Reid) Low; less than 0.01 mm.

n-DECYL ALCOHOL   REC. D:0700

SYNONYMS: AGENT 504; ALCOHOL C-10; ALCOHOL *n*-DECILICO (Spanish); ANTAK; C-10 ALCOHOL; CAPRIC ALCOHOL; CAPRINIC ALCOHOL; DECANAL DIMETHYL ACETAL; DECANOL; L-DECANOL; DYTOL S-91; *N*-DECATYL ALCOHOL; *N*-DECYL ALCOHOL; DECYLIC ALCOHOL; DYTOL S-91; EPAL 10; LOROL-22; NONYLCARBINOL; PRIMARY DECYL ALCOHOL; ROYALTAC; SIPOL L10

IDENTIFICATION
CAS Number: 112-30-1
Formula: $CH_3(CH_2)_8CH_2OH$
DOT ID Number: UN 1987; DOT Guide Number: 127
Proper Shipping Name: Alcohol, n.o.s.

DESCRIPTION: Colorless to pale yellow liquid. Faint alcohol odor. Floats on the surface of water.

Combustible • Containers may BLEVE when exposed to fire • Vapors may form explosive mixture with air • Vapors are heavier than air and will collect and stay in low areas • Vapors may travel long distances to ignition sources and flashback • Vapors in confined areas (e.g., tanks, sewers, buildings) may explode when exposed to fire • Irritating to the skin, eyes, and respiratory tract • Toxic products of combustion may include carbon monoxide.

Hazard Classification (based on NFPA-704 Rating System)
Health Hazards (Blue): 0; Flammability (Red): 2; Reactivity (Yellow): 0

EMERGENCY RESPONSE: See Appendix A (127)
Evacuation:
Public safety: Isolate spill area for at least 25 to 50 meters (80 to 160 feet) in all directions.
Spill: Large spill–Consider initial downwind evacuation for at least 300 meters (1000 feet).
Fire: Isolate for 800 meters (½ mile) in all directions, especially if tank, rail car, or tank truck is involved in fire.

EXPOSURE
Short-term effects: *SEEK MEDICAL ATTENTION.* The use of alcoholic beverages may enhance toxic effects. **Liquid:** Irritating to eyes. *IF IN EYES*, hold eyelids open and flush with plenty of water.

HEALTH HAZARDS
Personal protective equipment (PPE): Goggles or face shield. Natural rubber, neoprene, and nitrile rubber may offer some protection from alcohols. Chemical protective material(s) reported to offer minimal to poor protection: Viton®/neoprene, butyl rubber,/neoprene.
Toxicity by ingestion: Grade 1; LD_{50} = 5 to 15 g/kg (rat).
Liquid or solid irritant characteristics: Eye irritant. Practically harmless to the skin.

FIRE DATA
Flash point: 180°F/82°C (oc).
Autoignition temperature: 550°F/288°C/561°K.
Electrical hazard: Class I, Group D.

CHEMICAL REACTIVITY
Binary reactants: Oxidizers.
Reactivity group: 20
Compatibility class: Alcohols, glycols

ENVIRONMENTAL DATA
Food chain concentration potential: Negative; unlikely to accumulate.
Water pollution: DOT Appendix B, §172.101–marine pollutant. Effect of low concentrations on aquatic life is unknown. Fouling to shoreline. May be dangerous if it enters nearby water intakes; notify operators. Notify local health and pollution control officials.
Response to discharge: Mechanical containment. Should be removed. Chemical and physical treatment.

324 *n*-Decylbenzene

SHIPPING INFORMATION
Grades of purity: 95-99%; **Storage temperature:** Ambient; **Inert atmosphere:** None; **Venting:** Open (flame arrester); **Stability during transport:** Stable.

NAS HAZARD CLASSIFICATIONS FOR BULK WATER TRANSPORTATION
FIRE: 1
HEALTH: Vapor irritant: 0; Liquid or solid irritant: 0; Poisons: 0
WATER POLLUTION: Human toxicity: 0; Aquatic toxicity: 0; Aesthetic effect: 2
REACTIVITY: Other chemicals: 2; Water: 0; Self-reaction: 0

PHYSICAL AND CHEMICAL PROPERTIES
Physical state @ 59°F/15°C and 1 atm: Liquid.
Molecular weight: 158.29
Boiling point @ 1 atm: 446°F/230°C/503°K.
Melting/Freezing point: 44°F/6.9°C/280.1°K.
Critical temperature: 801°F/427°C/700°K.
Critical pressure: 320 psi = 22 atm = 2.2 MN/m^2.
Specific gravity (water = 1): 0.840 @ 68°F/20°C (liquid).
Liquid surface tension: 28.9 dynes/cm = 0.0289 N/m @ 68°F/20°C.
Liquid water interfacial tension: estimate 60 dynes/cm = 0.06 N/m at 10°C.
Relative vapor density (air = 1): 5.5
Ratio of specific heats of vapor (gas): 1.035
Latent heat of vaporization: estimate 130 Btu/lb = 74 cal/g = 3.1 x 10^5 J/kg.
Heat of combustion: –18,000 Btu/lb = –9980 cal/g = 418 x 10^5 J/kg.
Vapor pressure: (Reid) Very low.

n-DECYLBENZENE   REC. D:0750

SYNONYMS: DECILBENCENO (Spanish); DECYLBENZENE; *L*-PHENYLDECANE

IDENTIFICATION
CAS Number: 104-72-3
Formula: $C_6NO_5(CH_2)_9CH_3$

DESCRIPTION: White liquid. Floats on water surface; insoluble.

Combustible • Corrosive to the skin, eyes, and respiratory tract; contact with skin or eyes can cause burns and vision impairment; inhalation symptoms may be delayed • Containers may BLEVE when exposed to fire • Vapors in confined areas (e.g., tanks, sewers, buildings) may explode when exposed to fire • Toxic products of combustion may include carbon monoxide.

Hazard Classification (based on NFPA-704 Rating System)
Health Hazards (Blue): 2; Flammability (Red): 1; Reactivity (Yellow): 0

EMERGENCY RESPONSE: See Appendix A (171)
Evacuation:
Public safety: Isolate the area of spill or leak for at least 10 to 25 meters (30 to 80 feet) in all directions.
Spill: Increase, in the downwind direction, as necessary, the distance shown under "Public Safety."
Fire: If any large container is involved in fire, isolate for at least 800 meters (½ mile) in all directions; also, consider initial evacuation for 800 meters (½ mile) in all directions.

EXPOSURE
Short-term effects: *SEEK MEDICAL ATTENTION*. **Vapor:** Irritating to eyes, nose, and throat. *IF INHALED*, will, will cause coughing or difficult breathing. *IF IN EYES*, hold eyelids open and flush with plenty of water. *IF BREATHING HAS STOPPED*, give artificial respiration. If breathing is difficult, administer oxygen. May cause lung edema; physical exertion will aggravate this condition. **Liquid:** Irritating to skin and eyes. *IF SWALLOWED*, will cause nausea and vomiting. Remove contaminated clothing and shoes. Flush affected areas with plenty of water. *IF IN EYES*, hold eyelids open and flush with plenty of water. **Do NOT induce vomiting.**
Note to physician or authorized medical personnel: Medical observation is recommended for 24 to 48 hours after breathing overexposure, as pulmonary edema may be delayed. As first aid for pulmonary edema, consider administering a corticosteroid spray. Cigarette smoking may exacerbate pulmonary injury and should be discouraged for at least 72 hours following exposure.

HEALTH HAZARDS
Personal protective equipment (PPE): B-Level PPE. Goggles or face shield; rubber gloves.
Vapor (gas) irritant characteristics: Respiratory tract irritant.
Liquid or solid irritant characteristics: Eye and skin irritant.

FIRE DATA
Flash point: 225°F/107°C (cc).
Fire extinguishing agents not to be used: Water or foam may cause frothing.
Burning rate: 5.04 mm/min
Electrical hazard: Class I, Group D

CHEMICAL REACTIVITY
Binary reactants: Oxidizers, nitric acid. May attack some forms of plastics, rubber, and coatings.
Reactivity group: 32
Compatibility class: Aromatic hydrocarbons

ENVIRONMENTAL DATA
Food chain concentration potential: Negative; unlikely to accumulate.
Water pollution: Effect of low concentrations on aquatic life is unknown. Fouling to shoreline. May be dangerous if it enters nearby water intakes; notify operators. Notify local health and wildlife officials. **Response to discharge:** Mechanical containment. Should be removed. Chemical and physical treatment.

SHIPPING INFORMATION
Grades of purity: Commercial; **Storage temperature:** Ambient; **Inert atmosphere:** None; **Venting:** Open (flame arrester); **Stability during transport:** Stable.

PHYSICAL AND CHEMICAL PROPERTIES
Physical state @ 59°F/15°C and 1 atm: Liquid.
Molecular weight: 218
Boiling point @ 1 atm: 572°F/300°C/573°K.
Specific gravity (water = 1): 0.855 @ 68°F/20°C (liquid).
Liquid surface tension: 29.95 dynes/cm = 0.02995 N/m @ 68°F/20°C.
Latent heat of vaporization: 103.8 Btu/lb = 57.67 cal/g = 2.413 x 10^5 J/kg.
Heat of combustion: –18,400 Btu/lb = –10,200 cal/g = –427 x 10^5 J/kg.
Vapor pressure: Unknown.

DEMETON
REC. D:0800

SYNONYMS: BAY 10756; BAYER 8169; DEMETONA (Spanish); DEMETON O+DEMETON S; DEMOX®; DENOX®; O,O-DIETHYL-2-ETHYLMERCAPTOETHYL THIOPHOSPHATE, DIETHOXYTHIOPHOSPHORIC ACID; E-1059; MERCAPTOPHOS; PHOSPHOROTHIOIC ACID,O,O-DIETHYL O-2-(ETHYLTHIO)ETHYL ESTER, MIXED WITH O,O-DIETHYL-s-2-(ETHYLTHIO)ETHYL PHOSPHOROTHIOATE; SYSTOX®; SYSTEMOX; O,O[DIETHYL-O (AND S)-]2-(ETHYLTHIO)ETHYLPHOSPHOROTHIOATES; SYSTOX® AND ISOSYSTOX MIXTURE; ULV

IDENTIFICATION
CAS Number: 8065-48-3
Formula: $C_8H_{19}O_3PS_2$; $C_8H_{19}O_3PS_2 \cdot C_8H_{10}$ (mixture)
DOT ID Number: UN 3018; DOT Guide Number: 152
Proper Shipping Name: Organophosphorus pesticide, liquid, toxic, n.o.s.
Reportable Quantity (RQ): **(EHS)** 1 lb/0.045 kg

DESCRIPTION: Pale yellow to amber liquid. Unpleasant odor; like sulfur. Sinks in water; slightly soluble; mixes slowly.

Poison! (organophosphate)• Breathing the vapor, skin or eye contact, or swallowing the material can kill you • Firefighting gear (including SCBA) does not provide adequate protection. If exposure occurs, remove and isolate gear immediately and thoroughly decontaminate personnel • Vapors are heavier than air and will collect and stay in low areas • Containers may BLEVE when exposed to fire • Toxic products of combustion may include sulfur dioxide and phosphoric acid • Do not put yourself in danger by entering a contaminated area to rescue a victim.

Hazard Classification (based on NFPA-704 Rating System)
Health Hazards (Blue): 3; Flammability (Red): 2; Reactivity (Yellow): 0

EMERGENCY RESPONSE: See Appendix A (152)
Evacuation:
Public safety: Isolate the area of spill or leak for at least 25 to 50 meters (80 to 160 feet) in all directions.
Spill: Increase, in the downwind direction, as necessary, the distance shown under "Public Safety."
Fire: If tank, rail car, or tank truck is involved in fire, isolate for at least 800 meters (½ mile) in all directions; also, consider initial evacuation for 800 meters (½ mile) in all directions.

EXPOSURE
Short-term effects: *SEEK MEDICAL ATTENTION.* **Vapor: Vapor or Liquid:** *POISONOUS IF INHALED, IF SWALLOWED OR IF SKIN IS EXPOSED.* Eye pupils are small; blurred vision; runny nose; cough; shortness of breath; pain; diarrhea, nausea and vomiting; increased blood pressure, hypermotility, hallucinations; loss of consciousness; convulsions; breathing stops; death. *IF BREATHING HAS STOPPED,* give artificial respiration; *avoid mouth-to-mouth resuscitation; use bag/mask apparatus.* **Skin:** *Remove and double-bag contaminated clothing (including shoes) and leave them in the Hot Zone;* wash skin with soap and water and large volumes of water for at least 15 minutes. *Skin can also be decontaminated with diluted hypochlorite solution, U.S. Army M291 kit, and M258(A1) skin decontamination kit.* **Eye:** Rinse with large volumes of water or saline for at least 15 minutes. *IF SWALLOWED,* will cause nausea, vomiting, or loss of consciousness. *Remove and double-bag contaminated clothing and shoes and leave in Hot Zone for later incineration by hazardous materials experts.* Flush affected areas with plenty of water. **If swallowed** and victim is *CONSCIOUS AND ABLE TO SWALLOW,* have victim drink 4 to 8 ounces of water. **Do NOT induce vomiting but immediately administer slurry of activated charcoal.** *IF SWALLOWED* and victim is *UNCONSCIOUS OR HAVING CONVULSIONS,* do nothing except keep victim warm. *Note to physician and medical personnel:* If symptoms indicate, initial treatment for an adult includes atropine, 2 mg (1/30 gr) intramuscularly or intravenously as soon as any local or systemic signs or symptoms of an intoxication are noted; repeat the administration of atropine every 3 to 8 minutes until signs of atropinization (mydriasis, dry mouth, rapid pulse, hot and dry skin) occur; initiate treatment in children with 0.05 mg/kg of atropine; repeat at 5- to 10-minute intervals. Watch respiration, and remove bronchial secretions if they appear to be obstructing the airway; intubate if necessary. Pralidoxime must be administered within minutes to a few hours following exposure (depending on the specific agent) to be effective. Give 2-PAMCI (Pralidoxime: Protopam), 2.5 g in 100 mL of sterile water or in 5% dextrose and water, intravenously, slowly, in 15 to 30 minutes; if sufficient fluid is not available, give 1 g of 2-PAMCI in 3 mL of distilled water by deep intramuscular injection; repeat this every half hour if respiration weakens or if muscle fasciculation or convulsions recur. Also Diazepam, an anticonvulsant, might be needed. For adults, the dose is 5 to 10 mg (slow IV), repeated every 12 to 15 minutes up to three (3) doses maximum. For children, the dose is 0.2 to 0.5 mg/kg.
Note to physician or authorized medical personnel: If inhaled, medical observation is recommended for 24 to 48 hours after breathing overexposure, as pulmonary edema may be delayed. As first aid for pulmonary edema, consider administering a corticosteroid spray. Cigarette smoking may exacerbate pulmonary injury and should be discouraged for at least 72 hours following exposure.

HEALTH HAZARDS
Personal protective equipment (PPE): A-Level PPE. OSHA Table Z-1-A air contaminant.
Recommendations for respirator selection: NIOSH/OSHA
1 mg/g3: SA (any supplied-air respirator). *2.5 mg/m³:* SA:CF (any supplied-air respirator operated in a continuous-flow mode). *5 mg/m³:* SAT:CF (any supplied-air respirator that has a tight-fitting facepiece and is operated in a continuous-flow mode); or SCBAF (any self-contained breathing apparatus with a full facepiece); or SAF (any supplied-air respirator with a full facepiece). *10 mg/m³:* SA:PD,PP (any supplied-air respirator operated in a pressure-demand or other positive-pressure mode). *EMERGENCY OR PLANNED ENTRY INTO UNKNOWN CONCENTRATIONS OR IDLH CONDITIONS:* SCBAF:PD,PP (any self-contained breathing apparatus that has a full faceplate and is operated in a pressure-demand or other positive-pressure mode); or SAF:PD,PP:ASCBA (any supplied-air respirator that has a full facepiece and is operated in a pressure-demand or other positive-pressure mode in combination with an auxiliary self-contained breathing apparatus operated in a pressure-demand or other positive pressure mode). *ESCAPE:* GMFOVHiE [any air-purifying, full-facepiece respirator (gas mask) with a chin-style, front- or back-mounted organic vapor canister having a high-efficiency particulate filter]; or SCBAE (any appropriate escape-type, self-contained breathing apparatus).
Exposure limits (TWA unless otherwise noted): ACGIH TLV 0.01 ppm (0.11 mg/m³); NIOSH/OSHA 0.1 mg/m³; skin contact
Toxicity by ingestion: Grade 4; oral LD_{50} = 1.7 mg/kg (rat).

Liquid or solid irritant characteristics: Eye and skin irritant.
IDLH value: 10 mg/m³.

FIRE DATA
Flash point: 113°F/45°C (cc).
Flammable limits in air: LEL: 1.0%; UEL: 5.3%.
Fire extinguishing agents not to be used: Water may be ineffective on fire.
Behavior in fire: Compound may volatilize and form toxic fumes.
Autoignition temperature: 867°F/464 (xylene solvent).
Electrical hazard: Class I, Group D as xylene.
Burning rate: 5.8 mm/min

CHEMICAL REACTIVITY
Reactivity with water: Reacts with water.
Binary reactants: May attack some forms of plastics; strong oxidizers, alkalis, water.

ENVIRONMENTAL DATA
Food chain concentration potential: Negative; unlikely to accumulate.
Water pollution: Effect of low concentrations on aquatic life is unknown. May be dangerous if it enters nearby water intakes; notify operators. Notify local health and wildlife officials. For weapons testing, use an M272 Water Detection Kit. **Response to discharge:** Issue warning–poison, water contaminant, high flammability. Restrict access. Mechanical containment. Should be removed. Chemical and physical treatment.

SHIPPING INFORMATION
Grades of purity: 25–66% solution in xylenes, which are combustible solvents; **Storage temperature:** Ambient; **Inert atmosphere:** None; **Venting:** Open (flame arrester); **Stability during transport:** Stable.

PHYSICAL AND CHEMICAL PROPERTIES
Physical state @ 59°F/15°C and 1 atm: Liquid.
Molecular weight: 258.3
Boiling point @ 1 atm: More than 284°F/more than 140°C/more than 413°K.
Melting/Freezing point: Less than –13°F/–25°C/248°K.
Specific gravity (water = 1): 1.1 @ 68°F/20°C (liquid).
Vapor pressure: 0.0003 mm.

DFP   REC. D:0850

SYNONYMS: DIFLUPYL; DIFLUROPHATE; DIISOPROPOXYPHOSPHORYL FLUORIDE; *O,O*-DIISOPROPYL FLUOROPHOSPHATE; DIISOPROPYL FLUOROPHOSPHONATE; DIISOPROPYLFLUOROPHOSPHORIC ACID ESTER; DIISOPROPYLFLUORPHOSPHORSAEUREESTER (German); DIISOPROPYL PHOSPHOFLUORIDATE; *O,O'*-DIISOPROPYL PHOSPHORYL FLUORIDE; DYFLOS; FLOROPRYL; FLUOPHOSPHORIC ACID, DIISOPROPYL ESTER; FLUORODIISOPROPYL PHOSPHATE; FLUOROPRYL; FLUOSTIGMINE; ISOFLUROPHATE; ISOPROPYL FLUOPHOSPHATE; NEOGLAUCIT; PF-3; PHOSPHOROFLUORIDIC ACID, DIISOPROPYL ESTER; RCRA No. P043; T-1703; TL 466

IDENTIFICATION:
CAS Registry Number: 55-91-4
Formula: $C_6H_{14}FO_3P$

DOT ID Number: UN 2810; DOT Guide Number: 153
Proper Shipping Name: Poisonous liquid, organic, n.o.s. (Inhalation hazard Zone B)
Reportable Quantity (RQ): **(EHS/CERCLA):** 100 lb (45.4 kg)

DESCRIPTION: Colorless oily liquid. Odorless. Slightly soluble in water; reacts; forms hydrogen fluoride. An organophosphate insecticide. *Note:* This chemical has been used as a chemical warfare agent (nerve gas). If used as a weapon, notify U.S. Department of Defense: Army. Damage and/or death may occur before chemical detection can take place. Detection limit: 0.005 mg/m³ based on G-Agents.

Flammable • Poison! (organophosphate) • Combustible • Breathing the vapor, skin or eye contact, or swallowing the material can kill you; symptoms may be delayed for several hours • Firefighting gear (including SCBA) provide NO protection. If exposure occurs, remove and isolate gear immediately and thoroughly decontaminate personnel • Containers may BLEVE when exposed to fire • Vapors are heavier than air and will collect and stay in low areas • Vapors in confined areas (e.g., tanks, sewers, buildings) may explode when exposed to fire • *Combustion products are less deadly than the material itself.* Toxic products of combustion may include carbon monoxide, sulfur oxide, phosphorus oxide, and nitrogen oxides • Do NOT attempt rescue.

Hazard Classification (based on NFPA-704 Rating System)

Health Hazards (Blue): 4; Flammability (Red): 1; Reactivity (Yellow): 1

EMERGENCY RESPONSE: See Appendix A (153) with notes.
Evacuation:
Public safety: See below.
Spill: Small spill–First: Isolate in all directions 60 meters (200 feet); Then: Protect persons downwind, DAY: 0.2 km (0.1 mile); NIGHT: 1.1 km (0.7 mile). Large spill–First: Isolate in all directions 185 meters (600 feet); Then: Protect persons downwind, DAY: 1.6 km (1.0 mile); NIGHT: 4.0 km (2.5 miles).
Fire: If tank, rail car, or tank truck is involved in fire, isolate for at least 800 meters (½ mile) in all directions; also, consider initial evacuation for 800 meters (½ mile) in all directions.

EXPOSURE
Short-term effects: *SEEK MEDICAL ATTENTION.* **Vapor or Liquid:** *POISONOUS IF INHALED, IF SWALLOWED, OR IF SKIN IS EXPOSED.* Eye pupils are small; blurred vision; runny nose; cough; shortness of breath; pain; diarrhea, nausea and vomiting; increased blood pressure, hypermotility, hallucinations; loss of consciousness; convulsions; breathing stops; death. *IF BREATHING HAS STOPPED,* give artificial respiration; *avoid mouth-to-mouth resuscitation; use bag/mask apparatus.* **Skin:** *Remove and double-bag contaminated clothing (including shoes) and leave them in the Hot Zone*; wash skin with soap and water and large volumes of water for at least 15 minutes. *Skin can also be decontaminated with diluted hypochlorite solution, U.S. Army M291 kit, and M258(A1) skin decontamination kit.* **Eye:** Rinse with large volumes of water or saline for at least 15 minutes. *IF SWALLOWED,* will cause nausea, vomiting, or loss of consciousness. *Remove and double-bag contaminated clothing and shoes and leave in Hot Zone for later incineration by hazardous materials experts.* Flush affected areas with plenty of water. **If swallowed** and victim is *CONSCIOUS AND ABLE TO SWALLOW*, have victim drink 4 to 8 ounces of water. **Do NOT induce vomiting but immediately administer slurry of activated**

charcoal. *IF SWALLOWED and victim is UNCONSCIOUS OR HAVING CONVULSIONS,* do nothing except keep victim warm. *Note to physician and medical personnel:* If symptoms indicate, initial treatment for an adult includes atropine, 2 mg (1/30 gr) intramuscularly or intravenously as soon as any local or systemic signs or symptoms of an intoxication are noted; repeat the administration of atropine every 3 to 8 minutes until signs of atropinization (mydriasis, dry mouth, rapid pulse, hot and dry skin) occur; initiate treatment in children with 0.05 mg/kg of atropine; repeat at 5- to 10-minute intervals. Watch respiration, and remove bronchial secretions if they appear to be obstructing the airway; intubate if necessary. Pralidoxime must be administered within minutes to a few hours following exposure (depending on the specific agent) to be effective. Give 2-PAMCI (Pralidoxime; Protopam), 2.5 g in 100 mL of sterile water or in 5% dextrose and water, intravenously, slowly, in 15 to 30 minutes; if sufficient fluid is not available, give 1 g of 2-PAMCI in 3 mL of distilled water by deep intramuscular injection; repeat this every half hour if respiration weakens or if muscle fasciculation or convulsions recur. Also Diazepam, an anticonvulsant, might be needed. For adults, the dose is 5 to 10 mg (slow IV), repeated every 12 to 15 minutes up to three (3) doses maximum. For children, the dose is 0.2 to 0.5 mg/kg.
Note to physician or authorized medical personnel: If inhaled, medical observation is recommended for 24 to 48 hours after breathing overexposure, as pulmonary edema may be delayed. As first aid for pulmonary edema, consider administering a corticosteroid spray. Cigarette smoking may exacerbate pulmonary injury and should be discouraged for at least 72 hours following exposure.

HEALTH HAZARDS
Personal protective equipment (PPE): A-Level PPE. Rubber gloves, protective clothing, goggles, respirators. Butyl rubber gloves and Tyvek® "F"decontamination suit provide barrier protection against chemical warfare agents. Airtight, impermeable clothing was developed for personnel who must enter heavily contaminated areas. This clothing is made of butyl rubber or a coated fabric such as Tyvek "F" and provide barrier protection against liquid chemical warfare agents. Although resistant to liquid chemical agents, impermeable protective clothing may be penetrated after a few hours of exposure to heavy concentration of agent. Consequently, liquid contamination on the clothing must be neutralized or removed as soon as possible. If the proper equipment is not available, or if the rescuers have not been trained in its use, call for assistance from the U.S. Soldier and Biological Chemical Command, Edgewood Research Development and Engineering Center (from 0700-1630 EST call 410-671-4411, and from 1630-0700 EST call 410-278-5201; ask for the Staff Duty Officer).
Recommendations for respirator selection: *EMERGENCY OR PLANNED ENTRY INTO UNKNOWN CONCENTRATIONS OR IDLH CONDITIONS:* SCBAF:PD,PP (any self-contained breathing apparatus that has a full facepiece and is operated in a pressure-demand or other positive-pressure mode); or SAF:PD,PP:ASCBA (any supplied-air respirator that has a full facepiece and is operated in a pressure-demand or other positive-pressure mode in combination with an auxiliary self-contained breathing apparatus operated in a pressure-demand or other positive pressure mode). *ESCAPE:* GMFOVHiE [any air-purifying, full-facepiece respirator (gas mask) with a chin-style, front-or back-mounted canister having a high efficiency particulate filter); or SCBAE (any appropriate escape-type, self-contained breathing apparatus). The U.S. Army standard M17A1 (which is being replaced by the M40 protective mask) mask provides complete respiratory protection against all known military toxic chemical agents, but it cannot be used in an oxygen deficient environment and it does not afford protection against industrial toxics such as ammonia and carbon monoxide. *It is not approved for civilian use.*

CHEMICAL REACTIVITY
Reactivity with water: Forms hydrogen fluoride.
Reactivity with other materials: Oxidizers.

ENVIRONMENTAL DATA
Water pollution: If used as a weapon, utilize use an M272 Water Detection Kit (Detection limit: 0.02 mg/L). Dangerous to aquatic life in high concentrations. May be dangerous if it enters water intakes. Notify local health and pollution control officials. Notify operators of nearby water intakes.

PHYSICAL AND CHEMICAL PROPERTIES
Physical state @ 59°F/15°C and 1 atm: Liquid.
Molecular weight: 184.17
Boiling point @ 1 atm: 46°C @ 5 mm
Melting/Freezing point: –116°F/–82°C/191°K
Specific gravity (water = 1): 1.067
Liquid surface tension:
Vapor density: 5.24

DIACETONE ALCOHOL   REC. D:0900

SYNONYMS: ACETONYLDIMETHYLCARBINOL; DIACETON-ALCOHOL (Spanish); DIACETONALKOHOL (German); DIACETONE-ALCOOL (French); DIMETHYLACETONYLCARBINOL; DIKETONE ALCOHOL; EEC No. 603-016-00-1; 4-HYDROXY-2-KETO-4-METHYLPENTANE; 4-HYDROXY-4-METHYL-2-PENTANONE; 4-HYDROXY-4-METHYL-PENTAN-2-ON (German); 4-HYDROXY-4-METHYLPENTAN-2-ONE; 4-HYDROXYL-2-KETO-4-METHYLPENTANE; 4-METHYL-4-HYDROXY-2-PENTANONE; 2-METHYL-2-PENTANOL-4-ONE; 2-PENTANONE, 4-HYDROXY-4-METHYL-; TYRANTON; 4-HYDROXY-4-METHYL-2-PENTANONE; DIACETONE,4-HYDROXY-4-METHYL-2-PENTATONE,2-METHYL-2-PENTANOL-4-ONE

IDENTIFICATION
CAS Number: 123-42-2
Formula: $CH_3C(OH)(CH_3)CH_2COCH_3$
DOT ID Number: UN 1148; DOT Guide Number: 129
Proper Shipping Name: Diacetone alcohol

DESCRIPTION: Colorless to pale yellow watery liquid. Mild, pleasant odor. Floats on the surface of water; soluble.

Flammable • Containers may BLEVE when exposed to fire • Vapors may form explosive mixture with air • Vapors are heavier than air and will collect and stay in low areas • Vapors may travel long distances to ignition sources and flashback • Vapors in confined areas (e.g., tanks, sewers, buildings) may explode when exposed to fire • Irritating to the skin, eyes, and respiratory tract • Toxic products of combustion may include acetone and mesityl oxide.

Hazard Classification (based on NFPA-704 Rating System)
Health Hazards (Blue): 1; Flammability (Red): 2; Reactivity (Yellow): 0

328   Diammonium salt of zinc EDTA

EMERGENCY RESPONSE: See Appendix A (129)
Evacuation:
Public safety: Isolate spill area for at least 50 to 100 meters (160 to 330 feet) in all directions.
Spill: Large spill–Consider initial downwind evacuation for at least 300 meters (1000 feet).
Fire: Isolate for 800 meters (½ mile) in all directions, especially if tank, rail car, or tank truck is involved in fire.

EXPOSURE
Short-term effects: *SEEK MEDICAL ATTENTION.* **Liquid:** Irritating to skin and eyes. Harmful if swallowed. Remove contaminated clothing and shoes. Flush affected areas with plenty of water. *IF IN EYES,* hold eyelids open and flush with plenty of water. *IF SWALLOWED* and victim is *CONSCIOUS AND ABLE TO SWALLOW,* have victim drink 4 to 8 ounces of water.

HEALTH HAZARDS
Personal protective equipment (PPE): B-Level PPE. OSHA Table Z-1-A air contaminant. Gloves, goggles. Chemical protective material(s) reported to offer minimal to poor protection: neoprene. Also, butyl rubber, natural rubber, nitrile+PVC, PV alcohol, butyl rubber/neoprene, styrene-butadiene, styrene-butadiene/neoprene offers limited protection
Recommendations for respirator selection: NIOSH/OSHA
1250 ppm: SA:CF* (any supplied-air respirator operated in a continuous-flow mode); or PAPROV* [any powered, air-purifying respirator with organic vapor cartridge(s)]. *1800 ppm:* CCRFOV [any chemical cartridge respirator with a full facepiece and organic vapor cartridges(s)]; or GMFOV [any air-purifying, full-facepiece respirator (gas mask) with a chin-style, front- or back-mounted acid gas canister]; or PAPRTOV* [any powered, air-purifying respirator with a tight-fitting facepiece and organic vapor cartridges(s)]; or SCBAF (any self-contained breathing apparatus with a full facepiece); or SAF (any supplied-air respirator with a full facepiece). *EMERGENCY OR PLANNED ENTRY INTO UNKNOWN CONCENTRATIONS OR IDLH CONDITIONS:* SCBAF:PD,PP (any self-contained breathing apparatus that has a full facepiece and is operated in a pressure-demand or other positive-pressure mode); or SAF:PD,PP:ASCBA (any supplied-air respirator that has a full facepiece and is operated in a pressure-demand or other positive-pressure mode in combination with an auxiliary self-contained breathing apparatus operated in a pressure-demand or other positive pressure mode). *ESCAPE:* GMFOV[any air-purifying, full-facepiece respirator (gas mask) with a chin-style, front- or back-mounted organic vapor cannister]; or SCBAE (any appropriate escape-type, self-contained breathing apparatus). *Note:* Substance causes eye irritation or damage; eye protection needed.
Exposure limits (TWA unless otherwise noted): ACGIH TLV 50 ppm (238 mg/m^3); NIOSH/OSHA 50 ppm (240 mg/m^3).
Toxicity by ingestion: Grade 2; LD_{50} = 0.5 to 5 g/kg (rat).
Long-term health effects: Possible dermatitis. Liver and kidney damage may occur. May effect the central nervous system.
Vapor (gas) irritant characteristics: Eye and respiratory tract irritant. Vapors cause moderate irritation such that personnel will find high concentrations unpleasant. The effect is temporary.
Liquid or solid irritant characteristics: Eye and skin irritant. If spilled on clothing and allowed to remain, may cause smarting and reddening of the skin.
Odor threshold: 0.27 ppm.
IDLH value: 1800 ppm [10% LEL].

FIRE DATA
Flash point: 142°F/61°C (oc); 125°F/52°C (cc) (acetone-free). Technical grades containing acetone may have lower flash points varying from –4°F/–20°C to 320°F/160°C.
Flammable limits in air: LEL: 1.8%; UEL: 6.9%.
Autoignition temperature: 1118°F/603°C/876°K (acetone-free); 1190°F/643°C/916°K (technical grades).
Electrical hazard: Above 135°F; explosion protected electrical equipment required.

CHEMICAL REACTIVITY
Binary reactants: Strong oxidizers, strong alkalis and acids, alkali metals: lithium, sodium, potassium, rubidium, cesium, and francium.
Reactivity group: 20
Compatibility class: Alcohols, glycols

ENVIRONMENTAL DATA
Food chain concentration potential: Log P_{ow} = –0.10. Unlikely to accumulate.
Water pollution: Effect of low concentrations on aquatic life is unknown. May be dangerous if it enters nearby water intakes; notify operators. Notify local health and pollution control officials.
Response to discharge: Disperse and flush.

SHIPPING INFORMATION
Grades of purity: 99.0%; technical (containing acetone); **Storage temperature:** Ambient, cool; **Inert atmosphere:** None; **Venting:** Open (flame arrester); **Stability during transport:** Stable.

PHYSICAL AND CHEMICAL PROPERTIES
Physical state @ 59°F/15°C and 1 atm: Liquid.
Molecular weight: 116.16
Boiling point @ 1 atm: 333°F/167°C/440°K.
Melting/Freezing point: –45°F/–43°C/230°K.
Critical temperature: 633°F/334°C/607°K.
Critical pressure: 380 psi = 36 atm = 3.6 MN/m^2.
Specific gravity (water = 1): 0.938 @ 68°F/20°C (liquid).
Relative vapor density (air = 1): 4.0
Ratio of specific heats of vapor (gas): 1.052
Latent heat of vaporization: 150 Btu/lb = 85 cal/g = 3.6 x 10^5 J/kg.
Heat of combustion: estimate –13,000 Btu/lb = –7250 cal/g = 303 x 10^5 J/kg.
Vapor pressure: (Reid) 0.07 psi; 1 mm.

DIAMMONIUM SALT OF ZINC EDTA   REC. D:0950

SYNONYMS: ZINC ETHYLENEDIAMINETETRAACETATE; EDTA ZINC SALT; EDTA-ZINC COMPLEX; EDTA-ZINC

IDENTIFICATION
CAS Number: 12519-36-7
Formula: $C_{10}H_{12}N_2O_8Zn$

DESCRIPTION: No reaction with water; may float or sink.

Irritating to the skin, eyes, and respiratory tract • Toxic products of combustion may include carbon monoxide and oxides of nitrogen and zinc.

EMERGENCY RESPONSE: See Appendix A (171)
Evacuation:
Public safety: Isolate the area of spill or leak for at least 10 to 25 meters (30 to 80 feet) in all directions.

Spill: Increase, in the downwind direction, as necessary, the distance shown under "Public Safety."
Fire: If any large container is involved in fire, isolate for at least 800 meters (½ mile) in all directions; also, consider initial evacuation for 800 meters (½ mile) in all directions.

EXPOSURE
Short-term effects: *SEEK MEDICAL ATTENTION*. Irritating to skin and eyes. *IF IN EYES*, hold eyes open and flush with running water for at least 15 minutes. *IF ON SKIN*, wash with soap and water. Remove contaminated clothing and shoes.

HEALTH HAZARDS
Personal protective equipment (PPE): B-Level PPE.
Toxicity by ingestion: Grade 3; LD_{50} = 85 mg/kg as zinc
Liquid or solid irritant characteristics: Eye and skin irritant. If spilled on clothing and allowed to remain, may cause smarting and reddening of skin.

CHEMICAL REACTIVITY
Binary reactants: Sulfuric acid and isocyanates.
Reactivity group: 43
Compatibility class: Miscellaneous water solutions.

ENVIRONMENTAL DATA
Food chain concentration potential: Negative; unlikely to accumulate.
Water pollution: Effects of low concentration on aquatic life is unknown. May be dangerous if it enters nearby water intakes; notify operators. Notify local health and wild life officials.
Response to discharge: Restrict access. Issue warning. Should be removed.

SHIPPING INFORMATION
Storage temperature: Ambient; **Inert atmosphere:** None; **Stability during transport:** Stable.

PHYSICAL AND CHEMICAL PROPERTIES
Molecular weight: 302.811

DI-*n*-AMYL PHTHALATE   REC. D:0975

SYNONYMS: AMOIL; 1,2-BENZENEDICARBOXYLIC ACID, DIPENTYL ESTER; DIAMYL PHTHALATE; DIPENTYL PHTHALATE; DI-*n*-PENTYLPHTHALATE; DPP; FTALATO de DIAMILO (Spanish); PHTHALIC ACID, DIAMYL ESTER

IDENTIFICATION
CAS Number: 131-18-0
Formula: $C_{18}H_{26}O_4$; $C_6H_4(COOC_5H_{11})_2$

DESCRIPTION: Colorless, oily liquid. Nearly odorless. Floats on the surface of water; somewhat soluble.

Combustible • Containers may BLEVE when exposed to fire • Vapors are heavier than air and will collect and stay in low areas • Vapors in confined areas (e.g., tanks, sewers, buildings) may explode when exposed to fire • Irritating to the skin, eyes, and respiratory tract • Toxic products of combustion may include carbon monoxide.

Hazard Classification (based on NFPA-704 Rating System)
Health Hazards (Blue): 0; Flammability (Red): 1; Reactivity (Yellow): 0

EMERGENCY RESPONSE: See Appendix A (171)
Evacuation:
Public safety: Isolate the area of spill or leak for at least 10 to 25 meters (30 to 80 feet) in all directions.
Spill: Increase, in the downwind direction, as necessary, the distance shown under "Public Safety."
Fire: If any large container is involved in fire, isolate for at least 800 meters (½ mile) in all directions; also, consider initial evacuation for 800 meters (½ mile) in all directions.

EXPOSURE
Short-term effects: *SEEK MEDICAL ATTENTION*. **Vapor:** Irritating to eyes, nose, and throat. *IF INHALED*, will, will cause headache, coughing, or difficult breathing. *IF IN EYES*, hold eyelids open and flush with plenty of water. *IF BREATHING HAS STOPPED*, give artificial respiration. If breathing is difficult, administer oxygen. **Liquid:** Irritating to skin and eyes. Harmful if swallowed. Remove contaminated clothing and shoes. Flush affected areas with plenty of water. *IF IN EYES*, hold eyelids open and flush with plenty of water. *IF SWALLOWED* and victim is *CONSCIOUS AND ABLE TO SWALLOW*, have victim drink 4 to 8 ounces of water. *IF SWALLOWED* and victim is *UNCONSCIOUS OR HAVING CONVULSIONS*, do nothing except keep victim warm.

HEALTH HAZARDS
Personal protective equipment (PPE): B-Level PPE. Butyl rubber is generally suitable for carbooxylic acid compounds.
Toxicity by ingestion: Grade 2: LD_{LO} = 2.2 g/kg (rat).
Long-term health effects: Causes birth defects in rats (skeletal and gross abnormalities).
Liquid or solid irritant characteristics: May cause eye or skin irritation.

FIRE DATA
Flash point: 245°F/118°C (cc).
Fire extinguishing agents not to be used: Water or foam may cause frothing.

CHEMICAL REACTIVITY
Binary reactants: May attack some forms of plastics
Reactivity group: 34
Compatibility class: Esters.

ENVIRONMENTAL DATA
Food chain concentration potential: Negative; unlikely to accumulate.
Water pollution: Effect of low concentrations on aquatic life is unknown. Fouling to shoreline. May be dangerous if it enters nearby water intakes; notify operators. Notify local health and wildlife officials. **Response to discharge:** Mechanical containment. Should be removed. Chemical and physical treatment.

SHIPPING INFORMATION
Grades of purity: Technical; **Storage temperature:** Ambient; **Inert atmosphere:** None; **Venting:** Open; **Stability during transport:** Stable.

NAS HAZARD CLASSIFICATIONS FOR BULK WATER TRANSPORTATION
FIRE: 1
HEALTH: Vapor irritant: 1; Liquid or solid irritant: 0; Poisons: 0
WATER POLLUTION: Human toxicity: 2; Aquatic toxicity: 2; Aesthetic effect: 1
REACTIVITY: Other chemicals: 0; Water: 0; Self-reaction: 0

PHYSICAL AND CHEMICAL PROPERTIES
Physical state @ 59°F/15°C and 1 atm: Liquid.
Boiling Point: 475–490°F/246–254°C/519–527°K.
Molecular weight: 306
Relative vapor density (air = 1): 10.5
Specific gravity (water = 1): 0.82 @ 68°F/20°C (liquid)
Liquid surface tension: 31.5 dynes/cm = 0.0315 N/m @ 68°F/20°C.
Latent heat of vaporization: 140 Btu/lb = 76 cal/g = 3.2×10^5 J/kg.
Heat of combustion: $-13,900$ Btu/lb = -7720 cal/g = -323×10^5 J/kg.

DIAZINON   REC. D:1000

SYNONYMS: ALFA-TOX; BASUDIN®; BASUDIN® 10 G; BAZUDEN; DAZZEL; *O,O*-DIAETHYL-*O*-(2-ISOPROPYL-4-METHYL-PYRIMIDIN-6-YL)-MONOTHIOPHOSPHAT (German); DIANON; DIATERR-FOS; DIAZAJET; DIAZATOL; DIAZIDE; DIAZINONE; DIAZITOL; DIAZOL; DIMPYLATE; DIPOFENE; DIZINON; DYZOL; G 301; G-24480; GARDENTOX; GEIGY 24480; KAYAZINON; KAYAZOL; NECIDOL; NEOCIDOL; NIPSAN; NUCIDOL; PHOSPHORIC ACID, *O,O*-DIETHYL-*O*-6-METHYL-2-(1-METHYLETHYL)-4-PYRIMIDINYL ESTER; SAROLEX; SPECTRACIDE; *O,O*-DIETHYLO-(2-ISOPROPYL-6-METHYL-4-PYRIMIDINYL) PHOSPHOROTHIOATE

IDENTIFICATION
CAS Number: 333-41-5
Formula: $C_{12}H_{21}N_2O_3PS$
DOT ID Number: UN 2783; DOT Guide Number: 152
Proper Shipping Name: Diazinon
Reportable Quantity (RQ): **(CERCLA)** 1 lb/0.454 kg

DESCRIPTION: Colorless liquid. Technical grade is a light to dark brown liquid. Sinks in water; slightly soluble; hydrolyzes slowly.

Poison! (organophosphate) • Combustible • Breathing the dust, skin or eye contact, or swallowing the material can kill you • Firefighting gear (including SCBA) does not provide adequate protection. If exposure occurs, remove and isolate gear immediately and thoroughly decontaminate personnel • Containers may BLEVE when exposed to fire • Concentrated dust in confined areas (e.g., tanks, sewers, buildings) may explode when exposed to fire • Toxic products of combustion may include oxides of nitrogen, sulfur and phosphorus • Do not put yourself in danger by entering a contaminated area to rescue a victim.

Hazard Classification (based on NFPA-704 Rating System)
Health Hazards (Blue): 3; Flammability (Red): 1; Reactivity (Yellow): 0

EMERGENCY RESPONSE: See Appendix A (152)
Evacuation:
Public safety: Isolate the area of spill or leak for at least 25 to 50 meters (80 to 160 feet) in all directions.
Spill: Increase, in the downwind direction, as necessary, the distance shown under "Public Safety."
Fire: If tank, rail car, or tank truck is involved in fire, isolate for at least 800 meters (½ mile) in all directions; also, consider initial evacuation for 800 meters (½ mile) in all directions.

EXPOSURE
Vapor or Liquid: *POISONOUS IF INHALED, IF SWALLOWED, OR IF SKIN IS EXPOSED*. Eye pupils are small; blurred vision; runny nose; cough; shortness of breath; pain; diarrhea, nausea and vomiting; increased blood pressure, hypermotility, hallucinations; loss of consciousness; convulsions; breathing stops; death. *IF BREATHING HAS STOPPED*, give artificial respiration; *avoid mouth-to-mouth resuscitation; use bag/mask apparatus*. **Skin:** *Remove and double-bag contaminated clothing (including shoes) and leave them in the Hot Zone*; wash skin with soap and water and large volumes of water for at least 15 minutes. *Skin can also be decontaminated with diluted hypochlorite solution, U.S. Army M291 kit, and M258(A1) skin decontamination kit.* **Eye:** Rinse with large volumes of water or saline for at least 15 minutes. *IF SWALLOWED*, will cause nausea, vomiting, or loss of consciousness. *Remove and double-bag contaminated clothing and shoes and leave in Hot Zone for later incineration by hazardous materials experts.* Flush affected areas with plenty of water. **If swallowed and victim is *CONSCIOUS AND ABLE TO SWALLOW*, have victim drink 4 to 8 ounces of water. Do NOT induce vomiting but immediately administer slurry of activated charcoal**. *IF SWALLOWED* and victim is *UNCONSCIOUS OR HAVING CONVULSIONS,* do nothing except keep victim warm. *Note to physician and medical personnel:* If symptoms indicate, initial treatment for an adult includes atropine, 2 mg (1/30 gr) intramuscularly or intravenously as soon as any local or systemic signs or symptoms of an intoxication are noted; repeat the administration of atropine every 3 to 8 minutes until signs of atropinization (mydriasis, dry mouth, rapid pulse, hot and dry skin) occur; initiate treatment in children with 0.05 mg/kg of atropine; repeat at 5- to 10-minute intervals. Watch respiration, and remove bronchial secretions if they appear to be obstructing the airway; intubate if necessary. Pralidoxime must be administered within minutes to a few hours following exposure (depending on the specific agent) to be effective. Give 2-PAMCI (Pralidoxime; Protopam), 2.5 g in 100 mL of sterile water or in 5% dextrose and water, intravenously, slowly, in 15 to 30 minutes; if sufficient fluid is not available, give 1 g of 2-PAMCI in 3 mL of distilled water by deep intramuscular injection; repeat this every half hour if respiration weakens or if muscle fasciculation or convulsions recur. Also Diazepam, an anticonvulsant, might be needed. For adults, the dose is 5 to 10 mg (slow IV), repeated every 12 to 15 minutes up to three (3) doses maximum. For children, the dose is 0.2 to 0.5 mg/kg.
Note to physician or authorized medical personnel: If inhaled, medical observation is recommended for 24 to 48 hours after breathing overexposure, as pulmonary edema may be delayed. As first aid for pulmonary edema, consider administering a corticosteroid spray. Cigarette smoking may exacerbate pulmonary injury and should be discouraged for at least 72 hours following exposure.

Personal protective equipment (PPE): A-Level PPE. OSHA Table Z-1-A air contaminant. Chemical-protective clothing and butyl rubber gloves are recommended when skin contact is possible because liquid is rapidly absorbed through the skin and may cause systemic toxicity.
Exposure limits (TWA unless otherwise noted): ACGIH TLV 0.1 mg/m³; NIOSH 0.1 mg/m³; skin contact contributes significantly in overall exposure.
Toxicity by ingestion: Grade 3; oral LD_{50} = 76 mg/kg (rat).
Long-term health effects: May be mutagenic. Reproductive effects.
Liquid or solid irritant characteristics: Severe eye and skin irritant.

FIRE DATA
Flash point: 180°F/82°C (cc).
Fire extinguishing agents not to be used: Water may be ineffective.
Burning rate: (for solutions) 4 mm/min

CHEMICAL REACTIVITY
Reactivity with water: Hydrolyzes slowly in water and dilute acid.
Binary reactants: Strong acids, alkali, copper-containing compounds.

ENVIRONMENTAL DATA
Water pollution: DOT Appendix B, §172.101–marine pollutant. Harmful to aquatic life in very low concentrations. May be dangerous if it enters nearby water intakes; notify operators. Notify local health and wildlife officials. For weapons testing, use an M272 Water Detection Kit. **Response to discharge:** Issue warning–poison, water contaminant, high flammability (if solution). Restrict access. Should be removed. Chemical and physical treatment.

SHIPPING INFORMATION
Grades of purity: Technical; wettable powders; a variety of emulsifiable solutions in combustible solvents; **Storage temperature:** Ambient; **Inert atmosphere:** None; **Venting:** Open (flame arrester); **Stability during transport:** Stable.

PHYSICAL AND CHEMICAL PROPERTIES
Physical state @ 59°F/15°C and 1 atm: Liquid.
Molecular weight: 304.4
Boiling point @ 1 atm: Very high; decomposes.
Specific gravity (water = 1): 1.117 @ 68°F/20°C (liquid).
Liquid surface tension: estimate 35 dynes/cm = 0.035 N/m @ 68°F/20°C.
Liquid water interfacial tension: estimate 40 dynes/cm = 0.040 N/m @ 68°F/20°C.
Heat of combustion: estimate $-12,000$ Btu/lb = -6500 cal/g = -270×10^5 J/kg.
Vapor pressure: 0.0001 mm.

DIBENZOYL PEROXIDE REC. D:1050

SYNONYMS: BENZOPEROXIDE; BENZOYL PEROXIDE; BENZOYL SUPEROXIDE; BPO; BP; LUCIDOL; OXYLITE; PEROXIDO de DIBENZOILO (Spanish)

IDENTIFICATION
CAS Number: 94-36-0
Formula: $C_{13}H_{10}O_4$; $C_6H_5CO\cdot O\cdot O\cdot COC_6NO_5$
DOT ID Number: UN 2085; UN 2087; UN 2088; UN 2089; UN 2090; DOT Guide Number: 145 (UN 2089); 146 (UN 2085; UN; (2087; UN 2088; UN 2090)
Proper Shipping Name: Benzoyl Peroxide

DESCRIPTION: White powder or granules. Odorless. Sinks in water.

Extremely flammable • Containers may BLEVE when exposed to fire • May explode if subjected to heat or contamination • Strong oxidizer which may react spontaneously with low flash point organics or reducing agents. Heat forms oxygen; will increase the activity of an existing fire • May cause fire or explosion on contact with combustibles (wood, paper, oil, clothing, etc.) • Vapors are heavier than air and will collect and stay in low areas • Vapors may travel long distances to ignition sources and flashback • Vapors in confined areas (e.g., tanks, sewers, buildings) may explode when exposed to fire • Irritating to the skin, eyes, and respiratory tract • Suffocating smoke may be produced. Toxic products of combustion may include carbon monoxide.

Hazard Classification (based on NFPA-704 Rating System)
Health Hazards (Blue): 1; Flammability (Red): 4; Reactivity (Yellow): 4; Special Notice (White): OXY

EMERGENCY RESPONSE: See Appendix A (145, 146)
Evacuation:
Public safety: Isolate the area of spill or leak for at least 25 to 50 meters (80 to 160 feet) in all directions.
Spill: Consider evacuation for at least 250 meters (800 feet) in all directions.
Fire: If any large container is involved in fire, isolate for at least 800 meters (½ mile) in all directions; also, consider initial evacuation for 800 meters (½ mile) in all directions.

EXPOSURE
Short-term effects: *SEEK MEDICAL ATTENTION.* **Solid:** Irritating to skin and eyes. Harmful if swallowed. Remove contaminated clothing and shoes. Flush affected areas with plenty of water. *IF IN EYES*, hold eyelids open and flush with plenty of water. *IF SWALLOWED* and victim is *CONSCIOUS AND ABLE TO SWALLOW*, have victim drink 4 to 8 ounces of water and have victim induce vomiting. *IF SWALLOWED* and victim is *UNCONSCIOUS OR HAVING CONVULSIONS*, do nothing except keep victim warm.

HEALTH HAZARDS
Personal protective equipment (PPE): B-Level PPE. OSHA Table Z-1-A air contaminant. Butyl rubber is generally suitable for peroxide compounds.
Recommendations for respirator selection: NIOSH/OSHA
50 mg/m³: DMXSQ * (any dust and mist respirator except single-use and quarter mask respirators); or SA* (any supplied-air respirator). *125 mg/m³:* SA:CF (any supplied-air respirator operated in a continuous-flow mode); or PAPRDM* (any powered, air-purifying respirator with a dust and mist filter). *250 mg/m³:* HiEF (any air-purifying, full-facepiece respirator with a high-efficiency particulate filter); or PAPRTHiE (any powered, air-purifying respirator with a tight-fitting facepiece and a high-efficiency particulate filter); or SCBAF (any self-contained breathing apparatus with a full facepiece); or SAF (any supplied-air respirator with a full facepiece). *1500 mg/m³:* SAF:PD,PP (any supplied-air respirator that has a full facepiece and is operated in a pressure-demand or other positive-pressure mode). *EMERGENCY OR PLANNED ENTRY INTO UNKNOWN CONCENTRATIONS OR IDLH CONDITIONS:* SCBAF:PD,PP (any self-contained breathing apparatus that has a full facepiece and is operated in a pressure-demand or other positive-pressure mode); or SAF:PD,PP:ASCBA (any supplied-air respirator that has a full facepiece and is operated in a pressure-demand or other positive-pressure mode in combination with an auxiliary self-contained breathing apparatus operated in a pressure-demand or other positive pressure mode). *ESCAPE:* HiEF (any air-purifying, full-facepiece respirator with a high-efficiency particulate filter); or SCBAE (any appropriate escape-type, self-contained breathing apparatus). *Note*: Substance reported to cause eye irritation or damage; may require eye protection.

332 Dibenzyl ether

Exposure limits (TWA unless otherwise noted): ACGIH TLV 5 mg/m^3, not classifiable as a human carcinogen; NIOSH/OSHA 5 mg/m^3.
Toxicity by ingestion: Grade 2; LD$_{50}$ = 0.5 to 5 g/kg.
Long-term health effects: Eczema and/or dermatitis may develop.
Liquid or solid irritant characteristics: Eye irritant. If spilled on clothing and allowed to remain, may cause smarting and reddening of the skin.
IDLH value: 1500 mg/m^3.

FIRE DATA
Flash point: 176°F/80°C. Highly combustible solid; explosion-sensitive to shock, heat, and friction
Fire extinguishing agents not to be used: Do not use hand extinguishers
Behavior in fire: May explode.

CHEMICAL REACTIVITY
Binary reactants: Special care must be taken to avoid contamination with combustible materials (wood, paper, etc.), various inorganic and organic acids, alkalies, alcohols, amines, easily oxidizable materials such as ethers, or materials; used as accelerators in polymerization reactions. Has been reported to explode for apparently no specific reason. Self-reactive.

ENVIRONMENTAL DATA
Food chain concentration potential: Negative; unlikely to accumulate.
Water pollution: Effect of low concentrations on aquatic life is unknown. May be dangerous if it enters nearby water intakes; notify operators. Notify local health and pollution control officials.
Response to discharge: Issue warning–high flammability. Should be removed. Chemical and physical treatment.

SHIPPING INFORMATION
Grades of purity: 98% (dry); 70–78% (wet); various pastes of dibenzoyl peroxide and liquid plasticizers such as tricresyl phosphate, silicone oil; **Storage temperature:** 65–85°F/18-30°C; **Stability during transport:** Extremely explosion-sensitive to shock (impact, blows), heat, and friction.

PHYSICAL AND CHEMICAL PROPERTIES
Physical state @ 59°F/15°C and 1 atm: Solid.
Molecular weight: 242.22
Boiling point @ 1 atm: Decomposes explosively
Melting/Freezing point: 217°F/103°C/376°K.
Specific gravity (water = 1): 1.334 @ 15°C (solid).
Vapor pressure: Less than 1 mm.

DIBENZYL ETHER **REC. D:1100**

SYNONYMS: BENZYL ETHER; ETER DIBENCILICO (Spanish); 1,1'-[OXYBIS(METHYLENE)]BISBENZENE; BENZYL OXIDE

IDENTIFICATION
CAS Number: 103-50-4
Formula: $C_{13}H_{14}O$
DOT ID No: UN 3271; DOT ID Number: 127
Proper Shipping Name: Ethers, n.o.s.

DESCRIPTION: Colorless to pale yellow liquid. almond-like odor. Floats on the surface of water; insoluble.

Combustible • Containers may BLEVE when exposed to fire • Vapors may form explosive mixture with air • Vapors are heavier than air and will collect and stay in low areas • Vapors in confined areas (e.g., tanks, sewers, buildings) may explode when exposed to fire • Irritating to the skin, eyes, and respiratory tract • Toxic products of combustion may include carbon monoxide.

Hazard Classification (based on NFPA-704 Rating System)
Health Hazards (Blue): 0; Flammability (Red): 1; Reactivity (Yellow): 0

EMERGENCY RESPONSE: See Appendix A (171)
Evacuation:
Public safety: Isolate spill area for at least 25 to 50 meters (80 to 160 feet) in all directions.
Spill: Large spill–Consider initial downwind evacuation for at least 300 meters (1000 feet).
Fire: Isolate for 800 meters (½ mile) in all directions, especially if tank, rail car, or tank truck is involved in fire.

EXPOSURE
Short-term effects: *SEEK MEDICAL ATTENTION*. **Liquid:** Irritating to skin and eyes. Harmful if swallowed. Remove contaminated clothing and shoes. Flush affected areas with plenty of water. *IF IN EYES*, hold eyelids open and flush with plenty of water. *IF SWALLOWED* and victim is *CONSCIOUS AND ABLE TO SWALLOW*, have victim drink 4 to 8 ounces of water.

HEALTH HAZARDS
Personal protective equipment (PPE): B-Level PPE. Chemical protective material(s) reported to offer minimal to poor protection: neoprene, PVC. Also, butyl rubber, nitrile+PVC, nitrile, and butyl rubber/neoprene materials offers limited protection
Toxicity by ingestion: Grade 2: LD$_{50}$ = 2.5 g/kg (rat).
Liquid or solid irritant characteristics: Eye irritant. If spilled on clothing and allowed to remain, may cause smarting and reddening of the skin.

FIRE DATA
Flash point: 275°F/135°C (cc).
Fire extinguishing agents not to be used: Water or foam may cause frothing.
Electrical hazard: Flow or agitation of substance may generate electrostatic charges due to low conductivity.

CHEMICAL REACTIVITY
Binary reactants: Strong oxidizers.
Polymerization: Forms unstable peroxides. Capable of spontaneous explosion.

ENVIRONMENTAL DATA
Water pollution: Effects of low concentrations on aquatic life is unknown. Fouling to shoreline. May be dangerous if it enters nearby water intakes; notify operators. Notify local health and wildlife officials. Notify operators of local water intakes. **Response to discharge:** Mechanical containment. Should be removed. Chemical and physical treatment.

SHIPPING INFORMATION
Grades of purity: 99%; **Stability during transport:** Stable.

NAS HAZARD CLASSIFICATION FOR BULK WATER TRANSPORTATION
FIRE: 1
HEALTH: Vapor irritant: 2; Liquid or solid irritant: 1; Poisons: 1

WATER POLLUTION: Human toxicity: 2; Aquatic toxicity:; Aesthetic effect: 2
REACTIVITY: Other chemicals: 1; Water: 0; Self-reaction: 0

PHYSICAL AND CHEMICAL PROPERTIES
Physical state @ 59°F/15°C and 1 atm: Liquid.
Molecular weight: 198.26
Boiling point @ 1 atm: 568°F/298°C/571°K.
Melting/Freezing point: 38.5°F/3.6°C/277°K.
Specific gravity (water = 1): 1.0428 @ 68°F/20°C.
Relative vapor density (air = 1): 6.84

DIBROMOMETHANE   REC. D:1150

SYNONYMS: DIBROMOMETANO (Spanish); METHYLENE BROMIDE; METHYLENE DIBROMIDE; METHANE, DIBROMO-; RCRA No. U068

IDENTIFICATION
CAS Number: 74-95-3
Formula: CH_2Br_2
DOT ID Number: UN 2664; DOT Guide Number: 160
Proper Shipping Name: Dibromomethane
Reportable Quantity (RQ): **(CERCLA)** 1000 lb/454 kg

DESCRIPTION: Colorless watery liquid. Sweet, ether, or chloroform odor. Sinks in water; slightly soluble. Irritating vapor is produced.

Containers may BLEVE when exposed to fire • Vapors are heavier than air and will collect and stay in low areas • Vapors in confined areas (e.g., tanks, sewers, buildings) may explode when exposed to fire • Irritating to skin, eyes, and respiratory tract • Toxic products of combustion may include hydrogen bromide.

Hazard Classification (based on NFPA-704 Rating System)
Health Hazards (Blue): 2; Flammability (Red): 0; Reactivity (Yellow): 0

EMERGENCY RESPONSE: See Appendix A (160)
Evacuation:
Public safety: Isolate the area of spill or leak for at least 25 to 50 meters (80- 160 feet) in all directions.
Spill: Consider initial downwind evacuation for at least 100 meters (330 feet); increase, in the downwind direction, as necessary.
Fire: If tank, rail car, or tank truck is involved in fire, isolate for at least 800 meters (½ mile) in all directions; also, consider initial evacuation for 800 meters (½ mile) in all directions.

EXPOSURE
Short-term effects: *SEEK MEDICAL ATTENTION*. **Vapor:** Irritating to eyes, nose, and throat. *IF INHALED*, will, will cause nausea and dizziness. Move to fresh air. *IF BREATHING HAS STOPPED*, give artificial respiration. If breathing is difficult, administer oxygen. **Liquid:** Irritating to skin and eyes. Harmful if swallowed. Remove contaminated clothing and shoes. Flush affected areas with plenty of water. *IF IN EYES*, hold eyelids open and flush with plenty of water. *IF SWALLOWED* and victim is *CONSCIOUS AND ABLE TO SWALLOW*, have victim drink 4 to 8 ounces of water.

HEALTH HAZARDS
Personal protective equipment (PPE): B-Level PPE. Chemical protective material(s) reported to have good to excellent resistance: PV alcohol, Viton®, Viton®/butyl rubber.
Toxicity by ingestion: Grade 3; LD_{50} = 108 mg/kg (rat).
Long-term health effects: Mutation data.
Vapor (gas) irritant characteristics: Eye and respiratory tract irritant. Vapors cause moderate irritation such that personnel will find high concentrations unpleasant. The effect is temporary.
Liquid or solid irritant characteristics: Eye and skin irritant. If spilled on clothing and allowed to remain, may cause smarting and reddening of the skin.

CHEMICAL REACTIVITY
Binary reactants: Contact with potassium may explode in light.

ENVIRONMENTAL DATA
Water pollution: Effects of low concentrations on aquatic life is unknown. May be dangerous if it enters nearby water intakes; notify operators. Notify local health and pollution control officials.
Response to discharge: Disperse and flush.

SHIPPING INFORMATION
Grades of purity: Technical grade; **Inert atmosphere:** Inerted; **Stability during transport:** Stable.

NAS HAZARD CLASSIFICATION FOR BULK WATER TRANSPORTATION
FIRE: 0
HEALTH: Vapor irritant: 2; Liquid or solid irritant: 1; Poisons: 2
WATER POLLUTION: Human toxicity: 2; Aquatic toxicity: 1; Aesthetic effect: 2
REACTIVITY: Other chemicals: 2; Water: 1; Self-reaction: 0

PHYSICAL AND CHEMICAL PROPERTIES
Physical state @ 59°F/15°C and 1 atm: Liquid.
Molecular weight: 173.83
Boiling point @ 1 atm: 206.6°F/97.0°C/370.2°K.
Melting/Freezing point: −62.5°F/−52.5°C/220.7°K.
Specific gravity (water = 1): 2.4970 @ 68°F/20°C (liquid).
Relative vapor density (air = 1): 6.05 (estimate).
Latent heat of vaporization: 166 Btu/lb = 92.3 cal/g = 3.86×10^5 J/kg.
Vapor pressure: (Reid) 1.7 psia.

DI-*n*-BUTYLAMINE   REC. D:1160

SYNONYMS: 1-BUTANAMINE, *n*-BUTYL; *n*-BUTYL-1-BUTANAMINE; DBA; DI-*n*-BUTILAMINA (Spanish); DIBUTYLAMINE; *n*-DIBUTYLAMINE; DNBA; EEC No. 612-049-00-0

IDENTIFICATION
CAS Number: 111-92-2
Formula: $C_8H_{19}N$; $(C_4H_9)_2NH$
DOT ID Number: UN 2248; DOT Guide Number: 132
Proper Shipping Name: Di-*n*-butylamine

DESCRIPTION: Colorless liquid. Weak ammonia, fishy odor. Floats on the surface of water; soluble, forming a strong base.

Flammable • Corrosive to the skin, eyes, and respiratory tract; contact with skin or eyes can cause burns and vision impairment; inhalation symptoms may be delayed • Containers may BLEVE when exposed to fire •Vapors may form explosive mixture with air • Vapors are heavier than air and will collect and stay in low areas • Vapors may travel long distances to ignition sources and flashback

334 Di-*n*-butyl ether

• Vapors in confined areas (e.g., tanks, sewers, buildings) may explode when exposed to fire • Toxic products of combustion may include nitrogen oxide.

Hazard Classification (based on NFPA-704 Rating System)
Health Hazards (Blue): 3; Flammability (Red): 2; Reactivity (Yellow): 0

EMERGENCY RESPONSE: See Appendix A (132)
Evacuation:
Public safety: Isolate spill area for at least 50 to 100 meters (160 to 330 feet) in all directions.
Spill: Increase, as necessary, the isolation distance shown above, in "Public safety."
Fire: Isolate for 800 meters (½ mile) in all directions, especially if tank, rail car, or tank truck is involved in fire.

EXPOSURE
Short-term effects: *SEEK MEDICAL ATTENTION. ABSORBED THROUGH THE SKIN.* **Vapor:** Irritating to eyes, nose, and throat. *IF INHALED*, will, will cause headache, coughing or difficult breathing. *IF IN EYES*, hold eyelids open and flush. with plenty of water. *IF BREATHING HAS STOPPED*, give artificial respiration. If breathing is difficult, administer oxygen.
May cause lung edema; physical exertion will aggravate this condition. **Liquid:** *TOXIC WHEN ABSORBED THROUGH THE SKIN.* Irritating to skin and eyes. *IF SWALLOWED*, will cause nausea and vomiting. Remove contaminated clothing and shoes. Flush affected areas with plenty of water. *IF IN EYES*, hold eyelids open and flush with plenty of water. *IF SWALLOWED* and victim is *CONSCIOUS AND ABLE TO SWALLOW*, have victim drink 4 to 8 ounces of water. *IF SWALLOWED* and victim is *UNCONSCIOUS OR HAVING CONVULSIONS*, do nothing except keep victim warm.
Note to physician or authorized medical personnel: Medical observation is recommended for 24 to 48 hours after breathing overexposure, as pulmonary edema may be delayed. As first aid for pulmonary edema, consider administering a corticosteroid spray. Cigarette smoking may exacerbate pulmonary injury and should be discouraged for at least 72 hours following exposure.

HEALTH HAZARDS
Personal protective equipment (PPE): B-Level PPE. Full face organic vapor respirator. Chemical protective material(s) reported to have good to excellent resistance: butyl rubber, PV alcohol, Viton®. Also, nitrile is listed in literature, and may offers limited protection
Exposure limits (TWA unless otherwise noted): 5 ppm (AIHA WEEL)
Toxicity by ingestion: Grade 3; oral LD_{50} = 360 mg/kg (rat).
Vapor (gas) irritant characteristics: Vapors cause irritation such that personnel will find high concentrations unpleasant. May cause lung edema; physical exertion will aggravate this condition.
Liquid or solid irritant characteristics: Corrosive to the eyes. Severe skin irritant. Causes second-and third-degree burns on short contact and is very injurious to the eyes.

FIRE DATA
Flash point: 117°F/47°C (cc).
Flammable limits in air: LEL: 1.1%.
Fire extinguishing agents not to be used: Water may be ineffective.
Burning rate: 5.84 mm/min
Electrical hazard: Flow or agitation of substance may generate electrostatic charges due to low conductivity.

CHEMICAL REACTIVITY
Reactivity with water: Aqueous solution is a strongly alkaline.
Binary reactants: Reacts with acids. Violent reaction with oxidizers. Incompatible with oxidizers, acids. May corrode copper and copper alloys, and attack some forms of plastics.
Reactivity group: 7
Compatibility class: Aliphatic amines

ENVIRONMENTAL DATA
Food chain concentration potential: Log P_{ow} = 2.8. Unlikely to accumulate.
Water pollution: Effect of low concentrations on aquatic life is unknown. May be dangerous if it enters nearby water intakes; notify operators. Notify local health and wildlife officials. **Response to discharge:** Issue warning–water contaminant. Restrict access. Disperse and flush.

SHIPPING INFORMATION
Grades of purity: Technical; **Storage temperature:** Ambient; **Inert atmosphere:** None; **Venting:** Open (flame arrester); **Stability during transport:** Stable.

NAS HAZARD CLASSIFICATION FOR BULK WATER TRANSPORTATION
FIRE: 2
HEALTH: Vapor irritant: 2; Liquid or solid irritant: 4; Poisons: 2
WATER POLLUTION: Human toxicity: 2; Aquatic toxicity: 2; Aesthetic effect: 2
REACTIVITY: Other chemicals: 3; Water: 0; Self-reaction: 0

PHYSICAL AND CHEMICAL PROPERTIES
Physical state @ 59°F/15°C and 1 atm: Liquid.
Molecular weight: 129.25
Boiling point @ 1 atm: 319.3°F/159.6°C/432.8°K.
Melting/Freezing point: –80°F/–62°C/211°K.
Specific gravity (water = 1): 0.759 @ 68°F/20°C (liquid).
Liquid surface tension: 24.76 dynes/cm = 0.02476 N/m @ 68°F/20°C.
Relative vapor density (air = 1): 4.5
Latent heat of vaporization: 130 Btu/lb = 72.3 cal/g = 3.03×10^5 J/kg.
Heat of combustion: –18,800 Btu/lb = –10,440 cal/g = -436.8×10^5 J/kg.
Vapor pressure: 2.0 mm.

DI-*n*-BUTYL ETHER   REC. D:1170

SYNONYMS: *n*-DIBUTYL ETHER; *n*-BUTYL ETHER; BUTYL ETHER; 1-BUTOXY BUTANE; DIBUTYL ETHER; DIBUTYL OXIDE; EEC No. 603-054-00-9; ETER *n*-DIBUTILICO (Spanish); ETHER BUTYLIQUE (French); 1,1'-OXYBIS(BUTANE); EK5425000

IDENTIFICATION
CAS Number: 142-96-1
Formula: $C_8H_{18}O$; $C_4H_9OC_4H_9$
DOT ID Number: UN 1149; DOT Guide Number: 127
Proper Shipping Name: Dibutyl ethers

DESCRIPTION: Colorless liquid. Mild, ether-like odor. Floats on the surface of water. Flammable, irritating vapor is produced.

Highly flammable • Containers may BLEVE when exposed to fire • Vapors may form explosive mixture with air • Vapors are heavier than air and will collect and stay in low areas • Vapors may travel long distances to ignition sources and flashback • Vapors in confined areas (e.g., tanks, sewers, buildings) may explode when exposed to fire • Irritating to the skin, eyes, and respiratory tract • Toxic products of combustion may include carbon monoxide.

Hazard Classification (based on NFPA-704 Rating System)
Health Hazards (Blue): 2; Flammability (Red): 3; Reactivity (Yellow): 1

EMERGENCY RESPONSE: See Appendix A (127)
Evacuation:
Public safety: Isolate spill area for at least 25 to 50 meters (80 to 160 feet) in all directions.
Spill: Large spill–Consider initial downwind evacuation for at least 300 meters (1000 feet).
Fire: Isolate for 800 meters (½ mile) in all directions, especially if tank, rail car, or tank truck is involved in fire.

EXPOSURE
Short-term effects: *SEEK MEDICAL ATTENTION.* **Vapor:** Irritating to eyes, nose, and throat. Move victim to fresh air. If breathing is difficult, administer oxygen. **Liquid:** Irritating to skin and eyes. Remove contaminated clothing and shoes. Flush affected areas with plenty of water. *IF IN EYES*, hold eyelids open and flush with plenty of water. *IF SWALLOWED* and victim is *CONSCIOUS AND ABLE TO SWALLOW*, have victim drink 4 to 8 ounces of water.

HEALTH HAZARDS
Personal protective equipment (PPE): B-Level PPE. B-Level PPE. Chemical protective material(s) reported to offer minimal protection: Teflon®, butyl rubber.
Toxicity by ingestion: Grade 1; oral LD_{50} = 7400 mg/kg (rat).
Long-term health effects: May effect the nervous system.
Vapor (gas) irritant characteristics: Respiratory tract irritation.
Liquid or solid irritant characteristics: Eye and skin irritation.

FIRE DATA
Flash point: 77°F/25°C (cc); 92°F/33°C (oc).
Flammable limits in air: LEL: 1.5%; UEL: 7.6%.
Fire extinguishing agents not to be used: Water may be ineffective.
Autoignition temperature: 382°F/194°C/467°K.
Electrical hazard: Flow or agitation of substance may generate electrostatic charges due to low conductivity.

CHEMICAL REACTIVITY
Binary reactants: Violent reaction with strong oxidizers. Incompatible with strong acids, nitrogen trichloride.
Polymerization: Contact with air forms unstable peroxides. Test regularly or peroxides.
Reactivity group: 41
Compatibility class: Ethers

ENVIRONMENTAL DATA
Food chain concentration potential: Log P_{ow} = 3.1. Values > 3.0 are likely to bioconcentrate in aquatic organisms and other living tissue, especially in fats.
Water pollution: Effect of low concentrations on aquatic life is unknown. Fouling to shoreline. May be dangerous if it enters nearby water intakes; notify operators. Notify local health and wildlife officials. **Response to discharge:** Mechanical containment. Should be removed. Chemical and physical treatment.

SHIPPING INFORMATION
Grades of purity: Ambient; **Storage temperature:** None; **Venting:** Open (flame arrester); **Stability during transport:** Stable.

PHYSICAL AND CHEMICAL PROPERTIES
Physical state @ 59°F/15°C and 1 atm: Liquid.
Molecular weight: 130.2
Boiling point @ 1 atm: 288°F/142°C/415°K.
Melting/Freezing point: –140°F/–95°C/178°K.
Specific gravity (water = 1): 0.767 @ 68°F/20°C (liquid).
Liquid surface tension: 23 dynes/cm = 0.023 N/m @ 68°F/20°C.
Liquid water interfacial tension: (estimate) 30 dynes/cm = 0.030 N/m @ 68°F/20°C.
Relative vapor density (air = 1): 4.5
Ratio of specific heats of vapor (gas): 1.0434
Latent heat of vaporization: 120 Btu/lb = 68 cal/g = 2.8×10^5 J/kg.
Heat of combustion: –17,670 Btu/lb = –9,820 cal/g = $–411 \times 10^5$ J/kg.
Vapor pressure: 5.0 mm.

DI-*n*-BUTYL KETONE   REC. D:1180

SYNONYMS: BUTYL KETONE; DIBUTILCETONA (Spanish); DIBUTYLKETONE; 5-NORANONE; 5-NONANONE

IDENTIFICATION
CAS Number: 502-56-7
Formula: $C_9H_{18}O$; $CH_3(CH_2)_3CO(CH_2)_3CH_3$
DOT ID Number: UN 1224; DOT Guide Number: 127
Proper Shipping Name: Ketones, liquid, n.o.s.

DESCRIPTION: Colorless to pale yellow liquid. Floats on the surface of water.

Highly flammable • Containers may BLEVE when exposed to fire • Vapors may form explosive mixture with air • Vapors are heavier than air and will collect and stay in low areas • Vapors may travel long distances to ignition sources and flashback • Vapors in confined areas (e.g., tanks, sewers, buildings) may explode when exposed to fire • Irritating to the skin, eyes, and respiratory tract • Toxic products of combustion may include carbon monoxide.

Hazard Classification (based on NFPA-704 Rating System)
Health Hazards (Blue): 1; Flammability (Red): 2; Reactivity (Yellow): 0

EMERGENCY RESPONSE: See Appendix A (127)
Evacuation:
Public safety: Isolate spill area for at least 25 to 50 meters (80 to 160 feet) in all directions.
Spill: Large spill–Consider initial downwind evacuation for at least 300 meters (1000 feet).
Fire: Isolate for 800 meters (½ mile) in all directions, especially if tank, rail car, or tank truck is involved in fire.

EXPOSURE
Short-term effects: *SEEK MEDICAL ATTENTION.* **Vapor:** Irritating to eyes, nose, and throat. *IF INHALED*, will, will cause coughing or difficult breathing. *IF IN EYES*, hold eyelids open and flush with plenty of water. *IF BREATHING HAS STOPPED*, give artificial respiration. If breathing is difficult, administer oxygen. **Liquid:** Irritating to skin and eyes. *IF SWALLOWED*, will cause

nausea and vomiting. Remove contaminated clothing and shoes. Flush affected areas with plenty of water. *IF IN EYES*, hold eyelids open and flush with plenty of water. *IF SWALLOWED* and victim is *CONSCIOUS AND ABLE TO SWALLOW*, have victim drink 4 to 8 ounces of water. *IF SWALLOWED* and victim is *UNCONSCIOUS OR HAVING CONVULSIONS*, do nothing except keep victim warm.

HEALTH HAZARDS
Personal protective equipment (PPE): B-Level PPE.
Vapor (gas) irritant characteristics: Respiratory tract irritation.
Liquid or solid irritant characteristics: Eye and skin irritation.

FIRE DATA
Fire extinguishing agents not to be used: Water may be ineffective.

CHEMICAL REACTIVITY
Binary reactants: May attack some forms of plastics.

ENVIRONMENTAL DATA
Food chain concentration potential: Negative; unlikely to accumulate.
Water pollution: Effect of low concentrations on aquatic life is unknown. Fouling to shoreline. May be dangerous if it enters nearby water intakes; notify operators. Notify local health and wildlife officials. **Response to discharge:** Mechanical containment. Should be removed. Chemical and physical treatment.

SHIPPING INFORMATION
Grades of purity: 98+%; **Storage temperature:** Ambient; **Inert atmosphere:** None; **Venting:** Open (flame arrester); **Stability during transport:** Stable.

PHYSICAL AND CHEMICAL PROPERTIES
Physical state @ 59°F/15°C and 1 atm: Liquid.
Molecular weight: 142
Boiling point @ 1 atm: 370°F/188°C/461°K.
Melting/Freezing point: 21°F/–6°C/267°K.
Specific gravity (water = 1): 0.822 @ 68°F/20°C (liquid).
Liquid surface tension: 26.60 dynes/cm = 0.0266 N/m at 21.1°C.
Latent heat of vaporization: 161 Btu/lb = 89.6 cal/g = 3.75×10^5 J/kg.
Heat of combustion: –16,080 Btu/lb = –8,930 cal/g = -374×10^5 J/kg.

DIBUTYLPHENOL   REC. D:1200

SYNONYMS: DIBUTILFENOL (Spanish); 2,4-DI-*tert*-BUTYLPHENOL; 2,6-DI-*tert*-BUTYLPHENOL

IDENTIFICATION
CAS Number: 26746-38-3; 128-39-2 (2,6-*tert*-isomer); 96-76-4 (2,4-*tert*-isomer)
Formula: $C_{13}H_{22}O$; 2,6-$(t-C_4H_9)_2C_6H_3OH$; 2,4-$(t-C_4H_9)_2C_6H_3OH$

DESCRIPTION: Colorless to pale yellow liquid or tan crystalline solid. Odorless. Floats on the surface of water.

Combustible • Containers may BLEVE when exposed to fire • Vapors in confined areas (e.g., tanks, sewers, buildings) may explode when exposed to fire • Very Irritating to the skin, eyes, and respiratory tract; contact may cause skin and eye burns • Toxic products of combustion may include carbon monoxide.

Hazard Classification (based on NFPA-704 Rating System)
Health Hazards (Blue): 1; Flammability (Red): 2; Reactivity (Yellow): 0

EMERGENCY RESPONSE: See Appendix A (171)
Evacuation:
Public safety: Isolate the area of spill or leak for at least 10 to 25 meters (30 to 80 feet) in all directions.
Spill: Increase, in the downwind direction, as necessary, the distance shown under "Public Safety."
Fire: If any large container is involved in fire, isolate for at least 800 meters (½ mile) in all directions; also, consider initial evacuation for 800 meters (½ mile) in all directions.

EXPOSURE
Short-term effects: *SEEK MEDICAL ATTENTION*. **Liquid or solid:** Will burn skin and eyes. Harmful if swallowed. Remove contaminated clothing and shoes. Flush affected areas with plenty of water. *IF IN EYES*, hold eyelids open and flush with plenty of water. *IF SWALLOWED* and victim is *CONSCIOUS AND ABLE TO SWALLOW*, have victim drink 4 to 8 ounces of water. **Do NOT induce vomiting.**

HEALTH HAZARDS
Personal protective equipment (PPE): B-Level PPE.
Toxicity by ingestion: Grade 2; oral LD_{50} (2, 6-di-*sec*-butyl phenol) = 1.32 g/kg (rat).
Liquid or solid irritant characteristics: Eye irritant. If spilled on clothing and allowed to remain, may cause smarting and reddening of the skin.

FIRE DATA
Flash point: More than 200°F/93°C (oc).
Fire extinguishing agents not to be used: Water may be ineffective.

CHEMICAL REACTIVITY
Binary reactants: Oxidizers.

ENVIRONMENTAL DATA
Food chain concentration potential: Negative; unlikely to accumulate.
Water pollution: DOT Appendix B, §172.101–marine pollutant. Effect of low concentrations on aquatic life is unknown. Fouling to shoreline. May be dangerous if it enters nearby water intakes; notify operators. Notify local health and wildlife officials. **Response to discharge:** Issue warning–water contaminant. Mechanical containment. Should be removed. Chemical and physical treatment.

SHIPPING INFORMATION
Grades of purity: Commercial; **Storage temperature:** Ambient (solid); 100°F(liquid); **Inert atmosphere:** None; **Venting:** Open; **Stability during transport:** Stable.

PHYSICAL AND CHEMICAL PROPERTIES
Physical state @ 59°F/15°C and 1 atm: Solid.
Molecular weight: 206.3
Boiling point @ 1 atm: 487°F/253°C/526°K.
Melting/Freezing point: 97°F/36°C/309°K.
Specific gravity (water = 1): 0.914 @ 68°F/20°C (solid).
Heat of combustion: estimate –18,000 Btu/lb = –9800 cal/g = -410×10^5 J/kg.

DIBUTYL PHTHALATE
REC. D:1250

SYNONYMS: *o*-BENZENEDICARBOXYLIC ACID, DIBUTYL ESTER; BENZENE-*o*-DICARBOXYLIC ACID DI-*n*-BUTYL ESTER; 1,2-BENZENEDICARBOXYLIC ACID, DIBUTYL ESTER; DI-*n*-BUTYL PHTHALATE; *n*-BUTYLPHTHALATE; PHTHALIC ACID DIBUTYL ESTERCELLUFLEX DBP; DBP; DIBUTYL-1,2-BENZENEDICARBOXYLATE; DI-*n*-BUTYL PHTHALATE; ELAOL; FTALATO de *n*-BUTILO (Spanish); HEXAPLAS M/B; PALATINOL C; POLYCIZER DBP; Px 104; STAFLEX DBP; WITCIZER 300; BUTYL PHTHALATE; PHTHALIC ACID, DIBUTYL ESTER; RC PLASTICIZER DBP; RCRA No. U069

IDENTIFICATION
CAS Number: 84-74-2
Formula: $C_{16}H_{22}O_4$; *o*-$C_6H_4[COO(CH_2)_3CH_3]_2$
DOT ID Number: UN 3077; DOT Guide Number: 171
Proper Shipping Name: Environmentally hazardous substance, liquid, n.o.s.
Reportable Quantity (RQ): **(CERCLA)** 10 lb/4.54 kg

DESCRIPTION: Colorless oily liquid. Practically odorless, like esters. Sinks slowly in water.

Combustible • Containers may BLEVE when exposed to fire • Vapors may form explosive mixture with air • Vapors are heavier than air and will collect and stay in low areas • Vapors in confined areas (e.g., tanks, sewers, buildings) may explode when exposed to fire • Irritating to the skin, eyes, and respiratory tract • Toxic products of combustion may include carbon monoxide.

Hazard Classification (based on NFPA-704 Rating System)
Health Hazards (Blue): 0; Flammability (Red): 1; Reactivity (Yellow): 0

EMERGENCY RESPONSE: See Appendix A (171)
Evacuation:
Public safety: Isolate the area of spill or leak for at least 10 to 25 meters (30 to 80 feet) in all directions.
Spill: Increase, in the downwind direction, as necessary, the distance shown under "Public Safety."
Fire: If any large container is involved in fire, isolate for at least 800 meters (½ mile) in all directions; also, consider initial evacuation for 800 meters (½ mile) in all directions.

EXPOSURE
Short-term effects: Liquid: Irritates the eye, skin, and respiratory tract.

HEALTH HAZARDS
Personal protective equipment (PPE): B-Level PPE. OSHA Table Z-1-A air contaminant. Chemical protective material(s) reported to have good to excellent resistance: butyl rubber, nitrile, nitrile = PVC, PV alcohol, Silvershield®, Viton®. Also, neoprene, styrene-butadiene, Viton®/neoprene, styrene-butadiene/neoprene offers limited protection.
Recommendations for respirator selection: NIOSH/OSHA
50 mg/m³: DMF (any dust and mist respirator with a full facepiece). *125 mg/m³*: SA:CF* (any supplied-air respirator operated in a continuous-flow mode); or PAPRDM* (any powered, air-purifying respirator with a dust and mist filter). *250 mg/m³*: HiEF (any air-purifying, full-facepiece respirator with a high-efficiency particulate filter); or SCBAF (any self-contained breathing apparatus with a full facepiece); or SAF (any supplied-air respirator with a full facepiece). *4000 mg/m³*: SAF:PD,PP (any supplied-air respirator that has a full facepiece and is operated in a pressure-demand or other positive-pressure mode). *EMERGENCY OR PLANNED ENTRY INTO UNKNOWN CONCENTRATIONS OR IDLH CONDITIONS*: SCBAF:PD,PP (any self-contained breathing apparatus that has a full facepiece and is operated in a pressure-demand or other positive-pressure mode); or SAF:PD,PP:ASCBA (any supplied-air respirator that has a full facepiece and is operated in a pressure-demand or other positive-pressure mode in combination with an auxiliary self-contained breathing apparatus operated in a pressure-demand or other positive pressure mode). *ESCAPE* : HiEF (any air-purifying, full-facepiece respirator with a high-efficiency particulate filter); or SCBAE (any appropriate escape-type, self-contained breathing apparatus). *Note: Substance causes eye irritation or damage; eye protection needed.
Exposure limits (TWA unless otherwise noted): ACGIH TLV 5 mg/m³; NIOSH/OSHA 5 mg/m³.
Toxicity by ingestion: Grade 1; LD_{50} = 5 to 15 g/kg (rat).
Long-term health effects: Birth defects in rats; polyneuritis in humans; and kidney, ureter, or bladder effects.
Vapor (gas) irritant characteristics: Eye and respiratory tract irritant.
Liquid or solid irritant characteristics: Eye and skin irritant.
IDLH value: 4000 mg/m³.

FIRE DATA
Flash point: 355°F/179°C (oc); 315°F/157°C (cc)
Flammable limits in air: LEL: 0.5% @ 456°F/235°C; UEL: 2.5% (estimate)
Fire extinguishing agents not to be used: Water or foam may cause frothing.
Autoignition temperature: 757°F/402°C/675°K.

CHEMICAL REACTIVITY
Binary reactants: Reacts with nitrates, strong oxidizers, alkalis, acids, liquid chlorine.
Reactivity group: 34
Compatibility class: Ester

ENVIRONMENTAL DATA
Food chain concentration potential: Negative; unlikely to accumulate.
Water pollution: Dangerous to aquatic life in high concentrations. Fouling to shoreline. May be dangerous if it enters nearby water intakes; notify operators. Notify local health and pollution control officials. **Response to discharge:** Mechanical containment. Should be removed. Chemical and physical treatment.

SHIPPING INFORMATION
Grades of purity: 99.6%; **Stability during transport:** Stable.

NAS HAZARD CLASSIFICATION FOR BULK WATER TRANSPORTATION
FIRE: 1
HEALTH: Vapor irritant: 0; Liquid or solid irritant: 0; Poisons: 0
WATER POLLUTION: Human toxicity: 0; Aquatic toxicity: 1; Aesthetic effect: 3
REACTIVITY: Other chemicals: 1; Water: 0; Self-reaction: 0

PHYSICAL AND CHEMICAL PROPERTIES
Molecular weight: 278.35
Boiling point @ 1 atm: 644°F/340°C/613°K.
Melting/Freezing point: –31°F/–35°C/238°K.
Critical temperature: 932°F/500°C/773°K.
Critical pressure: 250 psia = 17 atm = 1.7 MN/m².

Specific gravity (water = 1): 1.049 @ 68°F/20°C (liquid).
Liquid surface tension: 34 dynes/cm = 0.034 N/m @ 68°F/20°C.
Liquid water interfacial tension: 27 dynes/cm = 0.027 N/m at 22.7°C.
Heat of combustion: $-13,300$ Btu/lb = -7400 cal/g = -310×10^5 J/kg.
Vapor pressure: 0.00007 mm.

DICAMBA   REC. D:1300

SYNONYMS: BANEX; BANLEN; BANVEL; BANVEL D; BANVEL HERBICIDE; BRUSH BUSTER; COMPOUND B DICAMBRA; DIANATE; 3,6-DICHLORO-*o*-ANISIC ACID; MBDA; MEDIBEN; VELSACOL COMPOUND "D"; VELSICOL 58-CS-11

IDENTIFICATION
CAS Number: 1918-00-9
Formula: $C_8H_6Cl_2O_3$
DOT ID Number: UN 2769; DOT Guide Number: 151
Proper Shipping Name: Benzoic derivative pesticides, solid, toxic, n.o.s.
Reportable Quantity (RQ): **(CERCLA)** 1000 lb/454 kg

DESCRIPTION: Colorless crystalline solid. Turns brown on exposure to air. Odorless. Sinks in water; slowly dissolved.

Poison! (benzoic acid herbicide) • Extremely irritating to the eyes; irritating to the skin and respiratory system; inhaling, swallowing the material, or absorption through the skin is toxic • Firefighting gear (including SCBA) does not provide adequate protection. If exposure occurs, remove and isolate gear immediately and thoroughly decontaminate personnel • Containers may BLEVE when exposed to fire • At 392°F/200°C dicamba is decomposed, forming 2,5-dichloroanisole. Toxic products of combustion may include and carbon monoxide and hydrogen chloride.

Hazard Classification (based on NFPA-704 Rating System)
Health Hazards (Blue): 1; Flammability (Red): 0; Reactivity (Yellow): 0

EMERGENCY RESPONSE: See Appendix A (151)
Evacuation:
Public safety: Isolate the area of spill or leak for at least 25 to 50 meters (80 to 160 feet) in all directions.
Spill: Increase, in the downwind direction, as necessary, the distance shown under "Public Safety."
Fire: If tank, rail car, or tank truck is involved in fire, isolate for at least 800 meters (½ mile) in all directions; also, consider initial evacuation for 800 meters (½ mile) in all directions.

EXPOSURE
Short-term effects: *CALL FOR MEDICAL AID.* **Solid:** Harmful if swallowed. *IF BREATHING HAS STOPPED*, give artificial respiration; *avoid mouth-to-mouth resuscitation; use bag/mask apparatus. IF SWALLOWED* and victim is *CONSCIOUS AND ABLE TO SWALLOW*, have victim drink 4 to 8 ounces of water and induce vomiting. Flush affected area with plenty of water. *IF IN EYES*, hold eyelids open and flush with plenty of water.

HEALTH HAZARDS
Toxicity by ingestion: Grade 2; LD_{50} = 0.5 to 5 g/kg.
Long-term health effects: Two-year feeding of 500 ppm to rats and dogs-no observable effects.

FIRE DATA
Behavior in fire: Water streams applied to adjacent fires will spread contamination of pesticide over wide area.

ENVIRONMENTAL DATA
Food chain concentration potential: Log K_{ow} = 2.2. Unlikely to accumulate.
Water pollution: Dangerous to aquatic life in high concentrations. May be dangerous if it enters nearby water intakes; notify operators. Notify local health and wildlife officials. **Response to discharge:** Disperse and flush.

SHIPPING INFORMATION
Stability during transport: Stable.

PHYSICAL AND CHEMICAL PROPERTIES
Physical state @ 59°F/15°C and 1 atm: Solid.
Molecular weight: 221
Melting/Freezing point: 237.2° to 240.8°F/114° to 116°C/387.2° to 389.2°K.
Relative vapor density (air = 1): 7.62

DICHLOBENIL   REC. D:1350

SYNONYMS: CASORON; 2,6-DICHLOROBENZONITRILE; DU-SPREX; 2,6-DBN; NIA 5996

IDENTIFICATION
CAS Number: 1194-65-6
Formula: $C_7H_3Cl_2N$
DOT ID Number: UN 2769; DOT Guide Number: 151
Proper Shipping Name: Benzoic derivative pesticide, solid, toxic
Reportable Quantity (RQ): **(CERCLA)** 100 lb/45.4 kg

DESCRIPTION: White crystalline solid or liquid. Aromatic odor. Mixes slowly with water.

Poison! (benzonitrile herbicide) • Breathing the vapor, skin or eye contact, or swallowing the material can kill you • Firefighting gear (including SCBA) does not provide adequate protection. If exposure occurs, remove and isolate gear immediately and thoroughly decontaminate personnel • Toxic products of combustion may include hydrogen cyanide and nitrogen oxides • Do not put yourself in danger by entering a contaminated area to rescue a victim.

EMERGENCY RESPONSE: See Appendix A (151)
Evacuation:
Public safety: Isolate the area of spill or leak for at least 25 to 50 meters (80 to 160 feet) in all directions.
Spill: Increase, in the downwind direction, as necessary, the distance shown under "Public Safety."
Fire: If tank, rail car, or tank truck is involved in fire, isolate for at least 800 meters (½ mile) in all directions; also, consider initial evacuation for 800 meters (½ mile) in all directions.

EXPOSURE
Short-term effects: *SEEK MEDICAL ATTENTION.* **Solid or Liquid:** Poisonous if absorbed through the skin; inhaled, or swallowed. Wash with soap and water; flush affected areas with plenty of water. *IF SWALLOWED* and victim is *CONSCIOUS* have victim drink 4 to 8 ounces of water and induce vomiting.

HEALTH HAZARDS
Personal protective equipment (PPE): B-Level PPE.
Toxicity by ingestion: Grade 2; LD_{50} = 0.5 to 5 g/kg.
Long-term health effects: At 50 ppm growth inhibition occurred in second generation rats and at higher levels hypertrophy of liver and kidneys was found.

ENVIRONMENTAL DATA
Food chain concentration potential: Accumulated by goldfish at a 15 to 20 fold level in 3 months.
Water pollution: Harmful to aquatic life in very low concentrations. May be dangerous if it enters nearby water intakes; notify operators. Notify local health and wildlife officials.
Response to discharge: Issue warning–water contaminant. Chemical and physical treatment.

SHIPPING INFORMATION
Storage temperature: Cool; **Stability during transport:** Stable.

PHYSICAL AND CHEMICAL PROPERTIES
Physical state @ 59°F/15°C and 1 atm: Solid.
Molecular weight: 172
Boiling point @ 1 atm: 518°F/270°C/543.2°K.
Melting/Freezing point: 293–295°F/145–146°C/418–419°K.
Heat of decomposition: Thermally stable.

DICHLONE   REC. D:1400

SYNONYMS: AGISTAT; COMPOUND 604; DICLONA (Spanish); 2,3-DICHLORO-1,4-NAPHTHO-QUINONE; PHYGON®; PHYGON®-XL; QUINTAR; QUINTAR 504F; SANQUINON; UNIROYAL 604; USR-604; U.S. RUBBER 604

IDENTIFICATION
CAS Number: 117-80-6
Formula: $C_{10}H_4Cl_2O_2$
DOT ID Number: UN 2811; DOT Guide Number: 151
Proper Shipping Name: Poisonous solid, organic, **n.o.s.**
Reportable Quantity (RQ): (CERCLA) 1 lb/0.454 kg

DESCRIPTION: Yellow crystalline solid. Sinks and mixes with water.

Poison! (substituted benzene fungicide) • Breathing the dust, skin or eye contact can cause illness; swallowing the material can kill you • Firefighting gear (including SCBA) does not provide adequate protection. If exposure occurs, remove and isolate gear immediately and thoroughly decontaminate personnel • Containers may BLEVE when exposed to fire • Toxic products of combustion may include carbon monoxide and hydrogen chloride fumes • Do not put yourself in danger by entering a contaminated area to rescue a victim.

Hazard Classification (based on NFPA-704 Rating System)
Health Hazards (Blue): 2; Flammability (Red): 0; Reactivity (Yellow): 0

EMERGENCY RESPONSE: See Appendix A (151)
Evacuation:
Public safety: Isolate the area of spill or leak for at least 25 to 50 meters (80 to 160 feet) in all directions.
Spill: Increase, in the downwind direction, as necessary, the distance shown under "Public Safety."
Fire: If tank, rail car, or tank truck is involved in fire, isolate for at least 800 meters (½ mile) in all directions; also, consider initial evacuation for 800 meters (½ mile) in all directions.

EXPOSURE
Short-term effects: *SEEK MEDICAL ATTENTION.* **Dust:** Irritating to eyes, skin, nose, and throat. Move to fresh air. *IF BREATHING HAS STOPPED,* give artificial respiration; *avoid mouth-to-mouth resuscitation; use bag/mask apparatus.* Flush affected areas with plenty of water. *IF IN EYES,* hold eyelids open and flush with plenty of water. **Solid:** Harmful if swallowed. *IF SWALLOWED* and victim is *CONSCIOUS AND ABLE TO SWALLOW,* have victim drink 4 to 8 ounces of water. **Do NOT induce vomiting.** Administer a slurry of activated charcoal.
Personal protective equipment (PPE): B-Level PPE. Self-contained breathing apparatus, rubber gloves, hats, suits, and boots.
Short-term exposure limits (15-minute TWA): Maximum emission concentration (MIC) 0.05 mg/m³ [20–30 minutes].
Toxicity by ingestion: Grade 2; LD_{50} = 0.5 to 5 g/kg.
Long-term health effects: Irritating to skin. May cause dermatitis.
Vapor (gas) irritant characteristics: Eye and respiratory tract irritant.
Liquid or solid irritant characteristics: Eye and skin irritant.

ENVIRONMENTAL DATA
Food chain concentration potential: Log P = 3.15 (estimate). Values above 3.0 are very likely to accumulate in living tissues and especially in fats.
Water pollution: Harmful to aquatic life in very low concentrations. May be dangerous if it enters nearby water intakes; notify operators. Notify local health and wildlife officials. **Response to discharge:** Issue warning–water contaminant. Chemical and physical treatment. Disperse and flush.

SHIPPING INFORMATION
Grades of purity: Technical grade, 95%; **Stability during transport:** Stable.

PHYSICAL AND CHEMICAL PROPERTIES
Physical state @ 59°F/15°C and 1 atm: Solid.
Molecular weight: 227.06
Boiling point @ 1 atm: Sublimes 527°F/275°C/548.15°K.
Melting/Freezing point: Pure: 379°F/193°C/466°K; Technical: 370.4°F/188°C/461.2°K.
Relative vapor density (air = 1): 7.84

o-DICHLOROBENZENE   REC. D:1450

SYNONYMS: BENZENE, 1,2-DICHLORO-; CHLOROBEN; CHLORODEN; CLOROBEN; ODCB; *o*-DICLOROBENCENO (Spanish); *o*-DICHLOR BENZOL; 1,2-DICHLOROBENZENE; DICHLORICIDE; *o*-DICHLOROBENZOL; DILANTIN DB; DILATIN DB; DIZENE; DOWTHERM E®; ODB; ODCB; ORTHODICHLOROBENZENE; ORTHODICHLOROBENZOL; TERMITKIL; RCRA No. UO7O

IDENTIFICATION
CAS Number: 95-50-1
Formula: $C_6H_4Cl_2$; *o*-$C_6H_4Cl_2$
DOT ID Number: UN 1591; DOT Guide Number: 152
Proper Shipping Name: *o*-Dichlorobenzene
Reportable Quantity (RQ): **(CERCLA)** 100 lb/45.4 kg

o-Dichlorobenzene

DESCRIPTION: Colorless liquid. Pleasant odor. Sinks in water.

Flammable • Containers may BLEVE when exposed to fire • Vapors may form explosive mixture with air • Vapors are heavier than air and will collect and stay in low areas • Vapors may travel long distances to ignition sources and flashback • Vapors in confined areas (e.g., tanks, sewers, buildings) may explode when exposed to fire • Irritating to the skin, eyes, and respiratory tract • Toxic products of combustion may include hydrogen chloride gas, chlorocarbons, chlorine.

Hazard Classification (based on NFPA-704 Rating System)
Health Hazards (Blue): 2; Flammability (Red): 2; Reactivity (Yellow): 0

EMERGENCY RESPONSE: See Appendix A (152)
Evacuation:
Public safety: Isolate the area of spill or leak for at least 25 to 50 meters (80 to 160 feet) in all directions.
Spill: Increase, in the downwind direction, as necessary, the distance shown under "Public Safety."
Fire: If tank, rail car, or tank truck is involved in fire, isolate for at least 800 meters (½ mile) in all directions; also, consider initial evacuation for 800 meters (½ mile) in all directions.

EXPOSURE
Short-term effects: *SEEK MEDICAL ATTENTION.* **Liquid:** Irritating to skin and eyes. Harmful if swallowed. Remove contaminated clothing and shoes. Flush affected areas with plenty of water. *IF BREATHING HAS STOPPED,* give artificial respiration; *avoid mouth-to-mouth resuscitation; use bag/mask apparatus. IF IN EYES,* hold eyelids open and flush with plenty of water. *IF SWALLOWED* and victim is *CONSCIOUS AND ABLE TO SWALLOW,* have victim drink 4 to 8 ounces of water and have victim induce vomiting. *IF SWALLOWED* and victim is *UNCONSCIOUS OR HAVING CONVULSIONS,* do nothing except keep victim warm.

HEALTH HAZARDS
Personal protective equipment (PPE): B-Level PPE. OSHA Table Z-1-A air contaminant. Chemical protective material(s) reported to have good to excellent resistance: Viton®. Also, natural rubber offers limited protection.
Recommendations for respirator selection: NIOSH/OSHA
200 ppm: CCRFOV [any air-purifying, full-facepiece respirator (gas mask) with a chin-style, front- or back-mounted acid gas canister]; or PAPROV* [any powered, air-purifying respirator with organic vapor cartridge(s)]; or SCBAF (any self-contained breathing apparatus with a full facepiece); or SAF (any supplied-air respirator with a full facepiece). *EMERGENCY OR PLANNED ENTRY INTO UNKNOWN CONCENTRATIONS OR IDLH CONDITIONS:* SCBAF:PD,PP (any self-contained breathing apparatus that has a full facepiece and is operated in a pressure-demand or other positive-pressure mode); or SAF:PD,PP:ASCBA (any supplied-air respirator that has a full facepiece and is operated in a pressure-demand or other positive-pressure mode in combination with an auxiliary self-contained breathing apparatus operated in a pressure-demand or other positive pressure mode). *ESCAPE:* GMFOV[any air-purifying, full-facepiece respirator (gas mask) with a chin-style, front- or back-mounted organic vapor canister]; or SCBAE (any appropriate escape-type, self-contained breathing apparatus). *Note: Substance causes eye irritation or damage; eye protection needed.
Exposure limits (TWA unless otherwise noted): ACGIH TLV 25 ppm (150 mg/m^3); NIOSH/OSHA ceiling 50 ppm (300 mg/m^3).
Short-term exposure limits (15-minute TWA): ACGIH STEL 50 ppm (301 mg/m^3)
Toxicity by ingestion: Grade 2; LD$_{50}$ = 0.5 to 5 g/kg.
Long-term health effects: May cause lung, liver and kidney damage. Causes kidney and liver damage in rats. IARC unclassified, rating 3; inadequate human and animal evidence.
Vapor (gas) irritant characteristics: Eye and respiratory tract irritant. Vapors cause irritation such that personnel will find high concentrations unpleasant.
Liquid or solid irritant characteristics: Eye and skin irritant. If spilled on clothing and allowed to remain, may cause smarting and reddening of the skin.
Odor threshold: 0.70 ppm.
IDLH value: 200 ppm.

FIRE DATA
Flash point: 165°F/74°F/oc); 151°F/66°C (cc).
Flammable limits in air: LEL: 2.2%; UEL: 9.2%.
Autoignition temperature: 1198°F/648°C/921°K.
Electrical hazard: Class I, Group D.
Burning rate: 1.3 mm/min

CHEMICAL REACTIVITY
Binary reactants: Strong oxidizers, aluminum, chlorides, acids, acid fumes, alkali metals. Most rubbers not compatible.
Reactivity group: 36
Compatibility class: Halogenated hydrocarbons

ENVIRONMENTAL DATA
Food chain concentration potential: Log K$_{ow}$ = 3.5 (estimate). Values > 3.0 are likely to bioconcentrate in aquatic organisms and other living tissue, especially in fats.
Water pollution: DOT Appendix B, §172.101–marine pollutant. Effect of low concentrations on aquatic life is unknown. May be dangerous if it enters nearby water intakes; notify operators. Notify local health and pollution control officials. **Response to discharge:** Issue warning–water contaminant. Should be removed. Chemical and physical treatment.

SHIPPING INFORMATION
Grades of purity: Technical: 99.5% min. dichlorobenzene (ratio–ortho + para/meta: 80 min.); Technical: 85% orthodichlorobenzene, 14.0% paradichlorobenzene; Technical: 80% ortho, 17% para, 2% meta; Pure: not less than 99.5% ortho, not more than 0.5% para; **Stability during transport:** Stable.

NAS HAZARD CLASSIFICATION FOR BULK WATER TRANSPORTATION
FIRE: 1
HEALTH: Vapor irritant: 2; Liquid or solid irritant: 1; Poisons: 1
WATER POLLUTION: Human toxicity: 1; Aquatic toxicity: 3; Aesthetic effect: 2
REACTIVITY: Other chemicals: 1; Water: 0; Self-reaction: 0

PHYSICAL AND CHEMICAL PROPERTIES
Physical state @ 59°F/15°C and 1 atm: Liquid.
Molecular weight: 147.01
Boiling point @ 1 atm: 356.9°F/180.5°C/453.7°K.
Melting/Freezing point: 1°F/–17°C/256°K.
Specific gravity (water = 1): 1.306 @ 68°F/20°C (liquid).
Liquid surface tension: 37 dynes/cm = 0.037 N/m @ 68°F/20°C.
Liquid water interfacial tension: estimate 40 dynes/cm = 0.04 N/m @ 68°F/20°C.
Relative vapor density (air = 1): 5.1
Ratio of specific heats of vapor (gas): 1.080

Latent heat of vaporization: 115 Btu/lb = 63.9 cal/g = 2.68 x 10^5 J/kg.
Heat of combustion: –7969 Btu/lb = –4427 cal/g = –185.4 x 10^5 J/kg.
Heat of fusion: 21.02 cal/g.
Vapor pressure: (Reid) 0.06 psia; 1 mm.

m-DICHLOROBENZENE   REC. D:1500

SYNONYMS: *m*-DICLOROBENCENO (Spanish); 1,3-DICHLOROBENZENE; meta-DICHLOROBENZENE; MDB; MDCB; RCRA No. U071

IDENTIFICATION
Formula: $C_6H_4Cl_2$
CAS Number: 541-73-1
DOT ID Number: NA 1993; DOT Guide Number: 128
Proper Shipping Name: Combustible liquid, n.o.s.
Reportable Quantity (RQ): **(CERCLA)** 100 lb/45.4 kg

DESCRIPTION: Colorless liquid. Sinks in water.

Flammable • Containers may BLEVE when exposed to fire • Vapors may form explosive mixture with air • Vapors are heavier than air and will collect and stay in low areas • Vapors may travel long distances to ignition sources and flashback • Vapors in confined areas (e.g., tanks, sewers, buildings) may explode when exposed to fire • Irritating to the skin, eyes, and respiratory tract • Toxic products of combustion may include carbon monoxide.

Hazard Classification (based on NFPA-704 Rating System)
Health Hazards (Blue): 2; Flammability (Red): 2; Reactivity (Yellow): 0

EMERGENCY RESPONSE: See Appendix A (128)
Evacuation:
Public safety: Isolate spill area for at least 25 to 50 meters (80 to 160 feet) in all directions.
Spill: Large spill–Consider initial downwind evacuation for at least 300 meters (1000 feet).
Fire: Isolate for 800 meters (½ mile) in all directions, especially if tank, rail car, or tank truck is involved in fire.

EXPOSURE
Short-term effects: *SEEK MEDICAL ATTENTION*. **Liquid:** Irritating to skin and eyes. Harmful if swallowed. Remove contaminated clothing and shoes. Flush affected areas with plenty of water. *IF IN EYES*, hold eyelids open and flush with plenty of water. *IF SWALLOWED* and victim is *CONSCIOUS AND ABLE TO SWALLOW*, have victim drink 4 to 8 ounces of water and have victim induce vomiting. *IF SWALLOWED* and victim is *UNCONSCIOUS OR HAVING CONVULSIONS*, do nothing except keep victim warm.

Personal protective equipment (PPE): B-Level PPE. Chemical protective material(s) reported to have good to excellent resistance: Viton®.
Toxicity by ingestion: Grade 2; LD_{50} = 500 to 5000 mg/kg.
Long-term health effects: May cause some liver and kidney damage.
Vapor (gas) irritant characteristics: Severe eye and respiratory tract irritant. Vapors cause moderate irritation such that personnel will find high concentrations unpleasant. The effect is temporary.
Liquid or solid irritant characteristics: Severe eye and skin irritant. If spilled on clothing and allowed to remain, may cause smarting and reddening of skin.
Odor threshold: 0.02 ppm in water.

FIRE DATA
Flash point: estimate 165°F/74°C (oc); 151°F/66°C (cc).
Flammable limits in air: estimate LEL: 2.02%; UEL: 9.2%.
Autoignition temperature: estimate 1198°F/647°C/920°K.
Electrical hazard: Class I, Group D.

CHEMICAL REACTIVITY
Binary reactants: Reacts with strong oxidizers, alkali metals.
Reactivity group: 36
Compatibility class: Halogenated hydrocarbon.

SHIPPING INFORMATION
Stability during transport: Stable.

ENVIRONMENTAL DATA
Food chain concentration potential: Log K_{ow} = 3.60. Values > 3.0 are likely to bioconcentrate in aquatic organisms and other living tissue, especially in fats.
Water pollution: DOT Appendix B, §172.101–marine pollutant. Harmful to aquatic life in very low concentrations. May be dangerous if it enters nearby water intakes; notify operators. Notify local health and pollution control officials. **Response to discharge:** Issue warning–water contaminant. Should be removed. Chemical and physical treatment.

PHYSICAL AND CHEMICAL PROPERTIES
Physical state @ 59°F/15°C and 1 atm: Liquid.
Molecular weight: 147.01.
Boiling point @ 1 atm: 343.4°F/173°C/446.15°K.
Melting/Freezing point: –12.5°F/24.7°C/248.45°K.
Critical temperature: estimate 771.44°F/410.8°C/683.95°K.
Critical pressure: 562.9 psia = 38.3 atm = 3.88 NM/m$_2$.
Specific gravity (water = 1): 1.2884 @ 68°F/20°C.
Liquid surface tension: 36.01 dynes/cm = 0.03601 N/m @ 68°F/20°C.
Relative vapor density (air = 1): 5.07.
Latent heat of vaporization: At boiling point. 113.02 Btu/lb = 62.79 cal/g = 2.63 x 10^5 J/kg.
Heat of combustion: (net) –8096 Btu/lb = 4498 cal/g = –1.88 x 10^7 J/kg.
Heat of fusion: 20.55 cal/g.
Vapor pressure: 1 mm.

p-DICHLOROBENZENE   REC. D:1550

SYNONYMS: BENZENE, 1,4-DICHLORO-; *p*-DCB; DICHLOROCIDE; *p*-CHLOROPHENYL CHLORIDE; *p*-DICHLORBENZOL (German); 1,4-DICHLOR-BENZOL (German); DI-CHLORICIDE; 1,4-DICHLOROBENZENE; *p*-DICHLOROBENZOL; *p*-DICLOROBENCENO (Spanish); DICHLOROCIDE; EEC No. 602-036-00-2; EVOLA; PARACIDE; PARA CRYSTALS; PARADI; PARADICHLOROBENZENE; PARADICHLOROBENZOL; PARADOW; PARAMOTH; PARANUGGETS; PARAZENE; PDB; PDCB; PERSIA-PERAZOL; SANTOCHLOR; RCRA U072

IDENTIFICATION
CAS Number: 106-46-7
Formula: $C_6H_4Cl_2$; *p*-$C_6H_4Cl_2$
DOT ID Number: UN 1592; DOT Guide Number: 152

342   Di-(*p*-chlorobenzoyl) peroxide

Proper Shipping Name: *p*-Dichlorobenzene
Reportable Quantity (RQ): 100 lb/45.4 kg

DESCRIPTION: White solid. Mothball-like odor. Sinks in water; insoluble.

Combustible solid • Heat or flame may cause explosion • Poison! • Vapors are very irritating to skin, eyes, node, and lungs • Containers may explode when exposed to fire • Vapors are heavier than air and will collect and stay in low areas • Toxic products of combustion may include hydrogen chloride and phosgene.

Hazard Classification (based on NFPA-704 Rating System)
Health Hazards (Blue): 2; Flammability (Red): 2; Reactivity (Yellow): 0

EMERGENCY RESPONSE: See Appendix A (152)
Evacuation:
Public safety: Isolate the area of spill or leak for at least 25 to 50 meters (80 to 160 feet) in all directions.
Spill: Increase, in the downwind direction, as necessary, the distance shown under "Public Safety."
Fire: If tank, rail car, or tank truck is involved in fire, isolate for at least 800 meters (½ mile) in all directions; also, consider initial evacuation for 800 meters (½ mile) in all directions.

EXPOSURE
Short-term effects: *SEEK MEDICAL ATTENTION*. **Solid:** Irritating to skin and eyes. Harmful if swallowed. Remove contaminated clothing and shoes. Flush affected areas with plenty of water. *IF BREATHING HAS STOPPED*, give artificial respiration; *avoid mouth-to-mouth resuscitation; use bag/mask apparatus*. *IF IN EYES*, hold eyelids open and flush with plenty of water. *IF SWALLOWED* and victim is *CONSCIOUS AND ABLE TO SWALLOW*, have victim drink 4 to 8 ounces of water.

HEALTH HAZARDS
Personal protective equipment (PPE): B-Level PPE. OSHA Table Z-1-A air contaminant. Chemical protective material(s) reported to offer minimal to poor protection: nitrile
Recommendations for respirator selection: NIOSH
At any concentrations above the NIOSH REL, or where there is no REL, at any detectable concentration: SCBAF:PD,PP (any self-contained breathing apparatus that has a full facepiece and is operated in a pressure-demand or other positive-pressure mode); or SAF:PD,PP:ASCBA (any supplied-air respirator that has a full facepiece and is operated in a pressure-demand or other positive-pressure mode in combination with an auxiliary self-contained breathing apparatus operated in a pressure-demand or other positive pressure mode). *ESCAPE:* GMFOV [any air-purifying, full-facepiece respirator (gas mask) with a chin-style, front-or back-mounted organic vapor canister]; or SCBAE (any appropriate escape-type, self-contained breathing apparatus).
Exposure limits (TWA unless otherwise noted): ACGIH TLV 10 ppm (60 mg/m^3) animal carcinogen; OSHA 75 ppm (450 mg/m^3); NIOSH potential human carcinogen; reduce exposure to lowest feasible level.
Toxicity by ingestion: Grade 2; LD_{50} = 0.5 to 5 g/kg.
Long-term health effects: Liver and kidney damage may occur. IARC possible carcinogen, rating 2B; human evidence: inadequate; animal evidence: sufficient.
Vapor (gas) irritant characteristics: Eye and respiratory tract irritant. Vapors cause irritation such that personnel will find high concentrations unpleasant. The effect is temporary.
Liquid or solid irritant characteristics: Eye irritant. If spilled on clothing and allowed to remain, may cause smarting and reddening of the skin.
Odor threshold: 0.12 ppm.
IDLH value: 150 ppm; potential human carcinogen.

FIRE DATA
Flash point: 165°F/74°C (oc); 150°F/66°C (cc).
Flammable limits in air: LEL: 1.5%; UEL: 5.1%.
Autoignition temperature: More than 900°F/482°C/755°K.
Burning rate: 1.3 mm/min (approximate).

CHEMICAL REACTIVITY
Binary reactants: Reacts vigorously with oxidizing materials.
Reactivity group: 36
Compatibility class: Halogenated hydrocarbons

ENVIRONMENTAL DATA
Food chain concentration potential: Log K_{ow} = 3.37. Values > 3.0 are likely to bioconcentrate in aquatic organisms and other living tissue, especially in fats.
Water pollution: DOT Appendix B, §172.101–marine pollutant. Harmful to aquatic life in very low concentrations. Fouling to shoreline. May be dangerous if it enters nearby water intakes; notify operators. Notify local health and wildlife officials. **Response to discharge:** Issue warning–water contaminant. Should be removed. Chemical and physical treatment. Federal Water pollution Control Act; Clean Water Act.

SHIPPING INFORMATION
Grades of purity: Solid: 5 grades, chemical purity close to 100%; Liquid: 1–2% orthodichlorobenzene; **Stability during transport:** Stable.

PHYSICAL AND CHEMICAL PROPERTIES
Physical state @ 59°F/15°C and 1 atm: Solid.
Molecular weight: 147.01
Boiling point @ 1 atm: 345.6°F/174.2°C/447.4°K.
Melting/Freezing point: 128°F/54°C/583°K.
Specific gravity (water = 1): 1.25 @ 68°F/20°C (solid).
Relative vapor density (air = 1): 5.09
Heat of fusion: 29.07 cal/g.
Vapor pressure: 1.3 mm.

DI-(*p*-CHLOROBENZOYL) PEROXIDE   REC. D:1600

SYNONYMS: CADOX PS; BIS-(*p*-CHLOROBENZOY) PEROXIDE; *p*-CHLOROBENZOYL PEROXIDE; *p,p*-CHLOROBENZOYL PEROXIDE; DI-(4-CHLOROBENZOYL) PEROXIDE; *p,p*´-DICHLOROBENZOYL PEROXIDE

IDENTIFICATION
CAS Number: 94-17-7
Formula: $C_{13}H_8Cl_2O_4$; (*p*-ClC$_6$H$_4$COO)$_2$
DOT ID Number: UN 3102; DOT Guide Number: 146
Proper Shipping Name: Organic peroxide, Type B, solid

DESCRIPTION: White solid or paste in silicone fluid and dibutyl phthalate. Odorless. Sinks in water.

Extremely flammable • Containers may BLEVE when exposed to fire • May explode if subjected to heat or flame or shock • Vapors are heavier than air and will collect and stay in low areas • Strong oxidizer which may react spontaneously with low flash point organics or reducing agents. Heat forms oxygen; will increase the activity of an existing fire • May cause fire or explosion on contact with combustibles (wood, paper, oil, clothing, etc.) • Vapors in confined areas (e.g., tanks, sewers, buildings) may explode when exposed to fire • Irritating to the skin, eyes, and respiratory tract • Toxic products of combustion may include chlorinated biphenyls.

EMERGENCY RESPONSE: See Appendix A (146)
Evacuation:
Public safety: Isolate the area of spill or leak for at least 25 to 50 meters (80 to 160 feet) in all directions.
Spill: Consider initial evacuation for at least 250 meters (800 feet).
Fire: If tank, rail car, or tank truck is involved in fire, isolate for at least 800 meters (½ mile) in all directions; also, consider initial evacuation for 800 meters (½ mile) in all directions.

EXPOSURE
Short-term effects: *SEEK MEDICAL ATTENTION.* **Solid:** Irritating to skin and eyes. Harmful if swallowed. Remove contaminated clothing and shoes. Flush affected areas with plenty of water. *IF IN EYES*, hold eyelids open and flush with plenty of water. *IF SWALLOWED and victim is CONSCIOUS AND ABLE TO SWALLOW*, have victim drink 4 to 8 ounces of water and have victim induce vomiting. *IF SWALLOWED and victim is UNCONSCIOUS OR HAVING CONVULSIONS*, do nothing except keep victim warm.

HEALTH HAZARDS
Personal protective equipment (PPE): B-Level PPE. Butyl rubber is generally suitable for peroxide compounds.
Liquid or solid irritant characteristics: Eye and skin irritant.

FIRE DATA
Behavior in fire: Solid may explode. Burns very rapidly when ignited. Smoke is unusually heavy when paste form is involved.

CHEMICAL REACTIVITY
Binary reactants: Vigorous reaction with combustible materials.

ENVIRONMENTAL DATA
Water pollution: Effect of low concentrations on aquatic life is unknown. May be dangerous if it enters nearby water intakes; notify operators. Notify local health and wildlife officials.
Response to discharge: Issue warning–oxidizing material, water contaminant. Should be removed. Chemical and physical treatment.

SHIPPING INFORMATION
Grades of purity: Dry; wet with more than 30% water; 50% paste with silicone fluid; **Storage temperature:** Below 80°F/27°C; **Inert atmosphere:** None; **Venting:** Pressure-vacuum; **Stability during transport:** Stable if below 80°F/27°C.

PHYSICAL AND CHEMICAL PROPERTIES
Physical state @ 59°F/15°C and 1 atm: Solid.
Molecular weight: 311.1
Boiling point @ 1 atm: Decomposes.
Specific gravity (water = 1): More than 1.1 @ 68°F/20°C (solid).
Heat of combustion: estimate -9000 Btu/lb $= -5000$ cal/g $= -210 \times 10^5$ J/kg.

DICHLORODIFLUOROMETHANE   REC. D:1650

SYNONYMS: ALGOFRENE TYPE 2; ARCTON 6; CFC-12; D I C L O R O D I F L U O M E T A N O (Spanish); DIFLUORODICHLOROMETHANE; ELECTRO-CF 12; ESKIMON 12; F 12; FC 12; FLUOROCARBON-12; FREON 12; FREON F-12; FRIGEN 12; GENETRON 12; HALON; HALON 122; ISCEON 122; ISOTRON 12; LEDON 12; METHANE, DICHLORODIFLUORO-; PROPELLANT 12; R 12; REFRIGERANT 12; UCON 12; UCON 12/HALOCARBON 12; RCRA No. U075

IDENTIFICATION
CAS Number: 75-71-8
Formula: CCl_2F_2
DOT ID Number: UN 1028; DOT Guide Number: 126
Proper Shipping Name: Dichlorodifluoromethane, R12
Reportable Quantity (RQ): **(CERCLA)** 5000 lb/2270 kg. Notify local health and pollution control agencies.

DESCRIPTION: Colorless gas. Shipped and stored as a liquefied compressed gas. Faint ether-like odor. Slightly soluble in water; Visible vapor cloud is produced.

Breathing the gas can kill you; it can cause the heart to beat irregularly or lead to cardiac arrest. Breathing the gas can irritate the mouth, nose and throat Contact with the liquid can cause severe eye and skin burns from frostbite • High levels can cause asphyxiation • Toxic products of combustion may include carbon monoxide and hydrogen chloride.

Hazard Classification (based on NFPA-704 M Rating System)
Health Hazards (Blue): 0; Flammability (Red): 0; Reactivity (Yellow): 0

EMERGENCY RESPONSE: See Appendix A (126)
Evacuation:
Public safety: Isolate spill area for at least 100 meters (330 feet) in all directions.
Spill: Large spill–Consider initial downwind evacuation for at least 500 meters (⅓ mile).
Fire: Isolate for 800 meters (½ mile) in all directions, especially if tank, rail car, or tank truck is involved in fire.

EXPOSURE
Short-term effects: *SEEK MEDICAL ATTENTION.* **Vapor:** Not irritating to eyes, nose or throat. *IF INHALED*, will, will cause dizziness, difficult breathing. Move to fresh air. *IF BREATHING HAS STOPPED,* give artificial respiration. If breathing is difficult, administer oxygen.

HEALTH HAZARDS
Personal protective equipment (PPE): B-Level PPE. OSHA Table Z-1-A air contaminant. Wear thermal protective clothing. Chemical protective material(s) reported to have good to excellent resistance: neoprene. Also, nitrile and Viton® offers limited protection.
Recommendations for respirator selection: NIOSH/OSHA
10,000 ppm: SA (any supplied-air respirator). *15,000 ppm:* SA:CF (any supplied-air respirator operated in a continuous-flow mode); or SCBAF (any self-contained breathing apparatus with a full facepiece); or SAF (any supplied-air respirator with a full facepiece). *EMERGENCY OR PLANNED ENTRY INTO UNKNOWN CONCENTRATIONS OR IDLH CONDITIONS:* SCBAF:PD,PP (any self-contained breathing apparatus that has a full facepiece and is operated in a pressure-demand or other positive-pressure mode); or SAF:PD,PP:ASCBA (any supplied-air respirator that has a full facepiece and is operated in a pressure-demand or other positive-pressure mode in combination with an

1,1-Dichloroethane

auxiliary self-contained breathing apparatus operated in a pressure-demand or other positive pressure mode). *ESCAPE:* GMFOV[any air-purifying, full-facepiece respirator (gas mask) with a chin-style, front- or back-mounted organic vapor canister]; or SCBAE (any appropriate escape-type, self-contained breathing apparatus).
Exposure limits (TWA unless otherwise noted): ACGIH TLV 1000 ppm (4950 mg/m^3); NIOSH/OSHA 1000 pm (4950 mg/m^3).
Vapor (gas) irritant characteristics: High concentrations may irritate lungs.
Liquid or solid irritant characteristics: Possible frostbite. Practically harmless to the skin because it is very volatile and evaporates quickly.
IDLH value: 15,000 ppm.

FIRE DATA
Behavior in fire: Material helps extinguish fire.

CHEMICAL REACTIVITY
Reactivity with water: Slow decomposition.
Binary reactants: chemically active metals such as sodium, potassium, calcium, powdered aluminum, zinc, magnesium.
Reactivity group: 36
Compatibility class: Halogenated hydrocarbons

ENVIRONMENTAL DATA
Food chain concentration potential: Negative; unlikely to accumulate.
Water pollution: Not harmful to aquatic life. **Response to discharge:** Disperse and flush.

SHIPPING INFORMATION
Grades of purity: 99.5% (vol.); **Storage temperature:** Ambient; **Inert atmosphere:** None; **Venting:** Safety relief; **Stability during transport:** Stable.

NAS HAZARD CLASSIFICATIONS FOR BULK WATER TRANSPORTATION
FIRE: 0
HEALTH: Vapor irritant: 0; Liquid or solid irritant: 0; Poisons: 1
WATER POLLUTION: Human toxicity: 0; Aquatic toxicity: 0; Aesthetic effect: 0
REACTIVITY: Other chemicals: 1; Water: 0; Self-reaction: 0

PHYSICAL AND CHEMICAL PROPERTIES
Physical state @ 59°F/15°C and 1 atm: Gas.
Molecular weight: 120.91
Boiling point @ 1 atm: –21.6°F/–29.8°C/243.4°K.
Melting/Freezing point: –251.9°F/157.7°C/115.5°K.
Critical temperature: 233.2°F/111.8°C/385.0°K.
Critical pressure: 598 psia = 40.7 atm = 4.12 MN/m^2.
Specific gravity (water = 1): 1.35 @ 15°C (liquid); RGasD: 4.2
Relative vapor density (air = 1): 4.2
Ratio of specific heats of vapor (gas): 1.129
Latent heat of vaporization: 140 Btu/lb = 77.9 cal/g = 3.26 x 10^5 J/kg.
Vapor pressure: (Reid) 132 psia; more than 4000 mm.

1,1-DICHLOROETHANE   **REC. D:1700**

SYNONYMS: AETHYLIDENCHLORID (German); ASYMMETRICAL DICHLOROETHANE; as-DICHLOROETHANE; CHLORINATED HYDROCHLORIC ETHER; CHLORURE D'ETHYLIDENE (French); 1,1-DICHLORAETHAN (German); 1,1-DICHLORETHANE; 1,1-DICLOROETANO (Spanish); EEC No. 602-011-00-1; ETHYLIDENE CHLORIDE; ETHYLIDENE DICHLORIDE; 1,1-ETHYLIDENE DICHLORIDE; ETHANE,1,1-DICHLORO-; DICHLOROMETHYLETHANE; RCRA No. U076

IDENTIFICATION
CAS Number: 75-34-3
Formula: $C_2H_4Cl_2$
DOT ID Number: UN 2362; DOT Guide Number: 128
Proper Shipping Name: 1,1-Dichloroethane
Reportable Quantity (RQ): **(CERCLA)** 1000 lb/454 kg

DESCRIPTION: Colorless oily liquid. Chloroform- or ether-like odor. Sinks and mixes with water.

Highly flammable • A tear gas • Containers may BLEVE when exposed to fire •Vapors may form explosive mixture with air • Vapors are heavier than air and will collect and stay in low areas • Vapors may travel long distances to ignition sources and flashback • Vapors in confined areas (e.g., tanks, sewers, buildings) may explode when exposed to fire • Irritating to the skin, eyes, and respiratory tract • Toxic products of combustion may include hydrogen chloride (1,1-isomer); phosgene and hydrogen chloride (1,2-isomer).

Hazard Classification (based on NFPA-704 Rating System)
Health Hazards (Blue): 2; Flammability (Red): 3; Reactivity (Yellow): 0

EMERGENCY RESPONSE: See Appendix A (128)
Evacuation:
Public safety: Isolate spill area for at least 25 to 50 meters (80 to 160 feet) in all directions.
Spill: Large spill–Consider initial downwind evacuation for at least 300 meters (1000 feet).
Fire: Isolate for 800 meters (½ mile) in all directions, especially if tank, rail car, or tank truck is involved in fire.

EXPOSURE
Short-term effects: *SEEK MEDICAL ATTENTION.* **Liquid:** If swallowed may cause nausea, vomiting and faintness. Irritating to skin and eyes. Flush affected areas with plenty of water. *IF IN EYES,* hold eyelids open and flush with plenty of water. *IF SWALLOWED* and victim is CONSCIOUS have victim drink 4 to 8 ounces of water and induce vomiting.

HEALTH HAZARDS
Personal protective equipment (PPE): B-Level PPE. OSHA Table Z-1-A air contaminant. Chemical protective material(s) reported to have good to excellent resistance: Teflon®. Also, PV alcohol, Viton®/neoprene, and Viton® offers limited protection
Recommendations for respirator selection: NIOSH/OSHA *1000 ppm:* SA (any supplied-air respirator). *2500 ppm:* SA:CF (any supplied-air respirator operated in a continuous-flow mode). *3000 ppm:* SCBAF (any self-contained breathing apparatus with a full facepiece); or SAF (any supplied-air respirator with a full facepiece). *EMERGENCY OR PLANNED ENTRY INTO UNKNOWN CONCENTRATIONS OR IDLH CONDITIONS:* SCBAF:PD,PP (any self-contained breathing apparatus that has a full facepiece and is operated in a pressure-demand or other positive-pressure mode); or SAF:PD,PP:ASCBA (any supplied-air respirator that has a full facepiece and is operated in a pressure-demand or other positive-pressure mode in combination with an auxiliary self-contained breathing apparatus operated in a pressure-demand or other positive pressure mode). *ESCAPE:* GMFOV[any

air-purifying, full-facepiece respirator (gas mask) with a chin-style, front- or back- mounted organic vapor canister]; or SCBAE (any appropriate escape-type, self-contained breathing apparatus).
Exposure limits (TWA unless otherwise noted): ACGIH TLV 100 ppm (405 mg/m^3); NIOSH/OSHA 100 ppm (400 mg/m^3) NIOSH suggests this chloroethane be treated as a potential carcinogen.
Toxicity by ingestion: Grade 2; LD$_{50}$ = 0.5 to 5 g/kg (rat).
Long-term health effects: Chronic exposure may cause organ damage; liver and kidney, and dermatitis. Animal experimentation has shown this compound to be slightly embryo-toxic and to retard fetal development. Skin problems from repeated or prolonged dermal exposure.
Vapor (gas) irritant characteristics: Eye and respiratory tract irritation. Vapors cause smarting of the eyes or respiratory system if present in high concentrations. The effect is temporary.
Liquid or solid irritant characteristics: Minimum hazard. Eye and skin irritation. If spilled on clothing and allowed to remain, may cause smarting and reddening of skin.
Odor threshold: 50–1375 ppm.
IDLH value: 3000 ppm.

FIRE DATA
Flash point: 2°F/–17°C (cc).
Flammable limits in air: LEL: 5.4%; UEL: 11.4%.
Fire extinguishing agents not to be used: Water may be ineffective.
Behavior in fire: Explosion hazard
Autoignition temperature: 856°F/458°C/731°K.
Electrical hazard: Class I, Group D.

CHEMICAL REACTIVITY
Binary reactants: Violent reaction with strong oxidizers, potassium. Incompatible with strong caustics, alkaline earth, and alkali metals.
Reactivity group: 36
Compatibility class: Halogenated hydrocarbons

ENVIRONMENTAL DATA
Food chain concentration potential: Log K$_{ow}$ = 1.89. Unlikely to accumulate.
Water pollution: Dangerous to aquatic life in high concentrations. May be dangerous if it enters nearby water intakes; notify operators. Notify local health and wildlife officials. **Response to discharge:** Issue warning–high flammability. Restrict access. Chemical and physical treatment.

SHIPPING INFORMATION
Storage temperature: Cool

PHYSICAL AND CHEMICAL PROPERTIES
Physical state @ 59°F/15°C and 1 atm: Liquid.
Molecular weight: 98.97
Boiling point @ 1 atm: 135.14°F/57.3°C/330.5°K.
Melting/Freezing point: –143.32°F/–97.4°C/175.75°K.
Critical temperature: 502.7°F/261.5°C/534.65°K.
Critical pressure: 734.8 psia = 50 atm = 5.065 MN/m^2.
Specific gravity (water = 1): 1.174 @ 68°F/20°C.
Liquid surface tension: 24.75 dynes/cm = 0.02475 N/m @ 68°F/20°C.
Relative vapor density (air = 1): 3.42
Ratio of specific heats of vapor (gas): 1.136 @ 68°F/20°C (68°F).
Latent heat of vaporization: 131.6 Btu/lb = 73.1 cal/g = 3.06 x 10^5 J/kg.
Heat of combustion: –4,774 Btu/lb = –2,652 cal/g = –111 x 10^5 J/kg.
Vapor pressure: (Reid) 7.35 psia; 182 mm; 230 mm @ 77°F/25°C.

DICHLOROETHYLENE   REC. D:1750

SYNONYMS: ACETYLENE DICHLORIDE; *cis*-ACETYLENE DICHLORIDE; *trans*-ACETYLENE DICHLORIDE; 1,2-DCE; 1,2-DICHLOR-AETHEN (German); 1,2-DICHLOROETHYLENE; DICHLORO-1,2-ETHYLENE (French); 1,2-DICLOROETENO (Spanish); DIOFORM; ETHENE, 1,2-DICHLORO-; *cis*-1,2-DICHLOROETHYLENE; *trans*-1,2-DICHLOROETHYLENE

IDENTIFICATION
CAS Number: 540-59-0; 156-59-2 (*cis*-isomer); 156-60-5 (*trans*-isomer); for 1,1-isomer (75-35-4) see vinylidene chloride
Formula: C$_2$H$_2$Cl$_2$; ClCH=CHCl Usually a mixture of the *cis*- and *trans*-isomers
DOT ID Number: UN 1150; DOT Guide Number: 132P
Proper Shipping Name: Dichloroethylene
Reportable Quantity (RQ): 1000 lb/454 kg

DESCRIPTION: Colorless liquid. Sweet pleasant odor; like ether or chloroform. Sinks in water. Flammable, irritating vapor is produced.

Highly flammable • Polymerization hazard • Heat can induce polymerization with rapid release of energy; sealed containers may rupture explosively • Containers may BLEVE when exposed to fire • Vapors may form explosive mixture with air • Vapors are heavier than air and will collect and stay in low areas • Vapors may travel long distances to ignition sources and flashback • Vapors in confined areas (e.g., tanks, sewers, buildings) may explode when exposed to fire • Irritating to the skin, eyes, and respiratory tract • Toxic products of combustion may include hydrogen chloride and phosgene.

Hazard Classification (based on NFPA-704 Rating System)
Health Hazards (Blue): 2; Flammability (Red): 3; Reactivity (Yellow): 2

EMERGENCY RESPONSE: See Appendix A (132)
Evacuation:
Public safety: Isolate spill area for at least 50 to 100 meters (160 to 330 feet) in all directions.
Spill: Increase, as necessary, the isolation distance shown above, in "Public safety."
Fire: Isolate for 800 meters (½ mile) in all directions, especially if tank, rail car, or tank truck is involved in fire.

EXPOSURE
Short-term effects: *SEEK MEDICAL ATTENTION*. **Vapor:** IF inhaled, will cause dizziness, nausea, vomiting, or difficult breathing. Move victim to fresh air. *IF BREATHING HAS STOPPED*, give artificial respiration. If breathing is difficult, administer oxygen. **Liquid:** Harmful if swallowed. *IF SWALLOWED* and victim is *CONSCIOUS AND ABLE TO SWALLOW*, have victim drink 4 to 8 ounces of water.

HEALTH HAZARDS
Personal protective equipment (PPE): B-Level PPE. OSHA Table Z-1-A air contaminant. Gloves; safety goggles; air supply mask or self-contained breathing apparatus. Chemical protective

2,2'-Dichloroethyl ether

material(s) reported to offer limited protection from all isomers: Viton®. Also, PV alcohol offers limited protection from the *trans*-isomer.

Recommendations for respirator selection: NIOSH/OSHA 1000 ppm: SA:CF* (any supplied-air respirator operated in a continuous-flow mode); or PAPROV* [any powered, air-purifying respirator with organic vapor cartridge(s)]; or CCRFOV [any chemical cartridge respirator with a full facepiece and organic vapor cartridges(s)]; or GMFOV [any air-purifying, full-facepiece respirator (gas mask) with a chin-style, front- or back-mounted acid gas canister]; or SCBAF (any self-contained breathing apparatus with a full facepiece); or SAF (any supplied-air respirator with a full facepiece). *EMERGENCY OR PLANNED ENTRY INTO UNKNOWN CONCENTRATIONS OR IDLH CONDITIONS*: SCBAF:PD,PP (any self-contained breathing apparatus that has a full facepiece and is operated in a pressure-demand or other positive-pressure mode); or SAF:PD,PP:ASCBA (any supplied-air respirator that has a full facepiece and is operated in a pressure-demand or other positive-pressure mode in combination with an auxiliary self-contained breathing apparatus operated in a pressure-demand or other positive pressure mode). *ESCAPE*: GMFOV[any air-purifying, full-facepiece respirator (gas mask) with a chin-style, front- or back-mounted organic vapor cannister]; or SCBAE (any appropriate escape-type, self-contained breathing apparatus). *Note: Substance causes eye irritation or damage; eye protection needed.

Exposure limits (TWA unless otherwise noted): ACGIH TLV 200 ppm (793 mg/m^3); NIOSH/OSHA 200 ppm (790 mg/m^3).

Toxicity by ingestion: Grade 2; oral LD_{50} = 770 mg/kg (rat).

Long-term health effects: Produced liver and kidney injury in experimental animals.

Liquid or solid irritant characteristics: Eye irritant. If spilled on clothing and allowed to remain, may cause smarting and reddening of the skin.

Odor threshold: 0.07–18 ppm.

IDLH value: 1000 ppm.

FIRE DATA

Flash point: 37°F/19°C (cc).
Flammable limits in air: LEL: 9.7%; UEL: 12.8%.
Fire extinguishing agents not to be used: Water may be ineffective.
Autoignition temperature: 860°F/460°C/733°K.
Electrical hazard: Class I, Group D.
Burning rate: 2.6 mm/min

CHEMICAL REACTIVITY

Binary reactants: Strong oxidizers, strong alkalis, potassium hydroxide, copper, aluminum.
Polymerization: May be caused by oxidizers, heat, peroxides or sunlight. May not occur under ordinary conditions of shipment. The reaction is not vigorous.

ENVIRONMENTAL DATA

Food chain concentration potential: Negative; unlikely to accumulate.
Water pollution: Effect of low concentrations on aquatic life is unknown. May be dangerous if it enters nearby water intakes; notify operators. Notify local health and wildlife officials.
Response to discharge: Issue warning–high flammability. Restrict access. Evacuate area. Should be removed. Chemical and physical treatment.

SHIPPING INFORMATION

Grades of purity: Commercial; **Storage temperature:** Ambient; **Inert atmosphere:** None; **Venting:** Pressure-vacuum; **Stability during transport:** Stable.

PHYSICAL AND CHEMICAL PROPERTIES

Physical state @ 59°F/15°C and 1 atm: Liquid.
Molecular weight: 97.0
Boiling point @ 1 atm: (*cis*-) 140°F/60°C/333°K; (*trans*-) 118°F/48°C/321°K.
Melting/Freezing point: (*cis*-) –114°F/–81°C/192°K; (*trans*-) –58°F/–50°C/223°K.
Specific gravity (water = 1): 1.27 at 77°F/liquid).
Liquid surface tension: 24 dynes/cm = 0.024 N/m @ 68°F/20°C.
Liquid water interfacial tension: estimate 30 dynes/cm = 0.030 N/m @ 68°F/20°C.
Relative vapor density (air = 1): 3.34
Ratio of specific heats of vapor (gas): 1.1468
Latent heat of vaporization: 130 Btu/lb = 72 cal/g = 3.0 x 10^5 J/kg.
Heat of combustion: –4,847.2 Btu/lb = –2,692.9 cal/g = –112.67 x 10^5 J/kg.
Vapor pressure: 180–262 mm.

2,2'-DICHLOROETHYL ETHER REC. D:1800

SYNONYMS: CHLOROETHYL ETHER; BIS(2-CHLOROETHYL) ETHER; DICHLOROETHER; DI-(2-CHLOROETHYL) ETHER; CHLOREX; 2,2'-DICHLORODIETHYL ETHER; DICHLORODIETHYL ETHER; ETER *sim*-DICLOROETILO (Spanish); DCEE; 2,2'-DCEE; DICHLORODIETHYL OXIDE; 1,1-OXYBIS(2-CHLOROETHANE); EEC No. 603-029-00-2; RCRA NO U025

IDENTIFICATION

CAS Number: 111-44-4
Formula: $C_4H_8Cl_2O$; $(ClCH_2CH_2)_2O$
DOT ID Number: UN 1916; DOT Guide Number: 152
Proper Shipping Name: 2,2'-Dichlorodiethyl ether
Reportable Quantity (RQ): **(CERCLA)** 10 lb/4.54 kg

DESCRIPTION: Colorless liquid. Sweet pleasant odor, like chloroform. Sinks and mixes slowly with water.

Combustible • May form explosive peroxides upon exposure to air • Containers may BLEVE when exposed to fire •Vapors may form explosive mixture with air • Vapors are heavier than air and will collect and stay in low areas • Vapors may travel long distances to ignition sources and flashback • Vapors in confined areas (e.g., tanks, sewers, buildings) may explode when exposed to fire • Irritating to the skin, eyes, and respiratory tract • Toxic products of combustion may include phosgene or hydrogen chloride.

Hazard Classification (based on NFPA-704 Rating System)
Health Hazards (Blue): 3; Flammability (Red): 2; Reactivity (Yellow): 1

EMERGENCY RESPONSE: See Appendix A (152)
Evacuation:
Public safety: Isolate the area of spill or leak for at least 25 to 50 meters (80 to 160 feet) in all directions.
Spill: Increase, in the downwind direction, as necessary, the distance shown under "Public Safety."
Fire: If tank, rail car, or tank truck is involved in fire, isolate for at least 800 meters (½ mile) in all directions; also, consider initial evacuation for 800 meters (½ mile) in all directions.

EXPOSURE
Short-term effects: *SEEK MEDICAL ATTENTION.* **Liquid:** *POISONOUS IF SWALLOWED OR IF SKIN IS EXPOSED.* Irritating to skin and eyes. Remove contaminated clothing and shoes. Flush affected areas with plenty of water. *IF BREATHING HAS STOPPED,* give artificial respiration; *avoid mouth-to-mouth resuscitation; use bag/mask apparatus. IF IN EYES,* hold eyelids open and flush with plenty of water. *IF SWALLOWED* and victim is *CONSCIOUS AND ABLE TO SWALLOW,* have victim drink 4 to 8 ounces of water and have victim induce vomiting. *IF SWALLOWED* and victim is *UNCONSCIOUS OR HAVING CONVULSIONS,* do nothing except keep victim warm. The use of alcoholic beverages may enhance the toxic effect.

HEALTH HAZARDS
Personal protective equipment (PPE): B-Level PPE. OSHA Table Z-1-A air contaminant. Chemical protective material(s) reported to have good to excellent resistance: Chlorinated polyethylene, Teflon®.
Recommendations for respirator selection: NIOSH
At any concentrations above the NIOSH REL, or where there is no REL, at any detectable concentration: SCBAF:PD,PP (any self-contained breathing apparatus that has a full facepiece and is operated in a pressure-demand or other positive-pressure mode); or SAF:PD,PP:ASCBA (any supplied-air respirator that has a full facepiece and is operated in a pressure-demand or other positive-pressure mode in combination with an auxiliary self-contained breathing apparatus operated in a pressure-demand or other positive pressure mode). *ESCAPE:* GMFOV [any air-purifying, full-facepiece respirator (gas mask) with a chin-style, front-or back-mounted organic vapor canister]; or SCBAE (any appropriate escape-type, self-contained breathing apparatus).
Exposure limits (TWA unless otherwise noted): ACGIH TLV 5 ppm (29 mg/m^3); OSHA PEL ceiling 15 ppm (90 mg/m^3); NIOSH potential human carcinogen, 5 ppm (30 mg/m^3); skin contact contributes significantly in overall exposure.
Short-term exposure limits (15-minute TWA): ACGIH STEL 10 ppm (58 mg/m^3); NIOSH STEL 10 ppm (60 mg/m^3), skin contact contributes significantly in overall exposure.
Toxicity by ingestion: Grade 3; oral LD$_{50}$ = 75 mg/kg (rat).
Long-term health effects: Potential carcinogen. IARC rating 3; limited animal evidence. Possible liver and kidney injury.
Vapor (gas) irritant characteristics: Eye and respiratory tract irritant. Vapors are irritating such that personnel will not usually tolerate moderate or high vapor concentrations.
Liquid or solid irritant characteristics: Severe eye and skin irritant. Causes smarting of the skin and first-degree burns on short exposure; may cause second-degree burns on long exposure.
Odor threshold: Less than 35 ppm.
IDLH value: Potential human carcinogen; 100 ppm.

FIRE DATA
Flash point: 131°F/55°C (cc).
Autoignition temperature: 696°F/369°C/642°K.
Electrical hazard: Class I, Group C.
Burning rate: 2.4 mm/min

CHEMICAL REACTIVITY
Reactivity with water: Decomposes in the presence of moisture to form hydrochloric acid gas.
Binary reactants: Vigorous reaction with oxidizers. Metal powders may cause fire or explosive hazards.
Reactivity group: 41
Compatibility class: Ethers

ENVIRONMENTAL DATA
Food chain concentration potential: Negative; unlikely to accumulate.
Water pollution: DOT Appendix B, §172.101–marine pollutant. Effect of low concentrations on aquatic life is unknown. May be dangerous if it enters nearby water intakes; notify operators. Notify local health and wildlife officials. **Response to discharge:** Issue warning–poison, water contaminant. Restrict access. Should be removed. Chemical and physical treatment.

SHIPPING INFORMATION
Grades of purity: Commercial; **Storage temperature:** Ambient; **Inert atmosphere:** None; **Venting:** Open (flame arrester); **Stability during transport:** Stable.

NAS HAZARD CLASSIFICATION FOR BULK WATER TRANSPORTATION
FIRE: 2
HEALTH: Vapor irritant: 3; Liquid or solid irritant: 2; Poisons: 3
WATER POLLUTION: Human toxicity: 3; Aquatic toxicity: 2; Aesthetic effect: 2
REACTIVITY: Other chemicals: 1; Water: 0; Self-reaction: 0

PHYSICAL AND CHEMICAL PROPERTIES
Molecular weight: 143.0
Boiling point @ 1 atm: 353°F/178°C/451°K.
Melting/Freezing point: –62°F/–52°C.
Specific gravity (water = 1): 1.22 @ 68°F/20°C (liquid).
Liquid surface tension: 37.9 dynes/cm = 0.0379 N/m @ 66°F/19°C/292°K.
Liquid water interfacial tension: estimate 40 dynes/cm = 0.040 N/m @ 68°F/20°C.
Relative vapor density (air = 1): 4.93
Ratio of specific heats of vapor (gas): 1.0743
Latent heat of vaporization: 143 Btu/lb = 79.5 cal/g = 3.33 x 10^5 J/kg.
Heat of combustion: estimate –7,530 Btu/lb = –4,180 cal/g = –175 x 10^5 J/kg.
Vapor pressure: 0.7 mm.

1,6-DICHLOROHEXANE   REC. D:1850

SYNONYMS: 1,6-DICLOROHEXANO (Spanish)

IDENTIFICATION
CAS Number: 2163-00-0
Formula: ClH$_2$C(C$_4$H$_8$)CH$_2$Cl
DOT ID Number: NA 1993; DOT Guide Number: 128
Proper Shipping Name: Combustible liquid, n.o.s.

DESCRIPTION: Colorless liquid. Mixes with water.

Combustible • Containers may BLEVE when exposed to fire • Vapors may form explosive mixture with air • Vapors are heavier than air and will collect and stay in low areas • Vapors may travel long distances to ignition sources and flashback • Vapors in confined areas (e.g., tanks, sewers, buildings) may explode when exposed to fire • Irritating to the skin, eyes, and respiratory tract • Toxic products of combustion may include carbon monoxide and hydrogen chloride.

Hazard Classification (based on NFPA-704 Rating System)
Health Hazards (Blue): 1; Flammability (Red): 2; Reactivity (Yellow): 0

EMERGENCY RESPONSE: See Appendix A (128)
Evacuation:
Public safety: Isolate spill area for at least 25 to 50 meters (80 to 160 feet) in all directions.
Spill: Large spill–Consider initial downwind evacuation for at least 300 meters (1000 feet).
Fire: Isolate for 800 meters (½ mile) in all directions, especially if tank, rail car, or tank truck is involved in fire.

EXPOSURE
Short-term effects: *SEEK MEDICAL ATTENTION*. **Vapor:** Move victim to fresh air. *IF BREATHING HAS STOPPED*, give artificial respiration. If breathing is difficult, administer oxygen. **Liquid:** Remove contaminated clothing and shoes. Wash affected areas with soap and water. *IF IN EYES*, hold eyelids open and flush with plenty of water.

HEALTH HAZARDS
Personal protective equipment (PPE): B-Level PPE.
Vapor (gas) irritant characteristics: Vapors cause a slight smarting of the eyes or respiratory system if present in high concentrations. The effect is temporary.
Liquid or solid irritant characteristics: Minimum hazard. If spilled on clothing and allowed to remain, may cause smarting and reddening of the skin.

FIRE DATA
Flash point: 165°F/74°C (cc).
Fire extinguishing agents not to be used: Water.

ENVIRONMENTAL DATA
Water pollution: DOT Appendix B, §172.101–marine pollutant. Effects of low concentrations on aquatic life is unknown. May be dangerous if it enters nearby water intakes; notify operators. Notify local health and wildlife officials. **Response to discharge:** Should be removed.

CHEMICAL REACTIVITY
Binary reactants: Vigorous reaction with oxidizers.

SHIPPING INFORMATION
Grades of purity: 98%; technical grades; **Storage temperature:** Ambient; **Inert atmosphere:** None; **Stability during transport:** Stable.

PHYSICAL AND CHEMICAL PROPERTIES
Physical state @ 59°F/15°C and 1 atm: Liquid.
Molecular weight: 155.07
Specific gravity (water = 1): 1.068 @ 20°C.

1,1-DICHLORO-1-NITROETHANE   **REC. D:1900**

SYNONYMS: DICHLORONITROETHANE; DICLORONITROETANO (Spanish); ETHIDE

IDENTIFICATION
CAS Number: 594-72-9
Formula: $C_2H_3Cl_2NO_2$; $CH_3CCl_2NO_2$
DOT ID Number: UN 2650; DOT Guide Number: 153
Proper Shipping Name: 1,1-Dichloro-1-nitroethane

DESCRIPTION: Colorless liquid. Unpleasant odor that causes tears.

Flammable • Containers may BLEVE when exposed to fire • Vapors may form explosive mixture with air • Vapors are heavier than air and will collect and stay in low areas • Vapors may travel long distances to ignition sources and flashback • Vapors in confined areas (e.g., tanks, sewers, buildings) may explode when exposed to fire • Irritating to the skin, eyes, and respiratory tract • Toxic products of combustion may include nitrogen oxides, hydrogen chloride, and carbon monoxide.

Hazard Classification (based on NFPA-704 Rating System)
Health Hazards (Blue): 2; Flammability (Red): 2; Reactivity (Yellow): 3

EMERGENCY RESPONSE: See Appendix A (153)
Evacuation:
Public safety: Isolate the area of spill or leak for at least 25 to 50 meters (80 to 160 feet) in all directions.
Spill: Increase, in the downwind direction, as necessary, the distance shown under "Public Safety."
Fire: If tank, rail car, or tank truck is involved in fire, isolate for at least 800 meters (½ mile) in all directions; also, consider initial evacuation for 800 meters (½ mile) in all directions.

EXPOSURE
Short-term effects: *SEEK MEDICAL ATTENTION*. Move victim to fresh air. Remove contaminated clothing and shoes. Wash affected areas with plenty of soap and water. *IF BREATHING HAS STOPPED*, give artificial respiration; *avoid mouth-to-mouth resuscitation; use bag/mask apparatus*. *IF IN EYES*, hold eyelids open and flush with plenty of water. *IF SWALLOWED* and victim is *CONSCIOUS AND ABLE TO SWALLOW*, have victim drink water, then induce vomiting. *IF BREATHING HAS STOPPED*, perform artificial respiration.

Personal protective equipment (PPE): B-Level PPE. OSHA Table Z-1-A air contaminant.
Recommendations for respirator selection: NIOSH/OSHA
20 ppm: SA (any supplied-air respirator). *25 ppm:* SA:CF (any supplied-air respirator operated in a continuous-flow mode); or SCBAF (any self-contained breathing apparatus with a full facepiece); or SAF (any supplied-air respirator with a full facepiece). *EMERGENCY OR PLANNED ENTRY INTO UNKNOWN CONCENTRATIONS OR IDLH CONDITIONS:* SCBAF:PD,PP (any self-contained breathing apparatus that has a full facepiece and is operated in a pressure-demand or other positive-pressure mode); or SAF:PD,PP:ASCBA (any supplied-air respirator that has a full facepiece and is operated in a pressure-demand or other positive-pressure mode in combination with an auxiliary self-contained breathing apparatus operated in a pressure-demand or other positive pressure mode). *ESCAPE:* GMFOV[any air-purifying, full-facepiece respirator (gas mask) with a chin-style, front- or back-mounted organic vapor canister]; or SCBAE (any appropriate escape-type, self-contained breathing apparatus].
Exposure limits (TWA unless otherwise noted): ACGIH TLV 2 ppm (12 mg/m³); OSHA PEL ceiling 10 ppm (60 mg/m³); NIOSH REL 2 ppm (10 mg/m³).
Long-term health effects: Exposure of animals produced severe irritation of lungs with severe breathing difficulties, which may be delayed in onset. Liver, heart, kidney, and blood vessel damage were also reported in animals.
Vapor (gas) irritant characteristics: Vapors cause irritation such that personnel will find high concentrations unpleasant. The effect is temporary.

Liquid or solid irritant characteristics: Eye and skin irritant. Causes smarting of the skin and first-degree burns on short exposure; may cause second-degree burns on long exposure.
IDLH value: 25 ppm.

FIRE DATA
Flash point: 136°F/58°C (cc).
Fire extinguishing agents not to be used: Water.
Stoichiometric air-to-fuel ratio: 8.3

CHEMICAL REACTIVITY
Binary reactants: Contact with strong oxidizers or alkyl metals may cause fires and explosions. Incompatible with caustics, ammonia, aliphatic amines, alkanolamines, aromatic amines. Corrosive to iron in the presence of moisture.
Reactivity group: 42
Compatibility class: Nitrocompounds

ENVIRONMENTAL DATA
Water pollution: DOT Appendix B, §172.101–marine pollutant. Effects of low concentrations on aquatic life is unknown. May be dangerous if it enters nearby water intakes; notify operators. Notify local health and wildlife officials. **Response to discharge:** Restrict access. Mechanical containment. Should be removed. Chemical and physical treatment.

SHIPPING INFORMATION
Grades of purity: Technical grades; **Storage temperature:** Ambient; **Inert atmosphere:** None.

PHYSICAL AND CHEMICAL PROPERTIES
Physical state @ 59°F/15°C and 1 atm: Liquid.
Molecular weight: 143.9
Boiling point @ 1 atm: 257°F/125°C/398°K.
Specific gravity (water = 1): 1.43
Vapor pressure: 15 mm.

2,2'-DICHLOROISOPROPYL ETHER REC. D:1950

SYNONYMS: BIS(2-CHLOROISOPROPYL) ETHER; BIS(2-CHLORO-1-METHYLETHYL ETHER); DICHLOROISOPROPYL ETHER; DICHLORODIISOPROPYL ETHER; ETER DICLOROISOPROPILICO (Spanish); ETHER, BIS(2-CHLORO-1-METHYLETHYL); RCRA No. U027

IDENTIFICATION
CAS Number: 108-60-1
Formula: $C_6H_{12}Cl_2O$; $(ClCH_2C(CH_3)H)_2O$
DOT ID Number: UN 2490; DOT Guide Number: 153
Proper Shipping Name: Dichloroisopropyl ether
Reportable Quantity (RQ): **(CERCLA)** 1000 lb/454 kg

DESCRIPTION: Colorless liquid. Sinks and mixes slowly with water.

Flammable • Containers may BLEVE when exposed to fire • Vapors may form explosive mixture with air • Vapors are heavier than air and will collect and stay in low areas • Vapors may travel long distances to ignition sources and flashback • Vapors in confined areas (e.g., tanks, sewers, buildings) may explode when exposed to fire • Irritating to the skin, eyes, and respiratory tract • Toxic products of combustion may include hydrochloric acid and phosgene gas.

Hazard Classification (based on NFPA-704 Rating System)
Health Hazards (Blue): 2; Flammability (Red): 2; Reactivity (Yellow): 0

EMERGENCY RESPONSE: See Appendix A (153)
Evacuation:
Public safety: Isolate the area of spill or leak for at least 25 to 50 meters (80 to 160 feet) in all directions.
Spill: Increase, in the downwind direction, as necessary, the distance shown under "Public Safety."
Fire: If tank, rail car, or tank truck is involved in fire, isolate for at least 800 meters (½ mile) in all directions; also, consider initial evacuation for 800 meters (½ mile) in all directions.

EXPOSURE
Short-term effects: *SEEK MEDICAL ATTENTION*. **Vapor:** May be fatal if inhaled. Highly irritating to upper respiratory tract. Move to fresh air. *IF BREATHING HAS STOPPED*, give artificial respiration; *avoid mouth-to-mouth resuscitation; use bag/mask apparatus*. If breathing is difficult, administer oxygen. **Liquid:** *POISONOUS IF SWALLOWED OR ABSORBED THROUGH SKIN. IF IN EYES OR ON SKIN*, flush with running water for at least 15 minutes; hold eyelids open if necessary. Remove contaminated clothing and shoes at the site. *IF SWALLOWED* and victim is *CONSCIOUS AND ABLE TO SWALLOW*, have victim drink 4 to 8 ounces of water and induce vomiting. *IF SWALLOWED* and victim is *UNCONSCIOUS OR HAVING CONVULSIONS*, do nothing except keep victim warm.

Personal protective equipment (PPE): B-Level PPE.
Toxicity by ingestion: Grade 3; LD_{50} = 240 mg/kg (rat).
Long-term health effects: May cause mutagenic effects and liver and kidney damage. IARC rating 3; limited animal evidence.
Vapor (gas) irritant characteristics: Strong respiratory tract irritant. Eye irritant. Vapor cause smarting of the eyes or respiratory system if present in high concentration.
Liquid or solid irritant characteristics: Eye and skin irritant. Can be absorbed through the skin.
Odor threshold: 0.32 ppm (detection in water).

FIRE DATA
Flash point: 185°F/85°C (oc); 170°F/77°C (cc).

CHEMICAL REACTIVITY
Binary reactants: Oxidizing materials. Incompatible with aluminum, copper, epoxy coatings. Avoid high heat and oxidizing materials. Subject to peroxide formation if not handled properly.
Reactivity group: 36
Compatibility class: Halogenated hydrocarbons

ENVIRONMENTAL DATA
Water pollution: Effect of low concentrations on aquatic life is unknown. May be dangerous if it enters nearby water intakes; notify operators. Notify local health and wildlife officials. **Response to discharge:** Issue warning–poison, corrosive Restrict access. Should be removed. Chemical and physical treatment. Disperse and flush.

SHIPPING INFORMATION
Grades of purity: 95% (mixed isomers); **Storage temperature:** Ambient; **Stability during transport:** Stable if kept cool and handled properly.

PHYSICAL AND CHEMICAL PROPERTIES
Physical state @ 59°F/15°C and 1 atm: Liquid.

Molecular weight: 171.07
Boiling point @ 1 atm: 369°F/187°C/461°K.
Melting/Freezing point: −142 to −151°F/−97 to −102°C/176 −171°K.
Critical temperature: 723°F/384°C/657°K (estimate).
Critical pressure: 413 psia = 28.1 atm = 2.85 MN/m^2 (estimate).
Specific gravity (water = 1): 1.1122 @ 68°F/20°C.
Relative vapor density (air = 1): 5.9
Latent heat of vaporization: 19.8 Btu/lb = 11.0 cal/g = 4.60 x 10^4 J/kg.

DICHLOROBUTENE   **REC. D:2000**

SYNONYMS: DCB; 1,4-DCB; 1,4-DICHLORO-2-BUTENE; 2-BUTYLENE DICHLORIDE; 1,4-DICHLORO-2-BUTYLENE; *cis*-1,4-DICHLORO-2-BUTENE; DICLOROBUTANO (Spanish); TETRAMETHYLENE DICHLORIDE, DCB; *trans*-1,4-DICHLORO-2-BUTENE; RCRA No. U074

IDENTIFICATION
CAS Number: 764-41-0
Formula: C$_4$H$_6$Cl$_2$; ClCH$_2$CH=CHCH$_2$Cl
DOT ID Number: UN 2920; DOT Guide Number: 132
Proper Shipping Name: Dichlorobutene
Reportable Quantity (RQ): (CERCLA) 1 lb/0.454 kg

DESCRIPTION: Colorless liquid. Sweet odor. Sinks and mixes slowly with water.

Flammable • Containers may BLEVE when exposed to fire • Vapors may form explosive mixture with air • Vapors are heavier than air and will collect and stay in low areas • Vapors may travel long distances to ignition sources and flashback • Vapors in confined areas (e.g., tanks, sewers, buildings) may explode when exposed to fire • Irritating to the skin, eyes, and respiratory tract • Toxic products of combustion may include phosgene and hydrogen chloride.

Hazard Classification (based on NFPA-704 Rating System)
Health Hazards (Blue): 3; Flammability (Red): 2; Reactivity (Yellow): 0

EMERGENCY RESPONSE: See Appendix A (132)
Evacuation:
Public safety: Isolate spill area for at least 50 to 100 meters (160 to 330 feet) in all directions.
Spill: Increase, as necessary, the isolation distance shown above, in "Public safety."
Fire: Isolate for 800 meters (½ mile) in all directions, especially if tank, rail car, or tank truck is involved in fire.

EXPOSURE
Short-term effects: *SEEK MEDICAL ATTENTION.* **Liquid:** *HARMFUL IF SWALLOWED OR ABSORBED THROUGH THE SKIN.* Irritating to skin and eyes. Harmful if swallowed or absorbed through the skin. Remove contaminated clothing and shoes. Flush affected areas with plenty of water. *IF IN EYES*, hold eyelids open and flush with plenty of water. *IF SWALLOWED* and victim is *CONSCIOUS AND ABLE TO SWALLOW*, have victim drink 4 to 8 ounces of water and have victim induce vomiting. *IF SWALLOWED* and victim is *UNCONSCIOUS OR HAVING CONVULSIONS,* do nothing except keep victim warm.

HEALTH HAZARDS
Personal protective equipment (PPE): B-Level PPE. Full face organic vapor respirator. Rubber gloves; rubber boots; barrier cream.
Exposure limits (TWA unless otherwise noted): ACGIH TLV 0.005 ppm. skin contact contributes significantly in overall exposure. suspected human carcinogen.
Toxicity by ingestion: Grade 3; oral LD$_{50}$ = 89 mg/kg (rat).
Vapor (gas) irritant characteristics: Eye and respiratory tract irritant.
Liquid or solid irritant characteristics: Severe eye and skin irritant.

FIRE DATA
Flash point: 126°F/52°C (cc).
Flammable limits in air: LEL: 1.5%; UEL: 4%.
Behavior in fire: Containers may explode.
Electrical hazard: Class I, Group D.
Burning rate: 2.6 mm/min

CHEMICAL REACTIVITY
Reactivity with water: Reacts slowly to form hydrochloric acid.
Binary reactants: Reacts with strong oxidizers and bases. Corrodes metal when wet.
Neutralizing agents for acids and caustics: Flood with water and rinse with sodium bicarbonate or lime solution.

ENVIRONMENTAL DATA
Food chain concentration potential: Negative; unlikely to accumulate.
Water pollution: Effect of low concentrations on aquatic life is unknown. May be dangerous if it enters nearby water intakes; notify operators. Notify local health and wildlife officials. **Response to discharge:** Issue warning–corrosive, water contaminant. Restrict access. Should be removed. Chemical and physical treatment.

SHIPPING INFORMATION
Grades of purity: *cis*-trans equilibrium mixture, 98+%; **Storage temperature:** Ambient; **Inert atmosphere:** None; **Venting:** Open (flame arrester); **Stability during transport:** Stable.

PHYSICAL AND CHEMICAL PROPERTIES
Physical state @ 59°F/15°C and 1 atm: Liquid.
Molecular weight: 125.0
Boiling point @ 1 atm: 313°F/156°C/429°K.
Melting/Freezing point: (*cis*-) −54°F/−48°C/225°K; (*trans*-) 37°F/3°C/276°K.
Specific gravity (water = 1): 1.112 @ 68°F/20°C (liquid).
Liquid surface tension: estimate 24 dynes/cm = 0.024 N/m @ 68°F/20°C.
Liquid water interfacial tension: estimate 30 dynes/cm = 0.030 N/m @ 68°F/20°C.
Relative vapor density (air = 1): 4
Ratio of specific heats of vapor (gas): 1.0874
Latent heat of vaporization: estimate 130 Btu/lb = 73 cal/g = 3.1 x 10^5 J/kg.
Heat of combustion: −17,500 Btu/lb = −9720 cal/g = −407 x 10^5 J/kg.

DICHLOROMETHANE   **REC. D:2050**

SYNONYMS: DICLOROMETANO (Spanish); EEC No. 602-004-00-3; METHYLENE CHLORIDE; METHYLENE DICHLORIDE; RCRA No. U080

Dichloromethane

IDENTIFICATION
CAS Number: 75-09-2
Formula: CH_2Cl_2
DOT ID Number: UN 1593; DOT Guide Number: 160
Proper Shipping Name: Dichloromethane
Reportable Quantity (RQ): **(CERCLA)** 1000 lb/454 kg

DESCRIPTION: Colorless, watery liquid. Sweet, pleasant odor, like chloroform or ether. Sinks in water; slightly soluble. Large amounts of irritating vapor is produced.

Combustible • Harmful or fatal if swallowed; produces carbon monoxide in the body • Containers may BLEVE when exposed to fire • Vapors may form explosive mixture with air • Vapors are heavier than air and will collect and stay in low areas • Vapors may travel long distances to ignition sources and flashback • Vapors in confined areas (e.g., tanks, sewers, buildings) may explode when exposed to fire • Irritating to the skin, eyes, and respiratory tract • Toxic products of combustion may include phosgene and hydrogen chloride which are more toxic than the material itself.

Hazard Classification (based on NFPA-704 Rating System)
Health Hazards (Blue): 1; Flammability (Red): 3; Reactivity (Yellow): 1

EMERGENCY RESPONSE: See Appendix A (160)
Evacuation:
Public safety: Isolate the area of spill or leak for at least 25 to 50 meters (80- 160 feet) in all directions.
Spill: Consider initial downwind evacuation for at least 100 meters (330 feet); increase, in the downwind direction, as necessary.
Fire: If tank, rail car, or tank truck is involved in fire, isolate for at least 800 meters (½ mile) in all directions; also, consider initial evacuation for 800 meters (½ mile) in all directions.

EXPOSURE
Short-term effects: *SEEK MEDICAL ATTENTION.* **Vapor:** Irritating to eyes, nose, and throat. *IF INHALED*, will, will cause nausea and dizziness. Move to fresh air. *IF BREATHING HAS STOPPED*, give artificial respiration, *avoid mouth-to-mouth resuscitation; use bag/mask apparatus*. If breathing is difficult, administer oxygen. **Liquid:** Irritating to skin and eyes. Harmful if swallowed. Remove contaminated clothing and shoes. Flush affected areas with plenty of water. *IF IN EYES*, hold eyelids open and flush with plenty of water. *IF SWALLOWED* and victim is *CONSCIOUS AND ABLE TO SWALLOW*, have victim drink 4 to 8 ounces of water. **Do NOT induce vomiting.** Administer a slurry of activated charcoal.

HEALTH HAZARDS
Personal protective equipment (PPE): OSHA Table Z-1-A air contaminant. OSHA Table Z-2 air contaminant. B-Level PPE. Chemical protective material(s) reported to have good to excellent resistance: PV alcohol, Silvershield®. Also, Viton®/neoprene and Teflon® offers limited protection
Recommendations for respirator selection: NIOSH
At any concentrations above the NIOSH REL, or where there is no REL, at any detectable concentration: SCBAF:PD,PP (any self-contained breathing apparatus that has a full facepiece and is operated in a pressure-demand or other positive-pressure mode); or SAF:PD,PP:ASCBA (any supplied-air respirator that has a full facepiece and is operated in a pressure-demand or other positive-pressure mode in combination with an auxiliary self-contained breathing apparatus operated in a pressure-demand or other positive pressure mode). *ESCAPE:* GMFOV [any air-purifying, full-facepiece respirator (gas mask) with a chin-style, front-or back-mounted organic vapor canister]; or SCBAE (any appropriate escape-type, self-contained breathing apparatus).
Exposure limits (TWA unless otherwise noted): ACGIH TLV 50 ppm (174 mg/m^3), suspected human carcinogen; OSHA PEL 25 ppm; NIOSH REL potential human carcinogen; reduce exposure to lowest feasible level.
Short-term exposure limits (15-minute TWA): OSHA STEL 125 ppm.
Toxicity by ingestion: Grade 2; LD_{50} = 0.5 to 5 g/kg.
Long-term health effects: IARC possible carcinogen; rating 2B; Adequate naimal evidence. Organ damage; liver.
Vapor (gas) irritant characteristics: Respiratory tract irritant. Eye and respiratory tract irritant. Vapors cause irritation such that personnel will find high concentrations unpleasant.
Liquid or solid irritant characteristics: Eye and skin irritant. If spilled on clothing and allowed to remain, may cause smarting and reddening of the skin.
Odor threshold: 160 ppm.
IDLH value: Potential human carcinogen; 2300 ppm.

FIRE DATA
Flash point: No flash point by conventional methods, but forms flammable vapor-air mixtures at 212°F/100°C and higher.
Flammable limits in air: LEL: 13%; UEL: 23%.
Autoignition temperature: 1033°F/556°C/829°K.
Electrical hazard: Class I, Group D. Due to low electric conductivity, this substance may generate electrostatic charges as a result of agitation and flow.

CHEMICAL REACTIVITY
Reactivity with water: Slow reaction forms methanol and formic acid
Binary reactants: Reacts with strong oxidizers, caustics; concentrated nitric acid; some plastics. Corrosive to aluminum and magnesium; corrosive to steel when wet.
Reactivity group: 36
Compatibility class: Halogenated hydrocarbons

ENVIRONMENTAL DATA
Food chain concentration potential: Log P_{ow} = 1.24. Unlikely to accumulate.
Food chain concentration potential: Negative; unlikely to accumulate.
Water pollution: Effect of low concentrations on aquatic life is unknown. May be dangerous if it enters nearby water intakes; notify operators. Notify local health and pollution control officials.
Response to discharge: Disperse and flush.

SHIPPING INFORMATION
Grades of purity: Aerosol grade; technical grade; **Inert atmosphere:** Inert; **Stability during transport:** Stable.

NAS HAZARD CLASSIFICATION FOR BULK WATER TRANSPORTATION
FIRE: 1
HEALTH: Vapor irritant: 2; Liquid or solid irritant: 1; Poisons: 2
WATER POLLUTION: Human toxicity: 1; Aquatic toxicity: 2; Aesthetic effect: 2
REACTIVITY: Other chemicals: 1; Water: 0; Self-reaction: 0

PHYSICAL AND CHEMICAL PROPERTIES
Physical state @ 59°F/15°C and 1 atm: Liquid.
Molecular weight: 84.93
Boiling point @ 1 atm: 104°F/39.8°C/313.0°K.

Melting/Freezing point: –139°F
Critical temperature: 473°F/245°C/518°K.
Critical pressure: 895 psia = 60.9 atm = 6.17 MN/m².
Specific gravity (water = 1): 1.33 @ 68°F/20°C (liquid).
Relative vapor density (air = 1): 2.9
Ratio of specific heats of vapor (gas): 1.199
Latent heat of vaporization: 142 Btu/lb = 78.7 cal/g = 3.30×10^5 J/kg.
Heat of fusion: 16.89 cal/g.
Vapor pressure: (Reid) 13.9 psia; 350 mm.

DICHLOROMONOFLUOROMETHANE   REC. D:2100

SYNONYMS: ALGOFRENE TYPE 5; ARCTON 7; DICHLOROFLUOROMETHANE; DICLOROMONOFLUOMETANO (Spanish); FLUORODICHLOROMETHANE; FREON 21; F-21; GENETRON 21; HALON 112; REFRIGERANT 21; R-21

IDENTIFICATION
CAS Number: 75-43-4
Formula: $CHCl_2F$
DOT ID Number: UN 1029; DOT Guide Number: 126
Proper Shipping Name: Dichlorofluoromethane, R21

DESCRIPTION: Colorless liquid or compressed gas. Slight ether-like odor. Insoluble in water; slowly decomposes.

Toxic substances formed on contact with a flame or hot metal surfaces • Gas from liquefied gas is initially heavier than air and spread along ground; in enclosed spaces it can cause oxygen deficiency • Contact with liquid can cause frostbite • Toxic products of combustion may include carbon monoxide, hydrogen fluoride, and hydrogen chloride.

Hazard Classification (based on NFPA-704 Rating System)
Health Hazards (Blue): 1; Flammability (Red): 0; Reactivity (Yellow): 0

EMERGENCY RESPONSE: See Appendix A (126)
Evacuation:
Public safety: Isolate spill area for at least 100 meters (330 feet) in all directions.
Spill: Large spill–Consider initial downwind evacuation for at least 500 meters (⅓ mile).
Fire: Isolate for 800 meters (½ mile) in all directions, especially if tank, rail car, or tank truck is involved in fire.

EXPOSURE
Short-term effects: *SEEK MEDICAL ATTENTION.* VAPORS: Vapors may cause dizziness or suffocation. Move victim to fresh air. If not breathing, give artificial respiration. If breathing is difficult, administer oxygen. **Liquid:** Contact with liquid may cause frostbite. Remove contaminated clothing and shoes. Flush affected areas with plenty of lukewarm water. *DO NOT USE HOT WATER.*

HEALTH HAZARDS
Personal protective equipment (PPE): OSHA Table Z-1-A air contaminant. B-Level PPE. Nitrile rubber is generally suitable for freons. Wear thermal protective clothing.
Recommendations for respirator selection: NIOSH/OSHA
100 ppm: SA (any supplied-air respirator). *250 ppm*: SA:CF (any supplied-air respirator operated in a continuous-flow mode). 500 ppm: SCBAF (any self-contained breathing apparatus with a full facepiece); or SAF (any supplied-air respirator with a full facepiece). *5000 ppm*: SA:PD,PP (any supplied-air respirator operated in a pressure-demand or other positive-pressure mode). *EMERGENCY OR PLANNED ENTRY INTO UNKNOWN CONCENTRATIONS OR IDLH CONDITIONS*: SCBAF:PD,PP (any self-contained breathing apparatus that has a full facepiece and is operated in a pressure-demand or other positive-pressure mode); or SAF:PD,PP:ASCBA (any supplied-air respirator that has a full facepiece and is operated in a pressure-demand or other positive-pressure mode in combination with an auxiliary self-contained breathing apparatus operated in a pressure-demand or other positive pressure mode). *ESCAPE*: GMFOV [any air-purifying, full-facepiece respirator (gas mask) with a chin-style, front- or back-mounted organic vapor canister]; or SCBAE (any appropriate escape-type, self-contained breathing apparatus).
Exposure limits (TWA unless otherwise noted): ACGIH TLV 10 ppm (42 mg/m³); OSHA PEL 1000 ppm (4200 mg/m³); NIOSH REL 10 ppm (40 mg/m³).
Vapor (gas) irritant characteristics: Although dangerous, vapors are nonirritating to eyes and throat.
Liquid or solid irritant characteristics: If spilled on clothing or skin, may cause frostbite.
IDLH value: 5000 ppm.

FIRE DATA
Autoignition temperature: 1022°F/550°C/823°K.

CHEMICAL REACTIVITY
Reactivity with water: Slow decomposition.
Binary reactants: Reacts with chemically active metals; acid, acid fumes.

ENVIRONMENTAL DATA
Food chain concentration potential: Log P_{ow} = 1.6. Unlikely to accumulate.
Water pollution: Effect of low concentrations on aquatic life is unknown. May be dangerous if it enters nearby water intakes; notify operators. Notify local health and pollution control officials.
Response to discharge: Disperse and flush.

SHIPPING INFORMATION
Grades of purity: 98%; **Stability during transport:** Stable.

NAS HAZARD CLASSIFICATION FOR BULK WATER TRANSPORTATION
FIRE: 0
HEALTH: Vapor irritant: 0; Liquid or solid irritant: 2; Poisons: 0
WATER POLLUTION: Human toxicity:–; Aquatic toxicity: 3; Aesthetic effect: 0
REACTIVITY: Other chemicals: 0; Water: 0; Self-reaction: 0

PHYSICAL AND CHEMICAL PROPERTIES
Physical state @ 59°F/15°C and 1 atm: Gas.
Molecular weight: 102.92
Boiling point @ 1 atm: 48°F/8.9°C/282°K.
Melting/Freezing point: –211°F/–135°F/138°K.
Critical temperature: 353.3°F/178.5°C/451.7°K.
Critical pressure: 749.5 psia = 51 atm = 5.2 MN/m².
Specific gravity (water = 1): 1.48 @ 68°F/20°C.
Liquid surface tension: 18 dyne/cm = 0.018 N/m @ 77°F/25°C.
Relative vapor density (air = 1): 3.57
Latent heat of vaporization: 104.2 Btu/lb = 57.9 cal/g = 2.42×10^5 J/kg.
Vapor pressure: (Reid) 42.9 psia; 1100 mm.

2,4-DICHLOROPHENOL   REC. D:2200

SYNONYMS: DCP; 2,4-DCP; DICLOROFENOL (Spanish); RCRA No. U081

IDENTIFICATION
CAS Number: 120-83-2; may also apply to 576-24-9 (2,3-isomer); 583-78-8 (2,5-isomer); 87-65-0 (2,6-isomer)
Formula: $HOC_6H_3Cl_2$-2,4
DOT ID Number: UN 2020; DOT Guide Number: 153
Proper Shipping Name: Chlorophenols, solid
Reportable Quantity (RQ): **(CERCLA)** 100 lb/45.4 kg

DESCRIPTION: Colorless, crystalline solid. Medicinal odor. Sinks in water.

Poison! • Combustible • Breathing the dust, skin or eye contact, or swallowing the material can kill you • Firefighting gear (including SCBA) does not provide adequate protection. If exposure occurs, remove and isolate gear immediately and thoroughly decontaminate personnel • Containers may BLEVE when exposed to fire • Concentrated dust in confined areas (e.g., tanks, sewers, buildings) may explode when exposed to fire • Toxic products of combustion may include toxic fumes of nitrogen oxide.

Hazard Classification (based on NFPA-704 Rating System)
Health Hazards (Blue): 1; Flammability (Red): 1; Reactivity (Yellow): 0

EMERGENCY RESPONSE: See Appendix A (153)
Evacuation:
Public safety: Isolate the area of spill or leak for at least 25 to 50 meters (80 to 160 feet) in all directions.
Spill: Increase, in the downwind direction, as necessary, the distance shown under "Public Safety."
Fire: If tank, rail car, or tank truck is involved in fire, isolate for at least 800 meters (½ mile) in all directions; also, consider initial evacuation for 800 meters (½ mile) in all directions.

EXPOSURE
Short-term effects: *SEEK MEDICAL ATTENTION. IF BREATHING HAS STOPPED*, give artificial respiration; *avoid mouth-to-mouth resuscitation; use bag/mask apparatus.* **Solid or dust:** Will burn skin and eyes. *POISONOUS IF SWALLOWED*. Remove contaminated clothing and shoes. Flush affected areas with plenty of water. *IF IN EYES*, hold eyelids open and flush with plenty of water. *IF SWALLOWED and victim is CONSCIOUS AND ABLE TO SWALLOW*, have victim drink 4 to 8 ounces of water.

HEALTH HAZARDS
Personal protective equipment (PPE): A-Level PPE. Bureau of Mines or NIOSH/OSHA approved respirator. Chemical protective material(s) reported to have good to excellent resistance: butyl rubber. Also polycarbonate offers good resistance.
Toxicity by ingestion: Grade 2; LD_{50} = 0.5 to 5 g/kg (rat).
Long-term health effects: May effect the nervous system. Possible liver and kidney damage; dermatitis. Metabolic effects. Some chlorophenols are suspected carcinogens.
Liquid or solid irritant characteristics: Fairly severe eye and skin irritant. May cause pain and second-degree burns after a few minutes of contact.

FIRE DATA
Flash point: 237°F/114°C (oc)
Fire extinguishing agents not to be used: Water or foam may cause frothing.
Electrical hazard: Class I, Group D.

CHEMICAL REACTIVITY
Binary reactants: May react vigorously with strong oxidizing materials. Incompatible with acids, alkali solutions. Rapidly corrodes aluminum. Slowly corrodes zinc, tin, brass, bronze, copper.

ENVIRONMENTAL DATA
Food chain concentration potential: Log P_{ow} = 3.0 (estimate). Log P_{ow} = 3.1 (2,3-isomer); Log P_{ow} = 2.6 (2,6-isomer). Values > 3.0 are likely to bioconcentrate in aquatic organisms and other living tissue (in fats).

SHIPPING INFORMATION
Stability during transport: Stable.

PHYSICAL AND CHEMICAL PROPERTIES
Physical state @ 59°F/15°C and 1 atm: Solid.
Molecular weight: 163.01
Boiling point @ 1 atm: 421°F/216°C/489°K.
Melting/Freezing point: 110°F/45°C/318°K.
Specific gravity (water = 1): 1.40 @ 15°C (solid)1.40 @ 140°F/liquid).
Relative vapor density (air = 1): 5.6
Vapor pressure: 0.10 mm.

1,1-DICHLOROPROPANE   REC. D:2250

SYNONYMS: 1,1-DICLOROPROPANO (Spanish); PROPYLIDENE CHLORIDE; PROPYLIDENE DICHLORIDE; PROPANE,1,1-DICHLORO-

IDENTIFICATION
CAS Number: 78-99-9
Formula: $C_3H_6Cl_2$; $C_2NO_5 \cdot CH \cdot Cl_2$
DOT ID Number: UN 1279; DOT Guide Number: 128
Proper Shipping Name: Propylene dichloride
Hazard class or division: 3
Reportable Quantity (RQ): **(CERCLA)** 1000 lb/454 kg

DESCRIPTION: Colorless liquid. Sweet, chloroform odor. Sinks in water; insoluble. Flammable, irritating vapor is produced.

Highly flammable • Containers may BLEVE when exposed to fire • Vapors may form explosive mixture with air • Vapors are heavier than air and will collect and stay in low areas • Vapors may travel long distances to ignition sources and flashback • Vapors in confined areas (e.g., tanks, sewers, buildings) may explode when exposed to fire • Irritating to the skin, eyes, and respiratory tract • Toxic products of combustion may include phosgene and hydrogen chloride.

Hazard Classification (based on NFPA-704 Rating System)
Health Hazards (Blue): 2; Flammability (Red): 3; Reactivity (Yellow): 0

EMERGENCY RESPONSE: See Appendix A (128)
Evacuation:
Public safety: Isolate spill area for at least 25 to 50 meters (80 to 160 feet) in all directions.

1,2-Dichloropropane

Spill: Large spill–Consider initial downwind evacuation for at least 300 meters (1000 feet).
Fire: Isolate for 800 meters (½ mile) in all directions, especially if tank, rail car, or tank truck is involved in fire.

EXPOSURE
Short-term effects: *SEEK MEDICAL ATTENTION.* **Vapor:** Irritating to eyes, nose, and throat. Move to fresh air. *IF BREATHING HAS STOPPED,* give artificial respiration. If breathing is difficult, administer oxygen. **Liquid:** Irritating to skin and eyes. Harmful if swallowed. Remove contaminated clothing and shoes. Flush affected areas with plenty of water. *IF IN EYES,* hold eyelids open and flush with plenty of water. *IF SWALLOWED* and victim is *CONSCIOUS AND ABLE TO SWALLOW,* have victim drink 4 to 8 ounces of water. The use of alcoholic beverages may enhance the toxic effect.

ENVIRONMENTAL DATA
Water pollution: DOT Appendix B, §172.101–marine pollutant. Effect of low concentrations on aquatic life is unknown. May be dangerous if it enters nearby water intakes; notify operators. Notify local health and wildlife officials. **Response to discharge:** Issue warning–high flammability. Evacuate area.

HEALTH HAZARDS
Personal protective equipment (PPE): B-Level PPE. Chemical protective material(s) reported to have good to excellent resistance: Teflon®, Viton®, PV alcohol. Also, chlorinated polyethylene (CPE) offers limited protection, breakthrough < 1 hour.
Toxicity by ingestion: Grade 1; LD_{50} = 5 to 15 g/kg.
Long-term health effects: Frequent or prolonged contact may result in organ damage to the kidney, liver or heart.
Vapor (gas) irritant characteristics: Vapors cause a slight smarting of the eyes or respiratory system if present in high concentrations. The effect is temporary.
Liquid or solid irritant characteristics: Eye and skin irritant. If spilled on clothing and allowed to remain may cause smarting and reddening of skin.

FIRE DATA
Flash point: estimate 158°F/70°C (cc).
Flammable limits in air: LEL: 3.4%; UEL: 14.5%.
Autoignition temperature: (estimate) 1035°F/557°C/830°K.
Electrical hazard: Class I, Group D.

CHEMICAL REACTIVITY
Binary reactants: Reacts with oxidizers. May attack some plastics and metals. Corrodes aluminum.
Reactivity group: 36
Compatibility class: Halogenated hydrocarbons

ENVIRONMENTAL DATA
Food chain concentration potential: Log K_{ow} = 2.34. Unlikely to accumulate.
Water pollution: DOT Appendix B, §172.101–marine pollutant. Effect of low concentrations on aquatic life is unknown. May be dangerous if it enters nearby water intakes; notify operators. Notify local health and wildlife officials. **Response to discharge:** Issue warning–water contaminant. Should be removed. Chemical and physical treatment.

SHIPPING INFORMATION
Storage temperature: Ambient; **Inert atmosphere:** None;
Venting: Pressure-vacuum; **Stability during transport:** Stable.

NAS HAZARD CLASSIFICATION FOR BULK WATER TRANSPORTATION
FIRE: 3
HEALTH: Vapor irritant: 1; Liquid or solid irritant: 1; Poisons: 3
WATER POLLUTION: Human toxicity: 1; Aquatic toxicity: 3; Aesthetic effect: 2
REACTIVITY: Other chemicals: 1; Water: 0; Self-reaction: 0

PHYSICAL AND CHEMICAL PROPERTIES
Physical state @ 59°F/15°C and 1 atm: Liquid.
Molecular weight: 112.99
Boiling point @ 1 atm: 191°F/88°C/361°K.
Critical temperature: (estimate) 511°F/266°C/539°K.
Critical pressure: (estimate) 563 psia = 38.3 atm = 3.88 MN/m².
Specific gravity (water = 1): 1.1321 @ 68°F/20°C.
Liquid surface tension: (estimate) 26.1 dynes/cm = 0.0261 N/m @ 68°F/20°C.
Liquid water interfacial tension: (estimate) 46.9 dynes/cm = 0.0469 N/m @ 68°F/20°C.
Relative vapor density (air = 1): 3.90
Ratio of specific heats of vapor (gas): (estimate) 1.094 @ 68°F/20°C (68°F).
Heat of combustion: (estimate) –6667 Btu/lb = –3704 cal/g = –155 x 10^5 J/kg.
Vapor pressure: 40 mm.

1,2-DICHLOROPROPANE REC. D:2300

SYNONYMS: BICHLORURE de PROPYLENE (French); DICHLORO-1,2-PROPANE; DICHLOROPROPANE; α,β-DICHLOROPROPANE; 1,2-DICLOROPROPANO (Spanish); EEC No. 602-020-00-0; PROPANE, 1,2-DICHLORO-; PROPYLENE CHLORIDE; α,β-PROPYLENE DICHLORIDE; PROPYLENE DICHLORIDE; RCRA No. U083

IDENTIFICATION
CAS Number: 78-87-5
Formula: $C_3H_6Cl_2$; $CH_3CHClCH_2Cl$
DOT ID Number: UN 1279; DOT Guide Number: 128
Proper Shipping Name: Propylene dichloride
Reportable Quantity (RQ): **(CERCLA)** 1000 lb/454 kg

DESCRIPTION: Colorless watery liquid. Sweet odor. Sinks in water; insoluble. Flammable, irritating vapor is produced.

Poison! • Highly flammable • Breathing the vapor, skin or eye contact, or swallowing the material can cause severe illness • Containers may BLEVE when exposed to fire •Vapors may form explosive mixture with air • Vapors are heavier than air and will collect and stay in low areas • Vapors may travel long distances to ignition sources and flashback • Vapors in confined areas (e.g., tanks, sewers, buildings) may explode when exposed to fire • Irritating to the skin, eyes, and respiratory tract • Toxic products of combustion may include phosgene and hydrogen chloride.

Hazard Classification (based on NFPA-704 Rating System)
Health Hazards (Blue): 2; Flammability (Red): 3; Reactivity (Yellow): 0

EMERGENCY RESPONSE: See Appendix A (128)
Evacuation:
Public safety: Isolate spill area for at least 25 to 50 meters (80 to 160 feet) in all directions.

Spill: Large spill–Consider initial downwind evacuation for at least 300 meters (1000 feet).
Fire: Isolate for 800 meters (½ mile) in all directions, especially if tank, rail car, or tank truck is involved in fire.

EXPOSURE
Short-term effects: *SEEK MEDICAL ATTENTION.* **Vapor:** Irritating to eyes, nose, and throat. Move to fresh air. *IF BREATHING HAS STOPPED,* give artificial respiration. If breathing is difficult, administer oxygen. **Liquid:** Irritating to skin and eyes. Harmful if swallowed. Remove contaminated clothing and shoes. Flush affected areas with plenty of water. *IF IN EYES,* hold eyelids open and flush with plenty of water. *IF SWALLOWED* and victim is *CONSCIOUS AND ABLE TO SWALLOW,* have victim drink 4 to 8 ounces of water. The use of alcoholic beverages may enhance the toxic effect.

HEALTH HAZARDS
Personal protective equipment (PPE): A-Level PPE. OSHA Table Z-1-A air contaminant. Chemical protective material(s) reported to have good to excellent resistance: Teflon®, Viton®, PV alcohol.
Recommendations for respirator selection: NIOSH
At any concentrations above the NIOSH REL, or where there is no REL, at any detectable concentration: SCBAF:PD,PP (any self-contained breathing apparatus that has a full facepiece and is operated in a pressure-demand or other positive-pressure mode); or SAF:PD,PP:ASCBA (any supplied-air respirator that has a full facepiece and is operated in a pressure-demand or other positive-pressure mode in combination with an auxiliary self-contained breathing apparatus operated in a pressure-demand or other positive pressure mode). *ESCAPE:* GMFOV [any air-purifying, full-facepiece respirator (gas mask) with a chin-style, front-or back-mounted organic vapor canister]; or SCBAE (any appropriate escape-type, self-contained breathing apparatus).
Exposure limits (TWA unless otherwise noted): ACGIH TLV 75 ppm (347 mg/m^3); OSHA PEL 75 ppm (350 mg/m^3); NIOSH REL potential human carcinogen; reduce exposure to lowest feasible level
Short-term exposure limits (15-minute TWA): AGCIH STEL 110 ppm (508 mg/m^3).
Toxicity by ingestion: Grade 2; LD$_{50}$ = 0.5 to 5 g/kg (guinea pig).
Long-term health effects: IARC rating 3, limited animal evidence. May effect the nervous system. Frequent or prolonged contact may cause organ damage to liver, kidney or heart.
Vapor (gas) irritant characteristics: Eye and respiratory tract irritant. Vapors cause a slight smarting of the eyes or respiratory system if present in high concentrations.
Liquid or solid irritant characteristics: Eye and skin irritant. If spilled on clothing and allowed to remain, may cause smarting and reddening of the skin.
Odor threshold: 0.25 ppm.
IDLH value: Potential human carcinogen; 400 ppm.

FIRE DATA
Flash point: 70°F/21°C (oc); 60°F/16°C (cc).
Flammable limits in air: LEL: 3.4%; UEL: 14.5%.
Autoignition temperature: 1035°F/557°C/830°K.
Electrical hazard: Class I, Group D.
Burning rate: (estimate) 3.2 mm/min

CHEMICAL REACTIVITY
Binary reactants: Strong oxidizers; strong acids; active metals. Corrodes aluminum.
Reactivity group: 33
Compatibility class: Halogenated hydrocarbons

ENVIRONMENTAL DATA
Food chain concentration potential: Log K$_{ow}$ = 2.0–2.3. Unlikely to accumulate.
Water pollution: Effect of low concentrations on aquatic life is unknown. May be dangerous if it enters nearby water intakes; notify operators. Notify local health and wildlife officials. **Response to discharge:** Issue warning–high flammability. Evacuate area.

SHIPPING INFORMATION
Grades of purity: Refined; **Storage temperature:** Ambient; **Inert atmosphere:** None; **Venting:** Pressure-vacuum; **Stability during transport:** Stable.

NAS HAZARD CLASSIFICATION FOR BULK WATER TRANSPORTATION
FIRE: 3
HEALTH: Vapor irritant: 1; Liquid or solid irritant: 1; Poisons: 3
WATER POLLUTION: Human toxicity: 1; Aquatic toxicity: 3; Aesthetic effect: 2
REACTIVITY: Other chemicals: 1; Water: 0; Self-reaction: 0

PHYSICAL AND CHEMICAL PROPERTIES
Physical state @ 59°F/15°C and 1 atm: Liquid.
Molecular weight: 113.0
Boiling point @ 1 atm: 206°F/96°C/370°K.
Melting/Freezing point: –148°F/–100°C/173°K.
Specific gravity (water = 1): 1.158 @ 68°F/20°C (liquid).
Liquid surface tension: 29 dynes/cm = 0.029 N/m @ 68°F/20°C.
Liquid water interfacial tension: 37.9 dynes/cm = 0.0379 N/m at 22.7°C.
Relative vapor density (air = 1): 3.7
Ratio of specific heats of vapor (gas): 1.094
Latent heat of vaporization: 122 Btu/lb = 67.7 cal/g = 2.83 x 10^5 J/kg.
Heat of combustion: (estimate) 7300 Btu/lb = 4100 cal/g = 170 x 10^5 J/kg.
Heat of fusion: 13.53 cal/g.
Vapor pressure: (Reid) 1.9 psia; 40 mm.

1,3-DICHLOROPROPANE REC. D:2350

SYNONYMS: 1,3-DICLOROPROPANO (Spanish); TRIMETHYLENE DICHLORIDE; TRIMETHYLENE CHLORIDE

IDENTIFICATION
CAS Number: 142-28-9
Formula: C$_3$H$_6$Cl$_2$; CH$_2$ClCH$_2$CH$_2$Cl
DOT ID Number: UN 1279; DOT Guide Number: 130
Proper Shipping Name: Dichloropropane
Reportable Quantity (RQ): **(CERCLA)** 5000 lb/2270 kg

DESCRIPTION: Colorless watery liquid. Sweet odor. Sinks in water. Flammable, irritating vapor is produced.

Poison! • Highly flammable • Breathing the vapor, skin or eye contact, or swallowing the material can cause severe illness • Containers may BLEVE when exposed to fire •Vapors may form explosive mixture with air • Vapors are heavier than air and will collect and stay in low areas • Vapors may travel long distances to ignition sources and flashback • Vapors in confined areas (e.g., tanks, sewers, buildings) may explode when exposed to fire • Irritating to the skin, eyes, and respiratory tract • Toxic products of combustion may include phosgene and hydrogen chloride.

Hazard Classification (based on NFPA-704 Rating System)
Health Hazards (Blue): 2; Flammability (Red): 3; Reactivity (Yellow): 0

EMERGENCY RESPONSE: See Appendix A (130)
Evacuation:
Public safety: Isolate spill area for at least 50 to 100 meters (160 to 330 feet) in all directions.
Spill: Large spill–Consider initial downwind evacuation for at least 300 meters (1000 feet).
Fire: Isolate for 800 meters (½ mile) in all directions, especially if tank, rail car, or tank truck is involved in fire.

EXPOSURE
Short-term effects: *SEEK MEDICAL ATTENTION*. **Vapor:** Irritating to eyes, nose, and throat. Move to fresh air. *IF BREATHING HAS STOPPED*, give artificial respiration. If breathing is difficult, administer oxygen. **Liquid:** Irritating to skin and eyes. Harmful if swallowed. Remove contaminated clothing and shoes. Flush affected areas with plenty of water. *IF IN EYES*, hold eyelids open and flush with plenty of water. *IF SWALLOWED* and victim is *CONSCIOUS AND ABLE TO SWALLOW*, have victim drink 4 to 8 ounces of water. The use of alcoholic beverages may enhance the toxic effect.

HEALTH HAZARDS
Personal protective equipment (PPE): B-Level PPE.
Long-term health effects: Alters pancreatic function. May effect the nervous system, liver, kidney and heart.
Vapor (gas) irritant characteristics: Eye and respiratory tract irritant. Vapors cause smarting of the eyes or respiratory system if present in high concentrations. The effect is temporary.
Liquid or solid irritant characteristics: Eye and skin irritant. If spilled on clothing and allowed to remain, may cause smarting and reddening of skin.

FIRE DATA
Flash point: (estimate) 60°F/16°C (oc)
Flammable limits in air: LEL: 3.4%; UEL: 14.5%.
Autoignition temperature: (approx) 1031°F/555°C.
Electrical hazard: Class I, Group D.

CHEMICAL REACTIVITY
Binary reactants: Reacts with oxidizers, light metals forming heat, and plastics.

ENVIRONMENTAL DATA
Food chain concentration potential: Log K_{ow} = 1.98 (estimate).
Water pollution: DOT Appendix B, §172.101–marine pollutant. Effect of low concentrations on aquatic life is unknown. May be dangerous if it enters nearby water intakes; notify operators. Notify local health and wildlife officials. **Response to discharge:** Issue warning–high flammability: Evacuate area.

SHIPPING INFORMATION
Storage temperature: Ambient; **Inert atmosphere:** None; **Venting:** Pressure-vacuum; **Stability during transport:** Stable.

NAS HAZARD CLASSIFICATION FOR BULK WATER TRANSPORTATION
FIRE: 3
HEALTH: Vapor irritant: 1; Liquid or solid irritant: 1; Poisons: 3
WATER POLLUTION: Human toxicity: 1; Aquatic toxicity: 3; Aesthetic effect: 2
REACTIVITY: Other chemicals: 1; Water: 0; Self-reaction: 0

PHYSICAL AND CHEMICAL PROPERTIES
Physical state @ 59°F/15°C and 1 atm: Liquid.
Molecular weight: 112.99
Boiling point @ 1 atm: 249°F/120°C/394°K.
Melting/Freezing point: –147°F/–100°C/174°K.
Critical temperature: (estimate) 598°F/314°C/588°K.
Critical pressure: (estimate) 613.7 psia = 41.75 atm = 4.23 MN/m².
Specific gravity (water = 1): 1.1878 @ 68°F/20°C.
Liquid surface tension: 33.93 dynes/cm = 0.03393 N/m @ 68°F/20°C.
Liquid water interfacial tension: 41.1 dynes/cm = 0.0411 N/m @ 68°F/20°C.
Relative vapor density (air = 1): 3.90.
Ratio of specific heats of vapor (gas): (estimate) 1.094 @ 68°F/20°C/68°F
Latent heat of vaporization: At boiling point, 129 Btu/lb = 71.71 cal/g = 3.0 x 10^5 J/kg.
Heat of combustion: –6676 Btu/lb = –3709 cal/g = –155 x 10^5 J/kg.
Vapor pressure: 40 mm.

1,3-DICHLOROPROPENE   REC. D:2400

SYNONYMS: α-CHLOROALLYL CHLORIDE; 3-CHLORO-ALLYL CHLORIDE; γ-CHLORO ALLYL CHLORIDE; 3-CHLOROPROPENYL CHLORIDE; α,γ-DICHLOROPROPYLENE; DICHLOROPROPENE; 1,3-DICHLORO-1-PROPENE; 1,3-DICHLOROPROPENE-1; 1,3-DICHLOROPROPYLENE; 1,3-DICLOROPROPENO (Spanish); EEC No. 602-030-00-5; NEMEX; TELONE; TELONE C; TELONE II SOIL FUMIGANT; VIDDEN D; VORLEX: VORLEX-201; RCRA No. U084

IDENTIFICATION
CAS Number: 542-75-6; 10061-01-5 (*cis*-isomer); 10061-02-6 (*trans*-isomer)
Formula: $C_3H_4Cl_2$; $ClCH_2CH=CHCl$
DOT ID Number: UN 2047; DOT Guide Number: 132
Proper Shipping Name: Dichloropropenes
Reportable Quantity (RQ): (CERCLA) 100 lb/45.4 kg

DESCRIPTION: Colorless to light yellow liquid. Sweet odor, like chloroform. Sinks in water; insoluble. Flammable, irritating vapor is produced.

Poison! (insecticide) • Highly flammable • Breathing the vapor, skin or eye contact, or swallowing the material can cause illness, and possible death • Firefighting gear (including SCBA) does not provide adequate protection. If exposure occurs, remove and isolate gear immediately and thoroughly decontaminate personnel • Containers may BLEVE when exposed to fire •Vapors may form explosive mixture with air • Vapors are heavier than air and will collect and stay in low areas • Vapors may travel long distances to ignition sources and flashback • Vapors in confined areas (e.g., tanks, sewers, buildings) may explode when exposed to fire • Toxic products of combustion may include hydrogen chloride.

Hazard Classification (based on NFPA-704 Rating System)
Health Hazards (Blue): 2; Flammability (Red): 3; Reactivity (Yellow): 0

EMERGENCY RESPONSE: See Appendix A (132)
Evacuation:
Public safety: Isolate spill area for at least 50 to 100 meters (160 to 330 feet) in all directions.
Spill: Increase, as necessary, the isolation distance shown above, in "Public safety."
Fire: Isolate for 800 meters (½ mile) in all directions, especially if tank, rail car, or tank truck is involved in fire.

EXPOSURE
Short-term effects: *SEEK MEDICAL ATTENTION.* **Vapor:** Irritating to eyes, nose, and throat. Move to fresh air. *IF BREATHING HAS STOPPED,* give artificial respiration. If breathing is difficult, administer oxygen. **Liquid:** *ABSORBED THROUGH THE SKIN.* Will burn skin and eyes. Harmful if swallowed or absorbed through the skin. Remove contaminated clothing and shoes. Flush affected areas with plenty of water. *IF IN EYES,* hold eyelids open and flush with plenty of water. *IF SWALLOWED* and victim is *CONSCIOUS AND ABLE TO SWALLOW,* have victim drink 4 to 8 ounces of water and have victim induce vomiting. *IF SWALLOWED* and victim is *UNCONSCIOUS OR HAVING CONVULSIONS,* do nothing, except keep victim warm. The use of alcoholic beverages may enhance the toxic effect.

HEALTH HAZARDS
Personal protective equipment (PPE): A-Level PPE. OSHA Table Z-1-A air contaminant. Full protective clothing and approved full-face organic vapor respirator (see below). Chemical protective material(s) reported to have good to excellent resistance: Teflon®, Viton®, PV alcohol.
Recommendations for respirator selection: NIOSH
At any concentrations above the NIOSH REL, or where there is no REL, at any detectable concentration: SCBAF:PD,PP (any self-contained breathing apparatus that has a full facepiece and is operated in a pressure-demand or other positive-pressure mode); or SAF:PD,PP:ASCBA (any supplied-air respirator that has a full facepiece and is operated in a pressure-demand or other positive-pressure mode in combination with an auxiliary self-contained breathing apparatus operated in a pressure-demand or other positive pressure mode). *ESCAPE:* GMFOV [any air-purifying, full-facepiece respirator (gas mask) with a chin-style, front-or back-mounted organic vapor canister]; or SCBAE (any appropriate escape-type, self-contained breathing apparatus).
Exposure limits (TWA unless otherwise noted): ACGIH TLV 1 ppm (4.5 mg/m^3); NIOSH REL 1 ppm (5 mg/m^3); potential human carcinogen; reduce exposure to lowest feasible level; skin contact contributes significantly in overall exposure.
Toxicity by ingestion: Grade 3; LD$_{50}$ = 50 to 500 mg/kg.
Long-term health effects: Possible nervous system, liver and kidney damage may occur. Suspected human carcinogen. NTP anticipated carcinogen. IARC possible carcinogen, rating 2B; sufficient animal evidence.
Vapor (gas) irritant characteristics: Eye and respiratory tract irritant. Vapors cause irritation such that personnel will find high concentrations unpleasant.
Liquid or solid irritant characteristics: Eye and skin irritant. Causes smarting of the skin and first-degree burns on short exposure and may cause secondary burns on long exposure.
Odor threshold: 1 ppm.

FIRE DATA
Flash point: 95°F/35°C (cc); 69°F/21°C (*cis-* or *trans-*isomer).
Flammable limits in air: LEL: 5.3%; UEL: 14.5%.
Behavior in fire: Containers may rupture or explode. Water streams applied to adjacent fires will spread contamination of pesticide over wide area.
Electrical hazard: Class I, Group D. Flow or agitation of substance may generate electrostatic charges due to low conductivity.
Burning rate: (estimate) 3.4 mm/min

CHEMICAL REACTIVITY
Binary reactants: Reacts with strong oxidizers, aluminum, magnesium, halogens, oxidizers. Will corrode steel if wet or at elevated temperatures; also attacks aluminum and rubber.
Reactivity group: 15
Compatibility class: Substituted allyl.

ENVIRONMENTAL DATA
Food chain concentration potential: Log K$_{ow}$ = 1.4–1.6. *Cis-* and *trans-*isomers are in the same range. Unlikely to accumulate.
Water pollution: Effect of low concentrations on aquatic life is unknown. May be dangerous if it enters nearby water intakes; notify operators. Notify local health and wildlife officials. **Response to discharge:** Disperse and flush.

SHIPPING INFORMATION
Grades of purity: Telone soil fumigant: 100%; Telone C soil fungicide: 85%; **Storage temperature:** Ambient; **Inert atmosphere:** None; **Venting:** Pressure-vacuum; **Stability during transport:** Stable.

NAS HAZARD CLASSIFICATION FOR BULK WATER TRANSPORTATION
FIRE: 3
HEALTH: Vapor irritant: 2; Liquid or solid irritant: 2; Poisons: 3
WATER POLLUTION: Human toxicity: 3; Aquatic toxicity: 3; Aesthetic effect: 2
REACTIVITY: Other chemicals: 1; Water: 0; Self-reaction: 1

PHYSICAL AND CHEMICAL PROPERTIES
Physical state @ 59°F/15°C and 1 atm: Liquid.
Molecular weight: 110.98
Boiling point @ 1 atm: 2196°F/104°C/377°K.
Melting/Freezing point: 119°F/48°C/321°K.
Specific gravity (water = 1): 1.21 @ 68°F/20°C (liquid).
Liquid surface tension: 31.2 dynes/cm = 0.0312 N/m at 24°C.
Liquid water interfacial tension: 23.8 dynes/cm = 0.0238 N/m at 24°C.
Relative vapor density (air = 1): 3.75
Ratio of specific heats of vapor (gas): (estimate) 1.116
Latent heat of vaporization: (estimate) 113 Btu/lb = 62.8 cal/g = 2.63 x 10^5 J/kg.
Heat of combustion: (estimate) 6900 Btu/lb = 3900 cal/g = 160 x 10^5 J/kg.
Vapor pressure: (Reid) 4.0 psia; 28 mm.

2,3-DICHLOROPROPENE   REC. D:2450

SYNONYMS: 2,3-DICHLORO-1-PROPENE; 2,3-DICHLOROPROPYLENE; 2-CHLOROALLYL CHLORIDE; 2,3-DICHLORO-1-PROPANE; 2,3-DICLOROPROPENO (Spanish)

IDENTIFICATION
CAS Number: 78-88-6

Dichloropropene–dichloropropane mixture

Formula: $C_3H_4Cl_2$; CH_2CClCH_2Cl
DOT ID Number: UN 2047; DOT Guide Number: 132
Proper Shipping Name: Dichloropropenes
Reportable Quantity (RQ): **(CERCLA)** 100 lb/45.4 kg

DESCRIPTION: Colorless to yellow liquid. Chloroform-like odor. Sinks in water. Flammable, irritating vapor is produced.

Poison! • Highly flammable • A tear gas • Breathing the vapor, skin or eye contact, or swallowing the material can cause illness, and possible death • Firefighting gear (including SCBA) does not provide adequate protection. If exposure occurs, remove and isolate gear immediately and thoroughly decontaminate personnel • Containers may BLEVE when exposed to fire • Vapors may form explosive mixture with air • Vapors are heavier than air and will collect and stay in low areas • Vapors may travel long distances to ignition sources and flashback • Vapors in confined areas (e.g., tanks, sewers, buildings) may explode when exposed to fire • Irritating to the skin, eyes, and respiratory tract • Toxic products of combustion may include phosgene and hydrogen chloride.

Hazard Classification (based on NFPA-704 Rating System)
Health Hazards (Blue): 3; Flammability (Red): 3; Reactivity (Yellow): 0

EMERGENCY RESPONSE: See Appendix A (132)
Evacuation:
Public safety: Isolate spill area for at least 50 to 100 meters (160 to 330 feet) in all directions.
Spill: Increase, as necessary, the isolation distance shown above, in "Public safety."
Fire: Isolate for 800 meters (½ mile) in all directions, especially if tank, rail car, or tank truck is involved in fire.

EXPOSURE
Short-term effects: *SEEK MEDICAL ATTENTION*. **Vapor:** Irritating to eyes, nose, and throat. Move to fresh air. *IF BREATHING HAS STOPPED*, give artificial respiration. If breathing is difficult, administer oxygen. **Liquid:** Will burn skin and eyes. Harmful if swallowed. Remove contaminated clothing and shoes. Flush affected areas with plenty of water. *IF IN EYES*, hold eyelids open and flush with plenty of water. *IF SWALLOWED* and victim is *CONSCIOUS AND ABLE TO SWALLOW*, have victim drink 4 to 8 ounces of water and have victim induce vomiting. *IF SWALLOWED* and victim is *UNCONSCIOUS OR HAVING CONVULSIONS*, do nothing, except keep victim warm.

HEALTH HAZARDS
Personal protective equipment (PPE): A-Level PPE. Chemical protective material(s) reported to have good to excellent resistance: Viton®, PV alcohol.
Toxicity by ingestion: Grade 3; LD_{50} = 50 to 500 mg/kg.
Long-term health effects: Mutagenic; can injure liver, kidneys and heart.
Vapor (gas) irritant characteristics: Eye and respiratory tract irritant and poison. Vapors are irritating such that personnel will not usually tolerate moderate or high concentrations.
Liquid or solid irritant characteristics: Severe eye and skin irritant. May cause pain and second-degree burns after a few minutes of contact.

FIRE DATA
Flash point: 59-62°F/15-17°C (cc).
Flammable limits in air: LEL: 2.6%; UEL: 7.8%.

Behavior in fire: Water streams applied to adjacent fires will spread contamination of pesticide over wide area. vapors.

CHEMICAL REACTIVITY
Binary reactants: Reacts with aluminum, amines, and ammonia.
Reactivity group: 15
Compatibility class: Substituted allyls.

ENVIRONMENTAL DATA
Food chain concentration potential: Log P_{ow} = 1.9. Unlikely to accumulate.
Water pollution: Harmful to aquatic life in very low concentrations. May be dangerous if it enters nearby water intakes; notify operators. Notify local health and wildlife officials. **Response to discharge:** Issue warning–high flammability, water contaminant. Restrict access. Should be removed. Chemical and physical treatment.

SHIPPING INFORMATION
Grades of purity: 98% minimum; **Storage temperature:** Ambient; **Inert atmosphere:** None; **Venting:** Pressure-vacuum; **Stability during transport:** Stable.

NAS HAZARD CLASSIFICATION FOR BULK WATER TRANSPORTATION
FIRE: 3
HEALTH: Vapor irritant: 2; Liquid or solid irritant: 2; Poisons: 3
WATER POLLUTION: Human toxicity: 3; Aquatic toxicity: 3; Aesthetic effect: 2
REACTIVITY: Other chemicals: 1; Water: 0; Self-reaction: 1

PHYSICAL AND CHEMICAL PROPERTIES
Physical state @ 59°F/15°C and 1 atm: Liquid.
Molecular weight: 110.98
Boiling point @ 1 atm: 196-201°F/92-94°C/364.2-366.2°K.
Melting/Freezing point: –115°F/–81.7°C/191.5°K.
Critical temperature: (estimate) 526.7°F/274.8°C/548°K.
Critical pressure: (estimate) 513 psia = 34.9 atm = 3.54 MN/m^2.
Specific gravity (water = 1): 1.211 @ 68°F/20°C/4°C.
Liquid surface tension: (calculated) 29.9 dynes/cm = 0.029 N/m @ 68°F/20°C.
Liquid water interfacial tension: (Calculated) 45.1 dynes/cm = 0.0451 N/m @ 68°F/20°C.
Relative vapor density (air = 1): 3.8
Ratio of specific heats of vapor (gas): (estimate) 1.116 @ 68°F/20°C.
Latent heat of vaporization: 137.1 Btu/lb = 76.1 cal/g = 3.19 x 10^5 J/kg.
Heat of combustion: (estimate) –6900 Btu/lb = –3900 cal/g = –160 x 10^5 J/kg.

DICHLOROPROPENE–DICHLOROPROPANE MIXTURE
REC. D:2500

SYNONYMS: DD MIXTURE; 1,3-DICHLOROPROPENE AND 1,2-DICHLOROPROPANE MIXTURE; DICHLORPROPAN-DICHLORPROPEN GEMISCH (German); D-D SOIL FUMIGANT; DOWFUME N; MEZCLA de DICLOROPROPENO y DICLOROPROPANO (Spanish); NEMAFENE; TELONE; VIDDEN D

IDENTIFICATION
CAS Number: 8003-19-8
Formula: $C_3H_4Cl_2$

DOT ID Number: UN 2047; DOT Guide Number: 132
Proper Shipping Name: Dichloropropenes
Reportable Quantity (RQ): **(CERCLA)** 100 lb/45.4 kg

DESCRIPTION: Straw to amber liquid. Pungent, garlic-like odor. Sinks and slowly mixes with water.

Poison! • Highly flammable • Breathing the vapor, skin or eye contact, or swallowing the material can cause illness, and possible death • Firefighting gear (including SCBA) does not provide adequate protection. If exposure occurs, remove and isolate gear immediately and thoroughly decontaminate personnel • Containers may BLEVE when exposed to fire • Vapors may form explosive mixture with air • Vapors are heavier than air and will collect and stay in low areas • Vapors may travel long distances to ignition sources and flashback • Vapors in confined areas (e.g., tanks, sewers, buildings) may explode when exposed to fire • Irritating to the skin, eyes, and respiratory tract • Toxic products of combustion may include phosgene and hydrogen chloride.

Hazard Classification (based on NFPA-704 Rating System)
Health Hazards (Blue): 3; Flammability (Red): 3; Reactivity (Yellow): 0

EMERGENCY RESPONSE: See Appendix A (132)
Evacuation:
Public safety: Isolate spill area for at least 50 to 100 meters (160 to 330 feet) in all directions.
Spill: Increase, as necessary, the isolation distance shown above, in "Public safety."
Fire: Isolate for 800 meters (½ mile) in all directions, especially if tank, rail car, or tank truck is involved in fire.

EXPOSURE
Short-term effects: *SEEK MEDICAL ATTENTION.* **Vapor:** Irritating to eyes, and respiratory tract, skin and digestive tract. Inhalation will cause gasping, refusal to breathe, and respiratory distress; may be fatal. Move victim to fresh air. *IF BREATHING HAS STOPPED,* give artificial respiration. If breathing is difficult, administer oxygen. **Liquid:** May be fatal if swallowed or absorbed through skin. Will burn exposed tissues. *IF IN EYES,* hold eyelids open and flush with water for 15 minutes. *IF ON SKIN,* flush with water for 15 minutes; wash with soap and water. Remove and double bag contaminated clothing and shoes at the site. *IF SWALLOWED* and victim is *CONSCIOUS AND ABLE TO SWALLOW,* have victim drink water and induce vomiting. *IF SWALLOWED* and victim is *UNCONSCIOUS OR HAVING CONVULSIONS,* do nothing except keep victim warm.

HEALTH HAZARDS
Personal protective equipment (PPE): A-Level PPE. This material will penetrate ordinary rubber protective equipment such as boots and gloves; therefore chemical resistant equipment must be worn. Chemical protective material(s) reported to have good to excellent resistance: Viton®, PV alcohol. Also, Teflon® may offers limited protection
Toxicity by ingestion: Grade 3; LD_{50} = 140 mg/kg (rat).
Long-term health effects: Has mutagenic effects. May cause liver and kidney damage.
Vapor (gas) irritant characteristics: Causes intense irritation of eyes, skin, and respiratory mucosa.
Liquid or solid irritant characteristics: Severe eye and skin irritant. Contact with liquid will cause burns to exposed surfaces.
Odor threshold: 1,3-Dichloropropene (a major component) 1–3 ppm; 1,2-dichloropropane (a major component) 0.1 ppm.

FIRE DATA
Flash point: 67°F/19°C (cc).
Flammable limits in air: LEL: 5.3%; UEL: 14.5%.
Fire extinguishing agents not to be used: Water streams applied to adjacent fires will spread contamination over wide area.
Electrical hazard: Class I, Group D. Due to low electric conductivity, this substance may generate electrostatic charges as a result of agitation and flow.

CHEMICAL REACTIVITY
Binary reactants: Reacts with strong oxidizers, aluminum, magnesium, and their alloys.
Neutralizing agents for acids and caustics: Sodium bicarbonate or sand and soda ash mixture.
Reactivity group: 15
Compatibility class: Substituted allyls

ENVIRONMENTAL DATA
Food chain concentration potential: Log P_{ow} = 2.0. Unlikely to accumulate.
Water pollution: Dangerous to aquatic life in high concentrations. May be dangerous if it enters nearby water intakes; notify operators. Notify local health and wildlife officials. **Response to discharge:** Issue warning–high flammability, water contaminant. Restrict access. Evacuate area. Should be removed. Chemical and physical treatment.

SHIPPING INFORMATION
Grades of purity: Mixture; **Storage temperature:** Ambient; **Inert atmosphere:** Nones; **Venting:** Pressure-vacuum valve; **Stability during transport:** Stable.

PHYSICAL AND CHEMICAL PROPERTIES
Physical state @ 59°F/15°C and 1 atm: Liquid.
Boiling point @ 1 atm: 217–340°F/103–171°C/376–444°K.
Melting/Freezing point: (estimated) –78°F/–61°C/212°K.
Specific gravity (water = 1): 1.2 (temperature unknown).
Relative vapor density (air = 1): 3.9

DICHLOROTETRAFLUOROETHANE REC. D:2550

SYNONYMS: 1,2-DICHLOROTETRAFLUOROETHANE; DICLOROTETRAFLUOETANO (Spanish); FLUORANE 114; FREON 114; F-114; GENETRON 114; HALON 242; PROPELLANT 14; R-14; REFRIGERANT 114

IDENTIFICATION
CAS Number: 76-14-2
Formula: $C_2Cl_2F_4$
DOT ID Number: UN 1958; DOT Guide Number: 126
Proper Shipping Name: Dichlorotetrafluoroethane

DESCRIPTION: Colorless gas. Shipped and stored as a liquefied, compressed gas. Slight, ether-like odor at high concentrations. Practically insoluble in water. A liquid below 38.8°F/3.8°C.

Toxic substances formed on contact with a flame or hot metal surfaces • Irritating to lungs • Contact with liquid may cause frostbite • Toxic products of combustion may include carbon monoxide, hydrogen fluoride, and hydrogen chloride.

Hazard Classification (based on NFPA-704 Rating System)
Health Hazards (Blue): 0; Flammability (Red): 0; Reactivity (Yellow): 0

360   4,4'-Dichloro-α-trichloromethylbenzhydrol

EMERGENCY RESPONSE: See Appendix A (126)
Evacuation:
Public safety: Isolate spill area for at least 100 meters (330 feet) in all directions.
Spill: Large spill–Consider initial downwind evacuation for at least 500 meters (⅓ mile).
Fire: Isolate for 800 meters (½ mile) in all directions, especially if tank, rail car, or tank truck is involved in fire.

EXPOSURE
Short-term effects: *SEEK MEDICAL ATTENTION.*
Vapor: Vapors may cause dizziness or suffocation. Move victim to fresh air; call emergency medical care. If not breathing, give artificial respiration. If breathing is difficult, administer oxygen.
Liquid: Contact with liquid may cause frostbite. Remove contaminated clothing and shoes. Flush affected areas with lukewarm water. *DO NOT USE HOT WATER.*

HEALTH HAZARDS
Personal protective equipment (PPE): B-Level PPE. OSHA Table Z-1-A air contaminant. Nitrile rubber is generally suitable for freons. Wear thermal protective clothing.
Recommendations for respirator selection: NIOSH/OSHA
10,000 ppm: SA (any supplied-air respirator). *15,000 ppm:* SA:CF (any supplied-air respirator operated in a continuous-flow mode); or SCBAF (any self-contained breathing apparatus with a full facepiece); or SAF (any supplied-air respirator with a full facepiece). *EMERGENCY OR PLANNED ENTRY INTO UNKNOWN CONCENTRATIONS OR IDLH CONDITIONS:* SCBAF:PD,PP (any self-contained breathing apparatus that has a full facepiece and is operated in a pressure-demand or other positive-pressure mode); or SAF:PD,PP:ASCBA (any supplied-air respirator that has a full facepiece and is operated in a pressure-demand or other positive-pressure mode in combination with an auxiliary self-contained breathing apparatus operated in a pressure-demand or other positive pressure mode). *ESCAPE:* GMFOV[any air-purifying, full-facepiece respirator (gas mask) with a chin-style, front- or back-mounted organic vapor cannister]; or SCBAE (any appropriate escape-type, self-contained breathing apparatus).
Exposure limits (TWA unless otherwise noted): ACGIH TLV 1000 ppm (6990 mg/m^3); NIOSH/OSHA 1000 ppm (7000 mg/m^3).
Vapor (gas) irritant characteristics: Vapors are nonirritating to eyes and throat.
Liquid or solid irritant characteristics: Contact may cause frostbite.
IDLH value: 15,000 ppm.

FIRE DATA
Electrical hazard: Not flammable.

CHEMICAL REACTIVITY
Binary reactants: Chemically active metals such as sodium, potassium, calcium, powdered aluminum, zinc and magnesium; acids; acid fumes.

SHIPPING INFORMATION
Grades of purity: 99+%; **Stability during transport:** Stable.

NAS HAZARD CLASSIFICATION FOR BULK WATER TRANSPORTATION
FIRE: 0
HEALTH: Vapor irritant: 1; Liquid or solid irritant: 0; Poisons: 0
WATER POLLUTION: Human toxicity: 0; Aquatic toxicity: 1; Aesthetic effect: 0
REACTIVITY: Other chemicals: 0; Water: 0; Self-reaction: 0

PHYSICAL AND CHEMICAL PROPERTIES
Physical state @ 59°F/15°C and 1 atm: Gas.
Molecular weight: 170.93
Boiling point @ 1 atm: 38.8°F/3.8°C/277°K.
Melting/Freezing point: –137°F/–94°C/179°K.
Critical temperature: 294.3°F/145.7°C/418.9°K.
Critical pressure: 473.2 psia = 32.2 atm = 3.3 MN/m^2.
Specific gravity (water = 1): 1.455
Liquid surface tension: 12 dyne/cm = 0.012 N/m @ 77°F/25°C.
Relative vapor density (air = 1): 5.93
Latent heat of vaporization: 58.5 Btu/lb = 32.5 cal/g = 1.36 x 10^5 J/kg.
Vapor pressure: (Reid) 50.6 psia; 1.9 atm @ 70°F

4,4'-DICHLORO-α-TRICHLOROMETHYLBENZHYDROL
REC. D:2600

SYNONYMS: ACARIN; CARBAX; DICOFOL; DTMC; KELTHANE; KELTHANETHANOL; p,p-KELTHANE; 1,1-BIS(*p*-CHLOROPHENYL)-2,2,2-TRICHLOROETHANOL; DI-(*p*-CHLOROPHENYL) TRICHLOROMETHYLCARBINOL

IDENTIFICATION
CAS Number: 115-32-2
Formula: $C_{14}H_9Cl_5O$
DOT ID Number: UN 3077; DOT Guide Number: 171
Proper Shipping Name: Environmentally hazardous substances, solid, n.o.s.
Reportable Quantity (RQ): **(CERCLA)** 10 lb/4.54 kg

DESCRIPTION: White to gray solid; red to black semi-solid; brown liquid solution. Solid is odorless; liquid has odor of the solvent xylene. Pure material will sink in water; formulated material may mix with water.

Combustible solid, flammable liquid • Containers may BLEVE when exposed to fire • Vapors of liquid may form explosive mixture with air • Vapors from liquid are heavier than air and will collect and stay in low areas • Vapors from xylene solvent may travel long distances to ignition sources and flashback • Vapors from liquid in confined areas (e.g., tanks, sewers, buildings) may explode when exposed to fire • Irritating to the skin, eyes, and respiratory tract • Toxic products of combustion may include carbon monoxide and hydrogen chloride.

EMERGENCY RESPONSE: See Appendix A (171)
Evacuation:
Public safety: Isolate the area of spill or leak for at least 10 to 25 meters (30 to 80 feet) in all directions.
Spill: Increase, in the downwind direction, as necessary, the distance shown under "Public Safety."
Fire: If any large container is involved in fire, isolate for at least 800 meters (½ mile) in all directions; also, consider initial evacuation for 800 meters (½ mile) in all directions.

EXPOSURE
Short-term effects: *SEEK MEDICAL ATTENTION.* **Dust:** Irritating to eyes, nose, and throat. *IF INHALED*, will, will cause headache or dizziness. *IF IN EYES*, hold eyelids open and flush with plenty of water. *IF BREATHING HAS STOPPED*, give artificial respiration. If breathing is difficult, administer oxygen.
Liquid or solid: Irritating to skin and eyes. *IF SWALLOWED*, will cause headache, nausea or dizziness. Remove contaminated clothing and shoes. Flush affected areas with plenty of water. *IF IN*

EYES, hold eyelids open and flush with plenty of water. *IF SWALLOWED* and victim is *CONSCIOUS AND ABLE TO SWALLOW*, have victim drink 4 to 8 ounces of water. *IF SWALLOWED* and victim is *UNCONSCIOUS OR HAVING CONVULSIONS*, do nothing except keep victim warm.

HEALTH HAZARDS
Personal protective equipment (PPE): B-Level PPE.
Toxicity by ingestion: Grade 2; oral LD_{50} = 575 mg/kg (rat), 1810 mg/kg (rabbit).
Long-term health effects: Suppresses immune reactions in rats. IARC rating 3; limited animal evidence. Organ damage to liver and kidneys.
Liquid or solid irritant characteristics: Eye irritant.

FIRE DATA
Flash point: 75°F/4°C (oc) (xylene).
Flammable limits in air: LEL: 1.1%; UEL: 7.0%.
Flammable limits in air: (for xylene solution).
Fire extinguishing agents not to be used: Water may be ineffective.
Autoignition temperature: 986°F/530°C.
Autoignition temperature: (for xylene solution).
Electrical hazard: (xylene) Class 1, Group D.
Burning rate: (xylene) 5.8 mm/min

CHEMICAL REACTIVITY
Binary reactants: Contact with steel at elevated temperature causes formation of toxic chlorine and hydrogen chloride gases. Hydrolyzes in alkali. Liquid may attack some forms of plastics.

ENVIRONMENTAL DATA
Food chain concentration potential: Log P_{ow} = 2.8. Values > 3.0 are likely to bioconcentrate in aquatic organisms and other living tissue, especially in fats.
Water pollution: Dangerous to aquatic life in high concentrations. Fouling to shoreline. May be dangerous if it enters nearby water intakes; notify operators. Notify local health and wildlife officials.
Response to discharge: Issue warning–water contaminant, high flammability (liquid only). Mechanical containment (liquid only) Should be removed. Chemical and physical treatment.

SHIPPING INFORMATION
Grades of purity: Technical; 82–88%; 35% plus 65% inert solids; 75% solution in xylene, a combustible solvent; **Storage temperature:** Ambient; **Inert atmosphere:** None; **Venting:** Open (flame arrester); **Stability during transport:** Stable.

PHYSICAL AND CHEMICAL PROPERTIES
Physical state @ 59°F/15°C and 1 atm: Solid or Liquid.
Molecular weight: 470.5
Boiling point @ 1 atm: (Data apply to xylene solution; solid decomposes) 282°F/139°C/412°K.
Specific gravity (water = 1): More than 1.1 @ 68°F/20°C (solid); less than 0.9 @ 68°F/20°C (liquid).

DICHLORVOS   REC. D:2650

SYNONYMS: APAVAP; ASTROBOT; ATGARD; BENFOS; BIBESOL; BREVINYL; CANOGARD; CEKUSAN; CYPONA; DDVP; DERABAN; DERRIBANTE; DEVIKOL; DIVIPAN; DICHLOROPHOS; DICLORVOS (Spanish); DUO-KILL; DURAVOS; EQUIGEL; ESTROSEL; ESTROSOL; FECAMA; FLY-DIE; FLY FIGHTER; HERKAL; KRECALVIN; MAFU; MARVEX; MOPARI; NERKOL; NOGOS; NO-PEST STRIP; NUVA; OKO; OMS 14; PHOSVIT; SD-1750; TAP 9VP; TASK; TASK TABS; TENAC; TETRAVOS; VAPONA; VAPONITE; VERDICAN; VERDIPOR; VINYLFOS; VINYLOPHOS; 2,2-DICHLOROVINYL *O,O*-DIMETHYL PHOSPHATE; 2,2-DICHLOROVINYL DIMETHYL PHOSPHATE

IDENTIFICATION
CAS Number: 62-73-7
Formula: $C_4H_7Cl_2O_4P$; $(CH_3O)_2P(O)OCH$: $CCl_2C_4H_7Cl_2O_4P$
DOT ID Number: UN 2783; DOT Guide Number: 152
Proper Shipping Name: Dichlorvos
Reportable Quantity (RQ): **(CERCLA)** 10 lb/4.54 kg

DESCRIPTION: Colorless to amber liquid. Aromatic, characteristic odor. Sinks and mixes with water.

Poison! (organophosphate) • Combustible • Breathing the vapor, skin or eye contact, or swallowing the material can kill you • Firefighting gear (including SCBA) does not provide adequate protection. If exposure occurs, remove and isolate gear immediately and thoroughly decontaminate personnel • Containers may BLEVE when exposed to fire • Toxic products of combustion may include carbon monoxide, chlorine and phosphorus oxides • Do not put yourself in danger by entering a contaminated area to rescue a victim.

Hazard Classification (based on NFPA-704 Rating System)
Health Hazards (Blue): 3; Flammability (Red): 1; Reactivity (Yellow): 0

EMERGENCY RESPONSE: See Appendix A (152)
Evacuation:
Public safety: Isolate the area of spill or leak for at least 25 to 50 meters (80 to 160 feet) in all directions.
Spill: Increase, in the downwind direction, as necessary, the distance shown under "Public Safety."
Fire: If tank, rail car, or tank truck is involved in fire, isolate for at least 800 meters (½ mile) in all directions; also, consider initial evacuation for 800 meters (½ mile) in all directions.

EXPOSURE
Short-term effects: *SEEK MEDICAL ATTENTION*. **Vapor:** *POISONOUS IF INHALED OR SKIN IS EXPOSED*. Move to fresh air. *IF BREATHING HAS STOPPED*, give artificial respiration; avoid mouth-to-mouth resuscitation; use bag/mask apparatus. **Liquid:** *POISONOUS IF SWALLOWED* OR SKIN IS EXPOSED. Remove contaminated clothing and shoes. Flush affected areas with plenty of water. *IF IN EYES*, hold eyelids open and flush with plenty of water. *IF SWALLOWED* and victim is *CONSCIOUS AND ABLE TO SWALLOW*, have victim drink 4 to 8 ounces of water and have victim induce vomiting.
Note to physician or authorized medical personnel: Administer atropine, 2 mg (1/30 gr) intramuscularly or intravenously as soon as any local or systemic signs or symptoms of an intoxication are noted; repeat the administration of atropine every 3 to 8 minutes until signs of atropinization (mydriasis, dry mouth, rapid pulse, hot and dry skin) occur; initiate treatment in children with 0.05 mg/kg of atropine; repeat at 5- to 10-minute intervals. Watch respiration, and remove bronchial secretions if they appear to be obstructing the airway; intubate if necessary. Pralidoxime must be administered within minutes to a few hours following exposure (depending on the specific agent) to be effective. Give 2-PAMCI (Pralidoxime; Protopam), 2.5 g in 100 mL of sterile water or in 5% dextrose and water, intravenously, slowly, in 15 to 30 minutes; if sufficient fluid

is not available, give 1 g of 2-PAMCI in 3 mL of distilled water by deep intramuscular injection; repeat this every half hour if respiration weakens or if muscle fasciculation or convulsions recur. Also Diazepam, an anticonvulsant, might be needed.

HEALTH HAZARDS
Personal protective equipment (PPE): A-Level PPE. OSHA Table Z-1-A air contaminant. Chemical-protective clothing and butyl rubber gloves are recommended when skin contact is possible because liquid is rapidly absorbed through the skin and may cause systemic toxicity.

Recommendations for respirator selection: NIOSH/OSHA *10 mg/m^3:* SA (any supplied-air respirator). *25 mg/m^3:* SA:CF (any supplied-air respirator operated in a continuous-flow mode). *50 mg/m^3:* SAT:CF (any supplied-air respirator that has a tight-fitting facepiece and is operated in a continuous-flow mode); or SCBAF (any self-contained breathing apparatus with a full facepiece); or SAF (any supplied-air respirator with a full facepiece). *100 mg/m^3:* SA:PD,PP (any supplied-air respirator operated in a pressure-demand or other positive-pressure mode). *EMERGENCY OR PLANNED ENTRY INTO UNKNOWN CONCENTRATIONS OR IDLH CONDITIONS:* SCBAF:PD,PP (any self-contained breathing apparatus that has a full facepiece and is operated in a pressure-demand or other positive-pressure mode); or SAF:PD,PP:ASCBA (any supplied-air respirator that has a full facepiece and is operated in a pressure-demand or other positive-pressure mode in combination with an auxiliary self-contained breathing apparatus operated in a pressure-demand or other positive pressure mode). *ESCAPE:* GMFOVHiE [any air-purifying, full-facepiece respirator (gas mask) with a chin-style, front- or back-mounted organic vapor canister having a high-efficiency particulate filter]; or SCBAE (any appropriate escape-type, self-contained breathing apparatus).

Exposure limits (TWA unless otherwise noted): ACGIH TLV 0.1 ppm (0.9 mg/m^3); NIOSH/OSHA 1 mg/m^3; skin contact contributes significantly in overall exposure.

Toxicity by ingestion: Grade 3; LD$_{50}$ = 50 to 500 mg/kg.

Long-term health effects: Teratogenic effects. Workers exposed to low levels of pesticide suffered a decrease in serum and red cell cholinesterase. These workers had more health complaints (frequent headaches, dizziness, sore throat, nausea, etc.) than nonexposed workers. IARC rating 3; inadequate animal evidence.

Liquid or solid irritant characteristics: Skin irritant from prolonged contact. If spilled on clothing and allowed to remain, may cause smarting and reddening of skin.

IDLH value: 100 mg/m^3.

FIRE DATA
Fire extinguishing agents not to be used: Water streams applied to adjacent fires will spread contamination over wide area.

CHEMICAL REACTIVITY
Binary reactants: Strong acids; strong alkalis. Corrosive to iron and mild steel.

ENVIRONMENTAL DATA
Food chain concentration potential: Prolonged exposure to organophosphorus pesticides at concentrations as low as 0.01 ppb are toxic to marine animals due to bioconcentration.

Water pollution: DOT Appendix B, §172.101- severe marine pollutant. Harmful to aquatic life in very low concentrations. May be dangerous if it enters nearby water intakes; notify operators. Notify local health and wildlife officials. **Response to discharge:** Issue warning–poison, water contaminant. Restrict access. Should be removed. Chemical and physical treatment.

SHIPPING INFORMATION
Stability during transport: Stable.

PHYSICAL AND CHEMICAL PROPERTIES
Physical state @ 59°F/15°C and 1 atm: Liquid.
Molecular weight: 220.98
Boiling point @ 1 atm: 284°F/140°C/413.2°K (decomposes).
Specific gravity (water = 1): 1.415 @ 77°F/25°C.
Vapor pressure: 0.01 mm.

DICYCLOPENTADIENE   REC. D:2675

SYNONYMS: BICYCLOPENTADIENE; BISCYCLOPENTADIENE; 1,3-CYCLOPENTADIENE, DIMER; DCPD; 1,3-DICYCLOPENTADIENE DIMER; DICICLOPENTADIENO (Spanish); 4,7-METHANO-1H-INDENE; 3*A*,4,7,7A-TETRAHYDRO-4,7-METHANOINDENE

IDENTIFICATION
CAS Number: 77-73-6
Formula: C$_{10}$H$_{12}$
DOT ID Number: UN 2048; DOT Guide Number: 129
Proper Shipping Name: Dicyclopentadiene

DESCRIPTION: Colorless liquid or crystalline solid. Sweet, camphor-like odor. Liquid floats on the surface of water; solid sinks; insoluble. Melts at 90°F/32°C.

Highly flammable • Containers may BLEVE when exposed to fire • Vapors may form explosive mixture with air • Vapors are heavier than air and will collect and stay in low areas • Vapors may travel long distances to ignition sources and flashback • Vapors in confined areas (e.g., tanks, sewers, buildings) may explode when exposed to fire • Irritating to the skin, eyes, and respiratory tract • Toxic products of combustion may include carbon monoxide.

Hazard Classification (based on NFPA-704 Rating System)
Health Hazards (Blue): 1; Flammability (Red): 3; Reactivity (Yellow): 1

EMERGENCY RESPONSE: See Appendix A (129)
Evacuation:
Public safety: Isolate spill area for at least 50 to 100 meters (160 to 330 feet) in all directions.
Spill: Large spill–Consider initial downwind evacuation for at least 300 meters (1000 feet).
Fire: Isolate for 800 meters (½ mile) in all directions, especially if tank, rail car, or tank truck is involved in fire.

EXPOSURE
Short-term effects: *SEEK MEDICAL ATTENTION.* **Liquid or solid:** Irritating to skin and eyes. Remove contaminated clothing and shoes. Flush affected areas with plenty of water. *IF IN EYES*, hold eyelids open and flush with plenty of water.

HEALTH HAZARDS
Personal protective equipment (PPE): B-Level PPE. OSHA Table Z-1-A air contaminant.
Exposure limits (TWA unless otherwise noted): ACGIH TLV 5 ppm (27 mg/m^3); NIOSH REL 5 ppm (30 mg/m^3).
Toxicity by ingestion: Grade 2; oral rat LD$_{50}$ = 0.82 g/kg.
Long-term health effects: Possible nervous system damage.

Vapor (gas) irritant characteristics: Respiratory tract irritant. Vapors cause smarting of the eyes and respiratory system if present in high concentrations. The effect is temporary.
Liquid or solid irritant characteristics: Skin irritant. If spilled on clothing and allowed to remain, may cause smarting and reddening of the skin.
Odor threshold: 0.011 ppm.

FIRE DATA
Flash point: 90°F/32°C (oc).
Flammable limits in air: LEL: 0.8%; UEL: 6.3%.
Autoignition temperature: 941°F/506°C/779°K.
Electrical hazard: Class I, Group C. Finely dispersed particles may be electrostaticly charged pouring and similar movement.

CHEMICAL REACTIVITY
Binary reactants: Reacts violently with oxidizers. Incompatible with nitric and organic acids.
Polymerization: May occur in presence of acids. Depolymerizes at boiling point and forms two molecules of cyclopentadiene. Must be inhibited and maintained under an inert atmosphere to prevent polymerization.
Reactivity group: 30
Compatibility class: Olefins

ENVIRONMENTAL DATA
Food chain concentration potential: Log P_{ow} = 2.9. Values > 3.0 are likely to bioconcentrate in aquatic organisms and other living tissue, especially in fats.
Water pollution: Effect of low concentrations on aquatic life is unknown. Fouling to shoreline. May be dangerous if it enters nearby water intakes; notify operators. Notify local health and wildlife officials. **Response to discharge:** Mechanical containment. Should be removed. Chemical and physical treatment.

SHIPPING INFORMATION
Grades of purity: 97%; **Storage temperature:** Ambient; **Inert atmosphere:** None; **Venting:** Open (flame arrester); **Stability during transport:** Stable.

NAS HAZARD CLASSIFICATION FOR BULK WATER TRANSPORTATION
FIRE: 3
HEALTH: Vapor irritant: 1; Liquid or solid irritant: 1; Poisons: 2
WATER POLLUTION: Human toxicity: 1; Aquatic toxicity: 1; Aesthetic effect: 2
REACTIVITY: Other chemicals: 1; Water: 0; Self-reaction: 1

PHYSICAL AND CHEMICAL PROPERTIES
Physical state @ 59°F/15°C and 1 atm: Liquid.
Molecular weight: 132.31
Boiling point @ 1 atm: 342°F/172°C/445°K.
Melting/Freezing point: 90°F/32°C/305°K.
Specific gravity (water = 1): 0.978 @ 68°F/20°C (liquid).
Heat of combustion: −18,800 Btu/lb = −10,400 cal/g = −437 x 10^5 J/kg.
Vapor pressure: (Reid) 0.16 psia; 1.4 mm.

DIELDRIN   REC. D:2700

SYNONYMS: ALVIT; COMPOUND 497; DIELDRINA (Spanish); DIELDREX; DIELDRINE (French); DIELDRITE; 2,7:3,6-DIMETHANONAPTH 2,3-*B* OXIRENE, 3,4,5,6,9,9-HEXACHLORO-1*A*,2,2*A*,3,6,6*A*,7,7*A*-OCTAHYDRO-(1*A*α,2 β-2*A*α, 3 β, 6β, 6*A*α, 7β, 7*A*α)-; HEOD; HEOD-*endo,exo*-1,2,3,4,10,10-; HEXACHLOROEPOXYOCTAHYDRO-*endo,exo*-DIMETHANONAPHTHALENE; ILLOXOL; KILLGERM DETHLAC INSECTICIDAL LAQUER; OCTALOX; OXRALOX; PANORAM; PANORAM D-31; QUINTOX; 1,2,3,4,10,10-HEXACHLORO-6,7-EPOXY-1,4,4*A*,5,6,7,8,8*a*-OCTAHYDRO-1,4-*endo-exo*-5,8-DI-METHANONAPHTHALENE; 3,4,5,6,9,9-HEXACHLORO-1*A*,2,2*A*,3,6,6*A*,7,7*a*-OCTAHYDRO-2,7:3,6-DIMETHANO; HEXACHLORO-6,7-EXPOXY-1,4,4*A*,5,6,7,8,8*A*-OCTAHYDRO-1,4:5,8- DIMETHANONAPHTHALENE; RCRA No. U037

IDENTIFICATION
CAS Number: 60-57-1
Formula: $C_{12}H_8Cl_6O$
DOT ID Number: 2761; DOT Guide Number: 151
Proper Shipping Name: Dieldrin
Reportable Quantity (RQ): **(CERCLA)** 1 lb/0.454 kg

DESCRIPTION: Tan to light brown flakes. Mild chemical (insecticide) odor. Sinks in water; insoluble.

Poison! (organochlorine) • Breathing the vapor, skin or eye contact, or swallowing the material can kill you • Firefighting gear (including SCBA) does not provide adequate protection. If exposure occurs, remove and isolate gear immediately and thoroughly decontaminate personnel • Containers may BLEVE when exposed to fire • Toxic products of combustion may include carbon monoxide and hydrogen chloride fumes • Do not put yourself in danger by entering a contaminated area to rescue a victim.

Hazard Classification (based on NFPA-704 Rating System)
Health Hazards (Blue): 3; Flammability (Red): 0; Reactivity (Yellow): 0

EMERGENCY RESPONSE: See Appendix A (151)
Evacuation:
Public safety: Isolate the area of spill or leak for at least 25 to 50 meters (80 to 160 feet) in all directions.
Spill: Increase, in the downwind direction, as necessary, the distance shown under "Public Safety."
Fire: If tank, rail car, or tank truck is involved in fire, isolate for at least 800 meters (½ mile) in all directions; also, consider initial evacuation for 800 meters (½ mile) in all directions.

EXPOSURE
Short-term effects: *SEEK MEDICAL ATTENTION*. **Dust:** *POISONOUS IF INHALED OR IF SKIN IS EXPOSED. IF INHALED*, will, will cause headache, dizziness, or loss of consciousness. *IF IN EYES*, hold eyelids open and flush. with plenty of water. *IF BREATHING HAS STOPPED*, give artificial respiration; *avoid mouth-to-mouth resuscitation; use bag/mask apparatus*. If breathing is difficult, administer oxygen. **Solid:** POISONOUS IF SWALLOWED OR IF SKIN IS EXPOSED. *IF SWALLOWED*, will cause headache, nausea, dizziness, vomiting, or loss of consciousness. Remove contaminated clothing and shoes. Flush affected areas with plenty of water. *IF IN EYES*, hold eyelids open and flush with plenty of water. *IF SWALLOWED* and victim is *CONSCIOUS AND ABLE TO SWALLOW*, have victim drink 4 to 8 ounces of water. **Do NOT induce vomiting.** Administer a slurry of activated charcoal. *IF SWALLOWED* and victim is *UNCONSCIOUS OR HAVING CONVULSIONS*, do nothing except keep victim warm.

HEALTH HAZARDS
Personal protective equipment (PPE): A-Level PPE. OSHA Table Z-1-A air contaminant.
Recommendations for respirator selection: NIOSH
NIOSH *At any concentrations above the NIOSH REL, or where there is no REL, at any detectable concentration:* SCBAF:PD,PP (any self-contained breathing apparatus that has a full facepiece and is operated in a pressure-demand or other positive-pressure mode); or SAF:PD,PP:ASCBA (any supplied-air respirator that has a full facepiece and is operated in a pressure-demand or other positive-pressure mode in combination with an auxiliary, self-contained breathing apparatus operated in a pressure-demand or other positive pressure mode). *ESCAPE:* GMFOVHiE [any air-purifying, full-facepiece respirator (gas mask) with a chin-style, front- or back-mounted organic vapor canister having a high-efficiency particulate filter]; or SCBAE (any appropriate escape-type, self-contained breathing apparatus). *At any concentrations above the NIOSH REL, or where there is no REL, at any detectable concentration:* SCBAF:PD,PP (any self-contained breathing apparatus that has a full facepiece and is operated in a pressure-demand or other positive-pressure mode); or SAF:PD,PP:ASCBA (any supplied-air respirator that has a full facepiece and is operated in a pressure-demand or other positive-pressure mode in combination with an auxiliary, self-contained breathing apparatus operated in a pressure-demand or other positive pressure mode). *ESCAPE:* GMFOVHiE [any air-purifying, full-facepiece respirator (gas mask) with a chin-style, front- or back-mounted organic vapor canister having a high-efficiency particulate filter]; or SCBAE (any appropriate escape-type, self-contained breathing apparatus).
Exposure limits (TWA unless otherwise noted): ACGIH TLV 0.25 mg/m^3; OSHA/NIOSH 0.25 mg/m^3; potential human carcinogen; reduce exposure to lowest feasible level; skin contact contributes significantly in overall exposure; possible carcinogen.
Toxicity by ingestion: Grade 4; oral LD$_{50}$ = 46 mg/kg (rat), 65 mg/kg (dog).
Long-term health effects: Cholinergic. Exposure to low levels of pesticide may suffer a decrease in serum and red cell cholinesterase. Exposed persons may experience health complaints (frequent headaches, dizziness, sore throat, nausea, etc.).
Liquid or solid irritant characteristics: Eye irritant. If spilled on clothing and allowed to remain, may cause smarting and reddening of the skin.
Odor threshold: 0.041 ppm.
IDLH value: Potential human carcinogen; 50 mg/m^3.

FIRE DATA
Behavior in fire: Water streams applied to adjacent fires will spread contamination of pesticide over wide area.

CHEMICAL REACTIVITY
Binary reactants: Reacts with strong oxidizers, strong acids, phenols; active metal such as sodium. Slightly corrosive to metals.

ENVIRONMENTAL DATA
Food chain concentration potential: High
Water pollution: DOT Appendix B, §172.101–marine pollutant. Harmful to aquatic life in very low concentrations. May be dangerous if it enters nearby water intakes; notify operators. Notify local health and wildlife officials. **Response to discharge:** Issue warning–water contaminant. Restrict access. Should be removed. Chemical and physical treatment. Banned by EPA (October 1974) because of alleged "imminent hazard to human health" as a potential carcinogen in man.

SHIPPING INFORMATION
Grades of purity: Technical, 85+% HEOD; 18% emulsifiable concentrates in petroleum hydrocarbons, which are combustible; **Storage temperature:** Ambient; **Inert atmosphere:** None; **Venting:** Open; (flame arrester for liquid form); **Stability during transport:** Stable.

PHYSICAL AND CHEMICAL PROPERTIES
Physical state @ 59°F/15°C and 1 atm: Solid.
Molecular weight: 380.93
Boiling point @ 1 atm: Decomposes.
Melting/Freezing point: 349°F/176°C/449°K.
Specific gravity (water = 1): 1.75 @ 68°F/20°C (solid).
Vapor pressure: 0.0000008 mm @ 77°F/25°C.

DIETHANOLAMINE   REC. D:2750

SYNONYMS: BIS(2-HYDROXYETHYL)AMINE; BUTANE, 1,2,3,4-DIEPOXY; BUTADIENE DIOXIDE; DEA; DIAETHANOLAMIN (German); *N,N*-DIETHANOLAMINE; DIETHYLAMINE, 2,2'-DIHYDROXY-; DIETANOLAMINA (Spanish); 2,2'-DIHYDROXYDIETHYLAMINE; DI(2-HYDROXYETHYL)AMINE; DIOLAMINE; ETHANOL, 2,2'-IMINOBIS-; 2,2'-IMINOBISETHANOL; 2,2'-IMINODIETHANOL; 2,2'-DIHYDROXYETHYLAMINE; DI(2-HYDROXYETHYL) AMINE

IDENTIFICATION
CAS Number: 111-42-2
Formula: $C_4H_{11}NO_2$; $(HOCH_2CH_2)_2NH$
DOT ID Number: UN 3077 (solid); UN 3082 (liquid); DOT Guide Number: 171
Proper Shipping Name: Environmentally hazardous substances, solid, n.o.s; Environmentally hazardous substances, liquid, n.o.s.
Reportable Quantity (RQ): **(CERCLA)** 1 lb/0.454 kg

DESCRIPTION: Colorless thick, oily liquid or white crystalline solid. Slight fishy or ammonia odor. Sinks and mixes with water.

Combustible • Vapor or dust cloud may explode if ignited in an enclosed area • Containers may BLEVE when exposed to fire • Irritating to the skin, eyes, and respiratory tract • Toxic products of combustion may include nitrogen oxides.

Hazard Classification (based on NFPA-704 Rating System)
Health Hazards (Blue): 1; Flammability (Red): 1; Reactivity (Yellow): 0

EMERGENCY RESPONSE: See Appendix A (171)
Evacuation:
Public safety: Isolate the area of spill or leak for at least 10 to 25 meters (30 to 80 feet) in all directions.
Spill: Increase, in the downwind direction, as necessary, the distance shown under "Public Safety."
Fire: If any large container is involved in fire, isolate for at least 800 meters (½ mile) in all directions; also, consider initial evacuation for 800 meters (½ mile) in all directions.

EXPOSURE
Short-term effects: *SEEK MEDICAL ATTENTION.* **Liquid:** *HARMFUL IF SWALLOWED OR ABSORBED THROUGH THE SKIN.* Irritating to skin and eyes. Remove contaminated clothing and shoes. Flush affected areas with plenty of water. *IF IN EYES,* hold eyelids open and flush with plenty of water. *IF SWALLOWED*

and victim is *CONSCIOUS AND ABLE TO SWALLOW*, have victim drink 4 to 8 ounces of water and have victim induce vomiting. *IF SWALLOWED* and victim is *UNCONSCIOUS OR HAVING CONVULSIONS,* do nothing except keep victim warm. The fatal dose in humans is estimated to be about 20 g.

HEALTH HAZARDS
Personal protective equipment (PPE): B-Level PPE. OSHA Table Z-1-A air contaminant. Organic vapor respirator. Chemical protective material(s) reported to have good to excellent resistance: butyl rubber, neoprene, nitrile, Teflon®, Viton®. Also, neoprene+ natural rubber, and PVC offers limited protection
Recommendations for respirator selection: *EMERGENCY OR PLANNED ENTRY INTO UNKNOWN CONCENTRATIONS OR IDLH CONDITIONS:* SCBAF:PD,PP (any self-contained breathing apparatus that has a full facepiece and is operated in a pressure-demand or other positive-pressure mode); or SAF:PD,PP:ASCBA (any supplied-air respirator that has a full facepiece and is operated in a pressure-demand or other positive-pressure mode in combination with an auxiliary self-contained breathing apparatus operated in a pressure-demand or other positive pressure mode). *ESCAPE:* GMFS [any air-purifying, full-facepiece respirator (gas mask) with a chin-style, front- or back-mounted canister providing protection against the compound of concern]; or SCBAE (any appropriate escape-type, self-contained breathing apparatus).
Exposure limits (TWA unless otherwise noted): ACGIH TLV 0.46 ppm (2 mg/m^3); NIOSH REL 3 ppm (15 mg/m^3).
Toxicity by ingestion: Grade 2; LD_{50} = 0.5 to 5 g/kg (rat).
Vapor (gas) irritant characteristics: Eye and respiratory tract irritant. Vapors cause moderate irritation such that personnel will find high concentrations unpleasant. The effect may be temporary. At room temperature the vapor pressure is so low that inhalation may be unlikely.
Liquid or solid irritant characteristics: Eye and skin irritant. Causes smarting of the skin and first-degree burns on short exposure and may cause secondary burns on long exposure.

FIRE DATA
Flash point: 342°F/172°C (oc); 297°F/147°C (cc).
Flammable limits in air: LEL: 1.6%; UEL: 9.8%.
Fire extinguishing agents not to be used: Water or foam may cause frothing.
Autoignition temperature: 1224°F/662°C/935°K.
Electrical hazard: Class I. Group D.
Burning rate: 0.74 mm/min

CHEMICAL REACTIVITY
Binary reactants: Oxidizers, acids, acid anhydrides, halides. Reacts with CO_2 in air. Hygroscopic. Corrosive to copper and its alloys, zinc, and galvanized iron
Neutralizing agents for acids and caustics: Flush with water.
Reactivity group: 8
Compatibility class: Alkanolamines

ENVIRONMENTAL DATA
Food chain concentration potential: Log P_{ow} = –1.4. Unlikely to accumulate.
Water pollution: Dangerous to aquatic life in high concentrations. May be dangerous if it enters nearby water intakes; notify operators. Notify local health and wildlife officials. **Response to discharge:** Disperse and flush.

SHIPPING INFORMATION
Grades of purity: 85–99.5%; **Storage temperature:** Ambient; **Inert atmosphere:** None; **Venting:** Open; **Stability during transport:** Stable.

NAS HAZARD CLASSIFICATION FOR BULK WATER TRANSPORTATION
FIRE: 1
HEALTH: Vapor irritant: 2; Liquid or solid irritant: 2; Poisons: 2
WATER POLLUTION: Human toxicity: 1; Aquatic toxicity: 1; Aesthetic effect: 2
REACTIVITY: Other chemicals: 3; Water: 0; Self-reaction: 0

PHYSICAL AND CHEMICAL PROPERTIES
Physical state @ 59°F/15°C and 1 atm: Solid.
Molecular weight: 105.2
Boiling point @ 1 atm: 515.1°F/268.4°C/541.6°K.
Melting/Freezing point: 82°F/28°C/301°K.
Critical temperature: 828°F/442°C/715°K.
Critical pressure: 470 psia = 32 atm = 3.2 MN/m^2.
Specific gravity (water = 1): 1.10 at 28°C (liquid).
Ratio of specific heats of vapor (gas): 1.053
Latent heat of vaporization: 266 Btu/lb = 148 cal/g = 6.20 x 10^5 J/kg.
Heat of combustion: –10,790 Btu/lb = –6000 cal/g = 0.251 x 10^5 J/kg.
Heat of solution: (estimate) –13 Btu/lb = –7 cal/g = –0.3 x 10^5 J/kg.
Vapor pressure: (Reid) 0.97 psia; less than 0.01 mm.

DIETHYLAMINE   REC. D:2800

SYNONYMS: 2-AMINOPENTANE; DEN; DIETHAMINE; *N,N*-DIETHYLAMINE; DIETILAMINA (Spanish); EEC No. 612-003-00-X; *N*-ETHYLETHANAMINE

IDENTIFICATION
CAS Number: 109-89-7
Formula: $C_4H_{11}N$; $(CH_3CH_2)_2NH$; $(C_2NO_5)_2C_6H_3NH_2$
DOT ID Number: UN 1154; DOT Guide Number: 132
Proper Shipping Name: Diethylamine
Reportable Quantity (RQ): **(CERCLA)** 1000 lb/454 kg

DESCRIPTION: Colorless watery liquid. Fishy, ammonia-like odor. Floats on water surface; soluble; flammable, large amounts of vapor is produced.

Highly flammable • Corrosive to skin, eyes, and respiratory tract; skin and eye contact causes severe burns and blindness; inhalation symptoms may be delayed • Firefighting gear (including SCBA) does not provide adequate protection. If exposure occurs, remove and isolate gear immediately and thoroughly decontaminate personnel • Containers may BLEVE when exposed to fire •Vapors may form explosive mixture with air • Vapors are heavier than air and will collect and stay in low areas • Vapors may travel long distances to ignition sources and flashback • Vapors in confined areas (e.g., tanks, sewers, buildings) may explode when exposed to fire • Toxic products of combustion may include nitrogen oxides • Do not put yourself in danger by entering a contaminated area to rescue a victim.

Hazard Classification (based on NFPA-704 Rating System)
Health Hazards (Blue): 3; Flammability (Red): 3; Reactivity (Yellow): 0

EMERGENCY RESPONSE: See Appendix A (132)
Evacuation:
Public safety: Isolate spill area for at least 50 to 100 meters (160 to 330 feet) in all directions.
Spill: Increase, as necessary, the isolation distance shown above, in "Public safety."
Fire: Isolate for 800 meters (½ mile) in all directions, especially if tank, rail car, or tank truck is involved in fire.

EXPOSURE
Short-term effects: *SEEK MEDICAL ATTENTION.* **Vapor:** Irritating to eyes, nose, and throat. Harmful if inhaled. Move to fresh air. *IF BREATHING HAS STOPPED,* give artificial respiration. If breathing is difficult, administer oxygen. May cause lung edema; physical exertion will aggravate this condition. **Liquid:** Organic vapor respirator. Will burn eyes. Harmful if swallowed. Remove contaminated clothing and shoes. Flush affected areas with plenty of water. *IF IN EYES,* hold eyelids open and flush with plenty of water. *IF SWALLOWED* and victim is *CONSCIOUS AND ABLE TO SWALLOW,* have victim drink 4 to 8 ounces of water.
Note to physician or authorized medical personnel: Medical observation is recommended for 24 to 48 hours after breathing overexposure, as pulmonary edema may be delayed. As first aid for pulmonary edema, consider administering a corticosteroid spray. Cigarette smoking may exacerbate pulmonary injury and should be discouraged for at least 72 hours following exposure.

HEALTH HAZARDS
Personal protective equipment (PPE): B-Level PPE. OSHA Table Z-1-A air contaminant. Full face organic vapor respirator. Chemical protective material(s) reported to have good to excellent resistance: Teflon®, Silvershield®. Also, Styrene-butadiene nitrile offers limited protection.
Recommendations for respirator selection: NIOSH/OSHA
200 ppm: SA:CF* (any supplied-air respirator operated in a continuous-flow mode); or PAPRS* (any powered, air-purifying respirator with cartridge(s) providing protection against the compound of concern; or CCRFS [any chemical cartridge respirator with a full facepiece and cartridge(s) providing protection against the compound of concern]; or GMFS [any air-purifying, full-facepiece respirator (gas mask) with a chin-style, front- or back-mounted canister providing protection against the compound of concern]; or SCBAF (any self-contained breathing apparatus with a full facepiece); or SAF (any supplied-air respirator with a full facepiece). *EMERGENCY OR PLANNED ENTRY INTO UNKNOWN CONCENTRATIONS OR IDLH CONDITIONS:* SCBAF:PD,PP (any self-contained breathing apparatus that has a full facepiece and is operated in a pressure-demand or other positive-pressure mode); or SAF:PD,PP:ASCBA (any supplied-air respirator that has a full facepiece and is operated in a pressure-demand or other positive-pressure mode in combination with an auxiliary self-contained breathing apparatus operated in a pressure-demand or other positive pressure mode). *ESCAPE:* GMFS [any air-purifying, full-facepiece respirator (gas mask) with a chin-style, front- or back-mounted canister providing protection against the compound of concern]; or SCBAE (any appropriate escape-type, self-contained breathing apparatus).
Note: Substance causes eye irritation or damage; eye protection needed.
Exposure limits (TWA unless otherwise noted): ACGIH TLV 5 ppm (15 mg/m^3) not classified as a human carcinogen; OSHA PEL 25 ppm (75 mg/m^3; NIOSH REL 10 ppm (30 mg/m^3).
Short-term exposure limits (15-minute TWA): ACGIH STEL 15 ppm (45 mg/m^3); NIOSH STEL 25 ppm (75 mg/m^3).

Toxicity by ingestion: Grade 2; LD_{50} = 0.5 to 5 g/kg (rat).
Vapor (gas) irritant characteristics: Eye and respiratory tract irritant. Vapor is irritating such that personnel will not usually tolerate moderate or high vapor concentrations.
Liquid or solid irritant characteristics: Severe eye and skin irritant. If spilled on clothing and allowed to remain, may cause smarting and reddening of the skin.
Odor threshold: 0.02–14 ppm.
IDLH value: 200 ppm.

FIRE DATA
Flash point: 5°F/oc); –9°F/–23°C (cc).
Flammable limits in air: LEL: 1.8%; UEL: 10.1%.
Autoignition temperature: 594°F/312°C/585°K.
Electrical hazard: Class I, Group C. Flow or agitation may generate electrostatic charges due to low conductivity.
Burning rate: 6.7 mm/min

CHEMICAL REACTIVITY
Binary reactants: A strong organic base. Incompatible with strong oxidizers, acids, mercury, cellulose nitrate. Corrosive to aluminum, copper, copper alloys, lead, tin, and zinc.
Neutralizing agents for acids and caustics: Flush with water.
Reactivity group: 7
Compatibility class: Aliphatic amines

ENVIRONMENTAL DATA
Food chain concentration potential: Log P_{ow} = 0.48. Unlikely to accumulate.
Water pollution: Harmful to aquatic life in very low concentrations. May be dangerous if it enters nearby water intakes; notify operators. Notify local health and wildlife officials. **Response to discharge:** Issue warning–high flammability. Evacuate area. Disperse and flush.

SHIPPING INFORMATION
Grades of purity: Technical: 99%; **Stability during transport:** Stable.

NAS HAZARD CLASSIFICATION FOR BULK WATER TRANSPORTATION
FIRE: 3
HEALTH: Vapor irritant: 3; Liquid or solid irritant: 1; Poisons: 2
WATER POLLUTION: Human toxicity: 2; Aquatic toxicity: 3; Aesthetic effect: 1
REACTIVITY: Other chemicals: 3; Water: 0; Self-reaction: 0

PHYSICAL AND CHEMICAL PROPERTIES
Physical state @ 59°F/15°C and 1 atm: Liquid.
Molecular weight: 73.14
Boiling point @ 1 atm: 132°F/55.5°C/328.7°K.
Melting/Freezing point: –57.6°F/–49.8°C/223.4°K.
Critical temperature: 434.3°F/223.5°C/496.7°K.
Critical pressure: 538 psia = 36.6 atm = 3.71 MN/m^2.
Specific gravity (water = 1): 0.708 @ 68°F/20°C (liquid).
Liquid surface tension: 20.05 dynes/cm = 0.02005 N/m @ 68°F/20°C.
Relative vapor density (air = 1): 2.5
Ratio of specific heats of vapor (gas): 1.079
Latent heat of vaporization: 170 Btu/lb = 93 cal/g = 3.9 x 10^5 J/kg.
Heat of combustion: –17,990 Btu/lb = –9994 cal/g = –418.4 x 10^5 J/kg.
Heat of solution: –202 Btu/lb = –112 cal/g = –4.69 x 10^5 J/kg.
Vapor pressure: (Reid) 0.7 psia; 192 mm; 20.4 mm.

2,6-DIETHYL ANILINE   REC. D:2850

SYNONYMS: ANILINE, 2,6-DIETHYL; BENZENEAMINE, 2,6-DIETHYL-; 2,6-DIETHYLBENZENAMINE; DIETILANILINA (Spanish)

IDENTIFICATION
CAS Number: 579-66-8; health information may apply to 91-66-7 (*N,N*-diethylaniline)
Formula: $C_{10}H_{15}N$
DOT ID Number: UN 2432; DOT Guide Number: 153
Proper Shipping Name: *N,N*-Diethylaniline

DESCRIPTION: Colorless to yellow liquid. Floats on water; sparingly soluble.

Flammable • Poison! Inhalation or absorbed through the skin; forms methemoglobin; may interfere with the body's ability to use oxygen •. Irritating to the skin, eyes, and respiratory tract • Containers may BLEVE when exposed to fire •Vapors may form explosive mixture with air • Vapors are heavier than air and will collect and stay in low areas • Vapors may travel long distances to ignition sources and flashback • Vapors in confined areas (e.g., tanks, sewers, buildings) may explode when exposed to fire • Toxic products of combustion may include nitrogen oxides• Toxic products of combustion may include nitrogen oxides.

Hazard Classification (based on NFPA-704 Rating System)
Health Hazards (Blue): 2; Flammability (Red): 2; Reactivity (Yellow): 0

EMERGENCY RESPONSE: See Appendix A (153)
Evacuation:
Public safety: Isolate the area of spill or leak for at least 25 to 50 meters (80 to 160 feet) in all directions.
Spill: Increase, in the downwind direction, as necessary, the distance shown under "Public Safety."
Fire: If tank, rail car, or tank truck is involved in fire, isolate for at least 800 meters (½ mile) in all directions; also, consider initial evacuation for 800 meters (½ mile) in all directions.

EXPOSURE
Short-term effects: *SEEK MEDICAL ATTENTION*. **Vapor/mist:** *POISONOUS IF INHALED*. Irritating to eyes, nose, and throat. Move the victim to fresh air. *IF IN EYES*, hold eyes open and flush with plenty of water. *IF BREATHING HAS STOPPED,* give artificial respiration; *avoid mouth-to-mouth resuscitation; use bag/mask apparatus*. If breathing is difficult, administer oxygen.
Liquid: Irritating to skin and eyes. *IF SWALLOWED*, will cause nausea, dizziness, and headaches. Remove contaminated clothing and shoes. Flush affected areas with plenty of water. *IF IN EYES*, hold eyelids open and flush with plenty of water.

Personal protective equipment (PPE): A-Level PPE.
Recommendations for respirator selection: *EMERGENCY OR PLANNED ENTRY INTO UNKNOWN CONCENTRATIONS OR IDLH CONDITIONS:* SCBAF:PD,PP (any self-contained breathing apparatus that has a full facepiece and is operated in a pressure-demand or other positive-pressure mode); or SAF:PD,PP:ASCBA (any supplied-air respirator that has a full facepiece and is operated in a pressure-demand or other positive-pressure mode in combination with an auxiliary self-contained breathing apparatus operated in a pressure-demand or other positive pressure mode). *ESCAPE:* GMFS [any air-purifying, full-facepiece respirator (gas mask) with a chin-style, front- or back-mounted canister providing protection against the compound of concern]; or SCBAE (any appropriate escape-type, self-contained breathing apparatus). Self-contained breathing apparatus, rubber boots, heavy rubber gloves and protective clothing.
Toxicity by ingestion: Grade 2: LD_{50} = 1.8 g/kg (rat).
Long-term health effects: May be mutagenic. May form methemoglobin.
Vapor (gas) irritant characteristics: Eye and respiratory tract irritant and poison. Vapors are irritating such that personnel will not usually tolerate moderate or high concentrations.
Liquid or solid irritant characteristics: Severe eye and skin irritant. Causes second- and third-degree burns on short contact and is very injurious to the eyes.

FIRE DATA
Flash point: 254°F/123°C (cc) (2,6-isomer); 185°F/85°C (*N,N*-isomer).

CHEMICAL REACTIVITY
Binary reactants: Oxidizers, acids, isocyanates.
Reactivity group: 9
Compatibility class: Aromatic amines

ENVIRONMENTAL DATA
Water pollution: Effects of low concentrations on aquatic life is unknown. Fouling to shoreline. May be dangerous if it enters nearby water intakes; notify operators. Notify local health and wildlife officials. **Response to discharge:** Restrict access. Evacuate areas. Mechanical containment. Should be removed. Chemical and physical treatment.

SHIPPING INFORMATION
Grades of purity: 99.5%; **Storage temperature:** Ambient; **Stability during transport:** Stable.

PHYSICAL AND CHEMICAL PROPERTIES
Physical state @ 59°F/15°C and 1 atm: Liquid.
Molecular weight: 149.24 (2,6-isomer); 149.26 (*N,N*-isomer).
Boiling point @ 1 atm: 237°F/114°C/387°K (2,6-isomer); 420°F/215°C/488°K (*N,N*-isomer).
Melting/Freezing point: 37.4–39.2°F/3–4°C/275.2–277.2°K (at 10mmHg = 0.0132 atm) (2,6-isomer); –36°F/–38°C/235°K (*N,N*-isomer).
Specific gravity (water = 1): 0.906
Relative vapor density (air = 1): 5.15

DIETHYLBENZENE   REC. D:2900

SYNONYMS: DIETHYLBENZOL; DIETILBENCENO (Spanish)

IDENTIFICATION
CAS Number: 25340-17-4 (mixed isomers); 1300-82-9 (*p*-isomer); 141-93-5 (*m*-isomer); 135-01-3 (*o*-isomer)
Formula: $C_{10}H_{14}$; $C_6H_4(C_2NO_5)_2$
DOT ID Number: UN 2049; DOT Guide Number: 130
Proper Shipping Name: Diethylbenzene

DESCRIPTION: Colorless liquid. Sweet odor, like benzene or toluene. Floats on the surface of water.

Hazard Classification (based on NFPA-704 Rating System)
Health Hazards (Blue): 1; Flammability (Red): 2; Reactivity (Yellow): 0

368 Diethyl carbonate

EMERGENCY RESPONSE: See Appendix A (130)
Evacuation:
Public safety: Isolate spill area for at least 50 to 100 meters (160 to 330 feet) in all directions.
Spill: Large spill–Consider initial downwind evacuation for at least 300 meters (1000 feet).
Fire: Isolate for 800 meters (½ mile) in all directions, especially if tank, rail car, or tank truck is involved in fire.

EXPOSURE
Short-term effects: *SEEK MEDICAL ATTENTION.* **Liquid:** Irritating to skin and eyes. Remove contaminated clothing and shoes. Flush affected areas with plenty of water. *IF IN EYES*, hold eyelids open and flush with plenty of water.

HEALTH HAZARDS
Personal protective equipment (PPE): B-Level PPE. Chemical protective material(s) reported to offer minimal to poor protection: Viton®/neoprene.
Toxicity by ingestion: Grade 2; oral rat LD_{50} = 1.2 g/kg.
Vapor (gas) irritant characteristics: Eye and respiratory tract irritant. Vapors cause smarting of the eyes and respiratory system if present in high concentrations.
Liquid or solid irritant characteristics: Eye and skin irritant. If spilled on clothing and allowed to remain, may cause smarting and reddening of the skin.

FIRE DATA
Flash point: 135°F/57°C (cc) (*o*-isomer); 132–133°F/55–56°C (cc) (*m*- and *p*-isomers).
Behavior in fire: Containers may explode.
Autoignition temperature: 743°F/395°C/668°K (*o*-isomer).
Electrical hazard: Class I, Group D.

CHEMICAL REACTIVITY
Binary reactants: Reacts with strong oxidizers. On long immersion, rubber will swell, then soften.
Reactivity group: 32
Compatibility class: Aromatic hydrocarbons

ENVIRONMENTAL DATA
Food chain concentration potential: Negative; unlikely to accumulate.
Water pollution: Effect of low concentrations on aquatic life is unknown. Fouling to shoreline. May be dangerous if its enters water intakes. Notify local health and wildlife officials. **Response to discharge:** Mechanical containment. Should be removed. Chemical and physical treatment.

SHIPPING INFORMATION
Grades of purity: Technical (mixture of isomers); **Storage temperature:** Ambient; **Inert atmosphere:** None; **Venting:** Open (flame arrester); **Stability during transport:** Stable.

NAS HAZARD CLASSIFICATION FOR BULK WATER TRANSPORTATION
FIRE: 2
HEALTH: Vapor irritant: 1; Liquid or solid irritant: 1; Poisons: 1
WATER POLLUTION: Human toxicity: 1; Aquatic toxicity: 3; Aesthetic effect: 2
REACTIVITY: Other chemicals: 1; Water: 0; Self-reaction: 0

PHYSICAL AND CHEMICAL PROPERTIES
Physical state @ 59°F/15°C and 1 atm: Liquid.
Molecular weight: 134.21
Boiling point @ 1 atm: 356°F/180°C/453°K.
Melting/Freezing point: Less than 160°F/70°C/343°K.
Specific gravity (water = 1): 0.86 @ 68°F/20°C (liquid).
Liquid surface tension: 30 dynes/cm = 0.030 N/m @ 68°F/20°C.
Relative vapor density (air = 1): 4.6
Latent heat of vaporization: 140 Btu/lb = 77 cal/g = 3.2×10^5 J/kg.
Heat of combustion: –17,800 Btu/lb = –9890 cal/g = -414×10^5 J/kg.
Vapor pressure: (Reid) 0.05 psia; 1 mm.

DIETHYL CARBONATE   REC. D:2950

SYNONYMS: CARBONATO de DIETILO (Spanish); CARBONIC ACID, DIETHYL ESTER; DIETHYLCARBONAT (German); ETHYL CARBONATE; ETHOXYFORMIC ANHYDRIDE; EUFIN

IDENTIFICATION
CAS Number: 105-58-8
Formula: $C_5H_{10}O_3$; $(CH_3CH_2)_2CO_3$
DOT ID Number: UN 2366; DOT Guide Number: 127
Proper Shipping Name: Diethyl carbonate

DESCRIPTION: Colorless, watery liquid. Pleasant, sweet, ether-like odor. Floats on the surface of water; insoluble. Flammable, irritating vapor is produced.

Highly flammable • Containers may BLEVE when exposed to fire • Vapors may form explosive mixture with air • Vapors are heavier than air and will collect and stay in low areas • Vapors may travel long distances to ignition sources and flashback • Vapors in confined areas (e.g., tanks, sewers, buildings) may explode when exposed to fire • Irritating to the skin, eyes, and respiratory tract • Toxic products of combustion may include carbon monoxide.

Hazard Classification (based on NFPA-704 Rating System)
Health Hazards (Blue): 2; Flammability (Red): 3; Reactivity (Yellow): 1

EMERGENCY RESPONSE: See Appendix A (127)
Evacuation:
Public safety: Isolate spill area for at least 25 to 50 meters (80 to 160 feet) in all directions.
Spill: Large spill–Consider initial downwind evacuation for at least 300 meters (1000 feet).
Fire: Isolate for 800 meters (½ mile) in all directions, especially if tank, rail car, or tank truck is involved in fire.

EXPOSURE
Short-term effects: *SEEK MEDICAL ATTENTION.* VAPORS: Irritating to eyes, nose, and throat. *IF INHALED*, will, will cause headache, dizziness, nausea, or loss of consciousness. Move to fresh air. *IF BREATHING HAS STOPPED,* give artificial respiration. If breathing is difficult, administer oxygen. **Liquid:** Irritating to skin and eyes. Harmful if swallowed. Remove contaminated clothing and shoes. Flush affected areas with plenty of water. *IF IN EYES*, hold eyelids open and flush with plenty of water. *IF SWALLOWED* and victim is *CONSCIOUS AND ABLE TO SWALLOW*, have victim drink 4 to 8 ounces of water.

HEALTH HAZARDS
Personal protective equipment (PPE): B-Level PPE.

Vapor (gas) irritant characteristics: Vapors may cause smarting of eyes.
Liquid or solid irritant characteristics: Irritant. Prolonged contact may cause redness and pain.

FIRE DATA
Flash point: 115°F/46°C (oc); 77°F/25°C (cc).
Fire extinguishing agents not to be used: Water.
Electrical hazard: Class I, Group C.
Burning rate: 3.4 mm/min

CHEMICAL REACTIVITY
Reactivity with water: Slow; may not be hazardous
Binary reactants: Reacts with oxidizers.

ENVIRONMENTAL DATA
Food chain concentration potential: Negative; unlikely to accumulate.
Water pollution: Effect of low concentrations on aquatic life is unknown. Fouling to shoreline. May be dangerous if it enters nearby water intakes; notify operators. Notify local health and wildlife officials. **Response to discharge:** Disperse and flush.

SHIPPING INFORMATION
Stability during transport: Stable.

SHIPPING INFORMATION
Grades of purity: 99–100%.
PHYSICAL AND CHEMICAL PROPERTIES
Physical state @ 59°F/15°C and 1 atm: Liquid.
Molecular weight: 118.13
Boiling point @ 1 atm: 260.2°F/126.8°C/400.0°K.
Melting/Freezing point: –45°F/–43°C/230°K.
Specific gravity (water = 1): 0.975 @ 68°F/20°C (liquid).
Liquid surface tension: 26.3 dynes/cm = 0.0263 N/m @ 68°F/20°C.
Liquid water interfacial tension: 12.86 dynes/cm = 0.01286 N/m @ 68°F/20°C.
Relative vapor density (air = 1): 4.1
Ratio of specific heats of vapor (gas): (estimate) 1.110
Latent heat of vaporization: 130 Btu/lb = 73 cal/g = 3.1×10^5 J/kg.
Heat of combustion: –9760 Btu/lb = –5420 cal/g = -227×10^5 J/kg.

DIETHYLENE GLYCOL   REC. D:3000

SYNONYMS: BIS(2-HYDROXYETHYL) ETHER; BRECOLANE NDG; CARBITOL; DIETHYLENE GLYCOL MONOMETHYL ETHER; DEG; DICOL; DIETHYLENE GLYCOL ETHYL ETHER; DIETILENGLICOL (Spanish); DIGLYCOL; DIGGE; β,β-DIHYDROXYDIETHYL ETHER; 2-(2-ETHOXYETHOXY)ETHANOL; ETHYLENE DIGLYCOL; GLYCOL ETHER; GLYCOL ETHER DE; GLYCOL ETHYL ETHER; 3-OXYPENTANE-1,5-DIOL; 2,2'OXYBISETHANOL; 3-OXA-1, 5-PENTANEDIOL; 2,2'-OXYDIETHANOL; TL4N

IDENTIFICATION
CAS Number: 111-46-6
Formula: $C_4H_{10}O_3$; $(HOCH_2CH_2)_2O$
DOT ID Number: UN 3082; DOT GUIDE NUMBER: 171
Proper Shipping Name: Environmentally hazardous substances, liquid, n.o.s.

DESCRIPTION: Colorless, oily liquid. Practically odorless. Sinks and mixes with water.

Combustible • Containers may BLEVE when exposed to fire • Vapors are heavier than air and will collect and stay in low areas • Vapors in confined areas (e.g., tanks, sewers, buildings) may explode when exposed to fire • Irritating to the skin, eyes, and respiratory tract • Toxic products of combustion may include carbon monoxide.

Hazard Classification (based on NFPA-704 Rating System)
Health Hazards (Blue): 1; Flammability (Red): 1; Reactivity (Yellow): 0

EMERGENCY RESPONSE: See Appendix A (171)
Evacuation:
Public safety: Isolate the area of spill or leak for at least 10 to 25 meters (30 to 80 feet) in all directions.
Spill: Increase, in the downwind direction, as necessary, the distance shown under "Public Safety."
Fire: If any large container is involved in fire, isolate for at least 800 meters (½ mile) in all directions; also, consider initial evacuation for 800 meters (½ mile) in all directions.

HEALTH HAZARDS
Personal protective equipment (PPE): B-Level PPE. Organic vapor respirator. Chemical protective material(s) reported to have good to excellent resistance: butyl rubber, nitrile. Also, neoprene, Viton/neoprene, PVC, butyl rubber/neoprene offers limited protection
Exposure limits (TWA unless otherwise noted): 25 ppm (AIHA WEEL); 25 ppm (AIHAWEEL).
Toxicity by ingestion: Grade 0; LD_{50} = > 15 g/kg (rat).
Long-term health effects: Kidney and liver damage.
Vapor (gas) irritant characteristics: Eye and respiratory tract irritant.
Liquid or solid irritant characteristics: Possible skin irritation from prolonged contact.

FIRE DATA
Flash point: 255°F/124°C (cc).
Flammable limits in air: LEL: 1.6%; UEL: 10.8%.
Fire extinguishing agents not to be used: Water or foam may cause frothing.
Autoignition temperature: 444°F/229°C/502°K.
Electrical hazard: Class I, Group C.
Burning rate: 1.5 mm/min

CHEMICAL REACTIVITY
Binary reactants: Violent reaction with strong oxidizers. Incompatible with strong acids (e.g., sulfuric acid, nitric acid), aliphatic amines, amides, caustics, and isocyanates.
Reactivity group: 40
Compatibility class: Glycol ethers

ENVIRONMENTAL DATA
Food chain concentration potential: Log P_{ow} = –2.1. Negative; unlikely to accumulate.
Water pollution: Dangerous to aquatic life in high concentrations. May be dangerous if it enters nearby water intakes; notify operators. Notify local health and wildlife officials. **Response to discharge:** Disperse and flush.

SHIPPING INFORMATION
Grades of purity: Technical grade, regular grade; polyester grade;

Storage temperature: Ambient; **Inert atmosphere:** None; **Venting:** Open (flame arrester); **Stability during transport:** Stable.

NAS HAZARD CLASSIFICATION FOR BULK WATER TRANSPORTATION
FIRE: 1
HEALTH: Vapor irritant: 0; Liquid or solid irritant: 0; Poisons: 1
WATER POLLUTION: Human toxicity: 1; Aquatic toxicity: 1; Aesthetic effect: 0
REACTIVITY: Other chemicals: 2; Water: 0; Self-reaction: 0

PHYSICAL AND CHEMICAL PROPERTIES
Physical state @ 59°F/15°C and 1 atm: Liquid.
Molecular weight: 106.12
Boiling point @ 1 atm: 473°F/245°C/518°K.
Melting/Freezing point: 20°F/–8°C/265°K.
Critical temperature: 766°F/408°C/681°K.
Critical pressure: 680 psia = 46 atm = 4.7 MN/m^2.
Specific gravity (water = 1): 1.118 @ 68°F/20°C (liquid).
Latent heat of vaporization: 270 Btu/lb = 150 cal/g = 6.28 x 10^5 J/kg.
Heat of combustion: –9617 Btu/lb = –5343 cal/g = –223.7 x 10^5 J/kg.
Vapor pressure: (Reid) Very low.

DIETHYLENE GLYCOL DIBUTYL ETHER   REC. D:3050

SYNONYMS: BIS(2-BUTOXYETHYL) ETHER; BUTYL DIGLYME; 2,2'-DIBUTOXYETHYL ETHER; DIBUTYL CARBITOL; DIETHYLENE GLYCOL DI-*n*-BUTYL ETHER; DIETILENGLICOLDIBUTIL ETER (Spanish); 5,8,11-TRIOXAPENTADECANE

IDENTIFICATION
CAS Number: 112-73-2
Formula: $C_{12}H_{26}O_3$; $[CH_3(CH_2)_3OCH_2CH_2]_2O$
DOT ID Number: UN 3082; DOT GUIDE NUMBER: 171
PROPER SHIPPING NAME: Environmentally hazardous substances, liquid, n.o.s.

DESCRIPTION: Colorless liquid. Floats on the surface of water.

Combustible • Able to form unstable peroxide; may be a polymerization hazard • Containers may BLEVE when exposed to fire • Vapors are heavier than air and will collect and stay in low areas • Vapors in confined areas (e.g., tanks, sewers, buildings) may explode when exposed to fire • Irritating to the skin, eyes, and respiratory tract; absorbed by the skin • Toxic products of combustion may include carbon monoxide.

Hazard Classification (based on NFPA-704 Rating System)
Health Hazards (Blue): 1; Flammability (Red): 1; Reactivity (Yellow): 0

EMERGENCY RESPONSE: See Appendix A (171)
Evacuation:
Public safety: Isolate the area of spill or leak for at least 10 to 25 meters (30 to 80 feet) in all directions.
Spill: Increase, in the downwind direction, as necessary, the distance shown under "Public Safety."
Fire: If any large container is involved in fire, isolate for at least 800 meters (½ mile) in all directions; also, consider initial evacuation for 800 meters (½ mile) in all directions.

EXPOSURE
Short-term effects: *SEEK MEDICAL ATTENTION*. **Vapor:** Vapor irritating to eyes, nose, and throat. Move victim to fresh air. **Liquid:** Irritating to skin and eyes. Harmful if swallowed. Remove contaminated clothing and shoes. Flush affected areas with plenty of water. *IF IN EYES*, hold eyelids open and flush with plenty of water.

Personal protective equipment (PPE): B-Level PPE. Wear self-contained breathing apparatus, rubber boots and heavy rubber gloves.
Toxicity by ingestion: Grade 2: LD_{50} = 3.9 g/kg rat
Vapor (gas) irritant characteristics: Eye and respiratory tract irritant. Vapors cause irritation such that personnel will find high concentrations unpleasant.
Liquid or solid irritant characteristics: Eye and skin irritant. Causes smarting of skin and first-degree burn on short exposure; may cause second-degree burn on long exposure.

FIRE DATA
Flash point: 245°F/118°C (oc).
Autoignition temperature: 590°F/310°C/583°K.
Electrical hazard: Class I, Group C.

CHEMICAL REACTIVITY
Binary reactants: Reacts with strong oxidizers.
Reactivity group: 40
Compatibility class: Glycol ethers

ENVIRONMENTAL DATA
Water pollution: Effect of low concentrations on aquatic life is unknown. Fouling to shoreline. May be dangerous if it enters nearby water intakes; notify operators. Notify local health and wildlife officials. **Response to discharge:** Restrict access. Evacuate area. Mechanical containment. Should be removed. Chemical and physical treatment.

SHIPPING INFORMATION
Grades of purity: 99+%; **Storage temperature:** Ambient; **Inert atmosphere:** Nitrogen atmosphere; **Stability during transport:** Stable.

PHYSICAL AND CHEMICAL PROPERTIES
Physical state @ 59°F/15°C and 1 atm: Liquid.
Molecular weight: 218.34
Boiling point @ 1 atm: 493°F/256°C/529°K.
Melting/Freezing point: –76°F/–60°C/213°K.
Specific gravity (water = 1): 0.885
Relative vapor density (air = 1): More than 1.0

DIETHYLENE GLYCOL DIMETHYL ETHER
   REC. D:3100

SYNONYMS: BIS(2-METHOXYETHYL)-ETHER; DIETHYL GLYCOL DIMETHYL ETHER; DIETILENGLICOLDIMETIL ETER (Spanish); DIGLYME; POLY SOLV

IDENTIFICATION
CAS Number: 111-96-6
Formula: $C_6H_{14}O_3$; $(CH_3OCH_2CH_2)_2O$
DOT ID Number: UN 3082; DOT GUIDE NUMBER: 171
PROPER SHIPPING NAME: Environmentally hazardous substances, liquid, n.o.s.

DESCRIPTION: Colorless watery liquid. Pleasant odor. Floats on the surface of water; soluble.

Highly flammable • Able to form unstable peroxide; may be a polymerization hazard • Containers may BLEVE when exposed to fire • Vapors may form explosive mixture with air • Vapors are heavier than air and will collect and stay in low areas • Vapors may travel long distances to ignition sources and flashback • Vapors in confined areas (e.g., tanks, sewers, buildings) may explode when exposed to fire • Irritating to the skin, eyes, and respiratory tract • Toxic products of combustion may include carbon monoxide.

Hazard Classification (based on NFPA-704 Rating System)
Health Hazards (Blue): 1; Flammability (Red): 2; Reactivity (Yellow): 1

EMERGENCY RESPONSE: See Appendix A (171)
Evacuation:
Public safety: Isolate the area of spill or leak for at least 10 to 25 meters (30 to 80 feet) in all directions.
Spill: Increase, in the downwind direction, as necessary, the distance shown under "Public Safety."
Fire: If any large container is involved in fire, isolate for at least 800 meters (½ mile) in all directions; also, consider initial evacuation for 800 meters (½ mile) in all directions.

EXPOSURE
Short-term effects: *SEEK MEDICAL ATTENTION*. **Liquid:** If swallowed, will cause nausea, vomiting, or loss of consciousness. *IF SWALLOWED* and victim is *CONSCIOUS AND ABLE TO SWALLOW*, have victim drink 4 to 8 ounces of water.

HEALTH HAZARDS
Personal protective equipment (PPE): B-Level PPE. Vinyl (not rubber) gloves; safety goggles.
Long-term health effects: Mutagen data; experimental reproductive effects. Affects kidneys.

FIRE DATA
Flash point: 158°F/70°C (oc).
Electrical hazard: Class I, Group C.

CHEMICAL REACTIVITY
Binary reactants: Forms explosive reaction upon exposure to air, heat and light. Reacts with oxidizers and metal hydrides.
Reactivity group: 40
Compatibility class: Glycol ethers

ENVIRONMENTAL DATA
Food chain concentration potential: Negative; unlikely to accumulate.
Water pollution: Effect of low concentrations on aquatic life is unknown. May be dangerous if it enters nearby water intakes; notify operators. Notify local health and wildlife officials.
Response to discharge: Disperse and flush

SHIPPING INFORMATION
Grades of purity: Technical; **Stability during transport:** Stable.

PHYSICAL AND CHEMICAL PROPERTIES
Physical state @ 59°F/15°C and 1 atm: Liquid.
Molecular weight: 134.12
Boiling point @ 1 atm: 324°F/162°C/435°K.
Melting/Freezing point: –94°F/–70°C/203°K.
Specific gravity (water = 1): 0.945 @ 68°F/20°C (liquid).
Latent heat of vaporization: 130 Btu/lb = 74 cal/g = 3.1×10^5 J/kg.
Heat of combustion: (estimate) –11,300 Btu/lb = –6260 cal/g = -262×10^5 J/kg.

DIETHYLENE GLYCOL ETHYL ETHER ACETATE
REC. D:3150

SYNONYMS: ACETATO de DIETILENGLICOL MONOETIL ETER (Spanish); CARBITOL ACETATE; DIGLYCOL MONOETHYL ETHER ACETATE; 2-(2-ETHOXYETHOXY)ETHANOL ACETATE

IDENTIFICATION
CAS Number: 112-15-2
Formula: $C_8H_{16}O_4$; $CH_3COOCH_2OCH_2CH_2OC_2NO_5$
DOT ID Number: UN 3082; DOT GUIDE NUMBER: 171
PROPER SHIPPING NAME: Environmentally hazardous substances, liquid, n.o.s.

DESCRIPTION: Colorless to light yellow liquid. Characteristic odor. Insoluble in water.

Combustible • Containers may BLEVE when exposed to fire • Vapors are heavier than air and will collect and stay in low areas • Vapors in confined areas (e.g., tanks, sewers, buildings) may explode when exposed to fire • Toxic products of combustion may include carbon monoxide • Toxic products of combustion may include carbon monoxide.
Health Hazards (Blue): 1; Flammability (Red): 1; Reactivity (Yellow): 0

EMERGENCY RESPONSE: See Appendix A (171)
Evacuation:
Public safety: Isolate the area of spill or leak for at least 10 to 25 meters (30 to 80 feet) in all directions.
Spill: Increase, in the downwind direction, as necessary, the distance shown under "Public Safety."
Fire: If any large container is involved in fire, isolate for at least 800 meters (½ mile) in all directions; also, consider initial evacuation for 800 meters (½ mile) in all directions.

EXPOSURE
Short-term effects: *SEEK MEDICAL ATTENTION*. **Vapor:** Move victim to fresh air. *IF BREATHING HAS STOPPED*, give artificial respiration. If breathing is difficult, administer oxygen. **Liquid:** Remove contaminated clothing and shoes. Wash affected areas with soap and water. *IF IN EYES*, hold eyelids open and flush with plenty of water.

HEALTH HAZARDS
Personal protective equipment (PPE): B-Level PPE.
Vapor (gas) irritant characteristics: Vapors cause a slight smarting of the eyes or respiratory system if present in high concentrations. The effect is temporary.
Liquid or solid irritant characteristics: If spilled on clothing and allowed to remain, may cause smarting and reddening of the skin.

FIRE DATA
Flash point: 230°F/110°C (oc).
Fire extinguishing agents not to be used: Water.
Electrical hazard: Class I, Group C.

372 Diethylene glycol methyl ether acetate

CHEMICAL REACTIVITY
Binary reactants: Oxidizers, strong acids.
Reactivity group: 40
Compatibility class: Glycol ethers

ENVIRONMENTAL DATA
Water pollution: Effects of low concentrations on aquatic life is unknown. May be dangerous if it enters nearby water intakes; notify operators. Notify local health and wildlife officials.
Response to discharge: Should be removed.

SHIPPING INFORMATION
Grades of purity: Technical grades; **Storage temperature:** Ambient; **Inert atmosphere:** None; **Stability during transport:** Stable.

PHYSICAL AND CHEMICAL PROPERTIES
Physical state @ 59°F/15°C and 1 atm: Liquid.
Molecular weight: 162.12
Boiling point @ 1 atm: 423.32°F/217.4°C/490.4°K.
Melting/Freezing point: –13°F/–25°C/248°K.
Specific gravity (water = 1): 1.0114 @ 20°C.

DIETHYLENE GLYCOL n-HEXYL ETHER REC. D:3200

SYNONYMS: DIETILENGLICOL n- HEXIL ETER (Spanish); HEXYL CARBITOL

IDENTIFICATION
CAS Number: 112-59-4
Formula: $C_{10}H_{22}O_3$; $CH_3(CH_2)_5OC_2H_4OC_2H_4OH$
DOT ID Number: UN 3082; DOT Guide Number: 171
Proper Shipping Name: Environmentally hazardous substances, liquid, n.o.s.

DESCRIPTION: Clear liquid. Soluble in water.

Combustible • Containers may BLEVE when exposed to fire • Vapors are heavier than air and will collect and stay in low areas • Vapors in confined areas (e.g., tanks, sewers, buildings) may explode when exposed to fire • Toxic products of combustion may include carbon monoxide.

Hazard Classification (based on NFPA-704 Rating System)
Health Hazards (Blue): 1; Flammability (Red): 1; Reactivity (Yellow): 0

EMERGENCY RESPONSE: See Appendix A (171)
Evacuation:
Public safety: Isolate the area of spill or leak for at least 10 to 25 meters (30 to 80 feet) in all directions.
Spill: Increase, in the downwind direction, as necessary, the distance shown under "Public Safety."
Fire: If any large container is involved in fire, isolate for at least 800 meters (½ mile) in all directions; also, consider initial evacuation for 800 meters (½ mile) in all directions.

EXPOSURE
Short-term effects: *SEEK MEDICAL ATTENTION.* **Vapor:** Move victim to fresh air. *IF BREATHING HAS STOPPED,* give artificial respiration. If breathing is difficult, administer oxygen. **Liquid:** Remove contaminated clothing and shoes. Wash affected areas with soap and water. *IF IN EYES,* hold eyelids open and flush with plenty of water.

HEALTH HAZARDS
Personal protective equipment (PPE): B-Level PPE.
Vapor (gas) irritant characteristics: Eye and respiratory tract irritant. Vapors cause smarting of the eyes or respiratory system if present in high concentrations. The effect is temporary.
Liquid or solid irritant characteristics: Eye irritant. If spilled on clothing and allowed to remain, may cause smarting and reddening of the skin.

FIRE DATA
Flash point: 285°F/141°C (oc).
Fire extinguishing agents not to be used: Water.
Electrical hazard: Class I, Group C.

CHEMICAL REACTIVITY
Binary reactants: Oxidizers, strong acids.
Reactivity group: 40
Compatibility class: Glycol ethers

ENVIRONMENTAL DATA
Water pollution: Effects of low concentrations on aquatic life is unknown. May be dangerous if it enters nearby water intakes; notify operators. Notify local health and wildlife officials. **Response to discharge:** Should be removed.

SHIPPING INFORMATION
Grades of purity: Technical grades; **Storage temperature:** Ambient; **Inert atmosphere:** None; **Stability during transport:** Stable.

PHYSICAL AND CHEMICAL PROPERTIES
Physical state @ 59°F/15°C and 1 atm: Liquid.
Molecular weight: 190.32
Boiling point @ 1 atm: 498°F/259°C/532°K.
Melting/Freezing point: –27°F/–33°C/240°K.
Specific gravity (water = 1): 0.9346 @ 20°C.

DIETHYLENE GLYCOL METHYL ETHER ACETATE
REC. D:3250

SYNONYMS: ACETATO de DIETILENGLICOL MONOMETIL ETER (Spanish); ACETIC ACID, 2-(2-METHOXYETHOXY)ETHYL ESTER; 2-(2-METHOXY-ETHOXY)ETHANOL ACETATE; METHYL CARBITOL ACETATE

IDENTIFICATION
CAS Number: 629-38-9
Formula: $C_7H_{14}O_4$; $CH_3COOC_2H_4OC_2H_4OCH_3$
DOT ID Number: UN 3082; DOT Guide Number: 171
Proper Shipping Name: Environmentally hazardous substances, liquid, n.o.s.

DESCRIPTION: Colorless liquid. Mixes with water.

Combustible • Containers may BLEVE when exposed to fire • Vapors are heavier than air and will collect and stay in low areas • Vapors in confined areas (e.g., tanks, sewers, buildings) may explode when exposed to fire • Toxic products of combustion may include carbon monoxide.

Hazard Classification (based on NFPA-704 Rating System)
Health Hazards (Blue): 1; Flammability (Red): 2; Reactivity (Yellow): 0

EMERGENCY RESPONSE: See Appendix A (171)
Evacuation:
Public safety: Isolate the area of spill or leak for at least 10 to 25 meters (30 to 80 feet) in all directions.
Spill: Increase, in the downwind direction, as necessary, the distance shown under "Public Safety."
Fire: If any large container is involved in fire, isolate for at least 800 meters (½ mile) in all directions; also, consider initial evacuation for 800 meters (½ mile) in all directions.

EXPOSURE
Short-term effects: *SEEK MEDICAL ATTENTION.* **Vapor:** Move victim to fresh air. *IF BREATHING HAS STOPPED,* give artificial respiration. If breathing is difficult, administer oxygen. **Liquid:** Remove contaminated clothing and shoes. Wash affected areas with soap and water. *IF IN EYES*, hold eyelids open and flush with plenty of water.

HEALTH HAZARDS
Personal protective equipment (PPE): B-Level PPE.
Vapor (gas) irritant characteristics: Vapors cause a slight smarting of the eyes or respiratory system if present in high concentrations. The effect is temporary.
Liquid or solid irritant characteristics: Eye irritant. If spilled on clothing and allowed to remain, may cause smarting and reddening of the skin.

FIRE DATA
Flash point: 180°F/82°C (oc).
Fire extinguishing agents not to be used: Water.
Electrical hazard: Class I, Group C.

CHEMICAL REACTIVITY
Binary reactants: Oxidizers, strong acids,
Reactivity group: 40
Compatibility class: Glycol ethers

ENVIRONMENTAL DATA
Water pollution: Effects of low concentrations on aquatic life is unknown. May be dangerous if it enters nearby water intakes; notify operators. Notify local health and wildlife officials.
Response to discharge: Should be removed.

SHIPPING INFORMATION
Grades of purity: Technical grades; **Storage temperature:** Ambient; **Inert atmosphere:** None; **Stability during transport:** Stable.

PHYSICAL AND CHEMICAL PROPERTIES
Physical state @ 59°F/15°C and 1 atm: Liquid.
Molecular weight: 162.18
Boiling point @ 1 atm: 408°F/209°C/482°K.
Specific gravity (water = 1): 1.04 @ 20°C.
Relative vapor density (air = 1): 5.6

DIETHYLENE GLYCOL MONOBUTYL ETHER
REC. D:3300

SYNONYMS: BUTOXYDIGLYCOL; 2-(2-BUTOXYETHOXY) ETHANOL; BUTOXYDIETHYLENE GLYCOL; BUTYL CARBITOL; DIETILENGLICOL MONOBUTIL ETER (Spanish); DIGLYCOL MONOBUTYL ETHER; DIETHYLENE GLYCOL BUTYL ETHER; DIETHYLENE GLYCOL *n*-BUTYL ETHER; DOWANOL DB; POLY-SOLV DB

IDENTIFICATION
CAS Number: 112-34-5
Formula: $C_8H_{18}O_3$; $C_4H_9OCH_2CH_2OCH_2CH_2OH$
DOT ID Number: UN 3082; DOT Guide Number: 171
Proper Shipping Name: Environmentally hazardous substances, liquid, n.o.s.

DESCRIPTION: Colorless liquid. Mild pleasant odor. Mixes with water.

Combustible • May form unstable peroxides after prolonged storage in air; peroxides may be detonated by heating, impact, or friction • Containers may BLEVE when exposed to fire • Vapors are heavier than air and will collect and stay in low areas • Vapors in confined areas (e.g., tanks, sewers, buildings) may explode when exposed to fire • Toxic products of combustion may include carbon monoxide.

Hazard Classification (based on NFPA-704 Rating System)
Health Hazards (Blue): 1; Flammability (Red): 2; Reactivity (Yellow): 0

EMERGENCY RESPONSE: See Appendix A (171)
Evacuation:
Public safety: Isolate the area of spill or leak for at least 10 to 25 meters (30 to 80 feet) in all directions.
Spill: Increase, in the downwind direction, as necessary, the distance shown under "Public Safety."
Fire: If any large container is involved in fire, isolate for at least 800 meters (½ mile) in all directions; also, consider initial evacuation for 800 meters (½ mile) in all directions.

EXPOSURE
Short-term effects: *SEEK MEDICAL ATTENTION.* **Liquid:** Irritating to skin and eyes. Remove contaminated clothing and shoes. Flush affected areas with plenty of water. *IF IN EYES*, hold eyelids open and flush with plenty of water. *IF SWALLOWED* and victim is *CONSCIOUS AND ABLE TO SWALLOW*, have victim drink 4 to 8 ounces of water.

HEALTH HAZARDS
Personal protective equipment (PPE): B-Level PPE. Sealed chemical protective materials of Viton®/neoprene and butyl/neoprene offers limited protection
Toxicity by ingestion: Grade 2; oral LD_{50} = 2 g/kg (guinea pig).
Vapor (gas) irritant characteristics: Eye irritant. Vapors cause smarting of the eyes or respiratory system if present in high concentrations. The effect is temporary.
Liquid or solid irritant characteristics: Eye irritant. If spilled on clothing and allowed to remain, may cause smarting and reddening of the skin. Practically harmless to the skin.

FIRE DATA
Flash point: 230°F/110°C (oc); 172°F/78°C (cc).
Flammable limits in air: LEL: 0.85%; UEL: 24.6%.
Autoignition temperature: 403°F/206°C/479°K.
Electrical hazard: Class I, Group C.
Burning rate: 3.3 mm/min

CHEMICAL REACTIVITY
Binary reactants: Strong oxidizers, strong bases. Attacks light metals forming flammable hydrogen.
Reactivity group: 40
Compatibility class: Glycol ethers
Polymerization: Unstable peroxides may accumulate upon prolonged storage in presence of air.

374 *Diethylene glycol monobutyl ether acetate*

ENVIRONMENTAL DATA
Food chain concentration potential: Log P_{ow} = 0.6. Unlikely to accumulate.
Water pollution: Effect of low concentrations on aquatic life is unknown. May be dangerous if it enters nearby water intakes; notify operators. Notify local health and wildlife officials.
Response to discharge: Issue warning–water contaminant. Disperse and flush.

SHIPPING INFORMATION
Grades of purity: Commercial; **Storage temperature:** Ambient; **Inert atmosphere:** None; **Venting:** Open; **Stability during transport:** Stable.

NAS HAZARD CLASSIFICATION FOR BULK WATER TRANSPORTATION
FIRE: 1
HEALTH: Vapor irritant: 1; Liquid or solid irritant: 0; Poisons: 1
WATER POLLUTION: Human toxicity: 1; Aquatic toxicity: 3; Aesthetic effect: 1
REACTIVITY: Other chemicals: 2; Water: 0; Self-reaction: 0

PHYSICAL AND CHEMICAL PROPERTIES
Physical state @ 59°F/15°C and 1 atm: Liquid.
Molecular weight: 162.2
Boiling point @ 1 atm: 448°F/231°C/504°K.
Melting/Freezing point: –90°F/–68°C/205°K.
Specific gravity (water = 1): 0.954 @ 68°F/20°C (liquid).
Liquid surface tension: 34 dynes/cm = 0.034 N/m @ 77°F/25°C.
Relative vapor density (air = 1): 5.58
Latent heat of vaporization: 130 Btu/lb = 74 cal/g = 3.1×10^5 J/kg.
Heat of combustion: (estimate) –14,000 Btu/lb = –7900 cal/g = -330×10^5 J/kg.
Heat of solution: (estimate) –36 Btu/lb = –20 cal/g = -0.84×10^5 J/kg.
Vapor pressure: 0.02 mm.

DIETHYLENE GLYCOL MONOBUTYL ETHER ACETATE
REC. D:3350

SYNONYMS: 2-(2-BUTOXYETHOXY)ETHANOL ACETATE; 2-(2-BUTOXYETHOXY) ETHYL ACETATE; BUTYL CARBITOL ACETATE; DIETHYLENE GLYCOL BUTYL ETHER ACETATE; ACETATO de DIETILENGLICOL MONOBUTIL ETER (Spanish); DIGLYCOL MONOBUTYL ETHER ACETATE; EKTASOLVE DB ACETATE; GLYCOL ETHER DB ACETATE

IDENTIFICATION
CAS Number: 124-17-4
Formula: $C_{10}H_{20}O_4$; $C_4H_9OCH_2CH_2OCH_2CH_2OCOCH_3$
DOT ID Number: UN 3082; DOT Guide Number: 171
Proper Shipping Name: Environmentally hazardous substances, liquid, n.o.s.

DESCRIPTION: Colorless liquid. Mild odor Floats and mixes slowly with water.

Combustible • Containers may BLEVE when exposed to fire • Vapors are heavier than air and will collect and stay in low areas • Vapors in confined areas (e.g., tanks, sewers, buildings) may explode when exposed to fire • Toxic products of combustion may include carbon monoxide.

Hazard Classification (based on NFPA-704 Rating System)
Health Hazards (Blue): 1; Flammability (Red): 1; Reactivity (Yellow): 0

EMERGENCY RESPONSE: See Appendix A (171)
Evacuation:
Public safety: Isolate the area of spill or leak for at least 10 to 25 meters (30 to 80 feet) in all directions.
Spill: Increase, in the downwind direction, as necessary, the distance shown under "Public Safety."
Fire: If any large container is involved in fire, isolate for at least 800 meters (½ mile) in all directions; also, consider initial evacuation for 800 meters (½ mile) in all directions.

EXPOSURE
Short-term effects: *SEEK MEDICAL ATTENTION*. **Liquid:** Irritating to skin and eyes. Remove contaminated clothing and shoes. Flush affected areas with plenty of water. *IF IN EYES*, hold eyelids open and flush with plenty of water. *IF SWALLOWED* and victim is *CONSCIOUS AND ABLE TO SWALLOW*, have victim drink 4 to 8 ounces of water.

HEALTH HAZARDS
Personal protective equipment (PPE): B-Level PPE.
Toxicity by ingestion: Grade 2; oral LD_{50} = 2.34 g/kg (guinea pig).
Long-term health effects: Kidney damage noted in animals following repeated contact with skin.
Vapor (gas) irritant characteristics: Respiratory tract irritant.
Liquid or solid irritant characteristics: Eye and skin irritant.

FIRE DATA
Flash point: 240°F/116°C (oc).
Flammable limits in air: LEL 0.8%; UEL: 11.%.
Autoignition temperature: 572°F/300°C/573°K.
Electrical hazard: Class I, Group C.
Burning rate: 3.8 mm/min

CHEMICAL REACTIVITY
Binary reactants: Oxidizers, strong acids.
Reactivity group: 40
Compatibility class: Glycol ethers

ENVIRONMENTAL DATA
Food chain concentration potential: Negative; unlikely to accumulate.
Water pollution: Effect of low concentrations on aquatic life is unknown. May be dangerous if it enters nearby water intakes; notify operators. Notify local health and wildlife officials. **Response to discharge:** Issue warning–water contaminant. Disperse and flush.

SHIPPING INFORMATION
Grades of purity: 98+%; **Storage temperature:** Ambient; **Inert atmosphere:** None; **Venting:** Open (flame arrester); **Stability during transport:** Stable.

PHYSICAL AND CHEMICAL PROPERTIES
Physical state @ 59°F/15°C and 1 atm: Liquid.
Molecular weight: 204.3
Boiling point @ 1 atm: 475°F/246°C/519°K.
Melting/Freezing point: –27°F/–33°C/240°K.
Specific gravity (water = 1): 0.985 @ 68°F/20°C (liquid).
Liquid surface tension: (estimate) 22 dynes/cm = 0.022 N/m @ 68°F/20°C.

Latent heat of vaporization: 106 Btu/lb = 59 cal/g = 2.5 x 10^5 J/kg.
Heat of combustion: (estimate) –13,000 Btu/lb = –7400 cal/g = –310 x 10^5 J/kg.
Heat of solution: (estimate) –27 Btu/lb = –15 cal/g = –0.63 x 10^5 J/kg.

DIETHYLENE GLYCOL MONOETHYL ETHER
REC. D:3400

SYNONYMS: CARBITOL; CARBITOL CELLOSOLVE; CARBITOL SOLVENT; DIETHYLENE GLYCOL ETHYL ETHER; DIOXITOL; DOWANOL; DOWANOL DE; DIETILENGLICOL MONOETIL ETER (Spanish); ETHOXY DIGLYCOL; 2-(2-ETHOXYETHOXY) ETHANOL; ETHYL CARBITOL; ETHYL DIETHYLENE GLYCOL; LOSUNGSMITTEL APV; POLY-SOLV; SOLVOSOL

IDENTIFICATION
CAS Number: 111-90-0
Formula: $C_6H_{14}O_3$; $CH_3CH_2OCH_2CH_2OCH_2CH_2OH$
DOT ID Number: UN 3082; DOT Guide Number: 171
Proper Shipping Name: Environmentally hazardous substances, liquid, n.o.s.

DESCRIPTION: Colorless liquid. Fruity odor. Floats on the surface of water; soluble.

Combustible • Containers may BLEVE when exposed to fire • Vapors are heavier than air and will collect and stay in low areas • Vapors in confined areas (e.g., tanks, sewers, buildings) may explode when exposed to fire • Irritates eyes • Toxic products of combustion may include carbon monoxide.

Hazard Classification (based on NFPA-704 Rating System)
Health Hazards (Blue): 1; Flammability (Red): 1; Reactivity (Yellow): 0

EMERGENCY RESPONSE: See Appendix A (171)
Evacuation:
Public safety: Isolate the area of spill or leak for at least 10 to 25 meters (30 to 80 feet) in all directions.
Spill: Increase, in the downwind direction, as necessary, the distance shown under "Public Safety."
Fire: If any large container is involved in fire, isolate for at least 800 meters (½ mile) in all directions; also, consider initial evacuation for 800 meters (½ mile) in all directions.

EXPOSURE
Short-term effects: *SEEK MEDICAL ATTENTION.*
Liquid: Harmful if swallowed. *IF SWALLOWED* and victim is *CONSCIOUS AND ABLE TO SWALLOW*, have victim drink 4 to 8 ounces of water.

HEALTH HAZARDS
Personal protective equipment (PPE): B-Level PPE. Neoprene is generally suitable for cellosolve.
Toxicity by ingestion: Grade 2; LD_{50} = 0.5 to 5 g/kg.
Long-term health effects: Organ damage.
Liquid or solid irritant characteristics: Eye irritant. Practically harmless to the skin.

FIRE DATA
Flash point: 205°F/96°C (oc); 201°F/94°C (cc).
Flammable limits in air: LEL: (estimate) 1.2%; UEL: 8.5%.
Autoignition temperature: 400°F/205°C/478°K.
Electrical hazard: Class I, Group C.
Burning rate: 2.5 mm/min

CHEMICAL REACTIVITY
Binary reactants: Reacts with strong oxidizers.
Reactivity group: 40
Compatibility class: Glycol ethers
Polymerization: Unstable peroxides may accumulate after prolonged storage in presence of air.

ENVIRONMENTAL DATA
Food chain concentration potential: Negative; unlikely to accumulate.
Water pollution: Effect of low concentrations on aquatic life is unknown. May be dangerous if it enters nearby water intakes; notify operators. Notify local health and wildlife officials. **Response to discharge:** Disperse and flush

SHIPPING INFORMATION
Grades of purity: Commercial; **Storage temperature:** Ambient; **Inert atmosphere:** None; **Venting:** Open (flame arrester); **Stability during transport:** Stable.

NAS HAZARD CLASSIFICATION FOR BULK WATER TRANSPORTATION
FIRE: 1
HEALTH: Vapor irritant: 0; Liquid or solid irritant: 0; Poisons: 0
WATER POLLUTION: Human toxicity: 1; Aquatic toxicity: 3; Aesthetic effect: 1
REACTIVITY: Other chemicals: 2; Water: 0; Self-reaction: 0

PHYSICAL AND CHEMICAL PROPERTIES
Physical state @ 59°F/15°C and 1 atm: Liquid.
Molecular weight: 134.17
Boiling point @ 1 atm: 396°F/202°C/475°K.
Melting/Freezing point: –105°F/–76°C/197°K.
Specific gravity (water = 1): 0.99 @ 68°F/20°C (liquid).
Relative vapor density (air = 1): 4.58
Latent heat of vaporization: 150 Btu/lb = 85 cal/g = 3.6 x 10^5 J/kg.
Heat of combustion: –11,390 Btu/lb = –6330 cal/g = –265 x 10^5 J/kg.
Vapor pressure: 1 mm.

DIETHYLENE GLYCOL MONOMETHYL ETHER
REC. D: 3415

SYNONYMS: DIETHYLENE GLYCOL METHYL ETHER; DIETILENGLICOL MONOMETIL ETER (Spanish); DIGLYCOL MONOMETHYL ETHER; ETHYLENE DIGLYCOL MONOMETHYL ETHER; MECB; METHOXYDIGLYCOL; DOWANOL DM; 2-(2-METHOXYETHOXY)-ETHANOL; METHYL CARBITOL; POLY-SOLV DM

IDENTIFICATION
CAS Number: 111-77-3
Formula: $C_5H_{12}O_3$; $CH_3OCHCHOCHCH_2OH$
DOT ID Number: UN 3082; DOT Guide Number: 171
Proper Shipping Name: Environmentally hazardous substances, liquid, n.o.s.

DESCRIPTION: Colorless liquid. Pleasant odor. Floats on the surface of water; soluble.

Combustible • Containers may BLEVE when exposed to fire • Vapors are heavier than air and will collect and stay in low areas • Vapors in confined areas (e.g., tanks, sewers, buildings) may explode when exposed to fire • Toxic products of combustion may include carbon monoxide.

Hazard Classification (based on NFPA-704 Rating System)
Health Hazards (Blue): 1; Flammability (Red): 1; Reactivity (Yellow): 0

EMERGENCY RESPONSE: See Appendix A (171)
Evacuation:
Public safety: Isolate the area of spill or leak for at least 10 to 25 meters (30 to 80 feet) in all directions.
Spill: Increase, in the downwind direction, as necessary, the distance shown under "Public Safety."
Fire: If any large container is involved in fire, isolate for at least 800 meters (½ mile) in all directions; also, consider initial evacuation for 800 meters (½ mile) in all directions.

EXPOSURE
Short-term effects: *SEEK MEDICAL ATTENTION.* **Liquid:** Irritating to eyes. Harmful if swallowed. *IF IN EYES*, hold eyelids open and flush with plenty of water. *IF SWALLOWED* and victim is *CONSCIOUS AND ABLE TO SWALLOW*, have victim drink 4 to 8 ounces of water.

HEALTH HAZARDS
Personal protective equipment (PPE): B-Level PPE.
Toxicity by ingestion: Grade 2; LD_{50} = 0.5 to 5 g/kg (guinea pig).
Liquid or solid irritant characteristics: May cause eye irritation.

FIRE DATA
Flash point: 205°F/96°C (oc).
Flammable limits in air: LEL: 1.2%; UEL: 24.0%.

Autoignition temperature: 400°F/204°C/477°K.
Electrical hazard: Class I, Group C.

CHEMICAL REACTIVITY
Binary reactants: Oxidizers, strong acids, isocyanates.
Reactivity group: 40
Compatibility class: Glycol ethers

ENVIRONMENTAL DATA
Food chain concentration potential: Negative; unlikely to accumulate.
Water pollution: Effect of low concentrations on aquatic life is unknown. May be dangerous if it enters nearby water intakes; notify operators. Notify local health and wildlife officials.
Response to discharge: Disperse and flush.

SHIPPING INFORMATION
Grades of purity: Commercial; **Storage temperature:** Ambient; **Inert atmosphere:** None; **Venting:** Open (flame arrester); **Stability during transport:** Stable.

NAS HAZARD CLASSIFICATION FOR BULK WATER TRANSPORTATION
FIRE: 1
HEALTH: Vapor irritant: 0; Liquid or solid irritant: 0; Poisons: 0
WATER POLLUTION: Human toxicity: 1; Aquatic toxicity: 3; Aesthetic effect: 1
REACTIVITY: Other chemicals: 2; Water: 0; Self-reaction: 0

PHYSICAL AND CHEMICAL PROPERTIES
Physical state @ 59°F/15°C and 1 atm: Liquid.
Molecular weight: 120.15
Boiling point @ 1 atm: 381°F/194°C/467°K.
Melting/Freezing point: –120°F/–85°C/188°K.
Specific gravity (water = 1): 1.025 @ 68°F/20°C (liquid).
Latent heat of vaporization: 160 Btu/lb = 90 cal/g = 3.8×10^5 J/kg.
Heat of combustion: –10,830 Btu/lb = –6020 cal/g = -252×10^5 J/kg.

DIETHYLENE TRIAMINE   REC. D:3450

SYNONYMS: AMINOETHANDIAMINE; AMINOETHYLETHANDIAMINE; 3-AZAPENTANE-1,5-DIAMINE; BIS(2-AMINOETHYL)AMINE; *N*-(2-AMINOETHYL)ETHYLENEDIAMINE; BIS(*β*-AMINOETHYL)AMINE; DETA; DIETILENTRIAMINA (Spanish); EEC No. 612-058-00-X; 1,2-ETHANEDIAMINE, 3-AZAPENTANE-1, 5-DIAMINE; 2,2'-DIAMINODIETHYLAMINE; 2,2'-IMINOBISETHYLAMINE; 1,4,7-TRIAZAHEPTANE

IDENTIFICATION
CAS Number: 111-40-0
Formula: $C_4H_{13}N_3$; $NH_2(CH)_2NH(CH)_2NH_2$
DOT ID Number: UN 2079; DOT Guide Number: 154
Proper Shipping Name: Diethylenetriamine

DESCRIPTION: Colorless to yellow liquid. Ammonia-like odor. Floats on the surface of water; soluble, forming a basic solution.

Combustible • Corrosive to the skin, eyes, and respiratory tract; contact with skin or eyes can cause burns and vision impairment; inhalation symptoms may be delayed • Do NOT use HALONS • Containers may BLEVE when exposed to fire • Vapors are heavier than air and will collect and stay in low areas • Vapors in confined areas (e.g., tanks, sewers, buildings) may explode when exposed to fire • Toxic products of combustion may include hydrogen cyanide, and nitrogen oxides.

Hazard Classification (based on NFPA-704 Rating System)

Health Hazards (Blue): 3; Flammability (Red): 1; Reactivity (Yellow): 0

EMERGENCY RESPONSE: See Appendix A (154)
Evacuation:
Public safety: Isolate the area of spill or leak for at least 25 to 50 meters (80 to 160 feet) in all directions.
Spill: Increase, in the downwind direction, as necessary, the distance shown under "Public Safety."
Fire: If tank, rail car, or tank truck is involved in fire, isolate for at least 800 meters (½ mile) in all directions; also, consider initial evacuation for 800 meters (½ mile) in all directions.

EXPOSURE
Short-term effects: *SEEK MEDICAL ATTENTION.* **Vapor:** May cause lung edema; physical exertion will aggravate this condition. *IF BREATHING HAS STOPPED*, give artificial respiration; *avoid*

mouth-to-mouth resuscitation; use bag/mask apparatus. **Liquid:** *HARMFUL IF SWALLOWED OR ABSORBED THROUGH THE SKIN.* Will burn skin and eyes. Harmful if swallowed. Remove contaminated clothing and shoes. Flush affected areas with plenty of water. *IF IN EYES*, hold eyelids open and flush with plenty of water. *IF SWALLOWED* and victim is *CONSCIOUS AND ABLE TO SWALLOW*, have victim drink 4 to 8 ounces of water. **Do NOT induce vomiting.**

Note to physician or authorized medical personnel: Medical observation is recommended for 24 to 48 hours after breathing overexposure, as pulmonary edema may be delayed. As first aid for pulmonary edema, consider administering a corticosteroid spray. Cigarette smoking may exacerbate pulmonary injury and should be discouraged for at least 72 hours following exposure.

HEALTH HAZARDS

Personal protective equipment (PPE): A-Level PPE. OSHA Table Z-1-A air contaminant. Full face organic vapor respirator. Chemical protective material(s) reported to have good to excellent resistance: butyl rubber.

Recommendations for respirator selection: *EMERGENCY OR PLANNED ENTRY INTO UNKNOWN CONCENTRATIONS OR IDLH CONDITIONS*: SCBAF:PD,PP (any self-contained breathing apparatus that has a full facepiece and is operated in a pressure-demand or other positive-pressure mode); or SAF:PD,PP:ASCBA (any supplied-air respirator that has a full facepiece and is operated in a pressure-demand or other positive-pressure mode in combination with an auxiliary self-contained breathing apparatus operated in a pressure-demand or other positive pressure mode). *ESCAPE*: GMFS [any air-purifying, full-facepiece respirator (gas mask) with a chin-style, front- or back-mounted canister providing protection against the compound of concern]; or SCBAE (any appropriate escape-type, self-contained breathing apparatus).

Exposure limits (TWA unless otherwise noted): ACGIH TLV 1 ppm (4.2 mg/m^3); NIOSH REL 1 ppm (4 mg/m^3); skin contact

Toxicity by ingestion: Grade 2; LD_{50} = 0.5 to 5 g/kg (rat).

Long-term health effects: Asthma, eczema and liver damage may occur.

Vapor (gas) irritant characteristics: Eye and respiratory tract irritant. Vapors cause irritation such that personnel will find high concentrations unpleasant.

Liquid or solid irritant characteristics: Severe eye and skin irritant. Causes smarting of the skin and first-degree burns on short exposure and may cause secondary burns on long exposure.

Odor threshold: 10 ppm.

FIRE DATA

Flash point: 208°F/98°C (oc).
Flammable limits in air: LEL: 2%; UEL: 6.7%.
Fire extinguishing agents not to be used: Water or foam may cause frothing.
Autoignition temperature: 676°F/358°C/631°K.
Electrical hazard: Class I, Group D.

CHEMICAL REACTIVITY

Binary reactants: Oxidizers, strong bases, cellulose nitrate. May form explosive complexes with silver, cobalt, chlorinated hydrocarbon and chromium compounds. Corrosive to aluminum, copper, brass and zinc.
Neutralizing agents for acids and caustics: Flush with water.
Reactivity group: 7
Compatibility class: Aliphatic amines

ENVIRONMENTAL DATA

Food chain concentration potential: 23% of theoretical in 5 days/freshwater.
Water pollution: Effect of low concentrations on aquatic life is unknown. May be dangerous if it enters nearby water intakes; notify operators. Notify local health and wildlife officials. **Response to discharge:** Disperse and flush.

SHIPPING INFORMATION

Grades of purity: 98–99%; **Storage temperature:** Ambient; **Inert atmosphere:** None; **Venting:** Open; **Stability during transport:** Stable.

NAS HAZARD CLASSIFICATION FOR BULK WATER TRANSPORTATION

FIRE: 1
HEALTH: Vapor irritant: 2; Liquid or solid irritant: 2; Poisons: 2
WATER POLLUTION: Human toxicity: 2; Aquatic toxicity: 2; Aesthetic effect: 3
REACTIVITY: Other chemicals: 3; Water: 0; Self-reaction: 0

PHYSICAL AND CHEMICAL PROPERTIES

Physical state @ 59°F/15°C and 1 atm: Liquid.
Molecular weight: 103.17
Boiling point @ 1 atm: 405°F/207°C/480°K.
Melting/Freezing point: –38°F/–39°C/234°K.
Specific gravity (water = 1): 0.954 @ 68°F/20°C (liquid).
Relative vapor density (air = 1): 3.48
Heat of combustion: (estimate) –13,300 Btu/lb = –7390 cal/g = –309 x 10^5 J/kg.
Heat of solution: (estimate) –13 Btu/lb = –7 cal/g = –0.3 x 10^5 J/kg.
Vapor pressure: (Reid) 0.02 psia; 0.4 mm.

DIETHYLETHANOLAMINE   **REC. D:3500**

SYNONYMS: DEAE; DIAETHYLAMINOAETHANOL (German); β-DIETHYLAMINOETHANOL; *N*-DIETHYLAMINOETHANOL; 2-DIETHYLAMINO-; 2-(DIETHYLAMINO) ETHYL ALCOHOL; 2-(DIETHYLAMINO)ETHANOL; 2-*N*-DIETHYLAMINOETHANOL; β-DIETHYLAMINOETHYL ALCOHOL; *N*,*N*-DIETHYLETHANOLAMINE; *N*,*N*-DIETHYL-*N*-(β-HYDROXYETHYL)AMINE; *N*,*N*-DIETILETANOLAMINA (Spanish); 2-HYDROXYTRIETHYLAMINE; N,11-DIETHYL-*N*-(8-HYDROXY-ETHYL)AMINE; EEC No. 603-048-00-6; ETHANOL,2-(DIETHYLAMINO)-; *N*,*N*-DIETHYL-2-HYDROXYETHYLAMINE

IDENTIFICATION

CAS Number: 100-37-8
Formula: $C_6H_{15}NO$; $(C_2NO_5)_2NC_2H_4OH$
DOT ID Number: UN 2686; DOT Guide Number: 132
Proper Shipping Name: Diethylaminoethanol

DESCRIPTION: Colorless liquid. Ammonia-like odor. Floats on the surface of water; soluble.

Flammable • Firefighting gear (including SCBA) does not provide adequate protection. If exposure occurs, remove and isolate gear immediately and thoroughly decontaminate personnel • Containers may BLEVE when exposed to fire • Severely irritating to skin, eyes, and respiratory tract; can cause burns and blindness. Containers may BLEVE when exposed to fire •Vapors may form explosive mixture with air • Vapors are heavier than air and will collect and stay in low areas • Vapors may travel long distances to ignition sources and flashback • Vapors in confined areas (e.g., tanks, sewers, buildings) may explode when exposed to fire • Toxic products of combustion may include carbon monoxide and nitrogen oxides • Do not put yourself in danger by entering a contaminated area to rescue a victim.

Hazard Classification (based on NFPA-704 Rating System)
Health Hazards (Blue): 3; Flammability (Red): 2; Reactivity (Yellow): 0

EMERGENCY RESPONSE: See Appendix A (132)
Evacuation:
Public safety: Isolate spill area for at least 50 to 100 meters (160 to 330 feet) in all directions.
Spill: Increase, as necessary, the isolation distance shown above, in "Public safety."
Fire: Isolate for 800 meters (½ mile) in all directions, especially if tank, rail car, or tank truck is involved in fire.

EXPOSURE
Short-term effects: *SEEK MEDICAL ATTENTION.* **Liquid:** *HARMFUL IF SWALLOWED OR ABSORBED THROUGH THE SKIN.* Will burn skin and eyes. Harmful if swallowed. Remove contaminated clothing and shoes. Flush affected areas with plenty of water. *IF IN EYES*, hold eyelids open and flush with plenty of water. *IF SWALLOWED* and victim is *CONSCIOUS AND ABLE TO SWALLOW*, have victim drink 4 to 8 ounces of water.

HEALTH HAZARDS
Personal protective equipment (PPE): B-Level PPE. OSHA Table Z-1-A air contaminant. Organic vapor respirator. Chemical protective material(s) reported to have good to excellent resistance: butyl rubber, nitrile, PV alcohol, Viton®. Also, natural rubber, and neoprene may offer some protection from alcohols.
Recommendations for respirator selection: NIOSH/OSHA
100 ppm: CCROV* [any chemical cartridge respirator with organic vapor cartridge(s)]; or GMFOV [any air-purifying, full-facepiece respirator (gas mask) with a chin-style, front- or back-mounted acid gas canister]; or PAPROV* [any powered, air-purifying respirator with organic vapor cartridge(s)]; or SA* [any supplied-air respirator]; or SCBAF (any self-contained breathing apparatus with a full facepiece). *EMERGENCY OR PLANNED ENTRY INTO UNKNOWN CONCENTRATIONS OR IDLH CONDITIONS:* SCBAF:PD,PP (any self-contained breathing apparatus that has a full facepiece and is operated in a pressure-demand or other positive-pressure mode); or SAF:PD,PP:ASCBA (any supplied-air respirator that has a full facepiece and is operated in a pressure-demand or other positive-pressure mode in combination with an auxiliary self-contained breathing apparatus operated in a pressure-demand or other positive pressure mode). *ESCAPE:* GMFOV[any air-purifying, full-facepiece respirator (gas mask) with a chin-style, front- or back-mounted organic vapor cannister]; or SCBAE (any appropriate escape-type, self-contained breathing apparatus). **Note*: Substance reported to cause eye irritation or damage; may require eye protection.
Exposure limits (TWA unless otherwise noted): ACGIH TLV 2 ppm (9.6 mg/m^3); NIOSH/OSHA 10 ppm (50 mg/m^3; skin absorption contributes to overall exposure.
Toxicity by ingestion: Grade 2; LD$_{50}$ = 0.5 to 5 g/kg.
Vapor (gas) irritant characteristics: Eye and respiratory tract irritant. Vapors cause a slight smarting of the eyes or respiratory system if present in high concentration. The effect is temporary.
Liquid or solid irritant characteristics: Severe eye and skin irritant.

Odor threshold: Abs. perception limit in air = 0.011 ppm. 100% recognition in air = 0.04 ppm.
IDLH value: 100 ppm.

FIRE DATA
Flash point: 140°F/60°C (oc); 126°F/52°C (cc).
Flammable limits in air: LEL: 0.9%; UEL: 4.0%.
Autoignition temperature: 608°F/320°C/593°K.
Electrical hazard: Class I, Group C.

CHEMICAL REACTIVITY
Binary reactants: Reacts with strong oxidizers, acids. Corrodes light metals.
Neutralizing agents for acids and caustics: Dilute with water.
Reactivity group: 8
Compatibility class: Alkanolamines

ENVIRONMENTAL DATA
Food chain concentration potential: Log P$_{ow}$ = 0.39. Unlikely to accumulate.
Water pollution: Dangerous to aquatic life in high concentrations. May be dangerous if it enters nearby water intakes; notify operators. Notify local health and wildlife officials. **Response to discharge:** Chemical and physical treatment. Disperse and flush.

SHIPPING INFORMATION
Grades of purity: 99.5%; **Stability during transport:** Stable.

PHYSICAL AND CHEMICAL PROPERTIES
Physical state @ 59°F/15°C and 1 atm: Liquid.
Molecular weight: 117.19
Boiling point @ 1 atm: 315.5–327.2°F/157.5–164°C/430.7–437.2°K.
Melting/Freezing point: –94°F/–70°C/203°K.
Critical temperature: 709.8°F/376.6°C/649.8°K.
Critical pressure: 457.3 psia = 31.11 atm = 3.15 MN/m^2.
Specific gravity (water = 1): 0.8921 @ 77°F/25°C.
Liquid surface tension: (estimate) 34.3 dynes/cm = 0.0343 N/m @ 68°F/20°C.
Relative vapor density (air = 1): 4.03
Ratio of specific heats of vapor (gas): More than 1 (estimate).
Latent heat of vaporization: 140.2 Btu/lb = 77.9 cal/g = 3.26 x 10^5 J/kg.
Heat of combustion: 964 K cal/mole
Vapor pressure: 1.4 mm @ 20°C.

DI-(2-ETHYLHEXYL)ADIPATE   **REC. D:3550**

SYNONYMS: ADIPATO de DI(2-ETILHEXILO) (Spanish); ADIPIC ACID, DIBUTYL ESTER; ADIPIC ACID, BIS(2-ETHYLHEXYL) ESTER; BIS(2-ETHYLHEXYL)ADIPATE; BUTYL ADIPATE; BIS(2-ETHYLHEXYL) ADIPATE; DIBUTYL ADIPINATE; DIBUTYL HEXANEDIOATE; HEXANEDIOIC ACID, DIBUTYL ESTER

IDENTIFICATION
CAS Number: 103-23-1
Formula: $C_{22}H_{42}O_4$; $C_8H_{17}OOC(CH)_4COOC_8H_{17}$

DESCRIPTION: Clear to straw colored, oily liquid. Mild odor. Floats on water surface; insoluble.

Combustible • Containers may BLEVE when exposed to fire • Vapors are heavier than air and will collect and stay in low areas •

Vapors in confined areas (e.g., tanks, sewers, buildings) may explode when exposed to fire • Irritating to the skin, eyes, and respiratory tract.

Hazard Classification (based on NFPA-704 Rating System)
Health Hazards (Blue): 1; Flammability (Red): 1; Reactivity (Yellow): 0

EMERGENCY RESPONSE: See Appendix A (171)
Evacuation:
Public safety: Isolate the area of spill or leak for at least 10 to 25 meters (30 to 80 feet) in all directions.
Spill: Increase, in the downwind direction, as necessary, the distance shown under "Public Safety."
Fire: If any large container is involved in fire, isolate for at least 800 meters (½ mile) in all directions; also, consider initial evacuation for 800 meters (½ mile) in all directions.

EXPOSURE
Short-term effects: *SEEK MEDICAL ATTENTION*. **Liquid:** Move to fresh air. Wash affected areas with plenty of soap and water. *IF IN EYES*, hold eyelids open and flush with plenty of water.

HEALTH HAZARDS
Personal protective equipment (PPE): B-Level PPE. Wear splash goggles, impervious apron, and impervious gloves.
Toxicity by ingestion: Grade 1; LD_{50} = 5 to 15 g/kg.
Long-term health effects: IARC rating 3; limited animal evidence.
Vapor (gas) irritant characteristics: Vapors cause a slight smarting of the eyes or respiratory system if present in high concentrations. The effect is temporary.
Liquid or solid irritant characteristics: May cause eye irritation. If spilled on clothing and allowed to remain, may cause skin irritation.

FIRE DATA
Flash point: 400°F/204°C (oc); 385°F/196°C (cc).
Autoignition temperature: 710°F/377°C/650°K.

CHEMICAL REACTIVITY
Binary reactants: Avoid contact with strong oxidizers.
Reactivity group: 34
Compatibility class: Esters.

ENVIRONMENTAL DATA
Food chain concentration potential: Negative; unlikely to accumulate.
EPCRA Section 313: Deleted 7/31/96 (FR Vol. 61, No. 148 p.39891-39892).
Water pollution: Regulated under Safe Drinking Water Act: MCL, 0.4 mg/L; MCLG, 0.4 mg/L. May be dangerous if it enters nearby water intakes; notify operators. Notify local health and wildlife officials. **Response to discharge:** Mechanical containment. Should be removed.

SHIPPING INFORMATION
Grades of purity: 99.6%; Technical grades; **Storage temperature:** Ambient; **Inert atmosphere:** None; **Stability during transport:** Stable.

PHYSICAL AND CHEMICAL PROPERTIES
Physical state @ 59°F/15°C and 1 atm: Liquid.
Molecular weight: 370.58
Boiling point @ 1 atm: 788°F/420°C/693°K.
Specific gravity (water = 1): 0.923
Relative vapor density (air = 1): 12.8
Vapor pressure: 2.60 mmHg

DI-(2-ETHYLHEXYL)PHOSPHORIC ACID   REC. D:3600

SYNONYMS: ACIDO DI(2-ETILHEXIL)FOSFORICO (Spanish); BIS(2-ETHYLHEXYL)HYDROGEN PHOSPHATE; BIS(2-ETHYLHEXYL)ORTHOPHOSPHORIC ACID; BIS(2-ETHYLHEXYL)PHOSPHORIC ACID; DEHPA; DEHPA EXTRACT; 2-ETHYL-1-HEXANOL HYDROGEN PHOSPHATE; DI-(2-ETHYLHEXYL)PHOSPHATE; DI-(2-ETHYLHEXYL)PHOSPHORIC ACID; 2-ETHYL-1-HEXANOL HYDROGEN PHOSPHATE; DIISOOCTYL ACID PHOSPHATE; HDEHP

IDENTIFICATION
CAS Number: 298-07-7
Formula: $C_{16}H_{35}O_4P$; $[CH_3CHCHCHCH(C_2H_5)CH_2O]_2POOH$
DOT ID Number: UN 1902; DOT Guide Number: 153
Proper Shipping Name: Diisooctyl acid phosphate or Di-(2-ethylhexyl)phosphoric acid

DESCRIPTION: Light yellow liquid. Odorless. Floats on water surface; practically insoluble.

Combustible • Containers may BLEVE when exposed to fire • Vapors are heavier than air and will collect and stay in low areas • Vapors in confined areas (e.g., tanks, sewers, buildings) may explode when exposed to fire • Irritating to the skin, eyes, and respiratory tract • Toxic products of combustion may include phosphorus oxides.

Hazard Classification (based on NFPA-704 Rating System)
Health Hazards (Blue): 1; Flammability (Red): 1; Reactivity (Yellow): 1

EMERGENCY RESPONSE: See Appendix A (153)
Evacuation:
Public safety: Isolate the area of spill or leak for at least 25 to 50 meters (80 to 160 feet) in all directions.
Spill: Increase, in the downwind direction, as necessary, the distance shown under "Public Safety."
Fire: If tank, rail car, or tank truck is involved in fire, isolate for at least 800 meters (½ mile) in all directions; also, consider initial evacuation for 800 meters (½ mile) in all directions.

EXPOSURE
Short-term effects: *SEEK MEDICAL ATTENTION*. **Liquid:** Irritating to skin and eyes. Harmful if swallowed. Remove contaminated clothing and shoes. Flush affected areas with plenty of water. *IF BREATHING HAS STOPPED*, give artificial respiration; *avoid mouth-to-mouth resuscitation; use bag/mask apparatus. IF IN EYES*, hold eyelids open and flush with plenty of water. *IF SWALLOWED* and victim is *CONSCIOUS AND ABLE TO SWALLOW*, have victim drink 4 to 8 ounces of water and have victim induce vomiting. *IF SWALLOWED* and victim is *UNCONSCIOUS OR HAVING CONVULSIONS*, do nothing except keep victim warm.

HEALTH HAZARDS
Personal protective equipment (PPE): B-Level PPE. Neoprene is generally suitable for mineral acids.
Toxicity by ingestion: Grade 2; LD_{50} = 0.5 to 5 g/kg.

Di-(2-ethylhexyl)phthalate

Vapor (gas) irritant characteristics: Vapors are nonirritating to the eyes and throat.
Liquid or solid irritant characteristics: Severe eye and skin irritant: may cause first-degree burns on short exposure and may cause second-degree burns on long exposure.

FIRE DATA
Flash point: 385°F/196°C (oc).
Fire extinguishing agents not to be used: Water or foam may cause frothing.

CHEMICAL REACTIVITY
Binary reactants: Mildly corrosive to most metals; may form flammable hydrogen gas.
Neutralizing agents for acids and caustics: Sodium bicarbonate or lime solution
Reactivity group: 1
Compatibility class: Nonoxidizing Mineral Acids

ENVIRONMENTAL DATA
Food chain concentration potential: Log P_{ow} = 2.94. Values > 3.0 are likely to bioconcentrate in aquatic organisms and other living tissue, especially in fats.
Water pollution: Effect of low concentrations on aquatic life is unknown. Fouling to shoreline. May be dangerous if it enters nearby water intakes; notify operators. Notify local health and wildlife officials. **Response to discharge:** Issue warning–corrosive, water contaminant. Restrict access. Mechanical containment. Should be removed. Chemical and physical treatment.

SHIPPING INFORMATION
Grades of purity: 92+%; **Storage temperature:** Ambient; **Inert atmosphere:** None; **Venting:** Open; **Stability during transport:** Stable.

NAS HAZARD CLASSIFICATION FOR BULK WATER TRANSPORTATION
FIRE: 1
HEALTH: Vapor irritant: 0; Liquid or solid irritant: 2; Poisons: 3
WATER POLLUTION: Human toxicity:–; Aquatic toxicity:–; Aesthetic effect: 2
REACTIVITY: Other chemicals: 1; Water: 0; Self-reaction: 0

PHYSICAL AND CHEMICAL PROPERTIES
Physical state @ 59°F/15°C and 1 atm: Liquid.
Molecular weight: 322.48
Boiling point @ 1 atm: Decomposes.
Melting/Freezing point: Less than –76°F/–60°C/213°K.
Specific gravity (water = 1): 0.977 @ 68°F/20°C (liquid).
Liquid surface tension: (estimate) 20 dynes/cm = 0.020 N/m @ 68°F/20°C.
Liquid water interfacial tension: (estimate) 30 dynes/cm = 0.030 N/m @ 68°F/20°C.
Heat of combustion: –13,970 Btu/lb = –7,760 cal/g = –325 x 10^5 J/kg.

DI-(2-ETHYLHEXYL)PHTHALATE   REC. D:3650

SYNONYMS: BEHP; BIS-(2-ETHYLHEXYL) PHTHALATE; BISOFLEX-81; BISOFLEX DOP; DAF 68; DEHP; DOP; DI-*sec*-OCTYL PHTHALATE; DIOCTYL PHTHALATE; FTALATO de DI(2-ETILHEXILO) (Spanish); PHTHALIC ACID, BIS(2-ETHYLHEXYL ESTER); RCRA No. U028; WITCIZER 312

IDENTIFICATION
CAS Number: 117-81-7
Formula: $C_{24}H_{38}O_4$; $C_6H_4[COOCHCH(C_2NO_5)(CH)_3CH_3]_2$
DOT ID Number: UN 3082; DOT Guide Number: 171
Proper Shipping Name: Environmentally hazardous liquid, n.o.s.
Reportable Quantity (RQ): **(CERCLA)** 100 lb/45.4 kg

DESCRIPTION: Colorless oily liquid. Slight odor. Floats on the surface of water; practically insoluble.

Combustible • Containers may BLEVE when exposed to fire • Vapors are heavier than air and will collect and stay in low areas • Vapors in confined areas (e.g., tanks, sewers, buildings) may explode when exposed to fire • Irritating to the skin, eyes, and respiratory tract • Reacts violently with strong oxidizers • Toxic products of combustion may include carbon monoxide.

Hazard Classification (based on NFPA-704 Rating System)
Health Hazards (Blue): 0; Flammability (Red): 1; Reactivity (Yellow): 0

EMERGENCY RESPONSE: See Appendix A (171)
Evacuation:
Public safety: Isolate the area of spill or leak for at least 10 to 25 meters (30 to 80 feet) in all directions.
Spill: Increase, in the downwind direction, as necessary, the distance shown under "Public Safety."
Fire: If any large container is involved in fire, isolate for at least 800 meters (½ mile) in all directions; also, consider initial evacuation for 800 meters (½ mile) in all directions.

EXPOSURE
Short-term effects: *SEEK MEDICAL ATTENTION*. Move to fresh air. Remove contaminated clothing and shoes. Wash affected areas with plenty of soap and water. *IF IN EYES*, hold eyelids open and flush with plenty of water.

HEALTH HAZARDS
Personal protective equipment (PPE): B-Level PPE. OSHA Table Z-1-A air contaminant. Chemical protective clothing, gloves, and other appropriate protective clothing to prevent skin contact. Contact lenses should not be worn when working with this material. Use self-contained breathing apparatus if vapor concentrations are present. Chemical protective material(s) reported to have good to excellent resistance: butyl rubber, Viton®.
Recommendations for respirator selection: NIOSH
At any concentrations above the NIOSH REL, or where there is no REL, at any detectable concentration: SCBAF:PD,PP (any self-contained breathing apparatus that has a full facepiece and is operated in a pressure-demand or other positive-pressure mode); or SAF:PD,PP:ASCBA (any supplied-air respirator that has a full facepiece and is operated in a pressure-demand or other positive-pressure mode in combination with an auxiliary self-contained breathing apparatus operated in a pressure-demand or other positive pressure mode). *ESCAPE:* HiEF (any air-purifying, full-facepiece respirator with a high-efficiency particulate filter); or SCBAE (any appropriate escape-type, self-contained breathing apparatus).
Exposure limits (TWA unless otherwise noted): ACGIL TLV 5 mg/m³; OSHA PEL 5 mg/m³; NIOSH REL 5 mg/m³; potential human carcinogen; reduce exposure to lowest feasible level
Short-term exposure limits (15-minute TWA): ACGIH STEL 10 mg/m³; NIOSH STEL 10 mg/m³.
Long-term health effects: Subchronic inhalation by rats or mice caused pulmonary irritation, congestion of liver and kidneys, renal cysts, bladder stones, and increased liver metabolism. Chronic

inhalation by mice or rats caused liver cancer. NTP anticipated carcinogen. IARC possible carcinogen, rating 2B; sufficient animal evidence.
Vapor (gas) irritant characteristics: Eye and respiratory tract irritant. Vapors cause irritation such that personnel will find high concentrations unpleasant. The effect is temporary.
Liquid or solid irritant characteristics: Eye and skin irritant. If spilled on clothing and allowed to remain, may cause poisoning.
IDLH value: 5000 mg/m^3; potential human carcinogen.

FIRE DATA
Flash point: 425°F/218°C (oc).
Fire extinguishing agents not to be used: Water or foam may cause frothing.
Autoignition temperature: 735°F/391°C/664°K.

CHEMICAL REACTIVITY
Binary reactants: Incompatible with nitrates, strong oxidizers, strong alkalis, or strong acids. Reactions with these compounds may cause fires and explosions.
Reactivity group: 34
Compatibility class: Esters.

ENVIRONMENTAL DATA
Food chain concentration potential: Negative; unlikely to accumulate.
Water pollution: Effects of low concentrations on aquatic life is unknown. May be dangerous if it enters nearby water intakes; notify operators. Notify local health and wildlife officials.
Response to discharge: Mechanical containment. Should be removed. CLEAN WATER ACT: Section 307 Priority Pollutants

SHIPPING INFORMATION
Grades of purity: 99%; Technical grades; **Storage temperature:** Ambient; **Inert atmosphere:** None; **Stability during transport:** Stable.

PHYSICAL AND CHEMICAL PROPERTIES
Physical state @ 59°F/15°C and 1 atm: Liquid.
Molecular weight: 390.5
Boiling point @ 1 atm: 727°F/386°C/659°K.
Melting/Freezing point: –58°F/–50°C/223°K.
Specific gravity (water = 1): 0.9861
Liquid surface tension: (estimate) 15 dynes/cm = 0.015 N/m @ 68°F/20°C.
Liquid water interfacial tension: (estimate) 30 dynes/cm = 0.03 N/m @ 68°F/20°C.
Relative vapor density (air = 1): 16
Vapor pressure: Less than 0.01 mm.

DIETHYL KETONE   REC. D:3700

SYNONYMS: DEK; DIETHYLCETONE (French); DIETILCETONA (Spanish); DIMETHYLACETONE; EEC No. 606-006-00-5; ETHYL KETONE; ETHYL PROPIONYL; METACETONE; METHACETONE; PENTANONE-3; 3-PENTANONE; PROPIONE; 3-PENTANONE DIMETHYL ACETONE

IDENTIFICATION
CAS Number: 96-22-0
Formula: $C_5H_{10}O$; $CH_3CHCOCH_2CH_3$
DOT ID Number: UN 1156; DOT Guide Number: 127
Proper Shipping Name: Diethyl ketone

DESCRIPTION: Colorless, clear liquid. Acetone-like odor. Floats on the surface of water; slightly soluble. Flammable, irritating vapor is produced.

Highly flammable • Containers may BLEVE when exposed to fire • Vapors may form explosive mixture with air • Vapors are heavier than air and will collect and stay in low areas • Vapors may travel long distances to ignition sources and flashback • Vapors in confined areas (e.g., tanks, sewers, buildings) may explode when exposed to fire • Irritating to the skin, eyes, and respiratory tract • Toxic products of combustion may include carbon monoxide.

Hazard Classification (based on NFPA-704 Rating System)
Health Hazards (Blue): 1; Flammability (Red): 3; Reactivity (Yellow): 0

EMERGENCY RESPONSE: See Appendix A (127)
Evacuation:
Public safety: Isolate spill area for at least 25 to 50 meters (80 to 160 feet) in all directions.
Spill: Large spill–Consider initial downwind evacuation for at least 300 meters (1000 feet).
Fire: Isolate for 800 meters (½ mile) in all directions, especially if tank, rail car, or tank truck is involved in fire.

EXPOSURE
Short-term effects: *SEEK MEDICAL ATTENTION.* **Vapor:** Irritating to eyes, nose, and throat. *IF INHALED*, will, will cause nausea, vomiting, headache, dizziness, difficult breathing, or loss of consciousness. Move to fresh air. *IF BREATHING HAS STOPPED,* give artificial respiration. If breathing is difficult, administer oxygen. **Liquid:** Will burn eyes. Harmful if swallowed. Absorbed by the skin. Remove contaminated clothing and shoes. Flush affected areas with plenty of water. *IF IN EYES*, hold eyelids open and flush with plenty of water. *IF SWALLOWED* and victim is *CONSCIOUS AND ABLE TO SWALLOW*, have victim drink 4 to 8 ounces of water.

HEALTH HAZARDS
Personal protective equipment (PPE): B-Level PPE. OSHA Table Z-1-A air contaminant. Organic canister); or air pack; plastic gloves; goggles or face shield.
Exposure limits (TWA unless otherwise noted): ACGIH TLV 200 ppm (705 mg/m^3); NIOSH REL 200 ppm (705 mg/m^3).
Short-term exposure limits (15-minute TWA): ACGIH STEL 300 ppm.
Toxicity by ingestion: Grade 2; LD_{50} = 2.14 g/kg (rat).
Vapor (gas) irritant characteristics: Eye and respiratory tract irritant. Vapors cause smarting of the eyes or respiratory system if present in high concentrations.
Liquid or solid irritant characteristics: Severe eye and skin irritant. If spilled on clothing and allowed to remain, may cause smarting and reddening of the skin.
Odor threshold: 0.85–9.2 ppm.

FIRE DATA
Flash point: 55°F/13°C (oc).
Flammable limits in air: LEL 1.6; UEL: 6.4%.
Fire extinguishing agents not to be used: Water may be ineffective due to low flash point.
Autoignition temperature: 842°F/450°C/723°K.
Electrical hazard: Class 1, Group D. Due to low electric conductivity, this substance may generate electrostatic charges as a result of agitation and flow.

382 Diethyl phthalate

CHEMICAL REACTIVITY
Binary reactants: Reacts with strong oxidizers, alkalis, mineral acids, plastics.

ENVIRONMENTAL DATA
Food chain concentration potential: Negative; unlikely to accumulate.
Water pollution: Dangerous to aquatic life in high concentrations. May be dangerous if it enters nearby water intakes; notify operators. Notify local health and wildlife officials. **Response to discharge:** Issue warning–high flammability. Disperse and flush.

SHIPPING INFORMATION
Grades of purity: 99.5+%; **Storage temperature:** Ambient; **Inert atmosphere:** None; **Venting:** Open (flame arrester) or pressure-vacuum; **Stability during transport:** Stable.

NAS HAZARD CLASSIFICATION FOR BULK WATER TRANSPORTATION
FIRE: 3
HEALTH: Vapor irritant: 1; Liquid or solid irritant: 1; Poisons: 2
WATER POLLUTION: Human toxicity: 2; Aquatic toxicity: 1; Aesthetic effect: 1
REACTIVITY: Other chemicals: 2; Water: 0; Self-reaction: 0

PHYSICAL AND CHEMICAL PROPERTIES
Physical state @ 59°F/15°C and 1 atm: Liquid.
Molecular weight: 86.13
Boiling point @ 1 atm: 216°F/102°C/375°K.
Melting/Freezing point: –44°F/–42°C/231°K.
Critical temperature: 550°F/288°C/561°K.
Critical pressure: 543 psia = 36.9 atm = 3.74 MN/m^2
Specific gravity (water = 1): 0.8159 @ 66°F/19°C/292°K (liquid).
Liquid surface tension: 25.33 @ 15°C.
Relative vapor density (air = 1): 2.96
Latent heat of vaporization: 163.4 Btu/lb = 90.8 cal/g = 3.8 x 10^5 J/kg.
Heat of combustion: –15373 Btu/lb = –8541 cal/g = –357 x 10^5 J/kg.
Heat of decomposition: (estimate) –1542 Btu/lb = –857 cal/g = –35.9 x 10^5 J/kg.
Vapor pressure: (Reid) 0.5 psia; 35.43 mmHg @ 77°F/25°C.

DIETHYL PHTHALATE   REC. D:3750

SYNONYMS: ANOZOL; 1,2-BENZENEDICARBOXYLIC ACID, DIETHYL ESTER; DEP; DIETHYL ESTER OF PHTHALIC ACID; p-DIETHYL PHTHALATE; ETHYL PHTHALATE; ESTOL-1550; FTALATO de DIETILO (Spanish); NEANTINE; PALATINOL A; PHTHALIC ACID, DIETHYL ESTER; PHTHALIC ACID, DIETHYL ESTER; PHTHALOL; PLACIDOL E; RCRA No. U088; SOLVANOL

IDENTIFICATION
CAS Number: 84-66-2
Formula: $C_{12}H_{14}O_4$; $C_6H_4(COOC_2NO_5)_2$
DOT ID Number: UN 3082; DOT Guide Number: 171
Proper Shipping Name: Environmentally hazardous substances, liquid, n.o.s.
Reportable Quantity (RQ): **(CERCLA)** 1000 lb/454 kg

DESCRIPTION: White liquid. Mild chemical odor. Sinks in water; insoluble.

Combustible • Containers may BLEVE when exposed to fire • Vapors are heavier than air and will collect and stay in low areas • Vapors in confined areas (e.g., tanks, sewers, buildings) may explode when exposed to fire • Irritating to the skin, eyes, and respiratory tract • Toxic products of combustion may include carbon monoxide.

Hazard Classification (based on NFPA-704 Rating System)
Health Hazards (Blue): 0; Flammability (Red): 1; Reactivity (Yellow): 0

EMERGENCY RESPONSE: See Appendix A (171)
Evacuation:
Public safety: Isolate the area of spill or leak for at least 10 to 25 meters (30 to 80 feet) in all directions.
Spill: Increase, in the downwind direction, as necessary, the distance shown under "Public Safety."
Fire: If any large container is involved in fire, isolate for at least 800 meters (½ mile) in all directions; also, consider initial evacuation for 800 meters (½ mile) in all directions.

EXPOSURE
Short-term effects:
Vapor: High concentration may affect the central nervous system; dizziness, drowsiness.

HEALTH HAZARDS
Personal protective equipment (PPE): B-Level PPE. OSHA Table Z-1-A air contaminant. Butyl rubber is generally suitable for carbooxylic acid compounds.
Exposure limits (TWA unless otherwise noted): ACGIH TLV 5 mg/m3; NIOSH REL 5 mg/m^3.
Toxicity by ingestion: Grade 2; oral LD_{50} = 1000 mg/kg (rabbit).
Long-term health effects: Prolonged inhalation of heated vapor produces irritation of upper respiratory tract in humans
Vapor (gas) irritant characteristics: May cause eye and respiratory tract irritation.
Liquid or solid irritant characteristics: Eye and skin irritant.

FIRE DATA
Flash point: 322°F/161°C (oc).
Flammable limits in air: LEL: 0.7% @ 368°F/187°C.
Fire extinguishing agents not to be used: Water or foam may cause frothing.
Autoignition temperature: 855°F/457°C/730°K.

CHEMICAL REACTIVITY
Reactivity with water: May react with water; danger unknown.
Binary reactants: Strong oxidizers, strong acids, nitric acid, permanganates. May attack some forms of plastics.

ENVIRONMENTAL DATA
Food chain concentration potential: Negative; unlikely to accumulate.
Water pollution: Harmful to aquatic life in very low concentrations. May be dangerous if it enters nearby water intakes; notify operators. Notify local health and wildlife officials. **Response to discharge:** Should be removed. Chemical and physical treatment. Clean water act: Section 307; 313.

SHIPPING INFORMATION
Grades of purity: Technical; 99+%; **Storage temperature:** Ambient; **Inert atmosphere:** None; **Venting:** Open; **Stability during transport:** Stable.

PHYSICAL AND CHEMICAL PROPERTIES
Physical state @ 59°F/15°C and 1 atm: Liquid.
Molecular weight: 222.3
Boiling point @ 1 atm: 563°F/294°C/567°K.
Melting/Freezing point: –41°F/–41°C/232°K.
Specific gravity (water = 1): 1.12 @ 68°F/20°C (liquid).
Liquid surface tension: 37.5 dynes/cm = 0.0375 N/m @ 68°F/20°C.
Liquid water interfacial tension: 16.27 dynes/cm = 0.01627 N/m at 20.5°C.
Relative vapor density (air = 1): 7.68
Ratio of specific heats of vapor (gas): 1.0 (approximate).
Latent heat of vaporization: 170 Btu/lb = 96 cal/g = 4.0×10^5 J/kg.
Heat of combustion: –10,920 Btu/lb = –6,070 cal/g = $–254 \times 10^5$ J/kg.
Vapor pressure: 0.002 mm @ 77°F/25°C.

DIETHYL SULFATE   REC. D:3800

SYNONYMS: DIAETHYLSULFAT (German); DIETHYL MONOSULFATE; DIETHYL SULPHATE; DS; DIETHYL ESTER SULFURIC ACID; EEC No. 016-027-00-6; ETHYL SULFATE; SULFATO de DIETILO (Spanish); SULFURIC ACID, DIETHYL ESTER

IDENTIFICATION
CAS Number: 64-67-5
Formula: $(C_2NO_5)_2SO_4$
DOT ID Number: UN 1594; DOT Guide Number: 152
Proper Shipping Name: Diethyl sulfate
Reportable Quantity (RQ): **(CERCLA)** 1 lb/0.454 kg

DESCRIPTION: Colorless liquid. Ether-like or peppermint odor. Sinks and very slowly dissolves in water; gradually decomposes to produce sulfuric acid, ethyl alcohol, and monoethyl sulfate.

Combustible • Poison! • Breathing the vapor can cause severe illness; skin or eye contact causes severe burns, impaired vision, or blindness • Firefighting gear (including SCBA) does not provide adequate protection. If exposure occurs, remove and isolate gear immediately and thoroughly decontaminate personnel • Containers may BLEVE when exposed to fire • Vapors are heavier than air and will collect and stay in low areas • Vapors in confined areas (e.g., tanks, sewers, buildings) may explode when exposed to fire • Irritating to the skin, eyes, and respiratory tract • Above 212°F/100°C, it undergoes thermal decomposition yielding ethyl ether, ethylene, and sulfur oxides which may cause an explosion in closed containers or confined spaces. Toxic products of combustion may also include carbon monoxide and sulfur oxides.

Hazard Classification (based on NFPA-704 Rating System)
Health Hazards (Blue): 3; Flammability (Red): 1; Reactivity (Yellow): 1

EMERGENCY RESPONSE: See Appendix A (152)
Evacuation:
Public safety: Isolate the area of spill or leak for at least 25 to 50 meters (80 to 160 feet) in all directions.
Spill: Increase, in the downwind direction, as necessary, the distance shown under "Public Safety."
Fire: If tank, rail car, or tank truck is involved in fire, isolate for at least 800 meters (½ mile) in all directions; also, consider initial evacuation for 800 meters (½ mile) in all directions.

EXPOSURE
Short-term effects: *SEEK MEDICAL ATTENTION*. **Vapor:** *POISONOUS IF INHALED*. Irritating to eyes, skin and mucous membranes. Move to fresh air. *IF BREATHING HAS STOPPED*, give artificial respiration; *avoid mouth-to-mouth resuscitation; use bag/mask apparatus*. If breathing is difficult, administer oxygen. **Liquid:** *POISONOUS IF SWALLOWED*. Will burn skin and eyes. Remove contaminated clothing and shoes. Flush contaminated areas with plenty of running water. *IF IN EYES*, hold eyelids open and flush with running water for at least 15 minutes. *IF ON SKIN:* Wash with soap and plenty of running water; speed in removing material from skin is of extreme importance. Remove and double bag contaminated clothing and shoes at the site. *IF SWALLOWED* and victim is *CONSCIOUS AND ABLE TO SWALLOW*, give two glasses of water and induce vomiting. *IF SWALLOWED* and victim is *UNCONSCIOUS,* do nothing but keep victim warm.

HEALTH HAZARDS
Personal protective equipment (PPE): A-Level PPE.
Toxicity by ingestion: Grade 2; LD_{50} = 647 mg/kg (mouse).
Long-term health effects: Causes mutagenic, tumorigenic and cancerogenic effects. IARC possible carcinogen, rating 2B; human evidence: limited; animal evidence: sufficient.
Vapor (gas) irritant characteristics: Vapor cause severe irritation of eyes and throat (mucous membranes) and can cause eye and lung injury.
Liquid or solid irritant characteristics: Severe eye and skin irritant. May cause pain and second-degree burns after a few minutes of contact.

FIRE DATA
Flash point: 220°F/104°C (cc).
Fire extinguishing agents not to be used: Alcohol foam, universal foam. Water or foam may cause frothing.
Autoignition temperature: 817°F/436°C/709°K.
Electrical hazard: Class I, Group D.
Stoichiometric air-to-fuel ratio: 4.95 (estimate).

CHEMICAL REACTIVITY
Reactivity with water: Reacts slowly with cold water about 0.05% per hour at 77°F/25°C, to yield monoethyl sulfate and ethyl alcohol. Reacts vigorously with water at temperature above 58°F/50°C. Sulfuric acid may be produced along with ethyl alcohol and monoethyl sulfate.
Binary reactants: Violent reaction with aluminum, magnesium. Avoid contact with aqueous alkali, concentrated nitric acid and strong oxidizing agents such as peroxides and peracids. Incompatible with zinc, galvanized iron, lead, nickel or copper and its alloys. Violent reaction occurs with potassium and *tert*-butoxide. It may react with moisture to yield sulfuric acid which subsequently may react with a metal container to liberate hydrogen gas resulting in an explosion.
Neutralizing agents for acids and caustics: Dilute aqueous sodium hydroxide.
Reactivity group: 34
Compatibility class: Esters.

ENVIRONMENTAL DATA
Water pollution: Harmful to aquatic life in very low concentrations. May be dangerous if it enters nearby water intakes; notify operators. Notify local health and wildlife officials. **Response to discharge:** Issue warning–poison, air contaminant, water contaminant. Restrict access. Should be removed. Chemical and physical treatment.

SHIPPING INFORMATION
Grades of purity: 100%; **Storage temperature:** Ambient; **Stability during transport:** Stable.

PHYSICAL AND CHEMICAL PROPERTIES
Physical state @ 59°F/15°C and 1 atm: Liquid.
Molecular weight: 154.18
Boiling point @ 1 atm: 409°F/210°C/483°K (decomposes).
Melting/Freezing point: −32°F/−25°C/248°K.
Specific gravity (water = 1): 1.1803 @ 20°C.
Liquid surface tension: 33.5 dynes/cm = 0.034 N/m @ 68°F/20°C.
Relative vapor density (air = 1): 5.3
Vapor pressure: 0.2 mm.

DIETHYL ZINC   REC. D:3850

SYNONYMS: DIETILZINC (Spanish); ETHYLZINC; ZINC ETHIDE; ZINC ETHYL; ZINC DIETHYL-

IDENTIFICATION
CAS Number: 557-20-0
Formula: $C_4H_{10}Zn$; $(C_2NO_5)_2Zn$
DOT ID Number: UN 1366; DOT Guide Number: 135
Proper Shipping Name: Diethylzinc

DESCRIPTION: Colorless watery liquid. Liquid is odorless but burns with a garlic odor. Pyrophoric (ignites spontaneously when exposed to air). Decomposes in water producing flammable ethane gas and irritating vapor.

Pyrophoric. Explosive decomposition at 245°F/120°C • Firefighting gear (including SCBA) does not provide adequate protection. If exposure occurs, remove and isolate gear immediately and thoroughly decontaminate personnel • Containers may BLEVE when exposed to fire • Vapors may form explosive mixture with air • Vapors are heavier than air and will collect and stay in low areas • Vapors may travel long distances to ignition sources and flashback • Vapors in confined areas (e.g., tanks, sewers, buildings) may explode when exposed to fire • Severely Irritating to the skin, eyes, and respiratory tract; skin and eye contact causes severe burns and blindness • Toxic products of combustion may include zinc oxide fumes • Violent reaction with water producing explosive ethane gas • May re-ignite after fire is extinguished • Do not put yourself in danger by entering a contaminated area to rescue a victim.

Hazard Classification (based on NFPA-704 Rating System)
Health Hazards (Blue): 3; Flammability (Red): 4; Reactivity (Yellow): 3; Special Notice (White): Water reactive

EMERGENCY RESPONSE: See Appendix A (135)
Evacuation:
Public safety: Isolate spill area for at least 100 to 150 meters (330 to 490 feet) in all directions.
Spill: Increase downwind, the distance shown above in "Public safety."
Fire: Isolate for 800 meters (½ mile) in all directions, especially if tank, rail car, or tank truck is involved in fire.

EXPOSURE
Short-term effects: SEEK MEDICAL ATTENTION. **Vapor or dust:** Irritating to eyes, nose, and throat. *IF INHALED*, will, will cause headache, or difficult breathing. Move victim to fresh air. *IF IN EYES*, hold eyelids open and flush with plenty of water. *IF BREATHING HAS STOPPED*, give artificial respiration. If breathing is difficult, administer oxygen. **Liquid:** Will burn skin and eyes. *IF SWALLOWED*, will cause nausea, and vomiting. Remove contaminated clothing and shoes. Flush affected areas with plenty of water. *IF IN EYES*, hold eyelids open and flush with plenty of water. *IF SWALLOWED* and victim is *CONSCIOUS AND ABLE TO SWALLOW*, have victim drink 4 to 8 ounces of water. **Do NOT induce vomiting.**

HEALTH HAZARDS
Personal protective equipment (PPE): B-Level PPE. Cartridge-type or fresh air mask for fumes or smoke; PVC fire-retardant or asbestos gloves; full face shield, safety glasses, or goggles; fire-retardant coveralls as standard wear; for special cases, use asbestos coat or rain suit.
Vapor (gas) irritant characteristics: Severe eye and respiratory tract irritant. Exposure may cause permanent damage.
Liquid or solid irritant characteristics: Extremely severe eye, skin and respiratory tract irritant. May cause burns and permanent damage.

FIRE DATA
Fire extinguishing agents not to be used: Water, foam, halogenated agents, CO_2. Contact with water applied to adjacent fires will intensify the fire.
Special hazards of combustion products: Zinc oxide fumes can cause metal fume fever.
Behavior in fire: Reacts spontaneously with air or oxygen, and violently with water, evolving flammable ethane gas.
Autoignition temperature: Less than 0°F/−18°C/255°K.

CHEMICAL REACTIVITY
Reactivity with water: Reacts violently to form flammable ethane gas.
Binary reactants: Spontaneously ignites in air or on contact with oxidizers. Reacts, often violently, with many compounds. Will react with surface moisture, generating flammable ethane gas.

ENVIRONMENTAL DATA
Food chain concentration potential: Negative; unlikely to accumulate.
Water pollution: Effect of low concentrations on aquatic life is unknown. May be dangerous if it enters nearby water intakes; notify operators. Notify local health and wildlife officials. **Response to discharge:** Issue warning–high flammability. Restrict access. Evacuate area.

SHIPPING INFORMATION
Grades of purity: 95–98%. Also, shipped as 15–25% by weight solutions in hydrocarbon solvents; **Storage temperature:** Ambient; **Inert atmosphere:** Inerted with dry nitrogen gas; **Venting:** Safety relief; **Stability during transport:** Stable.

PHYSICAL AND CHEMICAL PROPERTIES
Physical state @ 59°F/15°C and 1 atm: Liquid.
Molecular weight: 123.5
Boiling point @ 1 atm: 243°F/117°C/390°K.
Melting/Freezing point: −18°F/−28°C/245°K.
Specific gravity (water = 1): 1.207 @ 20°C (liquid).
Liquid surface tension: (estimate) 20 dynes/cm = 0.020 N/m @ 68°F/20°C.
Latent heat of vaporization: 120 Btu/lb = 68 cal/g = 2.8×10^5 J/kg.
Heat of combustion: −11,700 Btu/lb = −6495 cal/g = -272×10^5 J/kg.

1,1-DIFLUOROETHANE   REC. D:3900

SYNONYMS: ALGOFRENE TYPE 67; DIFLUROETHANE; 1,1-DIFLUROETHANE; 1,1-DIFLUORETANO (Spanish); ETHYLENE FLUORIDE; ETHYLIDENE DIFLUORIDE; ETHYLIDENE FLUORIDE; R 152A; REFRIGERANT 152a; FREON 152; GENETRON 100; HALOCARBON 152a

IDENTIFICATION
CAS Number: 75-37-6
Formula: $C_2H_4F_2$; CH_3CHF_2
DOT ID Number: UN 1030; DOT Guide Number: 115
Proper Shipping Name: Difluoroethane, R152

DESCRIPTION: Colorless gas. Shipped and stored as a liquefied compressed gas. Odorless. May boil on the surface of water; insoluble; reacts, forming hydrogen fluoride; produces a large vapor cloud. A liquid below –13°F/–25°C.

Extremely flammable • Forms explosive mixture with air • Reacts explosively with many materials • Gas from liquefied gas are initially heavier than air and spread along ground • Gas may travel to source of ignition and flashback • Exposure of cylinders to elevated temperatures, fire, and flame may cause cylinders to rupture or cause frangible disk to burst, releasing entire contents of cylinder • Ruptured or venting cylinders may rocket through buildings and/or travel a considerable distance • Irritating to the skin, eyes, and respiratory tract • Vapors may cause dizziness or asphyxiation without warning • Contact with liquid may cause frostbite • Toxic products of combustion may includes irritating hydrogen fluoride fumes.

Hazard Classification (based on NFPA-704 Rating System)
Health Hazards (Blue): 1; Flammability (Red): 4; Reactivity (Yellow): 0

EMERGENCY RESPONSE: See Appendix A (115)
Evacuation:
Public safety: Isolate spill area for at least 50 to 100 meters (160 to 330 feet) in all directions.
Spill: Consider initial downwind evacuation for at least 800 meters (½ mile).
Fire: Isolate for 1600 meters (1 mile) in all directions, especially if tank, rail car, or tank truck is involved in fire.

EXPOSURE
Short-term effects: *SEEK MEDICAL ATTENTION*. The use of alcoholic beverages may enhance toxic effects. **Vapor:** Irritating to eyes, nose, and throat. Harmful if inhaled. Move victim to fresh air. *IF BREATHING HAS STOPPED,* give artificial respiration. If breathing is difficult, administer oxygen. **Liquid:** Will burn eyes. Will cause frostbite. Harmful if swallowed. Remove contaminated clothing and shoes. Flush affected areas with plenty of water. *DO NOT RUB AFFECTED AREAS. IF SWALLOWED* and victim is *CONSCIOUS AND ABLE TO SWALLOW,* have victim drink 4 to 8 ounces of water. The use of alcoholic beverages may enhance the toxic effect.

HEALTH HAZARDS
Personal protective equipment (PPE): B-Level PPE. Individual breathing devices with air supply; gloves; protective clothing; eye protection Nitrile rubber is generally suitable for freons. Wear thermal protective clothing.
Long-term health effects: May effect the nervous system.
Vapor (gas) irritant characteristics: Eye irritant.
Liquid or solid irritant characteristics: Rapid evaporation of liquid may cause frostbite.

FIRE DATA
Flash point: Gas.
Flammable limits in air: LEL: 3.7%; UEL: 18%.
Behavior in fire: Containers may explode.
Electrical hazard: Due to low electric conductivity, this substance may generate electrostatic charges as a result of agitation and flow.
Stoichiometric air-to-fuel ratio: 5.196 (estimate).

CHEMICAL REACTIVITY
Binary reactants: Violent reaction with strong oxidizers. Attacks many metals in presence of water or vapor.

ENVIRONMENTAL DATA
Food chain concentration potential: Log P_{ow} = 0.72. Unlikely to accumulate.
Water pollution: Not harmful to aquatic life. **Response to discharge:** Issue warning–high flammability. Restrict access. Evacuate area.

SHIPPING INFORMATION
Grades of purity: Commercial; **Storage temperature:** Ambient; **Inert atmosphere:** None; **Venting:** Safety relief; **Stability during transport:** Stable.

PHYSICAL AND CHEMICAL PROPERTIES
Physical state @ 59°F/15°C and 1 atm: Gas.
Molecular weight: 66.05
Boiling point @ 1 atm: –12.46°F/–24.7°C/248.3°K.
Melting/Freezing point: –179°F/–117°C/156°K.
Critical temperature: 236.3°F/113.5°C/386.6°K.
Critical pressure: 652 psia = 44.37 atm = 4.50 MN/m².
Specific gravity (water = 1): 0.95 @ 68°F/20°C (liquid).
Liquid surface tension: 11.25 dynes/cm = 0.01125 N/m @ 68°F/20°C.
Relative vapor density (air = 1): 2.3
Ratio of specific heats of vapor (gas): 1.141
Latent heat of vaporization: 140.5 Btu/lb = 78.03 cal/g = 3.265 x 10^5 J/kg.
Heat of combustion: –7950 Btu/lb = –4420 cal/g = –185 x 10^5 J/kg.
Vapor pressure: 4 mm.

DIFLUOROPHOSPHORIC ACID   REC. D:3950

SYNONYMS: ACIDO DIFLUOROFOSFORICO (Spanish); PHOSPHORODIFLUORIDIC ACID

IDENTIFICATION
CAS Number: 13779-41-4
Formula: F_2HO_2P; $HOPOF_2$
DOT ID Number: UN 1768; DOT Guide Number: 154
Proper Shipping Name: Difluorophosphoric acid, anhydrous

DESCRIPTION: Colorless liquid; fumes in moist air. Sharp, irritating odor. Reacts violently with water forming hydrogen fluoride fumes.

Toxic! • Do not use water or foam • Corrosive to the skin, eyes, and respiratory tract; contact with skin or eyes can cause burns and vision impairment; inhalation symptoms may be delayed • Firefighting gear (including SCBA) does not provide adequate protection. If exposure occurs, remove and isolate gear immediately and thoroughly decontaminate personnel • Vapors are heavier than air and will collect and stay in low areas • Toxic products of combustion may include hydrogen fluoride and phosphoric acid fumes.

Hazard Classification (based on NFPA-704 Rating System)
Health Hazards (Blue): 3; Flammability (Red): 0; Reactivity (Yellow): 1

EMERGENCY RESPONSE: See Appendix A (154)
Evacuation:
Public safety: Isolate the area of spill or leak for at least 25 to 50 meters (80 to 160 feet) in all directions.
Spill: Increase, in the downwind direction, as necessary, the distance shown under "Public Safety."
Fire: If tank, rail car, or tank truck is involved in fire, isolate for at least 800 meters (½ mile) in all directions; also, consider initial evacuation for 800 meters (½ mile) in all directions.

EXPOSURE
Short-term effects: *SEEK MEDICAL ATTENTION.* **Vapor:** Irritating to eyes, nose, and throat. Move victim to fresh air. *IF IN EYES,* hold eyelids open and flush with plenty of water. *IF BREATHING HAS STOPPED,* give artificial respiration; *avoid mouth-to-mouth resuscitation; use bag/mask apparatus.* If breathing is difficult, administer oxygen. **Liquid:** Will burn skin and eyes. Harmful if swallowed. Remove contaminated clothing and shoes. Flush affected areas with plenty of water. *IF IN EYES,* hold eyelids open and flush with plenty of water. *IF SWALLOWED* and victim is *CONSCIOUS AND ABLE TO SWALLOW,* have victim drink 4 to 8 ounces of water. **Do NOT induce vomiting.**
Note to physician or authorized medical personnel: Medical observation is recommended for 24 to 48 hours after breathing overexposure, as pulmonary edema may be delayed. As first aid for pulmonary edema, consider administering a corticosteroid spray. Cigarette smoking may exacerbate pulmonary injury and should be discouraged for at least 72 hours following exposure.

HEALTH HAZARDS
Personal protective equipment (PPE): B-Level PPE. Air line mask or self-contained breathing apparatus; full protective clothing.
Vapor (gas) irritant characteristics: Severe eye and respiratory tract irritant. Vapors cause irritation such that personnel will find high concentrations unpleasant.
Liquid or solid irritant characteristics: Severe eye and skin irritant. Causes second- and third-degree burns on short contact and is very injurious to the eyes.

FIRE DATA
Fire extinguishing agents not to be used: Do not use water on adjacent fires.

CHEMICAL REACTIVITY
Reactivity with water: Reacts vigorously to form corrosive and toxic hydrofluoric acid.
Binary reactants: In the presence of moisture, it is corrosive to glass, other siliceous materials, and most metals.
Neutralizing agents for acids and caustics: Flush with water, rinse with sodium bicarbonate or lime solution.

ENVIRONMENTAL DATA
Food chain concentration potential: Negative; unlikely to accumulate.

Water pollution: Effect of low concentrations on aquatic life is unknown. May be dangerous if it enters nearby water intakes; notify operators. Notify local health and wildlife officials. **Response to discharge:** Issue warning–poison, corrosive, water contaminant. Restrict access. Evacuate area. Disperse and flush with care.

SHIPPING INFORMATION
Grades of purity: FP Acid No. 2, 90+%; Commercial, 96+% plus 3.5% monofluorophosphoric acid; **Storage temperature:** Ambient; **Inert atmosphere:** None; **Venting:** Pressure-vacuum; **Stability during transport:** Stable.

NAS HAZARD CLASSIFICATION FOR BULK WATER TRANSPORTATION
FIRE: 0
HEALTH: Vapor irritant: 2; Liquid or solid irritant: 4; Poisons: 2
WATER POLLUTION: Human toxicity:–; Aquatic toxicity:–; Aesthetic effect: 2
REACTIVITY: Other chemicals: 3; Water: 1; Self-reaction: 0

PHYSICAL AND CHEMICAL PROPERTIES
Physical state @ 59°F/15°C and 1 atm: Liquid.
Molecular weight: 103.0
Boiling point @ 1 atm: 241°F/116°C/389°K.
Melting/Freezing point: –139°F/–95°C/178°K.
Specific gravity (water = 1): 1.583 @ 77°F/25°C (liquid).
Latent heat of vaporization: 140 Btu/lb = 77 cal/g = 3.2×10^5 J/kg.

DIHEPTYL PHTHALATE REC. D:4000

SYNONYMS: FTALATO de DICLICOL (Spanish); HEPTYL PHTHALATE; DI-*n*-HEPTYL PHTHALATE; PHTHALIC ACID, DIHEPTYL ESTER

IDENTIFICATION
CAS Number: 3648-21-3
Formula: $C_{22}H_{34}O_4$; $C_6H_4(COOC_7H_{15})_2$

DESCRIPTION: White liquid. Odorless. May float or sink in water.

Combustible • Containers may BLEVE when exposed to fire • Vapors are heavier than air and will collect and stay in low areas • Vapors in confined areas (e.g., tanks, sewers, buildings) may explode when exposed to fire • Irritating to the skin, eyes, and respiratory tract • Toxic products of combustion may include carbon monoxide.

EMERGENCY RESPONSE: See Appendix A (171)
Evacuation:
Public safety: Isolate the area of spill or leak for at least 10 to 25 meters (30 to 80 feet) in all directions.
Spill: Increase, in the downwind direction, as necessary, the distance shown under "Public Safety."
Fire: If any large container is involved in fire, isolate for at least 800 meters (½ mile) in all directions; also, consider initial evacuation for 800 meters (½ mile) in all directions.

EXPOSURE
Short-term effects: *SEEK MEDICAL ATTENTION.* **Liquid:** Irritating to skin and eyes. Harmful if swallowed. Remove contaminated clothing and shoes. Flush affected areas with plenty of water. *IF IN EYES,* hold eyelids open and flush with plenty of

water. *IF SWALLOWED* and victim is *CONSCIOUS AND ABLE TO SWALLOW*, have victim drink 4 to 8 ounces of water. *IF SWALLOWED* and victim is *UNCONSCIOUS OR HAVING CONVULSIONS*, do nothing except keep victim warm.

HEALTH HAZARDS
Personal protective equipment (PPE): B-Level PPE. Goggles or face shield; rubber gloves.
Liquid or solid irritant characteristics: Eye irritant.

FIRE DATA
Fire extinguishing agents not to be used: Water may be ineffective.

CHEMICAL REACTIVITY
Binary reactants: May attack some forms of plastics.
Reactivity group: 34
Compatibility class: Ester

ENVIRONMENTAL DATA
Food chain concentration potential: Negative; unlikely to accumulate.
Water pollution: Effect of low concentrations on aquatic life is unknown. Fouling to shoreline. May be dangerous if it enters nearby water intakes; notify operators. Notify local health and wildlife officials. **Response to discharge:** Mechanical containment. Should be removed. Chemical and physical treatment.

SHIPPING INFORMATION
Grades of purity: Pure; **Storage temperature:** Ambient; **Inert atmosphere:** None; **Venting:** Open; **Stability during transport:** Stable.

PHYSICAL AND CHEMICAL PROPERTIES
Physical state @ 59°F/15°C and 1 atm: Liquid.
Molecular weight: 363
Boiling point @ 1 atm: Decomposes.
Specific gravity (water = 1): (estimate) 1.0 @ 68°F/20°C (liquid).
Heat of combustion: (estimate) $-16,850$ Btu/lb = $-9,370$ cal/g = -392×10^5 J/kg.

DI-*n*-HEXYL ADIPATE   REC. D:4050

SYNONYMS: ADIPTATO de DI-*n*-HEXILO (Spanish); ADIPIC ACID, DIHEXYL ESTER: DIHEXYL HEXANEDIOATE; HEXANEDIOIC ACID, DIHEXYL ESTER

IDENTIFICATION
CAS Number: 110-33-8
Formula: $C_{18}H_{34}O_4$; $(-CH_2CH_2COOC_6H_{13})_2$

DESCRIPTION: Colorless liquid. Floats on the surface of water.

Combustible • Containers may BLEVE when exposed to fire • Vapors are heavier than air and will collect and stay in low areas • Vapors in confined areas (e.g., tanks, sewers, buildings) may explode when exposed to fire • Irritating to the skin, eyes, and respiratory tract • Toxic products of combustion may include irritating vapors.

Hazard Classification (based on NFPA-704 M Rating System)
Health Hazards (Blue): 0; Flammability (Red): 1; Reactivity (Yellow): 0

EMERGENCY RESPONSE: See Appendix A (171)
Evacuation:
Public safety: Isolate the area of spill or leak for at least 10 to 25 meters (30 to 80 feet) in all directions.
Spill: Increase, in the downwind direction, as necessary, the distance shown under "Public Safety."
Fire: If any large container is involved in fire, isolate for at least 800 meters (½ mile) in all directions; also, consider initial evacuation for 800 meters (½ mile) in all directions.

EXPOSURE
Short-term effects: *SEEK MEDICAL ATTENTION*. **Vapor:** Move victim to fresh air. *IF BREATHING HAS STOPPED*, give artificial respiration. If breathing is difficult, administer oxygen. **Liquid:** Remove contaminated clothing and shoes. Wash affected areas with soap and water. *IF IN EYES*, hold eyelids open and flush with plenty of water.

HEALTH HAZARDS
Personal protective equipment (PPE): B-Level PPE. Full impervious protective clothing, including boots and gloves. Where splashing is possible wear full face shield or chemical safety goggles. Use approved respirator to protect against vapors.
Vapor (gas) irritant characteristics: Eye and respiratory tract irritant. Vapors cause a slight smarting of the eyes or respiratory system if present in high concentrations. The effect is temporary.
Liquid or solid irritant characteristics: Eye irritant. If spilled on clothing and allowed to remain, may cause smarting and reddening of the skin.

FIRE DATA
Flash point: 325°F/163°C (cc).
Fire extinguishing agents not to be used: Water.

CHEMICAL REACTIVITY
Reactivity group: 34
Compatibility class: Esters.
Water pollution: Effects of low concentrations on aquatic life is unknown. May be dangerous if it enters nearby water intakes; notify operators. Notify local health and wildlife officials. **Response to discharge:** Should be removed.

SHIPPING INFORMATION
Grades of purity: Technical grades; **Storage temperature:** Ambient; **Inert atmosphere:** None; **Venting:** Open; **Stability during transport:** Stable.

PHYSICAL AND CHEMICAL PROPERTIES
Physical state @ 59°F/15°C and 1 atm: Liquid
Molecular weight: 314.52
Boiling point @ 1 atm: (approximate) 376°F/190°C/463°K.
Specific gravity (water = 1): 0.939 @ 20°C.

DIISOBUTYLAMINE   REC. D:4100

SYNONYMS: *N,N*-BIS(2-METHYLPROPYL) AMINE; 1-PROPANAMINE, 2-METHYL-*N*-(2-METHYL PROPYL)-; DIISOBUTILAMINA (Spanish); 2 METHYL-*N*-(2-METHYLPROPYL)-1-PROPANAMINE

IDENTIFICATION
CAS Number: 110-96-3
Formula: $C_8H_{19}N$; $[(CH_3)_2CHCH_2]_2NH$

388 Diisobutyl carbinol

DOT ID Number: UN 2361; DOT Guide Number: 132
Proper Shipping Name: Diisobutylamine

DESCRIPTION: Colorless liquid. Ammonia odor. Floats on water surface; soluble.

Extremely flammable • Containers may BLEVE when exposed to fire •Vapors may form explosive mixture with air • Vapors are heavier than air and will collect and stay in low areas • Vapors may travel long distances to ignition sources and flashback • Vapors in confined areas (e.g., tanks, sewers, buildings) may explode when exposed to fire • Irritating to the skin, eyes, and respiratory tract • Toxic products of combustion may include carbon monoxide and nitrogen oxide fumes.

Hazard Classification (based on NFPA-704 Rating System)
Health Hazards (Blue): 3; Flammability (Red): 3; Reactivity (Yellow): 0

EMERGENCY RESPONSE: See Appendix A (132)
Evacuation:
Public safety: Isolate spill area for at least 50 to 100 meters (160 to 330 feet) in all directions.
Spill: Increase, as necessary, the isolation distance shown above, in "Public safety."
Fire: Isolate for 800 meters (½ mile) in all directions, especially if tank, rail car, or tank truck is involved in fire.

EXPOSURE
Short-term effects: *SEEK MEDICAL ATTENTION.* **Vapor:** Irritating to eyes, nose, throat. *IF IN EYES,* hold eyelids open and flush with running water for at least 15 minutes. *IF BREATHING HAS STOPPED,* give artificial respiration. If breathing is difficult, administer oxygen. **Liquid:** Irritating to skin and eyes. Harmful if swallowed. Remove and double bag contaminated clothing and shoes at the site. *IF IN EYES OR ON SKIN,* flush with running water for at least 15 minutes; hold eyelids open if necessary. *IF SWALLOWED* and victim is *CONSCIOUS AND ABLE TO SWALLOW,* have victim drink 4 to 8 ounces of water and induce vomiting. *IF SWALLOWED* and victim is *UNCONSCIOUS OR HAVING CONVULSIONS,* do nothing except keep victim warm.

HEALTH HAZARDS
Personal protective equipment (PPE): B-Level PPE. Wear self-contained positive pressure breathing apparatus and full protective clothing. Chemical protective material(s) reported to have good to excellent resistance: nitrile rubber, PV alcohol, Viton®.
Toxicity by ingestion: Grade 3; LD_{50} = 258 mg/kg (rat).
Vapor (gas) irritant characteristics: Eye and respiratory tract irritant. Vapors are irritating such that personnel will not usually tolerate moderate or high concentrations.
Liquid or solid irritant characteristics: Fairly severe eye and skin irritant. May cause pain and second-degree burns after a few minutes of contact.

FIRE DATA
Flash point: 69°F/21°C (cc); also listed at 85°F/29°C.
Fire extinguishing agents not to be used: Water may not be effective.
Autoignition temperature: 554°F/290°C/563°K.
Electrical hazard: Class 1; Group C.

CHEMICAL REACTIVITY
Binary reactants: Oxidizers. Reacts with aluminum or aluminum alloys; copper or copper alloys; zinc, galvanized steel or alloys having more than 10% zinc by weight; mercury. Will dissolve paint and most plastic material.
Reactivity group: 7
Compatibility class: Aliphatic amines

ENVIRONMENTAL DATA
Water pollution: Harmful to aquatic life in very low concentrations. May be dangerous if it enters nearby water intakes; notify operators. Notify local health and wildlife officials. **Response to discharge:** Issue warning–high flammability, water contaminant. Restrict access. Should be removed. Chemical and physical treatment.

SHIPPING INFORMATION
Grades of purity: 99%; 98%; **Storage temperature:** Ambient; **Inert atmosphere:** Nones; **Venting:** Pressure-vacuum; **Stability during transport:** Stable.

PHYSICAL AND CHEMICAL PROPERTIES
Physical state @ 59°F/15°C and 1 atm: Liquid.
Molecular weight: 129.25
Boiling point @ 1 atm: 283.1°F/139.5°C/412.7°K.
Melting/Freezing point: –94°F/–70°C/203.2°K.
Critical temperature: 538°F/281.3°C/554.5°K.
Critical pressure: 370 psia = 25.16 atm = 2.55 MN/m^2.
Specific gravity (water = 1): 0.745 @ 68°F/20°C.
Liquid surface tension: 22.58 dynes/cm = 0.02258 N/m @ 15.1°C.
Relative vapor density (air = 1): 4.46
Latent heat of vaporization: 140 Btu/lb = 77.8 cal/g = 3.26 x 10^5 J/kg.
Heat of solution: –73.31 +/- 1.49 Btu/lb = –40.73 +/- 0.83 cal/g = –1.71 +/- 0.03 x 10^5 J/kg.
Vapor pressure: 10 mm.

DIISOBUTYL CARBINOL   REC. D:4150

SYNONYMS: 2,6-DIMETHYL-4-HEPTANOL; DIISOBUTILCARBINOL (Spanish); *sec*-NONYL ALCOHOL

IDENTIFICATION
CAS Number: 108-82-7
Formula: $C_9H_{20}O$; $[(CH_3)_2CHCH_2]_2CHOH$
DOT ID Number: UN 1993; DOT Guide Number: 128
Proper Shipping Name: Combustible liquid, n.o.s.

DESCRIPTION: Colorless thick, oily liquid. Forms slick on water surface.

Combustible • Containers may BLEVE when exposed to fire •Vapors may form explosive mixture with air • Vapors are heavier than air and will collect and stay in low areas • Vapors may travel long distances to ignition sources and flashback • Vapors in confined areas (e.g., tanks, sewers, buildings) may explode when exposed to fire • Irritating to the skin, eyes, and respiratory tract • Toxic products of combustion may include carbon monoxide.

Hazard Classification (based on NFPA-704 Rating System)
Health Hazards (Blue): 1; Flammability (Red): 2; Reactivity (Yellow): 0

EMERGENCY RESPONSE: See Appendix A (128)
Evacuation:
Public safety: Isolate spill area for at least 25 to 50 meters (80 to 160 feet) in all directions.

Spill: Large spill–Consider initial downwind evacuation for at least 300 meters (1000 feet).
Fire: Isolate for 800 meters (½ mile) in all directions, especially if tank, rail car, or tank truck is involved in fire.

EXPOSURE
Short-term effects: *SEEK MEDICAL ATTENTION*.
Liquid: Harmful if swallowed. *IF SWALLOWED* and victim is *CONSCIOUS AND ABLE TO SWALLOW*, have victim drink 4 to 8 ounces of water.

HEALTH HAZARDS
Personal protective equipment (PPE): B-Level PPE. Air–supplied mask for prolonged exposure; gloves; goggles. Chemical protective material(s) reported to offer minimal to poor protection: Viton®/neoprene, butyl rubber,/neoprene.
Toxicity by ingestion: Grade 2; LD_{50} = 0.5 to 5 g/kg (rat). Possible central nervous system and liver damage.
Vapor (gas) irritant characteristics: Eye and respiratory tract irritant.
Liquid or solid irritant characteristics: Eye and skin irritant.

FIRE DATA
Flash point: 162°F/72°C (oc); 165°F/74°C (cc)
Flammable limits in air: LEL: 0.8%; UEL: 6.1% (both @ 212°F)
Behavior in fire: Containers may explode.
Autoignition temperature: 494°F/257°C (calculated)
Electrical hazard: Class I, Group D. Due to low electric conductivity, this substance may generate electrostatic charges.

CHEMICAL REACTIVITY
Binary reactants: Reacts with strong oxidizers. Incompatible with caustics, aliphatic amines, isocyanates, sulfuric acid, nitric acid.
Reactivity group: 20
Compatibility class: Alcohols, glycols

ENVIRONMENTAL DATA
Food chain concentration potential: Negative; unlikely to accumulate.
Water pollution: Effect of low concentrations on aquatic life is unknown. Fouling to shoreline. A danger to waterfowl. May be dangerous if it enters nearby water intakes; notify operators. Notify local health and wildlife officials. **Response to discharge:** Mechanical containment. Chemical and physical treatment. Oil-skimming equipment.

SHIPPING INFORMATION
Grades of purity: 98.0%; **Storage temperature:** Ambient; **Inert atmosphere:** None; **Venting:** Open (flame arrester); **Stability during transport:** Stable.

PHYSICAL AND CHEMICAL PROPERTIES
Physical state @ 59°F/15°C and 1 atm: Liquid.
Molecular weight: 144.26
Boiling point @ 1 atm: 353°F/178°C/451°K.
Melting/Freezing point: –85°F/–65°C/208°K.
Specific gravity (water = 1): 0.812 @ 68°F/20°C (liquid).
Relative vapor density (air = 1): 4.8
Latent heat of vaporization: 140 Btu/lb = 76 cal/g = 3.2×10^5 J/kg.
Heat of combustion: (estimate) –17,400 Btu/lb = –9680 cal/g = -405×10^5 J/kg.
Vapor pressure: (Reid) 0.06 psia; 0.21 mmHg @ 68°F/20°C (liquid).

DIISOBUTYLENE   REC. D:4200

SYNONYMS: DIISOBUTILENO (Spanish); EEC No. 601-031-00-8; ISOOCTENE; 2,4,4-TRIMETHYL-1-PENTENE; 2,4,4-TRIMETHYLPENTENE-1; 2,4,4-TRIMETHYLPENTENE-2

IDENTIFICATION
CAS Number: 25167-70-8
Formula: C_8H_{16}; $(CH_3)_3CCH_2C(CH_3)=CH_2$
DOT ID Number: UN 2050; DOT Guide Number: 127
Proper Shipping Name: Diisobutylene, isomeric compounds

DESCRIPTION: Colorless liquid. Gasoline-like odor. Floats on the surface of water. Flammable, irritating vapor is produced.

Highly flammable • Polymerization hazard; unstable peroxides may accumulate after prolonged storage in presence of air; peroxides may be detonated by heating, impact, or friction • Containers may BLEVE when exposed to fire • Vapors may form explosive mixture with air • Vapors are heavier than air and will collect and stay in low areas • Vapors may travel long distances to ignition sources and flashback • Vapors in confined areas (e.g., tanks, sewers, buildings) may explode when exposed to fire • Irritating to the skin, eyes, and respiratory tract • Toxic products of combustion may include carbon monoxide.

Hazard Classification (based on NFPA-704 Rating System)
Health Hazards (Blue): 2; Flammability (Red): 3; Reactivity (Yellow): 0

EMERGENCY RESPONSE: See Appendix A (127)
Evacuation:
Public safety: Isolate spill area for at least 25 to 50 meters (80 to 160 feet) in all directions.
Spill: Large spill–Consider initial downwind evacuation for at least 300 meters (1000 feet).
Fire: Isolate for 800 meters (½ mile) in all directions, especially if tank, rail car, or tank truck is involved in fire.

EXPOSURE
Short-term effects: *SEEK MEDICAL ATTENTION*. **Vapor:** Irritating to eyes, nose, and throat. *IF INHALED*, will, will cause dizziness, headache, difficult breathing or loss of consciousness. Move to fresh air. *IF BREATHING HAS STOPPED,* give artificial respiration. If breathing is difficult, administer oxygen. **Liquid:** Irritating to skin and eyes. *IF SWALLOWED*, will cause nausea or vomiting; possible pneumonia. Remove contaminated clothing and shoes. Flush affected areas with plenty of water. *IF IN EYES*, hold eyelids open and flush with plenty of water. *IF SWALLOWED* and victim is *CONSCIOUS AND ABLE TO SWALLOW*, have victim drink 4 to 8 ounces of water. **Do NOT induce vomiting**.

HEALTH HAZARDS
Personal protective equipment (PPE): B-Level PPE. Protective goggles. Chemical protective material(s) reported to offer minimal to poor protection: Viton®/neoprene.
Exposure limits (TWA unless otherwise noted): 300 ppm (AIHAWEEL).
Long-term health effects: Liver and kidney damage in experimental animals. May effect the nervous system.
Vapor (gas) irritant characteristics: May cause eye and respiratory tract irritation.
Liquid or solid irritant characteristics: may cause eye and skin irritation. If spilled on clothing and allowed to remain, may cause smarting and reddening of the skin.

FIRE DATA

Flash point: −20°F/−29°C (cc); also listed 23°F/−5°C (NFPA).
Flammable limits in air: LEL: 0.8%; UEL: 4.8%.
Fire extinguishing agents not to be used: Water may be ineffective
Autoignition temperature: 736°F/391°C/664°K.
Electrical hazard: Class I, Group D. Flow or agitation of substance may generate electrostatic charges due to low conductivity.
Burning rate: 7.9 mm/min

CHEMICAL REACTIVITY

Binary reactants: Reacts with strong oxidizers.
Polymerization: Peroxides may accumulate after prolonged storage in presence of air. Peroxides may be detonated by heating, impact, or friction
Reactivity group: 30
Compatibility class: Olefins

ENVIRONMENTAL DATA

Food chain concentration potential: Negative; unlikely to accumulate.
Water pollution: Effect of low concentrations on aquatic life is unknown. Fouling to shoreline. May be dangerous if it enters nearby water intakes; notify operators. Notify local health and wildlife officials. **Response to discharge:** Issue warning–high flammability. Evacuate area.

SHIPPING INFORMATION

Grades of purity: Research grade: 99.86%; Pure grade: 99.39%; Technical grade: 98.7%; **Storage temperature:** Ambient; **Inert atmosphere:** None; **Venting:** Open (flame arrester) or pressure-vacuum; **Stability during transport:** Stable.

NAS HAZARD CLASSIFICATION FOR BULK WATER TRANSPORTATION

FIRE: 3
HEALTH: Vapor irritant: 0; Liquid or solid irritant: 1; Poisons: 0
WATER POLLUTION: Human toxicity: 1; Aquatic toxicity: 1; Aesthetic effect: 2
REACTIVITY: Other chemicals: 1; Water: 0; Self-reaction: 1

PHYSICAL AND CHEMICAL PROPERTIES

Physical state @ 59°F/15°C and 1 atm: Liquid.
Molecular weight: 112.22
Boiling point @ 1 atm: 214.7°F/101.5°C/374.7°K.
Melting/Freezing point: −136.3°F/−93.5°C/179.7°K.
Critical temperature: 548°F/286.7°C/559.9°K.
Critical pressure: 380 psia = 25.85 atm = 2.619 MN/m².
Specific gravity (water = 1): 0.715 @ 68°F/20°C (liquid).
Liquid surface tension: 20.7 dynes/cm = 0.0207 N/m @ 68°F/20°C.
Relative vapor density (air = 1): 3.99
Ratio of specific heats of vapor (gas): (estimate) 1.049
Latent heat of vaporization: 110 Btu/lb = 60 cal/g = 2.5 x 10^5 J/kg.
Heat of combustion: −18,900 Btu/lb = −10,500 cal/g = −440 x 10^5 J/kg.
Vapor pressure: (Reid) 1.6 psia; 30 mm.

DIISOBUTYL KETONE   REC. D:4250

SYNONYMS: DiBk; DIBK; DI-ISOBUTYLCETONE (French); DIISOBUTILCETONA (Spanish); DIISOBUTYLKETON (German); DIISOPROPYLACETONE; 2,6-DIMETHYL-HEPTAN-4-ON (German); 2,6-DIMETHYLHEPTAN-4-ONE; 2,6-DIMETHYL-4-HEPTANONE; 2,6-DIMETHYLHEPTANONE; 2,6-DIMETHYL-4-HEPTANE; 5-DIISOPROPYLACETONE; DIISIPROPYL-ACETONE; EEC No. 606-005-00-X; 4-HEPTANONE,2,6-DIMETHYL-; ISOBUTYL KETONE; ISOVALERONE VALERONE

IDENTIFICATION

CAS Number: 108-83-8
Formula: $C_9H_{18}O$; $(CH_3)_2CHCH_2COCH_2CH(CH_3)_2$
DOT ID Number: UN 1157; DOT Guide Number: 127
Proper Shipping Name: Diisobutyl ketone

DESCRIPTION: Colorless liquid. Mild, sweet, ketonic odor. Forms slick on water surface.

Flammable • Containers may BLEVE when exposed to fire • Vapors may form explosive mixture with air • Vapors are heavier than air and will collect and stay in low areas • Vapors may travel long distances to ignition sources and flashback • Vapors in confined areas (e.g., tanks, sewers, buildings) may explode when exposed to fire • Irritating to the skin, eyes, and respiratory tract; absorbed by the skin. • Toxic products of combustion may include carbon monoxide.

Hazard Classification (based on NFPA-704 Rating System)
Health Hazards (Blue): 1; Flammability (Red): 2; Reactivity (Yellow): 0

EMERGENCY RESPONSE: See Appendix A (127)
Evacuation:
Public safety: Isolate spill area for at least 25 to 50 meters (80 to 160 feet) in all directions.
Spill: Large spill–Consider initial downwind evacuation for at least 300 meters (1000 feet).
Fire: Isolate for 800 meters (½ mile) in all directions, especially if tank, rail car, or tank truck is involved in fire.

EXPOSURE

Short-term effects: *SEEK MEDICAL ATTENTION.* **Vapor:** Irritating to eyes, nose, and throat. *IF INHALED*, will, will cause coughing or difficult breathing. *IF IN EYES*, hold eyelids open and flush with plenty of water. *IF BREATHING HAS STOPPED*, give artificial respiration. If breathing is difficult, administer oxygen. **Liquid:** Irritating to skin and eyes. *IF SWALLOWED*, will cause nausea and vomiting. Remove contaminated clothing and shoes. Flush affected areas with plenty of water. *IF IN EYES*, hold eyelids open and flush with plenty of water. *IF SWALLOWED* and victim is *CONSCIOUS AND ABLE TO SWALLOW*, have victim drink 4 to 8 ounces of water. *IF SWALLOWED* and victim is *UNCONSCIOUS OR HAVING CONVULSIONS*, do nothing except keep victim warm.

HEALTH HAZARDS

Personal protective equipment (PPE): B-Level PPE. OSHA Table Z-1-A air contaminant. Air–supplied mask in confined areas; gloves, faceshield, and safety glasses. Chemical protective material(s) reported to have good to excellent resistance: PV alcohol, Silvershield®. Also, butyl rubber and nitrile offers limited protection.
Recommendations for respirator selection: NIOSH
500 ppm: SA:CF* (any supplied-air respirator operated in a continuous-flow mode); or PAPROV* [any powered, air-purifying respirator with organic vapor cartridge(s)]; or CCRFOV [any air-

purifying, full-facepiece respirator (gas mask) with a chin-style, front- or back-mounted acid gas canister]; or GMFOV [any air-purifying, full-facepiece respirator (gas mask) with a chin-style, front- or back-mounted organic vapor canister]; or SCBAF (any self-contained breathing apparatus with a full facepiece); or SAF (any supplied-air respirator with a full facepiece). *EMERGENCY OR PLANNED ENTRY INTO UNKNOWN CONCENTRATIONS OR IDLH CONDITIONS*: SCBAF:PD,PP (any self-contained breathing apparatus that has a full facepiece and is operated in a pressure-demand or other positive-pressure mode); or SAF:PD,PP:ASCBA (any supplied-air respirator that has a full facepiece and is operated in a pressure-demand or other positive-pressure mode in combination with an auxiliary self-contained breathing apparatus operated in a pressure-demand or other positive pressure mode). *ESCAPE*: GMFOV[any air-purifying, full-facepiece respirator (gas mask) with a chin-style, front- or back-mounted organic vapor canister]; or SCBAE (any appropriate escape-type, self-contained breathing apparatus). *Note*: Substance causes eye irritation or damage; eye protection needed.
Exposure limits (TWA unless otherwise noted): ACGIH TLV 25 ppm (145 mg/m^3); OSHA PEL 50 ppm (290 mg/m^3); NIOSH REL 25 ppm (150 mg/m^3).
Toxicity by ingestion: Grade 2; oral LD$_{50}$ = 1.4 g/kg (mouse), 5.75 g/kg (rat).
Long-term health effects: Causes increased liver and kidney weights in rats, decreased liver weights in guinea pigs. Possible eczema.
Vapor (gas) irritant characteristics: Eye and respiratory tract irritant. Vapors cause moderate irritation such that personnel will find high concentrations unpleasant. The effect is temporary.
Liquid or solid irritant characteristics: Skin irritant. May cause eye irritation. If spilled on clothing and allowed to remain, may cause smarting and reddening of skin.
Odor threshold: 0.1–0.3 ppm.
IDLH value: 500 ppm.

FIRE DATA
Flash point: 131°F/55°C (oc); 120°F/49°C (cc).
Flammable limits in air: LEL: 0.8%; UEL: 7.1% both @ 200°F/93°C.
Fire extinguishing agents not to be used: Water may be ineffective.
Autoignition temperature: 745°F/396°C/669°K.
Electrical hazard: Class I, Group D. Due to low electric conductivity, this substance may generate electrostatic charges.

CHEMICAL REACTIVITY
Binary reactants: Reacts with strong oxidizers, strong acids, aliphatic amines. May attack various forms of plastics, coatings, and rubber.
Reactivity group: 18
Compatibility class: Ketone

ENVIRONMENTAL DATA
Food chain concentration potential: Negative; unlikely to accumulate.
Water pollution: Effect of low concentrations on aquatic life is unknown. Fouling to shoreline. May be dangerous if it enters nearby water intakes; notify operators. Can be a danger to waterfowl. Notify local health and wildlife officials. **Response to discharge:** Mechanical containment. Should be removed. Chemical and physical treatment.

SHIPPING INFORMATION
Grades of purity: Technical; **Storage temperature:** Ambient; **Inert atmosphere:** None; **Venting:** Open (flame arrester); **Stability during transport:** Stable.

NAS HAZARD CLASSIFICATION FOR BULK WATER TRANSPORTATION
FIRE: 2
HEALTH: Vapor irritant: 2; Liquid or solid irritant: 1; Poisons: 1
WATER POLLUTION: Human toxicity: 1; Aquatic toxicity: 2; Aesthetic effect: 2
REACTIVITY: Other chemicals: 2; Water: 0; Self-reaction: 0

PHYSICAL AND CHEMICAL PROPERTIES
Physical state @ 59°F/15°C and 1 atm: Liquid.
Molecular weight: 142.23
Boiling point @ 1 atm: 334°F 167°C/440°K.
Melting/Freezing point: –43°F/–42°C/231°K.
Specific gravity (water = 1): 0.806 @ 68°F/20°C (liquid).
Liquid surface tension: 23.92 dynes/cm = 0.02392 N/m at 71°F/22°C.
Relative vapor density (air = 1): 4.9
Latent heat of vaporization: 121 Btu/lb = 67 cal/g = 2.8 x 10^5 J/kg.
Heat of combustion: –16,040 Btu/lb = –8,910 cal/g = –373 x 10^5 J/kg.
Vapor pressure: (Reid) 0.21 psia; 2 mm.

DIISOBUTYL PHTHALATE REC. D:4300

SYNONYMS: 1,2-BENZENE DICARBOXYLIC ACID, DI-(2-METHYLPROPYL)ESTER; DIBP; DISB; FTALATO de DIISOBUTILO (Spanish); HEXAPLAS M/B; HEXAPLAS M-1B; PHTHALIC ACID, DI-ISOBUTYL ESTER; PALATINOL-IC; ISOBUTYL PHTHALATE

IDENTIFICATION
CAS Number: 84-69-5
Formula: $C_{16}H_{22}O_4$; $o\text{-}C_6H_4[COOCH_2CH(CH_3)_2]_2$

DESCRIPTION: Colorless oily liquid. Slight ester odor. Sinks slowly in water; mixes and slowly decomposes.

Combustible • Containers may BLEVE when exposed to fire • Vapors are heavier than air and will collect and stay in low areas • Vapors in confined areas (e.g., tanks, sewers, buildings) may explode when exposed to fire • Irritating to the skin, eyes, and respiratory tract; mildly toxic by skin and eye contact • Toxic products of combustion may include carbon monoxide.

Hazard Classification (based on NFPA-704 Rating System)
Health Hazards (Blue): 0; Flammability (Red): 1; Reactivity (Yellow): 0

EMERGENCY RESPONSE: See Appendix A (171)
Evacuation:
Public safety: Isolate the area of spill or leak for at least 10 to 25 meters (30 to 80 feet) in all directions.
Spill: Increase, in the downwind direction, as necessary, the distance shown under "Public Safety."
Fire: If any large container is involved in fire, isolate for at least 800 meters (½ mile) in all directions; also, consider initial evacuation for 800 meters (½ mile) in all directions.

EXPOSURE
Short-term effects: *SEEK MEDICAL ATTENTION*. **Mildly toxic**

392 Diisodecyl phthalate

if swallowed or upon skin or eye contact. **Vapor:** Irritating to eyes, nose, and throat.. *IF IN EYES*, hold eyelids open and flush with plenty of water. **Liquid:** Irritating to skin and eyes. Remove contaminated clothing and shoes. Flush affected areas with plenty of water. *IF IN EYES*, hold eyelids open and flush with plenty of water. *IF SWALLOWED* and victim is *CONSCIOUS AND ABLE TO SWALLOW*, have victim drink 4 to 8 ounces of water. *IF SWALLOWED* and victim is *UNCONSCIOUS OR HAVING CONVULSIONS,* do nothing except keep victim warm.

HEALTH HAZARDS
Personal protective equipment (PPE): B-Level PPE. Eye protection. **Butyl rubber is generally suitable for carbooxylic acid compounds.**
Toxicity by ingestion: Grade 0: LD_{50} = 20 g/kg (rat).
Long-term health effects: May be a reproduction hazard.
Vapor (gas) irritant characteristics: May cause eye irritation.
Liquid or solid irritant characteristics: May cause skin or eye irritation.
IDLH value: 9.3 g/m3

FIRE DATA
Flash point: 365°F/185°C (oc).
Flammable limits in air: LEL: 0.4% @ 448°F/231°C.
Fire extinguishing agents not to be used: Water or foam may cause frothing.
Autoignition temperature: 810°F/432°C/705°K.

CHEMICAL REACTIVITY
Binary reactants: Oxidizers, strong acids.
Reactivity group: 34
Compatibility class: Ester

ENVIRONMENTAL DATA
Food chain concentration potential: Negative; unlikely to accumulate.
Water pollution: Dangerous to aquatic life in high concentrations. Fouling to shoreline. May be dangerous if it enters nearby water intakes; notify operators. Notify local health and pollution control officials. **Response to discharge:** Mechanical containment. Should be removed. Chemical and physical treatment.

SHIPPING INFORMATION
Grades of purity: 99.6%; **Stability during transport:** Stable.

NAS HAZARD CLASSIFICATION FOR BULK WATER TRANSPORTATION
FIRE: 1
HEALTH: Vapor irritant: 0; Liquid or solid irritant: 0; Poisons: 0
WATER POLLUTION: Human toxicity: 1; Aquatic toxicity: 1; Aesthetic effect: 1
REACTIVITY: Other chemicals: 3; Water: 1; Self-reaction: 0

PHYSICAL AND CHEMICAL PROPERTIES
Physical state @ 59°F/15°C and 1 atm: Liquid.
Molecular weight: 278.35
Boiling point @ 1 atm: 621°F/327°C/600°K.
Melting/Freezing point: –83°F/–64°C/209°K.
Specific gravity (water = 1): 1.047 @ 68°F/20°C (liquid).
Relative vapor density (air = 1): 9.59

DIISODECYL PHTHALATE   REC. D:4350

SYNONYMS: FTALATO de DIISODECILO (Spanish); PHTHALIC ACID, BIS(8-METHYL-NONYL) ESTER; PHTHALIC ACID, DIISODECYL ESTER; PLASTICIZED DDP

IDENTIFICATION
CAS Number: 26761-40-0
Formula: $C_{28}H_{46}O_4$

DESCRIPTION: Colorless liquid. May float or sink in water.

Combustible • Containers may BLEVE when exposed to fire • Vapors are heavier than air and will collect and stay in low areas • Vapors in confined areas (e.g., tanks, sewers, buildings) may explode when exposed to fire • Irritating to the skin, eyes, and respiratory tract • Toxic products of combustion may include carbon monoxide.

Hazard Classification (based on NFPA-704 Rating System)
Health Hazards (Blue): 0; Flammability (Red): 1; Reactivity (Yellow): 0

EMERGENCY RESPONSE: See Appendix A (171)
Evacuation:
Public safety: Isolate the area of spill or leak for at least 10 to 25 meters (30 to 80 feet) in all directions.
Spill: Increase, in the downwind direction, as necessary, the distance shown under "Public Safety."
Fire: If any large container is involved in fire, isolate for at least 800 meters (½ mile) in all directions; also, consider initial evacuation for 800 meters (½ mile) in all directions.

HEALTH HAZARDS
Personal protective equipment (PPE): B-Level PPE. Goggles or face shield; rubber gloves.

FIRE DATA
Flash point: 450°F/232°C (cc).
Flammable limits in air: LEL: 0.3% @ 508°F/264°C.
Fire extinguishing agents not to be used: Water may be ineffective.
Autoignition temperature: 755°F/402°C/675°K.

CHEMICAL REACTIVITY
Binary reactants: Strong acids, strong alkalis, nitrates, oxidizers. May attack some forms of plastics
Reactivity group: 34
Compatibility class: Ester

ENVIRONMENTAL DATA
Food chain concentration potential: Negative; unlikely to accumulate.
Water pollution: Effect of low concentrations on aquatic life is unknown. Fouling to shoreline. May be dangerous if it enters nearby water intakes; notify operators. Notify local health and wildlife officials. **Response to discharge:** Mechanical containment. Should be removed. Chemical and physical treatment.

SHIPPING INFORMATION
Grades of purity: Technical; **Storage temperature:** Ambient; **Inert atmosphere:** None; **Venting:** Open; **Stability during transport:** Stable.

PHYSICAL AND CHEMICAL PROPERTIES
Physical state @ 59°F/15°C and 1 atm: Liquid.
Molecular weight: 446.7
Boiling point @ 1 atm: 482°F/250°C/523°K.
Melting/Freezing point: –58°F/–50°C/223°K.

Specific gravity (water = 1): 0.967 @ 68°F/20°C (liquid).
Heat of combustion: (estimate) –16,600 Btu/lb = –9220 cal/g = –386 x 10^5 J/kg.

DIISONONYL ADIPATE   REC. D:4400

SYNONYMS: ADIPATO de DIISONONILO (Spanish)

IDENTIFICATION
Formula: $C_9H_{19}OOC(CH_2)_4COOC_9H_{19}$

DESCRIPTION: Colorless liquid.

Combustible • Containers may BLEVE when exposed to fire • Vapors are heavier than air and will collect and stay in low areas • Vapors in confined areas (e.g., tanks, sewers, buildings) may explode when exposed to fire • Irritating to the skin, eyes, and respiratory tract • Toxic products of combustion may include carbon monoxide and CO_2.

EMERGENCY RESPONSE: See Appendix A (171)
Evacuation:
Public safety: Isolate the area of spill or leak for at least 10 to 25 meters (30 to 80 feet) in all directions.
Spill: Increase, in the downwind direction, as necessary, the distance shown under "Public Safety."
Fire: If any large container is involved in fire, isolate for at least 800 meters (½ mile) in all directions; also, consider initial evacuation for 800 meters (½ mile) in all directions.

EXPOSURE
Short-term effects: *SEEK MEDICAL ATTENTION*. **Vapor:** Move victim to fresh air. *IF BREATHING HAS STOPPED,* give artificial respiration. If breathing is difficult, administer oxygen. **Liquid:** Remove contaminated clothing and shoes. Wash affected areas with soap and water. *IF IN EYES*, hold eyelids open and flush with plenty of water.

HEALTH HAZARDS
Personal protective equipment (PPE): B-Level PPE. Full impervious protective clothing, including boots and gloves. Where splashing is possible wear full face shield or chemical safety goggles. Use approved respirator to protect against vapors.
Vapor (gas) irritant characteristics: Eye and respiratory tract irritant. Vapors cause a slight smarting of the eyes or respiratory system if present in high concentrations.
Liquid or solid irritant characteristics: Eye and skin irritant. If spilled on clothing and allowed to remain, may cause smarting and reddening of the skin.

FIRE DATA
Fire extinguishing agents not to be used: Water.

CHEMICAL REACTIVITY
Binary reactants: Oxidizers, strong acids
Reactivity group: 34
Compatibility class: Esters.

ENVIRONMENTAL DATA
Water pollution: Effects of low concentrations on aquatic life is unknown. May be dangerous if it enters nearby water intakes; notify operators. Notify local health and wildlife officials.
Response to discharge: Should be removed.

SHIPPING INFORMATION
Grades of purity: Technical grades; **Storage temperature:** Ambient; **Inert atmosphere:** None; **Stability during transport:** Stable.

PHYSICAL AND CHEMICAL PROPERTIES
Physical state @ 59°F/15°C and 1 atm: Liquid.
Molecular weight: 398.63

DIISONONYL PHTHALATE   REC. D:4450

SYNONYMS: DI(7-METHYLOCTYL) PHTHALATE; 1,2-BENZENEDICARBOXYLIC ACID, DI-ISONONYL ESTER; *o*-DICARBOXYBENZENE; FTALATO de DIISONONILO (Spanish); PHTHALIC ACID; PHTHALIC ACID, BIS-(7-METHYLOCTYL) ESTER

IDENTIFICATION
CAS Number: 88-99-3
Formula: *o*-$C_6H_4[COO(CH_2)_6CH(CH_3)_2]_2$

DESCRIPTION: Colorless liquid. Odorless. Floats on water surface; slightly soluble.

Combustible • Containers may BLEVE when exposed to fire • Vapors are heavier than air and will collect and stay in low areas • Vapors in confined areas (e.g., tanks, sewers, buildings) may explode when exposed to fire • Irritating to the skin, eyes, and respiratory tract • Toxic products of combustion may include carbon monoxide.

Hazard Classification (based on NFPA-704 Rating System)
Health Hazards (Blue): 0; Flammability (Red): 1; Reactivity (Yellow): 0

EMERGENCY RESPONSE: See Appendix A (171)
Evacuation:
Public safety: Isolate the area of spill or leak for at least 10 to 25 meters (30 to 80 feet) in all directions.
Spill: Increase, in the downwind direction, as necessary, the distance shown under "Public Safety."
Fire: If any large container is involved in fire, isolate for at least 800 meters (½ mile) in all directions; also, consider initial evacuation for 800 meters (½ mile) in all directions.

HEALTH HAZARDS
Personal protective equipment (PPE): Butyl rubber is generally suitable for carbooxylic acid compounds.
Vapor (gas) irritant characteristics: Nonirritating to the eyes and throat.
Liquid or solid irritant characteristics: Practically harmless to the skin.

CHEMICAL REACTIVITY
Binary reactants: Reacts with sodium nitrate; nitric acid.
Reactivity group: 34
Compatibility class: Esters.

ENVIRONMENTAL DATA
Water pollution: Effects of low concentrations on aquatic life not known. Fouling to shoreline. May be harmful if it enters water intakes. Notify local health and wildlife officials. Notify operators of local water intakes. **Response to discharge:** Mechanical containment. Chemical and physical treatment.

SHIPPING INFORMATION
Stability during transport: Stable.

NAS HAZARD CLASSIFICATION FOR BULK WATER TRANSPORTATION
FIRE: 0
HEALTH: Vapor irritant: 0; Liquid or solid irritant: 0; Poisons: 0
WATER POLLUTION: Human toxicity:–; Aquatic toxicity:–; Aesthetic effect: 1
REACTIVITY: Other chemicals: 2; Water: 0; Self-reaction: 0

PHYSICAL AND CHEMICAL PROPERTIES
Physical state @ 59°F/15°C and 1 atm: Liquid.
Molecular weight: 418.6
Boiling point @ 1 atm: 172°F/78°C/351°K.

DIISOOCTYL PHTHALATE   REC. D:4500

SYNONYMS: 1,2-BENZENEDICARBOXYLIC ACID; 1,2-BENZENEDICARBOXYLIC ACID, DIISOOCTYL ESTER; BIS(6-METHYLHEPTYL) PHTHALATE; BIS(6-METHYLHEPTYL) ESTER OF PHTHALIC ACID; CORFLEX 880; DIOP; DI(6-METHYLHEPTYL) PHTHALATE; FTALATO de DIISOCTILO (Spanish); FLEXOL PLASTICIZER P; HEXAPLAS DIOP; HEXAPLAS M/O; ISOOCTYL PHTHALATE

IDENTIFICATION
CAS Number: 27554-26-3
Formula: $C_{24}H_{38}O_4$; o-$C_6H_4[COO(CH_2)_5CH(CH_3)_2]_2$

DESCRIPTION: Nearly colorless, thick, oily liquid. Slight odor. Floats on water surface; insoluble.

Combustible • Containers may BLEVE when exposed to fire • Vapors are heavier than air and will collect and stay in low areas • Vapors in confined areas (e.g., tanks, sewers, buildings) may explode when exposed to fire • Irritating to the skin, eyes, and respiratory tract • Toxic products of combustion may include carbon monoxide.

Hazard Classification (based on NFPA-704 Rating System)
Health Hazards (Blue): 0; Flammability (Red): 1; Reactivity (Yellow): 0

EMERGENCY RESPONSE: See Appendix A (171)
Evacuation:
Public safety: Isolate the area of spill or leak for at least 10 to 25 meters (30 to 80 feet) in all directions.
Spill: Increase, in the downwind direction, as necessary, the distance shown under "Public Safety."
Fire: If any large container is involved in fire, isolate for at least 800 meters (½ mile) in all directions; also, consider initial evacuation for 800 meters (½ mile) in all directions.

EXPOSURE
Short-term effects: Irritating to the skin, eyes, and respiratory tract.

HEALTH HAZARDS
Personal protective equipment (PPE): Butyl rubber is generally suitable for carbooxylic acid compounds.
Toxicity by ingestion: Grade 0: LD_{50} = 22 g/kg (rat).

Long-term health effects: Repeated or prolonged dermal exposure may cause eczema or other skin problems.
Vapor (gas) irritant characteristics: Eye and respiratory tract irritant, especially at high temperatures.
Liquid or solid irritant characteristics: May cause eye and skin irritation. Practically harmless to the skin.

FIRE DATA
Flash point: 450°F/232°C (cc).
Fire extinguishing agents not to be used: Water or foam may cause frothing
Electrical hazard: Due to low electric conductivity, this substance may generate electrostatic charges as a result of agitation and flow.

CHEMICAL REACTIVITY
Binary reactants: Reacts with strong oxidizers, strong acids.
Reactivity group: 34
Compatibility class: Ester

ENVIRONMENTAL DATA
Food chain concentration potential: Negative; unlikely to accumulate.
Water pollution: Effects of low concentrations on aquatic life is unknown. Fouling to shoreline. May be dangerous if it enters nearby water intakes; notify operators. Notify local health and water life officials. **Response to discharge:** Mechanical containment. Chemical and physical treatment.

SHIPPING INFORMATION
Storage temperature: Ambient; **Inert atmosphere:** None; **Venting:** Open (flame arrester); **Stability during transport:** Stable.

NAS HAZARD CLASSIFICATION FOR BULK WATER TRANSPORTATION
FIRE: 1
HEALTH: Vapor irritant: 0; Liquid or solid irritant: 0; Poisons: 0
WATER POLLUTION: Human toxicity: 0; Aquatic toxicity: 0; Aesthetic effect: 3
REACTIVITY: Other chemicals: 0; Water: 0; Self-reaction: 0

PHYSICAL AND CHEMICAL PROPERTIES
Physical state @ 59°F/15°C and 1 atm: Liquid.
Molecular weight: 390.62
Boiling point @ 1 atm: 698°F/370°C/643°K.
Melting/Freezing point: –45°F/–43°C/230°K.
Relative vapor density (air = 1): 13.5
Ratio of specific heats of vapor (gas): 1.0
Vapor pressure: Less than 0.08 mm.

DIISOPROPANOLAMINE   REC. D:4550

SYNONYMS: 2,2'-DIHYDROXYDIPROPYLAMINE; DIISOPROPANOLAMINA (Spanish); DIPA; 1,1'-IMINODI-2-PROPANOL

IDENTIFICATION
CAS Number: 110-97-4
Formula: $[CH_3CH(OH)CH_2]_2NH$
DOT ID Number: UN 1158; DOT Guide Number: 132
Proper Shipping Name: Diisopropylamine

DESCRIPTION: Colorless liquid or white to off-white crystalline solid. Dead fish or ammonia odor. Liquid floats on the surface of water; slightly soluble. Solid sinks and mixes in water.

Highly flammable • Containers may BLEVE when exposed to fire •Vapors may form explosive mixture with air • Vapors are heavier than air and will collect and stay in low areas • Vapors may travel long distances to ignition sources and flashback • Vapors in confined areas (e.g., tanks, sewers, buildings) may explode when exposed to fire • Irritating to the skin, eyes, and respiratory tract • Toxic products of combustion may include carbon monoxide and nitrogen oxides • Do not put yourself in danger by entering a contaminated area to rescue a victim:

Hazard Classification (based on NFPA-704 Rating System)
Health Hazards (Blue): 3; Flammability (Red): 3; Reactivity (Yellow): 0

EMERGENCY RESPONSE: See Appendix A (132)
Evacuation:
Public safety: Isolate spill area for at least 50 to 100 meters (160 to 330 feet) in all directions.
Spill: Increase, as necessary, the isolation distance shown above, in "Public safety."
Fire: Isolate for 800 meters (½ mile) in all directions, especially if tank, rail car, or tank truck is involved in fire.

EXPOSURE
Short-term effects: *SEEK MEDICAL ATTENTION.* **Liquid or solid:** *ABSORBED THROUGH THE SKIN.* Will burn skin and eyes. Harmful if swallowed. Remove contaminated clothing and shoes. Flush affected areas with plenty of water. *IF IN EYES*, hold eyelids open and flush with plenty of water. *IF SWALLOWED* and victim is *CONSCIOUS AND ABLE TO SWALLOW*, have victim drink 4 to 8 ounces of water and have victim induce vomiting. *IF SWALLOWED* and victim is *UNCONSCIOUS OR HAVING CONVULSIONS*, do nothing except keep victim warm. **Vapor or liquid:** May cause blurred vision that usually improves in a few hours following exposure.

HEALTH HAZARDS
Personal protective equipment (PPE): B-Level PPE. Full face organic vapor respirator.
Toxicity by ingestion: Grade 2; LD_{50} = 0.5 to 5 g/kg (rat).
Vapor (gas) irritant characteristics: Vapors cause moderate irritation such that personnel will find high concentrations unpleasant. The effect is temporary.
Liquid or solid irritant characteristics: Eye and skin irritant. Causes smarting of the skin and first-degree burns on short exposure and may cause secondary burns on long exposure.

FIRE DATA
Flash point: 260°F/127°C (oc).
Flammable limits in air: LEL: 1.1%; UEL: 5.4%
Fire extinguishing agents not to be used: Water or foam may cause frothing.
Autoignition temperature: 705°F/374°C/647°K.
Electrical hazard: Class I, Group D.

CHEMICAL REACTIVITY
Binary reactants: Oxidizers, acids. Attacks copper and its alloys.
Reactivity group: 8
Compatibility class: Alkanolamines

ENVIRONMENTAL DATA
Food chain concentration potential: Negative.

Water pollution: Effect of low concentrations on aquatic life is unknown. May be dangerous if it enters nearby water intakes; notify operators. Notify local health and wildlife officials. **Response to discharge:** Disperse and flush

SHIPPING INFORMATION
Grades of purity: 97%; **Storage temperature:** Ambient; **Inert atmosphere:** None; **Venting:** Open; **Stability during transport:** Stable.

NAS HAZARD CLASSIFICATION FOR BULK WATER TRANSPORTATION
FIRE: 1
HEALTH: Vapor irritant: 2; Liquid or solid irritant: 2; Poisons: 2
WATER POLLUTION: Human toxicity: 2; Aquatic toxicity: 2; Aesthetic effect: 2
REACTIVITY: Other chemicals: 3; Water: 0; Self-reaction: 0

PHYSICAL AND CHEMICAL PROPERTIES
Physical state @ 59°F/15°C and 1 atm: Liquid or Solid.
Molecular weight: 133.19
Boiling point @ 1 atm: 479.7°F/248.7°C/521.9°K.
Melting/Freezing point: 108°F/42°C/315°K.
Critical temperature: 750°F/399°C/672°K.
Critical pressure: 529 psia = 36 atm = 3.6 MN/m².
Specific gravity (water = 1): 0.99 at 42°C (liquid).
Latent heat of vaporization: 185 Btu/lb = 103 cal/g = 4.31 x 10^5 J/kg.
Heat of combustion: (estimate) –12,300 Btu/lb = –6860 cal/g = –287 x 10^5 J/kg.
Heat of solution: (estimate) –13 Btu/lb = –7 cal/g = –0.3 x 10^5 J/kg.

DIISOPROPYLAMINE   REC. D:4600

SYNONYMS: DIPA; DIISOPROPILAMINA (Spanish); 2-PROPANAMINE, N-(1-METHYLETHYL)-; *N*-(1-METHYETHYL)-2-PROPANAMINE; BIS(ISOPROPYL)AMINEENTIFICATION:

CAS Number: 108-18-9
Formula: $C_6H_{15}N$; $[(CH_3)_2CH]_2NH$
DOT ID Number: UN 1158; DOT Guide Number: 132
Proper Shipping Name: Diisopropylamine

DESCRIPTION: Colorless liquid. Fishy odor. Floats on water surface; slightly soluble.

Extremely flammable • Containers may BLEVE when exposed to fire •Vapors may form explosive mixture with air • Vapors are heavier than air and will collect and stay in low areas • Vapors may travel long distances to ignition sources and flashback • Vapors in confined areas (e.g., tanks, sewers, buildings) may explode when exposed to fire • Irritating to the skin, eyes, and respiratory tract • Toxic products of combustion may include nitrogen oxide.

Hazard Classification (based on NFPA-704 Rating System)
Health Hazards (Blue): 3; Flammability (Red): 3; Reactivity (Yellow): 0

EMERGENCY RESPONSE: See Appendix A (132)
Evacuation:
Public safety: Isolate spill area for at least 50 to 100 meters (160 to 330 feet) in all directions.

Spill: Increase, as necessary, the isolation distance shown above, in "Public safety."
Fire: Isolate for 800 meters (½ mile) in all directions, especially if tank, rail car, or tank truck is involved in fire.

EXPOSURE
Short-term effects: *SEEK MEDICAL ATTENTION*. **Vapor:** Irritating to eyes, nose, and throat. *IF INHALED*, will, will cause coughing or difficult breathing. *IF IN EYES*, hold eyelids open and flush with plenty of water. *IF BREATHING HAS STOPPED*, give artificial respiration. If breathing is difficult, administer oxygen. **Liquid:** Will burn eyes. Irritating to eyes. *IF SWALLOWED*, will cause nausea and vomiting. Remove contaminated clothing and shoes. Flush affected areas with plenty of water. *IF IN EYES*, hold eyelids open and flush with plenty of water. *IF SWALLOWED* and victim is *CONSCIOUS AND ABLE TO SWALLOW*, have victim drink 4 to 8 ounces of water and have victim induce vomiting. *IF SWALLOWED* and victim is *UNCONSCIOUS OR HAVING CONVULSIONS*, do nothing except keep victim warm.

HEALTH HAZARDS
Personal protective equipment (PPE): B-Level PPE. OSHA Table Z-1-A air contaminant. Air–supplied mask, gloves, monogoggles, apron. Protective materials with good to excellent resistance: Teflon®, Viton®. Also, nitrile offers limited protection.
Recommendations for respirator selection: NIOSH/OSHA *125 ppm:* SA:CF* (any supplied-air respirator operated in a continuous-flow mode); or PAPROV* [any powered, air-purifying respirator with organic vapor cartridge(s)]. *200 ppm:* CCRFOV [any air-purifying, full-facepiece respirator (gas mask) with a chin-style, front- or back-mounted acid gas canister]; or GMFOV [any air-purifying, full-facepiece respirator (gas mask) with a chin-style, front- or back-mounted organic vapor canister]; or PAPRTOV [any powered, air-purifying respirator with a tight-fitting facepiece and organic vapor cartridges(s)]; or SCBAF (any self-contained breathing apparatus with a full facepiece); or SAF (any supplied-air respirator with a full facepiece). *EMERGENCY OR PLANNED ENTRY INTO UNKNOWN CONCENTRATIONS OR IDLH CONDITIONS:* SCBAF:PD,PP (any self-contained breathing apparatus that has a full facepiece and is operated in a pressure-demand or other positive-pressure mode); or SAF:PD,PP:ASCBA (any supplied-air respirator that has a full facepiece and is operated in a pressure-demand or other positive-pressure mode in combination with an auxiliary self-contained breathing apparatus operated in a pressure-demand or other positive pressure mode). *ESCAPE:* GMFOV [any air-purifying, full-facepiece respirator (gas mask) with a chin-style, front- or back-mounted organic vapor canister]; or SCBAE (any appropriate escape-type, self-contained breathing apparatus). **Note:* Substance causes eye irritation or damage; eye protection needed.
Exposure limits (TWA unless otherwise noted): ACGIH TLV 5 ppm (21 mg/m^3); NIOSH/OSHA 5 ppm (20 mg/m^3); skin contact contributes significantly in overall exposure.
Toxicity by ingestion: Grade 2; oral LD$_{50}$ = 0.7 g/kg (rat).
Vapor (gas) irritant characteristics: Very severe eye and respiratory tract irritant. Vapors are irritating such that personnel will not usually tolerate moderate or high concentrations.
Liquid or solid irritant characteristics: Severe eye and skin irritant. Causes smarting of the skin and first-degree burns on short exposure; may cause second-degree burns on long exposure.
Odor threshold: 0.13 ppm.
IDLH value: 200 ppm.

FIRE DATA
Flash point: 30°F/–1°C (oc); 19°F/–7°C(cc).
Flammable limits in air: LEL: 1.1%; UEL: 7.1%.
Fire extinguishing agents not to be used: Water may be ineffective.
Autoignition temperature: 600°F/316°C/589°K.
Electrical hazard: Class I, Group C.

CHEMICAL REACTIVITY
Binary reactants: Strong acids, strong oxidizers. May attack some forms of plastics; copper and its alloys.
Reactivity group: 7
Compatibility class: Aliphatic amines

ENVIRONMENTAL DATA
Food chain concentration potential: Negative; unlikely to accumulate.
Water pollution: Harmful to aquatic life in very low concentrations. May be dangerous if it enters nearby water intakes; notify operators. Notify local health and wildlife officials. **Response to discharge:** Issue warning–high flammability, water contaminant, air contaminant. Restrict access. Disperse and flush.

SHIPPING INFORMATION
Grades of purity: Commercial, 100%; **Storage temperature:** Ambient; **Inert atmosphere:** None; **Venting:** Open (flame arrester); **Stability during transport:** Stable.

NAS HAZARD CLASSIFICATION FOR BULK WATER TRANSPORTATION
FIRE: 3
HEALTH: Vapor irritant: 3; Liquid or solid irritant: 2; Poisons: 4
WATER POLLUTION: Human toxicity: 2; Aquatic toxicity: 3; Aesthetic effect: 2
REACTIVITY: Other chemicals: 3; Water: 0; Self-reaction: 0

PHYSICAL AND CHEMICAL PROPERTIES
Physical state @ 59°F/15°C and 1 atm: Liquid.
Molecular weight: 101.19
Boiling point @ 1 atm: 183.0°F/83.9°C/357.1°K.
Melting/Freezing point: –141.3°F/–96.3°C/176.9°K.
Critical temperature: 480.2°F/249.0°C/522.2°K.
Critical pressure: (estimate) 400 psia = 30 atm = 3 MN/m^2.
Specific gravity (water = 1): 0.717 @ 68°F/20°C (liquid).
Liquid surface tension: 19.64 dynes/cm = 0.01964 N/m @ 68°F/20°C.
Relative vapor density (air = 1): 3.5
Ratio of specific heats of vapor (gas): (estimate) 1.064
Latent heat of vaporization: 121 Btu/lb = 67.5 cal/g = 2.82 x 10^5 J/kg.
Heat of combustion: –19,800 Btu/lb = –11,000 cal/g = –460 x 10^5 J/kg.
Heat of solution: –140 Btu/lb = –76 cal/g = 3.2 x 10^5 J/kg.
Vapor pressure: (Reid) 2.5 psia; 70 mm

DIISOPROPYLBENZENE (all isomers)   REC. D:4650

SYNONYMS: BENZENE, DIISOPROPYL; CUMENE BOTTOMS; BIS(1-METHYLETHYL)-BENZENE; DIISOPROPILBENCENO (Spanish)

IDENTIFICATION
CAS Number: 25321-09-9; 577-55-9 (*o*-isomer); 99-62-7 (*m*-isomer); 100-18-5 (*p*-isomer)
Formula: C12H$_{18}$; (C$_6$H$_4$)[CH(CH$_3$)$_2$]$_2$

DOT ID Number: UN 1993; DOT Guide Number: 128
Proper Shipping Name: Combustible liquid, n.o.s.

DESCRIPTION: Clear amber liquid. Sharp penetrating and aromatic odor.

Flammable • Containers may BLEVE when exposed to fire •Vapors may form explosive mixture with air • Vapors are heavier than air and will collect and stay in low areas • Vapors may travel long distances to ignition sources and flashback • Vapors in confined areas (e.g., tanks, sewers, buildings) may explode when exposed to fire • Irritating to the skin, eyes, and respiratory tract • Toxic products of combustion may include carbon monoxide.

Hazard Classification (based on NFPA-704 Rating System)
Health Hazards (Blue): 0; Flammability (Red): 2; Reactivity (Yellow): 0

EMERGENCY RESPONSE: See Appendix A (128)
Evacuation:
Public safety: Isolate spill area for at least 25 to 50 meters (80 to 160 feet) in all directions.
Spill: Large spill–Consider initial downwind evacuation for at least 300 meters (1000 feet).
Fire: Isolate for 800 meters (½ mile) in all directions, especially if tank, rail car, or tank truck is involved in fire.

EXPOSURE
Short-term effects: *SEEK MEDICAL ATTENTION.* **Vapor:** Irritating to eyes, nose, and throat. *IF INHALED*, will, will cause dizziness or difficult breathing. Move to fresh air. *IF BREATHING HAS STOPPED*, give artificial respiration. If breathing is difficult, administer oxygen. **Liquid:** Will burn skin and eyes. Harmful if swallowed. Remove contaminated clothing and shoes. Flush affected areas with plenty of water. *IF IN EYES*, hold eyelids open and flush with plenty of water. *IF SWALLOWED*, **Do NOT induce vomiting.**

HEALTH HAZARDS
Personal protective equipment (PPE): B-Level PPE. Gloves and clothing impervious to aromatic hydrocarbon and splash-proof safety goggles.
Toxicity by ingestion: Grade 1: LD_{50} = 6.5 g/kg (rat).
Vapor (gas) irritant characteristics: Severe eye and respiratory tract irritant. Vapors cause smarting of the eyes or respiratory system if present in high concentrations. The effect is temporary.
Liquid or solid irritant characteristics: Eye and skin irritant. If spilled on clothing and allowed to remain, may cause smarting and reddening of skin.

FIRE DATA
Flash point: 170°F/77°C (oc) (*o*-isomer).
Flammable limits in air: LEL: 0.9%; UEL: 6.5%.
Fire extinguishing agents not to be used: Water may be ineffective.
Autoignition temperature: 840°F/449°C/722°K.

CHEMICAL REACTIVITY
Binary reactants: Oxidizers, nitric acids.
Reactivity group: 32
Compatibility class: Aromatic hydrocarbons

ENVIRONMENTAL DATA
Water pollution: DOT Appendix B, §172.101–marine pollutant. Effects of low concentrations on aquatic life is not known. Fouling to shoreline. May be dangerous if it enters nearby water intakes; notify operators. Notify local health and wildlife officials. **Response to discharge:** Restrict access. Mechanical containment. Should be removed. Chemical and physical treatment.

SHIPPING INFORMATION
Storage temperature: Ambient; **Venting:** Explosion proof type exhaust. Keep vapors below 100 ppm; **Stability during transport:** Stable.

PHYSICAL AND CHEMICAL PROPERTIES
Physical state @ 59°F/15°C and 1 atm: Liquid.
Molecular weight: 162.30
Boiling point @ 1 atm: 401°F/205°C/478°K.
Melting/Freezing point: Less than 15°F/–9°C/264°K.
Specific gravity (water = 1): 0.956 @ 15°C (liquid).
Relative vapor density (air = 1): 5.63

DIISOPROPYLBENZENE HYDROPEROXIDE
REC. D:4700

SYNONYMS: DIISOPROPYLPHENYLHYDROPEROXIDE; DIISOPROPYLBENZENE HYDROPEROXIDE, NOT MORE THAN 72% IN SOLUTION; HIDROPEROXIDO de DIISOPROPILBENCENO (Spanish); ISOPROPYLCUMYLHYDROPEROXIDE

IDENTIFICATION
CAS Number: 26762-93-6
Formula: $C_{12}H_{19}O_2$; $(CH_3)_2CHC_6H_4C(CH_3)_2OOH$ +$(CH_3)_2CHC_6H_4CH(CH_3)_2$
DOT ID Number: UN 2171; DOT Guide Number: 145
Proper Shipping Name: Diisopropylbenzene hydroperoxide

DESCRIPTION: Colorless to pale yellow liquid. Sharp, unpleasant odor. May float or sink in water.

Highly flammable • Heat or contamination may cause explosion • Strong oxidizer which may react spontaneously with low flash point organics or reducing agents. Heat forms oxygen; will increase the activity of an existing fire • May cause fire or explosion on contact with combustibles (wood, paper, oil, clothing, etc.) all fuels and most metals • Containers may BLEVE when exposed to fire •Vapors may form explosive mixture with air • Vapors are heavier than air and will collect and stay in low areas • Vapors may travel long distances to ignition sources and flashback • Vapors in confined areas (e.g., tanks, sewers, buildings) may explode when exposed to fire • Irritating to the skin, eyes, and respiratory tract • Toxic products of combustion may include acrid smoke and flammable alcohol and ketone gases.

Hazard Classification (based on NFPA-704 Rating System)
Health Hazards (Blue): 0; Flammability (Red): 2; Reactivity (Yellow): 0

EMERGENCY RESPONSE
Evacuation:
Public safety: Isolate the area of spill or leak for at least 25 to 50 meters (80 to 160 feet) in all directions.
Spill: Consider evacuation for at least 250 meters (800 feet) in all directions.
Fire: If any large container is involved in fire, isolate for at least 800 meters (½ mile) in all directions; also, consider initial evacuation for 800 meters (½ mile) in all directions.

EXPOSURE

Short-term effects: *SEEK MEDICAL ATTENTION.* **Vapor:** Irritating to eyes, nose, and throat. *IF INHALED*, will, will cause coughing or difficult breathing. *IF IN EYES*, hold eyelids open and flush with plenty of water. *IF BREATHING HAS STOPPED,* give artificial respiration. If breathing is difficult, administer oxygen. **Liquid:** Irritating to skin and eyes. Harmful if swallowed. Remove contaminated clothing and shoes. Flush affected areas with plenty of water. *IF IN EYES*, hold eyelids open and flush with plenty of water. *IF SWALLOWED* and victim is *CONSCIOUS AND ABLE TO SWALLOW*, have victim drink 4 to 8 ounces of water. *IF SWALLOWED* and victim is *UNCONSCIOUS OR HAVING CONVULSIONS,* do nothing except keep victim warm.

HEALTH HAZARDS

Personal protective equipment (PPE): B-Level PPE. solvent-resistant gloves; chemical-resistant apron; chemical goggles or face shield; self-contained breathing apparatus. Butyl rubber is generally suitable for peroxide compounds.
Vapor (gas) irritant characteristics: Eye and respiratory tract irritant.
Liquid or solid irritant characteristics: Severe eye and skin irritant.

FIRE DATA

Flash point: 175°F/79°C (cc).
Fire extinguishing agents not to be used: Water may be ineffective.
Behavior in fire: Burns with a flare effect. Containers may explode.

CHEMICAL REACTIVITY

Binary reactants: Aluminum, copper, brass, lead, zinc salts, mineral acids, and oxidizing or reducing agents all can cause rapid decomposition.

ENVIRONMENTAL DATA

Food chain concentration potential: Negative; unlikely to accumulate.
Water pollution: DOT Appendix B, §172.101–marine pollutant. Effect of low concentrations on aquatic life is unknown. Fouling to shoreline. May be dangerous if it enters nearby water intakes; notify operators. Notify local health and wildlife officials.
Response to discharge: Issue warning–oxidizing material. Restrict access. Mechanical containment. Should be removed. Chemical and physical treatment.

SHIPPING INFORMATION

Grades of purity: 54.06%, the balance is diisopropylbenzene, a combustible hydrocarbon; **Storage temperature:** Ambient; **Inert atmosphere:** None; **Venting:** Open (flame arrester); **Stability during transport:** Unstable; slowly evolves oxygen.

PHYSICAL AND CHEMICAL PROPERTIES

Physical state @ 59°F/15°C and 1 atm: Liquid.
Molecular weight: 194.26
Melting/Freezing point: Less than 15°F/less than –9°C/less than 264°K.
Specific gravity (water = 1): 0.956 @ 15°C (liquid).

DIISOPROPYL NAPHTHALENE   REC. D:4750

SYNONYMS: BIS(ISOPROPYL)NAPHTHALENE; 2,6-DIISOPROPYL NAPHTHALENE; 2,6-DIISOPROPIL NAFTALENO (Spanish); K 113; KMC 113; KMC-R 113; NAPHTHALENE, BIS(1-METHYLETHYL)-

IDENTIFICATION

CAS Number: 24157-81-1; 38640-62-9
Formula: $C_{16}H_{20}$; $(CH_3)_2CH(C_{10}H_8)CH(CH_3)_2$

DESCRIPTION: Clear yellowish brown liquid. Faint, sweet odor.

Combustible • Containers may BLEVE when exposed to fire • Vapors are heavier than air and will collect and stay in low areas • Vapors in confined areas (e.g., tanks, sewers, buildings) may explode when exposed to fire • Irritating to the skin, eyes, and respiratory tract • Toxic products of combustion may include carbon monoxide.

EMERGENCY RESPONSE: See Appendix A (145)
Evacuation:
Public safety: Isolate the area of spill or leak for at least 10 to 25 meters (30 to 80 feet) in all directions.
Spill: Increase, in the downwind direction, as necessary, the distance shown under "Public Safety."
Fire: If any large container is involved in fire, isolate for at least 800 meters (½ mile) in all directions; also, consider initial evacuation for 800 meters (½ mile) in all directions.

EXPOSURE

Short-term effects: *SEEK MEDICAL ATTENTION.* **Vapor:** Move victim to fresh air. *IF BREATHING HAS STOPPED,* give artificial respiration. If breathing is difficult, administer oxygen. **Liquid:** Remove contaminated clothing and shoes. Wash affected areas with soap and water. *IF IN EYES*, hold eyelids open and flush with plenty of water.

HEALTH HAZARDS

Personal protective equipment (PPE): B-Level PPE. Full impervious protective clothing, including boots and gloves. Where splashing is possible wear full face shield or chemical safety goggles. Use approved respirator to protect against vapors.
Vapor (gas) irritant characteristics: eye and respiratory tract irritant. Vapors cause smarting of the eyes or respiratory system if present in high concentrations. The effect is temporary.
Liquid or solid irritant characteristics: Eye and skin irritant. If spilled on clothing and allowed to remain, may cause poisoning.

FIRE DATA

Flash point: 284°F/140°C (cc).
Fire extinguishing agents not to be used: Water.

CHEMICAL REACTIVITY

Binary reactants: Oxidizers, nitric acid.
Reactivity group: 32
Compatibility class: Aromatic hydrocarbons

ENVIRONMENTAL DATA

Water pollution: Effects of low concentrations on aquatic life is unknown. May be dangerous if it enters nearby water intakes; notify operators. Notify local health and wildlife officials. **Response to discharge:** Should be removed.

SHIPPING INFORMATION

Grades of purity: Technical grades; **Storage temperature:** Ambient; **Inert atmosphere:** None; **Stability during transport:** Stable.

PHYSICAL AND CHEMICAL PROPERTIES
Physical state @ 59°F/15°C and 1 atm: Liquid.
Molecular weight: 212.34
Boiling point @ 1 atm: 554–563°F/290–295°C/302–307°K.
Specific gravity (water = 1): 0.95 @ 30°C.

DIMETHYL ACETAMIDE   REC. D:4800

SYNONYMS: ACETAMIDE *N,N*-DIMETHYL; ACETDIMETHYLAMIDE; ACETIC ACID, DIMETHYLAMIDE; ACETYLDIMETHYLAMINE; *N,N*-DIMETHYL ACETAMIDE; DIMETHYLACETAMIDE; DIMETHYLACETONE AMIDE; DIMETHYLAMIDE ACETATE; *N,N*-DIMETILACETAMIDA (Spanish); DMA; DMAC; EEC No. 616-011-00-4; U-5954

IDENTIFICATION
CAS Number: 127-19-5
Formula: C_4H_9NO; $CH_3CON(CH_3)_2$
DOT ID Number: UN 1993; DOT Guide Number: 128
Proper Shipping Name: Combustible liquid, n.o.s.

DESCRIPTION: Colorless, oily liquid. Weak ammonia or fish-like odor. Initially floats on water surface; soluble.

Flammable • Do Not use HALONS •Containers may BLEVE when exposed to fire •Vapors may form explosive mixture with air • Vapors are heavier than air and will collect and stay in low areas • Vapors may travel long distances to ignition sources and flashback • Vapors in confined areas (e.g., tanks, sewers, buildings) may explode when exposed to fire • Irritating to the skin, eyes, and respiratory tract; absorbed by the skin • Toxic products of combustion may include carbon monoxide and nitrogen oxides.

Hazard Classification (based on NFPA-704 Rating System)
Health Hazards (Blue): 2; Flammability (Red): 2; Reactivity (Yellow): 0

EMERGENCY RESPONSE: See Appendix A (128)
Evacuation:
Public safety: Isolate spill area for at least 25 to 50 meters (80 to 160 feet) in all directions.
Spill: Large spill–Consider initial downwind evacuation for at least 300 meters (1000 feet).
Fire: Isolate for 800 meters (½ mile) in all directions, especially if tank, rail car, or tank truck is involved in fire.

EXPOSURE
Short-term effects: *SEEK MEDICAL ATTENTION.* **Liquid:** *HARMFUL IF SWALLOWED OR ABSORBED THROUGH THE SKIN.* Irritating to skin and eyes. Harmful if swallowed. Remove contaminated clothing and shoes. Flush affected areas with plenty of water. *IF IN EYES*, hold eyelids open and flush with plenty of water. *IF SWALLOWED* and victim is *CONSCIOUS AND ABLE TO SWALLOW*, have victim drink 4 to 8 ounces of water and have victim induce vomiting. *IF SWALLOWED* and victim is *UNCONSCIOUS OR HAVING CONVULSIONS,* do nothing except keep victim warm.

HEALTH HAZARDS
Personal protective equipment (PPE): B-Level PPE. OSHA Table Z-1-A air contaminant. Full face organic vapor respirator. Chemical protective material(s) reported to offer minimal to poor protection: butyl rubber.
Recommendations for respirator selection: NIOSH/OSHA
100 ppm: SA (any supplied-air respirator). *250 ppm:* SA:CF (any supplied-air respirator operated in a continuous-flow mode). *300 ppm:* SCBAF (any self-contained breathing apparatus with a full facepiece); or SAF (any supplied-air respirator with a full facepiece). *EMERGENCY OR PLANNED ENTRY INTO UNKNOWN CONCENTRATIONS OR IDLH CONDITIONS:*SCBAF:PD,PP (any self-contained breathing apparatus that has a full facepiece and is operated in a pressure-demand or other positive-pressure mode); or SAF:PD,PP:ASCBA (any supplied-air respirator that has a full facepiece and is operated in a pressure-demand or other positive-pressure mode in combination with an auxiliary self-contained breathing apparatus operated in a pressure-demand or other positive pressure mode). *ESCAPE:* GMFOV[any air-purifying, full-facepiece respirator (gas mask) with a chin-style, front- or back-mounted organic vapor canister]; or SCBAE (any appropriate escape-type, self-contained breathing apparatus).
Exposure limits (TWA unless otherwise noted): ACGIH TLV 10 ppm (36 mg/m^3); NIOSH/OSHA 10 ppm (35 mg/m^3); skin contact contributes significantly in overall exposure.
Toxicity by ingestion: Grade 1; oral LD_{50} = 5.63 g/kg (rat).
Long-term health effects: May produce chronic liver, kidney, and brain damage.
Vapor (gas) irritant characteristics: Eye and respiratory tract irritant.
Liquid or solid irritant characteristics: Eye and skin irritant.
Odor threshold: 21–47 ppm.
IDLH value: 300 ppm.

FIRE DATA
Flash point: 158°F/70°C (oc).
Flammable limits in air: LEL: 1.8% @ 212°F/100°C; UEL: 11.5% @ 320°F/160°C.
Fire extinguishing agents not to be used: Do not use halons.
Autoignition temperature: 914°F/490°C/763°K.
Burning rate: 2.8 mm/min

CHEMICAL REACTIVITY
Reactivity with water: Becomes corrosive.
Binary reactants: Reacts with oxidizers, carbon tetrachloride, other halogenated compounds when in contact with iron. Attacks many forms of plastic materials.

ENVIRONMENTAL DATA
Food chain concentration potential: Log P_{ow} = 0.78. Unlikely to accumulate.
Water pollution: Effect of low concentrations on aquatic life is unknown. May be dangerous if it enters nearby water intakes; notify operators. Notify local health and wildlife officials. **Response to discharge:** Issue warning–water contaminant. Disperse and flush

SHIPPING INFORMATION
Grades of purity: Technical; **Storage temperature:** Ambient; **Inert atmosphere:** None; **Venting:** Open (flame arrester); **Stability during transport:** Stable.

PHYSICAL AND CHEMICAL PROPERTIES
Physical state @ 59°F/15°C and 1 atm: Liquid.
Molecular weight: 87.1
Boiling point @ 1 atm: 331°F/166°C/439°K.
Melting/Freezing point: –4°F/–20°C/253°K.
Specific gravity (water = 1): 0.943 @ 68°F/20°C (liquid).

400 Dimethylamine

Liquid surface tension: 34 dynes/cm = 0.034 N/m @ 68°F/20°C.
Relative vapor density (air = 1): 3.0
Latent heat of vaporization: 214 Btu/lb = 119 cal/g = 4.98 x 10^5 J/kg.
Heat of combustion: –12,560 Btu/lb = –6980 cal/g = –292 x 10^5 J/kg.
Vapor pressure: 2 mm.

DIMETHYLADIPATE   REC. D:4900

SYNONYMS: ADIPTATO de METILO (Spanish); ADIPIC ACID, DIMETHYL ESTER; DIMETHYL HEXANEDIOATE; HEXANEDIOIC ACID, DIMETHYL ESTER; METHYL ADIPATE

IDENTIFICATION
CAS Number: 627-93-0
Formula: $C_8H_{14}O_4$; $CH_3O_2C(CH_2)_4CO_2CH_3$

DESCRIPTION: Colorless liquid.

Combustible • Containers may BLEVE when exposed to fire • Vapors are heavier than air and will collect and stay in low areas • Irritating to the skin, eyes, and respiratory tract • Toxic products of combustion may include carbon monoxide.

Hazard Classification (based on NFPA-704 Rating System)
Health Hazards (Blue): 1; Flammability (Red): 1; Reactivity (Yellow): 0

EMERGENCY RESPONSE: See Appendix A (171)
Evacuation:
Public safety: Isolate the area of spill or leak for at least 10 to 25 meters (30 to 80 feet) in all directions.
Spill: Increase, in the downwind direction, as necessary, the distance shown under "Public Safety."
Fire: If any large container is involved in fire, isolate for at least 800 meters (½ mile) in all directions; also, consider initial evacuation for 800 meters (½ mile) in all directions.

EXPOSURE
Short-term effects: *SEEK MEDICAL ATTENTION.* **Vapor or mist:** Move to fresh air. *IF BREATHING HAS STOPPED,* give artificial respiration. If breathing is difficult, administer oxygen. **Liquid:** Flush affected areas with plenty of water. *IF IN EYES,* hold eyelids open and flush with plenty of water.

HEALTH HAZARDS
Personal protective equipment (PPE): B-Level PPE. Self-contained breathing apparatus, rubber boots and heavy rubber gloves.
Long-term health effects: Prolonged exposure may result in infertility or infant developmental abnormalities.
Vapor (gas) irritant characteristics: Vapors cause smarting of the eyes or respiratory system if present in high concentrations.
Liquid or solid irritant characteristics: Minimum hazard. Eye and skin irritant. If spilled on clothing and allowed to remain, may cause smarting and reddening of skin.

FIRE DATA
Flash point: 225°F/107°C (cc).
Flammable limits in air: LEL: 0.81%; UEL 8.1%.
Autoignition temperature: 680°F/360°C/633°K.

CHEMICAL REACTIVITY
Binary reactants: Oxidizers, strong acids
Reactivity group: 34

ENVIRONMENTAL DATA
Water pollution: Effects of low concentrations on aquatic life is unknown. Fouling to shoreline. May be dangerous if it enters nearby water intakes; notify operators. Notify local health and wildlife officials. **Response to discharge:** Should be removed. Mechanical containment. Chemical and physical treatment.

SHIPPING INFORMATION
Grades of purity: 99+%; **Storage temperature:** Ambient; **Inert atmosphere:** Not required; **Venting:** Not required; **Stability during transport:** Stable.

PHYSICAL AND CHEMICAL PROPERTIES
Physical state @ 59°F/15°C and 1 atm: Liquid.
Molecular weight: 174.20
Boiling point @ 1 atm: 228-230°F/109-110°C/382-283°K (at 14 mmHg = 0.0184 atm).
Melting/Freezing point: 46.4°F/8°C/281.2°K.
Specific gravity (water = 1): 1.063
Relative vapor density (air = 1): 6.0 (estimate).

DIMETHYLAMINE   REC. D:4950

SYNONYMS: DIMETHYLAMINE, AQUEOUS SOLUTION; *N,N*-DIMETHYLAMINE; DIMETHYLAMINE, ANHYDROUS; DIMETHYLAMINE SOLUTION; DIMETILAMINA (Spanish); DMA; EEC No. 612-001-00-9; METHANAMINE, *N*-METHYL-; *N*-METHYLMETHANAMINE; RCRA No. U092

IDENTIFICATION:
CAS Number: 124-40-3
Formula: C_2H_7N; $(CH_3)_2NH$
DOT ID Number: UN 1160 (solution); UN 1032 (anhydrous); DOT Guide Number: 129 (solution); 118 (anhydrous)
Proper Shipping Name: Dimethylamine solution; Dimethylamine, anhydrous
Reportable Quantity (RQ): **(CERCLA)** 1000 lb/454 kg

DESCRIPTION: Colorless gas. Shipped or stored as a liquefied compressed gas or in a water solution. Dead fish or ammonia-like odor. Floats and boils on water; highly soluble. Flammable, irritating vapor is produced. Become a liquid at 44°F/7°C.

Extremely flammable Corrosive to the skin, eyes, and respiratory tract; contact with skin or eyes can cause burns and vision impairment; inhalation symptoms may be delayed • Contact with liquid may cause frostbite • Firefighting gear (including SCBA) does not provide adequate protection. If exposure occurs, remove and isolate gear immediately and thoroughly decontaminate personnel •Containers may BLEVE when exposed to fire •Vapors may form explosive mixture with air • Vapors are heavier than air and will collect and stay in low areas • Vapors may travel long distances to ignition sources and flashback • Vapors in confined areas (e.g., tanks, sewers, buildings) may explode when exposed to fire • Irritating to the skin, eyes, and respiratory tract • Toxic products of combustion may include nitrogen oxide • Do not put yourself in danger by entering a contaminated area to rescue a victim.

Dimethylamine

Hazard Classification (based on NFPA-704 Rating System)
Health Hazards (Blue): 3; Flammability (Red): 4; Reactivity (Yellow): 0

EMERGENCY RESPONSE, gas: See Appendix A (118)
Evacuation:
Public safety: Isolate spill area for at least 100 to 200 meters (330 to 660 feet) in all directions.
Spill: Consider initial downwind evacuation for at least 800 meters (½ mile).
Fire: Isolate for 1600 meters (1 mile) in all directions, especially if tank, rail car, or tank truck is involved in fire.

EMERGENCY RESPONSE, solution: See Appendix A (129)
Evacuation:
Public safety: Isolate spill area for at least 50 to 100 meters (160 to 330 feet) in all directions.
Spill: Large spill–Consider initial downwind evacuation for at least 300 meters (1000 feet).
Fire: Isolate for 800 meters (½ mile) in all directions, especially if tank, rail car, or tank truck is involved in fire.

EXPOSURE
Short-term effects: *SEEK MEDICAL ATTENTION*. **Vapor:** Irritating to eyes, nose, and throat. *IF INHALED*, will, will cause difficult breathing. Move to fresh air. *IF BREATHING HAS STOPPED,* give artificial respiration. If breathing is difficult, administer oxygen. May cause lung edema; physical exertion will aggravate this condition. **Liquid:** Will burn skin and eyes. Harmful if swallowed. Remove contaminated clothing and shoes. Flush affected areas with plenty of water. *IF IN EYES*, hold eyelids open and flush with plenty of water. *IF SWALLOWED* and victim is *CONSCIOUS AND ABLE TO SWALLOW*, have victim drink 4 to 8 ounces of water.
Note to physician or authorized medical personnel: Medical observation is recommended for 24 to 48 hours after breathing overexposure, as pulmonary edema may be delayed. As first aid for pulmonary edema, consider administering a corticosteroid spray. Cigarette smoking may exacerbate pulmonary injury and should be discouraged for at least 72 hours following exposure.

HEALTH HAZARDS
Personal protective equipment (PPE): B-Level PPE. OSHA Table Z-1-A air contaminant. Chemical goggles and full face shield; molded acid gloves; self-contained breathing apparatus. Chemical protective material(s) reported to have good to excellent resistance: butyl rubber, neoprene.
Recommendations for respirator selection: NIOSH/OSHA
250 ppm: SA:CF* (any supplied-air respirator operated in a continuous-flow mode). *500 ppm:* SCBAF (any self-contained breathing apparatus with a full facepiece); or SAF (any supplied-air respirator with a full facepiece). *EMERGENCY OR PLANNED ENTRY INTO UNKNOWN CONCENTRATIONS OR IDLH CONDITIONS:* SCBAF:PD,PP (any self-contained breathing apparatus that has a full facepiece and is operated in a pressure-demand or other positive-pressure mode); or SAF:PD,PP:ASCBA (any supplied-air respirator that has a full facepiece and is operated in a pressure-demand or other positive-pressure mode in combination with an auxiliary self-contained breathing apparatus operated in a pressure-demand or other positive pressure mode). *ESCAPE:* GMFS [any air-purifying, full-facepiece respirator (gas mask) with a chin-style, front- or back-mounted canister providing protection against the compound of concern]; or SCBAE (any appropriate escape-type, self-contained breathing apparatus).
Note: Causes eye irritation or damage; eye protection needed.

Exposure limits (TWA unless otherwise noted): ACGIH TLV 5 ppm (9.2 mg/m^3); NIOSH/OSHA PEL 10 ppm (18 mg/m^3).
Short-term exposure limits (15-minute TWA): ACGIH STEL 15 ppm (27.6 mg/m^3).
Long-term health effects: May cause eczema or other skin disorders.
Vapor (gas) irritant characteristics: Eye and respiratory tract irritant. Vapors cause irritation such that personnel will find high concentrations unpleasant. The effect is temporary.
Liquid or solid irritant characteristics: Severe ye and skin irritant. Causes smarting of the skin and first-degree burns on short exposure and may cause secondary burns on long exposure.
Odor threshold: 0.001–1.58 ppm.
IDLH value: 500 ppm.

FIRE DATA
Flash point: 20°F/–7°C (cc) (liquid).
Flammable limits in air: LEL: 2.8%; UEL: 14.4% (gas).
Fire extinguishing agents not to be used: Do not use regular foam on small fires.
Autoignition temperature: 756°F/403°C/676°K.
Electrical hazard: Class I, Group C. Due to low electric conductivity, may generate electrostatic charges from agitation and flow.
Burning rate: 4.5 mm/min
Stoichiometric air-to-fuel ratio: 11.42 (estimate).

CHEMICAL REACTIVITY
Reactivity with water: Solution is a base.
Binary reactants: Strong oxidizers, acids, chlorine, mercury, acetaldehyde, fluorine, maleic anhydride. Attacks aluminum, brass, magnesium, copper, zinc, and galvanized metal.
Reactivity group: 7
Compatibility class: Aliphatic amines

ENVIRONMENTAL DATA
Food chain concentration potential: Log P_{ow} = –0.22. Unlikely to accumulate.
Water pollution: Harmful to aquatic life in very low concentrations. May be dangerous if it enters nearby water intakes; notify operators. Notify local health and wildlife officials. **Response to discharge:** Issue warning–high flammability. Restrict access. Evacuate area.

SHIPPING INFORMATION
Grades of purity: Anhydrous: 99.5%; Aqueous solutions: 25%, 40%, 50%, 60%; **Storage temperature:** Ambient; **Inert atmosphere:** None; **Venting:** Safety relief; **Stability during transport:** Stable. Highly reactive with other material.

NAS HAZARD CLASSIFICATION FOR BULK WATER TRANSPORTATION
FIRE: 4
HEALTH: Vapor irritant: 2; Liquid or solid irritant: 2; Poisons: 2
WATER POLLUTION: Human toxicity: 2; Aquatic toxicity: 3; Aesthetic effect: 2
REACTIVITY: Other chemicals: 3; Water: 0; Self-reaction: 0

PHYSICAL AND CHEMICAL PROPERTIES
Physical state @ 59°F/15°C and 1 atm: Gas.
Molecular weight: 45.08
Boiling point @ 1 atm: 44.42°F/6.9°C/280.1°K.
Melting/Freezing point: –134.0°F/–92.2°C/181.0°K.
Critical temperature: 328.3°F/164.6°C/437.8°K.
Critical pressure: 770 psia = 52.4 atm = 5.31 MN/m^2.

402 *N,N*-Dimethylcarbamoyl chloride

Specific gravity (water = 1): 0.671 at 44°F/liquid).
Relative vapor density (air = 1): 1.56
Ratio of specific heats of vapor (gas): 1.139
Latent heat of vaporization: 252.9 Btu/lb = 140.5 cal/g = 5.882 x 10^5 J/kg.
Heat of combustion: −16,800 Btu/lb = −9340 cal/g = −391.0 x 10^5 J/kg.
Heat of solution: −515 Btu/lb = −286 cal/g = −12.0 x 10^5 J/kg.
Heat of fusion: 31.51 cal/g.
Vapor pressure: (Reid) 45 psia; less than 220 mm.

N,N-DIMETHYLCARBAMOYL CHLORIDE REC. D:5050

SYNONYMS: CARBAMIC CHLORIDE, DIMETHYL-; CARBAMOYL CHLORIDE, DIMETHYL-; CLORURO de DIMETILCARBAMOILO (Spanish); DIMETHYLCARBAMYL CHLORIDE; DIMETHYL CARBAMOYL CHLORIDE; DDC; DIMETHYLAMINO CARBONYL CHLORIDE; CHLOROFORMIC ACID DIMETHYLAMIDE; DIMETHYLCARBAMIC ACID CHLORIDE; DIMETHYLCARBAMIC CHLORIDE; DIMETHYLCARBAMIDOYL CHLORIDE; *N,N*-DIMETHYLCARBAMYL CHLORIDE; DMCC; TL 389; *N,N*-DIMETHYLCHLOROFORMAMIDE; CHLOROFORMIC ACID DIMETHYLAMIDE; RCRA No. U097

IDENTIFICATION
CAS Number: 79-44-7
Formula: C_3H_6ClNO; $(CH_3)_2NCOCl$
DOT ID Number: UN 2262; DOT Guide Number: 156
Proper Shipping Name: Dimethylcarbamoyl chloride
Reportable Quantity (RQ): **(CERCLA)** 1 lb/0.454 kg

DESCRIPTION: Colorless liquid. Irritating odor. Sinks in water; rapidly decomposes forming extremely flammable dimethylamine and corrosive hydrochloric acid.

Combustible • Severely irritating to skin, eyes, and respiratory tract; contact with skin and eyes causes severe burns and blindness • Firefighting gear (including SCBA) does not provide adequate protection. If exposure occurs, remove and isolate gear immediately and thoroughly decontaminate personnel • Containers may BLEVE when exposed to fire • Vapors are heavier than air and will collect and stay in low areas • Vapors in confined areas (e.g., tanks, sewers, buildings) may explode when exposed to fire • Irritating to the skin, eyes, and respiratory tract • Toxic products of combustion may include carbon monoxide, nitrogen oxides, and hydrogen chloride • Do not put yourself in danger by entering a contaminated area to rescue a victim.

Hazard Classification (based on NFPA-704 Rating System)
Health Hazards (Blue): 3; Flammability (Red): 1; Reactivity (Yellow): 1; Special Notice (White): Water reactive

EMERGENCY RESPONSE: See Appendix A (156)
Evacuation:
Public safety: Isolate the area of spill or leak for at least 50 to 100 meters (160 to 330 feet) in all directions.
Spill: Increase, in the downwind direction, as necessary, the distance shown under "Public Safety."
Fire: If tank, rail car, or tank truck is involved in fire, isolate for at least 800 meters (½ mile) in all directions; also, consider initial evacuation for 800 meters (½ mile) in all directions.

EXPOSURE
Short-term effects: *SEEK MEDICAL ATTENTION.* **Vapor:** Harmful if inhaled or absorbed through the skin. Extremely irritating to the eyes, nose, and throat. Remove to fresh air. *IF BREATHING HAS STOPPED,* give artificial respiration; *avoid mouth-to-mouth resuscitation; use bag/mask apparatus.* **Liquid:** Corrosive to skin and eyes, and harmful if ingested or absorbed through the skin. Remove contaminated clothing and shoes. Flush affected areas with plenty of water. *IF IN EYES,* hold eyelids open, flush with plenty of water for at least 15 minutes. *IF SWALLOWED: CALL A DOCTOR* immediately.

HEALTH HAZARDS
Personal protective equipment (PPE): B-Level PPE. Approved respirator, chemical-resistant gloves, full protective clothing, safety goggles or 8-inch minimum face shield.
Recommendations for respirator selection: NIOSH
At any concentrations above the NIOSH REL, or where there is no REL, at any detectable concentration: SCBAF:PD,PP (any self-contained breathing apparatus that has a full facepiece and is operated in a pressure-demand or other positive-pressure mode); or SAF:PD,PP:ASCBA (any supplied-air respirator that has a full facepiece and is operated in a pressure-demand or other positive-pressure mode in combination with an auxiliary self-contained breathing apparatus operated in a pressure-demand or other positive pressure mode). *ESCAPE:* GMFOV[any air-purifying, full-facepiece respirator (gas mask) with a chin-style, front- or back-mounted organic vapor canister]; or SCBAE (any appropriate escape-type, self-contained breathing apparatus).
Exposure limits (TWA unless otherwise noted): ACGIH TLV suspected human carcinogen; NIOSH REL potential human carcinogen; reduce exposure to lowest feasible level
Toxicity by ingestion: Grade 2: LD_{50} = 1 g/kg (rat).
Long-term health effects: Suspected Human Carcinogen. NTP anticipated carcinogen; IARC probable carcinogen, rating 2A; animal evidence: sufficient. May cause lung cancer. Amides may cause liver, kidney, and brain damage.
Vapor (gas) irritant characteristics: Vapors cause severe irritation of eyes and throat and can cause eye and lung injury. They cannot be tolerated even at low concentrations.
Liquid or solid irritant characteristics: Severe eye and skin irritant. Causes second- and third-degree burns on short contact and is very injurious to the eyes.

FIRE DATA
Flash point: 155°F/68°C (cc).

CHEMICAL REACTIVITY
Reactivity with water: Rapidly hydrolyzed in water to dimethylamine, CO_2, and hydrogen chloride.
Binary reactants: Violent reaction with strong oxidizers. Contact with acid or acid fumes form toxic chloride fumes. Incompatible with alkalis, ammonia, amines, amides, vinyl acetate, epichlorohydrin. Attacks some steels, causing pitting and stress corrosion.

ENVIRONMENTAL DATA
Water pollution: Effects of low concentrations on aquatic life are not known. May be dangerous if it enters nearby water intakes; notify operators. Notify local health and wildlife officials. **Response to discharge:** Issue warning–corrosive. Restrict access. Mechanical containment. Should be removed.

SHIPPING INFORMATION
Grades of purity: 98%; **Stability during transport:** Stable.

NAS HAZARD CLASSIFICATION FOR BULK WATER TRANSPORTATION
FIRE: 1
HEALTH: Vapor irritant: 4; Liquid or solid irritant: 4; Poisons: 2
WATER POLLUTION: Human toxicity: 2; Aquatic toxicity:–; Aesthetic effect: 2
REACTIVITY: Other chemicals: 4; Water: 2; Self-reaction: 0

PHYSICAL AND CHEMICAL PROPERTIES
Physical state @ 59°F/15°C and 1 atm: Liquid.
Molecular weight: 107.54
Boiling point @ 1 atm: 329°F
Melting/Freezing point: –27°F
Specific gravity (water = 1): 1.168 @ 68°F/20°C.
Relative vapor density (air = 1): 3.71

N,N-DIMETHYLCYCLOHEXYLAMINE   REC. D:5100

SYNONYMS: CYCLOHEXYLAMINE, *N,N*-DIMETHYL; CYCLOHEXYLDIMETHYLAMINE; *N*-CYCLOHEXYLDIMETHYLAMINE; *N,N*-DIMETHYLAMINOCYCLOHEXANE; *N*-DIMETHYLCYCLOHEXANAMINE; *N,N*-DIMETILCICLOHEXILAMINA (Spanish); POLYCAT-8

IDENTIFICATION
CAS Number: 98-94-2
Formula: $C_8H_{17}N$; $(CH_3)_2NC_6H_{11}$
DOT ID Number: UN 2264; DOT Guide Number: 132
Proper Shipping Name: Dimethylcyclohexylamine

DESCRIPTION: Colorless liquid. Musky ammonia odor. Floats and mixes slowly with water.

Flammable • Containers may BLEVE when exposed to fire •Vapors may form explosive mixture with air • Vapors are heavier than air and will collect and stay in low areas • Vapors may travel long distances to ignition sources and flashback • Vapors in confined areas (e.g., tanks, sewers, buildings) may explode when exposed to fire • Irritating to the skin, eyes, and respiratory tract • Toxic products of combustion may include carbon monoxide and nitrogen oxides.

Hazard Classification (based on NFPA-704 Rating System)
Health Hazards (Blue): 3; Flammability (Red): 3; Reactivity (Yellow): 0

EMERGENCY RESPONSE: See Appendix A (132)
Evacuation:
Public safety: Isolate spill area for at least 50 to 100 meters (160 to 330 feet) in all directions.
Spill: Increase, as necessary, the isolation distance shown above, in "Public safety."
Fire: Isolate for 800 meters (½ mile) in all directions, especially if tank, rail car, or tank truck is involved in fire.

EXPOSURE
Short-term effects: *SEEK MEDICAL ATTENTION.* **Vapor:** Strong irritant to eyes, nose, and throat. Harmful if inhaled; could be fatal. Move victim to fresh air. *IF BREATHING HAS STOPPED,* give artificial respiration. If breathing is difficult, administer oxygen. **Liquid:** Will burn skin and eyes. Harmful if swallowed. *IF IN EYES,* flush with plenty of water for at least 15 minutes. Remove contaminated clothing and shoes. Flush affected areas with plenty of running water.

HEALTH HAZARDS
Personal protective equipment (PPE): B-Level PPE. Wear self-contained breathing apparatus, rubber boots, heavy rubber gloves. If entering spill area, wear self-contained breathing apparatus and full protective
clothing, including boots.
Toxicity by ingestion: Grade 3: LD_{50} = 348 mg/kg (rat).
Vapor (gas) irritant characteristics: Eye and respiratory tract irritant. Vapors are irritating such that personnel will not tolerate moderate or high concentrations.
Liquid or solid irritant characteristics: Severe eye and skin irritant. Causes second- and third-degree burns on short contact and is very injurious to eyes.

FIRE DATA
Flash point: 108°F/42°C (cc).

CHEMICAL REACTIVITY
Binary reactants: Can react vigorously with oxidizing materials. Incompatible with acids and isocyanates.
Reactivity group: 7
Compatibility class: Aliphatic amines

ENVIRONMENTAL DATA
Water pollution: Effects of low concentration on aquatic life is unknown. May be dangerous if it enters nearby water intakes; notify operators. Notify local health and wildlife officials. **Response to discharge:** Issue warning–high flammability, air contaminant. Restrict access. Should be removed.

SHIPPING INFORMATION
Grades of purity: 97%, 99%; **Storage temperature:** Ambient; **Stability during transport:** Stable.

NAS HAZARD CLASSIFICATIONS FOR BULK WATER TRANSPORTATION
FIRE: 2
HEALTH: Vapor irritant: 4; Liquid or solid irritant: 4; Poisons: 3
WATER POLLUTION: Human toxicity: 3; Aquatic toxicity:–; Aesthetic effect: 3
REACTIVITY: Other chemicals: 2; Water: 1; Self-reaction: 0

PHYSICAL AND CHEMICAL PROPERTIES
Physical state @ 59°F/15°C and 1 atm: Liquid.
Molecular weight: 127.23
Boiling point @ 1 atm: 324°F/162°C/435°K.
Specific gravity (water = 1): 0.849 @ 68°F/20°C.

DIMETHYLDICHLOROSILANE   REC. D:5150

SYNONYMS: DICHLORODIMETHYL SILANE; SILANE, DICHLORODIMETHYL-; DICHLOROETHYLSILANE; DIMETHYLDICHLOROSILAN; DIMETILDICLOROSILANO (Spanish); EEC No. 014-003-00-X; NCLC50704

IDENTIFICATION
CAS Number: 75-78-5
Formula: $(CH_3)_2SiCl_2$
DOT ID Number: UN 1162; DOT Guide Number: 155
Proper Shipping Name: Dimethyldichlorosilane
Reportable Quantity (RQ): **(EHS)** 1 lb/0.045 kg

Dimethyldichlorosilane

DESCRIPTION: Clear, colorless, fuming liquid. Sharp irritating odor; like hydrochloric acid. Sinks in water; decomposes violently forming hydrochloric acid and hydrogen chloride gas.

Poison! • Corrosive • Breathing the vapor can kill you; skin or eye contact causes severe burns, impaired vision, or blindness • Extremely flammable • Firefighting gear (including SCBA) does not provide adequate protection. If exposure occurs, remove and isolate gear immediately and thoroughly decontaminate personnel • Containers may explode when exposed to fire • Vapors may form explosive mixture with air • Vapors are heavier than air and will collect and stay in low areas • Vapors may travel long distances to ignition sources and flashback • Vapors in confined areas (e.g., tanks, sewers, buildings) may explode when exposed to fire • Toxic products of combustion may include carbon monoxide, hydrogen chloride, and phosgene • Do not put yourself in danger by entering a contaminated area to rescue a victim.

Hazard Classification (based on NFPA-704 Rating System)
Health Hazards (Blue): 3; Flammability (Red): 3; Reactivity (Yellow): 1; Special Notice (White): Water reactive (decomposes).

EMERGENCY RESPONSE: See Appendix A (155)
Evacuation:
Public safety: Isolate the area of spill or leak for at least 50 to 100 meters (160 to 330 feet) in all directions.
Spill: Increase, in the downwind direction, as necessary, the distance shown under "Public Safety."
IF SPILLED IN WATER. Small spill–First: Isolate in all directions 30 meters (100 feet); Then: Protect persons downwind, DAY: 0.2 km (0.1 mile); NIGHT: 0.3 km (0.2 mile). Large spill–First: Isolate in all directions 125 meters (400 feet); Then: Protect persons downwind, DAY: 1.1 km (0.7 mile); NIGHT: 2.9 km (1.8 miles).
Fire: If tank, rail car, or tank truck is involved in fire, isolate for at least 800 meters (½ mile) in all directions; also, consider initial evacuation for 800 meters (½ mile) in all directions.

EXPOSURE
Short-term effects: *SEEK MEDICAL ATTENTION.* **Vapor:** Irritating to eyes, nose, and throat. Move victim to fresh air. *IF BREATHING HAS STOPPED,* give artificial respiration; *avoid mouth-to-mouth resuscitation; use bag/mask apparatus.* If breathing is difficult, administer oxygen. May cause lung edema; physical exertion will aggravate this condition. **Liquid:** Will burn skin and eyes. Harmful if swallowed. Remove contaminated clothing and shoes. Flush affected areas with plenty of water. *IF IN EYES,* hold eyelids open and flush with plenty of water. *IF SWALLOWED* and victim is *CONSCIOUS AND ABLE TO SWALLOW,* have victim drink 4 to 8 ounces of water. **Do NOT induce vomiting.**
Note to physician or authorized medical personnel: Medical observation is recommended for 24 to 48 hours after breathing overexposure, as pulmonary edema may be delayed. As first aid for pulmonary edema, consider administering a corticosteroid spray. Cigarette smoking may exacerbate pulmonary injury and should be discouraged for at least 72 hours following exposure.

HEALTH HAZARDS
Personal protective equipment (PPE): B-Level PPE. Organic vapor respirator/acid-vapor respiratory protection; rubber gloves; chemical worker's goggles; other protective equipment as necessary to protect skin and eyes.
Short-term exposure limits (15-minute TWA): 1 ppm (ceiling) (AIHA WEEL).
Toxicity by ingestion: Grade 3; LD_{50} = 50 to 500 mg/kg.
Long-term health effects: Lung damage.
Vapor (gas) irritant characteristics: Vapors cause severe irritation of eyes and throat and can cause eye and lung injury. They cannot be tolerated even at low concentrations.
Liquid or solid irritant characteristics: Severe eye and skin irritant. Causes second-and third-degree burns on short contact and is very injurious to the eyes.

FIRE DATA
Flash point: 15°F/–9°C (oc).
Flammable limits in air: LEL: 3.4%; UEL: 9.5%.
Fire extinguishing agents not to be used: Water or foam
Behavior in fire: Difficult to extinguish. Re-ignition may occur.
Autoignition temperature: 707°F/375°C/648°K.
Electrical hazard: Due to low electric conductivity, this substance may generate electrostatic charges as a result of agitation and flow.
Burning rate: 3.3 mm/min

CHEMICAL REACTIVITY
Reactivity with water: Reacts vigorously with water to generate hydrogen chloride.
Binary reactants: Will react with surface moisture to generate hydrogen chloride, which is corrosive to common metals. Reacts violently with acetone, amines, ammonia, alcohols. Contact with air forms corrosive fumes of hydrochloric acid.
Neutralizing agents for acids and caustics: Sodium bicarbonate or lime.

ENVIRONMENTAL DATA
Food chain concentration potential: Negative; unlikely to accumulate.
Water pollution: Effect of low concentrations on aquatic life is unknown. May be dangerous if it enters nearby water intakes; notify operators. Notify local health and wildlife officials. **Response to discharge:** Issue warning–high flammability, corrosive. Restrict access. Evacuate area. Disperse and flush with care.

SHIPPING INFORMATION
Grades of purity: 99+%; **Storage temperature:** Ambient; **Inert atmosphere:** None; **Venting:** Pressure-vacuum; **Stability during transport:** Stable.

NAS HAZARD CLASSIFICATIONS FOR BULK WATER TRANSPORTATION
FIRE: 3
HEALTH: Vapor irritant: 4; Liquid or solid irritant: 4; Poisons: 3
WATER POLLUTION: Human toxicity: 3; Aquatic toxicity: 3; Aesthetic effect: 2
REACTIVITY: Other chemicals: 3; Water: 4; Self-reaction: 1

PHYSICAL AND CHEMICAL PROPERTIES
Physical state @ 59°F/15°C and 1 atm: Liquid.
Molecular weight: 129
Boiling point @ 1 atm: 158.8°F/70.5°C/343.7°K.
Melting/Freezing point: –122°F/–86°C/187°K.
Specific gravity (water = 1): 1.07 @ 77°F/25°C (liquid).
Liquid surface tension: 20.1 dynes/cm = 0.0201 N/m @ 68°F/20°C.
Relative vapor density (air = 1): 4.4
Latent heat of vaporization: 100 Btu/lb = 58 cal/g = 2.4×10^5 J/kg.
Heat of combustion: (estimate) –6000 Btu/lb = –3300 cal/g = –140 $\times 10^5$ J/kg.
Vapor pressure: 110 mm.

DIMETHYLETHANOLAMINE REC. D:5175

SYNONYMS: DEANOL; 2-(DIMETHYLAMINO)ETHANOL; β-DIMETHYLAMINOETHYL ALCOHOL; *N,N*-DIMETHYLETHANOLAMINE; DIMETILETANOAMINA (Spanish); DMAE; EEC No. 603-047-00-0; β-HYDROXYETHYLDIMETHYLAMINE; *N,N*-DIMETHYL-*N*-(2-HYDROXYETHYL) AMINE

IDENTIFICATION
CAS Number: 108-01-0
Formula: $C_4H_{11}NO$; $(CH_3)_2NCH_2CH_2OH$
DOT ID Number: UN 2051; DOT Guide Number: 132
Proper Shipping Name: Dimethylethanolamine

DESCRIPTION: Colorless liquid. Amine odor. Floats on the surface of water; soluble.

Flammable • Containers may BLEVE when exposed to fire • Vapors may form explosive mixture with air • Vapors are heavier than air and will collect and stay in low areas • Vapors may travel long distances to ignition sources and flashback • Vapors in confined areas (e.g., tanks, sewers, buildings) may explode when exposed to fire • Irritating to the skin, eyes, and respiratory tract • Toxic products of combustion may include nitrogen oxide.

Hazard Classification (based on NFPA-704 Rating System)
Health Hazards (Blue): 3; Flammability (Red): 2; Reactivity (Yellow): 0

EMERGENCY RESPONSE: See Appendix A (132)
Evacuation:
Public safety: Isolate spill area for at least 50 to 100 meters (160 to 330 feet) in all directions.
Spill: Increase, as necessary, the isolation distance shown above, in "Public safety."
Fire: Isolate for 800 meters (½ mile) in all directions, especially if tank, rail car, or tank truck is involved in fire.

EXPOSURE
Short-term effects: *SEEK MEDICAL ATTENTION.* **Vapor:** Irritating to eyes, nose, and throat. Possible lung edema. Harmful if inhaled. Move to fresh air. *IF BREATHING HAS STOPPED,* give artificial respiration. If breathing is difficult, administer oxygen. Vapor or liquid can cause blurred vision that can improve in several hours. **Liquid:** Will burn skin and eyes. Harmful if swallowed. *IF IN EYES OR ON SKIN,* flush with plenty of water for at least 15 minutes. Remove and double bag contaminated clothing and shoes at the site. *IF SWALLOWED* and victim is *CONSCIOUS AND ABLE TO SWALLOW,* have victim drink 4 to 8 ounces of water. **Do NOT induce vomiting.**
Note to physician or authorized medical personnel: Medical observation is recommended for 24 to 48 hours after breathing overexposure, as pulmonary edema may be delayed. As first aid for pulmonary edema, consider administering a corticosteroid spray. Cigarette smoking may exacerbate pulmonary injury and should be discouraged for at least 72 hours following exposure.

HEALTH HAZARDS
Personal protective equipment (PPE): B-Level PPE. Wear self-contained positive pressure breathing apparatus and full protective clothing. Chemical protective material(s) reported to have good to excellent resistance: butyl rubber, nitrile. Also, natural rubber, and neoprene may offer some protection from alcohols.
Toxicity by ingestion: Grade 2; LD_{50} = 2.34 g/kg, (rat).
Long-term health effects: Chronic exposure may cause asthma and grand mal epilepsy.
Vapor (gas) irritant characteristics: Eye and respiratory tract irritation. Vapors are irritating such that personnel will not usually tolerate moderate or high concentrations.
Liquid or solid irritant characteristics: Fairly severe skin irritant. May cause pain and second-degree burns after a few minutes of contact.
Odor threshold: 0.015 ppm detection; 0.045 ppm recognition

FIRE DATA
Flash point: 140°F/60°C (oc).
Autoignition temperature: 608°F/320°C/573°K.
Electrical hazard: Class 1, Group C.

CHEMICAL REACTIVITY
Binary reactants: Violent reaction with oxidizers, acids. Attacks copper, copper alloys, zinc, galvanized steel, or zinc alloys having more than 10% zinc by weight.
Neutralizing agents for acids and caustics: Sodium bisulfate
Reactivity group: 8
Compatibility class: Alkanolamines

ENVIRONMENTAL DATA
Water pollution: Harmful to aquatic life in very low concentrations. May be dangerous if it enters nearby water intakes; notify operators. Notify local health and wildlife officials. **Response to discharge:** Issue warning–high flammability, water contaminant. Restrict access. Mechanical containment. Should be removed. Chemical and physical treatment.

SHIPPING INFORMATION
Grades of purity: 99%; **Storage temperature:** Ambient temperature; **Venting:** Pressure-vacuum; **Stability during transport:** Stable.

PHYSICAL AND CHEMICAL PROPERTIES
Physical state @ 59°F/15°C and 1 atm: Liquid.
Molecular weight: 89.14
Boiling point @ 1 atm: 324°F/162°C/435°K.
Melting/Freezing point: –73.5°F/–58.6°C/214.6°K.
Critical temperature: 572°F/300°C/573°K (estimate).
Critical pressure: 600 psia = 40.8 atm = 4.13 MN/m^2.
Specific gravity (water = 1): 0.8870 @ 68°F/20°C.
Liquid surface tension: 27.1 dynes/cm = 0.0271 N/m at 24.5°C.
Relative vapor density (air = 1): 4.0
Latent heat of vaporization: 170.6 Btu/lb = 94.8 cal/g = 3.97 x 10^5 J/kg.
Heat of combustion: 15508 Btu/lb = 8616 cal/g = 360 x 10^5 J/kg.
Vapor pressure: 4.3 mm.

DIMETHYL ETHER REC. D:5200

SYNONYMS: EEC No. 603-019-00-8; ETER DIMETILICO (Spanish); METHYL ETHER; OXYBISMETHYANE; DME; WOOD ETHER

IDENTIFICATION
CAS Number: 115-10-6
Formula: CH_3OCH_3
DOT ID Number: UN 1033; DOT Guide Number: 115
Proper Shipping Name: Dimethyl ether

DESCRIPTION: Colorless compressed gas. Sweet odor, like ether. Floats and boils on water surface; slowly dissolves. Flammable, irritating vapor is produced.

Extremely flammable • Technical grades be able to form unstable peroxides after prolonged storage in air; peroxides may be detonated by heating, impact, or friction. • Forms explosive mixture with air • Reacts explosively with many materials • Vapors from liquefied gas are initially heavier than air and spread along ground • Gas may travel to source of ignition and flashback • Exposure of cylinders to elevated temperatures, fire, and flame may cause cylinders to rupture or cause frangible disk to burst, releasing entire contents of cylinder • Ruptured or venting cylinders may rocket through buildings and/or travel a considerable distance •Irritating to the skin, eyes, and respiratory tract • Vapors may cause dizziness or asphyxiation without warning • Contact with liquid may cause frostbite.

Hazard Classification (based on NFPA-704 Rating System)
Health Hazards (Blue): 1; Flammability (Red): 4; Reactivity (Yellow): 1

EMERGENCY RESPONSE: See Appendix A (115)
Evacuation:
Public safety: Isolate spill area for at least 50 to 100 meters (160 to 330 feet) in all directions.
Spill: Consider initial downwind evacuation for at least 800 meters (½ mile).
Fire: Isolate for 1600 meters (1 mile) in all directions, especially if tank, rail car, or tank truck is involved in fire.

EXPOSURE
Short-term effects: *SEEK MEDICAL ATTENTION*. **Vapor:** Irritating to eyes, nose, and throat. *IF INHALED*, will, will cause headache, dizziness, or loss of consciousness. Move victim to fresh air. *IF BREATHING HAS STOPPED*, give artificial respiration. If breathing is difficult, administer oxygen. **Liquid:** Irritating to skin and eyes. Will cause frostbite. Remove contaminated clothing and shoes. Flush affected areas with plenty of *water. DO NOT RUB AFFECTED AREAS. IF IN EYES*, hold eyelids open and flush with plenty of water. *IF SWALLOWED* and victim is *CONSCIOUS AND ABLE TO SWALLOW*, have victim drink 4 to 8 ounces of water.

HEALTH HAZARDS
Personal protective equipment (PPE): A-Level PPE. Mask for organic vapors; plastic or rubber gloves; safety glasses. Wear thermal protective clothing. Chemical protective material(s) reported to have good to excellent resistance: butyl rubber, neoprene, PV alcohol, Teflon®, Silvershield®.
Exposure limits (TWA unless otherwise noted): 1000 ppm (AIHA WEEL).
Long-term health effects: May effect the nervous system.
Vapor (gas) irritant characteristics: Eye and respiratory tract irritant. Anesthetic effects.
Liquid or solid irritant characteristics: Rapid evaporation of liquid may cause frostbite.

FIRE DATA
Flash point: Flammable gas.
Flammable limits in air: LEL: 3.4%; UEL: 27%.
Behavior in fire: Containers may explode. Vapors are heavier than air and may travel long distance to a source of ignition and flashback.
Autoignition temperature: 662°F/350°C/623°K.

Electrical hazard: Class I, Group C. Flow or agitation of substance may generate electrostatic charges due to low conductivity.
Burning rate: 6.6 mm/min
Stoichiometric air-to-fuel ratio: 8.934 (estimate).

CHEMICAL REACTIVITY
Binary reactants: Reacts with oxidizers. Technical grades may accumulate unstable peroxides after prolonged storage. Peroxides may be detonated by heating, impact, or friction.
Polymerization: Technical grades may be able to form unstable peroxides.

ENVIRONMENTAL DATA
Food chain concentration potential: Negative; unlikely to accumulate.
Water pollution: Effect of low concentrations on aquatic life is unknown. May be dangerous if it enters nearby water intakes; notify operators. Notify local health and wildlife officials. **Response to discharge:** Issue warning–high flammability. Restrict access. Evacuate area.

SHIPPING INFORMATION
Grades of purity: 99+%; Technical grade; **Storage temperature:** Ambient; **Inert atmosphere:** None; **Venting:** Safety relief; **Stability during transport:** Pure DME is stable.

PHYSICAL AND CHEMICAL PROPERTIES
Physical state @ 59°F/15°C and 1 atm: Gas.
Molecular weight: 46.1
Boiling point @ 1 atm: –13°F/–25°C/249°K.
Melting/Freezing point: –223°F/–142°C/132°K.
Critical temperature: 260°F/127°C/400°K.
Critical pressure: 780 psia = 53 atm = 5.4 MN/m^2.
Specific gravity (water = 1): 0.724 at –24.7°C (liquid).
Liquid surface tension: 21 dynes/cm = 0.021 N/m at –40°F/–40°C.
Liquid water interfacial tension: (estimate) 15 dynes/cm = 0.015 N/m at –40°F/–40°C.
Relative vapor density (air = 1): 1.6
Ratio of specific heats of vapor (gas): 1.1456
Latent heat of vaporization: 200 Btu/lb = 111 cal/g = 4.65 x 10^5 J/kg.
Heat of combustion: –13,450 Btu/lb = –7,480 cal/g = –313 x 10^5 J/kg.
Heat of fusion: 25.62 cal/g.
Vapor pressure: 4000 mm (approximate).

DIOCTYL PHTHALATE REC. D:5250

Synonyms: 1,2-BENZENEDICARBOXYLIC ACID, DI-*n*-OCTYL ESTER; BIS(2-ETHYLHEXYL) PHTHALATE; CELLULEX DOP; DI-*n*-OCTYL PHTHALATE; DNOP; DINOPOL NOP; DOP; PHTHALIC ACID, DIOCTYL ESTER; OCTOIL; PHTHALIC ACID,BIS(2-ETHYLHEXYL ESTER); VINICIZER; RCRA No. U017

IDENTIFICATION:

CAS Registry Number: 117-84-0
Formula: $C_{24}H_{38}O_4$; *o*-$C_6H_4[COOCH_2CH(C_2H_5)(CH_2)_3CH_3]_2$; $C_6H_4(COOC_5H_{11})_2$
DOT ID Number: UN 3082; DOT Guide Number: 171
Proper Shipping Name: Environmentally hazardous substances, liquid, n.o.s.
Reportable Quantity (RQ): **(CERCLA)** 5000 lb/2270 kg

DESCRIPTION: Colorless oily liquid or molten solid. Slight odor. Floats on water; soluble.

Irritates eyes, skin, and respiratory tract; Decomposes at more than 330°F/166°C; Toxic products of combustion may include carbon monoxide.

Hazard Classification (based on NFPA-704 Rating System)
Health Hazards (BLUE): 0; Flammability (RED): 1; Reactivity (YELLOW): 0

EMERGENCY RESPONSE: See Appendix A (171)
Evacuation:
Public safety: Isolate the area of spill or leak for at least 10 to 25 meters (30 to 80 feet) in all directions.
Spill: Increase, in the downwind direction, as necessary, the distance shown under "Public Safety."
Fire: If any large container is involved in fire, isolate for at least 800 meters (½ mile) in all directions; also, consider initial evacuation for 800 meters (½ mile) in all directions.

EXPOSURE
Short-term effects: Toxic if swallowed. *SEEK MEDICAL ATTENTION*. Move to fresh air. Remove contaminated clothing and shoes. Wash affected areas with plenty of soap and water. *IF IN EYES*, hold eyelids open and flush with plenty of water.

HEALTH HAZARDS
Personal protective equipment (PPE): B-Level PPE. Butyl rubber is generally suitable for carbooxylic acid compounds.
Toxicity by ingestion: Grade 0; LD_{50} = > 15 g/kg (rat).
Vapor (gas) irritant characteristics: Eye and respiratory tract irritant.

FIRE DATA
Flash point: 425°F/218°C (oc)
Fire Extinguishing Agents Not to be Used: Water or water-based foam may cause frothing
Autoignition Temperature: 725°F/385°C/658°K.

CHEMICAL REACTIVITY
Reactivity with water: Water may cause a foaming, frothing action.
Reactivity with other materials: Reacts with oxidizers, acids.
Reactivity group: 34
Compatibility class: Esters.

ENVIRONMENTAL DATA
Food chain concentration potential: High. The Log K_{ow} for dioctyl phthalate is in the range of 4.9–5.2. Values above 3.0 are very likely to accumulate in living tissues and especially in fats.
Water pollution: Effect of low concentrations on aquatic life is unknown. Fouling to shoreline. May be dangerous if it enters water intakes. Notify local health and wildlife officials. Notify operators of nearby water intakes. **Response to discharge:** Mechanical containment. Chemical and physical treatment.

SHIPPING INFORMATION
Storage temperature: Ambient; **Inert atmosphere:** No requirement; **Venting:** Open (flame arrester); **Stability during transport:** Stable

PHYSICAL AND CHEMICAL PROPERTIES
Physical State @ 59°F/15°C and 1 atm: Liquid.
Molecular Weight: 390.6
Boiling Point @ 1 atm: 727°F/386°C/659°K.
Melting/Freezing Point: –67°F/–55°C/218°K.
Specific Gravity (Water = 1): 0.980 at 25°C (liquid)
Liquid Surface Tension: (estimate) 15 dynes/cm = 0.015 N/m at 20°C.
Liquid Water Interfacial Tension: (estimate) 30 dynes/cm = 0.03 N/m at 20°C.
Relative Vapor Density (Air = 1): 13.48
Heat of Combustion: –15,130 Btu/lb = –8410 cal/g = –352 x 10^5 J/kg.
Vapor Pressure: (Reid) Low

DIMETHYLFORMAMIDE   REC. D:5550

SYNONYMS: DIMETHYLFORMAMID (German); DIMETHYL FORMAMIDE; *N,N*-DIMETHYL FORMAMIDE; *N,N*-DIMETHYLFORMAMIDE; *N,N*-DIMETILFORMAMIDA (Spanish); DMF; DMFA; EEC No. 616-001-00-X; *N*-FORMYLDIMETHYLAMINE; FORMAMIDE,*N,N*-DIMETHYL-; *N,N*-DIMETHYLMETHANIDE; U-4224

IDENTIFICATION
CAS Number: 68-12-2
Formula: C_3H_7NO; $HCON(CH_3)_2$
DOT ID Number: UN 2265; DOT Guide Number: 129
Proper Shipping Name: *N,N*-Dimethylformamide
Reportable Quantity (RQ): **(CERCLA)** 1 lb/0.45 kg

DESCRIPTION: Colorless to slightly yellow, watery liquid. Slight ammonia or fish-like odor. Floats on water surface; soluble.

Very Flammable • Containers may BLEVE when exposed to fire • Vapors may form explosive mixture with air • Vapors are heavier than air and will collect and stay in low areas • Vapors may travel long distances to ignition sources and flashback • Vapors in confined areas (e.g., tanks, sewers, buildings) may explode when exposed to fire • Irritating to the skin, eyes, and respiratory tract • Toxic products of combustion may include nitrogen oxides.

Hazard Classification (based on NFPA-704 Rating System)
Health Hazards (Blue): 1; Flammability (Red): 2; Reactivity (Yellow): 0

EMERGENCY RESPONSE: See Appendix A (129)
Evacuation:
Public safety: Isolate spill area for at least 50 to 100 meters (160 to 330 feet) in all directions.
Spill: Large spill–Consider initial downwind evacuation for at least 300 meters (1000 feet).
Fire: Isolate for 800 meters (½ mile) in all directions, especially if tank, rail car, or tank truck is involved in fire.

EXPOSURE
Short-term effects: *SEEK MEDICAL ATTENTION*. **Liquid:** *HARMFUL IF SWALLOWED OR ABSORBED THROUGH THE SKIN*. Will burn skin and eyes. Remove contaminated clothing and shoes. Flush affected areas with plenty of water. *IF IN EYES*, hold eyelids open and flush with plenty of water. *IF SWALLOWED* and victim is *CONSCIOUS AND ABLE TO SWALLOW*, have victim drink 4 to 8 ounces of water. **Do NOT induce vomiting**. The use of alcoholic beverages may enhance the toxic effect.

408 Dimethyl glutarate

HEALTH HAZARDS
Personal protective equipment (PPE): A-Level PPE. OSHA Table Z-1-A air contaminant. Full face organic vapor respirator. Chemical protective material(s) reported to have good to excellent resistance: supported neoprene, butyl rubber, Teflon®, Silvershield®.
Recommendations for respirator selection: NIOSH
100 ppm: SA* (any supplied-air respirator). *250 ppm:* SA:CF* (any supplied-air respirator operated in a continuous-flow mode). *500 ppm:* SAT:CF* (any supplied-air respirator that has a tight-fitting facepiece and is operated in a continuous-flow mode); or SCBAF (any self-contained breathing apparatus with a full facepiece); or SAF (any supplied-air respirator with a full facepiece). *EMERGENCY OR PLANNED ENTRY INTO UNKNOWN CONCENTRATIONS OR IDLH CONDITIONS:* SCBAF:PD,PP (any self-contained breathing apparatus that has a full facepiece and is operated in a pressure-demand or other positive-pressure mode); or SAF:PD,PP:ASCBA (any supplied-air respirator that has a full facepiece and is operated in a pressure-demand or other positive-pressure mode in combination with an auxiliary self-contained breathing apparatus operated in a pressure-demand or other positive pressure mode). *ESCAPE:* GMFOV[any air-purifying, full-facepiece respirator (gas mask) with a chin-style, front- or back-mounted organic vapor canister]; or SCBAE (any appropriate escape-type, self-contained breathing apparatus).
Note: Substance reported to cause eye irritation or damage; may require eye protection.
Exposure limits (TWA unless otherwise noted): ACGIH TLV 10 ppm (30 mg/m^3); NIOSH/OSHA 10 ppm (30 mg/m^3); skin
Toxicity by ingestion: Grade 1; LD_{50} = 5 to 15 g/kg (rat).
Long-term health effects: May cause liver, kidney, and brain damage. Frequent contact may cause dermatitis. Avoid exposure of pregnant women to this product. Causes abortions in pregnant rats, possibly in humans also. IARC possible carcinogen; rating 2B; human evidence: limited; animal evidence: inadequate.
Vapor (gas) irritant characteristics: Eye and respiratory tract irritation. Vapors cause irritation, such that personnel will find high concentrations unpleasant. The effect is temporary.
Liquid or solid irritant characteristics: Eye and skin irritation. Causes smarting of the skin and first-degree burns on short exposure and may cause secondary burns on long exposure.
Odor threshold: 0.047–100 ppm.
IDLH value: 500 ppm.

FIRE DATA
Flash point: 153°F/67°C (oc); 136°F/58°C (cc).
Flammable limits in air: LEL: 2.2% @ 212°F/100°C; UEL: 15.2%.
Autoignition temperature: 833°F/445°F/718°K.
Burning rate: 2.2 mm/min

CHEMICAL REACTIVITY
Binary reactants: Carbon tetrachloride; other halogenated compounds when in contact with iron; strong oxidizers; alkyl aluminums; inorganic nitrates. Attacks certain plastics.
Reactivity group: 10
Compatibility class: Amide

ENVIRONMENTAL DATA
Food chain concentration potential: Log P_{ow} = –0.7. Unlikely to accumulate.
Water pollution: Effect of low concentrations on aquatic life is unknown. May be dangerous if it enters nearby water intakes; notify operators. Notify local health and wildlife officials.
Response to discharge: Restrict access. Disperse and flush.

SHIPPING INFORMATION
Storage temperature: Ambient; **Inert atmosphere:** None; **Venting:** Open (flame arrester); **Stability during transport:** Stable.

NAS HAZARD CLASSIFICATION FOR BULK WATER TRANSPORTATION
FIRE: 2
HEALTH: Vapor irritant: 2; Liquid or solid irritant: 2; Poisons: 3
WATER POLLUTION: Human toxicity: 2; Aquatic toxicity: 2; Aesthetic effect: 0
REACTIVITY: Other chemicals: 1; Water: 0; Self-reaction: 0

PHYSICAL AND CHEMICAL PROPERTIES
Physical state @ 59°F/15°C and 1 atm: Liquid.
Molecular weight: 73.09
Boiling point @ 1 atm: 307°F/153°C/426°K.
Melting/Freezing point: –78°F/–61°C/212°K.
Specific gravity (water = 1): 0.950 @ 68°F/20°C (liquid).
Relative vapor density (air = 1): 2.48
Ratio of specific heats of vapor (gas): 1.101
Latent heat of vaporization: 248 Btu/lb = 138 cal/g = 5.78 x 10^5 J/kg.
Heat of combustion: –11,280 Btu/lb = –6267 cal/g = –262.4 x 10^5 J/kg.
Heat of decomposition: 348°C.
Heat of solution: –63 Btu/lb = –35 cal/g = –1.5 x 10^5 J/kg.
Vapor pressure: (Reid) 0.16 psia; 3 mm.

DIMETHYL GLUTARATE   REC. D:5600

IDENTIFICATION
CAS Number: 1119-40-0
Formula: $CH_3O_2OC(CH_2)_3CO_2CH_3$

DESCRIPTION: Colorless liquid.

Combustible • Containers may BLEVE when exposed to fire • Vapors are heavier than air and will collect and stay in low areas • Vapors in confined areas (e.g., tanks, sewers, buildings) may explode when exposed to fire • Irritating to the skin, eyes, and respiratory tract • Toxic products of combustion may include carbon monoxide.

Hazard Classification (based on NFPA-704 Rating System)
Health Hazards (Blue): 1; Flammability (Red): 1; Reactivity (Yellow): 0

EMERGENCY RESPONSE: See Appendix A (171)
Evacuation:
Public safety: Isolate the area of spill or leak for at least 10 to 25 meters (30 to 80 feet) in all directions.
Spill: Increase, in the downwind direction, as necessary, the distance shown under "Public Safety."
Fire: If any large container is involved in fire, isolate for at least 800 meters (½ mile) in all directions; also, consider initial evacuation for 800 meters (½ mile) in all directions.

EXPOSURE
Short-term effects: *SEEK MEDICAL ATTENTION.* **Vapor:** May be irritating. Move to fresh air. If breathing has stopped give artificial respiration. If breathing is difficult give oxygen. **Liquid:** May be harmful by ingestion or skin absorption. Remove

contaminated clothing and shoes. Flush affected areas with plenty of water. *IF IN EYES*, hold eyelids open and flush with plenty of water.

HEALTH HAZARDS
Personal protective equipment (PPE): B-Level PPE. Self-contained breathing apparatus, rubber boots and heavy rubber gloves.
Vapor (gas) irritant characteristics: Vapors cause smarting of the eyes or respiratory system if present in high concentrations. The effect is temporary.
Liquid or solid irritant characteristics: If spilled on clothing and allowed to remain, may cause smarting and reddening skin.

FIRE DATA
Flash point: 218°F/103°C (cc).

CHEMICAL REACTIVITY
Binary reactants: Oxidizers, strong acids.
Reactivity group: 34
Compatibility class: Esters.

ENVIRONMENTAL DATA
Water pollution: Effect of low concentrations on aquatic life is unknown. Fouling to shoreline. May be dangerous if it enters nearby water intakes; notify operators. Notify local health and wildlife officials. **Response to discharge:** Restrict access. Should be removed. Mechanical containment. Chemical and physical treatment.

SHIPPING INFORMATION
Grades of purity: 98%; **Storage temperature:** Ambient; **Stability during transport:** Stable.

PHYSICAL AND CHEMICAL PROPERTIES
Physical state @ 59°F/15°C and 1 atm: Liquid.
Molecular weight: 160.17
Boiling point @ 1 atm: 199.4–203°F/93–95°C/366.2°K–368.2°K (at 13 mmHg = 0.0171 atm).
Specific gravity (water = 1): 1.087
Relative vapor density (air = 1): 5.52

DIMETHYLHEXANE DIHYDROPEROXIDE
REC. D:5650

SYNONYMS: 2,5-DIHIDROPEROXIDO de 2,5-DIMETILHEXANO (Spanish); 2,5-DIMETHYL-2,5-DIHYDROPEROXY HEXANE, not more than 82% with water; DIMETHYLHEXANE DIHYDROPEROXIDE with 18% or more water; 2,5-DIMETHYLHEXANE-2,5-DIHYDROPEROXIDE; 2,5-DIHYDROPEROXY-2, 5-DIMETHYLHEXANE

IDENTIFICATION
CAS Number: 3025-88-5
Formula: $C_8H_{16}(OOH)_2 \cdot H_2O$
DOT ID Number: UN 2174; DOT Guide Number: 146
Proper Shipping Name: 2,5-Dimethyl-2,5-dihydroperoxy hexane, not more than 82% with water; Dimethylhexane dihydroperoxide with 18% or more water; Dimethylhexane dihydroperoxide, dry [Forbidden].

DESCRIPTION: White, wet solid (dry is forbidden and may not be transported). May float or sink in water; slightly soluble.

Extremely flammable • Containers may BLEVE when exposed to fire •Vapors may form explosive mixture with air • Vapors are heavier than air and will collect and stay in low areas • Vapors may travel long distances to ignition sources and flashback • Vapors in confined areas (e.g., tanks, sewers, buildings) may explode when exposed to fire • Irritating to the skin, eyes, and respiratory tract • May explode if subjected to heat or flame or shock • May cause fire and explode on contact with combustibles.

Hazard Classification (based on NFPA-704 Rating System)
Health Hazards (Blue): 1; Flammability (Red): 2; Reactivity (Yellow): 3; Special Notice (White): OXY

EMERGENCY RESPONSE: See Appendix A (146)
Evacuation:
Public safety: Isolate the area of spill or leak for at least 25 to 50 meters (80 to 160 feet) in all directions.
Spill: Consider initial evacuation for at least 250 meters (800 feet).
Fire: If tank, rail car, or tank truck is involved in fire, isolate for at least 800 meters (½ mile) in all directions; also, consider initial evacuation for 800 meters (½ mile) in all directions.

EXPOSURE
Short-term effects: *SEEK MEDICAL ATTENTION*. **Vapor:** Irritating to eyes, nose, and throat. *IF INHALED*, will, will cause coughing or difficult breathing. *IF IN EYES*, hold eyelids open and flush with plenty of water. *IF BREATHING HAS STOPPED*, give artificial respiration. If breathing is difficult, administer oxygen. **Solid:** Irritating to skin and eyes. Harmful if swallowed. Remove contaminated clothing and shoes. Flush affected areas with plenty of water. *IF IN EYES*, hold eyelids open and flush with plenty of water. *IF SWALLOWED* and victim is *CONSCIOUS AND ABLE TO SWALLOW*, have victim drink 4 to 8 ounces of water and have victim induce vomiting. *IF SWALLOWED* and victim is *UNCONSCIOUS OR HAVING CONVULSIONS*, do nothing except keep victim warm.

HEALTH HAZARDS
Personal protective equipment (PPE): B-Level PPE. Goggles or face shield; rubber gloves. Butyl rubber is generally suitable for peroxide compounds.
Liquid or solid irritant characteristics: Eye and skin irritation.

FIRE DATA
Fire extinguishing agents not to be used: Water may be ineffective on fire.
Behavior in fire: Decomposes violently when heated in fire. Can increase intensity of fire when in contact with combustible material. Containers may explode.

CHEMICAL REACTIVITY
Binary reactants: A powerful oxidizer. violent reaction with reducing agents, combustible materials, powdered metals, organic substances. Contact with transition metals, may cause explosive decomposition. Decomposes in contact with many metals and acids.

ENVIRONMENTAL DATA
Food chain concentration potential: Negative; unlikely to accumulate.
Water pollution: Effect of low concentrations on aquatic life is unknown. May be dangerous if it enters nearby water intakes; notify operators. Notify local health and wildlife officials. **Response to discharge:** Issue warning–oxidizing material. Restrict access. Mechanical containment. Should be removed. Chemical and physical treatment.

SHIPPING INFORMATION
Grades of purity: (approximately) 30+% water. The dry chemical is too hazardous to ship; **Storage temperature:** 40–100°F/4–38°C; **Inert atmosphere:** None; **Venting:** Open; **Stability during transport:** Stable below 100°F/32°C.

PHYSICAL AND CHEMICAL PROPERTIES
Physical state @ 59°F/15°C and 1 atm: Solid.
Molecular weight: 178.2
Melting/Freezing point: 221°F/105°C/378°K.
Specific gravity (water = 1): (estimate) 1.0 @ 68°F/20°C (solid).

1,1-DIMETHYLHYDRAZINE   REC. D:5700

SYNONYMS: DIMAZINE; DIMETHYLHYDRAZINE; 1,1-DIMETHYLHYDRAZIN (German); asym-DIMETHYLHYDRAZINE; unsym-DIMETHYLHYDRAZINE; 1,1-DIMETILHIDRAZINA (Spanish); N,N-DIMETHYLHYDRAZINE; DMH; EEC No. 007-012-00-5; HYDRAZINE, 1,1-DIMETHYL-; RCRA No. U098; UDMH

IDENTIFICATION
CAS Number: 57-14-7
Formula: $C_2H_8N_2$; $(CH_3)_2N \cdot N \cdot NH_2$
DOT ID Number: UN 1163; DOT Guide Number: 131
Proper Shipping Name: 1,1-Dimethylhydrazine; Dimethylhydrazine, unsymmetrical
Reportable Quantity (RQ): **(CERCLA)** 10 lb/4.54 kg

DESCRIPTION: Colorless watery liquid. Turns yellow and fumes upon contact with air. Fishy or ammonia-like odor. Floats on the surface of water; soluble. Used as a high energy propellant for liquid fueled rockets.

Poison! • Highly flammable • Corrosive • Breathing the vapor, skin or eye contact, or swallowing the material can kill you; may interfere with the body's ability to use oxygen • Firefighting gear (including SCBA) does not provide adequate protection. If exposure occurs, remove and isolate gear immediately and thoroughly decontaminate personnel • Containers may explode and rocket when exposed to fire • Vapors may form explosive mixture with air • Vapors are heavier than air and will collect and stay in low areas • Vapors may travel long distances to ignition sources and flashback • Vapors in confined areas (e.g., tanks, sewers, buildings) may explode when exposed to fire • Corrosive to plastics; causes disintegration and swelling • Toxic products of combustion may include nitrogen oxide • Do not put yourself in danger by entering a contaminated area to rescue a victim. *Warning:* Odor is not a reliable indicator of the presence of toxic amounts of this material. Odor detection is higher than the Exposure limits.

Hazard Classification (based on NFPA-704 Rating System)
Health Hazards (Blue): 4; Flammability (Red): 3; Reactivity (Yellow): 1

EMERGENCY RESPONSE: See Appendix A (131)
Evacuation:
Public safety: Isolate spill area for at least 100 to 200 meters (330 to 660 feet) in all directions.
Spill: Small spill–First: Isolate in all directions 30 meters (100 feet); Then: Protect persons downwind, DAY: 0.2 km (0.1 mile); NIGHT: 0.2 km (0.1 mile). Large spill–First: Isolate in all directions 60 meters (200 feet); Then: Protect persons downwind, DAY: 0.5 km (0.3 mile); NIGHT: 1.1 km (0.7 mile).
Fire: Isolate for 800 meters (½ mile) in all directions, especially if tank, rail car, or tank truck is involved in fire.

EXPOSURE
Short-term effects: *SEEK MEDICAL ATTENTION.* **Vapor:** *POISONOUS IF INHALED OR ABSORBED THROUGH THE SKIN.* May cause lung edema; physical exertion will aggravate this condition. Irritating to eyes. Move to fresh air. *IF BREATHING HAS STOPPED*, give artificial respiration. If breathing is difficult, administer oxygen. **Liquid:** *POISONOUS IF SWALLOWED OR ABSORBED THROUGH THE SKIN.* Will burn eyes. Remove contaminated clothing and shoes. Flush affected areas with plenty of water. *IF IN EYES*, hold eyelids open and flush with plenty of water. *IF SWALLOWED* and victim is *CONSCIOUS AND ABLE TO SWALLOW*, have victim drink 4 to 8 ounces of water. **Do NOT induce vomiting.**

Note to physician or authorized medical personnel: Medical observation is recommended for 24 to 48 hours after breathing overexposure, as pulmonary edema may be delayed. As first aid for pulmonary edema, consider administering a corticosteroid spray. Cigarette smoking may exacerbate pulmonary injury and should be discouraged for at least 72 hours following exposure.

HEALTH HAZARDS
Personal protective equipment (PPE): A-Level PPE. OSHA Table Z-1-A air contaminant. Chemical protective material(s) reported to have good to excellent resistance: butyl rubber, chlorobutyl rubber.
Recommendations for respirator selection: NIOSH
At any detectable concentration: SCBAF:PD,PP (any MSHA/NIOSH approved self-contained breathing apparatus that has a full facepiece and is operated in a pressure-demand or other positive-pressure mode); or SAF:PD,PP:ASCBA (any supplied-air respirator that has a full facepiece and is operated in a pressure-demand or other positive-pressure mode in combination with an auxiliary, self-contained breathing apparatus operated in a pressure-demand or other positive pressure mode). *ESCAPE:* GMFS [any air-purifying, full-facepiece respirator (gas mask) with a chin-style, front- or back-mounted canister providing protection against the compound of concern/any appropriate escape-type, self-contained breathing apparatus]; or SCBAE (any appropriate escape-type, self-contained breathing apparatus).
Exposure limits (TWA unless otherwise noted): ACGIH TLV 0.01 ppm (0.025 mg/m^3), suspected human carcinogen; OSHA 0.5 ppm (1 mg/m^3); NIOSH ceiling 0.06 ppm (0.15 mg/m^3)/2-hour; potential human carcinogen; exposure to lowest feasible level; skin contact contributes significantly in overall exposure; NIOSH REL ceiling 0.15 mg/m^3/2 hour.
Toxicity by ingestion: Grade 3; LD_{50} = 50 to 500 mg/kg (rat, mouse).
Long-term health effects: IARC possible carcinogen, rating 2B; sufficient animal evidence. NTP anticipated carcinogen. May effect the nervous system, liver and kidneys. Blood disorders may occur. Mild anemia, upper respiratory irritation, and muscle tremors in dogs following chronic exposure.
Vapor (gas) irritant characteristics: Vapor is irritating such that personnel will not usually tolerate moderate or high concentrations.
Liquid or solid irritant characteristics: Eyes, skin, and respiratory tract irritation. Severe skin irritant. Causes second- and third-degree burns on short contact and is very injurious to the eyes.
Odor threshold: 6.0–14 ppm. *Note:* These values are higher than the adopted Exposure limits.
IDLH value: Potential human carcinogen; 15 ppm.

FIRE DATA
Flash point: 5°F/15°C (cc).
Flammable limits in air: LEL: 2.0%; UEL: 95%.
Fire extinguishing agents not to be used: In large fires, water fog, CO_2, and bicarbonate types may allow flashback and explosive re-ignition.
Behavior in fire: Tends to re-ignite unless diluted with much water.
Autoignition temperature: 452-482°F/233-250°C/506-523°K.
Electrical hazard: Class I, Group C.
Burning rate: 3.8 mm/min

CHEMICAL REACTIVITY
Binary reactants: Dissolves, swells, and disintegrates many plastics. Reacts with oxidizers, halogens, metallic mercury, fuming nitric acid, hydrogen peroxide, nitrogen tetroxide (may be violent). May ignite spontaneously in contact with oxidizers.
Neutralizing agents for acids and caustics: Flush with water.

ENVIRONMENTAL DATA
Water pollution: DOT Appendix B, §172.101–marine pollutant. Effect of low concentrations on aquatic life is unknown. May be dangerous if it enters nearby water intakes; notify operators. Notify local health and wildlife officials. **Response to discharge:** Issue warning–high flammability. Restrict access. Evacuate area. Disperse and flush.

SHIPPING INFORMATION
Grades of purity: Propellant-grade: 98% min; **Storage temperature:** Below 120°F/49°C; **Inert atmosphere:** Inerted; **Stability during transport:** Stable below 1112°F/600°C.

NAS HAZARD CLASSIFICATION FOR BULK WATER TRANSPORTATION
FIRE: 4
HEALTH: Vapor irritant: 3; Liquid or solid irritant: 4; Poisons: 4
WATER POLLUTION: Human toxicity: 4; Aquatic toxicity: 2; Aesthetic effect: 2
REACTIVITY: Other chemicals: 4; Water: 0; Self-reaction: 4

PHYSICAL AND CHEMICAL PROPERTIES
Physical state @ 59°F/15°C and 1 atm: Liquid.
Molecular weight: 60.11
Boiling point @ 1 atm: 146°F/63°C/337°K.
Melting/Freezing point: –71°F/–57°C/216°K.
Critical temperature: 480°F/249°C/522°K.
Critical pressure: 865 psia = 53.5 atm = 5.40 MN/m^2.
Specific gravity (water = 1): 0.791 @ 68°F/20°C (liquid).
Liquid surface tension: 28 dynes/cm = 0.028 N/m @ 77°F/25°C.
Relative vapor density (air = 1): 2.1
Ratio of specific heats of vapor (gas): (estimate) 1.152
Latent heat of vaporization: 261 Btu/lb = 145 cal/g = 6.07 x 10^5 J/kg.
Heat of combustion: –14,170 Btu/lb = –7870 cal/g = –329.3 x 10^5 J/kg.
Heat of solution: (estimate) –30 Btu/lb = –10 cal/g = –0.6 x 10^5 J/kg.
Vapor pressure: 103 mm.

1,2-DIMETHYLHYDRAZINE   REC. D:5750

SYNONYMS: 1,2-DIMETHYLHYDRAZIN (German); N,N-DIMETHYLHYDRAZINE; sym-DIMETHYLHYDRAZINE; 1,2-DIMETILHIDRAZINA (Spanish); DMH; HYDRAZINE, 1,2-DIMETHYL-; RCRA No. U099; SDMH

IDENTIFICATION
CAS Number: 540-73-8
Formula: $C_2H_8N_2$; $(CH_3)HN:NHCH_3$
DOT ID Number: UN 2382; DOT Guide Number: 131
Proper Shipping Name: Dimethylhydrazine, symmetrical
Reportable Quantity (RQ): **(CERCLA)** 1 lb/0.454 kg

DESCRIPTION: Clear, colorless, watery liquid. Fishy or ammonia-like odor. Floats on the surface of water; soluble.

Toxic! • Highly flammable • Breathing the vapor, skin or eye contact can cause severe illness; swallowing the material can kill you; may interfere with the body's ability to use oxygen • Firefighting gear (including SCBA) does not provide adequate protection. If exposure occurs, remove and isolate gear immediately and thoroughly decontaminate personnel • Containers may explode and rocket when exposed to fire • Vapors may form explosive mixture with air • Vapors are heavier than air and will collect and stay in low areas • Vapors may travel long distances to ignition sources and flashback • Vapors in confined areas (e.g., tanks, sewers, buildings) may explode when exposed to fire • Corrosive to plastics; causes disintegration and swelling • Toxic products of combustion may include nitrogen oxide • Do not put yourself in danger by entering a contaminated area to rescue a victim. *Warning:* Odor is not a reliable indicator of the presence of toxic amounts of this material. Odor detection is higher than the Exposure limits.

Hazard Classification (based on NFPA-704 Rating System)
Health Hazards (Blue): 4; Flammability (Red): 3; Reactivity (Yellow): 1

EMERGENCY RESPONSE: See Appendix A (131)
Evacuation:
Public safety: Isolate spill area for at least 100 to 200 meters (330 to 660 feet) in all directions.
Spill: Small spill–First: Isolate in all directions 30 meters (100 feet); Then: Protect persons downwind, DAY: 0.2 km (0.1 mile); NIGHT: 0.3 km (0.2 mile). Large spill–First: Isolate in all directions 60 meters (200 feet); Then: Protect persons downwind, DAY: 0.5 km (0.3 mile); NIGHT: 1.1 km (0.7 mile).
Fire: Isolate for 800 meters (½ mile) in all directions, especially if tank, rail car, or tank truck is involved in fire.

EXPOSURE
Short-term effects: *SEEK MEDICAL ATTENTION.* **Vapor:** *POISONOUS IF INHALED OR IF SKIN IS EXPOSED.* May cause lung edema; physical exertion will aggravate this condition. Move to fresh air. *IF BREATHING HAS STOPPED,* give artificial respiration. If breathing is difficult, administer oxygen. **Liquid:** *POISONOUS IF SWALLOWED OR IF SKIN IS EXPOSED.* Will burn eyes. Remove contaminated clothing and shoes. Flush affected areas with plenty of water. *IF IN EYES*, hold eyelids open and flush with plenty of water. *IF SWALLOWED* and victim is *CONSCIOUS AND ABLE TO SWALLOW*, have victim drink 4 to 8 ounces of water. **Do NOT induce vomiting.**
Note to physician or authorized medical personnel: Medical observation is recommended for 24 to 48 hours after breathing overexposure, as pulmonary edema may be delayed. As first aid for pulmonary edema, consider administering a corticosteroid spray. Cigarette smoking may exacerbate pulmonary injury and should be discouraged for at least 72 hours following exposure.

HEALTH HAZARDS

Personal protective equipment (PPE): A-Level PPE. Chemical protective material(s) reported to have good to excellent resistance: butyl rubber, chlorobutyl rubber.

Recommendations for respirator selection: NIOSH as 1,1-dimethylhydrazine.

At any concentrations above the NIOSH REL, or where there is no REL, at any detectable concentration: SCBAF:PD,PP (any self-contained breathing apparatus that has a full facepiece and is operated in a pressure-demand or other positive-pressure mode); or SAF:PD,PP:ASCBA (any supplied-air respirator that has a full facepiece and is operated in a pressure-demand or other positive-pressure mode in combination with an auxiliary self-contained breathing apparatus operated in a pressure-demand or other positive pressure mode). *ESCAPE:* GMFS [any air-purifying, full-facepiece respirator (gas mask) with a chin-style, front- or back-mounted canister providing protection against the compound of concern]; or SCBAE (any appropriate escape-type, self-contained breathing apparatus).

Exposure limits (TWA unless otherwise noted): As 1,1-dimethylhydrazine. ACGIH TLV 0.01 ppm (0.025 mg/m^3), suspected human carcinogen; OSHA 0.5 ppm (1 mg/m^3); NIOSH ceiling 0.06 ppm (0.15 mg/m^3)/2-hour; potential human carcinogen; exposure to lowest feasible level; skin contact contributes significantly in overall exposure; NIOSH REL ceiling 0.15 mg/m^3/2 hour.

Toxicity by ingestion: Grade 3; LD$_{50}$ = 100 mg/kg (rat).

Long-term health effects: Mild anemia, upper respiratory irritation, and muscle tremors in dogs following chronic exposures. Has caused colon cancer in laboratory animals following a single exposure. IARC possible carcinogen, rating 2B; sufficient animal evidence.

Vapor (gas) irritant characteristics: Eye, skin and respiratory tract irritation. Vapor is irritating such that personnel will not usually tolerate moderate or high concentrations.

Liquid or solid irritant characteristics: Eye, skin and respiratory tract irritation. Severe skin irritant. Causes second- and third-degree burns on short contact and is very injurious to the eyes.

Odor threshold: 6.0-14 ppm as 1,1-dimethylhydrazine. *Note:* These values are higher than the adopted Exposure limits and are not adequate for prolonged exposure.

IDLH value: 15 ppm as 1,1-dimethylhydrazine.

FIRE DATA

Flash point: Less than 73°F/23°C.

Fire extinguishing agents not to be used: In large fires, water fog, CO$_2$, and bicarbonate types may allow flashback and explosive re-ignition.

Behavior in fire: Tends to re-ignite unless diluted with much water.

Electrical hazard: Class 1, Group C.

CHEMICAL REACTIVITY

Binary reactants: Dissolves, swells, and disintegrates many plastics

Neutralizing agents for acids and caustics: Flush with water.

ENVIRONMENTAL DATA

Water pollution: DOT Appendix B, §172.101–marine pollutant. Effects of low concentrations on aquatic life is unknown. May be dangerous if it enters nearby water intakes; notify operators. Notify local health and wildlife officials. **Response to discharge:** Issue warning–high flammability. Restrict access. Evacuate area. Disperse and flush.

SHIPPING INFORMATION

Grades of purity: 99+% as hydrochloride, 98% minimum; **Storage temperature:** Below 120°F; **Inert atmosphere:** Inerted

NAS HAZARD CLASSIFICATION FOR BULK WATER TRANSPORTATION

FIRE: 4
HEALTH: Vapor irritant: 3; Liquid or solid irritant: 4; Poisons: 4
WATER POLLUTION: Human toxicity: 3; Aquatic toxicity: 4; Aesthetic effect: 3
REACTIVITY: Other chemicals: 2; Water: 4; Self-reaction: 0

PHYSICAL AND CHEMICAL PROPERTIES

Physical state @ 59°F/15°C and 1 atm: Liquid.
Molecular weight: 60.10
Boiling point @ 1 atm: 177.8°F/81°C/354.2°K.
Melting/Freezing point: 16°F/–9°C/264°K.
Specific gravity (water = 1): 0.8274 @ 68°F/20°C (liquid).
Relative vapor density (air = 1): 2.07 (estimate).

DIMETHYL HYDROGEN PHOSPHITE   REC. D:5800

SYNONYMS: DIMETHYL PHOSPHITE; FOSFITO de DIMETILO (Spanish); PHOSPHONIC ACID, DIMETHYL ESTER; DIMETHYL PHOSPHONATE

IDENTIFICATION

CAS Number: 868-85-9
Formula: C$_2$H$_7$O$_3$P; (CH$_3$O)$_2$P(O)H
DOT ID Number: UN 1993; DOT Guide Number: 128
Proper Shipping Name: Flammable liquid, n.o.s.

DESCRIPTION: Colorless liquid. Odorless.

Highly flammable • Containers may BLEVE when exposed to fire • Vapors may form explosive mixture with air • Vapors are heavier than air and will collect and stay in low areas • Vapors may travel long distances to ignition sources and flashback • Vapors in confined areas (e.g., tanks, sewers, buildings) may explode when exposed to fire • Irritating to the skin, eyes, and respiratory tract • Toxic products of combustion may include phosphorus oxides.

Hazard Classification (based on NFPA-704 Rating System)
Health Hazards (Blue): 1; Flammability (Red): 3; Reactivity (Yellow): 0

EMERGENCY RESPONSE: See Appendix A (128)
Evacuation:
Public safety: Isolate spill area for at least 25 to 50 meters (80 to 160 feet) in all directions.
Spill: Large spill–Consider initial downwind evacuation for at least 300 meters (1000 feet).
Fire: Isolate for 800 meters (½ mile) in all directions, especially if tank, rail car, or tank truck is involved in fire.

EXPOSURE

Short-term effects: *CALL FOR MEDICAL HELP.* **Vapor:** Harmful if inhaled. Move to fresh air. If breathing has stopped give artificial respiration. If breathing is difficult, administer oxygen. **Liquid:** Harmful if absorbed through skin or swallowed. Remove contaminated clothing. Flush affected area with soap and plenty of water. *IF IN EYES,* hold eyelids open and flush with water for 15 minutes.

HEALTH HAZARDS

Personal protective equipment (PPE): B-Level PPE. Self-contained breathing apparatus, rubber boots, and heavy rubber gloves.
Toxicity by ingestion: Grade 2: LD_{50} = 3.05 g/kg rat
Long-term health effects: Suspected tumorigen and may be mutagen. IARC limited animal evidence of cancer.
Vapor (gas) irritant characteristics: Eye and respiratory tract irritation. Vapors are irritating such that personnel will not usually tolerate moderate or high concentrations.
Liquid or solid irritant characteristics: Causes smarting of the skin and first-degree burns on short exposure; may cause second-degree burns on long exposure.

FIRE DATA

Flash point: 85°F/30°C.
Fire extinguishing agents not to be used: Water may be ineffective.

CHEMICAL REACTIVITY

Binary reactants: Oxidizers, strong acids.
Neutralizing agents for acids and caustics: Dry lime, soda ash
Reactivity group: 34
Compatibility class: Esters.

ENVIRONMENTAL DATA

Water pollution: Effects of low concentration on aquatic life is unknown. May be dangerous if it enters nearby water intakes; notify operators. Notify health and wildlife officials. Notify operators of nearby water intakes. **Response to discharge:** Should be removed. Mechanical and physical treatment.

SHIPPING INFORMATION

Grades of purity: 99%; **Storage temperature:** Ambient; **Stability during transport:** Stable.

PHYSICAL AND CHEMICAL PROPERTIES

Physical state @ 59°F/15°C and 1 atm: Liquid.
Molecular weight: 110.05
Boiling point @ 1 atm: 338–340°F/170–171°C/443–444°K.
Specific gravity (water = 1): 1.200
Relative vapor density (air = 1): 3.79

2,2-DIMETHYL OCTANOIC ACID   REC. D:5850

SYNONYMS: ACIDO DIMETILOCTANOICO (Spanish); 2,2-DIMETHYLCAPRYLIC ACID; ISODECANOIC ACID

IDENTIFICATION

Formula: $C_{10}H_{20}O_2$; $CH_3(CH_2)_5C(CH_3)_2CO_2H$
DOT ID Number: UN 1760; DOT Guide Number: 154
Proper Shipping Name: Corrosive liquids, n.o.s.

DESCRIPTION: Colorless liquid. Burning, rancid odor. Floats on water surface; slightly soluble.

Flammable • Containers may BLEVE when exposed to fire • Vapors may form explosive mixture with air • Vapors are heavier than air and will collect and stay in low areas • Vapors may travel long distances to ignition sources and flashback • Vapors in confined areas (e.g., tanks, sewers, buildings) may explode when exposed to fire • Irritating to the skin, eyes, and respiratory tract • Toxic products of combustion may include carbon monoxide.

Hazard Classification (based on NFPA-704 Rating System)
Health Hazards (Blue): 1; Flammability (Red): 2; Reactivity (Yellow): 0

EMERGENCY RESPONSE: See Appendix A (154)

Evacuation:
Public safety: Isolate the area of spill or leak for at least 25 to 50 meters (80 to 160 feet) in all directions.
Spill: Increase, in the downwind direction, as necessary, the distance shown under "Public Safety."
Fire: If tank, rail car, or tank truck is involved in fire, isolate for at least 800 meters (½ mile) in all directions; also, consider initial evacuation for 800 meters (½ mile) in all directions.

EXPOSURE

Short-term effects: *SEEK MEDICAL ATTENTION.* **Vapor:** Harmful if inhaled or skin is exposed. Irritating to the eyes, nose, and throat. Move victim to fresh air. *IF BREATHING HAS STOPPED,* give artificial respiration; *avoid mouth-to-mouth resuscitation; use bag/mask apparatus.* If breathing is difficult, administer oxygen. **Liquid:** Harmful if swallowed or absorbed through the skin. Irritating to the eyes and skin. Remove contaminated clothing and shoes, flush affected areas with plenty of water. *IF IN EYES,* hold eyelids open, flush with plenty of water for at least 15 minutes. *IF SWALLOWED*: Do nothing except keep victim warm. *DO NOT INDUCE VOMITING*

HEALTH HAZARDS

Personal protective equipment (PPE): B-Level PPE. Approved respirator, chemical resistant gloves, chemical safety gloves, other protective clothing.
Vapor (gas) irritant characteristics: Vapors cause severe irritation of the eyes and throat and can cause eye and lung injury. They cannot be tolerated even at low concentrations.
Liquid or solid irritant characteristics: Severe skin irritant. Causes second- and third-degree burns on short contact and is very injurious to the eyes.

CHEMICAL REACTIVITY

Binary reactants: Oxidizers, ammonia, amines, sulfuric acid, caustics.
Neutralizing agents for acids and caustics: Caustic soda, soda ash, lime
Reactivity group: 4
Compatibility class: Organic acids

ENVIRONMENTAL DATA

Water pollution: Effects of low concentrations on aquatic life are not known. May be dangerous if it enters nearby water intakes; notify operators. Notify local health and wildlife officials. **Response to discharge:** Mechanical containment. Should be removed. Chemical and physical treatment.

SHIPPING INFORMATION

Storage temperature: Ambient; **Stability during transport:** Stable.

NAS HAZARD CLASSIFICATION FOR BULK WATER TRANSPORTATION

FIRE: 2
HEALTH: Vapor irritant: 4; Liquid or solid irritant: 4; Poisons:–
WATER POLLUTION: Human toxicity:–; Aquatic toxicity:–; Aesthetic effect: 3
REACTIVITY: Other chemicals: 2; Water: 0; Self-reaction: 0

PHYSICAL AND CHEMICAL PROPERTIES
Physical state @ 59°F/15°C and 1 atm: Liquid.
Boiling point: 489°F/254°C/527°K.
Molecular weight: 174.3
Relative vapor density (air = 1): 6.0

DIMETHYL PHTHALATE REC. D:5900

SYNONYMS: AVOLIN; 1,2-BENZENEDICARBOXYLIC ACID, DIMETHYL ESTER; DIMETHYL 1,2-BENZENEDICARBOXYLATE; DIMETHYLBENZENEORTHODICARBOXYLATE; DIMETHYL PHTHALATE; DMP; FERMINE; METHYL PHTHALATE; MIPAX; NTM; PALATINOL M; PHTHALSAEUREDIMETHYLESTER (German); PHTHALIC ACID DIMETHYL ESTER; SOLVANOM; SOLVARONE; RCRA No. U102

IDENTIFICATION
CAS Number: 131-11-3
Formula: $C_{10}H_{10}O_4$; $(C_6H_4)(COOCH_3)_2$
DOT ID Number: UN 3082; DOT Guide Number: 171
Proper Shipping Name: Environmentally hazardous substances, liquid, n.o.s.
Reportable Quantity (RQ): **(CERCLA)** 5000 lb/2270 kg

DESCRIPTION: Colorless to pale yellow liquid. Odorless. Sinks in water.

Combustible • Containers may BLEVE when exposed to fire • Vapors are heavier than air and will collect and stay in low areas • Vapors in confined areas (e.g., tanks, sewers, buildings) may explode when exposed to fire • Irritating to the skin, eyes, and respiratory tract • Toxic products of combustion may include carbon monoxide.

Hazard Classification (based on NFPA-704 Rating System)
Health Hazards (Blue): 0; Flammability (Red): 1; Reactivity (Yellow): 0

EMERGENCY RESPONSE: See Appendix A (171)
Evacuation:
Public safety: Isolate the area of spill or leak for at least 10 to 25 meters (30 to 80 feet) in all directions.
Spill: Increase, in the downwind direction, as necessary, the distance shown under "Public Safety."
Fire: If any large container is involved in fire, isolate for at least 800 meters (½ mile) in all directions; also, consider initial evacuation for 800 meters (½ mile) in all directions.

HEALTH HAZARDS
Personal protective equipment (PPE): B-Level PPE. OSHA Table Z-1-A air contaminant. Rubber gloves; goggles or face shield. Butyl rubber is generally suitable for carbooxylic acid compounds. Chemical protective material(s) reported to offer minimal to poor protection: Viton®/neoprene, butyl rubber/neoprene.
Recommendations for respirator selection: NIOSH/OSHA
50 mg/m³: DMF (any dust and mist respirator with a full facepiece). *125 mg/m³*: SA:CF* (any supplied-air respirator operated in a continuous-flow mode); or PAPRDM* (any powered, air-purifying respirator with a dust and mist filter). *250 mg/m³*: HiEF (any air-purifying, full-facepiece respirator with a high-efficiency particulate filter); or SCBAF (any self-contained breathing apparatus with a full facepiece); or SAF (any supplied-air respirator with a full facepiece). *2000 mg/m³*: SAF:PD,PP (any supplied-air respirator that has a full facepiece and is operated in a pressure-demand or other positive-pressure mode). *EMERGENCY OR PLANNED ENTRY INTO UNKNOWN CONCENTRATIONS OR IDLH CONDITIONS*: SCBAF:PD,PP (any self-contained breathing apparatus that has a full facepiece and is operated in a pressure-demand or other positive-pressure mode); or SAF:PD,PP:ASCBA (any supplied-air respirator that has a full facepiece and is operated in a pressure-demand or other positive-pressure mode in combination with an auxiliary self-contained breathing apparatus operated in a pressure-demand or other positive pressure mode). *ESCAPE*: HiEF (any air-purifying, full-facepiece respirator with a high-efficiency particulate filter); or SCBAE (any appropriate escape-type, self-contained breathing apparatus). *Note*: Substance causes eye irritation or damage; eye protection needed.
Exposure limits (TWA unless otherwise noted): ACGIH TLV 5 mg/m³; NIOSH/OSHA 5 mg/m³.
Toxicity by ingestion: Grade 1: LD_{50} = 6.8 g/kg (rat).
Long-term health effects: May effect the nervous system.
Vapor (gas) irritant characteristics: Eye and respiratory tract irritation. Vapors are nonirritating to eyes and throat.
Liquid or solid irritant characteristics: Eye, skin and respiratory tract irritation.
IDLH value: 2000 mg/m³.

FIRE DATA
Flash point: 295°F/146°C (cc).
Flammable limits in air: LEL: 0.9% @ 358°F/181°C; UEL: 8.0%.
Autoignition temperature: 915°F/490°C/763°K.
Electrical hazard: Flow or agitation of substance may generate electrostatic charges due to low conductivity.

CHEMICAL REACTIVITY
Binary reactants: Nitrates; strong oxidizers, alkalis, and acids.
Polymerization: None
Reactivity group: 34
Compatibility class: Esters.

ENVIRONMENTAL DATA
Water pollution: Effects of low concentrations on aquatic life are not known. May be dangerous if it enters nearby water intakes; notify operators. Notify local health and wildlife officials. Notify operators of local water intakes. **Response to discharge:** Should be removed. Chemical and physical treatment.

SHIPPING INFORMATION
Grades of purity: 99%; **Stability during transport:** Stable.

NAS HAZARD CLASSIFICATIONS FOR BULK WATER TRANSPORTATION
FIRE: 1
HEALTH: Vapor irritant: 0; Liquid or solid irritant: 0; Poisons: 0
WATER POLLUTION: Human toxicity: 1; Aquatic toxicity:–; Aesthetic effect: 2
REACTIVITY: Other chemicals: 1; Water: 0; Self-reaction: 0

PHYSICAL AND CHEMICAL PROPERTIES
Physical state @ 59°F/15°C and 1 atm: Liquid.
Molecular weight: 194.20
Boiling point @ 1 atm: 543°F/284°C/557°K.
Melting/Freezing point: 42°F/20°C/293°K.
Specific gravity (water = 1): 1.1905 @ 68°F/20°C.
Relative vapor density (air = 1): 6.69

Latent heat of vaporization: 284 Btu/lb = 158 cal/g = 6.6 x 10^5 J/kg.
Heat of combustion: –10379 Btu/lb = –5766 cal/g = –241 x 10^5 J/kg.
Vapor pressure: (Reid) Very low; 0.01 mm.

DIMETHYLPOLYSILOXANE   REC. D:5950

SYNONYMS: DIMETHYL SILICONE; DIMETHYL SILICONE FLUIDS; DIMETHYL SILICONE OIL; DIMETILPOLISILOXANO (Spanish); POLY(DIMETHYLSILOXANE); SILICONE FLUIDS

IDENTIFICATION
CAS Number: 63148-62-9
Formula: $(CH_3)_3Si\text{-}O\text{-}[Si(CH_3)_2O]_n\text{-}Si(CH_3)_3$

DESCRIPTION: Colorless, thick liquid. Odorless. Floats on water surface; insoluble.

Combustible • Containers may BLEVE when exposed to fire • Vapors are heavier than air and will collect and stay in low areas • Vapors in confined areas (e.g., tanks, sewers, buildings) may explode when exposed to fire • Irritating to the skin, eyes, and respiratory tract • Decomposes above 300°F/149°C. Toxic products of combustion may include CO_2, formaldehyde, and carbon monoxide.

Hazard Classification (based on NFPA-704 Rating System)
Health Hazards (Blue): 0; Flammability (Red): 1; Reactivity (Yellow): 0

EMERGENCY RESPONSE: See Appendix A (171)
Evacuation:
Public safety: Isolate the area of spill or leak for at least 10 to 25 meters (30 to 80 feet) in all directions.
Spill: Increase, in the downwind direction, as necessary, the distance shown under "Public Safety."
Fire: If any large container is involved in fire, isolate for at least 800 meters (½ mile) in all directions; also, consider initial evacuation for 800 meters (½ mile) in all directions.

EXPOSURE
Short-term effects: *SEEK MEDICAL ATTENTION.* **Liquid:** Irritating to eyes. *IF IN EYES,* hold eyelids open and flush with plenty of water. *IF SWALLOWED and victim is CONSCIOUS AND ABLE TO SWALLOW,* have victim drink 4 to 8 ounces of water.

HEALTH HAZARDS
Personal protective equipment (PPE): B-Level PPE. Safety goggles.
Liquid or solid irritant characteristics: Eye irritation.

FIRE DATA
Flash point: 275–635°F/135–335°C (oc).
Fire extinguishing agents not to be used: Water may be ineffective.
Behavior in fire: May emit irritating vapors and acrid smoke.
Autoignition temperature: 820–860°F/438–460°C/711–733°K.

CHEMICAL REACTIVITY
Binary reactants: Strong oxidizers, strong acids.

ENVIRONMENTAL DATA
Water pollution: Effect of low concentrations on aquatic life is unknown. Fouling to shoreline. May be dangerous if it enters nearby water intakes; notify operators. Notify local health and wildlife officials. **Response to discharge:** Mechanical containment. Should be removed. Chemical and physical treatment.

SHIPPING INFORMATION
Grades of purity: A series of compounds having viscosities of from 50 to 100,000 cp is available; **Storage temperature:** Ambient; **Inert atmosphere:** None; **Venting:** Open; **Stability during transport:** Stable.

PHYSICAL AND CHEMICAL PROPERTIES
Physical state @ 59°F/15°C and 1 atm: Liquid.
Boiling point @ 1 atm: More than 300°F/149°C/422°K.
Specific gravity (water = 1): 0.98 @ 68°F/20°C (liquid).
Liquid surface tension: 19-21 dynes/cm = 0.019-0.021 N/m @ 68°F/20°C.
Liquid water interfacial tension: (estimate) 30 dynes/cm = 0.030 N/m @ 68°F/20°C.
Heat of combustion: (estimate) –11,000 Btu/lb = –6200 cal/g = –260 x 10^5 J/kg.

2,2-DIMETHYLPROPANE-1,3-DIOL   REC. D:6000

SYNONYMS: 2,2-DIMETHYL-1,3-PROPANEDIOL; 1,3-PROPANEDIOL, 2,2-DIMETHYL; DIMETHYLOL PROPANE; DIMETHYLTRIMETHYLENE GLYCOL; 2,2-DIMETIL-1,3-PROPANDIOL (Spanish); NEOL; NEOPENTYL GLYCOL; NEOPENTYLENE GLYCOL

IDENTIFICATION
CAS Number: 126-30-7
Formula: $C_5H_{12}O_2$; $HOCH_2C(CH_3)_2CH_2OH$

DESCRIPTION: White crystalline solid.

Combustible • Containers may BLEVE when exposed to fire • Vapors are heavier than air and will collect and stay in low areas • Vapors in confined areas (e.g., tanks, sewers, buildings) may explode when exposed to fire • Irritating to the skin, eyes, and respiratory tract • Toxic products of combustion may include carbon monoxide.

Hazard Classification (based on NFPA-704 Rating System)
Health Hazards (Blue): 1; Flammability (Red): 1; Reactivity (Yellow): 0

EMERGENCY RESPONSE: See Appendix A (171)
Evacuation:
Public safety: Isolate the area of spill or leak for at least 10 to 25 meters (30 to 80 feet) in all directions.
Spill: Increase, in the downwind direction, as necessary, the distance shown under "Public Safety."
Fire: If any large container is involved in fire, isolate for at least 800 meters (½ mile) in all directions; also, consider initial evacuation for 800 meters (½ mile) in all directions.

EXPOSURE
Short-term effects: *SEEK MEDICAL ATTENTION.* **Liquid or solid:** Irritating to skin or eyes. Harmful if swallowed. Flush affected areas with plenty of water.

HEALTH HAZARDS
Personal protective equipment (PPE): B-Level PPE. Self-contained breathing apparatus, rubber boots, and heavy rubber gloves.
Toxicity by ingestion: Grade 1: LD_{50} = 6.4 g/kg (rat).
Vapor (gas) irritant characteristics: Vapors cause smarting of the eyes or respiratory system if present in high concentrations. The effect may be temporary.
Liquid or solid irritant characteristics: If spilled on clothing and allowed to remain, may cause smarting and reddening of skin.

FIRE DATA
Flash point: 225°F/107°C (cc).
Flammable limits in air: LEL: 1.37% @ 300°F/149°C; UEL: 18.8 @ 351°F/177°C.
Autoignition temperature: 730°F/388°C/661°K.

CHEMICAL REACTIVITY
Binary reactants: Incompatible with acids, caustics, amines.
Reactivity group: 20

ENVIRONMENTAL DATA
Water pollution: Effects of low concentrations on aquatic life is unknown. May be dangerous if it enters nearby water intakes; notify operators. Notify local health and wildlife officials.
Response to discharge: Should be removed.

SHIPPING INFORMATION
Grades of purity: 99%.

PHYSICAL AND CHEMICAL PROPERTIES
Physical state @ 59°F/15°C and 1 atm: Solid.
Molecular weight: 104.15
Boiling point @ 1 atm: 406.4°F/208°C/481.2°K.
Melting/Freezing point: 253–261°F/123–127°C/396–400°K.
Relative vapor density (air = 1): 3.6

DIMETHYL SUCCINATE   REC. D:6050

SYNONYMS: BUTANEDIOIC ACID, DIMETHYL ESTER; DIMETHYL BUTANEDIOATE; SUCCINIC ACID, DIMETHYL ESTER

IDENTIFICATION
CAS Number: 106-65-0
Formula: $C_6H_{10}O_4$; $CH_3O_2CCH_2CH_2CO_2CH_3$

DESCRIPTION: Colorless liquid.

Combustible • Containers may BLEVE when exposed to fire • Vapors are heavier than air and will collect and stay in low areas • Vapors in confined areas (e.g., tanks, sewers, buildings) may explode when exposed to fire • Irritating to the skin, eyes, and respiratory tract • Toxic products of combustion may include carbon monoxide.

Hazard Classification (based on NFPA-704 Rating System)
Health Hazards (Blue): 0; Flammability (Red): 2; Reactivity (Yellow): 0

EMERGENCY RESPONSE: See Appendix A (171)
Evacuation:
Public safety: Isolate the area of spill or leak for at least 10 to 25 meters (30 to 80 feet) in all directions.
Spill: Increase, in the downwind direction, as necessary, the distance shown under "Public Safety."
Fire: If any large container is involved in fire, isolate for at least 800 meters (½ mile) in all directions; also, consider initial evacuation for 800 meters (½ mile) in all directions.

EXPOSURE
Short-term effects: *SEEK MEDICAL ATTENTION*. **Vapor:** May be irritating. Move to fresh air. *IF BREATHING HAS STOPPED*, give artificial respiration. If breathing is difficult, administer oxygen. **Liquid:** May be irritating. Remove contaminated clothing and shoes. Flush affected areas with plenty of water. IF IN EYES, hold eyelids open and flush with plenty of water.

HEALTH HAZARDS
Personal protective equipment (PPE): B-Level PPE. Wear self-contained breathing apparatus, rubber boots and heavy rubber gloves.
Vapor (gas) irritant characteristics: Vapors cause smarting of the eyes or respiratory system if present in high concentrations. The effect may be temporary.
Liquid or solid irritant characteristics: If spilled on clothing and allowed to remain, may cause smarting and reddening of skin.

FIRE DATA
Flash point: 185°F/85°C (cc).
Flammable limits in air: LEL: 1%; UEL: 8.5%.
Autoignition temperature: 689°F/365°C/638°K.

CHEMICAL REACTIVITY
Binary reactants: Incompatible with oxidizers, strong acids.
Reactivity group: 34
Compatibility class: Esters.

ENVIRONMENTAL DATA
Water pollution: Effects of low concentrations on aquatic life is unknown. Fouling to shoreline. May be dangerous if it enters nearby water intakes; notify operators. Notify local health and wildlife officials. **Response to discharge:** Restrict access. Mechanical containment. Chemical and physical treatment.

SHIPPING INFORMATION
Grades of purity: 99%; **Storage temperature:** Ambient; **Stability during transport:** Stable.

PHYSICAL AND CHEMICAL PROPERTIES
Physical state @ 59°F/15°C and 1 atm: Solid.
Molecular weight: 146.14
Boiling point @ 1 atm: 392°F/200°C/473.2°K.
Melting/Freezing point: 64–66°F/18–19°C/291–292°K.
Specific gravity (water = 1): 1.117
Relative vapor density (air = 1): 5.04

DIMETHYL SULFATE   REC. D:6100

SYNONYMS: DIMETHYL ESTER OF SULFURIC ACID; DIMETHYL MONOSULFATE; DIMETHYL SULFATE; DIMETHYL SULPHATE; DMS; DMS (METHYL SULFATE); EEC No. 016-023-00-4; METHYL ESTER OF SULFURIC ACID; METHYL SULFATE; METHYLE (SULFATE de) (French); METHYL SULFATE; SULFATE de METHYLE (French); SULFATE DIMETHYLIQUE (French); SULFATO de DIMETILO (Spanish); SULFURIC ACID, DIMETHYL ESTER; RCRA No. U013

IDENTIFICATION
CAS Number: 77-78-1
Formula: $C_2H_6O_4S$; $(CH_3)_2SO_4$
DOT ID Number: UN 1595; DOT Guide Number: 156
Proper Shipping Name: Dimethyl sulfate
Reportable Quantity (RQ): **(CERCLA)** 100 lb/45.4 kg

DESCRIPTION: Colorless, oily liquid. Weak, onion-like odor. Very slightly soluble in water; decomposes forming sulfuric acid. If used as a weapon, utilize M8 Paper (detection: Red) or M256-A1 Detector Kit (Detection limits: 3.0 mg/m^3) if available.

Poison! • Highly flammable • Corrosive • Breathing the vapor, skin or eye contact, or swallowing the material can kill you; contact with the liquid will cause burns and blindness; inhalation symptoms may be delayed • Firefighting gear (including SCBA) does not provide adequate protection. If exposure occurs, remove and isolate gear immediately and thoroughly decontaminate personnel • Containers may BLEVE when exposed to fire • Vapors may form explosive mixture with air • Vapors are heavier than air and will collect and stay in low areas • Vapors may travel long distances to ignition sources and flashback • Vapors in confined areas (e.g., tanks, sewers, buildings) may explode when exposed to fire • Toxic products of combustion may include sulfur dioxide • Do not put yourself in danger by entering a contaminated area to rescue a victim.

Hazard Classification (based on NFPA-704 Rating System)
Health Hazards (Blue): 4; Flammability (Red): 2; Reactivity (Yellow): 0

EMERGENCY RESPONSE: See Appendix A (156)
Evacuation:
Public safety: Isolate the area of spill or leak for at least 50 to 100 meters (160 to 330 feet) in all directions.
Spill: Small spill–First: Isolate in all directions 30 meters (100 feet); Then: Protect persons downwind, DAY: 0.2 km (0.1 mile); NIGHT: 0.2 km (0.1 mile). Large spill–First: Isolate in all directions 30 meters (100 feet); Then: Protect persons downwind, DAY: 0.3 km (0.2 mile); NIGHT: 0.6 km (0.4 mile).
Fire: If tank, rail car, or tank truck is involved in fire, isolate for at least 800 meters (½ mile) in all directions; also, consider initial evacuation for 800 meters (½ mile) in all directions.

EXPOSURE
Short-term effects: *SEEK MEDICAL ATTENTION. THIS MATERIAL IS DANGEROUS WHEN ABSORBED THROUGH THE SKIN.* **Vapor/Fumes:** If artificial respiration is administered, *avoid mouth-to-mouth resuscitation; use bag/mask apparatus.* May cause lung edema; physical exertion will aggravate this condition. **Liquid:** *POISONOUS IF SWALLOWED OR IF SKIN IS EXPOSED.* Will burn eyes and skin. Remove contaminated clothing and shoes. Flush affected areas with plenty of water. *IF IN EYES*, hold eyelids open and flush with plenty of water. *IF SWALLOWED* and victim is *CONSCIOUS AND ABLE TO SWALLOW*, have victim drink 4 to 8 ounces of water. **Do NOT induce vomiting.** *IF SWALLOWED* and victim is *UNCONSCIOUS OR HAVING CONVULSIONS*, do nothing except keep victim warm.
Note to physician or authorized medical personnel: Medical observation is recommended for 24 to 48 hours after breathing overexposure, as pulmonary edema may be delayed. As first aid for pulmonary edema, consider administering a corticosteroid spray. Cigarette smoking may exacerbate pulmonary injury and should be discouraged for at least 72 hours following exposure.

HEALTH HAZARDS
Personal protective equipment (PPE): A-Level PPE. Full face organic vapor respirator. OSHA Table Z-1-A air contaminant.
Recommendations for respirator selection: NIOSH
At any detectable concentration: SCBAF:PD,PP (any MSHA/NIOSH approved self-contained breathing apparatus that has a full facepiece and is operated in a pressure-demand or other positive-pressure mode); or SAF:PD,PP:ASCBA (any supplied-air respirator that has a full facepiece and is operated in a pressure-demand or other positive-pressure mode in combination with an auxiliary, self-contained breathing apparatus operated in a pressure-demand or other positive pressure mode). *ESCAPE:* GMFS [any air-purifying, full-facepiece respirator (gas mask) with a chin-style, front- or back-mounted canister providing protection against the compound of concern/any appropriate escape-type, self-contained breathing apparatus]; or SCBAE (any appropriate escape-type, self-contained breathing apparatus).
Exposure limits (TWA unless otherwise noted): ACGIH TLV 0.1 ppm; OSHA PEL: 1 ppm (5 mg/m^3); NIOSH REL 0.1 ppm (5 mg/m^3); potential human carcinogen; reduce exposure to lowest feasible level; skin contact contributes significantly in overall exposure.
Toxicity by ingestion: Grade 3; LD_{50} = 50 to 500 mg/kg (rat).
Long-term health effects: OSHA specifically regulated carcinogen. IARC probable carcinogen; sufficient animal evidence. NTP anticipated carcinogen. Possible liver and kidney damage. Causes birth defects in rats (malignant tumors in nervous system).
Vapor (gas) irritant characteristics: Vapors cause severe irritation of eye and throat and can cause eye and lung injury. They cannot be tolerated even at low concentration.
Liquid or solid irritant characteristics: Severe skin irritant. Causes second-and third-degree burns on short contact; very injurious to the eyes.
IDLH value: Potential human carcinogen; 7 ppm.

FIRE DATA
Flash point: 182°F/83°C (oc); 180°F/82°C (cc).
Flammable limits in air: LEL: 3.6%; UEL: 23%.
Autoignition temperature: 370°F/188°C/461°K.
Electrical hazard: Class I, Group D.

CHEMICAL REACTIVITY
Reactivity with water: Decomposes in water forming sulfuric acid.
Binary reactants: Violent reaction with aluminum, magnesium. Strong oxidizers, bases, acids, ammonia solutions. Corrodes metal when wet.
Neutralizing agents for acids and caustics: Sodium bicarbonate or lime

ENVIRONMENTAL DATA
Food chain concentration potential: Negative; unlikely to accumulate.
Water pollution: If used as a weapon, utilize use an M272 Water Detection Kit (Detection limit: 2.0 mg/L). Effect of low concentrations on aquatic life is unknown. May be dangerous if it enters nearby water intakes; notify operators. Notify local health and wildlife officials. **Response to discharge:** Issue warning–poison, corrosive. Restrict access. Chemical and physical treatment.

SHIPPING INFORMATION
Grades of purity: Technical; **Storage temperature:** Ambient; **Inert atmosphere:** None; **Venting:** Pressure-vacuum; **Stability during transport:** Stable.

NAS HAZARD CLASSIFICATION FOR BULK WATER TRANSPORTATION
FIRE: 1
HEALTH: Vapor irritant: 4; Liquid or solid irritant: 4; Poisons: 4
WATER POLLUTION: Human toxicity: 4; Aquatic toxicity: 3; Aesthetic effect: 2
REACTIVITY: Other chemicals: 3; Water: 0; Self-reaction: 0

PHYSICAL AND CHEMICAL PROPERTIES
Physical state @ 59°F/15°C and 1 atm: Liquid.
Molecular weight: 126.13
Boiling point @ 1 atm: 370°F/188°C/461°K (decomposes).
Melting/Freezing point: −25°F/−32°C/241°K.
Specific gravity (water = 1): 1.33 @ 59°F/15°C (liquid).
Liquid surface tension: 40.1 dynes/cm = 0.0401 N/m at 18°C.
Liquid water interfacial tension: (estimate) 20 dynes/cm = 0.02 N/m @ 68°F/20°C.
Vapor pressure: 0.1 mm.

DIMETHYL SULFIDE   REC. D:6150

SYNONYMS: DIMETHYL SULPHIDE; DMS; EXTRACT-S; METHANETHIOMETHANE; METHYL SULFIDE; METHYL SULPHIDE; SULFURE DE METHYLE (French); SULFURO de DIMETILO (Spanish); 2-THIOPROPANE

IDENTIFICATION
CAS Number: 75-18-3
Formula: C_2H_6S; $(CH_3)_2S$; CH_3SOCH_3
DOT ID Number: UN 1164; DOT Guide Number: 128
Proper Shipping Name: Dimethyl sulfide

DESCRIPTION: Colorless to pale yellow liquid. Cabbage-like odor. Floats on the surface of water; slightly soluble. Irritating vapor is produced. Boils at 99°F/37°C.

Extremely flammable • Containers may BLEVE when exposed to fire •Vapors form explosive mixture with air • Vapors are heavier than air and will collect and stay in low areas • Vapors may travel long distances to ignition sources and flashback • Vapors in confined areas (e.g., tanks, sewers, buildings) may explode when exposed to fire • Irritating to the skin, eyes, and respiratory tract • Toxic products of combustion may include carbon monoxide and hydrogen sulfide gas.

Hazard Classification (based on NFPA-704 Rating System)
Health Hazards (Blue): 1; Flammability (Red): 4; Reactivity (Yellow): 0

EMERGENCY RESPONSE: See Appendix A (128)
Evacuation:
Public safety: Isolate spill area for at least 25 to 50 meters (80 to 160 feet) in all directions.
Spill: Large spill–Consider initial downwind evacuation for at least 300 meters (1000 feet).
Fire: Isolate for 800 meters (½ mile) in all directions, especially if tank, rail car, or tank truck is involved in fire.

EXPOSURE
Short-term effects: *SEEK MEDICAL ATTENTION.* **Vapor:** Irritating to eyes, nose, and throat. Move victim to fresh air. **Liquid:** Irritating to skin and eyes. Harmful if swallowed. Remove contaminated clothing and shoes. Flush affected areas with plenty of water. *IF IN EYES,* hold eyelids open and flush with plenty of water. *IF SWALLOWED* and victim is *CONSCIOUS AND ABLE TO SWALLOW,* have victim drink 4 to 8 ounces of water and have victim induce vomiting by tickling the back of the throat with the finger or by giving an emetic such as two tablespoons of common salt in a glass of warm water. *IF SWALLOWED* and victim is *UNCONSCIOUS OR HAVING CONVULSIONS,* do nothing except keep victim warm.

HEALTH HAZARDS
Personal protective equipment (PPE): B-Level PPE. Respirator with organic vapor canister; rubber or plastic gloves; goggles or face shield.
Toxicity by ingestion: Grade 2; oral LD_{50} = 535 mg/kg (rat).
Vapor (gas) irritant characteristics: Vapors are irritating such that personnel will not usually tolerate moderate or high vapor concentrations.
Liquid or solid irritant characteristics: Causes smarting of the skin and first-degree burns on short exposure and may cause second-degree burns on long exposure.
Odor threshold: 0.001 ppm.

FIRE DATA
Flash point: −36°F/−38°C (cc).
Flammable limits in air: LEL: 2.2%; UEL: 19.7%.
Fire extinguishing agents not to be used: Water may be ineffective.
Autoignition temperature: 403°F/206°C/479°K.
Electrical hazard: Class I, Group D.
Burning rate: 4.8 mm/min.

ENVIRONMENTAL DATA
Food chain concentration potential: Log K_{ow} = 0.85 (estimate). Unlikely to accumulate.
Water pollution: DOT Appendix B, §172.101–marine pollutant. Effect of low concentrations on aquatic life is unknown. May be dangerous if it enters nearby water intakes; notify operators. Notify local health and wildlife officials. **Response to discharge:** Issue warning–high flammability. Restrict access. Evacuate area. Disperse and flush.

SHIPPING INFORMATION
Grades of purity: 99.8%; **Storage temperature:** Ambient; **Inert atmosphere:** None; **Venting:** Pressure-vacuum; **Stability during transport:** Stable.

NAS HAZARD CLASSIFICATION FOR BULK WATER TRANSPORTATION
FIRE: 4
HEALTH: Vapor irritant: 3; Liquid or solid irritant: 2; Poisons: 2
WATER POLLUTION: Human toxicity: 3; Aquatic toxicity: 4; Aesthetic effect: 1
REACTIVITY: Other chemicals: 0; Water: 0; Self-reaction: 0

PHYSICAL AND CHEMICAL PROPERTIES
Physical state @ 59°F/15°C and 1 atm: Liquid.
Molecular weight: 62.1
Boiling point @ 1 atm: 99°F/37°C/310°K.
Melting/Freezing point: −144°F/−98°C/175°K.
Critical temperature: 444°F/229°C/502°K.
Critical pressure: 826 psia = 56.1 atm = 5.69 MN/m^2.
Specific gravity (water = 1): 0.85 @ 68°F/20°C (liquid).
Liquid surface tension: 26.5 dynes/cm = 0.0265 N/m at 11°C.
Liquid water interfacial tension: (estimate) 30 dynes/cm = 0.030 N/m @ 68°F/20°C.
Relative vapor density (air = 1): 2.14

Ratio of specific heats of vapor (gas): 1.1277 at 16°C.
Latent heat of vaporization: 194 Btu/lb = 108 cal/g = 4.52 x 10^5 J/kg.
Heat of combustion: –13,200 Btu/lb = –7,340 cal/g = –307 x 10^5 J/kg.
Heat of fusion: 30.73 cal/g.

DIMETHYL SULFOXIDE   REC. D:6200

SYNONYMS: A-10846; DELTAN; DEMASORB; DEMAVET; DEMESO; DEMSODROX; DERMASORB; DIMETHYL SULPHOXIDE; DIMEXIDE; DIPIRATRIL-TROPICO; DMS-70; DMS-90; DMSO; DOLICUR; DOLIGUR; DOMOSO; DURASORB; GAMASOL-90; HYADUR; M-176; INFILTRINA; METHYL SULFOXIDE; RIMSO-50; SOMI-PRONT; SQ 9453; SULFICYL BIS(METHANE); SULFOXIDO de DIMETILO (Spanish); SYNTEXAN; TOPSYM

IDENTIFICATION
CAS Number: 67-68-5
Formula: C_2H_6OS
DOT ID Number: NA 1993; DOT Guide Number: 128
Proper Shipping Name: Combustible liquid, n.o.s.

DESCRIPTION: Colorless liquid. Mild garlic odor. Sinks and mixes with water.

Combustible • Containers may BLEVE when exposed to fire • Vapors are heavier than air and will collect and stay in low areas • Vapors in confined areas (e.g., tanks, sewers, buildings) may explode when exposed to fire • Irritating to the skin, eyes, and respiratory tract • Toxic products of combustion may include sulfur dioxide and other sulfur oxides, formaldehyde, and methyl mercaptan.

Hazard Classification (based on NFPA-704 Rating System)
Health Hazards (Blue): 1; Flammability (Red): 1; Reactivity (Yellow): 0

EMERGENCY RESPONSE: See Appendix A (128)
Evacuation:
Public safety: Isolate spill area for at least 25 to 50 meters (80 to 160 feet) in all directions.
Spill: Large spill–Consider initial downwind evacuation for at least 300 meters (1000 feet).
Fire: Isolate for 800 meters (½ mile) in all directions, especially if tank, rail car, or tank truck is involved in fire.

EXPOSURE
Short-term effects: *SEEK MEDICAL ATTENTION.* **Liquid:** Irritating to skin and eyes. Flush affected areas with plenty of water. *IF IN EYES*, hold eyelids open and flush with plenty of water.

HEALTH HAZARDS
Personal protective equipment (PPE): B-Level PPE. Gloves, safety goggles. Respiratory filter if airborne sprays or drops are present. Chemical protective material(s) reported to have good to excellent resistance: butyl rubber, neoprene, neoprene, Silvershield®, natural rubber, nitrile.
Toxicity by ingestion: Grade 0; LD_{50} = > 15 g/kg.
Long-term health effects: May cause blood problems. Causes damage to eye in dogs, pigs, rats, and rabbits.
Vapor (gas) irritant characteristics: Vapors cause irritation such that personnel will find high concentrations unpleasant. The effect may be temporary.
Liquid or solid irritant characteristics: Eye, skin and respiratory tract irritation. If spilled on clothing and allowed to remain, may cause smarting and reddening of the skin.

FIRE DATA
Flash point: 203°F/95°C (oc); 190°F/102°C (cc).
Flammable limits in air: LEL: 2.6%; UEL: 41.8%.
Autoignition temperature: 419°F/215°C/488°K.
Burning rate: 2.0 mm/min

CHEMICAL REACTIVITY
Binary reactants: Violent reaction with strong oxidants, acetyl chloride, cyanuric chloride. Attacks plastics.

ENVIRONMENTAL DATA
Food chain concentration potential: Log P_{ow} = –2.3. Unlikely to accumulate.
Water pollution: Dangerous to aquatic life in high concentrations. May be dangerous if it enters nearby water intakes; notify operators. Notify local health and wildlife officials. **Response to discharge:** Disperse and flush.

SHIPPING INFORMATION
Grades of purity: 99%; **Storage temperature:** Ambient; **Inert atmosphere:** None; **Venting:** Open (flame arrester) or pressure-vacuum; **Stability during transport:** Stable.

PHYSICAL AND CHEMICAL PROPERTIES
Physical state @ 59°F/15°C and 1 atm: Liquid.
Molecular weight: 78.13
Boiling point @ 1 atm: 372°F/189°C/462°K.
Melting/Freezing point: 66°F/19°C/292°K.
Specific gravity (water = 1): 1.101 @ 68°F/20°C (liquid).
Latent heat of vaporization: 259 Btu/lb = 144 cal/g = 6.03 x 10^5 J/kg.
Heat of combustion: –10,890 Btu/lb = –6050 cal/g = 253.3 x 10^5 J/kg.
Heat of decomposition: 100°C.
Heat of solution: –97 Btu/lb = –54 cal/g = 2.3 x 10^5 J/kg.

DIMETHYL TEREPHTHALATE   REC. D:6250

SYNONYMS: 1,4-BENZENE DICARBOXYLIC ACID METHYL ESTER; DMT; METHYL-4-CARBOMETHOXY BENZOATE; TEREFTALATO de DIMETILO (Spanish); TEREPHTHALIC ACID, DIMETHYL ESTER; TEREPHTHALIC ACID, METHYL ESTER

IDENTIFICATION
CAS Number: 120-61-6
Formula: $C_{10}H_{10}O_4$; 1,4-$CH_3OOCC_6H_4COOCH_3$

DESCRIPTION: White crystalline solid or colorless, heated liquid. Odorless. Liquid solidifies. Solid and liquid sink in water; insoluble.

Combustible • Containers may BLEVE when exposed to fire • Vapors are heavier than air and will collect and stay in low areas • Vapors in confined areas (e.g., tanks, sewers, buildings) may explode when exposed to fire • Irritating to the skin, eyes, and respiratory tract • Toxic products of combustion may include carbon monoxide.

Hazard Classification (based on NFPA-704 Rating System)
Health Hazards (Blue): 1; Flammability (Red): 1; Reactivity (Yellow): 0

EMERGENCY RESPONSE: See Appendix A (171)
Evacuation:
Public safety: Isolate the area of spill or leak for at least 10 to 25 meters (30 to 80 feet) in all directions.
Spill: Increase, in the downwind direction, as necessary, the distance shown under "Public Safety."
Fire: If any large container is involved in fire, isolate for at least 800 meters (½ mile) in all directions; also, consider initial evacuation for 800 meters (½ mile) in all directions.

EXPOSURE
Short-term effects: *SEEK MEDICAL ATTENTION.* **Dust:** Not harmful. Move victim to fresh air. *IF IN EYES*, hold eyelids open and flush with plenty of water. **Liquid or solid:** Heated liquid will burn skin and eyes. Harmful if swallowed. Remove contaminated clothing and shoes. Flush affected areas with plenty of water. *IF IN EYES*, hold eyelids open and flush with plenty of water. *IF SWALLOWED* and victim is *CONSCIOUS AND ABLE TO SWALLOW*, have victim drink 4 to 8 ounces of water.

HEALTH HAZARDS
Personal protective equipment (PPE): B-Level PPE. Organic vapor respirator. Molten: goggles, face shield, gauntlets, and protective clothing. Butyl rubber is generally suitable for carbooxylic acid compounds.
Exposure limits (TWA unless otherwise noted): 5 mg/m^3 (respirable fraction) (AIHA WEEL).
Toxicity by ingestion: Grade 2; oral LD$_{50}$ = 4390 mg/kg (rat).
Liquid or solid irritant characteristics: Eye and respiratory tract irritation.

FIRE DATA
Flash point: 298°F/148°C (oc) (molten).
Flammable limits in air: LEL: 0.8%; UEL 11.8%.
Autoignition temperature: 965°F/518°C/791°K (dust).
Electrical hazard: Flow or agitation of substance may generate electrostatic charges due to low conductivity.

CHEMICAL REACTIVITY
Binary reactants: Strong oxidizers, strong acids (i.e., sulfuric acid, nitric acid), nitrates.

ENVIRONMENTAL DATA
Food chain concentration potential: Negative; unlikely to accumulate.
Water pollution: Effect of low concentrations on aquatic life is unknown. Fouling to shoreline. May be dangerous if it enters nearby water intakes; notify operators. Notify local health and wildlife officials. **Response to discharge:** Mechanical containment (if floating). Should be removed. Chemical and physical treatment.

SHIPPING INFORMATION
Grades of purity: 99.9%; **Storage temperature:** Ambient; **Inert atmosphere:** Inert gas blanket is; **Inert atmosphere:** Advisable; **Venting:** Pressure-vacuum; **Stability during transport:** Stable.

PHYSICAL AND CHEMICAL PROPERTIES
Physical state @ 59°F/15°C and 1 atm: Solid.
Molecular weight: 194.2
Boiling point @ 1 atm: 540°F/282°C/555°K.
Melting/Freezing point: 284°F/140°C/413°K.
Specific gravity (water = 1): 1.2 @ 68°F/20°C (solid); 1.08 at 145°C (liquid).
Latent heat of vaporization: 121 Btu/lb = 67.2 cal/g = 2.81 x 10^5 J/kg.
Heat of combustion: −10,310 Btu/lb = −5,727 cal/g = −239.6 x 10^5 J/kg.
Vapor pressure: 13 mm.

DIMETHYL ZINC   REC. D:6300

SYNONYMS: DIMETILZINC (Spanish); METHYLZINC; ZINC DIMETHYL; ZINC METHYL

IDENTIFICATION
CAS Number: 544-97-8
Formula: C_2H_6Zn; $(CH_3)_2Zn$
DOT ID Number: UN 1370; DOT Guide Number: 135
Proper Shipping Name: Dimethylzinc

DESCRIPTION: Colorless liquid. *Spontaneously ignites when exposed to air.* Reacts violently with water forming flammable methane gas.

Spontaneously combustible • Extremely flammable • Do not use water • Firefighting gear (including SCBA) does not provide adequate protection. If exposure occurs, remove and isolate gear immediately and thoroughly decontaminate personnel • Containers may BLEVE when exposed to fire • Vapors may form explosive mixture with air • Vapors are heavier than air and will collect and stay in low areas • Vapors may travel long distances to ignition sources and flashback • Vapors in confined areas (e.g., tanks, sewers, buildings) may explode when exposed to fire • Severely Irritating to the skin, eyes, and respiratory tract; skin and eye contact causes severe burns and blindness • Toxic products of combustion may include zinc oxide gas • May re-ignite after fire is extinguished • Do NOT attempt rescue.

Hazard Classification (based on NFPA-704 Rating System)
Health Hazards (Blue): 3; Flammability (Red): 4; Reactivity (Yellow): 3; Special Notice (White): Water reactive

EMERGENCY RESPONSE: See Appendix A (135)
Evacuation:
Public safety: Isolate spill area for at least 100 to 150 meters (330 to 490 feet) in all directions.
Spill: Increase downwind, the distance shown above in "Public safety."
Fire: Isolate for 800 meters (½ mile) in all directions, especially if tank, rail car, or tank truck is involved in fire.

EXPOSURE
Short-term effects: *SEEK MEDICAL ATTENTION.* **Vapor or mist:** Irritating to eyes, nose, and throat. *IF INHALED*, will, will cause headache, nausea, vomiting or difficult breathing. Move victim to fresh air. *IF IN EYES*, hold eyelids open and flush with plenty of water. *IF BREATHING HAS STOPPED*, give artificial respiration. If breathing is difficult, administer oxygen. **Liquid:** Will burn skin and eyes. *IF SWALLOWED*, will cause nausea, or vomiting. Remove contaminated clothing and shoes. Flush affected areas with plenty of water. *IF IN EYES*, hold eyelids open and flush with plenty of water. *IF SWALLOWED* and victim is *CONSCIOUS AND ABLE TO SWALLOW*, have victim drink 4 to 8 ounces of water. **Do NOT induce vomiting.**

HEALTH HAZARDS
Personal protective equipment (PPE): B-Level PPE. Cartridge-type or fresh air mask for fumes or smoke; PVC fire-retardant or asbestos gloves; full face shield, safety glasses, or goggles; fire-retardant coveralls as standard wear; for special cases, use asbestos coat or rain suit.

FIRE DATA
Flash point: Ignites spontaneously.
Fire extinguishing agents not to be used: Water, foam, halogenated agents, or CO_2
Behavior in fire: Water applied to adjacent fires will intensify fire.
Autoignition temperature: < 0°F/18°C.

CHEMICAL REACTIVITY
Reactivity with water: Reacts vigorously, generating flammable methane gas.
Binary reactants: Will react with surface moisture to generate flammable methane.

ENVIRONMENTAL DATA
Food chain concentration potential: Negative; unlikely to accumulate.
Water pollution: Effect of low concentrations on aquatic life is unknown. May be dangerous if it enters nearby water intakes; notify operators. Notify local health and wildlife officials.
Response to discharge: Issue warning–high flammability.

SHIPPING INFORMATION
Grades of purity: Technical; Electronic; **Storage temperature:** Ambient; **Inert atmosphere:** Dry nitrogen gas; **Venting:** Safety relief; **Stability during transport:** Stable.

PHYSICAL AND CHEMICAL PROPERTIES
Physical state @ 59°F/15°C and 1 atm: Liquid.
Molecular weight: 95.4
Boiling point @ 1 atm: 113°F/45°C/318°K.
Melting/Freezing point: –44°F/–42°C/231°K.
Specific gravity (water = 1): 1.39 at 10.5°C (liquid).
Liquid surface tension: (estimate) 18 dynes/cm = 0.018 N/m @ 68°F/20°C.
134.9 Btu/lb = 74.95 cal/g.
Latent heat of vaporization: = 3.138×10^5 J/kg.

2,4-DINITROANILINE   REC. D:6350

SYNONYMS: 2,4-DINITRANILINE; DINITROANILINE; 2,4-DINITROANILIN (German); 2,4-DINITROANILINA (Spanish); 2,4-DINITROBENZENAMIME; DNA; EEC No. 612-040-00-1

IDENTIFICATION
CAS Number: 97-02-9
Formula: $C_6H_5N_3O_4$; $NH_2C_6H_3(NO_2)_2$-2,4
DOT ID Number: UN 1596; DOT Guide Number: 153
Proper Shipping Name: Dinitroanilines

DESCRIPTION: Yellow crystals or powder. Slightly musty odor. Sinks in water; insoluble.

Explosive! • Heat can cause explosion • Poison! • Breathing the dust or vapor, or swallowing the material can kill you; can affect the body's ability to use oxygen. Irritating to the eyes, skin, and reapiratory tract • Firefighting gear (including SCBA) does not provide adequate protection. If exposure occurs, remove and isolate gear immediately and thoroughly decontaminate personnel • Vapors are heavier than air and will collect and stay in low areas • Vapors in confined areas (e.g., tanks, sewers, buildings) may explode when exposed to fire • Toxic products of combustion may include nitrogen oxides • Do not put yourself in danger by entering a contaminated area to rescue a victim.

Hazard Classification (based on NFPA-704 Rating System)
Health Hazards (Blue): 3; Flammability (Red): 1; Reactivity (Yellow): 3

EMERGENCY RESPONSE: See Appendix A (153)
Evacuation:
Public safety: Isolate the area of spill or leak for at least 25 to 50 meters (80 to 160 feet) in all directions.
Spill: Increase, in the downwind direction, as necessary, the distance shown under "Public Safety."
Fire: If tank, rail car, or tank truck is involved in fire, isolate for at least 800 meters (½ mile) in all directions; also, consider initial evacuation for 800 meters (½ mile) in all directions.

EXPOSURE
Short-term effects: *SEEK MEDICAL ATTENTION.* **Dust:** *POISONOUS IF INHALED OR ABSORBED THROUGH THE SKIN.* Forms methemoglobin. Move to fresh air. *IF BREATHING HAS STOPPED*, give artificial respiration; *avoid mouth-to-mouth resuscitation; use bag/mask apparatus.* **Solids:** *POISONOUS IF SWALLOWED OR IF SKIN IS EXPOSED.* Irritating to eyes. Remove contaminated clothing and shoes. Flush affected areas with plenty of water. *IF IN EYES*, hold eyelids open and flush with plenty of water. *IF SWALLOWED* and victim is *CONSCIOUS AND ABLE TO SWALLOW*, have victim drink 4 to 8 ounces of water and have victim induce vomiting. *IF SWALLOWED* and victim is *UNCONSCIOUS OR HAVING CONVULSIONS*, do nothing except keep victim warm.
Note to physician or authorized medical personnel: Give milk and demulcents, induce emesis or perform gastric lavage: Give fluids: observe for methemoglobinemia. If needed, give methylene blue as a 1% solution intravenously, 1–2 mg/kg; an oral dose of 3–5 mg/kg. If severe, consider exchange transfusion with whole blood.

HEALTH HAZARDS
Personal protective equipment (PPE): B-Level PPE. Self-contained breathing apparatus; butyl rubber gloves; eye goggles; plastic lab coat; protective shoes.
Toxicity by ingestion: Grade 3; oral rate LD_{50} = 418 mg/kg.
Long-term health effects: May effect the blood, liver, or kidneys.
Vapor (gas) irritant characteristics: Eye and respiratory irritation.
Liquid or solid irritant characteristics: Eye, skin and respiratory tract irritant. Causes smarting of the skin and first-degree burns on short exposure; may cause second-degree burns on long exposure.

FIRE DATA
Flash point: 435°F/224°C (cc).
Fire extinguishing agents not to be used: Water or foam may cause frothing
Behavior in fire: May explode in confined spaces. Fight advanced stage fires from explosion-protected positions.

CHEMICAL REACTIVITY
Binary reactants: Reacts violently with strong oxidizing materials.

ENVIRONMENTAL DATA
Food chain concentration potential: Log P_{ow} = 1.86. Unlikely to accumulate.

Water pollution: Effect of low concentrations on aquatic life is unknown. May be dangerous if it enters nearby water intakes; notify operators. Notify local health and wildlife officials. **Response to discharge:** Should be removed. Chemical and physical treatment.

SHIPPING INFORMATION
Grades of purity: Technical and Pure; **Storage temperature:** Ambient; **Inert atmosphere:** None; **Stability during transport:** May detonate when heated under confinement.

PHYSICAL AND CHEMICAL PROPERTIES
Physical state @ 59°F/15°C and 1 atm: Solid.
Molecular weight: 183.12
Melting/Freezing point: 368°F/187°C/460°K.
Specific gravity (water = 1): 1.615 @ 59°F/15°C (solid).
Vapor pressure: 9.7 mm.

---

m-DINITROBENZENE   REC. D:6400

SYNONYMS: BINITROBENZENE; *m*-DINITROBENCENO (Spanish); 1,3-DINITROBENZENE; DINITROBENZOL; 1,3-DINITROBENZOL; 2,4-DINITROBENZENE

IDENTIFICATION
CAS Number: 99-65-0; 25154-54-5 (mixed isomers)
Formula: $C_6H_4N_2O_4$; $1,3\text{-}C_6H_4(NO_2)_2$
DOT ID Number: UN 1597; DOT Guide Number: 152
Proper Shipping Name: Dinitrobenzenes
Reportable Quantity (RQ): **(CERCLA)** 100 lb/45.4 kg

DESCRIPTION: Yellow solid. Slight odor. Sinks in water; poor solubility.

Combustible solid • Heat, flame, or shock may cause explosion • Poison! • Breathing the vapor, skin or eye contact, or swallowing the material can kill you; may interfere with the body's ability to use oxygen • Firefighting gear (including SCBA) does not provide adequate protection. If exposure occurs, remove and isolate gear immediately and thoroughly decontaminate personnel • Containers may BLEVE when exposed to fire • Vapors are heavier than air and will collect and stay in low areas • Vapors in confined areas (e.g., tanks, sewers, buildings) may explode when exposed to fire • Irritating to the skin, eyes, and respiratory tract • Toxic products of combustion may include nitrogen oxides • Do not put yourself in danger by entering a contaminated area to rescue a victim.

Hazard Classification (based on NFPA-704 Rating System)
Health Hazards (Blue): 3; Flammability (Red): 1; Reactivity (Yellow): 4

EMERGENCY RESPONSE: See Appendix A (152)
Evacuation:
Public safety: Isolate the area of spill or leak for at least 25 to 50 meters (80 to 160 feet) in all directions.
Spill: Increase, in the downwind direction, as necessary, the distance shown under "Public Safety."
Fire: If tank, rail car, or tank truck is involved in fire, isolate for at least 800 meters (½ mile) in all directions; also, consider initial evacuation for 800 meters (½ mile) in all directions.

EXPOSURE
Short-term effects: *SEEK MEDICAL ATTENTION.* **Vapor or dust:** *POISONOUS IF INHALED OR IF SKIN IS EXPOSED.* Move victim to fresh air. *IF IN EYES,* hold eyelids open and flush with plenty of water. *IF BREATHING HAS STOPPED,* give artificial respiration; *avoid mouth-to-mouth resuscitation; use bag/mask apparatus.* If breathing is difficult, administer oxygen. **Solid:** *POISONOUS IF SWALLOWED OR IF SKIN IS EXPOSED.* Remove contaminated clothing and shoes. Flush affected areas with plenty of water. *IF IN EYES,* hold eyelids open and flush with plenty of water. *IF SWALLOWED* and victim is *CONSCIOUS AND ABLE TO SWALLOW,* have victim drink 4 to 8 ounces of water and have victim induce vomiting. *IF SWALLOWED* and victim is *UNCONSCIOUS OR HAVING CONVULSIONS,* do nothing except keep victim warm.

HEALTH HAZARDS
Personal protective equipment (PPE): A-Level PPE. OSHA Table Z-1-A air contaminant. Full face organic vapor respirator; rubber gloves; protective clothing.
Recommendations for respirator selection: NIOSH/OSHA
5 mg/m³: DM (any dust and mist respirator). *10 mg/m³*: DMXSQ (any dust and mist respirator except single-use and quarter mask respirators); or SA (any supplied-air respirator); or HiE (any air-purifying, respirator with a high-efficiency particulate filter. *25 mg/m³*: SA:CF (any supplied-air respirator operated in a continuous-flow mode); or PAPRDM (any powered, air-purifying respirator with a dust and mist filter). *50 mg/m³*: HiEF (any air-purifying, full-facepiece respirator with a high-efficiency particulate filter); or SAT:CF (any supplied-air respirator that has a tight-fitting facepiece and is operated in a continuous-flow mode); or PAPRTHiE (any powered, air-purifying respirator with a tight-fitting facepiece and a high-efficiency particulate filter); or SCBAF (any self-contained breathing apparatus with a full facepiece); or SAF (any supplied-air respirator with a full facepiece). *EMERGENCY OR PLANNED ENTRY INTO UNKNOWN CONCENTRATIONS OR IDLH CONDITIONS:* SCBAF:PD,PP (any self-contained breathing apparatus that has a full facepiece and is operated in a pressure-demand or other positive-pressure mode); or SAF:PD,PP:ASCBA (any supplied-air respirator that has a full facepiece and is operated in a pressure-demand or other positive-pressure mode in combination with an auxiliary self-contained breathing apparatus operated in a pressure-demand or other positive pressure mode). *ESCAPE:* HiEF (any air-purifying, full-facepiece respirator with a high-efficiency particulate filter); or SCBAE (any appropriate escape-type, self-contained breathing apparatus).
Exposure limits (TWA unless otherwise noted): ACGIH TLV 0.15 ppm (1.0 mg/m³); NIOSH/OSHA 1 mg/m³; skin contact contributes significantly in overall exposure.
Toxicity by ingestion: Grade 4; oral LD_{50} = 42 mg/kg (bird).
Long-term health effects: May cause liver damage, anemia, neuritis.
Liquid or solid irritant characteristics: Eye, skin and respiratory tract irritation.
IDLH value: 50 mg/m³.

FIRE DATA
Flash point: 302°F/150°C (cc).
Behavior in fire: May explode. Prolonged exposure to fire and heat may result in an explosion due to spontaneous combustion.

CHEMICAL REACTIVITY
Binary reactants: Strong oxidizers, caustics, metals such as tin and zinc. Prolonged exposure to fire and heat may result in explosion due to spontaneous decomposition
Reactivity group: 42
Compatibility class: Nitrocompounds.

ENVIRONMENTAL DATA
Food chain concentration potential: Log P_{ow} = 1.49. Unlikely to accumulate.
Water pollution: Harmful to aquatic life in very low concentrations. May be dangerous if it enters nearby water intakes; notify operators. Notify local health and wildlife officials.
Response to discharge: Issue warning–poison, water contaminant. Restrict access. Should be removed. Chemical and physical treatment.

SHIPPING INFORMATION
Grades of purity: Commercial; **Storage temperature:** Ambient; **Inert atmosphere:** None; **Venting:** Open (flame arrester); **Stability during transport:** Stable.

PHYSICAL AND CHEMICAL PROPERTIES
Physical state @ 59°F/15°C and 1 atm: Solid.
Molecular weight: 168.1
Boiling point @ 1 atm: 608°F/320°C/593°K.
Melting/Freezing point: 248°F/120°C/393°K.
Specific gravity (water = 1): 1.58 at 18°C (solid).
Heat of combustion: –7378 Btu/lb = –4099 cal/g = –171.5 x 10^5 J/kg.
Heat of fusion: 24.70 cal/g.
Vapor pressure: Less than 1 mm.

o-DINITROBENZENE REC. D:6450

SYNONYMS: o-DINITROBENCENO (Spanish); 1,2-DINITROBENZENE; o-DINITROBENZOL

IDENTIFICATION
CAS Number: 528-29-0; 25154-54-5 (mixed isomers)
Formula: $C_6H_4N_2O_4$; $C_6H_4(NO_2)_2$
DOT ID Number: UN 1597; DOT Guide Number: 152
Proper Shipping Name: Dinitrobenzenes
Reportable Quantity (RQ): **(CERCLA)** 100 lb/45.4 kg

DESCRIPTION: Colorless to yellow solid. Sinks in water; poor solubility.

Combustible solid • Heat, flame, or shock may cause explosion • Poison! • Breathing the vapor, skin or eye contact, or swallowing the material can kill you; may interfere with the body's ability to use oxygen • Firefighting gear (including SCBA) does not provide adequate protection. If exposure occurs, remove and isolate gear immediately and thoroughly decontaminate personnel • Containers may BLEVE when exposed to fire • Vapors are heavier than air and will collect and stay in low areas • Vapors in confined areas (e.g., tanks, sewers, buildings) may explode when exposed to fire • Irritating to the skin, eyes, and respiratory tract • Toxic products of combustion may include nitrogen oxides • Do not put yourself in danger by entering a contaminated area to rescue a victim.

Hazard Classification (based on NFPA-704 Rating System)
Health Hazards (Blue): 3; Flammability (Red): 1; Reactivity (Yellow): 4

EMERGENCY RESPONSE: See Appendix A (152)
Evacuation:
Public safety: Isolate the area of spill or leak for at least 25 to 50 meters (80 to 160 feet) in all directions.
Spill: Increase, in the downwind direction, as necessary, the distance shown under "Public Safety."
Fire: If tank, rail car, or tank truck is involved in fire, isolate for at least 800 meters (½ mile) in all directions; also, consider initial evacuation for 800 meters (½ mile) in all directions.

EXPOSURE
Short-term effects: *SEEK MEDICAL ATTENTION.* **Dust:** *POISONOUS IF INHALED, OR IF SKIN IS EXPOSED.* Move to fresh air. Headache, vertigo and vomiting followed by exhaustion, numbness of the legs, staggering, and collapse. Intense methemoglobinemia may lead to asphyxia severe enough to injure the central nervous system. Remove contaminated clothing and shoes. Flush affected areas with plenty of water. *IF BREATHING HAS STOPPED,* give artificial respiration; *avoid mouth-to-mouth resuscitation; use bag/mask apparatus.* **Solid:** *POISONOUS IF SWALLOWED. IF IN EYES,* hold eyelids open and flush with plenty of water. *IF SWALLOWED* and victim is *CONSCIOUS AND ABLE TO SWALLOW,* have victim drink 4 to 8 ounces of water. *Note to physician or authorized medical personnel:* Give milk and demulcents, induce emesis or perform gastric lavage: Give fluids: observe for methemoglobinemia. If needed, give methylene blue as a 1% solution intravenously, 1–2 mg/kg; an oral dose of 3–5 mg/kg. If severe, consider exchange transfusion with whole blood.

HEALTH HAZARDS
Personal protective equipment (PPE): A-Level PPE. OSHA Table Z-1-A air contaminant. Full face organic vapor respirator. Full protective gas-tight outerwear.
Recommendations for respirator selection: NIOSH/OSHA
5 mg/m³: DM (any dust and mist respirator). *10 mg/m³*: DMXSQ (any dust and mist respirator except single-use and quarter mask respirators); or HiE (any air-purifying, respirator with a high-efficiency particulate filter); or SA (any supplied-air respirator). *25 mg/m³*: SA:CF (any supplied-air respirator operated in a continuous-flow mode); or PAPRDM (any powered, air-purifying respirator with a dust and mist filter). *50 mg/m³*: HiEF (any air-purifying, full-facepiece respirator with a high-efficiency particulate filter); or SAT:CF (any supplied-air respirator that has a tight-fitting facepiece and is operated in a continuous-flow mode); or PAPRTHiE (any powered, air-purifying respirator with a tight-fitting facepiece and a high-efficiency particulate filter); or SCBAF (any self-contained breathing apparatus with a full facepiece); or SAF (any supplied-air respirator with a full facepiece) *EMERGENCY OR PLANNED ENTRY INTO UNKNOWN CONCENTRATIONS OR IDLH CONDITIONS:* SCBAF:PD,PP (any self-contained breathing apparatus that has a full facepiece and is operated in a pressure-demand or other positive-pressure mode); or SAF:PD,PP:ASCBA (any supplied-air respirator that has a full facepiece and is operated in a pressure-demand or other positive-pressure mode in combination with an auxiliary self-contained breathing apparatus operated in a pressure-demand or other positive pressure mode). *ESCAPE:* HiEF (any air-purifying, full-facepiece respirator with a high-efficiency particulate filter); or SCBAE (any appropriate escape-type, self-contained breathing apparatus).
Exposure limits (TWA unless otherwise noted): ACGIH TLV 0.15 ppm (1.0 mg/m³); NIOSH/OSHA 1 mg/m³; skin contact contributes significantly in overall exposure.
Toxicity by ingestion: Grade 4; LD_{50} less than 50 mg/kg.
Long-term health effects: Weight loss, anemia, weakness, irritability, and liver damage may occur. Skin may be discolored.
Liquid or solid irritant characteristics: Eye irritant.
IDLH value: 50 mg/m³.

FIRE DATA
Flash point: 302°F/150°C (cc).

Fire extinguishing agents not to be used: Water may cause frothing.
Behavior in fire: May explode when exposed to heat, flame, or when shocked.

CHEMICAL REACTIVITY
Binary reactants: Strong oxidizers, caustics, metals such as tin and zinc. Prolonged exposure to fire and heat may result in an explosion from spontaneous decomposition.
Reactivity group: 42
Compatibility class: Nitrocompounds

ENVIRONMENTAL DATA
Food chain concentration potential: Log P_{ow} = 1.6. Unlikely to accumulate.
Water pollution: Harmful to aquatic life in very low concentrations. May be dangerous if it enters nearby water intakes; notify operators. Notify local health and wildlife officials.
Response to discharge: Issue warning–poison, water contaminant. Restrict access. Evacuate area. Should be removed. Chemical and physical treatment.

SHIPPING INFORMATION
Grades of purity: 100%; **Storage temperature:** Cool; **Stability during transport:** May explode when shocked or heated under confinement.

PHYSICAL AND CHEMICAL PROPERTIES
Physical state @ 59°F/15°C and 1 atm: Solid.
Molecular weight: 168.11
Boiling point @ 1 atm: 606°F/319°C/592°K.
Melting/Freezing point: 244°F/118°C/391°K.
Specific gravity (water = 1): 1.31 @ 68°F/20°C.
Relative vapor density (air = 1): 5.8
Latent heat of vaporization: 145.8 Btu/lb = 81.0 cal/g = 3.39 x 10^5 J/kg = –167 x 10^5 J/kg.
Heat of combustion: –7187 Btu/lb = –3993 cal/g = –167 x 10^5 J/kg.
Heat of fusion: 32.25 cal/g.

---

p-DINITROBENZENE   REC. D:6500

SYNONYMS: *p*-DINITROBENCENO (Spanish); 1,4-DINITROBENZENE; *p*-DINITROBENZOL

IDENTIFICATION
CAS Number: 100-25-4; 25154-54-5 (mixed isomers)
Formula: $C_6H_4N_2O_4$; $C_6H_4(NO_2)_2$
DOT ID Number: UN 1597; DOT Guide Number: 152
Proper Shipping Name: Dinitrobenzenes
Reportable Quantity (RQ): **(CERCLA)** 100 lb/45.4 kg

DESCRIPTION: White to pale yellow solid. Sinks in water; insoluble.

Combustible solid • Heat, flame, or shock may cause explosion • Poison! • Breathing the vapor, skin or eye contact, or swallowing the material can kill you; may interfere with the body's ability to use oxygen • Firefighting gear (including SCBA) does not provide adequate protection. If exposure occurs, remove and isolate gear immediately and thoroughly decontaminate personnel • Containers may BLEVE when exposed to fire • Vapors are heavier than air and will collect and stay in low areas • Vapors in confined areas (e.g., tanks, sewers, buildings) may explode when exposed to fire • Irritating to the skin, eyes, and respiratory tract • Toxic products of combustion may include nitrogen oxides • Do not put yourself in danger by entering a contaminated area to rescue a victim.

Hazard Classification (based on NFPA-704 Rating System)
Health Hazards (Blue): 3; Flammability (Red): 1; Reactivity (Yellow): 4

EMERGENCY RESPONSE: See Appendix A (152)
Evacuation:
Public safety: Isolate the area of spill or leak for at least 25 to 50 meters (80 to 160 feet) in all directions.
Spill: Increase, in the downwind direction, as necessary, the distance shown under "Public Safety."
Fire: If tank, rail car, or tank truck is involved in fire, isolate for at least 800 meters (½ mile) in all directions; also, consider initial evacuation for 800 meters (½ mile) in all directions.

EXPOSURE
Short-term effects: *SEEK MEDICAL ATTENTION.* **Dust:** *POISONOUS IF INHALED, OR IF SKIN IS EXPOSED.* Move to fresh air. Remove contaminated clothing and shoes. Flush affected areas with plenty of water. *IF BREATHING HAS STOPPED*, give artificial respiration; *avoid mouth-to-mouth resuscitation; use bag/mask apparatus.* **Solid:** *POISONOUS IF SWALLOWED. IF IN EYES*, hold eyelids open and flush with plenty of water. *IF SWALLOWED* and victim is *CONSCIOUS AND ABLE TO SWALLOW*, have victim drink 4 to 8 ounces of water.

HEALTH HAZARDS
Personal protective equipment (PPE): A-Level PPE. OSHA Table Z-1-A air contaminant. Full face organic vapor respirator. Safety glasses, protective clothing and rubber gloves.
Recommendations for respirator selection: NIOSH/OSHA
5 mg/m³: DM (any dust and mist respirator). *10 mg/m³:* DMXSQ (any dust and mist respirator except single-use and quarter mask respirators); or HiE (any air-purifying, respirator with a high-efficiency particulate filter); or SA (any supplied-air respirator). *25 mg/m³:* SA:CF (any supplied-air respirator operated in a continuous-flow mode); or PAPRDM (any powered, air-purifying respirator with a dust and mist filter). *50 mg/m³:* HiEF (any air-purifying, full-facepiece respirator with a high-efficiency particulate filter); or SAT:CF (any supplied-air respirator that has a tight-fitting facepiece and is operated in a continuous-flow mode); or PAPRTHiE (any powered, air-purifying respirator with a tight-fitting facepiece and a high-efficiency particulate filter); or SCBAF (any self-contained breathing apparatus with a full facepiece); or SAF (any supplied-air respirator with a full facepiece). *EMERGENCY OR PLANNED ENTRY INTO UNKNOWN CONCENTRATIONS OR IDLH CONDITIONS:* SCBAF:PD,PP (any self-contained breathing apparatus that has a full facepiece and is operated in a pressure-demand or other positive-pressure mode); or SAF:PD,PP:ASCBA (any supplied-air respirator that has a full facepiece and is operated in a pressure-demand or other positive-pressure mode in combination with an auxiliary self-contained breathing apparatus operated in a pressure-demand or other positive pressure mode). *ESCAPE:* HiEF (any air-purifying, full-facepiece respirator with a high-efficiency particulate filter); or SCBAE (any appropriate escape-type, self-contained breathing apparatus).
Exposure limits (TWA unless otherwise noted): ACGIH TLV 0.15 ppm (1.0 mg/m³); NIOSH/OSHA 1 mg/m³; skin contact contributes significantly in overall exposure.
Toxicity by ingestion: Grade 4; LD_{50} below 50 mg/kg.

Long-term health effects: Secondary anemia, liver damage. Irritability, weakness, headache, anorexia, weight loss, nausea, vomiting, cyanosis, dyspnea and skin discoloration.
Liquid or solid irritant characteristics: Eye irritant.
IDLH value: 50 mg/m^3.

FIRE DATA
Flash point: 302°F/150°C (cc).
Fire extinguishing agents not to be used: Water may cause frothing.
Behavior in fire: Decomposes explosively. Can be detonated by shock or heat under confinement that will permit high pressure buildup.

CHEMICAL REACTIVITY
Binary reactants: Strong oxidizers, caustics, metals such as zinc and tin. Prolonged exposure to fire and heat may result in an explosion due to spontaneous decomposition.
Reactivity group: 42
Compatibility class: Nitrocompounds

ENVIRONMENTAL DATA
Water pollution: Harmful to aquatic life in very low concentrations. May be dangerous if it enters nearby water intakes; notify operators. Notify local health and wildlife officials.
Response to discharge: Issue warning–poison, water contaminant. Restrict access. Should be removed. Chemical and physical treatment.

SHIPPING INFORMATION
Grades of purity: Commercial; **Storage temperature:** Ambient; **Inert atmosphere:** None; **Venting:** Open (flame arrester); **Stability during transport:** Stable if kept cool. See also above (severe explosion hazard).

PHYSICAL AND CHEMICAL PROPERTIES
Physical state @ 59°F/15°C and 1 atm: Solid.
Molecular weight: 168.11
Boiling point @ 1 atm: 570°F/299°C/572°K.
Melting/Freezing point: 343°F/173°C/446°K.
Specific gravity (water = 1): 1.625 at 18°C.
Latent heat of vaporization: 142.8 btu/lb = 79.4 cal/g = 3.32 x 10^5 J/kg.
Heat of combustion: –7193 btu/lb = –3996 cal/g = –167.2 x 10^5 J/kg These values for *m*-isomer; use as estimate for *p*-isomer.
Heat of fusion: 39.99 cal/g.

o-DINITROCRESOL   REC. D:6550

SYNONYMS: DINITROCRESOL; DINITRO-*o*-CRESOL; 4,6-DINITRO-*o*-CRESOL; 3,5-DINITRO-*o*-CRESOL; 2,6-DINITRO-*o*-CRESOL; 3,5-DINITRO-2-HYDROXYTOLUENE; 4,6-DINITRO-2-METHYL PHENOL; DN; DNC; DNOC; NITROFAN; RAFEX; SINOX; TRIFRINA; RCRA No. PO47

IDENTIFICATION
CAS Number: 534-52-1
Formula: $C_7H_6N_2O_5$; $CH_3C_6H_2(NO_2)_2(OH)$
DOT ID Number: UN 1598; DOT Guide Number: 153
Proper Shipping Name: Dinitro-*o*-cresol
Reportable Quantity (RQ): **(CERCLA)** 10 lb/4.54 kg

DESCRIPTION: Yellow solid. Sinks in water. Combustible • Dilute material with water! Pure material (less than 10% water) may spontaneously explode if heated or shocked • Containers may BLEVE when exposed to fire • Vapors are heavier than air and will collect and stay in low areas • Vapors in confined areas (e.g., tanks, sewers, buildings) may explode when exposed to fire • Irritating to the skin, eyes, and respiratory tract • Toxic products of combustion may include nitrogen oxides • Do not put yourself in danger by entering a contaminated area to rescue a victim.

Hazard Classification (based on NFPA-704 Rating System)
Health Hazards (Blue): 3; Flammability (Red): 1; Reactivity (Yellow): 3

EMERGENCY RESPONSE: See Appendix A (153)
Evacuation:
Public safety: Isolate the area of spill or leak for at least 25 to 50 meters (80 to 160 feet) in all directions.
Spill: Increase, in the downwind direction, as necessary, the distance shown under "Public Safety."
Fire: If tank, rail car, or tank truck is involved in fire, isolate for at least 800 meters (½ mile) in all directions; also, consider initial evacuation for 800 meters (½ mile) in all directions.

EXPOSURE
Short-term effects: *SEEK MEDICAL ATTENTION.* **Dust:** *POISONOUS IF INHALED OR IF SKIN IS EXPOSED. IF INHALED,* will, will cause loss of consciousness. *IF IN EYES,* hold eyelids open and flush with plenty of water. *IF BREATHING HAS STOPPED,* give artificial respiration; *avoid mouth-to-mouth resuscitation; use bag/mask apparatus.* If breathing is difficult, administer oxygen. **Solid:** POISONOUS IF SWALLOWED OR IF SKIN IS EXPOSED. *IF SWALLOWED,* will cause nausea, vomiting, or loss of consciousness. Remove contaminated clothing and shoes. Flush affected areas with plenty of water. *IF IN EYES,* hold eyelids open and flush with plenty of water. *IF SWALLOWED* and victim is *CONSCIOUS AND ABLE TO SWALLOW,* have victim drink 4 to 8 ounces of water and have victim induce vomiting. *IF SWALLOWED* and victim is *UNCONSCIOUS OR HAVING CONVULSIONS,* do nothing except keep victim warm.
Medical note: Some authorities recommend that all exposed workers have blood tests regularly to determine the level of this substance. Further contact should be avoided if the level exceeds 20 micrograms per gram.

HEALTH HAZARDS
Personal protective equipment (PPE): B-Level PPE. OSHA Table Z-1-A air contaminant. Dust mask; goggles or face shield; protective clothing; rubber gloves.
Recommendations for respirator selection: NIOSH/OSHA
2 mg/m^3: DMF (any dust and mist respirator with a full facepiece). *5 mg/m^3:* HiEF (any air-purifying, full-facepiece respirator with a high-efficiency particulate filter); or SA:CF* (any supplied-air respirator operated in a continuous-flow mode); or PAPRDM* (any powered, air-purifying respirator with a dust and mist filter); or SCBAF (any self-contained breathing apparatus with a full facepiece); or SAF (any supplied-air respirator with a full facepiece). *EMERGENCY OR PLANNED ENTRY INTO UNKNOWN CONCENTRATIONS OR IDLH CONDITIONS:* SCBAF:PD,PP (any self-contained breathing apparatus that has a full facepiece and is operated in a pressure-demand or other positive-pressure mode); or SAF:PD,PP:ASCBA (any supplied-air respirator that has a full facepiece and is operated in a pressure-demand or other positive-pressure mode in combination with an auxiliary self-contained breathing apparatus operated in a pressure-

demand or other positive pressure mode). *ESCAPE:* HiEF (any air-purifying, full-facepiece respirator with a high-efficiency particulate filter); or SCBAE (any appropriate escape-type, self-contained breathing apparatus). *Note:* Substance causes eye irritation or damage; eye protection needed.
Exposure limits (TWA unless otherwise noted): ACGIH TLV 0.2 mg/m^3; NIOSH/OSHA 0.2 mg/m^3; skin contact contributes significantly in overall exposure.
Toxicity by ingestion: Grade 4; LD_{50} less than 50 mg/kg (rat).
IDLH value: 5 mg/m^3.

FIRE DATA
Behavior in fire: Containers may explode.

CHEMICAL REACTIVITY
Binary reactants: Strong oxidizers.

ENVIRONMENTAL DATA
Food chain concentration potential: Negative; unlikely to accumulate.
Water pollution: DOT Appendix B, §172.101–marine pollutant. Harmful to aquatic life in very low concentrations. May be dangerous if it enters nearby water intakes; notify operators. Notify local health and wildlife officials. **Response to discharge:** Issue warning–water contaminant. Restrict access. Should be removed. Chemical and physical treatment.

SHIPPING INFORMATION
Grades of purity: Technical, 90-95%; Paste containing 55-60% water; pure; **Storage temperature:** Ambient; **Inert atmosphere:** None; **Venting:** Open; **Stability during transport:** Stable.

PHYSICAL AND CHEMICAL PROPERTIES
Physical state @ 59°F/15°C and 1 atm: Solid.
Molecular weight: 198.1
Boiling point @ 1 atm: 594°F/312°C/585°K.
Melting/Freezing point: 190°F/88°C/361°K.
Specific gravity (water = 1): (estimate) more than 1.1 @ 68°F/20°C (solid).
Heat of combustion: -7050 Btu/lb = -3920 cal/g = -164×10^5 J/kg.
Vapor pressure: 0.00005 mm.

DINITROCYCLOHEXYLPHENOL   REC. D:6600

SYNONYMS: DINITRO-*o*-CYCLOHEXYLPHENOL; 2-CYCLOHEXYL-4,6-DINITROPHENOL; 2,4-DINITROCICLOHEXILFENOL (Spanish); 2,4-DINITRO-6-CYCLOHEXYLPHENOL; 4,6-DINITRO-*o*-CYCLOHEXYL PHENOL; RCRA No. P034

IDENTIFICATION
CAS Number: 131-89-5
Formula: $C_{12}H_{14}N_2O_5$; $C_6H_{11}C_6H_2(NO_2)_2OH$
DOT ID Number: UN 9026; DOT Guide Number: 153
Proper Shipping Name: Dinitrocyclohexylphenol
Reportable Quantity (RQ): **(CERCLA)** 100 lb/45.4 kg

DESCRIPTION: Yellow crystalline solid. Very slightly soluble in water.

Combustible • Containers may BLEVE when exposed to fire •Vapors may form explosive mixture with air • Vapors are heavier than air and will collect and stay in low areas • Vapors may travel long distances to ignition sources and flashback • Vapors in confined areas (e.g., tanks, sewers, buildings) may explode when exposed to fire • Irritating to the skin, eyes, and respiratory tract • Toxic products of combustion may include nitrogen oxides.

Hazard Classification (based on NFPA-704 Rating System)
Health Hazards (Blue): 2; Flammability (Red): 2; Reactivity (Yellow): 2

EMERGENCY RESPONSE: See Appendix A (153)
Evacuation:
Public safety: Isolate the area of spill or leak for at least 25 to 50 meters (80 to 160 feet) in all directions.
Spill: Increase, in the downwind direction, as necessary, the distance shown under "Public Safety."
Fire: If tank, rail car, or tank truck is involved in fire, isolate for at least 800 meters (½ mile) in all directions; also, consider initial evacuation for 800 meters (½ mile) in all directions.

EXPOSURE
Short-term effects: *SEEK MEDICAL ATTENTION.* **Dust:** *POISONOUS IF INHALED OR IF SKIN IS EXPOSED.* Move to fresh air. *IF BREATHING HAS STOPPED,* give artificial respiration; *avoid mouth-to-mouth resuscitation; use bag/mask apparatus.* If breathing is difficult, administer oxygen. **Solid:** *POISONOUS IF SWALLOWED.* Remove contaminated clothing and shoes. Flush affected areas with plenty of water. *IF IN EYES,* hold eyelids open and flush with plenty of water. *IF SWALLOWED and victim is CONSCIOUS AND ABLE TO SWALLOW,* have victim drink 4 to 8 ounces of water.

HEALTH HAZARDS
Personal protective equipment (PPE): B-Level PPE. Self-contained breathing apparatus; butyl rubber gloves; goggles; lab coat; protective shoes.
Toxicity by ingestion: Grade 4: LD_{50} = 50 mg/kg (mouse).
Liquid or solid irritant characteristics: Causes smarting of the skin and first-degree burns on short exposure; may cause second-degree burns on long exposure.

FIRE DATA
Special hazards of combustion products: Can detonate or explode when heated under confinement.

CHEMICAL REACTIVITY
Binary reactants: Reacts with oxidizing materials and combustibles; risk of fire and explosion. Incompatible with sulfuric acid, nitric acid, caustics, aliphatic amines, isocyanates.

ENVIRONMENTAL DATA
Water pollution: Harmful to aquatic life in very low concentrations. May be dangerous if it enters nearby water intakes; notify operators. Notify local health and wildlife officials. **Response to discharge:** Issue warning–poison. Restrict access. Should be removed. Chemical and physical treatment.

SHIPPING INFORMATION
Inert atmosphere: None; **Stability during transport:** May detonate when heated under confinement.

NAS HAZARD CLASSIFICATION FOR BULK WATER TRANSPORTATION
FIRE: 2
HEALTH: Vapor irritant:–; Liquid or solid irritant: 3; Poisons:–

WATER POLLUTION: Human toxicity: 4; Aquatic toxicity:–; Aesthetic effect: 1
REACTIVITY: Other chemicals: 3; Water: 1; Self-reaction: 0

PHYSICAL AND CHEMICAL PROPERTIES
Physical state @ 59°F/15°C and 1 atm: Solid.
Molecular weight: 266.25
Melting/Freezing point: 225°F/107°C/380°K.
Relative vapor density (air = 1): 9.2

2,4-DINITROPHENOL   **REC. D:6650**

SYNONYMS: ALDIFEN; CHEMOX PE; 2,4-DNP; 2,4-DINITROFENOL (Spanish); α-DINITROPHENOL; DINOFAN; EEC No. 609-016-00-8; FENOXYL CARBON N; 1-HYDROXY-2,4-DINITRO-BENZENE; MAROXOL-50; NITRO KLEENUP; RCRA No. P048; SOLFO BLACK B; SOLFO BLACK BB; SOLFO BLACK 2B SUPRA; SOLFO BLACK G; SOLFO BLACK SB; TERTROSULPHUR BLACK PB; TERTROSULSULPHUR PBR; RCRA No. PO48

IDENTIFICATION
CAS Number: 51-28-5; 25550-58-7 (mixed isomers)
Formula: $C_6H_4N_2O_5$; $HOC_6H_3(NO_2)_2$-2,4
DOT ID Number: UN 0076 (dry); UN 1320 (wet with > 15% water); UN 1599 (solution); DOT Guide Number: 113 (UN 1320); 153 (UN 1599)
Proper Shipping Name: Dinitrol solution; Dinitrophenol, wetted with not less than 15% water
Reportable Quantity (RQ): **(CERCLA)** 10 lb/4.54 kg

DESCRIPTION: Yellow crystalline solid (pure material). Stored and used as a paste in 15% water. Sweet, musty odor. Sinks and mixes slowly with water. Treat as an explosive; DRIED OUT MATERIAL may explode from heat, flame, friction.

Keep material wet. Severe explosion hazard when dry. Dry material (less than 15% water added) will spontaneously explode if heated or shocked • Poison! • Combustible solid • Very irritating to skin, eyes, nose, and lungs; skin and eye contact causes severe burns and blindness. • Firefighting gear (including SCBA) does not provide adequate protection. If exposure occurs, remove and isolate gear immediately and thoroughly decontaminate personnel • Containers may BLEVE when exposed to fire • Vapors are heavier than air and will collect and stay in low areas • Vapors in confined areas (e.g., tanks, sewers, buildings) may explode when exposed to fire • Irritating to the skin, eyes, and respiratory tract • Toxic products of combustion may include nitrogen oxides • Do not put yourself in danger by entering a contaminated area to rescue a victim.

Hazard Classification (based on NFPA-704 Rating System)
Health Hazards (Blue): 3; Flammability (Red): 1; Reactivity (Yellow): 3

EMERGENCY RESPONSE: See Appendix A (113)
Evacuation:
Public safety: Isolate spill area for at least 100 meters (330 feet) in all directions.
Spill: Consider initial downwind evacuation for at least 500 meters (⅓ mile).
Fire: Isolate for at least 800 meters (½ mile) in all directions, especially if tank, rail car, or tank truck is involved in fire.

EMERGENCY RESPONSE: See Appendix A (153)
Evacuation:
Public safety: Isolate the area of spill or leak for at least 25 to 50 meters (80 to 160 feet) in all directions.
Spill: Increase, in the downwind direction, as necessary, the distance shown under "Public Safety."
Fire: If tank, rail car, or tank truck is involved in fire, isolate for at least 800 meters (½ mile) in all directions; also, consider initial evacuation for 800 meters (½ mile) in all directions.

EXPOSURE
Short-term effects: *SEEK MEDICAL ATTENTION.* **Dust:** *POISONOUS IF INHALED OR IF SKIN IS EXPOSED.* Move to fresh air. *IF BREATHING HAS STOPPED,* give artificial respiration. If breathing is difficult, administer oxygen. **Solid:** *DEADLY POISONOUS IF SWALLOWED.* Remove contaminated clothing and shoes. Flush affected areas with plenty of water. *IF IN EYES,* hold eyelids open and flush with plenty of water. *IF SWALLOWED* and victim is *CONSCIOUS AND ABLE TO SWALLOW,* have victim drink 4 to 8 ounces of water. Have victim attempt to vomit. After vomiting, administer a slurry of activated charcoal.

HEALTH HAZARDS
Personal protective equipment (PPE): B-Level PPE. Self-contained breathing apparatus; butyl rubber gloves; goggles; lab coat; protective shoes.
Toxicity by ingestion: Grade 4; LD_{50} below 50 mg/kg.
Long-term health effects: May damage the nervous system. Possible liver, heart and kidney damage. Produces clouding of lens of eye (cataracts) in animals and humans, birth defects in chick embryos.
Liquid or solid irritant characteristics: Causes smarting of the skin and first-degree burns on short exposure; may cause second-degree burns on long exposure.
IDLH value: 5.0 mg/m^3.

FIRE DATA
Behavior in fire: Spontaneous explosion if heated or can detonate or explode when heated under confinement.
Electrical hazard: Flow or agitation of dry substance may generate electrostatic charges due to low conductivity.

CHEMICAL REACTIVITY
Binary reactants: Reacts with oxidizing materials and combustibles; bases and ammonia may form shock-sensitive materials.

ENVIRONMENTAL DATA
Food chain concentration potential: Log P_{ow} = 1.49. Unlikely to accumulate.
Water pollution: DOT Appendix B, §172.101–marine pollutant. Harmful to aquatic life in very low concentrations. May be dangerous if it enters nearby water intakes; notify operators. Notify local health and wildlife officials. **Response to discharge:** Issue warning–poison. Restrict access. Should be removed. Chemical and physical treatment.

SHIPPING INFORMATION
Storage temperature: Ambient; **Inert atmosphere:** None; **Stability during transport:** May detonate when heated under confinement.

PHYSICAL AND CHEMICAL PROPERTIES
Physical state @ 59°F/15°C and 1 atm: Solid.

Molecular weight: 184.1
Boiling point @ 1 atm: sublimes.
Melting/Freezing point: 235°F/113°C/386°K.
Specific gravity (water = 1): 1.68 @ 68°F/20°C (solid).

2,5-DINITROPHENOL REC. D:6700

SYNONYMS: γ-DINITROPHENOL; 2,5-DNP; 2,5-DINITROFENOL (Spanish)

IDENTIFICATION
CAS Number: 329-71-5
Formula: $C_6H_4N_2O_5$; $(NO_2)_2$–C_6H_3–OH
DOT ID Number: UN 1599; DOT Guide Number: 153
Proper Shipping Name: Dinitrophenol solution
Reportable Quantity (RQ): **(CERCLA)** 10 lb/4.54 kg

DESCRIPTION: Yellow crystalline solid or powder (pure material). May be stored and used as a paste in 15% water. Sweet, musty odor. Sinks and mixes slowly with water.

Severe explosion hazard when dry. Keep material wet. Dry material (less than 15% water added) will spontaneously explode if heated or shocked • Combustible solid • Very irritating to skin, eyes, nose, and lungs; skin and eye contact causes severe burns and blindness. • Firefighting gear (including SCBA) does not provide adequate protection. If exposure occurs, remove and isolate gear immediately and thoroughly decontaminate personnel • Containers may BLEVE when exposed to fire • Vapors are heavier than air and will collect and stay in low areas • Vapors in confined areas (e.g., tanks, sewers, buildings) may explode when exposed to fire • Irritating to the skin, eyes, and respiratory tract • Toxic products of combustion may include nitrogen oxides • Do not put yourself in danger by entering a contaminated area to rescue a victim.

Hazard Classification (based on NFPA-704 Rating System)
Health Hazards (Blue):3; Flammability (Red): 1; Reactivity (Yellow): 3

EMERGENCY RESPONSE: See Appendix A (153)
Evacuation:
Public safety: Isolate the area of spill or leak for at least 25 to 50 meters (80 to 160 feet) in all directions.
Spill: Increase, in the downwind direction, as necessary, the distance shown under "Public Safety."
Fire: If tank, rail car, or tank truck is involved in fire, isolate for at least 800 meters (½ mile) in all directions; also, consider initial evacuation for 800 meters (½ mile) in all directions.

EXPOSURE
Short-term effects: *SEEK MEDICAL ATTENTION.* **Dust:** *POISONOUS IF INHALED OR IF SKIN IS EXPOSED.* Move to fresh air. *IF BREATHING HAS STOPPED,* give artificial respiration; *avoid mouth-to-mouth resuscitation; use bag/mask apparatus.* If breathing is difficult, administer oxygen. **Solid:** *POISONOUS IF SWALLOWED.* Remove contaminated clothing and shoes. Flush affected areas with plenty of water. *IF IN EYES,* hold eyelids open and flush with plenty of water. *IF SWALLOWED* and victim is *CONSCIOUS AND ABLE TO SWALLOW,* have victim drink 4 to 8 ounces of water.

HEALTH HAZARDS
Personal protective equipment (PPE): B-Level PPE. Wear approved, self-contained breathing apparatus, butyl rubber gloves, goggles, protective shoes, and coat.
Toxicity by ingestion: Grade 4; LD_{50} below 50 mg/kg.
Long-term health effects: Liver and kidney damage, cataracts, skin lesions and peripheral neuritis.
Liquid or solid irritant characteristics: Causes smarting of the skin and first-degree burns on short exposure; may cause second-degree burns on long exposure.

FIRE DATA
Behavior in fire: Can detonate or explode when heated under confinement.

ENVIRONMENTAL DATA
Water pollution: DOT Appendix B, §172.101–marine pollutant. Harmful to aquatic life in very low concentrations. May be dangerous if it enters nearby water intakes; notify operators. Notify local health and wildlife officials. **Response to discharge:** Issue warning–poison, water contaminant. Restrict access. Should be removed. Chemical and physical treatment.

SHIPPING INFORMATION
Grades of purity: 35% water; **Storage temperature:** Cool; **Stability during transport:** Stable. Keep cool.

PHYSICAL AND CHEMICAL PROPERTIES
Physical state @ 59°F/15°C and 1 atm: Solid.
Molecular weight: 184.11.
Melting/Freezing point: 226.4°F/108°C/381.2°K.
Specific gravity (water = 1): 1.68
Relative vapor density (air = 1): 6.35.
Heat of combustion: (estimate) –6130 Btu/lb = –3406 cal/g = -142.5 x 10^5 J/kg.

2,6-DINITROPHENOL REC. D:6750

SYNONYMS: β-DINITROPHENOL; *O-O*-DINITROPHENOL; DNP; 2,6-DINITROFENOL (Spanish)

IDENTIFICATION
CAS Number: 573-56-8
Formula: $(NO_2)_2C_6H_3$·OH
DOT ID Number: UN 1599; DOT Guide Number: 153
Proper Shipping Name: Dinitrophenol solution
Reportable Quantity (RQ): **(CERCLA)** 10 lb/4.54 kg

DESCRIPTION: Yellow crystalline solid (pure material). Stored and used as a paste in 15% water. Sweet, musty odor. Sinks and mixes slowly with water.

Severe explosion hazard when dry. Keep material wet. Dry material (less than 15% water added) will spontaneously explode if heated or shocked • Combustible solid • Very irritating to skin, eyes, nose, and lungs; skin and eye contact causes severe burns and blindness. • Firefighting gear (including SCBA) does not provide adequate protection. If exposure occurs, remove and isolate gear immediately and thoroughly decontaminate personnel • Containers may BLEVE when exposed to fire • Vapors are heavier than air and will collect and stay in low areas • Vapors in confined areas (e.g., tanks, sewers, buildings) may explode when exposed to fire • Irritating to the skin, eyes, and respiratory tract • Toxic products of combustion may include nitrogen oxides • Do not put yourself in danger by entering a contaminated area to rescue a victim.

EMERGENCY RESPONSE: See Appendix A (153)
Evacuation:
Public safety: Isolate the area of spill or leak for at least 25 to 50 meters (80 to 160 feet) in all directions.
Spill: Increase, in the downwind direction, as necessary, the distance shown under "Public Safety."
Fire: If tank, rail car, or tank truck is involved in fire, isolate for at least 800 meters (½ mile) in all directions; also, consider initial evacuation for 800 meters (½ mile) in all directions.

EXPOSURE
Short-term effects: *SEEK MEDICAL ATTENTION.* **Dust:** *POISONOUS IF INHALED OR IF SKIN IS EXPOSED.* Move to fresh air. *IF BREATHING HAS STOPPED,* give artificial respiration; *avoid mouth-to-mouth resuscitation; use bag/mask apparatus.* If breathing is difficult, administer oxygen. **Solid:** *POISONOUS IF* Swallowed. Remove contaminated clothing and shoes. Flush affected areas with plenty of water. *IF IN EYES,* hold eyelids open and flush with plenty of water. *IF SWALLOWED* and victim is *CONSCIOUS AND ABLE TO SWALLOW*, have victim drink 4 to 8 ounces of water.

HEALTH HAZARDS
Personal protective equipment (PPE): B-Level PPE. Safety glasses, self-contained breathing apparatus, protective clothing, butyl rubber gloves and protective shoes. Natural rubber is generally suitable for phenols.
Toxicity by ingestion: Grade 4; LD_{50} below 50 mg/kg.
Long-term health effects: Liver and kidney damage as well as development of cataracts. May cause skin discoloration and dermatitis.
Liquid or solid irritant characteristics: Causes smarting of the skin and first-degree burns on short exposure; may cause second-degree burns on long exposure.
Odor threshold: 0.021 ppm.
IDLH value: 5.0 mg/m^3.

FIRE DATA
Behavior in fire: Can explode when heated under confinement.

CHEMICAL REACTIVITY
Binary reactants: Reducing agents
Reactivity group: 21
Compatibility class: Phenols, cresols

ENVIRONMENTAL DATA
Water pollution: DOT Appendix B, §172.101–marine pollutant. Harmful to aquatic life in very low concentrations. May be dangerous if it enters nearby water intakes; notify operators. Notify local health and wildlife officials. **Response to discharge:** Issue warning–poison. Restrict access. Should be removed. Chemical and physical treatment.

SHIPPING INFORMATION
Grades of purity: Mixture 2,3-, 2,4-, and 2,6-isomers; **Storage temperature:** Avoid heat; **Stability during transport:** Stable.

PHYSICAL AND CHEMICAL PROPERTIES
Physical state @ 59°F/15°C and 1 atm: Solid.
Molecular weight: 184.11
Melting/Freezing point: 145-147°F/63-64°C/336-337°K.
Specific gravity (water = 1): (estimate) 1.68
Relative vapor density (air = 1): 6.35
Heat of combustion: -6156 Btu/lb = -3420 cal/g = -143.1×10^5 J/kg.

2,4-DINITROTOLUENE   REC. D:6800

SYNONYMS: BENZENE, 1-METHYL-2,4-DINITRO-; 2,4-DINITROTOLUENO (Spanish); DINITROTOLUOL; 2,4-DNT; TOLUENE, 2,4-DINITRO-; 2,4-DINITROTOLUOL; 1-METHYL-2,4-DINITROBENZENE; RCRA No. U105

IDENTIFICATION
CAS Number: 121-14-2; 25321-14-6 (mixed isomers)
Formula: $C_7H_6N_2O_4$; 2,4-$(NO_2)_2C_6H_3CH_3$
DOT ID Number: UN 2038 (solid, liquid); UN 1600 (molten);
DOT Guide Number: 152
Proper Shipping Name: Dinitrotoluenes, liquid or dinitrotoluenes, solid; Dinitrotoluenes, molten
Reportable Quantity (RQ): **(CERCLA)** 10 lb/4.54 kg

DESCRIPTION: Orange-yellow to red crystalline solid or yellow molten liquid. Slight odor. Solid and liquid sink in water; practically insoluble.

Combustible solid • Spontaneously decomposes in elevated temperatures. Heat or flame may cause explosion • Poison! • Breathing the vapor, skin or eye contact, or swallowing the material can kill you; may interfere with the body's ability to use oxygen • Firefighting gear (including SCBA) does not provide adequate protection. If exposure occurs, remove and isolate gear immediately and thoroughly decontaminate personnel • Vapors are heavier than air and will collect and stay in low areas • Vapors in confined areas (e.g., tanks, sewers, buildings) may explode • Toxic products of combustion may include nitrogen oxides and black smoke.

Hazard Classification (based on NFPA-704 Rating System)
Health Hazards (Blue): 3; Flammability (Red): 1; Reactivity (Yellow): 3

EMERGENCY RESPONSE: See Appendix A (152)
Evacuation:
Public safety: Isolate the area of spill or leak for at least 25 to 50 meters (80 to 160 feet) in all directions.
Spill: Increase, in the downwind direction, as necessary, the distance shown under "Public Safety."
Fire: If tank, rail car, or tank truck is involved in fire, isolate for at least 800 meters (½ mile) in all directions; also, consider initial evacuation for 800 meters (½ mile) in all directions.

EXPOSURE
Short-term effects: *SEEK MEDICAL ATTENTION.* **Liquid or Solid:** *POISONOUS IF SWALLOWED OR IF SKIN IS EXPOSED.* Will burn skin and eyes. *IF SWALLOWED*, will cause nausea, vomiting, or loss of consciousness. *IF BREATHING HAS STOPPED,* give artificial respiration; *avoid mouth-to-mouth resuscitation; use bag/mask apparatus.* Remove contaminated clothing and shoes. Flush affected areas with plenty of water. *IF IN EYES*, hold eyelids open and flush with plenty of water. *IF SWALLOWED* and victim is *CONSCIOUS AND ABLE TO SWALLOW*, have victim drink 4 to 8 ounces of water and have victim induce vomiting. *IF SWALLOWED* and victim is *UNCONSCIOUS OR HAVING CONVULSIONS,* do nothing except keep victim warm.

HEALTH HAZARDS
Personal protective equipment (PPE): A-Level PPE. OSHA Table Z-1-A air contaminant. Organic vapor respirator. Chemical protective material(s) reported to have good to excellent resistance: Saranex®.

Recommendations for respirator selection: NIOSH
At any concentrations above the NIOSH REL, or where there is no REL, at any detectable concentration: SCBAF:PD,PP (any self-contained breathing apparatus that has a full facepiece and is operated in a pressure-demand or other positive-pressure mode); or SAF:PD,PP:ASCBA (any supplied-air respirator that has a full facepiece and is operated in a pressure-demand or other positive-pressure mode in combination with an auxiliary, self-contained breathing apparatus operated in a pressure-demand or other positive pressure mode). *ESCAPE:* GMFOVHiE [any air-purifying, full-facepiece respirator (gas mask) with a chin-style, front- or back-mounted organic vapor canister having a high-efficiency particulate filter]; or SCBAE (any appropriate escape-type, self-contained breathing apparatus).
Exposure limits (TWA unless otherwise noted): [as dinitrotoluene CAS 25321-14-6 (mixed isomers)] ACGIH TLV 0.2 mg/m^3 suspected human carcinogen; OSHA PEL 1.5 mg/m^3; NIOSH REL potential carcinogen; reduce exposure to lowest feasible level; skin contact contributes significantly in overall exposure.
Toxicity by ingestion: Grade 4; oral LD$_{50}$ = 30 mg/kg (rat).
Long-term health effects: May cause liver damage, methemoglobinemia, anemia, neuritis. Suspected human carcinogen.
IDLH value: 50 mg/m^3; potential human carcinogen

FIRE DATA
Flash point: 404°F/207°C (cc).
Behavior in fire: Decomposition begins at 428°F/250°C and is self-sustaining at 536°F/280°C (NFPA). Containers may explode in a fire.

CHEMICAL REACTIVITY
Binary reactants: Strong oxidizers, caustics, metals. Commercial grade of dinitrotoluene will decompose at 482°F/250°C, with self-sustaining decomposition at 536°F/280°C.
Reactivity group: 42
Compatibility class: Nitrocompounds

ENVIRONMENTAL DATA
Food chain concentration potential: Log P$_{ow}$ = 2.1. Unlikely to accumulate.
Water pollution: DOT Appendix B, §172.101–marine pollutant. Effect of low concentrations on aquatic life is unknown. May be dangerous if it enters nearby water intakes; notify operators. Notify local health and wildlife officials. **Response to discharge:** Issue warning–poison, water contaminant. Restrict access. Should be removed. Chemical and physical treatment.

SHIPPING INFORMATION
Grades of purity: Technical. Mixtures such as an 80:20 mixture of 2,4- and 2,6-isomers are also available. The hazard properties are similar; **Storage temperature:** Ambient (solid); more than 194°F/90°C (liquid); **Inert atmosphere:** None; **Venting:** Open (flame arrester); **Stability during transport:** Stable below 482°F/250°C.

PHYSICAL AND CHEMICAL PROPERTIES
Physical state @ 59°F/15°C and 1 atm: Solid.
Molecular weight: 182.2
Boiling point @ 1 atm: 572°F/300°C/573°K (decomposes).
Melting/Freezing point: 158°F/70°C/343°K.
Specific gravity (water = 1): 1.379 @ 68°F/20°C (liquid).
Relative vapor density (air = 1): 6.3

Latent heat of vaporization: 170 Btu/lb = 93 cal/g = 3.9 x 10^5 J/kg.
Heat of combustion: –8,305 Btu/lb = –4,614 cal/g = –193.0 x 10^5 J/kg.
Heat of fusion: 26.40 cal/g.
Vapor pressure: 1 mm.

2,6-DINITROTOLUENE REC. D:6850

SYNONYMS: 2,6-DNT; 2-METHYL-1,3-DINIROBENZENE; 2,6-DINITROTOLUENO (Spanish); 2,6-DNT; RCRA No. U106; TOLUENE, 2, 6-DINITRO-; RCRA No. U106

IDENTIFICATION
CAS Number: 606-20-2; 25321-14-6 (mixed isomers)
Formula: (NO$_2$)$_2$C$_6$H$_3$·CH$_3$
DOT ID Number: UN 2038 (liquid, solid); 1600 (molten); DOT Guide Number: 152; DOT Guide Number: 152
Proper Shipping Name: Dinitrotoluenes, liquid; Dinitrotoluenes, solid; Dinitrotoluenes, molten
Reportable Quantity (RQ): **(CERCLA)** 100 lb/45.4 kg

DESCRIPTION: Yellow to red solid or heated liquid. Liquid solidifies. Slight odor. Solid sinks in water; practically insouble.

Combustible • Heat or flame may cause explosion • Containers may BLEVE when exposed to fire • Vapors are heavier than air and will collect and stay in low areas • Vapors in confined areas (e.g., tanks, sewers, buildings) may explode when exposed to fire • Irritating to the skin, eyes, and respiratory tract • Toxic products of combustion may include nitrogen oxides.

Hazard Classification (based on NFPA-704 Rating System)
Health Hazards (Blue): 3; Flammability (Red): 1; Reactivity (Yellow): 3

EMERGENCY RESPONSE: See Appendix A (152)
Evacuation:
Public safety: Isolate the area of spill or leak for at least 25 to 50 meters (80 to 160 feet) in all directions.
Spill: Increase, in the downwind direction, as necessary, the distance shown under "Public Safety."
Fire: If tank, rail car, or tank truck is involved in fire, isolate for at least 800 meters (½ mile) in all directions; also, consider initial evacuation for 800 meters (½ mile) in all directions.

EXPOSURE
Short-term effects: *SEEK MEDICAL ATTENTION.* **Liquid or solid:** *POISONOUS IF SWALLOWED OR IF SKIN IS EXPOSED.* Will burn skin and eyes. Remove contaminated clothing and shoes. Flush affected areas with plenty of water. *IF BREATHING HAS STOPPED,* give artificial respiration; *avoid mouth-to-mouth resuscitation; use bag/mask apparatus. IF IN EYES*, hold eyelids open and flush with plenty of water. *IF SWALLOWED* and victim is *CONSCIOUS AND ABLE TO SWALLOW*, have victim drink 4 to 8 ounces of water and have victim induce vomiting. *IF SWALLOWED* and victim is *UNCONSCIOUS, OR HAVING CONVULSIONS*, do nothing except keep victim warm.

HEALTH HAZARDS
Personal protective equipment (PPE): A-Level PPE. OSHA Wear approved organic vapor respirator; butyl rubber gloves, boots, and protective clothing. Table Z-1-A air contaminant.

Recommendations for respirator selection: NIOSH for dintrotoluene
At any concentrations above the NIOSH REL, or where there is no REL, at any detectable concentration: SCBAF:PD,PP (any self-contained breathing apparatus that has a full facepiece and is operated in a pressure-demand or other positive-pressure mode); or SAF:PD,PP:ASCBA (any supplied-air respirator that has a full facepiece and is operated in a pressure-demand or other positive-pressure mode in combination with an auxiliary self-contained breathing apparatus operated in a pressure-demand or other positive pressure mode). *ESCAPE:* GMFOVHiE [any air-purifying, full-facepiece respirator (gas mask) with a chin-style, front- or back-mounted organic vapor canister having a high-efficiency particulate filter]; or SCBAE (any appropriate escape-type, self-contained breathing apparatus).
Exposure limits (TWA unless otherwise noted): [as dinitrotoluene CAS 25321-14-6 (mixed isomers)] ACGIH TLV 0.2 mg/m^3 suspected human carcinogen; OSHA PEL 1.5 mg/m^3; NIOSH REL potential carcinogen; reduce exposure to lowest feasible level; skin contact contributes significantly in overall exposure.
Toxicity by ingestion: Grade 3; LD_{50} = 50 to 500 mg/kg.
Long-term health effects: Causes methemoglobinemia and anemia. Damaged spleen and liver. Brain damage was observed as was testicular atrophy. Suspected human carcinogen.
Liquid or solid irritant characteristics: May cause eye and skin irritation.
Odor threshold: 0.1 ppm in water.
IDLH value: 50 mg/m^3; potential human carcinogen

FIRE DATA
Flash point: (estimate) 404°F/207°C (cc).
Behavior in fire: May explode.

CHEMICAL REACTIVITY
Binary reactants: Strong oxidizers, caustics, metals such as tin and zinc. Commercial grades of dinitrotoluene will decompose at 428°F, with self-sustaining decomposition at 536°F/280°C.
Reactivity group: 32
Compatibility class: Aromatic hydrocarbons

ENVIRONMENTAL DATA
Food chain concentration potential: Log P_{ow} = 1.7. Unlikely to accumulate.
Water pollution: DOT Appendix B, §172.101–marine pollutant. Harmful to aquatic life in very low concentrations. May be dangerous if it enters nearby water intakes; notify operators. Notify local health and wildlife officials. **Response to discharge:** Issue warning–poison, water contaminant. Restrict access. Should be removed. Chemical and physical treatment.

SHIPPING INFORMATION
Grades of purity: Technical mixtures such as 80-20 mix of 2, 4- and 2, 6-isomers available. Hazard properties are the same; **Storage temperature:** Ambient; **Inert atmosphere:** None; **Venting:** Open (flame arrester); **Stability during transport:** Stable.

PHYSICAL AND CHEMICAL PROPERTIES
Physical state @ 59°F/15°C and 1 atm: Solid.
Molecular weight: 182.13
Boiling point @ 1 atm: 482°F/250°C/523°K (decomposes).
Melting/Freezing point: 140.9°F/60.5°C/333.7°K.
Specific gravity (water = 1): 1.283 @ 111°C.
Relative vapor density (air = 1): 6.28
Heat of combustion: –8099 Btu/lb = –4499 cal/g = –188.3 x 10^5 J/kg.
Vapor pressure: 1 mm.

3,4-DINITROTOLUENE   REC. D:6900

SYNONYMS: 3,4-DINITROTOLUENO (Spanish); 3,4-DNT; 4-METHYL-1,2-DINITROBENZENE; 3,4-TOLUENE,3,4-DINITRO-

IDENTIFICATION
CAS Number: 610-39-9; 25321-14-6 (mixed isomers)
Formula: $C_7H_6N_2O_4$; $(NO_2)_2C_6H_3 \cdot CH_3$
Proper Shipping Name: Dinitrotoluenes, liquid; Dinitrotoluenes, solid; Dinitrotoluenes, molten
Reportable Quantity (RQ): **(CERCLA)** 10 lb/4.54 kg

DESCRIPTION: Yellow to red solid or heated liquid. Liquid solidifies. Slight odor. Solid sinks in water; practically insoluble.

Combustible • Heat or flame may cause explosion • Containers may BLEVE when exposed to fire • Vapors are heavier than air and will collect and stay in low areas • Vapors in confined areas (e.g., tanks, sewers, buildings) may explode when exposed to fire • Irritating to the skin, eyes, and respiratory tract • Toxic products of combustion may include nitrogen oxides.

Hazard Classification (based on NFPA-704 Rating System) (2,4- and 2,6-isomers)
Health Hazards (Blue): 3; Flammability (Red): 1; Reactivity (Yellow): 3

EMERGENCY RESPONSE: See Appendix A (171)
Evacuation:
Public safety: Isolate the area of spill or leak for at least 25 to 50 meters (80 to 160 feet) in all directions.
Spill: Increase, in the downwind direction, as necessary, the distance shown under "Public Safety."
Fire: If tank, rail car, or tank truck is involved in fire, isolate for at least 800 meters (½ mile) in all directions; also, consider initial evacuation for 800 meters (½ mile) in all directions.

EXPOSURE
Short-term effects: *SEEK MEDICAL ATTENTION.* **Liquid or solid:** *POISONOUS IF SWALLOWED OR IF SKIN IS EXPOSED.* Will burn skin and eyes. Remove contaminated clothing and shoes. Flush affected areas with plenty of water. *IF IN EYES*, hold eyelids open and flush with plenty of water. *IF SWALLOWED* and victim is *CONSCIOUS AND ABLE TO SWALLOW*, have victim drink 4 to 8 ounces of water and have victim induce vomiting. *IF SWALLOWED* and victim is *UNCONSCIOUS,* OR HAVING CONVULSIONS, do nothing except keep victim warm.

HEALTH HAZARDS
Personal protective equipment (PPE): A-Level PPE. Organic vapor respirator. Butyl rubber gloves, boots and protective clothing.
Recommendations for respirator selection: NIOSH for dintrotoluene
At any concentrations above the NIOSH REL, or where there is no REL, at any detectable concentration: SCBAF:PD,PP (any self-contained breathing apparatus that has a full facepiece and is operated in a pressure-demand or other positive-pressure mode); or SAF:PD,PP:ASCBA (any supplied-air respirator that has a full facepiece and is operated in a pressure-demand or other positive-

pressure mode in combination with an auxiliary self-contained breathing apparatus operated in a pressure-demand or other positive pressure mode). *ESCAPE:* GMFOVHiE [any air-purifying, full-facepiece respirator (gas mask) with a chin-style, front- or back-mounted organic vapor canister having a high-efficiency particulate filter]; or SCBAE (any appropriate escape-type, self-contained breathing apparatus.

Exposure limits (TWA unless otherwise noted): [as dinitrotoluene CAS 25321-14-6 (mixed isomers)] ACGIH TLV 0.2 mg/m^3 suspected human carcinogen; OSHA PEL 1.5 mg/m^3; NIOSH REL potential carcinogen; reduce exposure to lowest feasible level; skin contact contributes significantly in overall exposure.

Toxicity by ingestion: Grade 2; LD$_{50}$ = 0.5 to 5 g/kg.

Long-term health effects: Caused melhemoglobinenia and anemia. Damaged spleen and liver. Brain damage was observed as was testicular atrophy. suspected human carcinogen.

Liquid or solid irritant characteristics: Serious eye and skin irritant.

IDLH value: 50 mg/m^3; potential human carcinogen

FIRE DATA
Flash point: (estimate) 404°F/207°C (cc).

Behavior in fire: Commercial grades of dinitrotoluene will decompose at 428°F/220°C, with self-sustaining decomposition at 536°F/280°C. May explode.

CHEMICAL REACTIVITY
Binary reactants: Strong oxidizers, caustics, metals such as tin and zinc.

Reactivity group: 32

Compatibility class: Aromatic hydrocarbons

ENVIRONMENTAL DATA
Food chain concentration potential: Log P$_{ow}$ = 2.0. Unlikely to accumulate.

Water pollution: DOT Appendix B, §172.101–marine pollutant. Harmful to aquatic life in very low concentrations. May be dangerous if it enters nearby water intakes; notify operators. Notify local health and wildlife officials. **Response to discharge:** Issue warning–poison, water contaminant. Restrict access. Should be removed. Chemical and physical treatment.

SHIPPING INFORMATION
Storage temperature: Ambient; **Stability during transport:** Stable.

PHYSICAL AND CHEMICAL PROPERTIES
Physical state @ 59°F/15°C and 1 atm: Solid.
Molecular weight: 182.13
Boiling point @ 1 atm: Decomposes.
Melting/Freezing point: 140–142°F/60–61°C/333–334°K.
Specific gravity (water = 1): 1.2594 at 111°C.
Relative vapor density (air = 1): 6.28
Heat of combustion: –8075 Btu/lb = –4486 cal/g = –187.7 x 10^5 J/kg.
Vapor pressure: 1 mm.

DINONYL PHTHALATE   REC. D:6950

SYNONYMS: FTALATO de DINONILO (Spanish); PHTHALIC ACID, DINONYL ESTER; DINONYL 1,2-BENZENEDICARBOXYLATE; DI-*n*-NONYL PHTHALATE; BIOFLEX 91

IDENTIFICATION
CAS Number: 84-76-4
Formula: (C$_6$H$_4$)(COOC$_9$H$_{19}$)$_2$

DESCRIPTION: Colorless liquid. Odorless.

Combustible • Heat or flame may cause explosion • Containers may BLEVE when exposed to fire • Vapors are heavier than air and will collect and stay in low areas • Vapors in confined areas (e.g., tanks, sewers, buildings) may explode when exposed to fire • Irritating to the skin, eyes, and respiratory tract • Toxic products of combustion may include carbon monoxide.

Hazard Classification (based on NFPA-704 Rating System)
Health Hazards (Blue): 0; Flammability (Red): 1; Reactivity (Yellow): 0

EMERGENCY RESPONSE: See Appendix A (171)
Evacuation:
Public safety: Isolate the area of spill or leak for at least 10 to 25 meters (30 to 80 feet) in all directions.
Spill: Increase, in the downwind direction, as necessary, the distance shown under "Public Safety."
Fire: If any large container is involved in fire, isolate for at least 800 meters (½ mile) in all directions; also, consider initial evacuation for 800 meters (½ mile) in all directions.

EXPOSURE
Short-term effects: CALL FOR MEAL AID. **Vapor:** Remove contaminated clothing and shoes. Flush affected areas with plenty of water IF IN EYES, hold eyelids open and flush with plenty of water. *IF SWALLOWED* and victim is *CONSCIOUS AND ABLE TO SWALLOW*, have victim drink 1 to 2 glasses of water.

HEALTH HAZARDS
Personal protective equipment (PPE): B-Level PPE. Goggles or face shield; rubber gloves.
Toxicity by ingestion: Grade 2: LD5O = 2.0 g/kg (rat).

FIRE DATA
Flash point: 420°F/216°C (cc).
Fire extinguishing agents not to be used: Water may be ineffective.

CHEMICAL REACTIVITY
Binary reactants: Strong oxidizers. May attack some forms of plastics.
Reactivity group: 34
Compatibility class: Esters.

ENVIRONMENTAL DATA
Water pollution: Effects of low concentration on aquatic life is unknown. Fouling to shoreline. May be dangerous if it enters nearby water intakes; notify operators. Notify local health and wildlife officials. **Response to discharge:** Mechanical containment. Should be removed. Chemical and physical treatment.

SHIPPING INFORMATION
Storage temperature: Ambient; **Stability during transport:** Stable.

PHYSICAL AND CHEMICAL PROPERTIES
Physical state @ 59°F/15°C and 1 atm: Liquid.
Molecular weight: 418.68
Boiling point @ 1 atm: 775°F/413°C/686.2°K.

Specific gravity (water = 1): 0.97
Relative vapor density (air = 1): 14.44

DIOCTYL ADIPATE   REC. D:7000

SYNONYMS: ADIPATO de DIOCTILO (Spanish); ADIPOL 2EH; ADIPIC ACID, BIS(2-ETHYLHEXYL) ESTER; BEHA; BIS(2-ETHYLHEXYL)ADIPATE; BISOFLEX DOA; DEHA; DI (2-ETHYLHEXYL) ADIPATE; EFFEMOLL DOA; ERGOPLAST AdDO; FLEXOL A-26; KODAFLEX DOA; MONOPLEX DOA; OCTYL ADIPATE; PLASTOMOLL DOA; PX-238; REOMOL DOA; RUCOFLEX PLASTICIZER DOA; SICOL 250; TRUFLEX DOA; VESTINOL OA; WICKENOL 158; WITOMOL-320

IDENTIFICATION
CAS Number: 103-23-1
Formula: $C_{22}H_{42}O_4$; $C_8H_{17}OOC(CH_2)_4COOC_8H_{17}$

DESCRIPTION: Colorless oily liquid. Odorless. Floats on the surface of water.

Combustible • Heat or flame may cause explosion • Containers may BLEVE when exposed to fire • Vapors are heavier than air and will collect and stay in low areas • Vapors in confined areas (e.g., tanks, sewers, buildings) may explode when exposed to fire • Irritating to the skin, eyes, and respiratory tract • Toxic products of combustion may include carbon monoxide.

Hazard Classification (based on NFPA-704 Rating System)
Health Hazards (Blue): 0; Flammability (Red): 1; Reactivity (Yellow): 0

EMERGENCY RESPONSE: See Appendix A (171)
Evacuation:
Public safety: Isolate the area of spill or leak for at least 10 to 25 meters (30 to 80 feet) in all directions.
Spill: Increase, in the downwind direction, as necessary, the distance shown under "Public Safety."
Fire: If any large container is involved in fire, isolate for at least 800 meters (½ mile) in all directions; also, consider initial evacuation for 800 meters (½ mile) in all directions.

EXPOSURE
Short-term effects: *SEEK MEDICAL ATTENTION*. **Liquid:** Irritating to eyes. Flush affected areas with plenty of water. *IF IN EYES*, hold eyelids open and flush with plenty of water.

HEALTH HAZARDS
Toxicity by ingestion: Grade 1; LD_{50} = 5 to 15 g/kg.
Long-term health effects: IARC rating 3; animal evidence: limited.
Vapor (gas) irritant characteristics: Vapors are nonirritating to the eyes and throat.
Liquid or solid irritant characteristics: Minimum hazard. If spilled on clothing and allowed to remain, may cause smarting and reddening of the skin.

FIRE DATA
Flash point: 402°F/206°C (oc).
Fire extinguishing agents not to be used: Water or foam may cause frothing (NFPA).
Autoignition temperature: 710°F/377°C/650°K.

CHEMICAL REACTIVITY
Reactivity with water: May cause foaming.
Binary reactants: Strong oxidizers, strong acids.
Reactivity group: 34
Compatibility class: Ester

ENVIRONMENTAL DATA
Food chain concentration potential: Log P_{ow} = 6.1. Values > 3.0 are likely to bioconcentrate in aquatic organisms and other living tissue, especially in fats.
Water pollution: Effect of low concentrations on aquatic life is unknown. Fouling to shoreline. May be dangerous if it enters nearby water intakes; notify operators. Notify local health and wildlife officials. **Response to discharge:** Mechanical containment. Chemical and physical treatment.

SHIPPING INFORMATION
Grades of purity: 99.6%; **Storage temperature:** Ambient; **Inert atmosphere:** None; **Venting:** Open; **Stability during transport:** Stable.

PHYSICAL AND CHEMICAL PROPERTIES
Physical state @ 59°F/15°C and 1 atm: Liquid.
Molecular weight: 371
Specific gravity (water = 1): 0.928 @ 68°F/20°C (liquid).
Liquid surface tension: (estimate) 15 dynes/cm = 0.015 N/m @ 68°F/20°C.
Liquid water interfacial tension: (estimate) 30 dynes/cm = 0.03 N/m @ 68°F/20°C.
Heat of combustion: –15,430 Btu/lb = –8580 cal/g = –359 x 10^5 J/kg.

DIOCTYL SODIUM SULFOSUCCINATE   REC. D:7050

SYNONYMS: AEROSOL GPG; AEROSOL SURFACTANT; ALCOPOL O; ALPHASOL OT; ALROWET D65; BEROL 478; BIS(2-ETHYLHEXYL) SODIUM SULFOSUCCINATE; CELANOL DOS 75; CLESTOL; COLACE; COMPLEMIX; CONSTONATE; COPROL; DEFILIN; DI-(2-ETHYLHEXYL) SULFOSUCCINATE, SODIUM SALT; DIOCTYN; DIOCTYAL; DIOMEDICONE; DIOSUCCIN; DIOTILAN; DIOVAC; DOCUSATE SODIUM; DOXINATE; DOXOL; DSS; DULSIVAC; DUOSOL; HUMIFEN WT-27G; KONLAX; KOSATE; LAXINATE; MANOXAL OT; MERVAMINE; MODANE SOFT; MOLATOC; MOLCER; MOLOFAC; NEKTAL; WT-27; NEVAX; NIKKOL OTP 70; NORVAL; OBSTON; RAPISOL; REGUTOL; REQUTOL; REVAC; SANMORIN OT 70; SBO; SOBITOL; SOFTIL; SILIWAX; SOLUSOL–75%; SOLUSOL–100%; SULFIMEL DOS; SULFOSUCCINATO de DIOCTILO y SODIO (Spanish); TEX-WET; TEXWET 1001; TRITON GR-5; VATSOL OT; VELMOL; WAXOL; WETAID SR

IDENTIFICATION
CAS Number: 577-11-7
Formula: $C_{20}H_{38}O_7S·Na$; $C_8H_{17}OOCCH_2CH(SO_3Na)COOC_8H_{17}$

DESCRIPTION: Colorless or off-white waxy solid or watery solution. Odorless. Sinks and mixes slowly with water.

Irritating to the skin, eyes, and respiratory tract • Toxic products of combustion may include carbon monoxide, sulfur oxide and sodium oxide.

Hazard Classification (based on NFPA-704 Rating System)
Health Hazards (Blue): 0; Flammability (Red): 0; Reactivity (Yellow): 0

EMERGENCY RESPONSE: See Appendix A (171)
Evacuation:
Public safety: Isolate the area of spill or leak for at least 10 to 25 meters (30 to 80 feet) in all directions.
Spill: Increase, in the downwind direction, as necessary, the distance shown under "Public Safety."
Fire: If any large container is involved in fire, isolate for at least 800 meters (½ mile) in all directions; also, consider initial evacuation for 800 meters (½ mile) in all directions.

EXPOSURE
Short-term effects: *SEEK MEDICAL ATTENTION.* **Liquid or solid:** Irritating to skin and eyes. Harmful if swallowed. Remove contaminated clothing and shoes. Flush affected areas with plenty of water. *IF IN EYES*, hold eyelids open and flush with plenty of water. *IF SWALLOWED* and victim is *CONSCIOUS AND ABLE TO SWALLOW*, have victim drink 4 to 8 ounces of water.

HEALTH HAZARDS
Personal protective equipment (PPE): B-Level PPE. Chemical goggles; rubber gloves; dust respirator.
Toxicity by ingestion: Grade 2; oral LD_{50} = 1900 mg/kg (rat).
Liquid or solid irritant characteristics: Severe eye and skin irritant.

FIRE DATA
Behavior in fire: Causes foaming and spreading of water. Assists in putting out fires by water.

ENVIRONMENTAL DATA
Food chain concentration potential: Negative; unlikely to accumulate.
Water pollution: Effect of low concentrations on aquatic life is unknown. May be dangerous if it enters nearby water intakes; notify operators. Notify local health and wildlife officials.
Response to discharge: Issue warning–water contaminant. Disperse and flush.

SHIPPING INFORMATION
Grades of purity: Since many grades contain inert diluents, concentration of compound may be as low as 50% by weight. Many are water solutions; **Storage temperature:** Ambient; **Inert atmosphere:** None; **Venting:** None; **Stability during transport:** Stable.

PHYSICAL AND CHEMICAL PROPERTIES
Physical state @ 59°F/15°C and 1 atm: Solid or Liquid.
Molecular weight: 445.63
Melting/Freezing point: (solid) 311°F/155°C/428°K.
Specific gravity (water = 1): 1.1 @ 68°F/20°C (solid or liquid).

DIOXANE   **REC. D:7100**

SYNONYMS: DIETHYLENE DIOXIDE; 1,4-DIETHYLENEDIOXIDE; DIETHYLENE ETHER; DI(ETHYLENE OXIDE); DIOKAN; 1,4-DIOXACYCLOHEXANE; DIOXAN; DIOXAN-1,4 (German); 1,4-DIOXANE; DIOXANE-1,4; 1,4-DIOXANE; DIOXANNE (French); *p*-DIOXAN, TETRAHYDRO-; DIOXYETHYLENE ETHER; EEC No. 603-024-00-5; GLYCOL ETHYLENE ETHER; TETRAHYDRO-*p*-DIOXIN; TETRAHYDRO-1,4-DIOXIN; *p*-DIOXANE; DIETHYLENE-1,4-DIOXIDE; RCRA No. U108

IDENTIFICATION
CAS Number: 123-91-1
Formula: $C_4H_8O_2$; $CH_2CH_2OCH_2CH_2O$
DOT ID Number: UN 1165; DOT Guide Number: 127
Proper Shipping Name: Dioxane
Reportable Quantity (RQ): **(CERCLA)** 100 lb/45.4 kg

DESCRIPTION: Colorless liquid. Faint, ether-like odor. Sinks in water; soluble. Flammable, irritating vapor is produced. Freezes at 53°F/12°C.

Extremely flammable • Able to form unstable peroxides; polymerization may occur; danger of fire and explosions • Containers may BLEVE when exposed to fire •Vapors may form explosive mixture with air • Vapors are heavier than air and will collect and stay in low areas • Vapors may travel long distances to ignition sources and flashback • Vapors in confined areas (e.g., tanks, sewers, buildings) may explode when exposed to fire • Irritating to the skin, eyes, and respiratory tract • Toxic products of combustion may include carbon monoxide and vapors that may produce a narcotic effects.

Hazard Classification (based on NFPA-704 Rating System)
Health Hazards (Blue): 2; Flammability (Red): 3; Reactivity (Yellow): 1

EMERGENCY RESPONSE: See Appendix A (127)
Evacuation:
Public safety: Isolate spill area for at least 25 to 50 meters (80 to 160 feet) in all directions.
Spill: Large spill–Consider initial downwind evacuation for at least 300 meters (1000 feet).
Fire: Isolate for 800 meters (½ mile) in all directions, especially if tank, rail car, or tank truck is involved in fire.

EXPOSURE
Short-term effects: *SEEK MEDICAL ATTENTION.* **Vapor:** Irritating to eyes, nose, and throat.Harmful if inhaled. Move to fresh air. *IF BREATHING HAS STOPPED,* give artificial respiration. If breathing is difficult, administer oxygen. **Liquid:** Irritating to skin and eyes. Drops in the lungs may cause pneumonia.Harmful if swallowed. Remove contaminated clothing and shoes. Flush affected areas with plenty of water. *IF IN EYES*, hold eyelids open and flush with plenty of water. *IF SWALLOWED* and victim is *CONSCIOUS AND ABLE TO SWALLOW*, have victim drink 4 to 8 ounces of water, have victim induce vomiting. *IF SWALLOWED* and victim is *UNCONSCIOUS OR HAVING CONVULSIONS*, do nothing except keep victim warm. Chemical pneumonitis may develop if droplets enter the lungs. The use of alcoholic beverages may enhance the toxic effect.

HEALTH HAZARDS
Personal protective equipment (PPE): B-Level PPE. OSHA Table Z-1-A air contaminant. Fresh air mask; rubber gloves; goggles; safety shower and eye bath. Chemical protective material(s) reported to have good to excellent resistance: butyl rubber, PV alcohol, Teflon®, Silvershield®. Also, Chlorinated polyethylene (Chlorinated polyethylene), and natural rubber offers limited protection. For 1,3-dioxane butyl rubber only is recommended.
Recommendations for respirator selection: NIOSH
At any concentrations above the NIOSH REL, or where there is no

REL, at any detectable concentration: SCBAF:PD,PP (any self-contained breathing apparatus that has a full facepiece and is operated in a pressure-demand or other positive-pressure mode); or SAF:PD,PP:ASCBA (any supplied-air respirator that has a full facepiece and is operated in a pressure-demand or other positive-pressure mode in combination with an auxiliary self-contained breathing apparatus operated in a pressure-demand or other positive pressure mode). *ESCAPE:* GMFOV [any air-purifying, full-facepiece respirator (gas mask) with a chin-style, front-or back-mounted organic vapor canister]; or SCBAE (any appropriate escape-type, self-contained breathing apparatus).
Exposure limits (TWA unless otherwise noted): ACGIH TLV 20 ppm; OSHA PEL 100 mg/m^3 (360 mg/m^3); NIOSH REL ceiling 1 ppm (3.6 mg/m^3) [30 minutes]; potential human carcinogen; reduce exposure to lowest feasible level; skin contact
Toxicity by ingestion: Grade 2; LD$_{50}$ = 0.5 to 5 g/kg (guinea pig: 3.90 g/kg).
Long-term health effects: May cause nervous system, liver and kidney damage. Causes cancer in rats. IARC possible carcinogen, rating 2B; human evidence inadequate; animal evidence sufficient.
Vapor (gas) irritant characteristics: Vapors cause a slight smarting of the eyes or respiratory system if present in high concentration. The effect is temporary.
Liquid or solid irritant characteristics: Minimum hazard. If spilled on clothing and allowed to remain, may cause smarting and reddening of the skin.
Odor threshold: 0.82–175 ppm.
IDLH value: Potential human carcinogen; 500 ppm.

FIRE DATA
Flash point: 74°F/oc); 55°F/13°C (cc).
Flammable limits in air: LEL: 1.97%; UEL: 22.0%.
Autoignition temperature: 356°F/180°C/453°K.
Electrical hazard: Class I, Group C. Flow or agitation of substance may generate electrostatic charges due to low conductivity.

CHEMICAL REACTIVITY
Binary reactants: Anhydrous form forms explosive peroxides in air. May self-ignite on contact with air. Reacts violently with strong oxidizers, strong acids; decarborane; trithynyl aluminum. Attacks plastic. Copper and copper alloys may cause slight discoloration.
Polymerization: May occur slowly; **Inhibitor of polymerization:** To prevent peroxide formation use hydroquinone as an inhibitor.
Reactivity group: 41
Compatibility class: Ethers

ENVIRONMENTAL DATA
Food chain concentration potential: Log P$_{ow}$ = –0.33. Unlikely to accumulate.
Water pollution: Effect of low concentrations on aquatic life is unknown. May be dangerous if it enters nearby water intakes; notify operators. Notify local health and wildlife officials.
Response to discharge: Issue warning–high flammability. Disperse and flush.

SHIPPING INFORMATION
Storage temperature: Ambient; **Inert atmosphere:** None; **Venting:** Open (flame arrester) or pressure-vacuum; **Stability during transport:** Stable.

NAS HAZARD CLASSIFICATION FOR BULK WATER TRANSPORTATION
FIRE: 3
HEALTH: Vapor irritant: 1; Liquid or solid irritant: 1; Poisons: 3
WATER POLLUTION: Human toxicity: 1; Aquatic toxicity: 2; Aesthetic effect: 2
REACTIVITY: Other chemicals: 1; Water: 0; Self-reaction: 0

PHYSICAL AND CHEMICAL PROPERTIES
Physical state @ 59°F/15°C and 1 atm: Liquid.
Molecular weight: 88.11
Boiling point @ 1 atm: 214°F/101°C/375°K.
Melting/Freezing point: 53°F/12°C/285°K.
Critical temperature: 597°F/314°C/587°K.
Critical pressure: 755 psia = 51.4 atm = 5.21 MN/m^2.
Specific gravity (water = 1): 1.036 @ 68°F/20°C (liquid).
Relative vapor density (air = 1): 2.98
Ratio of specific heats of vapor (gas): 1.1
Latent heat of vaporization: 178 Btu/lb = 98.6 cal/g = 4.13 x 10^5 J/kg.
Heat of combustion: –11,590 Btu/lb = –6440 cal/g = –269.6 x 10^5 J/kg.
Heat of solution: (estimate) –9 Btu/lb = –5 cal/g = –0.2 x 10^5 J/kg.
Heat of fusion: 34.85 cal/g.
Vapor pressure: (Reid) 1.4 psia; 29 mm; 29 mmHg @ 20°C.

DIPENTENE   REC. D:7150

SYNONYMS: ACINTENE DP; ACINTENE DP DIPENTENE; CAJEPUTENE; CINENE; CYCLOHEXENE, 1-METHYL-4-(1-METHYLETHENYL)-; DIPANOL; DI-*p*-MENTHA-1,8-DIENE; DIPENTENO (Spanish); EEC No. 601-029-00-7; INACTIVE LIMONENE; KAUTSCHIN; LIMONENE; dl-LIMONENE; *p*-MENTHA-1,8-DIENE,DL-; 1,8(9)-*p*-MENTHADIENE; 1-METHYL-4-ISOPROPENYL-1-CYCLOHEXENE; NESOL; δ-1,8-TERPODIENE; *p*-MENTHA-1, 8-DIENE; TERPINENE; PHELLANDRENE

IDENTIFICATION
CAS Number: 138-86-3; may also apply to 5989-27-5 (d-); 5989-54-8
Formula: C$_{10}$H$_{16}$
DOT ID Number: UN 2052; DOT Guide Number: 128
Proper Shipping Name: Dipentene

DESCRIPTION: Colorless to pale yellow liquid. Pleasant lemon-like or pine-like odor. Floats on the surface of water.

Flammable • Containers may BLEVE when exposed to fire • Vapors may form explosive mixture with air • Vapors are heavier than air and will collect and stay in low areas • Vapors may travel long distances to ignition sources and flashback • Vapors in confined areas (e.g., tanks, sewers, buildings) may explode when exposed to fire • Irritating to the skin, eyes, and respiratory tract • Toxic products of combustion may include carbon monoxide.

Hazard Classification (based on NFPA-704 Rating System)
Health Hazards (Blue): 0; Flammability (Red): 2; Reactivity (Yellow): 0

EMERGENCY RESPONSE: See Appendix A (128)
Evacuation:
Public safety: Isolate spill area for at least 25 to 50 meters (80 to 160 feet) in all directions.
Spill: Large spill–Consider initial downwind evacuation for at least 300 meters (1000 feet).

Fire: Isolate for 800 meters (½ mile) in all directions, especially if tank, rail car, or tank truck is involved in fire.

EXPOSURE
Short-term effects: *SEEK MEDICAL ATTENTION.* **Vapor:** Irritating to eyes, nose, and throat. Move victim to fresh air. If breathing is difficult, administer oxygen. **Liquid:** Irritating to skin and eyes. Harmful if swallowed. Remove contaminated clothing and shoes. Flush affected areas with plenty of water. *IF IN EYES,* hold eyelids open and flush with plenty of water. *IF SWALLOWED* and victim is *CONSCIOUS AND ABLE TO SWALLOW,* have victim drink 4 to 8 ounces of water.

HEALTH HAZARDS
Personal protective equipment (PPE): B-Level PPE. Organic vapor respirator. Chemical protective material(s) reported to offer minimal to poor protection: Viton®/neoprene. Chemical protective material(s) reported to offer limited protection (d-isomer): butyl rubber, neoprene, nitrile, PV alcohol.
Exposure limits (TWA unless otherwise noted): 30 ppm (AIHAWEEL).
Toxicity by ingestion: Grade 2; oral LD_{LO} = 4600 mg/kg (rat).
Vapor (gas) irritant characteristics: Vapors cause smarting of the eyes or respiratory system if present in high concentrations. The effect may be temporary.
Liquid or solid irritant characteristics: Eye and skin irritant. If spilled on clothing and allowed to remain, may cause smarting and reddening of the skin.

FIRE DATA
Flash point: 113°F/45°C (cc).
Flammable limits in air: LEL; 0.7%; UEL: 6.1% (both @ 302°F/150°C.
Fire extinguishing agents not to be used: Water may be ineffective.
Behavior in fire: Containers may explode.
Autoignition temperature: 458°F/237°C/510°C.
Burning rate: 5.5 mm/min

ENVIRONMENTAL DATA
Food chain concentration potential: Negative; unlikely to accumulate.
Water pollution: DOT Appendix B, §172.101–marine pollutant. Effect of low concentrations on aquatic life is unknown. Fouling to shoreline. May be dangerous if it enters nearby water intakes; notify operators. Notify local health and wildlife officials.
Response to discharge: Mechanical containment. Should be removed. Chemical and physical treatment.

SHIPPING INFORMATION
Grades of purity: Several technical grades, all having same general properties; **Storage temperature:** Ambient; **Inert atmosphere:** None; **Venting:** Open (flame arrester); **Stability during transport:** Stable.

NAS HAZARD CLASSIFICATIONS FOR BULK WATER TRANSPORTATION
FIRE: 2
HEALTH: Vapor irritant: 1; Liquid or solid irritant: 1; Poisons: 0
WATER POLLUTION: Human toxicity: 1; Aquatic toxicity: 1; Aesthetic effect: 2
REACTIVITY: Other chemicals: 1; Water: 0; Self-reaction: 1

PHYSICAL AND CHEMICAL PROPERTIES
Physical state @ 59°F/15°C and 1 atm: Liquid.
Molecular weight: 136.2
Boiling point @ 1 atm: 352°F/178°C/451°K.
Specific gravity (water = 1): 0.842 at 21°C (liquid).
Liquid surface tension: (estimate) 26 dynes/cm = 0.026 N/m @ 68°F/20°C.
Liquid water interfacial tension: 27.45 dynes/cm = 0.02745 N/m at 33.5°C.
Relative vapor density (air = 1): 4.9
Latent heat of vaporization: 140 Btu/lb = 77 cal/g = 3.2×10^5 J/kg.
Heat of combustion: $-19,520$ Btu/lb = $-10,840$ cal/g = -454×10^5 J/kg.
Vapor pressure: 1.5 mm.

DIPHENYL   REC. D:7200

SYNONYMS: BIPHENYL; BIBENZENE; 1,1'-BIPHENYL; CAROLID AL; DIFENOL (Spanish); PHENADOR-X; LEMONENE; PHPH; PHENYL BENZENE; TETROSIN LY; XENENE

IDENTIFICATION
CAS Number: 92-52-4
Formula: $C_6H_5C_6NO_5$
DOT ID Number: UN 3077; DOT Guide Number: 171
Proper Shipping Name: Environmentally hazardous substance, solid, n.o.s.
Reportable Quantity (RQ): **(CERCLA)** 1 lb/0.454 kg

DESCRIPTION: Colorless to pale yellow solid. Characteristic aromatic odor.

Combustible solid • Heat or flame may cause explosion • Poison! • Breathing the vapor, skin or eye contact, or swallowing the material can kill you • Firefighting gear (including SCBA) does not provide adequate protection. If exposure occurs, remove and isolate gear immediately and thoroughly decontaminate personnel • Containers may BLEVE when exposed to fire • Vapors are heavier than air and will collect and stay in low areas • Vapors in confined areas (e.g., tanks, sewers, buildings) may explode when exposed to fire • Irritating to the skin, eyes, and respiratory tract • Toxic products of combustion may include nitrogen oxides • Do not put yourself in danger by entering a contaminated area to rescue a victim.

Hazard Classification (based on NFPA-704 Rating System)
Health Hazards (Blue): 2; Flammability (Red): 1; Reactivity (Yellow): 0

EMERGENCY RESPONSE: See Appendix A (171)
Evacuation:
Public safety: Isolate the area of spill or leak for at least 10 to 25 meters (30 to 80 feet) in all directions.
Spill: Increase, in the downwind direction, as necessary, the distance shown under "Public Safety."
Fire: If any large container is involved in fire, isolate for at least 800 meters (½ mile) in all directions; also, consider initial evacuation for 800 meters (½ mile) in all directions.

EXPOSURE
Short-term effects: *SEEK MEDICAL ATTENTION.* **Vapor, mist or dust:** Irritating to eyes, nose, throat and skin. Flush affected area with plenty of water. *IF IN EYES,* hold eyelids open and flush with plenty of water. **Liquid or solid:** Irritating to skin and eyes, nose,

and throat. Remove contaminated clothing and shoes. Flush affected areas with plenty water. *IF IN EYES*, hold eyelids open and flush with plenty of water. *IF SWALLOWED*, **Do NOT induce vomiting**.

HEALTH HAZARDS
Personal protective equipment (PPE): A-Level PPE. OSHA Table Z-1-A air contaminant. Self-contained breathing apparatus and protective equipment.
Recommendations for respirator selection: NIOSH/OSHA
10 mg/m³: CCROVDM [any chemical cartridge respirator with organic vapor cartridge(s) in combination with a dust and mist filter]; SA (any supplied-air respirator). *25 mg/m³:* SA:CF* (any supplied-air respirator operated in a continuous-flow mode); or PAPROVDM* (any powered, air-purifying respirator with organic vapor cartridge(s) in combination with a dust and mist filter]. *50 mg/m³:* CCRFOVHiE [any chemical cartridge respirator with a full facepiece and organic vapor cartridge(s) in combination with a high-efficiency particulate filter]; or GMFOVHiE [any air-purifying, full-facepiece respirator (gas mask) with a chin-style, front- or back-mounted organic vapor canister having a high-efficiency particulate filter]; or PAPRTOVHiE [any powered, air-purifying respirator with a tight-fitting facepiece and organic vapor cartridge(s) in combination with a high-efficiency particulate filter]; or SCBAF (any self-contained breathing apparatus with a full facepiece); or SAF (any supplied-air respirator with a full facepiece). *100 mg/m³:* SAF:PD,PP (any supplied-air respirator that has a full facepiece and is operated in a pressure-demand or other positive-pressure mode). *EMERGENCY OR PLANNED ENTRY INTO UNKNOWN CONCENTRATIONS OR IDLH CONDITIONS:* SCBAF:PD,PP (any self-contained breathing apparatus that has a full facepiece and is operated in a pressure-demand or other positive-pressure mode); or SAF:PD,PP:ASCBA (any supplied-air respirator that has a full facepiece and is operated in a pressure-demand or other positive-pressure mode in combination with an auxiliary self-contained breathing apparatus operated in a pressure-demand or other positive pressure mode). *ESCAPE:* GMFOVHiE [any air-purifying, full-facepiece respirator (gas mask) with a chin-style, front- or back-mounted organic vapor canister having a high-efficiency particulate filter]; or SCBAE (any appropriate escape-type, self-contained breathing apparatus).
Note: Substance reported to cause eye irritation or damage; may require eye protection.
Exposure limits (TWA unless otherwise noted): ACGIH TLV 0.2 ppm (1.3 mg/m³); NIOSH/OSHA 1 mg/m³ (0.2 ppm)
Toxicity by ingestion: Grade 2: LD_{50} = 1.9 g/kg (mouse).
Long-term health effects: May cause nervous system disturbance and damage to liver.
Vapor (gas) irritant characteristics: VAPORS cause smarting of the eyes or respiratory system if present in high concentrations. The effect is temporary.
Liquid or solid irritant characteristics: Eye, skin and respiratory tract irritant. Can be absorbed through the skin. If spilled on clothing and allowed to remain, may cause smarting and reddening of skin.
Odor threshold: 0.0095 ppm
IDLH value: 100 mg/m³.

FIRE DATA
Flash point: 235°F/113°C (cc).
Flammable limits in air: LEL: 0.6% @ 232°F/111°C; UEL: 5.8% @ 311°F/155°C.
Autoignition temperature: 1004°F/540°C/813°K.

CHEMICAL REACTIVITY
Binary reactants: Strong oxidizers.
Reactivity group: 32
Compatibility class: Aromatic hydrocarbons

ENVIRONMENTAL DATA
Water pollution: Harmful to aquatic life in very low concentrations. Fouling to shoreline. May be dangerous if it enters nearby water intakes; notify operators. Notify local health and wildlife officials. **Response to discharge:** Restrict access. Mechanical containment. Should be removed.

SHIPPING INFORMATION
Grades of purity: 99%; **Storage temperature:** Ambient; **Stability during transport:** Stable.

PHYSICAL AND CHEMICAL PROPERTIES
Physical state @ 59°F/15°C and 1 atm: Solid.
Molecular weight: 154.21
Boiling point @ 1 atm: 489°F/254°C/527°K.
Melting/Freezing point: 156°F/69°C/342°K.
Specific gravity (water = 1): 1.04
Relative vapor density (air = 1): 5.31
Vapor pressure: (Reid) Less than 0.001 psia; 0.005 mm.

DIPHENYLAMINE REC. D:7250

SYNONYMS: ANILINE, *N*-PHENYL; ANILINOBENZENE; BENZENE, ANILINO-; BIG DIPPER; C.I. 10355; DIFENILAMINA (Spanish); DFA; *N,N*-DIPHENYLAMINE; DPA; NO SCALD; PHENYLANILINE; *N*-PHENYLANILINE; *N*-PHENYLBENZENAMINE; SCALDIP

IDENTIFICATION
CAS Number: 122-39-4
Formula: $(C_6NO_5)_2NH$

DESCRIPTION: Light tan, amber, or brown solid. Pleasant floral odor. Sinks in water; slightly soluble.

Combustible solid • Heat or flame may cause explosion • Poison! • Breathing the vapor, skin or eye contact can cause severe irritation; swallowing the material can kill you • Firefighting gear (including SCBA) does not provide adequate protection. If exposure occurs, remove and isolate gear immediately and thoroughly decontaminate personnel • Containers may BLEVE when exposed to fire • Concentrated dust in confined areas (e.g., tanks, sewers, buildings) may explode when exposed to fire • Irritating to the skin, eyes, and respiratory tract • Toxic products of combustion may include carbon monoxide and nitrogen oxides • Do not put yourself in danger by entering a contaminated area to rescue a victim.

Hazard Classification (based on NFPA-704 Rating System)
Health Hazards (Blue): 3; Flammability (Red): 1; Reactivity (Yellow): 0

EMERGENCY RESPONSE: See Appendix A (171)
Evacuation:
Public safety: Isolate the area of spill or leak for at least 10 to 25 meters (30 to 80 feet) in all directions.
Spill: Increase, in the downwind direction, as necessary, the distance shown under "Public Safety."

Fire: If any large container is involved in fire, isolate for at least 800 meters (½ mile) in all directions; also, consider initial evacuation for 800 meters (½ mile) in all directions.

EXPOSURE
Short-term effects: *SEEK MEDICAL ATTENTION*. **Dust:** Irritating to eyes, nose, and throat. Harmful if inhaled. *IF IN EYES*, hold eyelids open and flush with plenty of water. *IF BREATHING HAS STOPPED*, give artificial respiration. If breathing is difficult, administer oxygen. **Liquid or solid:** Irritating to skin and eyes. Harmful if swallowed. Remove contaminated clothing and shoes. Flush affected areas with plenty of water. *IF IN EYES*, hold eyelids open and flush with plenty of water. *IF SWALLOWED* and victim is *CONSCIOUS AND ABLE TO SWALLOW*, have victim drink 4 to 8 ounces of water. *IF SWALLOWED* and victim is *UNCONSCIOUS OR HAVING CONVULSIONS*, do nothing except keep victim warm.

HEALTH HAZARDS
Personal protective equipment (PPE): B-Level PPE. OSHA Table Z-1-A air contaminant. Respirator; safety goggles or face shield; rubber gloves and clothing.
Exposure limits (TWA unless otherwise noted): ACGIH TLV 10 mg/m^3; NIOSH REL 10 mg/m^3; *Note:* The carcinogen, 4-aminodiphenyl, may be present as an impurity in the commercial grade.
Toxicity by ingestion: Grade 2; oral LD_{50} = 2000 mg/kg (rat).
Long-term health effects: Causes birth defects in rats (polycytic kidneys). Symptoms include anorexia, eczema, cyanosis, headache, methemoglobinemia, nausea, and vomiting, weight loss and emaciation, tachycardia, hypertension, bladder trouble, skin irritation, diarrhea, and general debility.
Liquid or solid irritant characteristics: Eye and skin irritant. May have the same effects as aniline, but is considered less toxic.
Odor threshold: 0.024 ppm.

FIRE DATA
Flash point: (liquid) 307°F/153°C (oc).
Fire extinguishing agents not to be used: Water or foam may cause frothing.
Autoignition temperature: 1175°F/635°C 908°K.

CHEMICAL REACTIVITY
Binary reactants: Oxidizers, hexachloromelamine, trichloromelamine

ENVIRONMENTAL DATA
Food chain concentration potential: Negative; unlikely to accumulate.
Water pollution: Effect of low concentrations on aquatic life is unknown. May be dangerous if it enters nearby water intakes; notify operators. Notify local health and wildlife officials.
Response to discharge: Issue warning–water contaminant. Should be removed. Chemical and physical treatment.

SHIPPING INFORMATION
Grades of purity: Technical; sometimes shipped as liquid;
Storage temperature: Ambient for solid, elevated for Liquid;
Venting: Open; **Stability during transport:** Stable.

PHYSICAL AND CHEMICAL PROPERTIES
Physical state @ 59°F/15°C and 1 atm: Solid.
Molecular weight: 169.2
Boiling point @ 1 atm: 576°F/302°C/575°K.
Melting/Freezing point: 127°F/53°C/326°K.
Specific gravity (water = 1): 1.16 at 68°F/liquid).
Liquid surface tension: 39.3 dynes/cm = 0.0393 N/m at 60°C.
Heat of combustion: –16,300 Btu/lb = –9060 cal/g = –379 x 10^5 J/kg.
Heat of fusion: 25.23 cal/g.
Vapor pressure: 1mm @ 227°F

DIPHENYLDICHLOROSILANE REC. D:7300

SYNONYMS: DICHLORODIPHENYLSILANE; SILANE, DICHLORODIPHENYL-; DICHLORODIPHENYLSILICANE; DIFENOL DICLOROSILANO (Spanish); DIPHENYLSILICON DICHLORIDE

IDENTIFICATION
CAS Number: 80-10-4
Formula: $(C_6NO_5)_2SiCl_2$
DOT ID Number: UN 1769; DOT Guide Number: 156
Proper Shipping Name: Diphenyldichlorosilane

DESCRIPTION: Colorless liquid. Sharp, irritating odor; like hydrochloric acid. Reacts with water; producing corrosive hydrogen chloride.

Combustible • Heat or flame may cause explosion • Containers may BLEVE when exposed to fire • Vapors are heavier than air and will collect and stay in low areas • Vapors in confined areas (e.g., tanks, sewers, buildings) may explode when exposed to fire • Irritating to the skin, eyes, and respiratory tract • Toxic products of combustion may include hydrogen chloride vapors.

Hazard Classification (based on NFPA-704 Rating System)
Health Hazards (Blue): 3; Flammability (Red): 1; Reactivity (Yellow): 0

EMERGENCY RESPONSE: See Appendix A (156)
Evacuation:
Public safety: Isolate the area of spill or leak for at least 50 to 100 meters (160 to 330 feet) in all directions.
Spill: Increase, in the downwind direction, as necessary, the distance shown under "Public Safety."
Fire: If tank, rail car, or tank truck is involved in fire, isolate for at least 800 meters (½ mile) in all directions; also, consider initial evacuation for 800 meters (½ mile) in all directions.

EXPOSURE
Short-term effects: *SEEK MEDICAL ATTENTION*. **Vapor:** Irritating to eyes, nose, and throat. Move victim to fresh air. *IF BREATHING HAS STOPPED*, give artificial respiration; *avoid mouth-to-mouth resuscitation; use bag/mask apparatus*. **Liquid:** Will burn skin and eyes. Harmful if swallowed. Remove contaminated clothing and shoes. Flush affected areas with plenty of water. *IF IN EYES*, hold eyelids open and flush with plenty of water. *IF SWALLOWED* and victim is *CONSCIOUS AND ABLE TO SWALLOW*, have victim drink 4 to 8 ounces of water. **Do NOT induce vomiting.**

HEALTH HAZARDS
Personal protective equipment (PPE): B-Level PPE. Acid-vapor-type respiratory protection; rubber gloves; chemical worker's goggles; other protective equipment as necessary to protect skin and eyes.
Toxicity by ingestion: Grade 3; LD_{50} = 50 to 500 mg/kg.

Vapor (gas) irritant characteristics: Respiratory tract and eye irritant. Vapors cause irritation such that personnel will find high concentrations unpleasant. The effect is temporary.
Liquid or solid irritant characteristics: Severe skin irritant. Causes second-and-third-degree burns on short contact and is very injurious to the eyes.

FIRE DATA
Flash point: 288°F/142°C (oc).
Fire extinguishing agents not to be used: Water and foam
Behavior in fire: Difficult to extinguish; re-ignition may occur.
Burning rate: 2.7 mm/min

CHEMICAL REACTIVITY
Reactivity with water: Reacts with water to generate hydrogen chloride (hydrochloric acid).
Binary reactants: Reacts with surface moisture to generate hydrogen chloride. Corrodes most common metals in the presence of moisture.
Neutralizing agents for acids and caustics: Flood with water, rinse with sodium bicarbonate or lime solution.

ENVIRONMENTAL DATA
Food chain concentration potential: Negative; unlikely to accumulate.
Water pollution: Effect of low concentrations on aquatic life is unknown. May be dangerous if it enters nearby water intakes; notify operators. Notify local health and wildlife officials.
Response to discharge: Issue warning–corrosive, water contaminant. Restrict access. Disperse and flush with care.

SHIPPING INFORMATION
Grades of purity: 96+%; **Storage temperature:** Ambient; **Inert atmosphere:** None; **Venting:** Pressure-vacuum; **Stability during transport:** Stable.

NAS HAZARD CLASSIFICATION FOR BULK WATER TRANSPORTATION
FIRE: 1
HEALTH: Vapor irritant: 2; Liquid or solid irritant: 4; Poisons: 3
WATER POLLUTION: Human toxicity: 3; Aquatic toxicity: 3; Aesthetic effect: 3
REACTIVITY: Other chemicals: 3; Water: 4; Self-reaction: 1

PHYSICAL AND CHEMICAL PROPERTIES
Physical state @ 59°F/15°C and 1 atm: Liquid.
Molecular weight: 253
Boiling point @ 1 atm: 579°F/304°C/577°K.
Melting/Freezing point: –8°F/–22°C/251°K.
Specific gravity (water = 1): 1.22 @ 77°F/25°C (liquid).
Liquid surface tension: (estimate) 26 dynes/cm = 0.026 N/m @ 68°F/20°C.
Latent heat of vaporization: 106 Btu/lb = 59 cal/g 2.5×10^5 J/kg.
Heat of combustion: (estimate) –11,000 Btu/lb = –6200 cal/g = -260×10^5 J/kg.

DIPHENYL ETHER
REC. D:7350

SYNONYMS: BENZENE, 1,1'-OXYBIS-; BIPHENYL ETHER; BIPHENYL OXIDE; DIPHENYL OXIDE; DOWTHERM®; ETER DIFENILICO (Spanish); GERANIUM CRYSTALS; 1,1-OXYBISBENZENE; PHENOXY BENZENE; PHENYL ETHER; PHENYL OXIDE

IDENTIFICATION
CAS Number: 101-84-8
Formula: $C_6H_5OC_6NO_5$

DESCRIPTION: Colorless liquid (above 82°F/28°C), or crystalline solid. Mild pleasant floral odor, like geranium. May float or sink in water.

Combustible • Heat or flame may cause explosion • Containers may BLEVE when exposed to fire • Vapors are heavier than air and will collect and stay in low areas • Vapors in confined areas (e.g., tanks, sewers, buildings) may explode when exposed to fire • Irritating to the skin, eyes, and respiratory tract • Toxic products of combustion may include irritating vapors.

Hazard Classification (based on NFPA-704 Rating System)
Health Hazards (Blue): 1; Flammability (Red): 1; Reactivity (Yellow): 0

EMERGENCY RESPONSE: See Appendix A (171)
Evacuation:
Public safety: Isolate the area of spill or leak for at least 10 to 25 meters (30 to 80 feet) in all directions.
Spill: Increase, in the downwind direction, as necessary, the distance shown under "Public Safety."
Fire: If any large container is involved in fire, isolate for at least 800 meters (½ mile) in all directions; also, consider initial evacuation for 800 meters (½ mile) in all directions.

EXPOSURE
Short-term effects: *SEEK MEDICAL ATTENTION*. **Liquid or solid:** Irritating to skin and eyes. Harmful if swallowed. Remove contaminated clothing and shoes. Flush affected areas with plenty of water. *IF IN EYES*, hold eyelids open and flush with plenty of water. *IF SWALLOWED* and victim is *CONSCIOUS AND ABLE TO SWALLOW*, have victim drink 4 to 8 ounces of water.

HEALTH HAZARDS
Personal protective equipment (PPE): B-Level PPE. OSHA Table Z-1-A air contaminant. Goggles or face shield; rubber gloves.
Recommendations for respirator selection: NIOSH/OSHA
25 ppm: SA:CF* (any supplied-air respirator operated in a continuous-flow mode); PAPROVDM* (any powered, air-purifying respirator with organic vapor cartridge(s) in combination with a dust and mist filter]. *50 ppm:* CCRFOVHiE [any chemical cartridge respirator with a full facepiece and organic vapor cartridge(s) in combination with a high-efficiency particulate filter]; or GMFOVHiE [any air-purifying, full-facepiece respirator (gas mask) with a chin-style, front- or back-mounted organic vapor canister having a high-efficiency particulate filter]; or SCBAF (any self-contained breathing apparatus with a full facepiece); or SAF (any supplied-air respirator with a full facepiece). *100 ppm:* SAF:PD,PP (any supplied-air respirator that has a full facepiece and is operated in a pressure-demand or other positive-pressure mode). *EMERGENCY OR PLANNED ENTRY INTO UNKNOWN CONCENTRATIONS OR IDLH CONDITIONS:* SCBAF:PD,PP (any self-contained breathing apparatus that has a full facepiece and is operated in a pressure-demand or other positive-pressure mode); or SAF:PD,PP:ASCBA (any supplied-air respirator that has a full facepiece and is operated in a pressure-demand or other positive-pressure mode in combination with an auxiliary self-contained breathing apparatus operated in a pressure-demand or other positive pressure mode). *ESCAPE:* GMFOVHiE [any air-purifying, full-facepiece respirator (gas mask) with a chin-style, front- or back-mounted organic vapor canister having a high-efficiency particulate

filter]; or SCBAE (any appropriate escape-type, self-contained).
Note: Substance causes eye irritation or damage; eye protection needed.
Exposure limits (TWA unless otherwise noted): ACGIH TLV 1 ppm (7 mg/m^3); NIOSH/OSHA 1 ppm (7 mg/m^3) as phenyl ether.
Short-term exposure limits (15-minute TWA): 2 ppm (14 mg/m^3).
Toxicity by ingestion: Grade 2; oral LD_{50} = 3370 mg/kg (rat).
Vapor (gas) irritant characteristics: May cause nausea.
Liquid or solid irritant characteristics: Eye, skin and respiratory tract irritant.
Odor threshold: 0.001–0.012 ppm.
IDLH value: 100 ppm.

FIRE DATA
Flash point: 239°F/115°C (cc).
Flammable limits in air: LEL: 0.7%; UEL: 6.0%.
Fire extinguishing agents not to be used: Water or foam may cause frothing.
Autoignition temperature: 1148°F/617°C/890°K.
Electrical hazard: Flow or agitation of substance may generate electrostatic charges due to low conductivity.
Burning rate: 3.2 mm/min

CHEMICAL REACTIVITY
Binary reactants: Strong oxydizers. May form unstable peroxides.

ENVIRONMENTAL DATA
Food chain concentration potential: Negative; unlikely to accumulate.
Water pollution: DOT Appendix B, §172.101–marine pollutant. Effect of low concentrations on aquatic life is unknown. Fouling to shoreline. May be dangerous if it enters nearby water intakes; notify operators. Notify local health and wildlife officials.
Response to discharge: Mechanical containment. Should be removed. Chemical and physical treatment.

SHIPPING INFORMATION
Grades of purity: Pure grade; Technical grade; Perfume grade; **Storage temperature:** Ambient; **Inert atmosphere:** None; **Venting:** Pressure-vacuum; **Stability during transport:** Stable.

PHYSICAL AND CHEMICAL PROPERTIES
Physical state @ 59°F/15°C and 1 atm: Liquid.
Molecular weight: 170.2
Boiling point @ 1 atm: 496°F/258°C/531°K.
Melting/Freezing point: 82°F/28°C/301°K.
Critical temperature: 921°F/494°C/767°K.
Critical pressure: 478 psia = 32.5 atm = 3.30 MN/m^2.
Specific gravity (water = 1): 1.08 at 68°F/liquid).
Liquid surface tension: 40.05 dynes/cm = 0.0401 N/m @ 68°F/20°C.
Liquid water interfacial tension: (estimate) 36 dynes/cm = 0.036 N/m @ 68°F/20°C.
Latent heat of vaporization: 130 Btu/lb = 72 cal/g = 3.0 x 10^5 J/kg.
Heat of combustion: –15,520 Btu/lb = –8620 cal/g = –361 x 10^5 J/kg.
Vapor pressure: (Reid) Low; 0.062 mm; 0.02 mm (77°F/25°C).

DIPHENYLMETHANE DIISOCYANATE   **REC. D:7400**

SYNONYMS: CARADATE 30; CARWINATE 125 M; DESMODUR-44; 4,4'-DIISOCIANAT de DIFENILMETANO (Spanish); 4,4'-DIPHENYLMETHANE DIISOCYANATE; DIPHENYLMETHANE-4,4'-DIISOCYANATE; EEC No. 615-005-01-6; HYLENE M50; ISONATE; MBI; METHYLENE BIS(PHENYLISOCYANATE); METHYLENE BISPHENYL ISOCYANATE; METHYLENE BIS(4-PHENYL ISOCYANATE); 4,4-METHYLENEDIPHENYL DIISOCYANATE; METHYLENE DI-*p*-PHENYLENE ESTER OF ISOCYANIC ACID; METILENBIS(FENILISOCIANATO) (Spanish); MULTRATHANE M; NACCONATE 300; RUBINATE; VILRATHANE 4300

IDENTIFICATION
CAS Number: 101-68-8
Formula: $C_{15}H_{10}N_2O_2$; (p-OCNC$_6$H$_4$)$_2$CH$_2$
DOT ID Number: UN 2489; DOT Guide Number: 156
Proper Shipping Name: Diphenylmethane-4,4'-diisocyanate
Reportable Quantity (RQ): **(CERCLA)** 1 lb/0.454 kg

DESCRIPTION: White to light yellow flakes. Odorless. Sinks in water; slightly soluble. A liquid above 99°F/37°C.

Combustible • Heat or flame may cause explosion • Containers may BLEVE when exposed to fire • Vapors are heavier than air and will collect and stay in low areas • Vapors in confined areas (e.g., tanks, sewers, buildings) may explode when exposed to fire • Irritating to the skin, eyes, and respiratory tract • Toxic products of combustion may include nitrogen oxides and sulfur oxides.

Hazard Classification (based on NFPA-704 Rating System)
Health Hazards (Blue): 2; Flammability (Red): 1; Reactivity (Yellow): 1

EMERGENCY RESPONSE: See Appendix A (156)
Evacuation:
Public safety: Isolate the area of spill or leak for at least 10 to 25 meters (30 to 80 feet) in all directions.
Spill: Increase, in the downwind direction, as necessary, the distance shown under "Public Safety."
Fire: If any large container is involved in fire, isolate for at least 800 meters (½ mile) in all directions; also, consider initial evacuation for 800 meters (½ mile) in all directions.

EXPOSURE
Short-term effects: *SEEK MEDICAL ATTENTION. IF BREATHING HAS STOPPED,* give artificial respiration; *avoid mouth-to-mouth resuscitation; use bag/mask apparatus.* **Solid:** Irritating to skin and eyes. Flush affected areas with plenty of water. *IF IN EYES,* hold eyelids open and flush with plenty of water.

HEALTH HAZARDS
Personal protective equipment (PPE): A-Level PPE. OSHA Table Z-1-A air contaminant. Approved mask or respirator; clean rubber gloves.
Recommendations for respirator selection: NIOSH
0.5 mg/m^3: SA* (any supplied-air respirator); or SCBA (any self-contained breathing apparatus). *1.25 mg/m^3:* SA:CF* (any supplied-air respirator operated in a continuous-flow mode). *2.5 mg/m^3:* SCBAF (any self-contained breathing apparatus with a full facepiece); or SAF (any supplied-air respirator with a full facepiece). *75 mg/m^3:* SAF:PD,PP (any supplied-air respirator that has a full facepiece and is operated in a pressure-demand or other positive-pressure mode). *EMERGENCY OR PLANNED ENTRY INTO UNKNOWN CONCENTRATIONS OR IDLH CONDITIONS:* SCBAF:PD,PP (any self-contained breathing apparatus that has a full facepiece and is operated in a pressure-demand or other

positive-pressure mode); or SAF:PD,PP:ASCBA (any supplied-air respirator that has a full facepiece and is operated in a pressure-demand or other positive-pressure mode in combination with an auxiliary self-contained breathing apparatus operated in a pressure-demand or other positive pressure mode). *ESCAPE:* GMFOVHiE [any air-purifying, full-facepiece respirator (gas mask) with a chin-style, front- or back-mounted organic vapor canister having a high-efficiency particulate filter]; or SCBAE (any appropriate escape-type, self-contained breathing apparatus). *Note*: Substance reported to cause eye irritation or damage; may require eye protection.

Exposure limits (TWA unless otherwise noted): ACGIH TLV 0.005 ppm (0.051 mg/m^3) as methylene bisphenyl isocyanate; OSHA PEL ceiling 0.2 mg/m^3 (0.02 ppm); NIOSH 0.05 mg/m^3 (0.005 ppm); as methylene bisphenyl isocyanate.

Short-term exposure limits (15-minute TWA): NIOSH ceiling 0.2 mg/m^3 (0.020 ppm) [10 min].

Long-term health effects: IARC rating 3; no human or animal data.

Vapor (gas) irritant characteristics: Severe irritation of eyes and throat; can cause eye and lung injury. Cannot be tolerated even at low concentrations.

Liquid or solid irritant characteristics: Eye, skin and respiratory tract irritant. If spilled on clothing and allowed to remain, may cause reddening of the skin.

IDLH value: 75 mg/m^3.

FIRE DATA
Flash point: 425°F/218°C (oc); 390°F/199°C (cc).

CHEMICAL REACTIVITY
Reactivity with water: Slow, nonhazardous. Forms CO_2 gas.
Binary reactants: Violent reaction with strong alkalis, acids, amines, and alcohols.
Polymerization: May occur slowly. May not be hazardous.
Reactivity group: 12
Compatibility class: Isocyanates

ENVIRONMENTAL DATA
Food chain concentration potential: Negative; unlikely to accumulate.
Water pollution: Effect of low concentrations on aquatic life is unknown. May be dangerous if it enters nearby water intakes; notify operators. Notify local health and wildlife officials.
Response to discharge: Issue warning–water contaminant. Should be removed. Chemical and physical treatment.

SHIPPING INFORMATION
Grades of purity: Solid grades: 91–99%; liquid grades may contain 50% *o*-dichlorobenzene; **Storage temperature:** 0 to 40°F/–18 to 40°C; **Venting:** Pressure-vacuum; **Stability during transport:** Stable.

NAS HAZARD CLASSIFICATION FOR BULK WATER TRANSPORTATION
FIRE: 1
HEALTH: Vapor irritant: 3; Liquid or solid irritant: 2; Poisons: 4
WATER POLLUTION: Human toxicity: 2; Aquatic toxicity: 1; Aesthetic effect: 2
REACTIVITY: Other chemicals: 1; Water: 3; Self-reaction: 1

PHYSICAL AND CHEMICAL PROPERTIES
Physical state @ 59°F/15°C and 1 atm: Solid.
Molecular weight: 250.3
Boiling point @ 1 atm: 597°F/313°C/586°K.
Melting/Freezing point: 99°F/37°C/310°K.
Vapor pressure: (Reid) Very low; 0.00014 mm.

DI-*n*-PROPYLAMINE REC. D:7450

SYNONYMS: DI-*n*-DIPROPILAMINA (Spanish); DIPROPYLAMINE; *n*-PROPYLAMINE; *n*-PROPYL-1-PROPANAMINE; RCRA No. U110

IDENTIFICATION
CAS Number: 142-84-7
Formula: $C_{10}H_{21}O_4S\cdot Na$; $(CH_3CH_2CH_2)_2NH$
DOT ID Number: UN 2383; DOT Guide Number: 132
Proper Shipping Name: Dipropylamine
Reportable Quantity (RQ): **(CERCLA)** 5000 lb/2270 kg

DESCRIPTION: Colorless liquid. Fish-like odor at low concentrations; ammonia-like odor at high concentrations. Floats on the surface of water; soluble.

Extremely flammable • Severely irritating to skin, eyes, and respiratory tract; can cause burns • Firefighting gear (including SCBA) does not provide adequate protection. If exposure occurs, remove and isolate gear immediately and thoroughly decontaminate personnel • Containers may BLEVE when exposed to fire •Vapors may form explosive mixture with air • Vapors are heavier than air and will collect and stay in low areas • Vapors may travel long distances to ignition sources and flashback • Vapors in confined areas (e.g., tanks, sewers, buildings) may explode when exposed to fire • Toxic products of combustion may include oxides of carbon monoxide, nitrogen, sulfur, and sodium.

Hazard Classification (based on NFPA-704 Rating System)
Health Hazards (Blue): 3; Flammability (Red): 3; Reactivity (Yellow): 0

EMERGENCY RESPONSE: See Appendix A (132)
Evacuation:
Public safety: Isolate spill area for at least 50 to 100 meters (160 to 330 feet) in all directions.
Spill: Increase, as necessary, the isolation distance shown above,in "Public safety."
Fire: Isolate for 800 meters (½ mile) in all directions, especially if tank, rail car, or tank truck is involved in fire.

EXPOSURE
Short-term effects: *SEEK MEDICAL ATTENTION*. **Vapor:** Irritating to eyes, nose, and throat. *IF INHALED*, will, will cause headache, dizziness, coughing, or difficult breathing. *IF IN EYES*, hold eyelids open and flush with plenty of water. *IF BREATHING HAS STOPPED*, give artificial respiration. If breathing is difficult, administer oxygen. **Liquid:** Will burn eyes. *IF SWALLOWED*, will cause nausea and vomiting. Remove contaminated clothing and shoes. Flush affected areas with plenty of water. *IF IN EYES*, hold eyelids open and flush with plenty of water. *IF SWALLOWED* and victim is *CONSCIOUS AND ABLE TO SWALLOW*, have victim drink 4 to 8 ounces of water. *IF SWALLOWED* and victim is *UNCONSCIOUS OR HAVING CONVULSIONS,* do nothing except keep victim warm.

HEALTH HAZARDS
Personal protective equipment (PPE): B-Level PPE. Self-contained breathing apparatus; butyl rubber gloves; butyl rubber

442   Dipropylene glycol

apron; face shield. Chemical protective material(s) reported to have good to excellent resistance: Teflon®, polycarbonate. Also, Viton® offers limited protection.
Toxicity by ingestion: Grade 3; oral rat LD_{50} = 200 mg/kg (rat), 800 mg/kg (mouse).
Long-term health effects: Causes degenerative changes in liver and kidneys of rats and rabbits
Liquid or solid irritant characteristics: Severe eye and skin irritant.

FIRE DATA
Flash point: 63°F/17°C (oc).
Fire extinguishing agents not to be used: Water may be ineffective.
Autoignition temperature: 572°F/300°C/573°K.
Electrical hazard: Class I, Group C.
Burning rate: 6.1 mm/min

CHEMICAL REACTIVITY
Binary reactants: Dissolves paint and most plastics; swells rubber. Attacks copper, zinc, brass, bronze, aluminum, magnesium.
Reactivity group: 7
Compatibility class: Aliphatic amines

ENVIRONMENTAL DATA
Food chain concentration potential: Log P_{ow} = 1.67–1.70. Unlikely to accumulate.
Water pollution: Effect of low concentrations on aquatic life is unknown. Fouling to shoreline. May be dangerous if it enters nearby water intakes; notify operators. Notify local health and wildlife officials. **Response to discharge:** Issue warning–water contaminant, air contaminant, high flammability. Restrict access. Disperse and flush.

SHIPPING INFORMATION
Grades of purity: Technical, 98%; Pure, 99+%; **Storage temperature:** Ambient; **Inert atmosphere:** None; **Venting:** Open (flame arrester); **Stability during transport:** Stable.

PHYSICAL AND CHEMICAL PROPERTIES
Physical state @ 59°F/15°C and 1 atm: Liquid.
Molecular weight: 101.19
Boiling point @ 1 atm: 229°F/109°C/383°K.
Melting/Freezing point: –81°F/–63°C/210°K.
Critical temperature: 531°F/277°C/550°K.
Critical pressure: 456 psia = 31.0 atm = 3.14 MN/m^2.
Specific gravity (water = 1): 0.738 @ 68°F/20°C (liquid).
Liquid surface tension: 6.58 dynes/cm = 0.00658 N/m @ 68°F/20°C.
Relative vapor density (air = 1): 3.5
Latent heat of vaporization: 143 Btu/lb = 79.5 cal/g = 3.33 x 10^5 J/kg.
Heat of combustion: –18,750 Btu/lb = –10,420 cal/g = –436.0 x 10^5 J/kg.

DIPROPYLENE GLYCOL   **REC. D:7500**

SYNONYMS: BIS(2HYDROXYPROPYL) ETHER; 2,2'-DIHYDROXYDIPROPYL ETHER; 1,1'-DIMETHYLDIETHYLENE GLYCOL; DIPROPILENGLICOL (Spanish); 1,1'-OXYDI-2-PROPANOL

IDENTIFICATION
CAS Number: 110-98-5
Formula: $C_6H_{14}O_3$; $(CH_3CHOHCH_2)_2O$

DESCRIPTION: Colorless, thick liquid. Odorless. Highly soluble in water.
Combustible • Heat or flame may cause explosion • Containers may BLEVE when exposed to fire • Vapors are heavier than air and will collect and stay in low areas • Vapors in confined areas (e.g., tanks, sewers, buildings) may explode when exposed to fire • Irritating to the skin, eyes, and respiratory tract • Toxic products of combustion may include carbon monoxide.

Hazard Classification (based on NFPA-704 Rating System)
Health Hazards (Blue): 0; Flammability (Red): 1; Reactivity (Yellow): 0

EMERGENCY RESPONSE: See Appendix A (171)
Evacuation:
Public safety: Isolate the area of spill or leak for at least 10 to 25 meters (30 to 80 feet) in all directions.
Spill: Increase, in the downwind direction, as necessary, the distance shown under "Public Safety."
Fire: If any large container is involved in fire, isolate for at least 800 meters (½ mile) in all directions; also, consider initial evacuation for 800 meters (½ mile) in all directions.

EXPOSURE
Short-term effects: *SEEK MEDICAL ATTENTION*. **Liquid:** Irritating to eyes. *IF IN EYES*, hold eyelids open and flush with plenty of water.

HEALTH HAZARDS
Personal protective equipment (PPE): B-Level PPE. Safety glasses with side shields or goggles; shower and eye bath.
Skin or Ingestion: If any ill effects develop, *GET MEDICAL ATTENTION*.
Toxicity by ingestion: Grade 1; LD_{50} = 5 to 15 g/kg (rat).
Vapor (gas) irritant characteristics: Eye and respiratory tract irritant.
Liquid or solid irritant characteristics: Eye and skin irritant.

FIRE DATA
Flash point: 250°F/121°C (oc).
Flammable limits in air: LEL: 2.9%; UEL: 12.7
Fire extinguishing agents not to be used: Water or foam may cause frothing.
Autoignition temperature: 594°F/312°C/585°K.
Electrical hazard: Class I, Group C.
Burning rate: 2.0 mm/min

CHEMICAL REACTIVITY
Binary reactants: Incompatible with sulfuric acid, perchloric acid, isocyanates, strong oxidizers.
Reactivity group: 40
Compatibility class: Glycol ethers

ENVIRONMENTAL DATA
Food chain concentration potential: Log P_{ow} = –1.21 (estimate). Negative; unlikely to accumulate.
Water pollution: Effect of low concentrations on aquatic life is unknown. May be dangerous if it enters nearby water intakes; notify operators. Notify local health and wildlife officials. **Response to discharge:** Disperse and flush.

SHIPPING INFORMATION
Grades of purity: Commercial: 99%; **Storage temperature:** Ambient; **Inert atmosphere:** None; **Venting:** Open (flame arrester); **Stability during transport:** Stable.

NAS HAZARD CLASSIFICATION FOR BULK WATER TRANSPORTATION
FIRE: 1
HEALTH: Vapor irritant: 0; Liquid or solid irritant: 0; Poisons: 1
WATER POLLUTION: Human toxicity: 0; Aquatic toxicity: 1; Aesthetic effect: 0
REACTIVITY: Other chemicals: 2; Water: 0; Self-reaction: 0

PHYSICAL AND CHEMICAL PROPERTIES
Physical state @ 59°F/15°C and 1 atm: Liquid.
Molecular weight: 134.17
Viscosity: 107 cP @ 68°F/20°C (liquid).
Boiling point @ 1 atm: 420°F/232°C/505°K.
Melting/Freezing point: More than −40°F/−40°C/233°K.
Critical temperature: 720°F/382°C/655°K.
Critical pressure: 529 psia = 36 atm = 3.6 MN/m^2.
Specific gravity (water = 1): 1.023 @ 68°F/20°C (liquid).
Relative vapor density (air = 1): 4.61
Ratio of specific heats of vapor (gas): 1.0
Latent heat of vaporization: 170 Btu/lb = 96 cal/g = 4.0 x 10^5 J/kg.
Heat of combustion: −11,650 Btu/lb = −6470 cal/g = −271 x 10^5 J/kg.
Heat of solution: (estimate) −13 Btu/lb = −7 cal/g = −0.3 x 10^5 J/kg.
Vapor pressure: (Reid) Very low; 0.008 mm.

DIPROPYLENE GLYCOL DIBENZOATE   REC. D:7550

SYNONYMS: BENZOFLEX 9-88; BENZOFLEX 9-98; BENZOFLEX 9-88 SG; DIBENSOATO de DIPROPILENGLICOL (Spanish); DIBENZOL DIPROPYLENE GLYCOL ESTER; DIPROPANEDIOL DIBENZOATE; K-FLEX DP

IDENTIFICATION
CAS Number: 94-51-9
Formula: $C_{20}H_{22}O_5$; $[(C_6NO_5)CO_2CH)(CH_3)CH_2]_2O$

DESCRIPTION: Slight yellow thick liquid. Faint aromatic odor.

Combustible • Heat or flame may cause explosion • Containers may BLEVE when exposed to fire • Vapors are heavier than air and will collect and stay in low areas • Vapors in confined areas (e.g., tanks, sewers, buildings) may explode when exposed to fire • Irritating to the skin, eyes, and respiratory tract • Toxic products of combustion may include carbon monoxide.

Hazard Classification (based on NFPA-704 Rating System)
Health Hazards (Blue): 0; Flammability (Red): 1; Reactivity (Yellow): 0

EMERGENCY RESPONSE: See Appendix A (171)
Evacuation:
Public safety: Isolate the area of spill or leak for at least 10 to 25 meters (30 to 80 feet) in all directions.
Spill: Increase, in the downwind direction, as necessary, the distance shown under "Public Safety."
Fire: If any large container is involved in fire, isolate for at least 800 meters (½ mile) in all directions; also, consider initial evacuation for 800 meters (½ mile) in all directions.

EXPOSURE
Short-term effects: *SEEK MEDICAL ATTENTION*. **Vapor, mist and liquid:** May irritate eyes, nose, and throat. *IF INHALED*, will, remove victim to fresh air. *IF BREATHING HAS STOPPED*, give artificial respiration. If breathing is difficult, administer oxygen. Flush affected areas with plenty of water. *IF IN EYES*, hold eyelids open and flush with plenty of water. *IF SWALLOWED* and victim is *CONSCIOUS AND ABLE TO SWALLOW*, have victim drink 2 glasses of water and induce vomiting.

HEALTH HAZARDS
Personal protective equipment (PPE): B-Level PPE. Self-contained breathing apparatus, rubber boots, rubber gloves, and rubber apron. If spill is small, a full facepiece air purifying cartridge respirator equipped with organic vapor cartridge may be satisfactory.
Toxicity by ingestion: Grade 2: LD_{50} = 9.80 g/kg (rat).
Long-term health effects: Defatting of skin
Vapor (gas) irritant characteristics: Vapors or mists cause smarting of the eyes or respiratory system if present in high concentrations. The effect may be temporary.
Liquid or solid irritant characteristics: If spilled on clothing and allowed to remain, may cause smarting and reddening of skin.

FIRE DATA
Flash point: More than 300°F/149°C (cc).

CHEMICAL REACTIVITY
Binary reactants: Oxidizers, strong acids, nitrates.
Reactivity group: 34
Compatibility class: Esters.

ENVIRONMENTAL DATA
Water pollution: Effects of low concentration on aquatic life is not known. May be dangerous if it enters nearby water intakes; notify operators. Notify local health and wildlife officials. **Response to discharge:** Restrict access. Mechanical containment. Chemical and physical treatment.

SHIPPING INFORMATION
Grades of purity: More than 99%; **Stability during transport:** Stable.

PHYSICAL AND CHEMICAL PROPERTIES
Physical state @ 59°F/15°C and 1 atm: Liquid.
Molecular weight: 342.42
Viscosity: 227 cP @ 68°F/20°C (liquid).
Boiling point @ 1 atm: 446°F/230°C/503.2°K.
Melting/Freezing point: −22°F/−30°C/243.2°K.
Specific gravity (water = 1): 1.13
Bulk density: 9.4 lb/gal @ 68°F/20°C (liquid).
Relative vapor density (air = 1): 11.8

DIPROPYLENE GLYCOL METHYL ETHER REC. D:7600

SYNONYMS: ARCOSOLV; DIPROPILENGLICOL MONOMETIL ETER (Spanish); DIPROPYLENE GLYCOL MONOMETHYL ETHER; DOWANOL-50B; DOWANOL DPM; PROPANOL, OXYBIS-, METHYL ETHER; UCAR SOLVENT 2LM

IDENTIFICATION
CAS Number: 34590-94-8
Formula: $C_7H_{16}O_3$; $CH_3OC_3H_6OC_3H_6OH$

DESCRIPTION: Colorless liquid. Weak odor. Miscible with water.

Combustible • Heat or flame may cause explosion • Containers may BLEVE when exposed to fire • Vapors are heavier than air and will collect and stay in low areas • Vapors in confined areas (e.g., tanks, sewers, buildings) may explode when exposed to fire • Irritating to the skin, eyes, and respiratory tract • Toxic products of combustion may include carbon monoxide.

Hazard Classification (based on NFPA-704 Rating System)
Health Hazards (Blue): 0; Flammability (Red): 2; Reactivity (Yellow): 0

Combustible • Heat or flame may cause explosion • Containers may BLEVE when exposed to fire • Vapors are heavier than air and will collect and stay in low areas • Vapors in confined areas (e.g., tanks, sewers, buildings) may explode when exposed to fire • Irritating to the skin, eyes, and respiratory tract • Toxic products of combustion may include carbon monoxide.

EXPOSURE
Short-term effects: *SEEK MEDICAL ATTENTION.* **Liquid:** *HARMFUL IF SWALLOWED OR ABSORBED THROUGH THE SKIN.* Irritating to skin and eyes. Harmful if swallowed. Remove contaminated clothing and shoes. Flush affected areas with plenty of water. *IF IN EYES*, hold eyelids open and flush with plenty of water.

HEALTH HAZARDS
Personal protective equipment (PPE): B-Level PPE. Organic vapor respirator. Wear rubber boots and heavy rubber gloves. OSHA Table Z-1-A air contaminant.
Recommendations for respirator selection: NIOSH/OSHA *600 ppm:* SA (any supplied-air respirator); or SCBAF (any self-contained breathing apparatus with a full facepiece). *EMERGENCY OR PLANNED ENTRY INTO UNKNOWN CONCENTRATIONS OR IDLH CONDITIONS:* SCBAF:PD,PP (any self-contained breathing apparatus that has a full facepiece and is operated in a pressure-demand or other positive-pressure mode); or SAF:PD,PP:ASCBA (any supplied-air respirator that has a full facepiece and is operated in a pressure-demand or other positive-pressure mode in combination with an auxiliary self-contained breathing apparatus operated in a pressure-demand or other positive pressure mode). *ESCAPE:* GMFOVHiE [any air-purifying, full-facepiece respirator (gas mask) with a chin-style, front- or back-mounted organic vapor canister having a high-efficiency particulate filter]; or SCBAE (any appropriate escape-type, self-contained breathing apparatus).
Exposure limits (TWA unless otherwise noted): NIOSH/OSHA 100 ppm (600 mg/m^3); skin contact contributes significantly in overall exposure.
Short-term exposure limits (15-minute TWA): NIOSH STEL 150 ppm (900 mg/m^3); skin contact contributes significantly in overall exposure.
Toxicity by ingestion: Grade 1: LD_{50} = 5.135 g/kg (rat).
Vapor (gas) irritant characteristics: Vapors cause smarting of the eyes or respiratory system if present in high concentrations. The effect may be temporary.
Liquid or solid irritant characteristics: If spilled on clothing and allowed to remain, may cause smarting and reddening of skin.

Odor threshold: 1000 ppm.
IDLH value: 600 ppm.

FIRE DATA
Flash point: 180°F/84°C (cc).

CHEMICAL REACTIVITY
Binary reactants: Strong oxidizers, sulfuric acids, isocyanates
Reactivity group: 40
Compatibility class: Glycol ethers

ENVIRONMENTAL DATA
Water pollution: Effects of low concentrations on aquatic life is unknown. May be dangerous if it enters nearby water intakes; notify operators. Notify local health and wildlife officials. **Response to discharge:** Evacuate areas. Should be removed. Chemical and physical treatment.

SHIPPING INFORMATION
Grades of purity: 97%; **Storage temperature:** Ambient; **Stability during transport:** Stable.

PHYSICAL AND CHEMICAL PROPERTIES
Physical state @ 59°F/15°C and 1 atm: Liquid.
Molecular weight: 148.2
Boiling point @ 1 atm: 408°F/209°C/482°K.
Specific gravity (water = 1): 0.951
Liquid surface tension: 28.8 dynes/cm = 0.029 N/m
Relative vapor density (air = 1): 5.11
Vapor pressure: 0.5 mm.

DIQUAT   REC. D:7650

SYNONYMS: AQUACIDE; DEIQUAT; DEXTRONE; 9,10-DIHYDRO-8*a*,10,-DIAZONIAPHENANTHRENE DIBROMIDE; 9,10-DIHYDRO-8*a*,10*a*,DIAZONIAPHENANTHRENE(1,1'-ETHYLENE-2,2'-BIPYRIDYLIUM)DIBROMIDE; 5,6-DIHYDRO-DIPYRIDO(1,2*a*:2,1*c*)PYRAZINIUM DIBROMIDE; 6,7-DIHYDROPYRIDO(1,2-*a*:2',1'-*c*)PYRAZINEDIUM DIBROMIDE; DIQUAT DIBROMIDE; 1,1'-ETHYLENE-2,2'-BIPYRIDYLIUMDIBROMIDE; ETHYLENE DIPYRIDYLIUM DIBROMIDE; 1,1-ETHYLENE 2,2-DIPYRIDYLIUM DIBROMIDE; 1,1'-ETHYLENE-2,2'-DIPYRIDYLIUMDIBROMIDE; FB/2; FEGLOX; PREEGLONE; REGLON; REGLONE; REGLOX; WEEDTRINE-D; REGALON; 6,7-DIHYDROPYRIDO[1,2*a*: 2',1'-c]PYRAZINEDINIUM ION

IDENTIFICATION
CAS Number: 2768-72-9; 2764-72-9; 85-00-7 (*o*-isomer)
Formula: $C_{12}H_{12}Br_2N_2$; $BrC_{12}H_{12}N_2 \cdot 2Br$
DOT ID Number: UN 2781 (solid); 2782 (liquid); DOT Guide Number: 151 (solid); 131 (liquid)
Proper Shipping Name: Bipyridilium pesticides, solid, toxic; Bipyridilium pesticides, liquid, flammable, toxic
Reportable Quantity (RQ): **(CERCLA)** 1000 lb/454 kg

DESCRIPTION: Yellow (pure salt monohydrate) solid or dark reddish-brown liquid. Sinks and mixes with water.

Poison! (bipyridylium) • Breathing the vapor, skin or eye contact can cause illness; swallowing the material can kill you • Firefighting gear (including SCBA) does not provide adequate protection. If exposure occurs, remove and isolate gear immediately and thoroughly decontaminate personnel • Ruptured or venting cylinders may rocket through buildings and/or travel a considerable distance •Toxic products of combustion may include carbon monoxide, nitrogen oxides, and hydrogen bromide.

Hazard Classification (based on NFPA-704 Rating System) (solid) Health Hazards (Blue): 2; Flammability (Red): 0; Reactivity (Yellow): 0

EMERGENCY RESPONSE, solid: See Appendix A (151)
Evacuation:
Public safety: Isolate the area of spill or leak for at least 25 to 50 meters (80 to 160 feet) in all directions.
Spill: Increase, in the downwind direction, as necessary, the distance shown under "Public Safety."
Fire: If tank, rail car, or tank truck is involved in fire, isolate for at least 800 meters (½ mile) in all directions; also, consider initial evacuation for 800 meters (½ mile) in all directions.
EMERGENCY RESPONSE, liquid: See Appendix A (131)
Evacuation:
Public safety: Isolate spill area for at least 100 to 200 meters (330 to 660 feet) in all directions.
Spill: Increase, as necessary, the isolation distance shown above, in "Public safety."
Fire: Isolate for 800 meters (½ mile) in all directions, especially if tank, rail car, or tank truck is involved in fire.

EXPOSURE
Short-term effects: *SEEK MEDICAL ATTENTION*. May be fatal if swallowed, inhaled, or absorbed through skin. **Liquid or solid:** *POISONOUS IF INHALED OR SWALLOWED. IF BREATHING HAS STOPPED*, give artificial respiration; *avoid mouth-to-mouth resuscitation; use bag/mask apparatus*. Remove contaminated clothing and shoes. Flush affected areas with plenty of water. *IF IN EYES*, hold eyelids open and flush with plenty of water. *IF SWALLOWED* and victim is *CONSCIOUS AND ABLE TO SWALLOW*, have victim drink 4 to 8 ounces of water and have victim induce vomiting.

HEALTH HAZARDS
Personal protective equipment (PPE): B-Level PPE. OSHA Table Z-1-A air contaminant. Wear face shield, rubber gloves, rubber apron when handling concentrate. When spraying, wear waterproof foot wear and clothing.
Exposure limits (TWA unless otherwise noted): ACGIH TLV 0.5 mg/m^3 (total dust), 0.1 mg/m^3 (respirable fraction); NIOSH 0.5 mg/m^3.
Toxicity by ingestion: Grade 3; LD_{50} = 50 to 500 mg/kg.
Long-term health effects: Possible kidney and liver damage. Prolonged feeding produced cataract in rats and dogs (In rat after 100 weeks at concentration 36 ppm; in dog after 15 months at concentration 150 ppm). 2.5 mg/kg for 24 months (oral-rat) caused no adverse effects.
Vapor (gas) irritant characteristics: Eye and skin irritant.
Liquid or solid irritant characteristics: Eye and skin irritant. If spilled on clothing and allowed to remain, may cause smarting and reddening of skin.

FIRE DATA
Flash point: Difficult to ignite solid.
Fire extinguishing agents not to be used: Water streams applied to adjacent fires will spread contamination of pesticide over wide area.
Behavior in fire: Decomposes at high temperature, charring rather than melting or boiling.

CHEMICAL REACTIVITY
Binary reactants: Concentrated solutions corrode aluminum rapidly. Should not be stored in contact with metals.

ENVIRONMENTAL DATA
Food chain concentration potential: Log K_{ow} = –3.0. Unlikely to accumulate. When present in fish, 50% of the residual diquat lost in less than 3 weeks.
Water pollution: Dangerous to aquatic life in high concentrations. May be dangerous if it enters nearby water intakes; notify operators. Notify local health and wildlife officials. **Response to discharge:** Issue warning–water contaminant. Chemical and physical treatment. Disperse and flush.

SHIPPING INFORMATION
Grades of purity: Technical aqueous solution, 2-lb cation/gal; **Storage temperature:** Ambient. Sensitive to UV light.; **Stability during transport:** Stable in original containers.

PHYSICAL AND CHEMICAL PROPERTIES
Physical state @ 59°F/15°C and 1 atm: Solid.
Molecular weight: 184.2 cation; 344.1 dibromide
at @ 1 atm: Salts decompose at high temperatures (above 572°F/300°C), charring rather than melting or boiling.
Melting/Freezing point: 636°F/336°C/609°K.
Specific gravity (water = 1): 1.22 to 1.27 at 20°C
Vapor pressure: Less than 0.00001 mm.

DISULFOTON   REC. D:7700

SYNONYMS: BAY 19639; BAYER 19639; *O,O*-DIAETHYL-*s*-(3-THIA-PENTYL)-DITHIOPHOSPHAT (German); *O,O*-DIAETHYL-*s*-(2-AETHYLTHIO-AETHYL)-DITHIOPHOSPHAT (German); *O,O*-DIETHYL-*s*-(2-ETHTHIOETHYL)PHOSPHORODITHIOATE; *O,O*-DIETHYL-*s*-(2-ETHTHIOETHYL)THIOTHIONOPHOSPHATE; *O,O*-DIETHYL-*s*-(2-ETHYLMERCAPTOETHYL)DITHIOPHOSPHATE; *O,O*-DIETHYL 2-ETHYLTHIOETHYLPHOSPHORODITHIOATE; *O,O*-DIETHYL *s*-2-(ETHYLTHIO)ETHYLPHOSPHORODITHIOATE; DIMAZ; DISULFATON; DI-SYSTON; DISYSTOX; DITHIODEMETON; DITHIOPHOSPHATE de *O,O*-DIETHYLE ETDE *s*-(2-ETHYLTHIO-ETHYLE) (French); DITHIOSYSTOX; *O,O*-ETHYL S-2(ETHYLTHIO)ETHYL PHOSPHORODITHIOATE; *s*-2-(ETHYLTHIO)ETHYL *O,O*-DIETHYL ESTER of PHOSPHORODITHIOIC ACID; FRUMIN AL; FRUMIN G; M-74; PHOSPHORODITHIONIC ACID, s-2-(ETHYLTHIO)ETHYL-*O,O*-DIETHYL ESTER; S 276; SOLVIREX; THIODEMETON; THIODEMETRON; *O,O*-DIETHYL-*s*-2-(ETHYLTHIO)ETHYL PHOSPHODITHIOATE; RCRA No. P039

IDENTIFICATION
CAS Number: 298-04-4
Formula: $C_8H_{19}O_2PS_3$
DOT ID Number: UN 2783; DOT Guide Number: 152
Proper Shipping Name: Disulfoton
Reportable Quantity (RQ): **(CERCLA)** 1 lb/0.454 kg

DESCRIPTION: Pale yellow, oily liquid (pure). Technical product is a brown liquid. Sulfur odor. Sinks and mixes slowly with water.

Poison! (organophosphate) • Combustible solid • Heat or flame may cause explosion • A tear gas • Breathing the vapor, skin or eye contact, or swallowing the material can kill you; cholinesterase inhibitor. Irritating to the skin, eyes, and respiratory tract • Firefighting gear (including SCBA) does not provide adequate protection. If exposure occurs, remove and isolate gear immediately and thoroughly decontaminate personnel • Containers may BLEVE when exposed to fire • Dust or vapors in confined areas (e.g., tanks, sewers, buildings) may explode when exposed to fire • Toxic products of combustion may include carbon monoxide, sulfur and phosphorus oxides.

Hazard Classification (based on NFPA-704 Rating System)
Health Hazards (Blue): 4; Flammability (Red): 1; Reactivity (Yellow): 0

EMERGENCY RESPONSE: See Appendix A (152)
Evacuation:
Public safety: Isolate the area of spill or leak for at least 25 to 50 meters (80 to 160 feet) in all directions.
Spill: Increase, in the downwind direction, as necessary, the distance shown under "Public Safety."
Fire: If tank, rail car, or tank truck is involved in fire, isolate for at least 800 meters (½ mile) in all directions; also, consider initial evacuation for 800 meters (½ mile) in all directions.

EXPOSURE
Short-term effects: *SEEK MEDICAL ATTENTION.* **Vapor:** *POISONOUS IF INHALED. Move to fresh air. IF BREATHING HAS STOPPED,* give artificial respiration; *avoid mouth-to-mouth resuscitation; use bag/mask apparatus.* If breathing is difficult, administer oxygen. **Liquid:** *POISONOUS IF SWALLOWED OR IF SKIN IS EXPOSED.* Remove contaminated clothing and shoes. Flush affected areas with plenty of water. *IF IN EYES,* hold eyelids open and flush with plenty of water. *IF SWALLOWED* and victim is *CONSCIOUS AND ABLE TO SWALLOW,* have victim drink 4 to 8 ounces of water and induce vomiting. **Inhalation and skin contact:** *SPEED IS ESSENTIAL.* Remove from exposure. Flood and wash exposed skin areas thoroughly with water. Remove contaminated clothing under a shower. In nonbreathing victim, immediately institute artificial respiration.
Note to physician or authorized medical personnel: Administer atropine, 2 mg (1/30 gr) intramuscularly or intravenously as soon as any local or systemic signs or symptoms of an intoxication are noted; repeat the administration of atropine every 3 to 8 minutes until signs of atropinization (mydriasis, dry mouth, rapid pulse, hot and dry skin) occur; initiate treatment in children with 0.05 mg/kg of atropine; repeat at 5- to 10-minute intervals. Watch respiration, and remove bronchial secretions if they appear to be obstructing the airway; intubate if necessary. Pralidoxime must be administered within minutes to a few hours following exposure (depending on the specific agent) to be effective. Give 2-PAMCl (Pralidoxime; Protopam), 2.5 g in 100 mL of sterile water or in 5% dextrose and water, intravenously, slowly, in 15 to 30 minutes; if sufficient fluid is not available, give 1 g of 2-PAMCl in 3 mL of distilled water by deep intramuscular injection; repeat this every half hour if respiration weakens or if muscle fasciculation or convulsions recur. Also Diazepam, an anticonvulsant, might be needed.

HEALTH HAZARDS
Personal protective equipment (PPE): A-Level PPE. OSHA Table Z-1-A air contaminant. Chemical-protective clothing and butyl rubber gloves are recommended when skin contact is possible because liquid is rapidly absorbed through the skin and may cause systemic toxicity.

Exposure limits (TWA unless otherwise noted): ACGIH TLV 0.1 mg/m^3; NIOSH REL 0.1 mg/m^3; skin contact contributes significantly in overall exposure.
Toxicity by ingestion: Grade 4; LD$_{50}$ less than 50 mg/kg.
Long-term health effects: Possible mutagen, positive in bacterial tests. Decrease in cholinesterase activity mainly in erythrocytes and mild abnormalities in liver enzyme activities in dogs.

FIRE DATA
Flash point: More than 180°F/82°C.
Fire extinguishing agents not to be used: Water streams applied to adjacent fires will spread contamination of pesticide over wide area.

CHEMICAL REACTIVITY
Binary reactants: Alkalis.

ENVIRONMENTAL DATA
Water pollution: DOT Appendix B, §172.101–marine pollutant. Harmful to aquatic life in very low concentrations. May be dangerous if it enters nearby water intakes; notify operators. Notify local health and wildlife officials. **Response to discharge:** Issue warning–poison, water contaminant. Restrict access. Should be removed. Chemical and physical treatment.

SHIPPING INFORMATION
Grades of purity: Technical purity–minimum 94%.

PHYSICAL AND CHEMICAL PROPERTIES
Physical state @ 59°F/15°C and 1 atm: Solid.
Molecular weight: 274.42
Boiling point @ 1 atm: (@ 0.01 mmHg) 144°F/62°C/335°K.
Melting/Freezing point: Less than –13°F/–25°C/248°K.
Specific gravity (water = 1): 1.144 @ 68°F/20°C.
Relative vapor density (air = 1): 9.45
Vapor pressure: 0.00002 mm.

DISTILLATES: FLASHED FEED STOCKS   REC. D:7750

SYNONYMS: ALIPHATIC PETROLEUM NAPHTHA; PETROLEUM DISTILLATE; PETROLEUM NAPHTHA

IDENTIFICATION
CAS Number: 8002-05-9
DOT ID Number: UN 1255; DOT Guide Number: 128; DOT Guide Number: 128
Proper Shipping Name: Petroleum naphtha

DESCRIPTION: Colorless liquid. Gasoline-like odor. Floats on the surface of water. Flammable, irritating vapor is produced.

Extremely flammable • Containers may BLEVE when exposed to fire •Vapors may form explosive mixture with air • Vapors are heavier than air and will collect and stay in low areas • Vapors may travel long distances to ignition sources and flashback • Vapors in confined areas (e.g., tanks, sewers, buildings) may explode when exposed to fire • Irritating to the skin, eyes, and respiratory tract • Toxic products of combustion may include irritating vapors.

Hazard Classification (based on NFPA-704 Rating System)
Health Hazards (Blue): 1; Flammability (Red): 4; Reactivity (Yellow): 0

EMERGENCY RESPONSE: See Appendix A (128)
Evacuation:
Public safety: Isolate spill area for at least 25 to 50 meters (80 to 160 feet) in all directions.
Spill: Large spill–Consider initial downwind evacuation for at least 300 meters (1000 feet).
Fire: Isolate for 800 meters (½ mile) in all directions, especially if tank, rail car, or tank truck is involved in fire.

EXPOSURE
Short-term effects: *SEEK MEDICAL ATTENTION.* **Vapor:** Irritating to eyes, nose, and throat. Move to fresh air. *IF BREATHING HAS STOPPED,* give artificial respiration. If breathing is difficult, administer oxygen. **Liquids:** Irritating to eyes. Harmful if swallowed. Remove contaminated clothing and shoes. Flush affected areas with plenty of water. *IF IN EYES*, hold eyelids open and flush with plenty of water. *IF SWALLOWED* and victim is *CONSCIOUS AND ABLE TO SWALLOW*, have victim drink 4 to 8 ounces of water. **Do NOT induce vomiting.**

HEALTH HAZARDS
Personal protective equipment (PPE): B-Level PPE.
Recommendations for respirator selection: NIOSH
850 ppm: SA (any supplied-air respirator). *1100 ppm*: SA:CF (any supplied-air respirator operated in a continuous-flow mode); or SCBAF (any self-contained breathing apparatus with a full facepiece); or SAF (any supplied-air respirator with a full facepiece). *EMERGENCY OR PLANNED ENTRY INTO UNKNOWN CONCENTRATIONS OR IDLH CONDITIONS:* SCBAF:PD,PP (any self-contained breathing apparatus that has a full facepiece and is operated in a pressure-demand or other positive-pressure mode); or SAF:PD,PP:ASCBA (any supplied-air respirator that has a full facepiece and is operated in a pressure-demand or other positive-pressure mode in combination with an auxiliary self-contained breathing apparatus operated in a pressure-demand or other positive pressure mode). *ESCAPE:* GMFOV[any air-purifying, full-facepiece respirator (gas mask) with a chin-style, front- or back-mounted organic vapor canister]; or SCBAE (any appropriate escape-type, self-contained breathing apparatus).
Exposure limits (TWA unless otherwise noted): NIOSH 350 mg/m^3; ceiling 1800 mg/m^3/15 minutes; OSHA PEL 500 ppm (2000 mg/m^3).
Toxicity by ingestion: Grade 2; LD_{50} = 0.5 to 5 g/kg.
Long-term health effects: IARC carcinogen.
Vapor (gas) irritant characteristics: Vapors cause smarting of the eyes or respiratory system if present in high concentrations. The effect is temporary.
Liquid or solid irritant characteristics: If spilled on clothing and allowed to remain, may cause smarting and reddening of the skin.
Odor threshold: 0.25 ppm.
IDLH value: 1100 ppm.

FIRE DATA
Flash point: –40 to –86°F/–40 to –66°C.
Flammable limits in air: LEL: 1.1%; UEL: 5.9%.
Fire extinguishing agents not to be used: Water may be ineffective.
Electrical hazard: Class I, Group D.
Burning rate: (approximately) 4 mm/min

CHEMICAL REACTIVITY
Binary reactants: Strong oxidizers
Reactivity group: 33
Compatibility class: Miscellaneous hydrocarbon mixtures

ENVIRONMENTAL DATA
Food chain concentration potential: Negative; unlikely to accumulate.
Water pollution: Harmful to aquatic life in very low concentrations. Fouling to shoreline. May be dangerous if it enters nearby water intakes; notify operators. Notify local health and wildlife officials. **Response to discharge:** Issue warning–high flammability. Evacuate area. Disperse and flush.

SHIPPING INFORMATION
Grades of purity: Composition varies with range of distillation temperatures used; **Storage temperature:** Ambient; **Inert atmosphere:** None; **Stability during transport:** Stable.

NAS HAZARD CLASSIFICATION FOR BULK WATER TRANSPORTATION
FIRE: 3
HEALTH: Vapor irritant: 1; Liquid or solid irritant: 1; Poisons: 2
WATER POLLUTION: Human toxicity: 1; Aquatic toxicity: 2; Aesthetic effect: 2
REACTIVITY: Other chemicals: 0; Water: 0; Self-reaction: 0

PHYSICAL AND CHEMICAL PROPERTIES
Physical state @ 59°F/15°C and 1 atm: Liquid.
Molecular weight: 99 (approximate).
Boiling point @ 1 atm: 58–275°F/14–135°C/287–408°K.
Melting/Freezing point: –99°F/–73°C/200°K.
Specific gravity (water = 1): 0.71–0.75 @ 15°C (liquid).
Liquid surface tension: 19-23 dynes/cm = 0.019-0.023 N/m @ 68°F/20°C.
Liquid water interfacial tension: 49–51 dynes/cm = 0.049-0.051 N/m @ 68°F/20°C.
Relative vapor density (air = 1): 3.4
Ratio of specific heats of vapor (gas): (estimate) 1.054
Latent heat of vaporization: 130–150 Btu/lb = 71-81 cal/g = 3.0-3.4 x 10^5 J/kg.
Heat of combustion: –18,720 Btu/lb = –10,400 cal/g = 435.4 x 10^5 J/kg.
Vapor pressure: 40 mm (approximate).

DISTILLATES: STRAIGHT RUN   REC. D:7800

SYNONYMS: STRAIGHT RUN GASOLINE; PETROLEUM DISTILLATE; GAS OILS (PETROLEUM), STRAIGHT RUN

IDENTIFICATION
CAS Number: 64741-43-1; 64741-42-0
Formula: $C_6H_4(COOC_{13}H_{27})_2$; $(C_6H_4)(COOC_{11}H_{23})_2$
Proper Shipping Name: Petroleum distillates, n.o.s.

DESCRIPTION: Colorless watery liquid. Gasoline-like odor. Floats on the surface of water. Flammable, irritating vapor is produced.

Highly flammable • Containers may BLEVE when exposed to fire •Vapors may form explosive mixture with air • Vapors are heavier than air and will collect and stay in low areas • Vapors may travel long distances to ignition sources and flashback • Vapors in confined areas (e.g., tanks, sewers, buildings) may explode when exposed to fire • Irritating to the skin, eyes, and respiratory tract • Toxic products of combustion may include carbon monoxide.

Ditridecyl phthalate

Hazard Classification (based on NFPA-704 Rating System)
Health Hazards (Blue): 1; Flammability (Red): 3; Reactivity (Yellow): 0

EMERGENCY RESPONSE: See Appendix A (171)
Evacuation:
Public safety: Isolate spill area for at least 25 to 50 meters (80 to 160 feet) in all directions.
Spill: Large spill–Consider initial downwind evacuation for at least 300 meters (1000 feet).
Fire: Isolate for 800 meters (½ mile) in all directions, especially if tank, rail car, or tank truck is involved in fire.

EXPOSURE
Short-term effects: *SEEK MEDICAL ATTENTION*. **Vapor:** Irritating to eyes, nose, and throat. Move to fresh air. *IF BREATHING HAS STOPPED*, give artificial respiration. If breathing is difficult, administer oxygen. **Liquids:** Irritating to eyes. Harmful if swallowed. Remove contaminated clothing and shoes. Flush affected areas with plenty of water. *IF IN EYES*, hold eyelids open and flush with plenty of water. *IF SWALLOWED* and victim is *CONSCIOUS AND ABLE TO SWALLOW*, have victim drink 4 to 8 ounces of water. **Do NOT induce vomiting.**

HEALTH HAZARDS
Personal protective equipment (PPE):
B-Level PPE. **Recommendations for respirator selection:** NIOSH as petroleum distillates
850 ppm: SA (any supplied-air respirator). *1100 ppm*: SA:CF (any supplied-air respirator operated in a continuous-flow mode); or SCBAF (any self-contained breathing apparatus with a full facepiece); or SAF (any supplied-air respirator with a full facepiece). *EMERGENCY OR PLANNED ENTRY INTO UNKNOWN CONCENTRATIONS OR IDLH CONDITIONS:* SCBAF:PD,PP (any self-contained breathing apparatus that has a full facepiece and is operated in a pressure-demand or other positive-pressure mode); or SAF:PD,PP:ASCBA (any supplied-air respirator that has a full facepiece and is operated in a pressure-demand or other positive-pressure mode in combination with an auxiliary self-contained breathing apparatus operated in a pressure-demand or other positive pressure mode). *ESCAPE:* GMFOV[any air-purifying, full-facepiece respirator (gas mask) with a chin-style, front- or back-mounted organic vapor canister]; or SCBAE (any appropriate escape-type, self-contained breathing apparatus).
Exposure limits (TWA unless otherwise noted): NIOSH 350 mg/m^3; ceiling 1800 mg/m^3/15 min; OSHA PEL 500 ppm (2000 mg/m^3) as petroleum distillates
Toxicity by ingestion: Grade 2; LD$_{50}$ = 0.5–5g/kg.
Vapor (gas) irritant characteristics: Vapors cause smarting of the eyes or respiratory system if present in high concentrations. The effect may be temporary.
Liquid or solid irritant characteristics: If spilled on clothing and allowed to remain, may cause smarting and reddening of the skin.
Odor threshold: 0.25 ppm.
IDLH value: 1100 as petroleum distillates

FIRE DATA
Flash point: (a) less than 0°F/–18°C (cc); (b) 0–73°F/–18–23°C (cc); (c) 73–141°F/23–61°C (cc).
Flammable limits in air: LEL: 1.1%; UEL: 8.7%.
Fire extinguishing agents not to be used: Water may be ineffective
Electrical hazard: Class I, Group D.
Burning rate: (approximately) 4 mm/min

CHEMICAL REACTIVITY
Binary reactants: Oxidizers, nitric acid
Reactivity group: 33
Compatibility class: Miscellaneous hydrocarbon mixtures

ENVIRONMENTAL DATA
Food chain concentration potential: Negative; unlikely to accumulate.
Water pollution: Harmful to aquatic life in very low concentrations. Fouling to shoreline. May be dangerous if it enters nearby water intakes; notify operators. Notify local health and wildlife officials. **Response to discharge:** Issue warning–high flammability. Evacuate area. Disperse and flush.

SHIPPING INFORMATION
Grades of purity: Composition varies with range of distillation temperatures used; **Storage temperature:** Ambient; **Inert atmosphere:** None; **Stability during transport:** Stable.

NAS HAZARD CLASSIFICATION FOR BULK WATER TRANSPORTATION
FIRE: 3
HEALTH: Vapor irritant: 1; Liquid or solid irritant: 1; Poisons: 2
WATER POLLUTION: Human toxicity: 1; Aquatic toxicity: 2; Aesthetic effect: 2
REACTIVITY: Other chemicals: 0; Water: 0; Self-reaction: 0

PHYSICAL AND CHEMICAL PROPERTIES
Physical state @ 59°F/15°C and 1 atm: Liquid.

Boiling point @ 1 atm: 58–275°F/14–135°C/287–408°K.
Specific gravity (water = 1): 0.731 at 16°C (liquid).
Liquid surface tension: 19-23 dynes/cm = 0.019–0.023 N/m @ 68°F/20°C.
Liquid water interfacial tension: 49-51 dynes/cm = 0.049-0.051 N/m @ 68°F/20°C.
Relative vapor density (air = 1): 3.4
Ratio of specific heats of vapor (gas): (estimate) 1.054
Latent heat of vaporization: 130–150 Btu/lb = 71–81 cal/g = 3.0-3.4 x 10^5 J/kg.
Heat of combustion: –18,720 Btu/lb = –10,400 cal/g = –435.4 x 10^5 J/kg.

DITRIDECYL PHTHALATE **REC. D:7850**

SYNONYMS: FTALATO de DITETRADECILO (Spanish); PHTHALIC ACID, DITRIDECYL ESTER; DTDP; NUOPLAZ; JAYFLEX DTDP; POLYCIZER 962-BPA; STAFLEX DTDP; 1-TRIDECANOL, PHTHALATE

IDENTIFICATION
CAS Number: 119-06-2
Formula: $C_{34}NO_{58}O_4$; $C_6H_4(COOC_{13}H_{27})_2$

DESCRIPTION: Colorless oily liquid. Nearly odorless. Floats on the surface of water.

Combustible • Heat or flame may cause explosion • Containers may BLEVE when exposed to fire • Vapors are heavier than air and will collect and stay in low areas • Vapors in confined areas (e.g., tanks, sewers, buildings) may explode when exposed to fire • Irritating to the skin, eyes, and respiratory tract • Toxic products of combustion may include carbon monoxide.

Hazard Classification (based on NFPA-704 Rating System)
Health Hazards (Blue): 0; Flammability (Red): 1; Reactivity (Yellow): 0

EMERGENCY RESPONSE: See Appendix A (171)
Evacuation:
Public safety: Isolate the area of spill or leak for at least 10 to 25 meters (30 to 80 feet) in all directions.
Spill: Increase, in the downwind direction, as necessary, the distance shown under "Public Safety."
Fire: If any large container is involved in fire, isolate for at least 800 meters (½ mile) in all directions; also, consider initial evacuation for 800 meters (½ mile) in all directions.

EXPOSURE
Short-term effects: *CALL FOR MEDICAL AID* **Liquid:** If ingested, induce vomiting. Remove contaminated clothes and shoes. Wash skin with soap and water. *IF IN EYES*, hold eyelids open and flush with plenty of water.

HEALTH HAZARDS
Personal protective equipment (PPE): B-Level PPE. Self-contained breathing apparatus and full turn out gear (Fire resistant helmet with full neck and face covers, self-contained breathing apparatus, flame resistant coat and pants, pair of boots). Chemical protective material(s) reported to offer minimal to poor protection: Viton®/neoprene.
Vapor (gas) irritant characteristics: Fumes may cause irritation of eyes and throat.
Liquid or solid irritant characteristics: May cause eye and skin irritation.

FIRE DATA
Flash point: 470°F/243°C (oc); 490°F/254°C (cc).

CHEMICAL REACTIVITY
Binary reactants: Strong oxidizers
Reactivity group: 34

ENVIRONMENTAL DATA
Water pollution: Effects of low concentration of aquatic life unknown. Fouling to shore line. May be dangerous if it enters nearby water intakes; notify operators. Notify local health and wildlife officials. **Response to discharge:** Issue warning-flammable. Should be removed. Mechanical containment. Chemical and physical treatment.

SHIPPING INFORMATION
Storage temperature: Ambient; **Stability during transport:** Stable.

PHYSICAL AND CHEMICAL PROPERTIES
Physical state @ 59°F/15°C and 1 atm: Liquid.
Molecular weight: 530.8
Boiling point @ 1 atm: 547°F/286°C/559°K.
Melting/Freezing point: Less than –34.6°F/–37°C/236.2°K (pour point).
Specific gravity (water = 1): 0.951
Relative vapor density (air = 1): 18.3

DIUNDECYL PHTHALATE   **REC. D:7900**

SYNONYMS: FTALATO de DITRIDECILO (Spanish); SANTICIZER 711; 1,2-BENZENEDICARBOXYLIC ACID, DI-UNDECYL ESTER; PHTHALIC ACID, DIUNDECYL ESTER

IDENTIFICATION
CAS Number: 3648-20-2
Formula: $C_{30}NO_{50}O_2$

DESCRIPTION: Colorless liquid. Odorless. Floats on the surface of water; slightly soluble.

Combustible • Heat or flame may cause explosion • Containers may BLEVE when exposed to fire • Vapors are heavier than air and will collect and stay in low areas • Vapors in confined areas (e.g., tanks, sewers, buildings) may explode when exposed to fire • Irritating to the skin, eyes, and respiratory tract • Toxic products of combustion may include irritating vapors.

Hazard Classification (based on NFPA-704 Rating System)
Health Hazards (Blue): 0; Flammability (Red): 1; Reactivity (Yellow): 0

EMERGENCY RESPONSE: See Appendix A (171)
Evacuation:
Public safety: Isolate the area of spill or leak for at least 10 to 25 meters (30 to 80 feet) in all directions.
Spill: Increase, in the downwind direction, as necessary, the distance shown under "Public Safety."
Fire: If any large container is involved in fire, isolate for at least 800 meters (½ mile) in all directions; also, consider initial evacuation for 800 meters (½ mile) in all directions.

HEALTH HAZARDS
Personal protective equipment (PPE): Butyl rubber is generally suitable for carbooxylic acid compounds.
Vapor (gas) irritant characteristics: Irritating at high temperatures.

FIRE DATA
Fire extinguishing agents not to be used: Water or foam may cause frothing

CHEMICAL REACTIVITY
Binary reactants: Strong acids, nitrates.
Reactivity group: 34
Compatibility class: Esters.

ENVIRONMENTAL DATA
Water pollution: Effects of low concentrations on aquatic life are not known. Fouling to shoreline. May be dangerous if it enters nearby water intakes; notify operators. Notify local health and wildlife officials. **Response to discharge:** Mechanical containment. Chemical and physical treatment.

SHIPPING INFORMATION
Storage temperature: Ambient; **Stability during transport:** Stable.

NAS HAZARD CLASSIFICATION FOR BULK WATER TRANSPORTATION
FIRE: 1
HEALTH: Vapor irritant: 0; Liquid or solid irritant: 0; Poisons: 0
WATER POLLUTION: Human toxicity:–; Aquatic toxicity:–; Aesthetic effect: 2
REACTIVITY: Other chemicals: 1; Water: 0; Self-reaction: 0

450 Dodecanol

PHYSICAL AND CHEMICAL PROPERTIES
Physical state @ 59°F/15°C and 1 atm: Liquid.
Molecular weight: 442.80
Relative vapor density (air = 1): 15.3

DIURON   REC. D:7950

SYNONYMS: AF 101; CEKIURON; CRISURON; DAILON; DCMU; DIATER; DICHLORFENIDIM; 3-(3,4-DICHLOROPHENOL)-1,1-DIMETHYL UREA; N'-(3,4-DICHLOROPHENYL)-N,N-DIMETHYLUREA; 3-(3,4-DICHLOROPHENYL)-1,1-DIMETHYLUREA; 1-(3,4-DICHLOROPHENYL)-3,3-DIMETHYLUREE (French); 3-(3,4-DICHLOR-PHENYL)-1,1-DIMETHYL-HARNSTOFF (German); 1,1-DIMETHYL-3-(3,4-DICHLOROPHENYL)UREA; DI-ON; DIUREX; DIUROL; DIURON 4L; DMU; DREXEL; DURAN; DYNEX; HERBATOX; HW 920; KARMEX; KARMEX DIURON HERBICIDE; KARMEX DW; MARMER; SUP'R FLO; TELVAR DIURON WEED KILLER; UNIDRON; UROX DUSAF XR-42; VONDURON

IDENTIFICATION
CAS Number: 330-54-1
Formula: $C_9H_{10}Cl_2N_2O$; $(C_6H_3Cl_2)NHCON(CH_3)_2$
DOT ID Number: UN 3077; DOT Guide Number: 171
Proper Shipping Name: Environmentally hazardous substances, solid, n.o.s.
Reportable Quantity (RQ): **(CERCLA)** 100 lb/45.4 kg

DESCRIPTION: White crystalline solid. May be dissolved in a hydrocarbon solvent such as benzene or acetone. Odorless.

Pesticide • Breathing the dust or swallowing the material can cause harm; skin or eye contact causes irritation • Toxic products of combustion may include carbon monoxide, hydrogen chloride, and nitrogen oxides.

Hazard Classification (based on NFPA-704 Rating System)
Health Hazards (Blue): 1; Flammability (Red): 0; Reactivity (Yellow): 0

EMERGENCY RESPONSE: See Appendix A (171)
Evacuation:
Public safety: Isolate the area of spill or leak for at least 10 to 25 meters (30 to 80 feet) in all directions.
Spill: Increase, in the downwind direction, as necessary, the distance shown under "Public Safety."
Fire: If any large container is involved in fire, isolate for at least 800 meters (½ mile) in all directions; also, consider initial evacuation for 800 meters (½ mile) in all directions.

EXPOSURE
Short-term effects: *SEEK MEDICAL ATTENTION.* **Solid:** Irritating to skin, eyes, nose, and throat. Harmful if swallowed. Remove contaminated clothing and shoes. Flush affected areas with plenty of water. *IF IN EYES*, hold eyelids open and flush with plenty of water. *IF SWALLOWED* and victim is *CONSCIOUS AND ABLE TO SWALLOW*, have victim drink 4 to 8 ounces of water.

HEALTH HAZARDS
Personal protective equipment (PPE): B-Level PPE. OSHA Table Z-1-A air contaminant. Self-contained breathing apparatus, rubber gloves, suits and boots. Chemical protective material(s) reported to have good to excellent resistance: butyl rubber, polycarbonate.
Exposure limits (TWA unless otherwise noted): ACGIH TLV 10 mg/m³; NIOSH REL 10 mg/m³.
Toxicity by ingestion: Grade 2; LD_{50} = 0.5 to 5 g/kg.
Long-term health effects: Suspected of affecting DNA (potential mutagen). Repeated doses produce anemia in rats and perhaps methemoglobinemia if the compound is hydrolyzed *in vivo* to dichloroaniline. At 2500 ppm for two years growth was retarded in both rats and dogs.
Vapor (gas) irritant characteristics: May cause respiratory tract irritation.
Liquid or solid irritant characteristics: Mild skin irritant.

FIRE DATA
Fire extinguishing agents not to be used: Water streams applied to adjacent fires will spread contamination of pesticide over wide area.
Behavior in fire: Decomposes above 356–374°F/180–190°C/453–463°K.

CHEMICAL REACTIVITY
Binary reactants: Strong acids.

ENVIRONMENTAL DATA
Food chain concentration potential: Log K_{ow} = 2.8. There is some evidence of bioaccumulation in fish tissues.
Water pollution: DOT Appendix B, §172.101–marine pollutant. Harmful to aquatic life in very low concentrations. May be harmful if it enters water intakes. Notify local health and wildlife officials.
Response to discharge: Issue warning–water contaminant. Chemical and physical treatment.

SHIPPING INFORMATION
Grades of purity: Wettable powder 80%; Granular 8%; **Storage temperature:** Ambient; **Stability during transport:** Stable.

PHYSICAL AND CHEMICAL PROPERTIES
Physical state @ 59°F/15°C and 1 atm: Solid.
Molecular weight: 233.1
Boiling point @ 1 atm: 356-374°F/180-190°C/453-463°K (decomposes).
Melting/Freezing point: 316-318°F/158-159°C/431-432°K.
Relative vapor density (air = 1): 8.04
Vapor pressure: 0.000000002 mm.

DODECANOL   REC. D:8000

SYNONYMS: ALCOHOL C-12; ALFOL 12; CO-12; N-DODECANOL; 1-DODECANOL; DODECYL ALCOHOL; N-DODECYL ALCOHOL; EPAL 12; LAURIC ALCOHOL; LAURYL ALCOHOL; n-LAURYL ALCOHOL; LOROL; MA 1214; SIPOL L12

IDENTIFICATION
CAS Number: 112-53-8
Formula: $C_{12}H_{26}O$; $CH_3(CH_2)_{10}CH_2OH$
DOT ID Number: UN 1986; DOT Guide Number: 131
Proper Shipping Name: Alcohols, n.o.s.

DESCRIPTION: Thick Colorless liquid. Sweet, floral or fatty alcohol odor. Floats on the surface of water. Melting point is 75°F/24°C.

Combustible • Contact with heat or flame may cause explosion • Containers may BLEVE when exposed to fire • Vapors are heavier than air and will collect and stay in low areas • Vapors in confined areas (e.g., tanks, sewers, buildings) may explode when exposed to fire • Irritating to the skin, eyes, and respiratory tract • Toxic products of combustion may include irritating vapors.

Hazard Classification (based on NFPA-704 Rating System)
Health Hazards (Blue): 0; Flammability (Red): 1; Reactivity (Yellow): 0

EMERGENCY RESPONSE: See Appendix A (131)
Evacuation:
Public safety: Isolate spill area for at least 100 to 200 meters (330 to 660 feet) in all directions.
Spill: Increase, as necessary, the isolation distance shown above, in "Public safety."
Fire: Isolate for 800 meters (½ mile) in all directions, especially if tank, rail car, or tank truck is involved in fire.
EXPOSURE
Short-term effects: *SEEK MEDICAL ATTENTION.* **Liquid:** Irritating to skin. Will burn eyes. Flush affected areas with plenty of water. *IF IN EYES*, hold eyelids open and flush with plenty of water.

HEALTH HAZARDS
Personal protective equipment (PPE): A-Level PPE. Chemical gloves; chemical goggles. Chemical protective material(s) reported to offer minimal to poor protection: Viton®/neoprene, butyl rubber,/neoprene. Also, natural rubber, neoprene, and nitrile rubber may offer some protection from alcohols.
Toxicity by ingestion: Grade 1; LD_{50} = 5 5 to 15 g/kg (humans).
Liquid or solid irritant characteristics: Irritating to the eyes and skin.

FIRE DATA
Flash point: 260°F/127°C (cc).
Fire extinguishing agents not to be used: Water or foam may cause frothing
Autoignition temperature: 527°F/275°C/548°K.

CHEMICAL REACTIVITY
Binary reactants: Oxidizing materials, strong acids, caustics, ammonia.
Reactivity group: 20
Compatibility class: Alcohols, glycols

ENVIRONMENTAL DATA
Food chain concentration potential: Negative; unlikely to accumulate.
Water pollution: Effect of low concentrations on aquatic life is unknown. Fouling to shoreline. May be dangerous if it enters nearby water intakes; notify operators. Notify local health and wildlife officials. **Response to discharge:** Disperse and flush

SHIPPING INFORMATION
Grades of purity: 98.5-99.5%; **Storage temperature:** Ambient; **Inert atmosphere:** None; **Venting:** Open (flame arrester); **Stability during transport:** Stable.

PHYSICAL AND CHEMICAL PROPERTIES
Physical state @ 59°F/15°C and 1 atm: Liquid.
Molecular weight: 186.33
Boiling point @ 1 atm: 498°F/259°C/532°K.
Melting/Freezing point: 75°F/24°C/297°K.
Critical temperature: 763°F/406°C/679°K.
Critical pressure: 280 psia = 19 atm = 1.9 MN/m^2.
Specific gravity (water = 1): 0.831 at 24°C (liquid).
Liquid surface tension: 27.4 dynes/cm = 0.0274 N/m @ 77°F/25°C.
Liquid water interfacial tension: (estimate) 30 dynes/cm = 0.03 N/m at 30°C.
Ratio of specific heats of vapor (gas): 1.030
Latent heat of vaporization: (estimate) 110 Btu/lb = 62 cal/g = 2.6 x 10^5 J/kg.
Heat of combustion: (estimate) –18,000 Btu/lb = –10,000 cal/g = –420 x 10^5 J/kg.

DODECENE   REC. D:8050

SYNONYMS: ADACENE-12; AMSCO TETRAMER; 1-DODECENE; DODECENE (NONLINEAR); DODECENO (Spanish); DODECYLENE; DODECYLENE-α; α-DODECYLENE; PROPENE TETRAMER; PROPYLENE TETRAMER; TETRAPROPYLENE

IDENTIFICATION
CAS Number: 6842-15-5
Formula: $C_{12}H_{24}$; $CH_3(CH_2)_9CH=CH_2$
DOT ID Number: UN 2850; DOT Guide Number: 128
Proper Shipping Name: Propylene tetramer

DESCRIPTION: Colorless watery liquid. Pleasant odor. Floats on the surface of water.

Combustible • Containers may BLEVE when exposed to fire • Vapors may form explosive mixture with air • Vapors are heavier than air and will collect and stay in low areas • Vapors may travel long distances to ignition sources and flashback • Vapors in confined areas (e.g., tanks, sewers, buildings) may explode when exposed to fire • Irritating to the skin, eyes, and respiratory tract • Toxic products of combustion may include irritating vapors.

Hazard Classification (based on NFPA-704 Rating System)
Health Hazards (Blue): 0; Flammability (Red): 1; Reactivity (Yellow): 0

EMERGENCY RESPONSE: See Appendix A (128)
Evacuation:
Public safety: Isolate spill area for at least 25 to 50 meters (80 to 160 feet) in all directions.
Spill: Large spill–Consider initial downwind evacuation for at least 300 meters (1000 feet).
Fire: Isolate for 800 meters (½ mile) in all directions, especially if tank, rail car, or tank truck is involved in fire.

EXPOSURE
Short-term effects: *SEEK MEDICAL ATTENTION.* **Liquid:** Irritating to skin and eyes. Harmful if swallowed. Remove contaminated clothing and shoes. Flush affected areas with plenty of water. *IF IN EYES*, hold eyelids open and flush with plenty of water. *IF SWALLOWED* and victim is *CONSCIOUS AND ABLE TO SWALLOW*, have victim drink 4 to 8 ounces of water. **Do NOT induce vomiting.**

HEALTH HAZARDS
Personal protective equipment (PPE): B-Level PPE. Protective gloves; no respiratory protection needed if ventilation is adequate.
Toxicity by ingestion: Grade 0; LD_{50} = > 15 g/kg.

452 Dodecylbenzene

Vapor (gas) irritant characteristics: Slight smarting of eyes and respiratory system at high concentrations. The effect may be temporary.
Liquid or solid irritant characteristics: Mild eye and skin irritant. If spilled on clothing and allowed to remain, may cause smarting and reddening of skin.

FIRE DATA
Flash point: <212°F/<100°C (cc) (NFPA).
Fire extinguishing agents not to be used: Water may be ineffective.
Autoignition temperature: (estimate) 491°F/255°C/528°K.
Burning rate: 5.8 mm/min

CHEMICAL REACTIVITY
Binary reactants: Strong acids.
Reactivity group: 30
Compatibility class: Olefins

ENVIRONMENTAL DATA
Food chain concentration potential: Log P_{ow} = 6.49. Values > 3.0 are likely to bioconcentrate in aquatic organisms and other living tissue, especially in fats.
Water pollution: Effect of low concentrations on aquatic life is unknown. Fouling to shoreline. May be dangerous if it enters nearby water intakes; notify operators. Notify local health and wildlife officials. **Response to discharge:** Mechanical containment. Chemical and physical treatment.

SHIPPING INFORMATION
Grades of purity: 98.5-99+% olefin; **Storage temperature:** Ambient; **Inert atmosphere:** None; **Venting:** Open (flame arrester); **Stability during transport:** Stable.

NAS HAZARD CLASSIFICATION FOR BULK WATER TRANSPORTATION
FIRE: 2
HEALTH: Vapor irritant: 1; Liquid or solid irritant: 1; Poisons: 1
WATER POLLUTION: Human toxicity: 0; Aquatic toxicity: 1; Aesthetic effect: 2
REACTIVITY: Other chemicals: 1; Water: 0; Self-reaction: 1

PHYSICAL AND CHEMICAL PROPERTIES
Physical state @ 59°F/15°C and 1 atm: Liquid.
Molecular weight: 168.36
Boiling point @ 1 atm: 406°F/208°C/481°K.
Melting/Freezing point: –25°F/–32°C/242°K.
Specific gravity (water = 1): 0.77 @ 68°F/20°C (liquid).
Liquid surface tension: 23.0 dynes/cm = 0.0230 N/m @ 68°F/20°C.
Relative vapor density (air = 1): 5.79
Latent heat of vaporization: 110 Btu/lb = 61.0 cal/g = 2.55 x 10^5 J/kg.
Heat of combustion: –19,100 Btu/lb = –10,600 cal/g = –444 x 10^5 J/kg.
Vapor pressure: (Reid) 0.01 psia; 1 mm @ 117°F/47°C/320°K.

DODECYLBENZENE REC. D:8100

SYNONYMS: DODECILBENCENO (Spanish); *n*-DODECYLBENZENE; LAURYLBENZENE; DETERGENT ALKYLATE No. 2; DETERGENT ALKYLATES; *n*-DODECYLBENZENE; PHENYLDODECAN; 1-PHENYLDODECANE; UCANE ALKYLATE 12

IDENTIFICATION
CAS Number: 123-01-3
Formula: $C_{18}H_{30}$; $C_6NO_5(CH_2)_{11}CH_3$

DESCRIPTION: Colorless liquid. Weak, oily odor. Floats on the surface of water.

Combustible • Heat or flame may cause explosion • Containers may BLEVE when exposed to fire • Vapors are heavier than air and will collect and stay in low areas • Vapors in confined areas (e.g., tanks, sewers, buildings) may explode when exposed to fire • Irritating to the skin, eyes, and respiratory tract • Toxic products of combustion may include carbon monoxide.

Hazard Classification (based on NFPA-704 Rating System)
Health Hazards (Blue): 1; Flammability (Red): 1; Reactivity (Yellow): 0

EMERGENCY RESPONSE: See Appendix A (171)
Evacuation:
Public safety: Isolate the area of spill or leak for at least 10 to 25 meters (30 to 80 feet) in all directions.
Spill: Increase, in the downwind direction, as necessary, the distance shown under "Public Safety."
Fire: If any large container is involved in fire, isolate for at least 800 meters (½ mile) in all directions; also, consider initial evacuation for 800 meters (½ mile) in all directions.

EXPOSURE
Short-term effects: *SEEK MEDICAL ATTENTION*. **Liquid:** Irritating to skin and eyes. Harmful if swallowed. Remove contaminated clothing and shoes. Flush affected areas with plenty of water. *IF IN EYES*, hold eyelids open and flush with plenty of water. *IF SWALLOWED* and victim is *CONSCIOUS AND ABLE TO SWALLOW*, have victim drink 4 to 8 ounces of water.

HEALTH HAZARDS
Personal protective equipment (PPE): B-Level PPE. Goggles or face shield; rubber gloves.
Toxicity by ingestion: Grade 1; LD_{50} = 5 to 15 g/kg.
Vapor (gas) irritant characteristics: May cause eye and respiratory tract irritation.
Liquid or solid irritant characteristics: May cause eye and skin irritation.

FIRE DATA
Flash point: 275°F/135°C (oc).
Fire extinguishing agents not to be used: Water may be ineffective.
Burning rate: 3.7 mm/min

CHEMICAL REACTIVITY
Binary reactants: Oxidizers, nitric acid.
Reactivity group: 32
Compatibility class: Aromatic hydrocarbons

ENVIRONMENTAL DATA
Food chain concentration potential: Negative; unlikely to accumulate.
Water pollution: Effect of low concentrations on aquatic life is unknown. Fouling to shoreline. May be dangerous if it enters nearby water intakes; notify operators. Notify local health and wildlife officials. **Response to discharge:** Mechanical containment. Should be removed. Chemical and physical treatment.

SHIPPING INFORMATION
Grades of purity: Various mixtures containing 70–80% of undecyl-plus dodecyl-benzene, along with 10% decylbenzene and other analogous hydrocarbons; **Storage temperature:** Ambient; **Inert atmosphere:** None; **Venting:** Open (flame arrester); **Stability during transport:** Stable.

NAS HAZARD CLASSIFICATION FOR BULK WATER TRANSPORTATION
FIRE: 1
HEALTH: Vapor irritant: 0; Liquid or solid irritant: 0; Poisons: 0
WATER POLLUTION: Human toxicity: 1; Aquatic toxicity: 3; Aesthetic effect: 3
REACTIVITY: Other chemicals: 1; Water: 0; Self-reaction: 0

PHYSICAL AND CHEMICAL PROPERTIES
Physical state @ 59°F/15°C and 1 atm: Liquid.
Molecular weight: 246.4
Boiling point @ 1 atm: 550°F/288°C/561°K.
Specific gravity (water = 1): 0.860 @ 68°F/20°C (liquid).
Liquid surface tension: 30.12 dynes/cm = 0.0301 N/m @ 68°F/20°C.
Liquid water interfacial tension: (estimate) 30 dynes/cm = 0.030 N/m @ 68°F/20°C.
Latent heat of vaporization: 150 Btu/lb = 82 cal/g = 3.4×10^5 J/kg.
Heat of combustion: $-18,100$ Btu/lb = $-10,000$ cal/g = -418×10^5 J/kg.
Vapor pressure: (Reid) 4.1 psia; less than 0.08 mm.

DODECYLBENZENESULFONIC ACID REC. D:8150

SYNONYMS: ACIDO DODECILBENCENOSULFONICO (Spanish); BENZENESULFONIC ACID, DODECYL-; BENZENE SULFONIC ACID, DODECYL ESTER; DDBSA; DODECYL BENZENESULFONATE; CONOCO SA 597; NACCONOL 988 A; LAURYLBENZENESULFONIC ACID

IDENTIFICATION
CAS Number: 27176-87-0
Formula: $C_{18}H_{30}O_3S$
DOT ID Number: UN 2584; DOT Guide Number: 153
Proper Shipping Name: Dodecylbenzenesulfonic acid
Reportable Quantity (RQ): **(CERCLA)** 1000 lb/454 kg

DESCRIPTION: Light yellow to brown liquid. Possible odor of SO_2. Mixes with water.

Combustible • Heat or flame may cause explosion • Containers may BLEVE when exposed to fire • Vapors are heavier than air and will collect and stay in low areas • Vapors in confined areas (e.g., tanks, sewers, buildings) may explode when exposed to fire • Corrosive • Irritating to the skin, eyes, and respiratory tract • Toxic products of combustion may include sulfur oxides (SO_3, SO_2) and hydrogen sulfide.

Hazard Classification (based on NFPA-704 Rating System)
Health Hazards (Blue): 1; Flammability (Red): 1; Reactivity (Yellow): 0

EMERGENCY RESPONSE: See Appendix A (153)
Evacuation:
Public safety: Isolate the area of spill or leak for at least 25 to 50 meters (80 to 160 feet) in all directions.
Spill: Increase, in the downwind direction, as necessary, the distance shown under "Public Safety."
Fire: If tank, rail car, or tank truck is involved in fire, isolate for at least 800 meters (½ mile) in all directions; also, consider initial evacuation for 800 meters (½ mile) in all directions.

EXPOSURE
Short-term effects: *SEEK MEDICAL ATTENTION*. **Liquid:** Will burn skin and eyes. Harmful if swallowed. Remove contaminated clothing and shoes. Flush exposed areas with plenty of water. *IF BREATHING HAS STOPPED*, give artificial respiration; *avoid mouth-to-mouth resuscitation; use bag/mask apparatus. IF IN EYES*, hold eyelids open and flush with plenty of water. *IF SWALLOWED* and victim is *CONSCIOUS AND ABLE TO SWALLOW*, have victim drink 4 to 8 ounces of water and induce vomiting.
Note to physician or authorized medical personnel: Medical observation is recommended for 24 to 48 hours after breathing overexposure, as pulmonary edema may be delayed. As first aid for pulmonary edema, consider administering a corticosteroid spray. Cigarette smoking may exacerbate pulmonary injury and should be discouraged for at least 72 hours following exposure.

HEALTH HAZARDS
Personal protective equipment (PPE): B-Level PPE. Goggles, rubber gloves, air pack for contact with fumes in unventilated area.
Toxicity by ingestion: Grade 3; LD_{50} = 50 to 500 mg/kg (mouse).
Long-term health effects: Affected serum enzymes and electrolytes. Decrease in fetal body weight, length and construction. May facilitate penetration of carcinogens into gastric mucosa.
Liquid or solid irritant characteristics: Fairly severe skin irritant. May cause pain and second-degree burns after a few minutes of contact.
Odor threshold: 200 mg/L affected odor of water.

FIRE DATA
Flash point: 300°F/149°C (oc).

CHEMICAL REACTIVITY
Binary reactants: Oxidizers (may be violent), alkalis. Reacts with acids or acid fumes forming toxic sulfur oxides. Corrosive to metals; do not store in carbon steel or aluminum.
Neutralizing agents for acids and caustics: Flush with water, rinse with dilute sodium bicarbonate or soda ash solution

ENVIRONMENTAL DATA
Food chain concentration potential: Negative; unlikely to accumulate.
Water pollution: Harmful to aquatic life in very low concentrations. May be dangerous if it enters nearby water intakes; notify operators. Notify local health and wildlife officials. **Response to discharge:** Issue warning–corrosive. Disperse and flush.

SHIPPING INFORMATION
Grades of purity: 95–97.5%; **Storage temperature:** Ambient; **Inert atmosphere:** None; **Venting:** Open; **Stability during transport:** Stable.

PHYSICAL AND CHEMICAL PROPERTIES
Physical state @ 59°F/15°C and 1 atm: Liquid.
Molecular weight: 316 (average); 322.
Boiling point @ 1 atm: More than 440°F/205°C/478°K.
Melting/Freezing point: 50°F/10°C/283°K.
Specific gravity (water = 1): 1.2 @ 77°F/25°C.

DODECYLBENZENESULFONIC ACID, CALCIUM SALT
REC. D:8200

SYNONYMS: BENZENESULFONIC ACID, DODECYL-; BENZENE SULFONIC ACID, DODECYL ESTER; DODECYL BENZENESULFONATE; CALCIUM ALKYLAROMATIC SULFONATE; CALCIUM ALKYLBENZENESULFONATE

IDENTIFICATION
CAS Number: 27176-87-0 (dodecylbenzenesulfonic acid)
Formula: $C_{18}H_{30}O_3S$; $(C_{12}H_{25}C_6H_4SO_3)_2Ca$-solvent
DOT ID Number: UN 2584; DOT Guide Number: 153
Proper Shipping Name: Dodecylbenzenesulfonic acid
Reportable Quantity (RQ): **(CERCLA)** 1000 lb/454 kg

DESCRIPTION: Light yellow to brown liquid. Solvent odor. Mixes with water; may cause foaming.

Flammable • Containers may BLEVE when exposed to fire • Vapors may form explosive mixture with air • Vapors are heavier than air and will collect and stay in low areas • Vapors may travel long distances to ignition sources and flashback • Vapors in confined areas (e.g., tanks, sewers, buildings) may explode when exposed to fire • Corrosive; Irritating to the skin, eyes, and respiratory tract • Toxic products of combustion may include hydrogen sulfide, and sulfur oxide vapors (SO_3, SO_2).

Hazard Classification (based on NFPA-704 Rating System)
Health Hazards (Blue): 1; Flammability (Red): 2; Reactivity (Yellow): 0

EMERGENCY RESPONSE: See Appendix A (153)
Evacuation:
Public safety: Isolate the area of spill or leak for at least 25 to 50 meters (80 to 160 feet) in all directions.
Spill: Increase, in the downwind direction, as necessary, the distance shown under "Public Safety."
Fire: If tank, rail car, or tank truck is involved in fire, isolate for at least 800 meters (½ mile) in all directions; also, consider initial evacuation for 800 meters (½ mile) in all directions.

EXPOSURE
Short-term effects: *SEEK MEDICAL ATTENTION*. **Liquid:** Irritating to skin and eyes. *IF SWALLOWED*, will cause nausea. Remove contaminated clothing and shoes. Flush affected areas with plenty of water. *IF BREATHING HAS STOPPED*, give artificial respiration; *avoid mouth-to-mouth resuscitation; use bag/mask apparatus.* IF IN EYES, hold eyelids open and flush with plenty of water. *IF SWALLOWED* and victim is *CONSCIOUS AND ABLE TO SWALLOW*, have victim drink 4 to 8 ounces of water. *IF SWALLOWED* and victim is *UNCONSCIOUS OR HAVING CONVULSIONS*, do nothing except keep victim warm. **Do NOT induce vomiting.**

HEALTH HAZARDS
Personal protective equipment (PPE): B-Level PPE. Goggles or face shield; rubber gloves.
Liquid or solid irritant characteristics: Eye irritant.

FIRE DATA
Flash point: Less than 100°F/38°C (cc).
Fire extinguishing agents not to be used: Water may be ineffective.
Burning rate: About 4 mm/min

CHEMICAL REACTIVITY
Binary reactants: Contact with acids or acid fumes form toxic sulfur oxides.

ENVIRONMENTAL DATA
Food chain concentration potential: Negative; unlikely to accumulate.
Water pollution: Effect of low concentrations on aquatic life is unknown. Fouling to shoreline. May be dangerous if it enters nearby water intakes; notify operators. Notify local health and wildlife officials. **Response to discharge:** Issue warning–high flammability. Mechanical containment. Should be removed. Chemical and physical treatment.

SHIPPING INFORMATION
Grades of purity: 59-67%; remainder is a hydrocarbon solvent that is combustible or flammable; **Storage temperature:** Ambient; **Inert atmosphere:** None; **Venting:** Open (flame arrester); **Stability during transport:** Stable.

PHYSICAL AND CHEMICAL PROPERTIES
Physical state @ 59°F/15°C and 1 atm: Liquid.
Specific gravity (water = 1): 1.04 @ 77°F/25°C (liquid solution) 0.9 @ 77°F/25°C (solvent).

DODECYLBENZENESULFONIC ACID, ISOPROPYLAMINE SALT
REC. D:8250

SYNONYMS: BENZENE SULFONIC ACID, DODECYL-, compd. with 1-AMINO-2-PROPANOL (1:1); ISOPROPYLAMINE DODECYLBENZENESULFONATE

IDENTIFICATION
CAS Number: 42504-46-1
Formula: $C_{12}H_{25}C_6H_4SO_3H \cdot NG_2CH(CH_3)_2$
DOT ID Number: UN 3077; DOT Guide Number: 171
Proper Shipping Name: Environmentally hazardous substances, n.o.s.
Reportable Quantity (RQ): **(CERCLA)** 1000 lb/454 kg

DESCRIPTION: Off-white solid or thick liquid. Sweet petroleum odor. May float or sink and mix with water.

Combustible • Heat or flame may cause explosion • Containers may BLEVE when exposed to fire • Vapors are heavier than air and will collect and stay in low areas • Vapors in confined areas (e.g., tanks, sewers, buildings) may explode when exposed to fire • Irritating to the skin, eyes, and respiratory tract • Toxic products of combustion may include oxides of nitrogen and sulfur.

EMERGENCY RESPONSE: See Appendix A (171)
Evacuation:
Public safety: Isolate the area of spill or leak for at least 10 to 25 meters (30 to 80 feet) in all directions.
Spill: Increase, in the downwind direction, as necessary, the distance shown under "Public Safety."
Fire: If any large container is involved in fire, isolate for at least 800 meters (½ mile) in all directions; also, consider initial evacuation for 800 meters (½ mile) in all directions.

EXPOSURE
Short-term effects: *SEEK MEDICAL ATTENTION*. **Solid:** Irritating to skin and eyes. *IF SWALLOWED*, will cause nausea and vomiting. Remove contaminated clothing and shoes. Flush affected

areas with plenty of water. *IF IN EYES*, hold eyelids open and flush with plenty of water. *IF SWALLOWED* and victim is *CONSCIOUS AND ABLE TO SWALLOW*, have victim drink 4 to 8 ounces of water. *IF SWALLOWED* and victim is *UNCONSCIOUS OR HAVING CONVULSIONS* do nothing except keep victim warm.

HEALTH HAZARDS
Personal protective equipment (PPE): B-Level PPE. Rubber gloves; goggles; mask (for liquid form).
Toxicity by ingestion: Grade 2; LD_{50} = 3.54 g/kg (rat).
Liquid or solid irritant characteristics: Eye irritant.

FIRE DATA
Flash point: (liquid) more than 300°F/149°C (cc).

CHEMICAL REACTIVITY
Binary reactants: Acids or acid fumes for toxic sulfur oxides. Oxidizers, strong acids.

ENVIRONMENTAL DATA
Food chain concentration potential: Negative; unlikely to accumulate.
Water pollution: Effect of low concentrations on aquatic life is unknown. May be dangerous if it enters nearby water intakes; notify operators. Notify local health and wildlife officials.
Response to discharge: Disperse and flush.

SHIPPING INFORMATION
Grades of purity: Technical, 96+%; may also be shipped as a concentrated solution in a combustible petroleum solvent; **Storage temperature:** Ambient; **Inert atmosphere:** None; **Venting:** Open; **Stability during transport:** Stable.

PHYSICAL AND CHEMICAL PROPERTIES
Physical state @ 59°F/15°C and 1 atm: Solid.
Molecular weight: 385.5
Boiling point @ 1 atm: Decomposes.
Specific gravity (water = 1): 1.03 @ 68°F/20°C (solid); 1.03 @ 77°F/25°C (liquid).

DODECYL BENZENE SULFONIC ACID, SODIUM SALT
REC. D:8300

SYNONYMS: AA-9; ABESON NAM; BIO-SOFT D-40; CALSOFT F-90; COCOCO C-50; DETERGENT HD-90; DODECYLBENZENESULFONATE SODIUM SALT; DETERGENT HD-90; NACCANOL NR; NACCANOL SW; MERCOL 25; PILOT HD-90; PILOT SF-40; RICHONATE 1850; SANTOMERSE-3; SOLAR-40; SODIUM DODECYLBENZENE-SULFONATE; SULFAPOL; SULFRAMIN 85; SULFRAMIN 40; SULFRAMIN 85; ULTRA WET K

IDENTIFICATION
CAS Number: 25155-30-0
Formula: $C_{18}H_{30}SO_3 \cdot Na$; $C_{18}H_{29}NaO_3S$
DOT ID Number: UN 3077; DOT Guide Number: 171
Proper Shipping Name: Environmentally hazardous substances, solid, n.o.s.
Reportable Quantity (RQ): **(CERCLA)** 1000 lb/454 kg

DESCRIPTION: White to light yellow flakes or powder; or, a pasty liquid (40% H_2O). Odorless to slight oily odor. Mixes with water; foaming may be produced.

Combustible • Heat or flame may cause explosion • Containers may BLEVE when exposed to fire • Vapors are heavier than air and will collect and stay in low areas • Vapors in confined areas (e.g., tanks, sewers, buildings) may explode when exposed to fire • Irritating to the skin, eyes, and respiratory tract • Toxic products of combustion may include oxides of sulfur and sodium.

Hazard Classification (based on NFPA-704 Rating System)
Health Hazards (Blue): 1; Flammability (Red): 1; Reactivity (Yellow): 0

EMERGENCY RESPONSE: See Appendix A (171)
Evacuation:
Public safety: Isolate the area of spill or leak for at least 10 to 25 meters (30 to 80 feet) in all directions.
Spill: Increase, in the downwind direction, as necessary, the distance shown under "Public Safety."
Fire: If any large container is involved in fire, isolate for at least 800 meters (½ mile) in all directions; also, consider initial evacuation for 800 meters (½ mile) in all directions.

EXPOSURE
Short-term effects: *SEEK MEDICAL ATTENTION*. **Liquid or solid:** Irritating to skin and eyes. *IF SWALLOWED*, will cause nausea, vomiting, and diarrhea. Flush affected areas with plenty of water. *IF IN EYES*, hold eyelids open and flush with plenty of water. *IF SWALLOWED* and victim is *CONSCIOUS AND ABLE TO SWALLOW*, have victim drink 4 to 8 ounces of water and have victim induce vomiting. *IF SWALLOWED* and victim is *UNCONSCIOUS OR HAVING CONVULSIONS*, do nothing except keep victim warm.

HEALTH HAZARDS
Personal protective equipment (PPE): B-Level PPE. Rubber gloves, safety glasses.
Toxicity by ingestion: Grade 2; LD_{50} = 0.5 to 5 g/kg.
Long-term health effects: Possible teratogenic effects. May cause skin rash after prolonged exposure. Nonvolatile (vapor is water).
Liquid or solid irritant characteristics: Severe eye and skin irritant. If spilled on clothing and allowed to remain, may cause smarting and reddening of skin.

FIRE DATA
Flash point: Burns with great difficulty.

CHEMICAL REACTIVITY
Binary reactants: Acids or acid fumes for toxic sulfur oxides.

ENVIRONMENTAL DATA
Food chain concentration potential: Negative; unlikely to accumulate.
Water pollution: Harmful to aquatic life in very low concentrations. May be dangerous if it enters nearby water intakes; notify operators. Notify local health and wildlife officials. **Response to discharge:** Issue warning–water contaminant. Should be removed. Chemical and physical treatment.

SHIPPING INFORMATION
Grades of purity: 40% H_2O. Varies with manufacturer and use; **Storage temperature:** Temperature above 32°F/0°C; **Inert atmosphere:** Not required; **Venting:** Open; **Stability during transport:** Stable.

PHYSICAL AND CHEMICAL PROPERTIES
Physical state @ 59°F/15°C and 1 atm: Solid or Liquid.

Molecular weight: 348.52
Boiling point @ 1 atm: Foams as water boils. 212°F/100°C/373°K.
Specific gravity (water = 1): 1.0 @ 68°F/20°C for 60% slurry
Relative vapor density (air = 1): Vapor is water.

DODECYLBENZENESULFONIC ACID, TRIETHANOLAMINE SALT   REC. D:8350

SYNONYMS: TRIETHANOLAMINE DODECEYL-BENZENESULFONATE

IDENTIFICATION
CAS Number: 27323-41-7
Formula: $C_{18}H_{20}O_3S \cdot C_6H_{15}NO_3$; $C_{12}H_{25}C_6H_4SO_3H \cdot N(CH_2CH_2OH)_3 \cdot H_2O$
DOT ID Number: UN 3082; DOT Guide Number: 171
Proper Shipping Name: Environmentally hazardous substances, liquid, n.o.s.
Reportable Quantity (RQ): **(CERCLA)** 1000 lb/454 kg

DESCRIPTION: Yellowish brown or amber liquid. Sinks and mixes with water.

Combustible • Heat or flame may cause explosion • Containers may BLEVE when exposed to fire • Vapors are heavier than air and will collect and stay in low areas • Vapors in confined areas (e.g., tanks, sewers, buildings) may explode when exposed to fire • Irritating to the skin, eyes, and respiratory tract • Toxic products of combustion may include nitrogen oxides and sulfur oxides.

Hazard Classification (based on NFPA-704 Rating System)
Health Hazards (Blue): 1; Flammability (Red): 1; Reactivity (Yellow): 0

EMERGENCY RESPONSE: See Appendix A (171)
Evacuation:
Public safety: Isolate the area of spill or leak for at least 10 to 25 meters (30 to 80 feet) in all directions.
Spill: Increase, in the downwind direction, as necessary, the distance shown under "Public Safety."
Fire: If any large container is involved in fire, isolate for at least 800 meters (½ mile) in all directions; also, consider initial evacuation for 800 meters (½ mile) in all directions.

EXPOSURE
Short-term effects: *SEEK MEDICAL ATTENTION.* **Liquid:** Irritating to skin and eyes. *IF SWALLOWED*, will cause nausea and vomiting. Remove contaminated clothing and shoes. Flush affected areas with plenty of water. *IF IN EYES*, hold eyelids open and flush with plenty of water. *IF SWALLOWED* and victim is *CONSCIOUS AND ABLE TO SWALLOW*, have victim drink 4 to 8 ounces of water. *IF SWALLOWED* and victim is *UNCONSCIOUS OR HAVING CONVULSIONS,* do nothing except keep victim warm.

HEALTH HAZARDS
Personal protective equipment (PPE): B-Level PPE. Rubber gloves; face mask or goggles
Liquid or solid irritant characteristics: Eye irritant.

FIRE DATA
Flash point: May burn; difficult to ignite.

CHEMICAL REACTIVITY
Binary reactants: Acids or acid fumes for toxic sulfur oxides.

ENVIRONMENTAL DATA
Food chain concentration potential: Negative; unlikely to accumulate.
Water pollution: Effect of low concentrations on aquatic life is unknown. May be dangerous if it enters nearby water intakes; notify operators. Notify local health and wildlife officials. **Response to discharge:** Disperse and flush.

SHIPPING INFORMATION
Grades of purity: 60% solution in water; **Storage temperature:** Ambient; **Inert atmosphere:** None; **Venting:** Open; **Stability during transport:** Stable.

PHYSICAL AND CHEMICAL PROPERTIES
Physical state @ 59°F/15°C and 1 atm: Liquid.
Molecular weight: 475.6 (solute).
Specific gravity (water = 1): (estimate) 1.2 @ 68°F/20°C (liquid).

DODECYL DIPHENYL ETHER DISULFONATE SOLUTION   REC. D:8400

SYNONYMS: DODECYL DIPHENYL ETHER SULFONATE, DISODIUM SALT, AQUEOUS SOLUTION; DOWFAX 2A1

IDENTIFICATION
CAS Number: 25167-32-2
Formula: $(Na^+)_2[C_{12}H_{25}C_6H_4(SO_3\text{-})]_2 \cdot O$ in water; $CH_2 = C(CH_3)CO_2(CH_2)_{11}CH_3$; $[CH_2 = C(CH_3)COO]C_{12}H_{25}$; $[CH_2 = C(CH_3)COO]C_{15}H_{31}$

DESCRIPTION: Light yellow to light brown solution. Disinfectant-type odor. Soluble in water.

Corrosive skin, eyes, and respiratory tract • Toxic products of combustion may include carbon monoxide and oxides of sulfur.

Hazard Classification (based on NFPA-704 Rating System)
Health Hazards (Blue): 1; Flammability (Red): 0; Reactivity (Yellow): 0

EMERGENCY RESPONSE: See Appendix A (171)
Evacuation:
Public safety: Isolate the area of spill or leak for at least 10 to 25 meters (30 to 80 feet) in all directions.
Spill: Increase, in the downwind direction, as necessary, the distance shown under "Public Safety."
Fire: If any large container is involved in fire, isolate for at least 800 meters (½ mile) in all directions; also, consider initial evacuation for 800 meters (½ mile) in all directions.

EXPOSURE
Short-term effects: *SEEK MEDICAL ATTENTION.* **Liquid:** Will burn skin and eyes. Harmful if swallowed. Remove contaminated clothing and shoes. Flush affected areas with water. *IF IN EYES*, hold eyelids open and flush with plenty of water. *IF SWALLOWED* and victim is CONSCIOUS: Have victim drink 4 to 8 ounces of water and induce vomiting. *IF SWALLOWED* and victim is *UNCONSCIOUS OR HAVING CONVULSIONS,* do nothing except keep victim warm.

HEALTH HAZARDS
Personal protective equipment (PPE): B-Level PPE. Goggles, rubber gloves, approved mist respirator, protective clothing.
Toxicity by ingestion: Grade 2: LD_{50} = 700 mg/kg (rat).
Long-term health effects: Excessive exposure may cause liver and kidney damage.
Liquid or solid irritant characteristics: Severe eye and skin irritant. May cause pain and second-degree burns after a few minutes of contact.

FIRE DATA
Behavior in fire: Organic portion may burn once water has evaporated. Use water to extinguish.

CHEMICAL REACTIVITY
Binary reactants: Acids or acid fumes for toxic sulfur oxides. Sulfuric acid, isocyantes.
Neutralizing agents for acids and caustics: Flush with water, rinse with dilute sodium bicarbonate or soda ash solution.
Reactivity group: 43
Compatibility class: Miscellaneous water solutions.

ENVIRONMENTAL DATA
Water pollution: DOT Appendix B, §172.101–marine pollutant. Harmful to aquatic life in low concentrations. Harmful if it enters local water intakes. Notify local health and wildlife officials. Notify operators of local water intakes. **Response to discharge:** Issue warning–corrosive. Should be removed. Physical and chemical treatment

SHIPPING INFORMATION
Grades of purity: Aqueous solution, 47% maximum concentration; **Inert atmosphere:** None; **Venting:** None; **Stability during transport:** Stable.

NAS HAZARD CLASSIFICATION FOR BULK WATER TRANSPORTATION
FIRE: 0
HEALTH: Vapor irritant: 0; Liquid or solid irritant: 3; Poisons: 1
WATER POLLUTION: Human toxicity:–; Aquatic toxicity:–; Aesthetic effect: 1
REACTIVITY: Other chemicals: 3; Water: 0; Self-reaction: 0

PHYSICAL AND CHEMICAL PROPERTIES
Physical state @ 59°F/15°C and 1 atm: Liquid (water solution).
Molecular weight: 712.95
Boiling point @ 1 atm: 212°F/100°C/373°K (approximate, varies with concentration).
Melting/Freezing point: 32°F/0°C/273°K (approximate, varies with concentration).
Specific gravity (water = 1): 1.161 @ 77°F/25°C.

DODECYLMETHACRYLATE   REC. D:8450

SYNONYMS: LAURYL METHACRYLATE; METHACRYLIC ACID, DODECYL ESTER; DODECYL-2-METHYL-2-PROPENOATE

IDENTIFICATION
CAS Number: 142-90-5
Formula: $C_{16}H_{30}O_2$
DOT ID Number: None; DOT Guide Number: 171P
Proper Shipping Name: Polymerizable material, stabilize with dry ice

DESCRIPTION: Liquid. Floats on water surface; insoluble.

Combustible • Polymerization hazard • Heat can induce polymerization with rapid release of energy; sealed containers may rupture explosively • Vapors are heavier than air and will collect and stay in low areas • Vapors in confined areas (e.g., tanks, sewers, buildings) may explode when exposed to fire • Irritating to the skin, eyes, and respiratory tract • Toxic products of combustion may include carbon monoxide.

Hazard Classification (based on NFPA-704 Rating System)
Health Hazards (Blue): 0; Flammability (Red): 1; Reactivity (Yellow): 1

HEALTH HAZARDS
EMERGENCY RESPONSE: See Appendix A (171)
Evacuation:
Public safety: Isolate the area of spill or leak for at least 10 to 25 meters (30 to 80 feet) in all directions.
Spill: Increase, in the downwind direction, as necessary, the distance shown under "Public Safety."
Fire: If any large container is involved in fire, isolate for at least 800 meters (½ mile) in all directions; also, consider initial evacuation for 800 meters (½ mile) in all directions.

EXPOSURE
Short-term effects: *SEEK MEDICAL ATTENTION.* **Liquid:** Irritating to skin and eyes. Remove and double bag contaminated clothing and shoes at the site. *IF IN EYES OR ON SKIN*, flush with running water for at least 15 minutes; hold eyelids open if necessary. Wash skin with soap and water. *IF SWALLOWED* and victim is *CONSCIOUS AND ABLE TO SWALLOW*, have victim drink 4 to 8 ounces of water and induce vomiting. *IF SWALLOWED* and victim is *UNCONSCIOUS OR HAVING CONVULSIONS,* do nothing except keep victim warm.

Personal protective equipment (PPE): B-Level PPE. Wear self-contained positive pressure breathing apparatus and full protective clothing.
Toxicity by ingestion: Grade 0; LD_{50} more than 87.25 g/Kg (mouse).
Liquid or solid irritant characteristics: May cause eye and skin irritation.

FIRE DATA
Flash point: More than 230°F/110°C (cc).
Behavior in fire: Heat can induce polymerization with rapid release of energy. Sealed containers may rupture explosively.
Stoichiometric air-to-fuel ratio: 12 (estimate).

CHEMICAL REACTIVITY
Binary reactants: Oxidizing and reducing agents may initiate polymerization.
Polymerization: Reaction may occur when heated; **Inhibitor of polymerization:** 90–120 ppm hydroquinone
Reactivity group: 14
Compatibility class: Acrylates

ENVIRONMENTAL DATA
Water pollution: Effect of low concentrations on aquatic life is unknown. Fouling to shoreline. May be dangerous if it enters nearby water intakes; notify operators. Notify local health and wildlife officials. Notify operators of local water intakes. **Response to discharge:** Mechanical containment. Should be removed. Chemical and physical treatment.

SHIPPING INFORMATION
Stability during transport: Stable.

PHYSICAL AND CHEMICAL PROPERTIES
Physical state @ 59°F/15°C and 1 atm: Liquid.
Molecular weight: 254.42
Melting/Freezing point: −8°F/−22°C/251°K.
Specific gravity (water = 1): 0.868 @ 77°F/25°C.
Relative vapor density (air = 1): 8.8 (estimate).

DODECYL/PENTADECYL METHACRYLATE
REC. D:8500

SYNONYMS: METHACRYLIC ACID, DODECYL AND PENTADECYL ESTER MIX; METHACRYLIC ACID, LAURYL AND PENTADECYL ESTER MIX

IDENTIFICATION
DOT ID Number: None; DOT Guide Number: 171P
Proper Shipping Name: Polymerizable material, stabilize with dry ice

DESCRIPTION: Colorless liquid. Floats on the surface of water.

Combustible • Polymerization hazard • Heat can induce polymerization with rapid release of energy; sealed containers may rupture explosively • Vapors are heavier than air and will collect and stay in low areas • Vapors in confined areas (e.g., tanks, sewers, buildings) may explode when exposed to fire • Irritating to the skin, eyes, and respiratory tract • Toxic products of combustion may include carbon monoxide.

Hazard Classification (based on NFPA-704 Rating System)
Health Hazards (Blue): 0; Flammability (Red): 1; Reactivity (Yellow): 1

EMERGENCY RESPONSE: See Appendix A (171)
Evacuation:
Public safety: Isolate the area of spill or leak for at least 10 to 25 meters (30 to 80 feet) in all directions.
Spill: Increase, in the downwind direction, as necessary, the distance shown under "Public Safety."
Fire: If any large container is involved in fire, isolate for at least 800 meters (½ mile) in all directions; also, consider initial evacuation for 800 meters (½ mile) in all directions.

EXPOSURE
Short-term effects: *SEEK MEDICAL ATTENTION.* **Liquid:** Irritating to skin and eyes. Vapors may cause dizziness or suffocation. Remove and double bag contaminated clothing and shoes at the site. *IF IN EYES OR ON SKIN,* flush with running water for at least 15 minutes, hold eyelids open if necessary. Wash skin with soap and water. *IF SWALLOWED* and victim is *CONSCIOUS AND ABLE TO SWALLOW,* have victim drink 4 to 8 ounces of water and induce vomiting. *IF SWALLOWED* and victim is *UNCONSCIOUS OR HAVING CONVULSIONS,* do nothing except keep victim warm.

HEALTH HAZARDS
Personal protective equipment (PPE): B-Level PPE. Wear self-contained positive pressure breathing apparatus and full protective clothing.
Liquid or solid irritant characteristics: May cause eye and skin irritation.

FIRE DATA
Behavior in fire: Heat can induce polymerization with rapid release of energy. Sealed containers may rupture explosively.

CHEMICAL REACTIVITY
Binary reactants: Oxidizing and reducing agents may initiate polymerization.
Polymerization: Reaction may occur when heated; **Inhibitor of polymerization:** 90–120 ppm hydroquinone
Reactivity group: 14
Compatibility class: Acrylates

ENVIRONMENTAL DATA
Water pollution: Effects of low concentrations on aquatic life is unknown. Fouling to shoreline. May be dangerous if it enters nearby water intakes; notify operators. Notify local health and wildlife officials. **Response to discharge:** Mechanical containment. Should be removed. Chemical and physical treatment.

SHIPPING INFORMATION
Stability during transport: Stable.

NAS HAZARD CLASSIFICATION FOR BULK WATER TRANSPORTATION
FIRE:–
HEALTH: Vapor irritant: 2; Liquid or solid irritant: 2; Poisons: 0
WATER POLLUTION: Human toxicity:–; Aquatic toxicity:–; Aesthetic effect: 3
REACTIVITY: Other chemicals: 3; Water: 1; Self-reaction: 3

PHYSICAL AND CHEMICAL PROPERTIES
Physical state @ 59°F/15°C and 1 atm: liquid (mixture).
Specific gravity (water = 1): 0.87 @ 77°F/25°C (approx).

DODECYL PHENOL
REC. D:8550

SYNONYMS: DODECILFENOL (Spanish)

IDENTIFICATION
CAS Number: 27193-86-8
Formula: $C_{18}H_{30}O$; $C_6H_4OHC_{12}H_{25}$
DOT ID Number: UN 2821; DOT Guide Number: 153
Proper Shipping Name: Phenol solutions.

DESCRIPTION: Straw colored liquid. Phenolic odor. Mixes with water.

Combustible • Heat or flame may cause explosion • Containers may BLEVE when exposed to fire • Vapors are heavier than air and will collect and stay in low areas • Vapors in confined areas (e.g., tanks, sewers, buildings) may explode when exposed to fire • Irritating to the skin, eyes, and respiratory tract • Toxic products of combustion may include acrid carbon monoxide.

Hazard Classification (based on NFPA-704 Rating System)
Health Hazards (Blue): 0; Flammability (Red): 1; Reactivity (Yellow): 0

EMERGENCY RESPONSE: See Appendix A (153)
Evacuation:
Public safety: Isolate the area of spill or leak for at least 25 to 50 meters (80 to 160 feet) in all directions.
Spill: Increase, in the downwind direction, as necessary, the distance shown under "Public Safety."

Fire: If tank, rail car, or tank truck is involved in fire, isolate for at least 800 meters (½ mile) in all directions; also, consider initial evacuation for 800 meters (½ mile) in all directions.

EXPOSURE
Short-term effects: *SEEK MEDICAL ATTENTION*. **Vapor:** Move victim to fresh air. *IF BREATHING HAS STOPPED*, give artificial respiration; *avoid mouth-to-mouth resuscitation; use bag/mask apparatus*. If breathing is difficult, administer oxygen. **Liquid:** *HARMFUL IF ABSORBED THROUGH THE SKIN*. Remove contaminated clothing and shoes. Wash affected areas with soap and water. *IF IN EYES*, hold eyelids open and flush with plenty of water.

HEALTH HAZARDS
Personal protective equipment (PPE): B-Level PPE. Full impervious protective clothing, including boots and natural rubber gloves. Wear full face shield or chemical safety goggles. Use approved respirator to protect against vapors.
Vapor (gas) irritant characteristics: Eye and respiratory irritant. Vapors are irritating such that personnel will not usually tolerate moderate or high concentrations.
Liquid or solid irritant characteristics: Eye and respiratory tract irritant. Causes smarting of the skin and first-degree burns on short exposure; may cause second-degree burns on long exposure.

FIRE DATA
Flash point: 325°F/163°C (oc).
Fire extinguishing agents not to be used: Water streams will spread contamination of pesticide over wide area.

CHEMICAL REACTIVITY
Binary reactants: Oxidizing materials
Reactivity group: 21
Compatibility class: Phenols, cresols

ENVIRONMENTAL DATA
Water pollution: Effects of low concentrations on aquatic life is unknown. May be dangerous if it enters nearby water intakes; notify operators. Notify local health and wildlife officials.
Response to discharge: Should be removed.

SHIPPING INFORMATION
Grades of purity: Technical grades; **Storage temperature:** Ambient; **Inert atmosphere:** None; **Venting:** Open; **Stability during transport:** Stable.

PHYSICAL AND CHEMICAL PROPERTIES
Physical state @ 59°F/15°C and 1 atm: Liquid.
Molecular weight: 262.48
Boiling point @ 1 atm: 597–633°F/314–334°C/587–607°K.
Specific gravity (water = 1): 0.94 @ 20°C.
Relative vapor density (air = 1): 9.0

DODECYL SULFATE, DIETHANOLAMINE SALT
REC. D:8600

SYNONYMS: DIETHANOLAMINE LAURYL SULFATE SOLUTION; LAURYL SULFATE, DIETHANOLAMINE SALT SOLUTION

IDENTIFICATION
CAS Number: 151-21-3
Formula: $C_{12}H_{25}OSO_3H \cdot (HOCH_2CH_2)_2NH \cdot H_2O$

DESCRIPTION: Clear to pale yellow liquid. Mild fatty odor. May float or sink and mix with water.

Combustible • Heat or flame may cause explosion • Containers may BLEVE when exposed to fire • Vapors are heavier than air and will collect and stay in low areas • Vapors in confined areas (e.g., tanks, sewers, buildings) may explode when exposed to fire • Irritating to the skin, eyes, and respiratory tract • Toxic products of combustion may include vapors of diethanolamine and oxides nitrogen.

Hazard Classification (based on NFPA-704 Rating System)
Health Hazards (Blue): 1; Flammability (Red): 1; Reactivity (Yellow): 0

EMERGENCY RESPONSE: See Appendix A (171)
Evacuation:
Public safety: Isolate the area of spill or leak for at least 10 to 25 meters (30 to 80 feet) in all directions.
Spill: Increase, in the downwind direction, as necessary, the distance shown under "Public Safety."
Fire: If any large container is involved in fire, isolate for at least 800 meters (½ mile) in all directions; also, consider initial evacuation for 800 meters (½ mile) in all directions.

EXPOSURE
Short-term effects: *SEEK MEDICAL ATTENTION*. **Liquid:** Irritating to skin and eyes. *IF SWALLOWED*, will cause nausea and vomiting. Remove contaminated clothing and shoes. Flush affected areas with plenty of water. *IF IN EYES*, hold eyelids open and flush with plenty of water. *IF SWALLOWED* and victim is *CONSCIOUS AND ABLE TO SWALLOW*, have victim drink 4 to 8 ounces of water. *IF SWALLOWED* and victim is *UNCONSCIOUS OR HAVING CONVULSIONS*, do nothing except keep victim warm.

HEALTH HAZARDS
Personal protective equipment (PPE): B-Level PPE. Chemical goggles or face shield; protective gloves.
Liquid or solid irritant characteristics: Eye and skin irritant.

FIRE DATA
Flash point: Combustible. More than 250°F/121°C.

CHEMICAL REACTIVITY
Binary reactants: Violent reaction with aluminum, magnesium.

ENVIRONMENTAL DATA
Food chain concentration potential: Negative; unlikely to accumulate.
Water pollution: Effect of low concentrations on aquatic life is unknown. May be dangerous if it enters nearby water intakes; notify operators. Notify local health and wildlife officials. **Response to discharge:** Disperse and flush.

SHIPPING INFORMATION
Grades of purity: 35–40% solution in water; **Storage temperature:** Below 100°F/38°C; **Inert atmosphere:** None; **Venting:** Open; **Stability during transport:** Stable.

PHYSICAL AND CHEMICAL PROPERTIES
Physical state @ 59°F/15°C and 1 atm: Solid.
Boiling point @ 1 atm: (decomposes).
Specific gravity (water = 1): 1.01 @ 68°F/20°C (liquid).

DODECYL SULFATE, MAGNESIUM SALT REC. D:8650

SYNONYMS: MAGNESIUM DODECYL SULFATE; MAGNESIUM LAURYL SULFATE; LAURYL SULFATE, MAGNESIUM SALT; LAURYL MAGNESIUM SULFATE

IDENTIFICATION
Formula: $(C_{12}H_{25}OSO_3)2\ mg \cdot H_2O$

DESCRIPTION: Light yellow liquid. Mild odor. May float or sink and mix with water.

Irritating to the skin, eyes, and respiratory tract • Toxic products of combustion may include carbon monoxide, sulfur oxide, and magnesium fumes.

Hazard Classification (based on NFPA-704 Rating System)
Health Hazards (Blue): 1; Flammability (Red): 0; Reactivity (Yellow): 0

EMERGENCY RESPONSE: See Appendix A (171)
Evacuation:
Public safety: Isolate the area of spill or leak for at least 10 to 25 meters (30 to 80 feet) in all directions.
Spill: Increase, in the downwind direction, as necessary, the distance shown under "Public Safety."
Fire: If any large container is involved in fire, isolate for at least 800 meters (½ mile) in all directions; also, consider initial evacuation for 800 meters (½ mile) in all directions.

EXPOSURE
Short-term effects: *SEEK MEDICAL ATTENTION.* **Liquid:** Irritating to skin and eyes. *IF SWALLOWED*, will cause nausea and vomiting. Remove contaminated clothing and shoes. Flush affected areas with plenty of water. *IF IN EYES*, hold eyelids open and flush with plenty of water. *IF SWALLOWED* and victim is *CONSCIOUS AND ABLE TO SWALLOW*, have victim drink 4 to 8 ounces of water. *IF SWALLOWED* and victim is *UNCONSCIOUS OR HAVING CONVULSIONS,* do nothing except keep victim warm.

HEALTH HAZARDS
Personal protective equipment (PPE): B-Level PPE. Goggles or face shield; rubber gloves.
Liquid or solid irritant characteristics: Eye and skin irritant.

CHEMICAL REACTIVITY
Binary reactants: Reacts with finely divided aluminum, magnesium.

ENVIRONMENTAL DATA
Food chain concentration potential: Negative; unlikely to accumulate.
Water pollution: Effect of low concentrations on aquatic life is unknown. May be dangerous if it enters nearby water intakes; notify operators. Notify local health and wildlife officials.
Response to discharge: Disperse and flush

SHIPPING INFORMATION
Grades of purity: 27–30% solution in water; **Storage temperature:** Ambient; **Inert atmosphere:** None; **Venting:** Open; **Stability during transport:** Stable.

PHYSICAL AND CHEMICAL PROPERTIES
Physical state @ 59°F/15°C and 1 atm: Liquid.
Molecular weight: 555 (solute).
Specific gravity (water = 1): 1.04 @ 68°F/20°C (liquid).

DODECYL SULFATE, SODIUM SALT REC. D:8700

SYNONYMS: AQUAREX METHYL; AVIROL 118 CONC; CONCO SULFATE WA; DETERGENT 66; DODECYL SODIUM SULFATE; DREFT; DUPONOL; EMERSAL 6400; HEXAMOL SLS; IRIUM; LANETTE WAX-S; LAURYL SULFATE, SODIUM SALT; MAPROFIX 563; NEUTRAZYME; PRODUCT 161; QUOLAC EX-UB; RICHONOL C; SIPEX OP; SIPON WD; SLS; SODIUM LAURYL SULFATE; LAURYL SODIUM SULFATE; SODIUM DODECYL SULFATE; SOLSOL NEEDLES; STANDAPOL 112 COND; STEPANOL WAQ; STERLING WAQ-COSMETIC; SULFOPON WA-1; SULFURIC ACID, MONODODECYL ESTER, SODIUM SALT; TARAPON K 12; TEXAPON ZHC; TREPENAOL WA; ULTRA SULFATE SL-1

IDENTIFICATION
CAS Number: 151-21-3
Formula: $C_{12}H_{26}O_4S \cdot Na$

DESCRIPTION: White to pale yellow crystals or powder, paste or liquid. Mild fatty odor. Sinks and mixes with water.

Toxic products of combustion may include carbon monoxide, sulfur oxides and sodium oxide.

Hazard Classification (based on NFPA-704 Rating System)
Health Hazards (Blue): 1; Flammability (Red): 0; Reactivity (Yellow): 0

EMERGENCY RESPONSE: See Appendix A (171)
Evacuation:
Public safety: Isolate the area of spill or leak for at least 10 to 25 meters (30 to 80 feet) in all directions.
Spill: Increase, in the downwind direction, as necessary, the distance shown under "Public Safety."
Fire: If any large container is involved in fire, isolate for at least 800 meters (½ mile) in all directions; also, consider initial evacuation for 800 meters (½ mile) in all directions.

EXPOSURE
Short-term effects: *SEEK MEDICAL ATTENTION.* **Dust:** Irritating to eyes, nose, and throat. *IF INHALED*, will, will cause coughing. *IF IN EYES*, hold eyelids open and flush with plenty of water. *IF BREATHING HAS STOPPED,* give artificial respiration. If breathing is difficult, administer oxygen. **Liquid or solid:** Will burn eyes. Irritating to eyes. *IF SWALLOWED*, will cause nausea and vomiting. Remove contaminated clothing and shoes. Flush affected areas with plenty of water. *IF IN EYES*, hold eyelids open and flush with plenty of water. *IF SWALLOWED* and victim is *UNCONSCIOUS OR HAVING CONVULSIONS,* do nothing except keep victim warm.

HEALTH HAZARDS
Personal protective equipment (PPE): B-Level PPE. Protective gloves; dust mask or face shield
Toxicity by ingestion: Grade 2; oral LD_{50} = 1 g/kg (rat).
Liquid or solid irritant characteristics: Eye and skin irritant.

CHEMICAL REACTIVITY
Binary reactants: Violent reaction with aluminum, magnesium.

ENVIRONMENTAL DATA
Food chain concentration potential: Negative; unlikely to accumulate.
Water pollution: Effect of low concentrations on aquatic life is unknown. May be dangerous if it enters nearby water intakes; notify operators. Notify local health and wildlife officials.
Response to discharge: Disperse and flush.

SHIPPING INFORMATION
Grades of purity: Technical, 89–96%; pharmaceutical grade; also shipped as 28–50% solutions in water; **Storage temperature:** Below 100°F/38°C; **Inert atmosphere:** None; **Venting:** Open; **Stability during transport:** Stable.

PHYSICAL AND CHEMICAL PROPERTIES
Physical state @ 59°F/15°C and 1 atm: Solid.
Molecular weight: 289.43
Boiling point @ 1 atm: (decomposes).
Specific gravity (water = 1): More than 1.1 @ 68°F/20°C (solid).

DODECYL SULFATE, TRIETHANOLAMINE SALT
REC. D:8750

SYNONYMS: TRIETHANOLAMINE LAURYL SULFATE; LAURYL SULFATE, TRIETHANOLAMINE SALT

IDENTIFICATION
CAS Number: 139-96-8
Formula: $C_{12}H_{26}O_4S \cdot C_6H_{15}NO_3$

DESCRIPTION: Colorless liquid or paste. Mild, fatty odor. Sinks and mixes with water.

Irritating to skin, eyes, and respiratory tract • Toxic products of combustion may include triethanolamine, carbon monoxide, and oxides of nitrogen and sulfur.

Hazard Classification (based on NFPA-704 Rating System)
Health Hazards (Blue): 1; Flammability (Red): 0; Reactivity (Yellow): 0

EMERGENCY RESPONSE: See Appendix A (171)
Evacuation:
Public safety: Isolate the area of spill or leak for at least 10 to 25 meters (30 to 80 feet) in all directions.
Spill: Increase, in the downwind direction, as necessary, the distance shown under "Public Safety."
Fire: If any large container is involved in fire, isolate for at least 800 meters (½ mile) in all directions; also, consider initial evacuation for 800 meters (½ mile) in all directions.

EXPOSURE
Short-term effects: *SEEK MEDICAL ATTENTION.* If artificial respiration is administered; *avoid mouth-to-mouth resuscitation; use bag/mask apparatus.* **Liquid:** Irritating to skin and eyes. *IF SWALLOWED,* will cause nausea and vomiting. Remove contaminated clothing and shoes. Flush affected areas with plenty of water. *IF IN EYES,* hold eyelids open and flush with plenty of water. *IF SWALLOWED* and victim is *CONSCIOUS AND ABLE TO SWALLOW,* have victim drink 4 to 8 ounces of water. *IF SWALLOWED* and victim is *UNCONSCIOUS OR HAVING CONVULSIONS,* do nothing except keep victim warm.
Note to physician or authorized medical personnel: Consider the use of amyl nitrite perles if symptoms of cyanide poisoning develop. If symptoms indicate, initial treatment includes the cyanide antidote kit. In all cases, break an amyl nitrite perle in a gauze pad and hold lightly under victim's nose for 15 seconds, repeating 5 times at about 15-second intervals; if necessary (and if sodium nitrite infusions will be delayed), repeat procedure every 3 minutes with fresh perles until 3 or 4 have been used. Avoid breathing the vapor while administering it to the victim. Administer sodium nitrite IV, ASAP. The usual adult dose is 10 to 20 mL of a 3% solution infused over no less than 5 minutes; the average child dose is 0.15 to 0.20 mL/kg. Monitor blood pressure during administration, and slow the rate of infusion if hypotention develops. Next, infuse sodium thiosulfate IV. The usual adult dose is 50 mL of a 25% solution infused over 10 to 20 minutes; the average child dose is 1.65 mL/kg. Repeat with nitrite and thiosulfate as required.

HEALTH HAZARDS
Personal protective equipment (PPE): B-Level PPE. Rubber gloves; goggles or face shield.
Liquid or solid irritant characteristics: Eye and skin irritant.

CHEMICAL REACTIVITY
Binary reactants: Sulfates react violently with aluminum, magnesium.

ENVIRONMENTAL DATA
Food chain concentration potential: Negative; unlikely to accumulate.
Water pollution: Effect of low concentrations on aquatic life is unknown. May be dangerous if it enters nearby water intakes; notify operators. Notify local health and wildlife officials. **Response to discharge:** Disperse and flush.

SHIPPING INFORMATION
Grades of purity: 40–45% solution in water; **Storage temperature:** Ambient; **Inert atmosphere:** None; **Venting:** Open; **Stability during transport:** Stable.

PHYSICAL AND CHEMICAL PROPERTIES
Physical state @ 59°F/15°C and 1 atm: Liquid.
Molecular weight: 415 (solute).
Specific gravity (water = 1): (estimate) more than 1.1 @ 68°F/20°C (liquid).

DODECYLTRICHLOROSILANE
REC. D:8800

SYNONYMS: DODECYL TRICHLOROSILANE; SILANE, TRICHLORODODECYL-; TRICHLORODODECYLSILANE; SILANE, DODECYLTRICHLORO-

IDENTIFICATION
CAS Number: 4484-72-4
Formula: $C_{12}H_{25}Cl_3Si$; $CH_3(CH_2)_{11}SiCl_3$
DOT ID Number: UN 1771; DOT Guide Number: 156
Proper Shipping Name: Dodecyltrichlorosilane

DESCRIPTION: Colorless to yellow liquid. Sharp, irritating odor; like hydrochloric acid. Reacts with water; hydrogen chloride gas is produced.

Poison! • Corrosive • Breathing the vapor can kill you • Firefighting gear (including SCBA) does not provide adequate protection. If exposure occurs, remove and isolate gear immediately and thoroughly decontaminate personnel • Combustible • Containers may BLEVE when exposed to fire •Vapors may form explosive mixture with air •Vapors are heavier than air and will collect and stay in low areas • Vapors may travel long distances to ignition sources and flashback • Vapors in confined areas (e.g., tanks, sewers, buildings) may explode when exposed to fire • Severely irritating to skin, eyes, and respiratory tract; prolonged contact with the skin causes burns • Toxic products of combustion may include hydrogen chloride and phosgene • Do not use water; reacts to form hydrogen chloride vapor • Do not put yourself in danger by entering a contaminated area to rescue a victim.

Hazard Classification (based on NFPA-704 Rating System)
Health Hazards (Blue): 3; Flammability (Red): 2; Reactivity (Yellow): 2; Special Notice (White): Water reactive

EMERGENCY RESPONSE: See Appendix A (156)
Evacuation:
Public safety: Isolate the area of spill or leak for at least 50 to 100 meters (160 to 330 feet) in all directions.
Spill: Increase, in the downwind direction, as necessary, the distance shown under "Public Safety."
Fire: If tank, rail car, or tank truck is involved in fire, isolate for at least 800 meters (½ mile) in all directions; also, consider initial evacuation for 800 meters (½ mile) in all directions.

EXPOSURE
Short-term effects: *SEEK MEDICAL ATTENTION.* **Vapor:** Irritating to eyes, nose, and throat. Move victim to fresh air. *IF BREATHING HAS STOPPED,* give artificial respiration; *avoid mouth-to-mouth resuscitation; use bag/mask apparatus.* If breathing is difficult, administer oxygen. **Liquid:** Will burn skin and eyes. Harmful if swallowed. Remove contaminated clothing and shoes. Flush affected areas with plenty of water. *IF IN EYES,* hold eyelids open and flush with plenty of water. *IF SWALLOWED* and victim is *CONSCIOUS AND ABLE TO SWALLOW,* have victim drink 4 to 8 ounces of water. **Do NOT induce vomiting.**
Note to physician or authorized medical personnel: Medical observation is recommended for 24 to 48 hours after breathing overexposure, as pulmonary edema may be delayed. As first aid for pulmonary edema, consider administering a corticosteroid spray. Cigarette smoking may exacerbate pulmonary injury and should be discouraged for at least 72 hours following exposure.

HEALTH HAZARDS
Personal protective equipment (PPE): B-Level PPE. Acid-vapor type respiratory protection; rubber gloves; chemical worker's goggles; other protective equipment as necessary to protect eyes and skin.
Toxicity by ingestion: Grade 3; LD_{50} = 50 to 500 mg/kg.
Vapor (gas) irritant characteristics: Respiratory tract irritation. Vapors cause irritation such that personnel will find high concentrations unpleasant. The effect is temporary.
Liquid or solid irritant characteristics: Severe eye and skin irritant. Causes second- and third-degree burns on short contact and is very injurious to the eyes.

FIRE DATA
Flash point: More than 150°F/66°C (oc).
Fire extinguishing agents not to be used: Water, foam
Behavior in fire: Difficult to extinguish; re-ignition may occur. Contact with water applied to adjacent fires produces irritating hydrogen chloride fumes.

CHEMICAL REACTIVITY
Reactivity with water: Hydrolyzes; hydrogen chloride formed.

Binary reactants: Corrosive to common metals in the presence of moisture.
Neutralizing agents for acids and caustics: Flush with water, rinse with sodium bicarbonate or lime solution.

ENVIRONMENTAL DATA
Food chain concentration potential: Negative; unlikely to accumulate.
Water pollution: Effect of low concentrations on aquatic life is unknown. May be dangerous if it enters nearby water intakes; notify operators. Notify local health and wildlife officials. **Response to discharge:** Issue warning–corrosive, water containment. Restrict access. Disperse and flush with care.

SHIPPING INFORMATION
Grades of purity: Commercial; **Storage temperature:** Ambient; **Inert atmosphere:** None; **Venting:** Pressure-vacuum; **Stability during transport:** Stable.

NAS HAZARD CLASSIFICATION FOR BULK WATER TRANSPORTATION
FIRE: 1
HEALTH: Vapor irritant: 2; Liquid or solid irritant: 4; Poisons: 3
WATER POLLUTION: Human toxicity: 3; Aquatic toxicity: 3; Aesthetic effect: 3
REACTIVITY: Other chemicals: 3; Water: 4; Self-reaction: 1

PHYSICAL AND CHEMICAL PROPERTIES
Physical state @ 59°F/15°C and 1 atm: Liquid.
Molecular weight: 303.8
Boiling point @ 1 atm: 550°F/288°C/561°K.
Specific gravity (water = 1): 1.03 @ 68°F/20°C (liquid).
Heat of combustion: (estimate) $-11{,}000$ Btu/lb = -6200 cal/g = -260×10^5 J/kg.

DOWTHERM®   **REC. D:8850**

SYNONYMS: BIPHENYL, mixed with BIPHENYL OXIDE; DIPHENYL-DIPHENYL ETHER MIXTURE; DINIL; DINYL; DIPHENYL mixed with DIPHENYL ETHER; DIPHYL; DOWTHERM® A; PHENYL ETHER-BIPHENYL MIXTURE

IDENTIFICATION
CAS Number: 8004-13-5; 101-84-8 (diphenyl oxide)
Formula: $C_{12}H_{10} \cdot C_{12}H_{10}O$; $(C_6NO_5\text{-O-}C_6NO_5)(C_6NO_5\text{-}C_6NO_5)$

DESCRIPTION: Colorless to straw colored liquid or solid below 54°F/12°C. Disagreeable aromatic odor. May float or sink:

Combustible • Heat or flame may cause explosion • Containers may BLEVE when exposed to fire • Vapors are heavier than air and will collect and stay in low areas • Vapors in confined areas (e.g., tanks, sewers, buildings) may explode when exposed to fire • Irritating to the skin, eyes, and respiratory tract • Toxic products of combustion may include carbon monoxide.

Hazard Classification (based on NFPA-704 Rating System)
Health Hazards (Blue): 1; Flammability (Red): 1; Reactivity (Yellow): 0

EMERGENCY RESPONSE: See Appendix A (171)
Evacuation:
Public safety: Isolate the area of spill or leak for at least 10 to 25 meters (30 to 80 feet) in all directions.

Spill: Increase, in the downwind direction, as necessary, the distance shown under "Public Safety."
Fire: If any large container is involved in fire, isolate for at least 800 meters (½ mile) in all directions; also, consider initial evacuation for 800 meters (½ mile) in all directions.

EXPOSURE
Short-term effects: *SEEK MEDICAL ATTENTION.* **Liquid:** Irritating to skin and eyes. Harmful if swallowed. Remove contaminated clothing and shoes. Flush affected areas with plenty of water. *IF IN EYES*, hold eyelids open and flush with plenty of water. *IF SWALLOWED* and victim is *CONSCIOUS AND ABLE TO SWALLOW*, have victim drink 4 to 8 ounces of water, have victim induce vomiting. *IF SWALLOWED* and victim is *UNCONSCIOUS OR HAVING CONVULSIONS*, do nothing except keep victim warm.

HEALTH HAZARDS
Personal protective equipment (PPE): B-Level PPE. Safety glasses
Recommendations for respirator selection: NIOSH/OSHA
10 ppm: SA:CF* (any supplied-air respirator operated in a continuous-flow mode); or CCRFOVHiE [any chemical cartridge respirator with a full facepiece and organic vapor cartridge(s) in combination with a high-efficiency particulate filter]; or GMFOVHiE [any air-purifying, full-facepiece respirator (gas mask) with a chin-style, front- or back-mounted organic vapor canister having a high-efficiency particulate filter]; or PAPROVDM* [any powered, air-purifying respirator with organic vapor cartridge(s) in combination with a dust and mist filter]; or SCBAF (any self-contained breathing apparatus with a full facepiece); or SAF (any supplied-air respirator with a full facepiece). *EMERGENCY OR PLANNED ENTRY INTO UNKNOWN CONCENTRATIONS OR IDLH CONDITIONS:* SCBAF:PD,PP (any self-contained breathing apparatus that has a full facepiece and is operated in a pressure-demand or other positive-pressure mode); or SAF:PD,PP:ASCBA (any supplied-air respirator that has a full facepiece and is operated in a pressure-demand or other positive-pressure mode in combination with an auxiliary self-contained breathing apparatus operated in a pressure-demand or other positive pressure mode). *ESCAPE:* GMFOVHiE [any air-purifying, full-facepiece respirator (gas mask) with a chin-style, front- or back-mounted organic vapor canister having a high-efficiency particulate filter]; or SCBAE (any appropriate escape-type, self-contained breathing apparatus). *Note:* Substance causes eye irritation or damage; eye protection needed.
Exposure limits (TWA unless otherwise noted): ACGIH TLV 1 ppm (7 mg/m^3) as phenyl ether; NIOSH/OSHA 1 ppm (7 mg/m^3).
Short-term exposure limits (15-minute TWA): ACGIH STEL 2 ppm (14 mg/m^3) as phenyl ether.
Toxicity by ingestion: Grade 2; LD$_{50}$ = 0.5 to 5 g/kg.
Vapor (gas) irritant characteristics: Vapors cause smarting of the eyes or respiratory system if present in high concentrations. The effect may be temporary.
Liquid or solid irritant characteristics: Eye and skin irritant. If spilled on clothing and allowed to remain, may cause smarting and reddening of the skin.
Odor threshold: 0.09–1.0 ppm.
IDLH value: 10 ppm.

FIRE DATA
Flash point: 255°F/124°C (oc); 239°F/115°C (cc).
Flammable limits in air: LEL: 0.5%; UEL: 6.2% @ 500°F/260°C; LEL: 0.8%; UEL: 3.3% @ 300°F/149°C.

CHEMICAL REACTIVITY
Binary reactants: Strong oxidizers

ENVIRONMENTAL DATA
Food chain concentration potential: Log P$_{ow}$ = 4.3. Values > 3.0 are likely to bioconcentrate in aquatic organisms and other living tissue, especially in fats.
Water pollution: DOT Appendix B, §172.101–marine pollutant. Effect of low concentrations on aquatic life is unknown. May be dangerous if it enters nearby water intakes; notify operators. Notify local health and wildlife officials. **Response to discharge:** Issue warning–water contaminant. Should be removed. Chemical and physical treatment.

SHIPPING INFORMATION
Grades of purity: 73.5% Diphenyl ether, 26.5% Diphenyl (eutectic); **Storage temperature:** Ambient; **Inert atmosphere:** None; **Venting:** Open (flame arrester); **Stability during transport:** Stable.

NAS HAZARD CLASSIFICATION FOR BULK WATER TRANSPORTATION
FIRE: 1
HEALTH: Vapor irritant: 1; Liquid or solid irritant: 1; Poisons: 1
WATER POLLUTION: Human toxicity: 0; Aquatic toxicity: 0; Aesthetic effect: 3
REACTIVITY: Other chemicals: 0; Water: 0; Self-reaction: 0

PHYSICAL AND CHEMICAL PROPERTIES
Physical state @ 59°F/15°C and 1 atm: Liquid.
Molecular weight: 166 (approximate).
Boiling point @ 1 atm: 494°F/257°C/530°K.
Melting/Freezing point: 54°F/12°C/285°K.
Critical temperature: 932°F/500°C/773°K.
Critical pressure: 456 psia = 31 atm = 3.1 MN/m^2.
Specific gravity (water = 1): 1.06 at 21°C (liquid).
Liquid surface tension: 40.1 dynes/cm = 0.0401 N/m @ 68°F/20°C.
Liquid water interfacial tension: (estimate) 30 dynes/cm = 0.03 N/m @ 68°F/20°C.
Ratio of specific heats of vapor (gas): 1.046
Heat of combustion: –14,000 Btu/lb = –7778 cal/g = –325.6 x 10^5 J/kg.
Vapor pressure: 0.08 mm @ 77°F/25°C.

DURSBAN   REC. D:8900

SYNONYMS: BRODAN; CHLORPYRIFOS; DOWCO 179; DURSBAN F; ERADEX; KILLMASTER; LORSBAN; OMS-0971; PYRINEX; 3,5,6-TRICHLORO-2-PYRIDOL-*O*-ESTER WITH *O,O*-DIETYLPHOSPHOROTHIOATE

IDENTIFICATION
CAS Number: 2921-88-2
Formula: C$_9$H$_{11}$Cl$_3$NO$_3$PS
DOT ID Number: UN 2783; DOT Guide Number: 152
Proper Shipping Name: Organophosphorus pesticides, solid, toxic, n.o.s.
Reportable Quantity (RQ): **(CERCLA)** 1 lb/0.454 kg

DESCRIPTION: White crystalline solid. Often shipped as a liquid in water or combustible petroleum products. Mild sulfur-like odor. Sinks in water; insoluble.

Poison! (organophosphate) • Breathing the vapor, skin or eye contact, or swallowing the material can kill you • Firefighting gear (including SCBA) does not provide adequate protection. If exposure occurs, remove and isolate gear immediately and thoroughly decontaminate personnel • Toxic products of combustion may include carbon monoxide, hydrogen chloride, and oxides of nitrogen, phosphorus, and sulfur • Do not put yourself in danger by entering a contaminated area to rescue a victim.

Hazard Classification (based on NFPA-704 Rating System)
Health Hazards (Blue): 3; Flammability (Red): 0; Reactivity (Yellow): 0

EMERGENCY RESPONSE: See Appendix A (152)
Evacuation:
Public safety: Isolate the area of spill or leak for at least 25 to 50 meters (80 to 160 feet) in all directions.
Spill: Increase, in the downwind direction, as necessary, the distance shown under "Public Safety."
Fire: If tank, rail car, or tank truck is involved in fire, isolate for at least 800 meters (½ mile) in all directions; also, consider initial evacuation for 800 meters (½ mile) in all directions.

EXPOSURE
Short-term effects: *SEEK MEDICAL ATTENTION.* **Solid or dust:** *POISONOUS IF SWALLOWED OR IF SKIN IS EXPOSED.* Eye pupils are small; blurred vision; runny nose; cough; shortness of breath; pain; diarrhea, nausea and vomiting; increased blood pressure, hypermotility, hallucinations; loss of consciousness; convulsions; breathing stops; death. *IF BREATHING HAS STOPPED,* give artificial respiration; *avoid mouth-to-mouth resuscitation; use bag/mask apparatus.* **Skin:** *Remove and double-bag contaminated clothing and shoes and leave in Hot Zone for later incineration by hazardous materials experts.* Flush affected areas with plenty of water. *Skin can also be decontaminated with diluted hypochlorite solution, U.S. Army M291 kit, and M258(A1) skin decontamination kit.* **Eyes:** hold eyelids open and flush with plenty of water. *IF SWALLOWED* and victim is *CONSCIOUS AND ABLE TO SWALLOW,* have victim drink 4 to 8 ounces of water. **Do NOT induce vomiting but immediately administer slurry of activated charcoal.** *IF SWALLOWED* and victim is *UNCONSCIOUS OR HAVING CONVULSIONS,* do nothing except keep victim warm.
Note to physician or authorized medical personnel: Administer atropine, 2 mg (1/30 gr) intramuscularly or intravenously as soon as any local or systemic signs or symptoms of an intoxication are noted; repeat the administration of atropine every 3 to 8 minutes until signs of atropinization (mydriasis, dry mouth, rapid pulse, hot and dry skin) occur; initiate treatment in children with 0.05 mg/kg of atropine; repeat at 5- to 10-minute intervals. Watch respiration, and remove bronchial secretions if they appear to be obstructing the airway; intubate if necessary. Pralidoxime must be administered within minutes to a few hours following exposure (depending on the specific agent) to be effective. Give 2-PAMCI (Pralidoxime; Protopam), 2.5 g in 100 mL of sterile water or in 5% dextrose and water, intravenously, slowly, in 15 to 30 minutes; if sufficient fluid is not available, give 1 g of 2-PAMCI in 3 mL of distilled water by deep intramuscular injection; repeat this every half hour if respiration weakens or if muscle fasciculation or convulsions recur. Also Diazepam, an anticonvulsant, might be needed.
Note to physician or authorized medical personnel: If inhaled, medical observation is recommended for 24 to 48 hours after breathing overexposure, as pulmonary edema may be delayed. As first aid for pulmonary edema, consider administering a corticosteroid spray. Cigarette smoking may exacerbate pulmonary injury and should be discouraged for at least 72 hours following exposure.

HEALTH HAZARDS
Personal protective equipment (PPE): A-Level PPE. OSHA Table Z-1-A air contaminant. Chemical protective material(s) reported to have good to excellent resistance: butyl rubber. Also polycarbonate has good chemical resistance.
Exposure limits (TWA unless otherwise noted): ACGIH TLV 0.2 mg/m^3; NIOSH REL 0.2 mg/m^3; skin contact contributes significantly in overall exposure.
Short-term exposure limits (15-minute TWA): NIOSH STEL 0.6 mg/m^3; skin contact contributes significantly in overall exposure.
Toxicity by ingestion: Grade 3; LD_{50} = 50 to 500 mg/kg.
Long-term health effects: Plasma, red cell, and brain cholinesterase activity was depressed.
Liquid or solid irritant characteristics: Eye and skin irritant. If spilled on clothing and allowed to remain may cause smarting and reddening of skin.

FIRE DATA
Fire extinguishing agents not to be used: Water streams applied to adjacent fires will spread contamination of pesticide over wide area.

CHEMICAL REACTIVITY
Binary reactants: Strong acids, caustics. Corrosive to copper and brass.

ENVIRONMENTAL DATA
Food chain concentration potential: Log K_{ow} = 5.0 (approx). Values > 3.0 are likely to bioconcentrate in aquatic organisms and other living tissue (in fats).
Water pollution: For weapons testing, use an M272 Water Detection Kit. Harmful to aquatic life in very low concentrations. May be dangerous if it enters nearby water intakes; notify operators. Notify local health and wildlife officials. **Response to discharge:** Issue warning–poison, water contaminant. Restrict access. Should be removed. Chemical and physical treatment.

SHIPPING INFORMATION
Grades of purity: Technical grade, minimum 98% purity; **Stability during transport:** Stable.

PHYSICAL AND CHEMICAL PROPERTIES
Molecular weight: 350.59
Boiling point @ 1 atm: Decomposes. 320°F/160°C/433°K.
Melting/Freezing point: 107–110.3°F/42–44°C/315–317°K.
Relative vapor density (air = 1): 12.09 (calculated).
Vapor pressure: 0.000002 mm.

ENDOSULFAN   REC. E:0100

SYNONYMS: BENZOEPIN; BEOSIT; BIO 5,462; CHLORTHEPIN; CRUISULFAN; CYCLODAN; ENDOCEL; ENDOSULPHAN; ENSURE; FMC 5462; 1,2,3,4,7,7-HEXACHLOROBICYCLO(2,2,1)HEPTEN-5,6-BIOXYMETHYLENESULFITE; HILDAN; 6,7,8,9,10-HEXACHLORO-1,5,5a,6,9,9a,-HEXAHYDRO-6,9-METHANO-2,4,3-BENZODIOXATHIEPIN-3-OXIDE; HOE-2671; KOP-THIODAN; MALIX; NIA 5462; NIAGRA 5462; OMS 570; RCRA No. P050; THIFOR; THIMUL; THIODAN; THIONEX; THISULFAN; TIOVEL

IDENTIFICATION
CAS Number: 115-29-7
Formula: $C_9H_5Cl_6O_3S$
DOT ID Number: UN 2761; DOT Guide Number: 151
Proper Shipping Name: Endosulfan
Reportable Quantity (RQ): **(CERCLA)** 1 lb/0.454 kg

DESCRIPTION: Brown crystalline solid (technical product is a tan, waxy, isomer mixture), dust, wettable powder, granule, or solution. Slight sulfur dioxide odor. Sinks in water; slightly soluble. Wettable form dissolves.

Poison! (organochlorine) • Combustible solid • Breathing the vapor, skin or eye contact, or swallowing the material can kill you • Firefighting gear (including SCBA) does not provide adequate protection. If exposure occurs, remove and isolate gear immediately and thoroughly decontaminate personnel • Containers may BLEVE when exposed to fire • Vapors are heavier than air and will collect and stay in low areas • Vapors in confined areas (e.g., tanks, sewers, buildings) may explode when exposed to fire • Irritating to the skin, eyes, and respiratory tract • Toxic products of combustion may include chlorine and sulfur oxides • Do not put yourself in danger by entering a contaminated area to rescue a victim.

Hazard Classification (based on NFPA-704 Rating System)
Health Hazards (Blue): 4; Flammability (Red): 1; Reactivity (Yellow): 0

EMERGENCY RESPONSE: See Appendix A (151)
Evacuation:
Public safety: Isolate the area of spill or leak for at least 25 to 50 meters (80 to 160 feet) in all directions.
Spill: Increase, in the downwind direction, as necessary, the distance shown under "Public Safety."
Fire: If tank, rail car, or tank truck is involved in fire, isolate for at least 800 meters (½ mile) in all directions; also, consider initial evacuation for 800 meters (½ mile) in all directions.

EXPOSURE
Short-term effects: *SEEK MEDICAL ATTENTION.* **Solid or solution:** *POISONOUS IF SWALLOWED OR IF SKIN IS EXPOSED.* Irritating to skin. *IF BREATHING HAS STOPPED,* give artificial respiration; *avoid mouth-to-mouth resuscitation; use bag/mask apparatus.* Remove contaminated clothing and shoes. Flush affected areas with plenty of water. *IF IN EYES*, hold eyelids open and flush. with plenty of water. *IF SWALLOWED* and victim is *CONSCIOUS AND ABLE TO SWALLOW*, have victim drink 4 to 8 ounces of water and have victim induce vomiting. *IF SWALLOWED* and victim is *UNCONSCIOUS OR HAVING CONVULSIONS,* do nothing except keep victim warm.
Note to physician or authorized medical personnel: CNS *SYMPTOMS*: Phenobarbital may be used.

HEALTH HAZARDS
Personal protective equipment (PPE): A-Level PPE. OSHA Table Z-1-A air contaminant. Rubber gloves, mask, or respirator.
Exposure limits (TWA unless otherwise noted): ACGIH TLV 0.1 mg/m³; NIOSH REL 0.1 mg/m³; skin contact contributes significantly in overall exposure.
Toxicity by ingestion: Grade 4; LD_{50} less than 50 mg/kg.
Long-term health effects: Occasional epileptiform convulsions of grand mal or petit mal type have occurred in workers from skin absorption. Neoplastic effects have been reported.
Liquid or solid irritant characteristics: As a solution incorporated in oily media or with surfactants or emulsifiers. Minimum hazard. If spilled on clothing and allowed to remain, may cause smarting and reddening of skin.

CHEMICAL REACTIVITY
Reactivity with water: Slowly hydrolyzes; decomposes more rapidly in presence of alkalis and acids to form sulfur dioxide.
Binary reactants: Alkalis, acids. Corrosive to iron.

ENVIRONMENTAL DATA
Food chain concentration potential: Log P_{ow} = 3.7 (endosulfan sulfate). Values > 3.0 are likely to bioconcentrate in aquatic organisms and other living tissue, especially in fats.
Water pollution: DOT Appendix B, §172.101- severe marine pollutant. Harmful to aquatic life in low concentrations. May be dangerous if it enters nearby water intakes; notify operators. Notify local health and wildlife officials. **Response to discharge:** Issue warning–poison, water contaminant. Should be removed. Chemical and physical treatment.

SHIPPING INFORMATION
Grades of purity: 35%, 50% (wettable powders); 17.5%, 35%, 50% (emulsifiable concentrates); 2 lb/gal; 1%, 2%, 3%, 4%, 5%, and 6% (dusts); **Storage temperature:** More than 20°F/–6°C (miscible); **Stability during transport:** Stable when dry.

PHYSICAL AND CHEMICAL PROPERTIES
Physical state @ 59°F/15°C and 1 atm: Solid.
Molecular weight: 406.9
Boiling point @ 1 atm: (decomposes).
Melting/Freezing point: Technical grade: 158–212°F/70–100°C/343.2–373.2°K; Pure p-isomer: 223–226°F/106–108°C/379–381.2°K; Pure o-isomer: 406–410°F/208–210°C/481–483°K.
Specific gravity (water = 1): 1.745 @ 68°F/20°C.
Relative vapor density (air = 1): 14.0
Vapor pressure: 0.00001 mm @ 77°F/25°C.

ENDRIN REC. E:0200

SYNONYMS: COMPOUND 269; ENDREX; ENDRINE (French); HEXACHLOROEPOXYOCTAHYDRO-*endo,endo*-DIMETHANONAPTHALENE; 1,2,3,4,10,10-HEXACHLORO-6,7-EPOXY-1,4,4A,5,6,7,8,8*A*-OCTAHYDRO-1,4-*endo-endo*-1,4,5,8-DIMETHANONAPHTHALENE; HEXADRIN; MENDRIN; NENDRIN; RCRA No. P051

IDENTIFICATION
CAS Number: 72-20-8
Formula: $C_{12}H_8Cl_6O$
DOT ID Number: UN 2761; DOT Guide Number: 151
Proper Shipping Name: Organochlorine pesticides, solid, toxic, n.o.s.
Reportable Quantity (RQ): 1 lb/0.454 kg

DESCRIPTION: Colorless to tan solid or solution. Sometimes shipped as an emulsifiable concentrate in xylene solution. Odorless. Sinks in water.

Poison! (organochlorine) • Combustible solid • Heat or flame may cause explosion • Breathing the vapor, skin or eye contact, or swallowing the material can kill you • Firefighting gear (including SCBA) does not provide adequate protection. If exposure occurs, remove and isolate gear immediately and thoroughly decontaminate personnel • Containers may BLEVE when exposed to fire • Vapors are heavier than air and will collect and stay in low areas • Vapors in confined areas (e.g., tanks, sewers, buildings) may explode when exposed to fire • Irritating to the skin, eyes, and respiratory tract • Toxic products of combustion may include hydrogen chloride and phosgene • Do not put yourself in danger by entering a contaminated area to rescue a victim. *Note:* Solution in xylene is highly flammable. See Xylene.

Hazard Classification (based on NFPA-704 Rating System)
Health Hazards (Blue): 4; Flammability (Red): 1; Reactivity (Yellow): 0

EMERGENCY RESPONSE: See Appendix A (151)
Evacuation:
Public safety: Isolate the area of spill or leak for at least 25 to 50 meters (80 to 160 feet) in all directions.
Spill: Increase, in the downwind direction, as necessary, the distance shown under "Public Safety."
Fire: If tank, rail car, or tank truck is involved in fire, isolate for at least 800 meters (½ mile) in all directions; also, consider initial evacuation for 800 meters (½ mile) in all directions.

EXPOSURE
Short-term effects: *SEEK MEDICAL ATTENTION.* **Dust:** *POISONOUS IF INHALED OR IF SKIN IS EXPOSED.* Irritating to eyes, nose, and throat. Move victim to fresh air. *IF IN EYES,* hold eyelids open and flush. with plenty of water. *IF BREATHING HAS STOPPED,* give artificial respiration; *avoid mouth-to-mouth resuscitation; use bag/mask apparatus.* IF breathing is difficult, administer oxygen. **Liquid or solid:** *POISONOUS IF SWALLOWED OR IF SKIN IS EXPOSED.* Irritating to skin and eyes. Remove contaminated clothing and shoes. Flush affected areas with plenty of water. *IF IN EYES,* hold eyelids open and flush with plenty of water. *IF SWALLOWED* and victim is *CONSCIOUS AND ABLE TO SWALLOW,* have victim drink 4 to 8 ounces of water. **Do NOT induce vomiting.** Administer a slurry of activated charcoal. *IF SWALLOWED* and victim is *UNCONSCIOUS OR HAVING CONVULSIONS,* do nothing except keep victim warm.

HEALTH HAZARDS
Personal protective equipment (PPE): A-Level PPE. OSHA Table Z-1-A air contaminant. Respirator for spray, fog, or dust; rubber gloves and boots.
Recommendations for respirator selection: NIOSH/OSHA *1 mg/m³*: CCROVDMFu [any chemical cartridge respirator with organic vapor cartridge(s) in combination with a dust, mist, and fume filter]; or SA (any supplied-air respirator). *2 mg/m³:* SA:CF (any supplied-air respirator operated in a continuous-flow mode); PAPROVDMFu [any powered, air purifying respirator with organic vapor cartridge (s) in combination with a dust, mist, and fume filter]; or CCRFOVHiE [any chemical cartridge respirator with a full facepiece and organic vapor cartridge(s) in combination with a high-efficiency particulate filter]; or GMFOVHiE [any air-purifying, full-facepiece respirator (gas mask) with a chin-style, front- or back-mounted organic vapor canister having a high-efficiency particulate filter]; or SCBAF (any self-contained breathing apparatus with full facepiece); or SAF (any supplied-air respirator with a full facepiece). *EMERGENCY OR PLANNED ENTRY INTO UNKNOWN CONCENTRATIONS OR IDLH CONDITIONS:* SCBAF:PD,PP (any self-contained breathing apparatus that has a full facepiece and is operated in a pressure-demand or other positive-pressure mode); or SAF:PD,PP:ASCBA (any supplied-air respirator that has a full facepiece and is operated in a pressure-demand or other positive-pressure mode in combination with an auxiliary self-contained breathing apparatus operated in a pressure-demand or other positive pressure mode). *ESCAPE:* GMFOVHiE [any air-purifying, full-facepiece respirator (gas mask) with a chin-style, front-or back-mounted canister having a high efficiency particulate filter); or SCBAE (any appropriate escape-type, self-contained breathing apparatus).
Exposure limits (TWA unless otherwise noted): ACGIH TLV 0.1 mg/m³; NIOSH/OSHA 0.1 mg/m³; skin contact contributes significantly in overall exposure.
Toxicity by ingestion: Grade 4; oral LD_{50} = 3 mg/kg (rat).
Vapor (gas) irritant characteristics: Eye and respiratory tract irritant.
Liquid or solid irritant characteristics: Eye and skin irritant (solid).
IDLH value: 2 mg/m³.

FIRE DATA
Flash point: Nonflammable solid or flammable solution in xylene [more than 80°F/27°C (oc)].
Flammable limits in air: LEL: 1.1%; UEL: 7% (solution in xylene).
Fire extinguishing agents not to be used: Water may be ineffective on solution fire.
Autoignition temperature: (solution in xylene) 850–900°F/454–482°C/737–755°K.
Electrical hazard: Class I, Group D (solution in xylene).
Burning rate: 4 mm/min (xylene).

CHEMICAL REACTIVITY
Binary reactants: Strong oxidizers, strong acids, parathion.

ENVIRONMENTAL DATA
Food chain concentration potential: Log P_{ow} = 4.7–5.0+. Values > 3.0 are likely to bioconcentrate in aquatic organisms and other living tissue, especially in fats.
Water pollution: DOT Appendix B, §172.101- severe marine pollutant. Harmful to aquatic life in very low concentrations. May be dangerous if it enters nearby water intakes; notify operators. Notify local health and wildlife officials. **Response to discharge:** Issue warning–poison, water contaminant. Restrict access. Should be removed. Chemical and physical treatment.

SHIPPING INFORMATION
Grades of purity: Technical, 95-98%; Dry formulations, up to 75% endrin; liquid formulations, up to 25% inflammable xylene;
Storage temperature: Ambient; **Inert atmosphere:** None;
Venting: Open; **Stability during transport:** Stable.

PHYSICAL AND CHEMICAL PROPERTIES
Physical state @ 59°F/15°C and 1 atm: Solid.
Molecular weight: 380.92
Boiling point @ 1 atm: (decomposes).
Melting/Freezing point: 392°F/200°C/573°K.
Specific gravity (water = 1): 1.70 @ 77°F/25°C (solid).
Vapor pressure: (Reid) Low

EPICHLOROHYDRIN REC. E:0300

SYNONYMS: 2-CHLOROPROPYLENE OXIDE; 3-CHLOROPROPYLENE OXIDE; CHLOROMETHYL OXIRANE; EEC No. 603-026-00-6; γ-CHLOROPROPYLENE OXIDE; GLYCEROL EPICHLOROHYDRIN; 1-CHLORO-2,3-EPOXYPROPANE; α-EPICHLOROHYDRIN; ECH;

Epichlorohydrin

EPICLORHIDRINA (Spanish); 1,2-EPOXY-3-CHLOROPROPANE; γ-CHLOROPROPYLENE OXIDE; GLYCEROL EPICHLORHYDRIN; GLYCIDYL-CHLORIDE; OXIRANE, 2-(CHLOROMETHYL); OXIRANE, 2-(CHLOROMETHYL)-; 3-CHLORO-1,2-PROPYLENE OXIDE; CHLOROMETHYLOXIRANE; RCRA No. U041

IDENTIFICATION
CAS Number: 106-89-8
Formula: C_3H_5ClO; $O \cdot CH_2 \cdot CH \cdot CH_2Cl$
DOT ID Number: UN 2023; DOT Guide Number: 131P
Proper Shipping Name: Epichlorohydrin
Reportable Quantity (RQ): **(CERCLA)** 100 lb/45.4 kg

DESCRIPTION: Colorless watery liquid. Irritating; chloroform-like odor. Sinks in water; moderately soluble; poisonous, flammable vapor is produced.

Highly flammable • Corrosive to skin, eyes, and respiratory tract; skin and eye contact can cause severe burns and blindness; inhalation symptoms may be delayed • Firefighting gear (including SCBA) does not provide adequate protection. If exposure occurs, remove and isolate gear immediately and thoroughly decontaminate personnel • Vapors form explosive mixture with air • Polymerization hazard • Heat can induce polymerization with rapid release of energy; sealed containers may rupture explosively • May react with itself blocking relief valves; leading to tank explosions • Containers may BLEVE when exposed to fire • Vapors are heavier than air and will collect and stay in low areas • Vapors may travel long distances to ignition sources and flashback • Vapors in confined areas (e.g., tanks, sewers, buildings) may explode when exposed to fire • Toxic products of combustion may include hydrogen chloride, and possibly phosgene • Do not put yourself in danger by entering a contaminated area to rescue a victim.

Hazard Classification (based on NFPA-704 Rating System)
Health Hazards (Blue): 3; Flammability (Red): 3; Reactivity (Yellow): 2

EMERGENCY RESPONSE: See Appendix A (131)
Evacuation:
Public safety: Isolate spill area for at least 100 to 200 meters (330 to 660 feet) in all directions.
Spill: Increase, as necessary, the isolation distance shown above, in "Public safety."
Fire: Isolate for 800 meters (½ mile) in all directions, especially if tank, rail car, or tank truck is involved in fire.

EXPOSURE
Short-term effects: *SEEK MEDICAL ATTENTION.* **Vapor:** *POISONOUS IF INHALED OR ABSORBED THROUGH THE SKIN.* Possible Lung edema. Irritating to eyes. Move to fresh air. *IF BREATHING HAS STOPPED,* give artificial respiration. IF breathing is difficult, administer oxygen. **Liquid:** *POISONOUS IF SWALLOWED OR ABSORBED THROUGH THE SKIN.* Will burn skin and eyes. Remove contaminated clothing and shoes. Flush affected areas with plenty of water. *IF IN EYES,* hold eyelids open and flush with plenty of water. *IF SWALLOWED* and victim is *CONSCIOUS AND ABLE TO SWALLOW,* have victim drink 4 to 8 ounces of water, and have victim induce vomiting. *IF SWALLOWED* and victim is *UNCONSCIOUS OR HAVING CONVULSIONS,* do nothing except keep victim warm.
Note to physician or authorized medical personnel: Medical observation is recommended for 24 to 48 hours after breathing overexposure, as pulmonary edema may be delayed. As first aid for pulmonary edema, consider administering a corticosteroid spray. Cigarette smoking may exacerbate pulmonary injury and should be discouraged for at least 72 hours following exposure.

HEALTH HAZARDS
Personal protective equipment (PPE): A-Level PPE. OSHA Table Z-1-A air contaminant. Full face organic vapor respirator; protective gloves and goggles. Chemical protective material(s) reported to have good to excellent resistance: butyl rubber, Teflon®, Viton®, PV alcohol.
Recommendations for respirator selection: NIOSH
At any detectable concentration: SCBAF:PD,PP (any MSHA/NIOSH approved self-contained breathing apparatus that has a full facepiece and is operated in a pressure-demand or other positive-pressure mode); or SAF:PD,PP:ASCBA (any supplied-air respirator that has a full facepiece and is operated in a pressure-demand or other positive-pressure mode in combination with an auxiliary, self-contained breathing apparatus operated in a pressure-demand or other positive pressure mode). *ESCAPE:* GMFOVAG (any air-purifying, full-facepiece respirator (gas mask) with a chin-style, front- or back-mounted organic vapor and acid gas canister); or SCBAE (any appropriate escape-type, self-contained breathing apparatus).
Exposure limits (TWA unless otherwise noted): ACGIH TLV 0.1 ppm (0.38 mg/m^3) suspected human carcinogen; OSHA PEL 5 ppm (19 mg/m^3); NIOSH REL potential human carcinogen; reduce exposure to lowest feasible level; skin contact contributes significantly in overall exposure
Toxicity by ingestion: Grade 3; LD_{50} = 50 to 500 mg/kg.
Long-term health effects: May cause liver, kidney and adrenal injury. Possible sterility (may be temporary). Intensive skin contact may cause dermatitis. Causes cancer in experimental animals. IARC probable carcinogen, rating 2A; sufficient animal evidence. NTP anticipated carcinogen.
Vapor (gas) irritant characteristics: Eye and respiratory tract irritant. Vapor is irritating such that personnel will not usually tolerate moderate or high vapor concentrations.
Liquid or solid irritant characteristics: Fairly severe eye and skin irritant. May cause pain and second-degree burns on the skin after a few minutes of contact.
Odor threshold: 0.082–11.9 ppm.
IDLH value: Potential human carcinogen; 75 ppm.

FIRE DATA
Flash point: 93°F/35°C (cc).
Flammable limits in air: LEL: 3.7%; UEL: 21.4%.
Fire extinguishing agents not to be used: Avoid use of dry chemical if fire occurs in container with confined vent.
Behavior in fire: Containers may explode in fire due to polymerization.
Autoignition temperature: 772°F/411°C/684°K.
Electrical hazard: Class I, Group C.
Burning rate: 2.6 mm/min

CHEMICAL REACTIVITY
Reactivity with water: Mild reaction; may not be hazardous.
Binary reactants: Strong oxidizers may cause fire and explosions. Reacts with acids, metallic halides, caustics, zinc, aluminum. The wet product will pit steel.
Polymerization: May polymerize in presence of fire, strong acids and bases, particularly when hot.
Reactivity group: 17
Compatibility class: Epichlorohydrins

ENVIRONMENTAL DATA
Food chain concentration potential: Log P_{ow} = 0.45; unlikely to accumulate.
Water pollution: DOT Appendix B, §172.101–marine pollutant. Effect of low concentrations on aquatic life is unknown. May be dangerous if it enters nearby water intakes; notify operators. Notify local health and wildlife officials. **Response to discharge:** Issue warning–water contaminant. Restrict access. Disperse and flush.

SHIPPING INFORMATION
Grades of purity: 99.0%; **Storage temperature:** Ambient; **Inert atmosphere:** None; **Venting:** Pressure-vacuum; **Stability during transport:** Stable.

NAS HAZARD CLASSIFICATION FOR BULK WATER TRANSPORTATION
FIRE: 3
HEALTH: Vapor irritant: 3; Liquid or solid irritant: 3; Poisons: 4
WATER POLLUTION: Human toxicity: 3; Aquatic toxicity: 3; Aesthetic effect: 2
REACTIVITY: Other chemicals: 3; Water: 1; Self-reaction: 2

PHYSICAL AND CHEMICAL PROPERTIES
Physical state @ 59°F/15°C and 1 atm: Liquid.
Molecular weight: 92.53
Boiling point @ 1 atm: 243°F/117°C/390°C.
Melting/Freezing point: –54°F/–48°C/225°K.
Specific gravity (water = 1): 1.18 @ 68°F/20°C (liquid).
Liquid surface tension: 37.0 dynes/cm = 0.037 N/m @ 68°F/20°C.
Relative vapor density (air = 1): 3.3
Ratio of specific heats of vapor (gas): 1.155
Latent heat of vaporization: 176 Btu/lb = 97.9 cal/g = 4.10×10^5 J/kg.
Heat of combustion: –8143 Btu/lb = –4524 cal/g = -189.4×10^5 J/kg.
Vapor pressure: (Reid) 0.67; 13 mm.

EPOXIDIZED VEGETABLE OILS   REC. E:0400

SYNONYMS: EPOXIDIZED DRYING OILS; EPOXIDIZED OILS; DRYING OIL EPOXIDES

IDENTIFICATION
CAS Number: 8013-07-8 and others
Formula: $(C_2H_4O)_n \cdot C_{18}H_{36}O$; $(O \cdot CRH \cdot CHR'COO)_3C_3NO_5$

DESCRIPTION: Colorless to pale yellow, oily liquid. Odorless. Floats on the surface of water.

Combustible • Heat or flame may cause explosion • Containers may BLEVE when exposed to fire • Vapors are heavier than air and will collect and stay in low areas • Vapors in confined areas (e.g., tanks, sewers, buildings) may explode when exposed to fire • Irritating to the skin, eyes, and respiratory tract • Toxic products of combustion may include carbon monoxide.

EMERGENCY RESPONSE: See Appendix A (171)
Evacuation:
Public safety: Isolate the area of spill or leak for at least 10 to 25 meters (30 to 80 feet) in all directions.
Spill: Increase, in the downwind direction, as necessary, the distance shown under "Public Safety."
Fire: If any large container is involved in fire, isolate for at least 800 meters (½ mile) in all directions; also, consider initial evacuation for 800 meters (½ mile) in all directions.

HEALTH HAZARDS
Toxicity by ingestion: Grade 0; LD_{50} = > 15 g/kg (rat).

FIRE DATA
Flash point: 585°F/307°C (oc).

ENVIRONMENTAL DATA
Water pollution: Effect of low concentrations on aquatic life is unknown. Fouling to shoreline. May be dangerous if it enters nearby water intakes; notify operators. Notify local health and wildlife officials. **Response to discharge:** Mechanical containment. Chemical and physical treatment.

SHIPPING INFORMATION
Grades of purity: Epoxidized vegetable oil; Epoxidized soybean oil; **Storage temperature:** Ambient; **Inert atmosphere:** None; **Venting:** Open; **Stability during transport:** Stable.

PHYSICAL AND CHEMICAL PROPERTIES
Physical state @ 59°F/15°C and 1 atm: Liquid.
Boiling point @ 1 atm: Very high.
Specific gravity (water = 1): 1.0 @ 68°F/20°C (liquid).
Liquid surface tension: 36.2 dynes/cm = 0.0362 N/m at 24°C.
Liquid water interfacial tension: 50 dynes/cm = 0.05 N/m at 22.7°C.
Heat of combustion: (estimate) –13,000 Btu/lb = –7.000 cal/g = -300×10^5 J/kg.

ETHANE   REC. E:0500

SYNONYMS: BIMETHYL; ETANO (Spanish); DIMETHYL; EEC No. 601-002-00-X; ETHYL HYDRIDE; METHYLMETHANE

IDENTIFICATION
CAS Number: 74-84-0
Formula: C_2H_6
DOT ID Number: UN 1035 (compressed gas); UN 1961 (refrigerated liquid); DOT Guide Number: 115
Proper Shipping Name: Ethane, compressed; Ethane, refrigerated liquid, or Ethane-propane mixture, refrigerated Liquid.

DESCRIPTION: Colorless gas. Shipped and stored as a liquefied compressed gas or cryogenic liquid. Mild gasoline-like odor. Floats and boils on water; insoluble; flammable visible vapor cloud is produced.

Extremely flammable • Vapors may cause oxygen deficiency; dizziness, suffocation, or asphyxiation without warning • Forms explosive mixture with air • Reacts explosively with many materials • Gas from liquefied gas are initially heavier than air and spread along ground • Exposure of cylinders to elevated temperatures, fire, and flame may cause cylinders to rupture or cause frangible disk to burst, releasing entire contents of cylinder • Ruptured or venting cylinders may rocket through buildings and/or travel a considerable distance • Contact with liquid may cause frostbite.

Hazard Classification (based on NFPA-704 Rating System)
Health Hazards (Blue): 1; Flammability (Red): 4; Reactivity (Yellow): 0

EMERGENCY RESPONSE: See Appendix A (115)
Evacuation:
Public safety: Isolate spill area for at least 50 to 100 meters (160 to 330 feet) in all directions.
Spill: Consider initial downwind evacuation for at least 800 meters (½ mile).
Fire: Isolate for 1600 meters (1 mile) in all directions, especially if tank, rail car, or tank truck is involved in fire.

EXPOSURE
Short-term effects: *SEEK MEDICAL ATTENTION.* **Vapor:** IF inhaled, will cause difficult breathing. Not irritating to eyes, nose or throat. Move to fresh air. *IF BREATHING HAS STOPPED,* give artificial respiration. IF breathing is difficult, administer oxygen.
Liquid: Will cause frostbite. Flush affected areas with plenty of water. *DO NOT RUB AFFECTED AREAS.*

HEALTH HAZARDS
Personal protective equipment (PPE): B-Level PPE. Self-contained breathing apparatus for high vapor concentrations. Wear thermal protective clothing. Chemical protective material(s) reported to have good to excellent resistance: polyethylene. Also, PVC and Chlorinated polyethylene offers limited protection.
Vapor (gas) irritant characteristics: Vapors are nonirritating to the eyes and throat but can cause difficult breathing.
Liquid or solid irritant characteristics: Practically harmless to the skin because it is very volatile and evaporates quickly. Possible frostbite from rapidly evaporating liquid.
Odor threshold: 899 ppm.

FIRE DATA
Flash point: –211°F/–135°C (flammable gas).
Flammable limits in air: LEL: 2.9%; UEL: 13.0%.
Autoignition temperature: 880°F/471°C/744°K.
Electrical hazard: Class I, Group D; Flow or agitation of substance may generate electrostatic charges due to low conductivity.
Burning rate: 7.3 mm/min
Adiabatic flame temperature: 2394°F/1312°C (estimate).
Stoichiometric air-to-fuel ratio: 15.98 (estimate).

CHEMICAL REACTIVITY
Binary reactants: Forms explosive mixture with air.
Reactivity group: 31
Compatibility class: Paraffins

ENVIRONMENTAL DATA
Food chain concentration potential: Negative; unlikely to accumulate.
Water pollution: Not harmful. **Response to discharge:** Issue warning–high flammability. Evacuate area.

SHIPPING INFORMATION
Grades of purity: Research; pure; **Storage temperature:** –128°F/–88°C; **Inert atmosphere:** None; **Venting:** Safety relief; **Stability during transport:** Stable.

PHYSICAL AND CHEMICAL PROPERTIES
Physical state @ 59°F/15°C and 1 atm: Gas.
Molecular weight: 30.07
Boiling point @ 1 atm: –127.5°F/–88.6°C/264.6°K.
Melting/Freezing point: –279.9°F/–183.3°C/89.9°K.
Critical temperature: 90.1°F/32.3°C/305.5°K.
Critical pressure: 708.0 psia = 48.16 atm = 4.879 MN/m².
Specific gravity (water = 1): 0.546 at –88.6°C (liquid).
Liquid surface tension: 16 dynes/cm = 0.016 N/m at –88°C.
Liquid water interfacial tension: (estimate) 45 dynes/cm = 0.045 N/m at –88°C.
Relative vapor density (air = 1): 1.1
Ratio of specific heats of vapor (gas): 1.191
Latent heat of vaporization: 211 Btu/lb = 117 cal/g = 4.90×10^5 J/kg.
Heat of combustion: –20,293 Btu/lb = –11,274 cal/g = $–472.02 \times 10^5$ J/kg.
Heat of fusion: 22.73 cal/g.
Vapor pressure: (Reid) Very high; 30 mm.

ETHION
REC. E:0600

SYNONYMS: AC 3422; BIS(s-(DIETHOXYPHOSPHINOTHIOYL)MERCAPTO)METHANE; BIS(DITHIOPHOSPHATEDE O,O-DIETHYLE) de s,s'-METHYLENE (French); BLADAN; DIETHION; EMBATHION; ETHIOL; ETION (Spanish); ETHODAN; ETHYL METHYLENE PHOSPHORODITHIOATE; FMC-1240; FOSFONO 50; HYLEMOX; ITOPAZ; KWIT; METHANEDITHIOL, S,S-DIESTER WITH O,O-DIETHYL PHOSPHORODITHIOATE ACID; METHYLENE-s,s'-BIS(O,O-DIAETHYL-DITHIOPHOSPHAT) (German); s,s'-METHYLENE O,O,O',O'-TETRAETHYL PHOSPHORODITHIOATE; NIAGARA 1240; NIALATE; PHOSPHORODITHIOIC ACID, O,O-DIETHYL ESTER, s,s-DIESTER WITH METHANEDITHIOL; PHOSPHOTOX E; RODOCID; RP 8167; SOPRATHION; O,O,O',O'-TETRAETHYL-BIS(DITHIOPHOSPHAT)(German); O, O,O,O-TETRAETHYL s,s'-METHYLENEBIS(DITHIOPHOSPHATE); O,O,O',O'-TETRAETHYL s,s'-METHYLENEBISPHOSPHORDITHIOATE; TETRAETHYL s,s'-METHYLENE BIS(PHOSPHOROTHIOLOTHIONATE); O,O,O',O'-TETRAETHYL s,s'-METHYLENE DI(PHOSPHORODITHIOATE); VEGFRUFOSMITE; s,s-METHYLENE O,O,O',O' TETRAETHYL ESTER PHOSPHORODITHIOIC ACID; VEGFRUFOSMITE

IDENTIFICATION
CAS Number: 563-12-2
Formula: $C_9H_{22}O_4P_2S_4$
DOT ID Number: UN 2783; DOT Guide Number: 152
Proper Shipping Name: Organophosphorus pesticides, solid, toxic, n.o.s.
Reportable Quantity (RQ): **(CERCLA)** 10 lb/4.54 kg

DESCRIPTION: Clear to amber liquid; mild dithiophosphate acid (technical grade). Pure is odorless. Sinks and mixes slowly with water.

Poison! (organophosphate) • Combustible • Breathing the vapor, skin or eye contact, or swallowing the material can kill you • Firefighting gear (including SCBA) does not provide adequate protection. If exposure occurs, remove and isolate gear immediately and thoroughly decontaminate personnel • Containers may BLEVE when exposed to fire • Vapors are heavier than air and will collect and stay in low areas • Vapors in confined areas (e.g., tanks, sewers, buildings) may explode when exposed to fire • Toxic products of combustion may include carbon monoxide, and oxides of phosphorus and sulfur • Do not put yourself in danger by entering a contaminated area to rescue a victim.

470 Ethoxy dihydropyran

Hazard Classification (based on NFPA-704 Rating System)
Health Hazards (Blue): 3; Flammability (Red): 1; Reactivity (Yellow): 0

EMERGENCY RESPONSE: See Appendix A (152)
Evacuation:
Public safety: Isolate the area of spill or leak for at least 25 to 50 meters (80 to 160 feet) in all directions.
Spill: Increase, in the downwind direction, as necessary, the distance shown under "Public Safety."
Fire: If tank, rail car, or tank truck is involved in fire, isolate for at least 800 meters (½ mile) in all directions; also, consider initial evacuation for 800 meters (½ mile) in all directions.

EXPOSURE
Short-term effects: *SEEK MEDICAL ATTENTION.* **Vapor or Liquid spray:** *POISONOUS IF INHALED, IF SWALLOWED, OR IF SKIN IS EXPOSED.* Eye pupils are small; blurred vision; runny nose; cough; shortness of breath; pain; diarrhea, nausea and vomiting; increased blood pressure, hypermotility, hallucinations; loss of consciousness; convulsions; breathing stops; death. *IF BREATHING HAS STOPPED,* give artificial respiration; *avoid mouth-to-mouth resuscitation; use bag/mask apparatus.* **Skin:** *Remove and double-bag contaminated clothing (including shoes) and leave them in the Hot Zone*; wash skin with soap and water and large volumes of water for at least 15 minutes. *Skin can also be decontaminated with diluted hypochlorite solution, U.S. Army M291 kit, and M258(A1) skin decontamination kit.* **Eye:** Rinse with large volumes of water or saline for at least 15 minutes. *IF SWALLOWED,* will cause nausea, vomiting, or loss of consciousness. *Remove and double-bag contaminated clothing and shoes and leave in Hot Zone for later incineration by hazardous materials experts.* Flush affected areas with plenty of water. **If swallowed** and victim is *CONSCIOUS AND ABLE TO SWALLOW,* have victim drink 4 to 8 ounces of water. **Do NOT induce vomiting but immediately administer slurry of activated charcoal**. *IF SWALLOWED* and victim is *UNCONSCIOUS OR HAVING CONVULSIONS,* do nothing except keep victim warm. *Note to physician or authorized medical personnel*: Administer atropine, 2 mg (1/30 gr) intramuscularly or intravenously as soon as any local or systemic signs or symptoms of an intoxication are noted; repeat the administration of atropine every 3 to 8 minutes until signs of atropinization (mydriasis, dry mouth, rapid pulse, hot and dry skin) occur; initiate treatment in children with 0.05 mg/kg of atropine; repeat at 5- to 10-minute intervals. Watch respiration, and remove bronchial secretions if they appear to be obstructing the airway; intubate if necessary. Pralidoxime must be administered within minutes to a few hours following exposure (depending on the specific agent) to be effective. Give 2-PAMCI (Pralidoxime; Protopam), 2.5 g in 100 mL of sterile water or in 5% dextrose and water, intravenously, slowly, in 15 to 30 minutes; if sufficient fluid is not available, give 1 g of 2-PAMCI in 3 mL of distilled water by deep intramuscular injection; repeat this every half hour if respiration weakens or if muscle fasciculation or convulsions recur. Also Diazepam, an anticonvulsant, might be needed.

HEALTH HAZARDS
Personal protective equipment (PPE): A-Level PPE. OSHA Table Z-1-A air contaminant. Wear safety glasses and gas masks. In fires full face masks of the oxygen-producing type should be worn. Chemical protective material(s) reported to have good to excellent resistance: Teflon®.

Exposure limits (TWA unless otherwise noted): AGCIH TLV 0.4 mg/m^3; NIOSH REL 0.4 mg/m^3; skin contact contributes significantly in overall exposure.
Toxicity by ingestion: Grade 3; LD_{50} = 50 to 500 mg/kg.
Long-term health effects: With chronic intoxication, minor additional exposure may lead to an acute serious episode due to asymptomatic cumulative depression of cholinesterase activity.
Vapor (gas) irritant characteristics: Spray not irritating to eyes and throat (vapor pressure low; eliminating vapor hazard).
Liquid or solid irritant characteristics: Poisonous, but no appreciable irritant hazard.
Odor threshold: Pure-odorless. Emulsifiable 0.6 mg/l.

FIRE DATA
Flash point: 349°F/176°C (oc).
Behavior in fire: Unstable at elevated temperatures. Violently decomposition above 302°F/150°C.

CHEMICAL REACTIVITY
Binary reactants: Acids, alkalis

ENVIRONMENTAL DATA
Water pollution: DOT Appendix B, §172.101- severe marine pollutant. Harmful to aquatic life in very low concentrations. May be dangerous if it enters nearby water intakes; notify operators. Notify local health and wildlife officials. For weapons testing, use an M272 Water Detection Kit. **Response to discharge:** Issue warning–poison, water contaminant. Restrict access. Should be removed. Chemical and physical treatment.

SHIPPING INFORMATION
Grades of purity: 47.7% emulsifiable concentrate; 2.2% emulsion; 8% granular; 25% wettable powder; **Storage temperature:** Do not store formulation 4EC below 0°F/–18°C and formulation 8EC below 20°F/–7°C. **Stability during transport:** Subject to slow oxidation in air.

PHYSICAL AND CHEMICAL PROPERTIES
Physical state @ 59°F/15°C and 1 atm: Liquid.
Molecular weight: 384.48
Boiling point @ 1 atm: Decomposes above 150°C.
Melting/Freezing point: Pure: 19.4°F/–7°C/266.2°K; Technical: 8.6°F/–13°C/260.2°K.
Specific gravity (water = 1): 1.220 @ 68°F/20°C (Pure); 1.215 to 1.230 @ 68°F/20°C (Technical).
Relative vapor density (air = 1): 13.26
Vapor pressure: 0.0000015 mm.

ETHOXY DIHYDROPYRAN   **REC. E:0700**

SYNONYMS: 2-ETHOXY DIHYDROPYRAN; 2-ETHOXY-2,3-DIHYDRO-γ-PYRAN; 2-ETHOXY-3,4-DIHYDRO-2-PYRAN; 2-ETHOXY-3,4-DIHYDRO-1,2-PYRAN; 2-ETHOXY-3,4-DIHYDRO-2H-PYRAN; 2-ETOXI-3,4-DIHIDRO-2H-PIRANO (Spanish)

IDENTIFICATION
CAS Number: 103-75-3
Formula: $C_7H_{12}O_2$; $OCH=CHCH_2CH_2CHOC_2NO_5$
DOT ID Number: UN 1993; DOT Guide Number: 128
Proper Shipping Name: Flammable liquid, n.o.s.

DESCRIPTION: Colorless liquid. Floats on the surface of water.

Flammable • Containers may BLEVE when exposed to fire • Vapors may form explosive mixture with air • Vapors are heavier than air and will collect and stay in low areas • Vapors may travel long distances to ignition sources and flashback • Vapors in confined areas (e.g., tanks, sewers, buildings) may explode when exposed to fire • Irritating to the skin, eyes, and respiratory tract • Toxic products of combustion may include carbon monoxide.

Hazard Classification (based on NFPA-704 Rating System)
Health Hazards (Blue): 2; Flammability (Red): 2; Reactivity (Yellow): 1

EMERGENCY RESPONSE: See Appendix A (128)
Evacuation:
Public safety: Isolate spill area for at least 25 to 50 meters (80 to 160 feet) in all directions.
Spill: Large spill–Consider initial downwind evacuation for at least 300 meters (1000 feet).
Fire: Isolate for 800 meters (½ mile) in all directions, especially if tank, rail car, or tank truck is involved in fire.

EXPOSURE
Short-term effects: *SEEK MEDICAL ATTENTION.*
HEALTH HAZARDS
Personal protective equipment (PPE): B-Level PPE. Goggles or face shield; rubber gloves.
Liquid or solid irritant characteristics: Eye and skin irritant.

FIRE DATA
Flash point: 108°F/42°C (oc).
Fire extinguishing agents not to be used: Water may be ineffective.
Burning rate: 4.8 mm/min

CHEMICAL REACTIVITY
Binary reactants: Oxidizers.

ENVIRONMENTAL DATA
Food chain concentration potential: Negative; unlikely to accumulate.
Water pollution: Effect of low concentrations on aquatic life is unknown. Fouling to shoreline. May be dangerous if it enters nearby water intakes; notify operators. Notify local health and wildlife officials. **Response to discharge:** Mechanical containment. Should be removed. Chemical and physical treatment.

SHIPPING INFORMATION
Grades of purity: Commercial; **Storage temperature:** Ambient; **Inert atmosphere:** None; **Venting:** Open (flame arrester); **Stability during transport:** Stable.

PHYSICAL AND CHEMICAL PROPERTIES
Physical state @ 59°F/15°C and 1 atm: Liquid.
Molecular weight: 128.17
Boiling point @ 1 atm: 289°F/143°C/416°K.
Melting/Freezing point: −148°F/−100°C/173°K.
Specific gravity (water = 1): 0.875 @ 68°F/20°C (liquid).
Liquid surface tension: (estimate) 25 dynes/cm = 0.025 N/m @ 68°F/20°C.
Liquid water interfacial tension: (estimate) 30 dynes/cm = 0.030 N/m @ 68°F/20°C.
Latent heat of vaporization: (estimate). 120 Btu/lb = 69 cal/g = 2.9×10^5 J/kg.
Heat of combustion: (estimate) −14,000 Btu/lb = −7900 cal/g = -330×10^5 J/kg.

2-ETHOXYETHANOL   REC. E:0800

SYNONYMS: ATHYLENGLYKOL-MONOATHYLATHER (German); CELLOSOLVE; CELLOSOLVE SOLVENT; DOWANOL E; DOWANOL EE; 2EE; EEC No. 603-012-00-x; EGEE; EKTASOLVE EE; ETHANOL,2-ETHOXY-; ETHER MONOETHYLIQUE de L'ETHYLENE-GLYCOL (French); ETHYL CELLOSOLVE; ETHYLENE GLYCOL ETHYL ETHER; 2-ETOXIETHANOL (Spanish); GLYCOL MONOETHYL ETHER; EGEE; ETHYLENE GLYCOL MONOETHYL ETHER; GLYCOL ETHYL ETHER; GLYCOL ETHER; GLYCOL MONOETHYL ETHER; HYDROXY ETHER; JEFFERSOL EE; OXITOL; POLY-SOLV E; POLY-SOLV EE; RCRA No. U359

IDENTIFICATION
CAS Number: 110-80-5
Formula: $C_4H_{10}O_2$; $HOCH_2CH_2OCH_2CH_3$
DOT ID Number: 1171; DOT Guide Number: 127
Proper Shipping Name: Ethylene glycol momoethyl ether
Reportable Quantity (RQ): **(CERCLA)** 1000 lb/454 kg

DESCRIPTION: Colorless oily liquid. Sweet, ether-like odor. Floats on the surface of water; soluble.

Flammable • Containers may BLEVE when exposed to fire • Vapors may form explosive mixture with air • Vapors are heavier than air and will collect and stay in low areas • Vapors may travel long distances to ignition sources and flashback • Vapors in confined areas (e.g., tanks, sewers, buildings) may explode when exposed to fire • Irritating to the skin, eyes, and respiratory tract • Toxic products of combustion may include carbon monoxide.

Hazard Classification (based on NFPA-704 Rating System)
Health Hazards (Blue): 2; Flammability (Red): 2; Reactivity (Yellow): 0

EMERGENCY RESPONSE: See Appendix A (127)
Evacuation:
Public safety: Isolate spill area for at least 25 to 50 meters (80 to 160 feet) in all directions.
Spill: Large spill–Consider initial downwind evacuation for at least 300 meters (1000 feet).
Fire: Isolate for 800 meters (½ mile) in all directions, especially if tank, rail car, or tank truck is involved in fire.

EXPOSURE
Short-term effects: *SEEK MEDICAL ATTENTION.* **Liquid:** *TOXIC WHEN ABSORBED THROUGH THE SKIN.* Irritating to skin and eyes. Harmful if swallowed. Remove contaminated clothing and shoes. Flush affected areas with plenty of water. *IF IN EYES,* hold eyelids open and flush. with plenty of water. *IF SWALLOWED* and victim is *CONSCIOUS AND ABLE TO SWALLOW,* have victim drink 4 to 8 ounces of water.

HEALTH HAZARDS
Personal protective equipment (PPE): B-Level PPE. OSHA Table Z-1-A air contaminant. Full face organic vapor respirator; rubber gloves. Neoprene is generally suitable for cellosolve.
Recommendations for respirator selection: NIOSH
5 ppm: SA* (any supplied-air respirator). *12.5 ppm:* SA:CF* (any supplied-air respirator operated in a continuous-flow mode). *25*

472 2-Ethoxyethyl acetate

ppm: SCBAF (any self-contained breathing apparatus with a full facepiece); or SAF (any supplied-air respirator with a full facepiece). *500 ppm:* SA:PD,PP* (any supplied-air respirator operated in a pressure-demand or other positive-pressure mode). *EMERGENCY OR PLANNED ENTRY INTO UNKNOWN CONCENTRATIONS OR IDLH CONDITIONS:* SCBAF:PD,PP (any self-contained breathing apparatus that has a full facepiece and is operated in a pressure-demand or other positive-pressure mode); or SAF:PD,PP:ASCBA (any supplied-air respirator that has a full facepiece and is operated in a pressure-demand or other positive-pressure mode in combination with an auxiliary self-contained breathing apparatus operated in a pressure-demand or other positive pressure mode). *ESCAPE:* GMFOV[any air-purifying, full-facepiece respirator (gas mask) with a chin-style, front- or back-mounted organic vapor canister]; or SCBAE (any appropriate escape-type, self-contained breathing apparatus). **Note:* Substance reported to cause eye irritation or damage; may require eye protection.

Exposure limits (TWA unless otherwise noted): ACGIH TLV 5 ppm (18 mg/m^3); OSHA PEL 200 ppm (740 mg/m^3); NIOSH REL 0.5 ppm (1.8 mg/m^3); skin contact contributes significantly in overall exposure; NIOSH REL reduce to lowest feasible level as glycol ethers

Toxicity by ingestion: Grade 2; LD$_{50}$ = 3.1 g/kg (rabbit).

Long-term health effects: Blood, liver and kidney damage. Animal studies have produced malformed offspring and morphological changes in the testes.

Vapor (gas) irritant characteristics: Eye and respiratory tract irritant. Vapors cause smarting of the eyes or respiratory system if present in high concentrations. The effect is temporary.

Liquid or solid irritant characteristics: Eye and skin irritant. If spilled on clothing and allowed to remain, may cause smarting and reddening of the skin.

IDLH value: 500 ppm.

FIRE DATA

Flash point: 110°F/43°C (cc).
Flammable limits in air: LEL: 1.7% @ 200°F; UEL: 15.6% @ 200°F
Fire extinguishing agents not to be used: Water.
Autoignition temperature: 455°F/235°C/508°K.
Burning rate: 2.4 mm/min
Stoichiometric air-to-fuel ratio: 35.7

CHEMICAL REACTIVITY

Binary reactants: Incompatible with strong oxidizers and alkalies, strong acids, copper. Attacks rubber.
Reactivity group: 40
Compatibility class: Glycol ethers

ENVIRONMENTAL DATA

Food chain concentration potential: Log P$_{ow}$ = –0.48. Unlikely to accumulate.
Water pollution: Effects of low concentrations on aquatic life is unknown. May be dangerous if it enters nearby water intakes; notify operators. Notify local health and wildlife officials.
Response to discharge: Disperse and flush.

SHIPPING INFORMATION

Grades of purity: Commercial; **Storage temperature:** Ambient; **Inert atmosphere:** None; **Stability during transport:** Stable.

NAS HAZARD CLASSIFICATION FOR BULK WATER TRANSPORTATION

FIRE: 2
HEALTH: Vapor irritant: 1; Liquid or solid irritant: 1; Poisons: 2
WATER POLLUTION: Human toxicity: 1; Aquatic toxicity: 2; Aesthetic effect: 1
REACTIVITY: Other chemicals: 2; Water: 0; Self-reaction: 0

PHYSICAL AND CHEMICAL PROPERTIES

Physical state @ 59°F/15°C and 1 atm: Liquid.
Molecular weight: 90.12
Boiling point @ 1 atm: 275.2°F/135.1°C/408.3°K.
Melting/Freezing point: –130°F
Specific gravity (water = 1): 0.931 @ 68°F/20°C (liquid).
Relative vapor density (air = 1): 3.0 (at boiling point).
Ratio of specific heats of vapor (gas): 1.064
Latent heat of vaporization: 191 Btu/lb = 106 cal/g = 4.44 x 10^5 J/kg
Heat of combustion: (estimate) –13,000 Btu/lb = –7400 cal/g = –310 x 10^5 J/kg.
Vapor pressure: (Reid) 0.1 psia; 4 mm.

2-ETHOXYETHYL ACETATE   REC. E:0900

SYNONYMS: ACETATO de 2-ETOXIETILO (Spanish); ACETIC ACID, 2-ETHOXYETHYL ESTER; CELLOSOLVE ACETATE; EEC No. 607-037-00-7; EGEEA; ETHOXYETHYL ACETATE; ETHYLENE GLYCOL ETHYL ETHER ACETATE; ETHYLENE GLYCOL MONOETHYL ETHER ACETATE; ETHYLENE GLYCOL MONOETHYL ETHER MONOACETATE; ETHYLOXITOL ACETATE; GLYCOL MONOETHYL ETHER ACETATE; HYDROXY ETHER; POLY-SOLV EE ACETATE

IDENTIFICATION

CAS Number: 111-15-9
Formula: C$_6$H$_{12}$O$_3$; CH$_3$COOCH$_2$CH$_2$OC$_2$NO$_5$
DOT ID Number: UN 1172; DOT Guide Number: 127
Proper Shipping Name: Ethylene glycol monoethyl ether acetate

DESCRIPTION: Colorless liquid. Mild ester-like odor. Floats and mixes slowly with water.

Extremely flammable • Containers may BLEVE when exposed to fire •Vapors may form explosive mixture with air • Vapors are heavier than air and will collect and stay in low areas • Vapors may travel long distances to ignition sources and flashback • Vapors in confined areas (e.g., tanks, sewers, buildings) may explode when exposed to fire • Irritating to the skin, eyes, and respiratory tract • Toxic products of combustion may include carbon monoxide.

Hazard Classification (based on NFPA-704 Rating System)
Health Hazards (Blue): 1; Flammability (Red): 2; Reactivity (Yellow): 0

EMERGENCY RESPONSE: See Appendix A (127)
Evacuation:
Public safety: Isolate spill area for at least 25 to 50 meters (80 to 160 feet) in all directions.
Spill: Large spill–Consider initial downwind evacuation for at least 300 meters (1000 feet).
Fire: Isolate for 800 meters (½ mile) in all directions, especially if tank, rail car, or tank truck is involved in fire.

EXPOSURE

Short-term effects: *SEEK MEDICAL ATTENTION. ABSORBED THROUGH THE SKIN.* **Vapor/fumes:** High concentrations can cause paralysis; tingling sensation in the arms, legs, facial muscles.

Liquid: Irritating to skin and eyes. Harmful if swallowed. Remove contaminated clothing and shoes. Flush affected areas with plenty of water. *IF IN EYES*, hold eyelids open and flush. with plenty of water. *IF SWALLOWED* and victim is *CONSCIOUS AND ABLE TO SWALLOW*, have victim drink 4 to 8 ounces of water. The use of alcoholic beverages may enhance the toxic effect.

HEALTH HAZARDS

Personal protective equipment (PPE): B-Level PPE. OSHA Table Z-1-A air contaminant. Full face organic vapor respirator. Neoprene is generally suitable for cellosolve.
Recommendations for respirator selection: NIOSH
5 ppm: CCROV* [any chemical cartridge respirator with organic vapor cartridge(s)]; SA* (any supplied-air respirator). *12.5 ppm:* SA:CF* (any supplied-air respirator operated in a continuous-flow mode); PAPROV* [any powered, air-purifying respirator with organic vapor cartridge(s)]. *25 ppm:* CCROV [any chemical cartridge respirator with organic vapor cartridge(s)]; GMFOV [any air-purifying, full-facepiece respirator (gas mask) with a chin-style, front- or back-mounted acid gas canister]; PAPRTOV* [any powered, air-purifying respirator with a tight-fitting facepiece and organic vapor cartridges(s)]; or SCBAF (any self-contained breathing apparatus with a full facepiece); or SAF (any supplied-air respirator with a full facepiece). *500 ppm:* SA:PD,PP* (any supplied-air respirator operated in a pressure-demand or other positive-pressure mode). *EMERGENCY OR PLANNED ENTRY INTO UNKNOWN CONCENTRATIONS OR IDLH CONDITIONS:* SCBAF:PD,PP (any self-contained breathing apparatus that has a full facepiece and is operated in a pressure-demand or other positive-pressure mode); or SAF:PD,PP:ASCBA (any supplied-air respirator that has a full facepiece and is operated in a pressure-demand or other positive-pressure mode in combination with an auxiliary self-contained breathing apparatus operated in a pressure-demand or other positive pressure mode). *ESCAPE:* GMFOV[any air-purifying, full-facepiece respirator (gas mask) with a chin-style, front- or back-mounted organic vapor canister]; or SCBAE (any appropriate escape-type, self-contained breathing apparatus).
Note: Substance reported to cause eye irritation or damage; may require eye protection.
Exposure limits (TWA unless otherwise noted): ACGIH TLV 5 ppm (27 mg/m^3); skin contact contributes significantly in overall exposure; OSHA PEL 100 ppm (540 mg/m^3); NIOSH REL 0.5 ppm (2.7 mg/m^3); skin contact contributes significantly in overall exposure
Toxicity by ingestion: Grade 2; LD_{50} = 0.5 to 5 g/kg (rabbit).
Long-term health effects: Kidney disorders.
Vapor (gas) irritant characteristics: Eye and respiratory tract irritant. Vapors cause smarting of the eyes and respiratory system if present in high concentrations. The effect may be temporary.
Liquid or solid irritant characteristics: Eye and skin irritant. If spilled on clothing and allowed to remain, may cause smarting and reddening of the skin.
Odor threshold: 0.056 ppm.
IDLH value: 500 ppm.

FIRE DATA

Flash point: 124°F/53°C (cc).
Flammable limits in air: LEL: 1.7%; UEL: 8.4%.
Fire extinguishing agents not to be used: Water.
Behavior in fire: Toxic gases, such as carbon monoxide, may be produced in fire.
Autoignition temperature: 720°F/382°C/655°K.
Stoichiometric air-to-fuel ratio: 35.7

CHEMICAL REACTIVITY

Binary reactants: Reacts with strong oxidizers, alkalis, and acids. Attacks rubber.
Reactivity group: 40
Compatibility class: Glycol ethers

ENVIRONMENTAL DATA

Food chain concentration potential: Negative; unlikely to accumulate.
Water pollution: Effects of low concentrations on aquatic life is unknown. May be dangerous if it enters nearby water intakes; notify operators. Notify local health and wildlife officials. **Response to discharge:** Disperse and flush.

SHIPPING INFORMATION

Grades of purity: Commercial; **Storage temperature:** Ambient; **Inert atmosphere:** None; **Venting:** Pressure vacuum valve; **Stability during transport:** Stable.

NAS HAZARD CLASSIFICATION FOR BULK WATER TRANSPORTATION

FIRE: 2
HEALTH: Vapor irritant: 1; Liquid or solid irritant: 1; Poisons: 1
WATER POLLUTION: Human toxicity: 1; Aquatic toxicity: 2; Aesthetic effect: 2
REACTIVITY: Other chemicals: 1; Water: 0; Self-reaction: 0

PHYSICAL AND CHEMICAL PROPERTIES

Physical state @ 59°F/15°C and 1 atm: Liquid.
Molecular weight: 132.16
Boiling point @ 1 atm: 313°F/156°C/429°K.
Melting/Freezing point: –79.1°F/–61.7°C/211.5°K.
Critical temperature: 633°F/334°C/607°K.
Critical pressure: 440 psia = 30 atm = 3.0 MN/m^2.
Specific gravity (water = 1): 0.974 @ 68°F/20°C (liquid).
Relative vapor density (air = 1): 4.7
Ratio of specific heats of vapor (gas): 1.054
Latent heat of vaporization: 130 Btu/lb = 74 cal/g = 3.1 x 10^5 J/kg.
Heat of combustion: (estimate) –10,700 Btu/lb = –6000 cal/g = –250 x 10^5 J/kg.
Vapor pressure: (Reid) 0.1 psia; 2 mm.

ETHOXYLATED DODECANOL   REC. E:1000

SYNONYMS: ETHOXYLATED DODECYL ALCOHOL; ETHOXYLATED LAURYL ALCOHOL; POLY(OXYETHYL) DODECYL ETHER; POLY(OXYETHYL) LAURYL ETHER

IDENTIFICATION

CAS Number: 9008-57-5
Formula: $C_{12}H_{25}O(CH_2CH_2O)_nCH_2CH_2OH$ where n = 6–10 (average)
DOT ID Number: UN 1987; DOT Guide Number: 127
Proper Shipping Name: Alcohols, n.o.s.

DESCRIPTION: Colorless to yellow oily liquid. Pleasant odor. Mixes slowly with water.

Combustible • Heat or flame may cause explosion • Containers may BLEVE when exposed to fire • Vapors are heavier than air and will collect and stay in low areas • Vapors in confined areas (e.g., tanks, sewers, buildings) may explode when exposed to fire • Irritating to the skin, eyes, and respiratory tract • Toxic products of combustion may include carbon monoxide.

474 Ethoxylated nonylphenol

Hazard Classification (based on NFPA-704 Rating System)
Health Hazards (Blue): 2; Flammability (Red): 1; Reactivity (Yellow): 0

EMERGENCY RESPONSE: See Appendix A (127)

EXPOSURE
Short-term effects: *SEEK MEDICAL ATTENTION.* **Liquid:** Irritating to skin. Will burn eyes. Harmful if swallowed. Flush affected areas with plenty of water. *IF IN EYES*, hold eyelids open and flush with plenty of water. *IF SWALLOWED* and victim is *CONSCIOUS AND ABLE TO SWALLOW*, have victim drink 4 to 8 ounces of water.

HEALTH HAZARDS
Personal protective equipment (PPE): B-Level PPE. Plastic gloves, goggles. Natural rubber, neoprene, and nitrile rubber may offer some protection from alcohols.
Toxicity by ingestion: Grade 1; LD_{50} = 5 to 15 g/kg (rat).
Liquid or solid irritant characteristics: Liquid causes eye injury. Contact with skin may cause irritation.

FIRE DATA
Flash point: 470°F/243°C (oc).

CHEMICAL REACTIVITY
Reactivity group: 20
Compatibility class: Alcohols, glycols

ENVIRONMENTAL DATA
Food chain concentration potential: Negative; unlikely to accumulate.
Water pollution: DOT Appendix B, §172.101–marine pollutant. Effect of low concentration on aquatic life is unknown. May be dangerous if it enters nearby water intakes; notify operators. Notify local health and wildlife officials. **Response to discharge:** Disperse and flush.

SHIPPING INFORMATION
Grades of purity: Usually 100%; **Storage temperature:** Ambient; **Inert atmosphere:** None; **Venting:** Open (flame arrester); **Stability during transport:** Stable.

PHYSICAL AND CHEMICAL PROPERTIES
Physical state @ 59°F/15°C and 1 atm: Liquid.
Molecular weight: 450-626
Boiling point @ 1 atm: Very high.
Melting/Freezing point: 61°F/16°C/289°K.
Specific gravity (water = 1): 1.02 @ 68°F/20°C (liquid).
Heat of combustion: (estimate) –11,00 Btu/lb = –6200 cal/g = –260 x 10^5 J/kg.

ETHOXYLATED NONYLPHENOL   **REC. E:1100**

IDENTIFICATION
Formula: $C_9H_{19}C_6H_4O(C_2H_4O)_nH$; n = 4 –100

DESCRIPTION: White liquid or solid. Mild aromatic odor. May float or sink in water.

Combustible • Heat or flame may cause explosion • Containers may BLEVE when exposed to fire • Vapors are heavier than air and will collect and stay in low areas • Vapors in confined areas (e.g., tanks, sewers, buildings) may explode when exposed to fire • Irritating to the skin, eyes, and respiratory tract • Toxic products of combustion may include carbon monoxide.

Hazard Classification (based on NFPA-704 Rating System)
Health Hazards (Blue): 0; Flammability (Red): 1; Reactivity (Yellow): 0

EMERGENCY RESPONSE: See Appendix A (171)
Evacuation:
Public safety: Isolate the area of spill or leak for at least 10 to 25 meters (30 to 80 feet) in all directions.
Spill: Increase, in the downwind direction, as necessary, the distance shown under "Public Safety."
Fire: If any large container is involved in fire, isolate for at least 800 meters (½ mile) in all directions; also, consider initial evacuation for 800 meters (½ mile) in all directions.

EXPOSURE
Short-term effects: *SEEK MEDICAL ATTENTION.* **Liquid or solid:** Irritating to skin and eyes. Harmful if swallowed. Remove contaminated clothing and shoes. Flush affected areas with plenty of water. *IF IN EYES*, hold eyelids open and flush. with plenty of water. *IF SWALLOWED* and victim is *CONSCIOUS AND ABLE TO SWALLOW*, have victim drink 4 to 8 ounces of water. *IF SWALLOWED* and victim is *UNCONSCIOUS OR HAVING CONVULSIONS*, do nothing except keep victim warm.

HEALTH HAZARDS
Personal protective equipment (PPE): B-Level PPE. Gloves and safety glasses
Toxicity by ingestion: Grade 2; oral LD_{50} = 1310 mg/kg (rat).
Liquid or solid irritant characteristics: Eye and skin irritant.

FIRE DATA
Flash point: 338–600°F/170–316°C (oc); > 140°F/60°C (cc).
Fire extinguishing agents not to be used: Water may be ineffective on fire.

ENVIRONMENTAL DATA
Food chain concentration potential: Negative; unlikely to accumulate.
Water pollution: DOT Appendix B, §172.101–marine pollutant. Effect of low concentrations on aquatic life is unknown. May be dangerous if it enters nearby water intakes; notify operators. Notify local health and wildlife officials. **Response to discharge:** Disperse and flush.

SHIPPING INFORMATION
Grades of purity: Commercial; **Storage temperature:** Ambient; **Inert atmosphere:** None; **Venting:** Open; **Stability during transport:** Stable.

NAS HAZARD CLASSIFICATION FOR BULK WATER TRANSPORTATION
FIRE: 1
HEALTH: Vapor irritant: 0; Liquid or solid irritant: 2; Poisons: 1
WATER POLLUTION: Human toxicity: 2; Aquatic toxicity: 1; Aesthetic effect: 0
REACTIVITY: Other chemicals: 1; Water: 0; Self-reaction: 0

PHYSICAL AND CHEMICAL PROPERTIES
Physical state @ 59°F/15°C and 1 atm: Solid or Liquid.
Molecular weight: More than 500.
Boiling point @ 1 atm: (decomposes).
Specific gravity (water = 1): 0.99–1.07 @ 77°F/25°C (liquid).

ETHOXYLATED PENTADECANOL REC. E:1200

SYNONYMS: ALCOHOLS, C11-15-secondary, ETHOXYLATED; ETHOXYLATED PENTADECYLALCOHOL; POLY(OXYETHYL) PENTADECYL ETHER

IDENTIFICATION
CAS Number: 68131-40-8
Formula: $C_{15}H_{31}O(CH_2CH_2O)_nCH_2CH_2OH$ where n = 10 (average)
DOT ID Number: UN 1987; DOT Guide Number: 127
Proper Shipping Name: Alcohols, n.o.s.

DESCRIPTION: Colorless to yellow liquid. Pleasant odor. Mixes slowly with water. Freezing point is 59°F/15°C.

Combustible • Heat or flame may cause explosion • Containers may BLEVE when exposed to fire • Vapors are heavier than air and will collect and stay in low areas • Vapors in confined areas (e.g., tanks, sewers, buildings) may explode when exposed to fire • Irritating to the skin, eyes, and respiratory tract • Toxic products of combustion may include carbon monoxide.

Hazard Classification (based on NFPA-704 Rating System)
Health Hazards (Blue): 2; Flammability (Red): 1; Reactivity (Yellow): 0

EMERGENCY RESPONSE: See Appendix A (127)
Evacuation:
Public safety: Isolate the area of spill or leak for at least 10 to 25 meters (30 to 80 feet) in all directions.
Spill: Increase, in the downwind direction, as necessary, the distance shown under "Public Safety."
Fire: If any large container is involved in fire, isolate for at least 800 meters (½ mile) in all directions; also, consider initial evacuation for 800 meters (½ mile) in all directions.

EXPOSURE
Short-term effects: *SEEK MEDICAL ATTENTION.* **Liquid:** Irritating to skin. Will burn eyes. Flush affected areas with plenty of water. *IF IN EYES*, hold eyelids open and flush with plenty of water.

HEALTH HAZARDS
Personal protective equipment (PPE): B-Level PPE. Gloves, goggles. Natural rubber, neoprene, and nitrile rubber may offer some protection from alcohols. Chemical protective material(s) reported to offer minimal to poor protection: Viton®/neoprene, butyl rubber,/neoprene.
Liquid or solid irritant characteristics: Liquid causes eye injury. Contact with skin may cause irritation.

FIRE DATA
Flash point: 470°F/243°C (oc).

CHEMICAL REACTIVITY
Binary reactants: Strong acids, caustics, isocyanates, amines.
Reactivity group: 20
Compatibility class: Alcohols, glycols

ENVIRONMENTAL DATA
Food chain concentration potential: Negative; unlikely to accumulate.
Water pollution: DOT Appendix B, §172.101–marine pollutant. Effect of low concentrations on aquatic life is unknown. May be dangerous if it enters nearby water intakes; notify operators. Notify local health and wildlife officials. **Response to discharge:** Disperse and flush.

SHIPPING INFORMATION
Grades of purity: Usually 100%; **Storage temperature:** Ambient; **Inert atmosphere:** None; **Venting:** Open (flame arrester); **Stability during transport:** Stable.

PHYSICAL AND CHEMICAL PROPERTIES
Physical state @ 59°F/15°C and 1 atm (liquid)
Molecular weight: 660
Boiling point @ 1 atm: Very high.
Melting/Freezing point: 59°F/15°C/288°K.
Specific gravity (water = 1): 1.007 @ 15°C (liquid).
Heat of combustion: (estimate) = $-11,000$ Btu/lb = -6200 cal/g = -260×10^5 J/kg.
Heat of solution: (estimate) -9 Btu/lb = -5 cal/g = -0.2×10^5 J/kg.

ETHOXYLATED TRIDECANOL REC. E:1300

SYNONYMS: ETHOXYLATED TRIDECYL ALCOHOL; TERGITOL NONIONIC 3-A-6; POLY(OXYETHYL) TRIDECYL ETHER

IDENTIFICATION
CAS Number: 9002-92-0
Formula: $C_{13}H_{27}O(CH_2CH_2O)_n CH_2CH_2OH$; n = 5 (average)
DOT ID Number: UN 1987; DOT Guide Number: 127
Proper Shipping Name: Alcohols, n.o.s.

DESCRIPTION: Colorless to yellow liquid. Mild, pleasant odor.

Combustible • Heat or flame may cause explosion • Containers may BLEVE when exposed to fire • Vapors are heavier than air and will collect and stay in low areas • Vapors in confined areas (e.g., tanks, sewers, buildings) may explode when exposed to fire • Irritating to the skin, eyes, and respiratory tract • Toxic products of combustion may include carbon monoxide.

Hazard Classification (based on NFPA-704 Rating System)
Health Hazards (Blue): 2; Flammability (Red): 1; Reactivity (Yellow): 0

EMERGENCY RESPONSE: See Appendix A (127)

EXPOSURE
Short-term effects: *SEEK MEDICAL ATTENTION.* **Liquid:** Irritating to skin. Will burn eyes. Harmful if swallowed. Flush affected areas with plenty of water. *IF IN EYES*, hold eyelids open and flush with plenty of water. *IF SWALLOWED* and victim is *CONSCIOUS AND ABLE TO SWALLOW*, have victim drink 4 to 8 ounces of water.

HEALTH HAZARDS
Personal protective equipment (PPE): B-Level PPE. Plastic gloves; goggles. Natural rubber, neoprene, and nitrile rubber may offer some protection from alcohols.
Toxicity by ingestion: Grade 2; LD_{50} = 0.5 to 5 g/kg (rat).
Liquid or solid irritant characteristics: Liquid causes eye injury. Contact with skin may cause irritation.

FIRE DATA
Flash point: 385°F/196°C (oc).

476 Ethoxy triglycol

CHEMICAL REACTIVITY
Reactivity group: 20
Compatibility class: Alcohols, glycols

ENVIRONMENTAL DATA
Water pollution: DOT Appendix B, §172.101–marine pollutant. Effect of low concentrations on aquatic life is unknown. May be dangerous if it enters nearby water intakes; notify operators. Notify local health and wildlife officials. **Response to discharge:** Disperse and flush.

SHIPPING INFORMATION
Grades of purity: Usually 100%; **Storage temperature:** Ambient; **Inert atmosphere:** None; **Venting:** Open (flame arrester); **Stability during transport:** Stable.

PHYSICAL AND CHEMICAL PROPERTIES
Physical state @ 59°F/15°C and 1 atm: Liquid.
Molecular weight: 464
Boiling point @ 1 atm: Very high.
Specific gravity (water = 1): 1.00 @ 15°C (liquid).
Heat of combustion: (estimate) –11,000 Btu/lb = –6200 cal/g = –260 x 10^5 J/kg.
Heat of solution: (estimate) –9 Btu/lb= –5 cal/g = –0.2 x 10^5 J/kg.

ETHOXYLATED TETRADECANOL REC. E:1400

SYNONYMS: ETHOXYLATED TETRADECYL ALCOHOL; ETHOXYLATED MYRISTYL ALCOHOL; POLY(OXYETHYL) MYRISTYL ETHER; POLY(OXYETHYL) TETRADECYL ETHER

IDENTIFICATION
Formula: $C_{13}H_{29}O(CH_2CH_2O)_n CH_2CH_2OH$
DOT ID Number: UN 1987; DOT Guide Number: 127
Proper Shipping Name: Alcohols, n.o.s.

DESCRIPTION: Colorless to yellow liquid. Mild, pleasant odor. Mixes slowly with water.

Combustible • Heat or flame may cause explosion • Containers may BLEVE when exposed to fire • Vapors are heavier than air and will collect and stay in low areas • Vapors in confined areas (e.g., tanks, sewers, buildings) may explode when exposed to fire • Irritating to the skin, eyes, and respiratory tract • Toxic products of combustion may include carbon monoxide.

Hazard Classification (based on NFPA-704 Rating System)
Health Hazards (Blue): 2; Flammability (Red): 1; Reactivity (Yellow): 0

EMERGENCY RESPONSE: See Appendix A (127)
Evacuation:
Public safety: Isolate spill area for at least 25 to 50 meters (80 to 160 feet) in all directions.
Spill: Large spill–Consider initial downwind evacuation for at least 300 meters (1000 feet).
Fire: Isolate for 800 meters (½ mile) in all directions, especially if tank, rail car, or tank truck is involved in fire.

EXPOSURE
Short-term effects: *SEEK MEDICAL ATTENTION.* **Liquid:** Irritating to skin. Will burn eyes. Flush affected areas with plenty of water. *IF IN EYES*, hold eyelids open and flush with plenty of water.

HEALTH HAZARDS
Personal protective equipment (PPE): B-Level PPE. Gloves; goggles. Natural rubber, neoprene, and nitrile rubber may offer some protection from alcohols. Chemical protective material(s) reported to offer minimal to poor protection: Viton®/neoprene, butyl rubber,/neoprene.
Liquid or solid irritant characteristics: Liquid causes eye injury. Contact with skin may cause irritation.

FIRE DATA
Flash point: 470°F/243°C (oc).

CHEMICAL REACTIVITY
Reactivity group: 20
Compatibility class: Alcohols, glycols

ENVIRONMENTAL DATA
Food chain concentration potential: Negative; unlikely to accumulate.
Water pollution: DOT Appendix B, §172.101–marine pollutant. Effect of low concentrations on aquatic life is unknown. May be dangerous if it enters nearby water intakes; notify operators. Notify local health and wildlife officials. **Response to discharge:** Disperse and flush.

SHIPPING INFORMATION
Grades of purity: Usually 100%; **Storage temperature:** Ambient; **Inert atmosphere:** None; **Venting:** Open (flame arrester); **Stability during transport:** Stable.

PHYSICAL AND CHEMICAL PROPERTIES
Physical state @ 59°F/15°C and 1 atm: Liquid.
Molecular weight: 660
Boiling point @ 1 atm: Very high.
Melting/Freezing point: 59°F/15°C/288°K.
Specific gravity (water = 1): 1.007 @ 15°C (liquid).
Heat of combustion: (estimate) –11,000 Btu/lb = –6200 cal/g = -260 x 10^5 J/kg.
Heat of solution: (estimate) –9 Btu/lb = –5 cal/g = –0.2 x 10^5 J/kg.

ETHOXY TRIGLYCOL REC. E:1500

SYNONYMS: DOWANOL TE; ETHOXYTRIETHYLENE GLYCOL; POLY-SOLV TE; TRIETHYLENE GLYCOL MONOETHYLETHER; TRIGLYCOL MONOETHYL ETHER

IDENTIFICATION
CAS Number: 112-50-5
Formula: $C_8H_{18}O_4$; $C_2H_5O(CH_2)_2O(CH_2)_2OCH_2CH_2OH$

DESCRIPTION: Colorless liquid. Odorless. Sinks and mixes with water.

Combustible • Heat or flame may cause explosion • Containers may BLEVE when exposed to fire • Vapors are heavier than air and will collect and stay in low areas • Vapors in confined areas (e.g., tanks, sewers, buildings) may explode when exposed to fire • Irritating to the skin, eyes, and respiratory tract • Toxic products of combustion may include carbon monoxide.

Hazard Classification (based on NFPA-704 Rating System)
Health Hazards (Blue): 0; Flammability (Red): 1; Reactivity (Yellow): 0

EMERGENCY RESPONSE: See Appendix A (171)
Evacuation:
Public safety: Isolate the area of spill or leak for at least 10 to 25 meters (30 to 80 feet) in all directions.
Spill: Increase, in the downwind direction, as necessary, the distance shown under "Public Safety."
Fire: If any large container is involved in fire, isolate for at least 800 meters (½ mile) in all directions; also, consider initial evacuation for 800 meters (½ mile) in all directions.

HEALTH HAZARDS
Personal protective equipment (PPE): B-Level PPE. Chemical safety goggles and adequate protective clothing.
Toxicity by ingestion: Grade 1; LD_{50} = 5 to 15 g/kg (rat).
Vapor (gas) irritant characteristics: Vapors may be mildly irritating to the eyes and throat.
Liquid or solid irritant characteristics: Mild skin irritant.

FIRE DATA
Flash point: 275°F/135°C (oc).
Fire extinguishing agents not to be used: Water or foam may cause frothing.

CHEMICAL REACTIVITY
Binary reactants: Sulfuric acid, oxidizers, isocyanates.
Reactivity group: 40
Compatibility class: Glycol ethers

ENVIRONMENTAL DATA
Food chain concentration potential: Negative; unlikely to accumulate.
Water pollution: Effect of low concentrations on aquatic life is unknown. May be dangerous if it enters nearby water intakes; notify operators. Notify local health and wildlife officials.
Response to discharge: Disperse and flush.

SHIPPING INFORMATION
Storage temperature: Ambient; **Inert atmosphere:** None; **Venting:** Open (flame arrester); **Stability during transport:** Stable.

NAS HAZARD CLASSIFICATIONS FOR BULK WATER TRANSPORTATION
FIRE: 1
HEALTH: Vapor irritant: 0; Liquid or solid irritant: 0; Poisons: 0
WATER POLLUTION: Human toxicity: 0; Aquatic toxicity: 1; Aesthetic effect: 1
REACTIVITY: Other chemicals: 2; Water: 0; Self-reaction: 0

PHYSICAL AND CHEMICAL PROPERTIES
Physical state @ 59°F/15°C and 1 atm: Liquid.
Molecular weight: 178
Boiling point @ 1 atm: 493°F/256°C/529°K.
Melting/Freezing point: –1.7°F/–18.7°C/254.5°K.
Specific gravity (water = 1): 1.020 @ 68°F/20°C (liquid).
Ratio of specific heats of vapor (gas): 1.033
Latent heat of vaporization: (estimate) 125 Btu/lb = 69 cal/g = 2.9 x 10^5 J/kg.
Heat of combustion: (estimate) = –11,000 Btu/lb = –6170 cal/g = -258 x 10^5 J/kg.
Vapor pressure: (Reid) Very low.

ETHYL ACETATE   REC. E:1600

SYNONYMS: ACETATO de 2-ETILBUTILO (Spanish); ACETIC ACID, ETHYL ESTER; ACETIC ESTER; ACETIC ETHER; ETHYL ACETIC ESTER; EEC No. 607-022-00-5; ETHYL ETHANOATE; ACETIDIN; ACETOXYETHANE; AETHYLACETAT (German); ESSIGESTER (German); ETHYL ACETIC ESTER; ETHYLE (ACETATE d') (French); ETHYL ETHANOATE; VINEGAR NAPHTHA; RCRA No. U112

IDENTIFICATION
CAS Number: 141-78-6
Formula: $C_4H_8O_2$; $CH_3COOCH_2CH_3$
DOT ID Number: UN 1173; DOT Guide Number: 127
Proper Shipping Name: Ethyl acetate
Reportable Quantity (RQ): **(CERCLA)** 5000 lb/2270 kg

DESCRIPTION: Colorless, watery liquid. Pleasant fruity odor. Floats on the surface of water; moderately soluble. Flammable, irritating vapor is produced.

Highly flammable • Containers may BLEVE when exposed to fire • Vapors may form explosive mixture with air • Vapors are heavier than air and will collect and stay in low areas • Vapors may travel long distances to ignition sources and flashback • Vapors in confined areas (e.g., tanks, sewers, buildings) may explode when exposed to fire • Irritating to the skin, eyes, and respiratory tract • Toxic products of combustion may include carbon monoxide.

Hazard Classification (based on NFPA-704 Rating System)
Health Hazards (Blue): 1; Flammability (Red): 3; Reactivity (Yellow): 0

EMERGENCY RESPONSE: See Appendix A (127)
Evacuation:
Public safety: Isolate spill area for at least 25 to 50 meters (80 to 160 feet) in all directions.
Spill: Large spill–Consider initial downwind evacuation for at least 300 meters (1000 feet).
Fire: Isolate for 800 meters (½ mile) in all directions, especially if tank, rail car, or tank truck is involved in fire.

EXPOSURE
Short-term effects: *SEEK MEDICAL ATTENTION*. **Vapor:** Irritating to eyes, nose, and throat. *IF INHALED*, will, will cause headache, dizziness, nausea, or loss of consciousness. Move to fresh air. *IF BREATHING HAS STOPPED*, give artificial respiration. IF breathing is difficult, administer oxygen. **Liquid:** Irritating to skin and eyes. Harmful if swallowed. Remove contaminated clothing and shoes. Flush affected areas with plenty of water. *IF IN EYES*, hold eyelids open and flush with plenty of water. *IF SWALLOWED* and victim is *CONSCIOUS AND ABLE TO SWALLOW*, have victim drink 4 to 8 ounces of water. The use of alcoholic beverages may enhance the toxic effect.

HEALTH HAZARDS
Personal protective equipment (PPE): B-Level PPE. OSHA Table Z-1-A air contaminant. Organic vapor canister); or air mask; goggles or face shield. Protective materials with good to excellent resistance: butyl rubber, PV alcohol, Teflon®, Silvershield®. Also, the following glove materials offer limited protection: supported neoprene, supported polyvinyl alcohol, natural rubber.
Recommendations for respirator selection: NIOSH/OSHA
2000 ppm: SA:CF* (any supplied-air respirator operated in a continuous-flow mode); or PAPROV* [any powered, air-purifying

Ethyl acetoacetate

respirator with organic vapor cartridge(s)]; or CCRFOV [any chemical cartridge respirator with a full facepiece and organic vapor cartridges(s)]; or GMFOV [any air-purifying, full-facepiece respirator (gas mask) with a chin-style, front- or back-mounted acid gas canister]; or SCBAF (any self-contained breathing apparatus with a full facepiece); or SAF (any supplied-air respirator with a full facepiece). *EMERGENCY OR PLANNED ENTRY INTO UNKNOWN CONCENTRATIONS OR IDLH CONDITIONS:* SCBAF:PD,PP (any self-contained breathing apparatus that has a full facepiece and is operated in a pressure-demand or other positive-pressure mode); or SAF:PD,PP:ASCBA (any supplied-air respirator that has a full facepiece and is operated in a pressure-demand or other positive-pressure mode in combination with an auxiliary self-contained breathing apparatus operated in a pressure-demand or other positive pressure mode). *ESCAPE:* GMFOV[any air-purifying, full-facepiece respirator (gas mask) with a chin-style, front- or back-mounted organic vapor canister]; or SCBAE (any appropriate escape-type, self-contained breathing apparatus). *Note*: Substance causes eye irritation or damage; eye protection needed.

Exposure limits (TWA unless otherwise noted): ACGIH TLV 400 ppm (1440 mg/m^3); NIOSH/OSHA 400 ppm (1400 mg/m^3).
Toxicity by ingestion: Grade 2; LD_{50} = 0.5 to 5 g/kg.
Long-term health effects: May effect the nervous system. Degreases the skin.
Vapor (gas) irritant characteristics: Vapors cause smarting of the eyes or respiratory system if present in high concentrations. The effect is temporary.
Liquid or solid irritant characteristics: Eye irritant. If spilled on clothing and allowed to remain, may cause smarting and reddening of the skin.
Odor threshold: 6.5–45 ppm.
IDLH value: 2000 ppm [10% LEL].

FIRE DATA
Flash point: 55°F/13°C (oc); 24°F/–4°C (cc).
Flammable limits in air: LEL: 2.0%; UEL: 11.5%.
Autoignition temperature: 800°F/426°C/699°K.
Electrical hazard: Class I, Group D; Flow or agitation of substance may generate electrostatic charges due to low conductivity.
Burning rate: 3.7 mm/min

CHEMICAL REACTIVITY
Binary reactants: Vigorous reaction with oxidizers. Nitrates, alkalis, and acids. Will hydrolyze on standing forming acetic acid and ethyl alcohol. This reaction is greatly accelerated by alkalies.
Reactivity group: 34
Compatibility class: Ester

ENVIRONMENTAL DATA
Food chain concentration potential: Log P_{ow} = 0.7. Unlikely to accumulate.
Water pollution: Effect of low concentrations on aquatic life is unknown. May be dangerous if it enters nearby water intakes; notify operators. Notify local health and wildlife officials.
Response to discharge: Issue warning–high flammability. Evacuate area. Disperse and flush.

SHIPPING INFORMATION
Grades of purity: 85–100%; **Storage temperature:** Ambient; **Inert atmosphere:** None; **Venting:** Open (flame arrester) or pressure-vacuum; **Stability during transport:** Stable.

NAS HAZARD CLASSIFICATION FOR BULK WATER TRANSPORTATION
FIRE: 3
HEALTH: Vapor irritant: 1; Liquid or solid irritant: 1; Poisons: 2
WATER POLLUTION: Human toxicity: 1; Aquatic toxicity: 2; Aesthetic effect: 2
REACTIVITY: Other chemicals: 1; Water: 0; Self-reaction: 0

PHYSICAL AND CHEMICAL PROPERTIES
Physical state @ 59°F/15°C and 1 atm: Liquid.
Molecular weight: 88.11
Boiling point @ 1 atm: 171°F/77°C/350°K.
Melting/Freezing point: –117°F/–83°C/190°K.
Critical temperature: 482°F/250°C/523°K.
Critical pressure: 558 psia = 38 atm = 3.8 MN/m^2.
Specific gravity (water = 1): 0.902 @ 68°F/20°C (liquid).
Liquid surface tension: 24 dynes/cm = 0.024 N/m @ 68°F/20°C.
Liquid water interfacial tension: 6.79 dynes/cm = 0.00679 N/m at 30°C.
Relative vapor density (air = 1): 3.0
Ratio of specific heats of vapor (gas): 1.080
Latent heat of vaporization: 158 Btu/lb = 87.6 cal/g = 3.67 x 10^5 J/kg.
Heat of combustion: –10,110 Btu/lb = –5616 cal/g = –235.1 x 10^5 J/kg.
Heat of fusion: 28.43 cal/g.
Vapor pressure: (Reid) 3.27 psia; 73 mm.

ETHYL ACETOACETATE   REC. E:1700

SYNONYMS: ACETATO de ETILO (Spanish); ACETOACETIC ACID, ETHYL ESTER; ACETOACETIC ESTER; ETHYL ACETYL ACETATE; ACTIVE ACETYL ACETATE; DIACETIC ETHER; EAA; ETHYL ACETYLACETATE; ETHYL 3-OXOBUTANOATE; 3-OXOBUTANOIC ACID ETHYL ESTER

IDENTIFICATION
CAS Number: 141-97-9
Formula: $C_6H_{10}O_3$; $CH_3CO \cdot CH_2CO \cdot OC_2NO_5$
DOT ID Number: UN 1993; DOT Guide Number: 128
Proper Shipping Name: Flammable liquids, n.o.s.

DESCRIPTION: Colorless liquid. Pleasant fruity odor. Mixes with water.

Flammable • Containers may BLEVE when exposed to fire • Vapors may form explosive mixture with air • Vapors are heavier than air and will collect and stay in low areas • Vapors may travel long distances to ignition sources and flashback • Vapors in confined areas (e.g., tanks, sewers, buildings) may explode when exposed to fire • Irritating to the skin, eyes, and respiratory tract • Toxic products of combustion may include carbon monoxide.

Hazard Classification (based on NFPA-704 Rating System)
Health Hazards (Blue): 2; Flammability (Red): 2; Reactivity (Yellow): 0

EMERGENCY RESPONSE: See Appendix A (128)
Evacuation:
Public safety: Isolate spill area for at least 25 to 50 meters (80 to 160 feet) in all directions.
Spill: Large spill–Consider initial downwind evacuation for at least 300 meters (1000 feet).

Fire: Isolate for 800 meters (½ mile) in all directions, especially if tank, rail car, or tank truck is involved in fire.

EXPOSURE
Short-term effects: *SEEK MEDICAL ATTENTION.* **Liquid:** Irritating to eyes. Harmful if swallowed. Remove contaminated clothing and shoes. Flush affected areas with plenty of water. *IF IN EYES*, hold eyelids open and flush with plenty of water. *IF SWALLOWED* and victim is *CONSCIOUS AND ABLE TO SWALLOW*, have victim drink 4 to 8 ounces of water.

HEALTH HAZARDS
Personal protective equipment (PPE): B-Level PPE. Goggles or face shield; rubber gloves.
Toxicity by ingestion: Grade 2; oral LD_{50} = 3,980 mg/kg (rat).
Vapor (gas) irritant characteristics: Eye and respiratory tract irritant.
Liquid or solid irritant characteristics: Eye irritant.

FIRE DATA
Flash point: 176°F/80°C (oc); 135°F/57°C (cc).
Flammable limits in air: LEL: 1.4% @ 200°F/93°C; UEL: 9.5% @ 350°F/177°C.
Fire extinguishing agents not to be used: Water may be ineffective.
Autoignition temperature: 563°F/295°C/568°K.
Burning rate: 2.4 mm/min

CHEMICAL REACTIVITY
Binary reactants: Reacts with strong oxidizers.

ENVIRONMENTAL DATA
Food chain concentration potential: Log P_{ow} = 1.21 (estimate). Unlikely to accumulate.
Water pollution: Effect of low concentrations on aquatic life is unknown. May be dangerous if it enters nearby water intakes; notify operators. Notify local health and wildlife officials.
Response to discharge: Issue warning–water contaminant. Disperse and flush.

SHIPPING INFORMATION
Grades of purity: 98+%; **Storage temperature:** Ambient; **Inert atmosphere:** None; **Venting:** Open (flame arrester); **Stability during transport:** Stable.

PHYSICAL AND CHEMICAL PROPERTIES
Physical state @ 59°F/15°C and 1 atm: Liquid.
Molecular weight: 130.1
Boiling point @ 1 atm: 363°F/184°C/457°K.
Melting/Freezing point: Less than –112°F/–80°C/193°K.
Specific gravity (water = 1): 1.028 @ 68°F/20°C (liquid).
Liquid surface tension: 32.5 dynes/cm = 0.035 N/m @ 68°F/20°C.
Liquid water interfacial tension: (estimate) 35 dynes/cm = 0.035 N/m @ 68°F/20°C.
Relative vapor density (air = 1): 4.48
Latent heat of vaporization: 160 Btu/lb = 91 cal/g = 3.8×10^5 J/kg.
Heat of combustion: –9349 Btu/lb = –5194 cal/g = $–217.3 \times 10^5$ J/kg.
Vapor pressure: 1 mm.

ETHYL ACRYLATE   REC. E:1800

SYNONYMS: ACRILATO de ETILO (Spanish); ACRYLATE d'ETHYLE (French); ACRYLIC ACID, ETHYL ESTER; A C R Y L S A E U R E A E T H Y L E S T E R (German); AETHYLACRYLAT (German); EEC No. 607-032-00-X; ETHOXYCARBONYLETHYLENE; ETHYL PROPENOATE; ETHYL 2-PROPENOATE; 2-PROPENOIC ACID, ETHYL ESTER; RCRA No. U113

IDENTIFICATION
CAS Number: 140-88-5
Formula: $C_5H_8O_2$; $CH_2=CHCOOCH_2CH_3$
DOT ID Number: UN 1917; DOT Guide Number: 128P
Proper Shipping Name: Ethyl acrylate, inhibited
Reportable Quantity (RQ): **(CERCLA)** 1000 lb/454 kg

DESCRIPTION: Colorless liquid. Fruity or acrylic odor; nauseating. Floats on the surface of water. Flammable, irritating vapor is produced.

Highly flammable • Corrosive to skin, eyes, and respiratory tract; inhalation symptoms may be delayed • Polymerization hazard • Heat can induce polymerization with rapid release of energy; sealed containers may rupture explosively • May react with itself blocking relief valves; leading to tank explosions • Containers may BLEVE when exposed to fire •Vapors may form explosive mixture with air • Vapors are heavier than air and will collect and stay in low areas • Vapors may travel long distances to ignition sources and flashback • Vapors in confined areas (e.g., tanks, sewers, buildings) may explode when exposed to fire • Toxic products of combustion may include carbon monoxide.

Hazard Classification (based on NFPA-704 Rating System)
Health Hazards (Blue): 2; Flammability (Red): 3; Reactivity (Yellow): 2

EMERGENCY RESPONSE: See Appendix A (128)
Evacuation:
Public safety: Isolate spill area for at least 25 to 50 meters (80 to 160 feet) in all directions.
Spill: Large spill–Consider initial downwind evacuation for at least 300 meters (1000 feet).
Fire: Isolate for 800 meters (½ mile) in all directions, especially if tank, rail car, or tank truck is involved in fire.

EXPOSURE
Short-term effects: *SEEK MEDICAL ATTENTION. ABSORBED THROUGH THE SKIN.* **Vapor:** Irritating to eyes, nose, and throat. May cause lung edema; physical exertion will aggravate this condition. *IF INHALED*, will, will cause headache or nausea. Move to fresh air. *IF BREATHING HAS STOPPED,* give artificial respiration. IF breathing is difficult, administer oxygen. **Liquid:** Will burn skin and eyes. Harmful if swallowed. Remove contaminated clothing and shoes. Flush affected areas with plenty of water. *IF IN EYES*, hold eyelids open and flush with plenty of water. *IF SWALLOWED* and victim is *CONSCIOUS AND ABLE TO SWALLOW*, have victim drink 4 to 8 ounces of water.
Note to physician or authorized medical personnel: Medical observation is recommended for 24 to 48 hours after breathing overexposure, as pulmonary edema may be delayed. As first aid for pulmonary edema, consider administering a corticosteroid spray. Cigarette smoking may exacerbate pulmonary injury and should be discouraged for at least 72 hours following exposure.

HEALTH HAZARDS
Personal protective equipment (PPE): A-Level PPE. OSHA Table Z-1-A air contaminant. Full face organic vapor respirator.

Chemical protective material(s) reported to have good to excellent resistance: butyl rubber, Teflon®, PV alcohol.
Recommendations for respirator selection: NIOSH
At any concentrations above the NIOSH REL, or where there is no REL, at any detectable concentration: SCBAF:PD,PP (any self-contained breathing apparatus that has a full facepiece and is operated in a pressure-demand or other positive-pressure mode); or SAF:PD,PP:ASCBA (any supplied-air respirator that has a full facepiece and is operated in a pressure-demand or other positive-pressure mode in combination with an auxiliary self-contained breathing apparatus operated in a pressure-demand or other positive pressure mode). *ESCAPE:* GMFOV [any air-purifying, full-facepiece respirator (gas mask) with a chin-style, front-or back-mounted organic vapor canister]; or SCBAE (any appropriate escape-type, self-contained breathing apparatus).
Exposure limits (TWA unless otherwise noted): ACGIH TLV 5 ppm (20 mg/m^3); OSHA PEL 25 ppm (100 mg/m^3); NIOSH REL potential human carcinogen; skin contact contributes significantly in overall exposure.
Short-term exposure limits (15-minute TWA): ACGIH STEL 15 ppm (61 mg/m^3).
Toxicity by ingestion: Grade 2; LD$_{50}$ = 0.5 to 5 g/kg (rat).
Long-term health effects: Possible liver and kidney injury. Repeated exposure may develop sensitivity. IARC possible carcinogen, rating 2B; sufficient animal data. NTP anticipated carcinogen. Intensive skin contact may cause dermatitis.
Vapor (gas) irritant characteristics: Eye and respiratory tract irritant. Vapor is irritating such that personnel will not usually tolerate moderate or high vapor concentrations.
Liquid or solid irritant characteristics: Severe eye irritant. Skin irritant. Causes smarting of the skin and first-degree burns on short exposure and may cause secondary burns on long exposure.
Odor threshold: 0.00024 ppm.
IDLH value: 300 ppm; potential human carcinogen.

FIRE DATA
Flash point: 44°F/7°C (oc); 48°F/9°C (cc).
Flammable limits in air: LEL: 1.4%; UEL: 14%.
Behavior in fire: May polymerize and cause container to rupture or explode. Liquid floats on water; may spread to ignition source.
Autoignition temperature: 721°F/383°C.
Electrical hazard: Class I, Group D.
Burning rate: 4.3 mm/min

CHEMICAL REACTIVITY
Binary reactants: Violent reaction with strong oxidizers, peroxides, polymerizers, strong alkalis, strong acids. amines, chlorosulfonic acid.
Polymerization: Readily polymerizes without inhibitor. Sunlight or UV may cause polymerization. Exclude moisture, light; avoid exposure to high temperatures; **Inhibitor of polymerization:** 13-17 ppm monomethyl ether of hydroquinone
Reactivity group: 14
Compatibility class: Acrylate

ENVIRONMENTAL DATA
Food chain concentration potential: Log P$_{ow}$ = 1.2. Unlikely to accumulate.
Water pollution: DOT Appendix B, §172.101–marine pollutant. Effect of low concentrations on aquatic life is unknown. Fouling to shoreline. May be dangerous if it enters nearby water intakes; notify operators. Notify local health and wildlife officials.
Response to discharge: Issue warning–high flammability. Restrict access. Evacuate area. Disperse and flush.

SHIPPING INFORMATION
Stability during transport: Stable.

NAS HAZARD CLASSIFICATION FOR BULK WATER TRANSPORTATION
FIRE: 3
HEALTH: Vapor irritant: 3; Liquid or solid irritant: 2; Poisons: 3
WATER POLLUTION: Human toxicity: 2; Aquatic toxicity: 2; Aesthetic effect: 2
REACTIVITY: Other chemicals: 2; Water: 0; Self-reaction: 3

PHYSICAL AND CHEMICAL PROPERTIES
Physical state @ 59°F/15°C and 1 atm: Liquid.
Molecular weight: 100.12
Boiling point @ 1 atm: 211.3°F/99.6°C/372.8°K.
Melting/Freezing point: –96°F/–71°C/202°K.
Critical temperature: 534°F/279°C/552°K.
Critical pressure: 544 psia = 37 atm = 3.7 MN/m^2.
Specific gravity (water = 1): 0.923 @ 68°F/20°C (liquid).
Liquid surface tension: 25 dynes/cm = 0.025 N/m @ 68°F/20°C.
Liquid water interfacial tension: (estimate) 40 dynes/cm = 0.04 N/m @ 68°F/20°C.
Relative vapor density (air = 1): 3.4
Ratio of specific heats of vapor (gas): 1.080
Latent heat of vaporization: 149 Btu/lb = 82.9 cal/g = 3.47 x 10^5 J/kg.
Heat of combustion: –11,880 Btu/lb = –6600 cal/g = –276.3 x 10^5 J/kg.
Heat of polymerization: –335 Btu/lb
Heat of polymerization: = –186 cal/g = –7.79 x 10^5 J/kg.
Vapor pressure: (Reid) 1.4 psia; 29 mm.

ETHYL ALCOHOL   REC. E:1900

SYNONYMS: ABSOLUTE ETHANOL; AETHANOL (German); AETHYLDLKOHOL (German); ALCOHOL; ALCOHOL C-2; ALCOHOL, ANHYDROUS; ALCOHOL ETILICO (Spanish); ALCOHOL, DEHYDRATED; ALCOOL ETHYLIQUE (French); ALGRAIN; ALKOHOL (German); ANHYDROL; COLOGNE SPIRIT; COLOGNE SPIRITS; EEC No. 603-002-00-5; ETHANOL; ETHANOL 200 PROOF; ETHYL ALCOHOL ANHYDRO-S; ETHYL HYDRATE; ETHYL HYDROXIDE; FERMENTATION ALCOHOL; GRAIN ALCOHOL; JAYSOL S; METHYL CARBINOL; MOLASSES ALCOHOL; POTATO ALCOHOL; PURE GRAIN ALCOHOL; SD ALCOHOL 23-HYDROGEN; SPIRITS OF WINE; SPIRIT; TESCOL

IDENTIFICATION
CAS Number: 64-17-5
Formula: CH$_3$CH$_2$OH
DOT ID Number: UN 1170; DOT Guide Number: 127
Proper Shipping Name: Ethanol; Ethyl alcohol; Ethanol, solution; Ethyl alcohol solutions.

DESCRIPTION: Colorless watery liquid. Alcohol odor, like whiskey or wine. Floats on the surface of water; soluble. Flammable, irritating vapor is produced.

Highly flammable • Containers may BLEVE when exposed to fire •Vapors may form explosive mixture with air • Vapors are heavier than air and will collect and stay in low areas • Vapors may travel long distances to ignition sources and flashback • Vapors in confined areas (e.g., tanks, sewers, buildings) may explode when exposed to fire • Irritating to the skin, eyes, and respiratory tract • Toxic products of combustion may include carbon monoxide.

Hazard Classification (based on NFPA-704 Rating System)
Health Hazards (Blue): 0; Flammability (Red): 3; Reactivity (Yellow): 0

EMERGENCY RESPONSE: See Appendix A (127)
Evacuation:
Public safety: Isolate spill area for at least 25 to 50 meters (80 to 160 feet) in all directions.
Spill: Large spill–Consider initial downwind evacuation for at least 300 meters (1000 feet).
Fire: Isolate for 800 meters (½ mile) in all directions, especially if tank, rail car, or tank truck is involved in fire.

EXPOSURE
Short-term effects: *SEEK MEDICAL ATTENTION.* **Vapor:** Irritating to eyes, nose, and throat. Move to fresh air. **Liquid:** Can cause inebriation.

HEALTH HAZARDS
Personal protective equipment (PPE): B-Level PPE. OSHA Table Z-1-A air contaminant. All-purpose canister; safety goggles. Chemical protective material(s) reported to have good to excellent resistance: butyl rubber, nitrile, Teflon®, Silvershield®, PVC, neoprene/natural rubber. Also, natural rubber and neoprene may offer some protection from alcohols.
Recommendations for respirator selection: NIOSH/OSHA
3300 ppm: SA (any supplied-air respirator); or SCBAF (any self-contained breathing apparatus with a full facepiece). *EMERGENCY OR PLANNED ENTRY INTO UNKNOWN CONCENTRATIONS OR IDLH CONDITIONS:* SCBAF:PD,PP (any self-contained breathing apparatus that has a full facepiece and is operated in a pressure-demand or other positive-pressure mode); or SAF:PD,PP:ASCBA (any supplied-air respirator that has a full facepiece and is operated in a pressure-demand or other positive-pressure mode in combination with an auxiliary self-contained breathing apparatus operated in a pressure-demand or other positive pressure mode). *ESCAPE:* SCBAE (any appropriate escape-type, self-contained breathing apparatus).
Exposure limits (TWA unless otherwise noted): ACGIH TLV 1000 ppm (1880 mg/m^3) as ethanol; OSHA PEL 1000 ppm (1900 mg/m^3).
Short-term exposure limits (15-minute TWA): 5000 ppm/15 min.
Toxicity by ingestion: Grade 1; LD$_{50}$ = 5 to 15 g/kg.
Long-term health effects: Possible liver damage.
Vapor (gas) irritant characteristics: Vapors cause smarting of the eyes or respiratory system if present in high concentrations. The effect may be temporary.
Liquid or solid irritant characteristics: Practically harmless to the skin.
Odor threshold: 50–115 ppm.
IDLH value: 3300 ppm [10% LEL].

FIRE DATA
Flash point: 64°F/18°C (oc); 55°F/13°C (cc).
Flammable limits in air: LEL: 3.3%; UEL: 19%.
Autoignition temperature: 689°F/365°C/638°K.
Electrical hazard: Class I, Group D.
Burning rate: 3.9 mm/min

CHEMICAL REACTIVITY
Binary reactants: Reacts with strong oxidizers, bases, potassium dioxide, bromine pentafluoride, acetyl bromide, acetyl chloride, platinum, sodium. Forms explosive mixtures with perchlorates, mercury nitrate(II).
Reactivity group: 20
Compatibility class: Alcohols, glycols

ENVIRONMENTAL DATA
Food chain concentration potential: Log P$_{ow}$ = –0.3 to –1.4. Unlikely to accumulate.
Water pollution: Dangerous to aquatic life in high concentrations. May be dangerous if it enters nearby water intakes; notify operators. Notify local health and wildlife officials. **Response to discharge:** Issue warning–high flammability. Disperse and flush.

SHIPPING INFORMATION
Grades of purity: Anhydrous (200 proof); 190 proof; specially denatured; completely denatured; **Storage temperature:** Ambient; **Inert atmosphere:** None; **Venting:** Open (flame arrester) or pressure-vacuum; **Stability during transport:** Stable.

NAS HAZARD CLASSIFICATION FOR BULK WATER TRANSPORTATION
FIRE: 3
HEALTH: Vapor irritant: 1; Liquid or solid irritant: 0; Poisons: 1
WATER POLLUTION: Human toxicity: 1; Aquatic toxicity: 1; Aesthetic effect: 1
REACTIVITY: Other chemicals: 2; Water: 0; Self-reaction: 0

PHYSICAL AND CHEMICAL PROPERTIES
Physical state @ 59°F/15°C and 1 atm: Liquid.
Molecular weight: 46.07
Boiling point @ 1 atm: 172.9°F/78.3°C/351.5°K.
Melting/Freezing point: –173°F/–114°C/159°K.
Critical temperature: 469.6°F/243.1°C/516.3°K.
Critical pressure: 926 psia = 63.0 atm = 6.38 MN/m^2.
Specific gravity (water = 1): 0.790 @ 68°F/20°C (liquid).
Relative vapor density (air = 1): 1.6
Ratio of specific heats of vapor (gas): 1.128
Latent heat of vaporization: 360 Btu/lb = 200 cal/g.
Heat of combustion: 8.37 x 10^5 J/kg –11,570 Btu/lb = 6425 cal/g = –268.8 x 10^5 J/kg.
Heat of solution: –99 Btu/lb = –55 cal/g = –2.3 x 10^5 J/kg.
Vapor pressure: (Reid) 2.3 psia; 44 mm.

ETHYL ALUMINUM DICHLORIDE   REC. E:2000

SYNONYMS: ALUMINUM ETHYL DICHLORIDE; DICLORURO de ETILALUMINIO (Spanish); EADC

IDENTIFICATION
CAS Number: 563-43-9
Formula: C$_2$NO$_5$AlCl$_2$
DOT ID Number: UN 3052; DOT Guide Number: 135
Proper Shipping Name: Aluminum alkyl halides

DESCRIPTION: Colorless to light yellow liquid (heated). *Ignites spontaneously when exposed to air.* Often shipped as a 15-30% solution in a hydrocarbon solvent such as hexane. Reacts violently with water; poisonous hydrogen chloride fumes and flammable ethane gas are produced.

Spontaneously combustible • Extremely flammable • Corrosive to skin, eyes, and respiratory tract; skin and eye contact causes severe burns, impaired vision, and blindness; inhalation symptoms may be

Ethylaluminum sesquichloride

delayed • Firefighting gear (including SCBA) does not provide adequate protection. If exposure occurs, remove and isolate gear immediately and thoroughly decontaminate personnel • Containers may BLEVE when exposed to fire • Vapors may form explosive mixture with air • Vapors are heavier than air and will collect and stay in low areas • Vapors may travel long distances to ignition sources and flashback • Vapors in confined areas (e.g., tanks, sewers, buildings) may explode when exposed to fire • Toxic products of combustion may include aluminum oxide (may cause metal fume fever) and hydrogen chlorine gas • May re-ignite after fire is extinguished • Do not put yourself in danger by entering a contaminated area to rescue a victim.

Hazard Classification (based on NFPA-704 Rating System)
Health Hazards (Blue): 3; Flammability (Red): 3; Reactivity (Yellow): 3; Special Notice (White): Water reactive

EMERGENCY RESPONSE: See Appendix A (135)
Evacuation:
Public safety: Isolate spill area for at least 100 to 150 meters (330 to 490 feet) in all directions.
Spill: Small spill–First: Isolate in all directions 30 meters (100 feet); Then: Protect persons downwind, DAY: 0.2 km (0.1 mile); NIGHT: 0.2 km (0.1 mile). Large spill–First: Isolate in all directions 30 meters (100 feet); Then: Protect persons downwind, DAY: 0.3 km (0.2 mile); NIGHT: 1.3 km (0.8 mile).
Fire: Isolate for 800 meters (½ mile) in all directions, especially if tank, rail car, or tank truck is involved in fire.

EXPOSURE
Short-term effects: *SEEK MEDICAL ATTENTION. GAS PRODUCED IN REACTION WITH WATER: POISONOUS IF INHALED.* Irritating to eyes, nose, and throat. Possible lung edema. Move to fresh air. *IF BREATHING HAS STOPPED,* give artificial respiration. IF breathing is difficult, administer oxygen.
Liquid (heated): Will burn skin and eyes. Harmful if swallowed. Remove contaminated clothing and shoes. Flush affected areas with plenty of water. *IF IN EYES,* hold eyelids open and flush with plenty of water. *IF SWALLOWED* and victim is *CONSCIOUS AND ABLE TO SWALLOW,* have victim drink 4 to 8 ounces of water. **Do NOT induce vomiting.**
Note to physician or authorized medical personnel: Medical observation is recommended for 24 to 48 hours after breathing overexposure, as pulmonary edema may be delayed. As first aid for pulmonary edema, consider administering a corticosteroid spray. Cigarette smoking may exacerbate pulmonary injury and should be discouraged for at least 72 hours following exposure.

HEALTH HAZARDS
Personal protective equipment (PPE): B-Level PPE. Full protective clothing, preferably of aluminized glass cloth; goggles, face shield, gloves; in case of fire, all-purpose canister); or self-contained breathing apparatus.
attention.
Long-term health effects: Metal-fume fever may develop after breathing smoke from fire.
Vapor (gas) irritant characteristics: Eye and respiratory tract irritant. May cause severe breathing difficulties.
Liquid or solid irritant characteristics: Severe eye and skin irritant. Causes second-and third-degree burns on short contact with skin; and, is very injurious to the eyes.

FIRE DATA
Flash point: Pure material may ignite spontaneously in air. F.P. of solution depends on solvent used.: −14°F/−26°C for hexane.
Fire extinguishing agents not to be used: Water, foam, dry chemicals, halons/halogenated agents, or CO_2.
Behavior in fire: Contact with water applied to adjacent fires will cause formation of irritating smoke containing aluminum oxide and hydrogen chloride.
Autoignition temperature: Ignites spontaneously in air at ambient temperature.

CHEMICAL REACTIVITY
Reactivity with water: Reacts violently to form hydrogen chloride fumes and flammable ethane gas.
Binary reactants: Strong reducing agent; violent reaction with oxidizing materials. Reacts violently with alcohols, amines, phenols, CO_2, sulfur oxides, halogens, halogenated hydrocarbons, nitrogen oxides. Reacts with surface moisture or moisture in air generating aluminum oxide and hydrogen chloride fumes. Attacks most common metals.
Neutralizing agents for acids and caustics: Rinse with sodium bicarbonate or lime solution.

ENVIRONMENTAL DATA
Food chain concentration potential: Negative; unlikely to accumulate.
Water pollution: Effect of low concentrations on aquatic life is unknown. May be dangerous if it enters nearby water intakes; notify operators. Notify local health and wildlife officials. **Response to discharge:** Issue warning–high flammability, corrosive. Restrict access. Evacuate area.

SHIPPING INFORMATION
Grades of purity: Pure (neat) 25% or less by weight in benzene, hexane, or heptane. Solutions are not pyrophoric; **Storage temperature:** 35-40°C; **Inert atmosphere:** Under inert gas; dry nitrogen at 5 psig; **Venting:** Safety relief with rupture disc; **Stability during transport:** Stable.

PHYSICAL AND CHEMICAL PROPERTIES
Physical state @ 59°F/15°C and 1 atm: Solid.
Molecular weight: 126.9
Boiling point @ 1 atm: 381°F/194°C/467°K.
Melting/Freezing point: 90°F/32°C/305°K.
Specific gravity (water = 1): 1.227 at 35°C (liquid).
Liquid surface tension: (estimate) 30 dynes/cm = 0.030 N/m @ 35°C.
Relative vapor density (air = 1): 4.6
Heat of combustion: (estimate) −5600 Btu/lb = −3100 cal/g = −130 x 10^5 J/kg.
Vapor pressure: 5 mm.

ETHYLALUMINUM SESQUICHLORIDE   REC. E:2100

SYNONYMS: SESQUICLORURO de ETILALUMINIO (Spanish)

IDENTIFICATION
CAS Number: 12075-68-2
Formula: $C_6H_{15}Al_2Cl_3$; $(C_2NO_5)_3Al_2Cl_3$
DOT ID Number: UN 3052; DOT Guide Number: 135
Proper Shipping Name: Aluminum alkyl halides

DESCRIPTION: Colorless to yellow liquid. *Ignites spontaneously when exposed to air.* May be shipped as a 15-30% solution in a hydrocarbon solvent such as hexane. Reacts violently with water; poisonous hydrogen chloride fumes and flammable ethane gas are produced.

Spontaneously combustible • Extremely flammable • Corrosive to skin, eyes, and respiratory tract; skin and eye contact causes severe burns, vision impairment, and blindness; inhalation symptoms may be delayed • Firefighting gear (including SCBA) does not provide adequate protection. If exposure occurs, remove and isolate gear immediately and thoroughly decontaminate personnel • Containers may BLEVE when exposed to fire •Vapors may form explosive mixture with air • Vapors are heavier than air and will collect and stay in low areas • Vapors may travel long distances to ignition sources and flashback • Vapors in confined areas (e.g., tanks, sewers, buildings) may explode when exposed to fire • Toxic products of combustion may include aluminum oxide and hydrogen chloride gas • Intense smoke may cause metal-fume fever. • May re-ignite after fire is extinguished • Do not put yourself in danger by entering a contaminated area to rescue a victim.

Hazard Classification (based on NFPA-704 Rating System)
Health Hazards (Blue): 3; Flammability (Red): 3; Reactivity (Yellow): 3; Special Notice (White): Water reactive

EMERGENCY RESPONSE: See Appendix A (135)
Evacuation:
Public safety: Isolate spill area for at least 100 to 150 meters (330 to 490 feet) in all directions.
Spill: Small spill–First: Isolate in all directions 30 meters (100 feet); Then: Protect persons downwind, DAY: 0.2 km (0.1 mile); NIGHT: 0.2 km (0.1 mile). Large spill–First: Isolate in all directions 30 meters (100 feet); Then: Protect persons downwind, DAY: 0.3 km (0.2 mile); NIGHT: 1.3 km (0.8 mile).
Fire: Isolate for 800 meters (½ mile) in all directions, especially if tank, rail car, or tank truck is involved in fire.

EXPOSURE
Short-term effects: *SEEK MEDICAL ATTENTION*. **Vapor:** Irritating to eyes, nose, and throat. May cause lung edema; physical exertion will aggravate this condition. Harmful if inhaled. Move victim to fresh air. If breathing is difficult, administer oxygen. **Liquid:** Will burn skin and eyes. Harmful if swallowed. Remove contaminated clothing and shoes. Flush affected areas with plenty of water. IF IN EYES, hold eyelids open and flush with plenty of water. *IF SWALLOWED* and victim is *CONSCIOUS AND ABLE TO SWALLOW*, have victim drink 4 to 8 ounces of water. **Do NOT induce vomiting**.
Note to physician or authorized medical personnel: Medical observation is recommended for 24 to 48 hours after breathing overexposure, as pulmonary edema may be delayed. As first aid for pulmonary edema, consider administering a corticosteroid spray. Cigarette smoking may exacerbate pulmonary injury and should be discouraged for at least 72 hours following exposure.

HEALTH HAZARDS
Personal protective equipment (PPE): B-Level PPE. Full protective clothing, preferably of aluminized glass cloth; goggles, face shield, gloves; in case of fire, all-purpose canister); or self-contained breathing apparatus.
Long-term health effects: Metal-fume fever with flu-like symptoms may develop after breathing smoke from fire.
Vapor (gas) irritant characteristics: Eye and respiratory tract irritant. May cause severe breathing difficulties.
Liquid or solid irritant characteristics: Severe eye and skin irritant. Causes second- and third-degree burns on short contact and is very injurious to the eyes.

FIRE DATA
Flash point: –4°F/–20°C (ignites spontaneously). F.P. of solution depends on solvent used.: –14°F/–26°C for hexane.
Fire extinguishing agents not to be used: Water, foam, dry chemicals, halons/halogenated agents, or CO_2
Behavior in fire: Water from adjacent fires will cause formation of irritating smoke containing aluminum oxide and hydrogen chloride.

CHEMICAL REACTIVITY
Reactivity with water: Reacts violently to form hydrogen chloride fumes and flammable ethane gas.
Binary reactants: Self reactive with air. Reacts with surface moisture to generate hydrogen chloride, which is corrosive to common metals. Reacts with carbon tetrachloride.
Neutralizing agents for acids and caustics: Rinse with sodium bicarbonate or lime solution.

ENVIRONMENTAL DATA
Food chain concentration potential: Negative; unlikely to accumulate.
Water pollution: Effect of low concentrations on aquatic life is unknown. May be dangerous if it enters nearby water intakes; notify operators. Notify local health and wildlife officials. **Response to discharge:** Issue warning–high flammability, corrosive. Restrict access. Evacuate area.

SHIPPING INFORMATION
Grades of purity: Pure (neat); 25% or less by weight in benzene, hexane, or heptane. Solutions are not pyrophoric; **Storage temperature:** Ambient; **Inert atmosphere:** Inerted; dry nitrogen at 5 psig; **Venting:** Safety relief with rupture disc **Stability during transport:** Stable.

PHYSICAL AND CHEMICAL PROPERTIES
Physical state @ 59°F/15°C and 1 atm: Liquid.
Molecular weight: 247.5
Boiling point @ 1 atm: 293°F/145°C/418°K.
Melting/Freezing point: –4°F/–20°C/253°K.
Specific gravity (water = 1): 1.092 @ 77°F/25°C (liquid).
Liquid surface tension: (estimate) 32 dynes/cm = 0.032 N/m @ 68°F/20°C.
Relative vapor density (air = 1): 8.49
Heat of combustion: (estimate) –8600 Btu/lb = –4800 cal/g = –200 x 10^5 J/kg.
Vapor pressure: 0.30 mm.

ETHYLAMINE   **REC. E:2200**

SYNONYMS: AETHYLAMINE (German); AMINOETHANE; 1-AMINOETHANE; EA; EEC No. 612-002-00-4; ETHANAMINE; ETILAMINA (Spanish); MONOETHYLAMINE

IDENTIFICATION
CAS Number: 75-04-7
Formula: $C_2H_5NH_2$
DOT ID Number: UN 1036 (gas); UN 2270 (aqueous solution, 50-70% ethylamine); DOT Guide Number: 118 (UN 1036); 132 (UN 2270)
Proper Shipping Name: Ethylamine; Ethylamine aqueous solution with not less than 50% but not more than 70% ethylamine

DESCRIPTION: Colorless gas which becomes a liquid below 62°F/16.7°C. Also shipped as a colorless water solution. Strong ammonia-like odor. Mixes with water forming a basic solution.

Ethylamine

Extremely flammable • Corrosive to skin, eyes, and respiratory tract; can cause burns, impaired vision, and blindness; inhalation symptoms may be delayed • Firefighting gear (including SCBA) does not provide adequate protection. If exposure occurs, remove and isolate gear immediately and thoroughly decontaminate personnel • Contact with liquid may cause frostbite • Containers may BLEVE when exposed to fire • Gas form s explosive mixture with air • Gas is heavier than air and will collect and stay in low areas • Gas may travel long distances to ignition sources and flashback • Gas in confined areas (e.g., tanks, sewers, buildings) may explode when exposed to fire • Toxic products of combustion may include nitrogen oxides • Do not put yourself in danger by entering a contaminated area to rescue a victim. *Warning:* Odor is not a reliable indicator of the presence of toxic amounts of ethylamine gas.

Hazard Classification (based on NFPA-704 Rating System)
Health Hazards (Blue): 3; Flammability (Red): 4; Reactivity (Yellow): 0

EMERGENCY RESPONSE, toxic gas: See Appendix A (118)
Evacuation:
Public safety: Isolate spill area for at least 100 to 200 meters (330 to 660 feet) in all directions.
Spill: Consider initial downwind evacuation for at least 800 meters (½ mile).
Fire: Isolate for 1600 meters (1 mile) in all directions, especially if tank, rail car, or tank truck is involved in fire.

EMERGENCY RESPONSE, liquid: See Appendix A (132)
Evacuation:
Public safety: Isolate spill area for at least 50 to 100 meters (160 to 330 feet) in all directions.
Spill: Increase, as necessary, the isolation distance shown above, in "Public Safety."
Fire: Isolate for 800 meters (½ mile) in all directions, especially if tank, rail car, or tank truck is involved in fire.

EXPOSURE
Short-term effects: *SEEK MEDICAL ATTENTION. ABSORBED THROUGH THE SKIN.* **Vapor:** Irritating to eyes, nose, and throat. Harmful if inhaled. May cause lung edema; physical exertion will aggravate this condition. Move victim to fresh air. If breathing is difficult, administer oxygen. **Liquid:** Irritating to skin and eyes. Harmful if swallowed. Remove contaminated clothing and shoes. Flush affected areas with plenty of water. *IF IN EYES*, hold eyelids open and flush with plenty of water. *IF SWALLOWED* and victim is *CONSCIOUS AND ABLE TO SWALLOW*, have victim drink 4 to 8 ounces of water.

Note to physician or authorized medical personnel: Medical observation is recommended for 24 to 48 hours after breathing overexposure, as pulmonary edema may be delayed. As first aid for pulmonary edema, consider administering a corticosteroid spray. Cigarette smoking may exacerbate pulmonary injury and should be discouraged for at least 72 hours following exposure.

HEALTH HAZARDS
Personal protective equipment (PPE): B-Level PPE. Full face organic vapor or ammonia-methylamine respirator. Wear thermal protective clothing. Chemical protective material(s) reported to have good to excellent resistance: butyl rubber, Teflon®, nitrile (30-70% solution). OSHA Table Z-1-A air contaminant.
Recommendations for respirator selection: NIOSH/OSHA
250 ppm: SA:CF (any supplied-air respirator operated in a continuous-flow mode); or PAPRS [any powered, air-purifying respirator with cartridge(s) providing protection against the compound of concern]. *500 ppm:* CCRFS [any chemical cartridge respirator with a full facepiece and cartridge(s) providing protection against the compound of concern]; or GMFS [any air-purifying, full-facepiece respirator (gas mask) with a chin-style, front- or back-mounted canister providing protection against the compound of concern]; or SCBAF (any self-contained breathing apparatus with a full facepiece); or SAF (any supplied-air respirator with a full facepiece). *600 ppm:* SAF:PD,PP (any supplied-air respirator that has a full facepiece and is operated in a pressure-demand or other positive-pressure mode). *EMERGENCY OR PLANNED ENTRY INTO UNKNOWN CONCENTRATIONS OR IDLH CONDITIONS:* SCBAF:PD,PP (any self-contained breathing apparatus that has a full facepiece and is operated in a pressure-demand or other positive-pressure mode); or SAF:PD,PP:ASCBA (any supplied-air respirator that has a full facepiece and is operated in a pressure-demand or other positive-pressure mode in combination with an auxiliary self-contained breathing apparatus operated in a pressure-demand or other positive pressure mode). *ESCAPE:* GMFS [any air-purifying, full-facepiece respirator (gas mask) with a chin-style, front- or back-mounted canister providing protection against the compound of concern]; or SCBAE (any appropriate escape-type, self-contained breathing apparatus). Note: Substance causes eye irritation or damage; eye protection needed.
Exposure limits (TWA unless otherwise noted): ACGIH TLV 5 ppm (9.2 mg/m^3); NIOSH/OSHA 10 ppm (18 mg/m^3).
Short-term exposure limits (15-minute TWA): ACGIH STEL 15 ppm (27.6 mg/m^3).
Toxicity by ingestion: Grade 3; oral LD$_{50}$ = 400 mg/kg (rat).
Vapor (gas) irritant characteristics: Eye and respiratory tract irritant. Vapors are irritating such that personnel will not usually tolerate moderate or high concentrations.
Liquid or solid irritant characteristics: Severe eye and skin irritant. Causes smarting of the skin and first-degree burns on short exposure and may cause second-degree burns on long exposure. Rapid evaporation of liquid may cause frostbite.
Odor threshold: 0.27 ppm.
IDLH value: 600 ppm.

FIRE DATA
Flash point: 0°F/–18°C (oc); 1°F/–17°C (cc).
Flammable limits in air: LEL: 3.5%; UEL: 14.0%.
Fire extinguishing agents not to be used: Water may be ineffective.
Special hazards of combustion products: Containers may explode.
Autoignition temperature: 724°F/384°C/657°K.
Electrical hazard: Class I, Group D.
Burning rate: 5.0 mm/min
Stoichiometric air-to-fuel ratio: 11.41 (estimate).

CHEMICAL REACTIVITY
Reactivity with water: Forms a strong base.
Binary reactants: Explosive reaction with mercury, acrolein (causes exothermic polymerization). Reacts with strong oxidizers and nonferrous metals–copper, aluminum, tin and zinc, and most of their alloys including brass–in the presence of moisture; cellulose nitrate; chlorine; hypochlorites. Will strip and dissolve paint; dissolves most plastic materials including polyethylene; can cause swelling of rubber by absorption. Reacts violently with strong acids.
Neutralizing agents for acids and caustics: Flush with water.
Reactivity group: 7
Compatibility class: Aliphatic amines

ENVIRONMENTAL DATA
Food chain concentration potential: Log P_{ow} = –0.17. Unlikely to accumulate.
Water pollution: Harmful to aquatic life in very low concentrations. May be dangerous if it enters nearby water intakes; notify operators. Notify local health and wildlife officials.
Response to discharge: Issue warning–high flammability, air contaminant, water contaminant. Restrict access. Evacuate area Disperse and flush.

SHIPPING INFORMATION
Grades of purity: Anhydrous (98.5+%); 70-72% in water; **Storage temperature:** Ambient; **Inert atmosphere:** None; **Venting:** Pressure-vacuum; **Stability during transport:** Stable.

NAS HAZARD CLASSIFICATION FOR BULK WATER TRANSPORTATION
FIRE: 4
HEALTH: Vapor irritant: 3; Liquid or solid irritant: 2; Poisons: 3
WATER POLLUTION: Human toxicity: 3; Aquatic toxicity: 3; Aesthetic effect: 3
REACTIVITY: Other chemicals: 2; Water: 3; Self-reaction: 0

PHYSICAL AND CHEMICAL PROPERTIES
Physical state @ 59°F/15°C and 1 atm: Liquid.
Molecular weight: 45.1
Boiling point @ 1 atm: 61.7°F/16.5°C/289.7°K.
Melting/Freezing point: –114°F/–81°C/192°K.
Critical temperature: 361°F/183°C/456°K.
Critical pressure: 827 psia = 56.2 atm = 5.70 MN/m².
Specific gravity (water = 1): 0.687 @ 15°C (liquid).
Liquid surface tension: 20.5 dynes/cm = 0.0205 N/m @ 15°C.
Relative vapor density (air = 1): 1.61
Ratio of specific heats of vapor (gas): 1.1181
Latent heat of vaporization: 253 Btu/lb = 146 cal/g = **Stoichiometric air-to-fuel ratio:** x 10^5 J/kg.
Heat of combustion: –16,180 Btu/lb = –8,990 cal/g = –376 x 10^5 J/kg.
Vapor pressure: (Reid) 29.8 psia; 874 mm.

ETHYL AMYL KETONE   REC. E:2300

SYNONYMS: AMYL ETHYL KETONE; EAK; EEC No. 606-020-00-1; ETIL AMIL CETONA (Spanish); 5-METHYL-3-HEPTANONE; 3-OCTANONE

IDENTIFICATION
CAS Number: 541-85-5; 106-68-3 (3-octanone)
Formula: $C_8H_{16}O$; $CH_3(CH_2)_4COC_2NO_5$
DOT ID Number: UN 2271; DOT Guide Number: 127
Proper Shipping Name: Ethyl amyl ketone

DESCRIPTION: Colorless liquid. Mild fruity odor. Practically insoluble in water.

Highly flammable • Narcotic effect at high concentrations •Containers may BLEVE when exposed to fire •Vapors may form explosive mixture with air • Vapors are heavier than air and will collect and stay in low areas • Vapors may travel long distances to ignition sources and flashback • Vapors in confined areas (e.g., tanks, sewers, buildings) may explode when exposed to fire • Irritating to the skin, eyes, and respiratory tract • Toxic products of combustion may include carbon monoxide.

Hazard Classification (based on NFPA-704 Rating System)
Health Hazards (Blue): 0; Flammability (Red): 2; Reactivity (Yellow): 0

EMERGENCY RESPONSE: See Appendix A (127)
Evacuation:
Public safety: Isolate spill area for at least 25 to 50 meters (80 to 160 feet) in all directions.
Spill: Large spill–Consider initial downwind evacuation for at least 300 meters (1000 feet).
Fire: Isolate for 800 meters (½ mile) in all directions, especially if tank, rail car, or tank truck is involved in fire.

EXPOSURE
Short-term effects: *SEEK MEDICAL ATTENTION.* **Vapor:** Irritating to eyes, nose, and throat. *IF INHALED*, will, will cause dizziness, headache, or difficult breathing. *IF IN EYES*, hold eyelids open and flush with plenty of water. *IF BREATHING HAS STOPPED*, give artificial respiration. IF breathing is difficult, administer oxygen. **Liquid:** Irritating to skin and eyes. *IF SWALLOWED*, will cause nausea or vomiting. Remove contaminated clothing and shoes. Flush affected areas with plenty of water. *IF IN EYES*, hold eyelids open and flush with plenty of water. The use of alcoholic beverages may enhance the toxic effect.

HEALTH HAZARDS
Personal protective equipment (PPE): B-Level PPE. OSHA Table Z-1-A air contaminant. Self-contained breathing apparatus, rubber boots and heavy rubber gloves.
Recommendations for respirator selection: NIOSH/OSHA
100 ppm: CCROV [any chemical cartridge respirator with organic vapor cartridge(s)]; or PAPROV [any powered, air-purifying respirator with organic vapor cartridge(s)]; or GMFOV[any air-purifying, full-facepiece respirator (gas mask) with a chin-style, front- or back-mounted acid gas canister]; or SA (any supplied-air respirator); or SCBAF (any self-contained breathing apparatus with a full facepiece). *EMERGENCY OR PLANNED ENTRY INTO UNKNOWN CONCENTRATIONS OR IDLH CONDITIONS:* SCBAF:PD,PP (any self-contained breathing apparatus that has a full facepiece and is operated in a pressure-demand or other positive-pressure mode); or SAF:PD,PP:ASCBA (any supplied-air respirator that has a full facepiece and is operated in a pressure-demand or other positive-pressure mode in combination with an auxiliary, self-contained breathing apparatus operated in a pressure-demand or other positive-pressure mode). *ESCAPE:* GMFOV [any air-purifying, full-facepiece respirator (gas mask) with a chin-style, front- or back-mounted organic vapor canister]; or SCBAE (any appropriate escape-type, self-contained breathing apparatus).
Exposure limits (TWA unless otherwise noted): ACGIH TLV 25 ppm (131 mg/m³); NIOSH/OSHA PEL 25 ppm (130 mg/m³).
Toxicity by ingestion: Grade 3: LD_{50} = 0.406 g/kg (mouse, intraperitoneal).
Long-term health effects: May effect the nervous system.
Vapor (gas) irritant characteristics: Eye and respiratory tract irritant. Vapors cause irritation such that personnel will find high concentrations unpleasant. The effect may be temporary. High concentration acts as a narcotic.
Liquid or solid irritant characteristics: Eye and skin irritant. If spilled on clothing and allowed to remain, may cause smarting and reddening of skin. Liquid degreases the skin.
Odor threshold: 6 ppm.
IDLH value: 3000 ppm.

FIRE DATA
Flash point: 115°F/46°C (cc).

CHEMICAL REACTIVITY

Binary reactants: Violent reaction with oxidizers.
Reactivity group: 18
Compatibility class: Ketones
Water pollution: Effect of low concentrations on aquatic life is unknown. Fouling to shoreline. May be dangerous if it enters nearby water intakes; notify operators. Notify local health and wildlife officials. **Response to discharge:** Issue warning–water contaminant. Restrict access. Evacuate areas. Mechanical containment. Should be removed. Chemical and physical treatment.

SHIPPING INFORMATION
Grades of purity: 99%; **Storage temperature:** Ambient.

PHYSICAL AND CHEMICAL PROPERTIES
Physical state @ 59°F/15°C and 1 atm: Liquid.
Molecular weight: 128.22
Boiling point @ 1 atm: 333–334°F/167–168°C/440–441°K.
Specific gravity (water = 1): 0.822
Relative vapor density (air = 1): 4.42
Vapor pressure: 2 mm.

ETHYLBENZENE REC. E:2400

SYNONYMS: AETHYLBENZOL (German); EB; EEC No. 601-023-00-4; ETHYLBENZOL; ETILBENCENO (Spanish); PHENYLETHANE

IDENTIFICATION
CAS Number: 100-41-4
Formula: C_8H_{10}; $C_6H_5CH_2CH_3$
DOT ID Number: UN 1175; DOT Guide Number: 129
Proper Shipping Name: Ethylbenzene
Reportable Quantity (RQ): **(CERCLA)** 1000 lb/454 kg

DESCRIPTION: Colorless liquid. Sweet, gasoline-like odor. Floats on the surface of water; practically insoluble. Flammable, irritating vapor is produced.

Highly flammable • Irritating to the skin, eyes, and respiratory tract; swallowing the liquid may cause chemical pneumonitis • Containers may BLEVE when exposed to fire •Vapors form explosive mixture with air • Vapors are heavier than air and will collect and stay in low areas • Vapors may travel long distances to ignition sources and flashback • Vapors in confined areas (e.g., tanks, sewers, buildings) may explode when exposed to fire • Toxic products of combustion may include carbon monoxide.

Hazard Classification (based on NFPA-704 Rating System)
Health Hazards (Blue): 2; Flammability (Red): 3; Reactivity (Yellow): 0

EMERGENCY RESPONSE: See Appendix A (129)
Evacuation:
Public safety: Isolate spill area for at least 50 to 100 meters (160 to 330 feet) in all directions.
Spill: Large spill–Consider initial downwind evacuation for at least 300 meters (1000 feet).
Fire: Isolate for 800 meters (½ mile) in all directions, especially if tank, rail car, or tank truck is involved in fire. **Vapor:** Irritating to eyes, nose, and throat. *IF INHALED*, will, will cause dizziness or difficult breathing. Move to fresh air. *IF BREATHING HAS STOPPED*, give artificial respiration. IF breathing is difficult, administer oxygen. **Liquid:** Will burn skin and eyes. Harmful if swallowed. Remove contaminated clothing and shoes. Flush affected areas with plenty of water. *IF IN EYES*, hold eyelids open and flush with plenty of water. *IF SWALLOWED* and victim is *CONSCIOUS AND ABLE TO SWALLOW*, have victim drink 4 to 8 ounces of water. **Do NOT induce vomiting.** Swallowing the liquid may cause chemical pneumonitis.

HEALTH HAZARDS
Personal protective equipment (PPE): B-Level PPE. OSHA Table Z-1-A air contaminant. SCBA; safety goggles. Chemical protective material(s) reported to have good to excellent resistance: Teflon®, Viton®. Also, Viton®/neoprene offers limited protection.
Recommendations for respirator selection: NIOSH/OSHA *800 ppm:* CCROV [any chemical cartridge respirator with organic vapor cartridge(s)]; or GMFOV[any air-purifying, full-facepiece respirator (gas mask) with a chin-style, front- or back-mounted acid gas canister]; or PAPROV [any powered, air-purifying respirator with organic vapor cartridge(s)]; or SA (any supplied-air respirator); or SCBAF (any self-contained breathing apparatus with a full facepiece). *EMERGENCY OR PLANNED ENTRY INTO UNKNOWN CONCENTRATIONS OR IDLH CONDITIONS:* SCBAF:PD,PP (any self-contained breathing apparatus that has a full facepiece and is operated in a pressure-demand or other positive-pressure mode); or SAF:PD,PP:ASCBA (any supplied-air respirator that has a full facepiece and is operated in a pressure-demand or other positive-pressure mode in combination with an auxiliary self-contained breathing apparatus operated in a pressure-demand or other positive pressure mode). *ESCAPE:* GMFOV[any air-purifying, full-facepiece respirator (gas mask) with a chin-style, front- or back-mounted organic vapor cannister]; or SCBAE (any appropriate escape-type, self-contained breathing apparatus). *Note:* Substance reported to cause eye irritation or damage; may require eye protection.
Exposure limits (TWA unless otherwise noted): ACGIH TLV 100 ppm (434 mg/m^3); NIOSH/OSHA 100 ppm (435 mg/m^3).
Short-term exposure limits (15-minute TWA): ACGIH STEL 125 ppm (543 mg/m^3); NIOSH STEL 125 ppm (545 mg/m^3).
Toxicity by ingestion: Grade 2; LD_{50} = 0.5 to 5 g/kg (rat).
Long-term health effects: Skin disorders. Absorbed into body fat; can be detected in the laboratory.
Vapor (gas) irritant characteristics: Eye and respiratory tract irritant. Vapors cause irritation such that personnel will find high concentrations unpleasant. The effect is temporary.
Liquid or solid irritant characteristics: Eye and skin irritant. Causes smarting of the skin and first-degree burns on short exposure; may cause secondary burns on long exposure.
Odor threshold: 0.09–0.6 ppm.
IDLH value: 800 ppm [10% LEL].

FIRE DATA
Flash point: 70°F/21°C (cc).
Flammable limits in air: LEL: 0.8%; UEL: 6.7%.
Fire extinguishing agents: Fight in the same manner as a Grade C petroleum fire.
Autoignition temperature: 806°F/430°C/703°K.
Electrical hazard: Class I, Group D. Flow or agitation of substance may generate electrostatic charges due to low conductivity.
Burning rate: 5.8 mm/min

CHEMICAL REACTIVITY
Binary reactants: Strong oxidizers may cause fire or explosion. Causes rubber to swell and soften.
Reactivity group: 32
Compatibility class: Aromatic hydrocarbons.

ENVIRONMENTAL DATA
Food chain concentration potential: Log P_{ow} = 3.2. Values > 3.0 are likely to bioconcentrate in aquatic organisms and other living tissue, especially in fats.
Food chain concentration potential: Negative; unlikely to accumulate.
Water pollution: Harmful to aquatic life in very low concentrations. Fouling to shoreline. May be dangerous if it enters nearby water intakes; notify operators. Notify local health and wildlife officials. **Response to discharge:** Mechanical containment. Should be removed. Chemical and physical treatment.

SHIPPING INFORMATION
Grades of purity: Research grade: 99.98%; pure grade: 99.5%; technical grade: 99.0%; **Storage temperature:** Ambient; **Inert atmosphere:** None; **Venting:** Open (flame arrester) or pressure-vacuum; **Stability during transport:** Stable.

NAS HAZARD CLASSIFICATION FOR BULK WATER TRANSPORTATION
FIRE: 3
HEALTH: Vapor irritant: 2; Liquid or solid irritant: 2; Poisons: 2
WATER POLLUTION: Human toxicity: 1; Aquatic toxicity: 3; Aesthetic effect: 2
REACTIVITY: Other chemicals: 1; Water: 0; Self-reaction: 0

PHYSICAL AND CHEMICAL PROPERTIES
Physical state @ 59°F/15°C and 1 atm: Liquid.
Molecular weight: 106.17
Boiling point @ 1 atm: 277.2°F/136.2°C/409.4°K.
Melting/Freezing point: –139°F/–95°C/178°K.
Critical temperature: 651.0°F/343.9°C/617.1°K.
Critical pressure: 523 psia = 35.6 atm = 3.61 MN/m².
Specific gravity (water = 1): 0.867 @ 68°F/20°C (liquid).
Liquid surface tension: 29.2 dynes/cm = 0.0292 N/m @ 68°F/20°C.
Liquid water interfacial tension: 35.48 dynes/cm = 0.03548 N/m @ 68°F/20°C.
Relative vapor density (air = 1): 3.69
Ratio of specific heats of vapor (gas): 1.071
Latent heat of vaporization: 144 Btu/lb = 80.1 cal/g = 3.35 x 10^5 J/kg.
Heat of combustion: –17,780 Btu/lb = –9877 cal/g = –413.5 x 10^5 J/kg.
Vapor pressure: (Reid) 0.4 psia; 7 mm.

ETHYLBUTANOL   REC. E:2500

SYNONYMS: 2-ETHYL BUTANOL; 2-ETHYL BUTANOL-1; 2-ETHYL-1-BUTANOL; 2-ETHYLBUTYL ALCOHOL; 2-ETILBUTINOL (Spanish); sec-HEXANOL; sec-HEXYL ALCOHOL; 2-METHYLOLPENTANE; sec-PENTYLCARBINOL; 3-PENTYLCARBINOL; PSEUDOHEXYL ALCOHOL

IDENTIFICATION
CAS Number: 97-95-0
Formula: $C_6H_{14}O$; $(C_2NO_5)_2CHCH_2OH$
DOT ID Number: UN 2275; DOT Guide Number: 129
Proper Shipping Name: 2-Ethylbutanol

DESCRIPTION: Colorless liquid. Mild alcohol odor. Floats on the surface of water.

Highly flammable • Containers may BLEVE when exposed to fire • Vapors may form explosive mixture with air • Vapors are heavier than air and will collect and stay in low areas • Vapors may travel long distances to ignition sources and flashback • Vapors in confined areas (e.g., tanks, sewers, buildings) may explode when exposed to fire • Irritating to the skin, eyes, and respiratory tract • Toxic products of combustion may include carbon monoxide.

Hazard Classification (based on NFPA-704 Rating System)
Health Hazards (Blue): 2; Flammability (Red): 3; Reactivity (Yellow): 1

EMERGENCY RESPONSE: See Appendix A (129)
Evacuation:
Public safety: Isolate spill area for at least 50 to 100 meters (160 to 330 feet) in all directions.
Spill: Large spill–Consider initial downwind evacuation for at least 300 meters (1000 feet).
Fire: Isolate for 800 meters (½ mile) in all directions, especially if tank, rail car, or tank truck is involved in fire.

EXPOSURE
Short-term effects: *SEEK MEDICAL ATTENTION*. **Liquid:** Will burn eyes. Harmful if swallowed. *IF IN EYES*, hold eyelids open and flush with plenty of water. *IF SWALLOWED* and victim is *CONSCIOUS AND ABLE TO SWALLOW*, have victim drink 4 to 8 ounces of water.

HEALTH HAZARDS
Personal protective equipment (PPE): B-Level PPE. Fresh-air mask; plastic gloves; coverall goggles; safety shower and eye bath. Also, natural rubber, neoprene, and nitrile rubber may offer some protection from alcohols. Chemical protective material(s) reported to offer minimal to poor protection: Viton®/neoprene, butyl rubber,/neoprene.
Toxicity by ingestion: Grade 2; LD_{50} = 0.5 to 5 g/kg (rat).
Vapor (gas) irritant characteristics: Vapors cause smarting of the eyes or respiratory system if present in high concentrations. The effect may be temporary.
Liquid or solid irritant characteristics: Irritates eyes; moderate irritation of skin.

FIRE DATA
Flash point: 70°F/21°C (cc).
Flammable limits in air: LEL: 1.2%; UEL: 7.7%.
Autoignition temperature: 580°F/304°C/577°K (calculated).
Electrical hazard: Class I, Group D.

CHEMICAL REACTIVITY
Binary reactants: Strong acids, caustics, isocyanates, amines.
Reactivity group: 20
Compatibility class: Alcohols, glycols

ENVIRONMENTAL DATA
Food chain concentration potential: Negative; unlikely to accumulate.
Water pollution: Effect of low concentrations on aquatic life is unknown. Fouling to shoreline. May be dangerous if it enters nearby water intakes; notify operators. Notify local health and wildlife officials. **Response to discharge:** Disperse and flush.

SHIPPING INFORMATION
Storage temperature: Ambient; **Inert atmosphere:** None; **Venting:** Open (flame arrester) or pressure-vacuum; **Stability during transport:** Stable.

PHYSICAL AND CHEMICAL PROPERTIES
Physical state @ 59°F/15°C and 1 atm: Liquid.
Molecular weight: 102.17
Boiling point @ 1 atm: 242°F/117°C/390°K.
Melting/Freezing point: –173°F/–114°C/159°K.
Specific gravity (water = 1): 0.834 @ 68°F/20°C (liquid).
Liquid surface tension: 24.3 dynes/cm = 0.0243 N/m @ 77°F/25°C.
Liquid water interfacial tension: (estimate) 40 dynes/cm = 0.04 N/m @ 68°F/20°C.
Latent heat of vaporization: 196.0 Btu/lb = 108.9 cal/g = 4.559 x 10^5 J/kg.
Heat of combustion: (estimate) = –16,600 Btu/lb = –9250 cal/g = –387 x 10^5 J/kg.
Vapor pressure: (Reid) 0.07 psia.

N-ETHYL-n-BUTYLAMINE   REC. E:2600

SYNONYMS: BUTYLETHYLAMINE; ETHYLBUTYLAMINE; *n*-ETIL BUTILAMINA (Spanish)

IDENTIFICATION
CAS Number: 13360-63-9; 617-79-8 (2-isomer)
Formula: $C_6H_{15}N$; $C_2H_5NHC_4H_9$
DOT ID Number: UN 2733; DOT Guide Number: 132
Proper Shipping Name: Amines, liquid, corrosive, flammable, n.o.s; Polyamines, flammable, corrosive, n.o.s

DESCRIPTION: Water-white liquid. Ammonia-like odor. Floats on the surface of water; soluble.

Poison! • Corrosive • Highly flammable • Severely irritating to eyes, skin, and respiratory tract; skin and eye contact can cause severe burns and blindness • Firefighting gear (including SCBA) does not provide adequate protection. If exposure occurs, remove and isolate gear immediately and thoroughly decontaminate personnel • Containers may BLEVE when exposed to fire • Vapors may form explosive mixture with air • Vapors are heavier than air and will collect and stay in low areas • Vapors may travel long distances to ignition sources and flashback • Vapors in confined areas (e.g., tanks, sewers, buildings) may explode when exposed to fire • Toxic products of combustion may include nitrogen oxide • Do not put yourself in danger by entering a contaminated area to rescue a victim.

Hazard Classification (based on NFPA-704 Rating System)
Health Hazards (Blue): 3; Flammability (Red): 3; Reactivity (Yellow): 0

EMERGENCY RESPONSE: See Appendix A (132)
Evacuation:
Public safety: Isolate spill area for at least 50 to 100 meters (160 to 330 feet) in all directions.
Spill: Increase, as necessary, the isolation distance shown above, in "Public safety."
Fire: Isolate for 800 meters (½ mile) in all directions, especially if tank, rail car, or tank truck is involved in fire.

EXPOSURE
Short-term effects: *SEEK MEDICAL ATTENTION.* **Vapor:** Irritating to eyes, nose, and throat. Harmful if inhaled. Move to fresh air. *IF BREATHING HAS STOPPED*, give artificial respiration. IF breathing is difficult, administer oxygen. **Liquid:** Will burn skin and eyes. Harmful if swallowed. Remove contaminated clothing and shoes. Flush affected areas with plenty of water. *IF IN EYES*, hold eyelids open and flush with plenty of water. *IF SWALLOWED* and victim is *CONSCIOUS AND ABLE TO SWALLOW*, have victim drink 4 to 8 ounces of water. **Do NOT induce vomiting.**
Note to physician or authorized medical personnel: Medical observation is recommended for 24 to 48 hours after breathing overexposure, as pulmonary edema may be delayed. As first aid for pulmonary edema, consider administering a corticosteroid spray. Cigarette smoking may exacerbate pulmonary injury and should be discouraged for at least 72 hours following exposure.

HEALTH HAZARDS
Personal protective equipment (PPE): B-Level PPE. Chemical protective material(s) reported to have good to excellent resistance: PV alcohol.
Toxicity by ingestion: Grade 3; LD_{50} = 50 to 500 mg/kg.
Vapor (gas) irritant characteristics: Respiratory tract and eye irritant.
Liquid or solid irritant characteristics: Severe eye and skin irritant. Causes second- and third-degree burns on short contact and is very injurious to the eyes.

FIRE DATA
Flash point: 56°F/13°C (cc); 64°F/18°C (oc) (2-isomer).
Fire extinguishing agents not to be used: Water may be ineffective.

CHEMICAL REACTIVITY
Neutralizing agents for acids and caustics: Dilute with water.

ENVIRONMENTAL DATA
Water pollution: Effect of low concentration on aquatic life is unknown. May be dangerous if it enters nearby water intakes; notify operators. Notify local health and wildlife officials. **Response to discharge:** Issue warning–high flammability, corrosive. Restrict access. Evacuate area. Disperse and flush.

SHIPPING INFORMATION
Grades of purity: 99%; **Storage temperature:** Ambient; **Stability during transport:** Stable.

PHYSICAL AND CHEMICAL PROPERTIES
Physical state @ 59°F/15°C and 1 atm: Liquid.
Molecular weight: 101.2
Boiling point @ 1 atm: 227.3°F/108.5°C/381.7°K.
Critical temperature: (estimate) 565.7°F/296.5°C/569.6°K.
Critical pressure: (estimate) 440.9 psia = 30 atm = 3.04 MN/m^2.
Specific gravity (water = 1): 0.7398 @ 68°F/20°C.
Liquid surface tension: (estimate) 21 dynes/cm = 0.021 N/m @ 68°F/20°C.
Liquid water interfacial tension: (estimate) 54 dynes/cm = 0.054 N/m @ 68°F/20°C.
Relative vapor density (air = 1): 3.5
Ratio of specific heats of vapor (gas): (estimate) more than 1 @ 68°F/20°C (68°F).
Latent heat of vaporization: 153 Btu/lb = 85.0 cal/g = 3.56 x 10^5 J/kg.
Heat of combustion: –17431 Btu/lb = –9684 cal/g = –405 x 10^5 J/kg.

ETHYL BUTYL KETONE   REC. E:2700

SYNONYMS: BUTYL ETHYL KETONE; *n*-BUTYL ETHYL

KETONE; ETIL BUTIL CETONA (Spanish); HEPTAN-3-ONE; 3-HEPTANONE; 3-HEPTANONE

IDENTIFICATION
CAS Number: 106-35-4
Formula: $C_7H_{14}O$; $C_2H_5COC_4H_9$
DOT ID Number: UN 1993; DOT Guide Number: 128
Proper Shipping Name: Flammable liquid, n.o.s.

DESCRIPTION: Colorless liquid. Mild, fruity odor. Insoluble in water.

Highly flammable • Containers may BLEVE when exposed to fire • Vapors may form explosive mixture with air • Vapors are heavier than air and will collect and stay in low areas • Vapors may travel long distances to ignition sources and flashback • Vapors in confined areas (e.g., tanks, sewers, buildings) may explode when exposed to fire • Irritating to the skin, eyes, and respiratory tract • Toxic products of combustion may include carbon monoxide and CO_2.

Hazard Classification (based on NFPA-704 Rating System)
Health Hazards (Blue): 1; Flammability (Red): 2; Reactivity (Yellow): 0

Highly flammable • Containers may BLEVE when exposed to fire • Vapors may form explosive mixture with air • Vapors are heavier than air and will collect and stay in low areas • Vapors may travel long distances to ignition sources and flashback • Vapors in confined areas (e.g., tanks, sewers, buildings) may explode when exposed to fire • Irritating to the skin, eyes, and respiratory tract • Toxic products of combustion may include carbon monoxide.

Hazard Classification (based on NFPA-704 Rating System)
Health Hazards (Blue): 1; Flammability (Red): 2; Reactivity (Yellow): 0

EMERGENCY RESPONSE: See Appendix A (128)
Evacuation:
Public safety: Isolate spill area for at least 25 to 50 meters (80 to 160 feet) in all directions.
Spill: Large spill–Consider initial downwind evacuation for at least 300 meters (1000 feet).
Fire: Isolate for 800 meters (½ mile) in all directions, especially if tank, rail car, or tank truck is involved in fire.

EXPOSURE
Short-term effects: *SEEK MEDICAL ATTENTION.* **Vapor:** Move victim to fresh air. *IF BREATHING HAS STOPPED,* give artificial respiration. IF breathing is difficult, administer oxygen. **Liquid:** Remove contaminated clothing and shoes. Wash affected areas with soap and water. *IF IN EYES,* hold eyelids open and flush with plenty of water. *IF SWALLOWED* and victim is *CONSCIOUS AND ABLE TO SWALLOW,* have victim drink 1 to 2 glasses of water and induce vomiting.

HEALTH HAZARDS
Personal protective equipment (PPE): B-Level PPE. OSHA Table Z-1-A air contaminant. Impervious clothing, gloves, face shield, and other appropriate protective clothing to prevent repeated or prolonged skin contact with liquid material. Use splash proof safety goggles where liquid may contact eyes. Use respiratory protection (approved respirator or self-contained breathing apparatus) where vapors may be encountered.
Recommendations for respirator selection: NIOSH/OSHA *500 ppm*: CCROV [any chemical cartridge respirator with organic vapor cartridge(s)]; or SA* (any supplied-air respirator). *1000 ppm:* SA:CF* (any supplied-air respirator operated in a continuous-flow mode); or PAPROV* [any powered, air-purifying respirator with organic vapor cartridge(s)]; or CCRFOV [any air-purifying, full-facepiece respirator (gas mask) with a chin-style, front- or back-mounted acid gas canister]; or GMFOV[any air-purifying, full-facepiece respirator (gas mask) with a chin-style, front- or back-mounted acid gas canister]; or SCBAF (any self-contained breathing apparatus with a full facepiece); or SAF (any supplied-air respirator with a full facepiece). *EMERGENCY OR PLANNED ENTRY INTO UNKNOWN CONCENTRATIONS OR IDLH CONDITIONS:* SCBAF:PD,PP (any self-contained breathing apparatus that has a full facepiece and is operated in a pressure-demand or other positive-pressure mode); or SAF:PD,PP:ASCBA (any supplied-air respirator that has a full facepiece and is operated in a pressure-demand or other positive-pressure mode in combination with an auxiliary self-contained breathing apparatus operated in a pressure-demand or other positive pressure mode). *ESCAPE:* GMFOV[any air-purifying, full-facepiece respirator (gas mask) with a chin-style, front- or back-mounted organic vapor canister]; or SCBAE (any appropriate escape-type, self-contained breathing apparatus). **Note*: Substance reported to cause eye irritation or damage; may require eye protection.*
Exposure limits (TWA unless otherwise noted): ACGIH TLV 50 ppm (234 mg/m^3); NIOSH/OSHA 50 ppm (230 mg/m^3).
Short-term exposure limits (15-minute TWA): ACGIH STEL 75 ppm.
Long-term health effects: Chemical is a defatting agent and can cause dermatitis on prolonged exposure. May exacerbate previously existing respiratory, liver, or skin ailments.
Vapor (gas) irritant characteristics: Vapors cause irritation such that personnel will find high concentrations unpleasant. The effect is temporary.
Liquid or solid irritant characteristics: Eye and skin irritant. If spilled on clothing and allowed to remain, may cause smarting and reddening of the skin.
IDLH value: 1000 ppm.

FIRE DATA
Flash point: 115°F/46°C (oc).
Flammable limits in air: LEL: 1.4%; UEL: 8.8%.
Fire extinguishing agents not to be used: Water.
Stoichiometric air-to-fuel ratio: 47.6

CHEMICAL REACTIVITY
Binary reactants: Contact with oxidizers can cause fires and explosions. Reacts with acetaldehyde, perchloric acid.

ENVIRONMENTAL DATA
Water pollution: Effect of low concentrations on aquatic life is unknown. May be dangerous if it enters nearby water intakes; notify operators. Notify local health and wildlife officials. **Response to discharge:** Restrict access. Should be removed.

SHIPPING INFORMATION
Grades of purity: CP; technical grades; 20% solution in hexane; **Storage temperature:** Ambient; **Inert atmosphere:** Store solution under nitrogen; **Stability during transport:** Stable.

PHYSICAL AND CHEMICAL PROPERTIES
Physical state @ 59°F/15°C and 1 atm: Liquid.
Molecular weight: 114.19
Boiling point @ 1 atm: 297°F/147°C/420°K.
Melting/Freezing point: –38°F/–39°C/234°K.

Specific gravity (water = 1): 0.818
Relative vapor density (air = 1): 3.93
Vapor pressure: 4 mm.

ETHYL BUTYRATE   REC. E:2800

SYNONYMS: BUTIRATO de EDTILO (Spanish); BUTONIC ACID ETHYL ESTER; BUTYRIC ACID, ETHYL ESTER; BUTYRIC ETHER; ETHYL BUTANOATE; ETHYL-*n*-BUTYRATE

IDENTIFICATION
CAS Number: 105-54-4
Formula: $C_6H_{12}O_2$; $CH_3CH_2CH_2COOC_2H_5$
DOT ID Number: UN 1180; DOT Guide Number: 129
Proper Shipping Name: Ethyl butyrate

DESCRIPTION: Colorless liquid. Fruity odor, like apple or pineapple. Floats on the surface of water; insoluble.

Highly flammable • Containers may BLEVE when exposed to fire • Vapors may form explosive mixture with air • Vapors are heavier than air and will collect and stay in low areas • Vapors may travel long distances to ignition sources and flashback • Vapors in confined areas (e.g., tanks, sewers, buildings) may explode when exposed to fire • Irritating to the skin, eyes, and respiratory tract • Toxic products of combustion may include carbon monoxide.

Hazard Classification (based on NFPA-704 Rating System)
Health Hazards (Blue): 0; Flammability (Red): 3; Reactivity (Yellow): 0

EMERGENCY RESPONSE: See Appendix A (129)
Evacuation:
Public safety: Isolate spill area for at least 50 to 100 meters (160 to 330 feet) in all directions.
Spill: Large spill–Consider initial downwind evacuation for at least 300 meters (1000 feet).
Fire: Isolate for 800 meters (½ mile) in all directions, especially if tank, rail car, or tank truck is involved in fire.

EXPOSURE
Short-term effects: *SEEK MEDICAL ATTENTION.* **Vapor:** Irritating to eyes, nose, and throat. *IF INHALED,* will, will cause headache or dizziness. *IF IN EYES,* hold eyelids open and flush with plenty of water. *IF BREATHING HAS STOPPED,* give artificial respiration. IF breathing is difficult, administer oxygen. **Liquid:** Irritating to skin and eyes. *IF SWALLOWED,* will cause nausea, vomiting, dizziness or headache. Remove contaminated clothing and shoes. Flush affected areas with plenty of water. *IF IN EYES,* hold eyelids open and flush with plenty of water. *IF SWALLOWED* and victim is *CONSCIOUS AND ABLE TO SWALLOW,* have victim drink 4 to 8 ounces of water and have victim induce vomiting. *IF SWALLOWED* and victim is *UNCONSCIOUS OR HAVING CONVULSIONS,* do nothing except keep victim warm.

HEALTH HAZARDS
Personal protective equipment (PPE): B-Level PPE. All-purpose canister mask or chemical cartridge respirator; glass or face shield; rubber gloves.
Toxicity by ingestion: Grade 1; oral LD_{50} = 13 g/kg (rat).
Liquid or solid irritant characteristics: Eye irritant.
Odor threshold: 0.015 ppm.

FIRE DATA
Flash point: 85°F/29°C (oc); 75°F/24°C (cc).
Fire extinguishing agents not to be used: Water may be ineffective
Autoignition temperature: 865°F/463°C/736°K.
Burning rate: 4.72 mm/min

CHEMICAL REACTIVITY
Binary reactants: Oxidizers. May attack some forms of plastics

ENVIRONMENTAL DATA
Food chain concentration potential: Log P_{ow} = 1.7. Unlikely to accumulate.
Water pollution: Effect of low concentrations on aquatic life is unknown. Fouling to shoreline. May be dangerous if it enters nearby water intakes; notify operators. Notify local health and wildlife officials. **Response to discharge:** Issue warning–high flammability. Restrict access. Mechanical containment. Should be removed. Chemical and physical treatment.

SHIPPING INFORMATION
Grades of purity: Commercial, 98+%; **Storage temperature:** Ambient; **Inert atmosphere:** None; **Venting:** Open (flame arrester); **Stability during transport:** Stable.

PHYSICAL AND CHEMICAL PROPERTIES
Physical state @ 59°F/15°C and 1 atm: Liquid.
Molecular weight: 116.16
Boiling point @ 1 atm: 250°F/121°C/394°K.
Melting/Freezing point: –135°F/–93°C/180°K.
Critical temperature: 559°F/293°C/566°K.
Critical pressure: 460 psia = 31 atm = 3.2 MN/m^2.
Specific gravity (water = 1): 0.879 @ 68°F/20°C (liquid).
Liquid surface tension: 24.5 dynes/cm = 0.0245 N/m @ 68°F/20°C.
Relative vapor density (air = 1): 4.0
Latent heat of vaporization: 128 Btu/lb = 71 cal/g = 3.0 x 10^5 J/kg.
Heat of combustion: –13,200 Btu/lb = –7330 cal/g = –306 x 10^5 J/kg.

ETHYL CHLORIDE   REC. E:2900

SYNONYMS: AETHYLCHLORID (German); AETHYLIS; AETHYLIS CHLORIDUM; ANODYNON; CHELEN; CHLORETHYL; CHLORIDUM; CHLOROAETHAN (German); CHLOROETHANE; CHLORURE D'ETHYLE (French); CLORURO de ETILO (Spanish); CHLORYL; EEC No. 602-009-00-0; ETHER CHLORATUS; ETHER HYDROCHLORIC; ETHER MURIATIC; HYDROCHLORIC ETHER; KELENE; MONOCHLORETHANE; MURIATIC ETHER; NARCOTILE

IDENTIFICATION
CAS Number: 75-00-3
Formula: C_2H_5Cl
DOT ID Number: UN 1037; DOT Guide Number: 115
Proper Shipping Name: Ethyl chloride
Reportable Quantity (RQ): **(CERCLA)** 100 lb/45.4 kg

DESCRIPTION: Colorless gas; shipped and stored as a colorless liquid under pressure. Ether-like odor. Burning taste. Slightly soluble in water; floats and may boil. Flammable, irritating vapor is produced.

Extremely flammable • Forms explosive mixture with air • Reacts explosively with many materials • Gas from liquefied gas are initially heavier than air and spread along ground • Gas may travel to source of ignition and flashback • Exposure of cylinders to elevated temperatures, fire, and flame may cause cylinders to rupture or cause frangible disk to burst, releasing entire contents of cylinder • Ruptured or venting cylinders may rocket through buildings and/or travel a considerable distance •Irritating to the skin, eyes, and respiratory tract • Vapors may cause dizziness or asphyxiation without warning • Contact with liquid may cause frostbite.

Hazard Classification (based on NFPA-704 Rating System)
Health Hazards (Blue): 1; Flammability (Red): 4; Reactivity (Yellow): 0

EMERGENCY RESPONSE: See Appendix A (115)
Evacuation:
Public safety: Isolate spill area for at least 50 to 100 meters (160 to 330 feet) in all directions.
Spill: Consider initial downwind evacuation for at least 800 meters (½ mile).
Fire: Isolate for 1600 meters (1 mile) in all directions, especially if tank, rail car, or tank truck is involved in fire.
EXPOSURE
Short-term effects: *SEEK MEDICAL ATTENTION. ABSORBED THROUGH THE SKIN.* **Vapor:** Irritating to eyes, nose, and throat. *IF INHALED*, will, will cause dizziness or loss of consciousness. Move to fresh air. *IF BREATHING HAS STOPPED,* give artificial respiration. IF breathing is difficult, administer oxygen. **Liquid:** Irritating to skin and eyes. Can cause frostbite. Flush affected areas with plenty of water. *IF IN EYES*, hold eyelids open and flush with plenty of water. *DO NOT RUB AFFECTED AREAS.*

HEALTH HAZARDS
Personal protective equipment (PPE): B-Level PPE. OSHA Table Z-1-A air contaminant. Neoprene rubber clothing where liquid contact is likely; chemical worker's goggles. Wear thermal protective clothing.
Recommendations for respirator selection: OSHA
3800 ppm: SA* (any supplied-air respirator); or SCBAF (any self-contained breathing apparatus with a full facepiece). *EMERGENCY OR PLANNED ENTRY INTO UNKNOWN CONCENTRATIONS OR IDLH CONDITIONS*: SCBAF:PD,PP (any self-contained breathing apparatus that has a full facepiece and is operated in a pressure-demand or other positive-pressure mode); or SAF:PD,PP:ASCBA (any supplied-air respirator that has a full facepiece and is operated in a pressure-demand or other positive-pressure mode in combination with an auxiliary self-contained breathing apparatus operated in a pressure-demand or other positive pressure mode). *ESCAPE*: GMFOV[any air-purifying, full-facepiece respirator (gas mask) with a chin-style, front- or back-mounted organic vapor canister]; or SCBAE (any appropriate escape-type, self-contained breathing apparatus).
Exposure limits (TWA unless otherwise noted): ACGIH TLV 100 ppm (264 mg/m^3) animal carcinogen; OSHA PEL 1000 ppm (2600 mg/m^3); NIOSH REL handle with caution in the workplace.
Long-term health effects: May effect the nervous system. Liver and kidney damage may occur. IARC rating 3; limited animal evidence.
Vapor (gas) irritant characteristics: Vapors cause smarting of the eyes or respiratory system if present in high concentrations.
Liquid or solid irritant characteristics: If spilled on clothing and allowed to remain, may cause smarting and reddening of the skin. Rapid evaporation may cause frostbite.

Odor threshold: 4.2 ppm.
IDLH value: 3800 ppm.

FIRE DATA
Flash point: –45°F/–43°C (oc); –58°F/–50°C (cc) (liquid).
Flammable limits in air: LEL: 3.6%; UEL: 15.3%.
Behavior in fire: Containers may explode.
Autoignition temperature: 966°F/519°C/792°K.
Electrical hazard: Class I, Group D. Flow or agitation of substance may generate electrostatic charges due to low conductivity.
Burning rate: 3.8 mm/min
Stoichiometric air-to-fuel ratio: 6.383°F/–14.232°C (estimate).

CHEMICAL REACTIVITY
Reactivity with water: Produces toxic and corrosive fumes; forms hydrogen chloride gas.
Binary reactants: Dangerous reaction with oxidizers. Reacts with chemically active metals such as sodium, potassium, calcium, powdered aluminum, zinc and magnesium.
Reactivity group: 36
Compatibility class: Halogenated hydrocarbons

ENVIRONMENTAL DATA
Food chain concentration potential: Log P_{ow} = 1.48. Unlikely to accumulate.
Water pollution: Not harmful to aquatic life. May be dangerous if it enters nearby water intakes; notify operators. **Response to discharge:** Issue warning–high flammability. Restrict access. Evacuate area.

SHIPPING INFORMATION
Grades of purity: Technical: 98–100%; USP: 100%; **Storage temperature:** Ambient; **Inert atmosphere:** None; **Venting:** Safety relief; **Stability during transport:** Stable.

NAS HAZARD CLASSIFICATION FOR BULK WATER TRANSPORTATION
FIRE: 4
HEALTH: Vapor irritant: 1; Liquid or solid irritant: 1; Poisons: 1
WATER POLLUTION: Human toxicity: 0; Aquatic toxicity: 1; Aesthetic effect: 1
REACTIVITY: Other chemicals: 1; Water: 0; Self-reaction: 0

PHYSICAL AND CHEMICAL PROPERTIES
Physical state @ 59°F/15°C and 1 atm: Gas.
Molecular weight: 64.52
Boiling point @ 1 atm: 54.0°F/12.2°C/285.4°K.
Melting/Freezing point: –218°F/–139°C/134°K.
Critical temperature: 369°F/187.2°C/460.4°K.
Critical pressure: 758 psia = 51.6 atm = 5.23 MN/m^2.
Specific gravity (water = 1): 0.92 @ 32°F/0°C (liquid).
Liquid surface tension: 19.5 dynes/cm = 0.0195 N/m @ 68°F/20°C.
Liquid water interfacial tension: (estimate) 40 dynes/cm = 0.04 N/m at @ 32°F/0°C.
Relative vapor density (air = 1): 2.23
Ratio of specific heats of vapor (gas): 1.155
Latent heat of vaporization: 163 Btu/lb = 90.6 cal/g = 3.79 x 10^5 J/kg.
Heat of combustion: –8100 Btu/lb = –4500 cal/g = –188.4 x 10^5 J/kg.
Heat of fusion: 16.49 cal/g.
Vapor pressure: (Reid) 34.5 psia; 1000 mm.

ETHYL CHLOROACETATE　　　　　REC. E:3000

SYNONYMS: CHLOROACETIC ACID, ETHYL ESTER; CLOROACETATO de ETILO (Spanish); EEC No. 607-070-00-7; ETHYL-α-CHLOROACETATE; ETHYL 2-CHLOROETHANOATE; ETHYL CHLOROETHANOATE; ETHYL MONOCHLOROACETATE; MONOCHLORETHANOIC ACID, ETHYL ESTER; AF9110000

IDENTIFICATION
CAS Number: 105-39-5
Formula: $C_4H_7ClO_2$; $ClCH_2COOC_2NO_5$
DOT ID Number: UN 1181; DOT Guide Number: 155
Proper Shipping Name: Ethyl chloroacetate

DESCRIPTION: Colorless to light brown liquid. Irritating, fruity odor. Sinks in water; insoluble.

Highly flammable • Corrosive skin, eyes, and respiratory tract; produces tearing; contact with skin and eyes can cause burns, impaired vision, and blindness; inhalation symptoms may be delayed • Firefighting gear (including SCBA) does not provide adequate protection. If exposure occurs, remove and isolate gear immediately and thoroughly decontaminate personnel • Containers may BLEVE when exposed to fire • Vapors may form explosive mixture with air • Vapors are heavier than air and will collect and stay in low areas • Vapors may travel long distances to ignition sources and flashback • Vapors in confined areas (e.g., tanks, sewers, buildings) may explode when exposed to fire • Toxic products of combustion may include hydrogen chloride and phosgene.

Hazard Classification (based on NFPA-704 Rating System)
Health Hazards (Blue): 3; Flammability (Red): 3; Reactivity (Yellow): 0

EMERGENCY RESPONSE: See Appendix A (155)
Evacuation:
Public safety: Isolate the area of spill or leak for at least 50 to 100 meters (160 to 330 feet) in all directions.
Spill: Increase, in the downwind direction, as necessary, the distance shown under "Public Safety."
Fire: If tank, rail car, or tank truck is involved in fire, isolate for at least 800 meters (½ mile) in all directions; also, consider initial evacuation for 800 meters (½ mile) in all directions.

EXPOSURE
Short-term effects: *SEEK MEDICAL ATTENTION.* **Vapor:** Irritating to eyes, nose, and throat. *IF INHALED*, will, will cause headache, or nausea. Move victim to fresh air. *IF BREATHING HAS STOPPED,* give artificial respiration; *avoid mouth-to-mouth resuscitation; use bag/mask apparatus.* If breathing is difficult, administer oxygen. **Liquid:** *POISONOUS IF SWALLOWED.* Irritating to skin and eyes. Remove contaminated clothing and shoes. Flush affected areas with plenty of water. *IF IN EYES*, hold eyelids open and flush with plenty of water. *IF SWALLOWED* and victim is *CONSCIOUS AND ABLE TO SWALLOW*, have victim drink 4 to 8 ounces of water and have victim induce vomiting. *IF SWALLOWED* and victim is *UNCONSCIOUS OR HAVING CONVULSIONS,* do nothing except keep victim warm.
Note to physician or authorized medical personnel: Medical observation is recommended for 24 to 48 hours after breathing overexposure, as pulmonary edema may be delayed. As first aid for pulmonary edema, consider administering a corticosteroid spray. Cigarette smoking may exacerbate pulmonary injury and should be discouraged for at least 72 hours following exposure.

HEALTH HAZARDS
Personal protective equipment (PPE): B-Level PPE. Organic canister mask; rubber gloves; chemical goggles.
Toxicity by ingestion: Grade 4; LD_{50} less than 50 mg/kg.
Vapor (gas) irritant characteristics: Eye and respiratory tract irritant.
Liquid or solid irritant characteristics: Severe eye and skin irritant. May cause permanent eye damage.

FIRE DATA
Flash point: 147°F/64°C (oc); 100°F/38°C (cc).
Burning rate: 2.3 mm/min.

CHEMICAL REACTIVITY
Reactivity with water: Cold water: Very slow, may not be hazardous. Warm water: Acetic acid may be formed; toxic and corrosive fumes are produced.
Binary reactants: Slow hydrolysis to acidic products will cause slow corrosion of common metals. Alkali metals can cause violent reaction.

ENVIRONMENTAL DATA
Food chain concentration potential: Negative; unlikely to accumulate.
Water pollution: Effect of low concentrations on aquatic life is unknown. May be dangerous if it enters nearby water intakes; notify operators. Notify local health and wildlife officials. **Response to discharge:** Issue warning–water contaminant. Should be removed. Chemical and physical treatment.

SHIPPING INFORMATION
Grades of purity: 99+1%; **Storage temperature:** Ambient; **Inert atmosphere:** None; **Venting:** Open (flame arrester); **Stability during transport:** Stable.

PHYSICAL AND CHEMICAL PROPERTIES
Physical state @ 59°F/15°C and 1 atm: Liquid.
Molecular weight: 122.6
Boiling point @ 1 atm: 289°F/143°C/416°K.
Melting/Freezing point: –14°F/–26°C/247°K.
Specific gravity (water = 1): 1.15 @ 68°F/20°C (liquid).
Liquid surface tension: (estimate) 26 dynes/cm = 0.026 N/m @ 68°F/20°C.
Liquid water interfacial tension: (estimate) 24 dynes/cm = 0.024 N/m @ 68°F/20°C.
Relative vapor density (air = 1): 4.3
Latent heat of vaporization: 155 Btu/lb = 86 cal/g = 3.6×10^5 J/kg.
Heat of combustion: –7250 Btu/lb = –4028 cal/g = $–168 \times 10^5$ J/kg.
Vapor pressure: 3.6 mm.

ETHYL CHLOROFORMATE　　　　　REC. E:3100

SYNONYMS: CHLOROFORMIC ACID, ETHYL ESTER; ECF; EEC No. 607-020-00-4; ETHYL CHLOROCARBONATE; CLOROFORMIATO de ETILO (Spanish); TL423

IDENTIFICATION
CAS Number: 541-41-3
Formula: $ClCOOC_2NO_5$

DOT ID Number: UN 1182; DOT Guide Number: 155
Proper Shipping Name: Ethyl chloroformate

DESCRIPTION: Colorless to light yellow liquid. Sharp, irritating odor; like hydrochloric acid. May be contaminated with small levels of phosgene gas. Reacts slowly with water forming ethyl alcohol, CO_2, and chloroformic acid; practically insoluble.

Poison! • Highly flammable • Corrosive • Breathing the vapor can kill you; skin and eye contact causes severe burns and blindness • Firefighting gear (including SCBA) provides NO protection. If exposure occurs, remove and isolate gear immediately and thoroughly decontaminate personnel • Containers may BLEVE when exposed to fire • Vapors may form explosive mixture with air • Vapors are heavier than air and will collect and stay in low areas • Vapors may travel long distances to ignition sources and flashback • Vapors in confined areas (e.g., tanks, sewers, buildings) may explode when exposed to fire • Toxic products of combustion may include hydrogen chloride and phosgene gas • Do not put yourself in danger by entering a contaminated area to rescue a victim.

Hazard Classification (based on NFPA-704 Rating System)
Health Hazards (Blue): 4; Flammability (Red): 3; Reactivity (Yellow): 1

EMERGENCY RESPONSE: See Appendix A (155)
Evacuation:
Public safety: Isolate the area of spill or leak for at least 50 to 100 meters (160 to 330 feet) in all directions.
Spill: Small spill–First: Isolate in all directions 30 meters (100 feet); Then: Protect persons downwind, DAY: 0.2 km (0.1 mile); NIGHT: 0.3 km (0.2 mile). Large spill–First: Isolate in all directions 60 meters (200 feet); Then: Protect persons downwind, DAY: 0.6 km (0.4 mile); NIGHT: 1.4 km (0.9 mile).
Fire: If tank, rail car, or tank truck is involved in fire, isolate for at least 800 meters (½ mile) in all directions; also, consider initial evacuation for 800 meters (½ mile) in all directions.

EXPOSURE
Short-term effects: *SEEK MEDICAL ATTENTION.* **Vapor:** *POISONOUS IF INHALED OR IF SKIN IS EXPOSED.* May cause lung edema; physical exertion will aggravate this condition. Irritating to eyes, nose, and throat. Move victim to fresh air. *IF IN EYES,* hold eyelids open and flush with plenty of water. *IF BREATHING HAS STOPPED,* give artificial respiration; *avoid mouth-to-mouth resuscitation; use bag/mask apparatus.* IF breathing is difficult, administer oxygen. **Liquid:** *POISONOUS IF SWALLOWED OR IF SKIN IS EXPOSED.* Remove contaminated clothing and shoes. Flush affected areas with plenty of water. *IF IN EYES,* hold eyelids open and flush with plenty of water. *IF SWALLOWED* and victim is *CONSCIOUS AND ABLE TO SWALLOW,* have victim drink 4 to 8 ounces of water. **Do NOT induce vomiting.**
Note to physician or authorized medical personnel: Medical observation is recommended for 24 to 48 hours after breathing overexposure, as pulmonary edema may be delayed. As first aid for pulmonary edema, consider administering a corticosteroid spray. Cigarette smoking may exacerbate pulmonary injury and should be discouraged for at least 72 hours following exposure.

HEALTH HAZARDS
Personal protective equipment (PPE): B-Level PPE. Air-line mask, self-contained breathing apparatus, or organic and acid canister mask; full protective clothing.
Toxicity by ingestion: Grade 4; oral LD_{50} less than 50 mg/kg (rat).
Vapor (gas) irritant characteristics: Eye and respiratory tract irritant. Vapors are irritating such that personnel will not usually tolerate moderate or high concentrations.
Liquid or solid irritant characteristics: Severe eyes and skin irritant. Causes smarting of the skin and first-degree burns on short exposure and may cause second-degree burns on long exposure.

FIRE DATA
Flash point: 82°F/28°C (oc); 61°F/16°C (cc).
Autoignition temperature: 932°F/500°C/773°K.
Electrical hazard: Class I, Group D.
Burning rate: 2.6 mm/min

CHEMICAL REACTIVITY
Reactivity with water: Slow decomposition, forming chloroformic acid, ethanol, and CO_2.
Binary reactants: Reacts violently with strong oxidizers. Slow evolution of hydrogen chloride from surface moisture reaction can cause slow corrosion of metals. Air contact may produce hydrochloric acid fumes.
Neutralizing agents for acids and caustics: Flush with water, rinse with sodium bicarbonate or lime solution.

ENVIRONMENTAL DATA
Food chain concentration potential: Log P_{ow} = 0.23.
Water pollution: Effect of low concentrations on aquatic life is unknown. May be dangerous if it enters nearby water intakes; notify operators. Notify local health and wildlife officials. **Response to discharge:** Issue warning–corrosive, high flammability, poison. Restrict access. Disperse and flush.

SHIPPING INFORMATION
Grades of purity: Technical: 94+%; **Storage temperature:** Ambient; **Inert atmosphere:** None; **Venting:** Pressure-vacuum; **Stability during transport:** Stable.

NAS HAZARD CLASSIFICATION FOR BULK WATER TRANSPORTATION
FIRE: 3
HEALTH: Vapor irritant: 3; Liquid or solid irritant: 3; Poisons: 3
WATER POLLUTION: Human toxicity: 3; Aquatic toxicity: 3; Aesthetic effect: 3
REACTIVITY: Other chemicals: 2; Water: 2; Self-reaction: 0

PHYSICAL AND CHEMICAL PROPERTIES
Physical state @ 59°F/15°C and 1 atm: Liquid.
Molecular weight: 108.5
Boiling point @ 1 atm: 201°F/94°C/367°K.
Melting/Freezing point: –114°F/–81°C/192°K.
Specific gravity (water = 1): 1.135 @ 68°F/20°C (liquid).
Liquid surface tension: 27.5 dynes/cm = 0.0275 N/m @ 15°C.
Relative vapor density (air = 1): 3.7
Ratio of specific heats of vapor (gas): 1.1044
Latent heat of vaporization: (estimate) 140 Btu/lb = 79 cal/g = 3.3 x 10^5 J/kg.
Heat of combustion: (estimate) –6900 Btu/lb = –3800 cal/g = –160 x 10^5 J/kg.
Vapor pressure: 40 mm.

ETHYL CHLOROTHIOFORMATE	REC. E:3200

SYNONYMS: ETHYL CHLOROTHIOLFORMATE; CLOROTIOFORMATO de ETILO (Spanish)

IDENTIFICATION
CAS Number: 2812-73-9
Formula: C_3H_5ClOS; $CH_3CH_2SC(O)Cl$
DOT ID Number: UN 2826; DOT Guide Number: 155
Proper Shipping Name: Ethyl chlorothioformate

DESCRIPTION: Colorless liquid. Pungent odor. Insoluble in water; decomposed slowly forming hydrogen chloride, CO_2, and ethanol.

Poison! • Corrosive • Combustible • Corrosive to the skin, eyes, and respiratory tract; inhalation symptoms may be delayed • Firefighting gear (including SCBA) does not provide adequate protection. If exposure occurs, remove and isolate gear immediately and thoroughly decontaminate personnel • Heat or flame may cause explosion • Containers may BLEVE when exposed to fire • Vapors are heavier than air and will collect and stay in low areas • Vapors can flow along surfaces to distant ignition source and flashback • Vapors in confined areas (e.g., tanks, sewers, buildings) may explode when exposed to fire • Toxic products of combustion may include sulfur oxides, hydrogen chloride, and carbon monoxide.

Hazard Classification (based on NFPA-704 Rating System)
Health Hazards (Blue): 0; Flammability (Red): 1; Reactivity (Yellow): 0

EMERGENCY RESPONSE: See Appendix A (155)
Evacuation:
Public safety: Isolate the area of spill or leak for at least 50 to 100 meters (160 to 330 feet) in all directions.
Spill: Small spill–First: Isolate in all directions 30 meters (100 feet); Then: Protect persons downwind, DAY: 0.2 km (0.1 mile); NIGHT: 0.2 km (0.1 mile). Large spill–First: Isolate in all directions 60 meters (200 feet); Then: Protect persons downwind, DAY: 0.5 km (0.3 mile); NIGHT: 0.8 km (0.5 mile).
Fire: If tank, rail car, or tank truck is involved in fire, isolate for at least 800 meters (½ mile) in all directions; also, consider initial evacuation for 800 meters (½ mile) in all directions.

EXPOSURE
Short-term effects: *SEEK MEDICAL ATTENTION.* **Vapor:** Move victim to fresh air. May cause lung edema; physical exertion will aggravate this condition. *IF BREATHING HAS STOPPED,* give artificial respiration; *avoid mouth-to-mouth resuscitation; use bag/mask apparatus.* If breathing is difficult, administer oxygen.
Liquid: Remove contaminated clothing and shoes. Wash affected areas with soap and water. *IF IN EYES,* hold eyelids open and flush with plenty of water.
Note to physician or authorized medical personnel: Medical observation is recommended for 24 to 48 hours after breathing overexposure, as pulmonary edema may be delayed. As first aid for pulmonary edema, consider administering a corticosteroid spray. Cigarette smoking may exacerbate pulmonary injury and should be discouraged for at least 72 hours following exposure.

HEALTH HAZARDS
Personal protective equipment (PPE): B-Level PPE. Full impervious protective clothing, including boots and gloves. Where splashing is possible wear full face shield or chemical safety goggles. Use approved respirator to protect against vapors.
Vapor (gas) irritant characteristics: Eye and respiratory tract irritant. Vapors cause smarting of the eyes or respiratory system if present in high concentrations. The effect is temporary.
Liquid or solid irritant characteristics: Eye and skin irritant. If spilled on clothing and allowed to remain, may cause smarting and reddening of the skin.

FIRE DATA
Flash point: 86°F/30°C (cc)
Autoignition temperature: 2725°F/1495°C/1748°K.
Fire extinguishing agents not to be used: Water may be ineffective.
Stoichiometric air-to-fuel ratio: 22.6

CHEMICAL REACTIVITY
Reactivity group: 0
Compatibility class: Unassigned cargoes

ENVIRONMENTAL DATA
Water pollution: DOT Appendix B, §172.101–marine pollutant. Effects of low concentrations on aquatic life is unknown. May be dangerous if it enters nearby water intakes; notify operators. Notify local health and wildlife officials. **Response to discharge:** Issue warning–high flammability. Should be removed.

SHIPPING INFORMATION
Grades of purity: Technical grades; **Storage temperature:** Ambient; **Inert atmosphere:** None; **Stability during transport:** Stable.

PHYSICAL AND CHEMICAL PROPERTIES
Physical state @ 59°F/15°C and 1 atm: Liquid.
Molecular weight: 124.59
Boiling point @ 1 atm: 269.6°F/132°C/405°K.
Specific gravity (water = 1): 1.195

ETHYL CYCLOHEXANE   REC. E:3300

SYNONYMS: CYCLOHEXYL ETHANE; ETIL CICLOHEXANO (Spanish)

IDENTIFICATION
CAS Number: 1678-91-7
Formula: $(C_6H_{11})C_2NO_5$
DOT ID Number: UN 1993; DOT Guide Number: 128
Proper Shipping Name: Flammable liquid, n.o.s.

DESCRIPTION: Colorless liquid. Soluble in water.

Highly flammable • Containers may BLEVE when exposed to fire • Vapors may form explosive mixture with air • Vapors are heavier than air and will collect and stay in low areas • Vapors may travel long distances to ignition sources and flashback • Vapors in confined areas (e.g., tanks, sewers, buildings) may explode when exposed to fire • Irritating to the skin, eyes, and respiratory tract • Toxic products of combustion may include carbon monoxide.

Hazard Classification (based on NFPA-704 Rating System)
Health Hazards (Blue): 1; Flammability (Red): 3; Reactivity (Yellow): 0

EMERGENCY RESPONSE: See Appendix A (128)
Evacuation:
Public safety: Isolate spill area for at least 25 to 50 meters (80 to 160 feet) in all directions.
Spill: Large spill–Consider initial downwind evacuation for at least 300 meters (1000 feet).

Fire: Isolate for 800 meters (½ mile) in all directions, especially if tank, rail car, or tank truck is involved in fire.

EXPOSURE
Short-term effects: *SEEK MEDICAL ATTENTION.* **Vapor:** Irritating to eyes, nose, and throat. *IF INHALED*, will, will cause dizziness, nausea, vomiting, or loss of consciousness. Move to fresh air. *IF BREATHING HAS STOPPED,* give artificial respiration. IF breathing is difficult, administer oxygen. **Liquid:** Irritating to skin and eyes. Harmful if swallowed. Remove contaminated clothing and shoes. Flush affected areas with plenty of water. *IF IN EYES*, hold eyelids open and flush with plenty of water. *IF SWALLOWED* and victim is *CONSCIOUS AND ABLE TO SWALLOW*, have victim drink 4 to 8 ounces of water.

HEALTH HAZARDS
Personal protective equipment (PPE): B-Level PPE. Hydrocarbon vapor canister, supplied-air or hose mask, hydrocarbon-insoluble rubber or plastic gloves, chemical goggles or face splash shield, hydrocarbon-insoluble rubber or plastic apron.

FIRE DATA
Flash point: 95°F/35°C (cc).
Flammable limits in air: LEL: 0.9; UEL: 6.6%.
Fire extinguishing agents not to be used: Water may be ineffective against fire.
Autoignition temperature: 460°F/238°C/511°K.

CHEMICAL REACTIVITY
Binary reactants: Oxidizers, acids, caustics.

ENVIRONMENTAL DATA
Water pollution: Dangerous to aquatic life in high concentrations. Fouling to shoreline. May be dangerous if it enters nearby water intakes; notify operators. Notify local health and wildlife officials.
Response to discharge: Issue warning–high flammability. Evacuate area. Disperse and flush.

SHIPPING INFORMATION
Grades of purity: 99+%; **Stability during transport:** Stable.

NAS HAZARD CLASSIFICATION FOR BULK WATER TRANSPORTATION
FIRE: 3
HEALTH: Vapor irritant: 1; Liquid or solid irritant: 0; Poisons:–
WATER POLLUTION: Human toxicity:–; Aquatic toxicity:–; Aesthetic effect: 1
REACTIVITY: Other chemicals: 1; Water: 0; Self-reaction: 0

PHYSICAL AND CHEMICAL PROPERTIES
Physical state @ 59°F/15°C and 1 atm: Liquid.
Molecular weight: 112.22
Boiling point @ 1 atm: 269°F/132°C/405°K.
Melting/Freezing point: –168°F/–111.3°C/162°K.
Specific gravity (water = 1): 0.7880 @ 20°C.
Relative vapor density (air = 1): 3.89
Latent heat of vaporization: 180 Btu/lb = 99.9 cal/g = 4.2 x 10^5 J/kg.
Heat of combustion: –20,024 Btu/lb = –11,124 cal/g = 466 x 10^5 J/kg.
Heat of decomposition: (estimate) –91,314 Btu/lb = –50,730 cal/g = –212 x 10^5 J/kg.
Heat of fusion: 17.75 cal/g.
Vapor pressure: (Reid) 0.6 psia.

N-ETHYL(CYCLO)HEXYLAMINE   REC. E:3325

SYNONYMS: ACCELERATOR HX; *N*-CICLOHEXILETILAMINA (Spanish); *N*-CYCLOHEXYLETHYLAMINE; CYCLOHEXYLAMINE, *N*-ETHYL; *N*-ETHYLCYCLOHEXANAMINE; VULKACIT HX

IDENTIFICATION
CAS Number: 5459-93-8
Formula: $C_8H_{17}N$; $C_2H_5NHC_6H_{11}$
DOT ID Number: UN 2920; DOT Guide Number: 132
Proper Shipping Name: Corrosive liquids, flammable, n.o.s.

DESCRIPTION: Colorless liquid. Mild, musky ammonia odor. Floats and mixes slowly with water.

Highly flammable • Corrosive • Containers may BLEVE when exposed to fire • Vapors may form explosive mixture with air • Vapors are heavier than air and will collect and stay in low areas • Vapors may travel long distances to ignition sources and flashback • Vapors in confined areas (e.g., tanks, sewers, buildings) may explode when exposed to fire • Severely Irritating to the skin, eyes, and respiratory tract • Toxic products of combustion may include nitrogen oxides.

Hazard Classification (based on NFPA-704 Rating System)
Health Hazards (Blue): 3; Flammability (Red): 3; Reactivity (Yellow): 0

EMERGENCY RESPONSE: See Appendix A (132)
Evacuation:
Public safety: Isolate spill area for at least 50 to 100 meters (160 to 330 feet) in all directions.
Spill: Increase, as necessary, the isolation distance shown above, in "Public safety."
Fire: Isolate for 800 meters (½ mile) in all directions, especially if tank, rail car, or tank truck is involved in fire.

EXPOSURE
Short-term effects: *SEEK MEDICAL ATTENTION.* **Vapor:** Irritating to eyes, nose, and throat. Possible lung edema. Harmful if inhaled. Move to fresh air. *IF BREATHING HAS STOPPED,* give artificial respiration. IF breathing is difficult, administer oxygen. **Liquid:** Will burn skin and eyes. Harmful if swallowed. Remove contaminated clothing and shoes. *IF IN EYES*, hold eyelids open and flush with plenty of water. *IF SWALLOWED* and victim is *CONSCIOUS AND ABLE TO SWALLOW*, have victim drink 4 to 8 ounces of water. **Do NOT induce vomiting.**
Note to physician or authorized medical personnel: Medical observation is recommended for 24 to 48 hours after breathing overexposure, as pulmonary edema may be delayed. As first aid for pulmonary edema, consider administering a corticosteroid spray. Cigarette smoking may exacerbate pulmonary injury and should be discouraged for at least 72 hours following exposure.

HEALTH HAZARDS
Personal protective equipment (PPE): B-Level PPE. Gas mask suitable for ammonia; face shield or splash proof goggles; rubber gloves. If entering spill area, wear self-contained breathing apparatus and full protective clothing, including boots.
Recommendations for respirator selection: See ammonia for NH_3 fumes
Toxicity by ingestion: Grade 2; LD_{50} = 590 mg/Kg (rat).
Long-term health effects: Possible liver and kidney injury.
Vapor (gas) irritant characteristics: Vapors are irritating such

496 Ethyl dichlorosilane

that personnel will not usually tolerate moderate or high concentrations.
Liquid or solid irritant characteristics: Fairly severe skin irritant. May cause pain and second-degree burns after a few minutes of contact.

FIRE DATA
Flash point: 86°F/30°C (oc); 115°F/46°C (cc).
Fire extinguishing agents not to be used: Water may be ineffective.
Behavior in fire: Dangerous when exposed to heat or flame. Can react vigorously with oxidizing materials.
Autoignition temperature: 545°F/285°C/558°K.

CHEMICAL REACTIVITY
Binary reactants: Strong oxidizers.
Reactivity group: 7
Compatibility class: Aliphatic amines

ENVIRONMENTAL DATA
Water pollution: Effect of low concentration on aquatic life is unknown. May be dangerous if it enters nearby water intakes; notify operators. Notify local health and wildlife officials.
Response to discharge: Issue warning–high flammability., air contaminant. Restrict access. Disperse and flush

SHIPPING INFORMATION
Grades of purity: 97%; 99%; **Storage temperature:** Ambient; **Stability during transport:** Stable.

PHYSICAL AND CHEMICAL PROPERTIES
Physical state @ 59°F/15°C and 1 atm: Liquid.
Molecular weight: 127.23
Boiling point @ 1 atm: 329°F/165°C/438.2°K.
Melting/Freezing point: –49°F/–45°C/228.2°K.
Critical temperature: 677.12°F/358.4°C/631.6°K.
Critical pressure: 446.91 psia = 30.39 atm = 3.04 MN/m^2.
Specific gravity (water = 1): 0.8527 @ 77°F/25°C.
Liquid surface tension: 29.52 dynes/cm = 0.02952 N/m @ 68°F/20°C.
Liquid water interfacial tension: 43.5 dynes/cm = 0.0435 N/m @ 68°F/20°C.
Relative vapor density (air = 1): 4.4

ETHYL DICHLOROSILANE   **REC. E:3400**

SYNONYMS: DICHLOROETHYLSILANE; ETHYLDICHLOROSILANE; ETILDICLOROSILANO (Spanish); SILANE, DICHLOROETHYL-

IDENTIFICATION
CAS Number: 1789-58-8
Formula: $C_2H_6Cl_2Si$; $C_2NO_5SiHCl_2$
DOT ID Number: UN 1183; DOT Guide Number: 139
Proper Shipping Name: Ethyldichlorosilane

DESCRIPTION: Colorless liquid. Sharp, irritating odor, hydrochloric acid. Sinlks in water; reacts, forming hydrochloric acid and explosive hydrogen gas.

Extremely flammable • Poison • Breathing the vapors can kill you; skin and eye contact can cause severe burns and blindness • Ignites spontaneously in air • Firefighting gear (including SCBA) provides NO protection. If exposure occurs, remove and isolate gear immediately and thoroughly decontaminate personnel • Vapors are heavier than air and will collect and stay in low areas • Vapors may travel long distances to source of ignition and flashback • Often shipped and stored dissolved in acetone • Exposure of cylinders to elevated temperatures, fire, and flame may cause cylinders to rupture or cause frangible disk to burst, releasing entire contents of cylinder • When combined with surface moisture this material is corrosive to most common metals • Toxic products of combustion may include hydrogen chloride • Do not put yourself in danger by entering a contaminated area to rescue a victim.

Hazard Classification (based on NFPA-704 Rating System)
Health Hazards (Blue): 3; Flammability (Red): 3; Reactivity (Yellow): 0

EMERGENCY RESPONSE: See Appendix A (139)
Evacuation:
Public safety: Isolate the area of spill or leak for at least 100 to 150 meters (330 to 490 feet) in all directions.
Spill: See Public Safety.
Fire: If any large container is involved in fire, isolate for at least 800 meters (½ mile) in all directions; also, consider initial evacuation for 800 meters (½ mile) in all directions.

EXPOSURE
Short-term effects: *SEEK MEDICAL ATTENTION*. **Vapor:** Irritating to eyes, nose, and throat. Move victim to fresh air. If breathing is difficult, administer oxygen. **Liquid:** Will burn skin and eyes. Harmful if swallowed. Remove contaminated clothing and shoes. Flush affected areas with plenty of water. *IF IN EYES*, hold eyelids open and flush with plenty of water. *IF SWALLOWED* and victim is *CONSCIOUS AND ABLE TO SWALLOW*, have victim drink 4 to 8 ounces of water. **Do NOT induce vomiting**.

HEALTH HAZARDS
Personal protective equipment (PPE): B-Level PPE. Acid-vapor-type respiratory protection; rubber gloves; chemical worker's goggles; other equipment as necessary to protect skin and eyes.
Toxicity by ingestion: Grade 3; LD_{50} = 50 to 500 mg/kg.
Vapor (gas) irritant characteristics: Vapors cause severe irritation of eyes and throat and can cause eye and lung injury. They cannot be tolerated even at low concentrations.
Liquid or solid irritant characteristics: Severe eye and skin irritant. Causes second-and third-degree burns on short contact and is very injurious to the eyes.

FIRE DATA
Flash point: 30°F/–1°C (oc).
Fire extinguishing agents not to be used: Water, foam
Behavior in fire: Difficult to extinguish; re-ignition may occur. Reacts with water.
Burning rate: 3.2 mm/min

CHEMICAL REACTIVITY
Reactivity with water: Reacts vigorously, evolving hydrogen chloride (hydrochloric acid) and hydrogen gas.
Binary reactants: Reaction with surface moisture will generate hydrogen chloride, which corrodes common metals.
Neutralizing agents for acids and caustics: Flood with water, rinse with sodium bicarbonate or lime solution.

ENVIRONMENTAL DATA
Food chain concentration potential: Negative; unlikely to accumulate.

Water pollution: Effect of low concentrations on aquatic life is unknown. May be dangerous if it enters nearby water intakes; notify operators. Notify local health and wildlife officials. **Response to discharge:** Issue warning–high flammability, corrosive. Restrict access. Evacuate area. Disperse and flush with care.

SHIPPING INFORMATION
Grades of purity: Commercial; **Storage temperature:** Ambient; **Inert atmosphere:** None; **Venting:** Pressure-vacuum; **Stability during transport:** Stable.

NAS HAZARD CLASSIFICATION FOR BULK WATER TRANSPORTATION
FIRE: 3
HEALTH: Vapor irritant: 4; Liquid or solid irritant: 4; Poisons: 3
WATER POLLUTION: Human toxicity: 3; Aquatic toxicity: 3; Aesthetic effect: 2
REACTIVITY: Other chemicals: 3; Water: 4; Self-reaction: 1

PHYSICAL AND CHEMICAL PROPERTIES
Physical state @ 59°F/15°C and 1 atm: Liquid.
Molecular weight: 129.1
Boiling point @ 1 atm: 165°F/74°C/347°K.
Specific gravity (water = 1): 1.092 @ 68°F/20°C (liquid).
Liquid surface tension: 21.7 dynes/cm = 0.0217 N/m @ 68°F/20°C.
Relative vapor density (air = 1): 4.5
Latent heat of vaporization: (estimate).
104 Btu/lb = 57.8 cal/g = 2.42 x 10^5 J/kg.
Heat of combustion: (estimate) –6500 Btu/lb = –3600 cal/g = –150 x 10^5 J/kg.

ETHYLENE   REC. E:3500

SYNONYMS: ACETENE; ATHYLEN (German); BICARBURRETTED HYDROGEN; ATHYLEN (German); DICARBURRETTED HYDROGEN; EEC No. 601-010-00-3; ELAYL; ETHENE; ETENO (Spanish); ETILENO (Spanish); LIQUID ETHELYNE; NCL-C5066; OLEFIANT GAS

IDENTIFICATION
CAS Number: 74-85-1
Formula: C_2H_4
DOT ID Number: UN 1962; DOT Guide Number: 116P
Proper Shipping Name: Ethylene, compressed

DESCRIPTION: Colorless gas which may appear white. Shipped and stored as a liquefied compressed gas or cryogenic liquid. Faint, slightly sweet, ether-like odor. Floats and boils on water; flammable visible vapor cloud is produced.

Extremely flammable • High concentrations can cause oxygen deficiency; suffocation • Forms explosive mixture with air • Reacts explosively with many materials • Vapors are anesthetic; may cause dizziness or asphyxiation without warning • Contact with liquid may cause frostbite • Cold gas is initially heavier than air and spread along ground • Gas may travel to source of ignition and flashback • Exposure of cylinders to elevated temperatures, fire, and flame may cause cylinders to rupture or cause frangible disk to burst, releasing entire contents of cylinder • Ruptured or venting cylinders may rocket through buildings and/or travel a considerable distance • Polymerization hazard • Heat can induce polymerization with rapid release of energy; sealed containers may rupture explosively • May react with itself blocking relief valves; leading to tank explosions • Do not put yourself in danger by entering a contaminated area to rescue a victim.
Warning: Odor is not a reliable indicator of the presence of toxic amounts of ethylene gas.

Hazard Classification (based on NFPA-704 Rating System)
Health Hazards (Blue): 1; Flammability (Red): 4; Reactivity (Yellow): 2

EMERGENCY RESPONSE: See Appendix A (116)
Evacuation:
Public safety: Isolate spill area for at least 100 meters (330 feet) in all directions.
Spill: Consider initial downwind evacuation for at least 800 meters (½ mile).
Fire: Isolate for 1600 meters (1 mile) in all directions, especially if tank, rail car, or tank truck is involved in fire.

EXPOSURE
Short-term effects: *SEEK MEDICAL ATTENTION.* **Vapor:** Not irritating to eyes, nose or throat. *IF INHALED*, will, will cause headache, dizziness, or loss of consciousness. Move to fresh air. *IF BREATHING HAS STOPPED,* give artificial respiration. IF breathing is difficult, administer oxygen. **Liquid:** Will cause frostbite. Flush affected areas with plenty of water. *DO NOT RUB AFFECTED AREAS.*

HEALTH HAZARDS
Personal protective equipment (PPE): B-Level PPE. Organic vapor canister); or air–supplied mask. Wear thermal protective clothing.
Exposure limits (TWA unless otherwise noted): Simple asphyxiant. High concentrations have an anesthetic effect.
Long-term health effects: IARC rating 3; no human or animal data.
Vapor (gas) irritant characteristics: Vapors are nonirritating to the eyes and throat.
Liquid or solid irritant characteristics: Practically harmless to the skin, but rapid evaporation of liquid may cause frostbite.

FIRE DATA
Flash point: –213°F/–136°C (cc) (approximate) Gas.
Flammable limits in air: LEL: 2.7%; UEL: 36.0%.
Behavior in fire: Containers may explode.
Autoignition temperature: 842°F/450°C/723°K.
Electrical hazard: Class I, Group C. Flow or agitation of substance may generate electrostatic charges due to low conductivity.
Burning rate: 7.4 mm/min
Adiabatic flame temperature: 2600°F/1427°C (estimate).
Stoichiometric air-to-fuel ratio: 14.68 (estimate).

CHEMICAL REACTIVITY
Binary reactants: Violent reaction with oxidizers, hydrogen bromide, halogen acids, and hydrochloric acid.
Polymerization: Reaction with oxidizers may cause polymerization.
Reactivity group: 30
Compatibility class: Olefins

ENVIRONMENTAL DATA
Food chain concentration potential: Negative; unlikely to accumulate.

Water pollution: Not harmful to aquatic life. **Response to discharge:** Issue warning–high flammability. Restrict access. Evacuate area.

SHIPPING INFORMATION
Grades of purity: 99–100%; **Storage temperature:** Ambient (gas); –155°F/–104°C (liquid); **Inert atmosphere:** None; **Venting:** Safety relief; **Stability during transport:** Stable. Although ethylene is highly reactive, it must be catalyzed before most reactions take place.

NAS HAZARD CLASSIFICATION FOR BULK WATER TRANSPORTATION
FIRE: 4
HEALTH: Vapor irritant: 0; Liquid or solid irritant: 0; Poisons: 1
WATER POLLUTION: Human toxicity: 0; Aquatic toxicity: 1; Aesthetic effect: 0
REACTIVITY: Other chemicals: 1; Water: 0; Self-reaction: 2

PHYSICAL AND CHEMICAL PROPERTIES
Physical state @ 59°F/15°C and 1 atm: Gas.
Molecular weight: 28.1
Boiling point @ 1 atm: –154.7°F/–103.7°C/169.5°K.
Melting/Freezing point: –272.4°F/–169.1°C/104.1°K.
Critical temperature: 49.8°F/9.9°C/283.1°K.
Critical pressure: 742 psia = 50.5 atm = 5.11 MN/m^2.
Specific gravity (water = 1): 0.569 at –103.8°C (liquid).
Liquid surface tension: 16 dynes/cm = 0.016 N/m at –104°C.
Liquid water interfacial tension: (estimate) 50 dynes/cm = 0.05 N/m at –104°C.
Relative vapor density (air = 1): 1.0
Ratio of specific heats of vapor (gas): 1.240
Latent heat of vaporization: 207.7 Btu/lb = 115.4 cal/g = 4.832 x 10^5 J/kg.
Heat of combustion: –20,290 Btu/lb = –11,272 cal/g = –471.94 x 10^5 J/kg.
Vapor pressure: 39,500 @ 42°F

ETHYLENE CHLOROHYDRIN   **REC. E:3600**

SYNONYMS: β-CHLORETHYL ALCOHOL; CHLOROETHANOL; 2-CHLOROETHANOL; 2-CHLORETHANOL; δ-CHLORETHANOL; 2-CHLOROETHYL ALCOHOL; CLORHIDRINA ETILENICA (Spanish); EEC No. 603-028-00-7; ETHYLENE CHLORHYDRIN; GLYCOL CHLOROHYDRIN; 2-MONOCHLOROETHANOL

IDENTIFICATION
CAS Number: 107-07-3
Formula: C$_2$H$_5$ClO; ClCH$_2$CH$_2$OH
DOT ID Number: UN 1135; DOT Guide Number: 131
Proper Shipping Name: Ethylene chlorohydrin
Reportable Quantity (RQ): **(CERCLA)** 1 lb/0.454 kg

DESCRIPTION: Colorless liquid. Faint, sweet, ether-like odor. Mixes with water; mild reaction. Irritating vapor is produced.

Poison! • Breathing the vapor, skin contact, or swallowing the material can kill you • Highly flammable • Containers may BLEVE when exposed to fire • Vapors may form explosive mixture with air • Vapors are heavier than air and will collect and stay in low areas • Vapors may travel long distances to ignition sources and flashback • Vapors in confined areas (e.g., tanks, sewers, buildings) may explode when exposed to fire • Irritating to the skin, eyes, and respiratory tract in low concentrations • Toxic products of combustion may include phosgene gas and hydrogen chloride gas.

Hazard Classification (based on NFPA-704 Rating System)
Health Hazards (Blue): 4; Flammability (Red): 2; Reactivity (Yellow): 0

EMERGENCY RESPONSE: See Appendix A (131)
Evacuation:
Public safety: Isolate spill area for at least 100 to 200 meters (330 to 660 feet) in all directions.
Spill: Small spill–First: Isolate in all directions 30 meters (100 feet); Then: Protect persons downwind, DAY: 0.2 km (0.1 mile); NIGHT: 0.3 km (0.2 mile). Large spill–First: Isolate in all directions 60 meters (200 feet); Then: Protect persons downwind, DAY: 0.6 km (0.4 mile); NIGHT: 1.3 km (0.8 mile).
Fire: Isolate for 800 meters (½ mile) in all directions, especially if tank, rail car, or tank truck is involved in fire.

EXPOSURE
Short-term effects: *SEEK MEDICAL ATTENTION. ABSORBED THROUGH THE SKIN.* **Vapor:** Irritating to eyes, nose, and throat. *IF INHALED*, will, will cause coughing or difficult breathing. Move victim to fresh air. *IF BREATHING HAS STOPPED*, give artificial respiration. IF breathing is difficult, administer oxygen. **Liquid:** Irritating to skin and eyes. Harmful if swallowed. Remove contaminated clothing and shoes. Flush affected areas with plenty of water. *IF IN EYES*, hold eyelids open and flush with plenty of water. *IF SWALLOWED* and victim is *CONSCIOUS AND ABLE TO SWALLOW*, have victim drink 4 to 8 ounces of water. **Do NOT induce vomiting.**

HEALTH HAZARDS
Personal protective equipment (PPE): A-Level PPE. OSHA Table Z-1-A air contaminant. Organic vapor respirator. Chemical protective material(s) reported to have good to excellent resistance: butyl rubber, PE, PV alcohol, Viton®. Also, Viton®/neoprene offers limited protection.
Recommendations for respirator selection: NIOSH/OSHA
7 ppm: SA* (any supplied-air respirator); or SCBAF (any self-contained breathing apparatus with a full facepiece). *EMERGENCY OR PLANNED ENTRY INTO UNKNOWN CONCENTRATIONS OR IDLH CODITIONS:* SCBAF:PD,PP (any self-contained breathing apparatus that has a full facepiece and is operated in a pressure-demand or other positive-pressure mode); or SAF:PD,PP:ASCBA (any supplied-air respirator that has a full facepiece and is operated in a pressure-demand or other positive-pressure mode in combination with an auxiliary self-contained breathing apparatus operated in a pressure-demand or other positive pressure mode). *ESCAPE:* GMFOV[any air-purifying, full-facepiece respirator (gas mask) with a chin-style, front- or back-mounted organic vapor cannister]; or SCBAE (any appropriate escape-type, self-contained breathing apparatus). **Note:* Substance reported to cause eye irritation or damage; may require eye protection.
Exposure limits (TWA unless otherwise noted): ACGIH TLV ceiling 1 ppm (3.3 mg/m^3); OSHA PEL 5 ppm (16 mg/m^3); NIOSH REL ceiling 1 ppm (3 mg/m^3); skin contact contributes significantly in overall exposure.
Toxicity by ingestion: Grade 3; oral LD$_{50}$ = 71 mg/kg (rat).
Long-term health effects: Damage to central nervous system. Brain, kidney and liver in humans.
Vapor (gas) irritant characteristics: Eye and respiratory tract irritant. May cause difficult breathing and lung edema.

Liquid or solid irritant characteristics: Eye and skin irritant.
Odor threshold: 0.4 ppm.
IDLH value: 7 ppm.

FIRE DATA
Flash point: 139°F/59°C (cc).
Flammable limits in air: LEL: 4.9%; UEL 15.9%.
Autoignition temperature: 797°F/425°C/698°K.
Electrical hazard: Class I, Group D.
Burning rate: 1.7 mm/min

CHEMICAL REACTIVITY
Binary reactants: Strong oxidizers, strong caustics.
Reactivity group: 20
Compatibility class: Alcohols, glycols

ENVIRONMENTAL DATA
Food chain concentration potential: Negative; unlikely to accumulate.
Water pollution: Effect of low concentrations on aquatic life is unknown. May be dangerous if it enters nearby water intakes; notify operators. Notify local health and wildlife officials.
Response to discharge: Issue warning–water contaminant. Restrict access. Disperse and flush.

SHIPPING INFORMATION
Grades of purity: 99+%; **Storage temperature:** Ambient. Dangerous decomposition at high temperatures; **Inert atmosphere:** None; **Venting:** Open (flame arrester); **Stability during transport:** Stable.

PHYSICAL AND CHEMICAL PROPERTIES
Physical state @ 59°F/15°C and 1 atm: Liquid.
Molecular weight: 80.51
Boiling point @ 1 atm: 264°F/129°C/402°K.
Melting/Freezing point: –90°F/68°C/341°K.
Specific gravity (water = 1): 1.197 @ 68°F/20°C (liquid).
Relative vapor density (air = 1): 2.8
Latent heat of vaporization: 221 Btu/lb = 123 cal/g = 5.15×10^5 J/kg.
Heat of combustion: –6,487 Btu/lb = –3,604 cal/g = -150.8×10^5 J/kg.
Vapor pressure: 6 mm.

ETHYLENE CYANOHYDRIN	REC. E:3700

SYNONYMS: CIANHIDRINA ETILENICA (Spanish); 2-CYANOETHANOL; 2-CYANOETHYL ALCOHOL; GLYCOL CYANOHYDRIN; β-HPN; HYDRACRYLONITRILE; 3-HYDROXYPROPANENITRILE; 1-HYDROXY-2-CYANOETHANE; METHANOLACETONITRILE

IDENTIFICATION
CAS Number: 109-78-4
Formula: C_3H_5NO; $HOCH_2CH_2CN$
DOT ID Number: UN 1935; DOT Guide Number: 157
Proper Shipping Name: Cyanide solution, n.o.s.
Reportable Quantity (RQ): CERCLA, 10 lb (4.54 kg) cyanides, soluble salts and complexes, n.o.s.

DESCRIPTION: Colorless to light yellow liquid. Weak odor to odorless. Sinks in water; soluble.

Combustible • Heat or flame may cause explosion • Polymerization hazard • May react with itself blocking relief valves; leading to tank explosions • Containers may BLEVE when exposed to fire • Vapors are heavier than air and will collect and stay in low areas • Vapors in confined areas (e.g., tanks, sewers, buildings) may explode when exposed to fire • Irritating to the skin, eyes, and respiratory tract • Toxic products of combustion may include hydrogen cyanide and nitrogen oxides.

Hazard Classification (based on NFPA-704 Rating System)
Health Hazards (Blue): 1; Flammability (Red): 1; Reactivity (Yellow): 2

EMERGENCY RESPONSE: See Appendix A (157)
Evacuation:
Public safety: Isolate the area of spill or leak for at least 50 to 100 meters (160 to 330 feet) in all directions.
Spill: Increase, in the downwind direction, as necessary, the distance shown under "Public Safety."
Fire: If tank, rail car, or tank truck is involved in fire, isolate for at least 800 meters (½ mile) in all directions; also, consider initial evacuation for 800 meters (½ mile) in all directions.

EXPOSURE
Short-term effects: *SEEK MEDICAL ATTENTION*. If artificial respiration is administered; *avoid mouth-to-mouth resuscitation; use bag/mask apparatus*. **Liquid:** Irritating to skin and eyes. Harmful if swallowed. Remove contaminated clothing and shoes. Flush affected areas with plenty of water. *IF IN EYES*, hold eyelids open and flush with plenty of water. *IF SWALLOWED* and victim is *CONSCIOUS AND ABLE TO SWALLOW*, have victim drink 4 to 8 ounces of water. Have victim induce vomiting. *IF SWALLOWED* and victim is *UNCONSCIOUS OR HAVING CONVULSIONS*, do nothing except keep victim warm.
Note to physician or authorized medical personnel: Consider the use of amyl nitrite perles if symptoms of cyanide poisoning develop. If symptoms indicate, initial treatment includes the cyanide antidote kit. In all cases, break an amyl nitrite perle in a gauze pad and hold lightly under victim's nose for 15 seconds, repeating 5 times at about 15-second intervals; if necessary (and if sodium nitrite infusions will be delayed), repeat procedure every 3 minutes with fresh perles until 3 or 4 have been used. Avoid breathing the vapor while administering it to the victim. Administer sodium nitrite IV, ASAP. The usual adult dose is 10 to 20 mL of a 3% solution infused over no less than 5 minutes; the average child dose is 0.15 to 0.20 mL/kg. Monitor blood pressure during administration, and slow the rate of infusion if hypotention develops. Next, infuse sodium thiosulfate IV. The usual adult dose is 50 mL of a 25% solution infused over 10 to 20 minutes; the average child dose is 1.65 mL/kg. Repeat with nitrite and thiosulfate as required.

HEALTH HAZARDS
Personal protective equipment (PPE): B-Level PPE. Cyanides are OSHA Table Z-1-A air contaminants. Air–supplied mask; plastic gloves; rubber clothing; vapor-proof goggles. Natural rubber, neoprene, and nitrile rubber may offer some protection from alcohols.
Recommendations for respirator selection: NIOSH as cyanides *25 mg/m³*: SA (any supplied-air respirator); or SCBAF (any self-contained breathing apparatus with full facepiece). *EMERGENCY OR PLANNED ENTRY INTO UNKNOWN CONCENTRATIONS OR IDLH CONDITIONS*: SCBAF:PD,PP (any self-contained breathing apparatus that has a full facepiece and is operated in a pressure-demand or other positive-pressure mode); or SAF:PD,PP:ASCBA (any supplied-air respirator that has a full

facepiece and is operated in a pressure-demand or other positive-pressure mode in combination with an auxiliary self-contained breathing apparatus operated in a pressure-demand or other positive pressure mode). *ESCAPE:* GMFSHiE [any air-purifying, full-facepiece respirator (gas mask) with a chin-style, front- or back-mounted canister providing protection against the compound of concern and having a high efficiency particulate filter); or SCBAE (any appropriate escape-type, self-contained breathing apparatus).

Exposure limits (TWA unless otherwise noted): ACGIH TLV 5 mg/m^3; OSHA PEL 5 mg/m^3; NIOSH ceiling 5 mg/m^3 (4.7 ppm)/10 minutes as cyanides; skin contact contributes significantly in overall exposure

Toxicity by ingestion: Grade 2; LD$_{50}$ = 0.5 to 5 g/kg.

Long-term health effects: Ingestion of liquid may cause severe kidney damage.

Vapor (gas) irritant characteristics: Eye and respiratory tract irritant.

Liquid or solid irritant characteristics: Eye and skin irritant.

IDLH value: 50 mg/m^3 as cyanides

FIRE DATA

Flash point: 265°F/129°C (oc) 22% solution in water.

Flammable limits in air: LEL: 2.3% (calculated); UEL: 12.1% (estimate).

Fire extinguishing agents not to be used: Water or foam may cause frothing.

Autoignition temperature: 922°F/494°C/767°K.

Electrical hazard: Class I, Group D.

CHEMICAL REACTIVITY

Reactivity with water: Hot water forms toxic cyanide gas.

Binary reactants: Reacts violently with strong oxidizers; sodium hydroxide. Cyanides react with acids, acid salts, chlorates, and nitrates. Attacks mild steel, copper, and copper alloys.

Polymerization: Polymerizes with inorganic bases and amines.

Reactivity group: 37

Compatibility class: Nitriles

ENVIRONMENTAL DATA

Food chain concentration potential: Negative; unlikely to accumulate.

Water pollution: DOT Appendix B, §172.101–marine pollutant.Effect of low concentrations on aquatic life is unknown. Fouling to shoreline. May be dangerous if it enters nearby water intakes; notify operators. Notify local health and wildlife officials.

Response to discharge: Issue warning–water contaminant. Should be removed.

SHIPPING INFORMATION

Grades of purity: 22% solution in water; **Storage temperature:** Ambient; **Inert atmosphere:** None; **Venting:** Open (flame arrester); **Stability during transport:** Stable.

NAS HAZARD CLASSIFICATION FOR BULK WATER TRANSPORTATION

FIRE: 1
HEALTH: Vapor irritant: 0; Liquid or solid irritant: 0; Poisons: 2
WATER POLLUTION: Human toxicity: 1; Aquatic toxicity: 2; Aesthetic effect: 2
REACTIVITY: Other chemicals: 2; Water: 0; Self-reaction: 0

PHYSICAL AND CHEMICAL PROPERTIES

Physical state @ 59°F/15°C and 1 atm: Liquid.

Molecular weight: 71.1

Boiling point @ 1 atm: 445.5°F/229.7°C/502.9°K (decomposes).

Melting/Freezing point: –51.2°F/–46.2°C/227.0°K.

Critical temperature: 804°F/429°C/702°K.

Critical pressure: 720 psia = 4.9 MN/m^2.

Specific gravity (water = 1): 1.047 @ 68°F/20°C (liquid).

Relative vapor density (air = 1): 2.6

Vapor pressure: (Reid) Very low; 0.08 mm.

ETHYLENE DIAMINE REC. E:3800

SYNONYMS: AETHALDIAMIN (German); AETHYLENEDIAMIN (German); β-AMINOETHYLAMINE; 1,2-ETHYLENEDIAMINE; 1,2-DIAMINOETHAN (German); 1,2-DIAMINOETHANE, ANHYDROUS; DIMETHYLENEDIAMINE; EEC No. 612-006-00-6; 1,2-ETHANEDIAMINE; ETILENDIAMINA (Spanish)

IDENTIFICATION

CAS Number: 107-15-3
Formula: $C_2H_8N_2$; $NH_2CH_2CH_2NH_2$
DOT ID Number: UN 1604; DOT Guide Number: 132
Proper Shipping Name: Ethylenediamine
Reportable Quantity (RQ): **(CERCLA)** 5000 lb/2270 kg

DESCRIPTION: Colorless to pale yellow, thick liquid. Ammonia-like odor. Initially floats on the surface of water; soluble. Freezes at 47°F/8.5°C.

Highly flammable • Corrosive to skin, eyes, and respiratory tract; skin and eye contact causes severe burns, impaired vision, and blindness; inhalation symptoms may be delayed • Do NOT use HALONS • Firefighting gear (including SCBA) does not provide adequate protection. If exposure occurs, remove and isolate gear immediately and thoroughly decontaminate personnel • Containers may BLEVE when exposed to fire •Vapors may form explosive mixture with air • Vapors are heavier than air and will collect and stay in low areas • Vapors may travel long distances to ignition sources and flashback • Vapors in confined areas (e.g., tanks, sewers, buildings) may explode when exposed to fire • Toxic products of combustion may include hydrogen cyanide and nitrogen oxides • Do not put yourself in danger by entering a contaminated area to rescue a victim.

Hazard Classification (based on NFPA-704 Rating System)
Health Hazards (Blue): 3; Flammability (Red): 2; Reactivity (Yellow): 0

EMERGENCY RESPONSE: See Appendix A (132)
Evacuation:
Public safety: Isolate spill area for at least 50 to 100 meters (160 to 330 feet) in all directions.
Spill: Increase, as necessary, the isolation distance shown above,in "Public safety."
Fire: Isolate for 800 meters (½ mile) in all directions, especially if tank, rail car, or tank truck is involved in fire.

EXPOSURE

Short-term effects: *SEEK MEDICAL ATTENTION.* **Vapor:** Move victim to fresh air. May cause lung edema; physical exertion will aggravate this condition. *IF BREATHING HAS STOPPED,* give artificial respiration. If breathing is difficult, administer oxygen.

Liquid or solid: Irritating to skin and eyes. Remove contaminated clothing and shoes. Flush affected areas with water. *IF IN EYES,* hold eyelids open and flush with plenty of water. *IF SWALLOWED*

and victim is *CONSCIOUS AND ABLE TO SWALLOW*, drink large quantity of water and induce vomiting.
Note to physician or authorized medical personnel: Medical observation is recommended for 24 to 48 hours after breathing overexposure, as pulmonary edema may be delayed. As first aid for pulmonary edema, consider administering a corticosteroid spray. Cigarette smoking may exacerbate pulmonary injury and should be discouraged for at least 72 hours following exposure.

HEALTH HAZARDS
Personal protective equipment (PPE): B-Level PPE. OSHA Table Z-1-A air contaminant. Full impervious protective clothing, including boots and gloves to prevent skin contact. Where splashing is possible wear full face shield or chemical safety goggles. Use approved respirator to protect against vapors.
Recommendations for respirator selection: NIOSH/OSHA *250 ppm*: SA:CF* (any supplied-air respirator operated in a continuous-flow mode); or PAPRS* [any powered, air-purifying respirator with cartridge(s) providing protection against the compound of concern]. *500 ppm*: CCRFS [any chemical cartridge respirator with a full facepiece and cartridge(s) providing protection against the compound of concern]; or GMFS [any air-purifying, full-facepiece respirator (gas mask) with a chin-style, front- or back-mounted canister providing protection against the compound of concern]; or PAPRTS* (any powered, air-purifying respirator with a tight-fitting facepiece and cartridge(s) providing protection against the compound of concern]; or SCBAF (any self-contained breathing apparatus with a full facepiece); or SAF (any supplied-air respirator with a full facepiece). *1000 ppm*: SAF:PD,PP (any supplied-air respirator that has a full facepiece and is operated in a pressure-demand or other positive-pressure mode). *EMERGENCY OR PLANNED ENTRY INTO UNKNOWN CONCENTRATIONS OR IDLH CONDITIONS*: SCBAF:PD,PP (any self-contained breathing apparatus that has a full facepiece and is operated in a pressure-demand or other positive-pressure mode); or SAF:PD,PP:ASCBA (any supplied-air respirator that has a full facepiece and is operated in a pressure-demand or other positive-pressure mode in combination with an auxiliary self-contained breathing apparatus operated in a pressure-demand or other positive pressure mode). *ESCAPE*: GMFS [any air-purifying, full-facepiece respirator (gas mask) with a chin-style, front- or back-mounted canister providing protection against the compound of concern]; or SCBAE (any appropriate escape-type, self-contained breathing apparatus). *Note*: Substance reported to cause eye irritation or damage; may require eye protection.
Exposure limits (TWA unless otherwise noted): ACGIH TLV 10 ppm (25 mg/m^3); NIOSH/OSHA 10 ppm (25 mg/m^3).
Toxicity by ingestion: Grade 2; oral rat LD$_{50}$ = 1.16 g/kg.
Long-term health effects: Repeated or prolonged overexposure may cause dermatitis. A person can become allergic (skin rash and asthma). Liver, kidney, and lung damage may occur.
Vapor (gas) irritant characteristics: Eye and respiratory tract irritant. Vapors are irritating such that a person will not usually tolerate moderate or high concentrations.
Liquid or solid irritant characteristics: Fairly severe skin irritant. May cause pain and second-degree burns after a few minutes of contact.
Odor threshold: 10 ppm.
IDLH value: 1000 ppm.

FIRE DATA
Flash point: 104°F/40°C (cc) (anhydrous); 150°F/66°C (75% solution).
Flammable limits in air: LEL: 2.7%; UEL: 17.0%; LEL: 2.5%; UEL: 12.0% @ 212°F/100°C (NFPA).
Fire extinguishing agents not to be used: Water.
Behavior in fire: Vapors can flow along surfaces to distant ignition source and flashback.
Autoignition temperature: 715°F/380°C/653°K.
Electrical hazard: Class I, Group D.
Stoichiometric air-to-fuel ratio: 19.1

CHEMICAL REACTIVITY
Binary reactants: Contact with strong acids will cause violent splattering; strong oxidizers and chlorinated hydrocarbons may cause fires and explosions. Reacts with carbon tetrachloride, and other chlorinated organic compounds, carbon disulfide. Corrosive to metals: copper, aluminum, nickel, cobalt and zinc.
Reactivity group: 7
Compatibility class: Aliphatic Amines

ENVIRONMENTAL DATA
Water pollution: Effect of low concentrations on aquatic life is unknown. Fouling to shoreline. May be dangerous if it enters nearby water intakes; notify operators. Notify local health and wildlife officials. **Response to discharge:** Issue warning–water contaminant. Should be removed.

SHIPPING INFORMATION
Grades of purity: 99%; technical; **Storage temperature:** Ambient; **Inert atmosphere:** None; **Venting:** Pressure vacuum valve; **Stability during transport:** Stable under ordinary conditions. Absorbs CO_2 from air.

NAS HAZARD CLASSIFICATION FOR BULK WATER TRANSPORTATION
FIRE: 3
HEALTH: Vapor irritant: 3; Liquid or solid irritant: 3; Poisons: 3
WATER POLLUTION: Human toxicity: 2; Aquatic toxicity: 3; Aesthetic effect: 2
REACTIVITY: Other chemicals: 3; Water: 0; Self-reaction: 0

PHYSICAL AND CHEMICAL PROPERTIES
Physical state @ 59°F/15°C and 1 atm: Liquid.
Molecular weight: 60.1
Boiling point @ 1 atm: 241°F/116°C/389°K.
Melting/Freezing point: 47.3°F/8.5°C/281.5°K.
Specific gravity (water = 1): 0.91
Relative vapor density (air = 1): 2.07
Ratio of specific heats of vapor (gas): 1.087
Latent heat of vaporization: 288 Btu/lb = 160 cal/g = 6.70 x 10^5 J/kg.
Heat of combustion: –12,290 Btu/lb = –6830 cal/g = –286.0 x 10^5 J/kg.
Heat of solution: (estimate) –9 Btu/lb = –5 cal/g = –0.2 x 10^5 J/kg.
Vapor pressure: (Reid) 0.6 psia; 11 mm.

ETHYLENEDIAMINETETRACETIC ACID (EDTA)
REC. E:3900

SYNONYMS: ACETIC ACID (ETHYLENEDINITRILO)TETRA-; ACIDE ETHYLENEDIAMINETE TRACETIQUE (French); ACIDO ETILENDIAMINOTETRAACETICO (Spanish); CELON A; CHEELOX; CHEMCOLOX 340; CELON ATH; COMPLEXON II; 3,6-DIAZAOCTANEDIOIC ACID,3,6-BIS(CARBOXYMETHYL)-; EDATHAMIL; EDETIC; EDETIC ACID; EDTA; EDTA ACID; ENDRATE; ETHYLENE BIS(IMINODIACETIC ACID); ETHYLENEDIAMINETETRAACETATE; ETHYLENEDIAMINE

502 Ethylene dibromide

TETRAACETIC ACID; ETHYLENEDIAMINE-N,N,N',N'-TETRAACETIC ACID; ETHYLENEDINITRILOTETRAACETIC ACID; (ETHYLENEDINITRILO)TETRAACETIC ACID; GLYCINE, N,N'-1,2-ETHANEDIYLBIS(N-(CARBOXYMETHYL)-; HAVIDOTE; METAQUEST A; NERVANAID B ACID; NULLAPON B ACID; NULLAPON BF ACID; PERMA KLEER 50 ACID; SEQ–100; SEQUESTRENE AA; SEQUESTRIC ACID; SEQUESTROL; TETRINE ACID; TITRIPLEX; TRICON BW; TRILON B; TRILON BW; VERSENE; VERSENE ACID; WARKEELATE ACID

IDENTIFICATION
CAS Number: 60-00-4
Formula: $C_{10}H_{16}N_2O_8$; $(HOOCCH_2)_2NCH_2CH_2N(CH_2COOH)_2$
DOT ID Number: UN 9117; DOT Guide Number: 171
Proper Shipping Name: EDTA; Ethylenediaminetetraacetic acid
Reportable Quantity (RQ): **(CERCLA)** 5000 lb/2270 kg

DESCRIPTION: Colorless crystals or white powder. Odorless. Floats on water surface; slightly soluble.

Corrosive to skin, eyes, and respiratory tract; skin and eye contact causes severe burns, impaired vision, and blindness; inhalation symptoms may be delayed •Toxic products of combustion may include hydrogen cyanide, and nitrogen oxides.

Hazard Classification (based on NFPA-704 Rating System)
Health Hazards (Blue): 1; Flammability (Red): 0; Reactivity (Yellow): 0

EMERGENCY RESPONSE: See Appendix A (171)
Evacuation:
Public safety: Isolate the area of spill or leak for at least 10 to 25 meters (30 to 80 feet) in all directions.
Spill: Increase, in the downwind direction, as necessary, the distance shown under "Public Safety."
Fire: If any large container is involved in fire, isolate for at least 800 meters (½ mile) in all directions; also, consider initial evacuation for 800 meters (½ mile) in all directions.

EXPOSURE
Short-term effects: *SEEK MEDICAL ATTENTION.* **Dust/fumes:** May cause lung edema; physical exertion will aggravate this condition. **Solid:** Irritating to skin and eyes. Remove contaminated clothing and shoes. Flush affected areas with plenty of water. *IF IN EYES,* hold eyelids open and flush with plenty of water.
Note to physician or authorized medical personnel: Medical observation is recommended for 24 to 48 hours after breathing overexposure, as pulmonary edema may be delayed. As first aid for pulmonary edema, consider administering a corticosteroid spray. Cigarette smoking may exacerbate pulmonary injury and should be discouraged for at least 72 hours following exposure.

HEALTH HAZARDS
Personal protective equipment (PPE): B-Level PPE. Gloves and goggles or face shield.
Toxicity by ingestion: Grade 1; LD_{50} = 5 to 15 g/kg (as sodium or calcium salt).
Long-term health effects: Possible kidney injury. Symptoms may include vomiting, depression, and bloody diarrhea.
Vapor (gas) irritant characteristics: Eye and respiratory tract irritant.
Liquid or solid irritant characteristics: Eye and skin irritant. If spilled on clothing and allowed to remain, may cause smarting and reddening of the skin.

ENVIRONMENTAL DATA
Food chain concentration potential: Negative; unlikely to accumulate.
Water pollution: Effect of low concentrations on aquatic life is unknown. Fouling to shoreline. May be dangerous if it enters nearby water intakes; notify operators. Notify local health and wildlife officials. **Response to discharge:** Issue warning–water contaminant. Should be removed. Designated as a hazardous substance under section 311 (b) (2) (A) of the Federal Water pollution Control Act and further regulated by the Clean Water Act Amendments of 1977 and 1978. These regulations apply to discharges of this substance.

SHIPPING INFORMATION
Storage temperature: Ambient; **Inert atmosphere:** None; **Venting:** Open; **Stability during transport:** Stable.

NAS HAZARD CLASSIFICATION FOR BULK WATER TRANSPORTATION
FIRE: 0
HEALTH: Vapor irritant: 0; Liquid or solid irritant: 1; Poisons: 0
WATER POLLUTION: Human toxicity: 0; Aquatic toxicity: 2; Aesthetic effect: 1
REACTIVITY: Other chemicals: 0; Water: 0; Self-reaction: 0

PHYSICAL AND CHEMICAL PROPERTIES
Physical state @ 59°F/15°C and 1 atm: Powder
Molecular weight: 292.2
Melting/Freezing point: 240°C; decomposes.
Specific gravity (water = 1): 0.86 @ 68°F/20°C (solid).
Heat of decomposition: Less than 300°F

ETHYLENE DIBROMIDE   REC. E:4000

SYNONYMS: AETHYLENBROMID (German); BROMOFUME; CELMIDE; DBE; 1,2-DIBROMAETHAN (German); DIBROMUO de ETILENO (Spanish); α,β-DIBROMOETHANE; SYM-DIBROMOETHANE; 1,2-DIBROMOETHANE; DIBROMURE D'ETHYLENE (French); DOWFUME 40; DOWFUME EDB; DOWFUME W-8; DOW-FUME W-10, DOW-FUME W-15, DOW-FUME W-40; DOWFUME W-85; EDB; EDB-85; E-D-BEE; EEC No. 602-010-00-6; ETHYLENE BROMIDE; 1,2-ETHYLENE DIBROMIDE; FUMO-GAS; GLYCOL BROMIDE; GLYCOL DIBROMIDE; ISCOBROME D; KOPFUME; NEPHIS; PESTMASTER EDB-85; SOILBROM-40, -85, -90EC; SOILFUME; UNIFUME; RCRA No. U067

IDENTIFICATION
CAS Number: 106-93-4
Formula: $C_2H_4Br_2$; $BrCH_2CH_2Br$
DOT ID Number: UN 1605; DOT Guide Number: 154
Proper Shipping Name: Ethylene dibromide
Reportable Quantity (RQ): **(CERCLA)** 1 lb/0.454 kg

DESCRIPTION: Colorless, heavy liquid. Mild, sweet odor, like chloroform. Sinks in water; insoluble; poisonous vapor is produced. Freezes at 50°F/10°C.

Very Irritating to the skin, eyes, and respiratory tract; prolonged skin contact will cause skin burns; affects the central nervous system • Firefighting gear (including SCBA) does not provide adequate protection. If exposure occurs, remove and isolate gear immediately and thoroughly decontaminate personnel • Containers may BLEVE when exposed to fire •Vapors are heavier than air and will collect and stay in low areas • Vapors in confined areas (e.g., tanks, sewers, buildings) may explode when exposed to fire • Toxic products of combustion or contact with hot surfaces include carbon monoxide and hydrogen bromide.

Hazard Classification (based on NFPA-704 Rating System)
Health Hazards (Blue): 3; Flammability (Red): 0; Reactivity (Yellow): 0

EMERGENCY RESPONSE: See Appendix A (154)
Evacuation:
Public safety: Isolate the area of spill or leak for at least 25 to 50 meters (80 to 160 feet) in all directions.
Spill: Small spill–First: Isolate in all directions 30 meters (100 feet); Then: Protect persons downwind, DAY: 0.2 km (0.1 mile); NIGHT: 0.2 km (0.1 mile). Large spill–First: Isolate in all directions 30 meters (100 feet); Then: Protect persons downwind, DAY: 0.3 km (0.2 mile); NIGHT: 0.5 km (0.3 mile).
Fire: If tank, rail car, or tank truck is involved in fire, isolate for at least 800 meters (½ mile) in all directions; also, consider initial evacuation for 800 meters (½ mile) in all directions.

EXPOSURE
Short-term effects: *SEEK MEDICAL ATTENTION. ABSORBED THROUGH THE SKIN.* **Vapor:** *POISONOUS IF INHALED.* Irritating to eyes, nose, and throat. Move to fresh air. *IF BREATHING HAS STOPPED, give artificial respiration; avoid mouth-to-mouth resuscitation; use bag/mask apparatus.* IF breathing is difficult, administer oxygen. **Liquid:** *POISONOUS IF SWALLOWED OR IF SKIN IS EXPOSED.* Irritating to skin and eyes. Remove contaminated clothing and shoes. Flush affected areas with plenty of water. *IF IN EYES,* hold eyelids open and flush with plenty of water. *IF SWALLOWED* and victim is *CONSCIOUS AND ABLE TO SWALLOW,* have victim drink 4 to 8 ounces of water.
Notes: Inhalation or Ingestion can cause Irritation in lungs and organic injury to liver and kidneys. 5000 ppm for several minutes might be fatal.

HEALTH HAZARDS
Personal protective equipment (PPE): B-Level PPE. OSHA Table Z-1-A air contaminant. OSHA Table Z-2 air contaminant. Full face organic vapor respirator. Chemical protective material(s) reported to have good to excellent resistance: PV alcohol, Teflon®, Viton®, Viton®/neoprene.
Recommendations for respirator selection: NIOSH
At any concentrations above the NIOSH REL, or where there is no REL, at any detectable concentration: SCBAF:PD,PP (any self-contained breathing apparatus that has a full facepiece and is operated in a pressure-demand or other positive-pressure mode); or SAF:PD,PP:ASCBA (any supplied-air respirator that has a full facepiece and is operated in a pressure-demand or other positive-pressure mode in combination with an auxiliary self-contained breathing apparatus operated in a pressure-demand or other positive pressure mode). *ESCAPE:* GMFOV [any air-purifying, full-facepiece respirator (gas mask) with a chin-style, front-or back-mounted organic vapor canister]; or SCBAE (any appropriate escape-type, self-contained breathing apparatus).
Exposure limits (TWA unless otherwise noted): ACGIH TLV suspected human carcinogen; personal protective clothing needed; OSHA PEL 20 ppm; NIOSH REL 0.045 ppm; potential human carcinogen; reduce exposure to lowest feasible level; skin contact contributes significantly in overall exposure.
Short-term exposure limits (15-minute TWA): NIOSH ceiling 0.13 ppm/15 min; OSHA ceiling 30 ppm; 50 ppm/5 min max peak.

Toxicity by ingestion: Grade 3; LD_{50} = 50 to 500 mg/kg.
Long-term health effects: IARC probable carcinogen, rating 2A; human evidence: inadequate; animal evidence: sufficient. NTP anticipated carcinogen. Liver damage.
Vapor (gas) irritant characteristics: Eye and respiratory tract irritant. Vapors cause smarting of the eyes or respiratory system if present in high concentrations. The effect may be temporary.
Liquid or solid irritant characteristics: Eye and skin irritant. If spilled on clothing and allowed to remain, may cause smarting and reddening of the skin.
IDLH value: Potential human carcinogen; 100 ppm.

FIRE DATA
Electrical hazard: Flow or agitation of substance may generate electrostatic charges due to low conductivity.

CHEMICAL REACTIVITY
Binary reactants: Heat, sunlight causes decomposition liberating hydrogen bromide. Reacts with chemically active metals such as sodium, potassium, calcium, hot aluminum, and magnesium; liquid ammonia; bases and strong oxidizers. Incompatible with organic anhydrides, isocyanates, acrylates, ketones, aldehydes, alcohols, glycols. Attacks some plastics and coatings.
Reactivity group: 36
Compatibility class: Halogenated hydrocarbons

ENVIRONMENTAL DATA
Food chain concentration potential: Negative; unlikely to accumulate.
Water pollution: Harmful to aquatic life in very low concentrations. May be dangerous if it enters nearby water intakes; notify operators. Notify local health and wildlife officials.
Response to discharge: Should be removed. Chemical and physical treatment.

SHIPPING INFORMATION
Grades of purity: Commercial; **Storage temperature:** Ambient; **Inert atmosphere:** None; **Venting:** Pressure-vacuum; **Stability during transport:** Stable.

NAS HAZARD CLASSIFICATION FOR BULK WATER TRANSPORTATION
FIRE: 0
HEALTH: Vapor irritant: 1; Liquid or solid irritant: 1; Poisons: 3
WATER POLLUTION: Human toxicity: 3; Aquatic toxicity: 3; Aesthetic effect: 2
REACTIVITY: Other chemicals: 1; Water: 0; Self-reaction: 0

PHYSICAL AND CHEMICAL PROPERTIES
Physical state @ 59°F/15°C and 1 atm: Liquid.
Molecular weight: 187.86
Boiling point @ 1 atm: 268°F/131°C/404°K.
Melting/Freezing point: 49.6°F/9.8°C/283.0°K.
Specific gravity (water = 1): 2.180 @ 68°F/20°C (liquid).
Liquid surface tension: 38.75 dynes/cm = 0.03875 N/m @ 68°F/20°C.
Liquid water interfacial tension: 36.54 dynes/cm = 0.03654 N/m @ 68°F/20°C.
Relative vapor density (air = 1): 6.5
Ratio of specific heats of vapor (gas): 1.109
Latent heat of vaporization: 82.1 Btu/lb = 45.6 cal/g = 1.91×10^5 J/kg.
Heat of fusion: 13.79 cal/g.
Vapor pressure: (Reid) 0.4 psia; 12 mm.

ETHYLENE DICHLORIDE   REC. E:4100

SYNONYMS: BROCIDE; 1,2-DICHLOROETHANE; DICLORURO de ETILENO (Spanish); DUTCH LIQUID; ETHANE DICHLORIDE; ETHYLENE CHLORIDE; 1,2-ETHYLENE DICHLORIDE; EDC; EEC No. 602-012-00-7; ETHANE DICHLORIDE; GLYCOL DICHLORIDE; RCRA No. U077

IDENTIFICATION
CAS Number: 107-06-2
Formula: $C_2H_4Cl_2$; $ClCH_2CH_2Cl$
DOT ID Number: UN 1184; DOT Guide Number: 129
Proper Shipping Name: Ethylene dichloride
Reportable Quantity (RQ): 100 lb/45.4 kg

DESCRIPTION: Colorless liquid. Sweet odor, like ether or chloroform. Sinks in water; practically insoluble; reacts slowly forming hydrogen chloride. Flammable, irritating vapor is produced.

Poison! Highly flammable • Containers may BLEVE when exposed to fire •Vapors may form explosive mixture with air • Vapors are heavier than air and will collect and stay in low areas • Vapors may travel long distances to ignition sources and flashback • Vapors in confined areas (e.g., tanks, sewers, buildings) may explode when exposed to fire • Irritating to the skin, eyes, and respiratory tract • Toxic products of combustion may include phosgene and hydrogen chloride.

Hazard Classification (based on NFPA-704 Rating System)
Health Hazards (Blue): 2; Flammability (Red): 3; Reactivity (Yellow): 0

EMERGENCY RESPONSE: See Appendix A (129)
Evacuation:
Public safety: Isolate spill area for at least 50 to 100 meters (160 to 330 feet) in all directions.
Spill: Large spill–Consider initial downwind evacuation for at least 300 meters (1000 feet).
Fire: Isolate for 800 meters (½ mile) in all directions, especially if tank, rail car, or tank truck is involved in fire.

EXPOSURE
Short-term effects: *SEEK MEDICAL ATTENTION.* **Vapor:** Irritating to eyes, nose, and throat. *IF INHALED*, will, will cause nausea, dizziness or difficult breathing. Move to fresh air. *IF BREATHING HAS STOPPED,* give artificial respiration. IF breathing is difficult, administer oxygen. **Liquid:** Will burn skin and eyes. Harmful if swallowed. Remove contaminated clothing and shoes. Flush affected areas with plenty of water. *IF IN EYES,* hold eyelids open and flush with plenty of water. *IF SWALLOWED* and victim is *CONSCIOUS AND ABLE TO SWALLOW*, have victim drink 4 to 8 ounces of water and have victim induce vomiting. *IF SWALLOWED* and victim is *UNCONSCIOUS OR HAVING CONVULSIONS,* do nothing except keep victim warm. The use of alcoholic beverages may enhance the toxic effect.

HEALTH HAZARDS
Personal protective equipment (PPE): B-Level PPE. OSHA Table Z-1-A air contaminant. OSHA Table Z-2 air contaminant. Clean, body-covering clothing and safety glasses with side shields. Respiratory protection: up to 50 ppm, none; 50 ppm to 2%, 1/2 hr or less, full face mask and canister; greater than 2%, self-contained breathing apparatus. Chemical protective material(s) reported to have good to excellent resistance: PV alcohol, Teflon®, Viton®, Silvershield®. Also, Viton®/neoprene offers limited protection.
Recommendations for respirator selection: NIOSH
At any concentrations above the NIOSH REL, or where there is no REL, at any detectable concentration: SCBAF:PD,PP (any self-contained breathing apparatus that has a full facepiece and is operated in a pressure-demand or other positive-pressure mode); or SAF:PD,PP:ASCBA (any supplied-air respirator that has a full facepiece and is operated in a pressure-demand or other positive-pressure mode in combination with an auxiliary self-contained breathing apparatus operated in a pressure-demand or other positive pressure mode). *ESCAPE:* GMFOV [any air-purifying, full-facepiece respirator (gas mask) with a chin-style, front-or back-mounted organic vapor canister]; or SCBAE (any appropriate escape-type, self-contained breathing apparatus).
Exposure limits (TWA unless otherwise noted): ACGIH TLV 10 ppm (40 mg/m^3); OSHA PEL 50 ppm; NIOSH REL 1 ppm (4 mg/m^3); potential human carcinogen; limit exposure to lowest feasible level.
Short-term exposure limits (15-minute TWA): NIOSH STEL 2 ppm (8 mg/m^3); OSHA ceiling 100 ppm; peak 200 ppm [5 minute maximum peak in any 3 hours]).
Toxicity by ingestion: Grade 2; LD_{50} = 0.5 to 5 g/kg (rat).
Long-term health effects: IARC possible carcinogen. NTP anticipated carcinogen. May damage liver, kidneys, and nervous system.
Vapor (gas) irritant characteristics: Eye, skin and respiratory tract irritant. Vapors cause irritation such that personnel will find high concentrations unpleasant. The effect may be temporary.
Liquid or solid irritant characteristics: Eye and skin irritant. Causes smarting of the skin and first-degree burns on short exposure; may cause secondary burns on long exposure. Degreases the skin.
Odor threshold: 6.0-180 ppm.
IDLH value: Potential human carcinogen; 50 ppm.

FIRE DATA
Flash point: 60°F/16°C (oc); 56°F/13°C (cc).
Flammable limits in air: LEL: 6.2%; UEL: 16%.
Fire extinguishing agents not to be used: Water may be ineffective.
Autoignition temperature: 775°F/413°C/686°K.
Electrical hazard: Class I, Group D; Flow or agitation of substance may generate electrostatic charges due to low conductivity.
Burning rate: 1.6 mm/min

CHEMICAL REACTIVITY
Reactivity with water: Contact with moisture forms hydrochloric acid.
Binary reactants: Strong oxidizers, caustics; chemically active metals such as aluminum (forms dangerous shock-sensitive compounds) or magnesium powder; sodium and potassium; liquid ammonia; alkyl amides. Decomposes to vinyl chloride and hydrochloric acid above 1112°F/600°C. Attacks plastics. Corrodes iron when contaminated with water at elevated temperatures.
Reactivity group: 36
Compatibility class: Halogenated hydrocarbons

ENVIRONMENTAL DATA
Food chain concentration potential: Negative; unlikely to accumulate.
Water pollution: Dangerous to aquatic life in high concentrations. May be dangerous if it enters nearby water intakes; notify operators. Notify local health and wildlife officials. **Response to discharge:** Issue warning–high flammability. Disperse and flush.

SHIPPING INFORMATION
Grades of purity: Commercial; **Storage temperature:** Ambient; **Inert atmosphere:** None; **Venting:** Pressure-vacuum; **Stability during transport:** Stable.

NAS HAZARD CLASSIFICATION FOR BULK WATER TRANSPORTATION
FIRE: 3
HEALTH: Vapor irritant: 2; Liquid or solid irritant: 2; Poisons: 3
WATER POLLUTION: Human toxicity: 3; Aquatic toxicity: 2; Aesthetic effect: 2
REACTIVITY: Other chemicals: 1; Water: 0; Self-reaction: 0

PHYSICAL AND CHEMICAL PROPERTIES
Physical state @ 59°F/15°C and 1 atm: Liquid.
Molecular weight: 98.96
Boiling point @ 1 atm: 182.3°F/83.5°C/356.7°K.
Melting/Freezing point: −32.3°F/−35.7°C/237.5°K.
Critical temperature: 550°F/288°C/561°K.
Critical pressure: 735 psia = 50 atm = 5.1 MN/m^2.
Specific gravity (water = 1): 1.253 @ 68°F/20°C (liquid).
Liquid surface tension: 32.2 dynes/cm = 0.0322 N/m @ 68°F/20°C.
Liquid water interfacial tension: (estimate) 30 dynes/cm = 0.03 N/m @ 77°F/25°C.
Relative vapor density (air = 1): 3.3
Ratio of specific heats of vapor (gas): 1.118
Latent heat of vaporization: 138 Btu/lb = 76.4 cal/g = 3.2 x 10^5 J/kg.
Heat of combustion: (estimate) 3400 Btu/lb
Heat of decomposition: 1112°F/600°C/873°K (approximate).
Heat of fusion: 21.12 cal/g.
Vapor pressure: (Reid) 2.7 psia; 64 mm.

ETHYLENE GLYCOL   REC. E:4200

SYNONYMS: 1,2-DIHYDROXYETHANE; DOWTHERM® SR-1; EEC No. 603-027-00-1; EG; 1,2-ETHANEDIOL; ETHYLENE ALCOHOL; ETILENGLICOL (Spanish); FRIDEX; GLYCOL; GLYCOL ALCOHOL; 2-HYDROXYETHANOL; LUTROL-9; MACROGOL 400 BPC; MEG; ETHYLENE DIHYDRATE; MONOETHYLENE GLYCOL; NORKOOL; RAMP; TESCOL; UCAR-17

IDENTIFICATION
CAS Number: 107-21-1
Formula: C$_2$H$_6$O$_2$; HOCH$_2$CH$_2$OH
DOT ID Number: UN 3077; DOT Guide Number: 171
Proper Shipping Name: Environmentally hazardous substances, liquid, n.o.s.
Reportable Quantity (RQ): **(CERCLA)** 1 lb/0.454 kg

DESCRIPTION: Colorless, thick liquid. Odorless. Sinks and mixes with water.

Combustible • Heat or flame may cause explosion • Containers may BLEVE when exposed to fire • Vapors are heavier than air and will collect and stay in low areas • Vapors in confined areas (e.g., tanks, sewers, buildings) may explode when exposed to fire • Irritating to the skin, eyes, and respiratory tract; poisonous if swallowed; affects central nervous system • Toxic products of combustion may include carbon monoxide.

Hazard Classification (based on NFPA-704 Rating System)
Health Hazards (Blue): 1; Flammability (Red): 1; Reactivity (Yellow): 0

EMERGENCY RESPONSE: See Appendix A (171)
Evacuation:
Public safety: Isolate the area of spill or leak for at least 10 to 25 meters (30 to 80 feet) in all directions.
Spill: Increase, in the downwind direction, as necessary, the distance shown under "Public Safety."
Fire: If any large container is involved in fire, isolate for at least 800 meters (½ mile) in all directions; also, consider initial evacuation for 800 meters (½ mile) in all directions.

EXPOSURE
Short-term effects: *SEEK MEDICAL ATTENTION.* **Liquid:** Irritating to skin and eyes. *IF SWALLOWED*, will cause loss of consciousness. Remove contaminated clothing and shoes. Flush affected areas with plenty of water. *IF IN EYES*, hold eyelids open and flush with plenty of water. *IF SWALLOWED* and victim is *CONSCIOUS AND ABLE TO SWALLOW*, have victim drink 4 to 8 ounces of water and have victim induce vomiting. *IF SWALLOWED* and victim is *UNCONSCIOUS OR HAVING CONVULSIONS,* do nothing except keep victim warm.

HEALTH HAZARDS
Personal protective equipment (PPE): B-Level PPE. OSHA Table Z-1-A air contaminant. Goggles; shower and eye bath. Chemical protective material(s) reported to have good to excellent resistance: butyl rubber, nitrile, neoprene, nitrile, nitrile+PVC, PVC, polyethylene, Teflon®, Viton®. Also, natural rubber, chlorinated polyethylene, Viton®/neoprene, Viton®, butyl rubber/neoprene, PV alcohol, neoprene+natural rubber offers limited protection.
Exposure limits (TWA unless otherwise noted): ACGIH TLV ceiling 100 ppm.
Toxicity by ingestion: Grade 1; LD$_{50}$ = 5 to 15 g/kg (rat, guinea pig, mouse). Fatal kidney injury may result if ingested.
Long-term health effects: Possible liver, kidney, and brain damage.
Liquid or solid irritant characteristics: Eye and skin irritant. Degreases the skin.
Odor threshold: More than 0.1 ppm.

FIRE DATA
Flash point: 240°F/116°C (oc); 232°F/111°C (cc).
Flammable limits in air: LEL: 3.2%; UEL: 15.3%.
Fire extinguishing agents not to be used: Water or foam may cause frothing
Autoignition temperature: 752°F/400°C/673°K (pure). 775°F/413°C/686°K (antifreeze).
Electrical hazard: Class I, Group D.
Burning rate: 1.0 mm/min

CHEMICAL REACTIVITY
Binary reactants: Strong oxidizers, chromium trioxide, potassium permanganate, sodium peroxide. Hygroscopic (absorbs moisture from the air).
Reactivity group: 20
Compatibility class: Alcohols, glycols

ENVIRONMENTAL DATA
Food chain concentration potential: Log P$_{ow}$ = −1.4. Unlikely to accumulate.
Water pollution: Effect of low concentrations on aquatic life is unknown. May be dangerous if it enters nearby water intakes; notify

operators. Notify local health and wildlife officials. **Response to discharge:** Disperse and flush.

SHIPPING INFORMATION
Grades of purity: Industrial grade; low-conductivity grade; **Storage temperature:** Ambient; **Inert atmosphere:** None; **Venting:** Open (flame arrester); **Stability during transport:** Stable.

NAS HAZARD CLASSIFICATION FOR BULK WATER TRANSPORTATION
FIRE: 1
HEALTH: Vapor irritant: 0; Liquid or solid irritant: 0; Poisons: 1
WATER POLLUTION: Human toxicity: 2; Aquatic toxicity: 1; Aesthetic effect: 1
REACTIVITY: Other chemicals: 2; Water: 0; Self-reaction: 0

PHYSICAL AND CHEMICAL PROPERTIES
Physical state @ 59°F/15°C and 1 atm: Liquid.
Molecular weight: 62.07.
Boiling point @ 1 atm: 387°F/197°C/470°K.
Melting/Freezing point: 8.6°F/13°C/260°K.
Specific gravity (water = 1): 1.115 @ 68°F/20°C (liquid).
Relative vapor density (air = 1): 2.1
Ratio of specific heats of vapor (gas): 1.095
Latent heat of vaporization: 344 Btu/lb = 191 cal/g = 8.00×10^5 J/kg.
Heat of combustion: −7259 Btu/lb = −4033 cal/g = -168.9×10^5 J/kg.
Heat of solution: (estimate) −20 Btu/lb = −12 cal/g = -0.5×10^5 J/kg.
Heat of fusion: 43.26 cal/g.
Vapor pressure: (Reid) 0.008 psia; −2 mm.

ETHYLENE GLYCOL ACETATE REC. E:4300

SYNONYMS: ACETATO de ETILENGLICOL (Spanish); ETHYLENE GLYCOL, MONOACETATE; 1,2-ETHANEDIOL, MONOACETATE; GLYCOL MONOACETATE; HYDROXYETHYLACETATE, 2-; 2-HYDROXYETHYL ACETATE; GLYCOL-MONOACETIN

IDENTIFICATION
CAS Number: 542-59-6
Formula: $C_4H_8O_3$; $CH_3CO_2CH_2CH_2OH$
DOT ID Number: NA 1993; DOT Guide Number: 128
Proper Shipping Name: Combustible liquid, n.o.s.

DESCRIPTION: Colorless liquid. Weak fruity odor. Sinks and mixes with water.

Combustible • Containers may BLEVE when exposed to fire • Vapors may form explosive mixture with air • Vapors are heavier than air and will collect and stay in low areas • Vapors may travel long distances to ignition sources and flashback • Vapors in confined areas (e.g., tanks, sewers, buildings) may explode when exposed to fire • Irritating to the skin, eyes, and respiratory tract • Toxic products of combustion may include carbon monoxide.

Hazard Classification (based on NFPA-704 Rating System)
Health Hazards (Blue): 0; Flammability (Red): 1; Reactivity (Yellow): 0

EMERGENCY RESPONSE: See Appendix A (128)
Evacuation:
Public safety: Isolate spill area for at least 25 to 50 meters (80 to 160 feet) in all directions.
Spill: Large spill–Consider initial downwind evacuation for at least 300 meters (1000 feet).
Fire: Isolate for 800 meters (½ mile) in all directions, especially if tank, rail car, or tank truck is involved in fire.

EXPOSURE
Short-term effects: *SEEK MEDICAL ATTENTION.* **Liquid:** Irritating to skin and eyes. Remove contaminated clothing and shoes. Flush affected areas with plenty of water. *IF IN EYES,* hold eyelids open and flush with plenty of water.

HEALTH HAZARDS
Personal protective equipment (PPE): B-Level PPE. Goggles or face shield, and rubber gloves.
Toxicity by ingestion: Grade 2: LD_{50} = 3.8 g/kg (guinea pig).
Vapor (gas) irritant characteristics: May cause eye irritation.
Liquid or solid irritant characteristics: Eye irritant; practically harmless to skin.

FIRE DATA
Flash point: 215°F/102°C (oc); 191°F/88°C (cc).

CHEMICAL REACTIVITY
Binary reactants: Oxidizers, strong acids.
Reactivity group: 34
Compatibility class: Esters.

ENVIRONMENTAL DATA
Food chain concentration potential: Log P_{ow} = −0.21. Unlikely to accumulate.
Water pollution: Effect of low concentrations on aquatic life is unknown. May be dangerous if it enters nearby water intakes; notify operators. Notify local health and wildlife officials. **Response to discharge:** Disperse and flush.

SHIPPING INFORMATION
Grades of purity: 50%; **Inert atmosphere:** Ambient; **Stability during transport:** Stable.

PHYSICAL AND CHEMICAL PROPERTIES
Physical state @ 59°F/15°C and 1 atm: Liquid.
Molecular weight: 104.10
Boiling point @ 1 atm: 368.6–372.2°F/187–189°C/460.2–462.2°K.
Specific gravity (water = 1): 1.110
Relative vapor density (air = 1): 3.59

ETHYLENE GLYCOL DIACETATE REC. E:4400

SYNONYMS: DIACETATO de ETILENGLICOL (Spanish); 1,2-ETHANEDIOL DIACETATE; ETHYLENE ACETATE; ETHYLENE DIACETATE; ETHYLENE GLYCOL ACETATE; GLYCOL DIACETATE

IDENTIFICATION
CAS Number: 111-55-7
Formula: $C_6H_{10}O_4$; $CH_3COOCH_2CH_2OCOCH_3$
DOT ID Number: NA 1993; DOT Guide Number: 128
Proper Shipping Name: Combustible liquid, n.o.s.

DESCRIPTION: Colorless liquid. Weak, fruity odor. Sinks and mixes with water.

Combustible • Containers may BLEVE when exposed to fire •Vapors may form explosive mixture with air • Vapors are heavier than air and will collect and stay in low areas • Vapors may travel long distances to ignition sources and flashback • Vapors in confined areas (e.g., tanks, sewers, buildings) may explode when exposed to fire • Irritating to the skin, eyes, and respiratory tract • Toxic products of combustion may include carbon monoxide.

Hazard Classification (based on NFPA-704 Rating System)
Health Hazards (Blue): 1; Flammability (Red): 1; Reactivity (Yellow): 0

EMERGENCY RESPONSE: See Appendix A (128)
Evacuation:
Public safety: Isolate spill area for at least 25 to 50 meters (80 to 160 feet) in all directions.
Spill: Large spill–Consider initial downwind evacuation for at least 300 meters (1000 feet).
Fire: Isolate for 800 meters (½ mile) in all directions, especially if tank, rail car, or tank truck is involved in fire.

EXPOSURE
Short-term effects: *SEEK MEDICAL ATTENTION.* **Liquid:** Irritating to skin and eyes. Remove contaminated clothing and shoes. Flush affected areas with plenty of water. *IF IN EYES*, hold eyelids open and flush with plenty of water. *IF SWALLOWED* and victim is *CONSCIOUS AND ABLE TO SWALLOW*, have victim drink 4 to 8 ounces of water.

HEALTH HAZARDS
Personal protective equipment (PPE): B-Level PPE. Goggles or face shield; rubber gloves.
EYES Liquid causes mild irritation of eyes. **Ingestion:** Ingestion causes stupor or coma.
Toxicity by ingestion: Grade 1; oral LD_{50} = 6860 mg/kg (rat).
Long-term health effects: Ingestion may cause severe injury to kidneys.
Vapor (gas) irritant characteristics: Vapors are nonirritating to eyes and throat.
Liquid or solid irritant characteristics: Mild eye irritant. Practically harmless to the skin.

FIRE DATA
Flash point: 205°F/96°C (oc); 191°F/88°C (cc).
Flammable limits in air: LEL: 1.6%; UEL 8.4%.
Autoignition temperature: 900°F/482°C/755°K.
Burning rate: 2.9 mm/min

ENVIRONMENTAL DATA
Food chain concentration potential: Negative; unlikely to accumulate.
Water pollution: Effect of low concentrations on aquatic life is unknown. May be dangerous if it enters nearby water intakes; notify operators. Notify local health and wildlife officials.
Response to discharge: Issue warning–water contaminant. Disperse and flush.

SHIPPING INFORMATION
Grades of purity: 98+%; **Storage temperature:** Ambient; **Inert atmosphere:** None; **Venting:** Open (flame arrester); **Stability during transport:** Stable.

NAS HAZARD CLASSIFICATION FOR BULK WATER TRANSPORTATION
FIRE: 1
HEALTH: Vapor irritant: 0; Liquid or solid irritant: 0; Poisons: 0
WATER POLLUTION: Human toxicity: 1; Aquatic toxicity: 2; Aesthetic effect: 2
REACTIVITY: Other chemicals: 1; Water: 0; Self-reaction: 0

PHYSICAL AND CHEMICAL PROPERTIES
Physical state @ 59°F/15°C and 1 atm: Liquid.
Molecular weight: 146.1
Boiling point @ 1 atm: 375.6°F/190.9°C/464.1°K.
Melting/Freezing point: –42.7°F/–41.5°C/231.7°K.
Specific gravity (water = 1): 1.104 @ 68°F/20°C (liquid).
Liquid surface tension: (estimate) 20 dynes/cm = 0.020 N/m @ 68°F/20°C.
Latent heat of vaporization: 133 Btu/lb = 74 cal/g = 3.1×10^5 J/kg.
Heat of combustion: (estimate) –11,000 Btu/lb = –6000 cal/g = -250×10^5 J/kg.

ETHYLENE GLYCOL DIBUTYL ETHER   REC. E:4500

SYNONYMS: ETHANE, 1,2-DIBUTOXY; 1,2-DIBUTOXYETHANE; ETHER ETHYLENE GLYCOL DIBUTYL; ETILENGLICOL DIBUTIL ETER (Spanish); DIBUTYL CELLOSOLVE

IDENTIFICATION
CAS Number: 112-48-1
Formula: $C_{10}H_{22}O_2$; $C_4H_9OC_2H_4OC_4H_9$
DOT ID Number: UN 1993; DOT Guide Number: 128
Proper Shipping Name: Combustible liquid, n.o.s.

DESCRIPTION: Colorless liquid. Insoluble in water.

Flammable • Containers may BLEVE when exposed to fire •Vapors may form explosive mixture with air • Vapors are heavier than air and will collect and stay in low areas • Vapors may travel long distances to ignition sources and flashback • Vapors in confined areas (e.g., tanks, sewers, buildings) may explode when exposed to fire • Irritating to the skin, eyes, and respiratory tract; absorbed by the skin • Toxic products of combustion may include carbon monoxide.

Hazard Classification (based on NFPA-704 Rating System)
Health Hazards (Blue): 1; Flammability (Red): 2; Reactivity (Yellow): 0

EMERGENCY RESPONSE: See Appendix A (128)
Evacuation:
Public safety: Isolate spill area for at least 25 to 50 meters (80 to 160 feet) in all directions.
Spill: Large spill–Consider initial downwind evacuation for at least 300 meters (1000 feet).
Fire: Isolate for 800 meters (½ mile) in all directions, especially if tank, rail car, or tank truck is involved in fire.

EXPOSURE
Short-term effects: *SEEK MEDICAL ATTENTION.* **Vapor:** Irritating to eyes, nose, and throat. Move victim to fresh air.
Liquid: Irritating to skin and eyes. Harmful if swallowed. Remove

contaminated clothing and shoes. Flush affected areas with plenty of water. *IF IN EYES*, hold eyelids open and flush with plenty of water.

HEALTH HAZARDS
Personal protective equipment (PPE): B-Level PPE. Protective goggles or face shield; rubber gloves. **Neoprene is generally suitable for cellosolve**
Toxicity by ingestion: Grade 2: LD_{50} = 3.25 g/kg (rat).
Vapor (gas) irritant characteristics: Vapors cause smarting of the eyes or respiratory system if present in high concentrations. The effect is temporary.
Liquid or solid irritant characteristics: Eye and skin irritant. If spilled on clothing and allowed to remain, may cause smarting and reddening of skin.

FIRE DATA
Flash point: 185°F/85°C (cc).
Fire extinguishing agents not to be used: Water may be ineffective.

CHEMICAL REACTIVITY
Binary reactants: Strong acids, oxidizers.
Reactivity group: 40
Compatibility class: Glycol ethers

ENVIRONMENTAL DATA
Water pollution: Effect of low concentrations on aquatic life is unknown. Fouling to shoreline. May be dangerous if it enters nearby water intakes; notify operators. Notify local health and wildlife officials. **Response to discharge:** Issue warning–water contaminant. Disperse and flush.

SHIPPING INFORMATION
Stability during transport: Stable.

PHYSICAL AND CHEMICAL PROPERTIES
Physical state @ 59°F/15°C and 1 atm: Liquid.
Molecular weight: 174.32
Boiling point @ 1 atm: 399°F/204°C/477.2°K.
Specific gravity (water = 1): 0.8
Relative vapor density (air = 1): 6.01

ETHYLENE GLYCOL DIETHYL ETHER   REC. E:4600

SYNONYMS: 1,2-DIETHOXYETHANE; DIETHYL CELLOSOLVE; ETHYL GLYME; ETILENGLICOL DIETIL ETER (Spanish)

IDENTIFICATION
CAS Number: 629-14-1
Formula: $C_6H_{14}O_2$; $C_2H_5OCH_2CH_2OC_2NO_5$
DOT ID Number: UN 1153; DOT Guide Number: 127
Proper Shipping Name: Ethylene glycol diethyl ether

DESCRIPTION: Colorless liquid. Mild, sweet odor. Floats on the surface of water; slightly soluble.

Highly flammable • Containers may BLEVE when exposed to fire • Vapors may form explosive mixture with air • Vapors are heavier than air and will collect and stay in low areas • Vapors may travel long distances to ignition sources and flashback • Vapors in confined areas (e.g., tanks, sewers, buildings) may explode when exposed to fire • Irritating to the skin, eyes, and respiratory tract • Toxic products of combustion may include carbon monoxide.

Hazard Classification (based on NFPA-704 Rating System)
Health Hazards (Blue): 1; Flammability (Red): 3; Reactivity (Yellow): 0

EMERGENCY RESPONSE: See Appendix A (127)
Evacuation:
Public safety: Isolate spill area for at least 25 to 50 meters (80 to 160 feet) in all directions.
Spill: Large spill–Consider initial downwind evacuation for at least 300 meters (1000 feet).
Fire: Isolate for 800 meters (½ mile) in all directions, especially if tank, rail car, or tank truck is involved in fire.

EXPOSURE
Short-term effects: *SEEK MEDICAL ATTENTION*. **Vapor:** Irritating to eyes, nose, and throat. Move victim to fresh air. **Liquid:** Irritating to skin and eyes. Harmful if swallowed. Remove contaminated clothing and shoes. Flush affected areas with plenty of water. *IF IN EYES*, hold eyelids open and flush with plenty of water. *IF SWALLOWED* and victim is *CONSCIOUS AND ABLE TO SWALLOW*, have victim drink 4 to 8 ounces of water.

HEALTH HAZARDS
Personal protective equipment (PPE): B-Level PPE. Protective goggles or face shield; rubber gloves. **Neoprene is generally suitable for cellosolve**
Toxicity by ingestion: Grade 2; oral LD_{50} = 4390 mg/kg (rat).
Vapor (gas) irritant characteristics: Respiratory tract irritant.
Liquid or solid irritant characteristics: Eye irritant.

FIRE DATA
Flash point: 95°F/35°C (oc).
Fire extinguishing agents not to be used: Water may be ineffective.
Autoignition temperature: 406°F/208°C/481°K.
Burning rate: 4.1 mm/min

CHEMICAL REACTIVITY
Reactivity group: 40
Compatibility class: Glycol ethers

ENVIRONMENTAL DATA
Food chain concentration potential: Negative; unlikely to accumulate.
Water pollution: Effect of low concentrations on aquatic life is unknown. Fouling to shoreline. May be dangerous if it enters nearby water intakes; notify operators. Notify local health and wildlife officials. **Response to discharge:** Issue warning–water contaminant. Disperse and flush.

SHIPPING INFORMATION
Grades of purity: Commercial; **Storage temperature:** Ambient; **Inert atmosphere:** None; **Venting:** Open (flame arrester); **Stability during transport:** Stable.

PHYSICAL AND CHEMICAL PROPERTIES
Physical state @ 59°F/15°C and 1 atm: Liquid.
Molecular weight: 118.2
Boiling point @ 1 atm: 252°F/122°C/395°K.
Melting/Freezing point: –101°F/–74°C/199°K.
Specific gravity (water = 1): 0.8484 @ 68°F/20°C (liquid).

Liquid surface tension: (estimate) 26 dynes/cm = 0.026 N/m @ 68°F/20°C.
Relative vapor density (air = 1): 4.1
Ratio of specific heats of vapor (gas): 1.0504
Latent heat of vaporization: 192 Btu/lb = 107 cal/g = 4.48 x 10^5 J/kg.
Heat of combustion: (estimate) 15,000 Btu/lb = –8100 cal/g = –340 x 10^5 J/kg.

ETHYLENE GLYCOL DIMETHYL ETHER   REC. E:4700

SYNONYMS: ANSUL ETHER 121; 1,2-DIMETHOXYETHANE; DIMETHYL CELLOSOLVE; EEC No. 603-031-00-3; EGDME; ETHYLENE DIMETHYL ETHER; ETILENGLICOL DIMETIL ETER (Spanish); GLYCOL DIMETHYL ETHER; GLYME; MONOETHYLENE GLYCOL ETHER; MONOGLYME

IDENTIFICATION
CAS Number: 110-71-4
Formula: $C_4H_{10}O_2$; $CH_3OCH_2CH_2OCH_3$
DOT ID Number: UN 1993; DOT Guide Number: 128
Proper Shipping Name: Flammable Liquid.

DESCRIPTION: Colorless liquid. Sharp, ether-like odor. Floats on water surface; slightly soluble. Irritating vapor is produced.

Flammable • Containers may BLEVE when exposed to fire • May be able to form unstable peroxides • Vapors may form explosive mixture with air • Vapors are heavier than air and will collect and stay in low areas • Vapors may travel long distances to ignition sources and flashback • Vapors in confined areas (e.g., tanks, sewers, buildings) may explode when exposed to fire • Irritating to the skin, eyes, and respiratory tract • Toxic products of combustion may include carbon monoxide.

Hazard Classification (based on NFPA-704 Rating System)
Health Hazards (Blue): 2; Flammability (Red): 2; Reactivity (Yellow): 0

EMERGENCY RESPONSE: See Appendix A (128)
Evacuation:
Public safety: Isolate spill area for at least 25 to 50 meters (80 to 160 feet) in all directions.
Spill: Large spill–Consider initial downwind evacuation for at least 300 meters (1000 feet).
Fire: Isolate for 800 meters (½ mile) in all directions, especially if tank, rail car, or tank truck is involved in fire.

EXPOSURE
Short-term effects: *SEEK MEDICAL ATTENTION.* **Vapor:** If inhaled, will cause dizziness or difficult breathing. Move to fresh air. *IF BREATHING HAS STOPPED,* give artificial respiration. *IF BREATHING IS DIFFICULT,* administer oxygen. **Liquid:** Not irritating to skin. *IF SWALLOWED*, will cause nausea, vomiting, or loss of consciousness. *IF SWALLOWED* and victim is *CONSCIOUS AND ABLE TO SWALLOW,* have victim drink 4 to 8 ounces of water. *IF SWALLOWED* and victim is *UNCONSCIOUS OR HAVING CONVULSIONS,* do nothing except keep victim warm.

HEALTH HAZARDS
Personal protective equipment (PPE): B-Level PPE. Protective gloves; safety glasses or goggles. Chemical protective material(s) reported to have good to excellent resistance: butyl rubber. also, neoprene is generally suitable for cellosolve.
Toxicity by ingestion: Grade 1; LD_{50} = 5 to 15 g/kg (adult albino rat).
Long-term health effects: Liver and kidney damage. May cause blood damage.
Vapor (gas) irritant characteristics: May cause eye and respiratory tract irritation.
Liquid or solid irritant characteristics: May cause eye irritation.

FIRE DATA
Flash point: 29°F/–2°C (cc).
Behavior in fire: Containers may explode in fires.
Autoignition temperature: 395°F/202°C/475°K.
Burning rate: 4.9 mm/min

CHEMICAL REACTIVITY
Reactivity group: 40
Compatibility class: Glycol ethers

ENVIRONMENTAL DATA
Food chain concentration potential: Negative; unlikely to accumulate.
Water pollution: Effect of low concentrations on aquatic life is unknown. May be dangerous if it enters nearby water intakes; notify operators. Notify local health and wildlife officials. **Response to discharge:** Issue warning–high flammability. Disperse and flush.

SHIPPING INFORMATION
Grades of purity: Commercial; **Storage temperature:** Ambient; **Inert atmosphere:** None; **Venting:** Pressure-vacuum; **Stability during transport:** Stable.

PHYSICAL AND CHEMICAL PROPERTIES
Physical state @ 59°F/15°C and 1 atm: Liquid.
Molecular weight: 90.12
Boiling point @ 1 atm: 185.4°F/85.2°C/358.4°K.
Melting/Freezing point: –92°F/–69°C/204°K.
Critical temperature: 505°F/263°C/536°K.
Critical pressure: 561 psia = 38.2 atm = 3.87 MN/m².
Specific gravity (water = 1): 0.868 @ 68°F/20°C (liquid).
Relative vapor density (air = 1): 3.1
Ratio of specific heats of vapor (gas): 1.071
Latent heat of vaporization: 134 Btu/lb = 74.6 cal/g = 3.12 x 10^5 J/kg.
Heat of combustion: –12,020 Btu/lb = –6680 cal/g = –279.7 x 10^5 J/kg.
Heat of solution: (estimate) –9 Btu/lb = –5 cal/g = –0.2 x 10^5 J/kg.
Vapor pressure: 5 mm.

ETHYLENE GLYCOL METHYL ETHER ACETATE   REC. E:4800

SYNONYMS: ACETATO de ETILENGLICOL METIL ETER (Spanish); EGMEA; ETHYLENE GLYCOL MONOMETHYL ETHER ACETATE; GLYCOL MONOMETHYL ETHER ACETATE; 2-METHOXYETHYL ACRYLATE; METHYL CELLOSOLVE ACETATE; METHYL GLYCOL ACETATE; 2-METHOXYETHYL ACETATE

IDENTIFICATION
CAS Number: 110-49-6
Formula: $C_5H_{10}O_3$; $CH_3COOCH_2CH_2OCH_3$

Ethylene glycol monobutyl ether

DOT ID Number: UN 1189; DOT Guide Number: 129
Proper Shipping Name: Ethylene glycol monomethyl ether acetate

DESCRIPTION: Colorless liquid. Mild, sweet, ether-like odor. Initailly flats on water; soluble.

Flammable • Containers may BLEVE when exposed to fire • Vapors may form explosive mixture with air • Vapors are heavier than air and will collect and stay in low areas • Vapors may travel long distances to ignition sources and flashback • Vapors in confined areas (e.g., tanks, sewers, buildings) may explode when exposed to fire • Irritating to the skin, eyes, and respiratory tract • Toxic products of combustion may include carbon monoxide.

Hazard Classification (based on NFPA-704 Rating System)
Health Hazards (Blue): 1; Flammability (Red): 2; Reactivity (Yellow): 0

EMERGENCY RESPONSE: See Appendix A (129)
Evacuation:
Public safety: Isolate spill area for at least 50 to 100 meters (160 to 330 feet) in all directions.
Spill: Large spill–Consider initial downwind evacuation for at least 300 meters (1000 feet).
Fire: Isolate for 800 meters (½ mile) in all directions, especially if tank, rail car, or tank truck is involved in fire.

EXPOSURE
Short-term effects: *SEEK MEDICAL ATTENTION.* **Vapor:** Move victim to fresh air. *IF BREATHING HAS STOPPED,* give artificial respiration. If breathing is difficult, administer oxygen. **Liquid:** *HARMFUL IF SWALLOWED OR ABSORBED THROUGH THE SKIN.* **Skin absorption or ingestion:** Swallowing a large single dose or absorbing large amount through skin could result in death. Remove contaminated clothing and shoes. Flush affected areas with water. *IF IN EYES,* hold eyelids open and flush with plenty of water. *IF SWALLOWED* and victim is *CONSCIOUS AND ABLE TO SWALLOW,* induce vomiting.

HEALTH HAZARDS
Personal protective equipment (PPE): B-Level PPE. Full face organic vapor respirator. OSHA Table Z-1-A air contaminant. Impervious clothing and gloves should be used to prevent skin contact. Neoprene is generally suitable for cellosolve
Recommendations for respirator selection: NIOSH
1 ppm: SA* (any supplied-air respirator). *2.5 ppm:* SA:CF* (any supplied-air respirator operated in a continuous-flow mode). *5 ppm:* SCBAF (any self-contained breathing apparatus with a full facepiece); SAF (any supplied-air respirator with a full facepiece). *100 ppm:* SA:PD,PP (any supplied-air respirator operated in a pressure-demand or other positive-pressure mode). *200 ppm:* SAF:PD,PP (any supplied-air respirator that has a full facepiece and is operated in a pressure-demand or other positive-pressure mode). *EMERGENCY OR PLANNED ENTRY INTO UNKNOWN CONCENTRATIONS OR IDLH CONDITIONS:* SCBAF:PD,PP (any self-contained breathing apparatus that has a full facepiece and is operated in a pressure-demand or other positive-pressure mode); or SAF:PD,PP:ASCBA (any supplied-air respirator that has a full facepiece and is operated in a pressure-demand or other positive-pressure mode in combination with an auxiliary self-contained breathing apparatus operated in a pressure-demand or other positive pressure mode). *ESCAPE:* GMFOV[any air-purifying, full-facepiece respirator (gas mask) with a chin-style, front- or back-mounted organic vapor canister]; or SCBAE (any appropriate escape-type, self-contained breathing apparatus).

Note: Substance reported to cause eye irritation or damage; may require eye protection.
Exposure limits (TWA unless otherwise noted): ACGIH TLV 5 ppm (24 mg/m^3); OSHA PEL 25 ppm (120 mg/m^3); NIOSH REL 0.1 ppm (0.5 mg/m^3); potential human carcinogen; reduce exposure to lowest feasible level; skin contact contributes significantly in overall exposure.
Toxicity by ingestion: Grade 2; oral rat LD$_{50}$ = 3.39 g/kg.
Long-term health effects: Repeated or prolonged overexposure may cause lung or kidney damage, brain damage, and death.
Vapor (gas) irritant characteristics: Vapors cause smarting of the eyes or respiratory system if present in high concentrations. The effect is temporary.
Liquid or solid irritant characteristics: Eye and skin irritant. If spilled on clothing and allowed to remain, may cause smarting and reddening of skin.
Odor threshold: 50 ppm.
IDLH value: Potential human carcinogen; 200 ppm.

FIRE DATA
Flash point: 120°F/49°C (cc).
Flammable limits in air: LEL: 1.7%; UEL: 8.2%.
Fire extinguishing agents not to be used: Water.
Autoignition temperature: 740°F/392°C/666°K.
Adiabatic flame temperature: 425°F/218°C.
Stoichiometric air-to-fuel ratio: 28.6

CHEMICAL REACTIVITY
Binary reactants: Contact with nitrates, strong oxidizers, strong alkalies, and strong acids may cause fires and explosions.
Reactivity group: 34
Compatibility class: Esters.

ENVIRONMENTAL DATA
Water pollution: Effects of low concentrations on aquatic life is unknown. May be dangerous if it enters nearby water intakes; notify operators. Notify local health and wildlife officials. **Response to discharge:** Should be removed.

SHIPPING INFORMATION
Grades of purity: 99%; technical; **Storage temperature:** Ambient; **Inert atmosphere:** None; **Stability during transport:** Stable.

PHYSICAL AND CHEMICAL PROPERTIES
Physical state @ 59°F/15°C and 1 atm: Liquid.
Molecular weight: 118.13
Boiling point @ 1 atm: 293°F/145°C/418°K.
Melting/Freezing point: –85°F/–65°C/208°K.
Specific gravity (water = 1): 1.006 @ 20°C.
Relative vapor density (air = 1): 4.1
Vapor pressure: 2 mm.

ETHYLENE GLYCOL MONOBUTYL ETHER
REC. E:4900

SYNONYMS: BUCS; 2-BUTOXYETHANOL; BUTYL CELLOSOLVE; BUTYL OXITOL; DOWANOL EB; EEC No. 603-014-00-0; EGBE; ETILENGLICOL MONOBUTIL ETER (Spanish); EKTASOLVE EB SOLVENT; JEFFERSOL EB; POLY-SOLV EB; GLYCOL BUTYL ETHER

IDENTIFICATION
CAS Number: 111-76-2
Formula: $C_6H_{14}O_2$; $CH_3(CH_2)_3OCH_2CH_2OH$

DOT ID Number: UN 2369; DOT Guide Number: 152
Proper Shipping Name: Ethylene glycol monobutyl ether

DESCRIPTION: Colorless oily liquid. Mild, rancid odor; ether-like. Floats on the surface of water; soluble.

Combustible • Containers may BLEVE when exposed to fire • Vapors may form explosive mixture with air • Vapors are heavier than air and will collect and stay in low areas • Vapors may travel long distances to ignition sources and flashback • Vapors in confined areas (e.g., tanks, sewers, buildings) may explode when exposed to fire • Irritating to the skin, eyes, and respiratory tract • Toxic products of combustion may include carbon monoxide.

Hazard Classification (based on NFPA-704 Rating System)
Health Hazards (Blue): 2; Flammability (Red): 2; Reactivity (Yellow): 0

EMERGENCY RESPONSE: See Appendix A (152)
Evacuation:
Public safety: Isolate the area of spill or leak for at least 25 to 50 meters (80 to 160 feet) in all directions.
Spill: Increase, in the downwind direction, as necessary, the distance shown under "Public Safety."
Fire: If tank, rail car, or tank truck is involved in fire, isolate for at least 800 meters (½ mile) in all directions; also, consider initial evacuation for 800 meters (½ mile) in all directions.

EXPOSURE
Short-term effects: *SEEK MEDICAL ATTENTION.* **Liquid:** *HARMFUL IF SWALLOWED OR ABSORBED THROUGH THE SKIN.* Irritating to skin and eyes. Harmful if swallowed. *IF BREATHING HAS STOPPED,* give artificial respiration; *avoid mouth-to-mouth resuscitation; use bag/mask apparatus.* Remove contaminated clothing and shoes. Flush affected areas with plenty of water. *IF IN EYES,* hold eyelids open and flush with plenty of water. *IF SWALLOWED* and victim is *CONSCIOUS AND ABLE TO SWALLOW,* have victim drink 4 to 8 ounces of water.

HEALTH HAZARDS
Personal protective equipment (PPE): B-Level PPE. OSHA Table Z-1-A air contaminant. Full face organic canister respirator. Chemical protective material(s) reported to offer protection: butyl rubber, Saranex®. Also, PV alcohol, nitrile, and neoprene offer limited protection.
Recommendations for respirator selection: NIOSH/OSHA
50 ppm: CCROV* [any chemical cartridge respirator with organic vapor cartridge(s)]; or SA* (any supplied-air respirator). *125 ppm:* SA:CF* (any supplied-air respirator operated in a continuous-flow mode); or PAPROV * [any powered, air-purifying respirator with organic vapor cartridge(s)]. *250 ppm:* CCRFOV [any chemical cartridge respirator with a full facepiece and organic vapor cartridge(s)]; or GMFOV [any air-purifying, full-facepiece respirator (gas mask) with a chin-style, front- or back-mounted acid gas canister]; or PAPRTOV* [any powered, air-purifying respirator with a tight-fitting facepiece and organic vapor cartridge(s)]; or SCBAF (any self-contained breathing apparatus with a full facepiece); or SAF (any supplied-air respirator with a full facepiece). *EMERGENCY OR PLANNED ENTRY INTO UNKNOWN CONCENTRATIONS OR IDLH CONDITIONS:* SCBAF:PD,PP (any self-contained breathing apparatus that has a full facepiece and is operated in a pressure-demand or other positive-pressure mode); or SAF:PD,PP:ASCBA (any supplied-air respirator that has a full facepiece and is operated in a pressure-demand or other positive-pressure mode in combination with an auxiliary self-contained breathing apparatus operated in a pressure-demand or other positive pressure mode). *ESCAPE:* GMFOV [any air-purifying, full-facepiece respirator (gas mask) with a chin-style, front- or back-mounted organic vapor canister]; or SCBAE (any appropriate escape-type, self-contained breathing apparatus). *Note:* Substance reported to cause eye irritation or damage; may require eye protection.
Exposure limits (TWA unless otherwise noted): ACGIH TLV 20 ppm; NIOSH REL 5 ppm (24 mg/m^3); OSHA PEL 50 ppm (240 mg/m^3); skin contact contributes significantly in overall exposure.
Toxicity by ingestion: Grade 2; LD_{50} = 0.5 to 5 g/kg (rat).
Vapor (gas) irritant characteristics: Vapors cause smarting of the eyes or respiratory system if present in high concentrations. The effect is temporary.
Liquid or solid irritant characteristics: If spilled on clothing and allowed to remain, may cause smarting and reddening of the skin.
Odor threshold: 0.10 ppm.
IDLH value: 700 ppm.

FIRE DATA
Flash point: 165°F/74°C (oc); 143°F/62°C (cc).
Flammable limits in air: LEL: 1.1% @ 200°F; UEL 12.7% @ 275°F
Autoignition temperature: 460°F/238°C/511°K.
Electrical hazard: Class I, Group C.
Burning rate: 6.7 mm/min

CHEMICAL REACTIVITY
Binary reactants: Strong oxidizers, strong caustics. Attacks plastic materials and rubber.
Reactivity group: 40
Compatibility class: Glycol ethers

ENVIRONMENTAL DATA
Food chain concentration potential: Negative; unlikely to accumulate.
Water pollution: Effect of low concentrations on aquatic life is unknown. May be dangerous if it enters nearby water intakes; notify operators. Notify local health and wildlife officials. **Response to discharge:** Disperse and flush.

SHIPPING INFORMATION
Grades of purity: Commercial; **Storage temperature:** Ambient; **Inert atmosphere:** None; **Venting:** Open (flame arrester); **Stability during transport:** Stable.

NAS HAZARD CLASSIFICATION FOR BULK WATER TRANSPORTATION
FIRE: 1
HEALTH: Vapor irritant: 1; Liquid or solid irritant: 1; Poisons: 2
WATER POLLUTION: Human toxicity: 2; Aquatic toxicity: 2; Aesthetic effect: 1
REACTIVITY: Other chemicals: 2; Water: 0; Self-reaction: 0

PHYSICAL AND CHEMICAL PROPERTIES
Physical state @ 59°F/15°C and 1 atm: Liquid.
Molecular weight: 118.18
Boiling point @ 1 atm: 340.2°F/171.2°C/444.4°K.
Melting/Freezing point: –107°F/–77°C/196°K.
Critical temperature: 694°F/368°C/641°K.
Critical pressure: 470 psia = 32 atm = 3.2 MN/m^2.
Specific gravity (water = 1): 0.902 @ 68°F/20°C (liquid).
Ratio of specific heats of vapor (gas): 1.047
Latent heat of vaporization: 157 Btu/lb = 87.1 cal/g = 3.65 x 10^5 J/kg.

Heat of combustion: −13,890 Btu/lb = −7720 cal/g = −323 x 10^5 J/kg.
Heat of solution: (estimate) −9 Btu/lb = −5 cal/g = −0.2 x 10^5 J/kg.
Vapor pressure: (Reid) 2.2 psia; 0.8 mm.

ETHYLENE GLYCOL MONOBUTYL ETHER ACETATE
REC. E:5000

SYNONYMS: ACETATO de ETILENGLICOL MONOBUTIL ETER (Spanish); ACETIC ACID,2-BUTOXYETHYL ESTER; 2-BUTOXYETHANOL ACETATE; 2-BUTOXYETHYL ACETATE; BUTOXYETHYL CELLOSOLVE ACETATE; BUTYL CELLOSOLVE ACETATE; BUTYL GLYCOL ACETATE; EGBEA; EKTASOLVE EB ACETATE; GLYCOL MONOBUTYL ETHER ACETATE

IDENTIFICATION
CAS Number: 112-07-2
Formula: $C_8H_{16}O_3$; n-$C_4H_9OCH_2CH_2OCOCH_3$
DOT ID Number: UN 1993; DOT Guide Number: 128
Proper Shipping Name: Flammable liquid, n.o.s.

DESCRIPTION: Colorless liquid. Weak, fruity odor. Floats and mixes slowly with water.

Flammable • Containers may BLEVE when exposed to fire • Vapors may form explosive mixture with air • May be able to form unstable peroxides • Vapors are heavier than air and will collect and stay in low areas • Vapors may travel long distances to ignition sources and flashback • Vapors in confined areas (e.g., tanks, sewers, buildings) may explode when exposed to fire • Irritating to the skin, eyes, and respiratory tract; absorbed by the skin • Toxic products of combustion may include carbon monoxide.

Hazard Classification (based on NFPA-704 Rating System)
Health Hazards (Blue): 1; Flammability (Red): 2; Reactivity (Yellow): 0

EMERGENCY RESPONSE: See Appendix A (128)
Evacuation:
Public safety: Isolate spill area for at least 25 to 50 meters (80 to 160 feet) in all directions.
Spill: Large spill–Consider initial downwind evacuation for at least 300 meters (1000 feet).
Fire: Isolate for 800 meters (½ mile) in all directions, especially if tank, rail car, or tank truck is involved in fire.

EXPOSURE
Short-term effects: *SEEK MEDICAL ATTENTION.* **Liquid:** Irritating to skin and eyes. Harmful if swallowed. Remove contaminated clothing and shoes. Flush affected areas with plenty of water. *IF IN EYES*, hold eyelids open and flush with plenty of water. *IF SWALLOWED* and victim is *CONSCIOUS AND ABLE TO SWALLOW*, have victim drink 4 to 8 ounces of water.

HEALTH HAZARDS
Personal protective equipment (PPE): B-Level PPE. Goggles or face shield; rubber gloves. Chemical protective material(s) reported to offer protection: natural rubber+neoprene+NBR. Neoprene is generally suitable for cellosolve
Recommendations for respirator selection: NIOSH/OSHA
50 ppm: CCROV* [any chemical cartridge respirator with organic vapor cartridge(s)]; or SA* (any supplied-air respirator). *125 ppm:* SA:CF (any supplied-air respirator operated in a continuous-flow mode); or PAPROV* [any powered, air-purifying respirator with organic vapor cartridge(s)]. *250 ppm:* CCRFOV [any chemical cartridge respirator with a full facepiece and organic vapor cartridge(s)]; or GMFOV [any air-purifying, full-facepiece respirator (gas mask) with a chin-style, front- or back-mounted acid gas canister]; or PAPRTOV [any powered, air-purifying respirator with a tight-fitting facepiece and organic vapor cartridge(s)]; or SCBAF (any self-contained breathing apparatus with a full facepiece); or SAF (any supplied-air respirator with a full facepiece). *EMERGENCY OR PLANNED ENTRY INTO UNKNOWN CONCENTRATIONS OR IDLH CONDITIONS:* SCBAF:PD,PP (any self-contained breathing apparatus that has a full facepiece and is operated in a pressure-demand or other positive-pressure mode); or SAF:PD,PP:ASCBA (any supplied-air respirator that has a full facepiece and is operated in a pressure-demand or other positive-pressure mode in combination with an auxiliary self-contained breathing apparatus operated in a pressure-demand or other positive pressure mode). *ESCAPE:* GMFOV [any air-purifying, full-facepiece respirator (gas mask) with a chin-style, front- or back-mounted organic vapor canister]; or SCBAE (any appropriate escape-type, self-contained breathing apparatus).
Note: Substance reported to cause eye irritation or damage; may require eye protection.
Exposure limits (TWA unless otherwise noted): NIOSH REL 5 ppm (33 mg/m^3).
Toxicity by ingestion: Grade 2; oral LD_{50} = 3200 mg/kg (mouse).
Long-term health effects: Kidney damage may occur.
Vapor (gas) irritant characteristics: Eye and respiratory tract irritant.
Liquid or solid irritant characteristics: Eye and skin irritant.

FIRE DATA
Flash point: 160°F/71°C (cc).
Flammable limits in air: LEL: 0.85% @ 200°F/93°C; UEL: 8.54% @ 275°F/135°C.
Fire extinguishing agents not to be used: Water may be ineffective.
Autoignition temperature: 645°F/340°C/613°K.
Electrical hazard: Class I, Group C.
Burning rate: 4.1 mm/min

CHEMICAL REACTIVITY
Binary reactants: Reacts with oxidizers.
Reactivity group: 34
Compatibility class: Esters.

ENVIRONMENTAL DATA
Food chain concentration potential: Negative; unlikely to accumulate.
Water pollution: Effect of low concentrations on aquatic life is unknown. Fouling to shoreline. May be dangerous if it enters nearby water intakes; notify operators. Notify local health and wildlife officials. **Response to discharge:** Issue warning–water contaminant. Disperse and flush.

SHIPPING INFORMATION
Grades of purity: 98+%; **Storage temperature:** Ambient; **Inert atmosphere:** None; **Venting:** Open (flame arrester); **Stability during transport:** Stable.

PHYSICAL AND CHEMICAL PROPERTIES
Physical state @ 59°F/15°C and 1 atm: Liquid.
Molecular weight: 160.21
Boiling point @ 1 atm: 378°F/192°C/465°K.

Melting/Freezing point: −82°F/−64°C/210°K.
Specific gravity (water = 1): 0.942 @ 68°F/20°C (liquid).
Liquid surface tension: (estimate) 26 dynes/cm = 0.026 N/m @ 68°F/20°C.
Relative vapor density (air = 1): 5.4
Latent heat of vaporization: 120 Btu/lb = 65 cal/g = 2.7 x 10^5 J/kg.
Heat of combustion: (estimate) 14,000 Btu/lb = −7700 cal/g = −320 x 10^5 J/kg.
Vapor pressure: 0.3 mm.

ETHYLENE GLYCOL MONOMETHYL ETHER
REC. E:5100

SYNONYMS: AETHYLENGYKOL-MONOMETHYLAETHER (German); DOWANOL EM; EEC No. 603-011-00-4; EGM; EGME; ETHER MONOMETHYLIQUE de L'ETHYLENE-GLYCIL (French); ETHYLENE GLYCOL METHYL ETHER; ETILENGLICOL MONOMETIL ETER (Spanish); GLYCOL ETHER EM; GLYCOL METHYL ETHER; GLYCOL MONOMETHYL ETHER; JEFFERSOL EM; MECS; 2-METHOXYETHANOL; METHYL CELLOSOLVE; METHYL ETHOXOL; METHYL GLYCOL; METHYL OXITOL; POLY-SOLV EM; PRIST

IDENTIFICATION
CAS Number: 109-86-4
Formula: $C_3H_8O_2$; $CH_3OCH_2CH_2OH$
DOT ID Number: UN 1188; DOT Guide Number: 127
Proper Shipping Name: Ethylene glycol monomethyl ether

DESCRIPTION: Colorless liquid. Mild, ether-like odor. Initially floats on the surface of water; soluble.

Highly flammable • Containers may BLEVE when exposed to fire • Vapors may form explosive mixture with air • Vapors are heavier than air and will collect and stay in low areas • Vapors may travel long distances to ignition sources and flashback • Vapors in confined areas (e.g., tanks, sewers, buildings) may explode when exposed to fire • Irritating to the skin, eyes, and respiratory tract • Toxic products of combustion may include carbon monoxide.

Hazard Classification (based on NFPA-704 Rating System)
Health Hazards (Blue): 2; Flammability (Red): 2; Reactivity (Yellow): 0

EMERGENCY RESPONSE: See Appendix A (127)
Evacuation:
Public safety: Isolate spill area for at least 25 to 50 meters (80 to 160 feet) in all directions.
Spill: Large spill–Consider initial downwind evacuation for at least 300 meters (1000 feet).
Fire: Isolate for 800 meters (½ mile) in all directions, especially if tank, rail car, or tank truck is involved in fire.

EXPOSURE
Short-term effects: *SEEK MEDICAL ATTENTION*. **Liquid:** *HARMFUL IF SWALLOWED OR ABSORBED THROUGH THE SKIN*. Irritating to skin and eyes. Harmful if swallowed. Remove contaminated clothing and shoes. Flush affected areas with plenty of water. *IF IN EYES*, hold eyelids open and flush with plenty of water. *IF SWALLOWED* and victim is *CONSCIOUS AND ABLE TO SWALLOW*, have victim drink 4 to 8 ounces of water.

HEALTH HAZARDS
Personal protective equipment (PPE): B-Level PPE. OSHA Table Z-1-A air contaminant. Organic vapor respirator. Chemical safety goggles; protective clothing; supplied-air respirator for high concentrations; safety shower and eye bath. Neoprene is generally suitable for cellosolve
Recommendations for respirator selection: NIOSH
1 ppm: SA* (any supplied-air respirator). *2.5 ppm:* SA:CF* (any supplied-air respirator operated in a continuous-flow mode). *5 ppm:* SCBAF (any self-contained breathing apparatus with a full facepiece); or SAF (any supplied-air respirator with a full facepiece). *100 ppm:* SA:PD,PP (any supplied-air respirator operated in a pressure-demand or other positive-pressure mode). *200 ppm:* SAF:PD,PP (any supplied-air respirator that has a full facepiece and is operated in a pressure-demand or other positive-pressure mode). *EMERGENCY OR PLANNED ENTRY INTO UNKNOWN CONCENTRATIONS OR IDLH CONDITIONS:* SCBAF:PD,PP (any self-contained breathing apparatus that has a full facepiece and is operated in a pressure-demand or other positive-pressure mode); or SAF:PD,PP:ASCBA (any supplied-air respirator that has a full facepiece and is operated in a pressure-demand or other positive-pressure mode in combination with an auxiliary self-contained breathing apparatus operated in a pressure-demand or other positive pressure mode). *ESCAPE:* GMFOV[any air-purifying, full-facepiece respirator (gas mask) with a chin-style, front- or back-mounted organic vapor canister]; or SCBAE (any appropriate escape-type, self-contained). *Note: Substance reported to cause eye irritation or damage; may require eye protection.
Exposure limits (TWA unless otherwise noted): ACGIH TLV 5 ppm (15 mg/m³) as methoxyethanol (EGME); OSHA PEL 25 ppm (80 mg/m³); NIOSH 0.1 ppm (0.3 mg/m³); skin contact contributes significantly in overall exposure.
Toxicity by ingestion: Grade 2; LD_{50} = 0.5 to 5g/kg (rat, rabbit, guinea pig).
Long-term health effects: Causes blood disorders and damage to central nervous system in humans.
Vapor (gas) irritant characteristics: Vapors cause smarting of the eyes or respiratory system if present in high concentrations. The effect may be temporary.
Liquid or solid irritant characteristics: Eye and skin irritant. If spilled on clothing and allowed to remain, may cause smarting and reddening of the skin.
Odor threshold: 2.4 ppm.
IDLH value: 200 ppm.

FIRE DATA
Flash point: 102°F/39°C (cc).
Flammable limits in air: LEL: LEL: 1.8%; UEL: 14%.
Autoignition temperature: 545°F/285°C/558°K.
Electrical hazard: Class I, Group C.
Burning rate: 1.8 mm/min

CHEMICAL REACTIVITY
Binary reactants: Strong oxidizers; caustics.
Reactivity group: 40
Compatibility class: Glycol ethers

ENVIRONMENTAL DATA
Food chain concentration potential: Log P_{ow} = −0.7. Unlikely to accumulate.
Water pollution: Effect of low concentrations on aquatic life is unknown. May be dangerous if it enters nearby water intakes; notify operators. Notify local health and wildlife officials. **Response to discharge:** Disperse and flush.

514 Ethylene glycol propyl ether

SHIPPING INFORMATION
Grades of purity: Commercial; **Storage temperature:** Ambient; **Inert atmosphere:** None; **Venting:** Open (flame arrester); **Stability during transport:** Stable.

NAS HAZARD CLASSIFICATION FOR BULK WATER TRANSPORTATION
FIRE: 2
HEALTH: Vapor irritant: 1; Liquid or solid irritant: 1; Poisons: 2
WATER POLLUTION: Human toxicity: 2; Aquatic toxicity: 2; Aesthetic effect: 1
REACTIVITY: Other chemicals: 2; Water: 0; Self-reaction: 0

PHYSICAL AND CHEMICAL PROPERTIES
Physical state @ 59°F/15°C and 1 atm: Liquid.
Molecular weight: 76.10
Boiling point @ 1 atm: 256.1°F/124.5°C/397.7°K.
Melting/Freezing point: −121.2°F/−85.1°C/188.1°K.
Critical temperature: 558°F/292°C/565°K.
Critical pressure: 735 psia = 50 atm = 5.1 MN/m^2.
Specific gravity (water = 1): 0.966 @ 68°F/20°C (liquid).
Liquid surface tension: 33 dynes/cm = 0.033 N/m @ 68°F/20°C.
Ratio of specific heats of vapor (gas): 1.079
Latent heat of vaporization: 223 Btu/lb = 124 cal/g = 5.19 x 10^5 J/kg.
Heat of combustion: −9460 Btu/lb = −5250 cal/g = −220 x 10^5 J/kg.
Vapor pressure: (Reid) 0.39 psia; 6 mm.

ETHYLENE GLYCOL PHENYL ETHER   REC. E:5200

SYNONYMS: AROSOL; DOWANOL EP; DOWANOL EPH; EMERSSENCE 1160; EMERY 6705; ETHYLENE GLYCOL MONOPHENYL ETHER; ETILENGLICOL FENIL ETER (Spanish); 1-HYDROXY-2-PHENOXYETHANE; PHENOOXY ALCOHOL; 2-PHENOXYETHANOL; PHENOXYTOL; PHENYL MONOGLYCOL ETHER; PHENYL CELLOSOLVE; ROSE ETHER

IDENTIFICATION
CAS Number: 122-99-6
Formula: $C_8H_{10}O_2$; $C_6H_5OCH_2CH_2OH$

DESCRIPTION: Colorless liquid. Pleasant odor.

Combustible • Heat or flame may cause explosion • Containers may BLEVE when exposed to fire • Vapors are heavier than air and will collect and stay in low areas • Vapors in confined areas (e.g., tanks, sewers, buildings) may explode when exposed to fire • Irritating to the skin, eyes, and respiratory tract • Toxic products of combustion may include carbon monoxide.

Hazard Classification (based on NFPA-704 Rating System)
Health Hazards (Blue): 0; Flammability (Red): 1; Reactivity (Yellow): 0

EMERGENCY RESPONSE: See Appendix A (171)
Evacuation:
Public safety: Isolate the area of spill or leak for at least 10 to 25 meters (30 to 80 feet) in all directions.
Spill: Increase, in the downwind direction, as necessary, the distance shown under "Public Safety."
Fire: If any large container is involved in fire, isolate for at least 800 meters (½ mile) in all directions; also, consider initial evacuation for 800 meters (½ mile) in all directions.

EXPOSURE
Short-term effects: *SEEK MEDICAL ATTENTION.* **Liquid:** Irritating to skin and eyes. Harmful if swallowed. Remove contaminated clothing and shoes. Flush affected areas with plenty of water. *IF IN EYES*, hold eyelids open and flush with plenty of water. *IF SWALLOWED*, induce vomiting.

HEALTH HAZARDS
Personal protective equipment (PPE):,B-Level PPE. protective clothing chemical goggles, gloves and boots. Butyl rubber is generally suitable for ethers. Natural rubber, neoprene, and nitrile rubber may offer some protection from alcohols.
Toxicity by ingestion: Grade 2: LD_{50} = 1.26 g/kg (rat).
Long-term health effects: Liver, kidney, thyroid and blood effects.
Vapor (gas) irritant characteristics: Vapors cause smarting of eyes and respiratory system if present in high concentration. The effect is temporary.
Liquid or solid irritant characteristics: If spilled on clothing and allowed to remain, may cause smarting and reddening of skin.

FIRE DATA
Flash point: 260°F/127°C (cc).
Electrical hazard: Class I, Group C.

CHEMICAL REACTIVITY
Binary reactants: Oxidizers, sulfuric acid, isocyanates.
Reactivity group: 40
Compatibility class: Glycol ethers

ENVIRONMENTAL DATA
Food chain concentration potential: Log P_{ow} = 1.17–1.2. Unlikely to accumulate.
Water pollution: Effect of low concentrations on aquatic life is unknown. May be dangerous if it enters nearby water intakes; notify operators. Notify local health and wildlife officials. **Response to discharge:** Should be removed. Chemical and physical treatment.

SHIPPING INFORMATION
Grades of purity: 90%; **Storage temperature:** Ambient; **Stability during transport:** Stable.

PHYSICAL AND CHEMICAL PROPERTIES
Physical state @ 59°F/15°C and 1 atm: Liquid.
Molecular weight: 138.2
Boiling point @ 1 atm: 474°F/246°C/519°K.
Melting/Freezing point: 51°F/11°C/284.2°K.
Specific gravity (water = 1): 1.104
Liquid surface tension: 42 dynes/cm = 0.042 N/m
Relative vapor density (air = 1): 4.8
Vapor pressure: 1.2 mm.

ETHYLENE GLYCOL PROPYL ETHER   REC. E:5300

SYNONYMS: ETHYLENE GLYCOL MONOPROPYL ETHER; ETILENGLICOL MONOPROPIL ETER (Spanish) EKTASOLVE EP; PROPYL CELLOSOLVE; 2-PROPOXYETHANOL

IDENTIFICATION
CAS Number: 2807-30-9
Formula: $C_5H_{12}O_2$; $CH_3CH_2CH_2OCH_2CH_2OH$

DOT ID Number: UN 1993; DOT Guide Number: 128
Proper Shipping Name: Flammable liquid, n.o.s.

DESCRIPTION: Colorless liquid. Mild rancid odor. Floats on the surface of water; soluble.

Highly flammable • Containers may BLEVE when exposed to fire • Vapors may form explosive mixture with air • Vapors are heavier than air and will collect and stay in low areas • Vapors may travel long distances to ignition sources and flashback • Vapors in confined areas (e.g., tanks, sewers, buildings) may explode when exposed to fire • Irritating to the skin, eyes, and respiratory tract • Toxic products of combustion may include carbon monoxide.

EMERGENCY RESPONSE: See Appendix A (128)
Evacuation:
Public safety: Isolate spill area for at least 25 to 50 meters (80 to 160 feet) in all directions.
Spill: Large spill–Consider initial downwind evacuation for at least 300 meters (1000 feet).
Fire: Isolate for 800 meters (½ mile) in all directions, especially if tank, rail car, or tank truck is involved in fire.

EXPOSURE
Short-term effects: *SEEK MEDICAL ATTENTION*. **Liquid:** Irritating to skin and eyes. Harmful if swallowed. Remove contaminated clothing and shoes. Flush affected areas with plenty of water. *IF IN EYES*, hold eyelids open and flush with plenty of water.

HEALTH HAZARDS
Personal protective equipment (PPE): B-Level PPE. Approved respirator; rubber gloves; goggles; clothing to prevent body contact with liquid. **Neoprene is generally suitable for cellosolve**
Long-term health effects: May cause anemia and kidney damage.
Vapor (gas) irritant characteristics: Vapors cause smarting of the eyes or respiratory system if present in high concentrations. The effect may be is temporary.
Liquid or solid irritant characteristics: If spilled on clothing and allowed to remain, may cause smarting and reddening of the skin.

FIRE DATA
Flash point: 120°F/49°C (cc).
Flammable limits in air: LEL: 1.26% @ 69°C; UEL: 15.8% @ 127°C
Fire extinguishing agents not to be used: Water.
Electrical hazard: Class I, Group C.
Stoichiometric air-to-fuel ratio: 33.3

CHEMICAL REACTIVITY
Binary reactants: Oxidizing materials, sulfuric acid, isocyanates.
Reactivity group: 40
Compatibility class: Glycol ethers

ENVIRONMENTAL DATA
Food chain concentration potential: Negative; unlikely to accumulate.
Water pollution: Effects of low concentrations on aquatic life is unknown. May be dangerous if it enters nearby water intakes; notify operators. Notify local health and wildlife officials.
Response to discharge: Should be removed.

SHIPPING INFORMATION
Grades of purity: Commercial; **Storage temperature:** Ambient; **Inert atmosphere:** None; **Stability during transport:** Stable.

PHYSICAL AND CHEMICAL PROPERTIES
Physical state @ 59°F/15°C and 1 atm: Liquid.
Molecular weight: 104.15
Boiling point @ 1 atm: 301°F/149.5°C/422.5°K.
Specific gravity (water = 1): 0.908 at 23°C/73°F
Relative vapor density (air = 1): 3.6

ETHYLENEIMINE   REC. E:5400

SYNONYMS: AMINOETHYLENE; AZACYCLOPROPANE; AZIRIDINA (Spanish); AZIRANE; AZIRIDINE; AZIRINE; 1H-AZIRINE, DIHYDRO-; DIHYDROAZIRINE; DIHYDRO-1-AZIRINE; DIMETHYLENEIMINE; DIMETHYLENIMINE; E-1; EEC No. 613-001-00-1; ETHYLIMENE; ETILENIMIDA (Spanish); TL 337; RCRA No. P054

IDENTIFICATION
CAS Number: 151-56-4
Formula: CH_2CH_2NH
DOT ID Number: UN 1185; DOT Guide Number: 131P
Proper Shipping Name: Ethyleneimine, inhibited
Reportable Quantity (RQ): **(CERCLA)** 1 lb/0.454 kg

DESCRIPTION: Colorless, oily liquid. Ammonia-like, fishy odor. Floats on water surface; soluble; poisonous, flammable vapor is produced.

Poison! • Highly flammable •Corrosive • Breathing the vapor, skin or eye contact, or swallowing the material can kill you • Polymerization hazard; may react with itself blocking relief valves; leading to tank explosions • Firefighting gear (including SCBA) does not provide adequate protection. If exposure occurs, remove and isolate gear immediately and thoroughly decontaminate personnel • Containers may BLEVE when exposed to fire • Vapors may form explosive mixture with air • Vapors are heavier than air and will collect and stay in low areas • Vapors may travel long distances to ignition sources and flashback • Vapors in confined areas (e.g., tanks, sewers, buildings) may explode when exposed to fire • Toxic products of combustion may include nitrogen oxide • Do not put yourself in danger by entering a contaminated area to rescue a victim.

Hazard Classification (based on NFPA-704 Rating System)
Health Hazards (Blue): 4; Flammability (Red): 3; Reactivity (Yellow): 3

EMERGENCY RESPONSE: See Appendix A (131)
Evacuation:
Public safety: Isolate spill area for at least 100 to 200 meters (330 to 660 feet) in all directions.
Spill: Increase, as necessary, the isolation distance shown above, in "Public safety."
Fire: Isolate for 800 meters (½ mile) in all directions, especially if tank, rail car, or tank truck is involved in fire.

EXPOSURE
Short-term effects: *SEEK MEDICAL ATTENTION. ABSORBED THROUGH THE SKIN.* **Vapor:** *POISONOUS IF INHALED OR IF SKIN IS EXPOSED.* Irritating to eyes. Move to fresh air. *IF BREATHING HAS STOPPED,* give artificial respiration. IF breathing is difficult, administer oxygen. **Liquid:** *POISONOUS IF SWALLOWED OR IF SKIN IS EXPOSED.* Will burn eyes. Remove contaminated clothing and shoes. Flush affected areas with plenty

of water. *IF IN EYES*, hold eyelids open and flush with plenty of water. *IF SWALLOWED* and victim is *CONSCIOUS AND ABLE TO SWALLOW*, have victim drink 4 to 8 ounces of water.

HEALTH HAZARDS
Personal protective equipment (PPE): A-Level PPE. OSHA Table Z-1-A air contaminant. Ful face respirator. Chemical protective material(s) reported to have good to excellent resistance: butyl rubber.
Recommendations for respirator selection: NIOSH
At any concentrations above the NIOSH REL, or where there is no REL, at any detectable concentration: SCBAF:PD,PP (any self-contained breathing apparatus that has a full facepiece and is operated in a pressure-demand or other positive-pressure mode); or SAF:PD,PP:ASCBA (any supplied-air respirator that has a full facepiece and is operated in a pressure-demand or other positive-pressure mode in combination with an auxiliary self-contained breathing apparatus operated in a pressure-demand or other positive pressure mode). *ESCAPE:* GMFOV [any air-purifying, full-facepiece respirator (gas mask) with a chin-style, front-or back-mounted organic vapor canister]; or SCBAE (any appropriate escape-type, self-contained breathing apparatus).
Exposure limits (TWA unless otherwise noted): ACGIH TLV 0.5 ppm (0.88 mg/m^3); OSHA PEL [1910.1012]; suspected human carcinogen; NIOSH REL potential human carcinogen; reduce exposure to lowest feasible level; skin contact contributes significantly in overall exposure.
Toxicity by ingestion: Grade 4; LD_{50} below 50 mg/kg (rat).
Long-term health effects: Causes cancer in mice. OSHA specifically regulated carcinogen. IARC rating 3; limited animal evidence. Kidney and skin damage.
Vapor (gas) irritant characteristics: Vapor is irritating such that personnel will not usually tolerate moderate or high concentrations.
Liquid or solid irritant characteristics: Corrosive to eyes. Causes smarting of the skin and first-degree burns on short exposure and may cause secondary burns on long exposure.
Odor threshold: 0.6–2.0 ppm.
IDLH value: 100 ppm; suspected human carcinogen.

FIRE DATA
Flash point: 12°F/–11°C (cc).
Flammable limits in air: LEL: 3.3%; UEL: 54.8%.
Behavior in fire: May polymerize in fires with evolution of heat and container rupture.
Autoignition temperature: 608°F/320°C/593°K.
Electrical hazard: Class I, Group C.

CHEMICAL REACTIVITY
Reactivity with water: Mild reaction; becomes alkaline.
Binary reactants: Polymerizes explosively in presence of acids; contact with silver, silver alloys (e.g., silver solder) or aluminum may cause polymerization. Attacks rubber, aluminum, zinc and other chemically active metals.
Neutralizing agents for acids and caustics: Flush with **water**.
Polymerization: Explosive polymerization can occur when in contact with acids, high temperatures, peroxides or sunlight.

ENVIRONMENTAL DATA
Food chain concentration potential: May be able to accumulate.
Water pollution: Effect of low concentrations on aquatic life is unknown. May be dangerous if it enters nearby water intakes; notify operators. Notify local health and wildlife officials.
Response to discharge: Issue warning–high flammability, poison. Restrict access. Evacuate area. Disperse and flush.

SHIPPING INFORMATION
Grades of purity: 99.0%; **Storage temperature:** Ambient; **Inert atmosphere:** Inerted; **Venting:** Safety relief; **Stability during transport:** Stable unless heated under pressure.

NAS HAZARD CLASSIFICATION FOR BULK WATER TRANSPORTATION
FIRE: 3
HEALTH: Vapor irritant: 3; Liquid or solid irritant: 2; Poisons: 4
WATER POLLUTION: Human toxicity: 4; Aquatic toxicity: 3; Aesthetic effect: 3
REACTIVITY: Other chemicals: 3; Water: 1; Self-reaction: 3

PHYSICAL AND CHEMICAL PROPERTIES
Physical state @ 59°F/15°C and 1 atm: Liquid.
Molecular weight: 43.07
Boiling point @ 1 atm: 133°F/56°C/329°K.
Melting/Freezing point: –97°F/–71°C/202°K.
Specific gravity (water = 1): 0.832 @ 68°F/20°C (liquid).
Liquid surface tension: 34.5 dynes/cm = 0.0345 N/m @ 68°F/20°C.
Relative vapor density (air = 1): 1.5
Ratio of specific heats of vapor (gas): 1.192
Latent heat of vaporization: 333 Btu/lb = 185 cal/g = 7.75 x 10^5 J/kg.
Heat of combustion: –15,930 Btu/lb = –8850 cal/g = –370.5 x 10^5 J/kg.
Heat of solution: (estimate) –26 Btu/lb = –14 cal/g = –0.6 x 10^5 J/kg.
Heat of polymerization: (estimate) –900 Btu/lb = –500 cal/g = –20 x 10^5 J/kg.
Vapor pressure: (Reid) 8.6 psia; 160 mm.

ETHYLENE OXIDE   REC. E:5500

SYNONYMS: AETHYLENOXID (German); ANPROLENE; DIHYDROOXIRENE; DIMETHYLENE OXIDE; EEC No. 603-023-00-X; E.O; 1,2-EPOXYETHANE; ETHYLENE (OXYDE d') (French); ETO; OXACYCLOPROPANE; OXANE; OXIDOETHANE; OXIDO de ETILENO (Spanish); OXIRAN; OXIRANE; OXIRENE, DIHYDRO-; RCRA No. U115

IDENTIFICATION
CAS Number: 75-21-8
Formula: C_2H_4O; CH_2CH_2O
DOT ID Number: UN 1040; DOT Guide Number: 119P
Proper Shipping Name: Ethylene oxide
Reportable Quantity (RQ): **(CERCLA)** 10 lb/4.54 kg

DESCRIPTION: Colorless gas. Shipped and stored as a liquefied gas. Sweet odor, like ether. Floats on the surface of water; soluble. *Warning:* May undergo a runaway reaction on contact with water. Becomes a liquid below 51°F/11°C.

Extremely flammable. Approach only after considering explosion danger. If water supply is inadequate or containers show signs of overheating, evacuate area. Does not need oxygen for combustion • Polymerization hazard when heated • This material has low ignition energy • May react with itself without warning; explosive decomposition may occur • Severe respiratory tract and skin irritant. Breathing the vapor, skin or eye contact, or swallowing the material can kill you • Firefighting gear (including SCBA) does not provide adequate protection. If exposure occurs, remove and isolate gear immediately and thoroughly decontaminate personnel • Containers may BLEVE or explode when exposed to fire • Gas forms explosive mixture with air • Gas is heavier than air and will collect and stay in low areas • Gas may travel long distances to ignition sources and flashback • Gas in confined areas (e.g., tanks, sewers, buildings) may explode when exposed to fire • Toxic products of combustion may include carbon monoxide • Do not put yourself in danger by entering a contaminated area to rescue a victim. *Warning:* Odor detection is much higher than Exposure limits.

Hazard Classification (based on NFPA-704 Rating System)
Health Hazards (Blue): 3; Flammability (Red): 4; Reactivity (Yellow): 3

EMERGENCY RESPONSE: See Appendix A (119)
Evacuation:
Public safety: See below.
Spill: Small spill–First: Isolate in all directions 30 meters (100 feet); Then: Protect persons downwind, DAY: 0.2 km (0.1 mile); NIGHT: 0.2 km (0.1 mile). Large spill–First: Isolate in all directions 60 meters (200 feet); Then: Protect persons downwind, DAY: 0.5 km (0.3 mile); NIGHT: 1.8 km (1.1 miles).
Fire: Isolate for 1600 meters (1 mile) in all directions, especially if tank, rail car, or tank truck is involved in fire.

EXPOSURE
Short-term effects: *SEEK MEDICAL ATTENTION.* **Vapor:** Irritating to eyes, nose, and throat. *IF INHALED*, will, will cause nausea, vomiting and difficult breathing. Move to fresh air. *IF BREATHING HAS STOPPED,* give artificial respiration; *avoid mouth-to-mouth resuscitation; use bag/mask apparatus.* IF breathing is difficult, administer oxygen. **Liquid:** Will burn skin and eyes. Harmful if swallowed. Remove contaminated clothing and shoes. Flush affected areas with plenty of water. *IF IN EYES*, hold eyelids open, and flush with plenty of water. *IF SWALLOWED* and victim is *CONSCIOUS AND ABLE TO SWALLOW*, have victim drink 4 to 8 ounces of water.
Note to physician or authorized medical personnel: There is no antidote for ethylene oxide. Observe patients who are in respiratory distress for up to 12 hours.

HEALTH HAZARDS
Personal protective equipment (PPE): A-Level PPE. OSHA Table Z-1-A air contaminant. Air–supplied mask; goggles or face shield; rubber shoes and coveralls. Tyvek® or Saranex® are recommended by ASTDR. Chemical protective material(s) reported to offer minimal to poor protection: butyl rubber, chlorinated polyethylene.
Recommendations for respirator selection: NIOSH
5 ppm: GMFS** [any air-purifying, full-facepiece respirator (gas mask) with a chin-style, front- or back-mounted canister providing protection against the compound of concern]; or SCBAF (any self-contained breathing apparatus with a full facepiece); or SAF (any supplied-air respirator with a full facepiece). *EMERGENCY OR PLANNED ENTRY INTO UNKNOWN CONCENTRATIONS OR IDLH CONDITIONS:* SCBAF:PD,PP (any self-contained breathing apparatus that has a full facepiece and is operated in a pressure-demand or other positive-pressure mode); or SAF:PD,PP:ASCBA (any supplied-air respirator that has a full facepiece and is operated in a pressure-demand or other positive-pressure mode in combination with an auxiliary self-contained breathing apparatus operated in a pressure-demand or other positive pressure mode). *ESCAPE:* GMFS** [any air-purifying, full-facepiece respirator (gas mask) with a chin-style, front- or back-mounted canister providing protection against the compound of concern]; or SCBAE (any appropriate escape-type, self-contained breathing apparatus). ***Note*: End of service life indicator (ESLI) required.**
Exposure limits (TWA unless otherwise noted): ACGIH TLV 1 ppm; OSHA 1 ppm [1910.1047]; NIOSH REL 0.1 ppm; less than 0.1 ppm (less than 0.18 mg/m^3) ceiling 5 ppm [10 min/day]; potential human carcinogen; reduce exposure to lowest feasible level.
Short-term exposure limits (15-minute TWA): OSHA 5 ppm [15 minute excursion]
Toxicity by ingestion: Grade 3; oral rat LD_{50} = 72 mg/kg.
Long-term health effects: Causes cancer in mice. OSHA specifically regulated carcinogen. IARC probable carcinogen, rating 2A. May effect the nervous system.
Vapor (gas) irritant characteristics: Eye and respiratory tract irritant. Vapor is irritating such that personnel will not usually tolerate moderate or high vapor concentrations.
Liquid or solid irritant characteristics: Fairly severe skin irritant; may cause pain and second-degree burns after a few minutes of contact. Possible frostbite.
Odor threshold: 255-700 ppm. *Warning:* This is above the exposure limit.
IDLH value: Potential human carcinogen; 800 ppm.

FIRE DATA
Flash point: –4°F/–20°C (cc) (gas); –20°F/–29°C (cc) (liquid).
Flammable limits in air: LEL: 3.0%; UEL: 100%.
Special hazards of combustion products: Decomposes when heated under the influence of metal salts and oxides causing fire and explosions. If local "hot spots" develop in container, the liquid in the tank may explode.
Autoignition temperature: 804°F/429°C/702°K.
Electrical hazard: Class I, Group B. Flow or agitation of substance may generate electrostatic charges due to low conductivity.
Burning rate: 3.5 mm/min
Stoichiometric air-to-fuel ratio: 7.790 (estimate).

CHEMICAL REACTIVITY
Reactivity with water: Slow reaction forming hydrate; not hazardous at normal temperatures. NFPA reports possible runaway reaction with water. Many materials may accelerate this reaction (NFPA 1994).
Binary reactants: Reacts, possibly violently with many substances including strong acids, alcohols, alkalis, amines and oxidizers; chlorides/oxides of iron, aluminum, and tin; mercury, magnesium. May polymerize violently when in contact with highly active catalytic surfaces such as anhydrous iron, tin and aluminum chlorides, pure iron and aluminum oxides and alkali metal hydroxides. Attacks copper, silver and their alloys.
Polymerization: May polymerize violently if contaminated with alkaline or acidic materials and metal oxides or chlorides.
Reactivity group: 16
Compatibility class: Alkylene oxides

ENVIRONMENTAL DATA
Food chain concentration potential: Negative; unlikely to accumulate.
Water pollution: Effect of low concentrations on aquatic life is unknown. May be dangerous if it enters nearby water intakes; notify operators. Notify local health and wildlife officials. **Response to discharge:** Issue warning–high flammability. Evacuate area.

SHIPPING INFORMATION
Grades of purity: Commercial: 100% Must contain no acetylene; **Storage temperature:** Ambient; **Inert atmosphere:** Inerted; **Venting:** Safety relief; **Stability during transport:** Stable.

NAS HAZARD CLASSIFICATION FOR BULK WATER TRANSPORTATION
FIRE: 4
HEALTH: Vapor irritant: 3; Liquid or solid irritant: 3; Poisons: 2
WATER POLLUTION: Human toxicity: 3; Aquatic toxicity: 2; Aesthetic effect: 1
REACTIVITY: Other chemicals: 3; Water: 1; Self-reaction: 4

EXPLOSIVE

PHYSICAL AND CHEMICAL PROPERTIES
Physical state @ 59°F/15°C and 1 atm: Gas.
Molecular weight: 44.05
Boiling point @ 1 atm: 51.1°F/10.6°C/283.8°K.
Melting/Freezing point: −170.7°F/−112.6°C/160.6°K.
Critical temperature: 385°F/196°C/469°K.
Critical pressure: 1040 psia = 71.0 atm = 7.2 MN/m².
Specific gravity (water = 1): 0.82 (liquid @ 50°F).
Liquid surface tension: 24.3 dynes/cm = 0.0243 N/m @ 68°F/20°C.
Relative vapor density (air = 1): 1.5
Ratio of specific heats of vapor (gas): 1.49
Latent heat of vaporization: 249.3 Btu/lb = 138.5 cal/g = 5.799×10^5 J/kg.
Heat of combustion: −11,480 Btu/lb = −6380 cal/g = -267.1×10^5 J/kg.
Heat of solution: −61 Btu/lb = −34 cal/g = -1.4×10^5 J/kg.
Heat of fusion: 28.07 cal/g.
Vapor pressure: (Reid) 38.5 psia; 1140 mm.

ETHYL ETHER　　　　　　　　　　　　　**REC. E:5600**

SYNONYMS: AETHER; ANAESTHETIC ETHER; ANESTHESIA ETHER; ANESTHETIC ETHER; DIAETHYLAETHER (German); DIETHYL ETHER; DIETHYL OXIDE; EEC No. 603-022-00-4; ETER ETILICO (Spanish); ETHANE, 1,1'-OXYBIS-; ETHER; ETHER, ETHYL; ETHER ETHYLIQUE (French); ETHOXYETHANE; ETHYL ETHER; OXYDE D'ETHYLE (French); SOLVENT ETHER; SULFURIC ETHER; RCRA No. U117

IDENTIFICATION
CAS Number: 60-29-7
Formula: $C_4H_{10}O$; $C_2H_5OC_2NO_5$
DOT ID Number: UN 1155; DOT Guide Number: 127
Proper Shipping Name: Diethyl ether, or Ethyl ether
Reportable Quantity (RQ): **(CERCLA)** 100 lb/45.4 kg

DESCRIPTION: Clear, colorless liquid. Sweet, pungent characteristic odor. Floats on the surface of water; moderately soluble. Large amounts of vapors produced. Boils at 94.3°F/34.6°C.

Extremely flammable • This material has low ignition energy • Vapors form s explosive mixture with air • Containers may BLEVE when exposed to fire • Vapors are heavier than air and will collect and stay in low areas • Vapors may travel long distances to ignition sources and flashback • Vapors in confined areas (e.g., tanks, sewers, buildings) may explode when exposed to fire • Irritating to the skin, eyes, and respiratory tract • Decomposes above 1022°F/550°C. Toxic products of combustion may include carbon monoxide and aldehydes.

Hazard Classification (based on NFPA-704 Rating System)
Health Hazards (Blue): 1; Flammability (Red): 4; Reactivity (Yellow): 1

EMERGENCY RESPONSE: See Appendix A (127)
Evacuation:
Public safety: Isolate spill area for at least 25 to 50 meters (80 to 160 feet) in all directions.
Spill: Large spill–Consider initial downwind evacuation for at least 300 meters (1000 feet).
Fire: Isolate for 800 meters (½ mile) in all directions, especially if tank, rail car, or tank truck is involved in fire.

EXPOSURE
Short-term effects: *SEEK MEDICAL ATTENTION.* **Vapor:** Irritating to eyes, nose, and throat. *IF INHALED*, will, will cause nausea, vomiting, headache, or loss of consciousness. Move to fresh air. *IF BREATHING HAS STOPPED,* give artificial respiration. IF breathing is difficult, administer oxygen. **Liquid:** Irritating to skin. Harmful if swallowed. Remove contaminated clothing and shoes. Flush affected areas with plenty of water. *IF IN EYES*, hold eyelids open and flush with plenty of water. *IF SWALLOWED* and victim is *CONSCIOUS AND ABLE TO SWALLOW*, have victim drink 4 to 8 ounces of water. The use of alcoholic beverages may enhance the toxic effect.

HEALTH HAZARDS
Personal protective equipment (PPE): B-Level PPE. OSHA Table Z-1-A air contaminant. Approved organic vapor canister mask; chemical goggles; gloves. Chemical protective material(s) reported to have good to excellent resistance: PV alcohol, styrene-butadiene, Teflon®, Silvershield®, nitrile. Also, chlorinated polyethylene offers limited protection.
Recommendations for respirator selection: OSHA
1900 ppm: CCROV * [any chemical cartridge respirator with organic vapor cartridge(s)]; or GMFOV[any air-purifying, full-facepiece respirator (gas mask) with a chin-style, front- or back-mounted acid gas canister]; or PAPROV* [any powered, air-purifying respirator with organic vapor cartridge(s)]; or SA* (any supplied-air respirator); or SCBAF (any self-contained breathing apparatus with a full facepiece *EMERGENCY OR PLANNED ENTRY INTO UNKNOWN CONCENTRATIONS OR IDLH CONDITIONS:* SCBAF:PD,PP (any self-contained breathing apparatus that has a full facepiece and is operated in a pressure-demand or other positive-pressure mode); or SAF:PD,PP:ASCBA (any supplied-air respirator that has a full facepiece and is operated in a pressure-demand or other positive-pressure mode in combination with an auxiliary self-contained breathing apparatus operated in a pressure-demand or other positive pressure mode). *ESCAPE:* GMFOV[any air-purifying, full-facepiece respirator (gas mask) with a chin-style, front- or back-mounted organic vapor cannister]; or SCBAE (any appropriate escape-type, self-contained breathing apparatus). *Note:* Substance reported to cause eye irritation or damage; may require eye protection.
Exposure limits (TWA unless otherwise noted): ACGIH TLV 400 ppm (1210 mg/m³); OSHA PEL 400 ppm (1200 mg/m³).
Short-term exposure limits (15-minute TWA): ACGIH STEL 500 ppm (1520 mg/m³).
Toxicity by ingestion: Grade 2; LD_{50} = 0.5 to 5 g/kg.
Long-term health effects: May effect the nervous system.
Vapor (gas) irritant characteristics: Vapors cause smarting of the eyes or respiratory system if present in high concentrations. The effect may be temporary.
Liquid or solid irritant characteristics: Practically harmless to the skin because it is very volatile and evaporates quickly.

Odor threshold: 0.1–9.0 ppm.
IDLH value: 1900 ppm.

FIRE DATA
Flash point: –49°F/–45°C (cc).
Flammable limits in air: LEL: 1.85%; UEL: 36.0%.
Behavior in fire: Decomposes violently.
Autoignition temperature: 356°F/180°C/453°K.
Electrical hazard: Class I, Group C; Flow or agitation of substance may generate electrostatic charges due to low conductivity.
Burning rate: 6.7 mm/min

CHEMICAL REACTIVITY
Binary reactants: Strong oxidizers, halogens, sulfur, sulfuryl chloride; sulfur compounds, zinc. Tends to form explosive peroxides under influence of air and light. Attacks some plastics and swells natural rubber.
Reactivity group: 41
Compatibility class: Ethers

ENVIRONMENTAL DATA
Food chain concentration potential: Log P_{ow} = 0.8. Unlikely to accumulate.
Water pollution: Effect of low concentrations on aquatic life is unknown. May be dangerous if it enters nearby water intakes; notify operators. Notify local health and wildlife officials.
Response to discharge: Issue warning–high flammability. Restrict access. Evacuate area.

SHIPPING INFORMATION
Grades of purity: Reagent; absolute; purified; anesthesia; USP; concentrated; **Storage temperature:** Ambient; **Inert atmosphere:** Inerted; **Venting:** Pressure-vacuum; **Stability during transport:** Stable.

NAS HAZARD CLASSIFICATION FOR BULK WATER TRANSPORTATION
FIRE: 4
HEALTH: Vapor irritant: 1; Liquid or solid irritant: 0; Poisons: 2
WATER POLLUTION: Human toxicity: 0; Aquatic toxicity: 1; Aesthetic effect: 1
REACTIVITY: Other chemicals: 1; Water: 0; Self-reaction: 0

PHYSICAL AND CHEMICAL PROPERTIES
Physical state @ 59°F/15°C and 1 atm: Liquid.
Molecular weight: 74.12
Boiling point @ 1 atm: 94.3°F/34.6°C/307.8°K.
Melting/Freezing point: –177.3°F/–116.3°C/156.9°K.
Critical temperature: 380.3°F/193.5°C/466.7°K.
Critical pressure: 527 psia = 35.9 atm = 3.64 MN/m^2.
Specific gravity (water = 1): 0.714 @ 68°F/20°C (liquid).
Liquid surface tension: 17.0 dynes/cm = 0.0170 N/m @ 68°F/20°C.
Relative vapor density (air = 1): 2.6
Ratio of specific heats of vapor (gas): 1.081
Latent heat of vaporization: 153 Btu/lb = 84.9 cal/g = 3.56 x 10^5 J/kg.
Heat of combustion: –14,550 Btu/lb = –8082 cal/g = –338.4 x 10^5 J/kg.
Heat of decomposition: 1022°F/550°C/823°K.
Heat of fusion: 23.45 cal/g.
Vapor pressure: (Reid) 16.0 psia; 440 mm.

ETHYL-3-ETHOXYPROPIONATE REC. E:5700

SYNONYMS: ETHOXY PROPIONIC ACID, ETHYL ESTER; ETHYL β-ETHOXYPROPIONATE; PROPIONIC ACID, 3-ETHOXYETHYL ESTER

IDENTIFICATION
CAS Number: 763-69-9
Formula: $C_7H_{14}O_3$; $C_2H_5OOCCH_2CH_2OC_2NO_5$
DOT ID Number: UN 1993; DOT Guide Number: 128
Proper Shipping Name: Flammable liquid, n.o.s.

DESCRIPTION: Water-white liquid. Ester-like odor. Floats on the surface of water.

Highly flammable • Containers may BLEVE when exposed to fire • Vapors may form explosive mixture with air • Vapors are heavier than air and will collect and stay in low areas • Vapors may travel long distances to ignition sources and flashback • Vapors in confined areas (e.g., tanks, sewers, buildings) may explode when exposed to fire • Irritating to the skin, eyes, and respiratory tract • Toxic products of combustion may include carbon monoxide.

Hazard Classification (based on NFPA-704 Rating System)
Health Hazards (Blue): 0; Flammability (Red): 2; Reactivity (Yellow): 0

EMERGENCY RESPONSE: See Appendix A (128)
Evacuation:
Public safety: Isolate spill area for at least 25 to 50 meters (80 to 160 feet) in all directions.
Spill: Large spill–Consider initial downwind evacuation for at least 300 meters (1000 feet).
Fire: Isolate for 800 meters (½ mile) in all directions, especially if tank, rail car, or tank truck is involved in fire.

EXPOSURE
Short-term effects: *SEEK MEDICAL ATTENTION.* **Vapor:** Irritating to eyes, nose, throat. *IF INHALED*, will, will cause headache or dizziness. *IF IN EYES*, hold eyelids open and flush with plenty of water. *IF BREATHING HAS STOPPED*, give artificial respiration. IF breathing is difficult, administer oxygen. **Liquid:** Irritating to skin and eyes. *IF SWALLOWED*, will cause nausea, vomiting, dizziness or headache. Remove contaminated clothing and shoes. Flush affected areas with plenty of water. *IF IN EYES*, hold eyelids open and flush with plenty of water. *IF SWALLOWED*, Do NOT induce vomiting.

HEALTH HAZARDS
Personal protective equipment (PPE): B-Level PPE. Rubber boots, rubber gloves, rubber apron, and self-contained breathing apparatus or a supplied air respirator.
Toxicity by ingestion: Grade 2: LD_{50} = 5 g/kg (rat).
Long-term health effects: Prolonged or repeated exposure may cause dermatitis or conjunctivitis.
Vapor (gas) irritant characteristics: Vapors cause smarting of the eyes or respiratory system if present in high concentrations. The effect may be temporary.
Liquid or solid irritant characteristics: If spilled on clothing and allowed to remain, may cause smarting and reddening of skin.

FIRE DATA
Flash point: 138°F/59°C (cc).
Flammable limits in air: LEL: 1.05%; UEL: unknown.

Fire extinguishing agents not to be used: Water may be ineffective.

CHEMICAL REACTIVITY
Reactivity group: 34
Compatibility class: Esters.

ENVIRONMENTAL DATA
Water pollution: Effect of low concentrations on aquatic life is unknown. Fouling to shoreline. May be dangerous if it enters nearby water intakes; notify operators. Notify local health and wildlife officials. **Response to discharge:** Restrict access. Mechanical containment. Should be removed. Chemical and physical treatment.

SHIPPING INFORMATION
Grades of purity: More than 99%; **Storage temperature:** Ambient; **Stability during transport:** Stable.

PHYSICAL AND CHEMICAL PROPERTIES
Physical state @ 59°F/15°C and 1 atm: Liquid.
Molecular weight: 146.21
Boiling point @ 1 atm: 338°F/170°C/443.2°K.
Melting/Freezing point: −148°F/−100°C/173.2°K.
Specific gravity (water = 1): 0.95
Relative vapor density (air = 1): 5.0

ETHYL FORMATE   REC. E:5800

SYNONYMS: EEC No. 607-015-00-7; ETHYL ESTER OF FORMIC ACID; ETHYL FORMIC ESTER; ETHYL METHANOATE; FORMIC ACID, ETHYL ESTER; FORMIATO de ETILO (Spanish); FORMIC ETHER

IDENTIFICATION
CAS Number: 109-94-4
Formula: $C_3H_6O_2$; $HCOOC_2NO_5$
DOT ID Number: UN 1190; DOT Guide Number: 129
Proper Shipping Name: Ethyl formate

DESCRIPTION: Colorless liquid. Pleasant odor, like rum. Floats on water surface; moderately soluble; slowly decomposes forming ethyl alcohol and formic acid. Flammable, irritating vapor is produced.

Highly flammable • Irritating to the skin, eyes, and respiratory tract; absorbed through the skin • Containers may BLEVE when exposed to fire • Vapors may form explosive mixture with air • Vapors are heavier than air and will collect and stay in low areas • Vapors may travel long distances to ignition sources and flashback • Vapors in confined areas (e.g., tanks, sewers, buildings) may explode when exposed to fire • Toxic products of combustion may include carbon monoxide.

Hazard Classification (based on NFPA-704 Rating System)
Health Hazards (Blue): 2; Flammability (Red): 3; Reactivity (Yellow): 0

EMERGENCY RESPONSE: See Appendix A (129)
Evacuation:
Public safety: Isolate spill area for at least 50 to 100 meters (160 to 330 feet) in all directions.
Spill: Large spill–Consider initial downwind evacuation for at least 300 meters (1000 feet).
Fire: Isolate for 800 meters (½ mile) in all directions, especially if tank, rail car, or tank truck is involved in fire.

EXPOSURE
Short-term effects: *SEEK MEDICAL ATTENTION*. **Vapor:** Irritating to eyes, nose, and throat. *IF INHALED*, will, will cause difficult breathing. Move victim to fresh air. *IF BREATHING HAS STOPPED*, give artificial respiration. IF breathing is difficult, administer oxygen. **Liquid:** Irritating to skin and eyes. Harmful if swallowed. Remove contaminated clothing and shoes. Flush affected areas with plenty of water. *IF IN EYES*, hold eyelids open and flush with plenty of water. *IF SWALLOWED* and victim is *CONSCIOUS AND ABLE TO SWALLOW*, have victim drink 4 to 8 ounces of water.

HEALTH HAZARDS
Personal protective equipment (PPE): B-Level PPE. OSHA Table Z-1-A air contaminant. Organic canister gas mask; goggles or face shield; rubber gloves. Chemical protective material(s) reported to have good to excellent resistance: neoprene, nitrile, styrene-butadiene. Also, butyl rubber, natural rubber, and styrene-butadiene/neoprene offer limited protection.
Recommendations for respirator selection: OSHA
1500 ppm: SA:CF* (any supplied-air respirator operated in a continuous-flow mode); or PAPROV* [any powered, air-purifying respirator with organic vapor cartridge(s)]; or CCRFOV [any chemical cartridge respirator with a full facepiece and organic vapor cartridges(s)]; or GMFOV [any air-purifying, full-facepiece respirator (gas mask) with a chin-style, front- or back-mounted acid gas canister]; or SCBAF (any self-contained breathing apparatus with a full facepiece; or SAF (any supplied-air respirator with a full facepiece). *EMERGENCY OR PLANNED ENTRY INTO UNKNOWN CONCENTRATIONS OR IDLH CONDITIONS*: SCBAF:PD,PP (any self-contained breathing apparatus that has a full facepiece and is operated in a pressure-demand or other positive-pressure mode); or SAF:PD,PP:ASCBA (any supplied-air respirator that has a full facepiece and is operated in a pressure-demand or other positive-pressure mode in combination with an auxiliary self-contained breathing apparatus operated in a pressure-demand or other positive pressure mode). *ESCAPE:* GMFOV [any air-purifying, full-facepiece respirator (gas mask) with a chin-style, front- or back-mounted organic vapor cannister]; or SCBAE (any appropriate escape-type, self-contained breathing apparatus).
**Note:* Substance causes eye irritation or damage; eye protection needed.
Exposure limits (TWA unless otherwise noted): ACGIH TLV 100 ppm (303 mg/m^3); NIOSH/OSHA 100 ppm (300 mg/m^3).
Toxicity by ingestion: Grade 2; oral LD_{50} = 1850 mg/kg (rat).
Long-term health effects: Nervous system damage may occur.
Vapor (gas) irritant characteristics: Eye and respiratory tract irritant. Vapors cause moderate irritation such that personnel will find high concentrations unpleasant. The effect is temporary.
Liquid or solid irritant characteristics: Eye and skin irritant. Fairly severe skin irritant. May cause pain and second-degree burns after a few minutes of contact.
Odor threshold: 19 ppm.
IDLH value: 1500 ppm.

FIRE DATA
Flash point: −4°F/−20°C (cc).
Flammable limits in air: LEL: 2.8%; UEL: 16.0%.
Fire extinguishing agents not to be used: Water may be ineffective.
Autoignition temperature: 851°F/455°C/728°K.
Burning rate: 3.6 mm/min

CHEMICAL REACTIVITY
Reactivity with water: Decomposes slowly in water to form ethyl alcohol and formic acid.
Binary reactants: Nitrates; strong oxidizers, alkalis, and acids.

ENVIRONMENTAL DATA
Food chain concentration potential: Negative; unlikely to accumulate.
Water pollution: Effect of low concentrations on aquatic life is unknown. May be dangerous if it enters nearby water intakes; notify operators. Notify local health and wildlife officials.
Response to discharge: Issue warning–high flammability, water contaminant. Restrict access. Evacuate area. Disperse and flush.

SHIPPING INFORMATION
Grades of purity: 95+%; **Storage temperature:** Ambient; **Inert atmosphere:** None; **Venting:** Pressure-vacuum; **Stability during transport:** Stable.

NAS HAZARD CLASSIFICATION FOR BULK WATER TRANSPORTATION
FIRE: 3
HEALTH: Vapor irritant: 2; Liquid or solid irritant: 3; Poisons: 3
WATER POLLUTION: Human toxicity: 2; Aquatic toxicity: 2; Aesthetic effect: 2
REACTIVITY: Other chemicals: 1; Water: 1; Self-reaction: 0

PHYSICAL AND CHEMICAL PROPERTIES
Physical state @ 59°F/15°C and 1 atm: Liquid.
Molecular weight: 74.1
Boiling point @ 1 atm: 129.6°F/54.2°C/327.4°K.
Melting/Freezing point: –113°F
Critical temperature: 455°F/235°C/508°K.
Critical pressure: 686 psia = 46.6 atm = 4.73 MN/m^2.
Specific gravity (water = 1): 0.922 @ 68°F/20°C (liquid).
Liquid surface tension: 24 dynes/cm = 0.024 N/m @ 68°F/20°C.
Liquid water interfacial tension: (estimate) 28 dynes/cm = 0.028 N/m @ 68°F/20°C.
Relative vapor density (air = 1): 2.6
Ratio of specific heats of vapor (gas): 1.1014
Latent heat of vaporization: 176 Btu/lb = 98 cal/g = 4.1 x 10^5 J/kg.
Heat of combustion: –9500 Btu/lb = –5300 cal/g = –220 x 10^5 J/kg.
Heat of solution: –50 Btu/lb = –28 cal/g = 1.2 x 10^5 J/kg.
Vapor pressure: 200 mm.

ETHYLHEXALDEHYDE   REC. E:5900

SYNONYMS: BUTYL ETHYL ACETALDEHYDE; ETHYL BUTYLACETALDEHYDE; 2-ETHYLCAPROALDEHYDE; α-ETHYLCAPROALDEHYDE; ETHYLHEXALDEHYDE; 2-ETHYLHEXALDEHYDE; 2-ETHYLHEXANAL; 2-ETHYL-1-HEXANAL; 2-ETILHEXALDEHIDO (Spanish); 3-FORMYLHEPTANE; HEXANAL, 2-ETHYL; OCTYL ALDEHYDE

IDENTIFICATION
CAS Number: 123-05-7
Formula: $C_8H_{16}O$; $C_4H_9CH(C_2NO_5)CHO$
DOT ID Number: UN 1191; DOT Guide Number: 129
Proper Shipping Name: Octyl aldehydes, flammable

DESCRIPTION: Colorless to yellow liquid. Mild odor. Floats on water surface; very slightly soluble.

Flammable • Containers may BLEVE when exposed to fire •Vapors may form explosive mixture with air • Vapors are heavier than air and will collect and stay in low areas • Vapors may travel long distances to ignition sources and flashback • Vapors in confined areas (e.g., tanks, sewers, buildings) may explode when exposed to fire • Highly irritating to the eyes and respiratory tract; this material has anesthetic properties and may be absorbed through the skin • Toxic products of combustion may include carbon monoxide.

Hazard Classification (based on NFPA-704 Rating System)
Health Hazards (Blue): 2; Flammability (Red): 2; Reactivity (Yellow): 1

EMERGENCY RESPONSE: See Appendix A (129)
Evacuation:
Public safety: Isolate spill area for at least 50 to 100 meters (160 to 330 feet) in all directions.
Spill: Large spill–Consider initial downwind evacuation for at least 300 meters (1000 feet).
Fire: Isolate for 800 meters (½ mile) in all directions, especially if tank, rail car, or tank truck is involved in fire.

EXPOSURE
Short-term effects: *SEEK MEDICAL ATTENTION*. **Vapor:** Irritating to eyes, nose, and throat. *IF INHALED*, will, will cause coughing or difficult breathing. *IF IN EYES*, hold eyelids open and flush with plenty of water. *IF BREATHING HAS STOPPED*, give artificial respiration. IF breathing is difficult, administer oxygen. **Liquid:** Irritating to skin and eyes. *IF SWALLOWED*, will cause nausea and vomiting. Remove contaminated clothing and shoes. Flush affected areas with plenty of water. *IF IN EYES*, hold eyelids open and flush with plenty of water. *IF SWALLOWED* and victim is *CONSCIOUS AND ABLE TO SWALLOW*, have victim drink 4 to 8 ounces of water. *IF SWALLOWED* and victim is *UNCONSCIOUS OR HAVING CONVULSIONS*, do nothing except keep victim warm.

HEALTH HAZARDS
Personal protective equipment (PPE): B-Level Butyl rubber is generally suitable for aldehydes.
Toxicity by ingestion: Grade 2; oral rat LD_{50} = 3,730 mg/kg.
Vapor (gas) irritant characteristics: Vapors cause smarting of the eyes or respiratory system if present in high concentrations. The effect may be temporary.
Liquid or solid irritant characteristics: Causes smarting of the skin and first-degree burns on short exposure; may cause second-degree burns on long exposure.

FIRE DATA
Flash point: 127°F/53°C (oc); 112°F/44°C (cc).
Flammable limits in air: LEL: 0.8%; UEL: 7.2%.
Fire extinguishing agents not to be used: Water may be ineffective
Autoignition temperature: 383°F/195°C/468°K.

CHEMICAL REACTIVITY
Binary reactants: Violent reaction with oxidants. May ignite spontaneously when spilled on clothing, paper, or other absorbent materials. May ignite spontaneously with air under certain conditions.
Reactivity group: 19
Compatibility class: Aldehyde

2-Ethylhexanoic acid

ENVIRONMENTAL DATA
Food chain concentration potential: Negative; unlikely to accumulate.
Water pollution: DOT Appendix B, §172.101–marine pollutant. Effect of low concentrations on aquatic life is unknown. Fouling to shoreline. May be dangerous if it enters nearby water intakes; notify operators. Notify local health and wildlife officials.
Response to discharge: Mechanical containment. Should be removed. Chemical and physical treatment.

SHIPPING INFORMATION
Grades of purity: Commercial, 95.0+%; **Storage temperature:** Ambient; **Inert atmosphere:** None; **Venting:** Pressure-vacuum; **Stability during transport:** Stable.

NAS HAZARD CLASSIFICATION FOR BULK WATER TRANSPORTATION
FIRE: 2
HEALTH: Vapor irritant: 1; Liquid or solid irritant: 2; Poisons: 1
WATER POLLUTION: Human toxicity: 2; Aquatic toxicity: 2; Aesthetic effect: 2
REACTIVITY: Other chemicals: 2; Water: 0; Self-reaction: 1

PHYSICAL AND CHEMICAL PROPERTIES
Physical state @ 59°F/15°C and 1 atm: Liquid.
Molecular weight: 128.22
Boiling point @ 1 atm: 327°F/164°C/437°K.
Specific gravity (water = 1): 0.820 @ 68°F/20°C (liquid).
Latent heat of vaporization: 164 Btu/lb = 91.2 cal/g = 3.82×10^5 J/kg.
Heat of combustion: $-15,860$ Btu/lb = $-8,810$ cal/g = -369×10^5 J/kg.
Vapor Pressure: 1.8 mm Hg @ 68°F/20°C.

2-ETHYLHEXANOIC ACID   REC. E:6000

SYNONYMS: ACIDO 2-ETILHEXANOICO (Spanish); 2-BUTYLBUTANOIC ACID; BUTYLETHYLACETIC ACID; α-ETHYLCAPROIC ACID; 2-ETYYLCAPROIC ACID; 2-ETHYLHEXOIC ACID; 3-HEPTANECARBOXYLIC ACID; HEXANOIC ACID, 2-ETHYL-

IDENTIFICATION
CAS Number: 149-57-5
Formula: $C_8H_{16}O_2$; $CH_3(CH_2)_3CH(C_2NO_5)CO_2H$
DOT ID Number: UN 1760; DOT Guide Number: 154
Proper Shipping Name: Corrosive liquids, n.o.s.

DESCRIPTION: Colorless liquid.

Combustible • Corrosive • Heat or flame may cause explosion • Severely irritating to eyes, skin, and respiratory tract; contact with material may cause severe burns and blindness; symptoms from inhalation may be delayed; absorbed through the skin • Containers may BLEVE when exposed to fire • Vapors are heavier than air and will collect and stay in low areas • Vapors in confined areas (e.g., tanks, sewers, buildings) may explode when exposed to fire • Toxic products of combustion may include carbon monoxide.

Hazard Classification (based on NFPA-704 Rating System)
Health Hazards (Blue): 1; Flammability (Red): 1; Reactivity (Yellow): 0

EMERGENCY RESPONSE: See Appendix A (154)
Evacuation:
Public safety: Isolate the area of spill or leak for at least 25 to 50 meters (80 to 160 feet) in all directions.
Spill: Increase, in the downwind direction, as necessary, the distance shown under "Public Safety."
Fire: If tank, rail car, or tank truck is involved in fire, isolate for at least 800 meters (½ mile) in all directions; also, consider initial evacuation for 800 meters (½ mile) in all directions.

EXPOSURE
Short-term effects: *SEEK MEDICAL ATTENTION.* **Vapor:** Irritating to eyes, nose, and throat. *IF INHALED*, will, will cause coughing or difficult breathing. *IF IN EYES*, hold eyelids open and flush with plenty of water. *IF BREATHING HAS STOPPED*, give artificial respiration; *avoid mouth-to-mouth resuscitation; use bag/mask apparatus*. IF breathing is difficult, administer oxygen. Lung edema may develop. **Liquid:** Will burn skin and eyes. *IF SWALLOWED*, will cause nausea and vomiting. Remove contaminated clothing and shoes. Flush affected areas with plenty of water. *IF IN EYES*, hold eyelids open and flush with plenty of water.
Note to physician or authorized medical personnel: Medical observation is recommended for 24 to 48 hours after breathing overexposure, as pulmonary edema may be delayed. As first aid for pulmonary edema, consider administering a corticosteroid spray. Cigarette smoking may exacerbate pulmonary injury and should be discouraged for at least 72 hours following exposure.

HEALTH HAZARDS
Personal protective equipment (PPE): B-Level PPE. Respirator, chemical safety goggles, boots, gloves, and impervious apron. Chemical protective material(s) reported to have good to excellent resistance: neoprene, nitrile, PVC. Also, butyl rubber is generally suitable for carbooxylic acid compounds.
Toxicity by ingestion: Grade 2: LD_{50} = 3.0 g/kg (rat).
Vapor (gas) irritant characteristics: Vapors cause severe irritation of eyes and throat and can cause eye and lung injury. They cannot be tolerated even at low concentrations.
Liquid or solid irritant characteristics: Severe eye and skin irritant. May cause pain and second-degree burns after a few minutes of contact.

FIRE DATA
Flash point: 245°F/118°C (oc); more than 230°F/118°C (cc).
Flammable limits in air: LEL 0.8%; UEL 6.0%.
Fire extinguishing agents not to be used: Water may not be effective
Autoignition temperature: 699°F/370°C/643°K.
Electrical hazard: Flow or agitation of substance may generate electrostatic charges due to low conductivity.

CHEMICAL REACTIVITY
Binary reactants: Corrosive, attacks most common metals. Reacts with oxidizers.
Neutralizing agents for acids and caustics: Sodium bicarbonate solution
Reactivity group: 4
Compatibility class: Organic acids

ENVIRONMENTAL DATA
Water pollution: May be dangerous to aquatic life in high concentrations. May be dangerous if it enters nearby water intakes; notify operators. Notify local health and wildlife officials.

Response to discharge: Issue warning–corrosive. Mechanical containment. Should be removed. Chemical and physical treatment.

SHIPPING INFORMATION
Grades of purity: 99+%; **Storage temperature:** Ambient; **Inert atmosphere:** Not required; **Venting:** Not required; **Stability during transport:** Stable.

PHYSICAL AND CHEMICAL PROPERTIES
Physical state @ 59°F/15°C and 1 atm: Liquid.
Molecular weight: 144.3
Boiling point @ 1 atm: 442.4°F/228°C/501.2°K.
Specific gravity (water = 1): 0.903
Relative vapor density (air = 1): 4.98

2-ETHYL HEXANOL   REC. E:6100

SYNONYMS: 2-ETHYL-1-HEXANOL; 2-ETHYLHEXYL ALCOHOL; 2-ETILHEXANOL (Spanish); OCTYL ALCOHOL

IDENTIFICATION
CAS Number: 104-76-7
Formula: $C_8H_{18}O$; $CH_3(CH_2)_3CH(C_2NO_5)CH_2OH$
DOT ID Number: NA 1993; DOT Guide Number: 128
Proper Shipping Name: Combustible liquid, n.o.s.

DESCRIPTION: Colorless oily liquid. Faint odor. Floats on water surface; slightly soluble.

Highly flammable • Containers may BLEVE when exposed to fire • Vapors may form explosive mixture with air • Vapors are heavier than air and will collect and stay in low areas • Vapors may travel long distances to ignition sources and flashback • Vapors in confined areas (e.g., tanks, sewers, buildings) may explode when exposed to fire • Irritating to the skin, eyes, and respiratory tract • Toxic products of combustion may include carbon monoxide.

Hazard Classification (based on NFPA-704 Rating System)
Health Hazards (Blue): 2; Flammability (Red): 2; Reactivity (Yellow): 0

EMERGENCY RESPONSE: See Appendix A (128)
Evacuation:
Public safety: Isolate spill area for at least 25 to 50 meters (80 to 160 feet) in all directions.
Spill: Large spill–Consider initial downwind evacuation for at least 300 meters (1000 feet).
Fire: Isolate for 800 meters (½ mile) in all directions, especially if tank, rail car, or tank truck is involved in fire.

EXPOSURE
Short-term effects: *SEEK MEDICAL ATTENTION*. The use of alcoholic beverages may enhance toxic effects. **Liquid:** Irritating to skin and eyes. Harmful if swallowed. Remove contaminated clothing and shoes. Flush affected areas with plenty of water. *IF IN EYES*, hold eyelids open and flush with plenty of water. *IF SWALLOWED* and victim is *CONSCIOUS AND ABLE TO SWALLOW*, have victim drink 4 to 8 ounces of water.

HEALTH HAZARDS
Personal protective equipment (PPE): B-Level PPE. Air pack or organic canister; goggles; gloves. Chemical protective material(s) reported to have good to excellent resistance: butyl rubber, neoprene, PV alcohol, Viton®. Also, natural rubber and nitrile rubber may offer some protection from alcohols.
Toxicity by ingestion: Grade 2; LD_{50} = 0.5 to 5 g/kg (lab animals).
Long-term health effects: Increased excitability of central nervous system in rats and rabbits.
Vapor (gas) irritant characteristics: Vapors cause smarting of the eyes or respiratory system if present in high concentrations. The effect may be temporary.
Liquid or solid irritant characteristics: Eye and skin irritant. If spilled on clothing and allowed to remain, may cause smarting and reddening of the skin.

FIRE DATA
Flash point: 164°F/73°C (cc).
Flammable limits in air: LEL: 0.88%; UEL: 9.7%.
Autoignition temperature: 446°F/230°C/503°K.
Electrical hazard: Class I, Group D.
Burning rate: 4.0 mm/min

CHEMICAL REACTIVITY
Binary reactants: Strong acids, caustics, isocyanates, amines.
Reactivity group: 20
Compatibility class: Alcohols, glycols

ENVIRONMENTAL DATA
Food chain concentration potential: Negative; unlikely to accumulate.
Water pollution: Effect of low concentrations on aquatic life is unknown. Fouling to shoreline. May be dangerous if it enters nearby water intakes; notify operators. Notify local health and wildlife officials. **Response to discharge:** Mechanical containment. Chemical and physical treatment.

SHIPPING INFORMATION
Grades of purity: 99–99.7%; **Storage temperature:** Ambient; **Inert atmosphere:** None; **Venting:** Open (flame arrester); **Stability during transport:** Stable.

NAS HAZARD CLASSIFICATION FOR BULK WATER TRANSPORTATION
FIRE: 1
HEALTH: Vapor irritant: 1; Liquid or solid irritant: 1; Poisons: 1
WATER POLLUTION: Human toxicity: 1; Aquatic toxicity: 1; Aesthetic effect: 2
REACTIVITY: Other chemicals: 2; Water: 0; Self-reaction: 0

PHYSICAL AND CHEMICAL PROPERTIES
Physical state @ 59°F/15°C and 1 atm: Liquid.
Molecular weight: 130.23
Boiling point @ 1 atm: 364.5°F/184.7°C/457.9°K.
Melting/Freezing point: Less than 158°F/less than 70°C/less than 343°K.
Critical temperature: 711°F/377°C/650°K.
Critical pressure: 512 psia = 34.8 atm = 3.53 MN/m^2.
Specific gravity (water = 1): 0.834 @ 68°F/20°C (liquid).
Liquid surface tension: 27.6 dynes/cm = 0.0276 N/m @ 68°F/20°C.
Liquid water interfacial tension: 22 dynes/cm = 0.022 N/m at 22.7°C.
Relative vapor density (air = 1): 4.5
Latent heat of vaporization: 167 Btu/lb = 92.8 cal/g = 3.89 x 10^5 J/kg.
Heat of combustion: –17,480 Btu/lb = –9710 cal/g = 406.5 x 10^5 J/kg.
Vapor pressure: (Reid) 0.01 psia.

2-ETHYLHEXYL ACETATE   REC. E:6200

SYNONYMS: ACETATO de 2-ETILHEXILO (Spanish); OCTYL ACETATE

IDENTIFICATION
CAS Number: 103-09-3
Formula: $C_{10}H_{20}O_2$; $CH_3COOCH_2CH(CH_2NO_5)C_4H_9$
DOT ID Number: NA 1993; DOT Guide Number: 128
Proper Shipping Name: Combustible liquid, n.o.s.

DESCRIPTION: Colorless liquid. Mild odor. Floats on water surface; insoluble. This material is lighter than water and virtually insoluble, in a fire situation it could easily be spread by use of water in an uncontained area.

Highly flammable • Containers may BLEVE when exposed to fire • Vapors may form explosive mixture with air • Vapors are heavier than air and will collect and stay in low areas • Vapors may travel long distances to ignition sources and flashback • Vapors in confined areas (e.g., tanks, sewers, buildings) may explode when exposed to fire • Irritating to the skin, eyes, and respiratory tract • Toxic products of combustion may include carbon monoxide.

Hazard Classification (based on NFPA-704 Rating System)
Health Hazards (Blue): 2; Flammability (Red): 2; Reactivity (Yellow): 0

EMERGENCY RESPONSE: See Appendix A (128)
Evacuation:
Public safety: Isolate spill area for at least 25 to 50 meters (80 to 160 feet) in all directions.
Spill: Large spill–Consider initial downwind evacuation for at least 300 meters (1000 feet).
Fire: Isolate for 800 meters (½ mile) in all directions, especially if tank, rail car, or tank truck is involved in fire.

EXPOSURE
Short-term effects: *SEEK MEDICAL ATTENTION*. **Vapor:** Move victim to fresh air. *IF BREATHING HAS STOPPED,* give artificial respiration. If breathing is difficult, administer oxygen. **Liquid:** Remove contaminated clothing and shoes.
Flush affected areas with water.
IF IN EYES, hold eyelids open and flush with plenty of water.
IF SWALLOWED and victim is *CONSCIOUS AND ABLE TO SWALLOW*, drink water or milk.

HEALTH HAZARDS
Personal protective equipment (PPE): B-Level PPE. Impervious clothing and gloves should be used to prevent skin contact. Where splashing is possible wear full face shield or chemical safety goggles. Use approved respirator to protect against vapors.
Vapor (gas) irritant characteristics: Vapors cause smarting of the eyes or respiratory system if present in high concentrations. The effect may be temporary.
Liquid or solid irritant characteristics: If spilled on clothing and allowed to remain, may cause smarting and reddening of skin.

FIRE DATA
Flash point: 160°F/71°C (cc).
Flammable limits in air: LEL: 0.76% @ 200°F; UEL: 8.14% @ 300°F.
Fire extinguishing agents not to be used: Use water with caution.
Autoignition temperature: 515°F/268°C/541°K.
Stoichiometric air-to-fuel ratio: 66.7

CHEMICAL REACTIVITY
Binary reactants: Contact with strong oxidizers may cause vigorous reaction.

ENVIRONMENTAL DATA
Water pollution: Effects of low concentrations on aquatic life is unknown. May be dangerous if it enters nearby water intakes; notify operators. Notify local health and wildlife officials. **Response to discharge:** Should be removed.

SHIPPING INFORMATION
Grades of purity: Technical grades; **Storage temperature:** Ambient; **Inert atmosphere:** None; **Stability during transport:** Stable.

PHYSICAL AND CHEMICAL PROPERTIES
Physical state @ 59°F/15°C and 1 atm: Liquid.
Molecular weight: 172.27
Boiling point @ 1 atm: 390°F/199°C/472°K.
Specific gravity (water = 1): 0.873 @ 20°C.
Relative vapor density (air = 1): 5.93

2-ETHYLHEXYL ACRYLATE   REC. E:6300

SYNONYMS: ACRILATO de 2-ETILHEXILO (Spanish); ACRYLIC ACID, 2-ETHYLHEXYLESTER; EEC No. 607-107-00-7; 2-ETHYLHEXYL, 2-PROPENOATE; OCTYL ACRYLATE

IDENTIFICATION
CAS Number: 103-11-7
Formula: $C_{11}H_{20}O_2$; $CH_2=CHCOOCH_2CH(C_2NO_5)(CH_2)_3CH_3$
DOT ID Number: NA 1993; DOT Guide Number: 128
Proper Shipping Name: Combustible liquid, n.o.s.

DESCRIPTION: Colorless liquid. Sharp odor; mild ester-type; inoffensive. Floats on the surface of water.

Highly flammable • Polymerization hazard • Heat can induce polymerization with rapid release of energy; sealed containers may rupture explosively • May react with itself blocking relief valves; leading to tank explosions • Containers may BLEVE when exposed to fire • Vapors may form explosive mixture with air • Vapors are heavier than air and will collect and stay in low areas • Vapors may travel long distances to ignition sources and flashback • Vapors in confined areas (e.g., tanks, sewers, buildings) may explode when exposed to fire • Irritating to the skin, eyes, and respiratory tract • Toxic products of combustion may include carbon monoxide.

Hazard Classification (based on NFPA-704 Rating System)
Health Hazards (Blue): 2; Flammability (Red): 2; Reactivity (Yellow): 2

EMERGENCY RESPONSE: See Appendix A (128)
Evacuation:
Public safety: Isolate spill area for at least 25 to 50 meters (80 to 160 feet) in all directions.
Spill: Large spill–Consider initial downwind evacuation for at least 300 meters (1000 feet).
Fire: Isolate for 800 meters (½ mile) in all directions, especially if tank, rail car, or tank truck is involved in fire.

EXPOSURE
Short-term effects: *SEEK MEDICAL ATTENTION*. **Liquid:** Irritating to skin and eyes. Harmful if swallowed. Remove

contaminated clothing and shoes. Flush affected areas with plenty of water. *IF IN EYES*, hold eyelids open and flush with plenty of water. *IF SWALLOWED* and victim is *CONSCIOUS AND ABLE TO SWALLOW*, have victim drink 4 to 8 ounces of water and have victim induce vomiting. *IF SWALLOWED* and victim is *UNCONSCIOUS OR HAVING CONVULSIONS*, do nothing except keep victim warm.

HEALTH HAZARDS
Personal protective equipment (PPE): B-Level PPE. Self-contained breathing apparatus; rubber gloves; vapor-proof chemical safety goggles; impervious apron and boots.
Toxicity by ingestion: Grade 2; oral rat LD_{50} = 1540 mg/kg.
Vapor (gas) irritant characteristics: Eye and respiratory tract irritant.
Liquid or solid irritant characteristics: Eye and skin irritant. If spilled on clothing and allowed to remain, may cause smarting and reddening of the skin.

FIRE DATA
Flash point: 180°F/82°C (oc).
Flammable limits in air: LEL: 0.8%; UEL: 6.4%.
Fire extinguishing agents not to be used: Water or foam may cause frothing.
Behavior in fire: Heat can result in a severe polymerization with rapid release of energy. Sealed containers may rupture explosively if hot.
Autoignition temperature: 482°F/250°C/523°K.
Electrical hazard: Class I, Group D.
Burning rate: 4.6 mm/min

CHEMICAL REACTIVITY
Binary reactants: Reacts with oxidizers.
Polymerization: Will polymerize unless inhibited and when heated; **Inhibitor of polymerization:** Monomethyl ether of hydroquinone, 13-120 ppm. Hydroquinone, 90–120 ppm.
Reactivity group: 14
Compatibility class: Acrylate

ENVIRONMENTAL DATA
Food chain concentration potential: Negative; unlikely to accumulate.
Water pollution: Effect of low concentrations on aquatic life is unknown. Fouling to shoreline. May be dangerous if it enters nearby water intakes; notify operators. Notify local health and wildlife officials. Notify operators or nearby water intakes.
Response to discharge: Issue warning–water contaminant. Restrict access. Mechanical containment. Should be removed. Chemical and physical treatment.

SHIPPING INFORMATION
Grades of purity: 99+%; **Storage temperature:** Less than 100°F/38°C; **Inert atmosphere:** None; **Venting:** Open (flame arrester); **Stability during transport:** Stable.

NAS HAZARD CLASSIFICATION FOR BULK WATER TRANSPORTATION
FIRE: 1
HEALTH: Vapor irritant: 0; Liquid or solid irritant: 1; Poisons: 1
WATER POLLUTION: Human toxicity: 1; Aquatic toxicity: 2; Aesthetic effect: 2
REACTIVITY: Other chemicals: 2; Water: 0; Self-reaction: 3

PHYSICAL AND CHEMICAL PROPERTIES
Physical state @ 59°F/15°C and 1 atm: Liquid.
Molecular weight: 184.2
Boiling point @ 1 atm: (polymerizes) 471°F/214°C/487°K.
Melting/Freezing point: –130°F/–90°C/183°K.
Specific gravity (water = 1): 0.885 @ 68°F/20°C (liquid).
Liquid surface tension: (estimate) 26 dynes/cm = 0.026 N/m @ 68°F/20°C.
Liquid water interfacial tension: (estimate) 30 dynes/cm = 0.030 N/m @ 68°F/20°C.
Relative vapor density (air = 1): 6.3
Latent heat of vaporization: 110 Btu/lb = 61 cal/g = 2.6 x 10^5 J/kg.
Heat of combustion: –15,500 Btu/lb = –8600 cal/g = 360 x 10^5 J/kg.
Heat of polymerization: –142 Btu/lb
Heat of polymerization: = –79 cal/g = –3.3 x 10^5 J/kg.
Vapor pressure: (Reid) 0.01 psia; 0.01 mm.

2-ETHYL HEXYLAMINE REC. E:6500

SYNONYMS: 1-AMINO-2-ETHYLHEXANE; β-ETHYLHEXYLAMINE; 2-ETHYL-1-HEXYLAMINE; 2-ETHYLHEXYLAMINE-1; 2-ETILHEXILAMINA (Spanish)

IDENTIFICATION
CAS Number: 104-75-6
Formula: $C_8H_{19}N$; $C_4H_9CH(C_2NO_5)CH_2NH_2$
DOT ID Number: UN 2276; DOT Guide Number: 132
Proper Shipping Name: 2-Ethylhexylamine

DESCRIPTION: Colorless liquid. Musky ammonia odor. Floats on water surface; soluble.

Flammable • Corrosive to the skin, eyes, and respiratory tract; contact with skin or eyes can cause burns and vision impairment; inhalation symptoms may be delayed • Containers may BLEVE when exposed to fire •Vapors may form explosive mixture with air • Vapors are heavier than air and will collect and stay in low areas • Vapors may travel long distances to ignition sources and flashback • Vapors in confined areas (e.g., tanks, sewers, buildings) may explode when exposed to fire • Toxic products of combustion may include nitrogen oxides.

Hazard Classification (based on NFPA-704 Rating System)
Health Hazards (Blue): 2; Flammability (Red): 2; Reactivity (Yellow): 0

EMERGENCY RESPONSE: See Appendix A (132)
Evacuation:
Public safety: Isolate spill area for at least 50 to 100 meters (160 to 330 feet) in all directions.
Spill: Increase, as necessary, the isolation distance shown above,in "Public safety."
Fire: Isolate for 800 meters (½ mile) in all directions, especially if tank, rail car, or tank truck is involved in fire.

EXPOSURE
Short-term effects: *SEEK MEDICAL ATTENTION*. **Vapor:** Irritating to eyes, nose, and throat. Harmful if inhaled. May cause lung edema; physical exertion will aggravate this condition. Move victim to fresh air. *IF BREATHING HAS STOPPED,* give artificial respiration. IF breathing is difficult, administer oxygen. **Liquid:** Harmful if swallowed. Chemical pneumonitis may develop. Will burn skin and eyes. Remove contaminated clothing and shoes. Flush affected areas with plenty of water. *IF IN EYES*, hold eyelids

open and flush with plenty of water. *IF SWALLOWED* and victim is *CONSCIOUS AND ABLE TO SWALLOW*, have victim drink 4 to 8 ounces of water. **Do NOT induce vomiting**.

Note to physician or authorized medical personnel: Medical observation is recommended for 24 to 48 hours after breathing overexposure, as pulmonary edema may be delayed. As first aid for pulmonary edema, consider administering a corticosteroid spray. Cigarette smoking may exacerbate pulmonary injury and should be discouraged for at least 72 hours following exposure.

HEALTH HAZARDS
Personal protective equipment (PPE): B-Level PPE. Air supplied or cartridge respirator, impermeable gloves, apron, and boots if condition warrants. Face shield or splash proof goggles and other protective equipment as necessary to prevent skin contact. Wash skin for at least 15 minutes while removing contaminated clothing and shoes. **Ingestion:** Drink water, lemon juice, milk or demulcents. **Do NOT induce vomiting**.
Toxicity by ingestion: Grade 3; LD_{50} = 50 to 500 mg/kg.
Long-term health effects: Prolonged exposure to vapor can result in systemic toxic effects. Liver an kidney damage can occur.
Vapor (gas) irritant characteristics: Vapors are moderately irritating such that personnel will not usually tolerate moderate or high concentrations.
Liquid or solid irritant characteristics: Severe skin irritant. May cause pain and second-degree burns after a few minutes of contact.

FIRE DATA
Flash point: 140°F/60°C (oc).
Autoignition temperature: 563°F/295°C/568°K.

CHEMICAL REACTIVITY
Binary reactants: Can react vigorously with oxidizing materials. A strong base; incompatible with acids; corrosive to copper and its alloys, aluminum and zinc.
Neutralizing agents for acids and caustics: Cover with a 90:10 mixture of sand/soda ash. Place in container. Flush area with water.
Reactivity group: 7
Compatibility class: Aliphatic amines

ENVIRONMENTAL DATA
Water pollution: Harmful to aquatic life in very low concentrations. May be dangerous if it enters nearby water intakes; notify operators. Notify local health and wildlife officials.
Response to discharge: Issue warning–corrosive. Chemical and physical treatment. Disperse and flush.

SHIPPING INFORMATION
Grades of purity: 98–99%; **Storage temperature:** Ambient; **Stability during transport:** Stable.

PHYSICAL AND CHEMICAL PROPERTIES
Physical state @ 59°F/15°C and 1 atm: Liquid.
Molecular weight: 129.25
Boiling point @ 1 atm: 337°F/169°C/442.2°K.
Melting/Freezing point: Less than –94°F/less than –70°C/less than 203.2°K.
Critical temperature: 620.6°F/327.0°C/600.2°K.
Critical pressure: 375 psia = 25.51 atm = 2.58 MN/m^2.
Specific gravity (water = 1): 0.79 @ 68°F/20°C.
Liquid surface tension: 27.85 dynes/cm = 0.02785 N/m @ 68°F/20°C.
Liquid water interfacial tension: (estimate) 45.15 dynes/cm = 0.04515 N/m @ 68°F/20°C.

Relative vapor density (air = 1): 4.45–4.5
Vapor pressure: 0.20 mm.

ETHYL HEXYL PHTHALATE   REC. E:6550

SYNONYMS: BEHP; DEHP; OCTOIL; DI-*sec*-OCTYL PHTHALATE; BIS(2-ETHYLHEXYL)PHTHALATE; DI(2-ETHYLHEXYL)PHTHALATE; DOP; FTALATO de ETILHEXILO (Spanish); OCTOIL; RCRA No. U028; SICOL-150; STAFLEX DOP; TRUFLEX DOP; WITCIZER 312

IDENTIFICATION
CAS Number: 117-81-7
Formula: $C_{24}H_{38}O_4$; $C_6H_4(COOCH_2CH[C_2NO_5]C_4H_9)_2$
DOT ID Number: UN 3082; DOT Guide Number: 171
Proper Shipping Name: Environmentally hazardous substances, liquid, n.o.s.
Reportable Quantity (RQ): **(CERCLA)** 100 lb/45.4 kg

DESCRIPTION: Colorless to pale yellow oily liquid. Slight odor to odorless. Insoluble in water.

Combustible • Heat or flame may cause explosion • Containers may BLEVE when exposed to fire • Vapors are heavier than air and will collect and stay in low areas • Vapors in confined areas (e.g., tanks, sewers, buildings) may explode when exposed to fire • Irritating to the skin, eyes, and respiratory tract • Toxic products of combustion may include CO_2 and carbon monoxide.

Hazard Classification (based on NFPA-704 Rating System)
Health Hazards (Blue): 2; Flammability (Red): 1; Reactivity (Yellow): 0

EMERGENCY RESPONSE: See Appendix A (171)
Evacuation:
Public safety: Isolate the area of spill or leak for at least 10 to 25 meters (30 to 80 feet) in all directions.
Spill: Increase, in the downwind direction, as necessary, the distance shown under "Public Safety."
Fire: If any large container is involved in fire, isolate for at least 800 meters (½ mile) in all directions; also, consider initial evacuation for 800 meters (½ mile) in all directions.

EXPOSURE
Short-term effects: *SEEK MEDICAL ATTENTION*. **Vapor:** Move victim to fresh air. If breathing is difficult, administer oxygen.
Liquid: Irritating to skin and eyes. Remove contaminated clothing and shoes. Wash affected areas with soap and water. *IF IN EYES*, hold eyelids open and flush with plenty of water.

HEALTH HAZARDS
Personal protective equipment (PPE): B-Level PPE. OSHA Z-1-A air contaminant. Safety goggles or full face shield; approved organic vapor respirator; chemical resistant gloves.
Recommendations for respirator selection: NIOSH
At any concentrations above the NIOSH REL, or where there is no REL, at any detectable concentration: SCBAF:PD,PP (any self-contained breathing apparatus that has a full facepiece and is operated in a pressure-demand or other positive-pressure mode); or SAF:PD,PP:ASCBA (any supplied-air respirator that has a full facepiece and is operated in a pressure-demand or other positive-pressure mode in combination with an auxiliary self-contained breathing apparatus operated in a pressure-demand or other positive

pressure mode). *ESCAPE:* HiEF (any air-purifying, full-facepiece respirator with a high-efficiency particulate filter); or SCBAE (any appropriate escape-type, self-contained breathing apparatus).
Exposure limits (TWA unless otherwise noted): ACGIH TLV 5 mg/m^3; OSHA PEL 5 mg/m3; NIOSH REL 5 mg/m^3, potential carcinogen, reduce to lowest feasible level.
Short-term exposure limits (15-minute TWA): ACGIH STEL 10 mg/m^3; NIOSH STEL 10 mg/m^3.
Toxicity by ingestion: Grade 1; oral rat LD$_{50}$ = 30.6 g/kg; oral-man TDLo-143 mg/kg.
Long-term health effects: Listed as a potential carcinogen based upon increased incidence of liver cancers in female rats and mice; and an increased incidence of liver cancers or neoplasms in male rats. NTP anticipated carcinogen. IARC possible carcinogen, evaluation 2B; sufficient animal evidence.
Vapor (gas) irritant characteristics: Vapors cause smarting of the eyes or respiratory system if present in high concentrations. The effect is temporary.
Liquid or solid irritant characteristics: If spilled on clothing and allowed to remain, may cause smarting and reddening of the skin.
IDLH value: potential carcinogen 5000 mg/m^3.

FIRE DATA
Flash point: 420°F/216°C (oc).
Flammable limits in air: LEL: 0.31% @ 493°F/256°C; 0.28% @ 507°F/264°C.
Fire extinguishing agents not to be used: Water.
Behavior in fire: Overheating of containers during fire can result in rupture.
Stoichiometric air-to-fuel ratio: 150.0

CHEMICAL REACTIVITY
Binary reactants: Nitrates, strong oxidizers (dangerous reaction), acids, and alkalis.
Reactivity group: 34
Compatibility class: Esters.

ENVIRONMENTAL DATA
Water pollution: Effects of low concentrations on aquatic life is unknown. May be dangerous if it enters nearby water intakes; notify operators. Notify local health and wildlife officials.
Response to discharge: Should be removed.

SHIPPING INFORMATION
Grades of purity: 99%; **Storage temperature:** Ambient; **Inert atmosphere:** None; **Venting:** Open; **Stability during transport:** Stable.

PHYSICAL AND CHEMICAL PROPERTIES
Physical state @ 59°F/15°C and 1 atm: Liquid.
Molecular weight: 390.56
Boiling point @ 1 atm: 724°F/384°C/657°K.
Melting/Freezing point: –58°F/–50°C/223°K.
Specific gravity (water = 1): 0.98 @ 77°F/25°C.
Relative vapor density (air = 1): 16

ETHYLIDENE NORBORNENE   REC. E:6600

SYNONYMS: BICYCLO 221 HEPT-2-ENE, 5-ETHYLIDENE-; ENB; 5-ETHYLIDENEBICYCLO(2.2.1)HEPT-2-ENE; 5-ETHYLIDENE-2-NORBORNENE; 5-ETILIDENO-2-NORBORNENO (Spanish); 2-NORBORNENE,5-ETHYLIDENE-; E T H Y L I D E N E N O R C A M P H E N E ; ETHYLIDENENORBORNYLENE; ENB

IDENTIFICATION
CAS Number: 16219-75-3
Formula: C_9H_{12}
DOT ID Number: UN 1993; DOT Guide Number: 128
Proper Shipping Name: Flammable liquid, n.o.s.

DESCRIPTION: Colorless to white liquid. Turpentine-like odor. Floats on the surface of water.

Highly flammable • Reacts with oxygen; forms unstable peroxides; polymerization hazard • Containers may BLEVE when exposed to fire •Vapors may form explosive mixture with air • Vapors are heavier than air and will collect and stay in low areas • Vapors may travel long distances to ignition sources and flashback • Vapors in confined areas (e.g., tanks, sewers, buildings) may explode when exposed to fire • Irritating to the skin, eyes, and respiratory tract • Toxic products of combustion may include carbon monoxide.

Hazard Classification (based on NFPA-704 Rating System)
Health Hazards (Blue): 2; Flammability (Red): 2; Reactivity (Yellow): 0

EMERGENCY RESPONSE: See Appendix A (128)
Evacuation:
Public safety: Isolate spill area for at least 25 to 50 meters (80 to 160 feet) in all directions.
Spill: Large spill–Consider initial downwind evacuation for at least 300 meters (1000 feet).
Fire: Isolate for 800 meters (½ mile) in all directions, especially if tank, rail car, or tank truck is involved in fire.

EXPOSURE
Short-term effects: *SEEK MEDICAL ATTENTION.* **Vapor:** Irritating to eyes, nose, and throat. *IF INHALED*, will, will cause headache, coughing or difficult breathing. *IF IN EYES*, hold eyelids open and flush with plenty of water. *IF BREATHING HAS STOPPED,* give artificial respiration. IF breathing is difficult, administer oxygen. **Liquid:** *POISONOUS IF SWALLOWED.* Irritating to skin and eyes. *IF SWALLOWED*, will cause nausea and vomiting. Remove contaminated clothing and shoes. Flush affected areas with plenty of water. *IF IN EYES*, hold eyelids open and flush with plenty of water. *IF SWALLOWED* and victim is *CONSCIOUS AND ABLE TO SWALLOW*, have victim drink 4 to 8 ounces of water and have victim induce vomiting. *IF SWALLOWED* and victim is *UNCONSCIOUS OR HAVING CONVULSIONS,* do nothing except keep victim warm.

HEALTH HAZARDS
Personal protective equipment (PPE): B-Level PPE. OSHA Table Z-1-A air contaminant. Organic canister; or air–supplied mask; goggles or face shield; rubber gloves.
Exposure limits (TWA unless otherwise noted): NIOSH REL ceiling 5 ppm (25 mg/m^3)
Short-term exposure limits (15-minute TWA): ACGIH TLV ceiling 5 ppm (25 mg/m^3)
Toxicity by ingestion: Grade 2; oral LD$_{50}$ = 2.83 g/kg (rat).
Long-term health effects: Causes kidney lesions and gain in kidney and liver weights in rats
Vapor (gas) irritant characteristics: Eye and respiratory tract irritant. Vapors are irritating such that personnel will not usually tolerate moderate or high concentrations.
Liquid or solid irritant characteristics: Eye irritant. If spilled on clothing and allowed to remain, may cause smarting and reddening of skin.
Odor threshold: 0.007–0.014 ppm.

528   Ethyl lactate

FIRE DATA
Flash point: 101°F/38°C (oc).
Flammable limits in air: LEL: 0.9%; UEL: 6.4%.
Fire extinguishing agents not to be used: Water may be ineffective
Electrical hazard: Class I, Group C. Flow or agitation of substance may generate electrostatic charges due to low conductivity.

CHEMICAL REACTIVITY
Binary reactants: Oxygen. Unstable above 350°F/177°C/450°K Reacts with oxidizers; may cause polymerization.
Polymerization: Able to polymerize; add inhibitor; use *tert*-butyl catechol to inhibit peroxide formation.

ENVIRONMENTAL DATA
Water pollution: Effect of low concentrations on aquatic life is unknown. Fouling to shoreline. May be dangerous if it enters nearby water intakes; notify operators. Notify local health and wildlife officials. **Response to discharge:** Mechanical containment. Should be removed. Chemical and physical treatment.

SHIPPING INFORMATION
Grades of purity: Commercial, 99+%; **Storage temperature:** Ambient; **Inert atmosphere:** Nitrogen atmosphere; reacts with oxygen; **Venting:** Open (flame arrester); **Stability during transport:** Stable.

NAS HAZARD CLASSIFICATION FOR BULK WATER TRANSPORTATION
FIRE: 3
HEALTH: Vapor irritant: 3; Liquid or solid irritant: 1; Poisons: 4
WATER POLLUTION: Human toxicity: 4; Aquatic toxicity: 3; Aesthetic effect: 2
REACTIVITY: Other chemicals: 0; Water: 3; Self-reaction: 1

PHYSICAL AND CHEMICAL PROPERTIES
Physical state @ 59°F/15°C and 1 atm: Liquid.
Molecular weight: 120.2
Boiling point @ 1 atm: 297.7°F/147.6°C/420.8°K.
Melting/Freezing point: −112°F/−80°C/193°K.
Specific gravity (water = 1): 0.896 @ 20°C (liquid).
Relative vapor density (air = 1): 4.1
Heat of combustion: (estimate) −18,000 Btu/lb = −10,450 cal/g = −437 x 10^5 J/kg.
Vapor pressure: (Reid) 0.23 psia; 4.4 mm.

ETHYL LACTATE   **REC. E:6700**

SYNONYMS: ACTYOL; ACYTOL; ETHYL 2-HYDROXYPROPIONATE; ETHYL 2-HYDROXYPROPANOATE; ETHYL A-HYDROXYPROPIONATE; ETHYL DL-LACTATE; LACTATO de ETILO (Spanish); LACTIC ACID, ETHYL ESTER; LACTATE D'ETHYLE (French); SOLACTOL

IDENTIFICATION
CAS Number: 97-64-3
Formula: $C_5H_{10}O_3$; $CH_3CH_2OCOO\ C_2H_5$
DOT ID Number: UN 1192; DOT Guide Number: 129
Proper Shipping Name: Ethyl lactate

DESCRIPTION: Colorless liquid. Mild odor. Mixes with water.

Highly flammable • Containers may BLEVE when exposed to fire • Vapors may form explosive mixture with air • Vapors are heavier than air and will collect and stay in low areas • Vapors may travel long distances to ignition sources and flashback • Vapors in confined areas (e.g., tanks, sewers, buildings) may explode when exposed to fire • Irritating to the skin, eyes, and respiratory tract • Toxic products of combustion may include carbon monoxide.

Hazard Classification (based on NFPA-704 Rating System)
Health Hazards (Blue): 2; Flammability (Red): 2; Reactivity (Yellow): 0

EMERGENCY RESPONSE: See Appendix A (129)
Evacuation:
Public safety: Isolate spill area for at least 50 to 100 meters (160 to 330 feet) in all directions.
Spill: Large spill–Consider initial downwind evacuation for at least 300 meters (1000 feet).
Fire: Isolate for 800 meters (½ mile) in all directions, especially if tank, rail car, or tank truck is involved in fire.

EXPOSURE
Short-term effects: *SEEK MEDICAL ATTENTION*. **Liquid:** Harmful if swallowed. *IF SWALLOWED* and victim is *CONSCIOUS AND ABLE TO SWALLOW*, have victim drink 4 to 8 ounces of water.

HEALTH HAZARDS
Personal protective equipment (PPE): B-Level PPE. Goggles or face shield; rubber gloves.
Toxicity by ingestion: Grade 2; oral LD_{50} = 2580 mg/kg (mouse).
Liquid or solid irritant characteristics: Eye and skin irritant

FIRE DATA
Flash point: 115°F/46°C (cc); 131°F/55°C (cc) (technical).
Flammable limits in air: LEL: 1.5% @ 212°F/100°C; UEL: 11.4%.
Autoignition temperature: 752°F/400°C/673°K.

ENVIRONMENTAL DATA
Food chain concentration potential: Negative; unlikely to accumulate.
Water pollution: Effect of low concentrations on aquatic life is unknown. May be dangerous if it enters nearby water intakes; notify operators. Notify local health and wildlife officials. **Response to discharge:** Issue warning–water contaminant. Disperse and flush.

SHIPPING INFORMATION
Grades of purity: Commercial; **Storage temperature:** Ambient; **Inert atmosphere:** None; **Venting:** Open (flame arrester); **Stability during transport:** Stable.

PHYSICAL AND CHEMICAL PROPERTIES
Physical state @ 59°F/15°C and 1 atm: Liquid.
Molecular weight: 118.1
Boiling point @ 1 atm: 309°F/154°C/427°K.
Specific gravity (water = 1): 1.03 @ 68°F/20°C (liquid).
Liquid surface tension: 29.20 dynes/cm = 0.0292 N/m @ 68°F/20°C.
Relative vapor density (air = 1): 4.08
Heat of combustion: (estimate) −11,600 Btu/lb = −6500 cal/g = 270 x 10^5 J/kg.

ETHYL MERCAPTAN   REC. E:6800

SYNONYMS: EEC No. 016-022-00-9; ETHANETHIOL; ETHYL HYDROSULFIDE; ETHYL SULFHYDRATE; ETHYL THIOALCOHOL; ETILMERCAPTANO (Spanish); LPG ETHYL MERCAPTAN 1010; MERCAPTOETHANE; THIOETHANOL; THIOETHYL ALCOHOL

IDENTIFICATION
CAS Number: 75-08-1
Formula: C_2H_6S; C_2NO_5SH
DOT ID Number: UN 2363; DOT Guide Number: 128
Proper Shipping Name: Ethyl mercaptan

DESCRIPTION: Colorless to yellow liquid. Strong skunk-like odor, or garlic-like odor at low concentrations. Floats and mixes slowly with water; moderately soluble; poisonous, flammable vapor is produced. Boils at 94°F/34°C.

Extremely flammable • Containers may BLEVE when exposed to fire • Vapors may form explosive mixture with air • Vapors are heavier than air and will collect and stay in low areas • Vapors may travel long distances to ignition sources and flashback • Vapors in confined areas (e.g., tanks, sewers, buildings) may explode when exposed to fire • Irritating to the skin, eyes, and respiratory tract • Toxic products of combustion may include carbon monoxide and sulfur oxides. *Warning:* Odor is not a reliable indicator of the presence of toxic amounts of gas.

Hazard Classification (based on NFPA-704 Rating System)
Health Hazards (Blue): 2; Flammability (Red): 4; Reactivity (Yellow): 0

EMERGENCY RESPONSE: See Appendix A (128)
Evacuation:
Public safety: Isolate spill area for at least 25 to 50 meters (80 to 160 feet) in all directions.
Spill: Large spill–Consider initial downwind evacuation for at least 300 meters (1000 feet).
Fire: Isolate for 800 meters (½ mile) in all directions, especially if tank, rail car, or tank truck is involved in fire.

EXPOSURE
Short-term effects: *SEEK MEDICAL ATTENTION.* **Vapor:** *POISONOUS IF INHALED.* Move victim to fresh air. *IF BREATHING HAS STOPPED,* give artificial respiration. IF breathing is difficult, administer oxygen. **Liquid:** *POISONOUS IF SWALLOWED. IF SWALLOWED* and victim is *CONSCIOUS AND ABLE TO SWALLOW,* have victim drink 4 to 8 ounces of water and have victim induce vomiting. *IF SWALLOWED* and victim is *UNCONSCIOUS OR HAVING CONVULSIONS,* do nothing except keep victim warm.

HEALTH HAZARDS
Personal protective equipment (PPE): B-Level PPE. OSHA Table Z-1-A air contaminant. Plastic gloves; goggles or face shield. Natural rubber, neoprene, and nitrile rubber may offer some protection from alcohols.
Recommendations for respirator selection: NIOSH/OSHA
5 ppm: CCRFOV [any chemical cartridge respirator with a full facepiece and organic vapor cartridges(s)]; or SA (any supplied-air respirator). *12.5 ppm*: SA:CF (any supplied-air respirator operated in a continuous-flow mode); or PAPROV [any powered, air-purifying respirator with organic vapor cartridge(s)]. *25 ppm*: CCRFOV [any chemical cartridge respirator with a full facepiece and organic vapor cartridges(s)]; or GMFOV [any air-purifying, full-facepiece respirator (gas mask) with a chin-style, front- or back-mounted acid gas canister]; or SAT:CF (any supplied-air respirator that has a tight-fitting facepiece and is operated in a continuous-flow mode); or PAPRTOV [any powered, air-purifying respirator with a tight-fitting facepiece and organic vapor cartridges(s)]; or SCBAF (any self-contained breathing apparatus with a full facepiece); or SAF (any supplied-air respirator with a full facepiece). *500 ppm:* SA:PD:PP (any supplied-air respirator operated in a pressure-demand or other positive-pressure mode). *EMERGENCY OR PLANNED ENTRY INTO UNKNOWN CONCENTRATIONS OR IDLH CONDITIONS:* SCBAF:PD,PP (any self-contained breathing apparatus that has a full facepiece and is operated in a pressure-demand or other positive-pressure mode); or SAF:PD,PP:ASCBA (any supplied-air respirator that has a full facepiece and is operated in a pressure-demand or other positive-pressure mode in combination with an auxiliary self-contained breathing apparatus operated in a pressure-demand or other positive pressure mode). *ESCAPE:* GMFOV [any air-purifying, full-facepiece respirator (gas mask) with a chin-style, front- or back-mounted organic vapor cannister]; or SCBAE (any appropriate escape-type, self-contained breathing apparatus).
Exposure limits (TWA unless otherwise noted): ACGIH TLV 0.5 ppm (1.3 mg/m^3); OSHA PEL ceiling 10 ppm (25 mg/m^3); NIOSH ceiling 0.5 ppm (1.3 mg/m^3)/15 min.
Toxicity by ingestion: Grade 2; oral LD_{50} = 682 mg/kg (rat).
Long-term health effects: May impair respiratory muscle function in warm-blooded experimental animals. May cause liver or nervous system damage.
Odor threshold: 0.0001–0.003 ppm.
IDLH value: 500 ppm.

FIRE DATA
Flash point: –55°F/–48°C (cc).
Flammable limits in air: LEL: 2.8%; UEL 18.2%.
Fire extinguishing agents not to be used: Water may be ineffective.
Autoignition temperature: 572°F/300°C = 573°K.
Burning rate: 5.7 mm/min

CHEMICAL REACTIVITY
Binary reactants: Strong oxidizers; reacts violently with calcium hypochlorite; acids (forming combustible hydrogen sulfide gas).

ENVIRONMENTAL DATA
Food chain concentration potential: Negative; unlikely to accumulate.
Water pollution: DOT Appendix B, §172.101–marine pollutant. Effect of low concentrations on aquatic life is unknown. May be dangerous if it enters nearby water intakes; notify operators. Notify local health and wildlife officials. **Response to discharge:** Issue warning–high flammability, water contaminant, air contaminant. Restrict access. Evacuate area. Disperse and flush.

SHIPPING INFORMATION
Grades of purity: 98.5+%; **Storage temperature:** Below 30°C; **Inert atmosphere:** None; **Venting:** Pressure-vacuum; **Stability during transport:** Stable.

PHYSICAL AND CHEMICAL PROPERTIES
Physical state @ 59°F/15°C and 1 atm: Liquid.
Molecular weight: 62.1
Boiling point @ 1 atm: 93.9°F/34.4°C/307.6°K.
Melting/Freezing point: –228°F
Critical temperature: 439°F/226°C/499°K.

Critical pressure: 798 psia = 54.2 atm = 5.50 MN/m^2.
Specific gravity (water = 1): 0.84 @ 68°F/20°C (liquid).
Liquid surface tension: 23.5 dynes/cm = 0.0235 N/m @ 68°F/20°C.
Liquid water interfacial tension: 25 dynes/cm = 0.025 N/m @ 68°F/20°C.
Relative vapor density (air = 1): 2.1
Ratio of specific heats of vapor (gas): 1.1308 at 16°C.
Latent heat of vaporization: 189 Btu/lb = 105 cal/g = 4.39 x 10^5 J/kg.
Heat of combustion: −15,000 Btu/lb = −8300 cal/g = −350 x 10^5 J/kg.
Heat of fusion: 19.14 cal/g.
Vapor pressure: 445 mm.

ETHYL METHACRYLATE   REC. E:6900

SYNONYMS: ETHYL 1-2-METHACRYLATE; ETHYL-α-METHYLACRYLATE; 1-2-METHACRYLIC ACID, ETHYL ESTER; 2-METHYLE-2-PROPANOIC ACID, ETHYL ESTER; ETHYL METHYL ACRYLATE; METACRILATO de ETILO (Spanish); 2-PROPANOIC ACID, 1-METHYL-, ETHYL ESTER; RHOPLEX AC-33; ETHYL 2-METHYL-2-PROPENOATE; RCRA No. U118

IDENTIFICATION
CAS Number: 97-63-2
Formula: $C_6H_{10}O_2$; $CH_2 = C(CH_3)COOC_2NO_5$
DOT ID Number: UN 2277; DOT Guide Number: 129P
Proper Shipping Name: Ethyl methacrylate
Reportable Quantity (RQ): **(CERCLA)** 1000 lb/454 kg

DESCRIPTION: Colorless liquid. Sharp unpleasant odor; acrylic. Floats on the surface of water.

Highly flammable • Polymerization hazard • Heat can induce polymerization with rapid release of energy; sealed containers may rupture explosively • May react with itself blocking relief valves; leading to tank explosions • Containers may BLEVE when exposed to fire •Vapors may form explosive mixture with air • Vapors are heavier than air and will collect and stay in low areas • Vapors may travel long distances to ignition sources and flashback • Vapors in confined areas (e.g., tanks, sewers, buildings) may explode when exposed to fire • Irritating to the skin, eyes, and respiratory tract • Toxic products of combustion may include smoke and irritating vapors.

Hazard Classification (based on NFPA-704 Rating System)
Health Hazards (Blue): 2; Flammability (Red): 3; Reactivity (Yellow): 0

EMERGENCY RESPONSE: See Appendix A (129)
Evacuation:
Public safety: Isolate spill area for at least 50 to 100 meters (160 to 330 feet) in all directions.
Spill: Large spill–Consider initial downwind evacuation for at least 300 meters (1000 feet).
Fire: Isolate for 800 meters (½ mile) in all directions, especially if tank, rail car, or tank truck is involved in fire.

EXPOSURE
Short-term effects: *SEEK MEDICAL ATTENTION.* **Vapor:** Irritating to eyes, nose, and throat. *IF INHALED*, will, will cause coughing or difficult breathing. *IF IN EYES*, hold eyelids open and flush with plenty of water. *IF BREATHING HAS STOPPED*, give artificial respiration. IF breathing is difficult, administer oxygen. **Liquid:** Irritating to skin and eyes. *IF SWALLOWED*, will cause nausea and vomiting. Remove contaminated clothing and shoes. Flush affected areas with plenty of water. *IF IN EYES*, hold eyelids open and flush with plenty of water. *IF SWALLOWED* and victim is *CONSCIOUS AND ABLE TO SWALLOW*, have victim drink 4 to 8 ounces of water and have victim induce vomiting. *IF SWALLOWED* and victim is *UNCONSCIOUS OR HAVING CONVULSIONS*, do nothing except keep victim warm.

HEALTH HAZARDS
Personal protective equipment (PPE): B-Level PPE. Impervious gloves; splash goggles; self-contained breathing apparatus if exposed to vapors; coveralls. Chemical protective material(s) reported to have good to excellent resistance: butyl rubber, PV alcohol.
Toxicity by ingestion: Grade 2; oral LD_{50} = 4 g/kg (rabbit).
Long-term health effects: Causes birth defects in experimental animals
Vapor (gas) irritant characteristics: Eye and respiratory tract irritant.
Liquid or solid irritant characteristics: Eye and skin irritant.

FIRE DATA
Flash point: 68°F/20°C (oc); 80°F/27°C (cc).
Flammable limits in air: LEL: 1.8% to saturation.
Fire extinguishing agents not to be used: Water may be ineffective
Behavior in fire: Sealed containers may rupture explosively if hot. Heat can cause violent polymerization; rapid release of energy.
Autoignition temperature: 734°F/390°C/663°K.
Electrical hazard: Class I, Group D. Ground storage tanks or drums to prevent accumulation of static electricity.
Burning rate: 4.56 mm/min

CHEMICAL REACTIVITY
Binary reactants: Easily forms explosive mixture with air in the presence of heat or flame. Incompatible with oxidizers, strong acids, amines. Corrosive to mild steel.
Polymerization: If proper level of inhibitor is not present; or, when material is hot, violent polymerization may occur.
Reactivity group: 14
Compatibility class: Acrylates

ENVIRONMENTAL DATA
Food chain concentration potential: Log P_{ow} = 1.9. Unlikely to accumulate.
Water pollution: Effect of low concentration on aquatic life is unknown. Fouling to shoreline. May be dangerous if it enters nearby water intakes; notify operators. Notify local health and wildlife officials. **Response to discharge:** Issue warning–high flammability. Restrict access. Mechanical containment. Should be removed. Chemical and physical treatment.

SHIPPING INFORMATION
Grades of purity: Technical; **Storage temperature:** Below 38°C (100°F); **Inert atmosphere:** Ventilated (natural); **Venting:** Open (flame arrester); **Stability during transport:** Stable.

PHYSICAL AND CHEMICAL PROPERTIES
Physical state @ 59°F/15°C and 1 atm: Liquid.
Molecular weight: 114
Boiling point @ 1 atm: 243°F/117°C/390°K.

Melting/Freezing point: Less than –58°F/less than –50°C/less than 223°K.
Specific gravity (water = 1): 0.9151 @ 68°F/20°C (liquid).
Relative vapor density (air = 1): 3.9
Ratio of specific heats of vapor (gas): 1.064
Latent heat of vaporization: 170 Btu/lb = 96 cal/g = 4.0 x 10^5 J/kg.
Heat of combustion: –12,670 Btu/lb = –7,040 cal/g = –294 x 10^5 J/kg.
Heat of polymerization: –218 Btu/lb = –121 cal/g = –5.06 x 10^5 J/kg.
Vapor pressure: (Reid) 0.77 psia.

N-ETHYLMORPHOLINE   REC. E:7000

SYNONYMS: 4-ETHYLMORPHOLINE; 4-ETILMORFOLINA (Spanish)

IDENTIFICATION
CAS Number: 100-74-3
Formula: $C_6H_{13}NO$; $C_4H_8ON(C_2NO_5)$
DOT ID Number: UN 1993; DOT Guide Number: 128
Proper Shipping Name: Flammable Liquid.

DESCRIPTION: Colorless to slight yellow liquid. Ammonia-like odor. Soluble in water.

Highly flammable • Containers may BLEVE when exposed to fire • Vapors may form explosive mixture with air • Vapors are heavier than air and will collect and stay in low areas • Vapors may travel long distances to ignition sources and flashback • Vapors in confined areas (e.g., tanks, sewers, buildings) may explode when exposed to fire • Irritating to the skin, eyes, and respiratory tract • Toxic products of combustion may include nitrogen oxides.

Hazard Classification (based on NFPA-704 Rating System)
Health Hazards (Blue): 2; Flammability (Red): 3; Reactivity (Yellow): 0

EMERGENCY RESPONSE: See Appendix A (128)
Evacuation:
Public safety: Isolate spill area for at least 25 to 50 meters (80 to 160 feet) in all directions.
Spill: Large spill–Consider initial downwind evacuation for at least 300 meters (1000 feet).
Fire: Isolate for 800 meters (½ mile) in all directions, especially if tank, rail car, or tank truck is involved in fire.

EXPOSURE
Short-term effects: *SEEK MEDICAL ATTENTION. ABSORBED THROUGH THE SKIN.* **Vapor:** Move victim to fresh air. *IF BREATHING HAS STOPPED,* give artificial respiration. If breathing is difficult, administer oxygen. **Liquid:** Irritating to skin and eyes. Remove contaminated clothing and shoes. Wash affected areas with soap and water. *IF IN EYES,* hold eyelids open and flush with plenty of water. If swallowed and victim is *CONSCIOUS AND ABLE TO SWALLOW,* give large quantities of water and induce vomiting.

HEALTH HAZARDS
Personal protective equipment (PPE): B-Level PPE. Full face organic vapor respirator. OSHA Table Z-1-A air contaminant. Impervious protective clothing and gloves.

Recommendations for respirator selection: NIOSH/OSHA
50 ppm: CCRFOV* [any chemical cartridge respirator with a full facepiece and organic vapor cartridges(s)]; or SA* (any supplied-air respirator). *100 ppm:* SA:CF* (any supplied-air respirator operated in a continuous-flow mode); or PAPROV* [any powered, air-purifying respirator with organic vapor cartridge(s)]; or CCRFOV [any chemical cartridge respirator with a full facepiece and organic vapor cartridges(s)]; or GMFOV [any air-purifying, full-facepiece respirator (gas mask) with a chin-style, front- or back-mounted acid gas canister]; or SCBAF (any self-contained breathing apparatus with a full facepiece); or SAF (any supplied-air respirator with a full facepiece). *EMERGENCY OR PLANNED ENTRY INTO UNKNOWN CONCENTRATIONS OR IDLH CODITIONS:* SCBAF:PD,PP (any self-contained breathing apparatus that has a full facepiece and is operated in a pressure-demand or other positive-pressure mode); or SAF:PD,PP:ASCBA (any supplied-air respirator that has a full facepiece and is operated in a pressure-demand or other positive-pressure mode in combination with an auxiliary self-contained breathing apparatus operated in a pressure-demand or other positive pressure mode). *ESCAPE:* GMFOV[any air-purifying, full-facepiece respirator (gas mask) with a chin-style, front- or back-mounted organic vapor cannister]; or SCBAE (any appropriate escape-type, self-contained breathing apparatus). **Note:* Substance reported to cause eye irritation or damage; may require eye protection.
Exposure limits (TWA unless otherwise noted): ACGIH TLV 5 ppm (24 mg/m^3); OSHA PEL 20 ppm (94 mg/m^3); NIOSH REL 5 ppm (23 mg/m^3); skin contact contributes significantly in overall exposure.
Toxicity by ingestion: Grade 2; oral rat LD_{50} = 1.78 g/kg.
Long-term health effects: Vapor causes visual disturbances and irritates mucous membranes. There is a possibility that eye damage could be permanent.
Vapor (gas) irritant characteristics: Vapors cause smarting of the eyes or respiratory system if present in high concentrations. The effect may be temporary.
Liquid or solid irritant characteristics: If spilled on clothing and allowed to remain, may cause smarting and reddening of the skin.
Odor threshold: 0.0277 ppm.
IDLH value: 100 ppm.

FIRE DATA
Flash point: 90°F/32°C (oc).
Flammable limits in air: LEL: 1.0%; UEL: 9.8%.
Fire extinguishing agents not to be used: Water.
Electrical hazard: Will attack some insulators.
Stoichiometric air-to-fuel ratio: 41.7

CHEMICAL REACTIVITY
Binary reactants: Oxidizing materials can cause a vigorous reaction; strong acids. Attacks some plastics.

ENVIRONMENTAL DATA
Water pollution: Effects of low concentrations on aquatic life is unknown. May be dangerous if it enters nearby water intakes; notify operators. Notify local health and wildlife officials. **Response to discharge:** Should be removed.

SHIPPING INFORMATION
Grades of purity: 99%; technical; **Storage temperature:** Ambient; **Inert atmosphere:** None; **Stability during transport:** Stable.

PHYSICAL AND CHEMICAL PROPERTIES
Physical state @ 59°F/15°C and 1 atm: Liquid.
Molecular weight: 115.2

Boiling point @ 1 atm: 281°F/138.6°C/411.6°K.
Melting/Freezing point: −81°F/−63°C/210°K.
Specific gravity (water = 1): 0.916 @ 68°F/20°C.
Relative vapor density (air = 1): 4.0
Vapor pressure: 6.1 mm.

ETHYL NITRITE REC. E:7100

SYNONYMS: ETHYL NITRITE SOLUTION; NITRITO de ETILO (Spanish); NITROUS ACID ETHYL ESTER; NITROUS ETHER; NITROUS ETHYL ETHER; NITROSYL ETHOXIDE; SWEET SPIRIT OF NITRE; SPIRIT OF ETHER NITRITE

IDENTIFICATION
CAS Number: 109-95-5
Formula: $C_2H_5NO_2$; C_2H_5ONO
DOT ID Number: UN 1194; DOT Guide Number: 131
Proper Shipping Name: Ethyl nitrite solutions.

DESCRIPTION: Colorless gas above 63°F/17°C; colorless to light yellow liquid below this temperature. May be shipped and stored in ethyl alcohol. Pleasant, rum- or ether-like odor. Floats on water surface; slightly soluble; may boil on water.

Extremely flammable • Decomposes explosively at 194°F/90°C; Decomposition does not require air or oxygen • Poison! • Corrosive • Breathing the gas can kill you • Firefighting gear (including SCBA) does not provide adequate protection. If exposure occurs, remove and isolate gear immediately and thoroughly decontaminate personnel • Containers may BLEVE when exposed to fire • Vapors may form explosive mixture with air • Vapors are heavier than air and will collect and stay in low areas • Vapors may travel long distances to ignition sources and flashback • Vapors in confined areas (e.g., tanks, sewers, buildings) may explode when exposed to fire • Toxic products of combustion may include nitrogen oxides • Do not put yourself in danger by entering a contaminated area to rescue a victim.

Hazard Classification (based on NFPA-704 Rating System)
Health Hazards (Blue): 3; Flammability (Red): 4; Reactivity (Yellow): 4

EMERGENCY RESPONSE: See Appendix A (131)
Evacuation:
Public safety: Isolate spill area for at least 100 to 200 meters (330 to 660 feet) in all directions.
Spill: Increase, as necessary, the isolation distance shown above, in "Public safety."
Fire: Isolate for 800 meters (½ mile) in all directions, especially if tank, rail car, or tank truck is involved in fire.

EXPOSURE
Short-term effects: *SEEK MEDICAL ATTENTION*. **Vapor:** *IF INHALED*, will cause headache, dizziness, or loss of consciousness. Move victim to fresh air. *IF BREATHING HAS STOPPED*, give artificial respiration. If breathing is difficult, administer oxygen. **Liquid:** *IF SWALLOWED*, will cause headache, or loss of consciousness. *IF SWALLOWED* and victim is *CONSCIOUS AND ABLE TO SWALLOW*, have victim drink 4 to 8 ounces of water.
Note to physician or authorized medical personnel: Medical observation is recommended for 24 to 48 hours after breathing overexposure, as pulmonary edema may be delayed. As first aid for pulmonary edema, consider administering a corticosteroid spray. Cigarette smoking may exacerbate pulmonary injury and should be discouraged for at least 72 hours following exposure.

HEALTH HAZARDS
Personal protective equipment (PPE): A-Level PPE. Self-contained breathing apparatus; goggles or face shield; rubber gloves. **Liquid or solid irritant characteristics:** Eye and skin irritant.

FIRE DATA
Flash point: −31°F/−35°C (cc).
Flammable limits in air: LEL: 3%; UEL: > 50%.
Behavior in fire: Air or oxygen not required for decomposition. Containers may explode in a fire.
Autoignition temperature: 194°F/90°C/363°K explosive decomposition.
Electrical hazard: Class I, Group C.
Burning rate: 2.6 mm/min
Stoichiometric air-to-fuel ratio: 4.113 (estimate).

CHEMICAL REACTIVITY
Binary reactants: violent reaction with oxidizers.

ENVIRONMENTAL DATA
Food chain concentration potential: Negative; unlikely to accumulate.
Water pollution: Effect of low concentrations on aquatic life is unknown. May be dangerous if it enters nearby water intakes; notify operators. Notify local health and wildlife officials. **Response to discharge:** Issue warning–high flammability. Restrict access. Evacuate area. Disperse and flush.

SHIPPING INFORMATION
Grades of purity: Often shipped as a 85-92% (by volume) solution in ethyl alcohol; **Storage temperature:** Cool ambient; **Inert atmosphere:** None; **Venting:** Safety relief; **Stability during transport:** Stable if stored in a cool place and not exposed to strong light.

NAS HAZARD CLASSIFICATION FOR BULK WATER TRANSPORTATION
FIRE: 4
HEALTH: Vapor irritant:–; Liquid or solid irritant:–; Poisons: 3
WATER POLLUTION: Human toxicity:–; Aquatic toxicity: 1; Aesthetic effect:–
REACTIVITY: Other chemicals:–; Water: 0; Self-reaction: 4

PHYSICAL AND CHEMICAL PROPERTIES
Physical state @ 59°F/15°C and 1 atm: Liquid.
Molecular weight: 75.1
Boiling point @ 1 atm: 63°F/17°C/290°K.
Melting/Freezing point: −58°F/−50°C/223°K.
Specific gravity (water = 1): 0.900 @ 15°C (liquid).
Liquid surface tension: (estimate) 30 dynes/cm = 0.030 N/m @ 68°F/20°C.
Liquid water interfacial tension: (estimate) 35 dynes/cm = 0.035 N/m @ 68°F/20°C.
Relative vapor density (air = 1): 2.6
Latent heat of vaporization: 229 Btu/lb = 127 cal/g = 5.32×10^5 J/kg.
Heat of combustion: (estimate) −7800 Btu/lb = −4300 cal/g = $−180 \times 10^5$ J/kg.
Vapor pressure: Negligible.

ETHYLPHENOL E:7200

SYNONYMS: 2-ETHYLPHENOL; o-ETHYLPHENOL; ETILFENOL (Spanish); PHLOROL; PHENOL, o-ETHYL

IDENTIFICATION
CAS Number: 90-00-6 (o-isomer); 123-07-9 (p-isomer)
Formula: $C_8H_{10}O$; $C_2H_5C_6H_4OH$
DOT ID Number: UN 2821; DOT Guide Number: 153
Proper Shipping Name: Phenol solutions.

DESCRIPTION: Yellow liquid. Phenol-like odor. Insoluble in water.

Combustible • Poison! • Heat or flame may cause explosion • Containers may BLEVE when exposed to fire • Vapors are heavier than air and will collect and stay in low areas • Vapors in confined areas (e.g., tanks, sewers, buildings) may explode when exposed to fire • Irritating to the skin, eyes, and respiratory tract • Toxic products of combustion may include carbon monoxide.

Hazard Classification (based on NFPA-704 Rating System) (p-isomer).
Health Hazards (Blue): 2; Flammability (Red): 1; Reactivity (Yellow): 0

EMERGENCY RESPONSE: See Appendix A (153)
Evacuation:
Public safety: Isolate the area of spill or leak for at least 25 to 50 meters (80 to 160 feet) in all directions.
Spill: Increase, in the downwind direction, as necessary, the distance shown under "Public Safety."
Fire: If tank, rail car, or tank truck is involved in fire, isolate for at least 800 meters (½ mile) in all directions; also, consider initial evacuation for 800 meters (½ mile) in all directions.

EXPOSURE
Short-term effects: *SEEK MEDICAL ATTENTION*. **Vapor:** Move to fresh air. *IF BREATHING HAS STOPPED*, give artificial respiration; *avoid mouth-to-mouth resuscitation; use bag/mask apparatus*. IF breathing is difficult, administer oxygen. **Liquid or solid:** *POISONOUS IF SWALLOWED*. Will burn skin and eyes. Remove contaminated clothing and shoes. Flush affected areas with plenty of water. *IF IN EYES*, hold eyelids open and flush with plenty of water.

HEALTH HAZARDS
Personal protective equipment (PPE): B-Level PPE. Wear self-contained breathing apparatus, rubber boots and heavy rubber gloves.
Toxicity by ingestion: Grade 2: LD_{50} = 0.6 g/kg (mouse).
Long-term health effects: Possible carcinogen.
Vapor (gas) irritant characteristics: Vapors cause severe irritation of eyes and throat and can cause eye and lung injury. They cannot be tolerated even at low concentrations.
Liquid or solid irritant characteristics: Severe skin irritant. Causes second- and third-degree burns on short contact and is very injurious to the eyes.

FIRE DATA
Flash point: 219°F/104°C (cc) (p-isomer).

CHEMICAL REACTIVITY
Neutralizing agents for acids and caustics: Dry lime or soda ash.

Reactivity group: 21
Compatibility class: Phenols, cresols

ENVIRONMENTAL DATA
Water pollution: Effect of low concentration on aquatic life is unknown. May be dangerous if it enters nearby water intakes; notify operators. Notify local health and wildlife officials. **Response to discharge:** Issue warning–poison. Restrict access. Should be removed. Chemical and physical treatment.

SHIPPING INFORMATION
Grades of purity: 99%; **Storage temperature:** Ambient

PHYSICAL AND CHEMICAL PROPERTIES
Physical state @ 59°F/15°C and 1 atm: Liquid.
Molecular weight: 122.17
Boiling point @ 1 atm: 383-387°F/195-197°C/468-470°K (o-isomer)
Melting/Freezing point: –0.4°F/–18°C/255°K (o-isomer)
Specific gravity (water = 1): 1.037
Relative vapor density (air = 1): 4.21
Vapor pressure: (Reid) Less than 0.01 psia.

ETHYLPHENYL DICHLOROSILANE REC. E:7300

SYNONYMS: DICHLOROETHYLPHENYLSILANE; ETILFENILDICLOROSILANO (Spanish); SILANE, DICHLOROETHYLPHENYL-; PHENYLETHYLDICHLOROSILANE

IDENTIFICATION
CAS Number: 1125-27-5
Formula: $C_8H_{10}Cl_2Si$; $(C_2NO_5)(C_6NO_5)SiCl_2$
DOT ID Number: UN 2435; DOT Guide Number: 156
Proper Shipping Name: Ethylphenyldichlorosilane

DESCRIPTION: Colorless liquid. Sharp irritating odor; like hydrochloric acid. Reacts violently with water; forming toxic hydrogen chloride.

Poison! • Highly flammable • Breathing the vapor can kill you! Skin or eye contact can cause severe burns and blindness • Firefighting gear (including SCBA) does not provide adequate protection. If exposure occurs, remove and isolate gear immediately and thoroughly decontaminate personnel • Containers may BLEVE when exposed to fire • Vapors may form explosive mixture with air • Vapors are heavier than air and will collect and stay in low areas • Vapors may travel long distances to ignition sources and flashback • Vapors in confined areas (e.g., tanks, sewers, buildings) may explode when exposed to fire • Toxic products of combustion may include hydrogen chloride and phosgene • Do not put yourself in danger by entering a contaminated area to rescue a victim:

Hazard Classification (based on NFPA-704 Rating System)
Health Hazards (Blue): 3; Flammability (Red): 2; Reactivity (Yellow): 2; Special Notice (White): Water reactive

EMERGENCY RESPONSE: See Appendix A (156)
Evacuation:
Public safety: Isolate the area of spill or leak for at least 50 to 100 meters (160 to 330 feet) in all directions.
Spill: Increase, in the downwind direction, as necessary, the distance shown under "Public Safety."

Fire: If tank, rail car, or tank truck is involved in fire, isolate for at least 800 meters (½ mile) in all directions; also, consider initial evacuation for 800 meters (½ mile) in all directions.

EXPOSURE
Short-term effects: *SEEK MEDICAL ATTENTION.* **Gas produced in reaction with water:** *POISONOUS IF INHALED.* Irritating to eyes, nose, and throat. Move to fresh air. *IF BREATHING HAS STOPPED,* give artificial respiration; *avoid mouth-to-mouth resuscitation; use bag/mask apparatus.* IF breathing is difficult, administer oxygen. **Liquid:** Will burn skin and eyes. Harmful if swallowed. Remove contaminated clothing and shoes. Flush affected areas with plenty of water. *IF IN EYES,* hold eyelids open and flush with plenty of water. *IF SWALLOWED* and victim is *CONSCIOUS AND ABLE TO SWALLOW,* have victim drink 4 to 8 ounces of water.

HEALTH HAZARDS
Personal protective equipment (PPE): B-Level PPE. Acid-vapor-type respiratory protection; rubber gloves; chemical worker's goggles; other equipment as necessary to protect skin and eyes.
Toxicity by ingestion: Grade 3; LD_{50} = 50 to 500 mg/kg.
Vapor (gas) irritant characteristics: Vapors cause irritation such that personnel will find high concentrations unpleasant. The effect is temporary.
Liquid or solid irritant characteristics: Severe skin irritant. Causes second- and third-degree burns on short contact and is very injurious to the eyes.

FIRE DATA
Flash point: More than 150°F/66°C (oc).
Fire extinguishing agents not to be used: Water, foam applied to adjacent fires will generate hydrogen chloride gas.
Behavior in fire: Difficult to extinguish; re-ignition may occur.
Burning rate: 3.7 mm/min

CHEMICAL REACTIVITY
Reactivity with water: Reacts with water to generate hydrogen chloride (hydrochloric acid).
Binary reactants: Will react with surface moisture to evolve hydrogen chloride, which is corrosive to common metals.
Neutralizing agents for acids and caustics: Flush with water, rinse with sodium bicarbonate or lime solution.

ENVIRONMENTAL DATA
Food chain concentration potential: Negative; unlikely to accumulate.
Water pollution: Effect of low concentration on aquatic life is unknown. May be dangerous if it enters nearby water intakes; notify operators. Notify local health and wildlife officials.
Response to discharge: Issue warning–corrosive, water contaminant. Restrict access. Disperse and flush with care

SHIPPING INFORMATION
Grades of purity: Commercial; **Storage temperature:** Ambient; **Inert atmosphere:** None; **Venting:** Pressure-vacuum; **Stability during transport:** Stable.

NAS HAZARD CLASSIFICATION FOR BULK WATER TRANSPORTATION
FIRE: 2
HEALTH: Vapor irritant: 2; Liquid or solid irritant: 4; Poisons: 3
WATER POLLUTION: Human toxicity: 3; Aquatic toxicity: 3; Aesthetic effect: 3
REACTIVITY: Other chemicals: 3; Water: 4; Self-reaction: 1

PHYSICAL AND CHEMICAL PROPERTIES
Physical state @ 59°F/15°C and 1 atm: Liquid.
Molecular weight: 205.1
Boiling point @ 1 atm: More than 300°F/149°C/422°K.
Specific gravity (water = 1): 1.159 @ 15°C (liquid).
Liquid surface tension: (estimate) 25 dynes/cm = 0.025 N/m @ 68°F/20°C.
Latent heat of vaporization: 103 Btu/lb = 57 cal/g = 2.4×10^5 J/kg.
Heat of combustion: (estimate) –9900 Btu/lb = –5500 cal/g = –230 $\times 10^5$ J/kg.

ETHYL PHOSPHONOTHIOIC DICHLORIDE REC. E:7400

SYNONYMS: ETHYL THIONOPHOSPHORYL DICHLORIDE; ETHYL PHOSPHORODICHLORIDOTHIONATE

IDENTIFICATION
CAS Number: 993-43-1
Formula: $C_2H_5Cl_2PS$; $CH_3CH_2PSCl_2$
DOT ID Number: UN 2927; DOT Guide Number: 154
Proper Shipping Name: Ethyl phosphonothioic dichloride, anhydrous

DESCRIPTION: Colorless liquid. Choking odor. Reacts with water; poisonous gas is produced on contact with water.

Poison! • Corrosive to the skin, eyes, and respiratory tract; contact with skin or eyes can cause burns and vision impairment; inhalation symptoms may be delayed • Firefighting gear (including SCBA) does not provide adequate protection. If exposure occurs, remove and isolate gear immediately and thoroughly decontaminate personnel • Combustible • Containers may BLEVE when exposed to fire • Toxic products of combustion may include oxides of sulfur and phosphorus; hydrogen chloride and phosgene • Do not put yourself in danger by entering a contaminated area to rescue a victim.

Hazard Classification (based on NFPA-704 Rating System)
Health Hazards (Blue): 3; Flammability (Red): 1; Reactivity (Yellow): 1; Special Notice (White): Water reactive

EMERGENCY RESPONSE: See Appendix A (154)
Evacuation:
Public safety: Isolate the area of spill or leak for at least 25 to 50 meters (80 to 160 feet) in all directions.
Spill: Small spill–First: Isolate in all directions 30 meters (100 feet); Then: Protect persons downwind, DAY: 0.3 km (0.2 mile); NIGHT: 0.2 km (0.1 mile). Large spill–First: Isolate in all directions 30 meters (100 feet); Then: Protect persons downwind, DAY: 0.2 km (0.1 mile); NIGHT: 0.3 km (0.2 mile).
Fire: If tank, rail car, or tank truck is involved in fire, isolate for at least 800 meters (½ mile) in all directions; also, consider initial evacuation for 800 meters (½ mile) in all directions.

EXPOSURE
Short-term effects: *SEEK MEDICAL ATTENTION.* **Gas produced in reaction with water:** *POISONOUS IF INHALED.* Irritating to eyes, nose, and throat. Move to fresh air. *IF BREATHING HAS STOPPED,* give artificial respiration; *avoid mouth-to-mouth resuscitation; use bag/mask apparatus.* IF breathing is difficult, administer oxygen. **Liquid:** Irritating to skin and eyes. Harmful if swallowed. Remove contaminated clothing and shoes. Flush affected areas with plenty of water. *IF IN EYES,* hold eyelids open

and flush with plenty of water. *IF SWALLOWED* and victim is *CONSCIOUS AND ABLE TO SWALLOW*, have victim drink 4 to 8 ounces of water.

Note to physician or authorized medical personnel: Medical observation is recommended for 24 to 48 hours after breathing overexposure, as pulmonary edema may be delayed. As first aid for pulmonary edema, consider administering a corticosteroid spray. Cigarette smoking may exacerbate pulmonary injury and should be discouraged for at least 72 hours following exposure.

HEALTH HAZARDS
Personal protective equipment (PPE): B-Level PPE. Air mask; rubber or neoprene gloves; vapor-tight goggles.
Toxicity by ingestion: Grade 2; LD_{50} = 0.5 to 5 g/kg.
Vapor (gas) irritant characteristics: Eye and respiratory tract irritant.
Liquid or solid irritant characteristics: Eye and skin irritant.

FIRE DATA
Flash point: 203°F/95°C (oc).
Fire extinguishing agents not to be used: Water or foam
Behavior in fire: Contact with water applied to adjacent fires will produce irritating vapors of hydrogen chloride.

CHEMICAL REACTIVITY
Reactivity with water: Reacts with water to evolve hydrogen chloride (hydrochloric acid).
Binary reactants: Will react with surface moisture to evolve hydrogen chloride, which is corrosive to common metals.
Neutralizing agents for acids and caustics: Flood with water, rinse with sodium bicarbonate or lime solution.

ENVIRONMENTAL DATA
Food chain concentration potential: Negative; unlikely to accumulate.
Water pollution: Effect of low concentrations on aquatic life is unknown. May be dangerous if it enters nearby water intakes; notify operators. Notify local health and wildlife officials.
Response to discharge: Issue warning–corrosive, water contaminant. Restrict access. Disperse and flush with care.

SHIPPING INFORMATION
Grades of purity: Commercial; **Storage temperature:** Ambient; **Inert atmosphere:** Inerted with dry nitrogen; **Venting:** Pressure-vacuum; **Stability during transport:** Stable.

PHYSICAL AND CHEMICAL PROPERTIES
Physical state @ 59°F/15°C and 1 atm: Liquid.
Molecular weight: 163
Boiling point @ 1 atm: 342°F/172°C/445°K.
Melting/Freezing point: Less than –58°F/–50°C/223°K.
Specific gravity (water = 1): 1.35 @ 68°F/20°C (liquid).
Liquid surface tension: (estimate) 28 dynes/cm = 0.028 N/m @ 68°F/20°C.
Heat of combustion: –7700 Btu/lb = –4,280 cal/g = –179 x 10^5 J/kg.

ETHYL PHOSPHORODICHLORIDATE   REC. E:7500

SYNONYMS: DICHLOROPHOSPHORIC ACID, ETHYL ESTER; ETHYL DICHLOROPHOSPHATE; PHOSPHORODICHLORIDIC ACID, ETHYL ESTER

IDENTIFICATION
CAS Number: 1498-51-7
Formula: $C_2H_5Cl_2O_2P$; $Cl_2(OC_2NO_5)PO$
DOT ID Number: UN 2927; DOT Guide Number: 154
Proper Shipping Name: Ethyl phosphorodichloridate

DESCRIPTION: Colorless liquid. Choking odor. Reacts with water. Irritating gas is produced on contact with water.

Poison! • Corrosive to the skin, eyes, and respiratory tract; contact with skin or eyes can cause burns and vision impairment; inhalation symptoms may be delayed • Firefighting gear (including SCBA) does not provide adequate protection. If exposure occurs, remove and isolate gear immediately and thoroughly decontaminate personnel • Containers may BLEVE when exposed to fire • Toxic products of combustion may include phosphorus oxide and chlorine • Do not put yourself in danger by entering a contaminated area to rescue a victim.

Hazard Classification (based on NFPA-704 Rating System)
Health Hazards (Blue): 3; Flammability (Red): 1; Reactivity (Yellow): 1

EMERGENCY RESPONSE: See Appendix A (154)
Evacuation:
Public safety: Isolate the area of spill or leak for at least 25 to 50 meters (80 to 160 feet) in all directions.
Spill: Small spill–First: Isolate in all directions 30 meters (100 feet); Then: Protect persons downwind, DAY: 0.3 km (0.2 mile); NIGHT: 0.2 km (0.1 mile). Large spill–First: Isolate in all directions 30 meters (100 feet); Then: Protect persons downwind, DAY: 0.2 km (0.1 mile); NIGHT: 0.3 km (0.2 mile).
Fire: If tank, rail car, or tank truck is involved in fire, isolate for at least 800 meters (½ mile) in all directions; also, consider initial evacuation for 800 meters (½ mile) in all directions.

EXPOSURE
Short-term effects: *SEEK MEDICAL ATTENTION*. **Vapor:** Irritating to eyes, nose, and throat. Harmful if inhaled. Move victim to fresh air. *IF BREATHING HAS STOPPED*, give artificial respiration; *avoid mouth-to-mouth resuscitation; use bag/mask apparatus*. If breathing is difficult, administer oxygen. **Liquid:** Will burn skin and eyes. Harmful if swallowed. Remove contaminated clothing and shoes. Flush affected areas with plenty of water. *IF IN EYES*, hold eyelids open and flush with plenty of water. *IF SWALLOWED* and victim is *CONSCIOUS AND ABLE TO SWALLOW*, have victim drink 4 to 8 ounces of water. **Do NOT induce vomiting.**

Note to physician or authorized medical personnel: Medical observation is recommended for 24 to 48 hours after breathing overexposure, as pulmonary edema may be delayed. As first aid for pulmonary edema, consider administering a corticosteroid spray. Cigarette smoking may exacerbate pulmonary injury and should be discouraged for at least 72 hours following exposure.

HEALTH HAZARDS
Personal protective equipment (PPE): B-Level PPE. Goggles and face shield; self-contained or air-line respirator; rubber gloves, boots, and clothing.
Vapor (gas) irritant characteristics: Eye and respiratory tract irritant.
Liquid or solid irritant characteristics: Severe eye and skin irritant.

FIRE DATA
Fire extinguishing agents not to be used: Water or foam applied to adjacent fires may produce vapors of hydrogen chloride.

CHEMICAL REACTIVITY
Reactivity with water: Reacts with water to evolve hydrogen chloride (hydrochloric acid).
Binary reactants: Will react with surface moisture to evolve hydrogen chloride, which is corrosive to common metals.
Neutralizing agents for acids and caustics: Flood with water, rinse with sodium bicarbonate or lime solution.

ENVIRONMENTAL DATA
Food chain concentration potential: Negative; unlikely to accumulate.
Water pollution: Effect of low concentrations on aquatic life is unknown. May be dangerous if it enters nearby water intakes; notify operators. Notify local health and wildlife officials.
Response to discharge: Issue warning–corrosive, water. contaminant. Restrict access. Disperse and flush with care.

SHIPPING INFORMATION
Grades of purity: 97%; **Storage temperature:** Ambient; **Inert atmosphere:** None; **Venting:** Pressure-vacuum; **Stability during transport:** Stable.

PHYSICAL AND CHEMICAL PROPERTIES
Physical state @ 59°F/15°C and 1 atm: Liquid.
Molecular weight: 162.9
Boiling point @ 1 atm: 333°F/167°C/440°K.
Specific gravity (water = 1): 1.35 @ 66°F/19°C/292°K (liquid).
Liquid surface tension: (estimate) 32.8 dynes/cm = 0.0328 N/m @ 68°F/20°C.
Heat of combustion: (estimate) –4700 Btu/lb = –2600 cal/g = –110 x 10^5 J/kg.

ETHYL PROPIONATE   REC. E:7600

SYNONYMS: ETHYL PROPANOATE; EEC No. 607-028-00-8; PROPANOIC ACID, ETHYL ESTER; PROPIONATO de ETILO (Spanish); PROPIONIC ETHER

IDENTIFICATION
CAS Number: 105-37-3
Formula: $C_5H_{10}O_2$; $C_2H_5COOC_2NO_5$
DOT ID Number: UN 1195; DOT Guide Number: 129
Proper Shipping Name: Ethyl propionate

DESCRIPTION: Colorless liquid. Pineapple- or rum-like odor. Floats on water surface; soluble.

Extremely flammable • Containers may BLEVE when exposed to fire •Vapors may form explosive mixture with air • Vapors are heavier than air and will collect and stay in low areas • Vapors may travel long distances to ignition sources and flashback • Vapors in confined areas (e.g., tanks, sewers, buildings) may explode when exposed to fire • Irritating to the skin, eyes, and respiratory tract • Toxic products of combustion may include carbon monoxide.

Hazard Classification (based on NFPA-704 Rating System)
Health Hazards (Blue): -; Flammability (Red): 3; Reactivity (Yellow): 0

EMERGENCY RESPONSE: See Appendix A (129)
Evacuation:
Public safety: Isolate spill area for at least 50 to 100 meters (160 to 330 feet) in all directions.
Spill: Large spill–Consider initial downwind evacuation for at least 300 meters (1000 feet).
Fire: Isolate for 800 meters (½ mile) in all directions, especially if tank, rail car, or tank truck is involved in fire.

EXPOSURE
Short-term effects: *SEEK MEDICAL ATTENTION.* **Vapor:** Move victim to fresh air. *IF BREATHING HAS STOPPED*, give artificial respiration. If breathing is difficult, administer oxygen. **Liquid:** Irritating to skin and eyes. Overexposure may have a narcotic effect. Remove contaminated clothing and shoes. Wash affected areas with soap and water. *IF IN EYES*, hold eyelids open and flush with plenty of water.

HEALTH HAZARDS
Personal protective equipment (PPE): B-Level PPE. Full impervious protective clothing, including boots and gloves. Where splashing is possible wear full face shield or chemical safety goggles. Use approved respirator to protect against vapors.
Vapor (gas) irritant characteristics: Eye and respiratory tract irritant. Vapors cause smarting of the eyes or respiratory system if present in high concentrations. The effect is temporary.
Liquid or solid irritant characteristics: Eye irritant. If spilled on clothing and allowed to remain, may cause smarting and reddening of the skin.

FIRE DATA
Flash point: 54°F/12°C (cc).
Flammable limits in air: LEL: 1.9%; UEL: 11.0%.
Fire extinguishing agents not to be used: Water may be ineffective.
Autoignition temperature: 887°F/475°C/748°K.
Stoichiometric air-to-fuel ratio: 31.0

CHEMICAL REACTIVITY
Binary reactants: Can react with oxidizing agents, bases, and acids.
Reactivity group: 34
Compatibility class: Esters.

ENVIRONMENTAL DATA
Food chain concentration potential: Log P_{ow} = 1.19. Unlikely to accumulate.
Water pollution: Effects of low concentrations on aquatic life is unknown. May be dangerous if it enters nearby water intakes; notify operators. Notify local health and wildlife officials. **Response to discharge:** Issue warning–high flammability. Should be removed.

SHIPPING INFORMATION
Grades of purity: 99%; technical; **Storage temperature:** Ambient; **Inert atmosphere:** None; **Stability during transport:** Stable.

PHYSICAL AND CHEMICAL PROPERTIES
Physical state @ 59°F/15°C and 1 atm: Liquid.
Molecular weight: 102.13
Boiling point @ 1 atm: 210°F/99°C/372°K.
Melting/Freezing point: –99°F/–73°C/200°K.
Specific gravity (water = 1): 0.891
Relative vapor density (air = 1): 3.52
Vapor pressure: 27 mm.

2-ETHYL-3-PROPYLACROLEIN   REC. E:7700

SYNONYMS: 2-ETHYL-2-HEXENAL; 2-ETHYL-3-PROPYL ACRYLALDEHYDE; 2-ETIL-3-PROPILACROLEINA (Spanish)

IDENTIFICATION
CAS Number: 645-62-5
Formula: $C_8H_{14}O$; $CH_3(CH_2)_2CH=C(C_2NO_5)CHO$
DOT ID Number: UN 1191; DOT Guide Number: 129
Proper Shipping Name: Octyl aldehydes, flammable

DESCRIPTION: Yellow liquid. Sharp, powerful, irritating odor. Floats on the surface of water.

Combustible • Containers may BLEVE when exposed to fire • Vapors may form explosive mixture with air • Vapors are heavier than air and will collect and stay in low areas • Vapors may travel long distances to ignition sources and flashback • Vapors in confined areas (e.g., tanks, sewers, buildings) may explode when exposed to fire • Highly irritating to the eyes and respiratory tract; this material has anesthetic properties • Toxic products of combustion may include carbon monoxide.

Hazard Classification (based on NFPA-704 Rating System)
Health Hazards (Blue): 2; Flammability (Red): 2; Reactivity (Yellow): 1

EMERGENCY RESPONSE: See Appendix A (129)
Evacuation:
Public safety: Isolate spill area for at least 50 to 100 meters (160 to 330 feet) in all directions.
Spill: Large spill–Consider initial downwind evacuation for at least 300 meters (1000 feet).
Fire: Isolate for 800 meters (½ mile) in all directions, especially if tank, rail car, or tank truck is involved in fire.

EXPOSURE
Short-term effects: *SEEK MEDICAL ATTENTION*. **Liquid:** Will burn skin and eyes. Harmful if swallowed. Remove contaminated clothing and shoes. Flush affected areas with plenty of water. *IF IN EYES*, hold eyelids open and flush with plenty of water. *IF SWALLOWED* and victim is *CONSCIOUS AND ABLE TO SWALLOW*, have victim drink 4 to 8 ounces of water.

HEALTH HAZARDS
Personal protective equipment (PPE): B-Level PPE. Protective clothing; eye protection; approved respirator for high vapor concentrations. **Butyl rubber is generally suitable for aldehydes.**
Toxicity by ingestion: Grade 2; LD_{50} = 0.5 to 5 g/kg (rat).
Vapor (gas) irritant characteristics: Vapor is irritating such that personnel will not usually tolerate moderate or high vapor concentrations.
Liquid or solid irritant characteristics: Eye and skin irritant. Causes smarting of the skin and first-degree burns on short exposure; may cause secondary burns on long exposure.

FIRE DATA
Flash point: 155°F/68°C (oc).
Autoignition temperature: 392°F/200°C/373°K.
Electrical hazard: Class I, Group C.

CHEMICAL REACTIVITY
Binary reactants: Violent reaction with strong oxidizers, ketones, bromine. Incompatible with strong acids, caustics, ammonia, amines.

Reactivity group: 19
Compatibility class: Aldehyde

ENVIRONMENTAL DATA
Food chain concentration potential: Negative; unlikely to accumulate.
Water pollution: DOT Appendix B, §172.101–marine pollutant. Effect of low concentrations on aquatic life is unknown. Fouling to shoreline. May be dangerous if it enters nearby water intakes; notify operators. Notify local health and wildlife officials. **Response to discharge:** Mechanical containment. Should be removed. Chemical and physical treatment.

SHIPPING INFORMATION
Grades of purity: Technical; **Storage temperature:** Ambient; **Inert atmosphere:** None; **Venting:** Pressure-vacuum; **Stability during transport:** Stable.

NAS HAZARD CLASSIFICATION FOR BULK WATER TRANSPORTATION
FIRE: 2
HEALTH: Vapor irritant: 3; Liquid or solid irritant: 2; Poisons: 3
WATER POLLUTION: Human toxicity: 3; Aquatic toxicity: 3; Aesthetic effect: 2
REACTIVITY: Other chemicals: 2; Water: 0; Self-reaction: 1

PHYSICAL AND CHEMICAL PROPERTIES
Physical state @ 59°F/15°C and 1 atm: Liquid.
Molecular weight: 126.2
Boiling point @ 1 atm: 283°F/175°C/448°K.
Specific gravity (water = 1): 0.857 @ 15°C (liquid).
Liquid surface tension: 28.2 dynes/cm = 0.0282 N/m @ 68°F/20°C.
Liquid water interfacial tension: (estimate) 40 dynes/cm = 0.0282 N/m @ 68°F/20°C.
Heat of combustion: $-15,610$ Btu/lb = -8670 cal/g = -363×10^5 J/kg.
Vapor pressure: (Reid) 0.07 psia.

ETHYL SILICATE   REC. E:7800

SYNONYMS: EEC No. 014-005-0; ETHYL ORTHOSILICATE; ETHYL SILICATE 40; EXTREMA; SILICATE d'ETHYLE (French); SILICATO de ETILO (Spanish); SILICIC ACID TETRAETHYL ESTER; TEOS; TETRAETHOXYSILANE; TETRAETHYL ORTHOSILICATE; TETRAETHYL ORTHOSILICATE; TETRAETHYL SILICATE; SILIBOND

IDENTIFICATION
CAS Number: 78-10-4
Formula: $C_8H_{20}O_4Si$; $(C_2H_5O)_4Si$
DOT ID Number: UN 1292; DOT Guide Number: 129
Proper Shipping Name: Ethyl silicate

DESCRIPTION: Colorless watery liquid. Mild, sharp odor. May float or sink in water; practically insoluble; reaction forms a milky-white mass of silicone adhesive.

Flammable • Corrosive to skin, eyes, and respiratory tract; Contact with the skin or eyes can cause burns, impaired vision; inhalation symptoms may be delayed • Containers may BLEVE when exposed to fire • Vapors may form explosive mixture with air • Vapors are heavier than air and will collect and stay in low areas • Vapors may travel long distances to ignition sources and flashback • Vapors in confined areas (e.g., tanks, sewers, buildings) may explode when exposed to fire • Toxic products of combustion may include carbon monoxide.

Hazard Classification (based on NFPA-704 Rating System)
Health Hazards (Blue): 2; Flammability (Red): 2; Reactivity (Yellow): 0

EMERGENCY RESPONSE: See Appendix A (129)
Evacuation:
Public safety: Isolate spill area for at least 50 to 100 meters (160 to 330 feet) in all directions.
Spill: Large spill–Consider initial downwind evacuation for at least 300 meters (1000 feet).
Fire: Isolate for 800 meters (½ mile) in all directions, especially if tank, rail car, or tank truck is involved in fire.

EXPOSURE
Short-term effects: *SEEK MEDICAL ATTENTION.*
Vaor/fumes: May cause lung edema; physical exertion will aggravate this condition. **Liquid:** Irritating to eyes. *IF SWALLOWED*, will cause nausea and vomiting. Remove contaminated clothing and shoes. Flush affected areas with plenty of water. *IF IN EYES*, hold eyelids open and flush with plenty of water. *IF SWALLOWED* and victim is *CONSCIOUS AND ABLE TO SWALLOW*, have victim drink 4 to 8 ounces of water and have victim induce vomiting. *IF SWALLOWED* and victim is *UNCONSCIOUS OR HAVING CONVULSIONS,* do nothing except keep victim warm.
Note to physician or authorized medical personnel: Medical observation is recommended for 24 to 48 hours after breathing overexposure, as pulmonary edema may be delayed. As first aid for pulmonary edema, consider administering a corticosteroid spray. Cigarette smoking may exacerbate pulmonary injury and should be discouraged for at least 72 hours following exposure.

HEALTH HAZARDS
Personal protective equipment (PPE): B-Level PPE. OSHA Table Z-1-A air contaminant. Gloves; safety glasses or other form of eye protection. Chemical protective material(s) reported to have good to excellent resistance: Viton®/neoprene, butyl rubber/neoprene.
Recommendations for respirator selection: NIOSH/OSHA
100 ppm: SA* (any supplied-air respirator). *250 ppm:* SA:CF* (any supplied-air respirator operated in a continuous-flow mode). *500 ppm:* SCBAF (any self-contained breathing apparatus with full facepiece); or SAF (any supplied-air respirator with a full facepiece). *700 ppm:* SAF:PD,PP (any supplied-air respirator that has a full facepiece and is operated in a pressure-demand or other positive-pressure mode). *EMERGENCY OR PLANNED ENTRY INTO UNKNOWN CONCENTRATIONS OR IDLH CONDITIONS:* SCBAF:PD,PP (any self-contained breathing apparatus that has a full facepiece and is operated in a pressure-demand or other positive-pressure mode); or SAF:PD,PP:ASCBA (any supplied-air respirator that has a full facepiece and is operated in a pressure-demand or other positive-pressure mode in combination with an auxiliary self-contained breathing apparatus operated in a pressure-demand or other positive pressure mode). *ESCAPE:* GMFOV [any air-purifying, full-facepiece respirator (gas mask) with a chin-style, front- or back-mounted organic vapor cannister]; or SCBAE (any appropriate escape-type, self-contained breathing apparatus).
Note: Substance reported to cause eye irritation or damage; may require eye protection.
Exposure limits (TWA unless otherwise noted): ACGIH TLV 10 ppm; OSHA 100 ppm (850 mg/m^3); NIOSH 10 ppm (85 mg/m^3).

Long-term health effects: Liver, kidney, and lung damage may result from overexposures by inhalation or ingestion.
Vapor (gas) irritant characteristics: Severe eye and respiratory tract irritant.
Liquid or solid irritant characteristics: Severe eye, skin and respiratory tract irritant.
Odor threshold: 3.6 ppm.
IDLH value: 700 ppm.

FIRE DATA
Flash point: 125°F/52°C (oc); 99°F/37°C (cc).
Flammable limits in air: LEL: 1.3%; UEL: 23%.
Burning rate: 4.4 mm/min

CHEMICAL REACTIVITY
Reactivity with water: Reacts slowly, forming nontoxic silica and ethyl alcohol. Forms a milky-white mass of silicone adhesive.
Binary reactants: Causes swelling and hardening of some plastics. Reacts with strong oxidizers and acids.

ENVIRONMENTAL DATA
Food chain concentration potential: Negative; unlikely to accumulate.
Water pollution: Effect of low concentrations on aquatic life is unknown. Fouling to shoreline. May be dangerous if it enters nearby water intakes; notify operators. Notify local health and wildlife officials. **Response to discharge:** Disperse and flush.

SHIPPING INFORMATION
Grades of purity: 90-99%; **Storage temperature:** Ambient; **Inert atmosphere:** None; **Venting:** Open (flame arrester); **Stability during transport:** Stable.

PHYSICAL AND CHEMICAL PROPERTIES
Physical state @ 59°F/15°C and 1 atm: Liquid.
Molecular weight: 208.3
Boiling point @ 1 atm: 336°F/169°C/442°K.
Melting/Freezing point: –117°F/–83°C/190°K.
Specific gravity (water = 1): 0.933 @ 68°F/20°C (liquid).
Liquid surface tension: 22.8 dynes/cm = 0.0228 N/m @ 68°F/20°C.
Relative vapor density (air = 1): 7.2
Latent heat of vaporization: 95 Btu/lb = 53 cal/g = 2.2 x 10^5 J/kg.
Heat of combustion: (estimate) –12,000 Btu/lb = –6700 cal/g = –280 x 10^5 J/kg.
Vapor pressure: 13 mm.

2-ETHYL TOLUENE REC. E:7900

SYNONYMS: 1-ETHYL-2-METHYLBENZENE; *o*-ETHYLTOLUENE; *o*-ETHYLMETHYLBENZENE; *o*-METHYLETHYLBENZENE; 2-ETIL TOLUENO (Spanish)

IDENTIFICATION
CAS Number: 611-14-3
Formula: C$_9$H$_{12}$; *o*-CH$_3$C$_6$H$_4$C$_2$H$_5$
DOT ID Number: NA 1993; DOT Guide Number: 128
Proper Shipping Name: Flammable liquid, n.o.s.

DESCRIPTION: Colorless watery liquid. Pungent, benzene-like, pleasant odor. Floats on the surface of water. Flammable, irritating vapor is produced.

Flammable • Containers may BLEVE when exposed to fire • Vapors may form explosive mixture with air • Vapors are heavier than air and will collect and stay in low areas • Vapors may travel long distances to ignition sources and flashback • Vapors in confined areas (e.g., tanks, sewers, buildings) may explode when exposed to fire • Irritating to the skin, eyes, and respiratory tract • Toxic products of combustion may include carbon monoxide.

Hazard Classification (based on NFPA-704 Rating System)
Health Hazards (Blue): 2; Flammability (Red): 2; Reactivity (Yellow): 0

EMERGENCY RESPONSE: See Appendix A (128)
Evacuation:
Public safety: Isolate spill area for at least 25 to 50 meters (80 to 160 feet) in all directions.
Spill: Large spill–Consider initial downwind evacuation for at least 300 meters (1000 feet).
Fire: Isolate for 800 meters (½ mile) in all directions, especially if tank, rail car, or tank truck is involved in fire.

EXPOSURE
Short-term effects: *SEEK MEDICAL ATTENTION.* **Vapor:** Irritating to eyes, nose, and throat. *IF INHALED*, will, will cause nausea, vomiting, headache, dizziness, difficult breathing, or loss of consciousness. Move to fresh air. *IF BREATHING HAS STOPPED*, give artificial respiration. IF breathing is difficult, administer oxygen. **Liquid:** Irritating to skin and eyes. *IF SWALLOWED*, will cause nausea, vomiting, or loss of consciousness. Remove contaminated clothing and shoes. Flush affected areas with plenty of water. *IF IN EYES*, hold eyelids open and flush with plenty of water. *IF SWALLOWED* and victim is *CONSCIOUS AND ABLE TO SWALLOW*, have victim drink 4 to 8 ounces of water. **Do NOT induce vomiting.**

HEALTH HAZARDS
Personal protective equipment (PPE): B-Level PPE. Air–supplied mask; goggles or face shield; plastic gloves.
Toxicity by ingestion: Grade 2: LD_{LO} = 5 g/kg (rat).
Long-term health effects: Kidney and liver damage may follow ingestion.
Vapor (gas) irritant characteristics: Eye and respiratory tract irritant. Vapors cause smarting of the eyes or respiratory system if present in high concentrations. The effect is temporary.
Liquid or solid irritant characteristics: Eye irritant. If spilled on clothing and allowed to remain, may cause smarting and reddening of the skin.

FIRE DATA
Flash point: 103°F/39°C (cc).
Fire extinguishing agents not to be used: Water may be ineffective.
Autoignition temperature: 824°F/440°C/713°K.
Electrical hazard: Flow or agitation may generate electrostatic charges due to low conductivity.

CHEMICAL REACTIVITY
Binary reactants: Violent reaction with strong oxidizers, nitric acid, perchloric acid.
Reactivity group: 32
Compatibility class: Aromatic hydrocarbons

ENVIRONMENTAL DATA
Water pollution: DOT Appendix B, §172.101–marine pollutant. Dangerous to aquatic life in high concentrations. Fouling to shoreline. May be dangerous if it enters nearby water intakes; notify operators. Notify local health and wildlife officials. **Response to discharge:** Issue warning–high flammability. Evacuate area.

SHIPPING INFORMATION
Grades of purity: Research, reagent, 99+%; **Storage temperature:** Ambient; **Inert atmosphere:** None; **Venting:** Open (flame arrester) or pressure-vacuum; **Stability during transport:** Stable.

NAS HAZARD CLASSIFICATION FOR BULK WATER TRANSPORTATION
FIRE: 2
HEALTH: Vapor irritant: 2; Liquid or solid irritant: 1; Poisons: 1
WATER POLLUTION: Human toxicity: 2; Aquatic toxicity: 2; Aesthetic effect: 2
REACTIVITY: Other chemicals: 1; Water: 0; Self-reaction: 0

PHYSICAL AND CHEMICAL PROPERTIES
Physical state @ 59°F/15°C and 1 atm: Liquid.
Molecular weight: 120.19
Boiling point @ 1 atm: 329°F/165°C/438°K.
Melting/Freezing point: –113°F/–81°C/192°K.
Specific gravity (water = 1): 0.8807 @ 68°F/20°C (liquid).
Liquid surface tension: 15.2 dyne/cm = 0.0152 N/m @ 161°C.
Relative vapor density (air = 1): 4.15
Latent heat of vaporization: 200 Btu/lb = 111 cal/g = 4.6×10^5 J/kg.
Heat of combustion: –18,650 Btu/lb = –10,361 cal/g = -434×10^5 J/kg.
Heat of decomposition: (estimate) –19,998 Btu/lb = –11,110 cal/g = -465×10^5 J/kg.
Vapor pressure: (Reid) 0.2 psia.

ETHYLTRICHLOROSILANE REC. E:8000

SYNONYMS: ETHYL SILICON TRICHLORIDE; ETHYL TRICHLOROSILANE; ETILTRICLOROSILANO (Spanish); SILANE, TRICHLOROETHYL-; SILICANE, TRICHLOROETHYL-; TRICHLOROETHYLSILANE; TRICHLOROETHYLSILICANE

IDENTIFICATION
CAS Number: 115-21-9
Formula: $C_2H_5Cl_3Si$
DOT ID Number: UN 1196; DOT Guide Number: 155
Proper Shipping Name: Ethyltrichlorosilane
Reportable Quantity (RQ): **(CERCLA)** 1 lb/0.454 kg

DESCRIPTION: Colorless fuming liquid. Intolerable, sharp, irritating odor; like hydrochloric acid. Reacts violently with water forming hydrogen chloride and heat.

Poison! • Do NOT use water • Breathing the vapor can kill you; severely irritating to skin, eyes, and respiratory tract; prolonged contact can cause severe burns and blindness • Firefighting gear (including SCBA) does not provide adequate protection. If exposure occurs, remove and isolate gear immediately and thoroughly decontaminate personnel • Highly flammable • Containers may BLEVE when exposed to fire • Vapors may form explosive mixture with air • Vapors are heavier than air and will collect and stay in low areas • Vapors may travel long distances to ignition sources and flashback • Vapors in confined areas (e.g., tanks, sewers, buildings) may explode when exposed to fire • Toxic

of combustion may include hydrogen chloride and other irritants
• Do not put yourself in danger by entering a contaminated area to rescue a victim.

Hazard Classification (based on NFPA-704 Rating System)
Health Hazards (Blue): 3; Flammability (Red): 2; Reactivity (Yellow): 2; Special Notice (White): Water reactive

EMERGENCY RESPONSE: See Appendix A (155)
Evacuation:
Public safety: Isolate the area of spill or leak for at least 50 to 100 meters (160 to 330 feet) in all directions.
Spill: Increase, in the downwind direction, as necessary, the distance shown under "Public Safety."
Fire: If tank, rail car, or tank truck is involved in fire, isolate for at least 800 meters (½ mile) in all directions; also, consider initial evacuation for 800 meters (½ mile) in all directions.

EXPOSURE
Short-term effects: *SEEK MEDICAL ATTENTION*. **Vapor:** Irritating to eyes, nose, and throat. *IF INHALED*, will, will cause difficult breathing. Move victim to fresh air. *IF BREATHING HAS STOPPED*, give artificial respiration; *avoid mouth-to-mouth resuscitation; use bag/mask apparatus*. IF breathing is difficult, administer oxygen. **Liquid:** Will burn skin and eyes. Harmful if swallowed. Remove contaminated clothing and shoes. Flush affected areas with plenty of water. *IF IN EYES*, hold eyelids open and flush with plenty of water. *IF SWALLOWED* and victim is *CONSCIOUS AND ABLE TO SWALLOW*, have victim drink 4 to 8 ounces of water. **Do NOT induce vomiting.**

HEALTH HAZARDS
Personal protective equipment (PPE): B-Level PPE. Full protective clothing; acid-vapor-type respiratory protection; rubber gloves; chemical worker's goggles; other equipment as necessary to protect skin and eyes.
Toxicity by ingestion: Grade 2; oral LD_{50} = 1330 mg/kg (rat).
Vapor (gas) irritant characteristics: Vapors cause severe irritation of the eyes and throat and can cause eye and lung injury. They cannot be tolerated even at low concentrations.
Liquid or solid irritant characteristics: Severe skin irritant. Causes second-and third-degree burns on short contact and is very injurious to the eyes.

FIRE DATA
Flash point: 71°F/22°C (cc).
Fire extinguishing agents not to be used: Water may be ineffective.
Behavior in fire: Difficult to extinguish; re-ignition may occur. Contact with water applied to adjacent fires will produce irritating hydrogen chloride fumes.
Burning rate: 2.0 mm/min

CHEMICAL REACTIVITY
Reactivity with water: Reacts vigorously, evolving hydrogen chloride (hydrochloric acid).
Binary reactants: Reacts with surface moisture to form hydrogen chloride, which is corrosive to common metals.
Neutralizing agents for acids and caustics: Flood with water, rinse with sodium bicarbonate or lime solution.

ENVIRONMENTAL DATA
Food chain concentration potential: Negative; unlikely to accumulate.
Water pollution: Effect of low concentrations on aquatic life is unknown. May be dangerous if it enters nearby water intakes; notify operators. Notify local health and wildlife officials. **Response to discharge:** Issue warning–high flammability, corrosive. Restrict access. Disperse and flush with care.

SHIPPING INFORMATION
Grades of purity: 98+%; **Storage temperature:** Ambient; **Inert atmosphere:** None; **Stability during transport:** Stable.

NAS HAZARD CLASSIFICATION FOR BULK WATER TRANSPORTATION
FIRE: 3
HEALTH: Vapor irritant: 4; Liquid or solid irritant: 4; Poisons: 4
WATER POLLUTION: Human toxicity:–; Aquatic toxicity:–; Aesthetic effect: 2
REACTIVITY: Other chemicals: 2; Water: 3; Self-reaction: 1

PHYSICAL AND CHEMICAL PROPERTIES
Physical state @ 59°F/15°C and 1 atm: Liquid.
Molecular weight: 163.5
Boiling point @ 1 atm: 210°F/99°C/372°K.
Specific gravity (water = 1): 1.24 @ 77°F/25°C (liquid).
Liquid surface tension: (estimate) 25 dynes/cm = 0.025 N/m @ 68°F/20°C.
Relative vapor density (air = 1): 5.6
Latent heat of vaporization: 104 Btu/lb = 58 cal/g = 2.4×10^5 J/kg.
Heat of combustion: (estimate) –4300 Btu/lb = –2400 cal/g = -100×10^5 J/kg.

FERRIC AMMONIUM CITRATE   REC. F:0100

SYNONYMS: AMMONIUM FERRIC CITRATE; CITRATO FERRICO AMONICO (Spanish); FERRIC AMMONIUM CITRATE, BROWN; FERRIC AMMONIUM CITRATE, GREEN

IDENTIFICATION
CAS Number: 1185-57-5
Formula: $C_6H_8O_7 \cdot xFe \cdot xH_4N$; Mixture of $FeC_6H_5O_7$, $(NH_4)_2HC_6H_5O_7$, and water of hydration
DOT ID Number: UN 9118; DOT Guide Number: 171
Proper Shipping Name: Ferric ammonium citrate
Reportable Quantity (RQ): **(CERCLA)** 1000 lb/454 kg

DESCRIPTION: Red, green or brown solid in the form of crystals, granules, scales, or powder. Odorless; the brown form has a slight odor of ammonia. Sinks and mixes with water; highly soluble.

Toxic products of combustion may include oxides of nitrogen or ammonia gas.

Hazard Classification (based on NFPA-704 Rating System)
Health Hazards (Blue): 0; Flammability (Red): 0; Reactivity (Yellow): 0

EMERGENCY RESPONSE: See Appendix A (171)
Evacuation:
Public safety: Isolate the area of spill or leak for at least 10 to 25 meters (30 to 80 feet) in all directions.
Spill: Increase, in the downwind direction, as necessary, the distance shown under "Public Safety."
Fire: If any large container is involved in fire, isolate for at least 800 meters (½ mile) in all directions; also, consider initial evacuation for 800 meters (½ mile) in all directions.

EXPOSURE

Short-term effects: *SEEK MEDICAL ATTENTION.* **Dust:** Irritating to eyes, nose, and throat. *IF INHALED,* will, will cause coughing or difficult breathing. *IF IN EYES,* hold eyelids open and flush with plenty of water. *IF BREATHING HAS STOPPED,* give artificial respiration. IF breathing is difficult, administer oxygen. **Solid:** Irritating to skin and eyes. *IF SWALLOWED,* will cause nausea and vomiting. Remove contaminated clothing and shoes. Flush affected areas with plenty of water. *IF IN EYES,* hold eyelids open and flush with plenty of water. *IF SWALLOWED* and victim is *CONSCIOUS AND ABLE TO SWALLOW,* have victim drink 4 to 8 ounces of water. *IF SWALLOWED* and victim is *UNCONSCIOUS OR HAVING CONVULSIONS,* do nothing except keep victim warm.

HEALTH HAZARDS

Personal protective equipment (PPE): B-Level PPE. Approved respirator for nuisance dust; chemical goggles or face shield.
Vapor (gas) irritant characteristics: Dust is a respiratory tract irritant.
Liquid or solid irritant characteristics: Eye and skin irritant.

FIRE DATA

Fire extinguishing agents not to be used: Water streams applied to adjacent fires will spread contamination of environmentally hazardous substance over wide area.

ENVIRONMENTAL DATA

Food chain concentration potential: Negative; unlikely to accumulate.
Water pollution: Effect of low concentrations on aquatic life is unknown. May be dangerous if it enters nearby water intakes; notify operators. Notify local health and wildlife officials.
Response to discharge: Disperse and flush.

SHIPPING INFORMATION

Grades of purity: Commercial; **Storage temperature:** Ambient; **Inert atmosphere:** None; **Venting:** Open; **Stability during transport:** Stable.

PHYSICAL AND CHEMICAL PROPERTIES

Physical state @ 59°F/15°C and 1 atm: Solid (mixture).
Boiling point @ 1 atm: Decomposes.
Specific gravity (water = 1): 1.8 @ 68°F/20°C (solid).

FERRIC AMMONIUM OXALATE REC. F:0150

SYNONYMS: AMMONIUM FERRIOXALATE; AMMONIUM FERRIC OXALATE TRIHYDRATE; AMMONIUM TRIOXALATOFERRATE(III); AMMONIUM TRIOXALATOFERRATE(III) TRIHYDRATE; OXALATO FERRICO AMONICO (Spanish)

IDENTIFICATION

CAS Number: 2944-67-4; 14221-47-7
Formula: $C_6FeO_{12} \cdot 3H_4N$; $Fe(NH_4)_3(C_2O_4)_3 \cdot 3H_2O$
DOT ID Number: UN 9119; DOT Guide Number: 171
Proper Shipping Name: Ferric ammonium oxalate
Reportable Quantity (RQ): **(CERCLA)** 1000 lb/454 kg

DESCRIPTION: Yellowish-green powder. Light burnt-sugar odor. Sinks and mixes with water.

Corrosive • Severe irritant to eyes and respiratory tract • Toxic products of combustion may include oxides of nitrogen, ammonia, and carbon monoxide.

Hazard Classification (based on NFPA-704 Rating System)
Health Hazards (Blue): 0; Flammability (Red): 0; Reactivity (Yellow): 0

EMERGENCY RESPONSE: See Appendix A (171)
Evacuation:
Public safety: Isolate the area of spill or leak for at least 10 to 25 meters (30 to 80 feet) in all directions.
Spill: Increase, in the downwind direction, as necessary, the distance shown under "Public Safety."
Fire: If any large container is involved in fire, isolate for at least 800 meters (½ mile) in all directions; also, consider initial evacuation for 800 meters (½ mile) in all directions.

EXPOSURE

Short-term effects: *SEEK MEDICAL ATTENTION.* **Dust:** Irritating to eyes, nose, and throat. *IF INHALED,* will, will cause coughing or difficult breathing. *IF IN EYES,* hold eyelids open and flush with plenty of water. *IF BREATHING HAS STOPPED,* give artificial respiration. IF breathing is difficult, administer oxygen. **Solid:** Will burn skin and eyes. *IF SWALLOWED,* will cause nausea and vomiting. Remove contaminated clothing and shoes. Flush affected areas with plenty of water. *IF IN EYES,* hold eyelids open and flush with plenty of water. *IF SWALLOWED* and victim is *CONSCIOUS AND ABLE TO SWALLOW,* have victim drink 4 to 8 ounces of water and have victim induce vomiting. *IF SWALLOWED* and victim is *UNCONSCIOUS OR HAVING CONVULSIONS,* do nothing except keep victim warm.

HEALTH HAZARDS

Personal protective equipment (PPE): B-Level PPE. Approved dust respirator; rubber or plastic-coated gloves; chemical goggles or face shield.
Vapor (gas) irritant characteristics: Dust may cause eye and respiratory tract irritation.
Liquid or solid irritant characteristics: Eye and skin irritant.

FIRE DATA

Fire extinguishing agents not to be used: Water streams applied to adjacent fires will spread contamination of environmentally hazardous substance over wide area.

ENVIRONMENTAL DATA

Food chain concentration potential: Negative; unlikely to accumulate.
Water pollution: Effect of low concentrations on aquatic life is unknown. May be dangerous if it enters nearby water intakes; notify operators. Notify local health and wildlife officials. **Response to discharge:** Issue warning–water contaminant. Disperse and flush.

SHIPPING INFORMATION

Grades of purity: Commercial; **Storage temperature:** Ambient; **Inert atmosphere:** None; **Venting:** Open; **Stability during transport:** Stable.

PHYSICAL AND CHEMICAL PROPERTIES

Physical state @ 59°F/15°C and 1 atm: Solid.
Molecular weight: 428
Boiling point @ 1 atm: (decomposes).
Specific gravity (water = 1): 1.78 @ 68°F/20°C (solid).

FERRIC CHLORIDE REC. F:0200

SYNONYMS: CLORURO FERRICO ANHIDRO (Spanish); FERRIC CHLORIDE, HEXAHYDRATE; FERROUS(III) CHLORIDE; FLORES MARTIS; IRON(III) CHLORIDE; IRON PERCHLORIDE; IRON TRICHLORIDE; PERCHLORURE DE FER (French).

IDENTIFICATION
CAS Number: 7705-08-0
Formula: $FeCl_3$; $FeCl_3 \cdot 6H_2O$
DOT ID Number: UN 1773; DOT Guide Number: 157
Proper Shipping Name: Ferric chloride, anhydrous
Reportable Quantity (RQ): **(CERCLA)** 1000 lb/454 kg

DESCRIPTION: Brown (hydrate) or greenish-black (anhydrous) crystalline solid. Odorless. Sinks and mixes with water forming a medium strong acid.

Corrosive to the skin and eyes; highly irritating to respiratory tract • Aqueous solution attacks metals forming explosive hydrogen gas • Toxic products of combustion may include chlorine gas hydrogen chloride.

Hazard Classification (based on NFPA-704 Rating System)
Health Hazards (Blue): 1; Flammability (Red): 0; Reactivity (Yellow): 0

EMERGENCY RESPONSE: See Appendix A (157)
Evacuation:
Public safety: Isolate the area of spill or leak for at least 50 to 100 meters (160 to 330 feet) in all directions.
Spill: Increase, in the downwind direction, as necessary, the distance shown under "Public Safety."
Fire: If tank, rail car, or tank truck is involved in fire, isolate for at least 800 meters (½ mile) in all directions; also, consider initial evacuation for 800 meters (½ mile) in all directions.

EXPOSURE
Short-term effects: *SEEK MEDICAL ATTENTION.* **Dust:** Irritating to eyes, nose, and throat. *IF INHALED*, will, will cause coughing or difficult breathing. *IF IN EYES,* hold eyelids open and flush with plenty of water. *IF BREATHING HAS STOPPED,* give artificial respiration. IF breathing is difficult, administer oxygen. **Solid:** Will burn skin and eyes. *IF SWALLOWED,* will cause nausea and vomiting. Remove contaminated clothing and shoes. Flush affected areas with plenty of water. *IF IN EYES,* hold eyelids open and flush with plenty of water. *IF SWALLOWED* and victim is *CONSCIOUS AND ABLE TO SWALLOW*, have victim drink or milk and have victim induce vomiting. *IF SWALLOWED* and victim is *UNCONSCIOUS OR HAVING CONVULSIONS,* do nothing except keep victim warm.

HEALTH HAZARDS
Personal protective equipment (PPE): B-Level PPE. Dust respirator if required; rubber apron and boots; chemical worker's goggles or face shield. Neoprene is generally suitable for mineral acids.
Exposure limits (TWA unless otherwise noted): ACGIH 1 mg/m^3; NIOSH REL 1 mg/m^3 as soluble iron salts
Toxicity by ingestion: Poison. Grade 2; LD_{50} = 0.5 to 5 g/kg (rat).
Long-term health effects: EPA Genetic Toxicology Program: mutation data.
Liquid or solid irritant characteristics: Severe eye irritant; skin irritant.

FIRE DATA
Fire extinguishing agents not to be used: Water streams applied to adjacent fires will spread contamination of environmentally hazardous substance over wide area.

CHEMICAL REACTIVITY
Reactivity with water: Reaction produces toxic, corrosive gas.
Binary reactants: Water solutions are acidic and slowly corrosive to most metals, evolving flammable hydrogen gas. May form shock-sensitive explosive product with potassium, sodium and other active metals. Strong reaction with strong bases and allyl chloride.
Neutralizing agents for acids and caustics: Flush with water, rinse with dilute sodium bicarbonate or soda ash solutions.
Reactivity group: 1 (solution)
Compatibility class: Nonoxidizing Mineral Acids

ENVIRONMENTAL DATA
Food chain concentration potential: Negative; unlikely to accumulate.
Water pollution: Harmful to aquatic life in very low concentrations. May be dangerous if it enters nearby water intakes; notify operators. Notify local health and wildlife officials. **Response to discharge:** Disperse and flush.

SHIPPING INFORMATION
Grades of purity: Anhydrous; Hydrate; Reagent; 46% solution in water; **Storage temperature:** Ambient; **Inert atmosphere:** None; **Venting:** Open; **Stability during transport:** Stable.

PHYSICAL AND CHEMICAL PROPERTIES
Physical state @ 59°F/15°C and 1 atm: Solid.
Molecular weight: 162.22 (anhydrous).
Boiling point @ 1 atm: 599°F/315°C/588°K (decomposes).
Melting/Freezing point: 583°F/306°C/579°K; (hydrated form only) 540°F/282°C/555°K.
Specific gravity (water = 1): 2.8 @ 68°F/20°C (anhydrous solid).
Heat of solution: (anhydrous) –360 Btu/lb = –200 cal/g = –8.4 x 10^5 J/kg.

FERRIC FLUORIDE REC. F:0250

SYNONYMS: FERRIC TRIFLUORIDE; FLUORURO FERRICO (Spanish); IRON FLUORIDE; IRON TRIFLUORIDE

IDENTIFICATION
CAS Number: 7783-50-8
Formula: FeF_3
DOT ID Number: UN 9120; DOT Guide Number: 171
Proper Shipping Name: Ferric fluoride
Reportable Quantity (RQ): **(CERCLA)** 100 lb/45.4 kg

DESCRIPTION: Green crystalline solid. Sinks and mixes slowly with water; very slightly soluble; forms acid solution.

Strong irritant to the skin, eyes, and respiratory tract • The fluoride ion is poisonous to human cells • Toxic products of combustion may include iron fume, fluorides, and hydrogen fluoride.

Hazard Classification (based on NFPA-704 Rating System)
Health Hazards (Blue): 3; Flammability (Red): 0; Reactivity (Yellow): 0

EMERGENCY RESPONSE: See Appendix A (171)
Evacuation:
Public safety: Isolate the area of spill or leak for at least 10 to 25 meters (30 to 80 feet) in all directions.
Spill: Increase, in the downwind direction, as necessary, the distance shown under "Public Safety."
Fire: If any large container is involved in fire, isolate for at least 800 meters (½ mile) in all directions; also, consider initial evacuation for 800 meters (½ mile) in all directions.

EXPOSURE
Short-term effects: *SEEK MEDICAL ATTENTION.* **Dust or Solid:** Irritating to eyes, nose, and throat. *IF SWALLOWED*, will cause lethargy, nausea, and vomiting. Move to fresh air. Flush affected areas with plenty of water. *IF IN EYES*, hold eyelids open and flush with plenty of water. *IF SWALLOWED* and victim is *CONSCIOUS AND ABLE TO SWALLOW*, have victim drink 4 to 8 ounces of water and have victim induce vomiting.

HEALTH HAZARDS
Personal protective equipment (PPE): B-Level PPE. OSHA Table Z-1-A air contaminant. Goggles, polyvinyl chloride or neoprene gloves, filter type respirator, chemical apron.
Recommendations for respirator selection: NIOSH/OSHA as F. *12.5 mg/m³:* DM (any dust and mist respirator). *25 mg/m³:* DMXSQ* (any dust and mist respirator except single-use and quarter-mask respirators); or SA* (any supplied-air respirator). *62.5 mg/m³:* SA:CF* [any supplied-air respirator operated in a continuous-flow mode)]; or PAPRDM*⁺ *if not present as a fume* (any powered, air-purifying respirator with a dust and mist filter). *125 mg/m³:* HiEF + (any air-purifying, full-facepiece respirator with a high-efficiency particulate filter); or SCBAF (any self-contained breathing apparatus with a full facepiece); or SAF (any supplied-air respirator with a full facepiece). *250 mg/m³:* SA:PD,PP (any supplied-air respirator operated in a pressure-demand or other positive-pressure mode). *EMERGENCY OR PLANNED ENTRY INTO UNKNOWN CONCENTRATIONS OR IDLH CONDITIONS:* SCBAF:PD,PP (any self-contained breathing apparatus that has a full faceplate and is operated in a pressure-demand or other positive-pressure mode); or SAF:PD,PP:ASCBA (any supplied-air respirator that has a full facepiece and is operated in a pressure-demand or other positive-pressure mode in combination with an auxiliary, self-contained breathing apparatus operated in a pressure-demand or other positive-pressure mode). *ESCAPE:* HiEF+ (any air-purifying, full-facepiece respirator with a high-efficiency particulate filter); or SCBAE (any appropriate escape-type, self-contained breathing apparatus). *Notes:* *Substance reported to cause eye irritation or damage; may require eye protection. ⁺May need acid gas sorbent.
Exposure limits (TWA unless otherwise noted): ACGIH TLV 2.5 mg/m³ as fluorides; NIOSH/OSHA 2.5 mg/m³ as inorganic fluorides.
Long-term health effects: IARC rating 3
Vapor (gas) irritant characteristics: May cause eye and respiratory tract irritation.
IDLH value: 250 mg/m³ as F.

CHEMICAL REACTIVITY
Binary reactants: Strong oxidizers.

ENVIRONMENTAL DATA
Water pollution: Dangerous to aquatic life in high concentrations. May be dangerous if it enters nearby water intakes; notify operators. Notify local health and pollution control officials.
Response to discharge: Disperse and flush.

SHIPPING INFORMATION
Storage temperature: Ambient; **Inert atmosphere:** None; **Venting:** Open; **Stability during transport:** Stable.

PHYSICAL AND CHEMICAL PROPERTIES
Physical state @ 59°F/15°C and 1 atm: Solid.
Molecular weight: 112.85
Boiling point @ 1 atm: Sublimes. more than 1832°F/1000°C.
Melting/Freezing point: More than 1832°F/1000°C/1273°K.
Specific gravity (water = 1): 3.87 at room temperature
Heat of solution: 159.5 Btu/lb = 88.6 cal/g = 3.7×10^5 J/kg.

FERRIC GLYCEROPHOSPHATE   **REC. F:0300**

SYNONYMS: GLICEROFOSFATO FERRICO (Spanish)

IDENTIFICATION
Formula: $C_9H_{21}Fe_2O_{18}P_3$; $Fe_2[C_3NO_5(OH)_2PO_4]_3 \cdot xH_2O$

DESCRIPTION: Orange to greenish-yellow scales or powder. Odorless. Sinks and mixes slowly with water.

Irritating to the skin, eyes, and respiratory tract • Toxic products of combustion may include iron fume, and phosphorus oxides or phosphoric acid.

Hazard Classification (based on NFPA-704 Rating System)
Health Hazards (Blue): 2; Flammability (Red): 0; Reactivity (Yellow): 0

EMERGENCY RESPONSE: See Appendix A (171)
Evacuation:
Public safety: Isolate the area of spill or leak for at least 10 to 25 meters (30 to 80 feet) in all directions.
Spill: Increase, in the downwind direction, as necessary, the distance shown under "Public Safety."
Fire: If any large container is involved in fire, isolate for at least 800 meters (½ mile) in all directions; also, consider initial evacuation for 800 meters (½ mile) in all directions.

EXPOSURE
Short-term effects: *SEEK MEDICAL ATTENTION.* **Dust:** Irritating to eyes, nose, and throat. *IF INHALED*, will, will cause coughing or difficult breathing. *IF IN EYES*, hold eyelids open and flush with plenty of water. *IF BREATHING HAS STOPPED*, give artificial respiration. IF breathing is difficult, administer oxygen. **Solid:** Irritating to skin and eyes. Harmful if swallowed. Remove contaminated clothing and shoes. Flush affected areas with plenty of water. *IF IN EYES*, hold eyelids open and flush with plenty of water. *IF SWALLOWED* and victim is *CONSCIOUS AND ABLE TO SWALLOW*, have victim drink 4 to 8 ounces of water and have victim induce vomiting. *IF SWALLOWED* and victim is *UNCONSCIOUS OR HAVING CONVULSIONS,* do nothing except keep victim warm.

HEALTH HAZARDS
Personal protective equipment (PPE): B-Level PPE. Goggles or face shield; dust mask; rubber gloves.
Liquid or solid irritant characteristics: May irritate eyes, skin, and respiratory tract.

ENVIRONMENTAL DATA
Water pollution: Effect of low concentrations on aquatic life is

unknown. May be dangerous if it enters nearby water intakes; notify operators. Notify local health and wildlife officials. **Response to discharge:** Disperse and flush.

SHIPPING INFORMATION
Grades of purity: Commercial; **Storage temperature:** Ambient; **Inert atmosphere:** None; **Venting:** Open; **Stability during transport:** Stable.

PHYSICAL AND CHEMICAL PROPERTIES
Physical state @ 59°F/15°C and 1 atm: Solid.
Molecular weight: 470 (approximate).
Specific gravity (water = 1): 1.5 @ 68°F/20°C (solid).

FERRIC NITRATE REC. F:0350

SYNONYMS: FERRIC NITRATE, NONHYDRATE; FERRIC NITRATE, NITRIC ACID SOLUTION; IRON NITRATE; IRON(III) NITRATE, ANHYDROUS; IRON TRINITRATE; NITRATO FERRICO (Spanish); NITRIC ACID, IRON(3+) SALT; NITRIC ACID, IRON(III) SALT

IDENTIFICATION
CAS Number: 10421-48-4
Formula: $N_3O_9 \cdot Fe$; $Fe(NO_3)_3 \cdot 9H_2O$
DOT ID Number: UN 1466; DOT Guide Number: 140
Proper Shipping Name: Ferric nitrate
Reportable Quantity (RQ): **(CERCLA)** 1000 lb/454 kg

DESCRIPTION: Green, colorless, to pale violet solid. Odorless. Sinks and mixes with water.

Strong oxidizer which may react spontaneously with low flash point organics or reducing agents. Heat forms oxygen; will increase the activity of an existing fire • May cause fire or explosion on contact with combustibles (wood, paper, oil, clothing, etc.) • Irritating to the skin, eyes, and respiratory tract • Toxic products of combustion may include oxides of nitrogen and nitric acid vapor.

Hazard Classification (based on NFPA-704 Rating System)
Health Hazards (Blue): 0; Flammability (Red): 0; Reactivity (Yellow): 0; Special Notice (White): OXY

EMERGENCY RESPONSE: See Appendix A (140)
Evacuation:
Public safety: Isolate the area of spill or leak for at least 10 to 25 meters (30 to 80 feet) in all directions.
Spill: Consider initial downwind evacuation for at least 100 meters (330 feet).
Fire: If any large container is involved in fire, isolate for at least 800 meters (½ mile) in all directions; also, consider initial evacuation for 800 meters (½ mile) in all directions.

EXPOSURE
Short-term effects: *SEEK MEDICAL ATTENTION.* **Dust:** Irritating to eyes, nose, and throat. *IF INHALED,* will, will cause coughing or difficult breathing. *IF IN EYES,* hold eyelids open and flush with plenty of water. *IF BREATHING HAS STOPPED,* give artificial respiration. IF breathing is difficult, administer oxygen. **Solid:** Irritating to skin and eyes. Harmful if swallowed. Remove contaminated clothing and shoes. Flush affected areas with plenty of water. *IF IN EYES,* hold eyelids open and flush with plenty of water. *IF SWALLOWED* and victim is *CONSCIOUS AND ABLE TO SWALLOW,* have victim drink 4 to 8 ounces of water and have victim induce vomiting. *IF SWALLOWED* and victim is *UNCONSCIOUS OR HAVING CONVULSIONS,* do nothing except keep victim warm.

HEALTH HAZARDS
Personal protective equipment (PPE): B-Level PPE. Dust mask; goggles or face shield; protective gloves.
Exposure limits (TWA unless otherwise noted): ACGIH 1 mg/m³; NIOSH REL 1 mg/m³ as soluble iron salts
Toxicity by ingestion: Grade 2; LD_{50} = 0.5 to 5 g/kg.
Liquid or solid irritant characteristics: May cause eye, skin and respiratory tract irritation.

CHEMICAL REACTIVITY
Binary reactants: Solutions are corrosive to most metals. Contact of solid with wood or paper may cause fire.
Reactivity group: 3
Compatibility class: Nitric Acid

ENVIRONMENTAL DATA
Food chain concentration potential: Negative; unlikely to accumulate.
Water pollution: Effect of low concentrations on aquatic life is unknown. May be dangerous if it enters nearby water intakes; notify operators. Notify local health and wildlife officials. **Response to discharge:** Disperse and flush.

SHIPPING INFORMATION
Grades of purity: Technical, 99.0%; Analytical; **Storage temperature:** Ambient; **Inert atmosphere:** None; **Venting:** Open; **Stability during transport:** Stable.

PHYSICAL AND CHEMICAL PROPERTIES
Physical state @ 59°F/15°C and 1 atm: Solid.
Molecular weight: 404.02
Boiling point @ 1 atm: (decomposes).
Melting/Freezing point: 117°F/47°C/320°K.
Specific gravity (water = 1): 1.7 @ 68°F/20°C (solid).
Heat of solution: 40 Btu/lb = 22 cal/g = 0.92 x 10⁵ J/kg.

FERRIC SULFATE REC. F:0400

SYNONYMS: DIIRONTRISULFATE; IRON PERSULFATE; IRON SESQUISULFATE; IRON(III) SULFATE; IRON SULFATE; IRON(3+) SULFATE; IRON TERSULFATE; SULFATO FERRICO (Spanish); SULFURIC ACID, IRON(3+) SALT

IDENTIFICATION
CAS Number: 10028-22-5
Formula: $Fe_2O_{12}S_3$; $Fe_2(SO_4)_3$
DOT ID Number: UN 9121; DOT Guide Number: 171
Proper Shipping Name: Ferric sulfate; Ferric sulphate
Reportable Quantity (RQ): **(CERCLA)** 1000 lb/454 kg

DESCRIPTION: Gray to white powder or yellow crystals. Odorless. Sinks and mixes slowly with water; slightly soluble; forms acid solution, precipitates hydroxide and some phosphate salts.

An oxidizer. Heat forms oxygen; will increase the activity of an existing fire • May cause fire or explosion contact with combustibles (wood, paper, oil, clothing, etc.) • Irritating to the skin, eyes, and respiratory tract •Toxic products of combustion may include sulfur oxides and iron fume.

Hazard Classification (based on NFPA-704 Rating System)
Health Hazards (Blue): 1; Flammability (Red): 0; Reactivity (Yellow): 0

EMERGENCY RESPONSE: See Appendix A (171)
Evacuation:
Public safety: Isolate the area of spill or leak for at least 10 to 25 meters (30 to 80 feet) in all directions.
Spill: Increase, in the downwind direction, as necessary, the distance shown under "Public Safety."
Fire: If any large container is involved in fire, isolate for at least 800 meters (½ mile) in all directions; also, consider initial evacuation for 800 meters (½ mile) in all directions.

EXPOSURE
Short-term effects: *SEEK MEDICAL ATTENTION.* **Dust:** Irritating to eyes, nose, and throat. *IF INHALED*, will, will cause coughing or difficult breathing. *IF IN EYES*, hold eyelids open and flush with plenty of water. *IF BREATHING HAS STOPPED,* give artificial respiration. IF breathing is difficult, administer oxygen.
Solid: Irritating to skin and eyes. *IF SWALLOWED*, will cause nausea and vomiting. Remove contaminated clothing and shoes. Flush affected areas with plenty of water. *IF IN EYES*, hold eyelids open and flush with plenty of water. *IF SWALLOWED* and victim is *CONSCIOUS AND ABLE TO SWALLOW*, have victim drink 4 to 8 ounces of water and have victim induce vomiting. *IF SWALLOWED* and victim is *UNCONSCIOUS OR HAVING CONVULSIONS,* do nothing except keep victim warm.

HEALTH HAZARDS
Personal protective equipment (PPE): B-Level PPE. Dust mask; goggles or face shield; protective gloves.
Exposure limits (TWA unless otherwise noted): ACGIH TLV 1 mg/m^3; NIOSH REL 1 mg/m^3 as iron soluble salts
Liquid or solid irritant characteristics: Eye, skin and respiratory tract irritant.

CHEMICAL REACTIVITY
Binary reactants: Violent reaction with aluminum, magnesium. Corrosive to copper, copper alloys, mild steel, and galvanized steel.
Neutralizing agents for acids and caustics: Flush with water.

ENVIRONMENTAL DATA
Food chain concentration potential: Negative; unlikely to accumulate.
Water pollution: Dangerous to aquatic life in high concentrations. May be dangerous if it enters nearby water intakes; notify operators. Notify local health and wildlife officials. **Response to discharge:** Disperse and flush.

SHIPPING INFORMATION
Grades of purity: Anhydrous; Hydrate, 73%; sometimes shipped as water solutions, which are acidic; **Storage temperature:** Ambient; **Inert atmosphere:** None; **Venting:** Open; **Stability during transport:** Stable.

PHYSICAL AND CHEMICAL PROPERTIES
Physical state @ 59°F/15°C and 1 atm: Solid.
Molecular weight: 399.88 (decomposes).
Specific gravity (water = 1): 3.1 @ 68°F/20°C (solid).
Heat of solution: -53.3 Btu/lb = -29.6 cal/g = -1.24×10^5 J/kg.

FERROPHOSPHORUS REC. F:0450

SYNONYMS: FERROFOSFORO (Spanish); FERROPHOS; IRON ALLOY, BASE; IRON–PHOSPHORUS CASTING ALLOY

IDENTIFICATION
CAS Number: 8049-19-2
Formula: FeP

DESCRIPTION: Solid alloy of iron and phosphorus (18-25%P). Insoluble in water.

Irritating to the skin, eyes, and respiratory tract • Toxic products of combustion may include iron fume, irritating vapors, and toxic gases of phosphorus oxides and/or phosphoric acid.

Hazard Classification (based on NFPA-704 Rating System)
Health Hazards (Blue): 0; Flammability (Red): 0; Reactivity (Yellow): 0

EMERGENCY RESPONSE: See Appendix A (171)
Evacuation:
Public safety: Isolate the area of spill or leak for at least 10 to 25 meters (30 to 80 feet) in all directions.
Spill: Increase, in the downwind direction, as necessary, the distance shown under "Public Safety."
Fire: If any large container is involved in fire, isolate for at least 800 meters (½ mile) in all directions; also, consider initial evacuation for 800 meters (½ mile) in all directions.

EXPOSURE
Short-term effects: *SEEK MEDICAL ATTENTION.* **Dust:** Move victim to fresh air. Remove contaminated clothing and shoes. Flush affected areas with water. IF IN EYES, hold eyelids open and flush with plenty of water. *IF BREATHING HAS STOPPED,* give artificial respiration. If breathing is difficult, administer oxygen.

HEALTH HAZARDS
Personal protective equipment (PPE): B-Level PPE. Wear protective clothing to prevent contact with dust. Use approved respirator to protect against dust.
Liquid or solid irritant characteristics: Eye and respiratory tract irritation. If spilled on clothing and allowed to remain, may cause smarting and reddening of skin.

FIRE DATA
Fire extinguishing agents not to be used: Water.

CHEMICAL REACTIVITY
Binary reactants: Powdered material reacts violently with strong oxidizers.

ENVIRONMENTAL DATA
Water pollution: Effects of low concentrations on aquatic life is unknown. May be dangerous if it enters nearby water intakes; notify operators. Notify local health and wildlife officials. **Response to discharge:** Should be removed.

SHIPPING INFORMATION
Grades of purity: Technical grades, 18% or 25% phosphorus; **Storage temperature:** Ambient; **Inert atmosphere:** None; **Stability during transport:** Stable.

PHYSICAL AND CHEMICAL PROPERTIES
Physical state @ 59°F/15°C and 1 atm: Solid.

FERROSILICON　　REC. F:0500

SYNONYMS: IRON-SILICON ALLOY; FERROSILICON, CONTAINING MORE THAN 30% BUT LESS THAN 90% SILICON; FERROSILICIO (Spanish)

IDENTIFICATION
CAS Number: 8049-17-0
Formula: FeSi
DOT ID Number: UN 1408; DOT Guide Number: 139
Proper Shipping Name: Ferrosilicon, with 30% or more but less than 90% silicon.

DESCRIPTION: An alloy of iron and silicon Insoluble in water; material containing 30–90% silicon evolves flammable hydrogen and acetylene gases in the presence of moisture.

Material containing 30–90% silicon is combustible • Heat or flame may cause explosion • Containers may BLEVE when exposed to fire • Vapors are heavier than air and will collect and stay in low areas • Vapors in confined areas (e.g., tanks, sewers, buildings) may explode when exposed to fire • Irritating to the skin, eyes, and respiratory tract • Toxic products of combustion may include iron fume and reactive or poisonous gases such as arsine and phosphine.

EMERGENCY RESPONSE: See Appendix A (139)
Evacuation:
Public safety: Isolate the area of spill or leak for at least 100 to 150 meters (330 to 490 feet) in all directions.
Spill: See Public Safety.
Fire: If any large container is involved in fire, isolate for at least 800 meters (½ mile) in all directions; also, consider initial evacuation for 800 meters (½ mile) in all directions.

EXPOSURE
Short-term effects: *SEEK MEDICAL ATTENTION*. **Dust:** Move victim to fresh air. Remove contaminated clothing and shoes. Flush affected areas with water. *IF IN EYES*, hold eyelids open and flush with plenty of water. *IF BREATHING HAS STOPPED*, give artificial respiration. If breathing is difficult, administer oxygen.

HEALTH HAZARDS
Personal protective equipment (PPE): B-Level PPE. Wear protective clothing to prevent contact with dust. Use approved respirator to protect against dust.
Liquid or solid irritant characteristics: Eye and respiratory tract irritant. If spilled on clothing and allowed to remain, may cause smarting and reddening of skin.

FIRE DATA
Fire extinguishing agents not to be used: Do not use water. Dangerous when wet.

CHEMICAL REACTIVITY
Reactivity with water: Dangerous when wet. Reacts with water to release hydrogen and acetylene gas.
Binary reactants: Incompatible with oxidizers, acid, sodium hydroxide.

ENVIRONMENTAL DATA
Water pollution: Effects of low concentrations on aquatic life is unknown. May be dangerous if it enters nearby water intakes; notify operators. Notify local health and wildlife officials.
Response to discharge: Should be removed.

SHIPPING INFORMATION
Grades of purity: Technical grades, silicon content ranges from 30–90%; **Storage temperature:** Ambient; **Inert atmosphere:** None; **Stability during transport:** Stable.

PHYSICAL AND CHEMICAL PROPERTIES
Physical state @ 59°F/15°C and 1 atm: Solid.
Specific gravity (water = 1): 5.4

FERROUS AMMONIUM SULFATE　　REC. F:0550

SYNONYMS: AMMONIUM IRON SULFATE; FERROUS AMMONIUM SULFATE HEXAHYDRATE; IRON AMMONIUM SULFATE; MOHR'S SALT; SULFURIC ACID, AMMONIUM IRON(2+) SALT (2:2:1); AMMONIUM FERROUS SULFATE; SULFATO FERROSO AMONICO (Spanish)

IDENTIFICATION
CAS Number: 10045-89-3; 7783-85-9 (hexahydrate)
Formula: $Fe(NH_4)_2(SO_4)_2 \cdot 6H_2O$
DOT ID Number: UN 9122; DOT Guide Number: 171
Proper Shipping Name: Ferrous Ammonium Sulfate; Ferrous Ammonium sulphate
Reportable Quantity (RQ): **(CERCLA)** 1000 lb/454 kg

DESCRIPTION: Pale blue-green solid. Odorless. Sinks and mixes slowly with water.

Toxic products of combustion may include iron nitrogen oxides and sulfur oxides.

Hazard Classification (based on NFPA-704 Rating System)
Health Hazards (Blue): 0; Flammability (Red): 0; Reactivity (Yellow): 0

EMERGENCY RESPONSE: See Appendix A (171)
Evacuation:
Public safety: Isolate the area of spill or leak for at least 10 to 25 meters (30 to 80 feet) in all directions.
Spill: Increase, in the downwind direction, as necessary, the distance shown under "Public Safety."
Fire: If any large container is involved in fire, isolate for at least 800 meters (½ mile) in all directions; also, consider initial evacuation for 800 meters (½ mile) in all directions.

EXPOSURE
Short-term effects: *SEEK MEDICAL ATTENTION*. **Dust:** Irritating to eyes, nose, and throat. *IF INHALED*, will, will cause coughing or difficult breathing. *IF IN EYES*, hold eyelids open and flush with plenty of water. *IF BREATHING HAS STOPPED*, give artificial respiration. IF breathing is difficult, administer oxygen.
Solid: Irritating to skin and eyes. *IF SWALLOWED*, will cause nausea and vomiting. Remove contaminated clothing and shoes. Flush affected areas with plenty of water. *IF IN EYES*, hold eyelids open and flush with plenty of water. *IF SWALLOWED* and victim is *CONSCIOUS AND ABLE TO SWALLOW*, have victim drink 4 to 8 ounces of water and have victim induce vomiting. *IF SWALLOWED* and victim is *UNCONSCIOUS OR HAVING CONVULSIONS*, do nothing except keep victim warm.

HEALTH HAZARDS
Personal protective equipment (PPE): B-Level PPE. Dust mask; goggles or face shield; protective gloves.

Exposure limits (TWA unless otherwise noted): ACGIH 1 mg/m^3; NIOSH REL 1 mg/m^3 as iron soluble salts
Toxicity by ingestion: Grade 2; LD$_{50}$ = 0.5 to 5 g/kg (rat).
Long-term health effects: May cause eye degeneration in rabbits
Liquid or solid irritant characteristics: Eye, skin and respiratory tract irritant.

FIRE DATA
Fire extinguishing agents not to be used: Water streams applied to adjacent fires will spread contamination of environmentally hazardous substance over wide area.

CHEMICAL REACTIVITY
Binary reactants: Violent reaction with aluminum, magnesium.

ENVIRONMENTAL DATA
Food chain concentration potential: Negative; unlikely to accumulate.
Water pollution: Effect of low concentrations on aquatic life is unknown. May be dangerous if it enters nearby water intakes; notify operators. Notify local health and wildlife officials.
Response to discharge: Disperse and flush.

SHIPPING INFORMATION
Grades of purity: Technical, 99–102%; Reagent; **Storage temperature:** Ambient; **Inert atmosphere:** None; **Venting:** Open; **Stability during transport:** Stable.

PHYSICAL AND CHEMICAL PROPERTIES
Physical state @ 59°F/15°C and 1 atm: Solid.
Molecular weight: 392.16
Boiling point @ 1 atm: (decomposes).
Specific gravity (water = 1): 1.86 @ 68°F/20°C (solid).

FERROUS CHLORIDE   **REC. F:0600**

SYNONYMS: CLORURO FERROSO (Spanish); IRON(II) CHLORIDE; IRON DICHLORIDE; IRON(II) CHLORIDE; IRON PROTOCHLORIDE; LAWRENCITE; FERROUS CHLORIDE TETRAHYDRATE

IDENTIFICATION
CAS Number: 7758-94-3
Formula: Cl$_2$Fe; FeCl$_2 \cdot$4H$_2$O
DOT ID Number: UN 1759; UN 1760; DOT Guide Number: 154
Proper Shipping Name: Ferrous chloride, solid; Ferrous chloride solution
Reportable Quantity (RQ): **(CERCLA)** 100 lb/45.4 kg

DESCRIPTION: Pale green to yellow crystalline solid. Odorless. Sinks and mixes slowly with water.

Combustible • Heat or flame may cause explosion • Containers may BLEVE when exposed to fire • Vapors are heavier than air and will collect and stay in low areas • Vapors in confined areas (e.g., tanks, sewers, buildings) may explode when exposed to fire • Corrosive; Irritating to the skin, eyes, and respiratory tract • Toxic products of combustion may include iron fume and hydrogen chloride.

Hazard Classification (based on NFPA-704 Rating System)
Health Hazards (Blue): 2; Flammability (Red): 1; Reactivity (Yellow): 1

EMERGENCY RESPONSE: See Appendix A (154)
Evacuation:
Public safety: Isolate the area of spill or leak for at least 25 to 50 meters (80 to 160 feet) in all directions.
Spill: Increase, in the downwind direction, as necessary, the distance shown under "Public Safety."
Fire: If tank, rail car, or tank truck is involved in fire, isolate for at least 800 meters (½ mile) in all directions; also, consider initial evacuation for 800 meters (½ mile) in all directions.

EXPOSURE
Short-term effects: *SEEK MEDICAL ATTENTION.* **Dust:** Irritating to eyes, nose, and throat. *IF INHALED*, will, will cause difficult breathing. *IF IN EYES*, hold eyelids open and flush with plenty of water. *IF BREATHING HAS STOPPED*, give artificial respiration; *avoid mouth-to-mouth resuscitation; use bag/mask apparatus.* IF breathing is difficult, administer oxygen. **Solid:** Irritating to skin and eyes. *IF SWALLOWED*, will cause nausea and vomiting. Remove contaminated clothing and shoes. Flush affected areas with plenty of water. *IF IN EYES*, hold eyelids open and flush with plenty of water. *IF SWALLOWED* and victim is *CONSCIOUS AND ABLE TO SWALLOW*, have victim drink 4 to 8 ounces of water and have victim induce vomiting. *IF SWALLOWED* and victim is *UNCONSCIOUS OR HAVING CONVULSIONS*, do nothing except keep victim warm.

HEALTH HAZARDS
Personal protective equipment (PPE): B-Level PPE. Dust mask; goggles or face shield; rubber gloves.
Exposure limits (TWA unless otherwise noted): ACGIH TLV 1 mg/m^3; NIOSH REL 1 mg/m^3 as iron soluble salts
Toxicity by ingestion: Grade 2; LD$_{50}$ = 0.5 to 5 g/kg (rat).
Long-term health effects: EPA Genetic Toxicology Program: mutagenic data
Liquid or solid irritant characteristics: Eye, skin and respiratory tract irritant.

FIRE DATA
Fire extinguishing agents not to be used: Water streams applied to adjacent fires will spread contamination of environmentally hazardous substance over wide area.

CHEMICAL REACTIVITY
Binary reactants: Solutions may corrode metals. Reacts violently with ethylene oxide; potassium, sodium.
Neutralizing agents for acids and caustics: Flush with water, rinse with dilute solution of sodium bicarbonate or soda ash.

ENVIRONMENTAL DATA
Food chain concentration potential: Negative; unlikely to accumulate.
Water pollution: Harmful to aquatic life in very low concentrations. May be dangerous if it enters nearby water intakes; notify operators. Notify local health and wildlife officials. **Response to discharge:** Disperse and flush.

SHIPPING INFORMATION
Grades of purity: Technical; 35% solution in water; **Storage temperature:** Ambient; **Inert atmosphere:** None; **Venting:** Open; **Stability during transport:** Stable.

PHYSICAL AND CHEMICAL PROPERTIES
Physical state @ 59°F/15°C and 1 atm: Solid.
Molecular weight: 198
Specific gravity (water = 1): 1.93 @ 68°F/20°C (solid).

Heat of solution: –18 Btu/lb = –10 cal/g = –0.42 x 10^5 J/kg.
Heat of fusion: 61.5 cal/g.

FERROUS FLUOROBORATE   REC. F:0650

SYNONYMS: FERROUS BOROFLUORIDE

IDENTIFICATION
CAS Number: 15283-51-9
Formula: $Fe(BF_4)_2·H_2O$

DESCRIPTION: Yellow-green liquid. Sinks and mixes with water.

Irritating to the skin, eyes, and respiratory tract • Toxic products of combustion may include irritating and toxic vapors including iron fume and fluorides.

Hazard Classification (based on NFPA-704 Rating System)
Health Hazards (Blue): 1; Flammability (Red): 0; Reactivity (Yellow): 0

EMERGENCY RESPONSE: See Appendix A (171)
Evacuation:
Public safety: Isolate the area of spill or leak for at least 10 to 25 meters (30 to 80 feet) in all directions.
Spill: Increase, in the downwind direction, as necessary, the distance shown under "Public Safety."
Fire: If any large container is involved in fire, isolate for at least 800 meters (½ mile) in all directions; also, consider initial evacuation for 800 meters (½ mile) in all directions.

EXPOSURE
Short-term effects: *SEEK MEDICAL ATTENTION.* **Vapor:** *POISONOUS IF INHALED. IF IN EYES*, hold eyelids open and flush with plenty of water. *IF BREATHING HAS STOPPED,* give artificial respiration. IF breathing is difficult, administer oxygen. **Liquid:** Irritating to skin and eyes. *IF SWALLOWED*, will cause nausea and vomiting. Remove contaminated clothing and shoes. Flush affected areas with plenty of water. *IF IN EYES*, hold eyelids open and flush with plenty of water. *IF SWALLOWED* and victim is *CONSCIOUS AND ABLE TO SWALLOW*, have victim drink 4 to 8 ounces of water and have victim induce vomiting. *IF SWALLOWED* and victim is *UNCONSCIOUS OR HAVING CONVULSIONS,* do nothing except keep victim warm.

HEALTH HAZARDS
Personal protective equipment (PPE): B-Level PPE. Goggles or face shield; rubber gloves.
Exposure limits (TWA unless otherwise noted): ACGIH TLV 1 mg/m³; NIOSH REL 1 mg/m³ as iron, soluble salts.
Liquid or solid irritant characteristics: Eye and skin irritant.

ENVIRONMENTAL DATA
Food chain concentration potential: Negative; unlikely to accumulate.
Water pollution: Effect of low concentrations on aquatic life is unknown. May be dangerous if it enters nearby water intakes; notify operators. Notify local health and wildlife officials.
Response to discharge: Issue warning–water contaminant. Disperse and flush.

SHIPPING INFORMATION
Grades of purity: 50% solution in water; **Storage temperature:** Ambient; **Inert atmosphere:** None; **Venting:** Open; **Stability during transport:** Stable.

PHYSICAL AND CHEMICAL PROPERTIES
Physical state @ 59°F/15°C and 1 atm: Liquid.
Molecular weight: 229.5 (solute only).
Boiling point @ 1 atm: (decomposes).
Specific gravity (water = 1): (estimate) more than 1.1 @ 68°F/20°C (liquid).

FERROUS OXALATE   REC. F:0700

SYNONYMS: FERROUS OXALATE DIHYDRATE; FERROX; IRON PROTOXALATE; OXALIC ACID, FERROUS SALT

IDENTIFICATION
CAS Number: 19469-07-9
Formula: $FeC_2O_4·2H_2O$

DESCRIPTION: Yellow solid. Odorless. Sinks in water.

Toxic products of combustion may include iron fume or iron oxide fume.

Hazard Classification (based on NFPA-704 Rating System)
Health Hazards (Blue): 0; Flammability (Red): 0; Reactivity (Yellow): 0

EMERGENCY RESPONSE: See Appendix A (171)
Evacuation:
Public safety: Isolate the area of spill or leak for at least 10 to 25 meters (30 to 80 feet) in all directions.
Spill: Increase, in the downwind direction, as necessary, the distance shown under "Public Safety."
Fire: If any large container is involved in fire, isolate for at least 800 meters (½ mile) in all directions; also, consider initial evacuation for 800 meters (½ mile) in all directions.

EXPOSURE
Short-term effects: *SEEK MEDICAL ATTENTION.* **Dust:** Irritating to eyes, nose, and throat. *IF INHALED*, will, will cause coughing or difficult breathing. *IF IN EYES*, hold eyelids open and flush with plenty of water. *IF BREATHING HAS STOPPED,* give artificial respiration. IF breathing is difficult, administer oxygen. **Solid:** Irritating to skin and eyes. *IF SWALLOWED*, will cause nausea, vomiting, or loss of consciousness. Remove contaminated clothing and shoes. Flush affected areas with plenty of water. *IF IN EYES*, hold eyelids open and flush with plenty of water. *IF SWALLOWED* and victim is *CONSCIOUS AND ABLE TO SWALLOW*, have victim drink 4 to 8 ounces of water. *IF SWALLOWED* and victim is *UNCONSCIOUS OR HAVING CONVULSIONS,* do nothing except keep victim warm.

HEALTH HAZARDS
Personal protective equipment (PPE): B-Level PPE. Dust mask; goggles or face shield; protective gloves.
Recommendations for respirator selection: NIOSH
50 mg/m³: DMFu (any dust, mist, and fume respirator); or SA (any supplied-air respirator). *125 mg/m³*: SA:CF (any supplied-air respirator operated in a continuous-flow mode); or PAPRDMFu (any powered, air-purifying respirator with a dust, mist, and fume filter). *250 mg/m³*: HiEF (any air-purifying, full-facepiece respirator with a high-efficiency particulate filter); or PAPRTHiE (any powered, air-purifying respirator with a tight-fitting facepiece and

a high-efficiency particulate filter); or SCBAF (any self-contained breathing apparatus with a full facepiece); or SAF (any supplied-air respirator with a full facepiece). *2500 mg/m³*: SA:PD,PP (any supplied-air respirator operated in a pressure-demand or other positive-pressure mode).*EMERGENCY OR PLANNED ENTRY INTO UNKNOWN CONCENTRATIONS OR IDLH CONDITIONS*: SCBAF:PD,PP (any self-contained breathing apparatus that has a full facepiece and is operated in a pressure-demand or other positive-pressure mode); or SAF:PD,PP:ASCBA (any supplied-air respirator that has a full facepiece and is operated in a pressure-demand or other positive-pressure mode in combination with an auxiliary, self-contained breathing apparatus operated in a pressure-demand or other positive-pressure mode). Escape: HiEF (any air-purifying, full-facepiece respirator with a high-efficiency particulate filter); or SCBAE (any appropriate escape-type, self-contained breathing apparatus).
Liquid or solid irritant characteristics: Eye and skin and respiratory tract irritant.

ENVIRONMENTAL DATA
Food chain concentration potential: Negative; unlikely to accumulate.
Water pollution: Effect of low concentrations on aquatic life is unknown. May be dangerous if it enters nearby water intakes; notify operators. Notify local health and wildlife officials.
Response to discharge: Should be removed. Chemical and physical treatment.

SHIPPING INFORMATION
Grades of purity: Commercial; **Storage temperature:** Ambient; **Inert atmosphere:** None; **Venting:** Open; **Stability during transport:** Stable.

PHYSICAL AND CHEMICAL PROPERTIES
Physical state @ 59°F/15°C and 1 atm: Solid.
Molecular weight: 179.9
Boiling point @ 1 atm: (decomposes).
Specific gravity (water = 1): 2.3 @ 68°F/20°C (solid)

FERROUS SULFATE REC. F:0750

SYNONYMS: COPPERAS; DURETTER; DUROFERON; EXSICCATED FERROUS SULFATE; EXSICCATED FERROUS SULPHATE; FEOSOL; FEOSPAN; FER-IN-SOL; FERRO-GRADUMET; FERRALYN; FERROSULFAT (German); FERROSULFATE; FERRO-THERON; FERROUS SULFATE; FERROUS SULPHATE; FERSOLATE; GREEN VITRIOL IRON MONOSULFATE; IRON PROTOSULFATE; IRON SULFATE; IRON(II) SULFATE; IRON(2+) SULFATE; IRON (OUS) SULFATE; IRON VITRIOL; IROSPAN; SLOW-FE; SULFERROUS; SULFURIC ACID IRON SALT; SULFURIC ACID, IRON(2+) SAL

IDENTIFICATION
CAS Number: 7720-78-7; 7782-63-0 (heptahydrate)
Formula: $O_4S \cdot Fe$; $FeSO_4 \cdot 7H_2O$ (heptahydrate)
DOT ID Number: UN 9125; DOT Guide Number: 171
Proper Shipping Name: Ferrous sulfate
Reportable Quantity (RQ): **(CERCLA)** 1000 lb/454 kg

DESCRIPTION: Bluish-green crystalline solid; turns brown on exposure to air. Odorless. Sinks and mixes slowly with water forming a medium strong acid.

Poison! •Toxic products of combustion may include sulfur dioxide • Do not put yourself in danger by entering a contaminated area to rescue a victim.

Hazard Classification (based on NFPA-704 Rating System)
Health Hazards (Blue): 2; Flammability (Red): 0; Reactivity (Yellow): 0

EMERGENCY RESPONSE: See Appendix A (171)
Evacuation:
Public safety: Isolate the area of spill or leak for at least 10 to 25 meters (30 to 80 feet) in all directions.
Spill: Increase, in the downwind direction, as necessary, the distance shown under "Public Safety."
Fire: If any large container is involved in fire, isolate for at least 800 meters (½ mile) in all directions; also, consider initial evacuation for 800 meters (½ mile) in all directions.

EXPOSURE
Short-term effects: *SEEK MEDICAL ATTENTION.* **Solid:** Poisonous. *IF SWALLOWED,* will cause nausea, vomiting, or loss of consciousness. *IF SWALLOWED* and victim is *CONSCIOUS AND ABLE TO SWALLOW,* have victim drink 4 to 8 ounces of water, and have victim induce vomiting. *IF SWALLOWED* and victim is *UNCONSCIOUS OR HAVING CONVULSIONS,* do nothing except keep victim warm.

HEALTH HAZARDS
Personal protective equipment (PPE): B-Level PPE.
Exposure limits (TWA unless otherwise noted): ACGIH TLV 1 mg/m³; NIOSH REL 1 mg/m³ as iron soluble salts
Toxicity by ingestion: Poison. Grade 2; LD_{50} = 0.5 to 5 g/kg (rat).
Long-term health effects: May cause tumors, reproductive problems, and cancer.

FIRE DATA
Fire extinguishing agents not to be used: Water streams applied to adjacent fires will spread contamination of environmentally hazardous substance over wide area.

CHEMICAL REACTIVITY Water solution is acidic.
Binary reactants: Violent reaction with aluminum, magnesium. Reacts with oxygen in air forming ferric sulfate. Reactions with bases form iron.

ENVIRONMENTAL DATA
Food chain concentration potential: Negative; unlikely to accumulate.
Water pollution: Harmful to aquatic life in very low concentrations. May be dangerous if it enters nearby water intakes; notify operators. Notify local health and wildlife officials. **Response to discharge:** Disperse and flush.

SHIPPING INFORMATION
Grades of purity: USP; Commercial; **Storage temperature:** Ambient; **Inert atmosphere:** None; **Venting:** Open; **Stability during transport:** Stable.

PHYSICAL AND CHEMICAL PROPERTIES
Physical state @ 59°F/15°C and 1 atm: Solid.
Molecular weight: 278.0
Melting/Freezing point: 149°F/65°C/338°K (decomposes).
Specific gravity (water = 1): 1.90 @ 15°C (solid).
Heat of decomposition: More than 752°F/400°C.

FLUORINE   REC. F:0800

SYNONYMS: EEC No. 009-001-00-0; FLUOR (French, German, Spanish); FLUORINE-19; FLUORURES ACIDE (French); SAEURE FLUORIDE (German); RCRA No. P056

IDENTIFICATION
CAS Number: 7782-41-4
Formula: F_2
DOT ID Number: UN 9192 (cryogenic liquid); UN 1045 (compressed gas); DOT Guide Number: 167 (cryogenic liquid); 124 (compressed gas)
Proper Shipping Name: Fluorine, compressed gas; Fluorine, refrigerated liquid (cryogenic liquid)
Reportable Quantity (RQ): **(CERCLA)** 10 lb/4.54 kg

DESCRIPTION: Pale yellow or greenish-yellow gas. Stored and shipped as a compressed gas or cryogenic liquid in special cylinders without relief valves. Intense irritating odor; choking. Liquid sinks and boils violently in water forming toxic hydrogen fluoride, ozone, and oxygen difluoride.

Poison! • Breathing the vapor can kill you; skin and eye contact causes severe burns and blindness • Contact with gas or liquefied gas may cause severe injury or frostbite • Firefighting gear (including SCBA) does not provide adequate protection. If exposure occurs, remove and isolate gear immediately and thoroughly decontaminate personnel • Containers may BLEVE when exposed to fire • Vapors are heavier than air and will collect and stay in low areas • Strong oxidizer. forms explosive or combustible mixtures with most materials including all fuels and most metals. Heat forms oxygen; will increase the activity of an existing fire • Toxic products of combustion may include hydrogen fluoride • Do not put yourself in danger by entering a contaminated area to rescue a victim.

Hazard Classification (based on NFPA-704 Rating System)
Health Hazards (Blue): 4; Flammability (Red): 0; Reactivity (Yellow): 4; Special Notice (White): Water reactive

EMERGENCY RESPONSE, gas: See Appendix A (124)
Evacuation:
Public safety: Isolate spill area for at least 100 to 200 meters (330 to 660 feet) in all directions.
Spill: Small spill–First: Isolate in all directions 30 meters (100 feet); Then: Protect persons downwind, DAY: 0.2 km (0.1 mile); NIGHT: 0.5 km (0.3 mile). Large spill–First: Isolate in all directions 185 meters (600 feet); Then: Protect persons downwind, DAY: 1.4 km (0.9 mile); NIGHT: 4.0 km (2.5 miles).
Fire: Isolate for 800 meters (½ mile) in all directions, especially if tank, rail car, or tank truck is involved in fire.
EMERGENCY RESPONSE, liquid: See Appendix A (167)
Evacuation:
Public safety: Isolate the area of spill or leak for at least 100 to 200 meters (330 to 660 feet) in all directions.
Spill: Increase, in the downwind direction, as necessary, the distance shown under "Public Safety."
Fire: If tank, rail car, or tank truck is involved in fire, isolate for at least 1600 meters (1-mile) in all directions; also, consider initial evacuation for 1600 meters (1-mile) in all directions.

EXPOSURE
Short-term effects: *SEEK MEDICAL ATTENTION.* **Vapor:** *POISONOUS IF INHALED.* May cause lung edema; physical exertion will aggravate this condition. Irritating to eyes. Move to fresh air. *IF BREATHING HAS STOPPED,* give artificial respiration. IF breathing is difficult, administer oxygen. **Liquid:** Will burn skin and eyes. Will cause frostbite. Flush affected areas with plenty of water. *IF IN EYES,* hold eyelids open and flush with plenty of water. *DO NOT RUB AFFECTED AREAS.*
Note to physician or authorized medical personnel: Medical observation is recommended for 24 to 48 hours after breathing overexposure, as pulmonary edema may be delayed. As first aid for pulmonary edema, consider administering a corticosteroid spray. Cigarette smoking may exacerbate pulmonary injury and should be discouraged for at least 72 hours following exposure.

HEALTH HAZARDS
Personal protective equipment (PPE): A-Level PPE. OSHA Table Z-1-A air contaminant. Tight-fitting chemical goggles; special clothing, not easily ignited by fluorine gas. Wear thermal protective clothing. Chemical protective material(s) reported to have good to excellent resistance: Viton®. Also, neoprene, Viton®/neoprene, and PVC offers limited protection.
Recommendations for respirator selection: NIOSH/OSHA
1 ppm: SA* (any supplied-air respirator). *2.5 ppm*: SA:CF* (any supplied-air respirator operated in a continuous-flow mode). *5 ppm:* SCBAF (any self-contained breathing apparatus with a full facepiece); or SAF (any supplied-air respirator with a full facepiece). *25 ppm:* SAF:PD,PP (any supplied-air respirator that has a full facepiece and is operated in a pressure-demand or other positive-pressure mode). *EMERGENCY OR PLANNED ENTRY INTO UNKNOWN CONCENTRATIONS OR IDLH CONDITIONS:* SCBAF:PD,PP (any self-contained breathing apparatus that has a full facepiece and is operated in a pressure-demand or other positive-pressure mode); or SAF:PD,PP:ASCBA (any supplied-air respirator that has a full facepiece and is operated in a pressure-demand or other positive-pressure mode in combination with an auxiliary self-contained breathing apparatus operated in a pressure-demand or other positive pressure mode). *ESCAPE:* GMFS** [any air-purifying, full-facepiece respirator (gas mask) with a chin-style, front- or back-mounted canister providing protection against the compound of concern]; or SCBAE (any appropriate escape-type, self-contained breathing apparatus). *Notes:* Substance reported to cause eye irritation or damage; may require eye protection.** End of service life indicator (ESLI) required.
Exposure limits (TWA unless otherwise noted): ACGIH TLV 1 ppm (1.6 mg/m^3); NIOSH/OSHA 0.1 ppm (0.2 mg/m^3).
Short-term exposure limits (15-minute TWA): ACGIH STEL 2 ppm (3.1 mg/m^3).
Long-term health effects: Severe burns may develop slowly after exposure.
Vapor (gas) irritant characteristics: Can be fatal if inhaled; possible respiratory paralysis. Vapors cause severe irritation of eye and throat and can cause eye and lung injury. They cannot be tolerated even at low concentrations.
Liquid or solid irritant characteristics: Severe skin irritant. Causes second- and third-degree burns on short contact and is very injurious to the eyes. May cause frostbite.
Odor threshold: 0.1–0.2 ppm.
IDLH value: 25 ppm.

FIRE DATA
Fire extinguishing agents not to be used: Water may be used to cool containers, but do not direct water onto fluorine leaks. Reaction may be violent with formation of hydrogen fluoride, ozone, and oxygen which will increase the activity of an existing fire.
Behavior in fire: Produces dangerously reactive gas. Ignites most combustibles.

CHEMICAL REACTIVITY
Reactivity with water: Reacts with water to form hydrofluoric acid, oxygen, and oxygen difluoride.
Binary reactants: Reacts violently with all combustible materials; with metals, except the metal cylinders in which it is shipped; with nitric acid. Reacts with nearly every known element. Corrosive to metals in presence of moisture.

ENVIRONMENTAL DATA
Food chain concentration potential: Negative; unlikely to accumulate.
Water pollution: Harmful to aquatic life in very low concentrations. May be dangerous if it enters nearby water intakes; notify operators. Notify local health and wildlife officials. **Response to discharge:** Issue warning–poison. Restrict access. *EVACUATE AREA.*

SHIPPING INFORMATION
Grades of purity: 98%; **Storage temperature:** Ambient, cool; **Inert atmosphere:** None; **Venting:** Safety relief; **Stability during transport:** Stable.

PHYSICAL AND CHEMICAL PROPERTIES
Physical state @ 59°F/15°C and 1 atm: Gas.
Molecular weight: 37.99
Boiling point @ 1 atm: –306°F/–188°C/85°K.
Melting/Freezing point: –362°F/–219°C/54°K.
Critical temperature: –199.5°F/–128.6°C/144.6°K.
Critical pressure: 809.7 psia = 55.08 atm = 5.58 MN/m^2.
Specific gravity (water = 1): 1.5 at –188°C (liquid).
Ratio of specific heats of vapor (gas): 1.362
Latent heat of vaporization: 71.6 Btu/lb = 39.8 cal/g = 1.67 x 10^5 J/kg.
Heat of fusion: 244.0 cal/g.

2-FLUOROANILINE   REC. F:0850

SYNONYMS: 1-AMINO-2-FLUOROBENZENE; 2-FLUOROPHENYLAMINE; FLUORANILINA (Spanish); *o*-FLUOROANILINE; 2-FLUOROBENZENAMINE

IDENTIFICATION
CAS Number: 348-54-9
Formula: C$_6$H$_6$FN; 2-FC$_6$H$_4$NH$_2$
DOT ID Number: UN 2941; DOT Guide Number: 153
Proper Shipping Name: Fluoroanilines

DESCRIPTION: Clear to pale yellow liquid. Mild, sweet, amine-like odor. Sinks and mixes slowly with water.

Flammable • Severely irritating to eyes, skin, and respiratory tract • Containers may BLEVE when exposed to fire • Vapors may form explosive mixture with air • Vapors are heavier than air and will collect and stay in low areas • Vapors may travel long distances to ignition sources and flashback • Vapors in confined areas (e.g., tanks, sewers, buildings) may explode when exposed to fire • Toxic products of combustion may include nitrogen oxides and hydrogen fluoride.

Hazard Classification (based on NFPA-704 Rating System)
Health Hazards (Blue): 2; Flammability (Red): 2; Reactivity (Yellow): 0

EMERGENCY RESPONSE: See Appendix A (153)
Evacuation:
Public safety: Isolate the area of spill or leak for at least 25 to 50 meters (80 to 160 feet) in all directions.
Spill: Increase, in the downwind direction, as necessary, the distance shown under "Public Safety."
Fire: If tank, rail car, or tank truck is involved in fire, isolate for at least 800 meters (½ mile) in all directions; also, consider initial evacuation for 800 meters (½ mile) in all directions.

EXPOSURE
Short-term effects: *SEEK MEDICAL ATTENTION.* **Vapor:** HARMFUL IF INHALED. Remove to fresh air. *IF BREATHING HAS STOPPED*, give artificial respiration; *avoid mouth-to-mouth resuscitation; use bag/mask apparatus.* IF breathing is difficult, administer oxygen. **Liquid:** *POISONOUS IF SWALLOWED OR IF SKIN IS EXPOSED.* Remove contaminated clothing and shoes. Flush affected areas with plenty of water. *IF IN EYES*, hold eyelids open and flush with plenty of water. *IF SWALLOWED* and victim is *CONSCIOUS AND ABLE TO SWALLOW*, have victim drink 4 to 8 ounces of water and induce vomiting. *IF SWALLOWED* and victim is *UNCONSCIOUS OR HAVING CONVULSIONS*, do nothing except keep victim warm.

HEALTH HAZARDS
Personal protective equipment (PPE): B-Level PPE. Rubber gloves; chemical goggles; protective clothing; approved respirator.
Toxicity by ingestion: Poison.
Liquid or solid irritant characteristics: Eye irritant.

FIRE DATA
Flash point: 140°F/60°C (cc).

ENVIRONMENTAL DATA
Water pollution: Effect of low concentrations on aquatic life is unknown. May be dangerous if it enter water intakes. Notify local health and wildlife officials. **Response to discharge:** Issue warning–poison. Restrict access. Should be removed. Chemical and physical treatment.

SHIPPING INFORMATION
Grades of purity: 99.0%, Technical; **Storage temperature:** Ambient; **Inert atmosphere:** None; **Venting:** Open (flame arrester); **Stability during transport:** Stable. Store containers in a well-ventilated area.

NAS HAZARD CLASSIFICATIONS FOR BULK WATER TRANSPORTATION
FIRE: 2
HEALTH: Vapor irritant: 2; Liquid or solid irritant: 1; Poisons:–
WATER POLLUTION: Human toxicity:–; Aquatic toxicity:–; Aesthetic effect: 1
REACTIVITY: Other chemicals: 2; Water: 0; Self-reaction: 0

PHYSICAL AND CHEMICAL PROPERTIES
Physical state @ 59°F/15°C and 1 atm: Liquid.
Molecular weight: 111.2
Boiling point @ 1 atm: 347°F/175°C/448°K.
Melting/Freezing point: –19°F/–28C = 245°K.
Specific gravity (water = 1): 1.1513 at 21°C.

4-FLUOROANILINE   REC. F:0900

SYNONYMS: 1-AMINO-4-FLUOROBENZENE;

BENZENEAMINE, *p*-FLUORANILINA (Spanish); 4-FLUORO-; 4-FLUOROPHENYLAMINE; *p*-FLUOROANILINE; 4-FLUOROBENZENAMINE

IDENTIFICATION
CAS Number: 371-40-4
Formula: C_6H_6FN; $4-FC_6H_4NH_2$
DOT ID Number: UN 2941; DOT Guide Number: 153
Proper Shipping Name: Fluoroanilines

DESCRIPTION: Clear to pale yellow liquid. Mild, sweet odor; amine. Sinks and mixes slowly with water.

Poison! • Combustible • Severely irritating to the eyes, skin, and respiratory tract • Containers may BLEVE when exposed to fire • Vapors may form explosive mixture with air • Vapors are heavier than air and will collect and stay in low areas • Vapors may travel long distances to ignition sources and flashback • Vapors in confined areas (e.g., tanks, sewers, buildings) may explode when exposed to fire • Toxic products of combustion may include nitrogen oxides, fluorine

Hazard Classification (based on NFPA-704 Rating System)
Health Hazards (Blue): 2; Flammability (Red): 2; Reactivity (Yellow): 0

EMERGENCY RESPONSE: See Appendix A (153)
Evacuation:
Public safety: Isolate the area of spill or leak for at least 25 to 50 meters (80 to 160 feet) in all directions.
Spill: Increase, in the downwind direction, as necessary, the distance shown under "Public Safety."
Fire: If tank, rail car, or tank truck is involved in fire, isolate for at least 800 meters (½ mile) in all directions; also, consider initial evacuation for 800 meters (½ mile) in all directions.

EXPOSURE
Short-term effects: *SEEK MEDICAL ATTENTION*. **Vapor:** *MAY BE HARMFUL IF INHALED.* Remove to fresh air. *IF BREATHING HAS STOPPED,* give artificial respiration; *avoid mouth-to-mouth resuscitation; use bag/mask apparatus.* IF breathing is difficult, administer oxygen. **Liquid:** *POISONOUS IF SWALLOWED OR IF SKIN IS EXPOSED.* Remove contaminated clothing and shoes. Flush affected areas with plenty of water. *IF IN EYES,* hold eyelids open and flush with plenty of water. *IF SWALLOWED* and victim is *CONSCIOUS AND ABLE TO SWALLOW,* have victim drink 4 to 8 ounces of water and induce vomiting. *IF SWALLOWED* and victim is *UNCONSCIOUS OR HAVING CONVULSIONS,* do nothing except keep victim warm.

HEALTH HAZARDS
Personal protective equipment (PPE): B-Level PPE. Rubber gloves; chemical goggles; protective clothing; dust respirator.
Toxicity by ingestion: Grade 3; oral LD_{50} = 417 mg/kg (rat).
Long-term health effects: EPA Genetic Toxicology Program; mutation data. Repeated exposure of skin may cause dermatitis due to defatting action. Chronic inhalation of vapors or mist may result to damage to lungs, liver and kidneys. Acute vapor exposures can cause symptoms ranging from coughing to transient anesthesia and central nervous system depression.
Liquid or solid irritant characteristics: Eye irritant.

FIRE DATA
Flash point: 165°F/74°C (cc).

ENVIRONMENTAL DATA
Water pollution: Effect of low concentrations on aquatic life is unknown. May be dangerous if it enters nearby water intakes; notify operators. Notify local health and wildlife officials. **Response to discharge:** Issue warning–poison. Restrict access. Should be removed. Chemical and physical treatment.

SHIPPING INFORMATION
Grades of purity: 99.0%, Technical; **Storage temperature:** Ambient; Store containers in a well-ventilated area; **Inert atmosphere:** None; **Venting:** Open (flame arrester); **Stability during transport:** Stable.

NAS HAZARD CLASSIFICATION FOR BULK WATER TRANSPORTATION
FIRE: 1
HEALTH: Vapor irritant: 2; Liquid or solid irritant: 1; Poisons:–
WATER POLLUTION: Human toxicity: 3; Aquatic toxicity:–; Aesthetic effect: 2
REACTIVITY: Other chemicals: 2; Water: 0; Self-reaction: 0

PHYSICAL AND CHEMICAL PROPERTIES
Physical state @ 59°F/15°C and 1 atm: Liquid.
Molecular weight: 111.2
Boiling point @ 1 atm: 358.7°F/181.5°C/454.7°K.
Melting/Freezing point: 31°F/–1°C/272°K.
Specific gravity (water = 1): 1.1725 @ 68°F/20°C.

FLUOROBENZENE REC. F:0950

SYNONYMS: BENZENE, FLUORO; FLUORBENCENO (Spanish); MONOFLUOROBENZENE; PHENYL FLUORIDE; BENZENE FLUORIDE; MFB

IDENTIFICATION
CAS Number: 462-06-6
Formula: C_6NO_5F
DOT ID Number: UN 2387; DOT Guide Number: 128
Proper Shipping Name: Fluorobenzene

DESCRIPTION: Colorless, watery liquid. Benzene-like odor. Sink in water; insoluble. Flammable vapor is produced.
Liquid.

Highly flammable • Containers may BLEVE when exposed to fire • Vapors may form explosive mixture with air • Vapors are heavier than air and will collect and stay in low areas • Vapors may travel long distances to ignition sources and flashback • Vapors in confined areas (e.g., tanks, sewers, buildings) may explode when exposed to fire • Irritating to the skin, eyes, and respiratory tract • Toxic products of combustion may include hydrogen fluoride.

Hazard Classification (based on NFPA-704 Rating System)
Health Hazards (Blue): 2; Flammability (Red): 3; Reactivity (Yellow): 0

EMERGENCY RESPONSE: See Appendix A (128)
Evacuation:
Public safety: Isolate spill area for at least 25 to 50 meters (80 to 160 feet) in all directions.
Spill: Large spill–Consider initial downwind evacuation for at least 300 meters (1000 feet).
Fire: Isolate for 800 meters (½ mile) in all directions, especially if tank, rail car, or tank truck is involved in fire.

EXPOSURE
Short-term effects: *SEEK MEDICAL ATTENTION.* **Vapor:** IF inhaled, will cause coughing or dizziness. Not irritating to eyes, nose, and throat. Move to fresh air. *IF BREATHING HAS STOPPED,* give artificial respiration. IF breathing is difficult, administer oxygen. **Liquid:** Irritating to skin and eyes. Harmful if swallowed. Remove contaminated clothing and shoes. Flush affected areas with plenty of water. *IF IN EYES,* hold eyelids open and flush with plenty of water. *IF SWALLOWED* and victim is *CONSCIOUS AND ABLE TO SWALLOW,* have victim drink 4 to 8 ounces of water.

HEALTH HAZARDS
Personal protective equipment (PPE): B-Level PPE. Organic vapor-acid gas respirator where appropriate; gloves; chemical safety spectacles, plus face shield where appropriate; rubber footwear; apron or impervious clothing for splash protection; hard hat. Chemical protective material(s) reported to offer minimal to poor protection: Viton®/neoprene.
Toxicity by ingestion: Grade 2: LD_{50} = 4.4 g/kg (rat).
Vapor (gas) irritant characteristics: Eye and respiratory tract irritant.
Liquid or solid irritant characteristics: Eye and skin irritant. If spilled on clothing and allowed to remain, may cause smarting and reddening of the skin.

FIRE DATA
Flash point: 5°F/–15°C (cc).

CHEMICAL REACTIVITY
Binary reactants: Reacts with oxidizers, ammonium nitrate, chromic acid, halogens, hydrogen peroxide, nitric acid.

ENVIRONMENTAL DATA
Water pollution: Harmful to aquatic life in very low concentrations. May be dangerous if it enters nearby water intakes; notify operators. Notify local health and wildlife officials. Notify operators of nearby intakes. **Response to discharge:** Should be removed. Chemical and physical treatment.

SHIPPING INFORMATION
Grades of purity: 99%; **Storage temperature:** Ambient; **Inert atmosphere:** None; **Venting:** Pressure-vacuum; **Stability during transport:** Stable.

NAS HAZARD CLASSIFICATION FOR BULK WATER TRANSPORTATION
FIRE: 4
HEALTH: Vapor irritant: 0; Liquid or solid irritant: 1; Poisons: 2
WATER POLLUTION: Human toxicity: 1; Aquatic toxicity: 3; Aesthetic effect: 2
REACTIVITY: Other chemicals: 1; Water: 0; Self-reaction: 0

PHYSICAL AND CHEMICAL PROPERTIES
Physical state @ 59°F/15°C and 1 atm: Liquid.
Molecular weight: 96.10
Boiling point @ 1 atm: 185.2°F/85.1°C/358.3°K.
Melting/Freezing point: –42.2°F/–41.2°C/232°K.
Critical temperature: 547°F/286°C/559°K.
Critical pressure: 656 psia = 44.6 atm = 4.52 MN/m^2.
Specific gravity (water = 1): 1.0225 @ 68°F/20°C (liquid).
Relative vapor density (air = 1): 3.31
Heat of combustion: (estimate) –13,995 Btu/lb = –7775 cal/g = –325 x 10^5 J/kg.
Vapor pressure: (Reid) 2.8 psia.

FLUOROSILICIC ACID   REC. F:1000

SYNONYMS: ACIDO FLUOSILICICO (Spanish); HYDROFLUOSILIC ACID; HYDROGEN HEXAFLUOROSILICATE; HEXAFLUOSILICIC ACID; SILICOFLUORIC ACID; FLUOSILIC ACID; SAND ACID

IDENTIFICATION
CAS Number: 16961-83-4
Formula: $F_6Si \cdot 2H$; $H_2SiF_6 \cdot H_2O$
DOT ID Number: UN 1778; DOT Guide Number: 154
Proper Shipping Name: Fluorosilicic acid

DESCRIPTION: Clear, colorless or light yellow, fuming liquid. Unpleasant, sour odor. Sinks and mixes with water. Freezes at 63°F/17°C.

Poison! • Corrosive • Breathing the vapor can kill you; skin or eye contact causes severe burns, impaired vision, or blindness • Firefighting gear (including SCBA) does not provide adequate protection. If exposure occurs, remove and isolate gear immediately and thoroughly decontaminate personnel • Containers may BLEVE when exposed to fire • Toxic products of combustion may include hydrogen fluoride • Do not put yourself in danger by entering a contaminated area to rescue a victim.

Hazard Classification (based on NFPA-704 Rating System)
Health Hazards (Blue): 3; Flammability (Red): 0; Reactivity (Yellow): 0

EMERGENCY RESPONSE: See Appendix A (154)
Evacuation:
Public safety: Isolate the area of spill or leak for at least 25 to 50 meters (80 to 160 feet) in all directions.
Spill: Increase, in the downwind direction, as necessary, the distance shown under "Public Safety."
Fire: If tank, rail car, or tank truck is involved in fire, isolate for at least 800 meters (½ mile) in all directions; also, consider initial evacuation for 800 meters (½ mile) in all directions.

EXPOSURE
Short-term effects: *SEEK MEDICAL ATTENTION.* **Vapor:** Irritating to eyes, nose, and throat. *IF INHALED*, will, will cause coughing or difficult breathing. *IF IN EYES*, hold eyelids open and flush with plenty of water. *IF BREATHING HAS STOPPED,* give artificial respiration; *avoid mouth-to-mouth resuscitation; use bag/mask apparatus.* IF breathing is difficult, administer oxygen. May cause lung edema; physical exertion will aggravate this condition. **Liquid:** Will burn skin and eyes. *IF SWALLOWED*, will cause nausea. Remove contaminated clothing and shoes. Flush affected areas with plenty of water. *IF IN EYES,* hold eyelids open and flush with plenty of water. *IF SWALLOWED* and victim is *CONSCIOUS AND ABLE TO SWALLOW,* have victim drink 4 to 8 ounces of water. *IF SWALLOWED* and victim is *UNCONSCIOUS OR HAVING CONVULSIONS,* do nothing except keep victim warm.
Note to physician or authorized medical personnel: Medical observation is recommended for 24 to 48 hours after breathing overexposure, as pulmonary edema may be delayed. As first aid for pulmonary edema, consider administering a corticosteroid spray. Cigarette smoking may exacerbate pulmonary injury and should be discouraged for at least 72 hours following exposure.

HEALTH HAZARDS
Personal protective equipment (PPE): B-Level PPE. Rubber

gloves; safety glasses; protective clothing. Chemical protective material(s) reported to offer minimal to poor protection: Viton®/neoprene, butyl rubber/neoprene.
Vapor (gas) irritant characteristics: Severe eye and respiratory tract irritant.
Liquid or solid irritant characteristics: Severe eye and skin irritant.

CHEMICAL REACTIVITY
Binary reactants: Will corrode most metals, producing flammable hydrogen gas, which may collect in enclosed spaces.
Neutralizing agents for acids and caustics: Flush with water, rinse with dilute solution of sodium carbonate or soda ash.

ENVIRONMENTAL DATA
Food chain concentration potential: Negative; unlikely to accumulate.
Water pollution: Effect of low concentrations on aquatic life is unknown. May be dangerous if it enters nearby water intakes; notify operators. Notify local health and wildlife officials.
Response to discharge: Issue warning–corrosive. Restrict access. Disperse and flush.

SHIPPING INFORMATION
Grades of purity: 22–30% solutions in water; **Storage temperature:** Ambient; **Inert atmosphere:** None; **Venting:** Open; **Stability during transport:** Stable.

NAS HAZARD CLASSIFICATION FOR BULK WATER TRANSPORTATION
FIRE: 0
HEALTH: Vapor irritant: 0; Liquid or solid irritant: 3; Poisons: 3
WATER POLLUTION: Human toxicity: 3; Aquatic toxicity: 0; Aesthetic effect: 1
REACTIVITY: Other chemicals: 0; Water: 0; Self-reaction: 0

PHYSICAL AND CHEMICAL PROPERTIES
Physical state @ 59°F/15°C and 1 atm: Liquid.
Molecular weight: 144.09 (solute only).
Boiling point @ 1 atm: Decomposes approximately 212°F/100°C/373°K.
Melting/Freezing point: (typical) –24-4°F/–31 to –20°C/242-253°K.
Specific gravity (water = 1): (approximate) 1.3 @ 77°F/25°C (liquid).

2-FLUOROTOLUENE   REC. F:1050

SYNONYMS: 2-FLUORO-1-METHYLBENZENE; FLUORTOLUENO (Spanish); o-FLUOROTOLUENE; 1-FLUORO-2-METHYLBENZENE; 1-METHYL-2-FLUOROBENZENE; SAND ACID; o-TOLYL FLUORIDE

IDENTIFICATION
CAS Number: 95-52-3
Formula: C_7H_7F; $CH_3-C_6H_4-F$
DOT ID Number: UN 2388; DOT Guide Number: 128
Proper Shipping Name: Fluorotoluenes

DESCRIPTION: Colorless liquid. Aromatic odor. May sink or float on water.

Highly flammable • Containers may BLEVE when exposed to fire • Vapors may form explosive mixture with air • Vapors are heavier than air and will collect and stay in low areas • Vapors may travel long distances to ignition sources and flashback • Vapors in confined areas (e.g., tanks, sewers, buildings) may explode when exposed to fire • Irritating to the skin, eyes, and respiratory tract • Toxic products of combustion may include hydrogen fluoride.

Hazard Classification (based on NFPA-704 Rating System)
Health Hazards (Blue): 2; Flammability (Red): 3; Reactivity (Yellow): 0

EMERGENCY RESPONSE: See Appendix A (128)
Evacuation:
Public safety: Isolate spill area for at least 25 to 50 meters (80 to 160 feet) in all directions.
Spill: Large spill–Consider initial downwind evacuation for at least 300 meters (1000 feet).
Fire: Isolate for 800 meters (½ mile) in all directions, especially if tank, rail car, or tank truck is involved in fire.

EXPOSURE
Short-term effects: *SEEK MEDICAL ATTENTION.* **Vapor:** May be harmful if inhaled or absorbed through the skin. Irritating to eyes, skin, nose and throat. Move to fresh air. *IF BREATHING HAS STOPPED,* give artificial respiration. IF breathing is difficult, administer oxygen. **Liquid:** Irritating to skin and eyes. Harmful if swallowed or absorbed through the skin. *IF IN EYES OR ON SKIN:* Flush with running water for at least 15 minutes; hold eyelids open if necessary. *IF SWALLOWED* and victim is *CONSCIOUS:* Have victim drink 4 to 8 ounces of water. **Do not induce vomiting.** *IF SWALLOWED* and victim is *UNCONSCIOUS OR HAVING CONVULSIONS:* Do nothing except keep victim warm. Remove and double bag contaminated clothing and shoes at the site.

HEALTH HAZARDS
Personal protective equipment (PPE): B-Level PPE. Wear self-contained positive pressure breathing apparatus and full protective clothing.
Toxicity by ingestion: Grade 3; LD_{50} = 100 mg/kg (BWD).
Long-term health effects: Prolonged and repeated vapor exposure may produce systemic effects.
Vapor (gas) irritant characteristics: Respiratory tract irritant. Vapors are irritating such that personnel will not usually tolerate moderate or high concentrations.
Liquid or solid irritant characteristics: Eye and skin irritant. If spilled on clothing and allowed to remain, may cause smarting and reddening of skin.

FIRE DATA
Flash point: 55°F/13°C (cc).
Behavior in fire: Container may explode in heat of fire.

CHEMICAL REACTIVITY
Binary reactants: Reacts with oxidizing agents.

ENVIRONMENTAL DATA
Water pollution: Effects of low concentration on aquatic life is unknown. May be dangerous if it enters nearby water intakes; notify operators. Notify local health and wildlife officials. Notify operators of local water intakes. **Response to discharge:** Issue warning–high flammability., water contaminant. Restrict access. Should be removed. Chemical and physical treatment.

SHIPPING INFORMATION
Grades of purity: 99+%; **Storage temperature:** Ambient; **Stability during transport:** Stable.

NAS HAZARD CLASSIFICATION FOR BULK WATER TRANSPORTATION
FIRE: 3
HEALTH: Vapor irritant: 2; Liquid or solid irritant: 1; Poisons: 2
WATER POLLUTION: Human toxicity: 3; Aquatic toxicity:–; Aesthetic effect: 1
REACTIVITY: Other chemicals: 2; Water: 0; Self-reaction: 0

PHYSICAL AND CHEMICAL PROPERTIES
Physical state @ 59°F/15°C and 1 atm: Liquid.
Molecular weight: 110.13
Boiling point @ 1 atm: 237.2°F/114°C/387.2°K.
Melting/Freezing point: –79.6°F/–62°C/211.2°K.
Specific gravity (water = 1): 1.0041 at 13°C.
Relative vapor density (air = 1): 3.8
Vapor pressure: (Reid) 0.91 psia.

3-FLUOROTOLUENE   REC. F:1100

SYNONYMS: *m*-FLUOROTOLUENE; *m*-TOLYL FLUORIDE; *m*-FLUORTOLUENO (Spanish); 1-FLUORO-3-METHYLBENZENE; 1-METHYL-3-FLUOROBENZENE

IDENTIFICATION
CAS Number: 352-70-5
Formula: C_7H_7F; 3-$FC_6H_4CH_3$
DOT ID Number: UN 2388; DOT Guide Number: 128
Proper Shipping Name: Fluorotoluenes

DESCRIPTION: Colorless liquid. Aromatic odor. May float or sink in water.

Highly flammable • Containers may BLEVE when exposed to fire • Vapors may form explosive mixture with air • Vapors are heavier than air and will collect and stay in low areas • Vapors may travel long distances to ignition sources and flashback • Vapors in confined areas (e.g., tanks, sewers, buildings) may explode when exposed to fire • Irritating to the skin, eyes, and respiratory tract • Toxic products of combustion may include hydrogen fluoride.

Hazard Classification (based on NFPA-704 Rating System)
Health Hazards (Blue): 2; Flammability (Red): 3; Reactivity (Yellow): 0

EMERGENCY RESPONSE: See Appendix A (128)
Evacuation:
Public safety: Isolate spill area for at least 25 to 50 meters (80 to 160 feet) in all directions.
Spill: Large spill–Consider initial downwind evacuation for at least 300 meters (1000 feet).
Fire: Isolate for 800 meters (½ mile) in all directions, especially if tank, rail car, or tank truck is involved in fire.

EXPOSURE
Short-term effects: *SEEK MEDICAL ATTENTION*. **Vapor:** May be harmful if inhaled or absorbed through the skin. Irritating to eyes, skin, nose and throat. Move to fresh air. *IF BREATHING HAS STOPPED,* give artificial respiration. IF breathing is difficult, administer oxygen. **Liquid:** Harmful if swallowed or absorbed through skin. Irritating to skin and eyes. *IF IN EYES OR ON SKIN:* Flush with running water for at least 15 minutes, hold eyelids open if necessary. Remove and double bag contaminated clothing and shoes at the site. *IF SWALLOWED* and victim is *CONSCIOUS*: Have victim drink 4 to 8 ounces of water. **Do NOT induce vomiting.** *IF SWALLOWED* and victim is *UNCONSCIOUS OR HAVING CONVULSIONS*: Do nothing except keep victim warm.

HEALTH HAZARDS
Personal protective equipment (PPE): B-Level PPE. Wear self-contained positive pressure breathing apparatus and full protective clothing.
Long-term health effects: Prolonged and repeated vapor exposure may result systemic toxic effects.
Vapor (gas) irritant characteristics: Eye and respiratory tract irritant. Vapors are irritating such that personnel will not usually tolerate moderate or high concentrations.
Liquid or solid irritant characteristics: Eye and skin irritant. If spilled on clothing and allowed to remain, may cause smarting and reddening of the skin.

FIRE DATA
Flash point: 49°F/9°C (cc).

CHEMICAL REACTIVITY
Binary reactants: Oxidizers

ENVIRONMENTAL DATA
Water pollution: Effect of low concentrations on aquatic life is unknown. May be dangerous if it enters nearby water intakes; notify operators. Notify local health and wildlife officials. Notify operators of local water intakes. **Response to discharge:** Issue warning–high flammability, water contaminant. Restrict access. Should be removed. Chemical and physical treatment.

SHIPPING INFORMATION
Storage temperature: Ambient; **Stability during transport:** Stable.

NAS HAZARD CLASSIFICATION FOR BULK WATER TRANSPORTATION
FIRE: 4
HEALTH: Vapor irritant: 2; Liquid or solid irritant: 1; Poisons:–
WATER POLLUTION: Human toxicity:–; Aquatic toxicity:–; Aesthetic effect: 1
REACTIVITY: Other chemicals: 2; Water: 0; Self-reaction: 0

PHYSICAL AND CHEMICAL PROPERTIES
Physical state @ 59°F/15°C and 1 atm: Liquid.
Molecular weight: 110.13
Boiling point @ 1 atm: 240.8°F/116°C/389.2°K.
Melting/Freezing point: –125.9°F/–87.7°C/185.5°K.
Specific gravity (water = 1): 0.9986 @ 68°F/20°C.
Relative vapor density (air = 1): 3.8
Vapor pressure: (Reid) 0.82 psia.

4-FLUOROTOLUENE   REC. F:1150

SYNONYMS: *p*-TOLYL FLUORIDE; 4-FLUORO-1-METHYLBENZENE; *p*-FLUORTOLUENO (Spanish); *p*-FLUOROTOLUENE; 1-FLUORO-4-METHYLBENZENE

IDENTIFICATION
CAS Number: 352-32-9
Formula: C_7H_7F; 4-$F(C_6H_4)CH_3$
DOT ID Number: UN 2388; DOT Guide Number: 128
Proper Shipping Name: Fluorotoluenes

DESCRIPTION: Colorless liquid. Aromatic odor. May sink or float on water.

Flammable • Containers may BLEVE when exposed to fire • Vapors may form explosive mixture with air • Vapors are heavier than air and will collect and stay in low areas • Vapors may travel long distances to ignition sources and flashback • Vapors in confined areas (e.g., tanks, sewers, buildings) may explode when exposed to fire • Irritating to the skin, eyes, and respiratory tract • Toxic products of combustion may include hydrogen fluoride.

Hazard Classification (based on NFPA-704 Rating System)
Health Hazards (Blue): 2; Flammability (Red): 2; Reactivity (Yellow): 0

EMERGENCY RESPONSE: See Appendix A (128)
Evacuation:
Public safety: Isolate spill area for at least 25 to 50 meters (80 to 160 feet) in all directions.
Spill: Large spill–Consider initial downwind evacuation for at least 300 meters (1000 feet).
Fire: Isolate for 800 meters (½ mile) in all directions, especially if tank, rail car, or tank truck is involved in fire.

EXPOSURE
Short-term effects: *SEEK MEDICAL ATTENTION*. **Vapor:** May be harmful if inhaled or absorbed through the skin. Irritating to eyes, skin, nose, and throat. Move to fresh air. *IF BREATHING HAS STOPPED*, give artificial respiration. IF breathing is difficult, administer oxygen. **Liquid:** Irritating to skin and eyes. Harmful if swallowed or absorbed through skin. Remove contaminated clothing and shoes. Flush affected areas with plenty of water. *IF IN EYES*, hold eyelids open and flush with plenty of water. *IF SWALLOWED* and victim in *CONSCIOUS*, have victim drink 4 to 8 ounces of water. *Do NOT induce vomiting*. IF SWALLOWED and victim is UNCONSCIOUS OR HAVING CONVULSIONS: Do nothing except keep victim warm.

HEALTH HAZARDS
Personal protective equipment (PPE): B-Level PPE. Respirator with proper filter, goggles. Chemical protective material(s) reported to have good to excellent resistance: PV alcohol, Teflon®, Viton®.
Toxicity by ingestion: Grade 3: LD_{LO} = 500 mg/kg.
Vapor (gas) irritant characteristics: Eye and respiratory tract irritant.
Liquid or solid irritant characteristics: Severe eye and skin irritant.

FIRE DATA
Flash point: 105°F/49°C (cc).

CHEMICAL REACTIVITY
Binary reactants: Oxidizing materials.

ENVIRONMENTAL DATA
Water pollution: Harmful to aquatic life in very low concentrations. May be dangerous if it enters nearby water intakes; notify operators. Notify local health and wildlife officials.
Response to discharge: Restrict access. Chemical and physical treatment. Disperse and flush.

SHIPPING INFORMATION
Grades of purity: 97%; **Storage temperature:** Keep cool.

NAS HAZARD CLASSIFICATION FOR BULK WATER TRANSPORTATION
FIRE: 2
HEALTH: Vapor irritant: 2; Liquid or solid irritant: 1; Poisons:–
WATER POLLUTION: Human toxicity: 3; Aquatic toxicity:–; Aesthetic effect: 1
REACTIVITY: Other chemicals: 2; Water: 0; Self-reaction: 0

PHYSICAL AND CHEMICAL PROPERTIES
Physical state @ 59°F/15°C and 1 atm: Liquid.
Molecular weight: 110.13
Boiling point @ 1 atm: 242°F/117°C/390°K.
Melting/Freezing point: –70°F/–57°C/217°K.
Specific gravity (water = 1): 1.0007 @ 68°F/20°C.
Relative vapor density (air = 1): 3.8 (estimate).
Vapor pressure: (Reid) 0.77 psia.

FLUOROSULFONIC ACID REC. F:1200

SYNONYMS: ACIDO FLUOSULFONICO (Spanish); FLUOSULFONIC ACID; FLUOROSULFURIC ACID

IDENTIFICATION
CAS Number: 7789-21-1
Formula: FHO_3S; FSO_3H
DOT ID Number: UN 1777; DOT Guide Number: 137
Proper Shipping Name: Fluorosulfonic acid

DESCRIPTION: Colorless, cloudy, to pale yellow liquid that fumes in moist air. Choking, irritating odor. Reacts violently with water; forming toxic hydrogen fluoride and sulfuric acid mist.

Poison! • Corrosive • Do not use water • Breathing the vapor can kill you; skin or eye contact causes severe burns, impaired vision, or blindness • Firefighting gear (including SCBA) does not provide adequate protection. If exposure occurs, remove and isolate gear immediately and thoroughly decontaminate personnel • Containers may BLEVE when exposed to fire • Vapors are heavier than air and will collect and stay in low areas • Reacts with metals generating flammable hydrogen gas • Toxic products of combustion may include hydrogen fluoride and sulfur oxide • Do not put yourself in danger by entering a contaminated area to rescue a victim.

Hazard Classification (based on NFPA-704 Rating System)
Health Hazards (Blue): 4; Flammability (Red): 0; Reactivity (Yellow): 3; Special Notice (White): Water reactive

EMERGENCY RESPONSE: See Appendix A (137)
Evacuation:
Public safety: Isolate the area of spill or leak for at least 50 to 100 meters (160 to 330 feet) in all directions.
Spill: Increase, in the downwind direction, as necessary, the distance shown under "Public Safety." FOR SPILL IN WATER: Small spill–First: Isolate in all directions 30 meters (100 feet); Then, Protect persons downwind, DAY: 0.2 km (0.1 mile); NIGHT: 0.2 km (0.1 mile). Large spill–First: Isolate in all directions 60 meters (200 feet); Then: Protect persons downwind, DAY: 0.5 km (0.3 mile); NIGHT: 2.3 km (1.4 miles).
Fire: If any large container is involved in fire, isolate for at least 800 meters (½ mile) in all directions; also, consider initial evacuation for 800 meters (½ mile) in all directions.

EXPOSURE
Short-term effects: *SEEK MEDICAL ATTENTION*. **Vapor or**

mist: Irritating to eyes, nose, and throat. Harmful if inhaled. Move victim to fresh air. If breathing is difficult, administer oxygen. *IF BREATHING HAS STOPPED,* give artificial respiration; *avoid mouth-to-mouth resuscitation; use bag/mask apparatus.* **Liquid:** Will burn skin and eyes. Harmful if swallowed. Remove contaminated clothing and shoes. Flush affected areas with plenty of water. *IF IN EYES,* hold eyelids open and flush with plenty of water. *IF SWALLOWED* and victim is *CONSCIOUS AND ABLE TO SWALLOW,* have victim drink 4 to 8 ounces of water. **Do NOT induce vomiting.**

Note to physician or authorized medical personnel: Medical observation is recommended for 24 to 48 hours after breathing overexposure, as pulmonary edema may be delayed. As first aid for pulmonary edema, consider administering a corticosteroid spray. Cigarette smoking may exacerbate pulmonary injury and should be discouraged for at least 72 hours following exposure.

HEALTH HAZARDS

Personal protective equipment (PPE): A-Level PPE. Chemical protective material(s) reported to have good to excellent resistance: Saranex®.

Recommendations for respirator selection: NIOSH/OSHA as F. *12.5 mg/m³:* DM (any dust and mist respirator). *25 mg/m³:* DMXSQ* (any dust and mist respirator except single-use and quarter-mask respirators); or SA* (any supplied-air respirator). *62.5 mg/m³:* SA:CF* [any supplied-air respirator operated in a continuous-flow mode)]; or PAPRDM*⁺ *if not present as a fume* (any powered, air-purifying respirator with a dust and mist filter). *125 mg/m³:* HiEF⁺ (any air-purifying, full-facepiece respirator with a high-efficiency particulate filter); or SCBAF (any self-contained breathing apparatus with a full facepiece); or SAF (any supplied-air respirator with a full facepiece). *250 mg/m³:* SA:PD,PP (any supplied-air respirator operated in a pressure-demand or other positive-pressure mode). *EMERGENCY OR PLANNED ENTRY INTO UNKNOWN CONCENTRATIONS OR IDLH CONDITIONS:* SCBAF:PD,PP (any self-contained breathing apparatus that has a full faceplate and is operated in a pressure-demand or other positive-pressure mode); or SAF:PD,PP:ASCBA (any supplied-air respirator that has a full facepiece and is operated in a pressure-demand or other positive-pressure mode in combination with an auxiliary, self-contained breathing apparatus operated in a pressure-demand or other positive-pressure mode). *ESCAPE:* HiEF⁺ (any air-purifying, full-facepiece respirator with a high-efficiency particulate filter); or SCBAE (any appropriate escape-type, self-contained breathing apparatus). *Notes:* *Substance reported to cause eye irritation or damage; may require eye protection. ⁺May need acid gas sorbent.

Exposure limits (TWA unless otherwise noted): ACGIH TLV 2.5 mg/m³ as fluorides; NIOSH/OSHA 2.5 mg/m³ as inorganic fluorides.

Vapor (gas) irritant characteristics: Vapors cause severe irritation of eyes and throat and can cause eye and lung injury. They cannot be tolerated even at low concentrations.

Liquid or solid irritant characteristics: Severe skin irritant. Causes second- and third-degree burns on short contact and is very injurious to the eyes.

FIRE DATA

Fire extinguishing agents not to be used: Do not use water or foam on adjacent fires.

CHEMICAL REACTIVITY

Reactivity with water: Reacts violently with water to generate hydrogen fluoride and sulfur oxides.

Binary reactants: Reacts with metals, generating hydrogen gas.

Neutralizing agents for acids and caustics: Flood with water, rinse with sodium bicarbonate or lime solution.

ENVIRONMENTAL DATA

Food chain concentration potential: Negative; unlikely to accumulate.

Water pollution: Effect of low concentrations on aquatic life is unknown. May be dangerous if it enters nearby water intakes; notify operators. Notify local health and wildlife officials. **Response to discharge:** Issue warning–corrosive, air contaminant, water contaminant. Restrict access. Evacuate area Disperse and flush with care.

SHIPPING INFORMATION

Grades of purity: 98.5%; **Storage temperature:** Ambient; **Inert atmosphere:** None; **Venting:** Pressure-vacuum, with protection from moisture in air; **Stability during transport:** Stable.

NAS HAZARD CLASSIFICATION FOR BULK WATER TRANSPORTATION

FIRE: 0
HEALTH: Vapor irritant: 4; Liquid or solid irritant: 4; Poisons: 3
WATER POLLUTION: Human toxicity:; Aquatic toxicity: 2; Aesthetic effect: 2
REACTIVITY: Other chemicals: 4; Water: 4; Self-reaction: 0

PHYSICAL AND CHEMICAL PROPERTIES

Physical state @ 59°F/15°C and 1 atm: Liquid.
Molecular weight: 100.07
Boiling point @ 1 atm: 325°F/163°C/436°K.
Melting/Freezing point: –125°F/–87°C/186°K.
Specific gravity (water = 1): 1.73 @ 77°F/25°C (liquid).
Latent heat of vaporization: 170 Btu/lb = 94 cal/g = 3.9×10^5 J/kg.

FORMALDEHYDE (SOLUTION)   REC. F:1250

SYNONYMS: ALDEHYDE FORMIQUE (French); BFV; EEC No. 605-001-00-5; FA; FANNOFORM; FORMALDEHIDO (Spanish); FORMALIN; FORMALIN 40; FORMALINE (German); FORMALIN-LOESUNGEN (German); FORMALITH; FORMIC ALDEHYDE; FORMOL; FYDE; IVALON; KARSAN; LYSOFORM; METHANAL; METHANAL SOLUTION; METHYL ALDEHYDE; METHYLENE GLYCOL; METHYLENE OXIDE; MORBICID; OXOMETHANE; OXYMETHYLENE; PARAFORM; POLYOXYMETHYLENE GLYCOLS; SUPERLYSOFORM; TETRAOXYMETHYLENE; TRIOXANE; RCRA No. U122

IDENTIFICATION

CAS Number: 50-00-0
Formula: CH_2O; HCHO
DOT ID Number: UN 1198 (flammable solutions); UN 2209 (corrosive solutions); DOT Guide Number: 132
Proper Shipping Name: Formaldehyde, solutions, flammable; Formaldehyde, solutions (formalin), flammable; Formaldehyde, solutions (formalin) (corrosive)
Reportable Quantity (RQ): **(CERCLA)** 100 lb/45.4 kg

DESCRIPTION: Colorless or cloudy watery liquid. Because the pure gas tends to polymerize, ths material is usually shipped as a solution in water (30–40%) or methanol (up to 15% as a stabilizer). Pungent, irritating odor. Sinks and mixes with water.

Formaldehyde (solution)

Extremely flammable • Firefighting gear (including SCBA) does not provide adequate protection. If exposure occurs, remove and isolate gear immediately and thoroughly decontaminate personnel • Containers may BLEVE when exposed to fire •Vapors may form explosive mixture with air • Vapors are heavier than air and will collect and stay in low areas • Vapors may travel long distances to ignition sources and flashback • Vapors in confined areas (e.g., tanks, sewers, buildings) may explode when exposed to fire • Highly irritating to the eyes and respiratory tract; this material has anesthetic properties • Toxic products of combustion may include formic acid and gaseous formaldehyde. *Note*: Aqueous formaldehyde solutions heated above their flash points are an explosion hazard. See flash points for high methanol content formalin.

Hazard Classification (based on NFPA-704 Rating System) (gas, solutions, flammable; UN 1198) Health Hazards (Blue): 3; Flammability (Red): 4; Reactivity (Yellow): 0

EMERGENCY RESPONSE: See Appendix A (132)
Evacuation:
Public safety: Isolate spill area for at least 50 to 100 meters (160 to 330 feet) in all directions.
Spill: Increase, as necessary, the isolation distance shown above,in "Public safety."
Fire: Isolate for 800 meters (½ mile) in all directions, especially if tank, rail car, or tank truck is involved in fire.

EXPOSURE
Short-term effects: *SEEK MEDICAL ATTENTION*. Vapor may cause lung edema; physical exertion will aggravate this condition.
Liquid: Will burn skin and eyes. *IF SWALLOWED*, will cause nausea, vomiting, or loss of consciousness. Remove contaminated clothing and shoes. Flush affected areas with plenty of water. *IF IN EYES*, hold eyelids open and flush with plenty of water. *IF SWALLOWED* and victim is *CONSCIOUS AND ABLE TO SWALLOW*, **do not induce emesis**; have victim drink 4 to 8 ounces of water or milk. The effectiveness of activated charcoal is unknown. *IF SWALLOWED* and victim is *UNCONSCIOUS OR HAVING CONVULSIONS*, do nothing except keep victim warm.

HEALTH HAZARDS
Personal protective equipment (PPE): B-Level PPE. OSHA Table Z-1-A; OSHA Table Z-2 air contaminant. air contaminant. Self-contained breathing apparatus; chemical goggles; protective clothing; synthetic rubber or plastic gloves. For <30% solutions, Protective materials with good to excellent resistance: butyl rubber, chlorinated polyethylene, polyethylene, Teflon®, Viton®, Silvershield®. Also, Viton®/neoprene and butyl rubber/neoprene offers limited protection. For 30-70% solutions, Protective materials with good to excellent resistance: butyl rubber, nitrile, polyethylene, Teflon®, Viton®.
Recommendations for respirator selection: NIOSH
At any detectable concentration: SCBAF:PD,PP (any MSHA/NIOSH approved self-contained breathing apparatus that has a full facepiece and is operated in a pressure-demand or other positive-pressure mode); or SAF:PD,PP:ASCBA (any supplied-air respirator that has a full facepiece and is operated in a pressure-demand or other positive-pressure mode in combination with an auxiliary, self-contained breathing apparatus operated in a pressure-demand or other positive pressure mode). *ESCAPE:*GMFS [any air-purifying, full-facepiece respirator (gas mask) with a chin-style, front- or back-mounted canister providing protection against the compound of concern]; or SCBAE (any appropriate escape-type, self-contained breathing apparatus).

Exposure limits (TWA unless otherwise noted): OSHA PEL 0.75 ppm; NIOSH REL 0.016 ppm.
Short-term exposure limits (15-minute TWA): ACGIH TLV ceiling 0.3 ppm (0.37 mg/m^3) suspected human carcinogen; OSHA STEL 2 ppm; NIOSH ceiling 0.1 ppm/15 min.
Toxicity by ingestion: (solution) Grade 2; LD_{50} = 0.5 to 5 g/kg; Liver and kidney injury may occur.
Long-term health effects: Possible liver and kidney damage. OSHA specifically regulated carcinogen. IARC probable carcinogen, rating 2A; NTP anticipated carcinogen (gas).
Vapor (gas) irritant characteristics: Eye and respiratory tract irritant. Vapor is irritating such that personnel will not usually tolerate moderate or high concentrations.
Liquid or solid irritant characteristics: Severe eye and skin irritant. Causes smarting of the skin and first-degree burns on short exposure. May cause secondary burns on long exposure.
Odor threshold: 0.025–9800 ppm.
IDLH value: Potential human carcinogen; 20 ppm.

FIRE DATA
Flash point: (37% methanol-free) 185°F/85°C (cc); (37%, 15% methanol) 122°F/50°C (cc).
Flammable limits in air: LEL: 7.0%; UEL: 73% (GAS).
Autoignition temperature: 806°F/430°C/703°K (solutions 37 to 56% formaldehyde); 572°F/300°C/573°K (gas).
Electrical hazard: Class I, Group B.

CHEMICAL REACTIVITY
Binary reactants: Strong oxidizers, alkalis, and acids; phenols; urea. Reacts with hydrochloric acid, forming very toxic bis-(chloromethyl)ether.
Polymerization: Pure formaldehyde polymerizes readily; **Inhibitor of polymerization:** Methanol up to 15 ppm.
Reactivity group: 19
Compatibility class: Aldehyde

ENVIRONMENTAL DATA
Food chain concentration potential: Log P_{ow} = 0 Unlikely to accumulate.
Water pollution: Harmful to aquatic life in very low concentrations. May be dangerous if it enters nearby water intakes; notify operators. Notify local health and wildlife officials. **Response to discharge:** Issue warning–water contaminant. Disperse and flush.

SHIPPING INFORMATION
Grades of purity: 37-50% formaldehyde by weight in water. 0-15% methyl alcohol; **Storage temperature:** Ambient; **Inert atmosphere:** None; **Venting:** Pressure-vacuum; **Stability during transport:** Stable.

NAS HAZARD CLASSIFICATION FOR BULK WATER TRANSPORTATION
FIRE: 2
HEALTH: Vapor irritant: 3; Liquid or solid irritant: 2; Poisons: 3
WATER POLLUTION: Human toxicity: 3; Aquatic toxicity: 3; Aesthetic effect: 2
REACTIVITY: Other chemicals: 2; Water: 0; Self-reaction: 1

PHYSICAL AND CHEMICAL PROPERTIES
Physical state @ 59°F/15°C and 1 atm: Liquid.
Molecular weight: 18-30
Boiling point @ 1 atm: 206-214°F/96-101°C/369-374°K (varies with concentration); −6°F/−21°C/252°K (gas).
Melting/Freezing point: −134°F/92°C/365°K (gas).

Specific gravity (water = 1): 1.1 @ 77°F/25°C (liquid); 0.82 @ –6°F/–21°C (GAS).
Relative vapor density (air = 1): 1.04
Heat of solution: (estimate) –9 Btu/lb = –5 cal/g = –0.2 x 10^5 J/kg.
Vapor pressure: (Reid) 0.09 psia; 1 mm; 17-20 mmHg @ 77°F/25°C (aqueous solutions)

FORMAMIDE   REC. F:1300

SYNONYMS: CARBAMALDEHYDE; FORMIMIDIC ACID; FORMAMIDA (Spanish); METHANAMIDE; FORMIC ACID, AMIDE; METHANOIC ACID, AMIDE

IDENTIFICATION
CAS Number: 75-12-7
Formula: CH_3NO; $HCONH_2$

DESCRIPTION: Colorless to pale yellow, thick liquid. Faint odor of ammonia. Sinks and mixes with water; reacts slowly forming ammonium formate. Freezing point is 36°F/2°C.

Poison! • Breathing the vapor, skin or eye contact, or swallowing the material can cause severe pain and illness; absdorbed by the skin • Firefighting gear (including SCBA) does not provide adequate protection. If exposure occurs, remove and isolate gear immediately and thoroughly decontaminate personnel • Combustible • Heat or flame may cause explosion • Containers may BLEVE when exposed to fire • Vapors are heavier than air and will collect and stay in low areas • Vapors in confined areas (e.g., tanks, sewers, buildings) may explode when exposed to fire • Highly irritating to the eyes and respiratory tract; this material has anesthetic properties • Decomposes at 356–410°F/180–210°C. Toxic products of combustion may include anhydrous ammonia, CO_2, prussic acid, and nitrogen oxides.

Hazard Classification (based on NFPA-704 Rating System)
Health Hazards (Blue): 2; Flammability (Red): 1; Reactivity (Yellow): 0

EMERGENCY RESPONSE: See Appendix A (171)
Evacuation:
Public safety: Isolate the area of spill or leak for at least 10 to 25 meters (30 to 80 feet) in all directions.
Spill: Increase, in the downwind direction, as necessary, the distance shown under "Public Safety."
Fire: If any large container is involved in fire, isolate for at least 800 meters (½ mile) in all directions; also, consider initial evacuation for 800 meters (½ mile) in all directions.

EXPOSURE
Short-term effects: *SEEK MEDICAL ATTENTION.* **Liquid:** *HARMFUL IF SWALLOWED OR ABSORBED THROUGH THE SKIN.* Remove contaminated clothing and shoes. Flush affected areas with plenty of water. *IF IN EYES*, hold eyelids open and flush with plenty of water.

HEALTH HAZARDS
Personal protective equipment (PPE): B-Level PPE. OSHA Table Z-1-A air contaminant. Organic vapor respirator. For 30% solution, Chemical protective material(s) reported to offer minimal to poor protection: neoprene+natural rubber.
Exposure limits (TWA unless otherwise noted): ACGIH TLV 10 ppm (18 mg/m^3); NIOSH REL 10 ppm (15 mg/m^3); skin contact contributes significantly in overall exposure.
Toxicity by ingestion: Grade 1; LD_{50} = 6.1 to 7.5 g/kg (rat).
Long-term health effects: EPA Genetic Toxicology Program; teratogenic and reproduction data. Amides may cause liver, kidney, and brain damage. Depending on level of exposure, regular medical checkups are advised.
Vapor (gas) irritant characteristics: Eye and respiratory tract irritant.
Liquid or solid irritant characteristics: Eye and skin irritant.

FIRE DATA
Flash point: 310°F/154°C (oc).
Autoignition temperature: 310°F/154°C/427°K.

CHEMICAL REACTIVITY
Reactivity with water: Hygroscopic, absorbs moisture from the air. Reacts slowly forming prussic acid. In an alkaline environment, reacts with moisture forming ammonium formate.
Binary reactants: Oxidizers, iodine, pyridine, sulfur trioxide, copper, brass, lead. Storage explosions reported.
Reactivity group: 10
Compatibility class: Amide

ENVIRONMENTAL DATA
Food chain concentration potential: Log P_{ow} = –1.6. Unlikely to accumulate.
Water pollution: Effect of low concentrations on aquatic life is unknown. May be dangerous if it enter water intakes. Notify local health and wildlife officials. **Response to discharge:** Restrict access. Disperse and flush.

SHIPPING INFORMATION
Grades of purity: Reagent Grade (98%), Technical Grade, Practical Grade, Commercial; **Storage temperature:** Ambient; **Stability during transport:** Stable in absence of high temperatures.

PHYSICAL AND CHEMICAL PROPERTIES
Physical state @ 59°F/15°C and 1 atm: Liquid.
Molecular weight: 45.1
Boiling point @ 1 atm: 412°F/211°C/484°K (decomposes).
Melting/Freezing point: 37°F/3°C/276°K.
Specific gravity (water = 1): 1.1334 at @ 68°F/20°C.
Liquid surface tension: 58.35 dynes/cm = 0.0584 N/m @ 68°F/20°C.
Liquid water interfacial tension: 14.40 dynes/cm = 0.0144 N/m @ 68°F/20°C.
Relative vapor density (air = 1): 1.6
Heat of combustion: 5,380 Btu/lb = 2,973.8 cal/g = 125.2 x 10^5 J/kg.
Vapor pressure: 0.1 mm @ 86°F/30°C.

FORMIC ACID   REC. F:1350

SYNONYMS: ACIDE FORMIQUE (French); ACIDO FORMICO (Spanish); AMEISENSAEURE (German); AMINIC ACID; BILORIN; EEC No. 607-001-00-1; FORMYLIC ACID; HYDROGEN CARBOXYLIC ACID; METHANOIC ACID; RCRA No. U123

IDENTIFICATION
CAS Number: 64-18-6
Formula: HCOOH
DOT ID Number: UN 1779; DOT Guide Number: 153
Proper Shipping Name: Formic acid
Reportable Quantity (RQ): **(CERCLA)** 5000 lb/2270 kg

Formic acid

DESCRIPTION: Colorless, fuming liquid. Irritating odor. Sinks in water; soluble. Freezing point is 47°F/8°C. *Note:* May deteriorate upon normal storage causing pressure buildup and container failure.

Highly flammable • Corrosive to skin, eyes, and respiratory tract; skin and eye contact can cause severe burns and blindness; inhalation symptoms may be delayed • Firefighting gear (including SCBA) does not provide adequate protection. If exposure occurs, remove and isolate gear immediately and thoroughly decontaminate personnel • Containers may BLEVE when exposed to fire • Vapors may form explosive mixture with air • Vapors are heavier than air and will collect and stay in low areas • Vapors may travel long distances to ignition sources and flashback • Vapors in confined areas (e.g., tanks, sewers, buildings) may explode when exposed to fire • Toxic products of combustion may include carbon monoxide, flammable hydrogen, and nitrogen oxides • Do not put yourself in danger by entering a contaminated area to rescue a victim. *Warning:* Odor is not a reliable indicator of the presence of toxic amounts of formic acid vapors.

Hazard Classification (based on NFPA-704 Rating System)
Health Hazards (Blue): 3; Flammability (Red): 2; Reactivity (Yellow): 0

EMERGENCY RESPONSE: See Appendix A (153)
Evacuation:
Public safety: Isolate the area of spill or leak for at least 25 to 50 meters (80 to 160 feet) in all directions.
Spill: Increase, in the downwind direction, as necessary, the distance shown under "Public Safety."
Fire: If tank, rail car, or tank truck is involved in fire, isolate for at least 800 meters (½ mile) in all directions; also, consider initial evacuation for 800 meters (½ mile) in all directions.

EXPOSURE
Short-term effects: *SEEK MEDICAL ATTENTION. IF BREATHING HAS STOPPED,* give artificial respiration; *avoid mouth-to-mouth resuscitation; use bag/mask apparatus.* Vapor may cause lung edema; physical exertion will aggravate this condition. **Liquid:** Will burn skin and eyes. Harmful if swallowed. Remove contaminated clothing and shoes. Flush affected areas with plenty of water. *IF IN EYES,* hold eyelids open and flush with plenty of water. *IF SWALLOWED* and victim is *CONSCIOUS AND ABLE TO SWALLOW,* have victim drink 4 to 8 ounces of water. *IF SWALLOWED* and victim is *UNCONSCIOUS OR HAVING CONVULSIONS,* do nothing except keep victim warm.
Do NOT induce vomiting.

Note to physician or authorized medical personnel: Medical observation is recommended for 24 to 48 hours after breathing overexposure, as pulmonary edema may be delayed. As first aid for pulmonary edema, consider administering a corticosteroid spray. Cigarette smoking may exacerbate pulmonary injury and should be discouraged for at least 72 hours following exposure.

HEALTH HAZARDS
Personal protective equipment (PPE): B-Level PPE. OSHA Table Z-1-A air contaminant. Self-contained breathing apparatus; chemical goggles or face shield; suit, gloves, and shoes. Chemical protective material(s) reported to have good to excellent resistance: butyl rubber, neoprene, neoprene+natural rubber, PVC, polyurethane, Saranex®. Also, natural rubber, polyethylene offers limited protection.
Recommendations for respirator selection: NIOSH/OSHA
30 ppm: SA (any supplied-air respirator); or SCBAF (any self-contained breathing apparatus with a full facepiece). *EMERGENCY OR PLANNED ENTRY INTO UNKNOWN CONCENTRATIONS OR IDLH CONDITIONS:* SCBAF:PD,PP (any self-contained breathing apparatus that has a full facepiece and is operated in a pressure-demand or other positive-pressure mode); or SAF:PD,PP:ASCBA (any supplied-air respirator that has a full facepiece and is operated in a pressure-demand or other positive-pressure mode in combination with an auxiliary self-contained breathing apparatus operated in a pressure-demand or other positive pressure mode). *ESCAPE:* GMFOVHiE [any air-purifying, full-facepiece respirator (gas mask) with a chin-style, front- or back-mounted organic vapor canister having a high-efficiency particulate filter]; or SCBAE (any appropriate escape-type, self-contained breathing apparatus). *Note:* Substance reported to cause eye irritation or damage; may require eye protection.
Exposure limits (TWA unless otherwise noted): ACGIH TLV 5 ppm (9.4 mg/m^3); NIOSH/OSHA 5 ppm (9 mg/m^3).
Short-term exposure limits (15-minute TWA): ACGIH STEL 10 ppm (19 mg/m^3).
Toxicity by ingestion: Toxic. Grade 2; oral rat LD_{50} = 1.21 g/kg.
Long-term health effects: Mutation data.
Vapor (gas) irritant characteristics: Eye and respiratory tract irritant. Vapor is irritating such that personnel will not usually tolerate moderate or high vapor concentrations.
Liquid or solid irritant characteristics: Eye and skin irritant. Fairly severe skin irritant; may cause pain and second-degree burns after a few minutes of contact.
Odor threshold: 28.2 ppm. *Note:* Far above exposure limits.
IDLH value: 30 ppm.

FIRE DATA
Flash point: 156°F/69°C (cc); 122°F/50°C (cc) (90% solution).
Flammable limits in air: LEL: 4%; UEL 33%; LEL: 18%; UEL: 57% (90% solution).
Autoignition temperature: 1004°F/539°F/812°K; 813°F/434°C/707°K (90% solution).
Electrical hazard: Class I, Group D.
Burning rate: 0.5 mm/min

CHEMICAL REACTIVITY
Binary reactants: A strong reducing agent. Vigorous reaction with strong oxidizers, furfuryl alcohol, isocyanates, strong caustics, peroxides, hydrogen peroxide, strong acids including concentrated sulfuric acid and chromic acid; furfuryl alcohol. Corrosive to cast iron, aluminum, and steel. Attacks some plastics, coatings and rubber.
Neutralizing agents for acids and caustics: Flush with water, then neutralize with lime.
Reactivity group: 4
Compatibility class: Organic acids

ENVIRONMENTAL DATA
Food chain concentration potential: Log P_{ow} = –0.51 (estimate). Unlikely to accumulate.
Water pollution: Dangerous to aquatic life in high concentrations. May be dangerous if it enters nearby water intakes; notify operators. Notify local health and wildlife officials. **Response to discharge:** Issue warning–corrosive. Restrict access. Disperse and flush.

SHIPPING INFORMATION
Grades of purity: Technical, pharmaceutical: 85-95%; **Storage temperature:** Ambient; **Inert atmosphere:** None; **Venting:** Pressure-vacuum; **Stability during transport:** Stable; May generate carbon monoxide during storage.

NAS HAZARD CLASSIFICATION FOR BULK WATER TRANSPORTATION
FIRE: 2
HEALTH: Vapor irritant: 3; Liquid or solid irritant: 3; Poisons: 3
WATER POLLUTION: Human toxicity: 3; Aquatic toxicity: 2; Aesthetic effect: 2
REACTIVITY: Other chemicals: 2; Water: 0; Self-reaction: 0

PHYSICAL AND CHEMICAL PROPERTIES
Physical state @ 59°F/15°C and 1 atm: Liquid.
Molecular weight: 46.03
Boiling point @ 1 atm: 224°F
Melting/Freezing point: 47°F/8°C/281°K; 20°F/–7°C/266°K (90% solution).
Specific gravity (water = 1): 1.22 @ 68°F/20°C (liquid, 90% solution).
Liquid surface tension: 38 dynes/cm = 0.038 N/m @ 15°C.
Relative vapor density (air = 1): 1.6
Ratio of specific heats of vapor (gas): 1.228
Latent heat of vaporization: 216 Btu/lb = 120 cal/g = 5.02×10^5 J/kg.
Heat of combustion: –2045 Btu/lb = –1136 cal/g = -47.56×10^5 J/kg.
Heat of solution: (estimate) –26 Btu/lb = –14 cal/g = -0.6×10^5 J/kg.
Heat of fusion: 66.05 cal/g.
Vapor pressure: (Reid) 1.5 psia; 35 mm.

FUMARIC ACID   REC. F:1400

SYNONYMS: ACIDO FUMARICO (Spanish); ALLOMALEIC ACID; BOLETIC ACID; 2-BUTENEDIOIC ACID (E); *trans*-BUTENEDIOIC ACID; (E)-BUTENEDIOIC ACID; BUTENEDIOIC ACID, (E)-; EEC No. 607-146-00-X; 1,2-ETHENEDICARBOXYLIC ACID, *trans*-; *trans*-1,2-ETHYLENEDICARBOXYLIC ACID; 1,2-ETHYLENEDICARBOXYLIC ACID, (E); LICHENIC ACID

IDENTIFICATION
CAS Number: 110-17-8
Formula: $C_4H_4O_4$; $HO_2CCH=CHCO_2H$
DOT ID Number: UN 3077; DOT Guide Number: 171
Proper Shipping Name: Environmentally hazardous substances, solid, n.o.s.
Reportable Quantity (RQ): 5000 lb/2270 kg

DESCRIPTION: White solid. Odorless. Sinks and mixes with water.

Combustible solid • Heat or flame may cause explosion • Containers may BLEVE when exposed to fire • Concentrated dust in confined areas (e.g., tanks, sewers, buildings) may explode when exposed to fire • Irritating to the skin, eyes, and respiratory tract • Toxic products of combustion may include carbon monoxide and maleic anhydride.

Hazard Classification (based on NFPA-704 Rating System)
Health Hazards (Blue): 0; Flammability (Red): 1; Reactivity (Yellow): 0

EMERGENCY RESPONSE: See Appendix A (171)
Evacuation:
Public safety: Isolate the area of spill or leak for at least 10 to 25 meters (30 to 80 feet) in all directions.
Spill: Increase, in the downwind direction, as necessary, the distance shown under "Public Safety."
Fire: If any large container is involved in fire, isolate for at least 800 meters (½ mile) in all directions; also, consider initial evacuation for 800 meters (½ mile) in all directions.

EXPOSURE
Short-term effects: *SEEK MEDICAL ATTENTION*. **Dust:** Irritating to eyes, nose, and throat. *IF INHALED*, will, will cause coughing or difficult breathing. *IF IN EYES*, hold eyelids open and flush with plenty of water. *IF BREATHING HAS STOPPED*, give artificial respiration. IF breathing is difficult, administer oxygen. **Solid:** Irritating to skin and eyes. Remove contaminated clothing and shoes. Flush affected areas with plenty of water. *IF IN EYES*, hold eyelids open and flush with plenty of water.

HEALTH HAZARDS
Personal protective equipment (PPE): B-Level PPE. Butyl rubber is generally suitable for carbooxylic acid compounds.
Toxicity by ingestion: Somewhat toxic.
Vapor (gas) irritant characteristics: Eye and respiratory tract irritant.
Liquid or solid irritant characteristics: Eye and skin irritant.

FIRE DATA
Behavior in fire: Dust presents explosion hazard; knock down dust with water fog.
Autoignition temperature: 1364°F/740°C/1013°K (powder).

CHEMICAL REACTIVITY
Binary reactants: Reacts with oxidizing materials, sulfuric acid, caustics, ammonia, amines, isocyanates, alkylene oxides.

ENVIRONMENTAL DATA
Food chain concentration potential: Log P_{ow} = 0.29 (estimate). Unlikely to accumulate.
Water pollution: Effect of low concentrations on aquatic life is unknown. May be dangerous if it enters nearby water intakes; notify operators. Notify local health and wildlife officials. **Response to discharge:** Disperse and flush.

SHIPPING INFORMATION
Grades of purity: Technical; Purified food grade; **Storage temperature:** Ambient; **Inert atmosphere:** None; **Venting:** Open; **Stability during transport:** Stable.

PHYSICAL AND CHEMICAL PROPERTIES
Physical state @ 59°F/15°C and 1 atm: Solid.
Molecular weight: 116.07
Boiling point @ 1 atm: Very high.
Melting/Freezing point: 549°F/287°C/560°K.
Specific gravity (water = 1): 1.635 @ 68°F/20°C (solid).
Heat of combustion: –4970 Btu/lb = –2760 cal/g = -116×10^5 J/kg.
Vapor pressure: <0.001 @ 20°C

FURAN   REC. F:1450

SYNONYMS: DIVINYLENE OXIDE; FURFURAN; FURANO (Spanish); OXACYCLOPENTADIENE; OXOLE; RCRA No. U124; TETROLE

IDENTIFICATION
CAS Number: 110-00-9

562 Furfural

Formula: C_4H_4O
DOT ID Number: UN 2389; DOT Guide Number: 127
Proper Shipping Name: Furan
Reportable Quantity (RQ): **(CERCLA)** 100 lb/45.4 kg

DESCRIPTION: Clear, colorless liquid; turns brown upon standing. Mild, pleasant odor. Floats on the surface of water; slightly soluble. Boils at 90°F/32°C. Produces large amounts of vapor.

Extremely flammable • Very Irritating to the skin, eyes, and respiratory tract; prolonged contact can cause burns to eyes • Containers may BLEVE when exposed to fire •Vapors may form explosive mixture with air • Vapors are heavier than air and will collect and stay in low areas • Vapors may travel long distances to ignition sources and flashback • Vapors in confined areas (e.g., tanks, sewers, buildings) may explode when exposed to fire • Runoff from fire control can cause pollution • Toxic products of combustion may include carbon monoxide.

Hazard Classification (based on NFPA-704 Rating System)
Health Hazards (Blue): 1; Flammability (Red): 4; Reactivity (Yellow): 1

EMERGENCY RESPONSE: See Appendix A (127)
Evacuation:
Public safety: Isolate spill area for at least 25 to 50 meters (80 to 160 feet) in all directions.
Spill: Large spill–Consider initial downwind evacuation for at least 300 meters (1000 feet).
Fire: Isolate for 800 meters (½ mile) in all directions, especially if tank, rail car, or tank truck is involved in fire.

EXPOSURE
Short-term effects: *CALL FOR MEDICAL HELP*. **Vapor:** May be harmful if inhaled. Narcotic; may cause dizziness or suffocation. Move victim to fresh air. If not breathing, give artificial respiration. *IF* breathing is difficult, administer oxygen. **Liquid:** May be harmful if swallowed or absorbed through skin. Contact may irritate or burn skin and eyes. *IF IN EYES OR ON SKIN* immediately flush with running water for at least 15 minutes; hold eyelids open if necessary. Remove and double bag contaminated clothing and shoes at the site. *IF SWALLOWED* and victim is *UNCONSCIOUS OR HAVING CONVULSIONS*, do nothing except keep victim warm.

HEALTH HAZARDS
Personal protective equipment (PPE): B-Level PPE. Wear self-contained positive pressure breathing apparatus and full protective clothing. Chemical protective material(s) reported to offer minimal to poor protection: PV alcohol
Long-term health effects: May cause mutagenic effects.
Liquid or solid irritant characteristics: Eye and skin irritant.

FIRE DATA
Flash point: –40°F/–40°C (oc); –58°F/–50 (cc).
Flammable limits in air: LEL: 2.3%; UEL: 14.3%.
Behavior in fire: Vapors may travel to a source of ignition and flashback. Container may explode in heat of fire. Vapor explosion hazard exists indoors, outdoors or in sewers.
Stoichiometric air-to-fuel ratio: 9.11

CHEMICAL REACTIVITY
Binary reactants: Violent reaction with acids; oxidizing materials. Exposure to air may form unstable peroxides.

ENVIRONMENTAL DATA
Food chain concentration potential: Log P_{ow} = 1.3. Unlikely to accumulate.
Water pollution: Effect of low concentrations on aquatic life is unknown. May be dangerous if it enters nearby water intakes; notify operators. Notify local health and wildlife officials. **Response to discharge:** Issue warning–poison., high flammability. Restrict access. Evacuate area. Should be removed. Chemical and physical treatment.

INFORMATION
Grades of purity: 99+% (Stabilized with 0.0254% 2,6-di-*tert*-butyl-4-Methylphenol to prevent formation of peroxide); **Storage temperature:** Keep cool; **Stability during transport:** Stable.

PHYSICAL AND CHEMICAL PROPERTIES
Physical state @ 59°F/15°C and 1 atm: Liquid.
Molecular weight: 68.08
Boiling point @ 1 atm: 88°F/31°C/304°K.
Melting/Freezing point: –122°F/–86°C/188°K.
Critical temperature: 417°F/214°C/487°K.
Critical pressure: 772 psia = 52.5 atm = 5.32 MN/m^2.
Specific gravity (water = 1): 0.9514 @ 68°F/20°C.
Liquid surface tension: 24.10 dynes/cm = 0.0241 N/m @ 68°F/20°C.
Relative vapor density (air = 1): 2.3
Latent heat of vaporization: 171.2 Btu/lb = 95.09 cal/g = 3.981 x 10^5 J/kg.
Heat of combustion: –12,599 Btu/lb = –7000 cal/g = –293 x 10^5 J/kg.

FURFURAL   **REC. F:1500**

SYNONYMS: ARTIFICIAL ANT OIL; EEC No. 605-010-00-4; 2-FURANCARBOXALDEHYDE; FURAL; FURALE; 2-FURALDEHYDE; 2-FURANALDEHYDE; 2-FURANCARBONAL; FURFURALDEHYDE; FURFUROL; 2-FURYL-METHANAL; FUROLE; α-FUROLE; PYROMUCIC ALDEHYDE; RCRA No. U125; FURFUROLE; FURAL/PYROMUCIC ALDEHYDE; QUAKERAL

IDENTIFICATION
CAS Number: 98-01-1
Formula: $C_5H_4O_2$; OCH=CHCH=CCO·H
DOT ID Number: UN 1199; DOT Guide Number: 132P
Proper Shipping Name: Furfural
Reportable Quantity (RQ): **(CERCLA)** 5000 lb/2270 kg

DESCRIPTION: Colorless to yellowish oily liquid; turns brown on exposure to air. Almond-like odor. Sinks in water; moderately soluble.

Very flammable • Corrosive to skin, eyes, and respiratory tract; inhalation symptoms may be delayed • Firefighting gear (including SCBA) does not provide adequate protection. If exposure occurs, remove and isolate gear immediately and thoroughly decontaminate personnel • Containers may BLEVE when exposed to fire •Vapors may form explosive mixture with air • Vapors are heavier than air and will collect and stay in low areas • Vapors may travel long distances to ignition sources and flashback • Vapors in confined areas (e.g., tanks, sewers, buildings) may explode when exposed to fire • Highly irritating to the eyes and respiratory tract; this material has anesthetic properties • Toxic products of combustion may include smoke and irritating fumes • Polymerization hazard on

Furfural

contact with alkalies or strong acids • May react with itself blocking relief valves; leading to tank explosions.

Hazard Classification (based on NFPA-704 Rating System)
Health Hazards (Blue): 3; Flammability (Red): 2; Reactivity (Yellow): 0

EMERGENCY RESPONSE: See Appendix A (132)
Evacuation:
Public safety: Isolate spill area for at least 50 to 100 meters (160 to 330 feet) in all directions.
Spill: Increase, as necessary, the isolation distance shown above, in "Public safety."
Fire: Isolate for 800 meters (½ mile) in all directions, especially if tank, rail car, or tank truck is involved in fire.

EXPOSURE
Short-term effects: *SEEK MEDICAL ATTENTION*. Vapor may cause lung edema; physical exertion will aggravate this condition.
Liquid: *HARMFUL IF SWALLOWED OR ABSORBED THROUGH THE SKIN*. Will burn skin and eyes. Harmful if swallowed. Remove contaminated clothing and shoes. Flush affected areas with plenty of water. *IF IN EYES*, hold eyelids open and flush with plenty of water. *IF SWALLOWED* and victim is *CONSCIOUS AND ABLE TO SWALLOW*, have victim drink 4 to 8 ounces of water, have victim induce vomiting. *IF SWALLOWED* and victim is *UNCONSCIOUS OR HAVING CONVULSIONS*, do nothing except keep victim warm.
Note to physician or authorized medical personnel: Medical observation is recommended for 24 to 48 hours after breathing overexposure, as pulmonary edema may be delayed. As first aid for pulmonary edema, consider administering a corticosteroid spray. Cigarette smoking may exacerbate pulmonary injury and should be discouraged for at least 72 hours following exposure.

HEALTH HAZARDS
Personal protective equipment (PPE): B-Level PPE. OSHA Table Z-1-A air contaminant. Full face organic vapor respirator. Chemical protective material(s) reported to have good to excellent resistance: butyl rubber, PV alcohol, Teflon®, Viton®, Silvershield®.
Recommendations for respirator selection: OSHA
50 ppm: CCROV* [any chemical cartridge respirator with organic vapor cartridge(s)]; or SA* (any supplied-air respirator). *100 ppm:* SA:CF* (any supplied-air respirator operated in a continuous-flow mode); or CCRFOV [any chemical cartridge respirator with a full facepiece and organic vapor cartridges(s)]; or PAPROV* [any powered, air-purifying respirator with organic vapor cartridge(s)]; or GMFOV [any air-purifying, full-facepiece respirator (gas mask) with a chin-style, front- or back-mounted acid gas canister]; or SCBAF (any self-contained breathing apparatus with a full facepiece); or SAF (any supplied-air respirator with a full facepiece). *EMERGENCY OR PLANNED ENTRY INTO UNKNOWN CONCENTRATIONS OR IDLH CONDITIONS:* SCBAF:PD,PP (any self-contained breathing apparatus that has a full facepiece and is operated in a pressure-demand or other positive-pressure mode); or SAF:PD,PP:ASCBA (any supplied-air respirator that has a full facepiece and is operated in a pressure-demand or other positive-pressure mode in combination with an auxiliary self-contained breathing apparatus operated in a pressure-demand or other positive pressure mode). *ESCAPE:* GMFOV [any air-purifying, full-facepiece respirator (gas mask) with a chin-style, front- or back-mounted organic vapor cannister]; or SCBAE (any appropriate escape-type, self-contained breathing apparatus).
Note: Reported to cause eye damage; may require eye protection.

Skin and mucous membranes: Flood affected tissues with water.
Exposure limits (TWA unless otherwise noted): ACGIH TLV 2 ppm (7.9 mg/m^3); OSHA PEL 5 mg/m^3 (20 mg/m^3); skin contact contributes significantly in overall exposure.
Toxicity by ingestion: Poison. Grade 3; LD_{50} = 50 to 500 mg/kg Liver cirrhosis in animal experiments.
Long-term health effects: May cause liver and kidney damage. EPA genetic Toxicology program; mutation data.
Vapor (gas) irritant characteristics: Vapors cause irritation such that personnel will find high concentrations unpleasant.
Liquid or solid irritant characteristics: Severe eye and skin irritant. Causes smarting of the skin and first-degree burns on short exposure; may cause secondary burns on long exposure.
Odor threshold: 0.002–0.64 ppm.
IDLH value: 100 ppm.

FIRE DATA
Flash point: 153°F/67°C (oc); 140°F/60°C (cc).
Flammable limits in air: LEL: 2.1%; UEL: 19.3%.
Autoignition temperature: 597°F/314°C/587°K.
Electrical hazard: Class I, Group C.
Burning rate: 2.6 mm/min

CHEMICAL REACTIVITY
Binary reactants: Violent reaction with acids, oxidizers, strong bases. Attacks many plastic materials.
Polymerization: May polymerize on contact with strong acids or strong alkalis
Reactivity group: 19
Compatibility class: Aldehyde

ENVIRONMENTAL DATA
Food chain concentration potential: Negative; unlikely to accumulate.
Water pollution: Harmful to aquatic life in very low concentrations. May be dangerous if it enters nearby water intakes; notify operators. Notify local health and wildlife officials. **Response to discharge:** Issue warning–water contaminant. Should be removed. Chemical and physical treatment.

SHIPPING INFORMATION
Grades of purity: Commercial; **Storage temperature:** Ambient; **Inert atmosphere:** None; **Venting:** Pressure-vacuum; **Stability during transport:** Stable.

NAS HAZARD CLASSIFICATION FOR BULK WATER TRANSPORTATION
FIRE: 2
HEALTH: Vapor irritant: 2; Liquid or solid irritant: 2; Poisons: 3
WATER POLLUTION: Human toxicity: 3; Aquatic toxicity: 3; Aesthetic effect: 2
REACTIVITY: Other chemicals: 2; Water: 0; Self-reaction: 1

PHYSICAL AND CHEMICAL PROPERTIES
Physical state @ 59°F/15°C and 1 atm: Liquid.
Molecular weight: 96.1
Boiling point @ 1 atm: 323°F/162°C/435°K.
Melting/Freezing point: –34°F/–37°C/237°K.
Critical temperature: 745°F/397°C/670°K.
Critical pressure: 798 psia = 54.3 atm = 5.50 MN/m^2.
Specific gravity (water = 1): 1.159 @ 68°F/20°C (liquid).
Liquid surface tension: 43.5 dynes/cm = 0.0435 N/m @ 68°F/20°C.
Liquid water interfacial tension: (estimate) 30 dynes/cm = 0.03 N/m @ 68°F/20°C.

Relative vapor density (air = 1): 3.2
Latent heat of vaporization: 191 Btu/lb = 106 cal/g = 4.44×10^5 J/kg.
Heat of combustion: $-10{,}490$ Btu/lb = -5830 cal/g = -244.1×10^5 J/kg.
Vapor pressure: (Reid) 0.1 psia; 2 mm.

FURFURYL ALCOHOL   REC. F:1550

SYNONYMS: ALCOHOL FURFURILICO (Spanish); EEC No. 603-018-00-2; 2-FURANCARBINOL; 2-FURANMETHANOL; FURFURAL ALCOHOL; FURYLALCOHOL; 2-FURYLCARBINOL; 2-HYDROXYMETHYLFURAN

IDENTIFICATION
CAS Number: 98-00-0
Formula: $C_5H_6O_2$
DOT ID Number: UN 2874; DOT Guide Number: 153
Proper Shipping Name: Furfuryl alcohol

DESCRIPTION: Colorless liquid; turning yellow to amber on exposure to air. Mildly irritating odor. Mixes with water.

Poison! • Flammable • Breathing the vapor or swallowing the material can cause severe illness; skin or eye contact can cause severe pain and burns • Firefighting gear (including SCBA) does not provide adequate protection. If exposure occurs, remove and isolate gear immediately and thoroughly decontaminate personnel • Containers may BLEVE when exposed to fire • Vapors may form explosive mixture with air • Vapors are heavier than air and will collect and stay in low areas • Vapors may travel long distances to ignition sources and flashback • Vapors in confined areas (e.g., tanks, sewers, buildings) may explode when exposed to fire • Toxic products of combustion may include nitrogen oxide • Do not put yourself in danger by entering a contaminated area to rescue a victim.

Hazard Classification (based on NFPA-704 Rating System)
Health Hazards (Blue): 1; Flammability (Red): 2; Reactivity (Yellow): 1

EMERGENCY RESPONSE: See Appendix A (153)
Evacuation:
Public safety: Isolate the area of spill or leak for at least 25 to 50 meters (80 to 160 feet) in all directions.
Spill: Increase, in the downwind direction, as necessary, the distance shown under "Public Safety."
Fire: If tank, rail car, or tank truck is involved in fire, isolate for at least 800 meters (½ mile) in all directions; also, consider initial evacuation for 800 meters (½ mile) in all directions.

EXPOSURE
Short-term effects: *SEEK MEDICAL ATTENTION.*
Vapor/fumes: Affects the nervous system. *IF BREATHING HAS STOPPED*, give artificial respiration; *avoid mouth-to-mouth resuscitation; use bag/mask apparatus.* **Liquid:** *HARMFUL IF SWALLOWED OR ABSORBED THROUGH THE SKIN.* Irritating to skin and eyes. Harmful if swallowed. Remove contaminated clothing and shoes. Flush affected areas with plenty of water. *IF IN EYES*, hold eyelids open and flush with plenty of water. *IF SWALLOWED* and victim is *CONSCIOUS AND ABLE TO SWALLOW*, have victim drink 4 to 8 ounces of water and have victim induce vomiting. *IF SWALLOWED* and victim is *UNCONSCIOUS OR HAVING CONVULSIONS,* do nothing except keep victim warm. The use of alcoholic beverages may enhance the toxic effect.

HEALTH HAZARDS
Personal protective equipment (PPE): B-Level PPE. OSHA Table Z-1-A air contaminant. Full face organic vapor respirator. Natural rubber, neoprene, and nitrile rubber may offer some protection from alcohols. Chemical protective material(s) reported to offer minimal to poor protection: Viton®/neoprene, butyl rubber/neoprene.
Recommendations for respirator selection: NIOSH/OSHA
75 ppm: CCROV* [any chemical cartridge respirator with organic vapor cartridge(s)]; or GMFOV[any air-purifying, full-facepiece respirator (gas mask) with a chin-style, front- or back-mounted acid gas canister]; or PAPROV* [any powered, air-purifying respirator with organic vapor cartridge(s)]; or SA* (any supplied-air respirator); or SCBAF (any self-contained breathing apparatus with a full facepiece). *EMERGENCY OR PLANNED ENTRY INTO UNKNOWN CONCENTRATIONS OR IDLH CONDITIONS:* SCBAF:PD,PP (any self-contained breathing apparatus that has a full facepiece and is operated in a pressure-demand or other positive-pressure mode); or SAF:PD,PP:ASCBA (any supplied-air respirator that has a full facepiece and is operated in a pressure-demand or other positive-pressure mode in combination with an auxiliary self-contained breathing apparatus operated in a pressure-demand or other positive pressure mode). *ESCAPE:* GMFOV[any air-purifying, full-facepiece respirator (gas mask) with a chin-style, front- or back-mounted organic vapor cannister]; or SCBAE (any appropriate escape-type, self-contained breathing apparatus). *Note:* Substance reported to cause eye irritation or damage; may require eye protection.
Exposure limits (TWA unless otherwise noted): ACGIH TLV 10 ppm (40 mg/m^3); OSHA PEL 50 (200 mg/m^3); NIOSH REL 10 ppm (40 mg/m^3); skin contact contributes significantly in overall exposure. Skin absorption very rapid.
Short-term exposure limits (15-minute TWA): ACGIH STEL 15 ppm (60 mg/m^3); NIOSH STEL 15 ppm (60 mg/m^3); skin contact contributes significantly in overall exposure.
Toxicity by ingestion: Poison. Grade 3; oral LD_{50} = 132 mg/kg (rat).
Long-term health effects: May cause nervous system damage.
Vapor (gas) irritant characteristics: Eye and respiratory tract irritant. Vapors cause irritation such that personnel will find high concentrations unpleasant. The effect is temporary.
Liquid or solid irritant characteristics: Severe eye irritant. Absorption through skin very rapid. Skin irritant. If spilled on clothing and allowed to remain, may cause smarting and reddening of the skin. Degreases the skin.
Odor threshold: 8 ppm. *Note:* This is very close to the TLV.
IDLH value: 75 ppm.

FIRE DATA
Flash point: 167°F/75°C (oc); 149°F/65°C (cc).
Flammable limits in air: LEL: 1.8%; UEL: 16.3%.
Autoignition temperature: 912°F/489°C/762°K.
Electrical hazard: Class I, Group C.
Burning rate: 2.3 mm/min

CHEMICAL REACTIVITY
Binary reactants: Violent reaction with strong oxidizers and acids: Explosive violence in contact with mineral acids or their vapors, or with strong organic acids or their vapors. Darkens and forms water-insoluble material on exposure to air or acids, particularly when hot.

Polymerization: Contact with organic acids may lead to polymerization.
Reactivity group: 20
Compatibility class: Alcohols, glycols

ENVIRONMENTAL DATA
Food chain concentration potential: Log K_{ow} = 0.30 (estimate). Unlikely to accumulate.
Water pollution: Effect of low concentrations on aquatic life is unknown. May be dangerous if it enters nearby water intakes; notify operators. Notify local health and wildlife officials.
Response to discharge: Issue warning–water contaminant. Disperse and flush.

SHIPPING INFORMATION
Grades of purity: Technical; **Storage temperature:** Ambient; **Inert atmosphere:** None; **Venting:** Open (flame arrester); **Stability during transport:** Stable.

NAS HAZARD CLASSIFICATION FOR BULK WATER TRANSPORTATION
FIRE: 1
HEALTH: Vapor irritant: 2; Liquid or solid irritant: 1; Poisons: 2
WATER POLLUTION: Human toxicity: 2; Aquatic toxicity: 3; Aesthetic effect: 1
REACTIVITY: Other chemicals: 2; Water: 0; Self-reaction: 0

PHYSICAL AND CHEMICAL PROPERTIES
Physical state @ 59°F/15°C and 1 atm: Liquid.
Molecular weight: 98.1
Boiling point @ 1 atm: 338°F/170°C/443°K.
Melting/Freezing point: 6°F/–14°C/259°K.
Specific gravity (water = 1): 1.13 @ 68°F/20°C (liquid).
Liquid surface tension: 38 dynes/cm = 0.038 N/m @ 68°F/20°C.
Relative vapor density (air = 1): 3.4
Latent heat of vaporization: 230 Btu/lb = 130 cal/g = 5.4 x 10^5 J/kg.
Heat of combustion: –11200 Btu/lb = –6200 cal/g = –260 x 10^5 J/kg.
Vapor pressure: (Reid) 0.07 psia; 0.6 mm @ 77°F/25°C.

GALLIC ACID   REC. G:0100

SYNONYMS: ACIDO GALLICO (Spanish); GALLIC ACID MONOHYDRATE; 3,4,5-TRIHYDROXYBENZOIC ACID

IDENTIFICATION
CAS Number: 149-91-7
Formula: $C_7H_6O_5$; 3,4,5-$(HO)_3C_6H_2COOH \cdot H_2O$

DESCRIPTION: White to slightly yellow crystalline solid. Odorless. Sinks in water; slightly soluble.

Combustible • Heat or flame may cause explosion • Containers may BLEVE when exposed to fire • Vapors are heavier than air and will collect and stay in low areas • Vapors in confined areas (e.g., tanks, sewers, buildings) may explode when exposed to fire • Irritating to the skin, eyes, and respiratory tract • Toxic products of combustion may include carbon monoxide.

Hazard Classification (based on NFPA-704 Rating System)
Health Hazards (Blue): 1; Flammability (Red): 1; Reactivity (Yellow): 0

EMERGENCY RESPONSE: See Appendix A (171)
Evacuation:
Public safety: Isolate the area of spill or leak for at least 10 to 25 meters (30 to 80 feet) in all directions.
Spill: Increase, in the downwind direction, as necessary, the distance shown under "Public Safety."
Fire: If any large container is involved in fire, isolate for at least 800 meters (½ mile) in all directions; also, consider initial evacuation for 800 meters (½ mile) in all directions.

EXPOSURE
Short-term effects: *SEEK MEDICAL ATTENTION.* **Dust:** Irritating to eyes, nose, and throat. *IF INHALED*, will, will cause coughing or difficult breathing. *IF IN EYES*, hold eyelids open and flush with plenty of water. *IF BREATHING HAS STOPPED*, give artificial respiration. IF breathing is difficult, administer oxygen. **Solid:** Irritating to skin and eyes. Harmful if swallowed. Remove contaminated clothing and shoes. Flush affected areas with plenty of water. *IF IN EYES*, hold eyelids open and flush with plenty of water. *IF SWALLOWED* and victim is *CONSCIOUS AND ABLE TO SWALLOW*, have victim drink 4 to 8 ounces of water and have victim induce vomiting. *IF SWALLOWED* and victim is *UNCONSCIOUS OR HAVING CONVULSIONS,* do nothing except keep victim warm.

HEALTH HAZARDS
Personal protective equipment (PPE): B-Level PPE. Bureau of Mines approved respirator; rubber gloves; safety goggles
Toxicity by ingestion: Grade 2; LD_{50} = 0.5 to 5 g/kg (rat).
Long-term health effects: EPA Genetic Toxicology program; mutation and reproductive data.
Vapor (gas) irritant characteristics: Eye and respiratory tract irritant.
Liquid or solid irritant characteristics: Eye and skin irritant.

CHEMICAL REACTIVITY
Binary reactants: Reacts with oxidizers.

ENVIRONMENTAL DATA
Food chain concentration potential: Log P_{ow} = –0.2. Unlikely to accumulate.
Water pollution: Harmful to aquatic life in very low concentrations. May be dangerous if it enters nearby water intakes; notify operators. Notify local health and wildlife officials. **Response to discharge:** Disperse and flush.

SHIPPING INFORMATION
Grades of purity: N.F; Practical; **Storage temperature:** Ambient; **Inert atmosphere:** None; **Venting:** Open; **Stability during transport:** Stable.

PHYSICAL AND CHEMICAL PROPERTIES
Physical state @ 59°F/15°C and 1 atm: Solid.
Molecular weight: 188
Melting/Freezing point: Decomposes 465°F/240°C/513°K.
Specific gravity (water = 1): 1.7 @ 68°F/20°C (solid).
Heat of combustion: –6,060 Btu/lb = –3,370 cal/g = –141 x 10^5 J/kg.

GAS OIL   REC. G:0150

SYNONYMS: GAS OIL, CRACKED; GAS OILS (PETROLEUM), LIGHT VACUUM

Gasolines: automotive (less than 4.23g lead/gal)

IDENTIFICATION
CAS Number: 64741-58-8
Formula: C_{12}-C_{25} mixture of hydrocarbons
DOT ID Number: UN 1202; DOT Guide Number: 128
Proper Shipping Name: Gas oil

DESCRIPTION: Colorless or dyed yellow to brown liquid. Gasoline- or petroleum-like odor. Floats on the surface of water; insoluble. Gas oil is often a mixture of hydrocarbons, used to describe non-road vehicle fuels.

Highly flammable • Containers may BLEVE when exposed to fire • Vapors may form explosive mixture with air • Vapors are heavier than air and will collect and stay in low areas • Vapors may travel long distances to ignition sources and flashback • Vapors in confined areas (e.g., tanks, sewers, buildings) may explode when exposed to fire • Irritating to the skin, eyes, and respiratory tract • Toxic products of combustion may include carbon monoxide.

Hazard Classification (based on NFPA-704 Rating System)
Health Hazards (Blue): 0; Flammability (Red): 2; Reactivity (Yellow): 0

EMERGENCY RESPONSE: See Appendix A (128)
Evacuation:
Public safety: Isolate spill area for at least 25 to 50 meters (80 to 160 feet) in all directions.
Spill: Large spill–Consider initial downwind evacuation for at least 300 meters (1000 feet).
Fire: Isolate for 800 meters (½ mile) in all directions, especially if tank, rail car, or tank truck is involved in fire.

EXPOSURE
Short-term effects: *SEEK MEDICAL ATTENTION*. **Liquid:** Harmful if swallowed. *IF SWALLOWED* and victim is *CONSCIOUS AND ABLE TO SWALLOW*, have victim drink 4 to 8 ounces of water. **Do NOT induce vomiting.**

HEALTH HAZARDS
Personal protective equipment (PPE): B-Level PPE. Protective goggles, gloves. Chemical protective material(s) reported to have good to excellent resistance: nitrile
Exposure limits (TWA unless otherwise noted): No single value applicable.
Toxicity by ingestion: Grade 2; LD_{50} = 0.5 to 5 g/kg.
Vapor (gas) irritant characteristics: Vapors cause a smarting of the eyes or respiratory system if present in high concentrations. The effect is temporary.
Liquid or solid irritant characteristics: If spilled on clothing and allowed to remain, may cause smarting and reddening of skin.
Odor threshold: 0.25 ppm.

FIRE DATA
Flash point: 150°F/66°C (cc) or higher
Flammable limits in air: LEL: 0.5%; UEL: 5.0%.
Autoignition temperature: 640°F/338°C/611°K.
Electrical hazard: Flow or agitation of substance may generate electrostatic charges due to low conductivity.
Burning rate: 4 mm/min

CHEMICAL REACTIVITY
Binary reactants: Keep away from strong oxidizers.
Reactivity group: 33
Compatibility class: Miscellaneous hydrocarbon mixtures

ENVIRONMENTAL DATA
Food chain concentration potential: Negative; unlikely to accumulate.
Water pollution: Harmful to aquatic life in very low concentrations. Fouling to shoreline. May be dangerous if it enters nearby water intakes; notify operators. Notify local health and wildlife officials. **Response to discharge:** Mechanical containment. Should be removed.
Physical and chemical treatment.

SHIPPING INFORMATION
Grades of purity: Composition varies widely with the refinery operation involved; **Storage temperature:** Ambient; **Inert atmosphere:** None; **Venting:** Open (flame arrester); **Stability during transport:** Stable.

NAS HAZARD CLASSIFICATION FOR BULK WATER TRANSPORTATION
FIRE: 3
HEALTH: Vapor irritant: 1; Liquid or solid irritant: 1; Poisons: 2
WATER POLLUTION: Human toxicity: 1; Aquatic toxicity: 2; Aesthetic effect: 2
REACTIVITY: Other chemicals: 0; Water: 0; Self-reaction: 0

PHYSICAL AND CHEMICAL PROPERTIES
Physical state @ 59°F/15°C and 1 atm: Liquid.
Boiling point @ 1 atm: 375–750°F/190–399°C/463–672°K.
Specific gravity (water = 1): 0.848 16°C (liquid).
Liquid surface tension: (estimate) 25 dynes/cm = 0.025 N/m @ 68°F/20°C.
Liquid water interfacial tension: (estimate) 50 dynes/cm = 0.05 N/m @ 68°F/20°C.
Relative vapor density (air = 1): 3.4
Heat of combustion: $-18{,}400$ Btu/lb = $-10{,}200$ cal/g = 428×10^5 J/kg.

GASOLINES: AUTOMOTIVE (less than 4.23g lead/gal)
REC. G:0200

SYNONYMS: BENZIN (German); EEC No. 650-001-01-8; GASOLINA (Spanish); ESSENCE (French); MOTOR FUEL; MOTOR SPIRIT; PETROL (Britain)

IDENTIFICATION
CAS Number: 8006-61-9
Formula: $C_{45}H_{12}$ to C_9H_{20}
DOT ID Number: UN 1203; DOT Guide Number: 128
Proper Shipping Name: Gasoline

DESCRIPTION: Colorless, pink to pale brown watery liquid. Dyes are usually added. Characteristic odor. Floats on the surface of water; insoluble. Flammable, irritating vapor is produced. As of 1996 lead has not been used in commercial gasoline sold in the United States.

Highly flammable • Containers may BLEVE when exposed to fire • Vapors may form explosive mixture with air • Vapors are heavier than air and will collect and stay in low areas • Vapors may travel long distances to ignition sources and flashback • Vapors in confined areas (e.g., tanks, sewers, buildings) may explode when exposed to fire • Irritating to the skin, eyes, and respiratory tract • Toxic products of combustion may include carbon monoxide.

Hazard Classification (based on NFPA-704 Rating System)
Health Hazards (Blue): 1; Flammability (Red): 3; Reactivity (Yellow): 0

EMERGENCY RESPONSE: See Appendix A (128)
Evacuation:
Public safety: Isolate spill area for at least 25 to 50 meters (80 to 160 feet) in all directions.
Spill: Large spill–Consider initial downwind evacuation for at least 300 meters (1000 feet).
Fire: Isolate for 800 meters (½ mile) in all directions, especially if tank, rail car, or tank truck is involved in fire.

EXPOSURE
Short-term effects: *SEEK MEDICAL ATTENTION*. **Vapor:** Irritating to eyes, nose, and throat. *IF INHALED*, will, will cause dizziness, headache, difficult breathing or loss of consciousness. Move to fresh air. *IF BREATHING HAS STOPPED*, give artificial respiration. *IF BREATHING IS DIFFICULT*, administer oxygen. **Liquid:** Irritating to skin and eyes. *IF SWALLOWED*, will cause nausea or vomiting. Remove contaminated clothing and shoes. Flush affected areas with plenty of water. *IF IN EYES*, hold eyelids open and flush with plenty of water. *IF SWALLOWED* and victim is *CONSCIOUS AND ABLE TO SWALLOW*, have victim drink 4 to 8 ounces of water. **Do NOT induce emesis and do NOT administer activated charcoal.** Observe for at least 6 hours for signs of chemical pneumonitis.
Note: Alcohol consumption exacerbates the toxic effects, especially in those products containing MTBE.

HEALTH HAZARDS
Personal protective equipment (PPE): a-Level PPE. OSHA Table Z-1-A air contaminant. Chemical protective material(s) reported to have good to excellent resistance: nitrile, PV alcohol, Silvershield®.
Recommendations for respirator selection: NIOSH
At any concentrations above the NIOSH REL, or where there is no REL, at any detectable concentration: SCBAF:PD,PP (any self-contained breathing apparatus that has a full facepiece and is operated in a pressure-demand or other positive-pressure mode); or SAF:PD,PP:ASCBA (any supplied-air respirator that has a full facepiece and is operated in a pressure-demand or other positive-pressure mode in combination with an auxiliary self-contained breathing apparatus operated in a pressure-demand or other positive pressure mode). *ESCAPE:* GMFOV [any air-purifying, full-facepiece respirator (gas mask) with a chin-style, front-or back-mounted organic vapor canister]; or SCBAE (any appropriate escape-type, self-contained breathing apparatus).
Exposure limits (TWA unless otherwise noted): ACGIH TLV 300 ppm (890 mg/m^3); NIOSH REL potential human carcinogen, reduce exposure to lowest feasible level.
Short-term exposure limits (15-minute TWA): ACGIH STEL 500 ppm (1480 mg/m^3).
Toxicity by ingestion: Grade 2; LD_{50} = 0.5 to 5 g/kg.
Long-term health effects: IARC possible human carcinogen, rating 2B
Vapor (gas) irritant characteristics: Vapors cause smarting of the eyes or respiratory system if present in high concentrations.
Liquid or solid irritant characteristics: If spilled on clothing and allowed to remain, may cause smarting and reddening of the skin.
Odor threshold: 0.25 ppm.

FIRE DATA
Flash point: −36 to −45°F/−38 to −43°C (cc) (varies with octane ratings and grades).
Flammable limits in air: LEL: 1.4 -1.5%; UEL: 7.4 -7.6%.
Fire extinguishing agents not to be used: Water may be ineffective
Autoignition temperature: 536–853°F/280–456°C/553–729°K (varies with octane rating and grade).
Electrical hazard: Class I, Group D. Flow or agitation of substance may generate electrostatic charges due to low conductivity.
Burning rate: 4 mm/min

CHEMICAL REACTIVITY
Binary reactants: Strong oxidizers, such as peroxides, nitric acid and perchlorates.
Reactivity group: 33
Compatibility class: Miscellaneous hydrocarbon mixtures

ENVIRONMENTAL DATA
Food chain concentration potential: Negative; unlikely to accumulate.
Water pollution: DOT Appendix B, §172.101–marine pollutant (leaded gasoline). Harmful to aquatic life in very low concentrations. Fouling to shoreline. May be dangerous if it enters nearby water intakes; notify operators. Notify local health and wildlife officials. **Response to discharge:** Issue warning–high flammability. Evacuate area. Disperse and flush.

SHIPPING INFORMATION
Grades of purity: Various octane ratings; military specifications;
Storage temperature: Ambient; **Inert atmosphere:** None; **Venting:** Open (flame arrester) or pressure-vacuum; **Stability during transport:** Stable.

NAS HAZARD CLASSIFICATION FOR BULK WATER TRANSPORTATION
FIRE: 3
HEALTH: Vapor irritant: 1; Liquid or solid irritant: 1; Poisons: 2
WATER POLLUTION: Human toxicity: 1; Aquatic toxicity: 2; Aesthetic effect: 2
REACTIVITY: Other chemicals: 0; Water: 0; Self-reaction: 0

PHYSICAL AND CHEMICAL PROPERTIES
Physical state @ 59°F/15°C and 1 atm: Liquid.
Molecular weight: 72 (approximate).
Boiling point @ 1 atm: 100–400°F/38–204°C/311–477°K.
Specific gravity (water = 1): 0.7321 @ 68°F/20°C (liquid).
Liquid surface tension: 19-23 dynes/cm = 0.019–0.023 N/m @ 68°F/20°C.
Liquid water interfacial tension: 49-51 dynes/cm = 0.049-0.051 N/m @ 68°F/20°C.
Relative vapor density (air = 1): 3–4
Ratio of specific heats of vapor (gas): (estimate) 1.054
Latent heat of vaporization: 130–150 Btu/lb = 71–81 cal/g = 3.0–3.4 x 10^5 J/kg.
Heat of combustion: −18,720 Btu/lb = −10,400 cal/g = 435.1 x 10^5 J/kg.
Vapor pressure: (Reid) 7.4 psia; 38–300 mm.

GASOLINES, AVIATION REC. G:0250

SYNONYMS: GASOLINE, AVIATION (less than 4.86g(Pb)/gal); GASOLINE, AVIATION GRADE (100–130 OCTANE); GASOLINE, AVIATION GRADE (115–145 OCTANE); GASOLINA de AVIACION (Spanish)

568 Gasolines, aviation

IDENTIFICATION
CAS Number: 8006-61-9
DOT ID Number: UN 1203; DOT Guide Number: 128
Proper Shipping Name: Gasoline

DESCRIPTION: Red, blue, green, brown or purple watery liquid. Characteristic odor. Floats on water surface; insoluble; . Flammable, irritating vapor is produced.

Highly flammable • Containers may BLEVE when exposed to fire •Vapors may form explosive mixture with air • Vapors are heavier than air and will collect and stay in low areas • Vapors may travel long distances to ignition sources and flashback • Vapors in confined areas (e.g., tanks, sewers, buildings) may explode when exposed to fire • Irritating to the skin, eyes, and respiratory tract • Toxic products of combustion may include carbon monoxide.

Hazard Classification (based on NFPA-704 Rating System)
Health Hazards (Blue): 1; Flammability (Red): 3; Reactivity (Yellow): 0

EMERGENCY RESPONSE: See Appendix A (128)
Evacuation:
Public safety: Isolate spill area for at least 25 to 50 meters (80 to 160 feet) in all directions.
Spill: Large spill–Consider initial downwind evacuation for at least 300 meters (1000 feet).
Fire: Isolate for 800 meters (½ mile) in all directions, especially if tank, rail car, or tank truck is involved in fire.

EXPOSURE
Short-term effects: *SEEK MEDICAL ATTENTION.* **Vapor:** Irritating to eyes, nose, and throat. *IF INHALED*, will, will cause dizziness, headache, difficult breathing or loss of consciousness. Move to fresh air. *IF BREATHING HAS STOPPED*, give artificial respiration. IF breathing is difficult, administer oxygen. **Liquid:** Irritating to skin and eyes. *IF SWALLOWED*, will cause nausea or vomiting. Remove contaminated clothing and shoes. Flush affected areas with plenty of water. *IF IN EYES*, hold eyelids open and flush with plenty of water. *IF SWALLOWED* and victim is *CONSCIOUS AND ABLE TO SWALLOW*, have victim drink 4 to 8 ounces of water. **Do NOT induce vomiting.**

HEALTH HAZARDS
Personal protective equipment (PPE): A-Level PPE. OSHA Table Z-1-A air contaminant. Protective goggles, gloves. Chemical protective material(s) reported to have good to excellent resistance: nitrile, Teflon®, Viton®, Silvershield®.
Recommendations for respirator selection: NIOSH
At any concentrations above the NIOSH REL, or where there is no REL, at any detectable concentration: SCBAF:PD,PP (any self-contained breathing apparatus that has a full facepiece and is operated in a pressure-demand or other positive-pressure mode); or SAF:PD,PP:ASCBA (any supplied-air respirator that has a full facepiece and is operated in a pressure-demand or other positive-pressure mode in combination with an auxiliary self-contained breathing apparatus operated in a pressure-demand or other positive pressure mode). *ESCAPE:* GMFOV [any air-purifying, full-facepiece respirator (gas mask) with a chin-style, front-or back-mounted organic vapor canister]; or SCBAE (any appropriate escape-type, self-contained breathing apparatus).
Exposure limits (TWA unless otherwise noted): ACGIH TLV 300 ppm (890 mg/m^3); NIOSH REL potential human carcinogen; reduce exposure to lowest feasible level.
Short-term exposure limits (15-minute TWA): ACGIH/OSHA STEL 500 ppm (1480 mg/m^3).
Toxicity by ingestion: Grade 2; LD$_{50}$ = 0.5 to 5 g/kg.
Long-term health effects: IARC possible human carcinogen, rating 2B
Vapor (gas) irritant characteristics: Vapors cause smarting of the eyes or respiratory system if present in high concentrations. The effect is temporary.
Liquid or solid irritant characteristics: If spilled on clothing and allowed to remain, may cause smarting and reddening of the skin.
Odor threshold: 0.25 ppm.

FIRE DATA
Flash point: –50°F/–46°C (cc) (approximate).
Flammable limits in air: LEL: 1.3%; UEL: 7.1% (100–130 octane); 1.2%; UEL: 7.1% (115–145 octane)
Fire extinguishing agents not to be used: Water may be ineffective
Autoignition temperature: 824–880°F/440–471°C/713–744°K.
Electrical hazard: Class I, Group D. Flow or agitation of substance may generate electrostatic charges due to low conductivity.
Burning rate: 4 mm/min

CHEMICAL REACTIVITY
Binary reactants: Strong oxidizers, nitric acid perchlorates.

ENVIRONMENTAL DATA
Food chain concentration potential: Negative; unlikely to accumulate.
Water pollution: Harmful to aquatic life in very low concentrations. Fouling to shoreline. May be dangerous if it enters nearby water intakes; notify operators. Notify local health and wildlife officials. **Response to discharge:** Issue warning–high flammability. Evacuate area. Disperse and flush.

SHIPPING INFORMATION
Grades of purity: Grades 80/87, 100/130, and 115/145: Specification MIL-G-5572e; **Storage temperature:** Ambient; **Inert atmosphere:** None; **Venting:** Open (flame arrester) or pressure-vacuum; **Stability during transport:** Stable.

NAS HAZARD CLASSIFICATION FOR BULK WATER TRANSPORTATION
FIRE: 3
HEALTH: Vapor irritant: 1; Liquid or solid irritant: 1; Poisons: 2
WATER POLLUTION: Human toxicity: 1; Aquatic toxicity: 2; Aesthetic effect: 2
REACTIVITY: Other chemicals: 0; Water: 0; Self-reaction: 0

PHYSICAL AND CHEMICAL PROPERTIES
Physical state @ 59°F/15°C and 1 atm: Liquid.
Boiling point @ 1 atm: 160–340°F/71–171°C/344–444°K.
Melting/Freezing point: Less than 76°F/24°C/298°K.
Specific gravity (water = 1): 0.711 @ 15°C (liquid).
Liquid surface tension: 19–23 dynes/cm = 0.019–0.023 N/m@ 68°F/20°C.
Liquid water interfacial tension: 49–51 dynes/cm = 0.049-0.051 N/m @ 68°F/20°C.
Relative vapor density (air = 1): 3.4
Ratio of specific heats of vapor (gas): (estimate) 1.054
Latent heat of vaporization: 130–150 Btu/lb = 71–81 cal/g = 3.0–3.4 x 10^5 J/kg.
Heat of combustion: –18,720 Btu/lb = –10,400 cal/g = –435.4 x 10^5 J/kg.

GASOLINE BLENDING STOCKS: ALKYLATES
REC. G:0300

SYNONYMS: ALKYLATES, GASOLINE BLENDING STOCKS; ALQUILADO GASOLINA (Spanish)

IDENTIFICATION
DOT ID Number: UN 1203; DOT Guide Number: 128
Proper Shipping Name: Gasoline

DESCRIPTION: Colorless watery liquid. May be dyed. Gasoline odor. Floats on water surface; insoluble.

Highly flammable • Containers may BLEVE when exposed to fire • Vapors may form explosive mixture with air • Vapors are heavier than air and will collect and stay in low areas • Vapors may travel long distances to ignition sources and flashback • Vapors in confined areas (e.g., tanks, sewers, buildings) may explode when exposed to fire • Irritating to the skin, eyes, and respiratory tract • Toxic products of combustion may include carbon monoxide.

Hazard Classification (based on NFPA-704 Rating System)
Health Hazards (Blue): 1; Flammability (Red): 3; Reactivity (Yellow): 0

EMERGENCY RESPONSE: See Appendix A (128)
Evacuation:
Public safety: Isolate spill area for at least 25 to 50 meters (80 to 160 feet) in all directions.
Spill: Large spill–Consider initial downwind evacuation for at least 300 meters (1000 feet).
Fire: Isolate for 800 meters (½ mile) in all directions, especially if tank, rail car, or tank truck is involved in fire.

EXPOSURE
Short-term effects: *SEEK MEDICAL ATTENTION*. **Vapor:** Irritating to eyes, nose, and throat. *IF INHALED*, will, will cause dizziness, headache, difficult breathing or loss of consciousness. Move to fresh air. *IF BREATHING HAS STOPPED*, give artificial respiration. IF breathing is difficult, administer oxygen. **Liquid:** Irritating to skin and eyes. *IF SWALLOWED*, will cause nausea and vomiting. Remove contaminated clothing and shoes. Flush affected areas with plenty of water. *IF IN EYES*, hold eyelids open and flush with plenty of water. *IF SWALLOWED* and victim is *CONSCIOUS AND ABLE TO SWALLOW*, have victim drink 4 to 8 ounces of water. **Do NOT induce vomiting.**

HEALTH HAZARDS
Personal protective equipment (PPE): A-Level PPE. Protective goggles, gloves. Chemical protective material(s) reported to have good to excellent resistance: nitrile, Teflon®, Viton®, Silvershield®.
Recommendations for respirator selection: NIOSH
At any concentrations above the NIOSH REL, or where there is no REL, at any detectable concentration: SCBAF:PD,PP (any self-contained breathing apparatus that has a full facepiece and is operated in a pressure-demand or other positive-pressure mode); or SAF:PD,PP:ASCBA (any supplied-air respirator that has a full facepiece and is operated in a pressure-demand or other positive-pressure mode in combination with an auxiliary self-contained breathing apparatus operated in a pressure-demand or other positive pressure mode). *ESCAPE:* GMFOV [any air-purifying, full-facepiece respirator (gas mask) with a chin-style, front-or back-mounted organic vapor canister]; or SCBAE (any appropriate escape-type, self-contained breathing apparatus).

Exposure limits (TWA unless otherwise noted): As gasoline: ACGIH TLV 300 ppm (890 mg/m^3); NIOSH REL potential human carcinogen; reduce exposure to lowest feasible level.
Short-term exposure limits (15-minute TWA): As gasoline: ACGIH/OSHA STEL 500 ppm (1480 mg/m^3).
Toxicity by ingestion: Grade 2; LD_{50} = 0.5 to 5 g/kg.
Long-term health effects: Possible human carcinogen
Vapor (gas) irritant characteristics: Vapors cause smarting of the eyes or respiratory system if present in high concentrations. The effect is temporary.
Liquid or solid irritant characteristics: If spilled on clothing and allowed to remain, may cause smarting and reddening of the skin.
Odor threshold: 0.25 ppm as gassoline.

FIRE DATA
Flash point: (a) less than 0°F/–18°C (cc); (b) 0–73°F/–18–23°C (cc).
Flammable limits in air: (a) LEL: 1.1%; UEL: 8.7%.
Fire extinguishing agents not to be used: Water may be ineffective
Electrical hazard: Class I, Group D.
Burning rate: 4 mm/min

CHEMICAL REACTIVITY
Binary reactants: Strong oxidizers, nitric acid, perchlorates
Reactivity group: 33
Compatibility class: Miscellaneous Hydrocarbon Mixtures

ENVIRONMENTAL DATA
Food chain concentration potential: Negative; unlikely to accumulate.
Water pollution: Harmful to aquatic life in very low concentrations. Fouling to shoreline. May be dangerous if it enters nearby water intakes; notify operators. Notify local health and wildlife officials. **Response to discharge:** Issue warning–high flammability. Evacuate area. Disperse and flush.

SHIPPING INFORMATION
Grades of purity: Composition varies with range of distillation temperatures used; **Storage temperature:** Ambient; **Inert atmosphere:** None; **Venting:** Open (flame arrester) or pressure-vacuum; **Stability during transport:** Stable.

NAS HAZARD CLASSIFICATION FOR BULK WATER TRANSPORTATION
FIRE: 3
HEALTH: Vapor irritant: 1; Liquid or solid irritant: 1; Poisons: 2
WATER POLLUTION: Human toxicity: 1; Aquatic toxicity: 2; Aesthetic effect: 2
REACTIVITY: Other chemicals: 0; Water: 0; Self-reaction: 0

PHYSICAL AND CHEMICAL PROPERTIES
Physical state @ 59°F/15°C and 1 atm: Liquid.
Boiling point @ 1 atm: 58–275°F/14–135°C/287–408°K.
Specific gravity (water = 1): 0.71–0.75 @ 15°C (liquid).
Liquid surface tension: 19–23 dynes/cm = 0.019–0.023 N/m @ 68°F/20°C.
Liquid water interfacial tension: 49–51 dynes/cm = 0.049-0.051 N/m @ 68°F/20°C.
Relative vapor density (air = 1): 3.4
Latent heat of vaporization: 130–150 Btu/lb = 71–81 cal/g = 3.0-3.4 x 10^5 J/kg.
Heat of combustion: –18,720 Btu/lb = –10,400 cal/g = –435.4 x 10^5 J/kg.

GASOLINE BLENDING STOCKS: REFORMATES
REC. G:0350

SYNONYMS: REFORMATES, GASOLINE BLENDING STOCKS

IDENTIFICATION
DOT ID Number: UN 1203; DOT Guide Number: 128
Proper Shipping Name: Gasoline

DESCRIPTION: Colorless watery liquid. Gasoline-like odor. Floats on the surface of water. Flammable, irritating vapor is produced.

Highly flammable • Containers may BLEVE when exposed to fire • Vapors may form explosive mixture with air • Vapors are heavier than air and will collect and stay in low areas • Vapors may travel long distances to ignition sources and flashback • Vapors in confined areas (e.g., tanks, sewers, buildings) may explode when exposed to fire • Irritating to the skin, eyes, and respiratory tract • Toxic products of combustion may include carbon monoxide.

Hazard Classification (based on NFPA-704 Rating System)
Health Hazards (Blue): 1; Flammability (Red): 3; Reactivity (Yellow): 0

EMERGENCY RESPONSE: See Appendix A (128)
Evacuation:
Public safety: Isolate spill area for at least 25 to 50 meters (80 to 160 feet) in all directions.
Spill: Large spill–Consider initial downwind evacuation for at least 300 meters (1000 feet).
Fire: Isolate for 800 meters (½ mile) in all directions, especially if tank, rail car, or tank truck is involved in fire.

EXPOSURE
Short-term effects: *SEEK MEDICAL ATTENTION.* **Vapor:** Irritating to eyes, nose, and throat. *IF INHALED*, will, will cause dizziness, headache, difficult breathing or loss of consciousness. Move to fresh air. *IF BREATHING HAS STOPPED,* give artificial respiration. IF breathing is difficult, administer oxygen. **Liquid:** Irritating to skin and eyes. *IF SWALLOWED*, will cause nausea or vomiting. Remove contaminated clothing and shoes. Flush affected areas with plenty of water. *IF IN EYES*, hold eyelids open and flush with plenty of water. *IF SWALLOWED* and victim is *CONSCIOUS AND ABLE TO SWALLOW*, have victim drink 4 to 8 ounces of water. **Do NOT induce vomiting.**

Fouling to shoreline. May be dangerous if it enters nearby water intakes; notify operators. Notify local health and wildlife officials.
Response to discharge: Issue warning–high flammability. Evacuate area. Disperse and flush.

HEALTH HAZARDS
Personal protective equipment (PPE): A-Level PPE. OSHA Table Z-1-A air contaminant. Protective goggles, gloves.
Recommendations for respirator selection: NIOSH *At any concentrations above the NIOSH REL, or where there is no REL, at any detectable concentration:* SCBAF:PD,PP (any self-contained breathing apparatus that has a full facepiece and is operated in a pressure-demand or other positive-pressure mode); or SAF:PD,PP:ASCBA (any supplied-air respirator that has a full facepiece and is operated in a pressure-demand or other positive-pressure mode in combination with an auxiliary self-contained breathing apparatus operated in a pressure-demand or other positive pressure mode). *ESCAPE:* GMFOV [any air-purifying, full-facepiece respirator (gas mask) with a chin-style, front-or back-mounted organic vapor canister]; or SCBAE (any appropriate escape-type, self-contained breathing apparatus).
Exposure limits (TWA unless otherwise noted): As gasoline: ACGIH TLV 300 ppm (890 mg/m^3); NIOSH REL potential human carcinogen; reduce exposure to lowest feasible level.
Short-term exposure limits (15-minute TWA): As gasoline: ACGIH/OSHA STEL 500 ppm (1480 mg/m^3).
Toxicity by ingestion: Grade 2; LD_{50} = 0.5 to 5 g/kg.
Long-term health effects: Possible human carcinogen
Vapor (gas) irritant characteristics: Vapors cause smarting of the eyes or respiratory system if present in high concentrations. The effect is temporary.
Liquid or solid irritant characteristics: If spilled on clothing and allowed to remain, may cause smarting and reddening of the skin.
Odor threshold: 0.25 ppm as gasoline.

FIRE DATA
Flash point: (a) less than 0°F/–18°C (cc); (b) 0-73°F/–18-23°C (cc).
Flammable limits in air: (a) LEL: 1.1%; UEL: 8.7%.
Fire extinguishing agents not to be used: Water may be ineffective.
Electrical hazard: Class I, Group D.
Burning rate: 4 mm/min

CHEMICAL REACTIVITY
Binary reactants: Strong oxidizers, nitric acid, perchlorates.
Reactivity group: 33
Compatibility class: Miscellaneous hydrocarbon mixtures

ENVIRONMENTAL DATA
Food chain concentration potential: Negative; unlikely to accumulate.
Harmful to aquatic life in very low concentrations.

SHIPPING INFORMATION
Grades of purity: Composition varies with range of distillation temperatures used; **Storage temperature:** Ambient; **Inert atmosphere:** None; **Venting:** Open (flame arrester) or pressure-vacuum; **Stability during transport:** Stable.

NAS HAZARD CLASSIFICATION FOR BULK WATER TRANSPORTATION
FIRE: 3
HEALTH: Vapor irritant: 1; Liquid or solid irritant: 1; Poisons: 2
WATER POLLUTION: Human toxicity: 1; Aquatic toxicity: 2; Aesthetic effect: 2
REACTIVITY: Other chemicals: 0; Water: 0; Self-reaction: 0

PHYSICAL AND CHEMICAL PROPERTIES
Physical state @ 59°F/15°C and 1 atm: Liquid.

Boiling point @ 1 atm: 58–275°F/14–135°C/287–408°K.
Specific gravity (water = 1): 0.7934 @ 68°F/20°C (liquid).
Liquid surface tension: 19–23 dynes/cm = 0.019–0.023 N/m @ 68°F/20°C.
Liquid water interfacial tension: 49–51 dynes/cm = 0.049-0.051 N/m @ 68°F/20°C.
Relative vapor density (air = 1): 3.4
Latent heat of vaporization: 130–150 Btu/lb = 71–81 cal/g = 3.0-3.4 x 10^5 J/kg.
Heat of combustion: –18,720 Btu/lb = –10,400 cal/g = –435.4 x 10^5 J/kg.

GASOLINES: CASINGHEAD   REC. G:0400

SYNONYMS: CASINGHEAD, GASOLINE; GASOLINA RECTIFICADA (Spanish); GASOLINE, NATURAL; GASOLINA RECTIFICADA (Spanish); NATURAL GASOLINE

IDENTIFICATION
CAS Number: 8006-61-9
DOT ID Number: UN 1203; DOT Guide Number: 128
Proper Shipping Name: Gasoline

DESCRIPTION: Colorless watery liquid. Gasoline-like odor. Floats on the surface of water; insoluble. Flammable, irritating vapor is produced.

Highly flammable • Containers may BLEVE when exposed to fire • Vapors may form explosive mixture with air • Vapors are heavier than air and will collect and stay in low areas • Vapors may travel long distances to ignition sources and flashback • Vapors in confined areas (e.g., tanks, sewers, buildings) may explode when exposed to fire • Irritating to the skin, eyes, and respiratory tract • Toxic products of combustion may include carbon monoxide.

Hazard Classification (based on NFPA-704 Rating System)
Health Hazards (Blue): 1; Flammability (Red): 4; Reactivity (Yellow): 0

EMERGENCY RESPONSE: See Appendix A (128)
Evacuation:
Public safety: Isolate spill area for at least 25 to 50 meters (80 to 160 feet) in all directions.
Spill: Large spill–Consider initial downwind evacuation for at least 300 meters (1000 feet).
Fire: Isolate for 800 meters (½ mile) in all directions, especially if tank, rail car, or tank truck is involved in fire.

EXPOSURE
Short-term effects: *SEEK MEDICAL ATTENTION.* **Vapor:** Irritating to eyes, nose, and throat. *IF INHALED*, will, will cause dizziness, headache, difficult breathing or loss of consciousness. Move to fresh air. *IF BREATHING HAS STOPPED*, give artificial respiration. IF breathing is difficult, administer oxygen. **Liquid:** Irritating to skin and eyes. *IF SWALLOWED*, will cause nausea or vomiting.
Flush affected areas with plenty of water. *IF IN EYES*, hold eyelids open and flush with plenty of water. *IF SWALLOWED* and victim is *CONSCIOUS AND ABLE TO SWALLOW*, have victim drink 4 to 8 ounces of water. **Do NOT induce vomiting**.

Fouling to shoreline. May be dangerous if it enters nearby water intakes; notify operators. Notify local health and wildlife officials. **Response to discharge:** Issue warning–high flammability. Evacuate area. Disperse and flush.

HEALTH HAZARDS
Personal protective equipment (PPE): A-Level PPE. OSHA Table Z-1-A air contaminant.
Recommendations for respirator selection: NIOSH
At any concentrations above the NIOSH REL, or where there is no REL, at any detectable concentration: SCBAF:PD,PP (any self-contained breathing apparatus that has a full facepiece and is operated in a pressure-demand or other positive-pressure mode); or SAF:PD,PP:ASCBA (any supplied-air respirator that has a full facepiece and is operated in a pressure-demand or other positive-pressure mode in combination with an auxiliary self-contained breathing apparatus operated in a pressure-demand or other positive pressure mode). *ESCAPE:* GMFOV [any air-purifying, full-facepiece respirator (gas mask) with a chin-style, front-or back-mounted organic vapor canister]; or SCBAE (any appropriate escape-type, self-contained breathing apparatus).
Exposure limits (TWA unless otherwise noted): As gasoline: ACGIH TLV 300 ppm (890 mg/m^3); NIOSH REL potential human carcinogen; reduce exposure to lowest feasible level.
Short-term exposure limits (15-minute TWA): As gasoline: ACGIH/OSHA STEL 500 ppm (1480 mg/m^3).
Toxicity by ingestion: Grade 2; LD_{50} = 0.5 to 5 g/kg.
Long-term health effects: Possible human carcinogen
Vapor (gas) irritant characteristics: Vapors cause smarting of the eyes or respiratory system if present in high concentrations. The effect is temporary.
Liquid or solid irritant characteristics: If spilled on clothing and allowed to remain, may cause smarting and reddening of the skin.
Odor threshold: 0.25 ppm as gasoline.

FIRE DATA
Flash point: Less than 0°F/–18°C (oc).
Flammable limits in air: LEL: 1.2%; UEL: 7.1% (gasoline, high octane).
Fire extinguishing agents not to be used: Water may be ineffective.
Electrical hazard: Class I, Group D.
Burning rate: 4 mm/min

CHEMICAL REACTIVITY
Binary reactants: Strong oxidizers, nitric acid, perchlorates
Reactivity group: 33
Compatibility class: Miscellaneous hydrocarbon mixtures

ENVIRONMENTAL DATA
Food chain concentration potential: Negative; unlikely to accumulate.
Water pollution: Harmful to aquatic life in very low concentrations. Fouling to shoreline. May be dangerous if it enters nearby water intakes; notify operators. Notify local health and wildlife officials. **Response to discharge:** Issue warning–high flammability. Evacuate area. Disperse and flush.

SHIPPING INFORMATION
Grades of purity: Composition
depends on location of oil well; **Storage temperature:** Ambient; **Inert atmosphere:** None; **Venting:** Open (flame arrester) or pressure-vacuum; **Stability during transport:** Stable.

NAS HAZARD CLASSIFICATION FOR BULK WATER TRANSPORTATION
FIRE: 4
HEALTH: Vapor irritant: 1; Liquid or solid irritant: 0; Poisons: 1
WATER POLLUTION: Human toxicity: 1; Aquatic toxicity: 2; Aesthetic effect: 1
REACTIVITY: Other chemicals: 0; Water: 0; Self-reaction: 0

PHYSICAL AND CHEMICAL PROPERTIES
Physical state @ 59°F/15°C and 1 atm: Liquid.
Boiling point @ 1 atm: 58–275°F/14–135°C/287–408°K.
Specific gravity (water = 1): 0.671 @ 15°C (liquid).
Liquid surface tension: 19-23 dynes/cm
Liquid surface tension: = 0.019-0.023 N/m @ 68°F/20°C.
Liquid water interfacial tension: 49–51 dynes/cm = 0.049-0.051 N/m @ 68°F/20°C.
Relative vapor density (air = 1): 3.4

Latent heat of vaporization: 130–150 Btu/lb = 71–81 cal/g = 3.0-3.4 x 10^5 J/kg.
Heat of combustion: –18,720 Btu/lb = –10,400 cal/g = –435.4 x 10^5 J/kg.

GASOLINES: POLYMER   REC. G:0450

SYNONYMS: POLYMER GASOLINES

IDENTIFICATION
DOT ID Number: UN 1215; DOT Guide Number: 128
Proper Shipping Name: Gasoline

DESCRIPTION: Colorless watery liquid. Gasoline odor. Floats on the surface of water. Flammable, irritating vapor is produced.

Highly flammable • Containers may BLEVE when exposed to fire • Vapors may form explosive mixture with air • Vapors are heavier than air and will collect and stay in low areas • Vapors may travel long distances to ignition sources and flashback • Vapors in confined areas (e.g., tanks, sewers, buildings) may explode when exposed to fire • Irritating to the skin, eyes, and respiratory tract • Toxic products of combustion may include carbon monoxide.

Hazard Classification (based on NFPA-704 Rating System)
Health Hazards (Blue): 1; Flammability (Red): 3; Reactivity (Yellow): 0

EMERGENCY RESPONSE: See Appendix A (128)
Evacuation:
Public safety: Isolate spill area for at least 25 to 50 meters (80 to 160 feet) in all directions.
Spill: Large spill–Consider initial downwind evacuation for at least 300 meters (1000 feet).
Fire: Isolate for 800 meters (½ mile) in all directions, especially if tank, rail car, or tank truck is involved in fire.

EXPOSURE
Short-term effects: *SEEK MEDICAL ATTENTION*. **Vapor:** Irritating to eyes, nose, and throat. *IF INHALED*, will, will cause dizziness, headaches, difficult breathing
or loss of consciousness. Move to fresh air. *IF BREATHING HAS STOPPED*, give artificial respiration. IF breathing is difficult, administer oxygen. **Liquid:** Irritating to skin and eyes. *IF SWALLOWED*, will cause nausea or vomiting. Remove contaminated clothing and shoes. Flush affected areas with plenty of water. *IF IN EYES*, hold eyelids open and flush with plenty of water. *IF SWALLOWED* and victim is *CONSCIOUS AND ABLE TO SWALLOW*, have victim drink 4 to 8 ounces of water. **Do NOT induce vomiting.**

HEALTH HAZARDS
Personal protective equipment (PPE): A-Level PPE. OSHA Table Z-1-A air contaminant. Protective goggles, gloves.
Recommendations for respirator selection: NIOSH
At any concentrations above the NIOSH REL, or where there is no REL, at any detectable concentration: SCBAF:PD,PP (any self-contained breathing apparatus that has a full facepiece and is operated in a pressure-demand or other positive-pressure mode); or SAF:PD,PP:ASCBA (any supplied-air respirator that has a full facepiece and is operated in a pressure-demand or other positive-pressure mode in combination with an auxiliary self-contained breathing apparatus operated in a pressure-demand or other positive pressure mode). *ESCAPE:* GMFOV [any air-purifying, full-facepiece respirator (gas mask) with a chin-style, front-or back-mounted organic vapor canister]; or SCBAE (any appropriate escape-type, self-contained breathing apparatus).
Exposure limits (TWA unless otherwise noted): As gasoline: ACGIH TLV 300 ppm (890 mg/m^3); NIOSH REL potential human carcinogen; reduce exposure to lowest feasible level.
Short-term exposure limits (15-minute TWA): As gasoline: ACGIH/OSHA STEL 500 ppm (1480 mg/m^3)
Toxicity by ingestion: Grade 2; LD_{50} = 0.5 to 5 g/kg.
Long-term health effects: Possible human carcinogen
Vapor (gas) irritant characteristics: Vapors cause smarting of the eyes or respiratory system if present in high concentrations. The effect is temporary.
Liquid or solid irritant characteristics: If spilled on clothing and allowed to remain, may cause smarting and reddening of the skin.
Odor threshold: 0.25 ppm as gasoline.

FIRE DATA
Flash point: 0 to 73°F/–18 to 23°C (cc).
Flammable limits in air: LEL: 1.3%; UEL: 7.1%.
Fire extinguishing agents not to be used: Water may be ineffective.
Electrical hazard: Class I, Group D.
Burning rate: 4 mm/min

CHEMICAL REACTIVITY
Binary reactants: Strong oxidizers, nitric acid and perchlorates
Reactivity group: 33
Compatibility class: Miscellaneous hydrocarbon mixtures

ENVIRONMENTAL DATA
Food chain concentration potential: Negative; unlikely to accumulate.
Water pollution: Notify local health and wildlife officials.
Response to discharge: Issue warning–high flammability. Evacuate area. Disperse and flush.

SHIPPING INFORMATION
Grades of purity: Composition varies with range of distillation temperatures used. Contains mostly isohexane-isooctane; **Storage temperature:** Ambient; **Inert atmosphere:** None; **Venting:** Open (flame arrester) or pressure-vacuum; **Stability during transport:** Stable.

NAS HAZARD CLASSIFICATION FOR BULK WATER TRANSPORTATION
FIRE: 3
HEALTH: Vapor irritant: 1; Liquid or solid irritant: 1; Poisons: 2
WATER POLLUTION: Human toxicity: 1; Aquatic toxicity: 2; Aesthetic effect: 2
REACTIVITY: Other chemicals: 0; Water: 0; Self-reaction: 0

PHYSICAL AND CHEMICAL PROPERTIES
Physical state @ 59°F/15°C and 1 atm: Liquid.
Boiling point @ 1 atm: 58–275°F/14–135°C/287–408°K.
Specific gravity (water = 1): 0.71–0.75 @ 15°C (liquid).
Liquid surface tension: 19–23 dynes/cm = 0.019–0.023 N/m @ 68°F/20°C.
Liquid water interfacial tension: 49–51 dynes/cm = 0.049–0.051 N/m @ 68°F/20°C.
Relative vapor density (air = 1): 3.4
Latent heat of vaporization: 130–150 Btu/lb = 71–81 cal/g = 3.0–3.4 x 10^5 J/kg.
Heat of combustion: –18,720 Btu/lb = –10,400 cal/g = –435.4 x 10^5 J/kg.

GASOLINES: STRAIGHT RUN   REC. G:0500

SYNONYMS: GASOLINE, STRAIGHT RUN, TOPPING-PLANT; STRAIGHT RUN GASOLINES

IDENTIFICATION
CAS Number: 68606-11-1
DOT ID Number: UN 1203; DOT Guide Number: 128
Proper Shipping Name: Gasoline

DESCRIPTION: Colorless watery liquid. Gasoline-like odor. Floats on the surface of water. Flammable, irritating vapor is produced.

Highly flammable • Containers may BLEVE when exposed to fire • Vapors may form explosive mixture with air • Vapors are heavier than air and will collect and stay in low areas • Vapors may travel long distances to ignition sources and flashback • Vapors in confined areas (e.g., tanks, sewers, buildings) may explode when exposed to fire • Irritating to the skin, eyes, and respiratory tract • Toxic products of combustion may include carbon monoxide.

Hazard Classification (based on NFPA-704 Rating System)
Health Hazards (Blue): 1; Flammability (Red): 3; Reactivity (Yellow): 0

EMERGENCY RESPONSE: See Appendix A (128)
Evacuation:
Public safety: Isolate spill area for at least 25 to 50 meters (80 to 160 feet) in all directions.
Spill: Large spill–Consider initial downwind evacuation for at least 300 meters (1000 feet).
Fire: Isolate for 800 meters (½ mile) in all directions, especially if tank, rail car, or tank truck is involved in fire.

EXPOSURE
Short-term effects: Vapor: Irritating to eyes, nose, and throat. *IF INHALED*, will, will cause dizziness, headache, difficult breathing or loss of consciousness. **Liquid:** Irritating to skin and eyes. *IF SWALLOWED*, will cause nausea or vomiting.

HEALTH HAZARDS
Personal protective equipment (PPE): A-Level PPE. OSHA Table Z-1-A air contaminant. Protective goggles, gloves.
Recommendations for respirator selection: NIOSH
At any concentrations above the NIOSH REL, or where there is no REL, at any detectable concentration: SCBAF:PD,PP (any self-contained breathing apparatus that has a full facepiece and is operated in a pressure-demand or other positive-pressure mode); or SAF:PD,PP:ASCBA (any supplied-air respirator that has a full facepiece and is operated in a pressure-demand or other positive-pressure mode in combination with an auxiliary self-contained breathing apparatus operated in a pressure-demand or other positive pressure mode). *ESCAPE:* GMFOV [any air-purifying, full-facepiece respirator (gas mask) with a chin-style, front-or back-mounted organic vapor canister]; or SCBAE (any appropriate escape-type, self-contained breathing apparatus).
Exposure limits (TWA unless otherwise noted): As gasoline: ACGIH TLV 300 ppm (890 mg/m^3); NIOSH REL potential human carcinogen; reduce exposure to lowest feasible level.
Short-term exposure limits (15-minute TWA): As gasoline: ACGIH/OSHA STEL 500 ppm (1480 mg/m^3).
Toxicity by ingestion: Grade 2; LD_{50} = 0.5 to 5 g/kg.
Long-term health effects: Possible human carcinogen
Vapor (gas) irritant characteristics: Vapors cause smarting of the eyes or respiratory system if present in high concentrations. The effect is temporary.
Liquid or solid irritant characteristics: If spilled on clothing and allowed to remain, may cause smarting and reddening of the skin.
Odor threshold: 0.25 ppm as gasoline.

FIRE DATA
Flash point: (a) less than 0°F/–18°C (cc); (b) 0-73°F/–8-23°C (cc).
Flammable limits in air: (a) LEL: 1.3%; UEL: 7.1%.
Fire extinguishing agents not to be used: Water may be ineffective.
Electrical hazard: Class I, Group D.
Burning rate: 4 mm/min

CHEMICAL REACTIVITY
Binary reactants: Strong oxidizers, nitric acid and perchlorates
Reactivity group: 33
Compatibility class: Miscellaneous hydrocarbon mixtures

ENVIRONMENTAL DATA
Food chain concentration potential: Negative; unlikely to accumulate.
Water pollution: Harmful to aquatic life in very low concentrations. Fouling to shoreline. May be dangerous if it enters nearby water intakes; notify operators. **Response to discharge:** Issue warning–high flammability. Evacuate area. Disperse and flush.

SHIPPING INFORMATION
Grades of purity: Composition varies with range of distillation temperatures used; **Storage temperature:** Ambient; **Inert atmosphere:** None; **Venting:** Open (flame arrester) or pressure-vacuum; **Stability during transport:** Stable.

NAS HAZARD CLASSIFICATION FOR BULK WATER TRANSPORTATION
FIRE: 3
HEALTH: Vapor irritant: 1; Liquid or solid irritant: 1; Poisons: 2
WATER POLLUTION: Human toxicity: 1; Aquatic toxicity: 2; Aesthetic effect: 2
REACTIVITY: Other chemicals: 0; Water: 0; Self-reaction: 0

PHYSICAL AND CHEMICAL PROPERTIES
Physical state @ 59°F/15°C and 1 atm: Liquid.
Boiling point @ 1 atm: 58–275°F/14–135°C/287–408°K.
Specific gravity (water = 1): 0.71–0.747 @ 15°C (liquid).
Liquid surface tension: 19–23 dynes/cm = 0.019–0.023 N/m @ 68°F/20°C.
Liquid water interfacial tension: 49–51 dynes/cm = 0.049–0.051 N/m @ 68°F/20°C.
Relative vapor density (air = 1): 3.4
Latent heat of vaporization: 130–150 Btu/lb = 71–81 cal/g = 3.0–3.4 x 10^5 J/kg.
Heat of combustion: –18,720 Btu/lb = –10,400 cal/g = –435.4 x 10^5 J/kg.

GLUTARALDEHYDE SOLUTION   REC. G:0600

SYNONYMS: CIDEX; CUDEX; 1,3-DIFORMAL PROPANE; GLUTAMIC DIALDEHYDE; GLUTARAL; GLUTARALDEHIDO (Spanish); GLUTARD DIALDEHYDE; GLUTARIC ACID DIALDEHYSE; GLUTARIC DIALDEHYDE; PENTANEDIAL; 1,5-PENTANEDIAL; 1,5-PENTANEDIONE; POTENTIATED ACID GLUTARALDEHYDE; SONACIDE

574   Glycerine

IDENTIFICATION
CAS Number: 111-30-8
Formula: $C_5H_8O_2$; $OHC·(CH_2)_3·CHO$ (in water).

DESCRIPTION: Colorless to pale yellow liquid. Usually sold as a 50% water solution. Pungent odor, like rotten apples. Soluble in water; reacts, forming soluble polymers.

Highly irritating to the eyes and respiratory tract; this material has anesthetic properties that affect the central nervous system and may cause unconsciousness; absorbed through the skin • Decomposes above 370°F/188°C giving off carbon monoxide and acid vapors.

Hazard Classification (based on NFPA-704 Rating System)
Health Hazards (Blue): 2; Flammability (Red): 0; Reactivity (Yellow): 0

EMERGENCY RESPONSE: See Appendix A (171)
Evacuation:
Public safety: Isolate the area of spill or leak for at least 10 to 25 meters (30 to 80 feet) in all directions.
Spill: Increase, in the downwind direction, as necessary, the distance shown under "Public Safety."
Fire: If any large container is involved in fire, isolate for at least 800 meters (½ mile) in all directions; also, consider initial evacuation for 800 meters (½ mile) in all directions.

EXPOSURE
Short-term effects: *SEEK MEDICAL ATTENTION.* **Liquid:** Irritating to skin and eyes. Harmful if swallowed. Remove contaminated clothing and shoes. Flush affected areas with plenty of water. *IF IN EYES*, hold eyelids open and flush with plenty of water. *IF SWALLOWED* and victim is *CONSCIOUS AND ABLE TO SWALLOW*, have victim drink 4 to 8 ounces of water.

HEALTH HAZARDS
Personal protective equipment (PPE): B-Level PPE. OSHA Table Z-1-A air contaminant. Goggles or face shield; rubber gloves. Chemical protective material(s) reported to have good to excellent resistance: butyl rubber, neoprene, Viton®. Also, PVC offers limited protection.
Exposure limits (TWA unless otherwise noted): NIOSH REL ceiling 0.2 ppm (0.8 mg/m^3); reduce exposure to lowest feasible level.
Short-term exposure limits (15-minute TWA): ACGIH TLV ceiling 0.2 ppm (0.82 mg/m^3)
Toxicity by ingestion: Grade 2; oral rat LD_{50} = 2380 mg/kg.
Long-term health effects: Induces contact dermatitis and possible skin sensitization in some people. Asthma from prolonged or repeated inhalation. Nervous system damage may occur. Genetic damage: Reproductive and mutation data.
Liquid or solid irritant characteristics: Eye and skin irritant.
Odor threshold: 0.04 ppm.

CHEMICAL REACTIVITY
Reactivity with water: Forms polymer solution.
Binary reactants: Strong oxidizers may cause a violent reaction. Incompatible with strong acids, caustics, ammonia, amines.
Polymerization: Liquid can polymerize.
Reactivity group: 19
Compatibility class: Aldehyde

ENVIRONMENTAL DATA
Food chain concentration potential: Negative.
Water pollution: Effect of low concentrations on aquatic life is unknown. May be dangerous if it enters nearby water intakes; notify operators. Notify local health and wildlife officials. **Response to discharge:** Issue warning–water contaminant. Restrict access. Disperse and flush.

SHIPPING INFORMATION
Grades of purity: 25% aqueous solution; 50% aqueous solution; **Storage temperature:** Ambient; **Inert atmosphere:** None; **Venting:** Open; **Stability during transport:** Stable.

PHYSICAL AND CHEMICAL PROPERTIES
Physical state @ 59°F/15°C and 1 atm: Liquid.
Molecular weight: 100.1
Boiling point @ 1 atm: (decomposes) 372°F/189°C/462°K.
Melting/Freezing point: 7°F/–14°C/259°K.
Specific gravity (water = 1): 1.10 @ 68°F/20°C (liquid).
Liquid surface tension: (estimate) less than 80 dynes/cm = less than 0.080 N/m @ 68°F/20°C.
Relative vapor density (air = 1): 3.5
Heat of decomposition: 372°F/190°C.
Vapor pressure: 17 mm.

GLYCERINE   REC. G:0650

SYNONYMS: GLICERINA (Spanish); GLYCERIN, ANHYDROUS; GLYCERIN, SYNTHETIC; GLYCERITOL; GLYCEROL; GLYCYL ALCOHOL; GROCOLENE; MOON; 1,2,3-PROPANETRIOL; SYNTHETIC GLYCERIN; 90 TECHNICAL GLYCERIN; TRIHYDROXYPROPANE; 1,2,3-TRIHYDROXYPROPANE

IDENTIFICATION
CAS Number: 56-81-5
Formula: $C_3H_8O_3$; $HOCH_2CH(OH)CH_2OH$
DOT ID Number: UN 1993; DOT Guide Number: 128
Proper Shipping Name: Combustible liquid, n.o.s.; Medicines, flammable liquid, n.o.s.

DESCRIPTION: Clear, colorless, oily, syrupy liquid or solid. The solid form melts above 64°F/18°C but the liquid freezes at a much lower temperature. Odorless. Sinks and mixes with water.

Combustible • Containers may BLEVE when exposed to fire • Polymerization hazard above 302°F/150°C. Vapors are heavier than air and will collect and stay in low areas • Vapors in confined areas (e.g., tanks, sewers, buildings) may explode when exposed to fire • Irritating to the skin, eyes, and respiratory tract • Toxic products of combustion may include carbon monoxide and corrosive acrolein gas.

Hazard Classification (based on NFPA-704 Rating System)
Health Hazards (Blue): 1; Flammability (Red): 1; Reactivity (Yellow): 0

EMERGENCY RESPONSE: See Appendix A (128)
Evacuation:
Public safety: Isolate spill area for at least 25 to 50 meters (80 to 160 feet) in all directions.
Spill: Large spill–Consider initial downwind evacuation for at least 300 meters (1000 feet).
Fire: Isolate for 800 meters (½ mile) in all directions, especially if tank, rail car, or tank truck is involved in fire.

EXPOSURE
Short-term effects: Can be mildly toxic if swallowed.

HEALTH HAZARDS
Personal protective equipment (PPE): B-Level PPE. OSHA Table Z-1-A air contaminant (mist, respirable fraction, total dust). Rubber gloves, goggles. Protective materials with good to excellent resistance: natural rubber, neoprene, nitrile. Also, butyl rubber, Viton®/neoprene, nitrile/PVC, butyl rubber,/neoprene, PVC offers limited protection.
Exposure limits (TWA unless otherwise noted): mist: ACGIH TLV (mist) 10 mg/m^3 (total); OSHA PEL (total) 15 mg/m^3; (respirable fraction) 5 mg/m^3).
Toxicity by ingestion: Mildly toxic. Grade 0; LD_{50} = > 15 g/kg.
Long-term health effects: Possible kidney damage.
Vapor (gas) irritant characteristics: Vapors are nonirritating to the eyes and throat.
Liquid or solid irritant characteristics: Eye and skin irritant. Practically harmless to the skin.

FIRE DATA
Flash point: 390°F/370°C (cc).
Fire extinguishing agents not to be used: Water or foam may cause frothing.
Autoignition temperature: 698°F/370°C/643°K.
Burning rate: 0.9 mm/min

CHEMICAL REACTIVITY
Binary reactants: Reacts with strong oxidizers, acetic anhydride, chromium oxide, ethylene oxide, sodium tetraborate. Hygroscopic, absorbs moisture from the air.
Reactivity group: 20
Compatibility class: Alcohols, glycols

ENVIRONMENTAL DATA
Food chain concentration potential: Log P_{ow} = –2.6. Negative; unlikely to accumulate.
Water pollution: Effect of low concentrations on aquatic life is unknown. May be dangerous if it enters nearby water intakes; notify operators. Notify local health and wildlife officials.
Response to discharge: Disperse and flush.

SHIPPING INFORMATION
Grades of purity: CP: 99.5%; USP: 96%; **Storage temperature:** Ambient; **Inert atmosphere:** None; **Venting:** Open (flame arrester) or pressure-vacuum; **Stability during transport:** Stable.

NAS HAZARD CLASSIFICATION FOR BULK WATER TRANSPORTATION
FIRE: 1
HEALTH: Vapor irritant: 0; Liquid or solid irritant: 0; Poisons: 0
WATER POLLUTION: Human toxicity: 0; Aquatic toxicity: 0; Aesthetic effect: 0
REACTIVITY: Other chemicals: 2; Water: 0; Self-reaction: 0

PHYSICAL AND CHEMICAL PROPERTIES
Physical state @ 59°F/15°C and 1 atm: Liquid.
Molecular weight: 92.1
Boiling point @ 1 atm: 340°F/171°C/444°K.
Melting/Freezing point: 64°F/18°C/291°K.
Specific gravity (water = 1): 1.261 @ 68°F/20°C (liquid).
Relative vapor density (air = 1): 3.2
Latent heat of vaporization: 288 Btu/lb = 160 cal/g = 6.70 x 10^5 J/kg.
Heat of combustion: –7758 Btu/lb = –4310 cal/g = –180.5 x 10^5 J/kg.
Heat of decomposition: 554°F/290°C/563°K.
Heat of solution: (estimate) –9 Btu/lb = –5 cal/g = –0.2 x 10^5 J/kg.
Heat of fusion: 47.95 cal/g.
Vapor pressure: (Reid) Very low: 0.002 mm.

GLYCIDYL METHACRYLATE REC. G:0675

SYNONYMS: GLYCIDYL α-METHYL ACRYLATE; METACRILATO de GLICIDILO (Spanish); METHACRYLIC ACID, 2, 3-EPOXY PROPYL ESTER

IDENTIFICATION
CAS Number: 106-91-2
Formula: $C_7H_{10}O_3$; CH_2=CH(CH_3)COOCH$_2$CHCH$_2$O
DOT ID Number: NA 1993; DOT Guide Number: 128
Proper Shipping Name: Combustible liquid, n.o.s.

DESCRIPTION: Colorless liquid. Fruity, rum-like odor. Soluble in water.

Combustible • May form unstable peroxides after prolonged storage; peroxides may be detonated by heating, impact, or friction. • Containers may BLEVE when exposed to fire •Vapors may form explosive mixture with air • Vapors are heavier than air and will collect and stay in low areas • Vapors may travel long distances to ignition sources and flashback • Vapors in confined areas (e.g., tanks, sewers, buildings) may explode when exposed to fire • Irritating to the skin, eyes, and respiratory tract • Toxic products of combustion may include carbon monoxide.

Hazard Classification (based on NFPA-704 Rating System)
Health Hazards (Blue): 2; Flammability (Red): 2; Reactivity (Yellow): 1

EMERGENCY RESPONSE: See Appendix A (128)
Evacuation:
Public safety: Isolate spill area for at least 25 to 50 meters (80 to 160 feet) in all directions.
Spill: Large spill–Consider initial downwind evacuation for at least 300 meters (1000 feet).
Fire: Isolate for 800 meters (½ mile) in all directions, especially if tank, rail car, or tank truck is involved in fire.

EXPOSURE
Short-term effects: *SEEK MEDICAL ATTENTION*. **Liquid:** Will burn skin and eyes. Harmful if swallowed. Remove contaminated clothing and shoes. Flush affected areas with plenty of water. *IF IN EYES*, hold eyelids open and flush with plenty of water. *IF SWALLOWED* and victim is *CONSCIOUS AND ABLE TO SWALLOW*, have victim drink 4 to 8 ounces of water.

HEALTH HAZARDS
Personal protective equipment (PPE): B-Level PPE. Polyethylene-coated apron and gloves and close-fitting goggles.
Toxicity by ingestion: Grade 2; LD_{50} = 0.5 to 5 g/kg (rat).
Vapor (gas) irritant characteristics: Vapor is irritating such that personnel will not usually tolerate moderate or high vapor concentrations.
Liquid or solid irritant characteristics: Causes smarting of the skin and first-degree burns on short exposure; may cause secondary burns on long exposure. In eyes the irritation is similar to that caused by ordinary soap.

FIRE DATA
Flash point: 183°F/84°C (oc).

CHEMICAL REACTIVITY
Binary reactants: Strong oxidizers, nitric acid and perchlorates
Polymerization: Heat, peroxides, and caustics all cause polymerization; the reaction is not considered hazardous; **Inhibitor of polymerization:** Hydroquinone monomethyl ether: 50 ppm.
Reactivity group: 14
Compatibility class: Acrylates

ENVIRONMENTAL DATA
Food chain concentration potential: Negative; unlikely to accumulate.
Water pollution: Effect of low concentrations on aquatic life is unknown. Fouling to shoreline. May be dangerous if it enters nearby water intakes; notify operators. Notify local health and wildlife officials. **Response to discharge:** Mechanical containment. Should be removed. Chemical and physical treatment.

SHIPPING INFORMATION
Grades of purity: Technical: 92%; **Storage temperature:** Ambient; **Inert atmosphere:** None; **Venting:** Open (flame arrester); **Stability during transport:** Stable.

PHYSICAL AND CHEMICAL PROPERTIES
Physical state @ 59°F/15°C and 1 atm: Liquid.
Molecular weight: 142.2
Boiling point @ 1 atm: Very high.
Specific gravity (water = 1): 1.073 @ 68°F/20°C (liquid).
Liquid surface tension: (estimate) 25 dynes/cm = 0.025 N/m @ 68°F/20°C.
Liquid water interfacial tension: (estimate) 40 dynes/cm = 0.04 N/m @ 68°F/20°C.
Ratio of specific heats of vapor (gas): (estimate) 1.043
Heat of combustion: (estimate) $-10,800$ Btu/lb = -5980 cal/g = -250×10^5 J/kg.
Heat of polymerization: (estimate) -900 Btu/lb = -500 cal/g = -20×10^5 J/kg.

GLYOXAL   REC. G:0700

SYNONYMS: AEROTEX GLYOXAL 40; BIFORMAL; BIFORMYL; DIFORMYL; ETHANDIAL; EEC No. 605-016-00-7; ETHANEDIAL; 1,2-ETHANEDIONE; GLIOXAL (Spanish); GLYOXYALDEHYDE; OXAL; OXALDEHYDE; OXALIC ALDEHYDE

IDENTIFICATION
CAS Number: 107-22-2
Formula: $C_2H_2O_2$; CHOCHO (in water)
DOT ID Number: None; DOT Guide Number: 171P
Proper Shipping Name: Polymerization material, stabilize with dry ice (*Note*: pure, solid material).

DESCRIPTION: Yellow crystals or light yellow liquid (usually a 40% inhibited solution). May not be available as a pure substance. Weak, sour odor. Dissolve readily in water; solid material may cause violent polymerization.

A strong reducing agent • Solid material is a polymerization hazard, especially above 122°F/50°C • May explode spontaneously, on contact with moist air, or after storage • Violent polymerization in presence of water; *aqueous solutions include inhibitors that are dissolved readily in water* • Do NOT use water-based extinguishers or foam on solid material • Highly irritating to the eyes and respiratory tract; this material has anesthetic properties; absorbed by the skin • Heat may cause polymerization to a combustible, viscous material. Toxic products of combustion may include carbon monoxide.

Hazard Classification (based on NFPA-704 Rating System)
Health Hazards (Blue): 1; Flammability (Red): 0; Reactivity (Yellow): 2; Special Notice (White): Water reactive

EMERGENCY RESPONSE: See Appendix A (171)
Evacuation:
Public safety: Isolate the area of spill or leak for at least 10 to 25 meters (30 to 80 feet) in all directions.
Spill: Increase, in the downwind direction, as necessary, the distance shown under "Public Safety."
Fire: If any large container is involved in fire, isolate for at least 800 meters (½ mile) in all directions; also, consider initial evacuation for 800 meters (½ mile) in all directions.

EXPOSURE
Short-term effects: *SEEK MEDICAL ATTENTION.* **Liquid:** Irritating to skin and eyes. Harmful if swallowed. Remove contaminated clothing and shoes. Flush affected areas with plenty of water. *IF IN EYES*, hold eyelids open and flush with plenty of water. *IF SWALLOWED* and victim is *CONSCIOUS AND ABLE TO SWALLOW*, have victim drink 4 to 8 ounces of water.

HEALTH HAZARDS
Personal protective equipment (PPE): B-Level PPE. Butyl rubber is generally suitable for aldehydes.
Skin or Ingestion: Skin contact and ingestion may be toxic.
Toxicity by ingestion: Grade, 40% solution 2; oral rat LD_{50} = 020 mg/kg.
Vapor (gas) irritant characteristics: Eye and respiratory tract irritant. Vapors cause smarting of the eyes or respiratory system if present in high concentrations. The effect is temporary.
Liquid or solid irritant characteristics: Eye and skin irritant. If spilled on clothing and allowed to remain, may cause smarting and reddening of the skin.

FIRE DATA
Fire extinguishing agents not to be used: Water, foam, or water-based extinguishing agents.
Behavior in fire: Containers may explode.

CHEMICAL REACTIVITY
Reactivity with water: Polymerizes on contact.
Binary reactants: A strong reducing agent; violent reaction with oxidizers. Contact with air may cause explosion. Violent reaction with nitric acid, oleum, sodium hydroxide, chlorosulfonic acid, ethylene amine. Aqueous solution is acidic; corrosive to most metals; the reaction is slow.
Polymerization: On contact with heat, bases, water; during storage may polymerize and ignite. Inhibitor may volatilize and allow polymerization of the monomer.
Reactivity group: 19
Compatibility class: Aldehyde

ENVIRONMENTAL DATA
Food chain concentration potential: Log P_{ow} = -1.12. Unlikely to accumulate.

Water pollution: Effect of low concentrations on aquatic life is unknown. May be dangerous if it enters nearby water intakes; notify operators. Notify local health and wildlife officials.
Response to discharge: Issue warning–water contaminant. Disperse and flush.

SHIPPING INFORMATION
Grades of purity: 40% in water; **Storage temperature:** 10-120°F; **Inert atmosphere:** None; **Venting:** Open; **Stability during transport:** Stable.

NAS HAZARD CLASSIFICATION FOR BULK WATER TRANSPORTATION
FIRE: 1
HEALTH: Vapor irritant: 1; Liquid or solid irritant: 1; Poisons: 1
WATER POLLUTION: Human toxicity: 1; Aquatic toxicity: 1; Aesthetic effect: 1
REACTIVITY: Other chemicals: 2; Water: 0; Self-reaction: 1

PHYSICAL AND CHEMICAL PROPERTIES
Physical state @ 59°F/15°C and 1 atm: Liquid, 40% solution
Molecular weight: Mixture
Boiling point @ 1 atm: 69°F/51°C/324°K.
Melting/Freezing point: 5°F/–15°C/258°K.
Specific gravity (water = 1): 1.29 @ 68°F/20°C (liquid, 40% solution).
Heat of polymerization: More than 120°F
Vapor pressure: 220 mmHg @ 68°F/20°C (approximate).

GLYOXYLIC ACID (50% or less)   REC. G:0750

SYNONYMS: FORMYLFORMIC ACID; OXACETIC ACID; OXOETHANOIC ACID

IDENTIFICATION
CAS Number: 298-12-4
Formula: $C_2H_2O_3$; OHCCOOH

DESCRIPTION: Colorless to yellow liquid. Soluble in water; aqueous solutions include inhibitors that are dissolved readily in water. Contact of water may cause splattering and overheating.

A reducing agent • Polymerization hazard • Toxic products of combustion may include carbon monoxide.

Hazard Classification (based on NFPA-704 Rating System)
Health Hazards (Blue): 1; Flammability (Red): 0; Reactivity (Yellow): 2

EMERGENCY RESPONSE: See Appendix A (171)
Evacuation:
Public safety: Isolate the area of spill or leak for at least 10 to 25 meters (30 to 80 feet) in all directions.
Spill: Increase, in the downwind direction, as necessary, the distance shown under "Public Safety."
Fire: If any large container is involved in fire, isolate for at least 800 meters (½ mile) in all directions; also, consider initial evacuation for 800 meters (½ mile) in all directions.

EXPOSURE
Short-term effects: *SEEK MEDICAL ATTENTION.* **Vapor:** Move victim to fresh air. If breathing is difficult, administer oxygen.
Liquid: Irritating to skin and eyes. Remove contaminated clothing and shoes. Flush skin with water. *IF IN EYES*, hold eyelids open and flush with plenty of water.

HEALTH HAZARDS
Personal protective equipment (PPE): B-Level PPE. Wear full impervious protective clothing and approved respirator. Where splashing is possible wear full face shield or chemical safety goggles. Use approved respirator to protect against vapors.
Vapor (gas) irritant characteristics: Vapors are irritating such that personnel will not usually tolerate moderate or high concentrations.
Liquid or solid irritant characteristics: Fairly severe skin irritant. May cause pain and second-degree burns after a few minutes of contact.

FIRE DATA
Fire extinguishing agents not to be used: Avoid direct contact between water and acid.

CHEMICAL REACTIVITY
Reactivity with water: May generate heat.
Binary reactants: Attacks aluminum, steel, and copper. Will react with base metals to release hydrogen gas.
Neutralizing agents for acids and caustics: Lime.
Polymerization: Inhibitor may volatilize allowing monomer to polymerize.

ENVIRONMENTAL DATA
Water pollution: Effects of low concentrations on aquatic life is unknown. May be dangerous if it enters nearby water intakes; notify operators. Notify local health and wildlife officials. **Response to discharge:** Issue warning–corrosive. Should be removed.

SHIPPING INFORMATION
Grades of purity: Varying concentrations available. Usually 40% water solution. Due to high reactivity, this material is not available in pure form; **Storage temperature:** Ambient. Use airtight packaging; **Inert atmosphere:** None; **Stability during transport:** Stable.

PHYSICAL AND CHEMICAL PROPERTIES
Physical state @ 59°F/15°C and 1 atm: Liquid.
Molecular weight: 74.04
Specific gravity (water = 1): 1.342

HEPTACHLOR   REC. H:0100

SYNONYMS: AGROCERES; 3-CHLOROCHLORDENE; DRINOX; DRINOX H-34; E 3314; GPKH; HEPTACHLOR; HEPTACHLORE (French); 3,4,5,6,7,8,8-HEPTACHLORODICYCLOPENTADIENE; 1,4,5,6,7,8,8-HEPTACHLORO-3*A*,4,7,7*A*-TETRAHYDRO-4,7-ENDOMETHANOINDENE; 1,4,5,6,7,8,8*A*-HEPTACHLORO-3*A*,4,7,7*A*-TETRAHYDRO-4,7-METHANOINDANE; 1,4,5,6,7,8,8-HEPTACHLORO-3*A*,4,7,7*A*-TETRAHYDRO-4,7-METHANOINDENE; 1(3*A*),4,5,6,7,8,8-HEPTACHLORO-3*A*(1),4,7,7*A*-TETRAHYDRO-4,7-METHANOINDENE; 1,4,5,6,7,8,8-HEPTACHLORO-3*A*,4,7,7*A*-TETRAHYDRO-4,7-METHANOL-1*H*-INDENE; 1,4,5,6,7,8,8-HEPTACHLORO-3*A*,4,7,7,7*A*-TETRAHYDRO-4,7-METHYLENE INDENE; 1,4,5,6,7,10,10-HEPTACHLORO-4,7,8,9-TETRAHYDRO-4,7-METHYLENEINDENE; 1,4,5,6,7,10,10-HEPTACHLORO-4,7,8,9-TETRAHYDRO-4,7-ENDOMETHYLENEINDENE; 1,4,5,6,7,8,8-HEPTACHLOR-3*A*,4,7,7*A*-TETRAHYDRO-4,7-

578 Heptane

endo-METHANO-INDEN (German); HEPTACLORO (Spanish); HEPTAGRAN; HEPTAMUL; RCRA No. P059; RHODIACHLOR; VELSICOL 104

IDENTIFICATION
CAS Number: 76-44-8
Formula: $C_{10}H_5Cl_7$
DOT ID Number: UN 2761; DOT Guide Number: 151
Proper Shipping Name: Organochlorine pesticides, solid, toxic, n.o.s.
Reportable Quantity (RQ): **(CERCLA)** 1 lb/0.454 kg

DESCRIPTION: White crystalline solid or light tan waxy solid; formulated as dust or granules. Camphor-like odor. Sinks in water; insoluble.

Poison! (organochlorine) • Breathing the vapor can kill you; skin or eye contact causes severe burns, impaired vision, or blindness • Firefighting gear (including SCBA) does not provide adequate protection. If exposure occurs, remove and isolate gear immediately and thoroughly decontaminate personnel • Toxic products of combustion may include hydrogen chloride • Do not put yourself in danger by entering a contaminated area to rescue a victim.

Hazard Classification (based on NFPA-704 Rating System)
Health Hazards (Blue): 3; Flammability (Red): 0; Reactivity (Yellow): 0

EMERGENCY RESPONSE: See Appendix A (151)
Evacuation:
Public safety: Isolate the area of spill or leak for at least 25 to 50 meters (80 to 160 feet) in all directions.
Spill: Increase, in the downwind direction, as necessary, the distance shown under "Public Safety."
Fire: If tank, rail car, or tank truck is involved in fire, isolate for at least 800 meters (½ mile) in all directions; also, consider initial evacuation for 800 meters (½ mile) in all directions.

EXPOSURE
Short-term effects: Dust: *POISONOUS IF INHALED. IF INHALED*, will, will cause headache or loss of consciousness. *IF IN EYES*, hold eyelids open and flush with plenty of water. *IF BREATHING HAS STOPPED*, give artificial respiration; *avoid mouth-to-mouth resuscitation; use bag/mask apparatus.* IF breathing is difficult, administer oxygen. **Solid:** *POISONOUS IF SWALLOWED.* Irritating to skin and eyes. *IF SWALLOWED*, will cause nausea and vomiting. Remove contaminated clothing and shoes. Flush affected areas with plenty of water. *IF IN EYES*, hold eyelids open and flush with plenty of water. *IF SWALLOWED* and victim is *CONSCIOUS AND ABLE TO SWALLOW*, have victim drink 4 to 8 ounces of water. **Do NOT induce vomiting.** Administer a slurry of activated charcoal. *IF SWALLOWED* and victim is *UNCONSCIOUS OR HAVING CONVULSIONS*, do nothing except keep victim warm.

HEALTH HAZARDS
Personal protective equipment (PPE): A-Level PPE. OSHA Table Z-1-A air contaminant. Protective respirator; rubber gloves; clean clothes.
Recommendations for respirator selection: NIOSH
At any concentrations above the NIOSH REL, or where there is no REL, at any detectable concentration: SCBAF:PD,PP (any self-contained breathing apparatus that has a full facepiece and is operated in a pressure-demand or other positive-pressure mode); or SAF:PD,PP:ASCBA (any supplied-air respirator that has a full facepiece and is operated in a pressure-demand or other positive-pressure mode in combination with an auxiliary, self-contained breathing apparatus operated in a pressure-demand or other positive pressure mode). *ESCAPE:* GMFOVHiE [any air-purifying, full-facepiece respirator (gas mask) with a chin-style, front- or back-mounted organic vapor canister having a high-efficiency particulate filter]; or SCBAE (any appropriate escape-type, self-contained breathing apparatus).
Exposure limits (TWA unless otherwise noted): ACGIH TLV 0.05 mg/m³ animal carcinogen; OSHA PEL 0.5 mg/m³; NIOSH REL 0.5 mg/m³; potential human carcinogen; reduce exposure to lowest feasible level; skin contact contributes significantly in overall exposure.
Toxicity by ingestion: Grade 4; oral LD_{50} = 40 mg/kg (rat).
Long-term health effects: Liver damage may develop. Potential carcinogen. IARC rating 3; limited animal evidence.
Liquid or solid irritant characteristics: Eye and skin irritant.
Odor threshold: 0.02 ppm.
IDLH value: Potential human carcinogen; 35 mg/m³.

CHEMICAL REACTIVITY
Binary reactants: Iron, rust.

ENVIRONMENTAL DATA
Food chain concentration potential: Bioconcentration of up to 17600 in oysters and 300 in bluegills. A spill could cause potential problem with shellfish.
Water pollution: DOT Appendix B, §172.101- severe marine pollutant. Harmful to aquatic life in very low concentrations. May be dangerous if it enters nearby water intakes; notify operators. Notify local health and wildlife officials. **Response to discharge:** Issue warning–water contaminant. Should be removed. Chemical and physical treatment.

SHIPPING INFORMATION
Grades of purity: Commercial, 72+%; **Storage temperature:** Ambient; **Inert atmosphere:** None; **Venting:** Open; **Stability during transport:** Stable.

PHYSICAL AND CHEMICAL PROPERTIES
Physical state @ 59°F/15°C and 1 atm: Solid.
Molecular weight: 373.4
Boiling point @ 1 atm: Decomposes. 293°F/145°C/418°K.
Melting/Freezing point: 203°F/95°C/368°C.
Specific gravity (water = 1): 1.66 @ 68°F/20°C (solid).
Vapor pressure: 0.0003 mm @ 77°F/25°C.

HEPTANE REC. H:0150

SYNONYMS: DIPROPYL METHANE; DIPROPYLMETHANE; EEC No. 601-008-00-2; *n*-HEPTANE; *n*-HEPTANO (Spanish); HEPTYL HYDRIDE; DIPROPAL METHANE; SKELLY-SOLVE C

IDENTIFICATION
CAS Number: 142-82-5
Formula: C_7H_{16}; $CH_3(CH_2)_5CH_3$
DOT ID Number: UN 1206; DOT Guide Number: 128
Proper Shipping Name: Heptanes

DESCRIPTION: Colorless watery liquid. Gasoline-like odor. Floats on water surface; insoluble. Flammable vapor is produced.

Heptane

Highly flammable • Containers may BLEVE when exposed to fire • Vapors may form explosive mixture with air • Vapors are heavier than air and will collect and stay in low areas • Vapors may travel long distances to ignition sources and flashback • Vapors in confined areas (e.g., tanks, sewers, buildings) may explode when exposed to fire • Irritating to the skin, eyes, and respiratory tract • Toxic products of combustion may include carbon monoxide.

Hazard Classification (based on NFPA-704 Rating System)
Health Hazards (Blue): 1; Flammability (Red): 3; Reactivity (Yellow): 0

EMERGENCY RESPONSE: See Appendix A (128)
Evacuation:
Public safety: Isolate spill area for at least 25 to 50 meters (80 to 160 feet) in all directions.
Spill: Large spill–Consider initial downwind evacuation for at least 300 meters (1000 feet).
Fire: Isolate for 800 meters (½ mile) in all directions, especially if tank, rail car, or tank truck is involved in fire.

EXPOSURE
Short-term effects: *SEEK MEDICAL ATTENTION.* **Vapor:** Not irritating to eyes, nose or throat. *IF INHALED*, will, will cause coughing or difficult breathing. Move to fresh air. *IF BREATHING HAS STOPPED,* give artificial respiration. IF breathing is difficult, administer oxygen. **Liquid:** Irritating to skin and eyes. *IF SWALLOWED*, will cause nausea or vomiting. Remove contaminated clothing and shoes. Flush affected areas with plenty of water. *IF IN EYES,* hold eyelids open and flush with plenty of water. *IF SWALLOWED* and victim is *CONSCIOUS AND ABLE TO SWALLOW*, have victim drink 4 to 8 ounces of water. **Do NOT induce vomiting.** Droplets entering the lungs may cause pneumonia.

HEALTH HAZARDS
Personal protective equipment (PPE): B-Level PPE. OSHA Table Z-1-A air contaminant. Safety glasses; gloves. Chemical protective material(s) reported to have good to excellent resistance: nitrile, nitrile+PVC, Viton®. Also, PV alcohol offers limited protection.
Recommendations for respirator selection: NIOSH/OSHA
750 ppm: CCROV [any chemical cartridge respirator with organic vapor cartridge(s)]; or GMFOV [any air-purifying, full-facepiece respirator (gas mask) with a chin-style, front- or back-mounted acid gas canister]; or PAPROV [any powered, air-purifying respirator with organic vapor cartridge(s)]; or SA (any supplied-air respirator); or SCBAF (any self-contained breathing apparatus with a full facepiece). *EMERGENCY OR PLANNED ENTRY INTO UNKNOWN CONCENTRATIONS OR IDLH CONDITIONS:* SCBAF:PD,PP (any self-contained breathing apparatus that has a full facepiece and is operated in a pressure-demand or other positive-pressure mode); or SAF:PD,PP:ASCBA (any supplied-air respirator that has a full facepiece and is operated in a pressure-demand or other positive-pressure mode in combination with an auxiliary self-contained breathing apparatus operated in a pressure-demand or other positive pressure mode). *ESCAPE:* GMFOV[any air-purifying, full-facepiece respirator (gas mask) with a chin-style, front- or back-mounted organic vapor cannister]; or SCBAE (any appropriate escape-type, self-contained breathing apparatus).

Exposure limits (TWA unless otherwise noted): ACGIH TLV 400 ppm (1640 mg/m^3); OSHA PEL 500 ppm (2000 mg/m^3); NIOSH 85 ppm (350 mg/m^3); ceiling 440 ppm (1800 mg/m^3)/15 min.

Short-term exposure limits (15-minute TWA): ACGIH STEL 500 ppm (2050 mg/m^3).
Toxicity by ingestion: Grade 0; LD$_{50}$ = > 15 g/kg.
Long-term health effects: May effect the nervous system. Constant use may cause degreasing of the skin.
Vapor (gas) irritant characteristics: High concentrations may have a narcotic effect.
Liquid or solid irritant characteristics: Eye and skin irritant. If spilled on clothing and allowed to remain, may cause smarting and reddening of the skin.
Odor threshold: 220–230 ppm
IDLH value: 750 ppm

FIRE DATA
Flash point: 25°F/–4°C (cc).
Flammable limits in air: LEL: 1.0%; UEL: 6.7%.
Autoignition temperature: 392°F/200°C/473°K.
Electrical hazard: Class I, Group D; Flow or agitation of substance may generate electrostatic charges due to low conductivity.
Burning rate: 6.8 mm/min

CHEMICAL REACTIVITY
Binary reactants: Strong oxidizers. Keep away from chlorine and phosphorus.
Reactivity group: 31
Compatibility class: Paraffins

ENVIRONMENTAL DATA
Food chain concentration potential: Negative; unlikely to accumulate.
Water pollution: Dangerous to aquatic life in high concentrations. Fouling to shoreline. May be dangerous if it enters nearby water intakes; notify operators. Notify local health and wildlife officials.
Response to discharge: Issue warning–high flammability. Evacuate area. Disperse and flush.

SHIPPING INFORMATION
Grades of purity: Various grades, all greater than 99%; **Storage temperature:** Ambient; **Inert atmosphere:** None; **Venting:** Open (flame arrester) or pressure-vacuum; **Stability during transport:** Stable.

NAS HAZARD CLASSIFICATION FOR BULK WATER TRANSPORTATION
FIRE: 3
HEALTH: Vapor irritant: 0; Liquid or solid irritant: 1; Poisons: 1
WATER POLLUTION: Human toxicity: 1; Aquatic toxicity: 1; Aesthetic effect: 2
REACTIVITY: Other chemicals: 0; Water: 0; Self-reaction: 0

PHYSICAL AND CHEMICAL PROPERTIES
Physical state @ 59°F/15°C and 1 atm: Liquid.
Molecular weight: 100.21
Boiling point @ 1 atm: 209.1°F/98.4°C/371.6°K.
Melting/Freezing point: –131°F/–90.6°C/182.6°K.
Critical temperature: 513°F/267°C/540°K.
Critical pressure: 400 psia = 27 atm = 2.7 MN/m^2.
Specific gravity (water = 1): 0.6838 @ 68°F/20°C (liquid).
Liquid surface tension: 19.3 dynes/cm = 0.0193 N/m @ 68°F/20°C.
Liquid water interfacial tension: 51 dynes/cm = 0.051 N/m @ 68°F/20°C.
Relative vapor density (air = 1): 3.5
Ratio of specific heats of vapor (gas): 1.054

Latent heat of vaporization: 136.1 Btu/lb = 75.61 cal/g = 3.166 x 10^5 J/kg.
Heat of combustion: –19,170 Btu/lb = –10,650 cal/g = –445.9 x 10^5 J/kg.
Heat of fusion: 33.78 cal/g.
Vapor pressure: (Reid) 1.8 psia; 40 mm @ 72°F

HEPTANOIC ACID   REC. H:0200

SYNONYMS: ACIDO *n*-HEPTANOICO (Spanish); ENANTHIC ACID; HEXANE CARBOXYLIC ACID; *n*-HEPTOIC ACID; HEPTHLIC ACID; *n*-HEPTYLIC ACID; 1-HEXANECARCOXYLIC ACID

IDENTIFICATION
CAS Number: 111-14-8
Formula: $C_7H_{14}O_2$
DOT ID Number: UN 3265; DOT Guide Number: 153
Proper Shipping Name: Corrosive liquid, acid, organic, n.o.s.

DESCRIPTION: Colorless, oily liquid. Unpleasant odor when pure; rancid odor as it ages. Floats on water surface; insoluble.

Combustible • Corrosive to skin, eyes, and lungs; contact can cause severe burns and blindness • Firefighting gear (including SCBA) does not provide adequate protection. If exposure occurs, remove and isolate gear immediately and thoroughly decontaminate personnel • Containers may BLEVE when exposed to fire • Vapors in confined areas (e.g., tanks, sewers, buildings) may explode when exposed to fire • Toxic products of combustion may include carbon monoxide.

Hazard Classification (based on NFPA-704 Rating System)
Health Hazards (Blue): 1; Flammability (Red): 1; Reactivity (Yellow): 0

EMERGENCY RESPONSE: See Appendix A (153)
Evacuation:
Public safety: Isolate the area of spill or leak for at least 25 to 50 meters (80 to 160 feet) in all directions.
Spill: Increase, in the downwind direction, as necessary, the distance shown under "Public Safety."
Fire: If tank, rail car, or tank truck is involved in fire, isolate for at least 800 meters (½ mile) in all directions; also, consider initial evacuation for 800 meters (½ mile) in all directions.

EXPOSURE
Short-term effects: *SEEK MEDICAL ATTENTION.* **Vapor:** Harmful if inhaled. Extremely irritating to eyes, nose, and throat. Remove to fresh air. *IF BREATHING HAS STOPPED,* give artificial respiration; *avoid mouth-to-mouth resuscitation; use bag/mask apparatus.* IF breathing is difficult, administer oxygen. **Liquid:** Harmful if swallowed or absorbed through skin. Will burn skin and eyes. Remove contaminated clothing and shoes. Flush affected areas with plenty of water. *IF IN EYES* hold eyelids open and flush with plenty of water. *IF SWALLOWED* and victim is *CONSCIOUS*: Have victim drink 4 to 8 ounces of water. **Do NOT induce vomiting.** *IF SWALLOWED* and victim is *UNCONSCIOUS OR HAVING CONVULSIONS*: Do nothing except keep victim warm. Chemical pneumonitis may develop.

HEALTH HAZARDS
Personal protective equipment (PPE): B-Level PPE. Approved respirator, rubber gloves, safety goggles. Butyl rubber is generally suitable for carbooxylic acid compounds.
Toxicity by ingestion: Grade 1: LD_{50} = 7 g/kg (rat).
Vapor (gas) irritant characteristics: Vapors cause severe irritation of eyes and throat and can cause eye and lung injury. They cannot be tolerated even at low concentrations.
Liquid or solid irritant characteristics: Severe eye and skin irritant. Causes second- and third-degree burns on short contact and is very injurious to the eyes.

FIRE DATA
Flash point: More than 230°F/110°C (cc).

CHEMICAL REACTIVITY
Binary reactants: Strong oxidizers, caustics, sulfuric acid, amines.
Neutralizing agents for acids and caustics: Caustic soda or lime.
Reactivity group: 4
Compatibility class: Organic acids

ENVIRONMENTAL DATA
Food chain concentration potential: Log P_{ow} = 2.7. Values > 3.0 are likely to bioconcentrate in aquatic organisms and other living tissue, especially in fats.
Water pollution: Effects of low concentrations on aquatic life is not known. May be dangerous if it enters nearby water intakes; notify operators. Notify local health and wildlife officials. Notify operators of local water intakes. **Response to discharge:** Should be removed. Chemical and physical treatment.

SHIPPING INFORMATION
Grades of purity: 96%; **Stability during transport:** Stable.

NAS HAZARD CLASSIFICATION FOR BULK WATER TRANSPORTATION
FIRE: 1
HEALTH: Vapor irritant: 4; Liquid or solid irritant: 4; Poisons: 1
WATER POLLUTION: Human toxicity: 1; Aquatic toxicity:–; Aesthetic effect: 4
REACTIVITY: Other chemicals: 3; Water: 1; Self-reaction: 0

PHYSICAL AND CHEMICAL PROPERTIES
Physical state @ 59°F/15°C and 1 atm: Liquid.
Molecular weight: 130.19
Boiling point @ 1 atm: 433°F/223°C/496°K.
Melting/Freezing point: 18°F/–8°C/266°K.
Specific gravity (water = 1): 0.9200 @ 68°F/20°C.
Relative vapor density (air = 1): 4.49
Latent heat of vaporization: 302.8 Btu/lb = 168.2 cal/g = 7.04 x 10^5 J/kg.
Heat of combustion: –13,634 Btu/lb = –7,574 cal/g = –317 x 10^5 J/kg.
Vapor pressure: (Reid) Low; less than 0.08 mm.

HEPTANOL   REC. H:0250

SYNONYMS: ENANTHIC ALCOHOL; *n*-HEPTANOL; *n*-HETANOL-1; 1-HYDROXYHEPTANE; HEPTANOL-1; 1-HEPTANOL; HEPTYL ALCOHOL

IDENTIFICATION
CAS Number: 111-70-6; 543-49-7 (2-isomer); 589-82-2 (3-isomer)
Formula: $C_7H_{16}O$; $CH_3(CH_2)_5CH_2OH$
DOT ID Number: NA 1993; DOT Guide Number: 128
Proper Shipping Name: Combustible liquid, n.o.s.

DESCRIPTION: Colorless watery liquid Weak alcohol odor. Floats on the surface of water.

Combustible • Containers may BLEVE when exposed to fire •Vapors may form explosive mixture with air • Vapors are heavier than air and will collect and stay in low areas • Vapors may travel long distances to ignition sources and flashback • Vapors in confined areas (e.g., tanks, sewers, buildings) may explode when exposed to fire • Irritating to the skin, eyes, and respiratory tract • Toxic products of combustion may include carbon monoxide.

Hazard Classification (based on NFPA-704 Rating System)
Health Hazards (Blue): 0; Flammability (Red): 2; Reactivity (Yellow): 0

EMERGENCY RESPONSE: See Appendix A (128)
Evacuation:
Public safety: Isolate spill area for at least 25 to 50 meters (80 to 160 feet) in all directions.
Spill: Large spill–Consider initial downwind evacuation for at least 300 meters (1000 feet).
Fire: Isolate for 800 meters (½ mile) in all directions, especially if tank, rail car, or tank truck is involved in fire.

EXPOSURE
Short-term effects: Irritates the eyes, skin, and respiratory system. The use of alcoholic beverages may enhance the toxic effects.

HEALTH HAZARDS
Personal protective equipment (PPE): B-Level PPE. Chemical goggles or face shield. Protective gloves. Natural rubber, neoprene, and nitrile rubber may offer some protection from alcohols.
Toxicity by ingestion: Grade 2; oral rat LD_{50} = 1.87 g/kg.
Long-term health effects: May effect the nervous system.
Vapor (gas) irritant characteristics: May irritate the eyes and respiratory system. At room temperature the vapor pressure is so low that inhalation is unlikely.
Liquid or solid irritant characteristics: Liquid may irritate eyes.
Odor threshold: 0.49 ppm.

FIRE DATA
Flash point: 170°F/77°C (oc); 160°F/71°C (2-isomer); 140°F/60°C (3-isomer)
Burning rate: 3.2 mm/min

CHEMICAL REACTIVITY
Binary reactants: Oxidants.

ENVIRONMENTAL DATA
Food chain concentration potential: Negative; unlikely to accumulate.
Water pollution: Effect of low concentrations on aquatic life is unknown. Fouling to shoreline. May be dangerous if it enters nearby water intakes; notify operators. Notify local health and wildlife officials. **Response to discharge:** Mechanical containment. Should be removed. Chemical and physical treatment.

SHIPPING INFORMATION
Storage temperature: Ambient; **Inert atmosphere:** None; **Venting:** Open (flame arrester); **Stability during transport:** Stable.

PHYSICAL AND CHEMICAL PROPERTIES
Physical state @ 59°F/15°C and 1 atm: Liquid.
Molecular weight: 116.2
Boiling point @ 1 atm: 349°F/176°C/449°K; 320°F/160°C (2-isomer); 313°F/156°C/429°K (3-isomer).
Melting/Freezing point: −29°F/−34°C/239°K; −31°F/−35°C/238°K (2-isomer); −94°F/−70°C/203°K (3-isomer).
Critical temperature: 680°F/360°C/633°K.
Critical pressure: 440 psia = 30 atm = 3.0 MN/m^2.
Specific gravity (water = 1): 0.822 @ 68°F/20°C (liquid).
Liquid surface tension: 26.2 dynes/cm = 0.0262 N/mm @ 15°C.
Liquid water interfacial tension: 7.7 dynes/cm = 0.0077 N/m @ 77°F/25°C
Relative vapor density (air = 1): 3.9
Ratio of specific heats of vapor (gas): 1.049
Latent heat of vaporization: 189 Btu/lb = 105 cal/g = 4.40 x 10^5 J/kg.
Heat of combustion: −15,810 Btu/lb = −8784 cal/g = −367.8 x 10^5 J/kg.
Vapor pressure: 0.1 mm.

1-HEPTENE   REC. H:0300

SYNONYMS: HEPTENE; *n*-HEPTENE; *n*-HEPTENO (Spanish); HEPTYLENE; 1-HEPTYLENE

IDENTIFICATION
CAS Number: 592-76-7
Formula: C_7H_{14}; $CH_3(CH_2)_4CH=CH_2$
DOT ID Number: UN 2278; DOT Guide Number: 128
Proper Shipping Name: *n*-Heptene

DESCRIPTION: Colorless, watery liquid. Gasoline-like odor. Floats on water surface; insoluble. Flammable, irritating vapor is produced.

Highly flammable • Containers may BLEVE when exposed to fire •Vapors may form explosive mixture with air • Vapors are heavier than air and will collect and stay in low areas • Vapors may travel long distances to ignition sources and flashback • Vapors in confined areas (e.g., tanks, sewers, buildings) may explode when exposed to fire • Mildly Irritating to the skin, eyes, and respiratory tract • Toxic products of combustion may include carbon monoxide.

Hazard Classification (based on NFPA-704 Rating System)
Health Hazards (Blue): 0; Flammability (Red): 3; Reactivity (Yellow): 0

EMERGENCY RESPONSE: See Appendix A (128)
Evacuation:
Public safety: Isolate spill area for at least 25 to 50 meters (80 to 160 feet) in all directions.
Spill: Large spill–Consider initial downwind evacuation for at least 300 meters (1000 feet).
Fire: Isolate for 800 meters (½ mile) in all directions, especially if tank, rail car, or tank truck is involved in fire.

EXPOSURE
Short-term effects: *SEEK MEDICAL ATTENTION*. **Vapor:** Irritating to eyes, nose, and throat. *IF INHALED*, will, will cause dizziness or difficult breathing. Move to fresh air. *IF BREATHING HAS STOPPED*, give artificial respiration. IF breathing is difficult, administer oxygen. **Liquid:** Irritating to skin and eyes. Harmful if swallowed. Remove contaminated clothing and shoes. Flush affected areas with plenty of water. *IF IN EYES*, hold eyelids open and flush with plenty of water. *IF SWALLOWED* and victim is

582   Heptyl acetate

CONSCIOUS AND ABLE TO SWALLOW, have victim drink 4 to 8 ounces of water. **Do NOT induce vomiting.**

HEALTH HAZARDS
Personal protective equipment (PPE): B-Level PPE. Safety goggles or face shield; similar to gasoline.
Vapor (gas) irritant characteristics: Vapors cause smarting of the eyes or respiratory system if present in high concentrations. The effect is temporary.
Liquid or solid irritant characteristics: If spilled on clothing and allowed to remain, may cause smarting and reddening of the skin.

FIRE DATA
Flash point: 25°F/–4°C (cc) (estimate).
Autoignition temperature: 500°F/260°C/533°K.
Burning rate: 6.4 mm/min
Electrical hazard: Class I, Group D; Flow or agitation of substance may generate electrostatic charges due to low conductivity.

CHEMICAL REACTIVITY
Binary reactants: Violent reaction with oxidizers. Incompatible with nitric acid.

ENVIRONMENTAL DATA
Food chain concentration potential: Negative; unlikely to accumulate.
Water pollution: Effect of low concentrations on aquatic life is unknown. Fouling to shoreline. May be dangerous if it enters nearby water intakes; notify operators. Notify local health and wildlife officials. **Response to discharge:** Issue warning–high flammability. Restrict access. Evacuate area.

SHIPPING INFORMATION
Grades of purity: Technical; **Storage temperature:** Ambient; **Inert atmosphere:** None; **Venting:** Open (flame arrester); **Stability during transport:** Stable.

NAS HAZARD CLASSIFICATION FOR BULK WATER TRANSPORTATION
FIRE: 3
HEALTH: Vapor irritant: 1; Liquid or solid irritant: 1; Poisons: 0
WATER POLLUTION: Human toxicity: 0; Aquatic toxicity: 1; Aesthetic effect: 2
REACTIVITY: Other chemicals: 1; Water: 0; Self-reaction: 1

PHYSICAL AND CHEMICAL PROPERTIES
Physical state @ 59°F/15°C and 1 atm: Liquid.
Molecular weight: 98.18
Boiling point @ 1 atm: 200.5°F/93.6°C/366.8°K.
Melting/Freezing point: –182°F/–119°C/154°K.
Critical temperature: 507.4°F/264.1°C/537.3°K.
Critical pressure: 420 psia = 28.57 atm = 2.89 MN/m^2.
Specific gravity (water = 1): 0.697 @ 68°F/20°C (liquid).
Liquid surface tension: 20.5 dynes/cm = 0.0205 N/m @ 68°F/20°C.
Liquid water interfacial tension: (estimate) 50 dynes/cm = 0.05 N/m @ 68°F/20°C.
Relative vapor density (air = 1): 3.4
Ratio of specific heats of vapor (gas): 1.057
Latent heat of vaporization: 137 Btu/lb = 76.3 cal/g = 3.20 x 10^5 J/kg.
Heat of combustion: –19,377 Btu/lb = –10,765 cal/g = –450.71 x 10^5 J/kg.
Heat of fusion: 30.82 cal/g.

HEPTYL ACETATE   REC. H:0350

SYNONYMS: ACETATO de 1-HEPTILO (Spanish); ACETIC ACID, HEPTYL ESTER; ACETATE C-7; HEPTANYL ACETATE; *n*-HEPTYL ACETATE; 1-HEPTYL ACETATE

IDENTIFICATION
CAS Number: 112-06-1
Formula: $C_9H_{18}O_2$; $CH_3(CH_2)_6OOCCH_3$
DOT ID Number: NA 1993; DOT Guide Number: 128
Proper Shipping Name: Combustible liquid, n.o.s.

DESCRIPTION: Colorless liquid. Fruity odor. Insoluble in water.

Flammable • Containers may BLEVE when exposed to fire • Vapors may form explosive mixture with air • Vapors are heavier than air and will collect and stay in low areas • Vapors in confined areas (e.g., tanks, sewers, buildings) may explode when exposed to fire • Irritating to the skin, eyes, and respiratory tract • Toxic products of combustion may include carbon monoxide.

Hazard Classification (based on NFPA-704 Rating System)
Health Hazards (Blue): 0; Flammability (Red): 2; Reactivity (Yellow): 0

EMERGENCY RESPONSE: See Appendix A (128)
Evacuation:
Public safety: Isolate spill area for at least 25 to 50 meters (80 to 160 feet) in all directions.
Spill: Large spill–Consider initial downwind evacuation for at least 300 meters (1000 feet).
Fire: Isolate for 800 meters (½ mile) in all directions, especially if tank, rail car, or tank truck is involved in fire.

EXPOSURE
Short-term effects: *SEEK MEDICAL ATTENTION.* **Liquid:** Irritating to skin and eyes. Remove contaminated clothing and shoes. Flush affected areas with plenty of water. *IF IN EYES,* hold eyelids open and flush with plenty of water.

HEALTH HAZARDS
Personal protective equipment (PPE): B-Level PPE. Self-contained breathing apparatus, rubber boots and rubber gloves.
Vapor (gas) irritant characteristics: Eye and respiratory tract irritant. Vapors cause irritation, such that personnel will find high concentrations unpleasant. The effect is temporary.
Liquid or solid irritant characteristics: Eye and skin irritant. If spilled on clothing and allowed to remain, may cause smarting and reddening of the skin.

FIRE DATA
Flash point: 154°F/68°C (cc).
Fire extinguishing agents not to be used: Water may be ineffective.

CHEMICAL REACTIVITY
Reactivity group: 34
Compatibility class: Esters.

ENVIRONMENTAL DATA
Food chain concentration potential: Log P_{ow} = 6.2. Values > 3.0 are likely to bioconcentrate in aquatic organisms and other living tissue, especially in fats.
Water pollution: Effect of low concentrations on aquatic life is unknown. Fouling to shoreline. May be dangerous if it enters

nearby water intakes; notify operators. Notify local health and wildlife officials. **Response to discharge:** Evacuate area. Mechanical containment. Chemical and physical treatment.

SHIPPING INFORMATION
Grades of purity: 98+%; **Inert atmosphere:** Not required; **Venting:** Not required; **Stability during transport:** Stable.

PHYSICAL AND CHEMICAL PROPERTIES
Physical state @ 59°F/15°C and 1 atm: Liquid.
Molecular weight: 158.27
Boiling point @ 1 atm: 378.5°F/192.5°C/465.7°K.
Melting/Freezing point: –58.4°F/–50.2°C/223°K.
Specific gravity (water = 1): 0.875
Relative vapor density (air = 1): 5.5

HEXACHLOROBENZENE   REC. H:0400

SYNONYMS: AMATIN; ANTICARIE; BENZENE, HEXACHLORO-; BENZENE HEXACHLORIDE; BUNT-CURE; BUNT-NO-MORE; CO-OP HEXA; GRANOX MN; HCB; HEXA C. B; HEXACLOROBENCENO (Spanish); HEXACHLOROBENZOL (German); JULIN'S CARBON CHLORIDE; NO BUNT LIQUID; PENTACHLOROPHENYL CHLORIDE; PERCHLOROBENZENE; PHENYL PERCHLORYL; SANOCIDE; SMUT-GO; SNIECIOTOX; BENZENE, HEXACHLORO-; RCRA No. U127

IDENTIFICATION
CAS Number: 118-74-1
Formula: C_6Cl_6
DOT ID Number: UN 2729; DOT Guide Number: 152
Proper Shipping Name: Hexachlorobenzene
Reportable Quantity (RQ): **(CERCLA)** 100 lb/45.4 kg

DESCRIPTION: White needle-like, crystalline solid. Faint, not unpleasant odor. Sinks in water.

Combustible • Heat or flame may cause explosion • Containers may BLEVE when exposed to fire • Vapors are heavier than air and will collect and stay in low areas • Vapors in confined areas (e.g., tanks, sewers, buildings) may explode when exposed to fire • Irritating to the skin, eyes, and respiratory tract • Toxic products of combustion may include hydrogen chloride.

Hazard Classification (based on NFPA-704 Rating System)
Health Hazards (Blue): 1; Flammability (Red): 1; Reactivity (Yellow): 0

EMERGENCY RESPONSE: See Appendix A (152)
Evacuation:
Public safety: Isolate the area of spill or leak for at least 25 to 50 meters (80 to 160 feet) in all directions.
Spill: Increase, in the downwind direction, as necessary, the distance shown under "Public Safety."
Fire: If tank, rail car, or tank truck is involved in fire, isolate for at least 800 meters (½ mile) in all directions; also, consider initial evacuation for 800 meters (½ mile) in all directions.

EXPOSURE
Short-term effects: *SEEK MEDICAL ATTENTION. ABSORBED THROUGH THE SKIN.* **Dust:** Harmful if inhaled. Possible lung edema. Irritating to eyes, skin and mucous membranes. Move victim to fresh air. *IF BREATHING HAS STOPPED,* give artificial respiration; *avoid mouth-to-mouth resuscitation; use bag/mask apparatus.* IF breathing is difficult, administer oxygen. *IF IN EYES OR ON SKIN,* flush with running water for at least 15 minutes; hold eyelids open if necessary. **Solid:** Harmful if swallowed. Irritating to eyes and skin. *IF IN EYES OR ON SKIN,* flush with running water for at least 15 minutes; hold eyelids open if necessary. Remove and double bag contaminated clothing and shoes at the site. *IF SWALLOWED* and victim is *UNCONSCIOUS OR HAVING CONVULSIONS,* do nothing except keep victim warm.
Note to physician or authorized medical personnel: Medical observation is recommended for 24 to 48 hours after breathing overexposure, as pulmonary edema may be delayed. As first aid for pulmonary edema, consider administering a corticosteroid spray. Cigarette smoking may exacerbate pulmonary injury and should be discouraged for at least 72 hours following exposure.

HEALTH HAZARDS
Personal protective equipment (PPE): B-Level PPE. Wear self-contained positive pressure breathing apparatus and full protective clothing.
Exposure limits (TWA unless otherwise noted): ACGIH 0.002 mg/m^3; skin contact contributes significantly in overall exposure; animal carcinogen.
Toxicity by ingestion: Grade 1; LD_{50} = 10.0 g/kg (rat).
Long-term health effects: IARC Group 2B: Animal carcinogen; possibly carcinogenic to humans. May cause liver, kidney, nerve, and lung damage. Mutagenic, reproductive and tumorigenic effects. Chronic ingestion has caused enlargement of the thyroid and lymph nodes. Skin photosensitization and abnormal growth of body hair.
Vapor (gas) irritant characteristics: Eye and respiratory tract irritant.
Liquid or solid irritant characteristics: Eye irritant. Solid may cause slight irritation of the skin.

FIRE DATA
Flash point: 468°F/242°C (cc).

CHEMICAL REACTIVITY
Binary reactants: Violent reaction with strong oxidizers, dimethylformamide, nitric acid, liquid oxygen. Incompatible with aluminum, sodium, potassium.

ENVIRONMENTAL DATA
Food chain concentration potential: Log P_{ow} = high readings variously reported as 5.3, 6.2, and 6.4. Values above 3.0 are likely to bioconcentrate in aquatic organisms and other living tissue, especially in fats.
Water pollution: Harmful to aquatic life in very low concentrations. May be dangerous if it enters nearby water intakes; notify operators. Notify local health and wildlife officials. **Response to discharge:** Issue warning–poison; water contaminant. Restrict access. Should be removed. Chemical and physical treatment.

SHIPPING INFORMATION
Grades of purity: 97%; **Storage temperature:** Ambient; **Stability during transport:** Stable.

PHYSICAL AND CHEMICAL PROPERTIES
Physical state @ 59°F/15°C and 1 atm: Solid.
Molecular weight: 284.78
Boiling point @ 1 atm: 589°F/309°C/583°K.
Melting/Freezing point: 446°F/230°C/503°K.
Critical temperature: 1025°F/552°C/825°K.
Critical pressure: 413 psia = 28.1 atm = 2.85 MN/m^2 (estimate).
Specific gravity (water = 1): 2.044 at 24°C.

Relative vapor density (air = 1): 9.8
Vapor pressure: 0.8 mm.

HEXACHLOROBUTADIENE REC. H:0450

SYNONYMS: DOLEN-PUR; GP-40-66:120; HCBD; 1,3-BUTADIENE, 1,1,2,3,4,4-HEXACHLORO-; HEXACHLORO-1,3-BUTADIENE; HEXACLOROBUTADIENO (Spanish); PERCHLOROBUTADIENE; RCRA No. U128

IDENTIFICATION
CAS Number: 87-68-3
Formula: C_4Cl_6; $CCl_2 = CCl–CCl = CCl_2$
DOT ID Number: UN 2279; DOT Guide Number: 151
Proper Shipping Name: Hexachlorobutadiene
Reportable Quantity (RQ): **(CERCLA)** 1 lb/0.454 kg

DESCRIPTION: Colorless liquid Mild, turpentine-like odor. Sinks in water; practically insoluble.

Poison! • Combustible • Breathing the vapor, skin or eye contact, or swallowing the material can kill you • Firefighting gear (including SCBA) does not provide adequate protection. If exposure occurs, remove and isolate gear immediately and thoroughly decontaminate personnel • Containers may BLEVE when exposed to fire • Vapors are heavier than air and will collect and stay in low areas • Irritating to the skin, eyes, and respiratory tract • Toxic products of combustion may include phosgene and hydrogen chloride.

Hazard Classification (based on NFPA-704 Rating System)
Health Hazards (Blue): 2; Flammability (Red): 1; Reactivity (Yellow): 1

EMERGENCY RESPONSE: See Appendix A (151)
Evacuation:
Public safety: Isolate the area of spill or leak for at least 25 to 50 meters (80 to 160 feet) in all directions.
Spill: Increase, in the downwind direction, as necessary, the distance shown under "Public Safety."
Fire: If tank, rail car, or tank truck is involved in fire, isolate for at least 800 meters (½ mile) in all directions; also, consider initial evacuation for 800 meters (½ mile) in all directions.

EXPOSURE
Short-term effects: *SEEK MEDICAL ATTENTION.* **Vapor:** *POISONOUS; MAY BE FATAL IF INHALED.* May cause respiratory difficulty and irritation of eyes, skin and mucous membranes. Remove to fresh air. *IF BREATHING HAS STOPPED,* give artificial respiration; *avoid mouth-to-mouth resuscitation; use bag/mask apparatus.* IF breathing is difficult, administer oxygen. **Liquid:** *POISONOUS; MAY BE FATAL IF SWALLOWED OR ABSORBED THROUGH SKIN.* May cause burns to skin and eyes. *IF IN EYES OR ON SKIN,* flush with running water for at least 15 minutes, hold eyelids open if necessary. *Speed in removing material from skin is of extreme importance.* Remove and double bag contaminated clothing and shoes at the site. Keep victim quiet and maintain normal body temperature. Effects may be delayed; keep victim under observation. *IF SWALLOWED* and victim is *UNCONSCIOUS OR HAVING CONVULSIONS,* do nothing except keep victim warm.

HEALTH HAZARDS
Personal protective equipment (PPE): A-Level PPE. OSHA Table Z-1-A air contaminant. Full face organic vapor respirator and special protective clothing.
Recommendations for respirator selection: NIOSH
At any concentrations above the NIOSH REL, or where there is no REL, at any detectable concentration: SCBAF:PD,PP (any self-contained breathing apparatus that has a full facepiece and is operated in a pressure-demand or other positive-pressure mode); or SAF:PD,PP:ASCBA (any supplied-air respirator that has a full facepiece and is operated in a pressure-demand or other positive-pressure mode in combination with an auxiliary self-contained breathing apparatus operated in a pressure-demand or other positive pressure mode). *ESCAPE:* GMFOV [any air-purifying, full-facepiece respirator (gas mask) with a chin-style, front-or back-mounted organic vapor canister]; or SCBAE (any appropriate escape-type, self-contained breathing apparatus).
Exposure limits (TWA unless otherwise noted): ACGIH TLV 0.02 ppm (0.21 mg/m^3); NIOSH REL 0.02 ppm (0.24 mg/m^3); potential human carcinogen; reduce exposure to lowest feasible level; skin contact contributes significantly in overall exposure.
Toxicity by ingestion: Grade: 3; LD_{50} = 90 mg/kg (rat).
Long-term health effects: Can cause mutagenic, teratogenic and tumorigenic effects. Has caused kidney cancer in animals; it is a suspect human carcinogen. May cause kidney and liver damage. May effect the nervous system.
Vapor (gas) irritant characteristics: Eye and respiratory tract irritant.
Liquid or solid irritant characteristics: Eye and skin irritant.
Odor threshold: 0.006 ppm Potential human carcinogen.

FIRE DATA
Flash point: 194°F/90°C.
Autoignition temperature: 1130°F/610°C/883°K.

CHEMICAL REACTIVITY
Binary reactants: Strong reaction with oxidizers, aluminum powder. Mixtures with bromine perchlorate form heat-, friction-, and shock-sensitive explosive compounds. Reacts with aluminum and other light metals, generating heat. May attack rubber, coatings, and certain plastics.
Polymerization: Able to develop unstable peroxides in storage; may cause polymerization.

ENVIRONMENTAL DATA
Food chain concentration potential: Log P_{ow} = 4.90. Values of more than 3.0 are likely to accumulate in living tissues and fats.
Water pollution: DOT Appendix B, §172.101- severe marine pollutant. Harmful to aquatic life in very low concentrations. May be dangerous if it enters nearby water intakes; notify operators. Notify local health and wildlife officials. **Response to discharge:** Issue warning–poison. Restrict access. Should be removed. Chemical and physical treatment.

SHIPPING INFORMATION
Grades of purity: 98%; **Storage temperature:** Ambient; **Stability during transport:** Stable.

PHYSICAL AND CHEMICAL PROPERTIES
Physical state @ 59°F/15°C and 1 atm: Liquid.
Molecular weight: 260.76
Boiling point @ 1 atm: 419°F/215°C/488°K.
Melting/Freezing point: –6°F/–21°C/252°K.
Critical temperature: (estimate) 315–342°F = 157–172°C = 430–445°K.
Critical pressure: 41 psia = 28 atm = 2.8 MN/m^2 (estimate).
Specific gravity (water = 1): 1.55 at 68°F.

Relative vapor density (air = 1): 9.0 (estimate)
Vapor pressure: 0.2 mm.

HEXACHLOROCYCLOPENTADIENE   REC. H:0500

SYNONYMS: C-56; HCCPD; HEXACHLORO-1,3-CYCLOPENTADIENE; 1,2,3,4,5,5-HEXACHLORO-1,3-CYCLOPENTADIENE; HEXACLOROCICLOPENTADIENO (Spanish); PCL; PERCHLOROCYCLOPENTADIENE; RCRA No. U130

IDENTIFICATION
CAS Number: 77-47-4
Formula: C_5Cl_6
DOT ID Number: UN 2646; DOT Guide Number: 151
Proper Shipping Name: Hexachlorocyclopentadiene
Reportable Quantity (RQ): **(CERCLA)** 10 lb/4.54 kg

DESCRIPTION: Greenish-yellow liquid. Harsh, unpleasant odor. Sinks in water.

Poison! • Breathing the vapor can kill you; skin or eye contact causes severe burns, impaired vision, or blindness • Firefighting gear (including SCBA) does not provide adequate protection. If exposure occurs, remove and isolate gear immediately and thoroughly decontaminate personnel • May be combustible but does not ignite readily • Containers may BLEVE when exposed to fire • In presence of moisture, explosive hydrogen gas may collect in enclosed space • Toxic products of combustion may include hydrogen chloride, chlorine, and phosgene • Do not put yourself in danger by entering a contaminated area to rescue a victim.

Hazard Classification (based on NFPA-704 Rating System)
Health Hazards (Blue): 2; Flammability (Red): 1; Reactivity (Yellow): 0

EMERGENCY RESPONSE: See Appendix A (151)
Evacuation:
Public safety: Isolate the area of spill or leak for at least 25 to 50 meters (80 to 160 feet) in all directions.
Spill: Small spill–First: Isolate in all directions 30 meters (100 feet); Then: Protect persons downwind, DAY: 0.2 km (0.1 mile); NIGHT: 0.2 km (0.1 mile). Large spill–First: Isolate in all directions 30 meters (100 feet); Then: Protect persons downwind, DAY: 0.2 km (0.1 mile); NIGHT: 0.3 km (0.2 mile).
Fire: If tank, rail car, or tank truck is involved in fire, isolate for at least 800 meters (½ mile) in all directions; also, consider initial evacuation for 430 meters (1400 ft) in all directions.

EXPOSURE
Short-term effects: *SEEK MEDICAL ATTENTION*. **Vapor:** *POISONOUS IF INHALED. IF INHALED*, will, will cause coughing or difficult breathing. *IF IN EYES*, hold eyelids open and flush with plenty of water. *IF BREATHING HAS STOPPED*, give artificial respiration; *avoid mouth-to-mouth resuscitation; use bag/mask apparatus*. IF breathing is difficult, administer oxygen. **Liquid:** Will burn skin and eyes. *IF SWALLOWED*, will cause nausea and vomiting. Remove contaminated clothing and shoes. Flush affected areas with plenty of water. *IF IN EYES*, hold eyelids open and flush with plenty of water. *IF SWALLOWED* and victim is *CONSCIOUS AND ABLE TO SWALLOW*, have victim drink 4 to 8 ounces of water and have victim induce vomiting. *IF SWALLOWED* and victim is *UNCONSCIOUS OR HAVING CONVULSIONS*, do nothing except keep victim warm.

HEALTH HAZARDS
Personal protective equipment (PPE): A-Level PPE. OSHA Table Z-1-A air contaminant. Full face organic vapor respirator. Protective clothing, including gloves and shoes or boots. Chemical protective material(s) reported to have good to excellent resistance: butyl rubber, nitrile, PV alcohol, butyl rubber/neoprene, Viton®.
Exposure limits (TWA unless otherwise noted): ACGIH TLV 0.01 ppm (0.11 mg/m^3); NIOSH REL 0.01 ppm (0.1 mg/m^3).
Toxicity by ingestion: Grade 4; oral LD$_{50}$ = 0.505 mg/kg (mouse), 113 mg/kg (rat).
Vapor (gas) irritant characteristics: Severe eye and respiratory tract irritant.
Liquid or solid irritant characteristics: Severe eye and skin irritant.
Odor threshold: 0.03–0.2 ppm.

FIRE DATA
Flash point: Combustible, but does not ignite readily.
Fire extinguishing agents not to be used: If water is used on adjacent fires, do not allow water to enter drums or storage tanks.

CHEMICAL REACTIVITY
Reactivity with water: Reacts slowly to form hydrochloric acid. The reaction may not be hazardous.
Binary reactants: Explosive reaction with sodium. In presence of moisture, will corrode iron and other metals. Flammable and explosive hydrogen gas may collect in enclosed space.
Neutralizing agents for acids and caustics: Rinse with dilute solution of sodium bicarbonate or soda ash.

ENVIRONMENTAL DATA
Food chain concentration potential: Possible accumulation of breakdown products
Water pollution: DOT Appendix B, §172.101–marine pollutant. Dangerous to aquatic life in high concentrations. May be dangerous if it enters nearby water intakes; notify operators. Notify local health and wildlife officials. **Response to discharge:** Issue warning–poison, water contaminant, air contaminant. Restrict access. Should be removed. Chemical and physical treatment.

SHIPPING INFORMATION
Grades of purity: Commercial, 97+%; Synthesis grade; **Storage temperature:** Ambient; **Inert atmosphere:** None; **Venting:** Open; **Stability during transport:** Stable.

PHYSICAL AND CHEMICAL PROPERTIES
Physical state @ 59°F/15°C and 1 atm: Liquid.
Molecular weight: 272.8
Boiling point @ 1 atm: 462°F/239°C/512°K.
Melting/Freezing point: 16°F/–9°C/264°C.
Specific gravity (water = 1): 1.71 @ 68°F/20°C (liquid).
Liquid surface tension: 37.5 dynes/cm = 0.0375 N/m @ 68°F/20°C.
Relative vapor density (air = 1): 9.42
Latent heat of vaporization: (estimate) 76 Btu/lb = 42 cal/g = 1.8 x 10^5 J/kg.
Vapor pressure: 0.08 mm @77°F

HEXACHLOROETHANE   REC. H:0550

SYNONYMS: AVLOTHANE; DISTOKAL; CARBON HEXACHLORIDE; DISTOPAN; EGITOL; ETHANE HEXACHLORIDE; ETHYLENE HEXACHLORIDE; 1,1,1,2,2,2-

Hexachloroethane

HEXACHLOROETHANE; HEXACLOROETANO (Spanish); FALKITOL, FASCIOLIN; PERCHLOROETHANE; RCRA No. U131

IDENTIFICATION
CAS Number: 67-72-1
Formula: C_2Cl_6; Cl_3CCCl_3
DOT ID Number: UN 9037; DOT Guide Number: 151
Proper Shipping Name: Hexachloroethane
Reportable Quantity (RQ): **(CERCLA)** 100 lb/45.4 kg

DESCRIPTION: Colorless to pale yellow crystalline solid. Camphor-like odor. Sinks in water; insoluble.

Poison! • Breathing the vapor can kill you; may be fatal if swallowed or absorbed through the skin; effects of contact or inhalation may be delayed • Firefighting gear (including SCBA) does not provide adequate protection. If exposure occurs, remove and isolate gear immediately and thoroughly decontaminate personnel • Decomposes above 572°F/300°C. Toxic products of combustion or decomposition include hydrogen chloride and phosgene • Do not put yourself in danger by entering a contaminated area to rescue a victim • Slightly explosive (by spontaneous chemical reaction).

Hazard Classification (based on NFPA-704 Rating System)
Health Hazards (Blue): 2; Flammability (Red): 0; Reactivity (Yellow): 0

EMERGENCY RESPONSE: See Appendix A (151)
Evacuation:
Public safety: Isolate the area of spill or leak for at least 25 to 50 meters (80 to 160 feet) in all directions.
Spill: Increase, in the downwind direction, as necessary, the distance shown under "Public Safety."
Fire: If tank, rail car, or tank truck is involved in fire, isolate for at least 800 meters (½ mile) in all directions; also, consider initial evacuation for 800 meters (½ mile) in all directions.

EXPOSURE
Short-term effects: *SEEK MEDICAL ATTENTION.* **Vapor:** Severely Irritating to eyes, nose, and throat. Harmful if inhaled. *IF IN EYES*, hold eyelids open and flush with plenty of water.*IF BREATHING HAS STOPPED,* give artificial respiration; *avoid mouth-to-mouth resuscitation; use bag/mask apparatus.* IF breathing is difficult, administer oxygen. **Liquid:** *POISONOUS IF SWALLOWED OR IF ABSORBED THROUGH THE SKIN.* Irritating to skin and eyes. *IF SWALLOWED*, will cause nausea and vomiting. Remove contaminated clothing and shoes. Flush affected areas with plenty of water. *IF IN EYES*, hold eyelids open and flush with plenty of water. *IF SWALLOWED* and victim is *CONSCIOUS AND ABLE TO SWALLOW*, have victim drink 4 to 8 ounces of water and induce vomiting. *IF SWALLOWED* and victim is *UNCONSCIOUS OR HAVING CONVULSIONS,* do nothing except keep victim warm.

HEALTH HAZARDS
Personal protective equipment (PPE): B-Level PPE. OSHA Table Z-1-A air contaminant. Full face organic vapor respirator. Close-fitting chemical safety goggles; plastic face shield; air- or oxygen- supplied mask; safety hat with brim; solvent-proof apron; synthetic rubber **gloves.**
Recommendations for respirator selection: NIOSH
At any concentrations above the NIOSH REL, or where there is no REL, at any detectable concentration: SCBAF:PD,PP (any self-contained breathing apparatus that has a full facepiece and is operated in a pressure-demand or other positive-pressure mode); or SAF:PD,PP:ASCBA (any supplied-air respirator that has a full facepiece and is operated in a pressure-demand or other positive-pressure mode in combination with an auxiliary self-contained breathing apparatus operated in a pressure-demand or other positive pressure mode). *ESCAPE:* GMFOV [any air-purifying, full-facepiece respirator (gas mask) with a chin-style, front-or back-mounted organic vapor canister]; or SCBAE (any appropriate escape-type, self-contained breathing apparatus).
Exposure limits (TWA unless otherwise noted): ACGIH TLV 1 ppm (9.7 mg/m^3) suspected human carcinogen; OSHA PEL 1 ppm (10 mg/m^3); NIOSH REL 1 ppm (10 mg/m^3); reduce exposure to lowest feasible level; potential human carcinogen; skin contact contributes significantly in overall exposure.
Short-term exposure limits (15-minute TWA): No data available
Toxicity by ingestion: Grade 3; oral LD_{50} = 4.46 g/kg (rat).
Long-term health effects: Liver poisoning; nervous system disorders; suspected carcinogen. Possible kidney damage.
Vapor (gas) irritant characteristics: Eye and respiratory tract irritant. Vapor is irritating such that personnel will not usually tolerate moderate or high vapor concentrations.
Liquid or solid irritant characteristics: Eye and skin irritant. If spilled on clothing and allowed to remain, may cause smarting and reddening of the skin.
Odor threshold: 0.15 ppm
IDLH value: 300 ppm; potential human carcinogen.

CHEMICAL REACTIVITY
Binary reactants: Alkalis, metals such as zinc, cadmium, aluminum, hot iron and mercury. Reaction with aluminum and zinc metals form combustible gas. May attack some forms of plastics.

ENVIRONMENTAL DATA
Food chain concentration potential: Log P_{ow} = 3.34 (estimate). Values > 3.0 are likely to bioconcentrate in aquatic organisms and other living tissue, especially in fats.
Water pollution: Effect of low concentrations on aquatic life is unknown. May be dangerous if it enters nearby water intakes; notify operators. Notify local health and wildlife officials. **Response to discharge:** Issue warning–poisonous, air contaminant. Restrict access. Should be removed. Chemical and physical treatment.

SHIPPING INFORMATION
Grades of purity: 98%, 99%; **Storage temperature:** Ambient; **Inert atmosphere:** None; **Venting:** Open; **Stability during transport:** Stable.

NAS HAZARD CLASSIFICATION FOR BULK WATER TRANSPORTATION
FIRE: 0
HEALTH: Vapor irritant: 2; Liquid or solid irritant: 1; Poisons: 1
WATER POLLUTION: Human toxicity: 2; Aquatic toxicity: 3; Aesthetic effect: 0
REACTIVITY: Other chemicals: 0; Water: 0; Self-reaction: 0

PHYSICAL AND CHEMICAL PROPERTIES
Physical state @ 59°F/15°C and 1 atm: Solid.
Molecular weight: 236.74
Boiling point @ 1 atm: Sublimes
Melting/Freezing point: 368°F/186°C/459°K (sublimes).
Specific gravity (water = 1): 2.091 @ 68°F/20°C (liquid).
Relative vapor density (air = 1): 8.16
Latent heat of vaporization: 92.7 Btu/lb = 51.5 cal/g = 2.15 x 10^5 J/kg.

Heat of combustion: –836.4 Btu/lb = –464.6 cal/g = –19.4 x 10^5 J/kg.
Heat of decomposition: 570°F/298°C.
Vapor pressure: (Reid) 0.04 psia; 0.2 mm.

HEXACHLOROPHENE REC. H:0600

SYNONYMS: ACIGENA; ALMEDERM; AT-7; AT-17; B32; BILEVON; BIS(2-HYDROXY-3,5,6-TRICHLOROPHENYL)METHANE; BIS(3,5,6-TRICHLORO-2-HYDROXYPHENYL)METHANE; COMPOUND G-11; COTOFILM; DERMADEX; 2,2'-DIHYDROXY-3,3',5,5',6,6'-HEXACHLORODIPHENYLMETHANE; 2,2'-DIHYDROXY-3,5,6,3',5',6'-HEXACHLORODIPHENYLMEXOFENE; EEC No. 604-015-00-9; EXOPHENE; FOMAC; FOSTRIL; G-11; GAMOPHEN; GAMOPHENE; G-ELEVEN COMPOUND; GERMA-MEDICA; HCP; HEXABALM; 2,2',3,3',5,5'-HEXACHLORO-6,6'-DIHYDROXYDIPHENYLMETHANE; HEXACHLOROPHANE; HEXAFEN; HEXIDE; HEXOPHENE; HEXOSAN; HEXACLOROFENO (Spanish); ISOBAC 20; METHANE,BIS(2,3,5-TRICHLORO-6-HYDROXYPHENYL); 2,2'-METHYLENEBIS(3,4,6-TRICHLOROPHENOL); NABAC; NEOSEPT; PHISODAN; PHISOHEX; RITOSEPT; SEPTISOL; SEPTOFEN; STERAL; STERASKIN; SURGI-CEN; SURGI-CIN; SUROFENE; TERSASEPTIC; TRICHLOROPHENE; TURGEX; RCRA No. U132

IDENTIFICATION
CAS Number: 70-30-4
Formula: $C_{13}H_6Cl_6O_2$; $(C_6HCl_3OH)_2CH_2$
DOT ID Number: UN 2875; DOT Guide Number: 151
Proper Shipping Name: Hexachlorophene
Reportable Quantity (RQ): **(CERCLA)** 100 lb/45.4 kg

DESCRIPTION: White crystalline powder. Odorless. Sinks in water; insoluble.

Combustible • Containers may BLEVE when exposed to fire • Toxic products of combustion may include hydrogen chloride fumes • Do not put yourself in danger by entering a contaminated area to rescue a victim.

Hazard Classification (based on NFPA-704 Rating System)
Health Hazards (Blue): 3; Flammability (Red): 1; Reactivity (Yellow): 0

EMERGENCY RESPONSE: See Appendix A (151)
Evacuation:
Public safety: Isolate the area of spill or leak for at least 25 to 50 meters (80 to 160 feet) in all directions.
Spill: Increase, in the downwind direction, as necessary, the distance shown under "Public Safety."
Fire: If tank, rail car, or tank truck is involved in fire, isolate for at least 800 meters (½ mile) in all directions; also, consider initial evacuation for 800 meters (½ mile) in all directions.

EXPOSURE
Short-term effects: *SEEK MEDICAL ATTENTION.* **Dust:** *POISONOUS IF INHALED.* Irritating to mucous membranes. Move victim to fresh air. *IF BREATHING HAS STOPPED,* give artificial respiration; *avoid mouth-to-mouth resuscitation; use bag/mask apparatus.* IF breathing is difficult, administer oxygen. **Solid:** *POISONOUS IF SWALLOWED.* Irritating to eyes and skin. *IF IN EYES OR ON SKIN,* flush with running water for at least 15 minutes; hold eyelids open if necessary. Wash skin with soap and water. Remove and double bag contaminated clothing and shoes at the site. *IF SWALLOWED* and victim is *CONSCIOUS AND ABLE TO SWALLOW,* induce vomiting with warm salt water or syrup of ipecac. *IF SWALLOWED* and victim is *UNCONSCIOUS OR HAVING CONVULSIONS,* do nothing except keep victim warm.

HEALTH HAZARDS
Personal protective equipment (PPE): B-Level PPE. Wear self-contained positive pressure breathing apparatus and full protective clothing. Chemical protective material(s) reported to have good to excellent resistance: butyl rubber, polycarbonate.
Eyes or skin. EYE and skin irritant. **Ingestion:** *POISONOUS IF SWALLOWED.* Symptoms following ingestion include anorexia, nausea, vomiting, abdominal cramps, and diarrhea. Dehydration may be severe and may be associated with shock.
Toxicity by ingestion: Grade: 3; LD_{50} = 60 mg/kg (rat).
Long-term health effects: May cause dermatitis. Causes reproductive/birth defects and tumorigenic effects; indefinite carcinogen, male infertility. Prolonged or repeated exposure causes asthma, nervous system effects, possible blindness.
Vapor (gas) irritant characteristics: Eye and respiratory tract irritant.
Liquid or solid irritant characteristics: Eye and skin irritant.

ENVIRONMENTAL DATA
Water pollution: Effects of low concentrations on aquatic life is unknown. May be dangerous if it enters nearby water intakes; notify operators. Notify local health and wildlife officials. **Response to discharge:** Issue warning–poison. Restrict access.

SHIPPING INFORMATION
Grades of purity: 99%; **Storage temperature:** Ambient; **Stability during transport:** Stable.

PHYSICAL AND CHEMICAL PROPERTIES
Physical state @ 59°F/15°C and 1 atm: Solid.
Molecular weight: 406.91
Melting/Freezing point: 327–329°F/164–165°C/437–438°K.

HEXADECYL SULFATE, SODIUM SALT REC. H:0650

SYNONYMS: CETYL SODIUM SULFATE; SODIUM CETYL SULFATE

IDENTIFICATION
CAS Number: 1120-01-0
Formula: $C_{16}H_{33}O_4S \cdot Na$; $CH_3(CH_2)_4CH_2OSO_3Na \cdot H_2O$

DESCRIPTION: White solid paste or liquid. Mild odor. May float or sink in water; soluble.

Irritating to the skin, eyes, and respiratory tract • Toxic products of combustion may include carbon monoxide and sulfur oxides.

Hazard Classification (based on NFPA-704 Rating System)
Health Hazards (Blue): 1; Flammability (Red): 0; Reactivity (Yellow): 0

EMERGENCY RESPONSE: See Appendix A (171)
Evacuation:
Public safety: Isolate the area of spill or leak for at least 10 to 25 meters (30 to 80 feet) in all directions.

588   Hexadecyl trimethylammonium chloride

Spill: Increase, in the downwind direction, as necessary, the distance shown under "Public Safety."
Fire: If any large container is involved in fire, isolate for at least 800 meters (½ mile) in all directions; also, consider initial evacuation for 800 meters (½ mile) in all directions.

EXPOSURE
Short-term effects: *SEEK MEDICAL ATTENTION.* **Liquid or solid:** Irritating to skin and eyes. Harmful if swallowed. Remove contaminated clothing and shoes. Flush affected areas with plenty of water. *IF IN EYES,* hold eyelids open and flush with plenty of water. *IF SWALLOWED* and victim is *CONSCIOUS AND ABLE TO SWALLOW,* have victim drink 4 to 8 ounces of water. *IF SWALLOWED* and victim is *UNCONSCIOUS OR HAVING CONVULSIONS,* do nothing except keep victim warm.

HEALTH HAZARDS
Personal protective equipment (PPE): B-Level PPE. Plastic or rubber gloves; goggles or face shield.
Vapor (gas) irritant characteristics: May cause eye irritation.
Liquid or solid irritant characteristics: May cause eye and skin irritation.

CHEMICAL REACTIVITY
Binary reactants: Sulfates react violently with aluminum, magnesium.

ENVIRONMENTAL DATA
Food chain concentration potential: Negative; unlikely to accumulate.
Water pollution: Effect of low concentrations on aquatic life is unknown. May be dangerous if it enters nearby water intakes; notify operators. Notify local health and wildlife officials.
Response to discharge: Disperse and flush.

SHIPPING INFORMATION
Grades of purity: Commercial; **Storage temperature:** Ambient; **Inert atmosphere:** None; **Venting:** Open; **Stability during transport:** Stable.

PHYSICAL AND CHEMICAL PROPERTIES
Physical state @ 59°F/15°C and 1 atm: Solid or liquid (mixture).
Specific gravity (water = 1): 1 @ 68°F/20°C (liquid).

HEXADECYL TRIMETHYLAMMONIUM CHLORIDE
REC. H:0700

SYNONYMS: CETYLTRIMETHYL AMMONIUM CHLORIDE; CLORURO de HEXADECILTRIMETILAMONIO (Spanish)

IDENTIFICATION
CAS Number: 112-02-7
Formula: $C_{16}H_{33}(CH_3)_3NCl \cdot H_2O-(CH_3)_2CHOH$

DESCRIPTION: White powder or clear to pale yellow liquid (solution in isopropyl alcohol). Rubbing alcohol odor. Floats or sinks in water.

Solution is highly flammable • Corrosive to skin, eyes, and lungs; contact with skin or eyes may cause burns, impaired vision, or blindness; inhalation symptoms may be delayed • Containers may BLEVE when exposed to fire • Vapors may form explosive mixture with air • Vapors are heavier than air and will collect and stay in low areas • Vapors may travel long distances to ignition sources and flashback • Vapors in confined areas (e.g., tanks, sewers, buildings) may explode when exposed to fire • Decomposes below 302°F/150°C. Toxic products of combustion may include carbon monoxide, CO_2, nitrogen oxides, and hydrogen chloride.

Hazard Classification (based on NFPA-704 Rating System)
powder: Health Hazards (Blue): 2; Flammability (Red): 0; Reactivity (Yellow): 0
solution: Health Hazards (Blue): 1; Flammability (Red): 2; Reactivity (Yellow): 0

EMERGENCY RESPONSE: See Appendix A (171)
Evacuation:
Public safety: Isolate the area of spill or leak for at least 10 to 25 meters (30 to 80 feet) in all directions.
Spill: Increase, in the downwind direction, as necessary, the distance shown under "Public Safety."
Fire: If any large container is involved in fire, isolate for at least 800 meters (½ mile) in all directions; also, consider initial evacuation for 800 meters (½ mile) in all directions.

EXPOSURE
Short-term effects: *SEEK MEDICAL ATTENTION.* **Vapor:** Irritating to eyes, nose, and throat. Harmful if inhaled. *IF IN EYES,* hold eyelids open and flush with plenty of water. *IF BREATHING HAS STOPPED,* give artificial respiration. Do NOT use mouth-to-mouth resuscitation. IF breathing is difficult, administer oxygen. Lung edema may develop. **Liquid:** May be absorbed through the skin. Irritating to skin and eyes. May result in corneal damage and chemical conjunctivitis. Harmful if swallowed. Remove contaminated clothing and shoes. Flush affected areas with plenty of water. *IF IN EYES,* hold eyelids open and flush with plenty of water. *IF SWALLOWED* and victim is *CONSCIOUS AND ABLE TO SWALLOW,* have victim drink 4 to 8 ounces of water. *IF SWALLOWED* and victim is *UNCONSCIOUS OR HAVING CONVULSIONS,* do nothing except keep victim warm. **Do NOT induce vomiting.**
Note to physician or authorized medical personnel: Medical observation is recommended for 24 to 48 hours after breathing overexposure, as pulmonary edema may be delayed. As first aid for pulmonary edema, consider administering a corticosteroid spray. Cigarette smoking may exacerbate pulmonary injury and should be discouraged for at least 72 hours following exposure.

HEALTH HAZARDS
Personal protective equipment (PPE): B-Level PPE. Goggles or face shield; rubber gloves.
Exposure limits (TWA unless otherwise noted): ACGIH TLV 400 ppm (983 mg/m^3); NIOSH/OSHA 400 ppm (980 mg/m^3) as isopropyl alcohol
Short-term exposure limits (15-minute TWA): ACGIH STEL 500 ppm (1230 mg/m^3); NIOSH STEL 500 ppm (1225 mg/m^3) as isopropyl alcohol
Long-term health effects: May cause skin sensitization; allergy.
Toxicity by ingestion: Grade 3; LD_{50} = 250 mg/kg (rat).
Liquid or solid irritant characteristics: Severe eye and skin irritant.
IDLH value: 2000 ppm as isopropyl alcohol.

FIRE DATA
Flash point: 69°F/21°C (cc) (for isopropyl alcohol solutions only).
Flammable limits in air: LEL: 2%; UEL: 12% (isopropyl alcohol).
Fire extinguishing agents not to be used: Water may be ineffective.

Autoignition temperature: 750°F/399°C/672°K (isopropyl alcohol).
Electrical hazard: (as isopropyl alcohol) Class 1, Group D.
Burning rate: (isopropyl alcohol solutions) 2.3 mm/min

CHEMICAL REACTIVITY
Binary reactants: May react with strong oxidizers, acetaldehyde, chlorine, ethylene oxide, acids, isocyanates (isopropyl alcohol).

ENVIRONMENTAL DATA
Food chain concentration potential: Negative; unlikely to accumulate.
Water pollution: Effect of low concentrations on aquatic life is unknown. May be dangerous if it enters nearby water intakes; notify operators. Notify local health and wildlife officials. **Response to discharge:** Issue warning–water contaminant. Disperse and flush. Should be removed. Chemical and physical treatment.

SHIPPING INFORMATION
Grades of purity: 25% solution in water; 50% solution in a mixture of isopropyl alcohol and water, which is flammable. If the chemical (or a solution with concentration greater than 50%) is shipped, contact with skin and eyes should be avoided; **Storage temperature:** Ambient; **Inert atmosphere:** None; **Venting:** Open; **Stability during transport:** Stable.

PHYSICAL AND CHEMICAL PROPERTIES
Physical state @ 59°F/15°C and 1 atm: Liquid.
Molecular weight: 319.75 (solute only).
Melting/Freezing point: 130°F/55°C/328°K (approximate)
Boiling point @ 1 atm: (isopropyl alcohol) 180°F/82.3°C/355.5°K.
Specific gravity (water = 1): (approximate) 0.9 @ 77°F/25°C (liquid).

n-HEXALDEHYDE   REC. H:0750

SYNONYMS: ALDEHYDE C-6; CAPRONIC ALDEHYDE; CAPROALDEHYDE; CAPRONALDEHYDE; *n*-CAPROYLALDEHYDE; HEXANAL; HEXALDEHYDE; *n*-HEXALDEHIDO (Spanish); 1-HEXANAL

IDENTIFICATION
CAS Number: 66-25-1
Formula: $C_6H_{12}O$; $CH_3(CH_2)_4CHO$
DOT ID Number: UN 1207; DOT Guide Number: 129
Proper Shipping Name: Hexaldehyde

DESCRIPTION: Colorless liquid. Sharp unpleasant odor. Floats on water surface; insoluble.

Highly flammable • Containers may BLEVE when exposed to fire •Vapors may form explosive mixture with air • Vapors are heavier than air and will collect and stay in low areas • Vapors may travel long distances to ignition sources and flashback • Vapors in confined areas (e.g., tanks, sewers, buildings) may explode when exposed to fire • Highly irritating to the eyes and respiratory tract; this material has anesthetic properties • Toxic products of combustion may include carbon monoxide.

Hazard Classification (based on NFPA-704 Rating System)
Health Hazards (Blue): 2; Flammability (Red): 3; Reactivity (Yellow): 1

EMERGENCY RESPONSE: See Appendix A (129)
Evacuation:
Public safety: Isolate spill area for at least 50 to 100 meters (160 to 330 feet) in all directions.
Spill: Large spill–Consider initial downwind evacuation for at least 300 meters (1000 feet).
Fire: Isolate for 800 meters (½ mile) in all directions, especially if tank, rail car, or tank truck is involved in fire.

EXPOSURE
Short-term effects: Vapor: Irritating to eyes, nose, and throat. Harmful if inhaled. *IF IN EYES*, hold eyelids open and flush with plenty of water. *IF BREATHING HAS STOPPED*, give artificial respiration. IF breathing is difficult, administer oxygen. **Liquid:** Irritating to skin and eyes. *IF SWALLOWED*, will cause nausea and vomiting. Remove contaminated clothing and shoes. Flush affected areas with plenty of water. *IF IN EYES*, hold eyelids open and flush with plenty of water. *IF SWALLOWED* and victim is *CONSCIOUS AND ABLE TO SWALLOW*, have victim drink 4 to 8 ounces of water and have victim induce vomiting. *IF SWALLOWED* and victim is *UNCONSCIOUS OR HAVING CONVULSIONS*, do nothing except keep victim warm.

HEALTH HAZARDS
Personal protective equipment (PPE): B-Level PPE. Goggles or face shield; rubber gloves.
Toxicity by ingestion: Grade 2; oral LD_{50} = 4,890 mg/kg (rat).
Vapor (gas) irritant characteristics: Eye irritant.
Liquid or solid irritant characteristics: Eye and skin irritant.

FIRE DATA
Flash point: 90°F/32°C (oc).
Fire extinguishing agents not to be used: Water may be ineffective.
Burning rate: 5.21 mm/min

CHEMICAL REACTIVITY
Binary reactants: Violent reaction with strong oxidizers, bromine, ketones. Incompatible with strong acids, strong caustics. May attack some forms of plastics and coatings.

ENVIRONMENTAL DATA
Food chain concentration potential: Negative; unlikely to accumulate.
Water pollution: Effect of low concentrations on aquatic life is unknown. Fouling to shoreline. May be dangerous if it enters nearby water intakes; notify operators. Notify local health and wildlife officials. **Response to discharge:** Issue warning–high flammability. Restrict access. Mechanical containment. Should be removed. Chemical and physical treatment.

SHIPPING INFORMATION
Grades of purity: 99+%; Commercial; **Storage temperature:** Ambient; **Inert atmosphere:** None; **Venting:** Open (flame arrester); **Stability during transport:** Stable.

PHYSICAL AND CHEMICAL PROPERTIES
Physical state @ 59°F/15°C and 1 atm: Liquid.
Molecular weight: 100
Boiling point @ 1 atm: 262°F/128°C/401°K.
Specific gravity (water = 1): 0.83 @ 68°F/20°C (liquid).
Relative vapor density (air = 1): 3.5
Ratio of specific heats of vapor (gas): (estimate) 1.061 at 68°F/20°C.

Latent heat of vaporization: (estimate) 153 Btu/lb = 85 cal/g = 3.6×10^5 J/kg.
Heat of combustion: (estimate) –17,000 Btu/lb = –9,430 cal/g = -394×10^5 J/kg.

HEXAMETHYLENEDIAMINE REC. H:0800

SYNONYMS: 1,6-DIAMINOHEXANE; 1,6-HEXANEDIAMINE; HEXAMETILENDIAMINA (Spanish); HMDA

IDENTIFICATION
CAS Number: 124-09-4
Formula: $C_6H_{16}N_2$; $NH_2(CH_2)_6NH_2$
DOT ID Number: UN 2280 (solid); UN 1783 (solution); DOT Guide Number: 153
Proper Shipping Name: Hexamethylenediamine solid; Hexamethylenediamine solution

DESCRIPTION: Colorless glassy solid or clear solution (70%). Weak, fishy, ammonia odor. Floats on the surface of water; soluble.

Combustible • Corrosive to the skin, eyes, and respiratory tract; contact with skin and eyes may cause burns, vision impairment, and blindness; inhalation symptoms may be delayed • May interfere with the body's ability to use oxygen • Containers may BLEVE when exposed to fire • Vapors may form explosive mixture with air • Vapors are heavier than air and will collect and stay in low areas • Vapors in confined areas (e.g., tanks, sewers, buildings) may explode when exposed to fire • Toxic products of combustion may include anhydrous ammonia and nitrogen oxides.

Hazard Classification (based on NFPA-704 Rating System)
Health Hazards (Blue): 1; Flammability (Red): 2; Reactivity (Yellow): 0

EMERGENCY RESPONSE: See Appendix A (153)
Evacuation:
Public safety: Isolate the area of spill or leak for at least 25 to 50 meters (80 to 160 feet) in all directions.
Spill: Increase, in the downwind direction, as necessary, the distance shown under "Public Safety."
Fire: If tank, rail car, or tank truck is involved in fire, isolate for at least 800 meters (½ mile) in all directions; also, consider initial evacuation for 800 meters (½ mile) in all directions.

EXPOSURE
Short-term effects: *SEEK MEDICAL ATTENTION*. **Inhalation:** *IF BREATHING HAS STOPPED*, give artificial respiration; *avoid mouth-to-mouth resuscitation; use bag/mask apparatus*. Vapor may cause lung edema; physical exertion will aggravate this condition. **Liquid or solid:** *POISONOUS IF SWALLOWED OR IF SKIN IS EXPOSED*. Will burn eyes. Remove contaminated clothing and shoes. Flush affected areas with plenty of water. *IF IN EYES*, hold eyelids open and flush with plenty of water. *IF SWALLOWED* and victim is *CONSCIOUS AND ABLE TO SWALLOW*, have victim drink 4 to 8 ounces of water.
Note to physician or authorized medical personnel: Medical observation is recommended for 24 to 48 hours after breathing overexposure, as pulmonary edema may be delayed. As first aid for pulmonary edema, consider administering a corticosteroid spray. Cigarette smoking may exacerbate pulmonary injury and should be discouraged for at least 72 hours following exposure.

HEALTH HAZARDS
Personal protective equipment (PPE): B-Level PPE. Organic vapor respirator. Protective clothing; eye protection.
Exposure limits (TWA unless otherwise noted: 5 mg/m³ (AIHAWEEL).
Long-term health effects: Repeated exposure can cause anemia and damage to the kidney and liver. Possible dermatitis from prolonged contact.
Vapor (gas) irritant characteristics: Eye and respiratory tract irritant. Vapors cause smarting of the eyes or respiratory system if present in high concentrations. May cause lung edema; physical exertion will aggravate this condition.
Liquid or solid irritant characteristics: Severe eye and skin irritant. Causes smarting of the skin and first-degree burns on short exposure; may cause secondary burns on long exposure.
Odor threshold: 0.0041 mg/m³.

FIRE DATA
Flash point: 160°F/71°C (oc).
Flammable limits in air: LEL: 0.7%; UEL: 6.3%.
Fire extinguishing agents not to be used: Water and CO_2.
Autoignition temperature: 585°F/307°C/580°K.

CHEMICAL REACTIVITY
Reactivity with water: Dissolved in water, it forms a strong base.
Binary reactants: Violent reaction with strong oxidizers, strong acids, maleic anhydride. In the presence of moisture corrodes chemically active metals such as aluminum, copper, lead, zinc, and their alloys.
Neutralizing agents for acids and caustics: Flush with water.

ENVIRONMENTAL DATA
Food chain concentration potential: Negative; unlikely to accumulate.
Water pollution: Effect of low concentrations on aquatic life is unknown. May be dangerous if it enters nearby water intakes; notify operators. Notify local health and wildlife officials. **Response to discharge:** Issue warning–corrosive, water contaminant. Disperse and flush.

SHIPPING INFORMATION
Grades of purity: Anhydrous: 99.8%; 70% solution; **Storage temperature:** Ambient; **Inert atmosphere:** Nitrogen; **Venting:** Pressure-vacuum; **Stability during transport:** Stable.

NAS HAZARD CLASSIFICATION FOR BULK WATER TRANSPORTATION
FIRE: 1
HEALTH: Vapor irritant: 1; Liquid or solid irritant: 2; Poisons: 3
WATER POLLUTION: Human toxicity: 2; Aquatic toxicity: 3; Aesthetic effect: 1
REACTIVITY: Other chemicals: 3; Water: 0; Self-reaction: 0

PHYSICAL AND CHEMICAL PROPERTIES
Physical state @ 59°F/15°C and 1 atm: Solid (anhydrous).
Molecular weight: 116.21
Boiling point @ 1 atm: 478°K = 205°C/401°F
Melting/Freezing point: (anhydrous) 104.9°F/40.5°C/313.7°K; (70% solution) 28°F/–2°C/269°K.
Specific gravity (water = 1): (anhydrous) 0.799 at 60°C (liquid); (70% solution) 0.933 @ 68°F/20°C (liquid).
Liquid surface tension: (anhydrous) 34.6 dynes/cm = 0.0346 N/m at 60°C.
Latent heat of vaporization: 203 Btu/lb = 113 cal/g = 4.73×10^5 J/kg.

Heat of combustion: (estimate) –12,200 Btu/lb = –6790 cal/g = –284 x 10^5 J/kg.
Heat of solution: (estimate) –9 Btu/lb = –5 cal/g = –0.2 x 10^5 J/kg.

HEXAMETHYLENEIMINE   REC. H:0850

SYNONYMS: AZACYCLOHEPTANE; 1-AZACYCLOHEPTANE; CYCLOHEXAMETHYENEIMINE; HEXAHYDROAZEPINE; HEXAMETILENIMINA (Spanish); HOMOPIPERIDINE; PERHYDROAZEPINE

IDENTIFICATION
CAS Number: 111-49-9
Formula: $C_6H_{13}N$; $CH_2CH_2CH_2CH_2CH_2CH_2NH$
DOT ID Number: UN 2493; DOT Guide Number: 132
Proper Shipping Name: Hexamethyleneimine

DESCRIPTION: Colorless to light yellow liquid. Ammonia-like odor. Floats and mixes slowly with water. Irritating vapor is produced.

Highly flammable • Corrosive to the skin, eyes, and respiratory tract; contact with skin and eyes may cause burns, vision impairment, and blindness; inhalation symptoms may be delayed • Vapors are heavier than air and will collect and stay in low areas • Vapors may travel long distances to ignition sources and flashback • Vapors in confined areas (e.g., tanks, sewers, buildings) may explode when exposed to fire • Toxic products of combustion may include nitrogen oxides.

Hazard Classification (based on NFPA-704 Rating System)
Health Hazards (Blue): 2; Flammability (Red): 3; Reactivity (Yellow): 0

EMERGENCY RESPONSE: See Appendix A (132)
Evacuation:
Public safety: Isolate spill area for at least 50 to 100 meters (160 to 330 feet) in all directions.
Spill: Increase, as necessary, the isolation distance shown above, in "Public safety."
Fire: Isolate for 800 meters (½ mile) in all directions, especially if tank, rail car, or tank truck is involved in fire.

EXPOSURE
Short-term effects: *SEEK MEDICAL ATTENTION*. **Vapor:** Irritating to eyes, nose, and throat. *IF INHALED*, will, will cause coughing, difficult breathing, or loss of consciousness. *IF IN EYES*, hold eyelids open and flush with plenty of water. *IF BREATHING HAS STOPPED*, give artificial respiration. IF breathing is difficult, administer oxygen. May cause lung edema; physical exertion will aggravate this condition. **Liquid:** *POISONOUS IF SWALLOWED*. Will burn skin and eyes. *IF SWALLOWED*, will cause nausea. Remove contaminated clothing and shoes. Flush affected areas with plenty of water. *IF IN EYES*, hold eyelids open and flush with plenty of water. *IF SWALLOWED* and victim is *CONSCIOUS AND ABLE TO SWALLOW*, have victim drink 4 to 8 ounces of water. *IF SWALLOWED* and victim is *UNCONSCIOUS OR HAVING CONVULSIONS*, do nothing except keep victim warm. **Do NOT induce vomiting.**
Note to physician or authorized medical personnel: Medical observation is recommended for 24 to 48 hours after breathing overexposure, as pulmonary edema may be delayed. As first aid for pulmonary edema, consider administering a corticosteroid spray. Cigarette smoking may exacerbate pulmonary injury and should be discouraged for at least 72 hours following exposure.

HEALTH HAZARDS
Personal protective equipment (PPE): B-Level PPE. Self-contained breathing apparatus; impervious gloves; chemical safety goggles; impervious apron and boots.
Toxicity by ingestion: Grade 4; oral LD_{50} = 32 mg/kg (rat).
Vapor (gas) irritant characteristics: Eye and respiratory tract irritant.
Liquid or solid irritant characteristics: Eye and skin irritant.

FIRE DATA
Flash point: 99°F/37°C (oc).
Flammable limits in air: LEL: 1.6%; UEL: 2.3%.
Fire extinguishing agents not to be used: Water may be ineffective.
Electrical hazard: Class I, Group C.

CHEMICAL REACTIVITY
Binary reactants: Strong acids, vinyl acetates, ketones, aldehydes, phenols, cresols, caprolactum solution. Corrodes copper and its alloys in air; aluminum, especially when wet. Removes paint, swells rubber.
Reactivity group: 7
Compatibility class: Aliphatic amines

ENVIRONMENTAL DATA
Food chain concentration potential: Negative; unlikely to accumulate.
Water pollution: Effect of low concentrations on aquatic life is unknown. May be dangerous if it enters nearby water intakes; notify operators. Notify local health and wildlife officials. **Response to discharge:** Issue warning–corrosive, air contaminant, water contaminant. Restrict access. Disperse and flush.

SHIPPING INFORMATION
Grades of purity: Commercial; Pure; **Storage temperature:** Ambient; **Inert atmosphere:** None; **Venting:** Open (flame arrester); **Stability during transport:** Stable.

PHYSICAL AND CHEMICAL PROPERTIES
Physical state @ 59°F/15°C and 1 atm: Liquid.
Molecular weight: 99
Boiling point @ 1 atm: 270°F/132°C/405°K.
Melting/Freezing point: –36°F/–38°C/235°K.
Specific gravity (water = 1): 0.880 @ 68°F/20°C (liquid).
Vapor pressure: (Reid) 4.2 psia; 5 mm.

HEXAMETHYLENE TETRAMINE   REC. H:0900

SYNONYMS: AMINOFORM; AMMONIOFORMALDEHYDE; HEXA; HEXAMETILENTETRAMINA (Spanish); UROTROPIN; HEXAMINE; METHENEAMINE

IDENTIFICATION
CAS Number: 100-97-0
Formula: $C_6H_{12}N_4$
DOT ID Number: UN 1328; DOT Guide Number: 133
Proper Shipping Name: Hexamine

DESCRIPTION: White crystalline solid or powder. Mild ammonia odor. Sinks and mixes with water.

Combustible • Heat or flame may cause explosion • Highly irritating to the eyes and respiratory tract; this material has anesthetic properties. Inhalation of heated fumes can cause lung edema • Containers may BLEVE when exposed to fire • Vapors are heavier than air and will collect and stay in low areas • Vapors in confined areas (e.g., tanks, sewers, buildings) may explode when exposed to fire • Toxic products of combustion may include formaldehyde and nitrogen oxides.

Hazard Classification (based on NFPA-704 Rating System)
Health Hazards (Blue): 2; Flammability (Red): 1; Reactivity (Yellow): 0

EMERGENCY RESPONSE: See Appendix A (133)
Evacuation:
Public safety: Isolate spill area for at least 10 to 25 meters (30 to 80 feet) in all directions.
Spill: Consider initial downwind evacuation of at least 100 meters (330 feet).
Fire: Isolate for 800 meters (½ mile) in all directions, especially if tank, rail car, or tank truck is involved in fire.

EXPOSURE
Short-term effects: *SEEK MEDICAL ATTENTION.*
Vapors/fumes (heated): May cause lung edema; physical exertion will aggravate this condition. **Solid:** Irritating to skin and eyes. Remove contaminated clothing and shoes. Flush affected areas with plenty of water. *IF IN EYES*, hold eyelids open and flush with plenty of water.
Note to physician or authorized medical personnel: Medical observation is recommended for 24 to 48 hours after breathing overexposure, as pulmonary edema may be delayed. As first aid for pulmonary edema, consider administering a corticosteroid spray. Cigarette smoking may exacerbate pulmonary injury and should be discouraged for at least 72 hours following exposure.

HEALTH HAZARDS
Personal protective equipment (PPE): B-Level PPE. Gloves; for dusty or spatter conditions, use dust filter respirator and goggles.
Toxicity by ingestion: Grade 2; LD_{50} = 0.5 to 5 g/kg (human).
Long-term health effects: EPA Genetic Toxicology Program; mutagenic data. possible dermatitis.
Vapor (gas) irritant characteristics: Vapors may cause irritation of the eyes, nose, and mucous membrane.
Liquid or solid irritant characteristics: Eye and skin irritant. If spilled on clothing and allowed to remain, may cause smarting and reddening of the skin.

FIRE DATA
Flash point: 482°F/250°C (cc).
Autoignition temperature: More than 700°F/371°C/644°K.

CHEMICAL REACTIVITY
Binary reactants: Oxidizers. Violent reaction with sodium peroxide.

ENVIRONMENTAL DATA
Food chain concentration potential: Log P_{ow} = –2.19. Negative; unlikely to accumulate.
Water pollution: Effect of low concentrations on aquatic life is unknown. May be dangerous if it enters nearby water intakes; notify operators. Notify local health and wildlife officials.
Response to discharge: Disperse and flush.

SHIPPING INFORMATION

Grades of purity: Technical; USP; **Storage temperature:** Ambient; **Inert atmosphere:** None; **Venting:** Open (flame arrester); **Stability during transport:** Stable.

PHYSICAL AND CHEMICAL PROPERTIES
Physical state @ 59°F/15°C and 1 atm: Solid.
Molecular weight: 140.19
Melting/Freezing point: 536°F/280°C/553°K (sublimates).
Specific gravity (water = 1): 1.35 @ 68°F/20°C (solid).
Relative vapor density (air = 1): 5.0
Heat of combustion: –13,300 Btu/lb = –7400 cal/g = –310 x 10^5 J/kg.

n-HEXANE   REC. H:0950

SYNONYMS: EEC No. 601-037-00-0; HEXANO (Spanish); HEXYL HYDRIDE; SKELLYSOLVE-B

IDENTIFICATION
CAS Number: 110-54-3; 107-83-5 (*iso*-isomer)
Formula: C_6H_{14}; $CH_3(CH_2)_4CH_3$
DOT ID Number: UN 1208; DOT Guide Number: 128
Proper Shipping Name: Hexanes
Reportable Quantity (RQ): **(CERCLA)** 1 lb/0.454 kg

DESCRIPTION: Colorless watery liquid. Gasoline-like odor. Floats on water surface; insoluble. Large amounts of flammable, irritating vapor is produced.

Highly flammable • Containers may BLEVE when exposed to fire •Vapors may form explosive mixture with air • Vapors are heavier than air and will collect and stay in low areas • Vapors may travel long distances to ignition sources and flashback • Vapors in confined areas (e.g., tanks, sewers, buildings) may explode when exposed to fire • Irritating to the skin, eyes, and respiratory tract • Toxic products of combustion may include carbon monoxide.

Hazard Classification (based on NFPA-704 Rating System)
Health Hazards (Blue): 1; Flammability (Red): 3; Reactivity (Yellow): 0

EMERGENCY RESPONSE: See Appendix A (128)
Evacuation:
Public safety: Isolate spill area for at least 25 to 50 meters (80 to 160 feet) in all directions.
Spill: Large spill–Consider initial downwind evacuation for at least 300 meters (1000 feet).
Fire: Isolate for 800 meters (½ mile) in all directions, especially if tank, rail car, or tank truck is involved in fire.

EXPOSURE
Short-term effects: *SEEK MEDICAL ATTENTION.* **Vapor:** Irritating to nose and throat. *IF INHALED*, will, will cause coughing or dizziness. Move to fresh air. *IF BREATHING HAS STOPPED*, give artificial respiration. IF breathing is difficult, administer oxygen. **Liquid:** *HARMFUL IF SWALLOWED OR ABSORBED THROUGH THE SKIN*. Irritating to skin and eyes. *IF SWALLOWED*, will cause nausea or vomiting. May cause lung edema or pneumonia. Remove contaminated clothing and shoes. Flush affected areas with plenty of water. *IF IN EYES*, hold eyelids open and flush with plenty of water. *IF SWALLOWED* and victim is *CONSCIOUS AND ABLE TO SWALLOW*, have victim drink 4 to 8 ounces of water. **Do NOT induce vomiting.** The use of alcoholic beverages may enhance the toxic effect.

HEALTH HAZARDS

Personal protective equipment (PPE): B-Level PPE. OSHA Table Z-1-A air contaminant. Organic vapor respirator. Chemical protective material(s) reported to have good to excellent resistance: Chlorinated polyethylene, PV alcohol, nitrile rubber, polyurethane, Viton®, Teflon®, Viton®/chlorobutyl rubber, Silvershield®.

Recommendations for respirator selection: NIOSH
500 ppm: SA* (any supplied-air respirator). *1100 ppm*: SA:CF* (any supplied-air respirator operated in a continuous-flow mode); or SCBAF (any self-contained breathing apparatus with a full facepiece); or SAF (any supplied-air respirator with a full facepiece). *EMERGENCY OR PLANNED ENTRY INTO UNKNOWN CONCENTRATIONS OR IDLH CONDITIONS:* SCBAF:PD,PP (any self-contained breathing apparatus that has a full facepiece and is operated in a pressure-demand or other positive-pressure mode); or SAF:PD,PP:ASCBA (any supplied-air respirator that has a full facepiece and is operated in a pressure-demand or other positive-pressure mode in combination with an auxiliary self-contained breathing apparatus operated in a pressure-demand or other positive pressure mode). *ESCAPE:* GMFOV[any air-purifying, full-facepiece respirator (gas mask) with a chin-style, front- or back-mounted organic vapor cannister]; or SCBAE (any appropriate escape-type, self-contained breathing apparatus).
**Note:* Substance reported to cause eye irritation or damage; may require eye protection.

Exposure limits (TWA unless otherwise noted): ACGIH TLV 50 ppm (176 mg/m^3); OSHA PEL 500 ppm (1800 mg/m^3); NIOSH 50 ppm (180 mg/m^3).

Toxicity by ingestion: Very slight

Long-term health effects: May effect the nervous system.

Vapor (gas) irritant characteristics: Vapors are irritating to the eyes.

Liquid or solid irritant characteristics: Eye irritant. Degreases the skin; dryness may cause irritation.

Odor threshold: 60–245 ppm.

IDLH value: 1100 ppm.

FIRE DATA

Flash point: –7°F/–22°C (cc).

Flammable limits in air: LEL: 1.1%; UEL: 7.5%.

Behavior in fire: Vapors may explode. Vapors are heavier than air and may travel considerable distance to a source of ignition and flashback.

Autoignition temperature: 437°F/225°C/498°K.

Electrical hazard: Class I, Group D; Flow or agitation of substance may generate electrostatic charges due to low conductivity.

Burning rate: 7.3 mm/min

CHEMICAL REACTIVITY

Binary reactants: Strong oxidizers
Reactivity group: 31
Compatibility class: Paraffins

ENVIRONMENTAL DATA

Food chain concentration potential: Negative; unlikely to accumulate.

Water pollution: Effect of low concentrations on aquatic life is unknown. Fouling to shoreline. May be dangerous if it enters nearby water intakes; notify operators. Notify local health and wildlife officials. **Response to discharge:** Issue warning–high flammability. Evacuate area. Disperse and flush.

SHIPPING INFORMATION

Grades of purity: Research grade; technical grade; **Storage temperature:** Ambient; **Inert atmosphere:** None; **Venting:** Open (flame arrester) or pressure-vacuum; **Stability during transport:** Stable.

NAS HAZARD CLASSIFICATION FOR BULK WATER TRANSPORTATION

FIRE: 3
HEALTH: Vapor irritant: 0; Liquid or solid irritant: 0; Poisons: 1
WATER POLLUTION: Human toxicity: 1; Aquatic toxicity: 1; Aesthetic effect: 1
REACTIVITY: Other chemicals: 0; Water: 0; Self-reaction: 0

PHYSICAL AND CHEMICAL PROPERTIES

Physical state @ 59°F/15°C and 1 atm: Liquid.
Molecular weight: 86.17
Boiling point @ 1 atm: 155.7°F/68.7°C/341.9°K.
Melting/Freezing point: –219.3°F/–139.6°C/133.6°K.
Critical temperature: 453.6°F/234.2°C/507.4°K.
Critical pressure: 436.6 psia = 29.7 atm = 3.01 MN/m^2.
Specific gravity (water = 1): 0.659 @ 68°F/20°C (liquid).
Liquid surface tension: 18.4 dynes/cm = 0.0184 N/m @ 68°F/20°C.
Liquid water interfacial tension: 51.1 dynes/cm = 0.0511 N/m @ 68°F/20°C.
Relative vapor density (air = 1): 3.0
Ratio of specific heats of vapor (gas): 1.063
Latent heat of vaporization: 144 Btu/lb = 80.0 cal/g = 3.35 x 10^5 J/kg.
Heat of combustion: –19,246 Btu/lb = –10,692 cal/g = –447.65 x 10^5 J/kg.
Heat of fusion: 36.27 cal/g.
Vapor pressure: (Reid) 5.0 psia; 124 mm.

HEXANOIC ACID   REC. H:1000

SYNONYMS: ACIDO HEXANOICO (Spanish); BUTYLACETIC ACID; *n*-CAPROIC ACID; CAPRONIC ACID; HEXACID 698; *n*-HEXANOIC ACID; *n*-HEXOIC ACID; PENTIFORMIC ACID; PENTYLFORMIC ACID

IDENTIFICATION

CAS Number: 142-62-1
Formula: $C_6H_{12}O_2$; $CH_3(CH_2)_4CO_2H$
DOT ID Number: UN 2829; DOT Guide Number: 153
Proper Shipping Name: Caproic acid

DESCRIPTION: Colorless or slightly yellow oily liquid. Limburger cheese odor. Floats; practically insoluble in water.

Combustible • Heat or flame may cause explosion • Containers may BLEVE when exposed to fire • Vapors are heavier than air and will collect and stay in low areas • Vapors in confined areas (e.g., tanks, sewers, buildings) may explode when exposed to fire • Corrosive to the skin, eyes, and respiratory tract • Toxic products of combustion may include carbon monoxide.

Hazard Classification (based on NFPA-704 Rating System)
Health Hazards (Blue): 2; Flammability (Red): 1; Reactivity (Yellow): 0

EMERGENCY RESPONSE: See Appendix A (153)
Evacuation:
Public safety: Isolate the area of spill or leak for at least 25 to 50 meters (80 to 160 feet) in all directions.

Spill: Increase, in the downwind direction, as necessary, the distance shown under "Public Safety."
Fire: If tank, rail car, or tank truck is involved in fire, isolate for at least 800 meters (½ mile) in all directions; also, consider initial evacuation for 800 meters (½ mile) in all directions.

EXPOSURE
Short-term effects: *SEEK MEDICAL ATTENTION.* **Vapor:** Irritating to eyes, nose, and throat. *IF INHALED*, will, will cause coughing or difficult breathing. *IF IN EYES*, hold eyelids open and flush with plenty of water. *IF BREATHING HAS STOPPED,* give artificial respiration; *avoid mouth-to-mouth resuscitation; use bag/mask apparatus.* IF breathing is difficult, administer oxygen. **Liquid:** Will burn skin and eyes. *IF SWALLOWED*, will cause nausea and vomiting. Remove contaminated clothing and shoes. Flush affected areas, with plenty of water. *IF IN EYES*, hold eyelids open and flush with plenty of water. The use of alcoholic beverages may enhance toxic effects.

HEALTH HAZARDS
Personal protective equipment (PPE): B-Level PPE. Respirator, chemical safety goggles, rubber boots and heavy rubber gloves.
Toxicity by ingestion: Grade 2: LD_{50} = 3 g/kg (rat).
Long-term health effects: Possibly mutagenic.
Vapor (gas) irritant characteristics: Vapors cause severe irritation of eyes and throat and can cause eye and lung injury. They cannot be tolerated even at low concentrations.
Liquid or solid irritant characteristics: Fairly severe eye and skin irritant. May cause pain and second-degree burns after a few minutes of contact.

FIRE DATA
Flash point: 215°F/102°C (oc).
Fire extinguishing agents not to be used: Water may not be effective.
Autoignition temperature: 716°F/380°C/650°K.

CHEMICAL REACTIVITY
Binary reactants: Reacts with oxidizing materials. Corrosive, attacks most common metals.
Neutralizing agents for acids and caustics: Sodium bicarbonate solution
Reactivity group: 4
Compatibility class: Organic acids

ENVIRONMENTAL DATA
Food chain concentration potential: Log P_{ow} = 1.91. Unlikely to accumulate.
Water pollution: May be dangerous to aquatic life in high concentrations. May be dangerous if it enters nearby water intakes; notify operators. Notify local health and wildlife officials.
Response to discharge: Issue warning–corrosive. Mechanical containment. Should be removed. Chemical and physical treatment.

SHIPPING INFORMATION
Grades of purity: 99.5+%; **Storage temperature:** Ambient; **Stability during transport:** Stable.

PHYSICAL AND CHEMICAL PROPERTIES
Physical state @ 59°F/15°C and 1 atm: Liquid.
Molecular weight: 110.16
Boiling point @ 1 atm: 395.6–397°F/202–203°C/475–476°K.
Melting/Freezing point: 27°F/–3°C/270°K.
Specific gravity (water = 1): 0.927
Relative vapor density (air = 1): 4.0
Vapor pressure: (Reid) Very low.

1-HEXANOL   REC. H:1050

SYNONYMS: ALCOHOL C-6; *n*-AMYL CARBINOL; AMYL CARBINOL; CAPROYL ALCOHOL; EEC No. 603-059-00-6; EPAL-6; HEXANOL; *n*-HEXANOL; HEXYL ALCOHOL; *n*-HEXYL ALCOHOL; 1-HYDROXYHEXANE; PENTYL CARBINOL

IDENTIFICATION
CAS Number: 111-27-3
Formula: $C_6H_{14}O$; $CH_3(CH_2)_4CH_2OH$
DOT ID Number: UN 2282; DOT Guide Number: 129
Proper Shipping Name: Hexanols

DESCRIPTION: Clear colorless liquid. Sweet odor. Floats on the surface of water.

Highly flammable • Containers may BLEVE when exposed to fire • Vapors may form explosive mixture with air • Vapors are heavier than air and will collect and stay in low areas • Vapors may travel long distances to ignition sources and flashback • Vapors in confined areas (e.g., tanks, sewers, buildings) may explode when exposed to fire • Irritating to the skin, eyes, and respiratory tract • Toxic products of combustion may include carbon monoxide.

Hazard Classification (based on NFPA-704 Rating System)
Health Hazards (Blue): 2; Flammability (Red): 3; Reactivity (Yellow): 1

EMERGENCY RESPONSE: See Appendix A (129)
Evacuation:
Public safety: Isolate spill area for at least 50 to 100 meters (160 to 330 feet) in all directions.
Spill: Large spill–Consider initial downwind evacuation for at least 300 meters (1000 feet).
Fire: Isolate for 800 meters (½ mile) in all directions, especially if tank, rail car, or tank truck is involved in fire.

EXPOSURE
Short-term effects: *SEEK MEDICAL ATTENTION.* **Liquid:** Will burn eyes. Harmful if swallowed. The use of alcoholic beverages may enhance the toxic effect. Remove contaminated clothing and shoes. Flush affected areas with plenty of water. *IF IN EYES*, hold eyelids open and flush with plenty of water. *IF SWALLOWED* and victim is *CONSCIOUS AND ABLE TO SWALLOW*, have victim drink 4 to 8 ounces of water. The use of alcoholic beverages may enhance toxic effects.

HEALTH HAZARDS
Personal protective equipment (PPE): B-Level PPE. Chemical gloves, chemical goggles; face shield. Natural rubber, neoprene, and nitrile rubber may offer some protection from alcohols. Chemical protective material(s) reported to offer minimal to poor protection: Viton®/neoprene, butyl rubber,/neoprene.
Exposure limits (TWA unless otherwise noted): ACGIH TLV 500 ppm (1760 mg/m³) as hexane, other isomers; NIOSH 100 ppm (350 mg/m³) as hexane isomers.
Short-term exposure limits (15-minute TWA): ACGIH STEL 1000 ppm (3500 mg/m³) as hexane, other isomers; NIOSH ceiling (15 minute) 510 ppm (1800 mg/m³) as hexane isomers.
Toxicity by ingestion: Grade 2; LD_{50} = 0.5 to 5 g/kg (rat).

Long-term health effects: May effect the nervous system.
Vapor (gas) irritant characteristics: Irritates the eyes, skin, and respiratory tract.
Liquid or solid irritant characteristics: Causes smarting of the skin and first-degree burns on short exposure; may cause second-degree burns on long exposure.

FIRE DATA
Flash point: 149°F/65°C (oc); 145°F/63°C (cc).
Flammable limits in air: LEL: 1.2%; UEL: 7.7% (calculated).
Autoignition temperature: 554°F/290°C/563°K.
Electrical hazard: Class I, Group D.

CHEMICAL REACTIVITY
Binary reactants: Violent reaction with strong oxidizers.
Reactivity group: 20
Compatibility class: Alcohols, glycols

ENVIRONMENTAL DATA
Food chain concentration potential: Log P_{ow} = 1.9. Unlikely to accumulate.
Water pollution: Effect of low concentrations on aquatic life is unknown. Fouling to shoreline. May be dangerous if it enters nearby water intakes; notify operators. Notify local health and wildlife officials. **Response to discharge:** Mechanical containment. Chemical and physical treatment.

SHIPPING INFORMATION
Grades of purity: 99+%; **Storage temperature:** Ambient; **Inert atmosphere:** None; **Venting:** Open (flame arrester); **Stability during transport:** Stable.

PHYSICAL AND CHEMICAL PROPERTIES
Physical state @ 59°F/15°C and 1 atm: Liquid.
Molecular weight: 102.18
Boiling point @ 1 atm: 314.8°F/157.1°C/430.3°K.
Melting/Freezing point: –48.3°F/–44.6°C/228.6°K.
Critical temperature: 638.6°F/337°C/610.2°K.
Critical pressure: 485 psia = 33 atm = 3.34 MN/m².
Specific gravity (water = 1): 0.850 @ 68°F/20°C (liquid).
Liquid surface tension: 24.5 dynes/cm = 0.0245 N/m @ 68°F/20°C.
Liquid water interfacial tension: 6.8 dynes/cm = 0.0068 N/m @ 77°F/25°C.
Relative vapor density (air = 1): 3.5
Ratio of specific heats of vapor (gas): 1.057
Latent heat of vaporization: 209 Btu/lb = 116 cal/g = 4.86 x 10^5 J/kg.
Heat of combustion: –16,810 Btu/lb = –9340 cal/g = –391.0 x 10^5 J/kg.
Vapor pressure: (Reid) 0.75 psia; 1 mm.

1-HEXENE   REC. H:1100

SYNONYMS: BUTYL ETHYLENE; HEXENE; α-HEXENE; 1-HEXENEO (Spanish); HEXYLENE

IDENTIFICATION
CAS Number: 592-41-6
Formula: C_6H_{12}; $CH_3(CH_2)_3CH=CH_2$
DOT ID Number: UN 2370; DOT Guide Number: 128
Proper Shipping Name: 1-Hexene

DESCRIPTION: Colorless watery liquid. Mild pleasant odor. Floats on the surface of water. Flammable, irritating vapor is produced.

Highly flammable • Containers may BLEVE when exposed to fire • Vapors may form explosive mixture with air • Vapors are heavier than air and will collect and stay in low areas • Vapors may travel long distances to ignition sources and flashback • Vapors in confined areas (e.g., tanks, sewers, buildings) may explode when exposed to fire • Irritating to the skin, eyes, and respiratory tract • Toxic products of combustion may include carbon monoxide.

Hazard Classification (based on NFPA-704 Rating System)
Health Hazards (Blue): 1; Flammability (Red): 3; Reactivity (Yellow): 0

EMERGENCY RESPONSE: See Appendix A (128)
Evacuation:
Public safety: Isolate spill area for at least 25 to 50 meters (80 to 160 feet) in all directions.
Spill: Large spill–Consider initial downwind evacuation for at least 300 meters (1000 feet).
Fire: Isolate for 800 meters (½ mile) in all directions, especially if tank, rail car, or tank truck is involved in fire.

EXPOSURE
Short-term effects: *SEEK MEDICAL ATTENTION.* **Vapor:** If inhaled, will cause dizziness, difficult breathing, or loss of consciousness. Move to fresh air. *IF BREATHING HAS STOPPED,* give artificial respiration. IF breathing is difficult, administer oxygen. **Liquid:** Irritating to skin and eyes. Harmful if swallowed. Remove contaminated clothing and shoes. Flush affected areas with plenty of water. *IF IN EYES,* hold eyelids open and flush with plenty of water. *IF SWALLOWED* and victim is *CONSCIOUS AND ABLE TO SWALLOW,* have victim drink 4 to 8 ounces of water. **Do NOT induce vomiting.** The use of alcoholic beverages may enhance toxic effects.

HEALTH HAZARDS
Personal protective equipment (PPE): B-Level PPE. Approved organic vapor respirator or air-line mask; protective goggles or face shield. Chemical protective material(s) reported to offer minimal to poor protection: Viton®/neoprene.
Exposure limits (TWA unless otherwise noted): ACGIH TLV 30 ppm.
Vapor (gas) irritant characteristics: Slight smarting of the eyes or respiratory system if present in high concentrations. Effect is temporary.

FIRE DATA
Flash point: –14°F/–26°C (cc).
Fire extinguishing agents not to be used: Water may be ineffective
Autoignition temperature: 487°F/253°C/526°K.
Burning rate: 8.1 mm/min

CHEMICAL REACTIVITY
Binary reactants: Violent reaction with oxidizers.
Reactivity group: 30
Compatibility class: Olefins

ENVIRONMENTAL DATA
Food chain concentration potential: Negative; unlikely to accumulate.
Water pollution: Effect of low concentrations on aquatic life is unknown. Fouling to shoreline. May be dangerous if it enters

nearby water intakes; notify operators. Notify local health and wildlife officials. **Response to discharge:** Issue warning–high flammability. Evacuate area. Disperse and flush.

SHIPPING INFORMATION
Grades of purity: Technical, 95-98%; Pure, 99+%; **Storage temperature:** Ambient; **Inert atmosphere:** None; **Venting:** Open (flame arrester) or pressure-vacuum; **Stability during transport:** Stable.

PHYSICAL AND CHEMICAL PROPERTIES
Physical state @ 59°F/15°C and 1 atm: Liquid.
Molecular weight: 84.16
Boiling point @ 1 atm: 146°F/64°C/337°K.
Melting/Freezing point: −220°F/−140°C/134°K.
Critical temperature: 447°F/231°C/504°K.
Critical pressure: 460 psia = 31.3 atm = 3.17 MN/m².
Specific gravity (water = 1): 0.673 @ 68°F/20°C (liquid).
Liquid surface tension: 18.8 dynes/cm = 0.0188 N/m @ 68°F/20°C.
Liquid water interfacial tension: 31.6 dynes/cm = 0.0316 N/m at 22.7°C.
Relative vapor density (air = 1): 2.9
Ratio of specific heats of vapor (gas): 1.068
Latent heat of vaporization: 140 Btu/lb = 80 cal/g = 3.3 x 10^5 J/kg.
Heat of combustion: −19,134 Btu/lb = −10,630 cal/g = −445.06 x 10^5 J/kg.

HEXYL ACETATE   REC. H:1150

SYNONYMS: ACETATO de *sec*-HEXILO (Spanish); ACETIC ACID, HEXYL ESTER; *n*-HEXYL ACETATE; 1-HEXYL ACETATE; HEXYL ALCOHOL, ACETATE; HEXYL ETHANOATE; METHYLAMYL ACETATE

IDENTIFICATION
CAS Number: 142-92-7
Formula: $C_8H_{16}O_2$; $CH_3CO_2C_6H_{13}$
DOT ID Number: NA 1993; DOT Guide Number: 128
Proper Shipping Name: Combustible liquid, n.o.s.

DESCRIPTION: Colorless, oily liquid. Fruity odor. Initially appears as a slick; Insoluble in water.

Flammable • Containers may BLEVE when exposed to fire • Vapors may form explosive mixture with air • Vapors are heavier than air and will collect and stay in low areas • Vapors may travel long distances to ignition sources and flashback • Vapors in confined areas (e.g., tanks, sewers, buildings) may explode when exposed to fire • Irritating to the skin, eyes, and respiratory tract • Toxic products of combustion may include carbon monoxide.

Hazard Classification (based on NFPA-704 Rating System)
Health Hazards (Blue): 1; Flammability (Red): 2; Reactivity (Yellow): 0

EMERGENCY RESPONSE: See Appendix A (128)
Evacuation:
Public safety: Isolate spill area for at least 25 to 50 meters (80 to 160 feet) in all directions.
Spill: Large spill–Consider initial downwind evacuation for at least 300 meters (1000 feet).

Fire: Isolate for 800 meters (½ mile) in all directions, especially if tank, rail car, or tank truck is involved in fire.

EXPOSURE
Short-term effects: *SEEK MEDICAL ATTENTION.* **Vapor:** May be harmful. Move to fresh air. *IF BREATHING HAS STOPPED,* give artificial respiration. IF breathing is difficult, administer oxygen. **Liquid:** Irritating to skin and eyes. Remove contaminated clothing and shoes. Flush affected areas with plenty of water. *IF IN EYES*, hold eyelids open and flush with plenty of water.

HEALTH HAZARDS
Personal protective equipment (PPE): B-Level PPE. Self-contained breathing apparatus, rubber boots and heavy rubber gloves.
Toxicity by ingestion: Mildly toxic. Grade 0: LD_{50} = 42 g/kg (rat).
Vapor (gas) irritant characteristics: Vapors cause smarting of the eyes or respiratory system if present in high concentrations. The effect is temporary.
Liquid or solid irritant characteristics: May cause eye irritation. If spilled on clothing and allowed to remain, may cause smarting and reddening of skin.

FIRE DATA
Flash point: 113°F/45°C (cc).
Fire extinguishing agents not to be used: Water may not be effective.
Behavior in fire: Container explosion may occur under fire conditions.

CHEMICAL REACTIVITY
Binary reactants: Strong acids, nitrates, strong oxidizers, strong caustics.
Reactivity group: 34
Compatibility class: Esters.

ENVIRONMENTAL DATA
Food chain concentration potential: Negative; unlikely to accumulate.
Water pollution: Effect of low concentrations on aquatic life is unknown. Fouling to shoreline May be dangerous if it enters nearby water intakes; notify operators. Notify local health and wildlife officials. **Response to discharge:** Restrict access. Should be removed. Mechanical containment. Chemical and physical treatment.

SHIPPING INFORMATION
Grades of purity: 99%; **Stability during transport:** Stable.

PHYSICAL AND CHEMICAL PROPERTIES
Physical state @ 59°F/15°C and 1 atm: Liquid.
Molecular weight: 144.21
Boiling point @ 1 atm: 285°F/141°C/414°K.
Melting/Freezing point: −112°F/−80°C/193.2°K.
Specific gravity (water = 1): 0.876
Relative vapor density (air = 1): 4.97

HEXYLENE GLYCOL   REC. H:1200

SYNONYMS: 2,4-DIHYDROXY-2-METHYLPENTANE; DIOLANE; EEC No. 603-053-00-3; 1,2-HEXANEDIOL; HEXALENGLICOL (Spanish); ISOL; 2-METHYL-2,4-PENTANEDIOL; 4-METHYL-2,4-PENTANEDIOL;

2-METHYLPENTANE-2,4-DIOL; 2,4-PENTANEDIOL, 2-METHYL-; PINAKON; α,α,α'-TRIMETHYLENE GLYCOL

IDENTIFICATION
CAS Number: 107-41-5
Formula: $C_6H_{14}O_2$; $(CH_3)_2OHCH_2CH(OH)CH_3$
DOT ID Number: NA 1993; DOT Guide Number: 128
Proper Shipping Name: Combustible liquid, n.o.s.

DESCRIPTION: Colorless, viscous liquid. Mild sweet odor. Floats and mixes slowly with water.

Combustible • Corrosive to skin, eyes, and lungs • Containers may BLEVE when exposed to fire • Vapors are heavier than air and will collect and stay in low areas • Vapors in confined areas (e.g., tanks, sewers, buildings) may explode when exposed to fire • Toxic products of combustion may include carbon monoxide.

Hazard Classification (based on NFPA-704 Rating System)
Health Hazards (Blue): 1; Flammability (Red): 1; Reactivity (Yellow): 0

EMERGENCY RESPONSE: See Appendix A (128)
Evacuation:
Public safety: Isolate spill area for at least 25 to 50 meters (80 to 160 feet) in all directions.
Spill: Large spill–Consider initial downwind evacuation for at least 300 meters (1000 feet).
Fire: Isolate for 800 meters (½ mile) in all directions, especially if tank, rail car, or tank truck is involved in fire.

EXPOSURE
Short-term effects: *SEEK MEDICAL ATTENTION*. **Liquid:** Irritating to skin and eyes. Remove contaminated clothing and shoes. Flush affected areas with plenty of water. *IF IN EYES*, hold eyelids open and flush with plenty of water.

HEALTH HAZARDS
Personal protective equipment (PPE): B-Level PPE. OSHA Table Z-1-A air contaminant. Organic canister; or air pack; rubber gloves; goggles. Chemical protective material(s) reported to offer minimal to poor protection: Viton®/neoprene, butyl rubber/neoprene.
Exposure limits (TWA unless otherwise noted): NIOSH REL ceiling 25 ppm (125 mg/m^3).
Short-term exposure limits (15-minute TWA): ACGIH TLV ceiling 25 ppm (121 mg/m^3).
Toxicity by ingestion: Grade 2; LD_{50} = 0.5 to 5 g/kg.
Long-term health effects: May effect the nervous system. Mutation data. Eye damage may be slow to heal.
Vapor (gas) irritant characteristics: Vapors cause a smarting of the eyes, skin, and respiratory system. May have systemic effects.
Liquid or solid irritant characteristics: Eye irritant; corneal damage. Degreases the skin.
Odor threshold: 50 ppm.

FIRE DATA
Flash point: 215°F/102°C (oc); 209°F/98°C (cc).
Flammable limits in air: LEL: 1.3% (calculated); UEL: 7.4% (estimated).
Fire extinguishing agents not to be used: Water or foam may cause frothing.
Autoignition temperature: 583°F/306°C/579°K (calculated).
Electrical hazard: Class I, Group D.

CHEMICAL REACTIVITY
Binary reactants: Strong oxidizers, strong acids. Hygroscopic (absorbs moisture from the air).
Reactivity group: 20
Compatibility class: Alcohols, glycols

ENVIRONMENTAL DATA
Food chain concentration potential: Log P_{ow} = –0.1. Negative; unlikely to accumulate.
Water pollution: Effect of low concentrations on aquatic life is unknown. May be dangerous if it enters nearby water intakes; notify operators. Notify local health and wildlife officials. **Response to discharge:** Disperse and flush.

SHIPPING INFORMATION
Grades of purity: 99%; **Storage temperature:** Ambient; **Inert atmosphere:** None; **Venting:** Open (flame arrester); **Stability during transport:** Stable.

NAS HAZARD CLASSIFICATION FOR BULK WATER TRANSPORTATION
FIRE: 1
HEALTH: Vapor irritant: 1; Liquid or solid irritant: 0; Poisons: 0
WATER POLLUTION: Human toxicity: 1; Aquatic toxicity: 2; Aesthetic effect: 1
REACTIVITY: Other chemicals: 2; Water: 0; Self-reaction: 0

PHYSICAL AND CHEMICAL PROPERTIES
Physical state @ 59°F/15°C and 1 atm: Liquid.
Molecular weight: 118.19
Boiling point @ 1 atm: 387°F/197°C/470°K.
Melting/Freezing point: –58°F/–50°C/223°K.
Critical temperature: 752°F/400°C/673°K.
Critical pressure: 497 psia = 33.8 atm = 3.42 MN/m^2.
Specific gravity (water = 1): 0.923 @ 68°F/20°C (liquid).
Latent heat of vaporization: 187 Btu/lb = 104 cal/g = 4.35 x 10^5 J/kg.
Heat of combustion: (estimate) –13,600 Btu/lb = –7550 cal/g = –316 x 10^5 J/kg.
Heat of solution: (estimate) –11 Btu/lb = –6 cal/g = –0.25 x 10^5 J/kg.
Vapor pressure: 0.05 mm.

HYDRAZINE   REC. H:1250

SYNONYMS: DIAMIDE; DIAMINE; DIAMINE, HYDRAZINE BASE; HIDRAZINA (Spanish); HYDRAZINE, ANHYDROUS; HYDRAZINE BASE; RCRA No. U133

IDENTIFICATION
CAS Number: 302-01-2; 7803-57-8 (15% solution)
Formula: H_4N_2; H_2NNH_2
DOT ID Number: UN 2029 (anhydrous or aqueous solution >64%); UN 2030 (hydrate or solutions with not less than 37%, but not; DOT Guide Number: 132 (UN 2029); 153 (UN 2030); 152 (UN 3293). >64%); UN 3293 (solution not >37% hydrazine)
Proper Shipping Name: Hydrazine, anhydrous; Hydrazine aqueous solutions with more than 64% hydrazine, by mass; Hydrazine aqueous solution, with not less than 37% but not more than 64% hydrazine; Hydrazine, aqueous solution with not more than 37% hydrazine
Reportable Quantity (RQ): **(CERCLA)** 1 lb/0.454 kg

Hydrazine

DESCRIPTION: Colorless to slightly yellow, oily, fuming liquid. Ammonia odor; fishy. Mixes with water; poisonous, flammable vapor is produced. freezes at 35°F/1.6°C.

Poison! • Highly flammable • Highly reactive; a powerful reducing agent. Ignites spontaneously upon contact with porous materials such as wood, cloth, soil, or rusting metals • Corrosive; breathing the vapor, skin or eye contact, or swallowing the material can kill you; Skin and eye contact can cause severe burns, vision impairment, and blindness • Firefighting gear (including SCBA) does not provide adequate protection. If exposure occurs, remove and isolate gear immediately and thoroughly decontaminate personnel • Containers may BLEVE when exposed to fire • Vapors may form explosive mixture with air • Vapors are heavier than air and will collect and stay in low areas • Vapors may travel long distances to ignition sources and flashback • Vapors in confined areas (e.g., tanks, sewers, buildings) may explode when exposed to fire • Toxic products of combustion may include nitrogen oxide • Do not put yourself in danger by entering a contaminated area to rescue a victim.

Hazard Classification (based on NFPA-704 Rating System)
Health Hazards (Blue): 3; Flammability (Red): 3; Reactivity (Yellow): 3

EMERGENCY RESPONSE, liquid: See Appendix A (132)
Evacuation:
Public safety: Isolate spill area for at least 50 to 100 meters (160 to 330 feet) in all directions.
Spill: Increase, as necessary, the isolation distance shown above, in "Public safety."
Fire: Isolate for 800 meters (½ mile) in all directions, especially if tank, rail car, or tank truck is involved in fire.

EMERGENCY RESPONSE: See Appendix A (152, 153)
Evacuation:
Public safety: Isolate the area of spill or leak for at least 25 to 50 meters (80 to 160 feet) in all directions.
Spill: Increase, in the downwind direction, as necessary, the distance shown under "Public Safety."
Fire: If tank, rail car, or tank truck is involved in fire, isolate for at least 800 meters (½ mile) in all directions; also, consider initial evacuation for 800 meters (½ mile) in all directions.

EXPOSURE
Short-term effects: *SEEK MEDICAL ATTENTION.* **Vapor:** *POISONOUS IF INHALED OR IF SKIN IS EXPOSED.* May cause lung edema; physical exertion will aggravate this condition. Irritating to eyes. Move to fresh air. *IF BREATHING HAS STOPPED,* give artificial respiration. IF breathing is difficult, administer oxygen. **Liquid:** *POISONOUS IF SWALLOWED OR IF SKIN IS EXPOSED.* Will burn eyes. Remove contaminated clothing and shoes. Flush affected areas with plenty of water. *IF IN EYES,* hold eyelids open and flush with plenty of water. *IF SWALLOWED* and victim is *CONSCIOUS AND ABLE TO SWALLOW,* have victim drink 4 to 8 ounces of water. **Do NOT induce vomiting. Ingestion or absorption:** Ingestion or absorption through skin causes nausea, dizziness, headache. Severe exposure may cause death.
Note to physician or authorized medical personnel: Medical observation is recommended for 24 to 48 hours after breathing overexposure, as pulmonary edema may be delayed. As first aid for pulmonary edema, consider administering a corticosteroid spray. Cigarette smoking may exacerbate pulmonary injury and should be discouraged for at least 72 hours following exposure.

HEALTH HAZARDS
Personal protective equipment (PPE): A-Level PPE. OSHA Table Z-1-A air contaminant. Chemical protective material(s) reported to have good to excellent resistance: butyl rubber, chlorobutyl rubber, neoprene, nitrile, PVC, nitrile, Teflon®, Saranex®. Also, natural rubber offers limited protection.
Recommendations for respirator selection: NIOSH
At any concentrations above the NIOSH REL, or where there is no REL, at any detectable concentration: SCBAF:PD,PP (any self-contained breathing apparatus that has a full facepiece and is operated in a pressure-demand or other positive-pressure mode); or SAF:PD,PP:ASCBA (any supplied-air respirator that has a full facepiece and is operated in a pressure-demand or other positive-pressure mode in combination with an auxiliary self-contained breathing apparatus operated in a pressure-demand or other positive pressure mode). *ESCAPE:* SCBAE (any appropriate escape-type, self-contained breathing apparatus).
Exposure limits (TWA unless otherwise noted): ACGIH TLV 0.01 ppm (0.013 mg/m^3) animal carcinogen; OSHA PEL 1 ppm (1.3 mg/m^3); NIOSH REL ceiling 0.03 ppm (0.04 mg/m^3)/2 hours; potential human carcinogen; reduce exposure to lowest feasible level; skin contact contributes significantly in overall exposure.
Toxicity by ingestion: Grade 3; LD_{50} = 50 to 500 mg/kg (rat).
Long-term health effects: Causes lung cancer in mice. Suspected human carcinogen. NTP anticipated carcinogen. May cause blood, liver, kidney, and brain damage. Allergic eczema may be caused by intensive contact.
Vapor (gas) irritant characteristics: Eye and respiratory tract irritant. Vapor is irritating such that personnel will not usually tolerate moderate or high vapor concentrations.
Liquid or solid irritant characteristics: Severe eye, skin and respiratory tract irritant. Severe skin irritant; causes second-and third-degree burns on short contact; very injurious to the eyes.
Odor threshold: 3.5 ppm. *Note*: Higher than the Exposure limits.
IDLH value: Potential human carcinogen. 50 ppm.

FIRE DATA
Flash point: 100°F/38°C (cc).
Flammable limits in air: LEL: 2.9%; UEL 98%.
Behavior in fire: May explode if confined.
Autoignition temperature: 518°F/270°C/543°K.
Electrical hazard: Class I, Group C.
Burning rate: 1 mm/min (estimate).

CHEMICAL REACTIVITY
Binary reactants: A strong reducing agent and a strong base. Severe explosion hazard; with many compounds. Reacts, possibly violently, with oxidizers, halogens, hydrogen peroxide, nitric acid, metallic oxides, alkali metals, chromates, acids. Once dry, this material can ignite spontaneously in porous materials such as wood, asbestos, cloth, earth and rusty metals. Corrosive to glass, rubber, and chemically active metals such as aluminum and zinc.
Neutralizing agents for acids and caustics: Flush with water. Neutralize the resulting solution with calcium hypochlorite (7 lb per pound of hydrazine).

ENVIRONMENTAL DATA
Food chain concentration potential: Log P_{ow} = –1.2. Unlikely to accumulate.
Water pollution: Harmful to aquatic life in very low concentrations. May be dangerous if it enters nearby water intakes; notify operators. Notify local health and wildlife officials. **Response to discharge:** Issue warning–high flammability, corrosive. Restrict access. Chemical and physical treatment.

SHIPPING INFORMATION
Grades of purity: Anhydrous; 35–64% water solutions; **Storage temperature:** Ambient; **Inert atmosphere:** Padded; **Venting:** Pressure-vacuum; **Stability during transport:** Keep away from heat; dangerous explosion hazard. Stable at ordinary temperatures. When heated, can decompose to nitrogen and ammonia gases, decomposition is extremely hazardous if material is confined.

NAS HAZARD CLASSIFICATION FOR BULK WATER TRANSPORTATION
FIRE: 4
HEALTH: Vapor irritant: 3; Liquid or solid irritant: 4; Poisons: 4
WATER POLLUTION: Human toxicity: 4; Aquatic toxicity: 3; Aesthetic effect: 2
REACTIVITY: Other chemicals: 4; Water: 0; Self-reaction: 4

PHYSICAL AND CHEMICAL PROPERTIES
Physical state @ 59°F/15°C and 1 atm: Liquid.
Molecular weight: 32.05; 50.1 (15% solution)
Boiling point @ 1 atm: 236.3°F/113.5°C/386.7°K; 215°F/102°C/375°K (15% solution)
Melting/Freezing point: 34.7°F/1.5°C/274.7°K; 7°F/–14°C/259°K (15% solution)
Critical temperature: 716°F/380°C/653°K.
Critical pressure: 2130 psia = 145 atm = 14.7 MN/m^2.
Specific gravity (water = 1): 1.008 @ 68°F/20°C (liquid).
Relative vapor density (air = 1): 1.1
Ratio of specific heats of vapor (gas): 1.191
Latent heat of vaporization: 538 Btu/lb = 299 cal/g = 12.5 x 10^5 J/kg.
Heat of combustion: –8345 Btu/lb = –4636 cal/g = –194.1 x 10^5 J/kg.
Heat of solution: –218 Btu/lb = –121 cal/g = –5.07 x 10^5 J/kg.
Vapor pressure: 10 mm.

HYDROFLUOROSILICIC ACID   REC. H:1300

SYNONYMS: ACIDO FLUOSILICICO (Spanish); FLUOROSILICIC ACID; FLUOSILICIC ACID; HEXAFLUOSILICIC ACID; SAND ACID; SILICOFLUORIC ACID

IDENTIFICATION
CAS Number: 16961-83-4
Formula: H_2SiF_6; $F_6Si \cdot 2H$
DOT ID Number: UN 1778; DOT Guide Number: 154
Proper Shipping Name: Fluorosilicic acid; Fluosilicic acid

DESCRIPTION: Colorless to straw yellow, fuming liquid. None to slight acrid odor. Reacts with water producing toxic and corrosive fumes.

Corrosive• Firefighting gear (including SCBA) does not provide adequate protection. If exposure occurs, remove and isolate gear immediately and thoroughly decontaminate personnel • Toxic products of combustion may include fluorides • Severely Irritating to the skin, eyes, and respiratory tract; effects of contact or inhalation may be delayed • Do not put yourself in danger by entering a contaminated area to rescue a victim.

Hazard Classification (based on NFPA-704 Rating System)
Health Hazards (Blue): 4; Flammability (Red): 0; Reactivity (Yellow): 2; Special Notice (White): Water reactive

EMERGENCY RESPONSE: See Appendix A (154)
Evacuation:
Public safety: Isolate the area of spill or leak for at least 25 to 50 meters (80 to 160 feet) in all directions.
Spill: Increase, in the downwind direction, as necessary, the distance shown under "Public Safety."
Fire: If tank, rail car, or tank truck is involved in fire, isolate for at least 800 meters (½ mile) in all directions; also, consider initial evacuation for 800 meters (½ mile) in all directions.

EXPOSURE
Short-term effects: *SEEK MEDICAL ATTENTION*. **Vapor:** Move victim to fresh air. *IF BREATHING HAS STOPPED*, give artificial respiration; *avoid mouth-to-mouth resuscitation; use bag/mask apparatus*. **Liquid:** Irritating to skin and eyes. Remove contaminated clothing and shoes. Flush skin with water. *IF IN EYES*, hold eyelids open and flush with plenty of water. If swallowed and victim is *CONSCIOUS AND ABLE TO SWALLOW*, give large quantity of water followed by milk of magnesia or milk.

HEALTH HAZARDS
Personal protective equipment (PPE): B-Level PPE. Wear full impervious protective clothing and approved respirator. Where splashing is possible wear full face shield or chemical safety goggles. Use approved respirator to protect against vapors. Neoprene is generally suitable for mineral acids.
Recommendations for respirator selection: NIOSH/OSHA as F. *12.5 mg/m^3:* DM (any dust and mist respirator). *25 mg/m^3:* DMXSQ* (any dust and mist respirator except single-use and quarter-mask respirators); or SA* (any supplied-air respirator). *62.5 mg/m^3:* SA:CF* [any supplied-air respirator operated in a continuous-flow mode)]; or PAPRDM*+ *if not present as a fume* (any powered, air-purifying respirator with a dust and mist filter). *125 mg/m^3:* HiEF + (any air-purifying, full-facepiece respirator with a high-efficiency particulate filter); or SCBAF (any self-contained breathing apparatus with a full facepiece); or SAF (any supplied-air respirator with a full facepiece). *250 mg/m^3:* SA:PD,PP (any supplied-air respirator operated in a pressure-demand or other positive-pressure mode). *EMERGENCY OR PLANNED ENTRY INTO UNKNOWN CONCENTRATIONS OR IDLH CONDITIONS:* SCBAF:PD,PP (any self-contained breathing apparatus that has a full faceplate and is operated in a pressure-demand or other positive-pressure mode); or SAF:PD,PP:ASCBA (any supplied-air respirator that has a full facepiece and is operated in a pressure-demand or other positive-pressure mode in combination with an auxiliary, self-contained breathing apparatus operated in a pressure-demand or other positive-pressure mode). *ESCAPE:* HiEF+ (any air-purifying, full-facepiece respirator with a high-efficiency particulate filter); or SCBAE (any appropriate escape-type, self-contained breathing apparatus). *Notes:* *Substance reported to cause eye irritation or damage; may require eye protection. +May need acid gas sorbent.
Exposure limits (TWA unless otherwise noted): ACGIH TLV 2.5 mg/m^3 as fluorides; NIOSH/OSHA 2.5 mg/m^3 as inorganic fluorides.
Toxicity by ingestion: Grade 3; oral guinea pig LD$_{50}$ = 200 mg/kg.
Long-term health effects: Kidneys, liver, and lungs may be affected by exposures. Osteofluorosis is softening of the bones.
Vapor (gas) irritant characteristics: Eye and respiratory tract irritant. Vapors are irritating such that personnel will not usually tolerate moderate or high concentrations.
Liquid or solid irritant characteristics: Fairly severe eye and skin irritant. May cause pain and second-degree burns after a few minutes of contact.

FIRE DATA
Fire extinguishing agents not to be used: Water. Avoid direct contact between water and acid.

CHEMICAL REACTIVITY
Reactivity with water: Reacts with generation of heat.
Binary reactants: Can react with strong acids to release hydrogen fluoride fumes. Will react with metals to release flammable hydrogen gas. Will attack glass and materials containing silica.
Neutralizing agents for acids and caustics: Lime.
Reactivity group: 1
Compatibility class: Nonoxidizing mineral acids

ENVIRONMENTAL DATA
Water pollution: Effects of low concentrations on aquatic life is unknown. May be dangerous if it enters nearby water intakes; notify operators. Notify local health and wildlife officials.
Response to discharge: Issue warning–corrosive. Chemical and physical treatment.

SHIPPING INFORMATION
Grades of purity: Varying concentrations available; **Storage temperature:** Ambient; **Inert atmosphere:** None; **Venting:** Pressure vacuum valve; **Stability during transport:** Stable.

NAS HAZARD CLASSIFICATIONS FOR BULK WATER TRANSPORTATION
FIRE: 0
HEALTH: Vapor irritant: 0; Liquid or solid irritant: 3; Poisons: 3
WATER POLLUTION: Human toxicity:–; Aquatic toxicity: 3; Aesthetic effect: 0
REACTIVITY: Other chemicals: 1; Water: 0; Self-reaction: 0

PHYSICAL AND CHEMICAL PROPERTIES
Physical state @ 59°F/15°C and 1 atm: Liquid.
Molecular weight: 144.08
Boiling point @ 1 atm: Decomposes.
Specific gravity (water = 1): 1.25

HYDROGEN   REC. H:1350

SYNONYMS: EEC No. 001-001-00-9; HIDROGENO (Spanish); HYDROGEN, COMPRESSED; HYDROGEN, REFRIGERATED LIQUID; LIQUID HYDROGEN; PARA HYDROGEN; PROTIUM

IDENTIFICATION
CAS Number: 1333-74-0
Formula: H_2
DOT ID Number: UN 1966 (Refrigerated liquid); UN 1049 (Compressed); DOT Guide Number: 115
Proper Shipping Name: Hydrogen, refrigerated liquid (cryogenic liquid); Hydrogen, compressed

DESCRIPTION: Colorless compressed gas or cryogenic liquid. Odorless. Floats and boils on water; insoluble. Flammable visible vapor cloud is produced.

Extremely flammable; burns with an invisible flame:even in the dark • Asphyxiant • Forms explosive mixture with air • Reacts explosively with many materials • Gas is much lighter than air, and will disperse slowly unless confined • Vapors from liquefied gas are initially heavier than air and spread along ground • Gas or vapors may travel to source of ignition and flashback •High pressure releases often ignite without any apparent source of ignition • Exposure of cylinders to elevated temperatures, fire, and flame may cause cylinders to rupture or cause frangible disk to burst, releasing entire contents of cylinder • Ruptured or venting cylinders may rocket through buildings and/or travel a considerable distance •Irritating to the skin, eyes, and respiratory tract • Vapors may cause dizziness or asphyxiation without warning • Contact with liquid may cause frostbite or burns.

Hazard Classification (based on NFPA-704 Rating System)
Health Hazards (Blue): 0; Flammability (Red): 4; Reactivity (Yellow): 0

EMERGENCY RESPONSE: See Appendix A (115)
Evacuation:
Public safety: Isolate spill area for at least 50 to 100 meters (160 to 330 feet) in all directions.
Spill: Consider initial downwind evacuation for at least 800 meters (½ mile).
Fire: Isolate for 1600 meters (1 mile) in all directions, especially if tank, rail car, or tank truck is involved in fire.

EXPOSURE
The only effect of exposure to liquid hydrogen is that caused by its unusually low temperature and its action as a simple asphyxiant.
Inhalation: If victim is unconscious (due to oxygen deficiency), move him to fresh air and apply resuscitation methods; *CALL A DOCTOR. IF ON SKIN OR IN THE EYES:* Treat for frostbite; soak in lukewarm water; get medical attention if burn is severe.

HEALTH HAZARDS
Personal protective equipment (PPE): B-Level PPE. Safety goggles or face shield; insulated gloves and long sleeves; cuffless trousers worn outside boots or over high-top shoes to shed spilled liquid; self-contained breathing apparatus containing air (never use oxygen). Wear thermal protective clothing.
Exposure limits (TWA unless otherwise noted): Gas is nonpoisonous but can act as a simple asphyxiant.
Liquid or solid irritant characteristics: Rapid evaporation of liquid may cause frostbite.

FIRE DATA
Flash point: Flammable gas.
Flammable limits in air: LEL: 4.0%; UEL: 74.2%.
Fire extinguishing agents not to be used: CO_2. Use water spray to cool containers.
Behavior in fire: Flashback along vapor trail may occur. Burns with an almost invisible flame. Containers may rupture and explode. Vapor may explode if ignited in an enclosed area.
Autoignition temperature: 932°F/500°C/723°K.
Electrical hazard: Class I, Group B; Flow or agitation of substance may generate electrostatic charges due to low conductivity.
Burning rate: 9.9 mm/min
Adiabatic flame temperature: 2497°F/1370°C (estimate).
Stoichiometric air-to-fuel ratio: 34.32 (estimate).

CHEMICAL REACTIVITY
Reactivity with water: Ambient temperature of water will cause vigorous vaporization of hydrogen.
Binary reactants: Violent reaction with strong oxidizers; acetylene, halogens, nitrous oxide and other gases. Attacks (makes brittle) mild steels and some iron alloys at high temperature and pressure.

ENVIRONMENTAL DATA
Food chain concentration potential: Negative; unlikely to accumulate.
Water pollution: Not harmful to aquatic life. **Response to discharge:** Issue warning–high flammability. Restrict access. Evacuate area.

SHIPPING INFORMATION
Grades of purity: Commercial; **Storage temperature:** –423°F/–253°C/20°K; **Inert atmosphere:** None; **Venting:** Safety relief; **Stability during transport:** Stable.

PHYSICAL AND CHEMICAL PROPERTIES
Physical state @ 59°F/15°C and 1 atm: Gas.
Molecular weight: 2.0
Boiling point @ 1 atm: –423°F/–253°C/20°K.
Melting/Freezing point: –434°F/–259°C/14°K.
Critical temperature: –400°F/–240°C/33°K.
Critical pressure: 188 psia = **Liquid surface tension:** Atm = 1.30 MN/m^2.
Specific gravity (water = 1): 0.071 @ –253°C (liquid).
Liquid surface tension: 2.3 dynes/cm = 0.023 N/m at –255°C.
Relative vapor density (air = 1): 0.067
Ratio of specific heats of vapor (gas): 1.3962
Latent heat of vaporization: 190.5 Btu/lb = 105.8 cal/g = 4.427 x 10^5 J/kg.
Heat of combustion: –50,080 Btu/lb = –27,823 cal/g = –1164.1 x 10^5 J/kg.
Heat of fusion: 13.8 cal/g.
Vapor pressure: (Reid) Very high.

HYDROGEN BROMIDE   REC. H:1400

SYNONYMS: ACIDE BROMHYDRIQUE (French); ANHYDROUS HYDROBROMIC ACID; BROMWASSERSTOFF (German); EEC No. 035-002-00-0; BROMURO de HIDROGENO (Spanish); HYDROBROMIC ACID; HYDROBROMIC ACID, ANHYDROUS; HYDROGEN BROMIDE, ANHYDROUS

IDENTIFICATION
CAS Number: 10035-10-6
Formula: HBr
DOT ID Number: UN 1048 (anhydrous); UN 1788 (hydrobromic acid); DOT Guide Number: 125 (anhydrous); 154
Proper Shipping Name: Hydrogen bromide, anhydrous; Hydrobromic acid

DESCRIPTION: Colorless compressed, liquefied gas. May be shipped as a light yellow liquid under pressure. Sharp, irritating odor. Sinks and mixes with water; dissolves; produces heat and toxic hydrobromic acid. Boiling liquid may produce poisonous white vapor cloud.

Poison! • Breathing the vapor can kill you [inhalation exposures of 500 ppm (10 minutes) can be fatal]; skin or eye contact causes severe burns, impaired vision, or blindness • Firefighting gear (including SCBA) does not provide adequate protection. If exposure occurs, remove and isolate gear immediately and thoroughly decontaminate personnel • Containers may BLEVE when exposed to fire • Toxic products of combustion may include bromine and hydrogen bromide • Contact with the liquid may cause severe injury or frostbite • Reacts with metals producing flammable hydrogen gas • Do not put yourself in danger by entering a contaminated area to rescue a victim.

Hazard Classification (based on NFPA-704 Rating System)
Health Hazards (Blue): 3; Flammability (Red): 0; Reactivity (Yellow): 0

EMERGENCY RESPONSE, anhydrous: See Appendix A (125)
Evacuation:
Public safety: See below.
Spill: Small spill–First: Isolate in all directions 30 meters (100 feet); Then: Protect persons downwind, DAY: 0.2 km (0.1 mile); NIGHT: 0.5 km (0.3 mile). Large spill–First: Isolate in all directions 125 meters (400 feet); Then: Protect persons downwind, DAY: 1.1 km (0.7 mile); NIGHT: 3.4 km (2.1 miles).
Fire: Isolate for 1600 meters (1 mile) in all directions, especially if tank, rail car, or tank truck is involved in fire.

EMERGENCY RESPONSE: See Appendix A (154)
Evacuation:
Public safety: Isolate the area of spill or leak for at least 25 to 50 meters (80 to 160 feet) in all directions.
Spill: Increase, in the downwind direction, as necessary, the distance shown under "Public Safety."
Fire: If tank, rail car, or tank truck is involved in fire, isolate for at least 800 meters (½ mile) in all directions; also, consider initial evacuation for 800 meters (½ mile) in all directions.

EXPOSURE
Short-term effects: *SEEK MEDICAL ATTENTION.* **Vapor:** *POISONOUS IF INHALED.* May cause lung edema; physical exertion will aggravate this condition. Irritating to eyes, nose, and throat. Move to fresh air. *IF BREATHING HAS STOPPED,* give artificial respiration; *avoid mouth-to-mouth resuscitation; use bag/mask apparatus.* IF breathing is difficult, administer oxygen. **Liquid:** *POISONOUS IF SWALLOWED.* Will burn skin and eyes. Will cause frostbite. Remove contaminated clothing and shoes. Flush affected areas with plenty of water. *IF IN EYES,* hold eyelids open and flush with plenty of water. *IF SWALLOWED* and victim is *CONSCIOUS AND ABLE TO SWALLOW,* have victim drink 4 to 8 ounces of water. **Do NOT induce vomiting.** *DO NOT RUB AFFECTED AREAS.*

Note to physician or authorized medical personnel: Medical observation is recommended for 24 to 48 hours after breathing overexposure, as pulmonary edema may be delayed. As first aid for pulmonary edema, consider administering a corticosteroid spray. Cigarette smoking may exacerbate pulmonary injury and should be discouraged for at least 72 hours following exposure.

HEALTH HAZARDS
Personal protective equipment (PPE): A-Level PPE. OSHA Table Z-1-A air contaminant. Chemical goggles; rubber apron and gloves; acid-proof clothing; safety shower. Wear thermal protective clothing. Chemical protective material(s) reported to have good to excellent resistance: butyl rubber, neoprene, Saranex®.
Recommendations for respirator selection: NIOSH/OSHA
30 ppm: SA:CF* (any supplied-air respirator operated in a continuous-flow mode); or PAPRAG* [any powered, air-purifying respirator with acid gas cartridge(s)]; or GMFAG [any air-purifying, full-facepiece respirator (gas mask) with a chin-style, front- or back-mounted organic vapor cannister]; or SCBAF (any self-contained breathing apparatus with a full facepiece); or SAF (any supplied-air respirator with a full facepiece). *EMERGENCY OR PLANNED ENTRY INTO UNKNOWN CONCENTRATIONS OR IDLH CONDITIONS:* SCBAF:PD,PP (any self-contained

breathing apparatus that has a full facepiece and is operated in a pressure-demand or other positive-pressure mode); or SAF:PD,PP:ASCBA (any supplied-air respirator that has a full facepiece and is operated in a pressure-demand or other positive-pressure mode in combination with an auxiliary self-contained breathing apparatus operated in a pressure-demand or other positive pressure mode). *ESCAPE:* GMFAG [any air-purifying, full-facepiece respirator (gas mask) with a chin-style, front- or back-mounted organic vapor cannister]; or SCBAE (any appropriate escape-type, self-contained breathing apparatus).
Note: Substance causes eye irritation or damage; eye protection needed.
Exposure limits (TWA unless otherwise noted): ACGIH TLV ceiling 3 ppm (9.9 mg/m^3); OSHA PEL 3 ppm; NIOSH REL ceiling 3 ppm (10 mg/m^3).
Long-term health effects: Prolonged contact or inhalation of inorganic bromides causes acne-like bromoderma (bromide rash), especially of the hands and face. Other effects include emaciation, depression; and, in severe cases, psychosis and mental deterioration.
Vapor (gas) irritant characteristics: Eye, skin and respiratory tract irritant. May cause difficult breathing and lung edema.
Liquid or solid irritant characteristics: Severe eye, skin and respiratory tract irritant.
Odor threshold: 2 ppm.
IDLH value: 30 ppm.

FIRE DATA
Fire extinguishing agents not to be used: Do not apply water to cryogenic liquid containers.
Behavior in fire: Pressurized container may explode and release toxic, irritating vapor.

CHEMICAL REACTIVITY
Reactivity with water: Moderate reaction with evolution of heat. In water it is a strong acid.
Binary reactants: On contact with air forms corrosive fumes which may travel along the ground. Reacts with strong oxidizers, strong caustics, ammonia, fluorine, ozone, copper, brass, zinc. Rapidly absorbs moisture, forming hydrobromic acid. In presence of moisture, highly corrosive to most metals, with evolution of flammable hydrogen gas.
Neutralizing agents for acids and caustics: Flush with water; apply powdered limestone, slaked lime, soda ash, or sodium bicarbonate.

ENVIRONMENTAL DATA
Food chain concentration potential: Negative; unlikely to accumulate.
Water pollution: Dangerous to aquatic life in high concentrations. May be dangerous if it enters nearby water intakes; notify operators. Notify local health and wildlife officials. **Response to discharge:** Issue warning–poison, corrosive, air contaminant. Restrict access. Evacuate area. Disperse and flush.

SHIPPING INFORMATION
Grades of purity: 99.8+%; **Storage temperature:** Ambient; **Storage temperature:** or lower; **Inert atmosphere:** None; **Venting:** Safety relief; **Stability during transport:** Stable.

PHYSICAL AND CHEMICAL PROPERTIES
Physical state @ 59°F/15°C and 1 atm: Gas.
Molecular weight: 80.92
Boiling point @ 1 atm: −88.2°F/−66.8°C/206.4°K.
Melting/Freezing point: −124°F/−87°C/186°K.
Critical temperature: 193.6°F/89.8°C/363.0°K.
Critical pressure: 1235 psia = 84 atm = 8.52 MN/m^2.
Specific gravity (water = 1): 2.14 @ −67°C (liquid); 3.5 (gas).
Liquid surface tension: 27.1 dynes/cm = 0.0271 N/m @ −67.1°C.
Relative vapor density (air = 1): 2.81
Ratio of specific heats of vapor (gas): 1.38
Latent heat of vaporization: 92.3 Btu/lb = 51.3 cal/g = 2.15 x 10^5 J/kg.
Heat of solution: 445 Btu/lb = 247 cal/g = 10.3 x 10^5 J/kg.
Heat of fusion: 7.1 cal/g.
Vapor pressure: 15 mm.

HYDROGEN CHLORIDE   REC. H:1450

SYNONYMS: ACIDO CLORHIDRICO (Spanish); ACIDE CHLORHYDRIQUE (French); ANHYDROUS HYDROCHLORIC ACID; AQUEOUS HYDROGEN CHLORIDE; CHLOROHYDRIC ACID; CHLORWASSERSTOFF (German); EEC No. 017-002-00-2; HCL; HYDROCHLORIC ACID; HYDROCHLORIC ACID, ANHYDROUS; HYDROCHLORIDE; HYDROGEN CHLORIDE; MURIATIC ACID; SPIRITS OF SALT

IDENTIFICATION
CAS Number: 7647-01-0
Formula: HCl·H$_2$O
DOT ID Number: UN 1789 (solution); UN 1050 (anhydrous, gas); UN 2186 (refrigerated liquefied gas); DOT Guide Number: 157 (solution); 125 (anhydrous and refrigerated)
Proper Shipping Name: Hydrochloric acid, solution; Hydrogen chloride, anhydrous; Hydrogen chloride, refrigerated Liquid
Reportable Quantity (RQ): **(CERCLA)** 5000 lb/2270 kg

DESCRIPTION: Colorless gas. Shipped and stored as a cryogenic liquid or a light yellow watery solution. Sharp, irritating odor. Sinks and mixes with water; produces heat and hydrochloric acid. Irritating vapor is produced.

Poison! • Breathing the vapor can kill you; skin or eye contact causes severe burns, impaired vision, or blindness • Firefighting gear (including SCBA) does not provide adequate protection. If exposure occurs, remove and isolate gear immediately and thoroughly decontaminate personnel • Containers may BLEVE when exposed to fire • Toxic products of combustion may include hydrogen chloride • Contact with liquid may cause frostbite • Corrosive to common metals, especially in the presence of moisture, forming flammable hydrogen gas • Do not put yourself in danger by entering a contaminated area to rescue a victim.

Hazard Classification (based on NFPA-704 Rating System)
Health Hazards (Blue): 3; Flammability (Red): 0; Reactivity (Yellow): 1

EMERGENCY RESPONSE, solution: See Appendix A (157)
Evacuation:
Public safety: Isolate the area of spill or leak for at least 50 to 100 meters (160 to 330 feet) in all directions.
Spill: Increase, in the downwind direction, as necessary, the distance shown under "Public Safety."
Fire: If tank, rail car, or tank truck is involved in fire, isolate for at least 800 meters (½ mile) in all directions; also, consider initial evacuation for 800 meters (½ mile) in all directions.

EMERGENCY RESPONSE, anhydrous and refrigerated): See Appendix A (125)
Evacuation:
Public safety: Isolate spill area for at least 100 to 200 meters (330 to 660 feet) in all directions.
Spill: Small spill–First: Isolate in all directions 30 meters (100 feet); Then: Protect persons downwind, DAY: 0.2 km (0.1 mile); NIGHT: 0.6 km (0.4 mile). Large spill–First: Isolate in all directions 185 meters (600 feet); Then: Protect persons downwind, DAY: 1.6 km (1.0 mile); NIGHT: 4.3 km (2.7 miles).
Fire: Isolate for 1600 meters (1 mile) in all directions, especially if tank, rail car, or tank truck is involved in fire.

EXPOSURE
Short-term effects: *SEEK MEDICAL ATTENTION*. **Vapor:** Irritating to eyes, nose, and throat. *IF INHALED*, will, will cause coughing or difficult breathing. May cause lung edema; physical exertion will aggravate this condition. Move to fresh air. *IF BREATHING HAS STOPPED*, give artificial respiration. If breathing is difficult, administer oxygen. **Liquid:** Will burn skin and eyes. Harmful if swallowed. Remove contaminated clothing and shoes. Flush affected areas with plenty of water. *IF IN EYES*, hold eyelids open and flush with plenty of water. *IF SWALLOWED* and victim is *CONSCIOUS AND ABLE TO SWALLOW*, have victim drink 4 to 8 ounces of water. **Do NOT induce emesis. Do NOT administer activated charcoal or attempt to neutralize stomach contents.**
Note to physician or authorized medical personnel: Medical observation is recommended for 24 to 48 hours after breathing overexposure, as pulmonary edema may be delayed. As first aid for pulmonary edema, consider administering a corticosteroid spray. Cigarette smoking may exacerbate pulmonary injury and should be discouraged for at least 72 hours following exposure.

HEALTH HAZARDS
Personal protective equipment (PPE): B-Level PPE. OSHA Table Z-1-A air contaminant. Self-contained breathing equipment, air-line mask, or industrial canister-type gas mask; rubber or rubber-coated gloves, apron, coat, overalls, shoes. Wear thermal protective clothing.
For <30% hydrochloric acid, Protective materials with good to excellent resistance: butyl rubber, natural rubber, nitrile+PVC, neoprene, nitrile, PVC, Viton®, Saranex®, Silvershield®, nitrile .
For 30-70% hydrochloric acid, Protective materials with good to excellent resistance: butyl rubber, natural rubber, neoprene, nitrile, Viton®, Saranex®, nitrile, Silvershield®. Also, chlorinated polyethylene, PVC, butyl rubber/natural rubber offers limited protection *For >70% hydrochloric acid*, Protective materials with good to excellent resistance: natural rubber, neoprene+natural rubber, neoprene/natural rubber, nitrile.
Recommendations for respirator selection: NIOSH/OSHA
50 ppm: CCRS* [any chemical cartridge respirator with cartridge(s) providing protection against the compound of concern)]; or GMFS [any air-purifying, full-facepiece respirator (gas mask) with a chin-style, front- or back-mounted canister providing protection against the compound of concern]; or PAPRS* [any powered, air-purifying respirator with cartridge(s) providing protection against the compound of concern]; or SA* (any supplied-air respirator); or SCBAF (any self-contained breathing apparatus with a full facepiece). *EMERGENCY OR PLANNED ENTRY INTO UNKNOWN CONCENTRATIONS OR IDLH CONDITIONS:* SCBAF:PD,PP (any self-contained breathing apparatus that has a full facepiece and is operated in a pressure-demand or other positive-pressure mode); or SAF:PD,PP:ASCBA (any supplied-air respirator that has a full facepiece and is operated in a pressure-demand or other positive-pressure mode in combination with an auxiliary self-contained breathing apparatus operated in a pressure-demand or other positive pressure mode). *ESCAPE:* GMFAG [any air-purifying, full-facepiece respirator (gas mask) with a chin-style, front- or back-mounted organic vapor cannister]; or SCBAE (any appropriate escape-type, self-contained breathing apparatus). *Note:* Substance reported to cause eye irritation or damage; may require eye protection.
Exposure limits (TWA unless otherwise noted): ACGIH TLV ceiling 5 ppm (7.5 mg/m^3); NIOSH/OSHA ceiling 5 ppm (7 mg/m^3).
Long-term health effects: Tooth erosion.
Vapor (gas) irritant characteristics: Eye and respiratory tract irritant. Vapor is irritating such that personnel will not usually tolerate moderate or high vapor concentrations.
Liquid or solid irritant characteristics: Fairly severe eye and skin irritant; may cause pain and second-degree burns after a few minutes of contact. Rapid evaporation of liquid may cause frostbite.
Odor threshold: 0.3–10 ppm.
IDLH value: 50 ppm.

FIRE DATA
Flash point: Not combustible.
Fire extinguishing agents not to be used: Do not apply water to cryogenic liquid containers.

CHEMICAL REACTIVITY
Reactivity with water: Anhydrous hydrogen chloride are easily absorbed in to water forming corrosive hydrochloric acid.
Binary reactants: A strong acid; violent reaction with bases; forms explosive hydrogen gas with base metals; oxidizers form toxic chlorine gas; reaction with sulfuric acid forms corrosive hydrochloric acid gas. Corrosive fume on contact with air. Highly corrosive to most metals with evolution of hydrogen gas, which may form explosive mixtures with air. Reacts with hydroxides, amines, alkalis, copper, brass, zinc.
Neutralizing agents for acids and caustics: Flush with water; apply powdered limestone, slaked lime, soda ash, or sodium bicarbonate.
Reactivity group: 1
Compatibility class: Nonoxidizing mineral acids

ENVIRONMENTAL DATA
Food chain concentration potential: Log P_{ow} = 0.32. Unlikely to accumulate.
Water pollution: Dangerous to aquatic life in high concentrations. May be dangerous if it enters nearby water intakes; notify operators. Notify local health and wildlife officials. **Response to discharge:** Issue warning–corrosive. Restrict access. Disperse and flush.

SHIPPING INFORMATION
Grades of purity: Food processing or technical: 18° Be–27.9%; 20 Be–31.5%; 22° Be–35.2%; Reagent, ACS, and USP: 23° Be–37.1%; **Storage temperature:** Ambient; **Inert atmosphere:** None; **Venting:** Open; **Stability during transport:** Stable.

NAS HAZARD CLASSIFICATION FOR BULK WATER TRANSPORTATION
FIRE: 0
HEALTH: Vapor irritant: 3; Liquid or solid irritant: 3; Poisons: 2
WATER POLLUTION: Human toxicity: 2; Aquatic toxicity: 2; Aesthetic effect: 2
REACTIVITY: Other chemicals: 3; Water: 0; Self-reaction: 0

Hydrogen cyanide

PHYSICAL PROPERTIES: (37% solution unless otherwise noted).
Physical state @ 59°F/15°C and 1 atm: Liquid.
Molecular weight: 36.46
Boiling point @ 1 atm: 125°F/52°C/325°K (liquid); −121°F/−139°C/134°K (gas); −245°F/−154°C = 119°K @ 1 mm.
Melting/Freezing point: −174°F/−113°C/160°K (gas).
Specific gravity (water = 1): 1.27 @ 68°F/20°C (liquid).
Relative vapor density (air = 1): 1.27 (gas).
Latent heat of vaporization: 178 Btu/lb = 98.6 cal/g = 4.13×10^5 J/kg.
Heat of solution: −860 Btu/lb = −480 cal/g = -20×10^5 J/kg.
Heat of fusion: 13.0 cal/g.
Vapor pressure: (Reid) 8.0 psia; 37 mm.

HYDROGEN CYANIDE   REC. H:1500

SYNONYMS: AC (weaponized); ACIDE CYANHYDRIQUE (French); ACIDO CIANHIDRICO (Spanish); AERO LIQUID HCN; BLAUSAEURE (German); CYANWASSERSTOFF (German); CYCLON; CYCLONE B; EEC No. 006-006-00-X; FORMONITRILE; HCN; HYDROCYANIC ACID; PRUSSIC ACID; ZACLON DISCOIDS; RCRA No. P063

IDENTIFICATION

CAS Number: 74-90-8
Formula: HCN; CNH
DOT ID Number: UN 1051 (anhydrous; ≥20% solution); UN 1613 (≤20% solution); UN 3294 (solution in alcohol ≤ 45%); DOT Guide Number: 117 (UN 1051); 131 (UN 3294); 154 (UN 1613)
Proper Shipping Name: AC; Hydrogen cyanide, anhydrous, stabilized; Hydrogen cyanide, stabilized; Hydrocyanic acid, liquefied; Hydrocyanic acid, aqueous solutions, with more than 20% Hydrogen cyanide (UN 1051); Hydrogen cyanide solution in alcohol with not more than 45% hydrogen cyanide (UN 3294)
Reportable Quantity (RQ): **(CERCLA)** 10 lb/4.54 kg

DESCRIPTION: Colorless or pale blue-white, watery liquid, or gas (above 78°F/26°C). At higher temperatures it is a colorless gas. Transported in red and white candy-striped containers. Often used as a 96% solution in water Sweet, bitter almond odor; some people cannot smell it. Mixes slowly with water. Poisonous, flammable vapor is produced and rises. Freezes at 8°F/−13°C. *Note*: Has been used as a chemical warfare agent. If used as a weapon use M256-A1 Detector Kit. Detection limits: 11 mg/m³. Notify U.S. Department of Defense: Army. *Warning:* Odor is not a reliable indicator of the presence of toxic amounts of HCN gas.

Highly volatile poison! CAUTION-Class A poison; asphyxiation can be caused by ingestion, inhalation, or absorption of liquid or vapor through skin (particularly eyes, mucous membranes, and feet). Extremely toxic vapors (unspecified) are generated even at ordinary temperatures (USCG) • Extremely hazardous and flammable • May react with itself without warning with explosive violence • Firefighting gear (including SCBA) does not provide adequate protection. If exposure occurs, remove and isolate gear immediately and thoroughly decontaminate personnel • Containers may BLEVE when exposed to fire • Vapors may form explosive mixture with air • Gas is slightly lighter than air, and will disperse slowly unless confined • Vapors from liquefied gas are initially heavier than air and spread along ground • Vapors may travel long distances to ignition sources and flashback • Vapors in confined areas (e.g., tanks, sewers, buildings) may explode when exposed to fire • Toxic products of combustion may include cyanide; combustion products are less toxic than the material itself • Do NOT attempt rescue • *Warning:* Initial odor may deaden your sense of smell, and is not a reliable indicator of the presence of toxic amounts of hydrogen cyanide gas. Exposure to concentrations of 110–181 ppm for periods of 10 to 30 minutes can be fatal.

Hazard Classification (based on NFPA-704 Rating System)
Health Hazards (Blue): 4; Flammability (Red): 4; Reactivity (Yellow): 2

EMERGENCY RESPONSE: See Appendix A (117)
Evacuation:
Public safety: Isolate spill area for at least 100 to 200 meters (330 to 660 feet) in all directions.
Spill: (UN1051) Small spill–First: Isolate in all directions 60 meters (200 feet); Then: Protect persons downwind, DAY: 0.2 km (0.1 mile); NIGHT: 0.5 km (0.3 mile). Large spill–First: Isolate in all directions 400 meters (1300 feet) [450 m (1500 ft) weaponized]; Then: Protect persons downwind, DAY: 1.3 km (0.8 mile) [1.6 km (1.0 mile) weaponized]; NIGHT: 3.4 km (2.1 miles) [3.9 km (2.4 miles) weaponized].
Fire: Isolate for 1600 meters (1 mile) in all directions, especially if tank, rail car, or tank truck is involved in fire.

EMERGENCY RESPONSE: See Appendix A (131)
Evacuation:
Public safety: Isolate spill area for at least 100 to 200 meters (330 to 660 feet) in all directions.
Spill: Increase, as necessary, the isolation distance shown above, in "Public safety."
Fire: Isolate for 800 meters (½ mile) in all directions, especially if tank, rail car, or tank truck is involved in fire.

EXPOSURE

Short-term effects: *SEEK MEDICAL ATTENTION.* **Vapor:** *POISONOUS IF INHALED OR IF SKIN IS EXPOSED.* Irritating to eyes. Move to fresh air. *IF BREATHING HAS STOPPED,* give artificial respiration; *avoid mouth-to-mouth resuscitation; use bag/mask apparatus.* IF breathing is difficult, administer oxygen.
Liquid: *POISONOUS IF SWALLOWED OR IF SKIN IS EXPOSED.* Irritating to eyes. Remove contaminated clothing and shoes. Flush affected areas with plenty of water. *IF IN EYES,* hold eyelids open and flush with plenty of water. *IF SWALLOWED* and victim is *CONSCIOUS AND ABLE TO SWALLOW,* have victim drink 4 to 8 ounces of water and have victim induce vomiting. *IF SWALLOWED* and victim is *UNCONSCIOUS OR HAVING CONVULSIONS,* do nothing except keep victim warm.
Note to physician or authorized medical personnel: Consider the use of amyl nitrite perles if symptoms of cyanide poisoning develop. If symptoms indicate, initial treatment includes the cyanide antidote kit. In all cases, break an amyl nitrite perle in a gauze pad and hold lightly under victim's nose for 15 seconds, repeating 5 times at about 15-second intervals; if necessary (and if sodium nitrite infusions will be delayed), repeat procedure every 3 minutes with fresh perles until 3 or 4 have been used. Avoid breathing the vapor while administering it to the victim. Administer sodium nitrite IV, ASAP. The usual adult dose is 10 to 20 mL of a 3% solution infused over no less than 5 minutes; the average child dose is 0.15 to 0.20 mL/kg. Monitor blood pressure during administration, and slow the rate of infusion if hypotention develops. Next, infuse sodium thiosulfate IV. The usual adult dose is 50 mL of a 25% solution infused over 10 to 20 minutes; the average child dose is 1.65 mL/kg. Repeat with nitrite and thiosulfate as required.

Note: Observe patients for at least 2 hours in the Emergency Department if they have ingested solutions or have had direct eye or skin contact.

HEALTH HAZARDS

Personal protective equipment (PPE): A-Level PPE. OSHA Table Z-1-A air contaminant. Chemical protective material(s) reported to have good to excellent resistance: Teflon®. Also, butyl rubber or polyethylene offers minimal to poor protection. If the proper equipment is not available, or if the rescuers have not been trained in its use, call for assistance from the U.S. Soldier and Biological Chemical Command–Edgewood Research Development and Engineering Center (from 0700-1630 EST call 410-671-4411, and from 1630-0700 EST call 410-278-5201; ask for the Staff Duty Officer).
Recommendations for respirator selection: NIOSH/OSHA
47 ppm: SA (any supplied-air respirator). *50 ppm*: SA:CF (any supplied-air respirator operated in a continuous-flow mode); or SCBAF (any self-contained breathing apparatus with a full facepiece); or SAF (any supplied-air respirator with a full facepiece). *EMERGENCY OR PLANNED ENTRY INTO UNKNOWN CONCENTRATIONS OR IDLH CONDITIONS:* SCBAF:PD,PP (any self-contained breathing apparatus that has a full facepiece and is operated in a pressure-demand or other positive-pressure mode); or SAF:PD,PP:ASCBA (any supplied-air respirator that has a full facepiece and is operated in a pressure-demand or other positive-pressure mode in combination with an auxiliary self-contained breathing apparatus operated in a pressure-demand or other positive pressure mode). *ESCAPE:* GMFS [any air-purifying, full-facepiece respirator (gas mask) with a chin-style, front- or back-mounted canister providing protection against the compound of concern]; or SCBAE (any appropriate escape-type, self-contained breathing apparatus).
Exposure limits (TWA unless otherwise noted): ACGIH TLV ceiling 10 ppm; OSHA PEL 10 ppm (11 mg/m^3); skin contact contributes significantly in overall exposure.
Short-term exposure limits (15-minute TWA): NIOSH STEL 4.7 ppm (5 mg/m^3); skin contact contributes significantly in overall exposure.
Toxicity by ingestion: Grade 4; LD$_{50}$ less than 50 mg/kg.
Long-term health effects: May cause respiratory problems.
Vapor (gas) irritant characteristics: Vapor is extremely poisonous. See Odor threshold, below.
Liquid or solid irritant characteristics: Liquid irritates the eyes, skin, and respiratory tract. It is extremely poisonous if absorbed through skin or eyes.
Odor threshold: 0.1–5.0 ppm Perception of the odor is a genetic trait (20–40% of the general population cannot detect HCN); also, rapid olfactory fatigue can occur.
IDLH value: 50 ppm.

FIRE DATA

Flash point: 0°F/–18°C (cc) (96%).
Flammable limits in air: LEL: 5.6%; UEL 40.0%.
foam, CO$_2$, dry chemical. Fight fire from protected area or from maximum possible distance.
Fire extinguishing agents not to be used: Water may be ineffective.
Behavior in fire: Containers may explode with ignition of contents.
Autoignition temperature: 1004°F/540°C/813°K.
Electrical hazard: Class I, Group C.
Burning rate: 1.8 mm/min
Stoichiometric air-to-fuel ratio: 6.350 (estimate).

CHEMICAL REACTIVITY

Reactivity with water: Dissolves with a moderate reaction.
Binary reactants: Reacts with amines, oxidizers, acids, sodium hydroxide, calcium hydroxide, sodium carbonate, caustics, ammonia.
Neutralizing agents for acids and caustics: The weak acidity can be neutralized by slaked lime, but this does not destroy the poisonous property.
Polymerization: Can polymerize in temperatures above 122-140°F, or in the presence of 2-5% water.

ENVIRONMENTAL DATA

Food chain concentration potential: Log P$_{ow}$ = 0.7. Unlikely to accumulate.
Water pollution: DOT Appendix B, §172.101–marine pollutant. If used as a weapon, utilize M272 Chemical Agent Water Detection Kit: Detection limits: 20 mg/L. Harmful to aquatic life in very low concentrations. May be dangerous if it enters nearby water intakes; notify operators. Notify local health and wildlife officials. **Response to discharge:** Issue warning–high flammability, water contaminant. Restrict access. Evacuate area.

SHIPPING INFORMATION

Grades of purity: 96%; sometimes shipped as a water solution, or absorbed on an inert solid. All forms are extremely toxic; **Inert atmosphere:** May be padded; **Stability during transport:** May become unstable and subject to explosion if stored for extended time or exposed to high temp. and pressure.

NAS HAZARD CLASSIFICATION FOR BULK WATER TRANSPORTATION

FIRE: 4
HEALTH: Vapor irritant: 2; Liquid or solid irritant: 1; Poisons: 4
WATER POLLUTION: Human toxicity: 4; Aquatic toxicity: 4; Aesthetic effect: 1
REACTIVITY: Other chemicals: 3; Water: 0; Self-reaction: 3

PHYSICAL AND CHEMICAL PROPERTIES

Physical state @ 59°F/15°C and 1 atm: Liquid.
Molecular weight: 27.03
Boiling point @ 1 atm: 78.3°F/25.7°C/298.9°K.
Melting/Freezing point: 8°F/–13°C/260°K; 7°F/–14°C/259°K (96%).
Critical temperature: 362°F/184°C/457°K.
Critical pressure: 735 psia = 50 atm = 5.07 MN/m^2.
Specific gravity (water = 1): 0.689 @ 68°F/20°C (liquid).
Relative vapor density (air = 1): 0.9
Ratio of specific heats of vapor (gas): 1.303
Latent heat of vaporization: 444 Btu/lb = 247 cal/g = 10.3 x 10^5 J/kg.
Heat of combustion: –10,560 Btu/lb = –5864 cal/g = –245.3 x 10^5 J/kg.
Heat of fusion: 74.38
Vapor pressure: 630 mm.

HYDROGEN FLUORIDE REC. H:1550

SYNONYMS: ACIDO FLUORHIDRICO (Spanish); ANHYDROUS HYDROGEN FLUORIDE; ANHYDROFLUORIC ACID; AQUEOUS HYDROGEN FLUORIDE; EEC No. 009-002-00-6; HYDROFLUORIDE; HYDROFLUORIC ACID; HF; HF-A; RCRA No. U134

Hydrogen fluoride

IDENTIFICATION
CAS Number: 7664-39-3
Formula: HF·H_2O
DOT ID Number: UN 1052 (anhydrous); 1790 (solution); DOT Guide Number: 125 (anhydrous); 157 (solution)
Proper Shipping Name: Hydrogen fluoride, anhydrous; Hydrofluoric acid; Hydrofluoric acid, solution
Reportable Quantity (RQ): **(CERCLA)** 100 lb/45.4 kg

DESCRIPTION: Colorless to slightly yellow gas, or fuming watery liquid (below 67°F/20°C). The pure compound is a crystalline solid below 12°F/–11°C. Often used in aqueous solution. Sharp, irritating odor. Sinks and mixes with water. Harmful vapor is produced.

Corrosive to the respiratory tract and a systemic poison! • Breathing the vapor can kill you; skin or eye contact causes blindness or severe burns which may not be immediately painful • Firefighting gear (including SCBA) provides NO protection. If exposure occurs, remove and isolate gear immediately and thoroughly decontaminate personnel • Containers may BLEVE when exposed to fire • Vapor from solution is lighter than air • Vapors from liquefied gas are initially heavier than air and spread along ground • Toxic products of combustion may include fluorides • reacts with some metals producing flammable hydrogen gas • Highly corrosive to rubber, leather, glass and other materials • Do not put yourself in danger by entering a contaminated area to rescue a victim.

Hazard Classification (based on NFPA-704 Rating System)
Health Hazards (Blue): 4; Flammability (Red): 0; Reactivity (Yellow): 1

EMERGENCY RESPONSE, anhydrous: See Appendix A (125)
Evacuation:
Public safety: See below.
Spill: Small spill–First: Isolate in all directions 30 meters (100 feet); Then: Protect persons downwind, DAY: 0.2 km (0.1 mile); NIGHT: 0.6 km (0.4 mile). Large spill–First: Isolate in all directions 125 meters (400 feet); Then: Protect persons downwind, DAY: 1.1 km (0.7 mile); NIGHT: 2.9 km (1.8 miles).
Fire: Isolate for 1600 meters (1 mile) in all directions, especially if tank, rail car, or tank truck is involved in fire.
EMERGENCY RESPONSE, solution: See Appendix A (157)
Evacuation:
Public safety: Isolate the area of spill or leak for at least 50 to 100 meters (160 to 330 feet) in all directions.
Spill: Increase, in the downwind direction, as necessary, the distance shown under "Public Safety."
Fire: If tank, rail car, or tank truck is involved in fire, isolate for at least 800 meters (½ mile) in all directions; also, consider initial evacuation for 800 meters (½ mile) in all directions.

EXPOSURE
Short-term effects: *SEEK MEDICAL ATTENTION.* **Vapor:** Will burn eyes, nose, and throat. Harmful if inhaled. May cause lung edema; physical exertion will aggravate this condition. Move to fresh air. *IF BREATHING HAS STOPPED,* give artificial respiration ; *avoid mouth-to-mouth resuscitation; use bag/mask apparatus.* IF breathing is difficult, administer oxygen. *IF IN EYES*, hold eyelids open and flush with plenty of water. **Liquid:** Will burn skin and eyes. Corrosive if swallowed. Remove contaminated clothing and shoes. Flush affected areas with plenty of water. Cover burns with magnesium-containing solution (e.g., Maalox, epsom salts). *IF IN EYES,* hold eyelids open and flush with plenty of water. *IF SWALLOWED* and victim is *CONSCIOUS AND ABLE TO SWALLOW*, **do NOT induce emesis and do not administer activated charcoal.** Have victim drink 4 to 12 ounces of water to dilute the acid. Orally administer a one-time dose of several ounces of antacid containing magnesia (Mylanta®, Maalox®, or milk of magnesia) or calcium (e.g., TUMS®).
Note to physician or authorized medical personnel: **Skin burns:** Do NOT inject calcium chloride to treat skin burns (it will cause extreme pain and further tissue damage). Cover skin with one of the following preparations: (a) calcium-containing slurry or gel (2.5 g calcium gluconate in 100 mL of water-soluble lubricant such as K-Y Jelly or one (1) ampule of 10% calcium gluconate per ounce of K-Y Jelly); or (b) Aqueosu quaternary ammonium salt (Zephiran, 0.13%. If using Zephiran concentrated solution [17%] be sure to properly dilute it by adding 30 mL [1 ounce] of concentrate to one (1) gallon of water. **Do NOT use on face or eyes**). Treat large burn areas with 5% calcium gluconate solution using small-gauge (#30) needle. Do not inject more than 0.5 mL/cm^2 of affected skin surface area. **Eyes:** Do NOT use salves, oils, or ointments for injured eyes. Do NOT use Zephiran or the gel form of calcium gluconate in the eyes, as described for skin treatment.
Medical observation is recommended for 24 to 48 hours after breathing overexposure, as pulmonary edema may be delayed. As first aid for pulmonary edema, consider administering a corticosteroid spray. Cigarette smoking may exacerbate pulmonary injury and should be discouraged for at least 72 hours following exposure.

HEALTH HAZARDS
Personal protective equipment (PPE): A-Level PPE. OSHA Table Z-1-A air contaminant. OSHA Table Z-2 air contaminant. All persons handling this product must be familiar with and must observe all the precautions contained in the latest version of Manufacturing Chemists' Association's Safety Data Sheet SD-25. A shower and an eye wash must be available. *For <30% hydrofluoric acid,* Protective materials with good to excellent resistance: neoprene. Also, Viton®/neoprene, nitrile, PVC, natural rubber+neoprene+NBR offers limited protection *For 30-70% hydrofluoric acid,* Chemical protective material(s) reported to offer minimal to poor protection: natural rubber [>3 hrs, 48% solution], neoprene, Silvershied®, nitrile (2 hrs, 48% solution). *For >70% hydrofluoric acid,* Chemical protective material(s) reported to offer minimal to poor protection: natural rubber, neoprene+natural rubber, neoprene/natural rubber.
Recommendations for respirator selection: NIOSH/OSHA
30 ppm: CCRS* [any chemical cartridge respirator with cartridge(s) providing protection against the compound of concern]; or PAPRS* [any powered, air-purifying respirator with cartridge(s) providing protection against the compound of concern]; or GMFS [any air-purifying, full-facepiece respirator (gas mask) with a chin-style, front- or back-mounted canister providing protection against the compound of concern]; or SA* (any supplied-air respirator); or SCBA (any self-contained breathing apparatus). *EMERGENCY OR PLANNED ENTRY INTO UNKNOWN CONCENTRATIONS OR IDLH CONDITIONS:* SCBAF:PD,PP (any self-contained breathing apparatus that has a full facepiece and is operated in a pressure-demand or other positive-pressure mode); or SAF:PD,PP:ASCBA (any supplied-air respirator that has a full facepiece and is operated in a pressure-demand or other positive-pressure mode in combination with an auxiliary self-contained breathing apparatus operated in a pressure-demand or other positive pressure mode). *ESCAPE:* GMFS [any air-purifying, full-facepiece respirator (gas mask) with a chin-style, front- or back-mounted canister providing protection against the compound of concern]; or SCBAE (any appropriate escape-type, self-contained breathing apparatus).

Note: Substance reported to cause eye irritation or damage; may require eye protection.
Exposure limits (TWA unless otherwise noted): ACGIH TLV ceiling 3 ppm (2.6 mg/m^3); OSHA PEL 3 ppm; NIOSH REL 3 ppm (2.5 mg/m^3).
Short-term exposure limits (15-minute TWA): OSHA STEL 6 ppm (5 mg/m^3). NIOSH ceiling 6 ppm (5 mg/m^3)/15 min.
Toxicity by ingestion: Poison.
Long-term health effects: Possible mutagen. Reparatory ulcers may occur. Skin or eye injuries may result in gangrene.
Vapor (gas) irritant characteristics: Vapors cause severe irritation of eye and throat and can cause eye and lung injury. Vapors cannot be tolerated even at low concentrations.
Liquid or solid irritant characteristics: Severe eye and skin irritant. Causes second- and third-degree burns on short contact; very injurious to the eyes.
Odor threshold: 0.04 ppm.
IDLH value: 30 ppm.

FIRE DATA
Fire extinguishing agents not to be used: Water evolves heat.
Behavior in fire: Containers may rupture and explode.

CHEMICAL REACTIVITY
Reactivity with water: Dissolved in water it becomes a strong acid.
Binary reactants: Reacts violently with many compounds and corrodes most substances (EXCEPT lead, polyethylene, platinum, Teflon®, wax). Even dilute solutions will attack glass and other silicon-containing compounds, concrete and certain metals containing silica, such as cast iron; natural rubber, leather, and many organic materials. May generate flammable hydrogen in contact with some metals.
Neutralizing agents for acids and caustics: Flush with water; apply powdered limestone, slaked lime, soda ash, or sodium bicarbonate.
Reactivity group: 1
Compatibility class: Nonoxidizing mineral acids

ENVIRONMENTAL DATA
Food chain concentration potential: Log P_{ow} = –0.93. Unlikely to accumulate.
Water pollution: Harmful to aquatic life in very low concentrations. May be dangerous if it enters nearby water intakes; notify operators. Notify local health and wildlife officials.
Response to discharge: Issue warning–corrosive. Restrict access. Disperse and flush. Designated as a hazardous substance under section 311 (b) (2) (A) of the Federal Water pollution: Control Act and further regulated by the Clean Water Act. These regulations apply to discharges of this substance

SHIPPING INFORMATION
Grades of purity: Reagent: 48–51%; technical: 52–55%; 70% grade; **Storage temperature:** Ambient; **Inert atmosphere:** None; **Venting:** Pressure-vacuum; **Stability during transport:** Stable.

NAS HAZARD CLASSIFICATION FOR BULK WATER TRANSPORTATION
FIRE: 0
HEALTH: Vapor irritant: 4; Liquid or solid irritant: 4; Poisons: 4
WATER POLLUTION: Human toxicity: 4; Aquatic toxicity: 3; Aesthetic effect: 2
REACTIVITY: Other chemicals: 3; Water: 0; Self-reaction: 0

PHYSICAL AND CHEMICAL PROPERTIES (applies to 70% of solution unless otherwise noted).

Physical state @ 59°F/15°C and 1 atm: Liquid.
Molecular weight: 20.0
Boiling point @ 1 atm: 67°F/20°C/293°K (gas); 152°F/67°C/340°K (solution).
Melting/Freezing point: –118°F/–83°C/190°K (gas); less than 95°F/35°C/308°K (solution).
Specific gravity (water = 1): 1.258 @ 77°F/25°C (liquid).
Relative vapor density (air = 1): 0.7 (solution); 2.5 (gas).
Latent heat of vaporization: 649 Btu/lb = 361 cal/g = 15.1 x 10^5 J/kg.
Heat of solution: –66.6 Btu/lb = –37.0 cal/g = –1.55 x 10^5 J/kg.
Heat of fusion: 54.7 cal/g.
Vapor pressure: (Reid) Varies; 0.8 (gas); 150 mmHg (solution).

HYDROGEN PEROXIDE   REC. H:1600

SYNONYMS: ALBONE; EEC No. 008-003-00-9; HIGH STRENGTH HYDROGEN PEROXIDE; HYOXYL; HYDROGEN DIOXIDE; HYDROPEROXIDE; INHIBINE; OXYDOL; PERHYDROL; PERONE; PEROXAN; PEROXIDO de HIDROGENO (Spanish); SUPEROXOL; T-STUFF; WASSERSTOSSPEROXIDE (German); PEROXIDE

IDENTIFICATION
CAS Number: 7722-84-1
Formula: H_2O_2
DOT ID Number: UN 2984 (8–20% solution); UN 2014 (20–60% solution); UN 2015 (>60% solution, stabilized); DOT Guide Number: 140 (UN 2984; UN 2014); 143 (UN 2015)
Proper Shipping Name: Hydrogen peroxide, aqueous solution, with not less than 8% but less than 20% hydrogen peroxide; Hydrogen peroxide, aqueous solution, with not less than 20% but not more than 60% hydrogen peroxide (stabilized if necessary); Hydrogen Peroxide aqueous solutions, stabilized, with more than 60% Hydrogen peroxide; Hydrogen peroxide, stabilized
Reportable Quantity (RQ): Concentrated; more than 52% **(CERCLA)** 1 lb/0.454 kg

DESCRIPTION: Colorless watery liquid. Shipped and stored in water solution. Slightly sharp, irritating odor, resembles ozone. Sinks and mixes with water. Irritating vapor is produced. *Warning:* Odor is not a reliable indicator of the presence of toxic amounts of hydrogen peroxide.

Corrosive; may cause severe eye and skin burns • A strong oxidizer • Strong oxidizer that may react spontaneously with low-flash-point organics or reducing agents. Heat forms oxygen; will increase the activity of an existing fire • Firefighting gear (including SCBA) does not provide adequate protection. If exposure occurs, remove and isolate gear immediately and thoroughly decontaminate personnel • Containers may BLEVE when exposed to heat or fire • Reacts with iron, copper, brass and many other metals • Vapor concentrations greater than 40% by weight may decompose explosively @ 1 atm pressure (ILO). *Warning*: Inhalation exposures of more than 400 ppm (10 minutes) can be fatal.

Hazard Classification (based on NFPA-704 Rating System)
Aqueous solution, 8-20%:   Health Hazards (Blue): 1; Flammability (Red): 0; Reactivity (Yellow): 1; Special Notice (White): OXY

Aqueous solution, 40-60%: Health Hazards (Blue): 2; Flammability (Red): 0; Reactivity (Yellow): 1; Special Notice (White): OXY
Aqueous solution, more than 60%: Health Hazards (Blue): 2; Flammability (Red): 0; Reactivity (Yellow): 3; Special Notice (White): OXY

EMERGENCY RESPONSE: See Appendix A (140)
Evacuation:
Public safety: Isolate the area of spill or leak for at least 10 to 25 meters (30 to 80 feet) in all directions.
Spill: Consider initial downwind evacuation for at least 100 meters (330 feet).
Fire: If any large container is involved in fire, isolate for at least 800 meters (½ mile) in all directions; also, consider initial evacuation for 800 meters (½ mile) in all directions.

EMERGENCY RESPONSE: See Appendix A (171)
Evacuation:
Public safety: Isolate the area of spill or leak for at least 50 to 100 meters (160 to 300 feet) in all directions.
Spill: Increase, in the downwind direction, as necessary, the distance shown under "Public Safety."
Fire: If any large container is involved in fire, isolate for at least 800 meters (½ mile) in all directions; also, consider initial evacuation for 800 meters (½ mile) in all directions.

EXPOSURE
Short-term effects: *SEEK MEDICAL ATTENTION.* **Vapor:** Irritating to eyes, nose, and throat. Harmful if inhaled. May cause lung edema; physical exertion will aggravate this condition. Move to fresh air. *IF BREATHING HAS STOPPED,* give artificial respiration. IF breathing is difficult, administer oxygen. **Liquid:** *POORLY ABSORBED THROUGH THE INTACT SKIN.* Will burn skin and eyes. Harmful if swallowed. Remove contaminated clothing and shoes. Flush affected areas with plenty of water. *IF IN EYES,* hold eyelids open and flush with plenty of water. *IF SWALLOWED* and victim is *CONSCIOUS AND ABLE TO SWALLOW,* have victim drink 4 to 8 ounces of water. **Do NOT induce vomiting.** The ability of activated charcoal to absorb hydrogen peroxide is unknown.
Note to physician or authorized medical personnel: Medical observation is recommended for 24 to 48 hours after breathing overexposure, as pulmonary edema may be delayed. As first aid for pulmonary edema, consider administering a corticosteroid spray. Cigarette smoking may exacerbate pulmonary injury and should be discouraged for at least 72 hours following exposure.

HEALTH HAZARDS
Personal protective equipment (PPE): B-Level PPE. OSHA Table Z-1-A air contaminant. Impermeable apron, gloves, and boots; goggles. *<30%,* Protective materials with good to excellent resistance: chlorobutyl rubber. *30-70%,* Protective materials with good to excellent resistance: nitrile, polyethylene, PVC, natural rubber, neoprene+natural rubber, neoprene/natural rubber. Also, neoprene offers limited protection. *>70%,* Chemical protective material(s) reported to offer minimal protection: Viton®/neoprene. Protective garments, both outer and inner, made of a woven polyester fabric or of modacrylic or polyvinylidene fabrics.
Recommendations for respirator selection: NIOSH/OSHA
10 ppm: SA* (any supplied-air respirator). *25 ppm:* SA:CF* (any supplied-air respirator operated in a continuous-flow mode). *50 ppm:* SCBAF (any self-contained breathing apparatus with a full facepiece); or SAF (any supplied-air respirator with a full facepiece). *75 ppm:* SAF:PD,PP (any supplied-air respirator that has a full facepiece and is operated in a pressure-demand or other positive-pressure mode). *EMERGENCY OR PLANNED ENTRY INTO UNKNOWN CONCENTRATIONS OR IDLH CONDITIONS:* SCBAF:PD,PP (any self-contained breathing apparatus that has a full facepiece and is operated in a pressure-demand or other positive-pressure mode); or SAF:PD,PP:ASCBA (any supplied-air respirator that has a full facepiece and is operated in a pressure-demand or other positive-pressure mode in combination with an auxiliary self-contained breathing apparatus operated in a pressure-demand or other positive pressure mode). *ESCAPE:* GMFS [any air-purifying, full-facepiece respirator (gas mask) with a chin-style, front- or back-mounted canister providing protection against the compound of concern]; or SCBAE (any appropriate escape-type, self-contained breathing apparatus). *Note:* Substance reported to cause eye irritation or damage; may require eye protection
Exposure limits (TWA unless otherwise noted): ACGIH TLV 1 ppm (1.4 mg/m^3); NIOSH/OSHA 1 ppm (1.4 mg/m^3).
Long-term health effects: IARC rating 3; limited animal evidence. Possible genetic damage.
Vapor (gas) irritant characteristics: Vapors cause irritation, such that personnel will find high concentrations unpleasant. The effect is temporary.
Liquid or solid irritant characteristics: Severe eye irritant. Fairly severe skin irritant. May cause pain and second-degree burns after a few minutes of contact.
IDLH value: 75 ppm.

FIRE DATA
Flash point: Noncombustible, but concentrated material can generate heat and decompose spontaneously from heat, shock, contamination, or if placed in a basic environment–with added risk of fire.
Behavior in fire: Containers may explode in fire

CHEMICAL REACTIVITY
Binary reactants: A powerful oxidizer. Reacts with oxidizable materials, iron, copper, brass, bronze, chromium, zinc, lead, manganese, silver. Contact with combustible materials may result in spontaneous combustion. Dirt, rough surfaces and many metals cause a rapid decomposition with liberation of oxygen gas; occurs particularly if concentration is above 40%. Attacks many materials and substances.

ENVIRONMENTAL DATA
Food chain concentration potential: Log P_{ow} = –1.2. Unlikely to accumulate.
Water pollution: Effect of low concentrations on aquatic life is unknown. May be dangerous if it enters nearby water intakes; notify operators. Notify local health and wildlife officials. **Response to discharge:** Issue warning–corrosive. Restrict access. Disperse and flush.

SHIPPING INFORMATION
Grades of purity: Common commercial strengths are 27.5%, 35%, 50%, 70%, 90%, and 98%. "High strength" means greater than 52%. Purity: Technical; Military Specification; ACS. The hazard increases with the strength. Commercial peroxide products contain a stabilizer (usually acetanilide) to slow the rate of spontaneous decomposition; **Storage temperature:** Ambient; **Inert atmosphere:** None; **Venting:** Safety relief or pressure-vacuum; **Stability during transport:** Pure grades are quite stable, but contamination with metals or dirt can cause rapid or violent decomposition. Hydrogen peroxide is a very unstable compound; it chemically degrades rapidly, especially in sunlight.

NAS HAZARD CLASSIFICATION FOR BULK WATER TRANSPORTATION
FIRE: 0
HEALTH: Vapor irritant: 2; Liquid or solid irritant: 3; Poisons: 1
WATER POLLUTION: Human toxicity: 1; Aquatic toxicity: 3; Aesthetic effect: 1
REACTIVITY: Other chemicals: 4; Water: 1; Self-reaction: 3

PHYSICAL AND CHEMICAL PROPERTIES (applies to 70% of solution).

Physical state @ 59°F/15°C and 1 atm: Liquid.
Molecular weight: 34.01
Boiling point @ 1 atm: 258°F/126°C/399°K (>60% solution).
Melting/Freezing point: 12°F/–11°C/262°K.
Specific gravity (water = 1): 1.39 @ 68°F/20°C (liquid).
Ratio of specific heats of vapor (gas): 1.241
Latent heat of vaporization: 542 Btu/lb = 301 cal/g = 12.6 x 10^5 J/kg.
Heat of decomposition: –1220 Btu/lb = –676 cal/g = –28.3 x 10^5 J/kg.
Heat of solution: –20.2 Btu/lb = –11.2 cal/g = –0.469 x 10^5 J/kg.
Heat of fusion: 8.58 cal/g.
Vapor pressure: (Reid) Varies; 18–23 mmHg @ 86°F/30°C/303°K (35–50%); 8 mmHg @ 77°F/25°C/298°K (>60%).

HYDROGEN SULFIDE   REC. H:1650

SYNONYMS: ACIDE SULFHYDRIQUE (French); EEC No. 016-001-00-4; HEPATIC GAS; HYDROGENE SULFURE (French); HYDROSULFURIC ACID; SCHWEFELWASSERSTOFF (German); SEWER GAS; STINK DAMP; SULFURETTED HYDROGEN; SULFUR HYDRIDE; SULFUR HYDROXIDE; SULPHURETTED HYDROGEN; SULFURO de HIDROGENO (Spanish); RCRA No. U135

IDENTIFICATION
CAS Number: 7783-06-4
Formula: H_2S
DOT ID Number: UN 1053; DOT Guide Number: 117
Proper Shipping Name: Hydrogen sulfide; Hydrogen sulfide, liquefied; Hydrogen sulphide; Hydrogen sulphide, liquefied
Reportable Quantity (RQ): **(CERCLA)** 100 lb/45.4 kg

DESCRIPTION: Colorless gas that produces a visible vapor cloud. Rotten egg odor, but odorless at poisonous concentrations. Sinks and boils in water; insoluble.

Poison! • Extremely dangerous and flammable • May be fatal if inhaled or absorbed through the skin • Contact with the liquid can cause frostbite • Firefighting gear (including SCBA) does not provide adequate protection. If exposure occurs, remove and isolate gear immediately and thoroughly decontaminate personnel • Containers may BLEVE when exposed to fire • Gas form s explosive mixture with air • Gas is heavier than air and will collect and stay in low areas • Gas may travel long distances to ignition sources and flashback • Vapors in confined areas (e.g., tanks, sewers, buildings) may explode when exposed to fire • Toxic products of combustion may include sulfur dioxides • Will burn or explode in the presence of metal oxides • Do NOT attempt rescue • *Warning*: Initial odor may deaden your sense of smell, and is not a reliable indicator of the presence of toxic amounts of hydrogen sulfide gas.

Hazard Classification (based on NFPA-704 Rating System)
Health Hazards (Blue): 4; Flammability (Red): 4; Reactivity (Yellow): 0

EMERGENCY RESPONSE: See Appendix A (117)
Evacuation:
Public safety: Isolate spill area for at least 100 to 200 meters (330 to 660 feet) in all directions.
Spill: Small spill–First: Isolate in all directions 30 meters (100 feet); Then: Protect persons downwind, DAY: 0.2 km (0.1 mile); NIGHT: 0.3 km (0.2 mile). Large spill–First: Isolate in all directions 215 meters (700 feet); Then: Protect persons downwind, DAY: 1.4 km (0.9 mile); NIGHT: 4.3 km (2.7 miles).
Fire: Isolate for 1600 meters (1 mile) in all directions, especially if tank, rail car, or tank truck is involved in fire.

EXPOSURE
Short-term effects: *SEEK MEDICAL ATTENTION*. **Vapor:** *POISONOUS IF INHALED*. May cause lung edema; physical exertion will aggravate this condition. Irritating to eyes. Move to fresh air. *IF BREATHING HAS STOPPED*, give artificial respiration; *avoid mouth-to-mouth resuscitation; use bag/mask apparatus*. IF breathing is difficult, administer oxygen. *IF IN EYES*, hold eyelids open and flush with plenty of water.
Note to physician or authorized medical personnel: Medical observation is recommended for 24 to 48 hours after breathing overexposure, as pulmonary edema may be delayed. As first aid for pulmonary edema, consider administering a corticosteroid spray. Cigarette smoking may exacerbate pulmonary injury and should be discouraged for at least 72 hours following exposure.

HEALTH HAZARDS
Personal protective equipment (PPE): B-Level PPE. OSHA Table Z-1-A air contaminant. OSHA Table Z-2 air contaminant. Rubber-framed goggles; approved respiratory protection. Wear thermal protective clothing.
Recommendations for respirator selection: NIOSH
100 ppm: PAPRS [any powered, air-purifying respirator with cartridge(s) providing protection against the compound of concern]; or GMFS [any air-purifying, full-facepiece respirator (gas mask) with a chin-style, front- or back-mounted canister providing protection against the compound of concern]; or SA* (any supplied-air respirator); or SCBAF (any self-contained breathing apparatus with a full facepiece). *EMERGENCY OR PLANNED ENTRY INTO UNKNOWN CONCENTRATIONS OR IDLH CONDITIONS:* SCBAF:PD,PP (any self-contained breathing apparatus that has a full facepiece and is operated in a pressure-demand or other positive-pressure mode); or SAF:PD,PP:ASCBA (any supplied-air respirator that has a full facepiece and is operated in a pressure-demand or other positive-pressure mode in combination with an auxiliary self-contained breathing apparatus operated in a pressure-demand or other positive pressure mode). *ESCAPE:* GMFS [any air-purifying, full-facepiece respirator (gas mask) with a chin-style, front- or back-mounted canister providing protection against the compound of concern]; or SCBAE (any appropriate escape-type, self-contained breathing apparatus). *Note*: Substance reported to cause eye irritation or damage; may require eye protection.
Exposure limits (TWA unless otherwise noted): ACGIH TLV 10 ppm (14 mg/m³)
Short-term exposure limits (15-minute TWA): ACGIH STEL 15 ppm (21 mg/m³).OSHA PEL ceiling 20 ppm; 50 ppm [10 minute maximum peak]; NIOSH REL ceiling 10 ppm (15 mg/m³) [10 minute].
Toxicity by ingestion: Hydrogen sulfide is present as a gas at room temperature; ingestion not likely.

610   Hydroquinone

Long-term health effects: May effect the central nervous system.
Vapor (gas) irritant characteristics: A severe eye and mucous membrane irritant. High concentrations may cause lung edema, or may be fatal.
Liquid or solid irritant characteristics: Eye and skin irritant. If spilled on clothing and allowed to remain, may cause smarting and reddening of the skin. Rapid evaporation of liquid may cause frostbite.
Odor threshold: 0.001–0.10 ppm.
IDLH value: 100 ppm.

FIRE DATA
Flash point: Flammable gas; low ignition energy.
Flammable limits in air: LEL: 4.3%; UEL: 45%.
Autoignition temperature: 500°F/260°C/533°K.
Electrical hazard: Class I, Group C. Flow or agitation of substance may generate electrostatic charges due to low conductivity.
Burning rate: 2.3 mm/min (liquid).
Stoichiometric air-to-fuel ratio: 6.040 (estimate).

CHEMICAL REACTIVITY
Binary reactants: Strong oxidizers, strong nitric acid, metals. Reacts with many compounds.

ENVIRONMENTAL DATA
Food chain concentration potential: Log P_{ow} = 1.2. Unlikely to accumulate.
Water pollution: Harmful to aquatic life in very low concentrations. May be dangerous if it enters nearby water intakes; notify operators. Notify local health and wildlife officials.
Response to discharge: Issue warning–high flammability, poison. Restrict access. Evacuate area.

SHIPPING INFORMATION
Grades of purity: Purified; technical; **Storage temperature:** Ambient; **Inert atmosphere:** None; **Venting:** Safety relief; **Stability during transport:** Stable.

PHYSICAL AND CHEMICAL PROPERTIES
Physical state @ 59°F/15°C and 1 atm: Gas.
Molecular weight: 34.08
Boiling point @ 1 atm: –76.7°F/–60.4°C/212.8°K.
Melting/Freezing point: –123°F/–86°C/187°K.
Critical temperature: 212.7°F/100.4°C/373.6°K.
Critical pressure: 1300 psia = 88.9 atm = 9.01 MN/m^2.
Specific gravity (water = 1): 1.54; 0.916@ –76°F/–60°C/213°K (liquid).
Liquid surface tension: (estimate) 30 dynes/cm = 0.03 N/m@ –61°C.
Relative vapor density (air = 1): 1.19
Ratio of specific heats of vapor (gas): 1.322
Latent heat of vaporization: 234 Btu/lb = 130 cal/g = 5.44 x 10^5 J/kg.
Heat of combustion: –6552 Btu/lb = –3640 cal/g = –152.4 x 10^5 J/kg.
Heat of fusion: 16.8 cal/g.
Vapor pressure: 17.6 atm; 13,700 mm.

HYDROQUINONE   **REC. H:1700**

SYNONYMS: ARCTUVIN; *p*-BENZENEDIOL; 1,4-BENZENEDIOL; BENZOHYDROQUINONE; BENZOQUINOL; BLACK AND WHITE BLEACHING CREAM; DIAK-S; DIHYDROXYBENZENE; *p*-DIHYDROXYBENZENE; 1,4-DIHYDROXYBENZENE; 1,4-DIHYDROXY-BENZOL (German); *p*-DIOXOBENZENE; EEC No. 604-005-00-4; ELDOPOQUE; ELDOQUIN; HIDROQUINONA (Spanish); HYDROQUINOL; HYDROQUINOLE; α-HYDROQUINONE; *p*-HYDROQUINONE; PARA-HYDROXYPHENOL; QUINOL; β-QUINOL; TECQUINOL; TENOX HQ6; PYROGENTISIC ACID

IDENTIFICATION
CAS Number: 123-31-9
Formula: $C_6H_6O_2$; 1, 4-$C_6H_4(OH)_2$
DOT ID Number: UN 2662; DOT Guide Number: 153
Proper Shipping Name: Hydroquinone
Reportable Quantity (RQ): **(CERCLA)** 1 lb/0.454 kg

DESCRIPTION: White, light tan to gray solid. Sinks in water; quickly dissolved

Poison! • Allergen • Breathing the dust can kill you • Firefighting gear (including SCBA) does not provide adequate protection. If exposure occurs, remove and isolate gear immediately and thoroughly decontaminate personnel • Combustible solid • Heat or flame may cause explosion • Containers may BLEVE when exposed to fire • Concentrated dust in confined areas may explode when exposed to fire • Severely Irritating to the skin, eyes, and respiratory tract; dermatitis and skin allergies can result from skin contact • Toxic products of combustion may include carbon monoxide. Heated vapors are extremely dangerous; may cause lung damage. • Do not put yourself in danger by entering a contaminated area to rescue a victim.

Hazard Classification (based on NFPA-704 Rating System)
Health Hazards (Blue): 2; Flammability (Red): 1; Reactivity (Yellow): 0

EMERGENCY RESPONSE: See Appendix A (153)
Evacuation:
Public safety: Isolate the area of spill or leak for at least 25 to 50 meters (80 to 160 feet) in all directions.
Spill: Increase, in the downwind direction, as necessary, the distance shown under "Public Safety."
Fire: If tank, rail car, or tank truck is involved in fire, isolate for at least 800 meters (½ mile) in all directions; also, consider initial evacuation for 800 meters (½ mile) in all directions.

EXPOSURE
Short-term effects: *SEEK MEDICAL ATTENTION.* **Dust:** Irritating to eyes, nose, and throat. Harmful if inhaled. *May cause lung edema; physical exertion will aggravate this condition. IF IN EYES*, hold eyelids open and flush with plenty of water. *IF BREATHING HAS STOPPED*, give artificial respiration; *avoid mouth-to-mouth resuscitation; use bag/mask apparatus.* IF breathing is difficult, administer oxygen. **Solid:** Will burn eyes. Irritating to eyes. *IF SWALLOWED*, will cause headache, dizziness, nausea, vomiting, or loss of consciousness. Remove contaminated clothing and shoes. Flush affected areas with plenty of water. *IF IN EYES*, hold eyelids open and flush with plenty of water. *IF SWALLOWED* and victim is *CONSCIOUS AND ABLE TO SWALLOW*, have victim drink 4 to 8 ounces of water and have victim induce vomiting. *IF SWALLOWED* and victim is *UNCONSCIOUS OR HAVING CONVULSIONS*, do nothing except keep victim warm.
Note to physician or authorized medical personnel: Medical observation is recommended for 24 to 48 hours after breathing overexposure, as pulmonary edema may be delayed. As first aid for

pulmonary edema, consider administering a corticosteroid spray. Cigarette smoking may exacerbate pulmonary injury and should be discouraged for at least 72 hours following exposure.

HEALTH HAZARDS
Personal protective equipment (PPE): B-Level PPE. OSHA Table Z-1-A air contaminant. Goggles; respiratory protection if dust is present. Chemical protective material(s) reported to have good to excellent resistance: natural rubber, neoprene, nitrile, PVC, nitrile+PVC, polyethylene, nitrile. Also, Viton® and styrene-butadiene offers limited protection
Recommendations for respirator selection: NIOSH/OSHA
50 mg/m³: PAPRD* (any powered, air-purifying respirator with a dust filter); or HiEF (any air-purifying, full-facepiece respirator with a high-efficiency particulate filter); or SAT:CF* (any supplied-air respirator that has a tight-fitting facepiece and is operated in a continuous-flow mode); or SCBAF (any self-contained breathing apparatus with a full facepiece); or SAF (any supplied-air respirator with a full facepiece). *EMERGENCY OR PLANNED ENTRY INTO UNKNOWN CONCENTRATIONS OR IDLH CONDITIONS:* SCBAF:PD,PP (any self-contained breathing apparatus that has a full faceplate and is operated in a pressure-demand or other positive-pressure mode); or SAF:PD,PP:ASCBA (any supplied-air respirator that has a full facepiece and is operated in a pressure-demand or other positive-pressure mode in combination with an auxiliary self-contained breathing apparatus operated in a pressure-demand or other positive pressure mode). *ESCAPE:* HiEF (any air-purifying, full-facepiece respirator with a high-efficiency particulate filter); or SCBAE (any appropriate escape-type, self-contained breathing apparatus). *Note:* Substance causes eye irritation or damage; eye protection needed.
Exposure limits (TWA unless otherwise noted): ACGIH TLV 2 mg/m³; OSHA PEL 2 mg/m³; NIOSH ceiling 2 mg/m³/15 min.
Toxicity by ingestion: Poison. Grade 3; LD_{50} = 370 mg/kg (rat).
Long-term health effects: Causes bladder cancer in mice, discoloration of eyelids and eye changes, possible vision loss; dermatitis. IARC rating 3; inadequate animal evidence. May effect the blood and lungs.
Vapor (gas) irritant characteristics: Respiratory tract and eye irritant. May cause difficult breathing and lung edema.
Liquid or solid irritant characteristics: Strong skin irritant.
IDLH value: 50 mg/m³.

FIRE DATA
Flash point: (molten) 329°F/165°C (oc).
Behavior in fire: Dust explosion is possible.
Autoignition temperature: 960°F/516°C/789°K.

CHEMICAL REACTIVITY
Binary reactants: Strong oxidizers, alkalis, especially sodium hydroxide. May explode on contact with oxygen under certain conditions.

ENVIRONMENTAL DATA
Food chain concentration potential: Log P_{ow} = 0.53. Unlikely to accumulate.
Water pollution: Effect of low concentrations on aquatic life is unknown. May be dangerous if it enters nearby water intakes; notify operators. Notify local health and wildlife officials.
Response to discharge: Issue warning–water contaminant. Disperse and flush.

SHIPPING INFORMATION
Grades of purity: Pure; Technical; **Storage temperature:** Ambient; **Inert atmosphere:** None; **Venting:** Open; **Stability during transport:** Stable.

PHYSICAL AND CHEMICAL PROPERTIES
Physical state @ 59°F/15°C and 1 atm: Solid.
Molecular weight: 110.11
Boiling point @ 1 atm: 545°F/285°C/558°K.
Melting/Freezing point: 338°F/170°C/443°K.
Specific gravity (water = 1): 1.33 @ 68°F/20°C (solid).
Relative vapor density (air = 1): 1.2
Heat of combustion: –11200 Btu/lb = –6,220 cal/g = –260 x 10^5 J/kg.
Heat of fusion: 58.84 cal/g.
Vapor pressure: 0.00001 mm.

2-HYDROXYETHYL ACRYLATE REC. H:1750

SYNONYMS: ACRYLIC ACID, 2-HYDROXYETHYL ESTER; HEA; β-HYDROXYETHYL ACRYLATE; 2-HIDROXIETILACRILATO (Spanish); 2-HYDROXYETHYL 2-PROPENOATE

IDENTIFICATION
CAS Number: 818-61-1
Formula: $C_5H_8O_3$; $CH_2=CHCOOCH_2CH_2OH$
DOT ID Number: None; DOT Guide Number: 171P
Proper Shipping Name: Polymerization material, stabilize with dry ice

DESCRIPTION: Colorless liquid (inhibited). Sweet, pleasant odor. Mixes with water.

Combustible • Heat or flame may cause explosion • Containers may BLEVE when exposed to fire • Vapors are heavier than air and will collect and stay in low areas • Vapors in confined areas (e.g., tanks, sewers, buildings) may explode when exposed to fire • Irritating to the skin, eyes, and respiratory tract • May be a polymerization hazard; unless inhibited, heat can induce polymerization • May react with itself blocking relief valves; leading to tank explosions • Toxic products of combustion may include carbon monoxide.

Hazard Classification (based on NFPA-704 Rating System)
Health Hazards (Blue): 2; Flammability (Red): 1; Reactivity (Yellow): 2

EMERGENCY RESPONSE: See Appendix A (171)
Evacuation:
Public safety: Isolate the area of spill or leak for at least 10 to 25 meters (30 to 80 feet) in all directions.
Spill: Increase, in the downwind direction, as necessary, the distance shown under "Public Safety."
Fire: If any large container is involved in fire, isolate for at least 800 meters (½ mile) in all directions; also, consider initial evacuation for 800 meters (½ mile) in all directions.

EXPOSURE
Short-term effects: *SEEK MEDICAL ATTENTION.* **Liquid:** Will burn skin and eyes. Harmful if swallowed. Remove contaminated clothing and shoes. Flush affected areas with plenty of water. *IF IN EYES*, hold eyelids open and flush with plenty of water. *IF SWALLOWED* and victim is *CONSCIOUS AND ABLE TO SWALLOW*, have victim drink 4 to 8 ounces of water. **Do NOT induce vomiting.**

Hydroxylamine

HEALTH HAZARDS
Personal protective equipment (PPE): B-Level PPE. Goggles or face shield; rubber gloves.
Toxicity by ingestion: Grade 2; oral LD_{50} = 1070 mg/kg (rat).
Vapor (gas) irritant characteristics: Vapors cause severe irritation of eyes and throat and can cause eye and lung injury. They cannot be tolerated even at low concentrations.
Liquid or solid irritant characteristics: Severe eye and skin irritant. Causes second- and third-degree burns on short contact and is very injurious to the eyes.

FIRE DATA
Flash point: 220°F/104°C (oc); 214°F/101°C (cc).
Behavior in fire: Containers may explode.
Electrical hazard: Class I, Group D.
Burning rate: 2.0 mm/min

CHEMICAL REACTIVITY
Binary reactants: Corrodes mild steel, tin plate.
Polymerization: Unless inhibited, polymerization will occur, especially when heated; **Inhibitor of polymerization:** Monomethyl ether of hydroquinone, 400 ppm.
Reactivity group: 34
Compatibility class: Esters.

ENVIRONMENTAL DATA
Food chain concentration potential: Negative; unlikely to accumulate.
Water pollution: Effect of low concentrations on aquatic life is unknown. May be dangerous if it enters nearby water intakes; notify operators. Notify local health and wildlife officials.
Response to discharge: Issue warning–water contaminant. Restrict access. Disperse and flush.

SHIPPING INFORMATION
Grades of purity: Commercial; **Storage temperature:** Ambient; **Inert atmosphere:** None; **Venting:** Open (flame arrester); **Stability during transport:** Stable.

NAS HAZARD CLASSIFICATIONS FOR BULK WATER TRANSPORTATION
FIRE: 1
HEALTH: Vapor irritant: 4; Liquid or solid irritant: 4; Poisons: 4
WATER POLLUTION: Human toxicity: 3; Aquatic toxicity: 3; Aesthetic effect: 1
REACTIVITY: Other chemicals: 1; Water: 0; Self-reaction: 3

PHYSICAL AND CHEMICAL PROPERTIES
Physical state @ 59°F/15°C and 1 atm: Liquid.
Molecular weight: 116.1
Boiling point @ 1 atm: 410°F/210°C/483°K.
Melting/Freezing point: –76°F/–60°C/213°K.
Specific gravity (water = 1): 1.10 @ 77°F/25°C (liquid).
Liquid surface tension: (estimate) 28 dynes/cm = 0.028 N/m @ 68°F/20°C.
Heat of combustion: (estimate) –10,800 Btu/lb = –6000 cal/g = –250 x 10^5 J/kg.
Heat of polymerization: (estimate) –218 Btu/lb = –121 cal/g = –5.06 x 10^5 J/kg.
Vapor pressure: (Reid) Low

HYDROXYLAMINE REC. H:1800

SYNONYMS: HIDROXILAMINA (Spanish); OXAMMONIUM

IDENTIFICATION
CAS Number: 7803-49-8
Formula: H_3NO; NH_2OH

DESCRIPTION: White crystalline solid; melts at 90°F/32°C, forming a colorless liquid. Usually shipped and stored in liquid form as hydroxylamine sulfate or hydroxylamine hydrochloride. Odorless. Sinks and mixes with water.

Thermally unstable EXPLOSIVE! • Container may explode when exposed to fire or heated above 265°F/129°C. May explode at lower temperatures if material is exposed to air • Hydroxylamine sulfate or hydroxylamine hydrochloride, if confined, will decompose and explode if exposed to contaminants or elevated temperatures above 265°F/129°C (contamination may lower this temperature to 158°F/70°C) • Extremely Irritating to the skin, eyes, and respiratory tract; prolonged contact with skin can cause burns; absorbed by the skin • May interfere with the body's ability to use oxygen • Toxic products of combustion may include nitrogen oxides.

Hazard Classification (based on NFPA-704 Rating System)
Health Hazards (Blue): 2; Flammability (Red): 0; Reactivity (Yellow): 3

EMERGENCY RESPONSE: See Appendix A (171)
Evacuation:
Public safety: Isolate the area of spill or leak for at least 50 to 100 meters (160 to 300 feet) in all directions.
Spill: Increase, in the downwind direction, as necessary, the distance shown under "Public Safety."
Fire: If any large container is involved in fire, isolate for at least 800 meters (½ mile) in all directions; also, consider initial evacuation for 800 meters (½ mile) in all directions.

EXPOSURE
Short-term effects: Following symptoms are possible; headache, vertigo, tinnitus, dyspnea, nausea and vomiting, cyanosis, proteinuria and hematuria, jaundice, restlessness, and convulsion. *SEEK MEDICAL ATTENTION.* **Liquid or solid:** Irritating to skin and eyes. *IF INHALED*, will or swallowed may cause headache, dizziness, ringing in ears, labored breathing, nausea and vomiting. Remove contaminated clothing and shoes. Flush affected areas with plenty of water. *IF IN EYES*, hold eyelids open and flush with plenty of water. *IF SWALLOWED* and victim is *CONSCIOUS AND ABLE TO SWALLOW*, have victim drink 4 to 8 ounces of water and have victim induce vomiting. *IF BREATHING HAS STOPPED*, give artificial respiration.
Note to physician or authorized medical personnel: Give milk and demulcents, induce emesis or perform gastric lavage: Give fluids: observe for methemoglobinemia. If needed, give methylene blue as a 1% solution intravenously, 1–2 mg/kg; an oral dose of 3–5 mg/kg. If severe, consider exchange transfusion with whole blood.

HEALTH HAZARDS
Personal protective equipment (PPE): B-Level PPE. Wear protective clothing, cap, gloves, goggles-canister type mask recommended. Butyl rubber is generally suitable for hydroxyl compounds.
Toxicity by ingestion: Grade 3; LD_{50} = 50 to 500 mg/kg.
Long-term health effects: Potential mutagenic and teratogenic effects. Repeated exposure may enhance allergic reaction of the back of hands and forearms. May cause blood disorders; asthma. Eczema following prolonged contact.

Vapor (gas) irritant characteristics: Irritates the eyes, skin, and respiratory tract.
Liquid or solid irritant characteristics: Causes smarting of the skin and first-degree burns on short exposure; may cause second-degree burns on long exposure.

FIRE DATA
Flash point: 265°F/129°C (cc) (may explode).
Behavior in fire: May explode when exposed to flame, or heat above 265°F/129°C.

CHEMICAL REACTIVITY
Reactivity with water: Forms alkaline solution. In high pH conditions, it decomposes to ammonium hydroxide, nitrogen and hydrogen.
Binary reactants: Reacts with strong oxidizers. May attack some metals.
Neutralizing agents for acids and caustics: Sodium bisulfate

ENVIRONMENTAL DATA
Food chain concentration potential: Log P_{ow} = –1.5. Negative; unlikely to accumulate.
Water pollution: Harmful to aquatic life in very low concentrations. May be dangerous if it enters nearby water intakes; notify operators. Notify local health and wildlife officials.
Response to discharge: Issue warning–high flammability. Evacuate area. Disperse and flush.

SHIPPING INFORMATION
Grades of purity: 100% pure; **Storage temperature:** Cool-noncombustible building; keep out of sunlight; **Venting:** Open occasionally to relieve decomposition products; **Stability during transport:** Unstable-hygroscopic. Decomposes even at room temperature, especially in presence of atmospheric moisture and CO_2.

PHYSICAL AND CHEMICAL PROPERTIES
Physical state @ 59°F/15°C and 1 atm: Solid.
Molecular weight: 33.03
Boiling point @ 1 atm: (decomposes)134°F/56°C/330°K @ 22 mm; 158°F/70°C/343.2°K @ 60 mm
Melting/Freezing point: 92°F/33°C/306°K.
Specific gravity (water = 1): 1.227 at room temperature
Relative vapor density (air = 1): 1.14 (calculated).
Latent heat of vaporization: 880 Btu/lb = 488.9 cal/g = 2.04 x 10^5 J/kg.
Heat of solution: 207 Btu/lb = 115 cal/g = 4.81 x 10^5 J/kg.
Vapor pressure: 10 mm @ 117°F/47°C/320°K.

HYDROXYLAMINE SULFATE   REC. H:1850

SYNONYMS: BIS(HYDROXYLAMINE) SULFATE; HYDROXYLAMINE NEUTRAL; HYDROXYLAMMONIUM SULFATE; OXAMMONIUM SULFATE; SULFATE; HS; SULFATO de HIDROXILAMINA (Spanish)

IDENTIFICATION
CAS Number: 10039-54-0
Formula: $H_6N_2O_2 \cdot H_2O_4S$; $(NH_2OH)_2 \cdot H_2SO_4$
DOT ID Number: UN 2865; DOT Guide Number: 154
Proper Shipping Name: Hydroxylamine sulfate; Hydroxylamine sulphate

DESCRIPTION: White solid. Slight ammonia odor. Sinks in water; soluble.

Poison! • Corrosive to the skin, eyes, and respiratory tract; contact may cause severe eye and skin burns; inhalation symptoms may be delayed • Can decompose explosively at temperatures as low as 194°F/90°C. Hydroxylamine sulfate, if confined, will decompose and explode if exposed to flame or temperatures above 284°F/140°C • May interfere with the body's ability to use oxygen • Firefighting gear (including SCBA) does not provide adequate protection. If exposure occurs, remove and isolate gear immediately and thoroughly decontaminate personnel • Containers may BLEVE when exposed to fire • Toxic products of combustion may include oxides of sulfur and nitrogen.

Hazard Classification (based on NFPA-704 Rating System)
Health Hazards (Blue): 1; Flammability (Red): 0; Reactivity (Yellow): 1

EMERGENCY RESPONSE: See Appendix A (154)
Evacuation:
Public safety: Isolate the area of spill or leak for at least 25 to 50 meters (80 to 160 feet) in all directions.
Spill: Increase, in the downwind direction, as necessary, the distance shown under "Public Safety."
Fire: If tank, rail car, or tank truck is involved in fire, isolate for at least 800 meters (½ mile) in all directions; also, consider initial evacuation for 800 meters (½ mile) in all directions.

EXPOSURE
Short-term effects: *SEEK MEDICAL ATTENTION.* **Dust:** Irritating to eyes, nose, and throat. *IF INHALED*, will, will cause difficult breathing or loss of consciousness. *IF IN EYES*, hold eyelids open and flush with plenty of water. *IF BREATHING HAS STOPPED*, give artificial respiration; *avoid mouth-to-mouth resuscitation; use bag/mask apparatus.* IF breathing is difficult, administer oxygen. **Solid:** *POISONOUS IF SWALLOWED.* Irritating to skin and eyes. *IF SWALLOWED*, will cause nausea or loss of consciousness. Remove contaminated clothing and shoes. Flush affected areas with plenty of water. *IF IN EYES*, hold eyelids open and flush with plenty of water. *IF SWALLOWED* and victim is *CONSCIOUS AND ABLE TO SWALLOW*, have victim drink 4 to 8 ounces of water and have victim induce vomiting. *IF SWALLOWED* and victim is *UNCONSCIOUS OR HAVING CONVULSIONS*, do nothing except keep victim warm.
Note to physician or authorized medical personnel: Medical observation is recommended for 24 to 48 hours after breathing overexposure, as pulmonary edema may be delayed. As first aid for pulmonary edema, consider administering a corticosteroid spray. Cigarette smoking may exacerbate pulmonary injury and should be discouraged for at least 72 hours following exposure.
Note to physician or authorized medical personnel: Observe for methemoglobinemia. If needed, give methylene blue as a 1% solution intravenously, 1–2 mg/kg; an oral dose of 3–5 mg/kg. If severe, consider exchange transfusion with whole blood.

HEALTH HAZARDS
Personal protective equipment (PPE): B-Level PPE. Acid-resistant protective clothing, including coveralls, wrist-length gloves, cap, goggles, and dust mask. Butyl rubber is generally suitable for hydroxyl compounds.
Toxicity by ingestion: Grade 3; LD_{50} = 50 to 500 mg/kg.
Long-term health effects: May effect the nervous system and blood.

Vapor (gas) irritant characteristics: Eye and respiratory tract irritant.
Liquid or solid irritant characteristics: Eye and skin irritant.

CHEMICAL REACTIVITY
Reactivity with water: Forms acid.
Binary reactants: Violent reaction with aluminum, magnesium, oxidants, and bases. Corrosive to metals in presence of moisture.
Neutralizing agents for acids and caustics: Flush with water.

ENVIRONMENTAL DATA
Food chain concentration potential: Negative; unlikely to accumulate.
Water pollution: Effect of low concentrations on aquatic life is unknown. May be dangerous if it enters nearby water intakes; notify operators. Notify local health and wildlife officials.
Response to discharge: Issue warning–water contaminant. Restrict access. Disperse and flush.

SHIPPING INFORMATION
Grades of purity: Commercial, 97–99%; **Storage temperature:** Ambient; **Inert atmosphere:** None; **Venting:** Open; **Stability during transport:** Stable.

PHYSICAL AND CHEMICAL PROPERTIES
Physical state @ 59°F/15°C and 1 atm: Solid.
Molecular weight: 164.14
Boiling point @ 1 atm: Decomposes.
Melting/Freezing point: Decomposes 250°F/121°C/394°K.
Specific gravity (water = 1): More than 1 @ 68°F/20°C (solid).

2-HYDROXY-4-(METHYLTHIO)-BUTANOIC ACID
REC. H:1900

SYNONYMS: ALIMET; BUTRYIC ACID, 2-HYDROXY-4-METHYLTHIO-; MHA ACID; MHA-FA; HSDB 5700; METHIONINE HYDROXY ANALOG

IDENTIFICATION
CAS Number: 583-91-5
Formula: $C_5H_{10}O_3S$; $CH_3S(CH_2)_2CHOHCOOH$
DOT ID Number: UN 1760; DOT Guide Number: 154
Proper Shipping Name: Corrosive liquid, n.o.s.

DESCRIPTION: Light brown liquid.

Combustible • Heat or flame may cause explosion • Containers may BLEVE when exposed to fire • Vapors are heavier than air and will collect and stay in low areas • Vapors in confined areas (e.g., tanks, sewers, buildings) may explode when exposed to fire • Decomposes above 160°C/320°F • Corrosive to the skin, eyes, and respiratory tract. May cause pulmonary edema (symptoms may be delayed) • Toxic products of combustion may include sulfur oxides.

Hazard Classification (based on NFPA-704 Rating System)
Health Hazards (Blue): 1; Flammability (Red): 1; Reactivity (Yellow): 0

EMERGENCY RESPONSE: See Appendix A (154)
Evacuation:
Public safety: Isolate the area of spill or leak for at least 25 to 50 meters (80 to 160 feet) in all directions.

Spill: Increase, in the downwind direction, as necessary, the distance shown under "Public Safety."
Fire: If tank, rail car, or tank truck is involved in fire, isolate for at least 800 meters (½ mile) in all directions; also, consider initial evacuation for 800 meters (½ mile) in all directions.

EXPOSURE
Short-term effects: *SEEK MEDICAL ATTENTION*. **Vapor:** Irritating to eyes, nose, and throat. *IF INHALED*, will, will cause coughing or difficult breathing. *IF IN EYES*, hold eyelids open and flush with plenty of water. *IF BREATHING HAS STOPPED*, give artificial respiration; *avoid mouth-to-mouth resuscitation; use bag/mask apparatus*. IF breathing is difficult, administer oxygen. **Liquid:** Will burn skin and eyes. *IF SWALLOWED*, may cause nausea and vomiting. Remove contaminated clothing and shoes. Flush affected areas with plenty of water. *IF IN EYES*, hold eyelids open and flush with plenty of water.

HEALTH HAZARDS
Personal protective equipment (PPE): B-Level PPE. Wear approved respirator with full face piece, chemical resistant gloves and clothing.
Toxicity by ingestion: Grade 2: LD_{50} = 3.478 g/kg (rat).
Vapor (gas) irritant characteristics: Vapors cause severe irritation of eyes and throat and can cause eye and lung injury. They cannot be tolerated even at low concentrations.
Liquid or solid irritant characteristics: Severe skin irritant. Causes second- and third-degree burns on short contact and is very injurious to the eyes.

FIRE DATA
Flash point: 250°F/121°C (cc).
Autoignition temperature: Decomposes above 160°C/320°F/593°K.

CHEMICAL REACTIVITY
Binary reactants: Reacts with sulfuric acid, ammonia, isocyanates, epichlorohydrin.
Neutralizing agents for acids and caustics: Sodium bicarbonate solution. Flush with water.
Reactivity group: 4
Compatibility class: Organic acids

ENVIRONMENTAL DATA
Water pollution: May be dangerous to aquatic life in high concentrations. May be dangerous if it enters nearby water intakes; notify operators. Notify local health and wildlife officials. **Response to discharge:** Issue warning–corrosive. Mechanical containment. Should be removed. Chemical and physical treatment.

SHIPPING INFORMATION
Stability during transport: Stable.

PHYSICAL AND CHEMICAL PROPERTIES
Physical state @ 59°F/15°C and 1 atm: Liquid.
Molecular weight: 150.2
Specific gravity (water = 1): 1.21–1.23
Relative vapor density (air = 1): 5.19
Vapor pressure: (Reid) 0.48–0.55 psia.

HYDROXYPROPYL ACRYLATE
REC. H:1950

SYNONYMS: ACRILATO de 2-HIDROXIPROPILO (Spanish); ACRYLIC ACID-2-HYDROXYPROPYL ESTER; *β*-

β-HYDROXYPROPYL ACRYLATE; HPA; 1,2-PROPANEDIOL-1-ACRYLATE; PROPYLENE GLYCOL MONOACRYLATE

IDENTIFICATION
CAS Number: 999-61-1
Formula: $C_6H_{10}O_3$; $CH_3CHOHCH_2OCOCH=CH_2$
DOT ID Number: UN 1760; DOT Guide Number: 154
Proper Shipping Name: Corrosive liquid, n.o.s.

DESCRIPTION: Colorless liquid Faint unpleasant odor; slightly acrylic. Mixes with water.

Combustible • Corrosive to the skin, eyes, and respiratory tract; inhalation symptoms may be delayed • Heat or flame may cause explosion • Containers may BLEVE when exposed to fire • Vapors are heavier than air and will collect and stay in low areas • Vapors in confined areas (e.g., tanks, sewers, buildings) may explode when exposed to fire • Toxic products of combustion may include carbon monoxide • Polymerization hazard • Heat can induce polymerization with rapid release of energy; sealed containers may rupture explosively • May react with itself blocking relief valves; leading to tank explosions.

Hazard Classification (based on NFPA-704 Rating System)
Health Hazards (Blue): 3; Flammability (Red): 1; Reactivity (Yellow): 2

EMERGENCY RESPONSE: See Appendix A (154)
Evacuation:
Public safety: Isolate the area of spill or leak for at least 25 to 50 meters (80 to 160 feet) in all directions.
Spill: Increase, in the downwind direction, as necessary, the distance shown under "Public Safety."
Fire: If tank, rail car, or tank truck is involved in fire, isolate for at least 800 meters (½ mile) in all directions; also, consider initial evacuation for 800 meters (½ mile) in all directions.

EXPOSURE
Short-term effects: *SEEK MEDICAL ATTENTION.* **Vapor:** Irritating to eyes, nose, and throat. *IF INHALED,* will, will cause coughing or difficult breathing. *IF IN EYES,* hold eyelids open and flush with plenty of water. *IF BREATHING HAS STOPPED,* give artificial respiration; *avoid mouth-to-mouth resuscitation; use bag/mask apparatus.* IF breathing is difficult, administer oxygen. May cause lung edema; physical exertion will aggravate this condition. **Liquid:** *HARMFUL IF SWALLOWED OR ABSORBED THROUGH THE SKIN.* Will burn skin and eyes. *IF SWALLOWED,* will cause nausea. Remove contaminated clothing and shoes. Flush affected areas with plenty of water. *IF IN EYES,* hold eyelids open and flush with plenty of water. *IF SWALLOWED* and victim is *CONSCIOUS AND ABLE TO SWALLOW,* have victim drink 4 to 8 ounces of water. *IF SWALLOWED* and victim is *UNCONSCIOUS OR HAVING CONVULSIONS,* do nothing except keep victim warm.
Note to physician or authorized medical personnel: Medical observation is recommended for 24 to 48 hours after breathing overexposure, as pulmonary edema may be delayed. As first aid for pulmonary edema, consider administering a corticosteroid spray. Cigarette smoking may exacerbate pulmonary injury and should be discouraged for at least 72 hours following exposure.

HEALTH HAZARDS
Personal protective equipment (PPE): B-Level PPE. OSHA Table Z-1-A air contaminant. Organic vapor respirator. Butyl rubber gloves, apron, and boots; worker's goggles or face shield. Also, polycarbonate offers good chemical resistance.
Exposure limits (TWA unless otherwise noted): ACGIH TLV 0.5 ppm (2.8 mg/m^3); NIOSH 0.5 ppm (3 mg/m^3), prevent skin contact.
Toxicity by ingestion: Poison. Grade 2; oral LD_{50} = 1230 mg/kg (rat).
Vapor (gas) irritant characteristics: Eye irritant. Vapors are moderately irritating such that personnel will not usually tolerate moderate or high concentrations.
Liquid or solid irritant characteristics: Severe eye and skin irritant. Causes second- and third-degree burns on short contact and is very injurious to the eyes.

FIRE DATA
Flash point: 212°F/100°C (oc); 207°F/97°C (cc).
Flammable limits in air: LEL: 1.4% @ 212°F/100°C.
Fire extinguishing agents not to be used: Water may be ineffective.
Behavior in fire: Can become unstable at high temperatures and pressure or may react with water with some release of energy, but not violently.

CHEMICAL REACTIVITY
Reactivity with water: Mild reaction.
Binary reactants: Reacts violently with strong oxidizers, sodium peroxide, uranium fluoride. Incompatible with strong acids, nitrates, sulfuric acid, nitric acid, caustics, aliphatic amines, isocyanates, boranes.
Polymerization: May occur; avoid exposure to high temperatures, ultraviolet light, free-radical initiators; **Inhibitor of polymerization:** 200 ppm hydroquinone.
Reactivity group: 14
Compatibility class: Acrylates

ENVIRONMENTAL DATA
Food chain concentration potential: Negative; unlikely to accumulate.
Water pollution: Effect of low concentrations on aquatic life is unknown. May be dangerous if it enters nearby water intakes; notify operators. Notify local health and wildlife officials. **Response to discharge:** Issue warning–water contaminant. Restrict access. Disperse and flush. Neutralize with crushed limestone, soda ash, or lime,

SHIPPING INFORMATION
Grades of purity: Commercial, 97%; **Storage temperature:** Ambient; **Inert atmosphere:** None; **Venting:** Open; **Stability during transport:** Stable.

NAS HAZARD CLASSIFICATION FOR BULK WATER TRANSPORTATION
FIRE: 1
HEALTH: Vapor irritant: 3; Liquid or solid irritant: 4; Poisons: 3
WATER POLLUTION: Human toxicity: 2; Aquatic toxicity: 2; Aesthetic effect: 1
REACTIVITY: Other chemicals: 2; Water: 1; Self-reaction: 3

PHYSICAL AND CHEMICAL PROPERTIES
Physical state @ 59°F/15°C and 1 atm: Liquid.
Molecular weight: 130.2
Boiling point @ 1 atm: 401°F/205°C/478°K.
Specific gravity (water = 1): 1.06 @ 77°F/25°C (liquid).
Relative vapor density (air = 1): 4.5
Heat of combustion: (estimate) –12,300 Btu/lb –6850 cal/g = –287 x 10^5 J/kg.

HYDROXYPROPYL METHACRYLATE   REC. H:2000

SYNONYMS: METACRILATO de HIDROXIPROPILO (Spanish); PROPYLENE GLYCOL MONOMETHACRYLATE; 1,2-PROPANEDIOL 1-METHACRYLATE

IDENTIFICATION
CAS Number: 27813-02-1
Formula: $C_3CHNOHCH_2OCOC(CH_3)=CH_2$

DESCRIPTION: Colorless, clear, mobile liquid. Slightly acrylic odor. Sinks slowly in water; slightly soluble.

Combustible • Heat or flame may cause explosion • Containers may BLEVE when exposed to fire • Vapors are heavier than air and will collect and stay in low areas • Vapors in confined areas (e.g., tanks, sewers, buildings) may explode when exposed to fire • Toxic products of combustion may include carbon monoxide • Polymerization hazard • Heat may induce polymerization with rapid release of energy; sealed containers may rupture explosively • May react with itself blocking relief valves; leading to tank explosions.
• Toxic products of combustion may include carbon monoxide.

EMERGENCY RESPONSE: See Appendix A (171)
Evacuation:
Public safety: Isolate the area of spill or leak for at least 10 to 25 meters (30 to 80 feet) in all directions.
Spill: Increase, in the downwind direction, as necessary, the distance shown under "Public Safety."
Fire: If any large container is involved in fire, isolate for at least 800 meters (½ mile) in all directions; also, consider initial evacuation for 800 meters (½ mile) in all directions.

EXPOSURE
Short-term effects: *SEEK MEDICAL ATTENTION.* **Vapor:** Irritating to eyes, nose, and throat. *IF INHALED*, will, will cause coughing or difficult breathing. *IF IN EYES*, hold eyelids open and flush with plenty of water. *IF BREATHING HAS STOPPED*, give artificial respiration. IF breathing is difficult, administer oxygen.
Liquid: Will burn eyes,
Irritating to eyes. *IF SWALLOWED*, will cause nausea and vomiting. Remove contaminated clothing and shoes. Flush affected areas with plenty of water. *IF IN EYES*, hold eyelids open and flush with plenty of water. *IF SWALLOWED* and victim is *CONSCIOUS AND ABLE TO SWALLOW*, have victim drink 4 to 8 ounces of water. *IF SWALLOWED* and victim is *UNCONSCIOUS OR HAVING CONVULSIONS*, do nothing except keep victim warm.

HEALTH HAZARDS
Personal protective equipment (PPE): B-Level PPE. Goggles or face shield; rubber gloves.
Toxicity by ingestion: Grade 1; LD_{50} = 5–500 g/kg (mouse).
Vapor (gas) irritant characteristics: Eye and respiratory tract irritant.
Liquid or solid irritant characteristics: Severe eye and skin irritant.

FIRE DATA
Flash point: 250°F/121°C (oc).
Fire extinguishing agents not to be used: Water may be ineffective.
Behavior in fire: Compound may polymerize when hot and burst container.

CHEMICAL REACTIVITY
Polymerization: May polymerize when hot or when exposed to ultraviolet light and free-radical catalysts; **Inhibitor of polymerization:** 200 ppm hydroquinone
Reactivity group: 14
Compatibility class: Acrylates

ENVIRONMENTAL DATA
Food chain concentration potential: Negative; unlikely to accumulate.
Water pollution: Effect of low concentrations on aquatic life is unknown. May be dangerous if it enters nearby water intakes; notify operators. Notify local health and wildlife officials. **Response to discharge:** Issue warning–water contaminant. Restrict access. Disperse and flush.

SHIPPING INFORMATION
Grades of purity: Commercial, 95+%; **Storage temperature:** Ambient; **Inert atmosphere:** None; **Venting:** Open; **Stability during transport:** Stable.

PHYSICAL AND CHEMICAL PROPERTIES
Physical state @ 59°F/15°C and 1 atm: Liquid.
Molecular weight: 144
Boiling point @ 1 atm: (decomposes).
Specific gravity (water = 1): 1.06 @ 68°F/20°C (liquid).

ISOAMYLACETATE   REC. I:0100

SYNONYMS: ACETATO de ISOAMILICO (Spanish); ACETIC ACID, ISOPENTYL ESTER; AMYL ACETATE, ISO; AMYLACETIC ESTER; BANANA OIL; ISOAMILACETATO (Spanish); ISOAMYL ETHANOATE; ISOPENTYL ACETATE; 3-METHYL-1-BUTYL ACETATE; 3-METHYL-1-BUTANOL ACETATE; 3-METHYLBUTYL ESTER OF ACETIC ACID; 3-METHYLBUTYL ETHANOATE; PEAR OIL

IDENTIFICATION
CAS Number: 123-92-2
Formula: $C_7H_{14}O_2$; $CH_3COOCH_2CH_2CH(CH_3)_2$
DOT ID Number: UN 1104; DOT Guide Number: 129
Proper Shipping Name: Amyl acetates
Reportable Quantity (RQ): **(CERCLA)** 5000 lb/2270 kg

DESCRIPTION: Colorless oily liquid. Banana odor. Floats and mixes slowly with water. Flammable, irritating vapor is produced.

Highly flammable • Containers may BLEVE when exposed to fire •Vapors may form explosive mixture with air • Vapors are heavier than air and will collect and stay in low areas • Vapors may travel long distances to ignition sources and flashback • Vapors in confined areas (e.g., tanks, sewers, buildings) may explode when exposed to fire • Irritating to the skin, eyes, and respiratory tract • Toxic products of combustion may include carbon monoxide.

Hazard Classification (based on NFPA-704 Rating System)
Health Hazards (Blue): 1; Flammability (Red): 3; Reactivity (Yellow): 0

EMERGENCY RESPONSE: See Appendix A (129)
Evacuation:
Public safety: Isolate spill area for at least 50 to 100 meters (160 to 330 feet) in all directions.

Spill: Large spill–Consider initial downwind evacuation for at least 300 meters (1000 feet).
Fire: Isolate for 800 meters (½ mile) in all directions, especially if tank, rail car, or tank truck is involved in fire.

EXPOSURE
Short-term effects: *SEEK MEDICAL ATTENTION.* **Vapor:** Irritating to eyes, nose, and throat. *IF INHALED*, will, will cause nausea, headache or dizziness. Move to fresh air. *IF BREATHING HAS STOPPED,* give artificial respiration. IF breathing is difficult, administer oxygen. Alcohol consumption will increase toxic effects. **Liquid:** Irritating to skin and eyes. Harmful if swallowed. Remove contaminated clothing and shoes. Flush affected areas with plenty of water. *IF IN EYES,* hold eyelids open and flush. with plenty of water. *IF SWALLOWED* and victim is *CONSCIOUS* have victim drink 4 to 8 ounces of water.

HEALTH HAZARDS
Personal protective equipment (PPE): B-Level PPE. Rubber gloves, chemical goggles or face shield, and lab coat.
Recommendations for respirator selection: NIOSH
1000 ppm: CCROV [any chemical cartridge respirator with organic vapor cartridge(s)]; or PAPROV [any powered, air-purifying respirator with organic vapor cartridge(s)]; or GMFOV[any air-purifying, full-facepiece respirator (gas mask) with a chin-style, front- or back-mounted organic vapor cannister]; or SA (any supplied-air respirator); or SCBAF (any self-contained breathing apparatus with a full facepiece). *EMERGENCY OR PLANNED ENTRY INTO UNKNOWN CONCENTRATIONS OR IDLH CONDITIONS:* SCBAF:PD,PP (any self-contained breathing apparatus that has a full facepiece and is operated in a pressure-demand or other positive-pressure mode); or SAF:PD,PP:ASCBA (any supplied-air respirator that has a full facepiece and is operated in a pressure-demand or other positive-pressure mode in combination with an auxiliary self-contained breathing apparatus operated in a pressure-demand or other positive pressure mode). *ESCAPE:* GMFOV[any air-purifying, full-facepiece respirator (gas mask) with a chin-style, front- or back-mounted organic vapor cannister]; or SCBAE (any appropriate escape-type, self-contained breathing apparatus).
Exposure limits (TWA unless otherwise noted): ACGIH TLV 50 ppm (266 mg/m^3); NIOSH/OSHA 100 ppm (525 mg/m^3).
Short-term exposure limits (15-minute TWA): ACGIH STEL 100 ppm (532 mg/m^3).
Toxicity by ingestion: Grade 1; LD_{50} = 5 to 15 g/kg.
Vapor (gas) irritant characteristics: Eye and respiratory tract irritant. Vapors cause smarting of the eyes or respiratory system if present in high concentration.
Liquid or solid irritant characteristics: Eye and skin irritant. If spilled on clothing and allowed to remain, may cause smarting and reddening of the skin.
Odor threshold: 0.22 ppm.
IDLH value: 1000 ppm.

FIRE DATA
Flash point: 77°F/25°C (cc).
Flammable limits in air: LEL: 1.0% @ 122°F/50°C; UEL: 7.5%.
Fire extinguishing agents not to be used: Water may be ineffective.
Behavior in fire: When exposed to flames can react vigorously with reducing materials.
Autoignition temperature: 680°F/360°C/633°K.
Electrical hazard: Class I, Group D.

CHEMICAL REACTIVITY
Binary reactants: Reacts with nitrates, strong oxidizers, alkalis, and acids. Attacks asbestos. Softens then dissolves many plastic materials, and rubber.
Reactivity group: 34
Compatibility class: Ester

ENVIRONMENTAL DATA
Food chain concentration potential: Log P_{ow} = 1.2. Unlikely to accumulate.
Water pollution: Harmful to aquatic life in very low concentrations. Fouling to shoreline. May be dangerous if it enters nearby water intakes; notify operators. Notify local health and pollution control official. **Response to discharge:** Issue warning–high flammability. Mechanical containment. Chemical and physical treatment.

SHIPPING INFORMATION
Storage temperature: Ambient; **Storage temperature:** (cool); **Stability during transport:** Stable.

PHYSICAL AND CHEMICAL PROPERTIES
Physical state @ 59°F/15°C and 1 atm: Liquid.
Molecular weight: 130.18
Boiling point @ 1 atm: 287.6°F/142°C/415.2°K.
Melting/Freezing point: –109.3°F/–78.5°C/194.7°K.
Critical temperature: 619°F/326.1°C/599.3°K.
Critical pressure: 411.5 psia = 28.0 atm = 2.84 MN/m^2.
Specific gravity (water = 1): 0.876@15°C.
Liquid surface tension: 24.77 dynes/cm = 0.02477 N/m @ 68°F/20°C.
Liquid water interfacial tension: 50.2 dynes/cm = 0.0502 N/m @ 15°C.
Relative vapor density (air = 1): 4.5
Ratio of specific heats of vapor (gas): (estimate) > 1-1.1
Latent heat of vaporization: (estimate) 132 Btu/lb = 73.3 cal/g = 3.07 x 10^5 J/kg.
Heat of combustion: –14,402 Btu/lb = –8000 cal/g = 334.9 x 10^5 J/kg.
Vapor pressure: 4 mm.

ISOAMYL ALCOHOL, PRIMARY   REC. I:0150

SYNONYMS: ALCOHOL ISOAMILICO PRIMARIO (Spanish); EEC No. 603-006-00-7; FERMENTATION AMYL ALCOHOL; FUSEL OIL; ISOAMYL ALCOHOL (primary); ISOAMYOL; ISOBUTYL CARBINOL; ISOPENTYL ALCOHOL; 3-METHYL-1-BUTANOL (primary); POTATO SPIRIT OIL

IDENTIFICATION
CAS Number: 123-51-3
Formula: $C_5H_{12}O$; $(CH_3)_2CHCH_2CH_2OH$
DOT ID Number: UN 1105; DOT Guide Number: 129
Proper Shipping Name: Amyl alcohols

DESCRIPTION: Colorless liquid Mild, choking alcohol odor. Floats on water surface; slightly soluble: Irritating vapor is produced.

Flammable • Containers may BLEVE when exposed to fire •Vapors may form explosive mixture with air • Vapors are heavier than air and will collect and stay in low areas •Vapors may travel long distances to ignition sources and flashback • Vapors in confined areas (e.g., tanks, sewers, buildings) may explode when exposed to

618 Isoamyl alcohol, secondary

fire • Irritating to the skin, eyes, and respiratory tract • Toxic products of combustion may include carbon monoxide.

Hazard Classification (based on NFPA-704 Rating System)
Health Hazards (Blue): 1; Flammability (Red): 2; Reactivity (Yellow): 0

EMERGENCY RESPONSE: See Appendix A (129)
Evacuation:
Public safety: Isolate spill area for at least 50 to 100 meters (160 to 330 feet) in all directions.
Spill: Large spill–Consider initial downwind evacuation for at least 300 meters (1000 feet).
Fire: Isolate for 800 meters (½ mile) in all directions, especially if tank, rail car, or tank truck is involved in fire.

EXPOSURE
Short-term effects: *SEEK MEDICAL ATTENTION*. **Vapor:** Irritating to eyes, nose, and throat. Harmful if inhaled. Move to fresh air. *IF BREATHING HAS STOPPED*, give artificial respiration. IF breathing is difficult, administer oxygen. **Liquid:** Irritating to eyes. *IF IN EYES*, hold eyelids open and flush. with plenty of water.

HEALTH HAZARDS
Personal protective equipment (PPE): B-Level PPE. OSHA Table Z-1 air contaminant. Face shield to avoid splash. Chemical protective material(s) reported to have good to excellent resistance: butyl rubber, natural rubber, neoprene, nitrile.
Recommendations for respirator selection: NIOSH/OSHA
500 ppm: SA:CF* (any supplied-air respirator operated in a continuous-flow mode); or CCRFOV [any air-purifying, full-facepiece respirator (gas mask) with a chin-style, front- or back-mounted acid gas canister]; or GMFOV[any air-purifying, full-facepiece respirator (gas mask) with a chin-style, front- or back-mounted organic vapor canister]; or PAPROV* [any powered, air-purifying respirator with organic vapor cartridge(s)]; or SCBAF (any self-contained breathing apparatus with a full facepiece); or SAF (any supplied-air respirator with a full facepiece). *EMERGENCY OR PLANNED ENTRY INTO UNKNOWN CONCENTRATIONS OR IDLH CONDITIONS*: SCBAF:PD,PP (any self-contained breathing apparatus that has a full facepiece and is operated in a pressure-demand or other positive-pressure mode); or SAF:PD,PP:ASCBA (any supplied-air respirator that has a full facepiece and is operated in a pressure-demand or other positive-pressure mode in combination with an auxiliary self-contained breathing apparatus operated in a pressure-demand or other positive pressure mode). *ESCAPE*: GMFOV[any air-purifying, full-facepiece respirator (gas mask) with a chin-style, front- or back-mounted organic vapor canister]; or SCBAE (any appropriate escape-type, self-contained breathing apparatus).
Note: Substance causes eye irritation or damage; eye protection needed.
Exposure limits (TWA unless otherwise noted): ACGIH TLV 100 ppm (361 mg/m^3); NIOSH/OSHA 100 ppm (360 mg/m^3).
Short-term exposure limits (15-minute TWA): ACGIH STEL 125 ppm (452 mg/m^3); NIOSH 125 ppm (450 mg/m^3).
Toxicity by ingestion: Grade 2; LD$_{50}$ = 0.5 to 5 g/kg.
Long-term health effects:
Vapor (gas) irritant characteristics: Eye and respiratory tract irritant. Vapor is irritating such that personnel will not usually tolerate moderate or high vapor concentrations.
Liquid or solid irritant characteristics: Liquid may irritate skin.
Odor threshold: 0.03–0.07 ppm.
IDLH value: 500 ppm.

FIRE DATA
Flash point: 114°F/46°C (oc); 109°F/43°C (cc).
Flammable limits in air: LEL: 1.2%; UEL: 9.0% @ 212°F/100°C.
Fire extinguishing agents not to be used: Water may be ineffective.
Ignition temperature: 644°F/340°C/613°K.
Electrical hazard: Class I, Group D.
Burning rate: 3.6 mm/min

CHEMICAL REACTIVITY
Binary reactants: Violent reaction with strong oxidizers.
Reactivity group: 20
Compatibility class: Alcohols, glycols

ENVIRONMENTAL DATA
Food chain concentration potential: Log P$_{ow}$ = 1.3. Unlikely to accumulate.
Water pollution: Dangerous to aquatic life in high concentrations. May be dangerous if it enters nearby water intakes; notify operators. Notify local health and wildlife officials. **Response to discharge:** Disperse and flush.

SHIPPING INFORMATION
Grades of purity: Pure; fusel oil; **Storage temperature:** Ambient; **Inert atmosphere:** None; **Venting:** Open (flame arrester); **Stability during transport:** Stable.

PHYSICAL AND CHEMICAL PROPERTIES
Physical state @ 59°F/15°C and 1 atm: Liquid.
Molecular weight: 88.15
Boiling point @ 1 atm: 280°F/138°C/411°K.
Melting/Freezing point: 179°F/–117°C/156°K.
Critical temperature: 585°F/307°C/580°K.
Specific gravity (water = 1): 0.82 @ 68°F/20°C (liquid).
Liquid surface tension: 23.8 dynes/cm = 0.0238 N/m @ 68°F/20°C.
Liquid water interfacial tension: 5 dynes/cm = 0.005 N/m at 18°C.
Relative vapor density (air = 1): 3.04
Ratio of specific heats of vapor (gas): (estimate) 1.062
Latent heat of vaporization: 215.6 Btu/lb = 119.8 cal/g = 5.016 x 10^5 J/kg.
Heat of combustion: –16,200 Btu/lb = –9000 cal/g = –376.8 x 10^5 J/kg.
Heat of solution: –57.1 Btu/lb = –31.7 cal/g = –1.33 x 10^5 J/kg.
Vapor pressure: 28 mm.

ISOAMYL ALCOHOL, SECONDARY   REC. I:0200

SYNONYMS: ALCOHOL ISOAMILICO SECUNDARIO (Spanish); 3-METHYL-2-BUTANOL; SECONDARY ISOAMYL ALCOHOL

IDENTIFICATION
CAS Number: 528-75-4
Formula: (CH$_3$)$_2$CHCH(OH)CH$_2$
DOT ID Number: UN 1105; DOT Guide Number: 129
Proper Shipping Name: Amyl alcohols

DESCRIPTION: Colorless liquid. Mild, alcoholic odor. Floats on the surface of water; soluble. Irritating vapor is produced.

Flammable • Containers may BLEVE when exposed to fire • Vapors may form explosive mixture with air • Vapors are heavier than air and will collect and stay in low areas • Vapors may travel long distances to ignition sources and flashback • Vapors in confined areas (e.g., tanks, sewers, buildings) may explode when exposed to fire • Irritating to the skin, eyes, and respiratory tract • Toxic products of combustion may include carbon monoxide.

Hazard Classification (based on NFPA-704 Rating System)
Health Hazards (Blue): 1; Flammability (Red): 2; Reactivity (Yellow): 0

EMERGENCY RESPONSE: See Appendix A (129)
Evacuation:
Public safety: Isolate spill area for at least 50 to 100 meters (160 to 330 feet) in all directions.
Spill: Large spill–Consider initial downwind evacuation for at least 300 meters (1000 feet).
Fire: Isolate for 800 meters (½ mile) in all directions, especially if tank, rail car, or tank truck is involved in fire.

EXPOSURE
Short-term effects: *SEEK MEDICAL ATTENTION.* **Vapor:** Irritating to eyes, nose, and throat. Harmful if inhaled. Move to fresh air. *IF BREATHING HAS STOPPED,* give artificial respiration. IF breathing is difficult, administer oxygen. **Liquid:** Irritating to eyes. *IF IN EYES,* hold eyelids open and flush. with plenty of water.

HEALTH HAZARDS
Personal protective equipment (PPE): B-Level PPE. OSHA Table Z-1 air contaminant. Face shield to avoid splash.
Recommendations for respirator selection: NIOSH/OSHA
500 ppm: SA:CF (any supplied-air respirator operated in a continuous-flow mode); or CCRFOV [any air-purifying, full-facepiece respirator (gas mask) with a chin-style, front- or back-mounted acid gas canister]; or GMFOV[any air-purifying, full-facepiece respirator (gas mask) with a chin-style, front- or back-mounted organic vapor canister]; or PAPROV [any powered, air-purifying respirator with organic vapor cartridge(s)]; or SCBAF (any self-contained breathing apparatus with a full facepiece); or SAF (any supplied-air respirator with a full facepiece). *EMERGENCY OR PLANNED ENTRY INTO UNKNOWN CONCENTRATIONS OR IDLH CONDITIONS:* SCBAF:PD,PP (any self-contained breathing apparatus that has a full facepiece and is operated in a pressure-demand or other positive-pressure mode); or SAF:PD,PP:ASCBA (any supplied-air respirator that has a full facepiece and is operated in a pressure-demand or other positive-pressure mode in combination with an auxiliary self-contained breathing apparatus operated in a pressure-demand or other positive pressure mode). *ESCAPE:* GMFOV[any air-purifying, full-facepiece respirator (gas mask) with a chin-style, front- or back-mounted organic vapor canister]; or SCBAE (any appropriate escape-type, self-contained breathing apparatus). *Note:* Substance causes eye irritation or damage; eye protection needed.
Exposure limits (TWA unless otherwise noted): ACGIH TLV 100 ppm (361 mg/m^3); NIOSH 100 ppm (360 mg/m^3); OSHA PEL 100 ppm (360 mg/m^3) as isoamyl alcohol (primary).
Short-term exposure limits (15-minute TWA): ACGIH STEL 125 ppm (452 mg/m^3); NIOSH 125 ppm (450 mg/m^3) as isoamyl alcohol (primary).
Toxicity by ingestion: Grade 2; LD_{50} = 0.5 to 5 g/kg.
Long-term health effects:
Vapor (gas) irritant characteristics: Eye and respiratory tract irritant. Vapor is irritating such that personnel will not usually tolerate moderate or high vapor concentrations.
Liquid or solid irritant characteristics: Liquid may irritate skin.

Odor threshold: 0.03-0.07 ppm.
IDLH value: 500 ppm.

FIRE DATA
Flash point: 95°F/35°C (oc).
Fire extinguishing agents not to be used: Water may be ineffective.
Electrical hazard: Class I, Group D.
Burning rate: 3.6 mm/min

CHEMICAL REACTIVITY
Binary reactants: Reacts with strong oxidizers, sulfuric acid, caustics, isocyanates, amines.
Reactivity group: 20
Compatibility class: Alcohols, glycols

ENVIRONMENTAL DATA
Food chain concentration potential: Negative; unlikely to accumulate.
Water pollution: Dangerous to aquatic life in high concentrations. May be dangerous if it enters nearby water intakes; notify operators. Notify local health and wildlife officials. **Response to discharge:** Disperse and flush.

SHIPPING INFORMATION
Grades of purity: Pure; fusel oil; **Storage temperature:** Ambient; **Inert atmosphere:** None; **Venting:** Open (flame arrester); **Stability during transport:** Stable.

PHYSICAL AND CHEMICAL PROPERTIES
Physical state @ 59°F/15°C and 1 atm: Liquid.
Molecular weight: 88.2
Boiling point @ 1 atm: 234°F/112°C/385°K.
Specific gravity (water = 1): 0.82 @ 68°F/20°C (liquid).
Relative vapor density (air = 1): 3.04
Vapor pressure: 1 mm.

ISOBUTANE   REC. I:0250

SYNONYMS: 1,1-DIMETHYLETHANE; EEC No. 601-004-00-0; ISOBUTANO (Spanish); 2-METHYLPROPANE; PROPANE, 2-METHYL; TRIMETHYLMETHANE

IDENTIFICATION
CAS Number: 75-28-5; can also apply to 463-82-1 (neopentane)
Formula: C_4H_{10}; $CH_3CH(CH_3)_2$; C_5H_{12} (neopentane)
DOT ID Number: UN 1969; DOT Guide Number: 115
Proper Shipping Name: Isobutane; Isobutane mixture

DESCRIPTION: Colorless, liquefied compressed gas. Gasoline-like odor. Liquid floats and boils on water surface; insoluble.

Extremely flammable • Forms explosive mixture with air • Reacts explosively with many materials • Gas is heavier than air and spread along ground • Gas may travel to source of ignition and flashback • Exposure of cylinders to elevated temperatures, fire, and flame may cause cylinders to rupture or cause frangible disk to burst, releasing entire contents of cylinder • Ruptured or venting cylinders may rocket through buildings and/or travel a considerable distance •Irritating to the skin, eyes, and respiratory tract • Vapors may cause dizziness or asphyxiation without warning • Contact with liquid may cause frostbite.

Hazard Classification (based on NFPA-704 Rating System)
Health Hazards (Blue): 1; Flammability (Red): 4; Reactivity (Yellow): 0

EMERGENCY RESPONSE: See Appendix A (115)
Evacuation:
Public safety: Isolate spill area for at least 50 to 100 meters (160 to 330 feet) in all directions.
Spill: Consider initial downwind evacuation for at least 800 meters (½ mile).
Fire: Isolate for 1600 meters (1 mile) in all directions, especially if tank, rail car, or tank truck is involved in fire.

HEALTH HAZARDS
Personal protective equipment (PPE): B-Level PPE. Self-contained breathing apparatus; safety goggles. Wear thermal protective clothing.
Exposure limits (TWA unless otherwise noted): ACGIH TLV 800 ppm (1900 mg/m^3) as n-butane; NIOSH 800 ppm (1900 mg/m^3).
Vapor (gas) irritant characteristics: None
Liquid or solid irritant characteristics: Frostbite is possible.

FIRE DATA
Flash point: –117°F/–83°C (cc).
Flammable limits in air: LEL: 1.8%; UEL: 8.4%; LEL: 1.4%; UEL: 7.5% (neopentane)
Behavior in fire: Containers may explode.
Autoignition temperature: 853°F/456°C/729°K.
Electrical hazard: Class I, Group D.
Burning rate: 9.3 mm/min
Stoichiometric air-to-fuel ratio: 15.35 (estimate).

CHEMICAL REACTIVITY
Binary reactants: Reacts with strong oxidizers, chlorine, fluorine.
Reactivity group: 31
Compatibility class: Paraffins

ENVIRONMENTAL DATA
Water pollution: Not harmful to aquatic life. **Response to discharge:** Issue warning–high flammability. Restrict access. Evacuate area.

SHIPPING INFORMATION
Grades of purity: Pure; technical; **Storage temperature:** Ambient; **Inert atmosphere:** None; **Venting:** Safety relief; **Stability during transport:** Stable.

PHYSICAL AND CHEMICAL PROPERTIES
Physical state @ 59°F/15°C and 1 atm: Gas.
Molecular weight: 58.12
Boiling point @ 1 atm: 10.8°F/–11.8°C/261.4°K.
Melting/Freezing point: –427.5°F/–255.3°C/17.9°K.
Critical temperature: 275°F/135°C/408°K.
Critical pressure: 529 psia = 36.0 atm = 3.65 MN/m^2.
Specific gravity (water = 1): 0.557 @ 68°F/20°C (liquid).
Liquid surface tension: 14 dynes/cm = 0.014 N/m at –10°C.
Liquid water interfacial tension: (estimate) 50 dynes/cm = 0.05 N/m at –10°C.
Relative vapor density (air = 1): 2.06
Ratio of specific heats of vapor (gas): 1.095
Latent heat of vaporization: 158 Btu/lb = 87.5 cal/g = 3.66 x 10^5 J/kg.
Heat of combustion: –19,458 Btu/lb = –10,810 cal/g = –452.59 x 10^5 J/kg.
Heat of fusion: 18.96 cal/g.
Vapor pressure: (Reid) 66 psia; 1530 mm; 3.1 atm @ 70°F

ISOBUTYL ACETATE   REC. I:0300

SYNONYMS: ACETATO de ISOBUTILO (Spanish); ACETIC ACID, ISOBUTYL ESTER; ACETIC ACID, 2-METHYLPROPYL ESTER; BUTYL ACETATE, ISO-; EEC No. 607-026-00-7; 2-METHYLPROPYL ACETATE; 2-METHYL-1-PROPYL ACETATE; β-METHYLPROPYL ETHANOATE

IDENTIFICATION
CAS Number: 110-19-0
Formula: $C_6H_{12}O_2$; $CH_3COOCH_2CH(CH_3)_2$
DOT ID Number: UN 1213; DOT Guide Number: 129
Proper Shipping Name: Isobutyl acetate
Reportable Quantity (RQ): **(CERCLA)** 5000 lb/2270 kg

DESCRIPTION: Colorless watery liquid Pleasant fruity odor at low concentrations; diagreeable at higher concentrations. Floats on the surface of water. Flammable, irritating vapor is produced.

Highly flammable • Containers may BLEVE when exposed to fire • Vapors may form explosive mixture with air • Vapors are heavier than air and will collect and stay in low areas • Vapors may travel long distances to ignition sources and flashback • Vapors in confined areas (e.g., tanks, sewers, buildings) may explode when exposed to fire • Irritating to the skin, eyes, and respiratory tract • Toxic products of combustion may include carbon monoxide.

Hazard Classification (based on NFPA-704 Rating System)
Health Hazards (Blue): 1; Flammability (Red): 3; Reactivity (Yellow): 0

EMERGENCY RESPONSE: See Appendix A (129)
Evacuation:
Public safety: Isolate spill area for at least 50 to 100 meters (160 to 330 feet) in all directions.
Spill: Large spill–Consider initial downwind evacuation for at least 300 meters (1000 feet).
Fire: Isolate for 800 meters (½ mile) in all directions, especially if tank, rail car, or tank truck is involved in fire.

EXPOSURE
Short-term effects: *SEEK MEDICAL ATTENTION.* **Vapor:** Irritating to eyes, nose, and throat. *IF INHALED*, will, will cause nausea, vomiting, dizziness, or loss of consciousness.
Move to fresh air. *IF BREATHING HAS STOPPED,* give artificial respiration. If breathing is difficult, administer oxygen. **Liquid:** Irritating to skin and eyes. Remove contaminated clothing and shoes. Flush affected areas with plenty of water. *IF IN EYES*, hold eyelids open and flush with plenty of water.

HEALTH HAZARDS
Personal protective equipment (PPE): B-Level PPE. Air pack or organic canister mask; chemical goggles.
Recommendations for respirator selection: NIOSH/OSHA
1300 ppm: SA:CF* (any supplied-air respirator operated in a continuous-flow mode); or CCRFOV [any air-purifying, full-facepiece respirator (gas mask) with a chin-style, front- or back-mounted acid gas canister]; or GMFOV[any air-purifying, full-facepiece respirator (gas mask) with a chin-style, front- or back-mounted organic vapor canister]; or PAPROV* [any powered, air-purifying respirator with organic vapor cartridge(s)]; or SCBAF

(any self-contained breathing apparatus with a full facepiece); or SAF (any supplied-air respirator with a full facepiece). *EMERGENCY OR PLANNED ENTRY INTO UNKNOWN CONCENTRATIONS OR IDLH CONDITIONS:* SCBAF:PD,PP (any self-contained breathing apparatus that has a full facepiece and is operated in a pressure-demand or other positive-pressure mode); or SAF:PD,PP:ASCBA (any supplied-air respirator that has a full facepiece and is operated in a pressure-demand or other positive-pressure mode in combination with an auxiliary self-contained breathing apparatus operated in a pressure-demand or other positive pressure mode). *ESCAPE:* GMFOV[any air-purifying, full-facepiece respirator (gas mask) with a chin-style, front- or back-mounted organic vapor canister]; or SCBAE (any appropriate escape-type, self-contained breathing apparatus).
Note: Substance causes eye irritation or damage; eye protection needed.
Exposure limits (TWA unless otherwise noted): ACGIH TLV 150 ppm (713 mg/m^3); NIOSH/OSHA 105 ppm (700 mg/m^3).
Vapor (gas) irritant characteristics: Vapors cause a slight smarting of the eyes or respiratory system if present in high concentrations. The effect is temporary.
Liquid or solid irritant characteristics: Minimum hazard. If spilled on clothing and allowed to remain, may cause smarting and reddening of the skin.
Odor threshold: 0.4–4.0
IDLH value: 1300 ppm [LEL]

FIRE DATA
Flash point: 85°F/30°C (oc); 62°F/16°C (cc).
Flammable limits in air: LEL: 1.3%; UEL: 10.5%.
Fire extinguishing agents not to be used: Water may be ineffective
Autoignition temperature: 793°F/422°C/695°K.
Electrical hazard: Class I, Group D.

CHEMICAL REACTIVITY
Reactivity with water: Slow decomposition to isobutanol and acetic acid.
Binary reactants: Mixture with air forms an explosive mixture. Reacts with nitrates, strong oxidizers, alkalis, and acids. Softens and dissolves many plastics.
Reactivity group: 34
Compatibility class: Ester

ENVIRONMENTAL DATA
Food chain concentration potential: Log K_{ow} = 1.62. Unlikely to accumulate.
Water pollution: Effect of low concentrations on aquatic life is unknown. Fouling to shoreline. May be dangerous if it enters nearby water intakes; notify operators. Notify local health and wildlife officials. **Response to discharge:** Issue warning–high flammability. Evacuate area. Disperse and flush.

SHIPPING INFORMATION
Grades of purity: 95-99+%; **Storage temperature:** Ambient; **Inert atmosphere:** None; **Venting:** Open (flame arrester); **Stability during transport:** Stable.

NAS HAZARD CLASSIFICATION FOR BULK WATER TRANSPORTATION
FIRE: 3
HEALTH: Vapor irritant: 1; Liquid or solid irritant: 1; Poisons: 2
WATER POLLUTION: Human toxicity: 1; Aquatic toxicity: 1; Aesthetic effect: 2
REACTIVITY: Other chemicals: 1; Water: 0; Self-reaction: 0

PHYSICAL AND CHEMICAL PROPERTIES
Physical state @ 59°F/15°C and 1 atm: Liquid.
Molecular weight: 116.16
Boiling point @ 1 atm: 243.1°F/117.3°C/390.5°K.
Melting/Freezing point: –143°F/–97°C/176°K.
Critical temperature: 565°F/296°C/569°K.
Critical pressure: 470 psia = 32 atm = 3.2 MN/m^2.
Specific gravity (water = 1): 0.871 @ 68°F/20°C (liquid).
Liquid surface tension: 23.7 dynes/cm = 0.0237 N/m @ 68°F/20°C.
Liquid water interfacial tension: (estimate) 40 dynes/cm = 0.04 N/m @ 68°F/20°C.
Relative vapor density (air = 1): 4.0
Latent heat of vaporization: 133 Btu/lb = 73.7 cal/g = 3.09 x 10^5 J/kg.
Heat of combustion: (estimate) –13,000 Btu/lb = –7220 cal/g = –302 x 10^5 J/kg.
Vapor pressure: (Reid) 0.4 psia; 13 mm.

ISOBUTYL ACRYLATE   REC. I:0325

SYNONYMS: ACRILATO de ISO-BUTILO (Spanish); ACRYLIC ACID, ISOBUTYL ESTER; ISOBUTYL PROPENOATE; ISOBUTYL-2-PROPENOATE; 2-METHYL-1-PROPYL ACRYLATE; ISOBUTYL 2-PROPENOATE

IDENTIFICATION
CAS Number: 106-63-8
Formula: $C_7H_{12}O_2$; CH_2=CHCOOCH$_2$CH(CH$_3$)$_2$
DOT ID Number: UN 2527; DOT Guide Number: 130P
Proper Shipping Name: Isobutyl acrylate

DESCRIPTION: Colorless watery liquid. Sharp fragrant odor. Floats on the surface of water. Irritating vapor is produced.

Corrosive to the skin, eyes, and respiratory tract; contact with skin or eyes can cause burns and vision impairment; inhalation symptoms may be delayed • Polymerization hazard • Heat can induce polymerization with rapid release of energy; sealed containers may rupture explosively • May react with itself blocking relief valves; leading to tank explosions • Containers may BLEVE when exposed to fire • Vapors may form explosive mixture with air • Vapors are heavier than air and will collect and stay in low areas • Vapors may travel long distances to ignition sources and flashback • Vapors in confined areas (e.g., tanks, sewers, buildings) may explode when exposed to fire • Toxic products of combustion may include carbon monoxide.

Hazard Classification (based on NFPA-704 Rating System)
Health Hazards (Blue): 2; Flammability (Red): 2; Reactivity (Yellow): 2

EMERGENCY RESPONSE: See Appendix A (130)
Evacuation:
Public safety: Isolate spill area for at least 50 to 100 meters (160 to 330 feet) in all directions.
Spill: Large spill–Consider initial downwind evacuation for at least 300 meters (1000 feet).
Fire: Isolate for 800 meters (½ mile) in all directions, especially if tank, rail car, or tank truck is involved in fire.

EXPOSURE
Short-term effects: *SEEK MEDICAL ATTENTION.* **Vapor:** Irritating to eyes, nose, and throat. Move to fresh air. *IF*

Isobutyl alcohol

BREATHING HAS STOPPED, give artificial respiration. IF breathing is difficult, administer oxygen. **Liquid:** Irritating to skin and eyes. Harmful if swallowed. Remove contaminated clothing and shoes. Flush affected areas with plenty of water. *IF IN EYES*, hold eyelids open and flush with plenty of water. *IF SWALLOWED* and victim is *CONSCIOUS AND ABLE TO SWALLOW*, have victim induce vomiting. *IF SWALLOWED* and victim is *UNCONSCIOUS OR HAVING CONVULSIONS*, do nothing except keep victim warm.

HEALTH HAZARDS
Personal protective equipment (PPE): B-Level PPE. Self-contained breathing apparatus, rubber gloves, chemical goggles. Chemical protective material(s) reported to offer protection: Teflon.
Vapor (gas) irritant characteristics: Vapors cause smarting of the eyes or respiratory system if present in high concentrations. The effect is temporary.
Liquid or solid irritant characteristics: If spilled on clothing and allowed to remain, may cause smarting and reddening of the skin.

FIRE DATA
Flash point: 94°F/34°C (oc).
Flammable limits in air: LEL: 1.9%; UEL: 8.0%.
Autoignition temperature: 644°F/340°C/613°K.
Electrical hazard: Class I, Group D.
Burning rate: 4.8 mm/min

CHEMICAL REACTIVITY
Binary reactants: Reacts with strong acids, aliphatic amines, alkanolamines.
Polymerization: Will polymerize when hot. Uncontrolled bulk polymerization can be explosive; **Inhibitor of polymerization:** Methyl ether of hydroquinone: 10–60 ppm; hydroquinone: 5 ppm.
Reactivity group: 14
Compatibility class: Acrylate

ENVIRONMENTAL DATA
Food chain concentration potential: Log P_{ow} = 2.2. Unlikely to accumulate.
Water pollution: Effect of low concentrations on aquatic life is unknown. Fouling to shoreline. May be dangerous if it enters nearby water intakes; notify operators. Notify local health and pollution control officials. **Response to discharge:** Mechanical containment. Chemical and physical treatment.

SHIPPING INFORMATION
Grades of purity: 99.0%; **Storage temperature:** Ambient; **Inert atmosphere:** None; **Venting:** Pressure-vacuum; **Stability during transport:** Stable.

NAS HAZARD CLASSIFICATION FOR BULK WATER TRANSPORTATION
FIRE: 2
HEALTH: Vapor irritant: 1; Liquid or solid irritant: 1; Poisons: 1
WATER POLLUTION: Human toxicity: 1; Aquatic toxicity: 2; Aesthetic effect: 2
REACTIVITY: Other chemicals: 2; Water: 0; Self-reaction: 3

PHYSICAL AND CHEMICAL PROPERTIES
Physical state @ 59°F/15°C and 1 atm: Liquid.
Molecular weight: 128.17
Boiling point @ 1 atm: 280.2°F/137.9°C/411.1°K.
Melting/Freezing point: –78.0°F/–61.1°C/212.1°K.
Critical temperature: 599°F/315°C/588°K.
Critical pressure: 440 psia = 30 atm = 3.0 MN/m².
Specific gravity (water = 1): 0.889 @ 68°F/20°C (liquid).
Liquid surface tension: 2.47 dynes/cm = 0.0247 N/m @ 77°F/25°C.
Liquid water interfacial tension: (estimate) 35 dynes/cm = 0.035 N/m at 27°C.
Ratio of specific heats of vapor (gas): 1.044
Latent heat of vaporization: 130 Btu/lb = 71 cal/g = 3.0 x 10^5 J/kg.
Heat of combustion: –13,500 Btu/lb = –7500 cal/g = –314 x 10^5 J/kg.
Heat of polymerization: –229 Btu/lb = –127 cal/g = –5.32 x 10^5 J/kg.
Vapor pressure: (Reid) 0.4 psia.

ISOBUTYL ALCOHOL   REC. I:0350

SYNONYMS: ALCOHOL ISOBUTILICO (Spanish); ALCOHOL C-4; EEC No. 603-004-00-6; IBA; ISOBUTANOL; ISOBUTYL CARBINOL; ISOPROPYLCARBINOL; 2-METHYL-1-PROPANOL; FERMENTATION BUTYL ALCOHOL; 1-HYDROXYMETHYLPROPANE; RCRA No. U140

IDENTIFICATION
CAS Number: 78-83-1
Formula: $C_4H_{10}O$; $(CH_3)CHCH_2OH$
DOT ID Number: UN 1212; DOT Guide Number: 129
Proper Shipping Name: Isobutanol; Isobutyl alcohol
Reportable Quantity (RQ): **(CERCLA)** 5000 lb/2270 kg

DESCRIPTION: Colorless oily liquid. Slightly suffocating, mild alcohol odor. Floats and mixes slowly with water. Irritating vapor is produced.

Highly flammable • Containers may BLEVE when exposed to fire • Vapors may form explosive mixture with air • Vapors are heavier than air and will collect and stay in low areas • Vapors may travel long distances to ignition sources and flashback • Vapors in confined areas (e.g., tanks, sewers, buildings) may explode when exposed to fire • Irritating to the skin, eyes, and respiratory tract • Toxic products of combustion may include carbon monoxide.

Hazard Classification (based on NFPA-704 Rating System)
Health Hazards (Blue): 1; Flammability (Red): 3; Reactivity (Yellow): 0

EMERGENCY RESPONSE: See Appendix A (129)
Evacuation:
Public safety: Isolate spill area for at least 50 to 100 meters (160 to 330 feet) in all directions.
Spill: Large spill–Consider initial downwind evacuation for at least 300 meters (1000 feet).
Fire: Isolate for 800 meters (½ mile) in all directions, especially if tank, rail car, or tank truck is involved in fire.

EXPOSURE
Short-term effects: *SEEK MEDICAL ATTENTION.* **Vapor:** Irritating to eyes, nose, and throat. *IF INHALED*, will, will cause nausea, dizziness, or headache. Move to fresh air. *IF BREATHING HAS STOPPED*, give artificial respiration. IF breathing is difficult, administer oxygen. **Liquid:** Irritating to eyes. Harmful if swallowed. *IF IN EYES*, hold eyelids open and flush. with plenty of water. *IF SWALLOWED* and victim is *CONSCIOUS AND ABLE TO SWALLOW*, have victim drink 4 to 8 ounces of water.

HEALTH HAZARDS

Personal protective equipment (PPE): B-Level PPE. OSHA table Z-1 air contaminant. Air pack or organic canister; chemical goggles. Chemical protective material(s) reported to have good to excellent resistance: butyl rubber, neoprene, nitrile, nitrile+PVC, Viton®. Also, Viton®/neoprene, butyl/neoprene, neoprene+natural rubber, styrene-butadiene offers limited protection

Recommendations for respirator selection: NIOSH
500 ppm: CCROV* [any chemical cartridge respirator with organic vapor cartridge(s)]; or SA* (any supplied-air respirator).*1250 ppm:* SA:CF* (any supplied-air respirator operated in a continuous-flow mode); or PAPROV* [any powered, air-purifying respirator with organic vapor cartridge(s)]. *1600 ppm:* GMFOV[any air-purifying, full-facepiece respirator (gas mask) with a chin-style, front- or back-mounted organic vapor canister]; or GMFOV [any air-purifying, full-facepiece respirator (gas mask) with a chin-style, front- or back-mounted acid gas canister]; or PAPRTOV [any powered, air-purifying respirator with a tight-fitting facepiece and organic vapor cartridges(s)]; or SCBAF (any self-contained breathing apparatus with a full facepiece); or SAF (any supplied-air respirator with a full facepiece). *EMERGENCY OR PLANNED ENTRY INTO UNKNOWN CONCENTRATIONS OR IDLH CONDITIONS:* SCBAF:PD,PP (any self-contained breathing apparatus that has a full facepiece and is operated in a pressure-demand or other positive-pressure mode); or SAF:PD,PP:ASCBA (any supplied-air respirator that has a full facepiece and is operated in a pressure-demand or other positive-pressure mode in combination with an auxiliary self-contained breathing apparatus operated in a pressure-demand or other positive pressure mode). *ESCAPE:* GMFOV[any air-purifying, full-facepiece respirator (gas mask) with a chin-style, front- or back-mounted organic vapor canister]; or SCBAE (any appropriate escape-type, self-contained breathing apparatus). **Note:* Substance causes eye irritation or damage; eye protection needed.

Exposure limits (TWA unless otherwise noted): ACGIH TLV 50 ppm (152 mg/m^3); NIOSH 50 ppm (150 mg/m^3) skin; OSHA PEL 100 ppm (300 mg/m^3).

Toxicity by ingestion: Grade 2; LD_{50} = 0.5 to 5 g/kg (rat).

Vapor (gas) irritant characteristics: Respiratory tract irritant. Vapors cause smarting of the eyes or respiratory system if present in high concentrations. The effect is temporary.

Liquid or solid irritant characteristics: Severe eye irritant. Practically harmless to the skin.

Odor threshold: 0.6–40 ppm.

IDLH value: 1600 ppm.

FIRE DATA

Flash point: 90°F/32°C (oc); 82°F/28°C (cc).

Flammable limits in air: LEL: 1.7% @ 123°F/51°C; UEL: 10.6% @ 202°F/95°C.

Fire extinguishing agents not to be used: Water may be ineffective

Autoignition temperature: 780°F/415°C/688°K.

Electrical hazard: Class I, Group D.

Burning rate: 3.5 mm/min

CHEMICAL REACTIVITY

Binary reactants: Reacts with strong oxidizers, strong mineral acids. Forms explosive hydrogen gas with alkali metals: lithium, sodium, potassium, rubidium, cesium, and francium. Water free isobutanol reacts with aluminum at temperatures above 120°F/49°C.

Reactivity group: 20

Compatibility class: Alcohols, glycols

ENVIRONMENTAL DATA

Food chain concentration potential: Log P_{ow} = 0.75. Unlikely to accumulate.

Water pollution: Effect of low concentrations on aquatic life is unknown. May be dangerous if it enters nearby water intakes; notify operators. Notify local health and wildlife officials. **Response to discharge:** Disperse and flush.

SHIPPING INFORMATION

Grades of purity: 99+%; **Storage temperature:** Ambient; **Inert atmosphere:** None; **Venting:** Open (flame arrester); **Stability during transport:** Stable.

NAS HAZARD CLASSIFICATION FOR BULK WATER TRANSPORTATION

FIRE:–
HEALTH: Vapor irritant: 1; Liquid or solid irritant: 0; Poisons: 1
WATER POLLUTION: Human toxicity: 2; Aquatic toxicity: 1; Aesthetic effect: 2
REACTIVITY: Other chemicals: 2; Water: 0; Self-reaction: 0

PHYSICAL AND CHEMICAL PROPERTIES

Physical state @ 59°F/15°C and 1 atm: Liquid.
Molecular weight: 74.12
Boiling point @ 1 atm: 226°F/108°C/381°K.
Melting/Freezing point: –162°F/–108°C/165°K.
Critical temperature: 526°F/275°C/548°K.
Critical pressure: 623 psia = 42.4 atm = 4.30 MN/m^2.
Specific gravity (water = 1): 0.802 @ 68°F/20°C (liquid).
Relative vapor density (air = 1): 2.6
Latent heat of vaporization: 248 Btu/lb = 138 cal/g = 5.78 x 10^5 J/kg.
Heat of combustion: –14,220 Btu/lb = –7900 cal/g = –330.8 x 10^5 J/kg.
Heat of solution: (estimate) –9 Btu/lb = –5 cal/g = –0.2 x 10^5 J/kg.
Vapor pressure: 9 mm.

ISOBUTYLAMINE   REC. I:0400

SYNONYMS: 1-AMINO-2-METHYLPROPANE; ISOBUTILAMINA (Spanish); MONOISOBUTYLAMINE; 2-METHYLPROPYLAMINE

IDENTIFICATION

CAS Number: 78-81-9
Formula: $C_4H_{11}N$; $(CH_3)_2CHCH_2NH_2$
DOT ID Number: UN 1214; DOT Guide Number: 132
Proper Shipping Name: Isobutylamine
Reportable Quantity (RQ): **(CERCLA)** 1000 lb/454 kg

DESCRIPTION: Colorless liquid. Strong ammonia odor. Floats on the surface of water; soluble.

Highly flammable • Containers may BLEVE when exposed to fire • Vapors may form explosive mixture with air • Vapors are heavier than air and will collect and stay in low areas • Vapors may travel long distances to ignition sources and flashback • Vapors in confined areas (e.g., tanks, sewers, buildings) may explode when exposed to fire • Irritating to the skin, eyes, and respiratory tract • Toxic products of combustion may include carbon monoxide.

Hazard Classification (based on NFPA-704 Rating System)
Health Hazards (Blue): 2; Flammability (Red): 3; Reactivity (Yellow): 0

Isobutylamine

EMERGENCY RESPONSE: See Appendix A (132)
Evacuation:
Public safety: Isolate spill area for at least 50 to 100 meters (160 to 330 feet) in all directions.
Spill: Increase, as necessary, the isolation distance shown above, in "Public safety."
Fire: Isolate for 800 meters (½ mile) in all directions, especially if tank, rail car, or tank truck is involved in fire.

EXPOSURE
Short-term effects: *SEEK MEDICAL ATTENTION.* **Vapor:** Irritating to eyes, nose, and throat. *IF INHALED*, will, will cause coughing, difficult breathing or loss of consciousness. *IF IN EYES*, hold eyelids open and flush. with plenty of water. *IF BREATHING HAS STOPPED*, give artificial respiration. If breathing is difficult, administer oxygen. **Liquid:** Will burn skin and eyes. *IF SWALLOWED*, will cause nausea or loss of consciousness. Remove contaminated clothing and shoes. Flush affected areas with plenty of water. *IF IN EYES*, hold eyelids open and flush. with plenty of water. *IF SWALLOWED and victim is CONSCIOUS AND ABLE TO SWALLOW*, have victim drink 4 to 8 ounces of water and have victim induce vomiting.

HEALTH HAZARDS
Personal protective equipment (PPE): B-Level PPE. Self-contained breathing apparatus; butyl rubber gloves; chemical face shield; butyl rubber apron.
Recommendations for respirator selection: NIOSH/OSHA AS butylamine
50 ppm: CCRS [any chemical cartridge respirator with cartridge(s) providing protection against the compound of concern]; or SA (any supplied-air respirator). *125 ppm:* SA:CF (any supplied-air respirator operated in a continuous-flow mode); or PAPRS [any powered, air-purifying respirator with cartridge(s) providing protection against the compound of concern]. *250 ppm:* CCRFS (any chemical cartridge respirator with a full facepiece and cartridge(s) providing protection against the compound of concern]; or GMFS [any air-purifying, full-facepiece respirator (gas mask) with a chin-style, front- or back-mounted canister providing protection against the compound of concern]; or PAPRTS (any powered, air-purifying respirator with a tight-fitting facepiece and cartridge(s) providing protection against the compound of concern]; or SCBAF (any self-contained breathing apparatus with a full facepiece); or SAF (any supplied-air respirator with a full facepiece). *300 ppm:* SAF:PD,PP (any supplied-air respirator that has a full facepiece and is operated in a pressure-demand or other positive-pressure mode). *EMERGENCY OR PLANNED ENTRY INTO UNKNOWN CONCENTRATIONS OR IDLH CONDITIONS:* SCBAF:PD,PP (any self-contained breathing apparatus that has a full facepiece and is operated in a pressure-demand or other positive-pressure mode); or SAF:PD,PP:ASCBA (any supplied-air respirator that has a full facepiece and is operated in a pressure-demand or other positive-pressure mode in combination with an auxiliary self-contained breathing apparatus operated in a pressure-demand or other positive pressure mode). *ESCAPE:* GMFS [any air-purifying, full-facepiece respirator (gas mask) with a chin-style, front- or back-mounted canister providing protection against the compound of concern]; or SCBAE (any appropriate escape-type, self-contained breathing apparatus). *Note:* Substance reported to cause eye irritation or damage; may require eye protection.
Toxicity by ingestion: Grade 3; oral LD_{50} = 120 mg/kg (rabbit), 250 mg/kg (rat).
Vapor (gas) irritant characteristics: Eye and respiratory tract irritant. Vapors are irritating such that personnel will not usually tolerate moderate or high concentrations.
Liquid or solid irritant characteristics: Severe skin and eye irritant. Causes second- and third-degree burns on short contact and is very injurious to the eyes.
Odor threshold: 0.8 ppm

FIRE DATA
Flash point: 15°F/–9°C (cc).
Flammable limits in air: LEL: 1.7%; UEL; 9.8%.
Fire extinguishing agents not to be used: Water may be ineffective, but can be used to keep containers cool.
Autoignition temperature: 712°F/378°C/651°K.
Electrical hazard: Class I, Group D.
Burning rate: 6.03 mm/min

CHEMICAL REACTIVITY
Binary reactants: Incompatible with organic anhydrides, acrylates, alcohols, aldehydes, alkali metals, alkylene oxides, cellulose nitrate, copper and its alloys, cresols, caprolactam, epichlorohydrin, ethylene dichloride, isocyanates, ketones, glycols, nitrates, phenols, vinyl acetate. Violent reaction with maleic anhydride.
Reactivity group: 7
Compatibility class: Aliphatic amines

ENVIRONMENTAL DATA
Food chain concentration potential: Log P_{ow} = 0.8. Unlikely to accumulate.
Water pollution: Effect of low concentrations on aquatic life is unknown. May be dangerous if it enters nearby water intakes; notify operators. Notify local health and wildlife officials. **Response to discharge:** Issue warning–high flammability, air contaminant, water contaminant. Restrict access. Evacuate area. Disperse and flush.

SHIPPING INFORMATION
Grades of purity: Technical; 99+%; **Storage temperature:** Ambient; **Inert atmosphere:** None; **Venting:** Open (flame arrester); **Stability during transport:** Stable.

NAS HAZARD CLASSIFICATION FOR BULK WATER TRANSPORTATION
FIRE: 3
HEALTH: Vapor irritant: 3; Liquid or solid irritant: 4; Poisons: 4
WATER POLLUTION: Human toxicity: 2; Aquatic toxicity: 2; Aesthetic effect: 1
REACTIVITY: Other chemicals: 3; Water: 0; Self-reaction: 0

PHYSICAL AND CHEMICAL PROPERTIES
Physical state @ 59°F/15°C and 1 atm: Liquid.
Molecular weight: 73.1
Boiling point @ 1 atm: 153.3°F/67.4°C/340.6°K.
Melting/Freezing point: –121.9°F/–85.5°C/187.7°K.
Critical temperature: 469.4°F/243.0°C/516.2°K.
Critical pressure: 620 psia = 42 atm = 4.3 MN/m².
Specific gravity (water = 1): 0.739 @ 68°F/20°C (liquid).
Liquid surface tension: 23.70 dynes/cm = 0.0237 N/m @ 68°F/20°C.
Relative vapor density (air = 1): 2.5
Ratio of specific heats of vapor (gas): 1.073 @ 68°F/20°C.
Latent heat of vaporization: 182 Btu/lb = 101 cal/g = 4.23 x 10^5 J/kg.
Heat of combustion: –17,550 Btu/lb = –9760 cal/g = –408 x 10^5 J/kg.
Heat of solution: –148 Btu/lb = –82 cal/g = 3.4 x 10^5 J/kg.
Vapor pressure: (Reid) 2.4 psia; 3.2 mm.

ISOBUTYLENE   REC. I:0450

SYNONYMS: γ-BUTYLENE; EEC No. 601-012-00-4; ISOBUTENE; ISOBUTILENO (Spanish); 2-METHYLPROPENE

IDENTIFICATION
CAS Number: 115-11-7
Formula: C_4H_8; $(CH_3)_2C=CH_2$
DOT ID Number: UN 1055; DOT Guide Number: 115
Proper Shipping Name: Isobutylene

DESCRIPTION: Colorless gas, liquefied compressed gas. Mild, gasoline-like odor. Liquid floats and boils on water; insoluble; flammable visible vapor cloud is produced. Becomes a liquid below 20°F/–7°C.

Extremely flammable • Do not spray liquid with water • Forms explosive mixture with air • Reacts explosively with many materials • Polymerization hazard; fire and explosion risk • Vapors from liquefied gas are initially heavier than air and spread along ground • Vapors may travel to source of ignition and flashback • Exposure of cylinders to elevated temperatures, fire, and flame may cause cylinders to rupture or cause frangible disk to burst, releasing entire contents of cylinder • Ruptured or venting cylinders may rocket through buildings and/or travel a considerable distance •Irritating to the skin, eyes, and respiratory tract • Vapors may cause dizziness or asphyxiation without warning • Seek assistance of an expert • Contact with liquid may cause frostbite.

Hazard Classification (based on NFPA-704 Rating System)
Health Hazards (Blue): 1; Flammability (Red): 4; Reactivity (Yellow): 0

EMERGENCY RESPONSE: See Appendix A (115)
Evacuation:
Public safety: Isolate spill area for at least 50 to 100 meters (160 to 330 feet) in all directions.
Spill: Consider initial downwind evacuation for at least 800 meters (½ mile).
Fire: Isolate for 1600 meters (1 mile) in all directions, especially if tank, rail car, or tank truck is involved in fire.

EXPOSURE
Short-term effects: *SEEK MEDICAL ATTENTION.* **Vapor:** Irritating to eyes, nose, and throat. *IF INHALED,* will, will cause dizziness, or loss of consciousness. Move to fresh air. *IF BREATHING HAS STOPPED,* give artificial respiration. IF breathing is difficult, administer oxygen. **Liquid:** Will cause frostbite. Flush affected areas with plenty of water. *DO NOT RUB AFFECTED AREAS.*

HEALTH HAZARDS
Personal protective equipment (PPE): B-Level PPE. Chemical gloves and eye protection; organic vapor canister); or self-contained breathing apparatus.
Wear thermal protective clothing. Chemical protective material(s) reported to have good to excellent resistance: polyethylene.
Vapor (gas) irritant characteristics: Vapors are nonirritating to eyes and throat.
Liquid or solid irritant characteristics: Eye and skin irritant. Rapid evaporation may cause frostbite.

FIRE DATA
Flash point: Gas.
Flammable limits in air: LEL: 1.8%; UEL: 9.6%.
Autoignition temperature: 869°F/465°C/738°K.
Electrical hazard: Class I, Group D.
Stoichiometric air-to-fuel ratio: 14.68 (estimate).

CHEMICAL REACTIVITY
Binary reactants: Violent reaction with oxidizers, peroxides (which may be formed by substance), hydrochloric acid, chlorine, fluorine, hydrogen bromide and oxides of nitrogen.
Polymerization: May be caused by the formation of peroxides.
Reactivity group: 30
Compatibility class: Olefins

ENVIRONMENTAL DATA
Food chain concentration potential: Negative; unlikely to accumulate.
Water pollution: Not harmful to aquatic life. **Response to discharge:** Issue warning–high flammability. Restrict access. Evacuate area.

SHIPPING INFORMATION
Grades of purity: Commercial; **Storage temperature:** Ambient; **Inert atmosphere:** None; **Venting:** Safety relief; **Stability during transport:** Stable.

PHYSICAL AND CHEMICAL PROPERTIES
Physical state @ 59°F/15°C and 1 atm: Gas.
Molecular weight: 56.10
Boiling point @ 1 atm: 19.6°F/–6.9°C/266.3°K.
Melting/Freezing point: –220°F/–140.3°C/132.9°K.
Critical temperature: 292.5°F/–144.7°C/417.9°K.
Critical pressure: 580 psia = 39.48 atm = 3.99 MN/m^2.
Specific gravity (water = 1): 0.59 @ 68°F/20°C (liquid).
Liquid surface tension: 15.8 dynes/cm = 0.0158 N/m @ 68°F/20°C.
Liquid water interfacial tension: (estimate) 40 dynes/cm = 0.04 N/m at –10°C.
Relative vapor density (air = 1): 1.9
Ratio of specific heats of vapor (gas): 1.061
Latent heat of vaporization: 170 Btu/lb = 94.3 cal/g = 3.95 x 10^5 J/kg.
Heat of combustion: –19,359 Btu/lb = –10,755 cal/g = –450.29 x 10^5 J/kg.
Heat of fusion: 25.25 cal/g.
Vapor pressure: 62.5 mm.

ISOBUTYL ISOBUTYRATE   REC. I:0475

SYNONYMS: ISOBUTYRIC ACID, ISOBUTYL ESTER; 2-METHYLPROPYLISOBUTYRATE; 2-METHYLPROPYL ESTER

IDENTIFICATION
CAS Number: 97-85-8
Formula: $C_8H_{16}O_2$; $(CH_3)_2CHCOOCH_2CH(CH_3)_2$
DOT ID Number: UN 2528; DOT Guide Number: 129
Proper Shipping Name: Isobutyl isobutyrate

DESCRIPTION: Colorless liquid. Fruity odor. Floats on the surface of water.

Highly flammable • Containers may BLEVE when exposed to fire •Vapors may form explosive mixture with air • Vapors are heavier than air and will collect and stay in low areas • Vapors may travel long distances to ignition sources and flashback • Vapors in confined areas (e.g., tanks, sewers, buildings) may explode when exposed to fire • Irritating to the skin, eyes, and respiratory tract • Toxic products of combustion may include CO_2 and carbon monoxide.

Hazard Classification (based on NFPA-704 Rating System)
Health Hazards (Blue): 1; Flammability (Red): 2; Reactivity (Yellow): 0

EMERGENCY RESPONSE: See Appendix A (147)
Evacuation:
Public safety: Isolate spill area for at least 50 to 100 meters (160 to 330 feet) in all directions.
Spill: Large spill–Consider initial downwind evacuation for at least 300 meters (1000 feet).
Fire: Isolate for 800 meters (½ mile) in all directions, especially if tank, rail car, or tank truck is involved in fire.

EXPOSURE
Short-term effects: *SEEK MEDICAL ATTENTION.* **Vapor:** Move victim to fresh air. *IF BREATHING HAS STOPPED,* give artificial respiration. IF breathing is difficult, administer oxygen. **Liquid:** Remove contaminated clothing and shoes. Wash affected areas with soap and water. *IF IN EYES*, hold eyelids open and flush with plenty of water.

HEALTH HAZARDS
Personal protective equipment (PPE): Full impervious protective clothing, including boots and gloves. Where splashing is possible wear full face shield or chemical safety goggles. Use approved respirator to protect against vapors.
Toxicity by ingestion: Grade 2; oral rat LD_{50} = 12.8 g/kg.
Vapor (gas) irritant characteristics: Toxic. Vapors cause smarting of the eyes or respiratory system if present in high concentrations. **Liquid or solid irritant characteristics:** If spilled on clothing and allowed to remain, may cause smarting and reddening of the skin.

FIRE DATA
Flash point: 99°F/37°C (cc).
Fire extinguishing agents not to be used: Water.
Stoichiometric air-to-fuel ratio: 52.4

CHEMICAL REACTIVITY
Reactivity group: 34
Compatibility class: Esters.

ENVIRONMENTAL DATA
Water pollution: Effect of low concentrations on aquatic life is unknown. May be dangerous if it enters nearby water intakes; notify operators. Notify local health and wildlife officials.
Response to discharge: Issue warning–high flammability.

SHIPPING INFORMATION
Grades of purity: 99%; technical; **Storage temperature:** Ambient; **Inert atmosphere:** None; **Venting:** Pressure vacuum valve.

PHYSICAL AND CHEMICAL PROPERTIES
Physical state @ 59°F/15°C and 1 atm: Liquid.
Molecular weight: 144.24
Boiling point @ 1 atm: 299.7°F/148.7°C/421.7°K.
Melting/Freezing point: –113.3°F/–80.7°C/192.3°K.
Specific gravity (water = 1): 0.853–0.857 @ 20°C.

ISOBUTYL METHACRYLATE REC. I:0485

SYNONYMS: METACRILATO de *iso*-BUTILO (Spanish); METHACRYLIC ACID, ISOBUTY ESTER; ISOBUTYL α-METHACRYLATE; ISOBUTYL 2-METHYL-2-PROPENOATE

IDENTIFICATION
CAS Number: 97-86-9
Formula: $C_8H_{14}O_2$; $CH_2=C(CH_3)CO_2CH_2CH(CH_3)_2$
DOT ID Number: UN 2283; DOT Guide Number: 130P
Proper Shipping Name: Isobutyl methacrylate, inhibited

DESCRIPTION: Colorless liquid. Acrylic odor. Floats on the surface of water.

Flammable • Containers may BLEVE when exposed to fire • Vapors may form explosive mixture with air • Vapors are heavier than air and will collect and stay in low areas • Vapors may travel long distances to ignition sources and flashback • Vapors in confined areas (e.g., tanks, sewers, buildings) may explode when exposed to fire • Polymerization hazard • Heat can induce polymerization with rapid release of energy; sealed containers may rupture explosively • May react with itself blocking relief valves; leading to tank explosions • Irritating to the skin, eyes, and respiratory tract; inhalation symptoms may be delayed • Toxic products of combustion may include carbon monoxide.

Hazard Classification (based on NFPA-704 Rating System)
Health Hazards (Blue): 2; Flammability (Red): 2; Reactivity (Yellow): 0

EMERGENCY RESPONSE: See Appendix A (130)
Evacuation:
Public safety: Isolate spill area for at least 50 to 100 meters (160 to 330 feet) in all directions.
Spill: Large spill–Consider initial downwind evacuation for at least 300 meters (1000 feet).
Fire: Isolate for 800 meters (½ mile) in all directions, especially if tank, rail car, or tank truck is involved in fire.

EXPOSURE
Short-term effects: *SEEK MEDICAL ATTENTION.* **Vapor:** May cause lung edema; physical exertion will aggravate this condition. **Liquid:** Irritating to skin and eyes. Harmful if swallowed. *IF IN EYES OR ON SKIN*, flush with running water for at least 15 minutes; hold eyelids open if necessary. Wash skin with soap and water. Remove and isolate contaminated clothing and shoes at the site. *IF SWALLOWED* and victim is *CONSCIOUS AND ABLE TO SWALLOW*, have victim drink 4 to 8 ounces of water and induce vomiting. *IF SWALLOWED* and victim is *UNCONSCIOUS OR HAVING CONVULSIONS,* do nothing except keep victim warm. *Note to physician or authorized medical personnel*: Medical observation is recommended for 24 to 48 hours after breathing overexposure, as pulmonary edema may be delayed. As first aid for pulmonary edema, consider administering a corticosteroid spray. Cigarette smoking may exacerbate pulmonary injury and should be discouraged for at least 72 hours following exposure.

HEALTH HAZARDS
Personal protective equipment (PPE): Wear self-contained positive pressure breathing apparatus and full protective clothing.
Toxicity by ingestion: Grade 1; LD_{50} = 11.99 g/kg (mouse).
Long-term health effects: Teratogen
Vapor (gas) irritant characteristics: Vapors cause smarting of the

eyes or respiratory system if present in high concentrations. The effect may be temporary.
Liquid or solid irritant characteristics: Causes smarting of the skin and first-degree burns on short exposure; may cause second-degree burns on long exposure.

FIRE DATA
Flash point: 120°F/88°C (oc); 112°F/44°C (cc).
alcohol foam.
Behavior in fire: Containers may rupture explosively.
Electrical hazard: Class I, Group D.
Stoichiometric air-to-fuel ratio: 10.2 (estimate).

CHEMICAL REACTIVITY
Binary reactants: Reacts with strong acids, amines.
Polymerization: Will occur with elevated temperatures or on contact with oxidizing agents; **Inhibitor of polymerization:** 25 ppm hydroquinone monomethyl ether, 10 ppm *p*-methoxy phenol (MEHQ).
Reactivity group: 14
Compatibility class: Acrylates

ENVIRONMENTAL DATA
Food chain concentration potential: Log P_{ow} = 2.7. Unlikely to accumulate.
Water pollution: Dangerous to aquatic life in high concentrations. May be dangerous if it enters nearby water intakes; notify operators. Fouling to shoreline. Notify local health and wildlife officials. **Response to discharge:** Issue warning–high flammability. Restrict access. Mechanical containment. Should be removed. Chemical and physical treatment.

SHIPPING INFORMATION
Grades of purity: 100%; **Storage temperature:** Ambient; Store away from heat, catalysts and strong oxidizing agents. **Inert atmosphere:** None; **Venting:** Pressure-Vacuum; **Stability during transport:** Stable.

PHYSICAL AND CHEMICAL PROPERTIES
Physical state @ 59°F/15°C and 1 atm: Liquid.
Molecular weight: 142.20
Boiling point @ 1 atm: 311°F/155°C/428°K.
Critical temperature: 642°F/339°C/612°K (estimate).
Critical pressure: 387 psia = 26.3 atm = 2.66 MN/m² (estimate).
Specific gravity (water = 1): 0.8858 @ 68°F/20°C.
Relative vapor density (air = 1): 4.9 (estimate).

ISOBUTYRIC ACID   REC. I:0500

SYNONYMS: ACIDO ISOBUTIRICO (Spanish); ISOBUTANOIC ACID; DIMETHYLACETIC ACID; ISOPROPYLFORMIC ACID; 2-METHYLPROPANOIC ACID; PROPANE-2-CARBOXYLIC ACID; α-METHYLPROPIONIC ACID

IDENTIFICATION
CAS Number: 79-31-2
Formula: $C_4H_8O_2$; $(CH_3)_2CHCOOH$
DOT ID Number: UN 2529; DOT Guide Number: 132
Proper Shipping Name: Isobutyric acid
Reportable Quantity (RQ): **(CERCLA)** 5000 lb/2270 kg

DESCRIPTION: Colorless liquid Unpleasant, acrid odor. Floats on the surface of water; soluble.

Flammable • Containers may BLEVE when exposed to fire • Vapors may form explosive mixture with air • Vapors are heavier than air and will collect and stay in low areas • Vapors may travel long distances to ignition sources and flashback • Vapors in confined areas (e.g., tanks, sewers, buildings) may explode when exposed to fire • Irritating to the skin, eyes, and respiratory tract • Toxic products of combustion may include carbon monoxide.

Hazard Classification (based on NFPA-704 Rating System)
Health Hazards (Blue): 1; Flammability (Red): 2; Reactivity (Yellow): 0

EMERGENCY RESPONSE: See Appendix A (132)
Evacuation:
Public safety: Isolate spill area for at least 50 to 100 meters (160 to 330 feet) in all directions.
Spill: Increase, as necessary, the isolation distance shown above, in "Public safety."
Fire: Isolate for 800 meters (½ mile) in all directions, especially if tank, rail car, or tank truck is involved in fire.

EXPOSURE
Short-term effects: *SEEK MEDICAL ATTENTION*. **Vapor:** Irritating to eyes, nose, and throat. Move to fresh air. *IF BREATHING HAS STOPPED,* give artificial respiration. IF breathing is difficult, administer oxygen. **Liquid:** Will burn skin and eyes. Harmful if swallowed. Remove contaminated clothing and shoes. Flush affected areas with plenty of water. *IF IN EYES,* hold eyelids open and flush. with plenty of water. *IF SWALLOWED* and victim is *CONSCIOUS AND ABLE TO SWALLOW,* have victim drink 4 to 8 ounces of water. **Do NOT induce vomiting**.

HEALTH HAZARDS
Personal protective equipment (PPE): B-Level PPE. Organic chemical respirator; goggles or face shield; rubber gloves. Butyl rubber is generally suitable for carbooxylic acid compounds.
Toxicity by ingestion: Grade 3; oral LD_{50} = 280 mg/kg (rat).

FIRE DATA
Flash point: 170°F/77°C (oc); 132°F/56°C (cc).
Flammable limits in air: LEL: 2.0%; UEL: 9.2%.
Fire extinguishing agents not to be used: Water may be ineffective.
Autoignition temperature: 935°F/502°C/775°C.
Burning rate: 2.6 mm/min

CHEMICAL REACTIVITY
Binary reactants: Reacts with strong oxidizers, bases. Corrosive to aluminum and other metals, producing hydrogen gas.
Neutralizing agents for acids and caustics: Flush with water.
Reactivity group: 4
Compatibility class: Organic acids

ENVIRONMENTAL DATA
Food chain concentration potential: Negative.
Water pollution: Effects of low concentrations on aquatic life is unknown. May be dangerous if it enters nearby water intakes; notify operators. Notify local health and pollution control officials.
Response to discharge: Issue warning–corrosive. Restrict access. Disperse and flush.

SHIPPING INFORMATION
Grades of purity: 99+%; **Storage temperature:** Ambient; **Inert atmosphere:** None; **Venting:** Open; **Stability during transport:** Stable.

PHYSICAL AND CHEMICAL PROPERTIES
Physical state @ 59°F/15°C and 1 atm: Liquid.
Molecular weight: 88
Boiling point @ 1 atm: 309°F/154°C/427°K.
Melting/Freezing point: –51°F/–46°C/227°K.
Critical temperature: 637°F/336°C/609°K.
Critical pressure: 588 psia = 40 atm = 4.06 MN/m^2.
Specific gravity (water = 1): 0.949 @ 68°F/20°C (liquid).
Liquid surface tension: 25.1 dynes/cm = 0.0251 N/m @ 68°F/20°C.
Relative vapor density (air = 1): 3.0
Latent heat of vaporization: 202 Btu/lb = 112 cal/g = 4.68 x 10^5 J/kg.
Heat of combustion: –10,600 Btu/lb = –5880 cal/g = –246 x 10^5 J/kg.
Heat of solution: –20.5 Btu/lb = –11.4 cal/g = –0.477 x 10^5 J/kg.

ISOBUTYRONITRILE   REC. I:0550

SYNONYMS: 2-CYANOPROPANE; ISOPROPYL CYANIDE; IBN; ISOBUTIRONITRILO (Spanish); 2-METHYLPROPANENITRILE; 2-METHYLPROPIONITRILE

IDENTIFICATION
CAS Number: 78-82-0
Formula: C$_4$H$_7$N; (CH$_3$)$_2$CHCN
DOT ID Number: UN 2284; DOT Guide Number: 131
Proper Shipping Name: Isobutyronitrile
Reportable Quantity (RQ): **(EHS)** 1 lb/0.454 kg

DESCRIPTION: Colorless, clear liquid Almond-like odor. Floats on water surface; slightly soluble. Flammable vapor is produced.

Poison! • Highly flammable • Breathing the vapor, skin or eye contact, or swallowing the material can kill you; converted to cyanide in the body • Firefighting gear (including SCBA) does not provide adequate protection. If exposure occurs, remove and isolate gear immediately and thoroughly decontaminate personnel • Containers may BLEVE when exposed to fire • Vapors may form explosive mixture with air • Vapors are heavier than air and will collect and stay in low areas • Vapors may travel long distances to ignition sources and flashback • Vapors in confined areas (e.g., tanks, sewers, buildings) may explode when exposed to fire • Toxic products of combustion may include nitrogen oxides • Do not put yourself in danger by entering a contaminated area to rescue a victim.

Hazard Classification (based on NFPA-704 Rating System)
Health Hazards (Blue): 3; Flammability (Red): 3; Reactivity (Yellow): 0

EMERGENCY RESPONSE: See Appendix A (131)
Evacuation:
Public safety: Isolate spill area for at least 100 to 200 meters (330 to 660 feet) in all directions.
Spill: Increase, as necessary, the isolation distance shown above, in "Public safety."
Fire: Isolate for 800 meters (½ mile) in all directions, especially if tank, rail car, or tank truck is involved in fire.

EXPOSURE
Short-term effects: *SEEK MEDICAL ATTENTION.* **Vapor:** *POISONOUS IF INHALED.* Irritating to eyes, nose, and throat. Move to fresh air. *IF BREATHING HAS STOPPED,* give artificial respiration; *avoid mouth-to-mouth resuscitation; use bag/mask apparatus.* IF breathing is difficult, administer oxygen. **Liquid:** *POISONOUS IF SWALLOWED OR IF SKIN IS EXPOSED.* Irritating to skin and eyes. Remove contaminated clothing and shoes. Flush affected areas with plenty of water. *IF IN EYES,* hold eyelids open and flush. with plenty of water. *IF SWALLOWED* and victim is *CONSCIOUS AND ABLE TO SWALLOW,* have victim drink 4 to 8 ounces of water.

Note to physician or authorized medical personnel: Consider the use of amyl nitrite perles if symptoms of cyanide poisoning develop. If symptoms indicate, initial treatment includes the cyanide antidote kit. In all cases, break an amyl nitrite perle in a gauze pad and hold lightly under victim's nose for 15 seconds, repeating 5 times at about 15-second intervals; if necessary (and if sodium nitrite infusions will be delayed), repeat procedure every 3 minutes with fresh perles until 3 or 4 have been used. Avoid breathing the vapor while administering it to the victim. Administer sodium nitrite IV, ASAP. The usual adult dose is 10 to 20 mL of a 3% solution infused over no less than 5 minutes; the average child dose is 0.15 to 0.20 mL/kg. Monitor blood pressure during administration, and slow the rate of infusion if hypotention develops. Next, infuse sodium thiosulfate IV. The usual adult dose is 50 mL of a 25% solution infused over 10 to 20 minutes; the average child dose is 1.65 mL/kg. Repeat with nitrite and thiosulfate as required.

HEALTH HAZARDS
Personal protective equipment (PPE): A-Level PPE. Self-contained breathing apparatus; goggles; rubber **gloves.**
Recommendations for respirator selection: NIOSH
80 ppm: CCROV [any chemical cartridge respirator with organic vapor cartridge(s)]; or SA (any supplied-air respirator). *200 ppm:* SA:CF (any supplied-air respirator operated in a continuous-flow mode); or PAPROV [any powered, air-purifying respirator with organic vapor cartridge(s)]. *400 ppm:* CCRFOV [any air-purifying, full-facepiece respirator (gas mask) with a chin-style, front- or back-mounted acid gas canister]; or GMFOV [any air-purifying, full-facepiece respirator (gas mask) with a chin-style, front- or back-mounted acid gas canister]; or PAPRTOV [any powered, air-purifying respirator with a tight-fitting facepiece and organic vapor cartridges(s)]; or SCBAF (any self-contained breathing apparatus with a full facepiece); or SAF (any supplied-air respirator with a full facepiece). *1000 ppm:* SAF:PD,PP (any supplied-air respirator that has a full facepiece and is operated in a pressure-demand or other positive-pressure mode). *EMERGENCY OR PLANNED ENTRY INTO UNKNOWN CONCENTRATIONS OR IDLH CONDITIONS:* SCBAF:PD,PP (any self-contained breathing apparatus that has a full facepiece and is operated in a pressure-demand or other positive-pressure mode); or SAF:PD,PP:ASCBA (any supplied-air respirator that has a full facepiece and is operated in a pressure-demand or other positive-pressure mode in combination with an auxiliary self-contained breathing apparatus operated in a pressure-demand or other positive pressure mode). *ESCAPE:* GMFOV[any air-purifying, full-facepiece respirator (gas mask) with a chin-style, front- or back-mounted organic vapor cannister]; or SCBAE (any appropriate escape-type, self-contained breathing apparatus).
Exposure limits (TWA unless otherwise noted): NIOSH 8 ppm (22 mg/m^3).
Toxicity by ingestion: Grade 3; oral LD$_{50}$ = 100 mg/kg (rat).

FIRE DATA
Flash point: 47°F/8°C (cc).
Fire extinguishing agents not to be used: Water may be ineffective.

CHEMICAL REACTIVITY
Reactivity group: 37
Compatibility class: Nitriles

ENVIRONMENTAL DATA
Food chain concentration potential: Log P_{ow} = 0.4. Unlikely to accumulate.
Water pollution: DOT Appendix B, §172.101–marine pollutant. Effect of low concentrations on aquatic life is unknown. May be dangerous if it enters nearby water intakes; notify operators. Notify local health and wildlife officials. **Response to discharge:** Issue warning–high flammability, poison. Restrict access. Evacuate area. Mechanical containment. Should be removed. Chemical and physical treatment.

SHIPPING INFORMATION
Grades of purity: Technical; Pure; **Storage temperature:** Ambient; **Inert atmosphere:** None; **Venting:** Open; **Stability during transport:** Stable.

PHYSICAL AND CHEMICAL PROPERTIES
Physical state @ 59°F/15°C and 1 atm: Liquid.
Molecular weight: 69.1
Boiling point @ 1 atm: 219°F/104°C/377°K.
Melting/Freezing point: –97°F/–72°C/201°K.
Specific gravity (water = 1): 0.774 @ 68°F/20°C (liquid).
Liquid surface tension: 24.9 dynes/cm = 0.0249 N/m @ 68°F/20°C.
Relative vapor density (air = 1): 2.4
Latent heat of vaporization: 200 Btu/lb = 110 cal/g = 4.7 x 10^5 J/kg.
Heat of combustion: –14,960 Btu/lb = –8,310 cal/g = –348 x 10^5 J/kg.
Vapor pressure: 100 mm @ 130°F

ISODECALDEHYDE   REC. I:0600

SYNONYMS: ISODECALDEHIDO (Spanish); ISODECALDEHYDE, mixed isomers; TRIMETHYLHEPTANALS

IDENTIFICATION
Formula: $C_9H_{19}CHO$
DOT ID Number: NA 1993; DOT Guide Number: 128
Proper Shipping Name: Combustible liquid, n.o.s.

DESCRIPTION: Colorless liquid. Fruity odor. Floats on water surface; insoluble.

Highly flammable • Containers may BLEVE when exposed to fire • Vapors may form explosive mixture with air • Vapors are heavier than air and will collect and stay in low areas • Vapors may travel long distances to ignition sources and flashback • Vapors in confined areas (e.g., tanks, sewers, buildings) may explode when exposed to fire • Highly irritating to the eyes and respiratory tract; this material has anesthetic properties • Toxic products of combustion may include carbon monoxide.

Hazard Classification (based on NFPA-704 Rating System)
Health Hazards (Blue): 0; Flammability (Red): 2; Reactivity (Yellow): 0

EMERGENCY RESPONSE: See Appendix A (128)

Evacuation:
Public safety: Isolate spill area for at least 25 to 50 meters (80 to 160 feet) in all directions.
Spill: Large spill–Consider initial downwind evacuation for at least 300 meters (1000 feet).
Fire: Isolate for 800 meters (½ mile) in all directions, especially if tank, rail car, or tank truck is involved in fire.

EXPOSURE
Short-term effects: *SEEK MEDICAL ATTENTION.* **Liquid:** Irritating to skin and eyes. Remove contaminated clothing and shoes. Flush affected areas with plenty of water. *IF IN EYES*, hold eyelids open and flush with plenty of water.

HEALTH HAZARDS
Personal protective equipment (PPE): B-Level PPE. Butyl rubber is generally suitable for aldehydes.
Vapor (gas) irritant characteristics: Vapors cause a slight smarting of the eyes or respiratory system if present in high concentrations. The effect is temporary.
Liquid or solid irritant characteristics: Eye and skin irritant. If spilled on clothing and allowed to remain, may cause smarting and reddening of the skin.

FIRE DATA
Flash point: 185°F/85°C (oc).
Fire extinguishing agents not to be used: Water may be ineffective. Cool containers with water.

CHEMICAL REACTIVITY
Binary reactants: Reacts with strong acids, caustics, ammonia, amines.
Reactivity group: 19
Compatibility class: Aldehyde

ENVIRONMENTAL DATA
Food chain concentration potential: Negative; unlikely to accumulate.
Water pollution: DOT Appendix B, §172.101–marine pollutant. Effect of low concentrations on aquatic life is unknown. Fouling to shoreline. May be dangerous if it enters nearby water intakes; notify operators. Notify local health and wildlife officials. **Response to discharge:** Mechanical containment. Should be removed. Chemical and physical treatment.

SHIPPING INFORMATION
Grades of purity: Commercial; **Storage temperature:** Ambient; **Inert atmosphere:** None; **Venting:** Open (flame arrester); **Stability during transport:** Stable.

NAS HAZARD CLASSIFICATION FOR BULK WATER TRANSPORTATION
FIRE: 1
HEALTH: Vapor irritant: 1; Liquid or solid irritant: 1; Poisons: 1
WATER POLLUTION: Human toxicity: 1; Aquatic toxicity: 1; Aesthetic effect: 2
REACTIVITY: Other chemicals: 2; Water: 0; Self-reaction: 0

PHYSICAL AND CHEMICAL PROPERTIES
Physical state @ 59°F/15°C and 1 atm: Liquid.
Molecular weight: 156.28
Boiling point @ 1 atm: 387°F/197°C/470°K.
Specific gravity (water = 1): (estimate) 0.84 at 15° (liquid).
Liquid surface tension: (estimate) 20 dynes/cm = 0.02 N/m @ 68°F/20°C.

Liquid water interfacial tension: (estimate) 40 dynes/cm = 0.04 N/m @ 68°F/20°C.
Relative vapor density (air = 1): 5.4
Vapor pressure: (Reid) 0.03 psia.

ISODECYL ACRYLATE   REC. I:0650

SYNONYMS: ACRILATO de ISODECILO (Spanish); ACRYLIC ACID, ISODECYL ESTER; ISODECYL PROPENOATE

IDENTIFICATION
CAS Number: 1330-61-6
Formula: $C_{13}H_{24}O_2$; CH_2=CHCOO$C_{10}H_{21}$

DESCRIPTION: Colorless liquid Weak acrylate odor. Floats on water surface; soluble.

Combustible • Heat or flame may cause explosion • Containers may BLEVE when exposed to fire • Vapors are heavier than air and will collect and stay in low areas • Vapors in confined areas (e.g., tanks, sewers, buildings) may explode when exposed to fire • Polymerization hazard • Heat can induce polymerization with rapid release of energy; sealed containers may rupture explosively • • May react with itself blocking relief valves; leading to tank explosions • Irritating to the skin, eyes, and respiratory tract • Toxic products of combustion may include carbon monoxide.

Hazard Classification (based on NFPA-704 Rating System)
Health Hazards (Blue): 0; Flammability (Red): 1; Reactivity (Yellow): 0

EMERGENCY RESPONSE: See Appendix A (171)
Evacuation:
Public safety: Isolate the area of spill or leak for at least 10 to 25 meters (30 to 80 feet) in all directions.
Spill: Increase, in the downwind direction, as necessary, the distance shown under "Public Safety."
Fire: If any large container is involved in fire, isolate for at least 800 meters (½ mile) in all directions; also, consider initial evacuation for 800 meters (½ mile) in all directions.

EXPOSURE
Short-term effects: *SEEK MEDICAL ATTENTION.* **Liquid:** Irritating to skin and eyes. Remove contaminated clothing and shoes. Flush affected areas with plenty of water. *IF IN EYES,* hold eyelids open and flush with plenty of water. *IF SWALLOWED* and victim is *CONSCIOUS AND ABLE TO SWALLOW*, have victim drink 4 to 8 ounces of water.

HEALTH HAZARDS
Personal protective equipment (PPE): B-Level PPE. Goggles or face shield; rubber gloves.
Toxicity by ingestion: Grade 1; LD_{50} = 5 to 15 g/kg.
Vapor (gas) irritant characteristics: Eye irritant. Vapors cause smarting of the eyes or respiratory system if present in high concentrations.
Liquid or solid irritant characteristics: Eye and skin irritant. If spilled on clothing and allowed to remain, may cause smarting and reddening of the skin.

FIRE DATA
Flash point: 240°F/116°C (oc).
Fire extinguishing agents not to be used: Water may be ineffective. Cool containers with water.
Behavior in fire: May polymerize to gummy solid.

CHEMICAL REACTIVITY
Binary reactants: Will swell and soften certain rubbers and remove certain paints. Reacts with strong acids, amines.
Polymerization: Unless inhibited, polymerization will occur, especially when heated; **Inhibitor of polymerization:** Monomethyl ether of hydroquinone, 25 ppm.
Reactivity group: 14
Compatibility class: Acrylates

ENVIRONMENTAL DATA
Food chain concentration potential: Negative; unlikely to accumulate.
Water pollution: DOT Appendix B, §172.101–marine pollutant. Effect of low concentrations on aquatic life is unknown. Fouling to shoreline. May be dangerous if it enters nearby water intakes; notify operators. Notify local health and wildlife officials. **Response to discharge:** Issue warning–water contaminant. Mechanical containment. Should be removed. Chemical and physical treatment.

SHIPPING INFORMATION
Grades of purity: 97.5+%; **Storage temperature:** Ambient; **Inert atmosphere:** None; **Venting:** Open (flame arrester); **Stability during transport:** Stable.

NAS HAZARD CLASSIFICATION FOR BULK WATER TRANSPORTATION
FIRE: 1
HEALTH: Vapor irritant: 1; Liquid or solid irritant: 1; Poisons: 1
WATER POLLUTION: Human toxicity: 1; Aquatic toxicity: 2; Aesthetic effect: 2
REACTIVITY: Other chemicals: 2; Water: 0; Self-reaction: 0

PHYSICAL AND CHEMICAL PROPERTIES
Physical state @ 59°F/15°C and 1 atm: Liquid.
Molecular weight: 212.4
Boiling point @ 1 atm: 250°F/121°C/394°K (polymerizes).
Melting/Freezing point: –148°F/–100°C/173°K.
Specific gravity (water = 1): 0.885 @ 68°F/20°C (liquid).
Liquid surface tension: (estimate) 30 dynes/cm = 0.030 N/m @ 68°F/20°C.
Liquid water interfacial tension: (estimate) 30 dynes/cm = 0.030 N/m @ 68°F/20°C.
Relative vapor density (air = 1): 7.3
Latent heat of vaporization: 110 Btu/lb = 61 cal/g = 2.6 x 10^5 J/kg.
Heat of combustion: (estimate) –16,300 Btu/lb = –9100 cal/g = –380 x 10^5 J/kg.
Heat of polymerization: (estimate) –119 Btu/lb = –66 cal/g = –2.8 x 10^5 J/kg.
Vapor pressure: (Reid) Low

ISODECYL ALCOHOL   REC. I:0700

SYNONYMS: ALCOHOL de ISODECILO (Spanish); ISODECANOL

IDENTIFICATION
CAS Number: 25339-17-7
Formula: $C_{10}H_{21}OH$

DOT ID Number: UN 1987; DOT Guide Number: 127
Proper Shipping Name: Alcohols, n.o.s.

DESCRIPTION: Colorless liquid. Mild alcohol odor. Floats on water surface; insoluble.

Highly flammable • Containers may BLEVE when exposed to fire • Vapors may form explosive mixture with air • Vapors are heavier than air and will collect and stay in low areas • Vapors may travel long distances to ignition sources and flashback • Vapors in confined areas (e.g., tanks, sewers, buildings) may explode when exposed to fire • Irritating to the skin, eyes, and respiratory tract • Toxic products of combustion may include carbon monoxide.

Hazard Classification (based on NFPA-704 Rating System)
Health Hazards (Blue): 0; Flammability (Red): 1; Reactivity (Yellow): 0

EMERGENCY RESPONSE: See Appendix A (127)
Evacuation:
Public safety: Isolate spill area for at least 25 to 50 meters (80 to 160 feet) in all directions.
Spill: Large spill–Consider initial downwind evacuation for at least 300 meters (1000 feet).
Fire: Isolate for 800 meters (½ mile) in all directions, especially if tank, rail car, or tank truck is involved in fire.

EXPOSURE
Short-term effects: *SEEK MEDICAL ATTENTION.* **Liquid:** Will burn skin and eyes. Remove contaminated clothing and shoes. Flush affected areas with plenty of water. *IF SWALLOWED* and victim is *CONSCIOUS AND ABLE TO SWALLOW*, have victim drink 4 to 8 ounces of water.

HEALTH HAZARDS
Personal protective equipment (PPE): B-Level PPE. Chemical goggles.
Vapor (gas) irritant characteristics: Vapors are nonirritating to the eyes and throat.
Liquid or solid irritant characteristics: Skin and eye irritant. Causes smarting of the skin and first-degree burns on short exposure; may cause secondary burns on long exposure.

FIRE DATA
Flash point: 220°F/104°C (oc).
Fire extinguishing agents not to be used: Water or foam may cause frothing
Autoignition temperature: 550°F/288°C/561°K.
Electrical hazard: Class I, Group D.

CHEMICAL REACTIVITY
Binary reactants: Reacts with strong oxidizers, acids, caustics, isocyanates, aliphatic amines.
Reactivity group: 20
Compatibility class: Alcohols, glycols

ENVIRONMENTAL DATA
Food chain concentration potential: Negative; unlikely to accumulate.
Water pollution: DOT Appendix B, §172.101–marine pollutant. Effect of low concentrations on aquatic life is unknown. Fouling to shoreline. May be dangerous if it enters nearby water intakes; notify operators. Notify local health and wildlife officials.
Response to discharge: Mechanical containment. Should be removed. Chemical and physical treatment.

SHIPPING INFORMATION
Grades of purity: Technical: mixed isomers; **Storage temperature:** Ambient; **Inert atmosphere:** None; **Venting:** Open (flame arrester); **Stability during transport:** Stable.

NAS HAZARD CLASSIFICATION FOR BULK WATER TRANSPORTATION
FIRE: 1
HEALTH: Vapor irritant: 0; Liquid or solid irritant: 2; Poisons: 0
WATER POLLUTION: Human toxicity: 0; Aquatic toxicity: 1; Aesthetic effect: 2
REACTIVITY: Other chemicals: 2; Water: 0; Self-reaction: 0

PHYSICAL AND CHEMICAL PROPERTIES
Physical state @ 59°F/15°C and 1 atm: Liquid.
Molecular weight: 158.29
Boiling point @ 1 atm: 428°F/220°C/493°K.
Melting/Freezing point: Less than 140°F/60°C/333°K.
Specific gravity (water = 1): 0.841 @ 68°F/20°C (liquid).
Relative vapor density (air = 1): 5.5
Ratio of specific heats of vapor (gas): (estimate) 1.032
Latent heat of vaporization: (estimate) 120 Btu/lb = 67 cal/g = 2.8 x 10^5 J/kg.

ISOHEXANE   REC. I:0750

SYNONYMS: ISOHEXANO (Spanish); 2-METHYL PENTANE

IDENTIFICATION
CAS Number: 107-83-5; 96-14-0 (3-isomer)
Formula: C_6H_{14}; $CH_3CH(CH_3)CH_2CH_2CH_3$
DOT ID Number: UN 2462; DOT Guide Number: 128
Proper Shipping Name: Methyl pentane

DESCRIPTION: Colorless watery liquid Gasoline-like odor. Floats on the surface of water. Flammable, irritating vapor is produced.

Highly flammable • Containers may BLEVE when exposed to fire • Vapors may form explosive mixture with air • Vapors are heavier than air and will collect and stay in low areas • Vapors may travel long distances to ignition sources and flashback • Vapors in confined areas (e.g., tanks, sewers, buildings) may explode when exposed to fire • Irritating to the skin, eyes, and respiratory tract • Toxic products of combustion may include carbon monoxide.

Hazard Classification (based on NFPA-704 Rating System)
Health Hazards (Blue): 1; Flammability (Red): 3; Reactivity (Yellow): 0

EMERGENCY RESPONSE: See Appendix A (128)
Evacuation:
Public safety: Isolate spill area for at least 25 to 50 meters (80 to 160 feet) in all directions.
Spill: Large spill–Consider initial downwind evacuation for at least 300 meters (1000 feet).
Fire: Isolate for 800 meters (½ mile) in all directions, especially if tank, rail car, or tank truck is involved in fire.

EXPOSURE
Short-term effects: *SEEK MEDICAL ATTENTION.* **Vapor:** Irritating to eyes, nose, and throat. *IF INHALED*, will, will cause dizziness, headache, difficult breathing or loss of consciousness. Move to fresh air. *IF BREATHING HAS STOPPED,* give artificial

respiration. IF breathing is difficult, administer oxygen. **Liquid:** Irritating to skin and eyes. *IF SWALLOWED*, will cause nausea or vomiting. Remove contaminated clothing and shoes. Flush affected areas with plenty of water. *IF IN EYES*, hold eyelids open and flush with plenty of water. *IF SWALLOWED* and victim is *CONSCIOUS AND ABLE TO SWALLOW*, have victim drink 4 to 8 ounces of water. **Do NOT induce vomiting.**

HEALTH HAZARDS
PERSONAL PROTECTIVE EQUIPMENT. EYE protection (as for gasoline).
Recommendations for respirator selection: NIOSH as hexane isomers.
1000 ppm: SA* (any supplied-air respirator). *2500 ppm:* SA:CF* (any supplied-air respirator operated in a continuous-flow mode). *500 ppm:* SA:PD:PP (any supplied-air respirator operated in a pressure-demand or other positive-pressure mode). *5000 ppm:* SAT:CF (any supplied-air respirator that has a tight-fitting facepiece and is operated in a continuous-flow mode); or SCBAF (any self-contained breathing apparatus with a full facepiece); or SAF (any supplied-air respirator with a full facepiece). *EMERGENCY OR PLANNED ENTRY INTO UNKNOWN CONCENTRATIONS OR IDLH CONDITIONS:* SCBAF:PD,PP (any self-contained breathing apparatus that has a full facepiece and is operated in a pressure-demand or other positive-pressure mode); or SAF:PD,PP:ASCBA (any supplied-air respirator that has a full facepiece and is operated in a pressure-demand or other positive-pressure mode in combination with an auxiliary self-contained breathing apparatus operated in a pressure-demand or other positive pressure mode). *ESCAPE:* GMFOV [any air-purifying, full-facepiece respirator (gas mask) with a chin-style, front- or back-mounted organic vapor canister]; or SCBAE (any appropriate escape-type, self-contained breathing apparatus. *Note:* Substance reported to cause eye irritation or damage; may require eye protection.
Exposure limits (TWA unless otherwise noted): ACGIH TLV 500 ppm (1760 mg/m^3); NIOSH 100 ppm (350 mg/m^3), ceiling 510 ppm (1800 mg/m^3)/15 minute.
Short-term exposure limits (15-minute TWA): ACGIH STEL 1000 ppm (3500 mg/m^3).
Long-term health effects: Possible central nervous system damage.
Vapor (gas) irritant characteristics: Respiratory tract irritant.

FIRE DATA
Flash point: <–20°F/<–29°C (cc).
Flammable limits in air: LEL: 1.2%; UEL: 7.7%.
Fire extinguishing agents not to be used: Water may be ineffective
Behavior in fire: Flashback along vapor trail may occur. vapor may explode in an enclosed area. Containers may explode in fire.
Autoignition temperature: 505°F/263°C/536°K.
Electrical hazard: Class I, Group D.
Burning rate: 8.2 mm/min

CHEMICAL REACTIVITY
Binary reactants: Vigorous reaction with strong oxidizers.
Reactivity group: 31
Compatibility class: Paraffins

ENVIRONMENTAL DATA
Food chain concentration potential: Log P_{ow} = 3.7. Values > 3.0 are likely to bioconcentrate in aquatic organisms and other living tissue, especially in fats.
Water pollution: Effect of low concentrations on aquatic life is unknown. May be dangerous if it enters nearby water intakes; notify operators. Notify local health and wildlife officials. **Response to discharge:** Issue warning–high flammability. Evacuate area. Disperse and flush.

SHIPPING INFORMATION
Grades of purity: Research: 99.95%; pure: 99.0%; technical: 95.0%; **Storage temperature:** Ambient; **Inert atmosphere:** None; **Venting:** Open (flame arrester) or pressure-vacuum; **Stability during transport:** Stable.

NAS HAZARD CLASSIFICATION FOR BULK WATER TRANSPORTATION
FIRE: 3
HEALTH: Vapor irritant: 0; Liquid or solid irritant: 0; Poisons: 1
WATER POLLUTION: Human toxicity: 1; Aquatic toxicity: 1; Aesthetic effect: 1
REACTIVITY: Other chemicals: 0; Water: 0; Self-reaction: 0

PHYSICAL AND CHEMICAL PROPERTIES
Physical state @ 59°F/15°C and 1 atm: Liquid.
Molecular weight: 86.18
Boiling point @ 1 atm: 140.5°F/60.3°C/333.5°K.
Melting/Freezing point: –244.6°F/–153.7°C/119.5°K.
Critical temperature: 435.7°F/224.3°C/497.5°K.
Critical pressure: 437 psia = 29.7 atm = 3.01 MN/m^2.
Specific gravity (water = 1): 0.653 @ 68°F/20°C (liquid).
Liquid surface tension: 17.38 dynes/cm = 0.01738 N/m @ 68°F/20°C.
Liquid water interfacial tension: (estimate) 40 dynes/cm = 0.04 N/m @ 68°F/20°C.
Relative vapor density (air = 1): 2.9
Ratio of specific heats of vapor (gas): 1.062
Latent heat of vaporization: 139 Btu/lb = 77.1 cal/g = 3.23 x 10^5 J/kg.
Heat of combustion: –19,147 Btu/lb = –10,637 cal/g = –445.35 x 10^5 J/kg.
Heat of fusion: 17.41 cal/g.
Vapor pressure: (Reid) 6.0 psia.

ISOOCTALDEHYDE　　　　　　　　REC. I:0800

SYNONYMS: DIMETHYLHEXANALS; ISOOCTYLALDEHYDE; 6-METHYL-1-HEPTANAL; OXO OCTALDEHYDE

IDENTIFICATION
Formula: $(CH_3)_2CH(CH_2)_4CHO$

DESCRIPTION: Colorless liquid Mild fruity odor. Floats on the surface of water.

Flammable • Containers may BLEVE when exposed to fire • Vapors may form explosive mixture with air • Vapors are heavier than air and will collect and stay in low areas • Vapors may travel long distances to ignition sources and flashback • Vapors in confined areas (e.g., tanks, sewers, buildings) may explode when exposed to fire • Highly irritating to the eyes and respiratory tract; this material has anesthetic properties • Toxic products of combustion may include carbon monoxide.

Hazard Classification (based on NFPA-704 Rating System)
Health Hazards (Blue): 0; Flammability (Red): 2; Reactivity (Yellow): 0

EMERGENCY RESPONSE: See Appendix A (171)
Evacuation:
Public safety: Isolate the area of spill or leak for at least 10 to 25 meters (30 to 80 feet) in all directions.
Spill: Increase, in the downwind direction, as necessary, the distance shown under "Public Safety."
Fire: If any large container is involved in fire, isolate for at least 800 meters (½ mile) in all directions; also, consider initial evacuation for 800 meters (½ mile) in all directions.

EXPOSURE
Short-term effects: *SEEK MEDICAL ATTENTION.* **Liquid:** Irritating to eyes. *IF IN EYES*, hold eyelids open and flush. with plenty of water.

HEALTH HAZARDS
Personal protective equipment (PPE): B-Level PPE. Chemical goggles. Butyl rubber is generally suitable for aldehydes.
Vapor (gas) irritant characteristics: Vapors cause a slight smarting of the eyes or respiratory system if present in high concentrations. The effect is temporary.
Liquid or solid irritant characteristics: Eye irritant. Practically harmless to the skin.

FIRE DATA
Flash point: 104°F/40°C (cc).
Autoignition temperature: 320°F/160°C/433°K.

CHEMICAL REACTIVITY
Binary reactants: Acids, caustics, ammonia, amines.
Reactivity group: 19
Compatibility class: Aldehyde

ENVIRONMENTAL DATA
Food chain concentration potential: Negative; unlikely to accumulate.
Water pollution: DOT Appendix B, §172.101–marine pollutant (n-octaldehyde). Effects of low concentrations on aquatic life is unknown. Fouling to shoreline. May be dangerous if it enters nearby water intakes; notify operators. Notify local health and wildlife officials. **Response to discharge:** Mechanical containment. Should be removed. Chemical and physical treatment.

SHIPPING INFORMATION
Storage temperature: Ambient; **Inert atmosphere:** None;
Venting: Open (flame arrester); **Stability during transport:** Stable.

NAS HAZARD CLASSIFICATION FOR BULK WATER TRANSPORTATION
FIRE: 2
HEALTH: Vapor irritant: 1; Liquid or solid irritant: 0; Poisons: 1
WATER POLLUTION: Human toxicity: 1; Aquatic toxicity: 1; Aesthetic effect: 2
REACTIVITY: Other chemicals: 2; Water: 0; Self-reaction: 1

PHYSICAL AND CHEMICAL PROPERTIES
Physical state @ 59°F/15°C and 1 atm: Liquid.
Molecular weight: 128.22
Boiling point @ 1 atm: 307–352°F/153–178°C/426–451°K.
Melting/Freezing point: –180°F/–118°C/155°K.
Specific gravity (water = 1): 0.825 @ 68°F/20°C (liquid).
Liquid surface tension: 26.9 dynes/cm = 0.0269 N/m @ 68°F/20°C.
Liquid water interfacial tension: (estimate) 40 dynes/cm = 0.04 N/m @ 68°F/20°C.
Ratio of specific heats of vapor (gas): (estimate) 1.040
Latent heat of vaporization: 140 Btu/lb = 77 cal/g = 3.2 x 10^5 J/kg.
Heat of combustion: (estimate) –17,000 Btu/lb = –9600 cal/g = –400 x 10^5 J/kg.

ISOOCTYL ALCOHOL   REC. I:0850

SYNONYMS: ALCOHOL ISOOCTILICO (Spanish); DIMETHYL-1-HEXANOLS; ISOOCTANOL; 6-METHYL-1-HEPTANOL; OXO OCTYL ALCOHOL

IDENTIFICATION
CAS Number: 26952-21-6
Formula: $C_8H_{18}O$; $(CH_3)_2CH(CH_2)_4CH_2OH$
DOT ID Number: NA 1993; DOT Guide Number: 128
Proper Shipping Name: Combustible liquid, n.o.s.

DESCRIPTION: Colorless liquid. Mild odor. Floats on water surface; insoluble.

Hazard Classification (based on NFPA-704 Rating System)
Health Hazards (Blue): 0; Flammability (Red): 2; Reactivity (Yellow): 0

EMERGENCY RESPONSE: See Appendix A (128)
Evacuation:
Public safety: Isolate spill area for at least 25 to 50 meters (80 to 160 feet) in all directions.
Spill: Large spill–Consider initial downwind evacuation for at least 300 meters (1000 feet).
Fire: Isolate for 800 meters (½ mile) in all directions, especially if tank, rail car, or tank truck is involved in fire.

EXPOSURE
Short-term effects: *SEEK MEDICAL ATTENTION.* **Liquid:** *HARMFUL IF SWALLOWED OR ABSORBED THROUGH THE SKIN.* Irritating to skin and eyes. Harmful if swallowed. Remove contaminated clothing. Flush affected areas with plenty of water. *IF IN EYES*, hold eyelids open and flush with plenty of water. *IF SWALLOWED* and victim is *CONSCIOUS AND ABLE TO SWALLOW*, have victim drink 4 to 8 ounces of water.

HEALTH HAZARDS
Personal protective equipment (PPE): B-Level PPE. OSHA Table Z-1 air contaminant. Organic vapor respirator.
Exposure limits (TWA unless otherwise noted): ACGIH TLV 50 ppm (266 mg/m³) skin; NIOSH 50 ppm (270 mg/m³); skin contact contributes to overall exposure.
Toxicity by ingestion: Grade 2; LD_{50} = 0.5 to 5 g/kg (lab animals).
Vapor (gas) irritant characteristics: Vapors are irritating to the eyes and throat.
Liquid or solid irritant characteristics: Severe eye irritant. Skin irritant.

FIRE DATA
Flash point: 180°F/82°C (oc).
Flammable limits in air: LEL: 0.9% (calculated); UEL: 5.7% (estimate).
Autoignition temperature: 530°F/277°C/550°K (estimate).
Electrical hazard: Class I, Group D.

634 Isopentane

CHEMICAL REACTIVITY
Reactivity group: 20
Compatibility class: Alcohols, glycols

ENVIRONMENTAL DATA
Food chain concentration potential: Negative; unlikely to accumulate.
Water pollution: DOT Appendix B, §172.101–marine pollutant. Effect of low concentrations on aquatic life is unknown. Fouling to shoreline. May be dangerous if it enters nearby water intakes; notify operators. Notify local health and wildlife officials.
Response to discharge: Mechanical containment. Should be removed. Chemical and physical treatment.

SHIPPING INFORMATION
Grades of purity: 99+% (mixed isomers); **Storage temperature:** Ambient; **Inert atmosphere:** None; **Venting:** Open (flame arrester); **Stability during transport:** Stable.

NAS HAZARD CLASSIFICATION FOR BULK WATER TRANSPORTATION
FIRE: 1
HEALTH: Vapor irritant: 0; Liquid or solid irritant: 0; Poisons: 1
WATER POLLUTION: Human toxicity: 1; Aquatic toxicity: 3; Aesthetic effect: 2
REACTIVITY: Other chemicals: 2; Water: 0; Self-reaction: 0

PHYSICAL AND CHEMICAL PROPERTIES
Physical state @ 59°F/15°C and 1 atm: Liquid.
Molecular weight: 130.22
Boiling point @ 1 atm: 359-383°F/182-195°C/455-468°K.
Melting/Freezing point: Less than 212°F/100°C/373°K.
Specific gravity (water = 1): 0.832 @ 68°F/20°C (liquid).
Liquid surface tension: 29.5 dynes/cm = 0.0295 N/m @ 68°F/20°C.
Liquid water interfacial tension: (estimate) 40 dynes/cm = 0.04 N/m @ 68°F/20°C.
Relative vapor density (air = 1): 4.5
Ratio of specific heats of vapor (gas): (estimate) 1.040
Latent heat of vaporization: (estimate) 140 Btu/lb = 77 cal/g = 3.2×10^5 J/kg.
Heat of combustion: (estimate) –17,400 Btu/lb = –9650 cal/g = -404×10^5 J/kg.
Vapor pressure: (Reid) 0.02 psia; 3.06 mm.

ISOPENTANE REC. I:0900

SYNONYMS: EEC No. 601-006-00-1; ETHYL DIMETHYL METHANE; ISOAMYL HYDRIDE; ISOPENTANO (Spanish); 2-METHYLBUTANE

IDENTIFICATION
CAS Number: 78-78-4
Formula: C_5H_{12}; $(CH_3)_2CHCH_2CH_3$
DOT ID Number: UN 1265; DOT Guide Number: 128
Proper Shipping Name: Isopentane

DESCRIPTION: Colorless watery liquid Gaseline-like odor. Floats on the surface of water. Flammable, irritating vapor is produced.

Extremely flammable • Containers may BLEVE when exposed to fire • Vapors may form explosive mixture with air • Vapors are heavier than air and will collect and stay in low areas • Vapors may travel long distances to ignition sources and flashback • Vapors in confined areas (e.g., tanks, sewers, buildings) may explode when exposed to fire • Irritating to the skin, eyes, and respiratory tract • Toxic products of combustion may include carbon monoxide.

Hazard Classification (based on NFPA-704 Rating System)
Health Hazards (Blue): 1; Flammability (Red): 4; Reactivity (Yellow): 0

EMERGENCY RESPONSE: See Appendix A (128)
Evacuation:
Public safety: Isolate spill area for at least 25 to 50 meters (80 to 160 feet) in all directions.
Spill: Large spill–Consider initial downwind evacuation for at least 300 meters (1000 feet).
Fire: Isolate for 800 meters (½ mile) in all directions, especially if tank, rail car, or tank truck is involved in fire.

EXPOSURE
Short-term effects: *SEEK MEDICAL ATTENTION.* **Vapor:** Irritating to nose and throat. *IF INHALED*, will, will cause coughing, difficult breathing, or
loss of consciousness. Move to fresh air. *IF BREATHING HAS STOPPED,* give artificial respiration. IF breathing is difficult, administer oxygen. **Liquid:** Irritating to skin and eyes. *IF SWALLOWED*, will cause nausea or vomiting. Remove contaminated clothing and shoes. Flush affected areas with plenty of water. *IF IN EYES*, hold eyelids open and flush with plenty of water. *IF SWALLOWED* and victim is *CONSCIOUS AND ABLE TO SWALLOW*, have victim drink 4 to 8 ounces of water. **Do NOT induce vomiting.**

HEALTH HAZARDS
Personal protective equipment. Eye protection (as for gasoline).
Recommendations for respirator selection: NIOSH as *n*-pentane *1200 ppm:* SA (any supplied-air respirator). *1500 ppm:* SA:CF (any supplied-air respirator operated in a continuous-flow mode); or SCBAF (any self-contained breathing apparatus with a full facepiece); or SAF (any supplied-air respirator with a full facepiece). *EMERGENCY OR PLANNED ENTRY INTO UNKNOWN CONCENTRATIONS OR IDLH CONDITIONS:* SCBAF:PD,PP (any self-contained breathing apparatus that has a full facepiece and is operated in a pressure-demand or other positive-pressure mode); or SAF:PD,PP:ASCBA (any supplied-air respirator that has a full facepiece and is operated in a pressure-demand or other positive-pressure mode in combination with an auxiliary self-contained breathing apparatus operated in a pressure-demand or other positive pressure mode). *ESCAPE:* GMFOV[any air-purifying, full-facepiece respirator (gas mask) with a chin-style, front- or back-mounted organic vapor canister]; or SCBAE (any appropriate escape-type, self-contained breathing apparatus).
Exposure limits (TWA unless otherwise noted): ACGIH TLV NIOSH 600 ppm (1770 mg/m^3) as pentane; NIOSH 120 ppm (350 mg/m^3); OSHA PEL 1000 ppm (2950 mg/m^3) as *n*-pentane.
Short-term exposure limits (15-minute TWA): ACGIH STEL 750 ppm (2210 mg/m^3) as pentane; NIOSH ceiling/15 min: 610 ppm (1800 mg/m^3) as *n*-pentane
Toxicity by ingestion: Grade 1; LD_{50} = 5 to 15 g/kg.
Vapor (gas) irritant characteristics: Eye and respiratory tract irritant.
Liquid or solid irritant characteristics: Eye irritant. Practically harmless to the skin.

FIRE DATA
Flash point: Less than –70°F/–57°C (cc).

Flammable limits in air: LEL: 1.4%; UEL: 8.0%.
Fire extinguishing agents not to be used: Water may be ineffective
Autoignition temperature: 797°F/425°C/698°K.
Electrical hazard: Class I, Group D.
Burning rate: 7.4 mm/min

CHEMICAL REACTIVITY
Binary reactants: Dangerous reaction with oxidizers. Highly volatile liquid; vapors may explode when mixed with air. May react with some plastics.
Reactivity group: 31
Compatibility class: Paraffins

ENVIRONMENTAL DATA
Food chain concentration potential: Log P_{ow} = 2.33. Unlikely to accumulate.
Water pollution: Effect of low concentrations on aquatic life is unknown. May be dangerous if it enters nearby water intakes; notify operators. Notify local health and wildlife officials.
Response to discharge: Issue warning–high flammability. Evacuate area. Disperse and flush.

SHIPPING INFORMATION
Grades of purity: Research: 99.99%; pure: 99.4%; technical: 97%; **Storage temperature:** Ambient; **Inert atmosphere:** None; **Venting:** Open (flame arrester) or pressure-vacuum; **Stability during transport:** Stable.

NAS HAZARD CLASSIFICATION FOR BULK WATER TRANSPORTATION
FIRE: 4
HEALTH: Vapor irritant: 0; Liquid or solid irritant: 0; Poisons: 1
WATER POLLUTION: Human toxicity: 1; Aquatic toxicity: 2; Aesthetic effect: 1
REACTIVITY: Other chemicals: 0; Water: 0; Self-reaction: 0

PHYSICAL AND CHEMICAL PROPERTIES
Physical state @ 59°F/15°C and 1 atm: Liquid.
Molecular weight: 72.15
Boiling point @ 1 atm: 82.2°F/27.9°C/301.1°K.
Melting/Freezing point: –255.8°F/–159.9°C/113.3°K.
Critical temperature: 369.0°F/187.2°C/460.4°K.
Critical pressure: 491.0 psia = 33.4 atm = 3.38 MN/m^2.
Specific gravity (water = 1): 0.620 @ 68°F/20°C (liquid).
Liquid surface tension: 16.05 dynes/cm = 0.01605 N/m @ 68°F/20°C.
Liquid water interfacial tension: 31 dynes/cm = 0.031 N/m at 22.7°C.
Relative vapor density (air = 1): 2.5
Ratio of specific heats of vapor (gas): 1.076
Latent heat of vaporization: 146 Btu/lb = 81.0 cal/g = 3.39 x 10^5 J/kg.
Heat of combustion: –19,314 Btu/lb = –10,730 cal/g = –449.24 x 10^5 J/kg.
Heat of fusion: 17.05 cal/g.
Vapor pressure: (Reid) 20 psia; 510 mm.

ISOPHORONE   **REC. I:0950**

SYNONYMS: EEC No. 606-012-00-8; ISOACETOPHORONE; ISOFORONA (Spanish); 1,1,3-TRIMETHYL-3-CYCLOHEXEN-5-ONE; 3,5,5-TRIMETHYL-2-CYCLOHEXANE-1-ONE; 3,5,5-TRIMETHYL-2-CYCLOHENONE

IDENTIFICATION
CAS Number: 78-59-1
Formula: $C_9H_{14}O$; $COCH=C(CH_3)CH_2C(CH_3)_2CH_2$
DOT ID Number: NA 1993; DOT Guide Number: 128
Proper Shipping Name: Combustible liquid, n.o.s.
Reportable Quantity (RQ): **(CERCLA)** 5000 lb/2270 kg

DESCRIPTION: Colorless to white liquid. Camphor- or peppermint-like odor. Floats and mixes slowly with water.

Poison! • Highly flammable • Severely irritates eyes, skin, and respiratory tract • Containers may BLEVE when exposed to fire • Vapors may form explosive mixture with air • Vapors are heavier than air and will collect and stay in low areas • Vapors may travel long distances to ignition sources and flashback • Vapors in confined areas (e.g., tanks, sewers, buildings) may explode when exposed to fire • Toxic products of combustion may include carbon monoxide.

Hazard Classification (based on NFPA-704 Rating System)
Health Hazards (Blue): 2; Flammability (Red): 2; Reactivity (Yellow): 0

EMERGENCY RESPONSE: See Appendix A (128)
Evacuation:
Public safety: Isolate spill area for at least 25 to 50 meters (80 to 160 feet) in all directions.
Spill: Large spill–Consider initial downwind evacuation for at least 300 meters (1000 feet).
Fire: Isolate for 800 meters (½ mile) in all directions, especially if tank, rail car, or tank truck is involved in fire.

EXPOSURE
Short-term effects: *SEEK MEDICAL ATTENTION.* **Liquid:** Irritating to skin and eyes. Harmful if swallowed. Remove contaminated clothing and shoes. Flush affected areas with plenty of water. *IF IN EYES*, hold eyelids open and flush with plenty of water. *IF SWALLOWED* and victim is *CONSCIOUS AND ABLE TO SWALLOW*, have victim drink 4 to 8 ounces of water. **Do NOT induce vomiting.**

HEALTH HAZARDS
Personal protective equipment (PPE): B-Level PPE. OSHA Table Z-1 air contaminant. Self-contained breathing apparatus with full face mask; gloves. Chemical protective material(s) reported to have good to excellent resistance: PV alcohol. Also, butyl rubber/deoprene offers limited protection.
Recommendations for respirator selection: NIOSH
up to 40 ppm: CCROV* [any chemical cartridge respirator with organic vapor cartridge(s)]; or SA* (any supplied-air respirator). *100 ppm:* SA:CF* (any supplied-air respirator operated in a continuous-flow mode); or PAPROV* [any powered, air-purifying respirator with organic vapor cartridge(s)]. *200 ppm:* CCRFOV [any chemical cartridge respirator with a full facepiece and organic vapor cartridge(s)]; or GMFOV (any air-purifying, full-facepiece respirator (gas mask) with a chin-style, front- or back-mounted organic vapor canister); or PAPRTOV* (any powered, air-purifying respirator with a tight-fitting facepiece and organic vapor cartridge(s); or SAT:CF* (any supplied-air respirator that has a tight-fitting facepiece and is operated in a continuous-flow mode); or SCBAF (any self-contained breathing apparatus with a full facepiece); or SAF (any supplied-air respirator with a full facepiece). *EMERGENCY OR PLANNED ENTRY INTO UNKNOWN CONCENTRATIONS OR IDLH CONDITIONS*: SCBAF:PD,PP (any self-contained breathing apparatus that has a

636 Isophorone diamine

full facepiece and is operated in a pressure-demand or other positive-pressure mode); or SAF:PD,PP:ASCBA (any supplied-air respirator that has a full facepiece and is operated in a pressure-demand or other positive-pressure mode in combination with an auxiliary self-contained breathing apparatus operated in a pressure-demand or other positive-pressure mode). *ESCAPE:* GMFOV [any air-purifying, full-facepiece respirator (gas mask) with a chin-style, front-or back-mounted organic vapor canister] or SCBAE (any appropriate escape-type, self-contained breathing apparatus).
Note: Substance reported to cause eye irritation or damage; may require eye protection.
Exposure limits (TWA unless otherwise noted): NIOSH 4 ppm (23 mg/m^3); OSHA 25 ppm (140 mg/m^3).
Short-term exposure limits (15-minute TWA): ACGIH TLV ceiling 5 ppm (28 mg/m^3)
Toxicity by ingestion: Grade 2; oral LD$_{50}$ = 2330 mg/kg (rat).
Long-term health effects: Possible kidney and lung damage; central nervous system effects; dermatitis.
Vapor (gas) irritant characteristics: Vapors cause moderate irritation such that personnel will find high concentrations unpleasant. The effect is temporary.
Liquid or solid irritant characteristics: Causes smarting of the skin and first-degree burns on short exposure; may cause second-degree burns on long exposure.
Odor threshold: 0.2 ppm
IDLH value: 200 ppm.

FIRE DATA
Flash point: 205°F/96°C (oc); 184°F/84°C (cc).
Flammable limits in air: LEL: 0.84%; UEL: 3.8%.
Fire extinguishing agents not to be used: Water may be ineffective
Autoignition temperature: 861°F/461°C/734°K.
Electrical hazard: Class I, Group D.
Burning rate: 4.0 mm/min

CHEMICAL REACTIVITY
Binary reactants: Reacts with oxidizers, strong alkalis, amines.
Reactivity group: 18
Compatibility class: Ketone

ENVIRONMENTAL DATA
Food chain concentration potential: Negative; unlikely to accumulate.
Water pollution: Effect of low concentrations on aquatic life is unknown. Fouling to shoreline. May be dangerous if it enters nearby water intakes; notify operators. Notify local health and wildlife officials. **Response to discharge:** Issue warning–water contaminant. Restrict access. Mechanical containment. Should be removed. Chemical and physical treatment.

SHIPPING INFORMATION
Grades of purity: 99+%; **Storage temperature:** Ambient; **Inert atmosphere:** None; **Venting:** Open (flame arrester); **Stability during transport:** Stable.

NAS HAZARD CLASSIFICATION FOR BULK WATER TRANSPORTATION
FIRE: 1
HEALTH: Vapor irritant: 2; Liquid or solid irritant: 2; Poisons: 2
WATER POLLUTION: Human toxicity: 1; Aquatic toxicity: 3; Aesthetic effect: 2
REACTIVITY: Other chemicals: 2; Water: 0; Self-reaction: 0

PHYSICAL AND CHEMICAL PROPERTIES

Physical state @ 59°F/15°C and 1 atm: Liquid.
Molecular weight: 138.2
Boiling point @ 1 atm: 419.5°F/215.3°C/488.5°K.
Melting/Freezing point: 17.4°F/–8.1°C/265.1°K.
Specific gravity (water = 1): 0.921 @ 77°F/25°C (liquid).
Liquid surface tension: 32.3 dynes/cm = 0.0323 N/m @ 68°F/20°C.
Relative vapor density (air = 1): 4.75
Latent heat of vaporization: 135 Btu/lb = 75 cal/g = 3.14 x 10^5 J/kg.
Heat of combustion: –16,170 Btu/lb = –8,980 cal/g = –376 x 10^5 J/kg.
Vapor pressure: (Reid) Low; 0.3 mm.

ISOPHORONE DIAMINE   REC. I:1000

SYNONYMS: 3-AMINOMETHYL-3,5,5- TRIMETHYLCYCLO-HEXYLAMINE; ISOFORONA DIAMINA (Spanish)

IDENTIFICATION
CAS Number: 2855-13-2
Formula: $(CH_3)_3C_6H_7(NH_2)CH_2NH_2$
DOT ID Number: UN 2289; DOT Guide Number: 153
Proper Shipping Name: Isophoronediamine

DESCRIPTION: Colorless liquid Faint amine odor. Floats on the surface of water; soluble.

Combustible • Heat or flame may cause explosion • Containers may BLEVE when exposed to fire • Vapors are heavier than air and will collect and stay in low areas • Vapors in confined areas (e.g., tanks, sewers, buildings) may explode when exposed to fire • Irritating to the skin, eyes, and respiratory tract • Toxic products of combustion may include nitrogen oxides • Do not put yourself in danger by entering a contaminated area to rescue a victim.

Hazard Classification (based on NFPA-704 Rating System)
Health Hazards (Blue): 1; Flammability (Red): 1; Reactivity (Yellow): 0

EMERGENCY RESPONSE: See Appendix A (153)
Evacuation:
Public safety: Isolate the area of spill or leak for at least 25 to 50 meters (80 to 160 feet) in all directions.
Spill: Increase, in the downwind direction, as necessary, the distance shown under "Public Safety."
Fire: If tank, rail car, or tank truck is involved in fire, isolate for at least 800 meters (½ mile) in all directions; also, consider initial evacuation for 800 meters (½ mile) in all directions.

EXPOSURE
Short-term effects: *SEEK MEDICAL ATTENTION.* **Vapor:** If inhaled, may cause irritation, coughing, and nausea. Move to fresh air. *IF BREATHING HAS STOPPED,* give artificial respiration; *avoid mouth-to-mouth resuscitation; use bag/mask apparatus.* **Liquid:** May cause inflammation or burns to eyes and skin. Harmful if swallowed. Remove contaminated clothing and shoes. Flush affected areas with plenty of water. *IF IN EYES*, hold eyelids open and flush with plenty of water. *IF SWALLOWED* and victim is *CONSCIOUS AND ABLE TO SWALLOW*, have victim drink 4 to 8 ounces of water. **Do NOT induce vomiting.**

HEALTH HAZARDS
Personal protective equipment (PPE): B-Level PPE. Wear rubber

overclothing, gloves, goggles, and self-contained breathing apparatus.
Toxicity by ingestion: Grade 2; LD_{50} = 1.03 g/kg (rat).

FIRE DATA
Flash point: 230°F/110°C (oc).

CHEMICAL REACTIVITY
Binary reactants: Reacts with acids, alcohol, glycol. Corrodes aluminum and steel in the presence of moisture and CO_2.
Neutralizing agents for acids and caustics: Flush with water.
Reactivity group: 7
Compatibility class: Aliphatic amines

ENVIRONMENTAL DATA
Water pollution: Effects of low concentrations on aquatic life is unknown. May be dangerous if it enters nearby water intakes; notify operators. Notify local health and wildlife officials.
Response to discharge: Flush and disperse with water.

SHIPPING INFORMATION
Grades of purity: 99.7%; **Storage temperature:** Ambient; **Stability during transport:** Stable.

PHYSICAL AND CHEMICAL PROPERTIES
Physical state @ 59°F/15°C and 1 atm: Liquid.
Molecular weight: 170.3
Boiling point @ 1 atm: 477°F/247°C/520°K.
Melting/Freezing point: 50°F/10°C/283°K.
Specific gravity (water = 1): 0.924 @ 68°F/20°C (liquid).
Liquid surface tension: 34.7 dynes/cm = 0.0347 N/m at 23°C.
Liquid water interfacial tension: 37.62 dynes/cm = 0.0376 N/m at 23°C.

ISOPHORONE DIISOCYANATE   REC. I:1050

SYNONYMS: IDI; IPDI; DIISOCIANATO de ISOFORONA (Spanish); ISOPHORONE DIAMINE DIISOCYANATE; 3-ISOCYANATOMETHYL-3,5,5-TRIMETHYLCYCLOHEXYL-ISOCYATE; ISOPHORONE DIAMINE DIISOCYANATE; TRIISOCYANATOISOCYANURATE

IDENTIFICATION
CAS Number: 4098-71-9
Formula: $C_{12}H_{18}N_2O_2$
DOT ID Number: UN 2290; UN 2906; DOT Guide Number: 156 (2290); 127 (UN 2906)
Proper Shipping Name: Isophorone diisocyanate (UN 2290); Triisocyanatoisocyanurate, solution (70%) (UN 2906)
Reportable Quantity (RQ): **(EHS)** 1 lb/0.454 kg

DESCRIPTION: Colorless or yellowish liquid. Sinks and reacts with water to produce gaseous carbon dioxide and the corresponding diamine.

Combustible • Toxic by skin absorption. Severely irritating to the skin, eyes, and respiratory tract • Heat or flame may cause explosion • Containers may BLEVE when exposed to fire • Vapors are heavier than air and will collect and stay in low areas • Vapors in confined areas (e.g., tanks, sewers, buildings) may explode when exposed to fire • Toxic products of combustion may include carbon monoxide, CO_2, and nitrogen oxides.

Hazard Classification (based on NFPA-704 Rating System)
Health Hazards (Blue): 2; Flammability (Red): 1; Reactivity (Yellow): 1; Special Notice (White): Water reactive

EMERGENCY RESPONSE, liquid: See Appendix A (129)
Evacuation:
Public safety: Isolate spill area for at least 50 to 100 meters (160 to 330 feet) in all directions.
Spill: Large spill–Consider initial downwind evacuation for at least 300 meters (1000 feet).
Fire: Isolate for 800 meters (½ mile) in all directions, especially if tank, rail car, or tank truck is involved in fire.

EMERGENCY RESPONSE: See Appendix A (156)
Evacuation:
Public safety: Isolate the area of spill or leak for at least 50 to 100 meters (160 to 330 feet) in all directions.
Spill: Increase, in the downwind direction, as necessary, the distance shown under "Public Safety."
Fire: If tank, rail car, or tank truck is involved in fire, isolate for at least 800 meters (½ mile) in all directions; also, consider initial evacuation for 800 meters (½ mile) in all directions.

EXPOSURE
Short-term effects: *SEEK MEDICAL ATTENTION.* **Vapor:** *POISONOUS. MAY BE FATAL IF INHALED OR ABSORBED THROUGH SKIN*: Contact may cause burns to skin and eyes. Move to fresh air. *IF BREATHING HAS STOPPED,* give artificial respiration; *avoid mouth-to-mouth resuscitation; use bag/mask apparatus.* IF breathing is difficult, administer oxygen. **Liquid:** *POISONOUS. MAY BE FATAL IF SWALLOWED OR ABSORBED THROUGH SKIN.* Contact may burn skin and eyes. Immediately flush skin or eyes with running water for at least 15 minutes. Hold eyelids open periodically while flushing eyes. Speed in removing material from skin is of extreme importance. Remove and double bag contaminated clothing and shoes at the site. Keep victim quiet and maintain normal body temperature. Effects may be delayed; keep victim under observation. *IF SWALLOWED* and victim is *UNCONSCIOUS OR HAVING CONVULSIONS,* do nothing except keep victim warm.

HEALTH HAZARDS
Personal protective equipment (PPE): B-Level PPE. OSHA Table Z-1 contaminant. Chemical protective material(s) reported to have good to excellent resistance: butyl rubber, PV alcohol, Viton®. Also, natural rubber offers limited protection.
Recommendations for respirator selection: NIOSH
0.05 ppm: SA* (any supplied-air respirator). *0.125 ppm:* SA:CF* (any supplied-air respirator operated in a continuous-flow mode).*0.25 ppm:* SCBAF (any self-contained breathing apparatus with a full facepiece); or SAF (any supplied-air respirator with a full facepiece). *EMERGENCY OR PLANNED ENTRY INTO UNKNOWN CONCENTRATIONS OR IDLH CONDITIONS:* SCBAF:PD,PP (any self-contained breathing apparatus that has a full facepiece and is operated in a pressure-demand or other positive-pressure mode); or SAF:PD,PP:ASCBA (any supplied-air respirator that has a full facepiece and is operated in a pressure-demand or other positive-pressure mode in combination with an auxiliary self-contained breathing apparatus operated in a pressure-demand or other positive pressure mode). *ESCAPE:* GMFOV [any air-purifying, full-facepiece respirator (gas mask) with a chin-style, front- or back-mounted organic vapor canister]; or SCBAE (any appropriate escape-type, self-contained breathing apparatus). *Note*: Substance reported to cause eye irritation or damage; may require eye protection.

Exposure limits (TWA unless otherwise noted): ACGIH TLV 0.005 ppm (0.045 mg/m^3); NIOSH 0.005 ppm (0.045 mg/m^3) skin contact contributes significantly in overall exposure.
Short-term exposure limits (15-minute TWA): NIOSH STEL 0.02 ppm (0.180 mg/m^3); skin contact contributes significantly in overall exposure.
Toxicity by ingestion: Grade 2; LD$_{50}$ more than 2.6 g/kg (rat).
Vapor (gas) irritant characteristics: Eye and respiratory tract irritant.
Liquid or solid irritant characteristics: Eye and skin irritant.

FIRE DATA
Flash point: 212°F/100°C.
Fire extinguishing agents not to be used: Although water is suitable for extinguishing open air fires, it should not be allowed to contaminate closed tanks containing this material due to the risk of hazardous gas generation.

CHEMICAL REACTIVITY
Reactivity with water: Reacts with water to produce water-soluble isophorone diamine (a toxic and corrosive compound) and CO$_2$.
Binary reactants: Reacts with substances containing active hydrogen; alcohols, amines, mercaptans, urethanes, ureas. Contact with aluminum, aluminum alloys, copper or copper alloys is prohibited.
Reactivity group: 12
Compatibility class: Isocyanates

ENVIRONMENTAL DATA
Water pollution: Effect of low concentrations on aquatic life is unknown. May be dangerous if it enters nearby water intakes; notify operators. Notify local health and wildlife officials.
Response to discharge: Issue warning–poison. Restrict access. Should be removed. Chemical and physical treatment.

SHIPPING INFORMATION
Inert atmosphere: Inerted; **Venting:** Pressure-vacuum

PHYSICAL AND CHEMICAL PROPERTIES
Physical state @ 59°F/15°C and 1 atm: Liquid.
Molecular weight: 222.32
Boiling point @ 1 atm: 316°F/158°C/431°K.
Melting/Freezing point: −76°F/−60°C/213°K.
Specific gravity (water = 1): 1.056 to 1.062 @ 68°F/20°C.
Relative vapor density (air = 1): 7.7 (calculated).
Vapor pressure: 0.0003 mm.

ISOPHTHALIC ACID   REC. I:1100

SYNONYMS: ACIDO ISOFTALICO (Spanish); BENZENE-1,3-DICARBOXYLIC ACID; *m*-PHTHALIC ACID

IDENTIFICATION
CAS Number: 121-91-5
Formula: $C_8H_6O_4$; 1,3-$C_6H_4(COOH)_2$

DESCRIPTION: White solid. Slightly acrid, unpleasant odor. Sinks in water; slightly soluble.

Combustible solid • Containers may BLEVE when exposed to fire • Concentrated dust in confined areas (e.g., tanks, sewers, buildings) may explode when exposed to fire • Irritating to the skin, eyes, and respiratory tract • Decomposes at approximately 375°F/191°C forming phthalic anhydride. Toxic products of combustion also include carbon monoxide.

Hazard Classification (based on NFPA-704 Rating System)
Health Hazards (Blue): 1; Flammability (Red): 1; Reactivity (Yellow): 0

EMERGENCY RESPONSE: See Appendix A (171)
Evacuation:
Public safety: Isolate the area of spill or leak for at least 10 to 25 meters (30 to 80 feet) in all directions.
Spill: Increase, in the downwind direction, as necessary, the distance shown under "Public Safety."
Fire: If any large container is involved in fire, isolate for at least 800 meters (½ mile) in all directions; also, consider initial evacuation for 800 meters (½ mile) in all directions.

EXPOSURE
Short-term effects: *SEEK MEDICAL ATTENTION.* **Dust:** Irritating to eyes, nose, and throat. *IF INHALED*, will, will cause coughing or difficult breathing. *IF IN EYES*, hold eyelids open and flush. with plenty of water. *IF BREATHING HAS STOPPED*, give artificial respiration. IF breathing is difficult, administer oxygen. **Solid:** Irritating to skin and eyes. *IF SWALLOWED*, will cause nausea. Remove contaminated clothing and shoes. Flush affected areas with plenty of water. *IF IN EYES*, hold eyelids open and flush with plenty of water. *IF SWALLOWED* and victim is *CONSCIOUS AND ABLE TO SWALLOW*, have victim drink 4 to 8 ounces of water. *IF SWALLOWED* and victim is *UNCONSCIOUS OR HAVING CONVULSIONS*, do nothing except keep victim warm.

HEALTH HAZARDS
Personal protective equipment (PPE): B-Level PPE. If without adequate ventilation, use respirator with dust filter, goggles, and gloves. Butyl rubber is generally suitable for carbooxylic acid compounds.
Exposure limits (TWA unless otherwise noted): 5 mg/m^3 (respirable fraction) (AIHA WEEL).
Toxicity by ingestion: Grade 1; LD$_{50}$ = 12.2 g/kg (rat).

FIRE DATA
Flash point: 334°F/168°C (combustible solid).

CHEMICAL REACTIVITY
Binary reactants: Reacts with strong oxidizers and sodium nitrite.

ENVIRONMENTAL DATA
Food chain concentration potential: Log P$_{ow}$ = 0.42. Unlikely to accumulate.
Water pollution: Effect of low concentrations on aquatic life is unknown. May be dangerous if it enters nearby water intakes; notify operators. Notify local health and wildlife officials. **Response to discharge:** Should be removed. Chemical and physical treatment.

SHIPPING INFORMATION
Grades of purity: Technical, 82–95%; **Storage temperature:** Ambient; **Inert atmosphere:** Inerted; **Venting:** Safety relief; **Stability during transport:** Stable.

PHYSICAL AND CHEMICAL PROPERTIES
Physical state @ 59°F/15°C and 1 atm: Solid.
Molecular weight: 166
Boiling point @ 1 atm: 554°F/290°C/563°K.
Melting/Freezing point: 376°F/191°C/464°K.

Specific gravity (water = 1): 1.54 @ 77°F/25°C (solid).
Relative vapor density (air = 1): 5.7
Heat of combustion: –8340 Btu/lb = –4630 cal/g = –194 x 10^5 J/kg.

ISOPRENE REC. I:1150

SYNONYMS: EEC No. 601-014-00-5; ISOPRENO (Spanish); β-METHYLBIVINYL; 2-METHYL-1,3-BUTADIENE; 3-METHYL-1,3-BUTADIENE

IDENTIFICATION
CAS Number: 78-79-5
Formula: C_5H_8; $CH_2 = C(CH_3)CH=CH_2$
DOT ID Number: UN 1218; DOT Guide Number: 128
Proper Shipping Name: Isoprene, inhibited

DESCRIPTION: Colorless watery liquid. Mild, petroleum-like odor. Floats on water surface; insoluble. Boils at 93°F/34°C.

Extremely flammable • Containers may BLEVE when exposed to fire •Vapors may form explosive mixture with air • Vapors are heavier than air and will collect and stay in low areas • Vapors may travel long distances to ignition sources and flashback • Vapors in confined areas (e.g., tanks, sewers, buildings) may explode when exposed to fire • Irritating to the skin, eyes, and respiratory tract • Able to form unstable peroxides; polymerization hazard • Heat can induce polymerization with rapid release of energy; sealed containers may rupture explosively • May react with itself blocking relief valves; leading to tank explosions • Toxic products of combustion may include carbon monoxide.

Hazard Classification (based on NFPA-704 Rating System)
Health Hazards (Blue): 1; Flammability (Red): 4; Reactivity (Yellow): 2

EMERGENCY RESPONSE: See Appendix A (128)
Evacuation:
Public safety: Isolate spill area for at least 25 to 50 meters (80 to 160 feet) in all directions.
Spill: Large spill–Consider initial downwind evacuation for at least 300 meters (1000 feet).
Fire: Isolate for 800 meters (½ mile) in all directions, especially if tank, rail car, or tank truck is involved in fire.

EXPOSURE
Short-term effects: *SEEK MEDICAL ATTENTION.* **Vapor:** Irritating to eyes, nose, and throat. Move to fresh air. *IF BREATHING HAS STOPPED*, give artificial respiration. IF breathing is difficult, administer oxygen. **Liquid:** Irritating to skin and eyes. Remove contaminated clothing and shoes. Flush affected areas with plenty of water. *IF IN EYES*, hold eyelids open and flush with plenty of water.

HEALTH HAZARDS
Personal protective equipment (PPE): B-Level PPE. Organic vapor respirator. Chemical protective material(s) reported to have good to excellent resistance: butyl rubber, PV alcohol, Viton®.
Exposure limits (TWA unless otherwise noted): 50 ppm (AIHAWEEL).
Vapor (gas) irritant characteristics: Eye and respiratory tract irritant. Vapors cause a slight smarting of the eyes or respiratory system if present in high concentrations.
Liquid or solid irritant characteristics: Eye and skin irritant. If spilled on clothing and allowed to remain, may cause smarting and reddening of the skin.
Odor threshold: 0.005 ppm.

FIRE DATA
Flash point: –65°F/–54°C (cc).
Flammable limits in air: LEL: 2.0%; UEL: 9.0%.
Fire extinguishing agents not to be used: Water may be ineffective.
Behavior in fire: May polymerize in containers and explode
Autoignition temperature: 743°F/395°C/668°K.
Electrical hazard: Class I, Group D.
Burning rate: 8.6 mm/min

CHEMICAL REACTIVITY
Binary reactants: Reacts with oxidizers. May soften some types of rubber or paint.
Polymerization: Polymerization is accelerated by heat and by oxygen, even by the presence of rusty iron. Iron surfaces should be treated with a suitable reducing agent, such as sodium nitrite, before they are placed into isoprene service; **Inhibitor of polymerization:** *tert*-butylcatechol (0.06%). Di-*n*-butylamine, phenyl-β-naphthylamine, and phenyl-α-naphthylamine are also used. Must be inhibited when transported interstate
Reactivity group: 30
Compatibility class: Olefins

ENVIRONMENTAL DATA
Food chain concentration potential: Log P_{ow} = 2.33. Unlikely to accumulate.
Water pollution: Harmful to aquatic life in very low concentrations. May be dangerous if it enters nearby water intakes; notify operators. Notify local health and wildlife officials. **Response to discharge:** Issue warning–high flammability. Restrict access. Evacuate area.

SHIPPING INFORMATION
Grades of purity: Research grade: 99.99%; polymerization grade: 99.8%; **Storage temperature:** Ambient; **Inert atmosphere:** None; **Venting:** Pressure-vacuum; **Stability during transport:** Stable.

NAS HAZARD CLASSIFICATION FOR BULK WATER TRANSPORTATION
FIRE: 4
HEALTH: Vapor irritant: 1; Liquid or solid irritant: 1; Poisons: 1
WATER POLLUTION: Human toxicity: 0; Aquatic toxicity: 1; Aesthetic effect: 1
REACTIVITY: Other chemicals: 2; Water: 0; Self-reaction: 3

PHYSICAL AND CHEMICAL PROPERTIES
Physical state @ 59°F/15°C and 1 atm: Liquid.
Molecular weight: 68.12
Boiling point @ 1 atm: 93.4°F/34.1°C/307.3°K.
Melting/Freezing point: –230.7°F/–145.9°C/127.3°K.
Critical temperature: 412°F/211.1°C/484.3°K.
Critical pressure: 550 psia = 37.4 atm = 3.79 MN/m^2.
Specific gravity (water = 1): 0.681 @ 68°F/20°C (liquid).
Liquid surface tension: 16.9 dynes/cm = 0.0169 N/m @ 68°F/20°C.
Liquid water interfacial tension: (estimate) 40 dynes/cm = 0.04 N/m @ 68°F/20°C.
Relative vapor density (air = 1): 2.3
Ratio of specific heats of vapor (gas): 1.091
Latent heat of vaporization: 150 Btu/lb = 85 cal/g = 3.6 x 10^5 J/kg.

Heat of combustion: –18,848 Btu/lb = –10,471 cal/g = –438.40 x 10^5 J/kg.
Heat of polymerization: –499 Btu/lb = –277 cal/g = –11.6 x 10^5 J/kg.
Heat of fusion: 16.80 cal/g.
Vapor pressure: (Reid) 15.0 psia; 400 mm.

2-ISOPROPOXYETHANOL   REC. I:1175

SYNONYMS: IPE; ETHANOL, 2-ISOPROPOXY; DOWANOL EIPAT; ETHYLENE GLYCOL ISOPROPYL ETHER; ETHYLENE GLYCOL MONOISOPROPYL ETHER; β-HYDROXYETHYL ISOPROPYL ETHER; ISOPROPOXIETANOL (Spanish); ISOPROPYL CELLOSOLVE; ISOPROPYL GLYCOL; ISOPROPYLOXITOL

IDENTIFICATION
CAS Number: 109-59-1
Formula: $C_5H_{12}O_2$; $(CH_3)_2CHOCH_2CH_2OH$
DOT ID Number: UN 1993; DOT Guide Number: 128
Proper Shipping Name: Flammable liquid, n.o.s.

DESCRIPTION: Colorless liquid Characteristic odor.

Highly flammable • Containers may BLEVE when exposed to fire • Vapors may form explosive mixture with air • Vapors are heavier than air and will collect and stay in low areas • Vapors may travel long distances to ignition sources and flashback • Vapors in confined areas (e.g., tanks, sewers, buildings) may explode when exposed to fire • Irritating to the skin, eyes, and respiratory tract • Toxic products of combustion may include carbon monoxide.

Hazard Classification (based on NFPA-704 Rating System)
Health Hazards (Blue): 1; Flammability (Red): 3; Reactivity (Yellow): 0

EMERGENCY RESPONSE: See Appendix A (128)
Evacuation:
Public safety: Isolate spill area for at least 25 to 50 meters (80 to 160 feet) in all directions.
Spill: Large spill–Consider initial downwind evacuation for at least 300 meters (1000 feet).
Fire: Isolate for 800 meters (½ mile) in all directions, especially if tank, rail car, or tank truck is involved in fire.

EXPOSURE
Short-term effects: *SEEK MEDICAL ATTENTION*. **Liquid:** Irritating to skin and eyes. Harmful if swallowed. Remove contaminated clothing and shoes. Flush affected areas with plenty of water. *IF IN EYES*, hold eyelids open and flush with plenty of water. *IF SWALLOWED* and victim is *CONSCIOUS AND ABLE TO SWALLOW*, immediately induce vomiting. The use of alcoholic beverages may enhance the toxic effect.

HEALTH HAZARDS
Personal protective equipment (PPE): B-Level PPE. OSHA Table Z-1-A air contaminant. Wear approved respirator, chemical resistant gloves, chemical safety goggles or full-face mask, boots, and apron.
Neoprene is generally suitable for cellosolve.
Exposure limits (TWA unless otherwise noted): ACGIH TLV 25 ppm; 25 ppm (AIHA WEEL).
Toxicity by ingestion: Grade 2: LD_{50} = 4.9 g/kg (mouse).

Vapor (gas) irritant characteristics: Vapors cause smarting of the eyes or respiratory system if present in high concentrations. The effect is temporary.
Liquid or solid irritant characteristics: Eye and skin irritant. If spilled on clothing and allowed to remain, may cause smarting and reddening of skin. Degreases the skin.

FIRE DATA
Flash point: 92°F/33°C (oc).
Flammable limits in air: LEL: 1.6%; UEL: 13.0%.

CHEMICAL REACTIVITY
Binary reactants: Reacts with oxidizers, sulfuric acid, isocyanates.
Reactivity group: 40
Compatibility class: Glycol ethers

ENVIRONMENTAL DATA
Water pollution: Effect of low concentrations on aquatic life is unknown. May be dangerous if it enters nearby water intakes; notify operators. Notify local health and wildlife officials. **Response to discharge:** Disperse and flush.

SHIPPING INFORMATION
Grades of purity: 99%; **Storage temperature:** Ambient; **Stability during transport:** Stable.

PHYSICAL AND CHEMICAL PROPERTIES
Physical state @ 59°F/15°C and 1 atm: Liquid.
Molecular weight: 104.2
Boiling point @ 1 atm: 283°F/139°C/412°K.
Melting/Freezing point: Less than –76°F/–60°C/213°K.
Specific gravity (water = 1): 0.90
Vapor (gas) Specific gravity (water = 1): 3.59
Vapor pressure: 3 mm; 3.5 mbar @ 20°C.

ISOPROPYL ACETATE   REC. I:1200

SYNONYMS: ACETATO de ISOPROPILO (Spanish); ACETIC ACID, ISOPROPYL ESTER; ACETIC ACID, 1-METHYLETHYL ESTER; 2-ACETOXYPROPANE; EEC No. 607-024-00-6; ISOPROPYL ESTER OF ACETIC ACID; 1-METHYLETHYL ACETATE; 2-PROPYL ACETATE; sec-PROPYL ACETATE

IDENTIFICATION
CAS Number: 108-21-4
Formula: $C_5H_{10}O_2$; $CH_3COOCH(CH_3)_2$
DOT ID Number: UN 1220; DOT Guide Number: 129
Proper Shipping Name: Isopropyl acetate

DESCRIPTION: Colorless watery liquid Pleasant, fruity odor. Floats on surface, and slowly mixes with water.

Highly flammable • Containers may BLEVE when exposed to fire • Vapors may form explosive mixture with air • Vapors are heavier than air and will collect and stay in low areas • Vapors may travel long distances to ignition sources and flashback • Vapors in confined areas (e.g., tanks, sewers, buildings) may explode when exposed to fire • Irritating to the skin, eyes, and respiratory tract • Toxic products of combustion may include carbon monoxide.

Hazard Classification (based on NFPA-704 Rating System)
Health Hazards (Blue): 1; Flammability (Red): 3; Reactivity (Yellow): 0

EMERGENCY RESPONSE: See Appendix A (129)
Evacuation:
Public safety: Isolate spill area for at least 50 to 100 meters (160 to 330 feet) in all directions.
Spill: Large spill–Consider initial downwind evacuation for at least 300 meters (1000 feet).
Fire: Isolate for 800 meters (½ mile) in all directions, especially if tank, rail car, or tank truck is involved in fire.

EXPOSURE
Short-term effects: *SEEK MEDICAL ATTENTION.* **Vapor:** Irritating to eyes, nose, and throat. Move to fresh air. *IF BREATHING HAS STOPPED*, give artificial respiration. IF breathing is difficult, administer oxygen. **Liquid:** Irritating to skin and eyes. Harmful if swallowed. Remove contaminated clothing. Flush affected areas with plenty of water. *IF IN EYES*, hold eyelids open and flush with plenty of water. *IF SWALLOWED* and victim is *CONSCIOUS AND ABLE TO SWALLOW*, have victim drink 4 to 8 ounces of water.

HEALTH HAZARDS
Personal protective equipment (PPE): B-Level PPE. OSHA Table Z-1 air contaminant. Organic vapor canister); or air–supplied mask; chemical goggles or face splash shield.
Recommendations for respirator selection: NIOSH/OSHA 1800 ppm: SA:CF (any supplied-air respirator operated in a continuous-flow mode); or SCBAF (any self-contained breathing apparatus with a full facepiece); or SAF (any supplied-air respirator with a full facepiece). *EMERGENCY OR PLANNED ENTRY INTO UNKNOWN CONCENTRATIONS OR IDLH CONDITIONS:* SCBAF:PD,PP (any self-contained breathing apparatus that has a full facepiece and is operated in a pressure-demand or other positive-pressure mode); or SAF:PD,PP:ASCBA (any supplied-air respirator that has a full facepiece and is operated in a pressure-demand or other positive-pressure mode in combination with an auxiliary self-contained breathing apparatus operated in a pressure-demand or other positive pressure mode). *ESCAPE:* GMFOV[any air-purifying, full-facepiece respirator (gas mask) with a chin-style, front- or back-mounted organic vapor canister]; or SCBAE (any appropriate escape-type, self-contained breathing apparatus). *Note:* Substance causes eye irritation or damage; eye protection needed.
Exposure limits (TWA unless otherwise noted): ACGIH TLV 250 ppm (1040 mg/m^3); OSHA 250 ppm (950 mg/m^3).
Short-term exposure limits (15-minute TWA): ACGIH STEL 310 ppm (1290 mg/m^3).
Toxicity by ingestion: Grade 2; LD$_{50}$ = 0.5 to 5 g/kg (rat).
Vapor (gas) irritant characteristics: Eye and respiratory tract irritant. Vapors cause smarting of the eyes or respiratory system if present in high concentrations.
Liquid or solid irritant characteristics: Eye and skin irritant. If spilled on clothing and allowed to remain, may cause smarting and reddening of the skin.
IDLH value: 1800 ppm.

FIRE DATA
Flash point: 60°F/15°C (oc); 36°F/2°C (cc).
Flammable limits in air: LEL: 1.8% @ 100°F/38°C; UEL: 8.0%.
Fire extinguishing agents not to be used: Use of dry chemical where it can get into a tank of isopropyl acetate is not recommended.
Autoignition temperature: 860°F/460°C/733°K.
Electrical hazard: Class I, Group D.

CHEMICAL REACTIVITY
Reactivity with water: Hydrolyzes on standing to form acetic acid and isopropyl alcohol; the presence of bases (alkalies) speeds up the reaction.
Binary reactants: Decomposes slowly in air forming acetic acid. Reacts vigorously with oxidizing agents. Reacts with nitrates, alkalis, and acids. Softens and dissolves many plastics. Contact with steel causes decomposition forming acetic acid and isopropyl alcohol. Do not store in an iron container.
Reactivity group: 34
Compatibility class: Ester

ENVIRONMENTAL DATA
Food chain concentration potential: Negative; unlikely to accumulate.
Water pollution: Effect of low concentrations on aquatic life is unknown. May be dangerous if it enters nearby water intakes; notify operators. Notify local health and wildlife officials. **Response to discharge:** Issue warning–high flammability. Evacuate area. Disperse and flush.

SHIPPING INFORMATION
Grades of purity: 95-99+%; **Storage temperature:** Ambient; **Inert atmosphere:** None; **Venting:** Open (flame arrester) or pressure-vacuum; **Stability during transport:** Stable.

NAS HAZARD CLASSIFICATION FOR BULK WATER TRANSPORTATION
FIRE: 3
HEALTH: Vapor irritant: 1; Liquid or solid irritant: 1; Poisons: 2
WATER POLLUTION: Human toxicity: 1; Aquatic toxicity: 1; Aesthetic effect: 2
REACTIVITY: Other chemicals: 1; Water: 0; Self-reaction: 0

PHYSICAL AND CHEMICAL PROPERTIES
Physical state @ 59°F/15°C and 1 atm: Liquid.
Molecular weight: 102.13
Boiling point @ 1 atm: 191.3°F/88.5°C/361.7°K.
Melting/Freezing point: –92.7°F/–69°C/204°K.
Critical temperature: 509°F/265°C/538°K.
Critical pressure: 529 psia = 36 atm = 3.65 MN/m^2.
Specific gravity (water = 1): 0.874 @ 68°F/20°C (liquid).
Liquid surface tension: 26 dynes/cm = 0.026 N/m @ 68°F/20°C.
Relative vapor density (air = 1): 3.5
Ratio of specific heats of vapor (gas): (estimate) 1.074
Latent heat of vaporization: 150 Btu/lb = 81 cal/g = 3.4 x 10^5 J/kg.
Heat of combustion: –9420 Btu/lb = –5230 cal/g = –219 x 10^5 J/kg.
Vapor pressure: (Reid) 2.0 psia; 42 mm.

ISOPROPYL ALCOHOL   REC. I:1250

SYNONYMS: ALCOHOL de ISOPROPILO (Spanish); DIMETHYLCARBINOL; EEC No. 603-003-00-0; IPA; ISOPROPANOL; 2-PROPANOL; sec-PROPYL ALCOHOL; RUBBING ALCOHOL; PETROHOL

IDENTIFICATION
CAS Number: 67-63-0
Formula: C$_3$H$_8$O; CH$_3$CH(OH)CH$_3$
DOT ID Number: UN 1219; DOT Guide Number: 129
Proper Shipping Name: Isopropanol; Isopropyl alcohol

Isopropyl alcohol

DESCRIPTION: Colorless watery liquid. Odor like rubbing alcohol. Soluble in water.

Highly flammable • Containers may BLEVE when exposed to fire • Vapors may form explosive mixture with air • Vapors are heavier than air and will collect and stay in low areas • Vapors may travel long distances to ignition sources and flashback • Vapors in confined areas (e.g., tanks, sewers, buildings) may explode when exposed to fire • Irritating to the skin, eyes, and respiratory tract • Toxic products of combustion may include irritating vapors.

Hazard Classification (based on NFPA-704 Rating System)
Health Hazards (Blue): 1; Flammability (Red): 3; Reactivity (Yellow): 0

EMERGENCY RESPONSE: See Appendix A (129)
Evacuation:
Public safety: Isolate spill area for at least 50 to 100 meters (160 to 330 feet) in all directions.
Spill: Large spill–Consider initial downwind evacuation for at least 300 meters (1000 feet).
Fire: Isolate for 800 meters (½ mile) in all directions, especially if tank, rail car, or tank truck is involved in fire.

EXPOSURE
Short-term effects: *SEEK MEDICAL ATTENTION.* **Vapor:** Irritating to eyes, nose, and throat. Move to fresh air. *IF BREATHING HAS STOPPED,* give artificial respiration. IF breathing is difficult, administer oxygen. **Liquid:** Irritating to eyes. Harmful if swallowed. *IF IN EYES,* hold eyelids open and flush with plenty of water. *IF SWALLOWED* and victim is *CONSCIOUS AND ABLE TO SWALLOW,* have victim drink 4 to 8 ounces of water.

HEALTH HAZARDS
Personal protective equipment (PPE): B-Level PPE. Organic vapor canister); or air–supplied mask; chemical goggles or face splash shield. Chemical protective material(s) reported to have good to excellent resistance: neoprene, nitrile, Teflon®, chlorobutyl rubber, chlorinated polyethylene, Silvershield®. Also, Viton®/neoprene, Viton® offers limited protection
Recommendations for respirator selection: NIOSH/OSHA 2000 ppm: SA:CF (any supplied-air respirator operated in a continuous-flow mode); or CCRFOV [any air-purifying, full-facepiece respirator (gas mask) with a chin-style, front- or back-mounted acid gas canister]; or GMFOV[any air-purifying, full-facepiece respirator (gas mask) with a chin-style, front- or back-mounted organic vapor canister]; or PAPROV [any powered, air-purifying respirator with organic vapor cartridge(s)]; or SCBAF (any self-contained breathing apparatus with a full facepiece); or SAF (any supplied-air respirator with a full facepiece). *EMERGENCY OR PLANNED ENTRY INTO UNKNOWN CONCENTRATIONS OR IDLH CONDITIONS:* SCBAF:PD,PP (any self-contained breathing apparatus that has a full facepiece and is operated in a pressure-demand or other positive-pressure mode); or SAF:PD,PP:ASCBA (any supplied-air respirator that has a full facepiece and is operated in a pressure-demand or other positive-pressure mode in combination with an auxiliary self-contained breathing apparatus operated in a pressure-demand or other positive pressure mode). *ESCAPE:* GMFOV[any air-purifying, full-facepiece respirator (gas mask) with a chin-style, front- or back-mounted organic vapor canister]; or SCBAE (any appropriate escape-type, self-contained breathing apparatus). *Note:* Substance causes eye irritation or damage; eye protection needed.

Exposure limits (TWA unless otherwise noted): ACGIH TLV 400 ppm (983 mg/m^3); NIOSH/OSHA 400 ppm (980 mg/m^3).
Short-term exposure limits (15-minute TWA): ACGIH STEL 500 ppm (1230 mg/m^3); NIOSH STEL 500 ppm (1225 mg/m^3).
Toxicity by ingestion: Grade 1; LD_{50} = 5 to 15 g/kg (rat: LD_{50}: 5.84 g/kg).
Vapor (gas) irritant characteristics: Eye and respiratory tract irritant. Vapors cause a slight smarting of the eyes or respiratory system if present in high concentrations. The effect is temporary.
Liquid or solid irritant characteristics: Eye irritant. Practically harmless to the skin.
Odor threshold: 40 ppm.
IDLH value: 2000 ppm [10% LEL].

FIRE DATA
Flash point: 65°F/18°C (oc); 53°F/12°C (cc).
Flammable limits in air: LEL: 2.0%; UEL: 12.7% @ 200°F
Fire extinguishing agents not to be used: Water may be ineffective. Use to cool exposed containers.
Behavior in fire: Containers may rupture and explode.
Autoignition temperature: 750°F/399°C/672°K.
Electrical hazard: Class I, Group D.
Burning rate: 2.3 mm/min

CHEMICAL REACTIVITY
Binary reactants: React with strong oxidizers, acetaldehyde, chlorine, crotonaldehyde, oleum, ethylene oxide, acids, phosgene, isocyanates, aluminum, and alkali metals.
Reactivity group: 20
Compatibility class: Alcohols, glycols

ENVIRONMENTAL DATA
Food chain concentration potential: Log P_{ow} = 0.07. Unlikely to accumulate.
Water pollution: Dangerous to aquatic life in high concentrations. May be dangerous if it enters nearby water intakes; notify operators. Notify local health and wildlife officials. **Response to discharge:** Issue warning–high flammability. Disperse and flush.

SHIPPING INFORMATION
Grades of purity: 91%, 95% Anhydrous; **Storage temperature:** Ambient; **Inert atmosphere:** None; **Venting:** Open (flame arrester) or pressure-vacuum; **Stability during transport:** Stable.

NAS HAZARD CLASSIFICATION FOR BULK WATER TRANSPORTATION
FIRE: 3
HEALTH: Vapor irritant: 1; Liquid or solid irritant: 0; Poisons: 2
WATER POLLUTION: Human toxicity: 2; Aquatic toxicity: 2; Aesthetic effect: 1
REACTIVITY: Other chemicals: 2; Water: 0; Self-reaction: 0

PHYSICAL AND CHEMICAL PROPERTIES
Physical state @ 59°F/15°C and 1 atm: Liquid.
Molecular weight: 60.10
Boiling point @ 1 atm: 180°F/82°C/356°K.
Melting/Freezing point: –127°F/–89°C/185°K.
Critical temperature: 455°F/235°C/508°K.
Critical pressure: 691 psia = 47.0 atm = 4.76 MN/m^2.
Specific gravity (water = 1): 0.785 @ 68°F/20°C (liquid).
Relative vapor density (air = 1): 2.1
Ratio of specific heats of vapor (gas): 1.105
Latent heat of vaporization: 286 Btu/lb = 159 cal/g = 6.66 x 10^5 J/kg.

Heat of combustion: –12,960 Btu/lb = –7,201 cal/g = –301.5 x 10^5 J/kg.
Heat of solution: (estimate) –9 Btu/lb = –5 cal/g = –0.2 x 10^5 J/kg.
Heat of fusion: 21.37 cal/g.
Vapor pressure: (Reid) 1.4 psia; 32.7 mm @ 68°F/20°C; 10 mm @ 59°F/14.7°C.

ISOPROPYLAMINE   REC. I:1300

SYNONYMS: 1-AMINO-2-HYDROXYPROPANE; 2-AMINOPROPANE; 1-AMINO-PROPANOL-2; EEC No. 612-007-00-1; 2-HYDROXYPROPYLAMINE; ISOPROPILAMINA (Spanish); MONOISOPROPYLAMINE; 2-PROPANAMINE; *sec*-PROPYLAMINE

IDENTIFICATION
CAS Number: 75-31-0
Formula: C_3H_9N; $(CH_3)_2CHNH_2$
DOT ID Number: UN 1221; DOT Guide Number: 132
Proper Shipping Name: Isopropylamine

DESCRIPTION: Colorless liquid. Strong ammonia odor. Floats on the surface of water; soluble. Flammable, irritating vapor is produced. Boils at 90°F/32°C.

Extremely flammable • Firefighting gear (including SCBA) does not provide adequate protection. If exposure occurs, remove and isolate gear immediately and thoroughly decontaminate personnel • Containers may BLEVE when exposed to fire • Vapors may form explosive mixture with air • Vapors are heavier than air and will collect and stay in low areas • Vapors may travel long distances to ignition sources and flashback • Vapors in confined areas (e.g., tanks, sewers, buildings) may explode when exposed to fire • Severely Irritating to the skin, eyes, and respiratory tract; skin and eye contact can cause severe burns and blindness • Toxic products of combustion may include nitrogen oxides.

Hazard Classification (based on NFPA-704 Rating System)
Health Hazards (Blue): 3; Flammability (Red): 4; Reactivity (Yellow): 0

EMERGENCY RESPONSE: See Appendix A (132)
Evacuation:
Public safety: Isolate spill area for at least 50 to 100 meters (160 to 330 feet) in all directions.
Spill: Increase, as necessary, the isolation distance shown above, in "Public safety."
Fire: Isolate for 800 meters (½ mile) in all directions, especially if tank, rail car, or tank truck is involved in fire.

EXPOSURE
Short-term effects: *SEEK MEDICAL ATTENTION.* **Vapor:** Irritating to eyes, nose, and throat. *IF INHALED*, will, will cause coughing, difficult breathing or loss of consciousness. *IF IN EYES*, hold eyelids open and flush. with plenty of water. *IF BREATHING HAS STOPPED*, give artificial respiration. IF breathing is difficult, administer oxygen. **Liquid:** Will burn skin and eyes. *IF SWALLOWED*, will cause nausea. Remove contaminated clothing and shoes. Flush affected areas with plenty of water. *IF IN EYES*, hold eyelids open and flush with plenty of water. *IF SWALLOWED* and victim is *CONSCIOUS AND ABLE TO SWALLOW*, have victim drink 4 to 8 ounces of water. *IF SWALLOWED* and victim is *UNCONSCIOUS OR HAVING CONVULSIONS*, do nothing except keep victim warm.

HEALTH HAZARDS
Personal protective equipment (PPE): B-Level PPE. OSHA Table Z-1 air contaminant. Self-contained breathing apparatus; gloves and apron; chemical face shield or safety goggles. Chemical protective material(s) reported to have good to excellent resistance: Teflon®, PV acetate. Alson, butyl rubber offers limited protection.
Recommendations for respirator selection: OSHA
125 ppm: SA:CF (any supplied-air respirator operated in a continuous-flow mode); or PAPRS [any powered, air-purifying respirator with cartridge(s) providing protection against the compound of concern]. *250 ppm:* CCRFS [any chemical cartridge respirator with a full facepiece and cartridge(s) providing protection against the compound of concern]; or GMFS [any air-purifying, full-facepiece respirator (gas mask) with a chin-style, front- or back-mounted canister protection against the compound of concern]; or PAPRTS (any powered, air-purifying respirator with a tight-fitting facepiece and cartridge(s) providing protection against the compound of concern]; or SCBAF (any self-contained breathing apparatus with a full facepiece); or SAF (any supplied-air respirator with a full facepiece). *750 ppm:* SAF:PD,PP (any supplied-air respirator that has a full facepiece and is operated in a pressure-demand or other positive-pressure mode). *EMERGENCY OR PLANNED ENTRY INTO UNKNOWN CONCENTRATIONS OR IDLH CONDITIONS:* SCBAF:PD,PP (any self-contained breathing apparatus that has a full facepiece and is operated in a pressure-demand or other positive-pressure mode); or SAF:PD,PP:ASCBA (any supplied-air respirator that has a full facepiece and is operated in a pressure-demand or other positive-pressure mode in combination with an auxiliary self-contained breathing apparatus operated in a pressure-demand or other positive pressure mode). *ESCAPE:* GMFS [any air-purifying, full-facepiece respirator (gas mask) with a chin-style, front- or back-mounted canister protection against the compound of concern]; or SCBAE (any appropriate escape-type, self-contained breathing apparatus). *Note:* Substance causes eye irritation or damage; eye protection needed.
Exposure limits (TWA unless otherwise noted): ACGIH TLV 5 ppm (12 mg/m³); OSHA 5 ppm (12 mg/m³).
Short-term exposure limits (15-minute TWA): ACGIH STEL 10 ppm (24 mg/m³).
Toxicity by ingestion: Grade 2; oral LD_{50} = 820 mg/kg (rat), 600 mg/kg (mouse).
Vapor (gas) irritant characteristics: Vapors are moderately irritating, such that personnel will
not usually tolerate moderate or high concentrations.
Liquid or solid irritant characteristics: Causes smarting of the skin and first-degree burns on short exposure; may cause second-degree burns on long exposure.
Odor threshold: 0.2 ppm.
IDLH value: 750 ppm.

FIRE DATA
Flash point: –35°F/–37°C (oc).
Flammable limits in air: LEL: 2.3%; UEL: 12%.
Fire extinguishing agents not to be used: Water may be ineffective.
Behavior in fire: Burning is difficult to control because of the ease of re-ignition of the vapor.
Autoignition temperature: 756°F/402°C/675°K.
Electrical hazard: Class I, Group D.
Burning rate: 6.33 mm/min

CHEMICAL REACTIVITY
Binary reactants: Reacts with strong acids, strong oxidizers,

644 Isopropyl cyclohexane

aldehydes, ketones, epoxides. Severely corrodes aluminum, copper, and copper-based alloys (except Monel).
Reactivity group: 7
Compatibility class: Aliphatic amines

ENVIRONMENTAL DATA
Food chain concentration potential: Log P_{ow} = 0.25. Unlikely to accumulate.
Water pollution: Harmful to aquatic life in very low concentrations. May be dangerous if it enters nearby water intakes; notify operators. Notify local health and wildlife officials.
Response to discharge: Issue warning–high flammability, air contaminant, water contaminant. Restrict access. Evacuate area. Disperse and flush.

SHIPPING INFORMATION
Grades of purity: Technical, 99.0%; **Storage temperature:** Ambient; **Inert atmosphere:** None; **Venting:** Open (flame arrester); **Stability during transport:** Stable.

NAS HAZARD CLASSIFICATION FOR BULK WATER TRANSPORTATION
FIRE: 4
HEALTH: Vapor irritant: 3; Liquid or solid irritant: 2; Poisons: 4
WATER POLLUTION: Human toxicity: 2; Aquatic toxicity: 3; Aesthetic effect: 2
REACTIVITY: Other chemicals: 3; Water: 0; Self-reaction: 0

PHYSICAL AND CHEMICAL PROPERTIES
Physical state @ 59°F/15°C and 1 atm: Liquid.
Molecular weight: 59.11
Boiling point @ 1 atm: 90.3°F/32.4°C/305.6°K.
Melting/Freezing point: –139°F/–95°C/178°K.
Critical temperature: 396°F/202°C/475°K.
Critical pressure: 740 psia = 50 atm = 5.1 MN/m^2.
Specific gravity (water = 1): (estimate) 0.691 @ 68°F/20°C (liquid).
Liquid surface tension: 16.8 dynes/cm = 0.0168 N/m @ 68°F/20°C.
Relative vapor density (air = 1): 2.04
Latent heat of vaporization: 193 Btu/lb = 107 cal/g = 4.48 x 10^5 J/kg.
Heat of combustion: –16,940 Btu/lb = –9,420 cal/g = –394 x 10^5 J/kg.
Heat of solution: –210 Btu/lb = –110 cal/g = –4.8 x 10^5 J/kg.
Vapor pressure: (Reid) 18.2 psia; 460 mm.

ISOPROPYL CYCLOHEXANE REC. I:1350

SYNONYMS: HEXAHYDROCUMENE; ISOPROPILCICLOHEXANO (Spanish); 1-METHYLETHYLCYCLOHEXANE; NORMENTHANE

IDENTIFICATION
CAS Number: 696-29-7
Formula: C_8H_{18}; $(CH_3)_2CHC_6H_{11}$

DESCRIPTION: Colorless Liquid.

Highly flammable • Containers may BLEVE when exposed to fire • Vapors may form explosive mixture with air • Vapors are heavier than air and will collect and stay in low areas • Vapors may travel long distances to ignition sources and flashback • Vapors in confined areas (e.g., tanks, sewers, buildings) may explode when exposed to fire • Irritating to the skin, eyes, and respiratory tract • Toxic products of combustion may include carbon monoxide.

Hazard Classification (based on NFPA-704 Rating System)
Health Hazards (Blue): 1; Flammability (Red): 3; Reactivity (Yellow): 0

EMERGENCY RESPONSE: See Appendix A (171)
Evacuation:
Public safety: Isolate the area of spill or leak for at least 10 to 25 meters (30 to 80 feet) in all directions.
Spill: Increase, in the downwind direction, as necessary, the distance shown under "Public Safety."
Fire: If any large container is involved in fire, isolate for at least 800 meters (½ mile) in all directions; also, consider initial evacuation for 800 meters (½ mile) in all directions.

EXPOSURE
Short-term effects: *SEEK MEDICAL ATTENTION.* **Vapor:** Irritating to eyes, nose, and throat. *IF INHALED,* will, will cause dizziness, nausea, vomiting, or loss of consciousness. Move to fresh air. *IF BREATHING HAS STOPPED,* give artificial respiration. IF breathing is difficult, administer oxygen. **Liquid:** Irritating to skin and eyes. Harmful if swallowed. Remove contaminated clothing and shoes. Flush affected areas with plenty of water. *IF IN EYES,* hold eyelids open and flush with plenty of water. *IF SWALLOWED* and victim is *CONSCIOUS AND ABLE TO SWALLOW,* have victim drink 4 to 8 ounces of water.

Fouling to shoreline May be dangerous if it enters nearby water intakes; notify operators. Notify local health and wildlife officials. Notify operators of local water intakes. **Response to discharge:** Issue warning–high flammability. Evacuate area. Disperse and flush.

HEALTH HAZARDS
Personal protective equipment (PPE): B-Level PPE. Hydrocarbon vapor canister, supplied-air, or hose mask, hydrocarbon-insoluble rubber or plastic gloves, chemical goggles or face splash shield, hydrocarbon- insoluble rubber or plastic apron.
Eyes or Skin contact: Remove contaminated clothing and gently flush affected areas with water for 15 minutes; *CALL A DOCTOR.*
Liquid or solid irritant characteristics: May cause eye and skin irritation.

FIRE DATA
Flash point: 96°F/36°C (cc).
Fire extinguishing agents not to be used: Water may not be effective on fire
Autoignition temperature: 541°F/283°C/556°K.

CHEMICAL REACTIVITY
Binary reactants: Reacts with strong oxidizers, acids, aldehydes. Reacts with sulfuric acid forming an explosive material. Heat can cause explosive decomposition; acidic materials can increase sensitivity.

ENVIRONMENTAL DATA
Water pollution: Dangerous to aquatic life in high concentrations.

SHIPPING INFORMATION
Grades of purity: 97%; **Storage temperature:** Ambient; **Inert atmosphere:** None; **Stability during transport:** Stable.

NAS HAZARD CLASSIFICATION FOR BULK WATER TRANSPORTATION
FIRE: 3
HEALTH: Vapor irritant: 1; Liquid or solid irritant: 1; Poisons:–
WATER POLLUTION: Human toxicity:–; Aquatic toxicity: 2; Aesthetic effect: 1
REACTIVITY: Other chemicals: 1; Water: 0; Self-reaction: 0

PHYSICAL AND CHEMICAL PROPERTIES
Physical state @ 59°F/15°C and 1 atm: Liquid.
Molecular weight: 126.24
Boiling point @ 1 atm: 310°F/154.5°C/428°K.
Melting/Freezing point: –129°F/–89.4°C/184°K.
Specific gravity (water = 1): 0.8023 @ 20°C.
Relative vapor density (air = 1): 4.35 (estimate).
Latent heat of vaporization: (estimate) 20,035 Btu/lb = 11,131 cal/g = 466 x 10^5 J/kg.

ISOPROPYL ETHER   REC. I:1400

SYNONYMS: DIISOPROPYL ETHER; DIISOPROPYL OXIDE; DI-(1-METHYLETHYL)ETHER; EEC No. 603-045-00-X; ETER ISOPROPILICO (Spanish); 2-ISOPROPOXY PROPANE; 2,2'-OXYBISPROPANE

IDENTIFICATION
CAS Number: 108-20-3
Formula: $C_6H_{14}O$; $(CH_3)_2CHOCH(CH_3)_2$
DOT ID Number: UN 1159; DOT Guide Number: 127
Proper Shipping Name: Diisopropyl ether

DESCRIPTION: Colorless liquid Sweet odor, like camphor or ethyl ether. Floats; practically insoluble in water. Flammable, irritating vapor is produced.

Highly flammable • Containers may BLEVE when exposed to fire • Vapors may form explosive mixture with air • Vapors are heavier than air and will collect and stay in low areas • Vapors may travel long distances to ignition sources and flashback • Vapors in confined areas (e.g., tanks, sewers, buildings) may explode when exposed to fire • Irritating to the skin, eyes, and respiratory tract • Toxic products of combustion may include carbon monoxide.

Hazard Classification (based on NFPA-704 Rating System)
Health Hazards (Blue): 1; Flammability (Red): 3; Reactivity (Yellow): 1

EMERGENCY RESPONSE: See Appendix A (127)
Evacuation:
Public safety: Isolate spill area for at least 25 to 50 meters (80 to 160 feet) in all directions.
Spill: Large spill–Consider initial downwind evacuation for at least 300 meters (1000 feet).
Fire: Isolate for 800 meters (½ mile) in all directions, especially if tank, rail car, or tank truck is involved in fire.

EXPOSURE
Short-term effects: *SEEK MEDICAL ATTENTION*. **Vapor:** Irritating to eyes, nose, and throat. *IF INHALED*, will, will cause headache, dizziness, or nausea. Move victim to fresh air. *IF BREATHING HAS STOPPED*, give artificial respiration. IF breathing is difficult, administer oxygen. **Liquid:** Irritating to skin and eyes. Remove contaminated clothing and shoes. Flush affected areas with plenty of water. *IF IN EYES*, hold eyelids open and flush with plenty of water. *IF SWALLOWED* and victim is *CONSCIOUS AND ABLE TO SWALLOW*, have victim drink 4 to 8 ounces of water.

HEALTH HAZARDS
Personal protective equipment (PPE): B-Level PPE. Sealed chemical protective materials offer limited protection: Chlorinated polyethylene, nitrile, PV alcohol, Viton®.
Recommendations for respirator selection: NIOSH/OSHA
1400 ppm: CCRFOV* [any air-purifying, full-facepiece respirator (gas mask) with a chin-style, front- or back-mounted acid gas canister]; or PAPROV* [any powered, air-purifying respirator with organic vapor cartridge(s)]; or GMFOV[any air-purifying, full-facepiece respirator (gas mask) with a chin-style, front- or back-mounted organic vapor canister]; or SA* (any supplied-air respirator); or SCBAF (any self-contained breathing apparatus with a full facepiece). *EMERGENCY OR PLANNED ENTRY INTO UNKNOWN CONCENTRATIONS OR IDLH CONDITIONS:* SCBAF:PD,PP (any self-contained breathing apparatus that has a full facepiece and is operated in a pressure-demand or other positive-pressure mode); or SAF:PD,PP:ASCBA (any supplied-air respirator that has a full facepiece and is operated in a pressure-demand or other positive-pressure mode in combination with an auxiliary self-contained breathing apparatus operated in a pressure-demand or other positive pressure mode). *ESCAPE:* GMFOV[any air-purifying, full-facepiece respirator (gas mask) with a chin-style, front- or back-mounted organic vapor canister]; or SCBAE (any appropriate escape-type, self-contained breathing apparatus).
*Note: Substance reported to cause eye irritation or damage; may require eye protection.
Exposure limits (TWA unless otherwise noted): ACGIH TLV 250 ppm (1040 mg/m³); NIOSH/OSHA 500 ppm (2100 mg/m³).
Short-term exposure limits (15-minute TWA): ACGIH STEL 310 ppm (1300 mg/m³).
Toxicity by ingestion: Grade 1; oral LD_{50} = 8470 mg/kg (rat).
Vapor (gas) irritant characteristics: Eye and respiratory tract. Vapors cause smarting of the eyes or respiratory system if present in high concentrations.
Liquid or solid irritant characteristics: Eye and skin irritant. If spilled on clothing and allowed to remain, may cause smarting and reddening of the skin.
IDLH value: 1400 ppm [LEL].

FIRE DATA
Flash point: –18°F/–28°C (cc).
Flammable limits in air: LEL: 1.4%; UEL: 7.9%.
Fire extinguishing agents not to be used: Water may be ineffective.
Behavior in fire: Containers may explode when heated.
Autoignition temperature: 830°F/443°C/716°K.
Burning rate: 5.0 mm/min

CHEMICAL REACTIVITY
Binary reactants: Reacts with strong oxidizers, acids. Contact with air form unstable peroxides.
Polymerization: Heat may cause explosive polymerization.
Inhibitor of polymerization: 0.01% hydroquinone.
Reactivity group: 41
Compatibility class: Ethers

ENVIRONMENTAL DATA
Food chain concentration potential: Log P_{ow} = 1.6. Unlikely to accumulate.

Water pollution: Effect of low concentrations on aquatic life is unknown. Fouling to shoreline. May be dangerous if it enters nearby water intakes; notify operators. Notify local health and wildlife officials. **Response to discharge:** Issue warning–high flammability. Restrict access. Evacuate area. Disperse and flush.

SHIPPING INFORMATION
Grades of purity: 94+% May contain 0.01% hydroquinone or other inhibitor to prevent peroxide formation; **Storage temperature:** Ambient; **Inert atmosphere:** None; **Venting:** Pressure-vacuum; **Stability during transport:** Stable if fresh. Unstable peroxides may form if containers are previously opened or on long standing in contact with air; these may explode spontaneously or when heated.

NAS HAZARD CLASSIFICATION FOR BULK WATER TRANSPORTATION
FIRE: 3
HEALTH: Vapor irritant: 1; Liquid or solid irritant: 1; Poisons: 1
WATER POLLUTION: Human toxicity: 1; Aquatic toxicity: 2; Aesthetic effect: 2
REACTIVITY: Other chemicals: 1; Water: 0; Self-reaction: 0

PHYSICAL AND CHEMICAL PROPERTIES
Physical state @ 59°F/15°C and 1 atm: Liquid.
Molecular weight: 102.2
Boiling point @ 1 atm: 156°F/69°C/342°K.
Melting/Freezing point: –76°F/–60°C/213°K.
Critical temperature: 440°F/227°C/500°K.
Critical pressure: 418 psia = 28.4 atm = 2.88 MN/m^2.
Specific gravity (water = 1): 0.724 @ 68°F/20°C (liquid).
Liquid surface tension: 17.1 dynes/cm = 0.0171 N/m @ 77°F/25°C.
Liquid water interfacial tension: 17.1 dynes/cm = 0.0171 N/m @ 77°F/25°C.
Relative vapor density (air = 1): 3.5
Ratio of specific heats of vapor (gas): 1.0590
Latent heat of vaporization: 131 Btu/lb = 73 cal/g = 3.1 x 10^5 J/kg.
Heat of combustion: –16900 Btu/lb = –9390 cal/g = –393 x 10^5 J/kg.
Heat of fusion: 25.79 cal/g.
Vapor pressure: (Reid) High; 119 mm.

ISOPROPYL GLYCIDYL ETHER   REC. I:1450

SYNONYMS: GLYCIDYL ISOPROPYL ETHER; 1,2-EPOXY-3-ISOPROPOXYPROPANE; IGE; ISOPROPIL GLICIDIL ETER (Spanish); ISOPROPYL EPOXYPROPYL ETHER; (ISOPROXYMETHYL)OXIRANE

IDENTIFICATION
CAS Number: 4016-14-2
Formula: $C_6H_{12}O_2$; $C_3H_7OCH_2CHOCH_2$
DOT ID Number: UN 1993; DOT Guide Number: 128
Proper Shipping Name: Flammable liquids, n.o.s.

DESCRIPTION: Colorless Liquid.

Flammable • Containers may BLEVE when exposed to fire • Vapors may form explosive mixture with air • Vapors are heavier than air and will collect and stay in low areas • Vapors may travel long distances to ignition sources and flashback • Vapors in confined areas (e.g., tanks, sewers, buildings) may explode when exposed to fire • Able to form unstable peroxides on contact with air or light; polymerization hazard • Irritating to the skin, eyes, and respiratory tract • Toxic products of combustion may include carbon monoxide.

Hazard Classification (based on NFPA-704 Rating System)
Health Hazards (Blue): 1; Flammability (Red): 2; Reactivity (Yellow): 1

EMERGENCY RESPONSE: See Appendix A (128)
Evacuation:
Public safety: Isolate spill area for at least 25 to 50 meters (80 to 160 feet) in all directions.
Spill: Large spill–Consider initial downwind evacuation for at least 300 meters (1000 feet).
Fire: Isolate for 800 meters (½ mile) in all directions, especially if tank, rail car, or tank truck is involved in fire.

EXPOSURE
Short-term effects: *SEEK MEDICAL ATTENTION*. **Vapor:** Move victim to fresh air. If breathing is difficult, administer oxygen.
Liquid: Remove contaminated clothing and shoes.
Wash skin with soap and water.
IF IN EYES, hold eyelids open and flush with plenty of water.

HEALTH HAZARDS
Personal protective equipment (PPE): B-Level PPE. OSHA Table Z-1-A air contaminant. Wear full impervious protective clothing and approved respirator. Where splashing is possible wear full face shield or chemical safety goggles.
Recommendations for respirator selection: NIOSH/OSHA
up to 400 ppm: SA:CF* (any supplied-air respirator operated in a continuous-flow mode); or SCBAF (any self-contained breathing apparatus with a full facepiece). *EMERGENCY OR PLANNED ENTRY INTO UNKNOWN CONCENTRATIONS OR IDLH CONDITIONS:* SCBAF:PD,PP (any self-contained breathing apparatus that has a full facepiece and is operated in a pressure-demand or other positive-pressure mode); or SAF:PD,PP:ASCBA (any supplied-air respirator that has a full facepiece and is operated in a pressure-demand or other positive-pressure mode in combination with an auxiliary self-contained breathing apparatus operated in a pressure-demand or other positive pressure mode). *ESCAPE:* GMFOV[any air-purifying, full-facepiece respirator (gas mask) with a chin-style, front- or back-mounted organic vapor canister]; or SCBAE (any appropriate escape-type, self-contained breathing apparatus). *Note:* Substance causes eye irritation or damage; eye protection needed.
Exposure limits (TWA unless otherwise noted): ACGIH TLV 50 ppm (238 mg/m^3); OSHA PEL 50 ppm (240 mg/m^3).
Short-term exposure limits (15-minute TWA): ACGIH STEL 75 ppm (356 mg/m^3); NIOSH ceiling 50 ppm (240 mg/m^3)/15 min.
Toxicity by ingestion: Moderately toxic.
Long-term health effects: Acute oral administration to mice, rats, and rabbits caused central nervous system depression. Subchronic inhalation by rats caused decreased weight gain, inflammation of the lungs, pneumonia, and respiratory distress. Mutation data. May cause dermatitis.
Vapor (gas) irritant characteristics: Eye and respiratory tract irritant. Vapors are irritating such that personnel will not usually tolerate moderate or high concentrations.
Liquid or solid irritant characteristics: Eye and skin irritant. If spilled on clothing and allowed to remain, may cause smarting and reddening of skin.
IDLH value: 400 ppm.

FIRE DATA
Flash point: 92°F/33°C (cc).
Fire extinguishing agents not to be used: Water.
Stoichiometric air-to-fuel ratio: 38.1

CHEMICAL REACTIVITY
Binary reactants: Contact with strong oxidizing agents can cause fires and explosions. Contact with strong caustics may cause polymerization. Exposure to air or light may cause formation of unstable peroxides.
Polymerization: May occur in contact with strong caustics.

ENVIRONMENTAL DATA
Water pollution: Effects of low concentrations on aquatic life is unknown. May be dangerous if it enters nearby water intakes; notify operators. Notify local health and wildlife officials.
Response to discharge: Issue warning–high flammability. Should be removed. Chemical and physical treatment.

SHIPPING INFORMATION
Grades of purity: Technical; 98%; **Stability during transport:** Stable.

PHYSICAL AND CHEMICAL PROPERTIES
Physical state @ 59°F/15°C and 1 atm: Liquid.
Molecular weight: 116.18
Boiling point @ 1 atm: 279°F/137°C/410°K.
Specific gravity (water = 1): 0.92
Vapor (gas) specific gravity: 4.0
Vapor pressure: 9 mm @77°F/25°C.

ISOPROPYL MERCAPTAN REC. I:1500

SYNONYMS: ISOPROPILMERCAPTANO (Spanish); ISOPROPYLTHIOL; 2-MERCAPTOPROPANE; 2-PROPANETHIOL; PROPANE-2-THIOL

IDENTIFICATION
CAS Number: 75-33-2
Formula: C_3H_8S; $(CH_3)_2CHSH$
DOT ID Number: UN 2402; DOT Guide Number: 128
Proper Shipping Name: Propanethiols

DESCRIPTION: White liquid. Strong skunk odor. Floats on the surface of water; soluble; may react forming flammable, irritating vapor.

Highly flammable • Containers may BLEVE when exposed to fire • Vapors may form explosive mixture with air • Vapors are heavier than air and will collect and stay in low areas • Vapors may travel long distances to ignition sources and flashback • Vapors in confined areas (e.g., tanks, sewers, buildings) may explode when exposed to fire • Irritating to the skin, eyes, and respiratory tract • Toxic products of combustion may include sulfur oxides.

Hazard Classification (based on NFPA-704 M Rating System)
Health Hazards (Blue): 1; Flammability (Red): 3; Reactivity (Yellow): 1

EMERGENCY RESPONSE: See Appendix A (128)
Evacuation:
Public safety: Isolate spill area for at least 25 to 50 meters (80 to 160 feet) in all directions.
Spill: Large spill–Consider initial downwind evacuation for at least 300 meters (1000 feet).
Fire: Isolate for 800 meters (½ mile) in all directions, especially if tank, rail car, or tank truck is involved in fire.

EXPOSURE
Short-term effects: *SEEK MEDICAL ATTENTION.* **Vapor:** Irritating to eyes, nose, and throat. *IF INHALED*, will, will cause difficult breathing, or loss of consciousness. *IF IN EYES*, hold eyelids open and flush with plenty of water. *IF BREATHING HAS STOPPED*, give artificial respiration. IF breathing is difficult, administer oxygen. **Liquid:** Irritating to skin and eyes. *IF SWALLOWED*, will cause nausea and vomiting. Remove contaminated clothing and shoes. Flush affected areas with plenty of water. *IF IN EYES*, hold eyelids open and flush with plenty of water. *IF SWALLOWED* and victim is *CONSCIOUS AND ABLE TO SWALLOW*, have victim drink 4 to 8 ounces of water and have victim induce vomiting. *IF SWALLOWED* and victim is *UNCONSCIOUS OR HAVING CONVULSIONS*, do nothing except keep victim warm.
Note to physician or authorized medical personnel: Medical observation is recommended for 24 to 48 hours after breathing overexposure, as pulmonary edema may be delayed. As first aid for pulmonary edema, consider administering a corticosteroid spray. Cigarette smoking may exacerbate pulmonary injury and should be discouraged for at least 72 hours following exposure.

HEALTH HAZARDS
Personal protective equipment (PPE): B-Level PPE. Self-contained breathing apparatus; goggles or face shield; rubber gloves.
Toxicity by ingestion: Grade 2; oral LD_{50} = 1790 mg/kg (rat).
Odor threshold: 0.25 ppm.

FIRE DATA
Flash point: –30°F/–34°C (oc).
Fire extinguishing agents not to be used: Water may be ineffective.

CHEMICAL REACTIVITY
Binary reactants: Reacts with oxidizers.

ENVIRONMENTAL DATA
Food chain concentration potential: Negative; unlikely to accumulate.
Water pollution: Effect of low concentrations on aquatic life is unknown. May be dangerous if it enters nearby water intakes; notify operators. Notify local health and wildlife officials. **Response to discharge:** Issue warning–high flammability, air contaminant, water contaminant. Restrict access. Evacuate area. Disperse and flush.

SHIPPING INFORMATION
Grades of purity: Technical, 98.0+%; **Storage temperature:** Ambient; **Inert atmosphere:** None; **Venting:** Pressure-vacuum; **Stability during transport:** Stable.

PHYSICAL AND CHEMICAL PROPERTIES
Physical state @ 59°F/15°C and 1 atm: Liquid.
Molecular weight: 76.2
Boiling point @ 1 atm: 126.6°F/52.5°C/325.8°K.
Melting/Freezing point: –202.8°F/–130.5°C/142.7°K.
Specific gravity (water = 1): 0.814 @ 68°F/20°C (liquid).
Liquid surface tension: 22.0 dynes/cm = 0.022 N/m @ 68°F/20°C.
Relative vapor density (air = 1): 2.6

Ratio of specific heats of vapor (gas): 1.0964 at 15.6°C.
Latent heat of vaporization: 165.7 Btu/lb = 92.1 cal/g = 3.83 x 10^5 J/kg.
Heat of combustion: –14,920 Btu/lb = –8,290 cal/g = –347 x 10^5 J/kg.

ISOPROPYL PERCARBONATE REC. I:1600

SYNONYMS: DIISOPROPYL PERCARBONATE; DIISOPROPYL PEROXYDICARBONATE; ISOPROPYL PEROXYDICARBONATE; PERCARBONATO de ISOPROPILO (Spanish); PEROXYDICARBONIC ACID, BIS(1-METHYLETHYL) ESTER

IDENTIFICATION
CAS Number: 105-64-6
Formula: $C_8H_{14}O_6$; $C_3H_7OOCOOCOOC_3H_7$
DOT ID Number: UN 3112; DOT Guide Number: 148
Proper Shipping Name: Organic peroxide type B, solid, temperature controlled.

DESCRIPTION: White solid (containers packed in dry ice). Sharp, unpleasant odor. Sinks in water. Unstable above 48°F/9°C.

Flammable • Thermally unstable; may explode from heat, shock, or contamination •Strong oxidizer which may react spontaneously with low flash point organics or reducing agents. Heat forms oxygen; will increase the activity of an existing fire • May explode on contact with combustibles (wood, paper, oil, clothing, etc.) • Containers may BLEVE when exposed to fire • Vapors are heavier than air and will collect and stay in low areas • Vapors may travel long distances to ignition sources and flashback • Irritating to the skin, eyes, and respiratory tract • Toxic products of combustion may include CO_2 and carbon monoxide, acetone, isopropyl alcohol, acetaldehyde, and ethane.

EMERGENCY RESPONSE: See Appendix A)
Evacuation:
Public safety: Isolate the area of spill or leak for at least 50 to 100 meters (160 to 330 feet) in all directions.
Spill: Consider initial evacuation for at least 250 meters (800 feet).
Fire: If tank, rail car, or tank truck is involved in fire, isolate for at least 800 meters (½ mile) in all directions; also, consider initial evacuation for 800 meters (½ mile) in all directions.

EXPOSURE
Short-term effects: *SEEK MEDICAL ATTENTION.* **Dust:** Irritating to eyes, nose, and throat. *IF INHALED*, will, will cause coughing or difficult breathing. *IF IN EYES*, hold eyelids open and flush with plenty of water. *IF BREATHING HAS STOPPED,* give artificial respiration. IF breathing is difficult, administer oxygen.
Solid: Irritating to skin and eyes. Harmful if swallowed. Remove contaminated clothing and shoes. Flush affected areas with plenty of water. *IF IN EYES*, hold eyelids open and flush with plenty of water. *IF SWALLOWED* and victim is *CONSCIOUS AND ABLE TO SWALLOW*, have victim drink 4 to 8 ounces of water. *IF SWALLOWED* and victim is *UNCONSCIOUS OR HAVING CONVULSIONS,* do nothing except keep victim warm.

HEALTH HAZARDS
Personal protective equipment (PPE): B-Level PPE. Rubber gloves and shoes; hard hat; chemical splash goggles; plastic apron; respirator (depending on solvent used).
Toxicity by ingestion: Grade 2; LD_{50} = 0.5 to 5 g/kg.

FIRE DATA
Flash point: Flammable solid.
Fire extinguishing agents not to be used: All extinguishing agents may be ineffective.
Behavior in fire: Undergoes autoaccelerative decomposition and may self-ignite. Confinement may lead to detonation. Fires very difficult to extinguish because air not needed

CHEMICAL REACTIVITY
Binary reactants: A powerful oxidizer. Concentrated solutions are usually in ethers or hydrocarbons and are unstable. Violent reaction with many materials including reducing agents, amines, potassium iodide, metal powders. May decompose with formation of oxygen when in contact with metals. Do not store in sealed containers.

ENVIRONMENTAL DATA
Food chain concentration potential: Negative; unlikely to accumulate.
Water pollution: Effect of low concentrations on aquatic life is unknown. May be dangerous if it enters nearby water intakes; notify operators. Notify local health and wildlife officials. **Response to discharge:** Issue warning–high flammability, oxidizing material. Restrict access. Should be removed. Chemical and physical treatment.

SHIPPING INFORMATION
Grades of purity: Technical, 98.5-99+%; **Storage temperature:** Below 0°F/–18°C; **Inert atmosphere:** None; **Venting:** Pressure-vacuum; **Stability during transport:** Unstable above 0°F/–18°C with formation of oxygen gas.

PHYSICAL AND CHEMICAL PROPERTIES
Physical state @ 59°F/15°C and 1 atm: Liquid.
Molecular weight: 206.2
Boiling point @ 1 atm: (decomposes).
Melting/Freezing point: 46-50°F/8-10°C/281-283°K.
Specific gravity (water = 1): 1.08@ 15°C (solid).
Heat of combustion: –8500 Btu/lb = –4,720 cal/g = –198 x 10^5 J/kg.
Heat of decomposition: –670 Btu/lb = –370 cal/g = 15.5 x 10^5 J/kg.

o-ISOPROPYL PHENOL REC. I:1650

SYNONYMS: *o*-ISOPROPILFENOL (Spanish); 2-ISOPROPYL PHENOL; PRODOX 131

IDENTIFICATION
CAS Number: 88-69-7
Formula: $C_9H_{12}O$; $(CH_3)_2CH–C_6H_4OH$

DESCRIPTION: Light yellow liquid. Floats on water surface; insoluble.

Poison! • Firefighting gear (including SCBA) does not provide adequate protection. If exposure occurs, remove and isolate gear immediately and thoroughly decontaminate personnel • Combustible • Containers may BLEVE when exposed to fire •Vapors may form explosive mixture with air • Vapors are heavier than air and will collect and stay in low areas • Vapors may travel long distances to ignition sources and flashback • Vapors in confined areas (e.g., tanks, sewers, buildings) may explode when exposed to fire • Severely Irritating to the skin, eyes, and respiratory tract • Toxic products of combustion may include CO_2 and carbon monoxide.

Hazard Classification (based on NFPA-704 Rating System)
Health Hazards (Blue): 2; Flammability (Red): 1; Reactivity (Yellow): 0

EMERGENCY RESPONSE: See Appendix A (148)
Evacuation:
Public safety: Isolate the area of spill or leak for at least 10 to 25 meters (30 to 80 feet) in all directions.
Spill: Increase, in the downwind direction, as necessary, the distance shown under "Public Safety."
Fire: If any large container is involved in fire, isolate for at least 800 meters (½ mile) in all directions; also, consider initial evacuation for 800 meters (½ mile) in all directions.

EXPOSURE
Short-term effects: *SEEK MEDICAL ATTENTION*. **Vapor:** Move victim to fresh air. *IF BREATHING HAS STOPPED*, give artificial respiration. If breathing is difficult, administer oxygen. **Liquid:** Irritating to skin and eyes.
Remove contaminated clothing and shoes.
Wash affected areas with soap and water.
IF IN EYES, hold eyelids open and flush with plenty of water.

HEALTH HAZARDS
Personal protective equipment (PPE): B-Level PPE. Full impervious protective clothing, including boots and gloves. Where splashing is possible wear full face shield or chemical safety goggles. Use approved respirator to protect against vapors.
Vapor (gas) irritant characteristics: Vapors are moderately irritating such that personnel will not usually tolerate moderate or high concentrations.
Liquid or solid irritant characteristics: Fairly severe eye and skin irritant. May cause pain and second-degree burns after a few minutes of contact.

FIRE DATA
Flash point: 220°F/104°C (oc).
Fire extinguishing agents not to be used: Water.
Stoichiometric air-to-fuel ratio: 54.8

CHEMICAL REACTIVITY
Binary reactants: Oxidizers, acids.

ENVIRONMENTAL DATA
Water pollution: Effects of low concentrations on aquatic life is unknown. May be dangerous if it enters nearby water intakes; notify operators. Notify local health and wildlife officials.
Response to discharge: Should be removed.

SHIPPING INFORMATION
Grades of purity: 98%; technical; **Storage temperature:** Ambient; **Inert atmosphere:** None; **Stability during transport:** Stable.

PHYSICAL AND CHEMICAL PROPERTIES
Physical state @ 59°F/15°C and 1 atm: Liquid.
Molecular weight: 136.21
Boiling point @ 1 atm: 417.2°F/214°C/487°K.
Melting/Freezing point: 62.6°F/17°C/290°K.
Specific gravity (water = 1): 0.995 @ 20°C.

ISOVALERALDEHYDE   REC. I:1700

SYNONYMS: ALDEHIDO ISOVALERIANICO (Spanish); ISOPENTALDEHYDE; ISOVALERIC ALDEHYDE; ISOVALERAL; 3-METHYLBUTANAL; 3-METHYLBUTYRALDEHYDE

IDENTIFICATION
CAS Number: 590-86-3
Formula: $C_5H_{10}O$; $(CH_3)_2CHCH_2CHO$
DOT ID Number: UN 1988; DOT Guide Number: 131
Proper Shipping Name: Aldehydes, flammable, toxic, n.o.s.

DESCRIPTION: Colorless liquid. Weak suffocating, apple-like odor. Floats on the surface of water. Flammable, irritating vapor is produced.

Highly flammable • Toxic! Breathing the vapor, skin or eye contact, or swallowing the material may kill you • Firefighting gear (including SCBA) does not provide adequate protection. If exposure occurs, remove and isolate gear immediately and thoroughly decontaminate personnel • Containers may BLEVE when exposed to fire •Vapors may form explosive mixture with air • Vapors are heavier than air and will collect and stay in low areas • Vapors may travel long distances to ignition sources and flashback • Vapors in confined areas (e.g., tanks, sewers, buildings) may explode when exposed to fire • Highly irritating to the eyes and respiratory tract; this material has anesthetic properties • Toxic products of combustion may include irritating vapors.

Hazard Classification (based on NFPA-704 Rating System)
Health Hazards (Blue): 2; Flammability (Red): 3; Reactivity (Yellow): 0

EMERGENCY RESPONSE: See Appendix A (131)
Evacuation:
Public safety: Isolate spill area for at least 100 to 200 meters (330 to 660 feet) in all directions.
Spill: Increase, as necessary, the isolation distance shown above, in "Public safety."
Fire: Isolate for 800 meters (½ mile) in all directions, especially if tank, rail car, or tank truck is involved in fire.

EXPOSURE
Short-term effects: *SEEK MEDICAL ATTENTION*. **Vapor:** Irritating to eyes, nose, and throat. *IF INHALED*, will, will cause headache, nausea, vomiting or difficult breathing. Move victim to fresh air. *IF BREATHING HAS STOPPED*, give artificial respiration. IF breathing is difficult, administer oxygen. **Liquid:** Irritating to skin and eyes. Harmful if swallowed. *IF SWALLOWED* and victim is *CONSCIOUS AND ABLE TO SWALLOW*, have victim drink 4 to 8 ounces of water.

HEALTH HAZARDS
Personal protective equipment (PPE): A-Level PPE. Goggles or face shield; rubber gloves; air mask or self-contained breathing apparatus for high vapor concentrations. Butyl rubber is generally suitable for aldehydes.
Exposure limits (TWA unless otherwise noted): ACGIH TLV 50 ppm (176 mg/m^3) as *n*-valeraldehyde.
Toxicity by ingestion: Grade 2; oral LD_{50} > 3200 mg/kg (rat).

FIRE DATA
Flash point: 23°F/–5°C (oc).
Fire extinguishing agents not to be used: Water may be ineffective.
Electrical hazard: Class I, Group C.
Burning rate: 5.3 mm/min

o-Isopropyl phenol

CHEMICAL REACTIVITY
Binary reactants: Violent reaction with strong oxidizers. Incompatible with nitric acid, sulfuric acid.
Reactivity group: 19
Compatibility class: Aldehyde

ENVIRONMENTAL DATA
Food chain concentration potential: Negative; unlikely to accumulate.
Water pollution: DOT Appendix B, §172.101–marine pollutant. Effect of low concentrations on aquatic life is unknown. Fouling to shoreline. May be dangerous if it enters nearby water intakes; notify operators. Notify local health and wildlife officials.
Response to discharge: Issue warning–high flammability, water contaminant. Restrict access. Mechanical containment. Should be removed. Chemical and physical treatment.

SHIPPING INFORMATION
Grades of purity: Commercial; **Storage temperature:** Ambient; **Inert atmosphere:** None; **Venting:** Open (flame arrester); **Stability during transport:** Stable.

PHYSICAL AND CHEMICAL PROPERTIES
Physical state @ 59°F/15°C and 1 atm: Liquid.
Molecular weight: 86.1
Boiling point @ 1 atm: 199°F/93°C/366°K.
Melting/Freezing point: –60°F/–51°C/222°K.
Specific gravity (water = 1): 0.785 @ 68°F/20°C (liquid).
Liquid surface tension: 23.7 dynes/cm = 0.0237 N/m @ 68°F/20°C.
Liquid water interfacial tension: (estimate) 30 dynes/cm = 0.030 N/m @ 68°F/20°C.
Relative vapor density (air = 1): 3
Ratio of specific heats of vapor (gas): (estimate) 1.0736
Latent heat of vaporization: (estimate) 167 Btu/lb = 93 cal/g = 3.9×10^5 J/kg.
Heat of combustion: –15,500 Btu/lb = –8620 cal/g = $–360 \times 10^5$ J/kg.
Vapor pressure: 50 mm @ 77°F/25°C (approximate).

JET FUELS: JP-1
REC. J:0100

SYNONYMS: COMBUSTIBLE de REACTOR, JP-1 (Spanish); EEC No. 650-001-02-5; JET A; JET A-1

IDENTIFICATION
CAS Number: 8008-20-6
Formula: C_nH_{2n+2}
DOT ID Number: UN 1863; **DOT Guide Number:** 128
Proper Shipping Name: Fuel, aviation, turbine engine

DESCRIPTION: Colorless to light brown watery liquid Fuel oil or petroleum odor. Floats on water surface; insoluble.

Vary Flammable • Containers may BLEVE when exposed to fire • Vapors form explosive mixture with air • Vapors are heavier than air and will collect and stay in low areas • Vapors may travel to ignition sources and flashback • Vapors in confined areas (e.g., tanks, sewers, buildings) may explode when exposed to fire • Irritating to the skin, eyes, and respiratory tract • Toxic products of combustion may include carbon monoxide and dark, heavy, smoke.

Hazard Classification (based on NFPA-704 Rating System)
Health Hazards (Blue): 0; Flammability (Red): 2; Reactivity (Yellow): 0

EMERGENCY RESPONSE: See Appendix A (128)
Evacuation:
Public safety: Isolate spill area for at least 25 to 50 meters (80 to 160 feet) in all directions.
Spill: Large spill–Consider initial downwind evacuation for at least 300 meters (1000 feet).
Fire: Isolate for 800 meters (½ mile) in all directions, especially if tank, rail car, or tank truck is involved in fire.

EXPOSURE
Short-term effects: *SEEK MEDICAL ATTENTION.* **Liquid:** Irritating to skin and eyes. Harmful if swallowed. Remove contaminated clothing and shoes. Flush affected areas with plenty of water. *IF IN EYES,* hold eyelids open and flush with plenty of water. *IF SWALLOWED* and victim is *CONSCIOUS AND ABLE TO SWALLOW,* have victim drink 4 to 8 ounces of water. **Do NOT induce vomiting.**

HEALTH HAZARDS
Personal protective equipment (PPE): B-Level PPE. Protective gloves; goggles or face shield. Chemical protective material(s) reported to have good to excellent resistance: nitrile, Saranex®, Viton®.
Recommendations for respirator selection: NIOSH as kerosene *1000 ppm*: CCROV [any chemical cartridge respirator with organic vapor cartridge(s)]; or SA (any supplied-air respirator). *2500 ppm:* SA:CF (any supplied-air respirator operated in a continuous-flow mode); or PAPROV [any powered, air-purifying respirator with organic vapor cartridge(s)]. *5000 ppm:* CCRFOV [any chemical cartridge respirator with a full facepiece and organic vapor cartridges(s)]; or GMFOV [any air-purifying, full-facepiece respirator (gas mask) with a chin-style, front- or back-mounted acid gas canister]; or PAPRTOV [any powered, air-purifying respirator with a tight-fitting facepiece and organic vapor cartridges(s)]; or SCBAF (any self-contained breathing apparatus with a full facepiece); or SAF (any supplied-air respirator with a full facepiece). *EMERGENCY OR PLANNED ENTRY INTO UNKNOWN CONCENTRATIONS OR IDLH CONDITIONS:* SCBAF:PD,PP (any self-contained breathing apparatus that has a full facepiece and is operated in a pressure-demand or other positive-pressure mode); or SAF:PD,PP:ASCBA (any supplied-air respirator that has a full facepiece and is operated in a pressure-demand or other positive-pressure mode in combination with an auxiliary self-contained breathing apparatus operated in a pressure-demand or other positive pressure mode). *ESCAPE:* GMFOV [any air-purifying, full-facepiece respirator (gas mask) with a chin-style, front- or back-mounted organic vapor canister]; or SCBAE (any appropriate escape-type, self-contained breathing apparatus).
Exposure limits (TWA unless otherwise noted): NIOSH REL 100 mg/m³ as kerosene
Toxicity by ingestion: Grade 1; LD_{50} = 5 to 15 g/kg.
Vapor (gas) irritant characteristics: Vapors cause smarting of the eyes or respiratory system if present in high concentrations. The effect is temporary.
Liquid or solid irritant characteristics: If spilled on clothing and allowed to remain, may cause smarting and reddening of the skin.
Odor threshold: 1 ppm.

FIRE DATA
Flash point: 110–150°F/43–66°C (cc) (JP-1); 100–150°F/37–66°C (JP-3)

Flammable limits in air: LEL: 0.7%; UEL: 5%.
containers.
Fire extinguishing agents not to be used: Water may be ineffective.
Autoignition temperature: 400–550°F/204–288°C/477–561°K.
Electrical hazard: Class I, Group D.
Burning rate: 4 mm/min

CHEMICAL REACTIVITY
Binary reactants: Reacts (possibly explosively) with oxidizing materials.
Reactivity group: 33
Compatibility class: Miscellaneous hydrocarbon mixtures

ENVIRONMENTAL DATA
Food chain concentration potential: Negative; unlikely to accumulate.
Water pollution: Dangerous to aquatic life in high concentrations. Fouling to shoreline. May be dangerous if it enters nearby water intakes; notify operators. Notify local health and wildlife officials.
Response to discharge: Mechanical containment. Should be removed. Chemical and physical treatment.

SHIPPING INFORMATION
Grades of purity: 100%; **Storage temperature:** Ambient; **Inert atmosphere:** None; **Venting:** Open (flame arrester); **Stability during transport:** Stable.

NAS HAZARD CLASSIFICATION FOR BULK WATER TRANSPORTATION
FIRE: 2
HEALTH: Vapor irritant: 1; Liquid or solid irritant: 1; Poisons: 1
WATER POLLUTION: Human toxicity: 1; Aquatic toxicity: 1; Aesthetic effect: 3
REACTIVITY: Other chemicals: 0; Water: 0; Self-reaction: 0

PHYSICAL AND CHEMICAL PROPERTIES
Physical state @ 59°F/15°C and 1 atm: Liquid.
Boiling point @ 1 atm: 392–500°F/200–260°C/473–533°K.
Melting/Freezing point: –45 to –55°F/–43 to –49°C/316 to 321°K.
Specific gravity (water = 1): 0.80 @ 15°C (liquid).
Liquid surface tension: 23 32 dynes/cm = 0.023 0.032 N/m @ 68°F/20°C.
Liquid water interfacial tension: 47–49 dynes/cm = 0.047–0.049 N/m @ 68°F/20°C.
Ratio of specific heats of vapor (gas): (estimate) 1.030
Latent heat of vaporization: 110 Btu/lb = 60 cal/g = 2.5×10^5 J/kg.
Heat of combustion: –18,540 Btu/lb = –10,300 cal/g = -431.24×10^5 J/kg.

JET FUELS: JP-4   **REC. J:0200**

SYNONYMS: COMBUSTIBLE de REACTOR, JP-4 (Spanish)

IDENTIFICATION
Formula: CNH_{2n+2}; a mixture of aromatic and aliphatic hydrocarbon compounds
DOT ID Number: UN 1863; DOT Guide Number: 128
Proper Shipping Name: Fuel, aviation, turbine engine

DESCRIPTION: Colorless watery liquid (65% gasoline, 35% light petroleum distillate). Fuel oil odor. Floats on water surface; insoluble.

Highly flammable • Containers may BLEVE when exposed to fire • Vapors may form explosive mixture with air • Vapors are heavier than air and will collect and stay in low areas • Vapors may travel long distances to ignition sources and flashback • Vapors in confined areas (e.g., tanks, sewers, buildings) may explode when exposed to fire • Irritating to the skin, eyes, and respiratory tract • Toxic products of combustion may include carbon monoxide, smoke, and irritating gases.

Hazard Classification (based on NFPA-704 Rating System)
Health Hazards (Blue): 1; Flammability (Red): 3; Reactivity (Yellow): 0

EMERGENCY RESPONSE: See Appendix A (128)
Evacuation:
Public safety: Isolate spill area for at least 25 to 50 meters (80 to 160 feet) in all directions.
Spill: Large spill–Consider initial downwind evacuation for at least 300 meters (1000 feet).
Fire: Isolate for 800 meters (½ mile) in all directions, especially if tank, rail car, or tank truck is involved in fire.

EXPOSURE
Short-term effects: *SEEK MEDICAL ATTENTION*. **Liquid:** Irritating to skin and eyes. Harmful if swallowed. Remove contaminated clothing and shoes. Flush affected areas with plenty of water. *IF IN EYES*, hold eyelids open and flush with plenty of water. *IF SWALLOWED* and victim is *CONSCIOUS AND ABLE TO SWALLOW*, have victim drink 4 to 8 ounces of water. **Do NOT induce vomiting.**

HEALTH HAZARDS
Personal protective equipment (PPE): B-Level PPE. Protective gloves; goggles or face shield. Chemical protective material(s) reported to have good to excellent resistance: nitrile, Viton®.
Ingestion and aspiration: Liquid irritates stomach; if taken into lungs, causes coughing, distress, and rapidly developing pulmonary edema.
Exposure limits (TWA unless otherwise noted): NIOSH REL 100 mg/m³ as kerosene
Toxicity by ingestion: Grade 2; LD_{50} = 0.5 to 5 g/kg.
Vapor (gas) irritant characteristics: Vapors cause smarting of the eyes or respiratory system if present in high concentrations. The effect is temporary.
Liquid or solid irritant characteristics: If spilled on clothing and allowed to remain, may cause smarting and reddening of the skin.
Odor threshold: 1 ppm.

FIRE DATA
Flash point: –10 to 30°F/–23 to –1°C (cc).
Flammable limits in air: LEL: 1.3%; UEL: 8.0%.
Autoignition temperature: 464°F/240°C/513°K.
Burning rate: 4 mm/min

CHEMICAL REACTIVITY
Binary reactants: Reacts with oxidizers, nitric acid.
Reactivity group: 33
Compatibility class: Miscellaneous hydrocarbon mixtures

ENVIRONMENTAL DATA
Food chain concentration potential: Negative; unlikely to accumulate.

Water pollution: Dangerous to aquatic life in high concentrations. Fouling to shoreline. May be dangerous if it enters nearby water intakes; notify operators. Notify local health and wildlife officials.
Response to discharge: Issue warning–high flammability. Mechanical containment. Should be removed. Chemical and physical treatment.

SHIPPING INFORMATION
Grades of purity: 100%; **Storage temperature:** Ambient; **Inert atmosphere:** None; **Venting:** Open (flame arrester) or pressure-vacuum; **Stability during transport:** Stable.

NAS HAZARD CLASSIFICATION FOR BULK WATER TRANSPORTATION
FIRE: 3
HEALTH: Vapor irritant: 1; Liquid or solid irritant: 1; Poisons: 1
WATER POLLUTION: Human toxicity: 1; Aquatic toxicity: 1; Aesthetic effect: 3
REACTIVITY: Other chemicals: 0; Water: 0; Self-reaction: 0

PHYSICAL AND CHEMICAL PROPERTIES
Physical state @ 59°F/15°C and 1 atm: Liquid.
Boiling point @ 1 atm: 349–549°F/176–287°C/449–560°K.
Melting/Freezing point: Less than –54°F/–48°C/225°K.
Specific gravity (water = 1): 0.81 @ 68°F/20°C (liquid).
Liquid surface tension: (estimate) 25 dynes/cm = 0.025 N/m @ 68°F/20°C.
Liquid water interfacial tension: (estimate) 50 dynes/cm = 0.05 N/m @ 68°F/20°C.
Ratio of specific heats of vapor (gas): (estimate) 1.030
Latent heat of vaporization: 140 Btu/lb = 78 cal/g = 3.3×10^5 J/kg.
Heat of combustion: –18,540 Btu/lb = –10,300 cal/g = $–431.24 \times 10^5$ J/kg.

JET FUELS: JP-5 REC. J:0250

SYNONYMS: COMBUSTIBLE de REACTOR, JP-5 (Spanish); KEROSENE, HEAVY

IDENTIFICATION
DOT ID Number: UN 1863; DOT Guide Number: 128
Proper Shipping Name: Fuel, aviation, turbine engine

DESCRIPTION: Colorless to light brown liquid. Fuel-oil odor. Floats on the surface of water.

Highly flammable • Containers may BLEVE when exposed to fire • Vapors may form explosive mixture with air • Vapors are heavier than air and will collect and stay in low areas • Vapors may travel long distances to ignition sources and flashback • Vapors in confined areas (e.g., tanks, sewers, buildings) may explode when exposed to fire • Irritating to the skin, eyes, and respiratory tract • Toxic products of combustion may include carbon monoxide.

Hazard Classification (based on NFPA-704 Rating System)
Health Hazards (Blue): 0; Flammability (Red): 2; Reactivity (Yellow): 0

EMERGENCY RESPONSE: See Appendix A (128)
Evacuation:
Public safety: Isolate spill area for at least 25 to 50 meters (80 to 160 feet) in all directions.
Spill: Large spill–Consider initial downwind evacuation for at least 300 meters (1000 feet).
Fire: Isolate for 800 meters (½ mile) in all directions, especially if tank, rail car, or tank truck is involved in fire.

EXPOSURE
Short-term effects: *SEEK MEDICAL ATTENTION*. **Liquid:** Irritating to skin and eyes. Harmful if swallowed. Remove contaminated clothing and shoes. Flush affected areas with plenty of water. *IF IN EYES*, hold eyelids open and flush with plenty of water. *IF SWALLOWED* and victim is *CONSCIOUS AND ABLE TO SWALLOW*, have victim drink 4 to 8 ounces of water. **Do NOT induce vomiting.**

HEALTH HAZARDS
Personal protective equipment (PPE): B-Level PPE. Protective gloves; goggles or face shield. Chemical protective material(s) reported to have good to excellent resistance: nitrile, Saranex®, Viton®.
Ingestion and aspiration: Liquid irritates stomach; if taken into lungs, causes coughing, distress, and rapidly developing pulmonary edema.
Exposure limits (TWA unless otherwise noted): NIOSH REL 100 mg/m^3.
Short-term exposure limits (15-minute TWA): CHRIS 2500 mg/m^3 for 60 minutes
Toxicity by ingestion: Grade 2; LD_{50} = 0.5 to 5 g/kg.
Vapor (gas) irritant characteristics: Vapors cause smarting of the eyes or respiratory system if present in high concentrations. The effect is temporary.
Liquid or solid irritant characteristics: If spilled on clothing and allowed to remain, may cause smarting and reddening of the skin.
Odor threshold: 1 ppm.

FIRE DATA
Flash point: 95-145°F/35-63°C (cc).
Flammable limits in air: LEL: 0.6%; UEL: 4.6%.
Fire extinguishing agents not to be used: Water may be ineffective
Autoignition temperature: 475°F/246°C/519°K (approximate).
Burning rate: 4 mm/min

CHEMICAL REACTIVITY
Binary reactants: Reacts with oxidizers, nitric acid.
Reactivity group: 33
Compatibility class: Miscellaneous hydrocarbon mixtures

ENVIRONMENTAL DATA
Food chain concentration potential: Negative; unlikely to accumulate.
Water pollution: Dangerous to aquatic life in high concentrations. Fouling to shoreline. May be dangerous if it enters nearby water intakes; notify operators. Notify local health and wildlife officials.
Response to discharge: Mechanical containment. Should be removed. Chemical and physical treatment.

SHIPPING INFORMATION
Grades of purity: 100%; **Storage temperature:** Ambient; **Inert atmosphere:** None; **Venting:** Open (flame arrester); **Stability during transport:** Stable.

NAS HAZARD CLASSIFICATION FOR BULK WATER TRANSPORTATION
FIRE: 1-2
HEALTH: Vapor irritant: 1; Liquid or solid irritant: 1; Poisons: 1

WATER POLLUTION: Human toxicity: 1; Aquatic toxicity: 1; Aesthetic effect: 3
REACTIVITY: Other chemicals: 0; Water: 0; Self-reaction: 0

PHYSICAL AND CHEMICAL PROPERTIES
Physical state @ 59°F/15°C and 1 atm: Liquid.
Boiling point @ 1 atm: 349-549°F/176-287°C/449-560°K.
Melting/Freezing point: Less than –54°F/–48°C/–225°K.
Specific gravity (water = 1): 0.82 @ 15°C (liquid).
Liquid surface tension: (estimate) 25 dynes/cm = 0.025 N/m @ 68°F/20°C.
Liquid water interfacial tension: (estimate) 50 dynes/cm = 0.05 N/m @ 68°F/20°C.
Latent heat of vaporization: 140 Btu/lb = 78 cal/g = 3.3×10^5 J/kg.
Heat of combustion: –18,540 Btu/lb = –10,300 cal/g = $–431.24 \times 10^5$ J/kg.

KEPONE REC. K:0100

SYNONYMS: CHLORDECONE; DECACHLOROKETONE; DECACHLOROOCTAHYDRO-KEPONE-2-ONE; DECACHLOROOCTAHYDRO-1,3,4-METHENO-2H-CYCLOBUTA(CD)-PENTALEN-2-ONE; DECACHLOROOCTAHYDRO-4,7-METHANOINDENEONE; MEREX; GENERAL CHEMICALS 1189; GC-1189; ENT-16391; RCRA No. U142

IDENTIFICATION
CAS Number: 143-50-0
Formula: $C_{10}Cl_{10}O$
DOT ID Number: UN 2761; DOT Guide Number: 151
Proper Shipping Name: Organochlorine, solid toxic
Reportable Quantity (RQ): **(CERCLA)** 1 lb/0.454 kg Insecticide (USEPA suspended registration).

DESCRIPTION: Colorless crystalline solid. Odorless. Soluble in water.

Poison! (organochlorine) • Breathing the vapor, skin or eye contact, or swallowing the material can kill you • Firefighting gear (including SCBA) does not provide adequate protection. If exposure occurs, remove and isolate gear immediately and thoroughly decontaminate personnel • Containers may BLEVE when exposed to fire • Toxic products of combustion may include carbon monoxide and chlorine gas • Do not put yourself in danger by entering a contaminated area to rescue a victim.

Hazard Classification (based on NFPA-704 Rating System)
Health Hazards (Blue): 2; Flammability (Red): 1; Reactivity (Yellow): 0

EMERGENCY RESPONSE: See Appendix A (151)
Evacuation:
Public safety: Isolate the area of spill or leak for at least 25 to 50 meters (80 to 160 feet) in all directions.
Spill: Increase, in the downwind direction, as necessary, the distance shown under "Public Safety."
Fire: If tank, rail car, or tank truck is involved in fire, isolate for at least 800 meters (½ mile) in all directions; also, consider initial evacuation for 800 meters (½ mile) in all directions.

EXPOSURE
Short-term effects: *SEEK MEDICAL ATTENTION.* Solid or dust: *POISONOUS IF SWALLOWED, INHALED, OR IF SKIN IS EXPOSED.* Flush affected areas with plenty of water. *IF IN EYES,* hold eyelids open and flush with plenty of water. *IF SWALLOWED* and victim is *CONSCIOUS AND ABLE TO SWALLOW,* have victim drink 4 to 8 ounces of water. **Do NOT induce vomiting**. Administer a slurry of activated charcoal. *IF SWALLOWED* and victim is *UNCONSCIOUS OR HAVING CONVULSIONS,* do nothing except keep victim warm. *IF BREATHING HAS STOPPED,* give artificial respiration; *avoid mouth-to-mouth resuscitation; use bag/mask apparatus.*

HEALTH HAZARDS
Personal protective equipment (PPE): A-Level PPE. Rubber gloves, self-contained breathing apparatus, and protective clothing.
Recommendations for respirator selection: NIOSH
At any concentrations above the NIOSH REL, or where there is no REL, at any detectable concentration: SCBAF:PD,PP (any self-contained breathing apparatus that has a full facepiece and is operated in a pressure-demand or other positive-pressure mode); or SAF:PD,PP:ASCBA (any supplied-air respirator that has a full facepiece and is operated in a pressure-demand or other positive-pressure mode in combination with an auxiliary, self-contained breathing apparatus operated in a pressure-demand or other positive pressure mode). *ESCAPE:* GMFOVHiE [any air-purifying, full-facepiece respirator (gas mask) with a chin-style, front- or back-mounted organic vapor canister having a high-efficiency particulate filter]; or SCBAE (any appropriate escape-type, self-contained breathing apparatus).
Exposure limits (TWA unless otherwise noted): NIOSH REL ceiling 0.001 mg/m^3; potential human carcinogen; limit exposure to lowest feasible level
Toxicity by ingestion: Grade 3; LD_{50} = 50 to 500 mg/kg.
Long-term health effects: Confirmed animal carcinogen. Possible liver, kidney, and nervous system damage. Mutagenic data.

CHEMICAL REACTIVITY
Binary reactants: Acids, acid fumes.

ENVIRONMENTAL DATA
Food chain concentration potential: Highly bioaccumulative (425 to 20,000) times
Water pollution: Harmful to aquatic life in very low concentrations. Dangerous if it enters water intakes. Notify local health and wildlife officials. **Response to discharge:** Issue warning–poison, water contaminant. Restrict access. Should be removed. Chemical and physical treatment.

SHIPPING INFORMATION
Grades of purity: 50% wettable powder; 2–4% baits.

PHYSICAL AND CHEMICAL PROPERTIES
Physical state @ 59°F/15°C and 1 atm: Solid.
Molecular weight: 490.68
Boiling point @ 1 atm: Sublimes at 350°C.
Melting/Freezing point: 662°F/349°C/623.2°K (sublimes).
Vapor pressure: Less than 3×10^{-7} mm @ 77°F/25°C.

KEROSENE REC. K:0150

SYNONYMS: COAL OIL; DEOBASE; EEC No. 650-001-02-5; KEROSENO (Spanish); STRAIGHT RUN KEROSENE; ILLUMINATING OIL; KEROSINE; NAPHTHA, 15-20% AROMATICS; RANGE OIL; JET FUEL: JP-1

Kerosene

IDENTIFICATION
CAS Number: 8008-20-6; may also apply to 68606-32-2 (partially sulfonized); 6474-81-0 (hydrosulfurized)
Formula: C_nH_{2n+2}
DOT ID Number: UN 1223; DOT Guide Number: 128
Proper Shipping Name: Kerosene

DESCRIPTION: Colorless to light brow watery liquid. Fuel oil odor. Floats on water surface; insoluble.

Highly flammable • Containers may BLEVE when exposed to fire • Vapors may form explosive mixture with air • Vapors are heavier than air and will collect and stay in low areas • Vapors may travel long distances to ignition sources and flashback • Vapors in confined areas (e.g., tanks, sewers, buildings) may explode when exposed to fire • Irritating to the skin, eyes, and respiratory tract • Toxic products of combustion may include carbon monoxide.

Hazard Classification (based on NFPA-704 Rating System)
Health Hazards (Blue): 0; Flammability (Red): 2; Reactivity (Yellow): 0

EMERGENCY RESPONSE: See Appendix A (128)
Evacuation:
Public safety: Isolate spill area for at least 25 to 50 meters (80 to 160 feet) in all directions.
Spill: Large spill–Consider initial downwind evacuation for at least 300 meters (1000 feet).
Fire: Isolate for 800 meters (½ mile) in all directions, especially if tank, rail car, or tank truck is involved in fire.

EXPOSURE
Short-term effects: *SEEK MEDICAL ATTENTION.* **Liquid:** Irritating to skin and eyes. Harmful if swallowed (droplets entering the lungs may cause pneumonia). Remove contaminated clothing and shoes. Flush affected areas with plenty of water. *IF IN EYES,* hold eyelids open and flush with plenty of water. *IF SWALLOWED* and victim is *CONSCIOUS AND ABLE TO SWALLOW,* have victim drink 4 to 8 ounces of water. Droplets in the lungs may cause pneumonia. **Do NOT induce vomiting.**
Note to physician or authorized medical personnel: Medical observation is recommended for 24 to 48 hours after breathing overexposure, as pulmonary edema may be delayed. As first aid for pulmonary edema, consider administering a corticosteroid spray. Cigarette smoking may exacerbate pulmonary injury and should be discouraged for at least 72 hours following exposure.

HEALTH HAZARDS
Personal protective equipment (PPE): B-Level PPE. Protective gloves; goggles or face shield. Chemical protective material(s) reported to have good to excellent resistance: nitrile, nitrile+PVC, neoprene, PV alcohol, PVC, Saranex®, Viton®. Also, Viton®/neopreneand polyurethane offers limited protection.
Recommendations for respirator selection: NIOSH
1000 ppm: CCROV [any chemical cartridge respirator with organic vapor cartridge(s)]; or SA (any supplied-air respirator). *2500 ppm:* SA:CF (any supplied-air respirator operated in a continuous-flow mode); or PAPROV [any powered, air-purifying respirator with organic vapor cartridge(s)]. *5000 ppm:* CCRFOV [any chemical cartridge respirator with a full facepiece and organic vapor cartridges(s)]; or GMFOV [any air-purifying, full-facepiece respirator (gas mask) with a chin-style, front- or back-mounted acid gas canister]; or PAPRTOV [any powered, air-purifying respirator with a tight-fitting facepiece and organic vapor cartridges(s)]; or SCBAF (any self-contained breathing apparatus with a full facepiece); or SAF (any supplied-air respirator with a full facepiece). *EMERGENCY OR PLANNED ENTRY INTO UNKNOWN CONCENTRATIONS OR IDLH CONDITIONS:* SCBAF:PD,PP (any self-contained breathing apparatus that has a full facepiece and is operated in a pressure-demand or other positive-pressure mode); or SAF:PD,PP:ASCBA (any supplied-air respirator that has a full facepiece and is operated in a pressure-demand or other positive-pressure mode in combination with an auxiliary self-contained breathing apparatus operated in a pressure-demand or other positive pressure mode). *ESCAPE:* GMFOV[any air-purifying, full-facepiece respirator (gas mask) with a chin-style, front- or back-mounted organic vapor canister]; or SCBAE (any appropriate escape-type, self-contained breathing apparatus).
Exposure limits (TWA unless otherwise noted): NIOSH REL 100 mg/m³.
Toxicity by ingestion: Grade 1; LD_{50} = 5 to 15 g/kg.
Long-term health effects: Mutation data. Suspected carcinogen.
Vapor (gas) irritant characteristics: Vapors cause smarting of the eyes or respiratory system if present in high concentrations. The effect is temporary.
Liquid or solid irritant characteristics: If spilled on clothing and allowed to remain, may cause smarting and reddening of the skin.
Odor threshold: 1 ppm.

FIRE DATA
Flash point: 100-162°F/38-72°C (cc).
Flammable limits in air: LEL: 0.7%; UEL: 5.0%.
Fire extinguishing agents not to be used: Water may be ineffective
Autoignition temperature: 410°F/210°C/483°K.
Electrical hazard: Flow or agitation of substance may generate electrostatic charges due to low conductivity.
Burning rate: 4 mm/min

CHEMICAL REACTIVITY
Binary reactants: Reacts with oxidizers, nitric acid. Kerosene causes rusting of steel.
Reactivity group: 33
Compatibility class: Miscellaneous hydrocarbon mixtures

ENVIRONMENTAL DATA
Food chain concentration potential: Negative; unlikely to accumulate.
Water pollution: Dangerous to aquatic life in high concentrations. Fouling to shoreline. May be dangerous if it enters nearby water intakes; notify operators. Notify local health and wildlife officials.
Response to discharge: Mechanical containment. Should be removed. Chemical and physical treatment.

SHIPPING INFORMATION
Grades of purity: Light hydrocarbon distillate: 100%; **Storage temperature:** Ambient; **Inert atmosphere:** None; **Venting:** Open (flame arrester); **Stability during transport:** Stable.

NAS HAZARD CLASSIFICATION FOR BULK WATER TRANSPORTATION
FIRE: 2
HEALTH: Vapor irritant: 1; Liquid or solid irritant: 1; Poisons: 1
WATER POLLUTION: Human toxicity: 1; Aquatic toxicity: 1; Aesthetic effect: 3
REACTIVITY: Other chemicals: 0; Water: 0; Self-reaction: 0

PHYSICAL AND CHEMICAL PROPERTIES
Physical state @ 59°F/15°C and 1 atm: Liquid.
Molecular weight: 170 (approximate).

Boiling point @ 1 atm: 304-574°F/151-301°C/424-574°K.
Melting/Freezing point: −50°F/−46°C/228°K.
Specific gravity (water = 1): 0.81 @ 15°C (liquid).
Liquid surface tension: 23-32 dynes/cm = 0.023-0.032 N/m @ 68°F/20°C.
Liquid water interfacial tension: 47-49 dynes/cm = 0.047–0.049 N/m @ 68°F/20°C.
Relative vapor density (air = 1): 4.5
Latent heat of vaporization: 110 Btu/lb = 60 cal/g = 2.5×10^5 J/kg.
Heat of combustion: −18,540 Btu/lb = −10,300 cal/g = $−431.24 \times 10^5$ J/kg.
Vapor pressure: (Reid) 0.1 psia; 5 mm @ 100°F/38°C.

LACTIC ACID   REC. L:0100

SYNONYMS: ACIDO LACTICO (Spanish); ACETONIC ACID; ETHYLIDENELACTIC ACID; α-HYDROXYPROPIONIC ACID; 2-HYDROXYPROPANOIC ACID; LD-LACTIC ACID; MILK ACID; ORDINARY LACTIC ACID; RACEMIC-LACTIC ACID

IDENTIFICATION
CAS Number: 50-21-5
Formula: $C_3H_6O_3$; $CH_3CHOHCOOH \cdot H_2O$

DESCRIPTION: Colorless to yellow, syrupy liquid or hygroscopic crystals. Slightly acrid. Sinks and mixes with water forming acid solution.

Concentrated solutions are caustic; corrosive to skin, eyes, and respiratory system • Toxic products of combustion may include carbon monoxide and acrid fumes.

Hazard Classification (based on NFPA-704 Rating System)
Health Hazards (Blue): 1; Flammability (Red): 1; Reactivity (Yellow): 0

EMERGENCY RESPONSE: See Appendix A (171)
Evacuation:
Public safety: Isolate the area of spill or leak for at least 10 to 25 meters (30 to 80 feet) in all directions.
Spill: Increase, in the downwind direction, as necessary, the distance shown under "Public Safety."
Fire: If any large container is involved in fire, isolate for at least 800 meters (½ mile) in all directions; also, consider initial evacuation for 800 meters (½ mile) in all directions.

EXPOSURE
Short-term effects: *SEEK MEDICAL ATTENTION.* **Vapor:** Irritating to eyes, nose, and throat. *IF INHALED*, will, will cause coughing or difficult breathing. *IF IN EYES*, hold eyelids open and flush with plenty of water. *IF BREATHING HAS STOPPED*, give artificial respiration. IF breathing is difficult, administer oxygen. **Liquid:** Will burn skin and eyes. *IF SWALLOWED*, will cause nausea. Remove contaminated clothing and shoes. Flush affected areas with plenty of water. *IF IN EYES*, hold eyelids open and flush with plenty of water. *IF SWALLOWED* and victim is *CONSCIOUS AND ABLE TO SWALLOW*, have victim drink 4 to 8 ounces of water. *IF SWALLOWED* and victim is *UNCONSCIOUS OR HAVING CONVULSIONS*, do nothing except keep victim warm.

HEALTH HAZARDS
Personal protective equipment (PPE): B-Level PPE. Gloves; goggles; self-contained breathing apparatus where high concentrations of mist are present. Chemical protective material(s) reported to have good to excellent resistance: butyl rubber, natural rubber, neoprene, nitrile, polyurethane, PVC, Viton®. Also, chlorinated polyethylene, Viton®/neoprene, PV alcohol, and butyl rubber/neoprene offers limited protection.
Toxicity by ingestion: Grade 2; oral LD_{50} = 1810 mg/kg (guinea pig).
Long-term health effects: Mutation data.
Vapor (gas) irritant characteristics: May cause respiratory irritation.
Liquid or solid irritant characteristics: Corrosive to eyes. Irritates the skin and respiratory tract.

FIRE DATA
Flash point: More than 160°F/71°C.

CHEMICAL REACTIVITY An acid and a corrosive when dissolved in water.
Binary reactants: Slowly corrodes most metals.
Neutralizing agents for acids and caustics: Dilute with water; rinse with sodium bicarbonate or lime solution.

ENVIRONMENTAL DATA
Food chain concentration potential: Log P_{ow} = −0.59. Unlikely to accumulate.
Water pollution: Dangerous to aquatic life in high concentrations. May be dangerous if it enters nearby water intakes; notify operators. Notify local health and wildlife officials. **Response to discharge:** Disperse and flush.

SHIPPING INFORMATION
Grades of purity: USP; Reagent; Technical, 88%; Food processing, 50%, 80%. The balance is water in all cases; **Storage temperature:** Ambient; **Inert atmosphere:** None; **Venting:** Open; **Stability during transport:** Stable.

PHYSICAL AND CHEMICAL PROPERTIES
Physical state @ 59°F/15°C and 1 atm: Liquid.
Molecular weight: 90.1
Boiling point @ 1 atm: Decomposes. 250°F/121°C/394°K @ 15 mm.
Melting/Freezing point: 64°F/18°C/291°K; L+ and D- lactic acid: 127°F/53°C/326°K.
Specific gravity (water = 1): 1.20 at 20° (liquid).
Heat of combustion: −6,520 Btu/lb = −3,620 cal/g = $−152 \times 10^5$ J/kg.

LATEX, LIQUID SYNTHETIC   REC. L:0200

SYNONYMS: DIMETHICONE 350; DOW CORNING 346; GEON; GOOD-RITE; GUM; HYCAR; LATEX; METHYL SILICONE; PLASTIC LATEX; POLYDIMETHYL SILYLENE; SYNTHETIC RUBBER LATEX

IDENTIFICATION
CAS Number: 9016-00-6
Formula: $(C_2H_6OSi)_x$

DESCRIPTION: Milky white liquid. Each type has a characteristic odor. Mixes with water.

If the latex dries out and then burns, irritating and poisonous hydrochloric acid, hydrogen cyanide, and styrene gases can be evolved.

EMERGENCY RESPONSE: See Appendix A (171)
Evacuation:
Public safety: Isolate the area of spill or leak for at least 10 to 25 meters (30 to 80 feet) in all directions.
Spill: Increase, in the downwind direction, as necessary, the distance shown under "Public Safety."
Fire: If any large container is involved in fire, isolate for at least 800 meters (½ mile) in all directions; also, consider initial evacuation for 800 meters (½ mile) in all directions.

EXPOSURE
Short-term effects: *SEEK MEDICAL ATTENTION.* **Liquid:** Irritating to eyes. *IF IN EYES*, hold eyelids open and flush with plenty of water.
Note to physician or authorized medical personnel: Consider the use of amyl nitrite perles if symptoms of cyanide poisoning develop. If symptoms indicate, initial treatment includes the cyanide antidote kit. In all cases, break an amyl nitrite perle in a gauze pad and hold lightly under victim's nose for 15 seconds, repeating 5 times at about 15-second intervals; if necessary (and if sodium nitrite infusions will be delayed), repeat procedure every 3 minutes with fresh perles until 3 or 4 have been used. Avoid breathing the vapor while administering it to the victim. Administer sodium nitrite IV, ASAP. The usual adult dose is 10 to 20 mL of a 3% solution infused over no less than 5 minutes; the average child dose is 0.15 to 0.20 mL/kg. Monitor blood pressure during administration, and slow the rate of infusion if hypotention develops. Next, infuse sodium thiosulfate IV. The usual adult dose is 50 mL of a 25% solution infused over 10 to 20 minutes; the average child dose is 1.65 mL/kg. Repeat with nitrite and thiosulfate as required.

HEALTH HAZARDS
Personal protective equipment (PPE): B-Level PPE. Chemical goggles or face shield.
Long-term health effects: Reproductive effects (experimental).
Liquid or solid irritant characteristics: Contact with eyes can cause irritation.

FIRE DATA
Behavior in fire: Heat may coagulate the latex and form sticky plastic lumps which may burn.

CHEMICAL REACTIVITY
Binary reactants: Sulfuric acid and isocyanates. Coagulated by heat and acids to gummy, flammable material.
Reactivity group: 43
Compatibility class: Water solutions.

ENVIRONMENTAL DATA
Water pollution: Effect of low concentrations on aquatic life is unknown. Fouling to shoreline. May be dangerous if it enters nearby water intakes; notify operators. Notify local health and wildlife officials. **Response to discharge:** Disperse and flush (if not coagulated). Should be removed (if coagulated). Chemical and physical treatment.

SHIPPING INFORMATION
Grades of purity: All commercial lattices are shipped in a variety of concentrations in water, depending on the particular polymer involved and the intended use of the latex. None are particularly hazardous except in fires, where all coagulate to gummy, flammable material; **Storage temperature:** Ambient; **Inert atmosphere:** None; **Venting:** Open; **Stability during transport:** Stable.

PHYSICAL AND CHEMICAL PROPERTIES
Physical state @ 59°F/15°C and 1 atm: Liquid.
Boiling point @ 1 atm: Very high.
Specific gravity (water = 1): 1.057 @ 77°F/25°C (liquid).

LAURIC ACID   REC. L:0300

SYNONYMS: ACIDO LAURICO (Spanish); C-1297; DODECANOIC ACID; *N*-DODECANOIC ACID; DODECONIC ACID; DUODECYLIC ACID; HYSTRENE 9512; HYDROFOL ACID 1255; HYDROFOL ACID 1295; LAUROSTEARIC ACID; NEOFAT 12; NEO-FAT 12-43; NINOL AA-62 EXTRA; WECOLINE 1295

IDENTIFICATION
CAS Number: 143-07-7
Formula: $C_{12}H_{24}O_2$; $CH_3(CH_2)_{10}CO_2H$

DESCRIPTION: Clear, colorless to white, crystalline solid. Slight odor of bay oil. Insoluble in water.
Combustible • Heat or flame may cause explosion • Containers may BLEVE when exposed to fire • Vapors are heavier than air and will collect and stay in low areas • Vapors in confined areas (e.g., tanks, sewers, buildings) may explode when exposed to fire • Irritating to the skin, eyes, and respiratory tract • Toxic products of combustion may include acid vapors and carbon monoxide.

Hazard Classification (based on NFPA-704 Rating System)
Health Hazards (Blue): 0; Flammability (Red): 1; Reactivity (Yellow): 0

EMERGENCY RESPONSE: See Appendix A (171)
Evacuation:
Public safety: Isolate the area of spill or leak for at least 10 to 25 meters (30 to 80 feet) in all directions.
Spill: Increase, in the downwind direction, as necessary, the distance shown under "Public Safety."
Fire: If any large container is involved in fire, isolate for at least 800 meters (½ mile) in all directions; also, consider initial evacuation for 800 meters (½ mile) in all directions.

EXPOSURE
Short-term effects: *CALL FOR MEDICAL AID*
Vapor, aerosol mist or dust: Irritating to eyes, mucous membranes, nose and throat. *IF INHALED*, will, will cause coughing or difficult breathing. *IF IN EYES*, hold eyelids open and flush with plenty of water. *IF BREATHING HAS STOPPED*, give artificial respiration. IF breathing is difficult, administer oxygen.
Liquid (solution): Will burn skin and eyes. *IF SWALLOWED*, will cause nausea and vomiting. Remove contaminated clothing and shoes. Flush affected areas with plenty of water. *IF IN EYES*, hold eyelids open and flush with plenty of water.

HEALTH HAZARDS
Personal protective equipment (PPE): B-Level PPE. Respirator, chemical safety goggles, boots and heavy gloves. Sealed chemical protective materials recommended (30 to 70%): Viton®/neoprene,

natural rubber, neoprene, nitrile. Also, PV alcohol offers limited protection.
Toxicity by ingestion: Grade 1: LD_{50} = 12 g/kg (rat).
Long-term health effects: Experimental carcinogen; mutation data.
Vapor (gas) irritant characteristics: Vapors or mists cause severe irritation and can cause eye and lung injury. They cannot be tolerated even at low concentrations.
Liquid or solid irritant characteristics: Severe skin irritant. Causes second- and third-degree burns on short contact and is very injurious to the eyes.

FIRE DATA
Flash point: More than 230°F/110°C (cc).
Behavior in fire: May cause dust explosion.

CHEMICAL REACTIVITY
Binary reactants: Strong oxidizing materials, strong acids, caustics, ammonia, isocyanates, epichlorohydrin.
Neutralizing agents for acids and caustics: Sodium bicarbonate solution; flush with water.
Reactivity group: 4
Compatibility class: Organic acids (carboxylic acids)

ENVIRONMENTAL DATA
Water pollution: May be dangerous to aquatic life in high concentrations. May be dangerous if it enters nearby water intakes; notify operators. Notify local health and wildlife officials.
Response to discharge: Restrict access. Mechanical containment. Should be removed.

PHYSICAL AND CHEMICAL PROPERTIES
Physical state @ 59°F/15°C and 1 atm: Solid.
Molecular weight: 200.32
Boiling point @ 1 atm: 437°F/225°C/498.2°K (at 100 mmHg = 0.132 atm).
Melting/Freezing point: 111–115°F/44–46°C/317–319°K.
Specific gravity (water = 1): 0.883
Relative vapor density (air = 1): 6.91
Vapor pressure: 1 mm @ 121°C.

LAUROYL PEROXIDE   REC. L:0400

SYNONYMS: ALPEROX C; DILAUROYL PEROXIDE; DPY-97 F; LAUROX; LAURYDOL; LYP; LYP 97; DODECANOYL PEROXIDE; PEROXIDO de LAUROILO (Spanish)

IDENTIFICATION
CAS Number: 105-74-8
Formula: $C_{24}H_{46}O_4$; $[CH_3(CH_2)_{10}COO]_2$
DOT ID Number: UN 2124; UN 2893 (42%); DOT Guide Number: 145
Proper Shipping Name: Lauroyl peroxide; Lauroyl peroxide, not more than 42%, stable dispersion, in water.

DESCRIPTION: White, coarse powder (oxidizing material). Faint, soapy odor. Sinks in water; insoluble.

Flammable • Poison! • Corrosive • Breathing the vapor can kill you; skin or eye contact causes severe burns, impaired vision, or blindness; inhalation symptoms may be delayed • Heat or contamination may cause explosion • Firefighting gear (including SCBA) does not provide adequate protection. If exposure occurs, remove and isolate gear immediately and thoroughly decontaminate personnel • Explosion hazard (less sensitive than most organic peroxides) • A powerful oxidizer which may react spontaneously with low flash point organics or reducing agents. Heat forms oxygen; will increase the activity of an existing fire • May cause fire or explosion on contact with combustibles (wood, paper, oil, clothing, etc.) • Containers may BLEVE when exposed to fire • Vapors may form explosive mixture with air • Vapors are heavier than air and will collect and stay in low areas • Vapors may travel long distances to ignition sources and flashback • Vapors in confined areas (e.g., tanks, sewers, buildings) may explode when exposed to fire • Toxic products of combustion may include carbon monoxide.

Hazard Classification (based on NFPA-704 Rating System)
Health Hazards (Blue): 3; Flammability (Red): 2; Reactivity (Yellow): 3; Special Notice (White): OXY

EMERGENCY RESPONSE: See Appendix A (145)
Evacuation:
Public safety: Isolate the area of spill or leak for at least 25 to 50 meters (80 to 160 feet) in all directions.
Spill: Consider evacuation for at least 250 meters (800 feet) in all directions.
Fire: If any large container is involved in fire, isolate for at least 800 meters (½ mile) in all directions; also, consider initial evacuation for 800 meters (½ mile) in all directions.

EXPOSURE
Short-term effects: *SEEK MEDICAL ATTENTION.* **Solid:** Inhalation of particles may cause lung edema; physical exertion will aggravate this condition. Irritating to skin and eyes. Harmful if swallowed. Remove contaminated clothing and shoes. Flush affected areas with plenty of water. *IF IN EYES*, hold eyelids open and flush with plenty of water. *IF SWALLOWED* and victim is *CONSCIOUS AND ABLE TO SWALLOW*, have victim drink 4 to 8 ounces of water and have victim induce vomiting. *IF SWALLOWED* and victim is *UNCONSCIOUS OR HAVING CONVULSIONS,* do nothing except keep victim warm.
Note to physician or authorized medical personnel: Medical observation is recommended for 24 to 48 hours after breathing overexposure, as pulmonary edema may be delayed. As first aid for pulmonary edema, consider administering a corticosteroid spray. *Insufficient data on harmful effects to humans.* Cigarette smoking may exacerbate pulmonary injury and should be discouraged for at least 72 hours following exposure.

HEALTH HAZARDS
Personal protective equipment (PPE): B-Level PPE. Protective gloves, goggles. Butyl rubber is generally suitable for peroxide compounds.
Long-term health effects: Weak carcinogen in mice. IARC rating 3.
Vapor (gas) irritant characteristics: Eye and respiratory irritation. May cause lung edema; physical exertion will aggravate this condition.
Liquid or solid irritant characteristics: Corrosive to the eyes, and respiratory tract; contact causes burns.

FIRE DATA
Flash point: Oxidizing combustible solid.
Behavior in fire: Becomes sensitive to shock when hot; burns explosively in fire. Can increase the severity of a fire. Containers may explode in a fire. May ignite or explode spontaneously if mixed with flammable materials.

CHEMICAL REACTIVITY
Binary reactants: A powerful oxidizer. May ignite or explode spontaneously when mixed with combustible materials.

ENVIRONMENTAL DATA
Food chain concentration potential: Negative; unlikely to accumulate.
Water pollution: Effect of low concentrations on aquatic life is unknown. May be dangerous if it enters nearby water intakes; notify operators. Notify local health and wildlife officials.
Response to discharge: Issue warning–oxidizing material, water contaminant. Mechanical containment. Should be removed. Chemical and physical treatment.

SHIPPING INFORMATION
Grades of purity: 97-98%; dry or wetted with water; **Storage temperature:** Less than 80°F/27°C; **Inert atmosphere:** None; **Venting:** Open; **Stability during transport:** Stable if not overheated.

PHYSICAL AND CHEMICAL PROPERTIES
Physical state @ 59°F/15°C and 1 atm: Solid.
Molecular weight: 399
Boiling point @ 1 atm: Decomposes.
Melting/Freezing point: 129°F/54°C/327°K.
Specific gravity (water = 1): 0.91 @ 77°F/25°C (solid).
Heat of combustion: (estimate) –16,300 btu/lb = –9100 cal/g = –380 x 10^5 J/kg.

n-LAURYL MERCAPTAN   REC. L:0500

SYNONYMS: 1-DODECANETHIOL; DODECYL MERCAPTAN; LAURILMERCAPTANO (Spanish); 1-MERCAPTODODECANE; PENNFLOAT M; PENNFLOAT S

IDENTIFICATION
CAS Number: 112-55-0
Formula: $C_{12}H_{26}S$; $CH_3(CH_2)_{10}CH_2SH$
DOT ID Number: UN 3071; UN 1228; DOT Guide Number: 131
Proper Shipping Name: Mercaptans, liquid, toxic, flammable, n.o.s; Mercaptan mixture, liquid, flammable, toxic, n.o.s.

DESCRIPTION: Colorless to pale yellow, oily liquid. Mild skunk odor. Floats on the surface of water; insoluble; reacts, forming toxic and flammable mercaptan vapors.

Combustible • Explosion hazard at high temperatures • Containers may BLEVE when exposed to fire • Vapors are heavier than air and will collect and stay in low areas • Vapors in confined areas (e.g., tanks, sewers, buildings) may explode when exposed to fire • Irritating to the skin, eyes, and respiratory tract; affects the central nervous system • Toxic products of combustion may include sulfur dioxide and hydrogen sulfide

Hazard Classification (based on NFPA-704 Rating System)
Health Hazards (Blue): 2; Flammability (Red): 1; Reactivity (Yellow): 0

EMERGENCY RESPONSE: See Appendix A (145)
Evacuation:
Public safety: Isolate spill area for at least 100 to 200 meters (330 to 660 feet) in all directions.
Spill: Increase, as necessary, the isolation distance shown above, in "Public safety."
Fire: Isolate for 800 meters (½ mile) in all directions, especially if tank, rail car, or tank truck is involved in fire.

EXPOSURE
Short-term effects: *SEEK MEDICAL ATTENTION.*
Vapors/fumes: Affects the central nervous system. Headache, nausea, shortness of breath. **Liquid or solid:** Irritating to skin and eyes. *IF SWALLOWED*, will cause nausea. Remove contaminated clothing and shoes. Flush affected areas with plenty of water. *IF IN EYES*, hold eyelids open and flush with plenty of water. *IF SWALLOWED* and victim is *CONSCIOUS AND ABLE TO SWALLOW*, have victim drink 4 to 8 ounces of water.

HEALTH HAZARDS
Personal protective equipment (PPE): A-Level PPE. Rubber or vinyl gloves; chemical goggles; rubber shoes and apron.
Recommendations for respirator selection: NIOSH
5 ppm: CCROV [any chemical cartridge respirator with organic vapor cartridge(s)]; SA (any supplied-air respirator). *12.5 ppm:* SA:CF (any supplied-air respirator operated in a continuous-flow mode); PAPROV [any powered, air-purifying respirator with organic vapor cartridge(s)]. *25 ppm:* CCRFOV [any chemical cartridge respirator with a full facepiece and organic vapor cartridges(s); GMFOV [any air-purifying, full-facepiece respirator (gas mask) with a chin-style, front- or back-mounted acid gas canister]; PAPRTOV [any powered, air-purifying respirator with a tight-fitting facepiece and organic vapor cartridges(s)]; or SCBAF (any self-contained breathing apparatus with a full facepiece); or SAF (any supplied-air respirator with a full facepiece). *EMERGENCY OR PLANNED ENTRY INTO UNKNOWN CONCENTRATIONS OR IDLH CONDITIONS:* SCBAF:PD,PP (any self-contained breathing apparatus that has a full facepiece and is operated in a pressure-demand or other positive-pressure mode); or SAF:PD,PP:ASCBA (any supplied-air respirator that has a full facepiece and is operated in a pressure-demand or other positive-pressure mode in combination with an auxiliary self-contained breathing apparatus operated in a pressure-demand or other positive pressure mode). *ESCAPE:* GMFOV [any air-purifying, full-facepiece respirator (gas mask) with a chin-style, front- or back-mounted organic vapor canister]; or SCBAE (any appropriate escape-type, self-contained breathing apparatus).
Exposure limits (TWA unless otherwise noted): NIOSH REL ceiling limit 0.5 ppm/15M
Long-term health effects: Causes decline in kidney and liver function in rats; mutation data. Possible nervous system damage.
Vapor (gas) irritant characteristics: Irritating concentrations of vapor unlikely, but mist can cause irritation of eyes and upper respiratory tract.
Liquid or solid irritant characteristics: If spilled on clothing and allowed to remain, may cause smarting and reddening of the skin.
Odor threshold: 4 mg/m^3.

FIRE DATA
Flash point: 262°F/128°C (oc).
Fire extinguishing agents not to be used: Water or foam may cause frothing.

CHEMICAL REACTIVITY
Reactivity with water: Reacts; forming flammable mercaptan vapors.
Binary reactants: Strong oxidizers.

ENVIRONMENTAL DATA
Food chain concentration potential: Negative; unlikely to accumulate.

Water pollution: Effect of low concentrations on aquatic life is unknown. Fouling to shoreline. May be dangerous if it enters nearby water intakes; notify operators. Notify local health and wildlife officials. **Response to discharge:** Mechanical containment. Should be removed. Chemical and physical treatment.

SHIPPING INFORMATION
Grades of purity: 95% minimum; **Storage temperature:** Ambient; **Inert atmosphere:** None; **Venting:** Open (flame arrester); **Stability during transport:** Stable.

PHYSICAL AND CHEMICAL PROPERTIES
Physical state @ 59°F/15°C and 1 atm: Liquid.
Molecular weight: 202
Boiling point @ 1 atm: 239–351°F/115–177°C/388–450°K.
Melting/Freezing point: 19°F/–7°C/266°K.
Specific gravity (water = 1): 0.85 @ 15°C (liquid).
Liquid surface tension: (estimate) 30 dynes/cm = 0.03 N/m @ 68°F/20°C.
Liquid water interfacial tension: (estimate) 30 dynes/cm = 0.03 N/m @ 68°F/20°C.
Latent heat of vaporization: (estimate) 110 Btu/lb = 60 cal/g = 2.5 x 10^5/kg.
Heat of combustion: (estimate) –18,200 Btu/lb = –10,100 cal/g = –422 x 10^5 J/kg.

LEAD ACETATE   REC. L:0600

SYNONYMS: ACETATO de PLOMO (Spanish); EEC No. 082-001-00-6; LEAD ACETATE TRIHYDRATE; LEAD ACETATE(II), TRIHYDRATE; LEAD DIACETATE; LEAD(II) ACETATE; LEAD(2+) ACETATE; NEUTRAL LEAD ACETATE; NORMAL LEAD ACETATE; SALT OF SATURN; SUGAR OF LEAD; RCRA No. U144

IDENTIFICATION
CAS Number: 301-04-2; 6080-56-4 (trihydrate)
Formula: $C_4H_6O_4 \cdot Pb$; $Pb(C_2H_3O_2)_2 \cdot 3H_2O$
DOT ID Number: UN 1616; DOT Guide Number: 151
Proper Shipping Name: Lead acetate
Reportable Quantity (RQ): **(CERCLA)** 5000 lb/2270 kg

DESCRIPTION: White crystalline solid or powder; commercial grades are frequently brown or gray lumps. Odorless. Sinks and mixes with water.

Poison! • Breathing the dust, skin or eye contact, or swallowing the material can cause illness • Firefighting gear (including SCBA) does not provide adequate protection. If exposure occurs, remove and isolate gear immediately and thoroughly decontaminate personnel • Toxic products of combustion may include lead.

Hazard Classification (based on NFPA-704 Rating System)
Health Hazards (Blue): 1; Flammability (Red): 0; Reactivity (Yellow): 0

EMERGENCY RESPONSE: See Appendix A (151)
Evacuation:
Public safety: Isolate the area of spill or leak for at least 25 to 50 meters (80 to 160 feet) in all directions.
Spill: Increase, in the downwind direction, as necessary, the distance shown under "Public Safety."
Fire: If tank, rail car, or tank truck is involved in fire, isolate for at least 800 meters (½ mile) in all directions; also, consider initial evacuation for 800 meters (½ mile) in all directions.

EXPOSURE
Short-term effects: *SEEK MEDICAL ATTENTION.* **Dust:** *POISONOUS IF INHALED. IF INHALED*, will, will cause dizziness or loss of consciousness. *IF IN EYES*, hold eyelids open and flush with plenty of water. *IF BREATHING HAS STOPPED*, give artificial respiration; *avoid mouth-to-mouth resuscitation; use bag/mask apparatus*. IF breathing is difficult, administer oxygen. **Solid:** Irritating to skin and eyes. *IF SWALLOWED*, will cause nausea, vomiting, or loss of consciousness. Remove contaminated clothing and shoes. Flush affected areas with plenty of water. *IF IN EYES*, hold eyelids open and flush with plenty of water. *IF SWALLOWED* and victim is *CONSCIOUS AND ABLE TO SWALLOW*, have victim drink 4 to 8 ounces of water and have victim induce vomiting. *IF SWALLOWED* and victim is *UNCONSCIOUS OR HAVING CONVULSIONS,* do nothing except keep victim warm.
Note to physician or authorized medical personnel: Atropine sulfate and other antispasmodics may relieve abdominal pain, but morphine may be necessary.

HEALTH HAZARDS
Personal protective equipment (PPE): B-Level PPE. OSHA Table Z-1-A air contaminant. Dust mask and protective gloves.
Recommendations for respirator selection: OSHA as lead.
0.5 mg/m^3: HiE (any air-purifying, respirator with a high-efficiency particulate filter); SA (any supplied-air respirator). *1.25 mg/m^3:* SA:CF (any supplied-air respirator operated in a continuous-flow mode); or PAPRHiE (any powered, air-purifying respirator with a high-efficiency particulate filter). *2.5 mg/m^3:* HiEF (any air-purifying, full-facepiece respirator with a high-efficiency particulate filter); or SAT:CF (any supplied-air respirator that has a tight-fitting facepiece and is operated in a continuous-flow mode); or PAPRTHiE (any powered, air-purifying respirator with a tight-fitting facepiece and a high-efficiency particulate filter); or SCBAF (any self-contained breathing apparatus with a full facepiece); or SAF (any supplied-air respirator with a full facepiece). *50 mg/m^3:* SA:PD,PP (any supplied-air respirator operated in a pressure-demand or other positive-pressure mode). *100 mg/m^3:* SAF:PD,PP (any supplied-air respirator that has a full facepiece and is operated in a pressure-demand or other positive-pressure mode). *EMERGENCY OR PLANNED ENTRY INTO UNKNOWN CONCENTRATIONS OR IDLH CONDITIONS:* SCBAF:PD,PP (any self-contained breathing apparatus that has a full facepiece and is operated in a pressure-demand or other positive-pressure mode); or SAF:PD,PP:ASCBA (any supplied-air respirator that has a full facepiece and is operated in a pressure-demand or other positive-pressure mode in combination with an auxiliary self-contained breathing apparatus operated in a pressure-demand or other positive pressure mode). *ESCAPE:* HiEF (any air-purifying, full-facepiece respirator with a high-efficiency particulate filter); or SCBAE (any appropriate escape-type, self-contained breathing apparatus).
Exposure limits (TWA unless otherwise noted): ACGIH TLV 0.15 mg/m^3; OSHA PEL [1910.1025] 0.050 mg/m^3; NIOSH REL 0.100 mg/m^3 potential human carcinogen; reduce exposure to lowest feasible level, **as lead.**
Toxicity by ingestion: Grade 2; LD_{50} = 0.5 to 5 g/kg.
Long-term health effects: IARC possible carcinogen, rating 2B; NTP anticipated carcinogen; possible mutagen. Possible blood disorders and nervous system damage; lead poisoning.
Vapor (gas) irritant characteristics: Respiratory tract irritation.

Lead arsenate

Liquid or solid irritant characteristics: Eye, skin and respiratory tract irritation.
IDLH value: 100 mg/m^3 as lead.

CHEMICAL REACTIVITY
Binary reactants: Decomposes on contact with strong acids forming acetic acid. Violent reaction with bromates. May react with strong oxidizers; hydrogen peroxide.

ENVIRONMENTAL DATA
Food chain concentration potential: Fish, waterfowl, and terrestrial animals are capable of concentrating lead. Certain fish species such as catfish, salmon, trout, and minnows are especially sensitive.
Water pollution: DOT Appendix B, §172.101–marine pollutant. Dangerous to aquatic life in high concentrations. May be dangerous if it enters nearby water intakes; notify operators. Notify local health and wildlife officials. **Response to discharge:** Issue warning–water contaminant. Restrict access. Disperse and flush.

SHIPPING INFORMATION
Grades of purity: NF; Reagent; Technical, 97%; **Storage temperature:** Ambient; **Inert atmosphere:** None; **Venting:** Open; **Stability during transport:** Stable.

PHYSICAL AND CHEMICAL PROPERTIES
Physical state @ 59°F/15°C and 1 atm: Solid.
Molecular weight: 379.35
Boiling point @ 1 atm: 596°F/200°C/473°C (decomposes).
Melting/Freezing point: 1387°F/753°C/1026°K.
Specific gravity (water = 1): 2.55 @ 68°F/20°C (solid).

LEAD ARSENATE   REC. L:0700

SYNONYMS: ARSENIATO de PLOMO (Spanish); ARSENATE OF LEAD; DIBASIC LEAD ARSENATE; LEAD ACID ARSENATE; PLUMBOUS ARSENATE; SOPRABEL

IDENTIFICATION
CAS Number: 7784-40-9; 3687-31-8; 10102-48-4 (dibasic)
Formula: AsHO$_4$·Pb; PbHAsO$_4$
DOT ID Number: UN 1617; DOT Guide Number: 151
Proper Shipping Name: Lead arsenates
Reportable Quantity (RQ): **(CERCLA)** 1 lb/0.454 kg

DESCRIPTION: White crystalline solid. Odorless. Sinks in water; insoluble.

Poison! • Breathing the dust, skin or eye contact, or swallowing the material can cause illness • Firefighting gear (including SCBA) does not provide adequate protection. If exposure occurs, remove and isolate gear immediately and thoroughly decontaminate personnel • Toxic products of combustion may include arsenic and lead.

Hazard Classification (based on NFPA-704 Rating System)
Health Hazards (Blue): 3; Flammability (Red): 0; Reactivity (Yellow): 0

EMERGENCY RESPONSE: See Appendix A (151)
Evacuation:
Public safety: Isolate the area of spill or leak for at least 25 to 50 meters (80 to 160 feet) in all directions.
Spill: Increase, in the downwind direction, as necessary, the distance shown under "Public Safety."
Fire: If tank, rail car, or tank truck is involved in fire, isolate for at least 800 meters (½ mile) in all directions; also, consider initial evacuation for 800 meters (½ mile) in all directions.

EXPOSURE
Short-term effects: A specific medical treatment is used for exposure to this chemical; *CALL A DOCTOR* immediately. If artificial respiration is administered, *avoid mouth-to-mouth resuscitation; use bag/mask apparatus*. **Solid:** *POISONOUS IF SWALLOWED. IF SWALLOWED* and victim is *CONSCIOUS AND ABLE TO SWALLOW*, give victim a tablespoon of salt in glass of warm water and repeat until vomit is clear. Then give two tablespoons of epsom salt or milk of magnesia in water, and plenty of milk and water. Have victim lie down and keep quiet. *IF SWALLOWED* and victim is *UNCONSCIOUS OR HAVING CONVULSIONS*, do nothing except keep victim warm.

HEALTH HAZARDS
Personal protective equipment (PPE): B-Level PPE. OSHA Table Z-1-A air contaminant. Dust respirator; protective clothing to prevent accidental inhalation or ingestion of dust.
Recommendations for respirator selection: OSHA as lead.
0.5 mg/m^3: HiE (any air-purifying respirator with a high-efficiency particulate filter); SA (any supplied-air respirator). *1.25 mg/m^3:* SA:CF (any supplied-air respirator operated in a continuous-flow mode); or PAPRHiE (any powered, air-purifying respirator with a high-efficiency particulate filter). *2.5 mg/m^3:* HiEF (any air-purifying, full-facepiece respirator with a high-efficiency particulate filter); or SAT:CF (any supplied-air respirator that has a tight-fitting facepiece and is operated in a continuous-flow mode); or PAPRTHiE (any powered, air-purifying respirator with a tight-fitting facepiece and a high-efficiency particulate filter); or SCBAF (any self-contained breathing apparatus with a full facepiece); or SAF (any supplied-air respirator with a full facepiece). *50 mg/m^3:* SA:PD,PP (any supplied-air respirator operated in a pressure-demand or other positive-pressure mode). *100 mg/m^3:* SAF:PD,PP (any supplied-air respirator that has a full facepiece and is operated in a pressure-demand or other positive-pressure mode). *EMERGENCY OR PLANNED ENTRY INTO UNKNOWN CONCENTRATIONS OR IDLH CONDITIONS:* SCBAF:PD,PP (any self-contained breathing apparatus that has a full facepiece and is operated in a pressure-demand or other positive-pressure mode); or SAF:PD,PP:ASCBA (any supplied-air respirator that has a full facepiece and is operated in a pressure-demand or other positive-pressure mode in combination with an auxiliary self-contained breathing apparatus operated in a pressure-demand or other positive pressure mode). *ESCAPE:* HiEF (any air-purifying, full-facepiece respirator with a high-efficiency particulate filter); or SCBAE (any appropriate escape-type, self-contained breathing apparatus).
NIOSH *(as inorganic arsenic compounds).*
AT ANY CONCENTRATIONS ABOVE THE NIOSH REL, OR WHERE THERE IS NO REL, AT ANY DETECTABLE CONCENTRATION: SCBAF:PD,PP (any self-contained breathing apparatus that has a full faceplate and is operated in a pressure-demand or other positive-pressure mode); or SAF:PD,PP:ASCBA (any supplied-air respirator that has a full facepiece and is operated in a pressure-demand or other positive-pressure mode in combination with an auxiliary self-contained breathing apparatus operated in a pressure-demand or other positive-pressure mode). *ESCAPE:* GMFAGHiE [any air-purifying, full-facepiece respirator (gas mask) with a chin-style, front-or back-mounted acid gas canister having a high-efficiency particulate filter]; or SCBAE (any appropriate escape-type, self-contained breathing apparatus).

Exposure limits (TWA unless otherwise noted): ACGIH TLV 0.15 mg/m^3; OSHA PEL [1910.1080] 0.010 mg/m^3; NIOSH REL ceiling 0.002 mg/m^3/15 min, as arsenic; potential human carcinogen; reduce exposure to lowest feasible level

Toxicity by ingestion: Grade 4; LD$_{50}$ below 50 mg/kg (rabbit, rat).

Long-term health effects: Lead poisoning; IARC possible carcinogen, rating 2B; NTP anticipated carcinogen. Arsenic compounds are allergens; skin irritants.

IDLH value: 100 mg/m^3 as lead; 5 mg/m^3 as inorganic arsenic; known human carcinogen.

CHEMICAL REACTIVITY
Binary reactants: Strong oxidizers.

ENVIRONMENTAL DATA
Food chain concentration potential: Fish, waterfowl, and terrestrial animals are capable of concentrating lead. Certain fish species such as catfish, salmon, trout, and minnows are especially sensitive.

Water pollution: DOT Appendix B, §172.101–marine pollutant. Harmful to aquatic life in very low concentrations. May be dangerous if it enters nearby water intakes; notify operators. Notify local health and wildlife officials. **Response to discharge:** Issue warning–poison, water contaminant. Should be removed. Chemical and physical treatment.

SHIPPING INFORMATION
Grades of purity: 94%; **Storage temperature:** Ambient; **Inert atmosphere:** None; **Venting:** Open; **Stability during transport:** Stable.

PHYSICAL AND CHEMICAL PROPERTIES
Physical state @ 59°F/15°C and 1 atm: Solid.
Molecular weight: 347.12
Boiling point @ 1 atm: Decomposes.
Specific gravity (water = 1): 5.79 @ 15°C (solid).

LEAD CHLORIDE   REC. L:0800

SYNONYMS: CLORURO de PLOMO (Spanish); LEAD(II) CHLORIDE; LEAD(2+) CHLORIDE; LEAD DICHLORIDE; PLUMBOUS CHLORIDE

IDENTIFICATION
CAS Number: 7758-95-4
Formula: Cl$_2$Pb
DOT ID Number: UN 2291; DOT Guide Number: 151
Proper Shipping Name: Lead chloride
Reportable Quantity (RQ): **(CERCLA)** 100 lb/45.4 kg

DESCRIPTION: White crystalline solid or powder. Sinks and dissolves slowly in water.

Poison! • Breathing the dust or swallowing the material can cause illness • Firefighting gear (including SCBA) does not provide adequate protection. If exposure occurs, remove and isolate gear immediately and thoroughly decontaminate personnel • Toxic products of combustion may include hydrogen chloride and lead.

Hazard Classification (based on NFPA-704 Rating System)
Health Hazards (Blue): 1; Flammability (Red): 0; Reactivity (Yellow): 0

EMERGENCY RESPONSE: See Appendix A (151)
Evacuation:
Public safety: Isolate the area of spill or leak for at least 25 to 50 meters (80 to 160 feet) in all directions.
Spill: Increase, in the downwind direction, as necessary, the distance shown under "Public Safety."
Fire: If tank, rail car, or tank truck is involved in fire, isolate for at least 800 meters (½ mile) in all directions; also, consider initial evacuation for 800 meters (½ mile) in all directions.

EXPOSURE
Short-term effects: *SEEK MEDICAL ATTENTION*. **Dust and fumes:** *POISONOUS IF INHALED*. Move to fresh air. Keep victim quiet and warm. *IF BREATHING HAS STOPPED,* give artificial respiration; *avoid mouth-to-mouth resuscitation; use bag/mask apparatus*. **Solid:** If swallowed, may cause metallic taste, abdominal pain, vomiting and diarrhea. Flush affected area with plenty of water. *IF IN EYES*, hold eyelids open and flush with plenty of water. *IF SWALLOWED* and victim is *CONSCIOUS AND ABLE TO SWALLOW*, have victim drink 4 to 8 ounces of water, have victim induce vomiting. *IF SWALLOWED* and victim is *UNCONSCIOUS,* do nothing except keep victim warm.

Note to physician and authorized medical personnel: Administer saline cathartic and an enema. Give antispasmodic (calcium gluconate, atropine, papaverine) for relief of colic. If pain is severe morphine sulfate may be considered.

HEALTH HAZARDS
Personal protective equipment (PPE): B-Level PPE. OSHA Table Z-1-A air contaminant. Wear approved filter mask, rubber gloves, and safety glasses.

Recommendations for respirator selection: OSHA as lead.
0.5 mg/m^3: HiE (any air-purifying, respirator with a high-efficiency particulate filter); SA (any supplied-air respirator). *1.25 mg/m^3:* SA:CF (any supplied-air respirator operated in a continuous-flow mode); or PAPRHiE (any powered, air-purifying respirator with a high-efficiency particulate filter). *2.5 mg/m^3:* HiEF (any air-purifying, full-facepiece respirator with a high-efficiency particulate filter); or SAT:CF (any supplied-air respirator that has a tight-fitting facepiece and is operated in a continuous-flow mode); or PAPRTHiE (any powered, air-purifying respirator with a tight-fitting facepiece and a high-efficiency particulate filter); or SCBAF (any self-contained breathing apparatus with a full facepiece); or SAF (any supplied-air respirator with a full facepiece). *50 mg/m^3:* SA:PD,PP (any supplied-air respirator operated in a pressure-demand or other positive-pressure mode). *100 mg/m^3:* SAF:PD,PP (any supplied-air respirator that has a full facepiece and is operated in a pressure-demand or other positive-pressure mode). *EMERGENCY OR PLANNED ENTRY INTO UNKNOWN CONCENTRATIONS OR IDLH CONDITIONS:* SCBAF:PD,PP (any self-contained breathing apparatus that has a full facepiece and is operated in a pressure-demand or other positive-pressure mode); or SAF:PD,PP:ASCBA (any supplied-air respirator that has a full facepiece and is operated in a pressure-demand or other positive-pressure mode in combination with an auxiliary self-contained breathing apparatus operated in a pressure-demand or other positive pressure mode). *ESCAPE:* HiEF (any air-purifying, full-facepiece respirator with a high-efficiency particulate filter); or SCBAE (any appropriate escape-type, self-contained breathing apparatus).

Exposure limits (TWA unless otherwise noted): ACGIH TLV 0.05 mg/m^3 as inorganic lead; OSHA PEL 0.05 mg/m^3; NIOSH REL 0.100 mg/m^3 as lead.

Toxicity by ingestion: Guinea pig minimum lethal dose 1500 to 2000 mg/kg.

Long-term health effects: Lead poisoning. In humans 6 mg/m³/day inhaled long term produces histological and pathological effects. 1.2 mg/day ingested long term produces central nervous system disorders. Teratogenic effects. Possible carcinogen; reproduction hazard.
IDLH value: 100 mg/m³ as lead.

CHEMICAL REACTIVITY
Binary reactants: Reacts explosively with calcium when heated.

ENVIRONMENTAL DATA
Food chain concentration potential: Fish, waterfowl, and terrestrial animals are capable of concentrating lead. Certain fish species such as catfish, salmon, trout, and minnows are especially sensitive.
Water pollution: DOT Appendix B, §172.101–marine pollutant. Harmful to aquatic life in very low concentrations. May be dangerous if it enters nearby water intakes; notify operators. Notify local health and wildlife officials. **Response to discharge:** Issue warning–water contaminant. Restrict access. Should be removed. Chemical and physical treatment.

PHYSICAL AND CHEMICAL PROPERTIES
Physical state @ 59°F/15°C and 1 atm: Solid.
Molecular weight: 278.12
Boiling point @ 1 atm: 1742°F/950°C/1223°K.
Melting/Freezing point: 934°F/501°C/774°K.
Specific gravity (water = 1): 5.85 at room temperature
Relative vapor density (air = 1): 9.59 (calculated).
Latent heat of vaporization: 191.5 Btu/lb = 106.4 cal/g = 4.45 x 10^5 J/kg.
Heat of solution: Endothermic 40.1 Btu/lb = 22.3 cal/g = 0.93 x 10^5 J/kg.
Heat of fusion: 20.3 cal/g.
Vapor pressure: 1 mm @ 547°C.

LEAD FLUORIDE
REC. L:0900

SYNONYMS: FLUORURO de PLOMO (Spanish); LEAD DIFLUORIDE; LEAD(II) FLUORIDE; LEAD(2+) FLUORIDE; PLOMB FLUORURE (French); PLUMBOUS FLUORIDE

IDENTIFICATION
CAS Number: 7783-46-2
Formula: F_2Pb
DOT ID Number: UN 2811; DOT Guide Number: 154
Proper Shipping Name: Lead fluoride
Reportable Quantity (RQ): **(CERCLA)** 10 lb/4.54 kg

DESCRIPTION: White solid. Odorless. Sinks in water; slightly soluble.

Poison! • Breathing the dust or swallowing the material can cause illness • Firefighting gear (including SCBA) does not provide adequate protection. If exposure occurs, remove and isolate gear immediately and thoroughly decontaminate personnel • Toxic products of combustion may include fluorides and lead fume.

Hazard Classification (based on NFPA-704 Rating System)
Health Hazards (Blue): 1; Flammability (Red): 0; Reactivity (Yellow): 0

EMERGENCY RESPONSE: See Appendix A (154)
Evacuation:
Public safety: Isolate the area of spill or leak for at least 25 to 50 meters (80 to 160 feet) in all directions.
Spill: Increase, in the downwind direction, as necessary, the distance shown under "Public Safety."
Fire: If tank, rail car, or tank truck is involved in fire, isolate for at least 800 meters (½ mile) in all directions; also, consider initial evacuation for 800 meters (½ mile) in all directions.

EXPOSURE
Short-term effects: *SEEK MEDICAL ATTENTION.* **Dust:** *POISONOUS IF INHALED. IF INHALED*, will, will cause dizziness or loss of consciousness. *IF IN EYES*, hold eyelids open and flush with plenty of water. *IF BREATHING HAS STOPPED*, give artificial respiration; *avoid mouth-to-mouth resuscitation; use bag/mask apparatus.* IF breathing is difficult, administer oxygen.
Solid: Irritating to skin and eyes. *IF SWALLOWED*, will cause nausea, vomiting, or loss of consciousness. Remove contaminated clothing and shoes. Flush affected areas with plenty of water. *IF IN EYES*, hold eyelids open and flush with plenty of water. *IF SWALLOWED* and victim is CONSCIOUS have victim drink 4 to 8 ounces of water and have victim induce vomiting. *IF SWALLOWED* and victim is *UNCONSCIOUS OR HAVING CONVULSIONS*, do nothing except keep victim warm.
Notes: Early symptoms of lead intoxication via inhalation or ingestion are most commonly gastrointestinal disorders, colic, constipation, etc; weakness, which may go on to paralysis chiefly of the extensor muscles of the wrists and less often the ankles, is noticeable in the most serious cases. Ingestion of a large amount causes local irritation of the alimentary tract; pain, leg cramps, muscle weakness, paresthesia, depression, coma, and death may follow in 1 or 2 days.
Note to physician or authorized medical personnel: Atropine sulfate and other antispasmodics may relieve abdominal pain, but morphine may be necessary.

HEALTH HAZARDS
Personal protective equipment (PPE): B-Level PPE. OSHA Table Z-1-A air contaminant. Dust and fumes of all but the most insoluble lead compounds are readily absorbed on inhalation and, to a lesser degree, after ingestion. Respirator for heavy dust exposure; safety goggles.
Recommendations for respirator selection: OSHA as lead.
0.5 mg/m³: HiE (any air-purifying, respirator with a high-efficiency particulate filter); SA (any supplied-air respirator). *1.25 mg/m³*: SA:CF (any supplied-air respirator operated in a continuous-flow mode); or PAPRHiE (any powered, air-purifying respirator with a high-efficiency particulate filter). *2.5 mg/m³*: HiEF (any air-purifying, full-facepiece respirator with a high-efficiency particulate filter); or SAT:CF (any supplied-air respirator that has a tight-fitting facepiece and is operated in a continuous-flow mode); or PAPRTHiE (any powered, air-purifying respirator with a tight-fitting facepiece and a high-efficiency particulate filter); or SCBAF (any self-contained breathing apparatus with a full facepiece); or SAF (any supplied-air respirator with a full facepiece). *50 mg/m³*: SA:PD,PP (any supplied-air respirator operated in a pressure-demand or other positive-pressure mode). *100 mg/m³*: SAF:PD,PP (any supplied-air respirator that has a full facepiece and is operated in a pressure-demand or other positive-pressure mode). *EMERGENCY OR PLANNED ENTRY INTO UNKNOWN CONCENTRATIONS OR IDLH CONDITIONS:* SCBAF:PD,PP (any self-contained breathing apparatus that has a full facepiece and is operated in a pressure-demand or other positive-pressure mode); or SAF:PD,PP:ASCBA (any supplied-air respirator that has a full

facepiece and is operated in a pressure-demand or other positive-pressure mode in combination with an auxiliary self-contained breathing apparatus operated in a pressure-demand or other positive pressure mode). *ESCAPE:* HiEF (any air-purifying, full-facepiece respirator with a high-efficiency particulate filter); or SCBAE (any appropriate escape-type, self-contained breathing apparatus).

Exposure limits (TWA unless otherwise noted): ACGIH TLV 0.05 mg/m^3 as inorganic lead; OSHA PEL 0.05 mg/m^3; NIOSH REL 0.100 mg/m^3 as lead.

Toxicity by ingestion: Grade 2; LD_{50} = 0.5 to 5 g/kg.

Long-term health effects: Lead poisoning. IARC possible carcinogen; rating 2B

IDLH value: 100 mg/m^3; OSHA; as lead.

CHEMICAL REACTIVITY

Binary reactants: Reacts violently with fluorine, boron, and calcium carbide.

ENVIRONMENTAL DATA

Food chain concentration potential: Fish and terrestrial animals are capable of concentrating lead.

Water pollution: DOT Appendix B, §172.101–marine pollutant. Dangerous to aquatic life in high concentrations. May be dangerous if it enters nearby water intakes; notify operators. Notify local health and wildlife officials. **Response to discharge:** Issue warning–water contaminant. Restrict access. Should be removed. Chemical and physical treatment.

SHIPPING INFORMATION

Grades of purity: Technical; C.P; Optical grade; **Storage temperature:** Ambient; **Inert atmosphere:** None; **Venting:** Open; **Stability during transport:** Stable.

PHYSICAL AND CHEMICAL PROPERTIES

Physical state @ 59°F/15°C and 1 atm: Solid.
Molecular weight: 245.19
Specific gravity (water = 1): 8.24 @ 68°F/20°C (solid).
Heat of fusion: 7.6 cal/g.

LEAD FLUOROBORATE REC. L:1000

SYNONYMS: FLUOBORATO de PLOMO (Spanish); LEAD FLUOROBORATE SOLUTION; TETRAFLUROBORATE(1–) LEAD(2+).

IDENTIFICATION

CAS Number: 13814-96-5
Formula: $B_2F_8 \cdot Pb$; $Pb(BF_4)_2 \cdot H_2O$
DOT ID Number: UN 2291; DOT Guide Number: 151
Proper Shipping Name: Lead fluoborate
Reportable Quantity (RQ): **(CERCLA)** 10 lb/4.54 kg

DESCRIPTION: Colorless liquid. Faint odor or odorless. Sinks and mixes with water; forms fluoboric acid.

Poison! • Breathing the vapor or swallowing the material can cause illness • Firefighting gear (including SCBA) does not provide adequate protection. If exposure occurs, remove and isolate gear immediately and thoroughly decontaminate personnel • Toxic products of combustion may include hydrogen fluoride, lead and oxides of boron.

Hazard Classification (based on NFPA-704 Rating System)
Health Hazards (Blue): 1; Flammability (Red): 0; Reactivity (Yellow): 0

EMERGENCY RESPONSE: See Appendix A (151)
Evacuation:
Public safety: Isolate the area of spill or leak for at least 25 to 50 meters (80 to 160 feet) in all directions.
Spill: Increase, in the downwind direction, as necessary, the distance shown under "Public Safety."
Fire: If tank, rail car, or tank truck is involved in fire, isolate for at least 800 meters (½ mile) in all directions; also, consider initial evacuation for 800 meters (½ mile) in all directions.

EXPOSURE

Short-term effects: *SEEK MEDICAL ATTENTION.* **Vapor:** *POISONOUS IF INHALED. IF INHALED,* will, will cause dizziness or loss of consciousness. *IF IN EYES*, hold eyelids open and flush with plenty of water. *IF BREATHING HAS STOPPED,* give artificial respiration; *avoid mouth-to-mouth resuscitation; use bag/mask apparatus.* IF breathing is difficult, administer oxygen. **Liquid:** Will burn skin and eyes. *IF SWALLOWED*, will cause nausea, vomiting, or loss of consciousness. Remove contaminated clothing and shoes. Flush affected areas with plenty of water. *IF IN EYES*, hold eyelids open and flush with plenty of water. *IF SWALLOWED* and victim is *CONSCIOUS AND ABLE TO SWALLOW*, have victim drink water or milk and have victim induce vomiting. Give gastric lavage using 1% solution of sodium or magnesium sulfate; leave 15 to 30 g magnesium sulfate in 6 to 8 ounces of water in the stomach as antidote and cathartic; egg white, milk, and tannin are useful demulcents. *IF SWALLOWED* and victim is *UNCONSCIOUS OR HAVING CONVULSIONS,* do nothing except keep victim warm.

Note to physician or authorized medical personnel: Atropine sulfate and other antispasmodics may relieve abdominal pain, but morphine may be necessary.

HEALTH HAZARDS

Personal protective equipment (PPE): B-Level PPE. OSHA Table Z-1-A air contaminant. Rubber gloves; face shield; rubber apron

Recommendations for respirator selection: OSHA as lead.
0.5 mg/m^3: HiE (any air-purifying, respirator with a high-efficiency particulate filter); SA (any supplied-air respirator). *1.25 mg/m^3:* SA:CF (any supplied-air respirator operated in a continuous-flow mode); or PAPRHiE (any powered, air-purifying respirator with a high-efficiency particulate filter). *2.5 mg/m^3:* HiEF (any air-purifying, full-facepiece respirator with a high-efficiency particulate filter); or SAT:CF (any supplied-air respirator that has a tight-fitting facepiece and is operated in a continuous-flow mode); or PAPRTHiE (any powered, air-purifying respirator with a tight-fitting facepiece and a high-efficiency particulate filter); or SCBAF (any self-contained breathing apparatus with a full facepiece); or SAF (any supplied-air respirator with a full facepiece). *50 mg/m^3:* SA:PD,PP (any supplied-air respirator operated in a pressure-demand or other positive-pressure mode). *100 mg/m^3:* SAF:PD,PP (any supplied-air respirator that has a full facepiece and is operated in a pressure-demand or other positive-pressure mode). *EMERGENCY OR PLANNED ENTRY INTO UNKNOWN CONCENTRATIONS OR IDLH CONDITIONS:* SCBAF:PD,PP (any self-contained breathing apparatus that has a full facepiece and is operated in a pressure-demand or other positive-pressure mode); or SAF:PD,PP:ASCBA (any supplied-air respirator that has a full facepiece and is operated in a pressure-demand or other positive-pressure mode in combination with an auxiliary self-contained

breathing apparatus operated in a pressure-demand or other positive pressure mode). *ESCAPE:* HiEF (any air-purifying, full-facepiece respirator with a high-efficiency particulate filter); or SCBAE (any appropriate escape-type, self-contained breathing apparatus).
Exposure limits (TWA unless otherwise noted): ACGIH TLV 0.05 mg/m³ as inorganic lead; OSHA PEL 0.05 mg/m³; NIOSH REL 0.100 mg/m³ as lead.
Toxicity by ingestion: Grade 2; LD_{50} = 0.5 to 5 g/kg.
Long-term health effects: IARC possible carcinogen, rating 2-B
IDLH value: 100 mg/m³; OSHA; as lead.

CHEMICAL REACTIVITY
Binary reactants: Solution is acidic and will corrode most metals.
Neutralizing agents for acids and caustics: Flush with water, rinse with dilute solution of sodium bicarbonate or soda ash.

ENVIRONMENTAL DATA
Food chain concentration potential: Fish, waterfowl, and terrestrial animals are capable of concentrating lead. Certain fish species such as catfish, salmon, trout, and minnows are especially sensitive.
Water pollution: DOT Appendix B, §172.101–marine pollutant. Effect of low concentrations on aquatic life is unknown. May be dangerous if it enters nearby water intakes; notify operators. Notify local health and wildlife officials. **Response to discharge:** Issue warning–water contaminant. Restrict access. Disperse and flush.

SHIPPING INFORMATION
Grades of purity: 50-62% solutions in water; **Storage temperature:** Ambient; **Inert atmosphere:** None; **Venting:** Open; **Stability during transport:** Stable.

PHYSICAL AND CHEMICAL PROPERTIES
Physical state @ 59°F/15°C and 1 atm: Liquid.
Molecular weight: 380.81
Specific gravity (water = 1): 1.75 @ 68°F/20°C (liquid).

LEAD IODIDE   REC. L:1100

SYNONYMS: YODURO de PLOMO (Spanish)

IDENTIFICATION
CAS Number: 10101-63-0
Formula: PbI2
DOT ID Number: UN 2291; DOT Guide Number: 151
Proper Shipping Name: Lead compound, soluble, n.o.s.
Reportable Quantity (RQ): **(CERCLA)** 10 lb/4.54 kg

DESCRIPTION: Bright yellow solid. Odorless. Sinks in water; soluble in warm water.

Poison! • Breathing the dust or swallowing the material can cause illness • Firefighting gear (including SCBA) does not provide adequate protection. If exposure occurs, remove and isolate gear immediately and thoroughly decontaminate personnel • Toxic products of combustion may include metal fumes of lead and iodine.

Hazard Classification (based on NFPA-704 Rating System)
Health Hazards (Blue): 1; Flammability (Red): 0; Reactivity (Yellow): 0

EMERGENCY RESPONSE: See Appendix A (151)
Evacuation:
Public safety: Isolate the area of spill or leak for at least 25 to 50 meters (80 to 160 feet) in all directions.
Spill: Increase, in the downwind direction, as necessary, the distance shown under "Public Safety."
Fire: If tank, rail car, or tank truck is involved in fire, isolate for at least 800 meters (½ mile) in all directions; also, consider initial evacuation for 800 meters (½ mile) in all directions.

EXPOSURE
Short-term effects: *SEEK MEDICAL ATTENTION.* **Dust:** *POISONOUS IF INHALED. IF INHALED,* will, will cause dizziness or loss of consciousness. *IF IN EYES,* hold eyelids open and flush with plenty of water. *IF BREATHING HAS STOPPED,* give artificial respiration; *avoid mouth-to-mouth resuscitation; use bag/mask apparatus.* IF breathing is difficult, administer oxygen. **Solid:** Irritating to skin and eyes. *IF SWALLOWED,* will cause nausea, vomiting, or loss of consciousness. Remove contaminated clothing and shoes. Flush affected areas with plenty of water. *IF IN EYES,* hold eyelids open and flush with plenty of water. *IF SWALLOWED* and victim is *CONSCIOUS AND ABLE TO SWALLOW,* have victim drink 4 to 8 ounces of water and have victim induce vomiting. *IF SWALLOWED* and victim is *UNCONSCIOUS OR HAVING CONVULSIONS,* do nothing except keep victim warm.
Note to physician or authorized medical personnel: Atropine sulfate and other antispasmodics may relieve abdominal pain, but morphine may be necessary.

HEALTH HAZARDS
Personal protective equipment (PPE): B-Level PPE. OSHA Table Z-1-A air contaminant. Dust mask and protective gloves.
Recommendations for respirator selection: OSHA as lead.
0.5 mg/m³: HiE (any air-purifying, respirator with a high-efficiency particulate filter); SA (any supplied-air respirator). *1.25 mg/m³:* SA:CF (any supplied-air respirator operated in a continuous-flow mode); or PAPRHiE (any powered, air-purifying respirator with a high-efficiency particulate filter). *2.5 mg/m³:* HiEF (any air-purifying, full-facepiece respirator with a high-efficiency particulate filter); or SAT:CF (any supplied-air respirator that has a tight-fitting facepiece and is operated in a continuous-flow mode); or PAPRTHiE (any powered, air-purifying respirator with a tight-fitting facepiece and a high-efficiency particulate filter); or SCBAF (any self-contained breathing apparatus with a full facepiece); or SAF (any supplied-air respirator with a full facepiece). *50 mg/m³:* SA:PD,PP (any supplied-air respirator operated in a pressure-demand or other positive-pressure mode). *100 mg/m³:* SAF:PD,PP (any supplied-air respirator that has a full facepiece and is operated in a pressure-demand or other positive-pressure mode). *EMERGENCY OR PLANNED ENTRY INTO UNKNOWN CONCENTRATIONS OR IDLH CONDITIONS:* SCBAF:PD,PP (any self-contained breathing apparatus that has a full facepiece and is operated in a pressure-demand or other positive-pressure mode); or SAF:PD,PP:ASCBA (any supplied-air respirator that has a full facepiece and is operated in a pressure-demand or other positive-pressure mode in combination with an auxiliary self-contained breathing apparatus operated in a pressure-demand or other positive pressure mode). *ESCAPE:* HiEF (any air-purifying, full-facepiece respirator with a high-efficiency particulate filter); or SCBAE (any appropriate escape-type, self-contained breathing apparatus).
Exposure limits (TWA unless otherwise noted): ACGIH TLV 0.05 mg/m³ as inorganic lead; OSHA PEL 0.05 mg/m³; NIOSH REL 0.100 mg/m³ as lead.

Toxicity by ingestion: Grade 2; LD_{50} = 0.5 to 5 g/kg.
Long-term health effects: Lead poisoning. IARC possible carcinogen, rating 2B
IDLH value: 100 mg/m^3 as lead.

ENVIRONMENTAL DATA
Food chain concentration potential: Fish, waterfowl, and terrestrial animals are capable of concentrating lead. Certain fish species such as catfish, salmon, trout, and minnows are especially sensitive.
Water pollution: DOT Appendix B, §172.101–marine pollutant. Dangerous to aquatic life in high concentrations. May be dangerous if it enters nearby water intakes; notify operators. Notify local health and wildlife officials. **Response to discharge:** Issue warning–water contaminant. Restrict access. Should be removed. Chemical and physical treatment.

SHIPPING INFORMATION
Grades of purity: 98.5+%; NF; **Storage temperature:** Ambient; **Inert atmosphere:** None; **Venting:** Open; **Stability during transport:** Stable.

PHYSICAL AND CHEMICAL PROPERTIES
Physical state @ 59°F/15°C and 1 atm: Solid.
Molecular weight: 461.03
Specific gravity (water = 1): 6.16 @ 68°F/20°C (solid).
Heat of fusion: 17.9 cal/g.

LEAD IOSULFATE   REC. L:1200

SYNONYMS: HIPOSULFITO de PLOMO (Spanish); LEAD HYPOSULFITE; THIOSULFURIC ACID, LEAD SALT

IDENTIFICATION
CAS Number: 13478-50-7
Formula: PbS_2O_3
DOT ID Number: UN 2291; DOT Guide Number: 151
Proper Shipping Name: Lead compounds, soluble, n.o.s.

DESCRIPTION: White crystalline solid. Sinks and mixes slowly with water.

Poison! • Breathing the dust or swallowing the material can cause illness • Firefighting gear (including SCBA) does not provide adequate protection. If exposure occurs, remove and isolate gear immediately and thoroughly decontaminate personnel • Toxic products of combustion may include metal fumes of lead and sulfur oxides.

Hazard Classification (based on NFPA-704 Rating System)
Health Hazards (Blue): 1; Flammability (Red): 0; Reactivity (Yellow): 0

EMERGENCY RESPONSE: See Appendix A (151)
Evacuation:
Public safety: Isolate the area of spill or leak for at least 25 to 50 meters (80 to 160 feet) in all directions.
Spill: Increase, in the downwind direction, as necessary, the distance shown under "Public Safety."
Fire: If tank, rail car, or tank truck is involved in fire, isolate for at least 800 meters (½ mile) in all directions; also, consider initial evacuation for 800 meters (½ mile) in all directions.

EXPOSURE
Short-term effects: *SEEK MEDICAL ATTENTION*. **Dust:** *POISONOUS IF INHALED*. Move to fresh air. **Solid:** If swallowed, will cause abdominal pain, diarrhea, weakness, nausea, and vomiting. *IF BREATHING HAS STOPPED,* give artificial respiration; *avoid mouth-to-mouth resuscitation; use bag/mask apparatus*. *IF IN EYES*, hold eyelids open and flush with plenty of water. *IF SWALLOWED* and victim is *CONSCIOUS AND ABLE TO SWALLOW*, have victim drink 4 to 8 ounces of water and have victim induce vomiting.

HEALTH HAZARDS
Personal protective equipment (PPE): B-Level PPE. OSHA Table Z-1-A air contaminant. Rubber gloves, safety glasses, respirator.
Recommendations for respirator selection: OSHA as lead.
0.5 mg/m^3: HiE (any air-purifying, respirator with a high-efficiency particulate filter); SA (any supplied-air respirator). *1.25 mg/m^3:* SA:CF (any supplied-air respirator operated in a continuous-flow mode); or PAPRHiE (any powered, air-purifying respirator with a high-efficiency particulate filter). *2.5 mg/m^3:* HiEF (any air-purifying, full-facepiece respirator with a high-efficiency particulate filter); or SAT:CF (any supplied-air respirator that has a tight-fitting facepiece and is operated in a continuous-flow mode); or PAPRTHiE (any powered, air-purifying respirator with a tight-fitting facepiece and a high-efficiency particulate filter); or SCBAF (any self-contained breathing apparatus with a full facepiece); or SAF (any supplied-air respirator with a full facepiece). *50 mg/m^3:* SA:PD,PP (any supplied-air respirator operated in a pressure-demand or other positive-pressure mode). *100 mg/m^3:* SAF:PD,PP (any supplied-air respirator that has a full facepiece and is operated in a pressure-demand or other positive-pressure mode). *EMERGENCY OR PLANNED ENTRY INTO UNKNOWN CONCENTRATIONS OR IDLH CONDITIONS:* SCBAF:PD,PP (any self-contained breathing apparatus that has a full facepiece and is operated in a pressure-demand or other positive-pressure mode); or SAF:PD,PP:ASCBA (any supplied-air respirator that has a full facepiece and is operated in a pressure-demand or other positive-pressure mode in combination with an auxiliary self-contained breathing apparatus operated in a pressure-demand or other positive pressure mode). *ESCAPE:* HiEF (any air-purifying, full-facepiece respirator with a high-efficiency particulate filter); or SCBAE (any appropriate escape-type, self-contained breathing apparatus).
Exposure limits (TWA unless otherwise noted): ACGIH TLV 0.05 mg/m^3 as inorganic lead; OSHA PEL 0.05 mg/m^3; NIOSH REL 0.100 mg/m^3 as lead.
Toxicity by ingestion: Poisonous.
Long-term health effects: Intermittent vomiting, irritability, nervousness, incoordination; vague pains in the arms, legs, joints, and abdomen. Sensory disturbances of extremities, paralysis of extensor muscles of arms and legs with wrist and foot drop. Disturbance of menstrual cycle, and abortion. Periods of stupor or lethargy, encephalopathy (with visual disturbances), elevated blood pressure, papilledema, cranial nerve paralysis, delirium, convulsions, and coma. IARC possible carcinogen, rating 2B.
IDLH value: 100 mg/m^3; OSHA; as lead.

CHEMICAL REACTIVITY
Binary reactants: Violent reaction with aluminum, magnesium.

ENVIRONMENTAL DATA
Food chain concentration potential: Fish, waterfowl, and terrestrial animals are capable of concentrating lead. Certain fish species such as catfish, salmon, trout, and minnows are especially sensitive.

666   Lead nitrate

Water pollution: DOT Appendix B, §172.101–marine pollutant. Harmful to aquatic life in very low concentrations. May be dangerous if it enters nearby water intakes; notify operators. Notify local health and wildlife officials. **Response to discharge:** Issue warning–poison, water contaminant. Restrict access. Should be removed. Chemical and physical treatment.

SHIPPING INFORMATION
Storage temperature: Cool, out of direct rays of sun; **Stability during transport:** Stable.

PHYSICAL AND CHEMICAL PROPERTIES
Physical state @ 59°F/15°C and 1 atm: Solid.
Molecular weight: 319.33
Melting/Freezing point: Decomposes at melting point.
Specific gravity (water = 1): 5.18 at room temperature
Relative vapor density (air = 1): 11.0 (calculated).

LEAD NITRATE   REC. L:1300

SYNONYMS: EEC No. 082-001-00-6; LEAD DINITRATE; LEAD(2+) NITRATE; NITROTO de PLOMO (Spanish); NITRIC ACID, LEAD(2+) SALT; NITRIC ACID, LEAD(II) SALT

IDENTIFICATION
CAS Number: 10099-74-8
Formula: $N_2O_6 \cdot Pb$; $Pb(NO_3)_2$
DOT ID Number: UN 1469; DOT Guide Number: 141
Proper Shipping Name: Lead nitrate
Reportable Quantity (RQ): **(CERCLA)** 10 lb/4.54 kg

DESCRIPTION: White crystalline solid. Odorless. Sinks and mixes with water.

Poison! • Corrosive to the eyes, skin, and respiratory tract; inhalation symptoms may be delayed • Breathing the dust or swallowing the material can cause illness • Firefighting gear (including SCBA) does not provide adequate protection. If exposure occurs, remove and isolate gear immediately and thoroughly decontaminate personnel • Strong oxidizer which may react spontaneously with low flash point organics or reducing agents. Heat forms oxygen; will increase the activity of an existing fire • May cause fire or explosion on contact with combustibles (wood, paper, oil, clothing, etc.) • Toxic products of combustion may include lead oxide and nitrogen oxides.

Hazard Classification (based on NFPA-704 Rating System)
Health Hazards (Blue): 1; Flammability (Red): 0; Reactivity (Yellow): 0; Special Notice (White): OXY

EMERGENCY RESPONSE: See Appendix A (141)
Evacuation:
Public safety: Isolate the area of spill or leak for at least 10 to 25 meters (30 to 80 feet) in all directions.
Spill: Consider initial downwind evacuation for at least 100 meters (330 feet).
Fire: If any large container is involved in fire, isolate for at least 800 meters (½ mile) in all directions; also, consider initial evacuation for 800 meters (½ mile) in all directions.
EXPOSURE
Short-term effects: *SEEK MEDICAL ATTENTION.* **Dust:** *POISONOUS IF INHALED.* Lung edema may develop. *IF INHALED*, will, will cause dizziness or loss of consciousness. *IF IN EYES*, hold eyelids open and flush with plenty of water. *IF BREATHING HAS STOPPED,* give artificial respiration. IF breathing is difficult, administer oxygen. **Solid:** Irritating to skin and eyes. *IF SWALLOWED*, will cause nausea, vomiting, or loss of consciousness. Remove contaminated clothing and shoes. Flush affected areas with plenty of water. *IF IN EYES*, hold eyelids open and flush with plenty of water. *IF SWALLOWED* and victim is *CONSCIOUS AND ABLE TO SWALLOW,* have victim drink 4 to 8 ounces of water and have victim induce vomiting. *IF SWALLOWED* and victim is *UNCONSCIOUS OR HAVING CONVULSIONS,* do nothing except keep victim warm.
Note to physician or authorized medical personnel: Atropine sulfate and other antispasmodics may relieve abdominal pain, but morphine may be necessary.
Note to physician or authorized medical personnel: Medical observation is recommended for 24 to 48 hours after breathing overexposure, as pulmonary edema may be delayed. As first aid for pulmonary edema, consider administering a corticosteroid spray. Cigarette smoking may exacerbate pulmonary injury and should be discouraged for at least 72 hours following exposure.

HEALTH HAZARDS
Personal protective equipment (PPE): B-Level PPE. OSHA Table Z-1-A air contaminant. Dust mask and protective **gloves**.
Recommendations for respirator selection: OSHA as lead.
0.5 mg/m³: HiE (any air-purifying, respirator with a high-efficiency particulate filter); SA (any supplied-air respirator). *1.25 mg/m³*: SA:CF (any supplied-air respirator operated in a continuous-flow mode); or PAPRHiE (any powered, air-purifying respirator with a high-efficiency particulate filter). *2.5 mg/m³*: HiEF (any air-purifying, full-facepiece respirator with a high-efficiency particulate filter); or SAT:CF (any supplied-air respirator that has a tight-fitting facepiece and is operated in a continuous-flow mode); or PAPRTHiE (any powered, air-purifying respirator with a tight-fitting facepiece and a high-efficiency particulate filter); or SCBAF (any self-contained breathing apparatus with a full facepiece); or SAF (any supplied-air respirator with a full facepiece). *50 mg/m³*: SA:PD,PP (any supplied-air respirator operated in a pressure-demand or other positive-pressure mode). *100 mg/m³*: SAF:PD,PP (any supplied-air respirator that has a full facepiece and is operated in a pressure-demand or other positive-pressure mode). *EMERGENCY OR PLANNED ENTRY INTO UNKNOWN CONCENTRATIONS OR IDLH CONDITIONS*: SCBAF:PD,PP (any self-contained breathing apparatus that has a full facepiece and is operated in a pressure-demand or other positive-pressure mode); or SAF:PD,PP:ASCBA (any supplied-air respirator that has a full facepiece and is operated in a pressure-demand or other positive-pressure mode in combination with an auxiliary self-contained breathing apparatus operated in a pressure-demand or other positive pressure mode). *ESCAPE*: HiEF (any air-purifying, full-facepiece respirator with a high-efficiency particulate filter); or SCBAE (any appropriate escape-type, self-contained breathing apparatus).
Exposure limits (TWA unless otherwise noted): ACGIH TLV 0.05 mg/m³ as inorganic lead; OSHA PEL 0.05 mg/m³; NIOSH REL 0.100 mg/m³ as lead.
Toxicity by ingestion: Grade 2; LD_{50} = 0.5 to 5 g/kg.
Long-term health effects: Lead poisoning. Possible blood, nervous system, kidney and liver damage. Possible limb paralysis. IARC possible carcinogen, rating 2B
Vapor (gas) irritant characteristics: May cause respiratory tract and eye irritation. Possible lung edema.
Liquid or solid irritant characteristics: Corrosive to the skin and eyes.
IDLH value: 100 mg/m³; OSHA; as lead.

FIRE DATA
Behavior in fire: Increases the intensity of a fire when in contact with burning material. Use plenty of water to cool containers or spilled material.

CHEMICAL REACTIVITY
Binary reactants: Strong oxidizer; contact with wood, paper and combustibles may cause fire. Violent reaction with reducing materials, carbon, lead hypophosphite, ammonium thiocyanate, potassium acetate.

ENVIRONMENTAL DATA
Food chain concentration potential: Fish, waterfowl, and terrestrial animals are capable of concentrating lead. Certain fish species such as catfish, salmon, trout, and minnows are especially sensitive.
Water pollution: DOT Appendix B, §172.101–marine pollutant. Dangerous to aquatic life in high concentrations. May be dangerous if it enters nearby water intakes; notify operators. Notify local health and wildlife officials. **Response to discharge:** Issue warning–water contaminant, oxidizing material. Restrict access. Disperse and flush.

SHIPPING INFORMATION
Grades of purity: Reagent; Technical, 98+%; **Storage temperature:** Ambient; **Inert atmosphere:** None; **Venting:** Open; **Stability during transport:** Stable.

PHYSICAL AND CHEMICAL PROPERTIES
Physical state @ 59°F/15°C and 1 atm: Solid.
Molecular weight: 331.2
Melting/Freezing point: 554°F/290°C/563°K.
Specific gravity (water = 1): 4.53 @ 68°F/20°C (solid).
Heat of solution: 41 Btu/lb = 23 cal/g = 0.96×10^5 J/kg.

LEAD STEARATE   REC. L:1400

SYNONYMS: BLEISTEARAT (German); ESTEARATO de PLOMO (Spanish); OCTADECANOIC ACID, LEAD SALT; STEARIC ACID, LEAD SALT; NEUTRAL LEAD STEARATE

IDENTIFICATION
CAS Number: 7428-48-0 (II); 52652-59-2 (dibasic); 56189-09-4
Formula: $C_{18}H_{36}O_{2-x}Pb$; $Pb(C_{18}H_{35}O_2)_2$
DOT ID Number: UN 2811; DOT Guide Number: 154
Proper Shipping Name: Poisonous solid, organic, n.o.s.
Reportable Quantity (RQ): **(CERCLA)** 5000 lb/2270 kg

DESCRIPTION: White powder. Slight fatty odor. Sinks in water; insoluble.

Poison! • Breathing the dust, skin or eye contact, or swallowing the material can cause illness • Firefighting gear (including SCBA) does not provide adequate protection. If exposure occurs, remove and isolate gear immediately and thoroughly decontaminate personnel • Dust may explode at high temperatures or source of ignition • Toxic products of combustion may include carbon monoxide and lead oxide.

Hazard Classification (based on NFPA-704 Rating System)
Health Hazards (Blue): 1; Flammability (Red): 0; Reactivity (Yellow): 0

EMERGENCY RESPONSE: See Appendix A (154)
Evacuation:
Public safety: Isolate the area of spill or leak for at least 25 to 50 meters (80 to 160 feet) in all directions.
Spill: Increase, in the downwind direction, as necessary, the distance shown under "Public Safety."
Fire: If tank, rail car, or tank truck is involved in fire, isolate for at least 800 meters (½ mile) in all directions; also, consider initial evacuation for 800 meters (½ mile) in all directions.

EXPOSURE
Short-term effects: *SEEK MEDICAL ATTENTION*. **Dust or solid:** *POISONOUS IF INHALED*. Move to fresh air. Keep victim quiet and warm. *IF BREATHING HAS STOPPED*, give artificial respiration; *avoid mouth-to-mouth resuscitation; use bag/mask apparatus. IF SWALLOWED*, will cause headache, abdominal pain, nausea and vomiting. Absorbed by the skin. Flush affected areas with water. *IF IN EYES*, hold eyelids open and flush with plenty of water. *IF SWALLOWED* and victim is *CONSCIOUS AND ABLE TO SWALLOW*, have victim drink milk or water, and have victim induce vomiting.

HEALTH HAZARDS
Personal protective equipment (PPE): B-Level PPE. OSHA Table Z-1-A air contaminant. Dust and/or fume respirator, gloves, goggles or safety glasses, coveralls, and cap.
Recommendations for respirator selection: OSHA as lead.
0.5 mg/m³: HiE (any air-purifying, respirator with a high-efficiency particulate filter); SA (any supplied-air respirator). *1.25 mg/m³*: SA:CF (any supplied-air respirator operated in a continuous-flow mode); or PAPRHiE (any powered, air-purifying respirator with a high-efficiency particulate filter). *2.5 mg/m³*: HiEF (any air-purifying, full-facepiece respirator with a high-efficiency particulate filter); or SAT:CF (any supplied-air respirator that has a tight-fitting facepiece and is operated in a continuous-flow mode); or PAPRTHiE (any powered, air-purifying respirator with a tight-fitting facepiece and a high-efficiency particulate filter); or SCBAF (any self-contained breathing apparatus with a full facepiece); or SAF (any supplied-air respirator with a full facepiece). *50 mg/m³*: SA:PD,PP (any supplied-air respirator operated in a pressure-demand or other positive-pressure mode). *100 mg/m³*: SAF:PD,PP (any supplied-air respirator that has a full facepiece and is operated in a pressure-demand or other positive-pressure mode). *EMERGENCY OR PLANNED ENTRY INTO UNKNOWN CONCENTRATIONS OR IDLH CONDITIONS*: SCBAF:PD,PP (any self-contained breathing apparatus that has a full facepiece and is operated in a pressure-demand or other positive-pressure mode); or SAF:PD,PP:ASCBA (any supplied-air respirator that has a full facepiece and is operated in a pressure-demand or other positive-pressure mode in combination with an auxiliary self-contained breathing apparatus operated in a pressure-demand or other positive pressure mode). *ESCAPE*: HiEF (any air-purifying, full-facepiece respirator with a high-efficiency particulate filter); or SCBAE (any appropriate escape-type, self-contained breathing apparatus).
and magnesium sulfate (epsom salts).
Exposure limits (TWA unless otherwise noted): OSHA PEL 0.05 mg/m³; NIOSH REL 0.10 mg/m³ as lead.
Toxicity by ingestion or skin absorption: Poison.
Long-term health effects: Possible lead poisoning. Intermittent vomiting, irritability, nervousness, incoordination, vague pains in the arms, legs, joints and abdomen. Sensory disturbances of extremities, paralysis of extensor muscles of arms and legs with wrist and foot drop. Disturbance of menstrual cycle, and abortion. Periods of stupor or lethargy, encephalopathy (with visual

disturbances), elevated blood pressure, papilledema, cranial nerve paralysis delirium, convulsions and coma. IARC possible carcinogen, rating 2B
IDLH value: 100 mg/m^3; OSHA; as lead.

FIRE DATA
Flash point: More than 450°F/232°C (oc).
Behavior in fire: Possibility of explosion exists under dusty conditions.

CHEMICAL REACTIVITY
Binary reactants: Strong oxidizers.

ENVIRONMENTAL DATA
Food chain concentration potential: Fish, waterfowl, and terrestrial animals are capable of concentrating lead. Certain fish species such as catfish, salmon, trout, and minnows are especially sensitive.
Water pollution: DOT Appendix B, §172.101–marine pollutant. Harmful to aquatic life in very low concentrations. May be dangerous if it enters nearby water intakes; notify operators. Notify local health and wildlife officials. **Response to discharge:** Issue warning–water contaminant. Restrict access. Should be removed. Chemical and physical treatment.

SHIPPING INFORMATION
Grades of purity: 29.5% PbO; **Storage temperature:** Cool; **Stability during transport:** Stable.

PHYSICAL AND CHEMICAL PROPERTIES
Physical state @ 59°F/15°C and 1 atm: Solid.
Molecular weight: 774.17

Melting/Freezing point: 240°F/116°C/389°K.
Specific gravity (water = 1): 1.34-1.4
Relative vapor density (air = 1): 26.7 (calculated).

LEAD SULFATE   REC. L:1500

SYNONYMS: ANGLISITE; BLEISULFAT (German); C.I. PIGMENT WHITE 3; FAST WHITE; FREEMANS WHITE LEAD; LANARKITE; LEAD BOTTOMS; LEAD(II) SULFATE; MILK WHITE; MULHOUSE WHITE; SULFATE de PLOMB (French); SULFATO de PLOMO (Spanish); SULFURIC ACID, LEAD(II) SALT; SULFURIC ACID, LEAD(2+) SALT; WHITE LEAD

IDENTIFICATION
CAS Number: 7446-14-2; 1573-98-07(II)
Formula: PbSO$_4$
DOT ID Number: UN 1794; DOT Guide Number: 154
Proper Shipping Name: Lead sulfate with more than 3% free acid
Reportable Quantity (RQ): **(CERCLA)** 10 lb/4.54 kg

DESCRIPTION: White powder. Odorless. Sinks in water.

Poison! • Corrosive • Breathing the dust, skin or eye contact, or swallowing the material can cause illness • Firefighting gear (including SCBA) does not provide adequate protection. If exposure occurs, remove and isolate gear immediately and thoroughly decontaminate personnel • Toxic products of combustion may include lead and oxides of sulfur.

Hazard Classification (based on NFPA-704 Rating System)
Health Hazards (Blue): 4; Flammability (Red): 0; Reactivity (Yellow): 0

EMERGENCY RESPONSE: See Appendix A (154)
Evacuation:
Public safety: Isolate the area of spill or leak for at least 25 to 50 meters (80 to 160 feet) in all directions.
Spill: Increase, in the downwind direction, as necessary, the distance shown under "Public Safety."
Fire: If tank, rail car, or tank truck is involved in fire, isolate for at least 800 meters (½ mile) in all directions; also, consider initial evacuation for 800 meters (½ mile) in all directions.

EXPOSURE
Short-term effects: *SEEK MEDICAL ATTENTION.* **Dust or Solid:** *POISONOUS IF INHALED.* Irritating to eyes. *IF BREATHING HAS STOPPED,* give artificial respiration; *avoid mouth-to-mouth resuscitation; use bag/mask apparatus. IF SWALLOWED,* will cause abdominal pain, nausea, vomiting, headache and muscular weakness. Move to fresh air. *IF IN EYES,* hold eyelids open and flush with plenty of water. Flush affected areas with plenty of water. *IF SWALLOWED* and victim is *CONSCIOUS AND ABLE TO SWALLOW,* have victim drink 4 to 8 ounces of water and induce vomiting.
Note to physician or authorized medical personnel: Medical observation is recommended for 24 to 48 hours after breathing overexposure, as pulmonary edema may be delayed. As first aid for pulmonary edema, consider administering a corticosteroid spray. Cigarette smoking may exacerbate pulmonary injury and should be discouraged for at least 72 hours following exposure.

HEALTH HAZARDS
Personal protective equipment (PPE): B-Level PPE. OSHA Table Z-1-A air contaminant. Wear approved filter mask, rubber gloves, and safety glasses.
Recommendations for respirator selection: OSHA as lead.
0.5 mg/m^3: HiE (any air-purifying, respirator with a high-efficiency particulate filter); SA (any supplied-air respirator). *1.25 mg/m^3*: SA:CF (any supplied-air respirator operated in a continuous-flow mode); or PAPRHiE (any powered, air-purifying respirator with a high-efficiency particulate filter). *2.5 mg/m^3*: HiEF (any air-purifying, full-facepiece respirator with a high-efficiency particulate filter); or SAT:CF (any supplied-air respirator that has a tight-fitting facepiece and is operated in a continuous-flow mode); or PAPRTHiE (any powered, air-purifying respirator with a tight-fitting facepiece and a high-efficiency particulate filter); or SCBAF (any self-contained breathing apparatus with a full facepiece); or SAF (any supplied-air respirator with a full facepiece). *50 mg/m^3*: SA:PD,PP (any supplied-air respirator operated in a pressure-demand or other positive-pressure mode). *100 mg/m^3*: SAF:PD,PP (any supplied-air respirator that has a full facepiece and is operated in a pressure-demand or other positive-pressure mode). *EMERGENCY OR PLANNED ENTRY INTO UNKNOWN CONCENTRATIONS OR IDLH CONDITIONS*: SCBAF:PD,PP (any self-contained breathing apparatus that has a full facepiece and is operated in a pressure-demand or other positive-pressure mode); or SAF:PD,PP:ASCBA (any supplied-air respirator that has a full facepiece and is operated in a pressure-demand or other positive-pressure mode in combination with an auxiliary self-contained breathing apparatus operated in a pressure-demand or other positive pressure mode). *ESCAPE*: HiEF (any air-purifying, full-facepiece respirator with a high-efficiency particulate filter); or SCBAE (any appropriate escape-type, self-contained breathing apparatus).

Exposure limits (TWA unless otherwise noted): ACGIH TLV 0.05 mg/m³ as inorganic lead; OSHA PEL 0.05 mg/m³; NIOSH REL 0.100 mg/m³ as lead.
Toxicity by ingestion: Poison. Grade 3 LD_{50} = 50 to 500 mg/kg.
Long-term health effects: possible lead poisoning. Intermittent vomiting, irritability, nervousness, incoordination, vague pains in the arms, legs, joints, and abdomen. Sensory disturbances of extremities, paralysis of extensor muscles of arms and legs with wrist and foot drop. Disturbance of menstrual cycle, and abortion. Periods of stupor or lethargy, encephalopathy (with visual disturbances), elevated blood pressure, papilledema, cranial nerve paralysis, delirium, convulsions, and coma. IARC possible carcinogen, rating 2B
Vapor (gas) irritant characteristics: May irritate eyes and respiratory tract.
Liquid or solid irritant characteristics: Corrosive irritant to eyes, skin, and respiratory tract.
IDLH value: 100 mg/m³; OSHA; as lead.

CHEMICAL REACTIVITY
Binary reactants: Violent reaction with aluminum, magnesium. Vigorous reaction with potassium.

ENVIRONMENTAL DATA
Food chain concentration potential: Fish, waterfowl, and terrestrial animals are capable of concentrating lead. Certain fish species such as catfish, salmon, trout, and minnows are especially sensitive.
Water pollution: DOT Appendix B, §172.101–marine pollutant. Harmful to aquatic life in very low concentrations. May be dangerous if it enters nearby water intakes; notify operators. Notify local health and wildlife officials. **Response to discharge:** Issue warning–water contaminant, corrosive. Restrict access. Should be removed. Chemical and physical treatment.

SHIPPING INFORMATION
Grades of purity: 83.2–88.7% PbO

PHYSICAL AND CHEMICAL PROPERTIES
Physical state @ 59°F/15°C and 1 atm: Solid.
Molecular weight: 303.28
Melting/Freezing point: 2138°F/1170°C/1443°K.
Specific gravity (water = 1): 6.2 at room temperature
Relative vapor density (air = 1): 10.46 (calculated).
Heat of fusion: 31.6 cal/g.

LEAD SULFIDE   REC. L:1600

SYNONYMS: C.I. 77640; GALENA; NATURAL LEAD SULFIDE; PLUMBOUS SULFIDE; SULFURO de PLOMO (Spanish)

IDENTIFICATION
CAS Number: 1314-87-0
Formula: PbS
DOT ID Number: UN 2811; DOT Guide Number: 154
Proper Shipping Name: Poisonous solid, n.o.s.
Reportable Quantity (RQ): **(CERCLA)** 10 lb/4.54 kg

DESCRIPTION: Black or silver-gray metallic crystals or powder. Sinks in water; practically insoluble.

Poison! • Corrosive • Breathing the dust, skin or eye contact, or swallowing the material can cause illness • Firefighting gear (including SCBA) does not provide adequate protection. If exposure occurs, remove and isolate gear immediately and thoroughly decontaminate personnel • Toxic products of combustion may include lead and oxides of sulfur.

Hazard Classification (based on NFPA-704 Rating System)
Health Hazards (Blue): 1; Flammability (Red): 0; Reactivity (Yellow): 0

EMERGENCY RESPONSE: See Appendix A (154)
Evacuation:
Public safety: Isolate the area of spill or leak for at least 25 to 50 meters (80 to 160 feet) in all directions.
Spill: Increase, in the downwind direction, as necessary, the distance shown under "Public Safety."
Fire: If tank, rail car, or tank truck is involved in fire, isolate for at least 800 meters (½ mile) in all directions; also, consider initial evacuation for 800 meters (½ mile) in all directions.

EXPOSURE
Short-term effects: *SEEK MEDICAL ATTENTION.* **Dust:** *POISONOUS IF INHALED.* Irritating to skin and eyes. Move to fresh air. *IF IN EYES*, hold eyelids open and flush with plenty of water. *IF BREATHING HAS STOPPED,* give artificial respiration; *avoid mouth-to-mouth resuscitation; use bag/mask apparatus.* IF breathing is difficult, administer oxygen. **Solid:** *POISONOUS IF SWALLOWED.* Flush affected areas with plenty of water. *IF SWALLOWED* and victim is *CONSCIOUS AND ABLE TO SWALLOW*, have victim drink 4 to 8 ounces of water and have victim induce vomiting. *IF SWALLOWED* and victim is *UNCONSCIOUS OR HAVING CONVULSIONS*, do nothing except keep victim warm.
Note to physician or authorized medical personnel: Medical observation is recommended for 24 to 48 hours after breathing overexposure, as pulmonary edema may be delayed. As first aid for pulmonary edema, consider administering a corticosteroid spray. Cigarette smoking may exacerbate pulmonary injury and should be discouraged for at least 72 hours following exposure.

HEALTH HAZARDS
Personal protective equipment (PPE): B-Level PPE. OSHA Table Z-1-A air contaminant. Protective clothing, rubber gloves, safety goggles, or face mask and an approved respirator.
Recommendations for respirator selection: OSHA as lead.
0.5 mg/m³: HiE (any air-purifying, respirator with a high-efficiency particulate filter); SA (any supplied-air respirator). *1.25 mg/m³:* SA:CF (any supplied-air respirator operated in a continuous-flow mode); or PAPRHiE (any powered, air-purifying respirator with a high-efficiency particulate filter). *2.5 mg/m³:* HiEF (any air-purifying, full-facepiece respirator with a high-efficiency particulate filter); or SAT:CF (any supplied-air respirator that has a tight-fitting facepiece and is operated in a continuous-flow mode); or PAPRTHiE (any powered, air-purifying respirator with a tight-fitting facepiece and a high-efficiency particulate filter); or SCBAF (any self-contained breathing apparatus with a full facepiece); or SAF (any supplied-air respirator with a full facepiece). *50 mg/m³:* SA:PD,PP (any supplied-air respirator operated in a pressure-demand or other positive-pressure mode). *100 mg/m³:* SAF:PD,PP (any supplied-air respirator that has a full facepiece and is operated in a pressure-demand or other positive-pressure mode). *EMERGENCY OR PLANNED ENTRY INTO UNKNOWN CONCENTRATIONS OR IDLH CONDITIONS:* SCBAF:PD,PP (any self-contained breathing apparatus that has a full facepiece and is operated in a pressure-demand or other positive-pressure mode); or SAF:PD,PP:ASCBA (any supplied-air respirator that has a full

facepiece and is operated in a pressure-demand or other positive-pressure mode in combination with an auxiliary self-contained breathing apparatus operated in a pressure-demand or other positive pressure mode). *ESCAPE:* HiEF (any air-purifying, full-facepiece respirator with a high-efficiency particulate filter); or SCBAE (any appropriate escape-type, self-contained breathing apparatus).
Exposure limits (TWA unless otherwise noted): ACGIH TLV 0.05 mg/m^3 as inorganic lead; OSHA PEL 0.05 mg/m^3; NIOSH REL 0.100 mg/m^3 as lead.
Toxicity by ingestion: Grade 1. LD$_{50}$ 5 to 15 g/kg.
Long-term health effects: Accumulative poison; repeated exposure can lead to damage to the liver, kidney, blood and nervous system. A suspected carcinogen of the lungs and kidney. Some evidence of teratogenic effects in laboratory animals. IARC possible carcinogen, rating 2B
IDLH value: 100 mg/m^3; OSHA; as lead.

CHEMICAL REACTIVITY
Binary reactants: Violent reaction with iodine monochloride; hydrogen peroxide.

ENVIRONMENTAL DATA
Food chain concentration potential: Fish, waterfowl, and terrestrial animals are capable of concentrating lead. Certain fish species such as catfish, salmon, trout, and minnows are especially sensitive.
Water pollution: DOT Appendix B, §172.101–marine pollutant. Dangerous to aquatic life in high concentrations. May be dangerous if it enters nearby water intakes; notify operators. Notify local health and wildlife officials. **Response to discharge:** Issue warning–water pollutant. Restrict access. Should be removed. Chemical and physical treatment.

SHIPPING INFORMATION
Stability during transport: Stable.

PHYSICAL AND CHEMICAL PROPERTIES
Physical state @ 59°F/15°C and 1 atm: Solid.
Molecular weight: 239.27.
Boiling point @ 1 atm: 2338°F/1281°C/1554°K.
Melting/Freezing point: 2037°F/1114°C/1387°K.
Specific gravity (water = 1): 7.5 @ 68°F/20°C.
Relative vapor density (air = 1): 8.25 (calculated).
Heat of fusion: 17.3 cal/g.

LEAD TETRAACETATE   REC. L:1700

SYNONYMS: LEAD(IV) ACETATE; LEAD(4+) ACETATE; TETRAACETATO de PLOMO (Spanish)

IDENTIFICATION
CAS Number: 546-67-8
Formula: Pb(C$_2$H$_3$O$_2$)$_4$·CH$_3$COOH
DOT ID Number: UN 2291; DOT Guide Number: 151
Proper Shipping Name: Lead compounds, soluble, n.o.s.
Reportable Quantity (RQ): **(CERCLA)** 10 lb/4.54 kg

DESCRIPTION: Light pink, wet (with glacial acetic acid) crystals. Vinegar-like odor; like acetic acid. Reacts with water.

Poison! • Breathing the dust, skin or eye contact, or swallowing the material can cause illness • Firefighting gear (including SCBA) does not provide adequate protection. If exposure occurs, remove and isolate gear immediately and thoroughly decontaminate personnel • Toxic products of combustion may include carbon monoxide and lead fume.

Hazard Classification (based on NFPA-704 Rating System)
Health Hazards (Blue): 2; Flammability (Red): 0; Reactivity (Yellow): 1

EMERGENCY RESPONSE: See Appendix A (151)
Evacuation:
Public safety: Isolate the area of spill or leak for at least 25 to 50 meters (80 to 160 feet) in all directions.
Spill: Increase, in the downwind direction, as necessary, the distance shown under "Public Safety."
Fire: If tank, rail car, or tank truck is involved in fire, isolate for at least 800 meters (½ mile) in all directions; also, consider initial evacuation for 800 meters (½ mile) in all directions.

EXPOSURE
Short-term effects: *SEEK MEDICAL ATTENTION.* **Liquid/solid:** Irritating to skin and eyes. *IF SWALLOWED,* will cause nausea, vomiting, or loss of consciousness. Remove contaminated clothing and shoes. Flush affected areas with plenty of water. *IF BREATHING HAS STOPPED,* give artificial respiration; *avoid mouth-to-mouth resuscitation; use bag/mask apparatus. IF IN EYES,* hold eyelids open and flush with plenty of water. IF swallowed and victim is *CONSCIOUS AND ABLE TO SWALLOW*, have victim drink 4 to 8 ounces of water and have victim induce vomiting. *IF SWALLOWED* and victim is *UNCONSCIOUS OR HAVING CONVULSIONS,* do nothing except keep victim warm.

HEALTH HAZARDS
Personal protective equipment (PPE): B-Level PPE. OSHA Table Z-1-A air contaminant. Goggles or face shield; rubber **gloves**.
Recommendations for respirator selection: OSHA as lead.
0.5 mg/m^3: HiE (any air-purifying, respirator with a high-efficiency particulate filter); SA (any supplied-air respirator). *1.25 mg/m^3:* SA:CF (any supplied-air respirator operated in a continuous-flow mode); or PAPRHiE (any powered, air-purifying respirator with a high-efficiency particulate filter). *2.5 mg/m^3:* HiEF (any air-purifying, full-facepiece respirator with a high-efficiency particulate filter); or SAT:CF (any supplied-air respirator that has a tight-fitting facepiece and is operated in a continuous-flow mode); or PAPRTHiE (any powered, air-purifying respirator with a tight-fitting facepiece and a high-efficiency particulate filter); or SCBAF (any self-contained breathing apparatus with a full facepiece); or SAF (any supplied-air respirator with a full facepiece). *50 mg/m^3:* SA:PD,PP (any supplied-air respirator operated in a pressure-demand or other positive-pressure mode). *100 mg/m^3:* SAF:PD,PP (any supplied-air respirator that has a full facepiece and is operated in a pressure-demand or other positive-pressure mode). *EMERGENCY OR PLANNED ENTRY INTO UNKNOWN CONCENTRATIONS OR IDLH CONDITIONS:* SCBAF:PD,PP (any self-contained breathing apparatus that has a full facepiece and is operated in a pressure-demand or other positive-pressure mode); or SAF:PD,PP:ASCBA (any supplied-air respirator that has a full facepiece and is operated in a pressure-demand or other positive-pressure mode in combination with an auxiliary self-contained breathing apparatus operated in a pressure-demand or other positive pressure mode). *ESCAPE:* HiEF (any air-purifying, full-facepiece respirator with a high-efficiency particulate filter); or SCBAE (any appropriate escape-type, self-contained breathing apparatus).
Exposure limits (TWA unless otherwise noted): OSHA PEL 0.05 mg/m^3; NIOSH REL 0.10 mg/m^3 as lead.
Toxicity by ingestion: Grade 2; LD$_{50}$ = 0.5 to 5 g/kg.

Long-term health effects: Lead poisoning. IARC possible carcinogen, rating 2B
IDLH value: 100 mg/m^3; OSHA; as lead.

CHEMICAL REACTIVITY
Reactivity with water: Forms lead dioxide and acetic acid in a reaction that is may not be violent.
Binary reactants: May corrode metals when moist.
Neutralizing agents for acids and caustics: Dilute with water, rinse with dilute sodium bicarbonate or lime solution.

ENVIRONMENTAL DATA
Food chain concentration potential: Fish, waterfowl, and terrestrial animals are capable of concentrating lead. Certain fish species such as catfish, salmon, trout, and minnows are especially sensitive.
Water pollution: DOT Appendix B, §172.101–marine pollutant. Effect of low concentrations on aquatic life is unknown. May be dangerous if it enters nearby water intakes; notify operators. Notify local health and wildlife officials. **Response to discharge:** Issue warning–oxidizing material, water contaminant. Restrict access. Disperse and flush.

SHIPPING INFORMATION
Grades of purity: Commercial, 80-90%; **Storage temperature:** Ambient; **Inert atmosphere:** None; **Venting:** Open; **Stability during transport:** Stable.

PHYSICAL AND CHEMICAL PROPERTIES
Physical state @ 59°F/15°C and 1 atm: Solid.
Molecular weight: 443.39

Melting/Freezing point: 347°F/175°C/448°K.
Specific gravity (water = 1): 2.2 @ 68°F/20°C (solid).

LEAD THIOCYANATE REC. L:1800

SYNONYMS: LEAD SULFOCYANATE; TIOCIANATO de PLOMO (Spanish)

IDENTIFICATION
CAS Number: 592-87-0
Formula: $C_2N_2PbS_2$; $Pb(SCN)_2$
DOT ID Number: UN 2291; DOT Guide Number: 151
Proper Shipping Name: Lead compounds, soluble, n.o.s.
Reportable Quantity (RQ): **(CERCLA)** 10 lb/4.54 kg

DESCRIPTION: White solid. Odorless. Sinks and mixes with water; decomposes in warm water.

Poison! • Breathing the dust, skin or eye contact, or swallowing the material can cause illness and possible death • Firefighting gear (including SCBA) does not provide adequate protection. If exposure occurs, remove and isolate gear immediately and thoroughly decontaminate personnel • Containers may explode in elevated temperatures• Toxic products of combustion may include lead metal fumes, cyanide fumes, and irritating sulfur dioxide gas.

Hazard Classification (based on NFPA-704 Rating System)
Health Hazards (Blue): 1; Flammability (Red): 0; Reactivity (Yellow): 0

EMERGENCY RESPONSE: See Appendix A (151)
Evacuation:
Public safety: Isolate the area of spill or leak for at least 25 to 50 meters (80 to 160 feet) in all directions.
Spill: Increase, in the downwind direction, as necessary, the distance shown under "Public Safety."
Fire: If tank, rail car, or tank truck is involved in fire, isolate for at least 800 meters (½ mile) in all directions; also, consider initial evacuation for 800 meters (½ mile) in all directions.

EXPOSURE
Short-term effects: *SEEK MEDICAL ATTENTION.* **Dust:** *POISONOUS IF INHALED. IF INHALED*, will, will cause dizziness or loss of consciousness. *IF IN EYES*, hold eyelids open and flush with plenty of water. *IF BREATHING HAS STOPPED*, give artificial respiration; *avoid mouth-to-mouth resuscitation; use bag/mask apparatus*. IF breathing is difficult, administer oxygen. **Solid:** Irritating to skin and eyes. *IF SWALLOWED*, will cause nausea, vomiting, or loss of consciousness. Remove contaminated clothing and shoes. Flush affected areas with plenty of water. *IF IN EYES*, hold eyelids open and flush with plenty of water. *IF SWALLOWED* victim is *CONSCIOUS AND ABLE TO SWALLOW*, have victim drink 4 to 8 ounces of water and have victim induce vomiting. *IF SWALLOWED* and victim is *UNCONSCIOUS OR HAVING CONVULSIONS*, do nothing except keep victim warm.
Note to physician or authorized medical personnel: Consider the use of amyl nitrite perles if symptoms of cyanide poisoning develop. If symptoms indicate, initial treatment includes the cyanide antidote kit. In all cases, break an amyl nitrite perle in a gauze pad and hold lightly under victim's nose for 15 seconds, repeating 5 times at about 15-second intervals; if necessary (and if sodium nitrite infusions will be delayed), repeat procedure every 3 minutes with fresh perles until 3 or 4 have been used. Avoid breathing the vapor while administering it to the victim. Administer sodium nitrite IV, ASAP. The usual adult dose is 10 to 20 mL of a 3% solution infused over no less than 5 minutes; the average child dose is 0.15 to 0.20 mL/kg. Monitor blood pressure during administration, and slow the rate of infusion if hypotention develops. Next, infuse sodium thiosulfate IV. The usual adult dose is 50 mL of a 25% solution infused over 10 to 20 minutes; the average child dose is 1.65 mL/kg. Repeat with nitrite and thiosulfate as required.
Note to physician or authorized medical personnel: Atropine sulfate and other antispasmodics may relieve abdominal pain, but morphine may be necessary.

HEALTH HAZARDS
Personal protective equipment (PPE): B-Level PPE. OSHA Table Z-1-A air contaminant.
Recommendations for respirator selection: OSHA as lead.
0.5 mg/m^3: HiE (any air-purifying, respirator with a high-efficiency particulate filter); SA (any supplied-air respirator). *1.25 mg/m^3:* SA:CF (any supplied-air respirator operated in a continuous-flow mode); or PAPRHiE (any powered, air-purifying respirator with a high-efficiency particulate filter). *2.5 mg/m^3*: HiEF (any air-purifying, full-facepiece respirator with a high-efficiency particulate filter); or SAT:CF (any supplied-air respirator that has a tight-fitting facepiece and is operated in a continuous-flow mode); or PAPRTHiE (any powered, air-purifying respirator with a tight-fitting facepiece and a high-efficiency particulate filter); or SCBAF (any self-contained breathing apparatus with a full facepiece); or SAF (any supplied-air respirator with a full facepiece). *50 mg/m^3:* SA:PD,PP (any supplied-air respirator operated in a pressure-demand or other positive-pressure mode). *100 mg/m^3:* SAF:PD,PP (any supplied-air respirator that has a full facepiece and is operated in a pressure-demand or other positive-pressure mode). *EMERGENCY OR PLANNED ENTRY INTO UNKNOWN*

672   Lead tungstate

CONCENTRATIONS OR IDLH CONDITIONS: SCBAF:PD,PP (any self-contained breathing apparatus that has a full facepiece and is operated in a pressure-demand or other positive-pressure mode); or SAF:PD,PP:ASCBA (any supplied-air respirator that has a full facepiece and is operated in a pressure-demand or other positive-pressure mode in combination with an auxiliary self-contained breathing apparatus operated in a pressure-demand or other positive pressure mode). *ESCAPE:* HiEF (any air-purifying, full-facepiece respirator with a high-efficiency particulate filter); or SCBAE (any appropriate escape-type, self-contained breathing apparatus).
Exposure limits (TWA unless otherwise noted): OSHA PEL 0.05 mg/m^3; NIOSH REL 0.10 mg/m^3 as lead.
Toxicity by ingestion: Grade 2; LD$_{50}$ = 0.5 to 5 g/kg.
Long-term health effects: Possible poisoning. IARC possible carcinogen, rating 2B
IDLH value: 100 mg/m^3; OSHA; as lead.

FIRE DATA
Behavior in fire: Containers may explode.

CHEMICAL REACTIVITY
Binary reactants: Strong oxidizers, strong acids.

ENVIRONMENTAL DATA
Food chain concentration potential: Fish, waterfowl, and terrestrial animals are capable of concentrating lead. Certain fish species such as catfish, salmon, trout, and minnows are especially sensitive.
Water pollution: DOT Appendix B, §172.101–marine pollutant. Effect of low concentrations on aquatic life is unknown. May be dangerous if it enters nearby water intakes; notify operators. Notify local health and wildlife officials. **Response to discharge:** Issue warning–water contaminant. Runoff from fire control water may cause pollution. Restrict access. Should be removed. Chemical and physical treatment.

SHIPPING INFORMATION
Grades of purity: Practical; Commercial; **Storage temperature:** Ambient; **Inert atmosphere:** None; **Venting:** Open; **Stability during transport:** Stable.

PHYSICAL AND CHEMICAL PROPERTIES
Physical state @ 59°F/15°C and 1 atm: Solid.
Molecular weight: 323.4
Specific gravity (water = 1): 3.82 @ 68°F/20°C (solid).

LEAD TUNGSTATE   REC. L:1900

SYNONYMS: LEAD WOLFRAMATE; RASPITE; SCHEELITE; STOLZITE; TUNGSTATO de PLOMO (Spanish)

IDENTIFICATION
CAS Number: 7759-01-5
Formula: PbWO$_4$
DOT ID Number: UN 3077; DOT Guide Number: 171
Proper Shipping Name: Environmentally hazardous substance, solid, n.o.s.
Reportable Quantity (RQ): **(CERCLA)** 10 lb/4.54 kg

DESCRIPTION: White to pale yellow powder. Sinks in water; insoluble.

Poison! • Breathing the dust, skin or eye contact, or swallowing the material can cause illness • Firefighting gear (including SCBA) does not provide adequate protection. If exposure occurs, remove and isolate gear immediately and thoroughly decontaminate personnel • Toxic products of combustion may include toxic lead metal fumes.

Hazard Classification (based on NFPA-704 Rating System)
Health Hazards (Blue): 1; Flammability (Red): 0; Reactivity (Yellow): 0

EMERGENCY RESPONSE: See Appendix A (171)
Evacuation:
Public safety: Isolate the area of spill or leak for at least 10 to 25 meters (30 to 80 feet) in all directions.
Spill: Increase, in the downwind direction, as necessary, the distance shown under "Public Safety."
Fire: If any large container is involved in fire, isolate for at least 800 meters (½ mile) in all directions; also, consider initial evacuation for 800 meters (½ mile) in all directions.

EXPOSURE
Short-term effects: *SEEK MEDICAL ATTENTION.* **Dust:** *POISONOUS IF INHALED.* Move to fresh air. **Solid:** *IF SWALLOWED,* will cause abdominal pain, diarrhea, weakness, nausea, and vomiting. Flush affected areas with plenty of water. *IF IN EYES,* hold eyelids open and flush with plenty of water. *IF SWALLOWED* and victim is *CONSCIOUS AND ABLE TO SWALLOW,* have victim drink 4 to 8 ounces of water and have victim induce vomiting.

HEALTH HAZARDS
Personal protective equipment (PPE): B-Level PPE. OSHA Table Z-1-A air contaminant. Rubber gloves, safety glasses, respirator.
Recommendations for respirator selection: OSHA as lead.
0.5 mg/m^3: HiE (any air-purifying, respirator with a high-efficiency particulate filter); SA (any supplied-air respirator). *1.25 mg/m^3:* SA:CF (any supplied-air respirator operated in a continuous-flow mode); or PAPRHiE (any powered, air-purifying respirator with a high-efficiency particulate filter). *2.5 mg/m^3:* HiEF (any air-purifying, full-facepiece respirator with a high-efficiency particulate filter); or SAT:CF (any supplied-air respirator that has a tight-fitting facepiece and is operated in a continuous-flow mode); or PAPRTHiE (any powered, air-purifying respirator with a tight-fitting facepiece and a high-efficiency particulate filter); or SCBAF (any self-contained breathing apparatus with a full facepiece); or SAF (any supplied-air respirator with a full facepiece). *50 mg/m^3:* SA:PD,PP (any supplied-air respirator operated in a pressure-demand or other positive-pressure mode). *100 mg/m^3:* SAF:PD,PP (any supplied-air respirator that has a full facepiece and is operated in a pressure-demand or other positive-pressure mode). *EMERGENCY OR PLANNED ENTRY INTO UNKNOWN CONCENTRATIONS OR IDLH CONDITIONS:* SCBAF:PD,PP (any self-contained breathing apparatus that has a full facepiece and is operated in a pressure-demand or other positive-pressure mode); or SAF:PD,PP:ASCBA (any supplied-air respirator that has a full facepiece and is operated in a pressure-demand or other positive-pressure mode in combination with an auxiliary self-contained breathing apparatus operated in a pressure-demand or other positive pressure mode). *ESCAPE:* HiEF (any air-purifying, full-facepiece respirator with a high-efficiency particulate filter); or SCBAE (any appropriate escape-type, self-contained breathing apparatus).
Exposure limits (TWA unless otherwise noted): ACGIH TLV 0.05 mg/m^3 as inorganic lead; OSHA PEL 0.05 mg/m^3; NIOSH REL 0.100 mg/m^3 as lead.

Long-term health effects: Intermittent vomiting, irritability, nervousness, incoordination; vague pains in the arms, legs, joints, and abdomen. Sensory disturbances of extremities, paralysis of extensor muscles of arms and legs with wrist and foot drop. Disturbance of menstrual cycle, and abortion. Periods of stupor or lethargy, encephalopathy (with visual disturbances), elevated blood pressure, papilledema, cranial nerve paralysis, delirium, convulsions, and coma. IARC possible carcinogen, rating 2B
IDLH value: 100 mg/m^3; OSHA; as lead.

ENVIRONMENTAL DATA
Food chain concentration potential: Fish, waterfowl, and terrestrial animals are capable of concentrating lead. Certain fish species such as catfish, salmon, trout, and minnows are especially sensitive.
Water pollution: DOT Appendix B, §172.101–marine pollutant. Harmful to aquatic life in very low concentrations. May be dangerous if it enters nearby water intakes; notify operators. Notify local health and wildlife officials. **Response to discharge:** Issue warning–poison, water contaminant. Restrict access. Should be removed. Chemical and physical treatment.

PHYSICAL AND CHEMICAL PROPERTIES
Physical state @ 59°F/15°C and 1 atm: Solid.
Molecular weight: 455.13
Melting/Freezing point: 2053°F/1123°C/1396°K.
Specific gravity (water = 1): 8.235 at room temperature (Stolzite) 8.46 at room temperature (Raspite).
Relative vapor density (air = 1): 15.7 (calculated).
Heat of fusion: 33.4 cal/g.

LINEAR ALCOHOLS   REC. L:2000

SYNONYMS: ALCOHOL C-12; ALCOHOL LINEAL (Spanish); ALFOL-12; CACALOT L-50; CO 12; CO-1214; *n*-DODECANOL; DODECANOL; 1-DODECANOL; DODECYL ALCOHOL; DYTOL J-68; EPAL 12; LAURIC ALCOHOL; LOROL; MA-1214; MYRISTYL ALCOHOL; MYRISTYL ALCOHOL, mixed isomers; TETRADECYL ALCOHOL PENTADECANOL; TRIDECANOL; TETRADECANOL; TETRADECANOL, mixed isomers

IDENTIFICATION
CAS Number: 112-53-8 (dodecyl alcohol); 112-72-1 (tetradecanol); 27196-00-5 (tetradecanol, mixed isomers)
Formula: $CH_3(CH_2)_{10-13}CH_2OH$
DOT ID Number: NA 1993; DOT Guide Number: 128
Proper Shipping Name: Combustible liquid, n.o.s.

DESCRIPTION: Colorless liquid or solid. Mild floral or alcohol odor. Floats on the surface of water.

Combustible • Containers may BLEVE when exposed to fire •Vapors may form explosive mixture with air • Vapors are heavier than air and will collect and stay in low areas • Vapors may travel long distances to ignition sources and flashback • Vapors in confined areas (e.g., tanks, sewers, buildings) may explode when exposed to fire • Irritating to the skin, eyes, and respiratory tract • Toxic products of combustion may include irritating vapors.

Hazard Classification (based on NFPA-704 Rating System)
Health Hazards (Blue): 0; Flammability (Red): 1; Reactivity (Yellow): 0

EMERGENCY RESPONSE: See Appendix A (128)
Evacuation:
Public safety: Isolate spill area for at least 25 to 50 meters (80 to 160 feet) in all directions.
Spill: Large spill–Consider initial downwind evacuation for at least 300 meters (1000 feet).
Fire: Isolate for 800 meters (½ mile) in all directions, especially if tank, rail car, or tank truck is involved in fire.

EXPOSURE
Short-term effects: *SEEK MEDICAL ATTENTION.* **Liquid:** Irritating to skin. Will burn eyes. Remove contaminated clothing and shoes. Flush affected areas with plenty of water. *IF IN EYES*, hold eyelids open and flush with plenty of water.

HEALTH HAZARDS
Personal protective equipment (PPE): B-Level PPE. Chemical protective material(s) reported to have good to excellent resistance: butyl rubber, natural rubber, neoprene, nitrile.
Toxicity by ingestion: Grade 1; LD_{50} = 5 to 15 g/kg (rat); LD_{50} = 12800 mg/kg (rat).
Long-term health effects: May be carcinogenic.
Liquid or solid irritant characteristics: Possible eye irritation. Severe skin irritant.

FIRE DATA
Flash point: 180°F–285°F/82–141°C (oc) (solid); 259°F/126°C (cc) (liquid).
Fire extinguishing agents not to be used: Water or foam may cause frothing
Autoignition temperature: 525°F/274°C/547°K.

CHEMICAL REACTIVITY
Binary reactants: Reacts with oxidizers, strong acids, caustics, aliphatic amines, and isocyanates.

ENVIRONMENTAL DATA
Food chain concentration potential: Negative; unlikely to accumulate.
Water pollution: Effect of low concentrations on aquatic life is unknown. Fouling to shoreline. May be dangerous if it enters nearby water intakes; notify operators. Notify local health and wildlife officials. **Response to discharge:** Mechanical containment. Should be removed. Chemical and physical treatment.

SHIPPING INFORMATION
Storage temperature: Ambient; **Inert atmosphere:** None; **Venting:** Open; **Stability during transport:** Stable (flame arrester); **Stability during transport:** Stable.

PHYSICAL AND CHEMICAL PROPERTIES
Physical state @ 59°F/15°C and 1 atm: Solid.
Molecular weight: More than 186
Boiling point @ 1 atm: More than 486°F/252°C/525°K.
Melting/Freezing point: More than 66°F/19°C/292°K.
Specific gravity (water = 1): 0.82 (liquid); 0.84 (solid) @ 68°F/20°C.
Liquid surface tension: (estimate) 30 dynes/cm = 0.03 N/m at 30°C.
Liquid water interfacial tension: (estimate) 30 dynes/cm = 0.03 N/m at 30°C.
Relative vapor density (air = 1): 7.4
Heat of combustion: (estimate) –18,500 Btu/lb = 10,300 cal/g = –429 x 10^5 J/kg.
Vapor pressure: 0.01 mm.

LIQUEFIED NATURAL GAS   REC. L:2100

SYNONYMS: GAS, NATURAL; GAS NATURAL LICUADO (Spanish); LNG; NATURAL GAS; NATURAL GAS, REFRIGERATED LIQUID; METHANE, REFRIGERATED Liquid.

IDENTIFICATION
CAS Number: 74-82-8
Formula: CH_4; $CH_4+C_2H_6$
DOT ID Number: UN 1972 (cryogenic); DOT Guide Number: 115
Proper Shipping Name: Liquefied natural gas (cryogenic liquid); LGN (cryogenic liquid); Natural gas, refrigerated liquid (cryogenic liquid).

DESCRIPTION: Colorless gas. Shipped and stored as a compressed gas or cryogenic liquid. Odorless or weak skunk odor. Floats and boils on the surface of water; insoluble. Flammable visible vapor cloud is produced.

Extremely flammable • Forms explosive mixture with air • Reacts explosively with many materials • Gas is lighter than air, and will disperse slowly unless confined • Vapors from liquefied gas are initially heavier than air and spread along ground • Gas may travel to source of ignition and flashback • Exposure of cylinders to elevated temperatures, fire, and flame may cause cylinders to rupture or cause frangible disk to burst, releasing entire contents of cylinder • Ruptured or venting cylinders may rocket through buildings and/or travel a considerable distance Irritating to the skin, eyes, and respiratory tract • Vapors may cause dizziness or asphyxiation without warning • Contact with liquid may cause frostbite. *Warning:* Odor is not a reliable indicator of the presence of toxic amounts of LNG.

Hazard Classification (based on NFPA-704 Rating System)
Health Hazards (Blue): 1; Flammability (Red): 4; Reactivity (Yellow): 0

EMERGENCY RESPONSE: See Appendix A (115)
Evacuation:
Public safety: Isolate spill area for at least 50 to 100 meters (160 to 330 feet) in all directions.
Spill: Consider initial downwind evacuation for at least 800 meters (½ mile).
Fire: Isolate for 1600 meters (1 mile) in all directions, especially if tank, rail car, or tank truck is involved in fire.

EXPOSURE
Short-term effects: *SEEK MEDICAL ATTENTION.* **Vapor:** Not irritating to eyes, nose or throat. *IF INHALED*, will, will cause dizziness, difficult breathing, or loss of consciousness. Move to fresh air. *IF BREATHING HAS STOPPED*, give artificial respiration. IF breathing is difficult, administer oxygen. **Liquid:** Will cause frostbite. Flush affected areas with plenty of water. *DO NOT RUB AFFECTED AREAS.*

HEALTH HAZARDS
Personal protective equipment (PPE): B-Level PPE. Self-contained breathing apparatus; face shield, protective clothing if exposed to liquid. Wear thermal protective clothing.
Vapor (gas) irritant characteristics: May act as a simple asphyxiant. Vapors are nonirritating to the eyes and throat.
Liquid or solid irritant characteristics: No appreciable hazard. Very volatile and evaporates quickly; may cause some frostbite.

FIRE DATA
Flash point: Flammable gas.
Flammable limits in air: LEL: 5.3%; UEL: 14.0%.
Shut off leak if possible. Extinguish small fires with dry chemicals.
Fire extinguishing agents not to be used: Water.
Autoignition temperature: 900°F/482°C/755°K.
Electrical hazard: Class I, Group D; Flow or agitation of substance may generate electrostatic charges due to low conductivity.
Burning rate: 12.5 mm/min
Adiabatic flame temperature: 2339°F/1282°C (estimate).
Stoichiometric air-to-fuel ratio: 17.16 (estimate).

CHEMICAL REACTIVITY
Binary reactants: Possible reaction with oxidizing materials.
Reactivity group: 31
Compatibility class: Paraffins

ENVIRONMENTAL DATA
Food chain concentration potential: Negative; unlikely to accumulate.
Water pollution: Not harmful to aquatic life. **Response to discharge:** Issue warning–high flammability. Restrict access. Evacuate area.

SHIPPING INFORMATION
Grades of purity: Varies with the point of origin. Usually contains at least 90% methane, with smaller quantities of ethane, propane, butanes and pentanes, CO_2 and nitrogen; **Storage temperature:** –260°F/–162°C; **Inert atmosphere:** None; **Venting:** Safety relief; **Stability during transport:** Stable.

NAS HAZARD CLASSIFICATION FOR BULK WATER TRANSPORTATION
FIRE: 4
HEALTH: Vapor irritant: 0; Liquid or solid irritant: 0; Poisons: 0
WATER POLLUTION: Human toxicity: 0; Aquatic toxicity: 0; Aesthetic effect: 0
REACTIVITY: Other chemicals: 0; Water: 0; Self-reaction: 0

PHYSICAL AND CHEMICAL PROPERTIES
Physical state @ 59°F/15°C and 1 atm: Gas.
Molecular weight: More than 16
Boiling point @ 1 atm: –258°F/–161°C/112°K.
Melting/Freezing point: –296°F/–182°C/91°K.
Critical temperature: –116°F/–82°C/191°K.
Critical pressure: 673 psia = 45.78 atm = 4.64 MN/m
Specific gravity (water = 1): (liquid) 0.415-0.45 at –162°C.
Liquid surface tension: 14 dynes/cm = 0.014 N/m at –161°C.
Relative vapor density (air = 1): 0.55-1.0
Ratio of specific heats of vapor (gas): 1.306
Latent heat of vaporization: (estimate) 220 Btu/lb = 120 cal/g = 5.1×10^5 J/kg.
Heat of combustion: –21,600 to –23,400 Btu/lb = –12,000 to –13,000 cal/g = –502.4 to –544.3 $\times 10^5$ J/kg.
Vapor pressure: (Reid) High (physical properties apply to methane; no "standard" LNG exists).

LIQUEFIED PETROLEUM GAS   REC. L:2200

SYNONYMS: BOTTLED GAS; COMPRESSED PETROLEUM GAS; GAS de PETROLEO LICUADO (Spanish); Liquefied HYDROCARBON GAS; LPG; PETROLEUM GAS, Liquefied; PROPANE-BUTANE-(PROPYLENE); PYROFAX

IDENTIFICATION
CAS Number: 68476-85-7
Formula: C_3H_6-C_3H_8-C_4H_{10} (mixture)
DOT ID Number: UN 1075; DOT Guide Number: 115
Proper Shipping Name: Liquefied petroleum gas; Petroleum gases, liquefied; LPG

DESCRIPTION: Colorless gas. May be shipped and stored as a compressed liquefied gas. Weak odor. May have a mercaptan or "skunk" fragrance added as a warning odorant. Floats and boils on water; insoluble; flammable vapor cloud is produced.

Extremely flammable • Forms explosive mixture with air • Reacts explosively with many materials • Vapors from liquefied gas are initially heavier than air and spread along ground • Gas may travel to source of ignition and flashback • Exposure of cylinders to elevated temperatures, fire, and flame may cause cylinders to rupture or cause frangible disk to burst, releasing entire contents of cylinder • Ruptured or venting cylinders may rocket through buildings and/or travel a considerable distance •Irritating to the skin, eyes, and respiratory tract • Vapors may cause dizziness or asphyxiation without warning • Contact with liquid may cause frostbite. *Warning:* Odor is not a reliable indicator of the presence of toxic amounts of LPG.

Hazard Classification (based on NFPA-704 Rating System)
Health Hazards (Blue): 1; Flammability (Red): 4; Reactivity (Yellow): 0

EMERGENCY RESPONSE: See Appendix A (115)
Evacuation:
Public safety: Isolate spill area for at least 50 to 100 meters (160 to 330 feet) in all directions.
Spill: Consider initial downwind evacuation for at least 800 meters (½ mile).
Fire: Isolate for 1600 meters (1 mile) in all directions, especially if tank, rail car, or tank truck is involved in fire.

EXPOSURE
Short-term effects: *SEEK MEDICAL ATTENTION.* **Vapor:** Not irritating to eyes, nose, and throat. *IF INHALED*, will, will cause dizziness, difficult breathing, or loss of consciousness. **Liquid:** Will cause frostbite. Move to fresh air. *IF BREATHING HAS STOPPED,* give artificial respiration. IF breathing is difficult, administer oxygen. Flush affected areas with plenty of water. *DO NOT RUB AFFECTED AREAS.*

HEALTH HAZARDS
Personal protective equipment (PPE): B-Level PPE. OSHA Table Z-1-A air contaminant. Self-contained breathing apparatus for high concentrations of gas.
Recommendations for respirator selection: NIOSH/OSHA *2000 ppm:* SA (any supplied-air respirator); or SCBAF (any self-contained breathing apparatus with a full facepiece). *EMERGENCY OR PLANNED ENTRY INTO UNKNOWN CONCENTRATIONS OR IDLH CONDITIONS:* SCBAF:PD,PP (any self-contained breathing apparatus that has a full facepiece and is operated in a pressure-demand or other positive-pressure mode); or SAF:PD,PP:ASCBA (any supplied-air respirator that has a full facepiece and is operated in a pressure-demand or other positive-pressure mode in combination with an auxiliary self-contained breathing apparatus operated in a pressure-demand or other positive pressure mode). *ESCAPE:* SCBAE (any appropriate escape-type, self-contained breathing apparatus).

Exposure limits (TWA unless otherwise noted): ACGIH TLV 1000 ppm (1800 mg/m^3); NIOSH/OSHA 1000 ppm (1800 mg/m^3).
Long-term health effects: Possible nervous system effects.
Vapor (gas) irritant characteristics: May act as a narcotic and simple asphyxiant.
Liquid or solid irritant characteristics: Very volatile and evaporates quickly; may cause frostbite.
Odor threshold: 5000-18,000 ppm.
IDLH value: 2000 ppm [10% LEL].

FIRE DATA
Flash point: Flammable gas (Propane: –156°F/–104°C (cc); Butane: –76°F/–60°C (cc).
Flammable limits in air: LEL: 1.5%; UEL 9.9% (Propane: LEL: 2.2%; UEL: 9.5%; Butane: LEL: 1.8%; UEL: 8.4%).
chemicals. Shut off leak if it can be done safely.
Fire extinguishing agents not to be used: Water (let fire burn).
Autoignition temperature: About 752°F/400°C/673°K; (Propane: 871°F/466°C/739°C; Butane: 761°F/405°C/678°K).
Electrical hazard: Class I, Group D; Flow or agitation of substance may generate electrostatic charges due to low conductivity.
Burning rate: 8.2 mm/min
Adiabatic flame temperature: 2419°F/1326°C/1599°K (estimate).
Stoichiometric air-to-fuel ratio: 17.16 (estimate).

CHEMICAL REACTIVITY
Binary reactants: Strong oxidizers; chlorine dioxide
Reactivity group: 31
Compatibility class: Paraffins or olefins

ENVIRONMENTAL DATA
Water pollution: Not harmful to aquatic life. **Response to discharge:** Issue warning–high flammability. Restrict access. Evacuate area.

SHIPPING INFORMATION
Grades of purity: Various grades, mostly propane. In some areas propylene may be included. The proportion may be varied with the season; **Storage temperature:** Ambient; **Inert atmosphere:** None; **Venting:** Safety relief; **Stability during transport:** Stable.

NAS HAZARD CLASSIFICATION FOR BULK WATER TRANSPORTATION
FIRE: 4
HEALTH: Vapor irritant: 0; Liquid or solid irritant: 0; Poisons: 0
WATER POLLUTION: Human toxicity: 0; Aquatic toxicity: 0; Aesthetic effect: 0
REACTIVITY: Other chemicals: 0; Water: 0; Self-reaction: 0

PHYSICAL AND CHEMICAL PROPERTIES
Physical state @ 59°F/15°C and 1 atm: Gas.
Molecular weight: More than 44
Boiling point @ 1 atm: More than –4°F/–20°C/253°K.
Melting/Freezing point: –256°F/–160°C/113°K (estimate).
Critical temperature: –142°F/–97°C/177°K.
Critical pressure: 616.5 psia = 41.94 atm = 4.249 MN/m^2.
Specific gravity (water = 1): 0.51 0.58 at –50°C (liquid).
Liquid surface tension: 16 dynes/cm = 0.016 N/m at –52.6°F/–47°C.
Liquid water interfacial tension: (estimate) 50 dynes/cm = 0.05 N/m at –38°C.
Relative vapor density (air = 1): 1.5
Ratio of specific heats of vapor (gas): 1.130
Latent heat of vaporization: 183.2 Btu/lb = 101.8 cal/g = 4.262 x 10^5 J/kg.

Heat of combustion: –19.782 Btu/lb = –10,990 cal/g = 460.13 x 10^5 J/kg.
Vapor pressure: (Reid) High (physical properties apply to propane. No "standard" LPG exists).

LITHARGE   REC. L:2300

SYNONYMS: C.I. 77577; C.I. PIGMENT YELLOW 46; LEAD MONOXIDE; LEAD OXIDE; LEAD (II) OXIDE; LEAD (2+) OXIDE; LEAD OXIDE YELLOW; PLUMBOUS OXIDE; LEAD PROTOXIDE; LITARGIRIO (Spanish); LITHARGE YELLOW L-28; MASSICOT; PLUMBOUS OXIDE; YELLOW LEAD OCHER

IDENTIFICATION
CAS Number: 1317-36-8
Formula: PbO
DOT ID Number: UN 3077; DOT Guide Number: 171
Proper Shipping Name: Environmentally hazardous substances, solid, n.o.s.

DESCRIPTION: An inorganic lead oxide compound. Exists in two crystalline forms: Red to reddish-yellow, tetragonal crystals or yellow, orthorhombic crystals. Odorless. Sinks in water; insoluble.

Poison! • Breathing the dust, skin or eye contact, or swallowing the material may cause chronic illness • Firefighting gear (including SCBA) does not provide adequate protection. If exposure occurs, remove and isolate gear immediately and thoroughly decontaminate personnel • Toxic products of combustion may include lead.

Hazard Classification (based on NFPA-704 Rating System)
Health Hazards (Blue): 1; Flammability (Red): 0; Reactivity (Yellow): 0

EMERGENCY RESPONSE: See Appendix A (171)
Evacuation:
Public safety: Isolate the area of spill or leak for at least 10 to 25 meters (30 to 80 feet) in all directions.
Spill: Increase, in the downwind direction, as necessary, the distance shown under "Public Safety."
Fire: If any large container is involved in fire, isolate for at least 800 meters (½ mile) in all directions; also, consider initial evacuation for 800 meters (½ mile) in all directions.

EXPOSURE:

Short-term effects: *SEEK MEDICAL ATTENTION*. **Solid or dust:** Irritating to eyes. Harmful if inhaled. Move victim to fresh air. *IF IN EYES*, hold eyelids open and flush with plenty of water. *IF SWALLOWED* and victim is *CONSCIOUS AND ABLE TO SWALLOW*, have victim drink 4 to 8 ounces of water and have victim induce vomiting. *IF SWALLOWED* and victim is *UNCONSCIOUS OR HAVING CONVULSIONS*, do nothing except keep victim warm.

HEALTH HAZARDS
Personal protective equipment (PPE): B-Level PPE.
Recommendations for respirator selection: OSHA as lead.
0.5 mg/m³: HiE (any air-purifying, respirator with a high-efficiency particulate filter); SA (any supplied-air respirator). *1.25 mg/m³*: SA:CF (any supplied-air respirator operated in a continuous-flow mode); or PAPRHiE (any powered, air-purifying respirator with a high-efficiency particulate filter). *2.5 mg/m³*: HiEF (any air-purifying, full-facepiece respirator with a high-efficiency particulate filter); or SAT:CF (any supplied-air respirator that has a tight-fitting facepiece and is operated in a continuous-flow mode); or PAPRTHiE (any powered, air-purifying respirator with a tight-fitting facepiece and a high-efficiency particulate filter); or SCBAF (any self-contained breathing apparatus with a full facepiece); or SAF (any supplied-air respirator with a full facepiece). *50 mg/m³*: SA:PD,PP (any supplied-air respirator operated in a pressure-demand or other positive-pressure mode). *100 mg/m³*: SAF:PD,PP (any supplied-air respirator that has a full facepiece and is operated in a pressure-demand or other positive-pressure mode). *EMERGENCY OR PLANNED ENTRY INTO UNKNOWN CONCENTRATIONS OR IDLH CONDITIONS*: SCBAF:PD,PP (any self-contained breathing apparatus that has a full facepiece and is operated in a pressure-demand or other positive-pressure mode); or SAF:PD,PP:ASCBA (any supplied-air respirator that has a full facepiece and is operated in a pressure-demand or other positive-pressure mode in combination with an auxiliary self-contained breathing apparatus operated in a pressure-demand or other positive pressure mode). *ESCAPE*: HiEF (any air-purifying, full-facepiece respirator with a high-efficiency particulate filter); or SCBAE (any appropriate escape-type, self-contained breathing apparatus).
Exposure limits (TWA unless otherwise noted): ACGIH TLV 0.05 mg/m³ as inorganic lead; OSHA PEL 0.05 mg/m³; NIOSH REL 0.100 mg/m³ as lead.
Toxicity by ingestion: Poison.
Long-term health effects: Lead poisoning. Impairs development of human fetal connective tissue cells
IDLH value: 100 mg/m³ as lead.

CHEMICAL REACTIVITY
Binary reactants: Strong oxidizers.

ENVIRONMENTAL DATA
Food chain concentration potential: Fish, waterfowl, and terrestrial animals are capable of concentrating lead. Certain fish species such as catfish, salmon, trout, and minnows are especially sensitive.
Water pollution: May be dangerous if it enters nearby water intakes; notify operators. Notify local health and wildlife officials.
Response to discharge: Should be removed. Chemical and physical treatment.

SHIPPING INFORMATION
Grades of purity: Low-metal-content oxides contain 98–99.8%. High metal or battery grades contain 50–95%. Reagent; purified. Most grades available in several particle sizes; **Storage temperature:** Ambient; **Inert atmosphere:** None; **Venting:** Open; **Stability during transport:** Stable; **Stability during transport:** Stable

PHYSICAL AND CHEMICAL PROPERTIES
Physical state @ 59°F/15°C and 1 atm: Solid.
Molecular weight: 223.2
Boiling point @ 1 atm: (decomposes).
Specific gravity (water = 1): 9.5 @ 68°F/20°C (solid).

LITHIUM   REC. L:2400

SYNONYMS: EEC No. 003-001-00-4; LITHIUM METAL; LITHIUM MONOHYDRIDE; LITIO (Spanish)

IDENTIFICATION
CAS Number: 7439-93-2
Formula: Li
DOT ID Number: UN 1415; DOT Guide Number: 138
Proper Shipping Name: Lithium

DESCRIPTION: Silvery white solid; turns yellow upon exposure to moisture. Shipped and stored under inert gas, mineral oil, or kerosene. Odorless. Reacts violently with water; producing flammable hydrogen gas, caustic fumes, and forming a caustic solution.

Very flammable • Do NOT use water, CO_2, bicarbonate, or halon extinguishers • Metal shavings or powder can ignite spontaneously in air • Skin and eye contact causes severe burns and blindness; inhalation symptoms may be delayed • Firefighting gear (including SCBA) does not provide adequate protection. If exposure occurs, remove and isolate gear immediately and thoroughly decontaminate personnel • Containers may BLEVE when exposed to fire •Vapors may form explosive mixture with air • Vapors are heavier than air and will collect and stay in low areas • Vapors may travel long distances to ignition sources and flashback • Vapors in confined areas (e.g., tanks, sewers, buildings) may explode when exposed to fire • Fumes from burning metal are extremely Irritating to the skin, eyes, and respiratory tract • Toxic products of combustion may include strong alkali fumes, lithium oxide, and lithium hydroxide.

Hazard Classification (based on NFPA-704 Rating System)
Health Hazards (Blue): 3; Flammability (Red): 2; Reactivity (Yellow): 2; Special Notice (White): Water reactive

EMERGENCY RESPONSE: See Appendix A (138)
Evacuation:
Public safety: Isolate the area of spill or leak for at least 50 to 100 meters (160 to 330 feet) in all directions.
Spill: Consider initial downwind evacuation for at least 250 meters (800 feet).
Fire: If any large container is involved in fire, isolate for at least 800 meters (½ mile) in all directions; also, consider initial evacuation for 800 meters (½ mile) in all directions.

EXPOSURE
Short-term effects: *SEEK MEDICAL ATTENTION.* **Fumes:** May cause lung edema to develop. **Solid:** Will burn skin and eyes. Harmful if swallowed. Remove contaminated clothing and shoes. Flush affected areas with plenty of water. *IF IN EYES,* hold eyelids open and flush with plenty of water. *IF SWALLOWED* and victim is *CONSCIOUS AND ABLE TO SWALLOW,* have victim drink 4 to 8 ounces of water. *IF SWALLOWED* and victim is *UNCONSCIOUS OR HAVING CONVULSIONS,* do nothing except keep victim warm.
Note to physician or authorized medical personnel: Pulmonary edema may be delayed. Medical observation is recommended for 24 to 48 hours after inhalation overexposure. As first aid for pulmonary edema, a physician or authorized medical personnel may consider administering a corticosteroid spray. Cigarette smoking may exacerbate pulmonary injury and should be discouraged for at least 72 hours following exposure.

HEALTH HAZARDS
Personal protective equipment (PPE): B-Level PPE. Rubber or plastic gloves; face shield; respirator; fire-retardant clothing.
Exposure limits (TWA unless otherwise noted): ACGIH TLV 0.025 mg/m^3 as lithium hydride; NIOSH/OSHA 0.025 mg/m^3 as lithium hydride.
Long-term health effects: Possible kidney damage.
Vapor (gas) irritant characteristics: Vapor may cause shortness of breath, irritation and lung edema.
Liquid or solid irritant characteristics: Corrosive to the eyes, skin, and respiratory tract.
IDLH value: 0.05 mg/m^3 as lithium hydride.

FIRE DATA
Flammable limits in air: (combustible solid) NA.
Fire extinguishing agents not to be used: Water, sand, halogenated hydrocarbons, CO_2, soda acid, or dry chemical.
Behavior in fire: Molten lithium is quite easily ignited and is then difficult to extinguish. Hot or burning lithium will react with all gases except those of the helium-argon group. It also reacts violently with concrete, wood, asphalt, sand, asbestos, and, in fact, nearly everything except metal. Do not apply water to adjacent fires. Hydrogen explosion may result.
Autoignition temperature: 354°F/179°C/452°K.

CHEMICAL REACTIVITY
Reactivity with water: Reacts violently to form flammable hydrogen gas, corrosive fumes and strong caustic solution. Ignition usually occurs.
Binary reactants: Reacts violently with oxidizers, acids and many other compounds including inert materials such as concrete and sand. A dangerous fire and explosive hazard. May ignite combustible materials if they are damp. Finely dispersed material may spontaneously ignite.
Neutralizing agents for acids and caustics: Residues should be flushed with water, then rinsed with dilute acetic acid.

ENVIRONMENTAL DATA
Food chain concentration potWater pollution: Effect of low concentrations on aquatic life is unknown. May be dangerous if it enters nearby water intakes; notify operators. Notify local health and wildlife officials. **Response to discharge:** Issue warning–high flammability, corrosive. Restrict access. Disperse and flush with care.

SHIPPING INFORMATION
Grades of purity: Pure, 99.9%; Powder, shot, wire, ribbon, rod;
Storage temperature: Ambient; **Inert atmosphere:** Inerted;
Venting: Safety relief; **Stability during transport:** Stable, if air and moisture are excluded.

PHYSICAL AND CHEMICAL PROPERTIES
Physical state @ 59°F/15°C and 1 atm: Solid.
Molecular weight: 6.939
Boiling point @ 1 atm: 2453°F/1345°C/1618°K.
Melting/Freezing point: 357°F/181°C/454°K.
Specific gravity (water = 1): 0.53 @ 68°F/20°C (solid).
Heat of combustion: −18,470 Btu/lb = −10,260 cal/g = −429.3 x 10^5 J/kg.
Heat of solution: −31,500 Btu/lb = −17,500 cal/g = −733 x 10^5 J/kg.
Heat of fusion: 158.5 cal/g.
Vapor pressure: 1 mm @ 1340°F

LITHIUM ALUMINUM HYDRIDE   REC. L:2500

SYNONYMS: ALUMINUM LITHIUM HYDRIDE; EEC No. 001-

Lithium aluminum hydride

-002-00-4; HIDRURO de LITIO y ALUMINIO (Spanish); LAH; LITHIUM TETRAHYDROALUMINATE

IDENTIFICATION
CAS Number: 1302-30-3; 16853-85-3
Formula: $AlH_4 \cdot Li$; $LiAlH_4$
DOT ID Number: UN 1410; UN 1411(etereal); DOT Guide Number: 138
Proper Shipping Name: Lithium aluminum hydride; Lithium aluminum hydride, etereal

DESCRIPTION: White crystalline powder. Turns gray on standing. Usually shipped and stored under nitrogen or argon gas. Odorless. Reacts violently with water; flammable hydrogen gas is produced.

Extremely flammable solid • Corrosive • Spontaneously combustible in moist air • Do not use water • A strong reducing agent • Firefighting gear (including SCBA) does not provide adequate protection. If exposure occurs, remove and isolate gear immediately and thoroughly decontaminate personnel • Containers may BLEVE when exposed to fire • Vapors in confined areas (e.g., tanks, sewers, buildings) may explode when exposed to fire • Severely Irritating to the skin, eyes, and respiratory tract; skin and eye contact causes severe burns and blindness • Decomposes at 257°F/125°C to form explosive hydrogen gas, aluminum, and lithium hydride • May re-ignite after fire is extinguished • Do not put yourself in danger by entering a contaminated area to rescue a victim.

Hazard Classification (based on NFPA-704 Rating System)
Health Hazards (Blue): 3; Flammability (Red): 2; Reactivity (Yellow): 2; Special Notice (White): Water reactive

EMERGENCY RESPONSE: See Appendix A (138)
Evacuation:
Public safety: Isolate the area of spill or leak for at least 50 to 100 meters (160 to 330 feet) in all directions.
Spill: Consider initial downwind evacuation for at least 250 meters (800 feet).
Fire: If any large container is involved in fire, isolate for at least 800 meters (½ mile) in all directions; also, consider initial evacuation for 800 meters (½ mile) in all directions.

EXPOSURE
Short-term effects: *SEEK MEDICAL ATTENTION.* **Inhalation:** Affects the nervous syste. Lung edema may develop. **Solid:** Will burn skin and eyes. Harmful if swallowed. Remove contaminated clothing and shoes. Flush affected areas with plenty of water. *IF IN EYES*, hold eyelids open and flush with plenty of water. *IF SWALLOWED* and victim is *CONSCIOUS AND ABLE TO SWALLOW*, have victim drink plenty of water.
Note to physician or authorized medical personnel: Medical observation is recommended for 24 to 48 hours after breathing overexposure, as pulmonary edema may be delayed. As first aid for pulmonary edema, consider administering a corticosteroid spray. Cigarette smoking may exacerbate pulmonary injury and should be discouraged for at least 72 hours following exposure.

HEALTH HAZARDS
Personal protective equipment (PPE): B-Level PPE. Rubberized gloves; full face shield.
Recommendations for respirator selection: NIOSH/OSHA as lithium hydride
0.25 mg/m³: SA (any supplied-air respirator); or SCBA (any self-contained breathing apparatus); or HiE (any air-purifying, respirator with a high-efficiency particulate filter). *0.5 mg/m³:* SA:CF (any supplied-air respirator operated in a continuous-flow mode); or HiEF (any air-purifying, full-facepiece respirator with a high-efficiency particulate filter); or PAPRHiE (any powered, air-purifying respirator with a high-efficiency particulate filter); or SCBAF (any self-contained breathing apparatus with a full facepiece); or SAF (any supplied-air respirator with a full facepiece). *EMERGENCY OR PLANNED ENTRY INTO UNKNOWN CONCENTRATIONS OR IDLH CONDITIONS:* SCBAF:PD,PP (any self-contained breathing apparatus that has a full facepiece and is operated in a pressure-demand or other positive-pressure mode); or SAF:PD,PP:ASCBA (any supplied-air respirator that has a full facepiece and is operated in a pressure-demand or other positive-pressure mode in combination with an auxiliary self-contained breathing apparatus operated in a pressure-demand or other positive pressure mode). *ESCAPE:* HiEF (any air-purifying, full-facepiece respirator with a high-efficiency particulate filter); or SCBAE (any appropriate escape-type, self-contained breathing apparatus). *Note:* Substance reported to cause eye irritation or damage; may require eye protection.
Exposure limits (TWA unless otherwise noted): ACGIH TLV 0.025 mg/m³; OSHA PEL 0.025 mg/m³.
Toxicity by ingestion: Corrosive.
Vapor (gas) irritant characteristics: May irritate eyes and respiratory tract.
Liquid or solid irritant characteristics: Moisture of skin causes caustic burns. Corrosive to skin, eyes and mucous membrane.

FIRE DATA
Fire extinguishing agents not to be used: Do NOT use water, soda acid, CO_2 or dry chemical.
Behavior in fire: Decomposes at about 250°F/121°C forming hydrogen gas. The heat generated may cause ignition and/or explosion.
Electrical hazard: Class II, Group Undesignated

CHEMICAL REACTIVITY
Reactivity with water: Reacts violently with water as a dry solid or when dissolved in ether. The hydrogen produced by the reaction with water is a major hazard and necessitates adequate ventilation.
Binary reactants: Possible spontaneous ignition on contact with air. Reacts violently (risk of explosion) with many materials including oxidizers, acids, alcohols, ethers. Can burn in heated or moist air.

ENVIRONMENTAL DATA
Food chain concentration potential: Negative; unlikely to accumulate.
Water pollution: Effect of low concentrations on aquatic life is unknown. May be dangerous if it enters nearby water intakes; notify operators. Notify local health and wildlife officials. **Response to discharge:** Issue warning–high flammability, corrosive. Restrict access. Disperse and flush with care.

SHIPPING INFORMATION
Grades of purity: 95–98%; **Storage temperature:** Ambient; **Inert atmosphere:** Dry air, nitrogen or argon gas; **Venting:** Store container in well-ventilated area; **Stability during transport:** Normally stable; unstable at high temperatures.

PHYSICAL AND CHEMICAL PROPERTIES
Physical state @ 59°F/15°C and 1 atm: Solid.
Molecular weight: 37.94
Boiling point @ 1 atm: Decomposes.

Melting/Freezing point: 257°F/125°C/398°K (decomposes).
Specific gravity (water = 1): 0.917 @ 15°C (solid).
Heat of decomposition: 125°C.

LITHIUM BICHROMATE
REC. L:2600

SYNONYMS: BICROMATO de LITIO (Spanish); LITHIUM BICHROMATE DIHYDRATE; LITHIUM DICHROMATE

IDENTIFICATION
CAS Number: 13843-81-7
Formula: $Li_2Cr_2O_7 \cdot 2H_2O$
DOT ID Number: UN 9134; DOT Guide Number: 171
Proper Shipping Name: Lithium chromate

DESCRIPTION: Yellowish-red to black brown crystalline solid or powder. Sinks and mixes with water.

Corrosive • Poison! • Breathing the vapor can kill you; skin or eye contact causes severe burns, impaired vision, or blindness • Firefighting gear (including SCBA) does not provide adequate protection. If exposure occurs, remove and isolate gear immediately and thoroughly decontaminate personnel • Containers may BLEVE when exposed to fire • Strong oxidizer which may react spontaneously with low flash point organics or reducing agents. Heat forms oxygen; will increase the activity of an existing fire • May cause fire or explosion on contact with combustibles (wood, paper, oil, clothing, etc.) • Toxic products of combustion may include chromium and lithium metal fume.

Hazard Classification (based on NFPA-704 Rating System)
Health Hazards (Blue): 1; Flammability (Red): 0; Reactivity (Yellow): 1; Special Notice (White): OXY

EMERGENCY RESPONSE: See Appendix A (171)
Evacuation:
Public safety: Isolate the area of spill or leak for at least 10 to 25 meters (30 to 80 feet) in all directions.
Spill: Increase, in the downwind direction, as necessary, the distance shown under "Public Safety."
Fire: If any large container is involved in fire, isolate for at least 800 meters (½ mile) in all directions; also, consider initial evacuation for 800 meters (½ mile) in all directions.

EXPOSURE
Short-term effects: *SEEK MEDICAL ATTENTION*. **Dust:** Irritating to eyes, nose, and throat. *IF INHALED*, will, will cause difficult breathing. Move to fresh air. *IF BREATHING HAS STOPPED*, give artificial respiration. IF breathing is difficult, administer oxygen. **Solid:** Will burn skin and eyes. *IF SWALLOWED* can cause dizziness, nausea, vomiting or coma. Remove contaminated clothing and shoes. Flush affected areas with plenty of water. *IF IN EYES*, hold eyelids open and flush with plenty of water. *IF SWALLOWED* and victim is *CONSCIOUS AND ABLE TO SWALLOW*, have victim drink 4 to 8 ounces of water. **Do NOT induce vomiting.**
Note to physician or authorized medical personnel: Medical observation is recommended for 24 to 48 hours after breathing overexposure, as pulmonary edema may be delayed. As first aid for pulmonary edema, consider administering a corticosteroid spray. Cigarette smoking may exacerbate pulmonary injury and should be discouraged for at least 72 hours following exposure.

HEALTH HAZARDS
Personal protective equipment (PPE): B-Level PPE. Approved dust mask, goggles or face shield, rubber gloves.
Recommendations for respirator selection: NIOSH As chromium(II) compounds
2.5 mg/m³: DM* (any dust and mist respirator). *5 mg/m³*: DMXSQ (any dust and mist respirator except single-use and quarter mask respirators); or SA* (any supplied-air respirator); *12.5 mg/m³*: SA:CF (any supplied-air respirator operated in a continuous-flow mode); or PAPRDM* (any powered, air-purifying respirator with a dust and mist filter). *25 mg/m³*: HiEF (any air-purifying, full-facepiece respirator with a high-efficiency particulate filter); or PAPRTHiE (any powered, air-purifying respirator with a tight-fitting facepiece and a high-efficiency particulate filter); or SCBAF (any self-contained breathing apparatus with a full facepiece); or SAF (any supplied-air respirator with a full facepiece). *250 mg/m³*: (any supplied-air respirator that has a full facepiece and is operated in a pressure-demand or other positive-pressure mode). *EMERGENCY OR PLANNED ENTRY INTO UNKNOWN CONCENTRATIONS OR IDLH CONDITIONS:* SCBAF:PD,PP (any self-contained breathing apparatus that has a full facepiece and is operated in a pressure-demand or other positive-pressure mode); or SAF:PD,PP:ASCBA (Any supplied-air respirator that has a full facepiece and is operated in a pressure-demand or other positive-pressure mode in combination with an auxiliary self-contained breathing apparatus operated in a pressure-demand or other positive pressure mode). *ESCAPE:* HiEF (Any air-purifying, full-facepiece respirator with a high-efficiency particulate filter); or SCBAE (Any appropriate escape-type, self-contained breathing apparatus). *Note:* Substance reported to cause eye irritation or damage; may require eye protection.
Exposure limits (TWA unless otherwise noted): ACGIH TLV 0.05 mg/m³ as water soluble Cr(VI) compounds; OSHA PEL 0.1 mg/m³ as CrO_3; NIOSH 0.001 mg/m³ as chromates; potential human carcinogen; reduce exposure to lowest feasible level.
Toxicity by ingestion: Grade 3; LD_{50} = 50 to 500 mg/kg.
Long-term health effects: A recognized carcinogen of the lungs, nasal cavity, and paranasal sinus. Lithium is teratogenic.
Liquid or solid irritant characteristics: Severe skin irritant; Causes second- and third-degree burns on short contact and is very injurious to the eye.
IDLH value: Potential human carcinogen; 15 mg/m³ as chromium(VI).

FIRE DATA
Behavior in fire: May give off oxygen with supports further combustion.

CHEMICAL REACTIVITY
Reactivity with water: Forms a caustic solution.
Binary reactants: An oxidizer, can react with combustibles. Vigorous reaction with reducing materials.

ENVIRONMENTAL DATA
Food chain concentration potential: Chromium can be accumulated and concentrated in fish.
Water pollution: Dangerous to aquatic life in high concentrations. May be dangerous if it enters nearby water intakes; notify operators. Notify local health and wildlife officials. **Response to discharge:** Issue warning–water contaminant, oxidizing material. Disperse and flush.

PHYSICAL AND CHEMICAL PROPERTIES
Physical state @ 59°F/15°C and 1 atm: Solid.
Molecular weight: 265.93

Boiling point @ 1 atm: Decomposes 369°F/187°C/460°K.
Melting/Freezing point: 230–266°F/110–130°C/283–403°K.
Specific gravity (water = 1): 2.34 at 30°C.

LITHIUM CHROMATE   REC. L:2700

SYNONYMS: CHROMIUM LITHIUM OXIDE; CHROMIUM LITHIUM OXIDE; CHROMIC ACID, DILITHIUM SALT; CROMATO de LITIO (Spanish); DILITHIUM CHROMATE

IDENTIFICATION
CAS Number: 14307-35-8 (Li_2CrO_4)
Formula: $Cr H_2O_4 \cdot 2Li$; $Li_2CrO_4 \cdot 2H_2O$
DOT ID Number: UN 9134; DOT Guide Number: 171
Proper Shipping Name: Lithium chromate
Reportable Quantity (RQ): **(CERCLA)** 10 lb/4.54 kg

DESCRIPTION: Yellow crystalline powder. Mixes with water; soluble; solution may be highly caustic.

Strong oxidizer which may react spontaneously with low flash point organics or reducing agents. Heat forms oxygen; will increase the activity of an existing fire • May cause fire or explosion on contact with combustibles (wood, paper, oil, clothing, etc.) • Irritating to the skin, eyes, and respiratory tract • Toxic products of combustion may include metal fumes and lithium oxide.

Hazard Classification (based on NFPA-704 Rating System)
Health Hazards (Blue): 1; Flammability (Red): 0; Reactivity (Yellow): 1; Special Notice (White): OXY

EMERGENCY RESPONSE: See Appendix A (171)
Evacuation:
Public safety: Isolate the area of spill or leak for at least 10 to 25 meters (30 to 80 feet) in all directions.
Spill: Increase, in the downwind direction, as necessary, the distance shown under "Public Safety."
Fire: If any large container is involved in fire, isolate for at least 800 meters (½ mile) in all directions; also, consider initial evacuation for 800 meters (½ mile) in all directions.

EXPOSURE
Short-term effects: *SEEK MEDICAL ATTENTION*. **Dust:** Irritating to eyes, nose, and throat. *IF INHALED*, will, will cause difficult breathing. Move to fresh air. *IF BREATHING HAS STOPPED*, give artificial respiration. IF breathing is difficult, administer oxygen. **Solid:** Will burn skin and eyes. *IF SWALLOWED* can cause dizziness, nausea, vomiting or coma. Remove contaminated clothing and shoes. Flush affected areas with plenty of water. *IF IN EYES*, hold eyelids open and flush with plenty of water. *IF SWALLOWED* and victim is *CONSCIOUS AND ABLE TO SWALLOW*, have victim drink 4 to 8 ounces of water. **Do NOT induce vomiting.**

HEALTH HAZARDS
Personal protective equipment (PPE): B-Level PPE. Approved filter-type respirator, close-fitting safety goggles, laboratory coat.
Recommendations for respirator selection: NIOSH as soluble chromic salt
2.5 mg/m³: DM* (any dust and mist respirator). *5 mg/m³*: DMXSQ (any dust and mist respirator except single-use and quarter mask respirators); or SA* (any supplied-air respirator); *12.5 mg/m³*: SA:CF (any supplied-air respirator operated in a continuous-flow mode); or PAPRDM* (any powered, air-purifying respirator with a dust and mist filter). *25 mg/m³*: HiEF (any air-purifying, full-facepiece respirator with a high-efficiency particulate filter); or PAPRTHiE (any powered, air-purifying respirator with a tight-fitting facepiece and a high-efficiency particulate filter); or SCBAF (any self-contained breathing apparatus with a full facepiece); or SAF (any supplied-air respirator with a full facepiece). *EMERGENCY OR PLANNED ENTRY INTO UNKNOWN CONCENTRATIONS OR IDLH CONDITIONS:* SCBAF:PD,PP (any self-contained breathing apparatus that has a full facepiece and is operated in a pressure-demand or other positive-pressure mode); or SAF:PD,PP:ASCBA (Any supplied-air respirator that has a full facepiece and is operated in a pressure-demand or other positive-pressure mode in combination with an auxiliary self-contained breathing apparatus operated in a pressure-demand or other positive pressure mode). *ESCAPE:* HiEF (Any air-purifying, full-facepiece respirator with a high-efficiency particulate filter); or SCBAE (Any appropriate escape-type, self-contained breathing apparatus). *Note:* Substance reported to cause eye irritation or damage; may require eye protection.
Exposure limits (TWA unless otherwise noted): ACGIH TLV 0.05 mg/m³ as water soluble Cr(VI) compounds; OSHA PEL 0.1 mg/m³ as CrO_3; NIOSH 0.001 mg/m³ as chromates; potential human carcinogen; reduce exposure to lowest feasible level.
Toxicity by ingestion: Grade 3; LD_{50} = 50 to 500 mg/kg.
Long-term health effects: A recognized carcinogen of the lungs, nasal cavity, and paranasal sinus. Lithium compounds may be teratogenic.
Liquid or solid irritant characteristics: Causes smarting of the skin and first-degree burns on short exposure; may cause second-degree burns on long exposure.
IDLH value: Potential human carcinogen; 15 mg/m³ as chromium(VI).

FIRE DATA
Behavior in fire: Formation of toxic metal fumes and lithium oxide.

CHEMICAL REACTIVITY
Binary reactants: Violent reaction with combustible materials, reducing agents.

ENVIRONMENTAL DATA
Food chain concentration potential: Cr can be accumulated and concentrated in fish.
Water pollution: Dangerous to aquatic life in high concentrations. May be dangerous if it enters nearby water intakes; notify operators. Notify local health and wildlife officials. **Response to discharge:** Disperse and flush.

SHIPPING INFORMATION
Storage temperature: Cool; **Venting:** Well-ventilated.

PHYSICAL AND CHEMICAL PROPERTIES
Physical state @ 59°F/15°C and 1 atm: Solid.
Molecular weight: 165.92
Relative vapor density (air = 1): 5.72 (calculated).

LITHIUM HYDRIDE   REC. L:2800

SYNONYMS: HYDRURE de LITHIUM (French); HYDRURO de LITIO (Spanish); LITHIUM MONOHYDRIDE

Lithium hydride

IDENTIFICATION
CAS Number: 7580-67-8
Formula: LiH
DOT ID Number: UN 1414; DOT Guide Number: 138
Proper Shipping Name: Lithium hydride
Reportable Quantity (RQ): **(EHS)** 1 lb/0.454 kg

DESCRIPTION: Gray or blue crystalline solid or white powder. Stored and transported under nitrogen or argon gas. Odorless. Reacts violently with water; produces a strong caustic solution, flammable hydrogen gas, and lithium hydroxide; ignition may occur, especially with powder.

Flammable • Thermally unstable; decomposes at 752°F/400°C forming explosive hydrogen gas • Corrosive to the skin, eyes, and respiratory tract; skin and eye contact causes severe burns and blindness • Firefighting gear (including SCBA) does not provide adequate protection. If exposure occurs, remove and isolate gear immediately and thoroughly decontaminate personnel • Containers may BLEVE when exposed to fire • Concentrated dust in confined areas (e.g., tanks, sewers, buildings) may explode when exposed to fire • Products of combustion may include alkali fumes and explosive hydrogen gas • Violet reaction with water producing explosive hydrogen gas • May re-ignite after fire is extinguished • Do not put yourself in danger by entering a contaminated area to rescue a victim.

Hazard Classification (based on NFPA-704 Rating System)
Health Hazards (Blue): 3; Flammability (Red): 2; Reactivity (Yellow): 2; Special Notice (White): Water reactive

EMERGENCY RESPONSE: See Appendix A (138)
Evacuation:
Public safety: Isolate the area of spill or leak for at least 50 to 100 meters (160 to 330 feet) in all directions.
Spill: Consider initial downwind evacuation for at least 250 meters (800 feet).
Fire: If any large container is involved in fire, isolate for at least 800 meters (½ mile) in all directions; also, consider initial evacuation for 800 meters (½ mile) in all directions.

EXPOSURE
Short-term effects: *SEEK MEDICAL ATTENTION.* **Dust:** *POISONOUS IF INHALED. IF INHALED,* will, will cause coughing or difficult breathing. *IF IN EYES,* hold eyelids open and flush with plenty of water. *IF BREATHING HAS STOPPED,* give artificial respiration. IF breathing is difficult, administer oxygen.
Solid: Will burn skin and eyes. *IF SWALLOWED,* will cause nausea or loss of consciousness. Remove contaminated clothing and shoes. Flush affected areas with plenty of water. *IF IN EYES,* hold eyelids open and flush with plenty of water. *IF SWALLOWED* and victim is *CONSCIOUS AND ABLE TO SWALLOW,* have victim drink 4 to 8 ounces of water. *IF SWALLOWED* and victim is *UNCONSCIOUS OR HAVING CONVULSIONS,* do nothing except keep victim warm.

HEALTH HAZARDS
Personal protective equipment (PPE): B-Level PPE. OSHA Table Z-1-A air contaminant. Goggles or face shield; rubberized gloves; flame proof outer clothing; respirator; high boots or shoes
Recommendations for respirator selection: NIOSH/OSHA *0.25 mg/m³*: SA (any supplied-air respirator); or SCBA (any self-contained breathing apparatus); or HiE (any air-purifying, respirator with a high-efficiency particulate filter). *0.5 mg/m³*: SA:CF (any supplied-air respirator operated in a continuous-flow mode); or HiEF (any air-purifying, full-facepiece respirator with a high-efficiency particulate filter); or PAPRHiE (any powered, air-purifying respirator with a high-efficiency particulate filter); or SCBAF (any self-contained breathing apparatus with a full facepiece); or SAF (any supplied-air respirator with a full facepiece). *EMERGENCY OR PLANNED ENTRY INTO UNKNOWN CONCENTRATIONS OR IDLH CONDITIONS:* SCBAF:PD,PP (any self-contained breathing apparatus that has a full facepiece and is operated in a pressure-demand or other positive-pressure mode); or SAF:PD,PP:ASCBA (any supplied-air respirator that has a full facepiece and is operated in a pressure-demand or other positive-pressure mode in combination with an auxiliary self-contained breathing apparatus operated in a pressure-demand or other positive pressure mode). *ESCAPE:* HiEF (any air-purifying, full-facepiece respirator with a high-efficiency particulate filter); or SCBAE (any appropriate escape-type, self-contained breathing apparatus). *Note:* Substance reported to cause eye irritation or damage; may require eye protection.
Exposure limits (TWA unless otherwise noted): ACGIH TLV 0.025 mg/m³; OSHA PEL 0.025 mg/m³.
Vapor (gas) irritant characteristics: Moisture causes the formation of corrosive lithium hydroxide making it particularly dangerous to eyes, and respiratory tract and lungs.
IDLH value: 0.5 mg/m³.

FIRE DATA
Flash point: Combustible solid; powder may ignite spontaneously in air.
Fire extinguishing agents not to be used: Never use water, foam, halogenated hydrocarbons, soda acid, dry chemical, or CO_2.
Behavior in fire: Thermally unstable. Decomposes at about 1000°F/538°C, forming explosive hydrogen gas.
Autoignition temperature: 392°F/200°C/473°K.
Electrical hazard: Class II, Group Undesignated.

CHEMICAL REACTIVITY
Reactivity with water: Dangerous when wet.
Binary reactants: May ignite combustible materials if they are damp. Reacts with strong oxidizers, halogenated hydrocarbons; acids, hydrazine. Keep away from liquid oxygen.
Neutralizing agents for acids and caustics: Residues should be washed well with water, then rinsed with dilute acetic acid.

ENVIRONMENTAL DATA
Food chain concentration potential: Negative; unlikely to accumulate.
Water pollution: Effect of low concentrations on aquatic life is unknown. May be dangerous if it enters nearby water intakes; notify operators. Notify local health and wildlife officials. **Response to discharge:** Issue warning–high flammability, corrosive. Restrict access. Disperse and flush with care

SHIPPING INFORMATION
Grades of purity: Commercial, 96.5%; **Storage temperature:** Ambient; **Inert atmosphere:** Inerted; **Venting:** Safety relief; **Stability during transport:** Stable, if air and moisture are excluded.

PHYSICAL AND CHEMICAL PROPERTIES
Physical state @ 59°F/15°C and 1 atm: Solid.
Molecular weight: 7.95
Boiling point @ 1 atm: (decomposes).
Melting/Freezing point: 1256°F/680°C/953°K.
Specific gravity (water = 1): 0.78 @ 68°F/20°C/293°K (solid).
Heat of solution: -7200 Btu/lb $= -4000$ cal/g $= -170 \times 10^5$ J/kg.

MAGNESIUM   REC. M:0100

SYNONYMS: EEC No. 012-002-00-9; MAGNESIO (Spanish)

IDENTIFICATION
CAS Number: 7439-95-4
Formula: Mg
DOT ID Number: UN 1418 (powder); UN 1869 (pellets); UN 2950 (granules, coated); DOT Guide Number: 138
Proper Shipping Name: Magnesium powder or Magnesium alloys, powder (UN 1418); Magnesium or Magnesium alloys, with more than 50% magnesium in pellets, turnings, or ribbons (UN 1869); Magnesium granules, coated

DESCRIPTION: Silvery-white (looks like aluminum, but is much lighter weight) solid. Odorless. Sinks in water; insoluble. Finely divided forms reacts with water producing explosive hydrogen gas.

Combustible solid • Do NOT use water or water-based extinguishers • Dust or other finely divided form (thin sheets, turnings, chips, pellets, or ribbons) are easily ignited and burn with intense white flame • Highly reactive with many materials • Containers may BLEVE when exposed to fire • Concentrated dust in confined areas (e.g., tanks, sewers, buildings) may explode when exposed to fire • Irritating to the skin, eyes, and respiratory tract • Toxic products of combustion may include magnesium metal fume. *Note:* Eye damage can be caused by looking directly at intense white flame.

Hazard Classification (based on NFPA-704M Rating System)
Health Hazards (Blue): 0; Flammability (Red): 1; Reactivity (Yellow): 1; Special Notice (White): Water reactive (finely divided forms)

EMERGENCY RESPONSE: See Appendix A (138)
Evacuation:
Public safety: Isolate the area of spill or leak for at least 50 to 100 meters (160 to 330 feet) in all directions.
Spill: Consider initial downwind evacuation for at least 250 meters (800 feet).
Fire: If any large container is involved in fire, isolate for at least 800 meters (½ mile) in all directions; also, consider initial evacuation for 800 meters (½ mile) in all directions.

EXPOSURE
Short-term effects: *SEEK MEDICAL ATTENTION.* **Solid:** Irritating to eyes; harmful if swallowed. *IF IN EYES*, hold eyelids open and flush with plenty of water. *IF SWALLOWED* and victim is *CONSCIOUS AND ABLE TO SWALLOW*, have victim drink 4 to 8 ounces of water.

HEALTH HAZARDS
Personal protective equipment (PPE): B-Level PPE. Eye and skin protection.
Exposure limits (TWA unless otherwise noted): ACGIH TL 10 gm/m3; OSHA PEL 15 mg/m^3 as magnesium oxide
Toxicity by ingestion: Poison. Oral $LD_{50\,Lo}$ = 230 mg/kg (dog).
Vapor (gas) irritant characteristics: Fumes can cause metal fume fever.
Liquid or solid irritant characteristics: Irritates the eyes and mucous membrane.

FIRE DATA
Fire extinguishing agents not to be used: Water, foam, halogenated agents, CO_2.
Behavior in fire: Forms dense white smoke. Flame is extremely bright white.
Autoignition temperature: 883°F/445°C/718°K.
Electrical hazard: Class II, Group E. Due to low electric conductivity, this substance may generate electrostatic charges as a result of agitation and flow.

CHEMICAL REACTIVITY
Reactivity with water: Can increase fire hazard, producing hydrogen.
Binary reactants: Finely divided form reacts violently with oxidizers. Solids forms react with acids producing explosive hydrogen gas. Reacts with chloromethane, carbonates, sulfates, cyanides, carbon tetrachloride, chlorinated hydrocarbons, oxides, chloroform; acetylene; ethylene oxide.

ENVIRONMENTAL DATA
Water pollution: Effect of low concentrations on aquatic life is unknown. May be dangerous if it enters nearby water intakes; notify operators. Notify local health and wildlife officials. Notify operators of nearby water intakes. **Response to discharge:** Issue warning–high flammability. Should be removed. Chemical and physical treatment.

SHIPPING INFORMATION:
Grades of purity: Pigs, ingots, turnings, sticks: All high purity; **Storage temperature:** Ambient; **Inert atmosphere:** None; **Venting:** Open; **Stability during transport:** Stable (flame arrester); **Stability during transport:** Stable.

PHYSICAL AND CHEMICAL PROPERTIES
Physical state @ 59°F/15°C and 1 atm: Solid.
Molecular weight: 24.3
Boiling point @ 1 atm: 2012°F/1100°C/1373°K.
Melting/Freezing point: 1202°F/650°C/923°K.
Specific gravity (water = 1): 1.74 @ 68°F/20°C (solid).
Relative vapor density (air = 1): 1.7
Heat of combustion: −11,950 Btu/lb = −6650 cal/g = −278 x 10^5 J/kg.
Heat of fusion: 88.9 cal/g.
Vapor pressure: 0.8 mm.

MAGNESIUM NITRATE   REC. M:0150

SYNONYMS: MAGNESIUM(II) NITRATE; MAGNESIUM(2+) NITRATE; HEXAHYDRATE; NITRATO MAGNESICO (Spanish); NITRIC ACID, MAGNESIUM SALT; NITROMAGNESITE

IDENTIFICATION
CAS Number: 10377-60-3; 10213-15-7 (hexahydrate)
Formula: $N_2O_6 \cdot Mg$; $Mg(NO_3)_2 \cdot 6H_2O$
DOT ID Number: UN 1474; DOT Guide Number: 140
Proper Shipping Name: Magnesium nitrate

DESCRIPTION: White crystalline solid. Odorless. Soluble in water.

Strong oxidizer which may react spontaneously with low flash point organics or reducing agents. Heat forms oxygen; will increase the activity of an existing fire • May cause fire or explosion on contact with combustibles (wood, paper, oil, clothing, etc.) • Irritating to the skin, eyes, and respiratory tract • Decomposes above 626°F/330°C. Toxic products of combustion may include nitrogen oxides

Hazard Classification (based on NFPA-704M Rating System)
Health Hazards (Blue): 1; Flammability (Red): 0; Reactivity (Yellow): 0; Special Notice (White): OXY

EMERGENCY RESPONSE: See Appendix A (140)
Evacuation:
Public safety: Isolate the area of spill or leak for at least 10 to 25 meters (30 to 80 feet) in all directions.
Spill: Consider initial downwind evacuation for at least 100 meters (330 feet).
Fire: If any large container is involved in fire, isolate for at least 800 meters (½ mile) in all directions; also, consider initial evacuation for 800 meters (½ mile) in all directions.

EXPOSURE
Short-term effects: *SEEK MEDICAL ATTENTION.* **Dust:** Move victim to fresh air. If not breathing, give artificial respiration. IF breathing is difficult, administer oxygen. **Liquid:** Remove contaminated clothing and shoes. Wash skin with soap and water. *IF IN EYES,* hold eyelids open and flush with plenty of water. *IF SWALLOWED* and victim is *CONSCIOUS AND ABLE TO SWALLOW,* have victim drink 4 to 8 ounces of water and induce vomiting.

HEALTH HAZARDS
Personal protective equipment (PPE): B-Level PPE. Wear protective gloves and clean body-covering clothing. If dust is encountered, use approved respirator.
Toxicity by ingestion: Blood effects.
Long-term health effects: Methemaglobin formation; may effect the blood.
Vapor (gas) irritant characteristics: May cause nose and lung irritation.
Liquid or solid irritant characteristics: Irritates the eyes, skin and mucous membrane. Spills on clothing may cause smarting and reddening of skin.

CHEMICAL REACTIVITY
Binary reactants: Violent reaction with dimethylformamide, combustibles, reducing materials, and oxidizable materials.

ENVIRONMENTAL DATA
Water pollution: Effects of low concentrations on aquatic life is unknown. May be dangerous if it enters nearby water intakes; notify operators. Notify local health and wildlife officials.
Response to discharge: Issue warning-oxidizer. Should be removed. Chemical and physical treatment.

SHIPPING INFORMATION
Grades of purity: Technical; 98%; **Storage temperature:** Ambient; **Inert atmosphere:** None required; **Stability during transport:** Stable

PHYSICAL AND CHEMICAL PROPERTIES
Physical state @ 59°F/15°C and 1 atm: Solid.
Molecular weight: 256.41
Boiling point @ 1 atm: Decomposes at 626°F/330°C/603°K.
Melting/Freezing point: 192°F/89°C/362°K.
Specific gravity (water = 1): 1.46
Heat of decomposition: 626°F

MAGNESIUM PERCHLORATE   REC. M:0200

SYNONYMS: ANHYDRONE; PERCHLORIC ACID, MAGNESIUM SALT; DEHYDRITE; PERCHLORATE de MAGNESIUM (French); MAGNESIUM PERCHLORATE, ANHYDROUS; MAGNESIUM PERCHLORATE HEXAHYDRATE; PERCLORATO MAGNESICO (Spanish)

IDENTIFICATION
CAS Number: 10034-81-8
Formula: $Cl_2O_8 \cdot Mg$; $Mg(ClO_4)_2$
DOT ID Number: UN 1475; DOT Guide Number: 140
Proper Shipping Name: Magnesium perchlorate

DESCRIPTION: White solid. Odorless. Sinks and mixes with water; moderately soluble; may cause heat and spattering.

Strong oxidizer which may react spontaneously with low flash point organics or reducing agents. Heat forms oxygen; will increase the activity of an existing fire • May cause fire or explosion on contact with combustibles (wood, paper, oil, clothing, etc.) • Irritating to the skin, eyes, and respiratory tract • Toxic products of combustion may include hydrogen chloride and magnesium oxide.

Hazard Classification (based on NFPA-704 Rating System)
Health Hazards (Blue): 1; Flammability (Red): 0; Reactivity (Yellow): 0; Special Notice (White): OXY

EMERGENCY RESPONSE: See Appendix A (140)
Note: Do not let spill area dry until it has been determined that there is no perchlorates left in the area. Continue cooling after fire has been extinguished.
Evacuation:
Public safety: Isolate the area of spill or leak for at least 10 to 25 meters (30 to 80 feet) in all directions.
Spill: Consider initial downwind evacuation for at least 100 meters (330 feet).
Fire: If any large container is involved in fire, isolate for at least 800 meters (½ mile) in all directions; also, consider initial evacuation for 800 meters (½ mile) in all directions.

EXPOSURE
Short-term effects: *SEEK MEDICAL ATTENTION.* **Dust:** Irritating to eyes, nose, and throat. *IF INHALED,* will, will cause difficult breathing. *IF IN EYES,* hold eyelids open and flush with plenty of water. *IF BREATHING HAS STOPPED,* give artificial respiration. IF breathing is difficult, administer oxygen. **Solid:** Irritating to skin and eyes. *IF SWALLOWED,* will cause nausea, vomiting, or loss of consciousness. Remove contaminated clothing and shoes. Flush affected areas with plenty of water. *IF IN EYES,* hold eyelids open and flush with plenty of water. *IF SWALLOWED* and victim is *CONSCIOUS AND ABLE TO SWALLOW,* have victim drink 4 to 8 ounces of water and have victim induce vomiting. *IF SWALLOWED* and victim is *UNCONSCIOUS OR HAVING CONVULSIONS,* do nothing except keep victim warm.

HEALTH HAZARDS
Personal protective equipment (PPE): B-Level PPE. U.S. Bureau of Mines approved respirator; chemical safety goggles; face shield.
Liquid or solid irritant characteristics: Eye, skin and respiratory tract irritant.

CHEMICAL REACTIVITY
Binary reactants: A powerful oxidizer. Violent reaction with reducing agents, alcohols, ammonia gas, argon (wet), butyl

fluorides, dimethyl sulfoxide, ethylene oxide, fluorobutane (wet), fuels, hydrazines. Contact with wood, paper, oils, grease, or finely divided metals may cause fires and explosions.

ENVIRONMENTAL DATA
Food chain concentration potential: Negative; unlikely to accumulate.
Water pollution: Effect of low concentrations on aquatic life is unknown. May be dangerous if it enters nearby water intakes; notify operators. Notify local health and wildlife officials. Response to discharge. Issue warning–oxidizing material. Restrict access. Disperse and flush.

SHIPPING INFORMATION
Grades of purity: Pure anhydrous; 65–68% solution of hexahydrate in water; **Storage temperature:** Ambient; **Inert atmosphere:** None; **Venting:** Safety relief; **Stability during transport:** Stable

PHYSICAL AND CHEMICAL PROPERTIES
Physical state @ 59°F/15°C and 1 atm: Solid.
Molecular weight: 223.2
Boiling point @ 1 atm: Decomposes above 482°F/250°C.
Specific gravity (water = 1): 2.21 @ 68°F/20°C (solid).
Heat of solution: -260 Btu/lb = -140 cal/g = -6.0×10^5 J/kg.

MALATHION　　　　　　　　　　　　　　**REC. M:0250**

SYNONYMS: AMERICAN CYANAMID 4,049; BAN-MITE; 1,2-BIS(ETHOXYCARBONYL)ETHYL; CALMATHION; COMPOUND 4049; CARBETHOXY MALATHION; CARBETOVUR; CARBETOX; CARBOFOS; CARBOPHOS; CHEMATHION; CIMEXAN; COMPOUND 4049; CYTHION; DETMOL MA; DETMOL MA 96%; O,O-DIMETHYL-PHOSPHORODITHIOATE; O,O-DIMETHYL DITHIOPHOSPHATE OF DIETHYL MERCAPTOSUCCINATE; DURAMITEX; EEC No. 015-014-00-X; EL 4049; EMMATOS; EMOTOS EXTRA; ETHIOLACAR; ETIOL; EXTERMATHION; s-[1,2-BIS(ETHOXYCARBONYL)ETHYL]O,O-DIMETHYL DITHIOPHOSPHATE OF DIETHYL MERCAPTOSUCCINATE; O,O-DIMETHYL S-(1,2-DICARBETHOXY-ETHYL)PHOSPHORO-DITHIOCITE; FORMAL; FORTHION; FOSFOTHION; HILTHION; HILTHION 25WDP; KARBOFOS; KOP-THION; KYPFOS; MALACIDE; MALAFOR; MALAGRAN; MALAKILL; MALAMAR; MALAMAR 50; MALAPHELE; MALAPHOS; MALASOL; MALASPRAY; MALATHIOZOO; MALATHON; MALATION (Spanish); MALATOL; MALATOX; MALDISON; MALMED; MALPHOS; MALTOX; MERCAPTOTHION; MLT; MOSCARDA; PBI CROP SAVER; PRIODERM; SADOFOS; SADOPHOS; SF 60; SIPTOX 1; SUMITOX; TK; TM-4049; VEGFRU MALATOX; VETIOL; ZITHIOL

IDENTIFICATION
CAS Number: 121-75-5
Formula: $C_{10}H_{19}O_6PS_2$
DOT ID Number: UN 3018 (liquid); UN 2783 (solid); DOT Guide Number: 152
Proper Shipping Name: Organophosphorus pesticides, liquid, toxic, n.o.s; organophosphorus pesticides, solid, toxic, n.o.s.
Reportable Quantity (RQ): **(CERCLA)** 100 lb/45.4 kg

DESCRIPTION: Clear colorless liquid when pure; becomes amber with time. Garlic- or skunk-like odor. Sinks in water; soluble. A solid below 37°F/3°C.

Poison! (organophosphate) • Combustible • Breathing the vapor, skin or eye contact, or swallowing the material can kill you • Firefighting gear (including SCBA) does not provide adequate protection. If exposure occurs, remove and isolate gear immediately and thoroughly decontaminate personnel • Containers may BLEVE when exposed to fire • Vapors in confined areas (e.g., tanks, sewers, buildings) may explode when exposed to fire • Toxic products of combustion may include carbon monoxide, sulfur oxide, phosphorus oxide, and phosphoric acid • Do not put yourself in danger by entering a contaminated area to rescue a victim.

Hazard Classification (based on NFPA-704 Rating System)
Health Hazards (Blue): 2; Flammability (Red): 1; Reactivity (Yellow): 0

EMERGENCY RESPONSE: See Appendix A (152)
Evacuation:
Public safety: Isolate the area of spill or leak for at least 25 to 50 meters (80 to 160 feet) in all directions.
Spill: Increase, in the downwind direction, as necessary, the distance shown under "Public Safety."
Fire: If tank, rail car, or tank truck is involved in fire, isolate for at least 800 meters (½ mile) in all directions; also, consider initial evacuation for 800 meters (½ mile) in all directions.

EXPOSURE
Short-term effects: *SEEK MEDICAL ATTENTION.* **Liquid:** *POISONOUS IF INHALED, IF SWALLOWED, OR IF SKIN IS EXPOSED.* Eye pupils are small; blurred vision; runny nose; cough; shortness of breath; pain; diarrhea, nausea and vomiting; increased blood pressure, hypermotility, hallucinations; loss of consciousness; convulsions; breathing stops; death. *IF BREATHING HAS STOPPED,* give artificial respiration; *avoid mouth-to-mouth resuscitation; use bag/mask apparatus.* **Skin:** *Remove and double-bag contaminated clothing (including shoes) and leave them in the Hot Zone*; wash skin with soap and water and large volumes of water for at least 15 minutes. *Skin can also be decontaminated with diluted hypochlorite solution, U.S. Army M291 kit, and M258(A1) skin decontamination kit.* **Eye:** Rinse with large volumes of water or saline for at least 15 minutes. *IF SWALLOWED,* will cause nausea, vomiting, or loss of consciousness. *Remove and double-bag contaminated clothing and shoes and leave in Hot Zone for later incineration by hazardous materials experts.* Flush affected areas with plenty of water. **If swallowed** and victim is *CONSCIOUS AND ABLE TO SWALLOW*, have victim drink 4 to 8 ounces of water. **Do NOT induce vomiting but immediately administer slurry of activated charcoal.** *IF SWALLOWED* and victim is *UNCONSCIOUS OR HAVING CONVULSIONS,* do nothing except keep victim warm. *Note to physician and medical personnel:* If symptoms indicate, initial treatment for an adult includes atropine, 2 mg (1/30 gr) intramuscularly or intravenously as soon as any local or systemic signs or symptoms of an intoxication are noted; repeat the administration of atropine every 3 to 8 minutes until signs of atropinization [mydriasis (abnormal dialation of the eye pupil), dry mouth, rapid pulse, hot and dry skin] occur; initiate treatment in children with 0.05 mg/kg of atropine; repeat at 5- to 10-minute intervals. Watch respiration, and remove bronchial secretions if they appear to be obstructing the airway; intubate if necessary. Pralidoxime must be administered within minutes to a few hours following exposure (depending on the specific agent) to be effective. Give 2-PAMCI (Pralidoxime; Protopam), 2.5 g in 100

mL of sterile water or in 5% dextrose and water, intravenously, slowly, in 15 to 30 minutes; if sufficient fluid is not available, give 1 g of 2-PAMCI in 3 mL of distilled water by deep intramuscular injection; repeat this every half hour if respiration weakens or if muscle fasciculation or convulsions recur. Also Diazepam, an anticonvulsant, might be needed. For adults, the dose is 5 to 10 mg (slow IV), repeated every 12 to 15 minutes up to three (3) doses maximum. For children, the dose is 0.2 to 0.5 mg/kg.

Note to physician or authorized medical personnel: If inhaled, medical observation is recommended for 24 to 48 hours after breathing overexposure, as pulmonary edema may be delayed. As first aid for pulmonary edema, consider administering a corticosteroid spray. Cigarette smoking may exacerbate pulmonary injury and should be discouraged for at least 72 hours following exposure.

HEALTH HAZARDS

Personal protective equipment (PPE): A-Level PPE. OSHA Table Z-1-A air contaminant. Wear respirator for organophosphate pesticides and rubber clothing while fighting fires of malathion with chlorine bleach solution. Chemical protective material(s) reported to have good to excellent resistance: nitrile, Teflon®. Wear sealed chemical suit of butyl rubber, polycarbonate, PVC, Nitrile, Neoprene. All clothing contaminated by fumes and vapors must be decontaminated.

Recommendations for respirator selection: NIOSH
100 mg/m³: CCROVDMFu [any chemical cartridge respirator with organic vapor cartridge(s) in combination with a dust, mist, and fume filter]; or SA (any supplied-air respirator); or SCBA (any self-contained breathing apparatus). *250 mg/m³*: SA:CF* (any supplied-air respirator operated in a continuous-flow mode); or CCRFOVHiE [any chemical cartridge respirator with a full facepiece and organic vapor cartridge(s) in combination with a high-efficiency particulate filter]; or GMFOVHiE [any air-purifying, full-facepiece respirator (gas mask) with a chin-style, front- or back-mounted organic vapor canister having a high-efficiency particulate filter]; or PAPROVDMFu* [any powered, air purifying respirator with organic vapor cartridge(s) in combination with a dust, mist, and fume filter]; SCBAF (any self-contained breathing apparatus with full facepiece); or SAF (any supplied-air respirator with a full facepiece). *EMERGENCY OR PLANNED ENTRY INTO UNKNOWN CONCENTRATIONS OR IDLH CONDITIONS*: SCBAF:PD,PP (any self-contained breathing apparatus that has a full facepiece and is operated in a pressure-demand or other positive-pressure mode); or SAF:PD,PP:ASCBA (any supplied-air respirator that has a full facepiece and is operated in a pressure-demand or other positive-pressure mode in combination with an auxiliary self-contained breathing apparatus operated in a pressure-demand or other positive pressure mode). *ESCAPE:* GMFOVHiE [any air-purifying, full-facepiece respirator (gas mask) with a chin-style, front-or back-mounted canister having a high efficiency particulate filter); or SCBAE (any appropriate escape-type, self-contained breathing apparatus).
Note: Substance reported to cause eye irritation or damage; may require eye protection.

Exposure limits (TWA unless otherwise noted): ACGIH TLV 10 mg/m³; OSHA PEL 15 mg/m³; NIOSH REL 10 mg/m³; skin contact contributes significantly in overall exposure.

Toxicity by ingestion: Poison. Grade 2; LD_{50} = 0.5 to 5g/kg(rat).

Long-term health effects: IARC rating 3; inadequate animal evidence. May cause skin allergy; nervous system effects. Reproductive effects; mutagen; possible teratogen.

Liquid or solid irritant characteristics: Liquid can be absorbed through the skin. If spilled on clothing and allowed to remain, may cause smarting and reddening of the skin.

Odor threshold: 10 mg/m³.
IDLH value: 250 mg/m³.

FIRE DATA
Flash point: More than 325°F/163°C.

CHEMICAL REACTIVITY
Binary reactants: Strong oxidizers, magnesium, alkaline pesticides. Corrosive to metals.
Neutralizing agents for acids and caustics: Liquid bleach solution for decontamination.

ENVIRONMENTAL DATA
Water pollution: DOT Appendix B, §172.101–marine pollutant. Harmful to aquatic life in very low concentrations. May be dangerous if it enters nearby water intakes; notify operators. Notify local health and wildlife officials. For weapons testing, use an M272 Water Detection Kit. **Response to discharge.** Issue warning–poison., water contaminant. Restrict access. Should be removed. Chemical and physical treatment.

SHIPPING INFORMATION
Grades of purity: Many powders, dusts, and spray solutions are sold under a variety of trade names; **Storage temperature:** Below 120°F/49°C. Decomposition (may not be hazardous) occurs at higher temperatures.

PHYSICAL AND CHEMICAL PROPERTIES
Physical state @ 59°F/15°C and 1 atm: Liquid.
Molecular weight: 330.36
Boiling point @ 1 atm: 140°F/60°C/333°K (decomposes).
Melting/Freezing point: 37°F/2.9°C/276°K.
Specific gravity (water = 1): 1.234 @ 77°F/25°C (liquid).
Liquid surface tension: 37.1 dynes/cm = 0.0371 N/m at 24°C.
Liquid water interfacial tension: 19 dynes/cm = 0.019 N/m at 24°C.
Vapor pressure: 0.00004 mm.

MALEIC ACID   REC. M:0300

SYNONYMS: ACIDO MALICO (Spanish); BUTENEDIOIC ACID, (Z)-; *cis*-BUTENEDIOIC ACID; (Z) BUTENEDIOIC ACID; *cis*-1,2-ETHYLENEDICARBOXYLIC ACID; EEC 607-095-00-3; 1,2-ETHYLENEDICARBOXYLIC ACID, (Z); MALEINIC ACID; MALENIC ACID; TOXILIC ACID

IDENTIFICATION
CAS Number: 110-16-7; may also apply to 141-82-2 (*cis*-isomer)
Formula: $C_4H_4O_4$; HOOC-CH=CH· COOH
DOT ID Number: UN 2215; DOT Guide Number: 156
Proper Shipping Name: Maleic acid
Reportable Quantity (RQ): **(CERCLA)** 5000 lb/2270 kg

DESCRIPTION: White crystalline solid. Faint acid odor. Sinks and mixes with water forming a medium-strong acid.

Combustible solid • Aqueous solution is corrosive to skin, eyes, and respiratory system • Heat or flame may cause explosion of dust cloud • Containers may BLEVE when exposed to fire • Concentrated dust in confined areas (e.g., tanks, sewers, buildings) may explode when exposed to fire • Toxic products of combustion may include fumaric acid and smoke containing maleic anhydride.

686   Maleic anhydride

Hazard Classification (based on NFPA-704 Rating System)
Health Hazards (Blue): 1; Flammability (Red): 1; Reactivity (Yellow): 0

EMERGENCY RESPONSE: See Appendix A (156)
Evacuation:
Public safety: Isolate the area of spill or leak for at least 50 to 100 meters (160 to 330 feet) in all directions.
Spill: Increase, in the downwind direction, as necessary, the distance shown under "Public Safety."
Fire: If tank, rail car, or tank truck is involved in fire, isolate for at least 800 meters (½ mile) in all directions; also, consider initial evacuation for 800 meters (½ mile) in all directions.

EXPOSURE
Short-term effects: *SEEK MEDICAL ATTENTION.* **Dust:** Irritating to eyes, nose, and throat. *IF INHALED*, will, will cause coughing or difficult breathing. *IF IN EYES*, hold eyelids open and flush with plenty of water. *IF BREATHING HAS STOPPED*, give artificial respiration; *avoid mouth-to-mouth resuscitation; use bag/mask apparatus.* IF breathing is difficult, administer oxygen. **Solid:** Irritating to skin and eyes. Harmful if swallowed. Remove contaminated clothing and shoes. Flush affected areas with plenty of water. *IF IN EYES*, hold eyelids open and flush with plenty of water. *IF SWALLOWED* and victim is *CONSCIOUS AND ABLE TO SWALLOW*, have victim drink 4 to 8 ounces of water. *IF SWALLOWED* and victim is *UNCONSCIOUS OR HAVING CONVULSIONS*, do nothing except keep victim warm.
Note to physician or authorized medical personnel: Medical observation is recommended for 24 to 48 hours after breathing overexposure, as pulmonary edema may be delayed. As first aid for pulmonary edema, consider administering a corticosteroid spray. Cigarette smoking may exacerbate pulmonary injury and should be discouraged for at least 72 hours following exposure.

HEALTH HAZARDS
Personal protective equipment (PPE): B-Level PPE. Dust mask; goggles or face shield; protective gloves. Chemical protective material(s) reported to have good to excellent resistance: natural rubber, neoprene, nitrile, nitrile+PVC, PVC. Also, Styrene-butadiene, styrene-butadiene/neoprene, Viton® offers limited protection.
Toxicity by ingestion: Grade 2; oral LD_{50} = 708 mg/kg (rat).
Vapor (gas) irritant characteristics: Eye and respiratory tract irritant.
Liquid or solid irritant characteristics: Severe skin and eye irritant. Passes through the skin which may increased the toxic effect.

FIRE DATA
Flash point: above 212°F/100°C.
Behavior in fire: May be converted to its isomer, fumaric acid.

CHEMICAL REACTIVITY
Reactivity with water: Solution is acetic.
Binary reactants: Violent reaction with oxidizers, bases. May corrode metals when wet.
Neutralizing agents for acids and caustics: Flush with water, rinse with dilute solution of sodium bicarbonate or soda ash.

ENVIRONMENTAL DATA
Food chain concentration potential: Log P_{ow} = –0.5 to –0.71 (estimate); Unlikely to accumulate. **Water pollution:** Effect of low concentrations on aquatic life is unknown. May be dangerous if it enters nearby water intakes; notify operators. Notify local health and wildlife officials. Response to discharge Disperse and flush.

SHIPPING INFORMATION
Grades of purity: Reagent; Technical; **Storage temperature:** Ambient; **Inert atmosphere:** None; **Venting:** Open; **Stability during transport:** Stable.

PHYSICAL AND CHEMICAL PROPERTIES
Physical state @ 59°F/15°C and 1 atm: Solid.
Molecular weight: 116.1
Boiling point @ 1 atm: 135°C/275°C/548°K (decomposes).
Melting/Freezing point: 266°F/130°C/403°K.
Specific gravity (water = 1): 1.59 @ 68°F/20°C (solid).
Heat of combustion: –5,000 Btu/lb = –2800 cal/g = –117 x 10^5 J/kg.

MALEIC ANHYDRIDE   **REC. M:0350**

SYNONYMS: ACIDO MALICO (Spanish); *cis*-BUTENEDIOIC ANHYDRIDE; DIHYDRO-2,5-DIOXOFURAN; EEC No. 607-096-00-9; 2,5-FURANDIONE; MAA; MALEIC ACID ANHYDRIDE; 2,5-FURANEDIONE; TOXILIC ANHYDRIDE; RCRA No. U147

IDENTIFICATION
CAS Number: 108-31-6
Formula: $C_4H_2O_3$; OCOCH=CHCO
DOT ID Number: UN 2215; DOT Guide Number: 156
Proper Shipping Name: Maleic anhydride
Reportable Quantity (RQ): **(CERCLA)** 5000 lb/2270 kg

DESCRIPTION: Colorless solid crystals, tablets, briquettes, pellets. Acrid, choking odor. Sinks and mixes slowly with water forming corrosive malic acid.

Combustible solid • Corrosive to the skin, eyes, skin, and respiratory tract; inhalation symptoms may be delayed • Heat or flame may cause explosion of dust cloud • Firefighting gear (including SCBA) does not provide adequate protection. If exposure occurs, remove and isolate gear immediately and thoroughly decontaminate personnel • Containers may BLEVE when exposed to fire • Concentrated dust in confined areas (e.g., tanks, sewers, buildings) may explode when exposed to fire • include smoke and irritating vapors.

Hazard Classification (based on NFPA-704 Rating System)
Health Hazards (Blue): 3; Flammability (Red): 1; Reactivity (Yellow): 1

EMERGENCY RESPONSE: See Appendix A (156)
Evacuation:
Public safety: Isolate the area of spill or leak for at least 50 to 100 meters (160 to 330 feet) in all directions.
Spill: Increase, in the downwind direction, as necessary, the distance shown under "Public Safety."
Fire: If tank, rail car, or tank truck is involved in fire, isolate for at least 800 meters (½ mile) in all directions; also, consider initial evacuation for 800 meters (½ mile) in all directions.

EXPOSURE
Short-term effects: *SEEK MEDICAL ATTENTION.* **Vapor:** *IF BREATHING HAS STOPPED*, give artificial respiration; *avoid*

mouth-to-mouth resuscitation; use bag/mask apparatus. **Liquid or solid:** Will burn skin and eyes. Harmful if swallowed. Remove contaminated clothing and shoes. Flush affected areas with plenty of water. *IF IN EYES*, hold eyelids open and flush with plenty of water. *IF SWALLOWED and victim is CONSCIOUS AND ABLE TO SWALLOW*, have victim drink 4 to 8 ounces of water.
Note to physician or authorized medical personnel: Medical observation is recommended for 24 to 48 hours after breathing overexposure, as pulmonary edema may be delayed. As first aid for pulmonary edema, consider administering a corticosteroid spray. Cigarette smoking may exacerbate pulmonary injury and should be discouraged for at least 72 hours following exposure.

HEALTH HAZARDS
Personal protective equipment (PPE): B-Level PPE. OSHA Table Z-1-A air contaminant. Approved respirator; chemical goggles and face shield; rubber gloves and boots; coveralls or rubber apron.
Recommendations for respirator selection: NIOSH/OSHA
10 mg/m³: SA:CF* (any supplied-air respirator operated in a continuous-flow mode); or SCBAF (any self-contained breathing apparatus with a full facepiece); or SAF (any supplied-air respirator with a full facepiece). *EMERGENCY OR PLANNED ENTRY INTO UNKNOWN CONCENTRATIONS OR IDLH CONDITIONS*: SCBAF:PD,PP (any self-contained breathing apparatus that has a full facepiece and is operated in a pressure-demand or other positive-pressure mode); or SAF:PD,PP:ASCBA (any supplied-air respirator that has a full facepiece and is operated in a pressure-demand or other positive-pressure mode in combination with an auxiliary self-contained breathing apparatus operated in a pressure-demand or other positive pressure mode). *ESCAPE*: GMFOVHiE [any air-purifying, full-facepiece respirator (gas mask) with a chin-style, front-or back-mounted canister having a high efficiency particulate filter) or SCBAE (any appropriate escape-type, self-contained breathing apparatus). *Note:* Substance causes eye irritation or damage; eye protection needed.
Exposure limits (TWA unless otherwise noted): ACGIH TLV 0.1; NIOSH/OSHA 0.25 ppm (1 mg/m³).
Toxicity by ingestion: Poison. Grade 2; LD_{50} = 0.5 to 5 g/kg.
Long-term health effects: Possible dermatitis.
Vapor (gas) irritant characteristics: eye and respiratory tract irritant. Vapors cause irritation, such that personnel will find high concentrations unpleasant. High concentrations can cause lung edema.
Liquid or solid irritant characteristics: Severe eye and skin irritant. Causes smarting of the skin and first-degree burns on short exposure; may cause secondary burns on long exposure.
Odor threshold: 0.25-0.3 ppm. 1 mg/m³ in air.
IDLH value: 10 mg/m³.

FIRE DATA
Flash point: (Liquid) 230°F (oc); 218°F (cc).
Flammable limits in air: LEL: 1.4%; UEL: 7.1%.
Fire extinguishing agents not to be used: Water or foam may cause frothing.
Behavior in fire: Above 300°F/149°C, in the presence of various materials, may generate CO_2. Will explode if confined.
Autoignition temperature: 883°F/473°C/746°K.
Electrical hazard: Class I, Group D.
Burning rate: 1.4 mm/min

CHEMICAL REACTIVITY
Reactivity with water: Hot water may cause frothing. Slow decomposition reaction with water (hydrolizes) to form malic acid. In water it becomes a strong acid.
Binary reactants: Strong oxidizers, alkalis, alkali metals; caustics and amines above 150°F/66°C. Corrosive to metals in the presence of water.
Neutralizing agents for acids and caustics: Solid spills can usually be recovered before any significant reaction with water occurs. Flush area of spill with water.

ENVIRONMENTAL DATA
Food chain concentration potential: Negative; unlikely to accumulate.
Water pollution: Dangerous to aquatic life in high concentrations. May be dangerous if it enters nearby water intakes; notify operators. Notify local health and wildlife officials. **Response to discharge:** Disperse and flush.

SHIPPING INFORMATION
Grades of purity: Commercial: 99.5%; **Storage temperature:** Ambient; **Inert atmosphere:** None; **Venting:** Open; **Stability during transport:** Stable; **Stability during transport:** Stable

NAS HAZARD CLASSIFICATION FOR BULK WATER TRANSPORTATION
FIRE: 1
HEALTH: Vapor irritant: 2; Liquid or solid irritant: 2; Poisons: 1
WATER POLLUTION: Human toxicity: 2; Aquatic toxicity: 2; Aesthetic effect: 1
REACTIVITY: Other chemicals: 3; Water: 2; Self-reaction: 0

PHYSICAL AND CHEMICAL PROPERTIES
Physical state @ 59°F/15°C and 1 atm: Solid.
Molecular weight: 98.06
Boiling point @ 1 atm: 396°F/202°C/475°K.
Melting/Freezing point: 127°F/53°C/326°K.
Specific gravity (water = 1): 1.43 @ 15°C (solid).
Relative vapor density (air = 1): 1.4
Heat of combustion: -5936 Btu/lb = -3298 cal/g = -138.1×10^5 J/kg.
Heat of decomposition: 150°C.
Heat of solution: -153 Btu/lb = -85.0 cal/g = -3.56×10^5 J/kg.
Vapor pressure: 0.2 mm.

MALEIC HYDRAZIDE   REC. M:0400

SYNONYMS: BH DOCK KILLER; BOS MH; *cis*-BUTENEDIOIC ANHYDRIDE; 1,2-DIHYDROPYRIDAZINE-3,6-DIONE; 1,2-DIHYDRO-3,6-PYRIDAZINEDIONE; EC 300; HYDRAZIDA MALEICA (Spanish); 6-HYDROXY-3(2H)-PYRIDAZINONE; MALAZIDE; MALEIC ACID HYDRAZIDE; MAZIDE; *N,N*-MALEOYLHYDRAZINE; MH; RCRA No. U14; REGULOX; ROYAL MH 30; SLO-GRO; STOP-GRO; STUNTMAN; SUPER-DE-SPROUT; VONDALDHYDE; VONDRAX; TOXILIC ANHYDRIDE

IDENTIFICATION
CAS Number: 123-33-1
Formula: $C_4H_4N_2O_2$
DOT ID Number: UN 2588; DOT Guide Number: 151
Proper Shipping Name: Pesticide, solid, poisonous, n.o.s.
Reportable Quantity (RQ): **(CERCLA)** 5000 lb/2270 kg

DESCRIPTION: White crystalline solid. Odorless. Sinks in water; soluble; sublimes, forming a strong acid solution. A restricted use pesticide (RUP).

Mercaptodimethur

Poison! (organic fungicide) • Combustible solid • Corrosive to skin, eyes, and respiratory tract; inhalation symptoms may be delayed • Heat or flame may cause explosion of dust cloud • Containers may BLEVE when exposed to fire • Concentrated dust in confined areas (e.g., tanks, sewers, buildings) may explode when exposed to fire • Toxic products of combustion may include carbon monoxide and nitrogen oxides.

Hazard Classification (based on NFPA-704 Rating System)
Health Hazards (Blue): 1; Flammability (Red): 1; Reactivity (Yellow): 0

EMERGENCY RESPONSE: See Appendix A (151)
Evacuation:
Public safety: Isolate the area of spill or leak for at least 25 to 50 meters (80 to 160 feet) in all directions.
Spill: Increase, in the downwind direction, as necessary, the distance shown under "Public Safety."
Fire: If tank, rail car, or tank truck is involved in fire, isolate for at least 800 meters (½ mile) in all directions; also, consider initial evacuation for 800 meters (½ mile) in all directions.

EXPOSURE
Short-term effects: *SEEK MEDICAL ATTENTION*. **Dust:** Irritating to eyes, nose, and throat. Move victim to fresh air. *IF IN EYES*, hold eyelids open and flush with plenty of water. IF breathing is difficult, administer oxygen. Lung edema may develop. **Solid:** Irritating to skin and eyes. Harmful if swallowed. Remove contaminated clothing and shoes. Flush affected areas with plenty of water. *IF IN EYES*, hold eyelids open and flush with plenty of water. *IF SWALLOWED* and victim is *CONSCIOUS AND ABLE TO SWALLOW*, have victim drink 4 to 8 ounces of water.
Note to physician or authorized medical personnel: Medical observation is recommended for 24 to 48 hours after breathing overexposure, as pulmonary edema may be delayed. As first aid for pulmonary edema, consider administering a corticosteroid spray. Cigarette smoking may exacerbate pulmonary injury and should be discouraged for at least 72 hours following exposure.

HEALTH HAZARDS
Personal protective equipment (PPE): B-Level PPE. Goggles or face shield; dust mask.
Toxicity by ingestion: Grade 2; oral LD_{50} = 3800 mg/kg (rat).
Long-term health effects: Causes cancer in rats. IARC rating 3; animal evidence inadequate.
Liquid or solid irritant characteristics: Eye, skin and respiratory tract irritant.

CHEMICAL REACTIVITY
Binary reactants: Reacts with strong oxidizers.

ENVIRONMENTAL DATA
Food chain concentration potential: Log P_{ow} = <1.0. Unlikely to accumulate.
Water pollution: Effect of low concentrations on aquatic life is unknown. May be dangerous if it enters nearby water intakes; notify operators. Notify local health and wildlife officials.
Response to discharge: Issue warning–water contaminant. Should be removed. Chemical and physical treatment.

SHIPPING INFORMATION
Grades of purity: Technical: 97+%; **Storage temperature:** Ambient; **Inert atmosphere:** None; **Venting:** Open; **Stability during transport:** Stable; **Stability during transport:** Stable

PHYSICAL AND CHEMICAL PROPERTIES
Physical state @ 59°F/15°C and 1 atm: Solid.
Molecular weight: 112.1
Boiling point @ 1 atm: Decomposes.
Melting/Freezing point: 558°F/292°C/565°K.
Specific gravity (water = 1): 1.60 @ 77°F/25°C (solid).
Heat of combustion: (estimate) –8200 Btu/lb = –4500 cal/g = –190 x 10^5 J/kg.

MERCAPTODIMETHUR   REC. M:0450

SYNONYMS: BAY 9026; BAY 37344; BAYER 37344; 3,5-DIMETHYL-4-(METHYLTHIO)-,METHYLCARBAMATE; 3,5-DIMETHYL-4-(METHYLTHIO) PHENOLMETHYL-CARBAMATE; 3,5-DIMETHYL-4-METHYLTHIOPHENYL *N*-METHYLCARBAMATE; DRAZA; H 321; MESUROL; METIOCARB (Spanish); METHYL CARBAMIC ACID 4-(METHYLTHIO)-3,5-XYLYL ESTER 4-METHYL-MERCAPTO-3,5-DIMETHYLPHENYL *N*-METHYL-CARBAMATE; 4-METHYLMERCAPTO-3,5-XYLYLMETHYL-CARBAMATE; 4-METHYLTHIO-3,5-DIMETHYLPHENYLMETHYL-CARBAMATE; 4-(METHYLTHIO)-3,5-XYLYLMETHYL-CARBAMATE; METMERCAPTURON; OMS-93; 3,5-XYLENOL, 4-(METHYLTHIO)-, METHYLCARBAMATE; METHMERCAPTURON; OMS-93

IDENTIFICATION
CAS Number: 2032-65-7
Formula: $C_{11}H_{15}NO_2S$
DOT ID Number: UN 2757; DOT Guide Number: 151
Proper Shipping Name: Carbamate pesticides, solid, toxic, n.o.s.
Reportable Quantity (RQ): **(CERCLA)** 10 lb/4.54 kg

DESCRIPTION: White crystalline solid. Mild odor. Sinks in water; insoluble.

Combustible solid • Poison! (carbamate) • Heat or flame may cause explosion of dust cloud • Breathing the dust, skin or eye contact, or swallowing the material can kill you • Firefighting gear (including SCBA) does not provide adequate protection. If exposure occurs, remove and isolate gear immediately and thoroughly decontaminate personnel • Containers may BLEVE when exposed to fire • Concentrated dust in confined areas (e.g., tanks, sewers, buildings) may explode when exposed to fire • Highly Irritating to the skin, eyes, and respiratory tract • Toxic products of combustion may include carbon monoxide, sulfur, and nitrogen oxides.

Hazard Classification (based on NFPA-704 Rating System)
Health Hazards (Blue): 3; Flammability (Red): 1; Reactivity (Yellow): 0

EMERGENCY RESPONSE: See Appendix A (151)
Evacuation:
Public safety: Isolate the area of spill or leak for at least 25 to 50 meters (80 to 160 feet) in all directions.
Spill: Increase, in the downwind direction, as necessary, the distance shown under "Public Safety."
Fire: If tank, rail car, or tank truck is involved in fire, isolate for at least 800 meters (½ mile) in all directions; also, consider initial evacuation for 800 meters (½ mile) in all directions.

EXPOSURE
Short-term effects: *SEEK MEDICAL ATTENTION*. **Dust:**

POISONOUS IF INHALED OR SKIN IS EXPOSED. **Inhalation, ingestion or absorption:** Inhalation of mist, dust, or vapor (or ingestion, or absorption through the skin) affects the nerves, muscles; causes salivation and other excessive respiratory tract secretion, nausea, stomach pain, headache, vomiting, diarrhea, headache, sweating, blurred vision, and pinpoint pupils of the eyes. High levels of exposure may cause convulsions, loss of reflexes, and loss of sphincter control unconsciousness, coma, and death. The symptoms may develop over a period of 8 hours. An increase in salivary and bronchial secretions may result which simulate severe pulmonary edema. Move to fresh air. *IF BREATHING HAS STOPPED,* give artificial respiration.*IF BREATHING HAS STOPPED,* give artificial respiration; *avoid mouth-to-mouth resuscitation; use bag/mask apparatus.* **Solid:** *POISONOUS IF SWALLOWED.* Remove contaminated clothing and shoes. Flush affected areas with plenty of water. *IF IN EYES*, hold eyelids open and flush with plenty of water. *IF SWALLOWED* and victim is *CONSCIOUS AND ABLE TO SWALLOW*, have victim drink 4 to 8 ounces of water and have victim induce vomiting.

Note to physician or authorized medical personnel: Administer atropine, 2 mg (1/30 gr) intramuscularly or intravenously as soon as any local or systemic signs or symptoms of an intoxication are noted; repeat the administration of atropine every 3 to 8 minutes until signs of atropinization (mydriasis, dry mouth, rapid pulse, hot and dry skin) occur; initiate treatment in children with 0.05 mg/kg of atropine; repeat at 5- to 10-minute intervals. Watch respiration, and remove bronchial secretions if they appear to be obstructing the airway; intubate if necessary. Pralidoxime must be administered within minutes to a few hours following exposure (depending on the specific agent) to be effective. Give 2-PAMCI (Pralidoxime; Protopam), 2.5 g in 100 mL of sterile water or in 5% dextrose and water, intravenously, slowly, in 15 to 30 minutes; if sufficient fluid is not available, give 1 g of 2-PAMCI in 3 mL of distilled water by deep intramuscular injection; repeat this every half hour if respiration weakens or if muscle fasciculation or convulsions recur. Also Diazepam, an anticonvulsant, might be needed.

Medical note: Due to the rapid regeneration of chlolinesterase and the fact that 2-PAMCI may be contraindicated in the case of some carbamate poisonings, 2-PAMCI (Pralidoxime; Protopam) may not be needed.

HEALTH HAZARDS
Personal protective equipment (PPE): A-Level PPE. Safety goggles or face mask, hooded suit, approved respirator or self-contained breathing apparatus, rubber gloves and boots.
Recommendations for respirator selection: SCBAF:PD,PP (any self-contained breathing apparatus that has a full facepiece and is operated in a pressure-demand or other positive-pressure mode); or SAF:PD,PP:ASCBA (any supplied-air respirator that has a full facepiece and is operated in a pressure-demand or other positive-pressure mode in combination with an auxiliary, self-contained breathing apparatus operated in a pressure-demand or other positive pressure mode). *ESCAPE:* GMFOVHiE [any air-purifying, full-facepiece respirator (gas mask) with a chin-style, front- or back-mounted organic vapor canister having a high-efficiency particulate filter]; or SCBAE (any appropriate escape-type, self-contained breathing apparatus).
Toxicity by ingestion: Poison. Grade 4; LD_{50} less than 50 mg/kg.
Long-term health effects: A suspected carcinogen.

CHEMICAL REACTIVITY
Binary reactants: Reacts with strong bases. May corrode some metals in the presence of moisture.

ENVIRONMENTAL DATA
Food chain concentration potential: Log P_{ow} = 2.9. Values > 3.0 are likely to bioconcentrate in aquatic organisms and other living tissue, especially in fats.
Water pollution: DOT Appendix B, §172.101–marine pollutant. Hazardous to aquatic life in very low concentrations. May be dangerous if it enters nearby water intakes; notify operators. Notify local health and wildlife officials. **Response to discharge:** Issue warning–poison, water contaminant. Restrict access. Should be removed. Will sink and may be almost completely recovered by physical treatment.

SHIPPING INFORMATION
Grades of purity: 50–75% wettable powder; 5% dust; 4% bait; **Storage temperature:** 0–100°F/18–38°C; **Stability during transport:** Stable.

PHYSICAL AND CHEMICAL PROPERTIES
Physical state @ 59°F/15°C and 1 atm: Solid.
Molecular weight: 225.305
Boiling point @ 1 atm: Very high.
Melting/Freezing point: 250.7°F/121.5°C/394.65°K.
Specific gravity (water = 1): More than 1
Relative vapor density (air = 1): 7.77

MERCURIC ACETATE   REC. M:0500

SYNONYMS: ACETATO MERCURIO (Spanish); ACETIC ACID, MERCURY(2+) SALT; BIS(ACETYLOXY)MERCURY; DIACETOXYMERCURY; EEC No. 080-002-00-6; MERCURIACETATE; MERCURIC DIACETATE; MERCURY ACETATE; MERCURY(2+) ACETATE; MERCURY(II) ACETATE; MERCURY DIACETATE; MERCURYL ACETATE

IDENTIFICATION
CAS Number: 1600-27-7
Formula: $C_4H_6O_4 \cdot Hg$; $(CH_3COO)_2Hg$
DOT ID Number: UN 1629; DOT Guide Number: 151
Proper Shipping Name: Mercury acetate
Reportable Quantity (RQ): **(EHS)** 1 lb/0.454 kg

DESCRIPTION: White crystals or powder. The aqueous solution is light-sensitive and yellows with age. Mild vinegar-like odor. Sinks and mixes with water.

Poison • Irritating to the skin, eyes, and respiratory tract; swallowing the material can kill you • Firefighting gear (including SCBA) does not provide adequate protection. If exposure occurs, remove and isolate gear immediately and thoroughly decontaminate personnel • Containers may BLEVE when exposed to fire • Toxic products of combustion may include carbon monoxide, mercury, and mercury oxide fumes.

Hazard Classification (based on NFPA-704 Rating System)
Health Hazards (Blue): 2; Flammability (Red): 0; Reactivity (Yellow): 0

EMERGENCY RESPONSE: See Appendix A (151)
Evacuation:
Public safety: Isolate the area of spill or leak for at least 25 to 50 meters (80 to 160 feet) in all directions.
Spill: Increase, in the downwind direction, as necessary, the distance shown under "Public Safety."

Mercuric ammonium chloride

Fire: If tank, rail car, or tank truck is involved in fire, isolate for at least 800 meters (½ mile) in all directions; also, consider initial evacuation for 800 meters (½ mile) in all directions.

EXPOSURE
Short-term effects: *SEEK MEDICAL ATTENTION.* **Dust:** *POISONOUS IF INHALED.* Inhalation may cause lung edema; physical exertion will aggravate this condition. Move victim to fresh air. *IF BREATHING HAS STOPPED,* give artificial respiration; *avoid mouth-to-mouth resuscitation; use bag/mask apparatus.* IF breathing is difficult, administer oxygen. *IF IN EYES,* hold eyelids open and flush with plenty of water. **Solid:** *POISONOUS IF SWALLOWED. IF SWALLOWED* and victim is *CONSCIOUS AND ABLE TO SWALLOW,* have victim drink 4 to 8 ounces of water and have victim induce vomiting. *IFSWALLOWED* and victim is *UNCONSCIOUS OR HAVING CONVULSIONS,* do nothing except keep victim warm.
Note to physician or authorized medical personnel: Pulmonary edema may be delayed. Medical observation is recommended for 24 to 48 hours after inhalation overexposure. As first aid for pulmonary edema, a physician or authorized medical personnel may consider administering a corticosteroid spray. Cigarette smoking may exacerbate pulmonary injury and should be discouraged for at least 72 hours following exposure.

HEALTH HAZARDS
Personal protective equipment (PPE): B-Level PPE. OSHA Table Z-1-A air contaminant. Rubber gloves, dust mask, goggles.
Recommendations for respirator selection: NIOSH/OSHA as other mercury compounds. See record M:1100 for mercury vapor. *up to 1 mg/m^3*: CCRS** [any chemical cartridge respirator with cartridge(s) providing protection against the compound of concern]; or SA (any supplied-air respirator). *up to 2.5 mg/m^3*: SA:CF (any supplied-air respirator operated in a continuous-flow mode); or PAPRS** [any powered, air-purifying respirator with cartridge(s) providing protection against the compound of concern]. *up to 5 mg/m^3*: CCRFS** [any chemical cartridge respirator with a full facepiece and cartridge(s) providing protection against the compound of concern]; or GMFS [any air-purifying, full-facepiece respirator (gas mask) with a chin-style, front- or back-mounted canister providing protection against the compound of concern]; or SAT:CF (any supplied-air respirator that has a tight-fitting facepiece and is operated in a continuous-flow mode); or PAPRTS [any powered, air-purifying respirator with a tight-fitting facepiece and cartridge(s) providing protection against the compound of concern]; or SCBAF (any self-contained breathing apparatus with a full facepiece); or SAF (any supplied-air respirator with a full facepiece). *up to 10 mg/m^3*: SA:PD,PP (any supplied-air respirator operated in a pressure-demand or other positive-pressure mode). *EMERGENCY OR PLANNED ENTRY INTO UNKNOWN CONCENTRATIONS OR IDLH CONDITIONS:* SCBAF:PD,PP (any self-contained breathing apparatus that has a full facepiece and is operated in a pressure-demand or other positive-pressure mode); or SAF:PD,PP:ASCBA (any supplied-air respirator that has a full facepiece and is operated in a pressure-demand or other positive-pressure mode in combination with an auxiliary, self-contained breathing apparatus operated in a pressure-demand or other positive-pressure mode). *ESCAPE:* GMFS** [any air-purifying, full-facepiece respirator (gas mask) with a chin-style, front- or back-mounted canister protection against the compound of concern]; or SCBAE (any appropriate escape-type, self-contained breathing apparatus). **Note:* End of service life indicator (ESLI) required.
Exposure limits (TWA unless otherwise noted): ACGIH TLV 0.1 mg/m^3 as aryl compounds; OSHA PEL ceiling 0.1 mg/m^3; NIOSH REL 0.05 mg/m^3 (vapor) as aryl and inorganic mercury compounds; skin contact contributes significantly in overall exposure.
Toxicity by ingestion: Grade 3; oral LD$_{50}$ = 76 mg/kg (rat).
Long-term health effects: Intestinal bleeding; kidney and liver damage may develop. A mutagen. Lethal blood level of inorganic mercury in blood (serum or plasma) in humans is 0.04–2.2 mg%; 0.4–22 mg/mL.
Vapor (gas) irritant characteristics: May cause eye and respiratory tract irritation.
Liquid or solid irritant characteristics: Severe eye, skin and respiratory system irritant.
IDLH value: 10 mg/m^3. as mercury.

CHEMICAL REACTIVITY
Binary reactants: Light and heat and can cause decomposition. May react violently or form sensitive, explosive compounds with 2-butyne-1,4-diol, fluoroacetylene, α-nitroguanidine, 5-nitrotetrazol, and others. Incompatible with ammonia, hydrozoic acid, methyl isocyanoacetate, sodium acetylide, sodium peroxyborate, trinitrobenzoic acid, urea nitrate. May also react with strong oxidizers, strong acids.

ENVIRONMENTAL DATA
Food chain concentration potential: Fish can accumulate mercury and transfer it to higher levels in the food chain
Water pollution: DOT Appendix B, §172.101- severe marine pollutant. Harmful to aquatic life in very low concentrations. May be dangerous if it enters nearby water intakes; notify operators. Notify local health and wildlife officials. **Response to discharge:** Issue warning–poison, water contaminant. Restrict access. Should be removed. Chemical and physical treatment.

SHIPPING INFORMATION
Grades of purity: C.P.: 99+%; **Storage temperature:** Ambient; **Inert atmosphere:** None; **Venting:** Open; **Stability during transport:** Stable; **Stability during transport:** Stable

PHYSICAL AND CHEMICAL PROPERTIES
Physical state @ 59°F/15°C and 1 atm: Solid.
Molecular weight: 318.7
Boiling point @ 1 atm: Decomposes.
Melting/Freezing point: 350°F/177°C/450°K (decomposes).
Specific gravity (water = 1): 3.27 @ 68°F/20°C (solid).

MERCURIC AMMONIUM CHLORIDE   REC. M:0550

SYNONYMS: ALBUS; AMINOMERCURIC CHLORIDE; AMMONIATED MERCURY; CLORURO MERCURICO AMONICAL (Spanish); MERCURIC CHLORIDE, AMMONIATED; MERCURY AMMONIUM CHLORIDE; MERCURY AMIDE CHLORIDE; WHITE MERCURY PRECIPITATE

IDENTIFICATION
CAS Number: 10124-48-8
Formula: ClH$_2$HgN; HgNH$_2$Cl
DOT ID Number: UN 1630; DOT Guide Number: 151
Proper Shipping Name: Mercury ammonium chloride

DESCRIPTION: White lumpy or powdered solid. Odorless. Sinks in water.

Containers may BLEVE when exposed to fire • Irritating to the skin, eyes, and respiratory tract • Toxic products of combustion may include hydrogen chloride, nitrogen oxides, and mercury fumes.

Hazard Classification (based on NFPA-704 Rating System)
Health Hazards (Blue): 3; Flammability (Red): 0; Reactivity (Yellow): 0

EMERGENCY RESPONSE: See Appendix A (151)
Evacuation:
Public safety: Isolate the area of spill or leak for at least 25 to 50 meters (80 to 160 feet) in all directions.
Spill: Increase, in the downwind direction, as necessary, the distance shown under "Public Safety."
Fire: If tank, rail car, or tank truck is involved in fire, isolate for at least 800 meters (½ mile) in all directions; also, consider initial evacuation for 800 meters (½ mile) in all directions.

EXPOSURE
Short-term effects: *SEEK MEDICAL ATTENTION.* **Dust:** *POISONOUS IF INHALED.* Move victim to fresh air. *IF IN EYES,* hold eyelids open and flush with plenty of water. IF breathing is difficult, administer oxygen. IF breathing is difficult, administer oxygen. **Solid:** *POISONOUS IF SWALLOWED. IF SWALLOWED* and victim is *CONSCIOUS AND ABLE TO SWALLOW,* have victim drink 4 to 8 ounces of water and have victim induce vomiting. *IF SWALLOWED* and victim is *UNCONSCIOUS OR HAVING CONVULSIONS,* do nothing except keep victim warm.

HEALTH HAZARDS
Personal protective equipment (PPE): B-Level PPE. OSHA Table Z-1-A air contaminant. Gloves, goggles, respirator.
Recommendations for respirator selection: NIOSH/OSHA as other mercury compounds. See record M:1100 for mercury vapor. *up to 1 mg/m³*: CCRS** [any chemical cartridge respirator with cartridge(s) providing protection against the compound of concern]; or SA (any supplied-air respirator). *up to 2.5 mg/m³*: SA:CF (any supplied-air respirator operated in a continuous-flow mode); or PAPRS** [any powered, air-purifying respirator with cartridge(s) providing protection against the compound of concern]. *up to 5 mg/m³*: CCRFS** [any chemical cartridge respirator with a full facepiece and cartridge(s) providing protection against the compound of concern]; or GMFS [any air-purifying, full-facepiece respirator (gas mask) with a chin-style, front- or back-mounted canister providing protection against the compound of concern]; or SAT:CF (any supplied-air respirator that has a tight-fitting facepiece and is operated in a continuous-flow mode); or PAPRTS [any powered, air-purifying respirator with a tight-fitting facepiece and cartridge(s) providing protection against the compound of concern]; or SCBAF (any self-contained breathing apparatus with a full facepiece); or SAF (any supplied-air respirator with a full facepiece). *up to 10 mg/m³*: SA:PD,PP (any supplied-air respirator operated in a pressure-demand or other positive-pressure mode). *EMERGENCY OR PLANNED ENTRY INTO UNKNOWN CONCENTRATIONS OR IDLH CONDITIONS:* SCBAF:PD,PP (any self-contained breathing apparatus that has a full facepiece and is operated in a pressure-demand or other positive-pressure mode); or SAF:PD,PP:ASCBA (any supplied-air respirator that has a full facepiece and is operated in a pressure-demand or other positive-pressure mode in combination with an auxiliary, self-contained breathing apparatus operated in a pressure-demand or other positive-pressure mode). *ESCAPE:* GMFS** [any air-purifying, full-facepiece respirator (gas mask) with a chin-style, front- or back-mounted canister protection against the compound of concern]; or SCBAE (any appropriate escape-type, self-contained breathing apparatus). ***Note:*** End of service life indicator (ESLI) required.
Exposure limits (TWA unless otherwise noted): ACGIH TLV 0.025 mg/m³; OSHA PEL 0.1 mg/m³; NIOSH REL 0.05 mg/m³; ceiling 0.1 mg/m³ as inorganic mercury; skin contact contributes significantly in overall exposure.
Toxicity by ingestion: Poisonous.
Long-term health effects: Intestinal bleeding and kidney damage may develop. Lethal blood level of inorganic mercury in blood (serum or plasma) in humans is 0.04–2.2 mg%; 0.4–22 mg/mL.
Vapor (gas) irritant characteristics: May cause mercury poisoning.
Liquid or solid irritant characteristics: Severe eye irritant. Skin irritant.
IDLH value: 10 mg/m³ as mercury.

CHEMICAL REACTIVITY
Binary reactants: Violent reaction with halogens and metal salts of amines.

ENVIRONMENTAL DATA
Food chain concentration potential: Log P_{ow} = 2.2. Unlikely to accumulate.
Water pollution: DOT Appendix B, §172.101- severe marine pollutant. Harmful to aquatic life in very low concentrations. May be dangerous if it enters nearby water intakes; notify operators. Notify local health and wildlife officials. Restrict access. Should be removed. Chemical and physical treatment. **Response to discharge:** Issue warning–poison, water contaminant.

SHIPPING INFORMATION
Grades of purity: U.S.P.: 98+%; **Storage temperature:** Ambient; **Inert atmosphere:** None; **Venting:** Open; **Stability during transport:** Stable; **Stability during transport:** Stable.

PHYSICAL AND CHEMICAL PROPERTIES
Physical state @ 59°F/15°C and 1 atm: Solid.
Molecular weight: 252.1
Boiling point @ 1 atm: (sublimes at red heat).
Melting/Freezing point: NA (infusible).
Specific gravity (water = 1): 5.7 @ 68°F/20°C (solid).

MERCURIC CHLORIDE REC. M:0600

SYNONYMS: BICHLORIDE OF MERCURY; BICHLORURE de MERCURE (French); CALOCHLOR; CHLORURE MERCURIQUE (French); CLORURO MERCURICO (Spanish); CORROSIVE MERCURY CHLORIDE; EEC No. 080-002-00-6; FUNGCHEX; MC; MERCURIC BICHLORIDE; MERCURY BICHLORIDE; MERCURY(II) CHLORIDE; MERCURY CHLORIDE; MERCURY PERCHLORIDE; QUECKSILBER CHLORID (German); PERCHLORIDE OF MERCURY; TL 898

IDENTIFICATION
CAS Number: 7487-94-7
Formula: Cl_2Hg; $HgCl_2$
DOT ID Number: UN 1624; DOT Guide Number: 154
Proper Shipping Name: Mercuric chloride
Reportable Quantity (RQ): 1 lb/0.454 kg

DESCRIPTION: White or colorless solid. Sinks and mixes slowly with water.

Irritating to the skin, eyes, and respiratory tract • Toxic products of combustion may include mercury and hydrogen chloride.

Hazard Classification (based on NFPA-704 Rating System)
Health Hazards (Blue): 3; Flammability (Red): 0; Reactivity (Yellow): 0

EMERGENCY RESPONSE: See Appendix A (154)
Evacuation:
Public safety: Isolate the area of spill or leak for at least 25 to 50 meters (80 to 160 feet) in all directions.
Spill: Increase, in the downwind direction, as necessary, the distance shown under "Public Safety."
Fire: If tank, rail car, or tank truck is involved in fire, isolate for at least 800 meters (½ mile) in all directions; also, consider initial evacuation for 800 meters (½ mile) in all directions.

EXPOSURE
Short-term effects: *SEEK MEDICAL ATTENTION.* **Dust:** *POISONOUS IF INHALED OR IF SKIN IS EXPOSED. IF INHALED*, will, will cause coughing or difficult breathing. *IF IN EYES*, hold eyelids open and flush with plenty of water. *IF BREATHING HAS STOPPED*, give artificial respiration; *avoid mouth-to-mouth resuscitation; use bag/mask apparatus.* IF breathing is difficult, administer oxygen. **Solid:** *POISONOUS IF SWALLOWED OR IF SKIN IS EXPOSED*. Irritating to skin and eyes. *IF SWALLOWED*, will cause nausea and vomiting. Remove contaminated clothing and shoes. Flush affected areas with plenty of water. *IF IN EYES*, hold eyelids open and flush with plenty of water. *IF SWALLOWED* and victim is *CONSCIOUS AND ABLE TO SWALLOW*, have victim drink 4 to 8 ounces of water and have victim induce vomiting. *IF SWALLOWED* and victim is *UNCONSCIOUS OR HAVING CONVULSIONS*, do nothing except keep victim warm.

HEALTH HAZARDS
Personal protective equipment (PPE): B-Level PPE. OSHA Table Z-1-A air contaminant. Bureau of Mines approved airline respirator; impervious suit; appropriate eye protection.
Recommendations for respirator selection: NIOSH/OSHA as other mercury compounds. See record M:1100 for mercury vapor. *up to 1 mg/m³*: CCRS** [any chemical cartridge respirator with cartridge(s) providing protection against the compound of concern]; or SA (any supplied-air respirator). *up to 2.5 mg/m³*: SA:CF (any supplied-air respirator operated in a continuous-flow mode); or PAPRS** [any powered, air-purifying respirator with cartridge(s) providing protection against the compound of concern]. *up to 5 mg/m³*: CCRFS** [any chemical cartridge respirator with a full facepiece and cartridge(s) providing protection against the compound of concern]; or GMFS [any air-purifying, full-facepiece respirator (gas mask) with a chin-style, front- or back-mounted canister providing protection against the compound of concern]; or SAT:CF (any supplied-air respirator that has a tight-fitting facepiece and is operated in a continuous-flow mode); or PAPRTS [any powered, air-purifying respirator with a tight-fitting facepiece and cartridge(s) providing protection against the compound of concern]; or SCBAF (any self-contained breathing apparatus with a full facepiece); or SAF (any supplied-air respirator with a full facepiece). *up to 10 mg/m³*: SA:PD,PP (any supplied-air respirator operated in a pressure-demand or other positive-pressure mode). *EMERGENCY OR PLANNED ENTRY INTO UNKNOWN CONCENTRATIONS OR IDLH CONDITIONS:* SCBAF:PD,PP (any self-contained breathing apparatus that has a full facepiece and is operated in a pressure-demand or other positive-pressure mode); or SAF:PD,PP:ASCBA (any supplied-air respirator that has a full facepiece and is operated in a pressure-demand or other positive-pressure mode in combination with an auxiliary, self-contained breathing apparatus operated in a pressure-demand or other positive-pressure mode). *ESCAPE:* GMFS** [any air-purifying, full-facepiece respirator (gas mask) with a chin-style, front- or back-mounted canister protection against the compound of concern]; or SCBAE (any appropriate escape-type, self-contained breathing apparatus). **Note:* End of service life indicator (ESLI) required.
Exposure limits (TWA unless otherwise noted): ACGIH TLV 0.025 mg/m³; OSHA PEL 0.1 mg/m³; NIOSH REL 0.05 mg/m³; ceiling 0.1 mg/m³ as mercury; skin contact contributes significantly in overall exposure.
Toxicity by ingestion: Poison. Grade 4; oral LD_{50} = 1 mg/kg (rat).
Long-term health effects: Mutagen. Reproduction hazard. Lethal blood level of inorganic mercury in blood (serum or plasma) in humans is 0.04–2.2 mg%; 0.4–22 mg/mL.
Vapor (gas) irritant characteristics: Highly poisonous.
Liquid or solid irritant characteristics: Severe eye and skin irritant.
IDLH value: 10 mg/m³ as mercury.

CHEMICAL REACTIVITY
Binary reactants: Violent reaction with potassium and sodium.

ENVIRONMENTAL DATA
Food chain concentration potential: Log P_{ow} = 0.1. Unlikely to accumulate.
Water pollution: DOT Appendix B, §172.101- severe marine pollutant. Harmful to aquatic life in very low concentrations. May be dangerous if it enters nearby water intakes; notify operators. Notify local health and wildlife officials. **Response to discharge:** Issue warning–poison, water contaminant. Restrict access. Disperse and flush.

SHIPPING INFORMATION
Grades of purity: Reagent; Analytical; **Storage temperature:** Ambient; **Inert atmosphere:** None; **Venting:** Open; **Stability during transport:** Stable; **Stability during transport:** Stable

PHYSICAL AND CHEMICAL PROPERTIES
Physical state @ 59°F/15°C and 1 atm: Solid.
Molecular weight: 271.50
Boiling point @ 1 atm: 576°F/302°C/575°K.
Melting/Freezing point: 531°F/277°C/550°K.
Specific gravity (water = 1): 5.4 @ 68°F/20°C (solid).
Heat of fusion: 15.3 cal/g.

MERCURIC CYANIDE   **REC. M:0650**

SYNONYMS: CIANURO MERCURICO (Spanish); CYANURE de MERCURE (French); MERCURY(II) CYANIDE; MERCURY(2+) CYANIDE; CIANURINA; MERCURY CYANIDE

IDENTIFICATION
CAS Number: 592-04-1
Formula: $Hg(CN)_2$
DOT ID Number: UN 1636; DOT Guide Number: 154
Proper Shipping Name: Mercury cyanide
Reportable Quantity (RQ): **(CERCLA)** 1 lb/0.454 kg

DESCRIPTION: Colorless or white crystals or powder. Odorless. Sinks and mixes slowly with water.

Poison • Breathing the vapor, skin or eye contact, or swallowing the material can cause illness and possible death • Firefighting gear (including SCBA) does not provide adequate protection. If exposure occurs, remove and isolate gear immediately and thoroughly decontaminate personnel • Containers may BLEVE when exposed to fire • Irritating to the skin, eyes, and respiratory tract • Friction-/impact-sensitive explosive • Toxic products of combustion may include mercury and hydrogen cyanide fumes.

Hazard Classification (based on NFPA-704 Rating System)
Health Hazards (Blue): 3; Flammability (Red): 0; Reactivity (Yellow): 0

EMERGENCY RESPONSE: See Appendix A (154)
Evacuation:
Public safety: Isolate the area of spill or leak for at least 25 to 50 meters (80 to 160 feet) in all directions.
Spill: Increase, in the downwind direction, as necessary, the distance shown under "Public Safety."
Fire: If tank, rail car, or tank truck is involved in fire, isolate for at least 800 meters (½ mile) in all directions; also, consider initial evacuation for 800 meters (½ mile) in all directions.

EXPOSURE
Short-term effects: *SEEK MEDICAL ATTENTION.* **Dust:** *POISONOUS IF INHALED OR IF SKIN IS EXPOSED. IF INHALED*, will, will cause coughing or difficult breathing. *IF IN EYES*, hold eyelids open and flush with plenty of water. *IF BREATHING HAS STOPPED*, give artificial respiration; *avoid mouth-to-mouth resuscitation; use bag/mask apparatus.* IF breathing is difficult, administer oxygen. **Solid:** *POISONOUS IF SWALLOWED OR IF SKIN IS EXPOSED.* Irritating to skin and eyes. *IF SWALLOWED*, will cause nausea and vomiting. Remove contaminated clothing and shoes. Flush affected areas with plenty of water. *IF IN EYES*, hold eyelids open and flush with plenty of water. *IF SWALLOWED* and victim is *CONSCIOUS:* Alimentary absorption is very rapid; action during first 10-15 minutes determines prognosis. Give egg whites, milk, or activated charcoal and induce vomiting; treat for cyanide poisoning *IF SWALLOWED* and victim is *UNCONSCIOUS OR HAVING CONVULSIONS*, do nothing except keep victim warm.
Note to physician or authorized medical personnel: Consider the use of amyl nitrite perles if symptoms of cyanide poisoning develop. If symptoms indicate, initial treatment includes the cyanide antidote kit. In all cases, break an amyl nitrite perle in a gauze pad and hold lightly under victim's nose for 15 seconds, repeating 5 times at about 15-second intervals; if necessary (and if sodium nitrite infusions will be delayed), repeat procedure every 3 minutes with fresh pearls until 3 or 4 have been used. Avoid breathing the vapor while administering it to the victim. Administer sodium nitrite IV, ASAP. The usual adult dose is 10 to 20 mL of a 3% solution infused over no less than 5 minutes; the average child dose is 0.15 to 0.20 mL/kg. Monitor blood pressure during administration, and slow the rate of infusion if hypotention develops. Next, infuse sodium thiosulfate IV. The usual adult dose is 50 mL of a 25% solution infused over 10 to 20 minutes; the average child dose is 1.65 mL/kg. Repeat with nitrite and thiosulfate as required.

HEALTH HAZARDS
Personal protective equipment (PPE): B-Level PPE. OSHA Table Z-1-A air contaminant. Dust mask; goggles or face shield; rubber gloves.

Recommendations for respirator selection: NIOSH/OSHA as other mercury compounds. See record M:1100 for mercury vapor. *up to 1 mg/m³:* CCRS** [any chemical cartridge respirator with cartridge(s) providing protection against the compound of concern]; or SA (any supplied-air respirator). *up to 2.5 mg/m³:* SA:CF (any supplied-air respirator operated in a continuous-flow mode); or PAPRS** [any powered, air-purifying respirator with cartridge(s) providing protection against the compound of concern]. *up to 5 mg/m³:* CCRFS** [any chemical cartridge respirator with a full facepiece and cartridge(s) providing protection against the compound of concern]; or GMFS [any air-purifying, full-facepiece respirator (gas mask) with a chin-style, front- or back-mounted canister providing protection against the compound of concern]; or SAT:CF (any supplied-air respirator that has a tight-fitting facepiece and is operated in a continuous-flow mode); or PAPRTS [any powered, air-purifying respirator with a tight-fitting facepiece and cartridge(s) providing protection against the compound of concern]; or SCBAF (any self-contained breathing apparatus with a full facepiece); or SAF (any supplied-air respirator with a full facepiece). *up to 10 mg/m³:* SA:PD,PP (any supplied-air respirator operated in a pressure-demand or other positive-pressure mode). *EMERGENCY OR PLANNED ENTRY INTO UNKNOWN CONCENTRATIONS OR IDLH CONDITIONS:* SCBAF:PD,PP (any self-contained breathing apparatus that has a full facepiece and is operated in a pressure-demand or other positive-pressure mode); or SAF:PD,PP:ASCBA (any supplied-air respirator that has a full facepiece and is operated in a pressure-demand or other positive-pressure mode in combination with an auxiliary, self-contained breathing apparatus operated in a pressure-demand or other positive-pressure mode). *ESCAPE:* GMFS** [any air-purifying, full-facepiece respirator (gas mask) with a chin-style, front- or back-mounted canister protection against the compound of concern]; or SCBAE (any appropriate escape-type, self-contained breathing apparatus). **Note:* End of service life indicator (ESLI) required.
Exposure limits (TWA unless otherwise noted): ACGIH TLV 0.025 mg/m³; OSHA PEL ceiling limit 0.1 mg/m³; NIOSH REL (vapor) 0.05 mg/m³; other: ceiling 0.1 mg/m³ as mercury; skin contact contributes significantly in overall exposure.
Toxicity by ingestion: Poison. Grade 4; oral LD_{50} = 25 mg/kg (rat). Kidney damage.
Long-term health effects: Possible kidney damage. Lethal blood level of inorganic mercury in blood (serum or plasma) in humans is 0.04–2.2 mg%; 0.4–22 mg/mL.
Vapor (gas) irritant characteristics: Poisonous.
Liquid or solid irritant characteristics: Severe eye and skin irritant.
IDLH value: 10 mg/m³ as mercury.

CHEMICAL REACTIVITY
Binary reactants: Contact with any acidic material will form poisonous hydrogen cyanide gas, which may collect in enclosed spaces. React with fluorine and magnesium.

ENVIRONMENTAL DATA
Food chain concentration potential: Possible bioaccumulation problem. Many organisms can accumulate mercury from water. Bioconcentrative up to 10,000-fold.
Water pollution: DOT Appendix B, §172.101- severe marine pollutant. Harmful to aquatic life in very low concentrations. May be dangerous if it enters nearby water intakes; notify operators. Notify local health and wildlife officials. **Response to discharge:** Issue warning–poison, water contaminant. Restrict access. Disperse and flush.

SHIPPING INFORMATION
Grades of purity: Reagent; **Storage temperature:** Ambient; **Inert atmosphere:** None; **Venting:** Open; **Stability during transport:** Stable; **Stability during transport:** Stable.

PHYSICAL AND CHEMICAL PROPERTIES
Physical state @ 59°F/15°C and 1 atm: Solid.
Molecular weight: 252.63
Boiling point @ 1 atm: (decomposes).
Specific gravity (water = 1): 4.0 @ 68°F/20°C (solid).

MERCURIC IODIDE   REC. M:0700

SYNONYMS: HYDRARGYRUM BIJODATUM (German); MERCURIC IODIDE, RED; MERCURY BINIODIDE; MERCURY(II) IODIDE; MERCURY(2+) IODIDE; RED MERCURIC IODIDE; YODURO MERCURICO (Spanish)

IDENTIFICATION
CAS Number: 7774-29-0
Formula: HgI_2
DOT ID Number: UN 1638; DOT Guide Number: 151
Proper Shipping Name: Mercury iodide

DESCRIPTION: Scarlet red powder. Solution is red. Odorless. Sinks in water; slightly soluble.

Poison! • Irritating to the skin, eyes, and respiratory tract • Swallowing the material can cause death • Firefighting gear (including SCBA) does not provide adequate protection. If exposure occurs, remove and isolate gear immediately and thoroughly decontaminate personnel • Containers may BLEVE when exposed to fire • Toxic products of combustion may include mercury and iodide fumes.

Hazard Classification (based on NFPA-704 Rating System)
Health Hazards (Blue): 3; Flammability (Red): 0; Reactivity (Yellow): 0

EMERGENCY RESPONSE: See Appendix A (151)
Evacuation:
Public safety: Isolate the area of spill or leak for at least 25 to 50 meters (80 to 160 feet) in all directions.
Spill: Increase, in the downwind direction, as necessary, the distance shown under "Public Safety."
Fire: If tank, rail car, or tank truck is involved in fire, isolate for at least 800 meters (½ mile) in all directions; also, consider initial evacuation for 800 meters (½ mile) in all directions.

EXPOSURE
Short-term effects: *SEEK MEDICAL ATTENTION.* **Dust:** *POISONOUS IF INHALED OR IF SKIN IS EXPOSED. IF INHALED,* will, will cause coughing or difficult breathing. *IF IN EYES,* hold eyelids open and flush with plenty of water. IF breathing is difficult, administer oxygen. IF breathing is difficult, administer oxygen. **Solid:** *POISONOUS IF SWALLOWED OR IF SKIN IS EXPOSED.* Irritating to skin and eyes. *IF SWALLOWED,* will cause nausea and vomiting. Remove contaminated clothing and shoes. Flush affected areas with plenty of water. *IF IN EYES,* hold eyelids open and flush with plenty of water. *IF SWALLOWED* and victim is *CONSCIOUS AND ABLE TO SWALLOW,* have victim drink 4 to 8 ounces of water and have victim induce vomiting. *IF SWALLOWED* and victim is *UNCONSCIOUS OR HAVING CONVULSIONS,* do nothing except keep victim warm.

HEALTH HAZARDS
Personal protective equipment (PPE): B-Level PPE. OSHA Table Z-1-A air contaminant. Dust mask; goggles or face shield; protective gloves.
Recommendations for respirator selection: NIOSH/OSHA as other mercury compounds. See record M:1100 for mercury vapor. *up to 1 mg/m³:* CCRS** [any chemical cartridge respirator with cartridge(s) providing protection against the compound of concern]; or SA (any supplied-air respirator). *up to 2.5 mg/m³:* SA:CF (any supplied-air respirator operated in a continuous-flow mode); or PAPRS** [any powered, air-purifying respirator with cartridge(s) providing protection against the compound of concern]. *up to 5 mg/m³:* CCRFS** [any chemical cartridge respirator with a full facepiece and cartridge(s) providing protection against the compound of concern]; or GMFS [any air-purifying, full-facepiece respirator (gas mask) with a chin-style, front- or back-mounted canister providing protection against the compound of concern]; or SAT:CF (any supplied-air respirator that has a tight-fitting facepiece and is operated in a continuous-flow mode); or PAPRTS [any powered, air-purifying respirator with a tight-fitting facepiece and cartridge(s) providing protection against the compound of concern]; or SCBAF (any self-contained breathing apparatus with a full facepiece); or SAF (any supplied-air respirator with a full facepiece). *up to 10 mg/m³:* SA:PD,PP (any supplied-air respirator operated in a pressure-demand or other positive-pressure mode). *EMERGENCY OR PLANNED ENTRY INTO UNKNOWN CONCENTRATIONS OR IDLH CONDITIONS:* SCBAF:PD,PP (any self-contained breathing apparatus that has a full facepiece and is operated in a pressure-demand or other positive-pressure mode); or SAF:PD,PP:ASCBA (any supplied-air respirator that has a full facepiece and is operated in a pressure-demand or other positive-pressure mode in combination with an auxiliary, self-contained breathing apparatus operated in a pressure-demand or other positive-pressure mode). *ESCAPE:* GMFS** [any air-purifying, full-facepiece respirator (gas mask) with a chin-style, front- or back-mounted canister protection against the compound of concern]; or SCBAE (any appropriate escape-type, self-contained breathing apparatus). **Note:* End of service life indicator (ESLI) required.
Exposure limits (TWA unless otherwise noted): ACGIH TLV 0.025 mg/m³; OSHA PEL ceiling limit 0.1 mg/m³; NIOSH REL (vapor) 0.05 mg/m³; other: ceiling 0.1 mg/m³ as mercury; ; skin contact contributes significantly in overall exposure.
Toxicity by ingestion: Poison. Grade 4; oral LD_{50} = 40 mg/kg (rat).
Long-term health effects: Lethal blood level of inorganic mercury in blood (serum or plasma) in humans is 0.04–2.2 mg%; 0.4–22 mg/mL.
Vapor (gas) irritant characteristics: Highly poisonous.
Liquid or solid irritant characteristics: Severe eye ans skin irritant. Can be absorbed through the skin.
IDLH value: 10 mg/m³ as mercury.

CHEMICAL REACTIVITY
Binary reactants: A Strong oxidizer. Violent reaction with reducing agents, acrolein, alcohols, hydrazine, metal powders.

ENVIRONMENTAL DATA
Food chain concentration potential: Many organisms can accumulate mercury from water. Bioconcentrative up to 10,000-fold. **Water pollution:** DOT Appendix B, §172.101–marine pollutant. Effect of low concentrations on aquatic life is unknown. May be dangerous if it enters nearby water intakes; notify operators. Notify local health and wildlife officials. **Response to discharge:** Issue warning–poison, water contaminant. Restrict access. Should be removed. Chemical and physical treatment.

SHIPPING INFORMATION

Grades of purity: Reagent, 99+%; **Storage temperature:** Ambient; **Inert atmosphere:** None; **Venting:** Open; **Stability during transport:** Stable; **Stability during transport:** Stable.

PHYSICAL AND CHEMICAL PROPERTIES

Physical state @ 59°F/15°C and 1 atm: Solid.
Molecular weight: 454.90
Boiling point @ 1 atm: 669°F/354°C/627°K.
Melting/Freezing point: 495°F/257°C/530°K.
Specific gravity (water = 1): 6.3 @ 68°F/20°C (solid).
Heat of fusion: 9.9 cal/g.

MERCURIC NITRATE   REC. M:0750

SYNONYMS: EEC No. 080-002-00-6; MERCURY(II) NITRATE; MERCURY PERNITRATE; NITRATE MERCURIQUE (French); NITRATO MERCURICO (Spanish); NITRIC ACID, MERCURY(II) SALT; MERCURY NITRATE MONOHYDRATE

IDENTIFICATION

CAS Number: 10045-94-0
Formula: $N_2O_6 \cdot Hg$; $Hg(NO_3)_2 \cdot H_2O$
DOT ID Number: UN 1625; DOT Guide Number: 141
Proper Shipping Name: Mercuric nitrate
Reportable Quantity (RQ): **(CERCLA)** 10 lb/4.54 kg

DESCRIPTION: Yellow hygroscopic crystals or yellow powder. Sharp odor of nitric acid. Sinks in water; soluble; forms a corrosive solution.

Poison • Corrosive to skin, eyes, and respiratory tract; inhalation symptoms may be delayed • Firefighting gear (including SCBA) does not provide adequate protection. If exposure occurs, remove and isolate gear immediately and thoroughly decontaminate personnel • Containers may BLEVE when exposed to fire • Strong oxidizer which may react spontaneously with low flash point organics or reducing agents. Heat forms oxygen; will increase the activity of an existing fire • May cause fire or explosion on contact with combustibles (wood, paper, oil, clothing, etc.) • Toxic products of combustion may include mercury and nitrogen oxides.

Hazard Classification (based on NFPA-704 Rating System)
Health Hazards (Blue): 3; Flammability (Red): 0; Reactivity (Yellow): 0; Special Notice (White): OXY

EMERGENCY RESPONSE: See Appendix A (141)
Evacuation:
Public safety: Isolate the area of spill or leak for at least 10 to 25 meters (30 to 80 feet) in all directions.
Spill: Consider initial downwind evacuation for at least 100 meters (330 feet).
Fire: If any large container is involved in fire, isolate for at least 800 meters (½ mile) in all directions; also, consider initial evacuation for 800 meters (½ mile) in all directions.

EXPOSURE

Short-term effects: *SEEK MEDICAL ATTENTION.* **Dust:** *POISONOUS IF INHALED OR IF SKIN IS EXPOSED. IF INHALED,* will, will cause coughing or difficult breathing. *IF IN EYES,* hold eyelids open and flush with plenty of water. *IF BREATHING HAS STOPPED,* give artificial respiration. IF breathing is difficult, administer oxygen. Lung edema may develop. **Solid:** *POISONOUS IF SWALLOWED OR IF SKIN IS EXPOSED.* Irritating to skin and eyes. *IF SWALLOWED,* will cause nausea and vomiting. Remove contaminated clothing and shoes. Flush affected areas with plenty of water. *IF IN EYES,* hold eyelids open and flush with plenty of water. *IF SWALLOWED* and victim is *CONSCIOUS AND ABLE TO SWALLOW,* have victim drink 4 to 8 ounces of water and have victim induce vomiting. *IF SWALLOWED* and victim is *UNCONSCIOUS OR HAVING CONVULSIONS,* do nothing except keep victim warm.

Note to physician or authorized medical personnel: Medical observation is recommended for 24 to 48 hours after breathing overexposure, as pulmonary edema may be delayed. As first aid for pulmonary edema, consider administering a corticosteroid spray. Cigarette smoking may exacerbate pulmonary injury and should be discouraged for at least 72 hours following exposure.

HEALTH HAZARDS

Personal protective equipment (PPE): B-Level PPE. OSHA Table Z-1-A air contaminant. Dust mask; goggles or face shield; protective gloves.

Recommendations for respirator selection: NIOSH/OSHA as other mercury compounds. See record M:1100 for mercury vapor. *up to 1 mg/m^3:* CCRS** [any chemical cartridge respirator with cartridge(s) providing protection against the compound of concern]; or SA (any supplied-air respirator). *up to 2.5 mg/m^3:* SA:CF (any supplied-air respirator operated in a continuous-flow mode); or PAPRS** [any powered, air-purifying respirator with cartridge(s) providing protection against the compound of concern]. *up to 5 mg/m^3:* CCRFS** [any chemical cartridge respirator with a full facepiece and cartridge(s) providing protection against the compound of concern]; or GMFS [any air-purifying, full-facepiece respirator (gas mask) with a chin-style, front- or back-mounted canister providing protection against the compound of concern]; or SAT:CF (any supplied-air respirator that has a tight-fitting facepiece and is operated in a continuous-flow mode); or PAPRTS [any powered, air-purifying respirator with a tight-fitting facepiece and cartridge(s) providing protection against the compound of concern]; or SCBAF (any self-contained breathing apparatus with a full facepiece); or SAF (any supplied-air respirator with a full facepiece). *up to 10 mg/m^3:* SA:PD,PP (any supplied-air respirator operated in a pressure-demand or other positive-pressure mode). *EMERGENCY OR PLANNED ENTRY INTO UNKNOWN CONCENTRATIONS OR IDLH CONDITIONS:* SCBAF:PD,PP (any self-contained breathing apparatus that has a full facepiece and is operated in a pressure-demand or other positive-pressure mode); or SAF:PD,PP:ASCBA (any supplied-air respirator that has a full facepiece and is operated in a pressure-demand or other positive-pressure mode in combination with an auxiliary, self-contained breathing apparatus operated in a pressure-demand or other positive-pressure mode). *ESCAPE:* GMFS** [any air-purifying, full-facepiece respirator (gas mask) with a chin-style, front- or back-mounted canister protection against the compound of concern]; or SCBAE (any appropriate escape-type, self-contained breathing apparatus). **Note:* End of service life indicator (ESLI) required.

Exposure limits (TWA unless otherwise noted): ACGIH TLV 0.025 mg/m^3; OSHA PEL ceiling limit 0.1 mg/m^3; NIOSH REL (vapor) 0.05 mg/m^3; other: ceiling 0.1 mg/m^3 as mercury; ; skin contact contributes significantly in overall exposure.

Toxicity by ingestion: Poison.
Long-term health effects: Lethal blood level of inorganic mercury in blood (serum or plasma) in humans is 0.04–2.2 mg%; 0.4–22 mg/mL.

Vapor (gas) irritant characteristics: Highly poisonous.
Liquid or solid irritant characteristics: Severe eye and skin irritant. Can be absorbed through the skin.
IDLH value: 10 mg/m^3 as mercury.

FIRE DATA
Behavior in fire: May increase intensity of fire if in contact with burning material.

CHEMICAL REACTIVITY
Reactivity with water: Dissolves, then forms cloudy acid solution.
Binary reactants: A powerful oxidizer; may cause fire when in contact with wood, paper or other combustibles. Solution corrodes most metals.
Neutralizing agents for acids and caustics: Flush well with water, rinse with dilute solution of sodium bicarbonate or soda ash.

ENVIRONMENTAL DATA
Food chain concentration potential: Possible bioaccumulation problem. Many organisms can accumulate mercury from water. Bioconcentrative up to 10,000-fold.
Water pollution: DOT Appendix B, §172.101- severe marine pollutant. Effect of low concentrations on aquatic life is unknown. May be dangerous if it enters nearby water intakes; notify operators. Notify local health and wildlife officials. **Response to discharge:** Issue warning–poison, water contaminant, oxidizing material. Restrict access. Disperse and flush.

SHIPPING INFORMATION
Grades of purity: Reagent, 99.0%; **Storage temperature:** Ambient; **Inert atmosphere:** None; **Venting:** Open; **Stability during transport:** Stable; **Stability during transport:** Stable

PHYSICAL AND CHEMICAL PROPERTIES
Physical state @ 59°F/15°C and 1 atm: Solid.
Molecular weight: 342.6
Boiling point @ 1 atm: (decomposes).
Melting/Freezing point: 175°F/79°C/352°K.
Specific gravity (water = 1): 4.3 @ 68°F/20°C (solid).

MERCURIC OXIDE **REC. M:0800**

SYNONYMS: EEC No. 080-002-00-6; MERCURIC OXIDE, RED; MERCURIC OXIDE, YELLOW; OXIDO MERCURICO ROJO (Spanish); RED OXIDE OF MERCURY; SANTAR; MERCURY OXIDE; YELLOW OXIDE OF MERCURY

IDENTIFICATION
CAS Number: 21908-53-2
Formula: HgO
DOT ID Number: UN 1641; DOT Guide Number: 151
Proper Shipping Name: Mercury oxide
Reportable Quantity (RQ): **(EHS)** 1 lb/0.454 kg

DESCRIPTION: Red or orange-yellow powder. Yellow powder is more reactive and finely divided; it turns dark on contact with air. Odorless. Sinks in water; red is insoluble; yellow is very slightly soluble.

Corrosive to the skin, eyes, and respiratory tract; symptoms of inhation my be delayed • Containers may BLEVE when exposed to fire • Strong oxidizer which may react spontaneously with low flash point organics or reducing agents. Heat forms oxygen; will increase the activity of an existing fire • May cause fire or explosion on contact with combustibles (wood, paper, oil, clothing, etc.) • Toxic products of combustion may include mercury vapors.

Hazard Classification (based on NFPA-704 Rating System)
Health Hazards (Blue): 3; Flammability (Red): 0; Reactivity (Yellow): 0; Special Notice (White): OXY

EMERGENCY RESPONSE: See Appendix A (151)
Evacuation:
Public safety: Isolate the area of spill or leak for at least 25 to 50 meters (80 to 160 feet) in all directions.
Spill: Increase, in the downwind direction, as necessary, the distance shown under "Public Safety."
Fire: If tank, rail car, or tank truck is involved in fire, isolate for at least 800 meters (½ mile) in all directions; also, consider initial evacuation for 800 meters (½ mile) in all directions.

EXPOSURE
Short-term effects: *SEEK MEDICAL ATTENTION.* **Dust:** *POISONOUS IF INHALED OR IF SKIN IS EXPOSED. IF INHALED*, will, will cause coughing or difficult breathing. *May cause lung edema; physical exertion will aggravate this condition. IF IN EYES,* hold eyelids open and flush with plenty of water. If breathing is difficult, administer oxygen. IF breathing is difficult, administer oxygen. **Solid:** *POISONOUS IF SWALLOWED OR IF SKIN IS EXPOSED.* Irritating to skin and eyes. *IF SWALLOWED*, will cause nausea or vomiting. Remove contaminated clothing and shoes. Flush affected areas with plenty of water. *IF IN EYES*, hold eyelids open and flush with plenty of water. *IF SWALLOWED* and victim is *CONSCIOUS AND ABLE TO SWALLOW,* have victim drink 4 to 8 ounces of water and have victim induce vomiting. *IF SWALLOWED* and victim is *UNCONSCIOUS OR HAVING CONVULSIONS,* do nothing except keep victim warm.
Note to physician or authorized medical personnel: Pulmonary edema may be delayed. Medical observation is recommended for 24 to 48 hours after inhalation overexposure. As first aid for pulmonary edema, a physician or authorized medical personnel may consider administering a corticosteroid spray. Cigarette smoking may exacerbate pulmonary injury and should be discouraged for at least 72 hours following exposure.

HEALTH HAZARDS
Personal protective equipment (PPE): B-Level PPE. OSHA Table Z-1-A air contaminant. Dust mask; goggles or face shield; protective gloves.
Recommendations for respirator selection: NIOSH/OSHA as other mercury compounds. See record M:1100 for mercury vapor. *up to 1 mg/m^3:* CCRS** [any chemical cartridge respirator with cartridge(s) providing protection against the compound of concern]; or SA (any supplied-air respirator). *up to 2.5 mg/m^3:* SA:CF (any supplied-air respirator operated in a continuous-flow mode); or PAPRS** [any powered, air-purifying respirator with cartridge(s) providing protection against the compound of concern]. *up to 5 mg/m^3:* CCRFS** [any chemical cartridge respirator with a full facepiece and cartridge(s) providing protection against the compound of concern]; or GMFS [any air-purifying, full-facepiece respirator (gas mask) with a chin-style, front- or back-mounted canister providing protection against the compound of concern]; or SAT:CF (any supplied-air respirator that has a tight-fitting facepiece and is operated in a continuous-flow mode); or PAPRTS [any powered, air-purifying respirator with a tight-fitting facepiece and cartridge(s) providing protection against the compound of concern]; or SCBAF (any self-contained breathing apparatus with a full facepiece); or SAF (any supplied-air respirator with a full

facepiece). *up to 10 mg/m³:* SA:PD,PP (any supplied-air respirator operated in a pressure-demand or other positive-pressure mode). *EMERGENCY OR PLANNED ENTRY INTO UNKNOWN CONCENTRATIONS OR IDLH CONDITIONS:* SCBAF:PD,PP (any self-contained breathing apparatus that has a full facepiece and is operated in a pressure-demand or other positive-pressure mode); or SAF:PD,PP:ASCBA (any supplied-air respirator that has a full facepiece and is operated in a pressure-demand or other positive-pressure mode in combination with an auxiliary, self-contained breathing apparatus operated in a pressure-demand or other positive-pressure mode). *ESCAPE:* GMFS** [any air-purifying, full-facepiece respirator (gas mask) with a chin-style, front- or back-mounted canister protection against the compound of concern]; or SCBAE (any appropriate escape-type, self-contained breathing apparatus). **Note:* End of service life indicator (ESLI) required.
Exposure limits (TWA unless otherwise noted): ACGIH TLV 0.025 mg/m³; OSHA PEL ceiling limit 0.1 mg/m³; NIOSH REL (vapor) 0.05 mg/m³; other: ceiling 0.1 mg/m³ as mercury; ; skin contact contributes significantly in overall exposure.
Toxicity by ingestion: Grade 4; oral LD_{50} = 18 mg/kg (rat).
Long-term health effects: Causes birth defects in rats. Possible teratogen. Kidney ans skin disorders. Lethal blood level of inorganic mercury in blood (serum or plasma) in humans is 0.04–2.2 mg%; 0.4–22 mg/mL.
Vapor (gas) irritant characteristics: May cause nose and respiratory irritation. Heavy exposure may cause lung edema; physical exertion will aggravate this condition.
Liquid or solid irritant characteristics: Corrosive to the eyes, skin, and respiratory tract.
IDLH value: 10 mg/m³ as mercury.

FIRE DATA
Behavior in fire: Decomposes at 932°F/500°C releasing oxygen, which can increase intensity of fire. Solid changes color when hot.

CHEMICAL REACTIVITY
Binary reactants: Reacts with combustible materials and alkali metals, reducing materials and many other compounds.

ENVIRONMENTAL DATA
Food chain concentration pot
Water pollution: DOT Appendix B, §172.101- severe marine pollutant. Effect of low concentrations on aquatic life is unknown. May be dangerous if it enters nearby water intakes; notify operators. Notify local health and wildlife officials. Restrict access. Should be removed. Chemical and physical treatment. **Response to discharge:** Issue warning–poison, water contaminant.

SHIPPING INFORMATION
Grades of purity: Red-technical; reagent; purified; yellow-technical; NF; reagent; **Storage temperature:** Ambient; **Inert atmosphere:** None; **Venting:** Open; **Stability during transport:** Stable; **Stability during transport:** Stable

PHYSICAL AND CHEMICAL PROPERTIES
Physical state @ 59°F/15°C and 1 atm: Solid.
Molecular weight: 216.61
Boiling point @ 1 atm: (decomposes).
Melting/Freezing point: 932°F/500°C/773°K (decomposes).
Specific gravity (water = 1): 11.1 @ 68°F/20°C (solid).
Heat of decomposition: 400°C.

MERCURIC SULFATE REC. M:0850

SYNONYMS: EEC No. 080-002-00-6; MERCURY(II)SULFATE; MERCURY(2+) SULFATE; MERCURY BISULFATE; MERCURY PERSULFATE; SULFATO MERCURICO (Spanish); SULFATE MERCURIQUE (French); SULFURIC ACID, MERCURY(2+) SALT

IDENTIFICATION
CAS Number: 7783-35-9
Formula: $HgSO_4$
DOT ID Number: UN 1645; DOT Guide Number: 151
Proper Shipping Name: Mercury sulfates
Reportable Quantity (RQ): **(CERCLA)** 10 lb/4.54 kg

DESCRIPTION: White crystalline powder. Odorless. Sinks in water; decomposes into insoluble, yellow basic mercury sulfate and sulfuric acid. Exposure to light causes material to decompose.

Poison! • Corrosive to skin, eyes, and respiratory tract; inhalation symptoms may be delayed • Persulfates are oxidizers. Heat forms oxygen; will increase the activity of an existing fire • May cause fire or explosion contact with combustibles (wood, paper, oil, clothing, etc.). • Containers may BLEVE when exposed to fire • Toxic products of combustion may include mercury and sulfur oxide fumes.

Hazard Classification (based on NFPA-704 Rating System)
Health Hazards (Blue): 3; Flammability (Red): 0; Reactivity (Yellow): 0

EMERGENCY RESPONSE: See Appendix A (151)
Evacuation:
Public safety: Isolate the area of spill or leak for at least 25 to 50 meters (80 to 160 feet) in all directions.
Spill: Increase, in the downwind direction, as necessary, the distance shown under "Public Safety."
Fire: If tank, rail car, or tank truck is involved in fire, isolate for at least 800 meters (½ mile) in all directions; also, consider initial evacuation for 800 meters (½ mile) in all directions.

EXPOSURE
Short-term effects: *SEEK MEDICAL ATTENTION.* **Dust:** Irritating to skin, eyes, and nose. *IF INHALED*, will, will cause coughing, pain, and breathing difficulty. May cause lung edema; physical exertion will aggravate this condition. Move to fresh air.*IF BREATHING HAS STOPPED,* give artificial respiration; *avoid mouth-to-mouth resuscitation; use bag/mask apparatus.* IF breathing is difficult, administer oxygen. **Solid:** *POISONOUS IF SWALLOWED.* Will burn skin and eyes. Remove contaminated clothing and shoes. Flush affected areas with plenty of water. *IF IN EYES,* hold eyelids open and flush with plenty of water. *IF SWALLOWED* and victim is *CONSCIOUS AND ABLE TO SWALLOW,* have victim drink 4 to 8 ounces of water and have victim induce vomiting. *IF SWALLOWED* and victim is *UNCONSCIOUS OR HAVING CONVULSIONS,* do nothing except keep victim warm.
Note to physician or authorized medical personnel: Pulmonary edema may be delayed. Medical observation is recommended for 24 to 48 hours after inhalation overexposure. As first aid for pulmonary edema, a physician or authorized medical personnel may consider administering a corticosteroid spray. Cigarette smoking may exacerbate pulmonary injury and should be discouraged for at least 72 hours following exposure.

HEALTH HAZARDS

Personal protective equipment (PPE): B-Level PPE. OSHA Table Z-1-A air contaminant. Self-contained breathing apparatus, rubber gloves, protective clothing, rubber apron, and safety goggles.

Recommendations for respirator selection: NIOSH/OSHA as other mercury compounds. See record M:1100 for mercury vapor. *up to 1 mg/m^3:* CCRS** [any chemical cartridge respirator with cartridge(s) providing protection against the compound of concern]; or SA (any supplied-air respirator). *up to 2.5 mg/m^3:* SA:CF (any supplied-air respirator operated in a continuous-flow mode); or PAPRS** [any powered, air-purifying respirator with cartridge(s) providing protection against the compound of concern]. *up to 5 mg/m^3:* CCRFS** [any chemical cartridge respirator with a full facepiece and cartridge(s) providing protection against the compound of concern]; or GMFS [any air-purifying, full-facepiece respirator (gas mask) with a chin-style, front- or back-mounted canister providing protection against the compound of concern]; or SAT:CF (any supplied-air respirator that has a tight-fitting facepiece and is operated in a continuous-flow mode); or PAPRTS [any powered, air-purifying respirator with a tight-fitting facepiece and cartridge(s) providing protection against the compound of concern]; or SCBAF (any self-contained breathing apparatus with a full facepiece); or SAF (any supplied-air respirator with a full facepiece). *up to 10 mg/m^3:* SA:PD,PP (any supplied-air respirator operated in a pressure-demand or other positive-pressure mode). *EMERGENCY OR PLANNED ENTRY INTO UNKNOWN CONCENTRATIONS OR IDLH CONDITIONS:* SCBAF:PD,PP (any self-contained breathing apparatus that has a full facepiece and is operated in a pressure-demand or other positive-pressure mode); or SAF:PD,PP:ASCBA (any supplied-air respirator that has a full facepiece and is operated in a pressure-demand or other positive-pressure mode in combination with an auxiliary, self-contained breathing apparatus operated in a pressure-demand or other positive-pressure mode). *ESCAPE:* GMFS** [any air-purifying, full-facepiece respirator (gas mask) with a chin-style, front- or back-mounted canister protection against the compound of concern]; or SCBAE (any appropriate escape-type, self-contained breathing apparatus). **Note:* End of service life indicator (ESLI) required.

Exposure limits (TWA unless otherwise noted): ACGIH TLV 0.025 mg/m^3; OSHA PEL ceiling limit 0.1 mg/m^3; NIOSH REL (vapor) 0.05 mg/m^3; other: ceiling 0.1 mg/m^3 as mercury; ; skin contact contributes significantly in overall exposure.

Toxicity by ingestion: Grade 4; LD$_{50}$ = 50 mg/kg.

Long-term health effects: Damage to kidney, heart, lung, and brain. Skin disorders. Psychic and emotional disturbances; fine tremors of hands, head, lips, tongue, or jaw. Salivation, gingivitis, and digestive disturbances are common. Stomatitis is sometimes severe. Lethal blood level of inorganic mercury in blood (serum or plasma) in humans is 0.04–2.2 mg%; 0.4–22 mg/mL.

Vapor (gas) irritant characteristics: Severe eye and respiratory tract irritant.

Liquid or solid irritant characteristics: Fairly severe skin and eye irritant. May cause pain and second-degree burns after a few minutes of contact.

IDLH value: 10 mg/m^3 as inorganic mercury

CHEMICAL REACTIVITY

Reactivity with water: Decomposes into yellow basic sulfate and sufuric acid. violent reaction with strong oxidizers, metal powders, organic materials, combustibles, etc.

Binary reactants: Violent reaction with aluminum, magnesium. Light causes decomposition and toxic fumes.

Neutralizing agents for acids and caustics: Flood with water and rinse with sodium bicarbonate or lime solution.

ENVIRONMENTAL DATA

Food chain concentration potential: Many organisms can bioaccumulate mercury from water. Bioconcentrative up to 10,000-fold.

Water pollution: DOT Appendix B, §172.101- severe marine pollutant. Harmful to aquatic life in very low concentrations. May be dangerous if it enters nearby water intakes; notify operators. Notify local health and wildlife officials. **Response to discharge:** Issue warning, water contaminant, poison. Restrict access. Should be removed. Chemical and physical treatment.

SHIPPING INFORMATION

Grades of purity: 100%; **Storage temperature:** Cool; **Stability during transport:** Stable.

PHYSICAL AND CHEMICAL PROPERTIES

Physical state @ 59°F/15°C and 1 atm: Solid.
Molecular weight: 296.68
Boiling point @ 1 atm: Decomposes.
Melting/Freezing point: Decomposes.
Specific gravity (water = 1): 6.47 at room temperature
Heat of fusion: 4.8 cal/g.

MERCURIC SULFIDE REC. M:0900

SYNONYMS: ARTIFICIAL CINNABAR; CHINESE RED; MERCURIC SULFIDE, BLACK; ETHIOPS MINERAL; MERCURIC SULFIDE, RED; SULFURO MERCURICO, NEGRO (Spanish); SULFURO MERCURICO, ROJO (Spanish); VERMILION

IDENTIFICATION
CAS Number: 1344-48-5
Formula: HgS
DOT ID Number: UN 2025; DOT Guide Number: 151
Proper Shipping Name: Mercury compounds, solid, n.o.s.

DESCRIPTION: Red or black powder. Odorless. Sinks in water; insoluble.

Irritating to the skin, eyes, and respiratory tract • Toxic products of combustion may include mercury and sulfur dioxide fumes.

Hazard Classification (based on NFPA-704 Rating System)
Health Hazards (Blue): 3; Flammability (Red): 0; Reactivity (Yellow): 0

EMERGENCY RESPONSE: See Appendix A (151)
Evacuation:
Public safety: Isolate the area of spill or leak for at least 25 to 50 meters (80 to 160 feet) in all directions.
Spill: Increase, in the downwind direction, as necessary, the distance shown under "Public Safety."
Fire: If tank, rail car, or tank truck is involved in fire, isolate for at least 800 meters (½ mile) in all directions; also, consider initial evacuation for 800 meters (½ mile) in all directions.

EXPOSURE

Short-term effects: *SEEK MEDICAL ATTENTION.* **Dust:** *POISONOUS IF INHALED OR IF SKIN IS EXPOSED. IF INHALED,* will, will cause coughing or difficult breathing. *IF IN*

EYES, hold eyelids open and flush with plenty of water. *IF BREATHING HAS STOPPED*, give artificial respiration; *avoid mouth-to-mouth resuscitation; use bag/mask apparatus.* IF breathing is difficult, administer oxygen. **Solid:** *POISONOUS IF SWALLOWED OR IF SKIN IS EXPOSED*. Irritating to skin and eyes. *IF SWALLOWED*, will cause coughing, nausea and vomiting. Remove contaminated clothing and shoes. Flush affected areas with plenty of water. *IF IN EYES*, hold eyelids open and flush with plenty of water. *IF SWALLOWED* and victim is *CONSCIOUS AND ABLE TO SWALLOW*, have victim drink 4 to 8 ounces of water and have victim induce vomiting. *IF SWALLOWED* and victim is *UNCONSCIOUS OR HAVING CONVULSIONS*, do nothing except keep victim warm.

HEALTH HAZARDS
Personal protective equipment (PPE): B-Level PPE. OSHA Table Z-1-A air contaminant. Dust mask; goggles or face shield; protective **gloves.**
Recommendations for respirator selection: NIOSH/OSHA as other mercury compounds. See record M:1100 for mercury vapor. *up to 1 mg/m³*: CCRS** [any chemical cartridge respirator with cartridge(s) providing protection against the compound of concern]; or SA (any supplied-air respirator). *up to 2.5 mg/m³*: SA:CF (any supplied-air respirator operated in a continuous-flow mode); or PAPRS** [any powered, air-purifying respirator with cartridge(s) providing protection against the compound of concern]. *up to 5 mg/m³*: CCRFS** [any chemical cartridge respirator with a full facepiece and cartridge(s) providing protection against the compound of concern]; or GMFS [any air-purifying, full-facepiece respirator (gas mask) with a chin-style, front- or back-mounted canister providing protection against the compound of concern]; or SAT:CF (any supplied-air respirator that has a tight-fitting facepiece and is operated in a continuous-flow mode); or PAPRTS [any powered, air-purifying respirator with a tight-fitting facepiece and cartridge(s) providing protection against the compound of concern]; or SCBAF (any self-contained breathing apparatus with a full facepiece); or SAF (any supplied-air respirator with a full facepiece). *up to 10 mg/m³*: SA:PD,PP (any supplied-air respirator operated in a pressure-demand or other positive-pressure mode). *EMERGENCY OR PLANNED ENTRY INTO UNKNOWN CONCENTRATIONS OR IDLH CONDITIONS*: SCBAF:PD,PP (any self-contained breathing apparatus that has a full facepiece and is operated in a pressure-demand or other positive-pressure mode); or SAF:PD,PP:ASCBA (any supplied-air respirator that has a full facepiece and is operated in a pressure-demand or other positive-pressure mode in combination with an auxiliary, self-contained breathing apparatus operated in a pressure-demand or other positive-pressure mode). *ESCAPE*: GMFS** [any air-purifying, full-facepiece respirator (gas mask) with a chin-style, front- or back-mounted canister protection against the compound of concern]; or SCBAE (any appropriate escape-type, self-contained breathing apparatus). **Note:* End of service life indicator (ESLI) required.
Exposure limits (TWA unless otherwise noted): ACGIH TLV 0.025 mg/m³; OSHA PEL ceiling limit 0.1 mg/m³; NIOSH REL (vapor) 0.05 mg/m³; other: ceiling 0.1 mg/m³ as mercury; ; skin contact contributes significantly in overall exposure.
Long-term health effects: Central nervous system affects, tremors, psychological disturbances in humans. Lethal blood level of inorganic mercury in blood (serum or plasma) in humans is 0.04–2.2 mg%; 0.4–22 mg/mL.
Vapor (gas) irritant characteristics: Poisonous.
Liquid or solid irritant characteristics: Eye and skin irritant. Possible dermatitis.
IDLH value: 10 mg/m³ as mercury.

FIRE DATA
Flash point: Combustible solid.
Behavior in fire: Changes color when hot. Decomposes at burning temperature. The black form may soften, and molten sulfur may flow out and burn.

ENVIRONMENTAL DATA
Food chain concentration potential: Many organisms can accumulate mercury from water. Bioconcentrative up to 10,000-fold.
Water pollution: DOT Appendix B, §172.101- severe marine pollutant. Effect of low concentrations on aquatic life is unknown. May be dangerous if it enters nearby water intakes; notify operators. Notify local health and wildlife officials. **Response to discharge:** Issue warning–water contaminant. Should be removed. Chemical and physical treatment.

SHIPPING INFORMATION
Grades of purity: The black form may contain up to 40% free sulfur; **Storage temperature:** Ambient; **Inert atmosphere:** None; **Venting:** Open; **Stability during transport:** Stable; **Stability during transport:** Stable

PHYSICAL AND CHEMICAL PROPERTIES
Physical state @ 59°F/15°C and 1 atm: Solid.
Molecular weight: 232.7
Specific gravity (water = 1): 8 @ 68°F/20°C (solid).
Heat of combustion: -1200 Btu/lb = -670 cal/g = -28×10^5 J/kg.

| MERCURIC THIOCYANATE | REC. M:0950 |

SYNONYMS: BIS(THIOCYANATO)-MERCURY; EEC No. 080-002-00-6; MERCURIC RHODANIDE; MERCURIC SULFOCYANATE; MERCURIC SULFOCYANIDE; MERCURY(II) THIOCYANATE; MERCURY RHODANIDE; TIOCIANATO MERCURICO (Spanish)

IDENTIFICATION
CAS Number: 592-85-8
Formula: $C_2HgN_2S_2$; $Hg(SCN)_2$
DOT ID Number: UN 1646; DOT Guide Number: 151
Proper Shipping Name: Mercury thiocyanate
Reportable Quantity (RQ): **(CERCLA)** 10 lb/4.54 kg

DESCRIPTION: White to tan powder. Odorless. Sinks and mixes slowly with water.

Thermally unstable combustible solid • Corrosive to skin, eyes, and respiratory tract; inhalation symptoms may be delayed • Breathing the vapor, skin or eye contact, or swallowing the material can kill you • Firefighting gear (including SCBA) does not provide adequate protection. If exposure occurs, remove and isolate gear immediately and thoroughly decontaminate personnel • Dust cloud may explode if ignited in an enclosed area • Containers may BLEVE when exposed to fire • Decomposes at about 329°F/165°C. Toxic products of combustion may include carbon monoxide, mercury, cyanide, sulfur, and nitrogen oxides. When heated, swells up to many times its original volume.

Hazard Classification (based on NFPA-704 Rating System)
Health Hazards (Blue): 3; Flammability (Red): 1; Reactivity (Yellow): 0

Mercurous chloride

EMERGENCY RESPONSE: See Appendix A (151)
Evacuation:
Public safety: Isolate the area of spill or leak for at least 25 to 50 meters (80 to 160 feet) in all directions.
Spill: Increase, in the downwind direction, as necessary, the distance shown under "Public Safety."
Fire: If tank, rail car, or tank truck is involved in fire, isolate for at least 800 meters (½ mile) in all directions; also, consider initial evacuation for 800 meters (½ mile) in all directions.

EXPOSURE

Short-term effects: *SEEK MEDICAL ATTENTION.* **Dust:** Irritating to skin, eyes, and nose. *IF INHALED,* will, will cause coughing, pain, and difficult breathing. Move to fresh air. *IF BREATHING HAS STOPPED,* give artificial respiration; *avoid mouth-to-mouth resuscitation; use bag/mask apparatus.* IF breathing is difficult, administer oxygen. **Solid:** *POISONOUS IF SWALLOWED.* Irritating to skin and eyes. Remove contaminated clothing and shoes. Flush affected areas with plenty of water. *IF IN EYES,* hold eyelids open and flush with plenty of water. *IF SWALLOWED* and victim is *CONSCIOUS AND ABLE TO SWALLOW,* have victim drink 4 to 8 ounces of water and have victim induce vomiting. *IF SWALLOWED* and victim is *UNCONSCIOUS OR HAVING CONVULSIONS,* do nothing except keep victim warm.
Note to physician or authorized medical personnel: Consider the use of amyl nitrite perles if symptoms of cyanide poisoning develop. If symptoms indicate, initial treatment includes the cyanide antidote kit. In all cases, break an amyl nitrite perle in a gauze pad and hold lightly under victim's nose for 15 seconds, repeating 5 times at about 15-second intervals; if necessary (and if sodium nitrite infusions will be delayed), repeat procedure every 3 minutes with fresh pearls until 3 or 4 have been used. Avoid breathing the vapor while administering it to the victim. Administer sodium nitrite IV, ASAP. The usual adult dose is 10 to 20 mL of a 3% solution infused over no less than 5 minutes; the average child dose is 0.15 to 0.20 mL/kg. Monitor blood pressure during administration, and slow the rate of infusion if hypotention develops. Next, infuse sodium thiosulfate IV. The usual adult dose is 50 mL of a 25% solution infused over 10 to 20 minutes; the average child dose is 1.65 mL/kg. Repeat with nitrite and thiosulfate as required.

HEALTH HAZARDS

Personal protective equipment (PPE): B-Level PPE. OSHA Table Z-1-A air contaminant. Dust mask or self-contained breathing apparatus, when required; rubber gloves, safety or chemical goggles, disposable clothing, or rubber apron.
Recommendations for respirator selection: NIOSH/OSHA as other mercury compounds. See record M:1100 for mercury vapor. *up to 1 mg/m^3:* CCRS** [any chemical cartridge respirator with cartridge(s) providing protection against the compound of concern]; or SA (any supplied-air respirator). *up to 2.5 mg/m^3:* SA:CF (any supplied-air respirator operated in a continuous-flow mode); or PAPRS** [any powered, air-purifying respirator with cartridge(s) providing protection against the compound of concern]. *up to 5 mg/m^3:* CCRFS** [any chemical cartridge respirator with a full facepiece and cartridge(s) providing protection against the compound of concern]; or GMFS [any air-purifying, full-facepiece respirator (gas mask) with a chin-style, front- or back-mounted canister providing protection against the compound of concern]; or SAT:CF (any supplied-air respirator that has a tight-fitting facepiece and is operated in a continuous-flow mode); or PAPRTS [any powered, air-purifying respirator with a tight-fitting facepiece and cartridge(s) providing protection against the compound of concern]; or SCBAF (any self-contained breathing apparatus with a full facepiece); or SAF (any supplied-air respirator with a full facepiece). *up to 10 mg/m^3:* SA:PD,PP (any supplied-air respirator operated in a pressure-demand or other positive-pressure mode). *EMERGENCY OR PLANNED ENTRY INTO UNKNOWN CONCENTRATIONS OR IDLH CONDITIONS:* SCBAF:PD,PP (any self-contained breathing apparatus that has a full facepiece and is operated in a pressure-demand or other positive-pressure mode); or SAF:PD,PP:ASCBA (any supplied-air respirator that has a full facepiece and is operated in a pressure-demand or other positive-pressure mode in combination with an auxiliary, self-contained breathing apparatus operated in a pressure-demand or other positive-pressure mode). *ESCAPE:* GMFS** [any air-purifying, full-facepiece respirator (gas mask) with a chin-style, front- or back-mounted canister protection against the compound of concern]; or SCBAE (any appropriate escape-type, self-contained breathing apparatus). **Note:* End of service life indicator (ESLI) required.
Exposure limits (TWA unless otherwise noted): ACGIH TLV 0.025 mg/m^3; OSHA PEL ceiling limit 0.1 mg/m^3; NIOSH REL (vapor) 0.05 mg/m^3; other: ceiling 0.1 mg/m^3 as mercury; ; skin contact contributes significantly in overall exposure.
Long-term health effects: Mercury vapor caused kidney, heart, lung, and brain damage in rabbits. In man-psychic and emotional disturbances. Fine tremors may affect hands, head, lips, tongue, or jaw. Salivation, gingivitis, and digestive disturbances. Liver and kidney damage. Lethal blood level of inorganic mercury in blood (serum or plasma) in humans is 0.04–2.2 mg%; 0.4–22 mg/mL.
Vapor (gas) irritant characteristics: May cause corrosive irritation of the nose and mucous membrane.
Liquid or solid irritant characteristics: Corrosive to the eyes, skin, and respiratory tract.
IDLH value: 10 mg/m^3 as mercury.

FIRE DATA
Flash point: More than 250°F/121°C.

ENVIRONMENTAL DATA
Food chain concentration potential: Many organisms can accumulate mercury from water. Bioconcentrative up to 10,000-fold.
Water pollution: DOT Appendix B, §172.101- severe marine pollutant. Harmful to aquatic life in very low concentrations. May be dangerous if it enters nearby water intakes; notify operators. Notify local health and wildlife officials. **Response to discharge:** Issue warning–poison, water contaminant. Restrict access. Should be removed. Chemical and physical treatment.

PHYSICAL AND CHEMICAL PROPERTIES
Physical state @ 59°F/15°C and 1 atm: Solid.
Molecular weight: 316.79
Boiling point @ 1 atm: 329°F/165°C/438°K (decomposes).
Specific gravity (water = 1): Approximately 4
Relative vapor density (air = 1): 10.9

MERCUROUS CHLORIDE   **REC. M:1000**

SYNONYMS: CALOMEL; CLORURO MERCURIOSO (Spanish); MILD MERCURY CHLORIDE; MERCURY MONOCHLORIDE; MERCURY PROTOCHLORIDE; MERCURY SUBCHLORIDE

IDENTIFICATION
CAS Number: 7546-30-7

Formula: HgCl; Hg_2Cl_2
DOT ID Number: UN 2025; DOT Guide Number: 151
Proper Shipping Name: Mercury compounds, solid, n.o.s.

DESCRIPTION: White crystalline solid or powder. Odorless. Sinks in water; insoluble.

Containers may BLEVE when exposed to fire • Irritating to the skin, eyes, and respiratory tract • Toxic products of combustion may include mercury and hydrogen chloride fumes.

Hazard Classification (based on NFPA-704 Rating System)
Health Hazards (Blue): 2; Flammability (Red): 0; Reactivity (Yellow): 0

EMERGENCY RESPONSE: See Appendix A (151)
Evacuation:
Public safety: Isolate the area of spill or leak for at least 25 to 50 meters (80 to 160 feet) in all directions.
Spill: Increase, in the downwind direction, as necessary, the distance shown under "Public Safety."
Fire: If tank, rail car, or tank truck is involved in fire, isolate for at least 800 meters (½ mile) in all directions; also, consider initial evacuation for 800 meters (½ mile) in all directions.

EXPOSURE
Short-term effects: *SEEK MEDICAL ATTENTION*. **Dust:** *POISONOUS IF INHALED. IF INHALED*, will, will cause coughing or difficult breathing. *IF IN EYES*, hold eyelids open and flush with plenty of water. *IF BREATHING HAS STOPPED*, give artificial respiration; *avoid mouth-to-mouth resuscitation; use bag/mask apparatus*. IF breathing is difficult, administer oxygen. **Solid:** *POISONOUS IF SWALLOWED*. Irritating to skin and eyes. *IF SWALLOWED*, will cause nausea and vomiting. Remove contaminated clothing and shoes. Flush affected areas with plenty of water. *IF IN EYES*, hold eyelids open and flush with plenty of water. *IF SWALLOWED* and victim is *CONSCIOUS AND ABLE TO SWALLOW*, have victim drink 4 to 8 ounces of water and have victim induce vomiting. *IF SWALLOWED* and victim is *UNCONSCIOUS OR HAVING CONVULSIONS*, do nothing except keep victim warm.

HEALTH HAZARDS
Personal protective equipment (PPE): B-Level PPE. OSHA Table Z-1-A air contaminant. Dust mask; goggles or face shield; protective gloves.
Recommendations for respirator selection: NIOSH/OSHA as other mercury compounds. See record M:1100 for mercury vapor. *up to 1 mg/m³*: CCRS** [any chemical cartridge respirator with cartridge(s) providing protection against the compound of concern]; or SA (any supplied-air respirator). *up to 2.5 mg/m³*: SA:CF (any supplied-air respirator operated in a continuous-flow mode); or PAPRS** [any powered, air-purifying respirator with cartridge(s) providing protection against the compound of concern]. *up to 5 mg/m³*: CCRFS** [any chemical cartridge respirator with a full facepiece and cartridge(s) providing protection against the compound of concern]; or GMFS [any air-purifying, full-facepiece respirator (gas mask) with a chin-style, front- or back-mounted canister providing protection against the compound of concern]; or SAT:CF (any supplied-air respirator that has a tight-fitting facepiece and is operated in a continuous-flow mode); or PAPRTS [any powered, air-purifying respirator with a tight-fitting facepiece and cartridge(s) providing protection against the compound of concern]; or SCBAF (any self-contained breathing apparatus with a full facepiece); or SAF (any supplied-air respirator with a full facepiece). *up to 10 mg/m³*: SA:PD,PP (any supplied-air respirator operated in a pressure-demand or other positive-pressure mode). *EMERGENCY OR PLANNED ENTRY INTO UNKNOWN CONCENTRATIONS OR IDLH CONDITIONS*: SCBAF:PD,PP (any self-contained breathing apparatus that has a full facepiece and is operated in a pressure-demand or other positive-pressure mode); or SAF:PD,PP:ASCBA (any supplied-air respirator that has a full facepiece and is operated in a pressure-demand or other positive-pressure mode in combination with an auxiliary, self-contained breathing apparatus operated in a pressure-demand or other positive-pressure mode). *ESCAPE:* GMFS** [any air-purifying, full-facepiece respirator (gas mask) with a chin-style, front- or back-mounted canister protection against the compound of concern]; or SCBAE (any appropriate escape-type, self-contained breathing apparatus). **Note:* End of service life indicator (ESLI) required.
Exposure limits (TWA unless otherwise noted): ACGIH TLV 0.025 mg/m³; OSHA PEL ceiling limit 0.1 mg/m³; NIOSH REL (vapor) 0.05 mg/m³; other: ceiling 0.1 mg/m³ as mercury; ; skin contact contributes significantly in overall exposure.
Toxicity by ingestion: Grade 3; oral LD_{50} = 210 mg/kg (rat).
Long-term health effects: Central nervous system effects, tremors, psychological disturbances in humans. Lethal blood level of inorganic mercury in blood (serum or plasma) in humans is 0.04–2.2 mg%; 0.4–22 mg/mL.
Vapor (gas) irritant characteristics: Respiratory tract irritant.
Liquid or solid irritant characteristics: Eye irritant.
IDLH value: 10 mg/m³ as inorganic mercury

FIRE DATA
Behavior in fire: Vaporizes and escapes as a sublimate.

ENVIRONMENTAL DATA
Food chain concentration potential: Many organisms can accumulate mercury from water. Bioconcentrative up to 10,000-fold.
Water pollution: DOT Appendix B, §172.101- severe marine pollutant. Effect of low concentrations on aquatic life is unknown. May be dangerous if it enters nearby water intakes; notify operators. Notify local health and wildlife officials. **Response to discharge:** Issue warning–water contaminant. Restrict access. Should be removed. Chemical and physical treatment.

SHIPPING INFORMATION
Grades of purity: NF; Technical, 99.6%; Reagent; **Storage temperature:** Ambient; **Inert atmosphere:** None; **Venting:** Open; **Stability during transport:** Stable; **Stability during transport:** Stable

PHYSICAL AND CHEMICAL PROPERTIES
Physical state @ 59°F/15°C and 1 atm: Solid.
Molecular weight: 236.1
Specific gravity (water = 1): 7.15 @ 68°F/20°C (solid).
Heat of fusion: 15.3 cal/g.

MERCUROUS NITRATE REC. M:1050

SYNONYMS: MERCURY NITRATE; NITRATE MERCUREUX (French); NITRIC ACID, MERCURY SALT; MERCURY PROTONITRATE; MERCUROUS NITRATE MONOHYDRATE

IDENTIFICATION
CAS Number: 10415-75-5
Formula: $HgNO_3 \cdot H_2O$

Mercurous nitrate

DOT ID Number: UN 1627; DOT Guide Number: 141
Proper Shipping Name: Mercurous nitrate
Reportable Quantity (RQ): **(CERCLA)** 10 lb/4.54 kg

DESCRIPTION: White, colorless, crystalline solid. Slight nitric acid odor. Soluble in water producing nitric acid.

Corrosive to skin, eyes, and respiratory tract; inhalation symptoms may be delayed • Containers may BLEVE when exposed to fire • Strong oxidizer which may react spontaneously with low flash point organics or reducing agents. Heat forms oxygen; will increase the activity of an existing fire • May cause fire or explosion on contact with combustibles (wood, paper, oil, clothing, etc.) • Toxic products of combustion may include mercury vapors and oxides of nitrogen.

Hazard Classification (based on NFPA-704 Rating System)
Health Hazards (Blue): 3; Flammability (Red): 0; Reactivity (Yellow): 0; Special Notice (White): OXY

EMERGENCY RESPONSE: See Appendix A (141)
Evacuation:
Public safety: Isolate the area of spill or leak for at least 10 to 25 meters (30 to 80 feet) in all directions.
Spill: Consider initial downwind evacuation for at least 100 meters (330 feet).
Fire: If any large container is involved in fire, isolate for at least 800 meters (½ mile) in all directions; also, consider initial evacuation for 800 meters (½ mile) in all directions.

EXPOSURE
Short-term effects: *SEEK MEDICAL ATTENTION.* **Dust:** *POISONOUS IF INHALED OR IF SKIN IS EXPOSED. IF INHALED,* will, will cause coughing or difficult breathing. *IF IN EYES,* hold eyelids open and flush with plenty of water. *IF BREATHING HAS STOPPED,* give artificial respiration. IF breathing is difficult, administer oxygen. Lung edema may develop. **Solid:** *POISONOUS IF SWALLOWED OR IF SKIN IS EXPOSED.* Irritating to skin and eyes. *IF SWALLOWED,* will cause nausea and vomiting. Remove contaminated clothing and shoes. Flush affected areas with plenty of water. *IF IN EYES*, hold eyelids open and flush with plenty of water. *IF SWALLOWED* and victim is *CONSCIOUS AND ABLE TO SWALLOW,* have victim drink 4 to 8 ounces of water and have victim induce vomiting. *IF SWALLOWED* and victim is *UNCONSCIOUS OR HAVING CONVULSIONS,* do nothing except keep victim warm.
Note to physician or authorized medical personnel: Medical observation is recommended for 24 to 48 hours after breathing overexposure, as pulmonary edema may be delayed. As first aid for pulmonary edema, consider administering a corticosteroid spray. Cigarette smoking may exacerbate pulmonary injury and should be discouraged for at least 72 hours following exposure.

HEALTH HAZARDS
Personal protective equipment (PPE): B-Level PPE. OSHA Table Z-1-A air contaminant. Dust mask; goggles or face shield; protective gloves.
Recommendations for respirator selection: NIOSH/OSHA as other mercury compounds; See record M:1100 for mercury vapor. *up to 1 mg/m^3:* CCRS** [any chemical cartridge respirator with cartridge(s) providing protection against the compound of concern]; or SA (any supplied-air respirator). *up to 2.5 mg/m^3:* SA:CF (any supplied-air respirator operated in a continuous-flow mode); or PAPRS** [any powered, air-purifying respirator with cartridge(s) providing protection against the compound of concern]. *up to 5 mg/m^3:* CCRFS** [any chemical cartridge respirator with a full facepiece and cartridge(s) providing protection against the compound of concern]; or GMFS [any air-purifying, full-facepiece respirator (gas mask) with a chin-style, front- or back-mounted canister providing protection against the compound of concern]; or SAT:CF (any supplied-air respirator that has a tight-fitting facepiece and is operated in a continuous-flow mode); or PAPRTS [any powered, air-purifying respirator with a tight-fitting facepiece and cartridge(s) providing protection against the compound of concern]; or SCBAF (any self-contained breathing apparatus with a full facepiece); or SAF (any supplied-air respirator with a full facepiece). *up to 10 mg/m^3:* SA:PD,PP (any supplied-air respirator operated in a pressure-demand or other positive-pressure mode). *EMERGENCY OR PLANNED ENTRY INTO UNKNOWN CONCENTRATIONS OR IDLH CONDITIONS:* SCBAF:PD,PP (any self-contained breathing apparatus that has a full facepiece and is operated in a pressure-demand or other positive-pressure mode); or SAF:PD,PP:ASCBA (any supplied-air respirator that has a full facepiece and is operated in a pressure-demand or other positive-pressure mode in combination with an auxiliary, self-contained breathing apparatus operated in a pressure-demand or other positive-pressure mode). *ESCAPE:* GMFS** [any air-purifying, full-facepiece respirator (gas mask) with a chin-style, front- or back-mounted canister protection against the compound of concern]; or SCBAE (any appropriate escape-type, self-contained breathing apparatus). **Note:* End of service life indicator (ESLI) required.
Exposure limits (TWA unless otherwise noted): ACGIH TLV 0.025 mg/m^3; OSHA PEL ceiling limit 0.1 mg/m^3; NIOSH REL (vapor) 0.05 mg/m^3; other: ceiling 0.1 mg/m^3 as mercury; ; skin contact contributes significantly in overall exposure.
Short-term exposure limits (15-minute TWA): Poisonous.
Long-term health effects: Lethal blood level of inorganic mercury in blood (serum or plasma) in humans is 0.04–2.2 mg%; 0.4–22 mg/mL.
Toxicity by ingestion: Grade 3; oral LD$_{50}$ = 297 mg/kg (rat).
Liquid or solid irritant characteristics: Severe eye ans skin irritant.
IDLH value: 10 mg/m^3 as inorganic mercury.

FIRE DATA
Behavior in fire: Not flammable but will increase intensity of fire.

CHEMICAL REACTIVITY
Reactivity with water: Dissolves, then forms cloudy acid solution. The reaction is not hazardous.
Binary reactants: A powerful oxidizer. violent reaction with reducing agents, acrolein, alcohols, ethers, fluorine, hydrazines, finely divided aluminum and magnesium, phosphinic acid, sodium acetylide. Aqueous solution may corrode most metals. Solid in contact with wood or paper may cause fire.
Neutralizing agents for acids and caustics: Flush with water, rinse with dilute solution of sodium bicarbonate or soda ash.

ENVIRONMENTAL DATA
Food chain concentration potential: Possible bioaccumulation problem. Many organisms can accumulate mercury from water. Bioconcentrative up to 10,000-fold.
Water pollution: DOT Appendix B, §172.101- severe marine pollutant. Effect of low concentrations on aquatic life is unknown. May be dangerous if it enters nearby water intakes; notify operators. Notify local health and wildlife officials. **Response to discharge:** Issue warning–poison, water contaminant, oxidizing material. Restrict access. Disperse and flush.

SHIPPING INFORMATION
Grades of purity: Reagent; Purified; **Storage temperature:** Ambient; **Inert atmosphere:** None; **Venting:** Open; **Stability during transport:** Stable; **Stability during transport:** Stable.

PHYSICAL AND CHEMICAL PROPERTIES
Physical state @ 59°F/15°C and 1 atm: Solid.
Molecular weight: 280.6
Boiling point @ 1 atm: (decomposes).
Specific gravity (water = 1): 4.78 @ 68°F/20°C (solid).

MERCURY   REC. M:1100

SYNONYMS: COLLOIDAL MERCURY; EEC No. 080-001-00-0; MERCURE (French); MERCURIO (Spanish); MERCURY, ELEMENTAL; MERCURY, METALLIC; METALLIC MERCURY; QUECKSILBER (German); QUICK SILVER; RCRA No. U151

IDENTIFICATION
CAS Number: 7439-97-6
Formula: Hg
DOT ID Number: UN 2809; DOT Guide Number: 172
Proper Shipping Name: Mercury
Reportable Quantity (RQ): **(CERCLA)** 1 lb/0.454 kg

DESCRIPTION: Silvery, heavy, mobile liquid metal. Odorless. Sinks in water; insoluble.

Corrosive •Highly Irritating to the skin, eyes, and respiratory tract; effects of exposure may be delayed • Toxic products of combustion may include highly toxi\c mercury vapors; reacts with oxygen when heated, forming mercury oxide.

Hazard Classification (based on NFPA-704 Rating System)
Health Hazards (Blue): 3; Flammability (Red): 0; Reactivity (Yellow): 0

EMERGENCY RESPONSE: See Appendix A (172)
Evacuation:
Public safety: Isolate the area of spill or leak for at least 10 to 25 meters (30 to 80 feet) in all directions.
Spill: Consider initial evacuation for at least 100 meters (330 feet).
Fire: If any large container is involved in fire, isolate for at least 500 meters (⅓ mile) in all directions; also, consider initial evacuation for 800 meters (½ mile) in all directions.

EXPOSURE
Short-term effects: *SEEK MEDICAL ATTENTION.* **Liquid:** *HARMFUL IF SWALLOWED OR ABSORBED THROUGH THE SKIN.* Effects of exposure may be delayed.
Note to physician or authorized medical personnel: Medical observation is recommended for 24 to 48 hours after breathing overexposure, as pulmonary edema may be delayed. As first aid for pulmonary edema, consider administering a corticosteroid spray. Cigarette smoking may exacerbate pulmonary injury and should be discouraged for at least 72 hours following exposure.

HEALTH HAZARDS
Personal protective equipment (PPE): B-Level PPE. OSHA Table Z-2 air contaminant. Avoid contact of liquid with skin. For vapor use chemical cartridge (Hopcalite®) respirator. Sealed chemical protective materials providing limited protection: Viton®/neoprene, neoprene, PVC, butyl rubber/neoprene.
Recommendations for respirator selection: NIOSH/OSHA as mercury vapor
NIOSH: *up to 0.5 mg/m³*: CCRS** [any chemical cartridge respirator with cartridge(s) providing protection against the compound of concern]; or SA (any supplied-air respirator). *up to 1.25 mg/m³*: SA:CF (any supplied-air respirator operated in a continuous-flow mode); or PAPRS** [any powered, air-purifying respirator with cartridge(s) providing protection against the compound of concern]. *up to 2.5 mg/m³*: CCRFS** [any chemical cartridge respirator with a full facepiece and cartridge(s) providing protection against the compound of concern]; or GMFS** [any air-purifying, full-facepiece respirator (gas mask) with a chin-style, front- or back-mounted canister providing protection against the compound of concern]; or SAT:CF (any supplied-air respirator that has a tight-fitting facepiece and is operated in a continuous-flow mode); or PAPRTS [any powered, air-purifying respirator with a tight-fitting facepiece and cartridge(s) providing protection against the compound of concern]; or SCBAF (any self-contained breathing apparatus with a full facepiece); or SAF (any supplied-air respirator with a full facepiece). *up to 10 mg/m³*: SA:PD,PP (any supplied-air respirator operated in a pressure-demand or other positive-pressure mode). *EMERGENCY OR PLANNED ENTRY INTO UNKNOWN CONCENTRATIONS OR IDLH CONDITIONS*: SCBAF:PD,PP (any self-contained breathing apparatus that has a full facepiece and is operated in a pressure-demand or other positive-pressure mode); or SAF:PD,PP:ASCBA (any supplied-air respirator that has a full facepiece and is operated in a pressure-demand or other positive-pressure mode in combination with an auxiliary, self-contained breathing apparatus operated in a pressure-demand or other positive-pressure mode). *ESCAPE:* GMFS ** (any air-purifying, full-facepiece respirator (gas mask) with a chin-style, front- or back-mounted canister protection against the compound of concern); or SCBAE (any appropriate escape-type, self-contained breathing apparatus). **Note:* End of service life indicator (ESLI) required.
Exposure limits (TWA unless otherwise noted): ACGIH TLV 0.025 mg/m³; OSHA PEL ceiling limit 0.1 mg/m³; NIOSH REL (vapor) 0.05 mg/m³; other: ceiling 0.1 mg/m³ as mercury; skin contact contributes significantly in overall exposure.
Toxicity by ingestion: No immediate toxicity
Long-term health effects: Development of mercury poisoning. Brain and kidney damage. Lethal blood level of inorganic mercury in blood (serum or plasma) in humans is 0.04–2.2 mg%; 0.4–22 mg/mL.
Vapor (gas) irritant characteristics: None
Liquid or solid irritant characteristics: None
IDLH value: 10 mg/m³.

CHEMICAL REACTIVITY
Binary reactants: Reacts with acetylene, ammonia, chlorine dioxide, azides, calcium (almagum formation), sodium carbide, lithium, rubidium, copper. May form shock-sensitive mixtures with some of these materials.

ENVIRONMENTAL DATA
Food chain concentration potential: Mercury concentrates in liver and kidneys of ducks and geese to levels above FDA limit of 0.5 ppm. Muscle tissue usually well below the limit.
Water pollution: Harmful to aquatic life in very low concentrations. May be dangerous if it enters nearby water intakes; notify operators. Notify local health and wildlife officials. **Response to discharge:** Should be removed. Chemical and physical treatment.

SHIPPING INFORMATION
Grades of purity: Pure; **Storage temperature:** Ambient; **Inert atmosphere:** None; **Venting:** Open; **Stability during transport:** Stable; **Stability during transport:** Stable

PHYSICAL AND CHEMICAL PROPERTIES
Physical state @ 59°F/15°C and 1 atm: Liquid.
Molecular weight: 200.59
Boiling point @ 1 atm: 675°F/357°C/630°K.
Melting/Freezing point: –38.0°F/–38.9°C/234.3°K.
Critical temperature: 2664°F/1462°C/1735°K.
Critical pressure: 23,300 psia = 1587 atm = 160.8 MN/m^2.
Specific gravity (water = 1): 13.55 @ 68°F/20°C (liquid).
Liquid surface tension: 470 dynes/cm = 0.470 N/m @ 68°F/20°C.
Liquid water interfacial tension: 375 dynes/cm = 0.375 N/m @ 68°F/20°C.
Relative vapor density (air = 1): 7.0
Heat of fusion: 2.7 cal/g.
Vapor pressure: 0.002 mm.

MESITYL OXIDE   REC. M:1150

SYNONYMS: EEC No. 606-009-00-1; ISOBUTENYL METHYL KETONE; ISOPROPYLIDENEACETONE; METHYL ISOBUTENYL KETONE; 4-METHYL-3-PENTEN-2-ONE; OXIDO de MESTILO (Spanish); OXYDE de MESITYLE (French).

IDENTIFICATION
CAS Number: 141-79-7
Formula: $C_6H_{10}O$; $CH_3COCH=C(CH_3)_2$
DOT ID Number: UN 1229; DOT Guide Number: 129
Proper Shipping Name: Mesityl oxide

DESCRIPTION: Colorless oily liquid. Strong, sweet, peppermint odor. Floats on the surface of water; moderately soluble. Flammable, irritating vapor is produced.

Highly flammable • Containers may BLEVE when exposed to fire • Vapors may form explosive mixture with air • Vapors are heavier than air and will collect and stay in low areas • Vapors may travel long distances to ignition sources and flashback • Vapors in confined areas (e.g., tanks, sewers, buildings) may explode when exposed to fire • Irritating to skin, eyes, and respiratory tract; can cause burns to the eyes • Toxic products of combustion may include carbon monoxide.

Hazard Classification (based on NFPA-704 Rating System)
Health Hazards (Blue): 2; Flammability (Red): 3; Reactivity (Yellow): 1

EMERGENCY RESPONSE: See Appendix A (129)
Evacuation:
Public safety: Isolate spill area for at least 50 to 100 meters (160 to 330 feet) in all directions.
Spill: Large spill–Consider initial downwind evacuation for at least 300 meters (1000 feet).
Fire: Isolate for 800 meters (½ mile) in all directions, especially if tank, rail car, or tank truck is involved in fire.

EXPOSURE
Short-term effects: *SEEK MEDICAL ATTENTION.* **Vapor:** Irritating to eyes, nose, and throat. *IF INHALED*, will, will cause headache, dizziness or difficult breathing. Move victim to fresh air. *IF BREATHING HAS STOPPED*, give artificial respiration. IF breathing is difficult, administer oxygen. **Liquid:** Irritating to skin and eyes. Harmful if swallowed. Remove contaminated clothing and shoes. Flush affected areas with plenty of water. *IF IN EYES*, hold eyelids open and flush with plenty of water. *IF SWALLOWED* and victim is *CONSCIOUS AND ABLE TO SWALLOW*, have victim drink 4 to 8 ounces of water.

HEALTH HAZARDS
Personal protective equipment (PPE): B-Level PPE. OSHA Table Z-1-A air contaminant. Air pack or organic canister mask; rubber gloves; goggles. Chemical protective material(s) reported to have good to excellent resistance: Viton®/chlorobutyl rubber. Also, Butyl rubber/neoprene offers limited protection.
Recommendations for respirator selection: NIOSH
250 ppm: SA:CF* (any supplied-air respirator operated in a continuous-flow mode); or PAPROV* [any powered, air-purifying respirator with organic vapor cartridge(s)]. *500 ppm*: CCRFOV [any chemical cartridge respirator with a full facepiece and organic vapor cartridges(s)]; or GMFOV [any air-purifying, full-facepiece respirator (gas mask) with a chin-style, front- or back-mounted acid gas canister]; or PAPRTOV* [any powered, air-purifying respirator with a tight-fitting facepiece and organic vapor cartridges(s)]; or SCBAF (any self-contained breathing apparatus with a full facepiece); or SAF (any supplied-air respirator with a full facepiece). *1400 ppm*: SAF:PD,PP (any supplied-air respirator that has a full facepiece and is operated in a pressure-demand or other positive-pressure mode). *EMERGENCY OR PLANNED ENTRY INTO UNKNOWN CONCENTRATIONS OR IDLH CONDITIONS*: SCBAF:PD,PP (any self-contained breathing apparatus that has a full facepiece and is operated in a pressure-demand or other positive-pressure mode); or SAF:PD,PP:ASCBA (any supplied-air respirator that has a full facepiece and is operated in a pressure-demand or other positive-pressure mode in combination with an auxiliary self-contained breathing apparatus operated in a pressure-demand or other positive pressure mode). *ESCAPE*: GMFOV [any air-purifying, full-facepiece respirator (gas mask) with a chin-style, front- or back-mounted organic vapor cannister]; or SCBAE (any appropriate escape-type, self-contained breathing apparatus). *Note*: Substance causes eye irritation or damage; eye protection needed.
Exposure limits (TWA unless otherwise noted): ACGIH TLV 15 ppm (60 mg/m^3); OSHA PEL 25 ppm (100 mg/m^3); NIOSH REL 10 ppm (40 mg/m^3).
Short-term exposure limits (15-minute TWA): ACGIH STEL 25 ppm (100 mg/m^3).
Toxicity by ingestion: Grade 2; oral LD_{50} = 1120 mg/kg (rat).
Vapor (gas) irritant characteristics: Eye and respiratory tract irritant. Vapors cause moderate irritation such that personnel will find high concentrations unpleasant.
Liquid or solid irritant characteristics: eye ans skin irritant. Causes smarting of the skin and first-degree burns on short exposure; may cause second-degree burns on long exposure.
Odor threshold: 0.017 ppm.
IDLH value: 1400 ppm [10% LEL].

FIRE DATA
Flash point: 87°F/31°C (cc).
Flammable limits in air: LEL: 1.4%; UEL: 7.4%.
Fire extinguishing agents not to be used: Water may be ineffective.
Autoignition temperature: 652°F/344°C/617°K.
Electrical hazard: Class I, Group D.
Burning rate: 4.2 mm/min

CHEMICAL REACTIVITY
Binary reactants: Oxidizers (when heated), acids, copper and its alloys, aliphatic amines. Attacks some plastic materials.
Reactivity group: 18
Compatibility class: Ketone

ENVIRONMENTAL DATA
Food chain concentration potential: Log P_{ow} = 1.3. Unlikely to accumulate.
Water pollution: Effect of low concentrations on aquatic life is unknown. Fouling to shoreline. May be dangerous if it enters nearby water intakes; notify operators. Notify local health and wildlife officials. **Response to discharge:** Issue warning–water contaminant. Restrict access. Disperse and flush.

SHIPPING INFORMATION
Grades of purity: 97+%; **Storage temperature:** Ambient; **Inert atmosphere:** None; **Venting:** Open; **Stability during transport:** Stable (flame arrester); **Stability during transport:** Stable.

FIRE: 3
HEALTH: Vapor irritant: 2; Liquid or solid irritant: 2; Poisons: 2
WATER POLLUTION: Human toxicity: 2; Aquatic toxicity: 3; Aesthetic effect: 2
REACTIVITY: Other chemicals: 2; Water: 0; Self-reaction: 1

PHYSICAL AND CHEMICAL PROPERTIES
Physical state @ 59°F/15°C and 1 atm: Liquid.
Molecular weight: 98.2
Boiling point @ 1 atm: 266°F/130°C/403°K.
Melting/Freezing point: –51°F/–46°C/227°K.
Specific gravity (water = 1): 0.853 @ 68°F/20°C (liquid).
Liquid surface tension: 22.9 dynes/cm = 0.0229 N/m @ 68°F/20°C.
Relative vapor density (air = 1): 3.4
Latent heat of vaporization: 157 Btu/lb = 87 cal/g = 3.7×10^5 J/kg.
Heat of combustion: –14,400 Btu/lb = –8000 cal/g = -330×10^5 J/kg.
Vapor pressure: 8 mm.

METHACRYLIC ACID   REC. M:1200

SYNONYMS: ACIDO METACRILICO (Spanish); ACIDO α-METACRILICO (Spanish); ACRYLIC ACID, 2-METHYL-; α-METHACRYLIC ACID; α-METHACRYLIC ACID; 2-METHACRYLIC ACID; METHACRYLIC ACID, glacial; EEC No. 607-088-00-5; 2-METHYLPROPIONIC ACID; PROPIONIC ACID, 2-METHYLENE-; 2-METHYL PROPENIC ACID

IDENTIFICATION
CAS Number: 79-41-4
Formula: $C_4H_6O_2$; $CH_2 = C(CH_3)COOH$
DOT ID Number: UN 2531; DOT Guide Number: 153P
Proper Shipping Name: Methacrylic acid, inhibited

DESCRIPTION: Colorless liquid. Acrid, repulsive odor. Sinks in water; moderately soluble. Freezes at 61°F/16°C.

Flammable • Containers may BLEVE when exposed to fire • Vapors may form explosive mixture with air • Vapors are heavier than air and will collect and stay in low areas • Vapors in confined areas (e.g., tanks, sewers, buildings) may explode when exposed to fire • Polymerization hazard • Heat can induce polymerization with rapid release of energy; sealed containers may rupture explosively • May react with itself blocking relief valves; leading to tank explosions • Irritating to the skin, eyes, and respiratory tract • Toxic products of combustion may include carbon monoxide and CO_2 • Do not put yourself in danger by entering a contaminated area to rescue a victim.

Hazard Classification (based on NFPA-704 Rating System)
Health Hazards (Blue): 3; Flammability (Red): 2; Reactivity (Yellow): 2

EMERGENCY RESPONSE: See Appendix A (153)
Evacuation:
Public safety: Isolate the area of spill or leak for at least 25 to 50 meters (80 to 160 feet) in all directions.
Spill: Increase, in the downwind direction, as necessary, the distance shown under "Public Safety."
Fire: If tank, rail car, or tank truck is involved in fire, isolate for at least 800 meters (½ mile) in all directions; also, consider initial evacuation for 800 meters (½ mile) in all directions.

EXPOSURE
Short-term effects: *SEEK MEDICAL ATTENTION*. **Vapor:** Irritating to skin, eyes and respiratory tract. Move to fresh air. *IF IN EYES*, hold eyelids open and flush with plenty of water. *IF BREATHING HAS STOPPED*, give artificial respiration; *avoid mouth-to-mouth resuscitation; use bag/mask apparatus*. IF breathing is difficult, administer oxygen. **Liquid:** *HARMFUL IF SWALLOWED OR ABSORBED THROUGH THE SKIN*. Severe irritant. Corrosive. Harmful if swallowed or absorbed through the skin. Remove contaminated clothing and shoes. Flush affected areas with water. *IF IN EYES*, hold eyelids open and flush with plenty of water. *IF SWALLOWED* and victim is *CONSCIOUS AND ABLE TO SWALLOW*, have victim drink 4 to 8 ounces of water. **Do NOT induce vomiting.**
Medical note: May affect blood pressure temporarily.

HEALTH HAZARDS
Personal protective equipment (PPE): B-Level PPE. OSHA Table Z-1-A air contaminant. Full face organic vapor respirator. Chemical protective material(s) reported to have good to excellent resistance: butyl rubber, Viton®.
Exposure limits (TWA unless otherwise noted): ACGIH TLV 20 ppm (70 mg/m³); NIOSH REL 20 ppm (70 mg/m3); skin contact contributes significantly in overall exposure.
Toxicity by ingestion: Grade 4; LD_{50} below 50 mg/kg (rats).
Long-term health effects: Prolonged exposure may damage lungs and kidneys.
Vapor (gas) irritant characteristics: Eye and respiratory tract irritant. Vapors are irritating such that personnel will not usually tolerate moderate or high concentrations. May cause difficult breathing and lung edema.
Liquid or solid irritant characteristics: Severe eye and skin irritant. May cause pain and second-degree burns after a few minutes of contact.

FIRE DATA
Flash point: 171°F/72°C (oc); 152°F/67°C (cc).
Flammable limits in air: LEL: 1.6%; UEL: 8.8%.
Behavior in fire: Sealed containers may rupture explosively (polymerization).
Autoignition temperature: 154°F/68°C/341°K.
Electrical hazard: Class I, Group D.

CHEMICAL REACTIVITY
Binary reactants: Oxidizers, hydrochloric acid, elevated temperatures. Corrodes steel, wood, cloth, and paint.
Neutralizing agents for acids and caustics: Sodium carbonate, dilute caustic solutions.
Polymerization: Heat, strong oxidizers, alkalies, or hydrogen chloride may cause rapid polymerization and release high energy rapidly; may cause explosion under confinement; **Inhibitor of polymerization:** 0.025% *p*-methoxyphenol; 1000 ppm hydroquinone + 250 ppm hydroquinone monomethyl ether.
Reactivity group: 4
Compatibility class: Organic acids

ENVIRONMENTAL DATA
Water pollution: Effects of low concentrations on aquatic life is unknown. May be dangerous if it enters nearby water intakes; notify operators. Notify local health and wildlife officials.
Response to discharge: Issue warning–corrosive, high health hazard. Restrict access. Disperse and flush.

SHIPPING INFORMATION
Grades of purity: 99% plus; 40% aqueous solution; crude monomer (85%); glacial (98% plus); **Storage temperature:** Ambient; **Stability during transport:** Stable if stored away from heat.

PHYSICAL AND CHEMICAL PROPERTIES
Physical state @ 59°F/15°C and 1 atm: Liquid.
Molecular weight: 86.09
Boiling point @ 1 atm: 320–325°F/160–163°C/433–436°K.
Melting/Freezing point: 61°F/16°C/289°K.
Specific gravity (water = 1): 1.015 @ 68°F/20°C.
Relative vapor density (air = 1): 2.97
Vapor pressure: 0.7 mm.

METHACRYLIC ACID, BUTYL, DECYL, CETYL, and EICOSYL ESTER MIX REC. M:1250

SYNONYMS: BUTYL, DECYL, CETYL, EICOSYL METHACRYLATE MIXTURE; BUTYL, DECYL, CETYL, EICOSYL 2-METHYL-2-PROPENOATE

IDENTIFICATION
CAS Number: 97-88-1 (butyl)
Formula: Mixture of RC_4H_9, $RC_{10}H_{21}$, $RC_{16}H_{31}$, and $RC_{20}H_{41}$ where R is $[CH_2 = C(CH_3)COO]$
DOT ID Number: None; DOT Guide Number: 171P
Proper Shipping Name: Polymerization material, stabilize with dry ice

DESCRIPTION: Colorless liquid. Mild acrylate odor. Floats on the surface of water; insoluble.

Flammable • Containers may BLEVE when exposed to fire • Vapors may form explosive mixture with air • Vapors are heavier than air and will collect and stay in low areas • Vapors may travel long distances to ignition sources and flashback • Vapors in confined areas (e.g., tanks, sewers, buildings) may explode when exposed to fire • Polymerization hazard • Heat can induce polymerization with rapid release of energy; sealed containers may rupture explosively • May react with itself blocking relief valves; leading to tank explosions • Irritating to the skin, eyes, and respiratory tract • Toxic products of combustion may include carbon monoxide and CO_2.

Hazard Classification (based on NFPA-704 Rating System)
Health Hazards (Blue): 3; Flammability (Red): 2; Reactivity (Yellow): 2

EMERGENCY RESPONSE: See Appendix A (171)
Evacuation:
Public safety: Isolate the area of spill or leak for at least 10 to 25 meters (30 to 80 feet) in all directions.
Spill: Increase, in the downwind direction, as necessary, the distance shown under "Public Safety."
Fire: If any large container is involved in fire, isolate for at least 800 meters (½ mile) in all directions; also, consider initial evacuation for 800 meters (½ mile) in all directions.

EXPOSURE
Short-term effects: *SEEK MEDICAL ATTENTION*. **Vapor:** May be harmful if inhaled or skin is exposed. Irritating to the eyes, nose, and throat. Move victim to fresh air. *IF BREATHING HAS STOPPED*, give artificial respiration. IF breathing is difficult, administer oxygen. **Liquid:** Irritating to skin and eyes. Harmful if swallowed. Remove contaminated clothing and shoes. Flush affected areas with plenty of water. *IF IN EYES*, hold eyelids open and flush with plenty of water. *IF SWALLOWED* and victim is *CONSCIOUS AND ABLE TO SWALLOW*, have victim drink 4 to 8 ounces of water and have victim induce vomiting. *IF SWALLOWED* and victim is *UNCONSCIOUS OR HAVING CONVULSIONS*, do nothing except keep victim warm.

HEALTH HAZARDS
Personal protective equipment (PPE): B-Level PPE. Self-contained respirator; impervious gloves; chemical splash goggles.
Long-term health effects: Birth defects in rats (gross and skeletal abnormalities).
Vapor (gas) irritant characteristics: Vapors are moderately irritating such that personnel will usually not tolerate moderate or high concentrations.
Liquid or solid irritant characteristics: Fairly severe eye and skin irritant. May cause pain and second-degree burns after a few minutes of contact.

FIRE DATA
Flash point: 126°F/52°C (oc).
Flammable limits in air: LEL: 2.0%; UEL: 8.0%.
Fire extinguishing agents not to be used: Water may be ineffective.
Behavior in fire: Containers may explode.
Autoignition temperature: 526°F/274°C/547°K.
Electrical hazard: Class I, Group D (metacrylic acid).

CHEMICAL REACTIVITY
Binary reactants: Strong oxidizers, strong acids, amines.
Polymerization: May occur when heated or exposed to light.
Inhibitor of polymerization: 10 ppm of hydroquinone monomethyl ether
Reactivity group: 14
Compatibility class: Acrylates

ENVIRONMENTAL DATA
Food chain concentration potential: Negative; unlikely to accumulate.
Water pollution: Effects of low concentrations on aquatic life is unknown. Fouling to shoreline. May be dangerous if it enters nearby water intakes; notify operators. Notify local health and waterlife officials. **Response to discharge:** Mechanical containment. Should be removed. Chemical and physical treatment.

SHIPPING INFORMATION
Storage temperature: Ambient; **Inert atmosphere:** None; **Venting:** Pressure-vacuum; **Stability during transport:** Stable.

NAS HAZARD CLASSIFICATION FOR BULK WATER TRANSPORTATION
FIRE:–
HEALTH: Vapor irritant: 3; Liquid or solid irritant: 3; Poisons:–
WATER POLLUTION: Human toxicity:–; Aquatic toxicity:–; Aesthetic effect: 2
REACTIVITY: Other chemicals: 3; Water: 1; Self-reaction: 3

PHYSICAL AND CHEMICAL PROPERTIES
Physical state @ 59°F/15°C and 1 atm: Liquid.
Vapor pressure: (Reid) Low

METHACRYLONITRILE   REC. M:1300

SYNONYMS: CLORURE de METALILO (Spanish); 2-CYANOPROPENE-1; 2-CYANO-1-PROPENE; EEC No. 608-010-00-2; ISOPROPENE CYANIDE; ISOPROPENYLNITRILE; METACRILONITRILO (Spanish); α-METHYLACRYLONITRILE; 2-METHYLACRYLONITRILE; 2-METHYLPROPENENITRILE; 2-METHYL-2-PROPENENITRILE; 2-PROPENENITRILE, 2-METHYL-; RCRA No. U152

IDENTIFICATION
CAS Number: 126-98-7
Formula: C_4H_5N; $H_2C = C(CH_3)CN$
DOT ID Number: UN 3079; DOT Guide Number: 131P
Proper Shipping Name: Methacryonitrile, inhibited
Reportable Quantity (RQ): **(CERCLA)** 1000 lb/454 kg

DESCRIPTION: Liquid. Colorless

Poison! • Highly flammable • Breathing the vapor, skin or eye contact, or swallowing the material can kill you • Firefighting gear (including SCBA) does not provide adequate protection. If exposure occurs, remove and isolate gear immediately and thoroughly decontaminate personnel • Polymerization hazard • Heat can induce polymerization with rapid release of energy; sealed containers may rupture explosively • May react with itself blocking relief valves; leading to tank explosions • Containers may BLEVE when exposed to fire • Vapors may form explosive mixture with air • Vapors are heavier than air and will collect and stay in low areas • Vapors may travel long distances to ignition sources and flashback • Vapors in confined areas (e.g., tanks, sewers, buildings) may explode when exposed to fire • Toxic products of combustion may include nitrogen oxide and cyanide • Do not put yourself in danger by entering a contaminated area to rescue a victim.

Hazard Classification (based on NFPA-704 Rating System)
Health Hazards (Blue): 2; Flammability (Red): 3; Reactivity (Yellow): 2

EMERGENCY RESPONSE: See Appendix A (131)
Evacuation:
Public safety: Isolate spill area for at least 100 to 200 meters (330 to 660 feet) in all directions.
Spill: Small spill–First: Isolate in all directions 30 meters (100 feet); Then: Protect persons downwind, DAY: 0.2 km (0.1 mile); NIGHT: 0.5 km (0.3 mile). Large spill–First: Isolate in all directions 60 meters (200 feet); Then: Protect persons downwind, DAY: 0.6 km (0.4 mile); NIGHT: 1.6 km (1.0 mile).
Fire: Isolate for 800 meters (½ mile) in all directions, especially if tank, rail car, or tank truck is involved in fire.

EXPOSURE
Short-term effects: *SEEK MEDICAL ATTENTION*. **Vapor:** Irritating to eyes, nose, and throat. *IF INHALED*, will, will cause headache or nausea. Move to fresh air. *IF BREATHING HAS STOPPED*, give artificial respiration; *avoid mouth-to-mouth resuscitation; use bag/mask apparatus*. If breathing is difficult, administer oxygen (100% if necessary). **Liquid:** *HARMFUL IF ABSORBED THROUGH THE SKIN*. Will burn skin and eyes. Harmful if swallowed. Remove contaminated clothing and shoes. Flush affected areas with plenty of water. *IF IN EYES*, hold eyelids open and flush with plenty of water.
Note to physician or authorized medical personnel: Consider the use of amyl nitrite perles if symptoms of cyanide poisoning develop. If symptoms indicate, initial treatment includes the cyanide antidote kit. In all cases, break an amyl nitrite perle in a gauze pad and hold lightly under victim's nose for 15 seconds, repeating 5 times at about 15-second intervals; if necessary (and if sodium nitrite infusions will be delayed), repeat procedure every 3 minutes with fresh pearls until 3 or 4 have been used. Avoid breathing the vapor while administering it to the victim. Administer sodium nitrite IV, ASAP. The usual adult dose is 10 to 20 mL of a 3% solution infused over no less than 5 minutes; the average child dose is 0.15 to 0.20 mL/kg. Monitor blood pressure during administration, and slow the rate of infusion if hypotention develops. Next, infuse sodium thiosulfate IV. The usual adult dose is 50 mL of a 25% solution infused over 10 to 20 minutes; the average child dose is 1.65 mL/kg. Repeat with nitrite and thiosulfate as required.

HEALTH HAZARDS
Personal protective equipment (PPE): A-Level PPE. OSHA Table Z-1-A air contaminant. Self contained breathing apparatus, rubber boots and heavy rubber gloves.
Exposure limits (TWA unless otherwise noted): ACGIH TLV 1 ppm (2.7 mg/m^3); NIOSH 1 ppm (3 mg/m^3); skin contact contributes significantly in overall exposure.
Toxicity by ingestion: Grade 4: LD_{50} = 16 mg/kg (rabbit).
Vapor (gas) irritant characteristics: Vapors cause severe irritation of eyes and throat and can cause eye and lung injury. They cannot be tolerated even at low concentrations.
Liquid or solid irritant characteristics: Severe eye and skin irritant. Causes second- and third-degree burns on short contact and is very injurious to the eyes.
Odor threshold: 6.9 ppm.

FIRE DATA
Flash point: 34°F/1°C (cc).
Flammable limits in air: LEL: 2%; UEL: 6.8%.
Behavior in fire: Container explosion may occur under fire conditions.

CHEMICAL REACTIVITY
Binary reactants: Strong oxidizers, acids, bases, light cause polymerization. Stable
Polymerization: Auto polymerization can occur; **Inhibitor of polymerization:** Stabilized with 50 ppm hydroquinone monomethyl ether.
Reactivity group: 15
Compatibility class: Substituted allyls

ENVIRONMENTAL DATA
Food chain concentration potential: Not expected to accumulate.
Water pollution: DOT Appendix B, §172.101–marine pollutant. Effects of low concentrations on aquatic life is unknown. Fouling to shoreline. May be dangerous if it enters nearby water intakes; notify operators. Notify local health and wildlife officials.
Response to discharge: Issue warning-flammable liquid. Restrict access. Evacuate area. Mechanical containment. Should be removed. Chemical and physical treatment.

SHIPPING INFORMATION
Storage temperature: Ambient; **Stability during transport:** Stable.

PHYSICAL AND CHEMICAL PROPERTIES
Physical state @ 59°F/15°C and 1 atm: Liquid.
Molecular weight: 67.09
Boiling point @ 1 atm: 194-197.6°F/90-92°C/365.2-365.2°K.
Melting/Freezing point: –32.4°F/–35.8°C/237.4°K.
Specific gravity (water = 1): 0.80 @ 68°F
Liquid surface tension: 24.45 dynes/cm = 0.024 N/m @ 68°F/20°C.
Relative vapor density (air = 1): 2.31
Vapor pressure: (Reid) 2.449 psia; 71 mm @ 77°F/25°C.

METHALLYL CHLORIDE   REC. M:1350

SYNONYMS: CLORURO de β-METALILO (Spanish); 3-CHLORO-2-METHYLPROPENE; γ-CHLOROISOBUTYLENE; β-METHALLYL CHLORIDE; 2-METHALLYL CHLORIDE; α-METHALLYL CHLORIDE

IDENTIFICATION
CAS Number: 563-47-3
Formula: C_4H_7Cl; $CH_2 = C(CH_3)CH_2Cl$
DOT ID Number: UN 2554; DOT Guide Number: 129P
Proper Shipping Name: Methallyl chloride

DESCRIPTION: Colorless to yellow liquid. Sharp penetrating odor. Floats on the surface of water. Slowly hydrolyzes in water forming methallyl alcohol and hydrochloric acid.

Highly flammable • Containers may BLEVE when exposed to fire • Vapors may form explosive mixture with air • Vapors are heavier than air and will collect and stay in low areas • Vapors may travel long distances to ignition sources and flashback • Vapors in confined areas (e.g., tanks, sewers, buildings) may explode when exposed to fire • Irritating to the skin, eyes, and respiratory tract • Polymerization hazard • Heat can induce polymerization with rapid release of energy; sealed containers may rupture explosively • May react with itself blocking relief valves; leading to tank explosions • Toxic products of combustion may include phosgene and hydrogen chloride.

Hazard Classification (based on NFPA-704 Rating System)
Health Hazards (Blue): 2; Flammability (Red): 3; Reactivity (Yellow): 1

EMERGENCY RESPONSE: See Appendix A (129)
Evacuation:
Public safety: Isolate spill area for at least 50 to 100 meters (160 to 330 feet) in all directions.
Spill: Large spill–Consider initial downwind evacuation for at least 300 meters (1000 feet).
Fire: Isolate for 800 meters (½ mile) in all directions, especially if tank, rail car, or tank truck is involved in fire.

EXPOSURE
Short-term effects: SEEK MEDICAL ATTENTION. **Vapor:** Harmful if inhaled. Move victim to fresh air. *IF BREATHING HAS STOPPED,* give artificial respiration. IF breathing is difficult, administer oxygen. **Liquid:** Irritating to skin and eyes. Harmful if swallowed. Remove Contaminated clothing and shoes. Flush affected areas with plenty of water. *IF IN EYES*, hold eyelids open and flush with plenty of water. *IF SWALLOWED* and victim is *CONSCIOUS AND ABLE TO SWALLOW*, have victim drink 4 to 8 ounces of water and have victim induce vomiting. *IF SWALLOWED* and victim is *UNCONSCIOUS OR HAVING CONVULSIONS,* do nothing except keep victim warm.

HEALTH HAZARDS
Personal protective equipment (PPE): B-Level PPE. Organic canister mask; goggles Protective materials with good to excellent resistance: Viton®.
Long-term health effects: NTP anticipated carcinogen.
Vapor (gas) irritant characteristics: Eye and respiratory tract irritant.
Liquid or solid irritant characteristics: Eye and skin irritant.

FIRE DATA
Flash point: 14°F/–10°C; (oc); 11°F/–12°C (cc).
Flammable limits in air: LEL: 3.2%; UEL: 8.1%.
Fire extinguishing agents not to be used: Water may be ineffective.
Burning rate: 4.4 mm/min

CHEMICAL REACTIVITY
Binary reactants: Violent reaction with strong oxidizers. Contact with acid or acid fumes forms toxic chloride fumes. Attacks metals in the presence of moisture.

ENVIRONMENTAL DATA
Food chain concentration potential: Negative; unlikely to accumulate.
Water pollution: Effect of low concentrations on aquatic life is unknown. Fouling to shoreline. May be dangerous if it enters nearby water intakes; notify operators. Notify local health and wildlife officials. **Response to discharge:** Issue warning–high flammability. Restrict access. Mechanical containment. Should be removed. Chemical and physical treatment.

SHIPPING INFORMATION
Grades of purity: 95+%; **Storage temperature:** Ambient; **Inert atmosphere:** None; **Venting:** Pressure-vacuum; **Stability during transport:** Stable

PHYSICAL AND CHEMICAL PROPERTIES
Physical state @ 59°F/15°C and 1 atm: Liquid.
Molecular weight: 90.55
Boiling point @ 1 atm: 162.0°F/72.2°C/345.4°K.
Melting/Freezing point: Less than –112°F/–80°C/193°K.
Specific gravity (water = 1): 0.928 @ 68°F/20°C (liquid).
Liquid surface tension: (estimate) 25 dynes/cm = 0.025 N/m @ 68°F/20°C.
Liquid water interfacial tension: (estimate) 32 dynes/cm = 0.032 N/m @ 68°F/20°C.
Relative vapor density (air = 1): 3.12
Ratio of specific heats of vapor (gas): 1.0893

Latent heat of vaporization: 160 Btu/lb = 89 cal/g = 3.7×10^5 J/kg.
Heat of combustion: (estimate) –11,600 Btu/lb = –6500 cal/g = -270×10^5 J/kg.

METHANE   REC. M:1400

SYNONYMS: BIOGAS; EEC No. 601-001-00-4; FIRE DAMP; MARSH GAS; METANO (Spanish); METHANE GAS; METHYL HYDRIDE; NATURAL GAS

IDENTIFICATION
CAS Number: 74-82-8
Formula: CH_4
DOT ID Number: UN 1971; DOT Guide Number: 115
Proper Shipping Name: Methane, compressed or natural gas, compressed (with high methane content).

DESCRIPTION: Colorless gas. Shipped and stored as a compressed gas or cryogenic liquid. Weak, skunk-like odor. Liquid floats floats and boils on water surface; insoluble. Flammable visible vapor cloud is produced.

Extremely flammable • Containers may BLEVE when exposed to fire • Gas forms explosive mixture with air • Gas is lighter than air, and will disperse slowly unless confined; it will also collect and stay in low areas • Gas may travel long distances to ignition sources and flashback • Gas in confined areas (e.g., tanks, sewers, buildings) may explode when exposed to fire • Irritating to the skin, eyes, and respiratory tract • Vapors may cause dizziness or asphyxiation without warning • Contact with liquid may cause frostbite • Ruptured or venting cylinders may rocket through buildings and/or travel a considerable distance.

Hazard Classification (based on NFPA-704 Rating System)
Health Hazards (Blue): 1; Flammability (Red): 4; Reactivity (Yellow): 0

EMERGENCY RESPONSE: See Appendix A (115)
Evacuation:
Public safety: Isolate spill area for at least 50 to 100 meters (160 to 330 feet) in all directions.
Spill: Consider initial downwind evacuation for at least 800 meters (½ mile).
Fire: Isolate for 1600 meters (1 mile) in all directions, especially if tank, rail car, or tank truck is involved in fire.

EXPOSURE
Short-term effects: *SEEK MEDICAL ATTENTION.* **Vapor:** Not irritating to eyes, nose or throat. *IF INHALED*, will, will cause dizziness, difficult breathing, and loss of consciousness. Move to fresh air. *IF BREATHING HAS STOPPED,* give artificial respiration. IF breathing is difficult, administer oxygen. **Liquid:** Contact will cause frostbite. Flush affected areas with plenty of water. *DO NOT RUB AFFECTED AREAS.*

HEALTH HAZARDS
Personal protective equipment (PPE): B-Level PPE. Self-contained breathing apparatus for high concentrations; protective clothing if exposed to liquid.
Wear thermal protective clothing. Chemical protective material(s) reported to have good to excellent resistance: PV alcohol.
Exposure limits (TWA unless otherwise noted): ACGIH TLV simple asphyxiant. **Inhalation:** Remove to fresh air. Support respiration. Methane is an asphyxiant, and limiting factor is available oxygen.
Vapor (gas) irritant characteristics: Vapors are nonirritating to the eyes and throat.
Liquid or solid irritant characteristics: Practically harmless to the skin; but, it evaporates quickly, but may cause frostbite.
Odor threshold: 200 ppm.

FIRE DATA
Flash point: Flammable gas.
Flammable limits in air: LEL: 5.0%; UEL: 15.0%.
Fire extinguishing agents not to be used: Water.
Autoignition temperature: 1004°F/540°C/813°K.
Electrical hazard: Class I, Group D.
Burning rate: 12.5 mm/min
Adiabatic flame temperature: 2339°F/1282°C (estimate).
Stoichiometric air-to-fuel ratio: 17.16 (estimate).

CHEMICAL REACTIVITY
Binary reactants: Strong oxidizers.
Reactivity group: 31
Compatibility class: Paraffins

ENVIRONMENTAL DATA
Water pollution: Not harmful to aquatic life. **Response to discharge:** Issue warning–high flammability. Restrict access. Evacuate area.

SHIPPING INFORMATION
Grades of purity: Research grade; pure grade; **Storage temperature:** –260°F/–162°C/111°K; **Inert atmosphere:** None; **Venting:** Safety relief; **Stability during transport:** Stable.

NAS HAZARD CLASSIFICATION FOR BULK WATER TRANSPORTATION
FIRE: 4
HEALTH: Vapor irritant: 0; Liquid or solid irritant: 0; Poisons: 0
WATER POLLUTION: Human toxicity: 0; Aquatic toxicity: 0; Aesthetic effect: 0
REACTIVITY: Other chemicals: 0; Water: 0; Self-reaction: 0

PHYSICAL AND CHEMICAL PROPERTIES
Physical state @ 59°F/15°C and 1 atm: Gas.
Molecular weight: 16.04
Boiling point @ 1 atm: –258.7°F/–161.5°C/111.7°K.
Melting/Freezing point: –296.5°F/–182.5°C/90.7°K.
Critical temperature: –116.5°F/–82.5°C/190.7°K.
Critical pressure: 668 psia = 45.44 atm = 4.60 MN/m^2.
Specific gravity (water = 1): 0.422 at –160°C (liquid).
Liquid surface tension: 14 dynes/cm = 0.014 N/m at –161°C.
Liquid water interfacial tension: (estimate) 50 dynes/cm = 0.050 N/m at –161°C.
Relative vapor density (air = 1): 0.55
Ratio of specific heats of vapor (gas): 1.306
Latent heat of vaporization: 219.4 Btu/lb = 121.9 cal/g = 5.100×10^5 J/kg.
Heat of combustion: –21,517 Btu/lb = –11,954 cal/g = -500.2×10^5 J/kg.
Heat of fusion: 13.96 cal/g.
Vapor pressure: 5 atm @ –138.3°C; 40 atm @ –86.3°C.

METHANEARSONIC ACID, SODIUM SALT   REC. M:1425

SYNONYMS: ANSAR 170; ARSONATE; ASAZOL; BUENO;

DICONATE 6; DAL-A-RAD; DISODIUM METHANE ARSONATE; DSMA; DISODIUM METHYL ARSONATE; HERB-ALL; HERBAN M; MERGE; MESAMATE; MONATE; MONOSODIUM METHANE ARSONATE; MSMA; MONOSODIUM METHYL ARSONATE; PHYBAN; SILVISAR; SILVISAR-550; TARGET MSMA; TRANS-VERT; WEED 108; WEED E RAD; WEED HOE

IDENTIFICATION
CAS Number: 2163-80-6
Formula: $CH_4AsO_3 \cdot Na$; $CH_3AsO(OH)(ONa)$ $CH_3AsO(ONa)_2 \cdot 6H_2O$
DOT ID Number: UN 1557; DOT Guide Number: 152
Proper Shipping Name: Arsenic compounds, solid, n.o.s.

DESCRIPTION: Colorless solid; solution may be dyed red or green. Odorless. Solid may float or sink in water; solid and solution mix with water.

Combustible • Containers may BLEVE when exposed to fire • Vapors are heavier than air and will collect and stay in low areas • Vapors in confined areas (e.g., tanks, sewers, buildings) may explode when exposed to fire • Toxic products of combustion may include arsenic and sodium oxide • Do not put yourself in danger by entering a contaminated area to rescue a victim.

Hazard Classification (based on NFPA-704 Rating System)
Health Hazards (Blue): 1; Flammability (Red): 1; Reactivity (Yellow): 0

EMERGENCY RESPONSE: See Appendix A (152)
Evacuation:
Public safety: Isolate the area of spill or leak for at least 25 to 50 meters (80 to 160 feet) in all directions.
Spill: Increase, in the downwind direction, as necessary, the distance shown under "Public Safety."
Fire: If tank, rail car, or tank truck is involved in fire, isolate for at least 800 meters (½ mile) in all directions; also, consider initial evacuation for 800 meters (½ mile) in all directions.

EXPOSURE
Short-term effects: *SEEK MEDICAL ATTENTION.* If artificial respiration is administered, *avoid mouth-to-mouth resuscitation; use bag/mask apparatus.*
Solid or solution: Irritating to skin and eyes. *IF SWALLOWED*, will cause nausea, vomiting, or loss of consciousness. Remove contaminated clothing and shoes. Flush affected areas with plenty of water. *IF IN EYES*, hold eyelids open and flush with plenty of water. *IF SWALLOWED* and victim is *CONSCIOUS AND ABLE TO SWALLOW*, have victim drink 4 to 8 ounces of water and have victim induce vomiting. *IF SWALLOWED* and victim is *UNCONSCIOUS OR HAVING CONVULSIONS,* do nothing except keep victim warm.

HEALTH HAZARDS
Personal protective equipment (PPE): A-Level PPE. Protective clothing to prevent contact with skin; chemical goggles.
Toxicity by ingestion: Grade 2; LD_{50} = 0.5 to 5 g/kg (rat).
Liquid or solid irritant characteristics: Repeated contact may cause skin sensitivity.
IDLH value: 5 mg/m^3 as inorganic arsenic; known human carcinogen. Not determined for organic arsenic.

FIRE DATA
Flash point: Combustible, but difficult to ignite.

ENVIRONMENTAL DATA
Water pollution: Dangerous to aquatic life in high concentrations. May be dangerous if it enters nearby water intakes; notify operators. Notify local health and wildlife officials. **Response to discharge:** Issue warning–poison, water contaminant. Should be removed. Chemical and physical treatment.

SHIPPING INFORMATION
Grades of purity: The solid disodium salt (DSMA) contains water crystallization. Salts are often shipped as solutions in water with concentrations up to about 50% solids; **Storage temperature:** Ambient; **Inert atmosphere:** None; **Venting:** Open; **Stability during transport:** Stable.

PHYSICAL AND CHEMICAL PROPERTIES
Physical state @ 59°F/15°C and 1 atm: Solid or water solution
Molecular weight: 162 (MSMA); 292 (DSMA hexahydrate).
Boiling point @ 1 atm: Decomposes.
Melting/Freezing point: (MSMA) 243°F/117°C/390°K; (DSMA) 137°F/58°C/332°K.
Specific gravity (water = 1): (DSMA) 1.0 @ 68°F/20°C (solid); (MSMA solutions) 1.4–1.6 @ 68°F/20°C (liquid).

3-METHOXYBUTYL ACETATE REC. M:1450

SYNONYMS: ACETIC ACID, 3-METHOXYBUTYL ESTER; BUTOXYL; 3-ETHOXYBUTYL ACETATE; METHYL-1,3-BUTYLENE GLYCOL ACETATE; 1-BUTANOL, 3-METHOXY-, ACETATE; 1-BUTANOL, 3-METHOXYACETATE; 3-ETHOXY-1-BUTANOL ACETATE

IDENTIFICATION
CAS Number: 4435-53-4
Formula: $C_7H_{14}O_3$
DOT ID Number: UN 2708; DOT Guide Number: 127
Proper Shipping Name: Butoxyl

DESCRIPTION: Liquid. Acrid odor.

Flammable • Containers may BLEVE when exposed to fire • Vapors may form explosive mixture with air • Vapors are heavier than air and will collect and stay in low areas • Vapors may travel long distances to ignition sources and flashback • Vapors in confined areas (e.g., tanks, sewers, buildings) may explode when exposed to fire • Irritating to the skin, eyes, and respiratory tract • Combustion products include smoke and irritating gases.

Hazard Classification (based on NFPA-704 Rating System)
Health Hazards (Blue): 1; Flammability (Red): 2; Reactivity (Yellow): 0

EMERGENCY RESPONSE: See Appendix A (127)
Evacuation:
Public safety: Isolate spill area for at least 25 to 50 meters (80 to 160 feet) in all directions.
Spill: Large spill–Consider initial downwind evacuation for at least 300 meters (1000 feet).
Fire: Isolate for 800 meters (½ mile) in all directions, especially if tank, rail car, or tank truck is involved in fire.

EXPOSURE
Short-term effects: *SEEK MEDICAL ATTENTION.* **Vapor:** May be harmful. Move to fresh air. *IF BREATHING HAS STOPPED,* give artificial respiration. IF breathing is difficult, administer

oxygen. **Liquid:** Irritating to skin and eyes. Remove contaminated clothing and shoes. Flush affected areas with plenty of water. *IF IN EYES*, hold eyelids open and flush with plenty of water. *IF SWALLOWED*, immediately induce vomiting.

HEALTH HAZARDS
Personal protective equipment (PPE): B-Level PPE. Self-contained breathing apparatus, rubber boots and heavy rubber gloves.
Toxicity by ingestion: Grade 2: LD_{50} = 4.2 g/kg.
Vapor (gas) irritant characteristics: Vapors cause smarting of the eyes or respiratory system if present in high concentrations. The effect is temporary.
Liquid or solid irritant characteristics: Eye ans skin irritant. If spilled on clothing and allowed to remain, may cause smarting and reddening of skin.

FIRE DATA
Flash point: 170°F/77°C (cc).

CHEMICAL REACTIVITY
Reactivity group: 34
Compatibility class: Esters.

ENVIRONMENTAL DATA
Water pollution: Effect of low concentrations on aquatic life is unknown. Fouling to shoreline. May be dangerous if it enters nearby water intakes; notify operators. Notify local health and wildlife officials. **Response to discharge:** Restrict access. Should be removed. Chemical and physical treatment.

SHIPPING INFORMATION
Stability during transport: Stable.

PHYSICAL AND CHEMICAL PROPERTIES
Physical state @ 59°F/15°C and 1 atm: Liquid.
Molecular weight: 146.19
Boiling point @ 1 atm: 275°F/135°C/408.2°K.
Specific gravity (water = 1): 0.96
Relative vapor density (air = 1): 5.05

METHOXYCHLOR   **REC. M:1500**

S Y N O N Y M S : C H E M F O R M ; *p , p '*-DIMETHOXYDIPHENYLTRICHLOROETHANE; DMDT; MARLATE 50; MARLATE; METOXICLORO (Spanish); METHOXY DDT; METOX; MOXIE; 2,2-BIS(*p*-METHOXYPHENYL)-1,1,1-TRICHLOROETHANE; 1,1,1-TRICHLORO-2,2-BIS(*p*-METHOXYPHENYL)ETHANE; RCRA No. U274

IDENTIFICATION
CAS Number: 72-43-5
Formula: $C_{16}H_{15}Cl_3O_2$
DOT ID Number: UN 2761; **DOT Guide Number:** 151
Proper Shipping Name: Organochlorine pesticides, solid, toxic, n.o.s.
Reportable Quantity (RQ): (CERCLA) 1 lb/0.454 kg

DESCRIPTION: White to light yellow solid (dust, granules, or wettable powder). Mild, slightly fruity odor. Sinks in water; wettable powder slowly dissolves. May be dissolved in kerosene or alcohol.

Poison! (organochlorine) • Breathing the dust, skin or eye contact can cause illness; swallowing the material can kill you • Firefighting gear (including SCBA) does not provide adequate protection. If exposure occurs, remove and isolate gear immediately and thoroughly decontaminate personnel • Containers may BLEVE when exposed to fire • Concentrated dust in confined areas (e.g., tanks, sewers, buildings) may explode when exposed to fire • Toxic products of combustion may include hydrogen chloride.

Hazard Classification (based on NFPA-704 Rating System)
Health Hazards (Blue): 2; Flammability (Red): 1; Reactivity (Yellow): 0

EMERGENCY RESPONSE: See Appendix A (151)
Evacuation:
Public safety: Isolate the area of spill or leak for at least 25 to 50 meters (80 to 160 feet) in all directions.
Spill: Increase, in the downwind direction, as necessary, the distance shown under "Public Safety."
Fire: If tank, rail car, or tank truck is involved in fire, isolate for at least 800 meters (½ mile) in all directions; also, consider initial evacuation for 800 meters (½ mile) in all directions.

EXPOSURE

Short-term effects: *SEEK MEDICAL ATTENTION.* **Dust:** *POISONOUS IF INHALED.* Move victim to fresh air. *IF IN EYES,* hold eyelids open and flush with plenty of water. *IF BREATHING HAS STOPPED,* give artificial respiration; *avoid mouth-to-mouth resuscitation; use bag/mask apparatus.* IF breathing is difficult, administer oxygen. **Solid:** *POISONOUS IF SWALLOWED.* Irritating to skin and eyes. Remove contaminated clothing and shoes. Flush affected areas with plenty of water. *IF IN EYES,* hold eyelids open and flush with plenty of water. *IF SWALLOWED* and victim is *CONSCIOUS AND ABLE TO SWALLOW,* have victim drink 4 to 8 ounces of water. **Do NOT induce vomiting.** Administer a slurry of activated charcoal.

HEALTH HAZARDS
Personal protective equipment (PPE): B-Level PPE. OSHA Table Z-1-A air contaminant. Dust respirator if needed; gloves and goggles.
Recommendations for respirator selection: NIOSH
At any concentrations above the NIOSH REL, or where there is no REL, at any detectable concentration: SCBAF:PD,PP (any self-contained breathing apparatus that has a full facepiece and is operated in a pressure-demand or other positive-pressure mode); or SAF:PD,PP:ASCBA (any supplied-air respirator that has a full facepiece and is operated in a pressure-demand or other positive-pressure mode in combination with an auxiliary, self-contained breathing apparatus operated in a pressure-demand or other positive pressure mode). *ESCAPE:* GMFOVHiE [any air-purifying, full-facepiece respirator (gas mask) with a chin-style, front- or back-mounted organic vapor canister having a high-efficiency particulate filter]; or SCBAE (any appropriate escape-type, self-contained breathing apparatus).
Exposure limits (TWA unless otherwise noted): ACGIH TLV 10 mg/m^3; OSHA PEL 15 mg/m^3; NIOSH potential human carcinogen; reduce exposure to lowest feasible level.
Toxicity by ingestion: Grade 1; LD_{50} = 5 to 15 g/kg.
Long-term health effects: Allergen.
IDLH value: 5000 mg/m^3; potential human carcinogen.

FIRE DATA
Flash point: Burns only at high temperatures.

CHEMICAL REACTIVITY
Binary reactants: Oxidizers

ENVIRONMENTAL DATA
Food chain concentration potential: Potential is high for chlorinated hydrocarbons.
Water pollution: Harmful to aquatic life in very low concentrations. May be dangerous if it enters nearby water intakes; notify operators. Notify local health and wildlife officials.
Response to discharge: Issue warning–poison, water contaminant. Restrict access. Should be removed. Chemical and physical treatment.

SHIPPING INFORMATION
Grades of purity: Technical flake or chip; 88% plus 12% isomers; **Wettable powders:** 50-75% Dust concentrate: 40% Emulsifiable concentrate (liquid): 25% solution in petroleum distillate; **Storage temperature:** Ambient; **Inert atmosphere:** None; **Venting:** Open; **Stability during transport:** Stable.

PHYSICAL AND CHEMICAL PROPERTIES
Physical state @ 59°F/15°C and 1 atm: Solid.
Molecular weight: 345.7
Boiling point @ 1 atm: (decomposes).
Melting/Freezing point: 171–192°F/77–89°C/350–362°K.
Specific gravity (water = 1): 1.41 @ 77°F/25°C (solid).
Vapor pressure: Very low.

METHYL ACETATE   REC. M:1550

SYNONYMS: ACETATE de METHYLE (French); ACETATO de METILO (Spanish); ACETIC ACID, METHYL ESTER; DEVOTON; EEC No. 607-021-00-X; METHYLACETAT (German); METHYL ACETIC ESTER; METHYLE (ACETATE de) (French); METHYL ESTER OF ACETIC ACID; METHYL ETHANOATE; TERETON

IDENTIFICATION
CAS Number: 79-20-9
Formula: $C_3H_6O_2$; CH_3COOCH_3
DOT ID Number: UN 1231; DOT Guide Number: 129
Proper Shipping Name: Methyl acetate

DESCRIPTION: Colorless liquid. Mild fruity odor. Initially floats on water surface; soluble. Produces large amounts of vapor.

Highly flammable • Containers may BLEVE when exposed to fire • Vapors may form explosive mixture with air • Vapors are heavier than air and will collect and stay in low areas • Vapors may travel long distances to ignition sources and flashback • Vapors in confined areas (e.g., tanks, sewers, buildings) may explode when exposed to fire • Irritating to the skin, eyes, and respiratory tract • Toxic products of combustion may include carbon monoxide.

Hazard Classification (based on NFPA-704 Rating System)
Health Hazards (Blue): 1; Flammability (Red): 3; Reactivity (Yellow): 0

EMERGENCY RESPONSE: See Appendix A (129)
Evacuation:
Public safety: Isolate spill area for at least 50 to 100 meters (160 to 330 feet) in all directions.
Spill: Large spill–Consider initial downwind evacuation for at least 300 meters (1000 feet).
Fire: Isolate for 800 meters (½ mile) in all directions, especially if tank, rail car, or tank truck is involved in fire.

EXPOSURE
Short-term effects: *SEEK MEDICAL ATTENTION.* **Vapor:** Irritating to eyes, nose, and throat. *IF INHALED*, will, will cause headache, or dizziness. Move victim to fresh air. *IF BREATHING HAS STOPPED*, give artificial respiration. IF breathing is difficult, administer oxygen. **Liquid:** Irritating to skin and eyes. Remove contaminated clothing and shoes. Flush affected areas with plenty of water. *IF IN EYES*, hold eyelids open and flush with plenty of water. *IF SWALLOWED* and victim is *CONSCIOUS AND ABLE TO SWALLOW*, have victim drink 4 to 8 ounces of water.

HEALTH HAZARDS
Personal protective equipment (PPE): B-Level PPE. OSHA Table Z-1-A air contaminant. Air mask or organic canister mask; goggles or face shield.
Protective materials with good to excellent resistance: butyl rubber, Silvershield®. Also, neoprene offers limited protection.
Recommendations for respirator selection: NIOSH/OSHA
2000 ppm: CCROV* (any chemical cartridge respirator with organic vapor cartridge(s); or SA* (any supplied-air respirator). *3100 ppm:* SA:CF* (any supplied-air respirator operated in a continuous-flow mode); or CCRFOV [any chemical cartridge respirator with a full facepiece and organic vapor cartridges(s)]; or GMFOV [any air-purifying, full-facepiece respirator (gas mask) with a chin-style, front- or back-mounted organic vapor cannister]; or PAPROV* [any powered, air-purifying respirator with organic vapor cartridge(s)]; or SCBAF (any self-contained breathing apparatus with a full facepiece); or SAF (any supplied-air respirator with a full facepiece). *EMERGENCY OR PLANNED ENTRY INTO UNKNOWN CONCENTRATIONS OR IDLH CONDITIONS:* SCBAF:PD,PP (any self-contained breathing apparatus that has a full facepiece and is operated in a pressure-demand or other positive-pressure mode); or SAF:PD,PP:ASCBA (any supplied-air respirator that has a full facepiece and is operated in a pressure-demand or other positive-pressure mode in combination with an auxiliary self-contained breathing apparatus operated in a pressure-demand or other positive pressure mode). *ESCAPE:* GMFOV [any air-purifying, full-facepiece respirator (gas mask) with a chin-style, front- or back-mounted organic vapor cannister]; or SCBAE (any appropriate escape-type, self-contained breathing apparatus).*Note:* Substance reported to cause eye irritation or damage; may require eye protection.
Exposure limits (TWA unless otherwise noted): ACGIH TLV 200 ppm (606 mg/m^3); OSHA PEL 200 ppm (610 mg/m^3); NIOSH REL 200 ppm (610 mg/m^3).
Short-term exposure limits (15-minute TWA): ACGIH STEL 250 ppm (757 mg/m^3); NIOSH STEL 250 ppm (760 mg/m^3).
Toxicity by ingestion: Grade 2; oral LD_{50} = 3700 mg/kg (rabbit).
Long-term health effects: Optic nerve may be damaged following overexposure to vapor or liquid.
Vapor (gas) irritant characteristics: Eye and respiratory tract irritant. Vapors cause irritation such that personnel will find high concentrations unpleasant. The effect is temporary.
Liquid or solid irritant characteristics: Eye irritant. Defatting of the skin may cause drying and irritation.
Odor threshold: 180 ppm.
IDLH value: 3100 ppm.

FIRE DATA
Flash point: 22°F/–6°C (oc), 14°F/–10°C (cc).
Flammable limits in air: LEL: 3.1%; UEL: 16%.

Fire extinguishing agents not to be used: Water may be ineffective.
Autoignition temperature: 850°F/454°C/727°K.
Electrical hazard: Class I, Group D.
Burning rate: 3.7 mm/min

CHEMICAL REACTIVITY
Reactivity with water: Reacts slowly to form acetic acid and methyl alcohol; the reaction is not violent.
Binary reactants: Reacts with nitrates, strong oxidizers, alkalis, and nitric and organic acids. May attack some plastic materials.
Neutralizing agents for acids and caustics: Mild acidic solution such as vinegar or 1-2% acetic acid.
Reactivity group: 34
Compatibility class: Esters.

ENVIRONMENTAL DATA
Food chain concentration potential: Log P_{ow} = 0.2. Unlikely to accumulate.
Water pollution: Effect of low concentrations on aquatic life is unknown. May be dangerous if it enters nearby water intakes; notify operators. Notify local health and wildlife officials.
Response to discharge: Issue warning–high flammability. Restrict access. Disperse and flush.

SHIPPING INFORMATION
Grades of purity: 78–82%; remainder is methyl alcohol; **Storage temperature:** Ambient; **Inert atmosphere:** None; **Venting:** Pressure-vacuum; **Stability during transport:** Stable.

NAS HAZARD CLASSIFICATION FOR BULK WATER TRANSPORTATION
FIRE: 3
HEALTH: Vapor irritant: 2; Liquid or solid irritant: 0; Poisons: 1
WATER POLLUTION: Human toxicity: 1; Aquatic toxicity: 2; Aesthetic effect: 1
REACTIVITY: Other chemicals: 1; Water: 0; Self-reaction: 0

PHYSICAL AND CHEMICAL PROPERTIES
Physical state @ 59°F/15°C and 1 atm: Liquid.
Molecular weight: 74.1
Boiling point @ 1 atm: 134.6°F/57.0°C/330.2°K.
Melting/Freezing point: –145.3°F/98.5°C/174.7°K.
Critical temperature: 452.7°F/233.7°C/506.9°K.
Critical pressure: 666 psia = 45.3 atm = 4.60 MN/m².
Specific gravity (water = 1): 0.927 @ 68°F/20°C (liquid).
Liquid surface tension: 24 dynes/cm = 0.024 N/m @ 68°F/20°C.
Liquid water interfacial tension: (estimate) 30 dynes/cm = 0.030 N/m @ 68°F/20°C.
Relative vapor density (air = 1): 2.8
Ratio of specific heats of vapor (gas): 1.1192
Latent heat of vaporization: 174 Btu/lb = 97 cal/g = 4.1 x 10^5 J/kg.
Heat of combustion: 9260 Btu/lb = 5150 cal/g = 215 x 10^5 J/kg.
Vapor pressure: (Reid) 4.6 psia; 173 mm.

METHYL ACETOACETATE   REC. M:1600

SYNONYMS: ACETOACETATO de METILO (Spanish); ACETIC METHYL ETHER; ACETOACETIC ACID, METHYL ESTER; BUTANOIC ACID, 3-oxo-METHYL ESTER; METHYL ACETYLACETONATE; 1-METHOXYBUTANE-1,3-DIONE; METHYL ACETYLACETATE; METHYL-3-OXOBUTYRATE; METHYL 3-OXOBUTYRATE; 3-OXOBUTANOIC ACID METHYL ESTER

IDENTIFICATION
CAS Number: 105-45-3
Formula: $C_5H_8O_3$; $CH_3COCH_2CO_2CH_3$
DOT ID Number: NA 1993; DOT Guide Number: 128
Proper Shipping Name: Combustible liquid, n.o.s.

DESCRIPTION: Colorless liquid. Soluble in water.

Combustible • Containers may BLEVE when exposed to fire • Vapors may form explosive mixture with air • Vapors are heavier than air and will collect and stay in low areas • Vapors may travel long distances to ignition sources and flashback • Vapors in confined areas (e.g., tanks, sewers, buildings) may explode when exposed to fire • Irritating to the skin, eyes, and respiratory tract • Toxic products of combustion may include carbon monoxide.

Hazard Classification (based on NFPA-704 Rating System)
Health Hazards (Blue): 2; Flammability (Red): 2; Reactivity (Yellow): 0

EMERGENCY RESPONSE: See Appendix A (128)
Evacuation:
Public safety: Isolate spill area for at least 25 to 50 meters (80 to 160 feet) in all directions.
Spill: Large spill–Consider initial downwind evacuation for at least 300 meters (1000 feet).
Fire: Isolate for 800 meters (½ mile) in all directions, especially if tank, rail car, or tank truck is involved in fire.

EXPOSURE
Short-term effects: *SEEK MEDICAL ATTENTION.* **Liquid:** Irritating to eyes. Harmful if swallowed. Remove contaminated clothing and shoes. Flush affected areas with plenty of water. *IF IN EYES*, hold eyelids open and flush with plenty of water.

HEALTH HAZARDS
Personal protective equipment (PPE): B-Level PPE. Respirator, chemical safety goggles, rubber boots and heavy rubber gloves.
Toxicity by ingestion: Grade 2: LD_{50} = 3.228 g/kg (rat).
Vapor (gas) irritant characteristics: Vapors are irritating such that personnel will not usually tolerate moderate or high concentrations.
Liquid or solid irritant characteristics: May cause eye and skin irritation. If spilled on clothing and allowed to remain, may cause smarting and reddening of skin.

FIRE DATA
Flash point: 170°F/77°C (cc).
Flammable limits in air: LEL: 3.1%; UEL: 16%.
Fire extinguishing agents not to be used: Water may be ineffective.
Autoignition temperature: 536°F/280°C/553°K.

CHEMICAL REACTIVITY
Binary reactants: Strong acids.
Reactivity group: 34
Compatibility class: Esters.

ENVIRONMENTAL DATA
Food chain concentration potential: Log K_{ow} = –0.26. Unlikely to accumulate. **Water pollution:** Effects of low concentrations on aquatic life is unknown. May be dangerous if it enters nearby water

intakes; notify operators. Notify local health and wildlife officials.
Response to discharge: Restrict access. Chemical and physical treatment.

SHIPPING INFORMATION
Grades of purity: 99+%; **Storage temperature:** Ambient; **Stability during transport:** Stable.

PHYSICAL AND CHEMICAL PROPERTIES
Physical state @ 59°F/15°C and 1 atm: Liquid.
Molecular weight: 116.12
Boiling point @ 1 atm: 336-338°F/169-170°C/442.2-443.2°K.
Melting/Freezing point: 82°F/28°C/301°K.
Specific gravity (water = 1): 1.076
Relative vapor density (air = 1): 4.0
Vapor pressure: 0.99 mmHg @ 68°F/20°C.

METHYLACETYLENE–PROPADIENE MIXTURE
REC. M:1650

SYNONYMS: ALLENE–METHYLACETYLENE MIXTURE; MAPP GAS; METHYL ACETYLENE–ALLENE MIXTURE; METILACETILENO–PROPADIENO, ESTABILIZADO (Spanish); PROPADIENE–METHYLACETYLENE MIXTURE; PROPADIENE–ALLENE MIXTURE; PROPYNE MIXED WITH PROPADIENE; PROPYNE-ALLENE mixture; PROPYNE-PROPADIENE mixture.

IDENTIFICATION
CAS Number: 59355-75-8
Formula: $CH_3CCH + CH_2=C=CH_2$
DOT ID Number: UN 1060; DOT Guide Number: 116P
Proper Shipping Name: Methylacetylene and propadiene mixtures, stabilized

DESCRIPTION: Colorless liquefied compressed gas. Garlic-like odor. Liquid floats and boils on water. Flammable visible vapor cloud is produced. A mixture containing 60-66.5% methylacetylene and propadiene, and smaller amounts of butane and propane.

Extremely flammable • Forms explosive mixture with air • Reacts explosively with many materials • Gas from liquefied gas are initially heavier than air and spread along ground • Gas may travel to source of ignition and flashback • Gas in confined areas (e.g., tanks, sewers, buildings) may explode when exposed to fire • Ruptured or venting cylinders may rocket through buildings and/or travel a considerable distance • Polymerization hazard • Heat can induce polymerization with rapid release of energy; sealed containers may rupture explosively • May react with itself blocking relief valves; leading to tank explosions • Moderately irritating to the skin, eyes, and respiratory tract • Vapors may cause dizziness or asphyxiation without warning • Contact with liquid may cause frostbite.

EMERGENCY RESPONSE: See Appendix A (116)
Caution: When in contact with refrigerated/cryogenic liquids, many materials become brittle and are likely to break without warning.
Evacuation:
Public safety: Isolate spill area for at least 100 meters (330 feet) in all directions.
Spill: Consider initial downwind evacuation for at least 800 meters (½ mile).
Fire: Isolate for 1600 meters (1 mile) in all directions, especially if tank, rail car, or tank truck is involved in fire.

EXPOSURE
Short-term effects: *SEEK MEDICAL ATTENTION.* **Vapor:** IF inhaled, will cause difficult breathing. Move victim to fresh air. If breathing is difficult, administer oxygen. **Liquid:** Will cause frostbite. Flush affected areas with plenty of water. *DO NOT RUB AFFECTED AREAS.*

HEALTH HAZARDS
Personal protective equipment (PPE): B-Level PPE. Safety goggles; protective gloves. Wear thermal protective clothing.
Recommendations for respirator selection: NIOSH/OSHA *3400 ppm:* SA (any supplied-air respirator); or SCBAF (any self-contained breathing apparatus with a full facepiece). *EMERGENCY OR PLANNED ENTRY INTO UNKNOWN CONCENTRATIONS OR IDLH CONDITIONS:* SCBAF:PD,PP (any self-contained breathing apparatus that has a full facepiece and is operated in a pressure-demand or other positive-pressure mode); or SAF:PD,PP:ASCBA (any supplied-air respirator that has a full facepiece and is operated in a pressure-demand or other positive-pressure mode in combination with an auxiliary self-contained breathing apparatus operated in a pressure-demand or other positive pressure mode). *ESCAPE:* GMFS [any air-purifying, full-facepiece respirator (gas mask) with a chin-style, front- or back-mounted canister providing protection against the compound of concern]; or SCBAE (any appropriate escape-type, self-contained breathing apparatus).
Exposure limits (TWA unless otherwise noted): ACGIH TLV 1000 ppm (1640 mg/m^3); OSHA PEL 1000 ppm (1800 mg/m^3); NIOSH 1000 ppm (1800 mg/m^3).
Short-term exposure limits (15-minute TWA): ACGIH STEL 1250 ppm (2050 mg/m^3); NIOSH STEL 1250 ppm (2250 mg/m^3).
Long-term health effects: Lung irritation in animals has been documented.
Liquid or solid irritant characteristics: Rapid evaporation of liquid causes frostbite.
Odor threshold: 100 ppm.
IDLH value: 3400 ppm [10% LEL].

FIRE DATA
Flammable limits in air: LEL: 3.4%; UEL: 10.8%.
Behavior in fire: Containers may explode.
Autoignition temperature: 850°F/454°C/727°K.
Stoichiometric air-to-fuel ratio: 13.69 (estimate).

CHEMICAL REACTIVITY
Binary reactants: Strong oxidizers, copper alloys. Forms explosive compounds at high pressure in contact with alloys containing more than 67% copper.
Reactivity group: 30
Compatibility class: Olefins

ENVIRONMENTAL DATA
Food chain concentration potential: Negative; unlikely to accumulate.
Water pollution: Not harmful to aquatic life. **Response to discharge:** Issue warning–high flammability. Restrict access. Evacuate area.

SHIPPING INFORMATION
Grades of purity: Commercial: 65% of a mixture of methylacetylene (85%) and propadiene (15%) plus 35% of a

mixture of C3 and C4 saturated and unsaturated hydrocarbons; **Storage temperature:** Ambient, but less than 125°F/52°C; **Inert atmosphere:** None; **Venting:** Safety relief; **Stability during transport:** Stable.

PHYSICAL AND CHEMICAL PROPERTIES
Physical state @ 59°F/15°C and 1 atm: Gas.
Molecular weight: 40.1
Boiling point @ 1 atm: –36 to –4°F/–38 to –20°C/235 to 253°K.
Melting/Freezing point: –213°F/136°C/409°K.
Specific gravity (water = 1): 0.576 @ 15°C (liquid).
Liquid surface tension: 18 dynes/cm = 0.018 N/m at –24°C.
Relative vapor density (air = 1): 1.48
Ratio of specific heats of vapor (gas): 1.1686
Latent heat of vaporization: 227 Btu/lb = 126 cal/g = 5.28×10^5 J/kg.
Heat of combustion: –19,800 Btu/lb = –11,000 cal/g = -460×10^5 J/kg.
Vapor pressure: (Reid) 165 psia; 1-7 mm.

METHYL ACRYLATE   REC. M:1700

SYNONYMS: ACRILATO de METILO (Spanish); ACRYLATE DE METHYLE (French); ACRYLIC ACID, METHYL ESTER; ACRYLSAEUREMETHYLESTER (German); CURITHANE 103; EEC No. 607-034-00-0; METHOXYCARBONYLETHYLENE; METHYL-ACRYLAT (German); METHYL ESTER OF ACRYLIC ACID; METHYL PROPENATE; METHYL-2-PROPENOATE; PROPENOIC ACID, METHYL ESTER; 2-PROPENOIC ACID, METHYL ESTER

IDENTIFICATION
CAS Number: 96-33-3
Formula: $C_4H_6O_2$; $CH_2=CHCOOCH_3$
DOT ID Number: UN 1919; DOT Guide Number: 129P
Proper Shipping Name: Methyl acrylate, inhibited

DESCRIPTION: Colorless watery liquid. Acrid odor. Floats and mixes slowly with water. Flammable, irritating vapor is produced.

Highly flammable • Corrosive to skin, eyes, and respiratory tract; skin or eye contact causes severe burns, impaired vision, or blindness; inhalation symptoms may be delayed • Firefighting gear (including SCBA) does not provide adequate protection. If exposure occurs, remove and isolate gear immediately and thoroughly decontaminate personnel • Containers may BLEVE when exposed to fire • Vapors may form explosive mixture with air • Vapors are heavier than air and will collect and stay in low areas • Vapors may travel long distances to ignition sources and flashback • Vapors in confined areas (e.g., tanks, sewers, buildings) may explode when exposed to fire • Polymerization hazard • Heat can induce polymerization with rapid release of energy; sealed containers may rupture explosively • May react with itself blocking relief valves; leading to tank explosions • Combustion products include smoke and irritating vapors that can cause tears.

Hazard Classification (based on NFPA-704 Rating System)
Health Hazards (Blue): 3; Flammability (Red): 3; Reactivity (Yellow): 2

EMERGENCY RESPONSE: See Appendix A (129)
Evacuation:
Public safety: Isolate spill area for at least 50 to 100 meters (160 to 330 feet) in all directions.
Spill: Large spill–Consider initial downwind evacuation for at least 300 meters (1000 feet).
Fire: Isolate for 800 meters (½ mile) in all directions, especially if tank, rail car, or tank truck is involved in fire.

EXPOSURE
Short-term effects: *SEEK MEDICAL ATTENTION.* **Vapor:** Irritating to eyes, nose, and throat. *IF INHALED*, will, will cause dizziness or difficult breathing. Move to fresh air. *IF BREATHING HAS STOPPED*, give artificial respiration. IF breathing is difficult, administer oxygen. **Liquid:** *HARMFUL IF ABSORBED THROUGH THE SKIN*. Will burn skin and eyes. Harmful if swallowed. Remove contaminated clothing. Flush affected areas with plenty of water. *IF IN EYES*, hold eyelids open and flush with plenty of water. *IF SWALLOWED* and victim is *CONSCIOUS AND ABLE TO SWALLOW*, have victim drink 4 to 8 ounces of water.
Note to physician or authorized medical personnel: Medical observation is recommended for 24 to 48 hours after breathing overexposure, as pulmonary edema may be delayed. As first aid for pulmonary edema, consider administering a corticosteroid spray. Cigarette smoking may exacerbate pulmonary injury and should be discouraged for at least 72 hours following exposure.

HEALTH HAZARDS
Personal protective equipment (PPE): B-Level PPE. OSHA Table Z-1-A air contaminant. Full face organic vapor respirator. Chemical protective material(s) reported to have good to excellent resistance: butyl rubber, PV alcohol. Also, Teflon® offers limited protection.
Recommendations for respirator selection: NIOSH/OSHA
100 ppm: SA* (any supplied-air respirator). *250 ppm:* SA:CF* (any supplied-air respirator operated in a continuous-flow mode); or SCBAF (any self-contained breathing apparatus with a full facepiece); or SAF (any supplied-air respirator with a full facepiece). *EMERGENCY OR PLANNED ENTRY INTO UNKNOWN CONCENTRATIONS OR IDLH CONDITIONS:* SCBAF:PD,PP (any self-contained breathing apparatus that has a full facepiece and is operated in a pressure-demand or other positive-pressure mode); or SAF:PD,PP:ASCBA (any supplied-air respirator that has a full facepiece and is operated in a pressure-demand or other positive-pressure mode in combination with an auxiliary self-contained breathing apparatus operated in a pressure-demand or other positive pressure mode). *ESCAPE:* GMFOV [any air-purifying, full-facepiece respirator (gas mask) with a chin-style, front- or back-mounted acid gas canister]; or SCBAE (any appropriate escape-type, self-contained breathing apparatus).
Note: Substance reported to cause eye irritation or damage; may require eye protection.
Exposure limits (TWA unless otherwise noted): ACGIH TLV 2 ppm (7 mg/m³) not classified as a human carcinogen; NIOSH/OSHA 10 ppm (35 mg/m³); skin contact contributes significantly in overall exposure.
Toxicity by ingestion: Grade 3; LD_{50} = 50 to 500 mg/kg (rabbit).
Long-term health effects: IARC rating 3; inadequate animal evidence. Liver and kidney damage.
Vapor (gas) irritant characteristics: Vapor is irritating such that personnel will not usually tolerate moderate or high vapor concentrations.
Liquid or solid irritant characteristics: Severe eye and skin irritant. Causes smarting of the skin and first-degree burns on short exposure; may cause secondary burns on long exposure.
Odor threshold: 0.003–0.02 ppm.
IDLH value: 250 ppm.

Methyl alcohol

FIRE DATA
Flash point: 27°F/–3°C (oc).
Flammable limits in air: LEL: 2.8%; UEL: 25%.
Fire extinguishing agents not to be used: Water may be ineffective
Behavior in fire: May polymerize.
Autoignition temperature: 865°F/463°C/736°K.

CHEMICAL REACTIVITY
Binary reactants: Violent reaction with strong oxidizers. Incompatible with strong acids, alkalis, aliphatic amines, alkanolamines, peroxides, nitrates. Usually stored in ambient air below 50°F/10°C.
Polymerization: Heat above 70°F/21°C may cause an explosive polymerization. Strong ultraviolet light, oxidizers and peroxides may also initiate polymerization; **Inhibitor of polymerization:** Hydroquinone and its methyl ether, in presence of air.
Reactivity group: 14
Compatibility class: Acrylates

ENVIRONMENTAL DATA
Food chain concentration potential: Log P_{ow} = 0.78. Unlikely to accumulate.
Water pollution: Effect of low concentrations on aquatic life is unknown. Fouling to shoreline. May be dangerous if it enters nearby water intakes; notify operators. Notify local health and wildlife officials. **Response to discharge:** Issue warning–high flammability. Evacuate area.

SHIPPING INFORMATION
Grades of purity: 99.9%; **Storage temperature:** Ambient if material is inhibited; under 40°F/40°C without inhibitor; **Inert atmosphere:** Air MUST be present; **Venting:** Open; **Stability during transport:** Stable (flame arrester)

NAS HAZARD CLASSIFICATION FOR BULK WATER TRANSPORTATION
FIRE: 3
HEALTH: Vapor irritant: 3; Liquid or solid irritant: 2; Poisons: 3
WATER POLLUTION: Human toxicity: 2; Aquatic toxicity: 2; Aesthetic effect: 2
REACTIVITY: Other chemicals: 2; Water: 0; Self-reaction: 3

PHYSICAL AND CHEMICAL PROPERTIES
Physical state @ 59°F/15°C and 1 atm: Liquid.
Molecular weight: 86.09
Boiling point @ 1 atm: 177°F/80.6°C/353.8°K.
Melting/Freezing point: –105.7°F/–76.5°C/196.7°K.
Critical temperature: 505°F/263°C/536°K.
Critical pressure: 630 psia = 43 atm = 4.3 MN/m².
Specific gravity (water = 1): 0.956 @ 68°F/20°C (liquid).
Liquid surface tension: 24.2 dynes/cm = 0.0242 N/m @ 68°F/20°C.
Liquid water interfacial tension: (estimate) 30 dynes/cm = 0.03 N/m @ 68°F/20°C.
Relative vapor density (air = 1): 3.0
Ratio of specific heats of vapor (gas): 1.102
Latent heat of vaporization: 160 Btu/lb = 90 cal/g = 3.8×10^5 J/kg.
Heat of combustion: (estimate) –9900 Btu/lb = –5500 cal/g = -230×10^5 J/kg.
Heat of polymerization: –392 Btu/lb = –218 cal/g = -9.13×10^5 J/kg.
Vapor pressure: (Reid) 3.1 psia; 65 mm.

METHYL ALCOHOL   REC. M:1750

SYNONYMS: ALCOHOL METILICO (Spanish); ALCOHOL C-1; ALCOOL METHYLIQUE (French); CARBINOL; COLONIAL SPIRIT; COLUMBIAN SPIRITS; EEC No. 603-001-00-X; METHANOL; METHYLOL; METHYLALKOHOL (German); METHYL HYDROXIDE; METHYL HYDRATE; METHYLOL; MONOHYDROXYMETHANE; PYROXYLIC SPIRIT; PYROLIGNEOUS SPIRIT; WOOD ALCOHOL; WOOD NAPHTHA; WOOD SPIRIT; RCRA No. U154

IDENTIFICATION
CAS Number: 67-56-1
Formula: CH_4O; CH_3OH
DOT ID Number: UN 1230; DOT Guide Number: 131
Proper Shipping Name: Methanol; Methyl alcohol
Reportable Quantity (RQ): **(CERCLA)** 5000 lb/2270 kg

DESCRIPTION: Colorless watery liquid. Sweet, fruity odor. Floats on the surface of water; soluble. Produces large amounts of vapor.

Highly flammable; burns with invisible flame • Containers may BLEVE when exposed to fire •Vapors may form explosive mixture with air • Vapors are heavier than air and will collect and stay in low areas • Vapors may travel long distances to ignition sources and flashback • Vapors in confined areas (e.g., tanks, sewers, buildings) may explode when exposed to fire • Irritating to the skin, eyes, and respiratory tract • Toxic products of combustion may include carbon monoxide.

Hazard Classification (based on NFPA-704 Rating System)
Health Hazards (Blue): 1; Flammability (Red): 3; Reactivity (Yellow): 0

EMERGENCY RESPONSE: See Appendix A (131)
Evacuation:
Public safety: Isolate spill area for at least 100 to 200 meters (330 to 660 feet) in all directions.
Spill: Increase, as necessary, the isolation distance shown above,in "Public safety."
Fire: Isolate for 800 meters (½ mile) in all directions, especially if tank, rail car, or tank truck is involved in fire.

EXPOSURE
Short-term effects: *SEEK MEDICAL ATTENTION.* **Vapor:** Irritating to eyes, nose, and throat. *IF INHALED*, will, will cause dizziness, headache, difficult breathing, or loss of consciousness. Move to fresh air. *IF BREATHING HAS STOPPED,* give artificial respiration. IF breathing is difficult, administer oxygen. **Liquid:** *POISONOUS IF SWALLOWED OR IF ABSORBED THROUGH THE SKIN.* Irritating to skin and eyes. Remove contaminated clothing and shoes. Flush affected areas with plenty of water. *IF IN EYES*, hold eyelids open and flush with plenty of water. *IF SWALLOWED* and victim is *CONSCIOUS AND ABLE TO SWALLOW*, have victim drink 4 to 8 ounces of water and have victim induce vomiting. *IF SWALLOWED* and victim is *UNCONSCIOUS OR HAVING CONVULSIONS,* do nothing except keep victim warm.

HEALTH HAZARDS
Personal protective equipment (PPE): A-Level PPE. OSHA Table Z-1-A air contaminant. Approved canister mask for high vapor concentrations; safety goggles; gloves. Chemical protective material(s) reported to have good to excellent resistance: butyl

rubber, butyl rubber/neoprene, chlorinated polyethylene, Viton®/neoprene, Viton®/chlorobutyl rubber, Teflon®, Viton®, Saranex®, styrene-butadiene, Silvershield®, PV acetate. Also, neoprene, PVC, and polyethylene offers limited protection.
Recommendations for respirator selection: NIOSH/OSHA
2000 ppm: SA (any supplied-air respirator). *5000 ppm*: SA:CF (any supplied-air respirator operated in a continuous-flow mode). *6000 ppm*: SAT:CF (any supplied-air respirator that has a tight-fitting facepiece and is operated in a continuous-flow mode); or SCBAF (any self-contained breathing apparatus with a full facepiece); or SAF (any supplied-air respirator with a full facepiece). *EMERGENCY OR PLANNED ENTRY INTO UNKNOWN CONCENTRATIONS OR IDLH CONDITIONS:* SCBAF:PD,PP (any self-contained breathing apparatus that has a full facepiece and is operated in a pressure-demand or other positive-pressure mode); or SAF:PD,PP:ASCBA (any supplied-air respirator that has a full facepiece and is operated in a pressure-demand or other positive-pressure mode in combination with an auxiliary self-contained breathing apparatus operated in a pressure-demand or other positive pressure mode). *ESCAPE:* SCBAE (any appropriate escape-type, self-contained breathing apparatus).
Exposure limits (TWA unless otherwise noted): ACGIH TLV 200 ppm; OSHA PEL 200 ppm (260 mg/m^3); NIOSH REL 200 ppm (260 mg/m^3); skin contact contributes significantly in overall exposure.
Short-term exposure limits (15-minute TWA): ACGIH STEL: 250 ppm (328 mg/m^3); NIOSH STEL 250 ppm (325 mg/m^3); skin contact contributes significantly in overall exposure.
Toxicity by ingestion: Grade 1; LD$_{50}$ = 5 to 15 g/kg (rat).
Long-term health effects: May effect the optic nerve.
Vapor (gas) irritant characteristics: Vapors cause smarting of the eyes or respiratory system if present in high concentrations.
Liquid or solid irritant characteristics: Eye irritant. If spilled on clothing and allowed to remain, may cause smarting and reddening of the skin.
Odor threshold: 141 ppm.
IDLH value: 6000 ppm.

FIRE DATA
Flash point: 61°F/oc); 52°F/11°C (cc).
Flammable limits in air: LEL: 6.0%; UEL: 36.5%.
Fire extinguishing agents not to be used: Water may be ineffective.
Behavior in fire: Containers may explode.
Autoignition temperature: 867°F/464°C/737°K.
Electrical hazard: Class I, Group D. Due to low electric conductivity, this substance may generate electrostatic charges as a result of agitation and flow.
Burning rate: 1.7 mm/min

CHEMICAL REACTIVITY
Binary reactants: Strong oxidizers, amines, isocyanates, strong acids.
Reactivity group: 20
Compatibility class: Alcohols, glycols

ENVIRONMENTAL DATA
Food chain concentration potential: Log P$_{ow}$ = –0.8. Unlikely to accumulate.
Water pollution: Dangerous to aquatic life in high concentrations. May be dangerous if it enters nearby water intakes; notify operators. Notify local health and wildlife officials. **Response to discharge:** Issue warning–high flammability. Restrict access. Evacuate area. Disperse and flush.

SHIPPING INFORMATION
Grades of purity: CP, Crude, ACS: All 99.9%; **Storage temperature:** Ambient; **Inert atmosphere:** None; **Venting:** Open (flame arrester) or pressure-vacuum; **Stability during transport:** Stable.

NAS HAZARD CLASSIFICATION FOR BULK WATER TRANSPORTATION
FIRE: 3
HEALTH: Vapor irritant: 1; Liquid or solid irritant: 1; Poisons: 2
WATER POLLUTION: Human toxicity: 1; Aquatic toxicity: 1; Aesthetic effect: 1
REACTIVITY: Other chemicals: 2; Water: 0; Self-reaction: 0

PHYSICAL AND CHEMICAL PROPERTIES
Physical state @ 59°F/15°C and 1 atm: Liquid.
Molecular weight: 32.1
Boiling point @ 1 atm: 147°F/64°C/337°K.
Melting/Freezing point: –144.0°F/–97.8°C/175.4°K.
Critical temperature: 464°F/240°C/513°K.
Critical pressure: 1142.0 psia = 77.7 atm = 7.87 MN/m^2.
Specific gravity (water = 1): 0.792 @ 68°F/20°C (liquid).
Relative vapor density (air = 1): 1.1
Ratio of specific heats of vapor (gas): 1.254
Latent heat of vaporization: 473.0 Btu/lb = 262.8 cal/g = 11.00 x 10^5 J/kg.
Heat of combustion: –8419 Btu/lb = –4677 cal/g = –195.8 x 10^5 J/kg.
Heat of solution: (estimate) –9 Btu/lb = –5 cal/g = –0.2 x 10^5 J/kg.
Heat of fusion: 23.70 cal/g.
Vapor pressure: (Reid) 4.5 psia; 97 mm.

METHYL ALLYL ALCOHOL   REC. M:1800

SYNONYMS: METHALLYL ALCOHOL; ISOPROPENYL CARBINOL; 2-METHYL-2-PROPEN-1-OL

IDENTIFICATION
CAS Number: 513-42-8
Formula: C$_4$H$_8$O; H$_2$C = C(CH$_3$)CH$_2$OH
DOT ID Number: UN 2614; DOT Guide Number: 129
Proper Shipping Name: Methallyl alcohol

DESCRIPTION: Colorless liquid. Pungent odor. Mixes with water.

Highly flammable • Containers may BLEVE when exposed to fire • Vapors may form explosive mixture with air • Vapors are heavier than air and will collect and stay in low areas • Vapors may travel long distances to ignition sources and flashback • Vapors in confined areas (e.g., tanks, sewers, buildings) may explode when exposed to fire • Irritating to the skin, eyes, and respiratory tract • Toxic products of combustion may include carbon monoxide.

Hazard Classification (based on NFPA-704 Rating System)
Health Hazards (Blue): 1; Flammability (Red): 2; Reactivity (Yellow): 0

EMERGENCY RESPONSE: See Appendix A (129)
Evacuation:
Public safety: Isolate spill area for at least 50 to 100 meters (160 to 330 feet) in all directions.
Spill: Large spill–Consider initial downwind evacuation for at least 300 meters (1000 feet).

Methylamine, anhydrous

Fire: Isolate for 800 meters (½ mile) in all directions, especially if tank, rail car, or tank truck is involved in fire.

EXPOSURE

Short-term effects: *SEEK MEDICAL ATTENTION.* **Vapor:** Move victim to fresh air. *IF BREATHING HAS STOPPED,* give artificial respiration. If breathing is difficult, administer oxygen. **Liquid:** Irritating to skin and eyes. Remove contaminated clothing and shoes. Flush affected areas with water. *IF IN EYES,* hold eyelids open and flush with plenty of water.

HEALTH HAZARDS

Personal protective equipment (PPE): B-Level PPE. Full impervious protective clothing, including boots and gloves. Where splashing is possible wear full face shield or chemical safety goggles. Use approved respirator to protect against vapors. Natural rubber, neoprene, and nitrile rubber may offer some protection from alcohols.

Vapor (gas) irritant characteristics: eye and respiratory tract irritant. Vapors are irriatating such that personnel will not usually tolerate moderate or high concentrations.

Liquid or solid irritant characteristics: Fairly severe eye and skin irritant. May cause pain and second-degree burns after a few minutes of contact.

FIRE DATA

Flash point: 92°F/33°C (cc).
Fire extinguishing agents not to be used: Water.
Stoichiometric air-to-fuel ratio: 26.2

CHEMICAL REACTIVITY

Reactivity group: 20
Compatibility class: Alcohols, glycols

ENVIRONMENTAL DATA

Water pollution: Effects of low concentrations on aquatic life is unknown. May be dangerous if it enters nearby water intakes; notify operators. Notify local health and wildlife officials.
Response to discharge: Issue warning–high flammability. Should be removed.

SHIPPING INFORMATION

Grades of purity: 98%; technical grades; **Storage temperature:** Ambient; **Inert atmosphere:** None; **Stability during transport:** Stable.

PHYSICAL AND CHEMICAL PROPERTIES

Physical state @ 59°F/15°C and 1 atm: Liquid.
Molecular weight: 72.10
Boiling point @ 1 atm: 239°F/115°C/388°K.
Specific gravity (water = 1): 0.8515 @ 20°C.

METHYLAMINE, anhydrous **REC. M:1850**

SYNONYMS: AMINOMETHANE; CARBINAMINE; EEC No. 612-001-00-9; MERCURIALIN; METHANAMINE; METILAMINA (Spanish); MONOMETHYLAMINE

IDENTIFICATION
CAS Number: 74-89-5
Formula: CH_5N; CH_3NH_2
DOT ID Number: UN 1061; DOT Guide Number: 118
Proper Shipping Name: Methylamine, anhydrous
Reportable Quantity (RQ): 100 lb/45.4 kg

DESCRIPTION: Colorless compressed gas. Shipped and stored as a refrigerated liquefied gas (a liquid below 21°F/–6°C). Ammonia-like odor; suffocating. Mixes with water and boils on surface.

Poison! • Extremely flammable • Breathing the vapors can kill you; skin and eye contact causes severe burns and blindness • Firefighting gear (including SCBA) does not provide adequate protection. If exposure occurs, remove and isolate gear immediately and thoroughly decontaminate personnel • Containers may BLEVE when exposed to fire • Gas from liquefied gas are initially heavier than air and spread along ground • Gas may travel to source of ignition and flashback • Gas in confined areas (e.g., tanks, sewers, buildings) may explode when exposed to fire • Ruptured or venting cylinders may rocket through buildings and/or travel a considerable distance • Contact with liquid may cause frostbite • Toxic products of combustion may include nitrogen oxides • Do not put yourself in danger by entering a contaminated area to rescue a victim. *Warning:* Odor is not a reliable indicator of the presence of toxic amounts of methylamine gas.

Hazard Classification (based on NFPA-704 Rating System)
Health Hazards (Blue): 3; Flammability (Red): 4; Reactivity (Yellow): 0

EMERGENCY RESPONSE: See Appendix A (118)
Evacuation:
Public safety: Isolate spill area for at least 100 to 200 meters (330 to 660 feet) in all directions.
Spill: Consider initial downwind evacuation for at least 800 meters (½ mile).
Fire: Isolate for 1600 meters (1 mile) in all directions, especially if tank, rail car, or tank truck is involved in fire.

EXPOSURE

Short-term effects: *SEEK MEDICAL ATTENTION.* **Vapor:** Irritating to eyes, nose, and throat. *IF INHALED,* will, will cause coughing or difficult breathing. Move victim to fresh air. *IF BREATHING HAS STOPPED,* give artificial respiration. IF breathing is difficult, administer oxygen. **Liquid:** Will burn skin and eyes. Remove contaminated clothing and shoes. Flush affected areas with plenty of water. *IF IN EYES,* hold eyelids open and flush with plenty of water. *IF SWALLOWED* and victim is *CONSCIOUS AND ABLE TO SWALLOW,* have victim drink 4 to 8 ounces of water. **Do NOT induce vomiting.**

HEALTH HAZARDS

Personal protective equipment (PPE): B-Level PPE. OSHA Table Z-1-A air contaminant. Goggles or face mask; rubber suit, apron, sleeves, and/or gloves; rubber or leather safety shoes; air-line mask, positive-pressure hose mask, self-contained breathing apparatus, or industrial canister-type gas mask. Wear thermal protective clothing. Chemical protective material(s) reported to have good to excellent resistance: butyl rubber, neoprene, nitrile, Viton®. Also, PVC, natural rubber, and Styrene-butadiene offers limited protection

Recommendations for respirator selection: NIOSH/OSHA
100 ppm: CCRFS [any chemical cartridge respirator with a full facepiece and cartridge(s) providing protection against the compound of concern]; or GMFS [any air-purifying, full-facepiece respirator (gas mask) with a chin-style, front- or back-mounted canister providing protection against the compound of concern]; or PAPRS* [any powered, air-purifying respirator with cartridge(s) providing protection against the compound of concern]; or SCBAF (any self-contained breathing apparatus with a full facepiece); or SAF (any supplied-air respirator with a full facepiece).

EMERGENCY OR PLANNED ENTRY INTO UNKNOWN CONCENTRATIONS OR IDLH CONDITIONS: SCBAF:PD,PP (any self-contained breathing apparatus that has a full facepiece and is operated in a pressure-demand or other positive-pressure mode); or SAF:PD,PP:ASCBA (any supplied-air respirator that has a full facepiece and is operated in a pressure-demand or other positive-pressure mode in combination with an auxiliary self-contained breathing apparatus operated in a pressure-demand or other positive pressure mode). *ESCAPE:* GMFS [any air-purifying, full-facepiece respirator (gas mask) with a chin-style, front- or back-mounted canister providing protection against the compound of concern]; or SCBAE (any appropriate escape-type, self-contained breathing apparatus). *Note:* Substance causes eye irritation or damage; eye protection needed.
Exposure limits (TWA unless otherwise noted): ACGIH TLV 5 ppm (6.4 mg/m^3); NIOSH/OSHA 10 ppm (12 mg/m^3).
Short-term exposure limits (15-minute TWA): ACGIH STEL 15 ppm (19 mg/m^3).
Toxicity by ingestion: Grade 2; LD$_{50}$ = 0.5 to 5 g/kg.
Vapor (gas) irritant characteristics: Vapors are irritating such that personnel will not usually tolerate moderate or high concentrations.
Liquid or solid irritant characteristics: Severe eye and skin irritant. Causes smarting of the skin and first-degree burns on short exposure and may cause second-degree burns on long exposure.
Odor threshold: 0.021 ppm.
IDLH value: 100 ppm.

FIRE DATA
Flash point: (flammable, liquefied compressed gas).
Flammable limits in air: LEL: 5%; UEL: 21%.
Autoignition temperature: 806°F/430°C/703°K.
Electrical hazard: Class I, Group D.
Stoichiometric air-to-fuel ratio: 9.932 (estimate).

CHEMICAL REACTIVITY
Reactivity with water: Solution is strongly alkaline.
Binary reactants: Strong oxidizers, acids, nitromethane; mercury (possible explosive reaction). Corrosive to copper and zinc, copper alloys, zinc alloys, aluminum, aluminum alloys and galvanized surfaces.
Neutralizing agents for acids and caustics: Mild acidic solution such as vinegar or 1-2% acetic acid.
Reactivity group: 7
Compatibility class: Aliphatic amines

ENVIRONMENTAL DATA
Food chain concentration potential: Log P$_{ow}$ = 2.3. Unlikely to accumulate.
Water pollution: Harmful to aquatic life in very low concentrations. May be dangerous if it enters nearby water intakes; notify operators. Notify local health and wildlife officials.
Response to discharge: Issue warning–high flammability, air contaminant, water contaminant. Restrict access. Evacuate area Disperse and flush.

SHIPPING INFORMATION
Grades of purity: Anhydrous, 99.3+%; Water solutions, 30–50% by weight; **Storage temperature:** Ambient; **Inert atmosphere:** None; **Venting:** Safety relief; **Stability during transport:** Stable.

NAS HAZARD CLASSIFICATION FOR BULK WATER TRANSPORTATION
FIRE: 4
HEALTH: Vapor irritant: 3; Liquid or solid irritant: 2; Poisons: 3
WATER POLLUTION: Human toxicity: 2; Aquatic toxicity:; Aesthetic effect: 2
REACTIVITY: Other chemicals: 3; Water: 0; Self-reaction: 0

PHYSICAL AND CHEMICAL PROPERTIES (Properties apply to anhydrous material unless other wise noted).

Physical state @ 59°F/15°C and 1 atm: Gas.
Molecular weight: 31.1
Boiling point @ 1 atm: 21°F/–6°C/267°K.
Melting/Freezing point: –134.5°F/–92.5°C/180.7°K.
Critical temperature: 318°F/159°C/432°K.
Critical pressure: 1080 psia = 73.6 atm = 7.47 MN/m^2.
Specific gravity (water = 1): 0.70 @ 13°F/liquid).
Liquid surface tension: 100.59 dynes/cm = 0.1006 N/m @ 68°F/20°C.
Relative vapor density (air = 1): 1.08
Ratio of specific heats of vapor (gas): 1.1946
Latent heat of vaporization: 358 Btu/lb = 199 cal/g = 8.33 x 10^5 J/kg.
Heat of combustion: –15,000 Btu/lb = –8340 cal/g = –34.9 x 10^6 J/kg.
Vapor pressure: 235 mm.

METHYLAMINE, AQUEOUS SOLUTION   REC. M:1900

SYNONYMS: AMINOMETHANE (cylinder); CARBINAMINE SOLUTION; EEC No. 612-001-00-9; MERCURIALIN SOLUTION; METILAMINA (Spanish); METHANAMINE SOLUTION; METHANAMINE SOLUTION; MONOMETHYLAMINE SOLUTION

IDENTIFICATION
CAS Number: 74-89-5
Formula: CH$_3$NH$_2$
DOT ID Number: UN 1235; DOT Guide Number: 132
Proper Shipping Name: Methylamine, aqeuous solution
Reportable Quantity (RQ): **(CERCLA)** 100 lb/45.4 kg

DESCRIPTION: Colorless liquid. Shipped as a 25-48% solution in water. Fishy odor at low concentrations; ammonia-like odor at higher concentrations. Mixes with water. Produces a large volume of vapor. Boils at 86–106°F/30–41°C.

Corrosive • Extremely flammable • Breathing the vapor can kill you; Skin or eye contact causes severe burns, impaired vision, or blindness; effects of contact or inhalation may be delayed • Firefighting gear (including SCBA) does not provide adequate protection. If exposure occurs, remove and isolate gear immediately and thoroughly decontaminate personnel • Containers may BLEVE when exposed to fire •Vapors may form explosive mixture with air • Vapors are heavier than air and will collect and stay in low areas • Vapors may travel long distances to ignition sources and flashback • Vapors in confined areas (e.g., tanks, sewers, buildings) may explode when exposed to fire • Toxic products of combustion may include nitrogen oxides • Do not put yourself in danger by entering a contaminated area to rescue a victim. *Warning:* Odor is not a reliable indicator of the presence of toxic amounts of methylamine gas.

Hazard Classification (based on NFPA-704 Rating System)
Health Hazards (Blue): 3; Flammability (Red): 4; Reactivity (Yellow): 0

Methylamine, aqueous solution

EMERGENCY RESPONSE: See Appendix A (132)
Evacuation:
Public safety: Isolate spill area for at least 50 to 100 meters (160 to 330 feet) in all directions.
Spill: Increase, as necessary, the isolation distance shown above, in "Public safety."
Fire: Isolate for 800 meters (½ mile) in all directions, especially if tank, rail car, or tank truck is involved in fire.

EXPOSURE
Short-term effects: *SEEK MEDICAL ATTENTION*. **Vapor:** Irritating to eyes, nose, and throat. *IF INHALED*, will, will causa coughing and difficult breathing. Move victim to fresh air. *IF BREATHING HAS STOPPED,* give artificial respiration. IF breathing is difficult, administer oxygen. **Liquid:** Will burn skin and eyes. Remove contaminated clothing and shoes. Wash affected area with soap and water. *IF IN EYES*, hold eyelids open and flush with plenty of water. *IF SWALLOWED* and victim is *CONSCIOUS AND ABLE TO SWALLOW*, have victim drink 4 to 8 ounces of water. **Do NOT induce vomiting**.
Note to physician or authorized medical personnel: Medical observation is recommended for 24 to 48 hours after breathing overexposure, as pulmonary edema may be delayed. As first aid for pulmonary edema, consider administering a corticosteroid spray. Cigarette smoking may exacerbate pulmonary injury and should be discouraged for at least 72 hours following exposure.

HEALTH HAZARDS
Personal protective equipment (PPE): B-Level PPE. OSHA Table Z-1-A air contaminant. Self-contained (positive pressure, if available) breathing apparatus and full protective clothing. No skin surface should be exposed. Chemical protective material(s) reported to have good to excellent resistance: butyl rubber, neoprene, nitrile, Viton®. Also, PVC, natural rubber, and Styrene-butadiene offers limited protection
Recommendations for respirator selection: NIOSH/OSHA
100 ppm: CCRFS [any chemical cartridge respirator with a full facepiece and cartridge(s) providing protection against the compound of concern]; or GMFS [any air-purifying, full-facepiece respirator (gas mask) with a chin-style, front- or back-mounted canister providing protection against the compound of concern]; or PAPRS [any powered, air-purifying respirator with cartridge(s) providing protection against the compound of concern]; or SCBAF (any self-contained breathing apparatus with a full facepiece); or SAF (any supplied-air respirator with a full facepiece). *EMERGENCY OR PLANNED ENTRY INTO UNKNOWN CONCENTRATIONS OR IDLH CONDITIONS:* SCBAF:PD,PP (any self-contained breathing apparatus that has a full facepiece and is operated in a pressure-demand or other positive-pressure mode); or SAF:PD,PP:ASCBA (any supplied-air respirator that has a full facepiece and is operated in a pressure-demand or other positive-pressure mode in combination with an auxiliary self-contained breathing apparatus operated in a pressure-demand or other positive pressure mode). *ESCAPE:* GMFS [any air-purifying, full-facepiece respirator (gas mask) with a chin-style, front- or back-mounted canister providing protection against the compound of concern]; or SCBAE (any appropriate escape-type, self-contained breathing apparatus). *Note:* Substance causes eye irritation or damage; eye protection needed.
Exposure limits (TWA unless otherwise noted): ACGIH TLV 5 ppm (6.4 mg/m^3); NIOSH/OSHA 10 ppm (12 mg/m^3).
Short-term exposure limits (15-minute TWA): ACGIH STEL 15 ppm (19 mg/m^3).
Toxicity by ingestion: Grade 3; LD$_{50}$ = 100-200 mg/kg (rat).
Long-term health effects: Dermatitis.

Vapor (gas) irritant characteristics: Eye and respiratory tract irritant. Vapors are irritating such that personnel will not usually tolerate moderate or high concentrations.
Liquid or solid irritant characteristics: Eye and skin irritant. Causes smarting of the skin and first-degree burns on short exposure and may cause second-degree burns on long exposure.
Odor threshold: 4.7
IDLH value: 100 ppm.

FIRE DATA
Flash point: 14°F/–10°C (cc) (liquid).
Flammable limits in air: LEL: 4.9%; UEL: 20.7%.
Autoignition temperature: 806°F/430°C/703°K.
Electrical hazard: Class I, Group D.
Stoichiometric air-to-fuel ratio: 9.932 (estimate)

CHEMICAL REACTIVITY
Reactivity with water: Dissolves completely.
Binary reactants: Strong oxidizers, acids, nitromethane. Corrosive to copper and zinc, copper alloys, zinc alloys, aluminum and galvanized surfaces. Contact with mercury may produce explosive reaction.
Neutralizing agents for acids and caustics: Mild acidic solution such as vinegar or 1–2% acetic acid.
Reactivity group: 7
Compatibility class: Aliphatic amines

ENVIRONMENTAL DATA
Food chain concentration potential: Log P$_{ow}$ = –0.6. Unlikely to accumulate.
Water pollution: Harmful to aquatic life in very low concentrations. May be dangerous if it enters nearby water intakes; notify operators. Notify local health and wildlife officials. **Response to discharge:** Issue warning. Restrict access. Evacuate area.

SHIPPING INFORMATION
Grades of purity: Anhydrous, 99.3+%; Water solutions, 25–50% by weight; **Storage temperature:** Ambient; **Inert atmosphere:** None; **Venting:** Safety relief; **Stability during transport:** Stable.

NAS HAZARD CLASSIFICATION FOR BULK WATER TRANSPORTATION
FIRE: 4
HEALTH: Vapor irritant: 3; Liquid or solid irritant: 2; Poisons: 3
WATER POLLUTION: Human toxicity: 2; Aquatic toxicity:–; Aesthetic effect: 2
REACTIVITY: Other chemicals: 3; Water: 0; Self-reaction: 0

PHYSICAL AND CHEMICAL PROPERTIES
Physical state @ 59°F/15°C and 1 atm: Gas or liquid solution.
Molecular weight: 31.06 (liquid); 31.1 (gas).
Boiling point @ 1 atm: 20.66°F/–6.3°C/266.7°K (liquid).
Melting/Freezing point: –136°F/–93°C/180°K.
Critical temperature: 314.42°F/156.9°C/429.9°K (liquid).
Critical pressure: 590 psia = 40.2 atm = 4.07 MN/m^2.
Specific gravity (water = 1): 0.70 @ 13°F/Liquid.
Liquid surface tension: 29.2 dynes/cm at –70°C.
Relative vapor density (air = 1): 1.08
Ratio of specific heats of vapor (gas): 1.1946
Latent heat of vaporization: 374.90 Btu/lb = 208.29 cal/g = 8.72 X 10^5 J/kg (liquid).
Heat of combustion: –15,000 Btu/lb = –8340 cal/g = –34.9 x 10^6 J/kg.
Vapor pressure: 236 mm.

METHYL AMYL ACETATE REC. M:1950

SYNONYMS: ACETATO de METILAMILO (Spanish); ACETIC ACID-1,3-DIMETHYLBUTYL ESTER; 1,3-DIMETHYLBUTYL ACETATE; HEXYL ACETATE; sec-HEXYL ACETATE; MAAC; METHYL ISOAMYL ACETATE; METHYLISOBUTYLCARBINOL ACETATE; METHYLISOBUTYLCARBINYL ACETATE; 4-METHYL-2-PENTANOL ACETATE; 4-METHYL-2-PENTYL ACETATE

IDENTIFICATION
CAS Number: 108-84-9
Formula: $C_8H_{16}O_2$; $CH_3COOCH(CH_3)CH_2CH(CH_3)_2$
DOT ID Number: UN 1233; DOT Guide Number: 129
Proper Shipping Name: Methyl amyl acetate

DESCRIPTION: Colorless, watery liquid. Mild, fruity odor. Floats on water surface; initially as a slick.

Flammable • Containers may BLEVE when exposed to fire • Vapors may form explosive mixture with air • Vapors are heavier than air and will collect and stay in low areas • Vapors may travel long distances to ignition sources and flashback • Vapors in confined areas (e.g., tanks, sewers, buildings) may explode when exposed to fire • Irritating to the skin, eyes, and respiratory tract • Toxic products of combustion may include carbon monoxide.

Hazard Classification (based on NFPA-704 Rating System)
Health Hazards (Blue): 1; Flammability (Red): 2; Reactivity (Yellow): 0

EMERGENCY RESPONSE: See Appendix A (129)
Evacuation:
Public safety: Isolate spill area for at least 50 to 100 meters (160 to 330 feet) in all directions.
Spill: Large spill–Consider initial downwind evacuation for at least 300 meters (1000 feet).
Fire: Isolate for 800 meters (½ mile) in all directions, especially if tank, rail car, or tank truck is involved in fire.

EXPOSURE
Short-term effects: *SEEK MEDICAL ATTENTION*. **Liquid:** Irritating to skin and eyes. Remove contaminated clothing and shoes. Flush affected areas with plenty of water. *IF IN EYES*, hold eyelids open and flush with plenty of water.

HEALTH HAZARDS
Personal protective equipment (PPE): B-Level PPE. OSHA Table Z-1-A air contaminant. Organic canister); or air pack; rubber gloves; goggles.
Recommendations for respirator selection: NIOSH/OSHA
500 ppm: CCROV* [any chemical cartridge respirator with organic vapor cartridge(s)]; or GMFOV [any air-purifying, full-facepiece respirator (gas mask) with a chin-style, front- or back-mounted acid gas canister]; or PAPROV* [any powered, air-purifying respirator with organic vapor cartridge(s)]; or SA* (any supplied-air respirator); or SCBAF (any self-contained breathing apparatus with a full facepiece). *EMERGENCY OR PLANNED ENTRY INTO UNKNOWN CONCENTRATIONS OR IDLH CONDITIONS:* SCBAF:PD,PP (any self-contained breathing apparatus that has a full facepiece and is operated in a pressure-demand or other positive-pressure mode); or SAF:PD,PP:ASCBA (any supplied-air respirator that has a full facepiece and is operated in a pressure-demand or other positive-pressure mode in combination with an auxiliary self-contained breathing apparatus operated in a pressure-demand or other positive pressure mode). *ESCAPE:* GMFOV [any air-purifying, full-facepiece respirator (gas mask) with a chin-style, front- or back-mounted organic vapor cannister]; or SCBAE (any appropriate escape-type, self-contained breathing apparatus). *Note:* Substance reported to cause eye irritation or damage; may require eye protection.
Exposure limits (TWA unless otherwise noted): ACGIH TLV 50 ppm (295 mg/m³); NIOSH/OSHA PEL 50 ppm (300 mg/m³).
Toxicity by ingestion: Grade 1; LD_{50} = 5 to 15 g/kg.
Vapor (gas) irritant characteristics: Vapors cause irritation, such that personnel will find high concentrations unpleasant. The effect is temporary.
Liquid or solid irritant characteristics: Eye irritant. If spilled on clothing and allowed to remain, may cause smarting and reddening of the skin.
Odor threshold: 0.39 ppm.
IDLH value: 500 ppm.

FIRE DATA
Flash point: 110°F/43°C (oc); 113°F/45°C (cc).
Flammable limits in air: LEL: 0.9%; UEL: 5.7% (calculated).
Fire extinguishing agents not to be used: Water may be ineffective.
Autoignition temperature: 510°F/266°C/539°K (calculated).
Electrical hazard: Class I, Group D.

CHEMICAL REACTIVITY
Binary reactants: Nitrates, strong oxidizers, alkalis, and acids. Will swell rubber and dissolve certain coatings.
Reactivity group: 34
Compatibility class: Esters.

ENVIRONMENTAL DATA
Food chain concentration potential: Negative; unlikely to accumulate.
Water pollution: Effect of low concentrations on aquatic life is unknown. Fouling to shoreline. May be dangerous if it enters nearby water intakes; notify operators. Notify local health and wildlife officials. **Response to discharge:** Mechanical containment. Chemical and physical treatment.

SHIPPING INFORMATION
Grades of purity: 95-99+%; **Storage temperature:** Ambient; **Inert atmosphere:** None; **Venting:** Open (flame arrester); **Stability during transport:** Stable.

NAS HAZARD CLASSIFICATION FOR BULK WATER TRANSPORTATION
FIRE: 2
HEALTH: Vapor irritant: 2; Liquid or solid irritant: 1; Poisons: 1
WATER POLLUTION: Human toxicity: 1; Aquatic toxicity: 1; Aesthetic effect: 2
REACTIVITY: Other chemicals: 1; Water: 0; Self-reaction: 0

PHYSICAL AND CHEMICAL PROPERTIES
Physical state @ 59°F/15°C and 1 atm: Liquid.
Molecular weight: 144.22
Boiling point @ 1 atm: 284°F/140°C/413°K.
Melting/Freezing point: –82.8°F/–63.8°C/209.4°K.
Critical temperature: 606°F/319°C/592°K.
Critical pressure: 382 psia = 26 atm = 2.6 MN/m².
Specific gravity (water = 1): 0.860 @ 68°F/20°C (liquid).
Liquid surface tension: (estimate) 25 dynes/cm = 0.025 N/m @ 77°F/25°C.

Liquid water interfacial tension: (estimate) 40 dynes/cm = 0.04 N/m @ 77°F/25°C.
Relative vapor density (air = 1): 5.0
Ratio of specific heats of vapor (gas): 1.046
Latent heat of vaporization: 225 Btu/lb = 125 cal/g = 5.23 x 10^5 J/kg.
Heat of combustion: (estimate) –14,400 Btu/lb = –8000 cal/g = –335 x 10^5 J/kg.
Vapor pressure: (Reid) 0.21 psia; 3 mm.

METHYL AMYL ALCOHOL   REC. M:2000

SYNONYMS: ALCOHOL METILAMILICO (Spanish); ALCOOL METHYL AMYLIQUE (French); 1,3-DIMETHYLBUTANOL; EEC No. 603-008-00-8; ISOBUTYLMETHYLCARBINOL; ISOBUTYLMETHYLMETHANOL; MAOH; METHYLAMYL ALCOHOL; METHYL ISOBUTYL CARBINOL; 2-METHYL-4-PENTANOL; 4-METHYLPENTANOL-2; 4-METHYL-2-PENTANOL; 4-METHY-2-PENTYL ALCOHOL; MIBC; MIC; 3-MIC; 2-PENTANOL, 4-METHYL-

IDENTIFICATION
CAS Number: 105-30-6
Formula: $C_6H_{14}O$; $(CH_3)_2CHCH_2CH(OH)CH_3$
DOT ID Number: UN 2053; DOT Guide Number: 129
Proper Shipping Name: Methylamyl alcohol

DESCRIPTION: Colorless oily liquid. Sharp, mild alcohol odor. Floats on the surface of water. Irritating vapor is produced.

Flammable • Containers may BLEVE when exposed to fire • Vapors may form explosive mixture with air • Vapors are heavier than air and will collect and stay in low areas • Vapors may travel long distances to ignition sources and flashback • Vapors in confined areas (e.g., tanks, sewers, buildings) may explode when exposed to fire • Irritating to the skin, eyes, and respiratory tract • Toxic products of combustion may include carbon monoxide.

Hazard Classification (based on NFPA-704 Rating System)
Health Hazards (Blue): 1; Flammability (Red): 2; Reactivity (Yellow): 0

EMERGENCY RESPONSE: See Appendix A (129)
Evacuation:
Public safety: Isolate spill area for at least 50 to 100 meters (160 to 330 feet) in all directions.
Spill: Large spill–Consider initial downwind evacuation for at least 300 meters (1000 feet).
Fire: Isolate for 800 meters (½ mile) in all directions, especially if tank, rail car, or tank truck is involved in fire.

EXPOSURE
Short-term effects: *SEEK MEDICAL ATTENTION.* **Vapor:** Irritating to eyes, nose, and throat. *IF INHALED*, will, will cause dizziness or difficult breathing. Harmful if skin is exposed. Move to fresh air. *IF BREATHING HAS STOPPED*, give artificial respiration. If breathing is difficult, administer oxygen. **Liquid:** *HARMFUL IF ABSORBED THROUGH THE SKIN.* Irritating to skin and eyes. Harmful if swallowed. Remove contaminated clothing and shoes. Flush affected areas with plenty of water. *IF IN EYES*, hold eyelids open and flush with plenty of water. *IF SWALLOWED* and victim is *CONSCIOUS AND ABLE TO SWALLOW*, have victim drink 4 to 8 ounces of water.

HEALTH HAZARDS
Personal protective equipment (PPE): B-Level PPE. OSHA Table Z-1-A air contaminant. Organic vapor respirator; rubber gloves; goggles or face shield. Natural rubber, neoprene, and nitrile rubber may offer some protection from alcohols.
Recommendations for respirator selection: NIOSH/OSHA *250 ppm:* SA (any supplied-air respirator). *400 ppm:* SA:CF (any supplied-air respirator operated in a continuous-flow mode); SCBAF (any self-contained breathing apparatus with a full facepiece); or SAF (any supplied-air respirator with a full facepiece). *EMERGENCY OR PLANNED ENTRY INTO UNKNOWN CONCENTRATIONS OR IDLH CONDITIONS:* SCBAF:PD,PP (any self-contained breathing apparatus that has a full facepiece and is operated in a pressure-demand or other positive-pressure mode); or SAF:PD,PP:ASCBA (any supplied-air respirator that has a full facepiece and is operated in a pressure-demand or other positive-pressure mode in combination with an auxiliary self-contained breathing apparatus operated in a pressure-demand or other positive pressure mode). *ESCAPE:* GMFOV [any air-purifying, full-facepiece respirator (gas mask) with a chin-style, front- or back-mounted organic vapor canister]; or SCBAE (any appropriate escape-type, self-contained breathing apparatus). *Note:* Substance reported to cause eye irritation or damage; may require eye protection.
Exposure limits (TWA unless otherwise noted): ACGIH TLV 25 ppm (104 mg/m^3); OSHA PEL 25 ppm (100 mg/m^3); NIOSH 25 ppm (100 mg/m^3); skin contact contributes significantly in overall exposure.
Short-term exposure limits (15-minute TWA): NIOSH STEL 40 ppm (1675 mg/m^3); skin contact contributes significantly in overall exposure.
Toxicity by ingestion: Grade 2; LD_{50} = 0.5 to 5 g/kg (rat).
Vapor (gas) irritant characteristics: Eye and respiratory tract irritant. Vapors cause moderate irritation such that personnel will find high concentrations unpleasant. The effect is temporary.
Liquid or solid irritant characteristics: Eye and skin irritant. If spilled on clothing and allowed to remain, may cause smarting and reddening of the skin.
Odor threshold: 0.01–1.0 ppm.
IDLH value: 400 ppm.

FIRE DATA
Flash point: 120–130°F/49–54°C (oc); 106°F/41°C (cc).
Flammable limits in air: LEL: 1.0%; UEL: 5.5%.
Autoignition temperature: 583°F/306°C/579°K.
Burning rate: 4.7 mm/min

CHEMICAL REACTIVITY
Binary reactants: Contact with alkali metals produces flammable hydrogen gas. Reacts, possibly violently, with oxidizers, acetaldehyde, alkaline earth metals, strong acids, strong caustics, aliphatic amines, benzoyl peroxide, chromic acid, chromium trioxide, dialkylzincs, dichlorine oxide, ethylene oxide, hypochlorous acid, isocyanates, isopropyl chlorocarbonate, lithium tetrahydroaluminate, nitric acid, nitrogen dioxide, pentafluoroguanidine, phosphorus pentasulfide, tangerine oil, triethylaluminum, triisobutylaluminum. Attacks some plastics, rubber, and coatings.
Reactivity group: 20
Compatibility class: Alcohols, glycols

ENVIRONMENTAL DATA
Food chain concentration potential: Log P_{ow} = 1.4 (estimate). Unlikely to accumulate. **Water pollution:** Effect of low concentrations on aquatic life is unknown. Fouling to shoreline.

May be dangerous if it enters nearby water intakes; notify operators. Notify local health and wildlife officials. **Response to discharge:** Mechanical containment. Chemical and physical treatment.

SHIPPING INFORMATION
Storage temperature: Ambient; **Inert atmosphere:** None; **Venting:** Open (flame arrester); **Stability during transport:** Stable.

NAS HAZARD CLASSIFICATION FOR BULK WATER TRANSPORTATION
FIRE: 2
HEALTH: Vapor irritant: 2; Liquid or solid irritant: 1; Poisons: 2
WATER POLLUTION: Human toxicity: 2; Aquatic toxicity: 2; Aesthetic effect: 2
REACTIVITY: Other chemicals: 2; Water: 0; Self-reaction: 0

PHYSICAL AND CHEMICAL PROPERTIES
Physical state @ 59°F/15°C and 1 atm: Liquid.
Molecular weight: 102.18
Boiling point @ 1 atm: 271°F/133°C/406°K.
Melting/Freezing point: Less than −130°F/−90°C/183°K.
Critical temperature: 556°F/291°C/564°K.
Specific gravity (water = 1): 0.807 @ 68°F/20°C (liquid).
Liquid surface tension: 22.8 dynes/cm = 0.0228 N/m @ 68°F/20°C.
Liquid water interfacial tension: (estimate) 40 dynes/cm = 0.04 N/m @ 68°F/20°C.
Relative vapor density (air = 1): 3.5
Ratio of specific heats of vapor (gas): 1.053
Latent heat of vaporization: 162 Btu/lb = 90.1 cal/g = 3.77×10^5 J/kg.
Heat of combustion: −16,640 Btu/lb = −9240 cal/g = -387×10^5 J/kg.
Vapor pressure: 5 mm.

N-METHYLANILINE   REC. M:2050

SYNONYMS: ANILINOMETHANE; BENZENENAMINE, *N*-METHYL-; EEC No. 612-015-00-5; MA; (METHYLAMINO)BENZENE; *N*-METILANILINA (Spanish); *N*-METHYLAMINOBENZENE; METHYL ANILINE; *N*-METHYLBENZENAMINE; METHYLPHENYL AMINE; *N*-METHYLPHENYLAMINE; MONOMETHYL ANILINE; *N*-MONOMETHYLANILINE; MA; MONOMETHYL ANILINE; *N*-PHENYLMETHYLAMINE; METHYLANILINE (MONO).

IDENTIFICATION
CAS Number: 100-61-8
Formula: C_7H_9N; $C_6H_5NHCH_3$
DOT ID Number: UN 2294; DOT Guide Number: 153
Proper Shipping Name: *N*-Methylaniline

DESCRIPTION: Yellow to light brown liquid. Chemical, aniline odor. May float or sink in water.

Combustible • Containers may BLEVE when exposed to fire • Vapors are heavier than air and will collect and stay in low areas • Vapors in confined areas (e.g., tanks, sewers, buildings) may explode when exposed to fire • Toxic products of combustion may include nitrogen oxides • Do not put yourself in danger by entering a contaminated area to rescue a victim.

Hazard Classification (based on NFPA-704 Rating System)
Health Hazards (Blue): 2; Flammability (Red): 2; Reactivity (Yellow): 0

EMERGENCY RESPONSE: See Appendix A (153)
Evacuation:
Public safety: Isolate the area of spill or leak for at least 25 to 50 meters (80 to 160 feet) in all directions.
Spill: Increase, in the downwind direction, as necessary, the distance shown under "Public Safety."
Fire: If tank, rail car, or tank truck is involved in fire, isolate for at least 800 meters (½ mile) in all directions; also, consider initial evacuation for 800 meters (½ mile) in all directions.

EXPOSURE
Short-term effects: *SEEK MEDICAL ATTENTION.* **Liquid:** *HARMFUL IF ABSORBED THROUGH THE SKIN.* Irritating to skin and eyes. Harmful if swallowed. Remove contaminated clothing and shoes. Flush affected areas with plenty of water. *IF BREATHING HAS STOPPED*, give artificial respiration; *avoid mouth-to-mouth resuscitation; use bag/mask apparatus. IF IN EYES*, hold eyelids open and flush with plenty of water. *IF SWALLOWED* and victim is *CONSCIOUS AND ABLE TO SWALLOW*, have victim drink 4 to 8 ounces of water. *IF SWALLOWED* and victim is *UNCONSCIOUS OR HAVING CONVULSIONS*, do nothing except keep victim warm.
Absorption through skin produces same symptoms as for ingestion.

HEALTH HAZARDS
Personal protective equipment (PPE): B-Level PPE. OSHA Table Z-1-A air contaminant. Organic vapor respirator; rubber gloves; splash proof goggles.
Recommendations for respirator selection: NIOSH
5 ppm: SA (any supplied-air respirator). *12.5 ppm*: SA:CF (any supplied-air respirator operated in a continuous-flow mode). *25 ppm*: SAT:CF (any supplied-air respirator that has a tight-fitting facepiece and is operated in a continuous-flow mode); or SCBAF (any self-contained breathing apparatus with a full facepiece); or SAF (any supplied-air respirator with a full facepiece). *100 ppm*: SAF:PD,PP (any supplied-air respirator that has a full facepiece and is operated in a pressure-demand or other positive-pressure mode). *EMERGENCY OR PLANNED ENTRY INTO UNKNOWN CONCENTRATIONS OR IDLH CONDITIONS*: SCBAF:PD,PP (any self-contained breathing apparatus that has a full facepiece and is operated in a pressure-demand or other positive-pressure mode); or SAF:PD,PP:ASCBA (any supplied-air respirator that has a full facepiece and is operated in a pressure-demand or other positive-pressure mode in combination with an auxiliary self-contained breathing apparatus operated in a pressure-demand or other positive pressure mode). *ESCAPE*: GMFS [any air-purifying, full-facepiece respirator (gas mask) with a chin-style, front- or back-mounted canister providing protection against the compound of concern]; or SCBAE (any appropriate escape-type, self-contained breathing apparatus).
Exposure limits (TWA unless otherwise noted): ACGIH TLV 0.5 ppm (2.2 mg/m³); OSHA PEL 2 ppm (9 mg/m³); NIOSH 0.5 ppm (2 mg/m³); skin contact contributes significantly in overall exposure.
Toxicity by ingestion: Poisonous.
Liquid or solid irritant characteristics: Eye irritant.
Odor threshold: 1.6–2.0 ppm.
IDLH value: 100 ppm.

FIRE DATA
Flash point: 175°F/79°C (cc).

Fire extinguishing agents not to be used: Water may be ineffective.
Burning rate: 3.65 mm/min

CHEMICAL REACTIVITY
Binary reactants: May attack some forms of plastics. Reacts with strong acids, strong oxidizers.

ENVIRONMENTAL DATA
Food chain concentration potential: Log P_{ow} = 1.7. Unlikely to accumulate.
Water pollution: Effect of low concentrations on aquatic life is unknown. Fouling to shoreline. May be dangerous if it enters nearby water intakes; notify operators. Notify local health and wildlife officials. **Response to discharge:** Issue warning–poison, water contaminant. Restrict access. Mechanical containment. Should be removed. Chemical and physical treatment.

SHIPPING INFORMATION
Grades of purity: Technical; Pure, 99+%; **Storage temperature:** Ambient; **Inert atmosphere:** None; **Venting:** Open; **Stability during transport:** Stable.

PHYSICAL AND CHEMICAL PROPERTIES
Physical state @ 59°F/15°C and 1 atm: Liquid.
Molecular weight: 107.2
Boiling point @ 1 atm: 384.6°F/195.9°C/469.1°K.
Melting/Freezing point: –71°F/–57°C/216°K.
Critical temperature: 802°F/428°C/701°K.
Critical pressure: 754 psia = 51.3 atm = 5.20 MN/m^2.
Specific gravity (water = 1): 0.989 @ 68°F/20°C (liquid).
Liquid surface tension: 39.6 dynes/cm = 0.0396 N/m @ 68°F/20°C.
Relative vapor density (air = 1): 3.70
Latent heat of vaporization: 180 Btu/lb = 100 cal/g = 4.20 x 10^5 J/kg.
Heat of combustion: –16,350 Btu/lb = –9085 cal/g = –380.1 x 10^5 J/kg.
Vapor pressure: 0.3 mm.

METHYL BENZOATE   REC. M:3000

SYNONYMS: BENZOATOATO de METILO (Spanish); BENZOIC ACID, METHYL ESTER; METHYL BENZENECARBOXYLATE; OXIDATE LE; NIOBE OIL; OIL OF NIOBE; ESSENCE OF NIOBE

IDENTIFICATION
CAS Number: 93-58-3
Formula: $C_8H_8O_2$; $C_6H_5CO_2CH_3$
DOT ID Number: UN 2938; DOT Guide Number: 171
Proper Shipping Name: Methyl benzoate

DESCRIPTION: Colorless liquid. Pleasant, fragrant odor. Sinks in water.

Combustible • Containers may BLEVE when exposed to fire • Vapors may form explosive mixture with air • Vapors are heavier than air and will collect and stay in low areas • Vapors may travel long distances to ignition sources and flashback • Vapors in confined areas (e.g., tanks, sewers, buildings) may explode when exposed to fire • Irritating to the skin, eyes, and respiratory tract • Toxic products of combustion may include carbon monoxide.

Hazard Classification (based on NFPA-704 Rating System)
Health Hazards (Blue): 0; Flammability (Red): 2; Reactivity (Yellow): 0

EMERGENCY RESPONSE: See Appendix A (171)
Evacuation:
Public safety: Isolate the area of spill or leak for at least 10 to 25 meters (30 to 80 feet) in all directions.
Spill: Increase, in the downwind direction, as necessary, the distance shown under "Public Safety."
Fire: If any large container is involved in fire, isolate for at least 800 meters (½ mile) in all directions; also, consider initial evacuation for 800 meters (½ mile) in all directions.

EXPOSURE
Short-term effects: *SEEK MEDICAL ATTENTION*. **Vapor:** Irritating to the eyes, nose, and throat. May be harmful if inhaled or absorbed through the skin. Move victim to fresh air *IF BREATHING HAS STOPPED,* give artificial respiration. IF breathing is difficult, administer oxygen. **Liquid:** *MAY BE HARMFUL IF SWALLOWED OR ABSORBED THROUGH THE SKIN*. Irritating to skin and eyes. *IF IN EYES:* immediately flush eyes with running water for at least 15 minutes. Remove and double bag contaminated clothing and shoes at the site. Wash skin with soap and water. *IF SWALLOWED*: Do nothing except keep victim warm. **Do not induce vomiting.**

HEALTH HAZARDS
Personal protective equipment (PPE): B-Level PPE. Approved respirator, chemical safety goggles, chemical-resistant gloves.
Toxicity by ingestion: Grade 2: LD_{50} = 1.35 g/kg (rat).
Vapor (gas) irritant characteristics: Eye and respiratory tract irritant. Vapors cause moderate irritation such that personnel will find high concentrations unpleasant. The effect is temporary.
Liquid or solid irritant characteristics: Eye and skin irritant. If spilled on clothing and allowed to remain, may cause smarting and reddening of skin.

FIRE DATA
Flash point: 181°F/83°C (cc).

CHEMICAL REACTIVITY
Neutralizing agents for acids and caustics: Dry lime, soda ash.

ENVIRONMENTAL DATA
Food chain concentration potential: Log P_{ow} = –0.31. Unlikely to accumulate.
Water pollution: Effects of low concentrations on aquatic life are not known. May be dangerous if it enters nearby water intakes; notify operators. Notify local health and wildlife officials. **Response to discharge:** Should be removed. Chemical and physical treatment.

SHIPPING INFORMATION
Grades of purity: 99%; **Inert atmosphere:** None; **Venting:** None; **Stability during transport:** Stable.

NAS HAZARD CLASSIFICATION FOR BULK WATER TRANSPORTATION
FIRE: 2
HEALTH: Vapor irritant: 2; Liquid or solid irritant: 1; Poisons: 1
WATER POLLUTION: Human toxicity: 2; Aquatic toxicity:–; Aesthetic effect: 1
REACTIVITY: Other chemicals: 2; Water: 0; Self-reaction: 0

PHYSICAL AND CHEMICAL PROPERTIES
Physical state @ 59°F/15°C and 1 atm: Liquid.
Molecular weight: 136.15
Boiling point @ 1 atm: 302°F/150°C/423°K.
Melting/Freezing point: 10°F/–12.3°C/261°K.
Specific gravity (water = 1): 1.0888 @ 68°F/20°C.
Liquid surface tension: 37.6 dynes/cm = 0.038 N/m @ 68°F/20°C.
Relative vapor density (air = 1): 4.7
Heat of combustion: –2432 Btu/lb = –1351 cal/g = –56 x 10^5 J/kg.
Vapor pressure: (Reid) 0.1 psia; 0.2 mm.

α-METHYLBENZYL ALCOHOL   REC. M:3050

SYNONYMS: (1-HYDROXYETHYL)BENZENE; α-METHYL BENZENE METHANOL; METHYLPHENYL METHANOL; STYRALLYL ALCOHOL; PHENYLMETHYL CARBINOL; α-PHENYL ETHYL ALCOHOL; 1-PHENYL ETHYL ALCOHOL; 1-PHENYLETHANOL; α-PHENETHYL ALCOHOL; α-PHENETHYL ALCOHOL; ETHANOL,1-PHENYL-; STYRALYL ALCOHOL

IDENTIFICATION
CAS Number: 98-85-1
Formula: $C_8H_{10}O$; $C_6H_5CH(OH)CH_3$
DOT ID Number: UN 2937; DOT Guide Number: 153
Proper Shipping Name: α-Methylbenzyl alcohol

DESCRIPTION: Colorless liquid. Flowery odor. Sinks and mixes with water.

Combustible • Containers may BLEVE when exposed to fire • Vapors may form explosive mixture with air • Vapors are heavier than air and will collect and stay in low areas • Vapors may travel long distances to ignition sources and flashback • Vapors in confined areas (e.g., tanks, sewers, buildings) may explode when exposed to fire • Irritating to the skin, eyes, and respiratory tract • Toxic products of combustion may include carbon monoxide.

Hazard Classification (based on NFPA-704 Rating System)
Health Hazards (Blue): 0; Flammability (Red): 2; Reactivity (Yellow): 0

EMERGENCY RESPONSE: See Appendix A (153)
Evacuation:
Public safety: Isolate the area of spill or leak for at least 25 to 50 meters (80 to 160 feet) in all directions.
Spill: Increase, in the downwind direction, as necessary, the distance shown under "Public Safety."
Fire: If tank, rail car, or tank truck is involved in fire, isolate for at least 800 meters (½ mile) in all directions; also, consider initial evacuation for 800 meters (½ mile) in all directions.

EXPOSURE
Short-term effects: *SEEK MEDICAL ATTENTION.* **Vapor:** Harmful if inhaled or absorbed through the skin. Move victim to fresh air. *IF BREATHING HAS STOPPED,* give artificial respiration; *avoid mouth-to-mouth resuscitation; use bag/mask apparatus.* IF breathing is difficult, administer oxygen. **Liquid or solid:** Harmful if swallowed or absorbed through the skin. *IF IN EYES:* Flush with water for at least 15 minutes. Remove contaminated clothing and shoes. Flush affected areas with water. *IF SWALLOWED*: Do nothing except keep victim warm. **Do not induce vomiting.**

HEALTH HAZARDS
Personal protective equipment (PPE): B-Level PPE. Approved respirator, chemical safety goggles, chemical resistant gloves, other protective clothing. Natural rubber, neoprene, and nitrile rubber may offer some protection from alcohols.
Toxicity by ingestion: Grade 3: LD_{50} = 400 mg/kg (rat).
Vapor (gas) irritant characteristics: Eye and respiratory tract irritant. Vapors cause moderate irritation such that personnel will find high concentrations unpleasant. The effect is temporary.
Liquid or solid irritant characteristics: Eye and skin irritant. Causes smarting of the skin and first-degree burns on short exposure; may cause second-degree burns on long exposure.

FIRE DATA
Flash point: 200°F/93°C (cc); 205°F/96°C (oc).

CHEMICAL REACTIVITY
Binary reactants: Strong oxidizers.

ENVIRONMENTAL DATA
Water pollution: Effects of low concentrations on aquatic life are not known. May be dangerous if it enters local water intakes. Notify local health and wildlife officials. **Response to discharge:** Should be removed. Chemical and physical treatment.

SHIPPING INFORMATION
Stability during transport: Stable.

NAS HAZARD CLASSIFICATION FOR BULK WATER TRANSPORTATION
FIRE: 2
HEALTH: Vapor irritant: 2; Liquid or solid irritant: 2; Poisons: 2
WATER POLLUTION: Human toxicity: 3; Aquatic toxicity:–; Aesthetic effect: 2
REACTIVITY: Other chemicals: 2; Water: 0; Self-reaction: 0

PHYSICAL AND CHEMICAL PROPERTIES
Physical state @ 59°F/15°C and 1 atm: Solid.
Molecular weight: 122.17
Boiling point @ 1 atm: 397°F/203°C/476°K.
Melting/Freezing point: 68°F/20°C/293°K.
Specific gravity (water = 1): 1.015 @ 68°F/20°C.
Relative vapor density (air = 1): 4.21
Vapor pressure: (Reid) Very low; 0.01 mm.

METHYL BROMIDE   REC. M:3100

SYNONYMS: BROMOMETHANE; BROM-O-GAS; DAWSON 100; DOWFUME; EDCO; EEC No. 602-002-00-3; EMBAFUME; HALON 1001; ISCOBROME; KAYAFUME; METHOGAS; M-B-C FUMIGANT; MONOBROMOMETHANE; R 40B1; RCRA No. U029; ROTOX; TERABOL; TERR-O-GAS 100; ZYTOX

IDENTIFICATION
CAS Number: 74-83-9
Formula: CH_3Br
DOT ID Number: UN 1062; DOT Guide Number: 123
Proper Shipping Name: Methyl bromide
Reportable Quantity (RQ): **(CERCLA)** 1000 lb/454 kg

Methyl bromide

DESCRIPTION: Colorless, liquefied gas. A liquid below 38°F/3°C. Odorless and nonirritating at low at low concentrations and has a musty or fruity odor at high concentrations (greater than 1000 ppm). Sinks and boils in water; poisonous vapor cloud is formed. *Warning:* Odor is not a reliable indicator of the presence of toxic amounts of methyl bromide gas. Because the material lacks adequate warning properties, up to 2% chloropicrin, a tear gas, is often added as a warning agent.

Poison! • Highly flammable • Breathing the gas can kill you • Firefighting gear (including SCBA) does not provide adequate protection. If exposure occurs, remove and isolate gear immediately and thoroughly decontaminate personnel • Containers may BLEVE when exposed to fire • Gas may form explosive mixture with air • Gas is heavier (3x) than air and will collect and stay in low areas • Gas may travel long distances to ignition sources and flashback • Gas in confined areas (e.g., tanks, sewers, buildings) may explode when exposed to fire • Ruptured or venting cylinders may rocket through buildings and/or travel a considerable distance • Toxic products of combustion may include bromine • Do not put yourself in danger by entering a contaminated area to rescue a victim.

Hazard Classification (based on NFPA-704 Rating System)
Health Hazards (Blue): 3; Flammability (Red): 1; Reactivity (Yellow): 0

EMERGENCY RESPONSE: See Appendix A (123)
Evacuation:
Public safety: See below.
Spill: Small spill–First: Isolate in all directions 30 meters (100 feet); Then: Protect persons downwind, DAY: 0.2 km (0.1 mile); NIGHT: 0.3 km (0.2 mile). Large spill–First: Isolate in all directions 95 meters (300 feet). Then: Protect persons downwind, DAY: 0.5 km (0.3 mile); NIGHT: 1.4 km (0.9 mile).
Fire: Isolate for 800 meters (½ mile) in all directions, especially if tank, rail car, or tank truck is involved in fire.

EXPOSURE
Short-term effects: *SEEK MEDICAL ATTENTION.* There is no antidote for methyl bromide poisoning, but its effects can be treated and most persons recover. **Vapor:** *POISONOUS. MAY BE FATAL IF INHALED.* Irritating to eyes. Move to fresh air. *IF BREATHING HAS STOPPED,* give artificial respiration; *avoid mouth-to-mouth resuscitation; use bag/mask apparatus.* IF breathing is difficult, administer oxygen. **Liquid:** *HARMFUL IF ABSORBED THROUGH THE SKIN.* Will burn skin and eyes. Harmful if absorbed by the gastrointestinal tract. *BECAUSE THIS CHEMICAL PERSISTS IN CLOTH, LEATHER, AND RUBBER, REMOVE AND DOUBLE-BAG CLOTHING AND SHOES AND LEAVE ALL CONTAMINATED CLOTHING IN THE HOT ZONE.* Flush affected areas with plenty of water. *DO NOT RUB AFFECTED AREAS. IF IN EYES*, hold eyelids open and flush with plenty of water. *IF SWALLOWED* and victim is *CONSCIOUS AND ABLE TO SWALLOW*, have victim drink 4 to 8 ounces of water. **Do NOT induce vomiting.**
Note to physician: There is no proven antidote for methyl bromide poisoning. Dimercaprol (BAL) or acetylcysteine (Mucomyst) have been suggested as antidotes based on the postulated mechanism of methyl bromide's toxicity. However, no adequate studies have tested the efficacy of these therapies, and they are NOT recommended for routine use (ATSDR).

HEALTH HAZARDS
Personal protective equipment (PPE): A-Level PPE. OSHA Table Z-1-A air contaminant. Acts as a narcotic if inhaled. Full face self-contained breathing apparatus; goggles. Methyl bromide gas easily penetrates most protective clothing (e.g., cloth, rubber, and leather). Chemical protective material(s) reported to have good to excellent resistance: PV alcohol, Saranex®. Also, styrene-butadiene, butyl rubber, neoprene, nitrile, and neoprene offers limited protection.
Recommendations for respirator selection: NIOSH
At any concentrations above the NIOSH REL, or where there is no REL, at any detectable concentration: SCBAF:PD,PP (any self-contained breathing apparatus that has a full facepiece and is operated in a pressure-demand or other positive-pressure mode); or SAF:PD,PP:ASCBA (any supplied-air respirator that has a full facepiece and is operated in a pressure-demand or other positive-pressure mode in combination with an auxiliary self-contained breathing apparatus operated in a pressure-demand or other positive pressure mode). *ESCAPE:* GMFOV [any air-purifying, full-facepiece respirator (gas mask) with a chin-style, front-or back-mounted organic vapor canister]; or SCBAE (any appropriate escape-type, self-contained breathing apparatus).
Exposure limits (TWA unless otherwise noted): ACGIH TLV 1 ppm; OSHA PEL ceiling 20 ppm (80 mg/m^3); NIOSH REL potential human carcinogen; reduce exposure to lowest feasible level; skin contact contributes significantly in overall exposure. AIHA ERPG-2 (the maximum airborne concentration below which it is believed that nearly all individuals could be exposed for up to 1 hour without experiencing or developing irreversible or other serious health effects or symptoms which could impair an individual's ability to take protective action) = 50 ppm.
Long-term health effects: IARC rating 3; animal evidence: limited. After a serious exposure that causes lung or nervous system-related problems, permanent brain or nerve damage can result.
Vapor (gas) irritant characteristics: Eye and respiratory tract irritant. Vapor is irritating such that personnel will not usually tolerate moderate or high vapor concentrations.
Liquid or solid irritant characteristics: Severe eye and skin irritant; may cause pain and second-degree burns after a few minutes of contact.
IDLH value: 250 ppm; potential human carcinogen.

FIRE DATA
Flash point: Practically not flammable.
Flammable limits in air: LEL: 10%; UEL: 15.9%.
Behavior in fire: Containers may explode.
Autoignition temperature: 999°F/537°C/810°K.
Electrical hazard: Class I, Group D.

CHEMICAL REACTIVITY
Reactivity with water: Reacts with water.
Binary reactants: Attacks aluminum to form pyrophoric alkyl aluminum salts. Incompatible with strong oxidizers, metals, dimethylsulfoxide, ethylene oxide, water. Attacks zinc, magnesium, alkali metals and their alloys. Attacks some plastics, rubber, and coatings.
Reactivity group: 36
Compatibility class: Halogenated hydrocarbons

ENVIRONMENTAL DATA
Food chain concentration potential: Log P_{ow} = 1.2. Unlikely to accumulate.
Water pollution: Not harmful to aquatic life. May be dangerous if it enters nearby water intakes; notify operators. Notify local health and wildlife officials. **Response to discharge:** Issue warning–poison. Restrict access.

SHIPPING INFORMATION
Grades of purity: Commercial: not less than 99.5%; **Storage temperature:** Ambient; **Inert atmosphere:** None; **Venting:** Safety relief; **Stability during transport:** Stable.

NAS HAZARD CLASSIFICATION FOR BULK WATER TRANSPORTATION
FIRE: 1
HEALTH: Vapor irritant: 3; Liquid or solid irritant: 3; Poisons: 4
WATER POLLUTION: Human toxicity: 0; Aquatic toxicity: 1; Aesthetic effect: 2
REACTIVITY: Other chemicals: 1; Water: 0; Self-reaction: 0

PHYSICAL AND CHEMICAL PROPERTIES
Physical state @ 59°F/15°C and 1 atm: Gas.
Molecular weight: 94.95
Boiling point @ 1 atm: 38.5°F/3.6°C/276.8°K.
Melting/Freezing point: −135°F/−93°C/180°K.
Critical temperature: 376°F/191°C/464°K.
Specific gravity (water = 1): 1.73 at 32°F/liquid).
Liquid surface tension: 24.5 dynes/cm = 0.0245 N/m @ 15°C.
Relative vapor density (air = 1): 3.36
Ratio of specific heats of vapor (gas): 1.247
Latent heat of vaporization: 108 Btu/lb = 59.7 cal/g = 2.50×10^5 J/kg.
Heat of combustion: −3188 Btu/lb = −1771 cal/g = 74.15×10^5 J/kg.
Heat of fusion: 15.05 cal/g.
Vapor pressure: (Reid) 45 psia; 685 mm.

3-METHYL-2-BUTANONE   REC. M:3150

SYNONYMS: 2-ACETYL PROPANE; ISOPROPYL METHYL KETONE; METIL BUTANONA (Spanish); 3-METHYLBUTAN-2-ONE; MIPK; METHYL ISOPROPYL KETONE

IDENTIFICATION
CAS Number: 563-80-4
Formula: $C_5H_{10}O$; $(CH_3)_2CH(CO)CH_3$
DOT ID Number: UN 2397; DOT Guide Number: 127
Proper Shipping Name: 3-Methylbutan-2-one

DESCRIPTION: Colorless liquid. Sweet, acetone odor. Floats on water surface; slightly soluble.

Extremely flammable • Containers may BLEVE when exposed to fire • Vapors may form explosive mixture with air • Vapors are heavier than air and will collect and stay in low areas • Vapors may travel long distances to ignition sources and flashback • Vapors in confined areas (e.g., tanks, sewers, buildings) may explode when exposed to fire • Irritating to the skin, eyes, and respiratory tract • Toxic products of combustion may include carbon monoxide.

Hazard Classification (based on NFPA-704 Rating System)
Health Hazards (Blue): 1; Flammability (Red): 3; Reactivity (Yellow): 0

EMERGENCY RESPONSE: See Appendix A (127)
Evacuation:
Public safety: Isolate spill area for at least 25 to 50 meters (80 to 160 feet) in all directions.
Spill: Large spill–Consider initial downwind evacuation for at least 300 meters (1000 feet).
Fire: Isolate for 800 meters (½ mile) in all directions, especially if tank, rail car, or tank truck is involved in fire.

EXPOSURE
Short-term effects: *SEEK MEDICAL ATTENTION.* **Vapor:** Skin and eye irritant. Harmful if inhaled or absorbed through the skin. Remove to fresh air. *IF BREATHING HAS STOPPED,* give artificial respiration. IF breathing is difficult, administer oxygen. **Liquid:** Harmful if swallowed or absorbed through the skin. Irritating to the skin and eyes. *IF IN EYES:* Flush with water for at least 15 minutes. Remove contaminated clothing and shoes, flush affected areas with plenty of water. *IF SWALLOWED,* do nothing except keep victim warm. **Do not induce vomiting.**

HEALTH HAZARDS
Personal protective equipment (PPE): B-Level PPE. OSHA Table Z-1-A air contaminant. Approved respirator, chemical resistant gloves, chemical safety goggles, other protective clothing.
Exposure limits (TWA unless otherwise noted): NIOSH REL: 200 ppm (705 mg/m^3) as methyl isopropyl ketone.
Toxicity by ingestion: Grade 3: LD_{50} = 148 mg/kg (rat).
Vapor (gas) irritant characteristics: Eye and respiratory tract irritant. Vapors cause smarting of the eyes or respiratory system if present in high concentrations. The effect is temporary.
Liquid or solid irritant characteristics: Eye and skin irritant. If spilled on clothing and allowed to remain, may cause smarting and reddening of skin.
Odor threshold: 4.3–5.0 ppm.

FIRE DATA
Flash point: 43°F/6°C (cc).

CHEMICAL REACTIVITY
Binary reactants: Oxidizers
Reactivity group: 18
Compatibility class: Ketones

ENVIRONMENTAL DATA
Water pollution: Effects of low concentration on aquatic life is not known. May be dangerous if it enters nearby water intakes; notify operators. Notify local health and wildlife officials. **Response to discharge:** Issue warning–high flammability. Mechanical containment. Should be removed. Chemical and physical treatment.

SHIPPING INFORMATION
Grades of purity: 99%; **Inert atmosphere:** None; **Venting:** None; **Stability during transport:** Stable.

NAS HAZARD CLASSIFICATION FOR BULK WATER TRANSPORTATION
FIRE: 4
HEALTH: Vapor irritant: 2; Liquid or solid irritant: 1; Poisons: 1
WATER POLLUTION: Human toxicity: 3; Aquatic toxicity: 2; Aesthetic effect: 1
REACTIVITY: Other chemicals: 1; Water: 0; Self-reaction: 0

PHYSICAL AND CHEMICAL PROPERTIES
Physical state @ 59°F/15°C and 1 atm: Liquid.
Molecular weight: 86.15
Boiling point @ 1 atm: 201°F/94°C/367°K.
Melting/Freezing point: −134°F/−92°C/181°K.
Specific gravity (water = 1): 0.8051 @ 68°F/20°C.
Relative vapor density (air = 1): 2.97
Latent heat of vaporization: 161.6 Btu/lb = 89.8 cal/g = 3.8×10^5 J/kg.

Heat of combustion: –15,334 Btu/lb = –8519 cal/g = –357 x 10^5 J/kg.
Vapor pressure: (Reid) 1.3 psia; 42 mm.

METHYL BUTENOL   REC. M:3200

SYNONYMS: 2-METHYL-3-BUTEN-2-OL; 3-METHYL-BUTEN-(1)-OL-(3); 3-METHYL-1-BUTEN-3-OL; 1-BUTEN-3-OL, 3-METHYL

IDENTIFICATION
CAS Number: 115-18-4
Formula: $C_5H_{10}O$
DOT ID Number: UN 1993; DOT Guide Number: 128
Proper Shipping Name: Flammable liquid, n.o.s.

DESCRIPTION: Colorless liquid. Slightly soluble in water.

Highly flammable • Containers may BLEVE when exposed to fire • Vapors may form explosive mixture with air • Vapors are heavier than air and will collect and stay in low areas • Vapors may travel long distances to ignition sources and flashback • Vapors in confined areas (e.g., tanks, sewers, buildings) may explode when exposed to fire • Irritating to the skin, eyes, and respiratory tract • Toxic products of combustion may include carbon monoxide.

Hazard Classification (based on NFPA-704 Rating System)
Health Hazards (Blue): 1; Flammability (Red): 3; Reactivity (Yellow): 0

EMERGENCY RESPONSE: See Appendix A (128)
Evacuation:
Public safety: Isolate spill area for at least 25 to 50 meters (80 to 160 feet) in all directions.
Spill: Large spill–Consider initial downwind evacuation for at least 300 meters (1000 feet).
Fire: Isolate for 800 meters (½ mile) in all directions, especially if tank, rail car, or tank truck is involved in fire.

EXPOSURE
Short-term effects: *SEEK MEDICAL ATTENTION.* **Vapor:** Harmful if inhaled or swallowed.
Material is irritating to mucous membrane and upper respiratory tract. Move to fresh air. *IF BREATHING HAS STOPPED,* give artificial respiration. IF breathing is difficult, administer oxygen. **Liquid:** Remove contaminated clothing and shoes. Flush affected areas with plenty of water. *IF IN EYES,* hold eyelids open and flush with plenty of water.

HEALTH HAZARDS
Personal protective equipment (PPE): B-Level PPE. Self-contained breathing apparatus, rubber boots and heavy rubber gloves.
Vapor (gas) irritant characteristics: Eye and respiratory tract irritant. Vapors cause smarting of the eyes or respiratory system if present in high concentrations. The effect is temporary.
Liquid or solid irritant characteristics: Eye and skin irritant. If spilled on clothing and allowed to remain, may cause smarting and reddening of skin.

FIRE DATA
Flash point: 56°F/13°C (cc).
Fire extinguishing agents not to be used: Water may be ineffective.

CHEMICAL REACTIVITY
Reactivity group: 20
Compatibility class: Alcohols, glycols

ENVIRONMENTAL DATA
Water pollution: Effects of low concentrations on aquatic life is unknown. May be dangerous if it enters nearby water intakes; notify operators. Notify local health and wildlife officials. **Response to discharge:** Issue warning-flammable liquid. Restrict access. Should be removed. Chemical and physical treatment.

SHIPPING INFORMATION
Grades of purity: 98%; **Storage temperature:** Ambient; **Stability during transport:** Stable.

PHYSICAL AND CHEMICAL PROPERTIES
Physical state @ 59°F/15°C and 1 atm: Liquid.
Molecular weight: 86.13
Boiling point @ 1 atm: 208.4–210.2°F/98–99°C/371.2–372.2°K.
Specific gravity (water = 1): 0.824
Relative vapor density (air = 1): 2.97

METHYL BUTYNOL   REC. M:3250

SYNONYMS: 2-HYDROXY-2-METHYL-3-BUTYNE; ETHYNYLDIMETHYLCARBINOL; METILBUTINOL (Spanish); 2-METHYL-3-BUTYN-2-OL

IDENTIFICATION
CAS Number: 115-19-5
Formula: C_5H_8O; $(CH_3)_2C(OH)CCH$
DOT ID Number: UN 1993; DOT Guide Number: 128
Proper Shipping Name: Flammable liquid, n.o.s.

DESCRIPTION: Colorless to straw yellow liquid.

Highly flammable • Containers may BLEVE when exposed to fire • Vapors may form explosive mixture with air • Vapors are heavier than air and will collect and stay in low areas • Vapors may travel long distances to ignition sources and flashback • Vapors in confined areas (e.g., tanks, sewers, buildings) may explode when exposed to fire • Irritating to the skin, eyes, and respiratory tract • Toxic products of combustion may include carbon monoxide.

Hazard Classification (based on NFPA-704 Rating System)
Health Hazards (Blue): 2; Flammability (Red): 3; Reactivity (Yellow): 0

EMERGENCY RESPONSE: See Appendix A (128)
Evacuation:
Public safety: Isolate spill area for at least 25 to 50 meters (80 to 160 feet) in all directions.
Spill: Large spill–Consider initial downwind evacuation for at least 300 meters (1000 feet).
Fire: Isolate for 800 meters (½ mile) in all directions, especially if tank, rail car, or tank truck is involved in fire.

EXPOSURE
Short-term effects: *SEEK MEDICAL ATTENTION.* **Vapor:** Move victim to fresh air. If breathing is difficult, administer oxygen. **Liquid:** Irritating to skin and eyes. Remove contaminated clothing and shoes. Flush skin with water. *IF IN EYES,* hold eyelids open and flush with plenty of water.

HEALTH HAZARDS

Personal protective equipment (PPE): B-Level PPE. Wear full impervious protective clothing and approved respirator. Where splashing is possible wear full face shield or chemical safety goggles. Use approved respirator to protect against vapors.
Vapor (gas) irritant characteristics: Eye and respiratory tract irritant. Vapors cause a slight smarting of the eyes or respiratory system if present in high concentrations. The effect is temporary.
Liquid or solid irritant characteristics: Eye and skin irritant. If spilled on clothing and allowed to remain, may cause smarting and reddening of the skin.

FIRE DATA

Flash point: Less than 70°F/21°C (cc).
Fire extinguishing agents not to be used: Water.
Behavior in fire: Containers may rupture and explode.
Stoichiometric air-to-fuel ratio: 31.0

CHEMICAL REACTIVITY

Binary reactants: Oxidizers, acids, caustics, ammonia, amines.
Reactivity group: 20
Compatibility class: Alcohols, glycols

ENVIRONMENTAL DATA

Water pollution: Effects of low concentrations on aquatic life is unknown. May be dangerous if it enters nearby water intakes; notify operators. Notify local health and wildlife officials.
Response to discharge: Issue warning-flammable. Should be removed.

SHIPPING INFORMATION

Grades of purity: Varying concentrations available; **Storage temperature:** Ambient; **Inert atmosphere:** None; **Stability during transport:** Stable.

PHYSICAL AND CHEMICAL PROPERTIES

Physical state @ 59°F/15°C and 1 atm: Liquid.
Molecular weight: 84.12
Boiling point @ 1 atm: 219.2–221°F/104–105°C/377–378°K.
Melting/Freezing point: 36.7°F/2.6°C/275.6°K.
Specific gravity (water = 1): 0.8672
Liquid surface tension: 23.8 dynes/cm @ 77°F/25°C.
Relative vapor density (air = 1): 2.9

METHYL *tert*-BUTYL ETHER   **REC. M:3300**

SYNONYMS: *tert*-BUTYL METHYL ETHER; METIL-*terc*-BUTIL ETER (Spanish); 2-METHOXY-2-METHYL PROPANE; 2-METHYL-2-METHOXY PROPANE; METHYL 1,1-DIMETHYLETHYL ETHER; PROPANE,2-METHOXY-2-METHYL(9CI).

IDENTIFICATION

CAS Number: 1634-04-4
Formula: $C_5H_{12}O$; $(CH_3)_3COCH_3(CH_3)_3COCH_3$
DOT ID Number: UN 2398; DOT Guide Number: 127
Proper Shipping Name: Methyl *tert*-butyl ether
Reportable Quantity (RQ): **(CERCLA)** 1 lb/0.454 kg

DESCRIPTION: Clear, colorless liquid. Turpentine-like odor. Floats and mixes slowly with water; moderately soluble. Produces large amounts of vapor.

Highly flammable • Containers may BLEVE when exposed to fire • Vapors may form explosive mixture with air • Vapors are heavier than air and will collect and stay in low areas • Vapors may travel long distances to ignition sources and flashback • Vapors in confined areas (e.g., tanks, sewers, buildings) may explode when exposed to fire • Irritating to the skin, eyes, and respiratory tract • Toxic products of combustion may include carbon monoxide.

Hazard Classification (based on NFPA-704 Rating System)
Health Hazards (Blue): 1; Flammability (Red): 3; Reactivity (Yellow): 0

EMERGENCY RESPONSE: See Appendix A (127)
Evacuation:
Public safety: Isolate spill area for at least 25 to 50 meters (80 to 160 feet) in all directions.
Spill: Large spill–Consider initial downwind evacuation for at least 300 meters (1000 feet).
Fire: Isolate for 800 meters (½ mile) in all directions, especially if tank, rail car, or tank truck is involved in fire.

EXPOSURE

Short-term effects: *SEEK MEDICAL ATTENTION*. **Vapor:** A mild irritant to eyes and skin. *IF INHALED*, will, may cause dizziness and/or suffocation. Move to fresh air. *IF BREATHING HAS STOPPED*, give artificial respiration. IF breathing is difficult, administer oxygen. **Liquid:** May irritate or burn skin and eyes. May be harmful if swallowed. *IF IN EYES OR ON SKIN*, flush with running water for at least 15 minutes; hold eyelids open if necessary. Wash skin with soap and water. Remove and double bag contaminated clothing and shoes at the site. *IF SWALLOWED* and victim is *UNCONSCIOUS OR HAVING CONVULSIONS*, do nothing except keep victim warm.

HEALTH HAZARDS

Personal protective equipment (PPE): B-Level PPE. Wear goggles, self-contained breathing apparatus, rubber gloves, boots and overclothing.
Toxicity by ingestion: Grade 2; LD_{50} = 2.96 g/kg (rat).
Vapor (gas) irritant characteristics: Vapors cause smarting of the eyes skin and respiratory system if present in high concentrations. The effect is temporary.
Liquid or solid irritant characteristics: Eye and skin irritant. If spilled on clothing and allowed to remain, may cause smarting and reddening of skin.

FIRE DATA

Flash point: –14°F/26°C (cc).
Behavior in fire: Containers may explode in heat of fire.
Stoichiometric air-to-fuel ratio: 11.9°F/24.4°C (estimate).

CHEMICAL REACTIVITY

Binary reactants: Reacts with strong acids; vigorously with oxidizing materials.
Reactivity group: 41
Compatibility class: Ethers

ENVIRONMENTAL DATA

Water pollution: Effects of low concentrations on aquatic life is unknown. May be dangerous if it enters nearby water intakes; notify operators. Notify local health and wildlife officials. **Response to discharge:** Issue warning–high flammability, air contaminant. Restrict access. Evacuate area. Should be removed. Chemical and physical treatment.

SHIPPING INFORMATION
Grades of purity: 97%; **Stability during transport:** May form explosive peroxides on standing.

PHYSICAL AND CHEMICAL PROPERTIES
Physical state @ 59°F/15°C and 1 atm: Liquid.
Molecular weight: 88.15
Boiling point @ 1 atm: 131.4°F/55.2°C/328.2°K.
Melting/Freezing point: −164,2°F/−109°C/164°K.
Critical temperature: 435.4°F/224.1°C/497.1°K.
Critical pressure: 520 psia = 35.4 atm = 3.59 MN/m^2.
Specific gravity (water = 1): 0.7405 @ 68°F/20°C.
Relative vapor density (air = 1): 3.0 (calculated).
Heat of combustion: 16,365 Btu/lb = 9092.4 cal/g = 380.7 x 10^5 J/kg.

METHYL *n*-BUTYL KETONE   REC. M:3350

SYNONYMS: BUTYL METHYL KETONE; *n*-BUTYL METHYL KETONE; EEC No. 606-030-00-6; 2-HEXANONE; HEXANONE-2; KETONE, BUTYL METHYL; METIL BUTIL CETONA (Spanish); MBK; METHYL BUTYL KETONE; MNBK; PROPYLACETONE

IDENTIFICATION
CAS Number: 591-78-6
Formula: $C_6H_{12}O$; $CH_3(CH_2)_3COCH_3$
DOT ID Number: UN 1993; DOT Guide Number: 128
Proper Shipping Name: Flammable liquids, n.o.s.

DESCRIPTION: Clear, colorless liquid. Acetone (finger nail polish remover)-like odor. Floats on surface of water; slightly soluble.

Highly flammable • Containers may BLEVE when exposed to fire • Vapors may form explosive mixture with air • Vapors are heavier than air and will collect and stay in low areas • Vapors may travel long distances to ignition sources and flashback • Vapors in confined areas (e.g., tanks, sewers, buildings) may explode when exposed to fire • Irritating to the skin, eyes, and respiratory tract • Toxic products of combustion may include carbon monoxide.

Hazard Classification (based on NFPA-704 Rating System)
Health Hazards (Blue): 2; Flammability (Red): 3; Reactivity (Yellow): 0

EMERGENCY RESPONSE: See Appendix A (128)
Evacuation:
Public safety: Isolate spill area for at least 25 to 50 meters (80 to 160 feet) in all directions.
Spill: Large spill–Consider initial downwind evacuation for at least 300 meters (1000 feet).
Fire: Isolate for 800 meters (½ mile) in all directions, especially if tank, rail car, or tank truck is involved in fire.

EXPOSURE
Short-term effects: *SEEK MEDICAL ATTENTION.* **Vapor:** Irritating to eyes, nose, and throat. Harmful if inhaled. *IF IN EYES*, hold eyelids open and flush with plenty of water. *IF BREATHING HAS STOPPED*, give artificial respiration. IF breathing is difficult, administer oxygen. **Liquid:** *HARMFUL IF ABSORBED THROUGH THE SKIN or SWALLOWED*. Irritating to skin and eyes. Remove contaminated clothing and shoes. Flush affected areas with plenty of water. *IF IN EYES*, hold eyelids open and flush with plenty of water. *IF SWALLOWED* and victim is *CONSCIOUS AND ABLE TO SWALLOW*, have victim drink water of milk and have victim induce vomiting. *IF SWALLOWED* and victim is *UNCONSCIOUS OR HAVING CONVULSIONS*, do nothing except keep victim warm. *Note:* The use of alcoholic beverages may enhance toxic effects.

HEALTH HAZARDS
Personal protective equipment (PPE): B-Level PPE. OSHA Table Z-1-A air contaminant. Organic vapor respirator. Protective gloves.
Recommendations for respirator selection: NIOSH
10 ppm: SA (any supplied-air respirator). *25 ppm:* SA:CF (any supplied-air respirator operated in a continuous-flow mode). *50 ppm:* SAT:CF (any supplied-air respirator that has a tight-fitting facepiece and is operated in a continuous-flow mode); or SCBAF (any self-contained breathing apparatus with a full facepiece); or SAF (any supplied-air respirator with a full facepiece). *1600 ppm:* SAF:PD,PP (any supplied-air respirator that has a full facepiece and is operated in a pressure-demand or other positive-pressure mode). *EMERGENCY OR PLANNED ENTRY INTO UNKNOWN CONCENTRATIONS OR IDLH CONDITIONS:* SCBAF:PD,PP (any self-contained breathing apparatus that has a full facepiece and is operated in a pressure-demand or other positive-pressure mode); or SAF:PD,PP:ASCBA (any supplied-air respirator that has a full facepiece and is operated in a pressure-demand or other positive-pressure mode in combination with an auxiliary self-contained breathing apparatus operated in a pressure-demand or other positive pressure mode). *ESCAPE:* GMFOV [any air-purifying, full-facepiece respirator (gas mask) with a chin-style, front- or back-mounted acid gas canister]; or SCBAE (any appropriate escape-type, self-contained breathing apparatus).
Exposure limits (TWA unless otherwise noted): ACGIH TLV 5 ppm (20 mg/m^3); OSHA PEL 100 ppm (410 mg/m^3); NIOSH REL 1 ppm (4 mg/m^3).
Toxicity by ingestion: Grade 2; oral LD$_{50}$ = 2590 mg/kg (rat).
Long-term health effects: Peripheral neuropathy in experimental animals and man (disease of motor and/or sensory nerves).
Vapor (gas) irritant characteristics: Eye and respiratory tract irritant.
Liquid or solid irritant characteristics: Eye and skin irritant.
Odor threshold: 0.065–0.09 ppm.
IDLH value: 1600 ppm.

FIRE DATA
Flash point: 83°F/28°C (oc); 77°F/25°C (cc).
Flammable limits in air: LEL: 1.3%; UEL: 8.0%.
Fire extinguishing agents not to be used: Water may be ineffective.
Autoignition temperature: 795°F/423°C/696°K.
Burning rate: 4.8 mm/min

CHEMICAL REACTIVITY
Binary reactants: Strong oxydizers. Attacks some metals and plastic materials.
Reactivity group: 18
Compatibility class: Ketone

ENVIRONMENTAL DATA
Food chain concentration potential: Log P_{ow} = 1.4. Unlikely to accumulate.
Water pollution: Effect of low concentrations on aquatic life is unknown. May be dangerous if it enters nearby water intakes; notify operators. Notify local health and wildlife officials. **Response to discharge:** Issue warning-air contaminant, water contaminant,

high flammability. Restrict access. Mechanical containment. Should be removed. Chemical and physical treatment.

SHIPPING INFORMATION
Grades of purity: Commercial, 95%; Pure, 99%; **Storage temperature:** Ambient; **Inert atmosphere:** None; **Venting:** Open (flame arrester); **Stability during transport:** Stable.

PHYSICAL AND CHEMICAL PROPERTIES
Physical state @ 59°F/15°C and 1 atm: Liquid.
Molecular weight: 100.16
Boiling point @ 1 atm: 261°F/127°C/400°K.
Melting/Freezing point: –70.4°F/–56.9°C/216.3°K.
Specific gravity (water = 1): 0.812 @ 68°F/20°C (liquid).
Liquid surface tension: 25.49 dynes/cm = 0.02549 N/m @ 68°F/20°C.
Liquid water interfacial tension: 9.73 dynes/cm = 0.00973 N/m @ 68°F/20°C.
Relative vapor density (air = 1): 3.5
Latent heat of vaporization: 148 Btu/lb = 82 cal/g = 3.4×10^5 J/kg.
Heat of combustion: $-16,100$ Btu/lb = -8940 cal/g = -374×10^5 J/kg.
Vapor pressure: 3 mm.

METHYL BUTYRATE   REC. M:3400

SYNONYMS: BUTANOIC ACID, METHYL ESTER; BUTIRATO de METILO (Spanish); BUTYRIC ACID, METHYL ESTER; METHYL, *n*-BUTYRATE; METHYL-*n*-BUTANOATE

IDENTIFICATION
CAS Number: 623-42-7
Formula: $C_5H_{10}O_2$; $C_4H_7OOCH_3$
DOT ID Number: UN 1237; DOT Guide Number: 129
Proper Shipping Name: Methyl butyrate

DESCRIPTION: Clear, colorless liquid. Apple-like odor. Floats on water surface; slightly soluble.

Highly flammable • Containers may BLEVE when exposed to fire • Vapors may form explosive mixture with air • Vapors are heavier than air and will collect and stay in low areas • Vapors may travel long distances to ignition sources and flashback • Vapors in confined areas (e.g., tanks, sewers, buildings) may explode when exposed to fire • Irritating to the skin, eyes, and respiratory tract • Toxic products of combustion may include carbon monoxide.

Hazard Classification (based on NFPA-704 Rating System)
Health Hazards (Blue): 2; Flammability (Red): 3; Reactivity (Yellow): 0

EMERGENCY RESPONSE: See Appendix A (129)
Evacuation:
Public safety: Isolate spill area for at least 50 to 100 meters (160 to 330 feet) in all directions.
Spill: Large spill–Consider initial downwind evacuation for at least 300 meters (1000 feet).
Fire: Isolate for 800 meters (½ mile) in all directions, especially if tank, rail car, or tank truck is involved in fire.

EXPOSURE
Short-term effects: *SEEK MEDICAL ATTENTION.* **Vapor:** Irritating to the eyes, nose, and throat. May be harmful if inhaled or absorbed through the skin. Remove to fresh air. *IF BREATHING HAS STOPPED,* give artificial respiration. IF breathing is difficult, administer oxygen. **Liquid:** Irritating to the eyes and skin. *MAY BE HARMFUL IF SWALLOWED OR ABSORBED THROUGH THE SKIN. IF IN EYES:* immediately flush with plenty of water for at least 15 minutes. Remove contaminated clothing and shoes. Wash affected areas with soap and water. *IF SWALLOWED:* Do nothing except keep victim warm. **Do not induce vomiting**.

HEALTH HAZARDS
Personal protective equipment (PPE): B-Level PPE. Approved respirator, chemical safety goggles, chemical-resistant gloves, other protective clothing.
Toxicity by ingestion: Grade 2: LD_{50} = 3.38 g/kg (rat).
Vapor (gas) irritant characteristics: Eye and respiratory tract irritant. Vapors cause a slight smarting of the eyes or respiratory system if present in high concentrations. The effect is temporary.
Liquid or solid irritant characteristics: Eye and skin irritant. If spilled on clothing and allowed to remain, may cause smarting and reddening of the skin.

FIRE DATA
Flash point: 57°F/14°C (cc).
Flammable limits in air: LEL: 0.9; UEL 3.5%.

CHEMICAL REACTIVITY
Binary reactants: Strong acids, oxidizers.
Neutralizing agents for acids and caustics: Dry lime, soda ash
Reactivity group: 34
Compatibility class: Esters.

ENVIRONMENTAL DATA
Water pollution: Effects of low concentrations on aquatic life are not known. May be dangerous if it enters nearby water intakes; notify operators. Notify local health and wildlife officials. **Response to discharge:** Issue warning–high flammability. Mechanical containment. Should be removed. Chemical and physical treatment.

SHIPPING INFORMATION
Grades of purity: 99%; **Inert atmosphere:** None; **Venting:** None; **Stability during transport:** Stable.

NAS HAZARD CLASSIFICATION FOR BULK WATER TRANSPORTATION
FIRE: 3
HEALTH: Vapor irritant: 2; Liquid or solid irritant: 1; Poisons: 1
WATER POLLUTION: Human toxicity: 2; Aquatic toxicity:–; Aesthetic effect: 1
REACTIVITY: Other chemicals: 1; Water: 0; Self-reaction: 0

PHYSICAL AND CHEMICAL PROPERTIES
Physical state @ 59°F/15°C and 1 atm: Liquid.
Molecular weight: 102.13
Boiling point @ 1 atm: 216°F/102.3°C/376°K.
Melting/Freezing point: –121°F/–84.8°C/188°K.
Critical temperature: 538.34°F/281.3°C/554.3°K.
Critical pressure: 34.2 atm.
Specific gravity (water = 1): 0.8984 @ 68°F/20°C.
Relative vapor density (air = 1): 3.53
Latent heat of vaporization: 143.6 Btu/lb = 79.8 cal/g = 3.3×10^5 J/kg.
Heat of combustion: -12209 Btu/lb = -6783 cal/g = -284×10^5 J/kg.
Vapor pressure: (Reid) 1.2 psia.

METHYL CHLORIDE   REC. M:3450

SYNONYMS: ARTIC; CHLOR-METHAN (German); CHLOROMETHANE; CHLORURE de METHYLE (French); CLORURO de METILO (Spanish); EEC No. 602-001-00-7; METHYLCHLORID (German); MONOCHLOROMETHANE; METHANE, CHLORO-; RCRA No. U045

IDENTIFICATION
CAS Number: 74-87-3
Formula: CH_3Cl
DOT ID Number: UN 1063; DOT Guide Number: 115
Proper Shipping Name: Methyl chloride
Reportable Quantity (RQ): 100 lb/45.4 kg

DESCRIPTION: Colorless gas. Shipped and stored as a liquefied compressed gas. Odorless or faint sweet odor, which may not be noticeably at dangerous concentrations. Floats and boils on the surface of water; slightly soluble; slowly hydrolyzed forming hydrochloric acid. Flammable, visible vapor cloud is formed. Becomes liquid below –11°F/–24°C.

Extremely flammable • Forms explosive mixture with air • Reacts explosively with many materials • Gas from liquefied gas are initially heavier than air and spread along ground • Gas may travel to source of ignition and flashback • Ruptured or venting cylinders may rocket through buildings and/or travel a considerable distance •Irritating to the skin, eyes, and respiratory tract • Vapors may cause dizziness or asphyxiation without warning • Contact with liquid may cause frostbite.

Hazard Classification (based on NFPA-704 Rating System)
Health Hazards (Blue): 1; Flammability (Red): 4; Reactivity (Yellow): 0

EMERGENCY RESPONSE: See Appendix A (115)
Evacuation:
Public safety: Isolate spill area for at least 50 to 100 meters (160 to 330 feet) in all directions.
Spill: Consider initial downwind evacuation for at least 800 meters (½ mile).
Fire: Isolate for 1600 meters (1 mile) in all directions, especially if tank, rail car, or tank truck is involved in fire.

EXPOSURE
Short-term effects: *SEEK MEDICAL ATTENTION.* **Vapor:** Not irritating to eyes, nose or throat. *IF INHALED*, will, will cause nausea, vomiting, headache, difficult breathing, or loss of consciousness. Move to fresh air. *IF BREATHING HAS STOPPED,* give artificial respiration. IF breathing is difficult, administer oxygen. **Liquid:** *HARMFUL IF ABSORBED THROUGH THE SKIN.* Will cause frostbite. Flush affected areas with plenty of water. *DO NOT RUB AFFECTED AREAS.*

HEALTH HAZARDS
Personal protective equipment (PPE): B-Level PPE. OSHA Table Z-1-A air contaminant. OSHA Table Z-2 air contaminant. Approved canister mask; leather or vinyl gloves; goggles or face shield. Wear thermal protective clothing.
Recommendations for respirator selection: NIOSH *At any concentrations above the NIOSH REL, or where there is no REL, at any detectable concentration:* SCBAF:PD,PP (any self-contained breathing apparatus that has a full facepiece and is operated in a pressure-demand or other positive-pressure mode); or SAF:PD,PP:ASCBA (any supplied-air respirator that has a full facepiece and is operated in a pressure-demand or other positive-pressure mode in combination with an auxiliary self-contained breathing apparatus operated in a pressure-demand or other positive pressure mode). *ESCAPE:* SCBAE (any appropriate escape-type, self-contained breathing apparatus).
Exposure limits (TWA unless otherwise noted): ACGIH TLV 50 ppm (103 mg/m^3); OSHA PEL 100 ppm; ceiling 200 ppm; 5 minute maximum peak in any 3 hours 300 ppm; NIOSH REL potental human carcinogen; reduce exposure to lowest feasible level; skin contact contributes significantly in overall exposure.
Short-term exposure limits (15-minute TWA): ACGIH TLV 100 ppm (207 mg/m^3).
Long-term health effects: IARC rating 3; inadequate human and animal evidence. Liver and kidney damage.
Vapor (gas) irritant characteristics: Vapors are nonirritating to the eyes and throat.
Liquid or solid irritant characteristics: No appreciable hazard. Practically harmless to the skin; however, liquid evaporates quickly, and may cause frostbite.
Odor threshold: More than 10 ppm.
IDLH value: Potential human carcinogen; 2000 ppm.

FIRE DATA
Flash point: –50°F/–46°C (cc).
Flammable limits in air: LEL: 8.1%; UEL: 17.4%.
may increase the danger by permitting the accumulation of an explosive mixture.
Behavior in fire: Containers may explode.
Autoignition temperature: 1170°F/632°C/905°K.
Electrical hazard: Class I, Group D.
Burning rate: 2.2 mm/min
Stoichiometric air-to-fuel ratio: 4.078 (estimate).

CHEMICAL REACTIVITY
Reactivity with water: Hydrolyzes to form hydrochloric acid.
Binary reactants: Reacts with fluorine, acetylene, zinc, aluminum, magnesium, zinc and their alloys. When in contact with aluminum, the product which forms may ignite spontaneously in air.
Neutralizing agents for acids and caustics: Flood with water and rinse with sodium bicarbonate or lime solution.
Reactivity group: 36
Compatibility class: Halogenated hydrocarbons

ENVIRONMENTAL DATA
Food chain concentration potential: Slowly hydrolyzes to hydrochloric acid. May volatilize rapidly and disperse.
Water pollution: Not harmful to aquatic life. **Response to discharge:** Issue warning–high flammability, air contaminant. Restrict access. Evacuate area.

SHIPPING INFORMATION
Grades of purity: Technical grade; ARTIC® refrigerant grade; **Storage temperature:** Ambient; **Inert atmosphere:** None; **Venting:** Safety relief; **Stability during transport:** Stable.

NAS HAZARD CLASSIFICATION FOR BULK WATER TRANSPORTATION
FIRE: 4
HEALTH: Vapor irritant: 0; Liquid or solid irritant: 0; Poisons: 2
WATER POLLUTION: Human toxicity: 0; Aquatic toxicity: 1; Aesthetic effect: 0
REACTIVITY: Other chemicals: 1; Water: 0; Self-reaction: 0

PHYSICAL AND CHEMICAL PROPERTIES
Physical state @ 59°F/15°C and 1 atm: Gas.

Molecular weight: 50.49
Boiling point @ 1 atm: –11.6°F/–24.2°C/249°K.
Melting/Freezing point: –143.9°F/97.7°C/175.5°K.
Critical temperature: 290.5°F/143.6°C/416.8°K.
Critical pressure: 969 psia = 65.9 atm = 6.68 MN/m^2.
Specific gravity (water = 1): 0.997 at –24°C (liquid).
Liquid surface tension: 16.2 dynes/cm = 0.0162 N/m @ 68°F/20°C.
Liquid water interfacial tension: (estimate) 50 dynes/cm = 0.05 N/m at –24°C.
Relative vapor density (air = 1): 1.78
Ratio of specific heats of vapor (gas): 1.259
Latent heat of vaporization: 182.3 Btu/lb = 101.3 cal/g = 4.241 x 10^5 J/kg.
Heat of combustion: –5290 Btu/lb = –2939 cal/g = –123.1 x 10^5 J/kg.
Vapor pressure: (Reid) 116.7 psia; 3800 mm.

METHYL CHLOROACETATE   REC. M:3500

SYNONYMS: CLOROACETATO de METILO (Spanish); CHLOROACETIC ACID, METHYL ESTER; METHYL MONOCHLOROACETATE; MONOCHLOROACETIC ACID, METHYL ESTER

IDENTIFICATION
CAS Number: 96-34-4
Formula: $C_3H_5ClO_2$; $ClCH_2CO_2CH_3$
DOT ID Number: UN 2295; DOT Guide Number: 155
Proper Shipping Name: Methyl chloroacetate

DESCRIPTION: Colorless liquid. Sweet, pungent odor. Sinks in water; insoluble. If water is hot acetic acid may be formed from decomposition. Freezes at –28°F/–33°C.

Highly flammable • Corrosive skin, eyes, and respiratory tract; produces tearing; contact with skin and eyes can cause burns, impaired vision, and blindness; inhalation symptoms may be delayed • Containers may BLEVE when exposed to fire •Vapors may form explosive mixture with air • Vapors are heavier than air and will collect and stay in low areas • Vapors may travel long distances to ignition sources and flashback • Vapors in confined areas (e.g., tanks, sewers, buildings) may explode when exposed to fire • Toxic products of combustion may include hydrogen chloride.

Hazard Classification (based on NFPA-704 Rating System)
Health Hazards (Blue): 2; Flammability (Red): 2; Reactivity (Yellow): 1

EMERGENCY RESPONSE: See Appendix A (155)
Evacuation:
Public safety: Isolate the area of spill or leak for at least 50 to 100 meters (160 to 330 feet) in all directions.
Spill: Increase, in the downwind direction, as necessary, the distance shown under "Public Safety."
Fire: If tank, rail car, or tank truck is involved in fire, isolate for at least 800 meters (½ mile) in all directions; also, consider initial evacuation for 800 meters (½ mile) in all directions.

EXPOSURE
Short-term effects: *SEEK MEDICAL ATTENTION*. **Vapor:** *HARMFUL IF INHALED OR ABSORBED THROUGH THE SKIN*. Irritating to the eyes, nose, and throat. May cause lung edema; physical exertion will aggravate this condition. Remove victim to fresh air. *IF BREATHING HAS STOPPED,* give artificial respiration; *avoid mouth-to-mouth resuscitation; use bag/mask apparatus.* IF breathing is difficult, administer oxygen. **Liquid:** Harmful if swallowed or absorbed through the skin. Corrosive to skin, eyes, nose, throat, and upper respiratory tract. Remove contaminated clothing and shoes. Flush affected areas with water. *IF IN EYES,* hold eyelids open, flush with running water for at least 15 minutes. *IF SWALLOWED*: Do nothing except keep victim warm. **Do not induce vomiting.**
Note to physician or authorized medical personnel: Medical observation is recommended for 24 to 48 hours after breathing overexposure, as pulmonary edema may be delayed. As first aid for pulmonary edema, consider administering a corticosteroid spray. Cigarette smoking may exacerbate pulmonary injury and should be discouraged for at least 72 hours following exposure.

HEALTH HAZARDS
Personal protective equipment (PPE): B-Level PPE. Approved respirator, chemical safety goggles, chemical-resistant gloves, other protective clothing.
Protective materials with good to excellent resistance: butyl rubber, neoprene, polyethylene, Viton®, Saranex®.
Toxicity by ingestion: Grade 3: LD_{50} = 240 mg/kg (mouse).
Vapor (gas) irritant characteristics: Vapors cause severe irritation of eyes and throat and can cause eye and lung injury. They cannot be tolerated even at low concentrations.
Liquid or solid irritant characteristics: Severe eye and skin irritant. Causes second- and third-degree burns on short contact and is very injurious to the eyes.

FIRE DATA
Flash point: 135°F/57°C (cc).
Flammable limits in air: LEL: 7.5; UEL: 18.5%.
Fire extinguishing agents not to be used: Water may be ineffective against fire.
Autoignition temperature: 869°F

CHEMICAL REACTIVITY
Reactivity group: 34
Compatibility class: Esters.

ENVIRONMENTAL DATA
Water pollution: Harmful to aquatic life. May be dangerous if it enters nearby water intakes; notify operators. Notify local health and wildlife officials. **Response to discharge:** Issue warning–corrosive. Should be removed. Chemical and physical treatment.

SHIPPING INFORMATION
Grades of purity: 99+%; **Inert atmosphere:** None; **Venting:** None; **Stability during transport:** Stable.

NAS HAZARD CLASSIFICATION FOR BULK WATER TRANSPORTATION
FIRE: 2
HEALTH: Vapor irritant: 3; Liquid or solid irritant: 2; Poisons: 2
WATER POLLUTION: Human toxicity: 3; Aquatic toxicity:–; Aesthetic effect: 1
REACTIVITY: Other chemicals: 1; Water: 0; Self-reaction: 0

PHYSICAL AND CHEMICAL PROPERTIES
Physical state @ 59°F/15°C and 1 atm: Liquid.
Molecular weight: 108.52
Boiling point @ 1 atm: 266°F/129.8°C/403°K.

Melting/Freezing point: −26°F/−32.1°C/241°K.
Specific gravity (water = 1): 1.2337 @ 68°F/20°C.
Relative vapor density (air = 1): 3.8
Vapor pressure: (Reid) 0.32 psia.

METHYL CHLOROFORMATE   REC. M:3550

SYNONYMS: CHLOROCARBONIC ACID, METHYL ESTER; CHLOROFORMIC ACID, METHYL ESTER; CLOROFORMIATO de METILO (Spanish); EEC No. 607-019-00-9; METHYL CHLOROCARBONATE; MCF; METHOXYCARBONYL CHLORIDE; RCRA No. U156

IDENTIFICATION
CAS Number: 79-22-1
Formula: $C_2H_3ClO_2$; $ClCOOCH_3$
DOT ID Number: UN 1238; DOT Guide Number: 155
Proper Shipping Name: Methyl chloroformate
Reportable Quantity (RQ): **(CERCLA)** 1000 lb/454 kg

DESCRIPTION: Liquid. Colorless to light yellow. Unpleasant, acrid odor. Sinks and reacts in water; slowly forms methanol, CO_2, and hydrochloric acid. Flammable, irritating vapor is produced.

Corrosive • Highly flammable • Reacts with moisture in air forming corrosive hydrochloric acid • Breathing the vapor can kill you; skin or eye contact causes severe burns, impaired vision, or blindness; effects of contact or inhalation may be delayed • Firefighting gear (including SCBA) does not provide adequate protection. If exposure occurs, remove and isolate gear immediately and thoroughly decontaminate personnel • Containers may BLEVE when exposed to fire •Vapors may form explosive mixture with air • Vapors are heavier than air and will collect and stay in low areas • Vapors may travel long distances to ignition sources and flashback • Vapors in confined areas (e.g., tanks, sewers, buildings) may explode when exposed to fire • Toxic products of combustion may include hydrogen chloride and phosgene.

Hazard Classification (based on NFPA-704 Rating System)
Health Hazards (Blue): 1; Flammability (Red): 3; Reactivity (Yellow): 1

EMERGENCY RESPONSE: See Appendix A (155)
Evacuation:
Public safety: Isolate the area of spill or leak for at least 50 to 100 meters (160 to 330 feet) in all directions.
Spill: Small spill–First: Isolate in all directions 30 meters (100 feet); Then: Protect persons downwind, DAY: 0.3 km (0.2 mile); NIGHT: 1.1 km (0.7 mile). Large spill–First: Isolate in all directions 155 meters (500 feet); Then: Protect persons downwind, DAY: 1.6 km (1.0 mile); NIGHT: 3.4 km (2.1 miles).
Fire: If tank, rail car, or tank truck is involved in fire, isolate for at least 800 meters (½ mile) in all directions; also, consider initial evacuation for 800 meters (½ mile) in all directions.

EXPOSURE
Short-term effects: *SEEK MEDICAL ATTENTION.* **Vapor:** Irritating to eyes, nose, and throat. *IF INHALED*, will, will cause difficult breathing. Move victim to fresh air. *IF BREATHING HAS STOPPED*, give artificial respiration; *avoid mouth-to-mouth resuscitation; use bag/mask apparatus.* IF breathing is difficult, administer oxygen. **Liquid:** *POISONOUS IF SWALLOWED.* Will burn skin and eyes. Remove contaminated clothing and shoes. Flush affected areas with plenty of water. *IF IN EYES*, hold eyelids open and flush with plenty of water. *IF SWALLOWED* and victim is *CONSCIOUS AND ABLE TO SWALLOW*, have victim drink 4 to 8 ounces of water. **Do NOT induce vomiting.**
Note to physician or authorized medical personnel: Medical observation is recommended for 24 to 48 hours after breathing overexposure, as pulmonary edema may be delayed. As first aid for pulmonary edema, consider administering a corticosteroid spray. Cigarette smoking may exacerbate pulmonary injury and should be discouraged for at least 72 hours following exposure.

HEALTH HAZARDS
Personal protective equipment (PPE): B-Level PPE. Acid- or organic-canister mask or self-contained breathing apparatus; goggles or face shield; plastic gloves.
Toxicity by ingestion: Grade 4; oral LD_{50} less than 50 mg/kg (rat).
Vapor (gas) irritant characteristics: Eye and respiratory tract irritant. Can lead to difficult breathing and lung edema.
Liquid or solid irritant characteristics: Severe eye and skin irritant.

FIRE DATA
Flash point: 76°F/24°C (oc); 73°F/23°C (cc).
Burning rate: 2.0 mm/min

CHEMICAL REACTIVITY
Reactivity with water: Reacts slowly, producing hydrochloric acid. Reaction can be hazardous if water is hot.
Binary reactants: Corrodes rubber and metals in presence of moisture.
Neutralizing agents for acids and caustics: Flood with water, rinse with sodium bicarbonate or lime solution.

ENVIRONMENTAL DATA
Food chain concentration potential: Hydrolyzes quickly in water; accumulation unlikely to be significant.
Water pollution: Effect of low concentrations on aquatic life is unknown. May be dangerous if it enters nearby water intakes; notify operators. Notify local health and wildlife officials.
Response to discharge: Issue warning–corrosive, poison, high flammability. Restrict access. Disperse and flush.

SHIPPING INFORMATION
Grades of purity: 97+%; **Storage temperature:** Ambient; **Inert atmosphere:** None; **Venting:** Pressure-vacuum; **Stability during transport:** Stable.

PHYSICAL AND CHEMICAL PROPERTIES
Physical state @ 59°F/15°C and 1 atm: Liquid.
Molecular weight: 94.5
Boiling point @ 1 atm: 160°F/71°C/344°K.
Melting/Freezing point: Less than −114°F/−81°C/192°K.
Specific gravity (water = 1): 1.22 @ 68°F/20°C (liquid).
Liquid surface tension: (estimate) 26 dynes/cm = 0.026 N/m @ 68°F/20°C.
Relative vapor density (air = 1): 3.25
Ratio of specific heats of vapor (gas): 1.1544
Latent heat of vaporization: (estimate) 153 Btu/lb = 85 cal/g = 3.6×10^5 J/kg.
Heat of combustion: −4690 Btu/lb = −2600 cal/g = -109×10^5 J/kg.

METHYL CYCLOHEXANE REC. M:3600

SYNONYMS: CYCLOHEXANE, METHYL-; CYCLOHEXYLMETHANE; EEC No. 601-018-00-7; HEPTANAPHTHENE; HEXAHYDROTOLUENE; METILCICLOHEXANO (Spanish); SEXTONE B; TOLUENE HEXAHYDRIDE

IDENTIFICATION
CAS Number: 108-87-2
Formula: C_7H_{14}; $C_6H_{11}CH_3$
DOT ID Number: UN 2296; DOT Guide Number: 128
Proper Shipping Name: Methyl cyclohexane

DESCRIPTION: Colorless liquid. Faint benzene-like odor. floats on water surface; insoluble.

Highly flammable • Containers may BLEVE when exposed to fire • Vapors may form explosive mixture with air • Vapors are heavier than air and will collect and stay in low areas • Vapors may travel long distances to ignition sources and flashback • Vapors in confined areas (e.g., tanks, sewers, buildings) may explode when exposed to fire • Irritating to the skin, eyes, and respiratory tract • Toxic products of combustion may include carbon monoxide.

Hazard Classification (based on NFPA-704 Rating System)
Health Hazards (Blue): 2; Flammability (Red): 3; Reactivity (Yellow): 0

EMERGENCY RESPONSE: See Appendix A (128)
Evacuation:
Public safety: Isolate spill area for at least 25 to 50 meters (80 to 160 feet) in all directions.
Spill: Large spill–Consider initial downwind evacuation for at least 300 meters (1000 feet).
Fire: Isolate for 800 meters (½ mile) in all directions, especially if tank, rail car, or tank truck is involved in fire.

EXPOSURE
Short-term effects: *SEEK MEDICAL ATTENTION*. **Vapor:** Irritating to eyes, nose, and throat. *IF INHALED*, will, will cause dizziness, lightheadedness, drowsiness, nausea, vomiting, or loss of consciousness. Move to fresh air. *IF BREATHING HAS STOPPED*, give artificial respiration. IF breathing is difficult, administer oxygen. **Liquid:** Irritating to skin and eyes. Harmful if swallowed. Remove contaminated clothing and shoes. Flush affected areas with plenty of water. *IF IN EYES*, hold eyelids open and flush with plenty of water.

HEALTH HAZARDS
Personal protective equipment (PPE): B-Level PPE. OSHA Table Z-1-A air contaminant.
Recommendations for respirator selection: NIOSH/OSHA
1200 ppm: SA (any supplied-air respirator); or SCBAF (any self-contained breathing apparatus with a full facepiece). *EMERGENCY OR PLANNED ENTRY INTO UNKNOWN CONCENTRATIONS OR IDLH CONDITIONS*: SCBAF:PD,PP (any self-contained breathing apparatus that has a full facepiece and is operated in a pressure-demand or other positive-pressure mode); or SAF:PD,PP:ASCBA (any supplied-air respirator that has a full facepiece and is operated in a pressure-demand or other positive-pressure mode in combination with an auxiliary self-contained breathing apparatus operated in a pressure-demand or other positive pressure mode). *ESCAPE*: GMFOV [any air-purifying, full-facepiece respirator (gas mask) with a chin-style, front- or back-mounted acid gas canister]; or SCBAE (any appropriate escape-type, self-contained breathing apparatus).
Exposure limits (TWA unless otherwise noted): ACGIH TLV 400 ppm (1610 mg/m^3); OSHS PEL 500 ppm (2000 mg/m^3); NIOSH 400 ppm (1600 mg/m^3).
Toxicity by ingestion: Grade 2: LD_{50} = 2.25 g/kg mouse
Vapor (gas) irritant characteristics: Eye and respiratory tract irritant. Vapors cause smarting of the eyes or respiratory system if present in high concentrations. The effect is temporary.
Liquid or solid irritant characteristics: Eye and skin irritant. If spilled on clothing and allowed to remain, may cause smarting and reddening of skin.
Odor threshold: 550–650 ppm.
IDLH value: 1200 ppm.

FIRE DATA
Flash point: 25°F/–4°C (cc).
Flammable limits in air: LEL 1.2%; UEL 6.7%.
Fire extinguishing agents not to be used: Water may be ineffective.
Autoignition temperature: 482°F/250°C/523°K.
Electrical hazard: Due to low electric conductivity, this substance may generate electrostatic charges as a result of agitation and flow.

CHEMICAL REACTIVITY
Binary reactants: Strong oxidizers.
Reactivity group: 31
Compatibility class: Paraffins.

ENVIRONMENTAL DATA
Food chain concentration potential: Negative; unlikely to accumulate.
Water pollution: Effects of low concentrations on aquatic life is unknown. Fouling to shoreline. May be dangerous if it enters nearby water intakes; notify operators. Notify local health and pollution control officials. **Response to discharge:** Issue warning-flammable. Restrict access. Evaluate area. Mechanical containment. Should be removed. Chemical and physical treatment.

SHIPPING INFORMATION
Grades of purity: 99+%; **Storage temperature:** Ambient; **Stability during transport:** Stable.

PHYSICAL AND CHEMICAL PROPERTIES
Physical state @ 59°F/15°C and 1 atm: Liquid.
Molecular weight: 98.21
Boiling point @ 1 atm: 213.8°F/101°C/374.2°K.
Melting/Freezing point: –194.8°F/–126°C/147.2°K.
Specific gravity (water = 1): 0.770
Relative vapor density (air = 1): 3.4
Vapor pressure: (Reid) 1.611 psia; 37 mm.

2-METHYLCYCLOHEXANOL REC. M:3650

SYNONYMS: HEXAHYDROCRESOL; HEXAHYDROMETHYLPHENOL; METHYLCYCLOHEXANOL; METHYLCYCLOHEXANE; METILCICLOHEXANOL (Spanish)

IDENTIFICATION
CAS Number: 25639-42-3
Formula: $C_7H_{14}O$; $C_6H_{10}(OH)(CH_3)$
DOT ID Number: UN 2617; DOT Guide Number: 129
Proper Shipping Name: Methyl cyclohexanols, flammable

736 *o*-Methylcyclohexanone

DESCRIPTION: Straw colored liquid. Weak methanol-like odor. Floats on water surface; slightly soluble.

Flammable • Containers may BLEVE when exposed to fire • Vapors may form explosive mixture with air • Vapors are heavier than air and will collect and stay in low areas • Vapors may travel long distances to ignition sources and flashback • Vapors in confined areas (e.g., tanks, sewers, buildings) may explode when exposed to fire • Irritating to the skin, eyes, and respiratory tract • Toxic products of combustion may include carbon monoxide.

Hazard Classification (based on NFPA-704 Rating System)
Health Hazards (Blue): -; Flammability (Red): 2; Reactivity (Yellow): 0

EMERGENCY RESPONSE: See Appendix A (129)
Evacuation:
Public safety: Isolate spill area for at least 50 to 100 meters (160 to 330 feet) in all directions.
Spill: Large spill–Consider initial downwind evacuation for at least 300 meters (1000 feet).
Fire: Isolate for 800 meters (½ mile) in all directions, especially if tank, rail car, or tank truck is involved in fire.

EXPOSURE
Short-term effects: *SEEK MEDICAL ATTENTION.* **Vapor:** Move victim to fresh air. *IF BREATHING HAS STOPPED,* give artificial respiration. If breathing is difficult, administer oxygen. **Liquid:** Remove contaminated clothing and shoes. Wash affected areas with soap and water. IF IN EYES, hold eyelids open and flush with plenty of water. IF SWALLOWED and victim is *CONSCIOUS AND ABLE TO SWALLOW,* give large quantity of water. After swallowing water, induce vomiting.

HEALTH HAZARDS
Personal protective equipment (PPE): B-Level PPE. OSHA Table Z-1-A air contaminant. Impervious clothing and gloves should be used to prevent skin contact. Where splashing is possible wear full face shield or chemical safety goggles. Use approved respirator to protect against vapors.
Recommendations for respirator selection: NIOSH/OSHA
500 ppm: SA* (any supplied-air respirator); or SCBAF (any self-contained breathing apparatus with a full facepiece). *EMERGENCY OR PLANNED ENTRY INTO UNKNOWN CONCENTRATIONS OR IDLH CONDITIONS:* SCBAF:PD,PP (any self-contained breathing apparatus that has a full facepiece and is operated in a pressure-demand or other positive-pressure mode); or SAF:PD,PP:ASCBA (any supplied-air respirator that has a full facepiece and is operated in a pressure-demand or other positive-pressure mode in combination with an auxiliary self-contained breathing apparatus operated in a pressure-demand or other positive pressure mode). *ESCAPE:* GMFOV [any air-purifying, full-facepiece respirator (gas mask) with a chin-style, front- or back-mounted acid gas canister]; or SCBAE (any appropriate escape-type, self-contained breathing apparatus).
Note: Substance reported to cause eye irritation or damage; may require eye protection.
Exposure limits (TWA unless otherwise noted): ACGIH TLV 50 ppm (234 mg/m^3); OSHA PEL 100 ppm (470 mg/m^3); NIOSH REL 50 ppm (235 mg/m^3).
Toxicity by ingestion: Grade 2; oral rat LD$_{50}$ = 2.0 g/kg.
Long-term health effects: Repeated or prolonged overexposure may cause a skin rash. In animals it has caused drowsiness, unconsciousness, and mild liver and kidney damage.

Vapor (gas) irritant characteristics: Vapors cause smarting of the eyes or respiratory system if present in high concentrations. The effect is temporary.
Liquid or solid irritant characteristics: Eye and skin irritant. If spilled on clothing and allowed to remain, may cause smarting and reddening of skin.
Odor threshold: 500 ppm.
IDLH value: 500 ppm.

FIRE DATA
Flash point: 149°F/70°C (cc).
Fire extinguishing agents not to be used: Water.
Autoignition temperature: 565°F/296°C/569°K.
Stoichiometric air-to-fuel ratio: 47.6

CHEMICAL REACTIVITY
Binary reactants: Contact with strong oxidizers may cause fires and explosions.

ENVIRONMENTAL DATA
Water pollution: Effect of low concentrations on aquatic life is unknown. May be dangerous if it enters nearby water intakes; notify operators. Notify local health and wildlife officials.
Response to discharge: Should be removed.

SHIPPING INFORMATION
Grades of purity: Technical grades; **Storage temperature:** Ambient; **Inert atmosphere:** None **Stability during transport:** Stable.

PHYSICAL AND CHEMICAL PROPERTIES
Physical state @ 59°F/15°C and 1 atm: Liquid.
Molecular weight: 114.2
Boiling point @ 1 atm: 325.4–330.8°F/163–166°C/436–439°K.
Melting/Freezing point: –58°F/–50°C/223°K.
Specific gravity (water = 1): 0.92
Relative vapor density (air = 1): 3.9
Vapor pressure: 2 mm @ 86°F

o-**METHYLCYCLOHEXANONE**   REC. M:3700

SYNONYMS: 2-METHYLCYCLOHEXANONE; 1-METHYLCYCLOHEXAN-2-ONE; METILCICLOHEXANONA (Spanish)

IDENTIFICATION
CAS Number: 583-60-8; may also apply to 591-24-2 (*m*-isomer); 589-92-4 (*p*-isomer); 1331-22-2 (mixed isomers)
Formula: $C_7H_{12}O$; $C_6H_9(O)(CH_3)$
DOT ID Number: UN 2297; DOT Guide Number: 127
Proper Shipping Name: Methyl cyclohexanone

DESCRIPTION: Colorless liquid. Weak peppermint odor. Floats on the surface of water.

Flammable • Containers may BLEVE when exposed to fire • Vapors may form explosive mixture with air • Vapors are heavier than air and will collect and stay in low areas • Vapors may travel long distances to ignition sources and flashback • Vapors in confined areas (e.g., tanks, sewers, buildings) may explode when exposed to fire • Irritating to the skin, eyes, and respiratory tract • Toxic products of combustion may include carbon monoxide.

Hazard Classification (based on NFPA-704 Rating System)
Health Hazards (Blue):–; Flammability (Red): 2; Reactivity (Yellow): 0

EMERGENCY RESPONSE: See Appendix A (127)
Evacuation:
Public safety: Isolate spill area for at least 25 to 50 meters (80 to 160 feet) in all directions.
Spill: Large spill–Consider initial downwind evacuation for at least 300 meters (1000 feet).
Fire: Isolate for 800 meters (½ mile) in all directions, especially if tank, rail car, or tank truck is involved in fire.

EXPOSURE
Short-term effects: *SEEK MEDICAL ATTENTION.* **Vapor:** Move victim to fresh air. *IF BREATHING HAS STOPPED,* give artificial respiration. If breathing is difficult, administer oxygen. **Liquid:** *HARMFUL IF ABSORBED THROUGH THE SKIN.* Remove contaminated clothing and shoes. Wash affected areas with soap and water. IF IN EYES, hold eyelids open and flush with plenty of water. IF SWALLOWED and victim is *CONSCIOUS AND ABLE TO SWALLOW*, give large quantity of water. After swallowing water, induce vomiting.

HEALTH HAZARDS
Personal protective equipment (PPE): B-Level PPE. OSHA Table Z-1-A air contaminant. Impervious clothing and gloves should be used to prevent skin contact. Where splashing is possible wear full face shield or chemical safety goggles. Use approved respirator to protect against vapors.
Recommendations for respirator selection: NIOSH/OSHA
500 ppm: SA* (any supplied-air respirator). *600 ppm*: SA:CF* (any supplied-air respirator operated in a continuous-flow mode); or SCBAF (any self-contained breathing apparatus with a full facepiece); or SAF (any supplied-air respirator with a full facepiece). *EMERGENCY OR PLANNED ENTRY INTO UNKNOWN CONCENTRATIONS OR IDLH CONDITIONS:* SCBAF:PD,PP (any self-contained breathing apparatus that has a full facepiece and is operated in a pressure-demand or other positive-pressure mode); or SAF:PD,PP:ASCBA (any supplied-air respirator that has a full facepiece and is operated in a pressure-demand or other positive-pressure mode in combination with an auxiliary self-contained breathing apparatus operated in a pressure-demand or other positive pressure mode). *ESCAPE:* GMFOV [any air-purifying, full-facepiece respirator (gas mask) with a chin-style, front- or back-mounted acid gas canister]; or SCBAE (any appropriate escape-type, self-contained breathing apparatus).
**Note:* Substance reported to cause eye irritation or damage; may require eye protection.
Exposure limits (TWA unless otherwise noted): ACGIH TLV 50 ppm (229 mg/m^3); OSHA PEL 100 ppm (460 mg/m^3); NIOSH 50 ppm (230 mg/m^3); skin contact contributes significantly in overall exposure.
Short-term exposure limits (15-minute TWA): ACGIH STEL 75 ppm (334 mg/m^3); NIOSH STEL 75 ppm (345 mg/m^3); skin contact contributes significantly in overall exposure.
Long-term health effects: Repeated or prolonged overexposure may cause dermatitis. In animals it has caused drowsiness, skin irritation, tremors, narcosis, and death.
Vapor (gas) irritant characteristics: Vapors cause smarting of the eyes or respiratory system if present in high concentrations. The effect is temporary.
Liquid or solid irritant characteristics: Eye and skin irritant. If spilled on clothing and allowed to remain, may cause smarting and reddening of skin.

IDLH value: 600 ppm.

FIRE DATA
Flash point: 118°F/48°C (cc).
Fire extinguishing agents not to be used: Water.
Stoichiometric air-to-fuel ratio: 45.2

CHEMICAL REACTIVITY
Binary reactants: Contact with strong oxidizers may cause fires and explosions.
Reactivity group: 18
Compatibility class: Ketones

ENVIRONMENTAL DATA
Water pollution: Effects of low concentrations on aquatic life is unknown. May be dangerous if it enters nearby water intakes; notify operators. Notify local health and wildlife officials. **Response to discharge:** Should be removed.

SHIPPING INFORMATION
Grades of purity: Technical grades; **Storage temperature:** Ambient; **Inert atmosphere:** None; **Stability during transport:** Stable.

PHYSICAL AND CHEMICAL PROPERTIES
Physical state @ 59°F/15°C and 1 atm: Liquid.
Molecular weight: 112.2
Boiling point @ 1 atm: 325°F/163°C/436°K.
Melting/Freezing point: –6.8°F/–14°C/259°K.
Specific gravity (water = 1): 0.93
Relative vapor density (air = 1): 3.9
Vapor pressure: 1 mm.

METHYLCYCLOPENTADIENE DIMER   REC. M:3750

SYNONYMS: BIS(METHYLCYCLOPENTADIENE); METILCICLOPENTADIENO DIMERO (Spanish); 3*A*,4,7,7*A*-TETRAHYDRODIMETHYL-4,7-METHANOINDENE; 4,7-METHANOINDENE, 3*A*,4,7,7*A*-TETRAHYDRODIMETHYL

IDENTIFICATION
CAS Number: 26472-00-4
Formula: $C_{12}H_{16}$
DOT ID Number: UN 1993; DOT Guide Number: 128
Proper Shipping Name: Flammable liquid, n.o.s.

DESCRIPTION: Colorless liquid.

Highly flammable • Containers may BLEVE when exposed to fire • Vapors may form explosive mixture with air • Gas is slightly lighter than air, and will disperse slowly unless confined • Vapors may travel long distances to ignition sources and flashback • Vapors in confined areas (e.g., tanks, sewers, buildings) may explode when exposed to fire • Irritating to the skin, eyes, and respiratory tract • Combustion products include smoke and irritating vapors.

Hazard Classification (based on NFPA-704 Rating System)
Health Hazards (Blue): 1; Flammability (Red): 3; Reactivity (Yellow): 0

EMERGENCY RESPONSE: See Appendix A (128)
Evacuation:
Public safety: Isolate spill area for at least 25 to 50 meters (80 to 160 feet) in all directions.

Spill: Large spill–Consider initial downwind evacuation for at least 300 meters (1000 feet).
Fire: Isolate for 800 meters (½ mile) in all directions, especially if tank, rail car, or tank truck is involved in fire.

EXPOSURE
Short-term effects: *SEEK MEDICAL ATTENTION.* **Vapor:** If inhaled, will cause dizziness, difficult breathing, or loss of consciousness. Move to fresh air. *IF BREATHING HAS STOPPED,* give artificial respiration. IF breathing is difficult, administer oxygen. **Liquid:** Irritating to skin and eyes. Harmful if swallowed. Remove contaminated clothing and shoes. Flush affected areas with plenty of water. *IF IN EYES,* hold eyelids open and flush with plenty of water.

HEALTH HAZARDS
Personal protective equipment (PPE): B-Level PPE. Self-contained breathing apparatus, protective clothing, rubber boots and heavy rubber gloves.
Eyes or Inhalation: Vapor or mist is irritating to the eyes, mucous membrane, upper respiratory tract. Exposure can cause nausea, headache, and vomiting. May contain 0.5% benzene, a known carcinogen.
Toxicity by ingestion: Grade 1: LD_{50} = 7.7 g/kg (mice).
Long-term health effects: Damage to liver, kidney and lung. Carcinogenic.
Vapor (gas) irritant characteristics: Eye and respiratory tract irritant. Vapors are irritating such that personnel will not usually tolerate moderate or high concentrations.
Liquid or solid irritant characteristics: Causes smarting of skin and first-degree burn on short exposure; may cause second-degree burn on long exposure.

FIRE DATA
Flash point: 80°F/27°C (cc).
Flammable limits in air: LEL: 1.0%; UEL: 10%.
Fire extinguishing agents not to be used: Water may be ineffective.

CHEMICAL REACTIVITY
Binary reactants: Strong oxidizers, nitric acid may cause fire or explosion.
Reactivity group: 30
Compatibility class: Olefins

ENVIRONMENTAL DATA
Water pollution: Effects of low concentrations on aquatic life is unknown. Fouling to shoreline. May be dangerous if it enters nearby water intakes; notify operators. Notify local health and wildlife officials. **Response to discharge:** Issue warning-flammable liquid. Evacuate area. Mechanical containment. Should be removed. Chemical and physical treatment.

SHIPPING INFORMATION
Grades of purity: 95%; **Storage temperature:** Ambient; **Stability during transport:** Stable.

PHYSICAL AND CHEMICAL PROPERTIES
Physical state @ 59°F/15°C and 1 atm: Liquid.
Molecular weight: 160.26
Boiling point @ 1 atm: 392°F/200°C/473.2°K.
Melting/Freezing point: –59.8°F/–51°C/222.2°K.
Specific gravity (water = 1): 0.941
Relative vapor density (air = 1): 0.93

METHYLCYCLOPENTADIENYL MANGANESE TRICARBONYL   REC. M:3800

SYNONYMS: AK-33X; ANTIKNOCK-33; COMBUSTION IMPROVER-2; COMBUSTION IMPROVER C-12; MANGANESE CYCLOPENTADIENYL TRICARBONYL; MANGANESE TRICARBONYL METHYLCYCLOPENTADIENYL; METILCICLOPENTADIENILO-MANGANESO TRICARBONILO (Spanish); 2- MMT

IDENTIFICATION
CAS Number: 12108-13-3; 12079-65-1
Formula: $C_9H_7MnO_3$; $C_9H_7O_3Mn$; $C_5H_5MnO_3$
Reportable Quantity (RQ): **(EHS)** 1 lb/0.454 kg

DESCRIPTION: Pale yellow to dark orange liquid. Faint pleasant herb-like odor. Sinks in water; practically insoluble; reaction may produce flammable and toxic vapors. A solid below 36°F/2°C.

Poison! • Combustible • Carbonyls are potential explosion hazards • Breathing the vapor or swallowing the material can kill you; skin and eye contact causes irritation; mildly narcotic if absorbed • Firefighting gear (including SCBA) does not provide adequate protection. If exposure occurs, remove and isolate gear immediately and thoroughly decontaminate personnel • Containers may BLEVE when exposed to fire • Vapors are heavier than air and will collect and stay in low areas • Vapors in confined areas (e.g., tanks, sewers, buildings) may explode when exposed to fire • Toxic products of combustion may include carbon monoxide and manganese fume • Do not put yourself in danger by entering a contaminated area to rescue a victim.

Hazard Classification (based on NFPA-704 Rating System)
Health Hazards (Blue): 3; Flammability (Red): 1; Reactivity (Yellow): 1

EMERGENCY RESPONSE: See Appendix A (171)
Evacuation:
Public safety: Isolate the area of spill or leak for at least 10 to 25 meters (30 to 80 feet) in all directions.
Spill: Increase, in the downwind direction, as necessary, the distance shown under "Public Safety."
Fire: If any large container is involved in fire, isolate for at least 800 meters (½ mile) in all directions; also, consider initial evacuation for 800 meters (½ mile) in all directions.

EXPOSURE
Short-term effects: *SEEK MEDICAL ATTENTION.* **Liquid:** POISONOUS IF SWALLOWED OR IF SKIN IS EXPOSED. *IF SWALLOWED,* will cause loss of consciousness. Remove contaminated clothing and shoes. Flush affected areas with plenty of water. *IF IN EYES,* hold eyelids open and flush with plenty of water. *IF SWALLOWED* and victim is *CONSCIOUS AND ABLE TO SWALLOW,* have victim drink 4 to 8 ounces of water and have victim induce vomiting. *IF SWALLOWED* and victim is *UNCONSCIOUS OR HAVING CONVULSIONS,* do nothing except keep victim warm.

HEALTH HAZARDS
Personal protective equipment (PPE): B-Level PPE. OSHA Table Z-1-A air contaminant. Organic vapor respirator. Rubber gloves and apron; protective goggles or face shield.
Recommendations for respirator selection:
10 mg/m³: DMXSQ, if not present as a fume (any dust and mist

respirator except single-use and quarter mask respirators) SA (any supplied-air respirator). *25 mg/m³:* SA:CF (any supplied-air respirator operated in a continuous-flow mode) PAPRDM, if not present as a fume (any powered, air-purifying respirator with a dust and mist filter). *50 mg/m³:* HiEF (any air-purifying, full-facepiece respirator with a high-efficiency particulate filter) SAT:CF (any supplied-air respirator that has a tight-fitting facepiece and is operated in a continuous-flow mode) PAPRTHiE (any powered, air-purifying respirator with a tight-fitting facepiece and a high-efficiency particulate filter) SCBAF (any self-contained breathing apparatus with a full facepiece) SAF (any supplied-air respirator with a full facepiece). *500 mg/m³:* SA:PD,PP (any supplied-air respirator operated in a pressure-demand or other positive-pressure mode).*EMERGENCY OR PLANNED ENTRY INTO UNKNOWN CONCENTRATIONS OR IDLH CONDITIONS:* SCBAF:PD,PP (any self-contained breathing apparatus that has a full facepiece and is operated in a pressure-demand or other positive-pressure mode) SAF:PD,PP:ASCBA (any supplied-air respirator that has a full facepiece and is operated in a pressure-demand or other positive-pressure mode in combination with an auxiliary self-contained breathing apparatus operated in a pressure-demand or other positive-pressure mode). *ESCAPE:* HiEF (any air-purifying, full-facepiece respirator with a high-efficiency particulate filter) SCBAE (any appropriate escape-type, self-contained breathing apparatus).

Exposure limits (TWA unless otherwise noted): ACGIH TLV 0.2 mg/m³; OSHA PEL ceiling 5 mg/m³; NIOSH 0.2 mg/m³; skin contact contributes significantly in overall exposure.
Toxicity by ingestion: Grade 4; oral LD_{50} = 23 mg/kg (rat).
Liquid or solid irritant characteristics: Eye and skin irritant. May have a narcotic effect.

FIRE DATA
Flash point: 230°F/110°C (cc).

CHEMICAL REACTIVITY
Reactivity with water: Some carbonyls react with moisture producing toxic and flammable vapors.
Binary reactants: Decomposes in light. Violent reaction with strong oxidizers.

ENVIRONMENTAL DATA
Water pollution: Effect of low concentrations on aquatic life is unknown. May be dangerous if it enters nearby water intakes; notify operators. Notify local health and wildlife officials.
Response to discharge: Issue warning–poison, water contaminant. Restrict access. Should be removed. Chemical and physical treatment.

SHIPPING INFORMATION
Grades of purity: 99.8%; **Storage temperature:** Ambient; **Inert atmosphere:** None; **Venting:** Pressure-vacuum; **Stability during transport:** Stable.

PHYSICAL AND CHEMICAL PROPERTIES
Physical state @ 59°F/15°C and 1 atm: Liquid.
Molecular weight: 218.1
Boiling point @ 1 atm: 451°F/233°C/506°K.
Melting/Freezing point: 36°F/2°C/275°K.
Specific gravity (water = 1): 1.39 @ 68°F/20°C (liquid).
Heat of combustion: (estimate) –9900 Btu/lb = –5500 cal/g = –230 x 10⁵ J/kg.
Vapor pressure: 7mm @ 212°F

METHYL CYCLOPENTANE   REC. M:3850

SYNONYMS: CYCLOPENTANE, METHYL-; METHYLCYCLOPENTANE; METILCICLOPENTANO (Spanish)

IDENTIFICATION
CAS Number: 96-37-7
Formula: C_6H_{12}
DOT ID Number: UN 2298; DOT Guide Number: 128
Proper Shipping Name: Methyl cyclopentane

DESCRIPTION: Colorless liquid. Gasoline-like odor. Floats on water surface; insoluble. Flammable, irritating vapor is produced.

Highly flammable • Containers may BLEVE when exposed to fire •Vapors may form explosive mixture with air • Vapors are heavier than air and will collect and stay in low areas • Vapors may travel long distances to ignition sources and flashback • Vapors in confined areas (e.g., tanks, sewers, buildings) may explode when exposed to fire • Irritating to the skin, eyes, and respiratory tract • Toxic products of combustion may include carbon monoxide.

Hazard Classification (based on NFPA-704 Rating System)
Health Hazards (Blue): 2; Flammability (Red): 3; Reactivity (Yellow): 0

EMERGENCY RESPONSE: See Appendix A (128)
Evacuation:
Public safety: Isolate spill area for at least 25 to 50 meters (80 to 160 feet) in all directions.
Spill: Large spill–Consider initial downwind evacuation for at least 300 meters (1000 feet).
Fire: Isolate for 800 meters (½ mile) in all directions, especially if tank, rail car, or tank truck is involved in fire.

EXPOSURE
Short-term effects: *SEEK MEDICAL ATTENTION.* **Vapor:** IF inhaled, will cause dizziness or difficult breathing. Move victim to fresh air. *IF BREATHING HAS STOPPED,* give artificial respiration. IF breathing is difficult, administer oxygen. The use of alcoholic beverages may enhance toxic effects. **Liquid:** Irritating to skin and eyes. Remove contaminated clothing and shoes. Flush affected areas with plenty of water. *IF IN EYES*, hold eyelids open and flush with plenty of water. *IF SWALLOWED* and victim is *CONSCIOUS AND ABLE TO SWALLOW,* have victim drink 4 to 8 ounces of water.

HEALTH HAZARDS
Personal protective equipment (PPE): B-Level PPE. Self-contained breathing apparatus; goggles or face shield; rubber gloves.
Toxicity by ingestion: Grade 1; LD_{50} = 5 to 15 g/kg.
Vapor (gas) irritant characteristics: Vapors are nonirritating to eyes and throat of most people.
Liquid or solid irritant characteristics: Eye and skin irritant. If spilled on clothing and allowed to remain, may cause smarting and reddening of skin.

FIRE DATA
Flash point: Less than 20°F/7°C (cc).
Flammable limits in air: LEL: 1.1%; UEL: 8.4%.
Fire extinguishing agents not to be used: Water may be ineffective.

Autoignition temperature: 496°F/258°C/531°K.
Burning rate: 7.1 mm/min

CHEMICAL REACTIVITY
Binary reactants: Oxidizers.

ENVIRONMENTAL DATA
Food chain concentration potential: Negative; unlikely to accumulate.
Water pollution: Effect of low concentrations on aquatic life is unknown. Fouling to shoreline. May be dangerous if it enters nearby water intakes; notify operators. Notify local health and wildlife officials. **Response to discharge:** Issue warning–high flammability. Evacuate area. Mechanical containment. Chemical and physical treatment.

SHIPPING INFORMATION
Grades of purity: Research: 99.94%; Pure: 99.5%; Technical: 96.5%; **Storage temperature:** Ambient; **Inert atmosphere:** None; **Venting:** Pressure-vacuum; **Stability during transport:** Stable.

NAS HAZARD CLASSIFICATION FOR BULK WATER TRANSPORTATION
FIRE: 3
HEALTH: Vapor irritant: 0; Liquid or solid irritant: 1; Poisons: 2
WATER POLLUTION: Human toxicity: 1; Aquatic toxicity: 2; Aesthetic effect: 2
REACTIVITY: Other chemicals: 0; Water: 0; Self-reaction: 0

PHYSICAL AND CHEMICAL PROPERTIES
Physical state @ 59°F/15°C and 1 atm: Liqui
Molecular weight: 84.2
Boiling point @ 1 atm: 161.3°F/71.8°C/345.0°K.
Melting/Freezing point: –224°F/–142°C/131°K.
Critical temperature: 499.3°F/259.6°C/532.8°K.
Critical pressure: 550 psia = 37.4 atm = 3.79 MN/m^2.
Specific gravity (water = 1): 0.749 @ 68°F/20°C (liquid).
Liquid surface tension: 21.60 dynes/cm = 0.0216 N/m @ 68°F/20°C.
Liquid water interfacial tension: (estimate) 35 dynes/cm = 0.035 N/m @ 68°F/20°C.
Relative vapor density (air = 1): 2.9
Ratio of specific heats of vapor (gas): 1.0834
Latent heat of vaporization: 162 Btu/lb = 90 cal/g = 3.8 x 10^5 J/kg.
Heat of combustion: (liquid) –18,900 Btu/lb = –10,500 cal/g = –440 x 10^5 J/kg.
Heat of fusion: 19.68 cal/g.

METHYL DICHLOROACETATE   REC. M:3900

SYNONYMS: DICHLOROACETIC ACID, METHYL ESTER; DICLOROACETATO de METILO (Spanish); METHYL DICHLOROETHANOATE

IDENTIFICATION
CAS Number: 116-54-1
Formula: $C_3H_4Cl_2O_2$; $Cl_2CHCO_2CH_3$
DOT ID Number: UN 2299; DOT Guide Number: 155
Proper Shipping Name: Methyl dichloroacetate

DESCRIPTION: Colorless liquid. Sweet, ether-like odor. Sinks in water; slightly soluble; reacts slowly forming hydrochloric acid. Combustible • Heat or flame may cause explosion • Highly Corrosive to the skin, eyes, and respiratory tract; even brief contact can cause burns and blindness • Firefighting gear (including SCBA) does not provide adequate protection. If exposure occurs, remove and isolate gear immediately and thoroughly decontaminate personnel • Containers may BLEVE when exposed to fire • Vapors are heavier than air and will collect and stay in low areas • Vapors in confined areas (e.g., tanks, sewers, buildings) may explode when exposed to fire • Toxic products of combustion may include hydrogen chloride.

Hazard Classification (based on NFPA-704 Rating System)
Health Hazards (Blue): 3; Flammability (Red): 2; Reactivity (Yellow): 0

EMERGENCY RESPONSE: See Appendix A (155)
Evacuation:
Public safety: Isolate the area of spill or leak for at least 50 to 100 meters (160 to 330 feet) in all directions.
Spill: Increase, in the downwind direction, as necessary, the distance shown under "Public Safety."
Fire: If tank, rail car, or tank truck is involved in fire, isolate for at least 800 meters (½ mile) in all directions; also, consider initial evacuation for 800 meters (½ mile) in all directions.

EXPOSURE
Short-term effects: *SEEK MEDICAL ATTENTION.* **Vapor:** *HARMFUL IF INHALED OR ABSORBED THROUGH THE SKIN.* Highly irritating to skin, eyes, and mucous membranes. Hydrolizes upon contact with moisture to form vapors corrosive to tissue. Remove victim to fresh air. *IF BREATHING HAS STOPPED,* give artificial respiration; *avoid mouth-to-mouth resuscitation; use bag/mask apparatus.* IF breathing is difficult, administer oxygen. **Liquid:** Harmful if swallowed or absorbed through the skin. Corrosive to eyes, skin, nose, throat, and upper respiratory tract. Hydrolyzes upon contact with moisture to form vapors corrosive to skin. *IF IN EYES,* hold eyelids open, flush with plenty of water for at least 15 minutes. Remove contaminated clothing and shoes; flush affected areas with water. *IF SWALLOWED:* Do nothing except keep victim warm. **Do NOT induce vomiting**.

HEALTH HAZARDS
Personal protective equipment (PPE): B-Level PPE. Approved respirator, chemical-resistant gloves, safety goggles or safety faceshield (8 inch minimum), other protective clothing.
Vapor (gas) irritant characteristics: Vapors cause severe irritation of eyes and throat and can cause eye and lung injury. They cannot be tolerated even at low concentrations.
Liquid or solid irritant characteristics: Severe skin irritant. Causes second- and third-degree burns on short contact and is very injurious to the eyes.

FIRE DATA
Flash point: 176°F/80°C (cc).

CHEMICAL REACTIVITY
Reactivity with water: Hydrolyzes to form corrosive products.
Reactivity group: 34
Compatibility class: Esters.

ENVIRONMENTAL DATA
Water pollution: Effects of low concentrations on aquatic life are not known. May be dangerous if it enters nearby water intakes; notify operators. Notify local health and wildlife officials. **Response**

to discharge: Issue warning–corrosive. Should be removed. Chemical and physical treatment.

SHIPPING INFORMATION
Grades of purity: 99+%; **Stability during transport:** Stable.

NAS HAZARD CLASSIFICATION FOR BULK WATER TRANSPORTATION
FIRE: 1
HEALTH: Vapor irritant: 4; Liquid or solid irritant: 3; Poisons: 3
WATER POLLUTION: Human toxicity:–; Aquatic toxicity:–; Aesthetic effect: 1
REACTIVITY: Other chemicals: 3; Water: 1; Self-reaction: 0

PHYSICAL AND CHEMICAL PROPERTIES
Physical state @ 59°F/15°C and 1 atm: Liquid.
Molecular weight: 142.97
Boiling point @ 1 atm: 289°F/143°C/416°K.
Melting/Freezing point: –62°F/–52°C/221°K.
Specific gravity (water = 1): 1.3774 @ 68°F/20°C.
Relative vapor density (air = 1): 4.93

METHYLDICHLOROSILANE   REC. M:3950

SYNONYMS: DICHLOROMETHYLSILANE; METILDICLOROSILANO (Spanish); SILANE, DICHLOROMETHYL-

IDENTIFICATION
CAS Number: 75-54-7
Formula: CH_4Cl_2Si; CH_3SiHCl_2
DOT ID Number: UN 1242; DOT Guide Number: 139
Proper Shipping Name: Methyldichlorsilane

DESCRIPTION: Colorless, fuming liquid. Sharp, irritating odor, like hydrochloric acid. Sinks in water; reacts violently, forming hydrochloric acid and explosive hydrogen gas. Boils at 107°F/41.6°C.

Extremely flammable • Poison • Breathing the vapors can kill you • Skin and eye contact causes severe burns and blindness • Ignites spontaneously in air • Firefighting gear (including SCBA) does not provide adequate protection. If exposure occurs, remove and isolate gear immediately and thoroughly decontaminate personnel • Vapors are heavier than air and will collect and stay in low areas • Vapors may travel long distances to source of ignition and flashback • Often shipped and stored dissolved in acetone • Ruptured or venting cylinders may rocket through buildings and/or travel a considerable distance • Toxic products of combustion may include hydrogen chloride • When combined with surface moisture this material is corrosive to most common metals • Do not put yourself in danger by entering a contaminated area to rescue a victim.

Hazard Classification (based on NFPA-704 Rating System)
Health Hazards (Blue): 3; Flammability (Red): 3; Reactivity (Yellow): 2; Special Notice (White): Water reactive

EMERGENCY RESPONSE (Chlorosilanes):
Evacuation:
Public safety: Isolate the area of spill or leak for at least 100 to 150 meters (330 to 490 feet) in all directions.
Spills: *IF SPILLED IN WATER*: Small spill–First: Isolate in all directions 30 meters (100 feet); Then: Protect persons downwind, DAY: 0.2 km (0.1 mile); NIGHT: 0.2 km (0.1 mile). Large spill–First: Isolate in all directions 60 meters (200 feet); Then: Protect persons downwind, DAY: 0.5 km (0.3 mile); NIGHT: 1.6 km (1.0 mile).
Fire: If any large container is involved in fire, isolate for at least 800 meters (½ mile) in all directions; also, consider initial evacuation for 800 meters (½ mile) in all directions.

EXPOSURE
Short-term effects: *SEEK MEDICAL ATTENTION.* **Vapor:** Irritating to eyes, nose, and throat. *IF INHALED*, will, will cause difficult breathing. Move victim to fresh air. *IF BREATHING HAS STOPPED,* give artificial respiration. IF breathing is difficult, administer oxygen. **Liquid:** Will burn skin and eyes. Harmful if swallowed. Remove contaminated clothing and shoes. Flush affected areas with plenty of water. *IF IN EYES*, hold eyelids open and flush with plenty of water. *IF SWALLOWED* and victim is *CONSCIOUS AND ABLE TO SWALLOW,* have victim drink 4 to 8 ounces of water. **Do NOT induce vomiting.**

HEALTH HAZARDS
Personal protective equipment (PPE): B-Level PPE. Full protective clothing; acid-vapor-type respiratory protection; rubber gloves; chemical worker's goggles; other protective equipment as necessary
to protect skin and eyes.
Toxicity by ingestion: Grade 3; LD_{50} = 50 to 500 mg/kg.
Vapor (gas) irritant characteristics: Vapors cause severe irritation of eyes and throat and can cause eye and lung injury. They cannot be tolerated even at low concentrations.
Liquid or solid irritant characteristics: Severe eye and skin irritant. Causes second-and third-degree burns on short contact and is very injurious to the eyes.

FIRE DATA
Flash point: 15°F/–9°C (cc).
Flammable limits in air: LEL: 6%; UEL: 55%.
Fire extinguishing agents not to be used: Water, foam
Behavior in fire: Difficult to extinguish; re-ignition may occur.
Autoignition temperature: More than 600°F/316°C/589°K.
Electrical hazard: Class I, Group C.
Burning rate: 3.0 mm/min

CHEMICAL REACTIVITY
Reactivity with water: Reacts violently to form hydrogen chloride and hydrogen gas.
Binary reactants: Reacts with surface moisture to evolve hydrogen chloride, which is corrosive to common metals.
Neutralizing agents for acids and caustics: Flood with water and rinse with sodium bicarbonate or lime solution.

ENVIRONMENTAL DATA
Food chain concentration potential: Negative; unlikely to accumulate.
Water pollution: Effect of low concentrations on aquatic life is unknown. May be dangerous if it enters nearby water intakes; notify operators. Notify local health and wildlife officials. **Response to discharge:** Issue warning–high flammability. Restrict access. Evacuate area. Disperse and flush with care.

SHIPPING INFORMATION
Grades of purity: 97%; **Storage temperature:** Ambient; **Inert atmosphere:** None; **Venting:** Pressure-vacuum; **Stability during transport:** Stable.

742   2-Methyl-6-ethyl aniline

NAS HAZARD CLASSIFICATION FOR BULK WATER TRANSPORTATION
FIRE: 3
HEALTH: Vapor irritant: 4; Liquid or solid irritant: 4; Poisons: 3
WATER POLLUTION: Human toxicity: 3; Aquatic toxicity: 3; Aesthetic effect: 2
REACTIVITY: Other chemicals: 3; Water: 4; Self-reaction: 1

PHYSICAL AND CHEMICAL PROPERTIES
Physical state @ 59°F/15°C and 1 atm: Liquid.
Molecular weight: 115
Boiling point @ 1 atm: 106.7°F/41.5°C/314.7°K.
Melting/Freezing point: –135°F/–93°C/180°K.
Specific gravity (water = 1): 1.11 @ 77°F/25°C (liquid).
Liquid surface tension: (estimate) 35 dynes/cm = 0.035 N/m @ 68°F/20°C.
Relative vapor density (air = 1): 4
Latent heat of vaporization: 106 Btu/lb = 59 cal/g = 2.5 x 10^5 J/kg.
Heat of combustion: (estimate) –4700 Btu/lb = –2600 cal/g = –110 x 10^5 J/kg.

METHYL DIETHANOLAMINE   REC. M:4000

SYNONYMS: *N*-METHYLDIETHANOLAMINE; 2,2'-METHYLIMINODIETHANOL; METILDIETANOLAMINA (Spanish)

IDENTIFICATION
CAS Number: 105-59-9
Formula: $C_5H_{13}NO_2$; $(HOCH_2CH_2)_2NCH_3$

DESCRIPTION: Colorless liquid. soluble in water.

Combustible • Containers may BLEVE when exposed to fire • Vapors are heavier than air and will collect and stay in low areas • Vapors in confined areas (e.g., tanks, sewers, buildings) may explode when exposed to fire • Irritating to the skin, eyes, and respiratory tract • Toxic products of combustion may include nitrogen oxides and carbon monoxide.

Hazard Classification (based on NFPA-704 Rating System)
Health Hazards (Blue): 1; Flammability (Red): 1; Reactivity (Yellow): 0

EMERGENCY RESPONSE: See Appendix A (139)
Evacuation:
Public safety: Isolate the area of spill or leak for at least 10 to 25 meters (30 to 80 feet) in all directions.
Spill: Increase, in the downwind direction, as necessary, the distance shown under "Public Safety."
Fire: If any large container is involved in fire, isolate for at least 800 meters (½ mile) in all directions; also, consider initial evacuation for 800 meters (½ mile) in all directions.

EXPOSURE
Short-term effects: *SEEK MEDICAL ATTENTION.* **Vapor:** Move victim to fresh air. *IF BREATHING HAS STOPPED,* give artificial respiration. If breathing is difficult, administer oxygen. **Liquid:** Irritating to skin and eyes. Remove contaminated clothing and shoes. Flush affected areas with water. *IF IN EYES,* hold eyelids open and flush with plenty of water.

HEALTH HAZARDS
Personal protective equipment (PPE): B-Level PPE. Full impervious protective clothing, including boots and gloves. Where splashing is possible wear full face shield or chemical safety goggles. Use approved respirator to protect against vapors.
Toxicity by ingestion: Grade 2; LD_{50} = 4.78 g/kg (rat).
Vapor (gas) irritant characteristics: Eye and respiratory tract irritant. Vapors are irriatating such that personnel will not usually tolerate moderate or high concentrations.
Liquid or solid irritant characteristics: Fairly severe eye and skin irritant. May cause pain and second-degree burns after a few minutes of contact.

FIRE DATA
Flash point: 259°F/126°C (cc).
Fire extinguishing agents not to be used: Water.
Stoichiometric air-to-fuel ratio: 34.5

CHEMICAL REACTIVITY
Binary reactants: Oxidizers, strong acids.
Reactivity group: 8
Compatibility class: Alkanolamines

ENVIRONMENTAL DATA
Water pollution: Effects of low concentrations on aquatic life is unknown. May be dangerous if it enters nearby water intakes; notify operators. Notify local health and wildlife officials.
Response to discharge: Should be removed.

SHIPPING INFORMATION
Grades of purity: 99%; technical grades; **Storage temperature:** Ambient; **Inert atmosphere:** None; **Venting:** Open; **Stability during transport:** Stable.

PHYSICAL AND CHEMICAL PROPERTIES
Physical state @ 59°F/15°C and 1 atm: Liquid.
Molecular weight: 119.16
Boiling point @ 1 atm: 474.8–478.4°F/246–248°C/519–521°K.
Specific gravity (water = 1): 1.0377 @ 20°C.

2-METHYL-6-ETHYL ANILINE   REC. M:4050

SYNONYMS: 6-ETHYL-2-METHYLANILINE; 6-ETHYL-*o*-TOLUIDINE; 2-METHYL-6-ETHYL BENZENEAMINE

IDENTIFICATION
CAS Number: 24549-06-2
Formula: $C_9H_{13}N$; $CH_3C_6H_4NHC_2H_5$

DESCRIPTION: Clear liquid. Pungent odor. Floats on water surface; insoluble.

Combustible • Containers may BLEVE when exposed to fire • Vapors are heavier than air and will collect and stay in low areas • Vapors in confined areas (e.g., tanks, sewers, buildings) may explode when exposed to fire • Irritating to the skin, eyes, and respiratory tract • Toxic products of combustion may include nitrogen oxides and CO_2.

Hazard Classification (based on NFPA-704 Rating System)
Health Hazards (Blue): 1; Flammability (Red): 1; Reactivity (Yellow): 0

EMERGENCY RESPONSE: See Appendix A (171)
Evacuation:
Public safety: Isolate the area of spill or leak for at least 10 to 25 meters (30 to 80 feet) in all directions.
Spill: Increase, in the downwind direction, as necessary, the distance shown under "Public Safety."
Fire: If any large container is involved in fire, isolate for at least 800 meters (½ mile) in all directions; also, consider initial evacuation for 800 meters (½ mile) in all directions.

EXPOSURE
Short-term effects: *SEEK MEDICAL ATTENTION.* **Liquid:** A severe eye irritant. *IF IN EYES*, hold eyelids open and flush with running water for at least 15 minutes. *IF SWALLOWED* and victim is *CONSCIOUS AND ABLE TO SWALLOW*, have victim drink 4 to 8 ounces of water and induce vomiting.

HEALTH HAZARDS
Personal protective equipment (PPE): B-Level PPE. Chemical goggles-face shield, protective gloves, organic vapor canister mask.
Toxicity by ingestion: Grade 2; LD_{50} = 1.18 g/kg (rat).
Liquid or solid irritant characteristics: Severe eye irritant.

FIRE DATA
Flash point: 232°F/111°C (oc); 215°F/102°C (cc).
Behavior in fire: Produces poisonous gases.

CHEMICAL REACTIVITY
Binary reactants: Oxidizers, strong acids, isocyanates, aldehydes.
Reactivity group: 9
Compatibility class: Aromatic amines

ENVIRONMENTAL DATA
Water pollution: Effects of low concentrations on aquatic life is unknown. May be dangerous if it enters nearby water intakes; notify operators. Notify local health and wildlife officials.
Response to discharge: Issue warning-air contaminant. Restrict access. Mechanical containment. Should be removed. Chemical and physical treatment.

SHIPPING INFORMATION
Storage temperature: Ambient; **Stability during transport:** Stable.

PHYSICAL AND CHEMICAL PROPERTIES
Physical state @ 59°F/15°C and 1 atm: Liquid.
Molecular weight: 135.2
Boiling point @ 1 atm: 447.8°F/231°C/504°K.
Melting/Freezing point: –27.4°F/–33°C/240°K.
Specific gravity (water = 1): 0.969 @ 68°F/20°C.

METHYL ETHYL KETONE REC. M:4100

SYNONYMS: ACETONE, METHYL-; AETHYLMETHYLKETON (German); BUTANONE; 2-BUTANONE; BUTANONE 2 (French); EEC No. 606-002-00-3; ETHYL METHYL CETONE (French); ETHYL METHYL KETONE; MEK; METHYL ACETONE; METIL ETIL CETONA (Spanish); KETONE, ETHYL METHYL; RCRA No. U159

IDENTIFICATION
CAS Number: 78-93-3
Formula: C_4H_8O; $CH_3COCH_2CH_3$
DOT ID Number: UN 1193; DOT Guide Number: 127
Proper Shipping Name: Ethyl methyl ketone; Methyl ethyl ketone
Reportable Quantity (RQ): **(CERCLA)** 5000 lb/2270 kg

DESCRIPTION: Colorless liquid. Sweet odor, like acetone. Initially floats on water surface; soluble. Flammable, irritating vapor is produced.

Highly flammable • Containers may BLEVE when exposed to fire • Vapors may form explosive mixture with air • Vapors are heavier than air and will collect and stay in low areas • Vapors may travel long distances to ignition sources and flashback • Vapors in confined areas (e.g., tanks, sewers, buildings) may explode when exposed to fire • Irritating to the skin, eyes, and respiratory tract • Toxic products of combustion may include carbon monoxide.

Hazard Classification (based on NFPA-704 Rating System)
Health Hazards (Blue): 1; Flammability (Red): 3; Reactivity (Yellow): 0

EMERGENCY RESPONSE: See Appendix A (127)
Evacuation:
Public safety: Isolate spill area for at least 25 to 50 meters (80 to 160 feet) in all directions.
Spill: Large spill–Consider initial downwind evacuation for at least 300 meters (1000 feet).
Fire: Isolate for 800 meters (½ mile) in all directions, especially if tank, rail car, or tank truck is involved in fire.

EXPOSURE
Short-term effects: *SEEK MEDICAL ATTENTION.* **Vapor:** Irritating to eyes, nose, and throat. *IF INHALED*, will, will cause nausea, vomiting, headache, dizziness, difficult breathing, or loss of consciousness. Move to fresh air. *IF BREATHING HAS STOPPED*, give artificial respiration. IF breathing is difficult, administer oxygen. **Liquid:** Will burn eyes. Harmful if swallowed. Remove contaminated clothing and shoes. Flush affected areas with plenty of water. *IF IN EYES*, hold eyelids open and flush with plenty of water. *IF SWALLOWED* and victim is *CONSCIOUS AND ABLE TO SWALLOW*, have victim drink 4 to 8 ounces of water.

HEALTH HAZARDS
Personal protective equipment (PPE): B-Level PPE. OSHA Table Z-1-A air contaminant. Chemical protective material(s) reported to have good to excellent resistance: butyl rubber, chlorobutyl rubber, Teflon®, Silvershield®. Also, natural rubber, PV alcohol, polyurethane, styrene-butadiene, offer limited protection
Recommendations for respirator selection: NIOSH/OSHA
3000 ppm: SA:CF$^£$ (any supplied-air respirator operated in a continuous-flow mode); or PAPROV$^£$ [any powered, air-purifying respirator with organic vapor cartridge(s)]; or CCRFOV [any air-purifying, full-facepiece respirator (gas mask) with a chin-style, front- or back-mounted acid gas canister]; or GMFOV [any air-purifying, full-facepiece respirator (gas mask) with a chin-style, front- or back-mounted acid gas canister]; SCBAF (any self-contained breathing apparatus with a full facepiece); or SAF (any supplied-air respirator with a full facepiece). *EMERGENCY OR PLANNED ENTRY INTO UNKNOWN CONCENTRATIONS OR IDLH CONDITIONS:* SCBAF:PD,PP (any self-contained breathing apparatus that has a full facepiece and is operated in a pressure-demand or other positive-pressure mode); or SAF:PD,PP:ASCBA (any supplied-air respirator that has a full facepiece and is operated in a pressure-demand or other positive-pressure mode in combination with an auxiliary self-contained

Methylethylpyridine

breathing apparatus operated in a pressure-demand or other positive pressure mode). *ESCAPE:* GMFOV [any air-purifying, full-facepiece respirator (gas mask) with a chin-style, front- or back-mounted organic vapor canister]; or SCBAE (any appropriate escape-type, self-contained breathing apparatus). *Note:* Substance causes eye irritation or damage; eye protection needed.
Exposure limits (TWA unless otherwise noted): ACGIH TLV 200 ppm (590 mg/m^3); OSHA PEL 200 ppm (590 mg/m^3); NIOSH REL 200 ppm (590 mg/m^3).
Short-term exposure limits (15-minute TWA): ACGIH STEL 300 ppm (885 mg/m^3); NIOSH STEL 300 ppm (885 mg/m^3).
Toxicity by ingestion: Grade 2; LD$_{50}$ = 0.5 to 5 g/kg (rat).
Vapor (gas) irritant characteristics: Eye and respiratory tract irritant. Vapors cause smarting of the eyes or respiratory system if present in high concentrations. The effect is temporary.
Liquid or solid irritant characteristics: Severe eye irritant. If spilled on clothing and allowed to remain, may cause smarting and reddening of the skin.
Odor threshold: 2–85 ppm.
IDLH value: 3000 ppm.

FIRE DATA
Flash point: 22°F/–6°C (oc); 16°F/–9°C (cc).
Flammable limits in air: LEL: 1.8%; UEL: 11.5%.
Fire extinguishing agents not to be used: Water may be ineffective.
Autoignition temperature: 960°F/515°C/788°K.
Electrical hazard: Class I, Group D. Due to low electric conductivity, this substance may generate electrostatic charges as a result of agitation and flow.
Burning rate: 4.1 mm/min

CHEMICAL REACTIVITY
Binary reactants: Strong oxidizers, amines, ammonia, inorganic acids, caustics, copper, isocyanates, pyridines
Reactivity group: 18
Compatibility class: Ketone

ENVIRONMENTAL DATA
Food chain concentration potential: Log P$_{ow}$ = 0.33. Unlikely to accumulate.
Water pollution: Dangerous to aquatic life in high concentrations. May be dangerous if it enters nearby water intakes; notify operators. Notify local health and wildlife officials. **Response to discharge:** Issue warning–high flammability. Disperse and flush.

SHIPPING INFORMATION
Grades of purity: 99.5+%; **Storage temperature:** Ambient; **Inert atmosphere:** None; **Venting:** Open (flame arrester) or pressure-vacuum; **Stability during transport:** Stable.

NAS HAZARD CLASSIFICATION FOR BULK WATER TRANSPORTATION
FIRE: 3
HEALTH: Vapor irritant: 1; Liquid or solid irritant: 1; Poisons: 2
WATER POLLUTION: Human toxicity: 2; Aquatic toxicity: 1; Aesthetic effect: 1
REACTIVITY: Other chemicals: 2; Water: 0; Self-reaction: 0

PHYSICAL AND CHEMICAL PROPERTIES
Physical state @ 59°F/15°C and 1 atm: Liquid.
Molecular weight: 72.11
Boiling point @ 1 atm: 175.3°F/79.6°C/352.8°K.
Melting/Freezing point: –123.3°F/–86.3°C/186.9°K.
Critical temperature: 504.5°F/262.5°C/535.7°K.
Critical pressure: 603 psia = 41.0 atm = 4.15 MN/m^2.
Specific gravity (water = 1): 0.806 @ 68°F/20°C (liquid).
Ratio of specific heats of vapor (gas): 1.075
Latent heat of vaporization: 191 Btu/lb = 106 cal/g = 4.44 x 10^5 J/kg.
Heat of combustion: –13,480 Btu/lb = –7491 cal/g = –313.6 x 10^5 J/kg.
Heat of solution: (estimate) –9 Btu/lb = –5 cal/g = –0.2 x 10^5 J/kg.
Vapor pressure: (Reid) 3.5 psia; 78 mm.

METHYLETHYLPYRIDINE   REC. M:4150

SYNONYMS: ALDEHYDE-COLLIDINE; ALDEHYDINE; COLLIDINE ALDEHYDECOLLIDINE; 5-ETHYL-2-METHYL PYRIDINE; 5-ETHYL-2-PICOLINE; MEP; 2-METHYL-5-ETHYLPYRIDINE; 2-METIL-5-ETILPIRIDINA (Spanish); 2-PICOLINE, 5-ETHYL

IDENTIFICATION
CAS Number: 104-90-5
Formula: C$_8$H$_{11}$N
DOT ID Number: UN 2300; DOT Guide Number: 153
Proper Shipping Name: 2-Methyl-5-ethylpyridine

DESCRIPTION: Colorless liquid. Sharp penetrating odor. Floats on the surface of water.

Poison! • Breathing the vapor, skin or eye contact, or swallowing the material can cause illness; toxic when absorbed by the skin • Corrosive to the eyes and respiratory tract; this material has anesthetic properties; inhalation symptoms may be delayed • Firefighting gear (including SCBA) does not provide adequate protection. If exposure occurs, remove and isolate gear immediately and thoroughly decontaminate personnel • Containers may BLEVE when exposed to fire • Vapors are heavier than air and will collect and stay in low areas • Vapors in confined areas (e.g., tanks, sewers, buildings) may explode when exposed to fire • Toxic products of combustion may include nitrogen oxides • Do not put yourself in danger by entering a contaminated area to rescue a victim.

Hazard Classification (based on NFPA-704 Rating System)
Health Hazards (Blue): 3; Flammability (Red): 2; Reactivity (Yellow): 0

EMERGENCY RESPONSE: See Appendix A (153)
Evacuation:
Public safety: Isolate the area of spill or leak for at least 25 to 50 meters (80 to 160 feet) in all directions.
Spill: Increase, in the downwind direction, as necessary, the distance shown under "Public Safety."
Fire: If tank, rail car, or tank truck is involved in fire, isolate for at least 800 meters (½ mile) in all directions; also, consider initial evacuation for 800 meters (½ mile) in all directions.

EXPOSURE
Short-term effects: *SEEK MEDICAL ATTENTION.* **Liquid:** Will burn skin and eyes. Harmful if swallowed. Remove contaminated clothing and shoes. Flush affected areas with plenty of water. *IF IN IF BREATHING HAS STOPPED,* give artificial respiration; *avoid mouth-to-mouth resuscitation; use bag/mask apparatus. EYES,* hold eyelids open and flush with plenty of water. *IF SWALLOWED* and victim is *CONSCIOUS AND ABLE TO SWALLOW,* have victim drink 4 to 8 ounces of water.

Note to physician or authorized medical personnel: Medical observation is recommended for 24 to 48 hours after breathing overexposure, as pulmonary edema may be delayed. As first aid for pulmonary edema, consider administering a corticosteroid spray. Cigarette smoking may exacerbate pulmonary injury and should be discouraged for at least 72 hours following exposure.

HEALTH HAZARDS
Personal protective equipment (PPE): B-Level PPE. Air–supplied mask for high vapor concentrations; plastic gloves; goggles or face shield.
Toxicity by ingestion: Grade 2; LD_{50} = 0.5 to 5 g/kg (rat).
Vapor (gas) irritant characteristics: Vapors cause irritation such that personnel will find high concentrations unpleasant. The effect is temporary.
Liquid or solid irritant characteristics: Severe eye and skin irritant. Causes smarting of the skin and first-degree burns on short exposure; may cause secondary burns on long exposure.

FIRE DATA
Flash point: 155°F/68°C (oc).
Flammable limits in air: LEL: 1.1%; UEL: 6.6%.
Autoignition temperature: 939°F/504°C/777°K.

CHEMICAL REACTIVITY
Binary reactants: Violent reaction with strong oxidizers. May forms heat-sensitive explosive materials with digold ketenide. Incompatible with alcohols, aldehydes, alkylene oxides, cresols, caprolactam solution, epichlorohydrin, organic anhydrides, glycols, maleic anhydride, phenols. Attacks copper and copper alloys.
Neutralizing agents for acids and caustics: Flush with water, neutralize with dilute acetic acid.
Reactivity group: 9
Compatibility class: Aromatic amines

ENVIRONMENTAL DATA
Food chain concentration potential: Log P_{ow} = 2.5. Unlikely to accumulate.
Water pollution: DOT Appendix B, §172.101–marine pollutant. Effect of low concentrations on aquatic life is unknown. Fouling to shoreline. May be dangerous if it enters nearby water intakes; notify operators. Notify local health and wildlife officials.
Response to discharge: Mechanical containment. Should be removed. Chemical and physical treatment.

SHIPPING INFORMATION
Grades of purity: 99.9%; **Storage temperature:** Ambient; **Inert atmosphere:** None; **Venting:** Open (flame arrester); **Stability during transport:** Stable.

NAS HAZARD CLASSIFICATION FOR BULK WATER TRANSPORTATION
FIRE: 1
HEALTH: Vapor irritant: 2; Liquid or solid irritant: 2; Poisons: 2
WATER POLLUTION: Human toxicity: 2; Aquatic toxicity: 2; Aesthetic effect: 3
REACTIVITY: Other chemicals: 3; Water: 0; Self-reaction: 0

PHYSICAL AND CHEMICAL PROPERTIES
Physical state @ 59°F/15°C and 1 atm: Liquid.
Molecular weight: 121.18
Boiling point @ 1 atm: 352°F/178°C/451°K.
Melting/Freezing point: –94.5°F/–70.3°C/202.9°K.
Specific gravity (water = 1): 0.922 @ 68°F/20°C (liquid).
Liquid surface tension: 36 dynes/cm = 0.036 N/m @ 68°F/20°C.
Vapor pressure: (Reid) 0.1 psia.

METHYL FORMAL   REC. M:4200

SYNONYMS: ANESTHENYL; DIMETHYLACETAL FORMALDEHYDE; DIMETHOXYMETHANE; FORMAL; FORMALDEHYDE DIMETHYLACETAL; METHANE, DIMETHOXY-; METHOXYMETHYL ETHER; METHYLENE DIMETHYL ETHER; METHYLAL; DIMETHYL FORMAL; METHYLAL

IDENTIFICATION
CAS Number: 109-87-5
Formula: $C_3H_8O_2$; $CH_2(OCH_3)_2$
DOT ID Number: UN 1234; DOT Guide Number: 127
Proper Shipping Name: Methylal

DESCRIPTION: Colorless liquid. Mild sweet odor, like chloroform. Mixes with water. Flammable, irritating vapor is produced.

Highly flammable • Containers may BLEVE when exposed to fire • Vapors may form explosive mixture with air • Vapors are heavier than air and will collect and stay in low areas • Vapors may travel long distances to ignition sources and flashback • Vapors in confined areas (e.g., tanks, sewers, buildings) may explode when exposed to fire • Highly irritating to the eyes and respiratory tract; this material has anesthetic properties • Toxic products of combustion may include formaldehyde gas and carbon monoxide.

Hazard Classification (based on NFPA-704 Rating System)
Health Hazards (Blue): 2; Flammability (Red): 3; Reactivity (Yellow): 2

EMERGENCY RESPONSE: See Appendix A (127)
Evacuation:
Public safety: Isolate spill area for at least 25 to 50 meters (80 to 160 feet) in all directions.
Spill: Large spill–Consider initial downwind evacuation for at least 300 meters (1000 feet).
Fire: Isolate for 800 meters (½ mile) in all directions, especially if tank, rail car, or tank truck is involved in fire.

EXPOSURE
Short-term effects: *SEEK MEDICAL ATTENTION.* **Vapor:** Irritating to eyes, nose, and throat. Harmful if inhaled. Move victim to fresh air. *IF BREATHING HAS STOPPED*, give artificial respiration. IF breathing is difficult, administer oxygen. **Liquid:** Irritating to skin and eyes. Harmful if swallowed. Remove contaminated clothing and shoes. Flush affected areas with plenty of water. *IF IN EYES*, hold eyelids open and flush with plenty of water. *IF SWALLOWED* and victim is *CONSCIOUS AND ABLE TO SWALLOW*, having victim drink 4 to 8 ounces of water and have victim induce vomiting. *IF SWALLOWED* and victim is *UNCONSCIOUS OR HAVING CONVULSIONS*, do nothing except keep victim warm.

HEALTH HAZARDS
Personal protective equipment (PPE): B-Level PPE. OSHA Table Z-1-A air contaminant. Self-contained breathing apparatus or all-purpose canister mask; rubber gloves; chemical safety goggles; impervious apron and boots.

Methyl formate

Recommendations for respirator selection: NIOSH/OSHA
2200 ppm: SA (any supplied-air respirator); or SCBAF (any self-contained breathing apparatus with a full facepiece); or SAF (any supplied-air respirator with a full facepiece). *EMERGENCY OR PLANNED ENTRY INTO UNKNOWN CONCENTRATIONS OR IDLH CONDITIONS*: SCBAF:PD,PP (any self-contained breathing apparatus that has a full facepiece and is operated in a pressure-demand or other positive-pressure mode); or SAF:PD,PP:ASCBA (any supplied-air respirator that has a full facepiece and is operated in a pressure-demand or other positive-pressure mode in combination with an auxiliary self-contained breathing apparatus operated in a pressure-demand or other positive pressure mode). *ESCAPE:* GMFOV [any air-purifying, full-facepiece respirator (gas mask) with a chin-style, front- or back-mounted organic vapor canister]; or SCBAE (any appropriate escape-type, self-contained breathing apparatus).
Exposure limits (TWA unless otherwise noted): ACGIH TLV 1000 ppm (3110 mg/m^3); NIOSH/OSHA 1000 ppm (3100 mg/m^3).
Toxicity by ingestion: Grade 1; LD_{50} = 5 to 15 g/kg.
Long-term health effects: Liver and kidney injury may follow high exposures. May effect the nervous system.
Vapor (gas) irritant characteristics: Vapors cause smarting of the eyes or respiratory system if present in high concentrations. The effect may be temporary.
Liquid or solid irritant characteristics: Irritates the skin, eyes and respiratory tract. If spilled on clothing and allowed to remain, may cause smarting and reddening of the skin.
IDLH value: 2200 ppm [10% LEL].

FIRE DATA
Flash point: –26°F/–32°C (oc).
Flammable limits in air: LEL: 2.2%; UEL: 13.8%.
Fire extinguishing agents not to be used: Water may be ineffective.
Autoignition temperature: 459°F/237°C/510°K.
Burning rate: 5.5 mm/min

CHEMICAL REACTIVITY
Binary reactants: Strong oxidizers (may be violent), acids, halogenated hydrocarbons. May form unstable peroxides in air.
Reactivity group: 41
Compatibility class: Ethers

ENVIRONMENTAL DATA
Food chain concentration potential: Log P_{ow} = 0.2. Unlikely to accumulate.
Water pollution: Effect of low concentrations on aquatic life is unknown. May be dangerous if it enters nearby water intakes; notify operators. Notify local health and wildlife officials.
Response to discharge: Issue warning–high flammability, water contaminant. Restrict access. Disperse and flush.

SHIPPING INFORMATION
Grades of purity: 97+%; **Storage temperature:** Ambient; **Inert atmosphere:** None; **Venting:** Pressure-vacuum; **Stability during transport:** Stable.

NAS HAZARD CLASSIFICATION FOR BULK WATER TRANSPORTATION
FIRE: 3
HEALTH: Vapor irritant: 1; Liquid or solid irritant: 1; Poisons: 1
WATER POLLUTION: Human toxicity: 1; Aquatic toxicity: 2; Aesthetic effect: 1
REACTIVITY: Other chemicals: 2; Water: 0; Self-reaction: 0

PHYSICAL AND CHEMICAL PROPERTIES
Physical state @ 59°F/15°C and 1 atm: Liquid.
Molecular weight: 76.1
Boiling point @ 1 atm: 111°F/44°C/317°K.
Melting/Freezing point: –157°F/–105°C/168°K.
Critical temperature: 419°F/215°C/488°K.
Specific gravity (water = 1): 0.861 @ 68°F/20°C (liquid).
Liquid surface tension: 21.1 dynes/cm = 0.0211 N/m @ 68°F/20°C.
Relative vapor density (air = 1): 2.6
Ratio of specific heats of vapor (gas): 1.0888
Latent heat of vaporization: 161.5 Btu/lb 89.8 cal/g = 3.76 x 10^5 J/kg.
Heat of combustion: –10,970 Btu/lb = –6100 cal/g = –255 x 10^5 J/kg.
Vapor pressure: 330 mm.

METHYL FORMATE   REC. M:4250

SYNONYMS: EEC No. 607-014-00-1; FORMIATE de METHYLE (French); FORMIATO de METILO (Spanish); FORMIC ACID, METHYL ESTER; METHYL ESTER OF FORMIC ACID; METHYLE FORMIATE de (French); METHYLFORMIAT (German); METHYL METHANOATE

IDENTIFICATION
CAS Number: 107-31-3
Formula: $C_2H_4O_2$; COOC3
DOT ID Number: UN 1243; DOT Guide Number: 129
Proper Shipping Name: Methyl formate

DESCRIPTION: Colorless liquid. Pleasant, fruity odor. Initially folats on water; soluble. Flammable, irritating vapor is produced. Boils at 89°F/32°C.

Extremely flammable • Containers may BLEVE when exposed to fire •Vapors may form explosive mixture with air • Vapors are heavier than air and will collect and stay in low areas • Vapors may travel long distances to ignition sources and flashback • Vapors in confined areas (e.g., tanks, sewers, buildings) may explode when exposed to fire • Highly Irritating to the skin, eyes, and respiratory tract • Toxic products of combustion may include carbon monoxide • Do not put yourself in danger by entering a contaminated area to rescue a victim.

Hazard Classification (based on NFPA-704 Rating System)
Health Hazards (Blue): 2; Flammability (Red): 4; Reactivity (Yellow): 0

EMERGENCY RESPONSE: See Appendix A (129)
Evacuation:
Public safety: Isolate spill area for at least 50 to 100 meters (160 to 330 feet) in all directions.
Spill: Large spill–Consider initial downwind evacuation for at least 300 meters (1000 feet).
Fire: Isolate for 800 meters (½ mile) in all directions, especially if tank, rail car, or tank truck is involved in fire.

EXPOSURE
Short-term effects: *SEEK MEDICAL ATTENTION.* **Vapor:** Irritating to eyes, nose, and throat. Harmful if inhaled. May cause lung edema; physical exertion will aggravate this condition. Move victim to fresh air. *IF BREATHING HAS STOPPED*, give artificial respiration. IF breathing is difficult, administer oxygen. **Liquid:**

Irritating to skin and eyes. Harmful if swallowed. Remove contaminated clothing and shoes. Flush affected areas with plenty of water. *IF IN EYES*, hold eyelids open and flush with plenty of water. *IF SWALLOWED* and victim is *CONSCIOUS AND ABLE TO SWALLOW*, have victim drink 4 to 8 ounces of water and have victim induce vomiting. *IF SWALLOWED* and victim is *UNCONSCIOUS OR HAVING CONVULSIONS*, do nothing except keep victim warm.

Note to physician or authorized medical personnel: Pulmonary edema may be delayed. Medical observation is recommended for 24 to 48 hours after inhalation overexposure. As first aid for pulmonary edema, a physician or authorized medical personnel may consider administering a corticosteroid spray. Cigarette smoking may exacerbate pulmonary injury and should be discouraged for at least 72 hours following exposure.

HEALTH HAZARDS

Personal protective equipment (PPE): B-Level PPE. OSHA Table Z-1-A air contaminant. Goggles or safety glasses; self-contained breathing apparatus; rubber gloves.

Recommendations for respirator selection: NIOSH/OSHA *1000 ppm*: SA* (any supplied-air respirator). *2500 ppm*: SA:CF* (any supplied-air respirator operated in a continuous-flow mode). *4500 ppm*: SCBAF (any self-contained breathing apparatus with a full facepiece); or SAF (any supplied-air respirator with a full facepiece). *EMERGENCY OR PLANNED ENTRY INTO UNKNOWN CONCENTRATIONS OR IDLH CONDITIONS*: SCBAF:PD,PP (any self-contained breathing apparatus that has a full facepiece and is operated in a pressure-demand or other positive-pressure mode); or SAF:PD,PP:ASCBA (any supplied-air respirator that has a full facepiece and is operated in a pressure-demand or other positive-pressure mode in combination with an auxiliary self-contained breathing apparatus operated in a pressure-demand or other positive pressure mode). *ESCAPE:* GMFOV [any air-purifying, full-facepiece respirator (gas mask) with a chin-style, front- or back-mounted acid gas canister]; or SCBAE (any appropriate escape-type, self-contained breathing apparatus). *Note*: Substance reported to cause eye irritation or damage; may require eye protection.

Exposure limits (TWA unless otherwise noted): ACGIH TLV 100 ppm (246 mg/m^3); NIOSH/OSHA 100 ppm (250 mg/m^3). SORT TERM Exposure limits: ACGIH STEL 150 ppm (368 mg/m^3); NIOSH STEL 150 ppm (375 mg/m^3).

Toxicity by ingestion: Grade 1; LD$_{50}$ = 5 to 15 g/kg.

Long-term health effects: The ocular nerve may be affected. Possible nervous system damage.

Vapor (gas) irritant characteristics: Vapors are irritating such that personnel will not usually tolerate moderate or high concentrations. May cause damage to the optic nerve or nervous system.

Liquid or solid irritant characteristics: Corrosive to the sin, eyes and respiratory tract. If spilled on clothing and allowed to remain, may cause smarting and reddening of the skin.

Odor threshold: 2000 ppm.

IDLH value: 4500 ppm.

FIRE DATA

Flash point: −2°F/−19°C (cc).
Flammable limits in air: LEL 4.5%; UEL 23%.
Fire extinguishing agents not to be used: Water may be ineffective.
Autoignition temperature: 840°F/449°C/722°K.
Electrical hazard: Class I, Group D.
Burning rate: 2.5 mm/min

CHEMICAL REACTIVITY

Reactivity with water: Slow reaction to form formic acid and methyl alcohol.
Binary reactants: Strong oxidizers; possible fire and explosion hazard. Decomposes on contact with acids + water; bases + water. May attack certain plastic materials.

ENVIRONMENTAL DATA

Food cain concentration potential: Negative; unlikely to accumulate.
Water pollution: Effect of low concentrations on aquatic life is unknown. May be dangerous if it enters nearby water intakes; notify operators. Notify local health and wildlife officials. **Response to discharge:** Issue warning–high flammability, water contaminant. Restrict access. Disperse and flush.

SIPPING INFORMATION

Grades of purity: Technical, practical, and spectro grades: All 97.5+%; **Storage temperature:** Less than 85°F; **Inert atmosphere:** None; **Venting:** Pressure-vacuum; **Stability during transport:** Stable.

NAS HAZARD CLASSIFICATION FOR BULK WATER TRANSPORTATION

FIRE: 4
HEALTH: Vapor irritant: 3; Liquid or solid irritant: 1; Poisons: 1
WATER POLLUTION: Human toxicity: 1; Aquatic toxicity: 3; Aesthetic effect: 1
REACTIVITY: Other chemicals: 1; Water: 0; Self-reaction: 0

PHYSICAL AND CHEMICAL PROPERTIES

Physical state @ 59°F/15°C and 1 atm: Liquid.
Molecular weight: 60.1
Boiling point @ 1 atm: 89.2°F/31.8°C/305°K.
Melting/Freezing point: −147.6°F/−99.8°C/173.4°K.
Critical temperature: 417°F/214°C/487°K.
Critical pressure: 870 psia = 59.2 atm = 6.00 MN/m^2.
Specific gravity (water = 1): 0.977 @ 68°F/20°C (liquid).
Liquid surface tension: 25 dynes/cm = 0.025 N/m @ 68°F/20°C.
Relative vapor density (air = 1): 2.07
Ratio of specific heats of vapor (gas): 1.1446
Latent heat of vaporization: 202 Btu/lb = 112 cal/g = 4.69 x 10^5 J/kg.
Heat of combustion: −6980 Btu/lb = −3880 cal/g = −162 x 10^5 J/kg.
Vapor pressure: 480 mm.

METHYL HEPTYL KETONE   REC. M:4300

SYNONYMS: KETONE, HEPTYL METHYL; NONAN-2-ONE; 2-NONANONE

IDENTIFICATION
CAS Number: 821-55-6
Formula: C$_9$H$_{18}$O; CH$_3$(CH$_2$)$_6$COCH$_3$
DOT ID Number: UN 1993; DOT Guide Number: 128
Proper Shipping Name: Flammable liquid, n.o.s.

DESCRIPTION: Colorless liquid.

Flammable • Containers may BLEVE when exposed to fire • Vapors may form explosive mixture with air • Vapors are heavier than air and will collect and stay in low areas • Vapors may travel long distances to ignition sources and flashback • Vapors in confined areas (e.g., tanks, sewers, buildings) may explode when exposed to fire • Irritating to the skin, eyes, and respiratory tract • Toxic products of combustion may include carbon monoxide.

Hazard Classification (based on NFPA-704 Rating System)
Health Hazards (Blue): 0; Flammability (Red): 2; Reactivity (Yellow): 0

EMERGENCY RESPONSE: See Appendix A (128)
Evacuation:
Public safety: Isolate spill area for at least 25 to 50 meters (80 to 160 feet) in all directions.
Spill: Large spill–Consider initial downwind evacuation for at least 300 meters (1000 feet).
Fire: Isolate for 800 meters (½ mile) in all directions, especially if tank, rail car, or tank truck is involved in fire.

EXPOSURE
Short-term effects: *SEEK MEDICAL ATTENTION.* **Vapor:** Irritating to eyes, nose, and throat. *IF INHALED*, will, will cause dizziness, headache, or difficult breathing. *IF IN EYES*, hold eyelids open and flush with plenty of water. *IF BREATHING HAS STOPPED*, give artificial respiration. IF breathing is difficult, administer oxygen. **Liquid:** Irritating to skin and eyes. *IF SWALLOWED*, will cause nausea or vomiting. Remove contaminated clothing and shoes. Flush affected areas with plenty of water. *IF IN EYES*, hold eyelids open and flush with plenty of water.

HEALTH HAZARDS
Personal protective equipment (PPE): B-Level PPE. Self-contained breathing apparatus, rubber boots and heavy rubber gloves.
Toxicity by ingestion: Grade 2: LD_{50} = 3.2 g/kg (rat).
Vapor (gas) irritant characteristics: Vapors cause moderate irritation such that personnel will find high concentrations unpleasant. The effect may be temporary.
Liquid or solid irritant characteristics: If spilled on clothing and allowed to remain, may cause smarting and reddening of skin.

FIRE DATA
Flash point: 140°F/60°C (cc).
Flammable limits in air: LEL: 0.9 @ 180°F; UEL: 5.9 @ 313°F
Fire extinguishing agents not to be used: Water may be ineffective.
Autoignition temperature: 680°F/360°C/633°K.

CHEMICAL REACTIVITY
Binary reactants: Reacts violently with aldehydes, nitric acid, strong oxidizers, perchloric acid. Incompatible with strong acids, aliphatic amines. Reaction with hydrogen peroxide may form unstable peroxides.
Reactivity group: 18
Compatibility class: Ketones

ENVIRONMENTAL DATA
Food chain concentration potential: Negative; unlikely to accumulate.
Water pollution: Effect of low concentrations on aquatic life is unknown. Fouling shoreline. May be dangerous if it enter water intakes. Notify local health and wildlife officials. **Response to discharge:** Restrict access. Evacuate areas. Mechanical containment. Should be removed. Chemical and physical treatment.

SHIPPING INFORMATION
Grades of purity: 99+%; **Storage temperature:** Ambient; **Stability during transport:** Stable.

PHYSICAL AND CHEMICAL PROPERTIES
Physical state @ 59°F/15°C and 1 atm: Liquid.
Molecular weight: 142.24
Boiling point @ 1 atm: 378°F/192°C/465°K (at 743 mmHg = 0.97 atm).
Melting/Freezing point: –5.8°F/–21°C/252.2°K.
Specific gravity (water = 1): 0.832
Relative vapor density (air = 1): 4.9
Vapor pressure: (Reid) 0.0309 psia.

METHYLHYDRAZINE REC. M:4350

SYNONYMS: HYDRAZINE, METHYL-; HYDRAZOMETHANE; 1-METHYL-HYDRAZINE; METILHIDRAZINA (Spanish); MMH; MONOMETHYL HYDRAZINE; RCRA No. P068

IDENTIFICATION
CAS Number: 60-34-4
Formula: CH_6N_2; CH_3NHNH_2
DOT ID Number: UN 1244; DOT Guide Number: 131
Proper Shipping Name: Methylhydrazine
Reportable Quantity (RQ): **(CERCLA)** 10 lb/4.54 kg

DESCRIPTION: Colorless, fuming liquid. Fishy, ammonia-like odor. Mixes with water forming a medium strong base. Poisonous, flammable vapor is produced.

Poison! • Highly flammable (Used as a rocket propellant) • May ignite spontaneously upon contact with porous materials (cloth, wood, earth, etc.) • Breathing the vapor, skin or eye contact, or swallowing the material can kill you • Firefighting gear (including SCBA) does not provide adequate protection. If exposure occurs, remove and isolate gear immediately and thoroughly decontaminate personnel • Containers may BLEVE when exposed to fire • Vapors are heavier than air and will collect and stay in low areas • Vapors in confined areas (e.g., tanks, sewers, buildings) may explode when exposed to fire • Toxic products of combustion may include nitrogen oxides • Do not put yourself in danger by entering a contaminated area to rescue a victim.

Hazard Classification (based on NFPA-704 Rating System)
Health Hazards (Blue): 4; Flammability (Red): 3; Reactivity (Yellow): 2

EMERGENCY RESPONSE: See Appendix A (131)
Evacuation:
Public safety: Isolate spill area for at least 100 to 200 meters (330 to 660 feet) in all directions.
Spill: Small spill–First: Isolate in all directions 30 meters (100 feet); Then: Protect persons downwind, DAY: 0.3 km (0.2 mile); NIGHT: 0.8 km (0.5 mile). Large spill–First: Isolate in all directions 125 meters (400 feet); Then: Protect persons downwind, DAY: 1.1 km (0.7 mile); NIGHT: 2.7 km (1.7 miles).
Fire: Isolate for 800 meters (½ mile) in all directions, especially if tank, rail car, or tank truck is involved in fire.

EXPOSURE
Short-term effects: *SEEK MEDICAL ATTENTION.* **Vapor:** *POISONOUS IF INHALED OR IF SKIN IS EXPOSED.* Irritating

to eyes, nose, and throat. Move victim to fresh air. *IF BREATHING HAS STOPPED,* give artificial respiration. IF breathing is difficult, administer oxygen. **Liquid:** *POISONOUS IF SWALLOWED OR IF SKIN IS EXPOSED.* Will burn skin and eyes. Remove contaminated clothing and shoes. Flush affected areas with plenty of water. *IF IN EYES,* hold eyelids open and flush with plenty of water. *IF SWALLOWED* and victim is *CONSCIOUS AND ABLE TO SWALLOW,* have victim drink 4 to 8 ounces of water. **Do NOT induce vomiting.**

HEALTH HAZARDS

Personal protective equipment (PPE): A-Level PPE. OSHA Table Z-1-A air contaminant. Chemical protective material(s) reported to have good to excellent resistance: Viton®, chlorobutyl rubber, CR-39. Also, butyl rubber offers limited protection.

Recommendations for respirator selection: NIOSH

At any concentrations above the NIOSH REL, or where there is no REL, at any detectable concentration: SCBAF:PD,PP (any self-contained breathing apparatus that has a full facepiece and is operated in a pressure-demand or other positive-pressure mode); or SAF:PD,PP:ASCBA (any supplied-air respirator that has a full facepiece and is operated in a pressure-demand or other positive-pressure mode in combination with an auxiliary self-contained breathing apparatus operated in a pressure-demand or other positive pressure mode). *ESCAPE:* SCBAE (any appropriate escape-type, self-contained breathing apparatus).

Exposure limits (TWA unless otherwise noted): ACGIH TLV 0.01 ppm (0.019 mg/m^3) animal carcinogen; OSHA PEL ceiling 0.2 ppm (0.35 mg/m^3); NIOSH REL ceiling 0.04 ppm (0.08 mg/m^3) [2 hours]; potential human carcinogen; skin contact contributes significantly in overall exposure.

Toxicity by ingestion: Poison. Grade 4; oral LD$_{50}$ = 33 mg/kg (rat).

Long-term health effects: May cause cancer. Hemolytic anemia may result from large doses by any route; suspected carcinogen and mutagen. May cause nervous system damage. Possible brain, kidney, liver and skin disorders.

Vapor (gas) irritant characteristics: Vapors are irritating such that personnel will not usually tolerate moderate or high concentrations.

Liquid or solid irritant characteristics: Corrosive the eyes, skin, and respiratory tract. Severe skin irritant. Causes second-and third-degree burns on short contact and is very injurious to the eyes.

Odor threshold: 1.7 ppm; also listed at 2.0–3.0 mg/m^3.

IDLH value: 20 ppm; potential human carcinogen.

FIRE DATA

Flash point: 17°F/–8°C (cc)
Flammable limits in air: LEL: 2.5%; UEL: 97%.
Behavior in fire: Vapors may explode.
Autoignition temperature: 382°F/194°C/467°K.
Electrical hazard: Class I, Group C.
Burning rate: 2.0 mm/min

CHEMICAL REACTIVITY

Binary reactants: Pyrophoric; may ignite spontaneously in air. A highly reactive reducing agent and medium-strong organic base. Violent reaction with strong oxidizers, dicyanofurazan, nitric acid, hydrogen peroxide, nitrogen tetroxide. Reacts, possibly violently, with organic anhydrides, acrylates, alcohols, aldehydes, alkylene oxides, substituted allyls, cellulose nitrate, cresols, caprolactam solution, epichlorohydrin, ethylene dichloride, isocyanates, ketones, glycols, maleic anhydride, nitrates, phenols, vinyl acetate. Contact with manganese, lead, copper, or their alloys may cause fire and explosions. Attacks some plastics, coatings, and rubber.

Neutralizing agents for acids and caustics: Flush with water.

ENVIRONMENTAL DATA

Food chain concentration potential: Negative; unlikely to accumulate.

Water pollution: Effect of low concentrations on aquatic life is unknown. May be dangerous if it enters nearby water intakes; notify operators. Notify local health and wildlife officials. **Response to discharge:** Issue warning–poison, high flammability, water contaminant, air contaminant. Restrict access. Evacuate area. Disperse and flush.

SHIPPING INFORMATION

Grades of purity: Propellant grade, 99+%; Laboratory grade, 98+%; **Storage temperature:** Ambient; **Inert atmosphere:** Padded with nitrogen; **Venting:** Safety relief; **Stability during transport:** Stable if not in contact with iron, copper, or their alloys.

NAS HAZARD CLASSIFICATION FOR BULK WATER TRANSPORTATION

FIRE: 4
HEALTH: Vapor irritant: 3; Liquid or solid irritant: 4; Poisons: 4
WATER POLLUTION: Human toxicity: 4; Aquatic toxicity:; Aesthetic effect: 2
REACTIVITY: Other chemicals: 4; Water: 0; Self-reaction: 4

PHYSICAL AND CHEMICAL PROPERTIES

Physical state @ 59°F/15°C and 1 atm: Liquid.
Molecular weight: 46.1
Boiling point @ 1 atm: 189.5°F/87.5°C/360.7°K.
Melting/Freezing point: –62.3°F/–52.4°C/220.8°K.
Critical temperature: 594°F/312°C/585°K.
Critical pressure: 1195 psia = 81.3 atm = 8.25 MN/m^2.
Specific gravity (water = 1): 0.878 @ 68°F/20°C (liquid).
Liquid surface tension: 34.3 dynes/cm = 0.0343 N/m @ 68°F/20°C.
Relative vapor density (air = 1): 1.59
Ratio of specific heats of vapor (gas): 1.1326
Latent heat of vaporization: 376 Btu/lb = 209 cal/g = 8.75 x 10^5 J/kg.
Heat of combustion: –12,178 Btu/lb = –6766 cal/g = –283.1 x 10^5 J/kg.
Vapor pressure: 38 mm.

2-METHYL-2-HYDROXY-3-BUTYNE   REC. M:4400

SYNONYMS: ALCOHOL *terc*-AMILICO (Spanish); *tert*-AMYL ALCOHOL; AMYLENE HYDRATE; DIMETHYLETHYLCARBINOL; DIMETHYLACETYLENECARBINOL; EEC No. 603-007-00-2; METHYL BUTYNOL; 2-METHYL-2-BUTANOL; 2-METHYL-2-BUTYNOL; *tert*-PENTANOL

IDENTIFICATION

CAS Number: 75-85-4
Formula: C$_5$H$_{12}$O; (CH$_3$)$_2$(OH)C–C≡CH
DOT ID Number: UN 1993; DOT Guide Number: 128
Proper Shipping Name: Flammable liquid, n.o.s.

DESCRIPTION: Colorless liquid. Characteristic. Floats on the surface of water; soluble. Flammable vapor is produced.

Highly flammable • Containers may BLEVE when exposed to fire •Vapors may form explosive mixture with air • Vapors are heavier

Methyl iodide

than air and will collect and stay in low areas • Vapors may travel long distances to ignition sources and flashback • Vapors in confined areas (e.g., tanks, sewers, buildings) may explode when exposed to fire • Irritating to the skin, eyes, and respiratory tract; high concentrations may act as a narcotic • Toxic products of combustion may include carbon monoxide.

Hazard Classification (based on NFPA-704 Rating System)
Health Hazards (Blue): 2; Flammability (Red): 3; Reactivity (Yellow): 0

EMERGENCY RESPONSE: See Appendix A (128)
Evacuation:
Public safety: Isolate spill area for at least 25 to 50 meters (80 to 160 feet) in all directions.
Spill: Large spill–Consider initial downwind evacuation for at least 300 meters (1000 feet).
Fire: Isolate for 800 meters (½ mile) in all directions, especially if tank, rail car, or tank truck is involved in fire.

EXPOSURE
Short-term effects: *SEEK MEDICAL ATTENTION*. **Liquid:** May cause eye, skin and respiratory tract irritation. High concentrations may cause unconsciousness. Flush affected area with plenty of water. The use of alcoholic beverages may enhance the toxic effect.

HEALTH HAZARDS
Personal protective equipment (PPE): B-Level PPE. Rubber gloves, face shield and laboratory coat. An all purpose canister mask should be available if needed. Natural rubber, neoprene, and nitrile rubber may offer some protection from alcohols.
Toxicity by ingestion: May be poisonous.
Long-term health effects: May cause blood and nervous system damage.
Vapor (gas) irritant characteristics: May be narcotic; hypnotic in high concentrations.
Liquid or solid irritant characteristics: Moderately toxic. May be absorbed through the skin.

FIRE DATA
Flash point: 77°F/25°C (oc).
Fire extinguishing agents not to be used: Water may be ineffective.

CHEMICAL REACTIVITY
Binary reactants: May self-ignite in air. Reacts with alkali metals and oxidizing materials (may be violent or cause fire).

ENVIRONMENTAL DATA
Food chain concentration potential: Log P_{ow} = 0.90. Unlikely to accumulate.
Water pollution: Effect of low concentration on aquatic life is unknown. May be dangerous if it enters nearby water intakes; notify operators. Notify local health and wildlife officials.
Response to discharge: Issue warning–high flammability. Restrict access. Disperse and flush.

SHIPPING INFORMATION
Stability during transport: Stable.

PHYSICAL AND CHEMICAL PROPERTIES
Physical state @ 59°F/15°C and 1 atm: Liquid.
Molecular weight: 84.11.
Boiling point @ 1 atm: 219.2°F/104°C/377.2°K.
Melting/Freezing point: 37.4°F/3.0°C/276.2°K.
Critical temperature: (estimate) 553.5°F/289.7°C/562.9°K.
Critical pressure: (estimate) 582 psia = 39.6 atm = 4.01 MN/m².
Specific gravity (water = 1): 0.8618 @ 68°F/20°C.
Liquid surface tension: 23.8 dynes/cm = 0.0238 N/m @ 77°F/25°C.
Relative vapor density (air = 1): 2.9.
Ratio of specific heats of vapor (gas): (estimate) Greater than one.
Latent heat of vaporization: (estimate) 165.6 Btu/lb = 92 cal/g = 3.8×10^5 J/kg.

METHYL IODIDE   REC. M:4450

SYNONYMS: EEC No. 602-005-00-9; HALON 10001; IODOMETHANE; IODURE DE METHYLE (French); METHANE, IODO-; METHYLJODID (German); MONOIODOMETHANE; RCRA No. U138; YODURO de METILO (Spanish)

IDENTIFICATION
CAS Number: 74-88-4
Formula: CH_3I
DOT ID Number: UN 2644; DOT Guide Number: 151
Proper Shipping Name: Methyl iodide
Reportable Quantity (RQ): **(CERCLA)** 100 lb/45.4 kg

DESCRIPTION: Colorless liquid; turns yellow, red or brown on exposure to light, air and moisture. Sweet, ether-like or chloroform odor. Sinks and decomposes slowly in water, forming hydrogen iodide; slightly soluble. Poisonous vapor cloud is formed.

Poison! • Breathing the vapors can kill you; prolonged contct with the skin or eyes can cause burns • Firefighting gear (including SCBA) does not provide adequate protection. If exposure occurs, remove and isolate gear immediately and thoroughly decontaminate personnel • Containers may BLEVE when exposed to fire • Decomposes above 518°F/270°C; toxic products of combustion may include hydrogen iodide • Do not put yourself in danger by entering a contaminated area to rescue a victim.

Hazard Classification (based on NFPA-704 Rating System)
Health Hazards (Blue): 3; Flammability (Red): 0; Reactivity (Yellow): 0

EMERGENCY RESPONSE: See Appendix A (151)
Evacuation:
Public safety: Isolate the area of spill or leak for at least 25 to 50 meters (80 to 160 feet) in all directions.
Spill: Small spill–First: Isolate in all directions 30 meters (100 feet); Then: Protect persons downwind, DAY: 0.2 km (0.1 mile); NIGHT: 0.3 km (0.2 mile). Large spill–First: Isolate in all directions 60 meters (200 feet); Then: Protect persons downwind, DAY: 0.3 km (0.2 mile); NIGHT: 1.0 km (0.6 mile).
Fire: If tank, rail car, or tank truck is involved in fire, isolate for at least 800 meters (½ mile) in all directions; also, consider initial evacuation for 800 meters (½ mile) in all directions.

EXPOSURE
Short-term effects: *SEEK MEDICAL ATTENTION*. **Vapor:** *POISONOUS IF INHALED*. Irritating to eyes. May cause lung edema; physical exertion will aggravate this condition. Move to fresh air. *IF BREATHING HAS STOPPED*, give artificial respiration; *avoid mouth-to-mouth resuscitation; use bag/mask apparatus*. IF breathing is difficult, administer oxygen. **Liquid:**

HARMFUL IF ABSORBED THROUGH THE SKIN. Corrosive. Will burn skin and eyes. Harmful if swallowed. Remove contaminated clothing and shoes. Flush affected areas with plenty of water. *DO NOT RUB AFFECTED AREAS. IF IN EYES*, hold eyelids open and flush with plenty of water. *IF SWALLOWED* and victim is *CONSCIOUS AND ABLE TO SWALLOW*, have victim drink 4 to 8 ounces of water. **Do NOT induce vomiting**.

Note to physician or authorized medical personnel: Pulmonary edema may be delayed. Medical observation is recommended for 24 to 48 hours after inhalation overexposure. As first aid for pulmonary edema, a physician or authorized medical personnel may consider administering a corticosteroid spray. Cigarette smoking may exacerbate pulmonary injury and should be discouraged for at least 72 hours following exposure.

HEALTH HAZARDS
Personal protective equipment (PPE): B-Level PPE. OSHA Table Z-1-A air contaminant. Full face self-contained breathing apparatus. Chemical protective material(s) reported to have good to excellent resistance: Viton®. Also, PV alcohol offers limited protection.
Recommendations for respirator selection: NIOSH
NIOSH *At any concentrations above the NIOSH REL, or where there is no REL, at any detectable concentration:* SCBAF:PD,PP (any self-contained breathing apparatus that has a full facepiece and is operated in a pressure-demand or other positive-pressure mode); or SAF:PD,PP:ASCBA (any supplied-air respirator that has a full facepiece and is operated in a pressure-demand or other positive-pressure mode in combination with an auxiliary self-contained breathing apparatus operated in a pressure-demand or other positive pressure mode). *ESCAPE:* GMFOV [any air-purifying, full-facepiece respirator (gas mask) with a chin-style, front-or back-mounted organic vapor canister]; or SCBAE (any appropriate escape-type, self-contained breathing apparatus). *At any concentrations above the NIOSH REL, or where there is no REL, at any detectable concentration:* SCBAF:PD,PP (any self-contained breathing apparatus that has a full facepiece and is operated in a pressure-demand or other positive-pressure mode); or SAF:PD,PP:ASCBA (any supplied-air respirator that has a full facepiece and is operated in a pressure-demand or other positive-pressure mode in combination with an auxiliary self-contained breathing apparatus operated in a pressure-demand or other positive pressure mode). *ESCAPE:* GMFOV [any air-purifying, full-facepiece respirator (gas mask) with a chin-style, front-or back-mounted organic vapor canister]; or SCBAE (any appropriate escape-type, self-contained breathing apparatus).
Exposure limits (TWA unless otherwise noted): ACGIH TLV 2 ppm (12 mg/m^3) suspected human carcinogen; OSHA PEL 5 ppm (28 mg/m^3); NIOSH REL 2 ppm (10 mg/m^3); potential human carcinogen; reduce exposure to lowest feasible level; skin contact contributes significantly in overall exposure.
Toxicity by ingestion: Poison. Grade 3: LD$_{50}$ = 51 mg/kg (guinea pig).
Long-term health effects: Possible human carcinogen. IARC rating 3; limited animal evidence. Mutation data.
Vapor (gas) irritant characteristics: Corrosive to skin, eyes and respiratory tract. Vapor is irritating such that personnel will not usually tolerate moderate or high vapor concentrations. May cause lung edema; physical exertion will aggravate this condition. affects the central nervous system.
Liquid or solid irritant characteristics: Fairly severe skin irritant; may cause pain and second-degree burns after a few minutes of contact. Toxic.
IDLH value: Suspected human carcinogen; 100 ppm.

FIRE DATA
Flash point: Practically not flammable.
Behavior in fire: Gives off corrosive and toxic hydrogen iodide gas. Containers may explode. Vapor heavier than air.

CHEMICAL REACTIVITY
Binary reactants: Strong oxidizers; oxygen at high temperatures; silver chlorite; sodium. Keep away from hot surfaces and welding opperations.

ENVIRONMENTAL DATA
Food chain concentration potential: Log P$_{ow}$ = 1.72. Unlikely to accumulate.
Water pollution: Not harmful to aquatic life May be dangerous if it enters nearby water intakes; notify operators. Notify local health and wildlife officials. **Response to discharge:** Issue warning–poison. Restrict access.

SHIPPING INFORMATION
Grades of purity: Commercial: not less than 99.5%; **Storage temperature:** Ambient; **Inert atmosphere:** None; **Venting:** Safety relief; **Stability during transport:** Stable.

NAS HAZARD CLASSIFICATION FOR BULK WATER TRANSPORTATION
FIRE: 0
HEALTH: Vapor irritant: 3; Liquid or solid irritant: 3; Poisons: 4
WATER POLLUTION: Human toxicity: 3; Aquatic toxicity: 3; Aesthetic effect: 3
REACTIVITY: Other chemicals: 1; Water: 1; Self-reaction: 0

PHYSICAL AND CHEMICAL PROPERTIES
Physical state @ 59°F/15°C and 1 atm: Liquid.
Molecular weight: 141.94
Boiling point @ 1 atm: 109°F/43°C/316°K.
Melting/Freezing point: –88°F/–65°C/208°K.
Critical temperature: 491°F/254.8°C/528°K.
Critical pressure: 1068 psia = 72.7 atm = 7.37 MN/m^2.
Specific gravity (water = 1): 2.279 @ 68°F/20°C (liquid).
Liquid surface tension: 25.8 dynes/cm = 0.026 N/m at 109°F/43°C/316°K.
Relative vapor density (air = 1): 4.89
Latent heat of vaporization: 82.6 Btu/lb = 45.9 cal/g = 1.9 x 10^5 J/kg.
Heat of combustion: –4793 Btu/lb = –2663 cal/g = –111 x 10^5 J/kg.
Heat of decomposition: 265°C.
Vapor pressure: (Reid) 12.8 psia; 450 mm.

METHYL ISOBUTYL CARBINOL   REC. M:4500

SYNONYMS: AMYL METHYL ALCOHOL; EEC No. 603-008-00-8; 1,3-DIMETHYL BUTANOL; ISOHEXYL ALCOHOL; METHYLAMYL ALCOHOL; 4-METHYL-2-PENTANOL; 4-METHY-2-PENTYL ALCOHOL; 2-METHYL-2-PROPYLETHANOL; METILIOSBUTILCARBINOL (Spanish); ISOBUTYLMETHYLCARBINOL; MAOH; MIBC; MIC; MAA

IDENTIFICATION
CAS Number: 108-11-2
Formula: C$_6$H$_{14}$O; (CH$_3$)$_2$CHCH$_2$CH(OH)CH$_3$
DOT ID Number: UN 2053; DOT Guide Number: 129
Proper Shipping Name: Methyl isobutyl carbinol

Methyl isobutyl carbinol

DESCRIPTION: Colorless, oily liquid. Sharp, irritating odor. Floats on water surface; slightly soluble. Irritating vapor is produced.

Flammable • Containers may BLEVE when exposed to fire • Vapors may form explosive mixture with air • Vapors are heavier than air and will collect and stay in low areas • Vapors may travel long distances to ignition sources and flashback • Vapors in confined areas (e.g., tanks, sewers, buildings) may explode when exposed to fire • Irritating to the skin, eyes, and respiratory tract • Toxic products of combustion may include carbon monoxide.

Hazard Classification (based on NFPA-704 Rating System)
Health Hazards (Blue): 2; Flammability (Red): 2; Reactivity (Yellow): 0

EMERGENCY RESPONSE: See Appendix A (129)
Evacuation:
Public safety: Isolate spill area for at least 50 to 100 meters (160 to 330 feet) in all directions.
Spill: Large spill–Consider initial downwind evacuation for at least 300 meters (1000 feet).
Fire: Isolate for 800 meters (½ mile) in all directions, especially if tank, rail car, or tank truck is involved in fire.

EXPOSURE
Short-term effects: *SEEK MEDICAL ATTENTION.* **Vapor:** Irritating to eyes, nose, and throat. Harmful if skin is exposed. *IF INHALED*, will, will cause dizziness or difficult breathing. Move to fresh air. *IF BREATHING HAS STOPPED,* give artificial respiration. IF breathing is difficult, administer oxygen. **Liquid:** Irritating to skin and eyes. Harmful if swallowed. Remove contaminated clothing and shoes. Flush affected areas with plenty of water. *IF IN EYES,* hold eyelids open and flush with plenty of water. *IF SWALLOWED* and victim is *CONSCIOUS AND ABLE TO SWALLOW,* have victim drink 4 to 8 ounces of water.

HEALTH HAZARDS
Personal protective equipment (PPE): B-Level PPE. OSHA Table Z-1-A air contaminant. Air pack or organic canister mask; rubber gloves; goggles or face shield. Also, natural rubber, neoprene, and nitrile rubber may offer some protection from alcohols.
Recommendations for respirator selection: NIOSH/OSHA *250 ppm:* SA* (any supplied-air respirator). *400 ppm:* SA:CF* (any supplied-air respirator operated in a continuous-flow mode); or SCBAF (any self-contained breathing apparatus with a full facepiece); or SAF (any supplied-air respirator with a full facepiece). *EMERGENCY OR PLANNED ENTRY INTO UNKNOWN CONCENTRATIONS OR IDLH CONDITIONS:* SCBAF:PD,PP (any self-contained breathing apparatus that has a full facepiece and is operated in a pressure-demand or other positive-pressure mode); or SAF:PD,PP:ASCBA (any supplied-air respirator that has a full facepiece and is operated in a pressure-demand or other positive-pressure mode in combination with an auxiliary self-contained breathing apparatus operated in a pressure-demand or other positive pressure mode). *ESCAPE:* GMFOV [any air-purifying, full-facepiece respirator (gas mask) with a chin-style, front- or back-mounted organic vapor canister]; or SCBAE (any appropriate escape-type, self-contained breathing apparatus).
Note: Substance reported to cause eye irritation or damage; may require eye protection.
Exposure limits (TWA unless otherwise noted): ACGIH TLV 25 ppm (104 mg/m^3); NIOSH/OSHA 25 ppm (100 mg/m^3); skin contact contributes significantly in overall exposure.

Short-term exposure limits (15-minute TWA): ACGIH STEL 40 ppm (167 mg/m^3); NIOSH STEL 40 ppm (165 mg/m^3); skin contact contributes significantly in overall exposure.
Toxicity by ingestion: Grade 2; LD$_{50}$ = 0.5 to 5 g/kg (rat).
Long-term health effects: Possible nervous system damage.
Vapor (gas) irritant characteristics: May cause irritation of the eyes, nose, and respiratory tract. Vapors cause irritation such that personnel will find high concentrations unpleasant. The effect may be temporary. High concentrations may have a narcotic effect.
Liquid or solid irritant characteristics: May enter body through the skin. May cause irritation of the eyes, skin and mucous membrane. If spilled on clothing and allowed to remain, may cause smarting and reddening of the skin. Prolonged or intensive contact may cause degreasing of the skin.
Odor threshold: 0.01–1.0 ppm.
IDLH value: 400 ppm.

FIRE DATA
Flash point: 120–130°F/49–54°C (oc); 106°F/41°C (cc).
Flammable limits in air: LEL: 1.0%; UEL: 5.5%.
Electrical hazard: Flow or agitation of substance may generate electrostatic charges due to low conductivity.

CHEMICAL REACTIVITY
Binary reactants: Violent reaction with strong oxidizers. contact with alkali metals may produce flammable hydrogen gas.
Reactivity group: 20
Compatibility class: Alcohols, glycols

ENVIRONMENTAL DATA
Food chain concentration potential: Negative; unlikely to accumulate.
Water pollution: Effect of low concentrations on aquatic life is unknown. Fouling to shoreline. May be dangerous if it enters nearby water intakes; notify operators. Notify local health and wildlife officials. **Response to discharge:** Mechanical containment. Chemical and physical treatment.

SHIPPING INFORMATION
Storage temperature: Ambient; **Inert atmosphere:** None; **Venting:** Open (flame arrester); **Stability during transport:** Stable.

NAS HAZARD CLASSIFICATION FOR BULK WATER TRANSPORTATION
FIRE: 2
HEALTH: Vapor irritant: 2; Liquid or solid irritant: 1; Poisons: 2
WATER POLLUTION: Human toxicity: 2; Aquatic toxicity: 2; Aesthetic effect: 2
REACTIVITY: Other chemicals: 2; Water: 0; Self-reaction: 0

PHYSICAL AND CHEMICAL PROPERTIES
Physical state @ 59°F/15°C and 1 atm: Liquid.
Molecular weight: 102.18
Boiling point @ 1 atm: 271°F/133°C/406°K.
Melting/Freezing point: –130°F/–90°C/183°K.
Critical temperature: 556°F/291°C/564°K.
Specific gravity (water = 1): 0.81 @ 68°F/20°C (liquid).
Liquid surface tension: 22.8 dynes/cm = 0.0228 N/m @ 68°F/20°C.
Liquid water interfacial tension: (estimate) 25 dynes/cm = 0.025 N/m @ 68°F/20°C.
Ratio of specific heats of vapor (gas): 1.053
Latent heat of vaporization: 162 Btu/lb = 90.1 cal/g = 3.77 x 10^5 J/kg.

Heat of combustion: (estimate) –16,600 Btu/lb = –9300 cal/g = –387 x 10^5 J/kg.
Vapor pressure: (Reid) 0.2 psia; 3 mm.

METHYL ISOBUTYL KETONE   REC. M:4550

SYNONYMS: EEC No. 603-008-00-8; HEXONE; ISOBUTYL METHYL KETONE; ISOPROPYLACETONE; METHYL-ISOBUTYL-CETONE (French); 2-METHYL-4-PENTANONE; 4-METHYL-2-PENTANONE; METIL ISOBUTIL CETONA (Spanish); MIBK; MIK; 2-PENTANONE, 4-METHYL-; SHELL MIBK; RCRA No. U161

IDENTIFICATION
CAS Number: 108-10-1
Formula: $C_6H_{12}O$; $(CH_3)_2CHCH_2COCH_3$
DOT ID Number: UN 1245; DOT Guide Number: 127
Proper Shipping Name: Methyl isobutyl ketone
Reportable Quantity (RQ): **(CERCLA)** 5000 lb/2270 kg

DESCRIPTION: Colorless watery liquid. Mild, pleasant, fruity odor. Floats on the surface of water; slightly soluble. Flammable, irritating vapor is produced.

Highly flammable • Containers may BLEVE when exposed to fire • Vapors may form explosive mixture with air • Vapors are heavier than air and will collect and stay in low areas • Vapors may travel long distances to ignition sources and flashback • Vapors in confined areas (e.g., tanks, sewers, buildings) may explode when exposed to fire • Irritating to the skin, eyes, and respiratory tract • Toxic products of combustion may include carbon monoxide.

Hazard Classification (based on NFPA-704 Rating System)
Health Hazards (Blue): 2; Flammability (Red): 3; Reactivity (Yellow): 1

EMERGENCY RESPONSE: See Appendix A (127)
Evacuation:
Public safety: Isolate spill area for at least 25 to 50 meters (80 to 160 feet) in all directions.
Spill: Large spill–Consider initial downwind evacuation for at least 300 meters (1000 feet).
Fire: Isolate for 800 meters (½ mile) in all directions, especially if tank, rail car, or tank truck is involved in fire.

EXPOSURE
Short-term effects: *SEEK MEDICAL ATTENTION.* **Vapor:** Irritating to eyes, nose, and throat. *IF INHALED*, will, will cause dizziness or loss of consciousness. Move to fresh air. *IF BREATHING HAS STOPPED,* give artificial respiration. IF breathing is difficult, administer oxygen. **Liquid:** Irritating to skin and eyes. Harmful if swallowed. Remove contaminated clothing and shoes. Flush affected areas with plenty of water. *IF IN EYES*, hold eyelids open and flush with plenty of water. *IF SWALLOWED* and victim is *CONSCIOUS AND ABLE TO SWALLOW*, have victim drink 4 to 8 ounces of water. The use of alcoholic beverages may enhance the toxic effect.

HEALTH HAZARDS
Personal protective equipment (PPE): B-Level PPE. OSHA Table Z-1-A air contaminant. Chemical protective material(s) reported to have good to excellent resistance: butyl rubber, PV alcohol, Teflon®, Silvershield®. Also, PVC and styrene-butadiene offers limited protection

Recommendations for respirator selection: NIOSH
500 ppm: CCROV* [any chemical cartridge respirator with organic vapor cartridge(s)]; or GMFOV [any air-purifying, full-facepiece respirator (gas mask) with a chin-style, front- or back-mounted organic vapor cannister]; or PAPROV* [any powered, air-purifying respirator with organic vapor cartridge(s)]; or SA* (any supplied-air respirator); or SCBAF (any self-contained breathing apparatus with a full facepiece). *EMERGENCY OR PLANNED ENTRY INTO UNKNOWN CONCENTRATIONS OR IDLH CONDITIONS:* SCBAF:PD,PP (any self-contained breathing apparatus that has a full facepiece and is operated in a pressure-demand or other positive-pressure mode); or SAF:PD,PP:ASCBA (any supplied-air respirator that has a full facepiece and is operated in a pressure-demand or other positive-pressure mode in combination with an auxiliary self-contained breathing apparatus operated in a pressure-demand or other positive pressure mode). *ESCAPE:* GMFOV [any air-purifying, full-facepiece respirator (gas mask) with a chin-style, front- or back-mounted organic vapor cannister]; or SCBAE (any appropriate escape-type, self-contained breathing apparatus).
**Note:* Substance reported to cause eye irritation or damage; may require eye protection.
Exposure limits (TWA unless otherwise noted): ACGIH TLV 50 ppm (205 mg/m^3); OSHA PEL 100 ppm (410 mg/m^3); NIOSH REL 50 mg/m^3 (205 mg/m^3).
Short-term exposure limits (15-minute TWA): ACGIH STEL 75 ppm (307 mg/m^3); NIOSH STEL 75 ppm (300 mg/m^3).
Toxicity by ingestion: Grade 2; LD_{50} = 0.5 to 5 g/kg (rat).
Vapor (gas) irritant characteristics: Vapors irritate the eyes and respiratory system. The effect may be temporary.
Liquid or solid irritant characteristics: Very irritating to the skin, eyes and respiratory system. If spilled on clothing and allowed to remain, may cause smarting and reddening of the skin. Causes degreasing of the skin.
Odor threshold: 0.08–8.3 ppm.
IDLH value: 500 ppm.

FIRE DATA
Flash point: 75°F/24°C (oc); 64°F/18°C (cc).
Flammable limits in air: LEL: 1.2%; UEL: 8.0%, both @ 200KF (93°C).
Fire extinguishing agents not to be used: Water may be ineffective.
Autoignition temperature: 846°F/452°C/725°K.
Electrical hazard: Class I, Group D.

CHEMICAL REACTIVITY
Binary reactants: May self-ignite in air. Violent reaction with strong oxidizers, potassium *tert*-butoxide. Softens many plastics.
Reactivity group: 18
Compatibility class: Ketone

ENVIRONMENTAL DATA
Food chain concentration potential: Log P_{ow} = 1.2. Unlikely to accumulate.
Water pollution: Effect of low concentrations on aquatic life is unknown. Fouling to shoreline. May be dangerous if it enters nearby water intakes; notify operators. Notify local health and wildlife officials. **Response to discharge:** Issue warning–high flammability. Evacuate area. Disperse and flush.

SHIPPING INFORMATION
Grades of purity: 99+%; **Storage temperature:** Ambient; **Inert atmosphere:** None; **Venting:** Open (flame arrester) or pressure-vacuum; **Stability during transport:** Stable.

NAS HAZARD CLASSIFICATION FOR BULK WATER TRANSPORTATION
FIRE: 3
HEALTH: Vapor irritant: 1; Liquid or solid irritant: 1; Poisons: 1
WATER POLLUTION: Human toxicity: 2; Aquatic toxicity: 1; Aesthetic effect: 2
REACTIVITY: Other chemicals: 2; Water: 0; Self-reaction: 0

PHYSICAL AND CHEMICAL PROPERTIES
Physical state @ 59°F/15°C and 1 atm: Liquid.
Molecular weight: 100.16
Boiling point @ 1 atm: 243°F/117°C/390°K.
Melting/Freezing point: –119°F/–120°C/153°K.
Critical temperature: 569°F/298°C/572°K.
Critical pressure: 475 psia = 32.3 atm = 3.27 MN/m^2.
Specific gravity (water = 1): 0.802 @ 68°F/20°C (liquid).
Liquid surface tension: 23.6 dynes/cm = 0.0236 N/m @ 68°F/20°C.
Liquid water interfacial tension: 15.7 dynes/cm = 0.0157 N/m at 22.7°C.
Ratio of specific heats of vapor (gas): 1.061
Latent heat of vaporization: 149 Btu/lb = 82.5 cal/g = 3.45 x 10^5 J/kg.
Heat of combustion: (estimate) –10,400 Btu/lb = –5800 cal/g = –242 x 10^5 J/kg.
Heat of solution: (estimate) –9 Btu/lb = –5 cal/g = –0.2 x 10^5 J/kg.
Vapor pressure: (Reid) 0.8 psia; 16 mm.

METHYL ISOCYANATE REC. M:4600

SYNONYMS: EEC No. 615-001-00-7; ISOCYANATE de METHYLE (French); ISOCIANATO de METILO (Spanish); ISO-CYANATOMETHANE; ISOCYANIC ACID, METHYL ESTER; ISOCYANATE METHANE; ISOCYANATOMETHANE; METHYLCARBAMYL AMINE; METHYL ESTER OF ISOCYANIC ACID; METHYL ISOCYANAT (German); METHYL CARBONIMIDE; MIC; TL 1450; METHANE, ISOCYANATO-; METHYL CARBONIMIDE; RCRA No. P064

IDENTIFICATION
CAS Number: 624-83-9
Formula: C$_2$H$_3$NO; CH$_3$NCO
DOT ID Number: UN 2480; DOT Guide Number: 155
Proper Shipping Name: Methyl isocyanate
Reportable Quantity (RQ): **(CERCLA)** 1 lb/0.454 kg

DESCRIPTION: Colorless liquid. Sharp, pungent odor. Floats on water surface; slowly mixes and reacts violently with water @ room temperature 68°F/20°C. Produces large amounts of toxic vapor. Boils at 100°F/38°C (decomposes giving off nitrous vapors).

Poison! • Highly flammable • Breathing the gas can kill you; skin and eye contact causes severe burns and blindness • Firefighting gear (including SCBA) does not provide adequate protection. If exposure occurs, remove and isolate gear immediately and thoroughly decontaminate personnel • Containers may BLEVE when exposed to fire • Vapors are heavier than air and will collect and stay in low areas • Vapors in confined areas (e.g., tanks, sewers, buildings) may explode when exposed to fire • Polymerization hazard • Heat can induce polymerization with rapid release of energy; sealed containers may rupture explosively • May react with itself blocking relief valves; leading to tank explosions • Toxic products of combustion may include hydrogen cyanide and nitrogen oxides • Do not put yourself in danger by entering a contaminated area to rescue a victim. *Warning:* Odor is not a reliable indicator of the presence of toxic amounts of MIC gas.

Hazard Classification (based on NFPA-704 Rating System)
Health Hazards (Blue): 4; Flammability (Red): 3; Reactivity (Yellow): 2; Special Notice (White): Water reactive

EMERGENCY RESPONSE: See Appendix A (155)
Evacuation:
Public safety: Isolate the area of spill or leak for at least 50 to 100 meters (160 to 330 feet) in all directions.
Spill: Small spill–First: Isolate in all directions 95 meters (300 feet). Then: Protect persons downwind, DAY: 0.8 km (0.5 mile); NIGHT: 2.7 km (1.7 miles). Large spill–First: Isolate in all directions: 490 meters (1600 feet); Then: Protect persons downwind, DAY: 0.5 km (0.3 mile); NIGHT: 9.8 km (6.1 mile).
Fire: If tank, rail car, or tank truck is involved in fire, isolate for at least 800 meters (½ mile) in all directions; also, consider initial evacuation for 800 meters (½ mile) in all directions.

EXPOSURE
Short-term effects: *SEEK MEDICAL ATTENTION.* **Vapor**: *POISONOUS IF INHALED OR IF SKIN EXPOSED.* May cause fatal pulmonary edema. Respiratory distress cited for most deaths. Severely irritating to eyes. High concentrations may cause blindness. Move to fresh air. *IF BREATHING HAS STOPPED,* give artificial respiration; *avoid mouth-to-mouth resuscitation; use bag/mask apparatus.* IF breathing is difficult, administer oxygen.
Liquid: *POISONOUS IF SWALLOWED OR IF SKIN EXPOSED.* Causes eye injury and skin burns. Remove contaminated clothing and shoes. Flush affected areas with plenty of running water for at least 15 minutes. *IF IN EYES,* hold eyelids open and flush with plenty of running water. *IF SWALLOWED* and victim is *CONSCIOUS AND ABLE TO SWALLOW,* have victim drink a large quantity of water and induce vomiting. *IF SWALLOWED* and victim is *UNCONSCIOUS OR HAVING CONVULSIONS,* do nothing except keep victim warm.
Note to physician or authorized medical personnel: Consider the use of amyl nitrite perles if symptoms of cyanide poisoning develop. If symptoms indicate, initial treatment includes the cyanide antidote kit. In all cases, break an amyl nitrite perle in a gauze pad and hold lightly under victim's nose for 15 seconds, repeating 5 times at about 15-second intervals; if necessary (and if sodium nitrite infusions will be delayed), repeat procedure every 3 minutes with fresh pearls until 3 or 4 have been used. Avoid breathing the vapor while administering it to the victim:

HEALTH HAZARDS
Personal protective equipment (PPE): A-Level PPE. OSHA Table Z-1-A air contaminant. Positive pressure breathing apparatus and special protective clothing. Chemical protective material(s) reported to have good to excellent resistance: PV alcohol.
Recommendations for respirator selection: NIOSH/OSHA
0.2 ppm: SA* (any supplied-air respirator). *0.5 ppm:* SA:CF* (any supplied-air respirator operated in a continuous-flow mode). 1 ppm: SCBAF (any self-contained breathing apparatus with a full facepiece); or SAF (any supplied-air respirator with a full facepiece). *3 ppm:* SAF:PD,PP (any supplied-air respirator that has a full facepiece and is operated in a pressure-demand or other positive-pressure mode). *EMERGENCY OR PLANNED ENTRY INTO UNKNOWN CONCENTRATIONS OR IDLH CONDITIONS:* SCBAF:PD,PP (any self-contained breathing apparatus that has a full facepiece and is operated in a pressure-demand or other positive-pressure mode); or SAF:PD,PP:ASCBA (any supplied-air

respirator that has a full facepiece and is operated in a pressure-demand or other positive-pressure mode in combination with an auxiliary self-contained breathing apparatus operated in a pressure-demand or other positive pressure mode). *ESCAPE*: GMFOV [any air-purifying, full-facepiece respirator (gas mask) with a chin-style, front- or back-mounted acid gas canister]; or SCBAE (any appropriate escape-type, self-contained breathing apparatus). *Note*: Substance reported to cause eye irritation or damage; may require eye protection.
Exposure limits (TWA unless otherwise noted): ACGIH TLV 0.02 ppm (0.047 mg/m^3); NIOSH/OSHA 0.02 ppm (0.05 mg/m^3); skin contact contributes significantly in overall exposure.
Toxicity by ingestion: Poison. Grade 3; LD_{50} = 71 mg/kg (rat).
Long-term health effects: Lung damage, bronchitis, fibrosis. Susceptible individuals may become sensitized so that subsequent exposure to extremely low concentrations provoke true asthma attacks. Cross sensitization to other isocyanates could also occur. Lung damage from frequent or intensive contact with vapor. Possible skin disorders.
Vapor (gas) irritant characteristics: Blindness can be caused by high concentrations. Vapors cause severe irritation of eyes and throat and can cause eye and lung injury.
Liquid or solid irritant characteristics: Severe skin irritant. Causes second- and third-degree burns on short contact and is very injurious to the eyes.
Odor threshold: 2.1 ppm. *Note*: This is above the exposure limits.
IDLH value: 3 ppm.

FIRE DATA
Flash point: 19°F/–7°C (cc).
Flammable limits in air: LEL: 5.3%; UEL: 26%.
Fire extinguishing agents not to be used: Reacts with water with release of heat (exothermic).
Behavior in fire: Container may explode violently.
Autoignition temperature: 995°F/535°C/808°K.
Electrical hazard: Class I, Group D.

CHEMICAL REACTIVITY
Reactivity with water: Reacts slowly and violently with water at room temperature (68°F/20°C) to produce gaseous CO_2, methylamine (BP: 21°F/–6°C/267°K) and heat (about 585 Btu per pound of methyl isocyanate or about 3700 Btu per pound of water). Resulting pressure increase may cause relief valves to open. Acids, alkalies and amides accelerate the reaction. Reactivity accelerates as temperature rises.
Binary reactants: Oxidizers, acids, alcohols, alkalis, amines, iron, steel, tin, copper, or their alloys. Avoid contact with all metals other than stainless steel, nickel, glass/ceramic. The metals may catalyze polymerization reactions. The heat of reaction can cause the polymerization to occur with explosive violence. Also attacks some plastics, rubber, and coatings; diffuses through polyethylene and attacks most elastomers. Fluorocarbon resins are resistant. Glass-lined containers (no pinholes) and fluorocarbon resin-lined transfer hoses are acceptable.
Neutralizing agents for acids and caustics: Caustic soda
Polymerization: Pure methyl isocyanate polymerizes spontaneously. Commercial product requires only heat or a trace of catalyst to initiate a potentially violent reaction; **Inhibitor of polymerization:** No inhibitor identified as such. Residual trace phosgene from production inhibits polymerization and reaction with water.

ENVIRONMENTAL DATA
Water pollution: Effect of low concentrations on aquatic life is unknown. May be dangerous if it enters nearby water intakes; notify operators. Notify local health and wildlife officials. **Response to discharge:** Issue warning-flammable liquid, poison. Restrict access. Evacuate area. Should be removed. Chemical and physical treatment.

SHIPPING INFORMATION
Grades of purity: Commercial (99%); **Storage temperature:** It is recommended that bulk quantities be cooled to approximately 32°F/0°C. Drums may be stored at ambient temperature out of direct sunlight. Storage temperature should not exceed 85°F/30°C; **Inert atmosphere:** Must be protected by a dry nitrogen (dew point –40°F/–40°C or lower) atmosphere; **Stability during transport:** Stable if kept as cool as practical and away from sources of heat, sparks, or flames. Protected from all contaminants. Cool bulk quantities to about 32°F/0°C.

PHYSICAL AND CHEMICAL PROPERTIES
Physical state @ 59°F/15°C and 1 atm: Liquid.
Molecular weight: 57.1
Boiling point @ 1 atm: 100°F/32°C/305°K.
Melting/Freezing point: –112°F/–80°C/193°K.
Critical temperature: 424°F/218°C/491°K.
Critical pressure: 808 psia = 55 atm = 5.6 MN/m^2.
Specific gravity (water = 1): 0.9599 @ 68°F/20°C (liquid).
Relative vapor density (air = 1): 2.0
Latent heat of vaporization: 223 Btu/lb = 124 cal/g = 5.19 x 10^5 J/kg.
Heat of combustion: 8041 Btu/lb = 4467 cal/g = 1.87 x 10^7 J/kg.
Heat of polymerization: –540 Btu/lb = –300 cal/g = –12.56 x 10^5 J/kg.
Vapor pressure: 348 mm.

METHYL ISOPROPENYL KETONE   REC. M:4650

SYNONYMS: ISOPROPENYL METHYL KETONE; 3-METHYL-3-BUTEN-2-ON (German); 2-METHYL-1-BUTENE-3-ONE; ISOPROPENYL METHYL KETONE; METIL ISOPROPIL CETONA (Spanish)

IDENTIFICATION
CAS Number: 814-78-8
Formula: C_5H_8O; $CH_3COC(CH_3)=CH_2$
DOT ID Number: UN 1246; DOT Guide Number: 127P
Proper Shipping Name: Methyl isopropenyl ketone, inhibited

DESCRIPTION: Clear, colorless liquid. Sweet pleasant odor. Floats on water surfacd; insoluble. Flammable, irritating vapor is produced.

Highly flammable • Containers may BLEVE when exposed to fire • Vapors may form explosive mixture with air • Vapors are heavier than air and will collect and stay in low areas • Vapors may travel long distances to ignition sources and flashback • Vapors in confined areas (e.g., tanks, sewers, buildings) may explode when exposed to fire • Irritating to the skin, eyes, and respiratory tract • Polymerization hazard • Heat can induce polymerization with rapid release of energy; sealed containers may rupture explosively • May react with itself blocking relief valves; leading to tank explosions • Toxic products of combustion may include carbon monoxide.

Hazard Classification (based on NFPA-704 Rating System)
Health Hazards (Blue): 2; Flammability (Red): 3; Reactivity (Yellow): 0

Methyl isothiocyanate

EMERGENCY RESPONSE: See Appendix A (127)
Evacuation:
Public safety: Isolate spill area for at least 25 to 50 meters (80 to 160 feet) in all directions.
Spill: Large spill–Consider initial downwind evacuation for at least 300 meters (1000 feet).
Fire: Isolate for 800 meters (½ mile) in all directions, especially if tank, rail car, or tank truck is involved in fire.

EXPOSURE
Short-term effects: *SEEK MEDICAL ATTENTION.* **Vapor:** Irritating to eyes, nose, and throat. Move victim to fresh air. *IF BREATHING HAS STOPPED,* give artificial respiration. IF breathing is difficult, administer oxygen. **Liquid:** Irritating to skin and eyes. Harmful if swallowed. Remove contaminated clothing and shoes. Flush affected areas with plenty of water. *IF IN EYES,* hold eyelids open and flush with plenty of water. *IF SWALLOWED* and victim is *CONSCIOUS AND ABLE TO SWALLOW,* have victim drink 4 to 8 ounces of water and have victim induce vomiting. *IF SWALLOWED* and victim is *UNCONSCIOUS OR HAVING CONVULSIONS,* do nothing except keep victim warm.

HEALTH HAZARDS
Personal protective equipment (PPE): B-Level PPE. Goggles or face shield; rubber gloves.
Toxicity by ingestion: Poison. Grade 3; oral LD_{50} = 180 mg/kg (rat).
Vapor (gas) irritant characteristics: Vapors are irritating such that personnel will not usually tolerate moderate or high concentrations.
Liquid or solid irritant characteristics: Severe irritant to eyes and skin. May cause smarting and first-degree burns on short exposure; may cause second-degree burns on long exposure.

FIRE DATA
Flash point: Less than 73°F/23°C (cc).
Flammable limits in air: LEL: 1.8%; UEL: 9.0%.
Fire extinguishing agents not to be used: Water may be ineffective.
Behavior in fire: May polymerize, resulting in explosion.
Burning rate: 4.7 mm/min

CHEMICAL REACTIVITY
Binary reactants: Violent reaction with aldehydes, nitric acid, perchloric acid, strong oxidizers, hydrogen peroxide
Polymerization: Will polymerize, unless inhibited, especially when heated; **Inhibitor of polymerization:** Up to 1% hydroquinone
Reactivity group: 18
Compatibility class: Ketone

ENVIRONMENTAL DATA
Food chain concentration potential: Negative; unlikely to accumulate.
Water pollution: Effect of low concentrations on aquatic life is unknown. Fouling to shoreline. May be dangerous if it enters nearby water intakes; notify operators. Notify local health and wildlife officials. **Response to discharge:** Issue warning–high flammability, water contaminant. Restrict access. Mechanical containment. Should be removed. Chemical and physical treatment.

SHIPPING INFORMATION
Grades of purity: Commercial; **Storage temperature:** Ambient; **Inert atmosphere:** None; **Venting:** Pressure-vacuum; **Stability during transport:** Stable.

NAS HAZARD CLASSIFICATION FOR BULK WATER TRANSPORTATION
FIRE: 3
HEALTH: Vapor irritant: 3; Liquid or solid irritant: 2; Poisons: 3
WATER POLLUTION: Human toxicity:–; Aquatic toxicity:–; Aesthetic effect: 2
REACTIVITY: Other chemicals: 3; Water: 0; Self-reaction: 3

PHYSICAL AND CHEMICAL PROPERTIES
Physical state @ 59°F/15°C and 1 atm: Liquid.
Molecular weight: 84.1
Boiling point @ 1 atm: 208°F/98°C/371°K.
Melting/Freezing point: –65°F/–54°C/219°K.
Specific gravity (water = 1): 0.85 @ 68°F/20°C (liquid).
Liquid surface tension: (estimate) 26 dynes/cm = 0.026 N/m @ 68°F/20°C.
Liquid water interfacial tension: (estimate) 30 dynes/cm = 0.030 N/m @ 68°F/20°C.
Relative vapor density (air = 1): 2.9
Ratio of specific heats of vapor (gas): 1.0796 @ 68°F/20°C (68°F).
Latent heat of vaporization: (estimate) 182 Btu/lb = 101 cal/g = 4.23 x 10^5 J/kg.
Heat of combustion: (estimate) –15,500 Btu/lb = –8600 cal/g = –360 x 10^5 J/kg.
Heat of polymerization: (estimate) –380 Btu/lb = –210 cal/g = –8.8 x 10^5 J/kg.

METHYL ISOTHIOCYANATE REC. M:4700

SYNONYMS: EP-161E; ISOTHIOCYANATE de METHYLE (French); ISOTHIOCYANIC ACID, METHYL ESTER; ISOTHIOCYANOMETHANE; METHANE, ISOTHIOCYANATO-; METHYL MUSTARD OIL; METHYLSENFOEL (German); METILISOTIOCIANATO (Spanish); MIC; MIT; MITC; MORTON WP-161-E; TRAPEX; TRAPEXIDE; VORLEX; VORTEX; WN 12

IDENTIFICATION
CAS Number: 556-61-6
Formula: C_2H_3NS; CH_3NCS
DOT ID Number: UN 2477; DOT Guide Number: 131
Proper Shipping Name: Methyl isothiocyanate
Reportable Quantity (RQ): **(CERCLA)** 1 lb/0.454 kg

DESCRIPTION: Colorless, crystalline solid. Horse raddish-like odor. Sinks and mixes with water; slightly soluble. Melts at 95°F/35°C.

Poison! • Breathing the vapor or dust or swallowing the dust can kill you • Firefighting gear (including SCBA) does not provide adequate protection. If exposure occurs, remove and isolate gear immediately and thoroughly decontaminate personnel • Containers may BLEVE when exposed to fire • Vapors are heavier than air and will collect and stay in low areas • Vapors in confined areas (e.g., tanks, sewers, buildings) may explode when exposed to fire • Toxic products of combustion may include cyanides and sulfur oxides • Do not put yourself in danger by entering a contaminated area to rescue a victim.

Hazard Classification (based on NFPA-704 Rating System)
Health Hazards (Blue): 4; Flammability (Red): 2; Reactivity (Yellow): 0

EMERGENCY RESPONSE: See Appendix A (131)
Evacuation:
Public safety: Isolate spill area for at least 100 to 200 meters (330 to 660 feet) in all directions.
Spill: Small spill–First: Isolate in all directions 155 meters (500 feet); Then: Protect persons downwind, DAY: 0.2 km (0.1 mile); NIGHT: 0.3 km (0.2 mile). Large spill–First: Isolate in all directions 155 meters (500 feet); Then: Protect persons downwind, DAY: 0.5 km (0.3 mile); NIGHT: 3.4 km (2.1 miles).
Fire: Isolate for 800 meters (½ mile) in all directions, especially if tank, rail car, or tank truck is involved in fire.

EXPOSURE
Short-term effects: *SEEK MEDICAL ATTENTION.* **Dust:** *POISONOUS IF INHALED OR ABSORBED THROUGH THE SKIN.* Move victim to fresh air. *IF BREATHING HAS STOPPED,* give artificial respiration. IF breathing is difficult, administer oxygen; *avoid mouth-to-mouth resuscitation; use bag/mask apparatus.* *IF IN EYES,* hold eyelids open and flush with water for at least 15 minutes. **Solid:** *POISONOUS IF SWALLOWED OR IF SKIN IS EXPOSED.* Remove contaminated clothing and shoes. Flush affected areas with water. *IF IN EYES,* hold eyelids open, flush with water for at least 15 minutes. *IF SWALLOWED:* Do nothing except keep victim warm. **Do NOT induce vomiting.**
Note to physician or authorized medical personnel: Consider the use of amyl nitrite perles if symptoms of cyanide poisoning develop. If symptoms indicate, initial treatment includes the cyanide antidote kit. In all cases, break an amyl nitrite perle in a gauze pad and hold lightly under victim's nose for 15 seconds, repeating 5 times at about 15-second intervals; if necessary (and if sodium nitrite infusions will be delayed), repeat procedure every 3 minutes with fresh pearls until 3 or 4 have been used. Avoid breathing the vapor while administering it to the victim. Administer sodium nitrite IV, ASAP. The usual adult dose is 10 to 20 mL of a 3% solution infused over no less than 5 minutes; the average child dose is 0.15 to 0.20 mL/kg. Monitor blood pressure during administration, and slow the rate of infusion if hypotention develops. Next, infuse sodium thiosulfate IV. The usual adult dose is 50 mL of a 25% solution infused over 10 to 20 minutes; the average child dose is 1.65 mL/kg. Repeat with nitrite and thiosulfate as required.

HEALTH HAZARDS
Personal protective equipment (PPE): A-Level PPE. Self-contained breathing apparatus.
Toxicity by ingestion: Grade 3: LD_{50} = 97 mg/kg (rat).
Long-term health effects: Prolonged contact may cause lung inflammation, chest pain, and edema, which may be fatal.
Vapor (gas) irritant characteristics: Vapors cause severe irritation of eyes and throat and can cause eye and lung injury. They cannot be tolerated even at low concentrations.
Liquid or solid irritant characteristics: Severe skin irritant. Causes second- and third-degree burns on short contact and is very injurious to the eyes.

FIRE DATA
Flash point: 90°F/32°C (cc).

CHEMICAL REACTIVITY
Binary reactants: Incompatible with strong acids, caustics, chlorates, ammonia, alcohols, glycols, strong oxidizers.

ENVIRONMENTAL DATA
Food chain concentration potential: Log P_{ow} = 0.9. Unlikely to accumulate.
Water pollution: DOT Appendix B, §172.101–marine pollutant. Effect of low concentrations on aquatic life is unknown. May be dangerous if it enters nearby water intakes; notify operators. Notify local health and wildlife officials. **Response to discharge:** Issue warning–water contaminant. Restrict access. Should be removed. Chemical and physical treatment.

NAS HAZARD CLASSIFICATION FOR BULK WATER TRANSPORTATION
FIRE: 0
HEALTH: Vapor irritant: 4; Liquid or solid irritant: 3; Poisons: 2
WATER POLLUTION: Human toxicity: 3; Aquatic toxicity:–; Aesthetic effect: 1
REACTIVITY: Other chemicals: 3; Water: 1; Self-reaction: 0

SHIPPING INFORMATION
Stability during transport: Stable.

PHYSICAL AND CHEMICAL PROPERTIES
Physical state @ 59°F/15°C and 1 atm: Solid.
Molecular weight: 73.11
Boiling point @ 1 atm: 246°F/119°C/392°K.
Melting/Freezing point: 97°F/36°C/309°K.
Specific gravity (water = 1): 1.069 @ 68°F/20°C.
Relative vapor density (air = 1): 2.5
Vapor pressure: (Reid) 0.76 psia.

METHYL MERCAPTAN   **REC. M:4750**

SYNONYMS: EEC No. 016-021-00-3; MERCAPTAN METHYLIQUE (French); MERCAPTOMETHANE; METHANETHIOL; METHANTHIOL (German); METHYL MERCAPTANE; METHYL SULFHYDRATE; METHYLTHIOALCOHOL; METILMERCAPTANO (Spanish); THIOMETHANOL; THIOMETHYL ALCOHOL; RCRA No. U153

IDENTIFICATION
CAS Number: 74-93-1
Formula: CH_3SH; CH_3–SH
DOT ID Number: UN 1064; DOT Guide Number: 117
Proper Shipping Name: Methyl mercaptan
Reportable Quantity (RQ): (CERCLA) 100 lb/45.4 kg

DESCRIPTION: Colorless gas. Shipped and stored as a liquefied gas under its own vapor pressure. Rotten-cabbage or strong garlic odor. Floats and boils on water; soluble; reacts, forming toxic and flammable mercaptan vapors. Poisonous, flammable vapor is produced. becomes liquid below 43°F/6°C.

Poison! • Breathing the gas can kill you; initial odor may deaden your sense of smell • Firefighting gear (including SCBA) provides NO protection. If exposure occurs, remove and isolate gear immediately and thoroughly decontaminate personnel • Containers may BLEVE when exposed to fire • Gas may travel long distances to ignition sources and flashback • Gas is heavier than air and will collect and stay in low areas • Gas in confined areas (e.g., tanks, sewers, buildings) may explode when exposed to fire • Contact with liquid may cause frostbite • Toxic products of combustion may include sulfur dioxide • Do not put yourself in danger by entering a contaminated area to rescue a victim.

Methyl mercaptan

Hazard Classification (based on NFPA-704 Rating System)
Health Hazards (Blue): 4; Flammability (Red): 4; Reactivity (Yellow): 0

EMERGENCY RESPONSE: See Appendix A (117)
Evacuation:
Public safety: Isolate spill area for at least 100 to 200 meters (330 to 660 feet) in all directions.
Spill: Increase, in a downwind direction, as necessary, the isolation distance shown in "Public safety."
Fire: Isolate for 1600 meters (1 mile) in all directions, especially if tank, rail car, or tank truck is involved in fire.

EXPOSURE
Short-term effects: *SEEK MEDICAL ATTENTION.* **Vapor:** POISONOUS IF INHALED. Irritating to eyes, nose, and throat. Move victim to fresh air. *IF IN EYES,* hold eyelids open and flush with plenty of water. *IF BREATHING HAS STOPPED,* give artificial respiration. IF breathing is difficult, administer oxygen. **Liquid:** *POISONOUS IF SWALLOWED.* Irritating to skin and eyes. Remove contaminated clothing and shoes. Flush affected areas with plenty of water. *IF IN EYES,* hold eyelids open and flush with plenty of water. *IF SWALLOWED* and victim is *CONSCIOUS AND ABLE TO SWALLOW*, have victim drink 4 to 8 ounces of water and have victim induce vomiting. *IF SWALLOWED* and victim is *UNCONSCIOUS OR HAVING CONVULSIONS,* do nothing except keep victim warm.

HEALTH HAZARDS
Personal protective equipment (PPE): B-Level PPE. OSHA Table Z-1-A air contaminant. Rubber gloves; goggles or face shield; air-line or self-contained breathing apparatus. Wear thermal protective clothing. Natural rubber, neoprene, and nitrile rubber may offer some protection from alcohols.
Recommendations for respirator selection: NIOSH/OSHA
5 ppm: CCROV [any chemical cartridge respirator with organic vapor cartridge(s)]; or SA (any supplied-air respirator). *12.5 ppm:* SA:CF (any supplied-air respirator operated in a continuous-flow mode); or PAPROV [any powered, air-purifying respirator with organic vapor cartridge(s)]. *25 ppm:* CCRFOV [any air-purifying, full-facepiece respirator (gas mask) with a chin-style, front- or back-mounted acid gas canister]; or GMFOV [any air-purifying, full-facepiece respirator (gas mask) with a chin-style, front- or back-mounted organic vapor cannister]; or PAPRTOV [any powered, air-purifying respirator with a tight-fitting facepiece and organic vapor cartridges(s)]; or SAT:CF (any supplied-air respirator that has a tight-fitting facepiece and is operated in a continuous-flow mode); or SCBAF (any self-contained breathing apparatus with a full facepiece); or SAF (any supplied-air respirator with a full facepiece). *150 ppm:* SA:PD,PP (any supplied-air respirator operated in a pressure-demand or other positive-pressure mode). *EMERGENCY OR PLANNED ENTRY INTO UNKNOWN CONCENTRATIONS OR IDLH CONDITIONS:* SCBAF:PD,PP (any self-contained breathing apparatus that has a full facepiece and is operated in a pressure-demand or other positive-pressure mode); or SAF:PD,PP:ASCBA (any supplied-air respirator that has a full facepiece and is operated in a pressure-demand or other positive-pressure mode in combination with an auxiliary self-contained breathing apparatus operated in a pressure-demand or other positive pressure mode). *ESCAPE:* GMFOV [any air-purifying, full-facepiece respirator (gas mask) with a chin-style, front- or back-mounted organic vapor cannister]; or SCBAE (any appropriate escape-type, self-contained breathing apparatus).
Exposure limits (TWA unless otherwise noted): ACGIH TLV 0.5 ppm (0.98 mg/m^3); OSHA ceiling 10 ppm (20 mg/m^3).
Short-term exposure limits (15-minute TWA): NIOSH ceiling 0.5 ppm (1 mg/m^3) [15 min].
Long-term health effects: May cause liver damage. Mutation data reported.
Vapor (gas) irritant characteristics: Irritates eyes and respiratory tract.
Liquid or solid irritant characteristics: May case irritation to the eyes, skin, and respiratory system. Rapid evaporation of the liquid may cause frostbite.
Odor threshold: Less than 0.0001–0.41 ppm.
IDLH value: 150 ppm.

FIRE DATA
Flash point: (Gas); 0°F/–18°C (oc) (liquid).
Flammable limits in air: LEL: 3.9%; UEL: 21.8%.
Fire extinguishing agents not to be used: Water may be ineffective.
Behavior in fire: Containers may explode.
Electrical hazard: Class I, Group C. Flow or agitation of substance may generate electrostatic charges due to low conductivity.
Burning rate: 3.8 mm/min
Stoichiometric air-to-fuel ratio: 8.562 (estimate).

CHEMICAL REACTIVITY
Binary reactants: Violent reaction with strong oxidizers. Reacts with water, steam, or acids forming hydrogen sulfide. Violent reaction with mercury(II) oxide. Potentially violent reaction with ethylene oxide. Reacts with light metals. Incompatible with caustics, aliphatic amines, isocyanates. Attacks some plastics, coatings, and rubber.

ENVIRONMENTAL DATA
Food chain concentration potential: Negative; unlikely to accumulate.
Water pollution: DOT Appendix B, §172.101–marine pollutant. Harmful to aquatic life in very low concentrations. May be dangerous if it enters nearby water intakes; notify operators. Notify local health and wildlife officials. **Response to discharge:** Issue warning–high flammability, water contaminant, air contaminant. Restrict access. Evacuate area. Disperse and flush.

SHIPPING INFORMATION
Grades of purity: 99.5+%; **Storage temperature:** Ambient; **Inert atmosphere:** None; **Venting:** Safety relief; **Stability during transport:** Stable.

PHYSICAL AND CHEMICAL PROPERTIES
Physical state @ 59°F/15°C and 1 atm: Gas.
Molecular weight: 48.1
Boiling point @ 1 atm: 43.2°F/6.2°C/279.4°K.
Melting/Freezing point: –186°F/–121°C/152°K.
Critical temperature: 386°F/197°C/470°K.
Critical pressure: 1050 psia = 71.4 atm = 7.25 MN/m^2.
Specific gravity (water = 1): 0.892 at 6°C (liquid).
Liquid surface tension: 31 dynes/cm = 0.031 N/m at 5°C.
Relative vapor density (air = 1): 1.66
Ratio of specific heats of vapor (gas): 1.1988
Latent heat of vaporization: 220 Btu/lb = 122 cal/g = 5.10×10^5 J/kg.
Heat of combustion: –11,054 Btu/lb = –6141 cal/g = $–257.0 \times 10^5$ J/kg.
Heat of fusion: 29.35 cal/g.
Vapor pressure: 1.7 atm.

METHYL METHACRYLATE REC. M:4800

SYNONYMS: DIAKON; EEC No. 607-035-00-6; METACRILATO de METILO (Spanish); METHACRYLATE DE METHYLE (French); METHACRYLIC ACID, METHYL ESTER; METHACRYLSAEUREMETHYL ESTER (German); METHYL-α-METHYLACRYLATE; METHYL 2-METHYLPROPENOATE; 2-METHYL PROPENIC ACID, METHYL ESTER; METHYL 2-METHYL-2-PROPENOATE; METIL MME; MONOCITE METHACRYLATE MONOMER; MER; 2-PROPENOIC ACID, 2-METHYL-, METHYL ESTER; METHYL α-METHYLACRYLATE; RCRA No. U162

IDENTIFICATION
CAS Number: 80-62-6
Formula: $C_5H_8O_2$; $CH_2 = C(CH_3)COOCH_3$
DOT ID Number: UN 1247; DOT Guide Number: 129P
Proper Shipping Name: Methyl methacrylate monomer, inhibited
Reportable Quantity (RQ): **(CERCLA)** 1000 lb/454 kg

DESCRIPTION: Colorless liquid. Pleasant, fruity odor. Floats on water surface; soluble. Flammable, irritating vapor is produced.

Highly flammable • Containers may BLEVE when exposed to fire • Vapors may form explosive mixture with air • Vapors are heavier than air and will collect and stay in low areas • Vapors may travel long distances to ignition sources and flashback • Vapors in confined areas (e.g., tanks, sewers, buildings) may explode when exposed to fire • Polymerization hazard • Heat can induce polymerization with rapid release of energy; sealed containers may rupture explosively • May react with itself blocking relief valves; leading to tank explosions • Irritating to the skin, eyes, and respiratory tract • Toxic products of combustion may include carbon monoxide.

Hazard Classification (based on NFPA-704 Rating System)
Health Hazards (Blue): 2; Flammability (Red): 3; Reactivity (Yellow): 2

EMERGENCY RESPONSE: See Appendix A (129)
Evacuation:
Public safety: Isolate spill area for at least 50 to 100 meters (160 to 330 feet) in all directions.
Spill: Large spill–Consider initial downwind evacuation for at least 300 meters (1000 feet).
Fire: Isolate for 800 meters (½ mile) in all directions, especially if tank, rail car, or tank truck is involved in fire.

EXPOSURE
Short-term effects: *SEEK MEDICAL ATTENTION.* **Vapor:** Irritating to eyes, nose, and throat. May cause lung edema; physical exertion will aggravate this condition. *IF INHALED*, will, will cause dizziness, headache, difficult breathing or loss of consciousness. May cause lung edema; physical exertion will aggravate this condition. Move to fresh air. *IF BREATHING HAS STOPPED*, give artificial respiration. IF breathing is difficult, administer oxygen. **Liquid:** Will burn skin and eyes. Harmful if swallowed. Remove contaminated clothing and shoes. Flush affected areas with plenty of water. *IF IN EYES*, hold eyelids open and flush with plenty of water. *IF SWALLOWED* and victim is *CONSCIOUS AND ABLE TO SWALLOW*, have victim drink 4 to 8 ounces of water.
Note to physician or authorized medical personnel: Pulmonary edema may be delayed. Medical observation is recommended for 24 to 48 hours after inhalation overexposure. As first aid for pulmonary edema, a physician or authorized medical personnel may consider administering a corticosteroid spray. Cigarette smoking may exacerbate pulmonary injury and should be discouraged for at least 72 hours following exposure.

HEALTH HAZARDS
Personal protective equipment (PPE): A-Level PPE. OSHA Table Z-1-A air contaminant. Air mask; protective gloves; face shield. Chemical protective material(s) reported to have good to excellent resistance: PV alcohol, Teflon®, Silvershield®.
Recommendations for respirator selection: NIOSH/OSHA
1000 ppm: SA:CF* (any supplied-air respirator operated in a continuous-flow mode); or CCRFOV [any chemical cartridge respirator with a full facepiece and organic vapor cartridges(s)]; or GMFOV [any air-purifying, full-facepiece respirator (gas mask) with a chin-style, front- or back-mounted acid gas canister]; or PAPROV* [any powered, air-purifying respirator with organic vapor cartridge(s)]; or SCBAF (any self-contained breathing apparatus with a full facepiece); or SAF (any supplied-air respirator with a full facepiece). *EMERGENCY OR PLANNED ENTRY INTO UNKNOWN CONCENTRATIONS OR IDLH CONDITIONS:* SCBAF:PD,PP (any self-contained breathing apparatus that has a full facepiece and is operated in a pressure-demand or other positive-pressure mode); or SAF:PD,PP:ASCBA (any supplied-air respirator that has a full facepiece and is operated in a pressure-demand or other positive-pressure mode in combination with an auxiliary self-contained breathing apparatus operated in a pressure-demand or other positive pressure mode). *ESCAPE:* GMFOV [any air-purifying, full-facepiece respirator (gas mask) with a chin-style, front- or back-mounted organic vapor cannister]; or SCBAE (any appropriate escape-type, self-contained breathing apparatus).
Note: Substance causes eye irritation or damage; eye protection needed.
Exposure limits (TWA unless otherwise noted): ACGIH TLV 50 ppm (205 mg/m^3); NIOSH/OSHA 100 ppm (410 mg/m^3).
Short-term exposure limits (15-minute TWA): ACGIH STEL 100 ppm (410 mg/m^3).
Toxicity by ingestion: Grade 1; LD_{50} = 5 to 15 g/kg (rat).
Long-term health effects: May cause allergic skin reaction; dermatitis; eczema.
Vapor (gas) irritant characteristics: May cause shortness of breath and lung edema. Vapor is irritating such that personnel will not usually tolerate moderate or high vapor concentrations.
Liquid or solid irritant characteristics: Causes irritation of the eyes, skin, and respiratory system. Causes smarting of the skin and first-degree burns on short exposure; may cause secondary burns on long exposure.
Odor threshold: 0.045 ppm.
IDLH value: 1000 ppm.

FIRE DATA
Flash point: 50°F/10°C (oc).
Flammable limits in air: LEL: 2.1%; UEL: 12.5%.
Fire extinguishing agents not to be used: Water may be ineffective.
Behavior in fire: Containers may explode due to polymerization.
Autoignition temperature: 790°F/421°C/694°K.
Electrical hazard: Class I, Group D. Flow or agitation of substance may generate electrostatic charges due to low conductivity.
Burning rate: 2.5 mm/min

CHEMICAL REACTIVITY
Reactivity with water: Mild reaction to moisture.
Binary reactants: Violent reaction with strong oxidizers; benzoyl peroxide or other polymerization initiators. Elevated temperatures,

light, contamination can cause spontaneous, explosive polymerization. Incompatible with caustics, nitrates, strong acids, aliphatic amines, alkanolamines, peroxides.
Polymerization: Heat, oxidizing agents, and ultraviolet light may cause polymerization; **Inhibitor of polymerization:** Hydroquinone, 22–65 ppm; hydroquinone methyl ether, 22–120 ppm; dimethyl *tert*-butylphenol, 45–65 ppm.
Reactivity group: 14
Compatibility class: Acrylates

ENVIRONMENTAL DATA
Food chain concentration potential: Log P_{ow} = 1.4. Unlikely to accumulate.
Water pollution: Dangerous to waterfowl and aquatic life in high concentrations. Fouling to shoreline. Biodegradation rate: moderate. May be dangerous if it enters nearby water intakes; notify operators. Notify local health and wildlife officials.
Response to discharge: Issue warning–high flammability. Evacuate area. Disperse and flush.

SHIPPING INFORMATION
Grades of purity: 99.8%; **Storage temperature:** Ambient; **Inert atmosphere:** None; **Venting:** Pressure-vacuum; **Stability during transport:** Stable.

NAS HAZARD CLASSIFICATION FOR BULK WATER TRANSPORTATION
FIRE: 3
HEALTH: Vapor irritant: 3; Liquid or solid irritant: 2; Poisons: 3
WATER POLLUTION: Human toxicity: 2; Aquatic toxicity: 2; Aesthetic effect: 2
REACTIVITY: Other chemicals: 2; Water: 0; Self-reaction: 3

PHYSICAL AND CHEMICAL PROPERTIES
Physical state @ 59°F/15°C and 1 atm: Liquid.
Molecular weight: 100.12
Boiling point @ 1 atm: 214°F/101°C/374°K.
Melting/Freezing point: –54°F/–48°C/225°K.
Critical temperature: 561°F/294°C/567°K.
Critical pressure: 485 psia = 33 atm = 3.3 MN/m².
Specific gravity (water = 1): 0.94 @ 68°F/20°C (liquid).
Liquid surface tension: 28 dynes/cm = 0.028 N/m @ 68°F/20°C.
Liquid water interfacial tension: 14.3 dynes/cm = 0.0143 N/m at 22.7°C.
Ratio of specific heats of vapor (gas): 1.059
Latent heat of vaporization: 140 Btu/lb = 77 cal/g = 3.2 x 10^5 J/kg.
Heat of combustion: (estimate) –11,400 Btu/lb = –6310 cal/g = –264 x 10^5 J/kg.
Heat of polymerization: –248 Btu/lb = –138 cal/g = –5.78 x 10^5 J/kg.
Vapor pressure: (Reid) 0.5 psia (approximate); 29 mm.

1-METHYLNAPHTHALENE   REC. M:4850

SYNONYMS: α-METHYLNAPHTHALENE; NAPHTHALENE, 1-METHYL-; 1-METILNAFTALENO (Spanish)

IDENTIFICATION
CAS Number: 90-12-0
Formula: $C_{11}H_{10}$; $C_{10}H_7CH_3$
DOT ID Number: NA 1993; DOT Guide Number: 128; DOT Guide Number: 152
Proper Shipping Name: Combustible liquid, n.o.s.

DESCRIPTION: Colorless oily liquid. Sinks slowly in water; practically insoluble.

Combustible • Containers may BLEVE when exposed to fire • Vapors are heavier than air and will collect and stay in low areas • Vapors may travel long distances to ignition sources and flashback • Vapors in confined areas (e.g., tanks, sewers, buildings) may explode when exposed to fire • Irritating to the skin, eyes, and respiratory tract • Toxic products of combustion may include carbon monoxide.

Hazard Classification (based on NFPA-704 Rating System)
Health Hazards (Blue): 2; Flammability (Red): 2; Reactivity (Yellow): 0

EMERGENCY RESPONSE: See Appendix A (128)
Evacuation:
Public safety: Isolate spill area for at least 25 to 50 meters (80 to 160 feet) in all directions.
Spill: Large spill–Consider initial downwind evacuation for at least 300 meters (1000 feet).
Fire: Isolate for 800 meters (½ mile) in all directions, especially if tank, rail car, or tank truck is involved in fire.

EXPOSURE
Short-term effects: *SEEK MEDICAL ATTENTION.* **Vapor:** Harmful if inhaled. May irritate the eyes and skin and photosensitize the skin. Move victim to fresh air. *IF IN EYES*, hold eyelids open and flush with running water. *IF BREATHING HAS STOPPED,* give artificial respiration; *avoid mouth-to-mouth resuscitation; use bag/mask apparatus.* IF breathing is difficult, administer oxygen. **Liquid:** Harmful if swallowed. May irritate the eyes and skin and photosensitize the skin. *IF IN EYES OR ON SKIN:* immediately flush with running water for at least 15 minutes; hold eyelids open if appropriate. Remove and double bag contaminated clothing and shoes at the site. *IF SWALLOWED* and victim is *CONSCIOUS AND ABLE TO SWALLOW*, have victim drink large volumes of warm water and induce vomiting. *IF SWALLOWED* and the victim is *UNCONSCIOUS OR HAVING CONVULSIONS,* do nothing except keep victim warm.

HEALTH HAZARDS
Personal protective equipment (PPE): B-Level PPE. Rubber gloves, safety goggles, coveralls, rubber shoes or boots, and hydrocarbon vapor canister mask.
Toxicity by ingestion: Grade 1; LD_{50} = 5000 mg/kg (rat).
Long-term health effects: Mutagenic, tumor promoting.
Vapor (gas) irritant characteristics: May be harmful.
Liquid or solid irritant characteristics: Causes irritation of the eyes, skin, and respiratory tract.
Odor threshold: 0.023 ppm.

FIRE DATA
Flash point: 180°F/82°C (cc).
Autoignition temperature: 984°F/529°C/802°K.

CHEMICAL REACTIVITY
Reactivity group: 32
Compatibility class: Aromatic hydrocarbons

ENVIRONMENTAL DATA
Food chain concentration potential: Moderate potential.
Water pollution: DOT Appendix B, §172.101–marine pollutant. Effect of low concentrations on aquatic life is unknown. May be dangerous if it enters nearby water intakes; notify operators. Notify

local health and wildlife officials. **Response to discharge:** Should be removed. Chemical and physical treatment.

SHIPPING INFORMATION
Storage temperature: Ambient

PHYSICAL AND CHEMICAL PROPERTIES
Physical state @ 59°F/15°C and 1 atm: Liquid.
Molecular weight: 142.20
Boiling point @ 1 atm: 464–469°F/240–243°C/513–516°K.
Melting/Freezing point: 25.6°F/–3.6°C/269.4°K.
Critical temperature: More than 923°F/495°C/768°K.
Specific gravity (water = 1): 1.0202 @ 68°F/20°C (liquid).
Relative vapor density (air = 1): 4.91
Heat of combustion: $-17,508.96$ Btu/lb = -9772.77 cal/g = -410.45×10^5 J/kg.

METHYL PARATHION   REC. M:4900

SYNONYMS: A-GRO; AZOFOS; AZOPHOS; BAY E-601; BAY 1145; BLADAN-M; CEKUMETHION; DALF; O,O-DIMETHYL 0-4 -NITROPHENYL PHOSPHOROTHIOATE; O,O-DIMETHYL O-p-NITROPHENYLPHOSPHOROTHIOATE; DEVITHION; FOLIDOL M; GEARPHOS; ME-PARATHION; MEPATON; MEPOX; METACIDE; METAFOS; METAPHOR; METAPHOS; METHYL-E 605; METHYL FOSFERNO; METHYL NIRAN; METHYLTHIOPHOS; METILPARATIONA (Spanish); METRON; NITROX; OLEOVOFOTOX; PARAPEST M-50; PARATAF; m-PARATHION; PARATHION METHYL; PATRON M; PARATOX; PENNCAP-M; PHOSPHOROTHIOIC ACID O,O-DIMETHYL O-(4-NITROPHENYL)ESTER; SINAFID M-48; TEKWAISA; THIOPHENIT; THYLFAR M-50; TOLL; VOFATOX; WOFATOS; WOFATOX; WOFOTOX; PARIDOL; ALKRON; NITRAN; RCRA No. P071

IDENTIFICATION
CAS Number: 298-00-0
Formula: $C_8H_{10}NO_5PS$; $(CH_3O)_2PSOC_6H_4NO_2$-p
DOT ID No. UN 3018 (liquid); NA 2783 (solid)
Proper Shipping Name: Methyl parathion, liquid or solid
Reportable Quantity (RQ): **(CERCLA)** 100 lb/45.4 kg

DESCRIPTION: White to tan solid; or, tan liquid (commercial product in xylene). Faint, rotten eggs or garlic odor. Solid and liquid sink in water, solution Floats on water surface; insoluble. Melts at 99°F/37°C. *Note:* If used as a weapon, notify U.S. Department of Defense: Army. Use M8 Paper (Detection: Yellow) or M256-A1 Detector Kit (Detection limit: 0.005 mg/m^3) if available. Damage and/or death may occur before chemical detection can take place.

Poison! (organophosphate) • Flammable • Breathing the vapor, skin or eye contact, or swallowing the material can kill you • Firefighting gear (including SCBA) does not provide adequate protection. If exposure occurs, remove and isolate gear immediately and thoroughly decontaminate personnel • Containers may BLEVE when exposed to fire • Toxic products of combustion may include nitrogen, sulfur, and phosphorus oxides • Do not put yourself in danger by entering a contaminated area to rescue a victim.

Hazard Classification (based on NFPA-704 Rating System)
Solutions: Health Hazards (Blue): 4; Flammability (Red): 3; Reactivity (Yellow): 2

Solids: Health Hazards (Blue): 4; Flammability (Red): 1; Reactivity (Yellow): 2

EMERGENCY RESPONSE: See Appendix A (171)
Evacuation:
Public safety: Isolate the area of spill or leak for at least 25 to 50 meters (80 to 160 feet) in all directions.
Spill: Increase, in the downwind direction, as necessary, the distance shown under "Public Safety."
Fire: If tank, rail car, or tank truck is involved in fire, isolate for at least 800 meters (½ mile) in all directions; also, consider initial evacuation for 800 meters (½ mile) in all directions.

EXPOSURE
Short-term effects: *SEEK MEDICAL ATTENTION*. **Vapor or Liquid:** *POISONOUS IF INHALED, IF SWALLOWED, OR IF SKIN IS EXPOSED.* Eye pupils are small; blurred vision; runny nose; cough; shortness of breath; pain; diarrhea, nausea and vomiting; increased blood pressure, hypermotility, hallucinations; loss of consciousness; convulsions; breathing stops; death. **Skin:** *Remove and double-bag contaminated clothing (including shoes) and leave them in the Hot Zone*; wash skin with soap and water and large volumes of water for at least 15 minutes. *Skin can also be decontaminated with diluted hypochlorite solution, U.S. Army M291 kit, and M258(A1) skin decontamination kit.* **Eye:** Rinse with large volumes of water or saline for at least 15 minutes. *IF SWALLOWED, will cause nausea, vomiting, or loss of consciousness. Remove and double-bag contaminated clothing and shoes and leave in Hot Zone for later incineration by hazardous materials experts.* Flush affected areas with plenty of water. **If swallowed** and victim is *CONSCIOUS AND ABLE TO SWALLOW*, have victim drink 4 to 8 ounces of water. **Do NOT induce vomiting but immediately administer slurry of activated charcoal**. *IF SWALLOWED* and victim is *UNCONSCIOUS OR HAVING CONVULSIONS,* do nothing except keep victim warm.
Note to physician and medical personnel: If symptoms indicate, initial treatment for an adult includes atropine, 2 mg (1/30 gr) intramuscularly or intravenously as soon as any local or systemic signs or symptoms of an intoxication are noted; repeat the administration of atropine every 3 to 8 minutes until signs of atropinization (mydriasis, dry mouth, rapid pulse, hot and dry skin) occur; initiate treatment in children with 0.05 mg/kg of atropine; repeat at 5- to 10-minute intervals. Watch respiration, and remove bronchial secretions if they appear to be obstructing the airway; intubate if necessary. Pralidoxime must be administered within minutes to a few hours following exposure (depending on the specific agent) to be effective. Give 2-PAMCI (Pralidoxime; Protopam), 2.5 g in 100 mL of sterile water or in 5% dextrose and water, intravenously, slowly, in 15 to 30 minutes; if sufficient fluid is not available, give 1 g of 2-PAMCI in 3 mL of distilled water by deep intramuscular injection; repeat this every half hour if respiration weakens or if muscle fasciculation or convulsions recur. Also Diazepam, an anticonvulsant, might be needed. For adults, the dose is 5 to 10 mg (slow IV), repeated every 12 to 15 minutes up to three (3) doses maximum. For children, the dose is 0.2 to 0.5 mg/kg.
Note to physician or authorized medical personnel: If inhaled, medical observation is recommended for 24 to 48 hours after breathing overexposure, as pulmonary edema may be delayed. As first aid for pulmonary edema, consider administering a corticosteroid spray. Cigarette smoking may exacerbate pulmonary injury and should be discouraged for at least 72 hours following exposure.

HEALTH HAZARDS

Personal protective equipment (PPE): A-Level PPE. OSHA Table Z-1-A air contaminant. Approved mask or respirator; natural rubber gloves, overshoes; protective clothing; goggles. For <30%, Protective materials with good to excellent resistance: Saranex®; for stronger solutions this material offers limited protection.

Recommendations for respirator selection: NIOSH
2 mg/m³: CCROVDMFu [any chemical cartridge respirator with organic vapor cartridge(s) in combination with a dust, mist, and fume filter]; or SA (any supplied-air respirator). *5 mg/m³:* SA:CF (any supplied-air respirator operated in a continuous-flow mode); or PAPROVDMFu [any powered, air purifying respirator with organic vapor cartridge (s) in combination with a dust, mist, and fume filter]. *10 mg/m³:* CCRFOVHiE [any chemical cartridge respirator with a full facepiece and organic vapor cartridge(s) in combination with a high-efficiency particulate filter]; or GMFOVHiE [any air-purifying, full-facepiece respirator (gas mask) with a chin-style, front- or back-mounted organic vapor canister having a high-efficiency particulate filter]; or PAPRTOVHiE [any powered, air-purifying respirator with a tight-fitting facepiece and organic vapor cartridge (s) in combination with a high-efficiency particulate filter]; or SAT:CF (any supplied-air respirator that has a tight-fitting facepiece and is operated in a continuous-flow mode); or SCBAF (any self-contained breathing apparatus with full facepiece); or SAF (any supplied-air respirator with a full facepiece). *200 mg/m³:* SAF:PD,PP (any supplied-air respirator that has a full facepiece and is operated in a pressure-demand or other positive-pressure mode). *EMERGENCY OR PLANNED ENTRY INTO UNKNOWN CONCENTRATIONS OR IDLH CONDITIONS:* SCBAF:PD,PP (any self-contained breathing apparatus that has a full facepiece and is operated in a pressure-demand or other positive-pressure mode); or SAF:PD,PP:ASCBA (any supplied-air respirator that has a full facepiece and is operated in a pressure-demand or other positive-pressure mode in combination with an auxiliary self-contained breathing apparatus operated in a pressure-demand or other positive pressure mode). *ESCAPE:* GMFOVHiE [any air-purifying, full-facepiece respirator (gas mask) with a chin-style, front-or back-mounted canister having a high efficiency particulate filter); or SCBAE (any appropriate escape-type, self-contained breathing apparatus).

Exposure limits (TWA unless otherwise noted): ACGIH TLV 0.2 mg/m³; NIOSH REL 0.2 mg/m³; skin contact contributes significantly in overall exposure.

Toxicity by ingestion: Poison. Grade 4; LD_{50} below 50 mg/kg (rat).

Long-term health effects: IARC rating 3; animal data: sufficient.

Vapor (gas) irritant characteristics: May cause poisoning.

Liquid or solid irritant characteristics: Poisonous when absorbed through skin. Concentrated solution can cause death by skin or eye contact.

FIRE DATA

Flash point: 115°F/46°C (oc).
Behavior in fire: Drums may rupture violently.

CHEMICAL REACTIVITY

Reactivity with water: Half decomposed in 8 days at 104°F/40°C.
Binary reactants: Incompatible with oxidizers, strong bases, perchloric acid, heat. Mixtures with magnesium, Endrin may be violent or explosive. Slightly decomposed by acid solutions. Rapidly decomposed by alkalis. The commercial product is a xylene solution; a storage hazard; an explosive risk; decomposes violently at 122°F/50°C. Attacks rubber and some plastics.

Neutralizing agents for acids and caustics: Apply caustic or soda ash slurry until yellow stains disappear.

ENVIRONMENTAL DATA

Water pollution: If used as a weapon, utilize use an M272 Water Detection Kit (Detection limit: 0.02 mg/L). DOT Appendix B, §172.101- severe marine pollutant. Harmful to aquatic life in very low concentrations. Solution is fouling to shoreline. May be dangerous if it enters nearby water intakes; notify operators. Notify local health and wildlife officials. **Response to discharge:** Issue warning–poison, water contaminant. Restrict access. Should be removed. Chemical and physical treatment.

SHIPPING INFORMATION

Grades of purity: Pure (solid); technical (liquid); 80% in xylene; **Storage temperature:** Below 50°F/10°C; **Inert atmosphere:** None; **Venting:** Pressure-vacuum; **Stability during transport:** Keep cool. Decomposes above 122°F/50°C with possible explosive force.

PHYSICAL AND CHEMICAL PROPERTIES

Physical state @ 59°F/15°C and 1 atm: Solid.
Molecular weight: 263.2
Boiling point @ 1 atm: 298°F/148°C/421°K.
Melting/Freezing point: 99°F/37°C/310°K.
Specific gravity (water = 1): 1.360 @ 68°F/20°C (liquid)
Vapor pressure: 0.00001 mm.

2-METHYL-1-PENTENE   REC. M:4950

SYNONYMS: *iso*-HEXENE; 2-METHYL PENTENE-1; 4-METHYL-4-PENTENE; 1-METHYL-1 PROPYLETHYLENE; 2-METIL-1-PENTENO

IDENTIFICATION

CAS Number: 763-29-1
Formula: C_6H_{12}; $CH_2C(CH_3)CH_2CH_2CH_3$
DOT ID Number: UN 2288; DOT Guide Number: 128
Proper Shipping Name: Isohexenes

DESCRIPTION: Colorless liquid. Gasoline-like odor. Floats on water surface; insoluble. Large amounts of flammable, irritating vapor is produced.

Highly flammable • Containers may BLEVE when exposed to fire • Vapors may form explosive mixture with air • Vapors are heavier than air and will collect and stay in low areas • Vapors may travel long distances to ignition sources and flashback • Vapors in confined areas (e.g., tanks, sewers, buildings) may explode when exposed to fire • Irritating to the skin, eyes, and respiratory tract • Toxic products of combustion may include carbon monoxide.

Hazard Classification (based on NFPA-704 Rating System)
Health Hazards (Blue): 1; Flammability (Red): 3; Reactivity (Yellow): 0

EMERGENCY RESPONSE: See Appendix A (128)
Evacuation:
Public safety: Isolate spill area for at least 25 to 50 meters (80 to 160 feet) in all directions.
Spill: Large spill–Consider initial downwind evacuation for at least 300 meters (1000 feet).

Fire: Isolate for 800 meters (½ mile) in all directions, especially if tank, rail car, or tank truck is involved in fire.

EXPOSURE
Short-term effects: *SEEK MEDICAL ATTENTION*. **Vapor:** If inhaled, will cause dizziness difficult breathing, or loss of consciousness. Move to fresh air. *IF BREATHING HAS STOPPED*, give artificial respiration. IF breathing is difficult, administer oxygen. **Liquid:** Irritating to skin and eyes. Harmful if swallowed. Remove contaminated clothing and shoes. Flush affected areas with plenty of water. *IF IN EYES*, hold eyelids open and flush with plenty of water. *IF SWALLOWED* and victim is *CONSCIOUS AND ABLE TO SWALLOW*, have victim drink 4 to 8 ounces of water. **Do NOT induce vomiting**.

HEALTH HAZARDS
Personal protective equipment (PPE): B-Level PPE. Rubber or neoprene gloves, splash goggles and NIOSH approved self-contained breathing apparatus).
Liquid or solid irritant characteristics: Eye irritant.

FIRE DATA
Flash point: Less than 20°F/–7°C.
Fire extinguishing agents not to be used: Water may be ineffective.
Behavior in fire: Can react vigorously with oxidizing materials.
Autoignition temperature: 579°F/304°C/577°K.

CHEMICAL REACTIVITY
Binary reactants: Oxidizers.

ENVIRONMENTAL DATA
Water pollution: Dangerous to aquatic life in high concentrations. Fouling to shoreline. May be dangerous if it enters nearby water intakes; notify operators. Notify local health and wildlife officials.
Response to discharge: Issue warning–high flammability. Restrict access.

SHIPPING INFORMATION
Grades of purity: 99.9% with 0.1% isoolefins (Research); 99.8% with.1% isoolefins and 0.1% *trans*-4-Methylpentene-2 (pure). 95.8% (technical). 95.0% Minimum; **Stability during transport:** Stable.

PHYSICAL AND CHEMICAL PROPERTIES
Physical state @ 59°F/15°C and 1 atm: Liquid.
Molecular weight: 84.156.
Boiling point @ 1 atm: 144°F/62°C/355°K.
Melting/Freezing point: –212°F/–136°C, 137°K.
Specific gravity (water = 1): 0.685 @ 15°C.
Relative vapor density (air = 1): 2.9.
Ratio of specific heats of vapor (gas): 1.067.
Latent heat of vaporization: 144.1 Btu/lb = 79.96 cal/g = 3.35 x 10^5 J/kg.

4-METHYL-1-PENTENE   REC. M:5000

SYNONYMS: ISOBUTYLETHENE; ISOHEXENE; 4-METHYL-PENTENE-1; 4-METIL-1-PENTENO (Spanish)

IDENTIFICATION
CAS Number: 691-37-2
Formula: C_6H_{12}; $(CH_3)_2CHCH_2CH=CH_2$
DOT ID Number: UN 2288; DOT Guide Number: 128
Proper Shipping Name: Isohexene

DESCRIPTION: Colorless liquid.
Highly flammable • Containers may BLEVE when exposed to fire • Vapors may form explosive mixture with air • Vapors are heavier than air and will collect and stay in low areas • Vapors may travel long distances to ignition sources and flashback • Vapors in confined areas (e.g., tanks, sewers, buildings) may explode when exposed to fire • Irritating to the skin, eyes, and respiratory tract • Toxic products of combustion may include carbon monoxide.

Hazard Classification (based on NFPA-704 Rating System)
Health Hazards (Blue): 1; Flammability (Red): 3; Reactivity (Yellow): 0

EMERGENCY RESPONSE: See Appendix A (128)
Evacuation:
Public safety: Isolate spill area for at least 25 to 50 meters (80 to 160 feet) in all directions.
Spill: Large spill–Consider initial downwind evacuation for at least 300 meters (1000 feet).
Fire: Isolate for 800 meters (½ mile) in all directions, especially if tank, rail car, or tank truck is involved in fire.

EXPOSURE
Short-term effects: *SEEK MEDICAL ATTENTION*. **Vapor:** If inhaled, will cause dizziness, difficult breathing, or loss of consciousness. Move to fresh air. *IF BREATHING HAS STOPPED*, give artificial respiration. IF breathing is difficult, administer oxygen. **Liquid:** Irritating to skin and eyes. Harmful if swallowed. Remove contaminated clothing and shoes. Flush affected areas with plenty of water. *IF IN EYES*, hold eyelids open and flush with plenty of water.

HEALTH HAZARDS
Personal protective equipment (PPE): B-Level PPE. Wear self-contained breathing apparatus, rubber boots, heavy rubber gloves and eye protection.
Vapor (gas) irritant characteristics: Vapors cause moderate irritation such that personnel will find high concentrations unpleasant. The effect is temporary.
Liquid or solid irritant characteristics: Causes smarting of skin and first-degree burn on short exposure. May cause second-degree burn on long exposure.

FIRE DATA
Flash point: –25°F/–32°C (cc).
Fire extinguishing agents not to be used: Water may be ineffective.
Autoignition temperature: 572°F/300°C/572°K.

CHEMICAL REACTIVITY
Binary reactants: Strong oxidizers.
Reactivity group: 30
Compatibility class: Olefins

ENVIRONMENTAL DATA
Water pollution: Effect of low concentrations on aquatic life is unknown. Fouling to shoreline. May be dangerous if it enters nearby water intakes; notify operators. Notify local health and wildlife officials. **Response to discharge:** Issue warning-flammable. Restrict access. Evacuate areas. Mechanical containment. Should be removed. Chemical and physical treatment.

SHIPPING INFORMATION
Grades of purity: 97%; **Storage temperature:** Ambient; **Stability during transport:** Stable.

PHYSICAL AND CHEMICAL PROPERTIES
Physical state @ 59°F/15°C and 1 atm: Liquid.
Molecular weight: 84.16
Boiling point @ 1 atm: 127-129°F/53-54°C/326-327°K.
Specific gravity (water = 1): 0.665
Relative vapor density (air = 1): 2.9
Vapor pressure: (Reid) 8.49 psia.

METHYL PHOSPHONOTHIOIC DICHLORIDE
REC. M:5050

SYNONYMS: METHYL PHOSPHONOUS DICHLORIDE; METHYL PHOSPHORUS DICHLORIDE; MPTD

IDENTIFICATION
CAS Number: 676-98-2
Formula: CH_3Cl_2PS; CH_3PSCl_2
DOT ID Number: NA 2845; DOT Guide Number: 135
Proper Shipping Name: Methyl phosphonous dichloride

DESCRIPTION: Clear, colorless liquid. Sharp, acrid, irritating odor. Sinks and mixes violently with water forming hydrochloric acid and hydrogen chloride vapors.

Poison! • Corrosive • Severely irritating to the skin; prolonged contact with the skin or eyes can cause burns • Firefighting gear (including SCBA) does not provide adequate protection. If exposure occurs, remove and isolate gear immediately and thoroughly decontaminate personnel • Containers may BLEVE when exposed to fire • Vapors are heavier than air and will collect and stay in low areas • Vapors in confined areas (e.g., tanks, sewers, buildings) may explode when exposed to fire • Toxic products of combustion may include hydrogen chloride and oxides of phosphorus and sulfur • Do not put yourself in danger by entering a contaminated area to rescue a victim.

Hazard Classification (based on NFPA-704 Rating System)
Health Hazards (Blue): 2; Flammability (Red): 2; Reactivity (Yellow): 2; Special Notice (White): Water reactive

EMERGENCY RESPONSE: See Appendix A (135)
Evacuation:
Public safety: Isolate spill area for at least 100 to 150 meters (330 to 490 feet) in all directions.
Spills: (anhydrous) Small spill–First: Isolate in all directions 60 meters (200 feet); Then: Protect persons downwind, DAY: 0.5 km (0.3 mile); NIGHT: 1.3 km (0.8 mile). Large spill–First: Isolate in all directions 155 meters (500 feet); Then: Protect persons downwind, DAY: 1.6 km (1.0 mile); NIGHT: 3.4 km (2.1 miles).
Fire: Isolate for 800 meters (½ mile) in all directions, especially if tank, rail car, or tank truck is involved in fire.

EXPOSURE
Short-term effects: *SEEK MEDICAL ATTENTION*. **Vapor:** Irritating to eyes, nose, and throat. *IF INHALED*, will, will cause coughing or difficult breathing. *IF IN EYES*, hold eyelids open and flush with plenty of water. *IF BREATHING HAS STOPPED*, give artificial respiration. IF breathing is difficult, administer oxygen.
Liquid: Will burn skin and eyes. *IF SWALLOWED*, will cause nausea and vomiting. Remove contaminated clothing and shoes. Flush affected areas with plenty of water. *IF IN EYES*, hold eyelids open and flush with plenty of water. *IF SWALLOWED* and victim is *CONSCIOUS AND ABLE TO SWALLOW*, have victim drink 4 to 8 ounces of water and have victim induce vomiting. *IF SWALLOWED* and victim is *UNCONSCIOUS OR HAVING CONVULSIONS*, do nothing except keep victim warm.
Note to physician or authorized medical personnel: Medical observation is recommended for 24 to 48 hours after breathing overexposure, as pulmonary edema may be delayed. As first aid for pulmonary edema, consider administering a corticosteroid spray. Cigarette smoking may exacerbate pulmonary injury and should be discouraged for at least 72 hours following exposure.

HEALTH HAZARDS
Personal protective equipment (PPE): A-Level PPE. Use extreme care when handling this compound. Avoid any contact with liquid or vapor. Rubber or neoprene gloves.
Vapor (gas) irritant characteristics: May cause irritation to the eyes and respiratory system.
Liquid or solid irritant characteristics: Corrosive. Extremely irritating to the eyes, skin, and respiratory tract.

FIRE DATA
Flash point: More than 122°F/50°C (oc).
Fire extinguishing agents not to be used: Water or foam

CHEMICAL REACTIVITY
Reactivity with water: Reacts with water to form hydrochloric acid and/or hydrogen chloride vapor. The reaction may be violent.
Binary reactants: Corrosive to metals.
Neutralizing agents for acids and caustics: Flush with water, rinse with dilute sodium bicarbonate or soda ash solution.

ENVIRONMENTAL DATA
Food chain concentration potential: Negative; unlikely to accumulate.
Water pollution: Effect of low concentrations on aquatic life is unknown. May be dangerous if it enters nearby water intakes; notify operators. Notify local health and wildlife officials.
Response to discharge: Issue warning–corrosive, water contaminant. Restrict access. Disperse and flush.

SHIPPING INFORMATION
Grades of purity: Technical; **Storage temperature:** Ambient; **Inert atmosphere:** None; **Venting:** Open; **Stability during transport:** Stable.

PHYSICAL AND CHEMICAL PROPERTIES
Physical state @ 59°F/15°C and 1 atm: Liquid.
Molecular weight: 149
Boiling point @ 1 atm: 352°F/178°C/451°K.
Melting/Freezing point: −14.1°F/−25.6°C/247.6°K.
Specific gravity (water = 1): 1.42 @ 68°F/20°C (liquid).
Latent heat of vaporization: (estimate) 110 Btu/lb = 60 cal/g = 2.5×10^5 J/kg.

METHYL PROPYL KETONE
REC. M:5100

SYNONYMS: ETHYL ACETONE; METHYL-*n*-PROPYL KETONE; MPK; METIL PROPIL CETONA (Spanish); 2-PENTANONE

IDENTIFICATION
CAS Number: 107-87-9
Formula: $C_5H_{10}O$; $CH_3CH_2CH_2COCH_3$; $CH_3C(O)CH_2CH_2CH_3$
DOT ID Number: UN 1249; DOT Guide Number: 127
Proper Shipping Name: Methyl propyl ketone

DESCRIPTION: Colorless to water white liquid. Characteristic acetone-like odor. Floats on the surface of water; slightly soluble.

Highly flammable • Containers may BLEVE when exposed to fire • Vapors may form explosive mixture with air • Vapors are heavier than air and will collect and stay in low areas • Vapors may travel long distances to ignition sources and flashback • Vapors in confined areas (e.g., tanks, sewers, buildings) may explode when exposed to fire • Irritating to the skin, eyes, and respiratory tract • Toxic products of combustion may include carbon monoxide.

Hazard Classification (based on NFPA-704 Rating System)
Health Hazards (Blue): 2; Flammability (Red): 3; Reactivity (Yellow): 0

EMERGENCY RESPONSE: See Appendix A (127)
Evacuation:
Public safety: Isolate spill area for at least 25 to 50 meters (80 to 160 feet) in all directions.
Spill: Large spill–Consider initial downwind evacuation for at least 300 meters (1000 feet).
Fire: Isolate for 800 meters (½ mile) in all directions, especially if tank, rail car, or tank truck is involved in fire.

EXPOSURE
Short-term effects: *SEEK MEDICAL ATTENTION*. **Vapor:** Move victim to fresh air. *IF BREATHING HAS STOPPED,* give artificial respiration. IF breathing is difficult, administer oxygen. **Liquid:** Irritating to skin and eyes. Remove contaminated clothing and shoes. Flush affected areas with water. *IF IN EYES*, hold eyelids open and flush with plenty of water. The use of alcoholic beverages may enhance the toxic effect.

HEALTH HAZARDS
Personal protective equipment (PPE): B-Level PPE. OSHA Table Z-1-A air contaminant. Full impervious protective clothing, including boots and gloves. Where splashing is possible wear full face shield or chemical safety goggles. Use approved respirator to protect against vapors.
Recommendations for respirator selection: NIOSH
1500 ppm: CCRFOV* [any air-purifying, full-facepiece respirator (gas mask) with a chin-style, front- or back-mounted acid gas canister]; or PAPROV* [any powered, air-purifying respirator with organic vapor cartridge(s)]; or GMFOV [any air-purifying, full-facepiece respirator (gas mask) with a chin-style, front- or back-mounted organic vapor cannister]; or SA* (any supplied-air respirator); or SCBAF (any self-contained breathing apparatus with a full facepiece). *EMERGENCY OR PLANNED ENTRY INTO UNKNOWN CONCENTRATIONS OR IDLH CONDITIONS:* SCBAF:PD,PP (any self-contained breathing apparatus that has a full facepiece and is operated in a pressure-demand or other positive-pressure mode); or SAF:PD,PP:ASCBA (any supplied-air respirator that has a full facepiece and is operated in a pressure-demand or other positive-pressure mode in combination with an auxiliary self-contained breathing apparatus operated in a pressure-demand or other positive pressure mode). *ESCAPE:* GMFOV [any air-purifying, full-facepiece respirator (gas mask) with a chin-style, front- or back-mounted organic vapor canister]; or SCBAE (any appropriate escape-type, self-contained breathing apparatus).

*Note: Substance reported to cause eye irritation or damage; may require eye protection.
Exposure limits (TWA unless otherwise noted): ACGIH TLV 200 ppm (705 mg/m^3); OSHA PEL 200 ppm (700 mg/m^3); NIOSH 150 ppm (530 mg/m^3).
Short-term exposure limits (15-minute TWA): ACGIH STEL 250 ppm (881 mg/m^3); OSHA STEL 250 ppm (875 mg/m^3).
Toxicity by ingestion: Grade 2; LD_{50} = 1.6 g/kg (rat).
Long-term health effects: Possible mutagen. May have an effect on the nervous system. Possible skin disorders.
Vapor (gas) irritant characteristics: May cause irritation of the eyes and respiratory system. Vapors are irritating such that personnel will not usually tolerate moderate or high concentrations.
Liquid or solid irritant characteristics: Irritates the eyes, skin, and respiratory tract. Fairly severe skin irritant. May cause pain and second-degree burns after a few minutes of contact. Degreases the skin.
Odor threshold: 11 ppm.
IDLH value: 1500 ppm.

FIRE DATA
Flash point: –20°F/–29°C (cc) (mixture of *cis*- and *trans*-); –45°F/–42°C (*trans*-).
Flammable limits in air: LEL: 1.5%; UEL: 8.2%.
Fire extinguishing agents not to be used: Use water with caution. The material is lighter than water and only moderately soluble, any fire could easily be spread by water in uncontained area.
Autoignition temperature: 846°F/452°C/725°K.
Electrical hazard: Flow or agitation of substance may generate electrostatic charges due to low conductivity.
Stoichiometric air-to-fuel ratio: 33.3

CHEMICAL REACTIVITY
Binary reactants: May self ignite in air. Violent reaction with strong oxidizers and bromine trifluoride
Reactivity group: 18
Compatibility class: Ketone
Water pollution: Effects of low concentrations on aquatic life is unknown. May be dangerous if it enters nearby water intakes; notify operators. Notify local health and wildlife officials. **Response to discharge:** Issue warning–high flammability. Should be removed.

SHIPPING INFORMATION
Grades of purity: 97%; 90%; technical; **Storage temperature:** Ambient; **Inert atmosphere:** None; **Stability during transport:** Stable.

PHYSICAL AND CHEMICAL PROPERTIES
Physical state @ 59°F/15°C and 1 atm: Liquid.
Molecular weight: 86.13
Boiling point @ 1 atm: 215°F/102°C/375°K.
Melting/Freezing point: –108°F/–78°C/196°K.
Specific gravity (water = 1): 0.81 @ 20°C.
Relative vapor density (air = 1): 3.0
Vapor pressure: 12 mm.

2-METHYLPYRIDINE　　　　　　　　　　　　　　**REC. M:5150**

SYNONYMS: EEC No. 613-036-00-2; *α*-METHYLPYRIDINE; METILPIRIDINA (Spanish); *α*-PICOLINE; 2-PICOLINE; RCRA No. U191

3-Methylpyridine

IDENTIFICATION
CAS Number: 109-06-8
Formula: C_6H_7N; $C_5H_4NCH_3$
DOT ID Number: UN 2313; DOT Guide Number: 128
Proper Shipping Name: Picolines
Reportable Quantity (RQ): **(CERCLA)** 5000 lb/2270 kg

DESCRIPTION: Colorless liquid. Strong, unpleasant odor. Floats on the surface of water. Poisonous, flammable vapor is produced.

Highly flammable • Containers may BLEVE when exposed to fire • Vapors may form explosive mixture with air • Vapors are heavier than air and will collect and stay in low areas • Vapors may travel long distances to ignition sources and flashback • Vapors in confined areas (e.g., tanks, sewers, buildings) may explode when exposed to fire • Irritating to the skin, eyes, and respiratory tract • Toxic products of combustion may include nitrogen oxides.

Hazard Classification (based on NFPA-704 Rating System)
Health Hazards (Blue): 2; Flammability (Red): 2; Reactivity (Yellow): 0

EMERGENCY RESPONSE: See Appendix A (128)
Evacuation:
Public safety: Isolate spill area for at least 25 to 50 meters (80 to 160 feet) in all directions.
Spill: Large spill–Consider initial downwind evacuation for at least 300 meters (1000 feet).
Fire: Isolate for 800 meters (½ mile) in all directions, especially if tank, rail car, or tank truck is involved in fire.

EXPOSURE
Short-term effects: *SEEK MEDICAL ATTENTION*. **Vapor:** Harmful if inhaled or if skin is exposed. Irritating to eyes, nose, and throat. Move to fresh air. *IF BREATHING HAS STOPPED*, give artificial respiration. IF breathing is difficult, administer oxygen. **Liquid:** *HARMFUL IF SWALLOWED OR IF SKIN IS EXPOSED*. Will burn eyes. Remove contaminated clothing and shoes. Flush affected areas with plenty of water. *IF IN EYES*, hold eyelids open and flush with plenty of water. *IF SWALLOWED* and victim is *CONSCIOUS AND ABLE TO SWALLOW*, have victim drink 4 to 8 ounces of water and have victim induce vomiting. *IF SWALLOWED* and victim in *UNCONSCIOUS OR HAVING CONVULSIONS*, do nothing except keep victim warm. The use of alcoholic beverages may enhance the toxic effect.

HEALTH HAZARDS
Personal protective equipment (PPE): B-Level PPE. Organic vapor respirator. Wear goggles, rubber gloves, self-contained breathing apparatus and protective overclothing.
Exposure limits (TWA unless otherwise noted): 2 ppm (AIHA WEEL).
Toxicity by ingestion: Grade 2; LD_{50} = 0.5 to 5 g/kg.
Long-term health effects: Chronic exposure may cause occasional vomiting and diarrhea; weight loss and anemia; ocular and facial paralysis. Kidney and liver injury have been reported. Possible nervous system damage.
Vapor (gas) irritant characteristics: May cause eye and respiratory system irritation. Vapors cause irritation such that personnel will find high concentrations unpleasant. The effect may be temporary.
Liquid or solid irritant characteristics: Irritates the eyes, skin, and respiratory system. Causes smarting of the skin and first-degree burns on short exposure; may cause second-degree burns on long exposure. Degreases the skin.

Odor threshold: 0.046 ppm–100% recognition in air. 0.023 ppm–50% recognition in air. 0.45–1.2 ppm; detectable in water.

FIRE DATA
Flash point: 102°F/39°C (oc); 79°F/26°C (cc).
Flammable limits in air: LEL: 1.4%; UEL: 8.6%.
Fire extinguishing agents not to be used: Water may be ineffective.
Behavior in fire: Heat may cause pressure buildup in closed containers. Use water to keep container cool.
Autoignition temperature: 1000°F/538°C/811°K.

CHEMICAL REACTIVITY
Binary reactants: Can react with strong oxidizers. Attacks copper and copper alloys
Neutralizing agents for acids and caustics: Flush with water.
Reactivity group: 9
Compatibility class: Aromatic amines

ENVIRONMENTAL DATA
Food chain concentration potential: Log P_{ow} = 1.1. Unlikely to accumulate.
Water pollution: Effect of low concentrations on aquatic life is unknown. May be dangerous if it enters nearby water intakes; notify operators. Notify local health and wildlife officials. **Response to discharge:** Issue warning–high flammability. Restrict access. Disperse and flush.

SHIPPING INFORMATION
Grades of purity: α-Picoline 98%, Water 0.2% max; **Storage temperature:** Ambient; **Inert atmosphere:** None; **Venting:** Pressure-vacuum; **Stability during transport:** Stable.

NAS HAZARD CLASSIFICATION FOR BULK WATER TRANSPORTATION
FIRE: 3
HEALTH: Vapor irritant: 2; Liquid or solid irritant: 2; Poisons: 1
WATER POLLUTION: Human toxicity: 2; Aquatic toxicity: 2; Aesthetic effect: 3
REACTIVITY: Other chemicals: 3; Water: 0; Self-reaction: 0

PHYSICAL AND CHEMICAL PROPERTIES
Physical state @ 59°F/15°C and 1 atm: Liquid.
Molecular weight: 93.13.
Boiling point @ 1 atm: 263.8°F/128.8°C/401.95°K.
Melting/Freezing point: –88.24°F/–66.8°C/206.35°K.
Critical temperature: 658.4°F/348°C/621.2°K.
Critical pressure: (estimate) 614.3 psia = 41.8 atm = 4.23 MN/m^2.
Specific gravity (water = 1): 0.944 @ 68°F/20°C.
Liquid surface tension: (estimate) 33.2 dynes/cm = 0.0332 N/m @ 68°F/20°C.
Relative vapor density (air = 1): 3.2.
Ratio of specific heats of vapor (gas): > 1, approximately 1.123.
Latent heat of vaporization: (est. at boiling point) 160.4 Btu/lb = 98.1 cal/g = 3.7 x 10^5 J/kg.
Heat of combustion: Net @ 77°F/25°C. –15,089 Btu/lb = –8383 cal/g = –350.7 x 10^5 J/kg.
Vapor pressure: 9.1 mm.

3-METHYLPYRIDINE   **REC. M:5200**

SYNONYMS: EEC No. 613-037-00-8; 3-METILPIRIDINA (Spanish); 3-PICOLINE; β-PICOLINE; m-PICOLINE; PYRIDINE, 3-METHYL

IDENTIFICATION
CAS Number: 108-99-6
Formula: C_6H_7N
DOT ID Number: UN 2313; DOT Guide Number: 128
Proper Shipping Name: Picolines

DESCRIPTION: Colorless liquid. Sweetish, not unpleasant, odor. Mixes with water.

Highly flammable • Containers may BLEVE when exposed to fire •Vapors may form explosive mixture with air • Vapors are heavier than air and will collect and stay in low areas • Vapors may travel long distances to ignition sources and flashback • Vapors in confined areas (e.g., tanks, sewers, buildings) may explode when exposed to fire • Irritating to the skin, eyes, and respiratory tract • Toxic products of combustion may include nitrogen oxides.

Hazard Classification (based on NFPA-704 Rating System)
Health Hazards (Blue): 2; Flammability (Red): 2; Reactivity (Yellow): 0

EMERGENCY RESPONSE: See Appendix A (128)
Evacuation:
Public safety: Isolate spill area for at least 25 to 50 meters (80 to 160 feet) in all directions.
Spill: Large spill–Consider initial downwind evacuation for at least 300 meters (1000 feet).
Fire: Isolate for 800 meters (½ mile) in all directions, especially if tank, rail car, or tank truck is involved in fire.

EXPOSURE
Short-term effects: *SEEK MEDICAL ATTENTION.* **Vapor:** *POISONOUS IF INHALED OR IF SKIN IS EXPOSED.* Irritating to eyes, nose, and throat. Move to fresh air. *IF BREATHING HAS STOPPED,* give artificial respiration. IF breathing is difficult, administer oxygen. **Liquid:** *POISONOUS IF SWALLOWED OR IF SKIN IS EXPOSED.* Will burn eyes. Remove contaminated clothing and shoes. Flush affected areas with plenty of water. *IF IN EYES,* hold eyelids open and flush with plenty of water. The use of alcoholic beverages may enhance the toxic effect.

HEALTH HAZARDS
Personal protective equipment (PPE): B-Level PPE. Organic vapor respirator. Self contained breathing apparatus, protective clothing, rubber boots, and heavy rubber gloves.
Exposure limits (TWA unless otherwise noted): 2 ppm (AIHAWEEL).
Toxicity by ingestion: Grade 3: LD_{50} = 400 mg/kg (rat).
Long-term health effects: Causes damage to liver and kidney. May effect the nervous system.
Vapor (gas) irritant characteristics: Vapor causes severe irritation to eyes and throat and can cause eye and lung injury. They cannot be tolerated even at low concentrations.
Liquid or solid irritant characteristics: Severe eye and skin irritant. Causes second- and third-degree burn on short contact and is very injurious to the eyes.

FIRE DATA
Flash point: 97°F/36°C (cc).
Fire extinguishing agents not to be used: Water may be ineffective.
Autoignition temperature: 1000°F/538°C/811°K.

CHEMICAL REACTIVITY
Binary reactants: Strong oxidizers may cause fire and explosions. Forms heat- and shock-sensitive explosive with digold ketenide. Incompatible with alcohols, aldehydes, alkylene oxides, cresols, caprolactam solution, epichlorohydrin, organic anhydrides, glycols, maleic anhydride, phenols.
Neutralizing agents for acids and caustics: Flush with water.
Reactivity group: 9
Compatibility class: Aromatic amines

ENVIRONMENTAL DATA
Food chain concentration potential: Log P_{ow} = 1.2. Unlikely to accumulate.
Water pollution: Effect of low concentration on aquatic life is unknown. May be dangerous if it enters nearby water intakes; notify operators. Notify local health and wildlife officials. **Response to discharge:** Issue warning–high flammability. Restrict access. Evacuate areas. Should be removed. Chemical and physical treatment.

SHIPPING INFORMATION
Grades of purity: 99.5%; **Storage temperature:** Ambient; **Stability during transport:** Stable.

PHYSICAL AND CHEMICAL PROPERTIES
Physical state @ 59°F/15°C and 1 atm: Liquid.
Molecular weight: 93.13
Boiling point @ 1 atm: 291.2°F/144°C/417.2°K.
Melting/Freezing point: 0.94°F/–18.3°C/254.9°K.
Specific gravity (water = 1): 0.957
Relative vapor density (air = 1): 3.2
Vapor pressure: 4.5 mm (approximate).

4-METHYLPYRIDINE   REC. M:5250

SYNONYMS: EEC No. 613-037-00-8; 4-METILPIRIDINA (Spanish); 4-PICOLINE; γ-PICOLINE

IDENTIFICATION
CAS Number: 108-89-4
Formula: C_6H_7N
DOT ID Number: UN 2313; DOT Guide Number: 128
Proper Shipping Name: Picolines

DESCRIPTION: Colorless to brown liquid. Obnoxious, sweetish odor. Water soluble liquid.

Highly flammable • Containers may BLEVE when exposed to fire •Vapors may form explosive mixture with air • Vapors are heavier than air and will collect and stay in low areas • Vapors may travel long distances to ignition sources and flashback • Vapors in confined areas (e.g., tanks, sewers, buildings) may explode when exposed to fire • Irritating to the skin, eyes, and respiratory tract • Toxic products of combustion may include nitrogen oxides.

Hazard Classification (based on NFPA-704 Rating System)
Health Hazards (Blue): 2; Flammability (Red): 2; Reactivity (Yellow): 0

EMERGENCY RESPONSE: See Appendix A (128)
Evacuation:
Public safety: Isolate spill area for at least 25 to 50 meters (80 to 160 feet) in all directions.
Spill: Large spill–Consider initial downwind evacuation for at least 300 meters (1000 feet).

768 1-Methylpyrrolidone

Fire: Isolate for 800 meters (½ mile) in all directions, especially if tank, rail car, or tank truck is involved in fire.

EXPOSURE
Short-term effects: *SEEK MEDICAL ATTENTION.* **Vapor:** *POISONOUS IF INHALED OR IF SKIN IS EXPOSED.* Irritating to eyes, nose, and throat. Move to fresh air. *IF BREATHING HAS STOPPED,* give artificial respiration. IF breathing is difficult, administer oxygen. **Liquid:** *POISONOUS IF SWALLOWED OR IF SKIN IS EXPOSED.* Will burn eyes. Remove contaminated clothing and shoes. Flush affected areas with plenty of water. *IF IN EYES,* hold eyelids open and flush with plenty of water.

HEALTH HAZARDS
Personal protective equipment (PPE): B-Level PPE. Wear self-contained breathing apparatus, rubber boots, heavy rubber gloves, and protective clothing.
Exposure limits (TWA unless otherwise noted): 2 ppm (AIHA WEEL).
Toxicity by ingestion: Grade 2: LD_{50} = 1.29 g/kg (rat).
Long-term health effects: Causes damage to the liver and kidney. May effect the nervous system.
Vapor (gas) irritant characteristics: Vapors cause severe irritation of eyes and throat and can cause eye and lung injury. They cannot be tolerated even at low concentrations.
Liquid or solid irritant characteristics: Severe skin and eye irritant. Causes second- and third-degree burns on short exposure and is very injurious to the eyes.

FIRE DATA
Flash point: 134°F/57°C (oc).
Flammable limits in air: LEL: 1.3%; UEL: 8.7%.
Fire extinguishing agents not to be used: Water may be ineffective.
Autoignition temperature: 1000°F/538°C/811°K.

CHEMICAL REACTIVITY
Binary reactants: Strong oxidizers may cause fire and explosions. Incompatible with alcohols, aldehydes, alkylene oxides, cresols, caprolactam solution, epichlorohydrin, organic anhydrides, glycols, maleic anhydride, phenols.
Neutralizing agents for acids and caustics: Flush with water.
Reactivity group: 9
Compatibility class: Aromatic amines

ENVIRONMENTAL DATA
Food chain concentration potential: Log P_{ow} = 1.23. Unlikely to accumulate.
Water pollution: Effect of low concentration on aquatic life is unknown. May be dangerous if it enters nearby water intakes; notify operators. Notify local health and wildlife officials.
Response to discharge: Restrict access. Evacuate areas. Should be removed. Chemical and physical treatment.

SHIPPING INFORMATION
Grades of purity: 99%; **Storage temperature:** Ambient; **Stability during transport:** Stable.

PHYSICAL AND CHEMICAL PROPERTIES
Physical state @ 59°F/15°C and 1 atm: Liquid.
Molecular weight: 93.13
Boiling point @ 1 atm: 293°F/145°C/418.2°K.
Melting/Freezing point: 36.3°F/2.4°C/275.6°K.
Specific gravity (water = 1): 0.957
Relative vapor density (air = 1): 3.2

1-METHYLPYRROLIDONE REC. M:5300

SYNONYMS: EEC No. 606-021-00-7; 1-METHYL-2-PYRROLIDINONE; *N*-METHYLPYRROLIDINONE; 1-METHYLPYRROLIDINONE; *N*-METHYL-α-PYRROLIDONE; *N*-METIL-2-PIRRIDONA (Spanish); M-PYROL; NMP

IDENTIFICATION
CAS Number: 872-50-4
Formula: C_5H_9NO
DOT ID Number: UN 1760; DOT Guide Number: 154
Proper Shipping Name: Corrosive liquids, n.o.s.

DESCRIPTION: Colorless to light yellow liquid. Mild, fishy odor. Soluble in water.

Highly flammable • Firefighting gear (including SCBA) does not provide adequate protection. If exposure occurs, remove and isolate gear immediately and thoroughly decontaminate personnel • Containers may BLEVE when exposed to fire • Vapors may form explosive mixture with air • Vapors are heavier than air and will collect and stay in low areas • Vapors may travel long distances to ignition sources and flashback • Vapors in confined areas (e.g., tanks, sewers, buildings) may explode when exposed to fire • Corrosive; highly irritating to the skin, eyes, and respiratory tract • Toxic products of combustion may include nitrogen oxides.

Hazard Classification (based on NFPA-704 Rating System)
Health Hazards (Blue): 2; Flammability (Red): 3; Reactivity (Yellow): 1

EMERGENCY RESPONSE: See Appendix A (154)
Evacuation:
Public safety: Isolate the area of spill or leak for at least 25 to 50 meters (80 to 160 feet) in all directions.
Spill: Increase, in the downwind direction, as necessary, the distance shown under "Public Safety."
Fire: If tank, rail car, or tank truck is involved in fire, isolate for at least 800 meters (½ mile) in all directions; also, consider initial evacuation for 800 meters (½ mile) in all directions.

EXPOSURE
Short-term effects: *SEEK MEDICAL ATTENTION.* **Liquid:** Irritating to skin and eyes. Remove contaminated clothing and shoes. Flush affected areas with plenty of water. *IF BREATHING HAS STOPPED,* give artificial respiration; *avoid mouth-to-mouth resuscitation; use bag/mask apparatus. IF IN EYES,* hold eyelids open and flush with plenty of water. *IF SWALLOWED,* will cause nausea and vomiting. *IF SWALLOWED* and victim is *CONSCIOUS AND ABLE TO SWALLOW,* have victim drink 4 to 8 ounces of water and have victim induce vomiting. *IF SWALLOWED* and victim is *UNCONSCIOUS OR HAVING CONVULSIONS,* do nothing except keep victim warm.

HEALTH HAZARDS
Personal protective equipment (PPE): B-Level PPE. Organic vapor respirator; rubber gloves. Chemical protective material(s) reported to have good to excellent resistance: butyl rubber, Silvershield®. Also, PV alcohol offers limited protection.
Exposure limits (TWA unless otherwise noted): 10 ppm [skin] (AIHA WEEL).
Toxicity by ingestion: Grade 2; oral LD_{50} = 3.5 mg/kg (rabbit).
Long-term health effects: Causes blood abnormalities in rats.
Vapor (gas) irritant characteristics: Corrosive to the eyes, skin, and respiratory system.

Liquid or solid irritant characteristics: Corrosive to the eyes, skin, and respiratory system.

FIRE DATA
Flash point: 7°F/–14°C (oc).
Fire extinguishing agents not to be used: Water may be ineffective.

CHEMICAL REACTIVITY
Binary reactants: Strong oxidizer; aluminum and other light metals. Strongly alkaline; keep away from acids, flammable and porous material

ENVIRONMENTAL DATA
Food chain concentration potential: Negative; unlikely to accumulate.
Water pollution: Effect of low concentrations on aquatic life is unknown. May be dangerous if it enters nearby water intakes; notify operators. Notify local health and wildlife officials.
Response to discharge: Issue warning–water contaminant. Disperse and flush.

SHIPPING INFORMATION
Grades of purity: Technical; **Storage temperature:** Ambient; **Inert atmosphere:** None; **Venting:** Open (flame arrester); **Stability during transport:** Stable.

PHYSICAL AND CHEMICAL PROPERTIES
Physical state @ 59°F/15°C and 1 atm: Liquid.
Molecular weight: 99
Boiling point @ 1 atm: 180°F/82°C/355°K.
Melting/Freezing point: 1°F/–17°C/256°K.
Specific gravity (water = 1): 1.03 @ 77°F/25°C (liquid).
Relative vapor density (air = 1): 2.9
Heat of combustion: –13,000 Btu/lb = –7220 cal/g = –302 x 10^5 J/kg.
Vapor pressure: 0.53 mm.

METHYL SALICYLATE   REC. M:5350

SYNONYMS: *o*-ANISIC ACID; BENZOIC ACID, 2-METHOXY-; *o*-METHOXYBENZOIC ACID; SWEET BIRCH OIL; BETULA OIL; GAULTHERIA OIL; SALICILATO de METILO (Spanish); TEABERRY OIL; WINTERGREEN OIL

IDENTIFICATION
CAS Number: 119-36-8
Formula: $C_8H_8O_3$; 2-(HO)$C_6H_4CO_2CH_3$

DESCRIPTION: Colorless, yellowish or reddish. liquid. Wintergreen odor. Mixes with water.

Combustible • Heat or flame may cause explosion • Breathing the vapor or swallowing the material can cause severe illness; skin or eye contact can cause severe burns • Firefighting gear (including SCBA) does not provide adequate protection. If exposure occurs, remove and isolate gear immediately and thoroughly decontaminate personnel • Containers may BLEVE when exposed to fire • Vapors are heavier than air and will collect and stay in low areas • Vapors in confined areas (e.g., tanks, sewers, buildings) may explode when exposed to fire • Irritating to the skin, eyes, and respiratory tract • Toxic products of combustion may include carbon monoxide.

Hazard Classification (based on NFPA-704 Rating System)
Health Hazards (Blue): 1; Flammability (Red): 1; Reactivity (Yellow): 0

EMERGENCY RESPONSE: See Appendix A (171)
Evacuation:
Public safety: Isolate the area of spill or leak for at least 10 to 25 meters (30 to 80 feet) in all directions.
Spill: Increase, in the downwind direction, as necessary, the distance shown under "Public Safety."
Fire: If any large container is involved in fire, isolate for at least 800 meters (½ mile) in all directions; also, consider initial evacuation for 800 meters (½ mile) in all directions.

EXPOSURE
Short-term effects: *CALL FOR MEDICAL AID*
Vapors: Irritating to eyes, nose, and throat. Move to fresh air. *IF BREATHING HAS STOPPED*, give artificial respiration. IF breathing is difficult, administer oxygen. **Liquid:** Irritating to skin, eyes, nose, and throat. Remove contaminated clothing and shoes. Flush affected areas with plenty of water. *IF IN EYES*, hold eyelids open and flush with plenty of water.

HEALTH HAZARDS
Personal protective equipment (PPE): B-Level PPE. Self contained breathing apparatus, rubber boots, and heavy rubber gloves. Chemical protective material(s) reported to offer minimal to poor protection: butyl rubber/neoprene.
Toxicity by ingestion: Poison. Grade 3: LD_{LO} = 101 mg/kg (man).
Long-term health effects: Possible reproductive damage. May effect the nervous system.
Vapor (gas) irritant characteristics: Vapors/mist cause moderate irritation such that personnel will find high concentrations unpleasant. The effect may be temporary.
Liquid or solid irritant characteristics: Causes smarting of the eyes and skin; first-degree burns on short exposure; may cause second-degree burns on long exposure.

FIRE DATA
Flash point: 205°F/96°C (cc).
Autoignition temperature: 847°F/453°C/726°K.

CHEMICAL REACTIVITY
Binary reactants: Reacts with strong oxidizers
Reactivity group: 34
Compatibility class: Esters.

ENVIRONMENTAL DATA
Food chain concentration potential: Log P_{ow} = 2.58. Unlikely to accumulate.
Water pollution: DOT Appendix B, §172.101–marine pollutant. Effects of low concentrations on aquatic life is unknown. Fouling to shoreline. May be dangerous if it enters nearby water intakes; notify operators. Notify local health and wildlife officials.
Response to discharge: Restrict access. Mechanical containment. Chemical and physical treatment.

SHIPPING INFORMATION
Grades of purity: 99 +%; **Stability during transport:** Stable.

PHYSICAL AND CHEMICAL PROPERTIES
Physical state @ 59°F/15°C and 1 atm: Liquid.
Molecular weight: 152.15
Boiling point @ 1 atm: 432°F/222°C/496°K.
Melting/Freezing point: 18-19°F/–8 to –7°C/265-266°K.

Specific gravity (water = 1): 1.174
Relative vapor density (air = 1): 5.25
Vapor pressure: (Reid) Less than 0.01 psia; 0.08 (approximate).

α-METHYL STYRENE   REC. M:5400

SYNONYMS: AMS; EEC No. 601-027-00-6; ISOPROPENYL BENZENE; 1-(METHYLETHYL) BENZENE; 1-METHYL-1-PHENYL-ETHENE; 1-METHYL-1-PHENYL-ETHYLENE; PHENYLPROPYLENE; α-METILESTIRENO (Spanish); 2-PHENYLPROPYLENE; β-PHENYLPROPYLENE

IDENTIFICATION
CAS Number: 98-83-9
Formula: C_9H_{10}; $C_6H_5C(CH_3){=}CH_2$
DOT ID Number: UN 1993; DOT Guide Number: 128
Proper Shipping Name: Flammable liquid, n.o.s.

DESCRIPTION: Colorless liquid. Sharp, aromatic, disagreeable odor. Floats on water surface; insoluble. Freezes at –9°F/–23°C.

Highly flammable • Containers may BLEVE when exposed to fire • Vapors may form explosive mixture with air • Vapors are heavier than air and will collect and stay in low areas • Vapors may travel long distances to ignition sources and flashback • Vapors in confined areas (e.g., tanks, sewers, buildings) may explode when exposed to fire • Polymerization hazard • Heat can induce polymerization with rapid release of energy; sealed containers may rupture explosively • May react with itself blocking relief valves; leading to tank explosions • Irritating to the skin, eyes, and respiratory tract • Toxic products of combustion may include carbon monoxide.

Hazard Classification (based on NFPA-704 Rating System)
Health Hazards (Blue): 2; Flammability (Red): 2; Reactivity (Yellow): 2

EMERGENCY RESPONSE: See Appendix A (128)
Evacuation:
Public safety: Isolate spill area for at least 25 to 50 meters (80 to 160 feet) in all directions.
Spill: Large spill–Consider initial downwind evacuation for at least 300 meters (1000 feet).
Fire: Isolate for 800 meters (½ mile) in all directions, especially if tank, rail car, or tank truck is involved in fire.

EXPOSURE
Short-term effects: *SEEK MEDICAL ATTENTION.* **Liquid:** Will burn skin and eyes. *IF SWALLOWED,* will cause nausea and vomiting. Remove contaminated clothing and shoes. Flush affected areas with plenty of water. *IF IN EYES*, hold eyelids open and flush with plenty of water. *IF SWALLOWED* and victim is *CONSCIOUS AND ABLE TO SWALLOW*, have victim drink 4 to 8 ounces of water. *IF SWALLOWED* and victim is *UNCONSCIOUS OR HAVING CONVULSIONS,* do nothing except keep victim warm. **Do NOT induce vomiting.**

HEALTH HAZARDS
Personal protective equipment (PPE): B-Level PPE. OSHA Table Z-1-A air contaminant. Gloves; splash proof goggles or face shield. Chemical protective material(s) reported to offer minimal to poor protection: butyl rubber/neoprene, natural rubber.
Recommendations for respirator selection: NIOSH/OSHA
500 ppm: CCROV* [any chemical cartridge respirator with organic vapor cartridge(s)]; SA* (any supplied-air respirator). *700 ppm:* SA:CF* (any supplied-air respirator operated in a continuous-flow mode); or CCRFOV [any air-purifying, full-facepiece respirator (gas mask) with a chin-style, front- or back-mounted acid gas canister]; or GMFOV [any air-purifying, full-facepiece respirator (gas mask) with a chin-style, front- or back-mounted organic vapor cannister]; or PAPROV [any powered, air-purifying respirator with organic vapor cartridge(s)]; or SCBAF (any self-contained breathing apparatus with a full facepiece); or SAF (any supplied-air respirator with a full facepiece). *EMERGENCY OR PLANNED ENTRY INTO UNKNOWN CONCENTRATIONS OR IDLH CONDITIONS:* SCBAF:PD,PP (any self-contained breathing apparatus that has a full facepiece and is operated in a pressure-demand or other positive-pressure mode); or SAF:PD,PP:ASCBA (any supplied-air respirator that has a full facepiece and is operated in a pressure-demand or other positive-pressure mode in combination with an auxiliary self-contained breathing apparatus operated in a pressure-demand or other positive pressure mode). *ESCAPE:* GMFOV [any air-purifying, full-facepiece respirator (gas mask) with a chin-style, front- or back-mounted organic vapor cannister]; or SCBAE (any appropriate escape-type, self-contained breathing apparatus). *Note:* Substance reported to cause eye irritation or damage; may require eye protection.
Exposure limits (TWA unless otherwise noted): ACGIH TLV 50 ppm (242 mg/m^3); OSHA PEL ceiling 100 ppm (480 mg/m^3); NIOSH REL 50 ppm (240 mg/m^3).
Short-term exposure limits (15-minute TWA): ACGIH STEL 100 ppm (483 mg/m^3); NIOSH STEL 100 ppm (485 mg/m^3).
Toxicity by ingestion: Grade 2; LD_{50} = 0.55 g/kg.
Long-term health effects: May cause liver and kidney damage; skin disorders.
Vapor (gas) irritant characteristics: May cause irritation of the eyes and respiratory tract.
Liquid or solid irritant characteristics: Irritates the eyes, skin, and respiratory system.
Odor threshold: 0.11–0.29 ppm.
IDLH value: 700 ppm.

FIRE DATA
Flash point: 129°F/54°C (cc).
Flammable limits in air: LEL: 1.9%; UEL: 6.1%.
Fire extinguishing agents not to be used: Water may be ineffective.
Autoignition temperature: 1066°F/574°C/847°K.
Electrical hazard: Class I, Group D. Flow or agitation of substance may generate electrostatic charges due to low conductivity.

CHEMICAL REACTIVITY
Binary reactants: Reacts with strong oxidizers, peroxides, halogens, catalysts for vinyl or ionic polymers; aluminum, iron chloride, copper. May attack some forms of plastics
Polymerization: Hazardous polymerization may occur when in contact with alkali metals or metalloorganic compounds, catalysts or heat; **Inhibitor of polymerization:** 10-20 ppm *tert*-butylcatechol
Reactivity group: 30
Compatibility class: Olefins

ENVIRONMENTAL DATA
Food chain concentration potential: Negative; unlikely to accumulate.
Water pollution: DOT Appendix B, §172.101–marine pollutant. Harmful to aquatic life in very low concentrations. Fouling to

shoreline. May be dangerous if it enters nearby water intakes; notify operators. Notify local health and wildlife officials.
Response to discharge: Mechanical containment. Should be removed. Chemical and physical treatment.

SHIPPING INFORMATION
Grades of purity: Commercial; **Storage temperature:** Ambient; **Inert atmosphere:** None; **Venting:** Open (flame arrester); **Stability during transport:** Stable.

PHYSICAL AND CHEMICAL PROPERTIES
Physical state @ 59°F/15°C and 1 atm: Liquid.
Molecular weight: 118.17
Boiling point @ 1 atm: 329°F/165°C/438°K.
Melting/Freezing point: –9.8°F/–23.2°C/250.0°K.
Critical temperature: 719.1°F/381.7°C/654.9°K.
Critical pressure: 494 psia = 33.6 atm = 3.41 MN/m^2.
Specific gravity (water = 1): 0.91 @ 68°F/20°C (liquid).
Liquid surface tension: 33.88 dynes/cm = 0.03388 N/m @ 68°F/20°C.
Relative vapor density (air = 1): 4.08
Ratio of specific heats of vapor (gas): 1.060 at 27°C.
Latent heat of vaporization: 140.4 Btu/lb = 78.0 cal/g = 3.26 x 10^5 J/kg.
Heat of combustion: –17,690 Btu/lb = –9,830 cal/g = –411 x 10^5 J/kg.
Vapor pressure: (Reid) 0.23 psia; 2.2 mm.

METHYL TRICHLOROSILANE   REC. M:5450

SYNONYMS: EEC No. 014-004-00-5; METILTRICLOROSILANO (Spanish); SILANE, TRICHLOROMETHYL-; SILANE,METHYLTRICHLORO-; METHYL CHLOROSILANE; TRICHLOROMETHYLSILANE

IDENTIFICATION
CAS Number: 75-79-6
Formula: CH$_3$Cl$_3$Si
DOT ID Number: UN 1250; DOT Guide Number: 155
Proper Shipping Name: Methyltrichlorosilane
Reportable Quantity (RQ): 1 lb/0.454 kg

DESCRIPTION: Colorless liquid. Sharp acrid odor, like hydrochloric acid. Reacts violently with water forming hydrochloric acid and hydrogen chloride vapors.

Poison! • Highly flammable • Do NOT use water • Breathing the vapor can kill you; skin and eye contact causes severe burns and blindness; inhalation symptoms may be delayed • Firefighting gear (including SCBA) does not provide adequate protection. If exposure occurs, remove and isolate gear immediately and thoroughly decontaminate personnel • Highly flammable • Containers may BLEVE when exposed to fire •Vapors may form explosive mixture with air •Vapors are heavier than air and will collect and stay in low areas • Vapors may travel long distances to ignition sources and flashback • Vapors in confined areas (e.g., tanks, sewers, buildings) may explode when exposed to fire • Severely irritating to skin, eyes, and respiratory tract; prolonged contact with the skin causes burns • Toxic products of combustion may include hydrogen chloride • Do not put yourself in danger by entering a contaminated area to rescue a victim.

Hazard Classification (based on NFPA-704 Rating System)
Health Hazards (Blue): 3; Flammability (Red): 3; Reactivity (Yellow): 2; Special Notice (White): Water reactive

EMERGENCY RESPONSE (Chlorosilanes): See Appendix A (155)
Evacuation:
Public safety: Isolate the area of spill or leak for at least 50 to 100 meters (160 to 330 feet) in all directions.
Spill: Increase, in the downwind direction, as necessary, the distance shown under "Public Safety."
IF SPILLED IN WATER: Small spill–First: Isolate in all directions 30 meters (100 feet); Then: Protect persons downwind, DAY: 0.2 km (0.1 mile); NIGHT: 0.3 km (0.2 mile). Large spill–First: Isolate in all directions 125 meters (400 feet); Then: Protect persons downwind, DAY: 1.1 km (0.7 mile); NIGHT: 2.9 km (1.8 miles).
Fire: If tank, rail car, or tank truck is involved in fire, isolate for at least 800 meters (½ mile) in all directions; also, consider initial evacuation for 800 meters (½ mile) in all directions.

EXPOSURE
Short-term effects: *SEEK MEDICAL ATTENTION*. **Vapor:** Irritating to eyes, nose, and throat. *IF INHALED*, will, will cause difficult breathing. Move victim to fresh air. *IF BREATHING HAS STOPPED*, give artificial respiration; *avoid mouth-to-mouth resuscitation; use bag/mask apparatus*. IF breathing is difficult, administer oxygen. Lung edema may develop. **Liquid:** Will burn skin and eyes. Harmful if swallowed. Drops may enter the lungs causing pneumonia. Remove contaminated clothing and shoes. Flush affected areas with plenty of water. *IF IN EYES*, hold eyelids open and flush with plenty of water. *IF SWALLOWED* and victim is *CONSCIOUS AND ABLE TO SWALLOW*, have victim drink 4 to 8 ounces of water. **Do NOT induce vomiting**.
Note to physician or authorized medical personnel: Medical observation is recommended for 24 to 48 hours after breathing overexposure, as pulmonary edema may be delayed. As first aid for pulmonary edema, consider administering a corticosteroid spray. Cigarette smoking may exacerbate pulmonary injury and should be discouraged for at least 72 hours following exposure.

HEALTH HAZARDS
Personal protective equipment (PPE): A-Level PPE. Full protective clothing; acid-vapor-type respiratory protection; rubber gloves; chemical worker's goggles; other protective equipment as necessary to protect skin and eyes.
Exposure limits (TWA unless otherwise noted): 1 ppm [ceiling] (AIHAWEEL).
Toxicity by ingestion: Grade 3; LD$_{50}$ = 50 to 500 mg/kg.
Vapor (gas) irritant characteristics: Vapors cause severe irritation of eyes and throat and can cause eye and lung injury. They cannot be tolerated even at low concentrations.
Liquid or solid irritant characteristics: Severe skin irritant. Causes second- and third-degree burns on short contact and is very injurious to the eyes.
Odor threshold: Decomposes in moist air, creating HCl with Odor threshold of 1 ppm.

FIRE DATA
Flash point: 45°F/7°C (oc); 15°F/–9°C (cc).
Flammable limits in air: LEL: 7.6%; UEL: More than 20%.
Fire extinguishing agents not to be used: Water, foam
Behavior in fire: Difficult to extinguish; re-ignition may occur.
Autoignition temperature: More than 760°F/404°C/1033°K.
Electrical hazard: Class I, Group Unassigned.
Burning rate: 1.9 mm/min

CHEMICAL REACTIVITY
Reactivity with water: Reacts violently to form hydrogen chloride gas.
Binary reactants: Vigorous reaction with strong oxidizers. Corrosive to metals in the presence of moisture.
Neutralizing agents for acids and caustics: Flood with water, rinse with sodium bicarbonate or lime solution.

ENVIRONMENTAL DATA
Food chain concentration potential: Negative; unlikely to accumulate.
Water pollution: Effect of low concentrations on aquatic life is unknown. May be dangerous if it enters nearby water intakes; notify operators. Notify local health and wildlife officials.
Response to discharge: Issue warning–high flammability, air contaminant. Restrict access. Evacuate area. Disperse and flush with care.

SHIPPING INFORMATION
Grades of purity: 98+%; **Storage temperature:** Ambient; **Inert atmosphere:** None; **Venting:** Safety relief; **Stability during transport:** Stable.

NAS HAZARD CLASSIFICATION FOR BULK WATER TRANSPORTATION
FIRE: 3
HEALTH: Vapor irritant: 4; Liquid or solid irritant: 4; Poisons: 3
WATER POLLUTION: Human toxicity: 3; Aquatic toxicity: 3; Aesthetic effect: 2
REACTIVITY: Other chemicals: 3; Water: 4; Self-reaction: 0

PHYSICAL AND CHEMICAL PROPERTIES
Physical state @ 59°F/15°C and 1 atm: Liquid.
Molecular weight: 149.5
Boiling point @ 1 atm: 152°F/66°C/340°K.
Melting/Freezing point: −130°F/−90°C/183°K.
Specific gravity (water = 1): 1.27 @ 77°F/25°C (liquid).
Liquid surface tension: 20.3 dynes/cm = 0.0203 N/m @ 68°F/20°C.
Relative vapor density (air = 1): 5.16
Latent heat of vaporization: 89.3 Btu/lb = 49.6 cal/g = 2.08×10^5 J/kg.
Heat of combustion: (estimate) −3000 Btu/lb = −1700 cal/g = -70×10^5 J/kg.
Vapor pressure: 137 mm.

METHYL VINYL KETONE   REC. M:5500

SYNONYMS: ACETYL ETHYLENE; 3-BUTEN-2-ONE; BUTENONE; 2-BUTENONE; METHYLENE ACETONE; METIL VINIL CETONA (Spanish); VINYL METHYL KETONE

IDENTIFICATION
CAS Number: 78-94-4
Formula: C_4H_6O; $CH_3COCH=CH_2$
DOT ID Number: UN 1251; DOT Guide Number: 131P
Proper Shipping Name: Methyl vinyl ketone, stabilized
Reportable Quantity (RQ): **(EHS)** 1 lb/0.454 kg

DESCRIPTION: Colorless to pale yellow liquid. Powerful, irritating odor. Mixes with water. Irritating vapor is produced.

Highly flammable • A tear gas • Firefighting gear (including SCBA) does not provide adequate protection. If exposure occurs, remove and isolate gear immediately and thoroughly decontaminate personnel • Containers may BLEVE when exposed to fire •Vapors may form explosive mixture with air • Vapors are heavier than air and will collect and stay in low areas • Vapors may travel long distances to ignition sources and flashback • Vapors in confined areas (e.g., tanks, sewers, buildings) may explode when exposed to fire • Polymerization hazard • Heat can induce polymerization with rapid release of energy; sealed containers may rupture explosively • May react with itself blocking relief valves; leading to tank explosions • Irritating to the skin, eyes, and respiratory tract • Toxic products of combustion may include carbon monoxide.

Hazard Classification (based on NFPA-704 Rating System)
Health Hazards (Blue): 4; Flammability (Red): 3; Reactivity (Yellow): 2

EMERGENCY RESPONSE: See Appendix A (131)
Evacuation:
Public safety: Isolate spill area for at least 100 to 200 meters (330 to 660 feet) in all directions.
Spill: Small spill–First: Isolate in all directions 155 meters (500 feet); Then: Protect persons downwind, DAY: 1.3 km (0.8 mile); NIGHT: 2.1, Large spill–First: Isolate in all directions: 915 meters (3000 feet); Then: Protect persons downwind, DAY: 8.7 km (5.4 mile); NIGHT: 11.0+ km (7.0+ miles).
Fire: Isolate for 800 meters (½ mile) in all directions, especially if tank, rail car, or tank truck is involved in fire.

EXPOSURE
Short-term effects: *SEEK MEDICAL ATTENTION.* **Vapor:** Irritating to eyes, nose, and throat. *IF INHALED*, will, will cause coughing or difficult breathing. May cause lung edema; physical exertion will aggravate this condition. Move victim to fresh air. *IF BREATHING HAS STOPPED,* give artificial respiration. IF breathing is difficult, administer oxygen. **Liquid:** *POISONOUS IF SWALLOWED or IF ABSORBED THROUGH THE SKIN.* Will burn skin and eyes. Remove contaminated clothing and shoes. Flush affected areas with plenty of water. *IF IN EYES*, hold eyelids open and flush with plenty of water. *IF SWALLOWED* and victim is *CONSCIOUS AND ABLE TO SWALLOW*, have victim drink 4 to 8 ounces of water. **Do NOT induce vomiting.**
Note to physician or authorized medical personnel: Pulmonary edema may be delayed. Medical observation is recommended for 24 to 48 hours after inhalation overexposure. As first aid for pulmonary edema, a physician or authorized medical personnel may consider administering a corticosteroid spray. Cigarette smoking may exacerbate pulmonary injury and should be discouraged for at least 72 hours following exposure.

HEALTH HAZARDS
Personal protective equipment (PPE): A-Level PPE. Organic vapor self-contained breathing apparatus with full face piece; rubber gloves; chemical goggles or face piece of breathing apparatus.
Toxicity by ingestion: Grade 4; LD_{50} less than 50 mg/kg.
Long-term health effects: Possible mutagen.
Vapor (gas) irritant characteristics: Vapors cause severe irritation of eyes and throat and can cause eye and lung injury. They cannot be tolerated even at low concentrations.
Liquid or solid irritant characteristics: Severe skin irritant. Causes second- and third-degree burns on short contact, and is very injurious to the eyes.
Odor threshold: 0.5 mg/m^3.

FIRE DATA
Flash point: 30°F/–1°C (oc); 20°F/–7°C (cc).
Flammable limits in air: LEL: 2.1%; UEL: 15.6%.
Fire extinguishing agents not to be used: Water may be ineffective.
Behavior in fire: distance to a source of ignition and flashback. At elevated temperatures (fire conditions) polymerization may take place in containers, causing violent rupture. Unburned vapors are very irritating.
Autoignition temperature: 915°F/491°C/764°K.
Electrical hazard: Class I, Group C.
Burning rate: 4.5 mm/min

CHEMICAL REACTIVITY
Binary reactants: Violent reaction with oxidizers. May form peroxides in air
Polymerization: Polymerizes spontaneously with generation of heat upon exposure to heat or sunlight; **Inhibitor of polymerization:** Up to 1% hydroquinone

ENVIRONMENTAL DATA
Food chain concentration potential: Log P_{ow} = 0.12. Unlikely to accumulate.
Water pollution: Effect of low concentrations on aquatic life is unknown. May be dangerous if it enters nearby water intakes; notify operators. Notify local health and wildlife officials.
Response to discharge: Issue warning–high flammability, water contaminant. Restrict access. Evacuate area. Disperse and flush.

SHIPPING INFORMATION
Grades of purity: 98.5+%; **Storage temperature:** Cool ambient; **Inert atmosphere:** None; **Venting:** Pressure-vacuum; **Stability during transport:** Stable.

NAS HAZARD CLASSIFICATION FOR BULK WATER TRANSPORTATION
FIRE: 3
HEALTH: Vapor irritant: 4; Liquid or solid irritant: 4; Poisons: 4
WATER POLLUTION: Human toxicity: 4; Aquatic toxicity: 2; Aesthetic effect: 3
REACTIVITY: Other chemicals: 3; Water: 0; Self-reaction: 3

PHYSICAL AND CHEMICAL PROPERTIES
Physical state @ 59°F/15°C and 1 atm: Liquid.
Molecular weight: 70.1
Boiling point @ 1 atm: 179°F/81°C/355°K.
Melting/Freezing point: 20°F/–7°C/266°K.
Specific gravity (water = 1): 0.864 @ 68°F/20°C (liquid).
Liquid surface tension: (estimate) 24 dynes/cm = 0.024 N/m @ 68°F/20°C.
Relative vapor density (air = 1): 2.4
Ratio of specific heats of vapor (gas): 1.1053
Latent heat of vaporization: (estimate) 203 Btu/lb = 113 cal/g = 4.73 x 10^5 J/kg.
Heat of combustion: (estimate) –14,600 Btu/lb = –8100 cal/g = –340 x 10^5 J/kg.
Heat of polymerization: –455 Btu/lb = –253 cal/g = –10.6 x 10^5 J/kg.
Vapor pressure: 75 mm.

METOLACHLOR   REC. M:5550

SYNONYMS: CODAL; DUAL; COTORAN MULTI; METELILACHLOR; MILOCEP; CGA-24705; ONTRACK 8E; METELILACHLOR; PRIMAGRAM; PRIMEXTRA

IDENTIFICATION
CAS Number: 51218-45-2
Formula: $C_{15}H_{22}ClNO_2$
DOT ID Number: UN 2996; DOT Guide Number: 151
Proper Shipping Name: Organochlorine pesticide, liquid, poisonous

DESCRIPTION: Tan to brown oily liquid. Slightly sweet odor. Slightly soluble; sinks in water.

Poison! (organochlorine; chloracetanilide herbicide) • Irritates the eye and skin; breathing the vapor can cause illness • Firefighting gear (including SCBA) does not provide adequate protection. If exposure occurs, remove and isolate gear immediately and thoroughly decontaminate personnel • Containers may BLEVE when exposed to fire • Vapors are heavier than air and will collect and stay in low areas • Vapors in confined areas (e.g., tanks, sewers, buildings) may explode when exposed to fire • Toxic products of combustion may include nitrogen oxides and hydrogen chloride • Do not put yourself in danger by entering a contaminated area to rescue a victim.

Hazard Classification (based on NFPA-704 Rating System)
Health Hazards (Blue): 1; Flammability (Red): 1; Reactivity (Yellow): 0

EMERGENCY RESPONSE: See Appendix A (151)
Evacuation:
Public safety: Isolate the area of spill or leak for at least 25 to 50 meters (80 to 160 feet) in all directions.
Spill: Increase, in the downwind direction, as necessary, the distance shown under "Public Safety."
Fire: If tank, rail car, or tank truck is involved in fire, isolate for at least 800 meters (½ mile) in all directions; also, consider initial evacuation for 800 meters (½ mile) in all directions.

EXPOSURE
Short-term effects: *CALL FOR MEDICAL AID.* **Liquid:** Irritating to skin and eyes; harmful if swallowed. Remove contaminated clothing and shoes. Flush affected areas with plenty of water. *IF BREATHING HAS STOPPED,* give artificial respiration; *avoid mouth-to-mouth resuscitation; use bag/mask apparatus. IF IN EYES,* hold eye lids open and flush with lenty of water. *IF SWALLOWED* and victim is *CONSCIOUS AND ABLE TO SWALLOW* have victim drink water. **Do NOT induce vomiting**. Administer a slurry of activated charcoal. *IF SWALLOWED* and victim is *UNCONSCIOUS OR HAVING CONVULSIONS,* do nothing except keep victim warm.

HEALTH HAZARDS
Personal protective equipment (PPE): B-Level PPE.
Toxicity by ingestion: Grade 2: LD_{50} = 2.75 g/kg (rat).
Long-term health effects: May cause tumors.
Vapor (gas) irritant characteristics: Vapor can cause slight smarting to eyes or respiratory system, if present in high concentration.
Liquid or solid irritant characteristics: Minimum hazard, if spilled on clothing and allowed to remain, may cause smarting and reddening of skin.

FIRE DATA
Flash point: More than 230°F/110°C (cc).
Autoignition temperature: 510°F/266°C/539°K.

CHEMICAL REACTIVITY
Binary reactants: Strong acids, oxidizers
Reactivity group: 34
Compatibility class: Esters.

ENVIRONMENTAL DATA
Water pollution: Effects of low concentration on aquatic life is unknown. May be dangerous if it enters nearby water intakes; notify operators. Notify local health and wildlife officials.
Response to discharge: Restrict access. Should be removed. Chemical and physical treatment.

SHIPPING INFORMATION
Grades of purity: 95%; **Storage temperature:** Ambient; **Stability during transport:** Stable.

PHYSICAL AND CHEMICAL PROPERTIES
Physical state @ 59°F/15°C and 1 atm: Liquid.
Molecular weight: 283.81
Boiling point @ 1 atm: 212°F/100°C/373°K (at 0.001 mmHg).
Specific gravity (water = 1): 1.12
Relative vapor density (air = 1): 9.79

MINERAL SPIRITS   REC. M:5600

SYNONYMS: LIGROIN; LIGROINA (Spanish); MINERAL THINNER; MINERAL TERPENTINE; PETROLEUM SOLVENT; TERPENTINE SUBSTITUTE

IDENTIFICATION
CAS Number: 8032-32-4 (ligroin); 64475-85-0
Formula: A refined petroleum solvent containing >65% C_{10} or higher hydrocarbons.*
DOT ID Number: UN 1268; DOT Guide Number: 128
Proper Shipping Name: Petroleum distillate, n.o.s.

DESCRIPTION: Colorless watery liquid. Gasoline-like odor. Floats on water surface; insoluble. *The term "mineral spirits" differs in the United States and the United Kingdom. In the United Kingdom this term refers to a volatile hydrocarbon mixture with a flash point below 32°F/0°C (Merck Index/9).

Flammable • Containers may BLEVE when exposed to fire • Vapors may form explosive mixture with air • Vapors are heavier than air and will collect and stay in low areas • Vapors may travel long distances to ignition sources and flashback • Vapors in confined areas (e.g., tanks, sewers, buildings) may explode when exposed to fire • Irritating to the skin, eyes, and respiratory tract • Toxic products of combustion may include carbon monoxide.

Hazard Classification (based on NFPA-704 Rating System)
Health Hazards (Blue): 0; Flammability (Red): 2; Reactivity (Yellow): 0

EMERGENCY RESPONSE: See Appendix A (128)
Evacuation:
Public safety: Isolate spill area for at least 25 to 50 meters (80 to 160 feet) in all directions.
Spill: Large spill–Consider initial downwind evacuation for at least 300 meters (1000 feet).
Fire: Isolate for 800 meters (½ mile) in all directions, especially if tank, rail car, or tank truck is involved in fire.

EXPOSURE
Short-term effects: *SEEK MEDICAL ATTENTION*. **Liquid:** Irritating to skin and eyes. Harmful if swallowed. Remove contaminated clothing and shoes. Flush affected areas with plenty of water. *IF IN EYES*, hold eyelids open and flush with plenty of water. *IF SWALLOWED* and victim is *CONSCIOUS AND ABLE TO SWALLOW*, have victim drink 4 to 8 ounces of water. **Do NOT induce vomiting.**

HEALTH HAZARDS
Personal protective equipment (PPE): B-Level PPE. Gloves; goggles or face shield (as for gasoline, naphtha, or stoddard solvent). Chemical protective material(s) reported to have good to excellent resistance: nitrile, PV alcohol, Viton®, Saranex®, Silvershield®. Also, neoprene and Styrene-butadiene offers limited protection
Toxicity by ingestion: Grade 2; LD_{50} = 0.5 to 5 g/kg.
Vapor (gas) irritant characteristics: Vapors may be nonirritating to the eyes and throat.
Liquid or solid irritant characteristics: If spilled on clothing and allowed to remain, may cause smarting and reddening of the skin.

FIRE DATA
Flash point: 104-140°F/40-60°C (cc), depending on grade, but are usually above 100°F/38°C.
Flammable limits in air: LEL: 0.8% @ 212°F/100°C; UEL: 5.0%.
Autoignition temperature: 470°F/244°C/517°K.
Electrical hazard: Class I, Group D.
Burning rate: 4 mm/min

CHEMICAL REACTIVITY
Binary reactants: Oxidizers
Reactivity group: 33
Compatibility class: Miscellaneous hydrocarbon mixtures

ENVIRONMENTAL DATA
Food chain concentration potential: Negative; unlikely to accumulate.
Water pollution: Effect of low concentrations on aquatic life is unknown. Fouling to shoreline. May be dangerous if it enters nearby water intakes; notify operators. Notify local health and wildlife officials. **Response to discharge:** Mechanical containment. Should be removed. Chemical and physical treatment.

SHIPPING INFORMATION
Grades of purity: Various grades available. 70–100% of the materials are derived from petroleum, and 0–30% are aromatic hydrocarbons like benzene and toluene; **Storage temperature:** Ambient; **Inert atmosphere:** None; **Venting:** Open (flame arrester); **Stability during transport:** Stable.

PHYSICAL AND CHEMICAL PROPERTIES
Boiling point @ 1 atm: 300–395°F/149–202°C/422–475°K.
Specific gravity (water = 1): 0.78 @ 68°F/20°C (liquid).
Ratio of specific heats of vapor (gas): (estimate) 1.030
Vapor pressure: (Reid) 0.13 psia.

MIREX   REC. M:5650

SYNONYMS: BICHLORENDO; CG-1283; DECHLORANE; DECHLORANE-4070; FERRIAMICIDE; HRS 1276; HEXACHLOROCYCLOPENTADIENE DIMER;

PERCHLORODIHOMOCUBANE; perchlordecone
PERCHLORDECONE;
PERCHLOROPENTACLYCLODECANE

IDENTIFICATION
CAS Number: 2385-85-5
Formula: $C_{10}Cl_{12}$
DOT ID Number: UN 2811; DOT Guide Number: 154
Proper Shipping Name: Poisonous solid, organic, n.o.s.

DESCRIPTION: Snow-white crystalline solid. Odorless. Practically insoluble in water. *Note*: Effective 12/1/77 all registered products containing Mirex were canceled and existing inventories within the continental U.S. were not to be sold, distributed, or used after 6/30/78 (U.S. EPA).

Poison! • Combustible solid • Corrosive • Heat or flame may cause explosion of dust • Breathing the dust, skin or eye contact, or swallowing the material can kill you • Firefighting gear (including SCBA) does not provide adequate protection. If exposure occurs, remove and isolate gear immediately and thoroughly decontaminate personnel • Containers may BLEVE when exposed to fire • Concentrated dust in confined areas (e.g., tanks, sewers, buildings) may explode when exposed to fire • Toxic products of combustion may include hydrogen chloride • Do not put yourself in danger by entering a contaminated area to rescue a victim.

Hazard Classification (based on NFPA-704 Rating System)
Health Hazards (Blue): 2; Flammability (Red): 1; Reactivity (Yellow): 0

EMERGENCY RESPONSE: See Appendix A (154)
Evacuation:
Public safety: Isolate the area of spill or leak for at least 25 to 50 meters (80 to 160 feet) in all directions.
Spill: Increase, in the downwind direction, as necessary, the distance shown under "Public Safety."
Fire: If tank, rail car, or tank truck is involved in fire, isolate for at least 800 meters (½ mile) in all directions; also, consider initial evacuation for 800 meters (½ mile) in all directions.

EXPOSURE
Short-term effects: *SEEK MEDICAL ATTENTION*. **Solid:** *POISONOUS IF SWALLOWED, INHALED, OR IF SKIN IS EXPOSED*. Remove contaminated clothing and shoes. Flush affected areas with plenty of water. *IF BREATHING HAS STOPPED,* give artificial respiration; *avoid mouth-to-mouth resuscitation; use bag/mask apparatus. IF IN EYES*, hold eyelids open and flush with plenty of water. *IF SWALLOWED* and victim is *CONSCIOUS AND ABLE TO SWALLOW*, have victim drink 4 to 8 ounces of water and have victim induce vomiting.
Note to physician or authorized medical personnel: Medical observation is recommended for 24 to 48 hours after breathing overexposure, as pulmonary edema may be delayed. As first aid for pulmonary edema, consider administering a corticosteroid spray. Cigarette smoking may exacerbate pulmonary injury and should be discouraged for at least 72 hours following exposure.

HEALTH HAZARDS
Toxicity by ingestion: Poison. Grade 3; LD_{50} = 50 to 500 mg/kg.
Long-term health effects: Chronic industrial exposure has caused apparently irreversible nerve damage. IARC possible human carcinogen, rating 2B; animal evidence: sufficient. Some teratogenic effects noted. Mutation data reported.

CHEMICAL REACTIVITY
Binary reactants: Oxidizers
Reactivity group: 36
Compatibility class: Halogenated hydrocarbons (Chlorinated hydrocarbons, aliphatic)

ENVIRONMENTAL DATA
Food chain concentration potential: High potential; bioaccumulates
Water pollution: DOT Appendix B, §172.101–marine pollutant. Harmful to aquatic life in very low concentrations. May be dangerous if it enters nearby water intakes; notify operators. Notify local health and wildlife officials. **Response to discharge:** Issue warning–poison, water contaminant. Restrict access. Should be removed. Chemical and physical treatment.

SHIPPING INFORMATION
Grades of purity: Pelleted bait "450" (0.45% Mirex); **Stability during transport:** Stable.

PHYSICAL AND CHEMICAL PROPERTIES
Physical state @ 59°F/15°C and 1 atm: Solid.
Molecular weight: 545.59
Boiling point @ 1 atm: Sublimes with decomposition 905°F/485°C/758°K.
Relative vapor density (air = 1): 18.8 (calculated).

MOLYBDIC TRIOXIDE   REC. M:5700

SYNONYMS: MOLYBDENUM(VI) OXIDE; MOLYBDIC ANHYDRIDE; MOLYBDENUM TRIOXIDE; TRIOXIDO de MOLIBDICO (Spanish)

IDENTIFICATION
CAS Number: 1313-27-5
Formula: MoO_3
DOT ID Number: UN 3077; DOT Guide Number: 171
Proper Shipping Name: Environmentally hazardous substances, solid, n.o.s.

DESCRIPTION: Colorless to slightly yellow solid. Odorless. Sinks in water; very slightly soluble; hydrolyzes to molybdic acid.

Poison! • Irritates the eyes; breathing the dust may cause illness (pulmonary fibrosis) • Firefighting gear (including SCBA) does not provide adequate protection. If exposure occurs, remove and isolate gear immediately and thoroughly decontaminate personnel • Containers may BLEVE when exposed to fire • Concentrated dust in confined areas (e.g., tanks, sewers, buildings) may explode when exposed to fire • Toxic products of combustion may include molybdenum.

Hazard Classification (based on NFPA-704 Rating System)
Health Hazards (Blue): 2; Flammability (Red): 0; Reactivity (Yellow): 0

EMERGENCY RESPONSE: See Appendix A (171)
Evacuation:
Public safety: Isolate the area of spill or leak for at least 10 to 25 meters (30 to 80 feet) in all directions.
Spill: Increase, in the downwind direction, as necessary, the distance shown under "Public Safety."

776   Monochloroacetic acid

Fire: If any large container is involved in fire, isolate for at least 800 meters (½ mile) in all directions; also, consider initial evacuation for 800 meters (½ mile) in all directions.

EXPOSURE
Short-term effects: *SEEK MEDICAL ATTENTION*. **Solid:** Irritating to skin and eyes. Harmful if swallowed. Remove contaminated clothing and shoes. Flush affected areas with plenty of water. *IF IN EYES*, hold eyelids open and flush with plenty of water. *IF SWALLOWED* and victim is *CONSCIOUS AND ABLE TO SWALLOW*, have victim drink 4 to 8 ounces of water. *IF SWALLOWED* and victim is *UNCONSCIOUS OR HAVING CONVULSIONS*, do nothing except keep victim warm.

HEALTH HAZARDS
Personal protective equipment (PPE): B-Level PPE.
Recommendations for respirator selection: OSHA for soluble compounds of molybdenum
$25\ mg/m^3$: DM* (any dust and mist respirator). $50\ mg/m^3$: DMXSQ* (any dust and mist respirator except single-use and quarter mask respirators); or SA* (any supplied-air respirator). $125\ mg/m^3$: SA:CF* (any supplied-air respirator operated in a continuous-flow mode); or PAPRDM* *If not present as a fume* (any powered, air-purifying respirator with a dust and mist filter). $250\ mg/m^3$: HiEF (any air-purifying, full-facepiece respirator with a high-efficiency particulate filter); or SAT:CF* (any supplied-air respirator that has a tight-fitting facepiece and is operated in a continuous-flow mode); or PAPRTHiE* (any powered, air-purifying respirator with a tight-fitting facepiece and a high-efficiency particulate filter); or SCBAF (any self-contained breathing apparatus with a full facepiece); or SAF (any supplied-air respirator with a full facepiece). $1000\ mg/m^3$: SAF:PD,PP (any supplied-air respirator that has a full facepiece and is operated in a pressure-demand or other positive-pressure mode). *EMERGENCY OR PLANNED ENTRY INTO UNKNOWN CONCENTRATIONS OR IDLH CONDITIONS*: SCBAF:PD,PP (any self-contained breathing apparatus that has a full facepiece and is operated in a pressure-demand or other positive-pressure mode); or SAF:PD,PP:ASCBA (any supplied-air respirator that has a full facepiece and is operated in a pressure-demand or other positive-pressure mode in combination with an auxiliary self-contained breathing apparatus operated in a pressure-demand or other positive pressure mode). *ESCAPE:* HiEF (any air-purifying, full-facepiece respirator with a high-efficiency particulate filter); or SCBAE (any appropriate escape-type, self-contained breathing apparatus). *Note:* Substance reported to cause eye irritation or damage; may require eye protection.
Exposure limits (TWA unless otherwise noted): ACGIH TLV $0.5\ mg/m^3$; OSHA PEL $5\ mg/m^3$ as molybdenum, soluble compounds.
Toxicity by ingestion: Grade 3; LD_{50} = 50 to 500 mg/kg.
Long-term health effects: May cause lung damage.
Vapor (gas) irritant characteristics: May cause irritation and cough; fibrosis of the lung.
Liquid or solid irritant characteristics: May cause irritation of the eyes and lungs.
IDLH value: $1000\ mg/m^3$ as molybdenum, soluble compounds.

CHEMICAL REACTIVITY
Binary reactants: Alkali metals, sodium, potassium, molten magnesium; bromine and chlorine compounds, lithium

ENVIRONMENTAL DATA
Water pollution: Harmful to aquatic life in very low concentrations. May be dangerous if it enters nearby water intakes; notify operators. Notify local health and wildlife officials. **Response to discharge:** Issue warning–water contaminant. Should be removed. Chemical and physical treatment.

SHIPPING INFORMATION
Grades of purity: Technical, 59.8-61.6%; Reagent; **Storage temperature:** Ambient; **Inert atmosphere:** None; **Venting:** Open; **Stability during transport:** Stable.

PHYSICAL AND CHEMICAL PROPERTIES
Physical state @ 59°F/15°C and 1 atm: Solid.
Molecular weight: 143.94
Specific gravity (water = 1): 4.69 @ 68°F/20°C (solid).

MONOCHLOROACETIC ACID   REC. M:5750

SYNONYMS: ACIDO MONOCLORACETICO (Spanish); CHLORACETIC ACID; α-CHLORACETIC ACID; CHLOROACETIC ACID; EEC No. 607-003-00-1; MCA; MONOCHLOROETHANOIC ACID

IDENTIFICATION
CAS Number: 79-11-8
Formula: $C_2H_3ClO_2$; $ClCH_2COOH$
DOT ID Number: UN 1751 (solid); UN 1750 (liquid or solution); UN 3250 (molten); DOT Guide Number: 153
Proper Shipping Name: Chloroacetic acid, solid; Chloroacetic acid, solution; Chloroacetic acid, liquid; Chloroacetic acid, molten
Reportable Quantity (RQ): **(CERCLA)** 1 lb/0.454 kg

DESCRIPTION: Cloudy translucent white solid; liquid is colorless to light yellow. Strong vinegar-like odor. Mixes with water.

Combustible • Heat or flame may cause explosion • Corrosive to skin, eyes, and respiratory tract; contact with skin or eyes can cause burns and blindness; inhalation symptoms may be delayed • Containers may BLEVE when exposed to fire • Vapors are heavier than air and will collect and stay in low areas • Vapors in confined areas (e.g., tanks, sewers, buildings) may explode when exposed to fire • Toxic products of combustion may include hydrogen chloride and phosgene.

Hazard Classification (based on NFPA-704 Rating System)
Health Hazards (Blue): 3; Flammability (Red): 1; Reactivity (Yellow): 0

EMERGENCY RESPONSE: See Appendix A (153)
Evacuation:
Public safety: Isolate the area of spill or leak for at least 25 to 50 meters (80 to 160 feet) in all directions.
Spill: Increase, in the downwind direction, as necessary, the distance shown under "Public Safety."
Fire: If tank, rail car, or tank truck is involved in fire, isolate for at least 800 meters (½ mile) in all directions; also, consider initial evacuation for 800 meters (½ mile) in all directions.

EXPOSURE
Short-term effects: *SEEK MEDICAL ATTENTION*. **Vapor:** Vapor may cause lung edema; physical exertion will aggravate this condition. *IF BREATHING HAS STOPPED*, give artificial respiration; *avoid mouth-to-mouth resuscitation; use bag/mask apparatus*. **Liquid or solid:** Will burn skin and eyes. Harmful if swallowed. Remove contaminated clothing and shoes. Flush

affected areas with plenty of water. *IF IN EYES*, hold eyelids open and flush with plenty of water. *IF SWALLOWED* and victim is *CONSCIOUS AND ABLE TO SWALLOW*, have victim drink 4 to 8 ounces of water. **Do NOT induce vomiting.**

Note to physician or authorized medical personnel: Medical observation is recommended for 24 to 48 hours after breathing overexposure, as pulmonary edema may be delayed. As first aid for pulmonary edema, consider administering a corticosteroid spray. Cigarette smoking may exacerbate pulmonary injury and should be discouraged for at least 72 hours following exposure.

HEALTH HAZARDS
Personal protective equipment (PPE): B-Level PPE. Self-contained breathing apparatus; vinyl or neoprene rubber gloves; goggles and protective face shield; rubberized or acid-resistant clothing. Chemical protective material(s) reported to offer protection: butyl rubber, PE, neoprene, Viton®.
Exposure limits (TWA unless otherwise noted): 1 mg/m^3 (AIHAWEEL).
Toxicity by ingestion: Poison. Grade 3; oral LD$_{50}$ = 76.2 mg/kg (rat).
Long-term health effects: Mutagen.
Vapor (gas) irritant characteristics: Vapors cause severe irritation of eyes and throat and can cause eye and lung injury. They cannot be tolerated even at low concentrations.
Liquid or solid irritant characteristics: Severe skin irritant. Causes second- and third-degree burns on short contact and is very injurious to the eyes.
Odor threshold: 0.15 mg/m^3.

FIRE DATA
Flash point: 259°F/126°C (cc).
Autoignition temperature: More than 932°F/500°C/773°C.

CHEMICAL REACTIVITY
Reactivity with water: Dissolved in water, substance becomes a strong acid.
Binary reactants: Solution reacts violently with bases and is corrosive. In the presence of moisture causes mild corrosion of common metals.
Neutralizing agents for acids and caustics: Flush with water, rinse with sodium bicarbonate or lime solution
Reactivity group: 4
Compatibility class: Organic acids

ENVIRONMENTAL DATA
Food chain concentration potential: Negative; unlikely to accumulate.
Water pollution: Effect of low concentrations on aquatic life is unknown. May be dangerous if it enters nearby water intakes; notify operators. Notify local health and wildlife officials.
Response to discharge: Issue warning–corrosive. Restrict access. Disperse and flush.

SHIPPING INFORMATION
Grades of purity: Commercial: 97.5+%; **Storage temperature: Solid:** Ambient; **Liquid:** 158°F/70°C; **Inert atmosphere:** None; **Venting:** Open (flame arrester); **Stability during transport:** Stable.

NAS HAZARD CLASSIFICATION FOR BULK WATER TRANSPORTATION
FIRE: 1
HEALTH: Vapor irritant: 4; Liquid or solid irritant: 4; Poisons: 4
WATER POLLUTION: Human toxicity: 3; Aquatic toxicity: 2; Aesthetic effect: 3
REACTIVITY: Other chemicals: 4; Water: 2; Self-reaction: 0

PHYSICAL AND CHEMICAL PROPERTIES
Physical state @ 59°F/15°C and 1 atm: Solid.
Molecular weight: 94.5
Boiling point @ 1 atm: 372°F/189°C/462°K.
Melting/Freezing point: 140°F/60°C/333°K.
Specific gravity (water = 1): 1.58 @ 68°F/20°C (solid).
Liquid surface tension: 33 dynes/cm = 0.033 N/m at 80°C.
Relative vapor density (air = 1): 3.3
Latent heat of vaporization: 250 Btu/lb = 139 cal/g = 5.82 x 10^5 J/kg.
Heat of combustion: (solid) –1814 Btu/lb = –1008 cal/g = –42.17 x 10^5 J/kg.
Heat of solution: –63 Btu/lb = –35 cal/g = –1.5 x 10^5 J/kg.
Heat of fusion: 31.06 cal/g.
Vapor pressure: 0.2 mm (approximate).

MONOCHLORODIFLUOROMETHANE   REC. M:5800

SYNONYMS: ALGOFRENE TYPE 6; ARCTON-4; CHLORODIFLUOROMETHANE (ACGIH); DIFLUROCHLOROMETHANE; ESKIMON-22; F 22; FREON-22; GENETRON-22; ICEON 22; ISOTRON-22; MONOCLORODIFLUOMETANO (Spanish); PROPELLENT-22; R22; UCONN-22; UCONN22/HALOCARBON 22

IDENTIFICATION
CAS Number: 75-45-6
Formula: CHClF$_2$
DOT ID Number: UN 1018; DOT Guide Number: 126
Proper Shipping Name: Chlorodifluoromethane; Refrigerant gas R-22

DESCRIPTION: Liquefied compressed gas. Colorless. Faint odor, like ether or carbon tetrachloride. Liquid sinks and boils in water; soluble. Visible vapor cloud is formed.

Gas from liquefied gas are initially heavier than air and spread along ground causing oxygen deficiency that may cause you to pass out • Do not spray liquid with water • Toxic products of combustion may include fluorine and hydrogen chloride • Contact with liquid may cause frostbite.

Hazard Classification (based on NFPA-704 Rating System)
Health Hazards (Blue): 0; Flammability (Red): 0; Reactivity (Yellow): 0

EMERGENCY RESPONSE: See Appendix A (126)
Evacuation:
Public safety: Isolate spill area for at least 100 meters (330 feet) in all directions.
Spill: Large spill–Consider initial downwind evacuation for at least 500 meters (⅓ mile).
Fire: Isolate for 800 meters (½ mile) in all directions, especially if tank, rail car, or tank truck is involved in fire.

EXPOSURE
Short-term effects: *SEEK MEDICAL ATTENTION.* **Vapor:** Not irritating to eyes, nose or throat. Vapors are heavier than air and may accumulate; may cause oxygen deficiency. *IF INHALED*, will, will cause dizziness or loss of consciousness. Move to fresh air. *IF*

Monochlorotetrafluoroethane

BREATHING HAS STOPPED, give artificial respiration. IF breathing is difficult, administer oxygen. **Liquid:** Will cause frostbite. Flush affected areas with plenty of water. *DO NOT RUB AFFECTED AREAS.*

HEALTH HAZARDS
Personal protective equipment (PPE): B-Level PPE. OSHA Table Z-1-A air contaminant. Rubber gloves; goggles. Wear thermal protective clothing. Chemical protective material(s) reported to offer minimal to poor protection: natural rubber, neoprene, neoprene+natural rubber. Nitrile rubber is generally suitable for freons.
Exposure limits (TWA unless otherwise noted): ACGIH TLV 1000 ppm (3,540 mg/m^3) as chlorodifluoromethane; OSHA PEL 1000 ppm (3600 mg/m^3).
Long-term health effects: Possible mutagenic and reproductive problems.
Vapor (gas) irritant characteristics: Vapors may be nonirritating to the eyes and throat. High concentrations may have a narcotic effect.
Liquid or solid irritant characteristics: No appreciable hazard. Practically harmless to the skin because it evaporates quickly. Evaporation of liquid may cause frostbite.

FIRE DATA
Electrical hazard: Flow or agitation of substance may generate electrostatic charges due to low conductivity.

CHEMICAL REACTIVITY
Reactivity with water: Decomposes slowly forming phosgene gas.
Binary reactants: Violent reaction with alkali metals; powdered chemically active metals such as aluminum, zinc, and magnesium
Reactivity group: 36
Compatibility class: Halogenated hydrocarbons

ENVIRONMENTAL DATA
Food chain concentration potential: Log P_{ow} = 1.1. Unlikely to accumulate.
Water pollution: Effect on aquatic life unknown. **Response to discharge:** Disperse and flush.

SHIPPING INFORMATION
Grades of purity: Propellant grade; **Storage temperature:** Ambient; **Inert atmosphere:** None; **Venting:** Safety relief; **Stability during transport:** Stable.

NAS HAZARD CLASSIFICATION FOR BULK WATER TRANSPORTATION
FIRE: 0
HEALTH: Vapor irritant: 0; Liquid or solid irritant: 0; Poisons: 1
WATER POLLUTION: Human toxicity: 0; Aquatic toxicity: 0; Aesthetic effect: 0
REACTIVITY: Other chemicals: 1; Water: 0; Self-reaction: 0

PHYSICAL AND CHEMICAL PROPERTIES
Physical state @ 59°F/15°C and 1 atm: Gas.
Molecular weight: 86.48
Boiling point @ 1 atm: –40.9°F/–40.5°C/232.7°K.
Melting/Freezing point: 231°F/–146°C/127°K.
Critical temperature: 205°F/96°C/369°K.
Critical pressure: 716 psia = 48.7 atm = 4.93 MN/m^2.
Specific gravity (water = 1): 1.41 at –40°F/–40°C (liquid).
Liquid surface tension: (estimate) 15 dynes/cm = 0.015 N/m @ –40°F/–40°C.
Relative vapor density (air = 1): 3.0
Ratio of specific heats of vapor (gas): (estimate) 1.13
Latent heat of vaporization: 101 Btu/lb = 55.9 cal/g = 2.34 x 10^5 J/kg.
Vapor pressure: (Reid) 212.6 psia; 6990 mm.

MONOCHLOROTETRAFLUOROETHANE   REC. M:5850

SYNONYMS: CHLOROTETRAFLUOROETHANE; 2-CHLORO-1,1,2,2-TETRAFLUOROETHANE; HALON 241; F-124; MONOCLOROTETRAFLUORETANO (Spanish); R-124

IDENTIFICATION
CAS Number: 63938-10-3
Formula: C$_2$HClF$_4$; CHF$_2$CHClF$_2$
DOT ID Number: UN 1021; DOT Guide Number: 126
Proper Shipping Name: Chlorotetrafluoroethane; 1-Chloro-1,1,2,2-tetrafluoroethane; Refrigerant gas R-124

DESCRIPTION: Colorless, compressed, liquefied gas. Odorless.

Gas from liquefied gas are initially heavier than air and spread along ground causing oxygen deficiency that may cause you to pass out • Do not spray liquid with water • Toxic products of combustion may include fluorine and hydrogen chloride • Contact with liquid may cause frostbite.

Hazard Classification (based on NFPA-704 Rating System)
Health Hazards (Blue): 1; Flammability (Red): 0; Reactivity (Yellow): 0

EMERGENCY RESPONSE: See Appendix A (126)
Evacuation:
Public safety: Isolate spill area for at least 100 meters (330 feet) in all directions.
Spill: Large spill–Consider initial downwind evacuation for at least 500 meters (⅓ mile).
Fire: Isolate for 800 meters (½ mile) in all directions, especially if tank, rail car, or tank truck is involved in fire.

EXPOSURE
Short-term effects: *SEEK MEDICAL ATTENTION.* **Vapor:** Vapors may cause dizziness or suffocation. Move victim to fresh air. If not breathing, give artificial respiration. IF breathing is difficult, administer oxygen. **Liquid:** Contact with liquid may cause frostbite. Remove contaminated clothing and shoes. Flush affected areas with lukewarm water. *DO NOT USE HOT WATER*

HEALTH HAZARDS
Personal protective equipment (PPE): B-Level PPE. Approved respirator, chemical safety goggles, chemical resistant gloves, other protective clothing. Wear thermal protective clothing.
Exposure limits (TWA unless otherwise noted): 1000 ppm (AIHA WEEL).
Vapor (gas) irritant characteristics: Vapors may not be irritating to the eyes and throat of most people.
Liquid or solid irritant characteristics: Minimum hazard. Contact with liquid may cause frostbite.

CHEMICAL REACTIVITY
Reactivity group: 36
Compatibility class: Halogenated hydrocarbons

ENVIRONMENTAL DATA
Ozone depleting chemical.
Food chain concentration potential: Unlikely to accumulate.

NAS HAZARD CLASSIFICATION FOR BULK WATER TRANSPORTATION
FIRE: 0
HEALTH: Vapor irritant: 0; Liquid or solid irritant: 1; Poisons: 0
WATER POLLUTION: Human toxicity:–; Aquatic toxicity:–; Aesthetic effect: 0
REACTIVITY: Other chemicals: 0; Water: 0; Self-reaction: 0

PHYSICAL AND CHEMICAL PROPERTIES
Physical state @ 59°F/15°C and 1 atm: Gas.
Molecular weight: 136.48
Boiling point @ 1 atm: 14°F/–10°C/263°K.
Melting/Freezing point: –179°F/–117°C/156°K.
Relative vapor density (air = 1): 4.71

MONOCHLOROTRIFLUOROMETHANE   REC. M:5900

SYNONYMS: ARCTON-3; CHLOROTRIFLUOROMETHANE; FREON 13; F-13; GENETRON 13; HALOCARBON 13/UCONN 13; R-13; TRIFLUOROCHLOROMETHANE; TRIFLUOROMETHYL CHLORIDE

IDENTIFICATION
CAS Number: 75-72-9
Formula: $CClF_3$
DOT ID Number: UN 1022; DOT Guide Number: 126
Proper Shipping Name: Chlorotrifluoromethane; Refrigerant gas R-13

DESCRIPTION: Colorless compressed, liquefied gas. Odorless. Insoluble in water.

Narcotic effect at high concentrations; gas from liquefied gas are initially heavier than air and spread along ground causing oxygen deficiency that may cause you to pass out • Do not spray liquid with water • Toxic products of combustion may include fluorine and hydrogen chloride • Contact with liquid may cause frostbite.

Hazard Classification (based on NFPA-704 Rating System)
Health Hazards (Blue): 1; Flammability (Red): 0; Reactivity (Yellow): 0

EMERGENCY RESPONSE: See Appendix A (126)
Evacuation:
Public safety: Isolate spill area for at least 100 meters (330 feet) in all directions.
Spill: Large spill–Consider initial downwind evacuation for at least 500 meters (⅓ mile).
Fire: Isolate for 800 meters (½ mile) in all directions, especially if tank, rail car, or tank truck is involved in fire.

EXPOSURE
Short-term effects: *SEEK MEDICAL ATTENTION.* **Vapors:** Vapors may be harmful if inhaled. Vapors may cause dizziness or suffocation. Move victim to fresh air. If not breathing, give artificial respiration. IF breathing is difficult, administer oxygen. Heavier than air, vapors may accumulate causing oxygen deficiency. **Liquid:** Contact with liquid may cause frostbite. Remove contaminated clothing and shoes. Flush affected areas with plenty of lukewarm water. *DO NOT USE HOT WATER.*

HEALTH HAZARDS
Personal protective equipment (PPE): B-Level PPE. Approved respirator, safety goggles, rubber gloves, safety shoes. Wear thermal protective clothing. Nitrile rubber is generally suitable for freons.
Long-term health effects: May cause nerve system damage.
Vapor (gas) irritant characteristics: Vapors are mildly irritating to eyes and throat. High concentrations may have a narcotic effect and may cause loss of consciousness.
Liquid or solid irritant characteristics: Mild irritant. Contact may cause frostbite.

CHEMICAL REACTIVITY
Binary reactants: Reacts with aluminum and its alloys; magnesium and its alloys; zinc and its alloys
Reactivity group: 36
Compatibility class: Halogenated hydrocarbons

ENVIRONMENTAL DATA
Ozone depleting chemical.
Food chain concentration potential: Unlikely to accumulate.

SHIPPING INFORMATION
Grades of purity: 99+%; **Storage temperature:** Below 130°F/54°C; **Stability during transport:** Stable.

NAS HAZARD CLASSIFICATION FOR BULK WATER TRANSPORTATION
FIRE: 0
HEALTH: Vapor irritant: 0; Liquid or solid irritant: 1; Poisons: 0
WATER POLLUTION: Human toxicity:–; Aquatic toxicity: 0; Aesthetic effect: 0
REACTIVITY: Other chemicals: 0; Water: 0; Self-reaction: 0

PHYSICAL AND CHEMICAL PROPERTIES
Physical state @ 59°F/15°C and 1 atm: Gas.
Molecular weight: 104.46
Boiling point @ 1 atm: –114°F/–81°C/192°K.
Melting/Freezing point: –294°F/–181°C/92°K.
Critical temperature: 83.93°F/28.85°C/302.05°K.
Critical pressure: 561.4 psia = 38.2 atm = 3.9 MN/m².
Specific gravity (water = 1): 1.298 at –30°C.
Liquid surface tension: 14 dyne/cm = 0.014 N/m at –73.3°C.
Relative vapor density (air = 1): 3.60
Latent heat of vaporization: 76.1 Btu/lb = 42.3 cal/g = 1.77×10^5 J/kg.
Vapor pressure: (Reid) 480 psia; 24,320 mm.

MONOETHANOLAMINE   REC. M:5950

SYNONYMS: 2-AMINOETHANOL; AMINOETHYL ALCOHOL; β-AMINOETHYL ALCOHOL; EEC No. 603-030-00-8; ETHANOLAMINE; GLYCINOL; ETANOLAMINA (Spanish); MONOETANOLAMINA (Spanish); 2-HYDROXYETHYLAMINE; OLAMINE

IDENTIFICATION
CAS Number: 141-43-5
Formula: $HOCH_2CH_2NH_2$
DOT ID Number: UN 2491; DOT Guide Number: 153
Proper Shipping Name: Ethanolamine; Ethanolamine solutions.

DESCRIPTION: Colorless, thick, oily liquid. Fishy odor. Initially sinks in water; soluble. Freezes at 51°F/11°C.

Monoethanolamine

Corrosive • Flammable • Severely Irritating to the skin, eyes, and respiratory tract; skin and eye contact causes severe burns and blindness • Firefighting gear (including SCBA) does not provide adequate protection. If exposure occurs, remove and isolate gear immediately and thoroughly decontaminate personnel • Containers may BLEVE when exposed to fire • Vapors are heavier than air and will collect and stay in low areas • Vapors in confined areas (e.g., tanks, sewers, buildings) may explode when exposed to fire • Toxic products of combustion may include carbon monoxide and nitrogen oxides.

Hazard Classification (based on NFPA-704 Rating System)
Health Hazards (Blue): 3; Flammability (Red): 2; Reactivity (Yellow): 0

EMERGENCY RESPONSE: See Appendix A (153)
Evacuation:
Public safety: Isolate the area of spill or leak for at least 25 to 50 meters (80 to 160 feet) in all directions.
Spill: Increase, in the downwind direction, as necessary, the distance shown under "Public Safety."
Fire: If tank, rail car, or tank truck is involved in fire, isolate for at least 800 meters (½ mile) in all directions; also, consider initial evacuation for 800 meters (½ mile) in all directions.

EXPOSURE
Short-term effects: *SEEK MEDICAL ATTENTION*. **Liquid or solid:** Irritating to skin and eyes. Harmful if swallowed. Remove contaminated clothing and shoes. Flush affected areas with plenty of water. *IF BREATHING HAS STOPPED*, give artificial respiration; *avoid mouth-to-mouth resuscitation; use bag/mask apparatus. IF IN EYES*, hold eyelids open and flush with plenty of water. *IF SWALLOWED* and victim is *CONSCIOUS AND ABLE TO SWALLOW*, have victim drink 4 to 8 ounces of water. The use of alcoholic beverages may enhance the toxic effects.
Note to physician or authorized medical personnel: Medical observation is recommended for 24 to 48 hours after breathing overexposure, as pulmonary edema may be delayed. As first aid for pulmonary edema, consider administering a corticosteroid spray. Cigarette smoking may exacerbate pulmonary injury and should be discouraged for at least 72 hours following exposure.

HEALTH HAZARDS
Personal protective equipment (PPE): B-Level PPE. OSHA Table Z-1-A air contaminant. Full face shield; goggles; eye wash facility. Chemical protective material(s) reported to have good to excellent resistance: butyl rubber, neoprene, nitrile, nitrile+PVC, PVC, Viton®. Also, chlorinated polyethylene offers limited protection.
Recommendations for respirator selection: NIOSH/OSHA
30 ppm: CCRS [any chemical cartridge respirator with cartridge(s) providing protection against the compound of concern]; or GMFS [any air-purifying, full-facepiece respirator (gas mask) with a chin-style, front- or back-mounted canister providing protection against the compound of concern]; or PAPRS [any powered, air-purifying respirator with cartridge(s) providing protection against the compound of concern]; or SA (any supplied-air respirator); or SCBAF (any self-contained breathing apparatus with a full facepiece). *EMERGENCY OR PLANNED ENTRY INTO UNKNOWN CONCENTRATIONS OR IDLH CONDITIONS*: SCBAF:PD,PP (any self-contained breathing apparatus that has a full facepiece and is operated in a pressure-demand or other positive-pressure mode); or SAF:PD,PP:ASCBA (any supplied-air respirator that has a full facepiece and is operated in a pressure-demand or other positive-pressure mode in combination with an auxiliary self-contained breathing apparatus operated in a pressure-demand or other positive pressure mode). *ESCAPE:* GMFS [any air-purifying, full-facepiece respirator (gas mask) with a chin-style, front- or back-mounted canister providing protection against the compound of concern]; or SCBAE (any appropriate escape-type, self-contained breathing apparatus). *Note:* Substance reported to cause eye irritation or damage; may require eye protection.
Exposure limits (TWA unless otherwise noted): ACGIH TLV 3 ppm (7.5 mg/m^3); NIOSH/OSHA 3 ppm (6 mg/m^3); NIOSH REL 3 ppm (8 mg/m3).
Short-term exposure limits (15-minute TWA): ACGIH STEL 6 ppm (15 mg/m^3); NIOSH STEL 6 ppm (15 mg/m^3).
Toxicity by ingestion: Grade 2; LD_{50} = 0.5 to 5 g/kg (rat).
Long-term health effects: Dermatitis may occur.
Vapor (gas) irritant characteristics: Irritates the eyes, skin and mucous membrane. Vapors cause irritation such that personnel will find high concentrations unpleasant. The effect may be temporary.
Liquid or solid irritant characteristics: Causes irritation of the eyes, skin, and respiratory tract. Causes smarting of the skin and first-degree burns on short exposure; may cause secondary burns on long exposure.
Odor threshold: 2.0–4.0 ppm.
IDLH value: 30 ppm.

FIRE DATA
Flash point: 200°F/93°C (oc); 186°F/86°C (cc).
Flammable limits in air: LEL: 3.0% @ 284°F/140°C; UEL 23.5% @ 140°F/60°C.
Autoignition temperature: 766°F/408°C/681°K.

CHEMICAL REACTIVITY
Reactivity with water: Forms alkaline solution.
Binary reactants: Reacts with strong oxidizers; strong acids, aluminum [at temperatures above 150°F/66°C], iron. Attack aluminum, copper, zinc, and rubber
Neutralizing agents for acids and caustics: Flush with water.
Reactivity group: 8
Compatibility class: Alkanolamines

ENVIRONMENTAL DATA
Food chain concentration potential: Log P_{ow} = –1.29. Unlikely to accumulate.
Water pollution: Dangerous to aquatic life in high concentrations. May be dangerous if it enters nearby water intakes; notify operators. Notify local health and wildlife officials. **Response to discharge:** Disperse and flush.

SHIPPING INFORMATION
Grades of purity: NF: 85% (15% water); commercial: 99+%; **Storage temperature:** Ambient; **Inert atmosphere:** None; **Venting:** Open; **Stability during transport:** Stable.

NAS HAZARD CLASSIFICATIONS FOR BULK WATER TRANSPORTATION
FIRE: 1
HEALTH: Vapor irritant: 2; Liquid or solid irritant: 2; Poisons: 2
WATER POLLUTION: Human toxicity: 2; Aquatic toxicity: 1; Aesthetic effect: 2
REACTIVITY: Other chemicals: 3; Water: 0; Self-reaction: 0

PHYSICAL AND CHEMICAL PROPERTIES
Physical state @ 59°F/15°C and 1 atm: Liquid.
Molecular weight: 61.08
Boiling point @ 1 atm: 338°F/170°C/443°K.
Melting/Freezing point: 50.5°F/10.3°C/283.5°K.

Critical temperature: 646°F/341°C/614°K.
Critical pressure: 647 psia = 44 atm = 4.45 MN/m².
Specific gravity (water = 1): 1.016 @ 68°F/20°C (liquid).
Relative vapor density (air = 1): 2.1
Latent heat of vaporization: 360 Btu/lb = 200 cal/g = 8.37 x 10⁵ J/kg.
Heat of combustion: –10,710 Btu/lb = –5950 cal/g = –249 x 10⁵ J/kg.
Heat of solution: (estimate) –17 Btu/lb = –10 cal/g = –0.4 x 10⁵ J/kg.
Vapor pressure: (Reid) 0.01 psia; 0.4 mm.

MONOISOPROPANOLAMINE   REC. M:6000

SYNONYMS: 1-AMINO-2-PROPANOL; 2-HYDROXYPROPYLAMINE; ISOPROPANOLAMINE; ISOPROPANOLAMINA (Spanish)

IDENTIFICATION
CAS Number: 78-96-6
Formula: C_3H_9NO; $CH_3CH(OH)CH_2NH_2$
DOT ID Number: UN 1760; DOT Guide Number: 154
Proper Shipping Name: Corrosive liquids, n.o.s.

DESCRIPTION: Colorless thick liquid. Ammonia-like odor. Initially floats on the surface of water; soluble; forms an alkaline solution. Freezes at 35°F/1.9°C.

Corrosive • Combustible • Severely Irritating to the skin, eyes, and respiratory tract; prolonged skin contact causes severe burns • Firefighting gear (including SCBA) does not provide adequate protection. If exposure occurs, remove and isolate gear immediately and thoroughly decontaminate personnel • Containers may BLEVE when exposed to fire • Vapors are heavier than air and will collect and stay in low areas • Vapors in confined areas (e.g., tanks, sewers, buildings) may explode when exposed to fire • Toxic products of combustion may include nitrogen oxides.

Hazard Classification (based on NFPA-704 Rating System)
Health Hazards (Blue): 2; Flammability (Red): 2; Reactivity (Yellow): 0

EMERGENCY RESPONSE: See Appendix A (154)
Evacuation:
Public safety: Isolate the area of spill or leak for at least 25 to 50 meters (80 to 160 feet) in all directions.
Spill: Increase, in the downwind direction, as necessary, the distance shown under "Public Safety."
Fire: If tank, rail car, or tank truck is involved in fire, isolate for at least 800 meters (½ mile) in all directions; also, consider initial evacuation for 800 meters (½ mile) in all directions.

EXPOSURE
Short-term effects: *SEEK MEDICAL ATTENTION.* **Liquid or solid:** Irritating to skin and eyes. Harmful if swallowed. Remove contaminated clothing and shoes. Flush affected areas with plenty of water. *IF BREATHING HAS STOPPED,* give artificial respiration; *avoid mouth-to-mouth resuscitation; use bag/mask apparatus. IF IN EYES,* hold eyelids open and flush with plenty of water. *IF SWALLOWED* and victim is *CONSCIOUS AND ABLE TO SWALLOW,* have victim drink 4 to 8 ounces of water and have victim induce vomiting. *IF SWALLOWED* and victim is *UNCONSCIOUS OR HAVING CONVULSIONS,* do nothing except keep victim warm.

Note to physician or authorized medical personnel: Medical observation is recommended for 24 to 48 hours after breathing overexposure, as pulmonary edema may be delayed. As first aid for pulmonary edema, consider administering a corticosteroid spray. Cigarette smoking may exacerbate pulmonary injury and should be discouraged for at least 72 hours following exposure.

HEALTH HAZARDS
Personal protective equipment (PPE): B-Level PPE. Full face shield; goggles; eye wash facility. Chemical protective material(s) reported to have good to excellent resistance: butyl rubber, neoprene, PVC, Viton®.
Toxicity by ingestion: Grade 2; LD_{50} = 0.5 to 5 g/kg (rat).
Vapor (gas) irritant characteristics: Vapors cause smarting of the eyes or respiratory system if present in high concentrations. The effect may be temporary.
Liquid or solid irritant characteristics: Causes smarting of the skin and first-degree burns on short exposure and may cause secondary burns on long exposure.

FIRE DATA
Flash point: 171°F/77°C (cc).
Flammable limits in air: LEL: 2.2% (calculated); UEL: 12% (estimate).
Autoignition temperature: 706°F/374°C/647°K (estimate).
Burning rate: 1.1 mm/min

CHEMICAL REACTIVITY
Reactivity with water: Forms alkaline solution
Neutralizing agents for acids and caustics: Flush with water.
Reactivity group: 8
Compatibility class: Alkanolamines

ENVIRONMENTAL DATA
Food chain concentration potential: Log P_{ow} = –1.1. Unlikely to accumulate.
Water pollution: Effect of low concentrations on aquatic life is unknown. May be dangerous if it enters nearby water intakes; notify operators. Notify local health and wildlife officials. **Response to discharge:** Disperse and flush.

SHIPPING INFORMATION
Grades of purity: 98.5+%; **Storage temperature:** Ambient; **Inert atmosphere:** None; **Venting:** Open; **Stability during transport:** Stable.

NAS HAZARD CLASSIFICATION FOR BULK WATER TRANSPORTATION
FIRE: 1
HEALTH: Vapor irritant: 1; Liquid or solid irritant: 2; Poisons: 1
WATER POLLUTION: Human toxicity: 2; Aquatic toxicity: 1; Aesthetic effect: 2
REACTIVITY: Other chemicals: 3; Water: 0; Self-reaction: 0

PHYSICAL AND CHEMICAL PROPERTIES
Physical state @ 59°F/15°C and 1 atm: Liquid.
Molecular weight: 75.11
Boiling point @ 1 atm: 320°F/160°C/433°K.
Melting/Freezing point: 35.4°F/1.9°C/275.1°K.
Critical temperature: 622°F/328°C/601°K.
Critical pressure: 850 psia = 58 atm = 5.9 MN/m².
Specific gravity (water = 1): 0.961 @ 68°F/20°C (liquid).
Latent heat of vaporization: 272 Btu/lb = 151 cal/g = 6.32 x 10⁵ J/kg.

Heat of combustion: (estimate) –13,900 Btu/lb = –7700 cal/g = –322 x 10^5 J/kg.
Heat of solution: (estimate) –17 Btu/lb = –10 cal/g = –0.4 x 10^5 J/kg.
Vapor pressure: (Reid) 0.05 psia.

MONOMETHYLETHANOLAMINE   REC. M:6050

SYNONYMS: EEC No. 603-080-00-0; 2-(HYDROXYETHYL) METHYLAMINE; *N*-METHYL ETHANOLAMINE; *N*-METHYLAMINOETHANOL; 2-(METHYLAMINO)ETHANOL; MONOMETHYLAMINOETHANOL

IDENTIFICATION
CAS Number: 109-83-1
Formula: C_3H_9NO; $CH_3NHCH_2CH_2OH$
DOT ID Number: NA 1993; DOT Guide Number: 128
Proper Shipping Name: Combustible liquid, n.o.s.

DESCRIPTION: Colorless liquid. Fishy odor. Soluble in water.

Combustible • Containers may BLEVE when exposed to fire • Vapors are heavier than air and will collect and stay in low areas • Vapors may travel long distances to ignition sources and flashback • Vapors in confined areas (e.g., tanks, sewers, buildings) may explode when exposed to fire • Irritating to the skin, eyes, and respiratory tract • Toxic products of combustion may include nitrogen oxides.

Hazard Classification (based on NFPA-704 Rating System)
Health Hazards (Blue): 2; Flammability (Red): 2; Reactivity (Yellow): 0

EMERGENCY RESPONSE: See Appendix A (128)
Evacuation:
Public safety: Isolate spill area for at least 25 to 50 meters (80 to 160 feet) in all directions.
Spill: Large spill–Consider initial downwind evacuation for at least 300 meters (1000 feet).
Fire: Isolate for 800 meters (½ mile) in all directions, especially if tank, rail car, or tank truck is involved in fire.

EXPOSURE
Short-term effects: *SEEK MEDICAL ATTENTION.* **Vapor:** Move victim to fresh air. *IF BREATHING HAS STOPPED,* give artificial respiration. IF breathing is difficult, administer oxygen. **Liquid:** Irritating to skin and eyes. Remove contaminated clothing and shoes. Flush affected areas with water. *IF IN EYES,* hold eyelids open and flush with plenty of water.

HEALTH HAZARDS
Personal protective equipment (PPE): B-Level PPE. Full impervious protective clothing, including boots and gloves. Where splashing is possible wear full face shield or chemical safety goggles. Use approved respirator to protect against vapors. Chemical protective material(s) reported to have good to excellent resistance: butyl rubber, neoprene, Viton®.
Toxicity by ingestion: Grade 2; LD_{50} = 2.34 g/kg (rat).
Vapor (gas) irritant characteristics: Vapors cause smarting of the eyes or respiratory system if present in high concentrations. The effect may be temporary.
Liquid or solid irritant characteristics: Irritates eyes, skin, and respiratory tract. If spilled on clothing and allowed to remain, may cause smarting and reddening of skin.

FIRE DATA
Flash point: 165°F/74°C (cc).
Flammable limits in air: LEL: 1.9%; UEL: 19.8%.
Fire extinguishing agents not to be used: Water.
Autoignition temperature: 491°F/255°C/528°K.
Electrical hazard: Flow or agitation of substance may generate electrostatic charges due to low conductivity.
Stoichiometric air-to-fuel ratio: 22.6

CHEMICAL REACTIVITY
Reactivity with water: Forms alkaline solution.
Binary reactants: Reaction with strong oxidizers may cause fire. Reacts with acids. Attacks metals such as aluminum, zinc, copper and copper alloys.
Reactivity group: 8
Compatibility class: Alkanolamines

ENVIRONMENTAL DATA
Water pollution: Effects of low concentrations on aquatic life is unknown. May be dangerous if it enters nearby water intakes; notify operators. Notify local health and wildlife officials. **Response to discharge:** Should be removed.

SHIPPING INFORMATION
Grades of purity: 99%; technical grades; **Storage temperature:** Ambient; **Inert atmosphere:** None; **Stability during transport:** Stable.

PHYSICAL PROPERTIES
Physical state @ 59°F/15°C and 1 atm: Liquid.
Molecular weight: 75.11
Boiling point @ 1 atm: 319°F/160°C/433°K.
Melting/Freezing point: 24°F/–5°C/269°K.
Specific gravity (water = 1): 0.9414
Vapor pressure: 0.71 mm.

MORPHOLINE   REC. M:6100

SYNONYMS: DIETHYLENEIMIDE OXIDE; DIETHYLENE IMIDOXIDE; DIETHYLENE OXIMIDE; EEC No. 613-028-00-9; *N,N*-DIMETHYLACETAMIDE; DIETHYLENIMIDE OXIDE; *p*-ISOXAZINE,TETRAHYDRO-; 1-OXA-4-AZACYCLOHEXANE; 2H-1,4-OXAZINE, TETRAHYDRO-; TETRAHYDRO-1,4-ISOXAZINE; TETRAHYDRO-1,4-OXAZINE; TETRAHYDRO-2H-1,4-OXAZINE; TETRAHYDRO-*p*-OXAZINE

IDENTIFICATION
CAS Number: 110-91-8
Formula: C_4H_9NO; $OCH_2CH_2NHCH_2CH_2$
DOT ID Number: UN 2054; DOT Guide Number: 132
Proper Shipping Name: Morpholine; Morpholine, aqueous mixture

DESCRIPTION: Colorless, oily liquid. Fishy, ammonia odor. Floats and mixes with water; forms alkaline solution. Irritating vapor is produced. A solid below 23°F/–5°C.

Poison! • Highly flammable • Corrosive to skin, eyes, and respiratory tract; skin and eye contact causes severe buns and possible blindness; can be absorbed through the skin at toxic levels • Firefighting gear (including SCBA) does not provide adequate protection. If exposure occurs, remove and isolate gear immediately and thoroughly decontaminate personnel • Containers may BLEVE when exposed to fire • Vapors are heavier than air and will collect and stay in low areas • Vapors in confined areas (e.g., tanks, sewers, buildings) may explode when exposed to fire • Toxic products of combustion may include nitrogen oxides • Do not put yourself in danger by entering a contaminated area to rescue a victim.

Hazard Classification (based on NFPA-704 Rating System)
Health Hazards (Blue): 3; Flammability (Red): 3; Reactivity (Yellow): 0

EMERGENCY RESPONSE: See Appendix A (132)
Evacuation:
Public safety: Isolate spill area for at least 50 to 100 meters (160 to 330 feet) in all directions.
Spill: Increase, as necessary, the isolation distance shown above, in "Public safety."
Fire: Isolate for 800 meters (½ mile) in all directions, especially if tank, rail car, or tank truck is involved in fire.

EXPOSURE
Short-term effects: *SEEK MEDICAL ATTENTION.* **Vapor:** Can be absorbed through the skin at toxic levels. Irritating to eyes, nose, and throat. *IF INHALED*, will, will cause nausea, headache, or difficult breathing. May cause lung edema; physical exertion will aggravate this condition. Move to fresh air. *IF BREATHING HAS STOPPED,* give artificial respiration. IF breathing is difficult, administer oxygen. **Liquid:** *HARMFUL IF ABSORBED THROUGH THE SKIN.* Irritating to skin and eyes. Remove contaminated clothing and shoes. Flush affected areas with plenty of water. *IF IN EYES*, hold eyelids open and flush with plenty of water. *IF SWALLOWED* and victim is *CONSCIOUS AND ABLE TO SWALLOW*, have victim drink 4 to 8 ounces of water and have victim induce vomiting. *IF SWALLOWED* and victim is *UNCONSCIOUS OR HAVING CONVULSIONS,* do nothing except keep victim warm.
Note to physician or authorized medical personnel: Pulmonary edema may be delayed. Medical observation is recommended for 24 to 48 hours after inhalation overexposure. As first aid for pulmonary edema, a physician or authorized medical personnel may consider administering a corticosteroid spray. Cigarette smoking may exacerbate pulmonary injury and should be discouraged for at least 72 hours following exposure.

HEALTH HAZARDS
Personal protective equipment (PPE): B-Level PPE. OSHA Table Z-1-A air contaminant. Respirator, boots and gloves; goggles or face shield. Chemical protective material(s) reported to have good to excellent resistance: butyl rubber, PV alcohol, Viton®. Also, Silvershield® and nitrile/PVC offers limited protection
Recommendations for respirator selection: NIOSH/OSHA
550 ppm: SA:CF* (any supplied-air respirator operated in a continuous-flow mode); or PAPROV* [any powered, air-purifying respirator with organic vapor cartridge(s)]. *1000 ppm:* CCRFOV [any chemical cartridge respirator with a full facepiece and organic vapor cartridges(s)]; or GMFOV [any air-purifying, full-facepiece respirator (gas mask) with a chin-style, front- or back-mounted acid gas canister]; or PAPRTOV* [any powered, air-purifying respirator with a tight-fitting facepiece and organic vapor cartridges(s)]; or SCBAF (any self-contained breathing apparatus with a full facepiece); or SAF (any supplied-air respirator with a full facepiece). *1400 ppm:* SAF:PD,PP (any supplied-air respirator that has a full facepiece and is operated in a pressure-demand or other positive-pressure mode). *EMERGENCY OR PLANNED ENTRY INTO UNKNOWN CONCENTRATIONS OR IDLH CONDITIONS:* SCBAF:PD,PP (any self-contained breathing apparatus that has a full facepiece and is operated in a pressure-demand or other positive-pressure mode); or SAF:PD,PP:ASCBA (any supplied-air respirator that has a full facepiece and is operated in a pressure-demand or other positive-pressure mode in combination with an auxiliary self-contained breathing apparatus operated in a pressure-demand or other positive pressure mode). *ESCAPE:* GMFOV [any air-purifying, full-facepiece respirator (gas mask) with a chin-style, front- or back-mounted organic vapor cannister]; or SCBAE (any appropriate escape-type, self-contained breathing apparatus). *Note:* Substance causes eye irritation or damage; eye protection needed.
Exposure limits (TWA unless otherwise noted): ACGIH TLV 20 ppm (71 mg/m^3); OSHA PEL 20 ppm (70 mg/m^3); NIOSH REL 20 ppm; skin contact contributes significantly in overall exposure.
Short-term exposure limits (15-minute TWA): NIOSH STEL 30 ppm (105 mg/m^3); skin contact contributes significantly in overall exposure.
Toxicity by ingestion: Grade 2; LD$_{50}$ = 0.5 to 5 g/kg (guinea pig, rat).
Long-term health effects: IARC rating 3; animal evidence: inadequate. May cause brain, kidney, and liver damage.
Vapor (gas) irritant characteristics: Corrosive irritant. Vapors cause smarting of the eyes or respiratory system if present in high concentrations. The effect may be temporary.
Liquid or solid irritant characteristics: If spilled on clothing and allowed to remain, may cause smarting and reddening of the skin.
Odor threshold: 0.01 ppm.
IDLH value: 1400 ppm [LEL]

FIRE DATA
Flash point: 98°F/37°C (oc).
Flammable limits in air: LEL: 1.4%; UEL: 11.2%.
Autoignition temperature: 590°F/310°C/583°K.
Burning rate: 1.9 mm/min

CHEMICAL REACTIVITY
Reactivity with water: Forms alkaline solution.
Binary reactants: Strong acids, strong oxidizers; nitro compounds; corrosive to aluminum, zinc, copper and copper alloys
Neutralizing agents for acids and caustics: Flush with water.
Reactivity group: 7
Compatibility class: Aliphatic amines

ENVIRONMENTAL DATA
Food chain concentration potential: Log P$_{ow}$ = –1.1. Unlikely to accumulate.
Water pollution: Effect of low concentrations on aquatic life is unknown. May be dangerous if it enters nearby water intakes; notify operators. Notify local health and wildlife officials. **Response to discharge:** Disperse and flush.

SHIPPING INFORMATION
Grades of purity: Several grades
available, most above 99%; **Storage temperature:** Ambient; **Inert atmosphere:** None; **Venting:** Open; **Stability during transport:** Stable.

NAS HAZARD CLASSIFICATION FOR BULK WATER TRANSPORTATION
FIRE: 3
HEALTH: Vapor irritant: 1; Liquid or solid irritant: 1; Poisons: 1
WATER POLLUTION: Human toxicity: 2; Aquatic toxicity: 2; Aesthetic effect: 2
REACTIVITY: Other chemicals: 3; Water: 0; Self-reaction: 0

PHYSICAL AND CHEMICAL PROPERTIES
Physical state @ 59°F/15°C and 1 atm: Liquid.

Molecular weight: 87.1
Boiling point @ 1 atm: 262.8°F/128.2°C/401.4°K.
Melting/Freezing point: 23.4°F/–4.8°C/268.4°K.
Critical temperature: 653°F/345°C/618°K.
Critical pressure: 794 psia = 54 atm = 5.47 MN/m².
Specific gravity (water = 1): 1.007 @ 68°F/20°C (liquid).
Ratio of specific heats of vapor (gas): (estimate) 1.091
Latent heat of vaporization: 182.9 Btu/lb = 101.6 cal/g = 4.254 x 10^5 J/kg.
Vapor pressure: (Reid) 0.55 psia; 8.0 mm.

MOTOR FUEL ANTI-KNOCK COMPOUNDS CONTAINING LEAD ALKYLS REC. M:6150

SYNONYMS: ANTI-KNOCK MIXTURE; MLA (mixed lead alkyls); LEAD ALKYL MIXTURE

IDENTIFICATION
Formula: Can be mixtures containing tetraethyl lead, tetramethyl lead, methyl triethyl lead, dimethyl diethyl lead, ethyl trimethyl lead
DOT ID Number: UN 1649; DOT Guide Number: 131
Proper Shipping Name: Motor fuel, anti-knock mixtures

DESCRIPTION: Dyed red, orange or blue oily liquid. Sweet fruity odor. Sinks in water.

May decompose explosively above 212–230°F/100–110°C. Highly flammable • Toxic! Breathing the vapor, skin or eye contact, or swallowing the material may kill you • Firefighting gear (including SCBA) does not provide adequate protection. If exposure occurs, remove and isolate gear immediately and thoroughly decontaminate personnel • Containers may BLEVE when exposed to fire • Vapors may form explosive mixture with air • Vapors are heavier than air and will collect and stay in low areas • Vapors may travel long distances to ignition sources and flashback • Vapors in confined areas (e.g., tanks, sewers, buildings) may explode when exposed to fire • Irritating to the skin, eyes, and respiratory tract • Toxic products of combustion may include fumes of lead.

Hazard Classification (based on NFPA-704 Rating System)

(tetraethyl lead): Health Hazards (Blue): 3; Flammability (Red): 2; Reactivity (Yellow): 3. See T: 0750.
(tetramethyl lead): Health Hazards (Blue): 3; Flammability (Red): 3; Reactivity (Yellow): 3. See T: 1050.

EMERGENCY RESPONSE: See Appendix A (131)
Evacuation:
Public safety: Isolate spill area for at least 100 to 200 meters (330 to 660 feet) in all directions.
Spill: Increase, as necessary, the isolation distance shown above, in "Public safety."
Fire: Isolate for 800 meters (½ mile) in all directions, especially if tank, rail car, or tank truck is involved in fire.

EXPOSURE
Short-term effects: *SEEK MEDICAL ATTENTION.* **Liquid:** *POISONOUS IF SWALLOWED OR IF SKIN IS EXPOSED.*
Will burn eyes. Remove contaminated clothes and shoes. Flush affected areas with plenty of water. *IF IN EYES*, hold eyelids open and flush with plenty of water. *IF SWALLOWED* and victim is *CONSCIOUS AND ABLE TO SWALLOW*, have victim drink 4 to 8 ounces of water. *IF SWALLOWED* and victim is *UNCONSCIOUS OR HAVING CONVULSIONS,* do nothing except keep victim warm. The use of alcoholic beverages may enhance the toxic effect.

HEALTH HAZARDS
Personal protective equipment (PPE): A-Level PPE. Organic vapor cartridge-type face mask for emergency or short duration; fresh air mask for longer duration; impervious protective gloves; goggles as required; boots and light-colored clothing.
Recommendations for respirator selection: OSHA as lead. *0.5 mg/m³*: HiE (any air-purifying, respirator with a high-efficiency particulate filter); SA (any supplied-air respirator). *1.25 mg/m³*: SA:CF (any supplied-air respirator operated in a continuous-flow mode); or PAPRHiE (any powered, air-purifying respirator with a high-efficiency particulate filter). *2.5 mg/m³*: HiEF (any air-purifying, full-facepiece respirator with a high-efficiency particulate filter); or SAT:CF (any supplied-air respirator that has a tight-fitting facepiece and is operated in a continuous-flow mode); or PAPRTHiE (any powered, air-purifying respirator with a tight-fitting facepiece and a high-efficiency particulate filter); or SCBAF (any self-contained breathing apparatus with a full facepiece); or SAF (any supplied-air respirator with a full facepiece). *50 mg/m³*: SA:PD,PP (any supplied-air respirator operated in a pressure-demand or other positive-pressure mode). *100 mg/m³*: SAF:PD,PP (any supplied-air respirator that has a full facepiece and is operated in a pressure-demand or other positive-pressure mode). *EMERGENCY OR PLANNED ENTRY INTO UNKNOWN CONCENTRATIONS OR IDLH CONDITIONS*: SCBAF:PD,PP (any self-contained breathing apparatus that has a full facepiece and is operated in a pressure-demand or other positive-pressure mode); or SAF:PD,PP:ASCBA (any supplied-air respirator that has a full facepiece and is operated in a pressure-demand or other positive-pressure mode in combination with an auxiliary self-contained breathing apparatus operated in a pressure-demand or other positive pressure mode). *ESCAPE:* HiEF (any air-purifying, full-facepiece respirator with a high-efficiency particulate filter); or SCBAE (any appropriate escape-type, self-contained breathing apparatus).
Exposure limits (TWA unless otherwise noted): ACGIH TLV 0.05 mg/m³ as inorganic lead; OSHA PEL 0.05 mg/m³; NIOSH REL 0.100 mg/m³ as lead.
Toxicity by ingestion: Poisonous; possible carcinogen.
Long-term health effects: Lead poisoning. May effect the nervous system.
Vapor (gas) irritant characteristics: Vapors cause a smarting of the eyes or respiratory system if present in high concentrations.
Liquid or solid irritant characteristics: If spilled on clothing and allowed to remain, may cause smarting and reddening of the skin. Toxic absorption through skin may occur.
IDLH value: 100 mg/m³ as lead.

FIRE DATA
Flash point: 89–265°F/32–129°C (oc).
Flammable limits in air: LEL: 1.4%; UEL: 11%.
Behavior in fire: Containers may explode. Vapors are heavier than air and may travel considerable distance to a source of ignition and flashback. Begins to decompose above 212°F/100°C.
Autoignition temperature: 559°F/293°C/566°K.
Electrical hazard: Flow or agitation of substance may generate electrostatic charges due to low conductivity.

CHEMICAL REACTIVITY
Reactivity with common materials: Reacts with oxidizing materials, active metals and rust. Attacks rubber.
Reactivity group: See Compatibility Guide
Compatibility class: Special class

ENVIRONMENTAL DATA
Water pollution: DOT Appendix B, §172.101–marine pollutant. Effect of low concentrations on aquatic life is unknown. May be dangerous if it enters nearby water intakes; notify operators. Notify local health and wildlife officials. **Response to discharge:** Issue warning–poison. Restrict access. Evacuate area. Should be removed. Chemical and physical treatment.

SHIPPING INFORMATION
Grades of purity: 50-60% mixed lead alkyls; 18-36% ethylene dibromide; 0-19% ethylene dichloride; 2-12% toluene, other solvents, dyes; **Storage temperature:** Ambient; **Inert atmosphere:** None; **Venting:** Pressure-vacuum; **Stability during transport:** A self-sustaining decomposition occurs if the temperature of the bulk liquid is above 212°F/100°C and a flame or hot metal surface serves to ignite the mass. The presence of ethylene dibromide renders the compound stable at 300°F/149°C for about 15 hours.

PHYSICAL AND CHEMICAL PROPERTIES
Physical state @ 59°F/15°C and 1 atm: Liquid.
Molecular weight: 265 (approximate).
Boiling point @ 1 atm: More than 200°F/93°C/367°K.
Melting/Freezing point: Less than –4°F/–20°C/253°K.
Specific gravity (water = 1): 1.5-1.7 @ 15°C (liquid).
Liquid surface tension: (estimate) 20 dynes/cm = 0.020 N/m @ 68°F/20°C.
Liquid water interfacial tension: (estimate) 45 dynes/cm = 0.045 N/m @ 68°F/20°C.
Ratio of specific heats of vapor (gas): (estimate) 1.030
Latent heat of vaporization: (estimate) 101 Btu/lb = 56.2 cal/g = 2.35 x 10^5 J/kg.
Heat of combustion: (estimate) –18,200 Btu/lb = –10,100 cal/g = –424 x 10^5 J/kg.
Vapor pressure: (Reid) 0.2 to 1.7 psia; 30 mm.

MYRCENE REC. M:6200

SYNONYMS: 3-METHYLENE-7-METHYL 1,6-OCTADIENE; 2-METHYL-6-METHYLENE-2,7-OCTADIENE; 1,6-OCTADIENE,7-METHYL-3-METHYLENE

IDENTIFICATION
CAS Number: 123-35-3
Formula: $C_{10}H_{16}$; $H_2CCHC(=CH_2)CH_2CH_2CH=C(CH_3)_2$
DOT ID Number: Un 1993; DOT Guide Number: 128
Proper Shipping Name: Flammable liquid, n.o.s.

DESCRIPTION: Yellow tinted, oily liquid. Pleasant odor, like balsamic vinegar. Insoluble in water.

Flammable • Containers may BLEVE when exposed to fire • Vapors may form explosive mixture with air • Vapors are heavier than air and will collect and stay in low areas • Vapors may travel long distances to ignition sources and flashback • Vapors in confined areas (e.g., tanks, sewers, buildings) may explode when exposed to fire • Irritating to the skin, eyes, and respiratory tract • Toxic products of combustion may include carbon monoxide.

Hazard Classification (based on NFPA-704 Rating System)
Health Hazards (Blue): 2; Flammability (Red): 2; Reactivity (Yellow): 0

EMERGENCY RESPONSE: See Appendix A (128)
Evacuation:
Public safety: Isolate spill area for at least 25 to 50 meters (80 to 160 feet) in all directions.
Spill: Large spill–Consider initial downwind evacuation for at least 300 meters (1000 feet).
Fire: Isolate for 800 meters (½ mile) in all directions, especially if tank, rail car, or tank truck is involved in fire.

EXPOSURE
Short-term effects: *SEEK MEDICAL ATTENTION.* **Liquid or solid:** Irritating to skin and eyes. Remove contaminated clothing and shoes. Flush affected areas with plenty of water. *IF IN EYES,* hold eyelids open and flush with plenty of water.

HEALTH HAZARDS
Personal protective equipment (PPE): B-Level PPE. Wear self-contained breathing apparatus, rubber boots and heavy rubber gloves.
Vapor (gas) irritant characteristics: Vapors cause moderate irritation such that personnel will find high concentrations unpleasant. The effect may be temporary.
Liquid or solid irritant characteristics: Causes smarting of the skin and first-degree burns on short exposure; may cause second-degree burns on long exposure.

FIRE DATA
Flash point: 103°F/40°C (cc).
Fire extinguishing agents not to be used: Water may be ineffective.

CHEMICAL REACTIVITY
Binary reactants: Strong oxidizers, strong acids.
Reactivity group: 30
Compatibility class: Olefins

ENVIRONMENTAL DATA
Water pollution: Effect of low concentrations on aquatic life is unknown. Fouling to shoreline. May be dangerous if it enters nearby water intakes; notify operators. Notify local health and wildlife officials. **Response to discharge:** Evacuate area. Mechanical containment. Should be removed. Chemical and physical treatment.

SHIPPING INFORMATION
Inert atmosphere: Ambient; **Stability during transport:** Stable.

PHYSICAL AND CHEMICAL PROPERTIES
Physical state @ 59°F/15°C and 1 atm: Liquid.
Molecular weight: 136.24
Boiling point @ 1 atm: 332.6°F/167°C/440.2°K.
Specific gravity (water = 1): 0.801
Relative vapor density (air = 1): 4.7
Vapor pressure: (Reid) 0.0891 psia.

NABAM REC. N:0100

SYNONYMS: CARBAMIC ACID, ETHYLENEBIS (DITHIO-), DISODIUM SALT; CHEM BAM; DITHANE; DITHANE A-40; DITHANE D-14; DSE; EBDC, SODIUM SALT; 1,2-ETHANEDIYLBISCARBAMODITHIOIC ACID, DISODIUM SALT; DISODIUM ETHYLENEBIS(DITHIOCARBAMATE); NABAME; NABASAM (obsolete); PARZATE; SPRING-BAK

IDENTIFICATION
CAS Number: 142-59-6
Formula: $C_4H_6N_2Na_2S_4$
DOT ID Number: UN 2757 (solid); UN 2992 (liquid); DOT Guide Number: 151
Proper Shipping Name: Carbamate pesticides, solid, toxic; Carbamate pesticides, liquid, toxic

DESCRIPTION: Colorless to light amber crystalline solid or 22% wettable powder solution. Slight odor of sulfide. Mixes with water.

Poison! (carbamate) • Irritating to skin, eyes, and respiratory tract; in high concentrations, it is narcotic • Firefighting gear (including SCBA) does not provide adequate protection. If exposure occurs, remove and isolate gear immediately and thoroughly decontaminate personnel • Containers may BLEVE when exposed to fire • Toxic products of combustion may include oxides of nitrogen, sulfur, and sodium; poisonous hydrogen sulfide and highly flammable carbon disulfide • Do not put yourself in danger by entering a contaminated area to rescue a victim.

Hazard Classification (based on NFPA-704 Rating System)
Health Hazards (Blue): 2; Flammability (Red): 0; Reactivity (Yellow): 0

EMERGENCY RESPONSE: See Appendix A (151)
Evacuation:
Public safety: Isolate the area of spill or leak for at least 25 to 50 meters (80 to 160 feet) in all directions.
Spill: Increase, in the downwind direction, as necessary, the distance shown under "Public Safety."
Fire: If tank, rail car, or tank truck is involved in fire, isolate for at least 800 meters (½ mile) in all directions; also, consider initial evacuation for 800 meters (½ mile) in all directions.

EXPOSURE
Short-term effects: *SEEK MEDICAL ATTENTION.* **Dust:** *POISONOUS IF INHALED.* Irritating to eyes, nose, and throat. Move victim to fresh air. *IF IN EYES*, hold eyelids open and flush with plenty of water.*IF BREATHING HAS STOPPED,* give artificial respiration; *avoid mouth-to-mouth resuscitation; use bag/mask apparatus.* IF breathing is difficult, administer oxygen.
Liquid or solid: *POISONOUS IF SWALLOWED.* Irritating to skin and eyes. Remove contaminated clothing and shoes. Flush affected areas with plenty of water. *IF IN EYES*, hold eyelids open and flush with plenty of water. *IF SWALLOWED* and victim is *CONSCIOUS AND ABLE TO SWALLOW*, have victim drink 4 to 8 ounces of water and have victim induce vomiting. *IF SWALLOWED* and victim is *UNCONSCIOUS OR HAVING CONVULSIONS,* do nothing except keep victim warm.
Note to physician or authorized medical personnel. Administer atropine, 2 mg (1/30 gr) intramuscularly or intravenously as soon as any local or systemic signs or symptoms of an intoxication are noted; repeat the administration of atropine every 3 to 8 minutes until signs of atropinization (mydriasis, dry mouth, rapid pulse, hot and dry skin) occur; initiate treatment in children with 0.05 mg/kg of atropine; repeat at 5- to 10-minute intervals. Watch respiration, and remove bronchial secretions if they appear to be obstructing the airway; intubate if necessary.
Medical note: Due to the rapid regeneration of chlolinesterase and the fact that 2-PAMCI may be contraindicated in the case of some carbamate poisonings, 2-PAMCI (Pralidoxime; Protopam) may not be needed.

HEALTH HAZARDS
Personal protective equipment (PPE): B-Level PPE. Dust mask; self-contained breathing apparatus if compound is hot; goggles; rubber gloves.
Recommendations for respirator selection: SCBAF:PD,PP (any self-contained breathing apparatus that has a full facepiece and is operated in a pressure-demand or other positive-pressure mode); or SAF:PD,PP:ASCBA (any supplied-air respirator that has a full facepiece and is operated in a pressure-demand or other positive-pressure mode in combination with an auxiliary, self-contained breathing apparatus operated in a pressure-demand or other positive pressure mode). *ESCAPE:* GMFOVHiE [any air-purifying, full-facepiece respirator (gas mask) with a chin-style, front- or back-mounted organic vapor canister having a high-efficiency particulate filter]; or SCBAE (any appropriate escape-type, self-contained breathing apparatus).
Toxicity by ingestion: Grade 3; oral LD = 395 mg/kg (rat).
Long-term health effects: Degrades to ethylenethiourea, which may affect thyroid gland of animals.
Liquid or solid irritant characteristics: Causes irritation of the eyes and skin.

ENVIRONMENTAL DATA
Food chain concentration potential: Unlikely to accumulate.
Water pollution: DOT Appendix B, §172.101–marine pollutant. Harmful to aquatic life in very low concentrations. May be dangerous if it enters nearby water intakes; notify operators. Notify local health and wildlife officials. **Response to discharge:** Issue warning–poison, water contaminant. Restrict access. Should be removed. Chemical and physical treatment.

SHIPPING INFORMATION
Grades of purity: Technical; 22% solution in water; **Storage temperature:** Ambient; **Inert atmosphere:** None; **Venting:** Open; **Stability during transport:** Stable.

PHYSICAL AND CHEMICAL PROPERTIES
Physical state @ 59°F/15°C and 1 atm: Solid.
Molecular weight: 256.3
Boiling point @ 1 atm: Decomposes.
Specific gravity (water = 1): 1.14 @ 68°F/20°C (solid).

NALED   **REC. N:0150**

SYNONYMS: ARTHODIBROM; BROMCHLOPHOS; BROMEX; DIBROM®; *O*-(1,2-DIBROM-2,2-DICHLOR-AETHYL)-*O,O*-DIMETHYL-PHOSPHAT (German); 1,2-DIBROMO-2,2-DICHLOROETHYL DIMETHYL PHOSPHATE; DIMETHYL 1,2-DIBROMO-2,2-DICHLOROETHYLPHOSPHATE; *O,O*-DIMETHYL-*O*-(1,2-DIBROMO-2,2-DICHLOROETHYL)PHOSPHATE; *O,O*-DIMETHYL *O*-2,2-DICHLORO-1,2-DIBROMOETHYL PHOSPHATE; ETHANOL,1,2-DIBROMO-2,2-DICHLORO-DIMETHYL PHOSPHATE; DIBROM; ORTHO 4355; ORTHODIBROM; ORTHODIBROMO; PHOSPHATE de *O,O*-DIMETHLE et de *O*-(1,2-DIBROMO-2,2-DICHLORETHYLE) (French); RE-4355

IDENTIFICATION
CAS Number: 300-76-5
Formula: $C_4H_7Br_2Cl_2O_4P$
DOT ID Number: UN 2783 (solid); UN 3018 (liquid); DOT Guide Number: 152

Naled

Proper Shipping Name: Organophosphorus pesticides, solid, toxic; organophosphorus pesticides, liquid, toxic
Reportable Quantity (RQ): **(CERCLA)** 10 lb/4.54 lb

DESCRIPTION: White solid; light yellow liquid. Slightly pungent odor. Sinks in water; practically insoluble; readily hydrolyzed.

Poison! (organophosphate). A cholinesterase inhibitor • Breathing the dust or vapor, skin or eye contact, or swallowing the material can kill you • Firefighting gear (including SCBA) does not provide adequate protection. If exposure occurs, remove and isolate gear immediately and thoroughly decontaminate personnel • Containers may BLEVE when exposed to fire • Toxic products of combustion may include bromine, chlorine, and oxides of sulfur and phosphorus • Do not put yourself in danger by entering a contaminated area to rescue a victim.

Hazard Classification (based on NFPA-704 Rating System)
Health Hazards (Blue): 2; Flammability (Red): 0; Reactivity (Yellow): 1

EMERGENCY RESPONSE: See Appendix A (152)
Evacuation:
Public safety: Isolate the area of spill or leak for at least 25 to 50 meters (80 to 160 feet) in all directions.
Spill: Increase, in the downwind direction, as necessary, the distance shown under "Public Safety."
Fire: If tank, rail car, or tank truck is involved in fire, isolate for at least 800 meters (½ mile) in all directions; also, consider initial evacuation for 800 meters (½ mile) in all directions.

EXPOSURE
Short-term effects: *SEEK MEDICAL ATTENTION.* **Spray or dust:** *POISONOUS IF INHALED, IF SWALLOWED, OR IF SKIN IS EXPOSED.* Eye pupils are small; blurred vision; runny nose; cough; shortness of breath; pain; diarrhea, nausea and vomiting; increased blood pressure, hypermotility, hallucinations; loss of consciousness; convulsions; breathing stops; death. *IF BREATHING HAS STOPPED,* give artificial respiration; *avoid mouth-to-mouth resuscitation; use bag/mask apparatus.* **Skin:** *Remove and double-bag contaminated clothing (including shoes) and leave them in the Hot Zone;* wash skin with soap and water and large volumes of water for at least 15 minutes. *Skin can also be decontaminated with diluted hypochlorite solution, U.S. Army M291 kit, and M258(A1) skin decontamination kit.* **Eye:** Rinse with large volumes of water or saline for at least 15 minutes. *IF SWALLOWED,* will cause nausea, vomiting, or loss of consciousness. *Remove and double-bag contaminated clothing and shoes and leave in Hot Zone for later incineration by hazardous materials experts.* Flush affected areas with plenty of water. **If swallowed** and victim is *CONSCIOUS AND ABLE TO SWALLOW,* have victim drink 4 to 8 ounces of water. **Do NOT induce vomiting but immediately administer slurry of activated charcoal.** *IF SWALLOWED* and victim is *UNCONSCIOUS OR HAVING CONVULSIONS,* do nothing except keep victim warm. *Note to physician or authorized medical personnel:* Administer atropine, 2 mg (1/30 gr) intramuscularly or intravenously as soon as any local or systemic signs or symptoms of an intoxication are noted; repeat the administration of atropine every 3 to 8 minutes until signs of atropinization (mydriasis, dry mouth, rapid pulse, hot and dry skin) occur; initiate treatment in children with 0.05 mg/kg of atropine; repeat at 5- to 10-minute intervals. Watch respiration, and remove bronchial secretions if they appear to be obstructing the airway; intubate if necessary. Pralidoxime must be administered within minutes to a few hours following exposure (depending on the specific agent) to be effective. Give 2-PAMCI (Pralidoxime; Protopam), 2.5 g in 100 mL of sterile water or in 5% dextrose and water, intravenously, slowly, in 15 to 30 minutes; if sufficient fluid is not available, give 1 g of 2-PAMCI in 3 mL of distilled water by deep intramuscular injection; repeat this every half hour if respiration weakens or if muscle fasciculation or convulsions recur. Also consider Diazepam, an anticonvulsant.

HEALTH HAZARDS
Personal protective equipment (PPE): A-Level PPE. Gloves, self-contained breathing apparatus, protective clothing. Chemical protective material(s) reported to have good to excellent resistance: Teflon®.
Recommendations for respirator selection: NIOSH/OSHA
$30\ mg/m^3$: DMFu (any dust, mist and fume respirator); or HiE (any air-purifying, respirator with a high-efficiency particulate filter); or SA (any supplied-air respirator). $75\ mg/m^3$: SA:CF (any supplied-air respirator operated in a continuous-flow mode); or PAPRDMFu (any powered, air-purifying respirator with a dust, mist, and fume filter). $150\ mg/m^3$: HiEF (any air-purifying, full-facepiece respirator with a high-efficiency particulate filter); or SAT:CF (any supplied-air respirator that has a tight-fitting facepiece and is operated in a continuous-flow mode); or PAPRTHiE (any powered, air-purifying respirator with a tight-fitting facepiece and a high-efficiency particulate filter); or SCBAF (any self-contained breathing apparatus with a full facepiece); or SAF (any supplied-air respirator with a full facepiece). $200\ mg/m^3$: SA:PD,PP (any supplied-air respirator operated in a pressure-demand or other positive-pressure mode). *EMERGENCY OR PLANNED ENTRY INTO UNKNOWN CONCENTRATIONS OR IDLH CONDITIONS:* SCBAF:PD,PP (any self-contained breathing apparatus that has a full facepiece and is operated in a pressure-demand or other positive-pressure mode); or SAF:PD,PP:ASCBA (any supplied-air respirator that has a full facepiece and is operated in a pressure-demand or other positive-pressure mode in combination with an auxiliary self-contained breathing apparatus operated in a pressure-demand or other positive pressure mode). *ESCAPE*: HiEF (any air-purifying, full-facepiece respirator with a high-efficiency particulate filter); or SCBAE (any appropriate escape-type, self-contained breathing apparatus).
Exposure limits (TWA unless otherwise noted): ACGIH TLV 3 mg/m³; OSHA PEL 3 mg/m³; NIOSH REL 3 mg/m³; skin contact contributes significantly in overall exposure.
Toxicity by ingestion: Grade 3; LD_{50} = 50 to 500 mg/kg.
Long-term health effects: Cholinesterase inhibition persists for several weeks making person more vulnerable in case of additional exposure. Exposure of rats at 0.3 to 2.5 mg/L for 4 hours daily for 6 months caused emphysema, interstitial pneumonia, bronchitis, and peribronchitis. Liver, spleen, and brain damage was noted.
Vapor (gas) irritant characteristics: Dangerous concentrations of vapor are not produced under normal conditions.
Liquid or solid irritant characteristics: If spilled on clothing and allowed to remain may cause smarting and reddening of skin.
IDLH value: 200 mg/m³.

CHEMICAL REACTIVITY
Reactivity with water: May hydrolyze.
Binary reactants: Unstable in presence of Iron. Reacts with strong oxidizers, acids; in sunlight. Corrosive to metals.

ENVIRONMENTAL DATA
Food chain concentration potential: Unlikely to accumulate.
Water pollution: DOT Appendix B, §172.101–marine pollutant. Harmful to aquatic life in very low concentrations. May be

dangerous if it enters nearby water intakes; notify operators. Notify local health and wildlife officials. For weapons testing, use an M272 Water Detection Kit. **Response to discharge:** Issue warning–poison, water contaminant. Restrict access. Should be removed. Chemical and physical treatment.

SHIPPING INFORMATION
Grades of purity: Technical, 93%; **Stability during transport:** Stable under anhydrous conditions. Unstable in alkaline conditions. Degraded by sunlight.

PHYSICAL AND CHEMICAL PROPERTIES
Physical state @ 59°F/15°C and 1 atm: Solid.
Molecular weight: 381
Boiling point @ 1 atm: 392°F/200°C/473°K.
Melting/Freezing point: Pure 80.6°F/27°C/300.2°K.
Specific gravity (water = 1): 1.97 @ 68°F/20°C.
Relative vapor density (air = 1): 13.1 (calculated).

NAPHTHA: VM&P   REC. N:0200

SYNONYMS: AROMATIC SOLVENT; BENZIN; BENZOLINE; LIGROIN (petroleum benzene); NAPHTHA, <3% AROMATICS, 120°-200°C; NAPHTHA PETROLEUM; NAPHTHA SOLVENT; PETROLEUM BENZIN; PETROLEUM ETHER; PETROLEUM SOLVENT; PETROLEUM SPIRITS; PAINTERS NAPHTHA; REFINED SOLVENT NAPHTHA; SOLVENT NAPHTHA; VARNISH MAKERS AND PAINTERS NAPHTHA; VARSOL; VM&P NAPHTHA; WHITE SPIRITS

IDENTIFICATION
CAS Number: 8032-32-4; 64475-85-0
DOT ID Number: UN 1271; DOT Guide Number: 128
Proper Shipping Name: Petroleum ether; Petroleum spirit

DESCRIPTION: Colorless to pale yellow watery liquid. Gasoline- or benzene-like odor. Floats on the surface of water. Irritating vapor is produced.

Highly flammable • Containers may BLEVE when exposed to fire • Vapors may form explosive mixture with air • Vapors are heavier than air and will collect and stay in low areas • Vapors may travel long distances to ignition sources and flashback • Vapors in confined areas (e.g., tanks, sewers, buildings) may explode when exposed to fire • Irritating to the skin, eyes, and respiratory tract • Toxic products of combustion may include carbon monoxide.

Hazard Classification (based on NFPA-704 Rating System)
Health Hazards (Blue): 1; Flammability (Red): 3; Reactivity (Yellow): 0

EMERGENCY RESPONSE: See Appendix A (128)
Evacuation:
Public safety: Isolate spill area for at least 25 to 50 meters (80 to 160 feet) in all directions.
Spill: Large spill–Consider initial downwind evacuation for at least 300 meters (1000 feet).
Fire: Isolate for 800 meters (½ mile) in all directions, especially if tank, rail car, or tank truck is involved in fire.

EXPOSURE
Short-term effects: *SEEK MEDICAL ATTENTION*. **Vapor:** Irritating to eyes, nose, and throat. *IF INHALED*, will, will cause dizziness, headache, difficult breathing or loss of consciousness. Move to fresh air. *IF BREATHING HAS STOPPED*, give artificial respiration. IF breathing is difficult, administer oxygen. **Liquid:** Irritating to skin and eyes. *IF SWALLOWED*, will cause nausea or vomiting. Remove contaminated clothing and shoes. Flush affected areas with plenty of water. *IF IN EYES*, hold eyelids open and flush with plenty of water. *IF SWALLOWED* and victim is *CONSCIOUS AND ABLE TO SWALLOW*, have victim drink 4 to 8 ounces of water. **Do NOT induce vomiting.**

HEALTH HAZARDS
Personal protective equipment (PPE): B-Level PPE. OSHA Table Z-1-A air contaminant. Hydrocarbon vapor canister); or air pack; plastic gloves; goggles or face shield. Chemical protective material(s) reported to have good to excellent resistance: Chlorinated polyethylene, nitrile, PV alcohol, polyurethane, Viton®. Also, PVC, Viton®/neoprene, and neoprene offers limited protection
Recommendations for respirator selection: NIOSH/OSHA
3500 mg/m³: CCROV [any chemical cartridge respirator with organic vapor cartridge(s)]; or SA (any supplied-air respirator). *8750 mg/m³:* SA:CF (any supplied-air respirator operated in a continuous-flow mode); or PAPROV [any powered, air-purifying respirator with organic vapor cartridge(s)]. *17,500 mg/m³:* CCRFOV [any air-purifying, full-facepiece respirator (gas mask) with a chin-style, front- or back-mounted acid gas canister]; or GMFOV [any air-purifying, full-facepiece respirator (gas mask) with a chin-style, front- or back-mounted organic vapor cannister]; or PAPRTOV [any powered, air-purifying respirator with a tight-fitting facepiece and organic vapor cartridges(s)]; or SCBAF (any self-contained breathing apparatus with a full facepiece); or SAF (any supplied-air respirator with a full facepiece). *EMERGENCY OR PLANNED ENTRY INTO UNKNOWN CONCENTRATIONS OR IDLH CONDITIONS:* SCBAF:PD,PP (any self-contained breathing apparatus that has a full facepiece and is operated in a pressure-demand or other positive-pressure mode); or SAF:PD,PP:ASCBA (any supplied-air respirator that has a full facepiece and is operated in a pressure-demand or other positive-pressure mode in combination with an auxiliary self-contained breathing apparatus operated in a pressure-demand or other positive pressure mode). *ESCAPE:* GMFOV [any air-purifying, full-facepiece respirator (gas mask) with a chin-style, front- or back-mounted organic vapor canister]; or SCBAE (any appropriate escape-type, self-contained breathing apparatus).
Exposure limits (TWA unless otherwise noted): ACGIH TLV: 300 ppm (1370 mg/m³); NIOSH: 350 mg/m³; ceiling 1800 mg/m³/ 15 minutes.
Toxicity by ingestion: Grade 1; LD_{50} = 5 to 15 mg/kg.
Long-term health effects: IARC substance, overall evaluation unassigned. May cause dermatitis from daily contact.
Vapor (gas) irritant characteristics: Eye and respiratory tract irritant. Vapors cause smarting of the eyes or respiratory system if present in high concentrations. The effect is temporary.
Liquid or solid irritant characteristics: If spilled on clothing and allowed to remain, may cause a smarting and reddening of the skin.
Odor threshold: 1–40 ppm.

FIRE DATA
Flash point: 20 to 55°F/–7 to 13°C (cc) (varies by manufacturer); 50°F/10°C (cc) (regular and flash); 5°F/29°C (cc) (high flash).
Flammable limits in air: LEL: 1.2%; UEL: 6.0%.
Fire extinguishing agents not to be used: Water may be ineffective.
Autoignition temperature: 450°F/232°C/505°K.
Electrical hazard: Class I, Group B
Burning rate: 4 mm/min

CHEMICAL REACTIVITY
Binary reactants: Strong oxidizers, nitric acid. Effects some paints and rubber. Corrosive presence in salt water.
Reactivity group: 33
Compatibility class: Miscellaneous hydrocarbon mixtures

ENVIRONMENTAL DATA
Food chain concentration potential: Negative; unlikely to accumulate.
Water pollution: Effect of low concentrations on aquatic life is unknown. Fouling to shoreline. May be dangerous if it enters nearby water intakes; notify operators. Notify local health and wildlife officials. **Response to discharge:** Mechanical containment. Should be removed. Chemical and physical treatment.

SHIPPING INFORMATION
Grades of purity: Petroleum hydrocarbons (90%) plus aromatic hydrocarbons such as benzene and toluene (10%); **Storage temperature:** Ambient; **Inert atmosphere:** None; **Venting:** Open (flame arrester) or pressure-vacuum; **Stability during transport:** Stable.

PHYSICAL AND CHEMICAL PROPERTIES
Physical state @ 59°F/15°C and 1 atm: Liquid.
Molecular weight: 87–114 (approximate).
Boiling point @ 1 atm: 212–350°F/100–177°C/373–450°K.
Specific gravity (water = 1): 0.73–0.76 at 60°F/16°C/289°K (liquid).
Liquid surface tension: 19–23 dynes/cm = 0.019-0.023 N/m @ 68°F/20°C.
Liquid water interfacial tension: 39–51 dynes/cm = 0.039–0.051 N/m @ 68°F/20°C.
Relative vapor density (air = 1): 4.2
Ratio of specific heats of vapor (gas): (estimate) 1.030
Latent heat of vaporization: 130–150 Btu/lb = 71-81 cal/g = 3.0–3.4 x 10^5 J/kg.
Heat of combustion: (estimate) –18,200 Btu/lb = –10,100 cal/g = –424 x 10^5 J/kg.
Vapor pressure: (Reid) 0.12 psia; 2–20 mm.

NAPHTHA: SOLVENT   REC. N:0250

SYNONYMS: LIGHT NAPHTHA; PETROLEUM SOLVENT

IDENTIFICATION
CAS Number: 8002-05-9
DOT ID Number: UN 1256; DOT Guide Number: 128
Proper Shipping Name: Naphtha solvent

DESCRIPTION: Colorless watery liquid. Gasoline- or kerosene-like odor. Floats on water surface; insoluble.

Highly flammable • Containers may BLEVE when exposed to fire • Vapors may form explosive mixture with air • Vapors are heavier than air and will collect and stay in low areas • Vapors may travel long distances to ignition sources and flashback • Vapors in confined areas (e.g., tanks, sewers, buildings) may explode when exposed to fire • Irritating to the skin, eyes, and respiratory tract • Toxic products of combustion may include carbon monoxide.

Hazard Classification (based on NFPA-704 Rating System)
Health Hazards (Blue): 0; Flammability (Red): 2; Reactivity (Yellow): 0

EMERGENCY RESPONSE: See Appendix A (128)
Evacuation:
Public safety: Isolate spill area for at least 25 to 50 meters (80 to 160 feet) in all directions.
Spill: Large spill–Consider initial downwind evacuation for at least 300 meters (1000 feet).
Fire: Isolate for 800 meters (½ mile) in all directions, especially if tank, rail car, or tank truck is involved in fire.

EXPOSURE
Short-term effects: *SEEK MEDICAL ATTENTION.* **Vapor:** Not irritating to eyes, nose or throat. *IF INHALED*, will, will cause dizziness or loss of consciousness. Move to fresh air. *IF BREATHING HAS STOPPED,* give artificial respiration. IF breathing is difficult, administer oxygen. **Liquid:** Irritating to skin and eyes. Harmful if swallowed. Remove contaminated clothing and shoes. Flush affected areas with plenty of water. *IF IN EYES*, hold eyelids open and flush with plenty of water. *IF SWALLOWED* and victim is *CONSCIOUS AND ABLE TO SWALLOW*, have victim drink 4 to 8 ounces of water. **Do NOT induce vomiting.**

HEALTH HAZARDS
Personal protective equipment (PPE): B-Level PPE. Goggles or face shield (as for gasoline). Chemical protective material(s) reported to have good to excellent resistance: nitrile, PV alcohol, Viton®, Saranex®, Silvershield®. Also, PVC, neoprene, and Styrene-butadiene offers limited protection
Recommendations for respirator selection: NIOSH/OSHA (Petroleum distillates (naphtha).
850 ppm: SA (any supplied-air respirator). *1100 ppm*: SA:CF* (any supplied-air respirator operated in a continuous-flow mode); or SCBAF (any self-contained breathing apparatus with a full facepiece); or SAF (any supplied-air respirator with a full facepiece). *ESCAPE:* GMFOV [any air-purifying, full-facepiece respirator (gas mask) with a chin-style, front- or back-mounted organic vapor canister]; or SCBAE (any appropriate escape-type, self-contained breathing apparatus). *Note:* Substance reported to cause eye irritation or damage; may require eye protection.
Exposure limits (TWA unless otherwise noted): 350 mg/m^3; ceiling 1800 mg/m^3 (15 minute).
Short-term exposure limits (15-minute TWA): See above
Toxicity by ingestion: Grade 2; LD_{50} = 0.5 to 5 g/kg.
Vapor (gas) irritant characteristics: Eye and respiratory tract irritant.
Liquid or solid irritant characteristics: If spilled on clothing and allowed to remain, may cause smarting and reddening of skin.
IDLH value: 1100 ppm.

FIRE DATA
Flash point: More than 100°F/38°C.
Flammable limits in air: LEL: 0.8%; UEL: 5.0%.
Autoignition temperature: 444°F/229°C/502°K.
Electrical hazard: Class I, Group D.
Burning rate: 4 mm/min

CHEMICAL REACTIVITY
Binary reactants: Reacts with strong oxidizers, nitric acid
Reactivity group: 33
Compatibility class: Miscellaneous hydrocarbon mixtures

ENVIRONMENTAL DATA
Food chain concentration potential: Negative; unlikely to accumulate.
Water pollution: Effect of low concentrations on aquatic life is unknown. Fouling to shoreline. May be dangerous if it enters

nearby water intakes; notify operators. Notify local health and wildlife officials. **Response to discharge:** Mechanical containment. Should be removed. Chemical and physical treatment.

SHIPPING INFORMATION
Grades of purity: Refined solvent; crude light solvent; crude heavy solvent; **Storage temperature:** Ambient; **Inert atmosphere:** None; **Venting:** Open (flame arrester) or pressure-vacuum; **Stability during transport:** Stable.

PHYSICAL AND CHEMICAL PROPERTIES
Physical state @ 59°F/15°C and 1 atm: Liquid.
Boiling point @ 1 atm: 266–311°F/130–155°C/403–428°K.
Specific gravity (water = 1): 0.85–0.87 @ 68°F/20°C (liquid).
Liquid surface tension: 19–23 dynes/cm = 0.019–0.023 N/m @ 68°F/20°C.
Liquid water interfacial tension: 39–51 dynes/cm = 0.039–0.051 N/m @ 68°F/20°C.
Ratio of specific heats of vapor (gas): (estimate) 1.030
Latent heat of vaporization: 130–150 Btu/lb = 71–81 cal/g = 3.0–3.4 x 10^5 J/kg.
Heat of combustion: (estimate) –18,200 Btu/lb = –10,100 cal/g = –424 x 10^5 J/kg.

NAPHTHA: STODDARD SOLVENT   REC. N:0300

SYNONYMS: CLEANING SOLVENT; DRYCLEANING SAFETY SOLVENT; DRYCLEANER NAPHTHA; EEC No. 650-001-02-5; MINERAL SPIRITS; NAPHTHA SAFETY SOLVENT; NONANE and TRIMETHYLBENZENE MIXTURE (85:15); PETROLEUM SOLVENT; PETROLEUM SPIRITS; PETROLEUM THINNER; SPOTTING NAPHTHA; VARNOLINE; WHITE SPIRIT

IDENTIFICATION
CAS Number: 8052-41-3
Formula: 85% Nonane (C_9H_{20}) + 15% Trimethyl benzene (C_9H_{12})
DOT ID Number: UN 1268; DOT Guide Number: 128
Proper Shipping Name: Petroleum distillates, n.o.s.

DESCRIPTION: Colorless to light yellow, watery liquid. Becomes gaseous at about 302°F/150°C. Gasoline-like odor. Floats on the surface of water.

Highly flammable • Containers may BLEVE when exposed to fire • Vapors may form explosive mixture with air • Vapors are heavier than air and will collect and stay in low areas • Vapors may travel long distances to ignition sources and flashback • Vapors in confined areas (e.g., tanks, sewers, buildings) may explode when exposed to fire • Irritating to the skin, eyes, and respiratory tract • Toxic products of combustion may include carbon monoxide.

Hazard Classification (based on NFPA-704 Rating System)
Health Hazards (Blue): 0; Flammability (Red): 2; Reactivity (Yellow): 0

EMERGENCY RESPONSE: See Appendix A (128)
Evacuation:
Public safety: Isolate spill area for at least 25 to 50 meters (80 to 160 feet) in all directions.
Spill: Large spill–Consider initial downwind evacuation for at least 300 meters (1000 feet).
Fire: Isolate for 800 meters (½ mile) in all directions, especially if tank, rail car, or tank truck is involved in fire.

EXPOSURE
Short-term effects: *SEEK MEDICAL ATTENTION*. **Liquid:** Irritating to skin and eyes. Harmful if swallowed. Remove contaminated clothing and shoes. Flush affected areas with plenty of water. *IF IN EYES*, hold eyelids open and flush with plenty of water. *IF SWALLOWED* and victim is *CONSCIOUS AND ABLE TO SWALLOW*, have victim drink 4 to 8 ounces of water. **Do NOT induce vomiting.**

HEALTH HAZARDS
Personal protective equipment (PPE): B-Level PPE. OSHA Table Z-1-A air contaminant. Goggles or face shield (as for gasoline). Chemical protective material(s) reported to have good to excellent resistance: nitrile, PV alcohol, Viton®, Saranex®, Silvershield®. Also, PVC, neoprene, and Styrene-butadiene offers limited protection
Recommendations for respirator selection: NIOSH/OSHA *3500 mg/m^3:* CCROV* [any chemical cartridge respirator with organic vapor cartridge(s)]; or SA* (any supplied-air respirator). *8750 mg/m^3:* SA:CF* (any supplied-air respirator operated in a continuous-flow mode); or *5900 mg/m^3:* PAPROV* [any powered, air-purifying respirator with organic vapor cartridge(s)]. *17,500 mg/m^3:* CCRFOV [any air-purifying, full-facepiece respirator (gas mask) with a chin-style, front- or back-mounted acid gas canister]; or GMFOV [any air-purifying, full-facepiece respirator (gas mask) with a chin-style, front- or back-mounted organic vapor cannister]; or PAPRTOV* [any powered, air-purifying respirator with a tight-fitting facepiece and organic vapor cartridges(s)]; or SCBAF (any self-contained breathing apparatus with a full facepiece); or SAF (any supplied-air respirator with a full facepiece). *20,000 mg/m^3:* SAF:PD,PP (any supplied-air respirator that has a full facepiece and is operated in a pressure-demand or other positive-pressure mode). *EMERGENCY OR PLANNED ENTRY INTO UNKNOWN CONCENTRATIONS OR IDLH CONDITIONS:* SCBAF:PD,PP (any self-contained breathing apparatus that has a full facepiece and is operated in a pressure-demand or other positive-pressure mode); or SAF:PD,PP:ASCBA (any supplied-air respirator that has a full facepiece and is operated in a pressure-demand or other positive-pressure mode in combination with an auxiliary self-contained breathing apparatus operated in a pressure-demand or other positive pressure mode). *ESCAPE:* GMFOV [any air-purifying, full-facepiece respirator (gas mask) with a chin-style, front- or back-mounted organic vapor cannister]; or SCBAE (any appropriate escape-type, self-contained breathing apparatus). *Note:* Substance reported to cause eye irritation or damage; may require eye protection.
Exposure limits (TWA unless otherwise noted): ACGIH TLV 100 ppm (525 mg/m^3); OSHA PEL 500 ppm (2900 mg/m^3); NIOSH 350 mg/m^3; ceiling 1800 mg/m^3/15 minutes.
Short-term exposure limits (15-minute TWA): OSHA STEL 400 ppm.
Toxicity by ingestion: Grade 2; LD_{50} = 0.5 to 5 g/kg.
Long-term health effects: Skin disorders from repeated or prolonged contact.
Vapor (gas) irritant characteristics: Vapors are mildly irritating to the eyes and throat.
Liquid or solid irritant characteristics: If spilled on clothing and allowed to remain, may cause smarting and reddening of the skin.
Odor threshold: 1–30 ppm.
IDLH value: 20,000 mg/m^3.

FIRE DATA
Flash point: 100-110°F/38-43°C (cc).

Flammable limits in air: LEL: 1.1%; UEL: 6.0%.
Behavior in fire: Closed containers may explode.
Autoignition temperature: 450°F/232°C/505°K.
Electrical hazard: Class I, Group D. Flow or agitation of substance may generate electrostatic charges due to low conductivity.
Burning rate: 4 mm/min

CHEMICAL REACTIVITY
Binary reactants: Strong oxidizers, nitric acid
Reactivity group: 33
Compatibility class: Miscellaneous hydrocarbon mixtures

ENVIRONMENTAL DATA
Food chain concentration potential: Negative; unlikely to accumulate.
Water pollution: Effect of low concentrations on aquatic life is unknown. Fouling to shoreline. May be dangerous if it enters nearby water intakes; notify operators. Notify local health and wildlife officials. **Response to discharge:** Mechanical containment. Should be removed. Chemical and physical treatment.

SHIPPING INFORMATION
Storage temperature: Ambient; **Inert atmosphere:** None; **Venting:** Open (flame arrester); **Stability during transport:** Stable.

PHYSICAL AND CHEMICAL PROPERTIES
Physical state @ 59°F/15°C and 1 atm: Liquid.
Boiling point @ 1 atm: 320–390°F/160–199°C/433–472°K.
Specific gravity (water = 1): 0.78 @ 68°F/20°C (liquid).
Liquid surface tension: 19–23 dynes/cm = 0.019–0.023 N/m @ 68°F/20°C.
Liquid water interfacial tension: 39–51 dynes/cm = 0.039–0.051 N/m @ 68°F/20°C.
Ratio of specific heats of vapor (gas): (estimate) 1.030
Latent heat of vaporization: 130–150 Btu/lb = 71–81 cal/g = 3.0–3.4 x 10^5 J/kg.
Heat of combustion: (estimate) –18,200 Btu/lb = –10,100 cal/g = –424 x 10^5 J/kg.
Vapor pressure: (Reid) 0.1 psia.

NAPHTHA   REC. N:0350

SYNONYMS: COAL TAR LIGHT OIL; COAL TAR NAPHTHA; COAL TAR OIL; CRUDE SOLVENT COAL TAR NAPHTHA; HIGH SOLVENT NAPHTHA; NAPHTHA; MIXTURE OF BENZENE, TOLUENE, XYLENES; MIXTURE OF PENTANE, HEXANE, HEPTANE; PETROLEUM NAPHTHA; SUPER VMP

IDENTIFICATION
CAS Number: 8030-30-6; MX8030-31-7
DOT ID Number: UN 1255; UN 1256; UN 1268; UN 2553; DOT Guide Number: 128
Proper Shipping Name: Naphtha, petroleum; Petroleum naphtha; Naphtha, solvent; Naphtha

DESCRIPTION: Colorless to pale yellow liquid. Gasoline-like odor. Floats on water surface; insoluble.

Highly flammable • Containers may BLEVE when exposed to fire • Vapors may form explosive mixture with air • Vapors are heavier than air and will collect and stay in low areas • Vapors may travel long distances to ignition sources and flashback • Vapors in confined areas (e.g., tanks, sewers, buildings) may explode when exposed to fire • Irritating to the skin, eyes, and respiratory tract • Toxic products of combustion may include carbon monoxide.

Hazard Classification (based on NFPA-704 Rating System)
Health Hazards (Blue): 2; Flammability (Red): 2; Reactivity (Yellow): 0

EMERGENCY RESPONSE: See Appendix A (128)
Evacuation:
Public safety: Isolate spill area for at least 25 to 50 meters (80 to 160 feet) in all directions.
Spill: Large spill–Consider initial downwind evacuation for at least 300 meters (1000 feet).
Fire: Isolate for 800 meters (½ mile) in all directions, especially if tank, rail car, or tank truck is involved in fire.

EXPOSURE
Short-term effects: *SEEK MEDICAL ATTENTION.* **Vapor:** Irritating to eyes, nose, and throat. *IF INHALED*, will, will cause dizziness or loss of consciousness. Move to fresh air. *IF BREATHING HAS STOPPED*, give artificial respiration. IF breathing is difficult, administer oxygen. **Liquid:** Irritating to skin and eyes. *IF SWALLOWED*, will cause nausea or vomiting. Remove contaminated clothing and shoes. Flush affected areas with plenty of water. *IF IN EYES*, hold eyelids open and flush with plenty of water. *IF SWALLOWED* and victim is *CONSCIOUS AND ABLE TO SWALLOW*, have victim drink 4 to 8 ounces of water. **Do NOT induce vomiting.**

HEALTH HAZARDS
Personal protective equipment (PPE): B-Level PPE. OSHA Table Z-1-A air contaminant. Goggles or face shield (as for gasoline). Chemical protective material(s) reported to have good to excellent resistance: nitrile, PV alcohol, Viton®, Saranex®, Silvershield®. Also, PVC, neoprene, and Styrene-butadiene offers limited protection
Recommendations for respirator selection: NIOSH/OSHA
1000 ppm: SA:CF* (any supplied-air respirator operated in a continuous-flow mode); or CCRFOV [any air-purifying, full-facepiece respirator (gas mask) with a chin-style, front- or back-mounted acid gas canister]; or GMFOV [any air-purifying, full-facepiece respirator (gas mask) with a chin-style, front- or back-mounted organic vapor cannister]; or PAPROV* [any powered, air-purifying respirator with organic vapor cartridge(s)]; or SCBAF (any self-contained breathing apparatus with a full facepiece); or SAF (any supplied-air respirator with a full facepiece). *EMERGENCY OR PLANNED ENTRY INTO UNKNOWN CONCENTRATIONS OR IDLH CONDITIONS*: SCBAF:PD,PP (any self-contained breathing apparatus that has a full facepiece and is operated in a pressure-demand or other positive-pressure mode); or SAF:PD,PP:ASCBA (any supplied-air respirator that has a full facepiece and is operated in a pressure-demand or other positive-pressure mode in combination with an auxiliary self-contained breathing apparatus operated in a pressure-demand or other positive pressure mode). *ESCAPE*: GMFOV [any air-purifying, full-facepiece respirator (gas mask) with a chin-style, front- or back-mounted organic vapor canister]; or SCBAE (any appropriate escape-type, self-contained breathing apparatus). **Note:* Substance causes eye irritation or damage; eye protection needed.
Exposure limits (TWA unless otherwise noted): NIOSH/OSHA PEL 100 ppm (400 mg/m³).
Toxicity by ingestion: Grade 3; LD_{50} = 50 to 500 g/kg.

Vapor (gas) irritant characteristics: Vapors cause smarting of the eyes or respiratory system if present in high concentrations. The effect is temporary.
Liquid or solid irritant characteristics: If spilled on clothing and allowed to remain, may cause smarting and reddening of the skin.
IDLH value: 1000 ppm.

FIRE DATA
Flash point: 107°F/42°C.
Fire extinguishing agents not to be used: Water may be ineffective
Autoignition temperature: 531°F/277°C/550°K. May vary by manufacturer.
Electrical hazard: Class I, Group D.
Burning rate: 4 mm/min

CHEMICAL REACTIVITY
Binary reactants: Strong oxidizers, nitric acid
Reactivity group: 33
Compatibility class: Miscellaneous hydrocarbon mixtures

ENVIRONMENTAL DATA
Food chain concentration potential: Negative; unlikely to accumulate.
Water pollution: Effect of low concentrations on aquatic life is unknown. Fouling to shoreline. May be dangerous if it enters nearby water intakes; notify operators. Notify local health and wildlife officials. **Response to discharge:** Issue warning–high flammability. Evacuate area. Disperse and flush.

SHIPPING INFORMATION
Grades of purity: Purity varies with coal used and distillation range taken; **Storage temperature:** Ambient; **Inert atmosphere:** None; **Venting:** Open (flame arrester); **Stability during transport:** Stable.

PHYSICAL AND CHEMICAL PROPERTIES
Physical state @ 59°F/15°C and 1 atm: Liquid.
Molecular weight: 110 (approximate).
Boiling point @ 1 atm: 320–428°F/160–220°C/433–493°K.
Specific gravity (water = 1): 0.89–0.97 @ 68°F/20°C (liquid).
Liquid surface tension: (estimate) 20 dynes/cm = 0.020 N/m @ 68°F/20°C.
Liquid water interfacial tension: (estimate) 45 dynes/cm = 0.045 N/m @ 68°F/20°C.
Relative vapor density (air = 1): 4.2
Ratio of specific heats of vapor (gas): (estimate) 1.030
Latent heat of vaporization: (estimate) 101 Btu/lb = 56.2 cal/g. = 2.35 x 10^5 J/kg.
Heat of combustion: (estimate) –18,200 Btu/lb = –10,100 cal/g = –424 x 10^5 J/kg.
Vapor pressure: (Reid) 0.13 psia; less than 5 mm.

NAPHTHALENE   REC. N:0400

SYNONYMS: CAMPHOR TAR; MIGHTY 150; MOTH BALLS; MOTH FLAKES; NAFTALENO (Spanish); NAPTHALIN; NAPTHALINE; RCRA No. U165; TAR CAMPHOR; WHITE TAR

IDENTIFICATION
CAS Number: 91-20-3
Formula: $C_{10}H_8$
DOT ID Number: UN 1334 (crude or refined); UN 2304 (molten);
DOT Guide Number: 133
Proper shipping name: Naphthalene, crude; or Naphthalene, refined; Naphthalene, molten
Reportable Quantity (RQ): **(CERCLA)** 100 lb/45.4 kg

DESCRIPTION: Colorless solid. Mothballs odor. Solidifies and floats or sinks in water.

Poison! • Combustible solid • Irritates the eyes and skin; poisonous and narcotic effects if inhaled or absorbed by the skin • Firefighting gear (including SCBA) does not provide adequate protection. If exposure occurs, remove and isolate gear immediately and thoroughly decontaminate personnel • Containers may BLEVE when exposed to fire • Concentrated dust in confined areas (e.g., tanks, sewers, buildings) may explode when exposed to fire • Toxic products of combustion may include smoke, irritants, and flammable vapors.

Hazard Classification (based on NFPA-704 Rating System)
Health Hazards (Blue): 2; Flammability (Red): 2; Reactivity (Yellow): 0

EMERGENCY RESPONSE: See Appendix A (133)
Evacuation:
Public safety: Isolate spill area for at least 10 to 25 meters (30 to 80 feet) in all directions.
Spill: Consider initial downwind evacuation of at least 100 meters (330 feet).
Fire: Isolate for 800 meters (½ mile) in all directions, especially if tank, rail car, or tank truck is involved in fire.

EXPOSURE
Short-term effects: *SEEK MEDICAL ATTENTION*. **Solid or liquid:** Irritating to skin and eyes. Remove contaminated clothing and shoes. Flush affected areas with plenty of water. *IF IN EYES*, hold eyelids open and flush with plenty of water.

HEALTH HAZARDS
Personal protective equipment (PPE): B-Level PPE. OSHA Table Z-1-A air contaminant. May act as a narcotic if inhaled. Approved organic vapor canister unit; rubber gloves; chemical safety goggles; face shield; coveralls and/or rubber apron; rubber shoes or boots. Chemical protective material(s) reported to have good to excellent resistance: Teflon®.
Recommendations for respirator selection: NIOSH/OSHA
100 ppm: CCROVDM * [any chemical cartridge respirator with organic vapor cartridge(s) in combination with a dust and mist filter]; or SA* (any supplied-air respirator). *250 ppm:* SA:CF* (any supplied-air respirator operated in a continuous-flow mode); or CCRFOVHiE [any chemical cartridge respirator with a full facepiece and organic vapor cartridge(s) in combination with a high-efficiency particulate filter]; or PAPROVDM* (any powered, air-purifying respirator with organic vapor cartridge(s) in combination with a dust and mist filter]; or SCBAF (any self-contained breathing apparatus with a full facepiece); or SAF (any supplied-air respirator with a full facepiece). *EMERGENCY OR PLANNED ENTRY INTO UNKNOWN CONCENTRATIONS OR IDLH CONDITIONS:* SCBAF:PD,PP (any self-contained breathing apparatus that has a full facepiece and is operated in a pressure-demand or other positive-pressure mode); or SAF:PD,PP:ASCBA (any supplied-air respirator that has a full facepiece and is operated in a pressure-demand or other positive-pressure mode in combination with an auxiliary self-contained breathing apparatus operated in a pressure-demand or other positive pressure mode). *ESCAPE:* GMFOVHiE [any air-purifying, full-facepiece respirator

(gas mask) with a chin-style, front- or back-mounted organic vapor canister having a high-efficiency particulate filter]; or SCBAE (any appropriate escape-type, self-contained breathing apparatus).
Note: Substance reported to cause eye irritation or damage; may require eye protection.
Exposure limits (TWA unless otherwise noted): ACGIH TLV 10 ppm (52 mg/m^3); OSHA PEL 10 ppm (50 mg/m^3).
Short-term exposure limits (15-minute TWA): ACGIH STEL 15 ppm (79 mg/m^3); NIOSH STEL 15 ppm (75 mg/m^3).
Toxicity by ingestion: Grade 2; oral rat LD_{50} = 1780 mg/kg.
Vapor (gas) irritant characteristics: Vapors cause irritation such that personnel will find high concentrations unpleasant. The effect is temporary.
Liquid or solid irritant characteristics: Hot liquid can cause severe burn. The solid may irritate the skin.
Odor threshold: 0.038 ppm.
IDLH value: 250 ppm.

FIRE DATA
Flash point: 190°F/88°C (oc); 174°F/79°C (cc).
Flammable limits in air: LEL: 0.9%; UEL: 5.9%.
Fire extinguishing agents not to be used: Water or water-based foam may cause frothing.
Autoignition temperature: 979°F/526°C/799°K.
Electrical hazard: Flow or agitation of substance may generate electrostatic charges due to low conductivity.
Burning rate: 4.3 mm/min

CHEMICAL REACTIVITY
Reactivity with water: Molten naphthalene spatters and foams in contact with water. No chemical reaction is involved.
Binary reactants: Strong oxidizers, chromic anhydride, chromium trioxide (may be violent).
Reactivity group: 32
Compatibility class: Aromatic hydrocarbons

ENVIRONMENTAL DATA
Food chain concentration potential: Log P_{ow} = 3.3 (estimate). Values > 3.0 are likely to bioconcentrate in aquatic organisms and other living tissue, especially in fats.
Water pollution: DOT Appendix B, §172.101–marine pollutant. Harmful to aquatic life in very low concentrations. Fouling to shoreline. May solidify into a foam and float. May be dangerous if it enters nearby water intakes; notify operators. Notify local health and wildlife officials. **Response to discharge:** Should be removed. Chemical and physical treatment.

SHIPPING INFORMATION
Grades of purity: Pure; Crude: 95%; Pure: mp = 176°F; Crude: mp = 165-176°F; **Storage temperature:** Elevated, may be stored under nitrogen gas; **Inert atmosphere:** None; **Venting:** Open (flame arrester) or pressure-vacuum; **Stability during transport:** Stable if kept reasonably cool. Solid is volatile forming flammable vapors when heated.

NAS HAZARD CLASSIFICATION FOR BULK WATER TRANSPORTATION
FIRE: 1
HEALTH: Vapor irritant: 2; Liquid or solid irritant: 1; Poisons: 2
WATER POLLUTION: Human toxicity: 1; Aquatic toxicity: 3; Aesthetic effect: 3
REACTIVITY: Other chemicals: 1; Water: 0; Self-reaction: 0

PHYSICAL AND CHEMICAL PROPERTIES
Physical state @ 59°F/15°C and 1 atm: Solid.
Molecular weight: 128.18
Boiling point @ 1 atm: 424°F/218°C/491°K.
Melting/Freezing point: 176.4°F/80.2°C/353.4°K.
Critical temperature: 887.4°F/475.2°C/748.4°K.
Critical pressure: 588 psia = 40.0 atm = 4.05 MN/m^2.
Specific gravity (water = 1): 1.145 @ 68°F/20°C (solid).
Liquid surface tension: 31.8 dynes/cm = 0.0318 N/m at 100°C.
Relative vapor density (air = 1): 4.4
Ratio of specific heats of vapor (gas): 1.068
Latent heat of vaporization: 145 Btu/lb = 80.7 cal/g = 3.38 x 10^5 J/kg.
Heat of combustion: –16,720 Btu/lb = –9287 cal/g = –388.8 x 10^5 J/kg.
Heat of fusion: 35.06 cal/g.
Vapor pressure: (Reid) Low; 0.030 mm.

NAPHTHENIC ACIDS   REC. N:0450

SYNONYMS: ACIDO NAFTENICO (Spanish); AGENAP; NAPHID; SUNATIPIC ACID-C

IDENTIFICATION
CAS Number: 1338-24-5
Formula: 1-$C_{10}H_7NH_2$; R_2C-Cr_2-Cr_2-Cr_2-CR-$(CH_2)_n$-COOH where n = 2–6
DOT ID Number: UN 3077; DOT Guide Number: 171
Proper Shipping Name: Environmentally hazardous substances, solid, n.o.s.
Reportable Quantity (RQ): **(CERCLA)** 100 lb/45.4 kg

DESCRIPTION: Golden to black liquid or crystalline solid. Odorless. Slightly soluble in water.

Combustible solid • Dust cloud may explode if ignited in an enclosed area • Containers may BLEVE when exposed to fire • Irritating to the skin, eyes, and respiratory tract • Toxic products of combustion may include carbon monoxide and possibly nitrogen oxides.

Hazard Classification (based on NFPA-704 Rating System)
Health Hazards (Blue): 0; Flammability (Red): 1; Reactivity (Yellow): 0

EMERGENCY RESPONSE: See Appendix A (171)
Evacuation:
Public safety: Isolate the area of spill or leak for at least 10 to 25 meters (30 to 80 feet) in all directions.
Spill: Increase, in the downwind direction, as necessary, the distance shown under "Public Safety."
Fire: If any large container is involved in fire, isolate for at least 800 meters (½ mile) in all directions; also, consider initial evacuation for 800 meters (½ mile) in all directions.

EXPOSURE
Short-term effects: *SEEK MEDICAL ATTENTION.* **Vapor:** Irritating to eyes, nose, and throat. *IF INHALED*, will, will cause coughing or difficult breathing. *IF IN EYES*, hold eyelids open and flush with plenty of water. *IF BREATHING HAS STOPPED*, give artificial respiration. IF breathing is difficult, administer oxygen. **Liquid:** Irritating to skin and eyes. *IF SWALLOWED*, will cause nausea. Remove contaminated clothing and shoes. Flush affected areas with plenty of water. *IF IN EYES*, hold eyelids open and flush with plenty of water. *IF SWALLOWED* and victim is *CONSCIOUS AND ABLE TO SWALLOW*, have victim drink 4 to 8 ounces of

of water. *IF SWALLOWED* and victim is *UNCONSCIOUS OR HAVING CONVULSIONS*, do nothing except keep victim warm.

HEALTH HAZARDS
Personal protective equipment (PPE): B-Level PPE. Safety glasses or face mask
Toxicity by ingestion: Grade 2; oral LD_{50} = 3000 mg/kg (rat).
Vapor (gas) irritant characteristics: Vapors cause smarting of the eyes or respiratory system if present in high concentrations. The effect is temporary.
Liquid or solid irritant characteristics: If spilled on clothing and allowed to remain, may cause smarting and reddening of skin.

FIRE DATA
Flash point: 300°F/149°C (oc).
Fire extinguishing agents not to be used: Water may be ineffective

CHEMICAL REACTIVITY
Binary reactants: Generally corrosive to metals.

ENVIRONMENTAL DATA
Food chain concentration potential: Negative; unlikely to accumulate.
Water pollution: DOT Appendix B, §172.101–marine pollutant. Harmful to aquatic life in very low concentrations. Fouling to shoreline. May be dangerous if it enters nearby water intakes; notify operators. Notify local health and wildlife officials.
Response to discharge: Prevent from reaching bodies of water, it fouls beaches and taints aquatic life. Mechanical containment. Should be removed. Chemical and physical treatment.

SHIPPING INFORMATION
Grades of purity: Commercial, 100%; **Storage temperature:** Ambient; **Inert atmosphere:** None; **Venting:** Open; **Stability during transport:** Stable.

NAS HAZARD CLASSIFICATION FOR BULK WATER TRANSPORTATION
FIRE: 1
HEALTH: Vapor irritant: 1; Liquid or solid irritant: 1; Poisons: 3
WATER POLLUTION: Human toxicity: 1; Aquatic toxicity: 3; Aesthetic effect: 4
REACTIVITY: Other chemicals: 3; Water: 0; Self-reaction: 0

PHYSICAL AND CHEMICAL PROPERTIES
Physical state @ 59°F/15°C and 1 atm: Liquid.
Molecular weight: 200-250 (mixture).
Boiling point @ 1 atm: 270–470°F/132–243°C/405–516°K.
Specific gravity (water = 1): 0.982 @ 68°F/20°C (liquid).

1-NAPHTHYLAMINE REC. N:0500

SYNONYMS: 1-AMINONAPHTHALENE; C.I. AZOIC DIAZO COMPONENT 114; FAST GARNET B BASE; FAST GARNET BASE B; 1-NAFTILAMINA (Spanish); 1-NAPHTHALENAMINE, TECHNICAL GRADE; NAPHTHALIDAM; NAPHTHALIDINE; α-NAPHTHYLAMINE; RCRA No. U167

IDENTIFICATION
CAS Number: 134-32-7
Formula: $C_{10}H_9N$
DOT ID Number: UN 2077; DOT Guide Number: 153
Proper Shipping Name: α-Naphthylamine
Reportable Quantity (RQ): **(CERCLA)** 100 lb/45.4 kg

DESCRIPTION: White crystalline solid; darkens on contact with air from tan to reddish to dark brown. Weak ammonia-like odor. Sinks in water; soluble.

Poison! • Combustible solid • Breathing the dust, skin or eye contact, or swallowing the material can cause illness • Firefighting gear (including SCBA) does not provide adequate protection. If exposure occurs, remove and isolate gear immediately and thoroughly decontaminate personnel • Containers may BLEVE when exposed to fire • Concentrated dust in confined areas (e.g., tanks, sewers, buildings) may explode when exposed to fire • Toxic products of combustion may include nitrogen oxides • Do not put yourself in danger by entering a contaminated area to rescue a victim.

Hazard Classification (based on NFPA-704 Rating System)
Health Hazards (Blue): 2; Flammability (Red): 1; Reactivity (Yellow): 0

EMERGENCY RESPONSE: See Appendix A (153)
Evacuation:
Public safety: Isolate the area of spill or leak for at least 25 to 50 meters (80 to 160 feet) in all directions.
Spill: Increase, in the downwind direction, as necessary, the distance shown under "Public Safety."
Fire: If tank, rail car, or tank truck is involved in fire, isolate for at least 800 meters (½ mile) in all directions; also, consider initial evacuation for 800 meters (½ mile) in all directions.

EXPOSURE
Short-term effects: *SEEK MEDICAL ATTENTION.* **Dust:** *POISONOUS IF INHALED OR IF SKIN IS EXPOSED.* Irritating to eyes. Move victim to fresh air. *IF IN EYES,* hold eyelids open and flush with plenty of water. *IF BREATHING HAS STOPPED,* give artificial respiration; *avoid mouth-to-mouth resuscitation; use bag/mask apparatus.* IF breathing is difficult, administer oxygen. **Solid:** *POISONOUS IF SWALLOWED OR IF SKIN IS EXPOSED.* Remove contaminated clothing and shoes. Flush affected areas with plenty of water. *IF IN EYES,* hold eyelids open and flush with plenty of water. *IF SWALLOWED* and victim is *CONSCIOUS AND ABLE TO SWALLOW,* have victim drink 4 to 8 ounces of water and have victim induce vomiting. Watch for signs fo cyanosis. *IF SWALLOWED* and victim is *UNCONSCIOUS OR HAVING CONVULSIONS,* do nothing except keep victim warm.
Note: Persons undergoing *severe* exposure to this compound should have continuing medical attention for possible development of cancer.

HEALTH HAZARDS
Personal protective equipment (PPE): B-Level PPE. OSHA Table Z-1-A air contaminant. Complete protection for respiratory system, eyes, and skin.
Recommendations for respirator selection: NIOSH
At any concentrations above the NIOSH REL, or where there is no REL, at any detectable concentration: SCBAF:PD,PP (any self-contained breathing apparatus that has a full facepiece and is operated in a pressure-demand or other positive-pressure mode); or SAF:PD,PP:ASCBA (any supplied-air respirator that has a full facepiece and is operated in a pressure-demand or other positive-pressure mode in combination with an auxiliary self-contained breathing apparatus operated in a pressure-demand or other positive pressure mode). ESCAPE: HiEF (any air-purifying, full-

facepiece respirator with a high-efficiency particulate filter); or SCBAE (any appropriate escape-type, self-contained breathing apparatus).
Exposure limits (TWA unless otherwise noted): OSHA PEL: Cancer suspect agent; NIOSH REL [1910.1009]
Toxicity by ingestion: Grade 2; oral LD_{50} = 779 mg/kg (rat), 4000 mg/kg (mammal).
Long-term health effects: Suspected human carcinogen; may cause bladder cancer.
Liquid or solid irritant characteristics: Causes eye irritation.

FIRE DATA
Flash point: 315°F/157°C (cc) (molten solid).
Fire extinguishing agents not to be used: Water or foam may cause frothing.

CHEMICAL REACTIVITY
Binary reactants: Nitrous acid, strong oxidizers

ENVIRONMENTAL DATA
Food chain concentration potential: Log K_{ow} = 2.25. No significant accumulation. expected
Water pollution: Effect of low concentrations on aquatic life is unknown. May be dangerous if it enters nearby water intakes; notify operators. Notify local health and wildlife officials.
Response to discharge: Issue warning–poison, water contaminant. Restrict access. Should be removed. Chemical and physical treatment.

SHIPPING INFORMATION
Grades of purity: Pure; Technical; **Storage temperature:** Cool ambient; **Inert atmosphere:** None; **Venting:** Open; **Stability during transport:** Stable.

PHYSICAL AND CHEMICAL PROPERTIES
Physical state @ 59°F/15°C and 1 atm: Solid.
Molecular weight: 143.2
Boiling point @ 1 atm: 572°F/300°C/573°K.
Melting/Freezing point: 118-122°F/48-50°C/321-323°K.
Specific gravity (water = 1): 1.12 @ 77°F/25°C (solid).
Heat of combustion: −15,290 Btu/lb = −8495 cal/g = −355.4 x 10^5 J/kg.

NEODECANOIC ACID   REC. N:0550

SYNONYMS: ACIDO NEODECANOICO (Spanish); 2,2-DIMETHYL OCTANOIC ACID; OCTANOIC ACID, DIMETHYL-; WILTZ 65

IDENTIFICATION
CAS Number: 26896-20-8
Formula: $C_{10}H_{20}O_2$; $CH_3(CH_2)_5C(CH_3)_2COOH$
DOT ID Number: UN 1760; DOT Guide Number: 154
Proper Shipping Name: Corrosive liquids, n.o.s.

DESCRIPTION: Clear, colorless liquid (when pure).

Combustible • Heat or flame may cause explosion • Containers may BLEVE when exposed to fire • Vapors are heavier than air and will collect and stay in low areas • Vapors in confined areas (e.g., tanks, sewers, buildings) may explode when exposed to fire • Irritating to the skin, eyes, and respiratory tract • Toxic products of combustion may include carbon monoxide.

Hazard Classification (based on NFPA-704 Rating System)
Health Hazards (Blue): 0; Flammability (Red): 1; Reactivity (Yellow): 0

EMERGENCY RESPONSE: See Appendix A (154)
Evacuation:
Public safety: Isolate the area of spill or leak for at least 25 to 50 meters (80 to 160 feet) in all directions.
Spill: Increase, in the downwind direction, as necessary, the distance shown under "Public Safety."
Fire: If tank, rail car, or tank truck is involved in fire, isolate for at least 800 meters (½ mile) in all directions; also, consider initial evacuation for 800 meters (½ mile) in all directions.

EXPOSURE
Short-term effects: *SEEK MEDICAL ATTENTION*. **Liquid or solid:** Will burn skin and eyes. Harmful if swallowed. Remove contaminated clothing and shoes. Flush affected areas with plenty of water. *IF BREATHING HAS STOPPED,* give artificial respiration; *avoid mouth-to-mouth resuscitation; use bag/mask apparatus. IF IN EYES,* hold eyelids open and flush with plenty of water. *IF SWALLOWED* and victim is *CONSCIOUS AND ABLE TO SWALLOW,* induce vomiting.

HEALTH HAZARDS
Personal protective equipment (PPE): B-Level PPE. When contact is likely wear long sleeves, chemical resistant gloves, and chemical goggles. Where contact may occur wear safety glasses with side shields. Where overexposure by inhalation may occur, wear approved respirator.
Vapor (gas) irritant characteristics: Vapors/mists cause severe irritation of eyes and throat and can cause eye and lung injury. They cannot be tolerated even at low concentrations.
Liquid or solid irritant characteristics: Severe skin irritant. Causes second- and third-degree burns on short contact and is very injurious to the eyes.

FIRE DATA
Flash point: 201°F/94°C (cc).
Electrical hazard: Flow or agitation of substance may generate electrostatic charges due to low conductivity.

CHEMICAL REACTIVITY
Binary reactants: Reacts with sulfuric acid, caustics, ammonia, amines, isocyanates. May corrode metals
Neutralizing agents for acids and caustics: Sodium bicarbonate or lime.
Reactivity group: 4
Compatibility class: Organic acids

ENVIRONMENTAL DATA
Water pollution: Effect of low concentrations on aquatic life is unknown. May be dangerous if it enters nearby water intakes; notify operators. Notify local health and wildlife officials.
Response to discharge: Restrict access. Mechanical containment. Chemical and physical treatment.

SHIPPING INFORMATION
Grades of purity: 79–100%; **Storage temperature:** Ambient; **Inert atmosphere:** None; **Venting:** Open; **Stability during transport:** Stable.

PHYSICAL AND CHEMICAL PROPERTIES
Physical state @ 59°F/15°C and 1 atm: Liquid.
Molecular weight: 172.27

Boiling point @ 1 atm: 482–494°F/250–257°C/523–530°K.
Melting/Freezing point: Less than 104°F/40°C/313°K.
Specific gravity (water = 1): 0.92
Relative vapor density (air = 1): 6.0

NEOHEXANE   REC. N:0600

SYNONYMS: 2,2-DIMETHYLBUTANE; NEOHEXANO (Spanish)

IDENTIFICATION
CAS Number: 75-83-2; 79-29-8; 594-84-3; 96-14-0 (all isomers of *n*-hexane)
Formula: C_6H_{14}; $CH_3C(CH_3)_2CH_2CH_3$
DOT ID Number: UN 1208; DOT Guide Number: 128
Proper Shipping Name: Hexanes

DESCRIPTION: Colorless liquid. Gasoline-like odor. Floats on water surface; insoluble. Flammable, irritating vapor is produced.

Highly flammable • Containers may BLEVE when exposed to fire • Vapors may form explosive mixture with air • Vapors are heavier than air and will collect and stay in low areas • Vapors may travel long distances to ignition sources and flashback • Vapors in confined areas (e.g., tanks, sewers, buildings) may explode when exposed to fire • Irritating to the skin, eyes, and respiratory tract • Toxic products of combustion may include carbon monoxide.

Hazard Classification (based on NFPA-704 Rating System)
Health Hazards (Blue): 1; Flammability (Red): 3; Reactivity (Yellow): 0

EMERGENCY RESPONSE: See Appendix A (128)
Evacuation:
Public safety: Isolate spill area for at least 25 to 50 meters (80 to 160 feet) in all directions.
Spill: Large spill–Consider initial downwind evacuation for at least 300 meters (1000 feet).
Fire: Isolate for 800 meters (½ mile) in all directions, especially if tank, rail car, or tank truck is involved in fire.

EXPOSURE
Short-term effects: *SEEK MEDICAL ATTENTION.* **Vapor:** Irritating to eyes, nose, and throat. *IF INHALED*, will, will cause dizziness, coughing, or difficult breathing. Move victim to fresh air. *IF BREATHING HAS STOPPED,* give artificial respiration. IF breathing is difficult, administer oxygen. **Liquid:** Irritating to skin and eyes. *IF SWALLOWED*, will cause nausea, or vomiting. Remove contaminated clothing and shoes. Flush affected areas with plenty of water. *IF IN EYES*, hold eyelids open and flush with plenty of water. *IF SWALLOWED* and victim is *CONSCIOUS AND ABLE TO SWALLOW*, have victim drink 4 to 8 ounces of water. **Do NOT induce vomiting.**

HEALTH HAZARDS
Personal protective equipment (PPE): B-Level PPE. Air–supplied apparatus or organic vapor cartridge; goggles or face shield; rubber gloves.
Recommendations for respirator selection: NIOSH as hexane isomers.
1000 ppm: SA* (any supplied-air respirator). *2500 ppm:* SA:CF* (any supplied-air respirator operated in a continuous-flow mode). *500 ppm:* SA:PD:PP (any supplied-air respirator operated in a pressure-demand or other positive-pressure mode). *5000 ppm:* SAT:CF (any supplied-air respirator that has a tight-fitting facepiece and is operated in a continuous-flow mode); or SCBAF (any self-contained breathing apparatus with a full facepiece); or SAF (any supplied-air respirator with a full facepiece). *EMERGENCY OR PLANNED ENTRY INTO UNKNOWN CONCENTRATIONS OR IDLH CONDITIONS:* SCBAF:PD,PP (any self-contained breathing apparatus that has a full facepiece and is operated in a pressure-demand or other positive-pressure mode); or SAF:PD,PP:ASCBA (any supplied-air respirator that has a full facepiece and is operated in a pressure-demand or other positive-pressure mode in combination with an auxiliary self-contained breathing apparatus operated in a pressure-demand or other positive pressure mode). *ESCAPE:* GMFOV [any air-purifying, full-facepiece respirator (gas mask) with a chin-style, front- or back-mounted organic vapor canister]; or SCBAE (any appropriate escape-type, self-contained breathing apparatus. **Note:* Substance reported to cause eye irritation or damage; may require eye protection.
Exposure limits (TWA unless otherwise noted): ACGIH TLV 500 ppm (1760 mg/m^3) as hexane, other isomers; OSHA PEL 500 ppm (1800 mg/m^3); NIOSH 50 ppm (180 mg/m^3).
Liquid or solid irritant characteristics: Causes eye irritation. Possible skin irritation.

FIRE DATA
Flash point: –54°F/–48°C (cc).
Flammable limits in air: LEL: 1.2%; UEL: 7.0%.
Fire extinguishing agents not to be used: Water may be ineffective.
Autoignition temperature: 770°F/410°C/683°K.
Burning rate: 9.2 mm/min

CHEMICAL REACTIVITY
Binary reactants: Strong oxidizers

ENVIRONMENTAL DATA
Food chain concentration potential: Negative; unlikely to accumulate.
Water pollution: Effect of low concentrations on aquatic life is unknown. Fouling to shoreline. May be dangerous if it enters nearby water intakes; notify operators. Notify local health and wildlife officials. Notify operators or nearby water intakes.
Response to discharge: Issue warning–high flammability. Restrict access. Evacuate area. Disperse and flush.

SHIPPING INFORMATION
Grades of purity: Research: 99.98%; Pure: 99.5%; Technical: 96.4%; **Storage temperature:** Ambient; **Inert atmosphere:** None; **Venting:** Open (flame arrester); **Stability during transport:** Stable.

PHYSICAL AND CHEMICAL PROPERTIES
Physical state @ 59°F/15°C and 1 atm: Liquid.
Molecular weight: 86.2
Boiling point @ 1 atm: 121.5°F/49.7°C/322.9°K.
Melting/Freezing point: –147.8°F/–99.9°C/173.3°K.
Critical temperature: 420°F/216°C/489°K.
Critical pressure: 447 psia = 30.4 atm = 3.08 MN/m^2.
Specific gravity (water = 1): 0.649 @ 68°F/20°C (liquid).
Liquid surface tension: 16.3 dynes/cm = 0.0163 N/m @ 68°F/20°C.
Liquid water interfacial tension: (estimate) 35 dynes/cm = 0.035 N/m @ 68°F/20°C.
Relative vapor density (air = 1): 3.0
Ratio of specific heats of vapor (gas): 1.064 at 16°C.

Latent heat of vaporization: 131 Btu/lb = 72.9 cal/g = 3.05 x 10^5 J/kg.
Heat of combustion: –19,310 Btu/lb = –10,730 cal/g = –448.9 x 10^5 J/kg.
Heat of fusion: 1.61 cal/g.

NICKEL ACETATE   REC. N:0650

SYNONYMS: ACETATO de NIQUEL (Spanish); ACETIC ACID,NICKEL(II) SALT; ACETIC ACID,NICKEL(2+) SALT; NICKEL ACETATE TETRAHYDRATE; NICKEL(II) ACETATE; NICKELOUS ACETATE

IDENTIFICATION
CAS Number: 373-02-4
Formula: $C_4H_6O_4 \cdot Ni$; $Ni(C_2H_3O_2)_2 \cdot 4H_2O$; $NiSO_4 \cdot (NH_4)_2SO_4 \cdot 6H_2O$
DOT ID Number: UN 3077; DOT Guide Number: 171
Proper Shipping Name: Environmentally hazardous substances, solid, n.o.s.

DESCRIPTION: Dull green solid. Odorless. Sinks and mixes slowly with water.

Poison! • Firefighting gear (including SCBA) does not provide adequate protection. If exposure occurs, remove and isolate gear immediately and thoroughly decontaminate personnel • Containers may BLEVE when exposed to fire • Toxic products of combustion may include nickel metal fumes and nickel carbonyl.

Hazard Classification (based on NFPA-704 Rating System)
Health Hazards (Blue): 2; Flammability (Red): 0; Reactivity (Yellow): 0

EMERGENCY RESPONSE: See Appendix A (171)
Evacuation:
Public safety: Isolate the area of spill or leak for at least 10 to 25 meters (30 to 80 feet) in all directions.
Spill: Increase, in the downwind direction, as necessary, the distance shown under "Public Safety."
Fire: If any large container is involved in fire, isolate for at least 800 meters (½ mile) in all directions; also, consider initial evacuation for 800 meters (½ mile) in all directions.

EXPOSURE
Short-term effects: *SEEK MEDICAL ATTENTION.* **Dust:** Irritating to eyes, nose, and throat. *IF INHALED,* will, will cause coughing or difficult breathing. *IF IN EYES,* hold eyelids open and flush with plenty of water. *IF BREATHING HAS STOPPED,* give artificial respiration. IF breathing is difficult, administer oxygen. **Solid:** Irritating to skin and eyes. *IF SWALLOWED,* will cause nausea and vomiting. Remove contaminated clothing and shoes. Flush affected areas with plenty of water. *IF IN EYES,* hold eyelids open and flush with plenty of water. *IF SWALLOWED* and victim is *CONSCIOUS AND ABLE TO SWALLOW,* have victim drink 4 to 8 ounces of water. *IF SWALLOWED* and victim is *UNCONSCIOUS OR HAVING CONVULSIONS,* do nothing except keep victim warm.

HEALTH HAZARDS
Personal protective equipment (PPE): B-Level PPE. OSHA Table Z-1-A air contaminant. Bureau of Mines approved respirator; rubber gloves; safety goggles; protective clothing
Recommendations for respirator selection: NIOSH *At any concentrations above the NIOSH REL, or where there is no REL, at any detectable concentration:* SCBAF:PD,PP (any self-contained breathing apparatus that has a full facepiece and is operated in a pressure-demand or other positive-pressure mode); or SAF:PD,PP:ASCBA (any supplied-air respirator that has a full facepiece and is operated in a pressure-demand or other positive-pressure mode in combination with an auxiliary self-contained breathing apparatus operated in a pressure-demand or other positive pressure mode). *ESCAPE:* HiEF (any air-purifying, full-facepiece respirator with a high-efficiency particulate filter); or SCBAE (any appropriate escape-type, self-contained breathing apparatus).
Exposure limits (TWA unless otherwise noted): ACGIH TLV 0.1 mg/m³ nickel, soluble compounds; OSHA PEL 1 mg/m³; NIOSH REL 0.015 mg/m³; reduce to lowest feasible level as nickel; potential human carcinogen.
Toxicity by ingestion: Grade 2; LD_{50} = 0.5–5g/kg.
Long-term health effects: Possible lung cancer. IARC listed with no overall value.
Liquid or solid irritant characteristics: Eye and skin irritation.
IDLH value: Potential human carcinogen; 10 mg/m³ as nickel.

CHEMICAL REACTIVITY
Binary reactants: Strong acids, sulfur, selenium

ENVIRONMENTAL DATA
Food chain concentration potential: Food chain concentration potential: May concentrate to a high degree in aquatic plants (103 to 104 mg/kg dry weight). Nat'l Research Council Canada, *Effects of Nickel in the Canadian Environment,* 1981, NRCC No.18568.
Water pollution: Effect of low concentrations on aquatic life is unknown; however, soluble nickel salts are of great concern to the environment. May be dangerous if it enters nearby water intakes; notify operators. Notify local health and wildlife officials. **Response to discharge:** Disperse and flush.

SHIPPING INFORMATION
Grades of purity: Commercial, 99%; Reagent; **Storage temperature:** Ambient; **Inert atmosphere:** None; **Venting:** Open; **Stability during transport:** Stable.

PHYSICAL AND CHEMICAL PROPERTIES
Physical state @ 59°F/15°C and 1 atm: Solid.
Molecular weight: 248.86
Boiling point @ 1 atm: Decomposes.
Specific gravity (water = 1): 1.74 @ 68°F/20°C (solid).

NICKEL AMMONIUM SULFATE   REC. N:0700

SYNONYMS: AMMONIUM NICKEL SULFATE; SULFURIC ACID, AMMONIUM NICKEL (2+) SALT (2:2:1); AMMONIUM NICKEL (II) SALT; AMMONIUM DISULFATONICKELATE (II); NICKEL AMMONIUM SULFATE HEXAHYDRATE; SULFATO de NIQUEL y AMONIO (Spanish)

IDENTIFICATION
CAS Number: 15699-18-0
Formula: $O_8S_2 \cdot Ni \cdot 2H_4N$
DOT ID Number: UN 9138; DOT Guide Number: 171
Proper Shipping Name: Nickel ammonium sulfate; Nickel ammonium sulphate
Reportable Quantity (RQ): **(CERCLA)** 100 lb/45.4 kg

DESCRIPTION: Dark green-blue crystalline solid. Odorless. Sinks and mixes slowly with water.

Poison! • Firefighting gear (including SCBA) does not provide adequate protection. If exposure occurs, remove and isolate gear immediately and thoroughly decontaminate personnel • Containers may BLEVE when exposed to fire • Toxic products of combustion may include nitrogen oxides, nickel metal fumes, and possibly nickel carbonyl fumes.

Hazard Classification (based on NFPA-704 Rating System)
Health Hazards (Blue): 2; Flammability (Red): 0; Reactivity (Yellow): 0

EMERGENCY RESPONSE: See Appendix A (171)
Evacuation:
Public safety: Isolate the area of spill or leak for at least 10 to 25 meters (30 to 80 feet) in all directions.
Spill: Increase, in the downwind direction, as necessary, the distance shown under "Public Safety."
Fire: If any large container is involved in fire, isolate for at least 800 meters (½ mile) in all directions; also, consider initial evacuation for 800 meters (½ mile) in all directions.

EXPOSURE
Short-term effects: *SEEK MEDICAL ATTENTION.* **Dust:** Irritating to eyes, nose, and throat. *IF INHALED*, will, will cause coughing or difficult breathing. *IF IN EYES*, hold eyelids open and flush with plenty of water. *IF BREATHING HAS STOPPED*, give artificial respiration. IF breathing is difficult, administer oxygen.
Solid: Irritating to skin and eyes. *IF SWALLOWED*, will cause nausea and vomiting. Remove contaminated clothing and shoes. Flush affected areas with plenty of water. *IF IN EYES*, hold eyelids open and flush with plenty of water. *IF SWALLOWED* and victim is *CONSCIOUS AND ABLE TO SWALLOW*, have victim drink 4 to 8 ounces of water. *IF SWALLOWED* and victim is *UNCONSCIOUS OR HAVING CONVULSIONS*, do nothing except keep victim warm.

HEALTH HAZARDS
Personal protective equipment (PPE): B-Level PPE. OSHA Table Z-1-A air contaminant. Bureau of Mines approved respirator; rubber gloves; face shield or safety goggles; protective clothing
Recommendations for respirator selection: NIOSH
At any concentrations above the NIOSH REL, or where there is no REL, at any detectable concentration: SCBAF:PD,PP (any self-contained breathing apparatus that has a full facepiece and is operated in a pressure-demand or other positive-pressure mode); or SAF:PD,PP:ASCBA (any supplied-air respirator that has a full facepiece and is operated in a pressure-demand or other positive-pressure mode in combination with an auxiliary self-contained breathing apparatus operated in a pressure-demand or other positive pressure mode). *ESCAPE:* HiEF (any air-purifying, full-facepiece respirator with a high-efficiency particulate filter); or SCBAE (any appropriate escape-type, self-contained breathing apparatus).
Exposure limits (TWA unless otherwise noted): ACGIH TLV 0.1 mg/m^3 as nickel, soluble compounds; OSHA PEL 1 mg/m^3; NIOSH REL 0.015 mg/m^3 as nickel; reduce exposure to lowest feasible level; potential human carcinogen.
Toxicity by ingestion: Grade 2; LD$_{50}$ = 0.5 to 5 g/kg.
Long-term health effects: Possible lung cancer. IARC listed, overall evaluation unassigned.
Liquid or solid irritant characteristics: Eye and skin irritation.
IDLH value: Potential human carcinogen; 10 mg/m^3 as nickel.

CHEMICAL REACTIVITY
Binary reactants: Violent reaction with aluminum, magnesium, strong acids, sulfur, selenium

ENVIRONMENTAL DATA
Food chain concentration potential: Food chain concentration potential: May concentrate to a high degree in aquatic plants (103 to 104 mg/kg dry weight). Nat'l Research Council Canada, *Effects of Nickel in the Canadian Environment*, 1981, NRCC No.18568.
Water pollution: Effect of low concentrations on aquatic life is unknown, but of great environmental concen. May be dangerous if it enters nearby water intakes; notify operators. Notify local health and wildlife officials. **Response to discharge:** Disperse and flush.

SHIPPING INFORMATION
Grades of purity: Reagent; Technical; **Storage temperature:** Ambient; **Inert atmosphere:** None; **Venting:** Open; **Stability during transport:** Stable.

PHYSICAL AND CHEMICAL PROPERTIES
Physical state @ 59°F/15°C and 1 atm: Solid.
Molecular weight: 395.00
Specific gravity (water = 1): 1.92 @ 68°F/20°C (solid).

NICKEL BROMIDE REC. N:0750

SYNONYMS: NICKEL BROMIDE TRIHYDRATE

IDENTIFICATION
CAS Number: 13462-88-9
Formula: NiBr$_2$; NiBr$_2$·3H$_2$O (trihydrate)
DOT ID Number: UN 3077; DOT Guide Number: 171
Proper Shipping Name: Environmentally hazardous substances, solid, n.o.s.

DESCRIPTION: Yellowish-green solid. Odorless. Sinks and mixes with water.

Poison! • Firefighting gear (including SCBA) does not provide adequate protection. If exposure occurs, remove and isolate gear immediately and thoroughly decontaminate personnel • Containers may BLEVE when exposed to fire • Toxic products of combustion may include hydrogen bromide and nickel.

Hazard Classification (based on NFPA-704 Rating System)
Health Hazards (Blue): 2; Flammability (Red): 0; Reactivity (Yellow): 0

EMERGENCY RESPONSE: See Appendix A (171)
Evacuation:
Public safety: Isolate the area of spill or leak for at least 10 to 25 meters (30 to 80 feet) in all directions.
Spill: Increase, in the downwind direction, as necessary, the distance shown under "Public Safety."
Fire: If any large container is involved in fire, isolate for at least 800 meters (½ mile) in all directions; also, consider initial evacuation for 800 meters (½ mile) in all directions.

EXPOSURE
Short-term effects: *SEEK MEDICAL ATTENTION.* **Dust:** Irritating to eyes, nose, and throat. *IF INHALED*, will, will cause coughing and difficult breathing. *IF IN EYES*, hold eyelids open and flush with plenty of water. *IF BREATHING HAS STOPPED*, give artificial respiration. IF breathing is difficult, administer

oxygen. **Solid:** Irritating to skin and eyes. *IF SWALLOWED*, will cause nausea and vomiting. Remove contaminated clothing and shoes. Flush affected areas with plenty of water. *IF IN EYES*, hold eyelids open and flush with plenty of water. *IF SWALLOWED* and victim is *CONSCIOUS AND ABLE TO SWALLOW*, have victim drink 4 to 8 ounces of water. *IF SWALLOWED* and victim is *UNCONSCIOUS OR HAVING CONVULSIONS*, do nothing except keep victim warm.

HEALTH HAZARDS
Personal protective equipment (PPE): B-Level PPE. OSHA Table Z-1-A air contaminant. Bureau of Mines approved respirator; rubber gloves; face shield or safety goggles; protective clothing.
Recommendations for respirator selection: NIOSH *At any concentrations above the NIOSH REL, or where there is no REL, at any detectable concentration:* SCBAF:PD,PP (any self-contained breathing apparatus that has a full facepiece and is operated in a pressure-demand or other positive-pressure mode); or SAF:PD,PP:ASCBA (any supplied-air respirator that has a full facepiece and is operated in a pressure-demand or other positive-pressure mode in combination with an auxiliary self-contained breathing apparatus operated in a pressure-demand or other positive pressure mode). *ESCAPE:* HiEF (any air-purifying, full-facepiece respirator with a high-efficiency particulate filter); or SCBAE (any appropriate escape-type, self-contained breathing apparatus).
Exposure limits (TWA unless otherwise noted): ACGIH TLV 0.1 mg/m^3 as nickel, soluble compounds; OSHA PEL 1 mg/m^3; NIOSH REL 0.015 mg/m^3; reduce to lowest feasible level as nickel; potential human carcinogen.
Toxicity by ingestion: Grade 2; LD$_{50}$ = 0.5 to 5 g/kg.
Long-term health effects: Prolonged contact or inhalation of inorganic bromides causes acne-like bromoderma (bromide rash), especially of the hands and face. Other effects include emaciation, depression; and, in severe cases, psychosis and mental deterioration. May cause lung cancer.
Liquid or solid irritant characteristics: Eye and skin irritation.
IDLH value: Potential human carcinogen; 10 mg/m^3 as nickel.

CHEMICAL REACTIVITY
Binary reactants: Strong acids, sulfur, selenium

ENVIRONMENTAL DATA
Food chain concentration potential: Food chain concentration potential: May concentrate to a high degree in aquatic plants (103 to 104 mg/kg dry weight). National Research Council Canada, *Effects of Nickel in the Canadian Environment*, 1981, NRCC No.18568.
Water pollution: Effect of low concentrations on aquatic life is unknown, but of great environmental concern. May be dangerous if it enters nearby water intakes; notify operators. Notify local health and wildlife officials. **Response to discharge:** Issue warning–water contaminant. Disperse and flush.

SHIPPING INFORMATION
Grades of purity: Reagent; Technical; **Storage temperature:** Ambient; **Inert atmosphere:** None; **Venting:** Open; **Stability during transport:** Stable.

PHYSICAL AND CHEMICAL PROPERTIES
Physical state @ 59°F/15°C and 1 atm: Solid.
Molecular weight: 272.6
Boiling point @ 1 atm: Decomposes.
Specific gravity (water = 1): (estimate) 4 @ 68°F/20°C (solid).

NICKEL CARBONYL   REC. N:0800

SYNONYMS: EEC No. 028-001-00-1; NIQUEL CARBONILO (Spanish); NICKEL CARBONYLE (French); NICKEL TETRACARBONYL; NICKEL TETRACARBONYLE (French); TETARCARBONYL NICKEL

IDENTIFICATION
CAS Number: 13463-39-3
Formula: Ni(CO)$_4$
DOT ID Number: UN 1259; DOT Guide Number: 131
Proper Shipping Name: Nickel carbonyl
Reportable Quantity (RQ): 10 lb/4.54 kg

DESCRIPTION: Colorless to yellow liquid at ambient temperature. Musty, stale odor. Sinks in water; slightly soluble. Poisonous, flammable vapor is produced.

Highly Poisonous! • Highly flammable • Explosive: Explosive decomposition at 140°F/60°C. [May explode spontaneously when mixed with air at temperatures as low as 68°F/20°C, even without ignition source (FEMA)] • Breathing the vapor or swallowing the material can kill you; can be absorbed through the skin; releases carbon monoxide in the body; inhalation symptoms may be delayed • Firefighting gear (including SCBA) does not provide adequate protection. If exposure occurs, remove and isolate gear immediately and thoroughly decontaminate personnel • Containers may BLEVE when exposed to fire • Vapors are heavier than air and will collect and stay in low areas • Vapors may travel long distances to ignition sources and flashback • Vapors in confined areas (e.g., tanks, sewers, buildings) may explode when exposed to fire • Toxic products of combustion may include nickel and carbon monoxide • Do not attempt rescue.

Hazard Classification (based on NFPA-704 Rating System)
Health Hazards (Blue): 4; Flammability (Red): 3; Reactivity (Yellow): 3

EMERGENCY RESPONSE: See Appendix A (131)
Evacuation:
Public safety: Isolate spill area for at least 100 to 200 meters (330 to 660 feet) in all directions.
Spill: Small spill–First: Isolate in all directions 60 meters (200 feet); Then: Protect persons downwind, DAY: 0.6 km (0.4 mile); NIGHT: 2.1 km (1.3 miles). Large spill–First: Isolate in all directions 215 meters (700 feet); Then: Protect persons downwind, DAY: 2.1 km (1.3 miles); NIGHT: 4.3 km (2.7 miles).
Fire: Isolate for 800 meters (½ mile) in all directions, especially if tank, rail car, or tank truck is involved in fire.

EXPOSURE
Short-term effects: *SEEK MEDICAL ATTENTION.* **Vapor:** *POISONOUS IF INHALED.* Irritating to eyes, nose, and throat. Move victim to fresh air. If breathing is difficult, administer oxygen. May cause lung edema; physical exertion will aggravate this condition. **Liquid:** *POISONOUS IF SWALLOWED OR IF SKIN IS EXPOSED.*
Will burn skin and eyes. Remove contaminated clothing and shoes. Flush affected areas with plenty of water. *IF IN EYES*, hold eyelids open and flush with plenty of water. *IF SWALLOWED* and victim is *CONSCIOUS AND ABLE TO SWALLOW*, have victim drink 4 to 8 ounces of water. **Do NOT induce vomiting.**
Note to physician or authorized medical personnel: Medical observation is recommended for 24 to 48 hours after breathing overexposure, as pulmonary edema may be delayed. As first aid for

Nickel chloride

pulmonary edema, consider administering a corticosteroid spray. Cigarette smoking may exacerbate pulmonary injury and should be discouraged for at least 72 hours following exposure.

HEALTH HAZARDS
Personal protective equipment (PPE): A-Level PPE. OSHA Table Z-1-A air contaminant. Sealed chemical protective materials pffering limited protection: butyl rubber, chlorinated polyethylene, PVC.
Recommendations for respirator selection: NIOSH
At any detectable concentration: SCBAF:PD,PP (any MSHA/NIOSH approved self-contained breathing apparatus that has a full facepiece and is operated in a pressure-demand or other positive-pressure mode); or SAF:PD,PP:ASCBA (any supplied-air respirator that has a full facepiece and is operated in a pressure-demand or other positive-pressure mode in combination with an auxiliary, self-contained breathing apparatus operated in a pressure-demand or other positive pressure mode). *ESCAPE:* GMFS [any air-purifying, full-facepiece respirator (gas mask) with a chin-style, front- or back-mounted canister providing protection against the compound of concern/any appropriate escape-type, self-contained breathing apparatus]; or SCBAE (any appropriate escape-type, self-contained breathing apparatus).
Exposure limits (TWA unless otherwise noted): ACGIH TLV 0.05 mg/m^3 (0.12 mg/m^3); NIOSH REL 0.001 ppm (0.007 mg/m^3), potential human carcinogen as nickel; OSHA 0.001 ppm (0.007 mg/m^3).
Long-term health effects: Potential human carcinogen.
Vapor (gas) irritant characteristics: Vapors cause severe irritation of eyes and throat and can cause eye and lung injury. They cannot be tolerated even at low concentrations.
Liquid or solid irritant characteristics: Severe skin irritant. Causes second-and third-degree burns on short contact and is very injurious to the eyes.
Odor threshold: 0.049–3.1 ppm.
IDLH value: Potential human carcinogen; 2 ppm.

FIRE DATA
Flash point: Less than –4°F/–24°C (cc).
Flammable limits in air: LEL: 2%; UEL: 34%.
Behavior in fire: Containers may explode when heated.
Autoignition temperature: 140°F/60°C/333°K.
Burning rate: 2.7 mm/min

CHEMICAL REACTIVITY
Binary reactants: Self igniting in air. Reacts with nitric acid; explosively with bromine, chlorine and other oxidizers; flammable materials.

ENVIRONMENTAL DATA
Food chain concentration potential: May concentrate to a high degree in aquatic plants (103 to 104 mg/kg dry weight). Nat'l Research Council Canada, *Effects of Nickel in the Canadian Environment*, 1981, NRCC No 18568.
Water pollution: DOT Appendix B, §172.101- severe marine pollutant. Effect of low concentrations on aquatic life is unknown. May be dangerous if it enters nearby water intakes; notify operators. Notify local health and wildlife officials. **Response to discharge:** Issue warning–poison and high flammability, air contaminant, water contaminant. Restrict access. Evacuate area. Should be removed. Chemical and physical treatment.

SHIPPING INFORMATION
Grades of purity: 99.9+%; **Storage temperature:** Cool ambient; **Inert atmosphere:** Carbon monoxide at 15 psi; CO_2; **Venting:** Cylinders must be stored in a well-ventilated area.

NAS HAZARD CLASSIFICATION FOR BULK WATER TRANSPORTATION
FIRE: 3
HEALTH: Vapor irritant: 4; Liquid or solid irritant: 4; Poisons: 4
WATER POLLUTION: Human toxicity:; Aquatic toxicity: 3; Aesthetic effect: 1
REACTIVITY: Other chemicals: 4; Water: 0; Self-reaction: 4

PHYSICAL AND CHEMICAL PROPERTIES
Physical state @ 59°F/15°C and 1 atm: Liquid.
Molecular weight: 170.7
Boiling point @ 1 atm: 109°F/43°C/316°K.
Melting/Freezing point: –13°F/25°C/248°K.
Specific gravity (water = 1): 1.322 at 17°C (liquid).
Liquid surface tension: 15.9 dynes/cm = 0.0159 N/m @ 68°F/20°C.
Relative vapor density (air = 1): 5.9
Latent heat of vaporization: 72 Btu/lb = 40 cal/g = 1.7×10^5 J/kg.
Heat of combustion: –2970 Btu/lb = –1650 cal/g = -69.0×10^5 J/kg.
Heat of decomposition: At 60°C decomposes explosively.
Vapor pressure: 325 mm.

NICKEL CHLORIDE   REC. N:0850

SYNONYMS: CLORURO de NIQUEL (Spanish); NICKELOUS CHLORIDE; NICKEL(II) CHLORIDE; NICKEL CHLORIDE HEXAHYDRATE

IDENTIFICATION
CAS Number: 7718-54-9
Formula: Cl_2Ni; $NiCl_2 \cdot 6H_2O$
DOT ID Number: UN 9136; DOT Guide Number: 151
Proper Shipping Name: Nickel chloride
Reportable Quantity (RQ): **(CERCLA)** 100 lb/45.4 kg

DESCRIPTION: Yellow or brown scales. Odorless. Sinks and mixes with water; solution is acid.

Poison! • Firefighting gear (including SCBA) does not provide adequate protection. If exposure occurs, remove and isolate gear immediately and thoroughly decontaminate personnel • Containers may BLEVE when exposed to fire • Toxic products of combustion may include nickel, hydrogen chloride, and possibly nickel carbonyl.

Hazard Classification (based on NFPA-704 Rating System)
Health Hazards (Blue): 2; Flammability (Red): 0; Reactivity (Yellow): 0

EMERGENCY RESPONSE: See Appendix A (151)
Evacuation:
Public safety: Isolate the area of spill or leak for at least 25 to 50 meters (80 to 160 feet) in all directions.
Spill: Increase, in the downwind direction, as necessary, the distance shown under "Public Safety."
Fire: If tank, rail car, or tank truck is involved in fire, isolate for at least 800 meters (½ mile) in all directions; also, consider initial evacuation for 800 meters (½ mile) in all directions.

EXPOSURE

Short-term effects: *SEEK MEDICAL ATTENTION.* **Dust:** Irritating to eyes, nose, and throat. *IF INHALED*, will, will cause coughing or difficult breathing. *IF IN EYES*, hold eyelids open and flush with plenty of water. *IF BREATHING HAS STOPPED,* give artificial respiration; *avoid mouth-to-mouth resuscitation; use bag/mask apparatus.* IF breathing is difficult, administer oxygen. **Solid:** Irritating to skin and eyes. *IF SWALLOWED*, will cause nausea and vomiting. Remove contaminated clothing and shoes. Flush affected areas with plenty of water. *IF IN EYES*, hold eyelids open and flush with plenty of water. *IF SWALLOWED* and victim is *CONSCIOUS AND ABLE TO SWALLOW*, have victim drink 4 to 8 ounces of water. *IF SWALLOWED* and victim is *UNCONSCIOUS OR HAVING CONVULSIONS,* do nothing except keep victim warm.

HEALTH HAZARDS

Personal protective equipment (PPE): B-Level PPE. Goggles or face shield; protective gloves; Bu. Mines approved respirator; protective clothing.

Recommendations for respirator selection: NIOSH
At any concentrations above the NIOSH REL, or where there is no REL, at any detectable concentration: SCBAF:PD,PP (any self-contained breathing apparatus that has a full facepiece and is operated in a pressure-demand or other positive-pressure mode); or SAF:PD,PP:ASCBA (any supplied-air respirator that has a full facepiece and is operated in a pressure-demand or other positive-pressure mode in combination with an auxiliary self-contained breathing apparatus operated in a pressure-demand or other positive pressure mode). *ESCAPE:* HiEF (any air-purifying, full-facepiece respirator with a high-efficiency particulate filter); or SCBAE (any appropriate escape-type, self-contained breathing apparatus).

Exposure limits (TWA unless otherwise noted): ACGIH TLV 0.1 mg/m^3 as nickel, soluble compounds; OSHA PEL 1 mg/m^3; NIOSH REL 0.015 mg/m^3; reduce to lowest feasible level as nickel; potential human carcinogen.

Toxicity by ingestion: Grade 2; LD$_{50}$ = 0.5 to 5 g/kg.

Long-term health effects: Possible lung cancer. IARC listed; overall evaluation unassigned.

Liquid or solid irritant characteristics: Eye and skin irritation.

IDLH value: Potential human carcinogen; 10 mg/m^3 as nickel.

ENVIRONMENTAL DATA

Food chain concentration potential: May concentrate to a high degree in aquatic plants (103 to 104 mg/kg dry weight). Nat'l Research Council Canada, *Effects of Nickel in the Canadian Environment,* 1981, NRCC No.18568.

Water pollution: Harmful to aquatic life in very low concentrations. May be dangerous if it enters nearby water intakes; notify operators. Notify local health and wildlife officials.

Response to discharge: Disperse and flush.

SHIPPING INFORMATION

Grades of purity: Technical, 99+%; **Storage temperature:** Ambient; **Inert atmosphere:** None; **Venting:** Open; **Stability during transport:** Stable.

PHYSICAL AND CHEMICAL PROPERTIES

Physical state @ 59°F/15°C and 1 atm: Solid.
Molecular weight: 237.7
Specific gravity (water = 1): 3.55 @ 15°C (solid).
Heat of solution: 8.8 Btu/lb = 4.9 cal/g = 0.21 x 10^5 J/kg.
Heat of fusion: 142.5 cal/g.

NICKEL CYANIDE   REC. N:0900

SYNONYMS: CIANURO de NIQUEL (Spanish); NICKEL(II) CYANIDE; RCRA No. P074

IDENTIFICATION
CAS Number: 557-19-7
Formula: C$_2$N$_2$Ni; Ni(CN)$_{2 \cdot n}$H$_2$O (where n = 0,2,3,4)
DOT ID Number: UN 1653; DOT Guide Number: 151
Proper Shipping Name: Nickel cyanide
Reportable Quantity (RQ): **(CERCLA)** 10 lb/4.54 kg

DESCRIPTION: Yellow-brown solid (anhydrous); pale green solid (hydrate). Weak almond odor. Sinks in water; insoluble.

Poison! • Firefighting gear (including SCBA) does not provide adequate protection. If exposure occurs, remove and isolate gear immediately and thoroughly decontaminate personnel • Containers may BLEVE when exposed to fire • Toxic products of combustion may include cyanide.

Hazard Classification (based on NFPA-704 Rating System)
Health Hazards (Blue): 2; Flammability (Red): 0; Reactivity (Yellow): 0

EMERGENCY RESPONSE: See Appendix A (151)
Evacuation:
Public safety: Isolate the area of spill or leak for at least 25 to 50 meters (80 to 160 feet) in all directions.
Spill: Increase, in the downwind direction, as necessary, the distance shown under "Public Safety."
Fire: If tank, rail car, or tank truck is involved in fire, isolate for at least 800 meters (½ mile) in all directions; also, consider initial evacuation for 800 meters (½ mile) in all directions.

EXPOSURE

Short-term effects: *SEEK MEDICAL ATTENTION.* Dust:*POISONOUS IF INHALED.* Irritating to eyes, nose, and throat. Move victim to fresh air. *IF IN EYES*, hold eyelids open and flush with plenty of water. *IF BREATHING HAS STOPPED,* give artificial respiration; *avoid mouth-to-mouth resuscitation; use bag/mask apparatus.* IF breathing is difficult, administer oxygen. **Solid:** *POISONOUS IF SWALLOWED.* Irritating to skin and eyes. Remove contaminated clothing and shoes. Flush affected areas with plenty of water. *IF IN EYES*, hold eyelids open and flush with plenty of water. *IF SWALLOWED* and victim is *CONSCIOUS AND ABLE TO SWALLOW*, have victim drink 4 to 8 ounces of water and have victim induce vomiting. *IF SWALLOWED* and victim is *UNCONSCIOUS OR HAVING CONVULSIONS,* do nothing except keep victim warm.

Note to physician or authorized medical personnel: Consider the use of amyl nitrite perles if symptoms of cyanide poisoning develop. If symptoms indicate, initial treatment includes the cyanide antidote kit. In all cases, break an amyl nitrite perle in a gauze pad and hold lightly under victim's nose for 15 seconds, repeating 5 times at about 15-second intervals; if necessary (and if sodium nitrite infusions will be delayed), repeat procedure every 3 minutes with fresh pearls until 3 or 4 have been used. Avoid breathing the vapor while administering it to the victim. Administer sodium nitrite IV, ASAP. The usual adult dose is 10 to 20 mL of a 3% solution infused over no less than 5 minutes; the average child dose is 0.15 to 0.20 mL/kg. Monitor blood pressure during administration, and slow the rate of infusion if hypotention develops. Next, infuse sodium thiosulfate IV. The usual adult dose is 50 mL of a 25% solution infused over 10 to 20 minutes; the average child dose is 1.65 mL/kg. Repeat with nitrite and thiosulfate as required.

HEALTH HAZARDS
Personal protective equipment (PPE): B-Level PPE. Full-body protective clothing is advisable.
Recommendations for respirator selection: NIOSH
At any concentrations above the NIOSH REL, or where there is no REL, at any detectable concentration: SCBAF:PD,PP (any self-contained breathing apparatus that has a full facepiece and is operated in a pressure-demand or other positive-pressure mode); or SAF:PD,PP:ASCBA (any supplied-air respirator that has a full facepiece and is operated in a pressure-demand or other positive-pressure mode in combination with an auxiliary self-contained breathing apparatus operated in a pressure-demand or other positive pressure mode). *ESCAPE:* HiEF (any air-purifying, full-facepiece respirator with a high-efficiency particulate filter); or SCBAE (any appropriate escape-type, self-contained breathing apparatus).
Exposure limits (TWA unless otherwise noted): ACGIH TLV 0.1 mg/m^3; OSHA PEL 1 mg/m^3; NIOSH REL 0.015 mg/m^3; reduce to lowest feasible level as nickel; potential human carcinogen.
Toxicity by ingestion: Poisonous.
Long-term health effects: Potential carcinogen.
IDLH value: Potential human carcinogen; 10 mg/m^3 as nickel.

CHEMICAL REACTIVITY
Binary reactants: Strong acids, sulfur, selenium

ENVIRONMENTAL DATA
Food chain concentration potential: Unlikely to accumulate.
Water pollution: DOT Appendix B, §172.101- severe marine pollutant. Dangerous to aquatic life in high concentrations. May be dangerous if it enters nearby water intakes; notify operators. Notify local health and wildlife officials. **Response to discharge:** Issue warning–poison, water contaminant. Restrict access. Should be removed. Chemical and physical treatment.

SHIPPING INFORMATION
Grades of purity: Commercial; **Storage temperature:** Ambient; **Inert atmosphere:** None; **Venting:** Sealed containers in well-ventilated area; **Stability during transport:** Stable.

PHYSICAL AND CHEMICAL PROPERTIES
Physical state @ 59°F/15°C and 1 atm: Solid.
Specific gravity (water = 1): 2.4 @ 77°F/25°C (solid).

NICKEL FLUOROBORATE   REC. N:0950

SYNONYMS: NICKEL BOROFLUORIDE; NICKELOUS TETRAFLUOROBORATE; NICKEL (II) FLUOBORATE; NICKEL FLUOROBORATE SOLUTION; TL-1091

IDENTIFICATION
CAS Number: 14708-14-6
Formula: $B_2F_8 \cdot Ni$; $Ni(BF_4)_2 \cdot H_2O$
DOT ID Number: UN 3077; DOT Guide Number: 171
Proper Shipping Name: Environmentally hazardous substances, solid, n.o.s.

DESCRIPTION: Green liquid. Sink and mixes with water.

Poison! • Firefighting gear (including SCBA) does not provide adequate protection. If exposure occurs, remove and isolate gear immediately and thoroughly decontaminate personnel • Containers may BLEVE when exposed to fire • Toxic products of combustion may include fluorides, boron, nickel, and other toxic fumes.

Hazard Classification (based on NFPA-704 Rating System)
Health Hazards (Blue): 2; Flammability (Red): 0; Reactivity (Yellow): 0

EMERGENCY RESPONSE: See Appendix A (171)
Evacuation:
Public safety: Isolate the area of spill or leak for at least 10 to 25 meters (30 to 80 feet) in all directions.
Spill: Increase, in the downwind direction, as necessary, the distance shown under "Public Safety."
Fire: If any large container is involved in fire, isolate for at least 800 meters (½ mile) in all directions; also, consider initial evacuation for 800 meters (½ mile) in all directions.

EXPOSURE
Short-term effects: *SEEK MEDICAL ATTENTION*. **Vapor:** Irritating to eyes, nose, and throat. Harmful if inhaled. *IF IN EYES*, hold eyelids open and flush with plenty of water. *IF BREATHING HAS STOPPED,* give artificial respiration. IF breathing is difficult, administer oxygen. **Liquid:** Irritating to skin and eyes. *IF SWALLOWED*, will cause nausea and vomiting. Remove contaminated clothing and shoes. Flush affected areas with plenty of water. *IF IN EYES*, hold eyelids open and flush with plenty of water. *IF SWALLOWED* and victim is *CONSCIOUS AND ABLE TO SWALLOW*, have victim drink 4 to 8 ounces of water and have victim induce vomiting. *IF SWALLOWED* and victim is *UNCONSCIOUS OR HAVING CONVULSIONS*, do nothing except keep victim warm.

HEALTH HAZARDS
Personal protective equipment (PPE): B-Level PPE. Safety glasses and face shield; rubber gloves; rubber apron.
Recommendations for respirator selection: NIOSH
At any concentrations above the NIOSH REL, or where there is no REL, at any detectable concentration: SCBAF:PD,PP (any self-contained breathing apparatus that has a full facepiece and is operated in a pressure-demand or other positive-pressure mode); or SAF:PD,PP:ASCBA (any supplied-air respirator that has a full facepiece and is operated in a pressure-demand or other positive-pressure mode in combination with an auxiliary self-contained breathing apparatus operated in a pressure-demand or other positive pressure mode). *ESCAPE:* HiEF (any air-purifying, full-facepiece respirator with a high-efficiency particulate filter); or SCBAE (any appropriate escape-type, self-contained breathing apparatus).
Exposure limits (TWA unless otherwise noted): ACGIH TLV 0.1 mg/m^3; OSHA PEL 1 mg/m^3; NIOSH REL 0.015 mg/m^3; reduce to lowest feasible level as nickel; potential human carcinogen.
Toxicity by ingestion: Grade 2; LD_{50} = 0.5 to 5 g/kg.
Long-term health effects: Possible lung cancer.
Liquid or solid irritant characteristics: Eye and skin irritation.
IDLH value: Potential human carcinogen; 10 mg/m^3 as nickel.

ENVIRONMENTAL DATA
Food chain concentration potential: May concentrate to a high degree in aquatic plants (103 to 104 mg/kg dry weight). Nat'l Research Council Canada, *Effects of Nickel in the Canadian Environment*, 1981, NRCC No.18568.
Water pollution: Effect of low concentrations on aquatic life is unknown. May be dangerous if it enters nearby water intakes;

notify operators. Notify local health and wildlife officials.
Response to discharge: Issue warning–water contaminant. Disperse and flush.

SHIPPING INFORMATION
Grades of purity: 44.2-45.2% in water; **Storage temperature:** Ambient; **Inert atmosphere:** None; **Venting:** Open; **Stability during transport:** Stable.

PHYSICAL AND CHEMICAL PROPERTIES
Physical state @ 59°F/15°C and 1 atm: Liquid.
Molecular weight: Mixture
Specific gravity (water = 1): 1.5 @ 68°F/20°C (liquid).

NICKEL FORMATE   REC. N:1000

SYNONYMS: FORMIATO de NIQUEL (Spanish); NICKEL FORMATE DIHYDRATE

IDENTIFICATION
CAS Number: 15694-70-9
Formula: $Ni(HCO_2)_2 \cdot 2H_2O$; $(HCOO)_2Ni \cdot 2H_2O$
DOT ID Number: UN 3077; DOT Guide Number: 171
Proper Shipping Name: Environmentally hazardous substances, solid, n.o.s.

DESCRIPTION: Green crystalline solid. Odorless. Sinks and mixes with water.

Poison! • Firefighting gear (including SCBA) does not provide adequate protection. If exposure occurs, remove and isolate gear immediately and thoroughly decontaminate personnel • Containers may BLEVE when exposed to fire • Toxic products of combustion may include nickel and formic acid fumes.

Hazard Classification (based on NFPA-704 Rating System)
Health Hazards (Blue): 2; Flammability (Red): 0; Reactivity (Yellow): 0

EMERGENCY RESPONSE: See Appendix A (171)
Evacuation:
Public safety: Isolate the area of spill or leak for at least 10 to 25 meters (30 to 80 feet) in all directions.
Spill: Increase, in the downwind direction, as necessary, the distance shown under "Public Safety."
Fire: If any large container is involved in fire, isolate for at least 800 meters (½ mile) in all directions; also, consider initial evacuation for 800 meters (½ mile) in all directions.

EXPOSURE
Short-term effects: *SEEK MEDICAL ATTENTION.* **Dust:** Irritating to eyes, nose, and throat. *IF INHALED,* will, will cause coughing or difficult breathing. *IF IN EYES,* hold eyelids open and flush with plenty of water. *IF BREATHING HAS STOPPED,* give artificial respiration. IF breathing is difficult, administer oxygen.
Solid: Irritating to skin and eyes. *IF SWALLOWED,* will cause nausea and vomiting. Remove contaminated clothing and shoes. Flush affected areas with plenty of water. *IF IN EYES,* hold eyelids open and flush with plenty of water. *IF SWALLOWED* and victim is *CONSCIOUS AND ABLE TO SWALLOW,* have victim drink 4 to 8 ounces of water. *IF SWALLOWED* and victim is *UNCONSCIOUS OR HAVING CONVULSIONS,* do nothing except keep victim warm.

HEALTH HAZARDS
Personal protective equipment (PPE): B-Level PPE. OSHA Table Z-1-A air contaminant. Bureau of Mines approved respirator; rubber gloves; face shield or safety goggles; full-body protective clothing
Recommendations for respirator selection: NIOSH
At any concentrations above the NIOSH REL, or where there is no REL, at any detectable concentration: SCBAF:PD,PP (any self-contained breathing apparatus that has a full facepiece and is operated in a pressure-demand or other positive-pressure mode); or SAF:PD,PP:ASCBA (any supplied-air respirator that has a full facepiece and is operated in a pressure-demand or other positive-pressure mode in combination with an auxiliary self-contained breathing apparatus operated in a pressure-demand or other positive pressure mode). *ESCAPE:* HiEF (any air-purifying, full-facepiece respirator with a high-efficiency particulate filter); or SCBAE (any appropriate escape-type, self-contained breathing apparatus).
Exposure limits (TWA unless otherwise noted): ACGIH TLV 0.1 mg/m^3; OSHA PEL 1 mg/m^3; NIOSH REL 0.015 mg/m^3; reduce to lowest feasible level as nickel; potential human carcinogen.
Toxicity by ingestion: Grade 2; LD_{50} = 0.5 to 5 g/kg.
Long-term health effects: Possible lung cancer
Liquid or solid irritant characteristics: Eye and skin irritation.
IDLH value: Potential human carcinogen; 10 mg/m^3 as nickel.

ENVIRONMENTAL DATA
Food chain concentration potential: May concentrate to a high degree in aquatic plants (103 to 104 mg/kg dry weight). Nat'l Research Council Canada, *Effects of Nickel in the Canadian Environment*, 1981, NRCC No.18568.
Water pollution: Effect of low concentrations on aquatic life is unknown. May be dangerous if it enters nearby water intakes; notify operators. Notify local health and wildlife officials. **Response to discharge:** Disperse and flush.

SHIPPING INFORMATION
Grades of purity: Commercial; **Storage temperature:** Ambient; **Inert atmosphere:** None; **Venting:** Open; **Stability during transport:** Stable.

PHYSICAL AND CHEMICAL PROPERTIES
Physical state @ 59°F/15°C and 1 atm: Solid.
Molecular weight: 184.8
Boiling point @ 1 atm: Decomposes.
Specific gravity (water = 1): 2.15 @ 68°F/20°C (solid).

NICKEL HYDROXIDE   REC. N:1050

SYNONYMS: GREEN NICKEL OXIDE; HIDROXIDO NIQUEL (Spanish); NICKELOUS HYDROXIDE; NICKEL DIHYDROXIDE; NICKEL(II) HYDROXIDE

IDENTIFICATION
CAS Number: 12054-48-7; 12125-56-3
Formula: H_2NiO_2; $Ni(OH)_2 \cdot H_2O$
DOT ID Number: UN 9140; DOT Guide Number: 154
Proper Shipping Name: Nickel hydroxide
Reportable Quantity (RQ): **(CERCLA)** 10 lb/4.54 kg

DESCRIPTION: Apple-green crystalline solid or powder. Sinks in water.

Poison! • Combustible; ignites spontaneously in air at approximately 752°F/400°C • Firefighting gear (including SCBA)

does not provide adequate protection. If exposure occurs, remove and isolate gear immediately and thoroughly decontaminate personnel • Containers may BLEVE when exposed to fire • Toxic products of combustion may include nickel fumes.

Hazard Classification (based on NFPA-704 Rating System)
Health Hazards (Blue): 2; Flammability (Red): 1; Reactivity (Yellow): 0

EMERGENCY RESPONSE: See Appendix A (154)
Evacuation:
Public safety: Isolate the area of spill or leak for at least 25 to 50 meters (80 to 160 feet) in all directions.
Spill: Increase, in the downwind direction, as necessary, the distance shown under "Public Safety."
Fire: If tank, rail car, or tank truck is involved in fire, isolate for at least 800 meters (½ mile) in all directions; also, consider initial evacuation for 800 meters (½ mile) in all directions.

EXPOSURE
Short-term effects: *SEEK MEDICAL ATTENTION.* **Dust or Solid:** Irritating to eyes, nose, and throat. Harmful if swallowed. Move to fresh air. Flush affected areas with plenty of water. *IF BREATHING HAS STOPPED, give artificial respiration; avoid mouth-to-mouth resuscitation; use bag/mask apparatus. IF IN EYES, hold eyelids open and flush with plenty of water. IF SWALLOWED and victim is CONSCIOUS AND ABLE TO SWALLOW,* have victim drink 4 to 8 ounces of water and have victim induce vomiting.

HEALTH HAZARDS
Personal protective equipment (PPE): B-Level PPE. NIOSH-approved respiratory protection, side shield safety glasses or goggles, and rubber gloves.
Recommendations for respirator selection: NIOSH
At any concentrations above the NIOSH REL, or where there is no REL, at any detectable concentration: SCBAF:PD,PP (any self-contained breathing apparatus that has a full facepiece and is operated in a pressure-demand or other positive-pressure mode); or SAF:PD,PP:ASCBA (any supplied-air respirator that has a full facepiece and is operated in a pressure-demand or other positive-pressure mode in combination with an auxiliary self-contained breathing apparatus operated in a pressure-demand or other positive pressure mode). *ESCAPE:* HiEF (any air-purifying, full-facepiece respirator with a high-efficiency particulate filter); or SCBAE (any appropriate escape-type, self-contained breathing apparatus).
Exposure limits (TWA unless otherwise noted): ACGIH TLV 0.2 mg/m^3 as nickel, confirmed human carcinogen; OSHA PEL 1 mg/m^3; NIOSH REL 0.015 mg/m^3; reduce to lowest feasible level as nickel; potential human carcinogen.
Toxicity by ingestion: Grade 2; LD_{50} = 0.5 to 5 g/kg (rat).
Long-term health effects: Carcinogen of nasal cavity, paranasal sinuses, and lungs.
Liquid or solid irritant characteristics: Eye and skin irritation.
IDLH value: Potential human carcinogen; 10 mg/m^3 as nickel.

FIRE DATA
Flash point: Combustible solid.
Behavior in fire: Converts to black nickelic oxide (Ni_2O_3).
Autoignition temperature: 752°F/400°C/773°K.

ENVIRONMENTAL DATA
Food chain concentration potential: May concentrate to a high degree in aquatic plants (103 to 104 mg/kg dry weight). Nat'l Research Council Canada, *Effects of Nickel in the Canadian Environment*, 1981, NRCC No.18568.
Water pollution: Harmful to aquatic life in very low concentrations. May be dangerous if it enters nearby water intakes; notify operators. Notify local health and wildlife officials.
Response to discharge: Issue warning–water contaminant. Should be removed. Chemical and physical treatment.

SHIPPING INFORMATION
Storage temperature: Ambient; **Stability during transport:** Stable.

PHYSICAL AND CHEMICAL PROPERTIES
Physical state @ 59°F/15°C and 1 atm: Solid.
Molecular weight: 92.72
Boiling point @ 1 atm: Decomposes.
Melting/Freezing point: 446°F/230°C/503°K.
Specific gravity (water = 1): 4.1 at room temperature
Relative vapor density (air = 1): 3.2 (calculated).

NICKEL(II) NITRATE, HEXAHYDRATE   REC. N:1100

SYNONYMS: NICKEL NITRATE HEXAHYDRATE; NICKEL(2+) NITRATE, HEXAHYDRATE; NITRATO de NIQUEL (Spanish); NITRIC ACID, NICKEL(2+) SALT, HEXAHYDRATE; NITRIC ACID, NICKEL(II) SALT, HEXAHYDRATE

IDENTIFICATION
CAS Number: 13478-00-7; 14216-75-2; 13138-45-9
Formula: $N_2O_6 \cdot Ni$; $Ni(NO_3)_2 \cdot 6H_2O$ (hexahydrate)
DOT ID Number: UN 2725; DOT Guide Number: 140
Proper Shipping Name: Nickel nitrate

DESCRIPTION: Yellow to pale green crystalline solid. Odorless. Sinks and mixes with water; aqueous solution is acidic.

Strong oxidizer which may react spontaneously with low-flash-point organics or reducing agents. Heat forms oxygen; will increase the activity of an existing fire and prolonged exposure to fire or heat may result in an explosion • May cause fire or explosion on contact with combustibles (wood, paper, oil, clothing, etc.) • Irritating to the skin, eyes, and respiratory tract • Toxic products of combustion may include nitrogen oxides and nickel fumes (possibly nickel carbonyl).

Hazard Classification (based on NFPA-704 Rating System)
Health Hazards (Blue): 2; Flammability (Red): 0; Reactivity (Yellow): 0; Special Notice (White): OXY

EMERGENCY RESPONSE: See Appendix A (140)
Evacuation:
Public safety: Isolate the area of spill or leak for at least 10 to 25 meters (30 to 80 feet) in all directions.
Spill: Consider initial downwind evacuation for at least 100 meters (330 feet).
Fire: If any large container is involved in fire, isolate for at least 800 meters (½ mile) in all directions; also, consider initial evacuation for 800 meters (½ mile) in all directions.

EXPOSURE
Short-term effects: *SEEK MEDICAL ATTENTION.* **Dust:** Irritating to eyes, nose, and throat. *IF INHALED,* will, will cause coughing or difficult breathing. *IF IN EYES,* hold eyelids open and

flush with plenty of water. *IF BREATHING HAS STOPPED*, give artificial respiration. IF breathing is difficult, administer oxygen.
Solid: Irritating to skin and eyes. *IF SWALLOWED*, will cause nausea and vomiting. Remove contaminated clothing and shoes. Flush affected areas with plenty of water. *IF IN EYES*, hold eyelids open and flush with plenty of water. *IF SWALLOWED* and victim is *CONSCIOUS AND ABLE TO SWALLOW*, have victim drink 4 to 8 ounces of water. *IF SWALLOWED* and victim is *UNCONSCIOUS OR HAVING CONVULSIONS*, do nothing except keep victim warm.

HEALTH HAZARDS
Personal protective equipment (PPE): B-Level PPE. OSHA Table Z-1-A air contaminant. Bureau of Mines approved respirator; rubber gloves; face shield or safety goggles; protective clothing
Recommendations for respirator selection: NIOSH
At any concentrations above the NIOSH REL, or where there is no REL, at any detectable concentration: SCBAF:PD,PP (any self-contained breathing apparatus that has a full facepiece and is operated in a pressure-demand or other positive-pressure mode); or SAF:PD,PP:ASCBA (any supplied-air respirator that has a full facepiece and is operated in a pressure-demand or other positive-pressure mode in combination with an auxiliary self-contained breathing apparatus operated in a pressure-demand or other positive pressure mode). *ESCAPE:* HiEF (any air-purifying, full-facepiece respirator with a high-efficiency particulate filter); or SCBAE (any appropriate escape-type, self-contained breathing apparatus).
Exposure limits (TWA unless otherwise noted): ACGIH TLV 0.1 mg/m^3; OSHA PEL 1 mg/m^3; NIOSH REL 0.015 mg/m^3; reduce to lowest feasible level as nickel; potential human carcinogen.
Toxicity by ingestion: Grade 2; LD$_{50}$ = 0.5 to 5 g/kg.
Long-term health effects: Possible lung cancer.
Liquid or solid irritant characteristics: Eye and skin irritation.
IDLH value: Potential human carcinogen; 10 mg/m^3 as nickel.

FIRE DATA
Behavior in fire: May increase intensity of fire if in contact with combustible material.

CHEMICAL REACTIVITY; forms acidic solution.
Binary reactants: Reacts with reducing agents, acids. Contact of solid with wood or paper may cause fires

ENVIRONMENTAL DATA
Food chain concentration potential: May concentrate to a high degree in aquatic plants (103 to 104 mg/kg dry weight). Nat'l Research Council Canada, *Effects of Nickel in the Canadian Environment*, 1981, NRCC No.18568.
Water pollution: Effect of low concentrations on aquatic life is unknown. May be dangerous if it enters nearby water intakes; notify operators. Notify local health and wildlife officials.
Response to discharge: Disperse and flush.

SHIPPING INFORMATION
Grades of purity: Purified, 99.1%; **Storage temperature:** Ambient; **Inert atmosphere:** None; **Venting:** None; **Stability during transport:** Stable.

PHYSICAL AND CHEMICAL PROPERTIES
Physical state @ 59°F/15°C and 1 atm: Solid.
Molecular weight: 290.8
Boiling point @ 1 atm: Decomposes; 279°F/137°C/410.2°K.
Melting Point: 134°F/57°C/330°K.
Specific gravity (water = 1): 2.05 @ 68°F/20°C (solid).
Heat of solution: 47 Btu/lb = 26 cal/g = 1.1 x 10^5 J/kg.

NICKEL SULFATE   REC. N:1150

SYNONYMS: NICKEL(II) SULFATE; NICKEL(2+) SULFATE; NICKELOUS SULFATE; SULFURIC ACID, NICKEL(2+) SALT; SULFURIC ACID, NICKEL(II) SALT; SULFATO de NIQUEL (Spanish)

IDENTIFICATION
CAS Number: 7786-81-4; 10101-98-1 (heptahydrate)
Formula: O$_4$S·Ni; NiSO$_4$; O$_4$NiS·7H$_2$O (heptahydrate)
DOT ID Number: UN 9141; DOT Guide Number: 154
Proper Shipping Name: Nickel sulfate; Nickel sulphate
Reportable Quantity (RQ): **(CERCLA)** 100 lb/45.4 kg

DESCRIPTION: Pale green crystalline solid; yellow in anhydrous state. Odorless. Sinks and mixes slowly with water.

Poison! • Firefighting gear (including SCBA) does not provide adequate protection. If exposure occurs, remove and isolate gear immediately and thoroughly decontaminate personnel • Containers may BLEVE when exposed to fire • Toxic products of combustion may include sulfur oxides and nickel fumes (possibly nickel carbonyl).

Hazard Classification (based on NFPA-704 Rating System)
Health Hazards (Blue): 2; Flammability (Red): 0; Reactivity (Yellow): 0

EMERGENCY RESPONSE: See Appendix A (154)
Evacuation:
Public safety: Isolate the area of spill or leak for at least 25 to 50 meters (80 to 160 feet) in all directions.
Spill: Increase, in the downwind direction, as necessary, the distance shown under "Public Safety."
Fire: If tank, rail car, or tank truck is involved in fire, isolate for at least 800 meters (½ mile) in all directions; also, consider initial evacuation for 800 meters (½ mile) in all directions.

EXPOSURE
Short-term effects: *SEEK MEDICAL ATTENTION*. **Solid:** Irritating to skin and eyes. Remove contaminated clothing and shoes. Flush affected areas with plenty of water. *IF BREATHING HAS STOPPED*, give artificial respiration; *avoid mouth-to-mouth resuscitation; use bag/mask apparatus*. *IF IN EYES*, hold eyelids open and flush with plenty of water.

HEALTH HAZARDS
Personal protective equipment (PPE): B-Level PPE. Full protective clothing.
Recommendations for respirator selection: NIOSH
At any concentrations above the NIOSH REL, or where there is no REL, at any detectable concentration: SCBAF:PD,PP (any self-contained breathing apparatus that has a full facepiece and is operated in a pressure-demand or other positive-pressure mode); or SAF:PD,PP:ASCBA (any supplied-air respirator that has a full facepiece and is operated in a pressure-demand or other positive-pressure mode in combination with an auxiliary self-contained breathing apparatus operated in a pressure-demand or other positive pressure mode). *ESCAPE:* HiEF (any air-purifying, full-facepiece respirator with a high-efficiency particulate filter); or SCBAE (any

806 Nicotine

appropriate escape-type, self-contained breathing apparatus).
Exposure limits (TWA unless otherwise noted): ACGIH TLV 0.1 mg/m³; OSHA PEL 1 mg/m³; NIOSH REL 0.015 mg/m³; reduce to lowest feasible level as nickel; potential human carcinogen.
Long-term health effects: Nose and lung cancer. IARC listed; overall evaluation unassigned.
Liquid or solid irritant characteristics: Repeated contact can cause dermatitis.
IDLH value: Potential human carcinogen; 10 mg/m³ as nickel.

CHEMICAL REACTIVITY
Binary reactants: Violent reaction with aluminum, magnesium

ENVIRONMENTAL DATA
Food chain concentration potential: May concentrate to a high degree in aquatic plants (103 to 104 mg/kg dry weight). Nat'l Research Council Canada, *Effects of Nickel in the Canadian Environment*, 1981, NRCC No.18568.
Water pollution: Dangerous to aquatic life in high concentrations. May be dangerous if it enters nearby water intakes; notify operators. Notify local health and wildlife officials. **Response to discharge:** Disperse and flush.

SHIPPING INFORMATION
Grades of purity: Technical; reagent; **Storage temperature:** Ambient; **Inert atmosphere:** None; **Venting:** Open; **Stability during transport:** Stable.

PHYSICAL AND CHEMICAL PROPERTIES
Physical state @ 59°F/15°C and 1 atm: Solid.
Molecular weight: 154.78
Boiling point @ 1 atm: Decomposes. 1544°F/840°C/1113°K.
Specific gravity (water = 1): 3.68 @ 68°F/20°C (solid).

NICOTINE   **REC. N:1200**

SYNONYMS: EEC No. 614-001-00-4; EMO-NIK; FLUX MAAY; MACH-NIC; NICODUST; NICOFUME; NICOCIDE; NICOTINA (Spanish); PYRIDINE, 3-(1-METHYL-2-PYRROLIDINYL)-; PYRIDINE, (S)-3-(1-METHYL-2-PYRROLIDINYL)-AND SALTS; 3-(1-METHYL-2-PYRROLIDYL) PYRIDINE; TENDUST; XL ALL INSECTICIDE; BLACK LEAF; 1-METHYL-2-(3-PYRIDYL)PYRROLIDINE; 3-(1-METHYL-2-PYRROLIDYL)PYRIDINE; RCRA No. P075

IDENTIFICATION
CAS Number: 54-11-5
Formula: $C_{10}H_{14}N_2$; $C_5H_4NC_4H_7NCH_3$
DOT ID Number: UN 1654; DOT Guide Number: 151
Proper Shipping Name: Nicotine
Reportable Quantity (RQ): **(CERCLA)** 100 lb/45.4 kg

DESCRIPTION: Colorless liquid; turns brown with age. Fishy odor; develops a pyridine or tobacco-like odor on exposure to air. Mixes with water.

Poison! • Combustible • Breathing the vapor, skin or eye contact, or swallowing the material can kill you • Firefighting gear (including SCBA) does not provide adequate protection. If exposure occurs, remove and isolate gear immediately and thoroughly decontaminate personnel • Heat or flame may cause explosion • Containers may BLEVE when exposed to fire • Vapors in confined areas (e.g., tanks, sewers, buildings) may explode when exposed to fire • Irritating to the skin, eyes, and respiratory tract • Toxic products of combustion may include nitrogen oxides, carbon monoxide, and vapors of unburned compound.

Hazard Classification (based on NFPA-704 Rating System)
Health Hazards (Blue): 4; Flammability (Red): 1; Reactivity (Yellow): 0

EMERGENCY RESPONSE: See Appendix A (151)
Evacuation:
Public safety: Isolate the area of spill or leak for at least 25 to 50 meters (80 to 160 feet) in all directions.
Spill: Increase, in the downwind direction, as necessary, the distance shown under "Public Safety."
Fire: If tank, rail car, or tank truck is involved in fire, isolate for at least 800 meters (½ mile) in all directions; also, consider initial evacuation for 800 meters (½ mile) in all directions.

EXPOSURE
Short-term effects: *SEEK MEDICAL ATTENTION.* **Liquid:** *POISONOUS IF SWALLOWED OR IF SKIN IS EXPOSED.* Remove contaminated clothing and shoes. Flush affected areas with plenty of water. *IF BREATHING HAS STOPPED,* give artificial respiration; *avoid mouth-to-mouth resuscitation; use bag/mask apparatus. IF IN EYES,* hold eyelids open and flush with plenty of water. *IF SWALLOWED* and victim is *CONSCIOUS AND ABLE TO SWALLOW,* have victim drink 4 to 8 ounces of water and have victim induce vomiting. *IF SWALLOWED* and victim is *UNCONSCIOUS OR HAVING CONVULSIONS,* do nothing except keep victim warm.

HEALTH HAZARDS
Personal protective equipment (PPE): A-Level PPE. OSHA Table Z-1-A air contaminant. Organic vapor respirator. Chemical protective material(s) reported to have good to excellent resistance: Silvershield®.
Recommendations for respirator selection: NIOSH/OSHA
5 mg/m³: SA (any supplied-air respirator); or SCBAF (any self-contained breathing apparatus with a full facepiece). *EMERGENCY OR PLANNED ENTRY INTO UNKNOWN CONCENTRATIONS OR IDLH CONDITIONS:* SCBAF:PD,PP (any self-contained breathing apparatus that has a full facepiece and is operated in a pressure-demand or other positive-pressure mode); or SAF:PD,PP:ASCBA (any supplied-air respirator that has a full facepiece and is operated in a pressure-demand or other positive-pressure mode in combination with an auxiliary self-contained breathing apparatus operated in a pressure-demand or other positive pressure mode). *ESCAPE:* GMFOV [any air-purifying, full-facepiece respirator (gas mask) with a chin-style, front- or back-mounted acid gas canister]; or SCBAE (any appropriate escape-type, self-contained breathing apparatus).
Exposure limits (TWA unless otherwise noted): ACGIH TLV 0.5 mg/m³; OSHA PEL 0.5 mg/m³; skin contact contributes significantly in overall exposure.
Toxicity by ingestion: Grade 4; oral LD_{50} = 53 mg/kg (rat), 1 mg/kg (human).
Long-term health effects: Birth defects (skeletal) in rats. Possible skin disorders.
Vapor (gas) irritant characteristics: Vapors are nonirritating to eyes and throat.
Liquid or solid irritant characteristics: May cause eye, skin and respiratory tract irritation.
IDLH value: 5 mg/m³.

FIRE DATA
Flash point: 203°F/95°C.
Flammable limits in air: LEL: 0.7%; UEL: 4.0%.
Fire extinguishing agents not to be used: Water or foam may cause frothing.
Autoignition temperature: 471°F/244°C/517°K.

CHEMICAL REACTIVITY
Binary reactants: Incompatible with strong acids, strong oxidizers. Mixtures with mineral acids forms salts. Attacks some plastics, rubber, and coatings.

ENVIRONMENTAL DATA
Food chain concentration potential: Log P_{ow} = 1.2. Unlikely to accumulate.
Water pollution: Harmful to aquatic life in very low concentrations. May be dangerous if it enters nearby water intakes; notify operators. Notify local health and wildlife officials.
Response to discharge: Issue warning–poison, water contaminant. Restrict access. Disperse and flush.

SHIPPING INFORMATION
Grades of purity: 93–98%; **Storage temperature:** Ambient; **Inert atmosphere:** None; **Venting:** Pressure-vacuum; **Stability during transport:** Stable.

NAS HAZARD CLASSIFICATION FOR BULK WATER TRANSPORTATION
FIRE: 1
HEALTH: Vapor irritant: 1; Liquid or solid irritant: 1; Poisons: 4
WATER POLLUTION: Human toxicity:; Aquatic toxicity: 3; Aesthetic effect: 4
REACTIVITY: Other chemicals: 3; Water: 0; Self-reaction: 0

PHYSICAL AND CHEMICAL PROPERTIES
Physical state @ 59°F/15°C and 1 atm: Liquid.
Molecular weight: 162.2
Boiling point @ 1 atm: Decomposes. 482°F/250°C/523°K.
Melting/Freezing point: –110°F/–79°C/192°K.
Specific gravity (water = 1): 1.016 @ 68°F/20°C (liquid).
Liquid surface tension: 38.61 dynes/cm = 0.03861 N/m @ 68°F/20°C.
Liquid water interfacial tension: (estimate) 20 dynes/cm = 0.020 N/m @ 68°F/20°C.
Relative vapor density (air = 1): 5.6
Heat of combustion: –15,836 Btu/lb = –8798 cal/g = –368.1 x 10^5 J/kg.
Vapor pressure: 0.08 mm.

NICOTINE SULFATE   REC. N:1250

SYNONYMS: 3-(1-METHYL-2-PYRROLIDYL)PYRIDINE; BLACK LEAF 40 (40% WATER SOLUTION); SULFATO de NICOTINA (Spanish)

IDENTIFICATION
CAS Number: 65-30-5
Formula: $C_{20}H_{26}N_4 \cdot O_4S$; $(C_{10}H_{14}N_2)_2 \cdot H_2SO_4$
DOT ID Number: UN 1658; DOT Guide Number: 151
Proper Shipping Name: Nicotine sulfate, solid; Nicotine sulfate, solution
Reportable Quantity (RQ): **(EHS)** 1 lb/0.454 kg

DESCRIPTION: White solid, hexagonal tablets or colorless water solution. Darkens on exposure to light. Odorless solid that develops a pyridine odor on standing; or, a solution with a tobacco odor. Mixes with water.

Poison! • Combustible • Toxic by all routes; breathing the material, skin or eye contact, or swallowing the material can cause illness and possible death; irritates the eyes and skin • Firefighting gear (including SCBA) does not provide adequate protection. If exposure occurs, remove and isolate gear immediately and thoroughly decontaminate personnel • Heat or flame may cause explosion • Containers may BLEVE when exposed to fire • Vapors in confined areas (e.g., tanks, sewers, buildings) may explode when exposed to fire • Toxic products of combustion may include carbon monoxide, sulfur oxide, nicotine, and nitrogen oxides.

Hazard Classification (based on NFPA-704 Rating System)
Health Hazards (Blue): 4; Flammability (Red): 1; Reactivity (Yellow): 0

EMERGENCY RESPONSE: See Appendix A (151)
Evacuation:
Public safety: Isolate the area of spill or leak for at least 25 to 50 meters (80 to 160 feet) in all directions.
Spill: Increase, in the downwind direction, as necessary, the distance shown under "Public Safety."
Fire: If tank, rail car, or tank truck is involved in fire, isolate for at least 800 meters (½ mile) in all directions; also, consider initial evacuation for 800 meters (½ mile) in all directions.

EXPOSURE
Short-term effects: *SEEK MEDICAL ATTENTION.* VAPOR OR **Dust:** *POISONOUS IF INHALED.* Move victim to fresh air. *IF BREATHING HAS STOPPED,* give artificial respiration; *avoid mouth-to-mouth resuscitation; use bag/mask apparatus. IF IN EYES,* hold eyelids open and flush with plenty of water. *IF BREATHING HAS STOPPED,* give artificial respiration. IF breathing is difficult, administer oxygen. **Liquid or solid:** *POISONOUS IF SWALLOWED OR IF SKIN IS EXPOSED.* Irritating to eyes. Remove contaminated clothing and shoes. Flush affected areas with plenty of water. *IF IN EYES,* hold eyelids open and flush with plenty of water. *IF SWALLOWED* and victim is *CONSCIOUS AND ABLE TO SWALLOW,* have victim drink 4 to 8 ounces of water and have victim induce vomiting. *IF SWALLOWED* and victim is *UNCONSCIOUS OR HAVING CONVULSIONS,* do nothing except keep victim warm.

HEALTH HAZARDS
Personal protective equipment (PPE): B-Level PPE. Dust mask; rubber gloves.
Recommendations for respirator selection: NIOSH/OSHA as nicotine (for reference)
5 mg/m³: SA (any supplied-air respirator); or SCBAF (any self-contained breathing apparatus with a full facepiece). *EMERGENCY OR PLANNED ENTRY INTO UNKNOWN CONCENTRATIONS OR IDLH CONDITIONS:* SCBAF:PD,PP (any self-contained breathing apparatus that has a full facepiece and is operated in a pressure-demand or other positive-pressure mode); or SAF:PD,PP:ASCBA (any supplied-air respirator that has a full facepiece and is operated in a pressure-demand or other positive-pressure mode in combination with an auxiliary self-contained breathing apparatus operated in a pressure-demand or other positive pressure mode). *ESCAPE:* GMFOV [any air-purifying, full-facepiece respirator (gas mask) with a chin-style, front- or back-mounted acid gas canister]; or SCBAE (any appropriate

escape-type, self-contained breathing apparatus). *Note*: Substance reported to cause eye irritation or damage; may require eye protection.
Toxicity by ingestion: Grade 3; oral LD_{50} = 55 mg/kg (rat).
Liquid or solid irritant characteristics: Eye and skin irritation.
IDLH value: 5 mg/m^3 as nicotine

FIRE DATA
Flash point: Nonflammable as solid or water solution.

CHEMICAL REACTIVITY
Binary reactants: Violent reaction with aluminum, magnesium

ENVIRONMENTAL DATA
Food chain concentration potential: Negative; unlikely to accumulate.
Water pollution: Harmful to aquatic life in very low concentrations. May be dangerous if it enters nearby water intakes; notify operators. Notify local health and wildlife officials.
Response to discharge: Issue warning–poison, water contaminant. Restrict access. Disperse and flush.

SHIPPING INFORMATION
Grades of purity: Solid: Commercial; Solution: 40%; **Storage temperature:** Ambient; **Inert atmosphere:** None; **Venting:** Open; **Stability during transport:** Stable.

PHYSICAL AND CHEMICAL PROPERTIES
Physical state @ 59°F/15°C and 1 atm: Solid.
Molecular weight: 422.5
Specific gravity (water = 1): 1.15 @ 68°F/20°C (solid).

NITRALIN   **REC. N:1300**

SYNONYMS: 4-(METHYLSULFONYL)-2,6-DINITRO; *N, N*-DIPROPYLANILINE; PLANAVIN

IDENTIFICATION
CAS Number: 472-61-4
Formula: $C_{13}H_{19}N_3O_6S$; $HNO_3 \cdot H_2O$
DOT ID Number: UN 2811; DOT Guide Number: 154
Proper Shipping Name: Poisonous solids, n.o.s.

DESCRIPTION: Light yellow to orange solid. Mild chemical odor. Sinks in water.

Poison! • Irritates the eyes, skin, and respiratory tract; toxic if inhaled or absorbed through the skin • Firefighting gear (including SCBA) does not provide adequate protection. If exposure occurs, remove and isolate gear immediately and thoroughly decontaminate personnel • Containers may BLEVE when exposed to fire • Decomposes vigorously in a self-sustaining reaction at or above 437°F/225°C. Toxic products of combustion may include nicotine and oxides of sulfur and nitrogen • Do not put yourself in danger by entering a contaminated area to rescue a victim.

Hazard Classification (based on NFPA-704 Rating System)
Health Hazards (Blue): 1; Flammability (Red): 1; Reactivity (Yellow): 0

EMERGENCY RESPONSE: See Appendix A (154)
Evacuation:
Public safety: Isolate the area of spill or leak for at least 25 to 50 meters (80 to 160 feet) in all directions.

Spill: Increase, in the downwind direction, as necessary, the distance shown under "Public Safety."
Fire: If tank, rail car, or tank truck is involved in fire, isolate for at least 800 meters (½ mile) in all directions; also, consider initial evacuation for 800 meters (½ mile) in all directions.

EXPOSURE
Short-term effects: *SEEK MEDICAL ATTENTION*. **Dust:** *POISONOUS IF INHALED OR ABSORBED THROUGH ATHE SKIN*. Move victim to fresh air. *IF BREATHING HAS STOPPED*, give artificial respiration; *avoid mouth-to-mouth resuscitation; use bag/mask apparatus*. If breathing is difficult, administer oxygen. **Solid:** *POISONOUS IF SWALLOWED*. Irritating to skin and eyes. Remove contaminated clothing and shoes. Flush affected areas with plenty of water. *IF IN EYES*, hold eyelids open and flush with plenty of water. *IF SWALLOWED* and victim is *CONSCIOUS AND ABLE TO SWALLOW*, have victim drink 4 to 8 ounces of water and have victim induce vomiting. *IF SWALLOWED* and victim is *UNCONSCIOUS OR HAVING CONVULSIONS*, do nothing except keep victim warm.

HEALTH HAZARDS
Personal protective equipment (PPE): B-Level PPE.
Toxicity by ingestion: Grade 2; oral LD_{50} more than 2000 mg/kg (rat).
Liquid or solid irritant characteristics: Eye irritation.

FIRE DATA
Flash point: Combustible solid.
Autoignition temperature: 435°F/224°C/497°K.

CHEMICAL REACTIVITY
Binary reactants: Strong oxidizers

ENVIRONMENTAL DATA
Food chain concentration potential: Negative.
Water pollution: Effect of low concentrations on aquatic life is unknown. May be dangerous if it enters nearby water intakes; notify operators. Notify local health and wildlife officials.
Response to discharge: Issue warning–poison, water contaminant. Restrict access. Should be removed. Chemical and physical treatment.

SHIPPING INFORMATION
Grades of purity: Technical: 94+%; Wettable powder: 75%; Emulsifiable concentrate; **Storage temperature:** Ambient; **Inert atmosphere:** None; **Venting:** Open; **Stability during transport:** Stable.

PHYSICAL AND CHEMICAL PROPERTIES
Physical state @ 59°F/15°C and 1 atm: Solid.
Molecular weight: 345.2
Boiling point @ 1 atm: Decomposes. more than 437°F/225°C/498°K.
Melting/Freezing point: 304°F/151°C/424°K.
Specific gravity (water = 1): (estimate) more than 1 @ 68°F/20°C (solid).
Heat of decomposition: –450 Btu/lb = –250 cal/g = –10.5 x 10^5 J/kg.

NITRIC ACID   **REC. N:1350**

SYNONYMS: ACIDE NITRIQUE (French); AQUA FORTIS; AZOTIC ACID; EEC No. 007-004-00-1; EEC No. 007-004-01-9;

Nitric acid

ENGRAVERS ACID; HYDROGEN NITRATE; NITAL; NITRIC ACID, RED FUMING; NITRIC ACID, WHITE FUMING; NITRYL HYDROXIDE; NITROUS FUMES; RED FUMING NITRIC ACID; RFNA; SALPETERSAURE (German); WHITE FUMING NITRIC ACID; WFNA (up to 70%) (more than 70%).

IDENTIFICATION
CAS Number: 7697-37-2
Formula: HNO_3
DOT ID Number: UN 1760; UN 2031; UN 2032; DOT Guide Number: 157
Proper Shipping Name: Nitric acid other than red fuming (UN 2031); Nitric acid, red fuming (UN 2032); Nitric acid, 40% or less (UN 1760)
Reportable Quantity (RQ): **(CERCLA)** 1000 lb/454 kg

DESCRIPTION: Pale yellow to reddish brown watery liquid. Choking, suffocating odor. Sinks and mixes with water; violent reaction producing heat.

Poison! • Breathing the vapor can kill you; Skin and eye contact causes severe burns and blindness • Firefighting gear (including SCBA) does not provide adequate protection. If exposure occurs, remove and isolate gear immediately and thoroughly decontaminate personnel • Containers may BLEVE when exposed to fire • Strong oxidizer which may react spontaneously with low flash point organics or reducing agents. Heat forms oxygen; will increase the activity of an existing fire • May cause fire or explosion on contact with combustibles (wood, paper, oil, clothing, etc.) • Corrosive to most metals forming flammable hydrogen gas • Toxic products of combustion may include nitrogen oxides and acid fumes • Do NOT attempt rescue.

Hazard Classification (based on NFPA-704 Rating System)
Health Hazards (Blue): 4; Flammability (Red): 0; Reactivity (Yellow): 1; Special Notice (White): OXY

EMERGENCY RESPONSE: See Appendix A (157)
Evacuation:
Public safety: Isolate the area of spill or leak for at least 50 to 100 meters (160 to 330 feet) in all directions.
Spill: Increase, in the downwind direction, as necessary, the distance shown under "Public Safety." NITRIC ACID FUMING AND RED FUMING: Small spill–First: Isolate in all directions 95 meters (300 feet). Then: Protect persons downwind, DAY: 0.3 km (0.2 mile); NIGHT: 0.5 km (0.3 mile). Large spill–First: Isolate in all directions 400 meters (1300 feet); Then: Protect persons downwind, DAY: 1.3 km (0.8 mile); NIGHT: 3.5 km (2.2 miles).
Fire: If tank, rail car, or tank truck is involved in fire, isolate for at least 800 meters (½ mile) in all directions; also, consider initial evacuation for 800 meters (½ mile) in all directions.

EXPOSURE
Short-term effects: *SEEK MEDICAL ATTENTION.* **Vapor:** Will burn eyes, nose, and throat. *IF INHALED*, will, will cause difficult breathing or loss of consciousness. Move to fresh air. *IF BREATHING HAS STOPPED*, give artificial respiration. IF breathing is difficult, administer oxygen. **Liquid:** Will burn skin and eyes. Harmful if swallowed. Remove contaminated clothing and shoes. Flush affected areas with plenty of water. *IF IN EYES*, hold eyelids open and flush with plenty of water. *IF SWALLOWED* and victim is *CONSCIOUS AND ABLE TO SWALLOW*, have victim drink 4 to 8 ounces of water. **Do NOT induce vomiting.**
Note to physician or authorized medical personnel: (Nitrous fumes) To relieve irritation of the respiratory tract consider a spray of 10% sodium thiosulfate. Keep the victim lying down. The slightest exertion, including walking, may result in cardiac arrest. Medical observation is recommended for 24 to 48 hours after breathing overexposure, as pulmonary edema or bronchopneumonia may be delayed. As first aid for pulmonary edema, consider administering a corticosteroid spray. Cigarette smoking may exacerbate pulmonary injury and should be discouraged for at least 72 hours following exposure.

HEALTH HAZARDS
Personal protective equipment (PPE): B-Level PPE. OSHA Table Z-1-A air contaminant. Air mask; rubber acid suit, hood, boots and gloves; chemical goggles; safety shower and eye bath. *For <30%*, Protective materials with good to excellent resistance: butyl rubber, natural rubber, nitrile+PVC, neoprene, neoprene+natural rubber, neoprene/natural rubber, nitrile, PVC, Saranex®, Silvershield®. Also, chlorinated polyethylene and Viton® offers limited protection *For 30-70%*, Protective materials with good to excellent resistance: natural rubber, neoprene, neoprene+natural rubber, neoprene/natural rubber, Saranex®, Silvershield®. Also, PVC and Viton® offers limited protection *For >70%*, Chemical protective material(s) reported to offer minimal to poor protection: neoprene/natural rubber, Saranex®. *For Red, fuming*, Chemical protective material(s) reported to offer minimal to poor protection: natural rubber, neoprene, nitrile, Viton®, chlorobutyl rubber, neoprene/natural rubber.
Recommendations for respirator selection: NIOSH/OSHA
25 ppm: SA:CF* (any supplied-air respirator operated in a continuous-flow mode); or CCRFS** [any chemical cartridge respirator with a full facepiece and cartridge(s) providing protection against the compound of concern]; or GMFS** [any air-purifying, full-facepiece respirator (gas mask) with a chin-style, front- or back-mounted canister providing protection against the compound of concern]; or SCBAF (any self-contained breathing apparatus with a full facepiece); or SAF (any supplied-air respirator with a full facepiece). *EMERGENCY OR PLANNED ENTRY INTO UNKNOWN CONCENTRATIONS OR IDLH CONDITIONS:* SCBAF:PD,PP (any self-contained breathing apparatus that has a full facepiece and is operated in a pressure-demand or other positive-pressure mode); or SAF:PD,PP:ASCBA (any supplied-air respirator that has a full facepiece and is operated in a pressure-demand or other positive-pressure mode in combination with an auxiliary self-contained breathing apparatus operated in a pressure-demand or other positive pressure mode). *ESCAPE:* GMFS** [any air-purifying, full-facepiece respirator (gas mask) with a chin-style, front- or back-mounted canister providing protection against the compound of concern]; or SCBAE (any appropriate escape-type, self-contained breathing apparatus). *Notes*: *Substance reported to cause eye irritation or damage; requires eye protection. **Only nonoxidizable sorbents are allowed (*not charcoal*).
Exposure limits (TWA unless otherwise noted): ACGIH TLV 2 ppm (5.2 mg/m^3); NIOSH OSHA 2 ppm (5 mg/m^3).
Short-term exposure limits (15-minute TWA): ACGIH, NIOSH/OSHA STEL 4 ppm (10 mg/m^3).
Toxicity by ingestion: Grade 3; LD_{50} = 50 to 500 mg/kg.
Vapor (gas) irritant characteristics: 58–68%; Vapor is moderately irritating such that personnel will not usually tolerate moderate or high vapor concentrations. 95%: Vapors cause severe irritation of eye and throat and can cause eye and lung injury. They cannot be tolerated even at low concentrations.
Liquid or solid irritant characteristics: Severe skin and eye irritant. Causes second- and third-degree burns on short contact.
Odor threshold: 0.27 ppm.
IDLH value: 25 ppm.

FIRE DATA
Electrical hazard: (56–68% solution) Class I, Group B (based on possible hydrogen gas generation should leak occur0.

CHEMICAL REACTIVITY
Reactivity with water: May heat up on mixing, but explosion or formation of steam unlikely.
Binary reactants: Very corrosive to wood, paper, cloth and most metals. Toxic red oxides of nitrogen are formed. Reacts with metallic powders, bases, hydrogen sulfide, carbides, alcohols, combustible materials. Reacts with many metals giving off flammable hydrogen gas. Attacks aluminum.
Neutralizing agents for acids and caustics: Flush with water and soda ash.
Reactivity group: 3
Compatibility class: Nitric acid

ENVIRONMENTAL DATA
330–1000 ppm/48 hr/cockle/LC50/salt water.
Food chain concentration potential: Negative; unlikely to accumulate.
Water pollution: Harmful to aquatic life in very low concentrations. May be dangerous if it enters nearby water intakes; notify operators. Notify local health and wildlife officials.
Response to discharge: Issue warning–corrosive. Restrict access. Evacuate area. Disperse and flush.

SHIPPING INFORMATION
Grades of purity: Various grades: 52–98%; **Storage temperature:** Ambient; **Inert atmosphere:** None; **Venting:** Open; **Stability during transport:** Stable. or pressure-vacuum; **Stability during transport:** Stable if kept cool.

NAS HAZARD CLASSIFICATION FOR BULK WATER TRANSPORTATION
FIRE: 0
HEALTH: Vapor irritant: 3; Liquid or solid irritant: 4; Poisons: 3
WATER POLLUTION: Human toxicity: 3; Aquatic toxicity: 3; Aesthetic effect: 2
REACTIVITY: Other chemicals: 4; Water: 0; Self-reaction: 0

PHYSICAL AND CHEMICAL PROPERTIES
Physical state @ 59°F/15°C and 1 atm: Liquid.
Molecular weight: 63.0
Boiling point @ 1 atm: 192°F/89°C/362°K; (less than 70%) 252°F/122°C/395°K.
Melting/Freezing point: –50°F/–46°C/228°K.
Specific gravity (water = 1): 1.49 @ 68°F/20°C (liquid).
Relative vapor density (air = 1): 2.2
Ratio of specific heats of vapor (gas): (estimate) 1.248
Latent heat of vaporization: 214 Btu/lb = 119 cal/g = 4.98×10^5 J/kg.
Heat of solution: –205 Btu/lb = –114 cal/g = -4.76×10^5 J/kg.
Vapor pressure: (Reid) 1.9 psia; 43 mm (100%); 7.2 (70% or less).

NITRIC OXIDE   REC. N:1400

SYNONYMS: MONONITROGEN MONOXIDE; NITROGEN MONOXIDE; NO; OXIDO NITRICO (Spanish); RCRA No. P076

IDENTIFICATION
CAS Number: 10102-43-9
Formula: NO
DOT ID Number: UN 1660; UN 1975 (mixture with dinitrogen tetroxide); DOT Guide Number: 124
Proper Shipping Name: Nitric oxide; Nitric oxide, compressed (UN 1660); Nitric oxide and dinitrogen tetroxide mixture; Nitric oxide and nitrogen dioxide mixture (UN 1975)
Reportable Quantity (RQ): **(CERCLA)** 10 lb/4.54 kg

DESCRIPTION: Colorless gas. Shipped and stored as liquefied compressed gas; becomes red-brown on contact with air due to formation of nitrogen tetroxide. Sharp, unpleasant odor. Sparingly soluble in water; reacts, producing corrosive nitric acid and nitrous acid. Rapidly converted in air to nitrogen dioxide.

Poison! • Breathing the vapor can kill you; skin and eye contact causes severe burns and blindness • May interfere with the body's ability to use oxygen • Contact with gas or liquefied gas may cause severe injury or frostbite • Firefighting gear (including SCBA) does not provide adequate protection. If exposure occurs, remove and isolate gear immediately and thoroughly decontaminate personnel • Containers may BLEVE when exposed to fire • Vapors are heavier than air and will collect and stay in low areas • Strong oxidizer which may react spontaneously with low flash point organics or reducing agents. Heat forms oxygen; will increase the activity of an existing fire • May cause fire or explosion on contact with combustibles (wood, paper, oil, clothing, etc.) all fuels and most metals • Reacts with O_2 in air to form nitrogen dioxide. Toxic products of combustion may include corrosive nitric acid and nitrous oxide• Do not put yourself in danger by entering a contaminated area to rescue a victim. *Warning:* Odor is not a reliable indicator of the presence of toxic amounts of nitric oxide gas.

Hazard Classification (based on NFPA-704 M Rating System)

Health Hazards (Blue): 3; Flammability (Red): 0; Reactivity (Yellow): 0; Special Notice (White): OXY

EMERGENCY RESPONSE: See Appendix A (124)
Evacuation:
Public safety: Isolate spill area for at least 100 to 200 meters (330 to 660 feet) in all directions.
Spill: Small spill–First: Isolate in all directions 30 meters (100 feet); Then: Protect persons downwind, DAY: 0.3 km (0.2 mile); NIGHT: 1.3 km (0.8 mile). Large spill–First: Isolate in all directions 155 meters (500 feet); Then: Protect persons downwind, DAY: 1.3 km (0.8 mile); NIGHT: 3.5 km (2.2 miles).
Fire: Isolate for 800 meters (½ mile) in all directions, especially if tank, rail car, or tank truck is involved in fire.

EXPOSURE
Short-term effects: *SEEK MEDICAL ATTENTION.* **Vapor:** *POISONOUS IF INHALED.* There is no antidote for nitrogen oxides. Primary treatment consists of respiratory and cardiovascular support. Methylene blue may be necessary to treat methemoglobinemia. Irritating to eyes, nose, and throat. Move victim to fresh air. *IF BREATHING HAS STOPPED,* give artificial respiration. IF breathing is difficult, administer oxygen. If the victim has ingested a solution of nitrogen oxides, Do not administer activated charcoal. **Do not induce vomiting**
Note to physician and medical personnel: Methylene blue (tetramethylthionine chloride) should be considered for patients who have signs and symptoms of hypoxia (other than cyanosis) or for patients who have methemoglobin levels >30%. Cyanosis alone does not require treatment. Methylene blue may not be effective in patients who have G6PD deficiency and may cause hemolysis. The

standard dose of methylene blue is 1 to 2 mg/kg body weight (0.1 to 0.2 mL/kg of a 1% solution) IV over 5 to 10 minutes, repeated in 1 hour if needed. The total initial dose should not exceed 7 mg/kg (Doses greater than 15 mg/kg may cause hemolysis). Clinical response to treatment is usually observed within 30 to 60 minutes. Side effects include nausea, vomiting, abdominal and chest pain, dizziness, diaphoresis, and dysuria. Consider exchange transfusion in severely poisoned patients who are deteriorating clinically in spite of methylene blue treatment.

HEALTH HAZARDS
Personal protective equipment (PPE): A-Level PPE. OSHA Table Z-1-A air contaminant. Self-contained breathing apparatus or gas mask with universal canister. Sealed chemical protective offering minimal to poor protection: butyl rubber, chlorinated polyethylene, PVC.
Recommendations for respirator selection: NIOSH/OSHA
1000 ppm: SA:CF* (any supplied-air respirator operated in a continuous-flow mode); or CCRFS** [any chemical cartridge respirator with a full facepiece and cartridge(s) providing protection against the compound of concern]; or *PAPRS* and*** [any powered, air-purifying respirator with cartridge(s) providing protection against the compound of concern]; or GMFS** [any air-purifying, full-facepiece respirator (gas mask) with a chin-style, front- or back-mounted canister providing protection against the compound of concern]; or SA* (any supplied-air respirator); or SCBAF (any self-contained breathing apparatus with a full facepiece). *EMERGENCY OR PLANNED ENTRY INTO UNKNOWN CONCENTRATIONS OR IDLH CONDITIONS:* SCBAF:PD,PP (any self-contained breathing apparatus that has a full facepiece and is operated in a pressure-demand or other positive-pressure mode); or SAF:PD,PP:ASCBA (any supplied-air respirator that has a full facepiece and is operated in a pressure-demand or other positive-pressure mode in combination with an auxiliary self-contained breathing apparatus operated in a pressure-demand or other positive pressure mode). *ESCAPE:* GMFS** [any air-purifying, full-facepiece respirator (gas mask) with a chin-style, front- or back-mounted canister providing protection against the compound of concern]; or SCBAE (any appropriate escape-type, self-contained breathing apparatus). *Notes*: *Substance reported to cause eye irritation or damage; requires eye protection. **Only nonoxidizable sorbents are allowed (*not charcoal*).
Exposure limits (TWA unless otherwise noted): ACGIH TLV 25 ppm (31 mg/m^3); NIOSH/OSHA 25 ppm (30 mg/m^3).
gas at normal temperatures).
Long-term health effects: May cause blood, central nervous system damage.
Vapor (gas) irritant characteristics: See symptoms, above.
Liquid or solid irritant characteristics: Eyes, skin, and respiratory tract irritation. See symptoms above.
IDLH value: 100 ppm.

FIRE DATA
Flash point: Nonflammable compressed gas.
Behavior in fire: Supports combustion; all fires burn more vigorously.

CHEMICAL REACTIVITY
Reactivity with water: Violent reaction with reducing agents, anhydrous ammonia, alcohols, butadiene, carbon disulfide, charcoal, chromium powders, dichlorine oxide, 1,3,5-cycloheptatriene, ethers, ethylene oxide, hydrogen, methanol, nitrogen chloride, oxygen, oxygen difluoride, oxygen, perchloryl fluoride, perfluoro-*tert*-nitrosobutane, phosphine, red phosphorus, rubidium acetylide, potassium sulfide, vinyl chloride, vinyl methyl ether. Forms an explosive product with propylene. Incompatible with combustible materials, calcium, chlorinated hydrocarbons, cyclopentadiene, fluorine, iron pentacarbonyl, metal powders, metal acetylides, metal carbides, potassium, ozone, tungsten carbide, uranium. Attacks some plastics, rubber, and coatings. Attacks metals in the presence of air and/or moisture.
Binary reactants: Reacts with O_2 in air to form nitrogen dioxide. Reacts with fluorine, combustible materials, ozone, ammonia, chlorinated hydrocarbons, metals, carbon disulfide. Reacts rapidly with air to form nitrogen tetroxide; see this compound.
Neutralizing agents for acids and caustics: Flood with water, rinse with sodium bicarbonate or lime solution.

ENVIRONMENTAL DATA
Food chain concentration potential: Negative; unlikely to accumulate.
Water pollution: Effect of low concentrations on aquatic life is unknown. May be dangerous if it enters nearby water intakes; notify operators. Notify local health and wildlife officials. **Response to discharge:** Issue warning–poison, air contaminant. Restrict access. Evacuate area.

SHIPPING INFORMATION
Grades of purity: C.P.: 99+%; **Storage temperature:** Cool ambient; **Inert atmosphere:** None; **Venting:** Safety relief. Containers must be in well-ventilated area; **Stability during transport:** Stable.

PHYSICAL AND CHEMICAL PROPERTIES
Physical state @ 59°F/15°C and 1 atm: Gas.
Molecular weight: 30.0
Boiling point @ 1 atm: –241.1°F/–151.7°C/121.5°K.
Melting/Freezing point: –262.5°F/–163.6°C/109.6°K.
Critical temperature: 847°F/453°C/180°K.
Critical pressure: 940 psia = 64 atm = 6.5 MN/m^2.
Relative vapor density (air = 1): 1.6 (nitrogen dioxide).
Ratio of specific heats of vapor (gas): 1.400 @ 15°C.
Heat of solution: –257 Btu/lb = –143 cal/g = 5.98 x 10^5 J/kg.
Heat of fusion: 18.3 cal/g.

o-NITROANILINE   REC. N:1450

SYNONYMS: 1-AMINO-2-NITROBENZENE; AZOENE FAST ORANGE GR SALT; AZOIC DIAZO COMPONENT 6; *o*-NITROANILINE; *o*-NITROANILINA (Spanish); ORANGE BASE CIBA 2; 2-NITRANILINE; ONA

IDENTIFICATION
CAS Number: 88-74-4 (*o*-isomer); 99-09-2 (*m*-isomer)
Formula: $C_6H_6N_2O_2$; 1,2-$C_6H_4NO_2NH_2$; 1,4-$C_6H_4NO_2NH_2$
DOT ID Number: UN 1661; DOT Guide Number: 153
Proper Shipping Name: Nitroanilines (*o*-; *m*-; *p*-).

DESCRIPTION: Orange-yellow crystalline solid. Odorless or slight musty odor. Sinks and mixes slowly with water.

Poison! • Combustible solid • Skin or eye contact, or swallowing the material can cause cyanosis or kill you • Firefighting gear (including SCBA) does not provide adequate protection. If exposure occurs, remove and isolate gear immediately and thoroughly decontaminate personnel • Containers may BLEVE when exposed to fire • Dust will collect and stay in low areas • Concentrated dust in confined areas (e.g., tanks, sewers, buildings) may explode when exposed to fire • Toxic products of combustion may include nitrogen oxide • Do not put yourself in danger by entering a contaminated area to rescue a victim.

Hazard Classification (based on NFPA-704 Rating System)
Health Hazards (Blue): 3; Flammability (Red): 1; Reactivity (Yellow): 2

EMERGENCY RESPONSE: See Appendix A (153)
Evacuation:
Public safety: Isolate the area of spill or leak for at least 25 to 50 meters (80 to 160 feet) in all directions.
Spill: Increase, in the downwind direction, as necessary, the distance shown under "Public Safety."
Fire: If tank, rail car, or tank truck is involved in fire, isolate for at least 800 meters (½ mile) in all directions; also, consider initial evacuation for 800 meters (½ mile) in all directions.

EXPOSURE
Short-term effects: *SEEK MEDICAL ATTENTION. HARMFUL IF ABSORBED THROUGH THE SKIN.* At ambient temperature vapor may not be as hazardous as dust because of low vapor pressure, but elevated temperature will increase the hazard and require SCBA. **Dust:** Irritating to eyes, nose, and throat. *IF INHALED*, will, will cause headache, dizziness, or loss of consciousness. *IF IN EYES*, hold eyelids open and flush with plenty of water. *IF BREATHING HAS STOPPED,* give artificial respiration; *avoid mouth-to-mouth resuscitation; use bag/mask apparatus.* IF breathing is difficult, administer oxygen. **Solid:** Irritating to skin and eyes. *IF SWALLOWED*, will cause headache, dizziness, nausea, vomiting, or loss of consciousness. Remove contaminated clothing and shoes. Flush affected areas with plenty of water. *IF IN EYES*, hold eyelids open and flush with plenty of water. *IF SWALLOWED* and victim is *CONSCIOUS AND ABLE TO SWALLOW*, have victim drink 4 to 8 ounces of water and have victim induce vomiting. *IF SWALLOWED* and victim is *UNCONSCIOUS OR HAVING CONVULSIONS,* do nothing except keep victim warm.

HEALTH HAZARDS
Personal protective equipment (PPE): B-Level PPE. Organic vapor breathing apparatus.
Toxicity by ingestion: Grade 2; LD_{50} = 0.5 to 5 g/kg.
Vapor (gas) irritant characteristics: See symptoms, above.
Liquid or solid irritant characteristics: Eye and skin irritation. See symptoms.

FIRE DATA
Flash point: Combustible solid.
Autoignition temperature: 970°F/521°C/794°K.

CHEMICAL REACTIVITY
Binary reactants: Sulfuric acid, strong oxidizers, strong reducers

ENVIRONMENTAL DATA
Food chain concentration potential: Log P_{ow} = 1.87. Unlikely to accumulate.
Water pollution: Harmful to aquatic life in very low concentrations. May be dangerous if it enters nearby water intakes; notify operators. Notify local health and wildlife officials.
Response to discharge: Issue warning–poison, water contaminant. Restrict access. Should be removed. Chemical and physical treatment.

SHIPPING INFORMATION
Grades of purity: Commercial, 100%; **Storage temperature:** Ambient; **Inert atmosphere:** None; **Venting:** Open; **Stability during transport:** Stable.

PHYSICAL AND CHEMICAL PROPERTIES
Physical state @ 59°F/15°C and 1 atm: Solid.
Molecular weight: 138.1
Boiling point @ 1 atm: 543°F/284°C/557°K.
Melting/Freezing point: 160°F/71°C/344°K.
Specific gravity (water = 1): 1.44 @ 68°F/20°C (solid).
Relative vapor density (air = 1): 4.8
Heat of combustion: –10,000 Btu/lb = –5550 cal/g = –232 x 10^5 J/kg.
Heat of fusion: 27.88 cal/g.

p-NITROANILINE   REC. N:1500

SYNONYMS: *p*-AMINONITROBENZENE; 1-AMINO-4-NITROBENZENE; AZOIC DIAZO COMPONENT 37; BENZENAMINE, 4-NITRO-; DEVELOPER P; EEC No. 612-012-00-9; FAST RED BASE; FAST RED 2G BASE; *p*-NITROANILINA (Spanish); 4-NITRANBINE; 4-NITROBENZENAMINE; *p*-NITROPHENYLAMINE; PNA; 4-NITROANILINE; FAST RED IG BASE; FAST RED GG BASE; RCRA No. P077

IDENTIFICATION
CAS Number: 100-01-6
Formula: $C_6H_6N_2O_2$
DOT ID Number: UN 1661; DOT Guide Number: 153
Proper Shipping Name: Nitroanilines (*o*-; *m*-; *p*-)
Reportable Quantity (RQ): **(CERCLA)** 5000 lb/2270 kg

DESCRIPTION: Bright yellow powder. Mild ammonia odor. Sinks in water; soluble.

Poison! • Combustible solid • Skin or eye contact, or swallowing the material can cause cyanosis or kill you • Firefighting gear (including SCBA) does not provide adequate protection. If exposure occurs, remove and isolate gear immediately and thoroughly decontaminate personnel • Containers may BLEVE when exposed to fire • Dust will collect and stay in low areas • Concentrated dust in confined areas (e.g., tanks, sewers, buildings) may explode when exposed to fire • Toxic products of combustion may include nitrogen oxide • Do not put yourself in danger by entering a contaminated area to rescue a victim.

Hazard Classification (based on NFPA-704 Rating System)
Health Hazards (Blue): 3; Flammability (Red): 1; Reactivity (Yellow): 2

EMERGENCY RESPONSE: See Appendix A (153)
Evacuation:
Public safety: Isolate the area of spill or leak for at least 25 to 50 meters (80 to 160 feet) in all directions.
Spill: Increase, in the downwind direction, as necessary, the distance shown under "Public Safety."
Fire: If tank, rail car, or tank truck is involved in fire, isolate for at least 800 meters (½ mile) in all directions; also, consider initial evacuation for 800 meters (½ mile) in all directions.

EXPOSURE
At ambient temperature vapor may not be as hazardous as dust because of low vapor pressure, but elevated temperature will increase the hazard and require SCBA. *IF BREATHING HAS*

STOPPED, give artificial respiration; *avoid mouth-to-mouth resuscitation; use bag/mask apparatus.*

Short-term effects: *SEEK MEDICAL ATTENTION.* **Dust:** *POISONOUS IF INHALED. IF INHALED*, will, will cause headache, coughing, difficult breathing, or loss of consciousness. *IF IN EYES*, hold eyelids open and flush with plenty of water. *IF BREATHING HAS STOPPED,* give artificial respiration. IF breathing is difficult, administer oxygen. **Solid:** Irritating to skin and eyes. *IF SWALLOWED*, will cause headache, coughing, or loss of consciousness. Remove contaminated clothing and shoes. Flush affected areas with plenty of water. *IF IN EYES*, hold eyelids open and flush with plenty of water. *IF SWALLOWED* and victim is *CONSCIOUS AND ABLE TO SWALLOW*, have victim drink 4 to 8 ounces of water and have victim induce vomiting. *IF SWALLOWED* and victim is *UNCONSCIOUS OR HAVING CONVULSIONS,* do nothing except keep victim warm.

HEALTH HAZARDS
Personal protective equipment (PPE): B-Level PPE. OSHA Table Z-1-A air contaminant. Bureau of Mines dust canister; rubber gloves; chemical safety goggles; face shield; rubber safety shoes.
Recommendations for respirator selection: NIOSH/OSHA
30 mg/m³: SA (any supplied-air respirator); or SCBA (any self-contained breathing apparatus). *75 mg/m³:* SA:CF (any supplied-air respirator operated in a continuous-flow mode). *150 mg/m³:* SCBAF (any self-contained breathing apparatus with a full facepiece); or SAF (any supplied-air respirator with a full facepiece). *300 mg/m³:* SAF:PD,PP (any supplied-air respirator that has a full facepiece and is operated in a pressure-demand or other positive-pressure mode). *EMERGENCY OR PLANNED ENTRY INTO UNKNOWN CONCENTRATIONS OR IDLH CONDITIONS:* SCBAF:PD,PP (any self-contained breathing apparatus that has a full facepiece and is operated in a pressure-demand or other positive-pressure mode); or SAF:PD,PP:ASCBA (any supplied-air respirator that has a full facepiece and is operated in a pressure-demand or other positive-pressure mode in combination with an auxiliary self-contained breathing apparatus operated in a pressure-demand or other positive pressure mode). *ESCAPE:* GMFOVDMFu [any air-purifying, full-facepiece respirator (gas mask) with a chin-style, front- or back-mounted organic vapor canister in combination with a dust, mist, and fume filter); or SCBAE (any appropriate escape-type, self-contained breathing apparatus). *Note:* Substance reported to cause eye irritation or damage; may require eye protection.
Exposure limits (TWA unless otherwise noted): ACGIH TLV 3 mg/m³; NIOSH REL 3 mg/m³; OSHA PEL 6 mg/m³ (1 ppm); skin contact contributes significantly in overall exposure.
Toxicity by ingestion: Grade 2; LD_{50} = 0.5 to 5 g/kg.
Liquid or solid irritant characteristics: Corrosive to eyes; irritating to skin.
IDLH value: 300 mg/m³.

FIRE DATA
Flash point: 329°F/165°C (oc); 390°F/199°C (cc) (molten solid).
Behavior in fire: Melting causes burning and slight explosions.
Electrical hazard: Flow or agitation of substance may generate electrostatic charges due to low conductivity.

CHEMICAL REACTIVITY
Binary reactants: Violent reaction with strong oxidizers, strong reducing agents, strong acids, acid anhydrides, acid chlorides. Forms explosive with hexanitroethane. Incompatible with nitrous acid, sulfuric acid; may form explosive with sodium hydroxide under certain conditions. Attacks some plastics, rubber, and coatings. May result in spontaneous heating of organic materials in the presence of moisture

ENVIRONMENTAL DATA
Food chain concentration potential: Log P_{ow} = 1.4. Unlikely to accumulate.
Water pollution: Harmful to aquatic life in very low concentrations. May be dangerous if it enters nearby water intakes; notify operators. Notify local health and wildlife officials. **Response to discharge:** Issue warning–poison, water contaminant. Restrict access. Should be removed. Chemical and physical treatment.

SHIPPING INFORMATION
Grades of purity: Technical, 100%; **Storage temperature:** Ambient; **Inert atmosphere:** None; **Venting:** Open; **Stability during transport:** Stable.

PHYSICAL AND CHEMICAL PROPERTIES
Physical state @ 59°F/15°C and 1 atm: Solid.
Molecular weight: 138.1
Boiling point @ 1 atm: 636°F/336°C/609°K.
Melting/Freezing point: 295°F/146°C/419°K.
Specific gravity (water = 1): 1.42 @ 68°F/20°C (solid).
Relative vapor density (air = 1): 4.8
Heat of combustion: –9,920 Btu/lb = –5510 cal/g = –231 x 10^5 J/kg.
Heat of fusion: 36.50 cal/g.

NITROBENZENE   REC. N:1550

SYNONYMS: BENZENE, NITRO-; EEC No. 609-003-00-7; ESSENCE OF MIRBANE; ESSENCE OF MYRBANE; MIRBANE OIL; NB; NITROBENCENO (Spanish); NITROBENZOL; NITROBENZOL,L; NITRO, LIQUID; OIL OF MIRBANE; OIL OF MYRBANE; RCRA No. U169

IDENTIFICATION
CAS Number: 98-95-3
Formula: $C_6H_5NO_2$; $C_6NO_5-NO_2$
DOT ID Number: UN 1662; DOT Guide Number: 152
Proper Shipping Name: Nitrobenzene
Reportable Quantity (RQ): **(CERCLA)** 1000 lb/454 kg

DESCRIPTION: Colorless to brownish oily liquid. Almond, or paste shoe polish odor. Sinks in water; dissolves very slowly.

Poison! • Breathing the vapor, skin or eye contact, or swallowing the material can kill you; may interfere with the body's ability to use oxygen • Firefighting gear (including SCBA) does not provide adequate protection. If exposure occurs, remove and isolate gear immediately and thoroughly decontaminate personnel • Containers may BLEVE when exposed to fire • Toxic products of combustion may include nitrogen oxides • Do not put yourself in danger by entering a contaminated area to rescue a victim.
Warning: Odor is not a reliable indicator of the presence of toxic amounts of nitrobenzene.

Hazard Classification (based on NFPA-704 Rating System)
Health Hazards (Blue): 3; Flammability (Red): 2; Reactivity (Yellow): 1

Nitrobenzene

EMERGENCY RESPONSE: See Appendix A (152)
Evacuation:
Public safety: Isolate the area of spill or leak for at least 25 to 50 meters (80 to 160 feet) in all directions.
Spill: Increase, in the downwind direction, as necessary, the distance shown under "Public Safety."
Fire: If tank, rail car, or tank truck is involved in fire, isolate for at least 800 meters (½ mile) in all directions; also, consider initial evacuation for 800 meters (½ mile) in all directions.

EXPOSURE
Short-term effects: *SEEK MEDICAL ATTENTION.* **Liquid:** *POISONOUS IF SWALLOWED OR IF SKIN IS EXPOSED.* Will burn eyes. Remove contaminated clothing and shoes. Flush affected areas with plenty of water. *IF BREATHING HAS STOPPED,* give artificial respiration; *avoid mouth-to-mouth resuscitation; use bag/mask apparatus. IF IN EYES,* hold eyelids open and flush with plenty of water. *IF SWALLOWED* and victim is *CONSCIOUS AND ABLE TO SWALLOW,* have victim drink 4 to 8 ounces of water.

HEALTH HAZARDS
Personal protective equipment (PPE): A-Level PPE. OSHA Table Z-1-A air contaminant. Organic vapor respirator. Chemical protective material(s) reported to have good to excellent resistance: butyl rubber, chlorinated polyethylene, PV alcohol, Teflon®, Viton®, Viton®/chlorobutyl rubber, Silvershield®.
Recommendations for respirator selection: NIOSH/OSHA
10 ppm: CCROV* [any chemical cartridge respirator with organic vapor cartridge(s)]; SA* (any supplied-air respirator). *25 ppm:* SA:CF (any supplied-air respirator operated in a continuous-flow mode); or PAPROV* [any powered, air-purifying respirator with organic vapor cartridge(s)]. *50 ppm:* CCRFOV [any air-purifying, full-facepiece respirator (gas mask) with a chin-style, front- or back-mounted acid gas canister]; or GMFOV [any air-purifying, full-facepiece respirator (gas mask) with a chin-style, front- or back-mounted organic vapor cannister]; or PAPRTOV* [any powered, air-purifying respirator with a tight-fitting facepiece and organic vapor cartridges(s)]; or SCBAF (any self-contained breathing apparatus with a full facepiece); or SAF (any supplied-air respirator with a full facepiece). *200 ppm:* SAF:PD,PP (any supplied-air respirator that has a full facepiece and is operated in a pressure-demand or other positive-pressure mode). *EMERGENCY OR PLANNED ENTRY INTO UNKNOWN CONCENTRATIONS OR IDLH CONDITIONS:* SCBAF:PD,PP (any self-contained breathing apparatus that has a full facepiece and is operated in a pressure-demand or other positive-pressure mode); or SAF:PD,PP:ASCBA (any supplied-air respirator that has a full facepiece and is operated in a pressure-demand or other positive-pressure mode in combination with an auxiliary self-contained breathing apparatus operated in a pressure-demand or other positive pressure mode). *ESCAPE:* GMFOV [any air-purifying, full-facepiece respirator (gas mask) with a chin-style, front- or back-mounted organic vapor cannister]; or SCBAE (any appropriate escape-type, self-contained breathing apparatus).
Note: Substance reported to cause eye irritation or damage; may require eye protection.
Exposure limits (TWA unless otherwise noted): ACGIH TLV 1 ppm (5 mg/m^3); NIOSH/OSHA 1 ppm (5 mg/m^3); skin contact contributes significantly in overall exposure.
Toxicity by ingestion: Grade 3; LD_{50} = 50 to 500 mg/kg (dog).
Long-term health effects: Blood damage. Confirmed carcinogen.
Vapor (gas) irritant characteristics: Light-sensitive eye irritation. Vapor is moderately irritating such that personnel will not usually tolerate moderate or high vapor concentrations.
Liquid or solid irritant characteristics: Causes smarting of the skin and first-degree burns on short exposure; may cause secondary burns on long exposure.
Odor threshold: 0.37 ppm.
IDLH value: 200 ppm.

FIRE DATA
Flash point: 171°F/77°C (oc); 190°F/88°C (cc).
Flammable limits in air: LEL: 1.8% @ 200°F/93°C; UEL 40%.
Autoignition temperature: 905°F/485°C/758°K.
Electrical hazard: Class I, Group D.
Burning rate: 2.9 mm/min

CHEMICAL REACTIVITY
Binary reactants: Strong oxidizers may form explosive mixtures; reacts with concentrated nitric acid, nitrogen tetroxide, caustics, phosphorus pentachloride, chemically active metals such as tin or zinc. Keep away from combustible materials. May attack plastics and promote stress cracks in polyethylene
Reactivity group: 42
Compatibility class: Nitrocompounds

ENVIRONMENTAL DATA
Food chain concentration potential: Log P_{ow} = 1.9. Unlikely to accumulate.
Water pollution: DOT Appendix B, §172.101–marine pollutant. Harmful to aquatic life in very low concentrations. May be dangerous if it enters nearby water intakes; notify operators. Notify local health and wildlife officials. **Response to discharge:** Issue warning–poison. Restrict access. Should be removed. Chemical and physical treatment.

SHIPPING INFORMATION
Grades of purity: Technical: 99.5–100%; **Storage temperature:** Ambient; **Inert atmosphere:** None; **Venting:** Open (flame arrester); **Stability during transport:** Stable.

NAS HAZARD CLASSIFICATION FOR BULK WATER TRANSPORTATION
FIRE: 1
HEALTH: Vapor irritant: 3; Liquid or solid irritant: 2; Poisons: 4
WATER POLLUTION: Human toxicity: 3; Aquatic toxicity: 3; Aesthetic effect: 3
REACTIVITY: Other chemicals: 2; Water: 0; Self-reaction: 1

PHYSICAL AND CHEMICAL PROPERTIES
Physical state @ 59°F/15°C and 1 atm: Liquid.
Molecular weight: 123.11
Boiling point @ 1 atm: 411.6°F/210.9°C/484.1°K.
Melting/Freezing point: 41.2°F/5.1°C/278.3°K.
Critical temperature: 836°F/447°C/720°K.
Critical pressure: 700 psia = 47.62 atm = 4.824 MN/m^2.
Specific gravity (water = 1): 1.204 @ 68°F/20°C (liquid).
Liquid surface tension: 43.9 dynes/cm = 0.0439 N/m @ 68°F/20°C.
Liquid water interfacial tension: 25.66 dynes/cm = 0.02566 N/m @ 68°F/20°C.
Relative vapor density (air = 1): 4.3
Latent heat of vaporization: 150 Btu/lb = 85 cal/g = 3.6 x 10^5 J/kg.
Heat of combustion: –10,420 Btu/lb = –5791 cal/g = –242.5 x 10^5 J/kg.
Heat of fusion: 22.50 cal/g.
Vapor pressure: (Reid) 0.01 psia; 0.14 mm.

NITROETHANE REC. N:1600

SYNONYMS: NITROETANO (Spanish); EEC No. 609-035-00-1

IDENTIFICATION
CAS Number: 79-24-3
Formula: $C_2H_5NO_2$; $CH_3CH_2NO_2$
DOT ID Number: UN 2842; DOT Guide Number: 129
Proper Shipping Name: Nitroethane

DESCRIPTION: Colorless oily liquid. Fruity, irritating, chloroform-like odor. May float and slowly sink in water; insoluble.

Highly flammable • Do NOT use dry chemical extinguishers • Elevated temperatures cause decomposition; may detonate in quickly elevating temperatures or under pressure or confinement • Containers may BLEVE when exposed to fire •Vapors may form explosive mixture with air • Vapors are heavier than air and will collect and stay in low areas • Vapors may travel long distances to ignition sources and flashback • Vapors in confined areas (e.g., tanks, sewers, buildings) may explode when exposed to fire • Irritating to the skin, eyes, and respiratory tract • Toxic products of combustion may include carbon monoxide CO_2, and nitrogen oxides.

Hazard Classification (based on NFPA-704 Rating System)
Health Hazards (Blue): 1; Flammability (Red): 3; Reactivity (Yellow): 3

EMERGENCY RESPONSE: See Appendix A (129)
Evacuation:
Public safety: Isolate spill area for at least 50 to 100 meters (160 to 330 feet) in all directions.
Spill: Large spill–Consider initial downwind evacuation for at least 300 meters (1000 feet).
Fire: Isolate for 800 meters (½ mile) in all directions, especially if tank, rail car, or tank truck is involved in fire.

EXPOSURE
Short-term effects: *SEEK MEDICAL ATTENTION.* **Vapor:** Irritating to eyes, nose, and throat. *IF INHALED*, will, will cause coughing or difficult breathing. *IF IN EYES*, hold eyelids open and flush with plenty of water. *IF BREATHING HAS STOPPED,* give artificial respiration. IF breathing is difficult, administer oxygen. **Liquid:** Irritating to skin and eyes. *IF SWALLOWED*, will cause nausea and vomiting. Remove contaminated clothing and shoes. Flush affected areas with plenty of water. *IF IN EYES*, hold eyelids open and flush with plenty of water. *IF SWALLOWED* and victim is *CONSCIOUS AND ABLE TO SWALLOW*, have victim drink 4 to 8 ounces of water. *IF SWALLOWED* and victim is *UNCONSCIOUS OR HAVING CONVULSIONS,* do nothing except keep victim warm.

HEALTH HAZARDS
Personal protective equipment (PPE): B-Level PPE. OSHA Table Z-1-A air contaminant. Supplied air or self-contained respirator; goggles or face shield; rubber gloves. Chemical protective material(s) reported to have good to excellent resistance: butyl rubber. Also. PV alcohol, butyl rubber/neoprene, neoprene+natural rubber offers limited protection
Recommendations for respirator selection: NIOSH/OSHA
1000 ppm: SCBAF (any self-contained breathing apparatus with a full facepiece); or SAF (any supplied-air respirator with a full facepiece). *EMERGENCY OR PLANNED ENTRY INTO UNKNOWN CONCENTRATIONS OR IDLH CONDITIONS:* SCBAF:PD,PP (any self-contained breathing apparatus that has a full facepiece and is operated in a pressure-demand or other positive-pressure mode); or SAF:PD,PP:ASCBA (any supplied-air respirator that has a full facepiece and is operated in a pressure-demand or other positive-pressure mode in combination with an auxiliary self-contained breathing apparatus operated in a pressure-demand or other positive pressure mode). *ESCAPE:* SCBAE (any appropriate escape-type, self-contained breathing apparatus).
Exposure limits (TWA unless otherwise noted): ACGIH TLV 100 ppm (307 mg/m^3); OSHA 100 ppm (310 mg/m^3).
Toxicity by ingestion: Grade 2; oral LD_{50} = 860 mg/kg (mouse).
Vapor (gas) irritant characteristics: Eye and respiratory irritation. Vapors cause irritation such that personnel will find high concentration unpleasant. The effect may be temporary.
Liquid or solid irritant characteristics: Eye and skin irritant. If spilled on clothing and allowed to remain, may cause smarting and reddening of skin.
Odor threshold: 2.0 ppm.
IDLH value: 750 ppm.

FIRE DATA
Flash point: 105°F/41°C (oc); 82°F/28°C (cc).
Fire extinguishing agents not to be used: Water or alcohol foam may be ineffective. Do NOT use chemical extinguishers.
Autoignition temperature: 778°F/414°C/687°K.
Electrical hazard: Class I, Group C.

CHEMICAL REACTIVITY
Binary reactants: Amines, strong acids, and oxidizers, hydrocarbons, combustibles, metal oxides. May attack some forms of plastics. Organic contamination may cause substance to become shock-sensitive
Reactivity group: 42
Compatibility class: Nitrocompounds

ENVIRONMENTAL DATA
Food chain concentration potential: Log P_{ow} = 0.2. Unlikely to accumulate.
Water pollution: DOT Appendix B, §172.101–marine pollutant. Effect of low concentrations on aquatic life is unknown. Fouling to shoreline. May be dangerous if it enters nearby water intakes; notify operators. Notify local health and wildlife officials. **Response to discharge:** Issue warning–high flammability. Restrict access. Disperse and flush.

SHIPPING INFORMATION
Grades of purity: Commercial, 92.5+%; **Storage temperature:** Ambient; **Inert atmosphere:** None; **Venting:** Open (flame arrester); **Stability during transport:** Stable.

NAS HAZARD CLASSIFICATION FOR BULK WATER TRANSPORTATION
FIRE: 3
HEALTH: Vapor irritant: 2; Liquid or solid irritant: 1; Poisons: 1
WATER POLLUTION: Human toxicity: 2; Aquatic toxicity: 2; Aesthetic effect: 1
REACTIVITY: Other chemicals: 0; Water: 4; Self-reaction: –

PHYSICAL AND CHEMICAL PROPERTIES
Physical state @ 59°F/15°C and 1 atm: Liquid.
Molecular weight: 75.07
Boiling point @ 1 atm: 237°F/114°C/387°K.
Melting/Freezing point: –130°F/–90°C/183°K.
Specific gravity (water = 1): 1.05 @ 68°F/20°C (liquid).

Liquid surface tension: 31.3 dynes/cm = 0.0313 N/m @ 68°F/20°C.
Relative vapor density (air = 1): 2.6
Ratio of specific heats of vapor (gas): (estimate) 1.115 @ 68°F/20°C.
Latent heat of vaporization: 211 Btu/lb = 117 cal/g = 4.90×10^5 J/kg.
Heat of combustion: $-7,720$ Btu/lb = -4290 cal/g = -179×10^5 J/kg.
Heat of decomposition: 599°F/315°C/588°K.
Vapor pressure: 15 mm.

NITROGEN   REC. N:1650

SYNONYMS: LIQUID NITROGEN; LN; NITROGEN, COMPRESSED; NITROGEN GAS; NXX

IDENTIFICATION
CAS Number: 7727-37-9
Formula: N_2
DOT ID Number: UN 1066 (compressed); UN 1977 (refrigerated liqid); DOT Guide Number: 121(compressed); 120 (liquid)
Proper Shipping Name: Nitrogen, compressed; Nitrogen, refrigerated liquid, cryogenic Liquid.

DESCRIPTION: Colorless gas or faint yellow liquid. Shipped and stored as a compressed gas or cryogenic liquid. Odorless. Liquid floats on water surface and boils; gas is insoluble in water.

Gas is lighter than air, and will disperse slowly unless confined •Contact with liquid may cause frostbite • Vapors may cause dizziness or asphyxiation without warning • Vapors from liquefied gas are initially heavier than air and spread along ground • Containers may BLEVE when exposed to fire • Contact with liquid may cause frostbite • *Warning:* Odor is not a reliable indicator of the presence of toxic amounts of nitrogen gas.

Hazard Classification (based on NFPA-704 Rating System)
Health Hazards (Blue): 3; Flammability (Red): 0; Reactivity (Yellow): 0

EMERGENCY RESPONSE, gas: See Appendix A (120)
Evacuation:
Public safety: Isolate spill area for at least 100 meters (330 feet) in all directions.
Spill: Consider initial downwind evacuation for at least 25 meters (80 feet).
Fire: Isolate for 800 meters (½ mile) in all directions, especially if tank, rail car, or tank truck is involved in fire.

EMERGENCY RESPONSE, liquid: See Appendix A (121)
Evacuation:
Public safety: Isolate spill area for at least 10 to 25 meters (30 to 80 feet) in all directions.
Spill: Large spill –Consider initial downwind evacuation for at least 100 meters (330 feet).
Fire: Isolate for 800 meters (½ mile) in all directions, especially if tank, rail car, or tank truck is involved in fire.

EXPOSURE
Short-term effects: *SEEK MEDICAL ATTENTION.* **Vapor:** Not harmful. In high concentrations may cause dizziness, difficult breathing or loss of consciousness. **Liquid:** Will cause frostbite.

Flush affected areas with plenty of water. *DO NOT RUB AFFECTED AREAS.*

HEALTH HAZARDS
Personal protective equipment (PPE): B-Level PPE. Safety glasses or face shield; insulated gloves; long sleeves; trousers worn outside boots or over high-top shoes to shed spilled liquid; self-contained breathing apparatus where insufficient air is present. Wear thermal protective clothing.
Exposure limits (TWA unless otherwise noted): Simple asphyxiant.
Vapor (gas) irritant characteristics: None.
Liquid or solid irritant characteristics: Rapid evaporation of liquid may cause frostbite.

FIRE DATA
Flash point: Nonflammable compressed gas.
Behavior in fire: Containers may explode when heated.

CHEMICAL REACTIVITY
Reactivity with water: Heat of water will vigorously vaporize liquid nitrogen.
Binary reactants: No chemical reaction. Low temperature may cause brittleness in rubber and plastics

SHIPPING INFORMATION
Grades of purity: 99.5+%; **Storage temperature:** –320°F/–196°C; **Inert atmosphere:** None; **Venting:** Open; **Stability during transport:** Stable.

PHYSICAL AND CHEMICAL PROPERTIES
Physical state @ 59°F/15°C and 1 atm: Gas.
Molecular weight: 28.0
Boiling point @ 1 atm: –320°F/–196°C/78°K.
Melting/Freezing point: –354°F/–215°C/58°K.
Critical temperature: –232.6°F/–147.0°C/126.2°K.
Critical pressure: 493 psia = 33.5 atm = 3.40 MN/m^2.
Specific gravity (water = 1): 0.807 at –195.5°C (liquid).
Liquid surface tension: 8.3 dynes/cm = 0.083 N/m at –193°C.
Relative vapor density (air = 1): 0.965
Ratio of specific heats of vapor (gas): 1.3962
Latent heat of vaporization: 95 Btu/lb = 53 cal/g = 2.2×10^5 J/kg.
Heat of fusion: 6.15 cal/g.

NITROGEN TETROXIDE (and other NITROGEN OXIDES)

REC. N:1700

SYNONYMS: DIMER OF NITROGEN DIOXIDE; DINITROGEN TETROXIDE; EEC No. 007-002-00-0; NTO; NOx; RED OXIDE OF NITROGEN; OXIDES OF NITROGEN; QW9800000

IDENTIFICATION
CAS Number: 10544-72-6 (nitrogen tetroxide); 10102-43-9 (nitric oxide); 10102-44-0 (nitrogen dioxide; nitrogen peroxide); 10544-73-7 (nitrogen trioxide); 10102-03-1 (nitrogen pentoxide)
Formula: NO (nitric oxide); NO_2 (nitrogen dioxide); N_2O_4 (nitrogen tetroxide); N_2O_3 (nitrogen trioxide); N_2O_5 (nitrogen pentoxide)
DOT ID Number: UN 1067; UN 1975; UN 2421; DOT Guide Number: 124
Proper Shipping Name: UN 1067 (nitrogen dioxide; nitrogen tetroxide, liquid; nitrogen peroxide, liquid); UN 1975 (nitrogen

dioxide and nitric oxide; nitrogen tetroxide and nitric oxide mixture); UN 2421(nitrogen trioxide)
Reportable Quantity (RQ): **(CERCLA)** 10 lb/4.54 kg

DESCRIPTION: Nitrogen monoxide is a colorless compressed, liquefied gas at ambient temperatures. Turns red-brown on contact with air, forming nitrogen dioxide, the principal component of nitrous vapors. Colorless or snow white below about 15°F/–9°C; or, yellow liquid. Sharp, irritating odor. Sinks and decomposes in water, releasing nitrogen oxide and forming nitric acid. Large amounts of poisonous brown vapor is produced. *Note*: In solid form, nitrogen dioxide is found structurally as nitrogen tetroxide. Nitrogen trioxide and nitrogen pentoxide are readily decomposed forming dinitrogen tetroxide and nitrogen dioxide. Boils at 70°F/21°C. *Warning:* Odor is not a reliable indicator of the presence of toxic amounts of nitrogen oxide gasses. Only one or two breaths of very high concentrations can cause severe toxicity. *Note:* The toxicity of nitrous oxide (N_2O) is different from that of other nitrogen oxides and is not discussed in this record.

Poison! • May be fatal if inhaled or absorbed through the skin • Corrosive; skin or eye contact can cause severe burns, impaired vision, or blindness • Strong oxidizer which may react spontaneously with low flash point organics or reducing agents. Heat forms oxygen; will increase the activity of an existing fire • May cause fire or explosion on contact with combustibles (wood, paper, oil, clothing, etc.) • Firefighting gear (including SCBA) does not provide adequate protection. If exposure occurs, remove and isolate gear immediately and thoroughly decontaminate personnel • Containers may BLEVE when exposed to fire • Contact with gas or liquefied gas may cause severe injury or frostbite • Toxic products of combustion may include fumes • Do not put yourself in danger by entering a contaminated area to rescue a victim.

Hazard Classification (based on NFPA-704 Rating System)
Health Hazards (Blue): 3; Flammability (Red): 0; Reactivity (Yellow): 0; Special Notice (White): OXY

EMERGENCY RESPONSE: See Appendix A (124)
Evacuation:
Public safety: Isolate spill area for at least 100 to 200 meters (330 to 660 feet) in all directions.
Spill: Small spill–First: Isolate in all directions 30 meters (100 feet) (UN 1067; UN 1975; UN 2421); Then: Protect persons downwind, DAY: 0.2 km (0.1 mile) (UN 1067; UN 2421); 0.3 km (0.2 mile) (UN 1975); NIGHT: 0.5 km (0.3 mile) (UN 1067); 1.3 km (0.8 mile) (UN 1975); 0.2 km (0.1 mile) (UN 2421). Large spill–First: Isolate in all directions: 305 meters (1000 feet) (UN 1067); 153 meters (500 feet) (UN 1975; UN 2421); Then: Protect persons downwind, DAY:1.3 km (0.8 mile) (UN 1067; UN 1975); 0.6 km (0.4 mile) (UN 2421); NIGHT: 3.9 km (2.4 miles) (UN 1067); 3.5 km (2.2 miles) (UN 1975); 2.1 km (1.3 miles) (UN 2421).
Fire: Isolate for 800 meters (½ mile) in all directions, especially if tank, rail car, or tank truck is involved in fire.

EXPOSURE
Short-term effects: *SEEK MEDICAL ATTENTION.* There is no antidote for nitrogen oxides. Primary treatment consists of respiratory and cardiovascular support. *Methylene blue may be necessary to treat methemoglobinemia.* **Vapor:** *POISONOUS IF INHALED.* Irritating to eyes. Move to fresh air. *IF BREATHING HAS STOPPED,* give artificial respiration. IF breathing is difficult, administer oxygen. Lung edema may develop. **Liquid:** Will burn skin and eyes. *POISONOUS IF SWALLOWED.* Remove contaminated clothing and shoes. Flush affected areas with plenty of water. *IF IN EYES,* hold eyelids open and flush with plenty of water. *IF SWALLOWED* and victim is *CONSCIOUS AND ABLE TO SWALLOW,* have victim drink 4 to 8 ounces of water or milk. Do NOT administer activated charcoal. **Do NOT induce vomiting.** *Note to physician and medical personnel:* Methylene blue (tetramethylthionine chloride) should be considered for patients who have signs and symptoms of hypoxia (other than cyanosis) or for patients who have methemoglobin levels >30%. Cyanosis alone does not require treatment. Methylene blue may not be effective in patients who have G6PD deficiency and may cause hemolysis. The standard dose of methylene blue is 1 to 2 mg/kg body weight (0.1 to 0.2 mL/kg of a 1% solution) IV over 5 to 10 minutes, repeated in 1 hour if needed. The total initial dose should not exceed 7 mg/kg (Doses greater than 15 mg/kg may cause hemolysis). Clinical response to treatment is usually observed within 30 to 60 minutes. Side effects include nausea, vomiting, abdominal and chest pain, dizziness, diaphoresis, and dysuria. Consider exchange transfusion in severely poisoned patients who are deteriorating clinically in spite of methylene blue treatment.
Note to physician or authorized medical personnel: Medical observation is recommended for 24 to 48 hours after breathing overexposure, as pulmonary edema may be delayed. As first aid for pulmonary edema, consider administering a corticosteroid spray. Cigarette smoking may exacerbate pulmonary injury and should be discouraged for at least 72 hours following exposure.

HEALTH HAZARDS
Personal protective equipment (PPE): A-Level PPE. Rubber gloves; safety goggles and face shield; protective clothing. The following materials are NOT recommended for service (nitrogen tetraoxide): natural rubber, nitrile rubber, PE, neoprene, PVC, Viton®.
Recommendations for respirator selection: NIOSH as NO_2.
20 ppm: SA:CF * (any supplied-air respirator operated in a continuous-flow mode); SCBAF (any self-contained breathing apparatus with a full facepiece); or SAF (any supplied-air respirator with a full facepiece). *EMERGENCY OR PLANNED ENTRY INTO UNKNOWN CONCENTRATIONS OR IDLH CONDITIONS:* SCBAF:PD,PP (any self-contained breathing apparatus that has a full facepiece and is operated in a pressure-demand or other positive-pressure mode); or SAF:PD,PP:ASCBA (any supplied-air respirator that has a full facepiece and is operated in a pressure-demand or other positive-pressure mode in combination with an auxiliary self-contained breathing apparatus operated in a pressure-demand or other positive pressure mode). *ESCAPE:* GMFS* [any air-purifying, full-facepiece respirator (gas mask) with a chin-style, front- or back-mounted canister protection against the compound of concern]; or SCBAE (any appropriate escape-type, self-contained breathing apparatus). *Notes:* *Substance reported to cause eye irritation or damage; may require eye protection. **Only nonoxidizable sorbents are allowed (*not charcoal*).
Exposure limits (TWA unless otherwise noted): ACGIH TLV 3 ppm (5.6 mg/m³) as nitrogen dioxide; OSHA ceiling 5 ppm (9 mg/m³) as nitrogen dioxide.
Short-term exposure limits (15-minute TWA): ACGIH STEL 5 ppm (9.4 mg/m³); NIOSH STEL 1 ppm (1.8 mg/m³) as nitrogen dioxide.
Vapor (gas) irritant characteristics: Vapors cause severe irritation of eyes and throat and can cause eye and lung injury. They cannot be tolerated even at low concentrations.
Liquid or solid irritant characteristics: Severe skin irritant. Causes second- and third-degree burns on short contact and is very

Causes second- and third-degree burns on short contact and is very injurious to the eyes.
Odor threshold: 0.06–0.15 ppm.
IDLH value: 20 ppm as nitrogen dioxide.

CHEMICAL REACTIVITY
Reactivity with water: Dissolves to form nitric acid and nitric oxide. Nitric oxide reacts with air to form more nitrogen tetroxide.
Binary reactants: Very corrosive to metals when wet. Violent reaction with alcohol, ammonia, barium oxide, carbon disulfide, cyclohexane, fluorine, formaldehyde, nitrobenzene, petroleum, toluene. Reacts vigorously with combustible materials such as wood, cloth, etc
Neutralizing agents for acids and caustics: Flush with water, then use soda ash or lime.

ENVIRONMENTAL DATA
Food chain concentration potential: Negative; unlikely to accumulate.
Water pollution: Harmful to aquatic life in very low concentrations. Nitrates persist for prolonged periods in natural waters. May be dangerous if it enters nearby water intakes; notify operators. Notify local health and wildlife officials. **Response to discharge:** Issue warning–poison, air contaminant, water contaminant, corrosive. Restrict access. Evacuate area.

SHIPPING INFORMATION
Storage temperature: Ambient. Storage and transfer structures shall be equipped with mechanical ventilation systems; **Inert atmosphere:** None; **Venting:** Pressure relief valves on containers; **Stability during transport:** Stable.

NAS HAZARD CLASSIFICATION FOR BULK WATER TRANSPORTATION
FIRE: 0
HEALTH: Vapor irritant: 4; Liquid or solid irritant: 4; Poisons: 4
WATER POLLUTION: Human toxicity: 3; Aquatic toxicity: 3; Aesthetic effect: 4
REACTIVITY: Other chemicals: 2; Water: 0; Self-reaction:

PHYSICAL AND CHEMICAL PROPERTIES
Physical state @ 59°F/15°C and 1 atm: Soluble
Molecular weight: 46.0 (nitrogen dioxide); 30.0 (nitric oxide); 92.0 (nitrogen tetroxide); 76.0 (nitrogen trioxide) 108.0 (nitrogen pentoxide).
Boiling point @ 1 atm: 70.1°F/21.2°C/294°K (nitrogen dioxide); –241°F/–52°C/221°K (nitric oxide); 117°F/47°C/320°K (notrogen pentoxide).
Melting/Freezing point: 11.8°F/–11.2°C/261.8°K (nitrogen tetroxide); –263°F/–164°C/209°K (nitric oxide); 86°F/30°C/303°K (nitrogen pentoxide); –15.26°F/–9.3°C/263.7°K (nitrogen dioxide).
Critical temperature: 317.0°F/158.2°C/431.4°K.
Critical pressure: 1470 psia = 100 atm = 10 MN/m².
Specific gravity (water = 1): 1.45 @ 68°F/20°C (liquid).
Relative vapor density (air = 1): 1.03-3.72; 3.2 (nitrogen tetroxide).
Ratio of specific heats of vapor (gas): (estimate) 1.262
Latent heat of vaporization: 178 Btu/lb = 99.1 cal/g = 4.15×10^5 J/kg.
Heat of decomposition: 320°F/160°C.
Heat of fusion: 60.2 cal/g.
Vapor pressure: (Reid) 30 psia; 720 mm.

NITRILOTRIACETIC ACID, AND SALTS   REC. N:1750

SYNONYMS: ACIDO NITRILOTRIACETICO (Spanish); AMINOTRIACETIC ACID; DISODIUM NITRILOTRIACETATE; NTA; TRIGLYCINE; TRIGLYCOLLAMIC ACID; TRISODIUM NITRILOTRIACETATE; VERSENE NTA ACID

IDENTIFICATION
CAS Number: 139-13-9; 15467-20-6 (disodium salt); 18994-66-6 (monosidium salt); 18662-53-8 (trisodium salt); 23255-03-0 (disodium salt, monohydrate); 10042-84-9 (sodium salt)
Formula: $C_6H_9NO_6$
DOT ID Number: 2811; DOT Guide Number: 154
Proper Shipping Name: Toxic solid, organic, n.o.s.

DESCRIPTION: White crystalline solid. Odorless. Sinks and mixes with water; forms an acid solution.

Toxic! • Combustible solid • Breathing the vapor, skin or eye contact, or swallowing the material can cause • Containers may BLEVE when exposed to fire • Concentrated dust in confined areas (e.g., tanks, sewers, buildings) may explode when exposed to fire • Toxic products of combustion may include nitrogen oxide.

Hazard Classification (based on NFPA-704 Rating System)
Health Hazards (Blue): 4; Flammability (Red): 1; Reactivity (Yellow): 0

EMERGENCY RESPONSE: See Appendix A (154)
Evacuation:
Public safety: Isolate the area of spill or leak for at least 25 to 50 meters (80 to 160 feet) in all directions.
Spill: Increase, in the downwind direction, as necessary, the distance shown under "Public Safety."
Fire: If tank, rail car, or tank truck is involved in fire, isolate for at least 800 meters (½ mile) in all directions; also, consider initial evacuation for 800 meters (½ mile) in all directions.

EXPOSURE
Short-term effects: *SEEK MEDICAL ATTENTION.* **Dust:** Irritating to eyes, nose, and throat. *IF IN EYES*, hold eyelids open and flush with plenty of water. *IF BREATHING HAS STOPPED*, give artificial respiration; *avoid mouth-to-mouth resuscitation; use bag/mask apparatus.* IF breathing is difficult, administer oxygen. **Solid:** Irritating to skin and eyes. Remove contaminated clothing and shoes. Flush affected areas with plenty of water. *IF IN EYES*, hold eyelids open and flush with plenty of water. *IF SWALLOWED* and victim is *CONSCIOUS AND ABLE TO SWALLOW*, have victim drink 4 to 8 ounces of water. *IF SWALLOWED* and victim is *UNCONSCIOUS OR HAVING CONVULSIONS*, do nothing except keep victim warm.

HEALTH HAZARDS
Personal protective equipment (PPE): B-Level PPE. Dust mask; rubber gloves; chemical safety goggles
Toxicity by ingestion: Grade 2; LD_{50} = 0.5 to 5 g/kg Disodium Grade 2; LD_{50} = 1.2 g/kg (rat). Trisodium Grade 2; LD_{50} = 4 g/kg (rat).
Long-term health effects: IARC confirmed animal carcinogen. Potential human carcinogen.
Liquid or solid irritant characteristics: Eye irritation.

FIRE DATA
Flash point: Combustible; may be difficult to ignite.

ENVIRONMENTAL DATA
Food chain concentration potential: Unlikely to accumulate.
Water pollution: Dangerous to aquatic life in high concentrations. May be dangerous if it enters nearby water intakes; notify operators. Notify local health and wildlife officials. **Response to discharge:** Disperse and flush.

SHIPPING INFORMATION
Grades of purity: Commercial, 99.5+%; a water solution of trisodium salt containing 43% solids is also shipped; **Storage temperature:** Ambient; **Inert atmosphere:** None; **Venting:** Open; **Stability during transport:** Stable.

PHYSICAL AND CHEMICAL PROPERTIES
Physical state @ 59°F/15°C and 1 atm: Solid.
Molecular weight: Acid 191; Disodium 253; Trisodium 275
Boiling point @ 1 atm: Decomposes.
Specific gravity (water = 1): More than 1 @ 68°F/20°C (solid).

NITROMETHANE   REC. N:1800

SYNONYMS: EEC No. 609-036-00-7; METHANE, NITRO-; NITROCARBOL; NITROMETANO (Spanish)

IDENTIFICATION
CAS Number: 75-52-5
Formula: CH_3NO_2
DOT ID Number: UN 1261; DOT Guide Number: 129
Proper Shipping Name: Nitromethane

DESCRIPTION: Clear, colorless or pale yellow, oily liquid. Disagreeable, slightly fruity odor. Sinks and mixes slowly with water; moderately soluble; forms acidic solution. Irritating vapor is produced.

Highly flammable • Do NOT use dry chemical extinguishers • Containers may BLEVE when exposed to fire •Vapors may form explosive mixture with air • Vapors are heavier than air and will collect and stay in low areas • Vapors may travel long distances to ignition sources and flashback • Vapors in confined areas (e.g., tanks, sewers, buildings) may explode when exposed to fire • Irritating to the skin, eyes, and respiratory tract • Explosive decomposition above 570°F/298°C (approximate). Toxic products of combustion may include nitrogen oxides.

Hazard Classification (based on NFPA-704 Rating System)
Health Hazards (Blue): 1; Flammability (Red): 3; Reactivity (Yellow): 4

EMERGENCY RESPONSE: See Appendix A (129)
Evacuation:
Public safety: Isolate spill area for at least 50 to 100 meters (160 to 330 feet) in all directions.
Spill: Large spill–Consider initial downwind evacuation for at least 300 meters (1000 feet).
Fire: Isolate for 800 meters (½ mile) in all directions, especially if tank, rail car, or tank truck is involved in fire.

EXPOSURE
Short-term effects: *SEEK MEDICAL ATTENTION*. **Vapor:** Irritating to eyes, nose, and throat. Harmful if inhaled. Move to fresh air. *IF BREATHING HAS STOPPED*, give artificial respiration. IF breathing is difficult, administer oxygen. **Liquid:** No appreciable harm to skin or eyes. Harmful if swallowed. Flush affected areas with plenty of water. *IF SWALLOWED* and victim is *CONSCIOUS AND ABLE TO SWALLOW*, have victim drink 4 to 8 ounces of water.

HEALTH HAZARDS
Personal protective equipment (PPE): B-Level PPE. OSHA Table Z-1-A air contaminant. Chemical protective material(s) reported to have good to excellent resistance: butyl rubber, neoprene, polyethylene. Also, PV alcohol and natural rubber offers limited protection
Recommendations for respirator selection: OSHA
750 ppm: SA:CF (any supplied-air respirator operated in a continuous-flow mode); or SCBAF (any self-contained breathing apparatus with a full facepiece); or SAF (any supplied-air respirator with a full facepiece). *EMERGENCY OR PLANNED ENTRY INTO UNKNOWN CONCENTRATIONS OR IDLH CONDITIONS*: SCBAF:PD,PP (any self-contained breathing apparatus that has a full facepiece and is operated in a pressure-demand or other positive-pressure mode); or SAF:PD,PP:ASCBA (any supplied-air respirator that has a full facepiece and is operated in a pressure-demand or other positive-pressure mode in combination with an auxiliary self-contained breathing apparatus operated in a pressure-demand or other positive pressure mode). *ESCAPE:* SCBAE (any appropriate escape-type, self-contained breathing apparatus). *Note*: Substance causes eye irritation or damage; eye protection needed.
Exposure limits (TWA unless otherwise noted): ACGIH TLV 20 ppm; OSHA PEL 100 ppm (250 mg/m^3).
Toxicity by ingestion: Grade 2; LD_{50} = 0.5 to 5 g/kg (rat).
Long-term health effects: Possible nervous system effects; dermatitis. Thyroid effects.
Vapor (gas) irritant characteristics: Respiratory tract and eye irritant. Vapors cause smarting of the eyes or respiratory system if present in high concentrations. The effect may be temporary.
Liquid or solid irritant characteristics: Eye irritant. Practically harmless to the skin.
Odor threshold: 3.5 ppm.
IDLH value: 750 ppm.

FIRE DATA
Flash point: 110°F/43°C (oc); 95°F/35°C (cc).
Flammable limits in air: LEL: 7.3%; UEL: 62%.
Fire extinguishing agents not to be used: Water may be ineffective; alcohol foam may not be effective. Do NOT use chemical extinguishers.
Behavior in fire: Containers may explode.
Autoignition temperature: 785°F/418°C/691°K.
Electrical hazard: Class I, Group C.
Burning rate: 1.1 mm/min

CHEMICAL REACTIVITY
Binary reactants: Reacts with amines, strong acids, strong bases; oxidizers, and other combustible materials; metallic oxides. Wet material corrodes steel and copper; the reaction is slow.

ENVIRONMENTAL DATA
Food chain concentration potential: Log P_{ow} = –0.1 to –0.4. Unlikely to accumulate.
Water pollution: DOT Appendix B, §172.101–marine pollutant. Effect of low concentrations on aquatic life is unknown. May be dangerous if it enters nearby water intakes; notify operators. Notify local health and wildlife officials. **Response to discharge:** Disperse and flush.

SHIPPING INFORMATION
Grades of purity: 95–99%; **Storage temperature:** Ambient; **Inert**

atmosphere: None; **Venting:** Open; **Stability during transport:** Stable. or pressure-vacuum; **Stability during transport:** Considered stable, but may become sensitized by organic bases (amines) and some metal oxides, such as lead pigments.

PHYSICAL AND CHEMICAL PROPERTIES
Physical state @ 59°F/15°C and 1 atm: Liquid.
Molecular weight: 61.04
Boiling point @ 1 atm: 214°F/101°C/374°K.
Melting/Freezing point: –20°F/–29°C/244°K.
Critical temperature: 599°F/315°C/588°K.
Critical pressure: 915.8 psia = 62.3 atm = 6.311 MN/m^2.
Specific gravity (water = 1): 1.139 @ 68°F/20°C (liquid).
Liquid surface tension: 37.0 dynes/cm = 0.0370 N/m @ 68°F/20°C.
Relative vapor density (air = 1): 2.1
Ratio of specific heats of vapor (gas): 1.172
Latent heat of vaporization: 241 Btu/lb = 134 cal/g = 5.61 x 10^5 J/kg.
Heat of combustion: –4531 Btu/lb = –2517 cal/g = –105.4 x 10^5 J/kg.
Heat of decomposition: 570°F/298°C (approximate).
Heat of solution: (estimate) –9 Btu/lb = –5 cal/g = –0.2 x 10^5 J/kg.
Vapor pressure: 28 mm.

2-NITROPHENOL REC. N:1850

SYNONYMS: 2-HYDROXYNITROBENZENE; *o*-NITROFENOL (Spanish); *o*-NITROPHENOL; ORTHONITROPHENOL; PHENOL, 2-NITRO-; PHENOL, *o*-NITRO; ONP

IDENTIFICATION
CAS Number: 88-75-5
Formula: $C_6H_5NO_3$; 1,2-$HOC_6H_4NO_2$
DOT ID Number: UN 1663; DOT Guide Number: 153
Proper Shipping Name: Nitrophenols (*o*-; *m*-; *p*-)
Reportable Quantity (RQ): **(CERCLA)** 100 lb/45.4 kg

DESCRIPTION: Yellow crystalline solid or powder. Peculiar aromatic odor. Sinks and mixes slowly with water.

Toxic! • Combustible solid • Containers may BLEVE when exposed to fire • Irritating to the eyes, skin, and respiratory tract; toxic if swallowed • Concentrated dust in confined areas (e.g., tanks, sewers, buildings) may explode when exposed to fire • Violent decomposition above 350°F/177°C. Toxic products of combustion may include vapors of unburned material, carbon monoxide, and nitrogen oxides.

Hazard Classification (based on NFPA-704 Rating System)
Health Hazards (Blue): 3; Flammability (Red): 1; Reactivity (Yellow): 2

EMERGENCY RESPONSE: See Appendix A (153)
Evacuation:
Public safety: Isolate the area of spill or leak for at least 25 to 50 meters (80 to 160 feet) in all directions.
Spill: Increase, in the downwind direction, as necessary, the distance shown under "Public Safety."
Fire: If tank, rail car, or tank truck is involved in fire, isolate for at least 800 meters (½ mile) in all directions; also, consider initial evacuation for 800 meters (½ mile) in all directions.

EXPOSURE
Short-term effects: *SEEK MEDICAL ATTENTION.* **Dust:** Irritating to eyes, nose, and throat. *IF INHALED*, will, will cause headache or loss of consciousness. *IF IN EYES*, hold eyelids open and flush with plenty of water. *IF BREATHING HAS STOPPED*, give artificial respiration; *avoid mouth-to-mouth resuscitation; use bag/mask apparatus.* IF breathing is difficult, administer oxygen. **Solid:** Irritating to skin and eyes. *IF SWALLOWED*, will cause headache, nausea, or loss of consciousness. Remove contaminated clothing and shoes. Flush affected areas with plenty of water. *IF IN EYES*, hold eyelids open and flush with plenty of water. *IF SWALLOWED* and victim is *CONSCIOUS AND ABLE TO SWALLOW*, have victim drink 4 to 8 ounces of water. *IF SWALLOWED* and victim is *UNCONSCIOUS OR HAVING CONVULSIONS,* do nothing except keep victim warm.

HEALTH HAZARDS
Personal protective equipment (PPE): B-Level PPE. Self-contained breathing apparatus for fumes; rubber gloves; goggles.
Toxicity by ingestion: Grade 2; oral LD_{50} = 1297 mg/kg (rat).
Vapor (gas) irritant characteristics: Eye and respiratory tract irritation.
Liquid or solid irritant characteristics: Eye and skin irritant.

CHEMICAL REACTIVITY
Binary reactants: Reducing agents and combustibles. Softens rubber and paint. Violent reaction with potassium hydroxide (caustic potash).

ENVIRONMENTAL DATA
Food chain concentration potential: Log K_{ow} = 1.1–1.6. Unlikely to accumulate.
Water pollution: Harmful to aquatic life in very low concentrations. May be dangerous if it enters nearby water intakes; notify operators. Notify local health and wildlife officials. **Response to discharge:** Issue warning–water contaminant. Should be removed. Chemical and physical treatment. Clean Water Act. RCRA Ground Water Monitoring List.

SHIPPING INFORMATION
Grades of purity: Commercial; Pure; **Storage temperature:** Ambient; **Inert atmosphere:** None; **Venting:** Open; **Stability during transport:** Stable.

PHYSICAL AND CHEMICAL PROPERTIES
Physical state @ 59°F/15°C and 1 atm: Solid.
Molecular weight: 139.1
Boiling point @ 1 atm: 417°F/214°C/487°K.
Melting/Freezing point: 111°F/44°C/313°K.
Specific gravity (water = 1): 1.49 @ 68°F/20°C (solid).
Relative vapor density (air = 1): 4.4
Heat of combustion: –8910 Btu/lb = –4950 cal/g = –207 x 10^5 J/kg.
Heat of fusion: 26.76 cal/g.
Vapor pressure: (Reid) Less than 0.1 psia; less than 1 mm.

3-NITROPHENOL REC. N:1900

SYNONYMS: EEC No. 609-015-00-2; *m*-NITROFENOL (Spanish); *m*-NITROPHENOL; 3-NITROTOLUENE; 3-METHYLNITROBENZENE; *m*-HYDROXYNITROBENZENE; 3-HYDROXYNITROBENZENE

IDENTIFICATION
CAS Number: 554-84-7
Formula: $C_6H_5NO_3$
DOT ID Number: UN 1663; DOT Guide Number: 153
Proper Shipping Name: Nitrophenols (*o-*; *m-*; *p-*)
Reportable Quantity (RQ): **(CERCLA)** 100 lb/45.4 kg

DESCRIPTION: Colorless to pale yellow crystalline solid. Sinks and mixes with water.

Toxic! • Combustible solid • Irritating to the eyes, skin, and respiratory tract; toxic if swallowed • Containers may BLEVE when exposed to fire • Concentrated dust in confined areas (e.g., tanks, sewers, buildings) may explode when exposed to fire • Toxic products of combustion may include carbon monoxide and nitrogen oxides • Do not put yourself in danger by entering a contaminated area to rescue a victim.

Hazard Classification (based on NFPA-704 Rating System)
Health Hazards (Blue): 3; Flammability (Red): 1; Reactivity (Yellow): 2

EMERGENCY RESPONSE: See Appendix A (153)
Evacuation:
Public safety: Isolate the area of spill or leak for at least 25 to 50 meters (80 to 160 feet) in all directions.
Spill: Increase, in the downwind direction, as necessary, the distance shown under "Public Safety."
Fire: If tank, rail car, or tank truck is involved in fire, isolate for at least 800 meters (½ mile) in all directions; also, consider initial evacuation for 800 meters (½ mile) in all directions.

EXPOSURE
Short-term effects: *SEEK MEDICAL ATTENTION.* **Solid or dust:** *HARMFUL IF ABSORBED THROUGH THE SKIN. IF INHALED,* will, swallowed, or if skin is exposed, may cause headache, lethargy, nausea, and cyanosis. Irritating to eyes. Move to fresh air. Remove contaminated clothing and shoes. Flush affected areas with plenty of water. *IF BREATHING HAS STOPPED,* give artificial respiration; *avoid mouth-to-mouth resuscitation; use bag/mask apparatus. IF IN EYES,* hold eyelids open and flush with plenty of water. *IF SWALLOWED* and victim is *CONSCIOUS AND ABLE TO SWALLOW,* have victim drink 4 to 8 ounces of water and have victim induce vomiting. *IF BREATHING HAS STOPPED,* give artificial respiration. IF breathing is difficult, administer oxygen.

HEALTH HAZARDS
Personal protective equipment (PPE): B-Level PPE. Wear butyl rubber gloves, protective clothing and shoes, and self-contained breathing apparatus).
Toxicity by ingestion: Grade 2; LD_{50} = 0.5 to 5 g/kg.
Long-term health effects: Chronic exposure adversely affects the neurohumoral regulation. Higher doses affect activity of all organs and systems such as gastritis, enteritis, colitis, hepatitis, neuritis, and
hyperplasia of the spleen, and hinders oxidation processes.
Vapor (gas) irritant characteristics: Eye and respiratory tract irritant.
Liquid or solid irritant characteristics: Eye and skin irritant.
Odor threshold: Limit concentration–acceptable odor concentration = 350.3 mg/l.

FIRE DATA
Flash point: Combustible solid.

CHEMICAL REACTIVITY
Binary reactants: A strong oxidizer which reacts with combustible, organic, or other readily oxidizable materials; reducing agents.

ENVIRONMENTAL DATA
Food chain concentration potential: Log K_{ow} = 2.0. Unlikely to accumulate.
Water pollution: Harmful to aquatic life in very low concentrations. May be dangerous if it enters nearby water intakes; notify operators. Notify local health and wildlife officials. **Response to discharge:** Issue warning–water contaminant. Should be removed. Chemical and physical treatment. Clean Water Act.

PHYSICAL AND CHEMICAL PROPERTIES
Physical state @ 59°F/15°C and 1 atm: Solid.
Molecular weight: 139.11
Boiling point @ 1 atm: 381.2°F/194°C/467.2°K @ 70 mmHg
Melting/Freezing point: 206.6°F/97°C/370.2
Specific gravity (water = 1): 1.485 @ 68°F/20°C; 1.2797 at 100°C (Liquid).
Relative vapor density (air = 1): 4.4
Heat of combustion: –8515 Btu/lb = –4731 cal/g = –198 x 10^5 J/kg.
Vapor pressure: 1 mm.

---

p-NITROPHENOL REC. N:1950

SYNONYMS: EEC No. 609-015-00-0; 4-HYDROXYNITROBENZENE; NIPHEN; *p*-NITROFENOL (Spanish); 4-NITROPHENOL; PARANITROPHENOL (French, German); PHENOL, 4-NITRO-; PHENOL, *p*-NITRO; PNP; PHENOL, 4-NITRO; RCRA No. U170

IDENTIFICATION
CAS Number: 100-02-7
Formula: 1,4-$HOC_6H_4NO_2$
DOT ID Number: UN 1663; DOT Guide Number: 153
Proper Shipping Name: Nitrophenols (*o-*; *m-*; *p-*)
Reportable Quantity (RQ): **(CERCLA)** 100 lb/45.4 kg

DESCRIPTION: Colorless to light yellow crystalline solid. Sweet odor. Sinks and mixes with water.

Poison! • Combustible solid; burns in the absence of air • Irritating to the eyes, skin, and respiratory tract; toxic if swallowed • Containers may BLEVE when exposed to fire • Concentrated dust in confined areas (e.g., tanks, sewers, buildings) may explode when exposed to fire • Decomposes violently at 555°F/291°C. Toxic products of combustion may include nitrogen oxides • Do not put yourself in danger by entering a contaminated area to rescue a victim.

Hazard Classification (based on NFPA-704 Rating System)
Health Hazards (Blue): 3; Flammability (Red): 1; Reactivity (Yellow): 2

EMERGENCY RESPONSE: See Appendix A (153)
Evacuation:
Public safety: Isolate the area of spill or leak for at least 25 to 50 meters (80 to 160 feet) in all directions.
Spill: Increase, in the downwind direction, as necessary, the distance shown under "Public Safety."

Fire: If tank, rail car, or tank truck is involved in fire, isolate for at least 800 meters (½ mile) in all directions; also, consider initial evacuation for 800 meters (½ mile) in all directions.

EXPOSURE
Short-term effects: *SEEK MEDICAL ATTENTION. HARMFUL IF ABSORBED THROUGH THE SKIN.* **Dust:** Irritating to eyes, nose, and throat. *IF INHALED*, will, will cause headache or loss of consciousness. *IF IN EYES*, hold eyelids open and flush with plenty of water. *IF BREATHING HAS STOPPED*, give artificial respiration; *avoid mouth-to-mouth resuscitation; use bag/mask apparatus*. IF breathing is difficult, administer oxygen. **Solid:** Irritating to skin and eyes. *IF SWALLOWED*, will cause headache, nausea, or loss of consciousness. Remove contaminated clothing and shoes. Flush affected areas with plenty of water. *IF IN EYES*, hold eyelids open and flush with plenty of water. *IF SWALLOWED* and victim is *CONSCIOUS AND ABLE TO SWALLOW*, have victim drink 4 to 8 ounces of water. *IF SWALLOWED* and victim is *UNCONSCIOUS OR HAVING CONVULSIONS,* do nothing except keep victim warm.

HEALTH HAZARDS
Personal protective equipment (PPE): B-Level PPE. Butyl rubber gloves; side-shield safety glasses; dust mask or self-contained breathing apparatus).
Toxicity by ingestion: Grade 3; LD_{50} = 50 to 500 mg/kg.
Vapor (gas) irritant characteristics: Eye and respiratory tract irritant.
Liquid or solid irritant characteristics: Eye, skin and respiratory tract irritant.

FIRE DATA
Flash point: Combustible solid.
Behavior in fire: Containers may explode.
Autoignition temperature: 541°F/283°C/556°K.

CHEMICAL REACTIVITY
Binary reactants: Reacts with oxidizers and alkalies

ENVIRONMENTAL DATA
Food chain concentration potential: Log P_{ow} = 1.9. Unlikely to accumulate.
Water pollution: Effect of low concentration on aquatic life is unknown. May be dangerous if it enters nearby water intakes; notify operators. Notify local health and wildlife officials.
Response to discharge: Issue warning–water contaminant. Should be removed. Chemical and physical treatment. Clean Water Act.

SHIPPING INFORMATION
Grades of purity: Technical; Pure; **Storage temperature:** Ambient; **Inert atmosphere:** None; **Venting:** Open; **Stability during transport:** Stable.

PHYSICAL AND CHEMICAL PROPERTIES
Physical state @ 59°F/15°C and 1 atm: Solid.
Molecular weight: 139.1
Boiling point @ 1 atm: 534°F/279°C/552°K.
Melting/Freezing point: 235°F/113°C/386°K.
Specific gravity (water = 1): 1.48 @ 68°F/20°C (solid).
Relative vapor density (air = 1): 4.4
Heat of combustion: −8,870 Btu/lb = −4,930 cal/g = −206 x 10^5 J/kg.
Heat of fusion: 41.70 cal/g.
Vapor pressure: 1 mm.

1-NITROPROPANE REC. N:2000

SYNONYMS: EEC No. 609-001-00-6; 1-NP; NITROPROPANE; 1-NITROPROPANO (Spanish); PROPANE, 1-NITRO-

IDENTIFICATION
CAS Number: 108-03-2
Formula: $C_3H_7NO_2$; CH_3–CH_2–CH_2–NO_2
DOT ID Number: UN 2608; DOT Guide Number: 129
Proper Shipping Name: Nitropropanes

DESCRIPTION: Colorless liquid. Mild, fruity odor. May float or sink in water, depending on temperature; slightly soluble.

Highly flammable • When heated, may decompose explosively • Containers may BLEVE when exposed to fire • Vapors may form explosive mixture with air • Vapors are heavier than air and will collect and stay in low areas • Vapors may travel long distances to ignition sources and flashback • Vapors in confined areas (e.g., tanks, sewers, buildings) may explode when exposed to fire • Irritating to the skin, eyes, and respiratory tract • Toxic products of combustion may include nitrogen oxides.

Hazard Classification (based on NFPA-704 Rating System)
Health Hazards (Blue): 1; Flammability (Red): 3; Reactivity (Yellow): 2

EMERGENCY RESPONSE: See Appendix A (129)
Evacuation:
Public safety: Isolate spill area for at least 50 to 100 meters (160 to 330 feet) in all directions.
Spill: Large spill–Consider initial downwind evacuation for at least 300 meters (1000 feet).
Fire: Isolate for 800 meters (½ mile) in all directions, especially if tank, rail car, or tank truck is involved in fire.

EXPOSURE
Short-term effects: *SEEK MEDICAL ATTENTION.* **Vapor:** Irritating to eyes and respiratory system. *IF INHALED*, will, will cause headache, nausea, vomiting, and diarrhea. Move victim to fresh air. *IF IN EYES*, hold eyelids open and flush with plenty of running water. *IF BREATHING HAS STOPPED,* give artificial respiration. IF breathing is difficult, administer oxygen. **Liquid:** Irritating to skin and eyes. *IF SWALLOWED*, will cause headache, dizziness, nausea, vomiting, diarrhea, restlessness and muscular incoordination. *IF SWALLOWED* and victim is *CONSCIOUS AND ABLE TO SWALLOW*, give large volumes of water and induce vomiting. *IF SWALLOWED* and the victim is *UNCONSCIOUS OR HAVING CONVULSIONS,* do nothing except keep victim warm. *IF IN EYES*, hold eyelids open and flush with plenty of running water for at least 15 minutes. *IF ON SKIN*, wash with soap or mild detergent under running water for at least 15 minutes. Remove and double bag contaminated clothing and shoes.

HEALTH HAZARDS
Personal protective equipment (PPE): B-Level PPE. OSHA Table Z-1-A air contaminant. Wear self-contained breathing apparatus and full protective clothing including helmet, coat and pants worn by fire fighters, rubber boots, gloves, bands around legs, arms and waist along with face mask and coverings for parts of neck and head not protected by other apparel. Carbon type respirators containing HOPCALITE, an oxide catalyst that converts carbon monoxide to CO_2, should not be used with high vapor concentrations of 1-nitropropane because the resulting reaction may cause a fire. Chemical protective material(s) reported to have good

to excellent resistance: butyl rubber, PV alcohol, Silvershield®. Also, neoprene and natural rubber offers limited protection
Recommendations for respirator selection: NIOSH/OSHA
250 ppm: SA* (any supplied-air respirator). *625 ppm:* SA:CF* (any supplied-air respirator operated in a continuous-flow mode). *1000 ppm*: SCBAF (any self-contained breathing apparatus with a full facepiece); or SAF (any supplied-air respirator with a full facepiece). *EMERGENCY OR PLANNED ENTRY INTO UNKNOWN CONCENTRATIONS OR IDLH CONDITIONS:* SCBAF:PD,PP (any self-contained breathing apparatus that has a full facepiece and is operated in a pressure-demand or other positive-pressure mode); or SAF:PD,PP:ASCBA (any supplied-air respirator that has a full facepiece and is operated in a pressure-demand or other positive-pressure mode in combination with an auxiliary self-contained breathing apparatus operated in a pressure-demand or other positive pressure mode). *ESCAPE:* SCBAE (any appropriate escape-type, self-contained breathing apparatus). **Note*: Substance reported to cause eye irritation or damage; may require eye protection.
Exposure limits (TWA unless otherwise noted): ACGIH TLV 25 ppm (91 mg/m^3); OSHA PEL 25 ppm (90 mg/m^3).
Toxicity by ingestion: GRADE 3; LD_{50} = 455 mg/kg (rat).
Long-term health effects: Causes liver, kidney, and heart damage. Blood damage.
Vapor (gas) irritant characteristics: Eye and respiratory tract irritation. Vapor causes moderate irritation such that personnel will find high concentrations unpleasant.
Liquid or solid irritant characteristics: Eye, skin and respiratory tract irritant. If spilled on clothing and allowed to remain, may cause smarting and reddening of skin.
Odor threshold: 140 ppm. Detectable odor is greater than the Exposure limits. Exposure to potentially dangerous vapor concentrations can occur before the vapor is detected by smell.
IDLH value: 1000 ppm.

FIRE DATA
Flash point: 120°F/49°C (oc); 96°F/36°C (cc)
Fire extinguishing agents not to be used: Do not use dry chemicals because some of them may react with water to make the fire worse. In the presence of water, inorganic bases react with nitropropane to produce salts which are explosive when dry. fight fire from protection or from maximum possible distance.
Behavior in fire: Rapid heating may cause explosion.
Autoignition temperature: 789°F/421°C/694°K.
Electrical hazard: Class I, Group C.

CHEMICAL REACTIVITY
Binary reactants: Contact with amines, strong acids, metal oxides; mercury and silver salts, and alkalies may cause 1-nitropropane to become unstable and lead to explosion. May react with strong oxidizers to produce fires or explosions. Highly flammable when mixed with hydrocarbons or other combustibles. Attacks some forms of plastics, rubber, and coatings
Neutralizing agents for acids and caustics: Small spills may be covered with soda ash then mixed with water and subsequently neutralized with 6 molar hydrochloric acid.
Reactivity group: 42
Compatibility class: Nitrocompounds

ENVIRONMENTAL DATA
Food chain concentration potential: Log P_{ow} = 0.7. Unlikely to accumulate. **Water pollution:** Effect of low concentrations on aquatic life is unknown. May be dangerous if it enters nearby water intakes; notify operators. Fouling to shoreline. Notify local health and wildlife officials. **Response to discharge:** Issue warning–high flammability. Restrict access. Evacuate area. Should be removed. Chemical and physical treatment.

SHIPPING INFORMATION
Grades of purity: 98%; **Storage temperature:** Ambient; **Inert atmosphere:** None; **Venting:** Pressure vacuum valve; **Stability during transport:** Stable.

PHYSICAL AND CHEMICAL PROPERTIES
Physical state @ 59°F/15°C and 1 atm: Liquid.
Molecular weight: 89.1
Boiling point @ 1 atm: 269°F/131.6°C/405°K.
Melting/Freezing point: –162°F/–108°C/165°K.
Specific gravity (water = 1): 0.9934 @ 77°F/25°C (liquid).
Liquid surface tension: 30.57 dynes/cm = 0.0306 N/m @ 77°F/25°C.
Relative vapor density (air = 1): 3.1
Latent heat of vaporization: 09.5 Btu/lb = 116.4 cal/g = 4.87 x 10^5 J/kg.
Heat of combustion: 9723 Btu/lb = 5402 cal/g = 22.62 x 10^6 J/kg.
Vapor pressure: 13 mm.

2-NITROPROPANE   REC. N:2050

SYNONYMS: DIMETHYLNITROMETHANE; EEC No. 609-002-00-1; ISONITROPROPANE; NIPAR S-20; NIPAR S-30 SOLVENT; 2-NITROPROPANO (Spanish); NITROISOPROPANE; β-NITROPROPANE; 2-NP; PROPANE, 2-NITRO; sec-NITROPROPANE; RCRA No. U171

IDENTIFICATION
CAS Number: 79-46-9
Formula: $C_3H_7NO_2$; $(CH_3)_2CHNO_2$
DOT ID Number: UN 2608; DOT Guide Number: 129
Proper Shipping Name: Nitropropanes
Reportable Quantity (RQ): **(CERCLA)** 10 lb/4.54 kg

DESCRIPTION: Colorless liquid. Mild, fruity odor. Floats on the surface of water; slightly soluble.

Highly flammable • When heated, may decompose explosively • Containers may BLEVE when exposed to fire •Vapors may form explosive mixture with air • Vapors are heavier than air and will collect and stay in low areas • Vapors may travel long distances to ignition sources and flashback • Vapors in confined areas (e.g., tanks, sewers, buildings) may explode when exposed to fire • Irritating to the skin, eyes, and respiratory tract • Toxic products of combustion may include carbon monoxide and nitrogen oxides.

Hazard Classification (based on NFPA-704 Rating System)
Health Hazards (Blue): 1; Flammability (Red): 3; Reactivity (Yellow): 2

EMERGENCY RESPONSE: See Appendix A (129)
Evacuation:
Public safety: Isolate spill area for at least 50 to 100 meters (160 to 330 feet) in all directions.
Spill: Large spill–Consider initial downwind evacuation for at least 300 meters (1000 feet).
Fire: Isolate for 800 meters (½ mile) in all directions, especially if tank, rail car, or tank truck is involved in fire.

EXPOSURE

Short-term effects: *SEEK MEDICAL ATTENTION.* **Vapor:** Irritating to eyes, nose, and throat. *IF INHALED*, will, will cause headache, dizziness, coughing, or difficult breathing. *IF IN EYES*, hold eyelids open and flush with plenty of water. *IF BREATHING HAS STOPPED*, give artificial respiration. IF breathing is difficult, administer oxygen. **Liquid:** Irritating to skin and eyes. *IF SWALLOWED*, will cause nausea, and vomiting. Remove contaminated clothing and shoes. Flush affected areas with plenty of water. *IF IN EYES*, hold eyelids open and flush with plenty of water. *IF SWALLOWED* and victim is *CONSCIOUS AND ABLE TO SWALLOW*, have victim drink 4 to 8 ounces of water and have victim induce vomiting. *IF SWALLOWED* and victim is *UNCONSCIOUS OR HAVING CONVULSIONS*, do nothing except keep victim warm.

HEALTH HAZARDS

Personal protective equipment (PPE): B-Level PPE. OSHA Table Z-1-A air contaminant. Self-contained breathing apparatus; goggles or face shield; rubber gloves. Chemical protective material(s) reported to have good to excellent resistance: butyl rubber, PV alcohol, Teflon®, Silvershield®. Also, neoprene and natural rubber offers limited protection

Recommendations for respirator selection: NIOSH

At any concentrations above the NIOSH REL, or where there is no REL, at any detectable concentration: SCBAF:PD,PP (any self-contained breathing apparatus that has a full facepiece and is operated in a pressure-demand or other positive-pressure mode); or SAF:PD,PP:ASCBA (any supplied-air respirator that has a full facepiece and is operated in a pressure-demand or other positive-pressure mode in combination with an auxiliary self-contained breathing apparatus operated in a pressure-demand or other positive pressure mode). *ESCAPE:* SCBAE (any appropriate escape-type, self-contained breathing apparatus).

Exposure limits (TWA unless otherwise noted): ACGIH TLV 10 ppm (36 mg/m^3) suspected human carcinogen; OSHA PEL 25 ppm (90 mg/m^3); NIOSH REL reduce to lowest feasible level; suspected human carcinogen.

Toxicity by ingestion: Grade 2; oral rat LD$_{50}$ = 720 mg/kg.

Long-term health effects: Causes liver cancer in rats; OSHA specifically regulated carcinogen; IARC possible carcinogen, rating 2B; NTP anticipated carcinogen. Liver and kidney damage.

Vapor (gas) irritant characteristics: Eye and respiratory tract irritant. Vapors cause smarting of the eyes or respiratory system if present in high concentrations. The effect may be temporary.

Liquid or solid irritant characteristics: Eye and skin irritant. If spilled on clothing and allowed to remain, may cause smarting and reddening of skin.

Odor threshold: 18–288 ppm Note: Higher than exposure limits.

IDLH value: Suspected human carcinogen; 100 ppm.

FIRE DATA

Flash point: 100°F/38°C (oc); 82°F/28°C (cc).
Flammable limits in air: LEL: 2.6%; UEL: 11%.
Fire extinguishing agents not to be used: Alcohol foam; water may be ineffective.
Behavior in fire: Containers may explode.
Autoignition temperature: 802°F/428°C/701°K.

CHEMICAL REACTIVITY

Binary reactants: Amines, strong acids, alkalis, and oxidizers, chlorosulfonic acid, metal oxides, oleum, combustible materials. Nitric acid and bases form shock-sensitive materials. May attack some forms of plastics.

Reactivity group: 42
Compatibility class: Nitrocompounds

ENVIRONMENTAL DATA

Food chain concentration potential: Negative; unlikely to accumulate.

Water pollution: DOT Appendix B, §172.101–marine pollutant. Effect of low concentrations on aquatic life is unknown. Fouling to shoreline. May be dangerous if it enters nearby water intakes; notify operators. Notify local health and wildlife officials. **Response to discharge:** Issue warning–high flammability. Restrict access. Disperse and flush.

SHIPPING INFORMATION

Grades of purity: Technical, 94+%; **Storage temperature:** Ambient; **Inert atmosphere:** None; **Venting:** Open (flame arrester); **Stability during transport:** Stable.

NAS HAZARD CLASSIFICATION FOR BULK WATER TRANSPORTATION

FIRE: 3
HEALTH: Vapor irritant: 1; Liquid or solid irritant: 1; Poisons: 1
WATER POLLUTION: Human toxicity: 2; Aquatic toxicity: 3; Aesthetic effect: 2
REACTIVITY: Other chemicals: 3; Water: 0; Self-reaction: 4

PHYSICAL AND CHEMICAL PROPERTIES

Physical state @ 59°F/15°C and 1 atm: Liquid.
Molecular weight: 89.09
Boiling point @ 1 atm: 248.5°F/120.3°C/393.5°K.
Melting/Freezing point: −132°F/−91°C/182°K.
Specific gravity (water = 1): 0.99 @ 68°F/20°C (liquid).
Liquid surface tension: 30 dynes/cm = 0.030 N/m @ 68°F/20°C.
Relative vapor density (air = 1): 3.1
Ratio of specific heats of vapor (gas): 1.090 @ 68°F/20°C.
Latent heat of vaporization: 178 Btu/lb = 99 cal/g = 4.1 x 10^5 J/kg.
Heat of combustion: −9650 Btu/lb = −5360 cal/g = −224 x 10^5 J/kg.
Vapor pressure: 13 mm.

NITROSYL CHLORIDE   REC. N:2100

SYNONYMS: CLORURO de NITROSILO (Spanish); NITROGEN CHLORIDE OXIDE; NITROGEN OXYCHLORIDE

IDENTIFICATION

CAS Number: 2696-92-6
Formula: NOCl
DOT ID Number: UN 1069; DOT Guide Number: 125
Proper Shipping Name: Nitrosyl chloride

DESCRIPTION: Yellow to reddish liquid or yellow gas. Choking odor. Liquid sinks and mixes with water; reacts, forming hydrochloric acid. Poisonous visible vapor cloud of hydrogen chloride is produced.

Poison! • Highly corrosive • May interfere with the body's ability to use oxygen • Breathing the vapor can kill you; skin or eye contact causes severe burns, impaired vision, or blindness • Firefighting gear (including SCBA) does not provide adequate protection. If exposure occurs, remove and isolate gear immediately and thoroughly decontaminate personnel • Strong oxidizer which may react spontaneously with low flash point organics or reducing agents. Heat forms oxygen; will increase the activity of an existing fire • May cause fire or explosion on contact with combustibles (wood, paper, oil, clothing, etc.) • Containers may BLEVE when exposed to fire • Toxic products of combustion may include hydrogen chloride and nitrogen oxides • Do not put yourself in danger by entering a contaminated area to rescue a victim.

Hazard Classification (based on NFPA-704 Rating System)
Health Hazards (Blue): 3; Flammability (Red): 0; Reactivity (Yellow): 1; Special Notice (White): OXY

EMERGENCY RESPONSE: See Appendix A (125)
Evacuation:
Public safety: Isolate spill area for at least 100 to 200 meters (330 to 660 feet) in all directions.
Spill: Small spill–First: Isolate in all directions 30 meters (100 feet); Then: Protect persons downwind, DAY: 0.3 km (0.2 mile); NIGHT: 0.9 mi, Large spill–First: Isolate in all directions 370 meters (1200 feet); 1200; Then: Protect persons downwind, DAY: 3.5 km (2.2 miles); NIGHT: 9.8 km (6.1miles).
Fire: Isolate for 1600 meters (1 mile) in all directions, especially if tank, rail car, or tank truck is involved in fire.

EXPOSURE
Short-term effects: *SEEK MEDICAL ATTENTION.* **Vapor:** *POISONOUS IF INHALED OR IF SKIN IS EXPOSED.* Irritating to eyes. Move to fresh air. *IF BREATHING HAS STOPPED,* give artificial respiration; *avoid mouth-to-mouth resuscitation; use bag/mask apparatus.* IF breathing is difficult, administer oxygen.
Liquid: *POISONOUS IF SWALLOWED OR IF SKIN IS EXPOSED.* Irritating to eyes. Remove contaminated clothing and shoes. Flush affected areas with plenty of water. *IF IN EYES*, hold eyelids open and flush with plenty of water.
Note to physician or authorized medical personnel: Medical observation is recommended for 24 to 48 hours after breathing overexposure, as pulmonary edema may be delayed. As first aid for pulmonary edema, consider administering a corticosteroid spray. Cigarette smoking may exacerbate pulmonary injury and should be discouraged for at least 72 hours following exposure.

HEALTH HAZARDS
Personal protective equipment (PPE): A-Level PPE. Self-contained breathing apparatus (approved mask may be used for short exposures only); rubberized clothing; gloves; shoes; chemical goggles.
Vapor (gas) irritant characteristics: Vapors cause severe irritation of eye and throat and can cause eye and lung injury. They cannot be tolerated even at low concentrations.
Liquid or solid irritant characteristics: Severe burns to eyes and skin.

CHEMICAL REACTIVITY
Reactivity with water: Dissolves and reacts to form acid solution and toxic red oxides of nitrogen.
Binary reactants: Corrosive to most metals, but reaction is not hazardous
Neutralizing agents for acids and caustics: Flush with water. Residual acid may be neutralized with soda ash.

ENVIRONMENTAL DATA
Food chain concentration potential: Unlikely to accumulate.
Water pollution: Harmful to aquatic life in very low concentrations. May be dangerous if it enters nearby water intakes; notify operators. Notify local health and wildlife officials.
Response to discharge: Issue warning–poison. Restrict access. Disperse and flush.

SHIPPING INFORMATION
Grades of purity: 97%; **Storage temperature:** Ambient; **Inert atmosphere:** None; **Venting:** Safety relief; **Stability during transport:** Stable.

PHYSICAL AND CHEMICAL PROPERTIES
Physical state @ 59°F/15°C and 1 atm: Gas.
Molecular weight: 65.46
Boiling point @ 1 atm: 21.6°F/–5.8°C/267.4°K.
Melting/Freezing point: –74°F/–59°C/214°K.
Critical temperature: 334°F/168°C/441°K.
Critical pressure: 1300 psia = 90 atm = 9.1 MN/m^2.
Specific gravity (water = 1): 1.36 at –5.7°C (liquid).
Relative vapor density (air = 1): 2.3
Ratio of specific heats of vapor (gas): 1.229
Latent heat of vaporization: 164 Btu/lb = 91.0 cal/g = 3.81 x 10^5 J/kg.

m-NITROTOLUENE   REC. N:2150

SYNONYMS: EEC No. 609-006-00-3; *m*-METHYLNITROBENZENE; 3-METHYLNITROBENZENE; MNT; 3-NITROTOLUENE; METANITROTOLUENE; *m*-NITROTOLUENO (Spanish); 3-NITROTOLUOL

IDENTIFICATION
CAS Number: 99-08-1
Formula: C$_7$H$_7$NO$_2$; C$_6$H$_4$CH$_3$NO$_2$
DOT ID Number: UN 1664; DOT Guide Number: 152
Proper Shipping Name: Nitrotoluenes, liquid or solid
Reportable Quantity (RQ): **(CERCLA)** 1000 lb/454 kg

DESCRIPTION: Yellow liquid. Weak, aromatic odor. Sinks in water; insoluble. Freezes at 60°F/15.6°C.

Poison! May interfere with the body's ability to use oxygen • Combustible • Containers may BLEVE when exposed to fire • Vapors are heavier than air and will collect and stay in low areas • Vapors in confined areas (e.g., tanks, sewers, buildings) may explode when exposed to fire • Irritating to the skin, eyes, and respiratory tract • Toxic products of combustion may include carbon monoxide and nitrogen oxides.

Hazard Classification (based on NFPA-704 Rating System)
Health Hazards (Blue): 3; Flammability (Red): 1; Reactivity (Yellow): 1

EMERGENCY RESPONSE: See Appendix A (152)
Evacuation:
Public safety: Isolate the area of spill or leak for at least 25 to 50 meters (80 to 160 feet) in all directions.
Spill: Increase, in the downwind direction, as necessary, the distance shown under "Public Safety."
Fire: If tank, rail car, or tank truck is involved in fire, isolate for at least 800 meters (½ mile) in all directions; also, consider initial evacuation for 800 meters (½ mile) in all directions.

EXPOSURE
Short-term effects: *SEEK MEDICAL ATTENTION.* **Vapor:** If inhaled may cause headache, dizziness, nausea, vomiting, and difficult breathing. Move to fresh air. *IF BREATHING HAS*

STOPPED, give artificial respiration; *avoid mouth-to-mouth resuscitation; use bag/mask apparatus.* IF breathing is difficult, administer oxygen. **Liquid:** *HARMFUL IF ABSORBED THROUGH THE SKIN.* If swallowed or skin is exposed, may cause headache, dizziness, nausea, vomiting, and difficult breathing. Remove contaminated clothing and shoes. Flush affected areas with plenty of water. *IF IN EYES*, hold eyelids open and flush with plenty of water. *IF SWALLOWED* and victim is *CONSCIOUS AND ABLE TO SWALLOW*, have victim drink 4 to 8 ounces of water and have victim induce vomiting.

HEALTH HAZARDS
Personal protective equipment (PPE): B-Level PPE. OSHA Table Z-1-A air contaminant. Protective clothing, including butyl rubber gloves and boots, safety goggles or face mask, respirator with approved canister); or self-contained breathing apparatus. Chemical protective material(s) reported to have good to excellent resistance: butyl rubber. Also, neoprene, PVC, styrene-butadiene, styrene-butadiene/neoprene offers limited protection.
Recommendations for respirator selection: NIOSH
20 ppm: SA* (any supplied-air respirator). *50 ppm:* SA:CF* (any supplied-air respirator operated in a continuous-flow mode). *100 ppm:* SAT:CF* (any supplied-air respirator that has a tight-fitting facepiece and is operated in a continuous-flow mode); or SCBAF (any self-contained breathing apparatus with a full facepiece); or SAF (any supplied-air respirator with a full facepiece). *200 ppm:* SAF:PD,PP (any supplied-air respirator that has a full facepiece and is operated in a pressure-demand or other positive-pressure mode). *EMERGENCY OR PLANNED ENTRY INTO UNKNOWN CONCENTRATIONS OR IDLH CONDITIONS:* SCBAF:PD,PP (any self-contained breathing apparatus that has a full facepiece and is operated in a pressure-demand or other positive-pressure mode); or SAF:PD,PP:ASCBA (any supplied-air respirator that has a full facepiece and is operated in a pressure-demand or other positive-pressure mode in combination with an auxiliary self-contained breathing apparatus operated in a pressure-demand or other positive pressure mode). *ESCAPE:* GMFOVHiE [any air-purifying, full-facepiece respirator (gas mask) with a chin-style, front- or back-mounted organic vapor canister having a high-efficiency particulate filter]; or SCBAE (any appropriate escape-type, self-contained breathing apparatus). *Note:* Substance reported to cause eye irritation or damage; may require eye protection:
Exposure limits (TWA unless otherwise noted): ACGIH TLV 2 ppm (11 mg/m^3); OSHA PEL 5 ppm (30 mg/m^3); NIOSH REL 2 ppm (11 mg/m^3); skin contact contributes significantly in overall exposure.
Toxicity by ingestion: Grade 2; LD$_{50}$ = 0.5 to 5 g/kg.
Long-term health effects: Poisonous through inhalation; has a cumulative effect. Chronic exposure can cause skin, eye, mucous membrane and respiratory irritation. Caused anemia and other blood changes in rats.
Vapor (gas) irritant characteristics: Eye and respiratory tract irritation.
Liquid or solid irritant characteristics: Eye and skin irritation.
Odor threshold: 0.05 ppm.
IDLH value: 200 ppm.

FIRE DATA
Flash point: 223°F/106°C (cc).
Fire extinguishing agents not to be used: Water or foam may cause frothing.

CHEMICAL REACTIVITY
Binary reactants: Violent reaction with strong acids, alkalis (e.g., sodium hydroxide), ammonia, amines, reducing agents, strong oxidizers. Elevated temperature may cause explosive decomposition. Attacks some plastics, rubber, and coatings.
Reactivity group: 42
Compatibility class: Nitrocompounds

ENVIRONMENTAL DATA
Food chain concentration potential: Log P$_{ow}$ = 2.5. Unlikely to accumulate.
Water pollution: DOT Appendix B, §172.101–marine pollutant. Harmful to aquatic life in very low concentrations. May be dangerous if it enters nearby water intakes; notify operators. Notify local health and wildlife officials. **Response to discharge:** Issue warning–water contaminant. Chemical and physical treatment.

SHIPPING INFORMATION
Storage temperature: Ambient; **Inert atmosphere:** None; **Venting:** Pressure-vacuum; **Stability during transport:** Stable.

PHYSICAL AND CHEMICAL PROPERTIES
Physical state @ 59°F/15°C and 1 atm: Liquid.
Molecular weight: 137.13
Boiling point @ 1 atm: 450°F/232.6°C/505.8°K.
Melting/Freezing point: 60.8°F/16.0°C/289.2°K.
Critical temperature: (estimate) 899.2°F/481.8°C/754.9°K.
Critical pressure: (estimate) 611.8 psia = 41.6 atm = 4.22 MN/m^2.
Specific gravity (water = 1): 1.1571 @ 68°F/20°C.
Liquid surface tension: 39.07 dynes/cm = 0.03907 N/m at 40°C.
Liquid water interfacial tension: 34.9 dynes/cm = 0.0349 N/m at 40°C.
Relative vapor density (air = 1): 4.73.
Ratio of specific heats of vapor (gas): (GAS): (estimate) more than 1.
Latent heat of vaporization: (estimate) at boiling point-155.3 Btu/lb = 86.3 cal/g = 3.61 x 10^5 J/kg.
Heat of combustion: (estimate) –11232 Btu/lb = –6240 cal/g = –261.1 x 10^5 J/kg.
Vapor pressure: 0.12 mm.

o-NITROTOLUENE   REC. N:2200

SYNONYMS: EEC No. 609-006-00-3; 2-METHYLNITROBENZENE; METHYL NITROBENZENE; O-METHYLNITROBENZENE; 2-NITROTOLUOL; 2-NITROTOLUENE; *o*-NITROTOLUENO (Spanish); ONT; TOLUENE, O-NITRO; NITROTOLUOL

IDENTIFICATION
CAS Number: 88-72-2
Formula: C$_7$H$_7$NO$_2$
DOT ID Number: UN 1664; DOT Guide Number: 152
Proper Shipping Name: Nitrololuenes, liquid or solid
Reportable Quantity (RQ): **(CERCLA)** 1000 lb/454 kg

DESCRIPTION: Yellow, oily liquid. Weak aromatic odor; like bitter almond. Sinks in water; insoluble. Freezes at 60°F/15.6°C.

Poison! May interfere with the body's ability to use oxygen • Combustible • Containers may BLEVE when exposed to fire • Vapors are heavier than air and will collect and stay in low areas • Vapors in confined areas (e.g., tanks, sewers, buildings) may explode when exposed to fire • Irritating to the skin, eyes, and respiratory tract • Toxic products of combustion may include carbon monoxide and nitrogen oxides.

Hazard Classification (based on NFPA-704 Rating System)
Health Hazards (Blue): 3; Flammability (Red): 1; Reactivity (Yellow): 1

EMERGENCY RESPONSE: See Appendix A (152)
Evacuation:
Public safety: Isolate the area of spill or leak for at least 25 to 50 meters (80 to 160 feet) in all directions.
Spill: Increase, in the downwind direction, as necessary, the distance shown under "Public Safety."
Fire: If tank, rail car, or tank truck is involved in fire, isolate for at least 800 meters (½ mile) in all directions; also, consider initial evacuation for 800 meters (½ mile) in all directions.

EXPOSURE
Short-term effects: *SEEK MEDICAL ATTENTION.* **Vapor:** If inhaled, may cause headache, dizziness, nausea, vomiting, and difficult breathing. Move to fresh air.*IF BREATHING HAS STOPPED*, give artificial respiration; *avoid mouth-to-mouth resuscitation; use bag/mask apparatus*. IF breathing is difficult, administer oxygen. **Liquid:** *HARMFUL IF ABSORBED THROUGH THE SKIN.* If swallowed or skin is exposed, may cause headache, dizziness, nausea, vomiting, and difficult breathing. Remove contaminated clothing and shoes. Flush affected areas with plenty of water. *IF IN EYES*, hold eyelids open and flush with plenty of water. *IF SWALLOWED* and victim is *CONSCIOUS AND ABLE TO SWALLOW*, have victim drink 4 to 8 ounces of water and have victim induce vomiting.

HEALTH HAZARDS
Personal protective equipment (PPE): B-Level PPE. OSHA Table Z-1-A air contaminant. Wear butyl rubber gloves, protective working clothes, self-contained breathing apparatus, and protective shoes. Chemical protective material(s) reported to have good to excellent resistance: butyl rubber.
Recommendations for respirator selection: NIOSH
20 ppm: SA* (any supplied-air respirator). *50 ppm:* SA:CF* (any supplied-air respirator operated in a continuous-flow mode). *100 ppm:* SAT:CF* (any supplied-air respirator that has a tight-fitting facepiece and is operated in a continuous-flow mode); or SCBAF (any self-contained breathing apparatus with a full facepiece); or SAF (any supplied-air respirator with a full facepiece). *200 ppm:* SAF:PD,PP (any supplied-air respirator that has a full facepiece and is operated in a pressure-demand or other positive-pressure mode). *EMERGENCY OR PLANNED ENTRY INTO UNKNOWN CONCENTRATIONS OR IDLH CONDITIONS:* SCBAF:PD,PP (any self-contained breathing apparatus that has a full facepiece and is operated in a pressure-demand or other positive-pressure mode); or SAF:PD,PP:ASCBA (any supplied-air respirator that has a full facepiece and is operated in a pressure-demand or other positive-pressure mode in combination with an auxiliary self-contained breathing apparatus operated in a pressure-demand or other positive pressure mode). *ESCAPE:* GMFOVHiE [any air-purifying, full-facepiece respirator (gas mask) with a chin-style, front- or back-mounted organic vapor canister having a high-efficiency particulate filter]; or SCBAE (any appropriate escape-type, self-contained breathing apparatus). *Note:* Substance reported to cause eye irritation or damage; may require eye protection.
Exposure limits (TWA unless otherwise noted): ACGIH 2 ppm (11 mg/m^3); OSHA PEL (transitional 5 ppm) 2 ppm; skin contact contributes significantly in overall exposure.
Toxicity by ingestion: Grade 2; LD$_{50}$ 0.5 to 5 g/kg.
Long-term health effects: Poisonous through inhalation; has a cumulative effect. Chronic exposure may produce a reversible anemia. Increased number of leukocytes and methemoglobin blood level, decreased number of erythrocytes and hemoglobin level. Impaired function of liver.
Vapor (gas) irritant characteristics: Eye and respiratory tract irritation.
Liquid or solid irritant characteristics: No appreciable hazard. Practically harmless to skin.
Odor threshold: 0.05 mg/l.
IDLH value: 200 ppm.

FIRE DATA
Flash point: 203°F/95°C (oc); 223°F/106°C (cc).
Fire extinguishing agents not to be used: Water or foam may cause frothing.
Autoignition temperature: 581°F/305°C/578°K.

CHEMICAL REACTIVITY
Binary reactants: Strong oxidizers, sulfuric acid, sodium hydroxide, strong acids and reducing agents
Reactivity group: 42
Compatibility class: Nitrocompounds

ENVIRONMENTAL DATA
Food chain concentration potential: Log P$_{ow}$ = 2.3. Unlikely to accumulate.
Water pollution: DOT Appendix B, §172.101–marine pollutant. Harmful to aquatic life in very low concentrations. May be dangerous if it enters nearby water intakes; notify operators. Notify local health and wildlife officials. **Response to discharge:** Issue warning–water contaminant. Chemical and physical treatment.

SHIPPING INFORMATION
Grades of purity: Technical, 99.5%; **Stability during transport:** Stable.

PHYSICAL AND CHEMICAL PROPERTIES
Physical state @ 59°F/15°C and 1 atm: Liquid.
Molecular weight: 137.13
Boiling point @ 1 atm: 431.6°F/222°C/495.2°K.
Melting/Freezing point: 15°F/–9°C/264°K.
Specific gravity (water = 1): 1.1622 @ 66°F/19°C/292°K.
Liquid surface tension: 42.29 dynes/cm = 0.04229 N/m @ 15°C; 41.67 dynes/cm = 0.04167 N/m @ 68°F/20°C; 40.50 dynes/cm = 0.04050 N/m at 30°C; 41.76 dynes/cm = 0.04176 N/m at 19.5°C.
Liquid water interfacial tension: 27.19 dynes/cm = 0.02719 N/m @ 68°F/20°C.
Relative vapor density (air = 1): 4.73
Latent heat of vaporization: 151 Btu/lb = 83.8 cal/g = 3.5 x 10^5 J/kg.
Heat of combustion: –11,290 Btu/lb = –6272 cal/g = –262 x 10^5 J/kg.
Vapor pressure: (Reid) Low; 0.12 mm.

p-NITROTOLUENE   REC. N:2250

SYNONYMS: EEC No. 609-006-00-3; 4-METHYLNITROBENZENE; *p*-METHYL NITROBENZENE; 4-NITROTOLUENE; 4-NITROTOLUOL; *p*-NITROTOLUENO (Spanish); para-NITROTOLUOL; TOLUENE, *p*-NITRO-; PNT

IDENTIFICATION
CAS Number: 99-99-0
Formula: C$_7$H$_7$NO$_2$
DOT ID Number: UN 1664; DOT Guide Number: 152

Proper Shipping Name: Nitrotoluenes, liquid or solid
Reportable Quantity (RQ): (CERCLA) 1000 lb/454 kg

DESCRIPTION: Yellow crystalline solid. Weak, bitter almond odor. Sinks in water; practically insoluble.

Poison! May interfere with the body's ability to use oxygen • Combustible solid • Containers may BLEVE when exposed to fire • Vapors are heavier than air and will collect and stay in low areas • Dust or vapors in confined areas (e.g., tanks, sewers, buildings) may explode when exposed to fire • Irritating to the skin, eyes, and respiratory tract • Toxic products of combustion may include carbon monoxide and nitrogen oxides. *Warning*: Unstable in heat, especially above 373°F/190°C. Material that has been standing may explode.

Hazard Classification (based on NFPA-704 Rating System)
Health Hazards (Blue): 3; Flammability (Red): 1; Reactivity (Yellow): 1

EMERGENCY RESPONSE: See Appendix A (152)
Evacuation:
Public safety: Isolate the area of spill or leak for at least 25 to 50 meters (80 to 160 feet) in all directions.
Spill: Increase, in the downwind direction, as necessary, the distance shown under "Public Safety."
Fire: If tank, rail car, or tank truck is involved in fire, isolate for at least 800 meters (½ mile) in all directions; also, consider initial evacuation for 800 meters (½ mile) in all directions.

EXPOSURE
Short-term effects: *SEEK MEDICAL ATTENTION. HARMFUL IF ABSORBED THROUGH THE SKIN.* **Dust or Solid:** If inhaled, swallowed, or skin is exposed, may cause headache, dizziness, nausea, vomiting, and difficult breathing. Move to fresh air. *IF BREATHING HAS STOPPED*, give artificial respiration; *avoid mouth-to-mouth resuscitation; use bag/mask apparatus*. IF breathing is difficult, administer oxygen. Remove contaminated clothing and shoes. Flush affected areas with plenty of water. *IF IN EYES*, hold eyelids open and flush with plenty of water. *IF SWALLOWED* and victim is *CONSCIOUS AND ABLE TO SWALLOW*, have victim drink 4 to 8 ounces of water and have victim induce vomiting.

HEALTH HAZARDS
Personal protective equipment (PPE): B-Level PPE. OSHA Table Z-1-A air contaminant. Organic vapor respirator. Wear butyl rubber gloves, protective clothing and shoes. Chemical protective material(s) reported to have good to excellent resistance: butyl rubber, polycarbonate.
Recommendations for respirator selection: NIOSH
20 ppm: SA* (any supplied-air respirator). *50 ppm*: SA:CF* (any supplied-air respirator operated in a continuous-flow mode). *100 ppm*: SAT:CF* (any supplied-air respirator that has a tight-fitting facepiece and is operated in a continuous-flow mode); or SCBAF (any self-contained breathing apparatus with a full facepiece); or SAF (any supplied-air respirator with a full facepiece). *200 ppm*: SAF:PD,PP (any supplied-air respirator that has a full facepiece and is operated in a pressure-demand or other positive-pressure mode). *EMERGENCY OR PLANNED ENTRY INTO UNKNOWN CONCENTRATIONS OR IDLH CONDITIONS*: SCBAF:PD,PP (any self-contained breathing apparatus that has a full facepiece and is operated in a pressure-demand or other positive-pressure mode); or SAF:PD,PP:ASCBA (any supplied-air respirator that has a full facepiece and is operated in a pressure-demand or other positive-pressure mode in combination with an auxiliary self-contained breathing apparatus operated in a pressure-demand or other positive pressure mode). *ESCAPE*: GMFOVHiE [any air-purifying, full-facepiece respirator (gas mask) with a chin-style, front- or back-mounted organic vapor canister having a high-efficiency particulate filter]; or SCBAE (any appropriate escape-type, self-contained breathing apparatus). **Note*: Substance reported to cause eye irritation or damage; may require eye protection:
Exposure limits (TWA unless otherwise noted): ACGIH TLV 2 ppm (11 mg/m^3); OSHA PEL 5 ppm (30 mg/m^3); NIOSH REL 2 ppm (11 mg/m^3); skin contact contributes significantly in overall exposure.
Toxicity by ingestion: Grade 2; LD_{50} = 0.5 to 5 g/kg.
Long-term health effects: Poisonous through inhalation; has a cumulative effect. Methemoglobin formation; kidney and liver damage. Changes in conditioned reflex activity with high dosage.
Odor threshold: 0.05 ppm.
IDLH value: 200 ppm.

FIRE DATA
Flash point: 223°F/106°C (cc).
Fire extinguishing agents not to be used: Water and water-based foam may cause frothing.
Autoignition temperature: 734°F/390°C/663°K.

CHEMICAL REACTIVITY
Binary reactants: Strong oxidizers, sulfuric acid or other strong acids, caustics, reducing agents, ammonia, amines.
Reactivity group: 42
Compatibility class: Nitrocompounds

ENVIRONMENTAL DATA
Food chain concentration potential: Log P_{ow} = 2.4. Unlikely to accumulate.
Water pollution: DOT Appendix B, §172.101–marine pollutant. Harmful to aquatic life in very low concentrations. May be dangerous if it enters nearby water intakes; notify operators. Notify local health and wildlife officials. **Response to discharge:** Issue warning–water contaminant. Chemical and physical treatment.

SHIPPING INFORMATION
Storage temperature: Cool

PHYSICAL AND CHEMICAL PROPERTIES
Physical state @ 59°F/15°C and 1 atm: Solid.
Molecular weight: 137.15
Boiling point @ 1 atm: 461°F/238.3°C/511.5°K.
Melting/Freezing point: 125°F/51.7°C/324.9°K.
Specific gravity (water = 1): 1.286 @ 68°F/20°C.
Liquid surface tension: 36.83 dynes/cm = 0.03683 N/m at 60°C; 35.64 dynes/cm = 0.03564 N/m at 158°F/70°C.
Relative vapor density (air = 1): 4.72
Latent heat of vaporization: 157 Btu/lb = 87 cal/g = 3.64 x 10^5 J/kg.
Heat of combustion: –11.181 Btu/lb = –6212 cal/g = –260 x 10^5 J/kg.
Vapor pressure: 1.0 mm.

NITROUS OXIDE REC. N:2300

SYNONYMS: DINITROGEN MONOXIDE; FACTITIOUS AIR; HYPONITROUS ACID ANHYDRIDE; LAUGHING GAS; NITROGEN OXIDE; STICKDIOXYD (German)

IDENTIFICATION
CAS Number: 10024-97-2
Formula: N_2O
DOT ID Number: UN 1070 (compressed gas); UN 2201 (refrigerated liquid); DOT Guide Number: 122
Proper Shipping Name: Nitrous oxide, compressed; Nitrous oxide, refrigerated Liquid.

DESCRIPTION: Colorless gas. Shipped and stored as a cryogenic liquid. Odorless or mild, sweet odor. Sinks and boils in water; soluble. Visible vapor cloud is produced.

An oxidizer; nonflammable but will support combustion and intensify fire • Vapors may cause dizziness or asphyxiation without warning; narcotic at high concentration • May form explosive mixture with air at elevated temperatures • Containers may BLEVE when exposed to fire • Vapors are heavier than air and will collect and stay in low areas • Vapors in confined areas (e.g., tanks, sewers, buildings) may explode when exposed to fire • Contact with liquid may cause frostbite • Toxic products of combustion may include nitrogen oxides.

EMERGENCY RESPONSE: See Appendix A (122)
Evacuation:
Public safety: Isolate spill area for at least 25 to 50 meters (80 to 160 feet) in all directions.
Spill: Large spill –Consider initial downwind evacuation for at least 500 meters (⅓ mile).
Fire: Isolate for 800 meters (½ mile) in all directions, especially if tank, rail car, or tank truck is involved in fire.

EXPOSURE
Short-term effects: *SEEK MEDICAL ATTENTION.* **Vapor:** IF inhaled, will cause dizziness, difficult breathing, or loss of consciousness. Move victim to fresh air. If breathing is difficult, administer oxygen. **Liquid:** Will cause frostbite. Flush affected areas with plenty of water. *DO NOT RUB AFFECTED AREAS.*

HEALTH HAZARDS
Personal protective equipment (PPE): B-Level PPE. Self-contained breathing apparatus for high vapor concentrations. Wear thermal protective clothing.
Exposure limits (TWA unless otherwise noted): ACGIH TLV 50 ppm (90 mg/m^3); NIOSH 25 ppm (46 mg/m^3).
Toxicity by ingestion: Grade 0; LD_{50} = > 15 g/kg.
Long-term health effects: Causes birth defects in rats; can cause lethal effects in chick eggs. IARC listed, overall evaluation unassigned.
Vapor (gas) irritant characteristics: Vapors are nonirritating to eyes and throat, but may affect the nervous system.
Liquid or solid irritant characteristics: No appreciable hazard; practically harmless to the skin.

FIRE DATA
Flash point: Nonflammable compressed gas.
Behavior in fire: Containers may explode.

CHEMICAL REACTIVITY
Binary reactants: Aluminum, boron, hydrazine, lithium hydride, phosphine, sodium; anhydrous ammonia (explosive). Supports combustion, but does not cause spontaneous ignition

ENVIRONMENTAL DATA
Food chain concentration potential: Negative; unlikely to accumulate.
Water pollution: Not harmful to aquatic life. **Response to discharge:** Restrict access.

SHIPPING INFORMATION
Grades of purity: 98.0+%; **Storage temperature:** Ambient; **Inert atmosphere:** None; **Venting:** Safety relief; **Stability during transport:** Stable.

NAS HAZARD CLASSIFICATION FOR BULK WATER TRANSPORTATION
FIRE: 0
HEALTH: Vapor irritant: 0; Liquid or solid irritant: 0; Poisons: 1
WATER POLLUTION: Human toxicity: 0; Aquatic toxicity: 0; Aesthetic effect: 0
REACTIVITY: Other chemicals: 3; Water: 0; Self-reaction: 0

PHYSICAL AND CHEMICAL PROPERTIES
Physical state @ 59°F/15°C and 1 atm: Gas.
Molecular weight: 44.0
Boiling point @ 1 atm: –129.1°F/–89.5°C/183.7°K.
Melting/Freezing point: –131.5°F/–90.8°C/182.4°K.
Critical temperature: 97.7°F/36.5°C/309.7°K.
Critical pressure: 1054 psia = 71.7 atm = 7.28 MN/m^2.
Specific gravity (water = 1): 1.266 at –89°C (liquid).
Liquid surface tension: 10 dynes/cm = 0.0101 N/m at –25°C.
Relative vapor density (air = 1): 1.53
Ratio of specific heats of vapor (gas): 1.303 @ 77°F/25°C.
Latent heat of vaporization: 161.7 Btu/lb = 89.9 cal/g = 3.76 x 10^5 J/kg.
Vapor pressure: 39 mm.

NONANE   REC. N:2350

SYNONYMS: *n*-NONANE; *n*-NONANO (Spanish); NONYLHYDRIDE; SHELLSOL 140

IDENTIFICATION
CAS Number: 111-84-2
Formula: C_9H_{20}
DOT ID Number: UN 1920; DOT Guide Number: 128
Proper Shipping Name: Nonanes

DESCRIPTION: Colorless liquid. Floral- or citronella-like odor. Floats on water surface; low solubility.

Highly flammable • Containers may BLEVE when exposed to fire • Vapors may form explosive mixture with air • Vapors are heavier than air and will collect and stay in low areas • Vapors may travel long distances to ignition sources and flashback • Vapors in confined areas (e.g., tanks, sewers, buildings) may explode when exposed to fire • Irritating to the skin, eyes, and respiratory tract • Toxic products of combustion may include carbon monoxide.

Hazard Classification (based on NFPA-704 Rating System)
Health Hazards (Blue): 0; Flammability (Red): 3; Reactivity (Yellow): 0

EMERGENCY RESPONSE: See Appendix A (128)
Evacuation:
Public safety: Isolate spill area for at least 25 to 50 meters (80 to 160 feet) in all directions.
Spill: Large spill–Consider initial downwind evacuation for at least 300 meters (1000 feet).

Fire: Isolate for 800 meters (½ mile) in all directions, especially if tank, rail car, or tank truck is involved in fire.

EXPOSURE
Short-term effects: *SEEK MEDICAL ATTENTION.* **Liquid:** Irritating to skin and eyes. *IF SWALLOWED*, will cause nausea, and vomiting. Remove contaminated clothing and shoes. Flush affected areas with plenty of water. *IF IN EYES*, hold eyelids open and flush with plenty of water. *IF SWALLOWED* and victim is *CONSCIOUS AND ABLE TO SWALLOW*, have victim drink 4 to 8 ounces of water. **Do NOT induce vomiting.**

HEALTH HAZARDS
Personal protective equipment (PPE): B-Level PPE. OSHA Table Z-1-A air contaminant. Self-contained breathing apparatus for high vapor concentrations; goggles or face shield; rubber gloves.
Exposure limits (TWA unless otherwise noted): ACGIH TLV 200 ppm (1050 mg/m^3); NIOSH REL 200 ppm (1050 mg/m^3).
Toxicity by ingestion: Grade 0; LD$_{50}$ = > 15 g/kg.
Vapor (gas) irritant characteristics: Eye and respiratory tract irritant.
Liquid or solid irritant characteristics: Practically harmless to the skin.
Odor threshold: 1–21 ppm.

FIRE DATA
Flash point: 88°F/31°C (cc).
Flammable limits in air: LEL: 0.87%; UEL: 2.9%.
Fire extinguishing agents not to be used: Water may be ineffective
Autoignition temperature: 401°F/205°C/474°K.
Electrical hazard: Class I, Group D.
Burning rate: 5.8 mm/min

CHEMICAL REACTIVITY
Binary reactants: Strong oxidizers.
Reactivity group: 31
Compatibility class: Paraffins

ENVIRONMENTAL DATA
Food chain concentration potential: Log P$_{ow}$ = 5.46 (Prager). Values > 3.0 are likely to bioconcentrate in aquatic organisms and other living tissue, especially in fats.
Water pollution: Effect of low concentrations on aquatic life is unknown. May be dangerous to waterfowl. Fouling to shoreline. May be dangerous if it enters nearby water intakes; notify operators. Notify local health and wildlife officials. **Response to discharge:** Mechanical containment. Should be removed. Chemical and physical treatment.

SHIPPING INFORMATION
Grades of purity: Research, Pure, Technical: All 99.5+%; **Storage temperature:** Ambient; **Inert atmosphere:** None; **Venting:** Open (flame arrester); **Stability during transport:** Stable.

NAS HAZARD CLASSIFICATION FOR BULK WATER TRANSPORTATION
FIRE: 3
HEALTH: Vapor irritant: 0; Liquid or solid irritant: 0; Poisons: 0
WATER POLLUTION: Human toxicity: 0; Aquatic toxicity: 2; Aesthetic effect: 2
REACTIVITY: Other chemicals: 0; Water: 0; Self-reaction: 0

PHYSICAL AND CHEMICAL PROPERTIES
Physical state @ 59°F/15°C and 1 atm: Liquid.
Molecular weight: 128.3
Boiling point @ 1 atm: 304°F/151°C/424°K.
Melting/Freezing point: −64.3°F/−53.5°C/219.7°K.
Critical temperature: 610.5°F/321.4°C/594.6°K.
Critical pressure: 335 psia = 22.8 atm = 2.31 MN/m^2.
Specific gravity (water = 1): 0.718 @ 68°F/20°C (liquid).
Liquid surface tension: 22.9 dynes/cm = 0.0229 N/m @ 68°F/20°C.
Liquid water interfacial tension: (estimate) 35 dynes/cm = 0.035 N/m @ 68°F/20°C.
Relative vapor density (air = 1): 4.4
Ratio of specific heats of vapor (gas): 1.042 at 16°C.
Latent heat of vaporization: 127 Btu/lb = 70.6 cal/g = 2.95 x 10^5 J/kg.
Heat of combustion: −19,067 Btu/lb = −10,593 cal/g = −443.21 x 10^5 J/kg.
Heat of fusion: 28.83 cal/g.
Vapor pressure: (Reid) 0.2 psia; 3 mm.

NONANOL REC. N:2400

SYNONYMS: ALCOHOL C-9; ALCOHOL *n*-NONILICO (Spanish); NONAN-1-OL; 1-NONANOL; *n*-NONYL ALCOHOL; NONYL ALCOHOL; OCTYL CARBINOL; PELARGONIC ALCOHOL

IDENTIFICATION
CAS Number: 143-08-8
Formula: C$_9$H$_{20}$O; CH$_3$(CH$_2$)7CH$_2$OH

DESCRIPTION: Colorless to yellowish liquid. Rose or citrus odor; like citronella. Floats on water as a slick; insoluble.

Flammable • Containers may BLEVE when exposed to fire • Vapors may form explosive mixture with air • Vapors are heavier than air and will collect and stay in low areas • Vapors may travel long distances to ignition sources and flashback • Vapors in confined areas (e.g., tanks, sewers, buildings) may explode when exposed to fire • Irritating to skin, eyes, and respiratory system; affects the nervous sytem • Toxic products of combustion may include carbon monoxide.

Hazard Classification (based on NFPA-704 Rating System)
Health Hazards (Blue): 1; Flammability (Red): 2; Reactivity (Yellow): 0

EMERGENCY RESPONSE: See Appendix A (171)
Evacuation:
Public safety: Isolate the area of spill or leak for at least 10 to 25 meters (30 to 80 feet) in all directions.
Spill: Increase, in the downwind direction, as necessary, the distance shown under "Public Safety."
Fire: If any large container is involved in fire, isolate for at least 800 meters (½ mile) in all directions; also, consider initial evacuation for 800 meters (½ mile) in all directions.

EXPOSURE
Short-term effects: *SEEK MEDICAL ATTENTION.* Vapors: Cause irritation. Affects the nervous system; high concentrations may cause dizziness, drowsiness; you may pass out. **Liquid:** Harmful if swallowed. *IF SWALLOWED* and victim is *CONSCIOUS AND ABLE TO SWALLOW*, have victim drink 4 to 8 ounces of water.

HEALTH HAZARDS

Personal protective equipment (PPE): B-Level PPE. Goggles or face shield; rubber gloves. **Natural rubber, neoprene, and nitrile rubber may offer some protection from alcohols.**
Toxicity by ingestion: Grade 2; LD_{50} = 0.5 to 5 g/kg (rat).
Long-term health effects: Liver and kidney damage.
Vapor (gas) irritant characteristics: Eye and respiratory tract irritant.
Liquid or solid irritant characteristics: Practically harmless to the skin.

FIRE DATA

Flash point: 210°F/99°C (oc): 165°F/74°C (cc).
Flammable limits in air: LEL: 0.8%; UEL: 6.1% (both @ 212°F).
Fire extinguishing agents not to be used: Water may be ineffective
Electrical hazard: Class I, Group D.

CHEMICAL REACTIVITY

Binary reactants: Strong oxidizers
Reactivity group: 20
Compatibility class: Alcohols, glycols

ENVIRONMENTAL DATA

Food chain concentration potential: Negative; unlikely to accumulate.
Water pollution: DOT Appendix B, §172.101–marine pollutant. Effect of low concentrations on aquatic life is unknown. Fouling to shoreline. May be dangerous if it enters nearby water intakes; notify operators. Notify local health and wildlife officials.
Response to discharge: Mechanical containment. Should be removed. Chemical and physical treatment.

SHIPPING INFORMATION

Grades of purity: 97%; **Storage temperature:** Ambient; **Inert atmosphere:** None; **Venting:** Open (flame arrester); **Stability during transport:** Stable.

NAS HAZARD CLASSIFICATION FOR BULK WATER TRANSPORTATION

FIRE: 1
HEALTH: Vapor irritant: 0; Liquid or solid irritant: 0; Poisons: 0
WATER POLLUTION: Human toxicity: 1; Aquatic toxicity: 3; Aesthetic effect: 2
REACTIVITY: Other chemicals: 2; Water: 0; Self-reaction: 0

PHYSICAL AND CHEMICAL PROPERTIES

Physical state @ 59°F/15°C and 1 atm: Liquid.
Molecular weight: 144.26
Boiling point @ 1 atm: 356°F/180°C/453°K.
Melting/Freezing point: 23°F/–5°C/268°K.
Critical temperature: 759°F/404°C/677°K.
Critical pressure: 350 psia = 24 atm = 2.4 MN/m^2.
Specific gravity (water = 1): 0.827 @ 68°F/20°C (liquid).
Liquid surface tension: 28 dynes/cm = 0.028 N/m at 24°C.
Liquid water interfacial tension: 9.0 dynes/cm = 0.0090 N/m at 21.3°C.
Relative vapor density (air = 1): 5.0
Ratio of specific heats of vapor (gas): 1.039
Latent heat of vaporization: 131 Btu/lb = 72.5 cal/g = 3.04 x 10^5 J/kg.
Heat of combustion: –17,800 Btu/lb = –9860 cal/g = –413 x 10^5 J/kg.
Vapor pressure: 0.76 mm.

1-NONENE
REC. N:2450

SYNONYMS: *n*-HEPTYLETHYLENE; NONENO (Spanish); NONYLENE; 1-NONYLENE; PROPENE TRIMER; PT 3; C-9 OLEFIN MIXTURE

IDENTIFICATION

CAS Number: 124-11-8; 27214-95-8
Formula: C_9H_{18}; $CH_3(CH_2)_6CH=CH_2$
DOT ID Number: UN 1993; DOT Guide Number: 131
Proper Shipping Name: Flammable liquid, n.o.s.

DESCRIPTION: Colorless liquid. Gasoline-like odor. Floats on water surface; insoluble.

Highly flammable • Containers may BLEVE when exposed to fire • Vapors may form explosive mixture with air • Vapors are heavier than air and will collect and stay in low areas • Vapors may travel long distances to ignition sources and flashback • Vapors in confined areas (e.g., tanks, sewers, buildings) may explode when exposed to fire • Irritating to the skin, eyes, and respiratory tract • Toxic products of combustion may include irritating vapors.

Hazard Classification (based on NFPA-704 Rating System)
Health Hazards (Blue): 0; Flammability (Red): 3; Reactivity (Yellow): 0

EMERGENCY RESPONSE: See Appendix A (131)
Evacuation:
Public safety: Isolate spill area for at least 100 to 200 meters (330 to 660 feet) in all directions.
Spill: Increase, as necessary, the isolation distance shown above, in "Public safety."
Fire: Isolate for 800 meters (½ mile) in all directions, especially if tank, rail car, or tank truck is involved in fire.

EXPOSURE

Short-term effects: *SEEK MEDICAL ATTENTION.* **Vapor:** Irritating to eyes, nose, and throat. *IF INHALED*, will, will cause dizziness, headache, difficult breathing or loss of consciousness. Move to fresh air. *IF BREATHING HAS STOPPED*, give artificial respiration. IF breathing is difficult, administer oxygen. **Liquid:** Irritating to skin and eyes. Remove contaminated clothing and shoes. Flush affected areas with plenty of water. *IF IN EYES*, hold eyelids open and flush with plenty of water. *IF SWALLOWED* and victim is *CONSCIOUS AND ABLE TO SWALLOW*, have victim drink 4 to 8 ounces of water. **Do NOT induce vomiting.**

HEALTH HAZARDS

Personal protective equipment (PPE): A-Level PPE. Respiratory organic vapor canister); or air–supplied mask; face splash shield.
Vapor (gas) irritant characteristics: Eye and respiratory tract irritant. Vapors cause smarting of the eyes or respiratory system if present at high concentrations. The effect may be temporary.
Liquid or solid irritant characteristics: If spilled on clothing and allowed to remain, may cause smarting and reddening of the skin.

FIRE DATA

Flash point: 78°F/26°C (oc).
Flammable limits in air: LEL: 0.8%; UEL 3.9% (estimate).
Fire extinguishing agents not to be used: Water may be ineffective
Behavior in fire: Containers may explode.
Electrical hazard: Class I, Group D.

Burning rate: 6.0 mm/min
Electrical hazards: Flow or agitation of substance may generate electrostatic charges due to low conductivity.

CHEMICAL REACTIVITY
Binary reactants: Incompatible with sulfuric and nitric acid. May soften some rubbers, paints, plastics
Reactivity group: 30
Compatibility class: Olefins

ENVIRONMENTAL DATA
Food chain concentration potential: Unlikely to accumulate.
Water pollution: Effect of low concentrations on aquatic life is unknown. Fouling to shoreline. May be dangerous if it enters nearby water intakes; notify operators. Notify local health and wildlife officials. **Response to discharge:** Mechanical containment. Should be removed. Chemical and physical treatment.

SHIPPING INFORMATION
Grades of purity: Technical; **Storage temperature:** Ambient; **Inert atmosphere:** None; **Venting:** Open (flame arrester); **Stability during transport:** Stable.

NAS HAZARD CLASSIFICATION FOR BULK WATER TRANSPORTATION
FIRE: 3
HEALTH: Vapor irritant: 1; Liquid or solid irritant: 1; Poisons: 0
WATER POLLUTION: Human toxicity: 0; Aquatic toxicity: 1; Aesthetic effect: 2
REACTIVITY: Other chemicals: 1; Water: 0; Self-reaction: 1

PHYSICAL AND CHEMICAL PROPERTIES
Physical state @ 59°F/15°C and 1 atm: Liquid.
Molecular weight: 126.24
Boiling point @ 1 atm: 297°F/147°C/420°K.
Melting/Freezing point: –115°F/–81.7°C/191.5°K.
Critical temperature: 622°F/327.8°C/601.0°K.
Critical pressure: 360 psia = 24.5 atm = 2.98 MN/m^2.
Specific gravity (water = 1): 0.733 g/cm3 @ 68°F/20°C (liquid).
Liquid surface tension: 23.0 dynes/cm = 0.0230 N/m @ 68°F/20°C.
Relative vapor density (air = 1): 4.36
Ratio of specific heats of vapor (gas): 1.044
Latent heat of vaporization: 124 Btu/lb = 68.8 cal/g = 2.88 x 10^5 J/kg.
Heat of combustion: –18,979 Btu/lb = –10,544 cal/g = –441.46 x 10^5 J/kg.
Vapor pressure: (Reid) 0.21 psia; 4.08 mm; 11 mmHg @ 100°F/38°C.

NONYL ACETATE   REC. N:2500

SYNONYMS: ACETATO de *n*-NONILO (Spanish); *n*-NONYL ACETATE; ACETIC ACID, *n*-NONYL ESTER; ACETATE C-9; NONANOL ACETATE

IDENTIFICATION
CAS Number: 143-13-5
Formula: $C_{11}H_{22}O_2$
DOT ID Number: UN1993; DOT Guide Number: 128
Proper Shipping Name: Flammable liquids, n.o.s.

DESCRIPTION: Colorless liquid. Fruity odor. Floats on water surface; very slightly soluble.

Highly flammable • Containers may BLEVE when exposed to fire • Vapors may form explosive mixture with air • Vapors are heavier than air and will collect and stay in low areas • Vapors may travel long distances to ignition sources and flashback • Vapors in confined areas (e.g., tanks, sewers, buildings) may explode when exposed to fire • Irritating to the skin, eyes, and respiratory tract • Toxic products of combustion may include irritating vapors and toxic gases, such as CO_2 and carbon monoxide.

Hazard Classification (based on NFPA-704 Rating System)
Health Hazards (Blue): 1; Flammability (Red): 2; Reactivity (Yellow): 0

EMERGENCY RESPONSE: See Appendix A (128)
Evacuation:
Public safety: Isolate spill area for at least 25 to 50 meters (80 to 160 feet) in all directions.
Spill: Large spill–Consider initial downwind evacuation for at least 300 meters (1000 feet).
Fire: Isolate for 800 meters (½ mile) in all directions, especially if tank, rail car, or tank truck is involved in fire.

EXPOSURE
Short-term effects: *SEEK MEDICAL ATTENTION*. **Vapor:** Move victim to fresh air. *IF BREATHING HAS STOPPED*, give artificial respiration. IF breathing is difficult, administer oxygen. **Liquid:** Remove contaminated clothing and shoes. Wash affected areas with soap and water. *IF IN EYES*, hold eyelids open and flush with plenty of water.

HEALTH HAZARDS
Personal protective equipment (PPE): B-Level PPE. Full impervious protective clothing, including boots and gloves. Where splashing is possible wear full face shield or chemical safety goggles. Use approved respirator to protect against vapors.
Vapor (gas) irritant characteristics: Eye and respiratory tract irritant.
Liquid or solid irritant characteristics: Eye and skin irritant.

FIRE DATA
Flash point: 155°F/68°C (cc).
Fire extinguishing agents not to be used: Water may be ineffective.
Electrical hazard: Class I, Group D. Flow or agitation of substance may generate electrostatic charges due to low conductivity.
Stoichiometric air-to-fuel ratio: 73.8

CHEMICAL REACTIVITY
Binary reactants: Oxidizers, strong acids, nitrates.

ENVIRONMENTAL DATA
Water pollution: Effects of low concentrations on aquatic life is unknown. May be dangerous if it enters nearby water intakes; notify operators. Notify local health and wildlife officials. **Response to discharge:** Should be removed.

SHIPPING INFORMATION
Grades of purity: Technical; **Storage temperature:** Ambient; **Inert atmosphere:** None; **Stability during transport:** Stable.

PHYSICAL AND CHEMICAL PROPERTIES
Physical state @ 59°F/15°C and 1 atm: Liquid.

Molecular weight: 186.29
Boiling point @ 1 atm: 378–413.6°F/192–212°C/465–485°K.
Specific gravity (water = 1): 0.8785 @ 15°C.
Relative vapor density (air = 1): 6.39

NONYL PHENOL   REC. N:2550

Synonyms: NONILFENOL (Spanish)

IDENTIFICATION
CAS Number: 25154-52-3
Formula: $C_{15}H_{24}O$; p-$HOC_6H_4(CH_2)_8CH_3$
DOT ID Number: 2922; DOT Guide Number: 154
Proper Shipping Name: Corrosive liquid, toxic, n.o.s.

DESCRIPTION: Pale yellow thick liquid. Medicinal odor; like disinfectant or phenol. Floats on the surface of water.

Corrosive to the skin, eyes, and respiratory tract • Breathing the vapor can cause lung edema and can be life threatening • Firefighting gear (including SCBA) does not provide adequate protection. If exposure occurs, remove and isolate gear immediately and thoroughly decontaminate personnel • Containers may BLEVE when exposed to fire • Vapors in confined areas (e.g., tanks, sewers, buildings) may explode when exposed to fire • Toxic products of combustion may include irritating vapors.

Hazard Classification (based on NFPA-704 Rating System)
Health Hazards (Blue): 2; Flammability (Red): 1; Reactivity (Yellow): 0

EMERGENCY RESPONSE: See Appendix A (154)
Evacuation:
Public safety: Isolate the area of spill or leak for at least 25 to 50 meters (80 to 160 feet) in all directions.
Spill: Increase, in the downwind direction, as necessary, the distance shown under "Public Safety."
Fire: If tank, rail car, or tank truck is involved in fire, isolate for at least 800 meters (½ mile) in all directions; also, consider initial evacuation for 800 meters (½ mile) in all directions.

EXPOSURE
Short-term effects: *SEEK MEDICAL ATTENTION*. **Vapor:** Corrosive. Lung edema may develop. **Liquid:** Will burn skin and eyes. Harmful if swallowed. Remove contaminated clothing and shoes. Flush affected areas with plenty of water. *IF BREATHING HAS STOPPED*, give artificial respiration; *avoid mouth-to-mouth resuscitation; use bag/mask apparatus. IF IN EYES*, hold eyelids open and flush with plenty of water. *IF SWALLOWED* and victim is *CONSCIOUS AND ABLE TO SWALLOW*, have victim drink 4 to 8 ounces of water.
Note to physician or authorized medical personnel: Medical observation is recommended for 24 to 48 hours after breathing overexposure, as pulmonary edema may be delayed. As first aid for pulmonary edema, consider administering a corticosteroid spray. Cigarette smoking may exacerbate pulmonary injury and should be discouraged for at least 72 hours following exposure.

HEALTH HAZARDS
Personal protective equipment (PPE): B-Level PPE. Chemical resistant gloves and splash-proof goggles. Chemical protective material(s) reported to have good to excellent resistance: neoprene, nitrile. Also, natural rubber may offer some protection from phenols.
Toxicity by ingestion: Grade 2; LD_{50} = 0.5 to 5 g/kg.
Vapor (gas) irritant characteristics: Vapors cause smarting of the eyes or respiratory system if present in high concentrations. The effect may be temporary.
Liquid or solid irritant characteristics: Causes smarting of the skin and first-degree burns on short exposure; may cause secondary burns on long exposure.

FIRE DATA
Flash point: 300°F/149°C (oc), 285°F/141°C (cc).
Fire extinguishing agents not to be used: Water or foam may cause frothing.

CHEMICAL REACTIVITY
Binary reactants: Reacts with strong acids, caustics, isocyanates, aliphatic amines
Reactivity group: 21
Compatibility class: Phenols, cresols

ENVIRONMENTAL DATA
Food chain concentration potential: Negative; unlikely to accumulate.
Water pollution: Effect of low concentrations on aquatic life is unknown. Fouling to shoreline. May be dangerous if it enters nearby water intakes; notify operators. Notify local health and wildlife officials. **Response to discharge:** Mechanical containment. Should be removed. Chemical and physical treatment.

SHIPPING INFORMATION
Grades of purity: 90% p-isomer plus 4% o-isomer, and 5% 2, 4-dinonylphenol; **Storage temperature:** Ambient; **Inert atmosphere:** None; **Venting:** Open (flame arrester); **Stability during transport:** Stable.

NAS HAZARD CLASSIFICATION FOR BULK WATER TRANSPORTATION
FIRE: 1
HEALTH: Vapor irritant: 1; Liquid or solid irritant: 2; Poisons: 1
WATER POLLUTION: Human toxicity: 2; Aquatic toxicity: 3; Aesthetic effect: 3
REACTIVITY: Other chemicals: 2; Water: 0; Self-reaction: 0

PHYSICAL AND CHEMICAL PROPERTIES
Physical state @ 59°F/15°C and 1 atm: Liquid.
Molecular weight: 220.36
Boiling point @ 1 atm: 579°F/304°C/577°K.
Critical temperature: 878°F/470°C/743°K.
Specific gravity (water = 1): 0.9494 @ 77°F/25°C (liquid).
Liquid surface tension: (estimate) 30 dynes/cm = 0.03 N/m @ 68°F/20°C.
Liquid water interfacial tension: (estimate) 30 dynes/cm = 0.03 N/m @ 68°F/20°C.
Heat of combustion: (estimate) –17,500 Btu/lb = –9730 cal/g = –407 x 10^5 J/kg.
Vapor pressure: (Reid) Low.

OCTANE   REC. O:0100

SYNONYMS: EEC No. 601-009-00-8; n-OCTANE; n-OCTANO (Spanish); 2,2,4-TRIMETHYLPENTANE

IDENTIFICATION
CAS Number: 111-65-9
Formula: C_8H_{18}

Octane

DOT ID Number: UN 1262; DOT Guide Number: 128
Proper Shipping Name: Octanes

DESCRIPTION: Clear, colorless liquid. Gasoline-like odor. Floats on water surface; insoluble. Flammable, irritating vapor is produced.

Highly flammable • Containers may BLEVE when exposed to fire • Vapors may form explosive mixture with air • Vapors are heavier than air and will collect and stay in low areas • Vapors may travel long distances to ignition sources and flashback • Vapors in confined areas (e.g., tanks, sewers, buildings) may explode when exposed to fire • Irritating to the skin, eyes, and respiratory tract • Toxic products of combustion may include acrid carbon monoxide.

Hazard Classification (based on NFPA-704 Rating System)
Health Hazards (Blue): 0; Flammability (Red): 3; Reactivity (Yellow): 0

EMERGENCY RESPONSE: See Appendix A (128)
Evacuation:
Public safety: Isolate spill area for at least 25 to 50 meters (80 to 160 feet) in all directions.
Spill: Large spill–Consider initial downwind evacuation for at least 300 meters (1000 feet).
Fire: Isolate for 800 meters (½ mile) in all directions, especially if tank, rail car, or tank truck is involved in fire.

EXPOSURE
Short-term effects: *SEEK MEDICAL ATTENTION.* **Vapor:** Irritating to eyes, nose, and throat. *IF INHALED*, will, will cause headache, dizziness, difficult breathing, or loss of consciousness. Move victim to fresh air. *IF BREATHING HAS STOPPED,* give artificial respiration. IF breathing is difficult, administer oxygen. **Liquid:** Irritating to skin and eyes. *IF SWALLOWED*, will cause nausea, and vomiting. Droplets in the lung may cause pneumonia. Remove contaminated clothing and shoes. Flush affected areas with plenty of water. *IF IN EYES*, hold eyelids open and flush with plenty of water. *IF SWALLOWED* and victim is *CONSCIOUS AND ABLE TO SWALLOW*, have victim drink 4 to 8 ounces of water. **Do NOT induce vomiting.**

HEALTH HAZARDS
Personal protective equipment (PPE): B-Level PPE. OSHA Table Z-1-A air contaminant. Goggles or face shield; chemical resistant gloves. Chemical protective material(s) reported to have good to excellent resistance: nitrile, Viton®, nitrile+PVC. Also, Viton/neoprene offers limited protection.
Recommendations for respirator selection: NIOSH
750 ppm: SA* (any supplied-air respirator). *1000 ppm:* SA:CF* (any supplied-air respirator operated in a continuous-flow mode); or SCBAF (any self-contained breathing apparatus with a full facepiece); or SAF (any supplied-air respirator with a full facepiece). *EMERGENCY OR PLANNED ENTRY INTO UNKNOWN CONCENTRATIONS OR IDLH CONDITIONS:* SCBAF:PD,PP (any self-contained breathing apparatus that has a full facepiece and is operated in a pressure-demand or other positive-pressure mode); or SAF:PD,PP:ASCBA (any supplied-air respirator that has a full facepiece and is operated in a pressure-demand or other positive-pressure mode in combination with an auxiliary self-contained breathing apparatus operated in a pressure-demand or other positive pressure mode). *ESCAPE:* GMFOV [any air-purifying, full-facepiece respirator (gas mask) with a chin-style, front- or back-mounted organic vapor canister]; or SCBAE (any appropriate escape-type, self-contained breathing apparatus).

Note: Substance reported to cause eye irritation or damage; may require eye protection.
Exposure limits (TWA unless otherwise noted): ACGIH TLV 300 ppm (1400 mg/m^3); OSHA PEL 500 ppm (2350 mg/m^3); NIOSH REL 75 ppm (350 mg/m^3); ceiling 385 ppm (1800 mg/m^3)/15 min.
Short-term exposure limits (15-minute TWA): ACGIH STEL 375 ppm (1750 mg/m^3).
Long-term health effects: May effect the nervous system.
Vapor (gas) irritant characteristics: Eye and respiratory tract irritant. High concentrations are a simple asphyxiant and narcotic.
Liquid or solid irritant characteristics: Degreases the skin; dryness may cause irritation.
Odor threshold: 14 ppm.
IDLH value: 1000 ppm [10% LEL].

FIRE DATA
Flash point: 56°F/13°C (cc).
Flammable limits in air: LEL: 0.8%; UEL: 6.5%.
Fire extinguishing agents not to be used: Water may be ineffective.
Autoignition temperature: 406°F/208°C/481°K.
Electrical hazard: Class I, Group D; Flow or agitation of substance may generate electrostatic charges due to low conductivity.
Burning rate: 6.3 mm/min

CHEMICAL REACTIVITY
Binary reactants: Strong oxidizers
Reactivity group: 31
Compatibility class: Paraffins

ENVIRONMENTAL DATA
Water pollution: Effect of low concentrations on aquatic life is unknown. Fouling to shoreline. May be dangerous if it enters nearby water intakes; notify operators. Notify local health and wildlife officials. **Response to discharge:** Issue warning–high flammability. Restrict access. Mechanical containment. Should be removed. Chemical and physical treatment.

SHIPPING INFORMATION
Grades of purity: Research: 99.92%; Pure: 99.6%; Technical: 98.7%; **Storage temperature:** Ambient; **Inert atmosphere:** None; **Venting:** Open (flame arrester); **Stability during transport:** Stable.

PHYSICAL AND CHEMICAL PROPERTIES
Physical state @ 59°F/15°C and 1 atm: Liquid.
Molecular weight: 114.2
Boiling point @ 1 atm: 258.1°F/125.6°C/398.9°K.
Melting/Freezing point: –70.2°F/–56.8°C/216.4°K.
Critical temperature: 563.7°F/295.4°C/568.6°K.
Critical pressure: 361 psia = 24.5 atm = 2.49 MN/m^2.
Specific gravity (water = 1): 0.703 @ 68°F/20°C (liquid).
Liquid surface tension: 21.7 dynes/cm = 0.0217 N/m @ 68°F/20°C.
Liquid water interfacial tension: (estimate) 35 dynes/cm = 0.035 N/m @ 68°F/20°C.
Relative vapor density (air = 1): 3.9
Ratio of specific heats of vapor (gas): 1.047 at 16°C.
Latent heat of vaporization: 130.4 Btu/lb = 72.5 cal/g = 3.03 x 10^5 J/kg.
Heat of combustion: –19,112 Btu/lb = –10,618 cal/g = –444.26 x 10^5 J/kg.
Heat of fusion: 43.21 cal/g.
Vapor pressure: 11 mm.

OCTANOIC ACID
REC. O:0200

SYNONYMS: ACIDO OCTANOICO (Spanish); C-8 ACID; *n*-CAPRYLIC ACID; HEXACID 898; 1-HEPTANECARBOXYLIC ACID; NEO-FAT 8; *n*-OCTOIC ACID

IDENTIFICATION
CAS Number: 124-07-2
Formula: $C_8H_{16}O_2$; $CH_3(CH_2)_6CO_2H$

DESCRIPTION: Colorless oily liquid. Slightly unpleasant odor. Practically insoluble in cold water; slightly soluble in hot.

Combustible • Containers may BLEVE when exposed to fire • Vapors in confined areas (e.g., tanks, sewers, buildings) may explode when exposed to fire • Irritating to the skin, eyes, and respiratory tract • Toxic products of combustion may include irritating vapors.

Hazard Classification (based on NFPA-704 Rating System)
Health Hazards (Blue): 1; Flammability (Red): 1; Reactivity (Yellow): 0

EMERGENCY RESPONSE: See Appendix A (171)
Evacuation:
Public safety: Isolate the area of spill or leak for at least 10 to 25 meters (30 to 80 feet) in all directions.
Spill: Increase, in the downwind direction, as necessary, the distance shown under "Public Safety."
Fire: If any large container is involved in fire, isolate for at least 800 meters (½ mile) in all directions; also, consider initial evacuation for 800 meters (½ mile) in all directions.

EXPOSURE
Short-term effects: *SEEK MEDICAL ATTENTION.* **Vapor:** Irritating to eyes, nose, and throat. *IF INHALED*, will, will cause coughing or difficult breathing. *IF IN EYES*, hold eyelids open and flush with plenty of water. *IF BREATHING HAS STOPPED*, give artificial respiration. IF breathing is difficult, administer oxygen. **Liquid:** Will burn skin and eyes. *IF SWALLOWED*, will cause nausea and vomiting. Remove contaminated clothing and shoes. Flush affected areas with plenty of water. *IF IN EYES*, hold eyelids open and flush with plenty of water.

HEALTH HAZARDS
Personal protective equipment (PPE): B-Level PPE. Full face self contained breathing apparatus, rubber boots, and heavy rubber gloves. Chemical protective material(s) reported to have good to excellent resistance: neoprene, nitrile, Viton®. Also, butyl rubber is generally suitable for carbooxylic acid compounds.
Toxicity by ingestion: Toxic. Grade 1: LD_{50} = 10.08 g/kg (rat).
Long-term health effects: Mutation data.
Vapor (gas) irritant characteristics: Vapors cause severe irritation of eyes and throat and can cause eye and lung injury. They cause coughing and cannot be tolerated even at low concentrations.
Liquid or solid irritant characteristics: Fairly severe eye and skin irritant. May cause pain and second-degree burns after a few minutes of contact.

FIRE DATA
Flash point: 230°F/110°C (cc).
Fire extinguishing agents not to be used: Water may be ineffective.

CHEMICAL REACTIVITY
Binary reactants: Reacts with sulfuric acid, caustics, ammonia, aliphatic amines, alkanol amines, alkylene oxides, epichlorohydrin. Attacks most common metals.
Neutralizing agents for acids and caustics: Sodium bicarbonate solution.
Reactivity group: 4
Compatibility class: Organic acids.

ENVIRONMENTAL DATA
Water pollution: May be dangerous to aquatic life in high concentrations. May be dangerous if it enters nearby water intakes; notify operators. Notify local health and wildlife officials. **Response to discharge:** Issue warning–corrosive. Mechanical containment. Should be removed. Chemical and physical treatment.

SHIPPING INFORMATION
Grades of purity: 99.5+%; **Storage temperature:** Ambient; **Stability during transport:** Stable.

PHYSICAL AND CHEMICAL PROPERTIES
Physical state @ 59°F/15°C and 1 atm: Solid.
Molecular weight: 144.21
Boiling point @ 1 atm: 458.6°F/237°C/510.2°K.
Melting/Freezing point: 60.8–61.7°F/16–16.5°C/289.2–289.7°K.
Specific gravity (water = 1): 0.910
Relative vapor density (air = 1): 5

OCTANOL
REC. O:0300

SYNONYMS: ALCOHOL C-8; ALFOL 8; DYTOL M-83; CAPRYL ALCOHOL; CAPRYLIC ALCOHOL; HEPTYL CARBINOL; 1-HYDROXY; LOROL 20; *n*-OCTANOL; 1-OCTANOL; OCTILIN; PRIMARY OCTYL ALCOHOL; SIPOL L8

IDENTIFICATION
CAS Number: 111-87-5 (1-*n*-isomer); also applies to 123-96-6 (2-*n*-isomer); 589-98-0 (3-*n*-isomer)
Formula: C_8H_{18}; $CH_3(CH_2)_6CH_2OH$
DOT ID Number: UN 1987; DOT Guide Number: 127
Proper Shipping Name: Alcohols, n.o.s.

DESCRIPTION: Colorless, thick liquid. Sweet odor. Floats on the surface of water.

Highly flammable • Containers may BLEVE when exposed to fire • Vapors may form explosive mixture with air • Vapors are heavier than air and will collect and stay in low areas • Vapors may travel long distances to ignition sources and flashback • Vapors in confined areas (e.g., tanks, sewers, buildings) may explode when exposed to fire • Irritating to the skin, eyes, and respiratory tract • Toxic products of combustion may include carbon monoxide.

Hazard Classification (based on NFPA-704 Rating System)
Health Hazards (Blue): 1; Flammability (Red): 2; Reactivity (Yellow): 0

EMERGENCY RESPONSE: See Appendix A (127)
Evacuation:
Public safety: Isolate spill area for at least 25 to 50 meters (80 to 160 feet) in all directions.
Spill: Large spill–Consider initial downwind evacuation for at least 300 meters (1000 feet).

Fire: Isolate for 800 meters (½ mile) in all directions, especially if tank, rail car, or tank truck is involved in fire.

EXPOSURE
Short-term effects: *SEEK MEDICAL ATTENTION.* **Liquid:** Irritating to skin. Will burn eyes. Remove contaminated clothing and shoes. Flush affected areas with plenty of water. *IF IN EYES,* hold eyelids open and flush with plenty of water. The use of alcoholic beverages may enhance any toxic effect.

HEALTH HAZARDS
Personal protective equipment (PPE): B-Level PPE. Chemical gloves and chemical goggles. Chemical protective material(s) reported to have good to excellent resistance: neoprene, nitrile, nitrile+PVC, PVC, styrene-butadiene, PV alcohol, PVC, nitrile+PVC. Also, natural rubber, styrene-butadiene/neoprene, Viton®/neoprene, and butyl rubber/neoprene offers limited protection
Exposure limits (TWA unless otherwise noted): AIHA WEEL 50 ppm.
Toxicity by ingestion: Grade 1; oral rat LD_{50} more than 3.2 g/kg.
Long-term health effects: Possible mutagen.
Vapor (gas) irritant characteristics: May irritate the eyes and mucous membrane.
Liquid or solid irritant characteristics: Eye, skin and respiratory tract irritant.
Odor threshold: 0.009–1.0 ppm.

FIRE DATA
Flash point: 178°F/81°C (cc) (1-*n*-isomer); 140°F/60°C (2-*n*-isomer).
Flammable limits in air: LEL: 0.3%; UEL: 31% (1-*n*-isomer); LEL: 0.8%; UEL: 7.4% (2-*n*-isomer).
Autoignition temperature: 487°F/253°C/526°K.
Burning rate: 3.7 mm/min (approximate).

CHEMICAL REACTIVITY
Binary reactants: Strong oxidizers
Reactivity group: 20
Compatibility class: Alcohols, glycols

ENVIRONMENTAL DATA
Food chain concentration potential: Log P_{ow} = listed at 2.97 and 3.2. Values > 3.0 are likely to bioconcentrate in aquatic organisms and other living tissue, especially in fats.
Water pollution: DOT Appendix B, §172.101–marine pollutant. Effect of low concentrations on aquatic life is unknown. Fouling to shoreline. May be dangerous if it enters nearby water intakes; notify operators. Notify local health and wildlife officials.
Response to discharge: Mechanical containment. Should be removed. Chemical and physical treatment.

SHIPPING INFORMATION
Storage temperature: Ambient; **Inert atmosphere:** None; **Venting:** Open (flame arrester); **Stability during transport:** Stable.

PHYSICAL AND CHEMICAL PROPERTIES
Physical state @ 59°F/15°C and 1 atm: Liquid.
Molecular weight: 130.23
Boiling point @ 1 atm: 335°F/168°C/441°K; 356°F/180°C/453°K (2-*n*-isomer); 349°F/176°C/449°K (3-*n*-isomer)
Melting/Freezing point: 5°F/–15°C/258°K; –36°F/–38°C/235°K (2-*n*-isomer).
Critical temperature: 725°F/385°C/658°K.
Critical pressure: 400 psia = 27 atm = 2.7 MN/m².
Specific gravity (water = 1): 0.829 @ 68°F/20°C (liquid).
Liquid surface tension: 27.5 dynes/cm = 0.0275 N/m @ 68°F/20°C.
Liquid water interfacial tension: 8.52 dynes/cm = 0.00852 N/m @ 68°F/20°C.
Relative vapor density (air = 1): 4.4
Ratio of specific heats of vapor (gas): 1.044
Latent heat of vaporization: 176 Btu/lb = 97.5 cal/g = 4.08 x 10^5 J/kg.
Heat of combustion: –16,130 Btu/lb = –8963 cal/g = –375.3 x 10^5 J/kg.
Vapor pressure: 0.02 mm.

1-OCTENE REC. O:0400

SYNONYMS: CAPRYLENE; 1-OCTENO (Spanish); α-OCTYLENE

IDENTIFICATION
CAS Number: 111-66-0
Formula: $CH_3(CH_2)_5CH=CH_2$
DOT ID Number: UN 3295; DOT Guide Number: 128
Proper Shipping Name: Hydrocarbons, liquid, n.o.s.

DESCRIPTION: Colorless liquid. Gasoline-like odor. Floats on water surface; flammable, harmful vapor is produced.

Highly flammable • Containers may BLEVE when exposed to fire • Vapors may form explosive mixture with air • Vapors are heavier than air and will collect and stay in low areas • Vapors may travel long distances to ignition sources and flashback • Vapors in confined areas (e.g., tanks, sewers, buildings) may explode when exposed to fire • Irritating to the skin, eyes, and respiratory tract • Toxic products of combustion may include irritating smoke and vapors.

Hazard Classification (based on NFPA-704 Rating System)
Health Hazards (Blue): 1; Flammability (Red): 3; Reactivity (Yellow): 0

EMERGENCY RESPONSE: See Appendix A (128)
Evacuation:
Public safety: Isolate spill area for at least 25 to 50 meters (80 to 160 feet) in all directions.
Spill: Large spill–Consider initial downwind evacuation for at least 300 meters (1000 feet).
Fire: Isolate for 800 meters (½ mile) in all directions, especially if tank, rail car, or tank truck is involved in fire.

EXPOSURE
Short-term effects: *SEEK MEDICAL ATTENTION.* **Vapor:** If inhaled, will cause dizziness. Move to fresh air. *IF BREATHING HAS STOPPED,* give artificial respiration. IF breathing is difficult, administer oxygen. **Liquid:** Irritating to skin and eyes. *IF SWALLOWED,* will cause nausea or vomiting. Remove contaminated clothing and shoes. Flush affected areas with plenty of water. *IF IN EYES,* hold eyelids open and flush with plenty of water. *IF SWALLOWED* and victim is *CONSCIOUS AND ABLE TO SWALLOW,* have victim drink 4 to 8 ounces of water. **Do NOT induce vomiting.**

HEALTH HAZARDS
Personal protective equipment (PPE): B-Level PPE. Organic vapor canister; goggles or face shield.
Vapor (gas) irritant characteristics: Vapors cause a slight smarting of the eyes or respiratory system if present at high concentrations. The effect is temporary.
Liquid or solid irritant characteristics: If spilled on clothing and allowed to remain, may cause smarting and reddening of the skin.

FIRE DATA
Flash point: 70°F/21°C (oc).
Fire extinguishing agents not to be used: Water may be ineffective
Autoignition temperature: 446°F/230°C/755°K.
Burning rate: 6.5 mm/min

CHEMICAL REACTIVITY
Binary reactants: Strong oxidizers
Reactivity group: 30
Compatibility class: Olefins

ENVIRONMENTAL DATA
Water pollution: Effect of low concentrations on aquatic life is unknown. Fouling to shoreline. May be dangerous if it enters nearby water intakes; notify operators. Notify local health and wildlife officials. **Response to discharge:** Issue warning–high flammability. Evacuate area. Disperse and flush.

SHIPPING INFORMATION
Grades of purity: Research: 99.7%; Pure: 99.3%; Technical: 95%; **Storage temperature:** Ambient; **Inert atmosphere:** None; **Venting:** Open (flame arrester) or pressure-vacuum; **Stability during transport:** Stable.

PHYSICAL AND CHEMICAL PROPERTIES
Physical state @ 59°F/15°C and 1 atm: Liquid.
Molecular weight: 112.22
Boiling point @ 1 atm: 250.3°F/121.3°C/394.5°K.
Melting/Freezing point: –151°F/–102°C/172°K.
Critical temperature: 560.1°F/293.4°C/566.6°K.
Critical pressure: 400 psia = 27.2 atm = 2.76 MN/m^2.
Specific gravity (water = 1): 0.715 @ 68°F/20°C (liquid).
Liquid surface tension: 21.76 dynes/cm = 0.02176 N/m @ 68°F/20°C.
Liquid water interfacial tension: (estimate) 50 dynes/cm = 0.05 N/m @ 68°F/20°C.
Relative vapor density (air = 1): 3.9
Ratio of specific heats of vapor (gas): 1.050
Latent heat of vaporization: 129 Btu/lb = 71.9 cal/g = 3.01 x 10^5 J/kg.
Heat of combustion: –19,170 Btu/lb = –10,650 cal/g = –445.89 x 10^5 J/kg.

OCTYL ALDEHYDES   REC. O:0500

SYNONYMS: ALDEHYDE C-8; C-8 ALDEHYDE; CAPRYLALDEHYDE; CAPRYLIC ALDEHYDE; 1-OCTANAL; OCTANALDEHYDE; *n*-OCTYL ALDEHYDE

IDENTIFICATION
CAS Number: 124-13-0
Formula: $C_8H_{16}O$; $C_7H_{15}CHO$
DOT ID Number: UN 1191; DOT Guide Number: 129
Proper Shipping Name: Octyl aldehydes; Ethylhexaldehydes

DESCRIPTION: Colorless to light yellow liquid. Strong, fatty, fruit odor. Floats on the surface of water.

Flammable • Containers may BLEVE when exposed to fire • Vapors may form explosive mixture with air • Vapors are heavier than air and will collect and stay in low areas • Vapors may travel long distances to ignition sources and flashback • Vapors in confined areas (e.g., tanks, sewers, buildings) may explode when exposed to fire • Highly irritating to the eyes and respiratory tract; this material has anesthetic properties • Toxic products of combustion may include carbon monoxide.

Hazard Classification (based on NFPA-704 Rating System)
Health Hazards (Blue): 2; Flammability (Red): 2; Reactivity (Yellow): 0

EMERGENCY RESPONSE: See Appendix A (129)
Evacuation:
Public safety: Isolate spill area for at least 50 to 100 meters (160 to 330 feet) in all directions.
Spill: Large spill–Consider initial downwind evacuation for at least 300 meters (1000 feet).
Fire: Isolate for 800 meters (½ mile) in all directions, especially if tank, rail car, or tank truck is involved in fire.

EXPOSURE
Short-term effects: *SEEK MEDICAL ATTENTION*. **Vapor:** Irritating to eyes, nose, and throat. *IF INHALED*, will, will cause coughing or difficult breathing. *IF IN EYES*, hold eyelids open and flush with plenty of water. *IF BREATHING HAS STOPPED*, give artificial respiration. IF breathing is difficult, administer oxygen. **Liquid:** Irritating to skin and eyes. *IF SWALLOWED*, will cause nausea and vomiting. Remove contaminated clothing and shoes. Flush affected areas with plenty of water. *IF IN EYES*, hold eyelids open and flush with plenty of water. *IF SWALLOWED* and victim is *CONSCIOUS AND ABLE TO SWALLOW*, have victim drink 4 to 8 ounces of water. *IF SWALLOWED* and victim is *UNCONSCIOUS OR HAVING CONVULSIONS*, do nothing except keep victim warm.

HEALTH HAZARDS
Personal protective equipment (PPE): B-Level PPE. Organic vapor respirator. Rubber gloves; safety goggles or face shield.
Butyl rubber is generally suitable for aldehydes.
Exposure limits (TWA unless otherwise noted): 50 ppm (AIHA WEEL).
Vapor (gas) irritant characteristics: Vapors cause smarting of the eyes or respiratory system if present in high concentrations. The effect is temporary.
Liquid or solid irritant characteristics: Eye and skin irritant. Causes smarting of the skin and first-degree burns on short exposure; may cause second-degree burns on long exposure.

FIRE DATA
Flash point: 125°F/52°C (cc).
Fire extinguishing agents not to be used: Water may be ineffective.
Stoichiometric air-to-fuel ratio: 54.8

CHEMICAL REACTIVITY
Binary reactants: Reacts violently with oxidizers. May ignite spontaneously when spilled on clothing, paper, wood or other absorbent materials. Incompatible with strong acids, caustics, ammonia, amines. Under certain conditions, it may be spontaneously flammable in air.

Polymerization: May form unstable peroxides in storage. May polymerize. .
Reactivity group: 19
Compatibility class: Aldehydes

ENVIRONMENTAL DATA
Food chain concentration potential: Log P_{ow} = 2.7. Low potential to accumulate.
Water pollution: Effects of low concentrations on aquatic life is unknown. Fouling to shoreline. May be dangerous if it enters nearby water intakes; notify operators. Notify local health and wildlife officials. **Response to discharge:** Mechanical containment. Should be removed. Chemical and physical treatment.

SHIPPING INFORMATION
Grades of purity: Commercial, 95.0+%; **Storage temperature:** Ambient; **Inert atmosphere:** None; **Venting:** Pressure vacuum valve; **Stability during transport:** Stable.

NAS HAZARD CLASSIFICATION FOR BULK WATER TRANSPORTATION
FIRE: 2
HEALTH: Vapor irritant: 1; Liquid or solid irritant: 2; Poisons: 1
WATER POLLUTION: Human toxicity: 2; Aquatic toxicity: 2; Aesthetic effect: 2
REACTIVITY: Other chemicals: 2; Water: 0; Self-reaction: 0

PHYSICAL AND CHEMICAL PROPERTIES
Physical state @ 59°F/15°C and 1 atm: Liquid.
Molecular weight: 128.22
Boiling point @ 1 atm: 335°F/168°C/441°K.
Specific gravity (water = 1): 0.82–0.83 Liquid.

OCTYL DECYL PHTHALATE   REC. O:0550

SYNONYMS: FTALATO de *n*-OCTIL-*n*-DECILO (Spanish); *n*-OCTYL-*n*-DECYL PHTHALATE

IDENTIFICATION
CAS Number: 119-07-3
Formula: $C_{26}H_{42}O_4$; $C_6H_4(COOC_8H_{17})(COOC_{10}H_{21})$

DESCRIPTION: Colorless liquid. Mild, characteristic odor. Floats on the surface of water.

Toxic products of combustion include CO_2 and carbon monoxide.

EMERGENCY RESPONSE: See Appendix A (171)
Evacuation:
Public safety: Isolate the area of spill or leak for at least 10 to 25 meters (30 to 80 feet) in all directions.
Spill: Increase, in the downwind direction, as necessary, the distance shown under "Public Safety."
Fire: If any large container is involved in fire, isolate for at least 800 meters (½ mile) in all directions; also, consider initial evacuation for 800 meters (½ mile) in all directions.

EXPOSURE
Short-term effects: *SEEK MEDICAL ATTENTION.* **Vapor:** Irritating to eyes, nose, and throat. *IF BREATHING HAS STOPPED,* give artificial respiration. IF breathing is difficult, administer oxygen. **Liquid:** Irritating to skin and eyes. Remove contaminated clothing and shoes. Flush affected areas with plenty of water. *IF IN EYES,* hold eyelids open and flush with plenty of water.

HEALTH HAZARDS
Personal protective equipment (PPE): B-Level PPE. Full impervious protective clothing, including boots and gloves. Where splashing is possible wear full face shield or chemical safety goggles. Use approved respirator to protect against vapors.
Vapor (gas) irritant characteristics: Vapors cause a slight smarting of the eyes or respiratory system if present in high concentrations. The effect is temporary.
Liquid or solid irritant characteristics: Minimum hazard. If spilled on clothing and allowed to remain, may cause smarting and reddening of skin.

FIRE DATA
Flash point: 455°F/235°C (cc).
Fire extinguishing agents not to be used: Water.
Stoichiometric air-to-fuel ratio: 160

ENVIRONMENTAL DATA
Water pollution: Effects of low concentrations on aquatic life is unknown. May be dangerous if it enters nearby water intakes; notify operators. Notify local health and wildlife officials. **Response to discharge:** Mechanical containment. Should be removed.

SHIPPING INFORMATION
Grades of purity: Technical grades; **Storage temperature:** Ambient; **Inert atmosphere:** None.

PHYSICAL AND CHEMICAL PROPERTIES
Physical state @ 59°F/15°C and 1 atm: Liquid.
Molecular weight: 418.68
Melting/Freezing point: –40°F/–40°C/233°K.
Specific gravity (water = 1): 0.972–0.976

OCTYL EPOXY TALLATE   REC. O:0600

SYNONYMS: EPOXIDIZED TALL OIL, OCTYL ESTER

IDENTIFICATION
Formula: Mixture
DOT ID Number: UN 9277; DOT Guide Number: 171
Proper Shipping Name: Oil, n.o.s., flash point not less than 93°C (200°F).

DESCRIPTION: Pale yellow liquid. Mild odor. Floats on the surface of water.

Flammable • Containers may BLEVE when exposed to fire • Vapors may form explosive mixture with air • Vapors are heavier than air and will collect and stay in low areas • Vapors may travel long distances to ignition sources and flashback • Vapors in confined areas (e.g., tanks, sewers, buildings) may explode when exposed to fire • Irritating to the skin, eyes, and respiratory tract • Toxic products of combustion may include acrid carbon monoxide.

Hazard Classification (based on NFPA-704 Rating System)
Health Hazards (Blue): 0; Flammability (Red): 2; Reactivity (Yellow): 0

Oils: clarified

EMERGENCY RESPONSE: See Appendix A (171)
Evacuation:
Public safety: Isolate the area of spill or leak for at least 10 to 25 meters (30 to 80 feet) in all directions.
Spill: Increase, in the downwind direction, as necessary, the distance shown under "Public Safety."
Fire: If any large container is involved in fire, isolate for at least 800 meters (½ mile) in all directions; also, consider initial evacuation for 800 meters (½ mile) in all directions.

EXPOSURE
Short-term effects: *SEEK MEDICAL ATTENTION.* **Liquid:** Irritating to skin and eyes. Harmful if swallowed. Remove contaminated clothing and shoes. Flush affected areas with plenty of water. *IF IN EYES,* hold eyelids open and flush with plenty of water. *IF SWALLOWED* and victim is *CONSCIOUS AND ABLE TO SWALLOW,* have victim drink 4 to 8 ounces of water. *IF SWALLOWED* and victim is *UNCONSCIOUS OR HAVING CONVULSIONS,* do nothing except keep victim warm.

HEALTH HAZARDS
Personal protective equipment (PPE): B-Level PPE. Chemical goggles; face shield; oil-resistant gloves.
Toxicity by ingestion: Grade 0; LD_{50} = > 15 g/kg.
Liquid or solid irritant characteristics: Eye and skin irritant.

FIRE DATA
Flash point: 450°F/232°C (oc).
Fire extinguishing agents not to be used: Water may be ineffective.

CHEMICAL REACTIVITY
Binary reactants: Strong acids, oxidizers. May attack some forms of plastics
Reactivity group: 34
Compatibility class: Esters.

ENVIRONMENTAL DATA
Water pollution: Effect of low concentrations on aquatic life is unknown. Fouling to shoreline. May be dangerous if it enters nearby water intakes; notify operators. Notify local health and wildlife officials. **Response to discharge:** Mechanical containment. Should be removed. Chemical and physical treatment.

SHIPPING INFORMATION
Grades of purity: Commercial; **Storage temperature:** Ambient; **Inert atmosphere:** None; **Venting:** Open (flame arrester); **Stability during transport:** Stable.

PHYSICAL AND CHEMICAL PROPERTIES
Physical state @ 59°F/15°C and 1 atm: Liquid.
Molecular weight: 420 (approximate).
Boiling point @ 1 atm: Decomposes.
Specific gravity (water = 1): (estimate) 1.002 @ 68°F/20°C (liquid).
Liquid surface tension: 40.1 dynes/cm = 0.0401 N/m @ 68°F/20°C.

OILS: CLARIFIED   **REC. O:0700**

SYNONYMS: CLARIFIED OILS (PETROLEUM), CATALYTIC CRACKED

IDENTIFICATION
CAS Number: 64741-62-4
DOT ID Number: UN 1270; DOT Guide Number: 128
Proper Shipping Name: Oil, petroleum, n.o.s.

DESCRIPTION: Oily Colorless liquid. Floats on water surface; insoluble.

Combustible • Containers may BLEVE when exposed to fire • Vapors are heavier than air and will collect and stay in low areas • Vapors in confined areas (e.g., tanks, sewers, buildings) may explode when exposed to fire • Irritating to the skin, eyes, and respiratory tract • Toxic products of combustion may include acrid carbon monoxide.

EMERGENCY RESPONSE: See Appendix A (128)
Evacuation:
Public safety: Isolate spill area for at least 25 to 50 meters (80 to 160 feet) in all directions.
Spill: Large spill–Consider initial downwind evacuation for at least 300 meters (1000 feet).
Fire: Isolate for 800 meters (½ mile) in all directions, especially if tank, rail car, or tank truck is involved in fire.

HEALTH HAZARDS
Personal protective equipment (PPE): B-Level PPE. Goggles or face shield
Toxicity by ingestion: Grade 1; LD_{50} = 5 to 15 g/kg.

FIRE DATA
Fire extinguishing agents not to be used: Water may be ineffective
Burning rate: 4 mm/min

CHEMICAL REACTIVITY
Binary reactants: Violent reaction with strong oxidizers, strong acids.
Reactivity group: 33
Compatibility class: Miscellaneous hydrocarbon mixtures

ENVIRONMENTAL DATA
Water pollution: Dangerous to waterfowl. Fouling to shoreline. May be dangerous if it enters nearby water intakes; notify operators. Notify local health and wildlife officials. **Response to discharge:** Mechanical containment. Should be removed. Chemical and physical treatment.

SHIPPING INFORMATION
Storage temperature: Ambient; **Inert atmosphere:** None; **Venting:** Open (flame arrester); **Stability during transport:** Stable.

PHYSICAL AND CHEMICAL PROPERTIES
Physical state @ 59°F/15°C and 1 atm: Liquid.
Specific gravity (water = 1): (estimate) 0.85 @ 68°F/20°C (liquid).
Liquid surface tension: (estimate) 25 dynes/cm = 0.025 N/m @ 68°F/20°C.
Liquid water interfacial tension: (estimate) 50 dynes/cm = 0.05 N/m @ 68°F/20°C.
Heat of combustion: (estimate) –18,000 Btu/lb = –10,000 cal/g = –420 x 10^5 J/kg.
Vapor pressure: Very low.

OILS: CRUDE REC. O:0800

SYNONYMS: COAL OIL; CRUDE OIL; PETROLEUM; PETROLEUM CRUDE OIL; ROCK OIL; SENECA OIL

IDENTIFICATION
CAS Number: 68308-34-9 (shale oil, crude)
Formula: Mixture of hydrocarbons
DOT ID Number: UN 1267; UN 1288 (shale oil); DOT Guide Number: 128
Proper Shipping Name: Petroleum crude oil (UN 1267); Shale oil (UN 1288)

DESCRIPTION: Dark oily liquid. Acrid, offensive, tarry odor. Floats on water surface; flammable vapor may be produced.

Highly flammable • Containers may BLEVE when exposed to fire • Vapors may form explosive mixture with air • Vapors are heavier than air and will collect and stay in low areas • Vapors may travel long distances to ignition sources and flashback • Vapors in confined areas (e.g., tanks, sewers, buildings) may explode when exposed to fire • Irritating to the skin, eyes, and respiratory tract • Toxic products of combustion may include acrid carbon monoxide.

Hazard Classification (based on NFPA-704 Rating System)
sour crude: Health Hazards (Blue): 2; Flammability (Red): 3; Reactivity (Yellow): 0
sweet crude: Health Hazards (Blue): 1; Flammability (Red): 3; Reactivity (Yellow): 0

EMERGENCY RESPONSE: See Appendix A (128)
Evacuation:
Public safety: Isolate spill area for at least 25 to 50 meters (80 to 160 feet) in all directions.
Spill: Large spill–Consider initial downwind evacuation for at least 300 meters (1000 feet).
Fire: Isolate for 800 meters (½ mile) in all directions, especially if tank, rail car, or tank truck is involved in fire.

EXPOSURE
Short-term effects: *SEEK MEDICAL ATTENTION.* **Vapor:** Not irritating to eyes, nose, or throat. **Liquid:** Irritating to skin and eyes. Remove contaminated clothing and shoes. Flush affected areas with plenty of water. *IF IN EYES*, hold eyelids open and flush with plenty of water.

HEALTH HAZARDS
Personal protective equipment (PPE): B-Level PPE. Goggles or face shield; rubber gloves and boots.
Recommendations for respirator selection: NIOSH/OSHA based on mineral oil mist
50 mg/m³: HiE (any air-purifying, respirator with a high-efficiency particulate filter); or SA (any supplied-air respirator). *125 mg/m³*: SA:CF (any supplied-air respirator operated in a continuous-flow mode); or PAPRHiE (any powered, air-purifying respirator with a high-efficiency particulate filter). *250 mg/m³*: HiEF (any air-purifying, full-facepiece respirator with a high-efficiency particulate filter); or SAT:CF (any supplied-air respirator that has a tight-fitting facepiece and is operated in a continuous-flow mode); or PAPRTHiE (any powered, air-purifying respirator with a tight-fitting facepiece and a high-efficiency particulate filter); or SCBAF (any self-contained breathing apparatus with a full facepiece); or SAF (any supplied-air respirator with a full facepiece). *EMERGENCY OR PLANNED ENTRY INTO UNKNOWN CONCENTRATIONS OR IDLH CONDITIONS*: SCBAF:PD,PP (any self-contained breathing apparatus that has a full facepiece and is operated in a pressure-demand or other positive-pressure mode); or SAF:PD,PP:ASCBA (any supplied-air respirator that has a full facepiece and is operated in a pressure-demand or other positive-pressure mode in combination with an auxiliary self-contained breathing apparatus operated in a pressure-demand or other positive pressure mode). *ESCAPE*: HiEF (any air-purifying, full-facepiece respirator with a high-efficiency particulate filter); or SCBAE (any appropriate escape-type, self-contained breathing apparatus).
Exposure limits (TWA unless otherwise noted): ACGIH TLV 5 mg/m³; OSHA PEL 5 mg/m³; NIOSH REL 5 mg/m³ as mineral oil mist
Short-term exposure limits (15-minute TWA): NIOSH STEL 10 mg/m³ as mineral oil mist
Vapor (gas) irritant characteristics: Vapors are nonirritating to the eyes and throat.
Liquid or solid irritant characteristics: Eye and skin irritant. If spilled on clothing and allowed to remain, may cause smarting and reddening of the skin.
IDLH value: 2500 mg/m³ as mineral oil mist.

FIRE DATA
Flash point: 20 to 90°F/–7 to 32°C (cc).
Fire extinguishing agents not to be used: Water may be ineffective
Electrical hazard: Class I, Group D.
Burning rate: 4 mm/min

CHEMICAL REACTIVITY
Binary reactants: Nitric acid, oxidizers
Reactivity group: 33
Compatibility class: Miscellaneous hydrocarbon mixtures

ENVIRONMENTAL DATA
Water pollution: Harmful to aquatic life in very low concentrations. Fouling to shoreline. May be dangerous if it enters nearby water intakes; notify operators. Notify local health and wildlife officials. **Response to discharge:** Mechanical containment. Should be removed. Chemical and physical treatment.

SHIPPING INFORMATION
Grades of purity: Wide variety, depending on oil field where produced; **Storage temperature:** Ambient; **Inert atmosphere:** None; **Venting:** Open (flame arrester); **Stability during transport:** Stable.

NAS HAZARD CLASSIFICATION FOR BULK WATER TRANSPORTATION
FIRE: 1
HEALTH: Vapor irritant: 0; Liquid or solid irritant: 1; Poisons: 1
WATER POLLUTION: Human toxicity: 1; Aquatic toxicity: 2; Aesthetic effect: 4
REACTIVITY: Other chemicals: 0; Water: 0; Self-reaction: 0

PHYSICAL AND CHEMICAL PROPERTIES
Physical state @ 59°F/15°C and 1 atm: Liquid.
Boiling point @ 1 atm: 90 to more than 750°F/32 to more than 400°C/305 to more than 673°K.
Melting/Freezing point: –50°F/–45°C/228°K.
Specific gravity (water = 1): 0.70-0.98 @ 15°C (liquid).
Liquid surface tension: 24-38 dynes/cm = 0.024-0.038 N/m @ 68°F/20°C.
Relative vapor density (air = 1): Above 1.0

Latent heat of vaporization: 140–150 Btu/lb = 76–86 cal/g = 3.2–3.6 x 10^5 J/kg.
Heat of combustion: –18,252 Btu/lb = –10,140 cal/g = –424.54 x 10^5 J/kg.
Vapor pressure: (Reid) 0.10 psia; 1-3 mm.

OILS: DIESEL   REC. O:0900

SYNONYMS: DIESEL FUEL; DIESEL OIL; FUEL OIL 1-D; FUEL OIL 2-D; PETROLEUM OIL

IDENTIFICATION
CAS Number: 68512-90-3; 68334-30-5 (automotive); 68476-30-2 (marine or Fuel oil No. 2)
Formula: Mixture of hydrocarbons
DOT ID Number: NA 1993; UN 1202; DOT Guide Number: 128
Proper Shipping Name: Diesel fuel

DESCRIPTION: Yellow to dark brown, oily liquid. Gasoline-like or fuel oil odor. Floats on water surface; insoluble.

Flammable • Containers may BLEVE when exposed to fire • Vapors are heavier than air and will collect and stay in low areas • Vapors in confined areas (e.g., tanks, sewers, buildings) may explode when exposed to fire • Irritating to the skin, eyes, and respiratory tract • Toxic products of combustion may include acrid carbon monoxide.

Hazard Classification (based on NFPA-704 Rating System)
Health Hazards (Blue): 0; Flammability (Red): 2; Reactivity (Yellow): 0

EMERGENCY RESPONSE: See Appendix A (128)
Evacuation:
Public safety: Isolate spill area for at least 25 to 50 meters (80 to 160 feet) in all directions.
Spill: Large spill–Consider initial downwind evacuation for at least 300 meters (1000 feet).
Fire: Isolate for 800 meters (½ mile) in all directions, especially if tank, rail car, or tank truck is involved in fire.

EXPOSURE
Short-term effects: *SEEK MEDICAL ATTENTION.* **Liquid:** Irritating to skin and eyes. Harmful if swallowed. Remove contaminated clothing and shoes. Flush affected areas with plenty of water. *IF IN EYES*, hold eyelids open and flush with plenty of water. *IF SWALLOWED* and victim is *CONSCIOUS AND ABLE TO SWALLOW*, have victim drink 4 to 8 ounces of water. **Do NOT induce vomiting.**

HEALTH HAZARDS
Personal protective equipment (PPE): B-Level PPE. Goggles or face shield. Chemical protective material(s) reported to have good to excellent resistance: nitrile
Recommendations for respirator selection: NIOSH *(diesel EXHAUST ONLY).*
At any concentrations above the NIOSH REL, or where there is no REL, at any detectable concentration: SCBAF:PD,PP (any self-contained breathing apparatus that has a full facepiece and is operated in a pressure-demand or other positive-pressure mode); or SAF:PD,PP:ASCBA (any supplied-air respirator that has a full facepiece and is operated in a pressure-demand or other positive-pressure mode in combination with an auxiliary, self-contained breathing apparatus operated in a pressure-demand or other positive pressure mode). *ESCAPE:* GMFOVHiE [any air-purifying, full-facepiece respirator (gas mask) with a chin-style, front- or back-mounted organic vapor canister having a high-efficiency particulate filter]; or SCBAE (any appropriate escape-type, self-contained breathing apparatus).
Exposure limits (TWA unless otherwise noted): ACGIH TLV 15 ppm.
Toxicity by ingestion: Grade 1; LD_{50} = 5 to 15 g/kg.
Long-term health effects: Dermatitis, heptoxic at high levels, lowpotency animal carcinogen.
Vapor (gas) irritant characteristics: Vapors cause a slight smarting of the eyes or respiratory system if present in high concentrations. The effect is temporary.
Liquid or solid irritant characteristics: Minimum hazard. If spilled on clothing and allowed to remain, may cause smarting and reddening of the skin.

FIRE DATA
Flash point: general range 100–190°F/38–88°C (cc); (1-D) 100°F/38°C (cc); (2-D) 125°F/52°C (cc); (4-D) 130°F/54°C (cc).
Flammable limits in air: LEL: 0.6–1.3%; UEL: 6.0–7.5%.
Fire extinguishing agents not to be used: Water may be ineffective
Autoignition temperature: (1-D) 350–625°F/177–329°C; (2-D) 490–545°F/254–285°C.
Electrical hazard: Class I, Group D. Flow or agitation of substance may generate electrostatic charges due to low conductivity.
Burning rate: 4 mm/min

CHEMICAL REACTIVITY
Binary reactants: Violent reaction with strong oxidizers, fluorine. Incompatible with nitric acid, ammonia, ammonium nitrate.
Reactivity group: 33
Compatibility class: Miscellaneous hydrocarbon mixtures

ENVIRONMENTAL DATA
Water pollution: Dangerous to aquatic life and waterfowl. Fouling to shoreline. May be dangerous if it enters nearby water intakes; notify operators. Notify local health and wildlife officials. **Response to discharge:** Mechanical containment. Should be removed. Chemical and physical treatment.

SHIPPING INFORMATION
Grades of purity: Diesel Fuel 1-D (ASTM); Diesel Fuel 2-D (ASTM); **Storage temperature:** Ambient; **Inert atmosphere:** None; **Venting:** Open (flame arrester); **Stability during transport:** Stable.

PHYSICAL AND CHEMICAL PROPERTIES
Physical state @ 59°F/15°C and 1 atm: Liquid.
Molecular weight: 190 (average).
Boiling point @ 1 atm: 450-800°F/232-425°C/505-698°K.
Melting/Freezing point: –29 to 18°F/–20 to –0.4°C/244 to 272.6°K.
Specific gravity (water = 1): 0.841 at 16°C (liquid); 0.87-0.95 g/mL @ 15°C to 20°C.
Liquid surface tension: (estimate) 25 dynes/cm – 0.025 N/m @ 68°F/20°C.
Liquid water interfacial tension: (estimate) 50 dynes/cm = 0.05 N/m @ 68°F/20°C.
Heat of combustion: –18,400 Btu/lb = –10,200 cal/g = 429 x 10^5 J/kg.
Vapor pressure: 2.6 mm @ 50°F; 2.1 to 2.6 torr @ 70°F/21°C/294°K.

OILS, DIESEL FUEL: 1-D REC. O:1000

SYNONYMS: DIESEL FUEL 1-D; DIESEL OIL (LIGHT); OILS, FUEL: 1-D

IDENTIFICATION
CAS Number: 68334-30-5
Formula: Mixture of hydrocarbons
DOT ID Number: NA 1993; UN 1202; DOT Guide Number: 128
Proper Shipping Name: Diesel fuel

DESCRIPTION: Yellow-brown, oily liquid. Kerosene or fuel oil odor. Floats on water surface; insoluble.

Flammable • Containers may BLEVE when exposed to fire • Vapors are heavier than air and will collect and stay in low areas • Vapors in confined areas (e.g., tanks, sewers, buildings) may explode when exposed to fire • Irritating to the skin, eyes, and respiratory tract • Toxic products of combustion may include acrid carbon monoxide.

Hazard Classification (based on NFPA-704 Rating System)
Health Hazards (Blue): 0; Flammability (Red): 2; Reactivity (Yellow): 0

EMERGENCY RESPONSE: See Appendix A (128)
Evacuation:
Public safety: Isolate spill area for at least 25 to 50 meters (80 to 160 feet) in all directions.
Spill: Large spill–Consider initial downwind evacuation for at least 300 meters (1000 feet).
Fire: Isolate for 800 meters (½ mile) in all directions, especially if tank, rail car, or tank truck is involved in fire.

EXPOSURE
Short-term effects: *SEEK MEDICAL ATTENTION*. **Liquid:** Irritating to skin and eyes. Harmful if swallowed. Remove contaminated clothing and shoes. Flush affected areas with plenty of water. *IF IN EYES*, hold eyelids open and flush with plenty of water. *IF SWALLOWED* and victim is *CONSCIOUS AND ABLE TO SWALLOW*, have victim drink 4 to 8 ounces of water. **Do NOT induce vomiting.**

HEALTH HAZARDS
Personal protective equipment (PPE): B-Level PPE. Protective gloves; goggles or face shield. Chemical protective material(s) reported to have good to excellent resistance: nitrile
Recommendations for respirator selection: NIOSH *(diesel EXHAUST ONLY). At any concentrations above the NIOSH REL, or where there is no REL, at any detectable concentration:* SCBAF:PD,PP (any self-contained breathing apparatus that has a full facepiece and is operated in a pressure-demand or other positive-pressure mode); or SAF:PD,PP:ASCBA (any supplied-air respirator that has a full facepiece and is operated in a pressure-demand or other positive-pressure mode in combination with an auxiliary, self-contained breathing apparatus operated in a pressure-demand or other positive pressure mode). *ESCAPE:* GMFOVHiE [any air-purifying, full-facepiece respirator (gas mask) with a chin-style, front- or back-mounted organic vapor canister having a high-efficiency particulate filter]; or SCBAE (any appropriate escape-type, self-contained breathing apparatus).
Exposure limits (TWA unless otherwise noted): ACGIH TLV 15 ppm (100 mg/m^3).
Toxicity by ingestion: Grade 1; LD_{50} = 5 to 15 g/kg.
Long-term health effects: Dermatitis, heptoxic at high levels, low potency animal carcinogen.
Vapor (gas) irritant characteristics: Slight smarting of eyes or respiratory system if present in high concentrations. The effect is temporary.
Liquid or solid irritant characteristics: Minimum hazard. If spilled on clothing and allowed to remain, may cause smarting and reddening of skin.
Odor threshold: 0.7 ppm.

FIRE DATA
Flash point: 100°F/38°C (cc).
Flammable limits in air: LEL: 1.3%; UEL: 6%.
Fire extinguishing agents not to be used: Water may be ineffective.
Autoignition temperature: 350–625°F/177–329°C/450–602°K.
Electrical hazard: Class I, Group D.
Burning rate: 4 mm/min

CHEMICAL REACTIVITY
Binary reactants: Strong oxidizers, nitric acid
Reactivity group: 33
Compatibility class: Miscellaneous hydrocarbon mixtures

ENVIRONMENTAL DATA
Water pollution: Dangerous to aquatic life and waterfowl. Fouling to shoreline. May be dangerous if it enters nearby water intakes; notify operators. Notify local health and wildlife officials. **Response to discharge:** Mechanical containment. Should be removed. Chemical and physical treatment.

SHIPPING INFORMATION
Grades of purity: Diesel fuel 1-D (ASTM); **Storage temperature:** Ambient; **Inert atmosphere:** None; **Venting:** Open (flame arrester); **Stability during transport:** Stable.

PHYSICAL AND CHEMICAL PROPERTIES
Physical state @ 59°F/15°C and 1 atm: Liquid.
Boiling point @ 1 atm: 380–560°F/193–293°C/466–566°K.
Melting/Freezing point: –30°F/–34°C/240°K.
Specific gravity (water = 1): 0.81–0.85 @ 15°C (liquid).
Liquid surface tension: 23–32 dynes/cm = 0.023–0.032 N/m @ 68°F/20°C.
Liquid water interfacial tension: 47-49 dynes/cm = 0.047–0.049 N/m @ 68°F/20°C.
Latent heat of vaporization: 110 Btu/lb = 60 cal/g = 2.5 x 10^5 J/kg.
Heat of combustion: –18,540 Btu/lb = –10,300 cal/g = –431.24 x 10^5 J/kg.

OILS, DIESEL FUEL: 2-D REC. O:1100

SYNONYMS: DIESEL OIL, MEDIUM; OIL, FUEL: 2-D

IDENTIFICATION
CAS Number: 68476-34-6; may also apply to 77650-28-3 (diesel No. 4, marine diesel)
DOT ID Number: UN 1202; DOT Guide Number: 128
Proper Shipping Name: Gas oil; Diesel oil; Heating oil, light

DESCRIPTION: Yellow-brown oily liquid. Lube or fuel oil odor. Floats on the surface of water.

Highly flammable • Containers may BLEVE when exposed to fire • Vapors may form explosive mixture with air • Vapors are heavier than air and will collect and stay in low areas • Vapors may travel long distances to ignition sources and flashback • Vapors in confined areas (e.g., tanks, sewers, buildings) may explode when exposed to fire • Irritating to the skin, eyes, and respiratory tract • Toxic products of combustion may include carbon monoxide.

Hazard Classification (based on NFPA-704 Rating System)
Health Hazards (Blue): 0; Flammability (Red): 2; Reactivity (Yellow): 0

EMERGENCY RESPONSE: See Appendix A (128)
Evacuation:
Public safety: Isolate spill area for at least 25 to 50 meters (80 to 160 feet) in all directions.
Spill: Large spill–Consider initial downwind evacuation for at least 300 meters (1000 feet).
Fire: Isolate for 800 meters (½ mile) in all directions, especially if tank, rail car, or tank truck is involved in fire.

EXPOSURE
Personal protective equipment (PPE): B-Level PPE. Chemical protective material(s) reported to have good to excellent resistance: nitrile
Recommendations for respirator selection: NIOSH *(diesel EXHAUST ONLY)*.
At any concentrations above the NIOSH REL, or where there is no REL, at any detectable concentration: SCBAF:PD,PP (any self-contained breathing apparatus that has a full facepiece and is operated in a pressure-demand or other positive-pressure mode); or SAF:PD,PP:ASCBA (any supplied-air respirator that has a full facepiece and is operated in a pressure-demand or other positive-pressure mode in combination with an auxiliary, self-contained breathing apparatus operated in a pressure-demand or other positive pressure mode). *ESCAPE:* GMFOVHiE [any air-purifying, full-facepiece respirator (gas mask) with a chin-style, front- or back-mounted organic vapor canister having a high-efficiency particulate filter]; or SCBAE (any appropriate escape-type, self-contained breathing apparatus). **Liquid:** Irritating to skin and eyes. Harmful if swallowed. Remove contaminated clothing and shoes. Flush affected areas with plenty of water. *IF IN EYES,* hold eyelids open and flush with plenty of water. *IF SWALLOWED* and victim is *CONSCIOUS AND ABLE TO SWALLOW,* have victim drink 4 to 8 ounces of water. **Do NOT induce vomiting.**

HEALTH HAZARDS
Personal protective equipment (PPE): B-Level PPE. Protective gloves; goggles or face shield. Chemical protective material(s) reported to have good to excellent resistance: nitrile
Exposure limits (TWA unless otherwise noted): ACGIH TLV 15 ppm (100 mg/m^3).
Toxicity by ingestion: Grade 1; LD$_{50}$ = 5 to 15 g/kg.
Long-term health effects: Dermatitis, heptoxic at high levels, low-potency animal carcinogen.
Vapor (gas) irritant characteristics: Slight smarting of eyes or respiratory system if present in high concentrations. The effect is temporary.
Liquid or solid irritant characteristics: Minimum hazard. If spilled on clothing and allowed to remain, may cause smarting and reddening of skin.

FIRE DATA
Flash point: 125°F/52°C (cc).
Flammable limits in air: LEL: 1.3%; UEL: 6.0%.
Fire extinguishing agents not to be used: Water may be ineffective
Autoignition temperature: 490–545°F/254–285°C/527–558°C.
Electrical hazard: Class I, Group D.
Burning rate: 4 mm/min

CHEMICAL REACTIVITY
Binary reactants: Strong oxidizers
Reactivity group: 33
Compatibility class: Miscellaneous hydrocarbon mixtures

ENVIRONMENTAL DATA
Water pollution: Dangerous to aquatic life and waterfowl. Fouling to shoreline. May be dangerous if it enters nearby water intakes; notify operators. Notify local health and wildlife officials. **Response to discharge:** Mechanical containment. Should be removed. Chemical and physical treatment.

SHIPPING INFORMATION
Grades of purity: Diesel fuel 2-D (ASTM); **Storage temperature:** Ambient; **Inert atmosphere:** None; **Venting:** Open (flame arrester); **Stability during transport:** Stable.

PHYSICAL AND CHEMICAL PROPERTIES
Physical state @ 59°F/15°C and 1 atm: Liquid.
Boiling point @ 1 atm: 540–640°F/282–338°C/555–611°K.
Melting/Freezing point: 0°F/18°C/255°K.
Specific gravity (water = 1): 0.87–0.90 @ 68°F/20°C (liquid).
Heat of combustion: –19,440 Btu/lb = –10,800 cal/g = –452.17 x 10^5 J/kg.

OILS, EDIBLE: CASTOR　　　　　　　　　　**REC. O:1200**

SYNONYMS: ACEITE de RICINO (Spanish); CASTER OIL; CASTOR OIL; CASTOR OIL, HYDROGENATED; NEOLID; OIL OF PALMA CHRISTI; RICINUS OIL; TANGANTANGAN OIL; TURKEY-RED OIL (SULFATED CASTOR OIL).

IDENTIFICATION
CAS Number: 8001-79-4
DOT ID Number: UN 9277; DOT Guide Number: 171
Proper Shipping Name: Oil, n.o.s., flash point not less than 93°C/200°F.

DESCRIPTION: Colorless to light yellow, oily liquid. Weak odor. Floats on water surface; insoluble.

Combustible • Containers may BLEVE when exposed to fire • Toxic products of combustion may include carbon monoxide.

Hazard Classification (based on NFPA-704 Rating System)
Health Hazards (Blue): 0; Flammability (Red): 1; Reactivity (Yellow): 0

EMERGENCY RESPONSE: See Appendix A (171)
Evacuation:
Public safety: Isolate the area of spill or leak for at least 10 to 25 meters (30 to 80 feet) in all directions.
Spill: Increase, in the downwind direction, as necessary, the distance shown under "Public Safety."
Fire: If any large container is involved in fire, isolate for at least 800 meters (½ mile) in all directions; also, consider initial evacuation for 800 meters (½ mile) in all directions.

HEALTH HAZARDS

Personal protective equipment (PPE): B-Level PPE. Goggles or face shield. Chemical protective material(s) reported to offer protection: neoprene.

Exposure limits (TWA unless otherwise noted): ACGIH TLV 10 mg/m^3 (vegetable oil mists); OSHA PEL TWA 15 mg/m^3 (total); 5 mg/m^3 (respirable fraction); NIOSH REL TWA 10 mg/m^3 (total); 5 mg/m^3 (respirable fraction) as vegetable oil mist.

Toxicity by ingestion: Grade 1; LD_{50} = 5 to 15 g/kg.

FIRE DATA

Flash point: 445°F/229°C (cc); 401°F/205°C (cc) (hydrogenated).
Fire extinguishing agents not to be used: Water or foam may cause frothing.
Autoignition temperature: 840°F/449°C/722°K.

CHEMICAL REACTIVITY

Binary reactants: Strong oxidizers, strong acids
Reactivity group: 34
Compatibility class: Esters.

ENVIRONMENTAL DATA

Water pollution: Effect of low concentrations on aquatic life is unknown. Fouling to shoreline. May be dangerous if it enters nearby water intakes; notify operators. Notify local health and wildlife officials. **Response to discharge:** Mechanical containment. Should be removed. Chemical and physical treatment.

SHIPPING INFORMATION

Grades of purity: Commercial: meets Military Specifications and ASTM; USP; USP Odorless; Technical. All grades differ only in color and acid values; **Storage temperature:** Ambient; **Inert atmosphere:** None; **Venting:** Open (flame arrester); **Stability during transport:** Stable.

PHYSICAL AND CHEMICAL PROPERTIES

Physical state @ 59°F/15°C and 1 atm: Liquid.
Boiling point @ 1 atm: Varies, depending on composition; 594°F/312°C/585°K.
Melting/Freezing point: –14°F/–10°C/263°K.
Specific gravity (water = 1): 0.96 @ 77°F/25°C (liquid).
Liquid surface tension: 39 dynes/cm = 0.039 N/m @ 68°F/20°C.
Liquid water interfacial tension: 19.2 dynes/cm = 0.0192 N/m at 71°F/22°C.
Relative vapor density (air = 1): 10
Heat of combustion: –15,950 Btu/lb = –8860 cal/g = –371.0 x 10^5 J/kg.
Vapor pressure: (Reid) 0.10 psia; 2.04 mm.

OILS, EDIBLE: COCONUT   REC. O:1300

SYNONYMS: ACEITE de NUEZ de COCO (Spanish); COCONUT BUTTER; COPRA OIL; COCONUT OIL; COCONUT OIL, REFINED; COCONUT OIL, CRUDE

IDENTIFICATION

CAS Number: 8001-31-8
DOT ID Number: UN 9277; DOT Guide Number: 171
Proper Shipping Name: Oil, n.o.s., flash point not less than 93°C (200°F).

DESCRIPTION: Light yellow to yellow-orange liquid or solid. Weak acid odor. Floats on the surface of water. Combustible • Containers may BLEVE when exposed to fire • Irritating to the skin, eyes, and respiratory tract • Toxic products of combustion may include irritating vapors.

Hazard Classification (based on NFPA-704 Rating System)
Health Hazards (Blue): 0; Flammability (Red): 1; Reactivity (Yellow): 0

EMERGENCY RESPONSE: See Appendix A (171)

Evacuation:
Public safety: Isolate the area of spill or leak for at least 10 to 25 meters (30 to 80 feet) in all directions.
Spill: Increase, in the downwind direction, as necessary, the distance shown under "Public Safety."
Fire: If any large container is involved in fire, isolate for at least 800 meters (½ mile) in all directions; also, consider initial evacuation for 800 meters (½ mile) in all directions.

EXPOSURE

Short-term effects: Liquid or solid: Not toxic. *IF SWALLOWED, do NOT induce vomiting.*

HEALTH HAZARDS

Personal protective equipment (PPE): B-Level PPE. OSHA Table Z-1-A air contaminant (vegetable oil). Goggles or face shield; rubber gloves.

Exposure limits (TWA unless otherwise noted): ACGIH TLV 10 mg/m^3 (vegetable oil mists); OSHA PEL TWA 15 mg/m^3 (total); 5 mg/m^3 (respirable fraction); NIOSH REL TWA 10 mg/m^3 (total); 5 mg/m^3 (respirable fraction) as vegetable oil mist.

Liquid or solid irritant characteristics: Mild eye irritant.

FIRE DATA

Flash point: 420°F/216°C (cc) (refined); 548°F/286°C (cc) (crude).
Fire extinguishing agents not to be used: Water or foam may cause frothing; water may be ineffective.
Burning rate: 4 mm/min

CHEMICAL REACTIVITY

Binary reactants: Strong oxidizers. If material is stored hot and wet it may spontaneously heat and ignite.
Reactivity group: 34
Compatibility class: Esters.

ENVIRONMENTAL DATA

Water pollution: Effect of low concentrations on aquatic life is unknown. Fouling to shoreline. May be dangerous if it enters nearby water intakes; notify operators. Notify local health and wildlife officials. **Response to discharge:** Mechanical containment. Should be removed. Chemical and physical treatment.

SHIPPING INFORMATION

Grades of purity: Crude; Cochin. All grades contain 3–5% free fatty acids; **Storage temperature:** Ambient; **Inert atmosphere:** None; **Venting:** Open (flame arrester); **Stability during transport:** Stable if dry and cool.

PHYSICAL AND CHEMICAL PROPERTIES

Physical state @ 59°F/15°C and 1 atm: Solid or Liquid.
Boiling point @ 1 atm: Very high.
Melting/Freezing point: (approximate) 76°F/24°C/297°K.
Specific gravity (water = 1): 0.922 @ 77°F/25°C (liquid).
Liquid surface tension: 33.4 dynes/cm = 0.0334 N/m @ 68°F/20°C.

Liquid water interfacial tension: (estimate) 50 dynes/cm = 0.050 N/m @ 77°F/25°C.
Heat of combustion: (estimate) = –15,500 Btu/lb = –8600 cal/g = –360 x 10^5 J/kg.

OILS, EDIBLE: COTTONSEED REC. O:1400

SYNONYMS: ACEITE de SEMILLA de ALGODON (Spanish); COTTONSEED OIL; COTTONSEED OIL, REFINED; COTTONSEED OIL, UNHYDROGENATED; DEODORIZED, WINTERIZED COTTONSEED OIL

IDENTIFICATION
CAS Number: 8001-29-4
Formula: Mixture of C–14 to C–16 fatty acids
DOT ID Number: UN 9277; DOT Guide Number: 171
Proper Shipping Name: Oil, n.o.s., flash point not less than 93°C/200°F

DESCRIPTION: Pale yellow, oily liquid. Odorless. Floats on water surface; insoluble.

Combustible • Containers may BLEVE when exposed to fire • Irritating to the skin, eyes, and respiratory tract • Toxic products of combustion may include irritating smoke and vapors.

Hazard Classification (based on NFPA-704 Rating System)
Health Hazards (Blue): 0; Flammability (Red): 1; Reactivity (Yellow): 0

EMERGENCY RESPONSE: See Appendix A (171)
Evacuation:
Public safety: Isolate the area of spill or leak for at least 10 to 25 meters (30 to 80 feet) in all directions.
Spill: Increase, in the downwind direction, as necessary, the distance shown under "Public Safety."
Fire: If any large container is involved in fire, isolate for at least 800 meters (½ mile) in all directions; also, consider initial evacuation for 800 meters (½ mile) in all directions.

HEALTH HAZARDS
Personal protective equipment (PPE): B-Level PPE. OSHA Table Z-1-A air contaminant (vegetable oil). Goggles or face shield.
Exposure limits (TWA unless otherwise specified): ACGIH TLV 10 mg/m^3 (vegetable oil mists); OSHA PEL TWA 15 mg/m^3 (total); 5 mg/m^3 (respirable fraction); NIOSH REL TWA 10 mg/m^3 (total); 5 mg/m^3 (respirable fraction) as vegetable oil mist.
Liquid or solid irritant characteristics: May cause eye irritation.

FIRE DATA
Flash point: 486°F/252°C (cc) (refined oil); 610°F/321°C (oc) (cooking oil).
Fire extinguishing agents not to be used: Water or foam may cause frothing.
Autoignition temperature: 650°F/343°C/616°K.
Electrical hazard: Class I, Group D.

CHEMICAL REACTIVITY
Binary reactants: Strong oxidizers
Reactivity group: 34
Compatibility class: Esters.

ENVIRONMENTAL DATA
Water pollution: Effect of low concentrations on aquatic life is unknown. Fouling to shoreline. May be dangerous if it enters nearby water intakes; notify operators. Notify local health and wildlife officials. **Response to discharge:** Mechanical containment. Should be removed. Chemical and physical treatment.

SHIPPING INFORMATION
Grades of purity: Refined; cooking; **Storage temperature:** Ambient; **Inert atmosphere:** None; **Venting:** Open (flame arrester); **Stability during transport:** Stable.

PHYSICAL AND CHEMICAL PROPERTIES
Physical state @ 59°F/15°C and 1 atm: Liquid.
Boiling point @ 1 atm: Very high.
Melting/Freezing point: 32°F/0°C/273°K.
Specific gravity (water = 1): 0.922 @ 68°F/20°C (liquid).
Liquid surface tension: 35 dynes/cm = 0.035 N/m @ 68°F/20°C.
Liquid water interfacial tension: (estimate) 50 dynes/cm = 0.05 N/m @ 68°F/20°C.
Heat of combustion: (estimate) = –16,000 Btu/lb = –8,870 cal/g = –371 x 10^5 J/kg.
Vapor pressure: (Reid) 0.1 psia; 2 mm.

OILS, EDIBLE: FISH REC. O:1500

SYNONYMS: ACEITE de PESCADO (Spanish); FISH OIL

IDENTIFICATION
DOT ID Number: UN 9277; DOT Guide Number: 171
Proper Shipping Name: Oil, n.o.s., flash point not less than 93°C (200°F).

DESCRIPTION: Pale yellow, oily liquid. Fishy odor. Floats on water surface; insoluble.

Combustible • Containers may BLEVE when exposed to fire.

Hazard Classification (based on NFPA-704 Rating System)
Health Hazards (Blue): 0; Flammability (Red): 1; Reactivity (Yellow): 0

EMERGENCY RESPONSE: See Appendix A (171)
Evacuation:
Public safety: Isolate the area of spill or leak for at least 10 to 25 meters (30 to 80 feet) in all directions.
Spill: Increase, in the downwind direction, as necessary, the distance shown under "Public Safety."
Fire: If any large container is involved in fire, isolate for at least 800 meters (½ mile) in all directions; also, consider initial evacuation for 800 meters (½ mile) in all directions.

HEALTH HAZARDS
Personal protective equipment (PPE): B-Level PPE. Goggles or face shield.
Liquid or solid irritant characteristics: May cause mild eye irritation.

FIRE DATA
Flash point: 420°F/216°C (cc).
Fire extinguishing agents not to be used: Water or foam may cause frothing.

CHEMICAL REACTIVITY
Binary reactants: Strong oxidizers, strong acids.
Reactivity group: 34
Compatibility class: Esters.

ENVIRONMENTAL DATA
Water pollution: Effect of low concentrations on aquatic life is unknown. Fouling to shoreline. May be dangerous if it enters nearby water intakes; notify operators. Notify local health and wildlife officials. **Response to discharge:** Mechanical containment. Should be removed. Chemical and physical treatment.

SHIPPING INFORMATION
Storage temperature: Ambient; **Inert atmosphere:** None; **Venting:** Open (flame arrester); **Stability during transport:** Stable.

PHYSICAL AND CHEMICAL PROPERTIES
Physical state @ 59°F/15°C and 1 atm: Liquid.
Boiling point @ 1 atm: Very high.
Specific gravity (water = 1): 0.93 @ 68°F/20°C (liquid).
Liquid surface tension: 38 dynes/cm = 0.038 N/m @ 68°F/20°C.
Liquid water interfacial tension: (estimate) 50 dynes/cm = 0.05 N/m @ 68°F/20°C.
Heat of combustion: (estimate) $-16{,}000$ Btu/lb = -8870 cal/g = -371×10^5 J/kg.
Vapor pressure: Very low.

OILS, EDIBLE: LARD　　　　　REC. O:1600

SYNONYMS: ACEITE de MANTECA de CERDO (Spanish); PRIME STEAM LARD; KETTLE RENDERED LARD; LEAF LARD; LARD; LARD OIL, ANIMAL; LARD OIL, COMMERCIAL; LARD OIL No. 1; LARD OIL No. 2; LARD OIL, MINERAL; LARD OIL, PURE

IDENTIFICATION
DOT ID Number: UN 9277; DOT Guide Number: 171
Proper Shipping Name: Oil, n.o.s., flash point not less than 93°C (200°F).

DESCRIPTION: Colorless to light yellow liquid or solid. Fatty odor. Floats on water surface; insoluble.

Combustible • Containers may BLEVE when exposed to fire • Toxic products of combustion may include acrid smoke.

Hazard Classification (based on NFPA-704 Rating System)
Health Hazards (Blue): 0; Flammability (Red): 1; Reactivity (Yellow): 0

EMERGENCY RESPONSE: See Appendix A (171)
Evacuation:
Public safety: Isolate the area of spill or leak for at least 10 to 25 meters (30 to 80 feet) in all directions.
Spill: Increase, in the downwind direction, as necessary, the distance shown under "Public Safety."
Fire: If any large container is involved in fire, isolate for at least 800 meters (½ mile) in all directions; also, consider initial evacuation for 800 meters (½ mile) in all directions.

EXPOSURE
Short-term effects: Liquid or solid: Not harmful.

HEALTH HAZARDS
Personal protective equipment (PPE): Goggles or face shield; rubber gloves.

FIRE DATA
Flash point: 395–500°F/202–260°C (cc).
Fire extinguishing agents not to be used: Water or foam may cause frothing; water may be ineffective.
Autoignition temperature: 833°F/445°C/718°K.
Burning rate: 4 mm/min

CHEMICAL REACTIVITY
Binary reactants: Strong oxidizers
Reactivity group: 34
Compatibility class: Esters.

ENVIRONMENTAL DATA
Water pollution: Effect of low concentrations on aquatic life is unknown. Fouling to shoreline. May be dangerous if it enters nearby water intakes; notify operators. Notify local health and wildlife officials. **Response to discharge:** Mechanical containment. Should be removed. Chemical and physical treatment.

SHIPPING INFORMATION
Grades of purity: Various grades, depending on source of animal fat used and method of rendering. Prime winter-edible, prime winter inedible, Antibiotic, Off prime, #1, #2, Extra #1; **Storage temperature:** Ambient, or elevated (for liquid); **Inert atmosphere:** None; **Venting:** Open (flame arrester); **Stability during transport:** Stable.

PHYSICAL AND CHEMICAL PROPERTIES
Physical state @ 59°F/15°C and 1 atm: Solid.
Melting/Freezing point: 66-99°F/19-37°C/292-310°K.
Specific gravity (water = 1): 0.861 @ 15°C (liquid).
Liquid surface tension: (estimate) 25 dynes/cm = 0.025 N/m at 30°C.
Liquid water interfacial tension: (estimate) 50 dynes/cm = 0.050 N/m at 30°C.
Heat of combustion: $-16{,}750$ Btu/lb = -9320 cal/g = -390×10^5 J/kg.

OILS, EDIBLE: OLIVE　　　　　REC. O:1700

SYNONYMS: ACEITE de OLIVA (Spanish); OLIVE OIL; SWEET OIL

IDENTIFICATION
CAS Number: 8001-25-0
DOT ID Number: UN 9277; DOT Guide Number: 171
Proper Shipping Name: Oil, n.o.s., flash point not less than 93°C (200°F).

DESCRIPTION: Pale yellow or greenish, oily liquid. Characteristic odor. Floats on water surface; insoluble.

Combustible • Containers may BLEVE when exposed to fire • Toxic products of combustion may include carbon monoxide.

Hazard Classification (based on NFPA-704 Rating System)
Health Hazards (Blue): 0; Flammability (Red): 1; Reactivity (Yellow): 0

EMERGENCY RESPONSE: See Appendix A (171)
Evacuation:
Public safety: Isolate the area of spill or leak for at least 10 to 25 meters (30 to 80 feet) in all directions.
Spill: Increase, in the downwind direction, as necessary, the distance shown under "Public Safety."
Fire: If any large container is involved in fire, isolate for at least 800 meters (½ mile) in all directions; also, consider initial evacuation for 800 meters (½ mile) in all directions.

HEALTH HAZARDS
Personal protective equipment (PPE): B-Level PPE. OSHA Table Z-1-A air contaminant (vegetable oil). Goggles or face shield
Exposure limits (TWA unless otherwise noted): ACGIH TLV 10 mg/m^3 (vegetable oil mists); OSHA PEL TWA 15 mg/m^3 (total); 5 mg/m^3 (respirable fraction); NIOSH REL TWA 10 mg/m^3 (total); 5 mg/m^3 (respirable fraction) as vegetable oil mist.
Liquid or solid irritant characteristics: Has caused skin irritation.

FIRE DATA
Flash point: 437°F/225°C (cc).
Fire extinguishing agents not to be used: Water or foam may cause frothing.
Autoignition temperature: 650°F/343°C/616°K.
Electrical hazard: Class I, Group D.

CHEMICAL REACTIVITY
Binary reactants: Oxidizing materials.
Reactivity group: 34
Compatibility class: Esters.

ENVIRONMENTAL DATA
Water pollution: Effect of low concentrations on aquatic life is unknown. Fouling to shoreline. May be dangerous if it enters nearby water intakes; notify operators. Notify local health and wildlife officials. **Response to discharge:** Mechanical containment. Should be removed. Chemical and physical treatment.

SHIPPING INFORMATION
Storage temperature: Ambient; **Inert atmosphere:** None; **Venting:** Open (flame arrester); **Stability during transport:** Stable.

PHYSICAL AND CHEMICAL PROPERTIES
Physical state @ 59°F/15°C and 1 atm: Liquid.
Boiling point @ 1 atm: Very high.
Specific gravity (water = 1): 0.915 @ 68°F/20°C (liquid).
Liquid surface tension: 36 dynes/cm = 0.036 N/m @ 68°F/20°C.
Liquid water interfacial tension: (estimate) 50 dynes/cm = 0.05 N/m @ 68°F/20°C.
Heat of combustion: (estimate) = −16,000 Btu/lb = −8870 cal/g = −371 x 10^5 J/kg.
Vapor pressure: (Reid) 0.1 psia; 2.04 mm.

OILS, EDIBLE: PALM   REC. O:1800

SYNONYMS: ACEITE de PALMA (Spanish); OIL, PALM; PALM BUTTER; PALM FRUIT OIL; PALM OIL

IDENTIFICATION
CAS Number: 8002-75-3
DOT ID Number: UN 9277; DOT Guide Number: 171
Proper Shipping Name: Oil, n.o.s., flash point not less than 93°C (200°F).

DESCRIPTION: Orange-red liquid or semi-solid. Pleasant odor. Floats on the surface of water.

Combustible • Containers may BLEVE when exposed to fire • Toxic products of combustion may include carbon monoxide.

Hazard Classification (based on NFPA-704 Rating System)
Health Hazards (Blue): 0; Flammability (Red): 1; Reactivity (Yellow): 0

EMERGENCY RESPONSE: See Appendix A (171)
Evacuation:
Public safety: Isolate the area of spill or leak for at least 10 to 25 meters (30 to 80 feet) in all directions.
Spill: Increase, in the downwind direction, as necessary, the distance shown under "Public Safety."
Fire: If any large container is involved in fire, isolate for at least 800 meters (½ mile) in all directions; also, consider initial evacuation for 800 meters (½ mile) in all directions.

HEALTH HAZARDS
Personal protective equipment (PPE): B-Level PPE. OSHA Table Z-1-A air contaminant (vegetable oil). Goggles or face shield; rubber gloves.
Exposure limits (TWA unless otherwise noted): ACGIH TLV 10 mg/m^3 (vegetable oil mists); OSHA PEL TWA 15 mg/m^3 (total); 5 mg/m^3 (respirable fraction); NIOSH REL TWA 10 mg/m^3 (total); 5 mg/m^3 (respirable fraction) as vegetable oil mist.
Liquid or solid irritant characteristics: May cause mild eye irritation.

FIRE DATA
Flash point: 323°F/162°C (cc).
Fire extinguishing agents not to be used: Water or foam may cause frothing; water may be ineffective.
Autoignition temperature: 600°F/316°C/589°K.
Burning rate: 4 mm/min

CHEMICAL REACTIVITY
Binary reactants: Oxidizers
Reactivity group: 34
Compatibility class: Esters.

ENVIRONMENTAL DATA
Water pollution: Effect of low concentrations on aquatic life is unknown. Fouling to shoreline. May be dangerous if it enters nearby water intakes; notify operators. Notify local health and wildlife officials. **Response to discharge:** Mechanical containment. Should be removed. Chemical and physical treatment.

SHIPPING INFORMATION
Grades of purity: Various grades, depending on source. Contains 3-45% fatty acids; **Storage temperature:** Ambient; **Inert atmosphere:** None; **Venting:** Open (flame arrester); **Stability during transport:** Stable.

PHYSICAL AND CHEMICAL PROPERTIES
Physical state @ 59°F/15°C and 1 atm: Solid to Liquid.
Melting/Freezing point: 70–110°F/21–43°C/294–316°K.
Specific gravity (water = 1): 0.906 at 38°C (liquid).
Liquid surface tension: (estimate) 25 dynes/cm = 0.025 N/m at 37°C.

Liquid water interfacial tension: (estimate) 50 dynes/cm = 0.050 N/m at 37°C.
Heat of combustion: (estimate) −15,500 Btu/lb = −8600 cal/g = −360 x 10^5 J/kg.

OILS, EDIBLE: PEANUT   REC. O:1900

SYNONYMS: ACEITE de CACAHUETE (Spanish); ARACHIS OIL; EARTHNUT OIL; GROUNDNUT OIL; KATCHUNG OIL; PEANUT OIL

IDENTIFICATION
CAS Number: 8002-03-7
DOT ID Number: UN 9277; DOT Guide Number: 171
Proper Shipping Name: Oil, n.o.s., flash point not less than 93°C (200°F).

DESCRIPTION: Pale yellow, oily liquid. Weak nutty odor. Floats on water surface; insoluble.

Combustible • Containers may BLEVE when exposed to fire • Toxic products of combustion may include carbon monoxide.

Hazard Classification (based on NFPA-704 Rating System)
Health Hazards (Blue): 0; Flammability (Red): 1; Reactivity (Yellow): 0

EMERGENCY RESPONSE: See Appendix A (171)
Evacuation:
Public safety: Isolate the area of spill or leak for at least 10 to 25 meters (30 to 80 feet) in all directions.
Spill: Increase, in the downwind direction, as necessary, the distance shown under "Public Safety."
Fire: If any large container is involved in fire, isolate for at least 800 meters (½ mile) in all directions; also, consider initial evacuation for 800 meters (½ mile) in all directions.

HEALTH HAZARDS
Personal protective equipment (PPE): B-Level PPE. OSHA Table Z-1-A air contaminant (vegetable oil). Goggles or face shield.
Exposure limits (TWA unless otherwise noted):
ACGIH TLV 10 mg/m^3 (vegetable oil mists); OSHA PEL TWA 15 mg/m^3 (total); 5 mg/m^3 (respirable fraction); NIOSH REL TWA 10 mg/m^3 (total); 5 mg/m^3 (respirable fraction) as vegetable oil mist.

FIRE DATA
Flash point: 640°F/338°C (oc); 540°F/282°C (cc).
Fire extinguishing agents not to be used: Water or foam may cause frothing.
Autoignition temperature: 833°F/445°C/718°K.
Electrical hazard: Class I, Group D.

CHEMICAL REACTIVITY
Binary reactants: Oxidizing materials
Reactivity group: 34
Compatibility class: Esters.

ENVIRONMENTAL DATA
Water pollution: Effect of low concentrations on aquatic life is unknown. Fouling to shoreline. May be dangerous if it enters nearby water intakes; notify operators. Notify local health and wildlife officials. **Response to discharge:** Mechanical containment. Should be removed. Chemical and physical treatment.

SHIPPING INFORMATION
Storage temperature: Ambient; **Inert atmosphere:** None; **Venting:** Open (flame arrester); **Stability during transport:** Stable.

PHYSICAL AND CHEMICAL PROPERTIES
Physical state @ 59°F/15°C and 1 atm: Liquid.
Boiling point @ 1 atm: Very high.
Melting/Freezing point: 28°F/−2°C/271°K.
Specific gravity (water = 1): 0.919 @ 68°F/20°C (liquid).
Liquid surface tension: 35.5 dynes/cm = 0.0355 N/m @ 68°F/20°C.
Liquid water interfacial tension: 30 dynes/cm = 0.030 N/m at 158°F/70°C.
Heat of combustion: (estimate) −16,000 Btu/lb = −8870 cal/g = −371 x 10^5 J/kg.
Vapor pressure: (Reid) 0.1 psia; 2.04 mm.

OILS, EDIBLE: SAFFLOWER   REC. O:2000

SYNONYMS: ACEITE de CARTAMO (Spanish); SAFFLOWER OIL; SAFFLOWER SEED OIL; CARTHAMUS TINCTORIUS OIL

IDENTIFICATION
CAS Number: 8001-23-8
DOT ID Number: UN 9277; DOT Guide Number: 171
Proper Shipping Name: Oil, n.o.s., flash point not less than 93°C (200°F).

DESCRIPTION: Light yellow liquid. Bland fatty odor. Floats on water surface; insoluble.

Combustible • Containers may BLEVE when exposed to fire • Toxic products of combustion may include carbon monoxide.

Hazard Classification (based on NFPA-704 Rating System)
Health Hazards (Blue): 0; Flammability (Red): 1; Reactivity (Yellow): 0

EMERGENCY RESPONSE: See Appendix A (171)
Evacuation:
Public safety: Isolate the area of spill or leak for at least 10 to 25 meters (30 to 80 feet) in all directions.
Spill: Increase, in the downwind direction, as necessary, the distance shown under "Public Safety."
Fire: If any large container is involved in fire, isolate for at least 800 meters (½ mile) in all directions; also, consider initial evacuation for 800 meters (½ mile) in all directions.

EXPOSURE
Short-term effects: Liquid: Not toxic. *IF SWALLOWED*, **do NOT induce vomiting**.

HEALTH HAZARDS
Personal protective equipment (PPE): B-Level PPE. OSHA Table Z-1-A air contaminant (vegetable oil). Goggles or face shield; rubber gloves.
Exposure limits (TWA unless otherwise noted): ACGIH TLV 10 mg/m^3 (vegetable oil mists); OSHA PEL TWA 15 mg/m^3 (total); 5

mg/m³ (respirable fraction); NIOSH REL TWA 10 mg/m³ (total); 5 mg/m³ (respirable fraction) as vegetable oil mist.

FIRE DATA
Flash point: More than 300°F/149°C.
Fire extinguishing agents not to be used: Water may be ineffective; water or foam may cause frothing.
Burning rate: 4 mm/min

CHEMICAL REACTIVITY
Binary reactants: Oxidizing materials
Reactivity group: 34
Compatibility class: Esters.

ENVIRONMENTAL DATA
Water pollution: Effect of low concentrations on aquatic life is unknown. Fouling to shoreline. May be dangerous if it enters nearby water intakes; notify operators. Notify local health and wildlife officials. **Response to discharge:** Mechanical containment. Should be removed. Chemical and physical treatment.

SHIPPING INFORMATION
Grades of purity: Food grade: contains 0.02% propyl gallate, 0.01% citric acid, or may contain no additives. Technical: non-break and alkali-refined; **Storage temperature:** Ambient; **Inert atmosphere:** None; **Venting:** Open (flame arrester); **Stability during transport:** Stable.

PHYSICAL AND CHEMICAL PROPERTIES
Physical state @ 59°F/15°C and 1 atm: Liquid.
Boiling point @ 1 atm: Very high.
Specific gravity (water = 1): 0.923 @ 77°F/25°C (liquid).
Liquid surface tension: (estimate) 25 dynes/cm = 0.025 N/m @ 68°F/20°C.
Liquid water interfacial tension: (estimate) 50 dynes/cm = 0.050 N/m @ 68°F/20°C.
Heat of combustion: (estimate) $-15,500$ Btu/lb = -8600 cal/g = -360×10^5 J/kg.

OILS, EDIBLE: SOYA BEAN **REC. O:2100**

SYNONYMS: ACEITE de SOJA (Spanish); CHINESE BEAN OIL; SOYABEAN OIL; SOY OIL; SOY BEAN OIL

IDENTIFICATION
DOT ID Number: UN 9277; DOT Guide Number: 171
Proper Shipping Name: Oil, n.o.s., flash point not less than 93°C (200°F).

DESCRIPTION: Pale yellow to dark brown, thick, oily liquid. Weak fruity odor. Floats on water surface; insoluble.

Combustible • Containers may BLEVE when exposed to fire • Toxic products of combustion may include carbon monoxide. If brominated combustion products may included bromine fumes.

Hazard Classification (based on NFPA-704 Rating System)
Health Hazards (Blue): 0; Flammability (Red): 1; Reactivity (Yellow): 0

EMERGENCY RESPONSE: See Appendix A (171)
Evacuation:
Public safety: Isolate the area of spill or leak for at least 10 to 25 meters (30 to 80 feet) in all directions.
Spill: Increase, in the downwind direction, as necessary, the distance shown under "Public Safety."
Fire: If any large container is involved in fire, isolate for at least 800 meters (½ mile) in all directions; also, consider initial evacuation for 800 meters (½ mile) in all directions.

HEALTH HAZARDS
Personal protective equipment (PPE): B-Level PPE. OSHA Table Z-1-A air contaminant (vegetable oil). Goggles or face shield.

FIRE DATA
Flash point: 540°F/282°C (cc).
Fire extinguishing agents not to be used: Water or foam may cause frothing.
Autoignition temperature: 833°F/445°C/718°K.
Electrical hazard: Class I, Group D.

CHEMICAL REACTIVITY
Binary reactants: Oxidizing materials. There is a chance of spontaneous heating in air.
Reactivity group: 34
Compatibility class: Esters.

ENVIRONMENTAL DATA
Water pollution: Effect of low concentrations on aquatic life is unknown. Fouling to shoreline. May be dangerous if it enters nearby water intakes; notify operators. Notify local health and wildlife officials. **Response to discharge:** Mechanical containment. Should be removed. Chemical and physical treatment.

SHIPPING INFORMATION
Grades of purity: Refined; crude; **Storage temperature:** Ambient; **Inert atmosphere:** None; **Venting:** Open (flame arrester); **Stability during transport:** Stable.

PHYSICAL AND CHEMICAL PROPERTIES
Physical state @ 59°F/15°C and 1 atm: Liquid.
Boiling point @ 1 atm: Very high.
Melting/Freezing point: -4°F/-20°C/253°K.
Specific gravity (water = 1): 0.922 @ 68°F/20°C (liquid).
Liquid surface tension: (estimate) 25 dynes/cm = 0.025 N/m @ 68°F/20°C.
Liquid water interfacial tension: (estimate) 50 dynes/cm = 0.05 N/m @ 68°F/20°C.
Heat of combustion: (estimate) $-16,000$ Btu/lb = -8870 cal/g = -371×10^5 J/kg.
Vapor pressure: (Reid) 0.10 psia.

OILS, EDIBLE: TUCUM **REC. O:2200**

SYNONYMS: AOUARA OIL; AMERICAN PALM KERNEL OIL; ARECA NUT OIL; PALM KERNEL OIL; PALM NUT OIL; PALM SEED OIL; TUCUM OIL; BETEL NUT OIL; SUPARI (India)

IDENTIFICATION
DOT ID Number: UN 9277; DOT Guide Number: 171
Proper Shipping Name: Oil, n.o.s., flash point not less than 93°C (200°F).

DESCRIPTION: Light yellow or orange (when ripe) liquid. Weak acid odor. Floats on water surface; insoluble.

850 Oils, edible: vegetable

Combustible • Containers may BLEVE when exposed to fire • Toxic products of combustion may include acrid carbon monoxide.

Hazard Classification (based on NFPA-704 Rating System)
Health Hazards (Blue): 0; Flammability (Red): 1; Reactivity (Yellow): 0

EMERGENCY RESPONSE: See Appendix A (171)
Evacuation:
Public safety: Isolate the area of spill or leak for at least 10 to 25 meters (30 to 80 feet) in all directions.
Spill: Increase, in the downwind direction, as necessary, the distance shown under "Public Safety."
Fire: If any large container is involved in fire, isolate for at least 800 meters (½ mile) in all directions; also, consider initial evacuation for 800 meters (½ mile) in all directions.

EXPOSURE
Short-term effects: Liquid: Not toxic. If swallowed, **Do NOT induce vomiting.**

HEALTH HAZARDS
Personal protective equipment (PPE): B-Level PPE. OSHA Table Z-1-A air contaminant (vegetable oil). Goggles or face shield; rubber gloves.
Exposure limits (TWA unless otherwise noted):
ACGIH TLV 10 mg/m^3 (vegetable oil mists); OSHA PEL TWA 15 mg/m^3 (total); 5 mg/m^3 (respirable fraction); NIOSH REL TWA 10 mg/m^3 (total); 5 mg/m^3 (respirable fraction) as vegetable oil mist.
Liquid or solid irritant characteristics: Mild eye irritant.

FIRE DATA
Flash point: 398°F/203°C (cc).
Fire extinguishing agents not to be used: Water or foam may cause frothing; water may be ineffective.
Burning rate: 4 mm/min

CHEMICAL REACTIVITY
Binary reactants: Oxidizing materials
Reactivity group: 34
Compatibility class: Esters.

ENVIRONMENTAL DATA
Water pollution: Effect of low concentrations on aquatic life is unknown. Fouling to shoreline. May be dangerous if it enters nearby water intakes; notify operators. Notify local health and wildlife officials. **Response to discharge:** Mechanical containment. Should be removed. Chemical and physical treatment.

SHIPPING INFORMATION
Grades of purity: Commercial; **Storage temperature:** Ambient; **Inert atmosphere:** None; **Venting:** Open (flame arrester); **Stability during transport:** Stable.

PHYSICAL AND CHEMICAL PROPERTIES
Physical state @ 59°F/15°C and 1 atm: Liquid.
Molecular weight: Not applicable
Boiling point @ 1 atm: Very high.
Melting/Freezing point: 86°F/30°C/303°K.
Specific gravity (water = 1): 0.908 at 60°C (liquid).
Liquid surface tension: (estimate) 25 dynes/cm = 0.025 N/m at 30°C.
Liquid water interfacial tension: (estimate) 50 dynes/cm = 0.050 N/m at 30°C.
Heat of combustion: (estimate) −15,500 Btu/lb = −8600 cal/g = −360 x 10^5 J/kg.

OILS, EDIBLE: VEGETABLE   REC. O:2300

SYNONYMS: ACEITE VEGETALES (Spanish); VEGETABLE MIST; VEGETABLE OIL MIST

IDENTIFICATION
CAS Number: 68956-68-3
DOT ID Number: UN 9277; DOT Guide Number: 171
Proper Shipping Name: Oil, n.o.s., flash point not less than 93°C/200°F.

DESCRIPTION: Pale yellow, oily liquid. Weak, fatty odor. Floats on water surface; insoluble.

Combustible • Containers may BLEVE when exposed to fire • Toxic products of combustion may include carbon monoxide.

Hazard Classification (based on NFPA-704 Rating System)
Health Hazards (Blue): 0; Flammability (Red): 1; Reactivity (Yellow): 0

EMERGENCY RESPONSE: See Appendix A (171)
Evacuation:
Public safety: Isolate the area of spill or leak for at least 10 to 25 meters (30 to 80 feet) in all directions.
Spill: Increase, in the downwind direction, as necessary, the distance shown under "Public Safety."
Fire: If any large container is involved in fire, isolate for at least 800 meters (½ mile) in all directions; also, consider initial evacuation for 800 meters (½ mile) in all directions.

HEALTH HAZARDS
Personal protective equipment (PPE): B-Level PPE. OSHA Table Z-1-A air contaminant (vegetable oil). Goggles or face shield.
Exposure limits (TWA unless otherwise noted): ACGIH TLV 10 mg/m^3 (vegetable oil mists); OSHA PEL TWA 15 mg/m^3 (total); 5 mg/m^3 (respirable fraction); NIOSH REL TWA 10 mg/m^3 (total); 5 mg/m^3 (respirable fraction) as vegetable oil mist.

FIRE DATA
Flash point: 323–540°F/162–282°C (cc).
Fire extinguishing agents not to be used: Water may be ineffective.

CHEMICAL REACTIVITY
Binary reactants: Oxidants
Reactivity group: 34
Compatibility class: Esters.

ENVIRONMENTAL DATA
Water pollution: Effect of low concentrations on aquatic life is unknown. Fouling to shoreline. May be dangerous if it enters nearby water intakes; notify operators. Notify local health and wildlife officials. **Response to discharge:** Mechanical containment. Should be removed. Chemical and physical treatment.

SHIPPING INFORMATION
Storage temperature: Ambient; **Inert atmosphere:** None; **Venting:** Open (flame arrester); **Stability during transport:** Stable.

PHYSICAL AND CHEMICAL PROPERTIES
Physical state @ 59°F/15°C and 1 atm: Liquid.
Boiling point @ 1 atm: Very high.
Specific gravity (water = 1): 0.923 @ 77°F/25°C (liquid).
Liquid surface tension: (estimate) 25 dynes/cm = 0.025 N/m @ 68°F/20°C.
Liquid water interfacial tension: (estimate) 50 dynes/cm = 0.05 N/m @ 68°F/20°C.
Heat of combustion: (estimate) –16,000 Btu/lb = –8870 cal/g = –371 x 10^5 J/kg.

OILS, FUEL: No. 1   **REC. O:2400**

SYNONYMS: COAL OIL; FUEL OIL No. 1; KEROSENE; KEROSINE; RANGE OIL; JET FUEL; JP-1

IDENTIFICATION
CAS Number: 8008-20-6
DOT ID Number: UN 1993; DOT Guide Number: 128
Proper Shipping Name: Fuel Oil

DESCRIPTION: Colorless to light brown watery liquid. Kerosene odor. Floats on water surface; insoluble.

Flammable • Containers may BLEVE when exposed to fire • Vapors may form explosive mixture with air • Vapors are heavier than air and will collect and stay in low areas • Vapors may travel long distances to ignition sources and flashback • Vapors in confined areas (e.g., tanks, sewers, buildings) may explode when exposed to fire • Irritating to the skin, eyes, and respiratory tract • Toxic products of combustion may include CO_2 and carbon monoxide.

Hazard Classification (based on NFPA-704 Rating System)
Health Hazards (Blue): 0; Flammability (Red): 2; Reactivity (Yellow): 0

EMERGENCY RESPONSE: See Appendix A (128)
Evacuation:
Public safety: Isolate spill area for at least 25 to 50 meters (80 to 160 feet) in all directions.
Spill: Large spill–Consider initial downwind evacuation for at least 300 meters (1000 feet).
Fire: Isolate for 800 meters (½ mile) in all directions, especially if tank, rail car, or tank truck is involved in fire.

EXPOSURE
Short-term effects: *SEEK MEDICAL ATTENTION.* **Liquid:** Irritating to skin and eyes. Harmful if swallowed. Remove contaminated clothing and shoes. Flush affected areas with plenty of water. *IF IN EYES*, hold eyelids open and flush with plenty of water. *IF SWALLOWED* and victim is *CONSCIOUS AND ABLE TO SWALLOW*, have victim drink 4 to 8 ounces of water. **Do NOT induce vomiting.**

HEALTH HAZARDS
Personal protective equipment (PPE): B-Level PPE. Protective gloves; goggles or face shield. Chemical protective material(s) reported to have good to excellent resistance: nitrile, nitrile+PVC, neoprene, PV alcohol, PVC, Saranex®, Viton®. Also, Viton®/neoprene and polyurethane offers limited protection.
Recommendations for respirator selection: NIOSH/OSHA as kerosene *1000 ppm:* CCROV [any chemical cartridge respirator with organic vapor cartridge(s)]; or SA (any supplied-air respirator). *2500 ppm:* SA:CF (any supplied-air respirator operated in a continuous-flow mode); or PAPROV [any powered, air-purifying respirator with organic vapor cartridge(s)]. *5000 ppm:* CCRFOV [any chemical cartridge respirator with a full facepiece and organic vapor cartridges(s)]; or GMFOV [any air-purifying, full-facepiece respirator (gas mask) with a chin-style, front- or back-mounted acid gas canister]; or PAPRTOV [any powered, air-purifying respirator with a tight-fitting facepiece and organic vapor cartridges(s)]; or SCBAF (any self-contained breathing apparatus with a full facepiece); or SAF (any supplied-air respirator with a full facepiece). *EMERGENCY OR PLANNED ENTRY INTO UNKNOWN CONCENTRATIONS OR IDLH CONDITIONS:* SCBAF:PD,PP (any self-contained breathing apparatus that has a full facepiece and is operated in a pressure-demand or other positive-pressure mode); or SAF:PD,PP:ASCBA (any supplied-air respirator that has a full facepiece and is operated in a pressure-demand or other positive-pressure mode in combination with an auxiliary self-contained breathing apparatus operated in a pressure-demand or other positive pressure mode). *ESCAPE:* GMFOV [any air-purifying, full-facepiece respirator (gas mask) with a chin-style, front- or back-mounted organic vapor cannister]; or SCBAE (any appropriate escape-type, self-contained breathing apparatus).
Exposure limits (TWA unless otherwise noted): NIOSH REL 100 mg/m³ as kerosene.
Toxicity by ingestion: Grade 1; LD_{50} = 5 to 15 g/kg.
Vapor (gas) irritant characteristics: Vapors cause smarting of the eyes or respiratory system if present in high concentrations. The effect is temporary.
Liquid or solid irritant characteristics: Eye irritant. If spilled on clothing and allowed to remain, may cause smarting and reddening of the skin.
Odor threshold: 1 ppm.

FIRE DATA
Flash point: 100–162°F/38–72°C.
Flammable limits in air: LEL: 0.7%; UEL: 5%.
Fire extinguishing agents not to be used: Water may be ineffective
Autoignition temperature: 410°F/210°C/483°K.
Burning rate: 4 mm/min

CHEMICAL REACTIVITY
Binary reactants: Oxidizers.
Reactivity group: 33
Compatibility class: Miscellaneous hydrocarbon mixtures

ENVIRONMENTAL DATA
Water pollution: Dangerous to aquatic life in high concentrations. Fouling to shoreline. May be dangerous if it enters nearby water intakes; notify operators. Notify local health and wildlife officials.
Response to discharge: Mechanical containment. Should be removed. Chemical and physical treatment.

SHIPPING INFORMATION
Grades of purity: Light hydrocarbon distillate: 100%; **Storage temperature:** Ambient; **Inert atmosphere:** None; **Venting:** Open (flame arrester); **Stability during transport:** Stable.

NAS HAZARD CLASSIFICATION FOR BULK WATER TRANSPORTATION
FIRE: 2
HEALTH: Vapor irritant: 1; Liquid or solid irritant: 1; Poisons: 1

WATER POLLUTION: Human toxicity: 1; Aquatic toxicity: 1; Aesthetic effect: 3
REACTIVITY: Other chemicals: 0; Water: 0; Self-reaction: 0

PHYSICAL AND CHEMICAL PROPERTIES
Physical state @ 59°F/15°C and 1 atm: Liquid.
Molecular weight: 170 (approximate).
Boiling point @ 1 atm: 304–574°F/151–301°C/424–574°K.
Melting/Freezing point: –45 to –55°F/–43 to –48°C/230-225°K.
Specific gravity (water = 1): 0.81 0.85 @ 15°C (liquid).
Liquid surface tension: 23 32 dynes/cm = 0.023-0.032 N/m @ 68°F/20°C.
Liquid water interfacial tension: 47-49 dynes/cm = 0.047 0.049 N/m @ 68°F/20°C.
Relative vapor density (air = 1): 1.0
Latent heat of vaporization: 110 Btu/lb = 60 cal/g = 2.5×10^5 J/kg.
Heat of combustion: –18,540 Btu/lb = –10,300 cal/g = -431.24×10^5 J/kg.
Vapor pressure: 3.0 mm.

OILS, FUEL: No. 2   REC. O:2500

SYNONYMS: HOME-HEATING OIL No. 2

IDENTIFICATION
CAS Number: 68476-30-2
DOT ID Number: UN 1993; DOT Guide Number: 128
Proper Shipping Name: Fuel oil

DESCRIPTION: Yellow-brown oily liquid. Lube or fuel oil odor. Floats on water surface; insoluble.

Highly flammable • Containers may BLEVE when exposed to fire • Vapors may form explosive mixture with air • Vapors are heavier than air and will collect and stay in low areas • Vapors may travel long distances to ignition sources and flashback • Vapors in confined areas (e.g., tanks, sewers, buildings) may explode when exposed to fire • Irritating to the skin, eyes, and respiratory tract • Toxic products of combustion may include carbon monoxide.

Hazard Classification (based on NFPA-704 Rating System)
Health Hazards (Blue): 0; Flammability (Red): 2; Reactivity (Yellow): 0

EMERGENCY RESPONSE: See Appendix A (128)
Evacuation:
Public safety: Isolate spill area for at least 25 to 50 meters (80 to 160 feet) in all directions.
Spill: Large spill–Consider initial downwind evacuation for at least 300 meters (1000 feet).
Fire: Isolate for 800 meters (½ mile) in all directions, especially if tank, rail car, or tank truck is involved in fire.

EXPOSURE
Short-term effects: *SEEK MEDICAL ATTENTION.* **Liquid:** Irritating to skin and eyes. *IF SWALLOWED*, will cause nausea, vomiting. Remove contaminated clothing and shoes. Flush affected areas with plenty of water. *IF IN EYES*, hold eyelids open and flush with plenty of water. *IF SWALLOWED* and victim is *CONSCIOUS AND ABLE TO SWALLOW*, have victim drink 4 to 8 ounces of water. **Do NOT induce vomiting.**

HEALTH HAZARDS
Personal protective equipment (PPE): B-Level PPE. Protective gloves; goggles or face shield. Chemical protective material(s) reported to have good to excellent resistance: nitrile, nitrile+PVC, neoprene, PV alcohol, PVC, Saranex®, Viton®. Also, Viton®/neoprene and polyurethane offers limited protection.
Exposure limits (TWA unless otherwise noted): ACGIH TLV 15 ppm (100 mg/m³).
Toxicity by ingestion: Grade 1; LD_{50} = 5 to 15 g/kg.
Vapor (gas) irritant characteristics: Slight smarting of eyes or respiratory system if present in high concentrations. The effect is temporary.
Liquid or solid irritant characteristics: Eye irritant. If spilled on clothing and allowed to remain, may cause smarting and reddening of skin.

FIRE DATA
Flash point: 136°F/58°C (cc).
Fire extinguishing agents not to be used: Water may be ineffective
Autoignition temperature: 494°F/257°C/530°K.
Burning rate: 4 mm/min

CHEMICAL REACTIVITY
Binary reactants: Oxidizers, nitric acid
Reactivity group: 33
Compatibility class: Miscellaneous hydrocarbon mixtures

ENVIRONMENTAL DATA
Water pollution: Dangerous to aquatic life in high concentrations. Fouling to shoreline. May be dangerous if it enters nearby water intakes; notify operators. Notify local health and wildlife officials.
Response to discharge: Should be removed. Mechanical containment. Chemical and physical treatment.

SHIPPING INFORMATION
Grades of purity: Commercial; **Storage temperature:** Ambient; **Inert atmosphere:** None; **Venting:** Open (flame arrester); **Stability during transport:** Stable.

PHYSICAL AND CHEMICAL PROPERTIES
Physical state @ 59°F/15°C and 1 atm: Liquid.
Boiling point @ 1 atm: 540–640°F/282–338°C/555–611°K.
Melting/Freezing point: –20°F/–29°C/244°K.
Specific gravity (water = 1): 0.879 @ 68°F/20°C (liquid).
Liquid surface tension: (estimate) 25 dynes/cm = 0.025 N/m @ 68°F/20°C.
Liquid water interfacial tension: (estimate) 50 dynes/cm = 0.05 N/m @ 68°F/20°C.
Heat of combustion: –19,440 Btu/lb = –10,800 cal/g = -452.17×10^5 J/kg.

OILS, FUEL: No. 4   REC. O:2600

SYNONYMS: RESIDUAL FUEL OIL No. 4

IDENTIFICATION
CAS Number: 68476-33-5
DOT ID Number: UN 1993; DOT Guide Number: 128
Proper Shipping Name: Fuel oil

DESCRIPTION: Brown, oily liquid. Lube or fuel oil odor. Floats on water surface; insoluble.

Highly flammable • Containers may BLEVE when exposed to fire • Vapors may form explosive mixture with air • Vapors are heavier than air and will collect and stay in low areas • Vapors may travel long distances to ignition sources and flashback • Vapors in confined areas (e.g., tanks, sewers, buildings) may explode when exposed to fire • Irritating to the skin, eyes, and respiratory tract • Toxic products of combustion may include acrid carbon monoxide.

Hazard Classification (based on NFPA-704 Rating System)
Health Hazards (Blue): 0; Flammability (Red): 2; Reactivity (Yellow): 0

EMERGENCY RESPONSE: See Appendix A (128)
Evacuation:
Public safety: Isolate spill area for at least 25 to 50 meters (80 to 160 feet) in all directions.
Spill: Large spill–Consider initial downwind evacuation for at least 300 meters (1000 feet).
Fire: Isolate for 800 meters (½ mile) in all directions, especially if tank, rail car, or tank truck is involved in fire.

EXPOSURE
Short-term effects: *SEEK MEDICAL ATTENTION*. **Liquid:** Irritating to skin and eyes. Harmful if swallowed. Remove contaminated clothing and shoes. Flush affected areas with plenty of water. *IF IN EYES*, hold eyelids open and flush with plenty of water. *IF SWALLOWED* and victim is *CONSCIOUS AND ABLE TO SWALLOW*, have victim drink 4 to 8 ounces of water. **Do NOT induce vomiting.**

HEALTH HAZARDS
Personal protective equipment (PPE): B-Level PPE. Protective gloves; goggles or face shield. Chemical protective material(s) reported to have good to excellent resistance: nitrile, nitrile+PVC, neoprene, PV alcohol, PVC, Saranex®, Viton®. Also, Viton®/neoprene and polyurethane offers limited protection.
Toxicity by ingestion: Grade 1; LD_{50} = 5 to 15 g/kg.
Long-term health effects: IARC possible carcinogen, rating 2B
Liquid or solid irritant characteristics: Eye irritant. If spilled on clothing and allowed to remain, may cause smarting and reddening of the skin.

FIRE DATA
Flash point: More than 140°F/60°C (cc).
Flammable limits in air: LEL: 1.0%; UEL: 5%.
Fire extinguishing agents not to be used: Water may be ineffective.
Autoignition temperature: 505°F/263°C/536°K.
Burning rate: 4 mm/min

CHEMICAL REACTIVITY
Binary reactants: Oxidizers
Reactivity group: 33
Compatibility class: Miscellaneous hydrocarbon mixtures

ENVIRONMENTAL DATA
Water pollution: Effect of low concentrations on aquatic life is unknown. Fouling to shoreline. May be dangerous if it enters nearby water intakes; notify operators. Notify local health and wildlife officials. **Response to discharge:** Mechanical containment. Should be removed. Chemical and physical treatment.

SHIPPING INFORMATION
Grades of purity: Commercial; **Storage temperature:** Ambient; **Inert atmosphere:** None; **Venting:** Open (flame arrester); **Stability during transport:** Stable.

PHYSICAL AND CHEMICAL PROPERTIES
Physical state @ 59°F/15°C and 1 atm: Liquid.
Boiling point @ 1 atm: 214–>1092°F/101–>588°C/374–>861°K.
Melting/Freezing point: –20 to +15°F/–29 to –9°C/244–264°K.
Specific gravity (water = 1): 0.904 @ 15°C (liquid).
Heat of combustion: –17,460 Btu/lb = –9700 cal/g = –406.1 x 10^5 J/kg.

OILS, FUEL: No. 5   **REC. O:2700**

SYNONYMS: RESIDUAL FUEL OIL No. 5

IDENTIFICATION
CAS Number: 68476-33-5
DOT ID Number: 1993; DOT Guide Number: 128
Proper Shipping Name: Fuel oil

DESCRIPTION: Brown, oily liquid. Strong lube oil odor. Floats on water surface; insoluble.

Combustible • Containers may BLEVE when exposed to fire • Vapors are heavier than air and will collect and stay in low areas • Vapors may travel long distances to ignition sources and flashback • Vapors in confined areas (e.g., tanks, sewers, buildings) may explode when exposed to fire • Irritating to the skin, eyes, and respiratory tract • Toxic products of combustion may include acrid carbon monoxide.

Hazard Classification (based on NFPA-704 Rating System)
Health Hazards (Blue): 0; Flammability (Red): 2; Reactivity (Yellow): 0

EMERGENCY RESPONSE: See Appendix A (128)
Evacuation:
Public safety: Isolate spill area for at least 25 to 50 meters (80 to 160 feet) in all directions.
Spill: Large spill–Consider initial downwind evacuation for at least 300 meters (1000 feet).
Fire: Isolate for 800 meters (½ mile) in all directions, especially if tank, rail car, or tank truck is involved in fire.

EXPOSURE
Short-term effects: *SEEK MEDICAL ATTENTION*. **Liquid:** Irritating to skin and eyes. Harmful if swallowed. Remove contaminated clothing and shoes. Flush affected areas with plenty of water. *IF IN EYES*, hold eyelids open and flush with plenty of water. *IF SWALLOWED* and victim is *CONSCIOUS AND ABLE TO SWALLOW*, have victim drink 4 to 8 ounces of water. **Do NOT induce vomiting.**

HEALTH HAZARDS
Personal protective equipment (PPE): B-Level PPE. Protective gloves; goggles or face shield. Chemical protective material(s) reported to have good to excellent resistance: nitrile, nitrile+PVC, neoprene, PV alcohol, PVC, Saranex®, Viton®. Also, Viton®/neoprene and polyurethane offers limited protection.
Toxicity by ingestion: Grade 1; LD_{50} = 5 to 15 g/kg.
Long-term health effects: IARC possible carcinogen, rating 2B
Liquid or solid irritant characteristics: Eye irritant. If spilled on clothing and allowed to remain, may cause smarting and reddening of the skin.

854   Oils, fuel: No. 6

FIRE DATA
Flash point: (light) more than 156°F/68°C (cc); (heavy) more than 160°F/71°C.
Flammable limits in air: LEL: 1%; UEL: 5%.
Fire extinguishing agents not to be used: Water may be ineffective.
Burning rate: 4 mm/min

CHEMICAL REACTIVITY
Binary reactants: Oxidizers
Reactivity group: 33
Compatibility class: Miscellaneous hydrocarbon mixtures

ENVIRONMENTAL DATA
Water pollution: Effect of low concentrations on aquatic life is unknown. Fouling to shoreline. May be dangerous if it enters nearby water intakes; notify operators. Notify local health and wildlife officials. **Response to discharge:** Mechanical containment. Should be removed. Chemical and physical treatment.

SHIPPING INFORMATION
Grades of purity: Fuel oil No. 5 (heavy); Fuel oil No. 5 (light); **Storage temperature:** Ambient; **Inert atmosphere:** None; **Venting:** Open (flame arrester); **Stability during transport:** Stable.

PHYSICAL AND CHEMICAL PROPERTIES
Physical state @ 59°F/15°C and 1 atm: Liquid.
Boiling point @ 1 atm: 426–>1062°F/218–>570°C/491–>843°K.
Melting/Freezing point: 0°F/–18°C/255°K.
Specific gravity (water = 1): 0.936 at 16°C (liquid).
Heat of combustion: –18,000 Btu/lb = –10,000 cal/g = –418.68 x 10^5 J/kg.

OILS, FUEL: No. 6   **REC. O:2800**

SYNONYMS: BUNKER C OIL; FUEL OIL No. 6; HEAVY INDUSTRIAL FUEL OIL; RESIDUAL FUEL OIL No. 6

IDENTIFICATION
CAS Number: 68476-33-5
DOT ID Number: UN 1993; DOT Guide Number: 128
Proper Shipping Name: Flammable liquid, n.o.s.

DESCRIPTION: Black, thick heated liquid. Black. Tarry odor; like kerosene. Floats on water surface; insoluble.

Flammable • Containers may BLEVE when exposed to fire • Vapors may form explosive mixture with air • Vapors are heavier than air and will collect and stay in low areas • Vapors may travel long distances to ignition sources and flashback • Vapors in confined areas (e.g., tanks, sewers, buildings) may explode when exposed to fire • Irritating to the skin, eyes, and respiratory tract • Toxic products of combustion may include acrid carbon monoxide.

Hazard Classification (based on NFPA-704 Rating System)
Health Hazards (Blue): 0; Flammability (Red): 2; Reactivity (Yellow): 0

EMERGENCY RESPONSE: See Appendix A (128)
Evacuation:
Public safety: Isolate spill area for at least 25 to 50 meters (80 to 160 feet) in all directions.
Spill: Large spill–Consider initial downwind evacuation for at least 300 meters (1000 feet).
Fire: Isolate for 800 meters (½ mile) in all directions, especially if tank, rail car, or tank truck is involved in fire.

EXPOSURE
Short-term effects: *SEEK MEDICAL ATTENTION.* **Liquid:** Irritating to skin and eyes. Harmful if swallowed. Remove contaminated clothing and shoes. Flush affected areas with plenty of water. *IF IN EYES*, hold eyelids open and flush with plenty of water. *IF SWALLOWED* and victim is *CONSCIOUS AND ABLE TO SWALLOW*, have victim drink 4 to 8 ounces of water. **Do NOT induce vomiting.**

HEALTH HAZARDS
Personal protective equipment (PPE): B-Level PPE. Protective gloves; goggles or face shield. Chemical protective material(s) reported to have good to excellent resistance: nitrile, nitrile+PVC, neoprene, PV alcohol, PVC, Saranex®, Viton®. Also, Viton®/neoprene and polyurethane offers limited protection.
Toxicity by ingestion: Grade 1; LD_{50} = 5 to 15 g/kg.
Long-term health effects: IARC possible carcinogen, rating 2B
Liquid or solid irritant characteristics: Minimum hazard. If spilled on clothing and allowed to remain, may cause smarting and reddening of the skin.

FIRE DATA
Flash point: More than 150°F/66°C (cc).
Flammable limits in air: LEL: 1%; UEL: 5%.
Fire extinguishing agents not to be used: Water may be ineffective
Autoignition temperature: 765°F/407°C/680°K.
Electrical hazard: Class I, Group D.
Burning rate: 4 mm/min

CHEMICAL REACTIVITY
Binary reactants: Oxidizers, nitric acid
Reactivity group: 33
Compatibility class: Miscellaneous hydrocarbon mixtures

ENVIRONMENTAL DATA
Water pollution: Dangerous to aquatic life in high concentrations. Fouling to shoreline. May be dangerous if it enters nearby water intakes; notify operators. Notify local health and wildlife officials.
Response to discharge: Mechanical containment. Should be removed. Chemical and physical treatment.

SHIPPING INFORMATION
Grades of purity: Commercial; **Storage temperature:** Elevated; **Inert atmosphere:** None; **Venting:** Open (flame arrester); **Stability during transport:** Stable.

PHYSICAL AND CHEMICAL PROPERTIES
Physical state @ 59°F/15°C and 1 atm: Liquid.
Boiling point @ 1 atm: 415–>1093°F/212–>588°C = 485–>861°K.
Melting/Freezing point: 25–55°F/–4 to +13°C/269–286°K.
Specific gravity (water = 1): 0.92-1.07 @ 68°F/20°C (liquid).
Liquid surface tension: (estimate) 25 dynes/cm = 0.025 N/m @ 68°F/20°C.
Liquid water interfacial tension: (estimate) 50 dynes/cm = 0.05 N/m @ 68°F/20°C.
Heat of combustion: –18,000 Btu/lb = –10,000 cal/g = –418.68 x 10^5 J/kg.
Vapor pressure: 0.042 mm.

OILS, MISCELLANEOUS: ABSORPTION   REC. O:2900

SYNONYMS: ACEITE ABSORBENTE (Spanish); ABSORBENT OIL; ABSORPTION OIL

IDENTIFICATION
DOT ID Number: UN 9277; DOT Guide Number: 171
Proper Shipping Name: Oil, n.o.s., flash point not less than 93°C (200°F).

DESCRIPTION: Colorless to pale yellow liquid. Fuel oil odor. Floats on water surface; insoluble.

Combustible • Containers may BLEVE when exposed to fire • Vapors are heavier than air and will collect and stay in low areas • Vapors in confined areas (e.g., tanks, sewers, buildings) may explode when exposed to fire • Irritating to the skin, eyes, and respiratory tract • Toxic products of combustion may include acrid carbon monoxide.

Hazard Classification (based on NFPA-704 Rating System)
Health Hazards (Blue): 0; Flammability (Red): 1; Reactivity (Yellow): 0

EMERGENCY RESPONSE: See Appendix A (171)
Evacuation:
Public safety: Isolate the area of spill or leak for at least 10 to 25 meters (30 to 80 feet) in all directions.
Spill: Increase, in the downwind direction, as necessary, the distance shown under "Public Safety."
Fire: If any large container is involved in fire, isolate for at least 800 meters (½ mile) in all directions; also, consider initial evacuation for 800 meters (½ mile) in all directions.

EXPOSURE
Short-term effects: *SEEK MEDICAL ATTENTION.* **Liquid:** Irritating to skin and eyes. Harmful if swallowed. Remove contaminated clothing and shoes. Flush affected areas with plenty of water. *IF IN EYES*, hold eyelids open and flush with plenty of water. *IF SWALLOWED* and victim is *CONSCIOUS AND ABLE TO SWALLOW*, have victim drink 4 to 8 ounces of water. **Do NOT induce vomiting.**

HEALTH HAZARDS
Personal protective equipment (PPE): B-Level PPE. Protective gloves; goggles or face shield.
Toxicity by ingestion: Grade 1; LD_{50} = 5 to 15 g/kg.
Liquid or solid irritant characteristics: If spilled on clothing and allowed to remain, may cause smarting and reddening of skin.

FIRE DATA
Flash point: 255°F/124°C (oc).
Fire extinguishing agents not to be used: Water may be ineffective
Autoignition temperature: 300°F/149°C/422°K.
Burning rate: 4 mm/min

CHEMICAL REACTIVITY
Binary reactants: Oxidizers
Reactivity group: 33
Compatibility class: Miscellaneous hydrocarbon mixtures

ENVIRONMENTAL DATA
Water pollution: Effect of low concentrations on aquatic life is unknown. Fouling to shoreline. May be dangerous if it enters nearby water intakes; notify operators. Notify local health and wildlife officials. **Response to discharge:** Mechanical containment. Should be removed. Chemical and physical treatment.

SHIPPING INFORMATION
Grades of purity: Commercial; **Storage temperature:** Ambient; **Inert atmosphere:** None; **Venting:** Open (flame arrester); **Stability during transport:** Stable.

PHYSICAL AND CHEMICAL PROPERTIES
Boiling point @ 1 atm: More than 500°F/260°C/533°K.
Specific gravity (water = 1): (estimate) 0.85 @ 68°F/20°C (liquid).
Liquid surface tension: (estimate) 25 dynes/cm = 0.025 N/m @ 68°F/20°C.
Liquid water interfacial tension: (estimate) 50 dynes/cm = 0.05 N/m @ 68°F/20°C.
Heat of combustion: (estimate) –18,000 Btu/lb = –10,000 cal/g = –420 x 10^5 J/kg.

OIL, MISCELLANEOUS: CASHEW NUT SHELL   REC. O:3100

SYNONYMS: ACEITE de CASCARA de NUEZ de ANACARDO (Spanish); CASHEW NUTSHELL LIQUID; CASHEW NUTSHELL OIL; OIL OF CASHEW NUTSHELL; ANACARDIC ACID; O-PENTADECADIENYL SALICYLIC ACID

IDENTIFICATION
CAS Number: 8001-24-7
DOT ID Number: UN 9277; DOT Guide Number: 171
Proper Shipping Name: Oil, n.o.s., flash point not less than 93°C (200°F).

DESCRIPTION: Black liquid. Floats on water surface; insoluble.

Combustible • Containers may BLEVE when exposed to fire • Toxic products of combustion may include acrid smoke and irritating vapors • May be a polymerization hazard.

Hazard Classification (based on NFPA-704 Rating System)
Health Hazards (Blue): 0; Flammability (Red): 1; Reactivity (Yellow): 0

EMERGENCY RESPONSE: See Appendix A (171)
Evacuation:
Public safety: Isolate the area of spill or leak for at least 10 to 25 meters (30 to 80 feet) in all directions.
Spill: Increase, in the downwind direction, as necessary, the distance shown under "Public Safety."
Fire: If any large container is involved in fire, isolate for at least 800 meters (½ mile) in all directions; also, consider initial evacuation for 800 meters (½ mile) in all directions.

EXPOSURE
Short-term effects: *SEEK MEDICAL ATTENTION.* **Vapor:** May be harmful if inhaled. May cause skin to blister (sensitive individuals). May irritate eyes, nose, and throat. Move to fresh air. *IF BREATHING HAS STOPPED*, give artificial respiration. IF breathing is difficult, administer oxygen. **Liquid:** Toxic and irritating. *IF SWALLOWED*, cardol, a principal constituent, produces severe gastroenteritis. Produces severe inflammation of the skin with subsequent blisters and desquamation–similar to poison ivy exposure. May burn eyes. Remove and double bag

contaminated clothing and shoes. Flush contaminated area with plenty of running water for at least 15 minutes. *IF IN EYES*, hold eyelids open and flush with water. *IF SWALLOWED* and victim is *UNCONSCIOUS OR HAVING CONVULSIONS*, do nothing except keep victim warm. Effects may be delayed; keep victim under observation.

HEALTH HAZARDS

Personal protective equipment (PPE): B-Level PPE. Wear positive pressure breathing apparatus and special protective clothing.
Exposure limits (TWA unless otherwise noted): ACGIH TLV 10 mg/m^3 (vegetable oil mists); OSHA PEL TWA 15 mg/m^3 (total); 5 mg/m^3 (respirable fraction); NIOSH REL TWA 10 mg/m^3 (total); 5 mg/m^3 (respirable fraction) as vegetable oil mist.
Vapor (gas) irritant characteristics: Similar to poison ivy.
Liquid or solid irritant characteristics: Similar to poison ivy exposures. Produces severe inflammation of the skin followed by blisters.

FIRE DATA

Flash point: More than 300°F/149°C.
Behavior in fire: The primary constituent, anacardic acid, decarboxylates produces CO$_2$ gas. Pressure will build up in heated, closed containers.

CHEMICAL REACTIVITY

Binary reactants: Oxidizers, sulfuric acid, caustics, ammonia, aliphatic amines, alkanolamines, isocyanates, alkylene oxides, epichlorohydrin.
Reactivity group: 33
Compatibility class: Miscellaneous hydrocarbon mixtures

ENVIRONMENTAL DATA

Water pollution: Effect of low concentrations on aquatic life is unknown. May be dangerous if it enters nearby water intakes; notify operators. Notify local health and wildlife officials.
Response to discharge: Issue warning-toxic irritant (similar to poison ivy), corrosive. Restrict access. Mechanical containment. Should be removed. Chemical and physical treatment.

SHIPPING INFORMATION

Grades of purity: Variable untreated mixture. Range of main C15 component phenols from six sources: Anacardic acid: 74.1 to 77.4%; cardol: 15.0 to 20.1%; 2-methylcardol: 1.7 to 2.6%; cardanol: 1.2 to 9.2%. Each component has four constituents because the C15 side chain for each component has 0, 1, 2, and 3 double bonds; **Inert atmosphere:** None; **Venting:** Pressure vacuum valve; **Stability during transport:** Stable.

PHYSICAL AND CHEMICAL PROPERTIES

Physical state @ 59°F/15°C and 1 atm: Liquid.

OILS, MISCELLANEOUS: COAL TAR   REC. O:3200

SYNONYMS: ACEITE de ALQUITRAN de HULLA (Spanish); COAL TAR OIL; CRUDE COAL TAR; ESTAR; LAVATAR; LIGHT OIL; TAR; TAR, COAL; TAR, LIQUID; ZETAR

IDENTIFICATION

CAS Number: 8007-45-2
DOT ID Number: UN 1136; DOT Guide Number: 128
Proper Shipping Name: Coal tar distillates, flammable

DESCRIPTION: Colorless to yellow liquid. Pleasant odor, like benzene or gasoline. Floats on water surface; insoluble.

Highly flammable • Containers may BLEVE when exposed to fire • Vapors may form explosive mixture with air • Vapors are heavier than air and will collect and stay in low areas • Vapors may travel long distances to ignition sources and flashback • Vapors in confined areas (e.g., tanks, sewers, buildings) may explode when exposed to fire • Irritating to the skin, eyes, and respiratory tract • Toxic products of combustion may include carbon monoxide.

Hazard Classification (based on NFPA-704 Rating System)
Health Hazards (Blue): 2; Flammability (Red): 3; Reactivity (Yellow): 0

EMERGENCY RESPONSE: See Appendix A (128)
Evacuation:
Public safety: Isolate spill area for at least 25 to 50 meters (80 to 160 feet) in all directions.
Spill: Large spill–Consider initial downwind evacuation for at least 300 meters (1000 feet).
Fire: Isolate for 800 meters (½ mile) in all directions, especially if tank, rail car, or tank truck is involved in fire.

EXPOSURE

Short-term effects: *SEEK MEDICAL ATTENTION*. **Vapor:** Irritating to eyes, nose, and throat. Move to fresh air. **Liquid:** Irritating to skin and eyes. Harmful if swallowed. Remove contaminated clothing and shoes. Flush affected areas with plenty of water. *IF IN EYES*, hold eyelids open and flush with plenty of water. *IF SWALLOWED* and victim is *CONSCIOUS AND ABLE TO SWALLOW*, have victim drink 4 to 8 ounces of water. **Do NOT induce vomiting**.

HEALTH HAZARDS

Personal protective equipment (PPE): B-Level PPE. Protective gloves; goggles or face shield. Sealed chemical protective materials offer limited protection (coal tar extract): nitrile, neoprene.
Recommendations for respirator selection: NIOSH
At any concentrations above the NIOSH REL, or where there is no REL, at any detectable concentration: SCBAF:PD,PP (any self-contained breathing apparatus that has a full facepiece and is operated in a pressure-demand or other positive-pressure mode); or SAF:PD,PP:ASCBA (any supplied-air respirator that has a full facepiece and is operated in a pressure-demand or other positive-pressure mode in combination with an auxiliary, self-contained breathing apparatus operated in a pressure-demand or other positive pressure mode). *ESCAPE:* GMFOVHiE [any air-purifying, full-facepiece respirator (gas mask) with a chin-style, front- or back-mounted organic vapor canister having a high-efficiency particulate filter]; or SCBAE (any appropriate escape-type, self-contained breathing apparatus).
Exposure limits (TWA unless otherwise noted): ACGIH TLV 0.2 mg/m^3 as benzene soluble aerosol (confirmed human carcinogen); OSHA PEL [1910.1002] 0.2 mg/m^3 (benzene-soluble fraction); NIOSH 0.1 mg/m^3 (cyclohexane-extractable fraction); carcinogen.
Long-term health effects: NTP anticipated carcinogen; IARC carcinogen; rating 1
Vapor (gas) irritant characteristics: Vapors cause a slight smarting of the eyes or respiratory system if present in high concentrations. The effect is temporary.
Liquid or solid irritant characteristics: Minimum hazard. If spilled on clothing and allowed to remain, may cause smarting and reddening of the skin.

IDLH value: 80 mg/m³ as coal tar pitch volatiles. Potential human carcinogen.

FIRE DATA
Flash point: 60–77°F/15–25°C (cc).
Flammable limits in air: LEL: 1.3%; UEL: 8%.
Fire extinguishing agents not to be used: Water may be ineffective.
Electrical hazard: Class I, Group D.
Burning rate: 4 mm/min

CHEMICAL REACTIVITY
Binary reactants: Strong oxidizers, nitric acid
Reactivity group: 33
Compatibility class: Miscellaneous hydrocarbon mixtures

ENVIRONMENTAL DATA
Water pollution: DOT Appendix B, §172.101–marine pollutant. Effect of low concentrations on aquatic life is unknown. Fouling to shoreline. May be dangerous if it enters nearby water intakes; notify operators. Notify local health and wildlife officials.
Response to discharge: Issue warning–high flammability. Evacuate area.

SHIPPING INFORMATION
Grades of purity: Various compositions, depending on type of coal used and boiling range taken; **Storage temperature:** Ambient; **Inert atmosphere:** None; **Venting:** Open (flame arrester) or pressure-vacuum; **Stability during transport:** Stable.

NAS HAZARD CLASSIFICATION FOR BULK WATER TRANSPORTATION
FIRE: 3
HEALTH: Vapor irritant: 1; Liquid or solid irritant: 1; Poisons: 2
WATER POLLUTION: Human toxicity: 2; Aquatic toxicity: 2; Aesthetic effect: 4
REACTIVITY: Other chemicals: 2; Water: 0; Self-reaction: 0

PHYSICAL AND CHEMICAL PROPERTIES
Physical state @ 59°F/15°C and 1 atm: Liquid.
Boiling point @ 1 atm: 223–333°F/106–167°C/379–440°K.
Specific gravity (water = 1): (estimate) 0.90 @ 68°F/20°C (liquid).
Liquid surface tension: (estimate) 25 dynes/cm = 0.025 N/m @ 68°F/20°C.
Ratio of specific heats of vapor (gas): (estimate) 1.071
Latent heat of vaporization: (estimate) 107 Btu/lb = 59.8 cal/g. = 2.5 x 10⁵ J/kg.
Heat of combustion: –17,440 Btu/lb = –9690 cal/g = –405.7 x 10⁵ J/kg.

OILS, MISCELLANEOUS: CROTON   REC. O:3300

SYNONYMS: ACEITE de CROTON (Spanish); CROTON TIGLIUM OIL; CROTON OIL; CROTON OEL (German)

IDENTIFICATION
CAS Number: 8001-28-3
DOT ID Number: UN 9277; DOT Guide Number: 171
Proper Shipping Name: Oil, n.o.s., flash point not less than 93°C (200°F).

DESCRIPTION: Brownish-yellow liquid. Unpleasant, acrid odor. Floats on water surface; insoluble.

Combustible • Containers may BLEVE when exposed to fire • Toxic products of combustion may include acrid carbon monoxide.

Hazard Classification (based on NFPA-704 Rating System)
Health Hazards (Blue): 0; Flammability (Red): 1; Reactivity (Yellow): 0

EMERGENCY RESPONSE: See Appendix A (171)
Evacuation:
Public safety: Isolate the area of spill or leak for at least 10 to 25 meters (30 to 80 feet) in all directions.
Spill: Increase, in the downwind direction, as necessary, the distance shown under "Public Safety."
Fire: If any large container is involved in fire, isolate for at least 800 meters (½ mile) in all directions; also, consider initial evacuation for 800 meters (½ mile) in all directions.

EXPOSURE
Short-term effects: SEEK MEDICAL ATTENTION. **Liquid:** POISONOUS IF SWALLOWED OR IF SKIN IS EXPOSED. Irritating to skin and eyes. Remove contaminated clothing and shoes. Flush affected areas with plenty of water. IF IN EYES, hold eyelids open and flush with plenty of water. IF SWALLOWED and victim is CONSCIOUS AND ABLE TO SWALLOW, have victim drink 4 to 8 ounces of water. **Do NOT induce vomiting.**

HEALTH HAZARDS
Personal protective equipment (PPE): B-Level PPE. Goggles or face shield; rubber gloves and any other protective clothing to prevent contact with skin.
ABSORPTION: Absorption through the skin may cause purging.
Ingestion: Causes burning of the mouth and stomach and drastic purging, possibly leading to collapse and death. Small doses have a strong laxative effect.
Toxicity by ingestion: Grade 4; LD_{50} less than 50 mg/kg.
Long-term health effects: Has been used in cancer research as a promoter for other compounds that cause skin cancer.
Liquid or solid irritant characteristics: Severe eye and skin irritant.

FIRE DATA
Fire extinguishing agents not to be used: Water may be ineffective.
Burning rate: 4 mm/min

CHEMICAL REACTIVITY
Binary reactants: Strong oxidizers.

ENVIRONMENTAL DATA
Water pollution: Effect of low concentrations on aquatic life is unknown. Fouling to shoreline. May be dangerous if it enters nearby water intakes; notify operators. Notify local health and wildlife officials. **Response to discharge:** Mechanical containment. Should be removed. Chemical and physical treatment.

SHIPPING INFORMATION
Grades of purity: Technical; **Storage temperature:** Ambient; **Inert atmosphere:** None; **Venting:** Open (flame arrester); **Stability during transport:** Stable.

PHYSICAL AND CHEMICAL PROPERTIES
Physical state @ 59°F/15°C and 1 atm: Liquid.
Boiling point @ 1 atm: Very high.
Melting/Freezing point: 0–18°F/–18 to –8°C/255–265°K.
Specific gravity (water = 1): (estimate) 0.946 @ 15°C (liquid).

Liquid surface tension: (estimate) 25 dynes/cm = 0.025 N/m @ 68°F/20°C.
Liquid water interfacial tension: (estimate) 50 dynes/cm = 0.050 N/m @ 68°F/20°C.
Heat of combustion: (estimate) –16,000 Btu/lb = –9300 cal/g = –390 x 10^5 J/kg.

OILS, MISCELLANEOUS: LINSEED   REC. O:3400

SYNONYMS: ACEITE de LINAZA (Spanish); BOILED LINSEED OIL; FLAXSEED OIL; GROCO; L-310; LINSEED OIL; RAW LINSEED OIL

IDENTIFICATION
CAS Number: 8001-26-1
DOT ID Number: UN 9277; DOT Guide Number: 171
Proper Shipping Name: Oil, n.o.s., flash point not less than 93°C (200°F).

DESCRIPTION: Light yellow to dark amber liquid. Oil base-paint-like odor. Floats on water surface; insoluble.

Combustible • Heat or flame may cause explosion • Containers may BLEVE when exposed to fire • Polymerization hazard on contact with air • Vapors are heavier than air and will collect and stay in low areas • Vapors in confined areas (e.g., tanks, sewers, buildings) may explode when exposed to fire • Irritating to the skin, eyes, and respiratory tract • Toxic products of combustion may include acrid carbon monoxide.

Hazard Classification (based on NFPA-704 Rating System)
Health Hazards (Blue): 0; Flammability (Red): 1; Reactivity (Yellow): 0

EMERGENCY RESPONSE: See Appendix A (171)
Evacuation:
Public safety: Isolate the area of spill or leak for at least 10 to 25 meters (30 to 80 feet) in all directions.
Spill: Increase, in the downwind direction, as necessary, the distance shown under "Public Safety."
Fire: If any large container is involved in fire, isolate for at least 800 meters (½ mile) in all directions; also, consider initial evacuation for 800 meters (½ mile) in all directions.

EXPOSURE
Short-term effects: Liquid: Not toxic. If swallowed, **Do NOT induce vomiting.**

HEALTH HAZARDS
Personal protective equipment (PPE): B-Level PPE. OSHA Table Z-1-A air contaminant (vegetable oil). Goggles or face shield; rubber gloves.
Toxicity by ingestion: Grade 0; LD_{50} = > 15 g/kg.
Long-term health effects: Liver damage in rats (from addition of oil to diet).
Liquid or solid irritant characteristics: Eye and skin irritant.

FIRE DATA
Flash point: 535°F/279°C (oc); 428°F/220°C (cc) (raw); 403°F/206°C (boiled).
Fire extinguishing agents not to be used: Water or foam may cause frothing; water may be ineffective
Autoignition temperature: 650°F/343°C/616°K.
Burning rate: 4 mm/min

CHEMICAL REACTIVITY
Binary reactants: Oxidizers, nitric acid. Spontaneous heating on contact with air.
Polymerization: May occur.
Reactivity group: 33
Compatibility class: Miscellaneous hydrocarbon mixtures

ENVIRONMENTAL DATA
Water pollution: Effect of low concentrations on aquatic life is unknown. Fouling to shoreline. May be dangerous if it enters nearby water intakes; notify operators. Notify local health and wildlife officials. **Response to discharge:** Mechanical containment. Should be removed. Chemical and physical treatment.

SHIPPING INFORMATION
Grades of purity: Raw grade; varnish makers grade; grinding grade; boiled; double-boiled; heat-bodied grade; blown grade; refined; **Storage temperature:** Ambient; **Inert atmosphere:** None; **Venting:** Open (flame arrester); **Stability during transport:** Stable.

PHYSICAL AND CHEMICAL PROPERTIES
Physical state @ 59°F/15°C and 1 atm: Liquid.
Boiling point @ 1 atm: More than 600°F/316°C/589°K.
Melting/Freezing point: –2°F/–19°C/254°K.
Specific gravity (water = 1): 0.932 @ 68°F/20°C (liquid).
Liquid surface tension: (estimate) 25 dynes/cm = 0.025 N/m @ 68°F/20°C.
Liquid water interfacial tension: (estimate) 50 dynes/cm = 0.050 N/m @ 68°F/20°C.
Heat of combustion: –16,800 Btu/lb = –9300 cal/g = –390 x 10^5 J/kg.

OILS, MISCELLANEOUS: LUBRICATING   REC. O:3500

SYNONYMS: ACEITE de LUBRICANTE (Spanish); CRANKCASE OIL; LUBRICATING OIL; PARAFFIN OIL; TRANSMISSION OIL

IDENTIFICATION
DOT ID Number: UN 1270; DOT Guide Number: 128
Proper Shipping Name: Petroleum oil

DESCRIPTION: Yellow-brown oily liquid. Lube oil odor. Floats on water surface; insoluble.

Combustible • Containers may BLEVE when exposed to fire • Vapors are heavier than air and will collect and stay in low areas • Vapors in confined areas (e.g., tanks, sewers, buildings) may explode when exposed to fire • Irritating to the skin, eyes, and respiratory tract • Toxic products of combustion may include carbon monoxide.

Hazard Classification (based on NFPA-704 Rating System)
Health Hazards (Blue): 0; Flammability (Red): 1; Reactivity (Yellow): 0

EMERGENCY RESPONSE: See Appendix A (128)
Evacuation:
Public safety: Isolate spill area for at least 25 to 50 meters (80 to 160 feet) in all directions.
Spill: Large spill–Consider initial downwind evacuation for at least 300 meters (1000 feet).

Fire: Isolate for 800 meters (½ mile) in all directions, especially if tank, rail car, or tank truck is involved in fire.

EXPOSURE
Short-term effects: *SEEK MEDICAL ATTENTION.* **Liquid:** Irritating to skin and eyes. Harmful if swallowed. Remove contaminated clothing and shoes. Flush affected areas with plenty of water. *IF IN EYES*, hold eyelids open and flush with plenty of water. *IF SWALLOWED* and victim is *CONSCIOUS AND ABLE TO SWALLOW*, have victim drink 4 to 8 ounces of water. **Do NOT induce vomiting.**

HEALTH HAZARDS
Personal protective equipment (PPE): B-Level PPE. Protective gloves; goggles or face shield.
Exposure limits (TWA unless otherwise noted): ACGIH TLV for oil mist, mineral is 5 mg/m^3.
Toxicity by ingestion: Grade 1; LD_{50} = 5 to 15 g/kg.
Vapor (gas) irritant characteristics: Vapors cause smarting of the eyes or respiratory system if present in high concentrations. The effect is temporary.
Liquid or solid irritant characteristics: Minimum hazard. If spilled on clothing and allowed to remain, may cause smarting and reddening of the skin.

FIRE DATA
Flash point: 300°F/149°C (cc).
Fire extinguishing agents not to be used: Water of foam may cause frothing.
Autoignition temperature: 500–700°F/260–371°C/533–644°K.
Electrical hazard: Class I, Group D. Flow or agitation of substance may generate electrostatic charges due to low conductivity.
Burning rate: 4 mm/min

CHEMICAL REACTIVITY
Binary reactants: Oxidizers, nitric acid.
Reactivity group: 33
Compatibility class: Miscellaneous hydrocarbon mixtures.

ENVIRONMENTAL DATA
Water pollution: Effect of low concentrations on aquatic life is unknown. Fouling to shoreline. May be dangerous if it enters nearby water intakes; notify operators. Notify local health and wildlife officials. **Response to discharge:** Mechanical containment. Should be removed. Chemical and physical treatment.

SHIPPING INFORMATION
Grades of purity: Various viscosities; **Storage temperature:** Ambient; **Inert atmosphere:** None; **Venting:** Open (flame arrester); **Stability during transport:** Stable.

PHYSICAL AND CHEMICAL PROPERTIES
Physical state @ 59°F/15°C and 1 atm: Liquid.
Boiling point @ 1 atm: 683°F/362°C/635°K.
Specific gravity (water = 1): (estimate) 0.902 @ 68°F/20°C (liquid).
Liquid surface tension: 36–37.5 dynes/cm = 0.036–0.0375 N/m @ 68°F/20°C.
Liquid water interfacial tension: 33–54 dynes/cm = 0.033–0.054 N/m @ 68°F/20°C.
Heat of combustion: –18,486 Btu/lb = –10,270 cal/g = –429.98 x 10^5 J/kg.

OILS, MISCELLANEOUS: MINERAL   REC. O:3600

SYNONYMS: ACEITE de MINERAL (Spanish); ALBOLINE; BAYOL F; BLANDLUBE; CABLE OIL; CRYSTOSOL; CUTTING OIL; DRAKEOL; FONOLINE; GLYMOL; HEAT TREATING OIL; HYDRAULIC OIL; LIQUID PETROLATUM; LUBRICATING OIL; MINERAL OIL; MOLOL; NUJOL; OIL MIST, MINERAL; PAROL; WHITE OIL; WHITE MINERAL OIL

IDENTIFICATION
CAS Number: 8012-95-1
DOT ID Number: UN 1270; DOT Guide Number: 128
Proper Shipping Name: Petroleum oil

DESCRIPTION: Colorless oily liquid. Very faint odor; nearly odorless. Floats on water surface; insoluble.

Combustible • Heat or flame may cause explosion • Containers may BLEVE when exposed to fire • Vapors are heavier than air and will collect and stay in low areas • Vapors in confined areas (e.g., tanks, sewers, buildings) may explode when exposed to fire • Irritating to the skin, eyes, and respiratory tract • Toxic products of combustion may include acrid carbon monoxide.

Hazard Classification (based on NFPA-704 Rating System)
Health Hazards (Blue): 0; Flammability (Red): 1; Reactivity (Yellow): 0

EMERGENCY RESPONSE: See Appendix A (128)
Evacuation:
Public safety: Isolate spill area for at least 25 to 50 meters (80 to 160 feet) in all directions.
Spill: Large spill–Consider initial downwind evacuation for at least 300 meters (1000 feet).
Fire: Isolate for 800 meters (½ mile) in all directions, especially if tank, rail car, or tank truck is involved in fire.

EXPOSURE
Short-term effects: *SEEK MEDICAL ATTENTION.* **Liquid:** Irritating to skin and eyes. Remove contaminated clothing and shoes. Flush affected areas with plenty of water. *IF IN EYES*, hold eyelids open and flush with plenty of water. *IF SWALLOWED* and victim is *CONSCIOUS AND ABLE TO SWALLOW*, have victim drink 4 to 8 ounces of water. **Do NOT induce vomiting.**

HEALTH HAZARDS
Personal protective equipment (PPE): B-Level PPE. OSHA Table Z-1-A air contaminant. Goggles or face shield.
Recommendations for respirator selection: NIOSH/OSHA as mineral oil mist
50 mg/m^3: HiE (any air-purifying, respirator with a high-efficiency particulate filter); or SA (any supplied-air respirator). *125 mg/m^3*: SA:CF (any supplied-air respirator operated in a continuous-flow mode); or PAPRHiE (any powered, air-purifying respirator with a high-efficiency particulate filter). *250 mg/m^3:* HiEF (any air-purifying, full-facepiece respirator with a high-efficiency particulate filter); or SAT:CF (any supplied-air respirator that has a tight-fitting facepiece and is operated in a continuous-flow mode); or PAPRTHiE (any powered, air-purifying respirator with a tight-fitting facepiece and a high-efficiency particulate filter); or SCBAF (any self-contained breathing apparatus with a full facepiece); or SAF (any supplied-air respirator with a full facepiece). *EMERGENCY OR PLANNED ENTRY INTO UNKNOWN CONCENTRATIONS OR IDLH CONDITIONS:* SCBAF:PD,PP

(any self-contained breathing apparatus that has a full facepiece and is operated in a pressure-demand or other positive-pressure mode); or SAF:PD,PP:ASCBA (any supplied-air respirator that has a full facepiece and is operated in a pressure-demand or other positive-pressure mode in combination with an auxiliary self-contained breathing apparatus operated in a pressure-demand or other positive pressure mode). *ESCAPE:* HiEF (any air-purifying, full-facepiece respirator with a high-efficiency particulate filter); or SCBAE (any appropriate escape-type, self-contained breathing apparatus).
Exposure limits (TWA unless otherwise noted): ACGIH TLV 5 mg/m^3; NIOSH/OSHA 5 mg/m^3 as mineral oil mist
Short-term exposure limits (15-minute TWA): ACGIH STEL 10 mg/m^3; NIOSH STEL 5 mg/m^3.
Toxicity by ingestion: Grade 1; LD$_{50}$ = 5 to 15 g/kg.
IDLH value: 2500 mg/m^3 as mineral oil mist

FIRE DATA
Flash point: 380°F/193°C (oc).
Fire extinguishing agents not to be used: Water or foam may cause frothing.
Autoignition temperature: 500–700°F/260–371°C/533–644°K.
Electrical hazard: Class I, Group D.
Burning rate: 4 mm/min

CHEMICAL REACTIVITY
Binary reactants: Oxidizers, nitric acid
Reactivity group: 33
Compatibility class: Miscellaneous hydrocarbon mixtures

ENVIRONMENTAL DATA
Food chain concentration potential: Negative; unlikely to accumulate.
Water pollution: Effect of low concentrations on aquatic life is unknown. Fouling to shoreline. May be dangerous if it enters nearby water intakes; notify operators. Notify local health and wildlife officials. **Response to discharge:** Mechanical containment. Should be removed. Chemical and physical treatment.

SHIPPING INFORMATION
Grades of purity: Commercial; refined; **Storage temperature:** Ambient; **Inert atmosphere:** None; **Venting:** Open (flame arrester); **Stability during transport:** Stable.

PHYSICAL AND CHEMICAL PROPERTIES
Physical state @ 59°F/15°C and 1 atm: Liquid.
Boiling point @ 1 atm: 684°F/362°C/635°K.
Specific gravity (water = 1): 0.822 @ 68°F/20°C (liquid).
Liquid surface tension: 27 dynes/cm = 0.027 N/m @ 68°F/20°C.
Liquid water interfacial tension: 47 dynes/cm = 0.047 N/m at 158°F/70°C.

OILS, MISCELLANEOUS: MINERAL SEAL REC. O:3700

SYNONYMS: LONG-TIME BURNING OIL; MINERAL COLZA OIL; MINERAL SEAL OIL; PETROLEUM DISTILLATES, HYDROTREATED MIDDLE; SIGNAL OIL

IDENTIFICATION
CAS Number: 64742-46-7 and others
DOT ID Number: UN 1270; DOT Guide Number: 128
Proper Shipping Name: Petroleum oil
Reportable Quantity (RQ):

DESCRIPTION: Colorless to yellow or light brown oily liquid. Kerosene odor. Floats on water surface; insoluble.

Flammable • Containers may BLEVE when exposed to fire • Vapors may form explosive mixture with air • Vapors are heavier than air and will collect and stay in low areas • Vapors may travel long distances to ignition sources and flashback • Vapors in confined areas (e.g., tanks, sewers, buildings) may explode when exposed to fire • Irritating to the skin, eyes, and respiratory tract • Toxic products of combustion may include acrid carbon monoxide.

Hazard Classification (based on NFPA-704 Rating System)
Health Hazards (Blue): 0; Flammability (Red): 2; Reactivity (Yellow): 0

EMERGENCY RESPONSE: See Appendix A (128)
Evacuation:
Public safety: Isolate spill area for at least 25 to 50 meters (80 to 160 feet) in all directions.
Spill: Large spill–Consider initial downwind evacuation for at least 300 meters (1000 feet).
Fire: Isolate for 800 meters (½ mile) in all directions, especially if tank, rail car, or tank truck is involved in fire.

EXPOSURE
Short-term effects: *SEEK MEDICAL ATTENTION.* **Liquid:** Irritating to skin and eyes. Harmful if swallowed. Remove contaminated clothing and shoes. Flush affected areas with plenty of water. *IF IN EYES,* hold eyelids open and flush with plenty of water. *IF SWALLOWED* and victim is *CONSCIOUS AND ABLE TO SWALLOW,* have victim drink 4 to 8 ounces of water. **Do NOT induce vomiting.**

HEALTH HAZARDS
Personal protective equipment (PPE): B-Level PPE. Protective gloves; goggles or face shield.
Exposure limits (TWA unless otherwise noted): ACGIH TLV for oil mist, mineral is 5 mg/m^3.
Toxicity by ingestion: Grade 2; LD$_{50}$ = 0.5 to 5 g/kg.
Vapor (gas) irritant characteristics: Vapors cause a slight smarting of the eyes or respiratory system if present in high concentrations. The effect is temporary.
Liquid or solid irritant characteristics: Eye and skin irritant. If spilled on clothing and allowed to remain, may cause smarting and reddening of the skin.
Odor threshold: 1 ppm.

FIRE DATA
Flash point: 170–275°F/77–135°C (oc).
Fire extinguishing agents not to be used: Water may be ineffective.
Electrical hazard: Class I, Group D.
Burning rate: 4 mm/min

CHEMICAL REACTIVITY
Binary reactants: Oxidizers, nitric acid
Reactivity group: 33
Compatibility class: Miscellaneous hydrocarbon mixtures

ENVIRONMENTAL DATA
Food chain concentration potential: Negative; unlikely to accumulate.
Water pollution: Effect of low concentrations on aquatic life is unknown. Fouling to shoreline. May be dangerous if it enters nearby water intakes; notify operators. Notify local health and

wildlife officials. **Response to discharge:** Mechanical containment. Should be removed. Chemical and physical treatment.

SHIPPING INFORMATION
Grades of purity: Several grades of varying pour points, all highly refined; **Storage temperature:** Ambient; **Inert atmosphere:** None; **Venting:** Open (flame arrester); **Stability during transport:** Stable.

PHYSICAL AND CHEMICAL PROPERTIES
Physical state @ 59°F/15°C and 1 atm: Liquid.
Boiling point @ 1 atm: More than 500°F/260°C/533°K.
Melting/Freezing point: 10°F/–12°C/261°K.
Specific gravity (water = 1): 0.811–0.825 @ 15°C (liquid).
Liquid surface tension: (estimate) 25 dynes/cm = 0.025 N/m @ 68°F/20°C.
Liquid water interfacial tension: 47–50 dynes/cm = 0.047–0.050 N/m @ 68°F/20°C.
Heat of combustion: (estimate) –18,000 Btu/lb = –10,000 cal/g = –420 x 10^5 J/kg.

OILS, MISCELLANEOUS: MOTOR   REC. O:3800

SYNONYMS: ACEITE de MOTOR (Spanish); CRANKCASE OIL; LUBRICATING OIL

IDENTIFICATION
DOT ID Number: UN 1270; DOT Guide Number: 128
Proper Shipping Name: Petroleum oil

DESCRIPTION: Yellow fluorescent to yellow-brown, oily liquid. Lube oil odor. Floats on water surface; insoluble.

Combustible • Heat or flame may cause explosion • Containers may BLEVE when exposed to fire • Vapors are heavier than air and will collect and stay in low areas • Vapors in confined areas (e.g., tanks, sewers, buildings) may explode when exposed to fire • Irritating to the skin, eyes, and respiratory tract • Toxic products of combustion may include acrid carbon monoxide.

Hazard Classification (based on NFPA-704 Rating System)
Health Hazards (Blue): 0; Flammability (Red): 1; Reactivity (Yellow): 0

EMERGENCY RESPONSE: See Appendix A (128)
Evacuation:
Public safety: Isolate spill area for at least 25 to 50 meters (80 to 160 feet) in all directions.
Spill: Large spill–Consider initial downwind evacuation for at least 300 meters (1000 feet).
Fire: Isolate for 800 meters (½ mile) in all directions, especially if tank, rail car, or tank truck is involved in fire.

EXPOSURE
Short-term effects: *SEEK MEDICAL ATTENTION.* **Liquid:** Irritating to skin and eyes. Harmful if swallowed. Remove contaminated clothing and shoes. Flush affected areas with plenty of water. *IF IN EYES*, hold eyelids open and flush with plenty of water. *IF SWALLOWED* and victim is *CONSCIOUS AND ABLE TO SWALLOW*, have victim drink 4 to 8 ounces of water. **Do NOT induce vomiting.**

HEALTH HAZARDS
Personal protective equipment (PPE): B-Level PPE. Protective gloves; goggles or face shield.
Toxicity by ingestion: Grade 1; LD_{50} = 5 to 15 g/kg.
Vapor (gas) irritant characteristics: Vapors cause smarting of the eyes or respiratory system if present in high concentrations. The effect is temporary.
Liquid or solid irritant characteristics: If spilled on clothing and allowed to remain, may cause smarting and reddening of the skin.

FIRE DATA
Flash point: 275–600°F/135–316°C (cc).
Fire extinguishing agents not to be used: Water may be ineffective
Autoignition temperature: 325–625°F/163–329°C/436–602°K.
Burning rate: 4 mm/min

CHEMICAL REACTIVITY
Binary reactants: Oxidizers, nitric acid
Reactivity group: 33
Compatibility class: Miscellaneous hydrocarbon mixtures

ENVIRONMENTAL DATA
Food chain concentration potential: Negative; unlikely to accumulate.
Water pollution: Effect of low concentrations on aquatic life is unknown. Fouling to shoreline. May be dangerous if it enters nearby water intakes; notify operators. Notify local health and wildlife officials. **Response to discharge:** Mechanical containment. Should be removed. Chemical and physical treatment.

SHIPPING INFORMATION
Grades of purity: Various viscosities; **Storage temperature:** Ambient; **Inert atmosphere:** None; **Venting:** Open (flame arrester); **Stability during transport:** Stable.

PHYSICAL AND CHEMICAL PROPERTIES
Physical state @ 59°F/15°C and 1 atm: Liquid.
Boiling point @ 1 atm: Very high.
Melting/Freezing point: –30°F/–34°C/239°K.
Specific gravity (water = 1): 0.84–0.96 @ 15°C (liquid).
Liquid surface tension: 36-37.5 dynes/cm = 0.036–0.0375 N/m @ 68°F/20°C.
Liquid water interfacial tension: 33–54 dynes/cm = 0.033–0.054 N/m @ 68°F/20°C.
Heat of combustion: –18,486 Btu/lb = –10,270 cal/g = –429.98 x 10^5 J/kg.

OILS, MISCELLANEOUS: NEATSFOOT   REC. O:3900

SYNONYMS: ACEITE de PIE de BUEY (Spanish); BABULUM OIL; HOOF OIL; NEATSFOOT OIL

IDENTIFICATION
DOT ID Number: UN 9277; DOT Guide Number: 171
Proper Shipping Name: Oil, n.o.s., flash point not less than 93°C (200°F).

DESCRIPTION: Pale yellow oily liquid. Peculiar, characteristic odor. Floats on water surface; insoluble.

Combustible • Containers may BLEVE when exposed to fire • Vapors in confined areas (e.g., tanks, sewers, buildings) may explode when exposed to fire • Irritating to the skin, eyes, and respiratory tract • Toxic products of combustion may include carbon monoxide.

Hazard Classification (based on NFPA-704 Rating System)
Health Hazards (Blue): 0; Flammability (Red): 1; Reactivity (Yellow): 0

EMERGENCY RESPONSE: See Appendix A (171)
Evacuation:
Public safety: Isolate the area of spill or leak for at least 10 to 25 meters (30 to 80 feet) in all directions.
Spill: Increase, in the downwind direction, as necessary, the distance shown under "Public Safety."
Fire: If any large container is involved in fire, isolate for at least 800 meters (½ mile) in all directions; also, consider initial evacuation for 800 meters (½ mile) in all directions.

HEALTH HAZARDS
Personal protective equipment (PPE): B-Level PPE. Wear rubber gloves, face shield and protective clothing.
Toxicity by ingestion: Grade 0; LD_{50} = > 15 g/kg.

FIRE DATA
Flash point: 430°F/221°C (cc).
Fire extinguishing agents not to be used: Water or foam may cause frothing.
Autoignition temperature: 828°F/442°C/715°K.
Electrical hazard: Class I, Group D.

CHEMICAL REACTIVITY
Binary reactants: Oxidizers, strong acids.
Reactivity group: 33
Compatibility class: Miscellaneous hydrocarbon mixtures

ENVIRONMENTAL DATA
Food chain concentration potential: Negative; unlikely to accumulate.
Water pollution: Effect of low concentrations on aquatic life is unknown. Fouling to shoreline. May be dangerous if it enters nearby water intakes; notify operators. Notify local health and wildlife officials. **Response to discharge:** Mechanical containment. Should be removed. Chemical and physical treatment.

SHIPPING INFORMATION
Grades of purity: Various grades designated by pour point (15–40°F); also various refined grades; **Storage temperature:** Ambient; **Inert atmosphere:** None; **Stability during transport:** Stable.

PHYSICAL AND CHEMICAL PROPERTIES
Physical state @ 59°F/15°C and 1 atm: Liquid.
Boiling point @ 1 atm: Very high.
Melting/Freezing point: 95°F/35°C/308°K.
Specific gravity (water = 1): 0.915 at 16°C (liquid).
Vapor pressure: (Reid) 0.1 psia.

OILS, MISCELLANEOUS: PENETRATING REC. O:4000

SYNONYMS: PENETRATING OIL; PRESERVATIVE OIL; WATER DISPLACING OIL

IDENTIFICATION
DOT ID Number: UN 1270; DOT Guide Number: 128
Proper Shipping Name: Petroleum oil

DESCRIPTION: Yellowish, oily liquid. Motor oil-like odor. Floats on water surface; insoluble.

Combustible • Heat or flame may cause explosion • Containers may BLEVE when exposed to fire • Vapors are heavier than air and will collect and stay in low areas • Vapors in confined areas (e.g., tanks, sewers, buildings) may explode when exposed to fire • Irritating to the skin, eyes, and respiratory tract • Toxic products of combustion may include carbon monoxide.

Hazard Classification (based on NFPA-704 Rating System)
Health Hazards (Blue): 0; Flammability (Red): 1; Reactivity (Yellow): 0

EMERGENCY RESPONSE: See Appendix A (128)
Evacuation:
Public safety: Isolate spill area for at least 25 to 50 meters (80 to 160 feet) in all directions.
Spill: Large spill–Consider initial downwind evacuation for at least 300 meters (1000 feet).
Fire: Isolate for 800 meters (½ mile) in all directions, especially if tank, rail car, or tank truck is involved in fire.

EXPOSURE
Short-term effects: *SEEK MEDICAL ATTENTION.* **Liquid:** Irritating to skin and eyes. Harmful if swallowed. Remove contaminated clothing and shoes. Flush affected areas with plenty of water. *IF IN EYES*, hold eyelids open and flush with plenty of water. *IF SWALLOWED* and victim is *CONSCIOUS AND ABLE TO SWALLOW*, have victim drink 4 to 8 ounces of water. **Do NOT induce vomiting**.

HEALTH HAZARDS
Personal protective equipment (PPE): B-Level PPE. Protective gloves; goggles or face shield.
Toxicity by ingestion: Grade 1; LD_{50} = 5 to 15 g/kg.
Vapor (gas) irritant characteristics: Vapors cause a slight smarting of the eyes or respiratory system if present in high concentrations. The effect is temporary.
Liquid or solid irritant characteristics: If spilled on clothing and allowed to remain, may cause smarting and reddening of skin.

FIRE DATA
Flash point: 295°F/146°C (oc).
Fire extinguishing agents not to be used: Water or foam may cause frothing.
Electrical hazard: Class I, Group D.

CHEMICAL REACTIVITY
Binary reactants: Oxidizers, nitric acid
Reactivity group: 33
Compatibility class: Miscellaneous hydrocarbon mixtures

ENVIRONMENTAL DATA
Food chain concentration potential: Negative; unlikely to accumulate.
Water pollution: Effect of low concentrations on aquatic life is unknown. Fouling to shoreline. May be dangerous if it enters nearby water intakes; notify operators. Notify local health and wildlife officials. **Response to discharge:** Mechanical containment. Should be removed. Chemical and physical treatment.

SHIPPING INFORMATION
Grades of purity: Commercial; **Storage temperature:** Ambient; **Inert atmosphere:** None; **Venting:** Open (flame arrester); **Stability during transport:** Stable.

PHYSICAL AND CHEMICAL PROPERTIES
Physical state @ 59°F/15°C and 1 atm: Liquid.
Boiling point @ 1 atm: Very high.
Specific gravity (water = 1): 0.8961 @ 68°F/20°C (liquid).
Liquid surface tension: 29.8 dynes/cm = 0.0298 N/m at 24°C.
Liquid water interfacial tension: 5.5 dynes/cm = 0.0055 N/m at 71°F/22°C.
Heat of combustion: (estimate) –18,000 Btu/lb = –10,000 cal/g = –420 x 10^5 J/kg.

OIL, MISCELLANEOUS: PINE   **REC. O:4100**

SYNONYMS: ACEITE de PINO (Spanish); ARIZOLE; OIL OF PINE; OLEUM ABIETIS; PINE OIL; PINE OIL, STEAM DISTILLED; TERPENTIN OEL (German); UNIPINE; YARMOR; YARMOR PINE OIL

IDENTIFICATION
CAS Number: 8002-09-3; 8021-27-0 (Silver); 8021-29-2 (Siberian); 8000-26-8 (Dwarf, Scotch, or Montana)
Formula: Mixture, primarily $C_{10}H_{17}OH$
DOT ID Number: UN 1272; DOT Guide Number: 129
Proper Shipping Name: Pine oil

DESCRIPTION: Colorless to pale yellow liquid. Pleasant, penetrating, pine odor. Floats on water surface; insoluble.

Flammable • Containers may BLEVE when exposed to fire • Vapors are heavier than air and will collect and stay in low areas • Vapors may travel long distances to ignition sources and flashback • Vapors in confined areas (e.g., tanks, sewers, buildings) may explode when exposed to fire • Irritating to the skin, eyes, and respiratory tract • Toxic products of combustion may include carbon monoxide.

Hazard Classification (based on NFPA-704 Rating System)
Health Hazards (Blue): 0; Flammability (Red): 2; Reactivity (Yellow): 0

EMERGENCY RESPONSE: See Appendix A (129)
Evacuation:
Public safety: Isolate spill area for at least 50 to 100 meters (160 to 330 feet) in all directions.
Spill: Large spill–Consider initial downwind evacuation for at least 300 meters (1000 feet).
Fire: Isolate for 800 meters (½ mile) in all directions, especially if tank, rail car, or tank truck is involved in fire.

EXPOSURE
Short-term effects: *SEEK MEDICAL ATTENTION.* **Vapor:** Irritating to eyes, nose, and throat. *IF INHALED*, will, will cause nausea, vomiting, headache, difficult breathing or loss of consciousness. Move to fresh air. *IF BREATHING HAS STOPPED*, give artificial respiration. IF breathing is difficult, administer oxygen. **Liquid:** *POISONOUS IF SWALLOWED.* Irritating to skin and eyes. Remove contaminated clothing and shoes. Flush affected areas with plenty of water. *IF IN EYES*, hold eyelids open and flush with plenty of water. *IF SWALLOWED*, Do NOT induce vomiting.

HEALTH HAZARDS
Personal protective equipment (PPE): B-Level PPE. Organic canister); or air–supplied mask; goggles or face shield; rubber gloves.
Toxicity by ingestion: Grade 2: LD_{50} = 3.2 g/kg (rat); TDLO = 4.78g/kg (human).
Vapor (gas) irritant characteristics: Vapors cause smarting of the eyes or respiratory system if present in high concentrations. The effect is temporary.
Liquid or solid irritant characteristics: Minimum hazard. If spilled on clothing and allowed to remain, may cause smarting and reddening of the skin.

FIRE DATA
Flash point: 138–172°F/59–78°C (cc).
Fire extinguishing agents not to be used: Water may be ineffective.
Behavior in fire: Forms heavy black smoke and soot.

CHEMICAL REACTIVITY
Binary reactants: Oxidizers, strong acids.
Reactivity group: 33

ENVIRONMENTAL DATA
Water pollution: Dangerous to aquatic life in high concentrations. Fouling to shoreline. May be dangerous if it enters nearby water intakes; notify operators. Notify local health and wildlife officials.
Response to discharge: Mechanical containment. Should be removed. Chemical and physical treatment.

SHIPPING INFORMATION
Storage temperature: Ambient; **Stability during transport:** Stable.

PHYSICAL AND CHEMICAL PROPERTIES
Physical state @ 59°F/15°C and 1 atm: Liquid.
Molecular weight: Alpha terpenol primary component 154.25
Boiling point @ 1 atm: More than 400°F/205°C/478°K.
Melting/Freezing point: Less than 50°F/10°C/283°K.
Specific gravity (water = 1): 0.95
Relative vapor density (air = 1): 5.3

OILS, MISCELLANEOUS: RESIN   **REC. O:4200**

SYNONYMS: ACEITE de RESINA (Spanish); CODOIL; RESIN OIL; RETINOL; ROSIN OIL; ROSINOL

IDENTIFICATION
DOT ID Number: UN 1286; DOT Guide Number: 127
Proper Shipping Name: Rosin oil

DESCRIPTION: Light amber to red to black liquid; depending on grade. Pine tree-pitch odor. Floats on water surface; insoluble.

Combustible • Containers may BLEVE when exposed to fire • Vapors may form explosive mixture with air • Vapors are heavier than air and will collect and stay in low areas • Vapors may travel long distances to ignition sources and flashback • Vapors in confined areas (e.g., tanks, sewers, buildings) may explode when exposed to fire • Irritating to the skin, eyes, and respiratory tract • Toxic products of combustion may include acrid carbon monoxide.

Hazard Classification (based on NFPA-704 Rating System)
Health Hazards (Blue): 0; Flammability (Red): 1; Reactivity (Yellow): 0

EMERGENCY RESPONSE: See Appendix A (127)
Evacuation:
Public safety: Isolate spill area for at least 25 to 50 meters (80 to 160 feet) in all directions.
Spill: Large spill–Consider initial downwind evacuation for at least 300 meters (1000 feet).
Fire: Isolate for 800 meters (½ mile) in all directions, especially if tank, rail car, or tank truck is involved in fire.

EXPOSURE
Short-term effects: *SEEK MEDICAL ATTENTION.* **Liquid:** Contact may cause irritation. Flush affected areas with plenty of water.

HEALTH HAZARDS
Personal protective equipment (PPE): B-Level PPE. Wear rubber gloves, face shield and protective clothing.

FIRE DATA
Flash point: 255–390°F/124–199°C (oc).
Fire extinguishing agents not to be used: Water may be ineffective
Autoignition temperature: 648°F/342°C/615°K.
Electrical hazard: Class I, Group D.

CHEMICAL REACTIVITY
Binary reactants: Oxidants
Reactivity group: 33
Compatibility class: Miscellaneous hydrocarbon mixtures

ENVIRONMENTAL DATA
Food chain concentration potential: Negative; unlikely to accumulate.
Water pollution: Effect of low concentrations on aquatic life is unknown. Fouling to shoreline. May be dangerous if it enters nearby water intakes; notify operators. Notify local health and wildlife officials. **Response to discharge:** Mechanical containment. Should be removed. Chemical and physical treatment.

SHIPPING INFORMATION
Grades of purity: A variety of grades that differ primarily in color and flash point; **Storage temperature:** Ambient; **Inert atmosphere:** None; **Venting:** Open (flame arrester); **Stability during transport:** Stable.

NAS HAZARD CLASSIFICATION FOR BULK WATER TRANSPORTATION
FIRE: 2
HEALTH: Vapor irritant: 2; Liquid or solid irritant: 2; Poisons: 2
WATER POLLUTION: Human toxicity: 1; Aquatic toxicity: 3; Aesthetic effect: 2
REACTIVITY: Other chemicals: 2; Water: 0; Self-reaction: 3

PHYSICAL AND CHEMICAL PROPERTIES
Physical state @ 59°F/15°C and 1 atm: Liquid.
Boiling point @ 1 atm: 572–750°F/300–400°C/573–673°K.
Specific gravity (water = 1): 0.96–1.02 @ 15°C (liquid).
Liquid surface tension: (estimate) 25 dynes/cm = 0.025 N/m @ 68°F/20°C.
Liquid water interfacial tension: (estimate) 50 dynes/cm = 0.05 N/m @ 68°F/20°C.
Heat of combustion: (estimate) –18,000 Btu/lb = –10,000 cal/g = –420 x 10^5 J/kg.
Vapor pressure: 0.04 mm.

OILS, MISCELLANEOUS: ROAD   **REC. O:4300**

SYNONYMS: ASPHALT, PETROLEUM; LIQUID ASPHALT; PETROLEUM ASPHALT; ROAD ASPHALT; PETROLEUM ROOFING TAR; SLOW CURING ASPHALT

IDENTIFICATION
CAS Number: 8052-42-4
DOT ID Number: UN 1999; DOT Guide Number: 128
Proper Shipping Name: Asphalt; Tars, liquid, including road asphalt and oils, bitumen and cutbacks.

DESCRIPTION: Black or dark brown oily liquid. Tar odor. Floats on the surface of water.

Combustible • Heat or flame may cause explosion • Containers may BLEVE when exposed to fire • Vapors are heavier than air and will collect and stay in low areas • Vapors in confined areas (e.g., tanks, sewers, buildings) may explode when exposed to fire • Irritating to the skin, eyes, and respiratory tract • Toxic products of combustion may include acrid carbon monoxide.

Hazard Classification (based on NFPA-704 Rating System)
Health Hazards (Blue): 0; Flammability (Red): 1; Reactivity (Yellow): 0

EMERGENCY RESPONSE: See Appendix A (128)
Evacuation:
Public safety: Isolate spill area for at least 25 to 50 meters (80 to 160 feet) in all directions.
Spill: Large spill–Consider initial downwind evacuation for at least 300 meters (1000 feet).
Fire: Isolate for 800 meters (½ mile) in all directions, especially if tank, rail car, or tank truck is involved in fire.

EXPOSURE
Short-term effects: *SEEK MEDICAL ATTENTION.* **Liquid:** Will burn skin and eyes. Harmful if swallowed. Remove contaminated clothing and shoes. Flush affected areas with plenty of water. *IF IN EYES*, hold eyelids open and flush with plenty of water. *IF SWALLOWED* and victim is *CONSCIOUS AND ABLE TO SWALLOW*, have victim drink 4 to 8 ounces of water. **Do NOT induce vomiting.**

HEALTH HAZARDS
Personal protective equipment (PPE): B-Level PPE. Protective clothing for hot asphalt; face and eye protection when hot.
Recommendations for respirator selection: NIOSH
At any concentrations above the NIOSH REL, or where there is no REL, at any detectable concentration: SCBAF:PD,PP (any self-contained breathing apparatus that has a full facepiece and is operated in a pressure-demand or other positive-pressure mode); or SAF:PD,PP:ASCBA (any supplied-air respirator that has a full facepiece and is operated in a pressure-demand or other positive-pressure mode in combination with an auxiliary, self-contained breathing apparatus operated in a pressure-demand or other positive pressure mode). *ESCAPE:* GMFOVHiE [any air-purifying, full-facepiece respirator (gas mask) with a chin-style, front- or back-

mounted organic vapor canister having a high-efficiency particulate filter]; or SCBAE (any appropriate escape-type, self-contained breathing apparatus).
Exposure limits (TWA unless otherwise noted): ACGIH TLV 5 mg/m³; NIOSH REL human carcinogen, ceiling 5 mg/m³/15 minutes, as asphalt fumes
Short-term exposure limits (15-minute TWA): See above.
Toxicity by ingestion: Grade 2; LD_{50} = 0.5 to 5 g/kg.
Long-term health effects: Suspected carcinogen
Vapor (gas) irritant characteristics: Respiratory tract irritant. Vapors cause irritation such that personnel will find high concentrations unpleasant. The effect is temporary.
Liquid or solid irritant characteristics: Causes smarting of the skin and first-degree burns on short exposure; may cause secondary burns on long exposure.

FIRE DATA
Flash point: 300–550°F/149–288°C.
Fire extinguishing agents not to be used: Water may be ineffective
Autoignition temperature: 400–700°F/204–142°C/477–415°K.
Electrical hazard: Class I, Group D.

CHEMICAL REACTIVITY
Binary reactants: Oxidizers, nitric acid
Reactivity group: 33
Compatibility class: Miscellaneous hydrocarbon mixtures

ENVIRONMENTAL DATA
Food chain concentration potential: Unlikely to accumulate.
Water pollution: Effect of low concentrations on aquatic life is unknown. Fouling to shoreline. May be dangerous if it enters nearby water intakes; notify operators. Notify local health and wildlife officials. **Response to discharge:** Mechanical containment. Chemical and physical treatment.

SHIPPING INFORMATION
Grades of purity: SC-0 to SC-5; **Storage temperature:** Ambient; **Inert atmosphere:** None; **Venting:** Open (flame arrester); **Stability during transport:** Stable.

NAS HAZARD CLASSIFICATION FOR BULK WATER TRANSPORTATION
FIRE: 1
HEALTH: Vapor irritant: 1; Liquid or solid irritant: 2; Poisons: 1
WATER POLLUTION: Human toxicity: 0; Aquatic toxicity: 1; Aesthetic effect: 4
REACTIVITY: Other chemicals: 0; Water: 0; Self-reaction: 0

PHYSICAL AND CHEMICAL PROPERTIES
Physical state @ 59°F/15°C and 1 atm: Liquid.
Boiling point @ 1 atm: Very high.
Specific gravity (water = 1): 1.0–1.2 @ 77°F/25°C (liquid).
Liquid surface tension: (estimate) 25 dynes/cm = 0.025 N/m @ 68°F/20°C.
Liquid water interfacial tension: (estimate) 50 dynes/cm = 0.05 N/m @ 68°F/20°C.
Heat of combustion: (estimate) –18,000 Btu/lb = –10,000 cal/g = –420 x 10^5 J/kg.

OILS, MISCELLANEOUS: SPERM   REC. O:4400

SYNONYMS: SPERM OIL; SPERM OIL No. 1; SPERM OIL No. 2

IDENTIFICATION
DOT ID Number: UN 9277; DOT Guide Number: 171
Proper Shipping Name: Oil, n.o.s., flash point not less than 93°C (200°F).

DESCRIPTION: Colorless to pale yellow, oily liquid. Characteristic odor. Floats on the surface of water.

Combustible • Heat or flame may cause explosion • Containers may BLEVE when exposed to fire • Irritating to the skin, eyes, and respiratory tract • Toxic products of combustion may include irritating vapors.

Hazard Classification (based on NFPA-704 Rating System)
Health Hazards (Blue): 0; Flammability (Red): 1; Reactivity (Yellow): 0

EMERGENCY RESPONSE: See Appendix A (171)
Evacuation:
Public safety: Isolate the area of spill or leak for at least 10 to 25 meters (30 to 80 feet) in all directions.
Spill: Increase, in the downwind direction, as necessary, the distance shown under "Public Safety."
Fire: If any large container is involved in fire, isolate for at least 800 meters (½ mile) in all directions; also, consider initial evacuation for 800 meters (½ mile) in all directions.

EXPOSURE
Short-term effects: *SEEK MEDICAL ATTENTION.* **Liquid:** Exposure data not available.
Flush affected areas with plenty of water.

HEALTH HAZARDS
Personal protective equipment (PPE): B-Level PPE. Wear rubber gloves, face shield and protective clothing.

FIRE DATA
Flash point: 500–510°F/260–266°C (oc); (No. 1) 428°F/220°C (cc); (No. 2) 460°F/238°C (cc) (these differ in purity and flash point).
Fire extinguishing agents not to be used: Water or foam may cause frothing.
Autoignition temperature: 586°F/308°C/581°K (No. 1).
Electrical hazard: Class I, Group D.

CHEMICAL REACTIVITY
Binary reactants: Oxidizers, nitric acid.
Reactivity group: 33
Compatibility class: Miscellaneous hydrocarbon mixtures

ENVIRONMENTAL DATA
Food chain concentration potential: Negative; unlikely to accumulate.
Water pollution: Effect of low concentrations on aquatic life is unknown. Fouling to shoreline. May be dangerous if it enters nearby water intakes; notify operators. Notify local health and wildlife officials. **Response to discharge:** Mechanical containment. Should be removed. Chemical and physical treatment.

SHIPPING INFORMATION
Grades of purity: No. 1, No. 2, Winterized (these differ in purity and flash point); **Storage temperature:** Ambient; **Inert atmosphere:** None; **Stability during transport:** Stable.

PHYSICAL AND CHEMICAL PROPERTIES
Physical state @ 59°F/15°C and 1 atm: Liquid.
Boiling point @ 1 atm: Very high.
Specific gravity (water = 1): 0.882 @ 68°F/20°C (liquid).
Liquid water interfacial tension: 5.7 dynes/cm = 0.0057 N/m @ 86°F/30°C/303°K.
Heat of combustion: $-17,900$ Btu/lb = -9943 cal/g = -416.3×10^5 J/kg.
Vapor pressure: (Reid) 0.1 psia; 2.0 mm.

OILS, MISCELLANEOUS: SPINDLE   REC. O:4500

SYNONYMS: ACEITE PARA HUSOS (Spanish); BEARING OIL; HIGH SPEED BEARING OIL; SPINDLE OIL

IDENTIFICATION
DOT ID Number: UN 1270; DOT Guide Number: 128
Proper Shipping Name: Petroleum oil

DESCRIPTION: Light brown, oily liquid. Weak kerosene-like odor. Floats on water surface; insoluble.

Combustible • Containers may BLEVE when exposed to fire • Vapors are heavier than air and will collect and stay in low areas • Vapors in confined areas (e.g., tanks, sewers, buildings) may explode when exposed to fire • Irritating to the skin, eyes, and respiratory tract • Toxic products of combustion may include carbon monoxide.

Hazard Classification (based on NFPA-704 Rating System)
Health Hazards (Blue): 0; Flammability (Red): 2; Reactivity (Yellow): 0

EMERGENCY RESPONSE: See Appendix A (128)
Evacuation:
Public safety: Isolate spill area for at least 25 to 50 meters (80 to 160 feet) in all directions.
Spill: Large spill–Consider initial downwind evacuation for at least 300 meters (1000 feet).
Fire: Isolate for 800 meters (½ mile) in all directions, especially if tank, rail car, or tank truck is involved in fire.

EXPOSURE
Short-term effects: *SEEK MEDICAL ATTENTION.* **Liquid:** Irritating to skin and eyes. Harmful if swallowed. Remove contaminated clothing and shoes. Flush affected areas with plenty of water. *IF IN EYES*, hold eyelids open and flush with plenty of water. *IF SWALLOWED* and victim is *CONSCIOUS AND ABLE TO SWALLOW*, have victim drink 4 to 8 ounces of water. **Do NOT induce vomiting.**

HEALTH HAZARDS
Personal protective equipment (PPE): B-Level PPE. Protective gloves; goggles or face shield.
Exposure limits (TWA unless otherwise noted): ACGIH TLV for oil mist, mineral is 5 mg/m³.
Toxicity by ingestion: Grade 1; LD_{50} = 5 to 15 g/kg.
Vapor (gas) irritant characteristics: Vapors cause a slight smarting of eyes or respiratory system if present in high concentrations.
Liquid or solid irritant characteristics: Minimum hazard. If spilled on clothing and allowed to remain, may cause smarting and reddening of skin.

FIRE DATA
Flash point: 169°F/76°C (cc).
Fire extinguishing agents not to be used: Water may be ineffective.
Autoignition temperature: 478°F/248°C/521°K.
Electrical hazard: Class I, Group D.

CHEMICAL REACTIVITY
Binary reactants: Oxidizers, nitric acid
Reactivity group: 33
Compatibility class: Miscellaneous hydrocarbon mixtures

ENVIRONMENTAL DATA
Food chain concentration potential: Negative; unlikely to accumulate.
Water pollution: Dangerous to aquatic life in high concentrations. Fouling to shoreline. May be dangerous if it enters nearby water intakes; notify operators. Notify local health and wildlife officials.
Response to discharge: Mechanical containment. Should be removed. Chemical and physical treatment.

SHIPPING INFORMATION
Grades of purity: Several grades, all with same hazard assessment; **Storage temperature:** Ambient; **Inert atmosphere:** None; **Venting:** Open (flame arrester); **Stability during transport:** Stable.

PHYSICAL AND CHEMICAL PROPERTIES
Physical state @ 59°F/15°C and 1 atm: Liquid.
Boiling point @ 1 atm: Very high.
Specific gravity (water = 1): 0.881 @ 15°C (liquid).

OILS, MISCELLANEOUS: SPRAY   REC. O:4600

SYNONYMS: PLANT SPRAY OIL; DORMANT OIL; FOLIAGE OIL; KEROSENE, HEAVY; SPRAY OIL

IDENTIFICATION
DOT ID Number: UN 1270; DOT Guide Number: 128
Proper Shipping Name: Petroleum oil

DESCRIPTION: Light brown oily liquid. Kerosene- or motor fuel-like odor.

Flammable • Containers may BLEVE when exposed to fire • Vapors may form explosive mixture with air • Vapors are heavier than air and will collect and stay in low areas • Vapors may travel long distances to ignition sources and flashback • Vapors in confined areas (e.g., tanks, sewers, buildings) may explode when exposed to fire • Irritating to the skin, eyes, and respiratory tract • Toxic products of combustion may include carbon monoxide.

Hazard Classification (based on NFPA-704 Rating System)
Health Hazards (Blue): 0; Flammability (Red): 2; Reactivity (Yellow): 0

EMERGENCY RESPONSE: See Appendix A (128)
Evacuation:
Public safety: Isolate spill area for at least 25 to 50 meters (80 to 160 feet) in all directions.
Spill: Large spill–Consider initial downwind evacuation for at least 300 meters (1000 feet).
Fire: Isolate for 800 meters (½ mile) in all directions, especially if tank, rail car, or tank truck is involved in fire.

EXPOSURE
Short-term effects: *SEEK MEDICAL ATTENTION.* **Liquid:** Irritating to skin and eyes. Harmful if swallowed. Remove contaminated clothing and shoes. Flush affected areas with plenty of water. *IF IN EYES*, hold eyelids open and flush with plenty of water. *IF SWALLOWED* and victim is *CONSCIOUS AND ABLE TO SWALLOW*, have victim drink 4 to 8 ounces of water. **Do NOT induce vomiting.**

HEALTH HAZARDS
Personal protective equipment (PPE): B-Level PPE. Protective gloves; goggles or face shield.
Recommendations for respirator selection: NIOSH as kerosene *1000 ppm:* CCROV [any chemical cartridge respirator with organic vapor cartridge(s)]; or SA (any supplied-air respirator). *2500 ppm:* SA:CF (any supplied-air respirator operated in a continuous-flow mode); or PAPROV [any powered, air-purifying respirator with organic vapor cartridge(s)]. *5000 ppm:* CCRFOV [any chemical cartridge respirator with a full facepiece and organic vapor cartridges(s)]; or GMFOV [any air-purifying, full-facepiece respirator (gas mask) with a chin-style, front- or back-mounted acid gas canister]; or PAPRTOV [any powered, air-purifying respirator with a tight-fitting facepiece and organic vapor cartridges(s)]; or SCBAF (any self-contained breathing apparatus with a full facepiece); or SAF (any supplied-air respirator with a full facepiece). *EMERGENCY OR PLANNED ENTRY INTO UNKNOWN CONCENTRATIONS OR IDLH CONDITIONS:* SCBAF:PD,PP (any self-contained breathing apparatus that has a full facepiece and is operated in a pressure-demand or other positive-pressure mode); or SAF:PD,PP:ASCBA (any supplied-air respirator that has a full facepiece and is operated in a pressure-demand or other positive-pressure mode in combination with an auxiliary self-contained breathing apparatus operated in a pressure-demand or other positive pressure mode). *ESCAPE:* GMFOV [any air-purifying, full-facepiece respirator (gas mask) with a chin-style, front- or back-mounted organic vapor canister]; or SCBAE (any appropriate escape-type, self-contained breathing apparatus).
Exposure limits (TWA unless otherwise noted): NIOSH 100 mg/m^3 as kerosene
Toxicity by ingestion: Grade 2; LD$_{50}$ = 0.5 to 5 g/kg.
Vapor (gas) irritant characteristics: Vapors cause a slight smarting of the eyes or respiratory system if present in high concentrations. The effect is temporary.
Liquid or solid irritant characteristics: Minimum hazard. If spilled on clothing and allowed to remain, may cause smarting and reddening of the skin.
Odor threshold: 1 ppm.

FIRE DATA
Flash point: 140°F/60°C (cc).
Flammable limits in air: LEL: 0.6%; UEL: 4.6%.
Fire extinguishing agents not to be used: Water may be ineffective
Autoignition temperature: 475°F/246°C/519°K.
Electrical hazard: Class I, Group D.
Burning rate: 4 mm/min

CHEMICAL REACTIVITY
Binary reactants: Oxidizers, nitric acid
Reactivity group: 33
Compatibility class: Miscellaneous hydrocarbon mixtures

ENVIRONMENTAL DATA
Food chain concentration potential: Negative; unlikely to accumulate.
Water pollution: Dangerous to aquatic life in high concentrations. Fouling to shoreline. May be dangerous if it enters nearby water intakes; notify operators. Notify local health and wildlife officials.
Response to discharge: Mechanical containment. Should be removed. Chemical and physical treatment.

SHIPPING INFORMATION
Grades of purity: Commercial; **Storage temperature:** Ambient; **Inert atmosphere:** None; **Venting:** Open (flame arrester); **Stability during transport:** Stable.

NAS HAZARD CLASSIFICATION FOR BULK WATER TRANSPORTATION
FIRE: 1/2
HEALTH: Vapor irritant: 1; Liquid or solid irritant: 1; Poisons: 1
WATER POLLUTION: Human toxicity: 1; Aquatic toxicity: 1; Aesthetic effect: 3
REACTIVITY: Other chemicals: 0; Water: 0; Self-reaction: 0

PHYSICAL AND CHEMICAL PROPERTIES
Physical state @ 59°F/15°C and 1 atm: Liquid.
Boiling point @ 1 atm: 590–700°F/310–371°C/583–644°K.
Specific gravity (water = 1): 0.82 @ 15°C (liquid).
Liquid surface tension: (estimate) 25 dynes/cm = 0.025 N/m @ 68°F/20°C.
Liquid water interfacial tension: (estimate) 50 dynes/cm = 0.05 N/m @ 68°F/20°C.
Heat of combustion: –18,540 Btu/lb = –10,300 cal/g = –431.24 x 10^5 J/kg.

OILS, MISCELLANEOUS: TALL   REC. O:4700

SYNONYMS: ACEITE de RESINA (Spanish); LIQUID ROSIN; TALL OIL; TALLEOL; TALL OIL; TALLOL

IDENTIFICATION
CAS Number: 8002-26-4
DOT ID Number: UN 9277; DOT Guide Number: 171
Proper Shipping Name: Oil, n.o.s., flash point not less than 93°C/ (200°F).

DESCRIPTION: Yellow to dark brown, oily liquid. Characteristic acrid odor. Floats on water surface; insoluble.

Combustible • Containers may BLEVE when exposed to fire • Irritating to the skin, eyes, and respiratory tract • Toxic products of combustion may include carbon monoxide.

Hazard Classification (based on NFPA-704 Rating System)
Health Hazards (Blue): 0; Flammability (Red): 1; Reactivity (Yellow): 0

EMERGENCY RESPONSE: See Appendix A (171)
Evacuation:
Public safety: Isolate the area of spill or leak for at least 10 to 25 meters (30 to 80 feet) in all directions.
Spill: Increase, in the downwind direction, as necessary, the distance shown under "Public Safety."
Fire: If any large container is involved in fire, isolate for at least 800 meters (½ mile) in all directions; also, consider initial evacuation for 800 meters (½ mile) in all directions.

EXPOSURE
Short-term effects: *SEEK MEDICAL ATTENTION.* **Liquid:** May cause a mild allergic reaction in some people. Flush affected areas with water.

HEALTH HAZARDS
Personal protective equipment (PPE): B-Level PPE. Butyl rubber gloves, face shield and protective clothing.

FIRE DATA
Flash point: 380°F/193°C (oc).
Fire extinguishing agents not to be used: Water may be ineffective.

CHEMICAL REACTIVITY
Binary reactants: Oxidizers, strong acids
Reactivity group: 34
Compatibility class: Esters.

ENVIRONMENTAL DATA
Food chain concentration potential: Negative; unlikely to accumulate.
Water pollution: Effect of low concentrations on aquatic life is unknown. Fouling to shoreline. May be dangerous if it enters nearby water intakes; notify operators. Notify local health and wildlife officials. **Response to discharge:** Mechanical containment. Should be removed. Chemical and physical treatment.

SHIPPING INFORMATION
Grades of purity: Various grades, which differ primarily in the relative content of fatty acids and rosin acids; **Storage temperature:** Ambient; **Inert atmosphere:** None; **Venting:** Open (flame arrester); **Stability during transport:** Stable.

PHYSICAL AND CHEMICAL PROPERTIES
Physical state @ 59°F/15°C and 1 atm: Liquid.
Boiling point @ 1 atm: Very high.
Specific gravity (water = 1): 0.951 at 16°C (liquid).
Liquid surface tension: 34.3 dynes/cm = 0.0343 N/m at 24°C.
Liquid water interfacial tension: 11 dynes/cm = 0.011 N/m at 22.5°C.
Heat of combustion: (estimate) $-18,000 = -10,000$ cal/g = -420×10^5 J/kg.
Vapor pressure: (Reid) 0.1 psia.

OILS, MISCELLANEOUS: TANNER'S   REC. O:4800

SYNONYMS: ACEITE de CURTIDO (Spanish); SULFATED NEATSFOOT OIL; TANNERS OIL

IDENTIFICATION
CAS Number: 8002-64-0
DOT ID Number: UN 9277; DOT Guide Number: 171
Proper Shipping Name: Oil, n.o.s., flash point not less than 93°C (200°F).

DESCRIPTION: Amber oily liquid. Peculiar, fatty odor. Floats on water surface; insoluble.

Combustible • Heat or flame may cause explosion • Containers may BLEVE when exposed to fire • Vapors are heavier than air and will collect and stay in low areas • Vapors in confined areas (e.g., tanks, sewers, buildings) may explode when exposed to fire • Irritating to the skin, eyes, and respiratory tract • Toxic products of combustion may include CO_2 and carbon monoxide.

Hazard Classification (based on NFPA-704 Rating System)
Health Hazards (Blue): 0; Flammability (Red): 1; Reactivity (Yellow): 0

EMERGENCY RESPONSE: See Appendix A (171)
Evacuation:
Public safety: Isolate the area of spill or leak for at least 10 to 25 meters (30 to 80 feet) in all directions.
Spill: Increase, in the downwind direction, as necessary, the distance shown under "Public Safety."
Fire: If any large container is involved in fire, isolate for at least 800 meters (½ mile) in all directions; also, consider initial evacuation for 800 meters (½ mile) in all directions.

EXPOSURE
Short-term effects: *SEEK MEDICAL ATTENTION.* **Liquid:** Exposure data not available. Flush affected areas with plenty of water.

HEALTH HAZARDS
Personal protective equipment (PPE): B-Level PPE. Rubber gloves, face shield and protective clothing.
Toxicity by ingestion: LD_{50} = > 15 g/kg.

FIRE DATA
Flash point: 430-620°F/221-327°C (approximate).
Fire extinguishing agents not to be used: Water may be ineffective.
Autoignition temperature: 828°F/442°C/715°K (approximate).
Electrical hazard: Class I, Group D.

CHEMICAL REACTIVITY
Binary reactants: Oxidizers, nitric acid, strong reducing agents.
Reactivity group: 33
Compatibility class: Miscellaneous hydrocarbon mixtures

ENVIRONMENTAL DATA
Water pollution: Effect of low concentrations on aquatic life is unknown. Fouling to shoreline. May be dangerous if it enters nearby water intakes; notify operators. Notify local health and wildlife officials. **Response to discharge:** Wear rubber gloves, face shield and protective clothing.

SHIPPING INFORMATION
Storage temperature: Ambient; **Inert atmosphere:** None; **Stability during transport:** Stable.

PHYSICAL AND CHEMICAL PROPERTIES
Physical state @ 59°F/15°C and 1 atm: Liquid.
Boiling point @ 1 atm: Very high.
Melting/Freezing point: 84-106°F/29-41°C/302-314°K (approximate).
Specific gravity (water = 1): 0.910-0.920 @ 68°F/20°C (liquid).
Liquid surface tension: (estimate) 25 dynes/cm = 0.025 N/m @ 68°F/20°C.
Liquid water interfacial tension: (estimate) 50 dynes/cm = 0.05 N/m @ 68°F/20°C.
Heat of combustion: (estimate) $-18,000$ Btu/lb = $-10,000$ cal/g = $-420 \cdot 10^5$ J/kg.
Vapor pressure: Low.

OILS, MISCELLANEOUS: TRANSFORMER   REC. O:4900

SYNONYMS: ACEITE de TRANSFORMADOR (Spanish); ELECTRICAL INSULATING OIL; INSULATING OIL; PETROLEUM INSULATING OIL; TRANSFORMER OIL

IDENTIFICATION
DOT ID Number: UN 1270; DOT Guide Number: 128
Proper Shipping Name: Petroleum oil

DESCRIPTION: Colorless to light brown oily liquid. Motor oil-like odor. Floats on the surface of water.
Note: There are two main types of transformer oils: (1) low viscosity mineral oils with having high chemical and oxidative stability properties (2) synthetics (also known as askarels) including silicone oils, chlorinated aromatics [eg, polychlorinated biphenyls (PCBs), trichlorobenzene], and liquid esters such as dibutyl sebacate.

Combustible • Heat or flame may cause explosion • Containers may BLEVE when exposed to fire • Vapors are heavier than air and will collect and stay in low areas • Vapors in confined areas (e.g., tanks, sewers, buildings) may explode when exposed to fire • Irritating to the skin, eyes, and respiratory tract • Toxic products of combustion may include acrid smoke and irritating and possibly toxic chloride fumes.

Hazard Classification (based on NFPA-704 Rating System)
Health Hazards (Blue): 0; Flammability (Red): 1; Reactivity (Yellow): 0

EMERGENCY RESPONSE: See Appendix A (128)
Evacuation:
Public safety: Isolate spill area for at least 25 to 50 meters (80 to 160 feet) in all directions.
Spill: Large spill–Consider initial downwind evacuation for at least 300 meters (1000 feet).
Fire: Isolate for 800 meters (½ mile) in all directions, especially if tank, rail car, or tank truck is involved in fire.

EXPOSURE
Short-term effects: *SEEK MEDICAL ATTENTION.* **Liquid:** Irritating to skin and eyes. Harmful if swallowed. Remove contaminated clothing.
Flush affected areas with plenty of water. *IF IN EYES*, hold eyelids open and flush with plenty of water. *IF SWALLOWED* and victim is *CONSCIOUS AND ABLE TO SWALLOW*, have victim drink 4 to 8 ounces of water. **Do NOT induce vomiting.**

HEALTH HAZARDS
Personal protective equipment (PPE): B-Level PPE. Protective gloves; goggles or face shield.
Exposure limits (TWA unless otherwise noted): ACGIH TLV for oil mist, mineral is 5 mg/m^3.
Toxicity by ingestion: Grade 1; LD$_{50}$ = 5 to 15 g/kg.
Vapor (gas) irritant characteristics: Vapors cause a slight smarting of the eyes or respiratory system if present in high concentrations. The effect is temporary.
Liquid or solid irritant characteristics: Minimum hazard. If spilled on clothing and allowed to remain, may cause smarting and reddening of the skin.

FIRE DATA
Flash point: 295°F/146°C (oc).
Fire extinguishing agents not to be used: Water may be ineffective.
Electrical hazard: Class I, Group D.

CHEMICAL REACTIVITY
Binary reactants: Oxidizers, nitric acid
Reactivity group: 33
Compatibility class: Miscellaneous hydrocarbon mixtures

ENVIRONMENTAL DATA
Food chain concentration potential: Negative; unlikely to accumulate.
Water pollution: Effect of low concentrations on aquatic life is unknown. Fouling to shoreline. May be dangerous if it enters nearby water intakes; notify operators. Notify local health and wildlife officials. **Response to discharge:** Mechanical containment. Should be removed. Chemical and physical treatment.

SHIPPING INFORMATION
Storage temperature: Ambient; **Inert atmosphere:** None; **Venting:** Open (flame arrester); **Stability during transport:** Stable.

PHYSICAL AND CHEMICAL PROPERTIES
Physical state @ 59°F/15°C and 1 atm: Liquid.
Boiling point @ 1 atm: Very high.
Melting/Freezing point: –75°F/–59°C/214°K.
Specific gravity (water = 1): 0.891 @ 15°C (liquid).
Liquid water interfacial tension: 49 dynes/cm = 0.049 N/m @ 77°F/25°C.

OILS, MISCELLANEOUS: TURBINE   REC. O:5000

SYNONYMS: LUBRICATING OIL, TURBINE; STEAM TURBINE OIL; STEAM TURBINE LUBE OIL; TURBINE OIL

IDENTIFICATION
DOT ID Number: UN 1270; DOT Guide Number: 128
Proper Shipping Name: Petroleum oil

DESCRIPTION: Colorless to light brown liquid. Kerosene or lube oil odor. Floats on the surface of water.

Combustible • Containers may BLEVE when exposed to fire • Irritating to the skin, eyes, and respiratory tract • Toxic products of combustion may include acrid carbon monoxide.

Hazard Classification (based on NFPA-704 Rating System)
Health Hazards (Blue): 0; Flammability (Red): 1; Reactivity (Yellow): 0

EMERGENCY RESPONSE: See Appendix A (128)
Evacuation:
Public safety: Isolate spill area for at least 25 to 50 meters (80 to 160 feet) in all directions.
Spill: Large spill–Consider initial downwind evacuation for at least 300 meters (1000 feet).
Fire: Isolate for 800 meters (½ mile) in all directions, especially if tank, rail car, or tank truck is involved in fire.

EXPOSURE
Short-term effects: *SEEK MEDICAL ATTENTION.* **Vapor:** Irritating to eyes, nose, and throat. Move victim to fresh air. If breathing is difficult, administer oxygen. **Liquid:** Irritating to skin

and eyes. Harmful if swallowed. Remove contaminated clothing and shoes. Flush affected areas with plenty of water. *IF IN EYES*, hold eyelids open and flush with plenty of water. *IF SWALLOWED* and victim is *CONSCIOUS AND ABLE TO SWALLOW*, have victim drink 4 to 8 ounces of water. **Do NOT induce vomiting.**

HEALTH HAZARDS
Personal protective equipment (PPE): B-Level PPE. Goggles or face shield; rubber gloves.
Exposure limits (TWA unless otherwise noted): ACGIH TLV for oil mist, mineral is 5 mg/m^3.
Toxicity by ingestion: Grade 0; LD_{50} = > 15 g/kg (rat).
Liquid or solid irritant characteristics: Mild eye and skin irritant.

FIRE DATA
Flash point: 390–485°F/199–252°C (oc).
Fire extinguishing agents not to be used: Water or foam may cause frothing; water may be ineffective.
Autoignition temperature: 700°F/371°C/644°K.
Electrical hazard: Class I, Group D.
Burning rate: (approximate) 4 mm/min

CHEMICAL REACTIVITY
Binary reactants: Oxidizers, nitric acid
Reactivity group: 33
Compatibility class: Miscellaneous hydrocarbon mixture

ENVIRONMENTAL DATA
Food chain concentration potential: Negative; unlikely to accumulate.
Water pollution: Effect of low concentrations on aquatic life is unknown. Fouling to shoreline. May be dangerous if it enters nearby water intakes; notify operators. Notify local health and wildlife officials. **Response to discharge:** Mechanical containment. Should be removed. Chemical and physical treatment.

SHIPPING INFORMATION
Grades of purity: Solvent refined paraffinic oils; 98.5+%. Grades vary in viscosity and flash point; **Storage temperature:** Ambient; **Inert atmosphere:** None; **Venting:** Open (flame arrester); **Stability during transport:** Stable.

PHYSICAL AND CHEMICAL PROPERTIES
Physical state @ 59°F/15°C and 1 atm: Liquid.
Specific gravity (water = 1): 0.87 @ 68°F/20°C (liquid).
Liquid surface tension: 25 dynes/cm = 0.025 N/m @ 68°F/20°C.
Liquid water interfacial tension: 50 dynes/cm = 0.050 N/m @ 68°F/20°C.
Heat of combustion: (estimate) –17,600 Btu/lb = –9800 cal/g = –410 x 10^5 J/kg.
Vapor pressure: Very low.

OLEIC ACID   REC. O:5100

SYNONYMS: ACIDO OLEICO (Spanish); EMERSOL 210; GLYCON RO; GLYCON WO; ELAIC ACID; *cis*-8-HEPTADECYLENECARBOXYLIC ACID; HY-PHY 1055; HY-PHI 2066; *cis*-9-OCTADECENOIC ACID; 9,10-OCTADECENOIC ACID K-52; METAUPON; NEO-FAT 90-04; OLEIC ACID, DISTILLED; OLEINIC ACID; PAMOLYN; RED OIL; WECOLINE OO; WOCHEM-320

IDENTIFICATION
CAS Number: 112-80-1
Formula: $C_{18}H_{34}O_2$; $CH_3(CH_2)_7CH=CH(CH_2)_7COOH$

DESCRIPTION: Colorless to pale yellow liquid. Turns red on exposure to air. Mild odor that becomes rancid on standing. Floats on water surface; insoluble. Becomes a solid at 39°F/4°C.

Combustible • Heat or flame may cause explosion • Containers may BLEVE when exposed to fire • Vapors in confined areas (e.g., tanks, sewers, buildings) may explode when exposed to fire • Irritating to the skin, eyes, and respiratory tract • Toxic products of combustion may include carbon monoxide.

Hazard Classification (based on NFPA-704 Rating System)
Health Hazards (Blue): 0; Flammability (Red): 1; Reactivity (Yellow): 0

EMERGENCY RESPONSE: See Appendix A (171)
Evacuation:
Public safety: Isolate the area of spill or leak for at least 10 to 25 meters (30 to 80 feet) in all directions.
Spill: Increase, in the downwind direction, as necessary, the distance shown under "Public Safety."
Fire: If any large container is involved in fire, isolate for at least 800 meters (½ mile) in all directions; also, consider initial evacuation for 800 meters (½ mile) in all directions.

EXPOSURE
Short-term effects: *SEEK MEDICAL ATTENTION.* **Liquid:** Irritating to skin and eyes. *IF SWALLOWED*, will cause nausea. Remove contaminated clothing and shoes. Flush affected areas with plenty of water. *IF IN EYES*, hold eyelids open and flush with plenty of water. *IF SWALLOWED* and victim is *CONSCIOUS AND ABLE TO SWALLOW*, have victim drink 4 to 8 ounces of water. *IF SWALLOWED* and victim is *UNCONSCIOUS OR HAVING CONVULSIONS,* do nothing except keep victim warm.

HEALTH HAZARDS
Personal protective equipment (PPE): B-Level PPE. Impervious chemical gloves; goggles or face shield; impervious apron. Chemical protective material(s) reported to have good to excellent resistance: nitrile. Also, butyl rubber is generally suitable for carbooxylic acid compounds.
Toxicity by ingestion: Grade 1; LD_{50} = > 15 g/kg.
Long-term health effects: Mutation and carcinogen data. May cause hemolytic anemia.
Vapor (gas) irritant characteristics: May cause respiratory irritation and related problems.
Liquid or solid irritant characteristics: Eye and skin irritant.

FIRE DATA
Flash point: 372°F/189°C (oc).
Fire extinguishing agents not to be used: Water or foam may cause frothing.
Autoignition temperature: 685°F/363°C/636°K.

CHEMICAL REACTIVITY
Binary reactants: Strong oxidants; perchloric acid. Corrodes aluminum

ENVIRONMENTAL DATA
Food chain concentration potential: Negative; unlikely to accumulate.

Water pollution: Effect of low concentrations on aquatic life is unknown. Fouling to shoreline. May be dangerous if it enters nearby water intakes; notify operators. Notify local health and wildlife officials. **Response to discharge:** Mechanical containment. Should be removed. Chemical and physical treatment.

SHIPPING INFORMATION
Grades of purity: Commercial, 79–83%; **Storage temperature:** Ambient; **Inert atmosphere:** None; **Venting:** Open (flame arrester); **Stability during transport:** Stable.

PHYSICAL AND CHEMICAL PROPERTIES
Physical state @ 59°F/15°C and 1 atm: Liquid.
Molecular weight: 277 (avg.).
Boiling point @ 1 atm: 536°F/280°C/553°K.
Melting/Freezing point: 39°F/4°C/277°K.
Specific gravity (water = 1): 0.89 @ 77°F/25°C (liquid).
Liquid surface tension: 32.8 dynes/cm = 0.0328 N/m @ 68°F/20°C.
Liquid water interfacial tension: 15.59 dynes/cm = 0.01559 N/m @ 68°F/20°C.
Relative vapor density (air = 1): 0.9
Latent heat of vaporization: 103 Btu/lb = 57 cal/g = 2.4×10^5 J/kg.
Vapor pressure: 1 mm.

OLEIC ACID, POTASSIUM SALT   REC. O:5200

SYNONYMS: POTASSIUM *cis*-9-OCTADECENOIC ACID; POTASSIUM OLEATE

IDENTIFICATION
CAS Number: 143-18-0
Formula: $C_{18}H_{34}O_2 \cdot K$; $C_{17}H_{33}COOK$
DOT ID Number: UN 1993; DOT Guide Number: 128
Proper Shipping Name: Flammable liquids, n.o.s.

DESCRIPTION: Brown solid or liquid. Faint, soapy odor. Sinks and mixes slowly with water.

Flammable • Containers may BLEVE when exposed to fire • Vapors are heavier than air and will collect and stay in low areas • Vapors in confined areas (e.g., tanks, sewers, buildings) may explode when exposed to fire • Irritating to the skin, eyes, and respiratory tract • Toxic products of combustion may include oxides of potassium.

Hazard Classification (based on NFPA-704 Rating System)
Health Hazards (Blue): 0; Flammability (Red): 2; Reactivity (Yellow): 0

EMERGENCY RESPONSE: See Appendix A (128)
Evacuation:
Public safety: Isolate spill area for at least 25 to 50 meters (80 to 160 feet) in all directions.
Spill: Large spill–Consider initial downwind evacuation for at least 300 meters (1000 feet).
Fire: Isolate for 800 meters (½ mile) in all directions, especially if tank, rail car, or tank truck is involved in fire.

EXPOSURE
Short-term effects: *SEEK MEDICAL ATTENTION.* **Liquid or solid:** Irritating to skin and eyes. IF SWALLOWED, will cause nausea and vomiting. Remove contaminated clothing and shoes. Flush affected areas with plenty of water. *IF IN EYES*, hold eyelids open and flush with plenty of water. *IF SWALLOWED* and victim is *CONSCIOUS AND ABLE TO SWALLOW*, have victim drink 4 to 8 ounces of water. *IF SWALLOWED* and victim is *UNCONSCIOUS OR HAVING CONVULSIONS,* do nothing except keep victim warm.

HEALTH HAZARDS
Personal protective equipment (PPE): B-Level PPE. Chemical goggles and gloves. Chemical protective material(s) reported to have good to excellent resistance: nitrile.
Liquid or solid irritant characteristics: Eye and respiratory tract irritant.

FIRE DATA
Flash point: 140°F/60°C (cc).
Fire extinguishing agents not to be used: Water may be ineffective.

CHEMICAL REACTIVITY
Binary reactants: Strong oxidizers

ENVIRONMENTAL DATA
Food chain concentration potential: Negative; unlikely to accumulate.
Water pollution: Effect of low concentrations on aquatic life is unknown. May be dangerous if it enters nearby water intakes; notify operators. Notify local health and wildlife officials. **Response to discharge:** Disperse and flush.

SHIPPING INFORMATION
Grades of purity: 19% solution in water; **Storage temperature:** Ambient; **Inert atmosphere:** None; **Venting:** Open; **Stability during transport:** Stable.

PHYSICAL AND CHEMICAL PROPERTIES
Physical state @ 59°F/15°C and 1 atm: Solid or Liquid.
Molecular weight: 320 (solute only).
Boiling point @ 1 atm: (decomposes).
Melting/Freezing point: 455–464°F/235–240°C/508–513°K.
Specific gravity (water = 1): More than 1.1 @ 68°F/20°C (solid or liquid).

OLEIC ACID, SODIUM SALT   REC. O:5300

SYNONYMS: EUNATROL; SODIUM OLEATE

IDENTIFICATION
CAS Number: 143-19-1
Formula: $C_{18}H_{33}O_2 \cdot Na$; $C_{17}H_{33}COONa$

DESCRIPTION: White crystalline solid or powder. Slight tallow-like odor. Sinks and mixes slowly with water; forms alkaline solution.

Combustible solid • Dust cloud may explode if ignited in an enclosed area • Containers may BLEVE when exposed to fire • Irritating to the skin, eyes, and respiratory tract • Toxic products of combustion may include oxides of sodium.

Hazard Classification (based on NFPA-704 Rating System)
Health Hazards (Blue): 0; Flammability (Red): 1; Reactivity (Yellow): 0

872 Oleum

EMERGENCY RESPONSE: See Appendix A (171)
Evacuation:
Public safety: Isolate the area of spill or leak for at least 10 to 25 meters (30 to 80 feet) in all directions.
Spill: Increase, in the downwind direction, as necessary, the distance shown under "Public Safety."
Fire: If any large container is involved in fire, isolate for at least 800 meters (½ mile) in all directions; also, consider initial evacuation for 800 meters (½ mile) in all directions.

EXPOSURE
Short-term effects: *SEEK MEDICAL ATTENTION.* **Solid:** Irritating to skin and eyes. *IF SWALLOWED*, will cause nausea and vomiting. Remove contaminated clothing and shoes. Flush affected areas with plenty of water. *IF IN EYES*, hold eyelids open and flush with plenty of water. *IF SWALLOWED* and victim is *CONSCIOUS AND ABLE TO SWALLOW*, have victim drink 4 to 8 ounces of water. *IF SWALLOWED* and victim is *UNCONSCIOUS OR HAVING CONVULSIONS*, do nothing except keep victim warm.

HEALTH HAZARDS
Personal protective equipment (PPE): B-Level PPE. Dust mask and gloves. Chemical protective material(s) reported to have good to excellent resistance: nitrile.
Liquid or solid irritant characteristics: Eye and respiratory tract irritant.

CHEMICAL REACTIVITY
Binary reactants: Strong oxidizers

ENVIRONMENTAL DATA
Food chain concentration potential: Negative; unlikely to accumulate.
Water pollution: Effect of low concentrations on aquatic life is unknown. May be dangerous if it enters nearby water intakes; notify operators. Notify local health and wildlife officials.
Response to discharge: Disperse and flush.

SHIPPING INFORMATION
Grades of purity: Commercial; **Storage temperature:** Ambient; **Inert atmosphere:** None; **Venting:** Open; **Stability during transport:** Stable.

PHYSICAL AND CHEMICAL PROPERTIES
Physical state @ 59°F/15°C and 1 atm: Solid.
Molecular weight: 304 (approximate)
Melting/Freezing point: 450–455°F/232–235°C/505–508°K.
Specific gravity (water = 1): More than 1.1 @ 68°F/20°C (solid).

OLEUM   REC. O:5400

SYNONYMS: DISULFURIC ACID; DITHIONIC ACID; EEC No. 016-019-00-2; FUMING SULFURIC ACID; PYROSULPHURIC ACID; SULFURIC ACID, FUMING; SULFURIC ACID MIXED WITH SULFUR TRIOXIDE

IDENTIFICATION
CAS Number: 8014-95-7
Formula: $H_2SO_4 \cdot O_3S$
DOT ID Number: NA 1831; DOT Guide Number: 137
Proper Shipping Name: Oleum; Sulfuric acid, fuming, with not less than 30% free sulfur trioxide; Sulfuric acid, fuming, with less than 30% free sulfur trioxide
Reportable Quantity (RQ): **(CERCLA)** 1000 lb/454 kg

DESCRIPTION: A solution of sulfur trioxide in concentrated sulfuric acid. Colorless to brown (depending on purity), oily liquid. Fuming in moist air. Sharp, choking odor. Mixes and reacts with water evolving heat and corrosive vapor.

Poison! • Corrosive • Breathing the vapor can kill you; skin or eye contact causes severe burns, impaired vision, or blindness; effects of contact or inhalation may be delayed • Firefighting gear (including SCBA) does not provide adequate protection. If exposure occurs, remove and isolate gear immediately and thoroughly decontaminate personnel • Containers may BLEVE when exposed to fire • Vapors in confined areas (e.g., tanks, sewers, buildings) may explode when exposed to fire • Strong oxidizer which may react spontaneously with low flash point organics or reducing agents. Heat forms oxygen; will increase the activity of an existing fire • May cause fire or explosion on contact with combustibles (wood, paper, oil, clothing, etc.) • Attacks most metals producing explosive hydrogen gas • Toxic products of combustion may include sulfur oxides.

Hazard Classification (based on NFPA-704 Rating System)
Health Hazards (Blue): 3; Flammability (Red): 0; Reactivity (Yellow): 2; Special Notice (White): OXY

EMERGENCY RESPONSE: See Appendix A (137)
Evacuation:
Public safety: Isolate the area of spill or leak for at least 50 to 100 meters (160 to 330 feet) in all directions.
Spill: Increase, in the downwind direction, as necessary, the distance shown under "Public Safety." For Sulfuric acid, fuming greater than or equal to 30% free sulfur trioxide. *FOR SPILL IN WATER*: Small spill–First: Isolate in all directions 60 meters (200 feet); Then: Protect persons downwind, DAY: 0.3 km (0.2 mile); NIGHT: 1.1 km (0.7 mile). Large spill–First: Isolate in all directions 305 meters (1000 feet); Then: Protect persons downwind, DAY: 2.1 km (1.3 miles); NIGHT: 5.6 km (3.5 mile).
Fire: If any large container is involved in fire, isolate for at least 800 meters (½ mile) in all directions; also, consider initial evacuation for 800 meters (½ mile) in all directions.

EXPOSURE
Short-term effects: *SEEK MEDICAL ATTENTION.* **Mist:** Irritating to eyes, nose, and throat. *IF INHALED*, will, will cause coughing or difficult breathing. May cause lung edema; physical exertion will aggravate this condition. IF breathing is difficult, administer oxygen. Move to fresh air. *IF BREATHING HAS STOPPED*, give artificial respiration; *avoid mouth-to-mouth resuscitation; use bag/mask apparatus.* **Liquid:** Will burn skin and eyes. Harmful if swallowed. Remove contaminated clothing and shoes. Flush affected areas with plenty of water. *IF IN EYES*, hold eyelids open and flush with plenty of water. *IF SWALLOWED* and victim is *CONSCIOUS AND ABLE TO SWALLOW*, have victim drink 4 to 8 ounces of water. **Do NOT induce vomiting.**
Note to physician or authorized medical personnel: Medical observation is recommended for 24 to 48 hours after breathing overexposure, as pulmonary edema may be delayed. As first aid for pulmonary edema, consider administering a corticosteroid spray. Cigarette smoking may exacerbate pulmonary injury and should be discouraged for at least 72 hours following exposure.

HEALTH HAZARDS

Personal protective equipment (PPE): A-Level PPE. Respirator approved for acid mists; rubber gloves; splash proof goggles; eyewash fountain and safety shower; rubber footwear; face shield. Chemical protective material(s) reported to have good to excellent resistance: Teflon®.

Recommendations for respirator selection: NIOSH/OSHA as sulfuric acid

15 mg/m³: SA:CF* (any supplied-air respirator operated in a continuous-flow mode); or PAPRAGHiE* (any powered, air-purifying respirator with acid gas cartridge(s) in combination with a high-efficiency particulate filter). *50 mg/m³:* CCRFAGHiE [any chemical cartridge respirator with a full facepiece and acid gas cartridge(s) in combination with a high-efficiency particulate filter]; or GMFAGHiE [any air-purifying, full-facepiece respirator (gas mask) with a chin-style, front- or back-mounted acid gas canister having a high-efficiency particulate filter]; or SCBAF (any self-contained breathing apparatus with a full facepiece); or SAF (any supplied-air respirator with a full facepiece). *EMERGENCY OR PLANNED ENTRY INTO UNKNOWN CONCENTRATIONS OR IDLH CONDITIONS:* SCBAF:PD,PP (any self-contained breathing apparatus that has a full facepiece and is operated in a pressure-demand or other positive-pressure mode); or SAF:PD,PP:ASCBA (any supplied-air respirator that has a full facepiece and is operated in a pressure-demand or other positive-pressure mode in combination with an auxiliary self-contained breathing apparatus operated in a pressure-demand or other positive pressure mode). *ESCAPE:* GMFAGHiE [any air-purifying, full-facepiece respirator (gas mask) with a chin-style, front- or back-mounted acid gas canister having a high-efficiency particulate filter]; or SCBAE (any appropriate escape-type, self-contained breathing apparatus). *Note:* Substance causes eye irritation or damage; eye protection needed.

Exposure limits (TWA unless otherwise noted): ACGIH TLV 1 mg/m³; NIOSH/OSHA 1 mg/m³ as sulfuric acid

Short-term exposure limits (15-minute TWA): ACGIH STEL 3 mg/m³ as sulfuric acid.

Toxicity by ingestion: Severe burns of mouth and stomach.

Vapor (gas) irritant characteristics: Vapors cause severe irritation of eye and throat and can cause eye and lung injury. Lung edema may occur. They cannot be tolerated even at low concentrations.

Liquid or solid irritant characteristics: Severe eye and skin irritant. Causes second- and third-degree burns on short contact; very injurious to the eyes.

Odor threshold: 1 mg/m³; 0.15 ppm as sulfuric acid

IDLH value: 15 mg/m³ as sulfuric acid

FIRE DATA

Fire extinguishing agents not to be used: Use water to cool closed containers *ONLY*.

Electrical hazard: Class I, Group B (based upon possible hydrogen gas generation should leak or spill occur.).

CHEMICAL REACTIVITY

Reactivity with water: Vigorous reaction with water; spatters.

Binary reactants: Strong oxidizer and acid. May react with cast iron with explosive violence. Attacks many metals, releasing flammable hydrogen gas. Capable of igniting finely divided combustible. Reacts with air and gives off corrosive, choking fumes that can are heavier than air and can travel along the ground material on contact. Extremely hazardous in contact with many materials including chlorates, carbides, fulminates, water and organic materials.

Neutralizing agents for acids and caustics: Cautious dilution with water, with protection against violent spattering. Diluted acid may be neutralized with lime or soda ash.

Reactivity group: Unassigned. Compatibility assistance available from F-MTH-1 (Telephone: 202-267-1577) or National Response Center (Telephone 800-424-8802).

Compatibility class: Special class

ENVIRONMENTAL DATA

Food chain concentration potential: Unlikely to accumulate.

Water pollution: Harmful to aquatic life in very low concentrations. Fouling to shoreline. May be dangerous if it enters nearby water intakes; notify operators. Notify local health and wildlife officials. **Response to discharge:** Issue warning–corrosive. Restrict access. Chemical and physical treatment. Disperse and flush.

SHIPPING INFORMATION

Grades of purity: 20% (104.5% sulfuric acid) to 65% (114.6% sulfuric acid); **Storage temperature:** Ambient; **Inert atmosphere:** None; **Venting:** Open; **Stability during transport:** Stable.

NAS HAZARD CLASSIFICATION FOR BULK WATER TRANSPORTATION

FIRE: 0
HEALTH: Vapor irritant: 4; Liquid or solid irritant: 4; Poisons: 3
WATER POLLUTION: Human toxicity: 2; Aquatic toxicity: 3; Aesthetic effect: 2
REACTIVITY: Other chemicals: 4; Water: 3; Self-reaction: 0

PHYSICAL AND CHEMICAL PROPERTIES

Physical state @ 59°F/15°C and 1 atm: Liquid.
Molecular weight: 90.1 (SO_3); 98.1 (H_2SO_4).
Boiling point @ 1 atm: (65% free SO_3) 131°F/55°C/328°K; (30% free SO_3) 241°F/116°C/389°K; (20% free SO_3) 280°F/138°C/411°K.
Melting/Freezing point: (65% free SO_3) 39°F/4°C/277°K; (30% free SO_3) 70°F/21°C/294°K; (20% free SO_3) 36°F/2°C/275°K.
Specific gravity (water = 1): 1.91–1.97 at 59°F/15°C (liquid).
Relative vapor density (air = 1): 2.76
Vapor pressure: (Reid) Low.

OXALIC ACID   REC. O:5500

SYNONYMS: ACIDE OXALIQUE (French); ACIDO OXALICO (Spanish); EEC No. 607-006-00-8; ETHANEDIOIC ACID; OXALSAEURE (German); OXALIC ACID DIHYDRATE; ETHANE DIOIC ACID

IDENTIFICATION

CAS Number: 144-62-7
Formula: $C_2H_2O_4$
DOT ID Number: UN 2449; DOT Guide Number: 154
Proper Shipping Name: Oxalates, water soluble

DESCRIPTION: Colorless crystalline solid or white powder. Odorless. Sinks and mixes with water forming a medium-strong acid.

Poison!• Combustible • Corrosive to the eyes, skin, and respiratory tract; inhalation symptoms may be delayed • Containers may BLEVE when exposed to fire • Dust may explode when exposed to fire • Toxic products of combustion may include carbon monoxide.

Hazard Classification (based on NFPA-704 Rating System)
Health Hazards (Blue): 3; Flammability (Red): 1; Reactivity (Yellow): 0

EMERGENCY RESPONSE: See Appendix A (154)
Evacuation:
Public safety: Isolate the area of spill or leak for at least 25 to 50 meters (80 to 160 feet) in all directions.
Spill: Increase, in the downwind direction, as necessary, the distance shown under "Public Safety."
Fire: If tank, rail car, or tank truck is involved in fire, isolate for at least 800 meters (½ mile) in all directions; also, consider initial evacuation for 800 meters (½ mile) in all directions.

EXPOSURE
Short-term effects: *SEEK MEDICAL ATTENTION.* **Dust:** Will burn eyes, nose, and throat. *IF INHALED*, will, will cause difficult breathing. Move to fresh air. *IF BREATHING HAS STOPPED,* give artificial respiration; *avoid mouth-to-mouth resuscitation; use bag/mask apparatus*. IF breathing is difficult, administer oxygen. **Solid:** Will burn skin and eyes. *IF SWALLOWED*, will cause nausea or loss of consciousness. Remove contaminated clothing and shoes. Flush affected areas with plenty of water. *IF IN EYES*, hold eyelids open and flush with plenty of water. *IF SWALLOWED* and victim is *CONSCIOUS AND ABLE TO SWALLOW*, have victim drink 4 to 8 ounces of water. **Do NOT induce vomiting**.
Note to physician or authorized medical personnel: Medical observation is recommended for 24 to 48 hours after breathing overexposure, as pulmonary edema may be delayed. As first aid for pulmonary edema, consider administering a corticosteroid spray. Cigarette smoking may exacerbate pulmonary injury and should be discouraged for at least 72 hours following exposure.

HEALTH HAZARDS
Personal protective equipment (PPE): B-Level PPE. OSHA Table Z-1-A air contaminant. Respirator for dust or mist protection; gloves; chemical safety glasses; rubbers over leather or rubber safety shoes; apron or impervious clothing for splash protection. Chemical protective material(s) reported to have good to excellent resistance: butyl rubber, natural rubber, neoprene, nitrile, PVC, nitrile/PVC, polyethylene, styrene-butadiene, Viton®.
Recommendations for respirator selection: NIOSH/OSHA
25 mg/m³: SA:CF* (any supplied-air respirator operated in a continuous-flow mode); or PAPRDM **If not present as a fume* (any powered, air-purifying respirator with a dust and mist filter). *50 mg/m³:* HiEF (any air-purifying, full-facepiece respirator with a high-efficiency particulate filter); or SCBAF (any self-contained breathing apparatus with a full facepiece); or SAF (any supplied-air respirator with a full facepiece). *500 mg/m³:* SAF:PD,PP (any supplied-air respirator that has a full facepiece and is operated in a pressure-demand or other positive-pressure mode). *EMERGENCY OR PLANNED ENTRY INTO UNKNOWN CONCENTRATIONS OR IDLH CONDITIONS:* SCBAF:PD,PP (any self-contained breathing apparatus that has a full facepiece and is operated in a pressure-demand or other positive-pressure mode); or SAF:PD,PP:ASCBA (any supplied-air respirator that has a full facepiece and is operated in a pressure-demand or other positive-pressure mode in combination with an auxiliary self-contained breathing apparatus operated in a pressure-demand or other positive pressure mode). *ESCAPE:* HiEF (any air-purifying, full-facepiece respirator with a high-efficiency particulate filter); or SCBAE (any appropriate escape-type, self-contained breathing apparatus). **Note:* Substance reported to cause eye irritation or damage; may require eye protection.

Exposure limits (TWA unless otherwise noted): ACGIH TLV 1 mg/m³; OSHA 1 mg/m³; NIOSH 1 mg/m3
Short-term exposure limits (15-minute TWA): ACGIH STEL 2 mg/m³; NIOSH STEL 2 mg/m³.
Toxicity by ingestion: Grade 3; LD_{50} = 50 to 500 mg/kg.
Long-term health effects: Kidney, liver and skin disorders. Gangrene has developed from repeated contact.
Vapor (gas) irritant characteristics: Eye and respiratory tract irritant.
Liquid or solid irritant characteristics: Severe eye and skin irritant. Causes second-and third-degree burns on short contact and is very injurious to the eyes.
IDLH value: 500 mg/m³.

FIRE DATA
Fire extinguishing agents not to be used: Water may not be effective; may cause foaming.

CHEMICAL REACTIVITY
Reactivity with water: Very slow reaction to form corrosive hydrofluoric acid.
Binary reactants: Fire and explosion hazard; reacts violently with strong oxidants, bases, chlorites, silver compounds. Combustible materials, chlorine, bromine, iodine, platinum, metal oxides, moist air, hydrogen, sulfide, hydrocarbons, water.
Neutralizing agents for acids and caustics: Lime or soda ash

ENVIRONMENTAL DATA
Food chain concentration potential: Log K_{ow} = –0.72 9 (estimate). Negative; unlikely to accumulate.
Water pollution: Dangerous to aquatic life in high concentrations. May be dangerous if it enters nearby water intakes; notify operators. Notify local health and wildlife officials. **Response to discharge:** Issue warning–poison. Should be removed. Chemical and physical treatment.

SHIPPING INFORMATION
Grades of purity: Technical: 99.8%; **Storage temperature:** Ambient; **Inert atmosphere:** None; **Venting:** Open; **Stability during transport:** Stable.

PHYSICAL AND CHEMICAL PROPERTIES
Physical state @ 59°F/15°C and 1 atm: Solid.
Molecular weight: 126.07
Boiling point @ 1 atm: (sublimes).
Melting/Freezing point: 373°F/190°C/463°K.
Specific gravity (water = 1): 1.90 @ 15°C (solid).
Vapor pressure: Less than 0.001 mm.

OXYGEN   **REC. O:5600**

SYNONYMS: LIQUID OXYGEN; LOX; OXIGENO (Spanish); OXYGEN, Liquid.

IDENTIFICATION
CAS Number: 7782-44-7
Formula: O_2
DOT ID Number: UN 1072; UN 1073; DOT Guide Number: 122
Proper Shipping Name: Oxygen, refrigerated liquid (cryogenic liquid) (UN 1073); Oxygen, compressed (UN 1072).

DESCRIPTION: Colorless gas or light blue liquid. Shipped and stored as a compressed gas or cryogenic liquid. Odorless. Sinks and boils in water.

At temperatures above 180°F/82°C LOX will flash into vapor. If unconfined the vapor will occupy about 860 times the volume of liquid • Vapors from liquefied gas are initially heavier than air and spread along ground • Containers may BLEVE when exposed to fire • Strong oxidizer which may react spontaneously with low flash point organics, fuels, or reducing agents. Will increase the activity of a fire • May cause fire or explosion on contact with combustibles (wood, paper, oil, clothing, etc.) • Contact with liquid may cause frostbite.

Hazard Classification (based on NFPA-704 Rating System)
Health Hazards (Blue): 3; Flammability (Red): 0; Reactivity (Yellow): 0; Special Notice (White): OXY

EMERGENCY RESPONSE: See Appendix A (122)
Evacuation:
Public safety: Isolate spill area for at least 25 to 50 meters (80 to 160 feet) in all directions.
Spill: Large spill –Consider initial downwind evacuation for at least 500 meters (⅓ mile).
Fire: Isolate for 800 meters (½ mile) in all directions, especially if tank, rail car, or tank truck is involved in fire.

EXPOSURE
Short-term effects: *SEEK MEDICAL ATTENTION.* **Vapor:** IF inhaled, will cause dizziness, or difficult breathing. **Liquid:** Will cause frostbite. Flush affected areas with plenty of water. *DO NOT RUB AFFECTED AREAS.*

HEALTH HAZARDS
Personal protective equipment (PPE): B-Level PPE. Safety goggles or face shield; insulated gloves; long sleeves; trousers worn outside boots or over high-top shoes to shed spilled liquid. Wear thermal protective clothing.
Liquid or solid irritant characteristics: Rapid evaporation of the liquid can cause frostbite.

FIRE DATA
Flash point: Nonflammable, but will support fire.
Behavior in fire: Mixtures of liquid oxygen and any fuel, or other combustible material, will burst into flame or explode if exposed to a spark source.

CHEMICAL REACTIVITY
Reactivity with water: Heat of water will vigorously vaporize liquid oxygen.
Binary reactants: Avoid organic and combustible materials, such as oil, grease, coal dust, etc. If ignited, such mixtures can explode. The low temperature may cause brittleness in some materials.
Reactivity group: Unassigned. LOX causes all combustible materials to burn vigorously. A spark is not always needed to ignite such a mixture.

ENVIRONMENTAL DATA
Food chain concentration potential: Negative; unlikely to accumulate.
Water pollution: Not harmful to aquatic life. **Response to discharge:** Restrict access.

SHIPPING INFORMATION
Grades of purity: 99.5+%; **Storage temperature:** –183°C; **Inert atmosphere:** None; **Venting:** Safety relief; **Stability during transport:** Stable.

PHYSICAL AND CHEMICAL PROPERTIES
Physical state @ 59°F/15°C and 1 atm: Gas.
Molecular weight: 32.0
Boiling point @ 1 atm: –297°F/–183°C/90°K.
Melting/Freezing point: –361°F/–218°C/55°K.
Critical temperature: –180°F/–118°C/155°K.
Critical pressure: 738 psia = 50.1 atm = 5.09 MN/m^2.
Specific gravity (water = 1): 1.14 at –297°F/–183°C/90°K.
Liquid surface tension: 13.47 dynes/cm = 0.01347 N/m at –183°C.
Relative vapor density (air = 1): 1.1
Ratio of specific heats of vapor (gas): 1.3962
Latent heat of vaporization: 91.6 Btu/lb = 50.9 cal/g = 2.13 x 10^5 J/kg.
Vapor pressure: High. 1 atm @ –297°F/–183°C/90°K; 2 atm @ –176°C; 20 atm @ –220°F/–140°C/134°K.

PARAFFIN   REC. P:0100

SYNONYMS: HARD WAX; PARAFFIN WAX; PARAFINA (Spanish); PETROLEUM WAX

IDENTIFICATION
CAS Number: 8002-74-2 (oil); 63449-39-8 (chlorinated); 108171-27-3 (43% chlorine); 108171-26-2 (60% chlorine)
Formula: $C_{(12-18)}H_{(26-38)}$

DESCRIPTION: Yellow to white thick liquid (heated) to hard solid. Very weak odor. Floats on water and solidifies.

Combustible; can ignite on hot surfaces at @ 392°F/200°C • Heat or flame may cause explosion • Containers may BLEVE when exposed to fire • Vapors in confined areas (e.g., tanks, sewers, buildings) may explode when exposed to fire • Irritating to the skin, eyes, and respiratory tract • Combustion products include smoke, irritating vapors, and possibly toxic hydrogen chloride from chlorinated paraffins.

Hazard Classification (based on NFPA-704 Rating System)
Health Hazards (Blue): 0; Flammability (Red): 1; Reactivity (Yellow): 0

EMERGENCY RESPONSE: See Appendix A (171)
Evacuation:
Public safety: Isolate the area of spill or leak for at least 10 to 25 meters (30 to 80 feet) in all directions.
Spill: Increase, in the downwind direction, as necessary, the distance shown under "Public Safety."
Fire: If any large container is involved in fire, isolate for at least 800 meters (½ mile) in all directions; also, consider initial evacuation for 800 meters (½ mile) in all directions.

EXPOSURE
Short-term effects: Liquid: Will burn skin and eyes. Remove wax. Flush affected areas with plenty of water.

HEALTH HAZARDS
Personal protective equipment (PPE): B-Level PPE. OSHA Table Z-1-A air contaminant. Goggles or face shield; protective gloves and clothing for hot liquid wax.
Exposure limits (TWA unless otherwise noted): ACGIH TLV 2 mg/m^3 (fume); NIOSH REL 2 mg/m^3.
Toxicity by ingestion: Grade 1; LD$_{50}$ = 5 to 15 g/kg.

Long-term health effects: Possible lung damage. Paraffin waxes, especially chlorinated, may contain carcinogens.
Liquid or solid irritant characteristics: None

FIRE DATA
Flash point: 390°F/199°C (cc); 380-465°F/193-241°C (oc).
Flammable limits in air: LEL: 0.5%; UEL: 8.0%.
Fire extinguishing agents not to be used: Water or foam may cause frothing.
Autoignition temperature: 390°F/199°C/472°K.
Electrical hazard: Class I, Group D.

CHEMICAL REACTIVITY
Binary reactants: Strong oxidizers

ENVIRONMENTAL DATA
Water pollution: Effect of low concentrations on aquatic life is unknown. Fouling to shoreline. May be dangerous if it enters nearby water intakes; notify operators. Notify local health and wildlife officials. **Response to discharge:** Mechanical containment. Should be removed. Chemical and physical treatment.

SHIPPING INFORMATION
Grades of purity: Crude-scale; refined; **Storage temperature:** Ambient; **Inert atmosphere:** None; **Venting:** Open (flame arrester); **Stability during transport:** Stable.

PHYSICAL AND CHEMICAL PROPERTIES
Physical state @ 59°F/15°C and 1 atm: Solid.
Molecular weight: 350-420 (solid); 165-250 (oil).
Boiling point @ 1 atm: Less than 460°F/238°F
Melting/Freezing point: 118-149°F/48-65°C/321-338°K.
Specific gravity (water = 1): 0.88-0.92 @ 68°F/20°C (liquid).
Liquid surface tension: 30.6 dynes/cm = 0.0306 N/m at 54°C.
Liquid water interfacial tension: 35–50 dynes/cm = 0.035-0.050 N/m at 54°C.
Relative vapor density (air = 1): 7.5
Heat of combustion: $-18,000$ Btu/lb = $-10,000$ cal/g = -430×10^5 J/kg.
Vapor pressure: (Reid) Very low; 0.24 mm.

PARAFORMALDEHYDE   **REC. P:0150**

SYNONYMS: FLO-MORE; FORMAGENE; PARAFORM; PARAFORM 3; PARAFORMALDEHIDO (Spanish); TRIFORMOL; TRIOXYMETHYLENE; FORMALDEHYDE POLYMER; PARAFORM; POLYFORMALDEHYDE; POLYOXYMETHYLENE; POLYOXYMETHYLENE GLYCOLENTIFICATION:

CAS Number: 30525-89-4
Formula: $HO(CH_2O)_nH$; n = 8–100
DOT ID Number: UN 2213; DOT Guide Number: 133
Proper Shipping Name: Paraformaldehyde
Reportable Quantity (RQ): **(CERCLA)** 1000 lb/454 kg

DESCRIPTION: White solid powder. A polymer of formaldehyde. Odor like formaldehyde. Sinks and mixes with water.

Poison! • Combustible • Breathing the vapor can kill you; skin or eye contact causes severe irritation • Firefighting gear (including SCBA) does not provide adequate protection. If exposure occurs, remove and isolate gear immediately and thoroughly decontaminate personnel • Vapor or dust may form explosive mixture with air • Containers may BLEVE when exposed to fire • Vapors and dust will collect and stay in low areas • Highly irritating to the eyes and respiratory tract; this material has anesthetic properties • Vapors or Concentrated dust in confined areas (e.g., tanks, sewers, buildings) may explode when exposed to fire • Toxic products of combustion may include carbon monoxide and formaldehyde gas which is highly flammable.

Hazard Classification (based on NFPA-704 Rating System)
Health Hazards (Blue): 3; Flammability (Red): 1; Reactivity (Yellow): 0

EMERGENCY RESPONSE: See Appendix A (133)
Evacuation:
Public safety: Isolate spill area for at least 10 to 25 meters (30 to 80 feet) in all directions.
Spill: Consider initial downwind evacuation of at least 100 meters (330 feet).
Fire: Isolate for 800 meters (½ mile) in all directions, especially if tank, rail car, or tank truck is involved in fire.

EXPOSURE
Short-term effects: *SEEK MEDICAL ATTENTION.* **Dust:** Irritating to eyes, nose, and throat. Harmful if inhaled. Lung edema may develop. Move to fresh air. *IF BREATHING HAS STOPPED,* give artificial respiration. IF breathing is difficult, administer oxygen. **Solid:** Irritating to skin and eyes. *IF SWALLOWED,* will cause nausea, vomiting, or loss of consciousness. Remove contaminated clothing and shoes. Flush affected areas with plenty of water. *IF IN EYES,* hold eyelids open and flush. with plenty of water. *IF SWALLOWED* and victim is *CONSCIOUS AND ABLE TO SWALLOW,* have victim drink 4 to 8 ounces of water.
Note to physician or authorized medical personnel: Medical observation is recommended for 24 to 48 hours after breathing overexposure, as pulmonary edema may be delayed. As first aid for pulmonary edema, consider administering a corticosteroid spray. Cigarette smoking may exacerbate pulmonary injury and should be discouraged for at least 72 hours following exposure

HEALTH HAZARDS
Personal protective equipment (PPE): B-Level PPE. Goggles or face shield; protective clothing.
Recommendations for respirator selection: NIOSH as formaldehyde
At any concentrations above the NIOSH REL, or where there is no REL, at any detectable concentration: SCBAF:PD,PP (any self-contained breathing apparatus that has a full facepiece and is operated in a pressure-demand or other positive-pressure mode); or SAF:PD,PP:ASCBA (any supplied-air respirator that has a full facepiece and is operated in a pressure-demand or other positive-pressure mode in combination with an auxiliary self-contained breathing apparatus operated in a pressure-demand or other positive pressure mode). *ESCAPE:* GMFS [any air-purifying, full-facepiece respirator (gas mask) with a chin-style, front- or back-mounted canister providing protection against the compound of concern]; or SCBAE (any appropriate escape-type, self-contained breathing apparatus).
Exposure limits (TWA unless otherwise noted): NIOSH 0.016 ppm, ceiling 0.1 ppm/15 min; OSHA 0.75 ppm as formaldehyde (product of decomposition).
Short-term exposure limits (15-minute TWA): ACGIH STEL 0.3 ppm (0.37 mg/m^3) suspected human carcinogen; OSHA STEL 2 ppm as formaldehyde (product of decomposition).

Toxicity by ingestion: Grade 3; LD_{50} = 50 to 500 mg/kg.
Long-term health effects: Lung and skin disorders.
Vapor (gas) irritant characteristics: Eye and respiratory tract irritant. Vapor is moderately irritating such that personnel will not usually tolerate moderate or high vapor concentrations.
Liquid or solid irritant characteristics: Severe eye irritant. Skin irritant. If spilled on clothing and allowed to remain, may cause smarting and reddening of the skin.

FIRE DATA
Flash point: 199°F/93°C (approximately) (oc); 160°F/71°C (approximate) (cc).
Flammable limits in air: LEL: 7.0%; UEL: 73.0%.
Behavior in fire: Containers may explode.
Autoignition temperature: 572°F/300°C/573°K.
Electrical hazard: Class I, Group C. Flow or agitation of substance may generate electrostatic charges due to low conductivity.

CHEMICAL REACTIVITY
Reactivity with water: Forms water solution of formaldehyde.
Binary reactants: Strong oxidizers may cause fire or explosion. Depolymerizes on contact with alkalies or acids.

ENVIRONMENTAL DATA
Food chain concentration potential: Negative; unlikely to accumulate.
Water pollution: Harmful to aquatic life in very low concentrations. May be dangerous if it enters nearby water intakes; notify operators. Notify local health and wildlife officials.
Response to discharge: Disperse and flush.

SHIPPING INFORMATION
Grades of purity: 91-99%, powder and flake; **Storage temperature:** Ambient; **Inert atmosphere:** None; **Venting:** Open (flame arrester); **Stability during transport:** Stable. Slowly decomposes to formaldehyde gas.

PHYSICAL AND CHEMICAL PROPERTIES
Physical state @ 59°F/15°C and 1 atm: Solid.
Molecular weight: 600 (approximate).
Boiling point @ 1 atm: Decomposes.
Melting/Freezing point: 246-342°F/119-172°C/392-445°K.
Specific gravity (water = 1): 1.46 @ 15°C (solid).
Heat of combustion: -6682 Btu/lb = -3712 cal/g = -155.4×10^5 J/kg.
Heat of solution: -150 Btu/lb = -83.5 cal/g = -3.50×10^5 J/kg.
Vapor pressure: 1.5 mm.

PARALDEHYDE REC. P:0175

SYNONYMS: *p*-ACETALDEHYDE; ACETALDEHYDE, TRIMER; EEC No. 605-004-00-1; ELALDEHYDE; PARAACETALDEHYDE; PARACETALDEHYDE; PARAL; PARALDEHYD (German); PARALDEHIDO (Spanish); PCHO; PORAL; TRIACETALDEHYDE (French); 2,4,6-TRIMETHYL-1,3,5-TRIOXANE; 2,4,6-TRIMETHYL-*s*-TRIOXANE; *s*-TRIMETHYLTRIOXYMETHYLENE; 1,3,5-TRIOXANE,2,4,6-TRIMETHYL-; RCRA No. U182; YK0525000

IDENTIFICATION
CAS Number: 123-63-7
Formula: $C_6H_{12}O_3$
DOT ID Number: UN 1264; DOT Guide Number: 129
Proper Shipping Name: Paraldehyde
Reportable Quantity (RQ): **(CERCLA)** 1000 lb/454 kg

DESCRIPTION: Colorless liquid. Agreeable, characteristic odor. Floats on the surface of water; soluble.

Highly flammable • Containers may BLEVE when exposed to fire •Vapors may form explosive mixture with air • Vapors are heavier than air and will collect and stay in low areas • Vapors may travel long distances to ignition sources and flashback • Vapors in confined areas (e.g., tanks, sewers, buildings) may explode when exposed to fire • Highly irritating to the eyes and respiratory tract; this material has anesthetic properties • Toxic products of combustion may include carbon monoxide.

Hazard Classification (based on NFPA-704 Rating System)
Health Hazards (Blue): 2; Flammability (Red): 3; Reactivity (Yellow): 1

EMERGENCY RESPONSE: See Appendix A (129)
Evacuation:
Public safety: Isolate spill area for at least 50 to 100 meters (160 to 330 feet) in all directions.
Spill: Large spill–Consider initial downwind evacuation for at least 300 meters (1000 feet).
Fire: Isolate for 800 meters (½ mile) in all directions, especially if tank, rail car, or tank truck is involved in fire.

EXPOSURE
Short-term effects: *SEEK MEDICAL ATTENTION.* **Vapor:** Harmful if inhaled. Move to fresh air. *IF BREATHING HAS STOPPED,* give artificial respiration. IF breathing is difficult, administer oxygen. **Liquid:** If swallowed, will cause headache, incoordination, drowsiness, or coma. Irritating to eyes and skin. Remove contaminated clothing and shoes. Flush affected areas with plenty of water. *IF IN EYES,* hold eyelids open and flush. with plenty of water. *IF SWALLOWED* and victim is *CONSCIOUS AND ABLE TO SWALLOW,* have victim drink 4 to 8 ounces of water.

HEALTH HAZARDS
Personal protective equipment (PPE): B-Level PPE. Wear rubber gloves, self-contained breathing apparatus. Butyl rubber is generally suitable for aldehydes.
Toxicity by ingestion: Grade 2; LD_{50} = 500 to 5000 mg/kg 4000 ppm was fatal to 3 of 6 rats in a 4-hour exposure.
Long-term health effects: Chronic intoxication–digestive disturbances, thirst, emaciation, muscular weakness, mental fatigue. Tremors of hands and tongue. Can cause skin eruptions. Substance is addictive.
Vapor (gas) irritant characteristics: Eye and respiratory tract irritant.
Liquid or solid irritant characteristics: Eye and skin irritant. If spilled on clothing and allowed to remain, may cause smarting and reddening of skin.

FIRE DATA
Flash point: 96°F/36°C (oc).
Fire extinguishing agents not to be used: Water may be ineffective.
Autoignition temperature: 460°F/238°C/511°K.
Electrical hazard: Class I, Group C. Flow or agitation of substance may generate electrostatic charges due to low conductivity.

CHEMICAL REACTIVITY
Binary reactants: Reacts with strong acids, caustics, ammonia,

amines, oxidizers. Decomposes on contact with acids or acid fumes forming acetaldehyde. Attacks rubber and copolymers of polystyrene and styrene-acrylonitrile.
Reactivity group: 19
Compatibility class: Aldehydes

ENVIRONMENTAL DATA
Food chain concentration potential: Log P_{ow} = 0.72. Unlikely to accumulate.
Water pollution: Effects of low concentrations on aquatic life are unknown. May be dangerous if it enters nearby water intakes; notify operators. Notify local health and wildlife officials.
Response to discharge: Issue warning–high flammability. Evacuate area. Chemical and physical treatment. Disperse and flush.

SHIPPING INFORMATION
Storage temperature: Ambient

PHYSICAL AND CHEMICAL PROPERTIES
Physical state @ 59°F/15°C and 1 atm: Liquid.
Molecular weight: 132.16
Boiling point @ 1 atm: 262.4°F/128°C/401.2°K.
Melting/Freezing point: 54.7°F/12.6°C/285.8°K.
Critical temperature: 500°F/260°C/533°K.
Specific gravity (water = 1): 0.9943 @ 68°F/20°C.
Liquid surface tension: 27.82 dynes/cm = 0.02782 N/m at 5°C.
Relative vapor density (air = 1): 4.55
Ratio of specific heats of vapor (gas): 1.1
Latent heat of vaporization: 135 Btu/lb = 75 cal/g = 3.1×10^5 J/kg.
Heat of combustion: @ 77°F/25°C –10,174 Btu/lb = –5,652 cal/g = 236×10^5 J/kg.
Vapor pressure: 26 mm.

PARATHION   REC. P:0200

SYNONYMS: AAT; AATP; AC 3422; ACC 3422; ALKRON; ALLERON; APHAMITE; ARALO; B 404; BAY E-605; BAYER E-605; BLADAN; BLADAN F; COMPOUND 3422; COROTHION; CORTHION; CORTHIONE; DANTHION; DIETHYL *p*-NITROPHENYL THIONOPHOSPHATE; DIETHYL 4-NITROPHENYL PHOSPHOROTHIONATE; *o,o*-DIETHYL-*O,p*-NITROPHENYL PHOSPHOROTHIOATE; DIETHYLPARATHION; DNTP; DPP; DREXEL PARATHION 8E; E 605; ECATOX; EEC No. 015-034-00-1; EKATIN WF & WF ULV; EKATOX; ETHLON; ETHYL PARATHION; FOLIDOL; FOLIDOL E605; FOLIDOL E&E 605; FOSFERMO; FOSFERNO; FOSFEX; FOSFIVE; FOSOVA; FOSTERN; FOSTOX; GEARPHOS; GENITHION; KALPHOS; KYPTHION; LETHALAIRE G-54; LIROTHION; MURFOS; NIRAN; NIRAN E-4; NITROSTIGMIN (German); NITROSTIGMINE; NIUIF–100; NOURITHION; OLEOFOS-20; PLEOPARAPHENE; OLEOPARATHION; ORTHOPHOS; PAC; PANTHION; PARADUST; PARAMAR; PARAMAR 50; PARAPHOS; PARATHENE; PARATHION-ETHYL; PARAWET; PARATIONA (Spanish); PESTOX PLUS; PETHION; PHOSKIL; PHOSPHOSTIGMINE; RB; PHOSPHOROTHIOIC ACID, *O,O,*-DIETHYLO-(*p*-NITROPHENYL)ESTER; RHODIASOL; RHODIATOX; RHODIATROX; SELEPHOS; SNP; SOPRATHION; STATHION; STRATHION; SULPHOS; SUPER RODIATOX; T-47; THIOPHOS; THIOPHOS 3422; TIOFOS; TOX 47; TOXOL (3); VAPOPHOS; VITREX; PHOSPHOROTHIOIC ACID, *O,O*-DIETHYL *O-p*-NITROPHENYL ESTER; RCRA No. P089

IDENTIFICATION
CAS Number: 56-38-2
Formula: $C_{10}H_{14}NO_5PS$; $(C_2H_5O)_2PSOC_6H_4NO_2$
DOT ID Number: UN 2783; DOT Guide Number: 152
Proper Shipping Name: Parathion; Parathion, mixture liquid or dry
Reportable Quantity (RQ): **(CERCLA)** 10 lb/4.54 kg

DESCRIPTION: Clear, colorless liquid when pure. Turns yellow to deep brown with age. Faint, garlic odor. Sinks in water; insoluble. *Note:* If used as a weapon, notify U.S. Department of Defense: Army. Use M8 Paper (Detection: Yellow) or M256-A1 Detector Kit (Detection limit: 0.005 mg/m^3) if available. Damage and/or death may occur before chemical detection can take place.

Poison! (organophosphate) • Breathing the vapor, skin or eye contact, or swallowing the material can kill you; symptoms may be delayed for several hours. *A single drop splashed in the eye can cause serious illness and possible death* • Firefighting gear (including SCBA) does not provide adequate protection. If exposure occurs, remove and isolate gear immediately and thoroughly decontaminate personnel • Vapors are heavier than air and will collect and stay in low areas • Containers may BLEVE when exposed to fire • Toxic products of combustion may include oxides of nitrogen, sulfur, and phosphorus • Do not put yourself in danger by entering a contaminated area to rescue a victim. *Note*: Has been used as a war nerve gas.

Hazard Classification (based on NFPA-704 Rating System)
Health Hazards (Blue): 4; Flammability (Red): 1; Reactivity (Yellow): 2

EMERGENCY RESPONSE: See Appendix A (152)
Evacuation:
Public safety: Isolate the area of spill or leak for at least 25 to 50 meters (80 to 160 feet) in all directions.
Spill: Increase, in the downwind direction, as necessary, the distance shown under "Public Safety."
Fire: If tank, rail car, or tank truck is involved in fire, isolate for at least 800 meters (½ mile) in all directions; also, consider initial evacuation for 800 meters (½ mile) in all directions.

EXPOSURE
Short-term effects: *SEEK MEDICAL ATTENTION.* **Vapor or Liquid:** *POISONOUS IF INHALED, IF SWALLOWED, OR IF SKIN IS EXPOSED.* Eye pupils are small; blurred vision; runny nose; cough; shortness of breath; pain; diarrhea, nausea and vomiting; increased blood pressure, hypermotility, hallucinations; loss of consciousness; convulsions; breathing stops; death. *IF BREATHING HAS STOPPED,* give artificial respiration; *avoid mouth-to-mouth resuscitation; use bag/mask apparatus.* **Skin:** *Remove and double-bag contaminated clothing (including shoes) and leave them in the Hot Zone*; wash skin with soap and water and large volumes of water for at least 15 minutes. *Skin can also be decontaminated with diluted hypochlorite solution, U.S. Army M291 kit, and M258(A1) skin decontamination kit.* **Eye:** Rinse with large volumes of water or saline for at least 15 minutes. *IF SWALLOWED,* will cause nausea, vomiting, or loss of consciousness. *Remove and double-bag contaminated clothing and shoes and leave in Hot Zone for later incineration by hazardous materials experts.* Flush affected areas with plenty of water. **If swallowed** and victim is *CONSCIOUS AND ABLE TO SWALLOW,* have victim drink 4 to 8 ounces of water. **Do NOT induce**

vomiting but immediately administer slurry of activated charcoal. *IF SWALLOWED* and victim is *UNCONSCIOUS OR HAVING CONVULSIONS,* do nothing except keep victim warm. *Note to physician or authorized medical personnel*: Do NOT induce emesis because of the risk of pulmonary aspiration of gastric contents. If the patient is alert and charcoal has not been given previously, administer a slurry of activated charcoal. Medical observation is recommended for 24 to 48 hours after breathing overexposure, as pulmonary edema may be delayed. As first aid for pulmonary edema, consider administering a corticosteroid spray. Cigarette smoking may exacerbate pulmonary injury and should be discouraged for at least 72 hours following exposure.

Note to physician or authorized medical personnel: Administer atropine, 2 mg (1/30 gr) intramuscularly or intravenously as soon as any local or systemic signs or symptoms of an intoxication are noted; repeat the administration of atropine every 3 to 8 minutes until signs of atropinization (mydriasis, dry mouth, rapid pulse, hot and dry skin) occur; initiate treatment in children with 0.05 mg/kg of atropine; repeat at 5- to 10-minute intervals. Watch respiration, and remove bronchial secretions if they appear to be obstructing the airway; intubate if necessary. Pralidoxime must be administered within minutes to a few hours following exposure (depending on the specific agent) to be effective. Give 2-PAMCl (Pralidoxime; Protopam), 2.5 g in 100 mL of sterile water or in 5% dextrose and water, intravenously, slowly, in 15 to 30 minutes; if sufficient fluid is not available, give 1 g of 2-PAMCl in 3 mL of distilled water by deep intramuscular injection; repeat this every half hour if respiration weakens or if muscle fasciculation or convulsions recur. Also Diazepam, an anticonvulsant, might be needed.

HEALTH HAZARDS
Personal protective equipment (PPE): A-Level PPE. Sealed chemical protective materials recommended (30-70%): Teflon®.
Recommendations for respirator selection: NIOSH
0.5 mg/m³: CCROVDMFu [any chemical cartridge respirator with organic vapor cartridge(s) in combination with a dust, mist, and fume filter]; or SA (any supplied-air respirator). *1.25 mg/m³*: SA:CF (any supplied-air respirator operated in a continuous-flow mode); or PAPROVDMFu [any powered, air purifying respirator with organic vapor cartridge (s) in combination with a dust, mist, and fume filter]. *2.5 mg/m³*: CCRFOVHiE [any chemical cartridge respirator with a full facepiece and organic vapor cartridge(s) in combination with a high-efficiency particulate filter]; or SAT:CF (any supplied-air respirator that has a tight-fitting facepiece and is operated in a continuous-flow mode); or PAPRTOVHiE [any powered, air-purifying respirator with a tight-fitting facepiece and organic vapor cartridge (s) in combination with a high-efficiency particulate filter]; or SCBAF (any self-contained breathing apparatus with full facepiece); or SAF (any supplied-air respirator with a full facepiece). *10 mg/m³*: SA:PD,PP (any supplied-air respirator operated in a pressure-demand or other positive-pressure mode). *EMERGENCY OR PLANNED ENTRY INTO UNKNOWN CONCENTRATIONS OR IDLH CONDITIONS*: SCBAF:PD,PP (any self-contained breathing apparatus that has a full facepiece and is operated in a pressure-demand or other positive-pressure mode); or SAF:PD,PP:ASCBA (any supplied-air respirator that has a full facepiece and is operated in a pressure-demand or other positive-pressure mode in combination with an auxiliary self-contained breathing apparatus operated in a pressure-demand or other positive pressure mode). *ESCAPE*: GMFOVHiE [any air-purifying, full-facepiece respirator (gas mask) with a chin-style, front-or back-mounted canister having a high efficiency particulate filter] or SCBAE (any appropriate escape-type, self-contained breathing apparatus).

Exposure limits (TWA unless otherwise noted): ACGIH TLV 0.1 mg/m³; NIOSH 0.05 mg/m³; OSHA PEL 0.1 mg/m³; skin contact contributes significantly in overall exposure.
Toxicity by ingestion: Deadly poison. Grade 4; oral LD_{50} = 2 mg/kg (rat).
Long-term health effects: Birth defects in chick embryos.
Liquid or solid irritant characteristics: Eye irritant.
Odor threshold: 0.04 ppm.
IDLH value: 10 mg/m³.

FIRE DATA
Flash point: 392°F/200°C.
Fire extinguishing agents not to be used: High-pressure water hoses may scatter parathion from broken containers, increasing contamination hazard.
Behavior in fire: Containers may explode.

CHEMICAL REACTIVITY
Reactivity with water: Slow reaction.
Binary reactants: Strong oxidizers, alkaline materials.

ENVIRONMENTAL DATA
Food chain concentration potential: Log P_{ow} = 3.8. Values > 3.0 are likely to bioconcentrate in aquatic organisms and other living tissue, especially in fats.
Water pollution: If used as a weapon, utilize use an M272 Water Detection Kit (Detection limit: 0.02 mg/L). DOT Appendix B, §172.101- severe marine pollutant. Harmful to aquatic life in very low concentrations. May be dangerous if it enters nearby water intakes; notify operators. Notify local health and wildlife officials. For weapons testing, use an M272 Water Detection Kit. **Response to discharge:** Issue warning–poison, water contaminant. Restrict access. Should be removed. Chemical and physical treatment.

SHIPPING INFORMATION
Grades of purity: 98.5+%. Sometimes distributed as solutions emulsifiable in water; **Storage temperature:** Ambient; **Inert atmosphere:** None; **Venting:** Pressure-vacuum; **Stability during transport:** Stable.

PHYSICAL AND CHEMICAL PROPERTIES
Physical state @ 59°F/15°C and 1 atm: Liquid.
Molecular weight: 291.3
Boiling point @ 1 atm: 315–323°F/157–162°C/430–435°K @ 6 mm (decomposes).
Melting/Freezing point: 37°F/3°C/276°K.
Specific gravity (water = 1): 1.269 @ 77°F/25°C (liquid).
Relative vapor density (air = 1): 10
Heat of combustion: -9240 Btu/lb = -5140 cal/g = -215×10^5 J/kg.
Vapor pressure: 0.00004 mmHg.

PENTABORANE   REC. P:0250

SYNONYMS: DIHYDROPENTABORANE(9); PENTABORANE(9); PENTABORANE UNDECAHYDRIDE; PENTABORON NONAHYDRIDE; PENTABORANO (Spanish); STABLE PENTABORANE; (9)-PENTABORON NONAHYDRIDE

IDENTIFICATION
CAS Number: 19624-22-7
Formula: B_5H_9
DOT ID Number: UN 1380; DOT Guide Number: 135

Pentaborane

Proper Shipping Name: Pentaborane
Reportable Quantity (RQ): **(EHS)** 1 lb/0.454 kg

DESCRIPTION: Colorless liquid. Strong sour milk odor. Floats on water surface; insoluble; slowly decomposes forming boric acid.

Poison! • Extremely flammable; spontaneously flammable when exposed to moist air • Violent reaction with halogenated (Halon) extinguishing agents • Breathing the vapor, skin or eye contact, or swallowing the material can kill you • Firefighting gear (including SCBA) does not provide adequate protection. If exposure occurs, remove and isolate gear immediately and thoroughly decontaminate personnel • Containers may BLEVE when exposed to fire • Corrosive to natural rubber • Toxic products of combustion may include boron • Do not attempt rescue.

Hazard Classification (based on NFPA-704 Rating System)
Health Hazards (Blue): 4; Flammability (Red): 4; Reactivity (Yellow): 2

EMERGENCY RESPONSE: See Appendix A (135)
Evacuation:
Public safety: Isolate spill area for at least 100 to 150 meters (330 to 490 feet) in all directions.
Spill: Small spill–First: Isolate in all directions 155 meters (500 feet); Then: Protect persons downwind, DAY: 1.3 km (0.8 mile); NIGHT: 3.7 km (2.3 mile). LARGE SPILL: First: Isolate in all directions 765 meters (2500 feet); Then: Protect persons downwind, DAY: 6.6 km (4.1 mile); NIGHT: 10.6 km (6.6 mile).
Fire: Isolate for 800 meters (½ mile) in all directions, especially if tank, rail car, or tank truck is involved in fire.

EXPOSURE
Short-term effects: *SEEK MEDICAL ATTENTION. HAZARD IS FROM PRODUCTS OF COMBUSTION.* **Vapor:** *POISONOUS IF INHALED.* Move victim to fresh air. If breathing is difficult, administer oxygen. **Liquid:** *POISONOUS IF SWALLOWED OR IF SKIN IS EXPOSED.* Irritating to eyes. Remove contaminated clothing and shoes. Flush affected areas with plenty of water. *IF IN EYES*, hold eyelids open and flush with plenty of water. *IF SWALLOWED* and victim is *CONSCIOUS AND ABLE TO SWALLOW*, have victim drink 4 to 8 ounces of water and have victim induce vomiting. *IF SWALLOWED* and victim is *UNCONSCIOUS OR HAVING CONVULSIONS*, do nothing except keep victim warm.

HEALTH HAZARDS
Personal protective equipment (PPE): A-Level PPE. Self-contained breathing apparatus or air-line mask; goggles or face shield; rubber gloves and protective clothing.
Recommendations for respirator selection: NIOSH/OSHA
0.05 ppm: SA (any supplied-air respirator). *0.125 ppm:* SA:CF (any supplied-air respirator operated in a continuous-flow mode). 0.25 ppm: SAT:CF (any supplied-air respirator that has a tight-fitting facepiece and is operated in a continuous-flow mode); or SCBAF (any self-contained breathing apparatus with a full facepiece); or SAF (any supplied-air respirator with a full facepiece). *1 ppm:* SA:PD:PP (any supplied-air respirator operated in a pressure-demand or other positive-pressure mode). *EMERGENCY OR PLANNED ENTRY INTO UNKNOWN CONCENTRATIONS OR IDLH CONDITIONS:* SCBAF:PD,PP (any self-contained breathing apparatus that has a full facepiece and is operated in a pressure-demand or other positive-pressure mode); or SAF:PD,PP:ASCBA (any supplied-air respirator that has a full facepiece and is operated in a pressure-demand or other positive-pressure mode in combination with an auxiliary self-contained breathing apparatus operated in a pressure-demand or other positive pressure mode). *ESCAPE:* GMFS [any air-purifying, full-facepiece respirator (gas mask) with a chin-style, front- or back-mounted canister providing protection against the compound of concern]; or SCBAE (any appropriate escape-type, self-contained breathing apparatus).
Exposure limits (TWA unless otherwise noted): ACGIH TLV 0.005 ppm (0.013 mg/m^3); OSHA PEL 0.005 ppm (0.01 mg/m^3); NIOSH 0.005 ppm (0.01 mg/m^3).
Short-term exposure limits (15-minute TWA): ACGIH STEL 0.015 ppm (0.039 mg/m^3); NIOSH STEL 0.015 ppm (0.03 mg/m^3).
Toxicity by ingestion: Grade 4; LD$_{50}$ less than 50 mg/kg.
Vapor (gas) irritant characteristics: Vapors cause moderate irritation such that personnel
will find high concentrations unpleasant. The effect is temporary.
Liquid or solid irritant characteristics: Eye and skin irritant. Can be absorbed through the skin.
Odor threshold: 0.8 ppm.
IDLH value: 1 ppm.

FIRE DATA
Flash point: 86°F/30°C (oc) (ignites spontaneously in moist air).
Flammable limits in air: LEL: 0.42%; UEL: 98%.
Fire extinguishing agents not to be used: Violent reaction with halogenated agents, water.
Behavior in fire: Tends to reignite. Contact with water applied to adjacent fires produces flammable hydrogen gas.
Autoignition temperature: Spontaneously flammable if impure. Approximately 95°F/35°C/308°K when pure.
Electrical hazards: Flow or agitation of substance may generate electrostatic charges due to low conductivity.

CHEMICAL REACTIVITY
Reactivity with water: Pyrophoric; may explode in moist air without a source of ignition. Hydrolyzes slowly with heat in water to form boric acid. Reacts slowly, forming flammable hydrogen gas.
Binary reactants: Reacts with ammonia forming violently explosive (at room temperature) pentaborane diammoniate. Violent reaction with strong oxidizers, halogens, and halogenated compounds, carbon tetrachloride. Incompatible with air, acetone, aldehydes, amines, carbon disulfide, chloroform, dioxane, hydrazines, ketones, trichloroethylene. Corrosive to natural rubber, some synthetic rubbers, some greases, and some lubricants.

ENVIRONMENTAL DATA
Food chain concentration potential: Negative; unlikely to accumulate.
Water pollution: Effect of low concentrations on aquatic life is unknown. May be dangerous if it enters nearby water intakes; notify operators. Notify local health and wildlife officials. **Response to discharge:** Issue warning–high flammability, water contaminant. Restrict access. Evacuate area. Disperse and flush.

SHIPPING INFORMATION
Grades of purity: Technical; 95+%; Hi-Purity: 99+%; **Storage temperature:** Cool ambient; **Inert atmosphere:** Inerted with dry nitrogen; **Venting:** Safety relief; **Stability during transport:** Stable below 302°F/150°C.

NAS HAZARD CLASSIFICATION FOR BULK WATER TRANSPORTATION
FIRE: 4
HEALTH: Vapor irritant: 2; Liquid or solid irritant:–; Poisons: 4

WATER POLLUTION: Human toxicity: 4; Aquatic toxicity:–; Aesthetic effect: 2
REACTIVITY: Other chemicals: 4; Water: 2; Self-reaction: 4

PHYSICAL AND CHEMICAL PROPERTIES
Physical state @ 59°F/15°C and 1 atm: Liquid.
Molecular weight: 63.2
Boiling point @ 1 atm: 137.1°F/58.4°C/331.5°K.
Melting/Freezing point: –52.2°F/–46.8°C/224.6°K.
Critical temperature: 441°F/227°C/500°K.
Critical pressure: 570 psia = 38 atm = 3.9 MN/m^2.
Specific gravity (water = 1): 0.623 @ 68°F/20°C (liquid).
Liquid surface tension: 20.8 dynes/cm = 0.0208 N/m @ 77°F/25°C.
Relative vapor density (air = 1): 2.2
Ratio of specific heats of vapor (gas): 1.0399
Latent heat of vaporization: 219 Btu/lb = 122 cal/g

PENTACHLOROETHANE   REC. P:0300

SYNONYMS: ETHANE PENTACHLORIDE; ETHANE, PENTACHLORO-; NCI-C53894; PENTACHLORAETHAN (German); PENTACHLORETHANE (French); PENTACLOROETANO (Spanish); PENTALINO

IDENTIFICATION
CAS Number: 76-01-7
Formula: C_2HCl_5; CCl_3CHCl_2
DOT ID Number: UN 1669; DOT Guide Number: 151
Proper Shipping Name: Pentachloroethane
Reportable Quantity (RQ): **(CERCLA)** 10 lb/4.54 kg

DESCRIPTION: Colorless liquid. Sweet, chloroform-like odor. Sinks in water; insoluble; reacts, forming dichloroacetic acid.

Poison! • Breathing the vapor can kill you • Firefighting gear (including SCBA) does not provide adequate protection. If exposure occurs, remove and isolate gear immediately and thoroughly decontaminate personnel • Flammable • Containers may BLEVE when exposed to fire • Vapors are heavier than air and will collect and stay in low areas • Vapors may travel long distances to ignition sources and flashback • Vapors in confined areas (e.g., tanks, sewers, buildings) may explode when exposed to fire • Severely irritating to skin, eyes, and respiratory tract; prolonged contact with the skin causes burns • Toxic products of combustion may include hydrogen chloride • Do not put yourself in danger by entering a contaminated area to rescue a victim.

Hazard Classification (based on NFPA-704 Rating System)
Health Hazards (Blue): 3; Flammability (Red): 2; Reactivity (Yellow): 0

EMERGENCY RESPONSE: See Appendix A (151)
Evacuation:
Public safety: Isolate the area of spill or leak for at least 25 to 50 meters (80 to 160 feet) in all directions.
Spill: Increase, in the downwind direction, as necessary, the distance shown under "Public Safety."
Fire: If tank, rail car, or tank truck is involved in fire, isolate for at least 800 meters (½ mile) in all directions; also, consider initial evacuation for 800 meters (½ mile) in all directions.

EXPOSURE
Short-term effects: *SEEK MEDICAL ATTENTION.* **Vapor:** *POISONOUS; MAY BE FATAL IF INHALED OR ABSORBED THROUGH SKIN.* Irritating to eyes, skin, lungs and mucous membrane. Narcotic effect greater than chloroform. More potent central nervous system depressant than chloroform or tetrachloroethane. *IF BREATHING HAS STOPPED,* give artificial respiration; *avoid mouth-to-mouth resuscitation; use bag/mask apparatus.* IF breathing is difficult, administer oxygen. *IF IN EYES OR ON SKIN,* flush with running water for at least 15 minutes; hold eyelids open if necessary. **Liquid:** *POISONOUS IF SWALLOWED OR IF SKIN IS EXPOSED.* Irritating to skin and eyes. *IF IN EYES OR ON SKIN,* flush with running water for at least 15 minutes; hold eyelids open if necessary.
Speed in removing material from skin is of extreme importance. Remove and double bag contaminated clothing and shoes at the site.
Effects may be delayed; keep victim under observation. *IF SWALLOWED* and victim is *UNCONSCIOUS OR HAVING CONVULSIONS,* do nothing except keep victim warm.

HEALTH HAZARDS
Personal protective equipment (PPE): A-Level PPE.
Exposure limits (TWA unless otherwise noted): NIOSH REL handle with caution in the workplace
Toxicity by ingestion: Toxic. Grade 3; LD_{50} = 500 g/kg (dog).
Long-term health effects: Exposure to 121 ppm 8 to 9 hours daily for 23 days caused significant pathological changes in the liver, lungs, and kidneys of cats. Exposure to vapor for 3 weeks caused fatty degeneration of the liver and injury to the kidneys and lungs of cats.
Vapor (gas) irritant characteristics: Causes mucous membrane, skin, lung, and cornea (eye) irritation.
Liquid or solid irritant characteristics: Eye and skin irritant.

FIRE DATA
Behavior in fire: Dangerous when heated to decomposition; dehalogenation will produce spontaneously explosive chloroacetylenes.

CHEMICAL REACTIVITY
Reactivity with water: Hydrolysis produces dichloroacetic acid.
Binary reactants: Alkalis, metals; reaction with alkalis and metals produce spontaneously explosive chloroacetylenes.
Reactivity group: 36
Compatibility class: Halogenated hydrocarbons

ENVIRONMENTAL DATA
Water pollution: DOT Appendix B, §172.101–marine pollutant. Harmful to aquatic life in very low concentrations. May be dangerous if it enters nearby water intakes; notify operators. Notify local health and wildlife officials. **Response to discharge:** Issue warning–poison, air contaminant. Restrict access. Should be removed. Chemical and physical treatment.

SHIPPING INFORMATION
Grades of purity: Technical; 96%; **Stability during transport:** Stable.

PHYSICAL AND CHEMICAL PROPERTIES
Physical state @ 59°F/15°C and 1 atm: Liquid.
Molecular weight: 202.30
Boiling point @ 1 atm: 321°F/160.5°C/434°K.
Melting/Freezing point: –7.6°F/–22°C/251.2°K.
Critical temperature: 740°F/390°C/663°K.
Critical pressure: 529 psia =36 atm = 3.6 MN/m^2.
Specific gravity (water = 1): 1.6728 @ 77°F/25°C.

Liquid surface tension: 34.55 dynes/cm = 0.0346 N/m @ 68°F/20°C.
Relative vapor density (air = 1): 7.0 (estimate).
Latent heat of vaporization: 78.5 Btu/lb = 43.6 cal/g = 1.83×10^5 J/kg.
Heat of decomposition: 1828 Btu/lb = 1015.8 cal/g = 4.25×10^6 J/kg.

PENTACHLOROPHENOL   REC. P:0350

SYNONYMS: CHEM-TOL; CHLOROPHEN; CRYPTOGIL OL; DOWCIDE 7; DOWICIDE 7; DOWICIDE EC-7; DOWICIDE G; DOW PENTACHLOROPHENOL DP-2 ANTIMICROBIAL; DUROTOX; EEC No. 604-002-00-8; EP 30; FUNGIFEN; GLAZD PENTA; GRUNDIER ARBEZOL; LAUXTOL; LAUXTOL A; LIROPREM; PCP; PENCHLOROL; PENTA; PENTACHLOROFENOL; PENTACLOROFENOL (Spanish); PENTACHLOROPHENATE; 2,3,4,5,6-PENTACHLOROPHENOL; PENTACHLOROPHENOL, DOWICIDE EC-7; PENTACHLOROPHENOL, DP-2; PENCHLOROL; PENTA; PENTACON; PENTA-KIL; PENTASOL; PENWAR; PERATOX; PERMACIDE; PERMAGARD; PERMASAN; PERMATOX DP-2; PERMATOX PENTA; PERMITE; PHENOL, PENTACHLORO-; PRILTOX; SANTOBRITE; SANTOPHEN; SANTOPHEN 20; SINITUHO; TERM-I-TROL; THOMPSONS® WOOD FIX; WEEDONE; RCRA No. U2420

IDENTIFICATION
CAS Number: 87-86-5
Formula: C_6HCl_5O
DOT ID Number: UN 3155; DOT Guide Number: 154
Proper Shipping Name: Pentachlorophenol
Reportable Quantity (RQ): **(CERCLA)** 10 lb/4.54 kg

DESCRIPTION: Colorless crystalline solid when pure; crude product is gray to light brown solid beads or flakes. May be shipped and stored as a solution in a flammable solvent. Sweet odor, like phenol. Sinks in water; insoluble; decomposes forming hydrochloric acid and hydrogen chloride gas.

Poison! (nitrophenol pesticide) • Irritating to skin, eyes, and respiratory tract; prolonged contact with the skin can cause burns • Containers may BLEVE when exposed to fire • Decomposes above 370°F/188°C; toxic products of combustion may include hydrogen chloride, chlorinated phenols, and carbon monoxide.

Hazard Classification (based on NFPA-704 Rating System)
Health Hazards (Blue): 3; Flammability (Red): 0 *note*: solutions of material are flammable; Reactivity (Yellow): 0

EMERGENCY RESPONSE: See Appendix A (154)
Evacuation:
Public safety: Isolate the area of spill or leak for at least 25 to 50 meters (80 to 160 feet) in all directions.
Spill: Increase, in the downwind direction, as necessary, the distance shown under "Public Safety."
Fire: If tank, rail car, or tank truck is involved in fire, isolate for at least 800 meters (½ mile) in all directions; also, consider initial evacuation for 800 meters (½ mile) in all directions.

EXPOSURE
Short-term effects: *SEEK MEDICAL ATTENTION.* **Dust:** Irritating to eyes, nose, and throat. *IF INHALED*, will, will cause coughing or difficult breathing. Move to fresh air. *IF BREATHING HAS STOPPED*, give artificial respiration; *avoid mouth-to-mouth resuscitation; use bag/mask apparatus.* IF breathing is difficult, administer oxygen. **Solid:** *POISONOUS IF SWALLOWED.* Will burn skin and eyes. Remove contaminated clothing and shoes. Flush affected areas with plenty of water. *IF IN EYES*, hold eyelids open and flush with plenty of water. *IF SWALLOWED* and victim is *CONSCIOUS AND ABLE TO SWALLOW*, have victim drink 4 to 8 ounces of water and have victim induce vomiting. *IF SWALLOWED* and victim is *UNCONSCIOUS OR HAVING CONVULSIONS*, do nothing except keep victim warm.

HEALTH HAZARDS
Personal protective equipment (PPE): B-Level PPE. Respirator for dust; goggles; protective clothing. Wear sealed chemical suit. Chemical protective material(s) reported to have good to excellent resistance: nitrile, Viton®. Also, PVC and polycarbonate offers limited protection
Recommendations for respirator selection: NIOSH/OSHA
$2.5\ mg/m^3$: CCROVDMFu* [any chemical cartridge respirator with organic vapor cartridge(s) in combination with a dust, mist, and fume filter]; or PAPROVDMFu* [any powered, air-purifying respirator with organic vapor cartridge(s) in combination with a dust, mist, and fume filter]; or SA* (any supplied-air respirator); or SCBA (any self-contained breathing apparatus). *EMERGENCY OR PLANNED ENTRY INTO UNKNOWN CONCENTRATIONS OR IDLH CONDITIONS:* SCBAF:PD,PP (any self-contained breathing apparatus that has a full facepiece and is operated in a pressure-demand or other positive-pressure mode); or SAF:PD,PP:ASCBA (any supplied-air respirator that has a full facepiece and is operated in a pressure-demand or other positive-pressure mode in combination with an auxiliary self-contained breathing apparatus operated in a pressure-demand or other positive pressure mode). *ESCAPE:* GMFOVHiE [any air-purifying, full-facepiece respirator (gas mask) with a chin-style, front- or back-mounted organic vapor canister having a high-efficiency particulate filter]; or SCBAE (any appropriate escape-type, self-contained breathing apparatus). *Note*: Substance reported to cause eye irritation or damage; may require eye protection.
Exposure limits (TWA unless otherwise noted): ACGIH TLV 0.5 mg/m^3; NIOSH/OSHA 0.5 mg/m^3; skin contact contributes significantly in overall exposure.
Toxicity by ingestion: Grade 3; LD_{50} = 50 to 500 mg/kg (rat).
Long-term health effects: Organ, liver and kidney damage. Confirmed carcinogen.
Vapor (gas) irritant characteristics: Eye and respiratory tract irritant. Vapor is irritating such that personnel will not usually tolerate moderate or high vapor concentrations.
Liquid or solid irritant characteristics: Severe eye and skin irritant. Causes smarting of the skin and first-degree burns on short exposure; may cause secondary burns on long exposure.
IDLH value: 2.5 mg/m^3.

CHEMICAL REACTIVITY
Reactivity with water: Decomposes forming hydrogen chloride.
Binary reactants: Strong oxidizers, acids, alkalis
Reactivity group: 21
Compatibility class: Phenols

ENVIRONMENTAL DATA
Food chain concentration potential: Log P_{ow} = 5.1. Able to bioconcentrate. Values > 3.0 are likely to bioconcentrate in aquatic organisms and other living tissue, especially in fats.
Water pollution: DOT Appendix B, §172.101- severe marine pollutant. Harmful to aquatic life in very low concentrations. May

be dangerous if it enters nearby water intakes; notify operators. Notify local health and wildlife officials. **Response to discharge:** Issue warning–poison. Restrict access. Should be removed.

SHIPPING INFORMATION
Grades of purity: 86–100%; **Storage temperature:** Ambient; **Inert atmosphere:** None; **Venting:** Open; **Stability during transport:** Stable.

PHYSICAL AND CHEMICAL PROPERTIES
Physical state @ 59°F/15°C and 1 atm: Solid.
Molecular weight: 266.35
Boiling point @ 1 atm: 590°F/310°C/583°K.
Melting/Freezing point: 370°F/188°C/461°K.
Specific gravity (water = 1): 1.98 @ 15°C (solid).
Vapor pressure: 0.0001 mm (77°F/25°C).

PENTADECANOL   REC. P:0400

SYNONYMS: 1-PENTADECANOL; PENTADECYL ALCOHOL

IDENTIFICATION
CAS Number: 629-76-5
Formula: $CH_3(CH_2)_{13}CH_2OH$
DOT ID Number: 1986; DOT Guide Number: 131
Proper Shipping Name: Alcohols, toxic, n.o.s.

DESCRIPTION: Colorless liquid. Faint alcohol odor. Floats on the surface of water.

Combustible • Toxic! Breathing the vapor, skin or eye contact, or swallowing the material may kill you • Firefighting gear (including SCBA) does not provide adequate protection. If exposure occurs, remove and isolate gear immediately and thoroughly decontaminate personnel • Containers may BLEVE when exposed to fire •Vapors may form explosive mixture with air • Vapors are heavier than air and will collect and stay in low areas • Vapors may travel long distances to ignition sources and flashback • Vapors in confined areas (e.g., tanks, sewers, buildings) may explode when exposed to fire • Irritating to the skin, eyes, and respiratory tract • Toxic products of combustion may include carbon monoxide.

Hazard Classification (based on NFPA-704 Rating System)
Health Hazards (Blue): 0; Flammability (Red): 1; Reactivity (Yellow): 0

EMERGENCY RESPONSE: See Appendix A (131)
Evacuation:
Public safety: Isolate spill area for at least 100 to 200 meters (330 to 660 feet) in all directions.
Spill: Increase, as necessary, the isolation distance shown above, in "Public safety."
Fire: Isolate for 800 meters (½ mile) in all directions, especially if tank, rail car, or tank truck is involved in fire.

EXPOSURE
Short-term effects: *SEEK MEDICAL ATTENTION.* **Liquid:** Irritating to skin and eyes. Remove contaminated clothing and shoes. Flush affected areas with plenty of water. *IF IN EYES*, hold eyelids open and flush with plenty of water.

HEALTH HAZARDS
Personal protective equipment (PPE): A-Level PPE.

FIRE DATA
Fire extinguishing agents not to be used: Water or foam may cause frothing.

CHEMICAL REACTIVITY
Reactivity group: 20
Compatibility class: Alcohols, glycols

ENVIRONMENTAL DATA
Food chain concentration potential: Negative; unlikely to accumulate.
Water pollution: Harmful to fish and water fowl. Fouling to shoreline. May be dangerous if it enters nearby water intakes; notify operators. Notify local health and pollution control officials. **Response to discharge:** Mechanical containment. Should be removed.

SHIPPING INFORMATION
Storage temperature: Ambient; **Inert atmosphere:** None; **Venting:** Open (flame arrester); **Stability during transport:** Stable.

PHYSICAL AND CHEMICAL PROPERTIES
Physical state @ 59°F/15°C and 1 atm: Liquid.
Molecular weight: 228.42
Boiling point @ 1 atm: 572°F/300°C/573°K.
Melting/Freezing point: 111°F/44°C/317°K.
Critical temperature: 824°F/440°C/713°K.
Specific gravity (water = 1): 0.829 at 50°C (liquid).
Liquid surface tension: (estimate) 25 dynes/cm = 0.025 N/m at 50°C.
Liquid water interfacial tension: (estimate) 35 dynes/cm = 0.035 N/m at 50°C.
Ratio of specific heats of vapor (gas): 1.024

1,3-PENTADIENE   REC. P:0450

SYNONYMS: α-METHYLBIVINYL; 1-METHYLBUTADIENE; *cis*-PENTADIENE-1,3; PIPERYLENE; *trans*-PENTADIENE-1,3; 1,3-PENTADIENO (Spanish)

IDENTIFICATION
CAS Number: 504-60-9; 2004-70-8 (*trans*-)
Formula: C_5H_8; $CH_2=CHCH=CHCH_3$
DOT ID Number: UN 3295; DOT Guide Number: 128
Proper Shipping Name: Hydrocarbons, liquid, n.o.s.

DESCRIPTION: Colorless liquid. Floats on water surface; slightly soluble. Boils at 108°F/42°C.

Extremely flammable • Containers may BLEVE when exposed to fire •Vapors may form explosive mixture with air • Vapors are heavier than air and will collect and stay in low areas • Vapors may travel long distances to ignition sources and flashback • Vapors in confined areas (e.g., tanks, sewers, buildings) may explode when exposed to fire • Irritating to the skin, eyes, and respiratory tract • Toxic products of combustion may include carbon monoxide and CO_2.

Hazard Classification (based on NFPA-704 Rating System)
Health Hazards (Blue): 0; Flammability (Red): 4; Reactivity (Yellow): 2

EMERGENCY RESPONSE: See Appendix A (128)
Evacuation:
Public safety: Isolate spill area for at least 25 to 50 meters (80 to 160 feet) in all directions.
Spill: Large spill–Consider initial downwind evacuation for at least 300 meters (1000 feet).
Fire: Isolate for 800 meters (½ mile) in all directions, especially if tank, rail car, or tank truck is involved in fire.

EXPOSURE
Short-term effects: *SEEK MEDICAL ATTENTION*. **Vapor:** May cause dizziness or suffocation. May irritate eyes and skin. Move to fresh air. *IF BREATHING HAS STOPPED*, give artificial respiration. IF breathing is difficult, administer oxygen. **Liquid:** May irritate skin and eyes. *IF IN EYES OR ON SKIN*, flush with running water for at least 15 minutes; hold eyelids open if necessary. Wash skin with soap and water. Remove and double bag contaminated clothing and shoes at the site.

HEALTH HAZARDS
Personal protective equipment (PPE): B-Level PPE. Wear self-contained positive pressure breathing apparatus and full protective clothing.
Vapor (gas) irritant characteristics: May irritate eyes and mucous membranes.
Liquid or solid irritant characteristics: Liquid may cause irritation of the eyes and skin.

FIRE DATA
Flash point: –20°F/–29°C (cc).
Flammable limits in air: LEL: 1.5%; UEL: 8.3%.
Fire extinguishing agents not to be used: Water may be ineffective; material floats on surface.
Behavior in fire: Will burn and produce irritating and poisonous gases. Container may explode in heat of fire. Vapor explosion hazard indoors, outdoors, or in sewers. Runoff to sewer may create fire or explosion hazard.
Electrical hazard: Class I, Group D.
Stoichiometric air-to-fuel ratio: 10.5

CHEMICAL REACTIVITY
Binary reactants: Violent reaction with oxidizers. Dissolves rubber and paint; **Inhibitor of polymerization:** 2,6-ditertiarybutyl-4-methylphenol
Reactivity group: 30
Compatibility class: Olefins

ENVIRONMENTAL DATA
Water pollution: Dangerous to aquatic life in high concentrations. May be dangerous if it enters nearby water intakes; notify operators. Notify local health and wildlife officials. **Response to discharge:** Issue warning–high flammability. Restrict access. Evacuate area. Mechanical containment. Should be removed. Chemical and physical treatment.

SHIPPING INFORMATION
Grades of purity: 90%.

PHYSICAL AND CHEMICAL PROPERTIES
Physical state @ 59°F/15°C and 1 atm: Liquid.
Molecular weight: 68.12
Boiling point @ 1 atm: –45°F/–43°C/230°K.
Melting/Freezing point: –222°F/–141°C/132°K.
Critical temperature: 411°F/211°C/484°K (estimate).
Critical pressure: 542 psia = 37 atm = 3.7 MN/m^2 (estimate).
Specific gravity (water = 1): 0.6834 @ 68°F/20°C (average value for cis and trans isomers).
Relative vapor density (air = 1): 2.4
Latent heat of vaporization: 193 Btu/lb = 107 cal/g = 4.50 x 10^5 J/kg.
Heat of combustion: –20,167 But/lb = –11,207 cal/g = –46.9 x 10^6 J/kg.
Vapor pressure: 345 mm.

1,4-PENTADIENE　　　　　　　　　　　　　　　　**REC. P:0500**

SYNONYMS: ALLYLETHYLENE; DIVINYMETHANE; PENTA-1,4-DIENE; 1,4-PENTADIENO (Spanish)

IDENTIFICATION
CAS Number: 591-93-5
Formula: H$_2$C=CHCH$_2$CH=CH$_2$
DOT ID Number: UN 1993; DOT Guide Number: 128
Proper Shipping Name: Flammable liquid, n.o.s.

DESCRIPTION: Colorless liquid. Floats on the surface of water.

Highly flammable • Containers may BLEVE when exposed to fire •Vapors may form explosive mixture with air • Vapors are heavier than air and will collect and stay in low areas • Vapors may travel long distances to ignition sources and flashback • Vapors in confined areas (e.g., tanks, sewers, buildings) may explode when exposed to fire • Irritating to the skin, eyes, and respiratory tract • Toxic products of combustion may include carbon monoxide.

Hazard Classification (based on NFPA-704 Rating System)
Health Hazards (Blue): 0; Flammability (Red): 4; Reactivity (Yellow): 2

EMERGENCY RESPONSE: See Appendix A (128)
Evacuation:
Public safety: Isolate spill area for at least 25 to 50 meters (80 to 160 feet) in all directions.
Spill: Large spill–Consider initial downwind evacuation for at least 300 meters (1000 feet).
Fire: Isolate for 800 meters (½ mile) in all directions, especially if tank, rail car, or tank truck is involved in fire.

EXPOSURE
Short-term effects: *SEEK MEDICAL ATTENTION*. **Vapor:** May cause dizziness or suffocation. May irritate eyes and skin. Move to fresh air. *IF BREATHING HAS STOPPED*, give artificial respiration. IF breathing is difficult, administer oxygen. **Liquid:** May irritate skin and eyes. *IF IN EYES OR ON SKIN*, flush with running water for at least 15 minutes; hold eyelids open if necessary. Wash skin with soap and water. Remove and double bag contaminated clothing and shoes at the site. *IF SWALLOWED* and victim is *UNCONSCIOUS OR HAVING CONVULSIONS*, do nothing except keep victim warm.

HEALTH HAZARDS
Personal protective equipment (PPE): B-Level PPE. Wear self-contained positive pressure breathing apparatus and full protective clothing.
Vapor (gas) irritant characteristics: Vapor may irritate eyes and respiratory tract.
Liquid or solid irritant characteristics: Liquid may irritate eyes and skin.

FIRE DATA
Flash point: 40°F/40°C (cc).
Fire extinguishing agents not to be used: Water may be ineffective since this material floats.
Behavior in fire: Container may explode.
Electrical hazard: Class I, Group D.
Stoichiometric air-to-fuel ratio: 10.5

ENVIRONMENTAL DATA
Water pollution: Effect of low concentration on aquatic life is unknown. May be dangerous if it enters nearby water intakes; notify operators. Notify local health and wildlife officials.
Response to discharge: Issue warning–high flammability. Restrict access. Evacuate area. Should be removed. Chemical and physical treatment.

SHIPPING INFORMATION
Grades of purity: 99%.

PHYSICAL AND CHEMICAL PROPERTIES
Physical state @ 59°F/15°C and 1 atm: Liquid.
Molecular weight: 68.13
Boiling point @ 1 atm: 78.8°F/26.0°C/299.2°K.
Melting/Freezing point: −234.9°F/−148.3°C/124.9°K.
Critical temperature: 369°F/187°C/460°K (estimate).
Critical pressure: 540 psia = 37 atm = 3.7 MN/m^2 (estimate).
Specific gravity (water = 1): 0.6608 @ 68°F/20°C.
Relative vapor density (air = 1): 2.3
Latent heat of vaporization: 180 Btu/lb = 100 cal/g = 4.20 x 10^5 J/kg.
Heat of combustion: -20,316 Btu/lb = -11,288 cal/g = -47.3 x 10^6 J/kg.

PENTAERYTHRITOL   REC. P:0550

SYNONYMS: 2,2-BIS(HYDROXYMETHYL)-1,3-PROPANEDIOL; MONO PE; PE; PENTEK; PENTAERITRITA (Spanish); PENTAERYTHRITE; TETRAHYDROXYMETHYLMETHANE; TETRAMETHYLOMETHANE

IDENTIFICATION
CAS Number: 115-77-5
Formula: C(CH$_2$OH)$_4$

DESCRIPTION: White solid. Odorless. Sinks and mixes slowly with water.

EMERGENCY RESPONSE: See Appendix A (171)
Evacuation:
Public safety: Isolate the area of spill or leak for at least 10 to 25 meters (30 to 80 feet) in all directions.
Spill: Increase, in the downwind direction, as necessary, the distance shown under "Public Safety."
Fire: If any large container is involved in fire, isolate for at least 800 meters (½ mile) in all directions; also, consider initial evacuation for 800 meters (½ mile) in all directions.

EXPOSURE
Short-term effects:
Vapor or dust: Not harmful. *IF IN EYES*, hold eyelids open and flush with plenty of water.

HEALTH HAZARDS
Personal protective equipment (PPE): B-Level PPE. Dust mask; goggles. nuisance dust.
Exposure limits (TWA unless otherwise noted): ACGIH TLV 10 mg/m^3; NIOSH 10 mg/m^3 (total), 5 mg/m^3 (respirable fraction); OSHA 15 mg/m^3 (total), 5 mg/m^3 (respirable fraction).
Toxicity by ingestion: Grade 0; LD$_{50}$ = > 15 g/kg.

FIRE DATA
Autoignition temperature: 842°F/450°C (dust cloud).

CHEMICAL REACTIVITY
Binary reactants: Organic acids, oxidizers. Explosive compound is formed when mixture of PE and thiophosphoryl chloride is heated

ENVIRONMENTAL DATA
Food chain concentration potential: Negative; unlikely to accumulate.
Water pollution: Effect of low concentrations on aquatic life is unknown. May be dangerous if it enters nearby water intakes; notify operators. Notify local health and wildlife officials. **Response to discharge:** Should be removed. Chemical and physical treatment.

SHIPPING INFORMATION
Grades of purity: Technical: 86–90% plus 10–14% dimer. Pure: 98+%; **Storage temperature:** Ambient; **Inert atmosphere:** None; **Venting:** Open; **Stability during transport:** Stable.

PHYSICAL AND CHEMICAL PROPERTIES
Physical state @ 59°F/15°C and 1 atm: Solid.
Molecular weight: 136.2
Boiling point @ 1 atm: Sublimes
Melting/Freezing point: 502°F/261°C/534°K.
Specific gravity (water = 1): 1.39 @ 77°F/25°C (solid).
Heat of combustion: −8730 Btu/lb = −4850 cal/g = −203 x 10^5 J/kg.

PENTAETHYLENEHEXAMINE   REC. P:0600

SYNONYMS: LEVEPOX HARDENER T3; PENTAETHYLENE HEXAMINE; 3,6,9,12-TETRAAZATETRADECANE-1,14-DIAMINE

IDENTIFICATION
CAS Number: 4067-16-7
Formula: H$_2$N(CH$_2$CH$_2$NH)$_4$CH$_2$CH$_2$NH$_2$

DESCRIPTION: Yellowish liquid. Ammonia-like odor.

Combustible • Heat or flame may cause explosion • Containers may BLEVE when exposed to fire • Vapors are heavier than air and will collect and stay in low areas • Vapors in confined areas (e.g., tanks, sewers, buildings) may explode when exposed to fire • Irritating to the skin, eyes, and respiratory tract • Toxic products of combustion may include amine vapors, nitrogen oxides, and carbon monoxide.

EMERGENCY RESPONSE: See Appendix A (171)
Evacuation:
Public safety: Isolate the area of spill or leak for at least 10 to 25 meters (30 to 80 feet) in all directions.
Spill: Increase, in the downwind direction, as necessary, the distance shown under "Public Safety."

Fire: If any large container is involved in fire, isolate for at least 800 meters (½ mile) in all directions; also, consider initial evacuation for 800 meters (½ mile) in all directions.

EXPOSURE
Short-term effects: *SEEK MEDICAL ATTENTION.* **Vapor:** Move victim to fresh air. If breathing is difficult, administer oxygen. **Liquid:** Remove contaminated clothing and shoes. Wash affected areas with soap and water. *IF IN EYES*, hold eyelids open and flush with plenty of water. *IF SWALLOWED* and victim is *CONSCIOUS AND ABLE TO SWALLOW*, administer large quantities of water.

HEALTH HAZARDS
Personal protective equipment (PPE): B-Level PPE. Wear chemical protective gloves. Wear full face shield or chemical safety goggles. Use approved respirator to protect against vapors.
Vapor (gas) irritant characteristics: Eye and respiratory tract irritant. Vapors cause irritation such that personnel will find high concentrations unpleasant. The effect is temporary.
Liquid or solid irritant characteristics: Eye and skin irritant. If spilled on clothing and allowed to remain, may cause smarting and reddening of skin.

FIRE DATA
Flash point: 347–405°F/175–207°C (cc).
Fire extinguishing agents not to be used: Do not use water.
Behavior in fire: Containers may rupture.
Autoignition temperature: 680°F/360°C/633°K.

Stoichiometric air-to-fuel ratio: 61.9

CHEMICAL REACTIVITY
Binary reactants: Incompatible with oxidizing materials, isocyanates, and acids.

ENVIRONMENTAL DATA
Water pollution: Effects of low concentrations on aquatic life is unknown. May be dangerous if it enters nearby water intakes; notify operators. Notify local health and wildlife officials.
Response to discharge: Should be removed.

SHIPPING INFORMATION
Grades of purity: Technical grades; **Storage temperature:** Ambient; **Inert atmosphere:** None; **Stability during transport:** Stable.

PHYSICAL AND CHEMICAL PROPERTIES
Physical state @ 59°F/15°C and 1 atm: Liquid.
Molecular weight: 232.38
Boiling point @ 1 atm: 662–734°F/350–390°C/623–663°K.
Specific gravity (water = 1): 1.0 @ 20°C.

PENTANE   REC. P:0650

SYNONYMS: AMYL HYDRIDE; EEC No. 601-006-00-1; PENTAN; *n*-PENTANE; *n*-PENTANO (Spanish)

IDENTIFICATION
CAS Number: 109-66-0
Formula: C_5H_{12}; *n*-C_5H_{12}
DOT ID Number: UN 1265; DOT Guide Number: 128
Proper Shipping Name: Pentanes

DESCRIPTION: Colorless liquid. Gasoline-like odor. Floats on the surface of water. Flammable vapor is produced.

Extremely flammable • Containers may BLEVE when exposed to fire • Vapors may form explosive mixture with air • Vapors are heavier than air and will collect and stay in low areas • Vapors may travel long distances to ignition sources and flashback • Vapors in confined areas (e.g., tanks, sewers, buildings) may explode when exposed to fire • Irritating to the skin, eyes, and respiratory tract • Toxic products of combustion may include carbon monoxide.

Hazard Classification (based on NFPA-704 Rating System)
Health Hazards (Blue): 1; Flammability (Red): 4; Reactivity (Yellow): 0

EMERGENCY RESPONSE: See Appendix A (128)
Evacuation:
Public safety: Isolate spill area for at least 25 to 50 meters (80 to 160 feet) in all directions.
Spill: Large spill–Consider initial downwind evacuation for at least 300 meters (1000 feet).
Fire: Isolate for 800 meters (½ mile) in all directions, especially if tank, rail car, or tank truck is involved in fire.

EXPOSURE
Short-term effects: *SEEK MEDICAL ATTENTION.* The use of alcoholic beverages may enhance toxic effects. **Vapor:** If inhaled, will cause dizziness or difficult breathing. Move to fresh air. *IF BREATHING HAS STOPPED,* give artificial respiration. IF breathing is difficult, administer oxygen. **Liquid:** Harmful if swallowed. *IF SWALLOWED* and victim is *CONSCIOUS AND ABLE TO SWALLOW,* have victim drink 4 to 8 ounces of water. **Do NOT induce vomiting.**

HEALTH HAZARDS
Personal protective equipment (PPE): B-Level PPE. Goggles or face shield (as for gasoline). Chemical protective material(s) reported to have good to excellent resistance: nitrile, PV alcohol, Viton®, Silvershield®, neoprene.
Recommendations for respirator selection: NIOSH
1200 ppm: SA (any supplied-air respirator). *1500 ppm:* SA:CF (any supplied-air respirator operated in a continuous-flow mode); or SCBAF (any self-contained breathing apparatus with a full facepiece); or SAF (any supplied-air respirator with a full facepiece). *EMERGENCY OR PLANNED ENTRY INTO UNKNOWN CONCENTRATIONS OR IDLH CONDITIONS:* SCBAF:PD,PP (any self-contained breathing apparatus that has a full facepiece and is operated in a pressure-demand or other positive-pressure mode); or SAF:PD,PP:ASCBA (any supplied-air respirator that has a full facepiece and is operated in a pressure-demand or other positive-pressure mode in combination with an auxiliary self-contained breathing apparatus operated in a pressure-demand or other positive pressure mode). *ESCAPE:* GMFOV [any air-purifying, full-facepiece respirator (gas mask) with a chin-style, front- or back-mounted organic vapor cannister]; or SCBAE (any appropriate escape-type, self-contained breathing apparatus).
Exposure limits (TWA unless otherwise noted): ACGIH TLV 600 ppm (1770 mg/m^3); OSHA 1000 ppm (2950 mg/m^3); NIOSH 120 ppm (350 mg/m^3), ceiling 610 ppm (1800 mg/m^3)/15 minutes
Short-term exposure limits (15-minute TWA): ACGIH STEL 750 ppm (2210 mg/m^3).
Vapor (gas) irritant characteristics: Eye and respiratory tract irritant.
Liquid or solid irritant characteristics: Liquid will defat the skin, causing dryness and possible irritation.

Odor threshold: 120–1150 ppm.
IDLH value: 1500 ppm (LEL).

FIRE DATA
Flash point: −57°F/−49°C (cc).
Flammable limits in air: LEL: 1.4%; UEL: 8.5%.
Fire extinguishing agents not to be used: Water may be ineffective.
Behavior in fire: Containers may explode.
Autoignition temperature: 544°F/284°C/557°K.
Electrical hazard: Class I, Group D. Due to low electric conductivity, this substance may generate electrostatic charges as a result of agitation and flow.
Burning rate: 8.6 mm/min

CHEMICAL REACTIVITY
Binary reactants: Strong oxidizers. Natural rubber will soften and deteriorate rapidly
Reactivity group: 31
Compatibility class: Paraffins

ENVIRONMENTAL DATA
Food chain concentration potential: Negative; unlikely to accumulate.
Water pollution: Effect of low concentrations on aquatic life is unknown. May be dangerous if it enters nearby water intakes; notify operators. Notify local health and wildlife officials.
Response to discharge: Issue warning–high flammability. Restrict access. Evacuate area.

SHIPPING INFORMATION
Grades of purity: Pure (99.2%); technical; research (99.98%);
Storage temperature: Ambient; Inert atmosphere: None;
Venting: Open (flame arrester) or pressure-vacuum; Stability during transport: Stable.

NAS HAZARD CLASSIFICATION FOR BULK WATER TRANSPORTATION
FIRE: 4
HEALTH: Vapor irritant: 0; Liquid or solid irritant: 0; Poisons: 1
WATER POLLUTION: Human toxicity: 1; Aquatic toxicity: 2; Aesthetic effect: 1
REACTIVITY: Other chemicals: 0; Water: 0; Self-reaction: 0

PHYSICAL AND CHEMICAL PROPERTIES
Physical state @ 59°F/15°C and 1 atm: Liquid.
Molecular weight: 72.15
Boiling point @ 1 atm: 97.0°F/36.1°C/309.3°K.
Melting/Freezing point: −201.0°F/129.4°C/143.8°K.
Critical temperature: 385.7°F/196.5°C/469.7°K.
Critical pressure: 490 psia = 33.3 atm = 3.37 MN/m^2.
Specific gravity (water = 1): 0.626 @ 68°F/20°C (liquid).
Liquid surface tension: 16 dynes/cm = 0.016 N/m @ 68°F/20°C.
Liquid water interfacial tension: 50.2 dynes/cm = 0.0502 N/m @ 68°F/20°C.
Relative vapor density (air = 1): 2.5
Ratio of specific heats of vapor (gas): 1.075
Latent heat of vaporization: 153.7 Btu/lb = 85.38 cal/g = 3.575 x 10^5 J/kg.
Heat of combustion: −19,352 Btu/lb = −10,751 cal/g = −450.12 x 10^5 J/kg.
Heat of fusion: 27.89 cal/g.
Vapor pressure: (Reid) 15.5 psia; 420 mm.

PENTANOIC ACID
REC. P:0700

SYNONYMS: ACIDO PENTNOICO (Spanish); 1-BUTANECARBOXYLIC ACID; n-PENTANOIC ACID; PROPYLACETIC ACID; VALERIC ACID

IDENTIFICATION
CAS Number: 109-52-4
Formula: $CH_3(CH_2)_3COOH$
DOT ID Number: UN 1993; DOT Guide Number: 128
Proper Shipping Name: Combustible liquid, n.o.s.

DESCRIPTION: Colorless liquid. Unpleasant odor.

Combustible • Heat or flame may cause explosion • Containers may BLEVE when exposed to fire • Vapors are heavier than air and will collect and stay in low areas • Vapors in confined areas (e.g., tanks, sewers, buildings) may explode when exposed to fire • Irritating to the skin, eyes, and respiratory tract • Toxic products of combustion may include CO_2 and carbon monoxide.

Hazard Classification (based on NFPA-704 Rating System)
Health Hazards (Blue): 2; Flammability (Red): 1; Reactivity (Yellow): 0

EMERGENCY RESPONSE: See Appendix A (128)
Evacuation:
Public safety: Isolate spill area for at least 25 to 50 meters (80 to 160 feet) in all directions.
Spill: Large spill–Consider initial downwind evacuation for at least 300 meters (1000 feet).
Fire: Isolate for 800 meters (½ mile) in all directions, especially if tank, rail car, or tank truck is involved in fire.

EXPOSURE
Short-term effects: *SEEK MEDICAL ATTENTION.* Vapor: Move victim to fresh air. *IF BREATHING HAS STOPPED,* give artificial respiration. IF breathing is difficult, administer oxygen. Liquid: Corrosive to skin and eyes. Remove contaminated clothing and shoes. Flush affected areas with water. *IF IN EYES,* hold eyelids open and flush with plenty of water.

HEALTH HAZARDS
Personal protective equipment (PPE): B-Level PPE. Full impervious protective clothing, including boots and gloves. Where splashing is possible wear full face shield or chemical safety goggles. Butyl rubber is generally suitable for carbooxylic acid compounds. Use approved respirator to protect against vapors.
Vapor (gas) irritant characteristics: Vapor cause severe irritation of eyes and throat and can cause eye and lung injury. They cannot be tolerated even at low concentrations.
Liquid or solid irritant characteristics: Severe eye and skin irritant. Causes second- and third-degree skin burns on short contact and is very injurious to the eyes.

FIRE DATA
Flash point: 205°F/96°C (cc).
Flammable limits in air: LEL: 1.6%; UEL: 7.6%.
Fire extinguishing agents not to be used: Water.
Autoignition temperature: 707°F/375°C/648°K.
Stoichiometric air-to-fuel ratio: 31.0

CHEMICAL REACTIVITY
Reactivity with water: May generate heat.
Binary reactants: Incompatible with strong oxidizers.

Neutralizing agents for acids and caustics: Lime.
Water pollution: Effects of low concentrations on aquatic life is unknown. May be dangerous if it enters nearby water intakes; notify operators. Notify local health and wildlife officials.
Response to discharge: Issue warning–corrosive. Should be removed.

SHIPPING INFORMATION
Grades of purity: 99%; technical; **Storage temperature:** Ambient; **Inert atmosphere:** None; **Stability during transport:** Stable.

PHYSICAL AND CHEMICAL PROPERTIES
Physical state @ 59°F/15°C and 1 atm: Liquid.
Molecular weight: 102.13
Boiling point @ 1 atm: 365°F/185°C/458°K.
Melting/Freezing point: 0°F/–18°C/255°K.
Specific gravity (water = 1): 0.939
Relative vapor density (air = 1): 3.5

1-PENTENE   REC. P:0800

SYNONYMS: AMYLENE; α-n-AMYLENE; PENTENO (Spanish); PENTYLENE; PROPYLETHYLENE

IDENTIFICATION
CAS Number: 109-67-1; 25377-72-4 (mixed isomers); also applies to 646-04-8 (2-isomer); 26760-64-5 (*tert*-isomer)
Formula: C_5H_{10}; $CH_3(CH_2)_2CH=CH_2$
DOT ID Number: UN 1108; DOT Guide Number: 127
Proper Shipping Name: *n*-Amylene

DESCRIPTION: Colorless liquid. Gasoline-like odor. Floats on the surface of water. Flammable vapor is produced. Can polymerize slowly.

Extremely flammable • Containers may BLEVE when exposed to fire •Vapors may form explosive mixture with air • Vapors are heavier than air and will collect and stay in low areas • Vapors may travel long distances to ignition sources and flashback • Vapors in confined areas (e.g., tanks, sewers, buildings) may explode when exposed to fire • Irritating to the skin, eyes, and respiratory tract • Toxic products of combustion may include carbon monoxide.

Hazard Classification (based on NFPA-704 Rating System)
Health Hazards (Blue): 1; Flammability (Red): 4; Reactivity (Yellow): 0

EMERGENCY RESPONSE: See Appendix A (127)
Evacuation:
Public safety: Isolate spill area for at least 25 to 50 meters (80 to 160 feet) in all directions.
Spill: Large spill–Consider initial downwind evacuation for at least 300 meters (1000 feet).
Fire: Isolate for 800 meters (½ mile) in all directions, especially if tank, rail car, or tank truck is involved in fire.

EXPOSURE
Short-term effects: *SEEK MEDICAL ATTENTION*. **Vapor:** If inhaled, will cause dizziness. Move to fresh air. *IF BREATHING HAS STOPPED*, give artificial respiration. IF breathing is difficult, administer oxygen. **Liquid:** Harmful if swallowed. *IF SWALLOWED* and victim is *CONSCIOUS AND ABLE TO SWALLOW*, have victim drink 4 to 8 ounces of water. **Do NOT induce vomiting**.

HEALTH HAZARDS
Personal protective equipment (PPE): B-Level PPE. Goggles or face shield (as for gasoline). Chemical protective materials offering limited protection: neoprene, nitrile, styrene-butadiene.
Long-term health effects: Possible heart damage.
Vapor (gas) irritant characteristics: Eye and respiratory tract irritant.
Liquid or solid irritant characteristics: Eye and skin irritant.

FIRE DATA
Flash point: –60°F/–51°C (cc); 0°F/–18°C (oc)
Flammable limits in air: LEL: 1.5%; UEL: 8.7%.
Fire extinguishing agents not to be used: Water may be ineffective.
Behavior in fire: Containers may explode.
Autoignition temperature: 527°F/275°C/548°K.
Electrical hazard: Flow or agitation of substance may generate electrostatic charges due to low conductivity.
Burning rate: 9.1 mm/min

CHEMICAL REACTIVITY
Binary reactants: Reacts with oxidizers, strong acids
Polymerization: May be able to polymerize on long standing. Heat may cause slow polymerization.
Reactivity group: 30
Compatibility class: Olefins

ENVIRONMENTAL DATA
Food chain concentration potential: Unlikely to accumulate.
Water pollution: Effect of low concentrations on aquatic life is unknown. May be dangerous if it enters nearby water intakes; notify operators. Notify local health and wildlife officials. **Response to discharge:** Issue warning–high flammability. Restrict access. Evacuate area.

SHIPPING INFORMATION
Grades of purity: Research: 99.9%; pure: 99.4%; technical: 97.0%; **Storage temperature:** Ambient; **Inert atmosphere:** None; **Venting:** Open (flame arrester) or pressure-vacuum; **Stability during transport:** Stable.

PHYSICAL AND CHEMICAL PROPERTIES
Physical state @ 59°F/15°C and 1 atm: Liquid.
Molecular weight: 70.13
Boiling point @ 1 atm: 85.8°F/29.9°C/303.1°K.
Melting/Freezing point: –265°F/–165°C/108°K.
Critical temperature: 376.9°F/191.6°C/464.8°K.
Critical pressure: 588 psia = 40 atm = 4.05 MN/m^2.
Specific gravity (water = 1): 0.641 @ 68°F/20°C (liquid).
Liquid surface tension: 16.5 dynes/cm = 0.0165 N/m @ 68°F/20°C.
Liquid water interfacial tension: (estimate) 50 dynes/cm = 0.05 N/m @ 68°F/20°C.
Relative vapor density (air = 1): 2.4
Ratio of specific heats of vapor (gas): 1.083
Latent heat of vaporization: 154.6 Btu/lb = 85.87 cal/g = 3.595 x 10^5 J/kg.
Heat of combustion: –19,359 Btu/lb = –10,755 cal/g = –450.29 x 10^5 J/kg.
Vapor pressure: 500 mm.

n-PENTYL PROPIONATE   REC. P:0850

SYNONYMS: n-AMYL PROPIONATE; PENTYL PROPIONATE

IDENTIFICATION
CAS Number: 624-54-4
Formula: $C_8H_{16}O_2$; $CH_3CH_2COO(CH_2)_4CH_3$
DOT ID Number: UN 1993; DOT Guide Number: 128
Proper Shipping Name: Flammable Liquid.

DESCRIPTION: Colorless liquid. Apple-like odor. Floats on the surface of water.

Highly flammable • Containers may BLEVE when exposed to fire •Vapors may form explosive mixture with air • Vapors are heavier than air and will collect and stay in low areas • Vapors may travel long distances to ignition sources and flashback • Vapors in confined areas (e.g., tanks, sewers, buildings) may explode when exposed to fire • Irritating to the skin, eyes, and respiratory tract • Toxic products of combustion may include carbon monoxide.

Hazard Classification (based on NFPA-704 Rating System)
Health Hazards (Blue): 0; Flammability (Red): 2; Reactivity (Yellow): 0

EMERGENCY RESPONSE: See Appendix A (128)
Evacuation:
Public safety: Isolate spill area for at least 25 to 50 meters (80 to 160 feet) in all directions.
Spill: Large spill–Consider initial downwind evacuation for at least 300 meters (1000 feet).
Fire: Isolate for 800 meters (½ mile) in all directions, especially if tank, rail car, or tank truck is involved in fire.

EXPOSURE
Short-term effects: *SEEK MEDICAL ATTENTION.* **Vapor:** Irritating to eyes, nose, and throat. *IF BREATHING HAS STOPPED,* give artificial respiration. If breathing is difficult, administer oxygen. **Liquid:** Irritating to skin and eyes. Remove contaminated clothing and shoes. Flush affected areas with plenty of water. *IF IN EYES,* hold eyelids open and flush with plenty of water.

HEALTH HAZARDS
Personal protective equipment (PPE): B-Level PPE. Full impervious protective clothing, including boots and gloves. Where splashing is possible wear full face shield or chemical safety goggles. Use approved respirator to protect against vapors.
Vapor (gas) irritant characteristics: Vapors cause a slight smarting of the eyes or respiratory system if present in high concentrations. The effect is temporary.
Liquid or solid irritant characteristics: If spilled on clothing and allowed to remain, may cause smarting and reddening of skin.

FIRE DATA
Flash point: 106°F/41°C (oc).
Fire extinguishing agents not to be used: Water.
Autoignition temperature: 712°F/378°C/651°K.
Stoichiometric air-to-fuel ratio: 52.4

CHEMICAL REACTIVITY
Binary reactants: Strong acids.
Reactivity group: 34
Compatibility class: Esters.

ENVIRONMENTAL DATA
Water pollution: Effects of low concentrations on aquatic life is unknown. May be dangerous if it enters nearby water intakes; notify operators. Notify local health and wildlife officials. **Response to discharge:** Mechanical containment. Should be removed.

SHIPPING INFORMATION
Grades of purity: Technical grades; **Storage temperature:** Ambient; **Inert atmosphere:** None; **Stability during transport:** Stable.

PHYSICAL AND CHEMICAL PROPERTIES
Physical state @ 59°F/15°C and 1 atm: Liquid.
Molecular weight: 144.22
Boiling point @ 1 atm: 275–347°F/135–175°C/408–448°K.
Specific gravity (water = 1): 0.869–0.873 @ 20°C.

PERACETIC ACID   REC. P:0900

SYNONYMS: ACIDO PERACETICO (Spanish); ACETIC PEROXIDE; ACETYL HYDROPEROXIDE; ACIDE PERACETIQUE (French); EEC No. 607-094-00-8; HYDROPEROXIDE, ACETYL; ETHANEPEROXIC ACID; PAA; PEROXYACETIC ACID

IDENTIFICATION
CAS Number: 79-21-0
Formula: $C_2H_4O_3$; $CH_3COOOH \cdot CH_3COOH$
DOT ID Number: UN 2131; DOT Guide Number: 147
Proper Shipping Name: Peracetic acid, solution; Peroxyacetic acid, solution
Reportable Quantity (RQ): **(EHS)** 1 lb/0.454 kg

DESCRIPTION: Colorless liquid. Strong odor. Mixes with water. Flammable, irritating vapor is produced.

Flammable • Heat, shock, or contamination may cause explosion. Liquid will detonate if concentration rises above 56% because of evaporation of acetic acid. • Strong oxidizer which may react spontaneously with low flash point organics or reducing agents. Decomposes at approximately 100°F/38°C producing oxygen and increased fire activity. Explosive at temperatures above 230°F/110°C • May cause fire or explosion on contact with combustibles (wood, paper, oil, clothing, etc.) • Containers may BLEVE when exposed to fire •Vapors may form explosive mixture with air • Vapors are heavier than air and will collect and stay in low areas • Vapors may travel long distances to ignition sources and flashback • Vapors in confined areas (e.g., tanks, sewers, buildings) may explode when exposed to fire • Irritating to the skin, eyes, and respiratory tract • Toxic products of combustion may include carbon monoxide.

Hazard Classification (based on NFPA-704 Rating System)
Health Hazards (Blue): 3; Flammability (Red): 2; Reactivity (Yellow): 4; Special Notice (White): OXY

EMERGENCY RESPONSE: See Appendix A (147)
Evacuation:
Public safety: Isolate the area of spill or leak for at least 25 to 50 meters (80 to 160 feet) in all directions.
Spill: Consider initial evacuation for at least 250 meters (800 feet).
Fire: If tank, rail car, or tank truck is involved in fire, isolate for at least 800 meters (½ mile) in all directions; also, consider initial evacuation for 800 meters (½ mile) in all directions.

EXPOSURE

Short-term effects: *SEEK MEDICAL ATTENTION*. **Vapor:** Irritating to eyes, nose, and throat. Move victim to fresh air. *IF BREATHING HAS STOPPED,* give artificial respiration. IF breathing is difficult, administer oxygen. **Liquid:** Irritating to skin and eyes. Harmful if swallowed. Remove contaminated clothing and shoes. Flush affected areas with plenty of water. *IF IN EYES,* hold eyelids open and flush with plenty of water. *IF SWALLOWED* and victim is *CONSCIOUS AND ABLE TO SWALLOW,* have victim drink 4 to 8 ounces of water.

HEALTH HAZARDS

Personal protective equipment (PPE): B-Level PPE. Self-contained breathing apparatus; full protective clothing (goggles, rubber gloves, etc.). Butyl rubber is generally suitable for peroxide compounds.
Toxicity by ingestion: Grade 4; oral LD_{50} = 10 mg/kg (guinea pig).
Vapor (gas) irritant characteristics: Severe eye and respiratory tract irritant.
Liquid or solid irritant characteristics: Severe eye and skin irritant. A strong oxidizer; saturated clothing may be a fire hazard.

FIRE DATA

Flash point: 104°F/40°C (oc).
Behavior in fire: Vapors are very flammable and explosive. Liquid will detonate if concentration rises above 56% because of evaporation of acetic acid.
Autoignition temperature: 392°F/200°C/573°K.

CHEMICAL REACTIVITY

Reactivity with water: Solution is acidic.
Binary reactants: Substance is a strong oxidizer; may cause fire in contact with organic materials such as wood, cotton or straw. Solution reacts with bases, thiocyanates. Corrosive to most metals, including aluminum.
Neutralizing agents for acids and caustics: Flush with water.

ENVIRONMENTAL DATA

Food chain concentration potential: Log P_{ow} = –0.9. Unlikely to accumulate.
Water pollution: Effect of low concentrations on aquatic life is unknown. May be dangerous if it enters nearby water intakes; notify operators. Notify local health and wildlife officials.
Response to discharge: Issue warning–oxidizing material, water contaminant. Restrict access. Disperse and flush.

SHIPPING INFORMATION

Grades of purity: 40% peracetic acid, 40% acetic acid, 5% hydrogen peroxide, 13% water, 500 ppm stabilizer; **Storage temperature:** 60–122°F; **Inert atmosphere:** None; **Venting:** Safety relief; **Stability during transport:** Stable if kept cool and out of contact with most metals. At 85°F/30°C concentration decreases about 0.4% each month.

PHYSICAL AND CHEMICAL PROPERTIES

Physical state @ 59°F/15°C and 1 atm: Liquid.
Boiling point @ 1 atm: 221°F/105°C/378°K.
Melting/Freezing point: (approximate) 40% solution in acetic acid/water –22°F/–30°C/243°K.
Specific gravity (water = 1): (estimate) 1.153 @ 77°F/25°C (liquid).
Relative vapor density (air = 1): 2.5
Vapor pressure: 20 mm.

PERCHLORIC ACID   REC. P:0950

SYNONYMS: ACIDO PERCLORICO (Spanish); DIOXONIUM PERCHLORATE SOLUTION; EEC No. 017-006-00-4; PERCHLORATE SOLUTION; PERCHLORIC ACID SOLUTION

IDENTIFICATION

CAS Number: 7601-90-3
Formula: $ClHO_4$; $HClO_4 \cdot H_2O$
DOT ID Number: UN 1802 (less than 50% acid); UN 1873 (50–72% acid); DOT Guide Number: 140 (less than 50% acid); 143 (50–72% acid)
Proper Shipping Name: Perchloric acid, with not more than 50% acid (UN 1802); Perchloric acid, with more than 50% but not more than 72% acid (UN 1873).

DESCRIPTION: Colorless, oily liquid. Odorless. Sinks and mixes with water (spatters) forming a strong acid; heat is evolved.

Can explode spontaneously as a result of dehydration; or, from shock or heat. Above 320°F/160°C will react with combustible material and increase intensity of fire • Corrosive! • Breathing the vapor can kill you; skin or eye contact causes severe burns, impaired vision, or blindness; inhalation symptoms may be delayed • Firefighting gear (including SCBA) does not provide adequate protection. If exposure occurs, remove and isolate gear immediately and thoroughly decontaminate personnel • Strong oxidizer which may react spontaneously with low flash point organics or reducing agents. Heat forms oxygen; will increase the activity of an existing fire • May cause fire or explosion contact with combustibles (wood, paper, oil, clothing, etc.). Containers may BLEVE when exposed to fire • Toxic products of combustion may include hydrogen chloride • Do not put yourself in danger by entering a contaminated area to rescue a victim.

Hazard Classification (based on NFPA-704 Rating System)
Health Hazards (Blue): 3; Flammability (Red): 0; Reactivity (Yellow): 3; Special Notice (White): OXY

EMERGENCY RESPONSE, less than 50% acid: See Appendix A (140)
Note: Do not let spill area dry until it has been determined that there is no perchlorates left in the area. Continue cooling after fire has been extinguished.
Evacuation:
Public safety: Isolate the area of spill or leak for at least 10 to 25 meters (30 to 80 feet) in all directions.
Spill: Consider initial downwind evacuation for at least 100 meters (330 feet).
Fire: If any large container is involved in fire, isolate for at least 800 meters (½ mile) in all directions; also, consider initial evacuation for 800 meters (½ mile) in all directions.
Note: Do not let spill area dry until it has been determined that there is no perchlorates left in the area. Continue cooling after fire has been extinguished.
EMERGENCY RESPONSE, 50–72% acid: See Appendix A (143)
Evacuation:
Public safety: Isolate the area of spill or leak for at least 50 to 100 meters (160 to 300 feet) in all directions.
Spill: Increase, in the downwind direction, as necessary, the distance shown under "Public Safety."
Fire: If any large container is involved in fire, isolate for at least 800 meters (½ mile) in all directions; also, consider initial evacuation for 800 meters (½ mile) in all directions.

EXPOSURE

Short-term effects: *SEEK MEDICAL ATTENTION.* **Vapor:** Irritating to eyes, nose, and throat. *IF INHALED*, will, will cause coughing or difficult breathing. *IF IN EYES*, hold eyelids open and flush with plenty of water. *IF BREATHING HAS STOPPED*, give artificial respiration. IF breathing is difficult, administer oxygen. **Liquid:** Will burn skin and eyes. *IF SWALLOWED*, will cause nausea and vomiting. Remove contaminated clothing and shoes. Flush affected areas with plenty of water. *IF IN EYES*, hold eyelids open and flush with plenty of water. *IF SWALLOWED* and victim is *CONSCIOUS AND ABLE TO SWALLOW*, have victim drink 4 to 8 ounces of water. *IF SWALLOWED* and victim is *UNCONSCIOUS OR HAVING CONVULSIONS*, do nothing except keep victim warm.

Note to physician or authorized medical personnel: Medical observation is recommended for 24 to 48 hours after breathing overexposure, as pulmonary edema may be delayed. As first aid for pulmonary edema, consider administering a corticosteroid spray. Cigarette smoking may exacerbate pulmonary injury and should be discouraged for at least 72 hours following exposure.

HEALTH HAZARDS

Personal protective equipment (PPE): B-Level PPE. Gloves; face shield or vapor-tight chemical-type safety goggles; apron; boots or shoes. Sealed chemical protective materials recommended (30-70%): natural rubber, neoprene, nitrile, PVC, nitrile/PVC. Also, butyl rubber, Viton®/neoprene, Viton®, butyl rubber/neoprene offers limited protection

Vapor (gas) irritant characteristics: Severe eye and respiratory tract irritant.

Liquid or solid irritant characteristics: severe eye and skin irritant. A strong oxidizer, saturated clothing may become a fire hazard.

FIRE DATA

Fire extinguishing agents not to be used: Do not use dry chemicals or CO_2. Use water spray only.

Behavior in fire: Containers may explode.

CHEMICAL REACTIVITY

Reactivity with water: Spatters and produces heat. Solution is a strong acid.

Binary reactants: Contact with dehydration agents such as alcohols, hydrogen fluoride, etc, can cause explosion. Contact with most combustible materials may cause fires and explosions. Reacts with bases. Corrosive to most metals with formation of flammable hydrogen gas, which may collect in enclosed spaces.

Neutralizing agents for acids and caustics: Flush with water and rinse with dilute sodium bicarbonate or soda ash solution.

ENVIRONMENTAL DATA

Food chain concentration potential: Negative; unlikely to accumulate.

Water pollution: Effect of low concentrations on aquatic life is unknown. May be dangerous if it enters nearby water intakes; notify operators. Notify local health and wildlife officials.

Response to discharge: Issue warning–corrosive, oxidizing material. Restrict access. Disperse and flush.

SHIPPING INFORMATION

Grades of purity: ACS, 60-72% solution in water; **Storage temperature:** Ambient; **Inert atmosphere:** None; **Venting:** Open; **Stability during transport:** Stable if cool.

PHYSICAL AND CHEMICAL PROPERTIES

Physical state @ 59°F/15°C and 1 atm: Liquid.
Molecular weight: 100.46 (solute only).
Boiling point @ 1 atm: 66°F/19°C/292°K (decomposes).
Melting/Freezing point: (anhydrous) –170°F/–112°C/161°K; (solution) 25°F/–4°C/269°K.
Specific gravity (water = 1): 1.6-1.7 @ 77°F/25°C (liquid).

PERCHLOROMETHYL MERCAPTAN   REC. P:1050

SYNONYMS: CLAIRSIT; PCM; PERCLOROMETILMERCAPTANO (Spanish); PMM; TRICHLORMETHYL SULFUR CHLORIDE; THIOCARBONYL TETRACHLORIDE; TRICHLOROMETHYL SULFOCHLORIDE; TRICHLOROMETHANE SULFURYL CHLORIDE; TRICHLOROMETHYL SULFUR CHLORIDE; TRICHLOROMETHANESULFENYL CHLORIDE; RCRA No. P118

IDENTIFICATION

CAS Number: 594-42-3
Formula: CCl_4S; Cl_3CSCl
DOT ID Number: UN 1670; DOT Guide Number: 157
Proper Shipping Name: Perchloromethylmercaptan
Reportable Quantity (RQ): **(CERCLA)** 100 lb/45.4 kg

DESCRIPTION: Bright yellow; pale yellow; orange-red liquid. Intensely strong, unbearable odor; like very strong acrid. Sinks in water; insoluble; reacts, forming toxic and flammable mercaptan vapors.

Poison! • Breathing the vapor can kill you; skin or eye contact causes severe burns, impaired vision, or blindness • Firefighting gear (including SCBA) does not provide adequate protection. If exposure occurs, remove and isolate gear immediately and thoroughly decontaminate personnel • Containers may BLEVE when exposed to fire • Toxic products of combustion may include phosgene gas, sulfur dioxide and hydrogen chloride • Do not put yourself in danger by entering a contaminated area to rescue a victim.

Hazard Classification (based on NFPA-704 Rating System)
Health Hazards (Blue): 2; Flammability (Red): 0; Reactivity (Yellow): 0

EMERGENCY RESPONSE: See Appendix A (157)
Evacuation:
Public safety: Isolate the area of spill or leak for at least 50 to 100 meters (160 to 330 feet) in all directions.
Spill: Small spill–First: Isolate in all directions 30 meters (100 feet); Then: Protect persons downwind, DAY: 0.2 km (0.1 mile); NIGHT: 0.3 km (0.2 mile). Large spill–First: Isolate in all directions 60 meters (200 feet); Then: Protect persons downwind, DAY: 0.5 km (0.3 mile); NIGHT: 1.1 km (0.7 mile).
Fire: If tank, rail car, or tank truck is involved in fire, isolate for at least 800 meters (½ mile) in all directions; also, consider initial evacuation for 800 meters (½ mile) in all directions.

EXPOSURE

Short-term effects: *SEEK MEDICAL ATTENTION.* **Vapor:** *POISONOUS IF INHALED.* Irritating to eyes, nose, and throat. Move victim to fresh air. *IF BREATHING HAS STOPPED*, give artificial respiration. IF breathing is difficult, administer oxygen.

Liquid: *POISONOUS IF SWALLOWED OR IF SKIN IS EXPOSED.* Irritating to eyes. Remove contaminated clothing and shoes. Flush affected areas with plenty of water. *IF IN EYES*, hold eyelids open and flush with plenty of water. *IF SWALLOWED* and victim is *CONSCIOUS AND ABLE TO SWALLOW*, have victim drink 4 to 8 ounces of water and have victim induce vomiting. *IF SWALLOWED* and victim is *UNCONSCIOUS OR HAVING CONVULSIONS*, do nothing except keep victim warm.

HEALTH HAZARDS

Personal protective equipment (PPE): B-Level PPE. Organic- and acid-type canister mask or self-contained breathing apparatus; goggles or face shield; rubber gloves.

Recommendations for respirator selection: NIOSH/OSHA
1 ppm: CCROV* [any chemical cartridge respirator with organic vapor cartridge(s)]; or SA* (any supplied-air respirator). *2.5 ppm:* SA:CF (any supplied-air respirator operated in a continuous-flow mode); or PAPROV* [any powered, air-purifying respirator with organic vapor cartridge(s)]. *5 ppm:* CCRFOV [any air-purifying, full-facepiece respirator (gas mask) with a chin-style, front- or back-mounted acid gas canister]; or GMFOV [any air-purifying, full-facepiece respirator (gas mask) with a chin-style, front- or back-mounted organic vapor cannister]; or PAPRTOV* [any powered, air-purifying respirator with a tight-fitting facepiece and organic vapor cartridges(s)]; or SAT:CF* (any supplied-air respirator that has a tight-fitting facepiece and is operated in a continuous-flow mode); or SCBAF (any self-contained breathing apparatus with a full facepiece); or SAF (any supplied-air respirator with a full facepiece). *10 ppm:* SAF:PD,PP (any supplied-air respirator that has a full facepiece and is operated in a pressure-demand or other positive-pressure mode). *EMERGENCY OR PLANNED ENTRY INTO UNKNOWN CONCENTRATIONS OR IDLH CONDITIONS:* SCBAF:PD,PP (any self-contained breathing apparatus that has a full facepiece and is operated in a pressure-demand or other positive-pressure mode); or SAF:PD,PP:ASCBA (any supplied-air respirator that has a full facepiece and is operated in a pressure-demand or other positive-pressure mode in combination with an auxiliary self-contained breathing apparatus operated in a pressure-demand or other positive pressure mode). *ESCAPE:* GMFOV [any air-purifying, full-facepiece respirator (gas mask) with a chin-style, front- or back-mounted organic vapor cannister]; or SCBAE (any appropriate escape-type, self-contained breathing apparatus).
Note: Substance reported to cause eye irritation or damage; may require eye protection.

Exposure limits (TWA unless otherwise noted): ACGIH TLV 0.1 ppm (0.76 mg/m^3); NIOSH/OSHA 0.1 ppm (0.8 mg/m^3).
Toxicity by ingestion: Grade 3; oral LD$_{50}$ = 83 mg/kg (rat).
Vapor (gas) irritant characteristics: Eye and respiratory tract irritant.
Liquid or solid irritant characteristics: Severe eye and skin irritant.
Odor threshold: 0.001 ppm.
IDLH value: 10 ppm.

FIRE DATA

Behavior in fire: At elevated temperatures will decompose to carbon tetrachloride, sulfur chloride, and heavy oily polymers.

CHEMICAL REACTIVITY

Reactivity with water: Forms CO_2, hydrochloric acid, and sulfur on contact.
Binary reactants: Reacts with alkalis, amines; iron or steel, evolving carbon tetrachloride. Corrosive to most metals.

Neutralizing agents for acids and caustics: Flood with water, rinse with dilute sodium bicarbonate or lime solution.

ENVIRONMENTAL DATA

Food chain concentration potential: Unlikely to accumulate.
Water pollution: DOT Appendix B, §172.101–marine pollutant. Effect of low concentrations on aquatic life is unknown. May be dangerous if it enters nearby water intakes; notify operators. Notify local health and wildlife officials. **Response to discharge:** Issue warning–poison, water contaminant, air contaminant. Restrict access. Disperse and flush.

SHIPPING INFORMATION

Grades of purity: Technical: 97+%; **Storage temperature:** Ambient; **Inert atmosphere:** None; **Venting:** Pressure-vacuum; **Stability during transport:** Stable.

PHYSICAL AND CHEMICAL PROPERTIES

Physical state @ 59°F/15°C and 1 atm: Liquid.
Molecular weight: 185.9
Boiling point @ 1 atm: 300°F/148°C/421°K (decomposes).
Specific gravity (water = 1): 1.706 at 11°C (liquid).
Liquid surface tension: 35.02 dynes/cm = 0.03502 N/m @ 68°F/20°C.
Latent heat of vaporization: (estimate) 94 Btu/lb = 52 cal/g = 2.2 x 10^5 J/kg.

PETROLATUM   REC. P:1100

SYNONYMS: PARAFFIN JELLY; PETROLEUM JELLY; PETROLATUM JELLY; VASELINE; VASELINA (Spanish); YELLOW PETROLATUM

IDENTIFICATION

CAS Number: 8012-95-1 (liquid).

DESCRIPTION: Colorless liquid. Practically odorless. Floats on water surface; insoluble.

Combustible • Heat or flame may cause explosion • Containers may BLEVE when exposed to fire • Vapors are heavier than air and will collect and stay in low areas • Vapors in confined areas (e.g., tanks, sewers, buildings) may explode when exposed to fire • Irritating to the skin, eyes, and respiratory tract • Toxic products of combustion may include carbon monoxide.

Hazard Classification (based on NFPA-704 Rating System)
Health Hazards (Blue): 0; Flammability (Red): 1; Reactivity (Yellow): 0

EMERGENCY RESPONSE: See Appendix A (171)
Evacuation:
Public safety: Isolate the area of spill or leak for at least 10 to 25 meters (30 to 80 feet) in all directions.
Spill: Increase, in the downwind direction, as necessary, the distance shown under "Public Safety."
Fire: If any large container is involved in fire, isolate for at least 800 meters (½ mile) in all directions; also, consider initial evacuation for 800 meters (½ mile) in all directions.

EXPOSURE

Short-term effects: *SEEK MEDICAL ATTENTION.* **Liquid:** Irritating to eyes. Flush affected areas with plenty of water. *IF IN EYES*, hold eyelids open and flush with plenty of water.

HEALTH HAZARDS
Personal protective equipment (PPE): B-Level PPE. Goggles or face shield
Toxicity by ingestion: Grade 1; LD_{50} = 5 to 15 g/kg.

FIRE DATA
Flash point: 360–430°F/182–221°C (cc).

CHEMICAL REACTIVITY
Binary reactants: Nitric acid, strong oxidizers.
Reactivity group: 33
Compatibility class: Miscellaneous hydrocarbon mixtures

ENVIRONMENTAL DATA
Food chain concentration potential: Negative; unlikely to accumulate.
Water pollution: Effect of low concentrations on aquatic life is unknown. Fouling to shoreline. May be dangerous if it enters nearby water intakes; notify operators. Notify local health and wildlife officials. **Response to discharge:** Mechanical containment. Should be removed. Chemical and physical treatment.

SHIPPING INFORMATION
Grades of purity: USP, NF, technical (these vary in color and in melting point); **Storage temperature:** Ambient; **Inert atmosphere:** None; **Venting:** Open (flame arrester); **Stability during transport:** Stable.

PHYSICAL AND CHEMICAL PROPERTIES
Physical state @ 59°F/15°C and 1 atm: Grease
Boiling point @ 1 atm: 302°F/150°C/423°K (approximate).
Melting/Freezing point: 100–135°F/38–57°C/311–330°K.
Specific gravity (water = 1): (estimate) 0.865 at 60°C (liquid).
Liquid water interfacial tension: 35–50 dynes/cm = 0.035–0.050 N/m
Latent heat of vaporization: 97–100 Btu/lb = 54–63 cal/g = 2.3–2.6 x 10^5 J/kg.
Vapor pressure: (Reid) Very low.

PETROLEUM NAPHTHA   REC. P:1150

SYNONYMS: NAFTA de PEROLEO (Spanish); PETROLEUM SOLVENT

IDENTIFICATION
CAS Number: 68955-35-1 (catalytic reformed); 64741-54-4 (heavy, catalytic cracked); 64741-68-0 (heavy, catalytic reformed); 64741-55-5 (light, catalytic cracked); 64741-63-5 (light catalytic reformed); 64741-46-4 (light straight run); 64741-87-3 (sweetened); 64741-83-9 (heavy, thermal-cracked); 64742-95-6 (high-flash, aromatic); 8030-30-6
DOT ID Number: UN 1255; DOT Guide Number: 128
Proper Shipping Name: Petroleum naphtha

DESCRIPTION: Colorless liquid. Gasoline- or kerosene-like odor. Floats on water surface; insoluble.

Extremely flammable • Containers may BLEVE when exposed to fire •Vapors may form explosive mixture with air • Vapors are heavier than air and will collect and stay in low areas • Vapors may travel long distances to ignition sources and flashback • Vapors in confined areas (e.g., tanks, sewers, buildings) may explode when exposed to fire • Irritating to the skin, eyes, and respiratory tract • Toxic products of combustion may include carbon monoxide.

Hazard Classification (based on NFPA-704 Rating System)
Health Hazards (Blue): 1; Flammability (Red): 4; Reactivity (Yellow): 0

EMERGENCY RESPONSE: See Appendix A (128)
Evacuation:
Public safety: Isolate spill area for at least 25 to 50 meters (80 to 160 feet) in all directions.
Spill: Large spill–Consider initial downwind evacuation for at least 300 meters (1000 feet).
Fire: Isolate for 800 meters (½ mile) in all directions, especially if tank, rail car, or tank truck is involved in fire.

EXPOSURE
Short-term effects: *SEEK MEDICAL ATTENTION*. The use of alcoholic beverages may enhance toxic effects. **Vapor:** Not irritating to eyes, nose, or throat. **Liquid:** Harmful if swallowed (droplets entering the lungs may cause pneumonia). *IF SWALLOWED* and victim is *CONSCIOUS AND ABLE TO SWALLOW*, have victim drink 4 to 8 ounces of water. **Do NOT induce vomiting.**

HEALTH HAZARDS
Personal protective equipment (PPE): B-Level PPE. Goggles or face shield. Chemical protective material(s) reported to have good to excellent resistance: nitrile, PV alcohol, Viton®, Saranex®, Silvershield®. Also, PVC, neoprene, and Styrene-butadiene offers limited protection
Recommendations for respirator selection: NIOSH as petroleum distillates (naphtha).
850 ppm: SA (any supplied-air respirator); *1100 ppm*: SA:CF* (any supplied-air respirator operated in a continuous-flow mode); or SCBAF (any self-contained breathing apparatus with a full facepiece); or SAF (any supplied-air respirator with a full facepiece). *EMERGENCY OR PLANNED ENTRY INTO UNKNOWN CONCENTRATIONS OR IDLH CONDITIONS:* SCBAF:PD,PP (any self-contained breathing apparatus that has a full facepiece and is operated in a pressure-demand or other positive-pressure mode); or SAF:PD,PP:ASCBA (any supplied-air respirator that has a full facepiece and is operated in a pressure-demand or other positive-pressure mode in combination with an auxiliary self-contained breathing apparatus operated in a pressure-demand or other positive pressure mode). *ESCAPE:* GMFOV [any air-purifying, full-facepiece respirator (gas mask) with a chin-style, front- or back-mounted organic vapor canister]; or SCBAE (any appropriate escape-type, self-contained breathing apparatus).
Note: Substance causes eye irritation or damage; eye protection needed.
Exposure limits (TWA unless otherwise noted): NIOSH 350 mg/m³, OSHA 500 ppm (2000 mg/m³).
Short-term exposure limits (15-minute TWA): NIOSH ceiling 1800 mg/m³/15 minutes.
Toxicity by ingestion: Grade 2; LD_{50} = 0.5 to 5 g/kg.
Vapor (gas) irritant characteristics: May cause eye and respiratory tract irritation.
Liquid or solid irritant characteristics: Can dry the skin resulting in irritation.
IDLH value: 1100 ppm [10% LEL] as petroleum distillates (naphtha).

FIRE DATA
Flash point: 20°F/–7°C (approximate) (cc).

Flammable limits in air: LEL: 0.9%; UEL: 6.0%.
Fire extinguishing agents not to be used: Water may be ineffective.
Autoignition temperature: 450°F/232°C/505°K (approximate).
Electrical hazard: Due to low electric conductivity, this substance may generate electrostatic charges as a result of agitation and flow.
Burning rate: 4 mm/min

CHEMICAL REACTIVITY
Binary reactants: Oxidizers, nitric acid
Reactivity group: 33
Compatibility class: Miscellaneous hydrocarbon mixtures

ENVIRONMENTAL DATA
Food chain concentration potential: Negative; unlikely to accumulate.
Water pollution: DOT Appendix B, §172.101–marine pollutant. Effect of low concentrations on aquatic life is unknown. Fouling to shoreline. May be dangerous if it enters nearby water intakes; notify operators. Notify local health and wildlife officials.
Response to discharge: Issue warning–high flammability. Restrict access. Evacuate area.

SHIPPING INFORMATION
Storage temperature: Ambient; **Inert atmosphere:** None; **Venting:** Open (flame arrester) or pressure-vacuum; **Stability during transport:** Stable.

PHYSICAL AND CHEMICAL PROPERTIES
Physical state @ 59°F/15°C and 1 atm: Liquid.
Boiling point @ 1 atm: 207°F/97.2°C/370.4°K.
Melting/Freezing point: Less than –40°F/–40°C/233°K.
Specific gravity (water = 1): 0.74 @ 68°F/20°C (liquid).
Liquid surface tension: 19–23 dynes/cm = 0.019–0.023 N/m @ 68°F/20°C.
Liquid water interfacial tension: 39–51 dynes/cm = 0.039–0.051 N/m @ 68°F/20°C.
Ratio of specific heats of vapor (gas): 3
Latent heat of vaporization: 130–150 Btu/lb = 71–81 cal/g = $3.0–3.4 \times 10^5$ J/kg.
Vapor pressure: 30–610 mm.

PHENOL   REC. P:1200

SYNONYMS: ACIDE CARBOLIQUE (French); BAKER'S P AND S LIQUID; BAKER'S P AND S OINTMENT; BENZENE,HYDROXY-; CARBOLIC ACID; CARBOLSAURE (German); EEC No. 604-001-00-2; HYDROXYBENZENE; MONOHYDROXY BENZENE; OXYBENZENE; PHENIC ACID; PHENOLE (German); PHENYL HYDRATE; PHENYL HYDROXIDE; PHENYLIC ACID; RCRA No. U188

IDENTIFICATION
CAS Number: 108-95-2
Formula: C_6H_6O; C_6H_5OH
DOT ID Number: UN 1671 (solid); UN 2312 (molten); UN 2821 (solution); DOT Guide Number: 153
Proper Shipping Name: Phenol, solid; Phenol, molten; Phenol, solution
Reportable Quantity (RQ): **(CERCLA)** 1000 lb/454 kg

DESCRIPTION: Colorless solid or liquid that darkens on exposure to light from light pink to red. Sweet, tarry odor; somewhat sickening sweet and acid. Sinks in water; moderately soluble. Often transported and processed im molten form (@ 140°F/60°C). Solid material melts at 106°F/41°C.

Poison! • Corrosive • Breathing the vapor or skin contact can kill you; skin or eye contact causes severe burns, impaired vision, or blindness • Firefighting gear (including SCBA) does not provide adequate protection. If exposure occurs, remove and isolate gear immediately and thoroughly decontaminate personnel • Containers may BLEVE when exposed to fire • Toxic products of combustion may include carbon monoxide • Do not put yourself in danger by entering a contaminated area to rescue a victim. *Warning:* Odor is not a reliable indicator of the presence of toxic amounts of phenol vapor.

Hazard Classification (based on NFPA-704 Rating System)
Health Hazards (Blue): 4 * (suspected carcinogen); Flammability (Red): 2; Reactivity (Yellow): 0

EMERGENCY RESPONSE: See Appendix A (153)
Evacuation:
Public safety: Isolate the area of spill or leak for at least 25 to 50 meters (80 to 160 feet) in all directions.
Spill: Increase, in the downwind direction, as necessary, the distance shown under "Public Safety."
Fire: If tank, rail car, or tank truck is involved in fire, isolate for at least 800 meters (½ mile) in all directions; also, consider initial evacuation for 800 meters (½ mile) in all directions.

EXPOSURE
Short-term effects: *SEEK MEDICAL ATTENTION.* **Liquid or solid:** *POISONOUS IF SWALLOWED OR RAPIDLY PASSING THROUGH THE SKIN.* Will burn skin and eyes. *IF BREATHING HAS STOPPED,* give artificial respiration; *avoid mouth-to-mouth resuscitation; use bag/mask apparatus.* Remove and double bag contaminated clothing and shoes. Flush affected areas with plenty of water. *IF IN EYES,* hold eyelids open and flush with plenty of water. *IF SWALLOWED* and victim is *CONSCIOUS AND ABLE TO SWALLOW,* have victim drink 4 to 8 ounces of water. **Do NOT induce vomiting.**
Note to physician or authorized medical personnel: Rapid skin decontamination is critical. If immediately available, use vegetable oil, polyethylene glycol (PEG 300 or 400), glycerol, or a water paste of polyvinyl pyrrolidone (PVP) to help remove the phenol from exposed skin. PVP may be superior to PEG for skin decontamination and detoxification. Isopropyl alcohol (rubbing alcohol) may be used for cleaning small burns, but caution is advised because isopropyl alcohol toxicity may occur form skin absorption.

HEALTH HAZARDS
Personal protective equipment (PPE): A-Level PPE. Full face organic vapor respirator. Fresh-air mask for confined areas; rubber gloves; protective clothing; full face shield. *For <30%,* Protective materials with good to excellent resistance: polyethylene. *For >70%,* Protective materials with good to excellent resistance: butyl rubber, neoprene, Teflon®, Viton®, silvershield. Also, polyethylene, and natural rubber offers limited protection.
Recommendations for respirator selection: NIOSH/OSHA
50 ppm: CCROVDM [any chemical cartridge respirator with organic vapor cartridge(s) in combination with a dust and mist filter]; SA (any supplied-air respirator). *125 ppm:* SA:CF (any supplied-air respirator operated in a continuous-flow mode); or PAPROVDM [any powered, air-purifying respirator with organic vapor cartridge(s) in combination with a dust and mist filter]. *250 ppm:* CCRFOVHiE [any chemical cartridge respirator with a full

facepiece and organic vapor cartridge(s) in combination with a high-efficiency particulate filter]; or GMFOVHiE [any air-purifying, full-facepiece respirator (gas mask) with a chin-style, front- or back-mounted organic vapor canister having a high-efficiency particulate filter]; or PAPRTOVHiE [any powered, air-purifying respirator with a tight-fitting facepiece and organic vapor cartridge(s) in combonation with a high-efficiency particulate filter]; or SCBAF (any self-contained breathing apparatus with a full facepiece); or SAF (any supplied-air respirator with a full facepiece). *EMERGENCY OR PLANNED ENTRY INTO UNKNOWN CONCENTRATIONS OR IDLH CONDITIONS*: SCBAF:PD,PP (any self-contained breathing apparatus that has a full facepiece and is operated in a pressure-demand or other positive-pressure mode); or SAF:PD,PP:ASCBA (any supplied-air respirator that has a full facepiece and is operated in a pressure-demand or other positive-pressure mode in combination with an auxiliary self-contained breathing apparatus operated in a pressure-demand or other positive pressure mode). *ESCAPE*: GMFOVHiE [any air-purifying, full-facepiece respirator (gas mask) with a chin-style, front- or back-mounted organic vapor canister having a high-efficiency particulate filter]; or SCBAE (any appropriate escape-type, self-contained breathing apparatus).
Exposure limits (TWA unless otherwise noted): ACGIH TLV 5 ppm (19 mg/m^3); OSHA 5 ppm (19 mg/m^3); NIOSH REL 5 ppm (19 mg/m^3); ceiling 15.6 ppm (60 mg/m^3)/15 minutes; skin contact contributes significantly in overall exposure.
Toxicity by ingestion: Grade 2; LD$_{50}$ = 0.5 to 5 g/kg (rat).
Long-term health effects: Carcinogenic in laboratory animals. Kidney and liver damage.
Vapor (gas) irritant characteristics: Eye and respiratory tract irritant. Vapors cause irritation such that personnel will find high concentrations unpleasant. The effect is temporary.
Liquid or solid irritant characteristics: Severe eye and skin irritant; may cause pain and second-degree burns after a few minutes of contact. May be absorbed through the skin.
Odor threshold: 0.06 ppm.
IDLH value: 250 ppm.

FIRE DATA
Flash point: 185°F (oc); 175°F/79°C (cc).
Flammable limits in air: LEL: 1.7%; UEL: 8.6%.
Autoignition temperature: 1319°F/715°F/988°K.
Electrical hazard: Class I, Group D.
Burning rate: 3.5 mm/min

CHEMICAL REACTIVITY
Binary reactants: Reacts with strong oxidizers, calcium hypochlorite, aluminum chloride, acids. Attacks rubber, aluminum, zinc and lead.
Reactivity group: 21
Compatibility class: Phenols, cresols

ENVIRONMENTAL DATA
Food chain concentration potential: Log P$_{ow}$ = 1.46–1.49. Unlikely to accumulate.
Water pollution: Harmful to aquatic life in very low concentrations. May be dangerous if it enters nearby water intakes; notify operators. Notify local health and wildlife officials.
Response to discharge: Issue warning–poison. Restrict access. Should be removed. Chemical and physical treatment.

SHIPPING INFORMATION
Grades of purity: 90–99% (solid), 60–85% (liquid). Technical: 82-92% (contains cresols); **Storage temperature:** Ambient; **Inert atmosphere:** None; **Venting:** Pressure-vacuum; **Stability during transport:** Stable.

NAS HAZARD CLASSIFICATION FOR BULK WATER TRANSPORTATION
FIRE: 1
HEALTH: Vapor irritant: 2; Liquid or solid irritant: 3; Poisons: 3
WATER POLLUTION: Human toxicity: 2; Aquatic toxicity: 3; Aesthetic effect: 3
REACTIVITY: Other chemicals: 2; Water: 0; Self-reaction: 0

PHYSICAL AND CHEMICAL PROPERTIES
Physical state @ 59°F/15°C and 1 atm: Solid or Liquid.
Molecular weight: 94.11
Boiling point @ 1 atm: 359.2°F/181.8°C/455.0°K.
Melting/Freezing point: 105.6°F/40.9°C/314.1°K.
Critical temperature: 790.0°F/421.1°C/694.3°K.
Critical pressure: 889 psia = 60.5 atm = 6.13 MN/m^2.
Specific gravity (water = 1): 1.058 at 41°C (liquid).
Liquid surface tension: 36.5 dynes/cm = 0.0365 N/m at 55°C.
Liquid water interfacial tension: (estimate) 20 dynes/cm = 0.02 N/m at 42°C.
Relative vapor density (air = 1): 3.2
Ratio of specific heats of vapor (gas): 1.089
Latent heat of vaporization: 130 Btu/lb = 72 cal/g = 3.0 x 10^5 J/kg.
Heat of combustion: –13,400 Btu/lb = –7445 cal/g = –311.7 x 10^5 J/kg.
Vapor pressure: (Reid) 0.3 psia; 0.2 mm.

PHENYLDICHLOROARSINE   REC. P:1250

SYNONYMS: DICHLOROPHENYLARSINE; FENILDICLOROARSINA (Spanish); FDA; PHENYLARSENIC DICHLORIDE; PHENYLARSENOUS DICHLORIDE; RCRA No. P036; TL 69

IDENTIFICATION
CAS Number: 696-28-6
Formula: $C_6NO_5AsCl_2$
DOT ID Number: UN 1556; DOT Guide Number: 152
Proper Shipping Name: Phenyldichloroarsine
Reportable Quantity (RQ): **(CERCLA)** 1 lb/0.454 kg

DESCRIPTION: Colorless to yellow liquid. Weak, but very unpleasant odor. Sinks in water; insoluble; reacts slowly forming hydrochloric acid. Freezes at 3.9°F/–15.6°C. Has been used in war as a poison gas and lacrimator.

Poison! • A tear gas • Breathing the vapor, swallowing the material or skin contact can kill you. Skin and eye contact causes severe burns and blindness • Firefighting gear (including SCBA) does not provide adequate protection. If exposure occurs, remove and isolate gear immediately and thoroughly decontaminate personnel • Containers may BLEVE when exposed to fire • Toxic products of combustion may include arsine and hydrogen chloride • Do not attempt rescue.

Hazard Classification (based on NFPA-704 Rating System)
Health Hazards (Blue): 4; Flammability (Red): 1; Reactivity (Yellow): 0

EMERGENCY RESPONSE: See Appendix A (152)
Evacuation:
Public safety: Isolate the area of spill or leak for at least 25 to 50 meters (80 to 160 feet) in all directions.
Spill: Increase, in the downwind direction, as necessary, the distance shown under "Public Safety."
Fire: If tank, rail car, or tank truck is involved in fire, isolate for at least 800 meters (½ mile) in all directions; also, consider initial evacuation for 800 meters (½ mile) in all directions.

EXPOSURE
Short-term effects: *SEEK MEDICAL ATTENTION*. If artificial respiration is administered, *avoid mouth-to-mouth resuscitation; use bag/mask apparatus*. **Liquid:** *POISONOUS IF SWALLOWED*. Will burn skin and eyes. Remove contaminated clothing and shoes. Flush affected areas with plenty of water. *IF IN EYES*, hold eyelids open and flush with plenty of water. *IF SWALLOWED* and victim is *CONSCIOUS AND ABLE TO SWALLOW*, have victim drink 4 to 8 ounces of water. **Do NOT induce vomiting.**

HEALTH HAZARDS
Personal protective equipment (PPE): B-Level PPE. Full protective clothing; gas mask or self-contained breathing apparatus.
Exposure limits (TWA unless otherwise noted): OSHA PEL 0.5 mg/m^3 as organic arsenic.
Long-term health effects: Arsenic compounds are allergens, irritants to skin.
Vapor (gas) irritant characteristics: Eye and respiratory tract irritant.
Liquid or solid irritant characteristics: Severe eye and skin irritant.

FIRE DATA
Flash point: 60°F/15°C (cc).
Burning rate: 1.8 mm/min

CHEMICAL REACTIVITY
Reactivity with water: Very slow reaction, considered nonhazardous. Hydrochloric acid is formed.
Binary reactants: Corrodes metals because of acid formed

ENVIRONMENTAL DATA
Food chain concentration potential: Negative; unlikely to accumulate.
Water pollution: Effect of low concentrations on aquatic life is unknown. May be dangerous if it enters nearby water intakes; notify operators. Notify local health and wildlife officials.
Response to discharge: Issue warning–poison, water containment, corrosive. Restrict access. Disperse and flush.

SHIPPING INFORMATION
Grades of purity: Commercial; **Storage temperature:** Ambient; **Inert atmosphere:** None; **Venting:** Pressure-vacuum; **Stability during transport:** Stable.

PHYSICAL AND CHEMICAL PROPERTIES
Physical state @ 59°F/15°C and 1 atm: Liquid.
Molecular weight: 222.9
Boiling point @ 1 atm: 495°F/257°C/530°K.
Melting/Freezing point: 3.9°F/–15.6°C/257.6°K.
Specific gravity (water = 1): 1.657 @ 68°F/20°C (liquid).
Liquid surface tension: 44.64 dynes/cm = 0.04464 N/m at 18°C.
Latent heat of vaporization: 99 Btu/lb = 55 cal/g = 2.3 x 10^5 J/kg.
Heat of combustion: (estimate) –6450 Btu/lb = –3600 cal/g = –150 x 10^5 J/kg.
Vapor pressure: 0.087 mmHg @ 77°F/25°C.

PHENYLHYDRAZINE   REC. P:1300

SYNONYMS: EEC No. 612-023-00-9; FENILHIDRIZINA (Spanish); HYDRAZINOBENZENE; HYDRAZINE-BENZENE; HYDRAZINOBENZENE; MONOPHENYLHYDRAZINE

IDENTIFICATION
CAS Number: 100-63-0
Formula: $C_6H_8N_2$; $C_6H_5NHNH_2$
DOT ID Number: UN 2572; DOT Guide Number: 153
Proper Shipping Name: Phenylhydrazine

DESCRIPTION: Colorless liquid or pale yellow crystalline solid; color darkens on exposure to light. Slight ammonia odor. Faint aromatic odor.

Corrosive • Combustible • Highly reactive; spontaneously ignites on contact with sand or cotton • Breathing the vapor, skin or eye contact, or swallowing the material can kill you; inhalation symptoms may be delayed • Firefighting gear (including SCBA) does not provide adequate protection. If exposure occurs, remove and isolate gear immediately and thoroughly decontaminate personnel • Containers may BLEVE when exposed to fire • Vapors are heavier than air and will collect and stay in low areas • Vapors in confined areas (e.g., tanks, sewers, buildings) may explode when exposed to fire • Toxic products of combustion may include nitrogen oxides and carbon monoxide • Do not put yourself in danger by entering a contaminated area to rescue a victim.

Hazard Classification (based on NFPA-704 Rating System)
Health Hazards (Blue): 3; Flammability (Red): 2; Reactivity (Yellow): 0

EMERGENCY RESPONSE: See Appendix A (153)
Evacuation:
Public safety: Isolate the area of spill or leak for at least 25 to 50 meters (80 to 160 feet) in all directions.
Spill: Increase, in the downwind direction, as necessary, the distance shown under "Public Safety."
Fire: If tank, rail car, or tank truck is involved in fire, isolate for at least 800 meters (½ mile) in all directions; also, consider initial evacuation for 800 meters (½ mile) in all directions.

EXPOSURE
Short-term effects: *SEEK MEDICAL ATTENTION*. **Vapor:** Move victim to fresh air. *IF BREATHING HAS STOPPED*, give artificial respiration; *avoid mouth-to-mouth resuscitation; use bag/mask apparatus*. If breathing is difficult, administer oxygen. Lung edema may develop. **Liquid or solid:** *POISONOUS IF ABSORBED THROUGH THE SKIN*. May be fatal if swallowed. Corrosive to skin and eyes. Remove contaminated clothing and shoes. Wash affected areas with soap and water. *IF IN EYES*, hold eyelids open and flush with plenty of water.
Note to physician or authorized medical personnel: Medical observation is recommended for 24 to 48 hours after breathing overexposure, as pulmonary edema may be delayed. As first aid for pulmonary edema, consider administering a corticosteroid spray. Cigarette smoking may exacerbate pulmonary injury and should be discouraged for at least 72 hours following exposure.

HEALTH HAZARDS

Personal protective equipment (PPE): A-Level PPE. Full face organic vapor respirator. Full impervious protective clothing, including boots and gloves. Where splashing is possible wear full face shield or chemical safety goggles.

Recommendations for respirator selection: NIOSH

At any concentrations above the NIOSH REL, or where there is no REL, at any detectable concentration: SCBAF:PD,PP (any self-contained breathing apparatus that has a full facepiece and is operated in a pressure-demand or other positive-pressure mode); or SAF:PD,PP:ASCBA (any supplied-air respirator that has a full facepiece and is operated in a pressure-demand or other positive-pressure mode in combination with an auxiliary self-contained breathing apparatus operated in a pressure-demand or other positive pressure mode). *ESCAPE:* SCBAE (any appropriate escape-type, self-contained breathing apparatus).

Exposure limits (TWA unless otherwise noted): ACGIH TLV 0.1 pp (0.44 mg/m^3) suspected human carcinogen; OSHA PEL 5 ppm (22 mg/m^3); NIOSH REL ceiling 0.14 ppm (0.6 mg/m^3)/2 hours; potential carcinogen-reduce to lowest feasible level; skin contact contributes significantly in overall exposure.

Toxicity by ingestion: Grade 3; oral rat LD_{50} = 188 mg/kg.

Long-term health effects: Has caused lung cancer in animal studies. Repeated skin contact can cause skin sensitization and eczematous dermatitis with redness, swelling, and rash. Chronic and acute exposures may produce blood effects, liver damage, and kidney damage.

Vapor (gas) irritant characteristics: Eye and respiratory tract irritant. Vapors are irritating such that personnel will not usually tolerate moderate or high concentrations.

Liquid or solid irritant characteristics: Eye and skin irritant. Causes smarting of the skin and first-degree burns on short exposure; may cause second-degree burns on long exposure.

IDLH value: Potential human carcinogen; 15 ppm.

FIRE DATA

Flash point: 190°F/88°C (cc).

Behavior in fire: May ignite spontaneously when spread on a large surface or when in air and in contact with porous materials such as soil, asbestos, wood, or cloth.

Autoignition temperature: 345°F/174°C/447°K.

Stoichiometric air-to-fuel ratio: 38.1

CHEMICAL REACTIVITY

Binary reactants: A strong reducing agent. May ignite spontaneously when in contact with oxidants such as hydrogen peroxide or nitric acid, oxides of iron or copper, or manganese, lead, copper or their alloys. Reacts with lead dioxide, ammonia, chlorine, fluorine, hydrogen peroxide, liquid oxygen, zinc diethyl. Will attack cork, some forms of plastics, coatings, and rubber (insulators).

ENVIRONMENTAL DATA

Food chain concentration potential: Log P_{ow} = 1.3. Unlikely to accumulate.

Water pollution: Effects of low concentrations on aquatic life is unknown. May be dangerous if it enters nearby water intakes; notify operators. Notify local health and wildlife officials.

Response to discharge: Should be removed.

SHIPPING INFORMATION

Grades of purity: 97%; technical; **Storage temperature:** Ambient; **Inert atmosphere:** None; **Stability during transport:** Stable.

PHYSICAL AND CHEMICAL PROPERTIES

Physical state @ 59°F/15°C and 1 atm: Solid.

Molecular weight: 108.16

Boiling point @ 1 atm: 471°F/243.5°C/516.5°K (with decomposition).

Melting/Freezing point: 68°F/19.6°C/292.6°K.

Specific gravity (water = 1): 1.0978

Relative vapor density (air = 1): 3.7

Vapor pressure: 0.8 mm.

PHENYLHYDRAZINE HYDROCHLORIDE   REC. P:1350

SYNONYMS: PHENYLHYDRAZINE MONOHYDROCHLORIDE; PHENYLHYDRAZINIUM CHLORIDE

IDENTIFICATION

CAS Number: 59-88-1

Formula: $C_6H_8N_2 \cdot ClH$; $C_6H_5NHNH_2 \cdot HCl$

DOT ID Number: UN 3077; DOT Guide Number: 171

Proper Shipping Name: Environmentally hazardous substances, solid, n.o.s.

Reportable Quantity (RQ): **(EHS)** 1lb/0.454 kg

DESCRIPTION: White to tan solid. Weak aromatic odor. Sinks and mixes with water.

Poison! • Combustible • Irritating to eyes, skin, and respiratory tract; poisonous if swallowed • Dust cloud may explode if ignited in an enclosed area • Containers may BLEVE when exposed to fire • Toxic products of combustion may include nitrogen oxides and hydrogen chloride • Do not put yourself in danger by entering a contaminated area to rescue a victim.

Hazard Classification (based on NFPA-704 Rating System)
Health Hazards (Blue): 3; Flammability (Red): 2; Reactivity (Yellow): 0

EMERGENCY RESPONSE: See Appendix A (171)

Evacuation:

Public safety: Isolate the area of spill or leak for at least 10 to 25 meters (30 to 80 feet) in all directions.

Spill: Increase, in the downwind direction, as necessary, the distance shown under "Public Safety."

Fire: If any large container is involved in fire, isolate for at least 800 meters (½ mile) in all directions; also, consider initial evacuation for 800 meters (½ mile) in all directions.

EXPOSURE

Short-term effects: *SEEK MEDICAL ATTENTION.* **Dust:** Irritating to eyes, nose, and throat. *IF INHALED*, will, will cause coughing or difficult breathing. *IF IN EYES*, hold eyelids open and flush with plenty of water. *IF BREATHING HAS STOPPED*, give artificial respiration. IF breathing is difficult, administer oxygen. **Solid:** POISONOUS IF SWALLOWED. *IF SWALLOWED*, will cause nausea. Remove contaminated clothing and shoes. Flush affected areas with plenty of water. *IF IN EYES*, hold eyelids open and flush with plenty of water. *IF SWALLOWED* and victim is *CONSCIOUS AND ABLE TO SWALLOW*, have victim drink 4 to 8 ounces of water and have victim induce vomiting. *IF SWALLOWED* and victim is *UNCONSCIOUS OR HAVING CONVULSIONS*, do nothing except keep victim warm.

HEALTH HAZARDS

Recommendations for respirator selection: NIOSH as phenylhydrazine

At any concentrations above the NIOSH REL, or where there is no REL, at any detectable concentration: SCBAF:PD,PP (any self-contained breathing apparatus that has a full facepiece and is operated in a pressure-demand or other positive-pressure mode); or SAF:PD,PP:ASCBA (any supplied-air respirator that has a full facepiece and is operated in a pressure-demand or other positive-pressure mode in combination with an auxiliary self-contained breathing apparatus operated in a pressure-demand or other positive pressure mode). *ESCAPE:* SCBAE (any appropriate escape-type, self-contained breathing apparatus).

Personal protective equipment (PPE): B-Level PPE. Dust respirator; rubber gloves; goggles

Long-term health effects: Causes tumors in mice; anemia and liver injury.

Vapor (gas) irritant characteristics: Chronic poison.

Liquid or solid irritant characteristics: Eye and skin irritant.

FIRE DATA

Flash point: Combustible solid.

Behavior in fire: The solid may sublime without melting and deposit on cool surfaces.

CHEMICAL REACTIVITY

Binary reactants: Reacts with alkali metals, ammonia, chlorine, fluorine, hydrogen peroxide. May be corrosive to metals, copper salts, nickel, chromates

ENVIRONMENTAL DATA

Food chain concentration potential: Unlikely to accumulate.

Water pollution: Effect of low concentrations on aquatic life is unknown. May be dangerous if it enters nearby water intakes; notify operators. Notify local health and wildlife officials.

Response to discharge: Issue warning–poison, water contaminant.

SHIPPING INFORMATION

Grades of purity: Commercial; Pure; **Storage temperature:** Ambient; **Inert atmosphere:** None; **Venting:** Open; **Stability during transport:** Stable.

PHYSICAL AND CHEMICAL PROPERTIES

Physical state @ 59°F/15°C and 1 atm: Solid.
Molecular weight: 144.6
Boiling point @ 1 atm: (decomposes).
Melting/Freezing point: 469°F/243°C/516°K.
Specific gravity (water = 1): More than 1 @ 68°F/20°C (solid).

PHENYLMERCURIC ACETATE REC. P:1400

SYNONYMS: ACETATE PHENYLMERCURIQUE (French); ACETATO FENILMERCURICO (Spanish); (ACETATO) PHENYL MERCURY; ACETIC ACID, PHENYLMERCURY DERIVATIVE; AGROSAN; ALGIMYCIN; ANTIMUCIN WDR; BUFEN; CEKUSIL; CELMER; CERESAN; CONTRA CREME CYANACIDE; FEMMA; FMA; FUNGITOX OR; GALLOTOX; HL-331; HONG KIEN; HOSTAQUICK; KWIKSAN; LEYTOSAN; LIQUIPHENE; NORFORMS; NYMERATE; PAMISAN; PHENMAD; PHENOMERCURIC ACETATE; PHENOMERCURY ACETATE; PMA; PMAC; PMACETATE; PMAL; PMAS; PURASAN-SC-10; PURATURF 10; QUICKSAN; SANITIZED SPG; SC-110; SEEDTOX; SPOR-KIL; TAG fungicide; ZIARNIK; ACETATOPHENYLMERCURY; COCURE 26; COSAN -100; RCRA No. P092

IDENTIFICATION

CAS Number: 62-38-4
Formula: $C_8H_8HgO_2$; $CH_3COOHgC_6NO_5$
DOT ID Number: UN 1674; DOT Guide Number: 151
Proper Shipping Name: Phenylmercuric acetate
Reportable Quantity (RQ): **(CERCLA)** 100 lb/45.4 kg

DESCRIPTION: Lustrous white solid. Slightly vinegary odor. Floats on and mixes with water.

Combustible solid • Heat or flame may cause explosion • Poison! • Irritating to the skin, eyes, and respiratory tract; absorption through the skin or swallowing the material can kill you • Firefighting gear (including SCBA) does not provide adequate protection. If exposure occurs, remove and isolate gear immediately and thoroughly decontaminate personnel • Containers may BLEVE when exposed to fire • Concentrated dust in confined areas (e.g., tanks, sewers, buildings) may explode when exposed to fire • Toxic products of combustion may include mercury fumes.

Hazard Classification (based on NFPA-704 Rating System)
Dry: Health Hazards (Blue): 3; Flammability (Red): 1; Reactivity (Yellow): 0
Organic solution: Health Hazards (Blue): 3; Flammability (Red): 2; Reactivity (Yellow): 0

EMERGENCY RESPONSE: See Appendix A (151)
Evacuation:
Public safety: Isolate the area of spill or leak for at least 25 to 50 meters (80 to 160 feet) in all directions.
Spill: Increase, in the downwind direction, as necessary, the distance shown under "Public Safety."
Fire: If tank, rail car, or tank truck is involved in fire, isolate for at least 800 meters (½ mile) in all directions; also, consider initial evacuation for 800 meters (½ mile) in all directions.

EXPOSURE

Short-term effects: *CALL FOR MEDICAL AID* **Dust:** *POISONOUS IF SWALLOWED OR ABSORBED THROUGH THE SKIN.* Will blister skin upon prolonged contact. *IF IN EYES*, hold eyelids open, flush with running water for at least 15 minutes. Move victim to fresh air. *IF BREATHING HAS STOPPED*, give artificial respiration; *avoid mouth-to-mouth resuscitation; use bag/mask apparatus.* If breathing is difficult, administer oxygen. **Solid:** *POISONOUS IF SWALLOWED OR ABSORBED THROUGH THE SKIN.* Will blister skin upon prolonged contact. *IF IN EYES*, hold eyelids open, flush with running water for at least 15 minutes. *IF SWALLOWED* and victim is *CONSCIOUS*: Have victim drink 4 to 8 ounces of water and induce vomiting. *IF SWALLOWED* and victim is *UNCONSCIOUS OR HAVING CONVULSIONS*: Do nothing except keep victim warm.

HEALTH HAZARDS

Personal protective equipment (PPE): B-Level PPE. Mercury vapor face mask or approved respirator, safety goggles, rubber gloves, other protective clothing.

Recommendations for respirator selection: NIOSH/OSHA
$0.5\ mg/m^3$: CCRS* (any chemical cartridge respirator with cartridge(s) providing protection against the compound of concern]; or SA (any supplied-air respirator). $1.25\ mg/m^3$: SA:CF (any supplied-air respirator operated in a continuous-flow mode); or PAPRS** [any powered, air-purifying respirator with cartridge(s)

providing protection against the compound of concern]. *2.5 mg/m³:* CCRFS* (any chemical cartridge respirator with a full facepiece and cartridge(s) providing protection against the compound of concern]; or GMFS* [any air-purifying, full-facepiece respirator (gas mask) with a chin-style, front- or back-mounted canister providing protection against the compound of concern]; or SAT:CF (any supplied-air respirator that has a tight-fitting facepiece and is operated in a continuous-flow mode); or PAPRTS (any powered, air-purifying respirator with a tight-fitting facepiece and cartridge(s) providing protection against the compound of concern]; or SCBAF (any self-contained breathing apparatus with a full facepiece); or SAF (any supplied-air respirator with a full facepiece). *10 mg/m³:* SA:PD,PP (any supplied-air respirator operated in a pressure-demand or other positive-pressure mode). *EMERGENCY OR PLANNED ENTRY INTO UNKNOWN CONCENTRATIONS OR IDLH CONDITIONS:* SCBAF:PD,PP (any self-contained breathing apparatus that has a full facepiece and is operated in a pressure-demand or other positive-pressure mode); or SAF:PD,PP:ASCBA (any supplied-air respirator that has a full facepiece and is operated in a pressure-demand or other positive-pressure mode in combination with an auxiliary self-contained breathing apparatus operated in a pressure-demand or other positive pressure mode). *ESCAPE:* GMFS [any air-purifying, full-facepiece respirator (gas mask) with a chin-style, front- or back-mounted canister providing protection against the compound of concern]; or SCBAE (any appropriate escape-type, self-contained breathing apparatus). ***Note:*** End of service life indicator (ESLI) required.
Exposure limits (TWA unless otherwise noted): ACGIH TLV TWA 0.01 mg/m³; NIOSH REL (organo) 0.01 mg/m³; OSHA PEL 0.01 mg/m³; ceiling 0.04 mg/m³.
Short-term exposure limits (15-minute TWA): ACGIH STEL 0.03 mg/m³.
Toxicity by ingestion: Grade 4: LD_{50} = 13 mg/kg (mouse).
Long-term health effects: May be a carcinogen.
Vapor (gas) irritant characteristics: See symptoms of exposure, above.
Liquid or solid irritant characteristics: Severe eye and skin irritant. May cause pain and second-degree burns after a few minutes of contact.
IDLH value: 2 mg/m³ as mercury.

FIRE DATA
Flash point: (approximate) 104°F/40°C.
Fire extinguishing agents not to be used: Water may be ineffective.
Behavior in fire: If solvent-based solution, fire characteristics may depend on solvent. Containers of solution may explode in fore.

CHEMICAL REACTIVITY
Binary reactants: Reacts with oxidizers, halides

ENVIRONMENTAL DATA
Food chain concentration potential: Unlikely to accumulate.
Water pollution: DOT Appendix B, §172.101- severe marine pollutant. Harmful to aquatic life in very low concentrations. Fouling to shoreline. Dangerous if it enters water intakes. Notify local health and wildlife officials. **Response to discharge:** Evacuate area. Mechanical containment. Should be removed. Chemical and physical treatment.

SHIPPING INFORMATION
Grades of purity: 97%; **Storage**; Ambient; stable for at least 2 years in unopened containers; **Stability during transport:** Stable.

NAS HAZARD CLASSIFICATION FOR BULK WATER TRANSPORTATION
FIRE: 0
HEALTH: Vapor irritant:–; Liquid or solid irritant: 3; Poisons: 4
WATER POLLUTION: Human toxicity: 4; Aquatic toxicity:–; Aesthetic effect: 1
REACTIVITY: Other chemicals: 1; Water: 1; Self-reaction: 0

PHYSICAL AND CHEMICAL PROPERTIES
Physical state @ 59°F/15°C and 1 atm: Solid.
Molecular weight: 336.7
Melting/Freezing point: 300°F/149°C/422°K.
Specific gravity (water = 1): 0.24 (estimate).

1-PHENYL-1-XYLYL ETHANE   REC. P:1450

SYNONYMS: PHENYL XYLYL ETHANE; HATCOL XP; 1-PHENIL-1-XILILO ETANO (Spanish)

IDENTIFICATION
CAS Number: 6196-95-8
Formula: $(C_6NO_5)[C_6H_3(CH_3)_2]CHCH_3$

DESCRIPTION: Clear, oily liquid. Practically odorless. Practically insoluble in water.

Combustible • Containers may BLEVE when exposed to fire • Vapors may form explosive mixture with air • Vapors are heavier than air and will collect and stay in low areas • Vapors may travel long distances to ignition sources and flashback • Vapors in confined areas (e.g., tanks, sewers, buildings) may explode when exposed to fire • Toxic products of combustion may include CO_2 and carbon monoxide.

Hazard Classification (based on NFPA-704 Rating System)
Health Hazards (Blue): 0; Flammability (Red): 1; Reactivity (Yellow): 0

EMERGENCY RESPONSE: See Appendix A (171)
Evacuation:
Public safety: Isolate the area of spill or leak for at least 10 to 25 meters (30 to 80 feet) in all directions.
Spill: Increase, in the downwind direction, as necessary, the distance shown under "Public Safety."
Fire: If any large container is involved in fire, isolate for at least 800 meters (½ mile) in all directions; also, consider initial evacuation for 800 meters (½ mile) in all directions.

EXPOSURE
Short-term effects: *SEEK MEDICAL ATTENTION.* **Vapor:** Move victim to fresh air. *IF BREATHING HAS STOPPED,* give artificial respiration. If breathing is difficult, administer oxygen. **Liquid or solid:** Remove contaminated clothing and shoes. Wash affected areas with soap and water. *IF IN EYES,* hold eyelids open and flush with plenty of water.

HEALTH HAZARDS
Personal protective equipment (PPE): B-Level PPE. Wear chemical protective goggles.
Use approved respirator to protect against vapors.
Vapor (gas) irritant characteristics: May cause eye and respiratory tract irritation.
Liquid or solid irritant characteristics: May cause eye and skin irritation.

FIRE DATA
Flash point: 290°F/143°C (oc).
Fire extinguishing agents not to be used: Water.
Stoichiometric air-to-fuel ratio: 97.6

CHEMICAL REACTIVITY
Binary reactants: Nitric acid, strong bases, oxidizers.
Reactivity group: 32
Compatibility class: Aromatic hydrocarbons

ENVIRONMENTAL DATA
Water pollution: Effects of low concentrations on aquatic life is unknown. May be dangerous if it enters nearby water intakes; notify operators. Notify local health and wildlife officials.
Response to discharge: Should be removed.

SHIPPING INFORMATION
Grades of purity: Technical; **Venting:** Open; **Stability during transport:** Stable.

PHYSICAL AND CHEMICAL PROPERTIES
Molecular weight: 210.3
Boiling Point: 390°F/199°C @ 35 mmHg
Specific gravity (water = 1): 0.983 @ 68°F/20°C.

PHOSDRIN®   REC. P:1500

SYNONYMS: APAVINPHOS; α-2-CARBOMETHOXY-1-METHYLVINYL DIMETHYL PHOSPHATE; 1-CARBOMETHOXY-1-PROPEN-2-Y PHOSPHATE; 2-CARBOMETHOXY-1-METHYLVINYLDIMETHYL PHOSPHATE; CMDP; COMPOUND 2046; O,O-DIMETHYL PHOSTENE; 3-[(DIMETHYLPHOSPHINYL)OXY]-2-BUTENOIC ACID METHYL ESTHER; DURAPHOS; FOSDRIN; GESFID; GESTID; 3-HYDROXYCROTONIC ACID METHYL ESTER DIMETHYL PHOSPHATE; MENIPHOS; MEVINPHOS; MENITE; 1-METHOXYCARBONYL-1-PROPEN-2-YLDIMETHYL PHOSPHATE; METHYL 3-(DIMETHOXYPHOSPHINYLOXYCROTONATE; MEVINFOS (Spanish); MEVINOX; MEVINPHOS; OS 2046; cis-PHOSDRIN; PHOSDRIN 24; PHOSFENE; PHOSPHENE (French)

IDENTIFICATION
CAS Number: 7786-34-7
Formula: $C_7H_{13}O_6P$
DOT ID Number: UN 2783; DOT Guide Number: 152
Proper Shipping Name: Mevinphos
Reportable Quantity (RQ): **(CERCLA)** 10 lb/4.54 kg

DESCRIPTION: Pale yellow to orange liquid. Mild to no odor. Sinks and mixes with water.

Poison! (organophosphate) • Combustible • Breathing the dust, skin or eye contact, or swallowing the material can kill you • Firefighting gear (including SCBA) does not provide adequate protection. If exposure occurs, remove and isolate gear immediately and thoroughly decontaminate personnel • Heat or flame may cause explosion • Containers may BLEVE when exposed to fire • Vapors are heavier than air and will collect and stay in low areas • Vapors in confined areas (e.g., tanks, sewers, buildings) may explode when exposed to fire • Irritating to the skin, eyes, and respiratory tract • Toxic products of combustion may include carbon monoxide and oxides of phosphorus.

Hazard Classification (based on NFPA-704 Rating System)
Health Hazards (Blue): 4; Flammability (Red): 1; Reactivity (Yellow): 0

EMERGENCY RESPONSE: See Appendix A (152)
Evacuation:
Public safety: Isolate the area of spill or leak for at least 25 to 50 meters (80 to 160 feet) in all directions.
Spill: Increase, in the downwind direction, as necessary, the distance shown under "Public Safety."
Fire: If tank, rail car, or tank truck is involved in fire, isolate for at least 800 meters (½ mile) in all directions; also, consider initial evacuation for 800 meters (½ mile) in all directions.

EXPOSURE
Short-term effects: Vapor or liquid: *POISONOUS IF INHALED, IF SWALLOWED, OR IF SKIN IS EXPOSED*. Eye pupils are small; blurred vision; runny nose; cough; shortness of breath; pain; diarrhea, nausea and vomiting; increased blood pressure, hypermotility, hallucinations; loss of consciousness; convulsions; breathing stops; death. *IF BREATHING HAS STOPPED*, give artificial respiration; *avoid mouth-to-mouth resuscitation; use bag/mask apparatus*. **Skin:** Remove and double-bag contaminated clothing (including shoes) and leave them in the Hot Zone; wash skin with soap and water and large volumes of water for at least 15 minutes. *Skin can also be decontaminated with diluted hypochlorite solution, U.S. Army M291 kit, and M258(A1) skin decontamination kit*. **Eye:** Rinse with large volumes of water or saline for at least 15 minutes. *IF SWALLOWED*, will cause nausea, vomiting, or loss of consciousness. *Remove and double-bag contaminated clothing and shoes and leave in Hot Zone for later incineration by hazardous materials experts*. Flush affected areas with plenty of water. **If swallowed** and victim is *CONSCIOUS AND ABLE TO SWALLOW*, have victim drink 4 to 8 ounces of water. **Do NOT induce vomiting but immediately administer slurry of activated charcoal**. *IF SWALLOWED* and victim is *UNCONSCIOUS OR HAVING CONVULSIONS*, do nothing except keep victim warm.
Note to physician and medical personnel: If symptoms indicate, initial treatment for an adult includes atropine, 2 mg (1/30 gr) intramuscularly or intravenously as soon as any local or systemic signs or symptoms of an intoxication are noted; repeat the administration of atropine every 3 to 8 minutes until signs of atropinization (mydriasis, dry mouth, rapid pulse, hot and dry skin) occur; initiate treatment in children with 0.05 mg/kg of atropine; repeat at 5- to 10-minute intervals. Watch respiration, and remove bronchial secretions if they appear to be obstructing the airway; intubate if necessary. Pralidoxime must be administered within minutes to a few hours following exposure (depending on the specific agent) to be effective. Give 2-PAMCl (Pralidoxime; Protopam), 2.5 g in 100 mL of sterile water or in 5% dextrose and water, intravenously, slowly, in 15 to 30 minutes; if sufficient fluid is not available, give 1 g of 2-PAMCl in 3 mL of distilled water by deep intramuscular injection; repeat this every half hour if respiration weakens or if muscle fasciculation or convulsions recur. Also Diazepam, an anticonvulsant, might be needed. For adults, the dose is 5 to 10 mg (slow IV), repeated every 12 to 15 minutes up to three (3) doses maximum. For children, the dose is 0.2 to 0.5 mg/kg.
Note to physician or authorized medical personnel: If inhaled, medical observation is recommended for 24 to 48 hours after breathing overexposure, as pulmonary edema may be delayed. As first aid for pulmonary edema, consider administering a corticosteroid spray. Cigarette smoking may exacerbate pulmonary injury and should be discouraged for at least 72 hours following exposure.

HEALTH HAZARDS

Personal protective equipment (PPE): A-Level PPE. Chemical-protective clothing and butyl rubber gloves are recommended when skin contact is possible because liquid is rapidly absorbed through the skin and may cause systemic toxicity.

Recommendations for respirator selection: NIOSH/OSHA
0.1 ppm: SA (any supplied-air respirator). *0.25 ppm:* SA:CF (any supplied-air respirator operated in a continuous-flow mode). *0.5 ppm:* SAT:CF (any supplied-air respirator that has a tight-fitting facepiece and is operated in a continuous-flow mode); or SCBAF (any self-contained breathing apparatus with a full facepiece); or SAF (any supplied-air respirator with a full facepiece). *4 ppm:* SA:PD:PP (any supplied-air respirator operated in a pressure-demand or other positive-pressure mode). *EMERGENCY OR PLANNED ENTRY INTO UNKNOWN CONCENTRATIONS OR IDLH CONDITIONS:* SCBAF:PD,PP (any self-contained breathing apparatus that has a full facepiece and is operated in a pressure-demand or othe positive-pressure mode); or SAF:PD,PP:ASCBA (any supplied-air respirator that has a full facepiece and is operated in a pressure-demand or other positive-pressure mode in combination with an auxiliary self-contained breathing apparatus operated in a pressure-demand or other positive pressure mode). *ESCAPE:* GMFOVHiE [any air-purifying, full-facepiece respirator (gas mask) with a chin-style, front- or back-mounted organic vapor canister having a high-efficiency particulate filter]; or SCBAE (any appropriate escape-type, self-contained breathing apparatus).

Exposure limits (TWA unless otherwise noted): ACGIH TLV 0.01 ppm (0.092 mg/m^3) as mevinphos; OSHA PEL 0.1 ppm; NIOSH REL 0.01 ppm (0.1 mg/m^3) as Phosdrin®; skin contact contributes significantly in overall exposure.

Short-term exposure limits (15-minute TWA): ACGIH STEL 0.03 ppm (0.27 mg/m^3); NIOSH STEL 0.03 ppm (0.3 mg/m^3); skin contact contributes significantly in overall exposure.

Toxicity by ingestion: Grade 4; LD 50 = below 50 mg/kg.

Long-term health effects: Positive teratogenicity at 10 mg in hen eggs. Cholinesterase inhibition persists for 2 to 6 weeks making subsequent exposures produce more severe symptoms.

IDLH value: 4 ppm.

FIRE DATA
Flash point: 347°F/175°C (oc).

CHEMICAL REACTIVITY
Reactivity with water: Hydrolyzes rapidly.
Binary reactants: Corrosive to many metals.

ENVIRONMENTAL DATA
Food chain concentration potential: Low, Highly soluble, Hydrolyzes rapidly, nonpersistent.
Water pollution: DOT Appendix B, §172.101–marine pollutant. Harmful to aquatic life in very low concentrations. May be dangerous if it enters nearby water intakes; notify operators. Notify local health and wildlife officials. For weapons testing, use an M272 Water Detection Kit. **Response to discharge:** Issue warning–poison, water contaminant. Restrict access. Should be removed. Chemical and physical treatment.

SHIPPING INFORMATION
Grades of purity: Technical >60% alpha isomer; 25% and 50% concentrates; 25% water-soluble solutions; 1% and 2% dusts and granules; **Stability during transport:** Stable when anhydrous.

PHYSICAL AND CHEMICAL PROPERTIES
Physical state @ 59°F/15°C and 1 atm: Liquid.
Molecular weight: 224.16
Boiling point @ 1 atm: 617°F/325°C/598.2°K.
Melting/Freezing point: –68.8°F/–56°C/217.2°K.
Specific gravity (water = 1): 1.26 @ 68°F/20°C.
Relative vapor density (air = 1): 7.73 (calculated).
Vapor pressure: 0.003 mm @ 20°C.

PHOSGENE REC. P:1550

SYNONYMS: CARBONE (OXYCHLORURE de) (French); CARBON OXYCHLORIDE; CARBONIC DICHLORIDE; CARBONYLCHLORID (German); CARBONYL CHLORIDE; CG; CHLOROFORMYL CHLORIDE; DIPHOSGENE; DP; EEC No. 006-002-00-8; FOSGENO (Spanish); NCI-C60219; PHOSGEN (German); RCRA No. P095

IDENTIFICATION
CAS Number: 75-44-5
Formula: CCl_2O
DOT ID Number: UN 1076; DOT Guide Number: 125
Proper Shipping Name: Phosgene; DP; Diphosgene
Reportable Quantity (RQ): **(CERCLA)** 10 lb/4.54 kg

DESCRIPTION: Colorless compressed, liquefied gas. A colorless to light yellow fuming liquid below 47°F/8.2°C. Musty odor like new-mown hay in low concentrations; may not be detectable at higher concentrations. Shipped as liquefied compressed gas in cylinders. Liquid sinks in water; slightly soluble; reacts slowly, forming hydrochloric acid and CO_2. *Note*: Has been used as a chemical warfare agent (diphosgene; DP). Notify U.S. Department of Defense: Army.

Corrosive • Breathing the vapor can kill you; skin or eye contact causes severe burns, impaired vision, or blindness; effects of contact or inhalation may be delayed • Firefighting gear (including SCBA) does not provide adequate protection. If exposure occurs, remove and isolate gear immediately and thoroughly decontaminate personnel • Gas is heavier than air and will collect and stay in low areas • Containers may BLEVE when exposed to fire • Contact with the liquid may cause frostbite • Decomposes above 572°F/300°C, forming toxic carbon monoxide, chlorine gas, and hydrogen chloride. Do NOT attempt rescue. *Warning:* Odor is not a reliable indicator of the presence of toxic amounts of phosgene gas.

Hazard Classification (based on NFPA-704 Rating System)
Health Hazards (Blue): 4; Flammability (Red): 0; Reactivity (Yellow): 1

EMERGENCY RESPONSE: See Appendix A (125)
Evacuation:
Public safety: See below.
Spill: Small spill–First: Isolate in all directions 95 meters (300 feet); Then: Protect persons downwind, DAY: 0.8 (0.5 mile); NIGHT: 2.7 km (1.7). Large spill–First: Isolate in all directions 765 meters (2500 feet); Then: Protect persons downwind, DAY: 6.6 km (4.1 mile); NIGHT: 11.0 km (6.9 mile).
Fire: Isolate for 1600 meters (1 mile) in all directions, especially if tank, rail car, or tank truck is involved in fire.

EMERGENCY RESPONSE: See Appendix A (125)
Public safety: Diphosgene (DP): See below.
Spill: Small spill–First: Isolate in all directions 60 meters (200 feet); Then: Protect persons downwind, DAY: 0.2 (0.1 mile), [0.3 km (0.2 mile) weaponized]; NIGHT: 0.5 km (0.3) [1.0 km (0.6

weaponized], Large spill–First: Isolate in all directions 95 meters (300 feet), [185 m (600 ft) weaponized]; Then: Protect persons downwind, DAY: 1.0 km (0.6 mile) [1.6 km (1.0 mile) weaponized]; NIGHT: 1.9 km (1.2 miles) [4.5 km (2.8 miles) weaponized].

Fire: Isolate for 1600 meters (1 mile) in all directions, especially if tank, rail car, or tank truck is involved in fire.

EXPOSURE

Short-term effects: *SEEK MEDICAL ATTENTION.* **Vapor:** *POISONOUS IF INHALED.* Irritating to eyes, nose, and throat. Effects may be delayed. Move to fresh air. *IF BREATHING HAS STOPPED,* give artificial respiration; *avoid mouth-to-mouth resuscitation; use bag/mask apparatus.* IF breathing is difficult, administer oxygen. Maintain absolute rest until medical aid arrives. If clothing is contaminated remove, double-bag and leave at incident site.

Note to physician or authorized medical personnel: There is no antidote for phosgene. Diuretics are contraindicated. Pulmonary edema due to phosgene inhalation is not hypervolemic in origin; patients tend to be hypovolemic and hypotensive. Dopamine may be required for treatment of hypotension, bradycardia, or renal failure. Initial fluid resuscitation as needed. To relieve irritation of the respiratory tract consider a spray of 10% sodium thiosulfate. Keep the victim lying down. The slightest exertion, including walking, may result in cardiac arrest. Medical observation is recommended for 24 to 48 hours after breathing overexposure, as pulmonary edema or bronchopneumonia may be delayed. As first aid for pulmonary edema, consider administering a corticosteroid spray. Cigarette smoking may exacerbate pulmonary injury and should be discouraged for at least 72 hours following exposure.

HEALTH HAZARDS

Personal protective equipment (PPE): A-Level PPE. If the proper equipment is not available, or if the rescuers have not been trained in its use, call for assistance from the U.S. Soldier and Biological Chemical Command–Edgewood Research Development and Engineering Center (from 0700-1630 EST call 410-671-4411, and from 1630-0700 EST call 410-278-5201; ask for the Staff Duty Officer).

Recommendations for respirator selection: NIOSH/OSHA

1 ppm: SA (any supplied-air respirator). *2 ppm:* SCBAF (any self-contained breathing apparatus with a full facepiece); or SAF (any supplied-air respirator with a full facepiece). *EMERGENCY OR PLANNED ENTRY INTO UNKNOWN CONCENTRATIONS OR IDLH CONDITIONS:* SCBAF:PD,PP (any self-contained breathing apparatus that has a full facepiece and is operated in a pressure-demand or other positive-pressure mode); or SAF:PD,PP:ASCBA (any supplied-air respirator that has a full facepiece and is operated in a pressure-demand or other positive-pressure mode in combination with an auxiliary self-contained breathing apparatus operated in a pressure-demand or other positive pressure mode). *ESCAPE:* GMFS [any air-purifying, full-facepiece respirator (gas mask) with a chin-style, front- or back-mounted canister providing protection against the compound of concern]; or SCBAE (any appropriate escape-type, self-contained breathing apparatus). *Note:* Substance reported to cause eye irritation or damage; may require eye protection.

Exposure limits (TWA unless otherwise noted): ACGIH TLV 0.1 ppm (0.40 mg/m^3); OSHA PEL 0.1 ppm (0.4 mg/m^3); NIOSH 0.1 ppm (0.4 mg/m^3).

Short-term exposure limits (15-minute TWA): NIOSH ceiling 0.2 ppm (0.8 mg/m^3)/15 minutes

Long-term health effects: Pneumonia or severe, delayed, and rapidly developing pulmonary edema.

Vapor (gas) irritant characteristics: Vapors cause severe irritation of eyes and throat and can cause eye and lung injury. They cannot be tolerated even at low concentrations.

Liquid or solid irritant characteristics: Severe irritant to all tissues.

Odor threshold: 0.1-5.9 ppm The Odor threshold is 5 times higher than the OSHA PEL. Thus, odor provides insufficient warning of hazardous concentrations.

IDLH value: 2 ppm.

CHEMICAL REACTIVITY

Reactivity with water: Decomposes, but not vigorously. Reacts slowly in water to form hydrochloric acid and CO_2.

Binary reactants: Oxidizers, alkalis, ammonia, alcohols, active metals including copper. Corrosive to metals in the presence of moisture.

Neutralizing agents for acids and caustics: Can be absorbed in caustic soda solution. One ton of phosgene requires 2480 pounds of caustic soda dissolved in 1000 gallons of water.

ENVIRONMENTAL DATA

Food chain concentration potential: Negative; unlikely to accumulate.

Water pollution: Hazardous to all aquatic life. Dangerous if it enters water intakes. Notify local health and wildlife officials.

Response to discharge: Issue warning–poison. Restrict access. Evacuate area.

SHIPPING INFORMATION

Grades of purity: Commercial; 100%; **Storage temperature:** Ambient; **Inert atmosphere:** None; **Venting:** Safety relief; **Stability during transport:** Stable.

PHYSICAL AND CHEMICAL PROPERTIES

Physical state @ 59°F/15°C and 1 atm: Gas.
Molecular weight: 98.92
Boiling point @ 1 atm: 44–46.8°F/7–8.2°C/280–281.4°K.
Melting/Freezing point: –195°F/–126°C/147°K.
Critical temperature: 360°F/182°C/455°K.
Critical pressure: 823 psia = 56.0 atm = 5.67 MN/m^2.
Specific gravity (water = 1): 1.38 @ 68°F/20°C (liquid).
Liquid surface tension: 22.8 dynes/cm = 0.0228 N/m at 0°C.
Relative vapor density (air = 1): 3.4
Ratio of specific heats of vapor (gas): 1.170
Latent heat of vaporization: 110 Btu/lb = 59 cal/g = 2.5 x 10^5 J/kg.
Vapor pressure: 1180 mm; 1.6 atm.

PHOSPHORIC ACID   REC. P:1600

SYNONYMS: ACIDE PHOSPHORIQUE (French); EEC No. 015-011-00-6; meta-PHOSPHORIC ACID; ORTHOPHOSPHORIC ACID; WHITE PHOSPHORIC ACID

IDENTIFICATION

CAS Number: 7664-38-2
Formula: H_3PO_4
DOT ID Number: UN 1805; DOT Guide Number: 154
Proper Shipping Name: Phosphoric acid
Reportable Quantity (RQ): **(CERCLA)** 5000 lb/2270 kg

DESCRIPTION: Colorless, thick liquid (25-85%) or crystalline solid (100%). Odorless. Sinks and mixes with water forming a strong acid.

Corrosive • Breathing the vapor can kill you; skin or eye contact causes severe burns, impaired vision, or blindness; effects of contact or inhalation may be delayed • Firefighting gear (including SCBA) does not provide adequate protection. If exposure occurs, remove and isolate gear immediately and thoroughly decontaminate personnel • Attacks many metals forming flammable hydrogen gas • Toxic products of combustion may include oxides of phosphorus.

Hazard Classification (based on NFPA-704 Rating System)
Health Hazards (Blue): 3; Flammability (Red): 0; Reactivity (Yellow): 0

EMERGENCY RESPONSE: See Appendix A (154)
Evacuation:
Public safety: Isolate the area of spill or leak for at least 25 to 50 meters (80 to 160 feet) in all directions.
Spill: Increase, in the downwind direction, as necessary, the distance shown under "Public Safety."
Fire: If tank, rail car, or tank truck is involved in fire, isolate for at least 800 meters (½ mile) in all directions; also, consider initial evacuation for 800 meters (½ mile) in all directions.

EXPOSURE
Short-term effects: *SEEK MEDICAL ATTENTION.* **Liquid:** Will burn skin and eyes. *IF SWALLOWED*, will cause nausea, vomiting, or loss of consciousness. Remove contaminated clothing and shoes. Flush affected areas with plenty of water. *IF BREATHING HAS STOPPED*, give artificial respiration; *avoid mouth-to-mouth resuscitation; use bag/mask apparatus*. *IF IN EYES*, hold eyelids open and flush with plenty of water. *IF SWALLOWED* and victim is *CONSCIOUS AND ABLE TO SWALLOW*, have victim drink 4 to 8 ounces of water. **Do NOT induce vomiting.**
Note to physician or authorized medical personnel: Medical observation is recommended for 24 to 48 hours after breathing overexposure, as pulmonary edema may be delayed. As first aid for pulmonary edema, consider administering a corticosteroid spray. Cigarette smoking may exacerbate pulmonary injury and should be discouraged for at least 72 hours following exposure.

HEALTH HAZARDS
Personal protective equipment (PPE): B-Level PPE. Goggles or face shield; rubber gloves and protective clothing. Sealed chemical protective materials recommended (>70%): natural rubber, neoprene, nitrile, nitrile+PVC, polyethylene, PV alcohol, PVC, Saranex®, neoprene+natural rubber, neoprene/natural rubber. Also, Viton® offers limited protection.
Recommendations for respirator selection: NIOSH/OSHA
$25\ mg/m^3$: SA:CF* (any supplied-air respirator operated in a continuous-flow mode). $50\ mg/m^3$: HiEF (any air-purifying, full-facepiece respirator with a high-efficiency particulate filter); or SCBAF (any self-contained breathing apparatus with a full facepiece); or SAF (any supplied-air respirator with a full facepiece). $1000\ mg/m^3$: SAF:PD,PP (any supplied-air respirator that has a full facepiece and is operated in a pressure-demand or other positive-pressure mode). *EMERGENCY OR PLANNED ENTRY INTO UNKNOWN CONCENTRATIONS OR IDLH CONDITIONS*: SCBAF:PD,PP (any self-contained breathing apparatus that has a full facepiece and is operated in a pressure-demand or other positive-pressure mode); or SAF:PD,PP:ASCBA (any supplied-air respirator that has a full facepiece and is operated in a pressure-demand or other positive-pressure mode in combination with an auxiliary self-contained breathing apparatus operated in a pressure-demand or other positive pressure mode). *ESCAPE:* HiEF (any air-purifying, full-facepiece respirator with a high-efficiency particulate filter); or SCBAE (any appropriate escape-type, self-contained breathing apparatus). *Note*: Substance reported to cause eye irritation or damage; may require eye protection.
Exposure limits (TWA unless otherwise noted): ACGIH TLV 1 mg/m^3; NIOSH/OSHA PEL 1.0 mg/m^3.
Short-term exposure limits (15-minute TWA): ACGIH STEL 3 mg/m^3; NIOSH STEL 3 mg/m^3.
Toxicity by ingestion: Grade 3; LD_{50} = 50 to 500 mg/kg.
Vapor (gas) irritant characteristics: Eye and respiratory tract irritant; difficult breathing and possible lung edema.
Liquid or solid irritant characteristics: Severe eye and skin irritant; may cause pain and second-degree burns after a few minutes of contact.
IDLH value: 1000 mg/m^3.

FIRE DATA
Behavior in fire: If trapped in a confined space, material can form an explosive mixture with air.
Electrical hazard: Class I, Group B. Based upon hydrogen gas generation should a leak occur.

CHEMICAL REACTIVITY
Reactivity with water: Mild evolution of heat; acidic.
Binary reactants: Reacts with strong caustics and most metals to liberate flammable hydrogen gas. Do not mix with solutions containing bleach or ammonia. Very corrosive to ferrous metals and alloys particularly at temperatures above 180°F/82°C. Does not attack red metals or stainless steel at ambient temperature. Corrosive to glass and pottery above 390°F/199°C.
Neutralizing agents for acids and caustics: Flush with water, neutralize with lime.
Reactivity group: 1
Compatibility class: Nonoxidizing mineral acids

ENVIRONMENTAL DATA
Food chain concentration potential: Negative; unlikely to accumulate.
Water pollution: Dangerous to aquatic life in high concentrations. May be dangerous if it enters nearby water intakes; notify operators. Notify local health and wildlife officials. **Response to discharge:** Issue warning–corrosive. Restrict access. Disperse and flush.

SHIPPING INFORMATION
Grades of purity: NF, food, fertilizer, commercial; all 75–85%, the balance is water; **Storage temperature:** Ambient; **Inert atmosphere:** None; **Venting:** Open; **Stability during transport:** Stable.

NAS HAZARD CLASSIFICATION FOR BULK WATER TRANSPORTATION
FIRE: 0
HEALTH: Vapor irritant: 0; Liquid or solid irritant: 3; Poisons: 1
WATER POLLUTION: Human toxicity: 2; Aquatic toxicity: 3; Aesthetic effect: 2
REACTIVITY: Other chemicals: 3; Water: 0; Self-reaction: 0

PHYSICAL AND CHEMICAL PROPERTIES
Physical state @ 59°F/15°C and 1 atm: Liquid.
Molecular weight: 98.00
Boiling point @ 1 atm: 415°F/213°C/486°K (decomposes).
Melting/Freezing point: 108°F/42°C/315°K.
Specific gravity (water = 1): 1.892 @ 77°F/25°C (liquid); 1.33 (50% solution).
Relative vapor density (air = 1): 3.4

Heat of decomposition: 390°F
Heat of solution: –52 Btu/lb = –29 cal/g = –1.2 x 10^5 J/kg.
Heat of fusion: 25.8 cal/g.
Vapor pressure: (Reid) Low; 0.2 mm.

PHOSPHORUS, BLACK　　　　　　　　REC. P:1650

SYNONYMS: AMORPHOUS PHOSPHORUS, BLACK; BLACK PHOSPHORUS; FOSFORO NEGRO (Spanish); PHOSPHORUS, AMORPHOUS, BLACK

IDENTIFICATION
CAS Number: 7723-14-0
Formula: Pn (an allotrope of white phosphorus)
DOT ID Number: UN 1338; DOT Guide Number: 133
Proper Shipping Name: Phosphorus, amorphous
Reportable Quantity (RQ): **(CERCLA)** 1 lb/0.454 kg

DESCRIPTION: Black (graphite-like) solid. Sinks in water; insoluble. An inorganic polymer produced from white phosphorus under pressure.

Combustible solid • Heat or flame may cause explosion • Containers may BLEVE when exposed to fire • Vapors are heavier than air and will collect and stay in low areas • Vapors in confined areas (e.g., tanks, sewers, buildings) may explode when exposed to fire • Irritating to the skin, eyes, and respiratory tract • At more than 550°F/288°C black phosphorus turns into red phosphorus; it may further decompose into very hazardous white phosphorus; may ignite spontaneously in the presence of air at this temperature • Toxic products of combustion may include oxides of phosphorus and phosphine (a highly toxic gas which ignites spontaneously).

Hazard Classification (based on NFPA-704 Rating System)
Health Hazards (Blue): 1; Flammability (Red): 1; Reactivity (Yellow): 0

EMERGENCY RESPONSE: See Appendix A (133)
Evacuation:
Public safety: Isolate spill area for at least 10 to 25 meters (30 to 80 feet) in all directions.
Spill: Consider initial downwind evacuation of at least 100 meters (330 feet).
Fire: Isolate for 800 meters (½ mile) in all directions, especially if tank, rail car, or tank truck is involved in fire.

EXPOSURE
Short-term effects: Harmless unless white phosphorus is present or if it is burning.

HEALTH HAZARDS
Personal protective equipment (PPE):
B-Level PPE. **Recommendations for respirator selection:** NIOSH/OSHA as yellow phosphorus
1 mg/m³: SA (any supplied-air respirator). *2.5 mg/m³:* SA:CF* (any supplied-air respirator operated in a continuous-flow mode). *5 mg/m³:* SCBAF (any self-contained breathing apparatus with a full facepiece); or SAF (any supplied-air respirator with a full facepiece). *EMERGENCY OR PLANNED ENTRY INTO UNKNOWN CONCENTRATIONS OR IDLH CONDITIONS:* SCBAF:PD,PP (any self-contained breathing apparatus that has a full facepiece and is operated in a pressure-demand or other positive-pressure mode); or SAF:PD,PP:ASCBA (any supplied-air respirator that has a full facepiece and is operated in a pressure-demand or other positive-pressure mode in combination with an auxiliary self-contained breathing apparatus operated in a pressure-demand or other positive pressure mode). *ESCAPE:* SCBAE (any appropriate escape-type, self-contained breathing apparatus). **Note:* Substance causes eye irritation or damage; eye protection needed
Exposure limits (TWA unless otherwise noted): ACGIH TLV 0.02 ppm (0.1 mg/m³) as yellow phosphorus; NIOSH/OSHA 0.1 mg/m³; DFG MAK (German) 0.1 mg/m³; as yellow phosphorus, a decomposition product.
Liquid or solid irritant characteristics: Eye irritant.
IDLH value: 5 mg/m³ as phosphorus

FIRE DATA
Behavior in fire: At more than 550°F/288°C black phosphorus may decompose into red phosphorus and then into very hazardous white phosphorus; may ignite spontaneously in the presence of air at this temperature. Also, white phosphorus may be present as a contaminant. Smoke from fire are highly irritating and corrosive to the eyes, skin, and respiratory tract.
Autoignition temperature: More than 752°F/400°C.

CHEMICAL REACTIVITY
Binary reactants: Reacts with oxidizers of all kinds, halogens, halides, metals sulfur and caustics (forms phosgene gas).
Reactivity group: See Compatibility Guide
Compatibility class: Special class

ENVIRONMENTAL DATA
Water pollution: Effects of low concentrations on aquatic life are unknown. May be dangerous if it enters nearby water intakes; notify operators. Notify local health and wildlife officials. **Response to discharge:** Disperse and flush.

SHIPPING INFORMATION
Stability during transport: Stable in air.

PHYSICAL AND CHEMICAL PROPERTIES
Physical state @ 59°F/15°C and 1 atm: Solid.
Molecular weight: 30.975
Boiling point @ 1 atm: 838.9°F/448.3°C/721.4°K.
Melting/Freezing point: 781°F/416°C/689°K.
Specific gravity (water = 1): 2.691 @ 68°F/20°C.
Relative vapor density (air = 1): 4.27 (vapor molecule is P4).
Heat of combustion: –9815 Btu/lb = –5453 cal/g = –228.2 x 10^5 J/kg.

PHOSPHORUS, RED　　　　　　　　REC. P:1700

SYNONYMS: AMORPHOUS PHOSPHORUS, RED; FOSFORO ROJO (Spanish); PHOSPHORUS, AMORPHOUS, RED; EEC No. 015-002-00-7; HITTORF'S PHOSPHORUS; RED PHOSPHORUS

IDENTIFICATION
CAS Number: 7723-14-0
Formula: Pn
DOT ID Number: UN 1338; DOT Guide Number: 133
Proper Shipping Name: Red phosphorus; Red phosphorus, amorphous; Phosphorus, amorphous, red
Reportable Quantity (RQ): **(CERCLA)** 1 lb/0.454 kg

DESCRIPTION: Reddish-brown powder or other forms. Odorless. Sinks in water; insoluble.

Combustible solid • Heat or flame may cause explosion • When material is cooling from the vapor or liquid stage, white phosphorus is formed; this material is toxic and spontaneously flammable upon contact with air • Containers may BLEVE when exposed to fire • Concentrated dust in confined areas (e.g., tanks, sewers, buildings) may explode when exposed to fire • Corrosive to the skin, eyes, and respiratory tract; inhalation symptoms may be delayed • Toxic products of combustion may include oxides of phosphorus pentoxide and phosphine.

Hazard Classification (based on NFPA-704 Rating System)
Health Hazards (Blue): 1; Flammability (Red): 1; Reactivity (Yellow): 1

EMERGENCY RESPONSE: See Appendix A (133)
Evacuation:
Public safety: Isolate spill area for at least 10 to 25 meters (30 to 80 feet) in all directions.
Spill: Consider initial downwind evacuation of at least 100 meters (330 feet).
Fire: Isolate for 800 meters (½ mile) in all directions, especially if tank, rail car, or tank truck is involved in fire.

EXPOSURE
Short-term effects: *SEEK MEDICAL ATTENTION.* **Vapor:** Lung edema may develop. **Solid:** Will burn eyes. Harmful if swallowed. Flush affected areas with plenty of water. *IF IN EYES*, hold eyelids open and flush with plenty of water. *IF SWALLOWED* and victim is *CONSCIOUS AND ABLE TO SWALLOW*, have victim drink 4 to 8 ounces of water.
Note to physician or authorized medical personnel: Medical observation is recommended for 24 to 48 hours after breathing overexposure, as pulmonary edema may be delayed. As first aid for pulmonary edema, consider administering a corticosteroid spray. Cigarette smoking may exacerbate pulmonary injury and should be discouraged for at least 72 hours following exposure.

HEALTH HAZARDS
Personal protective equipment (PPE): B-Level PPE. Dust mask; gloves of rubber or vinyl; chemical safety glasses; rubber shoes.
Recommendations for respirator selection: NIOSH/OSHA as phosphorus
1 mg/m^3: SA (any supplied-air respirator). 2.5 mg/m^3: SA:CF (any supplied-air respirator operated in a continuous-flow mode). 5 mg/m^3: SCBAF (any self-contained breathing apparatus with a full facepiece); or SAF (any supplied-air respirator with a full facepiece). *EMERGENCY OR PLANNED ENTRY INTO UNKNOWN CONCENTRATIONS OR IDLH CONDITIONS*: SCBAF:PD,PP (any self-contained breathing apparatus that has a full facepiece and is operated in a pressure-demand or other positive-pressure mode); or SAF:PD,PP:ASCBA (any supplied-air respirator that has a full facepiece and is operated in a pressure-demand or other positive-pressure mode in combination with an auxiliary self-contained breathing apparatus operated in a pressure-demand or other positive pressure mode). *ESCAPE:* SCBAE (any appropriate escape-type, self-contained breathing apparatus). *Note*: Substance causes eye irritation or damage; eye protection needed.
Exposure limits (TWA unless otherwise noted): ACGIH TLV 0.02 ppm (0.1 mg/m^3) as yellow phosphorus; NIOSH/OSHA 0.1 mg/m^3; DFG MAK (German) 0.1 mg/m^3; as phosphorus (Yellow): A decomposition product.
Vapor (gas) irritant characteristics: Nonvolatile, but smoke from fire is highly irritating and may cause eye damage.
Liquid or solid irritant characteristics: Eye irritant.
IDLH value: 5 mg/m^3 as yellow phosphorus.

FIRE DATA
Flash point: Combustible solid. Ignites in presence of air at approximately 500°F/260°C.
Fire extinguishing agents not to be used: Halogen based agents.
Behavior in fire: May cause decomposition to highly dangerous white/yellow phosphorus at more than 500°F/260°C which is toxic and spontaneously flammable upon contact with air.
Autoignition temperature: 500°F/260°C/533°K.

CHEMICAL REACTIVITY
Reactivity with water: May react spontaneously in moist air.
Binary reactants: Avoid contact with oxidizing agents (chlorates, nitrates, halogens, halides, etc.) or with strong alkaline hydroxides. Forms shock-sensitive mixtures with oxidizers. Can react violently with oxidizing agent in presence of air and moisture, liberating phosphorus acids and toxic, spontaneously flammable phosphine gas.
Reactivity group: See Compatibility Guide
Compatibility class: Special Class

ENVIRONMENTAL DATA
Food chain concentration potential: Unlikely to accumulate.
Water pollution: Harmful to aquatic life in very low concentrations. May be dangerous if it enters nearby water intakes; notify operators. Notify local health and wildlife officials. **Response to discharge:** Issue warning–high flammability. Should be removed. Chemical and physical treatment.

SHIPPING INFORMATION
Grades of purity: 99.9% Technical; **Storage temperature:** Ambient; **Inert atmosphere:** None; **Venting:** Open; **Stability during transport:** Stable.

PHYSICAL AND CHEMICAL PROPERTIES
Physical state @ 59°F/15°C and 1 atm: Solid.
Molecular weight: 123.89
Melting/Freezing point: 781°F/416°C/689°K.
Specific gravity (water = 1): 2.20 @ 68°F/20°C (solid).
Relative vapor density (air = 1): 2.3

PHOSPHORUS, YELLOW or WHITE REC. P:1750

SYNONYMS: EEC No. 015-001-00-1; ELEMENTAL WHITE PHOSPHORUS; FOSFORO BLANCO (Spanish); PHOSPHORIC CHLORIDE; PHOSPHORUS CHLORIDE OXIDE; PHOSPHORUS OXYCHLORIDE; PHOSPHORUS OXYTRICHLORIDE; PHOSPHORYL CHLORIDE; WHITE PHOSPHORUS; WP; YELLOW PHOSPHORUS

IDENTIFICATION
CAS Number: 7723-14-0
Formula: P$_4$
DOT ID Number: UN 1381; UN 2447 (molten); DOT Guide Number: 136
Proper Shipping Name: Phosphorus (dry, under water or in water solution); Phosphorus (white, molten)
Reportable Quantity (RQ): **(CERCLA)** 1 lb/0.454 kg

DESCRIPTION: Pale yellow to deep straw, waxy solid. Garlic odor. Fumes and burns in air, sinks in water. May ignite on contact with air.

Poison! • Corrosive • Extremely flammable; burns spontaneously upon contact with air; will reignite after fire is extinguished if material is still in contact with air • Do NOT use halogens • Breathing the vapor, skin or eye contact, or swallowing the material can kill you • Firefighting gear (including SCBA) does not provide adequate protection. If exposure occurs, remove and isolate gear immediately and thoroughly decontaminate personnel • Containers may BLEVE when exposed to fire • Toxic products of combustion may include highly toxic phosphorus pentoxide and phosphine (a highly toxic gas which ignites spontaneously) • Do not put yourself in danger by entering a contaminated area to rescue a victim

Hazard Classification (based on NFPA-704 Rating System)
Health Hazards (Blue): 4; Flammability (Red): 4; Reactivity (Yellow): 2

EMERGENCY RESPONSE: See Appendix A (136)
Evacuation:
Public safety: Isolate spill area for at least 100 to 150 meters (330 to 490 feet) in all directions.
Spill: Consider downwind evacuation for at least 300 meters (1000 feet).
Fire: Isolate for 800 meters (½ mile) in all directions, especially if tank, rail car, or tank truck is involved in fire.

EXPOSURE
Short-term effects: *SEEK MEDICAL ATTENTION*. **Solid:** Will burn skin and eyes; immediately flush with plenty of water for at least 15 minutes; keep skin area wet until medical attention is obtained. Large amounts of copper sulfate solution (10–15%) may be applied; continue for 3 minutes then try to was away the phosphorus particles. *IF SWALLOWED*, will cause nausea, vomiting, or loss of consciousness. Flush affected areas with plenty of water. *IF IN EYES*, hold eyelids open and flush with plenty of water. *IF SWALLOWED* and victim is *CONSCIOUS WITH NO CONVULSIONS*, have victim drink 4 to 8 ounces of water. **Do NOT induce vomiting.**

HEALTH HAZARDS
Personal protective equipment (PPE): A-Level PPE. Heavy rubber gloves and goggles or face shield.
Recommendations for respirator selection: NIOSH/OSHA as phosphorus
1 mg/m³: SA (any supplied-air respirator). *2.5 mg/m³*: SA:CF (any supplied-air respirator operated in a continuous-flow mode). *5 mg/m³*: SCBAF (any self-contained breathing apparatus with a full facepiece); or SAF (any supplied-air respirator with a full facepiece). *EMERGENCY OR PLANNED ENTRY INTO UNKNOWN CONCENTRATIONS OR IDLH CONDITIONS:* SCBAF:PD,PP (any self-contained breathing apparatus that has a full facepiece and is operated in a pressure-demand or other positive-pressure mode); or SAF:PD,PP:ASCBA (any supplied-air respirator that has a full facepiece and is operated in a pressure-demand or other positive-pressure mode in combination with an auxiliary self-contained breathing apparatus operated in a pressure-demand or other positive pressure mode). *ESCAPE:* SCBAE (any appropriate escape-type, self-contained breathing apparatus). *Note:* Substance causes eye irritation or damage; eye protection needed
Exposure limits (TWA unless otherwise noted): ACGIH TLV 0.02 ppm (0.1 mg/m³); NIOSH/OSHA 0.1 mg/m³.
Toxicity by ingestion: Grade 4; LD_{50} below 50 mg/kg.
Long-term health effects: Severe attack of liver and bones.
Vapor (gas) irritant characteristics: Combustion fumes are a severe eye, skin and respiratory tract irritant.
Liquid or solid irritant characteristics: Severe eye and skin irritant. Causes second-and third-degree burns on short contact, and is very injurious to the eyes.
IDLH value: 5 mg/m³.

FIRE DATA
Flash point: Ignites spontaneously in air, especially above 85°F/29°C.
Fire extinguishing agents not to be used: Halogen based agents.
Behavior in fire: Intense, acrid white smoke is formed.
Autoignition temperature: 86°F/30°C/303°K.

CHEMICAL REACTIVITY
Reactivity with water: Substance kept under water to prevent spontaneous combustion, but ignites spontaneously in moist air.
Binary reactants: May explode on contact with oxidizers (including elemental sulfur and strong caustics). Reacts with air, halogens, halides, alkali hydroxides (forms phosphine gas).
Polymerization: High heat (more than 450°F/232°C may cause polymerization to red phosphorus.
Reactivity group: See Compatibility Guide
Compatibility class: Special Class

ENVIRONMENTAL DATA
Food chain concentration potential: Unlikely to accumulate.
Water pollution: DOT Appendix B, §172.101- severe marine pollutant. Harmful to aquatic life in very low concentrations. May be dangerous if it enters nearby water intakes; notify operators. Notify local health and wildlife officials. **Response to discharge:** Issue warning–high flammability, poison. Restrict access. Evacuate area. Should be removed. Chemical and physical treatment.

SHIPPING INFORMATION
Grades of purity: 99.8–99.9%; **Storage temperature:** Elevated; **Inert atmosphere:** Padded; **Venting:** Pressure-vacuum; **Stability during transport:** Stable.

NAS HAZARD CLASSIFICATION FOR BULK WATER TRANSPORTATION
FIRE: 3
HEALTH: Vapor irritant:–; Liquid or solid irritant: 4; Poisons: 4
WATER POLLUTION: Human toxicity: 3; Aquatic toxicity: 4; Aesthetic effect: 1
REACTIVITY: Other chemicals: 4; Water: 0; Self-reaction: 0

PHYSICAL AND CHEMICAL PROPERTIES
Physical state @ 59°F/15°C and 1 atm: Solid.
Molecular weight: 123.89
Boiling point @ 1 atm: 535.5°F/279.7°C/552.9°K.
Melting/Freezing point: 111.4°F/44.1°C/317.3°K.
Specific gravity (water = 1): 1.82 @ 68°F/20°C (solid).
Relative vapor density (air = 1): 4.3
Heat of polymerization: More than 450°F
Heat of fusion: 4.8 cal/g.
Vapor pressure: (Reid) Very low; 0.03 mm.

PHOSPHORUS OXYCHLORIDE   **REC. P:1800**

SYNONYMS: EEC No. 015-009-00-5; OXICLUORO de FOSFORO (Spanish); PHOSPHORUS CHLORIDE; PHOSPHORUS OXYTRICHLORIDE; PHOSPHORYL CHLORIDE

IDENTIFICATION
CAS Number: 10025-87-3

Formula: Cl₃OP
DOT ID Number: UN 1810; DOT Guide Number: 137
Proper Shipping Name: Phosphorus oxychloride
Reportable Quantity (RQ): **(CERCLA)** 1000 lb/454 kg

DESCRIPTION: Colorless to pale yellow oily liquid that fumes in moist air. Musty odor. Sinks in water; reacts, forming heat, phosphoric acid, and hydrochloric acid. Freezes at 36°F/2°C.

Poison! • Corrosive • Breathing the vapor can kill you; skin or eye contact causes severe burns, impaired vision, or blindness • Firefighting gear (including SCBA) does not provide adequate protection. If exposure occurs, remove and isolate gear immediately and thoroughly decontaminate personnel • Vapors are heavier than air and will collect and stay in low areas • Containers may BLEVE when exposed to fire • Toxic products of combustion may include phosphine (a highly toxic gas which ignites spontaneously), hydrogen chloride and phosphoric acid • Do not put yourself in danger by entering a contaminated area to rescue a victim

Hazard Classification (based on NFPA-704 Rating System)
Health Hazards (Blue): 4; Flammability (Red): 0; Reactivity (Yellow): 2; Special Notice (White): Water reactive

EMERGENCY RESPONSE: See Appendix A (137)
Evacuation:
Public safety: See below.
Spill: Small spill–First: Isolate in all directions 30 meters (100 feet) (land or water); Then: Protect persons downwind, DAY: 0.2 km (0.1 mile) (land or water); NIGHT: 0.5 km (0.3 mile) (land); 0.3 km (0.2 mile) (water). Large spill–First: Isolate in all directions 95 meters (300 feet) (land or water); Then: Protect persons downwind, DAY: 0.8 km (0.5 mile) (land); 1.0 km (0.6 mile) (water); NIGHT: 1.8 km (1.1 miles) (land); 2.6 km (1.6 miles) (water).
Fire: If any large container is involved in fire, isolate for at least 800 meters (½ mile) in all directions; also, consider initial evacuation for 800 meters (½ mile) in all directions.

EXPOSURE
Short-term effects: *SEEK MEDICAL ATTENTION.* **Vapor:** Irritating to eyes, nose, and throat. Harmful if inhaled. Move to fresh air. *IF BREATHING HAS STOPPED,* give artificial respiration. IF breathing is difficult, administer oxygen. *IF BREATHING HAS STOPPED,* give artificial respiration; *avoid mouth-to-mouth resuscitation; use bag/mask apparatus.* Lung edema may develop. **Liquid:** Will burn skin and eyes. Harmful if swallowed. Remove to fresh air.
Flush affected areas with plenty of water. *IF IN EYES,* hold eyelids open and flush with plenty of water. *IF SWALLOWED* and victim is *CONSCIOUS AND ABLE TO SWALLOW,* have victim drink 4 to 8 ounces of water. **Do NOT induce vomiting.**
Note to physician or authorized medical personnel: Medical observation is recommended for 24 to 48 hours after breathing overexposure, as pulmonary edema may be delayed. As first aid for pulmonary edema, consider administering a corticosteroid spray. Cigarette smoking may exacerbate pulmonary injury and should be discouraged for at least 72 hours following exposure.

HEALTH HAZARDS
Personal protective equipment (PPE): A-Level PPE. Chemical protective material(s) reported to have good to excellent resistance: Teflon®.

Exposure limits (TWA unless otherwise noted): ACGIH TLV 0.1 ppm (0.63 mg/m³); NIOSH REL 0.1 ppm (0.6 mg/m3).
Short-term exposure limits (15-minute TWA): NIOSH STEL 0.5 (3 mg/m³).
Toxicity by ingestion: Grade 3; oral rat LD_{50} = 380 mg/kg.
Vapor (gas) irritant characteristics: Vapors cause severe irritation of eyes and throat and can cause eye and lung injury. They cannot be tolerated even at low concentrations.
Liquid or solid irritant characteristics: Severe eye and skin irritant. Causes second-and third-degree burns on short contact and is very injurious to the eyes.

FIRE DATA
Fire extinguishing agents not to be used: Water.

CHEMICAL REACTIVITY
Reactivity with water: Decomposes to hydrochloric and phosphoric acids.
Binary reactants: Reacts with acids, alcohols, alkali metals, caustics, combustible materials, carbon disulfide, dimethylformamide; most metals except nickel and lead. Products of its reaction with water rapidly corrodes wet metals producing flammable hydrogen gas.
Neutralizing agents for acids and caustics: Flush with water, neutralize acids formed with lime or soda ash.

ENVIRONMENTAL DATA
Food chain concentration potential: Negative; unlikely to accumulate.
Water pollution: Dangerous to aquatic life in high concentrations. May be dangerous if it enters nearby water intakes; notify operators. Notify local health and wildlife officials. **Response to discharge:** Issue warning–corrosive. Restrict access. Disperse and flush with care.

SHIPPING INFORMATION
Grades of purity: 99–99.9%; **Storage temperature:** Above 35°F/1.7°C; **Inert atmosphere:** None; **Venting:** Pressure-vacuum; **Stability during transport:** Stable.

NAS HAZARD CLASSIFICATION FOR BULK WATER TRANSPORTATION
FIRE: 0
HEALTH: Vapor irritant: 4; Liquid or solid irritant: 4; Poisons: 4
WATER POLLUTION: Human toxicity: 4; Aquatic toxicity: 3; Aesthetic effect: 2
REACTIVITY: Other chemicals: 4; Water: 4; Self-reaction: 0

PHYSICAL AND CHEMICAL PROPERTIES
Physical state @ 59°F/15°C and 1 atm: Liquid.
Molecular weight: 153.33
Boiling point @ 1 atm: 225°F/107°C/380°K.
Melting/Freezing point: 34°F/1°C/274°K.
Critical temperature: 630°F/332°C/605°K.
Specific gravity (water = 1): 1.675 @ 68°F/20°C (liquid).
Relative vapor density (air = 1): 5.3
Ratio of specific heats of vapor (gas): (estimate) 1.290
Latent heat of vaporization: 97 Btu/lb = 54 cal/g = 2.3 × 10⁵ J/kg.
Vapor pressure: 40 mm @ 81°F

PHOSPHORUS PENTASULFIDE   **REC. P:1850**

SYNONYMS: PENTASULFURE de PHOSPHORE (French); PENTASULFURO de FOSFORO (Spanish); PHOSPHORIC

SULFIDE; PHOSPHORUS PERSULFIDE; PHOSPHORUS SULFIDE; SULFUR PHOSPHIDE; THIOPHOSPHORIC ANHYDRIDE

IDENTIFICATION
CAS Number: 1314-80-3
Formula: P_2S_5
DOT ID Number: UN 1340; DOT Guide Number: 139
Proper Shipping Name: Phosphorus pentasulfide, free from yellow or white phosphorus
Reportable Quantity (RQ): **(CERCLA)** 1 lb/0.454 kg

DESCRIPTION: Greenish-yellow, greenish- gray or grayish-yellow solid flakes. Odor like rotten eggs. Sinks in water; reacts producing toxic hydrogen sulfide and phosphoric acid. *Warning:* Odor is not a reliable indicator of the presence of toxic amounts of phosphorus pentasulfide vapors. High (lethal) concentrations can paralyze the sense of smell.

Combustible solid; may produce heat and spontaneously ignite in the presence of moisture • Interferes with the body's ability to use oxygen • Heat or flame may cause explosion • Containers may BLEVE when exposed to fire • Concentrated dust in confined areas (e.g., tanks, sewers, buildings) may explode when exposed to fire • Corrosive to the skin, eyes, and respiratory tract; inhalation symptoms may be delayed • Toxic products of combustion may include sulfur oxide, phosphorus pentoxide, hydrogen sulfide gases, and phosphoric acid.

Hazard Classification (based on NFPA-704 Rating System)
Health Hazards (Blue): 2; Flammability (Red): 1; Reactivity (Yellow): 2; Special Notice (White): Water reactive

EMERGENCY RESPONSE: See Appendix A (139)
Evacuation:
Public safety: Isolate the area of spill or leak for at least 100 to 150 meters (330 to 490 feet) in all directions.
Spills: *IF SPILLED IN WATER*: Small spill–First: Isolate in all directions 30 meters (100 feet); Then: Protect persons downwind, DAY: 0.3 km (0.2 mile); NIGHT: 0.5 km (0.3 mile). Large spill–First: Isolate in all directions 155 meters (500 feet); Then: Protect persons downwind, DAY: 1.3 km (0.8 mile); NIGHT: 3.2 km (2.0 miles).
Fire: If any large container is involved in fire, isolate for at least 800 meters (½ mile) in all directions; also, consider initial evacuation for 800 meters (½ mile) in all directions.

EXPOSURE
Short-term effects: *SEEK MEDICAL ATTENTION.* **Dust:** Irritating to eyes, nose, and throat. Move to fresh air. *IF BREATHING HAS STOPPED,* give artificial respiration. IF breathing is difficult, administer oxygen. Inhalation may cause lung edema to develop. **Solid:** Irritating to skin and eyes. Harmful if swallowed. Remove contaminated clothing and shoes. Flush affected areas with plenty of water. *IF IN EYES,* hold eyelids open and flush with plenty of water. *IF SWALLOWED* and victim is *CONSCIOUS AND ABLE TO SWALLOW,* have victim drink 4 to 8 ounces of water and have victim induce vomiting. *IF SWALLOWED* and victim is *UNCONSCIOUS OR HAVING CONVULSIONS,* do nothing except keep victim warm.
Note to physician or authorized medical personnel: Medical observation is recommended for 24 to 48 hours after breathing overexposure, as pulmonary edema may be delayed. As first aid for pulmonary edema, consider administering a corticosteroid spray. Cigarette smoking may exacerbate pulmonary injury and should be discouraged for at least 72 hours following exposure.

HEALTH HAZARDS
Personal protective equipment (PPE): B-Level PPE. Chemical safety goggles; plastic face shield; self-contained or air-line respirator.
Recommendations for respirator selection: NIOSH/OSHA
$10\ mg/m^3$: SA* (any supplied-air respirator). $25\ mg/m^3$: SA:CF* (any supplied-air respirator operated in a continuous-flow mode). $50\ mg/m^3$: SCBAF (any self-contained breathing apparatus with a full facepiece); or SAF (any supplied-air respirator with a full facepiece). $250\ mg/m^3$: SAF:PD,PP (any supplied-air respirator that has a full facepiece and is operated in a pressure-demand or other positive-pressure mode). *EMERGENCY OR PLANNED ENTRY INTO UNKNOWN CONCENTRATIONS OR IDLH CONDITIONS:* SCBAF:PD,PP (any self-contained breathing apparatus that has a full facepiece and is operated in a pressure-demand or other positive-pressure mode); or SAF:PD,PP:ASCBA (any supplied-air respirator that has a full facepiece and is operated in a pressure-demand or other positive-pressure mode in combination with an auxiliary self-contained breathing apparatus operated in a pressure-demand or other positive pressure mode). *ESCAPE:* GMFSHiE [any air-purifying, full-facepiece respirator (gas mask) with a chin-style, front- or back-mounted canister providing protection against the compound of concern and having a high-efficiency particulate filter); or SCBAE (any appropriate escape-type, self-contained breathing apparatus). *Note*: Substance reported to cause eye irritation or damage; may require eye protection.
Exposure limits (TWA unless otherwise noted): ACGIH TLV 1 mg/m^3; NIOSH/OSHA 1 mg/m^3.
Short-term exposure limits (15-minute TWA): ACGIH & NIOSH STEL 3 mg/m^3.
Vapor (gas) irritant characteristics: Hydrogen sulfide gas, formed by reaction with moisture, causes severe irritation of eyes and throat and can cause eye and lung injury. It cannot be tolerated even at low concentrations.
Liquid or solid irritant characteristics: Severe eye and skin irritant. If spilled on clothing and allowed to remain, may cause smarting and reddening of the skin.
Odor threshold: 0.0047 ppm (hydrogen sulfide). High (lethal) concentrations can paralyze the sense of smell.
IDLH VALE: 250 mg/m^3.

FIRE DATA
Fire extinguishing agents not to be used: Water.
Autoignition temperature: 527°F/275°C/548°K (liquid).

CHEMICAL REACTIVITY
Reactivity with water: Reacts with liquid water or atmospheric moisture to form toxic hydrogen sulfide gas, sulfur dioxide and phosphoric acid.
Binary reactants: Reacts with acids, alcohols, strong oxidizers, amines, alkalis, organic compounds.

ENVIRONMENTAL DATA
Food chain concentration potential: Negative; unlikely to accumulate.
Water pollution: Harmful to aquatic life in very low concentrations. May be dangerous if it enters nearby water intakes; notify operators. Notify local health and wildlife officials. **Response to discharge:** Issue warning–high flammability, poison. Restrict access. Evacuate area. Should be removed.

SHIPPING INFORMATION
Grades of purity: "Regular" (low reactivity); "reactive" (high reactivity), distilled, undistilled: All 99+%; **Storage temperature:** Ambient; **Inert atmosphere:** None; **Venting:** Sealed containers must be stored in a well-ventilated area; **Stability during transport:** Stable, but cn be ignited by friction.

PHYSICAL AND CHEMICAL PROPERTIES
Physical state @ 59°F/15°C and 1 atm: Solid.
Molecular weight: 222.27/444.6
Boiling point @ 1 atm: 957°F/514°C/787°K.
Melting/Freezing point: 527°F/275°C/548°K.
Specific gravity (water = 1): 2.03 @ 68°F/20°C (solid).
Latent heat of vaporization: 184 Btu/lb = 102 cal/g = 4.27×10^5 J/kg.
Heat of combustion: –10,890 Btu/lb = –6,050 cal/g = $–253.3 \times 10^5$ J/kg.
Heat of solution: (estimate) –20 Btu/lb = –12 cal/g = $–0.5 \times 10^5$ J/kg.
Vapor pressure: 1 mm @ 527°F/275°C.

PHOSPHORUS TRIBROMIDE   REC. P:1900

SYNONYMS: EEC No. 015-103-00-6; PHOSPHORUS BROMIDE; TRIBROMO-PHOSPHINE; TRIBROMURO de FOSFORO (Spanish)

IDENTIFICATION
CAS Number: 7789-60-8
Formula: PBr_3
DOT ID Number: UN 1808; DOT Guide Number: 137
Proper Shipping Name: Phosphorus tribromide

DESCRIPTION: Colorless to pale yellow liquid that fumes in moist air (giving off corrosive hydrogen bromide fumes). Sharp, irritating. Sinks in water; reacts violently releasing toxic hydrobromic acid and phosphoric acid.

Poison! • Highly corrosive • Do NOT use water • Breathing the vapor can kill you; skin or eye contact causes severe burns, impaired vision, or blindness • Firefighting gear (including SCBA) does not provide adequate protection. If exposure occurs, remove and isolate gear immediately and thoroughly decontaminate personnel • Vapors are heavier than air and will collect and stay in low areas • Containers may BLEVE when exposed to fire • Toxic products of combustion may include hydrogen bromide, phosphoric acid, and phosphine. Phosphine is a highly toxic gas which ignites spontaneously • Do not put yourself in danger by entering a contaminated area to rescue a victim.

Hazard Classification (based on NFPA-704 Rating System)
Health Hazards (Blue): 3; Flammability (Red): 0; Reactivity (Yellow): 2; Special Notice (White): Water reactive

EMERGENCY RESPONSE: See Appendix A (137)
Evacuation:
Public safety: Isolate the area of spill or leak for at least 50 to 100 meters (160 to 330 feet) in all directions.
Spill: Increase, in the downwind direction, as necessary, the distance shown under "Public Safety."
Fire: If any large container is involved in fire, isolate for at least 800 meters (½ mile) in all directions; also, consider initial evacuation for 800 meters (½ mile) in all directions.

EXPOSURE
Short-term effects: SEEK MEDICAL ATTENTION. **Vapors:** IF BREATHING HAS STOPPED, give artificial respiration; *avoid mouth-to-mouth resuscitation; use bag/mask apparatus.* IF breathing is difficult, give oxygen. **Liquid:** Will burn skin and eyes. IF SWALLOWED, will cause nausea. Remove contaminated clothing and shoes. Flush affected areas with plenty of water. IF IN EYES, hold eyelids open and flush with plenty of water. IF SWALLOWED and victim is CONSCIOUS AND ABLE TO SWALLOW, have victim drink 4 to 8 ounces of water. IF SWALLOWED and victim is UNCONSCIOUS OR HAVING CONVULSIONS, do nothing except keep victim warm.
Note to physician or authorized medical personnel: Medical observation is recommended for 24 to 48 hours after breathing overexposure, as pulmonary edema may be delayed. As first aid for pulmonary edema, consider administering a corticosteroid spray. Cigarette smoking may exacerbate pulmonary injury and should be discouraged for at least 72 hours following exposure.

HEALTH HAZARDS
Personal protective equipment (PPE): A-Level PPE. Acid-gas canister-type mask (full face type for emergencies); chemical safety goggles; apron, gloves, clothing, and safety shoes all made from rubber
Vapor (gas) irritant characteristics: Eye and respiratory tract irritant.
Liquid or solid irritant characteristics: Eye and skin irritant.
Long-term health effects: Prolonged contact or inhalation of inorganic bromides causes acne-like bromoderma (bromide rash), especially of the hands and face. Other effects include emaciation, depression; and, in severe cases, psychosis and mental deterioration.

FIRE DATA
Fire extinguishing agents not to be used: Do not use water on adjacent fires.
Behavior in fire: Acids formed by reaction with water will attack metals and generate flammable hydrogen gas, which may form explosive mixtures in enclosed spaces.

CHEMICAL REACTIVITY
Reactivity with water: Reacts violently with water, evolving hydrogen bromide, an irritating and corrosive gas apparent as white fumes.
Binary reactants: Reacts violently acids, alkalis, alcohols, ammonia, oxidizers, alkali metals. In the presence of moisture, highly corrosive to most metals except lead and nickel.
Neutralizing agents for acids and caustics: Flush with water and rinse with dilute aqueous sodium bicarbonate or soda ash.

ENVIRONMENTAL DATA
Food chain concentration potential: Unlikely to accumulate.
Water pollution: Effect of low concentrations on aquatic life is unknown. May be dangerous if it enters nearby water intakes; notify operators. Notify local health and wildlife officials. **Response to discharge:** Issue warning–corrosive, air contaminant, water contaminant. Restrict access. Disperse and flush.

SHIPPING INFORMATION
Grades of purity: Purified, 88.55%; **Storage temperature:** Ambient; **Inert atmosphere:** None; **Venting:** Open; **Stability during transport:** Stable if cool; unstable if heated.

PHYSICAL AND CHEMICAL PROPERTIES
Physical state @ 59°F/15°C and 1 atm: Liquid.

Molecular weight: 270.73
Boiling point @ 1 atm: 343°F/173°C/446°K.
Melting/Freezing point: −42.9°F/−40.5°C/232.7°K.
Specific gravity (water = 1): 2.862 at 30°C (liquid).
Liquid surface tension: 45.8 dynes/cm = 0.0458 N/m at 24°C.
Relative vapor density (air = 1): 9.2
Latent heat of vaporization: 64.4 Btu/lb = 35.8 cal/g = 1.50×10^5 J/kg.
Heat of solution: −446 Btu/lb = −248 cal/g = $−10.4 \times 10^5$ J/kg.
Vapor pressure: 2 mm.

PHOSPHORUS TRICHLORIDE   REC. P:1950

SYNONYMS: CHLORIDE OF PHOSPHORUS; CLORURO de FOSFOILO (Spanish); EEC No. 015-007-00-4; PHOSPHORE (TRICHLORURE de) (French); Phosphorus CHLORIDE; PHOSPHORTRICHLORID (German); PHOSPHORUS CHLORIDE; TRICHLOROPHOSPHINE

IDENTIFICATION
CAS Number: 7719-12-2
Formula: PCl_3
DOT ID Number: UN 1809; DOT Guide Number: 137
Proper Shipping Name: Phosphorus trichloride
Reportable Quantity (RQ): **(CERCLA)** 1000 lb/454 kg

DESCRIPTION: Colorless to slightly yellow liquid that fumes in moist air. Sharp irritating odor; like hydrochloric acid. Sinks in water; reacts violently forming heat, phosphine, hydrochloric acid fumes, and phosphorus acid.

Poison! • Corrosive • Breathing the vapor can kill you; skin or eye contact causes severe burns, impaired vision, or blindness; inhalation symptoms may be delayed • Firefighting gear (including SCBA) does not provide adequate protection. If exposure occurs, remove and isolate gear immediately and thoroughly decontaminate personnel • Strong oxidizer which may react spontaneously with low flash point organics or reducing agents. Heat forms oxygen; will increase the activity of an existing fire • May cause fire or explosion contact with combustibles (wood, paper, oil, clothing, etc.) • Vapors are heavier than air and will collect and stay in low areas • Containers may BLEVE when exposed to fire • Toxic products of combustion may include hydrogen chloride, phosphoric acid, and phosphine fumes • Do not put yourself in danger by entering a contaminated area to rescue a victim.

Hazard Classification (based on NFPA-704 Rating System)
Health Hazards (Blue): 4; Flammability (Red): 0; Reactivity (Yellow): 2; Special Notice (White): Water reactive

EMERGENCY RESPONSE: See Appendix A (137)
Evacuation:
Public safety: See below.
Spill: Small spill–First: Isolate in all directions 30 meters (100 feet) (land or water); Then: Protect persons downwind, DAY: 0.2 km (0.1 mile) (land or water); NIGHT: 0.6 km (0.4 mile) (land); 0.3 km (0.2 mile) (water). Large spill–First: Isolate in all directions 125 meters (400 feet) (land or water); Then: Protect persons downwind, DAY: 1.1 km (0.7 mile) (land or water); NIGHT: 2.7 km (1.7 miles) (land); 2.6 km (1.6 miles) (water).
Fire: If any large container is involved in fire, isolate for at least 800 meters (½ mile) in all directions; also, consider initial evacuation for 800 meters (½ mile) in all directions.

EXPOSURE
Short-term effects: *SEEK MEDICAL ATTENTION. CAUTION: PERSONS DOING TREATMENT SHOULD PROTECT THEMSELVES.* **Vapor:** Irritating to eyes, nose, and throat. Harmful if inhaled. Move to fresh air. *IF BREATHING HAS STOPPED*, give artificial respiration; *avoid mouth-to-mouth resuscitation; use bag/mask apparatus.* IF breathing is difficult, administer oxygen.
Liquid: Will burn skin and eyes.
POISONOUS IF SWALLOWED. Remove contaminated clothing and shoes. Flush affected areas with plenty of water. *IF IN EYES*, hold eyelids open and flush with plenty of water. *IF SWALLOWED* and victim is *CONSCIOUS AND ABLE TO SWALLOW*, have victim drink 4 to 8 ounces of water. **Do NOT induce vomiting.**
Note to physician or authorized medical personnel: Medical observation is recommended for 24 to 48 hours after breathing overexposure, as pulmonary edema may be delayed. As first aid for pulmonary edema, consider administering a corticosteroid spray. Cigarette smoking may exacerbate pulmonary injury and should be discouraged for at least 72 hours following exposure.

HEALTH HAZARDS
Personal protective equipment (PPE): A-Level PPE. Chemical safety goggles; plastic face shield; self-contained or air-line respirator; safety hat; gloves and protective clothing. Chemical protective material(s) reported to have good to excellent resistance: Teflon®.
Recommendations for respirator selection: NIOSH
10 ppm: SCBAF (any self-contained breathing apparatus with a full facepiece); or SAF (any supplied-air respirator with a full facepiece). *25 ppm:* SAF:PD,PP (any supplied-air respirator that has a full facepiece and is operated in a pressure-demand or other positive-pressure mode). *EMERGENCY OR PLANNED ENTRY INTO UNKNOWN CONCENTRATIONS OR IDLH CONDITIONS:* SCBAF:PD,PP (any self-contained breathing apparatus that has a full facepiece and is operated in a pressure-demand or other positive-pressure mode); or SAF:PD,PP:ASCBA (any supplied-air respirator that has a full facepiece and is operated in a pressure-demand or other positive-pressure mode in combination with an auxiliary self-contained breathing apparatus operated in a pressure-demand or other positive pressure mode). *ESCAPE:* GMFS* [any air-purifying, full-facepiece respirator (gas mask) with a chin-style, front- or back-mounted canister providing protection against the compound of concern]; or SCBAE (any appropriate escape-type, self-contained breathing apparatus). *Note:* Substance causes eye irritation or damage; eye protection needed.
Exposure limits (TWA unless otherwise noted): ACGIH TLV 0.2 ppm (1.1 mg/m^3); OSHA PEL 0.5 ppm (3 mg/m^3); NIOSH REL 0.2 ppm (1.5 mg/m^3).
Short-term exposure limits (15-minute TWA): ACGIH 0.5 ppm (2.8 mg/m^3); NIOSH STEL 0.5 ppm (3 mg/m^3).
Toxicity by ingestion: Grade 2; oral rat LD_{50} = 550 mg/kg.
Vapor (gas) irritant characteristics: May be fatal if inhaled. Vapors cause severe irritation of eyes and throat and can cause eye and lung injury. They cannot be tolerated even at low concentrations.
Liquid or solid irritant characteristics: Severe eye and skin irritant. Causes second-and third-degree burns on short contact and is very injurious to the eyes.
IDLH value: 25 ppm.

FIRE DATA but will increase the activity of an existing fire.
Fire extinguishing agents not to be used: Water.

CHEMICAL REACTIVITY
Reactivity with water: Reacts violently and may cause flashes of

fire, hydrochloric acid (including gas) and phosphoric acid.
Binary reactants: Reacts with alkali metals, acids, especially strong nitric acid, acetic acid, alcohols, ammonia and amines, organics. Corrodes most common construction materials.
Neutralizing agents for acids and caustics: Flush with water; neutralize acids formed with lime or soda ash.

ENVIRONMENTAL DATA
Food chain concentration potential: Unlikely to accumulate.
Water pollution: Dangerous to aquatic life in high concentrations. May be dangerous if it enters nearby water intakes; notify operators. Notify local health and wildlife officials. **Response to discharge:** Issue warning–corrosive. Restrict access. Disperse and flush with care.

SHIPPING INFORMATION
Grades of purity: Pure: 99.5+%; technical: 98.5+%; **Storage temperature:** Ambient; **Inert atmosphere:** None; **Venting:** Pressure-vacuum; **Stability during transport:** Stable.

PHYSICAL AND CHEMICAL PROPERTIES
Physical state @ 59°F/15°C and 1 atm: Liquid.
Molecular weight: 137.33
Boiling point @ 1 atm: 169°F/76°C/349°K.
Melting/Freezing point: –170°F/–112°C/161°K.
Critical temperature: 547°F/286°C/559°K.
Specific gravity (water = 1): 1.575 @ 68°F/20°C (liquid).
Liquid surface tension: 25.6 dynes/cm = 0.0256 N/m @ 68°F/20°C.
Relative vapor density (air = 1): 4.7
Ratio of specific heats of vapor (gas): (estimate) 1.290
Latent heat of vaporization: 95 Btu/lb = 53 cal/g = 2.2×10^5 J/kg.
Vapor pressure: 100 mm.

PHTHALIC ANHYDRIDE   REC. P:2000

SYNONYMS: ANHIDRIDO FTALICO (Spanish); ANHYDRIDE PHTALIQUE (French); 1,2-BENZENEDICARBOXYLIC ANHYDRIDE; 1,2-BENZENEDICARBOXYLIC ACID ANHYDRIDE; 1,3-DIOXOPHTHALAN; EEC No. 607-009-00-4; ESEN; ISOBENZOFURAN,1,3-DIHYDRO-1,3-DIOXO-; 1,3-ISOBENZOFURANDIONE; PAN; PHTHALANDIONE; 1,3-PHTHALANDIONE; PHTHALIC ACID ANHYDRIDE; PHTHALSAEUREANHYDRID (German); RETARDER AK; RETARDER ESEN; RETARDER PD; RCRA No. U190

IDENTIFICATION
CAS Number: 85-44-9
Formula: $C_8H_4O_3$; $C_6H_4(CO)_2O$
DOT ID Number: UN 2214; DOT Guide Number: 156
Proper Shipping Name: Phthalic anhydride
Reportable Quantity (RQ): **(CERCLA)** 5000 lb/2270 kg

DESCRIPTION: White to pale yellow crystalline flakes/deedles/powder. Colorless liquid (molten) when heated. Choking odor. Solid sinks in water; soluble in hot water; liquid solidifies and sinks in cold water; decomposes to phthalic acid.

Combustible solid • Corrosive • Dust forms explosive mixtures with air • Breathing the dust can kill you; skin or eye contact causes severe burns, impaired vision, or blindness; contact or inhalation symptoms may be delayed • Firefighting gear (including SCBA) does not provide adequate protection. If exposure occurs, remove and isolate gear immediately and thoroughly decontaminate personnel • Containers may BLEVE when exposed to fire • Concentrated dust in confined areas (e.g., tanks, sewers, buildings) may explode when exposed to fire • Irritating to the skin, eyes, and respiratory tract • Toxic products of combustion may include irritating smoke and vapors.

Hazard Classification (based on NFPA-704 Rating System)
Health Hazards (Blue): 3; Flammability (Red): 1; Reactivity (Yellow): 0

EMERGENCY RESPONSE: See Appendix A (156)
Evacuation:
Public safety: Isolate the area of spill or leak for at least 50 to 100 meters (160 to 330 feet) in all directions.
Spill: Increase, in the downwind direction, as necessary, the distance shown under "Public Safety."
Fire: If tank, rail car, or tank truck is involved in fire, isolate for at least 800 meters (½ mile) in all directions; also, consider initial evacuation for 800 meters (½ mile) in all directions.

EXPOSURE
Short-term effects: *SEEK MEDICAL ATTENTION.* **Dust:** Irritating to eyes, nose, and throat. *IF INHALED*, will cause coughing. Move to fresh air. *IF BREATHING HAS STOPPED,* give artificial respiration; *avoid mouth-to-mouth resuscitation; use bag/mask apparatus.* IF breathing is difficult, administer oxygen.
Liquid or solid: Will burn skin or eyes. Harmful if swallowed. Remove contaminated clothing and shoes. Flush affected areas with plenty of water. *IF IN EYES*, hold eyelids open and flush with plenty of water. *IF SWALLOWED* and victim is *CONSCIOUS AND ABLE TO SWALLOW*, have victim drink 4 to 8 ounces of water and have victim induce vomiting. *IF SWALLOWED* and victim is *UNCONSCIOUS OR HAVING CONVULSIONS,* do nothing except keep victim warm.
Note to physician or authorized medical personnel: Medical observation is recommended for 24 to 48 hours after breathing overexposure, as pulmonary edema may be delayed. As first aid for pulmonary edema, consider administering a corticosteroid spray. Cigarette smoking may exacerbate pulmonary injury and should be discouraged for at least 72 hours following exposure.

HEALTH HAZARDS
Personal protective equipment (PPE): B-Level PPE. Coveralls and/or rubber apron; rubber shoes or boots; chemical goggles and/or face shield; gauntlet-type leather or rubber gloves. Butyl rubber is generally suitable for carbooxylic acid compounds.
Recommendations for respirator selection: NIOSH
$30\ mg/m^3$: DM* (any dust and mist respirator). $60\ mg/m^3$: DMXSQ * (any dust and mist respirator except single-use and quarter mask respirators); or HiEF (any air-purifying, full-facepiece respirator with a high-efficiency particulate filter); or PAPRDM * (any powered, air-purifying respirator with a dust and mist filter); or SA* (any supplied-air respirator); or SCBAF (any self-contained breathing apparatus with a full facepiece). *EMERGENCY OR PLANNED ENTRY INTO UNKNOWN CONCENTRATIONS OR IDLH CONDITIONS:* SCBAF:PD,PP (any self-contained breathing apparatus that has a full facepiece and is operated in a pressure-demand or other positive-pressure mode); or SAF:PD,PP:ASCBA (any supplied-air respirator that has a full facepiece and is operated in a pressure-demand or other positive-pressure mode in combination with an auxiliary self-contained breathing apparatus operated in a pressure-demand or other positive pressure mode). *ESCAPE:* HiEF (any air-purifying, full-facepiece respirator with a high-efficiency particulate filter); or SCBAE (any appropriate

escape-type, self-contained breathing apparatus). *Note*: Substance reported to cause eye irritation or damage; may require eye protection.
Exposure limits (TWA unless otherwise noted): ACGIH TLV 1 ppm (6.1 mg/m^3); OSHA PEL 2 ppm (12 mg/m^3); NIOSH 1 ppm (6 mg/m^3).
Toxicity by ingestion: Grade 2; LD$_{50}$ = 0.5 to 5 g/kg (rat).
Long-term health effects: Repeated or prolonged dermal exposure causes skin problems.
Vapor (gas) irritant characteristics: Eye and respiratory tract irritant. Vapor is irritating such that personnel will not usually tolerate moderate or high vapor concentrations.
Liquid or solid irritant characteristics: Eye and skin irritant. Causes smarting of the skin and first-degree burns on short exposure; may cause secondary burns on long exposure. Possible eczema.
Odor threshold: 0.053 mg/m^3.
IDLH value: 60 mg/m^3.

FIRE DATA
Flash point: 329°F/165°C (oc); 305°F/152°C (cc).
Flammable limits in air: LEL 1.7%; UEL 10.5%.
Fire extinguishing agents not to be used: Water or foam may cause frothing.
Behavior in fire: Dust may form explosive mixture in air.
Autoignition temperature: 1058°F/570°C/843°K.
Electrical hazard: Class II, Group G; Due to low electric conductivity, this substance may generate electrostatic charges as a result of agitation and flow.

CHEMICAL REACTIVITY
Reactivity with water: Solid has very slow reaction. Liquid spatters when in contact with water. Converted to phthalic acid in hot water.
Binary reactants: Strong oxidizers and acids. Can be self-reactive in air. Solution attacks ordinary iron and mild steel.
Neutralizing agents for acids and caustics: Water and sodium bicarbonate
Reactivity group: 11
Compatibility class: Organic anhydrides

ENVIRONMENTAL DATA
Food chain concentration potential: Unlikely to accumulate.
Water pollution: Effect of low concentrations on aquatic life is unknown. May be dangerous if it enters nearby water intakes; notify operators. Notify local health and wildlife officials.
Response to discharge: Disperse and flush.

SHIPPING INFORMATION
Grades of purity: Flake; molten; commercial: 99.8%; **Storage temperature:** 268–320°F/131–160°C (liquid); Ambient (solid); **Inert atmosphere:** None; **Venting:** Open (flame arrester); **Stability during transport:** Stable.

NAS HAZARD CLASSIFICATION FOR BULK WATER TRANSPORTATION
FIRE: 1
HEALTH: Vapor irritant: 2; Liquid or solid irritant: 3; Poisons: 1
WATER POLLUTION: Human toxicity: 2; Aquatic toxicity: 2; Aesthetic effect: 2
REACTIVITY: Other chemicals: 3; Water: 1; Self-reaction: 0

PHYSICAL AND CHEMICAL PROPERTIES
Physical state @ 59°F/15°C and 1 atm: Solid.
Molecular weight: 148.12
Boiling point @ 1 atm: 552°F/289°C/562°K.
Melting/Freezing point: 268°F/131°C/404°K.
Specific gravity (water = 1): 1.20 @ 135°C (molten); 1.53 @ 20°C (flakes).
Liquid surface tension: 35.5 dynes/cm = 0.0355 N/m at 155°C.
Liquid water interfacial tension: (estimate) 30 dynes/cm = 0.03 N/m at 155°C.
Relative vapor density (air = 1): 5.1
Ratio of specific heats of vapor (gas): 1.080
Latent heat of vaporization: 189 Btu/lb = 105 cal/g = 4.40 x 10^5 J/kg.
Heat of combustion: –9473 Btu/lb = –5263 cal/g = –220.4 x 10^5 J/kg.
Heat of solution: –127 Btu/lb = –70.8 cal/g = –2.96 x 10^5 J/kg.
Vapor pressure: (Reid) Low; 0.002 mm.

PINENE REC. P:2050

SYNONYMS: ACINTENE A; AUSTRALENE; 2-PINENE; α-PINENE; α-PINEO (Spanish); 2,6,6-TRIMETHYLBICYCLO[3.1.1]HEPT-2-ENE; 2,6,6-TRIMETHYLBICYCLO[3.1.1]-2-HEPT-2-ENE

IDENTIFICATION
CAS Number: 80-56-8
Formula: C$_{10}$H$_{16}$
DOT ID Number: UN 2368; DOT Guide Number: 127
Proper Shipping Name: Pinene (alpha)

DESCRIPTION: Colorless, oily liquid. Turpentine or pine-like odor. Floats on water surface; insoluble.

Highly flammable • Containers may BLEVE when exposed to fire • Vapors may form explosive mixture with air • Vapors are heavier than air and will collect and stay in low areas • Vapors may travel long distances to ignition sources and flashback • Vapors in confined areas (e.g., tanks, sewers, buildings) may explode when exposed to fire • Irritating to the skin, eyes, and respiratory tract • Toxic products of combustion may include carbon monoxide.

Hazard Classification (based on NFPA-704 Rating System)
Health Hazards (Blue): 1; Flammability (Red): 3; Reactivity (Yellow): 0

EMERGENCY RESPONSE: See Appendix A (127)
Evacuation:
Public safety: Isolate spill area for at least 25 to 50 meters (80 to 160 feet) in all directions.
Spill: Large spill–Consider initial downwind evacuation for at least 300 meters (1000 feet).
Fire: Isolate for 800 meters (½ mile) in all directions, especially if tank, rail car, or tank truck is involved in fire.

EXPOSURE
Short-term effects: *SEEK MEDICAL ATTENTION.* **Vapor:** Irritating to eyes, nose, and throat. *IF INHALED*, will, will cause nausea, vomiting, headache, difficult breathing, or loss of consciousness. Move to fresh air. *IF BREATHING HAS STOPPED*, give artificial respiration. IF breathing is difficult, administer oxygen. **Liquid:** *POISONOUS IF SWALLOWED*. Irritating to skin and eyes. Remove contaminated clothing and shoes. Flush affected areas with plenty of water. *IF IN EYES*, hold eyelids open and flush with plenty of water.

HEALTH HAZARDS

Personal protective equipment (PPE): B-Level PPE. Self-contained breathing apparatus and protective clothing, rubber boots, and heavy rubber gloves.
Vapor (gas) irritant characteristics: Vapors cause severe irritation of eyes and throat and can cause eye and lung injury. They cannot be tolerated even at low concentrations.
Liquid or solid irritant characteristics: Severe skin irritant. May cause pain and second-degree burns after a few minutes of contact.

FIRE DATA

Flash point: 91°F/33°C (cc).
Fire extinguishing agents not to be used: Water may be ineffective.
Autoignition temperature: 482°F/250°C/523°K.

CHEMICAL REACTIVITY

Binary reactants: Strong acids.
Reactivity group: 30
Compatibility class: Olefins

ENVIRONMENTAL DATA

Food chain concentration potential: Log P_{ow} = 4.12. Values > 3.0 are likely to bioconcentrate in aquatic organisms and other living tissue, especially in fats.
Water pollution: DOT Appendix B, §172.101–marine pollutant. May be dangerous to aquatic life. Fouling to shoreline. May be dangerous if it enters nearby water intakes; notify operators. Notify local health and wildlife officials. **Response to discharge:** Evacuate area. Mechanical containment. Should be removed. Chemical and physical treatment.

SHIPPING INFORMATION

Grades of purity: 98%; **Storage temperature:** Ambient; **Stability during transport:** Stable.

PHYSICAL AND CHEMICAL PROPERTIES

Physical state @ 59°F/15°C and 1 atm: Liquid.
Molecular weight: 136.26
Boiling point @ 1 atm: 311°F/155°C/428.16°K.
Melting/Freezing point: –58°F/–50°C/223.2°K.
Specific gravity (water = 1): 0.858
Relative vapor density (air = 1): 4.7
Vapor pressure: (Reid) 0.1991 psia.

PIPERAZINE   REC. P:2100

SYNONYMS: ANTIREN; 1,4-DIAZACYCLOHEXANE; 1,4-DIETHYLENEDIAMINE; DIETHYLENEIMINE; DISPERMINE; EEC No. 612-057-00-4; HEXAHYDRO-1,4-DIAZINE; HEXAHYDROPYRAZINE; LUMBRICAL; LUMBUCAL; PIPERAZIDINE; PIPERAZIN (German); PIPERAZINE, ANHYDROUS; PYRAZINE HEXAHYDRIDE; PYRAZINE, HEXAHYDRO-

IDENTIFICATION

CAS Number: 110-85-0; 142-63-2 (hexahydrate)
Formula: $C_4H_{10}N_2$; $NHCH_2CH_2NHCH_2CH_2$; $C_4H_{10}N_2 \cdot 6H_2O$ (hexahydrate)
DOT ID Number: UN 2579; DOT Guide Number: 153
Proper Shipping Name: Piperazine

DESCRIPTION: White solid. Mild, fishy, amine odor. Sinks and mixes with water.

Combustible • Corrosive to the skin, eyes, and respiratory tract; inhalation symptoms may be delayed • Containers may BLEVE when exposed to fire • Vapors may form explosive mixture with air • Vapors are heavier than air and will collect and stay in low areas • Vapors may travel long distances to ignition sources and flashback • Vapors in confined areas (e.g., tanks, sewers, buildings) may explode when exposed to fire • Toxic products of combustion may include nitrogen oxides.

Hazard Classification (based on NFPA-704 Rating System)
Health Hazards (Blue): 2; Flammability (Red): 2; Reactivity (Yellow): 0

EMERGENCY RESPONSE: See Appendix A (153)
Evacuation:
Public safety: Isolate the area of spill or leak for at least 25 to 50 meters (80 to 160 feet) in all directions.
Spill: Increase, in the downwind direction, as necessary, the distance shown under "Public Safety."
Fire: If tank, rail car, or tank truck is involved in fire, isolate for at least 800 meters (½ mile) in all directions; also, consider initial evacuation for 800 meters (½ mile) in all directions.

EXPOSURE

Short-term effects: *SEEK MEDICAL ATTENTION*. **Dust:** Irritating to eyes, nose, and throat. *IF INHALED*, will, will cause coughing or difficult breathing. *IF IN EYES*, hold eyelids open and flush with plenty of water. *IF BREATHING HAS STOPPED*, give artificial respiration; *avoid mouth-to-mouth resuscitation; use bag/mask apparatus*. IF breathing is difficult, administer oxygen. Lung edema may develop. **Solid:** Will burn eyes. Irritating to eyes. *IF SWALLOWED*, will cause nausea and vomiting. Remove contaminated clothing and shoes. Flush affected areas with plenty of water. *IF IN EYES*, hold eyelids open and flush with plenty of water. *IF SWALLOWED* and victim is *CONSCIOUS AND ABLE TO SWALLOW*, have victim drink 4 to 8 ounces of water and have victim induce vomiting. *IF SWALLOWED* and victim is *UNCONSCIOUS OR HAVING CONVULSIONS*, do nothing except keep victim warm.
Note to physician or authorized medical personnel: Medical observation is recommended for 24 to 48 hours after breathing overexposure, as pulmonary edema may be delayed. As first aid for pulmonary edema, consider administering a corticosteroid spray. Cigarette smoking may exacerbate pulmonary injury and should be discouraged for at least 72 hours following exposure.

HEALTH HAZARDS

Personal protective equipment (PPE): B-Level PPE. Monogoggles or face shield; rubber gloves; dust mask
Toxicity by ingestion: Grade 2; LD_{50} = 0.5 to 5 g/kg.
Vapor (gas) irritant characteristics: Respiratory tract irritant.
Liquid or solid irritant characteristics: Severe eye irritant; may cause burns. Skin irritant.

FIRE DATA

Flash point: 178°F/81°C (oc) (molten solid).
Flammable limits in air: LEL 1.6%; UEL 12.5%.
Fire extinguishing agents not to be used: Water may cause frothing.
Autoignition temperature: 851°F/455°C/728°K.

CHEMICAL REACTIVITY

Reaction with water: Aqueous solution forms a strong base.
Binary reactants: Violent or explosive reaction with strong oxidizers, strong acids, dicyanofurazan. Reacts with nitrogenous

compounds, carbon tetrachloride. Incompatible with alcohols, aldehydes, alkylene oxides, alkalis, cresols, caprolactam solution, epichlorohydrin, organic anhydrides, glycols, maleic anhydride, phenols. Attacks aluminum, cobalt, copper, nickel, magnesium, and zinc.
Neutralizing agents for acids and caustics: Flush with water.

ENVIRONMENTAL DATA
Food chain concentration potential: Log P_{ow} = –1.18. Unlikely to accumulate.
Water pollution: Effect of low concentrations on aquatic life is unknown. May be dangerous if it enters nearby water intakes; notify operators. Notify local health and wildlife officials.
Response to discharge: Issue warning–water contaminant. Disperse and flush.

SHIPPING INFORMATION
Grades of purity: Commercial, 99+%; may also be shipped as a solid hexahydrate, whose hazardous properties are similar; **Storage temperature:** Ambient; **Inert atmosphere:** None; **Venting:** Open (flame arrester); **Stability during transport:** Stable.

PHYSICAL AND CHEMICAL PROPERTIES
Physical state @ 59°F/15°C and 1 atm: Solid.
Molecular weight: 86
Boiling point @ 1 atm: 299°F/148°C/421°K.
Melting/Freezing point: 223°F/106°C/379°K.
Specific gravity (water = 1): 1.1 @ 68°F/20°C (solid).
Relative vapor density (air = 1): 3.0
Heat of combustion: –14,800 Btu/lb = –8200 cal/g = –343 x 10^5 J/kg.
Heat of solution: –34.9 Btu/lb = –19.4 cal/g = –0.812 x 10^5 J/kg.
Vapor pressure: 0.08 mm.

POLYBUTENE   REC. P:2150

SYNONYMS: BUTENE RESINS; INDOPOL L 50; POLYISOBUTYLENE PLASTICS, RESINS, and WAXES

IDENTIFICATION
CAS Number: 9003-29-6
Formula: $(C_4H_8)_x$; $[C(CH_3)_2CH_2]$

DESCRIPTION: Clear, colorless, oily liquid. Odorless. Floats on the surface of water.

Combustible • Containers may BLEVE when exposed to fire •Vapors may form explosive mixture with air • Vapors are heavier than air and will collect and stay in low areas • Vapors in confined areas (e.g., tanks, sewers, buildings) may explode when exposed to fire • Irritating to the skin, eyes, and respiratory tract • Toxic products of combustion may include carbon monoxide.

Hazard Classification (based on NFPA-704 Rating System)
Health Hazards (Blue): 0; Flammability (Red): 1; Reactivity (Yellow): 0

EMERGENCY RESPONSE: See Appendix A (171)
Evacuation:
Public safety: Isolate the area of spill or leak for at least 10 to 25 meters (30 to 80 feet) in all directions.
Spill: Increase, in the downwind direction, as necessary, the distance shown under "Public Safety."
Fire: If any large container is involved in fire, isolate for at least 800 meters (½ mile) in all directions; also, consider initial evacuation for 800 meters (½ mile) in all directions.

HEALTH HAZARDS
Personal protective equipment (PPE): B-Level PPE. Goggles or face shield.
Toxicity by ingestion: Grade 0; LD_{50} = > 15 g/kg (animals).
Vapor (gas) irritant characteristics: Vapors are nonirritating to the eyes and throat.
Liquid or solid irritant characteristics: Practically harmless to the skin.

FIRE DATA
Flash point: 215–470°F/102–243°C (oc) (polybutenes); 280°F/138°C (cc).
Fire extinguishing agents not to be used: Water may be ineffective.

CHEMICAL REACTIVITY
Binary reactants: Strong acids.
Reactivity group: 30
Compatibility class: Olefins

ENVIRONMENTAL DATA
Food chain concentration potential: Negative; unlikely to accumulate.
Water pollution: Effect of low concentrations on aquatic life is unknown. Fouling to shoreline. May be dangerous if it enters nearby water intakes; notify operators. Notify local health and wildlife officials. **Response to discharge:** Mechanical containment. Should be removed. Chemical and physical treatment.

SHIPPING INFORMATION
Grades of purity: 85–98%; **Storage temperature:** Ambient; **Inert atmosphere:** None; **Venting:** Open (flame arrester); **Stability during transport:** Stable.

PHYSICAL AND CHEMICAL PROPERTIES
Physical state @ 59°F/15°C and 1 atm: Liquid.
Molecular weight: 225–2300
Boiling point @ 1 atm: More than 90°F/32°C/305°K.
Melting/Freezing point: More than –40°F/–40°C/233°K.
Specific gravity (water = 1): 0.81–0.91 @ 15°C (liquid).
Liquid surface tension: (estimate) 25 dynes/cm = 0.025 N/m @ 68°F/20°C.
Liquid water interfacial tension: (estimate) 50 dynes/cm = 0.05 N/m @ 68°F/20°C.
Heat of combustion: (estimate) –20,000 Btu/lb = –11,000 cal/g = –470 x 10^5 J/kg.
Heat of solution: (estimate) –9 Btu/lb = –5 cal/g = 0.2 x 10^5 J/kg.

POLYCHLORINATED BIPHENYLS (PCBs)   REC. P:2200

SYNONYMS: AROCLOR®; AROCLOR® 1221; AROCLOR® 1232; AROCLOR® 1242; AROCLOR® 1248; AROCLOR® 1254; AROCLOR® 1260; AROCLOR® 1262; AROCLOR® 1268; AROCLOR® 2565; AROCLOR® 4465; BIPHENYL, POLYCHLORO-; CHLOPHEN; CHLOREXTOL; CHLORINATED BIPHENYLS; CHLORINATED DIPHENYL; CHLORINATED DIPHENYLENE; CHLORO BIPHENYLS; CHLORODIPHENYL (54% CHLORINE); CHLORO 1,1-BIPHENYL; CLOPHEN; DYKANOL; FENCLOR; INERTEEN; KANECHLOR; KANECHLOR 300; KANECHLOR 400;

KANECHLOR 500; MONTAR; NOFLAMOL; PCB; PHENOCHLOR; PHENOCLOR; POLYCHLOROBIPHENYL; PYRALENE; PYRANOL; SANTOTHERM; SANTOTHERM FR; SOVOL; THERMINOL FR-1; HALOGENATED WAXES; TQ1356000 (42% chlorine); TQ1360000 (54% chlorine)

IDENTIFICATION

CAS Number: 1336-36-3 (generic); 11096-82-5 (AROCLOR® 1260); 11097-69-1; 11100-14-4 (AROCLOR® 1268); 11104-28-2; 11120-29-9 (AROCLOR® 4465); 11141-16-5 (AROCLOR® 1232); 11097-69-1 (AROCLOR® 1254, 54% chlorine); 12672-29-6 (AROCLOR® 1248); 12737-87-0 (kanechlor 400); 37317-41-2 (kanechlor 500); 37324-24-6 (AROCLOR® 2565); 37324-23-5 (AROCLOR® 1262); 37353-63-2 (kanechlor 300); 37317-41-2; 53469-21-9 (arochlor 1242, 42% chlorine)

Formula: $(C_{12}H_{10}-10_x)Cl_x$

DOT ID Number: UN 2315; DOT Guide Number: 171

Proper Shipping Name: Polychlorinated biphenyls

Reportable Quantity (RQ): **(CERCLA)** 1 lb/0.454 kg

DESCRIPTION: Colorless mobile liquids (chlorinated AROCLOR® s 1221, 1232, 1016, 1242, and 1248); pale yellow, oily liquids (AROCLOR® 1254), or sticky resins (AROCLOR® 1260 and 1262) or white powder (AROCLOR® 1268 and 1270). Practically odorless. Sinks in water; insoluble.

Combustible • Poison! • Irritating to the skin, eyes, and respiratory tract • Breathing the vapor, skin or eye contact, or swallowing the material are poisonous and can cause long-term illness • Firefighting gear (including SCBA) does not provide adequate protection. If exposure occurs, remove and isolate gear immediately and thoroughly decontaminate personnel • Heat or flame may cause explosion • Containers may BLEVE when exposed to fire • Vapors are heavier than air and will collect and stay in low areas • Vapors in confined areas (e.g., tanks, sewers, buildings) may explode when exposed to fire • At high temperatures (over 1000°F/538°C) highly toxic derivatives of PCBs may be formed. These derivatives are more dangerous than PCBs. Toxic products of combustion may include the formation of a black soot containing PCBs, polychlorinated dibenzofurans, and chlorinated dibenzo-*p*-dioxins.

Hazard Classification (based on NFPA-704 Rating System)
Health Hazards (Blue): 2; Flammability (Red): 1; Reactivity (Yellow): 0

EMERGENCY RESPONSE: See Appendix A (171)
Evacuation:
Public safety: Isolate the area of spill or leak for at least 10 to 25 meters (30 to 80 feet) in all directions.
Spill: Increase, in the downwind direction, as necessary, the distance shown under "Public Safety."
Fire: If any large container is involved in fire, isolate for at least 800 meters (½ mile) in all directions; also, consider initial evacuation for 800 meters (½ mile) in all directions.

EXPOSURE

Short-term effects: *SEEK MEDICAL ATTENTION. POISONOUS OF ABSORBED THROUGH THE SKIN.* **Liquid or solid:** Irritating to skin and eyes. Flush affected areas with plenty of water. *IF IN EYES*, hold eyelids open and flush with plenty of water.

HEALTH HAZARDS

Personal protective equipment (PPE): A-Level PPE. Full face organic vapor respirator. Chemical protective material(s) reported to have good to excellent resistance: butyl rubber, Viton®/neoprene, neoprene, Teflon®, Viton®, Saranex®, Silvershield®. Also, nitrile and polyethylene offers limited protection

Recommendations for respirator selection: NIOSH

Any detectable concentration: SCBAF:PD,PP (any self-contained breathing apparatus that has a full facepiece and is operated in a pressure-demand or other positive-pressure mode); or SAF:PD,PP:ASCBA (any supplied-air respirator that has a full facepiece and is operated in a pressure-demand or other positive-pressure mode in combination with an auxiliary self-contained breathing apparatus operated in a pressure-demand or other positive pressure mode). *ESCAPE:* GMFOVHiE [any air-purifying, full-facepiece respirator (gas mask) with a chin-style, front- or back-mounted organic vapor canister having a high-efficiency particulate filter]; or SCBAE (any appropriate escape-type, self-contained breathing apparatus).

Exposure limits (TWA unless otherwise noted): ACGIH TLV 1 mg/m^3 (42% chlorine), 0.5 mg/m^3 (54% chlorine) as chlorodiphenyls; OSHA PEL 1 mg/m^3 (42% chlorine), 0.5 mg/m^3 (54% chlorine); NIOSH REL 0.001 mg/m^3 applies to all PCBs; skin contact contributes significantly in overall exposure.

Toxicity by ingestion: Grade 2; oral rat LD$_{50}$ = 3980 mg/kg.

Long-term health effects: Causes chromosomal abnormalities in rats, birth defects in birds. Confirmed carcinogen.

Vapor (gas) irritant characteristics: Vapors cause severe irritation of eyes and throat and cause eye and lung injury. They cannot be tolerated even at low concentrations.

Liquid or solid irritant characteristics: Skin irritant.

IDLH value: (42% and 54% chlorine) 5 mg/m^3.

FIRE DATA

Flash point: More than 286–385°F/141–196°C. The flash points of some of these materials are unknown.

CHEMICAL REACTIVITY

Binary reactants: Incompatible with strong oxidizers, strong acids. PCB's are generally chemically inert under normal conditions of temperature and pressure, and are stable to conditions of hydrolysis and oxidation in industrial use. However, strong ultraviolet or sunlight may cause the formation of phenolic materials, and traces of polychlorinated dibenzofurans. Attacks some plastics such as polyethylene and some rubbers: natural rubber and to a lesser degree nitrile rubber.

ENVIRONMENTAL DATA

Food chain concentration potential: High

Water pollution: DOT Appendix B, §172.101- severe marine pollutant. Harmful to aquatic life in very low concentrations. May be dangerous if it enters nearby water intakes; notify operators. Notify local health and wildlife officials. **Response to discharge:** Issue warning–water contaminant. Should be removed. Chemical and physical treatment.

SHIPPING INFORMATION

Grades of purity: 11 grades (some liquid, some solids) which differ primarily in their chlorine content (20–68% by weight). PCBs are no longer commercially produced in the United States, but may be found in various products (i.e., oil switches, capacitors, transformers) made prior to 1979; **Storage temperature:** Ambient; **Inert atmosphere:** None; **Venting:** Open; **Stability during transport:** Stable.

PHYSICAL AND CHEMICAL PROPERTIES

Physical state @ 59°F/15°C and 1 atm: Solid.

Molecular weight: 258 (approximate).
Boiling point @ 1 atm: 617–691°F/325–366°C/598–639°K.
Melting/Freezing point: –2°F/–18°C/255°K.
Specific gravity (water = 1): 1.3–1.8 @ 68°F/20°C (liquid).
Vapor pressure: 0.001 mm.

POLYETHYLENE POLYAMINES   P:2250

SYNONYMS: POLY(ETHYLENEIMINE); POLYETHYLENEIMINE

IDENTIFICATION
CAS Number: 26913-06-4
Formula: $(C_2H_5N_2)_n$; $(CH_2CH_2NH)_n$
DOT ID Number: UN 2735; DOT Guide Number: 153
Proper Shipping Name: Polyamines, liquid, corrosive, n.o.s.

DESCRIPTION: Yellowish to reddish-brown liquid. Fishy, amine odor. Mixes with water.

Combustible • Corrosive • Vapor may explode if ignited in an enclosed area • Containers may BLEVE when exposed to fire • Severely irritating to the skin, eyes, and respiratory tract • Toxic products of combustion may include nitrogen oxides.

Hazard Classification (based on NFPA-704 Rating System)
Health Hazards (Blue): 0; Flammability (Red): 1; Reactivity (Yellow): 0

EMERGENCY RESPONSE: See Appendix A (153)
Evacuation:
Public safety: Isolate the area of spill or leak for at least 25 to 50 meters (80 to 160 feet) in all directions.
Spill: Increase, in the downwind direction, as necessary, the distance shown under "Public Safety."
Fire: If tank, rail car, or tank truck is involved in fire, isolate for at least 800 meters (½ mile) in all directions; also, consider initial evacuation for 800 meters (½ mile) in all directions.

EXPOSURE
Short-term effects: *SEEK MEDICAL ATTENTION*. **Vapor:** May be harmful if inhaled. May irritate eyes and skin. *IF BREATHING HAS STOPPED*, give artificial respiration; *avoid mouth-to-mouth resuscitation; use bag/mask apparatus*. IF breathing is difficult, administer oxygen. Move to fresh air. **Liquid:** Irritating to eyes and skin. May be harmful if swallowed. *IF IN EYES OR ON SKIN*: Flush with plenty of running water for at least 15 minutes; hold eyelids open if necessary. Wash skin with soap and water. Remove and double bag contaminated clothing and shoes at the site. Keep victim quiet and maintain normal body temperature. *IF SWALLOWED* and victim is *CONSCIOUS AND ABLE TO SWALLOW*, have victim drink water and induce vomiting. *IF SWALLOWED* and victim is *UNCONSCIOUS OR HAVING CONVULSIONS*, do nothing except keep victim warm.
Note to physician or authorized medical personnel: Medical observation is recommended for 24 to 48 hours after breathing overexposure, as pulmonary edema may be delayed. As first aid for pulmonary edema, consider administering a corticosteroid spray. Cigarette smoking may exacerbate pulmonary injury and should be discouraged for at least 72 hours following exposure.

HEALTH HAZARDS
Personal protective equipment (PPE): B-Level PPE. Self-contained, positive-pressure breathing apparatus and full protective clothing.
Liquid or solid irritant characteristics: Severe eye and skin irritant.

FIRE DATA
Flash point: 207°F/97°C.
Autoignition temperature: 743°F/395°C/668°K.

CHEMICAL REACTIVITY
Binary reactants: Incompatible with aluminum, strong acids, organic acids, zinc, and other nonferrous metals.
Reactivity group: 7
Compatibility class: Aliphatic amines

ENVIRONMENTAL DATA
Water pollution: Effects of low concentration on aquatic life is unknown. May be dangerous if it enters nearby water intakes; notify operators. Notify local health and pollution control officials.
Response to discharge: Issue warning–water contaminant. Should be removed. Chemical and physical treatment.

SHIPPING INFORMATION
Storage temperature: Ambient; **Inert atmosphere:** None; **Venting:** Open; **Stability during transport:** Stable.

PHYSICAL AND CHEMICAL PROPERTIES
Physical state @ 59°F/15°C and 1 atm: Liquid.

Boiling point @ 1 atm: 401°F/205°C/478°K.
Specific gravity (water = 1): 0.99 (temperature unknown/liquid).

POLYMETHYLENEPOLYPHENYL ISOCYANATE   REC. P:2300

SYNONYMS: PAPI; METHYLENEDIPHENYL DIISOCYANATE URETHANE POLYMER

IDENTIFICATION
CAS Number: 9016-87-9
Formula: $C_6H_4(NCO)CH_2C_6H_4(NCO)$-and polymer
DOT ID Number: UN2206; DOT Guide Number: 155
Proper Shipping Name: Isocyanates, toxic, n.o.s.

DESCRIPTION: Dark brown liquid. Weak odor. Sinks in water.

Poison! •Irritating to eyes, skin, and respiratory tract; poisonous if absorbed by the skin or swallowed • Containers may BLEVE when exposed to fire • Vapors in confined areas (e.g., tanks, sewers, buildings) may explode when exposed to fire • Toxic products of combustion may include hydrogen cyanide • Do not put yourself in danger by entering a contaminated area to rescue a victim.

Hazard Classification (based on NFPA-704 Rating System)
Health Hazards (Blue): 1; Flammability (Red): 1; Reactivity (Yellow): 0

EMERGENCY RESPONSE: See Appendix A (155)
Evacuation:
Public safety: Isolate the area of spill or leak for at least 50 to 100 meters (160 to 330 feet) in all directions.
Spill: Increase, in the downwind direction, as necessary, the distance shown under "Public Safety."

Fire: If tank, rail car, or tank truck is involved in fire, isolate for at least 800 meters (½ mile) in all directions; also, consider initial evacuation for 800 meters (½ mile) in all directions.

EXPOSURE
Short-term effects: *SEEK MEDICAL ATTENTION.* If artificial respiration is administered; *avoid mouth-to-mouth resuscitation; use bag/mask apparatus.* **Liquid:** *POISONOUS IF SWALLOWED.* Irritating to skin and eyes. Remove contaminated clothing and shoes. Flush affected areas with plenty of water. *IF IN EYES*, hold eyelids open and flush with plenty of water. *IF SWALLOWED* and victim is *CONSCIOUS AND ABLE TO SWALLOW*, have victim drink 4 to 8 ounces of water and have victim induce vomiting. *IF SWALLOWED* and victim is *UNCONSCIOUS OR HAVING CONVULSIONS,* do nothing except keep victim warm.
Note to physician or authorized medical personnel: Consider the use of amyl nitrite perles if symptoms of cyanide poisoning develop. If symptoms indicate, initial treatment includes the cyanide antidote kit. In all cases, break an amyl nitrite perle in a gauze pad and hold lightly under victim's nose for 15 seconds, repeating 5 times at about 15-second intervals; if necessary (and if sodium nitrite infusions will be delayed), repeat procedure every 3 minutes with fresh pearls until 3 or 4 have been used. Avoid breathing the vapor while administering it to the victim. Administer sodium nitrite IV, ASAP. The usual adult dose is 10 to 20 mL of a 3% solution infused over no less than 5 minutes; the average child dose is 0.15 to 0.20 mL/kg. Monitor blood pressure during administration, and slow the rate of infusion if hypotention develops. Next, infuse sodium thiosulfate IV. The usual adult dose is 50 mL of a 25% solution infused over 10 to 20 minutes; the average child dose is 1.65 mL/kg. Repeat with nitrite and thiosulfate as required.

HEALTH HAZARDS
Personal protective equipment (PPE): B-Level PPE. Air-line or organic canister mask; goggles or face shield; rubber gloves and other protective clothing to prevent contact with skin.
Toxicity by ingestion: Grade 1; LD_{50} = 5 to 15 g/kg.
Vapor (gas) irritant characteristics: Respiratory tract irritant. Vapors are irritating such that personnel will not usually tolerate moderate or high concentrations.
Liquid or solid irritant characteristics: Causes smarting of the skin and first-degree burns on short exposure; may cause second-degree burns on long exposure.

FIRE DATA
Flash point: 425°F/218°C (oc).
Behavior in fire: Containers may explode.

CHEMICAL REACTIVITY
Reactivity with water: Reacts slowly, forming heavy scum and liberating CO_2 gas. Dangerous pressure can build up if container is sealed.
Binary reactants: Acids, caustic, ammonia. See reactivity group.
Reactivity group: 12
Compatibility class: Isocyanates

ENVIRONMENTAL DATA
Food chain concentration potential: Negative; unlikely to accumulate.
Water pollution: Effect of low concentrations on aquatic life is unknown. May be dangerous if it enters nearby water intakes; notify operators. Notify local health and wildlife officials.
Response to discharge: Issue warning–poison, water contaminant. Restrict access. Should be removed. Chemical and physical treatment.

SHIPPING INFORMATION
Grades of purity: 50% methylenebis-phenylisocyanate plus 50% polymer, 50% solution in monochlorobenzene; **Storage temperature:** 35-125°F/1.6-52°C; **Inert atmosphere:** Low-pressure dry; **Inert atmosphere:** nitrogen; **Venting:** Safety relief; **Stability during transport:** Stable if kept sealed and dry.

NAS HAZARD CLASSIFICATION FOR BULK WATER TRANSPORTATION
FIRE: 1
HEALTH: Vapor irritant: 3; Liquid or solid irritant: 2; Poisons: 4
WATER POLLUTION: Human toxicity: 1; Aquatic toxicity: 2; Aesthetic effect: 1
REACTIVITY: Other chemicals: 3; Water: 1; Self-reaction: –

PHYSICAL AND CHEMICAL PROPERTIES
Physical state @ 59°F/15°C and 1 atm: Liquid.
Molecular weight: 400 (approximate).
Boiling point @ 1 atm: 392°F/200°C/473°K.
Specific gravity (water = 1): 1.20 @ 68°F/20°C (liquid).
Heat of combustion: (estimate) –13,000 Btu/lb = –7200 cal/g = 300 x 10^5 J/kg.
Vapor pressure: (Reid) Very low.

POLYPHOSPHORIC ACID   REC. P:2350

SYNONYMS: ACID PHOSPHORIQUE (French); ACIDO POLIFOSFORICO (Spanish); ORTHOPHOSPHORIC ACID; METAPHOSPHORIC ACID; PHOSPHORIC ACID; PHORSAEURELOESUNGEN (German); TRIPHOSPHORIC ACID; WHITE PHOSPHORIC ACID; CONDENSED PHOSPHORIC ACID

IDENTIFICATION
CAS Number: 8017-16-1; 7664-38-2
Formula: H_3O_4P; $(P_2O_5)(H_2O)<3$; $H_{n2}P_nO_{3n+1}$, for n >1
DOT ID Number: UN 1805; DOT Guide Number: 154
Proper Shipping Name: Phosphoric acid
Reportable Quantity (RQ): **(CERCLA)** 5000 lb/2270 kg

DESCRIPTION: Clear, colorless, viscous liquid. Odorless. Sinks and mixes with water; slowly changes to orthophosphoric acid. A concentrated or high-content phosphoric acid.

Corrosive • Breathing the vapor can kill you; skin or eye contact causes severe burns, impaired vision, or blindness; effects of contact or inhalation may be delayed • Firefighting gear (including SCBA) does not provide adequate protection. If exposure occurs, remove and isolate gear immediately and thoroughly decontaminate personnel • Containers may BLEVE when exposed to fire • Toxic products of combustion may include oxides of phosphorus.

Hazard Classification (based on NFPA-704 Rating System)
Health Hazards (Blue): 3; Flammability (Red): 0; Reactivity (Yellow): 1

EMERGENCY RESPONSE: See Appendix A (154)
Evacuation:
Public safety: Isolate the area of spill or leak for at least 25 to 50 meters (80 to 160 feet) in all directions.

Spill: Increase, in the downwind direction, as necessary, the distance shown under "Public Safety."

Fire: If tank, rail car, or tank truck is involved in fire, isolate for at least 800 meters (½ mile) in all directions; also, consider initial evacuation for 800 meters (½ mile) in all directions.

EXPOSURE

Short-term effects: *SEEK MEDICAL ATTENTION.* **Liquid:** Will burn skin and eyes. Harmful if swallowed. Remove contaminated clothing and shoes. Flush affected areas with plenty of water. *IF BREATHING HAS STOPPED,* give artificial respiration; *avoid mouth-to-mouth resuscitation; use bag/mask apparatus. IF IN EYES,* hold eyelids open and flush with plenty of water. *IF SWALLOWED* and victim is *CONSCIOUS AND ABLE TO SWALLOW,* have victim drink 4 to 8 ounces of water. **Do NOT induce vomiting.**

Note to physician or authorized medical personnel: Medical observation is recommended for 24 to 48 hours after breathing overexposure, as pulmonary edema may be delayed. As first aid for pulmonary edema, consider administering a corticosteroid spray. Cigarette smoking may exacerbate pulmonary injury and should be discouraged for at least 72 hours following exposure.

HEALTH HAZARDS

Personal protective equipment (PPE): B-Level PPE. Goggles or face shield; rubber gloves or protective clothing. Sealed chemical protective materials recommended (>70%): natural rubber, neoprene, nitrile, nitrile+PVC, polyethylene, PV alcohol, PVC, Saranex®, neoprene+natural rubber, neoprene/natural rubber. Also, Viton® offers limited protection.

Recommendations for respirator selection: NIOSH/OSHA 25 mg/m^3: SA:CF (any supplied-air respirator operated in a continuous-flow mode). 50 mg/m^3: HiEF (any air-purifying, full-facepiece respirator with a high-efficiency particulate filter); or SCBAF (any self-contained breathing apparatus with a full facepiece); or SAF (any supplied-air respirator with a full facepiece). 1000 mg/m^3: SAF:PD,PP (any supplied-air respirator that has a full facepiece and is operated in a pressure-demand or other positive-pressure mode). *EMERGENCY OR PLANNED ENTRY INTO UNKNOWN CONCENTRATIONS OR IDLH CONDITIONS:* SCBAF:PD,PP (any self-contained breathing apparatus that has a full facepiece and is operated in a pressure-demand or other positive-pressure mode); or SAF:PD,PP:ASCBA (any supplied-air respirator that has a full facepiece and is operated in a pressure-demand or other positive-pressure mode in combination with an auxiliary self-contained breathing apparatus operated in a pressure-demand or other positive pressure mode). *ESCAPE:* HiEF (any air-purifying, full-facepiece respirator with a high-efficiency particulate filter); or SCBAE (any appropriate escape-type, self-contained breathing apparatus). *Note*: Substance reported to cause eye irritation or damage; may require eye protection.

Exposure limits (TWA unless otherwise noted): ACGIH TLV 1 mg/m^3; NIOSH/OSHA 1 mg/m^3.

Short-term exposure limits (15-minute TWA): ACGIH & NIOSH STEL 3 mg/m^3.

Toxicity by ingestion: Grade 3; LD$_{50}$ = 50 to 500 mg/kg.

Vapor (gas) irritant characteristics: Nonvolatile.

Liquid or solid irritant characteristics: Severe skin irritant; may cause pain and second-degree burns after a few minutes of contact.

IDLH value: 1000 mg/m^3.

CHEMICAL REACTIVITY

Reactivity with water: Reacts with water to generate heat and forms phosphoric acid. The reaction is not violent. Violent reaction if water is added to concentrated acid. To dilute, always add acid to water; heat will be generated.

Binary reactants: Violent reaction with strong bases. Reacts violently with solutions containing ammonia or bleach, azo compounds, epoxides and other polymerizable compounds. Reacts, possibly violently, with amines, aldehydes, alkanolamines, alcohols, alkylene oxides, amides, ammonium hydroxide, calcium oxide, cyanides, epichlorohydrin, esters, halogenated organics, isocyanates, ketones, oleum, organic anhydrides, sodium tetrahydroborate, sulfides, sulfuric acid, strong oxidizers, vinyl acetate. At elevated temperatures, attacks many metals, producing hydrogen gas. At room temperature does not attack stainless steel, copper or its alloys. Attacks glass, ceramics, and some plastics, rubber, and coatings.

Neutralizing agents for acids and caustics: Flush with water, neutralize acid with lime or soda ash.

Reactivity group: 1

Compatibility class: Nonoxidizing mineral acids

ENVIRONMENTAL DATA

Food chain concentration potential: Negative; unlikely to accumulate.

Water pollution: Dangerous to aquatic life in high concentrations. May be dangerous if it enters nearby water intakes; notify operators. Notify local health and wildlife officials. **Response to discharge:** Issue warning–corrosive. Restrict access. Disperse and flush.

SHIPPING INFORMATION

Grades of purity: 115% phosphoric acid; **Storage temperature:** Ambient; **Inert atmosphere:** None; **Venting:** Open or pressure-vacuum; **Stability during transport:** Stable.

PHYSICAL AND CHEMICAL PROPERTIES

Physical state @ 59°F/15°C and 1 atm: Liquid.
Molecular weight: 98.0
Boiling point @ 1 atm: 1022°F/550°C/823°K.
Melting/Freezing point: 100°F/38°C/311°K.
Specific gravity (water = 1): 2.05 at 38°C (liquid).
Vapor pressure: 0.03 mm.

POLYPROPYLENE　　　　　　　　　　　　　　REC. P:2400

SYNONYMS: AMCO; AZDEL; CLYSAR; ELPON; GERFIL; LAMBETH; NOBLEN; POLIPROPILENO (Spanish); POLY PROPENE; PROFAX; REXENE; SHELL 5520; PROPENE POLYMER; TENITE 423; TUFFLITE; ULSTRON; VISCOL 350P; W 101; WEX 1242

IDENTIFICATION

CAS Number: 9003-07-0
Formula: [CH(CH$_3$)-CH$_2$]n, where n is large

DESCRIPTION: Tan to white solid. Odorless. Floats on water surface; practically insoluble.

Combustible solid • Containers may BLEVE when exposed to fire • Concentrated dust in confined areas (e.g., tanks, sewers, buildings) may explode when exposed to fire • Toxic products of combustion may include carbon monoxide and CO$_2$.

Hazard Classification (based on NFPA-704 Rating System)
Health Hazards (Blue): 0; Flammability (Red): 2; Reactivity (Yellow): 0

EMERGENCY RESPONSE: See Appendix A (171)
Evacuation:
Public safety: Isolate the area of spill or leak for at least 10 to 25 meters (30 to 80 feet) in all directions.
Spill: Increase, in the downwind direction, as necessary, the distance shown under "Public Safety."
Fire: If any large container is involved in fire, isolate for at least 800 meters (½ mile) in all directions; also, consider initial evacuation for 800 meters (½ mile) in all directions.

EXPOSURE
Short-term effects:
Dust or Solid: Products of combustion are poisonous.

HEALTH HAZARDS
Personal protective equipment (PPE): B-Level PPE. Filter respirator
Long-term health effects: Causes central bronchitis and peripheral bronchitis in rats and rabbits

FIRE DATA
Flash point: 690°F/366°C.
Fire extinguishing agents not to be used: Water may be ineffective.

CHEMICAL REACTIVITY
Binary reactants: Keep away from strong oxidizers. No dangerous reaction, but this material is weakened by these materials.
Reactivity group: 30
Compatibility class: Olefins

ENVIRONMENTAL DATA
Food chain concentration potential: Negative; unlikely to accumulate.
Water pollution: Effect of low concentrations on aquatic life is unknown. May be dangerous if it enters nearby water intakes; notify operators. Notify local health and wildlife officials.
Response to discharge: Mechanical containment.

SHIPPING INFORMATION
Grades of purity: Commercial, 100%; **Storage temperature:** Ambient; **Inert atmosphere:** None; **Venting:** Open; **Stability during transport:** Stable.

PHYSICAL AND CHEMICAL PROPERTIES
Physical state @ 59°F/15°C and 1 atm: Solid.
Molecular weight: Mixture (decomposes).
Melting/Freezing point: 330°F/166°C/439°K
Specific gravity (water = 1): 0.90 @ 68°F/20°C (solid).
Heat of combustion: (estimate) $-19,600$ Btu/lb $= -10,900$ cal/g $= -456 \times 10^5$ J/kg.

POLYPROPYLENE GLYCOL REC. P:2450

SYNONYMS: ALKAPOL PPG-1200; JEFFOX; POLIPROPILENGLICOL (Spanish); POLYOXYPROPYLENE GLYCOL; POLYOXPROPYLENE GLYCOL; PLURACOL POLYOL; POLYPROPYLENE GLYCOL 750; THANOL PPG

IDENTIFICATION
CAS Number: 25322-69-4
Formula: $(C_3H_8O_2)_n$; $HOCH(CH_3)CH_2O[CH_2CH(CH_3)O]_n$-H (where n averages 2–34).

DESCRIPTION: Colorless liquid. Odorless or mild sweet odor, like ether. May float or sink in water; soluble.

Combustible • Heat or flame may cause explosion • Containers may BLEVE when exposed to fire • Vapors in confined areas (e.g., tanks, sewers, buildings) may explode when exposed to fire • Irritating to the skin, eyes, and respiratory tract • Toxic products of combustion may include CO_2 and carbon monoxide.

Hazard Classification (based on NFPA-704 Rating System)
Health Hazards (Blue): 0; Flammability (Red): 1; Reactivity (Yellow): 0

EMERGENCY RESPONSE: See Appendix A (171)
Evacuation:
Public safety: Isolate the area of spill or leak for at least 10 to 25 meters (30 to 80 feet) in all directions.
Spill: Increase, in the downwind direction, as necessary, the distance shown under "Public Safety."
Fire: If any large container is involved in fire, isolate for at least 800 meters (½ mile) in all directions; also, consider initial evacuation for 800 meters (½ mile) in all directions.

EXPOSURE
Short-term effects: *SEEK MEDICAL ATTENTION.* **Liquid:** Irritating to eyes. Harmful if swallowed. *IF IN EYES*, hold eyelids open and flush with plenty of water. *IF SWALLOWED* and victim is *CONSCIOUS AND ABLE TO SWALLOW*, have victim drink 4 to 8 ounces of water.

HEALTH HAZARDS
Personal protective equipment (PPE): B-Level PPE. Safety glasses or face shield; rubber gloves.
Exposure limits (TWA unless otherwise noted): 10 mg/m^3 (AIHAWEEL).
Toxicity by ingestion: (depends on molecular weight). Grade 2; oral LD_{50} = 2150 mg/kg (rat); Grade 1; LD_{50} 5 to 15 g/kg; Grade 0; LD_{50} = > 15 g/kg.
Liquid or solid irritant characteristics: Eye irritation and possible pain.

FIRE DATA
Flash point: 365–495°F/185–257°C (oc).

CHEMICAL REACTIVITY
Binary reactants: Incompatible with oxidizers, sulfuric acid, nitric acid, caustics, aliphatic amines, isocyanates, boranes.
Reactivity group: 40
Compatibility class: Glycol ethers

ENVIRONMENTAL DATA
Water pollution: Effect of low concentrations on aquatic life is unknown. Fouling to shoreline. May be dangerous if it enters nearby water intakes; notify operators. Notify local health and wildlife officials. **Response to discharge:** Issue warning–water contaminant. Mechanical containment. Should be removed. Disperse and flush.

SHIPPING INFORMATION
Grades of purity: Low molecular weight (miscible with water); medium molecular weight (2% soluble in water); high molecular weight (insoluble in water); **Storage temperature:** Below 140°F/60°C; **Inert atmosphere:** None; **Venting:** Open (flame arrester); **Stability during transport:** Stable.

PHYSICAL AND CHEMICAL PROPERTIES
Physical state @ 59°F/15°C and 1 atm: Liquid.
Molecular weight: Variable, 200 to 2000
Boiling point @ 1 atm: Decomposes.
Melting/Freezing point: −22 to −58°F/−30 to −50°C/243 to 223°K.
Specific gravity (water = 1): 1.012 @ 68°F/20°C (liquid).
Heat of combustion: −14,200 Btu/lb = −7900 cal/g = −330 x 10^5 J/kg.

POTASSIUM   REC. P:2500

SYNONYMS: EEC No. 019-001-00-2; KALIUM; POTASSIO (Spanish); POTASSIUM METAL; POTASSIUM, ELEMENTAL

IDENTIFICATION
CAS Number: 7440-09-7
Formula: K
DOT ID Number: UN 2257; DOT Guide Number: 138
Proper Shipping Name: Potassium

DESCRIPTION: Silver-white soft solid. Shipped and stored under oil or nitrogen. Odorless. Reacts violently with water; decomposes; flammable hydrogen gas is produced.

Highly flammable • Corrosive • Do not use water; violent reaction producing flammable and explosive hydrogen gas • Do not use CO_2 or halogenated extinguishing agents; these form explosive mixtures • May ignite spontaneously in air • Containers may explode when exposed to fire • Fumes from burning material are severely irritating to the skin, eyes, and respiratory tract • Toxic products of combustion may include oxides of potassium • May re-ignite after fire is extinguished • Do not put yourself in danger by entering a contaminated area to rescue a victim.

Hazard Classification (based on NFPA-704 Rating System)
Health Hazards (Blue): 3; Flammability (Red): 3; Reactivity (Yellow): 2; Special Notice (White): Water reactive

EMERGENCY RESPONSE: See Appendix A (138)
Evacuation:
Public safety: Isolate the area of spill or leak for at least 50 to 100 meters (160 to 330 feet) in all directions.
Spill: Consider initial downwind evacuation for at least 250 meters (800 feet).
Fire: If any large container is involved in fire, isolate for at least 800 meters (½ mile) in all directions; also, consider initial evacuation for 800 meters (½ mile) in all directions.

EXPOSURE
Short-term effects: *SEEK MEDICAL ATTENTION*. **Vapors or fumes:** Inhalation may cause lung edema to develop. **Solid:** Will burn skin and eyes. Flush affected areas with plenty of water. *IF IN EYES*, hold eyelids open and flush with plenty of water.
Note to physician or authorized medical personnel: Medical observation is recommended for 24 to 48 hours after breathing overexposure, as pulmonary edema may be delayed. As first aid for pulmonary edema, consider administering a corticosteroid spray. Cigarette smoking may exacerbate pulmonary injury and should be discouraged for at least 72 hours following exposure.

HEALTH HAZARDS
Personal protective equipment (PPE): B-Level PPE. Goggles or face shield; rubber gloves.

Vapor (gas) irritant characteristics: Eye and respiratory tract irritant.
Liquid or solid irritant characteristics: Severe eye and skin irritant.

FIRE DATA
Fire extinguishing agents not to be used: Water, foam, CO_2, or halogenated hydrocarbons
Behavior in fire: Reacts violently with water, forming flammable and explosive hydrogen gas. May ignite spontaneously in air.
Autoignition temperature: 825°F/441°C/714°K.

CHEMICAL REACTIVITY
Reactivity with water: Reacts violently to form flammable hydrogen gas and a strong caustic solution.
Binary reactants: May ignite combustible materials if they are damp. Reacts with oxidizers, acids, silicates, halogens, sulfates, nitrates, carbonates, phosphates, oxides, Teflon®, hydroxides of heavy metals, and a wide range of materials including air.
Neutralizing agents for acids and caustics: Caustic formed by reaction with water should be flushed with water, then area can be rinsed with dilute acetic acid.

ENVIRONMENTAL DATA
Food chain concentration potential: Negative.
Water pollution: Harmful to aquatic life in very low concentrations. May be dangerous if it enters nearby water intakes; notify operators. Notify local health and wildlife officials. **Response to discharge:** Issue warning–high flammability, corrosive. Restrict access. Disperse and flush.

SHIPPING INFORMATION
Grades of purity: Commercial, 99.9+% Shipped under oil or nitrogen; **Storage temperature:** Ambient; **Inert atmosphere:** Inerted; **Venting:** Pressure-vacuum; **Stability during transport:** Stable, if protected from air and moisture.

PHYSICAL AND CHEMICAL PROPERTIES
Physical state @ 59°F/15°C and 1 atm: Solid.
Molecular weight: 39
Boiling point @ 1 atm: 1425°F/774°C/1047°K.
Melting/Freezing point: 145°F/63°C/336°K.
Specific gravity (water = 1): 0.86 @ 68°F/20°C (solid).
Heat of combustion: −2003 Btu/lb = −1,113 cal/g = −46.57 x 10^5 J/kg.
Heat of solution: −2104 Btu/lb = −1,169 cal/g = −48.91 x 10^5 J/kg.
Heat of fusion: 14.6 cal/g.

POTASSIUM ARSENATE   REC. P:2550

SYNONYMS: ARSENIC ACID, MONOPOTASSIUM SALT; ARSENIATO POTASICO (Spanish); MACQUER'S SALT; MONOPOTASSIUM ARSENATE; MONOPOTASSIUM DIHYDROGEN ARSENATE; POTASSIUM ACID ARSENATE; POTASSIUM ARSENATE, MONOBASIC; POTASSIUM DIHYDROGEN ARSENATE; POTASSIUM HYDROGEN ARSENATE

IDENTIFICATION
CAS Number: 7784-41-0
Formula: $AsH_2O_4 \cdot K$; KH_2AsO_4
DOT ID Number: UN 1677; DOT Guide Number: 151
Proper Shipping Name: Potassium arsenate
Reportable Quantity (RQ): **(CERCLA)** 1 lb/0.454 kg

DESCRIPTION: White solid. Odorless. Mixes with water.

Poisonous! • Breathing the dust, skin or eye contact, or swallowing the material can cause illness • Firefighting gear (including SCBA) does not provide adequate protection. If exposure occurs, remove and isolate gear immediately and thoroughly decontaminate personnel • Containers may BLEVE when exposed to fire • Dust is heavier than air and will collect and stay in low areas • Concentrated dust in confined areas (e.g., tanks, sewers, buildings) may explode when exposed to fire • Toxic products of combustion may include oxides of potassium and arsenic.

Hazard Classification (based on NFPA-704 Rating System)
Health Hazards (Blue): 3; Flammability (Red): 0; Reactivity (Yellow): 0

EMERGENCY RESPONSE: See Appendix A (151)
Evacuation:
Public safety: Isolate the area of spill or leak for at least 25 to 50 meters (80 to 160 feet) in all directions.
Spill: Increase, in the downwind direction, as necessary, the distance shown under "Public Safety."
Fire: If tank, rail car, or tank truck is involved in fire, isolate for at least 800 meters (½ mile) in all directions; also, consider initial evacuation for 800 meters (½ mile) in all directions.

EXPOSURE
Short-term effects: *SEEK MEDICAL ATTENTION.* **Dust:** *POISONOUS IF INHALED OR SWALLOWED.* Irritating to eyes, nose, and throat. Move victim to fresh air. *IF IN EYES*, hold eyelids open and flush with plenty of water. If artificial respiration is administered, *avoid mouth-to-mouth resuscitation; use bag/mask apparatus.* IF breathing is difficult, administer oxygen.
Solid: *POISONOUS IF SWALLOWED.* Irritating to skin and eyes. Remove contaminated clothing and shoes. Flush affected areas with plenty of water. *IF IN EYES*, hold eyelids open and flush with plenty of water. *IF SWALLOWED* and victim is *CONSCIOUS AND ABLE TO SWALLOW*, have victim drink 4 to 8 ounces of water and have victim induce vomiting. *IF SWALLOWED* and victim is *UNCONSCIOUS OR HAVING CONVULSIONS*, do nothing except keep victim warm.

HEALTH HAZARDS
Personal protective equipment (PPE): A-Level PPE. Dust respirator; rubber **gloves.**
Recommendations for respirator selection: NIOSH
At any concentrations above the NIOSH REL, or where there is no REL, at any detectable concentration: SCBAF:PD,PP (any self-contained breathing apparatus that has a full faceplate and is operated in a pressure-demand or other positive-pressure mode); or SAF:PD,PP:ASCBA (any supplied-air respirator that has a full facepiece and is operated in a pressure-demand or other positive-pressure mode in combination with an auxiliary self-contained breathing apparatus operated in a pressure-demand or other positive-pressure mode). *ESCAPE:* GMFAGHiE [any air-purifying, full-facepiece respirator (gas mask) with a chin-style, front-or back-mounted acid gas canister having a high-efficiency particulate filter]; or SCBAE (any appropriate escape-type, self-contained breathing apparatus).
Exposure limits (TWA unless otherwise noted): ACGIH TLV 0.01 mg/m³; OSHA PEL [1910.1080] 0.010 mg/m³; NIOSH REL ceiling 0.002 mg/m³/15 min, as arsenic; potential human carcinogen; reduce exposure to lowest feasible level
Long-term health effects: Confirmed human carcinogen. Arsenic compounds are allergens, irritants to skin.

Vapor (gas) irritant characteristics: Respiratory tract irritant.
Liquid or solid irritant characteristics: May cause eye irritation.
IDLH value: 5 mg/m³ as arsenic; potential carcinogen

ENVIRONMENTAL DATA
Water pollution: Harmful to aquatic life in very low concentrations. May be dangerous if it enters nearby water intakes; notify operators. Notify local health and wildlife officials. **Response to discharge:** Issue warning–poison, water contaminant. Restrict access. Disperse and flush.

SHIPPING INFORMATION
Grades of purity: Commercial; **Storage temperature:** Ambient; **Inert atmosphere:** None; **Venting:** Open; **Stability during transport:** Stable.

PHYSICAL AND CHEMICAL PROPERTIES
Physical state @ 59°F/15°C and 1 atm: Solid.
Molecular weight: 180.0
Boiling point @ 1 atm: Decomposes.
Melting/Freezing point: 550°F/288°C/561°K.
Specific gravity (water = 1): 2.8 @ 68°F/20°C (solid).
Heat of solution: 49 Btu/lb = 27 cal/g = 1.1×10^5 J/kg.

POTASSIUM ARSENITE **REC. P:2600**

SYNONYMS: ARSENIOUS ACID, POTASSIUM SALT; ARSONIC ACID, POTASSIUM SALT; ARSENITO POTASICO (Spanish); FOWLER'S SOLUTION; POTASSIUM METAARSENITE

IDENTIFICATION
CAS Number: 13464-35-2; 10124-50-2
Formula: $KAsO_2 \cdot HAsO_2$
DOT ID Number: UN 1678; DOT Guide Number: 154
Proper Shipping Name: Potassium arsenite
Reportable Quantity (RQ): **(CERCLA)** 1 lb/0.454 kg

DESCRIPTION: White solid. Odorless. Mixes with water; forms an alkaline solution.

Poison! • Skin or eye contact, or swallowing the material can cause illness and possible death; irritating to the eyes, skin, and respiratory tract • Firefighting gear (including SCBA) does not provide adequate protection. If exposure occurs, remove and isolate gear immediately and thoroughly decontaminate personnel • Dust is heavier than air and will collect and stay in low areas • Toxic products of combustion may include oxides of potassium and arsenic.

Hazard Classification (based on NFPA-704 Rating System)
Health Hazards (Blue): 3; Flammability (Red): 0; Reactivity (Yellow): 0

EMERGENCY RESPONSE: See Appendix A (154)
Evacuation:
Public safety: Isolate the area of spill or leak for at least 25 to 50 meters (80 to 160 feet) in all directions.
Spill: Increase, in the downwind direction, as necessary, the distance shown under "Public Safety."
Fire: If tank, rail car, or tank truck is involved in fire, isolate for at least 800 meters (½ mile) in all directions; also, consider initial evacuation for 800 meters (½ mile) in all directions.

EXPOSURE

Short-term effects: *SEEK MEDICAL ATTENTION.* If artificial respiration is administered, *avoid mouth-to-mouth resuscitation; use bag/mask apparatus.* **Solid or dust:** *POISONOUS IF SWALLOWED OR SKIN CONTACT.* Irritating to skin, eyes, and nose. Move to fresh air. Remove contaminated clothing and shoes. Flush affected areas with plenty of water. *IF IN EYES*, hold eyelids open and flush with plenty of water. *IF SWALLOWED* and the victim is *CONSCIOUS AND ABLE TO SWALLOW*, have victim drink 4 to 8 ounces of water and have victim induce vomiting.

HEALTH HAZARDS

Personal protective equipment (PPE): A-Level PPE. Protective clothing, hand and arm protection, waterproof boots, respirator, and eye protection.
Recommendations for respirator selection: NIOSH
At any concentrations above the NIOSH REL, or where there is no REL, at any detectable concentration: SCBAF:PD,PP (any self-contained breathing apparatus that has a full faceplate and is operated in a pressure-demand or other positive-pressure mode); or SAF:PD,PP:ASCBA (any supplied-air respirator that has a full facepiece and is operated in a pressure-demand or other positive-pressure mode in combination with an auxiliary self-contained breathing apparatus operated in a pressure-demand or other positive-pressure mode). *ESCAPE:* GMFAGHiE [any air-purifying, full-facepiece respirator (gas mask) with a chin-style, front-or back-mounted acid gas canister having a high-efficiency particulate filter]; or SCBAE (any appropriate escape-type, self-contained breathing apparatus).
Exposure limits (TWA unless otherwise noted): ACGIH TLV 0.01 mg/m^3; OSHA PEL [1910.1080] 0.010 mg/m^3; NIOSH REL ceiling 0.002 mg/m^3/15 min, as arsenic; potential human carcinogen; reduce exposure to lowest feasible level
Toxicity by ingestion: Grade 4; LD$_{50}$ = 14 mg/kg.
Long-term health effects: Confirmed carcinogen. Can cause skin cancer. Other cancers, notably lung and liver have been reported. Chronic arsenic intoxication by ingestion may cause weakness, loss of appetite, gastrointestinal disturbances, peripheral neuritis, and skin disorders such as keratitis and pigmentation.
Vapor (gas) irritant characteristics: Severe eye and respiratory tract irritant.
Liquid or solid irritant characteristics: Eye and skin irritant.
IDLH value: 5 mg/m^3 as arsenic; cancer hazard

ENVIRONMENTAL DATA

Water pollution: Harmful to aquatic life in very low concentrations. May be dangerous if it enters nearby water intakes; notify operators. Notify local health and wildlife officials.
Response to discharge: Issue warning–poison, water contaminant. Restrict access. Should be removed. Chemical and physical treatment.

SHIPPING INFORMATION

Stability during transport: Stable but hygroscopic, gradually decomposes on exposure to air (by CO_2).

PHYSICAL AND CHEMICAL PROPERTIES

Physical state @ 59°F/15°C and 1 atm: Solid.
Molecular weight: 253.93
Relative vapor density (air = 1): 8.76 (calculated).

POTASSIUM BINOXALATE   REC. P:2650

SYNONYMS: BIOXALATO POTASICO (Spanish); POTASSIUM ACID OXALATE; SALT OF SORREL; SAL ACETOSELLA

IDENTIFICATION

CAS Number: 127-95-7
Formula: KHC_2O_4

DESCRIPTION: White solid. Odorless. Sinks in water; soluble.

Dust is irritating to skin, eyes, and respiratory tract • Toxic products of combustion may include carbon monoxide and potassium oxides

EMERGENCY RESPONSE: See Appendix A (171)
Evacuation:
Public safety: Isolate the area of spill or leak for at least 10 to 25 meters (30 to 80 feet) in all directions.
Spill: Increase, in the downwind direction, as necessary, the distance shown under "Public Safety."
Fire: If any large container is involved in fire, isolate for at least 800 meters (½ mile) in all directions; also, consider initial evacuation for 800 meters (½ mile) in all directions.

EXPOSURE

Short-term effects: *SEEK MEDICAL ATTENTION.* **Dust:** Irritating to eyes, nose, and throat. *IF INHALED*, will, will cause coughing, difficult breathing, or loss of consciousness. *IF IN EYES*, hold eyelids open and flush with plenty of water. *IF BREATHING HAS STOPPED*, give artificial respiration. IF breathing is difficult, administer oxygen. **Solid:** Irritating to skin and eyes. *IF SWALLOWED*, will cause nausea, vomiting, or loss of consciousness. Remove contaminated clothing and shoes. Flush affected areas with plenty of water. *IF IN EYES*, hold eyelids open and flush with plenty of water. *IF SWALLOWED* and victim is *CONSCIOUS AND ABLE TO SWALLOW*, have victim drink 4 to 8 ounces of water. *IF SWALLOWED* and victim is *UNCONSCIOUS OR HAVING CONVULSIONS*, do nothing except keep victim warm.

HEALTH HAZARDS

Personal protective equipment (PPE): B-Level PPE. Dust mask; goggles or face shield; protective gloves.
Toxicity by ingestion: Grade 3; LD$_{50}$ = 50 to 500 mg/kg.
Vapor (gas) irritant characteristics: Eye and respiratory tract irritant.
Liquid or solid irritant characteristics: Eye and skin irritant.

CHEMICAL REACTIVITY

Reactivity with water: Below 58°F/50°C dissolves in water and reacts to form the much less soluble potassium tetraoxalate, which separates out.

ENVIRONMENTAL DATA

Food chain concentration potential: Negative; unlikely to accumulate.
Water pollution: Effect of low concentrations on aquatic life if unknown. May be dangerous if it enters nearby water intakes; notify operators. Notify local health and wildlife officials. **Response to discharge:** Issue warning–poison, water contaminant. Restrict access. Disperse and flush.

SHIPPING INFORMATION

Grades of purity: Technical; **Storage temperature:** Ambient; **Inert atmosphere:** None; **Venting:** Open; **Stability during transport:** Stable.

PHYSICAL AND CHEMICAL PROPERTIES
Physical state @ 59°F/15°C and 1 atm: Solid.
Molecular weight: 128.11
Specific gravity (water = 1): 2.0 @ 68°F/20°C (solid).

POTASSIUM CHLORATE
REC. P:2700

SYNONYMS: BERTHOLLET'S SALT; CHLORATE de POTASSIUM (French); CHLORATE OF POTASH; CHLORIC ACID, POTASSIUM SALT; CLORATO POTASICO (Spanish); EEC No. 017-004-00-3; FEKABIT; KALIUMCHLORAT (German); OXYMURIATE OF POTASH; PEARL ASH; POTASH CHLORATE; POTASSIUM (CHLORATE DE) (French); POTASSIUM OXYMURIATE; POTCRATE; SALT OF TARTER

IDENTIFICATION
CAS Number: 3811-04-9
Formula: $KClO_3$
DOT ID Number: UN 1485; UN 2427; DOT Guide Number: 140
Proper Shipping Name: Potassium chlorate (UN 1485); Potassium chlorate, solution (UN 2427)

DESCRIPTION: White solid. Odorless. Mixes with water.

Poison! • Prolonged inhalation or ingestion interferes with the body's ability to use oxygen and is life threatening • Strong oxidizer which may react spontaneously with low flash point organics or reducing agents. Heat forms oxygen; will increase the activity of an existing fire • May cause fire or explosion on contact with combustibles (wood, paper, oil, clothing, etc.). Irritating to the skin, eyes, and respiratory tract • Explosive decomposition above 752°F/400°C giving off oxygen gas; toxic products of combustion may include hydrogen chloride, and oxides of potassium:

Hazard Classification (based on NFPA-704 Rating System)
Health Hazards (Blue): 1; Flammability (Red): 0; Reactivity (Yellow): 0; Special Notice (White): OXY

EMERGENCY RESPONSE: See Appendix A (140)
Note: Do not let spill area dry until it has been determined that there is no chlorates left in the area. Continue cooling after fire has been extinguished.
Evacuation:
Public safety: Isolate the area of spill or leak for at least 10 to 25 meters (30 to 80 feet) in all directions.
Spill: Consider initial downwind evacuation for at least 100 meters (330 feet).
Fire: If any large container is involved in fire, isolate for at least 800 meters (½ mile) in all directions; also, consider initial evacuation for 800 meters (½ mile) in all directions.

EXPOSURE
Short-term effects: *SEEK MEDICAL ATTENTION.* **Dust:** Irritating to eyes, nose, and throat. Move victim to fresh air. *IF IN EYES*, hold eyelids open and flush with plenty of water. **Solid:** Irritating to skin and eyes. *IF SWALLOWED*, will cause nausea, vomiting, or loss of consciousness. Remove contaminated clothing and shoes. Flush affected areas with plenty of water. *IF IN EYES*, hold eyelids open and flush with plenty of water. *IF SWALLOWED* and victim is *CONSCIOUS AND ABLE TO SWALLOW*, have victim drink 4 to 8 ounces of water and have victim induce vomiting. *IF SWALLOWED* and victim is *UNCONSCIOUS OR HAVING CONVULSIONS,* do nothing except keep victim warm. Methhemoglobinemia may develop.

HEALTH HAZARDS
Personal protective equipment (PPE): B-Level PPE. Dust mask; rubber gloves; goggles; protective clothing to prevent contact with skin.
Toxicity by ingestion: Grade 2; LD_{50} = 0.5 to 5 g/kg.
Long-term health effects: Can affect the blood. Liver and kidney damage. Skin discoloration.
Vapor (gas) irritant characteristics: Eye and skin irritant.
Liquid or solid irritant characteristics: eye and skin irritant.

CHEMICAL REACTIVITY
Binary reactants: Contact with combustible material or finely divided metals may cause fire or explosion. Reacts with strong acids.

ENVIRONMENTAL DATA
Food chain concentration potential: Negative; unlikely to accumulate.
Water pollution: Effect of low concentrations on aquatic life is unknown. May be dangerous if it enters nearby water intakes; notify operators. Notify local health and wildlife officials. **Response to discharge:** Issue warning–oxidizing material, water contaminant. Restrict access. Disperse and flush.

SHIPPING INFORMATION
Grades of purity: Commercial: 99.7+%; Reagent; Purified; **Storage temperature:** Ambient; **Inert atmosphere:** None; **Venting:** Open; **Stability during transport:** Stable.

PHYSICAL AND CHEMICAL PROPERTIES
Physical state @ 59°F/15°C and 1 atm: Solid.
Molecular weight: 122.6
Boiling point @ 1 atm: 752°F/400°C (decomposes).
Melting/Freezing point: 680°F/360°C/633°K.
Specific gravity (water = 1): 2.34 @ 68°F/20°C (solid).
Heat of decomposition: -176 Btu/lb = -98 cal/g = -4.1×10^5 J/kg.
Heat of solution: 147 Btu/lb = 81.9 cal/g = 3.43×10^5 J/kg.

POTASSIUM CHROMATE
REC. P:2750

SYNONYMS: BIPOTASSIUM CHROMATE; CHROMATO de POTASICO (Spanish); CHROMIC ACID, DIPOTASSIUM SALT; CHROMATE OF POTASSIUM; DIPOTASSIUM CHROMATE; DIPOTASSIUM MONOCHROMATE; EEC No. 024-006-00-8; NEUTRAL POTASSIUM CHROMATE; POTASSIUM CHROMATE(VI); TARAPACAITE

IDENTIFICATION
CAS Number: 7789-00-6
Formula: $CrO_4 \cdot 2K$; K_2CrO_4
DOT ID Number: UN 3077; DOT Guide Number: 171
Proper Shipping Name: Environmentally hazardous substances, solid, n.o.s.
Reportable Quantity (RQ): **(CERCLA)** 10 lb/4.54 kg

DESCRIPTION: Bright yellow crystalline solid. Odorless. Sinks and mixes with water forming an alkaline solution.

Corrosive to the skin, eyes, and respiratory tract; inhalation symptoms may be delayed • Strong oxidizer which may react spontaneously with low flash point organics or reducing agents.

Heat forms oxygen; will increase the activity of an existing fire • May cause fire or explosion on contact with combustibles (wood, paper, oil, clothing, etc.). Decomposition products upon heating may include chromium metal fumes and oxides of potassium.

Hazard Classification (based on NFPA-704 Rating System)
Health Hazards (Blue): 2; Flammability (Red): 0; Reactivity (Yellow): 0; Special Notice (White): OXY

EMERGENCY RESPONSE: See Appendix A (171)
Evacuation:
Public safety: Isolate the area of spill or leak for at least 10 to 25 meters (30 to 80 feet) in all directions.
Spill: Increase, in the downwind direction, as necessary, the distance shown under "Public Safety."
Fire: If any large container is involved in fire, isolate for at least 800 meters (½ mile) in all directions; also, consider initial evacuation for 800 meters (½ mile) in all directions.

EXPOSURE
Short-term effects: *SEEK MEDICAL ATTENTION.* **Dust:** Irritating to eyes, nose, and throat. *IF INHALED*, will, will cause coughing or difficult breathing. *IF IN EYES*, hold eyelids open and flush with plenty of water. *IF BREATHING HAS STOPPED*, give artificial respiration. IF breathing is difficult, administer oxygen. Lung edema may develop. **Solid:** *POISONOUS IF SWALLOWED.* Irritating to skin and eyes. *IF SWALLOWED*, will cause nausea, vomiting, or loss of consciousness. Remove contaminated clothing and shoes. Flush affected areas with plenty of water. *IF IN EYES*, hold eyelids open and flush with plenty of water. *IF SWALLOWED* and victim is *CONSCIOUS AND ABLE TO SWALLOW*, have victim drink 4 to 8 ounces of water and have victim induce vomiting. *IF SWALLOWED* and victim is *UNCONSCIOUS OR HAVING CONVULSIONS*, do nothing except keep victim warm. *Note to physician or authorized medical personnel*: Medical observation is recommended for 24 to 48 hours after breathing overexposure, as pulmonary edema may be delayed. As first aid for pulmonary edema, consider administering a corticosteroid spray. Cigarette smoking may exacerbate pulmonary injury and should be discouraged for at least 72 hours following exposure.

HEALTH HAZARDS
Personal protective equipment (PPE): B-Level PPE. Approved filter-type respirator; close-fitting safety goggles; rubber boots and apron; safety hat; face shield.
Recommendations for respirator selection: NIOSH *At any concentrations above the NIOSH REL, or where there is no REL, at any detectable concentration:* SCBAF:PD,PP (any self-contained breathing apparatus that has a full facepiece and is operated in a pressure-demand or other positive-pressure mode); or SAF:PD,PP:ASCBA (any supplied-air respirator that has a full facepiece and is operated in a pressure-demand or other positive-pressure mode in combination with an auxiliary self-contained breathing apparatus operated in a pressure-demand or other positive pressure mode). *ESCAPE:* HiEF (any air-purifying, full-facepiece respirator with a high-efficiency particulate filter); or SCBAE (any appropriate escape-type, self-contained breathing apparatus).
Exposure limits (TWA unless otherwise noted): ACGIH TLV 0.05 mg/m^3 as water soluble Cr(VI) compounds; OSHA PEL 0.1 mg/m^3 as CrO$_3$; NIOSH 0.001 mg/m^3 as chromates; potential human carcinogen; reduce exposure to lowest feasible level.
Toxicity by ingestion: Grade 3; LD$_{50}$ = 50 to 500 mg/kg.
Long-term health effects: Lung cancer may occur. Skin disorders; eczema, chromium ulcers; perforation of the nasal septum.

Vapor (gas) irritant characteristics: Eye and respiratory tract irritant.
Liquid or solid irritant characteristics: Eye and skin irritant.
IDLH value: 15 mg/m^3 as chromium(VI).

CHEMICAL REACTIVITY
Binary reactants: In contact with combustible materials or finely divided metals may cause fire or explosion.

ENVIRONMENTAL DATA
Food chain concentration potential: Plants can absorb compound from water and pass it on up the food chain. Bioconcentrative up to 2000-fold. Not likely to be a problem in a spill situation.
Water pollution: Harmful to aquatic life in very low concentrations. May be dangerous if it enters nearby water intakes; notify operators. Notify local health and wildlife officials. **Response to discharge:** Disperse and flush.

SHIPPING INFORMATION
Grades of purity: Reagent, 99%; CP; Technical; **Storage temperature:** Ambient; **Inert atmosphere:** None; **Venting:** Open; **Stability during transport:** Stable.

PHYSICAL AND CHEMICAL PROPERTIES
Physical state @ 59°F/15°C and 1 atm: Solid.
Molecular weight: 194.20
Boiling point @ 1 atm: (decomposes).
Specific gravity (water = 1): 2.73 at 18°C (solid).
Heat of fusion: 35.6 cal/g.

POTASSIUM CYANIDE   REC. P:2800

SYNONYMS: CYANIDE OF POTASSIUM; CYANURE de POTASSIUM (French); EEC No. 006-007-00-5; HYDROCYANIC ACID, POTASSIUM SALT; KALIUM-CYANID (German); KCN; POTASSIUM CYANIDE SOLUTION; POTASSIUM SALT OF HYDROCYANIC ACID; CYANIDE; RCRA No. P098

IDENTIFICATION
CAS Number: 151-50-8
Formula: KCN
DOT ID Number: UN 1680; DOT Guide Number: 157
Proper Shipping Name: Potassium cyanide
Reportable Quantity (RQ): **(CERCLA)** 10 lb/4.54 kg

DESCRIPTION: White lumps, crystals, or in solution. Faint, bitter almond odor. Sinks and mixes with water. When potassium cyanide dissolves in water, a mild reaction occurs and some poisonous hydrogen cyanide gas is released. This gas is hazardous in an enclosed space. The solution is a strong base; if the water is acidic, toxic amounts of gas will form at once.

Poison! • Corrosive • Breathing the vapor can kill you; skin or eye contact causes severe burns, impaired vision, or blindness • Firefighting gear (including SCBA) does not provide adequate protection. If exposure occurs, remove and isolate gear immediately and thoroughly decontaminate personnel • Toxic products of combustion may include oxides of potassium, nitrogen and cyanide fumes • Do not put yourself in danger by entering a contaminated area to rescue a victim.

Hazard Classification (based on NFPA-704 Rating System)
Health Hazards (Blue): 3; Flammability (Red): 0; Reactivity (Yellow): 0

EMERGENCY RESPONSE: See Appendix A (157)
Evacuation:
Public safety: Isolate the area of spill or leak for at least 50 to 100 meters (160 to 330 feet) in all directions.
Spill: *IF SPILLED IN WATER*: Small spill–First: Isolate in all directions 30 meters (100 feet); Then: Protect persons downwind, DAY: 0.2 km (0.1 mile); NIGHT: 0.3 km (0.2 mile). Large spill–First: Isolate in all directions 95 meters (300 feet). Then: Protect persons downwind, DAY: 0.8 km (0.5 mile); NIGHT: 2.6 km (1.6 miles).
Fire: If tank, rail car, or tank truck is involved in fire, isolate for at least 800 meters (½ mile) in all directions; also, consider initial evacuation for 800 meters (½ mile) in all directions.

EXPOSURE
Short-term effects: *SEEK MEDICAL ATTENTION*. **Dust:** *POISONOUS IF INHALED OR IF SKIN IS EXPOSED*. Move to fresh air. *IF BREATHING HAS STOPPED*, give artificial respiration; *avoid mouth-to-mouth resuscitation; use bag/mask apparatus*. IF breathing is difficult, administer oxygen. **Solid:** *A RAPIDLY FATAL POISON WHEN TAKEN INTO THE DIGESTIVE SYSTEM. A RAPIDLY FATAL POISON WHEN TAKEN INTO THE DIGESTIVE SYSTEM. POISONOUS IF SWALLOWED OR IF SKIN IS EXPOSED*. **Skin:** Remove and double bag contaminated clothing (including shoes) and leave them in the Hot Zone; wash skin with soap and water and large volumes of water for at least 15 minutes. *Skin can also be decontaminated with diluted hypochlorite solution, U.S. Army M291 kit, and M258(A1) skin decontamination kit*. **Eye:** Rinse with large volumes of water or saline for at least 15 minutes. **Swallowed:** Do NOT make victim vomit.
Note to physician or authorized medical personnel: Consider the use of amyl nitrite perles if symptoms of cyanide poisoning develop. If symptoms indicate, initial treatment includes the cyanide antidote kit. In all cases, break an amyl nitrite perle in a gauze pad and hold lightly under victim's nose for 15 seconds, repeating 5 times at about 15-second intervals; if necessary (and if sodium nitrite infusions will be delayed), repeat procedure every 3 minutes with fresh pearls until 3 or 4 have been used. Avoid breathing the vapor while administering it to the victim. Administer sodium nitrite IV, ASAP. The usual adult dose is 10 to 20 mL of a 3% solution infused over no less than 5 minutes; the average child dose is 0.15 to 0.20 mL/kg. Monitor blood pressure during administration, and slow the rate of infusion if hypotention develops. Next, infuse sodium thiosulfate IV. The usual adult dose is 50 mL of a 25% solution infused over 10 to 20 minutes; the average child dose is 1.65 mL/kg. Repeat with nitrite and thiosulfate as required.
Note to physician or authorized medical personnel: Medical observation is recommended for 24 to 48 hours after breathing overexposure, as pulmonary edema may be delayed. As first aid for pulmonary edema, consider administering a corticosteroid spray. Cigarette smoking may exacerbate pulmonary injury and should be discouraged for at least 72 hours following exposure.

HEALTH HAZARDS
Personal protective equipment (PPE): A-Level PPE. Sealed chemical protective materials offers limited protection: PVC. If the proper equipment is not available, or if the rescuers have not been trained in its use, call for assistance from the U.S. Soldier and Biological Chemical Command–Edgewood Research Development and Engineering Center (from 0700-1630 EST call 410-671-4411, and from 1630-0700 EST call 410-278-5201; ask for the Staff Duty Officer).

Recommendations for respirator selection: NIOSH/OSHA
25 mg/m³: SA (any supplied-air respirator); or SCBAF (any self-contained breathing apparatus with a full facepiece). *EMERGENCY OR PLANNED ENTRY INTO UNKNOWN CONCENTRATIONS OR IDLH CONDITIONS:* SCBAF:PD,PP (any self-contained breathing apparatus that has a full facepiece and is operated in a pressure-demand or other positive-pressure mode); or SAF:PD,PP:ASCBA (any supplied-air respirator that has a full facepiece and is operated in a pressure-demand or other positive-pressure mode in combination with an auxiliary self-contained breathing apparatus operated in a pressure-demand or other positive pressure mode). *ESCAPE:* GMFSHiE [any air-purifying, full-facepiece respirator (gas mask) with a chin-style, front- or back-mounted canister providing protection against the compound of concern and having a high-efficiency particulate filter); or SCBAE (any appropriate escape-type, self-contained breathing apparatus).
Exposure limits (TWA unless otherwise noted): ACGIH TLV 5 mg/m³ (4.7 ppm); OSHA PEL 5 mg/m³; NIOSH REL ceiling 5 mg/m³ (4.7 ppm)/10 minutes as cyanide; skin contact contributes significantly in overall exposure.
Short-term exposure limits (15-minute TWA): NIOSH ceiling 5 mg/m³ (4.7 ppm) 10-minutes.
Toxicity by ingestion: Grade 4; LD_{50} below 50 mg/kg (mice).
Vapor (gas) irritant characteristics: Nonvolatile, but moisture in air can liberate lethal hydrogen cyanide gas.
Liquid or solid irritant characteristics: Eye and skin irritant. Moist solid can cause caustic-type irritation of skin and formation of ulcers.
IDLH value: 25 mg/m³.

FIRE DATA
Fire extinguishing agents not to be used: CO_2, water.

CHEMICAL REACTIVITY
Reactivity with water: Reacts with moisture in air releasing flammable hydrogen cyanide gas; may ignite spontaneously. Aqueous solution is a strong base.
Binary reactants: Reacts with even weak acids, acid fumes, alcohols releasing flammable hydrogen cyanide gas. Violent reaction with oxidizers, fluorine, sodium chlorate. Aqueous solution is incompatible with organic anhydrides, isocyanates, alkylene oxides, epichlorohydrin, aldehydes, alcohols, glycols, phenols, cresols, caprolactam, strong oxidizers, sodium chlorate. Forms sensitive explosive mixtures with potassium chlorate. Incompatible with chlorates, gold, mercurous chloride, nitrates, nitrites. Attacks aluminum, copper, zinc in the presence of moisture.

ENVIRONMENTAL DATA
Food chain concentration potential: Negative; unlikely to accumulate.
Water pollution: DOT Appendix B, §172.101–marine pollutant. Harmful to aquatic life in very low concentrations. May be dangerous if it enters nearby water intakes; notify operators. Notify local health and wildlife officials. **Response to discharge:** Issue warning–poison. Restrict access. Evacuate area. Chemical and physical treatment.

SHIPPING INFORMATION
Grades of purity: 99.0%; **Storage temperature:** Ambient; **Inert atmosphere:** None; **Venting:** Sealed containers must be stored in a well-ventilated area; **Stability during transport:** Stable.

PHYSICAL AND CHEMICAL PROPERTIES
Physical state @ 59°F/15°C and 1 atm: Solid.
Molecular weight: 65.12

Boiling point @ 1 atm: 2957°F/1625°C/1898°K.
Melting/Freezing point: 1174.1°F/634.5°C/907.7°K.
Specific gravity (water = 1): 1.52 at 16°C (solid).
Heat of fusion: 53.7 cal/g.
Vapor pressure: 0 mm (approximate).

POTASSIUM DICHLORO-s-TRIAZINETRIONE
REC. P:2850

SYNONYMS: 1,3-DICHLORO-s-TRIAZINE-2,4,6(1H,3H,5H) TRIONE POTASSIUM SALT; DICHLOROISOCYANURIC ACID POTASSIUM SALT; DICHLORO-s-TRIAZINE-2,4,6(1H,3H,5H)-TRIONE POTASSIUM DERIV; DICHLOR-s-TRIAZIN-2,4,6(1H,3H,5H) TRIONE POTASSIUM; ISOCYANURIC ACID,DICHLORO-, POTASSIUM SALT; POTASSIUM DICHLORO ISOCYANURATE; POTASSIUM TROCLOSENE; 1,3,5-TRIAZINE-2,4,6(1H,3H,5H)-TRIONE, 1,3-DICHLORO-, POTASSIUM SALT; s-TRIAZINE-2,4,6(1H,3H,5H)-TRIONE,DICHLORO-, POTASSIUM DERIVATIVE; TROCLOSENE POTASSIUM; POTASSIUM DICHLOROISOCYANURATE

IDENTIFICATION
CAS Number: 2244-21-5
Formula: $C_3HCl_2N_3O_3 \cdot K$; $KCl_2(NCO)_3$
DOT ID Number: UN 2465; DOT Guide Number: 140
Proper Shipping Name: Potassium dichloro-s-triazinetrione

DESCRIPTION: White solid. Chlorine-like odor. Mixes with water.

Severely irritating to the skin, eyes, and respiratory tract • Strong oxidizer which may react spontaneously with low flash point organics or reducing agents. Heat forms oxygen; will increase the activity of an existing fire • May cause fire or explosion on contact with combustibles (wood, paper, oil, clothing, etc.) • Toxic products of combustion may include oxides of potassium and nitrogen and hydrogen chloride.

Hazard Classification (based on NFPA-704 Rating System)
Health Hazards (Blue): 3; Flammability (Red): 0; Reactivity (Yellow): 2; Special Notice (White): OXY

EMERGENCY RESPONSE: See Appendix A (140)
Evacuation:
Public safety: Isolate the area of spill or leak for at least 10 to 25 meters (30 to 80 feet) in all directions.
Spill: Consider initial downwind evacuation for at least 100 meters (330 feet).
Fire: If any large container is involved in fire, isolate for at least 800 meters (½ mile) in all directions; also, consider initial evacuation for 800 meters (½ mile) in all directions.

EXPOSURE
Short-term effects: *SEEK MEDICAL ATTENTION.* **Dust:** Severely irritating to eyes, nose, and throat. *IF INHALED,* will, will cause coughing or difficult breathing. Move victim to fresh air. *IF IN EYES,* hold eyelids open and flush with plenty of water. *IF BREATHING HAS STOPPED,* give artificial respiration. IF breathing is difficult, administer oxygen. **Solid:** Irritating to skin and eyes. Poisonous if swallowed. Remove contaminated clothing and shoes. Flush affected areas with plenty of water. *IF IN EYES,* hold eyelids open and flush with plenty of water. *IF SWALLOWED* and victim is *CONSCIOUS AND ABLE TO SWALLOW,* have victim drink 4 to 8 ounces of water and have victim induce vomiting. *IF SWALLOWED* and victim is *UNCONSCIOUS OR HAVING CONVULSIONS,* do nothing except keep victim warm.

HEALTH HAZARDS
Personal protective equipment (PPE): B-Level PPE. Dust mask or chlorine canister mask; goggles; rubber gloves and other protective clothing to prevent contact with skin.
Toxicity by ingestion: Grade 2; LD_{50} = 0.5 to 5 g/kg.
Vapor (gas) irritant characteristics: Eye and respiratory tract irritant.
Liquid or solid irritant characteristics: Eye and skin irritant.

FIRE DATA but may cause fire upon contact with ordinary combustibles.
Behavior in fire: Decomposition can be initiated with a heat source and can propagate throughout the mass with the evolution of dense fumes. Containers may explode.

CHEMICAL REACTIVITY
Reactivity with water: Forms a bleach solution; the reaction is not violent.
Binary reactants: Contact with most foreign materials, organic matter, or easily chlorinated or oxidized materials may result in fire. Avoid oil, grease, sawdust, floor sweepings, metal powders, hydrogen peroxide or other easily oxidized organic compounds.

ENVIRONMENTAL DATA
Food chain concentration potential: Negative; unlikely to accumulate.
Water pollution: Effect of low concentrations on aquatic life is unknown. May be dangerous if it enters nearby water intakes; notify operators. Notify local health and wildlife officials.
Response to discharge: Issue warning–oxidizing material, water contaminant. Restrict access. Disperse and flush.

SHIPPING INFORMATION
Grades of purity: Technical; 39-59% available chlorine; **Storage temperature:** Ambient; **Inert atmosphere:** None; **Venting:** Pressure-vacuum; **Stability during transport:** Stable if dry.

PHYSICAL AND CHEMICAL PROPERTIES
Physical state @ 59°F/15°C and 1 atm: Solid.
Molecular weight: 236.1
Boiling point @ 1 atm: 465°F/240°C/513°K (decomposes).
Specific gravity (water = 1): 0.96 @ 68°F/20°C (solid).

POTASSIUM DICHROMATE
REC. P:2900

SYNONYMS: BICHROMATE OF POTASH; CHROMIC ACID, DIPOTASSIUM SALT; HIDROXIDO POTASICO (Spanish) DICHROMIC ACID, DIPOTASSIUM SALT; DIPOTASSIUM DICHROMATE; DIPOTASSIUM DICHROMATE; EEC No. 024-002-00-6; IOPEZITE; KALIUMDICHROMAT (German); POTASSIUM BICHROMATE; POTASSIUM DICHROMATE(VI); RED CHROMATE OF POTASH; BICHROME

IDENTIFICATION
CAS Number: 7778-50-9
Formula: $K_2Cr_2O_7$; $K_2(OCrO_2OCrO_2O)$
DOT ID Number: UN 1479; DOT Guide Number: 140

Proper Shipping Name: Oxidizing substances, solid, n.o.s; Oxidizing substances, solid, n.o.s.
Reportable Quantity (RQ): **(CERCLA)** 10 lb/4.54 kg

DESCRIPTION: Red to orange crystalline solid. Odorless. Sinks and mixes with water; forms an acid solution.

Corrosive to the skin, eyes, and respiratory tract; contact with skin or eyes may cause severe burns; inhalation symptoms may be delayed • Firefighting gear (including SCBA) does not provide adequate protection. If exposure occurs, remove and isolate gear immediately and thoroughly decontaminate personnel • Containers may BLEVE when exposed to fire • Strong oxidizer which may react spontaneously with low flash point organics or reducing agents. Heat forms oxygen; will increase the activity of an existing fire • May cause fire or explosion on contact with combustibles (wood, paper, oil, clothing, etc.) • Toxic products of combustion may include potassium oxide.

Hazard Classification (based on NFPA-704 Rating System)
Health Hazards (Blue): 2; Flammability (Red): 0; Reactivity (Yellow): 2; Special Notice (White): OXY

EMERGENCY RESPONSE: See Appendix A (140)
Evacuation:
Public safety: Isolate the area of spill or leak for at least 10 to 25 meters (30 to 80 feet) in all directions.
Spill: Consider initial downwind evacuation for at least 100 meters (330 feet).
Fire: If any large container is involved in fire, isolate for at least 800 meters (½ mile) in all directions; also, consider initial evacuation for 800 meters (½ mile) in all directions.

EXPOSURE
Short-term effects: *SEEK MEDICAL ATTENTION.* **Dust:** Irritating to eyes, nose, and throat. *IF INHALED*, will, will cause difficult breathing. Move to fresh air. *IF BREATHING HAS STOPPED,* give artificial respiration. IF breathing is difficult, administer oxygen. Lung edema may develop.
SOLIDS: Will burn skin and eyes. *IF SWALLOWED*, will cause nausea, vomiting, or loss of consciousness. Remove contaminated clothing and shoes. Flush affected areas with plenty of water. *IF IN EYES*, hold eyelids open and flush with plenty of water. *IF SWALLOWED* and victim is *CONSCIOUS AND ABLE TO SWALLOW*, have victim drink 4 to 8 ounces of water. **Do NOT induce vomiting.**
Note to physician or authorized medical personnel: Medical observation is recommended for 24 to 48 hours after breathing overexposure, as pulmonary edema may be delayed. As first aid for pulmonary edema, consider administering a corticosteroid spray. Cigarette smoking may exacerbate pulmonary injury and should be discouraged for at least 72 hours following exposure.

HEALTH HAZARDS
Personal protective equipment (PPE): B-Level PPE. Approved dust mask; protective gloves; goggles or face shield. Sealed chemical protective materials offers limited protection: neoprene, PVC. Also, buty rubber, natural rubber, nitrile+PVC, styrene-butadiene, styrene-butadiene/neoprene offers limited protection
Recommendations for respirator selection: NIOSH
At any concentrations above the NIOSH REL, or where there is no REL, at any detectable concentration: SCBAF:PD,PP (any self-contained breathing apparatus that has a full facepiece and is operated in a pressure-demand or other positive-pressure mode); or SAF:PD,PP:ASCBA (any supplied-air respirator that has a full facepiece and is operated in a pressure-demand or other positive-pressure mode in combination with an auxiliary self-contained breathing apparatus operated in a pressure-demand or other positive pressure mode). *ESCAPE:* HiEF (any air-purifying, full-facepiece respirator with a high-efficiency particulate filter); or SCBAE (any appropriate escape-type, self-contained breathing apparatus).
Exposure limits (TWA unless otherwise noted): ACGIH TLV 0.05 ppm, confirmed human carcinogen, as chromium(VI) compounds; OSHA PEL ceiling 0.1 ppm; NIOSH 0.001 mg/m^3 as chromates.
Toxicity by ingestion: Grade 3; LD$_{50}$ to 500 mg/kg (human).
Long-term health effects: Possible lung cancer. Skin disorders; eczema.
Vapor (gas) irritant characteristics: Dusts or mists may cause severe irritation of eyes and throat and can cause eye and lung injury. They cannot be tolerated even at low concentrations.
Liquid or solid irritant characteristics: Severe eye and skin irritant; causes second-and third-degree burns on short contact and is very injurious to the eyes.
IDLH value: 15 mg/m^3 as chromates.

FIRE DATA
Behavior in fire: May decompose, generating oxygen. Supports the combustion of other materials.

CHEMICAL REACTIVITY
Binary reactants: Ignition may occur when in contact with finely divided combustibles, such as sawdust. Contact with nitric or hydrochloric acid form irritating and toxic vapors.

ENVIRONMENTAL DATA
Food chain concentration potential: Positive.
Water pollution: Dangerous to aquatic life in high concentrations. May be dangerous if it enters nearby water intakes; notify operators. Notify local health and wildlife officials. **Response to discharge:** Issue warning–water contaminant. Disperse and flush.

SHIPPING INFORMATION
Grades of purity: 99.9%; **Storage temperature:** Ambient; **Inert atmosphere:** None; **Venting:** Open; **Stability during transport:** Stable.

PHYSICAL AND CHEMICAL PROPERTIES
Physical state @ 59°F/15°C and 1 atm: Solid.
Molecular weight: 294.19
Boiling point @ 1 atm: 932°F/500°C/673°K (decomposes).
Melting/Freezing point: 748°F/398°C/671°K.
Specific gravity (water = 1): 2.676 @ 77°F/25°C (solid).
Heat of fusion: 29.8 cal/g.

POTASSIUM HYDROXIDE REC. P:2950

SYNONYMS: CAUSTIC POTASH; EEC No. 019-002-00-8; HIDROXIDO POTASICO (Spanish); HYDROXIDE de POTASSIUM (French); KALIUMHYDROXID (German); KALIUMHYDROXYDE; KOH; LYE; POTASSA; POTASSE CAUSTIQUE (French); POTASSIUM HYDRATE; POTASSIUM (HYDRIXYDE de) (French)

IDENTIFICATION
CAS Number: 1310-58-3
Formula: KOH
DOT ID Number: UN 1813 (dry, solid; flake); UN 1814 (solution);
DOT Guide Number: 154

Potassium iodide

Proper Shipping Name: Potassium hydroxide, solid; Potassium hydroxide, solution
Reportable Quantity (RQ): **(CERCLA)** 1000 lb/454 kg

DESCRIPTION: White solid crystals, or colorless, concentrated, watery liquid. Odorless. Solid sinks and mixes with water; violent reaction releasing large amounts of heat (that may ignite combustible materials) and spattering., forming a strong base.

Corrosive • Skin or eye contact causes severe burns, impaired vision, or blindness; poison if swallowed; effects of inhalation may be delayed • Firefighting gear (including SCBA) does not provide adequate protection. If exposure occurs, remove and isolate gear immediately and thoroughly decontaminate personnel • Reacts with metals forming explosive hydrogen gas • Toxic products of combustion may include potassium oxide.

Hazard Classification (based on NFPA-704 Rating System)
Health Hazards (Blue): 3; Flammability (Red): 0; Reactivity (Yellow): 1

EMERGENCY RESPONSE: See Appendix A (154)
Evacuation:
Public safety: Isolate the area of spill or leak for at least 25 to 50 meters (80 to 160 feet) in all directions.
Spill: Increase, in the downwind direction, as necessary, the distance shown under "Public Safety."
Fire: If tank, rail car, or tank truck is involved in fire, isolate for at least 800 meters (½ mile) in all directions; also, consider initial evacuation for 800 meters (½ mile) in all directions.

EXPOSURE
Short-term effects: *SEEK MEDICAL ATTENTION*. **Dust or mist:** Irritating to eyes, nose, and throat. Harmful if inhaled. Move to fresh air. *IF BREATHING HAS STOPPED,* give artificial respiration; *avoid mouth-to-mouth resuscitation; use bag/mask apparatus*. IF breathing is difficult, administer oxygen. *IF IN EYES,* hold eyelids open and flush with plenty of water. **Liquid or solid:** Will burn skin and eyes. Harmful if swallowed. Remove contaminated clothing or shoes. Flush affected areas with plenty of water. *IF IN EYES,* hold eyelids open and flush with plenty of water. *IF SWALLOWED* and victim is *CONSCIOUS AND ABLE TO SWALLOW,* have victim drink 4 to 8 ounces of water. **Do NOT induce vomiting.**
Note to physician or authorized medical personnel: Medical observation is recommended for 24 to 48 hours after breathing overexposure, as pulmonary edema may be delayed. As first aid for pulmonary edema, consider administering a corticosteroid spray. Cigarette smoking may exacerbate pulmonary injury and should be discouraged for at least 72 hours following exposure.

HEALTH HAZARDS
Personal protective equipment (PPE): B-Level PPE. Wide-brimmed hat and close-fitting safety goggles with rubber side shields; respirator for dust; long-sleeved cotton shirt or jacket with buttoned collar and buttoned sleeves; rubber or rubber-coated canvas gloves (shirt sleeves should be buttoned over the gloves); rubber shoes or boots; cotton coveralls (with trouser cuffs worn over boots); apron. Sealed chemical protective materials offers limited protection *(30–70% solution)*: butyl rubber, natural rubber, neoprene, nitrile, PV alcohol, PVC. Also, nitrile+PVC, styrene-butadiene, styrene-butadiene/neoprene, neoprene+styrene-butadiene offers limited protection
Exposure limits (TWA unless otherwise noted): ACGIH TLV ceiling 2 mg/m^3; NIOSH REL 2 mg/m^3.

Toxicity by ingestion: Grade 3; oral rat LD_{50} = 364 mg/kg.
Vapor (gas) irritant characteristics: Fumes can cause eye and respiratory tract irritation.
Liquid or solid irritant characteristics: Severe eye and skin irritant. Causes second-and third-degree burns on short contact; and is very injurious to the eyes.

FIRE DATA
Fire extinguishing agents not to be used: Water.

CHEMICAL REACTIVITY
Reactivity with water: Dissolves with liberation of much heat; may cause steam and spattering.
Binary reactants: Acids, halogenated hydrocarbons, maleic anhydride. When wet, attacks metals such as aluminum, tin, lead, and zinc to produce flammable hydrogen gas.
Neutralizing agents for acids and caustics: Flush with water, rinse with dilute acetic acid.
Reactivity group: 5
Compatibility class: Caustics

ENVIRONMENTAL DATA
Food chain concentration potential: Negative; unlikely to accumulate.
Water pollution: Harmful to aquatic life in very low concentrations. May be dangerous if it enters nearby water intakes; notify operators. Notify local health and wildlife officials. **Response to discharge:** Issue warning–corrosive. Restrict access. Disperse and flush.

SHIPPING INFORMATION
Grades of purity: Technical flake: 85-90%; USP pellets: 85-90%; **Storage temperature:** Ambient; **Inert atmosphere:** None; **Venting:** Open; **Stability during transport:** Stable.

PHYSICAL AND CHEMICAL PROPERTIES
Physical state @ 59°F/15°C and 1 atm: Solid.
Molecular weight: 56.11
Boiling point @ 1 atm: 2417°F/1325°C/1598°K.
Melting/Freezing point: 716°F/380°C/653°K.
Specific gravity (water = 1): 2.04 @ 15°C (solid).
Heat of fusion: 35.3 cal/g.
Vapor pressure: 1 mm @ 1317°F

POTASSIUM IODIDE REC. P:3000

SYNONYMS: IODURE DE POTASSIUM (French); POTASSIUM MONOIODIDE; TRIPOTASSIUM TRIIODIDE; YODURO POTASICO (Spanish)

IDENTIFICATION
CAS Number: 7681-11-0
Formula: KI

DESCRIPTION: White crystalline solid; tuns yellow with age. Odorless. Sinks and mixes with water; forms a corrosive solution.

Solution is corrosive • Breathing the vapor can kill you; skin or eye contact causes severe burns, impaired vision, or blindness; effects of contact or inhalation may be delayed • Firefighting gear (including SCBA) does not provide adequate protection. If exposure occurs, remove and isolate gear immediately and thoroughly decontaminate personnel • Toxic products of combustion may include potassium oxides and iodine fumes.

Hazard Classification (based on NFPA-704 Rating System)
Health Hazards (Blue): 1; Flammability (Red): 0; Reactivity (Yellow): 0

EMERGENCY RESPONSE: See Appendix A (171)
Evacuation:
Public safety: Isolate the area of spill or leak for at least 10 to 25 meters (30 to 80 feet) in all directions.
Spill: Increase, in the downwind direction, as necessary, the distance shown under "Public Safety."
Fire: If any large container is involved in fire, isolate for at least 800 meters (½ mile) in all directions; also, consider initial evacuation for 800 meters (½ mile) in all directions.

EXPOSURE
Short-term effects: *SEEK MEDICAL ATTENTION*. **Solid:** Harmful if swallowed.
Flush affected areas with plenty of water. *IF SWALLOWED* and victim is *CONSCIOUS AND ABLE TO SWALLOW*, have victim drink 4 to 8 ounces of water.
Note to physician or authorized medical personnel: Medical observation is recommended for 24 to 48 hours after breathing overexposure, as pulmonary edema may be delayed. As first aid for pulmonary edema, consider administering a corticosteroid spray. Cigarette smoking may exacerbate pulmonary injury and should be discouraged for at least 72 hours following exposure.

HEALTH HAZARDS
Personal protective equipment (PPE): B-Level PPE. Goggles or face shield.
Toxicity by ingestion: Grade 2; LD_{50} = 0.5 to 5 g/kg (human).
Vapor (gas) irritant characteristics: Nonvolatile
Liquid or solid irritant characteristics: Eye irritant.

CHEMICAL REACTIVITY
Reactivity with water: Solution is corrosive.
Binary reactants: Reacts violently with reducing agents, strong oxidizers, bromotrifluorides, bromopentafluoride, chlorotrifluorides, fluorine perchlorate, isopropyl peroxydicarbonate, metallic salts, mercurous chloride, diazonium salts, 3-toluenediazonium salts, trifluoroacetyl hypofluorite. Attacks metals in a moist environment.

ENVIRONMENTAL DATA
Food chain concentration potential: Negative; unlikely to accumulate.
Water pollution: Effect of low concentrations on aquatic life is unknown. May be dangerous if it enters nearby water intakes; notify operators. Notify local health and wildlife officials.
Response to discharge: Issue warning–water contaminant. Disperse and flush.

SHIPPING INFORMATION
Grades of purity: USP, ACS, CP (all 99+%); **Storage temperature:** Ambient; **Inert atmosphere:** None; **Venting:** Open; **Stability during transport:** Stable.

PHYSICAL AND CHEMICAL PROPERTIES
Physical state @ 59°F/15°C and 1 atm: Solid.
Molecular weight: 166.01
Boiling point @ 1 atm: 2584°F/1418°C/1691°K.
Melting/Freezing point: 1333°F/723°C/996°K.
Specific gravity (water = 1): 3.13 @ 15°C (solid).
Heat of fusion: 24.7 cal/g.

POTASSIUM NITRATE REC. P:3050

SYNONYMS: DEXOL STUMP REMOVER; NITER; NITRATO POTASICO (Spanish); NITRIC ACID, POTASSIUM SALT; POTASH NITRATE; SALTPETER; VICKNITE

IDENTIFICATION
CAS Number: 7757-79-1
Formula: KNO_3
DOT ID Number: UN 1486; DOT Guide Number: 140
Proper Shipping Name: Potassium nitrate

DESCRIPTION: Transparent or white crystalline solid or powder. Odorless. Mixes with water; very soluble.

Strong oxidizer which may react spontaneously with low flash point organics or reducing agents. Heat forms oxygen; will increase the activity of an existing fire • May cause fire or explosion on contact with combustibles (wood, paper, oil, clothing, etc.) • Irritating to the skin, eyes, and respiratory tract • Decomposes above 752°F/400°C with increased risk of fire; toxic products of decomposition and combustion may include potassium nitrite, oxides of nitrogen, and potassium • Elevated temperatures above 1652°F/900°C form potassium peroxide, a violently water reactive products.

Hazard Classification (based on NFPA-704 Rating System)
Health Hazards (Blue): 1; Flammability (Red): 0; Reactivity (Yellow): 1; Special Notice (White): OXY

EMERGENCY RESPONSE: See Appendix A (140)
Evacuation:
Public safety: Isolate the area of spill or leak for at least 10 to 25 meters (30 to 80 feet) in all directions.
Spill: Consider initial downwind evacuation for at least 100 meters (330 feet).
Fire: If any large container is involved in fire, isolate for at least 800 meters (½ mile) in all directions; also, consider initial evacuation for 800 meters (½ mile) in all directions.

EXPOSURE
Short-term effects: *SEEK MEDICAL ATTENTION*. **Dust:** Move victim to fresh air. **Solid:** Mildly irritating to skin, eyes and mucous membranes. Remove contaminated clothing and shoes. Flush affected areas with water. *IF IN EYES*, hold eyelids open and flush with plenty of water.

HEALTH HAZARDS
Personal protective equipment (PPE): B-Level PPE. Full cover clothing and chemical goggles. Use approved respirator to protect against dust.
Toxicity by ingestion: Grade 2; oral rabbit LD_{50} = 1.66 g/kg.
Long-term health effects: Methemoglobin formation; blood effects.
Vapor (gas) irritant characteristics: Not very volatile.
Liquid or solid irritant characteristics: Eye and skin irritant. Practically harmless to the skin.

FIRE DATA
Behavior in fire: Mixture may detonate by heat or shock. Increases the flammability of any combustible material. Cool containers with water spray.

CHEMICAL REACTIVITY
Binary reactants: Strong oxidizer. Incompatible with reducing

agents, combustible materials, acids and heat. Explosive when mixed with reducing agents.

ENVIRONMENTAL DATA
Water pollution: Effects of low concentrations on aquatic life is unknown. May be dangerous if it enters nearby water intakes; notify operators. Notify local health and wildlife officials.
Response to discharge: Issue warning-oxidizer. Should be removed. Disperse and flush.

SHIPPING INFORMATION
Grades of purity: 99.99%; CP; technical; **Storage temperature:** Ambient; **Inert atmosphere:** None; **Stability during transport:** Stable.

PHYSICAL AND CHEMICAL PROPERTIES
Physical state @ 59°F/15°C and 1 atm: Solid.
Molecular weight: 101.11
Boiling point @ 1 atm: 752°F/400°C/673°K (decomposes)
Melting/Freezing point: 633.2°F/334°C/607°K.
Specific gravity (water = 1): 2.109

POTASSIUM OLEATE   REC. P:3100

SYNONYMS: OLEIC ACID, POTASSIUM SALT; POTASH SOAP

IDENTIFICATION
CAS Number: 143-18-0; 61790-24-7 (18% solution in water)
Formula: $C_{18}H_{34}O_2 \cdot K$; $CH_3(CH_2)_7CH=CH(CH_2)_7COOK$

DESCRIPTION: Clear to hazy amber liquid or tan-gray paste. Soap-like odor. Mixes with water.

Material is Irritating to the skin, eyes, and respiratory tract • Toxic products of combustion may include potassium oxide.

Hazard Classification (based on NFPA-704 Rating System)
Health Hazards (Blue): 0; Flammability (Red): 0; Reactivity (Yellow): 0

EMERGENCY RESPONSE: See Appendix A (171)
Evacuation:
Public safety: Isolate the area of spill or leak for at least 10 to 25 meters (30 to 80 feet) in all directions.
Spill: Increase, in the downwind direction, as necessary, the distance shown under "Public Safety."
Fire: If any large container is involved in fire, isolate for at least 800 meters (½ mile) in all directions; also, consider initial evacuation for 800 meters (½ mile) in all directions.

EXPOSURE
Short-term effects: *SEEK MEDICAL ATTENTION.* **Vapor:** Move victim to fresh air. **Liquid:** Irritating to skin, eyes and mucous membranes. Remove contaminated clothing and shoes. Wash affected areas with soap and water. *IF IN EYES,* hold eyelids open and flush with plenty of water. *IF SWALLOWED* and victim is *CONSCIOUS AND ABLE TO SWALLOW,* give large amounts of water followed by milk. **Do NOT induce vomiting.**

HEALTH HAZARDS
Personal protective equipment (PPE): B-Level PPE. Wear full cover clothing, chemical safety glasses, and rubber or plastic gloves.

Vapor (gas) irritant characteristics: Eye and respiratory tract irritant. Vapors cause smarting of the eyes or respiratory system if present in high concentrations. The effect is temporary.
Liquid or solid irritant characteristics: Eye and skin irritant. If spilled on clothing and allowed to remain, may cause smarting and reddening of skin.

FIRE DATA
Fire extinguishing agents not to be used: Water.

CHEMICAL REACTIVITY
Binary reactants: Incompatible with strong oxidizing agents, strong acids and metallic salts.
Reactivity group: 34
Compatibility class: Esters.

ENVIRONMENTAL DATA
Water pollution: Effects of low concentrations on aquatic life is unknown. May be dangerous if it enters nearby water intakes; notify operators. Notify local health and wildlife officials. **Response to discharge:** Should be removed.

SHIPPING INFORMATION
Grades of purity: 40% paste; 18% solution in water; **Storage temperature:** Ambient; **Inert atmosphere:** None; **Stability during transport:** Stable.

PHYSICAL AND CHEMICAL PROPERTIES
Physical state @ 59°F/15°C and 1 atm: Liquid.
Molecular weight: 320.57
Specific gravity (water = 1): 1.0

POTASSIUM OXALATE   REC. P:3150

SYNONYMS: OXALATO POTASICO (Spanish); POTASSIUM OXALATE MONOHYDRATE

IDENTIFICATION
CAS Number: 583-52-8; 6487-48-5
Formula: $K_2C_2O_4 \cdot H_2O$

DESCRIPTION: White or colorless solid. Odorless. Sinks and mixes with water.

Dust irritates skin, eyes, and respiratory tract • Toxic products of combustion may include carbon monoxide.

Hazard Classification (based on NFPA-704 Rating System)
Health Hazards (Blue): 1; Flammability (Red): 0; Reactivity (Yellow): 0

EMERGENCY RESPONSE: See Appendix A (171)
Evacuation:
Public safety: Isolate the area of spill or leak for at least 10 to 25 meters (30 to 80 feet) in all directions.
Spill: Increase, in the downwind direction, as necessary, the distance shown under "Public Safety."
Fire: If any large container is involved in fire, isolate for at least 800 meters (½ mile) in all directions; also, consider initial evacuation for 800 meters (½ mile) in all directions.

EXPOSURE
Short-term effects: *SEEK MEDICAL ATTENTION.* **Dust:** Irritating to eyes, nose, and throat. *IF INHALED,* will, will cause

coughing or difficult breathing. *IF IN EYES*, hold eyelids open and flush with plenty of water. *IF BREATHING HAS STOPPED,* give artificial respiration. IF breathing is difficult, administer oxygen. **Solid:** *POISONOUS IF SWALLOWED.* Irritating to skin and eyes. *IF SWALLOWED*, will cause nausea, vomiting, or loss of consciousness. Remove contaminated clothing and shoes. Flush affected areas with plenty of water. *IF IN EYES*, hold eyelids open and flush with plenty of water. *IF SWALLOWED* and victim is *CONSCIOUS AND ABLE TO SWALLOW*, have victim drink 4 to 8 ounces of water. *IF SWALLOWED* and victim is *UNCONSCIOUS OR HAVING CONVULSIONS,* do nothing except keep victim warm.

HEALTH HAZARDS
Personal protective equipment (PPE): B-Level PPE. Approved dust respirator; chemical goggles; rubber or plastic-coated gloves.
Toxicity by ingestion: Grade 3; LD_{50} = 50 to 500 mg/kg.

FIRE DATA
Behavior in fire: Decomposes to carbonate with no charring. The reaction is not hazardous.

ENVIRONMENTAL DATA
Food chain concentration potential: Negative; unlikely to accumulate.
Water pollution: Effect of low concentrations on aquatic life is unknown. May be dangerous if it enters nearby water intakes; notify operators. Notify local health and wildlife officials.
Response to discharge: Issue warning–water contaminant. Disperse and flush.

SHIPPING INFORMATION
Grades of purity: Reagent, 99.0%; **Storage temperature:** Ambient; **Venting:** None; **Stability during transport:** Stable.

PHYSICAL AND CHEMICAL PROPERTIES
Physical state @ 59°F/15°C and 1 atm: Solid.
Molecular weight: 184.24
Boiling point @ 1 atm: (decomposes).
Specific gravity (water = 1): 2.13 at 18.5°C (solid).

POTASSIUM PERMANGANATE   REC. P:3200

SYNONYMS: CAIROX; CHAMELEON MINERAL; C.I. 77755; CONDY'S CRYSTALS; EEC No. 025-002-00-9; KALIUMPERMANGANAT (German); PERMANGANATE de POTASSIUM (French); PERMANGANATE OF POTASH; PERMANGANATO POTASICO (Spanish); POTASSIUM (PERMANGANATE de) (French); PURPLE SALT

IDENTIFICATION
CAS Number: 7722-64-7
Formula: $KMnO_4$
DOT ID Number: UN 1490; DOT Guide Number: 140
Proper Shipping Name: Potassium permanganate
Reportable Quantity (RQ): **(CERCLA)** 100 lb/45.4 kg

DESCRIPTION: Dark purple crystalline solid with a blue metallic sheen. Odorless. Sinks and mixes slowly with water.

Corrosive to the skin, eyes, and respiratory tract; contact with skin or eyes can cause burns, impaired vision or blindness; inhalation symptoms may be delayed • Strong oxidizer which may react spontaneously with low flash point organics or reducing agents.

Decomposes above 465°F/240°C forming oxygen that will increase the activity of an existing fire • May cause fire or explosion on contact with combustibles (wood, paper, oil, clothing, etc.) • Toxic products of combustion may include oxygen and potassium oxide.

Hazard Classification (based on NFPA-704 Rating System)
Health Hazards (Blue): 1; Flammability (Red): 0; Reactivity (Yellow): 1; Special Notice (White): OXY

EMERGENCY RESPONSE: See Appendix A (140)
Evacuation:
Public safety: Isolate the area of spill or leak for at least 10 to 25 meters (30 to 80 feet) in all directions.
Spill: Consider initial downwind evacuation for at least 100 meters (330 feet).
Fire: If any large container is involved in fire, isolate for at least 800 meters (½ mile) in all directions; also, consider initial evacuation for 800 meters (½ mile) in all directions.

EXPOSURE
Short-term effects: *SEEK MEDICAL ATTENTION.* **Vapors:** Lung edema may develop. **Solid:** Irritating to skin and eyes. *IF SWALLOWED,* will cause nausea, vomiting, or loss of consciousness.
Flush affected areas with plenty of water. *IF IN EYES*, hold eyelids open and flush with plenty of water. *IF SWALLOWED* and victim is *CONSCIOUS AND ABLE TO SWALLOW*, have victim drink 4 to 8 ounces of water and have victim induce vomiting. *IF SWALLOWED* and victim is *UNCONSCIOUS OR HAVING CONVULSIONS,* do nothing except keep victim warm.
Note to physician or authorized medical personnel: Medical observation is recommended for 24 to 48 hours after breathing overexposure, as pulmonary edema may be delayed. As first aid for pulmonary edema, consider administering a corticosteroid spray. Cigarette smoking may exacerbate pulmonary injury and should be discouraged for at least 72 hours following exposure.

HEALTH HAZARDS
Personal protective equipment (PPE): B-Level PPE. Goggles or face shield; rubber gloves.
Recommendations for respirator selection: NIOSH/OSHA as manganese
10 mg/m^3: DMXSQ *If not present as a fume* (any dust and mist respirator except single-use and quarter mask respirators); or SA (any supplied-air respirator). *25 mg/m^3*: SA:CF (any supplied-air respirator operated in a continuous-flow mode); or PAPRDM *If not present as a fume* (any powered, air-purifying respirator with a dust and mist filter). *50 mg/m^3*: HiEF (any air-purifying, full-facepiece respirator with a high-efficiency particulate filter); or SAT:CF (any supplied-air respirator that has a tight-fitting facepiece and is operated in a continuous-flow mode); or PAPRTHiE (any powered, air-purifying respirator with a tight-fitting facepiece and a high-efficiency particulate filter); or SCBAF (any self-contained breathing apparatus with a full facepiece); or SAF (any supplied-air respirator with a full facepiece). *500 mg/m^3*: SA:PD:PP (any supplied-air respirator operated in a pressure-demand or other positive-pressure mode). *EMERGENCY OR PLANNED ENTRY INTO UNKNOWN CONCENTRATIONS OR IDLH CONDITIONS*: SCBAF:PD,PP (any self-contained breathing apparatus that has a full facepiece and is operated in a pressure-demand or other positive-pressure mode); or SAF:PD,PP:ASCBA (any supplied-air respirator that has a full facepiece and is operated in a pressure-demand or other positive-pressure mode in combination with an auxiliary self-contained breathing apparatus operated in a pressure-demand or other positive pressure mode). *ESCAPE*: HiEF (any air-

purifying, full-facepiece respirator with a high-efficiency particulate filter); or SCBAE (any appropriate escape-type, self-contained breathing apparatus).
Exposure limits (TWA unless otherwise noted): ACGIH TLV 0.2 mg/m^3 as manganese; OSHA PEL ceiling limit 5 mg/m^3 as manganese; ACGIH TLV 5 mg/m^3; OSHA ceiling 5 mg/m^3; NIOSH 1 mg/m^3 as manganese
Short-term exposure limits (15-minute TWA): NIOSH STEL 3 mg/m^3 as manganese
Toxicity by ingestion: Grade 3; LD$_{50}$ = 50 to 500 mg/kg.
Vapor (gas) irritant characteristics: Nonvolatile
Liquid or solid irritant characteristics: Eye and skin irritant. Can burn skin if not flushed with water.
IDLH value: 500 mg/m^3 as manganese

FIRE DATA
Behavior in fire: Containers may explode.

CHEMICAL REACTIVITY
Binary reactants: Attacks rubber and most combustible materials. May cause ignition of wood. Some acids, such as sulfuric acid, may cause explosion. Violent reaction with ammonia, phosphorus, sulfur, hydrogen peroxide.

ENVIRONMENTAL DATA
Food chain concentration potential: Negative; unlikely to accumulate.
Water pollution: Harmful to aquatic life in very low concentrations. May be dangerous if it enters nearby water intakes; notify operators. **Response to discharge:** Issue warning–water contaminant. Disperse and flush

SHIPPING INFORMATION
Grades of purity: USP, Reagent (both 99+%); **Storage temperature:** Ambient; **Inert atmosphere:** None; **Venting:** Open; **Stability during transport:** Stable.

PHYSICAL AND CHEMICAL PROPERTIES
Physical state @ 59°F/15°C and 1 atm: Solid.
Molecular weight: 158.04
Boiling point @ 1 atm: Decomposes.
Melting/Freezing point: More than 464°F/240°C/513°K.
Specific gravity (water = 1): 2.70 @ 15°C (solid).

POTASSIUM PEROXIDE   REC. P:3250

SYNONYMS: PEROXIDO POTASICO (Spanish); POTASSIUM SUPEROXIDE

IDENTIFICATION
CAS Number: 17014-71-0
Formula: K$_2$O$_2$
DOT ID Number: UN 1491; DOT Guide Number: 144
Proper Shipping Name: Potassium peroxide

DESCRIPTION: Yellow powder. Odorless. Sinks and mixes violently (explodes) with water; evolving oxygen and forming hydrogen peroxide and potassium hydroxides.

Corrosive to the skin, eyes, and respiratory tract; inhalation symptoms may be delayed • Strong oxidizer which may react spontaneously with low flash point organics or reducing agents. Heat forms oxygen; will increase the activity of an existing fire • May cause fire or explosion on contact with combustibles (wood, paper, oil, clothing, etc.) • Toxic products of combustion may include oxygen and potassium oxide.

Hazard Classification (based on NFPA-704 Rating System)
Health Hazards (Blue): 3; Flammability (Red): 0; Reactivity (Yellow): 1; Special Notice (White): OXY

EMERGENCY RESPONSE: See Appendix A (144)
Evacuation:
Public safety: Isolate the area of spill or leak for at least 50 to 100 meters (160 to 300 feet) in all directions.
Spill: Increase, in the downwind direction, as necessary, the distance shown under "Public Safety."
Fire: If any large container is involved in fire, isolate for at least 800 meters (½ mile) in all directions; also, consider initial evacuation for 800 meters (½ mile) in all directions.

EXPOSURE
Short-term effects: *SEEK MEDICAL ATTENTION.* **Dust:** Irritating to eyes, nose, and throat. *IF INHALED*, will, will cause coughing or difficult breathing. *IF IN EYES*, hold eyelids open and flush with plenty of water. *IF BREATHING HAS STOPPED*, give artificial respiration; *avoid mouth-to-mouth resuscitation; use bag/mask apparatus*. IF breathing is difficult, administer oxygen. Lung edema may develop **Solid:** Will burn skin and eyes. *IF SWALLOWED*, will cause nausea. Remove contaminated clothing and shoes. Flush affected areas with plenty of water. *IF IN EYES*, hold eyelids open and flush with plenty of water. *IF SWALLOWED* and victim is *CONSCIOUS AND ABLE TO SWALLOW*, have victim drink 4 to 8 ounces of water. *IF SWALLOWED* and victim is *UNCONSCIOUS OR HAVING CONVULSIONS*, do nothing except keep victim warm. **Do NOT induce vomiting.**
Note to physician or authorized medical personnel: Medical observation is recommended for 24 to 48 hours after breathing overexposure, as pulmonary edema may be delayed. As first aid for pulmonary edema, consider administering a corticosteroid spray. Cigarette smoking may exacerbate pulmonary injury and should be discouraged for at least 72 hours following exposure.

HEALTH HAZARDS
Personal protective equipment (PPE): B-Level PPE. Dust mask; goggles or face shield; protective gloves.
Liquid or solid irritant characteristics: Severe eye and skin irritant. May cause burns.

FIRE DATA
Fire extinguishing agents not to be used: A small amount of water may cause explosions.
Behavior in fire: Increases intensity of fire and can start fires when in contact with organic combustibles.

CHEMICAL REACTIVITY
Reactivity with water: Reacts violently with liberation of heat, oxygen, hydrogen peroxide and caustic potassium hydroxide.
Binary reactants: Strong oxidizer; can form explosive and self-igniting mixtures with wood or other combustible materials. violent reaction with reducing agents.
Neutralizing agents for acids and caustics: Following reaction with water, caustic formed can be flushed away with water and area rinsed with dilute acetic acid.

ENVIRONMENTAL DATA
Food chain concentration potential: Negative; unlikely to accumulate.

Water pollution: Harmful to aquatic life in very low concentrations. May be dangerous if it enters nearby water intakes; notify operators. Notify local health and wildlife officials. **Response to discharge:** Issue warning–oxidizing material. Restrict access. Disperse and flush.

SHIPPING INFORMATION
Grades of purity: Commercial; Pure; **Storage temperature:** Ambient; **Inert atmosphere:** None; **Venting:** Pressure-vacuum; **Stability during transport:** Stable if kept dry.

PHYSICAL AND CHEMICAL PROPERTIES
Physical state @ 59°F/15°C and 1 atm: Solid.
Molecular weight: 110
Boiling point @ 1 atm: Decomposes.
Melting/Freezing point: 914°F/490°C/763°K.
Specific gravity (water = 1): More than 1 @ 68°F/20°C (solid).
Heat of fusion: 55.3 cal/g.

PROPANE   REC. P:3300

SYNONYMS: BOTTLED GAS; DIMETHYLMETHANE; EEC No. 601-003-00-5; Liquefied PETROLEUM GAS; *n*-PROPANE; PROPANO (Spanish); PROPYL HYDRATE; PROPYL HYDRIDE

IDENTIFICATION
CAS Number: 74-98-6
Formula: C_3H_8; $CH_3CH_2CH_3$
DOT ID Number: UN 1978; DOT Guide Number: 115
Proper Shipping Name: Propane; Propane mixtures

DESCRIPTION: Colorless gas. Shipped and stored as a liquefied gas under its own pressure. Faint petroleum odor or may have a foul smelling odorant added when used as a fuel. Liquid floats on the surface of water and boils; very slightly soluble; flammable visible vapor cloud is produced.

Extremely flammable • Forms explosive mixture with air • Reacts explosively with many materials • Gas from liquefied gas are initially heavier than air and spread along ground • Gas may travel to source of ignition and flashback • Ruptured or venting cylinders may rocket through buildings and/or travel a considerable distance • Vapors may cause dizziness or asphyxiation without warning • Contact with liquid may cause frostbite.

Hazard Classification (based on NFPA-704 Rating System)
Health Hazards (Blue): 1; Flammability (Red): 4; Reactivity (Yellow): 0

EMERGENCY RESPONSE: See Appendix A (115)
Evacuation:
Public safety: Isolate spill area for at least 50 to 100 meters (160 to 330 feet) in all directions.
Spill: Consider initial downwind evacuation for at least 800 meters (½ mile).
Fire: Isolate for 1600 meters (1 mile) in all directions, especially if tank, rail car, or tank truck is involved in fire.

EXPOSURE
Short-term effects: *SEEK MEDICAL ATTENTION.* **Vapor:** Not irritating to eyes, nose or throat. *IF INHALED*, will, will cause dizziness, difficult breathing, or loss of consciousness. Move to fresh air. *IF BREATHING HAS STOPPED,* give artificial respiration. IF breathing is difficult, administer oxygen. **Liquid:** May cause frostbite. Flush affected areas with plenty of water. *DO NOT RUB AFFECTED AREAS.*

HEALTH HAZARDS
Personal protective equipment (PPE): B-Level PPE. Self-contained breathing apparatus for high concentrations of gas. Wear thermal protective clothing. Chemical protective material(s) reported to have good to excellent resistance: neoprene, nitrile+PVC, polyethylene, polyurethane. Also, PVC, Viton® offers limited protection
Recommendations for respirator selection: NIOSH/OSHA *2100 ppm:* SA (any supplied-air respirator); or SCBAF (any self-contained breathing apparatus with a full facepiece). *EMERGENCY OR PLANNED ENTRY INTO UNKNOWN CONCENTRATIONS OR IDLH CONDITIONS:* SCBAF:PD,PP (any self-contained breathing apparatus that has a full facepiece and is operated in a pressure-demand or other positive-pressure mode); or SAF:PD,PP:ASCBA (any supplied-air respirator that has a full facepiece and is operated in a pressure-demand or other positive-pressure mode in combination with an auxiliary self-contained breathing apparatus operated in a pressure-demand or other positive pressure mode). *ESCAPE:* SCBAE (any appropriate escape-type, self-contained breathing apparatus).
Exposure limits (TWA unless otherwise noted): ACGIH 2500 ppm; NIOSH/OSHA 1000 ppm (1800 mg/m^3).
Vapor (gas) irritant characteristics: Vapors are nonirritating to the eyes and throat.
Liquid or solid irritant characteristics: No appreciable hazard. Fast evaporation of liquid may cause frostbite.
Odor threshold: 5,000–20,000 ppm.
IDLH value: 2100 ppm [10% LEL].

FIRE DATA
Flash point: Gas –156°F/–104°C (cc).
Flammable limits in air: LEL: 2.1%; UEL: 9.5%.
Fire extinguishing agents not to be used: Water.
Behavior in fire: Containers may explode. Unless the flow of gas can be stopped, extinguishing a propane fire will permit the accumulation of an explosive concentration of vapor, and subsequent explosion or reflash.
Autoignition temperature: 842°F/450°C/723°K.
Electrical hazard: Class I, Group D. Due to low electric conductivity, this substance may generate electrostatic charges as a result of agitation and flow.
Burning rate: 8.2 mm/min
Adiabatic flame temperature: 2419°F/1326°C (estimate).
Stoichiometric air-to-fuel ratio: 15.60 (estimate).

CHEMICAL REACTIVITY
Binary reactants: Strong oxidizers.
Reactivity group: 31
Compatibility class: Paraffins

ENVIRONMENTAL DATA
Food chain concentration potential: Negative; unlikely to accumulate.
Water pollution: Not harmful to aquatic life. **Response to discharge:** Issue warning–high flammability. Restrict access. Evacuate area.

SHIPPING INFORMATION
Grades of purity: Research; instrument, or Pure: 99.35+%; Technical: 97.50%; **Storage temperature:** Ambient; **Inert**

atmosphere: None; **Venting:** Safety relief; **Stability during transport:** Stable.

NAS HAZARD CLASSIFICATION FOR BULK WATER TRANSPORTATION
FIRE: 4
HEALTH: Vapor irritant: 0; Liquid or solid irritant: 0; Poisons: 0
WATER POLLUTION: Human toxicity: 0; Aquatic toxicity: 0; Aesthetic effect: 0
REACTIVITY: Other chemicals: 0; Water: 0; Self-reaction: 0

PHYSICAL AND CHEMICAL PROPERTIES
Physical state @ 59°F/15°C and 1 atm: Gas.
Molecular weight: 44.09
Boiling point @ 1 atm: –43.8°F/–42.1°C/231.1°K.
Melting/Freezing point: –305.9°F/–187.7°C/85.5°K.
Critical temperature: 206.0°F/96.67°C/369.67°K.
Critical pressure: 616.5 psia = 41.94 atm = 4.249 MN/m^2.
Specific gravity (water = 1): 0.590 at –50°C (liquid); 1.55 (gas).
Liquid surface tension: 16 dynes/cm = 0.016 N/m at –52.6°F/–47°C.
Liquid water interfacial tension: (estimate) 50 dynes/cm = 0.05 N/m at –58°F/–50°C/223.2°K.
Relative vapor density (air = 1): 1.5
Ratio of specific heats of vapor (gas): 1.130
Latent heat of vaporization: 183.2 Btu/lb = 101.8 cal/g = 4.262 x 10^5 J/kg.
Heat of combustion: –19,782 Btu/lb = –10,990 cal/g = –460.13 x 10^5 J/kg.
Vapor pressure: (Reid) 190 psia; 6850 mm.

N-PROPANOLAMINE   REC. P:3350

SYNONYMS: 3-AMINO-1-PROPANOL; β-ALANINOL; γ-AMINOPROPANOL; 3-HYDROXYPROPYLAMINE; 3-PROPANOLAMINE; 1-PROPANOL, 3-AMINO; 3-PROPANOLAMINA (Spanish); 1,3-PROPANOLAMINE

IDENTIFICATION
CAS Number: 156-87-6
Formula: C$_3$H$_9$NO; H$_2$NCH$_2$CH$_2$CH$_2$OH
DOT ID Number: NA 1993; DOT Guide Number: 128
Proper Shipping Name: Combustible liquid, n.o.s.

DESCRIPTION: Colorless to pale yellow liquid. Fishy odor. Liquid floats and mixes with water; forms alkaline solution.

Combustible • Containers may BLEVE when exposed to fire • Vapors may form explosive mixture with air • Vapors are heavier than air and will collect and stay in low areas • Vapors may travel long distances to ignition sources and flashback • Vapors in confined areas (e.g., tanks, sewers, buildings) may explode when exposed to fire • Irritating to the skin, eyes, and respiratory tract • Toxic products of combustion may include nitrogen oxides.

Hazard Classification (based on NFPA-704 Rating System)
Health Hazards (Blue): 3; Flammability (Red): 2; Reactivity (Yellow): 0

EMERGENCY RESPONSE: See Appendix A (128)
Evacuation:
Public safety: Isolate spill area for at least 25 to 50 meters (80 to 160 feet) in all directions.
Spill: Large spill–Consider initial downwind evacuation for at least 300 meters (1000 feet).
Fire: Isolate for 800 meters (½ mile) in all directions, especially if tank, rail car, or tank truck is involved in fire.

EXPOSURE
Short-term effects: *SEEK MEDICAL ATTENTION*. **Vapor:** May be harmful if inhaled. Move victim to fresh air. If not breathing, give artificial respiration. IF breathing is difficult, administer oxygen. **Liquid:** Irritating to eyes and skin. May cause burns. Remove and double bag contaminated clothing and shoes. *IF IN EYES*, hold eyelids open and flush with plenty of running water for at least 15 minutes.
Flush other affected areas for at least 15 minutes with plenty of running water.
Keep victim quiet and maintain normal body temperature.

HEALTH HAZARDS
Personal protective equipment (PPE): B-Level PPE. Wear butyl rubber gloves and face shield or all-purpose canister respirator for spills. Wear self-contained breathing apparatus and full protective clothing for fires.
Toxicity by ingestion: Grade 2: LD$_{50}$ = 2.8 g/kg (rat).
Vapor (gas) irritant characteristics: eye and respiratory tract irritant. Vapors are irritating such that personnel will usually not tolerate moderate or high concentrations
Liquid or solid irritant characteristics: Severe eye and skin irritant. May cause pain and second-degree burns after a few minutes of contact.

FIRE DATA
Flash point: 175°F/79°C (cc).
Electrical hazards: Flow or agitation of substance may generate electrostatic charges due to low conductivity.

CHEMICAL REACTIVITY
Binary reactants: Easily corrodes copper and its alloys.
Neutralizing agents for acids and caustics: Cover spilled material with sodium bisulfate. Flush with water.
Reactivity group: 8
Compatibility class: Alkanolamines

ENVIRONMENTAL DATA
Water pollution: Effect of low concentration on aquatic life is unknown. May be dangerous if it enters nearby water intakes; notify operators. Notify local health and wildlife officials. Notify operators of local water intakes. **Response to discharge:** Disperse and flush

SHIPPING INFORMATION
Grades of purity: 98%, 99+%.

NAS HAZARD CLASSIFICATION FOR BULK WATER TRANSPORTATION
FIRE: 1
HEALTH: Vapor irritant: 3; Liquid or solid irritant: 3; Poisons: 1
WATER POLLUTION: Human toxicity: 2; Aquatic toxicity:–; Aesthetic effect: 3
REACTIVITY: Other chemicals: 2; Water: 1; Self-reaction: 0

PHYSICAL AND CHEMICAL PROPERTIES
Physical state @ 59°F/15°C and 1 atm: Liquid.
Molecular weight: 75.11
Boiling point @ 1 atm: 370°F/188°C/461°K.
Melting/Freezing point: 52°F/11°C/284.2°K.

Specific gravity (water = 1): 0.982 @ 68°F/20°C.
Relative vapor density (air = 1): 2.6 (estimate).
Vapor pressure: 2.1 mm.

2-PROPANOLAMINE   REC. P:3400

SYNONYMS: ALANINOL; 2-AMINO-1-PROPANOL; 1-METHYL-2-HYDROXY-ETHYLAMINE; 2-PROPANOLAMINA (Spanish); 1-PROPANOL, 2-AMINO-

IDENTIFICATION
CAS Number: 78-91-1
Formula: C_3H_9NO; $CH_3CH(NH_2)CH_2OH$
DOT ID Number: NA 1993; DOT Guide Number: 128
Proper Shipping Name: Combustible liquid, n.o.s.

DESCRIPTION: Colorless to pale yellow liquid. Fishy odor. Liquid floats and mixes with water; forms alkaline solution.

Flammable • Containers may BLEVE when exposed to fire • Vapors may form explosive mixture with air • Vapors are heavier than air and will collect and stay in low areas • Vapors may travel long distances to ignition sources and flashback • Vapors in confined areas (e.g., tanks, sewers, buildings) may explode when exposed to fire • Irritating to the skin, eyes, and respiratory tract • Toxic products of combustion may include nitrogen oxides.

Hazard Classification (based on NFPA-704 Rating System)
Health Hazards (Blue): 3; Flammability (Red): 2; Reactivity (Yellow): 0

EMERGENCY RESPONSE: See Appendix A (128)
Evacuation:
Public safety: Isolate spill area for at least 25 to 50 meters (80 to 160 feet) in all directions.
Spill: Large spill–Consider initial downwind evacuation for at least 300 meters (1000 feet).
Fire: Isolate for 800 meters (½ mile) in all directions, especially if tank, rail car, or tank truck is involved in fire.

EXPOSURE
Short-term effects: *SEEK MEDICAL ATTENTION.* **Liquid:** Irritating to eyes and skin. May cause burns. Move victim to fresh air. If not breathing, give artificial respiration. IF breathing is difficult, administer oxygen. Remove and double bag contaminated clothing and shoes. *IF IN EYES*, hold eyelids open and flush with plenty of running water for at least 15 minutes. Flush other affected areas for at least 15 minutes with plenty of running water. Keep victim quiet and maintain normal body temperature. **Vapor:** May be harmful if inhaled.

HEALTH HAZARDS
Personal protective equipment (PPE): B-Level PPE. Wear butyl rubber gloves and face shield or all-purpose canister respirator for spills. Wear self-contained breathing apparatus and full protective clothing for fires.
Vapor (gas) irritant characteristics: May burn skin and eyes.
Liquid or solid irritant characteristics: Liquid may burn skin and eyes.

FIRE DATA
Flash point: 145°F/63°C (cc).
Autoignition temperature: (estimate) 706°F/374°C/647°K.
Electrical hazard: Class I, Group D.

CHEMICAL REACTIVITY
Binary reactants: All acids, isocyanates, organic anhydrides, acrylates, ketones.
Neutralizing agents for acids and caustics: Cover spilled material with sodium bisulfate. Flush with water.
Reactivity group: 8
Compatibility class: Alkanolamines

ENVIRONMENTAL DATA
Water pollution: Effect of low concentration on aquatic life is unknown. May be dangerous if it enters nearby water intakes; notify operators. Notify local health and wildlife officials. **Response to discharge:** Disperse and flush

SHIPPING INFORMATION
Grades of purity: 98%.

PHYSICAL AND CHEMICAL PROPERTIES
Physical state @ 59°F/15°C and 1 atm: Liquid.
Molecular weight: 75.11
Boiling point @ 1 atm: 311–317°F/173–176°C/446–449°K.
Melting/Freezing point: (estimate) 34°F/1.1°C/274°K.
Specific gravity (water = 1): 0.943 @ 68°F/20°C (*dl* mixture).
Relative vapor density (air = 1): 2.59 (approximated).

PROPANEDINITRILE   REC. P:3450

SYNONYMS: CYANOACETONITRILE; DIEYANOMETHANE; MALONONITRILE; MALONIC DINITRILE; METHYLENE CYANIDE; PROPANDINITRILO (Spanish); RCRA No. U1490

IDENTIFICATION
CAS Number: 109-77-3
Formula: $C_3H_2N_2$; $CH_2(CN)_2$
DOT ID Number: UN 2647; DOT Guide Number: 153
Proper Shipping Name: Malononitrile
Reportable Quantity (RQ): **(CERCLA)** 1000 lb/454 kg

DESCRIPTION: Colorless, white, crystalline solid or powder. Sinks in water; moderately soluble. Melts at 93°F/34°C.

Poison! • Breathing the vapor can kill you; skin or eye contact causes severe burns, impaired vision, or blindness; forms cyanide in the body • Firefighting gear (including SCBA) does not provide adequate protection. If exposure occurs, remove and isolate gear immediately and thoroughly decontaminate personnel • Polymerization hazard • Heat above 265°F/129°C (for a prolonged period) can induce polymerization with rapid release of energy; sealed containers may rupture explosively • May react with itself blocking relief valves; leading to tank explosions • Containers may BLEVE when exposed to fire • Toxic products of combustion may include oxides of nitrogen and cyanide and cyanogen gas • Do not put yourself in danger by entering a contaminated area to rescue a victim.

Hazard Classification (based on NFPA-704 Rating System)
Health Hazards (Blue): 2; Flammability (Red): 1; Reactivity (Yellow): 1

EMERGENCY RESPONSE: See Appendix A (153)
Evacuation:
Public safety: Isolate the area of spill or leak for at least 25 to 50 meters (80 to 160 feet) in all directions.

Spill: Increase, in the downwind direction, as necessary, the distance shown under "Public Safety."

Fire: If tank, rail car, or tank truck is involved in fire, isolate for at least 800 meters (½ mile) in all directions; also, consider initial evacuation for 800 meters (½ mile) in all directions.

EXPOSURE

Short-term effects: *SEEK MEDICAL ATTENTION.* **Dust:** *POISONOUS IF INHALED.* Move to fresh air. If not breathing, give artificial respiration; *avoid mouth-to-mouth resuscitation; use bag/mask apparatus*. IF breathing is difficult, administer oxygen. **Solid:** *POISONOUS IF SWALLOWED OR ABSORBED THROUGH SKIN.* Irritating to eyes. *IF IN EYES OR ON SKIN*, flush with running water for at least 15 minutes, hold eyelids open if necessary. Remove and double bag contaminated clothing and shoes at the site.

Note to physician or authorized medical personnel: Consider the use of amyl nitrite perles if symptoms of cyanide poisoning develop. If symptoms indicate, initial treatment includes the cyanide antidote kit. In all cases, break an amyl nitrite perle in a gauze pad and hold lightly under victim's nose for 15 seconds, repeating 5 times at about 15-second intervals; if necessary (and if sodium nitrite infusions will be delayed), repeat procedure every 3 minutes with fresh pearls until 3 or 4 have been used. Avoid breathing the vapor while administering it to the victim. Administer sodium nitrite IV, ASAP. The usual adult dose is 10 to 20 mL of a 3% solution infused over no less than 5 minutes; the average child dose is 0.15 to 0.20 mL/kg. Monitor blood pressure during administration, and slow the rate of infusion if hypotention develops. Next, infuse sodium thiosulfate IV. The usual adult dose is 50 mL of a 25% solution infused over 10 to 20 minutes; the average child dose is 1.65 mL/kg. Repeat with nitrite and thiosulfate as required.

HEALTH HAZARDS

Personal protective equipment (PPE): B-Level PPE. Wear self-contained positive pressure breathing apparatus and full protective clothing.

Recommendations for respirator selection: NIOSH
up to 80 mg/m³: SA (any supplied-air respirator). *up to 200 mg/m³:* SA:CF (any supplied-air respirator operated in a continuous-flow mode). *400 mg/m³:* SCBAF (any self-contained breathing apparatus with a full facepiece) SAF (any supplied-air respirator with a full facepiece). 667 mg/m³: SAF:PD,PP (any supplied-air respirator that has a full facepiece and is operated in a pressure-demand or other positive-pressure mode). *EMERGENCY OR PLANNED ENTRY INTO UNKNOWN CONCENTRATIONS OR IDLH CONDITIONS*: SCBAF:PD,PP (any self-contained breathing apparatus that has a full facepiece and is operated in a pressure-demand or other positive-pressure mode) SAF:PD,PP:ASCBA (any supplied-air respirator that has a full facepiece and is operated in a pressure-demand or other positive-pressure mode in combination with an auxiliary self-contained breathing apparatus operated in a pressure-demand or other positive-pressure mode). *ESCAPE:* GMFOV [any air-purifying, full-facepiece respirator (gas mask) with a chin-style, front-or back-mounted organic vapor canister] or SCBAE (any appropriate escape-type, self-contained breathing apparatus).

Exposure limits (TWA unless otherwise noted): NIOSH 3 ppm (8 mg/m³) as molononitrile.

Toxicity by ingestion: Grade 4; LD_{50} = 19 mg/kg (mouse).

Liquid or solid irritant characteristics: Eye irritant.

FIRE DATA

Flash point: 266°F/130°C (oc); 234°F/122°C (cc).

CHEMICAL REACTIVITY

Binary reactants: Incompatible with sulfuric acid.

Polymerization: Violent polymerization may occur if held at 266°F/130°C for an extended period of time, or at lower temperatures on contact with strong bases. Also, spontaneous explosion may occur (possibly with contamination) above 158°F/70°C.

Reactivity group: 37

Compatibility class: Nitriles

ENVIRONMENTAL DATA

Water pollution: DOT Appendix B, §172.101–marine pollutant. Harmful to aquatic life in very low concentrations. May be dangerous if it enters nearby water intakes; notify operators. Notify local health and wildlife officials. **Response to discharge:** Issue warning–poison, water contaminant. Restrict access. Should be removed. Chemical and physical treatment.

SHIPPING INFORMATION

Grades of purity: 99%; **Storage temperature:** Ambient. Keep as cool as reasonably practical; **Stability during transport:** Stable.

PHYSICAL AND CHEMICAL PROPERTIES

Physical state @ 59°F/15°C and 1 atm: Solid.
Molecular weight: 66.06
Boiling point @ 1 atm: 428°F/220°C/493°K.
Melting/Freezing point: 90–93°F/32–34°C/305–307°K.
Specific gravity (water = 1): 1.1910 @ 68°F/20°C.
Relative vapor density (air = 1): 2.3 (estimate).

PROPARGITE REC. P:3500

SYNONYMS: COMITE; DO 14; NAUGATUCK DO 14; NAUGATUCK DO-014; OMAIT; OMITE; PROPARGIL; PROPARGITA (Spanish); SULFUROUS ACID, 2-[4-(1,1-DIMETHYLETHYL] PHENOXY) CYCLOHEXYL 2-PROPYNYL ESTER; UNIROYAL D-014; U.S. RUBBER D-014

IDENTIFICATION

CAS Number: 2312-35-8
Formula: $C_{19}H_{26}O_4S$
DOT ID Number: UN 2902; DOT Guide Number: 151
Proper Shipping Name: Insecticide, liquid, poisonous, n.o.s.
Reportable Quantity (RQ): **(CERCLA)** 10 lb/4.54 kg

DESCRIPTION: Dark amber, thick, oily liquid. Odorless. Sinks in water; practically insoluble.

Highly flammable • Poison! (sulfur-containing insecticide) • Skin or eye contact or swallowing the material can cause illness and possible death • Firefighting gear (including SCBA) does not provide adequate protection. If exposure occurs, remove and isolate gear immediately and thoroughly decontaminate personnel • Containers may BLEVE when exposed to fire •Vapors may form explosive mixture with air • Vapors are heavier than air and will collect and stay in low areas • Vapors may travel long distances to ignition sources and flashback • Vapors in confined areas (e.g., tanks, sewers, buildings) may explode when exposed to fire • Irritating to the skin, eyes, and respiratory tract • Toxic products of combustion may include sulfur oxides.

Hazard Classification (based on NFPA-704 Rating System)
Health Hazards (Blue): 2; Flammability (Red): 3; Reactivity (Yellow): 1

EMERGENCY RESPONSE: See Appendix A (151)
Evacuation:
Public safety: Isolate the area of spill or leak for at least 25 to 50 meters (80 to 160 feet) in all directions.
Spill: Increase, in the downwind direction, as necessary, the distance shown under "Public Safety."
Fire: If tank, rail car, or tank truck is involved in fire, isolate for at least 800 meters (½ mile) in all directions; also, consider initial evacuation for 800 meters (½ mile) in all directions.

EXPOSURE
Short-term effects: *SEEK MEDICAL ATTENTION.* **Liquid:** Irritating to skin and eyes. Harmful if swallowed. *POISONOUS IF INHALED.* Move to fresh air. *IF BREATHING HAS STOPPED,* give artificial respiration. IF breathing is difficult, administer oxygen. Remove contaminated clothing and shoes. Flush affected areas with plenty of water. *IF IN EYES*, hold eyelids open and flush with plenty of water. *IF SWALLOWED* and victim is *CONSCIOUS AND ABLE TO SWALLOW*, have victim drink 4 to 8 ounces of water.

HEALTH HAZARDS
Personal protective equipment (PPE): B-Level PPE. Rubber gloves and boots, safety goggles or face mask, hooded suit and either a respirator with approved canister or a self-contained breathing apparatus.
Toxicity by ingestion: Grade 2; LD_{50} = 0.5 to 5 g/kg.
Vapor (gas) irritant characteristics: eye and respiratory tract irritant.
Liquid or solid irritant characteristics: Severe eye and skin irritant. If spilled on clothing and allowed to remain may cause smarting and reddening of skin.

FIRE DATA
Flash point: 82°F/28°C (oc).
Behavior in fire: Containers may rupture.

ENVIRONMENTAL DATA
Water pollution: Harmful to aquatic life in very low concentrations. May be dangerous if it enters nearby water intakes; notify operators. Notify local health and wildlife officials.
Response to discharge: Issue warning–water pollutant. Restrict access. Chemical and physical treatment.

SHIPPING INFORMATION
Grades of purity: Technical (at least 80%); 30% wettable powder; 4% dust

PHYSICAL AND CHEMICAL PROPERTIES
Physical state @ 59°F/15°C and 1 atm: Liquid.
Molecular weight: 350.472
Boiling point @ 1 atm: Decomposes.
Specific gravity (water = 1): 1.085-1.115 @ 68°F/20°C.

PROPARGYL ALCOHOL   **REC. P:3550**

SYNONYMS: ALCOHOL PROPARGILICO (Spanish); ETHYNYLCARBINOL; ETHYLMETHANOL; METHANOL, ETHYNYL; 1-PROPYNE-3-OL; 2-PROPYN-1-OL; 3-PROPYNOL; 2-PROPYNYL ALCOHOL; PROPIOLIC ALCOHOL; ETHYNYL CARBINOL; RCRA No. P102

IDENTIFICATION
CAS Number: 107-19-7
Formula: $HCCCH_2OH$
DOT ID Number: UN 1986; DOT Guide Number: 131
Proper Shipping Name: Propargyl alcohol
Reportable Quantity (RQ): **(CERCLA)** 1000 lb/454 kg

DESCRIPTION: Colorless to light yellow liquid. Mild, geranium-like odor. Initially floats on water surface; soluble.

Highly flammable • Poison! • Breathing the vapor, skin or eye contact, or swallowing the material can kill you • Firefighting gear (including SCBA) does not provide adequate protection. If exposure occurs, remove and isolate gear immediately and thoroughly decontaminate personnel •Polymerization hazard; May react spontaneously without warning with explosive violence • Containers may BLEVE when exposed to fire •Vapors may form explosive mixture with air • Vapors are heavier than air and will collect and stay in low areas • Vapors may travel long distances to ignition sources and flashback • Vapors in confined areas (e.g., tanks, sewers, buildings) may explode when exposed to fire • Irritating to the skin, eyes, and respiratory tract • Toxic products of combustion may include CO_2 and carbon monoxide • Do not put yourself in danger by entering a contaminated area to rescue a victim.

Hazard Classification (based on NFPA-704 Rating System)
Health Hazards (Blue): 4; Flammability (Red): 3; Reactivity (Yellow): 3

EMERGENCY RESPONSE: See Appendix A (131)
Evacuation:
Public safety: Isolate spill area for at least 100 to 200 meters (330 to 660 feet) in all directions.
Spill: Increase, as necessary, the isolation distance shown above,in "Public safety."
Fire: Isolate for 800 meters (½ mile) in all directions, especially if tank, rail car, or tank truck is involved in fire.

EXPOSURE
Short-term effects: *SEEK MEDICAL ATTENTION.* **Vapor:** May be harmful or fatal if inhaled or absorbed through the skin. Causes severe irritation of the eyes, nose, and throat. Move victim to fresh air. *IF BREATHING HAS STOPPED,* give artificial respiration. IF breathing is difficult, administer oxygen. **Liquid:** *MAY BE HARMFUL OR FATAL IF SWALLOWED OR ABSORBED THROUGH THE SKIN.* Extremely irritating to the eyes and skin. Remove contaminated clothing and shoes. Flush affected areas with plenty of running water. *IF IN EYES*, flush with running water for at least 15 minutes. *IF SWALLOWED,***Do NOT induce vomiting**.

HEALTH HAZARDS
Personal protective equipment (PPE): A-Level PPE. Approved organic vapor respirator, chemical resistant gloves, safety goggles, full protective clothing. **Natural rubber, neoprene, and nitrile** rubber may offer some protection from alcohols.
Exposure limits (TWA unless otherwise noted): ACGIH TLV 1 ppm (2.3 mg/m^3); NIOSH REL 1 ppm (2 mg/m^3); skin contact contributes significantly in overall exposure.
Toxicity by ingestion: Grade 4: LD_{50} = 20 mg/kg (rat).
Long-term health effects: Central nervous system depressant.
Vapor (gas) irritant characteristics: Severe eye and respiratory tract irritant. Vapors cause severe irritation of eyes and throat and can cause eye and lung injury. They cannot be tolerated even at low concentrations.
Liquid or solid irritant characteristics: Severe skin irritant. May cause pain and second-degree burns after a few minutes of contact.

β-Propiolactone

FIRE DATA
Flash point: 97°F/36°C (oc).
Fire extinguishing agents not to be used: Water may be ineffective against fire.
Behavior in fire: Containers may rupture and explode.
Electrical hazard: Class I. Group unassigned.

CHEMICAL REACTIVITY
Binary reactants: Reacts with oxidizers, phosphorus pentoxide, caustic materials.
Polymerization: May undergo autopolymerization from heat, sunlight, oxidizers, peroxides, or caustics.

ENVIRONMENTAL DATA
Water pollution: Effects of low concentrations on aquatic life are not known. Fouling to shoreline. May be dangerous if it enters local water intakes. Notify local health and wildlife officials.
Response to discharge: Issue warning–high flammability. Evacuate area. Should be removed. Chemical and physical treatment.

SHIPPING INFORMATION
Grades of purity: 99%; **Storage temperature:** Refrigerate; **Venting:** None; **Stability during transport:** Stable.

NAS HAZARD CLASSIFICATION FOR BULK WATER TRANSPORTATION
FIRE: 3
HEALTH: Vapor irritant: 4; Liquid or solid irritant: 3; Poisons: 4
WATER POLLUTION: Human toxicity: 4; Aquatic toxicity: 3; Aesthetic effect: 1
REACTIVITY: Other chemicals: 3; Water: 1; Self-reaction: 2

PHYSICAL AND CHEMICAL PROPERTIES
Physical state @ 59°F/15°C and 1 atm: Liquid.
Molecular weight: 56.07
Boiling point @ 1 atm: 237°F/114°C/387°K.
Melting/Freezing point: –54°F/–48°C/225°K.
Specific gravity (water = 1): 0.9485 @ 68°F/20°C.
Liquid surface tension: 36 dyne/cm = 0.036 N/m @ 68°F/20°C.
Relative vapor density (air = 1): 1.93
Vapor pressure: 12 mm.

β-PROPIOLACTONE   REC. P:3600

SYNONYMS: BPL; EEC No. 606-031-00-1; BETRAPRONE; HYDRACRYLIC ACID, β-LACTONE; 3-HYDROXY-β-PROPIOLACTONE; 2-OXETANONE; β-PROPIOLACTONA (Spanish); 3-PROPIONOLACTONE; PROPANOLIDE

IDENTIFICATION
CAS Number: 57-57-8
Formula: $C_3H_4O_2$; OCH_2CH_2CO
DOT ID Number: NA 1993; DOT Guide Number: 128
Proper Shipping Name: Combustible liquid, n.o.s.
Reportable Quantity (RQ): **(CERCLA)** 1 lb/0.454 kg

DESCRIPTION: Colorless liquid. Pungent odor Irritating odor (warning unknown); smells like acrylic. Mixes with water slowly forming a solution of 3-hydroxypropionic acid (a form of lactic acid).

Combustible • Corrosive to the skin, eyes, and respiratory tract; contact can cause burns and impaired vision; inhalation symptoms can be delayed • Firefighting gear (including SCBA) does not provide adequate protection. If exposure occurs, remove and isolate gear immediately and thoroughly decontaminate personnel •Polymerization hazard; heat can induce polymerization with rapid release of energy; sealed containers may rupture explosively • May react with itself blocking relief valves; leading to tank explosions • Containers may BLEVE when exposed to fire • Vapors are heavier than air and will collect and stay in low areas • Vapors in confined areas (e.g., tanks, sewers, buildings) may explode when exposed to fire • Toxic products of combustion may include unburned vapors, of this material, CO_2 and carbon monoxide • Do not put yourself in danger by entering a contaminated area to rescue a victim.

Hazard Classification (based on NFPA-704 Rating System)
Health Hazards (Blue): 0; Flammability (Red): 2; Reactivity (Yellow): 0

EMERGENCY RESPONSE: See Appendix A (128)
Evacuation:
Public safety: Isolate spill area for at least 25 to 50 meters (80 to 160 feet) in all directions.
Spill: Large spill–Consider initial downwind evacuation for at least 300 meters (1000 feet).
Fire: Isolate for 800 meters (½ mile) in all directions, especially if tank, rail car, or tank truck is involved in fire.

EXPOSURE
Short-term effects: *SEEK MEDICAL ATTENTION.* **Vapors:** Lung edema may develop. **Liquid:** *POISONOUS IF SWALLOWED OR IF SKIN IS EXPOSED.* Irritating to skin and eyes. Remove contaminated clothing and shoes. Flush affected areas with plenty of water. *IF IN EYES,* hold eyelids open and flush with plenty of water. *IF SWALLOWED* and victim is *CONSCIOUS AND ABLE TO SWALLOW*, have victim drink 4 to 8 ounces of water and have victim induce vomiting. *IF SWALLOWED* and victim is *UNCONSCIOUS OR HAVING CONVULSIONS,* do nothing except keep victim warm.
Note to physician or authorized medical personnel: Medical observation is recommended for 24 to 48 hours after breathing overexposure, as pulmonary edema may be delayed. As first aid for pulmonary edema, consider administering a corticosteroid spray. Cigarette smoking may exacerbate pulmonary injury and should be discouraged for at least 72 hours following exposure.

HEALTH HAZARDS
Personal protective equipment (PPE): B-Level PPE. Air mask or organic canister mask; goggles or face shield; gloves; protective clothing to prevent all contact with skin. Chemical protective material(s) reported to have good to excellent resistance: butyl rubber.
Recommendations for respirator selection: NIOSH
At any concentrations above the NIOSH REL, or where there is no REL, at any detectable concentration: SCBAF:PD,PP (any self-contained breathing apparatus that has a full facepiece and is operated in a pressure-demand or other positive-pressure mode); or SAF:PD,PP:ASCBA (any supplied-air respirator that has a full facepiece and is operated in a pressure-demand or other positive-pressure mode in combination with an auxiliary self-contained breathing apparatus operated in a pressure-demand or other positive pressure mode). *ESCAPE:* GMFOV [any air-purifying, full-facepiece respirator (gas mask) with a chin-style, front-or back-mounted organic vapor canister]; or SCBAE (any appropriate escape-type, self-contained breathing apparatus).

Exposure limits (TWA unless otherwise noted): ACGIH TLV 0.5 ppm (1.5 mg/m^3) suspected human carcinogen; OSHA PEL [1910.1013]; suspected human carcinogen.
Toxicity by ingestion: Grade 3; oral LD_{LO} = 50 mg/kg (rat).
Long-term health effects: Confirmed carcinogen. Because of the high incidence of cancer in animals, no exposure or contact by any route-respiratory, oral, or skin-should be permitted. Repeated or prolonged exposure may cause dermal problems.
Vapor (gas) irritant characteristics: Eye and respiratory tract irritant. Vapors are moderately irritating such that personnel will not usually tolerate moderate or high concentrations.
Liquid or solid irritant characteristics: Severe skin irritant. May cause pain and second-degree burns after a few minutes of contact.

FIRE DATA
Flash point: 165°F/73°C (cc).
Behavior in fire: Containers may explode.

CHEMICAL REACTIVITY
Reactivity with water: Slow, nonhazardous reaction to form β-hydroxypropionic acid.
Binary reactants: Acetates, halogens, thiocyanates, thiosulfates.
Polymerization: Can polymerize and rupture container, especially at elevated temperatures. At room temperature (71°F/22°C), 0.04% polymerizes each day.

ENVIRONMENTAL DATA
Food chain concentration potential: Negative; unlikely to accumulate.
Water pollution: Effect of low concentrations on aquatic life is unknown. May be dangerous if it enters nearby water intakes; notify operators. Notify local health and wildlife officials.
Response to discharge: Issue warning–poison, water contaminant. Restrict access. Disperse and flush.

SHIPPING INFORMATION
Grades of purity: 97+%; **Storage temperature:** Below 60°F/15°C; **Inert atmosphere:** None; **Venting:** Pressure-vacuum; **Stability during transport:** Stable.

NAS HAZARD CLASSIFICATION FOR BULK WATER TRANSPORTATION
FIRE: 1
HEALTH: Vapor irritant: 3; Liquid or solid irritant: 3; Poisons: 3
WATER POLLUTION: Human toxicity: 3; Aquatic toxicity: 2; Aesthetic effect: 1
REACTIVITY: Other chemicals: 2; Water: 0; Self-reaction: 1

PHYSICAL AND CHEMICAL PROPERTIES
Physical state @ 59°F/15°C and 1 atm: Liquid.
Molecular weight: 72.1
Boiling point @ 1 atm: 323°F/162°C/435°K (decomposes).
Melting/Freezing point: –28.1°F/–33.4°C/239.8°K.
Specific gravity (water = 1): 1.148 @ 68°F/20°C (liquid).
Liquid surface tension: (estimate) 22 dynes/cm = 0.022 N/m @ 68°F/20°C.
Liquid water interfacial tension: (estimate) 25 dynes/cm = 0.025 N/m @ 68°F/20°C.
Relative vapor density (air = 1): 2.51
Ratio of specific heats of vapor (gas): 1.1089
Heat of combustion: –8510 Btu/lb = –4730 cal/g = –198 x 10^5 J/kg.
Vapor pressure: 2 mm.

PROPIONALDEHYDE REC. P:3650

SYNONYMS: ALDEHYDE PROPIONIQUE (French); EEC No. 605-018-00-8; METHYLACETALDEHYDE; NCI-C61029; PROPALDEHYDE; PROPANAL; PROPIONIC ALDEHYDE; PROPIONALDEHIDO (Spanish); PROPYL ALDEHYDE; PROPYLIC ALDEHYDE

IDENTIFICATION
CAS Number: 123-38-6
Formula: C_3H_6O; CH_3CH_2CHO
DOT ID Number: UN 1275; DOT Guide Number: 129
Proper Shipping Name: Propionaldehyde
Reportable Quantity (RQ): **(CERCLA)** 1 lb/0.454 kg

DESCRIPTION: Colorless liquid. Fruity, suffocating odor. Floats and mixes slowly with water. Flammable, irritating vapor is produced.

Highly flammable • Corrosive to the eyes and respiratory tract; skin or eye contact can cause burns and impaired vision; inhalation symptoms may be delayed • Firefighting gear (including SCBA) does not provide adequate protection. If exposure occurs, remove and isolate gear immediately and thoroughly decontaminate personnel • Containers may BLEVE when exposed to fire •Vapors may form explosive mixture with air • Vapors are heavier than air and will collect and stay in low areas • Vapors may travel long distances to ignition sources and flashback • Vapors in confined areas (e.g., tanks, sewers, buildings) may explode when exposed to fire • Toxic products of combustion may include CO_2 and carbon monoxide • Do not put yourself in danger by entering a contaminated area to rescue a victim.

Hazard Classification (based on NFPA-704 Rating System)
Health Hazards (Blue): 2; Flammability (Red): 3; Reactivity (Yellow): 2

EMERGENCY RESPONSE: See Appendix A (129)
Evacuation:
Public safety: Isolate spill area for at least 50 to 100 meters (160 to 330 feet) in all directions.
Spill: Large spill–Consider initial downwind evacuation for at least 300 meters (1000 feet).
Fire: Isolate for 800 meters (½ mile) in all directions, especially if tank, rail car, or tank truck is involved in fire.

EXPOSURE
Short-term effects: *SEEK MEDICAL ATTENTION.* **Vapor:** Irritating to eyes, nose, and throat. *IF INHALED*, will, will cause nausea or vomiting. Move to fresh air. *IF BREATHING HAS STOPPED,* give artificial respiration. IF breathing is difficult, administer oxygen. **Liquid:** Irritating to skin and eyes. Harmful if swallowed. Remove contaminated clothing and shoes. Flush affected areas with plenty of water. *IF IN EYES*, hold eyelids open and flush with plenty of water. *IF SWALLOWED* and victim is *CONSCIOUS AND ABLE TO SWALLOW*, have victim drink 4 to 8 ounces of water.
Note to physician or authorized medical personnel: Medical observation is recommended for 24 to 48 hours after breathing overexposure, as pulmonary edema may be delayed. As first aid for pulmonary edema, consider administering a corticosteroid spray. Cigarette smoking may exacerbate pulmonary injury and should be discouraged for at least 72 hours following exposure.

HEALTH HAZARDS
Personal protective equipment (PPE): B-Level PPE. Air–supplied mask for high vapor concentrations; plastic gloves; goggles. Chemical protective material(s) reported to have good to excellent resistance: butyl rubber.
Toxicity by ingestion: Grade 2; $LD_{50} = 0.5$ to 5 g/kg (rat).
Long-term health effects: Liver damage.
Vapor (gas) irritant characteristics: Eye and respiratory tract irritant. Vapors cause irritation, such that personnel will find high concentrations unpleasant. The effect is temporary.
Liquid or solid irritant characteristics: Severe eye irritant. Skin irritant. If spilled on clothing and allowed to remain, may cause smarting and reddening of the skin.
Odor threshold: 1 ppm.

FIRE DATA
Flash point: –22°F/–30°C (oc).
Flammable limits in air: LEL: 2.2%; UEL: 21.1%.
Fire extinguishing agents not to be used: Water may be ineffective.
Autoignition temperature: 405°F/207°C/480°K.
Burning rate: 4.4 mm/min

CHEMICAL REACTIVITY
Binary reactants: Can ignite spontaneously when finely dispersed on porous materials. Reacts with oxidizers; strong acids, caustics, isocyantes, methyl methacrylate, amines.
Polymerization: Violent on contact with methyl methacrylate; may occur in presence of acids, caustics.
Reactivity group: 19
Compatibility class: Aldehyde

ENVIRONMENTAL DATA
Food chain concentration potential: Log $P_{ow} = 0.39$. Unlikely to accumulate.
Water pollution: DOT Appendix B, §172.101–marine pollutant. Effect of low concentrations on aquatic life is unknown. May be dangerous if it enters nearby water intakes; notify operators. Notify local health and wildlife officials. **Response to discharge:** Issue warning–high flammability. Evacuate area.

SHIPPING INFORMATION
Grades of purity: 97–99+%; **Storage temperature:** Ambient; **Inert atmosphere:** None; **Venting:** Open (flame arrester) or pressure-vacuum; **Stability during transport:** Stable.

NAS HAZARD CLASSIFICATION FOR BULK WATER TRANSPORTATION
FIRE: 3
HEALTH: Vapor irritant: 2; Liquid or solid irritant: 1; Poisons: 2
WATER POLLUTION: Human toxicity: 2; Aquatic toxicity: 2; Aesthetic effect: 2
REACTIVITY: Other chemicals: 2; Water: 0; Self-reaction: 1

PHYSICAL AND CHEMICAL PROPERTIES
Physical state @ 59°F/15°C and 1 atm: Liquid.
Molecular weight: 58.08
Boiling point @ 1 atm: 118.4°F/48.0°C/321.2°K.
Melting/Freezing point: –112°F/–80°C/193°K.
Critical temperature: 433°F/223°C/496°K.
Critical pressure: 690 psia = 47 atm = 4.8 MN/m^2.
Specific gravity (water = 1): 0.805 @ 68°F/20°C (liquid).
Liquid surface tension: 23.4 dynes/cm = 0.0234 N/m @ 68°F/20°C.
Liquid water interfacial tension: 29 dynes/cm = 0.029 N/m at 22.7°C.
Relative vapor density (air = 1): 2.0
Ratio of specific heats of vapor (gas): 1.120
Latent heat of vaporization: 211 Btu/lb = 117 cal/g = 4.9 x 10^5 J/kg.
Heat of combustion: –12,470 Btu/lb = –6930 cal/g = –290.1 x 10^5 J/kg.
Heat of solution: (estimate) –9 Btu/lb = –5 cal/g = 0.2 x 10^5 J/kg.
Vapor pressure: (Reid) 6.7 psia; 260 mm.

PROPIONIC ACID   REC. P:3700

SYNONYMS: ACIDO PROPRIONICO (Spanish); ACIDE PROPIONIQUE (French); CARBONYETHANE; CARBOXYETHANE; EEC No. 607-089-00-0; ETHANE CARBOXYLIC ACID; ETHYLFORMIC ACID; METACETONIC ACID; METHYLACETIC ACID; PROPANOIC ACID; PROPIONIC ACID GRAIN PRESERVER; PROZOIN; PSEUDOACETIC ACID; SENTRY GRAIN PRESERVER; TENOX P GRAIN PRESERVATIVE

IDENTIFICATION
CAS Number: 79-09-4
Formula: $C_3H_6O_2$; CH_3CH_2COOH
DOT ID Number: UN 1848; DOT Guide Number: 132
Proper Shipping Name: Propionic acid
Reportable Quantity (RQ): **(CERCLA)** 5000 lb/2270 kg

DESCRIPTION: Colorless, oily liquid. Sharp, rancid odor. Initially floats on the surface of water; soluble. Irritating vapor is produced.

Very flammable • Corrosive to the skin, eyes, and respiratory tract; skin or eye contact can cause severe burns, impaired vision, and blindness • Containers may BLEVE when exposed to fire • Vapors may form explosive mixture with air • Vapors are heavier than air and will collect and stay in low areas • Vapors may travel long distances to ignition sources and flashback • Vapors in confined areas (e.g., tanks, sewers, buildings) may explode when exposed to fire • Toxic products of combustion may include smoke, CO_2, and carbon monoxide vapors.

Hazard Classification (based on NFPA-704 Rating System)
Health Hazards (Blue): 3; Flammability (Red): 2; Reactivity (Yellow): 0

EMERGENCY RESPONSE: See Appendix A (132)
Evacuation:
Public safety: Isolate spill area for at least 50 to 100 meters (160 to 330 feet) in all directions.
Spill: Increase, as necessary, the isolation distance shown above, in "Public safety."
Fire: Isolate for 800 meters (½ mile) in all directions, especially if tank, rail car, or tank truck is involved in fire.

EXPOSURE
Short-term effects: *SEEK MEDICAL ATTENTION*. **Vapor:** Irritating to eyes, nose, and throat. Move to fresh air. **Liquid:** Will burn skin and eyes. Harmful if swallowed. Remove contaminated clothing and shoes. Flush affected areas with plenty of water. *IF IN EYES*, hold eyelids open and flush with plenty of water.

IF SWALLOWED and victim is CONSCIOUS AND ABLE TO SWALLOW, have victim drink 4 to 8 ounces of water. **Do NOT induce vomiting.**

HEALTH HAZARDS
Personal protective equipment (PPE): B-Level PPE. Air–supplied mask for high vapor concentrations; plastic gloves; goggles or face shield. Chemical protective material(s) reported to have good to excellent resistance: Teflon®. Also, butyl rubber is generally suitable for carbooxylic acid compounds.
Exposure limits (TWA unless otherwise noted): ACGIH TLV 10 ppm (30 mg/m^3); NIOSH REL 10 ppm (30 mg/m^3).
Short-term exposure limits (15-minute TWA): NIOSH STEL 15 ppm (45 mg/m^3)
Toxicity by ingestion: Grade 2; oral rat LD$_{50}$ = 2.6 g/kg.
Vapor (gas) irritant characteristics: Vapors cause irritation such that personnel will find high concentrations unpleasant. The effect is temporary.
Liquid or solid irritant characteristics: Severe skin irritant; may cause pain and second-degree burns after a few minutes of contact.
IDLH value: 0.02–0.2 ppm.

FIRE DATA
Flash point: 134°F/57°C (oc); 126°F/52°C (cc).
Flammable limits in air: LEL: 2.9%; UEL 12.1%.
Autoignition temperature: 864°F/462°C/735°K.
Electrical hazard: Class I, Group D.
Burning rate: 2.2 mm/min

CHEMICAL REACTIVITY
Binary reactants: Alkalis, strong oxidizers (e.g., chromium trioxide), amines, acids and bases. Corrodes ordinary steel and many other metals forming flammable hydrogen gas.
Neutralizing agents for acids and caustics: Dilute with water, then neutralize with lime or soda ash.
Reactivity group: 4
Compatibility class: Organic acids

ENVIRONMENTAL DATA
Food chain concentration potential: Log P$_{ow}$ = 0.3. Unlikely to accumulate.
Water pollution: Dangerous to aquatic life in high concentrations. May be dangerous if it enters nearby water intakes; notify operators. Notify local health and wildlife officials. **Response to discharge:** Disperse and flush.

SHIPPING INFORMATION
Grades of purity: 99+%; **Storage temperature:** Ambient; **Inert atmosphere:** None; **Venting:** Open (flame arrester); **Stability during transport:** Stable.

NAS HAZARD CLASSIFICATION FOR BULK WATER TRANSPORTATION
FIRE: 2
HEALTH: Vapor irritant: 2; Liquid or solid irritant: 3; Poisons: 2
WATER POLLUTION: Human toxicity: 2; Aquatic toxicity: 2; Aesthetic effect: 2
REACTIVITY: Other chemicals: 2; Water: 0; Self-reaction: 0

PHYSICAL AND CHEMICAL PROPERTIES
Physical state @ 59°F/15°C and 1 atm: Liquid.
Molecular weight: 74.08
Boiling point @ 1 atm: 285.4°F/140.8°C/414.0°K.
Melting/Freezing point: −5.3°F/−20.7°C/252.5°K.
Critical temperature: 642°F/339°C/612°K.
Critical pressure: 779 psia = 53 atm = 5.37 MN/m^2.
Specific gravity (water = 1): 0.995 @ 68°F/20°C (liquid).
Liquid surface tension: 26.2 dynes/cm = 0.0262 N/m @ 77°F/25°C.
Ratio of specific heats of vapor (gas): 1.103
Latent heat of vaporization: 248 Btu/lb = 138 cal/g = 5.78 x 10^5 J/kg.
Heat of combustion: −8883 Btu/lb = −4935 cal/g = 206.6 x 10^5 J/kg.
Vapor pressure: (Reid) 0.2 psia; 2.5 mm.

PROPIONIC ANHYDRIDE REC. P:3750

SYNONYMS: ANHIDRIDO PROPIONICO (Spanish); EEC No. 607-010-00-x; METHYLACETIC ACID ANHYDRIDE; PROPIONIC ACID ANHYDRIDE; PROPIONYL OXIDE

IDENTIFICATION
CAS Number: 123-62-6
Formula: C$_6$H$_{10}$O$_3$; (CH$_3$CH$_2$CO)$_2$O
DOT ID Number: UN 2496; DOT Guide Number: 156
Proper Shipping Name: Propionic anhydride
Reportable Quantity (RQ): **(CERCLA)** 5000 lb/2270 kg

DESCRIPTION: Colorless liquid. Sharp odor. Sinks and mixes slowly with water, forming propionic acid.

Combustible • Corrosive • Breathing the vapor can kill you; skin or eye contact causes severe burns, impaired vision, or blindness; effects of contact or inhalation may be delayed • Firefighting gear (including SCBA) does not provide adequate protection. If exposure occurs, remove and isolate gear immediately and thoroughly decontaminate personnel • Containers may BLEVE when exposed to fire • Vapors in confined areas (e.g., tanks, sewers, buildings) may explode when exposed to fire • Toxic products of combustion may include smoke, CO$_2$, and carbon monoxide.

Hazard Classification (based on NFPA-704 Rating System)
Health Hazards (Blue): 3; Flammability (Red): 2; Reactivity (Yellow): 1

EMERGENCY RESPONSE: See Appendix A (156)
Evacuation:
Public safety: Isolate the area of spill or leak for at least 50 to 100 meters (160 to 330 feet) in all directions.
Spill: Increase, in the downwind direction, as necessary, the distance shown under "Public Safety."
Fire: If tank, rail car, or tank truck is involved in fire, isolate for at least 800 meters (½ mile) in all directions; also, consider initial evacuation for 800 meters (½ mile) in all directions.
Evacuation:
Public safety: Isolate the area of spill or leak for at least 50 to 100 meters (160 to 330 feet) in all directions.
Spill: Increase, in the downwind direction, as necessary, the distance shown under "Public Safety."
Fire: If tank, rail car, or tank truck is involved in fire, isolate for at least 800 meters (½ mile) in all directions; also, consider initial evacuation for 800 meters (½ mile) in all directions.

EXPOSURE
Short-term effects: *SEEK MEDICAL ATTENTION.* **Vapor:** *IF BREATHING HAS STOPPED,* give artificial respiration; *avoid mouth-to-mouth resuscitation; use bag/mask apparatus.* **Liquid:** Will burn skin and eyes. Harmful if swallowed. Remove

contaminated clothing and shoes. Flush affected areas with plenty of water. *IF IN EYES*, hold eyelids open and flush with plenty of water. *IF SWALLOWED* and victim is *CONSCIOUS AND ABLE TO SWALLOW*, have victim drink 4 to 8 ounces of water. **Do NOT induce vomiting**.

Note to physician or authorized medical personnel: Medical observation is recommended for 24 to 48 hours after breathing overexposure, as pulmonary edema may be delayed. As first aid for pulmonary edema, consider administering a corticosteroid spray. Cigarette smoking may exacerbate pulmonary injury and should be discouraged for at least 72 hours following exposure.

HEALTH HAZARDS
Personal protective equipment (PPE): B-Level PPE. Organic canister mask; goggles or face shield; rubber gloves.
Exposure limits (TWA unless otherwise noted): ACGIH TLV 10 ppm (30 mg/m^3) as propionic acid; NIOSH 10 ppm (30 mg/m^3) as propionic acid.
Short-term exposure limits (15-minute TWA): NIOSH STEL 15 ppm (45 mg/m^3) as propionic acid.
Toxicity by ingestion: Grade 1; LD$_{50}$ = 5 to 15 g/kg; Grade 2; LD$_{50}$ = 0.5 to 5 g/kg.
Vapor (gas) irritant characteristics: Vapors are irritating such that personnel will not usually tolerate moderate or high concentrations.
Liquid or solid irritant characteristics: Severe eye and skin irritant. Causes smarting of the skin and first-degree burns on short exposure; may cause second-degree burns on long exposure.

FIRE DATA
Flash point: 156°F/69°C (oc); 145°F/63°C (cc).
Flammable limits in air: LEL: 1.48%; UEL: 11.9%.
Fire extinguishing agents not to be used: Water may cause a hazardous reaction unless used in large quantities to fully absorb the heat of reaction.
Autoignition temperature: 547°F/286°C/559°K.
Electrical hazard: Class I, Group D.
Burning rate: 3.0 mm/min

CHEMICAL REACTIVITY
Reactivity with water: Reacts to form weak propionic acid.
Binary reactants: Oxidizers, acids and bases. Slowly corrosive in the presence of moisture.
Neutralizing agents for acids and caustics: Flush with water, rinse with sodium bicarbonate or lime solution.
Reactivity group: 11
Compatibility class: Organic anhydrides

ENVIRONMENTAL DATA
Food chain concentration potential: Negative; unlikely to accumulate.
Water pollution: Dangerous to aquatic life in high concentrations. May be dangerous if it enters nearby water intakes; notify operators. Notify local health and wildlife officials. **Response to discharge:** Issue warning–corrosive. Restrict access. Disperse and flush.

SHIPPING INFORMATION
Grades of purity: 97+%; **Storage temperature:** Ambient; **Inert atmosphere:** None; **Venting:** Pressure-vacuum; **Stability during transport:** Stable.

NAS HAZARD CLASSIFICATION FOR BULK WATER TRANSPORTATION
FIRE: 1
HEALTH: Vapor irritant: 3; Liquid or solid irritant: 2; Poisons: 1
WATER POLLUTION: Human toxicity: 1; Aquatic toxicity: 3; Aesthetic effect: 2
REACTIVITY: Other chemicals: 3; Water: 1; Self-reaction: 0

PHYSICAL AND CHEMICAL PROPERTIES
Physical state @ 59°F/15°C and 1 atm: Liquid.
Boiling point @ 1 atm: 336°F/169°C/442°K.
Melting/Freezing point: –45°F/–43°C/230°K.
Critical temperature: 660°F/349°C/622°K.
Critical pressure: 490 psia = 33 atm = 3.3 MN/m^2.
Specific gravity (water = 1): 1.01 @ 68°F/20°C (liquid).
Liquid surface tension: 30 dynes/cm = 0.030 N/m @ 77°F/25°C.
Relative vapor density (air = 1): 4.5
Ratio of specific heats of vapor (gas): 1.0543
Latent heat of vaporization: 149 Btu/lb = 83 cal/g = 3.5 x 10^5 J/kg.
Heat of combustion: (@ 15°C) –10,320 Btu/lb = –5,740 cal/g = –240 x 10^5 J/kg.
Heat of solution: (estimate) –36 Btu/lb = –20 cal/g = –0.84 x 10^5 J/kg.
Vapor pressure: (Reid) Low; 1 mm.

PROPIONITRILE   REC. P:3800

SYNONYMS: CIANURO de ETILO (Spanish); CYANOETHANE; ETHER CYANATUS; ETHER CYANIDE; ETHYL CYANIDE; HYDROCYANIC ETHER; PROPANENITRILE; PROPIONIC NITRILE; PROPYLNITRILE; RCRA No. P101

IDENTIFICATION
CAS Number: 107-12-0
Formula: C$_2$H$_5$CN
DOT ID Number: UN 2404; DOT Guide Number: 131
Proper Shipping Name: Propionitrile
Reportable Quantity (RQ): **(CERCLA)** 10 lb/4.54 kg

DESCRIPTION: Colorless to brown liquid. Sweet, ether-like odor. Floats on water surface; moderately soluble.

Highly flammable • Poison! • Breathing the vapor, skin or eye contact, or swallowing the material can kill you; converted to cyanide in the body • Firefighting gear (including SCBA) does not provide adequate protection. If exposure occurs, remove and isolate gear immediately and thoroughly decontaminate personnel • Containers may BLEVE when exposed to fire •Vapors may form explosive mixture with air • Vapors are heavier than air and will collect and stay in low areas • Vapors may travel long distances to ignition sources and flashback • Vapors in confined areas (e.g., tanks, sewers, buildings) may explode when exposed to fire • Irritating to the skin, eyes, and respiratory tract • Toxic products of combustion may include hydrogen cyanide, carbon monoxide, and nitrogen oxides • Do not put yourself in danger by entering a contaminated area to rescue a victim.

Hazard Classification (based on NFPA-704 Rating System)
Health Hazards (Blue): 4; Flammability (Red): 3; Reactivity (Yellow): 1

EMERGENCY RESPONSE: See Appendix A (131)
Evacuation:
Public safety: Isolate spill area for at least 100 to 200 meters (330 to 660 feet) in all directions.

Spill: Increase, as necessary, the isolation distance shown above, in "Public safety."
Fire: Isolate for 800 meters (½ mile) in all directions, especially if tank, rail car, or tank truck is involved in fire.

EXPOSURE
Short-term effects: *SEEK MEDICAL ATTENTION.* **Vapor:** Irritating to eyes, nose, and throat. Harmful if inhaled. *IF INHALED*, will, move victim to fresh air. If not breathing, give artificial respiration; *avoid mouth-to-mouth resuscitation; use bag/mask apparatus*. IF breathing is difficult, administer oxygen. Toxic effects can be delayed. **Liquid:** Irritating to the skin and eyes. Harmful if swallowed or skin is exposed. Remove and double bag contaminated clothing and shoes at the site.
Flush affected areas immediately with plenty of water. *IF IN EYES*, hold eyelids open and flush with plenty of water. *IF SWALLOWED* and victim is *CONSCIOUS AND ABLE TO SWALLOW*, have victim drink 4 to 8 ounces of water and victim induce vomiting. *IF SWALLOWED* and victim is *UNCONSCIOUS OR HAVING CONVULSIONS*, do nothing except keep victim warm.
Effects may be delayed; keep victim under observation.
Note to physician or authorized medical personnel: Consider the use of amyl nitrite perles if symptoms of cyanide poisoning develop. If symptoms indicate, initial treatment includes the cyanide antidote kit. In all cases, break an amyl nitrite perle in a gauze pad and hold lightly under victim's nose for 15 seconds, repeating 5 times at about 15-second intervals; if necessary (and if sodium nitrite infusions will be delayed), repeat procedure every 3 minutes with fresh pearls until 3 or 4 have been used. Avoid breathing the vapor while administering it to the victim. Administer sodium nitrite IV, ASAP. The usual adult dose is 10 to 20 mL of a 3% solution infused over no less than 5 minutes; the average child dose is 0.15 to 0.20 mL/kg. Monitor blood pressure during administration, and slow the rate of infusion if hypotention develops. Next, infuse sodium thiosulfate IV. The usual adult dose is 50 mL of a 25% solution infused over 10 to 20 minutes; the average child dose is 1.65 mL/kg. Repeat with nitrite and thiosulfate as required.
Note to physician or authorized medical personnel: Antidote is amyl nitrite.

HEALTH HAZARDS
Personal protective equipment (PPE): A-Level PPE. Chemical protective material(s) reported to have good to excellent resistance: PV alcohol.
Recommendations for respirator selection: NIOSH/OSHA
60 ppm: CCROV [any chemical cartridge respirator with organic vapor cartridge(s)]; or SA (any supplied-air respirator). *150 ppm:* SA:CF (any supplied-air respirator operated in a continuous-flow mode); or PAPROV [any powered, air-purifying respirator with organic vapor cartridge(s)]. *300 ppm:* CCRFOV [any air-purifying, full-facepiece respirator (gas mask) with a chin-style, front- or back-mounted acid gas canister]; or GMFOV [any air-purifying, full-facepiece respirator (gas mask) with a chin-style, front- or back-mounted acid gas canister]; or PAPRTOV [any powered, air-purifying respirator with a tight-fitting facepiece and organic vapor cartridges(s)]; or SCBAF (any self-contained breathing apparatus with a full facepiece); or SAF (any supplied-air respirator with a full facepiece). *1000 ppm:* SAF:PD,PP (any supplied-air respirator that has a full facepiece and is operated in a pressure-demand or other positive-pressure mode). *EMERGENCY OR PLANNED ENTRY INTO UNKNOWN CONCENTRATIONS OR IDLH CONDITIONS:* SCBAF:PD,PP (any self-contained breathing apparatus that has a full facepiece and is operated in a pressure-demand or other positive-pressure mode); or SAF:PD,PP:ASCBA (any supplied-air respirator that has a full facepiece and is operated in a pressure-demand or other positive-pressure mode in combination with an auxiliary self-contained breathing apparatus operated in a pressure-demand or other positive pressure mode). *ESCAPE:* GMFOV [any air-purifying, full-facepiece respirator (gas mask) with a chin-style, front- or back-mounted organic vapor canister]; or SCBAE (any appropriate escape-type, self-contained breathing apparatus).
Exposure limits (TWA unless otherwise noted): NIOSH REL 6 ppm (14 mg/m^3).
Toxicity by ingestion: Grade 4: LD_{50} =36 mg/kg (mouse).
Long-term health effects: Repeated or prolonged exposure may cause skin problems. Teratogen.
Vapor (gas) irritant characteristics: Vapors cause sever irritation of eyes and throat and can cause eye and lung injury. They cannot be tolerated even at low concentrations.
Liquid or solid irritant characteristics: Minimum hazard. If spilled on clothing and allowed to remain, may cause smarting and reddening of skin.

FIRE DATA
Flash point: 36°F/2°C (cc).
Flammable limits in air: LEL: 3.1%.

CHEMICAL REACTIVITY
Reactivity with water: Reacts slowly in cold water to form hydrogen cyanide. Hot water may speed reaction.
Binary reactants: Strong oxidizers and reducing agents; strong acids and bases.
Reactivity group: 37
Compatibility class: Nitriles

ENVIRONMENTAL DATA
Food chain concentration potential: Log P_{ow} = 0.1. Unlikely to accumulate.
Water pollution: Harmful to aquatic life at low concentrations. May be dangerous if it enters nearby water intakes; notify operators. Fouling to shoreline. Notify local health and wildlife officials. Notify operators of local water intakes. **Response to discharge:** Issue warning–poison. Restrict access. Evacuate area. Mechanical containment. Should be removed. Chemical and physical treatment.

NAS HAZARD CLASSIFICATION FOR BULK WATER TRANSPORTATION
FIRE: 4
HEALTH: Vapor irritant: 4; Liquid or solid irritant: 2; Poisons: 4
WATER POLLUTION: Human toxicity: 4; Aquatic toxicity: 3; Aesthetic effect: 1
REACTIVITY: Other chemicals: 2; Water: 0; Self-reaction: 0

PHYSICAL AND CHEMICAL PROPERTIES
Physical state @ 59°F/15°C and 1 atm: Liquid.
Molecular weight: 55.08
Boiling point @ 1 atm: 207.1°F/97.3°C/370.5°K.
Melting/Freezing point: –135°F/–92.8°C/180°K.
Critical temperature: 555.4°F/290.8°C/564°K.
Critical pressure: 607.1 psia = 41.3 atm = 4.2 MN/m^2.
Specific gravity (water = 1): 0.702 @ 68°F/20°C.
Liquid surface tension: 27.2 dyne/cm @ 68°F/20°C.
Relative vapor density (air = 1): 1.9
Latent heat of vaporization: 241.7 Btu/lb = 134.3 cal/g = 5.6 x 10^5 J/kg.
Heat of combustion: –1,491.5 Btu/lb = –82.86 cal/g = –347 x 10^5 J/kg.
Vapor pressure: (Reid) 1.7 psia; 35 mm.

n-PROPOXYPROPANOL REC. P:3850

SYNONYMS: n-PROPOXIPROPANOL (Spanish); PROPYLENE GLYCOL MONPROPYL ETHER

IDENTIFICATION
CAS Number: 30136-13-1
Formula: $C_6H_{14}O_2$; $CH_3CH_2CH_2O(CH_2)_3OH$
DOT ID Number: UN 1993; DOT Guide Number: 128
Proper Shipping Name: Flammable liquid, n.o.s.

DESCRIPTION: Colorless liquid.

Flammable • Containers may BLEVE when exposed to fire •Vapors may form explosive mixture with air • Vapors are heavier than air and will collect and stay in low areas • Vapors may travel long distances to ignition sources and flashback • Vapors in confined areas (e.g., tanks, sewers, buildings) may explode when exposed to fire • Irritating to the skin, eyes, and respiratory tract • Toxic products of combustion may include carbon monoxide and CO_2.

Hazard Classification (based on NFPA-704 Rating System)
Health Hazards (Blue): 0; Flammability (Red): 2; Reactivity (Yellow): 0

EMERGENCY RESPONSE: See Appendix A (128)
Evacuation:
Public safety: Isolate spill area for at least 25 to 50 meters (80 to 160 feet) in all directions.
Spill: Large spill–Consider initial downwind evacuation for at least 300 meters (1000 feet).
Fire: Isolate for 800 meters (½ mile) in all directions, especially if tank, rail car, or tank truck is involved in fire.

EXPOSURE
Short-term effects: *SEEK MEDICAL ATTENTION*. **Vapor:** Irritating to eyes, nose, and throat. *IF BREATHING HAS STOPPED*, give artificial respiration. If breathing is difficult, administer oxygen. **Liquid:** Irritating to skin and eyes. Remove contaminated clothing and shoes. Flush affected areas with plenty of water. *IF IN EYES*, hold eyelids open and flush with plenty of water.

HEALTH HAZARDS
Personal protective equipment (PPE): B-Level PPE. Full impervious protective clothing, including boots and gloves. Where splashing is possible wear full face shield or chemical safety goggles. Use approved respirator to protect against vapors.
Vapor (gas) irritant characteristics: Vapors cause smarting of the eyes or respiratory system if present in high concentrations. The effect is temporary.
Liquid or solid irritant characteristics: Eye and skin irritant. IF spilled on clothing and allowed to remain, may cause smarting and reddening of skin.

FIRE DATA
Flash point: 128°F/53°C (cc).
Fire extinguishing agents not to be used: Water.
Stoichiometric air-to-fuel ratio: 40.5

CHEMICAL REACTIVITY
Binary reactants: Violent reaction with strong oxidizers, sulfuric acid, isocyanates, perchloric acid.

Reactivity group: 40
Compatibility class: Glycol ethers

ENVIRONMENTAL DATA
Water pollution: Effects of low concentrations on aquatic life is unknown. May be dangerous if it enters nearby water intakes; notify operators. Notify local health and wildlife officials. **Response to discharge:** Should be removed.

SHIPPING INFORMATION
Grades of purity: Technical grades; **Storage temperature:** Ambient; **Inert atmosphere:** None; **Stability during transport:** Stable.

PHYSICAL AND CHEMICAL PROPERTIES
Physical state @ 59°F/15°C and 1 atm: Liquid.
Molecular weight: 118.17
Boiling point @ 1 atm: 301.6°F/149.8°C/422.8°K.
Melting/Freezing point: –112°F/–80°C/193°K.
Specific gravity (water = 1): 0.8865 @ 20°C.

n-PROPYL ACETATE REC. P:3900

SYNONYMS: ACETATE de PROPYLE NORMAL (French); ACETIC ACID, PROPYL ESTER; ACETIC ACID, n-PROPYL ESTER; 1-ACETOXYPROPANE; EEC No. 607-024-00-6; PROPYL ACETATE; N-PROPYL ESTER of ACETIC ACID; 1-PROPYL ACETATE

IDENTIFICATION
CAS Number: 109-60-4
Formula: $C_5H_{10}O_2$; $CH_3COOCH_2CH_2CH_3$
DOT ID Number: UN 1276; DOT Guide Number: 129
Proper Shipping Name: n-Propyl acetate

DESCRIPTION: Colorless liquid. Mild, fruity odor. Floats on water surface; soluble. Flammable, irritating vapor is produced.

Highly flammable • Containers may BLEVE when exposed to fire •Vapors may form explosive mixture with air • Vapors are heavier than air and will collect and stay in low areas • Vapors may travel long distances to ignition sources and flashback • Vapors in confined areas (e.g., tanks, sewers, buildings) may explode when exposed to fire • Irritating to the skin, eyes, and respiratory tract • Toxic products of combustion may include CO_2 and carbon monoxide.

Hazard Classification (based on NFPA-704 Rating System)
Health Hazards (Blue): 1; Flammability (Red): 3; Reactivity (Yellow): 0

EMERGENCY RESPONSE: See Appendix A (129)
Evacuation:
Public safety: Isolate spill area for at least 50 to 100 meters (160 to 330 feet) in all directions.
Spill: Large spill–Consider initial downwind evacuation for at least 300 meters (1000 feet).
Fire: Isolate for 800 meters (½ mile) in all directions, especially if tank, rail car, or tank truck is involved in fire.

EXPOSURE
Short-term effects: *SEEK MEDICAL ATTENTION*. **Vapor:** Irritating to eyes, nose, and throat. *IF INHALED*, will, will cause nausea, vomiting, dizziness, or loss of consciousness. Move to fresh

air. *IF BREATHING HAS STOPPED,* give artificial respiration. IF breathing is difficult, administer oxygen. **Liquid:** Irritating to skin and eyes. Harmful if swallowed. Remove contaminated clothing and shoes. Flush affected areas with plenty of water. *IF IN EYES,* hold eyelids open and flush with plenty of water. *IF SWALLOWED* and victim is *CONSCIOUS AND ABLE TO SWALLOW,* have victim drink 4 to 8 ounces of water.

HEALTH HAZARDS
Personal protective equipment (PPE): B-Level PPE. Air–supplied mask or chemical canister; goggles or face shield; protective gloves. Chemical protective material(s) reported to have good to excellent resistance: Silvershield®. Also, butyl rubber, PV alcohol offers limited protection
Recommendations for respirator selection: NIOSH/OSHA
1700 ppm: SA:CF* (any supplied-air respirator operated in a continuous-flow mode); or CCRFOV [any air-purifying, full-facepiece respirator (gas mask) with a chin-style, front- or back-mounted acid gas canister]; or GMFOV [any air-purifying, full-facepiece respirator (gas mask) with a chin-style, front- or back-mounted organic vapor cannister]; or PAPROV* [any powered, air-purifying respirator with organic vapor cartridge(s)]; or SCBAF (any self-contained breathing apparatus with a full facepiece); or SAF (any supplied-air respirator with a full facepiece). *EMERGENCY OR PLANNED ENTRY INTO UNKNOWN CONCENTRATIONS OR IDLH CONDITIONS:* SCBAF:PD,PP (any self-contained breathing apparatus that has a full facepiece and is operated in a pressure-demand or other positive-pressure mode); or SAF:PD,PP:ASCBA (any supplied-air respirator that has a full facepiece and is operated in a pressure-demand or other positive-pressure mode in combination with an auxiliary self-contained breathing apparatus operated in a pressure-demand or other positive pressure mode). *ESCAPE:* GMFOV [any air-purifying, full-facepiece respirator (gas mask) with a chin-style, front- or back-mounted organic vapor cannister]; or SCBAE (any appropriate escape-type, self-contained breathing apparatus).*Note*: Substance causes eye irritation or damage; eye protection needed.
Exposure limits (TWA unless otherwise noted): ACGIH TLV 200 ppm (835 mg/m^3); NIOSH/OSHA PEL 200 ppm (840 mg/m^3).
Short-term exposure limits (15-minute TWA): ACGIH STEL 250 ppm (1040 mg/m^3); NIOSH STEL 250 ppm (1050 mg/m^3).
Toxicity by ingestion: Grade 2; LD$_{50}$ = 0.5 to 5 g/kg.
Vapor (gas) irritant characteristics: Eye and respiratory tract irritant. Vapors cause smarting of the eyes or respiratory system if present in high concentrations. The effect is temporary.
Liquid or solid irritant characteristics: Eye and skin irritant. If spilled on clothing and allowed to remain, may cause smarting and reddening of the skin.
Odor threshold: 70 mg/m^3.
IDLH value: 1700 ppm.

FIRE DATA
Flash point: 65°F/18°C (oc); 58°F/14°C (cc).
Flammable limits in air: LEL: 2.0%; UEL: 8.0%.
Fire extinguishing agents not to be used: Water may be ineffective.
Autoignition temperature: 842°F/450°C/723°K.
Electrical hazard: Class I, Group D.

CHEMICAL REACTIVITY
Reactivity with water: Hydrolyzes on standing to form acetic acid and *n*-propyl alcohol. The presence of bases speeds up the reaction.
Binary reactants: Reacts vigorously with oxidizers; nitrates, alkalis, and acids. Softens or dissolves many plastics.
Reactivity group: 34
Compatibility class: Esters.

ENVIRONMENTAL DATA
Food chain concentration potential: Log P$_{ow}$ = 1.49. Unlikely to accumulate.
Water pollution: Effect of low concentrations on aquatic life is unknown. May be dangerous if it enters nearby water intakes; notify operators. Notify local health and wildlife officials. **Response to discharge:** Issue warning–high flammability. Evacuate area.

SHIPPING INFORMATION
Grades of purity: 90–100%; **Storage temperature:** Ambient; **Inert atmosphere:** None; **Venting:** Open (flame arrester) or pressure-vacuum; **Stability during transport:** Stable.

NAS HAZARD CLASSIFICATION FOR BULK WATER TRANSPORTATION
FIRE: 3
HEALTH: Vapor irritant: 1; Liquid or solid irritant: 1; Poisons: 2
WATER POLLUTION: Human toxicity: 1; Aquatic toxicity: 2; Aesthetic effect: 2
REACTIVITY: Other chemicals: 1; Water: 0; Self-reaction: 0

PHYSICAL AND CHEMICAL PROPERTIES
Physical state @ 59°F/15°C and 1 atm: Liquid.
Molecular weight: 102.13
Boiling point @ 1 atm: 214.9°F/101.6°C/374.8°K.
Melting/Freezing point: –139°F/–95.0°C/178.2°K.
Critical temperature: 529°F/276°C/549°K.
Critical pressure: 485 psia = 33 atm = 3.3 MN/m^2.
Specific gravity (water = 1): 0.886 @ 68°F/20°C (liquid).
Liquid surface tension: 24.3 dynes/cm = 0.0243 N/m @ 68°F/20°C.
Relative vapor density (air = 1): 3.5
Ratio of specific heats of vapor (gas): 1.071
Latent heat of vaporization: 145 Btu/lb = 80.3 cal/g = 3.36 x 10^5 J/kg.
Vapor pressure: (Reid) 1.3 psia; 25 mm.

---

n-PROPYL ALCOHOL   REC. P:3950

SYNONYMS: ALCOHOL PROPILICO (Spanish); ALCOHOL C-3; ALCOOL PROPYLIQUE (French); EEC No. 603-003-00-0; ETHYL CARBINOL; 1-HYDROXYPROPANE; OPTAL; OSMOSOL EXTRA; PROPANOL-1; 1-PROPANOL; *n*-PROPANOL; PROPANOLE (German); PROPYL ALCOHOL; 1-PROPYL ALCOHOL; *n*-PROPYL ALKOHOL (German); PROPYLIC ALCOHOL

IDENTIFICATION
CAS Number: 71-23-8
Formula: C$_3$H$_8$O; CH$_3$CH$_2$CH$_2$OH
DOT ID Number: UN 1274; DOT Guide Number: 129
Proper Shipping Name: *n*-Propanol; normal Propyl alcohol; Propyl alcohol, normal

DESCRIPTION: Colorless liquid. Alcohol odor. Mixes with water. Flammable, irritating vapor is produced.

Highly flammable • Containers may BLEVE when exposed to fire • Vapors may form explosive mixture with air • Vapors are heavier than air and will collect and stay in low areas • Vapors may travel long distances to ignition sources and flashback • Vapors in confined areas (e.g., tanks, sewers, buildings) may explode when exposed to fire • Irritating to the skin, eyes, and respiratory tract • Toxic products of combustion may include CO_2 and carbon monoxide.

Hazard Classification (based on NFPA-704 Rating System)
Health Hazards (Blue): 1; Flammability (Red): 3; Reactivity (Yellow): 0

EMERGENCY RESPONSE: See Appendix A (129)
Evacuation:
Public safety: Isolate spill area for at least 50 to 100 meters (160 to 330 feet) in all directions.
Spill: Large spill–Consider initial downwind evacuation for at least 300 meters (1000 feet).
Fire: Isolate for 800 meters (½ mile) in all directions, especially if tank, rail car, or tank truck is involved in fire.

EXPOSURE
Short-term effects: *SEEK MEDICAL ATTENTION.* **Vapor:** Irritating to eyes, nose, and throat. *IF INHALED*, will, will cause nausea, dizziness, or headache. Move to fresh air. *IF BREATHING HAS STOPPED*, give artificial respiration. IF breathing is difficult, administer oxygen. **Liquid:** *HARMFUL IF ABSORBED THROUGH THE SKIN.* Will burn eyes. Harmful if swallowed. Flush affected areas with plenty of water. *IF IN EYES*, hold eyelids open and flush with plenty of water. *IF SWALLOWED* and victim is *CONSCIOUS AND ABLE TO SWALLOW*, have victim drink 4 to 8 ounces of water.

HEALTH HAZARDS
Personal protective equipment (PPE): B-Level PPE. Air–supplied respirator for high concentrations; goggles or face shield; plastic gloves. Chemical protective material(s) reported to have good to excellent resistance: neoprene, nitrile, Teflon®, Viton®. Also, butyl rubber, neoprene, butylrubber/neoprene, PVC, Viton®/neoprene offers limited protection
Recommendations for respirator selection: NIOSH/OSHA
800 ppm: CCROV* [any chemical cartridge respirator with organic vapor cartridge(s)]; or PAPROV* [any powered, air-purifying respirator with organic vapor cartridge(s)]; or GMFOV [any air-purifying, full-facepiece respirator (gas mask) with a chin-style, front- or back-mounted organic vapor cannister]; SA* (any supplied-air respirator); or SCBAF (any self-contained breathing apparatus with a full facepiece). *EMERGENCY OR PLANNED ENTRY INTO UNKNOWN CONCENTRATIONS OR IDLH CONDITIONS:* SCBAF:PD,PP (any self-contained breathing apparatus that has a full facepiece and is operated in a pressure-demand or other positive-pressure mode); or SAF:PD,PP:ASCBA (any supplied-air respirator that has a full facepiece and is operated in a pressure-demand or other positive-pressure mode in combination with an auxiliary self-contained breathing apparatus operated in a pressure-demand or other positive pressure mode). *ESCAPE:* GMFOV [any air-purifying, full-facepiece respirator (gas mask) with a chin-style, front- or back-mounted organic vapor cannister]; or SCBAE (any appropriate escape-type, self-contained breathing apparatus). *Note*: Substance reported to cause eye irritation or damage; may require eye protection.
Exposure limits (TWA unless otherwise noted): ACGIH TLV 200 ppm (492 mg/m^3); NIOSH/OSHA PEL 200 ppm (500 mg/m^3); skin contact contributes significantly in overall exposure.
Short-term exposure limits (15-minute TWA): ACGIH STEL 250 ppm (614 mg/m^3); NIOSH STEL 250 ppm (625 mg/m^3); skin contact contributes significantly in overall exposure.
Toxicity by ingestion: Grade 2; LD_{50} = 0.5 to 5 g/kg (rat).

Vapor (gas) irritant characteristics: Eye and respiratory tract irritant. Vapors cause smarting of the eyes or respiratory system if present in high concentrations. The effect is temporary.
Liquid or solid irritant characteristics: Eye and skin irritant. Practically harmless to the skin. Absorbed through the skin.
Odor threshold: 0.03–40 ppm.
IDLH value: 800 ppm.

FIRE DATA
Flash point: 81°F/27°C (oc); 77°F/25°C (cc).
Flammable limits in air: LEL: 2.2%; UEL: 13.7%.
Fire extinguishing agents not to be used: Water may be ineffective
Autoignition temperature: 770°F/410°C/683°K.
Electrical hazard: Class I, Group D.
Burning rate: 2.9 mm/min

CHEMICAL REACTIVITY
Binary reactants: Strong oxidizers.
Reactivity group: 20
Compatibility class: Alcohols, glycols

ENVIRONMENTAL DATA
Food chain concentration potential: Log P_{ow} = 0.3. Unlikely to accumulate.
Water pollution: Dangerous to aquatic life in high concentrations. May be dangerous if it enters nearby water intakes; notify operators. Notify local health and wildlife officials. **Response to discharge:** Issue warning–high flammability. Disperse and flush.

SHIPPING INFORMATION
Grades of purity: 99.8+%; **Storage temperature:** Ambient; **Inert atmosphere:** None; **Venting:** Open (flame arrester); **Stability during transport:** Stable.

NAS HAZARD CLASSIFICATION FOR BULK WATER TRANSPORTATION
FIRE: 3
HEALTH: Vapor irritant: 1; Liquid or solid irritant: 0; Poisons: 2
WATER POLLUTION: Human toxicity: 2; Aquatic toxicity: 2; Aesthetic effect: 1
REACTIVITY: Other chemicals: 2; Water: 0; Self-reaction: 0

PHYSICAL AND CHEMICAL PROPERTIES
Physical state @ 59°F/15°C and 1 atm: Liquid.
Molecular weight: 60.10
Boiling point @ 1 atm: 207.0°F/97.2°C/370.4°K.
Melting/Freezing point: −195.2°F/−126.2°C/147.0°K.
Critical temperature: 506.5°F/263.6°C/536.8°K.
Critical pressure: 750 psia = 51 atm = 5.2 MN/m^2.
Specific gravity (water = 1): 0.803 @ 68°F/20°C (liquid).
Relative vapor density (air = 1): 2.1
Ratio of specific heats of vapor (gas): 1.107
Latent heat of vaporization: 292.7 Btu/lb = 162.6 cal/g = 6.808 x 10^5 J/kg.
Heat of combustion: −13,130 Btu/lb = −7296 cal/g = −305.5 x 10^5 J/kg.
Heat of solution: (estimate) −9 Btu/lb = −5 cal/g = −0.2 x 10^5 J/kg.
Heat of fusion: 20.66 cal/g.
Vapor pressure: (Reid) 0.87 psia; 13 mm.

n-PROPYLAMINE　　　　　　　　　　　REC. P:4000

SYNONYMS: 1-AMINOPROPANE; MONO-*N*-

PROPYLAMINE; MONOPROPYLAMINE; PROPANAMINE; 1-PROPANAMINE; *n*-PROPILAMINA (Spanish); PROPYLAMINE; 2-PROPYLAMINE; TEA; RCRA No. U194

IDENTIFICATION
CAS Number: 107-10-8
Formula: C_3H_9N; $CH_3(CH_2)_2NH_2$
DOT ID Number: UN 1277; DOT Guide Number: 132
Proper Shipping Name: Propylamine
Reportable Quantity (RQ): **(CERCLA)** 5000 lb/2270 kg

DESCRIPTION: Colorless watery liquid. Strong ammonia odor. Floats and mixes with water forming an alkaline solution.

Highly flammable • Poison! • Corrosive to the skin, eyes, and respiratory tract; contact with eyes or skin can cause burns, impaired vision, or blindness; inhalation symptoms may be delayed • Firefighting gear (including SCBA) does not provide adequate protection. If exposure occurs, remove and isolate gear immediately and thoroughly decontaminate personnel • Containers may BLEVE when exposed to fire •Vapors may form explosive mixture with air • Vapors are heavier than air and will collect and stay in low areas • Vapors may travel long distances to ignition sources and flashback • Vapors in confined areas (e.g., tanks, sewers, buildings) may explode when exposed to fire • Toxic products of combustion may include nitrogen oxides, carbon monoxide, and CO_2 • Do not put yourself in danger by entering a contaminated area to rescue a victim.

Hazard Classification (based on NFPA-704 Rating System)
Health Hazards (Blue): 3; Flammability (Red): 3; Reactivity (Yellow): 0

EMERGENCY RESPONSE: See Appendix A (132)
Evacuation:
Public safety: Isolate spill area for at least 50 to 100 meters (160 to 330 feet) in all directions.
Spill: Increase, as necessary, the isolation distance shown above,in "Public safety."
Fire: Isolate for 800 meters (½ mile) in all directions, especially if tank, rail car, or tank truck is involved in fire.

EXPOSURE
Short-term effects: *SEEK MEDICAL ATTENTION.* **Vapor:** Irritating to eyes, nose, and throat. Harmful if inhaled. Move to fresh air. *IF BREATHING HAS STOPPED,* give artificial respiration. IF breathing is difficult, administer oxygen. Lung edema may develop. **Liquid:** Will burn eyes. Harmful if swallowed. Remove contaminated clothing and shoes. Flush affected areas with plenty of water. *IF IN EYES*, hold eyelids open and flush with plenty of water. *IF SWALLOWED* and victim is *CONSCIOUS AND ABLE TO SWALLOW*, have victim drink 4 to 8 ounces of water.
Note to physician or authorized medical personnel: Medical observation is recommended for 24 to 48 hours after breathing overexposure, as pulmonary edema may be delayed. As first aid for pulmonary edema, consider administering a corticosteroid spray. Cigarette smoking may exacerbate pulmonary injury and should be discouraged for at least 72 hours following exposure.

HEALTH HAZARDS
Personal protective equipment (PPE): B-Level PPE. With eyes, skin, and respiratory tract. Chemical protective material(s) reported to have good to excellent resistance: PV acetate, Teflon®.
Toxicity by ingestion: Grade 2; LD_{50} = 0.5 to 5 g/kg.

Long-term health effects: Weight loss, corneal opacities, and deaths occurred in laboratory animals exposed repeatedly to 800 ppm. A weak allergen.
Vapor (gas) irritant characteristics: Vapors cause severe irritation of eyes and throat and can cause eye and lung injury. They cannot be tolerated even at low concentrations.
Liquid or solid irritant characteristics: Severe eye and skin irritant. Causes second- and third-degree burns on short contact and is very injurious to the eyes.

FIRE DATA
Flash point: –35°F/–37°C (oc); –35°F/–37°C (cc).
Flammable limits in air: LEL: 2.0%; UEL: 10.4%.
Fire extinguishing agents not to be used: Water may be ineffective. Do not use halogenated agents.
Autoignition temperature: 604°F/318°C/591°K.

CHEMICAL REACTIVITY
Binary reactants: Strong oxidizers, chlorine, halogenated compounds and acids. Attacks chemically active metals, aluminum, zinc, copper and their alloys.
Reactivity group: 7
Compatibility class: Aliphatic amines

ENVIRONMENTAL DATA
Food chain concentration potential: Log P_{ow} = 0.2. Unlikely to accumulate.
Water pollution: Harmful to aquatic life in very low concentrations. May be dangerous if it enters nearby water intakes; notify operators. Notify local health and wildlife officials. **Response to discharge:** Issue warning–high flammability. Evacuate area. Disperse and flush.

SHIPPING INFORMATION
Grades of purity: 99% (minimum); **Storage temperature:** Cool; **Inert atmosphere:** Inert; **Venting:** PV; **Stability during transport:** Stable.

PHYSICAL AND CHEMICAL PROPERTIES
Physical state @ 59°F/15°C and 1 atm: Solid.
Molecular weight: 59.11
Boiling point @ 1 atm: 119.5°F/48.6°C/321.8°K.
Melting/Freezing point: –117.4°F/–83°C/190.2°K.
Critical temperature: 435°F/224°C/497°K.
Critical pressure: 687.8 psia = 46.8 atm = 4.74 MN/m².
Specific gravity (water = 1): 0.7182 @ 68°F/20°C.
Liquid surface tension: 57.72 dynes/cm = 0.05772 N/m @ 68°F/20°C.
Relative vapor density (air = 1): 2.04
Latent heat of vaporization: 219.4 Btu/lb = 121.9 cal/g = 5.1 x 10^5 J/kg.
Heat of combustion: –15,773 Btu/lb = –8763 cal/g = –366.6 x 10^5 J/kg.
Vapor pressure: (Reid) 10.4 psia; 250 mm.

n-**PROPYLBENZENE**   **REC. P:4050**

SYNONYMS: BENZENE, PROPYL; ISOCUMENE; 1-PHENYLPROPANE; *n*-PROPILBENCENO (Spanish); PROPYL BENZENE

IDENTIFICATION
CAS Number: 103-65-1
Formula: C_9H_{12}; $C_6H_5CH_2CH_2CH_3$

948 *n*-Propyl chloride

DOT ID Number: UN 2364; DOT Guide Number: 127
Proper Shipping Name: *n*-Propyl benzene

DESCRIPTION: Light yellow to colorless liquid. Petroleum-like odor. floats on the surface of water; insoluble.

Highly flammable • Containers may BLEVE when exposed to fire •Vapors may form explosive mixture with air • Vapors are heavier than air and will collect and stay in low areas • Vapors may travel long distances to ignition sources and flashback • Vapors in confined areas (e.g., tanks, sewers, buildings) may explode when exposed to fire • Irritating to the skin, eyes, and respiratory tract • Toxic products of combustion may include irritating vapors.

Hazard Classification (based on NFPA-704 Rating System)
Health Hazards (Blue): 2; Flammability (Red): 3; Reactivity (Yellow): 0

EMERGENCY RESPONSE: See Appendix A (127)
Evacuation:
Public safety: Isolate spill area for at least 25 to 50 meters (80 to 160 feet) in all directions.
Spill: Large spill–Consider initial downwind evacuation for at least 300 meters (1000 feet).
Fire: Isolate for 800 meters (½ mile) in all directions, especially if tank, rail car, or tank truck is involved in fire.

EXPOSURE
Short-term effects: *SEEK MEDICAL ATTENTION*. **Vapor:** May be irritating to eyes, nose, and throat. *IF INHALED*, will, will cause dizziness or difficult breathing. Move to fresh air. *IF BREATHING HAS STOPPED*, give artificial respiration. IF breathing is difficult, administer oxygen. **Liquid:** May be irritating to skin and eyes. May be harmful if swallowed. Remove contaminated clothing and shoes. Flush affected areas with plenty of water. *IF IN EYES*, hold eyelids open and flush with plenty of water.

HEALTH HAZARDS
Personal protective equipment (PPE): B-Level PPE. Self-contained breathing apparatus, rubber boots and heavy rubber gloves.
Toxicity by ingestion: Grade 1: LD_{50} = 6.04 g/kg rat
Vapor (gas) irritant characteristics: Vapors cause smarting of the eyes or respiratory system if present in high concentrations. The effect is temporary.
Liquid or solid irritant characteristics: If spilled on clothing and allowed to remain, may cause smarting and reddening of skin.

FIRE DATA
Flash point: 86°F/30°C (cc).
Flammable limits in air: LEL 0.8%; UEL 6%.
Autoignition temperature: 837°F/447°C/720°K.

CHEMICAL REACTIVITY
Binary reactants: Nitric acid.
Reactivity group: 32
Compatibility class: Aromatic hydrocarbons

ENVIRONMENTAL DATA
Water pollution: DOT Appendix B, §172.101–marine pollutant. Effect of low concentration on aquatic life is unknown. Fouling to shoreline. May be dangerous if it enters nearby water intakes; notify operators. Notify local health and wildlife officials.

Response to discharge: Evacuate area. Mechanical containment. Should be removed. Chemical and physical treatment.

SHIPPING INFORMATION
Grades of purity: 98%; **Storage temperature:** Ambient; **Stability during transport:** Stable.

PHYSICAL AND CHEMICAL PROPERTIES
Physical state @ 59°F/15°C and 1 atm: Liquid.
Molecular weight: 120.20
Boiling point @ 1 atm: 318.2°F/159°C/432.2°K.
Melting/Freezing point: –146.2°F/–99°C/174.2°K.
Specific gravity (water = 1): 0.862
Relative vapor density (air = 1): 4.14
Vapor pressure: (Reid) 0.1455 psia.

n-PROPYL CHLORIDE   REC. P:4100

SYNONYMS: 1-CHLOROPROPANE; CLORURO de PROPILO (Spanish); PROPANE, CHLORO-

IDENTIFICATION
CAS Number: 540-54-5
Formula: C_3H_7Cl; $ClCH_2CH_2CH_3$
DOT ID Number: UN 1278; DOT Guide Number: 129
Proper Shipping Name: Propyl chloride; 1-Chloropropane

DESCRIPTION: Colorless liquid. Chloroform-like odor. Floats on the surface of water.

Extremely flammable • Containers may BLEVE when exposed to fire •Vapors may form explosive mixture with air • Vapors are heavier than air and will collect and stay in low areas • Vapors may travel long distances to ignition sources and flashback • Vapors in confined areas (e.g., tanks, sewers, buildings) may explode when exposed to fire • Irritating to the skin, eyes, and respiratory tract • Toxic products of combustion may include hydrogen chloride and phosgene.

Hazard Classification (based on NFPA-704 Rating System)
Health Hazards (Blue): 2; Flammability (Red): 3; Reactivity (Yellow): 0

EMERGENCY RESPONSE: See Appendix A (129)
Evacuation:
Public safety: Isolate spill area for at least 50 to 100 meters (160 to 330 feet) in all directions.
Spill: Large spill–Consider initial downwind evacuation for at least 300 meters (1000 feet).
Fire: Isolate for 800 meters (½ mile) in all directions, especially if tank, rail car, or tank truck is involved in fire.

EXPOSURE
Short-term effects: *SEEK MEDICAL ATTENTION*. **Vapor:** Move victim to fresh air. *IF BREATHING HAS STOPPED*, give artificial respiration. If breathing is difficult, administer oxygen. **Liquid:** Irritating to skin and eyes. Remove contaminated clothing and shoes. Wash affected areas with soap and water. *IF IN EYES*, hold eyelids open and flush with plenty of water.

HEALTH HAZARDS
Personal protective equipment (PPE): B-Level PPE. Full impervious protective clothing, including boots and gloves. Where splashing is possible wear full face shield or chemical safety

goggles. Use approved respirator to protect against vapors.
Vapor (gas) irritant characteristics: eye and respiratory tract irritant. Vapors cause smarting of the eyes or respiratory system if present in high concentrations. The effect is temporary.
Liquid or solid irritant characteristics: If spilled on clothing and allowed to remain, may cause smarting and reddening of the skin.

FIRE DATA
Flash point: Less than 0°F/–18°C (cc).
Flammable limits in air: LEL: 2.6%; UEL: 11.0%.
Fire extinguishing agents not to be used: Water may be ineffective.
Autoignition temperature: 968°F/520°C/793°K.
Stoichiometric air-to-fuel ratio: 22.6

CHEMICAL REACTIVITY
Binary reactants: Incompatible with strong oxidizers.
Compatibility class: Aliphatic chlorinated hydrocarbon.

ENVIRONMENTAL DATA
Water pollution: Effects of low concentrations on aquatic life is unknown. May be dangerous if it enters nearby water intakes; notify operators. Notify local health and wildlife officials.
Response to discharge: Issue warning–high flammability. Should be removed.

SHIPPING INFORMATION
Grades of purity: 99%; technical; **Storage temperature:** Ambient; **Inert atmosphere:** None; **Stability during transport:** Stable.

PHYSICAL AND CHEMICAL PROPERTIES
Physical state @ 59°F/15°C and 1 atm: Liquid.
Molecular weight: 78.54
Boiling point @ 1 atm: 115–117°F/46–47°C/319–320°K
Melting/Freezing point: –189°F/–123°C/150°K.
Specific gravity (water = 1): 0.892
Relative vapor density (air = 1): 2.71

PROPYLENE   REC. P:4150

SYNONYMS: EEC No. 601-011-00-9; METHYLETHENE; METHYLETHYLENE; NCI-C50077; PROPENE: 1-PROPENE; PROPILENO (Spanish); 1-PROPYLENE

IDENTIFICATION
CAS Number: 115-07-1
Formula: C_3H_6; $CH_3CH=CH_2$
DOT ID Number: UN 1077; DOT Guide Number: 115
Proper Shipping Name: Propylene; Petroleum gases, liquefied

DESCRIPTION: Colorless gas. Shipped and stored as a liquefied compressed gas. Weak, petroleum odor. Floats and boils on water; insoluble; flammable, visible vapor cloud is produced.

Extremely flammable • Forms explosive mixture with air • Reacts explosively with many materials • Gas from liquefied gas are initially heavier than air and spread along ground • Gas may travel to source of ignition and flashback • Ruptured or venting cylinders may rocket through buildings and/or travel a considerable distance •Irritating to the skin, eyes, and respiratory tract • Vapors may cause dizziness or asphyxiation without warning • Contact with liquid may cause frostbite • Toxic products of combustion may include CO_2 and carbon monoxide.

Hazard Classification (based on NFPA-704 Rating System)
Health Hazards (Blue): 1; Flammability (Red): 4; Reactivity (Yellow): 1

EMERGENCY RESPONSE: See Appendix A (115)
Evacuation:
Public safety: Isolate spill area for at least 50 to 100 meters (160 to 330 feet) in all directions.
Spill: Consider initial downwind evacuation for at least 800 meters (½ mile).
Fire: Isolate for 1600 meters (1 mile) in all directions, especially if tank, rail car, or tank truck is involved in fire.

EXPOSURE
Short-term effects: *SEEK MEDICAL ATTENTION.* **Vapor:** If inhaled, will cause dizziness or loss of consciousness. Move to fresh air. *IF BREATHING HAS STOPPED,* give artificial respiration. IF breathing is difficult, administer oxygen. **Liquid:** Will cause frostbite. Flush affected areas with plenty of water. DO NOT RUB AFFECTED AREAS.

HEALTH HAZARDS
Personal protective equipment (PPE): B-Level PPE. Organic vapor canister or air–supplied mask; goggles or face shield (for liquid); protective clothing (for liquid). Wear thermal protective clothing.
Exposure limits (TWA unless otherwise noted): ACGIH simple asphyxiant
Vapor (gas) irritant characteristics: Vapors are nonirritating to the eyes and throat.
Liquid or solid irritant characteristics: Rapid evaporation of liquid may cause frostbite.

FIRE DATA
Flash point: Gas, –162°F/–108°C (cc).
Flammable limits in air: LEL: 2.0%; UEL: 11%.
Behavior in fire: Containers may explode.
Autoignition temperature: 843°F/451°C/724°K.
Electrical hazard: Class I, Group D. Flow or agitation of substance may generate electrostatic charges due to low conductivity.
Burning rate: 8 mm/min (liquid).
Adiabatic flame temperature: 2518°F/1381°C (estimate).
Stoichiometric air-to-fuel ratio: 14.68 (estimate).

CHEMICAL REACTIVITY
Binary reactants: Violent reaction with strong oxidizers, trifluoromethyl hypofluorite, fluoride, chlorine, and many other compounds. Incompatible with ammonium hydroxide. Forms explosive materials with nitrogen oxide compounds.
Polymerization: Production of peroxides may cause polymerization.
Reactivity group: 30
Compatibility class: Olefins

ENVIRONMENTAL DATA
Food chain concentration potential: Negative; unlikely to accumulate.
Water pollution: Not harmful to aquatic life. **Response to discharge:** Issue warning–high flammability. Evacuate area.

SHIPPING INFORMATION
Grades of purity: Chemical: 92+%; Polymerization: 99+%; research: 99+%; propylene concentrate: 80+%; **Storage temperature:** Ambient; **Inert atmosphere:** None; **Venting:** Safety relief; **Stability during transport:** Stable.

NAS HAZARD CLASSIFICATION FOR BULK WATER TRANSPORTATION
FIRE: 4
HEALTH: Vapor irritant: 0; Liquid or solid irritant: 0; Poisons: 1
WATER POLLUTION: Human toxicity: 0; Aquatic toxicity: 1; Aesthetic effect: 0
REACTIVITY: Other chemicals: 1; Water: 0; Self-reaction: 1

PHYSICAL AND CHEMICAL PROPERTIES
Physical state @ 59°F/15°C and 1 atm: Gas.
Molecular weight: 42.08
Boiling point @ 1 atm: −53.9°F/47.7°C/225.5°K.
Melting/Freezing point: −301.4°F/−185.2°C/88°K.
Critical temperature: 197°F/92°C/365°K.
Critical pressure: 670 psia = 45.6 atm = 4.62 MN/m².
Specific gravity (water = 1): 0.609 at −52.6°F/−47°C (liquid).
Liquid surface tension: 16.7 dynes/cm = 0.0167 N/m at −52.6°F/−47°C.
Relative vapor density (air = 1): 1.48
Ratio of specific heats of vapor (gas): 1.152
Latent heat of vaporization: 187 Btu/lb = 104 cal/g = 4.35×10^5 J/kg.
Heat of combustion: −19,692 Btu/lb = −10,940 cal/g = -458.04×10^5 J/kg.
Heat of fusion: 17.06 cal/g.
Vapor pressure: (Reid) 227.2 psia; 7600 mm.

PROPYLENE GLYCOL   REC. P:4200

SYNONYMS: DOWFROST; 1,2-DIHYDROXYPROPANE; 2-HYDROXYPROPANOL; METHYLETHYLENEGLYCOL; METHYLETHYLENE GLYCOL; METHYL GLYCOL; MONOPROPYLENE GLYCOL; PG12; PROPANE-1,2-DIOL; 1,2-PROPANEDIOL; PROPILENGLICOL (Spanish); 1,2-PROPYLENE GLYCOL; SIRLENE; SOLAR WINTER BAN; TRIMETHYL GLYCOL

IDENTIFICATION
CAS Number: 57-55-6
Formula: $C_3H_8O_2$; $CH_3CH(OH)CH_2OH$

DESCRIPTION: Colorless, thick liquid. Practically odorless. Mixes with water.

Combustible • Heat or flame may cause explosion • Containers may BLEVE when exposed to fire • Vapors in confined areas (e.g., tanks, sewers, buildings) may explode when exposed to fire • Irritating to the skin, eyes, and respiratory tract • Toxic products of combustion may include CO_2 and carbon monoxide.

Hazard Classification (based on NFPA-704 Rating System)
Health Hazards (Blue): 0; Flammability (Red): 1; Reactivity (Yellow): 0

EMERGENCY RESPONSE: See Appendix A (171)
Evacuation:
Public safety: Isolate the area of spill or leak for at least 10 to 25 meters (30 to 80 feet) in all directions.
Spill: Increase, in the downwind direction, as necessary, the distance shown under "Public Safety."
Fire: If any large container is involved in fire, isolate for at least 800 meters (½ mile) in all directions; also, consider initial evacuation for 800 meters (½ mile) in all directions.

EXPOSURE
Short-term effects: An eye and skin irritant. Slight anesthetic effect in high concentrations.

HEALTH HAZARDS
Personal protective equipment (PPE): B-Level PPE. Organic vapor respirator. Chemical protective material(s) reported to have good to excellent resistance: polyethylene, neoprene/natural rubber, nitrile+PVC, Silvershield®. Also, Natural rubber offers limited protection.
Exposure limits (TWA unless otherwise noted): 50 mg/m³ (AIHA WEEL). 10 mg/m³ [aerosols](AIHA WEEL).
Toxicity by ingestion: Grade 2; LD_{50} = 0.5 to 5 g/kg (mouse).
Vapor (gas) irritant characteristics: Vapors are nonirritating to the eyes and throat.
Liquid or solid irritant characteristics: May cause eye irritation. Practically harmless to the skin.

FIRE DATA
Flash point: 225°F; 107°C (oc); 210°F/99°C (cc).
Flammable limits in air: LEL: 2.6%; UEL: 12.5%.
Autoignition temperature: 707°F/375°C/648°K.
Burning rate: 1.5 mm/min

CHEMICAL REACTIVITY
Binary reactants: Violent reaction with strong oxidizers.
Reactivity group: 20
Compatibility class: Alcohols, glycols

ENVIRONMENTAL DATA
Food chain concentration potential: Log P_{ow} = −1.4. Unlikely to accumulate.
Water pollution: Effect of low concentrations on aquatic life is unknown. May be dangerous if it enters nearby water intakes; notify operators. Notify local health and wildlife officials. **Response to discharge:** Disperse and flush.

SHIPPING INFORMATION
Grades of purity: USP, industrial, food (all 99+%); **Storage temperature:** Ambient; **Inert atmosphere:** None; **Venting:** Open (flame arrester); **Stability during transport:** Stable.

NAS HAZARD CLASSIFICATION FOR BULK WATER TRANSPORTATION
FIRE: 1
HEALTH: Vapor irritant: 0; Liquid or solid irritant: 0; Poisons: 0
WATER POLLUTION: Human toxicity: 0; Aquatic toxicity: 1; Aesthetic effect: 1
REACTIVITY: Other chemicals: 2; Water: 0; Self-reaction: 0

PHYSICAL AND CHEMICAL PROPERTIES
Physical state @ 59°F/15°C and 1 atm: Liquid.
Molecular weight: 76.10
Boiling point @ 1 atm: 369.1°F/187.3°C/460.5°K.
Melting/Freezing point: Less than −76°F/−60°C/213°K.
Specific gravity (water = 1): 1.04 @ 68°F/20°C (liquid).
Liquid surface tension: 36 dynes/cm = 0.036 N/m @ 77°F/25°C.
Relative vapor density (air = 1): 2.6
Ratio of specific heats of vapor (gas): 1.073
Latent heat of vaporization: 306 Btu/lb = 170 cal/g = 7.12×10^5 J/kg.
Heat of combustion: −10,310 Btu/lb = −5,728 cal/g = -239.8×10^5 J/kg.
Vapor pressure: 0.1 mm.

PROPYLENE GLYCOL ETHYL ETHER REC. P:4250

SYNONYMS: 1-ETHOXY-2-PROPANOL; PROPYLENE GLYCOL MONO ETHYL ETHER; PROPILENGLICOL MONOETIL ETER (Spanish); 2-PROPANOL-1-ETHOXY

IDENTIFICATION
CAS Number: 1569-02-4
Formula: $C_5H_{12}O_2$; $C_2H_5OCH_2CHOHCH_3$
DOT ID Number: UN 1993; DOT Guide Number: 128
Proper Shipping Name: Flammable liquid, n.o.s.

DESCRIPTION: Colorless liquid. Mild odor. Mixes with water.

Flammable • Containers may BLEVE when exposed to fire • Vapors may form explosive mixture with air • May be able to form unstable peroxides • Vapors are heavier than air and will collect and stay in low areas • Vapors may travel long distances to ignition sources and flashback • Vapors in confined areas (e.g., tanks, sewers, buildings) may explode when exposed to fire • Irritating to the skin, eyes, and respiratory tract • Toxic products of combustion may include CO_2 and carbon monoxide.

Hazard Classification (based on NFPA-704 Rating System)
Health Hazards (Blue): 0; Flammability (Red): 2; Reactivity (Yellow): 0

EMERGENCY RESPONSE: See Appendix A (128)
Evacuation:
Public safety: Isolate spill area for at least 25 to 50 meters (80 to 160 feet) in all directions.
Spill: Large spill–Consider initial downwind evacuation for at least 300 meters (1000 feet).
Fire: Isolate for 800 meters (½ mile) in all directions, especially if tank, rail car, or tank truck is involved in fire.

EXPOSURE
Short-term effects: *SEEK MEDICAL ATTENTION*. **Vapor:** Irritating to eyes, nose, and throat. Move to fresh air. **Liquid:** Irritating to skin and eyes. Remove contaminated clothing and shoes. Flush affected areas with plenty of water. *IF IN EYES*, hold eyelids open and flush with plenty of water.

HEALTH HAZARDS
Personal protective equipment (PPE): B-Level PPE. Safety goggles, protective clothing.
Toxicity by ingestion: Grade 0: LD_{50} = 44g/kg (rat).
Vapor (gas) irritant characteristics: Vapors cause moderate irritation such that personnel will find high concentrations unpleasant. The effect is temporary.
Liquid or solid irritant characteristics: Eye and skin irritant. If spilled on clothing and allowed to remain, may cause smarting and reddening of skin.

FIRE DATA
Flash point: 109°F/43°C (cc).

CHEMICAL REACTIVITY
Binary reactants: Sulfuric acid, isocyanates. Hygroscopic; absorbs moisture form the air; may form unstable peroxides.
Reactivity group: 40
Compatibility class: Glycol ethers

ENVIRONMENTAL DATA
Water pollution: Effect of low concentrations on aquatic life is unknown. May be dangerous if it enters nearby water intakes; notify operators. Notify local health and wildlife officials. **Response to discharge:** Disperse and flush.

SHIPPING INFORMATION
Storage temperature: Ambient; **Stability during transport:** Stable.

PHYSICAL AND CHEMICAL PROPERTIES
Molecular weight: 104.15
Boiling point @ 1 atm: 518°F/270°C/543.2°K.
Specific gravity (water = 1): 0.895
Relative vapor density (air = 1): 3.59

POLYPROPYLENE GLYCOL METHYL ETHER REC. P:4300

SYNONYMS: DOWANOL 33B; DOWANOL PM; DOWTHERM® 209; EEC No. 603-064-00-3; METHOXY ETHER OF PROPYLENE GLYCOL; 1-METHOXY-2-HYDROXYPROPANE; 2-METHOXY-1-METHYLETHANOL; 1-METHOXY-2-PROPANOL; PROPILENGLICOL MONOMETIL ETER (Spanish); POLYOXYPROPYLENE GLYCOL METHYL ETHER; POLY (PROPYLENE GLYCOL) METHYL ETHER; PGME; POLY-SOLVE MPM; PROPASOL SOLVENT M; PROPYLENE GLYCOL MONOMETHYL ETHER; α-PROPYLENE GLYCOL MONOMETHYL ETHER; PROPYLENGLYKOL-MONOMETHYLAETHER (German)

IDENTIFICATION
CAS Number: 107-98-2; 37286-64-9
Formula: $C_4H_{10}O_2$; $CH_3CH(OH)CH_2OCH_3$
DOT ID Number: UN 3092; DOT Guide Number: 129
Proper Shipping Name: 1-Methoxy-2- propanol

DESCRIPTION: Colorless liquid. Mild, ether-like odor. Mixes with water; soluble.

Highly flammable • Containers may BLEVE when exposed to fire • Vapors may form explosive mixture with air • Vapors are heavier than air and will collect and stay in low areas • Vapors may travel long distances to ignition sources and flashback • Vapors in confined areas (e.g., tanks, sewers, buildings) may explode when exposed to fire • May be able to form unstable peroxides • Irritating to the skin, eyes, and respiratory tract • Toxic products of combustion may include CO_2 and carbon monoxide.

Hazard Classification (based on NFPA-704 Rating System)
Health Hazards (Blue): 0; Flammability (Red): 3; Reactivity (Yellow): 0

EMERGENCY RESPONSE: See Appendix A (129)
Evacuation:
Public safety: Isolate spill area for at least 50 to 100 meters (160 to 330 feet) in all directions.
Spill: Large spill–Consider initial downwind evacuation for at least 300 meters (1000 feet).
Fire: Isolate for 800 meters (½ mile) in all directions, especially if tank, rail car, or tank truck is involved in fire.

EXPOSURE
Short-term effects: *SEEK MEDICAL ATTENTION*. **Vapor:** Irritating to eyes, nose, and throat. Move to fresh air. **Liquid:** Irritating to skin and eyes. Remove contaminated clothing and

shoes. Flush affected areas with plenty of water. *IF IN EYES*, hold eyelids open and flush with plenty of water.

HEALTH HAZARDS
Personal protective equipment (PPE): B-Level PPE. Safety goggles, protective clothing. Chemical protective material(s) reported to have good to excellent resistance: butyl rubber, neoprene. Also, nitrile and PVC offers limited protection
Exposure limits (TWA unless otherwise noted): ACGIH TLV 100 ppm (369 mg/m^3); NIOSH REL 100 ppm (360 mg/m^3).
Short-term exposure limits (15-minute TWA): ACGIH STEL 150 ppm (553 mg/m^3); NIOSH STEL 150 ppm (540 mg/m^3).
Toxicity by ingestion: Grade 1; LD_{50} = 5 to 15 g/kg (rat).
Vapor (gas) irritant characteristics: Vapors cause irritation, such that personnel will find high concentrations unpleasant. The effect is temporary.
Liquid or solid irritant characteristics: Eye and skin irritation. If spilled on clothing and allowed to remain, may cause smarting and reddening of the skin.
Odor threshold: 10 ppm.

FIRE DATA
Flash point: 97°F/36°C (oc); 90°F/32°C (cc).
Flammable limits in air: LEL: 1.6%; UEL: 13.8%.
Fire extinguishing agents not to be used: Water may be ineffective.

CHEMICAL REACTIVITY
Binary reactants: Reacts with strong oxidizers. Hygroscopic; absorbs moisture form the air; may form unstable peroxides.

ENVIRONMENTAL DATA
Food chain concentration potential: Negative; unlikely to accumulate.
Water pollution: Effect of low concentrations on aquatic life is unknown. May be dangerous if it enters nearby water intakes; notify operators. Notify local health and wildlife officials.
Response to discharge: Disperse and flush.

SHIPPING INFORMATION
Grades of purity: Technical; **Storage temperature:** Ambient; **Inert atmosphere:** None; **Venting:** Open (flame arrester); **Stability during transport:** Stable.

PHYSICAL AND CHEMICAL PROPERTIES
Physical state @ 59°F/15°C and 1 atm: Liquid.
Molecular weight: 90.12
Boiling point @ 1 atm: 250°F/121°C/394°K.
Critical temperature: 538°F/281°C/554°K.
Specific gravity (water = 1): 0.924 @ 68°F/20°C (liquid).
Relative vapor density (air = 1): 3.09
Ratio of specific heats of vapor (gas): 1.066
Latent heat of vaporization: (estimate) 166 Btu/lb = 92.3 cal/g = 3.86 x 10^5 J/kg.
Heat of combustion: (estimate) –13,600 Btu/lb = –7580 cal/g = –317 x 10^5 J/kg.
Heat of solution: (estimate) –9 Btu/lb = –5 cal/g = –0.2 x 10^5 J/kg.
Vapor pressure: 10 mm.

PROPYLENE GLYCOL METHYL ETHER ACETATE
REC. P:4350

SYNONYMS: ACETATO de PROPILENGLICOL MONOMETIL ETER (Spanish); 1-METHOXY-2-PROPANOL ACETATE

IDENTIFICATION
CAS Number: 108-65-6
Formula: $C_6H_{12}O_3$; $CH_3COOCH(CH_3)CH_2OCH_3$
DOT ID Number: UN 1993; DOT Guide Number: 128
Proper Shipping Name: Flammable liquid, n.o.s.

DESCRIPTION: Colorless liquid. Sweet, ether-like odor. Slightly soluble in water.

Highly flammable • Containers may BLEVE when exposed to fire • Vapors may form explosive mixture with air • Vapors are heavier than air and will collect and stay in low areas • Vapors may travel long distances to ignition sources and flashback • Vapors in confined areas (e.g., tanks, sewers, buildings) may explode when exposed to fire • May be able to form unstable peroxides • Irritating to the skin, eyes, and respiratory tract • Toxic products of combustion may include CO_2 and carbon monoxide.

Hazard Classification (based on NFPA-704 Rating System)
Health Hazards (Blue): 0; Flammability (Red): 2; Reactivity (Yellow): 0

EMERGENCY RESPONSE: See Appendix A (128)
Evacuation:
Public safety: Isolate spill area for at least 25 to 50 meters (80 to 160 feet) in all directions.
Spill: Large spill–Consider initial downwind evacuation for at least 300 meters (1000 feet).
Fire: Isolate for 800 meters (½ mile) in all directions, especially if tank, rail car, or tank truck is involved in fire.

EXPOSURE Short-term effects: *SEEK MEDICAL ATTENTION.*
Vapor: Move to fresh air. **Liquid:** Remove contaminated clothing and shoes. Wash affected areas with soap and water. *IF IN EYES*, hold eyelids open and flush with plenty of water. *IF SWALLOWED* and victim is *CONSCIOUS AND ABLE TO SWALLOW*, induce vomiting after drinking two glasses of water.

HEALTH HAZARDS
Personal protective equipment (PPE): B-Level PPE. Safety goggles, protective clothing. Use approved organic vapor respirator.
Exposure limits (TWA unless otherwise noted): 100 ppm (AIHAWEEL).
Toxicity by ingestion: Grade 1; LD_{50} = more than 10 g/kg (rat).
Vapor (gas) irritant characteristics: Vapors may be irritating to eyes and throat.
Liquid or solid irritant characteristics: May cause eye and skin irritation.

FIRE DATA
Flash point: 114°F/46°C (oc); 108°F/42°C (cc).
Flammable limits in air: LEL: 1.3% @ 173°F/78°C; 1.5% @ 392°F/200°C; UEL: 13.1% @ 283°F/139°C; 7.0% @ 392°F/200°C.
Fire extinguishing agents not to be used: Water.
Autoignition temperature: 670°F/354°C.
Stoichiometric air-to-fuel ratio: 35.7

CHEMICAL REACTIVITY
Binary reactants: Incompatible with oxidizers. Hygroscopic; absorbs moisture form the air; may form unstable peroxides.

ENVIRONMENTAL DATA
Food chain concentration potential: Negative; unlikely to accumulate.
Water pollution: Effects of low concentrations on aquatic life is unknown. May be dangerous if it enters nearby water intakes; notify operators. Notify local health and wildlife officials.
Response to discharge: Should be removed.

SHIPPING INFORMATION
Grades of purity: 99%; technical; **Storage temperature:** Ambient; **Inert atmosphere:** None; **Stability during transport:** Stable.

PHYSICAL AND CHEMICAL PROPERTIES
Physical state @ 59°F/15°C and 1 atm: Liquid.
Molecular weight: 132.16
Boiling point @ 1 atm: 302°F/150°C/423°K.
Specific gravity (water = 1): 0.969 @ 68°F/20°C.
Relative vapor density (air = 1): 4.58

PROPYLENE IMINE   REC. P:4400

SYNONYMS: AZIRIDINA, 2-METIL (Spanish); AZIRIDINE, 2-METHYL-; 2-METHYLAZACLYCLOPROPANE; 2-METHYLAZIRIDINE; 1,2-PROPYLEN IMINE; 2-METHYLETHYLENIMINE; 1,2-PROPYLENIMINE; RCRA No. P067

IDENTIFICATION
CAS Number: 75-55-8
Formula: C_3H_7N; CH_3CHCH_2NH
DOT ID Number: UN 1921; DOT Guide Number: 131P
Proper Shipping Name: Propyleneimine, inhibited
Reportable Quantity (RQ): **(CERCLA)** 1 lb/0.454 kg

DESCRIPTION: Colorless oily liquid; fumes on contact with air. Strong ammonia-like odor. Mixes with water; hydrolyzes forming methylethanolamine. Flammable, irritating vapor is produced.

Highly flammable • Extremely irritating to skin, eyes, and respiratory tract; skin and eye contact causes severe burns and blindness • Firefighting gear (including SCBA) does not provide adequate protection. If exposure occurs, remove and isolate gear immediately and thoroughly decontaminate personnel • Polymerization hazard • Heat can induce polymerization with rapid release of energy; sealed containers may rupture explosively • May react with itself blocking relief valves; leading to tank explosions • Containers may BLEVE when exposed to fire • Vapors may form explosive mixture with air • Vapors are heavier than air and will collect and stay in low areas • Vapors may travel long distances to ignition sources and flashback • Vapors in confined areas (e.g., tanks, sewers, buildings) may explode when exposed to fire • Toxic products of combustion may include nitrogen oxides and carbon monoxide • Do not put yourself in danger by entering a contaminated area to rescue a victim.

Hazard Classification (based on NFPA-704 Rating System)
Health Hazards (Blue): 3; Flammability (Red): 3; Reactivity (Yellow): 2

EMERGENCY RESPONSE: See Appendix A (131)
Evacuation:
Public safety: Isolate spill area for at least 100 to 200 meters (330 to 660 feet) in all directions.
Spill: Increase, as necessary, the isolation distance shown above, in "Public safety."
Fire: Isolate for 800 meters (½ mile) in all directions, especially if tank, rail car, or tank truck is involved in fire.

EXPOSURE
Short-term effects: *SEEK MEDICAL ATTENTION*. **Vapor:** Irritating to eyes, nose, and throat. *IF INHALED*, will, will cause nausea, vomiting or difficult breathing. Move victim to fresh air. *IF BREATHING HAS STOPPED*, give artificial respiration. IF breathing is difficult, administer oxygen. **Liquid:** *POISONOUS IF SWALLOWED*. Will burn skin and eyes. Remove contaminated clothing and shoes. Flush affected areas with plenty of water. *IF IN EYES*, hold eyelids open and flush with plenty of water. *IF SWALLOWED* and victim is *CONSCIOUS AND ABLE TO SWALLOW*, have victim drink 4 to 8 ounces of water. **Do NOT induce vomiting.**

HEALTH HAZARDS
Personal protective equipment (PPE): A-Level PPE. Self-contained breathing apparatus; goggles or face shield; rubber gloves.
Recommendations for respirator selection: NIOSH
At any detectable concentration: SCBAF:PD,PP (any MSHA/NIOSH approved self-contained breathing apparatus that has a full facepiece and is operated in a pressure-demand or other positive-pressure mode); or SAF:PD,PP:ASCBA (any supplied-air respirator that has a full facepiece and is operated in a pressure-demand or other positive-pressure mode in combination with an auxiliary, self-contained breathing apparatus operated in a pressure-demand or other positive pressure mode). *ESCAPE:* GMFS [any air-purifying, full-facepiece respirator (gas mask) with a chin-style, front- or back-mounted canister providing protection against the compound of concern]; or SCBAE (any appropriate escape-type, self-contained breathing apparatus).
Exposure limits (TWA unless otherwise noted): ACGIH TLV 2 ppm (4.7 mg/m^3) suspected human carcinogen; OSHA PEL 2 ppm (5 mg/m^3); NIOSH 2 ppm (5 mg/m^3); skin contact contributes significantly in overall exposure; suspected human carcinogen.
Toxicity by ingestion: Grade 4; oral LD_{50} = 19 mg/kg (rat).
Long-term health effects: Suspected human carcinogen
Vapor (gas) irritant characteristics: Eye and respiratory tract irritant.
Liquid or solid irritant characteristics: Severe eye and skin irritant.
IDLH value: 100 ppm; suspected human carcinogen.

FIRE DATA
Flash point: 25°F/3.8°C (oc).
Fire extinguishing agents not to be used: Water or foam may be ineffective.
Behavior in fire: Containers may rupture and explode.
Burning rate: 4.1 mm/min

CHEMICAL REACTIVITY
Reactivity with water: Reacts slowly to form propanolamine. Hydrolyzes to form methylethanolamine.
Binary reactants: Acids (see polymerization, below), strong oxidizers, carbonyl compounds, quinone, sulfonyl halides.
Neutralizing agents for acids and caustics: Dilute with water, rinse with vinegar.
Polymerization: Polymerizes explosively when in contact with any acid; **Inhibitor of polymerization:** Solid sodium hydroxide (caustic soda).

ENVIRONMENTAL DATA
Water pollution: Effect of low concentrations on aquatic life is unknown. May be dangerous if it enters nearby water intakes; notify operators. Notify local health and wildlife officials.
Response to discharge: Issue warning–high flammability, water contaminant, air contaminant. Restrict access. Evacuate area. Disperse and flush.

SHIPPING INFORMATION
Grades of purity: Technical; **Storage temperature:** Ambient; **Inert atmosphere:** Exclude air; **Venting:** Pressure-vacuum; **Stability during transport:** Stable if kept in contact with solid caustic soda (sodium hydroxide).

PHYSICAL AND CHEMICAL PROPERTIES
Physical state @ 59°F/15°C and 1 atm: Liquid.
Molecular weight: 57.1
Boiling point @ 1 atm: 151°F/66°C/339°K.
Melting/Freezing point: –85°F/–65°C/208°K.
Specific gravity (water = 1): 0.802 @ 77°F/25°C (liquid).
Liquid surface tension: (estimate) 25 dynes/cm = 0.025 N/m @ 68°F/20°C.
Relative vapor density (air = 1): 2
Latent heat of vaporization: 250 Btu/lb = 139 cal/g = 5.82×10^5 J/kg.
Heat of combustion: (estimate) –15,500 Btu/lb = –8600 cal/g = -360×10^5 J/kg.
Heat of solution: –140 Btu/lb = –78 cal/g = -3.3×10^5 J/kg.
Heat of polymerization: (estimate) –720 Btu/lb = –400 cal/g = -17×10^5 J/kg.
Vapor pressure: 112 mm.

PROPYLENE OXIDE   REC. P:4450

SYNONYMS: EEC No. 603-055-00-4; EPOXYPROPANE; 1,2-EPOXYPROPANE; 2,3-EPOXYPROPANE; ETHYLENE OXIDE, METHYL-; METHYL ETHYLENE OXIDE; METHYL OXIRANE; NCI-C50099; OXIDO de PROPILENO (Spanish); OXIRANE, METHYL-; OXYDE de PROPYLENE (French); PROPANE, EPOXY-; PROPENE OXIDE; 1,2-PROPYLENE OXIDE; PROPENEOXIDE

IDENTIFICATION
CAS Number: 75-56-9
Formula: C_3H_6O; CH_3CHCH_2O
DOT ID Number: UN 1280; DOT Guide Number: 127P
Proper Shipping Name: Propylene oxide
Reportable Quantity (RQ): **(CERCLA)** 100 lb/45.4 kg

DESCRIPTION: Colorless liquid. Sweet, ether-like odor. Mixes with water forming a corrosive solution.

Extremely flammable • Corrosive to the skin, eyes, and respiratory tract; Contact with skin and eyes may cause burns and impaired vision; inhalation symptoms may be delayed • Polymerization hazard • Heat can induce polymerization with rapid release of energy; sealed containers may rupture explosively • May react with itself blocking relief valves; leading to tank explosions • Containers may BLEVE when exposed to fire • Vapors may form explosive mixture with air • Vapors are heavier than air and will collect and stay in low areas • Vapors may travel long distances to ignition sources and flashback • Vapors in confined areas (e.g., tanks, sewers, buildings) may explode when exposed to fire •

Toxic products of combustion may include carbon monoxide.
Warning: Odor is not a reliable indicator of the presence of toxic amounts of propylene oxide.

Hazard Classification (based on NFPA-704 Rating System)
Health Hazards (Blue): 3; Flammability (Red): 4; Reactivity (Yellow): 2

EMERGENCY RESPONSE: See Appendix A (127)
Evacuation:
Public safety: Isolate spill area for at least 25 to 50 meters (80 to 160 feet) in all directions.
Spill: Large spill–Consider initial downwind evacuation for at least 300 meters (1000 feet).
Fire: Isolate for 800 meters (½ mile) in all directions, especially if tank, rail car, or tank truck is involved in fire.

EXPOSURE
Short-term effects: *SEEK MEDICAL ATTENTION.* **Vapor:** Irritating to eyes, nose, and throat. *IF INHALED*, will, will cause headache, nausea, vomiting, or loss of consciousness. Move to fresh air. *IF BREATHING HAS STOPPED,* give artificial respiration. IF breathing is difficult, administer oxygen. Lung edema may develop.
Liquid: Will burn skin and eyes. Harmful if swallowed. Remove contaminated clothing and shoes. Flush affected areas with plenty of water. *IF IN EYES*, hold eyelids open and flush with plenty of water. *IF SWALLOWED* and victim is *CONSCIOUS AND ABLE TO SWALLOW*, have victim drink 4 to 8 ounces of water.
Note to physician or authorized medical personnel: Medical observation is recommended for 24 to 48 hours after breathing overexposure, as pulmonary edema may be delayed. As first aid for pulmonary edema, consider administering a corticosteroid spray. Cigarette smoking may exacerbate pulmonary injury and should be discouraged for at least 72 hours following exposure.

HEALTH HAZARDS
Personal protective equipment (PPE): A-Level PPE. Chemical protective material(s) reported to offer minimal to poor protection: butyl rubber, PV alcohol, Teflon®.
Recommendations for respirator selection: NIOSH
At any detectable concentration: SCBAF:PD,PP (any MSHA/NIOSH approved self-contained breathing apparatus that has a full facepiece and is operated in a pressure-demand or other positive-pressure mode); or SAF:PD,PP:ASCBA (any supplied-air respirator that has a full facepiece and is operated in a pressure-demand or other positive-pressure mode in combination with an auxiliary, self-contained breathing apparatus operated in a pressure-demand or other positive pressure mode). *ESCAPE:* GMFS [any air-purifying, full-facepiece respirator (gas mask) with a chin-style, front- or back-mounted canister providing protection against the compound of concern]; or SCBAE (any appropriate escape-type, self-contained breathing apparatus).

Exposure limits (TWA unless otherwise noted): ACGIH TLV 2 ppm (4.8 mg/m³); OSHA 100 ppm (240 mg/m³); NIOSH REL suspected carcinogen; reduce to lowest feasible level.
Toxicity by ingestion: Grade 2; LD_{50} = 0.5 to 5 g/kg (rat).
Long-term health effects: Repeated or prolonged exposure causes skin problems. Suspected human carcinogen.
Vapor (gas) irritant characteristics: Vapor is irritating such that personnel will not usually tolerate moderate or high vapor concentrations.
Liquid or solid irritant characteristics: Severe eye irritant. Causes smarting of the skin and first-degree burns on short exposure; may cause secondary burns on long exposure.

Odor threshold: 10–200 ppm.
IDLH value: 400 ppm; human carcinogen.

FIRE DATA
Flash point: Less than –20°F/29°C (oc); –35°F/–37°C (cc).
Flammable limits in air: LEL: 2.3%; UEL 36%.
Fire extinguishing agents not to be used: Water may cause a runaway reaction.
Autoignition temperature: 842°F/450°C/723°K.
Electrical hazard: Class I, Group B
Burning rate: 3.3 mm/min

CHEMICAL REACTIVITY
Reactivity with water: Reaction with water may be dangerous.
Binary reactants: Oxidizers, anhydrous metal hydrogen chloride; iron; strong acids, caustics, and peroxides. Reacts with copper and other acetylide-forming metals.
Polymerization: May occur due to high temperatures, contamination with alkalies, aqueous acids, metal halides, amines, and acidic alcohols.
Reactivity group: 16
Compatibility class: Alkylene oxides

ENVIRONMENTAL DATA
Food chain concentration potential: Negative; unlikely to accumulate.
Water pollution: Effect of low concentrations on aquatic life is unknown. May be dangerous if it enters nearby water intakes; notify operators. Notify local health and wildlife officials.
Response to discharge: Issue warning–high flammability. Restrict access. Evacuate area.

SHIPPING INFORMATION
Grades of purity: 99.99% (must contain no acetylene); **Storage temperature:** Ambient; **Inert atmosphere:** Inerted; **Venting:** Safety relief; **Stability during transport:** Stable.

NAS HAZARD CLASSIFICATION FOR BULK WATER TRANSPORTATION
FIRE: 4
HEALTH: Vapor irritant: 3; Liquid or solid irritant: 2; Poisons: 2
WATER POLLUTION: Human toxicity: 2; Aquatic toxicity: 1; Aesthetic effect: 1
REACTIVITY: Other chemicals: 3; Water: 1; Self-reaction: 3

PHYSICAL AND CHEMICAL PROPERTIES
Physical state @ 59°F/15°C and 1 atm: Liquid.
Molecular weight: 58.08
Boiling point @ 1 atm: 93.7°F/34.3°C/307.5°K.
Melting/Freezing point: –169.4°F/–111.9°C/161.3°K.
Critical temperature: 408.4°F/209.1°C/482.3°K.
Critical pressure: 714 psia = 48.6 atm = 4.92 MN/m^2.
Specific gravity (water = 1): 0.830 @ 68°F/20°C (liquid).
Liquid surface tension: 24.5 dynes/cm = 0.0245 N/m @ 15°C.
Relative vapor density (air = 1): 1.83
Ratio of specific heats of vapor (gas): 1.133
Latent heat of vaporization: 205 Btu/lb = 114 cal/g = 4.77 x 10^5 J/kg.
Heat of combustion: –13,000 Btu/lb
Latent heat of vaporization: 4.77 x 10^5 J/kg.
Heat of combustion: = –7221 cal/g = –302.3 x 10^5 J/kg.
Heat of solution: (estimate) –19 Btu/lb = –11 cal/g = –0.45 x 10^5 J/kg.
Vapor pressure: (Reid) 18.0 psia; 445 mm.

PROPYLENE TETRAMER   REC. P:4500

SYNONYMS: AMSCO TETRAMER; DODECENE (nonlinear); PROPENE TETRAMER; TETRAMERO de PROPILENO (Spanish); TETRAPROPYLENE

IDENTIFICATION
CAS Number: 6842-15-5
Formula: $C_{12}H_{14}$
DOT ID Number: UN 2850; DOT Guide Number: 128
Proper Shipping Name: Propylene tetramer

DESCRIPTION: Colorless liquid. Odorless. Floats on water surface; insoluble.

Flammable • Containers may BLEVE when exposed to fire • Vapors may form explosive mixture with air • Vapors are heavier than air and will collect and stay in low areas • Vapors may travel long distances to ignition sources and flashback • Vapors in confined areas (e.g., tanks, sewers, buildings) may explode when exposed to fire • Irritating to the skin, eyes, and respiratory tract; high concentrations are narcotic • Toxic products of combustion may include carbon monoxide.

EMERGENCY RESPONSE: See Appendix A (128)
Evacuation:
Public safety: Isolate spill area for at least 25 to 50 meters (80 to 160 feet) in all directions.
Spill: Large spill–Consider initial downwind evacuation for at least 300 meters (1000 feet).
Fire: Isolate for 800 meters (½ mile) in all directions, especially if tank, rail car, or tank truck is involved in fire.

EXPOSURE
Short-term effects: *SEEK MEDICAL ATTENTION.* **Liquid:** Irritating to skin and eyes. Harmful if swallowed.
Flush affected area with plenty of water. *IF IN EYES*, hold eyelids open and flush with plenty of water. *IF SWALLOWED* and victim is *CONSCIOUS AND ABLE TO SWALLOW*, have victim drink 4 to 8 ounces of water. **Do NOT induce vomiting.**

HEALTH HAZARDS
Personal protective equipment (PPE): B-Level PPE. Goggles or face shield.
Toxicity by ingestion: Grade 0; LD_{50} = > 15 g/kg.
Vapor (gas) irritant characteristics: Vapors cause smarting of the eyes or respiratory system if present in high concentrations. The effect is temporary.
Liquid or solid irritant characteristics: If spilled on clothing and allowed to remain, may cause smarting and reddening of the skin.

FIRE DATA
Flash point: 120°F/49°C (cc); 134°F/57°C (oc).
Autoignition temperature: 400°F/204°C/477°K (estimate).
Electrical hazards: Flow or agitation of substance may generate electrostatic charges due to low conductivity.

CHEMICAL REACTIVITY
Binary reactants: Reacts with inorganic acids, halogenated compounds, molten sulfur, and oxidizers.
Reactivity group: 30
Compatibility class: Olefins

ENVIRONMENTAL DATA
Food chain concentration potential: Negative.

Water pollution: Effect of low concentrations on aquatic life is unknown. Fouling to shoreline. May be dangerous if it enters nearby water intakes; notify operators. Notify local health and wildlife officials. **Response to discharge:** Mechanical containment. Should be removed. Chemical and physical treatment.

SHIPPING INFORMATION
Grades of purity: 98.5+%; **Storage temperature:** Ambient; **Inert atmosphere:** None; **Venting:** Open (flame arrester); **Stability during transport:** Stable.

NAS HAZARD CLASSIFICATION FOR BULK WATER TRANSPORTATION
FIRE: 2
HEALTH: Vapor irritant: 1; Liquid or solid irritant: 1; Poisons: 1
WATER POLLUTION: Human toxicity: 0; Aquatic toxicity: 1; Aesthetic effect: 2
REACTIVITY: Other chemicals: 1; Water: 0; Self-reaction: 1

PHYSICAL AND CHEMICAL PROPERTIES
Physical state @ 59°F/15°C and 1 atm: Liquid.
Molecular weight: 168.31
Boiling point @ 1 atm: 365–385°F/185–196°C/458–469°K.
Specific gravity (water = 1): 0.2937 @ 68°F/20°C (liquid).
Liquid surface tension: 23.9 dynes/cm = 0.0239 N/m at 24°C.
Liquid water interfacial tension: 44.5 dynes/cm = 0.0445 N/m at 22.7°C.
Latent heat of vaporization: (estimate) 154 Btu/lb = 58.6 cal/g = 2.45×10^5 J/kg.
Heat of combustion: $-19{,}100$ Btu/lb = $-10{,}600$ cal/g = -444×10^5 J/kg.

PROPYLENE TRIMER   REC. P:4550

SYNONYMS: PROPENE, TRIMER; NONENE (NONLINEAR); NONENE; TRIPROPYLENE; TRIPROPILENO (Spanish)

IDENTIFICATION
CAS Number: 13987-01-4
Formula: $(C_3H_6)_3$; C_9H_{18}
DOT ID Number: UN 2057; DOT Guide Number: 128
Proper Shipping Name: Tripropylene

DESCRIPTION: Colorless liquid. Odorless.

Highly flammable • Containers may BLEVE when exposed to fire • Vapors may form explosive mixture with air • Vapors are heavier than air and will collect and stay in low areas • Vapors may travel long distances to ignition sources and flashback • Vapors in confined areas (e.g., tanks, sewers, buildings) may explode when exposed to fire • Irritating to the skin, eyes, and respiratory tract • Toxic products of combustion may include carbon monoxide.

Hazard Classification (based on NFPA-704 Rating System)
Health Hazards (Blue): 0; Flammability (Red): 3; Reactivity (Yellow): 0

EMERGENCY RESPONSE: See Appendix A (128)
Evacuation:
Public safety: Isolate spill area for at least 25 to 50 meters (80 to 160 feet) in all directions.
Spill: Large spill–Consider initial downwind evacuation for at least 300 meters (1000 feet).
Fire: Isolate for 800 meters (½ mile) in all directions, especially if tank, rail car, or tank truck is involved in fire.

EXPOSURE
Short-term effects: *SEEK MEDICAL ATTENTION*. **Vapor:** Irritating to eyes, nose, and throat. *IF INHALED*, will, will cause dizziness, headache, difficult breathing or loss of consciousness. Move to fresh air. *IF BREATHING HAS STOPPED*, give artificial respiration. IF breathing is difficult, administer oxygen. **Liquid:** Irritating to skin and eyes. Remove contaminated clothing and shoes. Flush affected areas with plenty of water. *IF IN EYES*, hold eyelids open and flush with plenty of water.

HEALTH HAZARDS
Personal protective equipment (PPE): B-Level PPE. Respiratory and eye protection required for fire fighters. Supplied air breathing apparatus for large spill.
Toxicity by ingestion: Low toxicity
Long-term health effects: Possible lung damage.
Vapor (gas) irritant characteristics: Vapors cause moderate irritation such that personnel will not usually tolerate moderate or high concentration.
Liquid or solid irritant characteristics: If spilled on clothing and allowed to remain, may cause smarting and reddening.

FIRE DATA
Flash point: 75°F/24°C (oc).
Fire extinguishing agents not to be used: Water may be ineffective.

CHEMICAL REACTIVITY
Binary reactants: Strong acids, oxidizers.
Reactivity group: 30
Compatibility class: Olefins

ENVIRONMENTAL DATA
Water pollution: Effect of low concentrations on aquatic life is unknown. Fouling to shoreline. May be dangerous if it enters nearby water intakes; notify operators. Notify local health and wildlife officials. **Response to discharge:** Should be removed. Mechanical containment. Chemical and physical treatment. Do not use combustible adsorbents.

SHIPPING INFORMATION
Storage temperature: Ambient; **Stability during transport:** Stable.

PHYSICAL AND CHEMICAL PROPERTIES
Physical state @ 59°F/15°C and 1 atm: Liquid.
Molecular weight: 126.24
Boiling point @ 1 atm: 271.94–287.06°F/133.3–141.7°C/406.5–414.9°K.
Specific gravity (water = 1): 0.738
Relative vapor density (air = 1): 4.35

n-PROPYL ETHER   REC. P:4600

SYNONYMS: DIPROPYL ETHER; DIPROPYL OXIDE; DI-n-PROPYL ETHER; PROPYL ETHER; PROPIL ETER (Spanish)

IDENTIFICATION
CAS Number: 111-43-3
Formula: $C_6H_{14}O$; $(CH_3CH_2CH_2)_2O$
DOT ID Number: UN 238; DOT Guide Number: 127
Proper Shipping Name: Dipropyl ether

DESCRIPTION: Colorless liquid. Floats on water surface; slightly soluble.

Highly flammable • Containers may BLEVE when exposed to fire • Vapors may form explosive mixture with air • Vapors are heavier than air and will collect and stay in low areas • Vapors may travel long distances to ignition sources and flashback • Vapors in confined areas (e.g., tanks, sewers, buildings) may explode when exposed to fire • Irritating to the skin, eyes, and respiratory tract • May be able to form unstable peroxides in storage • Toxic products of combustion may include carbon monoxide.

Hazard Classification (based on NFPA-704 Rating System)
Health Hazards (Blue): -; Flammability (Red): 3; Reactivity (Yellow): 0

EMERGENCY RESPONSE: See Appendix A (127)
Evacuation:
Public safety: Isolate spill area for at least 25 to 50 meters (80 to 160 feet) in all directions.
Spill: Large spill–Consider initial downwind evacuation for at least 300 meters (1000 feet).
Fire: Isolate for 800 meters (½ mile) in all directions, especially if tank, rail car, or tank truck is involved in fire.

EXPOSURE Short-term effects: *SEEK MEDICAL ATTENTION*. **Vapor:** Move victim to fresh air. *IF BREATHING HAS STOPPED*, give artificial respiration. If breathing is difficult, administer oxygen. **Liquid:** Remove contaminated clothing and shoes. Wash affected areas with soap and water. *IF IN EYES*, hold eyelids open and flush with plenty of water.

HEALTH HAZARDS
Personal protective equipment (PPE): B-Level PPE. Full impervious protective clothing, including boots and gloves. Where splashing is possible wear full face shield or chemical safety goggles. Use approved respirator to protect against vapors.
FIRE DATA
Flash point: 70°F/21°C.
Flammable limits in air: LEL: 1.3%; UEL: 7.0%.
Fire extinguishing agents not to be used: Water.
Autoignition temperature: 374°F/190°C/463°K.
Stoichiometric air-to-fuel ratio: 42.9

CHEMICAL REACTIVITY
Binary reactants: Tends to form explosive peroxides, especially when anhydrous. May be able to form unstable peroxides.
Reactivity group: 41
Compatibility class: Ethers

ENVIRONMENTAL DATA
Water pollution: Effects of low concentrations on aquatic life is unknown. May be dangerous if it enters nearby water intakes; notify operators. Notify local health and wildlife officials.
Response to discharge: Issue warning–high flammability. Should be removed.

SHIPPING INFORMATION
Grades of purity: 99%; technical; **Storage temperature:** Ambient; **Inert atmosphere:** None; **Stability during transport:** Stable.

PHYSICAL AND CHEMICAL PROPERTIES
Physical state @ 59°F/15°C and 1 atm: Liquid.
Molecular weight: 102.17
Boiling point @ 1 atm: 192-196°F/89-91°C/362-364°K.
Melting/Freezing point: –187.6°F/–122°C/151°K.
Specific gravity (water = 1): 0.7360
Relative vapor density (air = 1): 3.5

n-PROPYL MERCAPTAN   REC. P:4650

SYNONYMS: 3-MERCAPTOPROPANE; 3-MERCAPTOPROPANOL; *n*-PROPILMERCAPTANO (Spanish); ROPYL MERCAPTAN; PROPANETHIOL; 1-PROPANETHIOL; PROPANE-1-THIOL

IDENTIFICATION
CAS Number: 107-03-9
Formula: C_3H_8S; $CH_3CH_2CH_2SH$
DOT ID Number: UN 2402; DOT Guide Number: 130
Proper Shipping Name: Propanethiols; Propyl mercaptan

DESCRIPTION: Colorless liquid. Skunk-like odor. Floats on water surface; slightly soluble; reacts, forming toxic mercaptan vapors.

Highly flammable • Containers may BLEVE when exposed to fire • Vapors may form explosive mixture with air • Vapors are heavier than air and will collect and stay in low areas • Vapors may travel long distances to ignition sources and flashback • Vapors in confined areas (e.g., tanks, sewers, buildings) may explode when exposed to fire • Irritating to the skin, eyes, and respiratory tract • Toxic products of combustion may include sulfur oxides.

Hazard Classification (based on NFPA-704 Rating System)
Health Hazards (Blue): 2; Flammability (Red): 4; Reactivity (Yellow): 0

EMERGENCY RESPONSE: See Appendix A (130)
Evacuation:
Public safety: Isolate spill area for at least 50 to 100 meters (160 to 330 feet) in all directions.
Spill: Large spill–Consider initial downwind evacuation for at least 300 meters (1000 feet).
Fire: Isolate for 800 meters (½ mile) in all directions, especially if tank, rail car, or tank truck is involved in fire.

EXPOSURE
Short-term effects: *SEEK MEDICAL ATTENTION*. **Vapor:** IF inhaled, will cause difficult breathing. Move victim to fresh air. *IF BREATHING HAS STOPPED*, give artificial respiration. IF breathing is difficult, administer oxygen. **Liquid:** Irritating to skin and eyes. Harmful if swallowed. Remove contaminated clothing and shoes. Flush affected area with plenty of water. *IF IN EYES*, hold eyelids open and flush with plenty of water. *IF SWALLOWED* and victim is *CONSCIOUS AND ABLE TO SWALLOW*, have victim drink 4 to 8 ounces of water and have victim induce vomiting. *IF SWALLOWED* and victim is *UNCONSCIOUS OR HAVING CONVULSIONS*, do nothing except keep victim warm. *Note to physician or authorized medical personnel*: Medical observation is recommended for 24 to 48 hours after breathing overexposure, as pulmonary edema may be delayed. As first aid for pulmonary edema, consider administering a corticosteroid spray. Cigarette smoking may exacerbate pulmonary injury and should be discouraged for at least 72 hours following exposure.

HEALTH HAZARDS

Personal protective equipment (PPE): B-Level PPE. Goggles or face shield; rubber gloves; self-contained breathing apparatus or organic canister mask

Recommendations for respirator selection: NIOSH
5 ppm: CCROV [any chemical cartridge respirator with organic vapor cartridge(s)]; or SA (any supplied-air respirator). *12.5 ppm:* SA:CF (any supplied-air respirator operated in a continuous-flow mode); or PAPROV [any powered, air-purifying respirator with organic vapor cartridge(s)]. *25 ppm:* CCROV [any chemical cartridge respirator with organic vapor cartridge(s)]; or GMFOV [any air-purifying, full-facepiece respirator (gas mask) with a chin-style, front- or back-mounted acid gas canister]; or PAPRTOV [any powered, air-purifying respirator with a tight-fitting facepiece and organic vapor cartridges(s)]; or SCBAF (any self-contained breathing apparatus with a full facepiece); or SAF (any supplied-air respirator with a full facepiece). *EMERGENCY OR PLANNED ENTRY INTO UNKNOWN CONCENTRATIONS OR IDLH CONDITIONS:* SCBAF:PD,PP (any self-contained breathing apparatus that has a full facepiece and is operated in a pressure-demand or other positive-pressure mode); or SAF:PD,PP:ASCBA (any supplied-air respirator that has a full facepiece and is operated in a pressure-demand or other positive-pressure mode in combination with an auxiliary self-contained breathing apparatus operated in a pressure-demand or other positive pressure mode). *ESCAPE:* GMFOV [any air-purifying, full-facepiece respirator (gas mask) with a chin-style, front- or back-mounted organic vapor cannister]; or SCBAE (any appropriate escape-type, self-contained breathing apparatus).

Toxicity by ingestion: Grade 2; oral LD_{50} = 1790 mg/kg (rat).
Liquid or solid irritant characteristics: Eye and skin irritation.
Odor threshold: 0.00075 ppm.

FIRE DATA

Flash point: –5°F/21°C (oc).
Fire extinguishing agents not to be used: Water may be ineffective.
Burning rate: 5.1 mm/min

CHEMICAL REACTIVITY

Binary reactants: Oxidizers, reducing agents, strong acids and bases, alkali metals, calcium hypochlorite.

ENVIRONMENTAL DATA

Food chain concentration potential: Log P_{ow} = 1.7. Unlikely to accumulate.
Water pollution: Effect of low concentrations on aquatic life is unknown. Fouling to shoreline. May be dangerous if it enters nearby water intakes; notify operators. Notify local health and wildlife officials. **Response to discharge:** Issue warning–high flammability, water contaminant, air contaminant. Restrict access. Evacuate area. Mechanical containment. Should be removed. Chemical and physical treatment.

SHIPPING INFORMATION

Grades of purity: 98+%; **Storage temperature:** Ambient; **Inert atmosphere:** None; **Venting:** Pressure-vacuum; **Stability during transport:** Stable.

PHYSICAL AND CHEMICAL PROPERTIES

Physical state @ 59°F/15°C and 1 atm: Liquid.
Molecular weight: 76.2
Boiling point @ 1 atm: 153°F/67°C/340°K.
Melting/Freezing point: –171°F/–113°C/160°K.
Critical temperature: (estimate) 495°F/257°C/530°K.
Critical pressure: (estimate) 667 psia = 45.3 atm = 4.60 MN/m^2.
Specific gravity (water = 1): 0.841 @ 68°F/20°C (liquid).
Liquid surface tension: 24.7 dynes/cm = 0.0247 N/m @ 68°F/20°C.
Liquid water interfacial tension: (estimate) 18 dynes/cm = 0.018 N/m @ 68°F/20°C.
Relative vapor density (air = 1): 2.6
Ratio of specific heats of vapor (gas): 1.0984
Latent heat of vaporization: 179 Btu/lb = 99 cal/g = 4.16 x 10^5 J/kg.
Heat of combustion: –15,990 Btu/lb = –8,890 cal/g = 372 x 10^5 J/kg.
Vapor pressure: 155 mm @ 77°F/25°C.

n-PROPYL NITRATE   REC. P:4700

SYNONYMS: NITRATO de *n*-PROPILO (Spanish); NITRIC ACID, PROPYL ESTER; PROPYL NITRATE; PROPYL ESTER OF NITRIC ACID

IDENTIFICATION

CAS Number: 627-13-4
Formula: $C_3H_7NO_3$; $CH_3CH_2CH_2NO_3$
DOT ID Number: UN 1865; DOT Guide Number: 131
Proper Shipping Name: *n*-Propyl nitrate

DESCRIPTION: Colorless to pale yellow liquid. Ether-like odor. Slightly soluble in water.

Thermally unstable and shock-sensitive. Decomposes at 347°F/175°C • Strong oxidizer which may react spontaneously with low flash point organics or reducing agents. Heat forms oxygen; will increase the activity of an existing fire • May cause fire or explosion on contact with combustibles (wood, paper, oil, clothing, etc.) • Irritating to the skin, eyes, and respiratory tract • Toxic products of combustion may include nitrogen oxides and carbon monoxide.

Hazard Classification (based on NFPA-704 Rating System)
Health Hazards (Blue): 2; Flammability (Red): 3; Reactivity (Yellow): 3; Special Notice (White): OXY

EMERGENCY RESPONSE: See Appendix A (131)
Evacuation:
Public safety: Isolate spill area for at least 100 to 200 meters (330 to 660 feet) in all directions.
Spill: Increase, as necessary, the isolation distance shown above, in "Public safety."
Fire: Isolate for 800 meters (½ mile) in all directions, especially if tank, rail car, or tank truck is involved in fire.

EXPOSURE

Short-term effects: *SEEK MEDICAL ATTENTION.* **Vapor:** Move victim to fresh air. *IF BREATHING HAS STOPPED,* give artificial respiration. If breathing is difficult, administer oxygen. **Liquid:** Remove contaminated clothing and shoes. Wash affected areas with soap and water. *IF IN EYES,* hold eyelids open and flush with plenty of water. *IF SWALLOWED* and victim is *CONSCIOUS AND ABLE TO SWALLOW,* give two glasses of water and induce vomiting.

HEALTH HAZARDS

Personal protective equipment (PPE): A-Level PPE. Full,

impervious chemical protective clothing and gloves, goggles, and approved respirator.
Recommendations for respirator selection: NIOSH/OSHA
250 ppm: SA (any supplied-air respirator). *500 ppm*: SA:CF (any supplied-air respirator operated in a continuous-flow mode); or SCBAF (any self-contained breathing apparatus with a full facepiece); or SAF (any supplied-air respirator with a full facepiece). *EMERGENCY OR PLANNED ENTRY INTO UNKNOWN CONCENTRATIONS OR IDLH CONDITIONS:* SCBAF:PD,PP (any self-contained breathing apparatus that has a full facepiece and is operated in a pressure-demand or other positive-pressure mode); or SAF:PD,PP:ASCBA (any supplied-air respirator that has a full facepiece and is operated in a pressure-demand or other positive-pressure mode in combination with an auxiliary self-contained breathing apparatus operated in a pressure-demand or other positive pressure mode). *ESCAPE:* GMFS* [any air-purifying, full-facepiece respirator (gas mask) with a chin-style, front- or back-mounted canister providing protection against the compound of concern]; or SCBAE (any appropriate escape-type, self-contained breathing apparatus). *Note*: End of service life indicator (ESLI) required.
Exposure limits (TWA unless otherwise noted): ACGIH TLV 25 ppm (107 mg/m^3); OSHA PEL 25 ppm (110 mg/m^3); NIOSH REL 25 ppm (110 mg/m^3).
Short-term exposure limits (15-minute TWA): ACGIH STEL 40 ppm (172 mg/m^3); NIOSH STEL 40 ppm (170 mg/m^3).
Vapor (gas) irritant characteristics: Vapors cause smarting of the eyes or respiratory system if present in high concentrations. The effect is temporary.
Liquid or solid irritant characteristics: If spilled on clothing and allowed to remain, may cause smarting and reddening of the skin.
Odor threshold: 50 ppm.
IDLH value: 500 ppm.

FIRE DATA
Flash point: 68°F/20°C (cc).
Flammable limits in air: LEL: 2%; UEL: 100%.
Fire extinguishing agents not to be used: Water may be ineffective; use to keep exposed containers cool.
Behavior in fire: Containers may rupture and explode.
Autoignition temperature: 347°F/175°C/448°K (explosive decomposition).
Electrical hazard: Class I, Group B
Stoichiometric air-to-fuel ratio: 15.5

CHEMICAL REACTIVITY
Binary reactants: Contact with either strong oxidizers or with combustibles may cause fires and explosions. Will attack some forms of plastics, rubber, and coatings (insulators).

ENVIRONMENTAL DATA
Water pollution: Effects of low concentrations on aquatic life is unknown. May be dangerous if it enters nearby water intakes; notify operators. Notify local health and wildlife officials.
Response to discharge: Should be removed.

SHIPPING INFORMATION
Grades of purity: Technical; **Storage temperature:** Ambient; **Inert atmosphere:** None; **Stability during transport:** Stable.

PHYSICAL AND CHEMICAL PROPERTIES
Physical state @ 59°F/15°C and 1 atm: Liquid.
Molecular weight: 105.1
Boiling point @ 1 atm: 231°F/111°C/384°K.
Melting/Freezing point: Less than –150°F/–101°C/172°K.
Specific gravity (water = 1): 1.07
Relative vapor density (air = 1): 3.6

PYRETHRINS REC. P:4750

SYNONYMS: BUHACH; CINERIN I; CINERIN II; FIRMOTOX; DALMATION-INSECT POWDER; JASMOLIN I; JASMOLIN II; PILITRE (Spanish); PERSIAN-INSECT POWDER; PIRETINA (Spanish); PYRETHRUM; PYRETHRUM I; PYRETHRUM II; PYRETHRUM FLOWERS; TRIESTE FLOWERS

IDENTIFICATION
CAS Number: 8003-34-7; 8003-73-7; 97-11-0; 121-21-1; 121-29-9; 584-79-2 (synthetic pyrethrins)
Formula: $C_{21}H_{28}O_3$ (pyrethrin I); $C_{22}H_{28}O_5$ (pyrethrin II); $C_{20}H_{28}O_3$ (cinerin I); $C_{21}H_{28}O_5$ (cinerin II); pyrethrum is a variable mixture of cinerin, jasmolin, and pyrethrin)
DOT ID Number: UN 2902; DOT Guide Number: 151
Proper Shipping Name: Pesticides, liquid, toxic, n.o.s.
Reportable Quantity (RQ): **(CERCLA)** 1 lb/0.454 kg

DESCRIPTION: Yellow to brown thick liquid or brown resin. Characteristic odor of carrier which may be alcohol or petroleum distillate. Sinks in water; insoluble.

Poison! • Irritating to the skin, eyes, and respiratory tract; ingestion may cause chronic sickness and death • Combustible • Heat or flame may cause explosion • Firefighting gear (including SCBA) does not provide adequate protection. If exposure occurs, remove and isolate gear immediately and thoroughly decontaminate personnel • Containers may BLEVE when exposed to fire • Vapors are heavier than air and will collect and stay in low areas • Vapors in confined areas (e.g., tanks, sewers, buildings) may explode when exposed to fire • Toxic products of combustion may include CO_2 and carbon monoxide.

Hazard Classification (based on NFPA-704 Rating System)
Health Hazards (Blue): 2; Flammability (Red): 1; Reactivity (Yellow): 0

EMERGENCY RESPONSE: See Appendix A (151)
Evacuation:
Public safety: Isolate the area of spill or leak for at least 25 to 50 meters (80 to 160 feet) in all directions.
Spill: Increase, in the downwind direction, as necessary, the distance shown under "Public Safety."
Fire: If tank, rail car, or tank truck is involved in fire, isolate for at least 800 meters (½ mile) in all directions; also, consider initial evacuation for 800 meters (½ mile) in all directions.

EXPOSURE
Short-term effects: *SEEK MEDICAL ATTENTION*. **Vapor:** Irritating to skin and eyes. *IF INHALED*, will, may cause sneezing, nasal discharge, and nasal stuffiness. Move to fresh air.*IF BREATHING HAS STOPPED*, give artificial respiration; *avoid mouth-to-mouth resuscitation; use bag/mask apparatus*. IF breathing is difficult, administer oxygen. **Liquid:** Irritating to skin and eyes. *IF SWALLOWED* may cause nausea, vomiting, headache, and other central nervous system disturbances. Remove contaminated clothing and shoes. Flush affected areas with plenty of water. *IF IN EYES*, hold eyelids open and flush with plenty of water. *IF SWALLOWED* and victim is *CONSCIOUS AND ABLE TO SWALLOW*, have victim drink 4 to 8 ounces of water.

Note to physician or authorized medical personnel: Diarrhea can be controlled with atropine sulfate.

HEALTH HAZARDS
Personal protective equipment (PPE): B-Level PPE. Protective clothing and filter mask recommended.
Recommendations for respirator selection: NIOSH/OSHA *50 mg/m³*: CCROVDMFu* [any chemical cartridge respirator with organic vapor cartridge(s) in combination with a dust, mist, and fume filter]; or SA* (any supplied-air respirator). *125 mg/m³*: SA:CF* (any supplied-air respirator operated in a continuous-flow mode); or PAPROVDMFu* [any powered, air purifying respirator with organic vapor cartridge (s) in combination with a dust, mist, and fume filter]. *250 mg/m³*: CCRFOVHiE [any chemical cartridge respirator with a full facepiece and organic vapor cartridge(s) in combination with a high-efficiency particulate filter]; or PAPRTOVHiE* [any powered, air-purifying respirator with a tight-fitting facepiece and organic vapor cartridge (s) in combination with a high-efficiency particulate filter]; or SCBAF (any self-contained breathing apparatus with full facepiece); or SAF (any supplied-air respirator with a full facepiece). *5000 mg/m³*: SAF:PD,PP (any supplied-air respirator that has a full facepiece and is operated in a pressure-demand or other positive-pressure mode). *EMERGENCY OR PLANNED ENTRY INTO UNKNOWN CONCENTRATIONS OR IDLH CONDITIONS*: SCBAF:PD,PP (any self-contained breathing apparatus that has a full facepiece and is operated in a pressure-demand or other positive-pressure mode); or SAF:PD,PP:ASCBA (any supplied-air respirator that has a full facepiece and is operated in a pressure-demand or other positive-pressure mode in combination with an auxiliary self-contained breathing apparatus operated in a pressure-demand or other positive pressure mode). *ESCAPE:* GMFOVHiE [any air-purifying, full-facepiece respirator (gas mask) with a chin-style, front-or back-mounted canister having a high efficiency particulate filter) or SCBAE (any appropriate escape-type, self-contained breathing apparatus). **Note*: Substance reported to cause eye irritation or damage; may require eye protection.
Exposure limits (TWA unless otherwise noted): ACGIH TLV 5 mg/m³; NIOSH/OSHA 5 mg/m³, as pyrethrum
Toxicity by ingestion: Grade 2; LD_{50} = 0.5 to 5 g/kg.
Long-term health effects: May cause hypersensitive reaction, especially following previous sensitizing exposure. May cause eczematous dermatitis.
Liquid or solid irritant characteristics: Eye and skin irritant. If spilled on clothing and allowed to remain, may cause smarting and reddening of skin.
IDLH value: 5000 mg/m³.

FIRE DATA
Flash point: 180–190°F/82–88°C (oc)

CHEMICAL REACTIVITY
Binary reactants: Strong oxidizers may cause fire and explosions. Incompatible with alkalis. The presence of light, heat, moisture, and air contribute to instability.

ENVIRONMENTAL DATA
Food chain concentration potential: Log P_{ow} = 3.6 (II). Values > 3.0 are likely to bioconcentrate in aquatic organisms and other living tissue, especially in fats.
Water pollution: Harmful to aquatic life in very low concentrations. May be dangerous if it enters nearby water intakes; notify operators. Notify local health and wildlife officials.
Response to discharge: Issue warning–water contaminant. Should be removed. Chemical and physical treatment. Disperse and flush

SHIPPING INFORMATION
Storage temperature: Cool; **Stability during transport:** Unstable in the presence of light, moisture, and air.

PHYSICAL AND CHEMICAL PROPERTIES
Physical state @ 59°F/15°C and 1 atm: Liquid.
Molecular weight: 328.4 Pyrethrin I; 372.4 Pyrethrin II; 316.4 Cinerin I; 360.4 Cinerin II
Boiling point @ 1 atm: 338°F/170°C/443.2°K (Pyrethrin I); 392°F/200°C/473.2°K (Pyrethrin II); 279°F/137°C/410.2°K (Cinerin I); 361°F/183°C/456.2°K (Cinerin II)
Specific gravity (water = 1): 1 (approximate).
Vapor pressure: (Reid) low.

PYRIDINE   REC. P:4800

SYNONYMS: AZABENZENE; AZINE; NCL-C55301; EEC No. 613-002-00-7; PYRIDIN (German); PRINDINA (Spanish); RCRA No. U196

IDENTIFICATION
CAS Number: 110-86-1
Formula: C_5H_5N
DOT ID Number: UN 1282; DOT Guide Number: 129
Proper Shipping Name: Pyridine
Reportable Quantity (RQ): **(CERCLA)** 1000 lb/454 kg

DESCRIPTION: Colorless to slightly yellow liquid. Sharp, nauseating, fish-like odor. Initially floats on water surface; soluble. Poisonous, flammable vapor is produced.

Highly flammable •Irritating to the skin, eyes, and respiratory tract; prolonged contact can cause burns; poisonous by skin contact or if swallowed • Containers may BLEVE when exposed to fire •Vapors may form explosive mixture with air • Vapors are heavier than air and will collect and stay in low areas • Vapors may travel long distances to ignition sources and flashback • Vapors in confined areas (e.g., tanks, sewers, buildings) may explode when exposed to fire • Toxic products of combustion may include nitrogen oxides • Do not put yourself in danger by entering a contaminated area to rescue a victim.

Hazard Classification (based on NFPA-704 Rating System)
Health Hazards (Blue): 3; Flammability (Red): 3; Reactivity (Yellow): 0

EMERGENCY RESPONSE: See Appendix A (129)
Evacuation:
Public safety: Isolate spill area for at least 50 to 100 meters (160 to 330 feet) in all directions.
Spill: Large spill–Consider initial downwind evacuation for at least 300 meters (1000 feet).
Fire: Isolate for 800 meters (½ mile) in all directions, especially if tank, rail car, or tank truck is involved in fire.

EXPOSURE
Short-term effects: *SEEK MEDICAL ATTENTION*. **Vapor:** *POISONOUS IF INHALED OR IF SKIN IS EXPOSED*. Irritating to eyes, nose, and throat. Move to fresh air. *IF BREATHING HAS STOPPED*, give artificial respiration. IF breathing is difficult, administer oxygen. **Liquid:** *POISONOUS IF SWALLOWED OR IF SKIN IS EXPOSED*. Will burn eyes. Remove contaminated clothing and shoes. Flush affected areas with plenty of water. *IF IN EYES*, hold eyelids open and flush with plenty of water. *IF SWALLOWED*

and victim is CONSCIOUS AND ABLE TO SWALLOW, have victim drink 4 to 8 ounces of water and have victim induce vomiting. *IF SWALLOWED and victim is UNCONSCIOUS OR HAVING CONVULSIONS,* do nothing except keep victim warm.

HEALTH HAZARDS
Personal protective equipment (PPE): B-Level PPE. Air-supplied mask or organic canister; vapor-proof goggles; rubber gloves and protective clothing. Protective materials with good to excellent resistance: butyl rubber, polyethylene. Also, PV alcohol and natural rubber offers limited protection.

Recommendations for respirator selection: NIOSH/OSHA *125 ppm:* SA:CF* (any supplied-air respirator operated in a continuous-flow mode); or PAPROV* [any powered, air-purifying respirator with organic vapor cartridge(s)]. *250 ppm:* CCRFOV [any air-purifying, full-facepiece respirator (gas mask) with a chin-style, front- or back-mounted acid gas canister]; or GMFOV [any air-purifying, full-facepiece respirator (gas mask) with a chin-style, front- or back-mounted organic vapor canister]; or PAPRTOV [any powered, air-purifying respirator with a tight-fitting facepiece and organic vapor cartridges(s)]; or SCBAF (any self-contained breathing apparatus with a full facepiece); or SAF (any supplied-air respirator with a full facepiece). *1000 ppm:* SAF:PD,PP (any supplied-air respirator that has a full facepiece and is operated in a pressure-demand or other positive-pressure mode). *EMERGENCY OR PLANNED ENTRY INTO UNKNOWN CONCENTRATIONS OR IDLH CONDITIONS:* SCBAF:PD,PP (any self-contained breathing apparatus that has a full facepiece and is operated in a pressure-demand or other positive-pressure mode); or SAF:PD,PP:ASCBA (any supplied-air respirator that has a full facepiece and is operated in a pressure-demand or other positive-pressure mode in combination with an auxiliary self-contained breathing apparatus operated in a pressure-demand or other positive pressure mode). *ESCAPE:* GMFOV [any air-purifying, full-facepiece respirator (gas mask) with a chin-style, front- or back-mounted organic vapor canister]; or SCBAE (any appropriate escape-type, self-contained breathing apparatus). **Note:* Substance causes eye irritation or damage; eye protection needed.
Exposure limits (TWA unless otherwise noted): ACGIH TLV 5 ppm (16 mg/m^3); NIOSH/OSHA 5 ppm (15 mg/m^3).
Toxicity by ingestion: Grade 2; LD$_{50}$ = 0.5 to 5 g/kg (rat).
Long-term health effects: Liver and kidney damage after ingestion.
Vapor (gas) irritant characteristics: Eye and respiratory tract irritant. Vapors cause irritation such that personnel will find high concentrations unpleasant. The effect is temporary.
Liquid or solid irritant characteristics: Eye and skin irritant. Causes smarting of the skin and first-degree burns on short exposure; may cause secondary burns on long exposure.
Odor threshold: 0.021 ppm.
IDLH value: 1000 ppm.

FIRE DATA
Flash point: 68°F/20°C (cc).
Flammable limits in air: LEL: 1.8%; UEL: 12.4%.
Fire extinguishing agents not to be used: Water may be ineffective.
Autoignition temperature: 900°F/482°C/755°K.
Electrical hazard: Class I, Group D. Due to low electric conductivity, this substance may generate electrostatic charges as a result of agitation and flow.
Burning rate: 4.3 mm/min

CHEMICAL REACTIVITY
Binary reactants: Strong oxidizers, strong acids. Corrosive to some forms of rubber, plastics, and coatings.
Neutralizing agents for acids and caustics: Flush with water.
Reactivity group: 9
Compatibility class: Aromatic amines

ENVIRONMENTAL DATA
Food chain concentration potential: Log P$_{ow}$ = 0.72. Unlikely to accumulate.
Water pollution: Dangerous to aquatic life in high concentrations. May be dangerous if it enters nearby water intakes; notify operators. Notify local health and wildlife officials. **Response to discharge:** Issue warning–high flammability. Restrict access. Evacuate area. Disperse and flush.

SHIPPING INFORMATION
Grades of purity: Technical; Pure; **Storage temperature:** Ambient; **Inert atmosphere:** None; **Venting:** Pressure-vacuum; **Stability during transport:** Stable.

NAS HAZARD CLASSIFICATION FOR BULK WATER TRANSPORTATION
FIRE: 3
HEALTH: Vapor irritant: 2; Liquid or solid irritant: 2; Poisons: 1
WATER POLLUTION: Human toxicity: 2; Aquatic toxicity: 2; Aesthetic effect: 3
REACTIVITY: Other chemicals: 3; Water: 0; Self-reaction: 0

PHYSICAL AND CHEMICAL PROPERTIES
Physical state @ 59°F/15°C and 1 atm: Liquid.
Molecular weight: 79.10
Boiling point @ 1 atm: 239.5°F/115.3°C/388.5°K.
Melting/Freezing point: –44°F/–42°C/231°K.
Critical temperature: 656.2°F/346.8°C/620°K.
Critical pressure: 817.3 psia = 55.6 atm = 5.63 MN/m^2.
Specific gravity (water = 1): 0.983 @ 68°F/20°C (liquid).
Liquid surface tension: 38.0 dynes/cm = 0.038 N/m @ 68°F/20°C.
Relative vapor density (air = 1): 2.73
Ratio of specific heats of vapor (gas): 1.123
Latent heat of vaporization: 193 Btu/lb = 107 cal/g = 4.48 x 10^5 J/kg.
Heat of combustion: –14,390 Btu/lb = –7992 cal/g = –334.6 x 10^5 J/kg.
Heat of solution: (estimate) –13 Btu/lb = –7 cal/g = –0.3 x 10^5 J/kg.
Vapor pressure: (Reid) 0.77 psia; 16 mm.

PYROGALLIC ACID   REC. P:4850

SYNONYMS: 1,2,3-BENZENETRIOL; C.I. 76515; C.I. OXIDATION BASE-32; EEC No. 604-009-00-6; FOURRINE PG; PYROGALLOL; 1,2,3-TRIHYDROXYBENZENE; UX2800000

IDENTIFICATION
CAS Number: 87-66-1
Formula: C$_6$H$_6$O$_3$; 1,2,3-C$_6$H$_3$(OH)$_3$

DESCRIPTION: White crystalline solid; turns gray on exposure to light and air. Odorless. Sinks and mixes with water.

Combustible • Heat or flame may cause explosion • Containers may BLEVE when exposed to fire • Vapors are heavier than air and will collect and stay in low areas • Vapors in confined areas (e.g., tanks, sewers, buildings) may explode when exposed to fire • Highly Irritating to the skin, eyes, and respiratory tract • Toxic products of combustion may include CO_2 and carbon monoxide.

Hazard Classification (based on NFPA-704 Rating System)
Health Hazards (Blue): 2; Flammability (Red): 1; Reactivity (Yellow): 0

EMERGENCY RESPONSE: See Appendix A (171)
Evacuation:
Public safety: Isolate the area of spill or leak for at least 10 to 25 meters (30 to 80 feet) in all directions.
Spill: Increase, in the downwind direction, as necessary, the distance shown under "Public Safety."
Fire: If any large container is involved in fire, isolate for at least 800 meters (½ mile) in all directions; also, consider initial evacuation for 800 meters (½ mile) in all directions.

EXPOSURE
Short-term effects: *SEEK MEDICAL ATTENTION*. Symptoms similar to phenol. **Dust:** *POISONOUS IF ABSORBED THROUGH THE SKIN OR IF SWALLOWED*. Irritating to eyes, nose, and throat. *IF INHALED*, will or absorbed through the skin causes nausea, vomiting; will cause coughing or difficult breathing. *IF IN EYES*, hold eyelids open and flush with plenty of water. *IF BREATHING HAS STOPPED*, give artificial respiration. IF breathing is difficult, administer oxygen. **Solid:** Irritating to skin and eyes. *IF SWALLOWED*, will cause nausea, vomiting, or loss of consciousness. Remove contaminated clothing and shoes. Flush affected areas with plenty of water. *IF IN EYES*, hold eyelids open and flush with plenty of water. *IF SWALLOWED* and victim is *CONSCIOUS AND ABLE TO SWALLOW*, have victim drink 4 to 8 ounces of water and have victim induce vomiting. *IF SWALLOWED* and victim is *UNCONSCIOUS OR HAVING CONVULSIONS*, do nothing except keep victim warm.

HEALTH HAZARDS
Personal protective equipment (PPE): B-Level PPE. Full face organic vapor respirator. Rubber gloves; safety goggles; dust mask.
Toxicity by ingestion: High dose by inhalation or ingestion of more than 2.5 grams can cause death in humans. Grade 2; oral LD_{50} = 789 mg/kg (rat); 300 mg/kg (mouse).
Long-term health effects: Methemoglobinemia, blood effects. Causes kidney and liver damage.
Vapor (gas) irritant characteristics: Respiratory tract irritation.
Liquid or solid irritant characteristics: Eye and skin irritation.

FIRE DATA
Flash point: Volatile, combustible solid.
Electrical hazard: Due to low electric conductivity, this substance may generate electrostatic charges as a result of agitation and flow.

CHEMICAL REACTIVITY
Binary reactants: Incompatible with strong oxidizers, alkalis, alkylene oxides, ammonia, amines, antipyrine, epichlorohydrin, iodine, isocyanates, menthol, peroxyfuroic acid, phenol, potassium permanganate, metallic salts of lead and iron.

ENVIRONMENTAL DATA
Food chain concentration potential: Log P_{ow} = 0.3.
Water pollution: Harmful to aquatic life in very low concentrations. May be dangerous if it enters nearby water intakes; notify operators. Notify local health and wildlife officials.
Response to discharge: Issue warning–water contaminant. Disperse and flush.

SHIPPING INFORMATION
Grades of purity: N.F; Reagent; **Storage temperature:** Ambient; **Inert atmosphere:** None; **Venting:** Open; **Stability during transport:** Stable.

PHYSICAL AND CHEMICAL PROPERTIES
Physical state @ 59°F/15°C and 1 atm: Solid.
Molecular weight: 126
Boiling point @ 1 atm: 588°F/309°C/582°K.
Melting/Freezing point: 268°F/131°C/404°K.
Specific gravity (water = 1): 1.45 @ 68°F/20°C (solid).
Heat of combustion: –9,130 Btu/lb = –5,070 cal/g = –212 x 10^5 J/kg.
Vapor pressure: 10 mm.

QUINOLINE REC. Q:0100

SYNONYMS: 1-AZANAPHTHALENE; 1-BENZAZINE; 1-BENZINE; BENZO(B) PYRIDINE; CHINOLEINE; CHINOLINE; LEUCOL; LEUCOLINE; LEUKOL8; QUINOLEINA (Spanish)

IDENTIFICATION
CAS Number: 91-22-5
Formula: C_9H_7N
DOT ID Number: UN 2656; DOT Guide Number: 154
Proper Shipping Name: Quinoline
Reportable Quantity (RQ): **(CERCLA)** 5000 lb/2270 kg

DESCRIPTION: Colorless liquid. Turns brown on exposure to air. Strong unpleasant odor, somewhat like pyridine. Sinks in water; dissolves slowly.

Combustible • Poison! • Swallowing the material can kill you; toxic if inhaled or skin contact; vapors are irritating to the skin, eyes, and respiratory tract • Heat may cause violent explosion • Vapors are heavier than air and will collect and stay in low areas • Vapors in confined areas (e.g., tanks, sewers, buildings) may explode when exposed to fire • Toxic products of combustion may include nitrogen oxides.

Hazard Classification (based on NFPA-704 Rating System)
Health Hazards (Blue): 2; Flammability (Red): 1; Reactivity (Yellow): 0

EMERGENCY RESPONSE: See Appendix A (154)
Evacuation:
Public safety: Isolate the area of spill or leak for at least 25 to 50 meters (80 to 160 feet) in all directions.
Spill: Increase, in the downwind direction, as necessary, the distance shown under "Public Safety."
Fire: If tank, rail car, or tank truck is involved in fire, isolate for at least 800 meters (½ mile) in all directions; also, consider initial evacuation for 800 meters (½ mile) in all directions.

EXPOSURE
Short-term effects: *SEEK MEDICAL ATTENTION*. **Liquid:** Irritating to skin and eyes. *IF SWALLOWED*, will cause nausea and vomiting. Remove contaminated clothing and shoes. Flush affected areas with plenty of water. *IF BREATHING HAS STOPPED*, give artificial respiration; *avoid mouth-to-mouth resuscitation; use bag/mask apparatus*. *IF IN EYES*, hold eyelids open and flush with plenty of water. *IF SWALLOWED* and victim is *CONSCIOUS AND ABLE TO SWALLOW*, have victim drink 4 to 8 ounces of water and have victim induce vomiting. *IF SWALLOWED* and victim is

UNCONSCIOUS OR HAVING CONVULSIONS, do nothing except keep victim warm.

HEALTH HAZARDS
Personal protective equipment (PPE): A-Level PPE. Full face organic vapor respirator; rubber gloves; coveralls and/or rubber apron; rubber shoes and boots.
Exposure limits (TWA unless otherwise noted): 0.1 ppm (AIHAWEEL).
Toxicity by ingestion: Grade 3; oral LD_{50} = 460 mg/kg (rat).
Vapor (gas) irritant characteristics: Respiratory tract irritant.
Liquid or solid irritant characteristics: Eye and skin irritant.
Odor threshold: 71 ppm.

FIRE DATA
Flash point: 225°F/107°C (cc).
Behavior in fire: Containers may rupture and explode.
Autoignition temperature: 896°F/480°C/753°K.
Burning rate: 4.06 mm/min

CHEMICAL REACTIVITY
Binary reactants: May attack some forms of plastics

ENVIRONMENTAL DATA
Food chain concentration potential: Log P_{ow} = 2.03. Unlikely to accumulate.
Water pollution: Harmful to aquatic life in very low concentrations. Fouling to shoreline. May be dangerous if it enters nearby water intakes; notify operators. Notify local health and wildlife officials. **Response to discharge:** Issue warning–water contaminant. Restrict access. Mechanical containment. Should be removed. Chemical and physical treatment.

SHIPPING INFORMATION
Grades of purity: Reagent; Technical; **Storage temperature:** Ambient; **Inert atmosphere:** None; **Venting:** Open (flame arrester); **Stability during transport:** Stable.

PHYSICAL AND CHEMICAL PROPERTIES
Physical state @ 59°F/15°C and 1 atm: Liquid.
Molecular weight: 129
Boiling point @ 1 atm: 459°F/237°C/510°K.
Melting/Freezing point: 5°F/–15°C/258°K.
Critical temperature: 948°F/509°C/782°K.
Specific gravity (water = 1): 1.095 @ 68°F/20°C (liquid).
Liquid surface tension: 45.0 dynes/cm = 0.0450 N/m @ 68°F/20°C.
Relative vapor density (air = 1): 4.5
Latent heat of vaporization: (estimate)155 Btu/lb = 86 cal/g = 3.6 x 10^5 J/kg.
Heat of combustion: –15,700 Btu/lb = –8710 cal/g = –365 x 10^5 J/kg.

RESORCINOL   REC. R:0100

SYNONYMS: BENZENE, *m*-DIHYDROXY-; *m*-BENZENEDIOL; 1,3-BENZENEDIOL; C.I. 76505; C.I. DEVELOPER 4; C.I. OXIDATION BASE 31; DEVELOPER O; DEVELOPER R; DEVELOPER RS; *m*-DIHYDROXYBENZENE; 1,3-DIHYDROXYBENZENE; *m*-DIOXYBENZENE; DURAFUR DEVELOPER G; EEC No. 604-010-00-1; FOURAMINE RS; FOURRINE 79; FOURRINE EW; *m*-HYDROQUINONE; 3-HYDROXYCYCLOHEXADIEN-1-ONE; *m*-HYDROXYPHENOL; 3-HYDROXYPHENOL; NAKO TGG; NCI-C05970; PELAGOL GREY RS; PELAGOL RS; PHENOL, *m*-HYDROXY-; RESORCIN; RESORCINA (Spanish); RESORCINE; *m*-DIHYDROXYBENZOL; RCRA No. U021

IDENTIFICATION
CAS Number: 108-46-3
Formula: $C_6H_6O_2$; 1,3-$C_6H_4(OH)_2$
DOT ID Number: UN 2876; DOT Guide Number: 153
Proper Shipping Name: Resorcinol
Reportable Quantity (RQ): **(CERCLA)** 5000 lb/2270 kg

DESCRIPTION: White or off-white solid when pure; turn pink on exposure to air. Faint odor. Sinks and mixes easily with water.

Combustible solid • Very Irritating to the skin, eyes, and respiratory tract; poisonous if absorbed through the skin or swallowed • Containers may BLEVE when exposed to fire • Concentrated dust in confined areas (e.g., tanks, sewers, buildings) may explode when exposed to fire • Toxic products of combustion may include CO_2 and carbon monoxide.

Hazard Classification (based on NFPA-704 Rating System)
Health Hazards (Blue): 3; Flammability (Red): 1; Reactivity (Yellow): 0

EMERGENCY RESPONSE: See Appendix A (153)
Evacuation:
Public safety: Isolate the area of spill or leak for at least 25 to 50 meters (80 to 160 feet) in all directions.
Spill: Increase, in the downwind direction, as necessary, the distance shown under "Public Safety."
Fire: If tank, rail car, or tank truck is involved in fire, isolate for at least 800 meters (½ mile) in all directions; also, consider initial evacuation for 800 meters (½ mile) in all directions.

EXPOSURE
Short-term effects: *SEEK MEDICAL ATTENTION.* **Dust:** Irritating to eyes, nose, and throat. *IF INHALED*, will, will cause coughing or difficult breathing. *IF IN EYES*, hold eyelids open and flush. with plenty of water. *IF BREATHING HAS STOPPED*, give artificial respiration; *avoid mouth-to-mouth resuscitation; use bag/mask apparatus.* IF breathing is difficult, administer oxygen. **Solid:** Irritating to skin and eyes. *IF SWALLOWED*, will cause nausea or loss of consciousness. Remove contaminated clothing and shoes. Flush affected areas with plenty of water. *IF IN EYES*, hold eyelids open and flush. with plenty of water. *IF SWALLOWED* and victim is *CONSCIOUS AND ABLE TO SWALLOW*, have victim drink 4 to 8 ounces of water. *IF SWALLOWED* and victim is *UNCONSCIOUS OR HAVING CONVULSIONS,* do nothing except keep victim warm.

HEALTH HAZARDS
Personal protective equipment (PPE): B-Level PPE. OSHA Table Z-1-A air contaminant. U.S. Bureau of Mines approved respirator; rubber gloves; safety glasses with side shields or chemical goggles; coveralls or rubber apron
Exposure limits (TWA unless otherwise noted): ACGIH TLV 10 ppm (45 mg/m3); NIOSH REL 10 ppm (45 mg/m³).
Short-term exposure limits (15-minute TWA): ACGIH STEL 20 ppm (90 mg/m³); NIOSH STEL 20 ppm (90 mg/m³).
Toxicity by ingestion: Grade 2; LD_{50} = 0.5 to 5.0 g/kg.
Long-term health effects: Can cause skin and blood disorders; heart, liver and kidney damage. Produces goiters in rats. IARC rating 3, human data: none; animal evidence: inadequate. May effect the nervous system.

Salicylaldehyde

Vapor (gas) irritant characteristics: May cause irritation of the eyes and respiratory tract.
Liquid or solid irritant characteristics: May cause irritation of the eyes, skin, and respiratory tract.

FIRE DATA
Flash point: 261°F/127°C.
Fire extinguishing agents not to be used: Water may cause frothing.
Behavior in fire: Containers may explode.
Autoignition temperature: 1125°F/607°C/880°K.
Electrical hazard: Flow or agitation of substance may generate electrostatic charges.

CHEMICAL REACTIVITY
Binary reactants: Acetanillide, albumin, alkalis, antipyrine, camphor, ferric salts, menthol, spirit nitrous ether, strong oxidizers (possible violent reaction), and bases. Hygroscopic (absorbs moisture from the air.).

ENVIRONMENTAL DATA
Food chain concentration potential: Log P_{ow} = 0.8. Unlikely to accumulate.
Water pollution: Harmful to aquatic life in very low concentrations. May be dangerous if it enters nearby water intakes; notify operators. Notify local health and wildlife officials.
Response to discharge: Issue warning–water contaminant. Disperse and flush.

SHIPPING INFORMATION
Grades of purity: USP, 99.5+%; Technical, 99%; **Storage temperature:** Ambient; **Inert atmosphere:** None; **Venting:** Open; **Stability during transport:** Stable.

PHYSICAL AND CHEMICAL PROPERTIES
Physical state @ 59°F/15°C and 1 atm: Solid.
Molecular weight: 110.1
Boiling point @ 1 atm: (sublimes) 537°F/277°C/550°K.
Melting/Freezing point: 228°F/109°C/382°K.
Specific gravity (water = 1): 1.2 @ 68°F/20°C (solid).
Heat of combustion: –11200 Btu/lb = –6200 cal/g = –259 x 10^5 J/kg.
Vapor pressure: 0.0075 mm.

SALICYLALDEHYDE   REC. S:0100

SYNONYMS: 2-FORMYLPHENOL; α-FORMYLPHENOL; o-HYDROXYBENZALDEHYDE; 2-HYDROXYBENZALDEHYDE; SAH; SALICILALDEHIDO (Spanish); SALICYLAL; SALICYLIC ALDEHYDE

IDENTIFICATION
CAS Number: 90-02-8
Formula: $C_7H_6O_2$; HOC_6H_4CHO
DOT ID Number: NA 1993; DOT Guide Number: 128
Proper Shipping Name: Combustible liquid, n.o.s.

DESCRIPTION: Colorless to pale yellow liquid or dark red oil. Bitter almond odor. Sinks and mixes slowly in water.

Combustible • Containers may BLEVE when exposed to fire • Vapors in confined areas (e.g., tanks, sewers, buildings) may explode when exposed to fire • Highly irritating to the eyes and respiratory tract; this material has anesthetic properties • Toxic products of combustion may include CO_2 and carbon monoxide.

Hazard Classification (based on NFPA-704 Rating System)
Health Hazards (Blue): 0; Flammability (Red): 2; Reactivity (Yellow): 0

EMERGENCY RESPONSE: See Appendix A (128)
Evacuation:
Public safety: Isolate spill area for at least 25 to 50 meters (80 to 160 feet) in all directions.
Spill: Large spill–Consider initial downwind evacuation for at least 300 meters (1000 feet).
Fire: Isolate for 800 meters (½ mile) in all directions, especially if tank, rail car, or tank truck is involved in fire.

EXPOSURE
Short-term effects: *SEEK MEDICAL ATTENTION*. **Liquid:** Irritating to skin and eyes. Harmful if swallowed. Remove contaminated clothing and shoes. Flush affected areas with plenty of water. *IF IN EYES* hold eyelids open and flush with plenty of water. *IF SWALLOWED* and victim is *CONSCIOUS AND ABLE TO SWALLOW*, have victim drink 4 to 8 ounces of water and have victim induce vomiting.

HEALTH HAZARDS
Personal protective equipment (PPE): B-Level PPE. Rubber gloves, self-contained breathing apparatus, laboratory coat and chemical work goggles.
Exposure limits (TWA unless otherwise noted): TLV not established-Dow Industrial Hygiene Guide is 5 mg/m³.
Toxicity by ingestion: Grade 2; LD_{50} = 500 to 5000 mg/kg.
Vapor (gas) irritant characteristics: May cause eye and respiratory tract irritation.
Liquid or solid irritant characteristics: Eye and skin irritant. Causes smarting of the skin and first-degree burns on short exposure; may cause second-degree burns on long exposure.

FIRE DATA
Flash point: 172°F/78°C (cc).

CHEMICAL REACTIVITY
Binary reactants: Reacts violently with strong oxidizers, bromine, ketones. Incompatible with strong acids, caustics, ammonia, amines.

ENVIRONMENTAL DATA
Food chain concentration potential: Unlikely to accumulate.
Water pollution: Harmful to aquatic life in low concentrations. May be dangerous if it enters nearby water intakes; notify operators. Notify local health and wildlife officials. **Response to discharge:** Chemical and physical treatment. Disperse and flush.

SHIPPING INFORMATION
Grades of purity: 98.5%; **Storage temperature:** Cool; **Stability during transport:** Stable.

PHYSICAL AND CHEMICAL PROPERTIES
Physical state @ 59°F/15°C and 1 atm: Liquid.
Molecular weight: 122.12.
Boiling point @ 1 atm: 386°F/197°C/470°K.
Melting/Freezing point: 19.4°F/–7°C/266.2°K.
Critical temperature: (estimate) 802.6°F/428.1°C/701.3°K.
Critical pressure: (estimate) 590.8 psia = 40.2 atm = 4.07 MN/m².
Specific gravity (water = 1): 1.1674 @ 68°F/20°C.

Liquid surface tension: 42.90 dynes/cm = 0.0490 N/m @ 68°F/20°C.
Relative vapor density (air = 1): 4.2
Latent heat of vaporization: (estimate) 145.2 Btu/lb = 80.7 cal/g = 3.37 x 10^5 J/kg.
Heat of combustion: @ 77°F/25°C –11,273 Btu/lb = –6263 cal/g = –262 x 10^5 J/kg.

SALICYLIC ACID   REC. S:0200

SYNONYMS: o-HYDROXYBENZOIC ACID; 2-HYDROXYBENZOIC ACID; KERALYT; RETARDER W; SA; SAX

IDENTIFICATION
CAS Number: 69-72-7
Formula: $C_7H_6O_3$; 1,2-HOC_6H_4COOH

DESCRIPTION: White crystalline solid. Turns yellow on exposure to air or sunlight. Odorless. Sinks and mixes slowly with water.

Combustible solid • Heat or flame may cause explosion • Containers may BLEVE when exposed to fire • Concentrated dust in confined areas (e.g., tanks, sewers, buildings) may explode when exposed to fire • Irritating to the skin, eyes, and respiratory tract • Toxic products of combustion may include fumes of carbon monoxide, CO_2, and phenol.

Hazard Classification (based on NFPA-704 Rating System)
Health Hazards (Blue): 0; Flammability (Red): 1; Reactivity (Yellow): 0

EMERGENCY RESPONSE: See Appendix A (171)
Evacuation:
Public safety: Isolate the area of spill or leak for at least 10 to 25 meters (30 to 80 feet) in all directions.
Spill: Increase, in the downwind direction, as necessary, the distance shown under "Public Safety."
Fire: If any large container is involved in fire, isolate for at least 800 meters (½ mile) in all directions; also, consider initial evacuation for 800 meters (½ mile) in all directions.

EXPOSURE
Short-term effects: *SEEK MEDICAL ATTENTION.* **Dust:** Irritating to eyes, nose, and throat. *IF INHALED,* will, will cause coughing or difficult breathing. *IF IN EYES,* hold eyelids open and flush with plenty of water. *IF BREATHING HAS STOPPED,* give artificial respiration. IF breathing is difficult, administer oxygen. **Solid:** Irritating to skin and eyes. *IF SWALLOWED,* will cause vomiting. Remove contaminated clothing and shoes. Flush affected areas with plenty of water. *IF IN EYES,* hold eyelids open and flush with plenty of water. *IF SWALLOWED* and victim is *CONSCIOUS AND ABLE TO SWALLOW,* have victim drink 4 to 8 ounces of water and have victim induce vomiting. *IF SWALLOWED* and victim is *UNCONSCIOUS OR HAVING CONVULSIONS,* do nothing except keep victim warm.

HEALTH HAZARDS
Personal protective equipment (PPE): B-Level PPE. Gloves; goggles; respirator for dust; clean body-covering clothing.
Toxicity by ingestion: Grade 2; LD_{50} = 0.5 to 5 g/kg.
Long-term health effects: Repeated or prolonged dermal exposure may cause skin problems; eczema.

Vapor (gas) irritant characteristics: Eye and respiratory tract irritant.
Liquid or solid irritant characteristics: Severe eye and skin irritant.

FIRE DATA
Flash point: 315°F/157°C.
Flammable limits in air: LEL: 1.1 @ 392°F/200°C (NFPA).
Fire extinguishing agents not to be used: Water or foam may cause frothing.
Behavior in fire: Sublimes and forms vapor or dust that may explode.
Autoignition temperature: 1008°F/542°C/815°K.
Electrical hazard: Due to low electric conductivity, this substance may generate electrostatic charges as a result of agitation and flow.

CHEMICAL REACTIVITY
Binary reactants: Strong oxidizers, iodine, salts of iron.

ENVIRONMENTAL DATA
Food chain concentration potential: Log K_{ow} = 2.26. Unlikely to accumulate.
Water pollution: Dangerous to aquatic life in high concentrations. May be dangerous if it enters nearby water intakes; notify operators. Notify local health and wildlife officials. **Response to discharge:** Should be removed. Chemical and physical treatment.

SHIPPING INFORMATION
Grades of purity: Technical, 98+%; Pure, 99+%; **Storage temperature:** Ambient; **Inert atmosphere:** None; **Venting:** Open; **Stability during transport:** Stable.

PHYSICAL AND CHEMICAL PROPERTIES
Physical state @ 59°F/15°C and 1 atm: Solid.
Molecular weight: 138.13
Boiling point @ 1 atm: 169°F/76°C/349°K (sublimes).
Melting/Freezing point: 315°F/157°C/430°K.
Specific gravity (water = 1): 1.44 @ 68°F/20°C (solid).
Heat of combustion: –9,420 Btu/lb = –5,230 cal/g = –219 x 10^5 J/kg.
Vapor pressure: 1 mm.

SARIN (GB)   REC. S:0250

SYNONYMS: FLUOROISOPROPOXYMETHYLPHOSPHINE OXIDE; GB; IMPF; ISOPROPYL METHANEFLUOROPHOSPHONATE ISOPROPHYL; METHYLPHOSPHONOFLUORIDATE; ISOPROPOXYMETHYLPHORYL, FLUORIDE; ISOPROPYL METHYLFLUOROPHOSPHATE; ISOPROPYL METHYLPHOSPHONOFLUORIDATE; O-ISOPROPYL METHYLPHOSPHONOFLUORIDATE; ISOPROPYL-METHYL-PHOSPHORYL FLUORIDE; METHYLFLUOROPHOSPHORIC ACID, ISOPROPYL ESTER; METHYLFLUORPHOSPHORSAEUREISOPROPYLESTER (German); METHYLPHOSPHONOFLUORIDIC ACID ISOPROPYL ESTER; METHYLPHOSPHONOFLUORIDIC ACID-1-METHYLETHYL ESTER; MFI; SARIN II; T-144; T-2106; TL 1618; TRILONE 46

IDENTIFICATION:
CAS Registry Number: 107-44-8
Formula: $C_4H_{10}FO_2P$; $[(CH_3)_2CHO](CH_3)FPO$
DOT ID Number: UN 2810; DOT Guide Number: 153

966   **Sarin (GB)**

Proper Shipping Name: Sarin
Reportable Quantity (RQ): **(EHS):** 10 lb (4.54 kg)

DESCRIPTION: Clear, colorless liquid. *No warning properties.* Odorless and tasteless. Sinks in water; soluble; hydrolyzes. The hydrolysis products are considerably less toxic than the material. *Note:* If used as a weapon, notify U.S. Department of Defense: Army. Use M8 Paper (Detection: Yellow) or M256-A1 Detector Kit (Detection limit: 0.005 mg/m^3) if available. Damage and/or death may occur before chemical detection can take place.

Highly volatile poison! (organophosphate) • Breathing the vapor, skin* or eye contact, or swallowing the material can kill you; symptoms may be delayed for several hours • Firefighting gear (including SCBA) provide NO protection. If exposure occurs, remove and isolate gear immediately and thoroughly decontaminate personnel • Containers may BLEVE when exposed to fire • Vapors are heavier than air and will collect and stay in low areas • Vapors in confined areas (e.g., tanks, sewers, buildings) may explode when exposed to fire • *Combustion products are less deadly than the material itself.* Toxic products of combustion may include carbon monoxide, fluorine, and phosphorus oxide • Do NOT attempt rescue. *Note: A single drop on the skin can be fatal.*

Hazard Classification (based on NFPA-704 Rating System)
Health Hazards (Blue): 4; Flammability (Red): 1; Reactivity (Yellow): 0

EMERGENCY RESPONSE: See Appendix A (153)
Evacuation:
Public safety: See below.
Spill: For Sarin [weaponized]: Small spill–First: Isolate in all directions 155 meters (500 feet); Then: Protect persons downwind, DAY: 1.6 km (1.0 mile); NIGHT: 3.4 km (2.1 miles). Large spill–First: Isolate in all directions 915 meters (3000 feet); Then: Protect persons downwind, DAY: 11.0+ km (7.0+ miles); NIGHT: 11.0+ km (7.0+ miles).
Fire: If tank, rail car, or tank truck is involved in fire, isolate for at least 800 meters (½ mile) in all directions; also, consider initial evacuation for 800 meters (½ mile) in all directions.

EXPOSURE
Short-term effects: *SEEK MEDICAL ATTENTION.* **Liquid:** *POISONOUS BY SKIN CONTACT OR IF SWALLOWED.* Will remain as a liquid for more than 24 hours. **Vapor or Liquid:** *POISONOUS IF INHALED, IF SWALLOWED, OR IF SKIN IS EXPOSED.* Eye pupils are small; blurred vision; runny nose; cough; shortness of breath; pain; diarrhea, nausea and vomiting; increased blood pressure, hypermotility, hallucinations; loss of consciousness; convulsions; breathing stops; death. *IF BREATHING HAS STOPPED,* give artificial respiration; *avoid mouth-to-mouth resuscitation; use bag/mask apparatus.* **Skin:** *Remove and double-bag contaminated clothing (including shoes) and leave them in the Hot Zone;* wash skin with soap and water and large volumes of water for at least 15 minutes. *Skin can also be decontaminated with diluted hypochlorite solution, U.S. Army M291 kit, and M258(A1) skin decontamination kit.* **Eye:** Rinse with large volumes of water or saline for at least 15 minutes. *IF SWALLOWED,* will cause nausea, vomiting, or loss of consciousness. *Remove and double-bag contaminated clothing and shoes and leave in Hot Zone for later incineration by hazardous materials experts.* Flush affected areas with plenty of water. **If swallowed** and victim is *CONSCIOUS AND ABLE TO SWALLOW,* have victim drink 4 to 8 ounces of water. **Do NOT induce vomiting but immediately administer slurry of activated charcoal.** *IF SWALLOWED* and victim is *UNCONSCIOUS OR HAVING CONVULSIONS,* do nothing except keep victim warm. *Note to physician and medical personnel:* If symptoms indicate, initial treatment for an adult includes atropine, 2 mg (1/30 gr) intramuscularly or intravenously as soon as any local or systemic signs or symptoms of an intoxication are noted; repeat the administration of atropine every 3 to 8 minutes until signs of atropinization (mydriasis, dry mouth, rapid pulse, hot and dry skin) occur; initiate treatment in children with 0.05 mg/kg of atropine; repeat at 5- to 10-minute intervals. Watch respiration, and remove bronchial secretions if they appear to be obstructing the airway; intubate if necessary. Pralidoxime must be administered within minutes to a few hours following exposure (depending on the specific agent) to be effective. Give 2-PAMCI (Pralidoxime; Protopam), 2.5 g in 100 mL of sterile water or in 5% dextrose and water, intravenously, slowly, in 15 to 30 minutes; if sufficient fluid is not available, give 1 g of 2-PAMCI in 3 mL of distilled water by deep intramuscular injection; repeat this every half hour if respiration weakens or if muscle fasciculation or convulsions recur. Also Diazepam, an anticonvulsant, might be needed. For adults, the dose is 5 to 10 mg (slow IV), repeated every 12 to 15 minutes up to three (3) doses maximum. For children, the dose is 0.2 to 0.5 mg/kg.
Note to physician or authorized medical personnel: Administer atropine. *IF INHALED,* medical observation is recommended for 24 to 48 hours after breathing overexposure, as pulmonary edema may be delayed. As first aid for pulmonary edema, consider administering a corticosteroid spray. Cigarette smoking may exacerbate pulmonary injury and should be discouraged for at least 72 hours following exposure.

HEALTH HAZARDS
Personal protective equipment (PPE): A-Level PPE. Rubber gloves, protective clothing, goggles, respirators. Butyl rubber gloves and Tyvek® "F" decontamination suit provide barrier protection against chemical warfare agents. Airtight, impermeable clothing was developed for personnel who must enter heavily contaminated areas. This clothing is made of butyl rubber or a coated fabric such as Tyvek "F" and provide barrier protection against liquid chemical warfare agents. Although resistant to liquid chemical agents, impermeable protective clothing may be penetrated after a few hours of exposure to heavy concentration of agent. Consequently, liquid contamination on the clothing must be neutralized or removed as soon as possible. If the proper equipment is not available, or if the rescuers have not been trained in its use, call for assistance from the U.S. Soldier and Biological Chemical Command–Edgewood Research Development and Engineering Center (from 0700-1630 EST call 410-671-4411, and from 1630-0700 EST call 410-278-5201; ask for the Staff Duty Officer).
Recommendations for respirator selection: *EMERGENCY OR PLANNED ENTRY INTO UNKNOWN CONCENTRATIONS OR IDLH CONDITIONS:* SCBAF:PD,PP (any self-contained breathing apparatus that has a full facepiece and is operated in a pressure-demand or other positive-pressure mode); or SAF:PD,PP:ASCBA (any supplied-air respirator that has a full facepiece and is operated in a pressure-demand or other positive-pressure mode in combination with an auxiliary self-contained breathing apparatus operated in a pressure-demand or other positive pressure mode). *ESCAPE:* GMFOVHiE [any air-purifying, full-facepiece respirator (gas mask) with a chin-style, front-or back-mounted canister having a high efficiency particulate filter); or SCBAE (any appropriate escape-type, self-contained breathing apparatus). The U.S. Army standard M17A1 (which is being replaced by the M40 protective mask) mask provides

complete respiratory protection against all known military toxic chemical agents, but it cannot be used in an oxygen deficient environment and it does not afford protection against industrial toxics such as ammonia and carbon monoxide. *It is not approved for civilian use.*

Exposure limits (TWA unless otherwise noted): Workplace: 0.0001 mg/m³; General population limits (as recommended by the Surgeon General's Working Group, U.S. Department of Health): 0.000003 mg/m³.

Human Toxicity: The human lethal dose (man) is approximately 0.01 mg/kg. LCt_{50} = 100 mg-min/m3. LD_{50} [skin] = 1.7 g/70 kg [man] (*Medical Sspects of Chemical and Biological Warfare, Part I*, Walter Reed Medical Center, 1997)

FIRE DATA
Flash point: Nonflammable.

CHEMICAL REACTIVITY
Reactivity with water: Hydrolyzed by water. Under acid conditions GB form s hydrogen fluoride (HF). Decomposes tin, magnesium, cadmium plated steel, and aluminum.
Reactivity with other materials: Slightly corrosive to steel.

ENVIRONMENTAL DATA
Food chain concentration potential: Sarin is soluble in water; therefore. bioconcentration in aquatic organisms is not expected to be an important fate process [U.S. Army Corps of Engineers. Special Report 86–38, Britton, K. B., *Low Temperature Effects on Sorption, Hydrolysis, and Photolysis of Organophosphates: A Literature Review, p. 23.* Washington DC, 1986]. Soil: 2 to 24 hr @ 41°–77°F/5°–25°C.
Water pollution: If used as a weapon, utilize use an M272 Water Detection Kit (Detection limit: 0.02 mg/L). Dangerous to aquatic life in high concentrations. May be dangerous if it enters water intakes. Notify local health and pollution control officials. Notify operators of nearby water intakes. This material will be broken down in water quickly, but small amounts may evaporate. This material will be broken down in moist soil quickly. Small amounts may evaporate into the air or travel below the soil surface and contaminated groundwater.

SHIPPING INFORMATION
Stability during transport: Unstable in the presence of moisture.

PHYSICAL AND CHEMICAL PROPERTIES
Physical state @ 59°F/15°C and 1 atm: Liquid.
Molecular weight: 140.09
Boiling point @ 1 atm: 316°F/158°C/431°K
Melting/Freezing point: –69°F/–56°C/217°K.
Liquid density: 1.10 g/mL @ 68°F/20°C.
Relative vapor density (air = 1): 4.86
Specific Gravity: 1.080
Vapor density: 4.9
Vapor pressure: 2.9 mm at 77°F/25°C. 2.1 mm @ 68°F/20°C.
Volatility: 22,000 mg/m³ @ 77°F/25°C.

SELENIUM DIOXIDE   REC. S:0300

SYNONYMS: DIOXIDO de SELENIO (Spanish); EEC No. 034-002-00-8; RCRA No. U204; SELENIOUS ANHYDRIDE; SELENIUM OXIDE

IDENTIFICATION
CAS Number: 7446-08-4
Formula: SeO_2; O_2Se
DOT ID Number: UN 2811; DOT Guide Number: 154
Proper Shipping Name: Selenium oxide
Reportable Quantity (RQ): **(CERCLA)** 10 lb/4.54 kg

DESCRIPTION: White crystalline solid. Green vapor. Sour odor. Sinks and mixes with water.

Poison! • Protect against inhalation and absorption through the skin. • Strong oxidizer which may react spontaneously with low flash point organics or reducing agents. Heat forms oxygen; will increase the activity of an existing fire • Do not put yourself in danger by entering a contaminated area to rescue a victim • Toxic products of combustion may include selenium oxide.

Hazard Classification (based on NFPA-704 Rating System)
Health Hazards (Blue): 1; Flammability (Red): 0; Reactivity (Yellow): 0

EMERGENCY RESPONSE: See Appendix A (154)
Evacuation:
Public safety: Isolate the area of spill or leak for at least 25 to 50 meters (80 to 160 feet) in all directions.
Spill: Increase, in the downwind direction, as necessary, the distance shown under "Public Safety."
Fire: If tank, rail car, or tank truck is involved in fire, isolate for at least 800 meters (½ mile) in all directions; also, consider initial evacuation for 800 meters (½ mile) in all directions.

EXPOSURE
Short-term effects: *SEEK MEDICAL ATTENTION.* **Dust:** *POISONOUS IF INHALED OR IF SKIN IS EXPOSED. IF INHALED*, will, will cause coughing or difficult breathing. *IF IN EYES*, hold eyelids open and flush with plenty of water.*IF BREATHING HAS STOPPED*, give artificial respiration; *avoid mouth-to-mouth resuscitation; use bag/mask apparatus.* IF breathing is difficult, administer oxygen. **Solid:** Irritating to skin and eyes. *IF SWALLOWED*, will cause coughing, nausea, or vomiting. Remove contaminated clothing and shoes. Flush affected areas with plenty of water. *IF IN EYES*, hold eyelids open and flush with plenty of water. *IF SWALLOWED* and victim is *CONSCIOUS AND ABLE TO SWALLOW*, have victim drink 4 to 8 ounces of water and have victim induce vomiting. *IF SWALLOWED* and victim is *UNCONSCIOUS OR HAVING CONVULSIONS,* do nothing except keep victim warm.

HEALTH HAZARDS
Personal protective equipment (PPE): B-Level PPE. OSHA Table Z-1-A air contaminant. This compound is highly toxic if inhaled or ingested. Dust mask; rubber gloves; protective clothing.
Recommendations for respirator selection: NIOSH/OSHA
1 mg/m³: DM* *if not present as a fume* (any dust and mist respirator); or DMFu* (any dust, mist and fume respirator); or HiEF (any air-purifying, full-facepiece respirator with a high-efficiency particulate filter); or PAPRDM* *if not present as a fume* (any powered, air-purifying respirator with a dust and mist filter); or PAPRDMFu* (any powered, air-purifying respirator with a dust, mist, and fume filter); or SA* (any supplied-air respirator); or SCBAF (any self-contained breathing apparatus with a full facepiece); or SAF (any supplied-air respirator with a full facepiece).*EMERGENCY OR PLANNED ENTRY INTO UNKNOWN CONCENTRATIONS OR IDLH CONDITIONS:* SCBAF:PD,PP (any self-contained breathing apparatus that has a full facepiece and is operated in a pressure-demand or other positive-pressure mode); or SAF:PD,PP:ASCBA (any supplied-air

respirator that has a full facepiece and is operated in a pressure-demand or other positive-pressure mode in combination with an auxiliary self-contained breathing apparatus operated in a pressure-demand or other positive-pressure mode). *ESCAPE:* HiEF (any air-purifying, full-facepiece respirator with a high-efficiency particulate filter); or SCBAE (any appropriate escape-type, self-contained breathing apparatus). *Note:* Substance reported to cause eye irritation or damage; may require eye protection.
Exposure limits (TWA unless otherwise noted): ACGIH TLV 0.2 mg/m^3; NIOSH/OSHA 0.2 mg/m^3 as selenium compounds.
Toxicity by ingestion: Poison.
Long-term health effects: IARC rating 3; inadequate human and animal evidence. Kidney damage. Organ damage: liver and kidney.
Vapor (gas) irritant characteristics: Severe eye and respiratory tract irritant.
Liquid or solid irritant characteristics: Severe eye and skin irritant.
IDLH value: 1 mg/m^3 as selenium

CHEMICAL REACTIVITY
Binary reactants: In presence of water corrodes most metals. An oxidizer. Reacts with reducing agents, acids and other chemical substances.

ENVIRONMENTAL DATA
Food chain concentration potential: Positive. Log P_{ow} likely to be = >3.0. Values > 3.0 are likely to bioconcentrate in aquatic organisms and other living tissue (in fats).
Water pollution: Threatening to recreation, fisheries, potable water supplies; industrial plants. Harmful to aquatic life in very low concentrations. May be dangerous if it enters nearby water intakes; notify operators. Notify local health and wildlife officials. Potentially corrosive. **Response to discharge:** Issue warning–poison, water contaminant. Restrict access. Disperse and flush.

SHIPPING INFORMATION
Grades of purity: Commercial, 99.5+%; **Storage temperature:** Cool ambient; **Inert atmosphere:** None; **Venting:** Open; **Stability during transport:** Stable.

PHYSICAL AND CHEMICAL PROPERTIES
Physical state @ 59°F/15°C and 1 atm: Solid.
Molecular weight: 111
Boiling point @ 1 atm: 599°F/315°C/583°K (sublimes).
Melting/Freezing point: 653°F/345°C/618°K.
Specific gravity (water = 1): 3.95 @ 68°F/20°C (solid).
Heat of solution: 12.1 Btu/lb = 6.7 cal/g = 0.28 x 10^5 J/kg.
Vapor pressure: 1 mm @ 315°F/157°C.

SELENIUM TRIOXIDE   REC. S:0400

SYNONYMS: EEC No. 034-002-00-8; SELENIC ANHYDRIDE; SELENIUM(VI) OXIDE; SELENIC ACID SOLUTION; TRIOXIDO de SELENIO (Spanish)

IDENTIFICATION
CAS Number: 13768-86-0
Formula: SeO$_3$; O$_3$Se
DOT ID Number: UN 3290; UN 1905 (solution); DOT Guide Number: 154
Proper Shipping Name: Toxic, solid, corrosive, inorganic, n.o.s; Selenic acid

DESCRIPTION: White solid. Sinks and mixes with water; reacts vigorously forming selenic acid.

Poison! • Corrosive to skin or eye, and respiratory tract; contact with skin or eyes causes severe burns, impaired vision, or blindness; effects of contact or inhalation may be delayed • Firefighting gear (including SCBA) does not provide adequate protection. If exposure occurs, remove and isolate gear immediately and thoroughly decontaminate personnel • Strong oxidizer which may react spontaneously with low flash point organics or reducing agents. Heat forms oxygen; will increase the activity of an existing fire • Do not put yourself in danger by entering a contaminated area to rescue a victim. Toxic products of combustion may include selenium oxide.

Hazard Classification (based on NFPA-704 Rating System)
Health Hazards (Blue): 1; Flammability (Red): 0; Reactivity (Yellow): 0

EMERGENCY RESPONSE: See Appendix A (154)
Evacuation:
Public safety: Isolate the area of spill or leak for at least 25 to 50 meters (80 to 160 feet) in all directions.
Spill: Increase, in the downwind direction, as necessary, the distance shown under "Public Safety."
Fire: If tank, rail car, or tank truck is involved in fire, isolate for at least 800 meters (½ mile) in all directions; also, consider initial evacuation for 800 meters (½ mile) in all directions.

EXPOSURE
Short-term effects: *SEEK MEDICAL ATTENTION.* **Dust:** *POISONOUS IF INHALED OR IF SKIN IS EXPOSED. IF INHALED,* will, will cause coughing or difficult breathing. *IF IN EYES,* hold eyelids open and flush with plenty of water. *IF BREATHING HAS STOPPED,* give artificial respiration; *avoid mouth-to-mouth resuscitation; use bag/mask apparatus.* IF breathing is difficult, administer oxygen. **Solid:** Irritating to skin and eyes. *IF SWALLOWED,* will cause coughing, nausea, and vomiting. Remove contaminated clothing and shoes. Flush affected areas with plenty of water. *IF IN EYES,* hold eyelids open and flush with plenty of water. *IF SWALLOWED* and victim is *CONSCIOUS AND ABLE TO SWALLOW,* have victim drink 4 to 8 ounces of water and have victim induce vomiting. *IF SWALLOWED* and victim is *UNCONSCIOUS OR HAVING CONVULSIONS,* do nothing except keep victim warm.
Note to physician or authorized medical personnel: Medical observation is recommended for 24 to 48 hours after breathing overexposure, as pulmonary edema may be delayed. As first aid for pulmonary edema, consider administering a corticosteroid spray. Cigarette smoking may exacerbate pulmonary injury and should be discouraged for at least 72 hours following exposure.

HEALTH HAZARDS
Personal protective equipment (PPE): B-Level PPE. OSHA Table Z-1-A air contaminant. This compound is highly toxic if ingested or inhaled. Dust mask; goggles or face shield; rubber gloves.
Recommendations for respirator selection: NIOSH/OSHA
1 mg/m^3: DM* *if not present as a fume* (any dust and mist respirator); or DMFu* (any dust, mist and fume respirator); or HiEF (any air-purifying, full-facepiece respirator with a high-efficiency particulate filter); or PAPRDM* *if not present as a fume* (any powered, air-purifying respirator with a dust and mist filter); or PAPRDMFu* (any powered, air-purifying respirator with a dust, mist, and fume filter); or SA* (any supplied-air respirator); or

Silicon tetrachloride

SCBAF (any self-contained breathing apparatus with a full facepiece); or SAF (any supplied-air respirator with a full facepiece).*EMERGENCY OR PLANNED ENTRY INTO UNKNOWN CONCENTRATIONS OR IDLH CONDITIONS:* SCBAF:PD,PP (any self-contained breathing apparatus that has a full facepiece and is operated in a pressure-demand or other positive-pressure mode); or SAF:PD,PP:ASCBA (any supplied-air respirator that has a full facepiece and is operated in a pressure-demand or other positive-pressure mode in combination with an auxiliary self-contained breathing apparatus operated in a pressure-demand or other positive-pressure mode).*ESCAPE:* HiEF (any air-purifying, full-facepiece respirator with a high-efficiency particulate filter); or SCBAE (any appropriate escape-type, self-contained breathing apparatus). **Note:* Substance reported to cause eye irritation or damage; may require eye protection.
Exposure limits (TWA unless otherwise noted): ACGIH TLV 0.2 mg/m^3; NIOSH/OSHA 0.2 mg/m^3 as selenium compounds.
Toxicity by ingestion: Poison.
Long-term health effects: IARC rating 3; inadequate human and animal evidence. Organ damage; liver and kidney.
Vapor (gas) irritant characteristics: Eye and respiratory tract irritant.
Liquid or solid irritant characteristics: Eye and skin irritant.
IDLH value: 1 mg/m^3 as selenium

CHEMICAL REACTIVITY
Reactivity with water: Reacts vigorously with water to form selenic acid solution
Binary reactants: A strong oxidizer. Violent reaction with many substances including reducing agents, combustible materials, organic substances, aldehydes, alkenes, carboxylic acids, isocyanates. Attacks most metals in the presence of moisture.
Neutralizing agents for acids and caustics: Flush with water, rinse with dilute solution of sodium bicarbonate or soda ash.

ENVIRONMENTAL DATA
Food chain concentration potential: Positive. Likely to accumulate in the food chain. Marine fish are efficient concentrators of selenium. (Prager).
Water pollution: Threatening to recreation, fisheries, potable water supplies; industrial plants. Harmful to aquatic life in very low concentrations. May be dangerous if it enters nearby water intakes; notify operators. Notify local health and wildlife officials. Potentially corrosive. **Response to discharge:** Issue warning–corrosive, water contaminant. Restrict access. Disperse and flush.

SHIPPING INFORMATION
Grades of purity: Commercial; also shipped as a 40% solution in water as selenic acid; **Storage temperature:** Ambient; **Inert atmosphere:** None; **Venting:** Open; **Stability during transport:** Stable.

PHYSICAL AND CHEMICAL PROPERTIES
Physical state @ 59°F/15°C and 1 atm: Solid.
Molecular weight: 126.9
Boiling point @ 1 atm: Decomposes.
Melting/Freezing point: 244°F/118°C/391°K.
Specific gravity (water = 1): 3.6 @ 68°F/20°C (solid).

SILICON TETRACHLORIDE   REC. S:0500

SYNONYMS: EEC No. 014-002-00-4; EXTREMA; SILICON CHLORIDE; SILICON CHLORIDE; TETRACHLOROSILANE

IDENTIFICATION
CAS Number: 10026-04-7
Formula: SiCl$_4$; Cl$_4$Si
DOT ID Number: UN 1818; DOT Guide Number: 156
Proper Shipping Name: Silicon tetrachloride

DESCRIPTION: Colorless to pale yellow, mobile, fuming liquid. Suffocating odor. Decomposes in water; reacts violently, forming silicic acid, hydrochloric acid, and hydrogen chloride gas. Produces large volumes of vapor.

Poison! • Breathing the vapors can kill you • Corrosive to skin, eyes, and respiratory tract; contact with skin or eyes causes severe burns, impaired vision, or blindness; inhalation symptoms may be delayed • Do NOT use water • Firefighting gear (including SCBA) does not provide adequate protection. If exposure occurs, remove and isolate gear immediately and thoroughly decontaminate personnel • Vapors are heavier than air and will collect and stay in low areas • Containers may BLEVE when exposed to fire • Toxic products of combustion may include hydrogen chloride • Do not put yourself in danger by entering a contaminated area to rescue a victim.

Hazard Classification (based on NFPA-704 Rating System)
Health Hazards (Blue): 3; Flammability (Red): 0; Reactivity (Yellow): 2; Special Notice (White): Water reactive

EMERGENCY RESPONSE: See Appendix A (156)
Evacuation:
Public safety: Isolate the area of spill or leak for at least 50 to 100 meters (160 to 330 feet) in all directions.
Spill: Increase, in the downwind direction, as necessary, the distance shown under "Public Safety."
Fire: If tank, rail car, or tank truck is involved in fire, isolate for at least 800 meters (½ mile) in all directions; also, consider initial evacuation for 800 meters (½ mile) in all directions.

EXPOSURE
Short-term effects: *SEEK MEDICAL ATTENTION.* **Vapor:** Irritating to eyes, nose, and throat. *IF INHALED*, will, will cause difficult breathing. Move victim to fresh air. *IF BREATHING HAS STOPPED*, give artificial respiration; *avoid mouth-to-mouth resuscitation; use bag/mask apparatus.* IF breathing is difficult, administer oxygen. Lung edema may develop. **Liquid:** *POISONOUS IF SWALLOWED.* Will burn skin and eyes. Remove contaminated clothing and shoes. Flush affected areas with plenty of water. *IF IN EYES*, hold eyelids open and flush. with plenty of water. *IF SWALLOWED* and victim is *CONSCIOUS AND ABLE TO SWALLOW*, have victim drink 4 to 8 ounces of water. **Do NOT induce vomiting**.
Note to physician or authorized medical personnel: Medical observation is recommended for 24 to 48 hours after breathing overexposure, as pulmonary edema may be delayed. As first aid for pulmonary edema, consider administering a corticosteroid spray. Cigarette smoking may exacerbate pulmonary injury and should be discouraged for at least 72 hours following exposure.

HEALTH HAZARDS
Personal protective equipment (PPE): B-Level PPE. Acid-canister-type gas mask or self-contained breathing apparatus; goggles or face shield; gloves; other protective clothing to prevent contact with skin. Chemical protective material(s) reported to have good to excellent resistance: Teflon®.
Exposure limits (TWA unless otherwise noted): 1 ppm [ceiling] (AIHAWEEL).

Toxicity by ingestion: Grade 4; LD_{50} less than 50 mg/kg.
Vapor (gas) irritant characteristics: Vapors cause severe irritation of eyes and throat and can cause eye and lung injury. They cannot be tolerated even at low concentrations.
Liquid or solid irritant characteristics: Severe eye and skin irritant.

FIRE DATA
Fire extinguishing agents not to be used: Water or water-based foam on adjacent fires *ONLY*.

CHEMICAL REACTIVITY
Reactivity with water: Reacts vigorously with water forming silicic and hydrochloric acid.
Binary reactants: In presence of moisture corrodes metals; flammable hydrogen gas may be formed. Violent reaction with sodium or potassium.
Neutralizing agents for acids and caustics: Flood with water, rinse with sodium bicarbonate or lime solution.

ENVIRONMENTAL DATA
Food chain concentration potential: Negative; unlikely to accumulate.
Water pollution: Effect of low concentrations on aquatic life is unknown. May be dangerous if it enters nearby water intakes; notify operators. Notify local health and wildlife officials.
Response to discharge: Issue warning–corrosive, air contaminant.

SHIPPING INFORMATION
Grades of purity: Technical: 99.7+%; C.P.: 99.9+%; **Storage temperature:** Ambient; **Inert atmosphere:** Dry air. Usually shipped under nitrogen blanket; **Venting:** Pressure-vacuum; **Stability during transport:** Stable.

NAS HAZARD CLASSIFICATIONS FOR BULK WATER TRANSPORTATION
FIRE: 0
HEALTH: Vapor irritant: 4; Liquid or solid irritant: 4; Poisons: 4
WATER POLLUTION: Human toxicity: 4; Aquatic toxicity: 2; Aesthetic effect: 2
REACTIVITY: Other chemicals: 4; Water: 4; Self-reaction: 0

PHYSICAL AND CHEMICAL PROPERTIES
Physical state @ 59°F/15°C and 1 atm: Liquid.
Molecular weight: 169.9
Boiling point @ 1 atm: 135.7°F/57.6°C/330.8°K.
Melting/Freezing point: –94°F/–70°C/203°K.
Critical temperature: 472.5°F/233.6°C/506.8°K.
Critical pressure: 542 psia = 36.8 atm = 3.74 MN/m^2.
Specific gravity (water = 1): 1.48 @ 68°F/20°C (liquid).
Liquid surface tension: 19.6 dynes/cm = 0.0196 N/m @ 68°F/20°C.
Relative vapor density (air = 1): 5.86
Latent heat of vaporization: 74.2 Btu/lb = 41.2 cal/g = 1.73×10^5 J/kg.
Heat of solution: –742 Btu/lb = –412 cal/g = -17.3×10^5 J/kg.
Vapor pressure: 195 mm.

SILVER ACETATE   REC. S:0600

SYNONYMS: ACETATO de PLATA (Spanish); ACETIC ACID, SILVER(I) SALT; SILVER(I) ACETATE; SILVER MONOACETATE

IDENTIFICATION
CAS Number: 563-63-3
Formula: $C_2H_3O_2 \cdot Ag$; CH_3COOAg

DESCRIPTION: White to gray solid. Odorless. Sinks in water; moderately soluble in hot water.

Used commercially as an oxidizing agent; may react spontaneously with low flash point organics or reducing agents. Strong oxidizer which may react spontaneously with low flash point organics or reducing agents. Heat forms oxygen; will increase the activity of an existing fire.

Hazard Classification (based on NFPA-704 Rating System)
Health Hazards (Blue): 0; Flammability (Red): 0; Reactivity (Yellow): 0

EMERGENCY RESPONSE: See Appendix A (171)
Evacuation:
Public safety: Isolate the area of spill or leak for at least 10 to 25 meters (30 to 80 feet) in all directions.
Spill: Increase, in the downwind direction, as necessary, the distance shown under "Public Safety."
Fire: If any large container is involved in fire, isolate for at least 800 meters (½ mile) in all directions; also, consider initial evacuation for 800 meters (½ mile) in all directions.

EXPOSURE
Short-term effects: *SEEK MEDICAL ATTENTION*. **Dust:** Irritating to eyes, nose, and throat. *IF INHALED*, will, will cause coughing or difficult breathing. *IF IN EYES*, hold eyelids open and flush with plenty of water. *IF BREATHING HAS STOPPED*, give artificial respiration. IF breathing is difficult, administer oxygen. **Solid:** Irritating to skin and eyes. Harmful if swallowed. Remove contaminated clothing and shoes. Flush affected areas with plenty of water. *IF IN EYES*, hold eyelids open and flush with plenty of water. *IF SWALLOWED* and victim is *CONSCIOUS AND ABLE TO SWALLOW*, have victim drink 4 to 8 ounces of water and have victim induce vomiting. *IF SWALLOWED* and victim is *UNCONSCIOUS OR HAVING CONVULSIONS*, do nothing except keep victim warm.

HEALTH HAZARDS
Personal protective equipment (PPE): B-Level PPE. OSHA Table Z-1-A air contaminant. Dust mask; goggles or face shield; protective **gloves.**
Recommendations for respirator selection: NIOSH/OSHA as silver soluble compounds
0.25 mg/m^3: SA:CF* (any supplied-air respirator operated in a continuous-flow mode); or PAPRHiE* (any powered, air-purifying respirator with a high-efficiency particulate filter). 0.5 mg/m^3: HiEF (any air-purifying, full-facepiece respirator with a high-efficiency particulate filter); or SCBAF (any self-contained breathing apparatus with a full facepiece); or SAF (any supplied-air respirator with a full facepiece). 10 mg/m^3: SAF:PD,PP (any supplied-air respirator that has a full facepiece and is operated in a pressure-demand or other positive-pressure mode). *EMERGENCY OR PLANNED ENTRY INTO UNKNOWN CONCENTRATIONS OR IDLH CONDITIONS:* SCBAF:PD,PP (any self-contained breathing apparatus that has a full facepiece and is operated in a pressure-demand or other positive-pressure mode); or SAF:PD,PP:ASCBA (any supplied-air respirator that has a full facepiece and is operated in a pressure-demand or other positive-pressure mode in combination with an auxiliary self-contained breathing apparatus operated in a pressure-demand or other positive pressure mode).

ESCAPE: HiEF (any air-purifying, full-facepiece respirator with a high-efficiency particulate filter); or SCBAE (any appropriate escape-type, self-contained breathing apparatus). *Note*: Substance causes eye irritation or damage; eye protection needed.
Exposure limits (TWA unless otherwise noted): ACGIH TLV 0.01 mg/m^3; OSHA PEL 0.01 mg/m^3 as silver soluble compounds.
Long-term health effects: If continued for a long period, ingestion or inhalation of silver compounds can cause permanent discoloration of the skin (argyria).
Vapor (gas) irritant characteristics: Dust irritates eyes and respiratory tract.
Liquid or solid irritant characteristics: Eye and skin irritant.

FIRE DATA
Behavior in fire: An oxidizer; increases flammability of combustibles.

CHEMICAL REACTIVITY
Binary reactants: Reacts with reducing agents, acids and combustible materials.

ENVIRONMENTAL DATA
Food chain concentration potential: Negative; unlikely to accumulate.
Water pollution: Effect of low concentrations on aquatic life is unknown. May be dangerous if it enters nearby water intakes; notify operators. Notify local health and wildlife officials.
Response to discharge: Issue warning–water contaminant. Restrict access. Disperse and flush.

SHIPPING INFORMATION
Grades of purity: Commercial; Purified; **Storage temperature:** Ambient; **Inert atmosphere:** None; **Venting:** Open; **Stability during transport:** Stable.

PHYSICAL AND CHEMICAL PROPERTIES
Physical state @ 59°F/15°C and 1 atm: Solid.
Molecular weight: 166.9
Boiling point @ 1 atm: Decomposes.
Specific gravity (water = 1): 3.26 @ 68°F/20°C (solid).

SILVER CARBONATE   REC. S:0700

SYNONYMS: CARBONATO de PLATA (Spanish)

IDENTIFICATION
CAS Number: 534-16-7
Formula: Ag_2CO_3

DESCRIPTION: Yellow to yellowish-gray to brown powder of high silver content (75–78%). Odorless. Sinks in water; insoluble.

Noncombustible solid but may be combustible in the form of dust or powder. Toxic products of combustion may include silver oxide, silver, and CO_2.

Hazard Classification (based on NFPA-704 Rating System)
Health Hazards (Blue): 1; Flammability (Red): 0; Reactivity (Yellow): 0

EMERGENCY RESPONSE: See Appendix A (171)
Evacuation:
Public safety: Isolate the area of spill or leak for at least 10 to 25 meters (30 to 80 feet) in all directions.
Spill: Increase, in the downwind direction, as necessary, the distance shown under "Public Safety."
Fire: If any large container is involved in fire, isolate for at least 800 meters (½ mile) in all directions; also, consider initial evacuation for 800 meters (½ mile) in all directions.

EXPOSURE
Short-term effects: *SEEK MEDICAL ATTENTION*. **Dust:** Irritating to eyes, nose, and throat. *IF INHALED*, will, will cause coughing or difficult breathing. *IF IN EYES*, hold eyelids open and flush with plenty of water. *IF BREATHING HAS STOPPED*, give artificial respiration. IF breathing is difficult, administer oxygen. **Solid:** Irritating to skin and eyes. Harmful if swallowed. Remove contaminated clothing and shoes. Flush affected areas with plenty of water. *IF IN EYES*, hold eyelids open and flush with plenty of water. *IF SWALLOWED* and victim is *CONSCIOUS AND ABLE TO SWALLOW*, have victim drink 4 to 8 ounces of water and have victim induce vomiting. *IF SWALLOWED* and victim is *UNCONSCIOUS OR HAVING CONVULSIONS*, do nothing except keep victim warm.

HEALTH HAZARDS
Personal protective equipment (PPE): B-Level PPE. OSHA Table Z-1-A air contaminant. Dust mask; goggles or face shield; rubber gloves.
Recommendations for respirator selection: NIOSH/OSHA as silver (insoluble, but contains 75–78% silver).
0.25 mg/m^3: SA:CF* (any supplied-air respirator operated in a continuous-flow mode); or PAPRHiE* (any powered, air-purifying respirator with a high-efficiency particulate filter). *0.5 mg/m^3:* HiEF (any air-purifying, full-facepiece respirator with a high-efficiency particulate filter); or SCBAF (any self-contained breathing apparatus with a full facepiece); or SAF (any supplied-air respirator with a full facepiece). *10 mg/m^3:* SAF:PD,PP (any supplied-air respirator that has a full facepiece and is operated in a pressure-demand or other positive-pressure mode). *EMERGENCY OR PLANNED ENTRY INTO UNKNOWN CONCENTRATIONS OR IDLH CONDITIONS:* SCBAF:PD,PP (any self-contained breathing apparatus that has a full facepiece and is operated in a pressure-demand or other positive-pressure mode); or SAF:PD,PP:ASCBA (any supplied-air respirator that has a full facepiece and is operated in a pressure-demand or other positive-pressure mode in combination with an auxiliary self-contained breathing apparatus operated in a pressure-demand or other positive pressure mode). *ESCAPE:* HiEF (any air-purifying, full-facepiece respirator with a high-efficiency particulate filter); or SCBAE (any appropriate escape-type, self-contained breathing apparatus). *Note:* Substance causes eye irritation or damage; eye protection needed.
Exposure limits (TWA unless otherwise noted): ACGIH TLV 0.01 mg/m^3; OSHA PEL 0.01 mg/m^3 as silver soluble compounds.
Long-term health effects: If continued for a long period, ingestion or inhalation of silver compounds can cause permanent discoloration of the skin (argyria).
Vapor (gas) irritant characteristics: May cause respiratory tract irritation.
Liquid or solid irritant characteristics: Eye irritant.

ENVIRONMENTAL DATA
Food chain concentration potential: Negative; unlikely to accumulate.
Water pollution: Effect of low concentrations on aquatic life is unknown. May be dangerous if it enters nearby water intakes; notify operators. Notify local health and wildlife officials. **Response to discharge:** Issue warning–water contaminant. Should be removed. Chemical and physical treatment.

SHIPPING INFORMATION
Grades of purity: Reagent, 98+%; **Storage temperature:** Ambient; **Inert atmosphere:** None; **Venting:** Open; **Stability during transport:** Stable.

PHYSICAL AND CHEMICAL PROPERTIES
Physical state @ 59°F/15°C and 1 atm: Solid.
Molecular weight: 275.75
Boiling point @ 1 atm: Decomposes.
Specific gravity (water = 1): 6.1 @ 68°F/20°C (solid).

SILVER FLUORIDE   REC. S:0800

SYNONYMS: ARGENTIC FLUORIDE; ARGENTOUS FLUORIDE; FLUORURO de PLATA (Spanish); SILVER(II) FLUORIDE; SILVER(2+) FLUORIDE; SILVER DIFLUORIDE

IDENTIFICATION
CAS Number: 7783-95-1 (II); 7775-41-9 (I)
Formula: AgF
DOT ID Number: UN 1479; DOT Guide Number: 140
Proper Shipping Name: Oxidizing solid, n.o.s.

DESCRIPTION: Light sensitive, yellow (I) or white (II) crystalline solid when pure; darkens to gray or brown on exposure to light. Odorless. Sinks and mixes with water.

Noncombustible solid but may be flammable in the form of dust or powder • Incandescent reaction in air at 608°F/320°C • Strong oxidizer which may react spontaneously with low flash point organics or reducing agents. Heat forms oxygen; will increase the activity of an existing fire • May cause fire or explosion contact with combustibles (wood, paper, oil, clothing, etc.) • Highly Irritating to the skin, eyes, and respiratory tract • Toxic products of combustion may include fluorides.

Hazard Classification (based on NFPA-704 Rating System)
Health Hazards (Blue): 2; Flammability (Red): 0; Reactivity (Yellow): 1; Special Notice (White): OXY

EMERGENCY RESPONSE: See Appendix A (140)
Evacuation:
Public safety: Isolate the area of spill or leak for at least 10 to 25 meters (30 to 80 feet) in all directions.
Spill: Consider initial downwind evacuation for at least 100 meters (330 feet).
Fire: If any large container is involved in fire, isolate for at least 800 meters (½ mile) in all directions; also, consider initial evacuation for 800 meters (½ mile) in all directions.

EXPOSURE
Short-term effects: *SEEK MEDICAL ATTENTION.* **Dust:** Irritating to eyes, nose, and throat. *IF INHALED*, will, will cause coughing or difficult breathing. *IF IN EYES*, hold eyelids open and flush with plenty of water. *IF BREATHING HAS STOPPED*, give artificial respiration. IF breathing is difficult, administer oxygen. **Solid:** Irritating to skin and eyes. *IF SWALLOWED*, will cause nausea, vomiting, or loss of consciousness. Remove contaminated clothing and shoes. Flush affected areas with plenty of water. *IF IN EYES*, hold eyelids open and flush with plenty of water. *IF SWALLOWED* and victim is *CONSCIOUS AND ABLE TO SWALLOW*, have victim drink 4 to 8 ounces of water and have victim induce vomiting. *IF SWALLOWED* and victim is *UNCONSCIOUS OR HAVING CONVULSIONS*, do nothing except keep victim warm.

HEALTH HAZARDS
Personal protective equipment (PPE): B-Level PPE. OSHA Table Z-1-A air contaminant. Dust mask; goggles or face shield; protective gloves.
Recommendations for respirator selection: NIOSH/OSHA as silver soluble compounds
0.25 mg/m³: SA:CF* (any supplied-air respirator operated in a continuous-flow mode); or PAPRHiE* (any powered, air-purifying respirator with a high-efficiency particulate filter). *0.5 mg/m³*: HiEF (any air-purifying, full-facepiece respirator with a high-efficiency particulate filter); or SCBAF (any self-contained breathing apparatus with a full facepiece); or SAF (any supplied-air respirator with a full facepiece). *10 mg/m³:* SAF:PD,PP (any supplied-air respirator that has a full facepiece and is operated in a pressure-demand or other positive-pressure mode). *EMERGENCY OR PLANNED ENTRY INTO UNKNOWN CONCENTRATIONS OR IDLH CONDITIONS:* SCBAF:PD,PP (any self-contained breathing apparatus that has a full facepiece and is operated in a pressure-demand or other positive-pressure mode); or SAF:PD,PP:ASCBA (any supplied-air respirator that has a full facepiece and is operated in a pressure-demand or other positive-pressure mode in combination with an auxiliary self-contained breathing apparatus operated in a pressure-demand or other positive pressure mode). *ESCAPE:* HiEF (any air-purifying, full-facepiece respirator with a high-efficiency particulate filter); or SCBAE (any appropriate escape-type, self-contained breathing apparatus). *Note*: Substance causes eye irritation or damage; eye protection needed.
Exposure limits (TWA unless otherwise noted): ACGIH TLV 0.01 mg/m³; OSHA PEL 0.01 mg/m³ as silver soluble compounds.
Vapor (gas) irritant characteristics: Eye and respiratory tract irritant.
Liquid or solid irritant characteristics: Eye and skin irritant.
IDLH value: 10 mg/m³ as silver, soluble compounds.

CHEMICAL REACTIVITY
Binary reactants: Contact with acetylene produces shock-sensitive material. Ammonia contact produces compounds that are explosive when dry. Hydrogen peroxide causes violent decomposition to oxygen gas. Soluble silver compounds attack some forms of plastics, rubber, and coatings.

ENVIRONMENTAL DATA
Food chain concentration potential: Negative; unlikely to accumulate.
Water pollution: Effect of low concentrations on aquatic life is unknown. May be dangerous if it enters nearby water intakes; notify operators. Notify local health and wildlife officials. **Response to discharge:** Issue warning–water contaminant. Disperse and flush.

SHIPPING INFORMATION
Grades of purity: Commercial; Pure, 99.9+%; **Storage temperature:** Ambient; **Inert atmosphere:** None; **Venting:** Open; **Stability during transport:** Stable.

PHYSICAL AND CHEMICAL PROPERTIES
Physical state @ 59°F/15°C and 1 atm: Solid.
Molecular weight: 126.9
Boiling point @ 1 atm: 2118°F/1,159°C/1,432°K.
Specific gravity (water = 1): 5.82 @ 68°F/20°C (solid).

SILVER IODATE
REC. S:0900

SYNONYMS: YODATO de PLATA (Spanish)

IDENTIFICATION
CAS Number: 7783-97-3
Formula: $AgIO_3$
DOT ID Number: UN 1479; DOT Guide Number: 140
Proper Shipping Name: Oxidizing solid, n.o.s.

DESCRIPTION: White solid. Odorless. Sinks in water.

Noncombustible solid but may be flammable in the form of dust or powder • Strong oxidizer which may react spontaneously with low flash point organics or reducing agents. Heat forms oxygen; will increase the activity of an existing fire • May cause fire or explosion contact with combustibles (wood, paper, oil, clothing, etc.) • Highly Irritating to the skin, eyes, and respiratory tract • Toxic products of combustion may include hydrogen iodide.

Hazard Classification (based on NFPA-704 Rating System)
Health Hazards (Blue): 1; Flammability (Red): 0; Reactivity (Yellow): 1; Special Notice (White): OXY

EMERGENCY RESPONSE: See Appendix A (140)
Evacuation:
Public safety: Isolate the area of spill or leak for at least 10 to 25 meters (30 to 80 feet) in all directions.
Spill: Consider initial downwind evacuation for at least 100 meters (330 feet).
Fire: If any large container is involved in fire, isolate for at least 800 meters (½ mile) in all directions; also, consider initial evacuation for 800 meters (½ mile) in all directions.

EXPOSURE
Short-term effects: *SEEK MEDICAL ATTENTION. CAN BE ABSORBED THROUGH THE SKIN.* **Dust or Skin:** Irritating to eyes, nose, and throat. *IF INHALED*, will, will cause coughing or difficult breathing. *IF IN EYES*, hold eyelids open and flush with plenty of water. *IF BREATHING HAS STOPPED*, give artificial respiration. IF breathing is difficult, administer oxygen. *IF ON SKIN*, remove contaminated clothing and shoes. Flush affected areas with plenty of water. Harmful if swallowed. *IF SWALLOWED* and victim is *CONSCIOUS AND ABLE TO SWALLOW*, have victim drink 4 to 8 ounces of water and have victim induce vomiting. *IF SWALLOWED* and victim is *UNCONSCIOUS OR HAVING CONVULSIONS*, do nothing except keep victim warm.

HEALTH HAZARDS
Personal protective equipment (PPE): B-Level PPE. OSHA Table Z-1-A air contaminant. Dust mask; goggles or face shield; protective **gloves.**
Recommendations for respirator selection: NIOSH/OSHA as silver soluble compounds.
0.25 mg/m³: SA:CF* (any supplied-air respirator operated in a continuous-flow mode); or PAPRHiE* (any powered, air-purifying respirator with a high-efficiency particulate filter). *0.5 mg/m³*: HiEF (any air-purifying, full-facepiece respirator with a high-efficiency particulate filter); or SCBAF (any self-contained breathing apparatus with a full facepiece); or SAF (any supplied-air respirator with a full facepiece). *10 mg/m³*: SAF:PD,PP (any supplied-air respirator that has a full facepiece and is operated in a pressure-demand or other positive-pressure mode). *EMERGENCY OR PLANNED ENTRY INTO UNKNOWN CONCENTRATIONS OR IDLH CONDITIONS:* SCBAF:PD,PP (any self-contained breathing apparatus that has a full facepiece and is operated in a pressure-demand or other positive-pressure mode); or SAF:PD,PP:ASCBA (any supplied-air respirator that has a full facepiece and is operated in a pressure-demand or other positive-pressure mode in combination with an auxiliary self-contained breathing apparatus operated in a pressure-demand or other positive pressure mode). *ESCAPE:* HiEF (any air-purifying, full-facepiece respirator with a high-efficiency particulate filter); or SCBAE (any appropriate escape-type, self-contained breathing apparatus). *Note*: Substance causes eye irritation or damage; eye protection needed.
Exposure limits (TWA unless otherwise noted): ACGIH TLV 0.1 mg/m³; OSHA PEL 0.01 mg/m³ as silver
Toxicity by ingestion: May be poisonous.
Long-term health effects: If continued for a long period, ingestion or inhalation of silver compounds can cause permanent discoloration of the skin (argyria).
Liquid or solid irritant characteristics: Eye irritant.
IDLH value: 10 mg/m³ as silver, soluble compounds.

CHEMICAL REACTIVITY
Binary reactants: Material is a strong oxidizer. Reacts with reducing agents, acids and combustible materials. Contamination with organic materials may form explosive compounds.

ENVIRONMENTAL DATA
Food chain concentration potential: Negative; unlikely to accumulate.
Water pollution: Effect of low concentrations on aquatic life is unknown. May be dangerous if it enters nearby water intakes; notify operators. Notify local health and wildlife officials. **Response to discharge:** Issue warning–water contaminant. Should be removed. Chemical and physical treatment.

SHIPPING INFORMATION
Grades of purity: Commercial, 99+%; **Storage temperature:** Ambient; **Inert atmosphere:** None; **Venting:** Open; **Stability during transport:** Stable.

PHYSICAL AND CHEMICAL PROPERTIES
Physical state @ 59°F/15°C and 1 atm: Solid.
Molecular weight: 282.1
Boiling point @ 1 atm: Decomposes.
Specific gravity (water = 1): 5.53 @ 68°F/20°C (solid).

SILVER NITRATE
REC. S:1000

SYNONYMS: EEC No. 047-001-00-2; LAPIS INFERNALIS; LUNAR CAUSTIC; NITRATE d'ARGENT (French); NITRATO de PLATA (Spanish); SILBERNITRAT (German)

IDENTIFICATION
CAS Number: 7761-88-8
Formula: $AgNO_3$
DOT ID Number: UN 1493; DOT Guide Number: 140
Proper Shipping Name: Silver nitrate
Reportable Quantity (RQ): **(CERCLA)** 1 lb/0454 kg

DESCRIPTION: Colorless to grayish black crystalline solid. Odorless. Sinks and mixes with water.

Poison! • Severely Irritating to the skin, eyes, and respiratory tract; toxic if ingested or inhaled; affects the blood • Firefighting gear (including SCBA) does not provide adequate protection. If exposure occurs, remove and isolate gear immediately and thoroughly decontaminate personnel • Noncombustible solid but may be flammable in the form of dust or powder • Strong oxidizer which may react spontaneously with low flash point organics or reducing agents. Heat forms oxygen; will increase the activity of an existing fire • May cause fire or explosion on contact with combustibles (wood, paper, oil, clothing, etc.) • Toxic products of combustion may include nitrogen oxides

Hazard Classification (based on NFPA-704 Rating System)
Health Hazards (Blue): 2; Flammability (Red): 0; Reactivity (Yellow): 1; Special Notice (White): OXY

EMERGENCY RESPONSE: See Appendix A (140)
Evacuation:
Public safety: Isolate the area of spill or leak for at least 10 to 25 meters (30 to 80 feet) in all directions.
Spill: Consider initial downwind evacuation for at least 100 meters (330 feet).
Fire: If any large container is involved in fire, isolate for at least 800 meters (½ mile) in all directions; also, consider initial evacuation for 800 meters (½ mile) in all directions.

EXPOSURE
Short-term effects: *SEEK MEDICAL ATTENTION*. **Solids:** Irritating to skin and eyes. Harmful if swallowed. Flush affected areas with plenty of water. *IF IN EYES*, hold eyelids open and flush with plenty of water. *IF SWALLOWED* and victim is *CONSCIOUS AND ABLE TO SWALLOW*, have victim drink 4 to 8 ounces of water. Methemoglobin formation may develop.

HEALTH HAZARDS
Personal protective equipment (PPE): B-Level PPE. OSHA Table Z-1-A air contaminant. Goggles or face shield; rubber gloves.
Recommendations for respirator selection: NIOSH/OSHA as silver soluble compounds.
0.25 mg/m³: SA:CF* (any supplied-air respirator operated in a continuous-flow mode); or PAPRHiE* (any powered, air-purifying respirator with a high-efficiency particulate filter). *0.5 mg/m³*: HiEF (any air-purifying, full-facepiece respirator with a high-efficiency particulate filter); or SCBAF (any self-contained breathing apparatus with a full facepiece); or SAF (any supplied-air respirator with a full facepiece). *10 mg/m³*: SAF:PD,PP (any supplied-air respirator that has a full facepiece and is operated in a pressure-demand or other positive-pressure mode). *EMERGENCY OR PLANNED ENTRY INTO UNKNOWN CONCENTRATIONS OR IDLH CONDITIONS:* SCBAF:PD,PP (any self-contained breathing apparatus that has a full facepiece and is operated in a pressure-demand or other positive-pressure mode); or SAF:PD,PP:ASCBA (any supplied-air respirator that has a full facepiece and is operated in a pressure-demand or other positive-pressure mode in combination with an auxiliary self-contained breathing apparatus operated in a pressure-demand or other positive pressure mode). *ESCAPE:* HiEF (any air-purifying, full-facepiece respirator with a high-efficiency particulate filter); or SCBAE (any appropriate escape-type, self-contained breathing apparatus). *Note:* Substance causes eye irritation or damage; eye protection needed.
Exposure limits (TWA unless otherwise noted): ACGIH TLV 0.01 mg/m³; OSHA PEL 0.01 mg/m³ as silver soluble compounds.
Toxicity by ingestion: May cause blood effects. Grade 3; LD_{50} = 50 to 500 mg/kg.
Long-term health effects: Mutagen. If continued for a long period, ingestion or inhalation of silver compounds can cause permanent discoloration of the skin (argyria).

Vapor (gas) irritant characteristics: Eye and respiratory tract irritant.
Liquid or solid irritant characteristics: Eye, skin and respiratory tract irritant. Burns skin on prolonged contact.
IDLH value: 10 mg/m³ as silver, soluble compounds.

CHEMICAL REACTIVITY
Binary reactants: Reacts with reducing agents; combustibles, amines, acetylene, arsenic compounds, creosote, alcohols, and many other compounds.

ENVIRONMENTAL DATA
Water pollution: Harmful to aquatic life in very low concentrations. May be dangerous if it enters nearby water intakes; notify operators. Notify local health and wildlife officials. **Response to discharge:** Issue warning–water contaminant. Should be removed. Clean Water Act. Water pollution Control Act.

SHIPPING INFORMATION
Grades of purity: Reagent 99.8+%; **Storage temperature:** Ambient; **Inert atmosphere:** None; **Venting:** Open; **Stability during transport:** Stable.

PHYSICAL AND CHEMICAL PROPERTIES
Physical state @ 59°F/15°C and 1 atm: Solid.
Molecular weight: 169.87
Boiling point @ 1 atm: Decomposes 830°F/443°C/758°K.
Melting/Freezing point: 414°F/212°C/485°K.
Specific gravity (water = 1): 4.35 @ 66°F/19°C/292°K (solid).
Heat of fusion: 16.2 cal/g.

SILVER OXIDE REC. S:1100

SYNONYMS: ARGENTOUS OXIDE; DISILVER OXIDE; OXIDO de PLATA (Spanish)

IDENTIFICATION
CAS Number: 20667-12-3
Formula: Ag2O
DOT ID Number: UN 1479; DOT Guide Number: 140
Proper Shipping Name: Oxidizing solid, n.o.s.

DESCRIPTION: Heavy, brown-black powder. Odorless. Sinks in water; slightly soluble.

Poison! • Toxic if ingested or inhaled • Firefighting gear (including SCBA) does not provide adequate protection. If exposure occurs, remove and isolate gear immediately and thoroughly decontaminate personnel • Noncombustible solid but may be flammable in the form of dust or powder • Strong oxidizer which may react spontaneously with low flash point organics or reducing agents. Heat forms oxygen; will increase the activity of an existing fire • May cause fire or explosion on contact with combustibles (wood, paper, oil, clothing, etc.) • Toxic products of combustion may include oxygen and toxic metal fumes.

Hazard Classification (based on NFPA-704 Rating System)
Health Hazards (Blue): 0; Flammability (Red): 0; Reactivity (Yellow): 1; Special Notice (White): OXY

EMERGENCY RESPONSE: See Appendix A (140)
Evacuation:
Public safety: Isolate the area of spill or leak for at least 10 to 25 meters (30 to 80 feet) in all directions.

Spill: Consider initial downwind evacuation for at least 100 meters (330 feet).
Fire: If any large container is involved in fire, isolate for at least 800 meters (½ mile) in all directions; also, consider initial evacuation for 800 meters (½ mile) in all directions.

EXPOSURE
Short-term effects: *SEEK MEDICAL ATTENTION.* **Dust:** Irritating to eyes, nose, and throat. *IF INHALED*, will, will cause coughing or difficult breathing. *IF IN EYES*, hold eyelids open and flush with plenty of water. *IF BREATHING HAS STOPPED*, give artificial respiration. IF breathing is difficult, administer oxygen. **Solid:** Irritating to skin and eyes. Harmful if swallowed. Remove contaminated clothing and shoes. Flush affected areas with plenty of water. *IF IN EYES*, hold eyelids open and flush with plenty of water. *IF SWALLOWED* and victim is *CONSCIOUS AND ABLE TO SWALLOW*, have victim drink 4 to 8 ounces of water and have victim induce vomiting. *IF SWALLOWED* and victim is *UNCONSCIOUS OR HAVING CONVULSIONS*, do nothing except keep victim warm.

HEALTH HAZARDS
Personal protective equipment (PPE): B-Level PPE. OSHA Table Z-1-A air contaminant. Dust mask; goggles or face shield; protective gloves.
Recommendations for respirator selection: NIOSH/OSHA as silver soluble compounds.
0.25 mg/m³: SA:CF* (any supplied-air respirator operated in a continuous-flow mode); or PAPRHiE* (any powered, air-purifying respirator with a high-efficiency particulate filter). *0.5 mg/m³*: HiEF (any air-purifying, full-facepiece respirator with a high-efficiency particulate filter); or SCBAF (any self-contained breathing apparatus with a full facepiece); or SAF (any supplied-air respirator with a full facepiece). *10 mg/m³*: SAF:PD,PP (any supplied-air respirator that has a full facepiece and is operated in a pressure-demand or other positive-pressure mode). *EMERGENCY OR PLANNED ENTRY INTO UNKNOWN CONCENTRATIONS OR IDLH CONDITIONS*: SCBAF:PD,PP (any self-contained breathing apparatus that has a full facepiece and is operated in a pressure-demand or other positive-pressure mode); or SAF:PD,PP:ASCBA (any supplied-air respirator that has a full facepiece and is operated in a pressure-demand or other positive-pressure mode in combination with an auxiliary self-contained breathing apparatus operated in a pressure-demand or other positive pressure mode). *ESCAPE:* HiEF (any air-purifying, full-facepiece respirator with a high-efficiency particulate filter); or SCBAE (any appropriate escape-type, self-contained breathing apparatus). *Note*: Substance causes eye irritation or damage; eye protection needed.
Exposure limits (TWA unless otherwise noted): ACGIH TLV 0.1 mg/m³; OSHA PEL 0.01 mg/m³ as silver, soluble compounds
Toxicity by ingestion: Grade 2; LD_{50} = 0.5 to 5 g/kg.
Long-term health effects: If continued for a long period, ingestion or inhalation of silver compounds can cause permanent discoloration of the skin (argyria).
Liquid or solid irritant characteristics: Eye irritant.
IDLH value: 10 mg/m³ as silver, soluble compounds

FIRE DATA
Behavior in fire: Decomposes into metallic silver and oxygen.

CHEMICAL REACTIVITY
Binary reactants: An oxidizer. Reacts with reducing agents, acids, amines, and combustible materials. Violent reaction with ammonia.

ENVIRONMENTAL DATA
Food chain concentration potential: Negative; unlikely to accumulate.
Water pollution: Effect of low concentrations on aquatic life is unknown. May be dangerous if it enters nearby water intakes; notify operators. Notify local health and wildlife officials. **Response to discharge:** Should be removed. Chemical and physical treatment.

SHIPPING INFORMATION
Grades of purity: Commercial, 99+%; **Storage temperature:** Ambient; **Inert atmosphere:** None; **Venting:** Open; **Stability during transport:** Stable.

PHYSICAL AND CHEMICAL PROPERTIES
Physical state @ 59°F/15°C and 1 atm: Solid.
Molecular weight: 231.8
Boiling point @ 1 atm: Decomposes @ approximately 390°F/199°C/472°K.
Specific gravity (water = 1): 7.14 @ 68°F/20°C (solid).

SILVER SULFATE   REC. S:1200

SYNONYMS: SULFATO de PLATA (Spanish)

IDENTIFICATION
CAS Number: 19287-89-9
Formula: Ag_2SO_4; $AgNO_3$

DESCRIPTION: Colorless crystalline solid when pure; turns gray on exposure to light. Odorless. Sinks and mixes with water.

Noncombustible solid, but may be combustible in the form of dust or powder • Irritating to the skin, eyes, and respiratory tract • Toxic products of combustion may include sulfur oxides.

Hazard Classification (based on NFPA-704 Rating System)
Health Hazards (Blue): 1; Flammability (Red): 0; Reactivity (Yellow): 0

EMERGENCY RESPONSE: See Appendix A (171)
Evacuation:
Public safety: Isolate the area of spill or leak for at least 10 to 25 meters (30 to 80 feet) in all directions.
Spill: Increase, in the downwind direction, as necessary, the distance shown under "Public Safety."
Fire: If any large container is involved in fire, isolate for at least 800 meters (½ mile) in all directions; also, consider initial evacuation for 800 meters (½ mile) in all directions.

EXPOSURE
Short-term effects: *SEEK MEDICAL ATTENTION.* **Dust:** Irritating to eyes, nose, and throat. *IF INHALED*, will, will cause coughing or difficult breathing. *IF IN EYES*, hold eyelids open and flush with plenty of water. *IF BREATHING HAS STOPPED*, give artificial respiration. IF breathing is difficult, administer oxygen. **Solid:** Irritating to skin and eyes. Harmful if swallowed. Remove contaminated clothing and shoes. Flush affected areas with plenty of water. *IF IN EYES*, hold eyelids open and flush with plenty of water. *IF SWALLOWED* and victim is *CONSCIOUS AND ABLE TO SWALLOW*, have victim drink 4 to 8 ounces of water and have victim induce vomiting. *IF SWALLOWED* and victim is *UNCONSCIOUS OR HAVING CONVULSIONS*, do nothing except keep victim warm.

HEALTH HAZARDS

Personal protective equipment (PPE): B-Level PPE. OSHA Table Z-1-A air contaminant. Dust mask; goggles or face shield; protective gloves.

Recommendations for respirator selection: NIOSH/OSHA as silver soluble compounds.

0.25 mg/m^3: SA:CF* (any supplied-air respirator operated in a continuous-flow mode); or PAPRHiE* (any powered, air-purifying respirator with a high-efficiency particulate filter). *0.5 mg/m^3:* HiEF (any air-purifying, full-facepiece respirator with a high-efficiency particulate filter); or SCBAF (any self-contained breathing apparatus with a full facepiece); or SAF (any supplied-air respirator with a full facepiece). *10 mg/m^3:* SAF:PD,PP (any supplied-air respirator that has a full facepiece and is operated in a pressure-demand or other positive-pressure mode). *EMERGENCY OR PLANNED ENTRY INTO UNKNOWN CONCENTRATIONS OR IDLH CONDITIONS:* SCBAF:PD,PP (any self-contained breathing apparatus that has a full facepiece and is operated in a pressure-demand or other positive-pressure mode); or SAF:PD,PP:ASCBA (any supplied-air respirator that has a full facepiece and is operated in a pressure-demand or other positive-pressure mode in combination with an auxiliary self-contained breathing apparatus operated in a pressure-demand or other positive pressure mode). *ESCAPE:* HiEF (any air-purifying, full-facepiece respirator with a high-efficiency particulate filter); or SCBAE (any appropriate escape-type, self-contained breathing apparatus). **Note*: Substance causes eye irritation or damage; eye protection needed.

Exposure limits (TWA unless otherwise noted): ACGIH TLV 0.01 mg/m^3; OSHA PEL 0.01 mg/m^3 as silver soluble compounds

Toxicity by ingestion: Poisonous.

Long-term health effects: If continued for a long period, ingestion or inhalation of silver compounds can cause permanent discoloration of the skin (argyria).

Liquid or solid irritant characteristics: Eye irritant.

IDLH value: 10 mg/m^3 as silver, soluble compounds.

CHEMICAL REACTIVITY

Binary reactants: Violent reaction with aluminum, magnesium.

ENVIRONMENTAL DATA

Food chain concentration potential: Negative; unlikely to accumulate.

Water pollution: Harmful to aquatic life in very low concentrations. May be dangerous if it enters nearby water intakes; notify operators. Notify local health and wildlife officials.

Response to discharge: Issue warning–water contaminant. Restrict access. Should be removed. Chemical and physical treatment.

SHIPPING INFORMATION

Grades of purity: Reagent; Commercial; **Storage temperature:** Ambient; **Inert atmosphere:** None; **Venting:** Open; **Stability during transport:** Stable.

PHYSICAL AND CHEMICAL PROPERTIES

Physical state @ 59°F/15°C and 1 atm: Solid.
Molecular weight: 311.80
Specific gravity (water = 1): 5.45 @ 68°F/20°C (solid).
Heat of fusion: 13.7 cal/g.

SODIUM   **REC. S:1300**

SYNONYMS: EEC No. 011-001-00-0; NATRIUM; SODIO (Spanish); SODIUM, METAL LIQUID ALLOY; SODIUM METAL

IDENTIFICATION

CAS Number: 7440-23-5
DOT ID Number: UN 1428; DOT Guide Number: 138
Proper Shipping Name: Sodium
Reportable Quantity (RQ): **(CERCLA)** 10 lb/4.54 kg

DESCRIPTION: Silver to grayish-white, soft, solid, or liquid. Shipped in hermetically sealed cans and shipped and stored under inert gas or mineral oil. Odorless. Floats on the surface of water; reacts violently producing flammable and explosive hydrogen gas.

Thermally unstable, highly flammable solid, especially when heated • Ignites spontaneously in air • Skin and eye contact causes severe burns and blindness • Do NOT use water, CO_2, or halogenated extinguishing agents • Firefighting gear (including SCBA) may not provide adequate protection. If exposure occurs, remove and isolate gear immediately and thoroughly decontaminate personnel • Toxic products of combustion may include extremely irritating vapors of Na • May re-ignite after fire is extinguished • Do not put yourself in danger by entering a contaminated area to rescue a victim.

Hazard Classification (based on NFPA-704 Rating System)
Health Hazards (Blue): 3; Flammability (Red): 1; Reactivity (Yellow): 2; Special Notice (White): Water reactive

EMERGENCY RESPONSE: See Appendix A (138)
Evacuation:
Public safety: Isolate the area of spill or leak for at least 50 to 100 meters (160 to 330 feet) in all directions.
Spill: Consider initial downwind evacuation for at least 250 meters (800 feet).
Fire: If any large container is involved in fire, isolate for at least 800 meters (½ mile) in all directions; also, consider initial evacuation for 800 meters (½ mile) in all directions.

EXPOSURE

Short-term effects: *SEEK MEDICAL ATTENTION.* **Solid:** Will burn skin and eyes. Remove contaminated clothing and shoes. Flush affected areas with plenty of water. *IF IN EYES*, hold eyelids open and flush with plenty of water.

HEALTH HAZARDS

Personal protective equipment (PPE): B-Level PPE. Maximum protective clothing; goggles and face shield.

Vapor (gas) irritant characteristics: Nonvolatile. Contact with water forms corrosive fumes that can irritate eyes and respiratory tract.

Liquid or solid irritant characteristics: Severe eye and skin irritant. Causes second- and third-degree burns on short contact and is very injurious to the eyes.

FIRE DATA

Fire extinguishing agents not to be used: Water, water-based foams, CO_2 or halogenated extinguishing agents.
Autoignition temperature: 250°F/121°C/394°K.

CHEMICAL REACTIVITY

Reactivity with water: Reacts violently, with formation of flammable hydrogen gas and caustic soda solution; fire often occurs.

Binary reactants: Self-reaction may occur in moist air. Strong

reaction with oxidizers; halogens, CO_2, and many hydrocarbon substances (especially halogenated).
Neutralizing agents for acids and caustics: After reaction with water, caustic soda formed can be diluted with water and/or neutralized with acetic acid.

ENVIRONMENTAL DATA
Food chain concentration potential: Negative; unlikely to accumulate.
Water pollution: Dangerous to aquatic life in high concentrations. May be dangerous if it enters nearby water intakes; notify operators. Notify local health and wildlife officials. **Response to discharge:** Issue warning–high flammability. Restrict access. Evacuate area. Chemical and physical treatment.

SHIPPING INFORMATION
Grades of purity: Commercial grade: 99.95%; **Storage temperature:** 230–250°F/110–121°C (liquid); ambient (solid); **Inert atmosphere:** Dry nitrogen or argon (for liquid); under kerosene (for solid); **Venting:** Pressure-vacuum; **Stability during transport:** Stable.

PHYSICAL AND CHEMICAL PROPERTIES
Physical state @ 59°F/15°C and 1 atm: Solid.
Molecular weight: 22.49
Boiling point @ 1 atm: 1621°F/883°C/1156°K.
Melting/Freezing point: 208°F/98°C/371°K.
Critical temperature: 3632°F/2000°C/2273°K.
Critical pressure: 5040 psia = 343 atm = 34.8 MN/m^2.
Specific gravity (water = 1): 0.971 @ 68°F/20°C (solid).
Heat of fusion: 27.4 cal/g.
Vapor pressure: 1 mm.

SODIUM ALKYLBENZENESULFONATES REC. S:1400

SYNONYMS: ABS-Na; ALKYLBENZENESULFONIC ACID, SODIUM SALT; *p-n*-ALKYLBENZENESULFONIC ACID DERIVITIVE, SODIUM SALT; ALKYLBENZENE, SODIUM SALT

IDENTIFICATION
CAS Number: 68411-30-3
Formula: $C_nH_{2n+1}C_6H_4SO_3Na$

DESCRIPTION: Pale yellow powder or thick liquid. Faint detergent odor. Mixes with water. Soap bubbles may be produced.

Very Irritating to the skin, eyes, and respiratory tract • Toxic products of combustion may include sulfur oxides.

Hazard Classification (based on NFPA-704 Rating System)
Health Hazards (Blue): 2; Flammability (Red): 0; Reactivity (Yellow): 0

EMERGENCY RESPONSE: See Appendix A (171)
Evacuation:
Public safety: Isolate the area of spill or leak for at least 10 to 25 meters (30 to 80 feet) in all directions.
Spill: Increase, in the downwind direction, as necessary, the distance shown under "Public Safety."
Fire: If any large container is involved in fire, isolate for at least 800 meters (½ mile) in all directions; also, consider initial evacuation for 800 meters (½ mile) in all directions.

EXPOSURE
Short-term effects: *SEEK MEDICAL ATTENTION*. **Liquid:** Irritating to skin and eyes. *IF SWALLOWED*, will cause nausea or vomiting. Remove contaminated clothing and shoes. Flush affected areas with plenty of water. *IF IN EYES*, hold eyelids open and flush with plenty of water. *IF SWALLOWED* and victim is *CONSCIOUS AND ABLE TO SWALLOW*, have victim drink 4 to 8 ounces of water and have victim induce vomiting. *IF SWALLOWED* and victim is *UNCONSCIOUS OR HAVING CONVULSIONS*, do nothing except keep victim warm.

HEALTH HAZARDS
Personal protective equipment (PPE): B-Level PPE. Goggles or face shield; rubber gloves.
Toxicity by ingestion: Grade 2; LD_{50} = 0.5 to 5 g/kg.
Vapor (gas) irritant characteristics: Nonvolatile
Liquid or solid irritant characteristics: If spilled on clothing and allowed to remain, may cause smarting and reddening of the skin.

CHEMICAL REACTIVITY
Binary reactants: Acids or acid fumes form toxic sulfur oxides.

ENVIRONMENTAL DATA
Food chain concentration potential: Negative; unlikely to accumulate.
Water pollution: Harmful to aquatic life in very low concentrations. May be dangerous if it enters nearby water intakes; notify operators. Notify local health and wildlife officials. **Response to discharge:** Issue warning–water contaminant. Should be removed. Chemical and physical treatment.

SHIPPING INFORMATION
Grades of purity: Vary with each manufacturer and with intended use. Some is shipped as a thick, concentrated water solution, some as a solid, often mixed with other solids such as sodium phosphate. Ordinary household detergents are good examples of this substance; **Storage temperature:** Ambient; **Inert atmosphere:** Not required; **Venting:** Open; **Stability during transport:** Stable.

PHYSICAL AND CHEMICAL PROPERTIES
Physical state @ 59°F/15°C and 1 atm: Liquid or solid.
Boiling point @ 1 atm: Decomposes.
Specific gravity (water = 1): 1.0 @ 68°F/20°C (liquid).

SODIUM ALKYL SULFATES REC. S:1500

SYNONYMS: SODIUM HYDROGEN ALKYL SULFATE

IDENTIFICATION
Formula: $CH_{2+1}OSO_2ONa$

DESCRIPTION: Colorless to pale yellow thick liquid or solid. Faint detergent odor. Mixes with water. Soap bubbles may be produced.

Hazard Classification (based on NFPA-704 Rating System)
Health Hazards (Blue): 1; Flammability (Red): 0; Reactivity (Yellow): 0

EMERGENCY RESPONSE: See Appendix A (171)
Evacuation:
Public safety: Isolate the area of spill or leak for at least 10 to 25 meters (30 to 80 feet) in all directions.

Spill: Increase, in the downwind direction, as necessary, the distance shown under "Public Safety."
Fire: If any large container is involved in fire, isolate for at least 800 meters (½ mile) in all directions; also, consider initial evacuation for 800 meters (½ mile) in all directions.

EXPOSURE
Short-term effects: *SEEK MEDICAL ATTENTION.* **Liquid or solid:** Irritating to skin and eyes. *IF SWALLOWED*, will cause nausea or vomiting. Flush affected areas with plenty of water. *IF IN EYES*, hold eyelids open and flush with plenty of water. *IF SWALLOWED* and victim is *CONSCIOUS AND ABLE TO SWALLOW*, have victim drink 4 to 8 ounces of water and have victim induce vomiting. *IF SWALLOWED* and victim is *UNCONSCIOUS OR HAVING CONVULSIONS*, do nothing except keep victim warm.

HEALTH HAZARDS
Personal protective equipment (PPE): B-Level PPE. Goggles or face shield; rubber gloves.
Toxicity by ingestion: Grade 2; LD_{50} = 0.5 to 5 g/kg.
Vapor (gas) irritant characteristics: Nonvolatile
Liquid or solid irritant characteristics: Minimum hazard. If spilled on clothing and allowed to remain, may cause smarting and reddening of the skin.

FIRE DATA
Behavior in fire: May produce irritating vapors.

CHEMICAL REACTIVITY
Binary reactants: Violent reaction with aluminum, magnesium.

ENVIRONMENTAL DATA
Food chain concentration potential: Negative; unlikely to accumulate.
Water pollution: Harmful to aquatic life in very low concentrations. May be dangerous if it enters nearby water intakes; notify operators. Notify local health and wildlife officials.
Response to discharge: Issue warning–water contaminant. Should be removed. Chemical and physical treatment.

SHIPPING INFORMATION
Grades of purity: Vary with each manufacturer and with intended use. Some is shipped as a thick concentrated water solution, some as a solid often mixed with other solids such as sodium phosphate. Ordinary household detergents are good examples of this substance; **Storage temperature:** Ambient; **Inert atmosphere:** None; **Venting:** Open; **Stability during transport:** Stable.

PHYSICAL AND CHEMICAL PROPERTIES
Physical state @ 59°F/15°C and 1 atm: Liquid or Solid.
Boiling point @ 1 atm: Decomposes.

SODIUM ALUMINATE SOLUTION (45% or less)
REC. S:1600

SYNONYMS: ALUMINATO SODICO (Spanish); SODIUM m-ALUMINATE SOLUTION

IDENTIFICATION
CAS Number: 1302-42-7; 11138-49-1
Formula: $NaAlO_2$; $AlNaO_2$
DOT ID Number: UN 1819; DOT Guide Number: 154
Proper Shipping Name: Sodium aluminate, solution

DESCRIPTION: Colorless to amber liquid. Odorless. Mixes with water.

Corrosive • Skin or eye contact causes severe burns, impaired vision, or blindness; effects of contact or inhalation may be delayed • Firefighting gear (including SCBA) may not provide adequate protection. If exposure occurs, remove and isolate gear immediately and thoroughly decontaminate personnel • Toxic products of combustion may include sodium oxide.

Hazard Classification (based on NFPA-704 Rating System)
Health Hazards (Blue): 2; Flammability (Red): 0; Reactivity (Yellow): 0

EMERGENCY RESPONSE: See Appendix A (154)
Evacuation:
Public safety: Isolate the area of spill or leak for at least 25 to 50 meters (80 to 160 feet) in all directions.
Spill: Increase, in the downwind direction, as necessary, the distance shown under "Public Safety."
Fire: If tank, rail car, or tank truck is involved in fire, isolate for at least 800 meters (½ mile) in all directions; also, consider initial evacuation for 800 meters (½ mile) in all directions.

EXPOSURE
Short-term effects: *SEEK MEDICAL ATTENTION.* **Liquid:** Remove contaminated clothing and shoes. Flush affected areas with water. *IF BREATHING HAS STOPPED*, give artificial respiration; *avoid mouth-to-mouth resuscitation; use bag/mask apparatus.* IF IN EYES, hold eyelids open and flush with plenty of water. *IF SWALLOWED* and victim is *CONSCIOUS AND ABLE TO SWALLOW*, dilute by drinking water or milk. **Do NOT induce vomiting.**
Neutralize with fruit juice.
Note to physician or authorized medical personnel: Medical observation is recommended for 24 to 48 hours after breathing overexposure, as pulmonary edema may be delayed. As first aid for pulmonary edema, consider administering a corticosteroid spray. Cigarette smoking may exacerbate pulmonary injury and should be discouraged for at least 72 hours following exposure.

HEALTH HAZARDS
Personal protective equipment (PPE): B-Level PPE. Full, impervious chemical protective clothing and gloves, goggles, and approved respirator.
Vapor (gas) irritant characteristics: Vapors are moderately irritating such that personnel will not usually tolerate moderate or high concentrations.
Liquid or solid irritant characteristics: Eye and skin irritant. Causes smarting of the skin and first-degree burns on contact.

FIRE DATA
Fire extinguishing agents not to be used: Irritating vapors may be formed.

CHEMICAL REACTIVITY
Binary reactants: Reacts with acids, copper, tin, zinc, aluminum, acids, phosphorus, or chlorocarbons.
Neutralizing agents for acids and caustics: Weak acid.
Reactivity group: 5
Compatibility class: Caustics

ENVIRONMENTAL DATA
Water pollution: Aqueous solution is a strong base. However,

effect of low concentrations on aquatic life is unknown. May be dangerous if it enters nearby water intakes; notify operators. Notify local health and wildlife officials. **Response to discharge:** Chemical and physical treatment.

SHIPPING INFORMATION
Grades of purity: Technical grades of varying concentrations; **Storage temperature:** Ambient; **Inert atmosphere:** None; **Venting:** Open; **Stability during transport:** Stable.

PHYSICAL AND CHEMICAL PROPERTIES
Physical state @ 59°F/15°C and 1 atm: Liquid.
Molecular weight: 81.97
Boiling point @ 1 atm: 239°F/115°C/388°K.
Melting/Freezing point: 32°F/0°C/273°K.
Specific gravity (water = 1): 1.55 @ 77°F/25°C.

SODIUM AMIDE   REC. S:1700

SYNONYMS: AMIDA SODICO (Spanish); SODAMIDE

IDENTIFICATION
CAS Number: 7782-92-5
Formula: $NaNH_2$
DOT ID Number: UN 1390; DOT Guide Number: 139
Proper Shipping Name: Alkali metal amides

DESCRIPTION: White to gray crystalline solid. Ammonia odor. Sinks in water; violent decomposition producing heat and forming yellowish brown vapors of sodium hydroxide and ammonia; may cause combustion.

Combustible solid • Forms unstable peroxides after prolonged storage; may become shock-sensitive and explosive • Corrosive to the skin, eyes, and respiratory tract; skin and eye contact causes severe burns, vision impairment, and blindness; inhalation symptoms may be delayed • Firefighting gear (including SCBA) does not provide adequate protection. If exposure occurs, remove and isolate gear immediately and thoroughly decontaminate personnel • Containers may BLEVE when exposed to fire • Vapors may form explosive mixture with air • Concentrated dust in confined areas (e.g., tanks, sewers, buildings) may explode when exposed to fire • Toxic products of combustion may include ammonia and sodium oxide • May re-ignite after fire is extinguished • Do not put yourself in danger by entering a contaminated area to rescue a victim. *Warning*: This material reacts with air forming ammonia and a highly explosive oxidation product that is yellow-brown in color. See also note under chemical reactivity.

Hazard Classification (based on NFPA-704 Rating System)
Health Hazards (Blue): 3; Flammability (Red): 3; Reactivity (Yellow): 2; Special Notice (White): Water reactive

EMERGENCY RESPONSE: See Appendix A (139)
Evacuation:
Public safety: Isolate the area of spill or leak for at least 100 to 150 meters (330 to 490 feet) in all directions.
Spill: Increase, in the downwind direction, as necessary, the distance shown under "Public Safety."
Fire: If any large container is involved in fire, isolate for at least 800 meters (½ mile) in all directions; also, consider initial evacuation for 800 meters (½ mile) in all directions.

EXPOSURE
Short-term effects: *SEEK MEDICAL ATTENTION*. **Solid:** Will burn skin and eyes. Remove contaminated clothing and shoes. Flush affected areas with plenty of water. *IF IN EYES*, hold eyelids open and flush with plenty of water.
Note to physician or authorized medical personnel: Medical observation is recommended for 24 to 48 hours after breathing overexposure, as pulmonary edema may be delayed. As first aid for pulmonary edema, consider administering a corticosteroid spray. Cigarette smoking may exacerbate pulmonary injury and should be discouraged for at least 72 hours following exposure.

HEALTH HAZARDS
Personal protective equipment (PPE): B-Level PPE. Goggles or face shield; dust respirator; rubber gloves and shoes.
Vapor (gas) irritant characteristics: Eye and reparatory tract irritant.
Liquid or solid irritant characteristics: Severe eye and skin irritant. Burns skin and eyes just like caustic soda.
Long-term health effects: Amides may cause liver, kidney, and brain damage.

FIRE DATA
Flash point: Combustible solid.
Fire extinguishing agents not to be used: Water.
Behavior in fire: Containers may explode.
Electrical hazard: Due to low electric conductivity, this substance may generate electrostatic charges as a result of agitation and flow.

CHEMICAL REACTIVITY
Reactivity with water: Reacts violently and frequently bursts into flames. Forms caustic soda solution.
Binary reactants: Reacts with oxidizers, halogenated hydrocarbons; acids, air. Reaction with air forms corrosive and irritating hydrogen chloride, ammonia vapors and explosive oxidation product.*
Neutralizing agents for acids and caustics: Caustic solution formed by reaction with water can be diluted with water and/or neutralized by acetic acid.
Safety Note: When explosive products develop on exposure to air (telltale sign is yellow-brown air) the material should be destroyed while wear protective clothing and approved respirator as follows: Cover material with toluene; then slowly stir in anhydrous ethanol; and finally, neutralizing mixture.

ENVIRONMENTAL DATA
Water pollution: Dangerous to aquatic life in high concentrations. May be dangerous if it enters nearby water intakes; notify operators. Notify local health and wildlife officials. **Response to discharge:** Issue warning–corrosive, flammable. Restrict access. Should be removed. Disperse and flush.

SHIPPING INFORMATION
Grades of purity: Pure; technical; **Storage temperature:** Ambient; **Inert atmosphere:** Must be dry.; **Venting:** Sealed, airtight containers must be stored in well-ventilated area; **Stability during transport:** Stable if dry. May be stored or shipped under inert gas or under dry pentane, toluene, or xylene.

PHYSICAL AND CHEMICAL PROPERTIES
Physical state @ 59°F/15°C and 1 atm: Solid.
Molecular weight: 39.01
Boiling point @ 1 atm: 752°F/400°C/673°K.
Melting/Freezing point: 410°F/210°C/483°K.
Specific gravity (water = 1): 1.39 @ 68°F/20°C (solid).

SODIUM ARSENATE   REC. S:1800

SYNONYMS: ARSENIATO SODICO (Spanish); ARSENIC ACID, SODIUM SALT; DISODIUM ARSENATE HEPTAHYDRATE; FATSCO ANT POISON; SODIUM METAARSENATE; SODIUM ORTHOARSENATE; SODIUM ARSENATE, DIBASIC; SWEENEY'S ANT-GO

IDENTIFICATION
CAS Number: 7631-89-2
Formula: $AsH_3O_4 \cdot 7Na$; $HAsO_4 \cdot 7H_2O$
DOT ID Number: UN 1685; DOT Guide Number: 151
Proper Shipping Name: Sodium arsenate
Reportable Quantity (RQ): **(CERCLA)** 1 lb/0.454 kg

DESCRIPTION: Clear crystalline solid. Odorless. Sinks and mixes with water.

Poison! • Irritates eyes, skin, or respiratory tract; poisonous if swallowed • Toxic products of combustion may include arsenic and sodium oxide.

Hazard Classification (based on NFPA-704 Rating System)
Health Hazards (Blue): 3; Flammability (Red): 0; Reactivity (Yellow): 0

Evacuation:
Public safety: Isolate the area of spill or leak for at least 25 to 50 meters (80 to 160 feet) in all directions.
Spill: Increase, in the downwind direction, as necessary, the distance shown under "Public Safety."
Fire: If tank, rail car, or tank truck is involved in fire, isolate for at least 800 meters (½ mile) in all directions; also, consider initial evacuation for 800 meters (½ mile) in all directions.

EXPOSURE
Short-term effects: *SEEK MEDICAL ATTENTION.* **Dust:** Irritating to eyes, nose, and throat. *IF INHALED*, will, will cause difficult breathing. *IF IN EYES*, hold eyelids open and flush with plenty of water. *IF BREATHING HAS STOPPED*, give artificial respiration; *avoid mouth-to-mouth resuscitation; use bag/mask apparatus*. IF breathing is difficult, administer oxygen. **Solid:** Irritating to skin and eyes. *IF SWALLOWED*, will cause coughing, nausea, vomiting, or loss of consciousness. Remove contaminated clothing and shoes. Flush affected areas with plenty of water. *IF IN EYES*, hold eyelids open and flush with plenty of water. *IF SWALLOWED* and victim is *CONSCIOUS AND ABLE TO SWALLOW*, have victim drink 4 to 8 ounces of water. *IF SWALLOWED* and victim is *UNCONSCIOUS OR HAVING CONVULSIONS*, do nothing except keep victim warm.

HEALTH HAZARDS
Personal protective equipment (PPE): B-Level PPE. Dust mask; goggles or face shield; protective gloves.
Recommendations for respirator selection: NIOSH *as arsenic. At any concentrations above the NIOSH REL, or where there is no REL, at any detectable concentration:* SCBAF:PD,PP (any self-contained breathing apparatus that has a full faceplate and is operated in a pressure-demand or other positive-pressure mode); or SAF:PD,PP:ASCBA (any supplied-air respirator that has a full facepiece and is operated in a pressure-demand or other positive-pressure mode in combination with an auxiliary self-contained breathing apparatus operated in a pressure-demand or other positive-pressure mode). *ESCAPE:* GMFAGHiE [any air-purifying, full-facepiece respirator (gas mask) with a chin-style, front-or back-mounted acid gas canister having a high-efficiency particulate filter]; or SCBAE (any appropriate escape-type, self-contained breathing apparatus).
Exposure limits (TWA unless otherwise noted): ACGIH TLV 0.01 mg/m³; OSHA PEL [1910.1080] 0.010 mg/m³; NIOSH REL ceiling 0.002 mg/m³/15 min, as arsenic; potential human carcinogen; reduce exposure to lowest feasible level
Toxicity by ingestion: Grade 4; LD_{50} less than 50 mg/kg.
Long-term health effects: NTP anticipated carcinogen. Possible carcinogenic effects on skin and lungs.
Liquid or solid irritant characteristics: Eye irritant.
IDLH value: Possible human carcinogen; 5 mg/m³ as arsenic.

ENVIRONMENTAL DATA
Food chain concentration potential: Bioconcentrative 300 fold. Solubility increases with acidity.
Water pollution: Dangerous to aquatic life in high concentrations. May be dangerous if it enters nearby water intakes; notify operators. Notify local health and wildlife officials. **Response to discharge:** Issue warning–poison, water contaminant. Restrict access. Disperse and flush.

SHIPPING INFORMATION
Grades of purity: Reagent; Technical, 98+%; **Storage temperature:** Ambient; **Inert atmosphere:** None; **Venting:** Open; **Stability during transport:** Stable.

PHYSICAL AND CHEMICAL PROPERTIES
Physical state @ 59°F/15°C and 1 atm: Solid.
Molecular weight: 312
Boiling point @ 1 atm: Decomposes. 356°F/180°C/453°K.
Melting/Freezing point: 135°F/57°C/330°K.
Specific gravity (water = 1): 1.87 @ 68°F/20°C (solid).

SODIUM ARSENITE   REC. S:1900

SYNONYMS: ARSENITO SODICO (Spanish); ARSENOUS ACID, SODIUM SALT; ATLAS-A; KILL-ALL; PENITE; SODANIT; SODIUM META ARSENITE; SODIUM ORTHO ARSENITE

IDENTIFICATION
CAS Number: 7784-46-5
Formula: $AsO_3 \cdot NaAsO_2$; $AsO_2 \cdot Na$
DOT ID Number: UN 1686 (aqueous solution); UN 2027 (solid);
DOT Guide Number: 154 (aqueous solution); 151 (solid)
Proper Shipping Name: Sodium arsenite, aqueous solutions; Sodium arsenite, solid
Reportable Quantity (RQ): **(CERCLA)** 1 lb/0.454 kg

DESCRIPTION: White to gray crystalline solid. Odorless. Mixes with water.

Poison! • Irritates eyes, skin, or respiratory tract; poisonous if swallowed • Toxic products of combustion may include arsenic and sodium oxide.

Hazard Classification (based on NFPA-704 Rating System)
Health Hazards (Blue): 3; Flammability (Red): 0; Reactivity (Yellow): 0

EMERGENCY RESPONSE aqueous solution: See Appendix A (154)

Evacuation:
Public safety: Isolate the area of spill or leak for at least 25 to 50 meters (80 to 160 feet) in all directions.
Spill: Increase, in the downwind direction, as necessary, the distance shown under "Public Safety."
Fire: If tank, rail car, or tank truck is involved in fire, isolate for at least 800 meters (½ mile) in all directions; also, consider initial evacuation for 800 meters (½ mile) in all directions.
EMERGENCY RESPONSE solid: See Appendix A (151)
Evacuation:
Public safety: Isolate the area of spill or leak for at least 25 to 50 meters (80 to 160 feet) in all directions.
Spill: Increase, in the downwind direction, as necessary, the distance shown under "Public Safety."
Fire: If tank, rail car, or tank truck is involved in fire, isolate for at least 800 meters (½ mile) in all directions; also, consider initial evacuation for 800 meters (½ mile) in all directions.

EXPOSURE
Short-term effects: *SEEK MEDICAL ATTENTION*. If artificial respiration is administered, *avoid mouth-to-mouth resuscitation; use bag/mask apparatus*. **Dust:** *POISONOUS IF INHALED*. Irritating to eyes, nose, and throat. Ingestion or excessive inhalation of dust causes irritation of stomach and intestines with nausea, vomiting, and diarrhea; bloody stools, shock, rapid pulse, coma. Move victim to fresh air. *IF IN EYES*, hold eyelids open and flush with plenty of water. IF breathing is difficult, administer oxygen. **Solid:** *POISONOUS IF SWALLOWED*. Irritating to skin and eyes. Remove contaminated clothing and shoes. Flush affected areas with plenty of water. *IF IN EYES*, hold eyelids open and flush with plenty of water. *IF SWALLOWED* and victim is *CONSCIOUS AND ABLE TO SWALLOW*, have victim drink 4 to 8 ounces of water and have victim induce vomiting. *IF SWALLOWED* and victim is *UNCONSCIOUS OR HAVING CONVULSIONS*, do nothing except keep victim warm.

HEALTH HAZARDS
Personal protective equipment (PPE): B-Level PPE. Dust mask; rubber gloves; goggles or face shield.
Recommendations for respirator selection: NIOSH
At any concentrations above the NIOSH REL, or where there is no REL, at any detectable concentration: SCBAF:PD,PP (any self-contained breathing apparatus that has a full faceplate and is operated in a pressure-demand or other positive-pressure mode); or SAF:PD,PP:ASCBA (any supplied-air respirator that has a full facepiece and is operated in a pressure-demand or other positive-pressure mode in combination with an auxiliary self-contained breathing apparatus operated in a pressure-demand or other positive-pressure mode). *ESCAPE:* GMFAGHiE [any air-purifying, full-facepiece respirator (gas mask) with a chin-style, front-or back-mounted acid gas canister having a high-efficiency particulate filter]; or SCBAE (any appropriate escape-type, self-contained breathing apparatus).
Exposure limits (TWA unless otherwise noted): ACGIH TLV 0.01 mg/m^3; OSHA PEL [1910.1080] 0.010 mg/m^3; NIOSH REL ceiling 0.002 mg/m^3/15 min, as arsenic; potential human carcinogen; reduce exposure to lowest feasible level
Toxicity by ingestion: Grade 4; oral LD$_{50}$ = 42 mg/kg (rat).
Long-term health effects: Carcinogen. Arsenic poisoning may develop.
Liquid or solid irritant characteristics: Eye irritant.
IDLH value: 5 mg/m^3.

ENVIRONMENTAL DATA
Water pollution: Harmful to aquatic life in very low concentrations. May be dangerous if it enters nearby water intakes; notify operators. Notify local health and wildlife officials. **Response to discharge:** Issue warning–poison, water contaminant.

SHIPPING INFORMATION
Grades of purity: Pure; technical (55–98%); **Storage temperature:** Ambient; **Inert atmosphere:** None; **Venting:** Pressure-vacuum; **Stability during transport:** Stable.

PHYSICAL AND CHEMICAL PROPERTIES
Physical state @ 59°F/15°C and 1 atm: Solid.
Boiling point @ 1 atm: Decomposes.
Melting/Freezing point: 1139°F/615°C/888°K.
Specific gravity (water = 1): 1.87 @ 68°F/20°C (solid).

SODIUM AZIDE REC. S:2000

SYNONYMS: AZIDA SODICO (Spanish); AZIDE; AZIUM; AZOTURE de SODIUM (French); EEC No. 011-004-00-7; HYDRAZOIC ACID, SODIUM SALT; KAZOE; NATRIUMAZID (German); SODIUM, AZOTURE de (French); SODIUM SALT OF HYDRAZOIC ACID; U-3886; RCRA No. P105

IDENTIFICATION
CAS Number: 26628-22-8
Formula: N$_3$Na
DOT ID Number: UN 1687; DOT Guide Number: 153
Proper Shipping Name: Sodium azide
Reportable Quantity (RQ): **(CERCLA)** 1000 lb/454 kg

DESCRIPTION: Colorless solid. Odorless. Initially sinks in water; soluble forming a weak base.

Explosive! Shock or heat may cause material to explode • Poison! • Breathing the dust, swallowing the material, or absorption through the skin can cause serious illness • Firefighting gear (including SCBA) does not provide adequate protection. If exposure occurs, remove and isolate gear immediately and thoroughly decontaminate personnel • Decomposes above 525°F/300°C. Toxic products of combustion may include hydrazoic acid and oxides of nitrogen • Do not put yourself in danger by entering a contaminated area to rescue a victim.

Hazard Classification (based on NFPA-704 Rating System)
Health Hazards (Blue): 3; Flammability (Red): 0; Reactivity (Yellow): 3

EMERGENCY RESPONSE: See Appendix A (153)
Evacuation:
Public safety: Isolate the area of spill or leak for at least 25 to 50 meters (80 to 160 feet) in all directions.
Spill: Increase, in the downwind direction, as necessary, the distance shown under "Public Safety."
Fire: If tank, rail car, or tank truck is involved in fire, isolate for at least 800 meters (½ mile) in all directions; also, consider initial evacuation for 800 meters (½ mile) in all directions.

EXPOSURE
Short-term effects: *SEEK MEDICAL ATTENTION*. **Dust:** *POISONOUS IF INHALED*. Move victim to fresh air. *IF BREATHING HAS STOPPED*, give artificial respiration; *avoid mouth-to-mouth resuscitation; use bag/mask apparatus*. IF breathing is difficult, administer oxygen. **Solid:** *POISONOUS IF SWALLOWED*. Remove contaminated clothing and shoes. Flush

affected areas with plenty of water. *IF IN EYES*, hold eyelids open and flush with plenty of water. *IF SWALLOWED* and victim is *CONSCIOUS AND ABLE TO SWALLOW*, have victim drink 4 to 8 ounces of water and have victim induce vomiting. *IF SWALLOWED* and victim is *UNCONSCIOUS OR HAVING CONVULSIONS*, do nothing except keep victim warm.

HEALTH HAZARDS
Personal protective equipment (PPE): A-Level PPE. OSHA Table Z-1-A air contaminant. Dust mask; protective clothing; goggles.
Exposure limits (TWA unless otherwise noted): NIOSH REL ceiling 0.1 ppm as hydrazoic acid; ceiling 0.3 mg/m^3 as NaN$_3$; skin contact contributes significantly in overall exposure.
Short-term exposure limits (15-minute TWA): ACGIH ceiling 0.29 mg/m^3.
Toxicity by ingestion: Grade 4; oral rat LD$_{50}$ = 27 mg/kg (technical).
Long-term health effects: Potent mutagen of salmon-sperm DNA. Possible damage of kidneys or cardiovascular system (blood vessels).
Liquid or solid irritant characteristics: Eye and skin irritant.

FIRE DATA
Flash point: Combustible solid. Approach with extreme caution; consider letting fire burn.
Behavior in fire: Containers may explode.

CHEMICAL REACTIVITY
Reactivity with water: Dissolves to form an alkaline solution.
Binary reactants: Reacts with acids and heavy metals (copper, brass, lead, etc.). Forms explosion-sensitive compounds of lead azide and copper azide with some metals such as lead, solder in plumbing systems, silver, mercury, and copper.

ENVIRONMENTAL DATA
Food chain concentration potential: Negative; unlikely to accumulate.
Water pollution: Harmful to aquatic life in very low concentrations. May be dangerous if it enters nearby water intakes; notify operators. Notify local health and wildlife officials.
Response to discharge: Issue warning–poison, water contaminant. Restrict access. Disperse and flush.

SHIPPING INFORMATION
Grades of purity: Pure: 99+%; Practical grade; **Storage temperature:** Ambient; **Inert atmosphere:** None; **Venting:** Open; **Stability during transport:** Stable unless in contact with acids.

PHYSICAL AND CHEMICAL PROPERTIES
Physical state @ 59°F/15°C and 1 atm: Solid.
Molecular weight: 65
Boiling point @ 1 atm: Decomposes.
Melting/Freezing point: Decomposes. 527°F/275°C/548°K may explode above this point.
Specific gravity (water = 1): 1.85 @ 68°F/20°C (solid).

SODIUM BIFLUORIDE   **REC. S:2100**

SYNONYMS: BIFLUORO SODICO (Spanish); SODIUM ACID FLUORIDE; SODIUM BIFLUORIDE (VAN); SODIUM DIFLUORIDE; SODIUM HYDROGEN FLUORIDE; SODIUM HYDROGEN DIFLUORIDE

IDENTIFICATION
CAS Number: 1333-83-1
Formula: F$_2$HNa
DOT ID Number: UN 2439; DOT Guide Number: 154
Proper Shipping Name: Sodium hydrogendifluoride; Sodium hydrogen fluoride
Reportable Quantity (RQ): **(CERCLA)** 100 lb/45.4 kg

DESCRIPTION: White crystalline powder. Sinks and mixes with water.

Corrosive • Breathing the vapor can kill you; skin or eye contact causes severe burns, impaired vision, or blindness; effects of contact or inhalation may be delayed • Firefighting gear (including SCBA) does not provide adequate protection. If exposure occurs, remove and isolate gear immediately and thoroughly decontaminate personnel • Toxic products of combustion may include fluorides and sodium oxide.

Hazard Classification (based on NFPA-704 Rating System)
Health Hazards (Blue): 3; Flammability (Red): 0; Reactivity (Yellow): 1

EMERGENCY RESPONSE: See Appendix A (154)
Evacuation:
Public safety: Isolate the area of spill or leak for at least 25 to 50 meters (80 to 160 feet) in all directions.
Spill: Increase, in the downwind direction, as necessary, the distance shown under "Public Safety."
Fire: If tank, rail car, or tank truck is involved in fire, isolate for at least 800 meters (½ mile) in all directions; also, consider initial evacuation for 800 meters (½ mile) in all directions.

EXPOSURE
Short-term effects: *SEEK MEDICAL ATTENTION*. **Dust or Solid:** Irritating to eyes, nose, and throat. *IF SWALLOWED*, will cause nausea, vomiting, abdominal pain, and diarrhea. Move to fresh air. Flush affected areas with plenty of water. *IF BREATHING HAS STOPPED*, give artificial respiration; *avoid mouth-to-mouth resuscitation; use bag/mask apparatus*. *IF IN EYES*, hold eyelids open and flush with plenty of water. *IF SWALLOWED* and victim is *CONSCIOUS AND ABLE TO SWALLOW*, have victim drink 4 to 8 ounces of water.
Note to physician or authorized medical personnel: Medical observation is recommended for 24 to 48 hours after breathing overexposure, as pulmonary edema may be delayed. As first aid for pulmonary edema, consider administering a corticosteroid spray. Cigarette smoking may exacerbate pulmonary injury and should be discouraged for at least 72 hours following exposure.

HEALTH HAZARDS
Personal protective equipment (PPE): B-Level PPE. Rubber gloves, safety glasses, self-contained breathing apparatus.
Recommendations for respirator selection: NIOSH/OSHA as F. *12.5 mg/m^3*: DM (any dust and mist respirator). *25 mg/m^3*: DMXSQ* (any dust and mist respirator except single-use and quarter-mask respirators); or SA* (any supplied-air respirator). *62.5 mg/m^3*: SA:CF* [any supplied-air respirator operated in a continuous-flow mode)]; or PAPRDM*$^+$ *if not present as a fume* (any powered, air-purifying respirator with a dust and mist filter). *125 mg/m^3*: HiEF + (any air-purifying, full-facepiece respirator with a high-efficiency particulate filter); or SCBAF (any self-contained breathing apparatus with a full facepiece); or SAF (any supplied-air respirator with a full facepiece). *250 mg/m^3*: SA:PD,PP (any supplied-air respirator operated in a pressure-demand or other

positive-pressure mode). *EMERGENCY OR PLANNED ENTRY INTO UNKNOWN CONCENTRATIONS OR IDLH CONDITIONS:* SCBAF:PD,PP (any self-contained breathing apparatus that has a full faceplate and is operated in a pressure-demand or other positive-pressure mode); or SAF:PD,PP:ASCBA (any supplied-air respirator that has a full facepiece and is operated in a pressure-demand or other positive-pressure mode in combination with an auxiliary, self-contained breathing apparatus operated in a pressure-demand or other positive-pressure mode). *ESCAPE:* HiEF+ (any air-purifying, full-facepiece respirator with a high-efficiency particulate filter); or SCBAE (any appropriate escape-type, self-contained breathing apparatus). *Notes:* *Substance reported to cause eye irritation or damage; may require eye protection. ⁺May need acid gas sorbent.
Exposure limits (TWA unless otherwise noted): ACGIH TLV 2.5 mg/m^3 as fluorides; NIOSH/OSHA 2.5 mg/m^3 as inorganic fluorides.
Toxicity by ingestion: Grade 3; LD$_{50}$ = 50 to 500 mg/kg.
Long-term health effects: Chronic exposure results in fluorosis. Symptoms are weight loss, brittleness of bones, anemia, weakness, stiffness of joints, and discoloration of teeth when exposure occurs during tooth development.
Vapor (gas) irritant characteristics: Dust may be eye and respiratory tract irritant.
Liquid or solid irritant characteristics: Eye and skin irritant.
IDLH value: 250 mg/m^3 as F.

FIRE DATA
Fire extinguishing agents not to be used: Water.

CHEMICAL REACTIVITY
Reactivity with water: Reacts with water liberating heat and forming a corrosive solution.
Binary reactants: Aqueous solution corrodes glass, concrete, and certain metals, especially those containing silica such as cast iron. Will attack natural rubber, leather, and many organic materials. May generate hydrogen gas on contact with some metals.
Neutralizing agents for acids and caustics: Dilution action will slowly neutralize the acid while the presence of calcium will precipitate excess fluoride. Apply powdered limestone, slaked lime, soda ash, or sodium bicarbonate.

ENVIRONMENTAL DATA
Food chain concentration potential: Negative; unlikely to accumulate.
Water pollution: Dangerous to aquatic life in high concentrations. May be dangerous if it enters nearby water intakes; notify operators. Notify local health and wildlife officials. **Response to discharge:** Disperse and flush.

PHYSICAL AND CHEMICAL PROPERTIES
Physical state @ 59°F/15°C and 1 atm: Solid.
Molecular weight: 61.99
Boiling point @ 1 atm: Decomposes.
Melting/Freezing point: Decomposes in melting
Specific gravity (water = 1): 2.08 at room temperature
Relative vapor density (air = 1): 2.14 (calculated).
Heat of combustion: Not flammable
Heat of solution: Absorbs heat @ 77°F/25°C. 156.8 Btu/lb = 87.1 cal/g = 3.64 x 10^5 J/kg.

SODIUM BISULFITE (SOLID) REC. S:2200

SYNONYMS: BISULFITE de SODIUM (French); BISULFITO SODICO (Spanish); HYDROGEN SULFITE SODIUM; MONOSODIUM SALT OF SULFUROUS ACID; SODIUM ACID BISULFITE; SODIUM ACID SULFITE; SODIUM BISULFITE, SOLUTION; SODIUM HYDROGEN SULFITE; SODIUM SULFHYDRATE; SULFUROUS ACID, MONOSODIUM SALT; SODIUM METABISULFITE; SODIUM PYROSULFITE

IDENTIFICATION
CAS Number: 7631-90-5
Formula: HO$_3$S·Na
DOT ID Number: UN 3077; DOT Guide Number: 171
Proper Shipping Name: Environmentally hazardous substances, solid, n.o.s.
Reportable Quantity (RQ): **(CERCLA)** 5000 lb/2270 kg

DESCRIPTION: White powder or granular solid. Odorless when dry; irritating sulfur odor when moist. Sinks and mixes with water form an acid; turns yellow in solution.

Powder or dust is extremely irritating to skin, eyes, and respiratory tract • Toxic products of combustion may include oxides of sulfur and sodium.

Hazard Classification (based on NFPA-704 Rating System)
Health Hazards (Blue): 3; Flammability (Red): 0; Reactivity (Yellow): 1

EMERGENCY RESPONSE: See Appendix A (171)
Evacuation:
Public safety: Isolate the area of spill or leak for at least 10 to 25 meters (30 to 80 feet) in all directions.
Spill: Increase, in the downwind direction, as necessary, the distance shown under "Public Safety."
Fire: If any large container is involved in fire, isolate for at least 800 meters (½ mile) in all directions; also, consider initial evacuation for 800 meters (½ mile) in all directions.

EXPOSURE
Short-term effects: *SEEK MEDICAL ATTENTION.* **Dust:** Harmful if inhaled. Move to fresh air. *IF BREATHING HAS STOPPED,* give artificial respiration. IF breathing is difficult, administer oxygen.
Solid: Irritating to skin and eyes. Harmful if swallowed.
Flush affected areas with plenty of water. *IF IN EYES,* hold eyelids open and flush with plenty of water. *IF SWALLOWED* and victim is *CONSCIOUS AND ABLE TO SWALLOW,* have victim drink 4 to 8 ounces of water.
Note to physician or authorized medical personnel: Keep under observation. This material is converted to sulfurous acid in stomach and acute obstruction of alimentary canal may occur up to 3 weeks following ingestion.

HEALTH HAZARDS
Personal protective equipment (PPE): B-Level PPE. OSHA Table Z-1-A air contaminant. Dust mask; goggles or face shield.
Exposure limits (TWA unless otherwise noted): ACGIH TLV 5 mg/m^3; NIOSH REL 5 mg/m^3.
Toxicity by ingestion: Grade 2; LD$_{50}$ = 0.5 to 5 g/kg.
Long-term health effects: Allergen, mutagen. Large doses have been shown to cause depression, nerve irritation, bone marrow atrophy, paralysis, and retarded growth.
Vapor (gas) irritant characteristics: Nonvolatile. Dust may cause eye and respiratory tract irritation.
Liquid or solid irritant characteristics: Irritates eyes, skin and mucous membranes.

CHEMICAL REACTIVITY
Binary reactants: Slowly oxidized to the sulfate on exposure to air. Reacts with oxidizers and acids forming sulfur dioxide gas.

ENVIRONMENTAL DATA
Food chain concentration potential: Negative; unlikely to accumulate.
Water pollution: Dangerous to aquatic life in high concentrations. May be dangerous if it enters nearby water intakes; notify operators. Notify local health and wildlife officials. **Response to discharge:** Issue warning–water contaminant. Disperse and flush.

SHIPPING INFORMATION
Grades of purity: Various grades (87–100%); **Storage temperature:** Ambient; **Inert atmosphere:** None; **Venting:** Open; **Stability during transport:** Stable.

PHYSICAL AND CHEMICAL PROPERTIES
Physical state @ 59°F/15°C and 1 atm: Solid.
Molecular weight: 104.06
Boiling point @ 1 atm: Decomposes.
Melting/Freezing point: Decomposes.
Specific gravity (water = 1): 1.48 @ 68°F/20°C (solid).

SODIUM BISULFITE SOLUTION REC. S:2300

SYNONYMS: BISULFITO SODICO (Spanish); SODIUM ACID BISULFITE SOLUTION; SODIUM HYDROGEN SULFITE SOLUTION (35% OR LESS); SODIUM HYDROGEN SULFITE SOLUTION

IDENTIFICATION
CAS Number: 7631-90-5
DOT ID Number: UN 2693; DOT Guide Number: 154
Proper Shipping Name: Bisulfites, aqueous solution
Reportable Quantity (RQ): **(CERCLA)** 5000 lb/2270 kg

DESCRIPTION: Pale yellow liquid. Pungent odor of sulfur dioxide. Mixes with water forming a weak acid.

Solution is irritating to skin, eyes, and respiratory tract • Toxic products of combustion may include oxides of sulfur and sodium.

Hazard Classification (based on NFPA-704 Rating System)
Health Hazards (Blue): 1; Flammability (Red): 0; Reactivity (Yellow): 0

EMERGENCY RESPONSE: See Appendix A (154)
Evacuation:
Public safety: Isolate the area of spill or leak for at least 25 to 50 meters (80 to 160 feet) in all directions.
Spill: Increase, in the downwind direction, as necessary, the distance shown under "Public Safety."
Fire: If tank, rail car, or tank truck is involved in fire, isolate for at least 800 meters (½ mile) in all directions; also, consider initial evacuation for 800 meters (½ mile) in all directions.

EXPOSURE
Short-term effects: *SEEK MEDICAL ATTENTION.* **Liquid:** Irritating to skin and eyes. Flush affected areas with plenty of water. *IF BREATHING HAS STOPPED,* give artificial respiration; *avoid mouth-to-mouth resuscitation; use bag/mask apparatus.* IF IN EYES, hold eyelids open and flush with plenty of water.

HEALTH HAZARDS
Personal protective equipment (PPE): B-Level PPE. OSHA Table Z-1-A air contaminant. Wear chemical splash goggles or face shield and chemical protective gloves.
Exposure limits (TWA unless otherwise noted): ACGIH TLV 5 mg/m^3; NIOSH REL 5 mg/m^3 as sodium bisulfite.
Toxicity by ingestion: Grade 2; LD_{50} = 2.0 g/kg (rat).
Long-term health effects: Allergen, mutagen.
Vapor (gas) irritant characteristics: Vapors cause smarting of the eyes or respiratory system if present in high concentrations. The effect is temporary.
Liquid or solid irritant characteristics: Eye and skin irritant. If spilled on clothing and allowed to remain, may cause smarting and reddening of the skin.

CHEMICAL REACTIVITY
Binary reactants: Slowly oxidized to the sulfate on contact with air. Reactions with oxidizers or acids produces toxic sulfur dioxide gas. Attacks many metals.
Reactivity group: 43
Compatibility class: Miscellaneous water solutions.

ENVIRONMENTAL DATA
Food chain concentration potential: Negative; unlikely to accumulate.
Water pollution: Effect of low concentrations on aquatic life is unknown. May be dangerous if it enters nearby water intakes; notify operators. Notify local health and wildlife officials. **Response to discharge:** Chemical and physical treatment. Disperse and flush.

SHIPPING INFORMATION
Grades of purity: Technical and photographic grades. **Storage temperature:** Ambient; **Inert atmosphere:** None; **Venting:** Open; **Stability during transport:** Stable.

PHYSICAL AND CHEMICAL PROPERTIES
Physical state @ 59°F/15°C and 1 atm: Liquid.
Molecular weight: 104.06
Boiling point @ 1 atm: More than 212°F/100°C/373°K.
Melting/Freezing point: Less than 32°F/0°C/273°K.
Specific gravity (water = 1): 1.36 @ 77°F/25°C.

SODIUM BORATE REC. S:2400

SYNONYMS: BORATO SODICO (Spanish); BORAX; SODIUM BIBORATE; SODIUM PYROBORATE; SODIUM TETRABORATE; SODIUM TETRABORATE DECAHYDRATE

IDENTIFICATION
CAS Number: 1303-96-4
Formula: $B_4O_7 \cdot 2Na$

DESCRIPTION: White crystalline solid. Odorless. Sinks and mixes slowly with water.

Irritating to skin, eyes, and respiratory tract • Toxic products of combustion may include sodium oxide and boron.

Hazard Classification (based on NFPA-704 Rating System)
Health Hazards (Blue): 0; Flammability (Red): 0; Reactivity (Yellow): 0

EMERGENCY RESPONSE: See Appendix A (171)

Evacuation:
Public safety: Isolate the area of spill or leak for at least 10 to 25 meters (30 to 80 feet) in all directions.
Spill: Increase, in the downwind direction, as necessary, the distance shown under "Public Safety."
Fire: If any large container is involved in fire, isolate for at least 800 meters (½ mile) in all directions; also, consider initial evacuation for 800 meters (½ mile) in all directions.

EXPOSURE
Short-term effects: *SEEK MEDICAL ATTENTION.* **Dust:** Irritating to eyes, nose, and throat. *IF IN EYES*, hold eyelids open and flush with plenty of water. **Solid:** Irritating to skin and eyes. *IF SWALLOWED*, will cause headache, dizziness, nausea, or vomiting. Remove contaminated clothing and shoes. Flush affected areas with plenty of water. *IF IN EYES*, hold eyelids open and flush with plenty of water. *IF SWALLOWED* and victim is *CONSCIOUS AND ABLE TO SWALLOW*, have victim drink 4 to 8 ounces of water and have victim induce vomiting. *IF SWALLOWED* and victim is *UNCONSCIOUS OR HAVING CONVULSIONS*, do nothing except keep victim warm.

HEALTH HAZARDS
Personal protective equipment (PPE): B-Level PPE. Dust mask and goggles or face shield.
Toxicity by ingestion: Grade 2; LD_{50} = 0.5 to 5 g/kg.
Liquid or solid irritant characteristics: May cause eye and skin irritation.

FIRE DATA
Behavior in fire: Compound melts to a glassy material that may flow in large quantities and ignite combustibles elsewhere.

CHEMICAL REACTIVITY
Binary reactants: Reacts with acids, metallic salts.

ENVIRONMENTAL DATA
Food chain concentration potential: Negative; unlikely to accumulate.
Water pollution: Dangerous to aquatic life in high concentrations. May be dangerous if it enters nearby water intakes; notify operators. Notify local health and wildlife officials. **Response to discharge:** Disperse and flush.

SHIPPING INFORMATION
Grades of purity: In addition to anhydrous sodium borate, both pentahydrate and decahydrate are commercially available, all in technical grade. The decahydrate is also available in Radio and U.S.P. grades; **Storage temperature:** Ambient; **Inert atmosphere:** None; **Venting:** Open; **Stability during transport:** Stable.

PHYSICAL AND CHEMICAL PROPERTIES
Physical state @ 59°F/15°C and 1 atm: Solid.
Molecular weight: 201.26
Boiling point @ 1 atm: 2867°F/1575°C/1848°K.
Melting/Freezing point: 1364°F/740°C/1013°K.
Specific gravity (water = 1): 2.367 @ 68°F/20°C (solid).
Heat of solution: −92 Btu/lb = −51 cal/g = −2.1 × 10^5 J/kg.

SODIUM BOROHYDRIDE   REC. S:2500

SYNONYMS: BOROHIDRURO SODICO (Spanish); BOROHYDRURE de SODIUM (French); SODIUM TETRAHYDROBORATE(1-).

IDENTIFICATION
CAS Number: 16940-66-2
Formula: $BH_4 \cdot Na$
DOT ID Number: UN 1426; DOT Guide Number: 138
Proper Shipping Name: Sodium borohydride

DESCRIPTION: White powder or pellets. Odorless. Sinks and mixes with water; reacts, forming flammable hydrogen gas.

Corrosive to the eyes, skin, and respiratory tract; inhalation symptoms may be delayed • Can explode spontaneously in moist air • Containers may BLEVE when exposed to fire • Toxic products of combustion may include sodium oxide. Also hydrogen gas may be released.

Hazard Classification (based on NFPA-704 Rating System)
Health Hazards (Blue): 3; Flammability (Red): 2; Reactivity (Yellow): 1; Special Notice (White): Water reactive

EMERGENCY RESPONSE: See Appendix A (138)
Evacuation:
Public safety: Isolate the area of spill or leak for at least 50 to 100 meters (160 to 330 feet) in all directions.
Spill: Consider initial downwind evacuation for at least 250 meters (800 feet).
Fire: If any large container is involved in fire, isolate for at least 800 meters (½ mile) in all directions; also, consider initial evacuation for 800 meters (½ mile) in all directions.

EXPOSURE
Short-term effects: *SEEK MEDICAL ATTENTION.* **Dust:** Harmful if inhaled. Move to fresh air. *IF BREATHING HAS STOPPED*, give artificial respiration. IF breathing is difficult, administer oxygen. **Solid:** Irritating to skin and eyes. Harmful if swallowed or if skin is exposed. Affects the nervous system and high exposure may result in death.
Flush affected areas with plenty of water. *IF IN EYES*, hold eyelids open and flush with plenty of water. *IF SWALLOWED* and victim is *CONSCIOUS AND ABLE TO SWALLOW*, have victim drink 4 to 8 ounces of water. **Do NOT induce vomiting**.
Note to physician or authorized medical personnel: Medical observation is recommended for 24 to 48 hours after breathing overexposure, as pulmonary edema may be delayed. As first aid for pulmonary edema, consider administering a corticosteroid spray. Cigarette smoking may exacerbate pulmonary injury and should be discouraged for at least 72 hours following exposure.

HEALTH HAZARDS
Personal protective equipment (PPE): B-Level PPE. Goggles, rubber gloves, and protective clothing.
Toxicity by ingestion: Violent reaction with acid in stomach. Considered toxic because of boron content.
Long-term health effects: May cause brain damage.
Vapor (gas) irritant characteristics: Nonvolatile.
Liquid or solid irritant characteristics: Severe eye, skin and respiratory tract irritant.

FIRE DATA
Flash point: Solid ignites in air above 525°F/274°C.
Fire extinguishing agents not to be used: Water, CO_2, or halogenated extinguishing agents.

Behavior in fire: Containers may explode.
Autoignition temperature: 480°F/249°C/522°K.

CHEMICAL REACTIVITY
Reactivity with water: Reacts to form flammable hydrogen gas.
Binary reactants: Reacts with many substances to form explosive mixtures; with acids to form toxic, flammable diborane gas; moist air causes decomposition; strong oxidizers; aldehydes. Slowly corrodes glass.
Neutralizing agents for acids and caustics: Caustic formed by reaction with water can be diluted and/or neutralized with acetic acid.

ENVIRONMENTAL DATA
Food chain concentration potential: Negative; unlikely to accumulate.
Water pollution: Effect of low concentrations on aquatic life is unknown. May be dangerous if it enters nearby water intakes; notify operators. Notify local health and wildlife officials.
Response to discharge: Issue warning-flammable, corrosive. Restrict access. Should be removed.

SHIPPING INFORMATION
Grades of purity: 95-98% minimum purity; dry powder; pellets; 12% solution in 43% aqueous sodium hydroxide; **Storage temperature:** Ambient; **Inert atmosphere:** None; **Venting:** Sealed containers must be stored in well-ventilated area; **Stability during transport:** Stable unless mixed with acids or overheated, when flammable hydrogen gas is formed.

PHYSICAL AND CHEMICAL PROPERTIES
Physical state @ 59°F/15°C and 1 atm: Solid.
Molecular weight: 37.83
Melting/Freezing point: Decomposes at more than 750°F/399°C/672°K.
Boiling point @ 1 atm: Decomposes below 572°F/300°C/573°K.
Specific gravity (water = 1): 1.074 @ 68°F/20°C (solid).

SODIUM BOROHYDRIDE (15% or less) and SODIUM HYDROXIDE SOLUTION   REC. S:2600

SYNONYMS: BOROHIDRURO SODICO (Spanish); BOROHYDRIDE SOLUTION

IDENTIFICATION
CAS Number: 16940-66-2
Formula: BH_4Na and $NaOH$ in aqueous solution
DOT ID Number: UN 3320; DOT Guide Number: 157
Proper Shipping Name: Sodium borohydride; Sodium hydroxide solution, with not more than 12% sodium borohydride and not more than 40% sodium hydroxide

DESCRIPTION: Colorless liquid. Odorless. Soluble in water.

Corrosive • Skin or eye contact causes severe burns, impaired vision, or blindness; effects of contact or inhalation may be delayed • Firefighting gear (including SCBA) does not provide adequate protection. If exposure occurs, remove and isolate gear immediately and thoroughly decontaminate personnel • Toxic products of combustion may include fumes and highly flammable hydrogen gas.

Hazard Classification (based on NFPA-704 Rating System) (15% or less)
Health Hazards (Blue): 3; Flammability (Red): 0; Reactivity (Yellow): 1

EMERGENCY RESPONSE: See Appendix A (157)
Evacuation:
Public safety: Isolate the area of spill or leak for at least 50 to 100 meters (160 to 330 feet) in all directions.
Spill: Increase, in the downwind direction, as necessary, the distance shown under "Public Safety."
Fire: If tank, rail car, or tank truck is involved in fire, isolate for at least 800 meters (½ mile) in all directions; also, consider initial evacuation for 800 meters (½ mile) in all directions.

EXPOSURE
Short-term effects: *SEEK MEDICAL ATTENTION.* **Liquid:** *POISONOUS IF SWALLOWED.* Extremely corrosive to eyes, skin, nose, throat, and upper respiratory tract. *IF IN EYES*, hold eyelids open, flush with running water for at least 15 minutes. Remove contaminated clothing and shoes, flush affected areas with plenty of running water for at least 15 minutes. *IF SWALLOWED* and victim is CONSCIOUS: Have victim drink water, milk, dilute vinegar, lemon juice, or olive oil to dilute the material. *IF SWALLOWED* and victim is *UNCONSCIOUS OR HAVING CONVULSIONS,* do nothing except keep victim warm. **Do not induce vomiting.**
Note to physician or authorized medical personnel: Medical observation is recommended for 24 to 48 hours after breathing overexposure, as pulmonary edema may be delayed. As first aid for pulmonary edema, consider administering a corticosteroid spray. Cigarette smoking may exacerbate pulmonary injury and should be discouraged for at least 72 hours following exposure.

HEALTH HAZARDS
Personal protective equipment (PPE): B-Level PPE. Goggles, rubber gloves, and protective clothing.
Toxicity by ingestion: Grade 4: LD_{50} = 18 mg/kg (rat) Violent reaction with acid in stomach. Toxic because of boron content.
Long-term health effects: Possible brain damage.
Vapor (gas) irritant characteristics: Nonvolatile.
Liquid or solid irritant characteristics: Severe skin irritant. Causes second- and third-degree burns on short contact and is very injurious to the eyes.

FIRE DATA
Behavior in fire: May decompose and produce highly flammable hydrogen gas.

CHEMICAL REACTIVITY
Binary reactants: Reacts with acid to form toxic, flammable diborane gas. Slowly corrodes glass.
Neutralizing agents for acids and caustics: Flush with water, rinse with dilute acetic acid or vinegar.
Reactivity group: 5
Compatibility class: Caustics

ENVIRONMENTAL DATA
Water pollution: Effect of low concentrations on aquatic life is unknown. May be dangerous if it enters nearby water intakes; notify operators. Notify local health and wildlife officials. **Response to discharge:** Issue warning–corrosive. Restrict access. Should be removed.

SHIPPING INFORMATION
Grades of purity: 12% solution in 43% aqueous sodium hydroxide; **Storage temperature:** Ambient; **Inert atmosphere:** None; **Venting:** Sealed containers must be stored in well-ventilated area. **Stability during transport:** Stable unless mixed with acids or overheated.

NAS HAZARD CLASSIFICATION FOR BULK WATER TRANSPORTATION
FIRE: 0
HEALTH: Vapor irritant: 0; Liquid or solid irritant: 4; Poisons: 4
WATER POLLUTION: Human toxicity: 4; Aquatic toxicity:–; Aesthetic effect: 1
REACTIVITY: Other chemicals: 3; Water: 0; Self-reaction: 0

PHYSICAL AND CHEMICAL PROPERTIES
Physical state @ 59°F/15°C and 1 atm: Liquid.

SODIUM CACODYLATE REC. S:2700

SYNONYMS: ALKARSODL; ANSAR 160; ARSICODILE; ARSYCODILE; CACODILATO SODICO (Spanish); CHEMAID; DUTCH TREAT; HYDROXYDIMETHYLARSINE OXIDE, SODIUM SALT; PHYTAR; PHYTAR-560; RAD-E-CATE 16; SILVISAR; SODIUM DIMETHYLARSENATE; SODIUM OF CACODYLIC ACID

IDENTIFICATION
CAS Number: 124-65-2
Formula: $(CH_3)_2AsOONa$; $C_2H_6AsO_2 \cdot Na$
DOT ID Number: UN 1688; DOT Guide Number: 152
Proper Shipping Name: Sodium cacodylate
Reportable Quantity (RQ): **(EHS)** 1 lb/0.454 kg

DESCRIPTION: White solid or colorless to light yellow solution. Odorless. Mixes with water.

Poison! • Breathing the vapor, skin or eye contact, or swallowing the material can kill you • Firefighting gear (including SCBA) does not provide adequate protection. If exposure occurs, remove and isolate gear immediately and thoroughly decontaminate personnel • Toxic products of combustion may include oxides of arsenic and sodium oxide • Do not put yourself in danger by entering a contaminated area to rescue a victim.

Hazard Classification (based on NFPA-704 Rating System)
Health Hazards (Blue): 4; Flammability (Red): 0; Reactivity (Yellow): 0

EMERGENCY RESPONSE: See Appendix A (152)
Evacuation:
Public safety: Isolate the area of spill or leak for at least 25 to 50 meters (80 to 160 feet) in all directions.
Spill: Increase, in the downwind direction, as necessary, the distance shown under "Public Safety."
Fire: If tank, rail car, or tank truck is involved in fire, isolate for at least 800 meters (½ mile) in all directions; also, consider initial evacuation for 800 meters (½ mile) in all directions.

EXPOSURE
Short-term effects: *SEEK MEDICAL ATTENTION*. **Vapor or dust:** *POISONOUS IF INHALED*. Irritating to eyes. Move victim to fresh air. *IF IN EYES*, hold eyelids open and flush with plenty of water. If breathing is difficult, administer oxygen. If artificial respiration is administered, *avoid mouth-to-mouth resuscitation; use bag/mask apparatus*. **Liquid or solid:** *POISONOUS IF SWALLOWED*. Irritating to skin and eyes. Remove contaminated clothing and shoes. Flush affected areas with plenty of water. *IF IN EYES*, hold eyelids open and flush with plenty of water. *IF SWALLOWED* and victim is *CONSCIOUS AND ABLE TO SWALLOW*, have victim drink 4 to 8 ounces of water and have victim induce vomiting. *IF SWALLOWED* and victim is *UNCONSCIOUS OR HAVING CONVULSIONS*, do nothing except keep victim warm.

HEALTH HAZARDS
Personal protective equipment (PPE): A-Level PPE.
Recommendations for respirator selection: NIOSH *(as inorganic arsenic, for reference only)*.
At any concentrations above the NIOSH REL, or where there is no REL, at any detectable concentration: SCBAF:PD,PP (any self-contained breathing apparatus that has a full faceplate and is operated in a pressure-demand or other positive-pressure mode); or SAF:PD,PP:ASCBA (any supplied-air respirator that has a full facepiece and is operated in a pressure-demand or other positive-pressure mode in combination with an auxiliary self-contained breathing apparatus operated in a pressure-demand or other positive-pressure mode). *ESCAPE:* GMFAGHiE [any air-purifying, full-facepiece respirator (gas mask) with a chin-style, front-or back-mounted acid gas canister having a high-efficiency particulate filter]; or SCBAE (any appropriate escape-type, self-contained breathing apparatus).
Exposure limits (TWA unless otherwise noted): OSHA PEL 0.5 mg/m^3 as organic arsenic.
Toxicity by ingestion: Grade 2; oral LD_{50} = 2600 mg/kg (rat).
Liquid or solid irritant characteristics: May cause eye irritation.
IDLH value: 5 mg/m^3 as inorganic arsenic; known human carcinogen. Not determined for organic arsenic.

CHEMICAL REACTIVITY
Binary reactants: Corrodes common metals.

ENVIRONMENTAL DATA
Water pollution: Effect of low concentrations on aquatic life is unknown. May be dangerous if it enters nearby water intakes; notify operators. Notify local health and wildlife officials. **Response to discharge:** Issue warning–poison, water contaminant. Restrict access. Disperse and flush.

SHIPPING INFORMATION
Grades of purity: 22–28% sodium cacodylate, 3–5% cacodylic acid,

SHIPPING INFORMATION
Grades of purity: balance inert solid (or water); **Storage temperature:** Ambient; **Inert atmosphere:** None; **Venting:** Open; **Stability during transport:** Stable.

PHYSICAL AND CHEMICAL PROPERTIES
Physical state @ 59°F/15°C and 1 atm: Solid.
Molecular weight: 160.0
Boiling point @ 1 atm: Decomposes.
Specific gravity (water = 1): More than 1 @ 68°F/20°C (solid).

SODIUM CHLORATE REC. S:2800

SYNONYMS: ASEX; ATLACIDE; ATRATOL B-HERBATOX; CHLORATE OF SODA; CHLORATE SALT OF SODIUM;

Sodium chlorate solution

CLORATO SODICO (Spanish); CHLORAX; CHLORIC ACID, SODIUM SALT; CHLORSAURE (German); DE-FOL-ATE; DESOLET; DREXEL DEFOL; DROP LEAF; EEC No. 017-005-00-9; EVAU-SUPERFALL; GRAIN SORGHUM HARVEST AID; GRANEX OK; HARVEST-AID; KLOREX; KUSA-TOHRUKUSATOL; LOREX; NATRIUMCHLORAT (German); ORTHO C-1 DEFOLIANT & WEED KILLER; OXYCIL; RASIKAL; SHED-A-LEAF; SHED-A-LEAF "L"; SODA CHLORATE; SODIUM (CHLORATE DE) (French); TRAVEX; TUMBLEAF; UNITED CHEMICAL DEFOLIANT No. 1; VAL-DROP

IDENTIFICATION
CAS Number: 7775-09-9
Formula: $ClNaO_3$; $ClO_3 \cdot Na$
DOT ID Number: UN 1495; UN 2428 (solution); DOT Guide Number: 140
Proper Shipping Name: Sodium chlorate; Sodium chlorate, aqueous solution

DESCRIPTION: Colorless to pale yellow crystals or powder. Odorless. Sinks and mixes with water.

Strong oxidizer which may react spontaneously with low flash point organics or reducing agents. Decomposes above 572°F/300°C. Heat forms oxygen; will increase the activity of an existing fire • May cause fire or explosion on contact with combustibles (wood, paper, oil, clothing, etc.) • Irritating to the skin, eyes, and respiratory tract • Decomposes above Toxic products of combustion or decomposition include oxygen, oxides of sodium, and hydrogen chloride.

Hazard Classification (based on NFPA-704 Rating System)
Health Hazards (Blue): 1; Flammability (Red): 0; Reactivity (Yellow): 2; Special Notice (White): OXY

EMERGENCY RESPONSE: See Appendix A (140)
Note: Do not let spill area dry until it has been determined that there is no chlorates left in the area. Continue cooling after fire has been extinguished.
Evacuation:
Public safety: Isolate the area of spill or leak for at least 10 to 25 meters (30 to 80 feet) in all directions.
Spill: Consider initial downwind evacuation for at least 100 meters (330 feet).
Fire: If any large container is involved in fire, isolate for at least 800 meters (½ mile) in all directions; also, consider initial evacuation for 800 meters (½ mile) in all directions.

EXPOSURE
Short-term effects: *SEEK MEDICAL ATTENTION*. **Solid:** Irritating to skin and eyes. Harmful if swallowed.
Flush affected areas with plenty of water. *IF IN EYES*, hold eyelids open and flush with plenty of water. *IF SWALLOWED* and victim is *CONSCIOUS AND ABLE TO SWALLOW*, have victim drink 4 to 8 ounces of water and have victim induce vomiting. *IF SWALLOWED* and victim is *UNCONSCIOUS OR HAVING CONVULSIONS*, do nothing except keep victim warm.

HEALTH HAZARDS
Personal protective equipment (PPE): B-Level PPE. Clean work clothing (must be washed well with water after each exposure); rubber gloves and shoes; where dusty, goggles and an approved dust respirator.

Toxicity by ingestion: Grade 3; LD_{50} = 50 to 500 mg/kg; ingestion of 15 to 30 g may be fatal.
Long-term health effects: Liver, kidney and blood damage.
Vapor (gas) irritant characteristics: Nonvolatile. Dust irritates eyes and respiratory tract.
Liquid or solid irritant characteristics: Eye and skin irritant. Prolonged exposure to solid or dust may irritate skin.

FIRE DATA
Behavior in fire: In fire situations oxygen may be liberated and increase the intensity of the fire.

CHEMICAL REACTIVITY
Binary reactants: Reacts explosively, either as a solid or a liquid, with all organic matter and some metals. Chlorates are powerful oxidizing agents and can cause explosions when heated or rubbed with wood, organic matter, sulfur, sulfides, powdered metals, ammonia compounds, phosphorus compounds. Reacts with acids. Even water solutions react in this way if stronger than 30%, especially when warm.

ENVIRONMENTAL DATA
Food chain concentration potential: Negative; unlikely to accumulate.
Water pollution: Dangerous to aquatic life in high concentrations. May be dangerous if it enters nearby water intakes; notify operators. Notify local health and wildlife officials. **Response to discharge:** Issue warning–high flammability. Should be removed. Disperse and flush.

SHIPPING INFORMATION
Grades of purity: Technical (99.5% minimum); treated (99.0% minimum); **Storage temperature:** Ambient; **Inert atmosphere:** None; **Venting:** Open; **Stability during transport:** Stable if cool. Decomposition starts at 572°F/300°C with evolution of oxygen gas. Decomposition may be self-sustaining.

PHYSICAL AND CHEMICAL PROPERTIES
Physical state @ 59°F/15°C and 1 atm: Solid.
Molecular weight: 106.45
Boiling point @ 1 atm: Decomposes 572°F/300°C/573°K.
Melting/Freezing point: 478°F/248°C/521°K.
Specific gravity (water = 1): 2.49 @ 15°C (solid).

SODIUM CHLORATE SOLUTION   REC. S:2900

SYNONYMS: CHLORATE OF SODA; CLORATO SODICO (Spanish); EEC No. 017-005-00-9; SODA CHLORIC ACID, SODIUM SALT

IDENTIFICATION
CAS Number: 7775-09-9
Formula: $ClNaO_3$; ClO_3
DOT ID Number: UN 2428; DOT Guide Number: 171
Proper Shipping Name: Sodium chlorate, aqueous solution

DESCRIPTION: Colorless to pale yellow liquid. Odorless. Mixes with water.

Strong oxidizer which may react spontaneously with low flash point organics or reducing agents. Heat forms oxygen; will increase the activity of an existing fire • May cause fire or explosion contact with combustibles (wood, paper, oil, clothing, etc.) • Toxic products of combustion may include chloride and sodium oxide fumes.

EMERGENCY RESPONSE: See Appendix A (171)
Note: Do not let spill area dry until it has been determined that there is no chlorates left in the area. Continue cooling after fire has been extinguished.
Evacuation:
Public safety: Isolate the area of spill or leak for at least 10 to 25 meters (30 to 80 feet) in all directions.
Spill: Increase, in the downwind direction, as necessary, the distance shown under "Public Safety."
Fire: If any large container is involved in fire, isolate for at least 800 meters (½ mile) in all directions; also, consider initial evacuation for 800 meters (½ mile) in all directions.
Note: Do not let spill area dry until it has been determined that there is no chlorates left in the area. Continue cooling after fire has been extinguished.

EXPOSURE
Short-term effects: *SEEK MEDICAL ATTENTION.* **Liquid:** Irritating to skin, eyes and mucous membranes. Poisonous if swallowed. *IF IN EYES OR ON SKIN*, flush with running water for at least 15 minutes; hold eyelids open if necessary. Wash skin with soap and water. *IF SWALLOWED* and victim is *CONSCIOUS AND ABLE TO SWALLOW*, have victim drink 4 to 8 ounces of water and induce vomiting. *IF SWALLOWED* and victim is *UNCONSCIOUS OR HAVING CONVULSIONS*, do nothing except keep victim warm.

HEALTH HAZARDS
Personal protective equipment (PPE): B-Level PPE. Wear self-contained positive pressure breathing apparatus and full protective clothing.
Toxicity by ingestion: Grade 3; LD_{50} = 1.2 g/kg (rat).
Long-term health effects: May cause mutagenic effects. Liver, kidney and blood damage reported.
Vapor (gas) irritant characteristics: Eye and respiratory tract irritant.
Liquid or solid irritant characteristics: Irritating to eyes and skin.

FIRE DATA but can support combustion, especially if dried.
Behavior in fire: Evaporation of water produces concentrated solutions or the dry salt. They can decompose to produce oxygen gas which increases fire intensity, and they can form explosive mixtures with organic matter and other easily oxidizable materials that are readily ignited by heat.

CHEMICAL REACTIVITY
Binary reactants: Chlorates are powerful oxidizing agents and can cause explosions when mixed or heated with organic matter and many metals. Reacts with acids, reducing agents, sulfur, sulfides, powdered metals, phosphorus or ammonia compounds, combustible materials.
Reactivity group: (50% or less) is not compatible with groups 1, 2, 3, 5, 7, 8, 10, 12, 13, 17, and 20.
Compatibility class: Special case.

ENVIRONMENTAL DATA
Food chain concentration potential: Negative; unlikely to accumulate.
Water pollution: Dangerous to aquatic life in high concentrations. May be dangerous if it enters nearby water intakes; notify operators. Notify local health and wildlife officials. **Response to discharge:** Issue warning–water contaminant, oxidizing agent. Restrict access. Should be removed. Chemical and physical treatment.

SHIPPING INFORMATION
Grades of purity: 50% or less; **Storage temperature:** Ambient; **Inert atmosphere:** None; **Venting:** Open; **Stability during transport:** Stable.

PHYSICAL AND CHEMICAL PROPERTIES
Physical state @ 59°F/15°C and 1 atm: Liquid.

SODIUM CHROMATE   REC. S:3000

SYNONYMS: CHROMATE OF SODA; CROMATO SODICO (Spanish); CHROMIUM DISODIUM OXIDE; CHROMIUM SODIUM OXIDE; DISODIUM CHROMATE; NEUTRAL SODIUM CHROMATE; SODIUM CHROMATE(VI); SODIUM CHROMATE, ANHYDROUS

IDENTIFICATION
CAS Number: 7775-11-3
Formula: $CrO_4 \cdot 2Na$
DOT ID Number: UN 3077; DOT Guide Number: 171
Proper Shipping Name: Environmentally hazardous substances, solid, n.o.s.
Reportable Quantity (RQ): **(CERCLA)** 10 lb/4.54 kg

DESCRIPTION: Orange to yellow crystalline solid. Odorless. Sinks and mixes with water; forms an alkaline solution.

Poisonous • Corrosive • Breathing the vapor can kill you; skin or eye contact causes severe burns, impaired vision, or blindness; effects of contact or inhalation may be delayed • Firefighting gear (including SCBA) does not provide adequate protection. If exposure occurs, remove and isolate gear immediately and thoroughly decontaminate personnel • Strong oxidizer which may react spontaneously with low flash point organics or reducing agents. Heat forms oxygen; will increase the activity of an existing fire • May cause fire or explosion contact with combustibles (wood, paper, oil, clothing, etc.) • Toxic products of combustion may include chromium oxide.

Hazard Classification (based on NFPA-704 Rating System)
Health Hazards (Blue): 2; Flammability (Red): 0; Reactivity (Yellow): 0; Special Notice (White): OXY

EMERGENCY RESPONSE: See Appendix A (171)
Evacuation:
Public safety: Isolate the area of spill or leak for at least 10 to 25 meters (30 to 80 feet) in all directions.
Spill: Increase, in the downwind direction, as necessary, the distance shown under "Public Safety."
Fire: If any large container is involved in fire, isolate for at least 800 meters (½ mile) in all directions; also, consider initial evacuation for 800 meters (½ mile) in all directions.

EXPOSURE
Short-term effects: *SEEK MEDICAL ATTENTION.* **Dust:** Irritating to eyes, nose, and throat. *IF INHALED*, will, will cause coughing or difficult breathing. *IF IN EYES*, hold eyelids open and flush with plenty of water. *IF BREATHING HAS STOPPED*, give artificial respiration. IF breathing is difficult, administer oxygen. **Solid:** *POISONOUS IF SWALLOWED.* Irritating to skin and eyes. *IF SWALLOWED*, will cause nausea or loss of consciousness. Remove contaminated clothing and shoes. Flush affected areas with plenty of water. *IF IN EYES*, hold eyelids open and flush with plenty of water. *IF SWALLOWED* and victim is *CONSCIOUS AND*

ABLE TO SWALLOW, have victim drink 4 to 8 ounces of water and have victim induce vomiting. *IF SWALLOWED* and victim is *UNCONSCIOUS OR HAVING CONVULSIONS*, do nothing except keep victim warm.

Note to physician or authorized medical personnel: Medical observation is recommended for 24 to 48 hours after breathing overexposure, as pulmonary edema may be delayed. As first aid for pulmonary edema, consider administering a corticosteroid spray. Cigarette smoking may exacerbate pulmonary injury and should be discouraged for at least 72 hours following exposure.

HEALTH HAZARDS
Personal protective equipment (PPE): B-Level PPE. Rubber gloves; chemical safety goggles; rubber apron and sleeves, face shield, rubber shoes, protective clothing.
Recommendations for respirator selection: NIOSH
At any concentrations above the NIOSH REL, or where there is no REL, at any detectable concentration: SCBAF:PD,PP (any self-contained breathing apparatus that has a full facepiece and is operated in a pressure-demand or other positive-pressure mode); or SAF:PD,PP:ASCBA (any supplied-air respirator that has a full facepiece and is operated in a pressure-demand or other positive-pressure mode in combination with an auxiliary self-contained breathing apparatus operated in a pressure-demand or other positive pressure mode). *ESCAPE:* HiEF (any air-purifying, full-facepiece respirator with a high-efficiency particulate filter); or SCBAE (any appropriate escape-type, self-contained breathing apparatus).
Exposure limits (TWA unless otherwise noted): ACGIH TLV 0.05 mg/m^3 as water soluble Cr(VI) compounds; OSHA PEL 0.1 mg/m^3 as CrO$_3$; NIOSH 0.001 mg/m^3 as chromates; potential human carcinogen; reduce exposure to lowest feasible level.
Toxicity by ingestion: Grade 3; LD$_{50}$ = 50 to 500 mg/kg.
Long-term health effects: NTP listed carcinogen. IARC rating 3. Possible lung cancer.
Vapor (gas) irritant characteristics: Respiratory tract irritant.
Liquid or solid irritant characteristics: Skin and eye irritant.
IDLH value: 15 mg/m^3, as hexavalent chromium.

CHEMICAL REACTIVITY
Binary reactants: A strong oxidizer. Reacts violently with reducing agents, acids, acetic anhydride, hydrazine, combustible materials, organic substances, metal powders. Reacts with acrolein, antimony trisulfide, antimony tritelluride, arsenic pentasulfide, 1,1-dichloro-1-nitroethane, 1,3-dichloropropene, diethylamine, fluorine, hydrazine, potassium iodide, sodium tetraborate, sodium tetraborate decahydrate, sodium borohydride, zirconium dusts. s-trioxane. Incompatible with hydroxylamine. Aqueous solution is caustic. Attacks aluminum copper, brass, bronze, tin, zinc, especially in the presence of moisture.

ENVIRONMENTAL DATA
Food chain concentration potential: High.
Water pollution: Dangerous to aquatic life in high concentrations. May be dangerous if it enters nearby water intakes; notify operators. Notify local health and wildlife officials. **Response to discharge:** Disperse and flush.

SHIPPING INFORMATION
Grades of purity: Reagent; Commercial; Tetrahydrate grade; **Storage temperature:** Ambient; **Inert atmosphere:** None; **Venting:** Open; **Stability during transport:** Stable.

PHYSICAL AND CHEMICAL PROPERTIES
Physical state @ 59°F/15°C and 1 atm: Solid.

Molecular weight: 162
Boiling point @ 1 atm: Decomposes.
Specific gravity (water = 1): 2.723 @ 77°F/25°C (solid).
Heat of solution: –24.5 Btu/lb = –13.6 cal/g = –0.57 x 10^5 J/kg.

SODIUM CYANIDE   REC. S:3100

SYNONYMS: CIANURO SODICO (Spanish); CYANIDE OF SODIUM; CYANOBRIK; CYANOGRAN; CYANURE de SODIUM (French); CYMAG; HYDROCYANIC ACID, SODIUM SALT; EEC No. 006-007-00-5; RCRA No. P106

IDENTIFICATION
CAS Number: 143-33-9
Formula: CNNa
DOT ID Number: UN 1689; UN 1935 (solutions); DOT Guide Number: 157
Proper Shipping Name: Sodium cyanide; Cyanide solutions, n.o.s.
Reportable Quantity (RQ): **(CERCLA)** 10 lb/4.54 kg

DESCRIPTION: Colorless solid; flakes, lumps, or "eggs." Odorless when pure and dry; slight odor of almond or hydrocyanic acid (hydrogen cyanide) when moist. Sinks and mixes with water forming a strong base; a small amount of cyanide gas is slowly released.

Poison! • *Warning*: Do not use CO$_2$ extinguishers; releases toxic hydrogen cyanide gas • Corrosive • Breathing the vapor, skin contact, or swallowing the material can kill you; Skin or eye contact can cause burns and impaired vision or blindness; inhalation symptoms may be delayed • Firefighting gear (including SCBA) does not provide adequate protection. If exposure occurs, remove and isolate gear immediately and thoroughly decontaminate personnel • Toxic products of combustion may include cyanide and oxides of sodium and nitrogen • Do not put yourself in danger by entering a contaminated area to rescue a victim.

Hazard Classification (based on NFPA-704 Rating System)
Health Hazards (Blue): 3; Flammability (Red): 0; Reactivity (Yellow): 0

EMERGENCY RESPONSE: See Appendix A (157)
Evacuation:
Public safety: Isolate the area of spill or leak for at least 50 to 100 meters (160 to 330 feet) in all directions.
Spill: Increase, in the downwind direction, as necessary, the distance shown under "Public Safety."
Fire: If tank, rail car, or tank truck is involved in fire, isolate for at least 800 meters (½ mile) in all directions; also, consider initial evacuation for 800 meters (½ mile) in all directions.
EXPOSURE
Short-term effects: *SEEK MEDICAL ATTENTION*. **Dust:** *POISONOUS IF INHALED OR IF SKIN IS EXPOSED*. Move to fresh air. *IF BREATHING HAS STOPPED*, give artificial respiration; *avoid mouth-to-mouth resuscitation; use bag/mask apparatus*. IF breathing is difficult, administer oxygen. Lung edema may develop. **Solid:** *POISONOUS IF SWALLOWED OR IF SKIN IS EXPOSED*.
Will burn eyes. Remove contaminated clothing and shoes. Flush affected areas with plenty of water. *IF IN EYES*, hold eyelids open and flush with plenty of water. *IF SWALLOWED* and victim is *CONSCIOUS AND ABLE TO SWALLOW*, have victim drink 4 to 8 ounces of water and have victim induce vomiting. *IF SWALLOWED* and victim is *UNCONSCIOUS OR HAVING*

CONVULSIONS, do nothing except keep victim warm.
Note to physician or authorized medical personnel: Consider the use of amyl nitrite perles if symptoms of cyanide poisoning develop. If symptoms indicate, initial treatment includes the cyanide antidote kit. In all cases, break an amyl nitrite perle in a gauze pad and hold lightly under victim's nose for 15 seconds, repeating 5 times at about 15-second intervals; if necessary (and if sodium nitrite infusions will be delayed), repeat procedure every 3 minutes with fresh pearls until 3 or 4 have been used. Avoid breathing the vapor while administering it to the victim. Administer sodium nitrite IV, ASAP. The usual adult dose is 10 to 20 mL of a 3% solution infused over no less than 5 minutes; the average child dose is 0.15 to 0.20 mL/kg. Monitor blood pressure during administration, and slow the rate of infusion if hypotention develops. Next, infuse sodium thiosulfate IV. The usual adult dose is 50 mL of a 25% solution infused over 10 to 20 minutes; the average child dose is 1.65 mL/kg. Repeat with nitrite and thiosulfate as required.
Note to physician or authorized medical personnel: Medical observation is recommended for 24 to 48 hours after breathing overexposure, as pulmonary edema may be delayed. As first aid for pulmonary edema, consider administering a corticosteroid spray. Cigarette smoking may exacerbate pulmonary injury and should be discouraged for at least 72 hours following exposure.

HEALTH HAZARDS
Personal protective equipment (PPE): A-Level PPE. Protective gloves when handling solid sodium cyanide; gloves when handling cyanide solutions; U.S. Bureau of Mines or NIOSH/OSHA approved dust respirator; approved chemical safety goggles. *For solid sodium cyanide,* Chemical protective material(s) reported to offer limited protection: natural rubber, neoprene, nitrile, PVC. *For <30% solution of sodium cyanide*, Chemical protective material(s) reported to offer limited protection: PV alcohol. Also, polyethylene offers limited protection.
Recommendations for respirator selection: NIOSH/OSHA
25 mg/m^3: SA (any supplied-air respirator); or SCBA (any self-contained breathing apparatus). *EMERGENCY OR PLANNED ENTRY INTO UNKNOWN CONCENTRATIONS OR IDLH CONDITIONS:* SCBAF:PD,PP (any self-contained breathing apparatus that has a full facepiece and is operated in a pressure-demand or other positive-pressure mode); or SAF:PD,PP:ASCBA (any supplied-air respirator that has a full facepiece and is operated in a pressure-demand or other positive-pressure mode in combination with an auxiliary self-contained breathing apparatus operated in a pressure-demand or other positive pressure mode). *ESCAPE:* GMFSHiE [any air-purifying, full-facepiece respirator (gas mask) with a chin-style, front- or back-mounted canister providing protection against the compound of concern and having a high efficiency particulate filter) or SCBAE (any appropriate escape-type, self-contained breathing apparatus).
Exposure limits (TWA unless otherwise noted): OSHA PEL 5 mg/m^3, as cyanides; skin contact contributes significantly in overall exposure.
Short-term exposure limits (15-minute TWA): NIOSH ceiling 5 mg/m^3 (4.7 ppm) [10 minutes] as cyanides
Toxicity by ingestion: Grade 4; LD$_{50}$ below 50 mg/kg.
Vapor (gas) irritant characteristics: Nonvolatile, but moisture in air can liberate lethal hydrogen cyanide gas.
Liquid or solid irritant characteristics: Severe eye and skin irritant; may cause pain and second-degree burns after a few minutes of contact.
IDLH value: 25 mg/m^3 as cyanide

FIRE DATA
Fire extinguishing agents not to be used: CO_2.

CHEMICAL REACTIVITY
Reactivity with water: When sodium cyanide dissolves in water, a mild reaction occurs; poisonous hydrogen cyanide gas is released. Solution is a strong base; if the water is acidic, however, toxic amounts of the gas will instantly form.
Binary reactants: Dangerous reaction with CO_2, acids, acid salts, forming hydrogen cyanide gas. Reacts with strong oxidizers, chlorates, and nitrates. Corrodes metals such as aluminum, zinc, copper and their alloys.

ENVIRONMENTAL DATA
Food chain concentration potential: Negative; unlikely to accumulate.
Water pollution: DOT Appendix B, §172.101–marine pollutant. Harmful to aquatic life in very low concentrations. May be dangerous if it enters nearby water intakes; notify operators. Notify local health and wildlife officials. **Response to discharge:** Issue warning–poison. Restrict access. Evacuate area. Chemical and physical treatment.

SHIPPING INFORMATION
Grades of purity: 99+%; **Storage temperature:** Ambient; **Inert atmosphere:** None; **Venting:** Sealed containers must be stored in well-ventilated area; **Stability during transport:** Stable.

PHYSICAL AND CHEMICAL PROPERTIES
Physical state @ 59°F/15°C and 1 atm: Solid.
Molecular weight: 49.01
Boiling point @ 1 atm: 2725°F/1495°C/1748°K.
Melting/Freezing point: 1047°F/564°C/837°K.
Specific gravity (water = 1): 1.60 @ 77°F/25°C (solid).
Heat of fusion: 88.9 cal/g.
Vapor pressure: 0 mm (approximate).

SODIUM DICHLORO-*s*-TRIAZINETRIONE DIHYDRATE
REC. S:3200

SYNONYMS: SODIUM DICHLOROISOCYANURATE; SODIUM DICHLORO-*s*-TRIAZINETRIONE

IDENTIFICATION
CAS Number: 2893-78-9; 10119-30-9
Formula: $C_3HCl_2N_3O_3 \cdot Na$
DOT ID Number: UN 2465; DOT Guide Number: 140
Proper Shipping Name: Dichloroisocyanuric acid, dry; Dichloroisocyanuric acid salts; Sodium dichloroisocyanurate; Sodium dichloro-*s*-triazinetrione

DESCRIPTION: White solid. Smells like bleach or chlorine. Mixes with water.

Thermally unstable. Decomposition can be initiated with a heat source and can propagate throughout the mass with the evolution of dense fumes • Strong oxidizer which may react spontaneously with low flash point organics or reducing agents. Heat forms oxygen; will increase the activity of an existing fire • May cause fire or explosion on contact with combustibles (wood, paper, oil, clothing, etc.) • Irritating to the skin, eyes, and respiratory tract • Toxic products of combustion may include hydrogen chloride and oxides of nitrogen and sodium.

Hazard Classification (based on NFPA-704 Rating System)
(anhydrous, molecular weight 219.9) Health Hazards (Blue): 2; Flammability (Red): 0; Reactivity (Yellow): 2; Special Notice (White): OXY
(molecular weight 255.9) Health Hazards (Blue): 2; Flammability (Red): 0; Reactivity (Yellow): 1; Special Notice (White): OXY

EMERGENCY RESPONSE: See Appendix A (140)
Evacuation:
Public safety: Isolate the area of spill or leak for at least 10 to 25 meters (30 to 80 feet) in all directions.
Spill: Consider initial downwind evacuation for at least 100 meters (330 feet).
Fire: If any large container is involved in fire, isolate for at least 800 meters (½ mile) in all directions; also, consider initial evacuation for 800 meters (½ mile) in all directions.

EXPOSURE
Short-term effects: *SEEK MEDICAL ATTENTION.* **Dust:** Irritating to eyes, nose, and throat. *IF INHALED*, will, will cause coughing or difficult breathing. Move victim to fresh air. *IF BREATHING HAS STOPPED*, give artificial respiration. *IF IN EYES*, hold eyelids open and flush with plenty of water. IF breathing is difficult, administer oxygen. **Solid:** Irritating to skin and eyes. Harmful if swallowed. Remove contaminated clothing and shoes. Flush affected areas with plenty of water. *IF IN EYES*, hold eyelids open and flush with plenty of water. *IF SWALLOWED* and victim is *CONSCIOUS AND ABLE TO SWALLOW*, have victim drink 4 to 8 ounces of water and have victim induce vomiting. *IF SWALLOWED* and victim is *UNCONSCIOUS OR HAVING CONVULSIONS*, do nothing except keep victim warm.

HEALTH HAZARDS
Personal protective equipment (PPE): B-Level PPE. Dust mask or chlorine-canister mask; goggles; rubber gloves and other protective clothing to prevent contact with skin.
Toxicity by ingestion: Grade 2; oral LD_{50} = 1670 mg/kg (rat).
Long-term health effects: Effects unknown in experimental animals
Liquid or solid irritant characteristics: Eye and skin irritant.

FIRE DATA, but contact with ordinary combustibles may cause fire.
Fire extinguishing agents not to be used: Ammonia-containing dry chemicals.
Behavior in fire: Containers may rupture and explode when heated.

CHEMICAL REACTIVITY
Reactivity with water: Reaction releases chlorine gas.
Binary reactants: Contact with most foreign materials, organic matter, or easily chlorinated or oxidized materials may result in fire. Explosive reaction with sodium hypochlorite in the presence of moisture. Avoid contact with ammonium salts, amines, oil, grease, sawdust, floor sweepings, easily oxidized organics.

ENVIRONMENTAL DATA
Food chain concentration potential: Negative; unlikely to accumulate.
Water pollution: Effect of low concentrations on aquatic life is unknown. May be dangerous if it enters nearby water intakes; notify operators. Notify local health and wildlife officials.
Response to discharge: Issue warning–oxidizing material, water contaminant. Restrict access. Disperse and flush.

SHIPPING INFORMATION
Grades of purity: Technical: 39-60% available chlorine; **Storage temperature:** Cool ambient; **Inert atmosphere:** None; **Venting:** Pressure-vacuum; **Stability during transport:** Stable if dry.

PHYSICAL AND CHEMICAL PROPERTIES
Physical state @ 59°F/15°C and 1 atm: Solid.
Molecular weight: 220.0
Boiling point @ 1 atm: Decomposes.
Specific gravity (water = 1): 0.96 @ 68°F/20°C (solid).

SODIUM DICHROMATE   REC. S:3300

SYNONYMS: BICHROMATE de SODIUM (French); BICHROMATE OF SODA; CHROMIC ACID,DISODIUM SALT; CHROMIUM SODIUM OXIDE; DICROMATO SODICO (Spanish); DISODIUM DICHROMATE; EEC No. 024-004-00-7; NATRIUMDICHROMAT (German); SODIUM BICHROMATE; SODIUM CHROMATE; SODIUM DICHROMATE(VI); SODIUM(DICHROMATE de) (French).

IDENTIFICATION
CAS Number: 10588-01-9
Formula: $Cr_2O_7 \cdot 2Na$; Cr_2O_7; $Cr_2Na_2O_7 \cdot 2H_2O$
DOT ID Number: UN 3077; DOT Guide Number: 171
Proper Shipping Name: Environmentally hazardous substances, solid, n.o.s.
Reportable Quantity (RQ): **(CERCLA)** 10 lb/4.54 kg

DESCRIPTION: Red to orange crystalline solid. Odorless. Sinks and dissolves in water; forms a corrosive solution.

Corrosive to skin, eyes, and respiratory tract; skin or eye contact causes severe burns, impaired vision, or blindness; effects of contact or inhalation may be delayed • Firefighting gear (including SCBA) does not provide adequate protection. If exposure occurs, remove and isolate gear immediately and thoroughly decontaminate personnel • Strong oxidizer which may react spontaneously with low flash point organics or reducing agents. Heat forms oxygen; will increase the activity of an existing fire • May cause fire or explosion on contact with combustibles (wood, paper, oil, clothing, etc.) • Irritating to the skin, eyes, and respiratory tract • Toxic products of combustion may include sodium oxide.

Hazard Classification (based on NFPA-704 Rating System)
Health Hazards (Blue): 1; Flammability (Red): 0; Reactivity (Yellow): 1; Special Notice (White): OXY

EMERGENCY RESPONSE: See Appendix A (171)
Evacuation:
Public safety: Isolate the area of spill or leak for at least 10 to 25 meters (30 to 80 feet) in all directions.
Spill: Increase, in the downwind direction, as necessary, the distance shown under "Public Safety."
Fire: If any large container is involved in fire, isolate for at least 800 meters (½ mile) in all directions; also, consider initial evacuation for 800 meters (½ mile) in all directions.

EXPOSURE
Short-term effects: *SEEK MEDICAL ATTENTION.* **Dust:** Irritating to eyes, nose, and throat. *IF INHALED*, will, will cause difficult breathing. Move to fresh air. *IF BREATHING HAS STOPPED*, give artificial respiration. IF breathing is difficult, administer oxygen. Lung edema may develop. **Solid:** Will burn skin

and eyes. *IF SWALLOWED*, will cause nausea and vomiting. Remove contaminated clothing and shoes. Flush affected areas with plenty of water. *IF ON SKIN*, treat like acid burns; external lesions can be scrubbed with a 2% solution of sodium thiosulfate. *IF IN EYES*, hold eyelids open and flush with plenty of water. *IF SWALLOWED* and victim is *CONSCIOUS AND ABLE TO SWALLOW*, have victim drink 4 to 8 ounces of water. **Do NOT induce vomiting.**

Note to physician or authorized medical personnel: Medical observation is recommended for 24 to 48 hours after breathing overexposure, as pulmonary edema may be delayed. As first aid for pulmonary edema, consider administering a corticosteroid spray. Cigarette smoking may exacerbate pulmonary injury and should be discouraged for at least 72 hours following exposure.

HEALTH HAZARDS
Personal protective equipment (PPE): B-Level PPE. Approved dust mask; protective gloves; goggles or face shield. Chemical protective material(s) reported to have good to excellent resistance: polyethylene, Saranex®.
Recommendations for respirator selection: NIOSH
At any concentrations above the NIOSH REL, or where there is no REL, at any detectable concentration: SCBAF:PD,PP (any self-contained breathing apparatus that has a full facepiece and is operated in a pressure-demand or other positive-pressure mode); or SAF:PD,PP:ASCBA (any supplied-air respirator that has a full facepiece and is operated in a pressure-demand or other positive-pressure mode in combination with an auxiliary self-contained breathing apparatus operated in a pressure-demand or other positive pressure mode). *ESCAPE:* HiEF (any air-purifying, full-facepiece respirator with a high-efficiency particulate filter); or SCBAE (any appropriate escape-type, self-contained breathing apparatus).
Exposure limits (TWA unless otherwise noted): ACGIH TLV 0.05 mg/m^3 as water soluble Cr(VI) compounds; OSHA PEL 0.1 mg/m^3 as CrO$_3$; NIOSH 0.001 mg/m^3 as chromates; potential human carcinogen; reduce exposure to lowest feasible level.
Toxicity by ingestion: Poison. Grade 3; LD$_{50}$ = 50 to 500 mg/kg.
Long-term health effects: Confirmed carcinogen. IARC substance; no overall evaluation. Possible mutagen. May cause skin disorders; lung cancer.
Vapor (gas) irritant characteristics: Dusts or mists may cause severe irritation of eye and throat and can cause eye and lung injury. They cannot be tolerated even at low concentrations.
Liquid or solid irritant characteristics: Severe skin irritant. Causes second- and third-degree burns on short contact and is very injurious to the eyes.
IDLH value: 15 mg/m^3 as chromates.

CHEMICAL REACTIVITY
Reactivity with water: Aqueous solution is corrosive.
Binary reactants: Strong oxidizer; reacts with reducing agents, acids. Violent reaction with acetic anhydride. On contact with finely divided combustibles, such as sawdust, ignition may occur. Attacks copper, zinc, tin, brass, bronze, organic linings. To acidify dichromate do not use nitric or hydrochloric acid (toxic fumes of chlorine and oxides of nitrogen are formed); sulfuric acid may be used.

ENVIRONMENTAL DATA
Food chain concentration potential: Positive; likely to accumulate.
Water pollution: Dangerous to aquatic life in high concentrations. May be dangerous if it enters nearby water intakes; notify operators. Notify local health and wildlife officials. **Response to discharge:** Issue warning–water contaminant. Disperse and flush.

SHIPPING INFORMATION
Grades of purity: Technical grades: 98.8–99.9%; High-purity grades: 99.3–99.9%; **Storage temperature:** Ambient; **Inert atmosphere:** None; **Venting:** Open; **Stability during transport:** Stable.

PHYSICAL AND CHEMICAL PROPERTIES
Physical state @ 59°F/15°C and 1 atm: Solid.
Molecular weight: 262.01
Boiling point @ 1 atm: Decomposes. 752°F/400°C/673°K.
Melting/Freezing point: 675°F/357°C/630°K.
Specific gravity (water = 1): 2.35 @ 77°F/25°C (solid).

SODIUM FLUORIDE   REC. S:3400

SYNONYMS: ALCOA SODIUM FLUORIDE; ANTIBULIT; CAVI-TROL; CHEMIFLUOR; CREDO; DISODIUM DIFLUORIDE; EEC No. 009-004-00-7; FLORIDINE; FLOROCID; FLOZENGES; FLUORURO SODICO (Spanish); FLUORAL; ROACH SALT; VILLIAUMITE

IDENTIFICATION
CAS Number: 7681-49-4
Formula: FNa
DOT ID Number: UN 1690; DOT Guide Number: 154
Proper Shipping Name: Sodium fluoride
Reportable Quantity (RQ): **(CERCLA)** 1000 lb/454 kg

DESCRIPTION: White crystalline solid or powder. May be tinted Nile blue for identification purposes. Odorless. Sinks in water.

Poison! • Corrosive to skin, eyes, and respiratory tract; skin or eye contact causes severe burns, impaired vision, or blindness; effects of contact or inhalation may be delayed • Firefighting gear (including SCBA) does not provide adequate protection. If exposure occurs, remove and isolate gear immediately and thoroughly decontaminate personnel • Containers may BLEVE when exposed to fire • Toxic products of combustion may include hydrogen fluoride gas, fumes of fluorine, and sodium oxide • Do not put yourself in danger by entering a contaminated area to rescue a victim.

Hazard Classification (based on NFPA-704 Rating System)
Health Hazards (Blue): 3; Flammability (Red): 0; Reactivity (Yellow): 0

EMERGENCY RESPONSE: See Appendix A (154)
Evacuation:
Public safety: Isolate the area of spill or leak for at least 25 to 50 meters (80 to 160 feet) in all directions.
Spill: Increase, in the downwind direction, as necessary, the distance shown under "Public Safety."
Fire: If tank, rail car, or tank truck is involved in fire, isolate for at least 800 meters (½ mile) in all directions; also, consider initial evacuation for 800 meters (½ mile) in all directions.

EXPOSURE
Short-term effects: *SEEK MEDICAL ATTENTION.* **Dust:** Lung edema may develop. *IF BREATHING HAS STOPPED*, give artificial respiration; *avoid mouth-to-mouth resuscitation; use bag/mask apparatus.* **Solid:** *POISONOUS IF SWALLOWED. IF*

SWALLOWED and victim is *CONSCIOUS AND ABLE TO SWALLOW*, have victim drink 4 to 8 ounces of water.
Note to physician or authorized medical personnel: Medical observation is recommended for 24 to 48 hours after breathing overexposure, as pulmonary edema may be delayed. As first aid for pulmonary edema, consider administering a corticosteroid spray. Cigarette smoking may exacerbate pulmonary injury and should be discouraged for at least 72 hours following exposure.

HEALTH HAZARDS
Personal protective equipment (PPE): B-Level PPE. Goggles or face shield. Chemical protective material(s) reported to have good to excellent resistance: natural rubber, neoprene, nitrile, PV alcohol.
Recommendations for respirator selection: NIOSH/OSHA as F. *12.5 mg/m³:* DM (any dust and mist respirator). *25 mg/m³:* DMXSQ* (any dust and mist respirator except single-use and quarter-mask respirators); or SA* (any supplied-air respirator). *62.5 mg/m³:* SA:CF* [any supplied-air respirator operated in a continuous-flow mode)]; or PAPRDM*⁺ *if not present as a fume* (any powered, air-purifying respirator with a dust and mist filter). *125 mg/m³:* HiEF + (any air-purifying, full-facepiece respirator with a high-efficiency particulate filter); or SCBAF (any self-contained breathing apparatus with a full facepiece); or SAF (any supplied-air respirator with a full facepiece). *250 mg/m³:* SA:PD,PP (any supplied-air respirator operated in a pressure-demand or other positive-pressure mode). *EMERGENCY OR PLANNED ENTRY INTO UNKNOWN CONCENTRATIONS OR IDLH CONDITIONS:* SCBAF:PD,PP (any self-contained breathing apparatus that has a full faceplate and is operated in a pressure-demand or other positive-pressure mode); or SAF:PD,PP:ASCBA (any supplied-air respirator that has a full facepiece and is operated in a pressure-demand or other positive-pressure mode in combination with an auxiliary, self-contained breathing apparatus operated in a pressure-demand or other positive-pressure mode). *ESCAPE:* HiEF+ (any air-purifying, full-facepiece respirator with a high-efficiency particulate filter); or SCBAE (any appropriate escape-type, self-contained breathing apparatus). *Notes:* *Substance reported to cause eye irritation or damage; may require eye protection. ⁺May need acid gas sorbent.
Exposure limits (TWA unless otherwise noted): ACGIH TLV 2.5 mg/m³ as fluorides; NIOSH/OSHA 2.5 mg/m³ as inorganic fluorides.
Toxicity by ingestion: Grade 3; LD_{50} = 50 to 500 mg/kg (rabbit, rat).
Long-term health effects: Organ damage; kidney.
Vapor (gas) irritant characteristics: Nonvolatile, but dust can cause eye and respiratory tract irritation.
Liquid or solid irritant characteristics: Severe eye and skin irritant.
IDLH value: 250 mg/m³ as F.

CHEMICAL REACTIVITY
Binary reactants: Reacts with acids.

ENVIRONMENTAL DATA
Water pollution: Dangerous to aquatic life in high concentrations. May be dangerous if it enters nearby water intakes; notify operators. Notify local health and wildlife officials. **Response to discharge:** Disperse and flush.

SHIPPING INFORMATION
Grades of purity: 97.5+%; **Storage temperature:** Ambient; **Inert atmosphere:** None; **Venting:** Open; **Stability during transport:** Stable.

PHYSICAL AND CHEMICAL PROPERTIES
Physical state @ 59°F/15°C and 1 atm: Solid.
Molecular weight: 41.99
Boiling point @ 1 atm: 3083°F/1695°C/1968°K.
Melting/Freezing point: 1819°F/993°C/1266°K.
Specific gravity (water = 1): 2.79 @ 68°F/20°C (solid).
Heat of fusion: 166.7 cal/g.
Vapor pressure: 1 mm @ 1971°F

SODIUM FLUOROACETATE   REC. S:3500

SYNONYMS: 1080; ACETIC ACID, FLUORO-, SODIUM SALT; COMPOUND 1080; FLUOROACETIC ACID, SODIUM SALT; FLUOACETATO SODICO (Spanish); FRATOL; FURATOL; RATBANE-1080; SFA; SODIUM MONOFLUOROACETATE; SODIUM FLUOROACETIC ACID; TL 869; YASO-KNOCK; RCRA No. P058

IDENTIFICATION
CAS Number: 62-74-8
Formula: $CH_2FCOONa$; $C_2H_2FO_2 \cdot Na$
DOT ID Number: UN 2629; DOT Guide Number: 151
Proper Shipping Name: Sodium fluoroacetate
Reportable Quantity (RQ): **(CERCLA)** 10 lb/4.54 kg

DESCRIPTION: Fluffy, colorless to white powder. May be dyed black or yellow. Faint, vinegar-like odor. Sinks and mixes with water. A liquid above 95°F.

Hazard Classification (based on NFPA-704 Rating System)
Health Hazards (Blue): 3; Flammability (Red): -; Reactivity (Yellow): -

EMERGENCY RESPONSE: See Appendix A (151)
Evacuation:
Public safety: Isolate the area of spill or leak for at least 25 to 50 meters (80 to 160 feet) in all directions.
Spill: Increase, in the downwind direction, as necessary, the distance shown under "Public Safety."
Fire: If tank, rail car, or tank truck is involved in fire, isolate for at least 800 meters (½ mile) in all directions; also, consider initial evacuation for 800 meters (½ mile) in all directions.

EXPOSURE
Short-term effects: *SEEK MEDICAL ATTENTION.* **Dust:** Very high acute toxicity. May be fatal if inhaled or absorbed through skin; harmful to eyes. *IF IN EYES OR ON SKIN*, flush with running water for at least 15 minutes; hold eyelids open periodically if appropriate. Remove and double bag contaminated clothing and shoes at the site. *IF BREATHING HAS STOPPED*, give artificial respiration; *avoid mouth-to-mouth resuscitation; use bag/mask apparatus.* IF breathing is difficult, administer oxygen. **Solid:** Very high acute toxicity. May be fatal if swallowed or absorbed through skin; harmful to eyes. *IF IN EYES OR ON SKIN*, flush with running water for at least 15 minutes; hold eyelids open periodically if appropriate. Remove and double bag contaminated clothing and shoes at the site. *IF SWALLOWED* and victim is *CONSCIOUS AND ABLE TO SWALLOW*, have victim drink water and touch back of throat to induce vomiting. *IF SWALLOWED* and victim is *UNCONSCIOUS OR HAVING CONVULSIONS*, do nothing except keep victim warm.
Note: The lethal dose is essentially the same for exposure via inhalation, ingestion or skin contact. Contact with the eyes can also affect the body. Absorption is very rapid by the gastrointestinal

tract; but skin absorption is slow unless the skin is cut or abraded. If ingested, smptoms may be delayed 30 minutes to 2 hours and may include vomiting, apprehension, auditory hallucinations, nystagmus, tingling sensations of the nose and face, facial numbness and twitching, and epileptiform convulsions. Several hours later, there may be pulsus alternans, ectopic heartbeats, tachycardia, ventricular fibrillation, and death. Autopsy findings pursuant to a lethal ingestion included hemorrhagic pulmonary edema and degeneration of renal tubules.

HEALTH HAZARDS
Personal protective equipment (PPE): A-Level PPE. OSHA Table Z-1-A air contaminant. Wear approved breathing apparatus and full protective clothing.
Recommendations for respirator selection: NIOSH/OSHA *0.25 mg/m³*: DM (any dust and mist respirator). *0.5 mg/m³*: DMXSQ (any dust and mist respirator except single-use and quarter mask respirators); or SA (any supplied-air respirator). *1.25 mg/m³*: SA:CF (any supplied-air respirator operated in a continuous-flow mode); or PAPRDM (any powered, air-purifying respirator with a dust and mist filter). *2.5 mg/m³*: HiEF (any air-purifying, full-facepiece respirator with a high-efficiency particulate filter); or SAT:CF (any supplied-air respirator that has a tight-fitting facepiece and is operated in a continuous-flow mode); or PAPRTHiE (any powered, air-purifying respirator with a tight-fitting facepiece and a high-efficiency particulate filter); or SCBAF (any self-contained breathing apparatus with a full facepiece); or SAF (any supplied-air respirator with a full facepiece). *EMERGENCY OR PLANNED ENTRY INTO UNKNOWN CONCENTRATIONS OR IDLH CONDITIONS:* SCBAF:PD,PP (any self-contained breathing apparatus that has a full facepiece and is operated in a pressure-demand or other positive-pressure mode); or SAF:PD,PP:ASCBA (any supplied-air respirator that has a full facepiece and is operated in a pressure-demand or other positive-pressure mode in combination with an auxiliary self-contained breathing apparatus operated in a pressure-demand or other positive pressure mode). *ESCAPE:* HiEF (any air-purifying, full-facepiece respirator with a high-efficiency particulate filter); or SCBAE (any appropriate escape-type, self-contained breathing apparatus).
Exposure limits (TWA unless otherwise noted): ACGIH TLV 0.05 mg/m³; NIOSH/OSHA 0.05 mg/m³; skin contact contributes significantly in overall exposure.
Short-term exposure limits (15-minute TWA): NIOSH STEL 0.15 mg/m³; skin contact contributes significantly in overall exposure.
Toxicity by ingestion: Highly poisonous. Grade 4; LD_{50} = 0.220 mg/kg (rat).
Long-term health effects: Symptoms included severe and progressive lesions of the renal tubular epithelium along with milder hepatic, neurologic and thyroid dysfunctions. Reproductive effects were observed in the rat. Organ damage; heart, liver kidneys.
IDLH value: 2.5 mg/m³.

CHEMICAL REACTIVITY
Binary reactants: In the presence of moisture, it may react with aluminum to produce highly flammable hydrogen gas. Forms hydrogen fluoride on contact with acids.

ENVIRONMENTAL DATA
Water pollution: Effect of low concentration on aquatic life is unknown. May be dangerous if it enters nearby water intakes; notify operators. Notify local health and wildlife officials.

Response to discharge: Issue warning–poison, water. contaminant. Restrict access. Should be removed. Chemical and physical treatment.

SHIPPING INFORMATION
Grades of purity: 98%; **Storage temperature:** Ambient; **Stability during transport:** Stable.

PHYSICAL AND CHEMICAL PROPERTIES
Physical state @ 59°F/15°C and 1 atm: Solid.
Molecular weight: 100.03
Boiling point @ 1 atm: Decomposes.
Melting/Freezing point: 392°F/200°C/473°K.
Vapor pressure: Low.

SODIUM HYDRIDE   REC. S:3600

SYNONYMS: HIDRURO SODICO (Spanish); NAH 80; SODIUM MONOHYDRIDE

IDENTIFICATION
CAS Number: 7646-69-7
Formula: HNa
DOT ID Number: UN 1427; DOT Guide Number: 138
Proper Shipping Name: Sodium hydride

DESCRIPTION: Gray powder in oil. Kerosene odor. Reacts violently with water; yields corrosive sodium hydroxide; flammable hydrogen gas is produced.

Extremely flammable solid • Corrosive • Severely Irritating to the skin, eyes, and respiratory tract; skin and eye contact causes severe burns and blindness • Do not use water or foam • The powder spontaneously ignites on contact with air • Firefighting gear (including SCBA) does not provide adequate protection. If exposure occurs, remove and isolate gear immediately and thoroughly decontaminate personnel • Containers may BLEVE when exposed to fire •Vapors may form explosive mixture with air • Vapors are heavier than air and will collect and stay in low areas • Vapors may travel long distances to ignition sources and flashback • Vapors in confined areas (e.g., tanks, sewers, buildings) may explode when exposed to fire • Toxic products of combustion may include sodium oxide • Violent reaction with water producing explosive hydrogen gas • May re-ignite after fire is extinguished • Do not put yourself in danger by entering a contaminated area to rescue a victim.

Hazard Classification (based on NFPA-704 Rating System)
Health Hazards (Blue): 3; Flammability (Red): 3; Reactivity (Yellow): 2; Special Notice (White): Water reactive

EMERGENCY RESPONSE: See Appendix A (138)
Evacuation:
Public safety: Isolate the area of spill or leak for at least 50 to 100 meters (160 to 330 feet) in all directions.
Spill: Consider initial downwind evacuation for at least 250 meters (800 feet).
Fire: If any large container is involved in fire, isolate for at least 800 meters (½ mile) in all directions; also, consider initial evacuation for 800 meters (½ mile) in all directions.

Evacuation:
Public safety: Isolate the area of spill or leak for at least 50 to 100 meters (160 to 330 feet) in all directions.

Spill: Consider initial downwind evacuation for at least 250 meters (800 feet).
Fire: If any large container is involved in fire, isolate for at least 800 meters (½ mile) in all directions; also, consider initial evacuation for 800 meters (½ mile) in all directions.

EXPOSURE
Short-term effects: *SEEK MEDICAL ATTENTION*. **Solid:** Will burn skin and eyes. Harmful if swallowed. Remove contaminated clothing and shoes. Flush affected areas with plenty of water. *IF IN EYES*, hold eyelids open and flush with plenty of water. *IF SWALLOWED* and victim is *CONSCIOUS AND ABLE TO SWALLOW*, have victim drink 4 to 8 ounces of water. **Do NOT induce vomiting.**

HEALTH HAZARDS
Personal protective equipment (PPE): B-Level PPE. Face shield; rubber gloves. Chemical protective material(s) reported to have good to excellent resistance: butyl rubber, polycarbonate.
Skin contact: brush off all particles at once and flood the affected area with water.
Vapor (gas) irritant characteristics: Nonvolatile
Liquid or solid irritant characteristics: Severe skin irritant. Causes second-and third-degree burns on short contact and is very injurious to the eyes.

FIRE DATA
Flash point: Flammable solid. May ignite spontaneously in presence of moisture.
Fire extinguishing agents not to be used: Water, soda acid, dry chemical, CO_2, or foam.
Behavior in fire: Accidental contact with water used to extinguish surrounding fire will result in the release of hydrogen gas and possible explosion

CHEMICAL REACTIVITY
Reactivity with water: Dangerous, potentially explosive, reaction with release of flammable hydrogen gas.
Binary reactants: Reacts with oxygen, acetylene, halogens, sulfur.
Neutralizing agents for acids and caustics: Neutralize only when accidental reaction with water is complete. Do not neutralize the flammable solid with aqueous solutions. Spent reaction solution may be neutralized with dilute solutions of acetic acid.

ENVIRONMENTAL DATA
Food chain concentration potential: Negative; unlikely to accumulate.
Water pollution: Dangerous to aquatic life in high concentrations. May be dangerous if it enters nearby water intakes; notify operators. Notify local health and wildlife officials. **Response to discharge:** Issue warning–high flammability. Restrict access. Should be removed. Chemical and physical treatment.

SHIPPING INFORMATION
Grades of purity: 55% plus 45% mineral oil; **Storage temperature:** Ambient; **Inert atmosphere:** Must be dry. Argon or nitrogen gas; **Venting:** Pressure-vacuum; **Stability during transport:** Stable below 435°F/225°C.

PHYSICAL AND CHEMICAL PROPERTIES
Physical state @ 59°F/15°C and 1 atm: Solid.
Molecular weight: 23.9
Boiling point @ 1 atm: Decomposes, Very high.
Melting/Freezing point: Decomposes 1470°F/800°C/973°K.

SODIUM HYDROSULFIDE   **REC. S:3700**

SYNONYMS: HIDROSULFURO SODICO (Spanish); SODIUM BISULFIDE; SODIUM HYDROGEN SULFIDE; SODIUM HYDROSULFIDE; SODIUM MERCAPTAN; SODIUM SULFHYDRATE

IDENTIFICATION
CAS Number: 16721-80-5
Formula: HNaS
DOT ID Number: NA 2922; UN 2318; NA 2949; DOT Guide Number: 154 (NA 2922; NA 2949); 135 (UN 2318)
Proper Shipping Name: Sodium hydrosulfide, solution (NA 2922); Sodium hydrosulfide, with less than 25% water of crystallization (UN 2318); Sodium hydrosulfide, with not less than 25% water of crystallization (UN 2949)
Reportable Quantity (RQ): **(CERCLA)** 5000 lb/2270 kg

DESCRIPTION: Light yellow; amber to dark red liquid. Hydrogen sulfide (rotten egg) odor. Mixes with water; emits sodium hydroxide and sodium sulfide. Moisture may cause self ignition or produce heat sufficient to ignite combustible materials.

Corrosive • Sodium hydrosulfide (<25% H_2O) may ignite spontaneously in air • Breathing the vapor can kill you; skin or eye contact causes severe burns, impaired vision, or blindness; effects of contact or inhalation may be delayed • Firefighting gear (including SCBA) does not provide adequate protection. If exposure occurs, remove and isolate gear immediately and thoroughly decontaminate personnel • Decomposition may start at 120°F/49°C. Toxic products of combustion may include sodium oxide and sulfur oxides.

Hazard Classification (based on NFPA-704 Rating System) (solution)
Health Hazards (Blue): 3; Flammability (Red): 0; Reactivity (Yellow): 0

EMERGENCY RESPONSE, spontaneously combustible: See Appendix A (135)
Evacuation:
Public safety: Isolate spill area for at least 100 to 150 meters (330 to 490 feet) in all directions.
Spill: Increase downwind, the distance shown above in "Public safety."
Fire: Isolate for 800 meters (½ mile) in all directions, especially if tank, rail car, or tank truck is involved in fire.

EMERGENCY RESPONSE: See Appendix A (154)
Evacuation:
Public safety: Isolate the area of spill or leak for at least 25 to 50 meters (80 to 160 feet) in all directions.
Spill: Increase, in the downwind direction, as necessary, the distance shown under "Public Safety."
Fire: If tank, rail car, or tank truck is involved in fire, isolate for at least 800 meters (½ mile) in all directions; also, consider initial evacuation for 800 meters (½ mile) in all directions.

EXPOSURE
Short-term effects: *SEEK MEDICAL ATTENTION*. **Liquid:** Irritating to skin and eyes. *IF SWALLOWED*, will cause nausea, vomiting, or loss of consciousness. Remove contaminated clothing and shoes. Flush affected areas with plenty of water. *IF BREATHING HAS STOPPED*, give artificial respiration; *avoid mouth-to-mouth resuscitation; use bag/mask apparatus. IF IN*

EYES, hold eyelids open and flush with plenty of water. *IF SWALLOWED* and victim is *CONSCIOUS AND ABLE TO SWALLOW*, have victim drink 4 to 8 ounces of water and have victim induce vomiting. *IF SWALLOWED* and victim is *UNCONSCIOUS OR HAVING CONVULSIONS*, do nothing except keep victim warm.

Note to physician or authorized medical personnel: Medical observation is recommended for 24 to 48 hours after breathing overexposure, as pulmonary edema may be delayed. As first aid for pulmonary edema, consider administering a corticosteroid spray. Cigarette smoking may exacerbate pulmonary injury and should be discouraged for at least 72 hours following exposure.

HEALTH HAZARDS

Personal protective equipment (PPE): A-Level PPE. Wear self-contained positive pressure breathing apparatus and full protective clothing.
Toxicity by ingestion: Grade 2; LD_{50} = 0.5 to 5 g/kg.
Long-term health effects: Mutation data. Allergen.
Vapor (gas) irritant characteristics: Vapors cause irritation such that personnel will find high concentrations unpleasant.
Liquid or solid irritant characteristics: Severe eye and skin irritant. May cause pain and second-degree burns after a few minutes of contact.
Odor threshold: 0.0047 ppm.

FIRE DATA

Flash point: 73°F/23°C (45% solution). Combustible solid (spontaneously combustible in the presence of moisture).
Flammable limits in air: LEL: 4.3%; UEL: 45.5%.
Behavior in fire: Containers may rupture and explode.

CHEMICAL REACTIVITY

Reactivity with water: Moisture may cause self ignition or produce heat sufficient to ignite combustible materials.
Binary reactants: Reacts with oxidizers, acids combustible materials. Hydrolyzed in moist air to sodium hydroxide and sodium sulfide. Corrodes steel above 150°F/66°C. Corrosive to aluminum and other metals.
Neutralizing agents for acids and caustics: Flood with water.
Reactivity group: 5
Compatibility class: Caustics

ENVIRONMENTAL DATA

Food chain concentration potential: Negative; unlikely to accumulate.
Water pollution: Dangerous to aquatic life in high concentrations. May be dangerous if it enters nearby water intakes; notify operators. Notify local health and wildlife officials. **Response to discharge:** Issue warning–water contaminant, corrosive. Restrict access. Disperse and flush.

SHIPPING INFORMATION

Grades of purity: 45% or less; **Storage temperature**: More than 63°F; **Inert atmosphere:** None; **Venting:** Pressure-vacuum; **Stability during transport:** Stable.

NAS HAZARD CLASSIFICATION FOR BULK WATER TRANSPORTATION

FIRE: 0
HEALTH: Vapor irritant: 2; Liquid or solid irritant: 3; Poisons: 4
WATER POLLUTION: Human toxicity: 2; Aquatic toxicity: 3; Aesthetic effect: 3
REACTIVITY: Other chemicals: 3; Water: 0; Self-reaction: 0

PHYSICAL AND CHEMICAL PROPERTIES

Physical state @ 59°F/15°C and 1 atm: Liquid.
Molecular weight: 56.05
Boiling point @ 1 atm: (approximate) 212°F/100°C/373°K.
Melting/Freezing point: (approximate) 63°F/17°C/290°K.
Specific gravity (water = 1): 1.3 @ 15°C (liquid).
Relative vapor density (air = 1): 1.17
Vapor pressure: (Reid) 0.95 psia.

SODIUM HYDROXIDE REC. S:3800

SYNONYMS: CAUSTIC SODA; EEC No. 011-002-00-6 (solid); EEC No. 011-002-01-3 (solution); HIDROXIDO SODICO (Spanish); HYDROXYDE de SODIUM (French); LEWIS-RED DEVIL® LYE; LYE; NATRIUMHYDROXID (German); SODA LYE; SODIUM HYDRATE; SODIUM HYDRATE; SODIUM(HYDROXYDE de) (French); WHITE CAUSTIC; SODIUM HYDROXIDE CAUSTIC SODA SOLUTION; SODA LYE; SODIUM HYDRATE SOLUTION

IDENTIFICATION

CAS Number: 1310-73-2
DOT ID Number: UN 1823 (dry solid); UN 1824 (solution); DOT Guide Number: 154
Proper Shipping Name: Sodium hydroxide, solid; Sodium hydroxide solution
Reportable Quantity (RQ): **(CERCLA)** 1000 lb/454 kg

DESCRIPTION: White solid crystals or pellets or clear to colorless solution in water. Odorless. Solid sinks and mixes with water; violent reaction releasing large amounts of heat (that may ignite combustible materials) and spattering.

Corrosive to skin, eyes, and respiratory tract; skin or eye contact causes severe burns, impaired vision, or blindness; poison if swallowed; effects of contact or inhalation may be delayed • Firefighting gear (including SCBA) does not provide adequate protection. If exposure occurs, remove and isolate gear immediately and thoroughly decontaminate personnel • Reacts with metals forming explosive hydrogen gas • Toxic products of combustion may include corrosive fumes and oxides of sodium • Do not put yourself in danger by entering a contaminated area to rescue a victim.

Hazard Classification (based on NFPA-704 Rating System)
Health Hazards (Blue): 3; Flammability (Red): 0; Reactivity (Yellow): 1

EMERGENCY RESPONSE: See Appendix A (154)
Evacuation:
Public safety: Isolate the area of spill or leak for at least 25 to 50 meters (80 to 160 feet) in all directions.
Spill: Increase, in the downwind direction, as necessary, the distance shown under "Public Safety."
Fire: If tank, rail car, or tank truck is involved in fire, isolate for at least 800 meters (½ mile) in all directions; also, consider initial evacuation for 800 meters (½ mile) in all directions.

EXPOSURE

Short-term effects: *SEEK MEDICAL ATTENTION*. **Dust:** Irritating to eyes, nose, and throat. Move to fresh air *IF BREATHING HAS STOPPED*, give artificial respiration; *avoid mouth-to-mouth resuscitation; use bag/mask apparatus.* IF breathing is difficult, administer oxygen. *IF IN EYES*, hold eyelids

Sodium hypochlorite

open and flush with plenty of water. **Solid:** Will burn skin and eyes. Harmful if swallowed. Remove contaminated clothing and shoes. Flush affected areas with plenty of water. *IF IN EYES*, hold eyelids open and flush with plenty of water. *IF SWALLOWED* and victim is *CONSCIOUS AND ABLE TO SWALLOW*, have victim drink 4 to 8 ounces of water. **Do NOT induce emesis. Do NOT administer activated charcoal or attempt to neutralize stomach contants.**

Note to physician or authorized authorized paramedic: Medical observation is recommended for 24 to 48 hours after breathing overexposure, as pulmonary edema may be delayed. As first aid for pulmonary edema, consider administering a corticosteroid spray. Cigarette smoking may exacerbate pulmonary injury and should be discouraged for at least 72 hours following exposure.

HEALTH HAZARDS
Personal protective equipment (PPE): B-Level PPE. OSHA Table Z-1-A air contaminant. Chemical safety goggles; face shield; filter or dust-type respirator; rubber boots; rubber gloves. *For 30–70% solution*, Protective materials with good to excellent resistance: butyl rubber, natural rubber, neoprene, nitrile rubber, polyethylene, PVC, Teflon®, Viton®, Viton®/chlorobutyl rubber, Saranex®, Silvershield®, butyl rubber/neoprene.
Recommendations for respirator selection: NIOSH/OSHA
10 mg/m³: SA:CF (any supplied-air respirator operated in a continuous-flow mode); or HiEF (any air-purifying, full-facepiece respirator with a high-efficiency particulate filter); PAPRDM (any powered, air-purifying respirator with a dust and mist filter); or SCBAF (any self-contained breathing apparatus with a full facepiece); or SAF (any supplied-air respirator with a full facepiece). *EMERGENCY OR PLANNED ENTRY INTO UNKNOWN CONCENTRATIONS OR IDLH CONDITIONS:* SCBAF:PD,PP (any self-contained breathing apparatus that has a full facepiece and is operated in a pressure-demand or other positive-pressure mode); or SAF:PD,PP:ASCBA (any supplied-air respirator that has a full facepiece and is operated in a pressure-demand or other positive-pressure mode in combination with an auxiliary self-contained breathing apparatus operated in a pressure-demand or other positive pressure mode). *ESCAPE:* HiEF (any air-purifying, full-facepiece respirator with a high-efficiency particulate filter); or SCBAE (any appropriate escape-type, self-contained breathing apparatus). *Note*: Substance causes eye irritation or damage; eye protection needed.
Exposure limits (TWA unless otherwise noted): OSHA PEL 2 mg/m³.
Short-term exposure limits (15-minute TWA): ACGIH STEL ceiling 2 mg/m³; NIOSH REL ceiling 2 mg/m³.
Toxicity by ingestion: (10% solution) oral rabbit LD_{Lo} = 500 mg/kg.
Long-term health effects: Possible respiratory tract/lung damage.
Vapor (gas) irritant characteristics: Severe eye and respiratory tract irritant. May cause lung edema; physical exertion will aggravate this condition.
Liquid or solid irritant characteristics: Severe eye and skin irritant. Causes second-and third-degree burns on short contact and is very injurious to the eyes.
IDLH value: 10 mg/m³.

CHEMICAL REACTIVITY
Binary reactants: Acids, flammable liquids, organics, halogens, metals, nitromethane. When wet, attacks chemically active metals such as aluminum, tin, lead, and zinc to produce flammable hydrogen gas.
Neutralizing agents for acids and caustics: Flush with water, rinse with dilute acetic acid.

Reactivity group: 5
Compatibility class: Caustics

ENVIRONMENTAL DATA
Food chain concentration potential: Negative; unlikely to accumulate.
Water pollution: Dangerous to aquatic life in high concentrations. May be dangerous if it enters nearby water intakes; notify operators. Notify local health and wildlife officials. **Response to discharge:** Issue warning–corrosive. Restrict access. Disperse and flush.

SHIPPING INFORMATION
Grades of purity: Technical flakes; USP pellets; **Storage temperature:** Ambient; **Inert atmosphere:** None; **Venting:** Open; **Stability during transport:** Stable.

PHYSICAL AND CHEMICAL PROPERTIES
Physical state @ 59°F/15°C and 1 atm: Solid.
Molecular weight: 40.00
Boiling point @ 1 atm: 2534°F/1390°C/1663°K; 248°F/120°C/393°K (33% solution); 54°F/12°C/285°K (50% solution).
Melting/Freezing point: 604°F/318°C/591°K; 46°F/8°C/281°K (33% solution)
Specific gravity (water = 1): 2.13 @ 68°F/20°C (solid); 1.3 (33% solution).
Heat of fusion: 50.0 cal/g.
Vapor pressure: 1 mm (approximate).

SODIUM HYPOCHLORITE　　　　　　　　　　　　**REC. S:3900**

SYNONYMS: ANTIFORMIN; CHLOROS; EEC No. 017-011-00-1; HIPOCLORITO SODICO (Spanish); HYCLORITE; HYPOCHLOROUS ACID, SODIUM SALT; MILTON; SODIUM OXYCHLORIDE; SURCHLOR (liquid): BLEACH; B-K LIQUID; CLOROX®; CARREL-DAKINS SOLUTION; DAKINS SOLUTION; EEC No. 017-011-01-9; JAVELLE WATER (50g/l); LIQUID BLEACH

IDENTIFICATION
CAS Number: 7681-52-9
Formula: $NaOCl \cdot 5H_2O$; $ClO \cdot Na$
DOT ID Number: UN 1791; DOT Guide Number: 154
Proper Shipping Name: Hypochlorite solutions with more than 5% available chlorine; Hypochlorite solution
Reportable Quantity (RQ): **(CERCLA)** 100 lb/45.4 kg

DESCRIPTION: Greenish-yellow solid but usually found as colorless to yellow watery solution. Faint chlorine bleach odor. Sinks and mixes with water.

Anhydrous sodium hypochlorite is explosive • Corrosive to skin, eyes, and respiratory tract; skin or eye contact causes severe burns, impaired vision, or blindness; effects of contact or inhalation may be delayed • Firefighting gear (including SCBA) does not provide adequate protection. If exposure occurs, remove and isolate gear immediately and thoroughly decontaminate personnel • Heat forms oxygen; will increase the activity of an existing fire • May cause fire or explosion contact with combustibles (wood, paper, oil, clothing, etc.) • Containers may BLEVE when exposed to fire • Toxic products of combustion may include sodium oxide and chlorine gas (releases chlorine gas when heated above 95°F/35°C).

Hazard Classification (based on NFPA-704 Rating System)
solution: Health Hazards (Blue): 1; Flammability (Red): 0; Reactivity (Yellow): 0
solid: Health Hazards (BLUE): 2; Flammability (RED): 0; Reactivity (YELLOW): 1

EMERGENCY RESPONSE: See Appendix A (154)
Evacuation:
Public safety: Isolate the area of spill or leak for at least 25 to 50 meters (80 to 160 feet) in all directions.
Spill: Increase, in the downwind direction, as necessary, the distance shown under "Public Safety."
Fire: If tank, rail car, or tank truck is involved in fire, isolate for at least 800 meters (½ mile) in all directions; also, consider initial evacuation for 800 meters (½ mile) in all directions.

EXPOSURE
Short-term effects: *SEEK MEDICAL ATTENTION.*
Vapor/fumes: Lung edema may develop. **Liquid:** Irritating to skin and eyes. Harmful if swallowed. Remove contaminated clothing and shoes. Flush affected areas with plenty of water. *IF BREATHING HAS STOPPED, give artificial respiration; avoid mouth-to-mouth resuscitation; use bag/mask apparatus. IF IN EYES,* hold eyelids open and flush with plenty of water. *IF SWALLOWED* and victim is *CONSCIOUS AND ABLE TO SWALLOW,* have victim drink 4 to 8 ounces of water, and have victim induce vomiting. *IF SWALLOWED* and victim is *UNCONSCIOUS OR HAVING CONVULSIONS,* do nothing except keep victim warm.
Note to physician or authorized medical personnel: Medical observation is recommended for 24 to 48 hours after breathing overexposure, as pulmonary edema may be delayed. As first aid for pulmonary edema, consider administering a corticosteroid spray. Cigarette smoking may exacerbate pulmonary injury and should be discouraged for at least 72 hours following exposure.

HEALTH HAZARDS
Personal protective equipment (PPE): B-Level PPE. Rubber gloves; goggles. Chemical protective material(s) reported to have good to excellent resistance: natural rubber, neoprene, nitrile+PVC, nitrile, PVC.
Exposure limits (TWA unless otherwise noted): 2 mg/m^3 (AIHA WEEL).
Toxicity by ingestion: Grade 1; oral rat LD$_{50}$ = 8.91 g/kg.
Long-term health effects: IARC rating 3; inadequate animal evidence; mutagen.
Vapor (gas) irritant characteristics: Eye and respiratory tract irritant.
Liquid or solid irritant characteristics: Liquid causes eye and skin irritant.

FIRE DATA
Behavior in fire: Containers may rupture and explode.

CHEMICAL REACTIVITY
Reactivity with water: Solution is a strong base.
Binary reactants: Strong oxidizer; reacts with reducing agents, acids and combustible materials. Corrosive to aluminum. Incompatible with steel, cast iron, 12% and 17% chromium steel, Monel, nickel, Inconel, brass, silicon, bronze, zinc.
Neutralizing agents for acids and caustics: Destroy with sodium bisulfite or hypo and water, then neutralize with soda ash.

ENVIRONMENTAL DATA
Water pollution: Harmful to aquatic life in very low concentrations. May be dangerous if it enters nearby water intakes; notify operators. Notify local health and wildlife officials. **Response to discharge:** Issue warning–corrosive. Disperse and flush.

SHIPPING INFORMATION
Grades of purity: Several grades and concentrations (50 g/L, 100 g/L, 150 g/L, etc.) depending on active chlorine; **Storage temperature:** Ambient to cool; **Inert atmosphere:** None; **Venting:** Pressure-vacuum; **Stability during transport:** Stable.

PHYSICAL AND CHEMICAL PROPERTIES
Physical state @ 59°F/15°C and 1 atm: Solid or Liquid.
Molecular weight: 74.44
Boiling point @ 1 atm: Decomposes above 230°F/110°C/383°K.
Melting/Freezing point: 19.5°F/–7°C/266°K (solution).
Specific gravity (water = 1): 1.093 for 5% solution
Heat of solution: (estimate) –90 Btu/lb = –50 cal/g = –2 x 10^5 J/kg.

SODIUM 2-MERCAPTOBENZOTHIAZOL   REC. S:4000

SYNONYMS: 2-BENZOTHIAZOLETHIOL, SODIUM SALT; 2-(3H)-BENZOTHIAZOLETHIONE, SODIUM SALT; DUODEX; SODIUM; 2-MERCAPTOBENZOTIAZOL SODICO (Spanish); 2-MERCAPTOBENZOTHIAZOL SODIUM SALT; 2-MERCAPTOBENZOTHIAZOL SOLUTION; NACAP; SODIUM 2-BENZOTHIAZOLETHIOATE

IDENTIFICATION
CAS Number: 2492-26-4
Formula: C$_6$H$_4$SC(SNa)$_n$; C$_7$H$_4$NS$_2$·Na
DOT ID Number: UN 1760; DOT Guide Number: 154
Proper Shipping Name: Corrosive, liquids, n.o.s.

DESCRIPTION: 50% aqueous solution. Amber. Old rubber odor. Sinks and mixes with water; reacts, forming toxic and flammable mercaptan vapors.

Poison! • Breathing the vapor can kill you; skin or eye contact causes severe burns, impaired vision, or blindness • Firefighting gear (including SCBA) does not provide adequate protection. If exposure occurs, remove and isolate gear immediately and thoroughly decontaminate personnel • Toxic products of combustion may include fumes of oxides of sulfur and nitrogen, sodium oxide, and sodium peroxide • Do not put yourself in danger by entering a contaminated area to rescue a victim.

Hazard Classification (based on NFPA-704 Rating System)
Health Hazards (Blue): 3; Flammability (Red): 0; Reactivity (Yellow): 0

EMERGENCY RESPONSE: See Appendix A (154)
Evacuation:
Public safety: Isolate the area of spill or leak for at least 25 to 50 meters (80 to 160 feet) in all directions.
Spill: Increase, in the downwind direction, as necessary, the distance shown under "Public Safety."
Fire: If tank, rail car, or tank truck is involved in fire, isolate for at least 800 meters (½ mile) in all directions; also, consider initial evacuation for 800 meters (½ mile) in all directions.

EXPOSURE
Short-term effects: *SEEK MEDICAL ATTENTION.* **Liquid:** *HARMFUL IF SWALLOWED OR ABSORBED THROUGH THE SKIN.* May cause irreversible damage to eyes or skin. Large amount

Sodium methylate

on skin may cause severe depression and/or death. *IF BREATHING HAS STOPPED*, give artificial respiration; *avoid mouth-to-mouth resuscitation; use bag/mask apparatus*. *IF SWALLOWED* may cause tremors, convulsions, severe depression, hematuria or death. Remove and double bag contaminated clothing and shoes at the site. Wash contaminated areas with soap and running water. *IF IN EYES OR ON SKIN*, flush with running water for at least 15 minutes; hold eyelids open if necessary. Clean skin with soap and water. *IF SWALLOWED* and victim is *UNCONSCIOUS OR HAVING CONVULSIONS,* do nothing but keep victim warm.

HEALTH HAZARDS

Personal protective equipment (PPE): B-Level PPE. For spills of aqueous solutions-wear splash-proof goggles, respirator, and protective rubber clothing. Fire fighters should wear full protective clothing and self-contained positive pressure breathing apparatus).
Exposure limits (TWA unless otherwise noted): 5 mg/m^3 [skin] (AIHA WEEL).
Toxicity by ingestion: Moderately toxic. Grade 2; LD_{50} = 3.12 g/kg (rat).
Vapor (gas) irritant characteristics: May cause eye and skin irritation.
Liquid or solid irritant characteristics: Severe eye and skin irritant. A 50% solution in water caused a severe degree of skin injury and destruction or irreversible eye damage in 24 hours or less (rabbit).
Odor threshold: 16 ppm.

CHEMICAL REACTIVITY

Reactivity with water: Concentrated aqueous solutions are corrosive.

ENVIRONMENTAL DATA

Food chain concentration potential: Log P_{ow} = 2.4. Unlikely to accumulate.
Water pollution: Effect of low concentrations on aquatic life is unknown. May be dangerous if it enters nearby water intakes; notify operators. Notify local health and wildlife officials.
Response to discharge: Issue warning–corrosive; water contaminant. Restrict access. Should be removed. Chemical and physical treatment.

SHIPPING INFORMATION

Grades of purity: 50% aqueous solution; **Stability during transport:** Stable.

PHYSICAL AND CHEMICAL PROPERTIES

Physical state @ 59°F/15°C and 1 atm: Liquid.
Molecular weight: 189.2
Specific gravity (water = 1): 1.255 @ 77°F/25°C.

SODIUM METHYLATE   REC. S:4200

SYNONYMS: EEC No. 603-040-00-2; METILATO SODICO (Spanish); METHANOL, SODIUM SALT; METHYL ALCOHOL, SODIUM SALT; SODIUM METHOXIDE

IDENTIFICATION
CAS Number: 124-41-4
Formula: CH_3ONa; CH_3NaO
DOT ID Number: UN 1431 (dry); UN 1289 (solution in alcohol);
DOT Guide Number: 138; 132 (UN 1289)

Proper Shipping Name: Sodium methylate, dry; Sodium methylate, solution in alcohol
Reportable Quantity (RQ): **(CERCLA)** 1000 lb/454 kg

DESCRIPTION: White powder. Odorless. Mixes with water; ignites on contact; decomposes producing sodium hydroxide and methyl alcohol.

Extremely flammable • Spontaneously combustible in moist air • Decomposes in air above 259°F/126°C. Corrosive to the skin, eyes, and respiratory tract; skin and eye contact causes severe burns and blindness; inhalation symptoms may be delayed • Firefighting gear (including SCBA) does not provide adequate protection. If exposure occurs, remove and isolate gear immediately and thoroughly decontaminate personnel • Containers may BLEVE when exposed to fire •Vapors may form explosive mixture with air • Concentrated dust in confined areas (e.g., tanks, sewers, buildings) may explode when exposed to fire • Toxic products of combustion may include sodium oxide • May re-ignite after fire is extinguished • Do not put yourself in danger by entering a contaminated area to rescue a victim.

Hazard Classification (based on NFPA-704 Rating System)
Health Hazards (Blue): 2; Flammability (Red): 4; Reactivity (Yellow): 1; Special Notice (White): Water reactive

EMERGENCY RESPONSE, dry: See Appendix A (138)
Evacuation:
Public safety: Isolate the area of spill or leak for at least 50 to 100 meters (160 to 330 feet) in all directions.
Spill: Consider initial downwind evacuation for at least 250 meters (800 feet).
Fire: Isolate for 800 meters (½ mile) in all directions, especially if tank, rail car, or tank truck is involved in fire.

EMERGENCY RESPONSE, liquid: See Appendix A (132)
Evacuation:
Public safety: Isolate spill area for at least 50 to 100 meters (160 to 330 feet) in all directions.
Spill: Increase, as necessary, the isolation distance shown in "Public safety."
Fire: Isolate for 800 meters (½ mile) in all directions, especially if tank, rail car, or tank truck is involved in fire.

EXPOSURE

Short-term effects: *SEEK MEDICAL ATTENTION*. **Dust:** Irritating to eyes, nose, and throat. Move victim to fresh air. *IF IN EYES*, hold eyelids open and flush with plenty of water. *IF BREATHING HAS STOPPED,* give artificial respiration. IF breathing is difficult, administer oxygen. Lung edema may develop. **Solid:** Irritating to skin and eyes. Harmful if swallowed. Remove contaminated clothing and shoes. Flush affected areas with plenty of water. *IF IN EYES*, hold eyelids open and flush with plenty of water. *IF SWALLOWED* and victim is *CONSCIOUS AND ABLE TO SWALLOW*, have victim drink 4 to 8 ounces of water and have victim induce vomiting. *IF SWALLOWED* and victim is *UNCONSCIOUS OR HAVING CONVULSIONS*, do nothing except keep victim warm.
Note to physician or authorized medical personnel: Medical observation is recommended for 24 to 48 hours after breathing overexposure, as pulmonary edema may be delayed. As first aid for pulmonary edema, consider administering a corticosteroid spray. Cigarette smoking may exacerbate pulmonary injury and should be discouraged for at least 72 hours following exposure.

HEALTH HAZARDS
Personal protective equipment (PPE): A-Level PPE (dry material or solution). Wear self-contained positive pressure breathing apparatus and full protective clothing.
Vapor (gas) irritant characteristics: Dust is an eye and respiratory tract irritant.
Liquid or solid irritant characteristics: Severe eye and skin irritant.

FIRE DATA
Flash point: Flammable solid that may ignite in the presence of moisture. Solution, in methyl alcohol, approximately 70°F/24°C.
Fire extinguishing agents not to be used: Water, water-based foam
Autoignition temperature: 158°F/70°C/343°K.

CHEMICAL REACTIVITY
Reactivity with water: Ignites on contact. Produces a caustic sodium hydroxide solution and flammable methyl alcohol.
Binary reactants: Self-ignition in air. Reacts with oxidizers. Solution reacts with acids. Attacks certain plastics such as nylon and polyesters. reacts with metals producing flammable and potentially explosive hydrogen gas.
Neutralizing agents for acids and caustics: Water, followed by dilute acetic acid or vinegar

ENVIRONMENTAL DATA
Food chain concentration potential: Negative; unlikely to accumulate.
Water pollution: Effect of low concentrations on aquatic life is unknown. May be dangerous if it enters nearby water intakes; notify operators. Notify local health and wildlife officials.
Response to discharge: Issue warning–high flammability, corrosive. Restrict access. Disperse and flush.

SHIPPING INFORMATION
Grades of purity: 97+%; **Storage temperature:** Ambient; **Inert atmosphere:** Padded, dry nitrogen; **Venting:** Safety relief; **Stability during transport:** Stable if kept dry.

PHYSICAL AND CHEMICAL PROPERTIES
Physical state @ 59°F/15°C and 1 atm: Solid.
Molecular weight: 54.0
Boiling point @ 1 atm: Decomposes.
Melting/Freezing point: Decomposes. 257°F/125°C/398°K.
Specific gravity (water = 1): More than 1 @ 68°F/20°C (solid).

SODIUM NITRATE   REC. S:4300

SYNONYMS: CHILE SALTPETER; CUBIC NITER; NITEROX®; NITRATINE; NITRATO SODICO (Spanish); NITRATE de SODIUM (French); NITRIC ACID, SODIUM SALT; SODA NITER; SODIUM NITRATE

IDENTIFICATION
CAS Number: 7631-99-4
Formula: $NO_3 \cdot Na$
DOT ID Number: UN 1498; DOT Guide Number: 140
Proper Shipping Name: Sodium nitrate

DESCRIPTION: Colorless to white crystalline solid. Odorless. Sinks and mixes with water; solution is neutral.

Corrosive • Strong oxidizer which may react spontaneously with low flash point organics or reducing agents. Heat forms oxygen; will increase the activity of an existing fire • May cause fire or explosion on contact with combustibles (wood, paper, oil, clothing, etc.) • Irritating to the skin, eyes, and respiratory tract • Explodes at temperatures above 716°F/380°C • Toxic products of combustion may include oxygen and oxides or nitrogen and sodium.

Hazard Classification (based on NFPA-704 Rating System)
Health Hazards (Blue): 1; Flammability (Red): 0; Reactivity (Yellow): 1; Special Notice (White): OXY

EMERGENCY RESPONSE: See Appendix A (140)
Evacuation:
Public safety: Isolate the area of spill or leak for at least 10 to 25 meters (30 to 80 feet) in all directions.
Spill: Consider initial downwind evacuation for at least 100 meters (330 feet).
Fire: If any large container is involved in fire, isolate for at least 800 meters (½ mile) in all directions; also, consider initial evacuation for 800 meters (½ mile) in all directions.

EXPOSURE
Short-term effects: *SEEK MEDICAL ATTENTION.* **Solid:** If swallowed, may cause dizziness, abdominal cramps, vomiting, convulsions, and collapse.
Flush exposed areas with plenty of water. *IF IN EYES*, hold eyelids open and flush with plenty of water. *IF SWALLOWED* and victim is *CONSCIOUS AND ABLE TO SWALLOW,* have victim drink 4 to 8 ounces of water and have victim induce vomiting. *IF SWALLOWED* and victim is *UNCONSCIOUS OR HAVING CONVULSIONS,* do nothing except keep victim warm.
Note to physician or authorized medical personnel: Medical observation is recommended for 24 to 48 hours after breathing overexposure, as pulmonary edema may be delayed. As first aid for pulmonary edema, consider administering a corticosteroid spray. Cigarette smoking may exacerbate pulmonary injury and should be discouraged for at least 72 hours following exposure.

HEALTH HAZARDS
Personal protective equipment (PPE): B-Level PPE. Rubber gloves, goggles, laboratory coat.
Toxicity by ingestion: Grade 1; LD_{50} = 5 to 15 g/kg.
Long-term health effects: Ingestion of nitrates has been implicated with cancer increase. Methemoglobin formation; possible blood and blood vessel damage. Small repeated doses may cause recurring headache and mental damage.
Vapor (gas) irritant characteristics: Eye and respiratory tract irritant.
Liquid or solid irritant characteristics: Eye and skin irritant.

FIRE DATA
Behavior in fire: Explodes when heated over 716°F/380°C. Will increase fire activity.

CHEMICAL REACTIVITY
Binary reactants: Powerful oxidizer. Reacts with oxidizable substances, powdered aluminum, gunpowder, bitumens, organic materials (i.e., wood, etc.), metals, combustibles, acids.

ENVIRONMENTAL DATA
Water pollution: Dangerous to aquatic life in very high concentrations. May be dangerous if it enters nearby water intakes; notify operators. Notify local health and wildlife officials. **Response to discharge:** Issue warning–oxidizing material. Chemical and physical treatment. Disperse and flush.

SHIPPING INFORMATION
Grades of purity: Purified-at least 99% NaNO$_3$; **Storage temperature:** Cool; **Stability during transport:** Stable.

PHYSICAL AND CHEMICAL PROPERTIES
Physical state @ 59°F/15°C and 1 atm: Solid.
Molecular weight: 84.99
Boiling point @ 1 atm: Decomposes (with possible explosion) 716°F/380°C/653.2°K.
Melting/Freezing point: 584.2°F/306.8°C/580.0°K.
Specific gravity (water = 1): 2.26
Relative vapor density (air = 1): 2.93 (calculated).
Heat of solution: @ 77°F/25°C −108 Btu/lb = −60.1 cal/g = −2.52 x 10^5 J/kg.
Heat of fusion: 44.2 cal/g.

SODIUM NITRITE   REC. S:4400

SYNONYMS: ANTI-RUST; EEC No. 007-010-00-4; ECRINITRIT; FILMERINE; NITRITO SODICO (Spanish); NITROUS ACID, SODIUM SALTo

IDENTIFICATION
CAS Number: 7632-00-0
Formula: NO$_2$·Na
DOT ID Number: UN 1500; DOT Guide Number: 140
Proper Shipping Name: Sodium nitrite
Reportable Quantity (RQ): **(CERCLA)** 100 lb/45.4 kg

DESCRIPTION: White or faint yellow crystalline solid. Odorless. Sinks and mixes with water.

Poison! • Breathing the vapor, skin or eye contact, or swallowing the material can kill you • Firefighting gear (including SCBA) does not provide adequate protection. If exposure occurs, remove and isolate gear immediately and thoroughly decontaminate personnel • Strong oxidizer which may react spontaneously with low flash point organics or reducing agents. Heat forms oxygen; will increase the activity of an existing fire • May cause fire or explosion on contact with combustibles (wood, paper, oil, clothing, etc.). Irritating to the skin, eyes, and respiratory tract • Explodes when heated to over 1000°F/538°C (CHRIS) • Toxic products of combustion may include oxygen and oxides or nitrogen and sodium.

Hazard Classification (based on NFPA-704 Rating System)
Health Hazards (Blue): 2; Flammability (Red): 0; Reactivity (Yellow): 1; Special Notice (White): OXY

EMERGENCY RESPONSE: See Appendix A (140)
Evacuation:
Public safety: Isolate the area of spill or leak for at least 10 to 25 meters (30 to 80 feet) in all directions.
Spill: Consider initial downwind evacuation for at least 100 meters (330 feet).
Fire: If any large container is involved in fire, isolate for at least 800 meters (½ mile) in all directions; also, consider initial evacuation for 800 meters (½ mile) in all directions.

EXPOSURE
Short-term effects: *SEEK MEDICAL ATTENTION.* **Dust:** Irritating to eyes, nose, and throat. *IF INHALED*, will, will cause headache, difficult breathing, or loss of consciousness. *IF IN EYES*, hold eyelids open and flush with plenty of water. *IF BREATHING HAS STOPPED*, give artificial respiration; *avoid mouth-to-mouth resuscitation; use bag/mask apparatus.* IF breathing is difficult, administer oxygen. **Solid:** *POISONOUS IF SWALLOWED.* Irritating to skin and eyes. *IF SWALLOWED*, will cause headache, nausea, vomiting, or loss of consciousness. Remove contaminated clothing and shoes. Flush affected areas with plenty of water. *IF IN EYES*, hold eyelids open and flush with plenty of water. *IF SWALLOWED* and victim is *CONSCIOUS AND ABLE TO SWALLOW*, have victim drink 4 to 8 ounces of water and have victim induce vomiting. *IF SWALLOWED* and victim is *UNCONSCIOUS OR HAVING CONVULSIONS*, do nothing except keep victim warm.

HEALTH HAZARDS
Personal protective equipment (PPE): A-Level PPE. Wear self-contained positive pressure breathing apparatus and full protective clothing.
Toxicity by ingestion: Grade 3; LD$_{50}$ = 50 to 500 mg/kg.
Long-term health effects: Methemoglobin formation; blood and blood vessel damage. It may form carcinogenic nitrosamines by reacting with organic amines in the body.
Liquid or solid irritant characteristics: Eye irritant. Can be absorbed through the skin.

FIRE DATA
Behavior in fire: May increase intensity of fire if in contact with combustible material. May melt and flow at elevated temperatures.

CHEMICAL REACTIVITY
Binary reactants: An oxidizer. Reacts with acids, phthalic anhydride, cyanides, sodium amide, sodium disulphite, sodium thiocyanate, wood urea, butadiene; ammonia salts, aminoguanidine salts. Slowly turns to nitrate in air.

ENVIRONMENTAL DATA
Food chain concentration potential: Unlikely to accumulate.
Water pollution: Harmful to aquatic life and highly toxic to livestock in very low concentrations. May be dangerous if it enters nearby water intakes; notify operators. Notify local health and wildlife officials. **Response to discharge:** Issue warning–oxidizing material and poison. Restrict access. Disperse and flush.

SHIPPING INFORMATION
Grades of purity: USP; Reagent; **Storage temperature:** Ambient; **Inert atmosphere:** None; **Venting:** Open; **Stability during transport:** Stable.

PHYSICAL AND CHEMICAL PROPERTIES
Physical state @ 59°F/15°C and 1 atm: Solid.
Molecular weight: 69.0
Boiling point @ 1 atm: Decomposes. More than 608°F/320°C/593°K.
Melting/Freezing point: 520°F/271°C/544°K.
Specific gravity (water = 1): 2.17 @ 68°F/20°C (solid).

SODIUM NITRITE SOLUTION   REC. S:4500

SYNONYMS: NITRITO SODICO (Spanish); SODIUM NITRITE LIQUOR

IDENTIFICATION
CAS Number: 7632-00-0
Formula: NO$_2$·Na
DOT ID Number: UN 1500; DOT Guide Number: 140

Proper Shipping Name: Sodium nitrite
Reportable Quantity (RQ): **(CERCLA)** 100 lb/45.4 kg

DESCRIPTION: Pale yellow liquid. Odorless. Soluble in water.

Strong oxidizer which may react spontaneously with low flash point organics or reducing agents. Heat forms oxygen; will increase the activity of an existing fire • May cause fire or explosion on contact with combustibles (wood, paper, oil, clothing, etc.) • Toxic products of combustion may include oxygen and oxides or nitrogen and sodium. See also previous entry.

Hazard Classification (based on NFPA-704 Rating System)
Health Hazards (Blue): 2; Flammability (Red): 0; Reactivity (Yellow): 1; Special Notice (White): OXY

EMERGENCY RESPONSE: See Appendix A (140)
Evacuation:
Public safety: Isolate the area of spill or leak for at least 10 to 25 meters (30 to 80 feet) in all directions.
Spill: Consider initial downwind evacuation for at least 100 meters (330 feet).
Fire: If any large container is involved in fire, isolate for at least 800 meters (½ mile) in all directions; also, consider initial evacuation for 800 meters (½ mile) in all directions.

EXPOSURE
Short-term effects: *SEEK MEDICAL ATTENTION*. **Liquid:** Irritating to eyes, nose, and throat. Remove contaminated clothing and shoes. Flush affected areas with plenty of water. *IF IN EYES*, hold eyelids open and flush with plenty of water. *IF BREATHING HAS STOPPED*, give artificial respiration. If breathing is difficult, administer oxygen. *IF SWALLOWED* and victim is *CONSCIOUS AND ABLE TO SWALLOW*, have victim drink 4 to 8 ounces of water and have victim induce vomiting.

HEALTH HAZARDS
Personal protective equipment (PPE): B-Level PPE. Wear impervious protective clothing and goggles.
Toxicity by ingestion: Grade 3; LD_{50} = 80–185 mg/kg (rat).
Long-term health effects: Methemoglobin formation: possible blood and blood vessel effects. It may form carcinogenic nitrosamines by reacting with organic amines in the body.
Vapor (gas) irritant characteristics: Vapors cause irritation such that personnel will find high concentrations unpleasant. The effect is temporary.
Liquid or solid irritant characteristics: Severe eye and skin irritant. Causes smarting of the skin and first-degree burns on short exposure; may cause second-degree burns on long exposure.

FIRE DATA, but may intensify fire.
Fire extinguishing agents not to be used: Do not use ammonium phosphate dry chemical.
Behavior in fire: May increase intensity of fire if water evaporates.

CHEMICAL REACTIVITY
Binary reactants: Strong oxidizer. Reacts with acids, ammonium salts, amines, cyanides, and reducing agents; organic materials, combustibles.
Reactivity group: 5
Compatibility class: Caustics

ENVIRONMENTAL DATA
Food chain concentration potential: Unlikely to accumulate.
Water pollution: Effect of low concentrations on aquatic life is unknown. May be dangerous if it enters nearby water intakes; notify operators. Notify local health and wildlife officials. **Response to discharge:** Issue warning–oxidizing material. Disperse and flush.

SHIPPING INFORMATION
Grades of purity: Technical grade solutions of varying concentrations; **Storage temperature:** Ambient; **Inert atmosphere:** None; **Venting:** Open; **Stability during transport:** Stable.

PHYSICAL AND CHEMICAL PROPERTIES
Physical state @ 59°F/15°C and 1 atm: Liquid.
Molecular weight: 69.0
Boiling point @ 1 atm: 239°F/115°C/388°K.
Melting/Freezing point: 30°F/–1°C/272°K.
Specific gravity (water = 1): 1.32 at 16°C (solid).
Relative vapor density (air = 1): Less than 1

SODIUM OXALATE   REC. S:4600

SYNONYMS: ETHANEDIOIC ACID, DISODIUM SALT

IDENTIFICATION
CAS Number: 62-76-0
Formula: $C_2O_4 \cdot 2Na$; $Na_2C_2O_4$

DESCRIPTION: White crystalline powder. Odorless. Sinks and mixes slowly with water.

Poison. Toxic products of combustion may include carbon monoxide, sodium oxide, and sodium carbonate (soda ash).

Hazard Classification (based on NFPA-704 Rating System)
Health Hazards (Blue): 1; Flammability (Red): 0; Reactivity (Yellow): 0

EMERGENCY RESPONSE: See Appendix A (171)
Evacuation:
Public safety: Isolate the area of spill or leak for at least 10 to 25 meters (30 to 80 feet) in all directions.
Spill: Increase, in the downwind direction, as necessary, the distance shown under "Public Safety."
Fire: If any large container is involved in fire, isolate for at least 800 meters (½ mile) in all directions; also, consider initial evacuation for 800 meters (½ mile) in all directions.

EXPOSURE
Short-term effects: *SEEK MEDICAL ATTENTION*. **Dust:** Irritating to eyes, nose, and throat. *IF INHALED*, will, will cause difficult breathing or loss of consciousness. *IF IN EYES*, hold eyelids open and flush with plenty of water. *IF BREATHING HAS STOPPED*, give artificial respiration. IF breathing is difficult, administer oxygen. **Solid:** *POISONOUS IF SWALLOWED*. Irritating to skin and eyes. *IF SWALLOWED*, will cause nausea, vomiting, or loss of consciousness. Remove contaminated clothing and shoes. Flush affected areas with plenty of water. *IF IN EYES*, hold eyelids open and flush with plenty of water. *IF SWALLOWED* and victim is *CONSCIOUS AND ABLE TO SWALLOW*, have victim drink 4 to 8 ounces of water and have victim induce vomiting. *IF SWALLOWED* and victim is *UNCONSCIOUS OR HAVING CONVULSIONS*, do nothing except keep victim warm.

HEALTH HAZARDS
Personal protective equipment (PPE): B-Level PPE. Dust mask; goggles or face shield; rubber gloves.
Toxicity by ingestion: Grade 3; LD_{50} = 50 to 500 mg/kg.
Vapor (gas) irritant characteristics: eye and respiratory tract irritant.
Liquid or solid irritant characteristics: Eye and skin irritant.

ENVIRONMENTAL DATA
Food chain concentration potential: Unlikely to accumulate.
Water pollution: Dangerous to aquatic life in high concentrations. May be dangerous if it enters nearby water intakes; notify operators. Notify local health and wildlife officials. **Response to discharge:** Issue warning–water contaminant. Disperse and flush.

SHIPPING INFORMATION
Grades of purity: Reagent; Primary standard grade; **Storage temperature:** Ambient; **Inert atmosphere:** None; **Venting:** Open; **Stability during transport:** Stable.

PHYSICAL AND CHEMICAL PROPERTIES
Physical state @ 59°F/15°C and 1 atm: Solid.
Molecular weight: 134.0
Boiling point @ 1 atm: (decomposes).
Specific gravity (water = 1): 2.27 @ 68°F/20°C (solid).

SODIUM PENTACHLOROPHENATE   REC. S:4700

SYNONYMS: DORMANT® FUNGICIDE (DOW); DOWCIDE G-ST; NAPCHLOR-G; PENTACLOROFENATO SODICO (Spanish); PHENOL, PENTACHLORO-,SODIUM SALT; SODIUM PCP; SODIUM PENTALCHLOROPHENOL; WEEDBEADS

IDENTIFICATION
CAS Number: 131-52-2
Formula: $C_6Cl_5 \cdot Na$
DOT ID Number: UN 2567; DOT Guide Number: 154
Proper Shipping Name: Sodium pentachlorophenate

DESCRIPTION: Tan colored beads or powder. Phenolic odor. Mixes with water.

Poison! • Corrosive to the eyes and skin • Breathing the dust, skin or eye contact, or swallowing the material can kill you • Firefighting gear (including SCBA) does not provide adequate protection. If exposure occurs, remove and isolate gear immediately and thoroughly decontaminate personnel • Toxic products of combustion may include hydrogen chloride, polychlorodibenzodioxins, carbon monoxide, phenol, and sodium oxide • Do not put yourself in danger by entering a contaminated area to rescue a victim.

Hazard Classification (based on NFPA-704 Rating System)
Health Hazards (Blue): 3; Flammability (Red): 0; Reactivity (Yellow): 0

EMERGENCY RESPONSE: See Appendix A (154)
Evacuation:
Public safety: Isolate the area of spill or leak for at least 25 to 50 meters (80 to 160 feet) in all directions.
Spill: Increase, in the downwind direction, as necessary, the distance shown under "Public Safety."
Fire: If tank, rail car, or tank truck is involved in fire, isolate for at least 800 meters (½ mile) in all directions; also, consider initial evacuation for 800 meters (½ mile) in all directions.

EXPOSURE
Short-term effects: *SEEK MEDICAL ATTENTION.* **Dust:** Move victim to fresh air. *IF BREATHING HAS STOPPED,* give artificial respiration; *avoid mouth-to-mouth resuscitation; use bag/mask apparatus.* If breathing is difficult, administer oxygen. **Solid:** Irritating to skin and eyes. May be fatal if absorbed through skin. Remove contaminated clothing and shoes. Wash affected areas with soap and water. *IF IN EYES,* hold eyelids open and flush with plenty of water. *IF SWALLOWED* and victim is *CONSCIOUS AND ABLE TO SWALLOW,* drink 1 or 2 glasses of water and induce vomiting.
Note to physician or authorized medical personnel: Medical observation is recommended for 24 to 48 hours after breathing overexposure, as pulmonary edema may be delayed. As first aid for pulmonary edema, consider administering a corticosteroid spray. Cigarette smoking may exacerbate pulmonary injury and should be discouraged for at least 72 hours following exposure.

HEALTH HAZARDS
Personal protective equipment (PPE): A-Level PPE. Full covering clothing, chemical protective gloves and goggles. Wear sealed chemical suit of butyl rubber. Also polycarbonate offers chemical resistance.
Toxicity by ingestion: A chlorophenol compound; poisonous, carcinogen.
Long-term health effects: Mutagen.
Liquid or solid irritant characteristics: Fairly severe skin irritant. May cause pain and second-degree burns after a few minutes of contact.

CHEMICAL REACTIVITY
Binary reactants: Incompatible with strong oxidizing agents.

ENVIRONMENTAL DATA
Note: A chlorophenol compound; may be contaminated with TCDD (2,3,7,8-tetrachlorodibenzo-1,4-dioxin).
Water pollution: DOT Appendix B, §172.101–marine pollutant. Dangerous to aquatic life in low concentrations. May be dangerous if it enters nearby water intakes; notify operators. Notify local health and wildlife officials. **Response to discharge:** Should be removed.

SHIPPING INFORMATION
Grades of purity: Technical grades, 85-95%; **Storage temperature:** Ambient; **Inert atmosphere:** None; **Stability during transport:** Stable.

PHYSICAL AND CHEMICAL PROPERTIES
Physical state @ 59°F/15°C and 1 atm: Solid.
Molecular weight: 288.35
Boiling point @ 1 atm: Decomposes.

SODIUM PHOSPHATE   REC. S:4800

SYNONYMS: DISODIUM DIHYDROGEN PYROPHOSPHATE; FOSFATO SODICO (Spanish); MONOSODIUM DIHYDROGEN PHOSPHATE; MONOSODIUM PHOSPHATE; MSP; SODIUM PHOSPHATE, MONOBASIC; SODIUM PHOSPHATE DIBASIC; SODIUM PHOSPHATE, TRIBASIC; SODIUM DIHYDROGEN PHOSPHATE; SODIUM ACID PYROPHOSPHATE

IDENTIFICATION
CAS Number: 7558-79-4 (dibasic); 7758-80-7 (monobasic); 7601-54-9 (tribasic)
Formula: (1) NaH_2PO_4; (2) Na_2HPO_4; (3) Na_3PO_4; (4) $Na_2H_2P_2O_7$; (5) $Na_4P_2O_7$; (6) $(NaPO_3)_n$; (7) $(NaPO_3)_3$; (8) $(NaPO_3)_n \cdot NaO$; (9) $Na5P_3O_{10}$
DOT ID Number: UN 9147; DOT Guide Number: 171
Proper Shipping Name: Sodium phosphate, dibasic
Reportable Quantity (RQ): **(CERCLA)** 5000 lb/2270 kg

DESCRIPTION: White granular or powdered solid; some may appear glassy. Odorless. Sinks and mixes with water forming a weak base.

Caustic; irritates skin, eyes, and respiratory tract •Toxic products of combustion may include oxides of phosphorus and sodium.

Hazard Classification (based on NFPA-704 Rating System) (dibasic) Health Hazards (Blue): 0; Flammability (Red): 0; Reactivity (Yellow): 0

EMERGENCY RESPONSE: See Appendix A (171)
Evacuation:
Public safety: Isolate the area of spill or leak for at least 10 to 25 meters (30 to 80 feet) in all directions.
Spill: Increase, in the downwind direction, as necessary, the distance shown under "Public Safety."
Fire: If any large container is involved in fire, isolate for at least 800 meters (½ mile) in all directions; also, consider initial evacuation for 800 meters (½ mile) in all directions.

EXPOSURE
Short-term effects: *SEEK MEDICAL ATTENTION.* **Dust:** Irritating to eyes, nose, and throat. *IF INHALED*, will, will cause coughing or difficult breathing. *IF IN EYES*, hold eyelids open and flush with plenty of water. *IF BREATHING HAS STOPPED*, give artificial respiration. IF breathing is difficult, administer oxygen. **Solid:** Irritating to skin and eyes. *IF SWALLOWED*, will cause nausea and vomiting. Remove contaminated clothing and shoes. Flush affected areas with plenty of water. *IF IN EYES*, hold eyelids open and flush with plenty of water. *IF SWALLOWED* and victim is *CONSCIOUS AND ABLE TO SWALLOW*, have victim drink 4 to 8 ounces of water and have victim induce vomiting. *IF SWALLOWED* and victim is *UNCONSCIOUS OR HAVING CONVULSIONS*, do nothing except keep victim warm.
Note to physician or authorized medical personnel: Medical observation is recommended for 24 to 48 hours after breathing overexposure, as pulmonary edema may be delayed. As first aid for pulmonary edema, consider administering a corticosteroid spray. Cigarette smoking may exacerbate pulmonary injury and should be discouraged for at least 72 hours following exposure.

HEALTH HAZARDS
Personal protective equipment (PPE): B-Level PPE. U.S. Bureau of Mines toxic dust mask; protective gloves; chemical-type goggles; full-cover clothing
Liquid or solid irritant characteristics: Eye and skin irritant.

FIRE DATA
Behavior in fire: May melt with formation of steam.

CHEMICAL REACTIVITY
Reactivity with water: All dissolve readily. MSP and A forms weakly acidic solutions; TSP forms strong caustic solution, similar to soda lye; T forms weakly alkaline solution.
Binary reactants: When wet, mild steel or brass may be corroded by MSP, A, and TSP. The others are not considered corrosive.
Neutralizing agents for acids and caustics: For those sodium phosphates that form acidic or basic solutions, dilution with water removes hazard.

ENVIRONMENTAL DATA
Food chain concentration potential: Negative; unlikely to accumulate.
Water pollution: Dangerous to aquatic life in high concentrations. May be dangerous if it enters nearby water intakes; notify operators. Notify local health and wildlife officials. **Response to discharge:** Disperse and flush.

SHIPPING INFORMATION
Grades of purity: All are available in Technical Grade, some in Food Grade and Reagent Grade. Some are available as hydrates as well as anhydrous forms; **Storage temperature:** Ambient; **Inert atmosphere:** None; **Venting:** Open; **Stability during transport:** All forms of sodium phosphate are stable. TSP tends to pick up moisture from air and form a hard cake.

PHYSICAL AND CHEMICAL PROPERTIES
Physical state @ 59°F/15°C and 1 atm: Solid.
Molecular weight: 142.0 (DSP); 120.0 (MSP); 163.9 (TSP). Values for anhydrous salts run from 120 to high polymer values.
Specific gravity (water = 1): 1.8-2.5 @ 77°F/25°C (solid).
Heat of solution: +83 to –81 Btu/lb = +46 to –45 cal/g = +1.93 to –1.88 x 10^5 J/kg.
Heat of fusion: 84.4 cal/g.

SODIUM PHOSPHATE, TRIBASIC   REC. S:4900

SYNONYMS: DRI-TRI; NUTRIFOS STP; FOSFATO TRIBASICO SODICO (Spanish); PHOSPHORIC ACID, TRISODIUM SALT; TRISODIUM PHOSPHATE; TRISODIUM ORTHOPHOSPHATE; TROMETE; TSP

IDENTIFICATION
CAS Number: 7601-54-9
Formula: $O_4P \cdot 3Na$
DOT ID Number: UN 9148; DOT Guide Number: 171
Proper Shipping Name: Sodium phosphate, tribasic
Reportable Quantity (RQ): **(CERCLA)** 5000 lb/2270 kg

DESCRIPTION: White crystalline powder. Odorless. Sinks and mixes with water.

Corrosive to the eyes, skin or respiratory tract; inhalation symptoms may be delayed • Toxic products of combustion may include oxides of phosphorus and sodium.

Hazard Classification (based on NFPA-704 Rating System)
Health Hazards (Blue): 1; Flammability (Red): 0; Reactivity (Yellow): 0

EMERGENCY RESPONSE: See Appendix A (171)
Evacuation:
Public safety: Isolate the area of spill or leak for at least 10 to 25 meters (30 to 80 feet) in all directions.
Spill: Increase, in the downwind direction, as necessary, the distance shown under "Public Safety."

Fire: If any large container is involved in fire, isolate for at least 800 meters (½ mile) in all directions; also, consider initial evacuation for 800 meters (½ mile) in all directions.

EXPOSURE

Short-term effects: *SEEK MEDICAL ATTENTION.* **Dust:** Irritating to eyes, nose, throat and skin. Move to fresh air. *IF BREATHING HAS STOPPED,* give artificial respiration. IF breathing is difficult, administer oxygen. *IF IN EYES,* hold eyelids open and flush with plenty of water. **Solid:** Will burn skin and eyes. Harmful if swallowed. Remove contaminated clothing or shoes. Flush affected areas with plenty of water. *IF IN EYES,* hold eyelids open and flush with plenty of water. *IF SWALLOWED* and victim is *CONSCIOUS AND ABLE TO SWALLOW,* have victim drink 4 to 8 ounces of water and have victim induce vomiting.

Note to physician or authorized medical personnel: Medical observation is recommended for 24 to 48 hours after breathing overexposure, as pulmonary edema may be delayed. As first aid for pulmonary edema, consider administering a corticosteroid spray. Cigarette smoking may exacerbate pulmonary injury and should be discouraged for at least 72 hours following exposure.

HEALTH HAZARDS

Personal protective equipment (PPE): B-Level PPE. Rubber gloves and boots, safety goggles or face mask, hooded suit, and respirator with approved canister, or a self-contained breathing apparatus.
Exposure limits (TWA unless otherwise noted): 5 mg/m^3 (AIHA WEEL).
Toxicity by ingestion: Grade 1; LD_{50} = 0 to 15 g/kg.
Long-term health effects: Mutagen
Vapor (gas) irritant characteristics: Eye and respiratory tract irritant.
Liquid or solid irritant characteristics: Severe eye and skin irritant. Causes second- and third-degree burns on short contact and is very injurious to the eyes.

CHEMICAL REACTIVITY

Reactivity with water: Aqueous solution is a strong caustic.
Binary reactants: Violent reaction with acids. Incompatible with organic anhydrides, acrylates, alcohols, aldehydes, alkylene oxides, substituted allyls, cresols, caprolactam solution, epichlorohydrin, ethylene dichloride, glycols, isocyanates, ketones, maleic anhydride, nitrates, nitromethane, phenols, vinyl acetate. Attacks aluminum, copper, zinc, and related alloys in the presence of moisture.

ENVIRONMENTAL DATA

Food chain concentration potential: Negative; unlikely to accumulate.
Water pollution: Dangerous to aquatic life in high concentrations. May be dangerous if it enters nearby water intakes; notify operators. Notify local health and wildlife officials. **Response to discharge:** Issue warning–corrosive. Restrict access. Chemical and physical treatment. Disperse and flush

SHIPPING INFORMATION

Grades of purity: Technical grade–92%; Commercial; NF–1 and NF–3; Food grade; Reagent Grade. Hydrated. Anhydrous; FCC–1 and FCC–2.; **Storage temperature:** Ambient; **Stability during transport:** Stable.

PHYSICAL AND CHEMICAL PROPERTIES

Physical state @ 59°F/15°C and 1 atm: Solid.
Molecular weight: 163.95 (anhydrous); 380.16 (hydrate).
Boiling point @ 1 atm: The hydrate loses 11 moles of water at 212°F/100°C/373°K.
Melting/Freezing point: 164–170°F/73–77°C/347–350°K.
Specific gravity (water = 1): 1.62 @ 68°F/20°C (Hydrate); 2.52 anhydrous.
Heat of solution: Hydrate absorbs heat. 75.2 Btu/lb = 40.3 cal/g = 1.69 x 10^5 J/kg.

SODIUM SELENITE   REC. S:5000

SYNONYMS: DISODIUM SELENITE; EEC No. 034-002-00-8; SELENIOUS ACID, DISODIUM SALT; SELENIATO SODICO (Spanish)

IDENTIFICATION

CAS Number: 10102-18-8
Formula: $O_3Se \cdot 2Na$
DOT ID Number: UN 2630; DOT Guide Number: 151
Proper Shipping Name: Sodium selenite
Reportable Quantity (RQ): **(CERCLA)** 100 lb/45.4 kg

DESCRIPTION: White to pink crystalline solid. Odorless. Mixes with water; forms slightly alkaline solution.

Poison! • Breathing the vapor, skin or eye contact, or swallowing the material can kill you • Firefighting gear (including SCBA) does not provide adequate protection. If exposure occurs, remove and isolate gear immediately and thoroughly decontaminate personnel • Containers may BLEVE when exposed to fire • Toxic products of combustion may include oxides of selenium and sodium • Do not put yourself in danger by entering a contaminated area to rescue a victim.

Hazard Classification (based on NFPA-704 Rating System)
Health Hazards (Blue): 4; Flammability (Red): 0; Reactivity (Yellow): 0

EMERGENCY RESPONSE: See Appendix A (151)
Evacuation:
Public safety: Isolate the area of spill or leak for at least 25 to 50 meters (80 to 160 feet) in all directions.
Spill: Increase, in the downwind direction, as necessary, the distance shown under "Public Safety."
Fire: If tank, rail car, or tank truck is involved in fire, isolate for at least 800 meters (½ mile) in all directions; also, consider initial evacuation for 800 meters (½ mile) in all directions.

EXPOSURE

Short-term effects: *SEEK MEDICAL ATTENTION.* **Dust or Solid:** *POISONOUS IF INHALED OR SWALLOWED.* Irritating to skin and eyes. Move to fresh air. *IF BREATHING HAS STOPPED,* give artificial respiration; *avoid mouth-to-mouth resuscitation; use bag/mask apparatus.* IF breathing is difficult, administer oxygen. Remove contaminated clothing and shoes. Flush affected areas with plenty of water. *IF SWALLOWED* and victim is *CONSCIOUS AND ABLE TO SWALLOW,* have victim drink 4 to 8 ounces of water and have victim induce vomiting. *IF SWALLOWED* and victim is *UNCONSCIOUS OR HAVING CONVULSIONS,* do nothing except keep victim warm.

HEALTH HAZARDS

Personal protective equipment (PPE): A-Level PPE. Wear self-contained positive pressure breathing apparatus and full protective clothing.

Recommendations for respirator selection: NIOSH/OSHA as selenium
1 mg/m³: DM-*If not present as a fume-* (any dust and mist respirator); or DMFu (any dust, mist and fume respirator); or HiEF (any air-purifying, full-facepiece respirator with a high-efficiency particulate filter); or PAPRDM-*If not present as a fume-*(any powered, air-purifying respirator with a dust and mist filter); or PAPRDMFu (any powered, air-purifying respirator with a dust, mist, and fume filter); or SA (any supplied-air respirator); or SCBAF (any self-contained breathing apparatus with a full facepiece). *EMERGENCY OR PLANNED ENTRY INTO UNKNOWN CONCENTRATIONS OR IDLH CONDITIONS:* SCBAF:PD,PP (any self-contained breathing apparatus that has a full facepiece and is operated in a pressure-demand or other positive-pressure mode); or SAF:PD,PP:ASCBA (any supplied-air respirator that has a full facepiece and is operated in a pressure-demand or other positive-pressure mode in combination with an auxiliary self-contained breathing apparatus operated in a pressure-demand or other positive pressure mode). *ESCAPE:* HiEF (any air-purifying, full-facepiece respirator with a high-efficiency particulate filter); or SCBAE (any appropriate escape-type, self-contained breathing apparatus). *Note*: Substance reported to cause eye irritation or damage; may require eye protection.
Exposure limits (TWA unless otherwise noted): ACGIH TLV 0.2 mg/m³; NIOSH/OSHA 0.2 mg/m³ as selenium
Toxicity by ingestion: Grade 4; LD_{50} below 50 mg/kg.
Long-term health effects: A suspected carcinogen. Chronic exposure can cause gastrointestinal disorders, nervousness, pallor, coated tongue, and a garlicky odor of the breath. Liver and spleen damage, emaciation, apathy, and progressive anemia. Mutagen.
Vapor (gas) irritant characteristics: Poisonous. Eye and respiratory tract irritant.
Liquid or solid irritant characteristics: Fairly severe eye and skin irritant. May cause pain and second-degree burns after a few minutes of contact.
IDLH value: 1 mg/m³ as selenium

FIRE DATA
Behavior in fire: Decomposes at more than 604°F/318°C.

ENVIRONMENTAL DATA
Food chain concentration potential: Positive. Concentration factors for Se-marine and freshwater plants 800, invertebrates and fish 400.
Water pollution: Harmful to aquatic life in very low concentrations. May be dangerous if it enters nearby water intakes; notify operators. Notify local health and wildlife officials. Indefinite persistence. **Response to discharge:** Issue warning–water contamination. Should be removed. Chemical and physical treatment.

SHIPPING INFORMATION
Grades of purity: Se: 44–46.5%, total metallic impurities: 0.02–0.1%; **Storage temperature:** Cool; **Stability during transport:** Stable.

PHYSICAL AND CHEMICAL PROPERTIES
Physical state @ 59°F/15°C and 1 atm: Solid.
Molecular weight: 172.95 (anhydrous salt); 263.01 (pentahydrate).
Boiling point @ 1 atm: Decomposes.
Melting/Freezing point: 1310°F/710°C/983°K (decomposes).
Relative vapor density (air = 1): (calculated) 5.96 (anhydrous salt); 9.07 (pentahydrate).

SODIUM SILICATE   REC. S:5100

SYNONYMS: B-W; CRYSTAMET; DISODIUM METASILICATE; DISODIUM MONOSILICATE; METSO 20; METSO PENTABEAD 20; ORTHOSIL; SILICATO SODICO (Spanish); SODIUM METASILICATE, ANHYDROUS; WATER GLASS; SOLUBLE GLASS

IDENTIFICATION
CAS Number: 6834-92-0 (anhydrous); 1344-09-8 (solution)
Formula: Na_2SiO_3; $O_3Si \cdot 2Na$

DESCRIPTION: White crystalline granules or colorless, thick liquid. Odorless. Sinks and mixes slowly with water forming a corrosive basic solution.

Corrosive • Skin or eye contact causes severe burns, impaired vision, or blindness; effects of contact or inhalation may be delayed • Firefighting gear (including SCBA) may not provide adequate protection. If exposure occurs, remove and isolate gear immediately and thoroughly decontaminate personnel • Toxic products of combustion may include sodium oxide.

Hazard Classification (based on NFPA-704 Rating System)
Health Hazards (Blue): 0; Flammability (Red): 0; Reactivity (Yellow): 0

EMERGENCY RESPONSE: See Appendix A (171)
Evacuation:
Public safety: Isolate the area of spill or leak for at least 10 to 25 meters (30 to 80 feet) in all directions.
Spill: Increase, in the downwind direction, as necessary, the distance shown under "Public Safety."
Fire: If any large container is involved in fire, isolate for at least 800 meters (½ mile) in all directions; also, consider initial evacuation for 800 meters (½ mile) in all directions.

EXPOSURE
Short-term effects: *SEEK MEDICAL ATTENTION.* **Liquid:** Harmful if swallowed. *IF SWALLOWED* and victim is *CONSCIOUS AND ABLE TO SWALLOW*, have victim drink 4 to 8 ounces of water. **Do NOT induce vomiting.**
Note to physician or authorized medical personnel: Medical observation is recommended for 24 to 48 hours after breathing overexposure, as pulmonary edema may be delayed. As first aid for pulmonary edema, consider administering a corticosteroid spray. Cigarette smoking may exacerbate pulmonary injury and should be discouraged for at least 72 hours following exposure.

HEALTH HAZARDS
Personal protective equipment (PPE): B-Level PPE. Goggles or face shield.
Toxicity by ingestion: Grade 2; LD_{50} = 0.5 to 5 g/kg (human).
Long-term health effects: Possible mutagen.
Vapor (gas) irritant characteristics: May cause eye and respiratory tract irritation.
Liquid or solid irritant characteristics: Severe eye and skin irritant.

CHEMICAL REACTIVITY
Reactivity with water: A strong base in solution.
Binary reactants: Solution reacts with acids. Attacks chemically active metals such as zinc and aluminum. Violent reaction with fluorine.

ENVIRONMENTAL DATA
Food chain concentration potential: Negative; unlikely to accumulate.
Water pollution: Dangerous to aquatic life in high concentrations. May be dangerous if it enters nearby water intakes; notify operators. Notify local health and wildlife officials. **Response to discharge:** Disperse and flush.

SHIPPING INFORMATION
Grades of purity: A wide variety of grades, which differ in concentration of sodium silicate in water, in specific gravity, and in viscosity; **Storage temperature:** Inerted; **Inert atmosphere:** None; **Venting:** Open; **Stability during transport:** Stable.

PHYSICAL AND CHEMICAL PROPERTIES
Physical state @ 59°F/15°C and 1 atm: Liquid.
Molecular weight: 122.1
Boiling point @ 1 atm: Decomposes.
Melting/Freezing point: 1990°F/1088°C/1361°K.
Specific gravity (water = 1): 1.1–1.7 @ 68°F/20°C (liquid); 2.4 (anhydrous).
Heat of solution: (estimate) -20 Btu/lb = -10 cal/g = -0.4×10^5 J/kg.
Heat of fusion: 84.4 cal/g.

SODIUM SILICOFLUORIDE   REC. S:5200

SYNONYMS: SILICOFLUORURO SODICO (Spanish); SODIUM FLUOSILICATE; SODIUM HEXAFLUOROSILICATE; SALUFER

IDENTIFICATION
CAS Number: 1310-02-7; 16893-85-9
Formula: SiF_6; $F_6Si \cdot 2Na$; Na_2SiF_6
DOT ID Number: UN 2674; DOT Guide Number: 154
Proper Shipping Name: Sodium fluorosilicate; Sodium silicofluoride

DESCRIPTION: White to yellow solid. Odorless. Sinks and slowly mixes in water.

Corrosive • Breathing the dust can kill you; skin or eye contact causes severe burns and may cause blindness; effects of contact or inhalation may be delayed • Firefighting gear (including SCBA) does not provide adequate protection. If exposure occurs, remove and isolate gear immediately and thoroughly decontaminate personnel • Toxic products of combustion may include sodium oxide and fluorine.

Hazard Classification (based on NFPA-704 Rating System)
Health Hazards (Blue): 2; Flammability (Red): 0; Reactivity (Yellow): 0

EMERGENCY RESPONSE: See Appendix A (154)
Evacuation:
Public safety: Isolate the area of spill or leak for at least 25 to 50 meters (80 to 160 feet) in all directions.
Spill: Increase, in the downwind direction, as necessary, the distance shown under "Public Safety."
Fire: If tank, rail car, or tank truck is involved in fire, isolate for at least 800 meters (½ mile) in all directions; also, consider initial evacuation for 800 meters (½ mile) in all directions.

EXPOSURE
Short-term effects: *SEEK MEDICAL ATTENTION. POISONOUS IF SWALLOWED.* **Dust:** Irritating to eyes, nose, and throat. *IF INHALED*, will, will cause coughing, or difficult breathing. *IF BREATHING HAS STOPPED,* give artificial respiration; *avoid mouth-to-mouth resuscitation; use bag/mask apparatus.* IF breathing is difficult, administer oxygen. *IF IN EYES*, hold eyelids open and flush with plenty of water. **Solid:** Will burn skin and eyes. *IF SWALLOWED*, will cause nausea, vomiting, or loss of consciousness. Remove contaminated clothing and shoes. Flush affected areas with plenty of water. *IF IN EYES*, hold eyelids open and flush with plenty of water. *IF SWALLOWED* and victim is *CONSCIOUS AND ABLE TO SWALLOW*, have victim drink 4 to 8 ounces of water. *IF SWALLOWED* and victim is *UNCONSCIOUS OR HAVING CONVULSIONS,* do nothing except keep victim warm.
Note to physician or authorized medical personnel: Medical observation is recommended for 24 to 48 hours after breathing overexposure, as pulmonary edema may be delayed. As first aid for pulmonary edema, consider administering a corticosteroid spray. Cigarette smoking may exacerbate pulmonary injury and should be discouraged for at least 72 hours following exposure.

HEALTH HAZARDS
Personal protective equipment (PPE): B-Level PPE. Dust respirator; goggles or face shield; protective **gloves.**
Recommendations for respirator selection: NIOSH/OSHA as F. *12.5 mg/m³:* DM (any dust and mist respirator). *25 mg/m³:* DMXSQ* (any dust and mist respirator except single-use and quarter-mask respirators); or SA* (any supplied-air respirator). *62.5 mg/m³:* SA:CF* [any supplied-air respirator operated in a continuous-flow mode)]; or PAPRDM*+ *if not present as a fume* (any powered, air-purifying respirator with a dust and mist filter). *125 mg/m³:* HiEF + (any air-purifying, full-facepiece respirator with a high-efficiency particulate filter); or SCBAF (any self-contained breathing apparatus with a full facepiece); or SAF (any supplied-air respirator with a full facepiece). *250 mg/m³:* SA:PD,PP (any supplied-air respirator operated in a pressure-demand or other positive-pressure mode). *EMERGENCY OR PLANNED ENTRY INTO UNKNOWN CONCENTRATIONS OR IDLH CONDITIONS:* SCBAF:PD,PP (any self-contained breathing apparatus that has a full faceplate and is operated in a pressure-demand or other positive-pressure mode); or SAF:PD,PP:ASCBA (any supplied-air respirator that has a full facepiece and is operated in a pressure-demand or other positive-pressure mode in combination with an auxiliary, self-contained breathing apparatus operated in a pressure-demand or other positive-pressure mode). *ESCAPE:* HiEF+ (any air-purifying, full-facepiece respirator with a high-efficiency particulate filter); or SCBAE (any appropriate escape-type, self-contained breathing apparatus). *Notes:* *Substance reported to cause eye irritation or damage; may require eye protection. +May need acid gas sorbent.
Exposure limits (TWA unless otherwise noted): ACGIH TLV 2.5 mg/m³ as fluorides; NIOSH/OSHA 2.5 mg/m³ as inorganic fluorides.
Toxicity by ingestion: Grade 3; LD_{50} = 50 to 500 mg/kg.
Vapor (gas) irritant characteristics: Dust is an eye and skin irritant.
Liquid or solid irritant characteristics: Severe eye and skin irritant.
IDLH value: 250 mg/m³ as F.

FIRE DATA
Behavior in fire: Decomposes at red heat, forming toxic fumes of oxides and fluorine.

CHEMICAL REACTIVITY
Reactivity with water: Solution may be corrosive.

ENVIRONMENTAL DATA
Food chain concentration potential: Negative; unlikely to accumulate.
Water pollution: Effect of low concentrations on aquatic life is unknown. May be dangerous if it enters nearby water intakes; notify operators. Notify local health and wildlife officials.
Response to discharge: Issue warning–water contaminant. Disperse and flush.

SHIPPING INFORMATION
Grades of purity: Technical, 98+%; **Storage temperature:** Ambient; **Inert atmosphere:** None; **Venting:** Open; **Stability during transport:** Stable.

PHYSICAL AND CHEMICAL PROPERTIES
Physical state @ 59°F/15°C and 1 atm: Solid.
Molecular weight: 188
Boiling point @ 1 atm: (decomposes).
Specific gravity (water = 1): 2.68 @ 68°F/20°C (solid).

SODIUM SULFIDE   REC. S:5300

SYNONYMS: EEC No. 016-009-00-8; DISODIUM MONOSULFIDE; DISODIUM SULFIDE; SODIUM MONOSULFIDE; SODIUM SULFIDE, ANHYDROUS; SODIUM SULFURET; SODIUM SULPHIDE; SULFURO SODICO (Spanish)

IDENTIFICATION
CAS Number: 1313-82-2
Formula: Na_2S
DOT ID Number: UN 1385; DOT Guide Number: 135
Proper Shipping Name: Sodium sulfide, anhydrous; Sodium sulfide with less than 30% water of crystallization; Sodium sulphide, anhydrous; Sodium sulphide with less than 30% water of crystallization

DESCRIPTION: White crystalline solid when pure. Turns yellow then brown in air. Rotten eggs odor. Sinks and mixes with water; reacts, forming toxic fumes of hydrogen sulfide.

Dust is spontaneously combustible in air • Heat or flame may cause explosion • Corrosive to skin, eyes and respiratory tract; skin or eye contact causes burns, impaired vision, or blindness; symptoms of contact or inhalation may be delayed • Firefighting gear (including SCBA) does not provide adequate protection. If exposure occurs, remove and isolate gear immediately and thoroughly decontaminate personnel • Containers may BLEVE when exposed to fire • Dust confined areas (e.g., tanks, sewers, buildings) may explode when exposed to fire • Irritating to the skin, eyes, and respiratory tract • Toxic products of combustion may include sulfur dioxide and sodium oxide.

Hazard Classification (based on NFPA-704 Rating System)
Health Hazards (Blue): 3; Flammability (Red): 1; Reactivity (Yellow): 1

EMERGENCY RESPONSE: See Appendix A (135)
Evacuation:
Public safety: Isolate spill area for at least 100 to 150 meters (330 to 490 feet) in all directions.

Spill: Increase downwind, the distance shown above in "Public safety."
Fire: Isolate for 800 meters (½ mile) in all directions, especially if tank, rail car, or tank truck is involved in fire.

EXPOSURE
Short-term effects: *SEEK MEDICAL ATTENTION.* **Dust:** Irritating to eyes, nose, and throat. Move to fresh air. *IF IN EYES*, hold eyelids open and flush with plenty of water. **Solid:** Harmful if swallowed. *IF SWALLOWED* and victim is *CONSCIOUS AND ABLE TO SWALLOW*, have victim drink 4 to 8 ounces of water.
Note to physician or authorized medical personnel: Medical observation is recommended for 24 to 48 hours after breathing overexposure, as pulmonary edema may be delayed. As first aid for pulmonary edema, consider administering a corticosteroid spray. Cigarette smoking may exacerbate pulmonary injury and should be discouraged for at least 72 hours following exposure.

HEALTH HAZARDS
Personal protective equipment (PPE): B-Level PPE. Goggles or face shield.
Eye or Skin contact: Wash with water for at least 15 minutes.
Toxicity by ingestion: Grade 3; LD_{50} = 50 to 500 mg/kg (human).
Vapor (gas) irritant characteristics: Nonvolatile, but dust irritates eyes and mucous membrane.
Liquid or solid irritant characteristics: Severe eye and skin irritant.

FIRE DATA
Flash point: Combustible.
Fire extinguishing agents not to be used: CO_2.
Behavior in fire: Unstable: Heat may cause container rupture and explosion.
Electrical hazard: Class II, Group G

CHEMICAL REACTIVITY
Reactivity with water: Forms corrosive hydrogen sulfide solution with water or acids.
Binary reactants: Self-reactive in air (finely divided material). Reacts with acids, carbon and oxidizers. Attacks metals.
Neutralizing agents for acids and caustics: Mild acidic solution such as vinegar or 1-2% acetic acid.

ENVIRONMENTAL DATA
Food chain concentration potential: Negative; unlikely to accumulate.
Water pollution: Harmful to aquatic life in very low concentrations. May be dangerous if it enters nearby water intakes; notify operators. Notify local health and wildlife officials. **Response to discharge:** Issue warning–corrosive. Restrict access. Disperse and flush.

SHIPPING INFORMATION
Grades of purity: Crystals, 60-62%, plus water; fused chips; **Storage temperature:** Ambient; **Inert atmosphere:** None; **Venting:** Open; **Stability during transport:** Stable.

PHYSICAL AND CHEMICAL PROPERTIES
Physical state @ 59°F/15°C and 1 atm: Solid.
Molecular weight: 78.0
Boiling point @ 1 atm: Very high.
Melting/Freezing point: 1706°F/930°C/1203°K.
Specific gravity (water = 1): 1.856 @ 68°F/20°C (solid).
Heat of fusion: 15.4 cal/g.

SODIUM SULFITE   REC. S:5400

SYNONYMS: DISODIUM SULFITE; SULFTECH; SULFUROUS ACID, SODIUM SALT; SULFITO SODICO (Spanish)

IDENTIFICATION
CAS Number: 7757-83-7
Formula: $O_3S \cdot 2Na$

DESCRIPTION: White crystalline solid. Odorless. Sinks and mixes slowly with water.

Corrosive to skin, eyes, and respiratory tract; skin or eye contact causes severe burns, impaired vision, or blindness; effects of contact or inhalation may be delayed • Decomposes above 1112°F/600°C. Toxic products of combustion or decomposition include oxides of sodium and sulfur.

Hazard Classification (based on NFPA-704 Rating System)
Health Hazards (Blue): 1; Flammability (Red): 0; Reactivity (Yellow): 1

EMERGENCY RESPONSE: See Appendix A (171)
Evacuation:
Public safety: Isolate the area of spill or leak for at least 10 to 25 meters (30 to 80 feet) in all directions.
Spill: Increase, in the downwind direction, as necessary, the distance shown under "Public Safety."
Fire: If any large container is involved in fire, isolate for at least 800 meters (½ mile) in all directions; also, consider initial evacuation for 800 meters (½ mile) in all directions.

EXPOSURE
Short-term effects: *SEEK MEDICAL ATTENTION.* **Dust or powder:** Lung edema may develop.
Solids: If swallowed, may cause loss of consciousness. *IF SWALLOWED* and victim is *CONSCIOUS AND ABLE TO SWALLOW*, have victim drink 4 to 8 ounces of water.
Note to physician or authorized medical personnel: Medical observation is recommended for 24 to 48 hours after breathing overexposure, as pulmonary edema may be delayed. As first aid for pulmonary edema, consider administering a corticosteroid spray. Cigarette smoking may exacerbate pulmonary injury and should be discouraged for at least 72 hours following exposure.

HEALTH HAZARDS
Personal protective equipment (PPE): B-Level PPE. Dust mask; goggles or face shield.
Toxicity by ingestion: Grade 2; LD_{50} = 0.5 to 5 g/kg.
Long-term health effects: Mutagen.
Vapor (gas) irritant characteristics: Nonvolatile, but dust can cause serious eye and respiratory tract irritation.
Liquid or solid irritant characteristics: Severe eye and skin irritant.

CHEMICAL REACTIVITY
Binary reactants: A reducing agent; reacts, possibly violently, with oxidizers and acids.

ENVIRONMENTAL DATA
Food chain concentration potential: Negative; unlikely to accumulate.
Water pollution: Dangerous to aquatic life in high concentrations. May be dangerous if it enters nearby water intakes; notify operators. Notify local health and wildlife officials. **Response to discharge:** Issue warning–water contaminant. Disperse and flush.

SHIPPING INFORMATION
Grades of purity: Technical anhydrous: 89–91%; **Storage temperature:** Ambient; **Inert atmosphere:** None; **Venting:** Open; **Stability during transport:** Stable.

PHYSICAL AND CHEMICAL PROPERTIES
Physical state @ 59°F/15°C and 1 atm: Solid.
Molecular weight: 126.04
Boiling point @ 1 atm: Decomposes.
Melting/Freezing point: Decomposes @ less than 1100°F/593°C/866°K.
Specific gravity (water = 1): 2.633 @ 15°C (solid).

SODIUM THIOCYANATE   REC. S:5500

SYNONYMS: HAIMASED; NATRIUMRHODANID (German); SCYAN; SODIUM ISOTHIOCYANATE; SODIUM SULFOCYANATE; SODIUM RHODANIDE; SODIUM THIOCYANIDE; RHODANATE434; TIOCIANATO SODICO (Spanish)

IDENTIFICATION
CAS Number: 540-72-7
Formula: $CNS \cdot Na$; NaSCN

DESCRIPTION: Colorless crystalline solid or white powder. Odorless. Sinks and mixes with water.

Finely divided material is combustible • Generally, thiocyanates have low acute toxicity. However, prolonged absorption may affect central nervous system • Concentrated dust in confined areas (e.g., tanks, sewers, buildings) may explode when exposed to fire • Toxic products of combustion may include oxides of sulfur, nitrogen and sodium.

Hazard Classification (based on NFPA-704 Rating System)
Health Hazards (Blue): 2; Flammability (Red): 0; Reactivity (Yellow): 0

EMERGENCY RESPONSE: See Appendix A (171)
Evacuation:
Public safety: Isolate the area of spill or leak for at least 10 to 25 meters (30 to 80 feet) in all directions.
Spill: Increase, in the downwind direction, as necessary, the distance shown under "Public Safety."
Fire: If any large container is involved in fire, isolate for at least 800 meters (½ mile) in all directions; also, consider initial evacuation for 800 meters (½ mile) in all directions.

EXPOSURE
Short-term effects: *SEEK MEDICAL ATTENTION.* **Dust:** Irritating to eyes, nose, and throat. *IF INHALED*, will, will cause coughing or difficult breathing. *IF IN EYES*, hold eyelids open and flush with plenty of water. *IF BREATHING HAS STOPPED*, give artificial respiration. IF breathing is difficult, administer oxygen.
Solid: Irritating to skin and eyes. *IF SWALLOWED*, will cause nausea, vomiting, or loss of consciousness. Remove contaminated clothing and shoes. Flush affected areas with plenty of water. *IF IN EYES*, hold eyelids open and flush with plenty of water. *IF SWALLOWED* and victim is *CONSCIOUS AND ABLE TO SWALLOW*, have victim drink 4 to 8 ounces of water. *IF*

SWALLOWED and victim is *UNCONSCIOUS OR HAVING CONVULSIONS,* do nothing except keep victim warm.

HEALTH HAZARDS
Personal protective equipment (PPE): A-Level PPE. Rubber or plastic gloves; standard goggles; rubber or plastic apron.
Toxicity by ingestion: Grade 2; LD_{50} = 0.5 to 5 g/kg.
Long-term health effects: Causes birth defects in chick embryos.
Liquid or solid irritant characteristics: Eye, skin and respiratory tract irritant.

ENVIRONMENTAL DATA
Food chain concentration potential: Negative; unlikely to accumulate.
Water pollution: DOT Appendix B, §172.101–marine pollutant. Dangerous to aquatic life in high concentrations. May be dangerous if it enters nearby water intakes; notify operators. Notify local health and wildlife officials. **Response to discharge:** Issue warning–water contaminant. Disperse and flush.

SHIPPING INFORMATION
Grades of purity: Commercial, 98%; Reagent, 99%; 50–60% solutions in water; **Storage temperature:** Ambient; **Inert atmosphere:** None; **Venting:** Open; **Stability during transport:** Stable.

PHYSICAL AND CHEMICAL PROPERTIES
Physical state @ 59°F/15°C and 1 atm: Solid.
Molecular weight: 81.08
Boiling point @ 1 atm: Decomposes.
Melting/Freezing point: 572°F/300°C/573°K.
Specific gravity (water = 1): More than 1 @ 68°F/20°C (solid).
Heat of solution: 34.9 Btu/lb = 19.4 cal/g = 0.812×10^5 J/kg.
Heat of fusion: 54.8 cal/g.

SODIUM THIOCYANATE SOLUTION (56% or less)
REC. S:5600

SYNONYMS: RHODANATE; SCYAN; SODIUM RHODANIDE; SODIUM SULFOCYANATE; THIOCYANATE SODIUM-434; TIOCIANATO SODICO (Spanish)

IDENTIFICATION
CAS Number: 540-72-7

DESCRIPTION: Clear to pale yellow liquid. Odorless.

Generally, thiocyanates have low acute toxicity. However, prolonged absorption may affect central nervous system • Toxic products of combustion may include oxides of sulfur, nitrogen and sodium.

Hazard Classification (based on NFPA-704 Rating System)
Health Hazards (Blue): 2; Flammability (Red): 0; Reactivity (Yellow): 0

EMERGENCY RESPONSE: See Appendix A (171)
Evacuation:
Public safety: Isolate the area of spill or leak for at least 10 to 25 meters (30 to 80 feet) in all directions.
Spill: Increase, in the downwind direction, as necessary, the distance shown under "Public Safety."
Fire: If any large container is involved in fire, isolate for at least 800 meters (½ mile) in all directions; also, consider initial evacuation for 800 meters (½ mile) in all directions.

EXPOSURE
Short-term effects: *SEEK MEDICAL ATTENTION.* **Vapor:** Irritating to eyes, nose, and throat. *IF IN EYES*, hold eyelids open and flush with plenty of water. *IF BREATHING HAS STOPPED*, give artificial respiration. *IF* breathing is difficult, administer oxygen. **Liquid:** Irritating to skin and eyes. *IF SWALLOWED*, will cause nausea, vomiting, or loss of consciousness. Remove contaminated clothing and shoes. Flush affected areas with plenty of water. *IF IN EYES*, hold eyelids open and flush with plenty of water. *IF SWALLOWED* and victim is *CONSCIOUS AND ABLE TO SWALLOW*, have victim drink 4 to 8 ounces of water. *IF SWALLOWED* and victim is *UNCONSCIOUS OR HAVING CONVULSIONS,* do nothing except keep victim warm.

HEALTH HAZARDS
Personal protective equipment (PPE): A-Level PPE. Rubber or plastic gloves; splash-proof goggles; rubber or plastic apron.
Toxicity by ingestion: Grade 3; LD_{50} = 764 mg/kg (rat).
Long-term health effects: Causes birth defects in chick.

CHEMICAL REACTIVITY
Reactivity group: 0
Compatibility class: Unassigned cargoes

ENVIRONMENTAL DATA
Food chain concentration potential: Negative; unlikely to accumulate.
Water pollution: Dangerous to aquatic life in high concentrations. May be dangerous if it enters nearby water intakes; notify operators. Notify local health and wildlife officials. **Response to discharge:** Issue warning–water contaminant.

SHIPPING INFORMATION
Grades of purity: 50-60% solutions in water; **Storage temperature:** Ambient; **Inert atmosphere:** None; **Venting:** Open; **Stability during transport:** Stable.

PHYSICAL AND CHEMICAL PROPERTIES
Physical state @ 59°F/15°C and 1 atm: Liquid.
Molecular weight: 81.08

SOMAN (GD)
REC. S:5650

SYNONYMS: 3,3-DIMETHYL-2-BUTANOL METHYLPHOSPHONOFLUORIDATE; 3,3-DIMETHYL-*n*-BUT-2-YL METHYLPHOSPHONOFLUORIDATE; 3,3-DIMETHYL-2-BUTYL METHYLPHOSPHONOFLUORIDATE; FLUOROMETHYL(1,2,2-TRIMETHYLPROPOXY)PHOSPHINE OXIDE; GD; METHYLFLUORPHOSPHORSAEUREPINAKOLYLESTER (GERMAN); METHYLPHOSPHONOFLUORIDIC ACID, 3,3-DIMETHYL-2-BUTYL ESTER; METHYLPHOSPHONOFLUORIDIC ACID 1,2,2-TRIMETHYLPROPYL ESTER; METHYL PINACOLYLOXY PHOSPHORYL FLUORIDE; PINACOLOXYMETHYLPHOSPHORYL FLUORIDE; PINACOLYL METHYLFLUOROPHOSPHONATE; PINACOLYL METHYLPHOSPHONOFLUORIDATE; PINACOLYL METHYLPHOSPHONOFLUORIDE; PINACOLYLOXY METHYLPHOSPHORYL FLUORIDE;

PINACOLYLOXY METHYLPHOSPHORYL FLUORIDE; PMFP; PYNACOLYL METHYLFLUOROPHOSPHONATE; 1,2,2-TRIMETHYLPROPYL METHYLPHOSPHONOFLUORIDATE

IDENTIFICATION:
CAS Registry Number: 96-64-0
Formula: $C_7H_{16}FO_2P$; $(CH_3)_3CCH(CH_3)OPF(O)CH_3$
DOT ID Number: UN 2810; DOT Guide Number: 153
Proper Shipping Name: Soman

DESCRIPTION: Clear, colorless liquid when pure. Turns yellow to brown with age. Faint, fruity, camphor-like odor. Sinks in water; soluble; hydrolyzes and forms less toxic materials. An organophosphate insecticide, has been used as a chemical warfare agent; *it is the most toxic of all the nerve agents. Warning:* Odor is not a reliable indicator of the presence of toxic amounts of *Note:* If used as a weapon, notify U.S. Department of Defense: Army. Damage and/or death may occur before chemical detection can take place. Use M8 Paper (Detection: Yellow) or M256-A1 Detector Kit (Detection limit: 0.005 mg/m^3) if available.

Highly volatile poison! (organophosphate) • Combustible • Breathing the vapor, skin or eye contact, or swallowing the material can kill you; symptoms may be delayed for several hours • Firefighting gear (including SCBA) provide NO protection. If exposure occurs, remove and isolate gear immediately and thoroughly decontaminate personnel • Containers may BLEVE when exposed to fire • Vapors are heavier than air and will collect and stay in low areas • Vapors in confined areas (e.g., tanks, sewers, buildings) may explode when exposed to fire • *Combustion products are less deadly than the material itself.* Toxic products of combustion may include carbon monoxide, fluorine, and phosphorus oxide • Do NOT attempt rescue.

Hazard Classification (based on NFPA-704 Rating System)
Health Hazards (Blue): 4; Flammability (Red): 1; Reactivity (Yellow): 0

EMERGENCY RESPONSE: See Appendix A (153) with notes.
Evacuation:
Public safety: See below.
Spill: For Soman [weaponized]: Small spill–First: Isolate in all directions 95 meters (300 feet); Then: Protect persons downwind, DAY: 0.8 km (0.5 mile); NIGHT: 1.8 km (1.1 miles). Large spill–First: Isolate in all directions 765 meters (2500 feet); Then: Protect persons downwind, DAY: 6.8 km (4.2 miles); NIGHT: 10.5 km (6.5 miles).
Fire: If tank, rail car, or tank truck is involved in fire, isolate for at least 800 meters (½ mile) in all directions; also, consider initial evacuation for 800 meters (½ mile) in all directions.

EXPOSURE
Short-term effects: *SEEK MEDICAL ATTENTION.* Will remain a liquid for more than 24 hours. **Vapor or Liquid:** *POISONOUS IF INHALED, IF SWALLOWED, OR IF SKIN IS EXPOSED.* Eye pupils are small; blurred vision; runny nose; cough; shortness of breath; pain; diarrhea, nausea and vomiting; increased blood pressure, hypermotility, hallucinations; loss of consciousness; convulsions; breathing stops; death. *IF BREATHING HAS STOPPED,* give artificial respiration; *avoid mouth-to-mouth resuscitation; use bag/mask apparatus.* **Skin:** *Remove and double-bag contaminated clothing (including shoes) and leave them in the Hot Zone*; wash skin with soap and water and large volumes of water for at least 15 minutes. *Skin can also be decontaminated with diluted hypochlorite solution, U.S. Army M291 kit, and M258(A1) skin decontamination kit.* **Eye:** Rinse with large volumes of water or saline for at least 15 minutes. *IF SWALLOWED,* will cause nausea, vomiting, or loss of consciousness. *Remove and double-bag contaminated clothing and shoes and leave in Hot Zone for later incineration by hazardous materials experts.* Flush affected areas with plenty of water. **If swallowed** and victim is *CONSCIOUS AND ABLE TO SWALLOW,* have victim drink 4 to 8 ounces of water. **Do NOT induce vomiting but immediately administer slurry of activated charcoal.** *IF SWALLOWED* and victim is *UNCONSCIOUS OR HAVING CONVULSIONS,* do nothing except keep victim warm. *Note to physician and medical personnel:* If symptoms indicate, initial treatment for an adult includes atropine, 2 mg (1/30 gr) intramuscularly or intravenously as soon as any local or systemic signs or symptoms of an intoxication are noted; repeat the administration of atropine every 3 to 8 minutes until signs of atropinization (mydriasis, dry mouth, rapid pulse, hot and dry skin) occur; initiate treatment in children with 0.05 mg/kg of atropine; repeat at 5- to 10-minute intervals. Watch respiration, and remove bronchial secretions if they appear to be obstructing the airway; intubate if necessary. Pralidoxime must be administered within minutes to a few hours following exposure (depending on the specific agent) to be effective. Give 2-PAMCI (Pralidoxime; Protopam), 2.5 g in 100 mL of sterile water or in 5% dextrose and water, intravenously, slowly, in 15 to 30 minutes; if sufficient fluid is not available, give 1 g of 2-PAMCI in 3 mL of distilled water by deep intramuscular injection; repeat this every half hour if respiration weakens or if muscle fasciculation or convulsions recur. Also Diazepam, an anticonvulsant, might be needed. For adults, the dose is 5 to 10 mg (slow IV), repeated every 12 to 15 minutes up to three (3) doses maximum. For children, the dose is 0.2 to 0.5 mg/kg.
Note to physician or authorized medical personnel: Administer atropine. *IF INHALED,* will, medical observation is recommended for 24 to 48 hours after breathing overexposure, as pulmonary edema may be delayed. As first aid for pulmonary edema, consider administering a corticosteroid spray. Cigarette smoking may exacerbate pulmonary injury and should be discouraged for at least 72 hours following exposure.

HEALTH HAZARDS
Personal protective equipment (PPE): A-Level PPE. Rubber gloves, protective clothing, goggles, respirators. Butyl rubber gloves and Tyvek® "F" decontamination suit provide barrier protection against chemical warfare agents. Airtight, impermeable clothing was developed for personnel who must enter heavily contaminated areas. This clothing is made of butyl rubber or a coated fabric such as Tyvek "F" and provide barrier protection against liquid chemical warfare agents. Although resistant to liquid chemical agents, impermeable protective clothing may be penetrated after a few hours of exposure to heavy concentration of agent. Consequently, liquid contamination on the clothing must be neutralized or removed as soon as possible. If the proper equipment is not available, or if the rescuers have not been trained in its use, call for assistance from the U.S. Soldier and Biological Chemical Command–Edgewood Research Development and Engineering Center (from 0700-1630 EST call 410-671-4411, and from 1630-0700 EST call 410-278-5201; ask for the Staff Duty Officer).

Recommendations for respirator selection: *EMERGENCY OR PLANNED ENTRY INTO UNKNOWN CONCENTRATIONS OR IDLH CONDITIONS:* SCBAF:PD,PP (any self-contained breathing apparatus that has a full facepiece and is operated in a pressure-demand or other positive-pressure mode); or

SAF:PD,PP:ASCBA (any supplied-air respirator that has a full facepiece and is operated in a pressure-demand or other positive-pressure mode in combination with an auxiliary self-contained breathing apparatus operated in a pressure-demand or other positive pressure mode). *ESCAPE:* GMFOVHiE [any air-purifying, full-facepiece respirator (gas mask) with a chin-style, front-or back-mounted canister having a high efficiency particulate filter); or SCBAE (any appropriate escape-type, self-contained breathing apparatus). The U.S. Army standard M17A1 (which is being replaced by the M40 protective mask) mask provides complete respiratory protection against all known military toxic chemical agents, but it cannot be used in an oxygen deficient environment and it does not afford protection against industrial toxics such as ammonia and carbon monoxide. *It is not approved for civilian use.*

Exposure limits (TWA unless otherwise noted): Workplace: 0.00003 mg/m^3; General population limits (as recommended by the Surgeon General's Working Group, U.S. Department of Health): 0.000003 mg/m^3.

Toxicity: The median lethal dosage (vapor/respiratory) LCt_{50} = 400 mg-minute/m^3 for humans; the median incapacitating dosage is 50 mg-minute/m^3. The LD$_{50}$ [skin] = 350 mg/70 kg/156.8 lb [man] (*Medical Sspects of Chemical and Biological Warfare, Part I*, Walter Reed Medical Center, 1997).

FIRE DATA
Flash point: 249.8°F/121°C.

CHEMICAL REACTIVITY
Binary reaction: Under acid conditions GD hydrolyze to form hydrogen fluoride (HF).

ENVIRONMENTAL DATA

Food chain concentration potential: Broken down quickly in water and moist soil. Not expected to accumulate in the food chain (ATSDR).
Water pollution: If used as a weapon, utilize use an M272 Water Detection Kit (Detection limit: 0.02 mg/L). Dangerous to aquatic life in high concentrations. May be dangerous if it enters water intakes. Notify local health and pollution control officials. Notify operators of nearby water intakes. This material will be broken down in water quickly, but small amounts may evaporate. This material will be broken down in moist soil quickly. Small amounts may evaporate into the air or travel below the soil surface and contaminated groundwater.

PHYSICAL AND CHEMICAL PROPERTIES
Physical state @ 59°F/15°C and 1 atm: Liquid.
Molecular weight: 182.19
Boiling point @ 1 atm: 332.6 to 392°F/167 to 200°C.
Melting/Freezing point: –43.6°F/–42°C/231°K.
Solubility in water: 2.1 g/100 g @ 68°F/20°C.
Relative vapor density (air = 1): 6.33
Specific gravity (water = 1): 1.022

Liquid density: 1.02g/mL @ 77°F/25°C.
Vapor pressure: 0.40 mm @ 77°F/25°C.
Volatility: 3,850 mg/m^3 @ 77°F/25°C.

SORBITOL   REC. S:5700

SYNONYMS: CHOLAXINE; DIAKARMON; GLUCITOL; D-GLUCITOL; 1,2,3,4,5,6,-HEXANNEHEXOL; HEXAHYDRIC ALCOHOL; SIONIT; SIONON; SORBOL; SORBITE; SORBO; SORBOL

IDENTIFICATION
CAS Number: 50-70-4
Formula: $C_6H_{14}O_6$; $CH_2OH(CHOH)_4CH_2OH$

DESCRIPTION: Colorless liquid. Odorless. Sinks and mixes with water.

Combustible • Containers may BLEVE when exposed to fire • Vapors in confined areas (e.g., tanks, sewers, buildings) may explode when exposed to fire • Irritating to the skin, eyes, and respiratory tract • Toxic products of combustion may include carbon monoxide.

Hazard Classification (based on NFPA-704 Rating System)
Health Hazards (Blue): 0; Flammability (Red): 1; Reactivity (Yellow): 0

EMERGENCY RESPONSE: See Appendix A (171)
Evacuation:
Public safety: Isolate the area of spill or leak for at least 10 to 25 meters (30 to 80 feet) in all directions.
Spill: Increase, in the downwind direction, as necessary, the distance shown under "Public Safety."
Fire: If any large container is involved in fire, isolate for at least 800 meters (½ mile) in all directions; also, consider initial evacuation for 800 meters (½ mile) in all directions.

EXPOSURE
Short-term effects: *SEEK MEDICAL ATTENTION.* **Liquid:** Will burn skin and. Flush affected areas with plenty of water. *IF IN EYES*, hold eyelids open and flush with plenty of water.

HEALTH HAZARDS
Personal protective equipment (PPE): B-Level PPE. Goggles or face shield; protective clothing for hot liquid. Natural rubber, neoprene, and nitrile rubber may offer some protection from alcohols.
Vapor (gas) irritant characteristics: Volatility is very low.

FIRE DATA
Flash point: 542°F/283°C (oc).
Electrical hazard: Class I, Group D.

CHEMICAL REACTIVITY
Reactivity group: 20
Compatibility class: Alcohols, glycols

ENVIRONMENTAL DATA
Food chain concentration potential: Negative; unlikely to accumulate.
Water pollution: Dangerous to aquatic life in high concentrations. May be dangerous if it enters nearby water intakes; notify operators. Notify local health and wildlife officials. **Response to discharge:** Disperse and flush.

SHIPPING INFORMATION
Grades of purity: USP crystalline: 93+%; **Storage temperature:** Elevated; **Inert atmosphere:** None; **Venting:** Open (flame arrester); **Stability during transport:** Stable.

PHYSICAL AND CHEMICAL PROPERTIES
Physical state @ 59°F/15°C and 1 atm: Solid.

Molecular weight: 182.17
Boiling point @ 1 atm: 221°F/105°C/378°K.
Melting/Freezing point: 230°F/110°C/383°K.
Specific gravity (water = 1): 1.49 @ 15°C (liquid).
Heat of combustion: (estimate) -6750 Btu/lb = -3750 cal/g = -157×10^5 J/kg.
Heat of solution: (estimate) -22 Btu/lb = -12 cal/g = -0.5×10^5 J/kg.
Vapor pressure: (Reid) Low.

STANNOUS FLUORIDE   REC. S:5800

SYNONYMS: AIM; FLUORISTAN; FLUORURO ESTANNOSO (Spanish); IRADICAR; IRADICAV; STANCARE; STANIDE; TIN BIFLUORIDE; TIN DIFLUORIDE; TIN FLUORIDE

IDENTIFICATION
CAS Number: 7783-47-3
Formula: SnF_2

DESCRIPTION: White crystalline powder. Sinks and mixes with water.

Toxic products of combustion may include hydrogen fluoride and tin oxides.

Hazard Classification (based on NFPA-704 Rating System)
Health Hazards (Blue): 2; Flammability (Red): 0; Reactivity (Yellow): 0

EMERGENCY RESPONSE: See Appendix A (171)
Evacuation:
Public safety: Isolate the area of spill or leak for at least 10 to 25 meters (30 to 80 feet) in all directions.
Spill: Increase, in the downwind direction, as necessary, the distance shown under "Public Safety."
Fire: If any large container is involved in fire, isolate for at least 800 meters (½ mile) in all directions; also, consider initial evacuation for 800 meters (½ mile) in all directions.

EXPOSURE
Short-term effects: *SEEK MEDICAL ATTENTION.* **Solid:** Will burn eyes. Harmful if swallowed.
Flush affected areas with plenty of water. *IF IN EYES*, hold eyelids open and flush with plenty of water. *IF SWALLOWED* and victim is *CONSCIOUS AND ABLE TO SWALLOW*, have victim drink 4 to 8 ounces of water and have victim induce vomiting.

HEALTH HAZARDS
Personal protective equipment (PPE): B-Level PPE. Goggles and dust mask.
Exposure limits (TWA unless otherwise noted): ACGIH TLV 2.5 mg/m³ as fluorides; NIOSH/OSHA 2.5 mg/m³ as inorganic fluorides. NIOSH 2 mg/m³ as tin.
Toxicity by ingestion: Grade 3; LD_{50} = 50 to 500 mg/kg.
Long-term health effects: Long exposure to large amounts of the fluoride may cause loss of weight, anorexia, anemia, wasting, and cachexia, and dental defects. An increase in bone density and discoloration of teeth may occur. A possible mutagen.
Liquid or solid irritant characteristics: Severe eye irritant. See above; practically harmless to intact skin.
IDLH value: 250 mg/m³ as F.

CHEMICAL REACTIVITY
Binary reactants: Avoid contact with acids–hydrogen fluoride (HF) fumes may be produced.

ENVIRONMENTAL DATA
Food chain concentration potential: Fluorine is concentrated by aquatic animals.
Water pollution: HARMFUL TO AQUATIC LIFE IN LOW Concentrations. May be dangerous if it enters nearby water intakes; notify operators. Notify local health and wildlife officials. **Response to discharge:** Disperse and flush.

SHIPPING INFORMATION
Grades of purity: 97.5%; **Stability during transport:** Stable.

PHYSICAL AND CHEMICAL PROPERTIES
Physical state @ 59°F/15°C and 1 atm: Solid.
Molecular weight: 156.70
Boiling point @ 1 atm: 1052°F/850°C/1123°K.
Melting/Freezing point: 419°F/215°C/488°K.
Specific gravity (water = 1): 2.79

STEARIC ACID   REC. S:5900

SYNONYMS: ACIDO ESTEARICO (Spanish); CENTURY-1240; DAR-CHEM; EMERSOL-120; GLYCON DP; GLYCON TP; GROCO-54; 1-HEPTADECANECARBOXYLIC ACID; KAM 1000; KAM 2000; NEO-FAT 18-61; OCTADECANOIC ACID; *N*-OCTADECYLIC ACID; PEARL STEARIC; STEAREX; STEAROPHANIC ACID

IDENTIFICATION
CAS Number: 57-11-4
Formula: $CH_3(CH_2)_{16}CO_2H$; $C_{18}H_{36}O_2$

DESCRIPTION: White to slightly yellow crystalline solid or white powder. Mild, tallow-like odor. Floats on water surface; insoluble.

Combustible • Containers may BLEVE when exposed to fire • Concentrated dust in confined areas (e.g., tanks, sewers, buildings) may explode when exposed to fire • Irritating to the skin, eyes, and respiratory tract • Toxic products of combustion may include CO_2 and carbon monoxide.

Hazard Classification (based on NFPA-704 Rating System)
Health Hazards (Blue): 1; Flammability (Red): 1; Reactivity (Yellow): 0

EMERGENCY RESPONSE: See Appendix A (171)
Evacuation:
Public safety: Isolate the area of spill or leak for at least 10 to 25 meters (30 to 80 feet) in all directions.
Spill: Increase, in the downwind direction, as necessary, the distance shown under "Public Safety."
Fire: If any large container is involved in fire, isolate for at least 800 meters (½ mile) in all directions; also, consider initial evacuation for 800 meters (½ mile) in all directions.

EXPOSURE
Short-term effects: *SEEK MEDICAL ATTENTION.* **Dust:** Irritating to eyes, nose, and throat. *IF INHALED*, will, will cause coughing or difficult breathing. *IF IN EYES*, hold eyelids open and flush with plenty of water. *IF BREATHING HAS STOPPED*, give

artificial respiration. IF breathing is difficult, administer oxygen.
Solid: Irritating to skin and eyes. Harmful if swallowed. Remove contaminated clothing and shoes. Flush affected areas with plenty of water. *IF IN EYES*, hold eyelids open and flush with plenty of water. *IF SWALLOWED and victim is CONSCIOUS AND ABLE TO SWALLOW*, have victim drink 4 to 8 ounces of water and have victim induce vomiting. *IF SWALLOWED and victim is UNCONSCIOUS OR HAVING CONVULSIONS,* do nothing except keep victim warm.

HEALTH HAZARDS
Personal protective equipment (PPE): B-Level PPE. For prolonged exposure to vapors, use air–supplied mask or chemical cartridge respirator; impervious gloves; goggles; impervious apron. Butyl rubber is generally suitable for carbooxylic acid compounds. Chemical protective material(s) reported to offer minimal to poor protection: natural rubber, neoprene, nitrile+PVC, nitrile, PVC, polyurethane, styrene-butadiene.
Toxicity by ingestion: Grade 0; LD_{50} = > 15 g/kg.
Long-term health effects: Possible carcinogen.
Liquid or solid irritant characteristics: Eye and skin irritant.
Odor threshold: 20 ppm.

FIRE DATA
Flash point: (molten solid) 410–435°F/210–224°C (oc); 385°F/196°C (cc).
Fire extinguishing agents not to be used: Water or foam may cause frothing.
Behavior in fire: Sealed containers may rupture and explode.
Autoignition temperature: 743°F/395°C/668°K.

CHEMICAL REACTIVITY
Binary reactants: Reacts with strong oxidizers; bases. Attacks chemically active metals such as aluminum and zinc.
Neutralizing agents for acids and caustics: Flood with water and rinse with sodium bicarbonate or lime solution.

ENVIRONMENTAL DATA
Food chain concentration potential: Negative; unlikely to accumulate.
Water pollution: May be toxic to fish; slicks threaten waterfowl. Frothing may occur and May be dangerous if it enters nearby water intakes; notify operators. Notify local health and wildlife officials.
Response to discharge: Mechanical containment. Should be removed. Chemical and physical treatment.

SHIPPING INFORMATION
Grades of purity: USP; Commercial; Triple pressed; Double pressed; **Storage temperature:** Ambient; **Inert atmosphere:** None; **Venting:** Open; **Stability during transport:** Stable.

PHYSICAL AND CHEMICAL PROPERTIES
Physical state @ 59°F/15°C and 1 atm: Solid.
Molecular weight: (avg.) 282
Boiling point @ 1 atm: 726°F/386°C/659°K.
Melting/Freezing point: 158°F/70°C/343°K.
Specific gravity (water = 1): 0.86 @ 68°F/20°C (solid).
Heat of combustion: $-17{,}310$ Btu/lb $= -9616$ cal/g $= -402.3 \times 10^5$ J/kg.

STRONTIUM CHROMATE REC. S:6000

SYNONYMS: CHROMIC ACID, STRONTIUM SALT; CLORURO de ESTRONICO (Spanish); C.I. PIGMENT YELLOW 32; DEEP LEMON YELLOW; STRONTIUM CHROMATE 12170; STRONTIUM CHROMATE A; STRONTIUM CHROMATE(VI); STRONTIUM CHROMATE X-2396; STRONTIUM YELLOW

IDENTIFICATION
CAS Number: 7789-06-2
Formula: $SrCrO_4$
DOT ID Number: UN 3077; DOT Guide Number: 171
Proper Shipping Name: Environmentally hazardous substances, solid, n.o.s.
Reportable Quantity (RQ): **(CERCLA)** 10 lb/4.54 kg

DESCRIPTION: Yellow crystalline solid or powder. Odorless. Sinks and mixes slowly with water.

Eye and respiratory tract irritant • Toxic products of combustion may include hydrogen chloride and chromium

Hazard Classification (based on NFPA-704 Rating System)
Health Hazards (Blue): 1; Flammability (Red): 0; Reactivity (Yellow): 1

EMERGENCY RESPONSE: See Appendix A (171)
Evacuation:
Public safety: Isolate the area of spill or leak for at least 10 to 25 meters (30 to 80 feet) in all directions.
Spill: Increase, in the downwind direction, as necessary, the distance shown under "Public Safety."
Fire: If any large container is involved in fire, isolate for at least 800 meters (½ mile) in all directions; also, consider initial evacuation for 800 meters (½ mile) in all directions.

EXPOSURE
Short-term effects: *SEEK MEDICAL ATTENTION*. **Solid or dust:** Irritating to skin, nose, and throat. Harmful if swallowed. Move to fresh air.
Flush affected areas with plenty of water. *IF IN EYES*, hold eyelids open and flush with plenty of water. *IF SWALLOWED* and victim is *CONSCIOUS AND ABLE TO SWALLOW*, have victim drink 4 to 8 ounces of water and have victim induce vomiting.

HEALTH HAZARDS
Personal protective equipment (PPE): B-Level PPE. Chemical safety goggles. Full-cover work clothes, work gloves, and toxic dust respirator.
Recommendations for respirator selection: NIOSH
At any concentrations above the NIOSH REL, or where there is no REL, at any detectable concentration: SCBAF:PD,PP (any self-contained breathing apparatus that has a full facepiece and is operated in a pressure-demand or other positive-pressure mode); or SAF:PD,PP:ASCBA (any supplied-air respirator that has a full facepiece and is operated in a pressure-demand or other positive-pressure mode in combination with an auxiliary self-contained breathing apparatus operated in a pressure-demand or other positive pressure mode). *ESCAPE:* HiEF (any air-purifying, full-facepiece respirator with a high-efficiency particulate filter); or SCBAE (any appropriate escape-type, self-contained breathing apparatus).
Exposure limits (TWA unless otherwise noted): ACGIH TLV 0.05 mg/m³ as water soluble Cr(VI) compounds; OSHA PEL 0.1 mg/m³ as CrO_3; NIOSH 0.001 mg/m³ as chromates; potential human carcinogen; reduce exposure to lowest feasible level.
Toxicity by ingestion: Grade 3; LD_{50} = 50 to 500 mg/kg.
Long-term health effects: Confirmed human carcinogen.

Vapor (gas) irritant characteristics: Eye and respiratory tract irritant.
Liquid or solid irritant characteristics: May cause eye irritation. Little or no irritation to skin on short-term contact (not true for repeated or prolonged contact).
IDLH value: 15 mg/m^3 as chromium(VI) compound

CHEMICAL REACTIVITY
Reactivity with water: Produces hazardous solution.
Binary reactants: Avoid contact with water, acids, and bases.

ENVIRONMENTAL DATA
Food chain concentration potential: Positive. Constant exposure can cause bioconcentration up to 2000-fold. Cr will concentrate. Probable Sr accumulation factor = 150 for goldfish.
Water pollution: Dangerous to aquatic life in high concentrations. May be dangerous if it enters nearby water intakes; notify operators. Notify local health and wildlife officials. **Response to discharge:** Issue warning–water contaminant. Chemical and physical treatment. Disperse and flush.

PHYSICAL AND CHEMICAL PROPERTIES
Physical state @ 59°F/15°C and 1 atm: Solid.
Molecular weight: 203.64
Specific gravity (water = 1): 3.895 @ 15°C.
Heat of combustion: Not flammable

STRONTIUM NITRATE   REC. S:6100

SYNONYMS: STRONTIUM(II) NITRATE; NITRIC ACID, STRONTIUM SALT; NITRATO de ESTRONICO (Spanish)

IDENTIFICATION
CAS Number: 10042-76-9
Formula: $Sr(NO_3)_2$
DOT ID Number: UN 1507; DOT Guide Number: 140
Proper Shipping Name: Strontium nitrate

DESCRIPTION: Colorless crystalline solid or white powder. Odorless. Sinks and mixes with water.

Strong oxidizer which may react spontaneously with low flash point organics or reducing agents. Heat forms oxygen; will increase the activity of an existing fire • May cause fire or explosion on contact with combustibles (wood, paper, oil, clothing, etc.). Irritating to the skin, eyes, and respiratory tract • Toxic products of combustion may include nitrogen oxides.

Hazard Classification (based on NFPA-704 Rating System)
Health Hazards (Blue): 1; Flammability (Red): 0; Reactivity (Yellow): 1; Special Notice: OXY

EMERGENCY RESPONSE: See Appendix A (140)
Evacuation:
Public safety: Isolate the area of spill or leak for at least 10 to 25 meters (30 to 80 feet) in all directions.
Spill: Consider initial downwind evacuation for at least 100 meters (330 feet).
Fire: If any large container is involved in fire, isolate for at least 800 meters (½ mile) in all directions; also, consider initial evacuation for 800 meters (½ mile) in all directions.

EXPOSURE
Short-term effects: *SEEK MEDICAL ATTENTION.* **Solid:** Flush exposed areas with plenty of water. *IF IN EYES,* hold eyelids open and flush with plenty of water. *IF SWALLOWED* and victim is *CONSCIOUS AND ABLE TO SWALLOW,* have victim drink 4 to 8 ounces of water and have victim induce vomiting. *IF SWALLOWED* and victim is *UNCONSCIOUS OR HAVING CONVULSIONS,* do nothing except keep victim warm.

HEALTH HAZARDS
Personal protective equipment (PPE): B-Level PPE. Rubber gloves, goggles, laboratory coat.
Toxicity by ingestion: Grade 2; LD_{50} = 2.75 g/kg (rat).
Long-term health effects: Ingestion of nitrates has been implicated with cancer increase.
Liquid or solid irritant characteristics: Minimum hazard. If spilled on clothing and allowed to remain, may cause smarting and reddening of the skin.

FIRE DATA
Behavior in fire: Containers may explode.

CHEMICAL REACTIVITY
Binary reactants: A strong oxidizer. Reacts with reducing agents, acids, organic materials, or combustible materials.

ENVIRONMENTAL DATA
Water pollution: Effect of low concentrations on aquatic life is unknown. May be dangerous if it enters nearby water intakes; notify operators. Notify local health and wildlife officials. **Response to discharge:** Issue warning–oxidizing material. Should be removed.

SHIPPING INFORMATION
Grades of purity: 99%; technical grades; **Storage temperature:** Cool; **Inert atmosphere:** None required; **Stability during transport:** Stable.

PHYSICAL AND CHEMICAL PROPERTIES
Physical state @ 59°F/15°C and 1 atm: Solid.
Molecular weight: 211.63
Boiling point @ 1 atm: 1193°F/645°C/918°K **Melting/Freezing point:** 1058°F/570°C/843°K CRITICAL
Specific gravity (water = 1): 2.98

STRYCHNINE   REC. S:6200

SYNONYMS: DOLCO MOUSE CEREAL; ESTRICNINA (Spanish); KWIK-KIL; MOLE-DEATH; MOUSE-TOX; NUX-VOMICA; RCRA No. P108; STRYCHNOS

IDENTIFICATION
CAS Number: 57-24-9
Formula: $C_{21}H_{22}N_2O_2$
DOT ID Number: UN 1692; DOT Guide Number: 151
Proper Shipping Name: Strychnine or strychnine salts
Reportable Quantity (RQ): **(CERCLA)** 10 lb/4.54 kg

DESCRIPTION: Colorless to white crystalline solid. Odorless. Sinks and mixes slowly in water.

Poison! • Breathing the dust or swallowing the material can kill you • Firefighting gear (including SCBA) does not provide adequate protection. If exposure occurs, remove and isolate gear immediately and thoroughly decontaminate personnel • Containers may BLEVE when exposed to fire • Toxic products of combustion may include nitrogen oxides • Do not put yourself in danger by entering a contaminated area to rescue a victim.

Hazard Classification (based on NFPA-704 Rating System)
Health Hazards (Blue): 3; Flammability (Red): 1; Reactivity (Yellow): 0

EMERGENCY RESPONSE: See Appendix A (151)
Evacuation:
Public safety: Isolate the area of spill or leak for at least 25 to 50 meters (80 to 160 feet) in all directions.
Spill: Increase, in the downwind direction, as necessary, the distance shown under "Public Safety."
Fire: If tank, rail car, or tank truck is involved in fire, isolate for at least 800 meters (½ mile) in all directions; also, consider initial evacuation for 800 meters (½ mile) in all directions.

EXPOSURE
Short-term effects: *SEEK MEDICAL ATTENTION*. If artificial respiration is administered, *avoid mouth-to-mouth resuscitation; use bag/mask apparatus*. **Solid:** *POISONOUS IF SWALLOWED*. Remove contaminated clothing and shoes. Flush affected areas with plenty of water. *IF SWALLOWED* and victim is *CONSCIOUS AND ABLE TO SWALLOW*, have victim ingest a slurry of activated charcoal. *IF SWALLOWED* and victim is *UNCONSCIOUS OR HAVING CONVULSIONS*, do nothing except keep victim warm.

HEALTH HAZARDS
Personal protective equipment (PPE): A-Level PPE. OSHA Table Z-1-A air contaminant. Butyl rubber may offer protection.
Recommendations for respirator selection: NIOSH/OSHA
0.75 mg/m^3: DM (any dust and mist respirator). *1.5 mg/m^3*: DMXSQ *If not present as a fume* (any dust and mist respirator except single-use and quarter mask respirators); or SA (any supplied-air respirator). *3 mg/m^3*: SA:CF (any supplied-air respirator operated in a continuous-flow mode); or PAPRDM (any powered, air-purifying respirator with a dust and mist filter); or HiEF (any air-purifying, full-facepiece respirator with a high-efficiency particulate filter); or SCBAF (any self-contained breathing apparatus with a full facepiece); or SAF (any supplied-air respirator with a full facepiece). *EMERGENCY OR PLANNED ENTRY INTO UNKNOWN CONCENTRATIONS OR IDLH CONDITIONS*: SCBAF:PD,PP (any self-contained breathing apparatus that has a full facepiece and is operated in a pressure-demand or other positive-pressure mode); or SAF:PD,PP:ASCBA (any supplied-air respirator that has a full facepiece and is operated in a pressure-demand or other positive-pressure mode in combination with an auxiliary self-contained breathing apparatus operated in a pressure-demand or other positive pressure mode). *ESCAPE*: HiEF (any air-purifying, full-facepiece respirator with a high-efficiency particulate filter); or SCBAE (any appropriate escape-type, self-contained breathing apparatus).
Exposure limits (TWA unless otherwise noted): ACGIH TLV 0.15 mg/m^3; NIOSH/OSHA 0.15 mg/m^3.
Toxicity by ingestion: Grade 4; LD$_{50}$ below 50 mg/kg.
Long-term health effects: Chronic skin allergen.
IDLH value: 3 mg/m^3.

ENVIRONMENTAL DATA
Food chain concentration potential: Log K$_{ow}$ = 1.7. Unlikely to accumulate.
Water pollution: DOT Appendix B, §172.101–marine pollutant. Harmful to aquatic life in very low concentrations. May be dangerous if it enters nearby water intakes; notify operators. Notify local health and wildlife officials. **Response to discharge:** Issue warning–poison, water contaminant. Restrict access. Should be removed. Chemical and physical treatment.

SHIPPING INFORMATION
Stability during transport: Stable, but sensitive to light.

PHYSICAL AND CHEMICAL PROPERTIES
Physical state @ 59°F/15°C and 1 atm: Solid.
Molecular weight: 334.40
Boiling point @ 1 atm: 518°F/270°C/543.2°K.
Melting/Freezing point: 514-558°F/268-290°C/541-563°K.
Specific gravity (water = 1): 1.36 @ 68°F/20°C/293°K.
Relative vapor density (air = 1): 11.5 (calculated).
Vapor pressure: (Reid) Low; 0 torr at 68°F/20°C/293°K.

STYRENE **REC. S:6300**

SYNONYMS: ANNAMENE; BENZENE, VINYL-; CINNAMENE; CINNAMENOL; CINNAMOL; DIAREX HF 77; EEC No. 601-026-00-0; ESTIRENO (Spanish); ETHYLBENZENE; ETHYLENE, PHENYL-; PHENYLETHENE; STYRENE MONOMER; STYROL (German); STYROLE; STYROLENE; STYRON; STYROPOL; STYROPOR; VINYL BENZENE; VINYLBENZOL

IDENTIFICATION
CAS Number: 100-42-5
Formula: C_8H_8; $C_6H_5CH=CH_2$
DOT ID Number: UN 2055; DOT Guide Number: 128
Proper Shipping Name: Styrene monomer, inhibited
Reportable Quantity (RQ): **(CERCLA)** 1000 lb/454 kg

DESCRIPTION: Colorless to pale yellow, watery liquid. Sharp, disagreeable odor that is sweet and floral at low concentrations. Floats on the surface of water; insoluble. Flammable, irritating vapor is produced.

Highly flammable • Polymerization hazard • Heat can induce polymerization with rapid release of energy; sealed containers may rupture explosively • May react with itself blocking relief valves; leading to tank explosions • Containers may BLEVE when exposed to fire •Vapors may form explosive mixture with air • Vapors are heavier than air and will collect and stay in low areas • Vapors may travel long distances to ignition sources and flashback • Vapors in confined areas (e.g., tanks, sewers, buildings) may explode when exposed to fire • Very Irritating to the skin, eyes, and respiratory tract • Toxic products of combustion may include very acrid fumes, carbon monoxide, and CO_2.

Hazard Classification (based on NFPA-704 Rating System)
Health Hazards (Blue): 2; Flammability (Red): 3; Reactivity (Yellow): 2

EMERGENCY RESPONSE: See Appendix A (128)
Evacuation:
Public safety: Isolate spill area for at least 25 to 50 meters (80 to 160 feet) in all directions.
Spill: Large spill–Consider initial downwind evacuation for at least 300 meters (1000 feet).
Fire: Isolate for 800 meters (½ mile) in all directions, especially if tank, rail car, or tank truck is involved in fire.

Styrene

EXPOSURE
Short-term effects: *SEEK MEDICAL ATTENTION.* **Vapor:** Irritating to eyes, nose, and throat. *IF INHALED*, will, will cause dizziness or loss of consciousness. Move to fresh air. *IF BREATHING HAS STOPPED*, give artificial respiration. IF breathing is difficult, administer oxygen. The use of alcoholic beverages may enhance toxic effects. **Liquid:** Will burn skin and eyes. Harmful if swallowed. Remove contaminated clothing and shoes. Flush affected areas with plenty of water. *IF IN EYES*, hold eyelids open and flush with plenty of water. *IF SWALLOWED* and victim is *CONSCIOUS AND ABLE TO SWALLOW*, have victim drink 4 to 8 ounces of water. **Do NOT induce vomiting.**

HEALTH HAZARDS
Personal protective equipment (PPE): B-Level PPE. OSHA Table Z-1-A air contaminant. OSHA Table Z-2 air contaminant. Air–supplied mask or approved canister; rubber or plastic gloves; boots; goggles or face shield. Chemical protective material(s) reported to offer minimal to poor protection: PV alcohol, Teflon®, Viton/chlorobutyl rubber, Silvershield®. Also, Viton®, Viton®/neoprene offers limited protection.
Recommendations for respirator selection: NIOSH
500 ppm: CCROV* [any chemical cartridge respirator with organic vapor cartridge(s)]; or SA* (any supplied-air respirator).
700 ppm: SA:CF* (any supplied-air respirator operated in a continuous-flow mode); or CCRFOV [any chemical cartridge respirator with a full facepiece and organic vapor cartridges(s)]; or GMFOV [any air-purifying, full-facepiece respirator (gas mask) with a chin-style, front- or back-mounted organic vapor canister]; or PAPROV* [any powered, air-purifying respirator with organic vapor cartridge(s)]; or SCBAF (any self-contained breathing apparatus with a full facepiece); or SAF (any supplied-air respirator with a full facepiece). *EMERGENCY OR PLANNED ENTRY INTO UNKNOWN CONCENTRATIONS OR IDLH CONDITIONS:* SCBAF:PD,PP (any self-contained breathing apparatus that has a full facepiece and is operated in a pressure-demand or other positive-pressure mode); or SAF:PD,PP:ASCBA (any supplied-air respirator that has a full facepiece and is operated in a pressure-demand or other positive-pressure mode in combination with an auxiliary self-contained breathing apparatus operated in a pressure-demand or other positive pressure mode). *ESCAPE:* GMFOV [any air-purifying, full-facepiece respirator (gas mask) with a chin-style, front- or back-mounted organic vapor cannister]; or SCBAE (any appropriate escape-type, self-contained breathing apparatus). *Note:* Substance reported to cause eye irritation or damage; may require eye protection.
Exposure limits (TWA unless otherwise noted): ACGIH TLV 50 ppm (213 mg/m^3) skin; OSHA PEL 100 ppm, ceiling 200 ppm, 600 ppm/5 minute maximum peak; NIOSH 50 ppm (215 mg/m^3) in any 3 hours
Short-term exposure limits (15-minute TWA): ACGIH STEL 100 ppm (426 mg/m^3); NIOSH STEL 100 ppm (425 mg/m^3).
Toxicity by ingestion: Grade 2; LD$_{50}$ = 0.5 to 5 g/kg.
Long-term health effects: IARC possible carcinogen; rating 2B; human evidence: Inadequate. Possible central nervous system damage: muscular and nervous system.
Vapor (gas) irritant characteristics: Vapors cause irritation such that personnel will find high concentrations unpleasant. The effect is temporary.
Liquid or solid irritant characteristics: Eye and skin irritant. Causes smarting of the skin and first-degree burns on short exposure; may cause secondary burns on long exposure.
Odor threshold: 0.02–2.0 ppm.
IDLH value: 700 ppm; At 10,000 ppm could be fatal in 30–60 minutes.

FIRE DATA
Flash point: 93°F/34°C (oc); 88°F/31°C (cc).
Flammable limits in air: LEL: 0.9%; UEL 6.8%.
Fire extinguishing agents not to be used: Water may be ineffective.
Behavior in fire: Polymerization may take place, leading to container explosion.
Autoignition temperature: 914°F/490°C/763°K.
Electrical hazard: Class I, Group D. Flow or agitation of substance may generate electrostatic charges due to low conductivity.
Burning rate: 5.2 mm/min

CHEMICAL REACTIVITY
Binary reactants: Reacts with oxidizers, catalysts for vinyl polymers, peroxides, strong acids, aluminum chloride, and copper and its alloys. May be polymerized at explosive rates by certain contaminants. See below.
Polymerization: May occur if heated above 125°F/52°C. Can cause rupture of container. Heat, light, metal salts, peroxides, and strong acids may also cause, or add to, polymerization; **Inhibitor of polymerization:** *tert*-butylcatechol, 10–15 ppm. *Note:* Even the inhibited product, when heated above 125°F/52°C, can polymerize with the generation of so much heat that ignition is possible.
Reactivity group: 30
Compatibility class: Olefins

ENVIRONMENTAL DATA
Food chain concentration potential: Negative; unlikely to accumulate.
Water pollution: DOT Appendix B, §172.101–marine pollutant. Harmful to aquatic life in very low concentrations. Fouling to shoreline. May be dangerous if it enters nearby water intakes; notify operators. Notify local health and wildlife officials. **Response to discharge:** Issue warning-air contaminant. Mechanical containment. Should be removed. Chemical and physical treatment.

SHIPPING INFORMATION
Grades of purity: 99.5+%; **Storage temperature:** Ambient; **Inert atmosphere:** None; **Venting:** Open (flame arrester); **Stability during transport:** Stable.

NAS HAZARD CLASSIFICATION FOR BULK WATER TRANSPORTATION
FIRE: 3
HEALTH: Vapor irritant: 2; Liquid or solid irritant: 2; Poisons: 2
WATER POLLUTION: Human toxicity: 1; Aquatic toxicity: 3; Aesthetic effect: 2
REACTIVITY: Other chemicals: 2; Water: 0; Self-reaction: 3

PHYSICAL AND CHEMICAL PROPERTIES
Physical state @ 59°F/15°C and 1 atm: Liquid.
Molecular weight: 104.15
Boiling point @ 1 atm: 293.4°F/145.2°C/418.4°K.
Melting/Freezing point: –23.1°F/–30.6°C/242.6°K.
Critical temperature: 703°F/373°C/646°K.
Critical pressure: 580 psia = 39.46 atm = 4.00 MN/m^2.
Specific gravity (water = 1): 0.906 @ 68°F/20°C (liquid).
Liquid surface tension: 32.14 dynes/cm = 0.03214 N/m @ 66°F/19°C/292°K.
Liquid water interfacial tension: 35.48 dynes/cm = 0.03548 N/m at 19°C.
Relative vapor density (air = 1): 3.59
Ratio of specific heats of vapor (gas): 1.074
Latent heat of vaporization: 156 Btu/lb = 86.8 cal/g = 3.63 x 10^5 J/kg.

Heat of polymerization: –277 Btu/lb = –154 cal/g = –6.45 x 10^5 J/kg.
Vapor pressure: (Reid) 0.27 psia; 5 mm.

SUCROSE   REC. S:6400

SYNONYMS: AZUCAR (Spanish); BEET SUGAR; CANE SUGAR; CONFECTIONER'S SUGAR; GRANULATED SUGAR; ROCK CANDY; SACCHAROSE; SACCHARUM; SACAROSA (Spanish); SUGAR; TABLE SUGAR

IDENTIFICATION
CAS Number: 57-50-1
Formula: $C_{12}H_{22}O11$

DESCRIPTION: Hard, white crystalline solid. Odorless; caramel odor when heated. Sinks in water; dissolves.

Fine airborne dust may explode.

Hazard Classification (based on NFPA-704 Rating System)
Health Hazards (Blue): 0; Flammability (Red): 1; Reactivity (Yellow): 0

EMERGENCY RESPONSE: See Appendix A (171)
Evacuation:
Public safety: Isolate the area of spill or leak for at least 10 to 25 meters (30 to 80 feet) in all directions.
Spill: Increase, in the downwind direction, as necessary, the distance shown under "Public Safety."
Fire: If any large container is involved in fire, isolate for at least 800 meters (½ mile) in all directions; also, consider initial evacuation for 800 meters (½ mile) in all directions.

HEALTH HAZARDS
Personal protective equipment (PPE): B-Level PPE. Dust mask and goggles or face shield.
Exposure limits (TWA unless otherwise noted): ACGIH TLV 10 mg/m^3; OSHA Table Z-1-A air contaminant. OSHA PEL 15 mg/m^3 (total), 5 mg/m^3 (respirable fraction); NIOSH 10 mg/m^3 (total), 5 mg/m^3 (respirable fraction).
Toxicity by ingestion: Grade 0; oral LD_{50} (100 days) = 28,500 mg/kg/day (rat).

FIRE DATA
Behavior in fire: Melts and chars.

CHEMICAL REACTIVITY
Binary reactants: Oxidizers, sulfuric acid; nitric acid

ENVIRONMENTAL DATA
Food chain concentration potential: Negative; unlikely to accumulate.
Water pollution: Effect of low concentrations on aquatic life is unknown. May be dangerous if it enters nearby water intakes; notify operators. Notify local health and wildlife officials.
Response to discharge: Disperse and flush.

SHIPPING INFORMATION
Grades of purity: Food grade; Technical; Storage temperature: Ambient; Inert atmosphere: None; Venting: Open; Stability during transport: Stable.

PHYSICAL AND CHEMICAL PROPERTIES
Physical state @ 59°F/15°C and 1 atm: Solid.
Molecular weight: 342.3
Boiling point @ 1 atm: Decomposes.
Melting/Freezing point: Decomposes. 320-367°F/160-186°C/433-459°K.
Specific gravity (water = 1): 1.59 @ 68°F/20°C (solid).
Heat of combustion: 6400 Btu/lb = –3600 cal/g = –150 x 10^5 J/kg.
Vapor pressure: 0 mm (approximate).

SULFOLANE   REC. S:6500

SYNONYMS: SULFOLANE-W; SULFOLANO (Spanish); TETRAHYDROTHIOPHENE-1,1-DIOXIDE; TETRAMETHYLENE SULFONE

IDENTIFICATION
CAS Number: 126-33-0
Formula: $C_4H_8O_2S$; $CH_2CH_2CH_2CH_2SO_2$

DESCRIPTION: White solid or colorless liquid. Solid is odorless; liquid has a weak oily odor. Solidifies and sinks and mixes with water. Freezing point is 79°F/26°C.

Combustible solid • Heat or flame may cause dust to explode • Containers may BLEVE when exposed to fire • Concentrated dust in confined areas (e.g., tanks, sewers, buildings) may explode when exposed to fire • Irritating to the skin, eyes, and respiratory tract • Toxic products of combustion may include sulfur oxides.

Hazard Classification (based on NFPA-704 Rating System)
Health Hazards (Blue): 2; Flammability (Red): 1; Reactivity (Yellow): 0

EMERGENCY RESPONSE: See Appendix A (171)
Evacuation:
Public safety: Isolate the area of spill or leak for at least 10 to 25 meters (30 to 80 feet) in all directions.
Spill: Increase, in the downwind direction, as necessary, the distance shown under "Public Safety."
Fire: If any large container is involved in fire, isolate for at least 800 meters (½ mile) in all directions; also, consider initial evacuation for 800 meters (½ mile) in all directions.

EXPOSURE
Short-term effects: *SEEK MEDICAL ATTENTION.* **Liquid:** Not irritating to skin. Irritating to eyes. Harmful if swallowed. *IF IN EYES,* hold eyelids open and flush with plenty of water. *IF SWALLOWED* and victim is *CONSCIOUS AND ABLE TO SWALLOW,* have victim drink 4 to 8 ounces of water, and have victim induce vomiting. *IF SWALLOWED* and victim is *UNCONSCIOUS OR HAVING CONVULSIONS,* do nothing except keep victim warm.

HEALTH HAZARDS
Personal protective equipment (PPE): B-Level PPE. Goggles or face shield; rubber gloves.
Skin or eye contact: Flush with water.
Toxicity by ingestion: Grade 2; LD_{50} = 0.5 to 5 g/kg (rat, mouse).
Liquid or solid irritant characteristics: Eye irritant. Practically harmless to the skin.

FIRE DATA
Flash point: 350°F/177°C (oc); 330°F/166°C (cc).

1020 Sulfur

CHEMICAL REACTIVITY
Binary reactants: Oxidizers, sulfuric acid
Reactivity group: 39
Compatibility class: Sulfolane

ENVIRONMENTAL DATA
Food chain concentration potential: Log P_{ow} = –0.8. Unlikely to accumulate.
Water pollution: Effect of low concentrations on aquatic life is unknown. May be dangerous if it enters nearby water intakes; notify operators. Notify local health and wildlife officials.
Response to discharge: Disperse and flush.

SHIPPING INFORMATION
Grades of purity: Anhydrous: 99+%; standard water blend: 97% plus 3% water; **Storage temperature:** Ambient; **Inert atmosphere:** None; **Venting:** Open (flame arrester); **Stability during transport:** Stable.

NAS HAZARD CLASSIFICATION FOR BULK WATER TRANSPORTATION
FIRE: 1
HEALTH: Vapor irritant: 0; Liquid or solid irritant: 0; Poisons: 1
WATER POLLUTION: Human toxicity: 2; Aquatic toxicity: 3; Aesthetic effect: 2
REACTIVITY: Other chemicals: 1; Water: 0; Self-reaction: 0

PHYSICAL AND CHEMICAL PROPERTIES
Physical state @ 59°F/15°C and 1 atm: Solid.
Molecular weight: 120.17
Boiling point @ 1 atm: 545°F/285°C/558°K.
Melting/Freezing point: 79°F/26°C/299°K.
Specific gravity (water = 1): 1.26 at 30°C (liquid).
Heat of combustion: (estimate) –9500 Btu/lb = –5300 cal/g = –220 x 10^5 J/kg.
Heat of solution: (estimate) –22 Btu/lb = –12 cal/g = –0.5 x 10^5 J/kg.

SULFUR   REC. S:6600

SYNONYMS: AZUFRE (Spanish); BENSULFOID; BRIMSTONE; COLLOIDAL-S; COLLOIDAL SULFUR; COLLOKIT; COLSUL; COROSUL D; COROSUL S; COSAN; COSAN 80; CRYSTEX; FLOUR SULPHUR; FLOWERS OF SULPHUR; GROUND VOCLE SULPHUR; HEXASUL; KOLOFOG; KOLOSPRAY; KUMULUS; MAGNETIC 70; MAGNETIC 90; MAGNETIC 95; MICROFLOTOX; PRECIPITATED SULFUR; SOFRIL; SPERLOX-S; SPERSUL; SPERSUL THIOVIT; SUBLIMED SULFUR; SULFIDAL; SULFORON; SULFUR, MOLTEN; SULKOL; SUPER COSAN; SULPHUR; SULSOL; TECHNETIUM TC 99M SULFUR COLLOID; TESULOID; THIOLUX; THIOVIT

IDENTIFICATION
CAS Number: 7704-34-9
Formula: S
DOT ID Number: UN 1350 (solid, powder); UN 2448 (molten);
DOT Guide Number:133
Proper Shipping Name: Sulfur; Sulfur, molten; Sulphur; Sulphur, molten

DESCRIPTION: Yellow, orange, tan, brown, or gray (depending upon amount and type of hydrocarbon impurity) powder or molten liquid. Pure sulfur is odorless, but traces of hydrocarbon impurity may impart a rotten egg odor. Thickens and sinks in water; insoluble. may be shipped in molten form at temperatures between 280–300°F/138–149°C.

Tanks of molten sulfur may contain toxic and flammable hydrogen sulfide gas under pressure • Combustible solid; flames of sulfur may be difficult to see in daylight • Fires may be difficult to extinguish • Hot molten sulfur will form a solid crust as it cools, liquid below will remain hot for an extended period of time • Containers may BLEVE when exposed to fire • Vapors are heavier than air and will collect and stay in low areas • Vapors in confined areas (e.g., tanks, sewers, buildings) may explode when exposed to fire • Irritating to the skin, eyes, and respiratory tract • Toxic products of combustion may include large amounts of sulfur dioxide gas.

Hazard Classification (based on NFPA-704 Rating System)
NON-FIRE SITUATIONS: Health Hazards (Blue): 1; Flammability (Red): 1; Reactivity (Yellow): 0
FIRE SITUATIONS: Health Hazards (Blue): 2; Flammability (Red): 1; Reactivity (Yellow): 0

EMERGENCY RESPONSE: See Appendix A (133)
Evacuation:
Public safety: Isolate spill area for at least 10 to 25 meters (30 to 80 feet) in all directions.
Spill: Consider initial downwind evacuation of at least 100 meters (330 feet).
Fire: Isolate for 800 meters (½ mile) in all directions, especially if tank, rail car, or tank truck is involved in fire.

EXPOSURE
Short-term effects: *SEEK MEDICAL ATTENTION.* **Liquid:** Will burn skin and eyes. Harmful if swallowed.
Flush affected areas with plenty of water. *IF IN EYES*, hold eyelids open and flush with plenty of water. *IF SWALLOWED* and victim is *CONSCIOUS AND ABLE TO SWALLOW*, have victim drink 4 to 8 ounces of water.

HEALTH HAZARDS
Personal protective equipment (PPE): B-Level PPE. Safety goggles with side shields; approved respirator; heat-resistant gloves; leather heat-resistant clothing.
Exposure limits (TWA unless otherwise noted): If recovered sulfur, use hydrogen sulfide*: ACGIH TLV 10 ppm (14 mg/m^3); NIOSH ceiling 10 ppm (15 mg/m^3) 10-minutes; OSHA ceiling 20 ppm; 50 ppm 10-minute maximum peak as hydrogen sulfide.
Short-term exposure limits (15-minute TWA): If recovered sulfur, use hydrogen sulfide*: ACGIH STEL 15 ppm (21 mg/m^3) as hydrogen sulfide.
Toxicity by ingestion: Grade 2; LD_{50} = 0.5 to 5 g/kg.
Vapor (gas) irritant characteristics: Nonvolatile
Liquid or solid irritant characteristics: Eye and skin irritant. If spilled on clothing and allowed to remain, may cause smarting and reddening of the skin.
Odor threshold: If recovered sulfur, see hydrogen sulfide.*
Note: Significant amounts of hydrogen sulfide, a very poisonous gas, may collect in poorly ventilated containers of liquid sulfur that has been recovered from hydrogen sulfide.

FIRE DATA
Flash point: 405°F/207°C (cc).
Behavior in fire: Burns with a pale blue flame that may be difficult to see in daylight.

Autoignition temperature: 450°F/232°C/505°K.
Electrical hazard: Flow or agitation of substance may generate electrostatic charges due to low conductivity.

CHEMICAL REACTIVITY
Binary reactants: Reacts with strong oxidizers, air, halogens, ammonia. Attacks zinc, iron, nickel, copper, and steel (in the presence of moisture).
Reactivity group: See Compatibility Guide
Compatibility class: Special class

ENVIRONMENTAL DATA
Food chain concentration potential: Negative; unlikely to accumulate.
Water pollution: Dangerous to aquatic life in high concentrations. May be dangerous if it enters nearby water intakes; notify operators. Notify local health and wildlife officials. **Response to discharge:** Should be removed. Chemical and physical treatment.

SHIPPING INFORMATION
Grades of purity: Frasch liquid sulfur: 99.8+%; solid sulfur is sold in many varieties and grades; these are not presently covered in this manual; **Storage temperature:** 270°F/132°C; **Inert atmosphere:** Ventilated (natural); **Venting:** Open; **Stability during transport:** Stable.

NAS HAZARD CLASSIFICATION FOR BULK WATER TRANSPORTATION
FIRE: 1-3
HEALTH: Vapor irritant: 1; Liquid or solid irritant: 1; Poisons: 1
WATER POLLUTION: Human toxicity: 0; Aquatic toxicity: 1; Aesthetic effect: 1
REACTIVITY: Other chemicals: 4; Water: 0; Self-reaction: 0

PHYSICAL AND CHEMICAL PROPERTIES
Physical state @ 59°F/15°C and 1 atm: Solid.
Molecular weight: 256.51
Boiling point @ 1 atm: 832°F/445°C/718°K.
Melting/Freezing point: 251°F/122°C/395°K.
Specific gravity (water = 1): 1.80 at 120°C (liquid).
Liquid surface tension: 60.8 dynes/cm = 0.0608 N/m at 120°C.
Liquid water interfacial tension: (estimate) 50 dynes/cm = 0.05 N/m at 127°C.
Ratio of specific heats of vapor (gas): 1.582 (estimate).
Latent heat of vaporization: 120 Btu/lb = 69 cal/g = 2.9×10^5 J/kg.
Heat of combustion: -4.741 Btu/lb = -2634 cal/g = -110.3×10^5 J/kg.
Heat of fusion: 9.2 cal/g.
Vapor pressure: (Reid) Very low.

SULFUR DIOXIDE   REC. S:6700

SYNONYMS: BISULFITE; DIOXIDO de AZUFRE (Spanish); EEC No. 016-011-00-9; FERMENICIDE; SCHWEFELDDIOXYD (German); SULFUROUS ACID ANHYDRIDE; SULFUROUS ANHYDRIDE; SULFUROUS OXIDE; SULFUR OXIDE; SULFUR DIOXIDE, Liquefied

IDENTIFICATION
CAS Number: 7446-09-5
Formula: SO_2
DOT ID Number: UN 1079; DOT Guide Number: 125
Proper Shipping Name: Sulfur dioxide; Sulfur dioxide, liquefied
Reportable Quantity (RQ): **(EHS)** 1 lb/0.454 kg

DESCRIPTION: Colorless gas. Shipped and stored as a compressed liquefied gas. Sharp, irritating odor, like burning sulfur. Liquid sinks in water, forming sulfurous acid. Becomes a liquid below 14°F/–10°C.

Poison! • Breathing the gas can kill you; corrosive to skin eyes, and respiratory tract; skin or eye contact causes severe burns, impaired vision, or blindness; symptoms of inhalation may be delayed • Firefighting gear (including SCBA) does not provide adequate protection. If exposure occurs, remove and isolate gear immediately and thoroughly decontaminate personnel • Containers may BLEVE when exposed to fire •Contact with liquid may cause severe injury or frostbite • Do not put yourself in danger by entering a contaminated area to rescue a victim. *Warning*: Inhalation exposures of 500 ppm (10 minutes) can be fatal.

Hazard Classification (based on NFPA-704 Rating System)
Health Hazards (Blue): 3; Flammability (Red): 0; Reactivity (Yellow): 0

EMERGENCY RESPONSE: See Appendix A (125)
Evacuation:
Public safety: Isolate spill area for at least 100 to 200 meters (330 to 660 feet) in all directions.
Spill: Small spill–First: Isolate in all directions 30 meters (100 feet); Then: Protect persons downwind, DAY: 0.3 km (0.2 mile); NIGHT: 1.1 km (0.7 mile). Large spill–First: Isolate in all directions 185 meters (600 feet); Then: Protect persons downwind, DAY: 3.1 km (1.9 miles); NIGHT: 7.2 km (4.5 miles).
Fire: Isolate for 1600 meters (1 mile) in all directions, especially if tank, rail car, or tank truck is involved in fire.

EXPOSURE
Short-term effects: *SEEK MEDICAL ATTENTION.* **Vapor:** *POISONOUS IF INHALED.* Move to fresh air. *IF BREATHING HAS STOPPED,* give artificial respiration; *avoid mouth-to-mouth resuscitation; use bag/mask apparatus.* IF breathing is difficult, administer oxygen. Lung edema may develop. **Liquid:** Will cause frostbite. Flush affected areas with plenty of water. *IF IN EYES*, hold eyelids open and flush with plenty of water. *DO NOT RUB AFFECTED AREAS.*

Note to physician or authorized medical personnel: Medical observation is recommended for 24 to 48 hours after breathing overexposure, as pulmonary edema may be delayed. As first aid for pulmonary edema, consider administering a corticosteroid spray. Cigarette smoking may exacerbate pulmonary injury and should be discouraged for at least 72 hours following exposure.

HEALTH HAZARDS
Personal protective equipment (PPE): A-Level PPE. OSHA Table Z-1-A air contaminant.
Recommendations for respirator selection: NIOSH
20 ppm: CCRS* [any chemical cartridge respirator with cartridge(s) providing protection against the compound of concern]; or SA* (any supplied-air respirator). *50: ppm:* SA:CF* (any supplied-air respirator operated in a continuous-flow mode); or *PAPRS** [any powered, air-purifying respirator with cartridge(s) providing protection against the compound of concern]. *100 ppm:* CCRFS [any chemical cartridge respirator with a full facepiece and cartridge(s) providing protection against the compound of concern]; or GMFS [any air-purifying, full-facepiece respirator (gas mask) with a chin-style, front- or back-mounted canister providing

1022   Sulfuric acid

protection against the compound of concern]; or PAPRTS (any powered, air-purifying respirator with a tight-fitting facepiece and cartridge(s) providing protection against the compound of concern]; or SAT:CF (any supplied-air respirator that has a tight-fitting facepiece and is operated in a continuous-flow mode); or SCBAF (any self-contained breathing apparatus with a full facepiece); or SAF (any supplied-air respirator with a full facepiece). *EMERGENCY OR PLANNED ENTRY INTO UNKNOWN CONCENTRATIONS OR IDLH CONDITIONS:* SCBAF:PD,PP (any self-contained breathing apparatus that has a full facepiece and is operated in a pressure-demand or other positive-pressure mode); or SAF:PD,PP:ASCBA (any supplied-air respirator that has a full facepiece and is operated in a pressure-demand or other positive-pressure mode in combination with an auxiliary self-contained breathing apparatus operated in a pressure-demand or other positive pressure mode). *ESCAPE:* GMFS [any air-purifying, full-facepiece respirator (gas mask) with a chin-style, front- or back-mounted canister providing protection against the compound of concern]; or SCBAE (any appropriate escape-type, self-contained breathing apparatus). *Note:* Substance reported to cause eye irritation or damage; may require eye protection.
Exposure limits (TWA unless otherwise noted): ACGIH TLV 2 ppm (5.2 mg/m^3); OSHA PEL 5 ppm (13 mg/m^3); NIOSH REL 2 ppm (5 mg/m^3).
Short-term exposure limits (15-minute TWA): ACGIH & NIOSH STEL 5 ppm (13 mg/m^3).
Long-term health effects: Mutagen. Lung damage.
Vapor (gas) irritant characteristics: Vapors cause severe irritation of eyes and throat and can cause eye and lung injury. They cannot be tolerated even at low concentrations.
Liquid or solid irritant characteristics: Liquid can cause frostbite.
Odor threshold: 2.7 ppm.
IDLH value: 100 ppm.

FIRE DATA
Behavior in fire: Containers may rupture and release toxic and irritating sulfur dioxide

CHEMICAL REACTIVITY
Reactivity with water: Reacts with water to form corrosive sulfurous acid.
Binary reactants: Reacts with oxidizers, alkaline materials, powdered alkali metals such as sodium and potassium, ammonia, zinc, brass, and copper. Reacts violently with acrolein, alcohols, aluminum powder, alkali metals, amines, bromine pentafluoride, caustics, cesium acetylene carbide, chlorates, chlorine trifluoride, chromium powder, copper or copper alloy powders, diethylzinc, fluorine, lead dioxide, lithium acetylene carbide, diamino-, metal powders, monolithium acetylide-ammonia, nitryl chloride, potassium acetylene carbide, potassium acetylide, potassium chlorate, rubidium carbide, silver azide, sodium, sodium acetylide, stannous oxide. Incompatible with alkalis, alkylene oxides, ammonia, aliphatic amines, alkanolamines, amides, organic anhydrides, cesium monoxide, epichlorohydrin, ferrous oxide, halogens, interhalogens, isocyanates, lithium nitrate, manganese, metal acetylides, metal oxides, perbromyl fluoride, red phosphorus, potassium azide, rubidium acetylyde, sodium hydride, sulfuric acid. Attacks some plastics, coatings, and rubber. Attacks metals, especially in the presence of moisture.
Neutralizing agents for acids and caustics: Mild acidity of water solution may be neutralized by dilute caustic soda

ENVIRONMENTAL DATA
Food chain concentration potential: May not accumulate.

Water pollution: Harmful to aquatic life in very low concentrations. May be dangerous if it enters nearby water intakes; notify operators. Notify local health and wildlife officials. **Response to discharge:** Issue warning-air contaminant. Restrict access. Evacuate area.

SHIPPING INFORMATION
Grades of purity: Refrigeration grade 99.98%; commercial grade 99.90%; **Storage temperature:** Less than 130°F/54°C; **Inert atmosphere:** None; **Venting:** Safety relief; **Stability during transport:** Stable.

NAS HAZARD CLASSIFICATION FOR BULK WATER TRANSPORTATION
FIRE: 0
HEALTH: Vapor irritant: 4; Liquid or solid irritant: 1; Poisons: 4
WATER POLLUTION: Human toxicity: 0; Aquatic toxicity: 3; Aesthetic effect: 1
REACTIVITY: Other chemicals: 1; Water: 1; Self-reaction: 0

PHYSICAL AND CHEMICAL PROPERTIES
Physical state @ 59°F/15°C and 1 atm: Gas.
Molecular weight: 64.06
Boiling point @ 1 atm: 14°F/–10°C/263.2°K.
Melting/Freezing point: –103.9°F/–75.5°C/197.7°K.
Critical temperature: 315°F/157°C/430°K.
Critical pressure: 1142 psia = 77.69 atm = 7.870 MN/m^2.
Specific gravity (water = 1): 1.45 at –10°C (liquid).
Relative vapor density (air = 1): 2.2
Ratio of specific heats of vapor (gas): 1.265
Latent heat of vaporization: 171 Btu/lb = 94.8 cal/g = 3.97 x 10^5 J/kg.
Heat of solution: –94.1 Btu/lb = –52.3 cal/g = –2.19 x 10^5 J/kg.
Heat of fusion: 32.2 cal/g.
Vapor pressure: (Reid) 84 psia; 2500 mm.

SULFURIC ACID   REC. S:6800

SYNONYMS: ACIDO SLFURICO (Spanish); BATTERY ACID; BOV; CHAMBER ACID; DIPPING ACID; EEC No. 016-020-00-8; FERTILIZER ACID; MATTING ACID; OIL OF VITRIOL; VITRIOL BROWN ACID

IDENTIFICATION
Formula: H_2SO_4
CAS Number: 7664-93-9
DOT ID Number: UN 1830; DOT Guide Number: 137
Proper Shipping Name: Sulfuric acid
Reportable Quantity (RQ): **(CERCLA)** 1000 lb/454 kg

DESCRIPTION: Colorless (pure) to dark brown liquid. Odorless; choking when hot. Sinks and mixes violently with water; soluble. Irritating mist is produced.

Poison! • Breathing the vapor can kill you; skin or eye contact causes severe burns, impaired vision, or blindness • Firefighting gear (including SCBA) does not provide adequate protection. If exposure occurs, remove and isolate gear immediately and thoroughly decontaminate personnel • Containers may BLEVE when exposed to fire • Strong oxidizer which may react spontaneously with low flash point organics or reducing agents. Heat forms oxygen; will increase the activity of an existing fire • May cause fire or explosion contact with combustibles (wood, paper, oil, clothing, etc.) • Contact with most metals produce flammable and potentially explosive hydrogen gas • Toxic products of combustion may include sulfur oxides • Do not put yourself in danger by entering a contaminated area to rescue a victim.

Hazard Classification (based on NFPA-704 Rating System)
Health Hazards (Blue): 3; Flammability (Red): 0; Reactivity (Yellow): 2; Special Notice (White): Water reactive

EMERGENCY RESPONSE: See Appendix A (137)
Evacuation:
Public safety: Isolate the area of spill or leak for at least 50 to 100 meters (160 to 330 feet) in all directions.
Spill: Increase, in the downwind direction, as necessary, the distance shown under "Public Safety."
Fire: If any large container is involved in fire, isolate for at least 800 meters (½ mile) in all directions; also, consider initial evacuation for 800 meters (½ mile) in all directions.

EXPOSURE
Short-term effects: *SEEK MEDICAL ATTENTION.* **Mist:** Irritating to eyes, nose, and throat. *IF INHALED*, will, will cause coughing, difficult breathing, or loss of consciousness. Move to fresh air. *IF IN EYES*, hold eyelids open and flush with plenty of water. *IF BREATHING HAS STOPPED,* give artificial respiration; *avoid mouth-to-mouth resuscitation; use bag/mask apparatus.* IF breathing is difficult, administer oxygen. **Liquid:** Will burn skin and eyes. Harmful if swallowed. Remove contaminated clothing and shoes. Flush affected areas with plenty of water. *IF IN EYES,* hold eyelids open and flush with plenty of water. *IF SWALLOWED* and victim is *CONSCIOUS AND ABLE TO SWALLOW*, have victim drink 4 to 8 ounces of water. **Do NOT induce vomiting.**

HEALTH HAZARDS
Personal protective equipment (PPE): B-Level PPE. OSHA Table Z-1-A air contaminant. Safety shower; eyewash fountain; safety goggles; face shield; approved respirator (self-contained or air-line); rubber safety shoes; rubber apron. *For <30%,* butyl rubber, natural rubber, neoprene, polyethylene, PVC, nitrile+PVC, Teflon®, Saranex®, Silvershield®. Also, Viton® offers limited protection. *For 30-70%,* butyl rubber, natural rubber, neoprene, polyethylene, PVC, Teflon®, Saranex®, Silvershield®. Also, nitrile offers limited protection. Also, Viton® offers limited protection. *For >70%,* butyl rubber, chlorinated polyethylene, polyethylene, neoprene/natural rubber, Teflon®, Saranex®, Silvershield®. Also, chlorinated polyethylene, neoprene, PVC, and Viton® offers limited protection
Fuming, see oleum.
Recommendations for respirator selection: NIOSH/OSHA 15 mg/m^3: SA:CF* (any supplied-air respirator operated in a continuous-flow mode); or PAPRAGHiE* (any powered, air-purifying respirator with acid gas cartridge(s) in combination with a high-efficiency particulate filter). 50 mg/m^3: CCRFAGHiE [any chemical cartridge respirator with a full facepiece and acid gas cartridge(s) in combination with a high-efficiency particulate filter]; or GMFAGHiE [any air-purifying, full-facepiece respirator (gas mask) with a chin-style, front- or back-mounted acid gas canister having a high-efficiency particulate filter]; or SCBAF (any self-contained breathing apparatus with a full facepiece); or SAF (any supplied-air respirator with a full facepiece). *EMERGENCY OR PLANNED ENTRY INTO UNKNOWN CONCENTRATIONS OR IDLH CONDITIONS:* SCBAF:PD,PP (any self-contained breathing apparatus that has a full facepiece and is operated in a pressure-demand or other positive-pressure mode); or SAF:PD,PP:ASCBA (any supplied-air respirator that has a full facepiece and is operated in a pressure-demand or other positive-pressure mode in combination with an auxiliary self-contained breathing apparatus operated in a pressure-demand or other positive pressure mode). *ESCAPE:* GMFAGHiE [any air-purifying, full-facepiece respirator (gas mask) with a chin-style, front- or back-mounted acid gas canister having a high-efficiency particulate filter]; or SCBAE (any appropriate escape-type, self-contained breathing apparatus). **Note:* Substance causes eye irritation or damage; eye protection needed.
Exposure limits (TWA unless otherwise noted): ACGIH TLV 1 mg/m^3; NIOSH/OSHA 1 mg/m^3.
Short-term exposure limits (15-minute TWA): ACGIH STEL 3 mg/m^3.
Toxicity by ingestion: No effects except those secondary to tissue damage.
Vapor (gas) irritant characteristics: Vapors from hot acid (77–98%) cause irritation of eyes and respiratory system. Effect is temporary.
Liquid or solid irritant characteristics: 77–98% acid causes severe second- and third-degree burns of skin on short contact and is very injurious to the eyes.
Odor threshold: 1 mg/m^3; 0.15 ppm.
IDLH value: 15 mg/m^3.

FIRE DATA
Fire extinguishing agents not to be used: Water used on adjacent fires should be carefully handled.

CHEMICAL REACTIVITY
Reactivity with water: Reacts violently with evolution of heat. Dangerous spattering occurs when water is added to the compound.
Binary reactants: Extremely hazardous in contact with many materials, particularly metals and combustibles. Dilute acid reacts with most metals, releasing hydrogen which can form explosive mixtures with air in confined spaces.
Neutralizing agents for acids and caustics: Dilute with water, then neutralize with lime, limestone, or soda ash.
Reactivity group: 2
Compatibility class: Sulfuric acid

ENVIRONMENTAL DATA
Food chain concentration potential: Negative; unlikely to accumulate.
Water pollution: Harmful to aquatic life in very low concentrations. May be dangerous if it enters nearby water intakes; notify operators. Notify local health and wildlife officials. **Response to discharge:** Issue warning–corrosive. Restrict access. Disperse and flush with care.

SHIPPING INFORMATION
Grades of purity: CP; USP; Technical, at 33–98% (50° Be to 66° Be); **Storage temperature:** Ambient; **Inert atmosphere:** None; **Venting:** Open; **Stability during transport:** Stable.

NAS HAZARD CLASSIFICATIONS FOR BULK WATER TRANSPORTATION
FIRE: 0
HEALTH: Vapor irritant: 2; Liquid or solid irritant: 4; Poisons: 2
WATER POLLUTION: Human toxicity: 2; Aquatic toxicity: 3; Aesthetic effect: 2
REACTIVITY: Other chemicals: 4; Water: 3; Self-reaction: 0

PHYSICAL AND CHEMICAL PROPERTIES (apply to concentrated (98%) acid unless otherwise stated. More dilute acid is more water-like).

Physical state @ 59°F/15°C and 1 atm: Liquid.
Molecular weight: 98.08
Boiling point @ 1 atm: 554°F/290°C/563°K.
Melting/Freezing point: 52°F/11°C/284°K.
Specific gravity (water = 1): 1.84 @ 68°F/20°C (liquid).
Relative vapor density (air = 1): 3.4
Heat of solution: –418.0 Btu/lb = –232.2 cal/g = –9.715 x 10^5 J/kg.
Vapor pressure: (Reid) Low; 0.001 mm.

SULFURIC ACID, SPENT   REC. S:6900

SYNONYMS: ACIDO SLFURICO (Spanish); DILUTE SULFURIC ACID; EEC No. 016-020-00-8

IDENTIFICATION
CAS Number: 7664-93-9
Formula: $H_2SO_4 \cdot H_2O$
DOT ID Number: UN 1832; DOT Guide Number: 137
Proper Shipping Name: Sulfuric acid, spent
Reportable Quantity (RQ): **(CERCLA)** 1000 lb/454 kg

DESCRIPTION: Colorless to dark brown oily liquid. Odorless. Sinks and mixes with water.

Hazard Classification (based on NFPA-704 Rating System)
Health Hazards (Blue): 3; Flammability (Red): 0; Reactivity (Yellow): 2; Special Notice (White): Water reactive

See S:6900. Isolate and remove discharged material. If possible, cover spill with sodium bicarbonate or soda ash-slaked lime mixture (50-50). Mix and add water to form a slurry. Scoop up slurry. wash site with soda ash solution. Otherwise flush cautiously with water. avoid directing stream into larger pools or pockets of concentrated acid.
NOTE of CAUTION: Never add water to the acid, otherwise spattering will occur. If dilution is required, always add the acid very carefully to the water. The acid is heavier than water. Thus, the heat of solution will be more uniformly dissipated, and spattering will be avoided.

EMERGENCY RESPONSE: See Appendix A (137)
Evacuation:
Public safety: Isolate the area of spill or leak for at least 50 to 100 meters (160 to 330 feet) in all directions.
Spill: Increase, in the downwind direction, as necessary, the distance shown under "Public Safety."
Fire: If any large container is involved in fire, isolate for at least 800 meters (½ mile) in all directions; also, consider initial evacuation for 800 meters (½ mile) in all directions.

EXPOSURE
Short-term effects: *SEEK MEDICAL ATTENTION. IF INHALED, will,* Move to fresh air. *IF BREATHING HAS STOPPED,* give artificial respiration; *avoid mouth-to-mouth resuscitation; use bag/mask apparatus.* IF breathing is difficult, give oxygen. **Liquid:** Will burn skin and eyes. Harmful if swallowed. Remove contaminated clothing and shoes. Flush affected areas with plenty of water. *IF IN EYES,* hold eyelids open and flush with plenty of water; do NOT use oils or ointments in eyes; treat burns. *IF SWALLOWED* and victim is *CONSCIOUS AND ABLE TO SWALLOW,* have victim drink 4 to 8 ounces of water. **Do NOT induce vomiting.**

HEALTH HAZARDS
Personal protective equipment (PPE): B-Level PPE. OSHA Table Z-1-A air contaminant. Chemical safety goggles and face shield; rubber gloves, boots, and apron. For <30%, butyl rubber, natural rubber, neoprene, polyethylene, PVC, nitrile+PVC, Teflon®, Saranex®, Silvershield®. Also, Viton® offers limited protection. For 30-70%, butyl rubber, natural rubber, neoprene, polyethylene, PVC, Teflon®, Saranex®, Silvershield®. Also, nitrile offers limited protection. Also, Viton® offers limited protection.
For >70%, butyl rubber, chlorinated polyethylene, polyethylene, neoprene/natural rubber, Teflon®, Saranex®, Silvershield®. Also, chlorinated polyethylene, neoprene, PVC, and Viton® offers limited protection *Fuming,* see oleum
Recommendations for respirator selection: NIOSH/OSHA as sulfuric acid
25 mg/m^3: SA:CF (any supplied-air respirator operated in a continuous-flow mode); or PAPRAGHiE (any powered, air-purifying respirator with acid gas cartridge(s) in combination with a high-efficiency particulate filter). 50 mg/m^3: CCRFAGHiE [any chemical cartridge respirator with a full facepiece and acid gas cartridge(s) in combination with a high-efficiency particulate filter]; or GMFAGHiE [any air-purifying, full-facepiece respirator (gas mask) with a chin-style, front- or back-mounted acid gas canister having a high-efficiency particulate filter]; or SCBAF (any self-contained breathing apparatus with a full facepiece); or SAF (any supplied-air respirator with a full facepiece). 80 mg/m^3: SAF:PD,PP (any supplied-air respirator that has a full facepiece and is operated in a pressure-demand or other positive-pressure mode). *EMERGENCY OR PLANNED ENTRY INTO UNKNOWN CONCENTRATIONS OR IDLH CONDITIONS:* SCBAF:PD,PP (any self-contained breathing apparatus that has a full facepiece and is operated in a pressure-demand or other positive-pressure mode); or SAF:PD,PP:ASCBA (any supplied-air respirator that has a full facepiece and is operated in a pressure-demand or other positive-pressure mode in combination with an auxiliary self-contained breathing apparatus operated in a pressure-demand or other positive pressure mode). *ESCAPE:* GMFAGHiE [any air-purifying, full-facepiece respirator (gas mask) with a chin-style, front- or back-mounted acid gas canister having a high-efficiency particulate filter]; or SCBAE (any appropriate escape-type, self-contained breathing apparatus). *Note:* Substance causes eye irritation or damage; eye protection needed.
Exposure limits (TWA unless otherwise noted): ACGIH TLV 1 mg/m^3; NIOSH/OSHA 1 mg/m^3 as sulfuric acid
Short-term exposure limits (15-minute TWA): ACGIH STEL 3 mg/m^3.
Toxicity by ingestion: No effects except those stemming from tissue damage.
Vapor (gas) irritant characteristics: Nonvolatile
Liquid or solid irritant characteristics: Severe skin irritant. Causes second-and third-degree burns on short contact and is very injurious to the eyes.
Odor threshold: 1 mg/m^3; 0.15 ppm as sulfuric acid
IDLH value: 15 mg/m^3.

FIRE DATA
Fire extinguishing agents not to be used: Be careful using water on adjacent fires.

CHEMICAL REACTIVITY
Reactivity with water: Dangerous if strength is above 80–90%, in which case heat is liberated, and spattering may occur.
Binary reactants: Bases, reducing agents, organics. Attacks many metals, releasing flammable hydrogen gas.

Neutralizing agents for acids and caustics: Limestone, lime, or soda ash.
Reactivity group: 2
Compatibility class: Sulfuric acid

ENVIRONMENTAL DATA
Food chain concentration potential: Negative; unlikely to accumulate.
Water pollution: Harmful to aquatic life in very low concentrations. May be dangerous if it enters nearby water intakes; notify operators. Notify local health and wildlife officials.
Response to discharge: Issue warning–corrosive. Restrict access. Disperse and flush.

SHIPPING INFORMATION
Grades of purity: Purity depends on the process in which the original acid is used. The strength (in water) is probably below 80%, and the solution may contain a wide variety of metals and organic compounds in solution; **Storage temperature:** Ambient; **Inert atmosphere:** None; **Venting:** Open; **Stability during transport:** Stable.

PHYSICAL AND CHEMICAL PROPERTIES
Physical state @ 59°F/15°C and 1 atm: Liquid.
Molecular weight: 98.1
Boiling point @ 1 atm: 212°F/100°C/373°K.
Specific gravity (water = 1): 1.39 @ 68°F/20°C (liquid).
Relative vapor density (air = 1): 3.4
Heat of solution: Less than –418 Btu/lb = less than –232 cal/g = less than 9.71×10^5 J/kg.
Vapor pressure: 0.001 mm.

SULFUR MONOCHLORIDE **REC. S:7000**

SYNONYMS: MONOCLORURO de AZUFRE (Spanish); DISULFUR DICHLORIDE; EEC No. 016-012-00-4; SULFUR CHLORIDE; SULFUR SUBCHLORIDE; THIOSULFUROUS DICHLORIDE0

IDENTIFICATION
CAS Number: 10025-67-9
Formula: S_2Cl_2
DOT ID Number: UN 1828; DOT Guide Number: 137
Proper Shipping Name: Sulfur chlorides
Reportable Quantity (RQ): 1000 lb/454 kg

DESCRIPTION: Light amber to yellowish-red oily fuming liquid. Irritating, nauseating, odor. Mixes and reacts with water forming hydrochloric acid; poisonous vapor is produced.

Combustible • Poison! • Breathing the vapor can kill you; skin or eye contact causes severe burns, impaired vision, or blindness • Firefighting gear (including SCBA) does not provide adequate protection. If exposure occurs, remove and isolate gear immediately and thoroughly decontaminate personnel • Strong oxidizer which may react spontaneously with low flash point organics or reducing agents. Heat forms oxygen; will increase the activity of an existing fire • May cause fire or explosion contact with combustibles (wood, paper, oil, clothing, etc.) • Containers may BLEVE when exposed to fire • Toxic products of combustion may include sulfur oxides and hydrogen chloride • Corrosive to metals and some plastics when wet • Do not put yourself in danger by entering a contaminated area to rescue a victim.

Hazard Classification (based on NFPA-704 Rating System)
Health Hazards (Blue): 3; Flammability (Red): 1; Reactivity (Yellow): 1

EMERGENCY RESPONSE: See Appendix A (137)
Evacuation:
Public safety: See below.
Spill: Small spill–First: Isolate in all directions 30 meters (100 feet) (land or water); Then: Protect persons downwind, DAY: 0.2 km (0.1 mile) (land or water); NIGHT: 0.3 km (0.2 mile) (land); 0.2 km (0.1 mile) (water). Large spill–First: Isolate in all directions 60 meters (200 feet) (land or water); Then: Protect persons downwind, DAY: 0.5 km (0.3 mile) (land); 0.6 km (0.4 mile) (water); NIGHT: 1.0 km (0.6 mile) (land); 2.3 km (1.4 miles)(water).
Fire: If any large container is involved in fire, isolate for at least 800 meters (½ mile) in all directions; also, consider initial evacuation for 800 meters (½ mile) in all directions.

EXPOSURE
Short-term effects: *SEEK MEDICAL ATTENTION*. **Vapor:** Irritating to eyes. *POISONOUS IF INHALED*. Move to fresh air. *IF BREATHING HAS STOPPED*, give artificial respiration respiration; *avoid mouth-to-mouth resuscitation; use bag/mask apparatus*. IF breathing is difficult, administer oxygen. Lung edema may develop. **Liquid:** Will burn skin and eyes. *POISONOUS IF SWALLOWED*. Remove contaminated clothing and shoes. Flush affected areas with plenty of water. *IF IN EYES*, hold eyelids open and flush with plenty of water. *IF SWALLOWED* and victim is *CONSCIOUS AND ABLE TO SWALLOW*, have victim drink 4 to 8 ounces of water. **Do NOT induce vomiting.**
Note to physician or authorized medical personnel: Medical observation is recommended for 24 to 48 hours after breathing overexposure, as pulmonary edema may be delayed. As first aid for pulmonary edema, consider administering a corticosteroid spray. Cigarette smoking may exacerbate pulmonary injury and should be discouraged for at least 72 hours following exposure.

HEALTH HAZARDS
Personal protective equipment (PPE): A-Level PPE. OSHA Table Z-1-A air contaminant. Chemical protective material(s) reported to offer minimal to poor protection: Viton®/neoprene.
Recommendations for respirator selection: NIOSH/OSHA
5 ppm: CCRFS [any chemical cartridge respirator with a full facepiece and cartridge(s) providing protection against the compound of concern]; or GMFS [any air-purifying, full-facepiece respirator (gas mask) with a chin-style, front- or back-mounted canister providing protection against the compound of concern]; or *PAPRS** [any powered, air-purifying respirator with cartridge(s) providing protection against the compound of concern]; or SCBAF (any self-contained breathing apparatus with a full facepiece); or SAF (any supplied-air respirator with a full facepiece). *EMERGENCY OR PLANNED ENTRY INTO UNKNOWN CONCENTRATIONS OR IDLH CONDITIONS:* SCBAF:PD,PP (any self-contained breathing apparatus that has a full facepiece and is operated in a pressure-demand or other positive-pressure mode); or SAF:PD,PP:ASCBA (any supplied-air respirator that has a full facepiece and is operated in a pressure-demand or other positive-pressure mode in combination with an auxiliary self-contained breathing apparatus operated in a pressure-demand or other positive pressure mode). *ESCAPE:* GMFS [any air-purifying, full-facepiece respirator (gas mask) with a chin-style, front- or back-mounted canister providing protection against the compound of concern]; or SCBAE (any appropriate escape-type, self-contained breathing apparatus). **Note*: Substance reported to cause eye irritation or damage; may require eye protection.

Exposure limits (TWA unless otherwise noted): ACGIH TLV ceiling limit 1 ppm (5.5 mg/m^3); OSHA PEL 1 ppm (6 mg/m^3); NIOSH ceiling 1 ppm (6 mg/m^3).
Vapor (gas) irritant characteristics: Vapors cause severe irritation of eye and throat and can cause eye and lung injury. They cannot be tolerated even at low concentrations.
Liquid or solid irritant characteristics: Severe eye and skin irritant. Causes second-and third-degree burns on short contact and is very injurious to the eyes.
IDLH value: 5 ppm.

FIRE DATA
Flash point: 266°F/130°C (oc); 245°F/118°C (cc).
Fire extinguishing agents not to be used: Water reacts violently with compound.
Autoignition temperature: 453°F/243°C/516°K.

CHEMICAL REACTIVITY
Reactivity with water: Reacts violently with water to produce heat and hydrochloric acid fumes, sulfur dioxide, sulfur, sulfur thiosulfate and hydrogen sulfide. The solution is a strong acid.
Binary reactants: Violent reaction with strong oxidizers, acetone, aluminum powders, ammonia, amines, dimethyl sulfoxide, lead dioxide, metal powders, nitric acid, perchloryl fluoride, red phosphorus, potassium, sodium, sodium peroxide, toluene. Incompatible with strong bases, oxides of phosphorus, organic substances. Incompatible with sulfuric acid, caustics, ammonia, aliphatic amines, alkanolamines, amides, organic anhydrides, isocyanates, phosphorus trichloride, vinyl acetate, alkylene oxides, epichlorohydrin. Attacks some plastics, rubber, and coatings. Attacks metals in the presence of moisture, forming flammable hydrogen gas.
Neutralizing agents for acids and caustics: After reaction with water, the acid formed can be neutralized with lime or soda ash.

ENVIRONMENTAL DATA
Food chain concentration potential: Negative; unlikely to accumulate.
Water pollution: Dangerous to aquatic life in high concentrations. May be dangerous if it enters nearby water intakes; notify operators. Notify local health and wildlife officials. **Response to discharge:** Issue warning–corrosive, air contaminant, water contaminant. Restrict access. Chemical and physical treatment.

SHIPPING INFORMATION
Grades of purity: Commercial material may contain 0–5% free sulfur; **Storage temperature:** Ambient; **Inert atmosphere:** None; **Venting:** Pressure-vacuum; **Stability during transport:** Stable.

NAS HAZARD CLASSIFICATION FOR BULK WATER TRANSPORTATION
FIRE: 1
HEALTH: Vapor irritant: 4; Liquid or solid irritant: 4; Poisons: 4
WATER POLLUTION: Human toxicity: 3; Aquatic toxicity: 2; Aesthetic effect: 4
REACTIVITY: Other chemicals: 3; Water: 4; Self-reaction: 0

PHYSICAL AND CHEMICAL PROPERTIES
Physical state @ 59°F/15°C and 1 atm: Liquid.
Molecular weight: 135.03
Boiling point @ 1 atm: 280°F/138°C/411°K.
Melting/Freezing point: –112°F/–80°C/193°K.
Specific gravity (water = 1): 1.68 @ 68°F/20°C (liquid).
Relative vapor density (air = 1): 4.7
Ratio of specific heats of vapor (gas): 1.129
Latent heat of vaporization: 115 Btu/lb = 63.8 cal/g = 2.67 x 10^5 J/kg.
Heat of solution: –502.2 Btu/lb = –279.0 cal/g = 11.67 x 10^5 J/kg.
Vapor pressure: 7 mm.

SULFURYL CHLORIDE   REC. S:7100

SYNONYMS: CLORURO de SULFURILO (Spanish); EEC No. 016-016-00-6; SULFONYL CHLORIDE; SULFURIC OXYCHLORIDE

IDENTIFICATION
CAS Number: 7791-25-5
Formula: SO$_2$Cl$_2$
DOT ID Number: UN 1834; DOT Guide Number: 137
Proper Shipping Name: Sulfuryl chloride

DESCRIPTION: Colorless to pale yellow watery liquid. Acrid, choking odor. Sinks in water and reacts violently producing toxic hydrochloric and sulfuric acids. Poisonous hydrogen chloride and sulfur oxide vapors are produced.

Poison! • Corrosive • Breathing the vapor can kill you; skin or eye contact causes severe burns, impaired vision, or blindness; symptoms of inhalation may be delayed • Firefighting gear (including SCBA) does not provide adequate protection. If exposure occurs, remove and isolate gear immediately and thoroughly decontaminate personnel • Do not use water • Decomposes above 320°F/160°C. Toxic products of combustion and decomposition include chlorine gas, sulfur oxides, and hydrogen chloride • Reacts with metals in the presence of moisture to release flammable hydrogen gas • Do not put yourself in danger by entering a contaminated area to rescue a victim.

Hazard Classification (based on NFPA-704 Rating System)
Health Hazards (Blue): 3; Flammability (Red): 0; Reactivity (Yellow): 2; Special Notice (White): Water reactive

EMERGENCY RESPONSE: See Appendix A (137)
Evacuation:
Public safety: See below.
Spill: Small spill–First: Isolate in all directions 30 meters (100 feet) (land or water); Then: Protect persons downwind, DAY: 0.2 km (0.1 mile) (land or water); NIGHT: 0.2 km (0.1 mile) (land or water). Large spill–First: Isolate in all directions 30 meters (100 feet) (land); 125 meters (400 feet) (water); Then: Protect persons downwind, DAY: 0.3 km (0.2 mile) (land); 1.1 km (0.7 mile) (water); NIGHT: 0.6 km (0.4 mile) (land); 2.4 km (1.5 miles) (water).
Fire: If any large container is involved in fire, isolate for at least 800 meters (½ mile) in all directions; also, consider initial evacuation for 800 meters (½ mile) in all directions.

EXPOSURE
Short-term effects: *SEEK MEDICAL ATTENTION.* **Vapor:** Irritating to eyes, nose, and throat. *IF INHALED*, will, will cause coughing, difficult breathing, or loss of consciousness. Move to fresh air. *IF IN EYES*, hold eyelids open and flush with plenty of water. *IF BREATHING HAS STOPPED*, give artificial respiration respiration; *avoid mouth-to-mouth resuscitation; use bag/mask apparatus.* IF breathing is difficult, administer oxygen. **Liquid:** Will burn skin and eyes. Harmful if swallowed. Remove contaminated clothing and shoes. Flush affected areas with plenty of water. *IF IN EYES*, hold eyelids open and flush with plenty of

water. *IF SWALLOWED* and victim is *CONSCIOUS AND ABLE TO SWALLOW*, have victim drink 4 to 8 ounces of water. **Do NOT induce vomiting.**

Note to physician or authorized medical personnel: Medical observation is recommended for 24 to 48 hours after breathing overexposure, as pulmonary edema may be delayed. As first aid for pulmonary edema, consider administering a corticosteroid spray. Cigarette smoking may exacerbate pulmonary injury and should be discouraged for at least 72 hours following exposure.

HEALTH HAZARDS
Personal protective equipment (PPE): A-Level PPE.
Vapor (gas) irritant characteristics: Vapors cause severe irritation of eyes and throat and can cause eye and lung injury. They cannot be tolerated even at low concentrations.
Liquid or solid irritant characteristics: Severe eye and skin irritant; may cause pain and second-degree burns after a few minutes of contact.

FIRE DATA
Fire extinguishing agents not to be used: Water applied to adjacent fires should be handled carefully. Do not use water if the material has been released.

CHEMICAL REACTIVITY
Reactivity with water: Reacts vigorously with water, releasing hydrogen chloride fumes and forming sulfuric acid.
Binary reactants: Reacts violently with alkalis, alkali metals, dimethyl sulfoxide, dinitrogen pentoxide, lead dioxide, N-methylformamide, red phosphorus. Reacts, possibly violently, with organic substances, strong acids, alcohols, amines, diethyl ether, glycols, peroxides. Attacks metals in the presence of moisture forming hydrogen gas.
Neutralizing agents for acids and caustics: Acid formed by reaction with water can be neutralized by limestone, lime, or soda ash.

ENVIRONMENTAL DATA
Food chain concentration potential: Negative; unlikely to accumulate.
Water pollution: Harmful to aquatic life in very low concentrations. May be dangerous if it enters nearby water intakes; notify operators. Notify local health and wildlife officials.
Response to discharge: Issue warning–corrosive, air contaminant. Restrict access. Evacuate area. Disperse and flush with care.

SHIPPING INFORMATION
Grades of purity: 99%; **Storage temperature:** Ambient; **Inert atmosphere:** None; **Venting:** Pressure-vacuum; **Stability during transport:** Stable.

NAS HAZARD CLASSIFICATION FOR BULK WATER TRANSPORTATION
FIRE: 0
HEALTH: Vapor irritant: 4; Liquid or solid irritant: 3; Poisons: 2
WATER POLLUTION: Human toxicity:; Aquatic toxicity: 3; Aesthetic effect: 2
REACTIVITY: Other chemicals: 4; Water: 4; Self-reaction: 0

PHYSICAL AND CHEMICAL PROPERTIES
Physical state @ 59°F/15°C and 1 atm: Liquid.
Molecular weight: 134.97
Boiling point @ 1 atm: 156°F/69°C/342°K.
Melting/Freezing point: –65°F/–54°C/219°K.
Specific gravity (water = 1): 1.67 @ 68°F/20°C (liquid).
Relative vapor density (air = 1): 4.6
Ratio of specific heats of vapor (gas): 1.122
Latent heat of vaporization: 89.1 Btu/lb = 49.5 cal/g = 2.07×10^5 J/kg.
Heat of solution: –885.5 Btu/lb = –491.9 cal/g = $–20.58 \times 10^5$ J/kg.
Vapor pressure: 105 mm.

TABUN (GA)   REC. T:0050

SYNONYMS: DIMETHYLAMIDOETHOXYPHOSPHORYL CYANIDE; DIMETHYLAMINOCYANPHOSPHORSAEURE-AETHYLESTER (German); DIMETHYLPHOSPHORAMIDO-CYANIDIC ACID, ETHYL ESTER; ETHYL *N,N*-DIMETHYLAMINO CYANOPHOSPHATE; ETHYL DIMETHYLPHOSPHORAMIDOCYANIDATE; ETHYL DIMETHYLAMIDOCYANOPHOSPHATE; ETHYL-*N,N*-DIMETHYLPHOSPHORAMIDOCYANIDATE; GA; GELAN I; Le-100; MCE; T-2104; TABOON A; TRILON 83

IDENTIFICATION:
CAS Registry Number: 77-81-6
Formula: $C_5H_{11}N_2O_2P$; $(CH_3)_2NPO(OC_2H_5)CN$
DOT ID Number: UN 2810; DOT Guide Number: 153
Proper Shipping Name: Tabun
Reportable Quantity (RQ): **(EHS):** 10 lb (4.54 kg)

DESCRIPTION: Clear, colorless to brown liquid. Fruit-like odor, or like bitter almonds. Tasteless. Slightly soluble in water; reacts quickly forming toxic hydrogen cyanide gas. *Warning:* Odor is not a reliable indicator of the presence of toxic amounts of Tabun.
Note: This chemical has been used as a chemical warfare agent (nerve gas). If used as a weapon, notify U.S. Department of Defense: Army. Damage and/or death may occur before chemical detection can take place. Use M8 Paper (Detection: Yellow) or M256-A1 Detector Kit (Detection limit: 0.005 mg/m^3) if available.

Combustible • Poison! (organophosphate) • Highly volatile • Breathing the vapor, skin or eye contact, or swallowing the material can kill you; symptoms may be delayed for several hours • Firefighting gear (including SCBA) provide NO protection. If exposure occurs, remove and isolate gear immediately and thoroughly decontaminate personnel • Containers may BLEVE when exposed to fire • Vapors are heavier than air and will collect and stay in low areas • Vapors in confined areas (e.g., tanks, sewers, buildings) may explode when exposed to fire • *Combustion products are less deadly than the material itself.* Toxic products of combustion may include cyanide, carbon monoxide, phosphorus oxide, and nitrogen oxides • Do NOT attempt rescue.

Hazard Classification (based on NFPA-704 Rating System)

Health Hazards (Blue): 4; Flammability (Red): 1; Reactivity (Yellow): 0

EMERGENCY RESPONSE: See Appendix A (153) with notes. Have available decontaminants (bleach, alkali). Bleaching powder (chlorinated lime) destroys tabun but gives rise to cyanogen chloride. Also have atropine available.
Evacuation:
Public safety: See below.

DAY: 0.3 km (0.2 mile); NIGHT: 0.6 km (0.4 mile). Large spill–First: Isolate in all directions 155 meters (500 feet); Then: Protect persons downwind, DAY: 1.6 km (1.0 mile); NIGHT: 3.1 km (1.9 miles).
Fire: If tank, rail car, or tank truck is involved in fire, isolate for at least 800 meters (½ mile) in all directions; also, consider initial evacuation for 800 meters (½ mile) in all directions.

EXPOSURE
Short-term effects: *SEEK MEDICAL ATTENTION.* Tabun is a nerve agent; it acts as a cholinesterase inhibitor. *Will remain as a liquid for more than 24 hours. Skin absorption great enough to cause death may occur in 1 to 2 minutes, but may be delayed for 1 to 2 hours. IF BREATHING HAS STOPPED,* give artificial respiration; *avoid mouth-to-mouth resuscitation; use bag/mask apparatus.* **Vapor or Liquid:** *POISONOUS IF INHALED, IF SWALLOWED, OR IF SKIN IS EXPOSED.* Eye pupils are small; blurred vision; runny nose; cough; shortness of breath; pain; diarrhea, nausea and vomiting; increased blood pressure, hypermotility, hallucinations; loss of consciousness; convulsions; breathing stops; death. **Skin:** *Remove and double-bag contaminated clothing (including shoes) and leave them in the Hot Zone;* wash skin with soap and water and large volumes of water for at least 15 minutes. *Skin can also be decontaminated with diluted hypochlorite solution, U.S. Army M291 kit, and M258(A1) skin decontamination kit.* **Eye:** Rinse with large volumes of water or saline for at least 15 minutes. *IF SWALLOWED,* will cause nausea, vomiting, or loss of consciousness. *Remove and double-bag contaminated clothing and shoes and leave in Hot Zone for later incineration by hazardous materials experts.* Flush affected areas with plenty of water. **If swallowed** and victim is *CONSCIOUS AND ABLE TO SWALLOW,* have victim drink 4 to 8 ounces of water. **Do NOT induce vomiting but immediately administer slurry of activated charcoal.** *IF SWALLOWED* and victim is *UNCONSCIOUS OR HAVING CONVULSIONS,* do nothing except keep victim warm.
Note to physician or authorized medical personnel: Administer atropine. *IF INHALED,* will, medical observation is recommended for 24 to 48 hours after breathing overexposure, as pulmonary edema may be delayed. As first aid for pulmonary edema, consider administering a corticosteroid spray. Cigarette smoking may exacerbate pulmonary injury and should be discouraged for at least 72 hours following exposure.
Note to physician and medical personnel: If symptoms indicate, initial treatment for an adult includes atropine, 2 mg (1/30 gr) intramuscularly or intravenously as soon as any local or systemic signs or symptoms of an intoxication are noted; repeat the administration of atropine every 3 to 8 minutes until signs of atropinization (mydriasis, dry mouth, rapid pulse, hot and dry skin) occur; initiate treatment in children with 0.05 mg/kg of atropine; repeat at 5- to 10-minute intervals. Watch respiration, and remove bronchial secretions if they appear to be obstructing the airway; intubate if necessary. Pralidoxime must be administered within minutes to a few hours following exposure (depending on the specific agent) to be effective. Give 2-PAMCI (Pralidoxime; Protopam), 2.5 g in 100 mL of sterile water or in 5% dextrose and water, intravenously, slowly, in 15 to 30 minutes; if sufficient fluid is not available, give 1 g of 2-PAMCI in 3 mL of distilled water by deep intramuscular injection; repeat this every half hour if respiration weakens or if muscle fasciculation or convulsions recur. Also Diazepam, an anticonvulsant, might be needed. For adults, the dose is 5 to 10 mg (slow IV), repeated every 12 to 15 minutes up to three (3) doses maximum. For children, the dose is 0.2 to 0.5 mg/kg.

HEALTH HAZARDS
Personal protective equipment (PPE): A-Level PPE. Wear positive pressure self-contained breathing apparatus (SCBA). Rubber gloves, protective clothing with full facepiece. Butyl rubber gloves and Tyvek® "F"decontamination suit provide barrier protection against chemical warfare agents. Airtight, impermeable clothing was developed for personnel who must enter heavily contaminated areas. This clothing is made of butyl rubber or a coated fabric such as Tyvek "F" and provide barrier protection against liquid chemical warfare agents. Although resistant to liquid chemical agents, impermeable protective clothing may be penetrated after a few hours of exposure to heavy concentration of agent. Consequently, liquid contamination on the clothing must be neutralized or removed as soon as possible. If the proper equipment is not available, or if the rescuers have not been trained in its use, call for assistance from the U.S. Soldier and Biological Chemical Command–Edgewood Research Development and Engineering Center (from 0700-1630 EST call 410-671-4411, and from 1630-0700 EST call 410-278-5201; ask for the Staff Duty Officer).
Recommendations for respirator selection: *EMERGENCY OR PLANNED ENTRY INTO UNKNOWN CONCENTRATIONS OR IDLH CONDITIONS:* SCBAF:PD,PP (any self-contained breathing apparatus that has a full facepiece and is operated in a pressure-demand or other positive-pressure mode); or SAF:PD,PP:ASCBA (any supplied-air respirator that has a full facepiece and is operated in a pressure-demand or other positive-pressure mode in combination with an auxiliary self-contained breathing apparatus operated in a pressure-demand or other positive pressure mode). *ESCAPE:* GMFOVHiE [any air-purifying, full-facepiece respirator (gas mask) with a chin-style, front-or back-mounted canister having a high efficiency particulate filter); or SCBAE (any appropriate escape-type, self-contained breathing apparatus).
The U.S. Army standard M17A1 (which is being replaced by the M40 protective mask) mask provides complete respiratory protection against all known military toxic chemical agents, but it cannot be used in an oxygen deficient environment and it does not afford protection against industrial toxics such as ammonia and carbon monoxide. *It is not approved for civilian use.*
Exposure limits (TWA unless otherwise noted): Workplace: 0.0001 mg/m^3; General population limits (as recommended by the Surgeon General's Working Group, U.S. Department of Health): 0.000003 mg/m^3.
Toxicity: The median lethal dosage (vapor/respiratory) LCt$_{50}$ = 400 mg-minute/m^3 for humans; the median incapacitating dosage is 300 mg-minute/m^3. Respiratory lethal dosages kill in 1 to 10 minutes; liquid in the eye kills nearly as rapidly. The LD$_{50}$ (skin) = 1.0g/70 kg /156.8 lb [man] (*Medical Sspects of Chemical and Biological Warfare, Part I,* Walter Reed Medical Center, 1997).
Long-term health effects: Repeated exposures may damage the nervous system, resulting in convulsions, respiratory failure. Liver and kidney damage. Cholinesterase inhibitor; cumulative effect is possible.

FIRE DATA
Flash point: 172.4°F/78°C.

CHEMICAL REACTIVITY
Reactivity with water: Produces hydrogen cyanide.
Binary reactants: Contact with acids produce hydrogen cyanide.

ENVIRONMENTAL DATA
Food chain concentration potential: Quickly hydrolyzed in water. Not expected to accumulate.
Water pollution: If used as a weapon, utilize use an M272 Water

Detection Kit (Detection limit: 0.02 mg/L). Dangerous to aquatic life in high concentrations. May be dangerous if it enters water intakes. Notify local health and pollution control officials. Notify operators of nearby water intakes. This material will be broken down in water quickly, but small amounts may evaporate. This material will be broken down in moist soil quickly. Small amounts may evaporate into the air or travel below the soil surface and contaminated groundwater. Persists 1½ to 2 days in soil. Bleaching powder (chlorinated line) destroys tabun but gives rise to cyanogen chloride.

PHYSICAL AND CHEMICAL PROPERTIES
Physical state @ 59°F/15°C and 1 atm: Liquid.
Molecular weight: 162.15
Boiling point @ 1 atm: 428 to 475°F/220 to 246°C/493 to 519°K a@ 760 mmHg.
Melting/Freezing point: –58°F/–50°C/223.2°K.
Solubility in water: 9.8g/100g @ 77°F/25°**C**.
Specific gravity (water = 1): 1.073 @ 77°F/25°**C**.
Vapor density: 5.63
Liquid density: 1.08 g/mL @ 77°F/25°**C**.
Vapor pressure: 0.037 mmHg @ 68°F/20°C; 0.057 mmHg @ 77°F/25°**C**.
Volatility: 490 mg/m^3 @ 77°F/25°**C**.

TALL OIL, FATTY ACID   REC. T:0100

SYNONYMS: ACEITE de RESINA (Spanish); LIQUID ROSIN; TALLEOL; TALLOL

IDENTIFICATION
CAS Number: 8002-26-4
Formula: Mixture

DESCRIPTION: Clear amber to dark brown thick liquid. Acrid odor. Floats on water surface; insoluble.

Combustible • Toxic products of combustion may include CO_2.

Hazard Classification (based on NFPA-704 Rating System)
Health Hazards (Blue): 0; Flammability (Red): 1; Reactivity (Yellow): 0

EMERGENCY RESPONSE: See Appendix A (171)
Evacuation:
Public safety: Isolate the area of spill or leak for at least 10 to 25 meters (30 to 80 feet) in all directions.
Spill: Increase, in the downwind direction, as necessary, the distance shown under "Public Safety."
Fire: If any large container is involved in fire, isolate for at least 800 meters (½ mile) in all directions; also, consider initial evacuation for 800 meters (½ mile) in all directions.

EXPOSURE
Short-term effects: *SEEK MEDICAL ATTENTION*. **Liquid:** Remove contaminated clothing and shoes. Wash affected areas with soap and water. *IF IN EYES*, hold eyelids open and flush with plenty of water.

HEALTH HAZARDS
Personal protective equipment (PPE): B-Level PPE. Full covering clothing and chemical protective butyl rubber gloves. Where splashing is possible wear full face shield or chemical safety goggles.

Long-term health effects: May cause allergic reaction.
Liquid or solid irritant characteristics: Mild skin irritant. If spilled on clothing and allowed to remain, may cause smarting and reddening of the skin.

FIRE DATA
Flash point: 365°F/185°C (cc).
Fire extinguishing agents not to be used: Water may be ineffective.
Electrical hazard: Class I, Group D.

CHEMICAL REACTIVITY
Binary reactants: Reacts with oxidizers.

ENVIRONMENTAL DATA
Water pollution: Effect of low concentrations on aquatic life is unknown. May be dangerous if it enters nearby water intakes; notify operators. Notify local health and wildlife officials. **Response to discharge:** Should be removed.

SHIPPING INFORMATION
Grades of purity: Technical; **Storage temperature:** Ambient; **Inert atmosphere:** None; **Venting:** Open; **Stability during transport:** Stable.

PHYSICAL AND CHEMICAL PROPERTIES
Physical state @ 59°F/15°C and 1 atm: Liquid.
Boiling point @ 1 atm: More than 662°F/350°C/623°K.
Specific gravity (water = 1): 0.91 to 1.00
Vapor pressure: (Reid) 0.1 psia; 1.5 mm.

TALLOW   REC. T:0150

SYNONYMS: EDIBLE TALLOW; INEDIBLE TALLOW; SEBO (Spanish); TALLOW OIL

IDENTIFICATION
Formula: Animal fats containing C_{16} to C18

DESCRIPTION: Dark yellow oily liquid. Waxy odor. Floats on the surface of water.

Combustible • Containers may BLEVE when exposed to fire • Toxic products of combustion may include acrid carbon monoxide.

Hazard Classification (based on NFPA-704 Rating System)
Health Hazards (Blue): 0; Flammability (Red): 1; Reactivity (Yellow): 0

EMERGENCY RESPONSE: See Appendix A (171)
Evacuation:
Public safety: Isolate the area of spill or leak for at least 10 to 25 meters (30 to 80 feet) in all directions.
Spill: Increase, in the downwind direction, as necessary, the distance shown under "Public Safety."
Fire: If any large container is involved in fire, isolate for at least 800 meters (½ mile) in all directions; also, consider initial evacuation for 800 meters (½ mile) in all directions.

HEALTH HAZARDS
Personal protective equipment (PPE): B-Level PPE. Goggles or face shield; protective clothing, if exposure to hot liquid is possible.
Toxicity by ingestion: Grade 0; LD_{50} = > 15 g/kg.
Long-term health effects: None.

1030 Tannic acid

Vapor (gas) irritant characteristics: Nonvolatile
Liquid or solid irritant characteristics: None

FIRE DATA
Flash point: 509°F/265°C (cc) (solid).
Fire extinguishing agents not to be used: Water or foam may cause frothing.
Electrical hazard: Class I, Group D.

CHEMICAL REACTIVITY
Reactivity group: 34
Compatibility class: Esters.

ENVIRONMENTAL DATA
Food chain concentration potential: Negative; unlikely to accumulate.
Water pollution: Effect of low concentrations on aquatic life is unknown. Fouling to shoreline. May be dangerous if it enters nearby water intakes; notify operators. Notify local health and wildlife officials. **Response to discharge:** Mechanical containment. Should be removed. Chemical and physical treatment.

SHIPPING INFORMATION
Grades of purity: Acidless; buffing; industrial fancy; edible; inedible; **Storage temperature:** Ambient; **Inert atmosphere:** None; **Venting:** Open (flame arrester); **Stability during transport:** Stable.

PHYSICAL AND CHEMICAL PROPERTIES
Physical state @ 59°F/15°C and 1 atm: Solid.
Boiling point @ 1 atm: Very high.
Melting/Freezing point: 88–100°F/31–38°C/304–311°K.
Specific gravity (water = 1): 0.85 to 0.89 at 158°F/70°C (liquid).
Heat of combustion: (estimate) $-18,000$ Btu/lb = $-10,000$ cal/g = -420×10^5 J/kg.
Vapor pressure: (Reid) 0.1 psia.

TALLOW FATTY ALCOHOL   **REC. T:0200**

SYNONYMS: HIGHER FATTY ALCOHOL; 1-OCTADECANOL, CRUDE; STEARYL ALCOHOL, CRUDE

IDENTIFICATION
CAS Number: 112-92-5 (steryl alcohol)
Formula: $C_{18}H_{37}OH$ (approximate).

DESCRIPTION: White, waxy solid. Mild, soapy odor. Floats on the surface of water.

Combustible solid • Containers may BLEVE when exposed to fire • Contact can cause skin or eye irritation • Toxic products of combustion may include CO_2, acrid carbon monoxide.

Hazard Classification (based on NFPA-704 Rating System)
Health Hazards (Blue): 0; Flammability (Red): 1; Reactivity (Yellow): 0

EMERGENCY RESPONSE: See Appendix A (171)
Evacuation:
Public safety: Isolate the area of spill or leak for at least 10 to 25 meters (30 to 80 feet) in all directions.
Spill: Increase, in the downwind direction, as necessary, the distance shown under "Public Safety."
Fire: If any large container is involved in fire, isolate for at least 800 meters (½ mile) in all directions; also, consider initial evacuation for 800 meters (½ mile) in all directions.

EXPOSURE
Short-term effects: Solid: Non-toxic.

HEALTH HAZARDS
Personal protective equipment (PPE): B-Level PPE. Goggles or face shield; rubber gloves. **Natural rubber, neoprene, and nitrile rubber may offer some protection from alcohols.**
Toxicity by ingestion: Grade 2; oral LD_{50} = 1900 mg/kg (rat).
Liquid or solid irritant characteristics: Eye irritant.

FIRE DATA
Flash point: More than 270°F/132°C (cc).
Fire extinguishing agents not to be used: Water may be ineffective.
Autoignition temperature: 842°C/450°C/723°K.

CHEMICAL REACTIVITY
Binary reactants: Strong acids, caustics, amines, isocyanates.
Reactivity group: 20
Compatibility class: Alcohols, glycols

ENVIRONMENTAL DATA
Food chain concentration potential: Negative; unlikely to accumulate.
Water pollution: Effect of low concentrations on aquatic life is unknown. Fouling to shoreline. May be dangerous if it enters nearby water intakes; notify operators. Notify local health and wildlife officials. **Response to discharge:** Mechanical containment. Should be removed. Chemical and physical treatment.

SHIPPING INFORMATION
Grades of purity: Technical; **Storage temperature:** Ambient; **Inert atmosphere:** None; **Venting:** Open (flame arrester); **Stability during transport:** Stable.

PHYSICAL AND CHEMICAL PROPERTIES
Physical state @ 59°F/15°C and 1 atm: Solid.
Molecular weight: 262 (avg.).
Boiling point @ 1 atm: More than 480°F/249°C/522°K.
Melting/Freezing point: 127°F/53°C/326°K.
Specific gravity (water = 1): 0.810 @ 77°F/25°C (solid).
Heat of combustion: (estimate) $-18,500$ Btu/lb = $-10,300$ cal/g = -430×10^5 J/kg.

TANNIC ACID   **REC. T:0250**

SYNONYMS: ACIDO TANICO (Spanish); CHINESE TANNIN; CHESTNUT TANNIN; DIGALLIC ACID; GALLOTANNIC ACID; GALLOTANNIN; GLYCERITE; TANNIN

IDENTIFICATION
CAS Number: 1401-55-4
Formula: $C_{76}NO_{52}O_{46}$

DESCRIPTION: Yellow to light tan solid. Darkens on contact with air. Faint odor. Sinks and mixes with water forming a weak acid.

Combustible solid • Containers may BLEVE when exposed to fire • Irritating to the skin, eyes, and respiratory tract • Concentrated dust in confined areas (e.g., tanks, sewers, buildings) may explode when exposed to fire • Toxic products of combustion may include carbon monoxide, CO_2, and pyrogallol.

Hazard Classification (based on NFPA-704 Rating System)
Health Hazards (Blue): 0; Flammability (Red): 1; Reactivity (Yellow): 0

EMERGENCY RESPONSE: See Appendix A (171)
Evacuation:
Public safety: Isolate the area of spill or leak for at least 10 to 25 meters (30 to 80 feet) in all directions.
Spill: Increase, in the downwind direction, as necessary, the distance shown under "Public Safety."
Fire: If any large container is involved in fire, isolate for at least 800 meters (½ mile) in all directions; also, consider initial evacuation for 800 meters (½ mile) in all directions.

EXPOSURE
Short-term effects: *SEEK MEDICAL ATTENTION.* **Dust:** Irritating to eyes, nose, and throat. *IF INHALED*, will, will cause coughing. *IF IN EYES*, hold eyelids open and flush with plenty of water. *IF BREATHING HAS STOPPED*, give artificial respiration. IF breathing is difficult, administer oxygen. **Solid:** Irritating to skin and eyes. *IF SWALLOWED*, will cause nausea and vomiting. Remove contaminated clothing and shoes. Flush affected areas with plenty of water. *IF IN EYES*, hold eyelids open and flush with plenty of water. *IF SWALLOWED* and victim is *CONSCIOUS AND ABLE TO SWALLOW*, have victim drink 4 to 8 ounces of water and have victim induce vomiting. *IF SWALLOWED* and victim is *UNCONSCIOUS OR HAVING CONVULSIONS*, do nothing except keep victim warm.

HEALTH HAZARDS
Personal protective equipment (PPE): B-Level PPE. Dust mask; goggles or face shield; protective gloves. Chemical protective material(s) reported to have good to excellent resistance: butyl rubber, natural rubber, neoprene, nitrile, PVC, nitrile+PVC, polyethylene, styrene-butadiene. Also, chlorinated polyethylene, Viton®, and Viton®/neoprene offers limited protection.
Toxicity by ingestion: Grade 2; oral LD_{50} = 2300 mg/kg (rat).
Long-term health effects: Causes cancer of liver in rats
Vapor (gas) irritant characteristics: Dust is an eye and respiratory tract irritant.
Liquid or solid irritant characteristics: Eye irritant.

FIRE DATA
Flash point: 390°F/199°C.
Fire extinguishing agents not to be used: Water or foam may cause frothing.
Autoignition temperature: 980°F/527°C/800°K.

ENVIRONMENTAL DATA
Food chain concentration potential: Negative; unlikely to accumulate.
Water pollution: Harmful to aquatic life in very low concentrations. May be dangerous if it enter water intakes. Notify local health and wildlife officials. **Response to discharge:** Disperse and flush.

SHIPPING INFORMATION
Grades of purity: Commercial, 84%; Reagent; **Storage temperature:** Ambient; **Inert atmosphere:** None; **Venting:** Open; **Stability during transport:** Stable.

PHYSICAL AND CHEMICAL PROPERTIES
Physical state @ 59°F/15°C and 1 atm: Solid.
Molecular weight: 1701
Boiling point @ 1 atm: Decomposes. 392°F/300°C/573°K.
Melting/Freezing point: 410°F/210°C/483°K.
Specific gravity (water = 1): More than 1 @ 68°F/20°C (solid).
Heat of combustion: (estimate) $-9,810$ Btu/lb = $-5,450$ cal/g = 228×10^5 J/kg.

TETRABUTYL TITANATE   REC. T:0300

SYNONYMS: TITANATO de TETRABUTILO (Spanish); TITANIUM BUTOXIDE; BUTYL TITANATE; TITANIUM TETRABUTOXIDE; BUTYL TITANATE MONOMER; ORTHOTITANIC ACID, TETRABUTYL ESTER

IDENTIFICATION
CAS Number: 5593-70-4
Formula: $Ti(OC_4H_9)_4$; $C_{16}H_{36}O_4 \cdot Ti$
DOT ID Number: UN 1993; DOT Guide Number: 128
Proper Shipping Name: Combustible liquid, n.o.s.

DESCRIPTION: Colorless to pale yellow liquid. Weak alcohol-like odor. May float or sink in water; reacts formine butanol and titanium dioxide; the reaction may not be hazardous.

Combustible • Containers may BLEVE when exposed to fire • Vapors in confined areas (e.g., tanks, sewers, buildings) may explode when exposed to fire • Irritating to the skin, eyes, and respiratory tract • Toxic products of combustion may include acrid carbon monoxide.

Hazard Classification (based on NFPA-704 Rating System)
Health Hazards (Blue): 0; Flammability (Red): 2; Reactivity (Yellow): 0

EMERGENCY RESPONSE: See Appendix A (128)
Evacuation:
Public safety: Isolate spill area for at least 25 to 50 meters (80 to 160 feet) in all directions.
Spill: Large spill–Consider initial downwind evacuation for at least 300 meters (1000 feet).
Fire: Isolate for 800 meters (½ mile) in all directions, especially if tank, rail car, or tank truck is involved in fire.

EXPOSURE
Short-term effects: *SEEK MEDICAL ATTENTION.* **Liquid:** Irritating to skin and eyes. *IF SWALLOWED*, will cause nausea and vomiting. Remove contaminated clothing and shoes. Flush affected areas with plenty of water. *IF IN EYES*, hold eyelids open and flush with plenty of water. *IF SWALLOWED* and victim is *CONSCIOUS AND ABLE TO SWALLOW*, have victim drink 4 to 8 ounces of water and have victim induce vomiting. *IF SWALLOWED* and victim is *UNCONSCIOUS OR HAVING CONVULSIONS*, do nothing except keep victim warm.

HEALTH HAZARDS
Personal protective equipment (PPE): B-Level PPE. Self-contained breathing apparatus or organic canister mask; goggles or face shield; rubber gloves.
Toxicity by ingestion: Moderately toxic.
Long-term health effects: Reproduction, mutagenic and teratogenic data.

1032   Tetrachloroethane

Liquid or solid irritant characteristics: Severe eye irritant. Skin irritant.

FIRE DATA
Flash point: 170°F/77°C (cc).
Flammable limits in air: LEL: 2%; UEL: 12%.
Fire extinguishing agents not to be used: Water.
Behavior in fire: May give off dense white smoke. Containers may explode.
Burning rate: 3.4 mm/min

CHEMICAL REACTIVITY
Binary reactants: Reacts violently with oxidizers. Reacts with water producing butanol and titanium dioxide. Incompatible with sulfuric acid, nitric acid, caustics, aliphatic amines, isocyanates, boranes. React violently with sodium peroxide, uranium fluoride.
Reactivity group: 34
Compatibility class: Esters.

ENVIRONMENTAL DATA
Water pollution: Effect of low concentrations on aquatic life is unknown. Fouling to shoreline. May be dangerous if it enters nearby water intakes; notify operators. Notify local health and wildlife officials. **Response to discharge:** Issue warning–water contaminant. Restrict access. Mechanical containment. Disperse and flush.

SHIPPING INFORMATION
Grades of purity: Technical; **Storage temperature:** Ambient; **Inert atmosphere:** None; **Venting:** Open (flame arrester); **Stability during transport:** Stable.

PHYSICAL AND CHEMICAL PROPERTIES
Physical state @ 59°F/15°C and 1 atm: Liquid.
Molecular weight: 340
Boiling point @ 1 atm: 593°F/312°C/585°K.
Melting/Freezing point: –67°F/–55°C/218°K.
Specific gravity (water = 1): 0.998 @ 77°F/25°C (liquid).
Latent heat of vaporization: 142 Btu/lb = 79 cal/g = 3.3×10^5 J/kg.
Heat of combustion: (estimate) –14,600 Btu/lb = –8100 cal/g = -340×10^5 J/kg.

TETRACHLOROETHANE   **REC. T:0350**

SYNONYMS: ACETYLENE TETRACHLORIDE; BONOFORM; CELLON; 1,1-DICHLORO-2,2-DICHLOROETHANE; EEC No. 602-015-00-3; ETHANE,1,1,2,2-TETRACHLORO-; 1,1,2,2-TETRACHLORAETHAN (German); 1,1,2,2-TETRACHLORETHANE; sym-TETRACHLOROETHANE; TETRACHLORURE d'ACETYLENE (French); TETRACLOROETANO (Spanish); 1,1,2,2-TETRACLOROETANO (Spanish); WESTRON; RCRA No. U209

IDENTIFICATION
CAS Number: 79-34-5
Formula: $C_2H_2Cl_4$; $Cl_2CHCHCl_2$
DOT ID Number: UN 1702; DOT Guide Number: 151
Proper Shipping Name: Tetrachloroethane
Reportable Quantity (RQ): **(CERCLA)** 100 lb/45.4 kg

DESCRIPTION: Colorless (pure) to light yellow, or yellowish-green, heavy, mobile liquid. Sweet odor, like chloroform. Sinks in water; insoluble; slowly forming hydrochloric acid and hydrogen chloride gas.

Corrosive; Irritates the skin, eyes, and respiratory tract • Containers may BLEVE when exposed to fire • Vapors are heavier than air and will collect and stay in low areas • Toxic products of combustion may include phosgene and hydrogen chloride gas.

Hazard Classification (based on NFPA-704 Rating System)
Health Hazards (Blue): 2; Flammability (Red): 0; Reactivity (Yellow): 0

EMERGENCY RESPONSE: See Appendix A (151)
Evacuation:
Public safety: Isolate the area of spill or leak for at least 25 to 50 meters (80 to 160 feet) in all directions.
Spill: Increase, in the downwind direction, as necessary, the distance shown under "Public Safety."
Fire: If tank, rail car, or tank truck is involved in fire, isolate for at least 800 meters (½ mile) in all directions; also, consider initial evacuation for 800 meters (½ mile) in all directions.

EXPOSURE
Short-term effects: *SEEK MEDICAL ATTENTION*. The use of alcoholic beverages may enhance toxic effects. **Vapor:** Harmful if inhaled. *IF IN EYES*, hold eyelids open and flush with plenty of water. *IF BREATHING HAS STOPPED,* give artificial respiration; *avoid mouth-to-mouth resuscitation; use bag/mask apparatus*. IF breathing is difficult, administer oxygen. **Liquid:** *POISONOUS IF SWALLOWED OR IF SKIN IS EXPOSED*. Irritating to skin and eyes. *IF SWALLOWED*, will cause nausea and vomiting. Remove contaminated clothing and shoes. Flush affected areas with plenty of water. *IF IN EYES*, hold eyelids open and flush with plenty of water. *IF SWALLOWED* and victim is *CONSCIOUS AND ABLE TO SWALLOW*, have victim drink 4 to 8 ounces of water and have victim induce vomiting. *IF SWALLOWED* and victim is *UNCONSCIOUS OR HAVING CONVULSIONS,* do nothing except keep victim warm.

HEALTH HAZARDS
Personal protective equipment (PPE): A-Level PPE. OSHA Table Z-1-A air contaminant. Chemical safety goggles; plastic face shield; air- or oxygen-supplied mask; safety hat with brim; solvent-proof apron and gloves. Chemical protective material(s) reported to have good to excellent resistance: PV alcohol, Viton®.
Recommendations for respirator selection: NIOSH
At any concentrations above the NIOSH REL, or where there is no REL, at any detectable concentration: SCBAF:PD,PP (any self-contained breathing apparatus that has a full facepiece and is operated in a pressure-demand or other positive-pressure mode); or SAF:PD,PP:ASCBA (any supplied-air respirator that has a full facepiece and is operated in a pressure-demand or other positive-pressure mode in combination with an auxiliary self-contained breathing apparatus operated in a pressure-demand or other positive pressure mode). *ESCAPE:* GMFOV [any air-purifying, full-facepiece respirator (gas mask) with a chin-style, front-or back-mounted organic vapor canister]; or SCBAE (any appropriate escape-type, self-contained breathing apparatus).
Exposure limits (TWA unless otherwise noted): ACGIH TLV 1 ppm (6.9 mg/m³); OSHA PEL 5 ppm (35 mg/m³); NIOSH REL potential human carcinogen; 1 ppm (7 mg/m³); skin contact contributes significantly in overall exposure.
Toxicity by ingestion: Grade 3; oral LD_{50} = 200 mg/kg (rat).
Long-term health effects: A powerful narcotic. Liver poison,

nervous disorders. Suspected carcinogen. IARC rating 3; animal evidence: limited. NCI–animal carcinogen. Kidney damage.
Vapor (gas) irritant characteristics: Eye and respiratory tract irritant. Vapor is irritating such that personnel will not usually tolerate moderate or high vapor concentrations.
Liquid or solid irritant characteristics: Eye and skin irritant. If spilled on clothing and allowed to remain, may cause smarting and reddening of the skin.
Odor threshold: 0.5 ppm.
IDLH value: Potential human carcinogen; 100 ppm.

FIRE DATA
Electrical hazard: Class I, Group C.

CHEMICAL REACTIVITY
Reactivity with water: Forms hydrochloric acid with moisture.
Binary reactants: Reacts with chemically active metals, strong bases, fuming sulfuric acid. Degrades slowly when exposed to air forming hydrochloric acid. May attack some forms of plastics and rubbers.
Reactivity group: 36
Compatibility class: Halogenated hydrocarbons

ENVIRONMENTAL DATA
Food chain concentration potential: Log K_{ow} = 2.39. Unlikely to accumulate.
Water pollution: DOT Appendix B, §172.101–marine pollutant. Effect of low concentrations on aquatic life is unknown. May be dangerous if it enters nearby water intakes; notify operators. Notify local health and wildlife officials. **Response to discharge:** Issue warning–poison, air contaminant. Restrict access. Should be removed. Chemical and physical treatment.

SHIPPING INFORMATION
Grades of purity: Technical, 98%; **Storage temperature:** Ambient; **Inert atmosphere:** None; **Venting:** Open; **Stability during transport:** Stable.

PHYSICAL AND CHEMICAL PROPERTIES
Physical state @ 59°F/15°C and 1 atm: Liquid.
Molecular weight: 167.85
Boiling point @ 1 atm: 295.3°F/146.3°C/419.5°K.
Melting/Freezing point: –33°F/–36°C/237°K.
Specific gravity (water = 1): 1.595 at 68°F/20°C (liquid).
Liquid surface tension: 37.85 dynes/cm = 0.03785 N/m @ 68°F/20°C.
Relative vapor density (air = 1): 5.79
Ratio of specific heats of vapor (gas): 1.090 at 77°F/25°C.
Latent heat of vaporization: 99.2 Btu/lb = 55.1 cal/g = 2.30 x 10^5 J/kg.
Vapor pressure: (Reid) 0.5 psia; 5 mm.

TETRACHLOROETHYLENE   **REC. T:0400**

SYNONYMS: CARBON BICHLORIDE; CARBON BICHLORIDE; DOW-PER; EEC No. 602-028-00-4; ETHYLENE TETRACHLORIDE; PERCHLOR; PERCLENE; PERCHLOROETHYLENE; PERCLOROETILENO (Spanish); PERK; PERKLONE; PERSEC; TETRACLOROETILENO (Spanish); TETRACAP; TETRALENO; TETRALEX; TETROPIL; RCRA No. U2100

IDENTIFICATION
CAS Number: 127-18-4
Formula: $Cl_2C = CCl_2$; C_2Cl_4
DOT ID Number: UN 1897; DOT Guide Number: 160
Proper Shipping Name: Tetrachloroethylene
Reportable Quantity (RQ): **(CERCLA)** 100 lb/45.4 kg

DESCRIPTION: Colorless watery liquid. Sweet odor, like ether or chloroform. Sinks in water; insoluble. Irritating vapor is produced.

Containers may BLEVE when exposed to fire • Vapors are heavier than air and will collect and stay in low areas • Irritating to the skin, eyes, and respiratory tract • Decomposes in heat above 302°F/150°C; toxic products of combustion or decomposition include hydrogen chloride, carbon monoxide, and phosgene.

Hazard Classification (based on NFPA-704 Rating System)
Health Hazards (Blue): 2; Flammability (Red): 0; Reactivity (Yellow): 0

EMERGENCY RESPONSE: See Appendix A (160)
Evacuation:
Public safety: Isolate the area of spill or leak for at least 10 to 25 meters (30 to 80 feet) in all directions.
Spill: Increase, in the downwind direction, as necessary, the distance shown under "Public Safety."
Fire: If any large container is involved in fire, isolate for at least 800 meters (½ mile) in all directions; also, consider initial evacuation for 800 meters (½ mile) in all directions.

EXPOSURE
Short-term effects: *SEEK MEDICAL ATTENTION*. The use of alcoholic beverages may enhance toxic effects. **Vapor:** Irritating to eyes, nose, and throat. *IF INHALED*, will, will cause difficult breathing, or loss of consciousness. Move to fresh air. *IF BREATHING HAS STOPPED*, give artificial respiration. IF breathing is difficult, administer oxygen. **Liquid:** Irritating to skin and eyes. Harmful if swallowed. Remove contaminated clothing and shoes. Flush affected areas with plenty of water. *IF IN EYES*, hold eyelids open and flush with plenty of water. *IF SWALLOWED* and victim is *CONSCIOUS AND ABLE TO SWALLOW*, have victim drink 4 to 8 ounces of water.

HEALTH HAZARDS
Personal protective equipment (PPE): B-Level PPE. OSHA Table Z-1-A air contaminant. For high vapor concentrations use approved canister); or air–supplied mask; chemical goggles or face shield; gloves. Chemical protective material(s) reported to have good to excellent resistance: PV alcohol, Teflon®, Viton®. Also, natural rubber may offer some protection from perchloroethylene.
Recommendations for respirator selection: NIOSH
At any concentrations above the NIOSH REL, or where there is no REL, at any detectable concentration: SCBAF:PD,PP (any self-contained breathing apparatus that has a full facepiece and is operated in a pressure-demand or other positive-pressure mode); or SAF:PD,PP:ASCBA (any supplied-air respirator that has a full facepiece and is operated in a pressure-demand or other positive-pressure mode in combination with an auxiliary self-contained breathing apparatus operated in a pressure-demand or other positive pressure mode). *ESCAPE:* GMFOV [any air-purifying, full-facepiece respirator (gas mask) with a chin-style, front-or back-mounted organic vapor canister]; or SCBAE (any appropriate escape-type, self-contained breathing apparatus).
Exposure limits (TWA unless otherwise noted): ACGIH TLV 25 ppm; OSHA PEL 100 ppm; ceiling 200 ppm; NIOSH REL

potential human carcinogen; minimize workplace odor exposure limit number of workers exposed.
Short-term exposure limits (15-minute TWA): ACGIH STEL 100 ppm; OSHA STEL 300 ppm [5-minute peak in any 3 hours] as tetrachloroethylene.
Toxicity by ingestion: Grade 2; LD_{50} = 0.5 to 5 g/kg.
Long-term health effects: OSHA specifically regulated carcinogen. ACGIH animal carcinogen. NTP anticipated carcinogen. IARC possible carcinogen, rating 2B; sufficient animal evidence. NCI-animal carcinogen. Causes liver damage.
Vapor (gas) irritant characteristics: Eye and respiratory tract irritant. Vapors cause a smarting of the eyes or throat if present in high concentrations. The effect is temporary.
Liquid or solid irritant characteristics: Eye and skin irritant. If spilled on clothing and allowed to remain, may cause smarting and reddening of the skin.
Odor threshold: 5 ppm.
IDLH value: 150 ppm; potential human carcinogen.

CHEMICAL REACTIVITY
Reactivity with water: Decomposes slowly in water yielding trichloroacetic and hydrochloric acids. Accellerated hydrolysis in the presence of O_2.
Binary reactants: Violent reaction with concentrated nitric acid (produces carbon dioxide), strong oxidizers, strong alkalis; powdered, chemically-active metals such as aluminum, barium, beryllium, lithium, and zinc. Incompatible with nitrogen tetroxide, finely divided metals. Decomposes in UV light, on contact with red- hot metals, and in temperatures above 302°F/150°C, releasing hydrogen chloride, carbon monoxide, and phosgene. Corrodes metals in the presence of moisture.

ENVIRONMENTAL DATA
Food chain concentration potential: Log K_{ow} = 2.6. Unlikely to accumulate. Values > 3.0 are likely to bioconcentrate in aquatic organisms and other living tissue, especially in fats.
Water pollution: DOT Appendix B, §172.101–marine pollutant. Not harmful to aquatic life. May be dangerous if it enters nearby water intakes; notify operators. Notify local health and wildlife officials. **Response to discharge:** Should be removed. Chemical and physical treatment. Clean Water Act.

SHIPPING INFORMATION
Grades of purity: Dry cleaning and industrial grades: 95+%; **Storage temperature:** Ambient; **Inert atmosphere:** None; **Venting:** Pressure-vacuum; **Stability during transport:** Stable.

NAS HAZARD CLASSIFICATION FOR BULK WATER TRANSPORTATION
FIRE: 0
HEALTH: Vapor irritant: 1; Liquid or solid irritant: 1; Poisons: 2
WATER POLLUTION: Human toxicity: 1; Aquatic toxicity: 3; Aesthetic effect: 2
REACTIVITY: Other chemicals: 1; Water: 0; Self-reaction: 1

PHYSICAL AND CHEMICAL PROPERTIES
Physical state @ 59°F/15°C and 1 atm: Liquid.
Molecular weight: 165.83
Boiling point @ 1 atm: 250°F/121°C/394°K.
Melting/Freezing point: –8°F/–22°C/251°K.
Critical temperature: 657°F/347°C/620°K.
Specific gravity (water = 1): 1.63 @ 68°F/20°C (liquid).
Liquid surface tension: 31.3 dynes/cm = 0.0313 N/m @ 68°F/20°C.
Liquid water interfacial tension: 44.4 dynes/cm = 0.0444 N/m @ 77°F/25°C.
Relative vapor density (air = 1): 5.78
Ratio of specific heats of vapor (gas): 1.116
Latent heat of vaporization: 90.2 Btu/lb = 50.1 cal/g = 2.10×10^5 J/kg.
Vapor pressure: 14 mm.

TETRADECANOL   REC. T:0450

SYNONYMS: ALCOHOL MIRISTILICO (Spanish); MYRISTIC ALCOHOL; MYRISTYL ALCOHOL; 1-TETRADECANOL; TETRADECANOL, mixed isomers; n-TETRADECYL ALCOHOL

IDENTIFICATION
CAS Number: 112-72-1; 27196-00-5 (mixed isomers)
Formula: $C_{13}H_{30}O$; $CH_3(CH_2)_{12}CH_2OH$

DESCRIPTION: White solid or colorless, thick liquid (heated). Faint, fatty alcohol odor. Solidifies and Floats on water surface; insoluble.

Combustible solid • Containers may BLEVE when exposed to fire • Vapors or Concentrated dust in confined areas (e.g., tanks, sewers, buildings) may explode when exposed to fire • Irritating to the skin, eyes, and respiratory tract • Toxic products of combustion may include carbon monoxide and CO_2.

Hazard Classification (based on NFPA-704 Rating System)
Health Hazards (Blue): 0; Flammability (Red): 1; Reactivity (Yellow): 0

EMERGENCY RESPONSE: See Appendix A (171)
Evacuation:
Public safety: Isolate the area of spill or leak for at least 10 to 25 meters (30 to 80 feet) in all directions.
Spill: Increase, in the downwind direction, as necessary, the distance shown under "Public Safety."
Fire: If any large container is involved in fire, isolate for at least 800 meters (½ mile) in all directions; also, consider initial evacuation for 800 meters (½ mile) in all directions.

EXPOSURE
Short-term effects: *SEEK MEDICAL ATTENTION.* **Liquid:** Irritating to skin and eyes. Remove contaminated clothing and shoes. Flush affected areas with plenty of water. *IF IN EYES*, hold eyelids open and flush with plenty of water.

HEALTH HAZARDS
Personal protective equipment (PPE): B-Level PPE. Goggles or face shield. **Natural rubber, neoprene, and nitrile rubber may** offer some protection from alcohols.
Toxicity by ingestion: Grade 1; LD_{50} = 5 to 15 g/kg.
Vapor (gas) irritant characteristics: Nonvolatile
Liquid or solid irritant characteristics: Prolonged contact with skin may cause irritation.

FIRE DATA
Flash point: 285°F/141°C (oc).
Fire extinguishing agents not to be used: Water or foam may cause frothing.

CHEMICAL REACTIVITY
Binary reactants: React, possibly violently, with oxidizers,

acetaldehyde, alkalis, alkaline-earth and alkali metals, ammonium persulfate, strong acids, aliphatic amines, benzoyl peroxide, boranes, bromine dioxide, chromic acid, chromium trioxide, dialkylzincs, dichlorine oxide, ethylene oxide, hypochlorous acid, isocyanates, isopropyl chlorocarbonate, lithium tetrahydroaluminate, nitric acid, nitrogen dioxide, pentafluoroguanidine, phosphorus pentasulfide, perchlorates, permanganates, peroxides, sodium peroxide, sulfuric acid, tangerine oil, thionyl chloride, triethylaluminum, triisobutylaluminum, uranium fluoride.
Reactivity group: 20
Compatibility class: Alcohol

ENVIRONMENTAL DATA
Food chain concentration potential: Negative; unlikely to accumulate.
Water pollution: Effect of low concentrations on aquatic life is unknown. Fouling to shoreline. May be dangerous if it enters nearby water intakes; notify operators. Notify local health and wildlife officials. **Response to discharge:** Mechanical containment. Should be removed. Chemical and physical treatment.

SHIPPING INFORMATION
Grades of purity: 96–99+%; **Storage temperature:** Ambient; **Inert atmosphere:** None; **Venting:** Open (flame arrester); **Stability during transport:** Stable.

PHYSICAL AND CHEMICAL PROPERTIES
Physical state @ 59°F/15°C and 1 atm: Solid.
Molecular weight: 214.38
Boiling point @ 1 atm: 505.8°F/263.2°C/536.4°K.
Melting/Freezing point: 99.7°F/37.6°C/310.8°K.
Critical temperature: 804°F/429°C/702°K.
Specific gravity (water = 1): 0.824 at 38°C (liquid).
Liquid surface tension: 23.7 dynes/cm = 0.0237 N/m @ 68°F/20°C.
Ratio of specific heats of vapor (gas): 1.026

1-TETRADECENE   REC. T:0500

SYNONYMS: DODECYLETHYLENE; TETRADECENO (Spanish)

IDENTIFICATION
CAS Number: 1120-36-1
Formula: $CH_3(CH_2)11CH=CH_2$
DOT ID Number: UN 1993; DOT Guide Number: 128
Proper Shipping Name: Combustible Liquid.

DESCRIPTION: Colorless watery liquid. Mild, pleasant odor. Floats on the surface of water.

Combustible • Heat or flame may cause explosion • Containers may BLEVE when exposed to fire • Vapors are heavier than air and will collect and stay in low areas • Vapors in confined areas (e.g., tanks, sewers, buildings) may explode when exposed to fire • Irritating to the skin, eyes, and respiratory tract • Toxic products of combustion may include carbon monoxide and CO_2.

Hazard Classification (based on NFPA-704 Rating System)
Health Hazards (Blue): 0; Flammability (Red): 1; Reactivity (Yellow): 0

EMERGENCY RESPONSE: See Appendix A (128)
Evacuation:
Public safety: Isolate spill area for at least 25 to 50 meters (80 to 160 feet) in all directions.
Spill: Large spill–Consider initial downwind evacuation for at least 300 meters (1000 feet).
Fire: Isolate for 800 meters (½ mile) in all directions, especially if tank, rail car, or tank truck is involved in fire.

EXPOSURE
Short-term effects: *SEEK MEDICAL ATTENTION.* **Liquid:** Irritating to eyes. *IF IN EYES*, hold eyelids open and flush with plenty of water.

HEALTH HAZARDS
Personal protective equipment (PPE): B-Level PPE. Goggles or face shield.
Vapor (gas) irritant characteristics: Nonvolatile
Liquid or solid irritant characteristics: May cause eye irritation. *Caution!* The toxicological properties of this material have not been fully investigated. May cause respiratory and digestive tract irritation.

FIRE DATA
Flash point: 230°F/110°C (cc); 284°F/140°C (92%).
Fire extinguishing agents not to be used: Water or foam may cause frothing.
Behavior in fire: Containers my explode.
Autoignition temperature: 455°F/235°C/508°K.

CHEMICAL REACTIVITY
Binary reactants: Nitric and organic acids.
Reactivity group: 30
Compatibility class: Olefins

ENVIRONMENTAL DATA
Food chain concentration potential: Negative; unlikely to accumulate.
Water pollution: Effect of low concentrations on aquatic life is unknown. Fouling to shoreline. May be dangerous if it enters nearby water intakes; notify operators. Notify local health and wildlife officials. **Response to discharge:** Mechanical containment. Should be removed. Chemical and physical treatment.

SHIPPING INFORMATION
Grades of purity: Technical: 96-99.6%; **Storage temperature:** Ambient; **Inert atmosphere:** None; **Venting:** Open (flame arrester); **Stability during transport:** Stable.

PHYSICAL AND CHEMICAL PROPERTIES
Physical state @ 59°F/15°C and 1 atm: Liquid.
Molecular weight: 196.38
Boiling point @ 1 atm: 484.0°F/251.1°C/524.3°K.
Melting/Freezing point: 8.8°F/–12.9°C/260.3°K.
Specific gravity (water = 1): 0.771 @ 68°F/20°C (liquid).
Liquid surface tension: 25.0 dynes/cm = 0.025 N/m @ 68°F/20°C.
Liquid water interfacial tension: 32.8 dynes/cm = 0.0328 N/m at 22.7°C.
Relative vapor density (air = 1): 6.8
Ratio of specific heats of vapor (gas): 1.027
Latent heat of vaporization: 103 Btu/lb = 57.1 cal/g = 2.39×10^5 J/kg.
Heat of combustion: –17,600 Btu/lb = –9779 cal/g = $–409.4 \times 10^5$ J/kg.

TETRADECYLBENZENE REC. T:0550

SYNONYMS: 1-PHENYLTETRADECANE; TETRADECILBENCENO (Spanish)

IDENTIFICATION
Formula: $C_6NO_5(CH_2)_{13}CH_3$

DESCRIPTION: Colorless liquid. Mild odor. Floats on the surface of water.

Combustible • Containers may BLEVE when exposed to fire • Vapors in confined areas (e.g., tanks, sewers, buildings) may explode when exposed to fire • Irritating to the skin, eyes, and respiratory tract • Toxic products of combustion may include carbon monoxide and CO_2.

Hazard Classification (based on NFPA-704 Rating System)
Health Hazards (Blue): 0; Flammability (Red): 1; Reactivity (Yellow): 0

EMERGENCY RESPONSE: See Appendix A (171)
Evacuation:
Public safety: Isolate the area of spill or leak for at least 10 to 25 meters (30 to 80 feet) in all directions.
Spill: Increase, in the downwind direction, as necessary, the distance shown under "Public Safety."
Fire: If any large container is involved in fire, isolate for at least 800 meters (½ mile) in all directions; also, consider initial evacuation for 800 meters (½ mile) in all directions.

EXPOSURE
Short-term effects: *SEEK MEDICAL ATTENTION*. **Liquid:** Irritating to skin and eyes. *IF SWALLOWED*, will cause nausea and vomiting. Remove contaminated clothing and shoes. Flush affected areas with plenty of water. *IF IN EYES*, hold eyelids open and flush with plenty of water. *IF SWALLOWED* and victim is *CONSCIOUS AND ABLE TO SWALLOW*, have victim drink 4 to 8 ounces of water and have victim induce vomiting. *IF SWALLOWED* and victim is *UNCONSCIOUS OR HAVING CONVULSIONS*, do nothing except keep victim warm.

HEALTH HAZARDS
Personal protective equipment (PPE): B-Level PPE. Goggles or face shield; rubber gloves.
Liquid or solid irritant characteristics: May cause eye irritation.

FIRE DATA
Fire extinguishing agents not to be used: Water may be ineffective.
Burning rate: 4.42 mm/min
Electrical hazards: Flow or agitation of substance may generate electrostatic charges due to low conductivity.

CHEMICAL REACTIVITY
Binary reactants: Strong oxidizers. May attack some forms of plastics
Reactivity group: 32
Compatibility class: Aromatic hydrocarbons

ENVIRONMENTAL DATA
Food chain concentration potential: Negative; unlikely to accumulate.
Water pollution: Effect of low concentrations on aquatic life is unknown. Fouling to shoreline. May be dangerous if it enters nearby water intakes; notify operators. Notify local health and wildlife officials. **Response to discharge:** Mechanical containment. Should be removed. Chemical and physical treatment.

SHIPPING INFORMATION
Grades of purity: Commercial; **Storage temperature:** Ambient; **Inert atmosphere:** None; **Venting:** Open (flame arrester); **Stability during transport:** Stable.

PHYSICAL AND CHEMICAL PROPERTIES
Physical state @ 59°F/15°C and 1 atm: Solid.
Molecular weight: 274.47
Boiling point @ 1 atm: 678°F/359°C/632°K.
Melting/Freezing point: 61°F/16°C/289°K.
Specific gravity (water = 1): 0.855 @ 68°F/20°C (liquid).
Liquid surface tension: 30.27 dynes/cm = 0.03027 N/m @ 68°F/20°C.
Latent heat of vaporization: 95.18 Btu/lb = 52.88 cal/g = 2.212×10^5 J/kg.
Heat of combustion: –18,430 Btu/lb = –10,240 cal/g = $–428.4 \times 10^5$ J/kg.

TETRAETHYL DITHIOPYROPHOSPHATE REC. T:0600

SYNONYMS: ASP 47; BAYER-E-393; BIS-*O,O*-DIETHYLPHOSPHOROTHIONIC ANHYDRIDE; BLADAFUME; BLADAFUN; DITHIO; DITHIODIPHOSPHORIC ACID, TETRAETHYL ESTER; DITHIOPIROFOSFATO de TETRAETILO (Spanish); DITHIOFOS; DITHION; DITHIONE; DITHIOPHOS; DI(THIOPHOSPHORIC) ACID, TETRAETHYL ESTER; DITHIOPYROPHOSPHATE de TETRAETHYLE (French); DITHIOTEP; E393; ETHYL THIOPYROPHOSPHATE; LETHALAIRE G-57; PIROFOS; PLANT DITHIO AEROSOL; PLANTFUME 103 SMOKE GENERATOR; PYROPHOSPHORODITHIOIC ACID, TETRAETHYL ESTER; PYROPHOSPHORODITHIOIC ACID,*O,O,O,O*-TETRAETHYL ESTER; SULFOTEP; SULFATEP; SULFOTEPP; TEDP; TEDTP; TETRAETHYL DITHIOPYROPHOSPHATE; *O,O,O,O*-TETRAETHYLDITHIOPYROPHOSPHATE; THIOTEPP; DITHIOPYROPHOSPHORIC ACID,*O,O,O,O*-TETRAETHYL ESTER; RCRA No. P109

IDENTIFICATION
CAS Number: 3689-24-5
Formula: $C_8H_{20}O_5P_2S_2$; $(C_2H_5O)_2PSOPS(OC_2NO_5)_2$
DOT ID Number: UN 1704; DOT Guide Number: 153
Proper Shipping Name: Tetraethyl dithiopyrophosphate
Reportable Quantity (RQ): **(CERCLA)** 100 lb/45.4 kg

DESCRIPTION: Yellow to dark colored liquid. May be mixed with organic solvents. Sinks in water; slightly soluble; hydrolyzes very slowly.

Poison! • Breathing the vapor can kill you; skin or eye contact causes severe burns, impaired vision, or blindness • Firefighting gear (including SCBA) does not provide adequate protection. If exposure occurs, remove and isolate gear immediately and thoroughly decontaminate personnel • Containers may BLEVE when exposed to fire • Toxic products of combustion may include oxides of phosphorus and sulfur • Do not put yourself in danger by entering a contaminated area to rescue a victim.

Hazard Classification (based on NFPA-704 Rating System)
Health Hazards (Blue): 4; Flammability (Red): 1; Reactivity (Yellow): 1

EMERGENCY RESPONSE: See Appendix A (153)
Evacuation:
Public safety: Isolate the area of spill or leak for at least 25 to 50 meters (80 to 160 feet) in all directions.
Spill: Increase, in the downwind direction, as necessary, the distance shown under "Public Safety."
Fire: If tank, rail car, or tank truck is involved in fire, isolate for at least 800 meters (½ mile) in all directions; also, consider initial evacuation for 800 meters (½ mile) in all directions.

EXPOSURE
Short-term effects: *SEEK MEDICAL ATTENTION*. **Liquid:** *POISONOUS IF SWALLOWED OR IF SKIN IS EXPOSED*. Remove contaminated clothing and shoes. Flush affected areas with plenty of water. *IF IN EYES*, hold eyelids open and flush with plenty of water. *IF BREATHING HAS STOPPED*, give artificial respiration; *avoid mouth-to-mouth resuscitation; use bag/mask apparatus*. *IF SWALLOWED* and victim is *CONSCIOUS AND ABLE TO SWALLOW*, have victim drink 4 to 8 ounces of water and have victim induce vomiting. *IF SWALLOWED* and victim is *UNCONSCIOUS OR HAVING CONVULSIONS*, do nothing except keep victim warm.
Notes: Ingestion of liquid or inhalation of mist causes nausea, vomiting, mental confusion, abdominal pain, sweating, giddiness, apprehension, and restlessness; later, muscular twitching of the eyelids and tongue begin, then other muscles of face and neck become involved; generalized twitching and muscle weakness may occur; pulmonary edema, ataxia, tremor, and convulsions may advance to coma.
Note to physician or authorized medical personnel: Medical observation is recommended for 24 to 48 hours after breathing overexposure, as pulmonary edema may be delayed. As first aid for pulmonary edema, consider administering a corticosteroid spray. Cigarette smoking may exacerbate pulmonary injury and should be discouraged for at least 72 hours following exposure.

HEALTH HAZARDS
Personal protective equipment (PPE): B-Level PPE. OSHA Table Z-1-A air contaminant. Mask with canister approved for organic phosphate pesticides; goggles or face shield; rubber gloves and other protective clothing to prevent contact with skin.
Recommendations for respirator selection: NIOSH/OSHA
2 mg/m³: SA (any supplied-air respirator); or SCBA (any self-contained breathing apparatus). *5 mg/m³:* SA:CF (any supplied-air respirator operated in a continuous-flow mode). *10 mg/m³:* SCBAF (any self-contained breathing apparatus with a full facepiece); or SAF (any supplied-air respirator with a full facepiece). *EMERGENCY OR PLANNED ENTRY INTO UNKNOWN CONCENTRATIONS OR IDLH CONDITIONS:* SCBAF:PD,PP (any self-contained breathing apparatus that has a full facepiece and is operated in a pressure-demand or other positive-pressure mode); or SAF:PD,PP:ASCBA (any supplied-air respirator that has a full facepiece and is operated in a pressure-demand or other positive-pressure mode in combination with an auxiliary self-contained breathing apparatus operated in a pressure-demand or other positive pressure mode). *ESCAPE:* GMFOVHiE [any air-purifying, full-facepiece respirator (gas mask) with a chin-style, front- or back-mounted organic vapor canister having a high-efficiency particulate filter]; or SCBAE (any appropriate escape-type, self-contained breathing apparatus).

Exposure limits (TWA unless otherwise noted): ACGIH TLV 0.2 mg/m³; NIOSH/OSHA 0.2 mg/m³; skin contact contributes significantly in overall exposure.
Toxicity by ingestion: Grade 4; oral LD_{50} = 5 mg/kg (rat).
Long-term health effects: Causes chromosomal damage in mice.
Liquid or solid irritant characteristics: Eye and skin irritant. Absorbed through the skin.
IDLH value: 10 mg/m³.

CHEMICAL REACTIVITY
Binary reactants: Incompatible with antimony(V) pentafluoride, lead diacetate, magnesium, silver nitrate. Corrodes iron and possibly other metals, especially in the presence of moisture.

ENVIRONMENTAL DATA
Food chain concentration potential: Unlikely to accumulate.
Water pollution: DOT Appendix B, §172.101–marine pollutant. Low concentrations may be toxic to aquatic life. May be dangerous if it enters nearby water intakes; notify operators. Notify local health and wildlife officials. **Response to discharge:** Issue warning–poison, water contaminant. Restrict access. Should be removed. Chemical and physical treatment. Disperse and flush (small discharges).

SHIPPING INFORMATION
Grades of purity: Technical; Emulsifiable concentrate; Dry mixtures with inert solid, more than 2%; **Storage temperature:** Ambient; **Inert atmosphere:** None; **Venting:** Pressure-vacuum; **Stability during transport:** Stable.

PHYSICAL AND CHEMICAL PROPERTIES
Physical state @ 59°F/15°C and 1 atm: Liquid.
Molecular weight: 322.3
Boiling point @ 1 atm: Decomposes. Very high.
Specific gravity (water = 1): 1.19 @ 77°F/25°C (liquid).
Vapor pressure: 0.0002 mm.

TETRAETHYLENE GLYCOL　　　　**REC. T:0650**

SYNONYMS: HI-DRY; BIS-[2-(2-HYDROXYETHOXY) ETHYL] ETHER; 2,2'-[OXYBIS(ETHYLENEOXY)]DIETHANOL; TEG; TETRAETILENGLICOL (Spanish); 3,6,9-TRIOXAUNDECAN-1,11-DIOL; 3,6,9-TRIOXAUNDECANOL,11-DIOL

IDENTIFICATION
CAS Number: 112-60-7
Formula: $C_8H_{18}O_5$; $HO(C_2H_4O)_4H$

DESCRIPTION: Colorless to straw colored liquid. Mild odor. Sinks and mixes with water.

Combustible • Heat or flame may cause explosion • Containers may BLEVE when exposed to fire • Vapors in confined areas (e.g., tanks, sewers, buildings) may explode when exposed to fire • May be Irritating to the skin, eyes, and respiratory tract • Toxic products of combustion may include carbon monoxide and CO_2.

Hazard Classification (based on NFPA-704 Rating System)
Health Hazards (Blue): 1; Flammability (Red): 1; Reactivity (Yellow): 0

Tetraethylene pentamine

EMERGENCY RESPONSE: See Appendix A (171)
Evacuation:
Public safety: Isolate the area of spill or leak for at least 10 to 25 meters (30 to 80 feet) in all directions.
Spill: Increase, in the downwind direction, as necessary, the distance shown under "Public Safety."
Fire: If any large container is involved in fire, isolate for at least 800 meters (½ mile) in all directions; also, consider initial evacuation for 800 meters (½ mile) in all directions.

HEALTH HAZARDS
Personal protective equipment (PPE): B-Level PPE. Goggles or face shield and rubber gloves.
Toxicity by ingestion: Grade 0; oral rat LD_{50} = 28-34 g/kg.
Vapor (gas) irritant characteristics: Vapors are nonirritating to eyes and throat.
Liquid or solid irritant characteristics: No known hazard.

FIRE DATA
Flash point: 360°F/182°C (oc).
Fire extinguishing agents not to be used: Water or foam may cause frothing.
Electrical hazard: Class I, Group C.

CHEMICAL REACTIVITY
Binary reactants: Reacts violently with strong oxidizers, sodium peroxide, uranium fluoride. Incompatible with aliphatic amines, boranes, caustics, isocyanates, nitric acid, perchloric acid, sulfuric acid. Attacks some plastics, rubber, and coatings
Reactivity group: 40
Compatibility class: Glycol ethers

ENVIRONMENTAL DATA
Food chain concentration potential: Log K_{ow} = –1.7. Negative; unlikely to accumulate.
Water pollution: Effect of low concentrations on aquatic life is unknown. May be dangerous if it enters nearby water intakes; notify operators. Notify local health and wildlife officials.
Response to discharge: Disperse and flush.

SHIPPING INFORMATION
Grades of purity: Commercial, 99+%; **Storage temperature:** Ambient; **Inert atmosphere:** None; **Venting:** Open (flame arrester); **Stability during transport:** Stable.

NAS HAZARD CLASSIFICATION FOR BULK WATER TRANSPORTATION
FIRE: 1
HEALTH: Vapor irritant: 0; Liquid or solid irritant: 0; Poisons: 0
WATER POLLUTION: Human toxicity: 0; Aquatic toxicity: 1; Aesthetic effect: 1
REACTIVITY: Other chemicals: 2; Water: 0; Self-reaction: 0

PHYSICAL AND CHEMICAL PROPERTIES
Physical state @ 59°F/15°C and 1 atm: Liquid.
Molecular weight: 194.23
Boiling point @ 1 atm: Decomposes. 621°F/327°C/600°K.
Melting/Freezing point: 24.8°F/4.0°C/269.0°K.
Specific gravity (water = 1): 1.12 @ 68°F/20°C (liquid).
Liquid surface tension: 18.81 dynes/cm = 0.01881 N/m at 327°C.
Relative vapor density (air = 1): 6.7
Heat of combustion: –10,530 Btu/lb = –5,850 cal/g = –245 x 10^5 J/kg.
Vapor pressure: (Reid) Low

TETRAETHYLENE PENTAMINE REC. T:0700

SYNONYMS: AMINO ETHYL-1,2-ETHANEDIAMINE, 1,4,7,10,13,-PENTAAZATRIDECANE; D.E.H. 26; EEC No. 612-060-00-0; 1,2-ETHANEDIAMINE, *N*-(2-AMINOETHYL)-*N'*-[2-(2-AMINOETHYL)ETHYL]-; 1,2-ETHANEDIAMINE, *N*-(2-AMINOETHYL)-*N'*-(2-AMINOETHYL) AMINOETHYL-; 1,4,7,10,13-PENTAAZATRIDECANE; 1,11-DIAMINO-3,6,9 TRIAZAUNDECANE; TETRAETILENPENTAMINA (Spanish)

IDENTIFICATION
CAS Number: 112-57-2
Formula: $C_8H_{23}N_5$; $H_2N(C_2H_4NH)_4H$
DOT ID Number: UN 2320; DOT Guide Number: 153
Proper Shipping Name: Tetraethylenepentamine

DESCRIPTION: Yellow, thick liquid. Penetrating, ammonia odor. May float or sink in water; highly soluble.

Poison • Do NOT use Halons • Corrosive to skin, eyes, and respiratory tract; skin or eye contact causes severe burns, impaired vision, or blindness; effects of contact or inhalation may be delayed • Firefighting gear (including SCBA) does not provide adequate protection. If exposure occurs, remove and isolate gear immediately and thoroughly decontaminate personnel • Combustible • Containers may BLEVE when exposed to fire • Vapors in confined areas (e.g., tanks, sewers, buildings) may explode when exposed to fire • Do NOT use Halons • Toxic products of combustion may include nitrogen oxides.

Hazard Classification (based on NFPA-704 Rating System)
Health Hazards (Blue): 2; Flammability (Red): 1; Reactivity (Yellow): 0

EMERGENCY RESPONSE: See Appendix A (153)
Evacuation:
Public safety: Isolate the area of spill or leak for at least 25 to 50 meters (80 to 160 feet) in all directions.
Spill: Increase, in the downwind direction, as necessary, the distance shown under "Public Safety."
Fire: If tank, rail car, or tank truck is involved in fire, isolate for at least 800 meters (½ mile) in all directions; also, consider initial evacuation for 800 meters (½ mile) in all directions.

EXPOSURE
Short-term effects: *SEEK MEDICAL ATTENTION*. **Liquid:** Will burn skin and eyes. *IF SWALLOWED*, will cause nausea. Remove contaminated clothing and shoes. Flush affected areas with plenty of water. *IF BREATHING HAS STOPPED*, give artificial respiration; *avoid mouth-to-mouth resuscitation; use bag/mask apparatus*. *IF IN EYES*, hold eyelids open and flush with plenty of water. *IF SWALLOWED* and victim is *CONSCIOUS AND ABLE TO SWALLOW*, have victim drink 4 to 8 ounces of water. *IF SWALLOWED* and victim is *UNCONSCIOUS OR HAVING CONVULSIONS*, do nothing except keep victim warm. **Do NOT induce vomiting.**
Note to physician or authorized medical personnel: Medical observation is recommended for 24 to 48 hours after breathing overexposure, as pulmonary edema may be delayed. As first aid for pulmonary edema, consider administering a corticosteroid spray. Cigarette smoking may exacerbate pulmonary injury and should be discouraged for at least 72 hours following exposure.

HEALTH HAZARDS
Personal protective equipment (PPE): A-Level PPE. Chemical

protective material(s) reported to have good to excellent resistance: butyl rubber, neoprene, Viton®.
Toxicity by ingestion: Grade 2; oral LD_{50} = 3,990 mg/kg (rat).
Vapor (gas) irritant characteristics: Eye and respiratory tract irritant. Vapors cause smarting of the eyes or respiratory system if present in high concentrations. The effect is temporary.
Liquid or solid irritant characteristics: Severe eye and skin irritant. Causes smarting of the skin and first-degree burns on short exposure; may cause second-degree burns on long exposure.

FIRE DATA
Flash point: 320°F/160°C (oc).
Flammable limits in air: (estimated) LEL: 0.8%; UEL: 4.6%.
Fire extinguishing agents not to be used: Water or foam may cause frothing.
Autoignition temperature: 606°F/319°C/592°K.

CHEMICAL REACTIVITY
Binary reactants: Violent reaction with Halons. Reacts with acids, nitrogen compounds, oxidizers. May attack some forms of plastics. Attacks copper and copper alloys.
Reactivity group: 7
Compatibility class: Aliphatic amines

ENVIRONMENTAL DATA
Food chain concentration potential: Log K_{ow} = –1.49. Negative; unlikely to accumulate.
Water pollution: Effect of low concentrations on aquatic life is unknown. May be fouling to shoreline. May be dangerous if it enters nearby water intakes; notify operators. Notify local health and wildlife officials. **Response to discharge:** Issue warning–water contaminant. Restrict access. Disperse and flush.

SHIPPING INFORMATION
Grades of purity: Commercial; **Storage temperature:** Ambient; **Inert atmosphere:** None; **Venting:** Open (flame arrester); **Stability during transport:** Stable.

NAS HAZARD CLASSIFICATION FOR BULK WATER TRANSPORTATION
FIRE: 1
HEALTH: Vapor irritant: 1; Liquid or solid irritant: 2; Poisons: 2
WATER POLLUTION: Human toxicity: 2; Aquatic toxicity: 2; Aesthetic effect: 2
REACTIVITY: Other chemicals: 3; Water: 0; Self-reaction: 0

PHYSICAL AND CHEMICAL PROPERTIES
Physical state @ 59°F/15°C and 1 atm: Liquid.
Molecular weight: 189
Boiling point @ 1 atm: 631°F/333°C/606°K.
Melting/Freezing point: –22°F/–30°C/243°K.
Specific gravity (water = 1): 0.998 @ 68°F/20°C (liquid).
Relative vapor density (air = 1): 6.48
Vapor pressure: (Reid) Low; less than 0.08 mm.

TETRAETHYL LEAD REC. T:0750

SYNONYMS: EEC No. 082-002-00-3; LEAD TETRAETHYL; MOTOR FUEL ANTI-KNOCK COMPOUND; PLUMBANE, TETRAETHYL-; TEL; TETRAETHYLE PLUMB; TETRAETHYLPLUMBANE; RCRA No. P110

IDENTIFICATION
CAS Number: 78-00-2
Formula: $C_8H_{20}Pb$; $Pb(C_2NO_5)_4$
DOT ID Number: UN 1649; DOT Guide Number: 131
Proper Shipping Name: Tetraethyl lead, liquid
Reportable Quantity (RQ): **(CERCLA)** 10 lb/4.54 kg

DESCRIPTION: Colorless oily liquid. Generally dyed red, orange, blue. Sweet, fruity odor. Sinks in water; insoluble. Poisonous, flammable vapor is produced. Slowly decomposes at room temperature; more rapidly at elevated temperatures.

Poison! • Breathing the vapor, absorption through the skin can cause severe lead poisoning • Firefighting gear (including SCBA) does not provide adequate protection. If exposure occurs, remove and isolate gear immediately and thoroughly decontaminate personnel • Containers may BLEVE when exposed to fire • Vapors are heavier than air and will collect and stay in low areas • Vapors in confined areas (e.g., tanks, sewers, buildings) may explode when exposed to fire • Decomposes above 257°F/125°C, and explosively, if rapidly heated to 608°F/320°C. Toxic products of combustion may include lead • Do not put yourself in danger by entering a contaminated area to rescue a victim.

Hazard Classification (based on NFPA-704 Rating System)
Health Hazards (Blue): 3; Flammability (Red): 2; Reactivity (Yellow): 3

EMERGENCY RESPONSE: See Appendix A (131)
Evacuation:
Public safety: Isolate spill area for at least 100 to 200 meters (330 to 660 feet) in all directions.
Spill: Increase, as necessary, the isolation distance shown above, in "Public safety."
Fire: Isolate for 800 meters (½ mile) in all directions, especially if tank, rail car, or tank truck is involved in fire.

EXPOSURE
Short-term effects: *SEEK MEDICAL ATTENTION.* **Vapor:** *POISONOUS IF INHALED OR IF SKIN IS EXPOSED.* Irritating to eyes. Move to fresh air. *IF BREATHING HAS STOPPED,* give artificial respiration. IF breathing is difficult, administer oxygen.
Liquid: *POISONOUS IF SWALLOWED OR IF SKIN IS EXPOSED.*
Will burn eyes. Remove contaminated clothing and shoes. Flush affected areas with plenty of water. *IF IN EYES,* hold eyelids open and flush with plenty of water. *IF SWALLOWED* and victim is *CONSCIOUS AND ABLE TO SWALLOW,* have victim drink 4 to 8 ounces of water and have victim induce vomiting. *IF SWALLOWED* and victim is *UNCONSCIOUS OR HAVING CONVULSIONS,* do nothing except keep victim warm.

HEALTH HAZARDS
Personal protective equipment (PPE): A-Level PPE. OSHA Table Z-1-A air contaminant. Neoprene-coated, liquid-proof gloves.
Recommendations for respirator selection: NIOSH/OSHA
0.75 ppm: SA (any supplied-air respirator). *1.875 ppm:* SA:CF (any supplied-air respirator operated in a continuous-flow mode). *3.75 ppm:* SAT:CF (any supplied-air respirator that has a tight-fitting facepiece and is operated in a continuous-flow mode); or SCBAF (any self-contained breathing apparatus with a full facepiece); or SAF (any supplied-air respirator with a full facepiece). *40 ppm:* SA:PD,PP (any supplied-air respirator operated in a pressure-demand or other positive-pressure mode). *EMERGENCY OR PLANNED ENTRY INTO UNKNOWN CONCENTRATIONS OR IDLH CONDITIONS:* SCBAF:PD,PP (any self-contained breathing apparatus that has a full facepiece and is operated in a pressure-

demand or other positive-pressure mode); or SAF:PD,PP:ASCBA (any supplied-air respirator that has a full facepiece and is operated in a pressure-demand or other positive-pressure mode in combination with an auxiliary self-contained breathing apparatus operated in a pressure-demand or other positive pressure mode). *ESCAPE:* GMFOV [any air-purifying, full-facepiece respirator (gas mask) with a chin-style, front- or back-mounted organic vapor cannister]; or SCBAE (any appropriate escape-type, self-contained breathing apparatus).

Exposure limits (TWA unless otherwise noted): ACGIH TLV 0.1 mg/m^3 as lead; NIOSH/OSHA 0.075 mg/m^3; skin contact contributes significantly in overall exposure.

Toxicity by ingestion: Oral rat LD$_{Lo}$ = 17 mg/kg.

Long-term health effects: Lead poisoning.

Vapor (gas) irritant characteristics: Vapors cause smarting of the eyes or respiratory system if present in high concentrations.

Liquid or solid irritant characteristics: Eye and skin irritant. Causes smarting of the skin and first-degree burns on short exposure; may cause secondary burns on long exposure.

IDLH value: 100 mg/m^3 as lead.

FIRE DATA

Flash point: 185°F/85°C (oc); 200°F/93°C (cc).
Behavior in fire: Containers may explode.
Autoignition temperature: Decomposes.
Autoignition temperature: More than 230°F/110°C.

CHEMICAL REACTIVITY

Binary reactants: Decomposes in sunlight and elevated temperatures above 230°F/110°C forming triethyl lead; explodes above 608°F/320°C. violent reaction with strong oxidizers, concentrated acids. Corrodes rubber. Attacks some plastics and coatings. Rust and some metals cause decomposition. Decomposes slowly at room temperature and more rapidly at higher temperatures.

ENVIRONMENTAL DATA

Water pollution: DOT Appendix B, §172.101–marine pollutant. Harmful to aquatic life in very low concentrations. May be dangerous if it enters nearby water intakes; notify operators. Notify local health and wildlife officials. **Response to discharge:** Issue warning–poison, water contaminant. Restrict access. Should be removed. Chemical and physical treatment.

SHIPPING INFORMATION

Grades of purity: Technical; **Storage temperature:** Ambient; **Inert atmosphere:** None; **Venting:** Pressure-vacuum; **Stability during transport:** Stable below 230°F/110°C. At higher temperatures, may detonate or explode when confined.

PHYSICAL AND CHEMICAL PROPERTIES

Physical state @ 59°F/15°C and 1 atm: Liquid.
Molecular weight: 323.44
Boiling point @ 1 atm: Decomposes. 228°F/109°C/382°K.
Melting/Freezing point: –215°F/–137°C/136°K.
Specific gravity (water = 1): 1.633 @ 68°F/20°C (liquid).
Liquid surface tension: 28.5 dynes/cm = 0.0285 N/m at (estimate) 77°F/25°C.
Liquid water interfacial tension: (estimate) 40 dynes/cm = 0.04 N/m @ 68°F/20°C.
Relative vapor density (air = 1): 8.6
Heat of combustion: (estimate) –7,870 Btu/lb = –4,380 cal/g = –183 × 10^5 J/kg.
Vapor pressure: 0.2 mm.

TETRAETHYL PYROPHOSPHATE   REC. T:0800

SYNONYMS: BIS-*O,O*-DIETHYLPHOSPHORIC ANHYDRIDE; BLADAN; BLADON; DIPHOSPHORIC ACID, TETRAETHYL ESTER; ETHYL PYROPHOSPHATE, TETRA-; FOSVEX; GRISOL; HEPT; HEXAMITE; KILLAX; KILMITE 40; LETHALAIRE G-52; LIROHEX; MORTOPAL; MOTOPAL; NIFOS; NIFOS T; NIFROST; PIROFOSFATO de TETRAETILO (Spanish; PHOSPHORIC ACID, TETRAETHYL ESTER; COMMERCIAL 40%; PYROPHOSPHATE de TETRAETHYLE (French); TEP; TEPP; BIS(*O,O*-DIAETHYLPHOSPHORSAEURE-ANHYDRID (German); TETRAETHYLPYROPHOSPHATE; TETRASTIGMINE; TETRON; TETRON-100; VAPOTONE; RCRA No. P111

IDENTIFICATION

CAS Number: 107-49-3
Formula: $(C_2H_5O)_2POOPO(OC_2NO_5)_2$; $(C_2NO_5)_4P_2O_7$
DOT ID Number: UN 3018 (liquid); UN 2783 (solid); UN 1705 (compressed gas mixtures); DOT Guide Number: 152 (liquid and solid); 123 (compressed gas mixtures).
PROPER SHIPPING NAMES: Tetraethyl pyrophosphate, liquid; Tetraethyl pyrophosphate, solid; Tetraethyl pyrophosphate and compressed gas mixtures LC$_{50}$ less than or equal to 200 ppm; Tetraethyl pyrophosphate and compressed gas mixtures LC$_{50}$ over 200 ppm but not greater than 5000 ppm
Reportable Quantity (RQ): **(CERCLA)** 10 lb/4.54 kg

DESCRIPTION: Colorless to yellow to dark amber liquid, depending on purity. Faint fruity odor. Mixes with water; forming corrosive phosphoric acid solution and toxic ethylene gas. Freezes at 32°F/0°C.

Poison! (organophosphate) • Breathing the vapor, skin or eye contact, or swallowing the material can kill you • Firefighting gear (including SCBA) does not provide adequate protection. If exposure occurs, remove and isolate gear immediately and thoroughly decontaminate personnel • Containers may BLEVE when exposed to fire • Toxic products of combustion may include phosphoric acid or phosphorus oxide, or ethylene gas • Do not put yourself in danger by entering a contaminated area to rescue a victim.

Hazard Classification (based on NFPA-704 Rating System)
Health Hazards (Blue): 4; Flammability (Red): 1; Reactivity (Yellow): 1

EMERGENCY RESPONSE, liquid/solid: see Appendix A (152)
Evacuation:
Public safety: Isolate the area of spill or leak for at least 25 to 50 meters (80 to 160 feet) in all directions.
Spill: Increase, in the downwind direction, as necessary, the distance shown under "Public Safety."
Fire: If tank, rail car, or tank truck is involved in fire, isolate for at DAY: 4.0 km (2.5 miles); NIGHT: 7.2 km (4.5 miles).
Fire: Isolate for 800 meters (½ mile) in all directions, especially if tank, rail car, or tank truck is involved in fire.
EMERGENCY RESPONSE, gas: See Appendix A (123)
Evacuation:
Public safety: See below (UN 1703 worst case)
Spill: Small spill–First: Isolate in all directions 30 meters (100 feet); Then: Protect persons downwind, DAY: 0.3 km (0.2 mile); NIGHT: 1.3 km (0.8 mile). Large spill–First: Isolate in all directions 400 meter (1300 feet); Then: Protect persons downwind,

DAY: 4.0 km (2.5 miles); NIGHT: 7.2 km (4.5 miles).
Fire: Isolate for 800 meters (½ mile) in all directions, especially if tank, rail car, or tank truck is involved in fire.

EXPOSURE
Short-term effects: *SEEK MEDICAL ATTENTION.* **Inhalation, ingestion or absorption:** Inhalation of mist, dust, or vapor (or ingestion, or absorption through the skin) affects the nerves, muscles; causes salivation and other excessive respiratory tract secretion, nausea, stomach pain, headache, vomiting, diarrhea, headache, sweating, blurred vision, and pinpoint pupils of the eyes. High levels of exposure may cause convulsions, loss of reflexes, and loss of sphincter control unconsciousness, coma, and death. The symptoms may develop over a period of 8 hours. An increase in salivary and bronchial secretions may result which simulate severe pulmonary edema. A cholinesterase inhibitor. The action is similar to that of parathion: causing an irreversible inhibition of the cholinesterase molecules and the consequent accumulation of large amounts of acetylcholine. *IF BREATHING HAS STOPPED,* give artificial respiration; *avoid mouth-to-mouth resuscitation: use bag/mask apparatus.* **Vapor or Liquid:** *POISONOUS IF INHALED, IF SWALLOWED, OR IF SKIN IS EXPOSED.* **Skin:** *Remove and double-bag contaminated clothing (including shoes) and leave them in the Hot Zone;* wash skin with soap and water and large volumes of water for at least 15 minutes. *Skin can also be decontaminated with diluted hypochlorite solution, U.S. Army M291 kit, and M258(A1) skin decontamination kit.* **Eye:** Rinse with large volumes of water or saline for at least 15 minutes. *IF SWALLOWED,* will cause nausea, vomiting, or loss of consciousness. *Remove and double-bag contaminated clothing and shoes and leave in Hot Zone for later incineration by hazardous materials experts.* Flush affected areas with plenty of water. **If swallowed** and victim is *CONSCIOUS AND ABLE TO SWALLOW,* have victim drink 4 to 8 ounces of water. **Do NOT induce vomiting but immediately administer slurry of activated charcoal.** *IF SWALLOWED* and victim is *UNCONSCIOUS OR HAVING CONVULSIONS,* do nothing except keep victim warm. *Note to physician and medical personnel:* If symptoms indicate, initial treatment for an adult includes atropine, 2 mg (1/30 gr) intramuscularly or intravenously as soon as any local or systemic signs or symptoms of an intoxication are noted; repeat the administration of atropine every 3 to 8 minutes until signs of atropinization (mydriasis, dry mouth, rapid pulse, hot and dry skin) occur; initiate treatment in children with 0.05 mg/kg of atropine; repeat at 5- to 10-minute intervals. Watch respiration, and remove bronchial secretions if they appear to be obstructing the airway; intubate if necessary. Pralidoxime must be administered within minutes to a few hours following exposure (depending on the specific agent) to be effective. Give 2-PAMCI (Pralidoxime; Protopam), 2.5 g in 100 mL of sterile water or in 5% dextrose and water, intravenously, slowly, in 15 to 30 minutes; if sufficient fluid is not available, give 1 g of 2-PAMCI in 3 mL of distilled water by deep intramuscular injection; repeat this every half hour if respiration weakens or if muscle fasciculation or convulsions recur. Also Diazepam, an anticonvulsant, might be needed. For adults, the dose is 5 to 10 mg (slow IV), repeated every 12 to 15 minutes up to three (3) doses maximum. For children, the dose is 0.2 to 0.5 mg/kg. *Note to physician or authorized medical personnel:* Medical observation is recommended for 24 to 48 hours after breathing overexposure, as pulmonary edema may be delayed. As first aid for pulmonary edema, consider administering a corticosteroid spray. Cigarette smoking may exacerbate pulmonary injury and should be discouraged for at least 72 hours following exposure.

HEALTH HAZARDS
Personal protective equipment (PPE): A-Level PPE. OSHA Table Z-1-A air contaminant. Mask with canister approved for organic phosphate pesticides; goggles or face shield; rubber gloves and other protective clothing to prevent contact with skin. Wear sealed chemical suit of butyl rubber. Also, polycarbonate offers chemical resistance.
Recommendations for respirator selection: NIOSH/OSHA
0.5 mg/m³: SA (any supplied-air respirator). *1.25 mg/m³:* SA:CF (any supplied-air respirator operated in a continuous-flow mode). *2.5 mg/m³:* SAT:CF (any supplied-air respirator that has a tight-fitting facepiece and is operated in a continuous-flow mode); or SCBAF (any self-contained breathing apparatus with a full facepiece); or SAF (any supplied-air respirator with a full facepiece). *5 mg/m³:* SA:PD,PP (any supplied-air respirator operated in a pressure-demand or other positive-pressure mode). *EMERGENCY OR PLANNED ENTRY INTO UNKNOWN CONCENTRATIONS OR IDLH CONDITIONS:* SCBAF:PD,PP (any self-contained breathing apparatus that has a full facepiece and is operated in a pressure-demand or other positive-pressure mode); or SAF:PD,PP:ASCBA (any supplied-air respirator that has a full facepiece and is operated in a pressure-demand or other positive-pressure mode in combination with an auxiliary self-contained breathing apparatus operated in a pressure-demand or other positive pressure mode). *ESCAPE:* GMFOVHiE [any air-purifying, full-facepiece respirator (gas mask) with a chin-style, front- or back-mounted organic vapor canister having a high-efficiency particulate filter]; or SCBAE (any appropriate escape-type, self-contained breathing apparatus).
Exposure limits (TWA unless otherwise noted): ACGIH TLV 0.04 mg/m³; NIOSH/OSHA 0.05 mg/m³; skin contact contributes significantly in overall exposure.
Short-term exposure limits (15-minute TWA): CHRIS 0.25 mg/m³ for 30 minutes
Toxicity by ingestion: Poison. Grade 4; LD_{50} less than 50 mg/kg.
Long-term health effects: Poison. Small doses may accumulate.
Liquid or solid irritant characteristics: Eye and skin irritant.
IDLH value: 5 mg/m³.

FIRE DATA
Behavior in fire: Water streams applied to adjacent fires will spread contamination of pesticide over wide area.

CHEMICAL REACTIVITY
Reactivity with water: Reacts slowly to form phosphoric acid.
Binary reactants: Corrosive to aluminum, slowly corrosive to copper, brass, zinc, and tin
Neutralizing agents for acids and caustics: Flush with water, rinse with sodium bicarbonate or lime solution.

ENVIRONMENTAL DATA
Food chain concentration potential: Negative; unlikely to accumulate.
Water pollution: DOT Appendix B, §172.101–marine pollutant. Harmful to aquatic life in very low concentrations. May be dangerous if it enters nearby water intakes; notify operators. Notify local health and wildlife officials. For weapons testing, use an M272 Water Detection Kit. **Response to discharge:** Issue warning–poison, water contaminant. Restrict access. Should be removed. Chemical and physical treatment. Disperse and flush (small discharges).

SHIPPING INFORMATION
Grades of purity: Technical: 40% plus 60% related ethyl phosphates; Aerosols (5–10%) (Class A poisons); Dusts

(0.66–1.2%); Sprays 10–40%; **Storage temperature:** Ambient; **Inert atmosphere:** None; **Venting:** Open; **Stability during transport:** Stable.

PHYSICAL AND CHEMICAL PROPERTIES
Physical state @ 59°F/15°C and 1 atm: Liquid.
Molecular weight: 290.2
Melting/Freezing point: 32°F/0°C/273°K.
Specific gravity (water = 1): 1.18 @ 77°F/25°C (liquid).
Vapor pressure: 0.00015 mm.

TETRAFLUOROETHYLENE   REC. T:0850

SYNONYMS: ETHYLENE, TETRAFLUORO-; FLUROPLAST 4; PERFLUROETHYLENE; 1,1,2,2-TETRAFLUOROETHYLENE; TETRAFLUROETHENE; TETRAFLUOETILENO (Spanish); TFE; TEFLON MONOMER

IDENTIFICATION
CAS Number: 116-14-3
Formula: C_2F_4; $F_2C = CF_2$
DOT ID Number: UN 1081; DOT Guide Number: 116P
Proper Shipping Name: Tetrafluoroethylene, inhibited

DESCRIPTION: Colorless compressed gas. Odorless or faint odor. Reacts with water; insoluble; visible vapor cloud is produced.

Extremely flammable • Containers may explode when exposed to fire • Gas form explosive mixture with air • Gas is heavier than air and will collect and stay in low areas • Gas may travel long distances to ignition sources and flashback • Gas in confined areas (e.g., tanks, sewers, buildings) may explode when exposed to fire • Vapors may cause dizziness or asphyxiation without warning • Contact with liquid may cause frostbite • Polymerization hazard • May react with itself without warning blocking relief valves leading to container explosion • Irritating to the skin, eyes, and respiratory tract • Toxic products of combustion may include carbonyl fluoride and hydrogen fluoride.

Hazard Classification (based on NFPA-704 Rating System)
Health Hazards (Blue): 2; Flammability (Red): 4; Reactivity (Yellow): 3

EMERGENCY RESPONSE: See Appendix A (116)
Evacuation:
Public safety: Isolate spill area for at least 100 meters (330 feet) in all directions.
Spill: Consider initial downwind evacuation for at least 800 meters (½ mile).
Fire: Isolate for 1600 meters (1 mile) in all directions, especially if tank, rail car, or tank truck is involved in fire.

EXPOSURE
Short-term effects: *SEEK MEDICAL ATTENTION.* **Vapor:** Irritating to eyes, nose, and throat. Move victim to fresh air. *IF BREATHING HAS STOPPED,* give artificial respiration. IF breathing is difficult, administer oxygen.

HEALTH HAZARDS
Personal protective equipment (PPE): B-Level PPE. Self-contained breathing apparatus for high gas concentrations. Chemical protective material(s) reported to have good to excellent resistance: Viton®. Butyl rubber, neoprene, PV alcohol, Viton®.

Long-term health effects: Causes possible impairment of immunological defense system in rats; IARC rating 3; no human or animal data.
Vapor (gas) irritant characteristics: Eye and respiratory tract irritant.
Liquid or solid irritant characteristics: Eye irritant.

FIRE DATA
Flash point: Extremely flammable gas.
Flammable limits in air: LEL: 10%; UEL: 50%.
Behavior in fire: Containers may explode.
Autoignition temperature: 383°F/195°C/468°K.
Electrical hazard: C_2F_4-air mixtures produced explosions which propagated through the smallest clearance in the standard test conducted by Underwriters Laboratories. It does not meet any group classification.
Stoichiometric air-to-fuel ratio: 2.746 (estimate).

CHEMICAL REACTIVITY
Binary reactants: Reacts with air, oxygen; oxidizers.
Polymerization: Can polymerize unless inhibited, especially when heated or in presence of oxygen; **Inhibitor of polymerization:** D-limonene; pinene; tetrahydronaphthalene; 1-octene; methyl methacrylate

ENVIRONMENTAL DATA
Water pollution: Not harmful to aquatic life. **Response to discharge:** Issue warning–high flammability. Restrict access.

SHIPPING INFORMATION
Grades of purity: 98+%; **Storage temperature:** Cool ambient; **Inert atmosphere:** None; **Venting:** Safety relief; **Stability during transport:** Stable.

PHYSICAL AND CHEMICAL PROPERTIES
Physical state @ 59°F/15°C and 1 atm: Gas.
Molecular weight: 100.0
Boiling point @ 1 atm: –105°F/–76°C/197°K.
Melting/Freezing point: –224°F/–142°C/131°K.
Critical temperature: (estimate) 92°F/33°C/306°K.
Critical pressure: (estimate) 573 psia = 38.9 atm = 3.95 MN/m².
Relative vapor density (air = 1): 3.9
Ratio of specific heats of vapor (gas): (estimate) 1.1261
Heat of combustion: (estimate) –4000 Btu/lb = –2000 cal/g = –90 x 10^5 J/kg.
Heat of polymerization: –450 Btu/lb = –250 cal/g = –10.5 x 10^5 J/kg.

TETRAHYDROFURAN   REC. T:0900

SYNONYMS: BUTANE, 1,4-EPOXY-; CYCLOTETRAMETHYLENE OXIDE; DIETHYLENE OXIDE; EEC No. 603-025-00-0; 1,4-EPOXYBUTANE; FURANIDINE; FURAN, TETRAHYDRO-; HYDROFURAN; OXACYCLOPENTANE; OXOLANE; TETRAHYDROFURANNE (French); TETRHIDROFURANO (Spanish); TETRAMETHYLENE OXIDE; THF; RCRA No. U213

IDENTIFICATION
CAS Number: 109-99-9
Formula: C_4H_8O; $CH_2CH_2CH_2CH_2-O$
DOT ID Number: UN 2056; DOT Guide Number: 127
Proper Shipping Name: Tetrahydrofuran
Reportable Quantity (RQ): **(CERCLA)** 1000 lb/454 kg

Tetrahydrofuran

DESCRIPTION: Colorless liquid. Faint fruity odor, like acetone. Initially floats on water surface; soluble. Large amounts of flammable, irritating vapor is produced. *Warning*:

Highly flammable • Forms unstable peroxides after prolonged exposure to air; may become shock- or heat-sensitive. Containers may BLEVE when exposed to fire • Vapors may form explosive mixture with air • Vapors are heavier than air and will collect and stay in low areas • Vapors may travel long distances to ignition sources and flashback • Vapors in confined areas (e.g., tanks, sewers, buildings) may explode when exposed to fire • Irritating to the skin, eyes, and respiratory tract • Toxic products of combustion may include carbon monoxide.

Hazard Classification (based on NFPA-704 Rating System)
Health Hazards (Blue): 2; Flammability (Red): 3; Reactivity (Yellow): 1

EMERGENCY RESPONSE: See Appendix A (127)
Evacuation:
Public safety: Isolate spill area for at least 25 to 50 meters (80 to 160 feet) in all directions.
Spill: Large spill–Consider initial downwind evacuation for at least 300 meters (1000 feet).
Fire: Isolate for 800 meters (½ mile) in all directions, especially if tank, rail car, or tank truck is involved in fire.

EXPOSURE
Short-term effects: *SEEK MEDICAL ATTENTION*. The use of alcoholic beverages may enhance toxic effects. **Vapor:** Irritating to eyes, nose, and throat. *IF INHALED*, will, will cause nausea, headache, loss of consciousness. May affect the nervous system.Move to fresh air. *IF BREATHING HAS STOPPED*, give artificial respiration. IF breathing is difficult, administer oxygen. **Liquid:** Irritating to skin and eyes. Harmful if swallowed. Remove contaminated clothing and shoes. Flush affected areas with plenty of water. *IF IN EYES*, hold eyelids open and flush with plenty of water. *IF SWALLOWED* and victim is *CONSCIOUS AND ABLE TO SWALLOW*, have victim drink 4 to 8 ounces of water.

HEALTH HAZARDS
Personal protective equipment (PPE): B-Level PPE. OSHA Table Z-1-A air contaminant. Self-contained breathing apparatus; goggles or face shield; gloves. Chemical protective material(s) reported to have good to excellent resistance: Teflon®, Silvershield®. Also, PV alcohol offers limited protection.
Recommendations for respirator selection: NIOSH/OSHA
2000 ppm: SA:CF* (any supplied-air respirator operated in a continuous-flow mode); CCRFOV [any chemical cartridge respirator with a full facepiece and organic vapor cartridges(s)]; or GMFOV [any air-purifying, full-facepiece respirator (gas mask) with a chin-style, front- or back-mounted acid gas canister]; or PAPROV* [any powered, air-purifying respirator with organic vapor cartridge(s)]; or SCBAF (any self-contained breathing apparatus with a full facepiece); or SAF (any supplied-air respirator with a full facepiece). *EMERGENCY OR PLANNED ENTRY INTO UNKNOWN CONCENTRATIONS OR IDLH CONDITIONS*: SCBAF:PD,PP (any self-contained breathing apparatus that has a full facepiece and is operated in a pressure-demand or other positive-pressure mode); or SAF:PD,PP:ASCBA (any supplied-air respirator that has a full facepiece and is operated in a pressure-demand or other positive-pressure mode in combination with an auxiliary self-contained breathing apparatus operated in a pressure-demand or other positive pressure mode). *ESCAPE:* GMFOV [any air-purifying, full-facepiece respirator (gas mask) with a chin-style, front- or back-mounted organic vapor canister]; or SCBAE (any appropriate escape-type, self-contained breathing apparatus). *Note*: Substance causes eye irritation or damage; eye protection needed.
Exposure limits (TWA unless otherwise noted): ACGIH TLV 200 ppm (590 mg/m^3); NIOSH/OSHA 200 ppm (590 mg/m^3).
Short-term exposure limits (15-minute TWA): ACGIH STEL 250 ppm (737 mg/m^3); NIOSH STEL 250 ppm (737 mg/m^3).
Toxicity by ingestion: Grade 3; LD_{50} = 50 to 500 mg/kg.
Vapor (gas) irritant characteristics: Vapors cause smarting of the eyes or respiratory system if present in high concentrations.
Liquid or solid irritant characteristics: Eye and skin irritant. If spilled on clothing and allowed to remain, may cause smarting and reddening of the skin.
Odor threshold: 19–50 ppm.
IDLH value: 2000 ppm [10% LEL].

FIRE DATA
Flash point: –4°F/–20°C (oc); 6°F/–14°C (cc).
Flammable limits in air: LEL: 1.8%; UEL: 11.8%.
Fire extinguishing agents not to be used: Water may be ineffective.
Minimum ignition energy, milliJoules: 0.54
Autoignition temperature: 610°F/321°C/594°K.
Electrical hazard: Class I, Group C.
Burning rate: 4.7 mm/min

CHEMICAL REACTIVITY
Binary reactants: Reacts with strong oxidizers, lithium–aluminum alloys. Attacks some plastic materials. Dissolves rubber. Peroxides may accumulate after prolonged storage in presence of air, or exposure to light.
Polymerization: Peroxide formation may cause explosion of uninhibited substance.; **Inhibitor of polymerization:** 0.025% butylated hydroxytoluene (BHT) present to prevent peroxide formation. Inhibitor may be depleted when THF comes in contact with highly alkaline or caustic substances.
Reactivity group: 41
Compatibility class: Ethers

ENVIRONMENTAL DATA
Food chain concentration potential: Log K_{ow} = 0.46. Negative; unlikely to accumulate.
Water pollution: Effect of low concentrations on aquatic life is unknown. May be dangerous if it enters nearby water intakes; notify operators. Notify local health and wildlife officials. **Response to discharge:** Issue warning–high flammability. Disperse and flush.

SHIPPING INFORMATION
Storage temperature: Ambient; **Inert atmosphere:** Padded; **Venting:** Pressure-vacuum; **Stability during transport:** Stable unless 0.1% of peroxides has accumulated because of prolonged storage in presence of air. When concentrated by evaporation of solution, they explode.

NAS HAZARD CLASSIFICATION FOR BULK WATER TRANSPORTATION
FIRE: 3
HEALTH: Vapor irritant: 1; Liquid or solid irritant: 1; Poisons: 2
WATER POLLUTION: Human toxicity: 2; Aquatic toxicity: 2; Aesthetic effect: 0
REACTIVITY: Other chemicals: 1; Water: 0; Self-reaction: 3

PHYSICAL AND CHEMICAL PROPERTIES
Physical state @ 59°F/15°C and 1 atm: Liquid.

Molecular weight: 72.10
Boiling point @ 1 atm: 151°F/66°C/339°K.
Melting/Freezing point: −163.3°F/−108.5°C/164.7°K.
Critical temperature: 512.6°F/267.0°C/540.2°K.
Critical pressure: 753 psia = 51.2 atm = 5.19 MN/m².
Specific gravity (water = 1): 0.888 @ 68°F/20°C (liquid).
Liquid surface tension: 28 dynes/cm = 0.028 N/m @ 68°F/20°C.
Relative vapor density (air = 1): 2.49
Ratio of specific heats of vapor (gas): (estimate) 1.083
Latent heat of vaporization: 180 Btu/lb = 98 cal/g = 4.1×10^5 J/kg.
Heat of combustion: −14,990 Btu/lb = −8330 cal/g = -348.8×10^5 J/kg.
Vapor pressure: (Reid) 7.7 psia; 132 mm.

TETRAHYDRONAPHTHALENE　　　　　　**REC. T:0950**

SYNONYMS: TETRAHIDRONAFTALEN O (Spanish); 1,2,3,4-TETRAHYDRONAPHTHALENE; TETRALIN; TETRALINE; TETRAMP; TETRANAP

IDENTIFICATION
CAS Number: 119-64-2
Formula: $C_{10}H_{12}$
DOT ID Number: NA 1993; DOT Guide Number: 128
Proper Shipping Name: Combustible liquid, n.o.s.

DESCRIPTION: Colorless watery liquid. Mild, turpentine or naphthalene odor. Floats on water surface; insoluble.

Highly flammable • Forms unstable peroxides after prolonged exposure to air; may become shock- or heat-sensitive; may explode spontaneously • Containers may BLEVE when exposed to fire • Vapors may form explosive mixture with air • Vapors are heavier than air and will collect and stay in low areas • Vapors may travel long distances to ignition sources and flashback • Vapors in confined areas (e.g., tanks, sewers, buildings) may explode when exposed to fire • Irritating to the skin, eyes, and respiratory tract • Toxic products of combustion may include carbon monoxide.

Hazard Classification (based on NFPA-704 Rating System)
Health Hazards (Blue): 1; Flammability (Red): 2; Reactivity (Yellow): 0

EMERGENCY RESPONSE: See Appendix A (128)
Evacuation:
Public safety: Isolate spill area for at least 25 to 50 meters (80 to 160 feet) in all directions.
Spill: Large spill—Consider initial downwind evacuation for at least 300 meters (1000 feet).
Fire: Isolate for 800 meters (½ mile) in all directions, especially if tank, rail car, or tank truck is involved in fire.

EXPOSURE
Short-term effects: *SEEK MEDICAL ATTENTION*. **Vapor:** Causes headache, dizziness. High levels can cause unconsciousness. The use of alcoholic beverages may enhance any toxic effects. **Liquid:** Irritating to skin and eyes. Harmful if swallowed. Remove contaminated clothing and shoes. Flush affected areas with plenty of water. *IF IN EYES*, hold eyelids open and flush with plenty of water. *IF SWALLOWED* and victim is *CONSCIOUS AND ABLE TO SWALLOW*, have victim drink 4 to 8 ounces of water.

HEALTH HAZARDS
Personal protective equipment (PPE): B-Level PPE. Air–supplied mask in closed tanks; goggles or face shield; gloves. Chemical protective material(s) reported to offer minimal to poor protection: Viton®/neoprene.
Toxicity by ingestion: Grade 2; LD_{50} = 0.5 to 5 g/kg.
Long-term health effects: Prolonged contact or high dose may cause liver and kidney damage, and skin problems.
Vapor (gas) irritant characteristics: Vapors cause a slight smarting of the eyes or respiratory system if present in high concentrations. The effect is temporary.
Liquid or solid irritant characteristics: Eye and skin irritant. If spilled on clothing and allowed to remain, may cause smarting and reddening of the skin.

FIRE DATA
Flash point: 190°F/88°C (oc); 160°F/71°C (cc).
Flammable limits in air: LEL: 0.8% @ 212°F/100°C; UEL: 5% @ 302°F/150°C.
Fire extinguishing agents not to be used: Water may be ineffective.
Autoignition temperature: 725°F/385°C/658°K.
Electrical hazard: Class I, Group D.

CHEMICAL REACTIVITY
Binary reactants: Reaction with some rubber and plastics.
Polymerization: Tetralin peroxide formation may cause explosion of uninhibited substance.
Polymerization: May polymerize forming resinous material and discoloration.
Reactivity group: 32
Compatibility class: Aromatic hydrocarbons

ENVIRONMENTAL DATA
Food chain concentration potential: Negative; unlikely to accumulate.
Water pollution: Effect of low concentrations on aquatic life is unknown. Fouling to shoreline. May be dangerous if it enters nearby water intakes; notify operators. Notify local health and wildlife officials. **Response to discharge:** Mechanical containment. Should be removed. Chemical and physical treatment.

SHIPPING INFORMATION
Grades of purity: 90+%; **Storage temperature:** Ambient; **Inert atmosphere:** None; **Venting:** Open (flame arrester); **Stability during transport:** Stable.

NAS HAZARD CLASSIFICATION FOR BULK WATER TRANSPORTATION
FIRE: 1
HEALTH: Vapor irritant: 1; Liquid or solid irritant: 1; Poisons: 2
WATER POLLUTION: Human toxicity: 1; Aquatic toxicity: 3; Aesthetic effect: 2
REACTIVITY: Other chemicals: 0; Water: 0; Self-reaction: 0

PHYSICAL AND CHEMICAL PROPERTIES
Physical state @ 59°F/15°C and 1 atm: Liquid.
Molecular weight: 132.21
Boiling point @ 1 atm: 406°F/208°C/481°K.
Melting/Freezing point: −23.1°F/30.6°C/242.6°K.
Specific gravity (water = 1): 0.974 @ 68°F/20°C (Liquid.
Liquid surface tension: 35.5 dynes/cm = 0.0355 N/m @ 68°F/20°C.
Relative vapor density (air = 1): 4.59

Latent heat of vaporization: 138 Btu/lb = 76.5 cal/lb = 3.20 x 10^5 J/kg.
Heat of combustion: –18,400 Btu/lb = –10,200 cal/g = –429 x 10^5 J/kg.
Vapor pressure: (Reid) 0.02 psia; 0.3 mm.

1,2,3,5-TETRAMETHYLBENZENE   REC. T:1000

SYNONYMS: BENZENE 1,2,3,5-TETRAMETHYL; 1,2,3,5-TETRAMETIL BENCENO (Spanish); ISODURENE

IDENTIFICATION
CAS Number: 527-53-7
Formula: $C_{10}H_{14}$; $C_6H_2(CH_3)_4$
DOT ID Number: NA 1993; DOT Guide Number: 128
Proper Shipping Name: Combustible liquid, n.o.s.

DESCRIPTION: Colorless liquid. Insoluble in water.

Flammable • Containers may BLEVE when exposed to fire • Vapors may form explosive mixture with air • Vapors are heavier than air and will collect and stay in low areas • Vapors may travel long distances to ignition sources and flashback • Vapors in confined areas (e.g., tanks, sewers, buildings) may explode when exposed to fire • Irritating to the skin, eyes, and respiratory tract • Toxic products of combustion may include carbon monoxide.

Hazard Classification (based on NFPA-704 Rating System)
Health Hazards (Blue): 0; Flammability (Red): 2; Reactivity (Yellow): 0

EMERGENCY RESPONSE: See Appendix A (128)
Evacuation:
Public safety: Isolate spill area for at least 25 to 50 meters (80 to 160 feet) in all directions.
Spill: Large spill–Consider initial downwind evacuation for at least 300 meters (1000 feet).
Fire: Isolate for 800 meters (½ mile) in all directions, especially if tank, rail car, or tank truck is involved in fire.

EXPOSURE
Short-term effects: *SEEK MEDICAL ATTENTION.* **Liquid:** Irritating to skin, eyes, and respiratory tract. Remove contaminated clothing and shoes. Flush affected areas with plenty of water. *IF IN EYES,* hold eyelids open and flush with plenty of water.
Vapors: Irritating to skin, eyes and respiratory tract. Move to fresh air. *IF BREATHING HAS STOPPED,* give artificial respiration. IF breathing is difficult, administer oxygen. *IF IN EYES,* hold eyelids open and flush with plenty of water.

HEALTH HAZARDS
Personal protective equipment (PPE): B-Level PPE. Wear self contained breathing apparatus, rubber boots and heavy rubber gloves.
Toxicity by ingestion: Grade 1: LD_{50} = 5.157 g/kg (rat).
Long-term health effects: Skin problems; eczema.
Vapor (gas) irritant characteristics: Vapors cause smarting of the eyes or respiratory system if present in high concentrations. The effect is temporary.
Liquid or solid irritant characteristics: Eye and skin irritant. If spilled on clothing and allowed to remain, may cause smarting and reddening of skin.

FIRE DATA
Flash point: 160°F/71°C (cc).
Autoignition temperature: 797°F/425°C/698°K (estimated).

CHEMICAL REACTIVITY
Binary reactants: Contact with strong oxidizers may cause fire and explosions.
Reactivity group: 32
Compatibility class: Aromatic hydrocarbons

ENVIRONMENTAL DATA
Water pollution: Effect of low concentrations on aquatic life is unknown. Fouling to shoreline. May be dangerous if it enters nearby water intakes; notify operators. Notify local health and wildlife officials. **Response to discharge:** Restrict access. Mechanical containment. Should be removed. Chemical and physical treatment.

SHIPPING INFORMATION
Grades of purity: 85+%; **Storage temperature:** Ambient; **Stability during transport:** Stable.

PHYSICAL AND CHEMICAL PROPERTIES
Physical state @ 59°F/15°C and 1 atm: Liquid.
Molecular weight: 134.22
Boiling point @ 1 atm: 388.4°F/198°C/471.2°K.
Melting/Freezing point: –10.6°F/–23.7°C/249.5°K.
Specific gravity (water = 1): 0.891
Relative vapor density (air = 1): 4.63
Vapor pressure: (Reid) 0.0147 psia.

TETRAMETHYL LEAD   REC. T:1050

SYNONYMS: EEC No. 082-002-00-1; LEAD, TETRAMETHYL-; MOTOR FUEL ANTI-KNOCK COMPOUND; PLUMBANE, TETRAMETHYL-; TETRAMETHYL PLUMBANE; TETRAMETILPLOMO (Spanish); TML

IDENTIFICATION
CAS Number: 75-74-1
Formula: $C_4H_{12}Pb$; $Pb(CH_3)_4$
DOT ID Number: UN 1649; DOT Guide Number: 131
Proper Shipping Name: Motor fuel anti-knock mixtures or compounds
Reportable Quantity (RQ): **(EHS)** 1 lb/0.454 kg

DESCRIPTION: Colorless, oily liquid. May be dyed red, orange, or blue. Faint, fruity odor. Sinks in water. Poisonous, flammable vapor is produced.

Highly flammable • Containers may BLEVE when exposed to fire • Vapors are heavier than air and will collect and stay in low areas • Vapors may travel long distances to ignition sources and flashback • Vapors in confined areas (e.g., tanks, sewers, buildings) may explode when exposed to fire • Irritating to the skin, eyes, and respiratory tract • Starts to decomposes above 212°F/100°C, forming explosive and toxic gases; explosive decomposition when heated rapidly above 608°F/320°C. Toxic products of combustion or decomposition include lead fumes.

Hazard Classification (based on NFPA-704 Rating System)
Health Hazards (Blue): 3; Flammability (Red): 3; Reactivity (Yellow): 3

EMERGENCY RESPONSE: See Appendix A (131)
Evacuation:
Public safety: Isolate spill area for at least 100 to 200 meters (330 to 660 feet) in all directions.
Spill: Increase, as necessary, the isolation distance shown above, in "Public safety."
Fire: Isolate for 800 meters (½ mile) in all directions, especially if tank, rail car, or tank truck is involved in fire.

EXPOSURE
Short-term effects: *SEEK MEDICAL ATTENTION*. **Vapor:** *POISONOUS IF INHALED*. Irritating to eyes. Move to fresh air. *IF BREATHING HAS STOPPED*, give artificial respiration. IF breathing is difficult, administer oxygen. **Liquid:** *POISONOUS IF SWALLOWED OR IF SKIN IS EXPOSED*. Will burn eyes. Remove contaminated clothing and shoes. Flush affected areas with plenty of water. *IF IN EYES*, hold eyelids open and flush with plenty of water. *IF SWALLOWED* and victim is *CONSCIOUS AND ABLE TO SWALLOW*, have victim drink 4 to 8 ounces of water.

HEALTH HAZARDS
Personal protective equipment (PPE): A-Level PPE. OSHA Table Z-1-A air contaminant. Neoprene-coated protective gloves.
Recommendations for respirator selection: NIOSH/OSHA
0.75 mg/m³: SA (any supplied-air respirator). *1.875 mg/m³:* SA:CF (any supplied-air respirator operated in a continuous-flow mode). *3.75 mg/m³:* SAT:CF (any supplied-air respirator that has a tight-fitting facepiece and is operated in a continuous-flow mode); or SCBAF (any self-contained breathing apparatus with a full facepiece); or SAF (any supplied-air respirator with a full facepiece). *40 mg/m³:* SA:PD,PP (any supplied-air respirator operated in a pressure-demand or other positive-pressure mode). *EMERGENCY OR PLANNED ENTRY INTO UNKNOWN CONCENTRATIONS OR IDLH CONDITIONS:* SCBAF:PD,PP (any self-contained breathing apparatus that has a full facepiece and is operated in a pressure-demand or other positive-pressure mode); or SAF:PD,PP:ASCBA (any supplied-air respirator that has a full facepiece and is operated in a pressure-demand or other positive-pressure mode in combination with an auxiliary self-contained breathing apparatus operated in a pressure-demand or other positive pressure mode). *ESCAPE:* GMFOV [any air-purifying, full-facepiece respirator (gas mask) with a chin-style, front- or back-mounted organic vapor canister]; or SCBAE (any appropriate escape-type, self-contained breathing apparatus).

Exposure limits (TWA unless otherwise noted): ACGIH TLV 0.15 mg/m³; NIOSH/OSHA 0.075 mg/m³; skin contact contributes significantly in overall exposure.
Toxicity by ingestion: Grade 3; oral rat LD_{50} = 109 mg/kg.
Long-term health effects: Lead poisoning; possible teratogen.
Vapor (gas) irritant characteristics: Vapors cause smarting of the eyes or respiratory system if present in high concentrations. The effect is temporary.
Liquid or solid irritant characteristics: Eye and skin irritant. Causes smarting of the skin and first-degree burns on short exposure; may cause secondary burns on long exposure.
IDLH value: 40 mg/m³ as lead.

FIRE DATA
Flash point: 100°F/38°C (cc).

CHEMICAL REACTIVITY
Reaction with water: Decomposes in water to forming trimethyl salt→diethyl salt→inorganic lead salt.
Binary reactants: Violent reaction with oxidizers, sulfuryl chloride, potassium permanganate, tetrachlorotrifluoromethylphosphorane. Decomposes in slunlight and in elevated temperatures above 212°F/100°C; explosion above 608°F/320°C. Strong oxidizers and strong acids cause fire and explosions. Attacks some plastics, rubber and coatings. More stable than tetraethyllead.

ENVIRONMENTAL DATA
Water pollution: DOT Appendix B, §172.101–marine pollutant. Effect of low concentrations on aquatic life is unknown. May be dangerous if it enters nearby water intakes; notify operators. Notify local health and wildlife officials. **Response to discharge:** Issue warning–poison, water contaminant. Restrict access. Should be removed. Chemical and physical treatment.

SHIPPING INFORMATION
Grades of purity: Technical grades (usually contain dibromoethane and dichloroethane); **Storage temperature:** Ambient; **Inert atmosphere:** None; **Venting:** Pressure-vacuum; **Stability during transport:** Stable unless 0.1% of peroxides has accumulated because of prolonged storage in presence of air. When concentrated by evaporation of solution, they explode.

PHYSICAL AND CHEMICAL PROPERTIES
Physical state @ 59°F/15°C and 1 atm: Liquid.
Molecular weight: 267.33
Boiling point @ 1 atm: Decomposes. 212°F/100°C/373°K.
Melting/Freezing point: –17.5°F/–27.5°C/245.7°K.
Specific gravity (water = 1): 1.999 @ 68°F/20°C (liquid).
Relative vapor density (air = 1): 9.2
Latent heat of vaporization: (estimate) 55.5 Btu/lb = 30.8 cal/g = 1.29×10^5 J/kg.
Heat of combustion: (estimate) –5290 Btu/lb = –2940 cal/g = –123 $\times 10^5$ J/kg.
Vapor pressure: 23 mm.

TETRANITROMETHANE REC. T:1100

SYNONYMS: METHANE, TETRANITRO-; RCRA No. P112; TNM; TETRANITROMETANO (Spanish); TETAN0

IDENTIFICATION
CAS Number: 509-14-8
Formula: $C(NO_2)_4$
DOT ID Number: UN 1510; DOT Guide Number: 143
Proper Shipping Name: Tetranitromethane
Reportable Quantity (RQ): **(CERCLA)** 10 lb/4.54 kg

DESCRIPTION: Colorless or yellow oily liquid. Pungent to acrid, biting odor. Sinks in water; insoluble. Freezes below 57°F/14°C.

Explosive! May explode when heated or shocked • Combustible • Strong oxidizer which may react spontaneously with low flash point organics or reducing agents. Heat forms oxygen; will increase the activity of an existing fire • May cause fire or explosion contact with combustibles (wood, paper, oil, clothing, etc.) • Do not use dry chemical extinguishers • Gas is lighter than air, and will disperse slowly unless confined • Vapors in confined areas (e.g., tanks, sewers, buildings) may explode when exposed to fire • Severely Irritating to the skin, eyes, and respiratory tract; prolonged contact with skin or eyes may cause burns • Interferes with the body's ability to use oxygen • Toxic products of combustion may include oxygen and toxic nitrogen oxides • Do not put yourself in danger by entering a contaminated area to rescue a victim.

Hazard Classification (based on NFPA-704 Rating System)
Health Hazards (Blue): 3; Flammability (Red): 1; Reactivity (Yellow): 3; Special Notice (White): OXY

EMERGENCY RESPONSE: See Appendix A (143)
Evacuation:
Public safety: Isolate the area of spill or leak for at least 50 to 100 meters (160 to 300 feet) in all directions.
Spills: Small spill–First: Isolate in all directions 30 meters (100 feet); Then: Protect persons downwind, DAY: 0.3 km (0.2 mile); NIGHT: 0.5 km (0.3 mile). Large spill–First: Isolate in all directions 60 meters (200 feet); Then: Protect persons downwind, DAY: 0.6 km (0.4 mile); NIGHT: 1.3 km (0.8 mile).
Fire: If any large container is involved in fire, isolate for at least 800 meters (½ mile) in all directions; also, consider initial evacuation for 800 meters (½ mile) in all directions.

EXPOSURE
Short-term effects: *SEEK MEDICAL ATTENTION.* **Vapor:** May be fatal if inhaled or absorbed through the skin. Irritating to the eyes, nose, throat, and lungs. Remove to fresh air. *IF BREATHING HAS STOPPED,* give artificial respiration. IF breathing is difficult, administer oxygen. **Liquid:** May be fatal if ingested or absorbed through the skin. Effects may be delayed. Causes eye and skin irritation. *IF IN EYES,* hold eyelids open, flush with running water for at least 15 minutes. Remove contaminated clothing and shoes. Flush affected areas with plenty of water. *IF SWALLOWED,* **Do NOT induce vomiting.** Keep victim quiet and maintain normal body temperature.
Notes: Central nervous system depressant. Skin absorption can lead to the formation of methemoglobin which may lead to cyanosis. Onset may be delayed 2 to 4 hours or longer.

HEALTH HAZARDS
Personal protective equipment (PPE): A-Level PPE. OSHA Table Z-1-A air contaminant. Approved respirator, safety goggles, chemical resistant gloves, other protective clothing.
Recommendations for respirator selection: NIOSH/OSHA
4 ppm: SA:CF* (any supplied-air respirator operated in a continuous-flow mode); or CCRFS* [any chemical cartridge respirator with a full facepiece and cartridge(s) providing protection against the compound of concern]; or GMFS* [any air-purifying, full-facepiece respirator (gas mask) with a chin-style, front- or back-mounted canister providing protection against the compound of concern]; or PAPRS* [any powered, air-purifying respirator with cartridge(s) providing protection against the compound of concern]; or SCBAF (any self-contained breathing apparatus with a full facepiece); or SAF (any supplied-air respirator with a full facepiece). *EMERGENCY OR PLANNED ENTRY INTO UNKNOWN CONCENTRATIONS OR IDLH CONDITIONS:* SCBAF:PD,PP (any self-contained breathing apparatus that has a full facepiece and is operated in a pressure-demand or other positive-pressure mode); or SAF:PD,PP:ASCBA (any supplied-air respirator that has a full facepiece and is operated in a pressure-demand or other positive-pressure mode in combination with an auxiliary self-contained breathing apparatus operated in a pressure-demand or other positive pressure mode). *ESCAPE:* GMFS** [any air-purifying, full-facepiece respirator (gas mask) with a chin-style, front- or back-mounted canister providing protection against the compound of concern]; or SCBAE (any appropriate escape-type, self-contained breathing apparatus).

Notes: **Substance causes eye irritation or damage; eye protection needed. **Only nonoxidizable sorbents are allowed (not charcoal).
Exposure limits (TWA unless otherwise noted): ACGIH TLV 0.005 ppm (0.04 mg/m^3); NIOSH/OSHA 1 ppm (8 mg/m^3).
Toxicity by ingestion: Grade 3: LD_{50} = 130 mg/kg (rat).
Long-term health effects: ACGIH suspected human carcinogen. May cause cyanosis due to formation of methemoglobin. Damage to liver, kidney heart and eyes. Methemoglobin formation; blood problems.
Vapor (gas) irritant characteristics: Eye and mucous membrane irritant. Vapors are moderately irritating such that personnel will not usually tolerate
moderate or high concentrations.
Liquid or solid irritant characteristics: eye and skin irritant. If spilled on clothing and allowed to remain, may cause smarting and reddening of skin.
IDLH value: 4 ppm.

FIRE DATA
Flash point: More than 230°F/110°C (cc).
Behavior in fire: Containers may explode.

CHEMICAL REACTIVITY
Binary reactants: Powerful oxidizer and high explosive. May react explosively with many substances. Incompatible with finely divided metals, iron and iron salts, copper, brass, zinc, or rubber.

ENVIRONMENTAL DATA
Food chain concentration potential: Log P_{ow} = –0.8. Unlikely to accumulate.
Water pollution: DOT Appendix B, §172.101–marine pollutant. Effects of low concentrations on aquatic life are not known. May be dangerous if it enters nearby water intakes; notify operators. Notify local health and wildlife officials. **Response to discharge:** Restrict access. Mechanical containment. Should be removed. Chemical and physical treatment.

SHIPPING INFORMATION
Grades of purity: 98%; **Storage temperature:** Refrigerate; **Venting:** None; **Stability during transport:** Stable.

NAS HAZARD CLASSIFICATION FOR BULK WATER TRANSPORTATION
FIRE: 1
HEALTH: Vapor irritant: 3; Liquid or solid irritant: 1; Poisons: 3
WATER POLLUTION: Human toxicity: 3; Aquatic toxicity: 3; Aesthetic effect: 2
REACTIVITY: Other chemicals: 4; Water: 0; Self-reaction: 0

PHYSICAL AND CHEMICAL PROPERTIES
Physical state @ 59°F/15°C and 1 atm: Liquid.
Molecular weight: 196.03
Boiling point @ 1 atm: 259°F/126°C/399°K.
Melting/Freezing point: 57°F/14°C/287°K.
Specific gravity (water = 1): 1.6380 @ 68°F/20°C.
Relative vapor density (air = 1): 6.76
Latent heat of vaporization: 188 Btu/lb = 104 cal/g = 4.4 x 10^5 J/kg.
Vapor pressure: (Reid) 0.5 psia; 8 mm.

THALLIUM ACETATE   **REC. T:1150**

SYNONYMS: ACETATO de TALIO (Spanish); ACETIC ACID, THALLOUS SALT; ACETIC ACID, THALLIUM SALT;

Thallium acetate

THALLIUM(I) ACETATE; THALLIUM(1+) ACETATE; THALLIUM MONOACETATE; THALLOUS ACETATE; RCRA No. U214

IDENTIFICATION
CAS Number: 563-68-8
Formula: $C_2H_3O_2 \cdot Tl$; CH_3COTl
DOT ID Number: UN 1707; DOT Guide Number: 151
Proper Shipping Name: Thallium compounds, n.o.s.
Reportable Quantity (RQ): **(CERCLA)** 100 lb/45.4 kg

DESCRIPTION: White crystalline solid. Odorless. Sinks and mixes with water.

Irritates eyes, skin, and respiratory tract; swallowing the material or inhaling large quantities of the dust can kill you • Toxic products of combustion may include thallium fumes.

Hazard Classification (based on NFPA-704 Rating System)
Health Hazards (Blue): 3; Flammability (Red): 0; Reactivity (Yellow): 0

EMERGENCY RESPONSE: See Appendix A (151)
Evacuation:
Public safety: Isolate the area of spill or leak for at least 25 to 50 meters (80 to 160 feet) in all directions.
Spill: Increase, in the downwind direction, as necessary, the distance shown under "Public Safety."
Fire: If tank, rail car, or tank truck is involved in fire, isolate for at least 800 meters (½ mile) in all directions; also, consider initial evacuation for 800 meters (½ mile) in all directions.

EXPOSURE
Short-term effects: *SEEK MEDICAL ATTENTION.* **Dust:** *POISONOUS. MAY BE FATAL IF INHALED OR ABSORBED THROUGH SKIN. ONSET OF SYMPTOMS MAY BE DELAYED SEVERAL HOURS. IF IN EYES OR ON SKIN,* flush with running water for at least 15 minutes holding eyelids open periodically, if appropriate. Remove and double bag contaminated clothing and shoes at the site. *IF BREATHING HAS STOPPED,* give artificial respiration; *avoid mouth-to-mouth resuscitation; use bag/mask apparatus.* IF breathing is difficult, administer oxygen. **Solid:** *POISONOUS. MAY BE FATAL IF SWALLOWED OR ABSORBED THROUGH SKIN.* Onset of symptoms may be delayed for several hours, perhaps as much as 12 to 24 hours. *IF SWALLOWED,* may cause nausea, vomiting, diarrhea, and abdominal pain. *IF IN EYES OR ON SKIN,* flush with running water for at least 15 minutes, hold eyelids open periodically if appropriate. Remove and double bag contaminated clothing and shoes at the site. *IF SWALLOWED and victim is CONSCIOUS AND ABLE TO SWALLOW,* have victim drink water and induce vomiting by touching a finger to the back of the throat. *IF SWALLOWED and victim is UNCONSCIOUS OR HAVING CONVULSIONS,* do nothing except keep victim quiet and maintain body temperature.
Notes: Speed in removing material from skin is important. Readily absorbed through the skin and digestive tract. Ingestion of soluble thallium compounds has caused many deaths. Ingestion of sublethal quantities may cause nausea, vomiting, diarrhea, abdominal pain, and bleeding from the gut accompanied or followed by drooping eyelids, crossed eyes, weakness, numbness, tingling of arms and legs, trembling, tightness and pain in the chest. Loss of hair may occur in two to three weeks. Severe intoxication may cause prostration, rapid heartbeat, convulsions, and psychosis. Some effects may be permanent.

HEALTH HAZARDS
Personal protective equipment (PPE): B-Level PPE. OSHA Table Z-1-A air contaminant. Wear self-contained positive pressure breathing apparatus and full protective clothing.
Recommendations for respirator selection: NIOSH/OSHA as thallium soluble compounds
0.5 mg/m³: DM *If not present as a fume* (any dust and mist respirator). *1 mg/m³:* DMXSQ *If not present as a fume* (any dust and mist respirator except single-use and quarter mask respirators); or SA (any supplied-air respirator). *2.5 mg/m³:* SA:CF (any supplied-air respirator operated in a continuous-flow mode); or PAPRDM *If not present as a fume* (any powered, air-purifying respirator with a dust and mist filter). *5 mg/m³:* HiEF (any air-purifying, full-facepiece respirator with a high-efficiency particulate filter); or SAT:CF (any supplied-air respirator that has a tight-fitting facepiece and is operated in a continuous-flow mode); or PAPRTHiE (any powered, air-purifying respirator with a tight-fitting facepiece and a high-efficiency particulate filter); or SCBAF (any self-contained breathing apparatus with a full facepiece); or SAF (any supplied-air respirator with a full facepiece). *15 mg/m³:* SAF:PD,PP (any supplied-air respirator that has a full facepiece and is operated in a pressure-demand or other positive-pressure mode). *EMERGENCY OR PLANNED ENTRY INTO UNKNOWN CONCENTRATIONS OR IDLH CONDITIONS:* SCBAF:PD,PP (any self-contained breathing apparatus that has a full facepiece and is operated in a pressure-demand or other positive-pressure mode); or SAF:PD,PP:ASCBA (any supplied-air respirator that has a full facepiece and is operated in a pressure-demand or other positive-pressure mode in combination with an auxiliary self-contained breathing apparatus operated in a pressure-demand or other positive pressure mode). *ESCAPE:* HiEF (any air-purifying, full-facepiece respirator with a high-efficiency particulate filter); or SCBAE (any appropriate escape-type, self-contained breathing apparatus).
Exposure limits (TWA unless otherwise noted): ACGIH TLV 0.1 mg/m³ as thallium; NIOSH/OSHA 0.1 mg/m³ as thallium; skin contact contributes significantly in overall exposure
Toxicity by ingestion: Grade 4; LD_{50} = 35 mg/kg (mouse).
Long-term health effects: Thallous ion causes mutagenic effects (chromosomal aberrations) in animals and plants, and teratogenic effects (detrimental to the sexual behavior, reproductive organs, egg and fetal development, and survival of the chicken). It also causes liver and kidney damage, hair loss and permanent effects such as staggering, visual difficulties, trembling, and mental abnormalities. Chronic oral or cutaneous exposure of mice to thallium caused cancer of the female genital tract.
Vapor (gas) irritant characteristics: Not volatile but toxic.
Liquid or solid irritant characteristics: Skin and eye irritant.
IDLH value: 15 mg/m³ as thallium.

ENVIRONMENTAL DATA
Food chain concentration potential: Thallium is a cumulative poison four times as toxic as arsenious oxide. Positive. Expected to accumulate. Plants growing in soils or water with very high thallium content may accumulate sufficient thallium to be toxic to organisms that feed on them. Algae from contaminated water exhibited thallium bioconcentration factor of more than 430. Other bioconcentration factors that have been reported include 130 for Atlantic salmon mussel and 18 for softshell clams.
Water pollution: DOT Appendix B, §172.101–marine pollutant. Harmful to aquatic life in very low concentrations. May be dangerous if it enters nearby water intakes; notify operators. Notify local health and wildlife officials. **Response to discharge:** Issue warning–poison, water contaminant. Restrict access. Should be removed. Chemical and physical treatment. Clean Water Act.

SHIPPING INFORMATION
Grades of purity: 99.99%; **Storage temperature:** Ambient; **Stability during transport:** Stable.

PHYSICAL AND CHEMICAL PROPERTIES
Physical state @ 59°F/15°C and 1 atm: Solid.
Molecular weight: 263.42
Boiling point @ 1 atm: (decomposes).
Melting/Freezing point: 267.8°F/131°C/404.2°K.
Specific gravity (water = 1): 3.765 at 137°C.

THALLIUM CARBONATE REC. T:1200

SYNONYMS: ARBONIC ACID, THALLIUM SALT; CARBONATO de TALIO (Spanish); DITHALLIUM CARBONATE; THALLOUS CARBONATE; RCRA No. U215

IDENTIFICATION
CAS Number: 6533-73-9
Formula: $T_{12}CO_3$
DOT ID Number: UN 1707; DOT Guide Number: 151
Proper Shipping Name: Thallium compound, n.o.s.
Reportable Quantity (RQ): **(CERCLA)** 100 lb/45.4 kg

DESCRIPTION: White crystalline solid. Sinks and mixes with water.

Irritates eyes, skin, and respiratory tract; swallowing the material or inhaling large quantities of the dust can kill you • Toxic products of combustion may include carbon monoxide and thallium fumes.

Hazard Classification (based on NFPA-704 Rating System)
Health Hazards (Blue): 3; Flammability (Red): 0; Reactivity (Yellow): 0

EMERGENCY RESPONSE: See Appendix A (151)
Evacuation:
Public safety: Isolate the area of spill or leak for at least 25 to 50 meters (80 to 160 feet) in all directions.
Spill: Increase, in the downwind direction, as necessary, the distance shown under "Public Safety."
Fire: If tank, rail car, or tank truck is involved in fire, isolate for at least 800 meters (½ mile) in all directions; also, consider initial evacuation for 800 meters (½ mile) in all directions.

EXPOSURE
Short-term effects: *SEEK MEDICAL ATTENTION.* **Dust:** *POISONOUS. MAY BE FATAL IF INHALED, IF SWALLOWED, OR ABSORBED THROUGH SKIN.* Onset of symptoms may be delayed for several hours, perhaps as much as 12 to 24 hours. Irritating to eyes and skin. *IF IN EYES OR ON SKIN,* flush with running water for at least 15 minutes; hold eyelids open periodically, if appropriate. Remove and double bag contaminated clothing and shoes at the site. *IF BREATHING HAS STOPPED,* give artificial respiration; *avoid mouth-to-mouth resuscitation; use bag/mask apparatus.* IF breathing is difficult, administer oxygen.
Solid: If swallowed, may cause nausea, vomiting, diarrhea, abdominal pain and bleeding from the gut. *IF IN EYES OR ON SKIN,* flush with running water for at least 15 minutes, hold eyelids open periodically, if appropriate. Remove and double bag contaminated clothing and shoes at the site. *IF SWALLOWED* and victim is *CONSCIOUS AND ABLE TO SWALLOW,* have victim drink water and induce vomiting. *IF SWALLOWED* and victim is *UNCONSCIOUS OR HAVING CONVULSIONS,* do nothing except keep victim warm.
Notes: Speed in removing material from skin is important. May be fatal if inhaled, ingested or absorbed through the skin. Readily absorbed through the skin and digestive tract. Digestion of soluble thallium compounds has caused many deaths. Ingestion of sublethal quantities may cause nausea, vomiting, diarrhea, abdominal pain, and bleeding from the gut accompanied or followed by drooping eyelids, crossed eyes, weakness, numbness, tingling of arms and legs, trembling, tightness and pain in the chest. Loss of hair may occur in two to three weeks. Severe intoxication may cause prostration, rapid heartbeat, convulsions, and psychosis. Some effects may be permanent.

HEALTH HAZARDS
Personal protective equipment (PPE): B-Level PPE. OSHA Table Z-1-A air contaminant. Wear self-contained positive pressure breathing apparatus and full protective clothing.
Recommendations for respirator selection: NIOSH/OSHA as thallium soluble compounds
0.5 mg/m³: DM *If not present as a fume* (any dust and mist respirator). *1 mg/m³:* DMXSQ *If not present as a fume* (any dust and mist respirator except single-use and quarter mask respirators); or SA (any supplied-air respirator). *2.5 mg/m³:* SA:CF (any supplied-air respirator operated in a continuous-flow mode); or PAPRDM *If not present as a fume* (any powered, air-purifying respirator with a dust and mist filter). *5 mg/m³:* HiEF (any air-purifying, full-facepiece respirator with a high-efficiency particulate filter); or SAT:CF (any supplied-air respirator that has a tight-fitting facepiece and is operated in a continuous-flow mode); or PAPRTHiE (any powered, air-purifying respirator with a tight-fitting facepiece and a high-efficiency particulate filter); or SCBAF (any self-contained breathing apparatus with a full facepiece); or SAF (any supplied-air respirator with a full facepiece). *15 mg/m³:* SAF:PD,PP (any supplied-air respirator that has a full facepiece and is operated in a pressure-demand or other positive-pressure mode). *EMERGENCY OR PLANNED ENTRY INTO UNKNOWN CONCENTRATIONS OR IDLH CONDITIONS:* SCBAF:PD,PP (any self-contained breathing apparatus that has a full facepiece and is operated in a pressure-demand or other positive-pressure mode); or SAF:PD,PP:ASCBA (any supplied-air respirator that has a full facepiece and is operated in a pressure-demand or other positive-pressure mode in combination with an auxiliary self-contained breathing apparatus operated in a pressure-demand or other positive pressure mode). *ESCAPE:* HiEF (any air-purifying, full-facepiece respirator with a high-efficiency particulate filter); or SCBAE (any appropriate escape-type, self-contained breathing apparatus).
Exposure limits (TWA unless otherwise noted): ACGIH TLV 0.1 mg/m³ as thallium; NIOSH/OSHA 0.1 mg/m³ as thallium; skin contact contributes significantly in overall exposure
Toxicity by ingestion: Grade 4; LD_{50} = 21 mg/kg (mouse).
Long-term health effects: Thallous ion causes mutagenic effects (chromosomal aberrations) in animals and plants, and teratogenic effects (detrimental to the sexual behavior, reproductive organs, egg and fetal development, and survival of the chicken). It also causes liver and kidney damage, hair loss and permanent effects such as staggering, visual difficulties, trembling, and mental abnormalities. Chronic oral or cutaneous exposure of mice to thallium caused cancer of the female genital tract.
Liquid or solid irritant characteristics: May cause eye and skin irritation.
IDLH value: 15 mg/m³ as thallium.

ENVIRONMENTAL DATA
Food chain concentration potential: Thallium is a cumulative

poison four times as toxic as arsenious oxide. Positive. Expected to accumulate. Plants growing in soils or water with very high thallium content may accumulate sufficient thallium to be toxic to organisms that feed on them.
Water pollution: DOT Appendix B, §172.101–marine pollutant. Harmful to aquatic life in very low concentrations. May be dangerous if it enters nearby water intakes; notify operators. Notify local health and wildlife officials. **Response to discharge:** Issue warning–poison, water contaminant. Restrict access. Should be removed. Chemical and physical treatment. Clean Water Act.

SHIPPING INFORMATION
Grades of purity: 99.99%; **Storage temperature:** Ambient

PHYSICAL AND CHEMICAL PROPERTIES
Physical state @ 59°F/15°C and 1 atm: Solid.
Molecular weight: 468.75
Melting/Freezing point: 521.6°F/272°C/545°K.
Specific gravity (water = 1): 7.11 @ 77°F/25°C (room temperature).

THALLIUM NITRATE   REC. T:1250

SYNONYMS: NITRATO de TALIO (Spanish); NITRIC ACID, THALLOUS SALT; THALLIUM MONONITRATE; THALLOUS NITRATE; RCRA No. U217

IDENTIFICATION
CAS Number: 10102-45-1
Formula: $NO_3 \cdot Tl$; $TlNO_3$
DOT ID Number: UN 2727; DOT Guide Number: 141
Proper Shipping Name: Thallium nitrate
Reportable Quantity (RQ): **(CERCLA)** 100 lb/45.4 kg

DESCRIPTION: White crystalline solid. Odorless. Sinks and mixes with water.

Irritates eyes, skin, and respiratory tract; swallowing the material or inhaling large quantities of the dust can kill you • Toxic products of combustion may include thallium and nitrogen oxides.

Hazard Classification (based on NFPA-704 Rating System)
Health Hazards (Blue): 3; Flammability (Red): 0; Reactivity (Yellow): 0

EMERGENCY RESPONSE: See Appendix A (141)
Evacuation:
Public safety: Isolate the area of spill or leak for at least 10 to 25 meters (30 to 80 feet) in all directions.
Spill: Consider initial downwind evacuation for at least 100 meters (330 feet).
Fire: If any large container is involved in fire, isolate for at least 800 meters (½ mile) in all directions; also, consider initial evacuation for 800 meters (½ mile) in all directions.

EXPOSURE
Short-term effects: *CALL FOR MEDICAL AID* **Dust:** *POISONOUS. MAY BE FATAL IF INHALED OR ABSORBED THROUGH SKIN.* Onset of symptoms may be delayed for several hours, perhaps as much as 12 to 24 hours. *IF IN EYES OR ON SKIN,* flush with running water for at least 15 minutes, holding eyelids open periodically, if appropriate. Remove and double bag contaminated clothing and shoes at the site. *IF BREATHING HAS STOPPED,* give artificial respiration; *avoid mouth-to-mouth resuscitation; use bag/mask apparatus*: IF breathing is difficult, administer oxygen. **Solid:** *POISONOUS. MAY BE FATAL IF SWALLOWED OR ABSORBED THROUGH SKIN. ONSET OF SYMPTOMS DELAYED 12 TO 24 HOURS AFTER INGESTION. IF SWALLOWED,* may cause nausea, vomiting, diarrhea, and abdominal pain. *IF IN EYES OR ON SKIN:* Flush with running water for at least 15 minutes, holding eyelids open periodically, if appropriate. Remove and double bag contaminated clothing and shoes at the site. *IF SWALLOWED* and victim is CONSCIOUS: Have victim drink water and induce vomiting by touching finger to back of throat. *IF SWALLOWED* and victim is UNCONSCIOUS OR HAVING CONVULSIONS: Do nothing except keep victim warm.
Notes: Speed in removing material from skin is important. May be fatal if inhaled, ingested or absorbed through the skin. Readily absorbed through the skin and digestive tract. Digestion of soluble thallium compounds has caused many deaths. Ingestion of sublethal quantities may cause nausea, vomiting, diarrhea, abdominal pain, and bleeding from the gut accompanied or followed by drooping eyelids, crossed eyes, weakness, numbness, tingling of arms and legs, trembling, tightness and pain in the chest. Loss of hair may occur in two to three weeks. Severe intoxication may cause prostration, rapid heartbeat, convulsions, and psychosis. Some effects may be permanent.

HEALTH HAZARDS
Personal protective equipment (PPE): B-Level PPE. OSHA Table Z-1-A air contaminant. Wear self-contained positive pressure breathing apparatus and full protective clothing.
Recommendations for respirator selection: NIOSH/OSHA as thallium soluble compounds
0.5 mg/m³: DM *If not present as a fume* (any dust and mist respirator). *1 mg/m³:* DMXSQ *If not present as a fume* (any dust and mist respirator except single-use and quarter mask respirators); or SA (any supplied-air respirator). *2.5 mg/m³:* SA:CF (any supplied-air respirator operated in a continuous-flow mode); or PAPRDM *If not present as a fume* (any powered, air-purifying respirator with a dust and mist filter). *5 mg/m³*: HiEF (any air-purifying, full-facepiece respirator with a high-efficiency particulate filter); or SAT:CF (any supplied-air respirator that has a tight-fitting facepiece and is operated in a continuous-flow mode); or PAPRTHiE (any powered, air-purifying respirator with a tight-fitting facepiece and a high-efficiency particulate filter); or SCBAF (any self-contained breathing apparatus with a full facepiece); or SAF (any supplied-air respirator with a full facepiece). *15 mg/m³:* SAF:PD,PP (any supplied-air respirator that has a full facepiece and is operated in a pressure-demand or other positive-pressure mode). *EMERGENCY OR PLANNED ENTRY INTO UNKNOWN CONCENTRATIONS OR IDLH CONDITIONS:* SCBAF:PD,PP (any self-contained breathing apparatus that has a full facepiece and is operated in a pressure-demand or other positive-pressure mode); or SAF:PD,PP:ASCBA (any supplied-air respirator that has a full facepiece and is operated in a pressure-demand or other positive-pressure mode in combination with an auxiliary self-contained breathing apparatus operated in a pressure-demand or other positive pressure mode). *ESCAPE:* HiEF (any air-purifying, full-facepiece respirator with a high-efficiency particulate filter); or SCBAE (any appropriate escape-type, self-contained breathing apparatus).
Exposure limits (TWA unless otherwise noted): ACGIH TLV 0.1 mg/m³ as thallium; NIOSH/OSHA 0.1 mg/m³ as thallium; skin contact contributes significantly in overall exposure.
Toxicity by ingestion: Grade 4; LD_{50} = 15 mg/kg (mouse).
Long-term health effects: Thallous ion causes mutagenic effects (chromosomal aberrations) in animals and plants, and teratogenic effects (detrimental to the sexual behavior, reproductive organs, egg

and fetal development, and survival of the chicken). It also causes liver and kidney damage, hair loss and permanent effects such as staggering, visual difficulties, trembling, and mental abnormalities. Chronic oral or cutaneous exposure of mice to thallium caused cancer of the female genital tract.
Liquid or solid irritant characteristics: Causes skin and eye irritation.
IDLH value: 15 mg/m^3 as thallium.

ENVIRONMENTAL DATA
Food chain concentration potential: Thallium is a cumulative poison four times as toxic as arsenious oxide. Positive. Expected to accumulate. Plants growing in soils or water with very high thallium content may accumulate sufficient thallium to be toxic to organisms that feed on them. Algae from contaminated water exhibited thallium bioconcentration factor of more than 430. Other bioconcentration factors that have been reported include 130 for atlantic salmon mussel and 18 for the edible portion of softshell clams.
Water pollution: DOT Appendix B, §172.101–marine pollutant. Harmful to aquatic life in very low concentrations. May be dangerous if it enters nearby water intakes; notify operators. Notify local health and wildlife officials. Notify operators of local water intakes. **Response to discharge:** Issue warning–poison, water contaminant. Restrict access. Should be removed. Chemical and physical treatment. Clean Water Act.

SHIPPING INFORMATION
Grades of purity: 99.999%; **Storage temperature:** Ambient; **Stability during transport:** Stable.

NAS HAZARD CLASSIFICATION FOR BULK WATER TRANSPORTATION
FIRE: 0
HEALTH: Vapor irritant:–; Liquid or solid irritant: 1; Poisons: 4
WATER POLLUTION: Human toxicity: 4; Aquatic toxicity: 4; Aesthetic effect: 1
REACTIVITY: Other chemicals: 1; Water: 1; Self-reaction: 0

PHYSICAL AND CHEMICAL PROPERTIES
Physical state @ 59°F/15°C and 1 atm: Solid.
Molecular weight: 266.39
Boiling point @ 1 atm: 806°F/430°C/703°K.
Melting/Freezing point: 403°F/206°C/479°K.
Specific gravity (water = 1): 5.556 at 21°C.
Heat of decomposition: (estimate) –393 Btu/lb = –218 cal/g = –9.1 x 10^5 J/kg.
Heat of fusion: 15.5 Btu/lb = 8.6 cal/g = 3.6 x 10^4 J/kg.

THALLIUM SULFATE REC. T:1300

SYNONYMS: RATOX; SULFATO de TALIO (Spanish); SULFURIC ACID, THALLIUM SALT; THALLOUS SULFATE; ZELIO

IDENTIFICATION
CAS Number: 10031-59-1; 7446-18-6
Formula: $T_{l2}SO_4$
DOT ID Number: UN 1707; DOT Guide Number: 151
Proper Shipping Name: Thallium sulfate, solid
Reportable Quantity (RQ): **(CERCLA)** 100 lb/45.4 kg

DESCRIPTION: Colorless to white crystalline solid. Odorless. Sinks and mixes with water. Irritates eyes, skin, and respiratory tract; swallowing the material or inhaling large quantities of the dust can kill you •Toxic products of combustion may include thallium and sulfur oxides.

Hazard Classification (based on NFPA-704 Rating System)
Health Hazards (Blue): 3; Flammability (Red): 0; Reactivity (Yellow): 0

EMERGENCY RESPONSE: See Appendix A (151)
Evacuation:
Public safety: Isolate the area of spill or leak for at least 25 to 50 meters (80 to 160 feet) in all directions.
Spill: Increase, in the downwind direction, as necessary, the distance shown under "Public Safety."
Fire: If tank, rail car, or tank truck is involved in fire, isolate for at least 800 meters (½ mile) in all directions; also, consider initial evacuation for 800 meters (½ mile) in all directions.

EXPOSURE
Short-term effects: *SEEK MEDICAL ATTENTION*. If artificial respiration is administered, *avoid mouth-to-mouth resuscitation; use bag/mask apparatus*. **Solid:** *POISONOUS IF SWALLOWED OR IF SKIN IS EXPOSED*. Remove contaminated clothing and shoes. Flush affected areas with plenty of water. *IF SWALLOWED* and victim is *CONSCIOUS AND ABLE TO SWALLOW*, have victim drink 4 to 8 ounces of water and have victim induce vomiting.
Notes: Ingestion or skin absorption can cause pain and tingling or numbnessof the extremities, drooping eyelids, incoordination of muscular action, loss of hair, fever, inflamed and runny nose, conjunctivitis, abdominal pain, nausea and vomiting. Lethargy, jumbled speech, tremors, convulsions and cyanosis may follow. Pulmonary edema and pneumonia may precede death from respiratory failure.
Note to physician or authorized medical personnel: Medical observation is recommended for 24 to 48 hours after breathing overexposure, as pulmonary edema may be delayed. As first aid for pulmonary edema, consider administering a corticosteroid spray. Cigarette smoking may exacerbate pulmonary injury and should be discouraged for at least 72 hours following exposure.

HEALTH HAZARDS
Personal protective equipment (PPE): B-Level PPE. OSHA Table Z-1-A air contaminant.
Recommendations for respirator selection: NIOSH/OSHA as thallium soluble compounds
0.5 mg/m^3: DM *If not present as a fume* (any dust and mist respirator). *1 mg/m^3*: DMXSQ *If not present as a fume* (any dust and mist respirator except single-use and quarter mask respirators); or SA (any supplied-air respirator). *2.5 mg/m^3*: SA:CF (any supplied-air respirator operated in a continuous-flow mode); or PAPRDM *If not present as a fume* (any powered, air-purifying respirator with a dust and mist filter). *5 mg/m^3*: HiEF (any air-purifying, full-facepiece respirator with a high-efficiency particulate filter); or SAT:CF (any supplied-air respirator that has a tight-fitting facepiece and is operated in a continuous-flow mode); or PAPRTHiE (any powered, air-purifying respirator with a tight-fitting facepiece and a high-efficiency particulate filter); or SCBAF (any self-contained breathing apparatus with a full facepiece); or SAF (any supplied-air respirator with a full facepiece). *15 mg/m^3*: SAF:PD,PP (any supplied-air respirator that has a full facepiece and is operated in a pressure-demand or other positive-pressure mode). *EMERGENCY OR PLANNED ENTRY INTO UNKNOWN CONCENTRATIONS OR IDLH CONDITIONS*: SCBAF:PD,PP (any self-contained breathing apparatus that has a full facepiece and

is operated in a pressure-demand or other positive-pressure mode); or SAF:PD,PP:ASCBA (any supplied-air respirator that has a full facepiece and is operated in a pressure-demand or other positive-pressure mode in combination with an auxiliary self-contained breathing apparatus operated in a pressure-demand or other positive pressure mode). *ESCAPE:* HiEF (any air-purifying, full-facepiece respirator with a high-efficiency particulate filter); or SCBAE (any appropriate escape-type, self-contained breathing apparatus).
Exposure limits (TWA unless otherwise noted): ACGIH TLV 0.1 mg/m^3; NIOSH/OSHA 0.1 mg/m^3 as thallium; skin contact contributes significantly in overall exposure.
Toxicity by ingestion: Grade 4; LD$_{50}$ less than 50 mg/kg.
Long-term health effects: Chronic exposure may cause hair loss, atrophic changes in skin and nails, salivation, pigmentation of the gums, and renal damage. Psychotic symptoms such as nervousness, anxiety, depression, impaired memory, sloppiness and deteriorating work performance indicate organic brain damage. Teratogenic effects in laboratory animals.
IDLH value: 15 mg/m^3 as thallium.

CHEMICAL REACTIVITY
Binary reactants: Violent reaction with aluminum, magnesium.

ENVIRONMENTAL DATA
Food chain concentration potential: Thallium is a cumulative poison four times as toxic as arsenious oxide. Positive. Expected to accumulate. Plants growing in soils or water with very high thallium content may accumulate sufficient thallium to be toxic to organisms that feed on them. Algae from contaminated water exhibited thallium bioconcentration factor of more than 430. Other bioconcentration factors that have been reported include 130 for atlantic salmon mussel and 18 for the edible portion of softshell clams.
Water pollution: DOT Appendix B, §172.101–marine pollutant. Harmful to aquatic life in very low concentrations. May be dangerous if it enters nearby water intakes; notify operators. Notify local health and wildlife officials. **Response to discharge:** Issue warning–poison; water contaminant. Should be removed. Clean Water Act.

PHYSICAL AND CHEMICAL PROPERTIES
Physical state @ 59°F/15°C and 1 atm: Solid.
Molecular weight: 504.85.
Boiling point @ 1 atm: Decomposes.
Melting/Freezing point: 1170°F/632°C/905°K.
Specific gravity (water = 1): 6.77 @ 68°F/20°C.
Heat of solution: (Absorbs heat). 29.5 Btu/lb = 16.4 cal/g = 6.86 x 10^5 J/kg.
Heat of fusion: 10.9 cal/g.

THIOCARBAMIDE REC. T:1350

SYNONYMS: ISOTHIOUREA; PSEUDOTHIOUREA; SULOUREA; THIOCARBAMATE; THIOUREA; 2-THIOUREA; THU; TIOCARBAMIDA (Spanish); UREA,THIO-7; RCRA No. U219

IDENTIFICATION
CAS Number: 62-56-6
Formula: CH_4N_2S; NH_2C5NH_2
DOT ID Number: UN 2757 (solid); UN 2992 (liquid); DOT Guide Number: 151
Proper Shipping Name: Carbamate pesticides, solid, toxic; Carbamate pesticides, liquid, toxic
Reportable Quantity (RQ): **(CERCLA)** 10 lb/4.54 kg

DESCRIPTION: White to off-white crystalline solid or powder. Odorless. Sinks and mixes with water.

Poison! (carbamate) • Breathing the dust or vapor, skin or eye contact, or swallowing the material can kill you • Firefighting gear (including SCBA) does not provide adequate protection. If exposure occurs, remove and isolate gear immediately and thoroughly decontaminate personnel • Toxic products of combustion may include oxides of sulfur and nitrogen.

Hazard Classification (based on NFPA-704 Rating System)
Health Hazards (Blue): 1; Flammability (Red): 0; Reactivity (Yellow): 0

EMERGENCY RESPONSE: See Appendix A (151)
Evacuation:
Public safety: Isolate the area of spill or leak for at least 25 to 50 meters (80 to 160 feet) in all directions.
Spill: Increase, in the downwind direction, as necessary, the distance shown under "Public Safety."
Fire: If tank, rail car, or tank truck is involved in fire, isolate for at least 800 meters (½ mile) in all directions; also, consider initial evacuation for 800 meters (½ mile) in all directions.

EXPOSURE
Short-term effects: *SEEK MEDICAL ATTENTION.* **Dust:** *POISONOUS IF INHALED.* May irritate skin. Move to fresh air. *IF BREATHING HAS STOPPED,* give artificial respiration; *avoid mouth-to-mouth resuscitation; use bag/mask apparatus.* IF breathing is difficult, administer oxygen. **Solid:** *POISONOUS IF SWALLOWED.* Irritating to skin. *IF IN EYES OR ON SKIN,* flush with running water for at least 15 minutes; hold eyelids open if necessary. Wash skin with soap and water. Remove and double bag contaminated clothing and shoes at the site. *IF SWALLOWED* and victim is *CONSCIOUS AND ABLE TO SWALLOW,* induce vomiting. *IF SWALLOWED* and victim is *UNCONSCIOUS OR HAVING CONVULSIONS,* do nothing except keep victim warm. *Note to physician or authorized medical personnel.* Administer atropine, 2 mg (1/30 gr) intramuscularly or intravenously as soon as any local or systemic signs or symptoms of an intoxication are noted; repeat the administration of atropine every 3 to 8 minutes until signs of atropinization (mydriasis, dry mouth, rapid pulse, hot and dry skin) occur; initiate treatment in children with 0.05 mg/kg of atropine; repeat at 5- to 10-minute intervals. Watch respiration, and remove bronchial secretions if they appear to be obstructing the airway; intubate if necessary.
Medical note: Due to the rapid regeneration of chlolinesterase and the fact that 2-PAMCI may be contraindicated in the case of some carbamate poisonings, 2-PAMCI (Pralidoxime; Protopam) may not be needed.

HEALTH HAZARDS
Personal protective equipment (PPE): A-Level PPE. Wear self-contained positive pressure breathing apparatus and full protective clothing. Sealed chemical protective materials recommended (<30%): Silvershield®.
Toxicity by ingestion: Grade: 3; LD$_{50}$ = 125 mg/kg (rat); varies with different stains of rats; less toxic to some strains.
Long-term health effects: Can cause cancer; mutagenic, teratogenic and tumorigenic effects. Confirmed animal carcinogen. IARC rating 2B; animal data: sufficient. Amides may cause liver, kidney, and brain damage.

Liquid or solid irritant characteristics: Skin irritant. May cause eye irritation.

CHEMICAL REACTIVITY
Binary reactants: Incompatible with metals.
Neutralizing agents for acids and caustics: Neutralize with six normal hydrochloric acid.

ENVIRONMENTAL DATA
Food chain concentration potential: Log K_{ow} = –1.0. Negative. Unlikely to accumulate.
Water pollution: Harmful to aquatic life in very low concentrations. May be dangerous if it inters water intakes. Notify local health and wildlife officials. **Response to discharge:** Issue warning–poison, water contaminant. Restrict access. Should be removed. Chemical and physical treatment.

SHIPPING INFORMATION
Grades of purity: 99%; **Storage temperature:** Ambient; **Stability during transport:** Stable.

PHYSICAL AND CHEMICAL PROPERTIES
Physical state @ 59°F/15°C and 1 atm: Solid.
Molecular weight: 76.12
Boiling point @ 1 atm: Decomposes.
Melting/Freezing point: 347–351°F/175–177°C/448–450°K.
Specific gravity (water = 1): 1.405 @ 68°F/20°C.
Relative vapor density (air = 1): 2.6 (estimate).
Heat of solution: –126.0 Btu/lb = –70.02 cal/g = –2.932 x 10^5 J/kg.

THIOPHOSGENE   REC. T:1400

SYNONYMS: DICHLOROTHIOCARBONYL; THIOCARBONYL CHLORIDE; THIOCARBONYL DICHLORIDE; THIOFOSGEN; TIOFOSGENO (Spanish)

IDENTIFICATION
CAS Number: 463-71-8
Formula: CCl_2S; $CSCl_2$
DOT ID Number: UN 2474; DOT Guide Number: 157
Proper Shipping Name: Thiophosgene

DESCRIPTION: Red liquid. Sharp, irritating, choking odor. Sinks in water. Reacts slowly with water; produces hydrochloric acid and poisonous fumes of sulfur oxides.

Poison! • Irritates eyes, skin, and respiratory tract; poisonous if inhaled or swallowed • Containers may BLEVE when exposed to fire • Decomposes above 392°F/300°C to carbon bisulfide (very flammable) and carbon tetrachloride. Toxic products of combustion may also include phosgene, hydrogen chloride, and sulfur oxides.

Hazard Classification (based on NFPA-704 Rating System)
Health Hazards (Blue): 1; Flammability (Red): 1; Reactivity (Yellow): 1
Special Notice: Water reactive

EMERGENCY RESPONSE: See Appendix A (157)
Evacuation:
Public safety: Isolate the area of spill or leak for at least 50 to 100 meters (160 to 330 feet) in all directions.
Spill: Small spill–First: Isolate in all directions 60 meters (200 feet); Then: Protect persons downwind, DAY: 0.6 km (0.4 mile); NIGHT: 1.8 km (1.1 miles). Large spill–First: Isolate in all directions 275 meters (900 feet); Then: Protect persons downwind, DAY: 2.6 km (1.6 miles); NIGHT: 5.0 km (3.1 mile).
Fire: If tank, rail car, or tank truck is involved in fire, isolate for at least 800 meters (½ mile) in all directions; also, consider initial evacuation for 800 meters (½ mile) in all directions.

EXPOSURE
Short-term effects: *SEEK MEDICAL ATTENTION.* **Vapor:** *POISONOUS IF INHALED.* Irritating to eyes, nose, and throat. Move victim to fresh air. *IF BREATHING HAS STOPPED,* give artificial respiration; *avoid mouth-to-mouth resuscitation; use bag/mask apparatus.* IF breathing is difficult, administer oxygen. **Liquid:** *POISONOUS IF SWALLOWED.* Irritating to skin and eyes. Remove contaminated clothing and shoes. Flush affected areas with plenty of water. *IF IN EYES*, hold eyelids open and flush with plenty of water. *IF SWALLOWED* and victim is *CONSCIOUS AND ABLE TO SWALLOW*, have victim drink 4 to 8 ounces of water and have victim induce vomiting. *IF SWALLOWED* and victim is *UNCONSCIOUS OR HAVING CONVULSIONS*, do nothing except keep victim warm.

HEALTH HAZARDS
Personal protective equipment (PPE): B-Level PPE. Self-contained breathing apparatus or organic canister mask; goggles or face shield; rubber gloves.
Toxicity by ingestion: Grade 2; oral LD_{50} = 929 mg/kg (rat).
Vapor (gas) irritant characteristics: Eye and respiratory tract irritant.
Liquid or solid irritant characteristics: Severe eye and skin irritant.

FIRE DATA
Fire extinguishing agents not to be used: Water, foam

CHEMICAL REACTIVITY
Reactivity with water: Evolves hydrogen chloride, carbon disulfide, and CO_2. Reaction is slow unless water is hot.
Binary reactants: Corrodes metals in presence of moisture.
Neutralizing agents for acids and caustics: Flush with water, rinse with sodium bicarbonate or lime solution

ENVIRONMENTAL DATA
Food chain concentration potential: Negative; unlikely to accumulate.
Water pollution: Effect of low concentrations on aquatic life is unknown. May be dangerous if it enters nearby water intakes; notify operators. Notify local health and wildlife officials. **Response to discharge:** Issue warning–poison, air contaminant, water contaminant. Restrict access. Evacuate area. Disperse and flush.

SHIPPING INFORMATION
Grades of purity: Commercial; **Storage temperature:** Ambient; **Inert atmosphere:** None; **Venting:** Pressure-vacuum; **Stability during transport:** Stable.

PHYSICAL AND CHEMICAL PROPERTIES
Physical state @ 59°F/15°C and 1 atm: Liquid.
Molecular weight: 115.0
Boiling point @ 1 atm: 163°F/73°C/346°K.
Specific gravity (water = 1): 1.513 @ 68°F/20°C.
Liquid surface tension: (estimate) 25 dynes/cm = 0.025 N/m @ 68°F/20°C.
Relative vapor density (air = 1): 4

Latent heat of vaporization: (estimate) 128 Btu/lb = 71 cal/g = 3.0×10^5 J/kg.
Heat of combustion: (estimate) -3400 Btu/lb = -1900 cal/g = -80×10^5 J/kg..

THIRAM REC. T:1450

SYNONYMS: AATACK; ACETO TETD; ARASAN; AULES; BIS(DIMETHYLAMINO)CARBONOTHIOYL) DISULPHIDE; BIS(DIMETHYL-THIOCARBAMOYL) DISULFID (German); α,α'-DITHIOBIS(DIMETHYLTHIO)FORMAMIDE; CHIPCO THIRAM 75; CYURAM DS; DISULFURE de TETRAMETHYLTHIOURAME (French); DURALYN; EBRAGOM TB; EKAGOM TB; FALITIRAM; FERNIDE; FERNACOL; FERNASAN; FERNIDE; FLO PRO T SEED PROTECTANT; HCDB; HERMAL; HERMAT TMT; HERYL; HEXATHIR; KREGASAN; MERCURAM; N,N'-(DITHIODICARBOROTHIOYL)BIS(N-METHYL-METHANAMINE); METHYL THIRAM; METHYL THIURAMDISULFIDE; METHYL TUADS; NOBECUTAN; NOMERSAN; NORMERSAN; NOWERGAN; PANORAM 75; PEZIFILM; POLYRAM ULTRA; POMARSOL; POMASOL; PURALIN; REZIFILM; ROYAL TMTD; SADOPLON; SPOTRETE; SQ 1489; TERSAN; TETRAMETHYL THIURAM DISULFIDE; TETRAMETHYLDIURANE SULPHITE; TETRAMETHYLENETHIURAM DISULPHIDE; TETRAMETHYLTHIOCARBAMOYLDISULPHIDE; TETRAMETHYLTHIOPEROXYDICARBONIC DIAMIDE; TETRAMETHYL-THIRAM DISULFID (German); TETRAMETHYLTHIURAM BISULFIDE; TETRAMETHYLTHIURAM DISULFIDE; N,N,N',N'-TETRAMETHYL THIURAM DISULFIDE; TETRAMETHYL THIURANE DISULFIDE; TETRAMETHYLTHIURAM DISULFIDE; THILLATE; THIMER; THIOSAN; THIOTOX; THIRAMAD; THIRAME (French); THIURAM; THIURAMYL; THYLATE; TIRAM (Spanish); TIURAMYL; TMTD; TTD; TUADS; TUEX; TULISAN USAF EK-2089VANCIDA TM-95; VANCIDE TM; VUAGT-1-4; VULCAFOR TMTD; VULKACIT MTIC; VULKACIT THIURAD; THIURAM; VULKACIT THIURAM/C; VULKAUT THIRAM; RCRA No. U244

IDENTIFICATION
CAS Number: 137-26-8
Formula: $C_6H_{12}N_2S_4$; $(CH_3)_2NC(S)SSC(S)N(CH_3)_2$
DOT ID Number: UN 2771; DOT Guide Number: 151
Proper Shipping Name: Thiram
Reportable Quantity (RQ): **(CERCLA)** 10 lb/4.54 kg

DESCRIPTION: White to light yellow solid. Commercial pesticide may be dyed blue. Sinks in water; insoluble.

Poison! • Irritating to eyes, skin, and respiratory tract; poisonous if inhaled, swallowed, or absorbed through the skin • Firefighting gear (including SCBA) does not provide adequate protection. If exposure occurs, remove and isolate gear immediately and thoroughly decontaminate personnel • Containers may BLEVE when exposed to fire • *Dangerous if in a fire:* Toxic products of combustion may include carbon disulfide, nitrogen monoxide, and sulfur dioxide • Do not put yourself in danger by entering a contaminated area to rescue a victim.

Hazard Classification (based on NFPA-704 Rating System)
Health Hazards (Blue): 2; Flammability (Red): 1; Reactivity (Yellow): 0

EMERGENCY RESPONSE: See Appendix A (151)
Evacuation:
Public safety: Isolate the area of spill or leak for at least 25 to 50 meters (80 to 160 feet) in all directions.
Spill: Increase, in the downwind direction, as necessary, the distance shown under "Public Safety."
Fire: If tank, rail car, or tank truck is involved in fire, isolate for at least 800 meters (½ mile) in all directions; also, consider initial evacuation for 800 meters (½ mile) in all directions.

EXPOSURE
Short-term effects: *SEEK MEDICAL ATTENTION.* **Dust:** *POISONOUS IF INHALED.* Irritating to eyes, nose, and throat. Move victim to fresh air. *IF IN EYES*, hold eyelids open and flush with plenty of water. *IF BREATHING HAS STOPPED,* give artificial respiration; *avoid mouth-to-mouth resuscitation; use bag/mask apparatus.* IF breathing is difficult, administer oxygen. **Solid:** *POISONOUS IF SWALLOWED.* Irritating to skin and eyes. Remove contaminated clothing and shoes. Flush affected areas with plenty of water. *IF IN EYES*, hold eyelids open and flush with plenty of water. *IF SWALLOWED* and victim is *CONSCIOUS AND ABLE TO SWALLOW*, have victim drink 4 to 8 ounces of water and have victim induce vomiting. *IF SWALLOWED* and victim is *UNCONSCIOUS OR HAVING CONVULSIONS*, do nothing except keep victim warm. The use of alcohol will exacerbate the toxic effects of Thiram.

HEALTH HAZARDS
Personal protective equipment (PPE): B-Level PPE. Rubber gloves; goggles; dust mask
Recommendations for respirator selection: NIOSH/OSHA
$50\ mg/m^3$: CCROVDMFu* [any chemical cartridge respirator with organic vapor cartridge(s) in combination with a dust, mist, and fume filter]; or SA* (any supplied-air respirator). $100\ mg/m^3$: SA:CF* (any supplied-air respirator operated in a continuous-flow mode); or CCRFOVHiE [any chemical cartridge respirator with a full facepiece and organic vapor cartridge(s) in combination with a high-efficiency particulate filter]; or GMFOVHiE [any air-purifying, full-facepiece respirator (gas mask) with a chin-style, front- or back-mounted organic vapor canister having a high-efficiency particulate filter]; or PAPROVDMFu* [any powered, air-purifying respirator with organic vapor cartridge(s) in combination with a dust, mist, and fume filter]; or SCBAF (any self-contained breathing apparatus with a full facepiece); or SAF (any supplied-air respirator with a full facepiece). *EMERGENCY OR PLANNED ENTRY INTO UNKNOWN CONCENTRATIONS OR IDLH CONDITIONS:* SCBAF:PD,PP (any self-contained breathing apparatus that has a full facepiece and is operated in a pressure-demand or other positive-pressure mode); or SAF:PD,PP:ASCBA (any supplied-air respirator that has a full facepiece and is operated in a pressure-demand or other positive-pressure mode in combination with an auxiliary self-contained breathing apparatus operated in a pressure-demand or other positive pressure mode). *ESCAPE:* GMFOVHiE [any air-purifying, full-facepiece respirator (gas mask) with a chin-style, front- or back-mounted organic vapor canister having a high-efficiency particulate filter]; or SCBAE (any appropriate escape-type, self-contained breathing apparatus). **Note:* Substance reported to cause eye irritation or damage; may require eye protection.
Exposure limits (TWA unless otherwise noted): ACGIH TLV 1 mg/m³; NIOSH/OSHA 5 mg/m³.
Toxicity by ingestion: Poison. Grade 2; oral LD_{50} = 560 mg/kg (rat); 375–865 mg/kg (mammal).
Long-term health effects: Human pulmonary problems. Causes birth defects, liver, kidney and brain damage in lab animals. IARC

rating 3; inadequate animal evidence. Possible allergen; skin problems; eczema.
Vapor (gas) irritant characteristics: Eye and respiratory tract irritation.
Liquid or solid irritant characteristics: Eye and skin irritant. Prolonged skin contact may cause skin problems.
IDLH value: 100 mg/m^3.

FIRE DATA
Flash point: Combustible solid.
Behavior in fire: Water streams applied to adjacent fires may spread contamination over wide area.

CHEMICAL REACTIVITY
Binary reactants: Strong oxidizers may cause fire and explosions; contact with strong acid or oxidizable materials produces toxic gases.

ENVIRONMENTAL DATA
Food chain concentration potential: Unlikely to accumulate.
Water pollution: Harmful to aquatic life in very low concentrations. May be dangerous if it enters nearby water intakes; notify operators. Notify local health and wildlife officials.
Response to discharge: Issue warning–water contaminant. Restrict access. Should be removed. Chemical and physical treatment.

SHIPPING INFORMATION
Grades of purity: 98% plus 2% oil; **Storage temperature:** Ambient; **Inert atmosphere:** None; **Venting:** Pressure-vacuum; **Stability during transport:** Stable.

PHYSICAL AND CHEMICAL PROPERTIES
Physical state @ 59°F/15°C and 1 atm: Solid.
Molecular weight: 240.4
Boiling point @ 1 atm: Decomposes. 264°F/129°C/402°K.
Melting/Freezing point: 288–313°F/142–156°C/415–429°K.
Specific gravity (water = 1): 1.43 @ 68°F/20°C (solid).
Vapor pressure: 0.000008 mm.

THORIUM NITRATE   **REC. T:1500**

SYNONYMS: NITRATO de TORIO (Spanish); NITRIC ACID, THORIUM(4+) SALT; NITRIC ACID, THORIUM(IV) SALT; THORIUM(4+) NITRATE; THORIUM(IV) NITRATE; THORIUM TETRANITRATE; THORIUM NITRATE TETRAHYDRATE

IDENTIFICATION
CAS Number: 13823-29-5
Formula: $N_4O_{12} \cdot Th$; $Th(NO_3)_4 \cdot 4H_2O$ (approximate)
DOT ID Number: UN 2976; DOT Guide Number: 162
Proper Shipping Name: Thorium nitrate, solid.

DESCRIPTION: White crystalline solid. Odorless. Mixes with water.

Radioactive ["priorities for rescue: lifesaving, first aid, and control of fire and other hazards are higher than the priority for measuring radiation levels" (ERG 2000)] • Strong oxidizer which may react spontaneously with low flash point organics or reducing agents. Heat forms oxygen; will increase the activity of an existing fire • May cause fire or explosion on contact with combustibles (wood, paper, oil, clothing, etc.) • Irritating to the skin, eyes, and respiratory tract • Toxic products of combustion may include nitrogen oxides.

Hazard Classification (based on NFPA-704 Rating System)
Health Hazards (Blue): 1; Flammability (Red): 0; Reactivity (Yellow): 0; Special Notice (White): OXY, Radioactive

EMERGENCY RESPONSE: See Appendix A (162)
Evacuation:
Public safety: Isolate the area of spill or leak for at least 25 to 50 meters (80 to 160 feet) in all directions. Stay upwind.
Spill: Consider initial downwind evacuation for at least 100 meters (330 feet); increase, in the downwind direction, as necessary.
Fire: When a large quantity of this material is involved in a major fire consider evacuation distance of 300 meters (1000 feet) in all directions.

EXPOSURE
Short-term effects: *SEEK MEDICAL ATTENTION*. Medical problems take priority over radiological concerns. **Dust:** Irritating to eyes, nose, and throat. Harmful if inhaled. Move victim to fresh air. *IF IN EYES*, hold eyelids open and flush with plenty of water. **Solid:** Irritating to skin and eyes. Harmful if swallowed. Remove contaminated clothing and shoes. Flush affected areas with plenty of water. *IF IN EYES*, hold eyelids open and flush with plenty of water. *IF SWALLOWED* and victim is *CONSCIOUS AND ABLE TO SWALLOW*, have victim drink 4 to 8 ounces of water.

HEALTH HAZARDS
Personal protective equipment (PPE): B-Level PPE. Dust respirator; gloves; rubber shoes or boots.
Long-term health effects: Genetic effects of long exposure to low level radiation are suspected to be harmful.
Vapor (gas) irritant characteristics: Dust may cause eye and respiratory tract irritation.
Liquid or solid irritant characteristics: Eye and skin irritant.

FIRE DATA, but may cause fire on contact with ordinary combustibles.
Behavior in fire: When large quantities are involved in fire, nitrate may fuse or melt, in which condition application of
water may result in extensive scattering of molten material. An oxidizer; will increase the intensity of a fire.
Note: Contact the local, state, or United States Department of Energy radiological response team.

CHEMICAL REACTIVITY
Reactivity with water: Forms a weak solution of nitric acid; the reaction is not hazardous.
Binary reactants: In contact with easily oxidizable substances, may react rapidly enough to cause ignition, violent combustion, or explosion. Solutions in water are acidic and can react with bases and corrode metals.
Neutralizing agents for acids and caustics: Flood with water and rinse with sodium bicarbonate or lime solution.

ENVIRONMENTAL DATA
Water pollution: Dangerous to aquatic life in high concentrations. May be dangerous if it enters nearby water intakes; notify operators. Notify local health and wildlife officials. **Response to discharge:** Issue warning-radioactive, oxidizing material, water contaminant. Restrict access. Should be removed. Chemical and physical treatment.
Disposal of wastes containing thorium: Follow guidelines set forth by the Nuclear Regulatory Commission (thorium and

compounds). Notify local and state health authorities, local solid waste disposal authorities, supplier, shipper, and United States Department of Energy.

SHIPPING INFORMATION
Grades of purity: Reagent; **Storage temperature:** Ambient; **Inert atmosphere:** None; **Venting:** Open; **Stability during transport:** Stable.

PHYSICAL AND CHEMICAL PROPERTIES
Physical state @ 59°F/15°C and 1 atm: Solid.
Molecular weight: 484.08
Boiling point @ 1 atm: Decomposes.
Specific gravity (water = 1): (estimate) more than 1 @ 68°F/20°C (solid).

TITANIUM TETRACHLORIDE   REC. T:1550

SYNONYMS: EEC No. 022-001-00-5; TETRACLORURO de TITANIO (Spanish); TITANIUM CHLORIDE; TITANIUM(IV) CHLORIDE; TITANIUM(4+) CHLORIDE

IDENTIFICATION
CAS Number: 7550-45-0
Formula: Cl_4Ti; $TiCl_4$
DOT ID Number: UN 1838; DOT Guide Number: 137
Proper Shipping Name: Titanium tetrachloride
Reportable Quantity (RQ): (CERCLA 1 lb/0.454 kg).

DESCRIPTION: Colorless to pale yellow watery liquid. Irritating odor; acrid, choking. Reacts violently with water, producing titanic acid and fumes as hydrogen chloride.

Corrosive • Breathing the vapor can kill you; skin or eye contact causes severe burns, impaired vision, or blindness; effects of contact or inhalation may be delayed • Firefighting gear (including SCBA) does not provide adequate protection. If exposure occurs, remove and isolate gear immediately and thoroughly decontaminate personnel • Forms hydrogen chloride fumes in moist air • Toxic products of combustion may include hydrogen chloride.

Hazard Classification (based on NFPA-704 Rating System)
Health Hazards (Blue): 3; Flammability (Red): 0; Reactivity (Yellow): 2; Special Notice (White): Water reactive

EMERGENCY RESPONSE: See Appendix A (137)
Evacuation:
Public safety: See below.
Spill: Small spill–First: Isolate in all directions 30 meters (100 feet) (land or water); Then: Protect persons downwind, DAY: 0.2 km (0.1 mile) (land or water); NIGHT: 0.5 km (0.3 mile) (land); 1.0 km (0.6 mile) (water). Large spill–First: Isolate in all directions 60 meters (200 feet) (land); 335 meyers (1100 feet) (water); Then: Protect persons downwind, DAY: 0.5 km (0.3 mile) (land); 0.3 km (0.2 mile) (water); NIGHT: 1.1 km (0.7 mile) (land); 7.1 km (4.4 mile) (water).
Fire: If any large container is involved in fire, isolate for at least 800 meters (½ mile) in all directions; also, consider initial evacuation for 800 meters (½ mile) in all directions.

EXPOSURE
Short-term effects: *SEEK MEDICAL ATTENTION.* **Vapor:** Irritating to eyes, nose, and throat. *IF INHALED*, will, will cause coughing or headache. Move to fresh air. *IF BREATHING HAS STOPPED*, give artificial respiration; *avoid mouth-to-mouth resuscitation; use bag/mask apparatus.* IF breathing is difficult, administer oxygen. **Liquid:** Will burn skin and eyes. *IF SWALLOWED*, will cause nausea and vomiting. Remove contaminated clothing and shoes. Flush affected areas with plenty of water. *IF IN EYES*, hold eyelids open and flush with plenty of water or milk and have victim induce vomiting. *IF SWALLOWED* and victim is *UNCONSCIOUS OR HAVING CONVULSIONS*, do nothing except keep victim warm.

Note to physician or authorized medical personnel: Medical observation is recommended for 24 to 48 hours after breathing overexposure, as pulmonary edema may be delayed. As first aid for pulmonary edema, consider administering a corticosteroid spray. Cigarette smoking may exacerbate pulmonary injury and should be discouraged for at least 72 hours following exposure.

HEALTH HAZARDS
Personal protective equipment (PPE): A-Level PPE. Acid gas respirator. Chemical protective material(s) reported to have good to excellent resistance: Saranex®. Also, Viton®/neoprene offers limited protection.
Recommendations for respirator selection: NIOSH/OSHA as hydrogen chloride
50 ppm: CCRS* [any chemical cartridge respirator with cartridge(s) providing protection against the compound of concern]; or GMFS [any air-purifying, full-facepiece respirator (gas mask) with a chin-style, front- or back-mounted canister providing protection against the compound of concern]; or PAPRS* [any powered, air-purifying respirator with cartridge(s) providing protection against the compound of concern]; or SA* (any supplied-air respirator); or SCBAF (any self-contained breathing apparatus with a full facepiece). *EMERGENCY OR PLANNED ENTRY INTO UNKNOWN CONCENTRATIONS OR IDLH CONDITIONS*: SCBAF:PD,PP (any self-contained breathing apparatus that has a full facepiece and is operated in a pressure-demand or other positive-pressure mode); or SAF:PD,PP:ASCBA (any supplied-air respirator that has a full facepiece and is operated in a pressure-demand or other positive-pressure mode in combination with an auxiliary self-contained breathing apparatus operated in a pressure-demand or other positive pressure mode). *ESCAPE*: GMFAG [any air-purifying, full-facepiece respirator (gas mask) with a chin-style, front- or back-mounted organic vapor cannister]; or SCBAE (any appropriate escape-type, self-contained breathing apparatus).
**Note:* Substance reported to cause eye irritation or damage; may require eye protection.
Exposure limits (TWA unless otherwise noted): 0.5 mg/m³ (AIHAWEEL).
Long-term health effects: Disturbances of upper respiratory and nervous system in man.
Vapor (gas) irritant characteristics: Eye and respiratory tract irritant. Vapor is moderately irritating such that personnel will not tolerate moderate or high vapor concentrations.
Liquid or solid irritant characteristics: Severe eye and skin irritant; may cause pain and second-degree burns after a few minutes of contact.

FIRE DATA
Fire extinguishing agents not to be used: Do not use water if it can contact titanium tetrachloride.
Behavior in fire: If container leaks, a very dense white fume forms and can obscure operations.

CHEMICAL REACTIVITY
Reactivity with water: Reacts with moisture in air, forming dense

white fume. Reaction with liquid water gives off heat and forms hydrochloric acid.
Binary reactants: Reacts with air forming hydrochloric acid fumes. The acid formed by reaction with
moisture attacks metals, forming flammable hydrogen gas.
Neutralizing agents for acids and caustics: Acid formed by reaction with water can be neutralized by limestone, lime, or soda ash.
Water pollution: Dangerous to aquatic life in high concentrations. May be dangerous if it enters nearby water intakes; notify operators. Notify local health and wildlife officials. **Response to discharge:** Issue warning–corrosive, air contaminant. Restrict access. Evacuate area.

SHIPPING INFORMATION
Grades of purity: Technical; **Storage temperature:** Ambient; **Inert atmosphere:** None; **Venting:** Pressure-vacuum; **Stability during transport:** Stable.

NAS HAZARD CLASSIFICATION FOR BULK WATER TRANSPORTATION
FIRE: 0
HEALTH: Vapor irritant: 3; Liquid or solid irritant: 3; Poisons: 2
WATER POLLUTION: Human toxicity: 2; Aquatic toxicity: 2; Aesthetic effect: 2
REACTIVITY: Other chemicals: 3; Water: 4; Self-reaction: 0

PHYSICAL AND CHEMICAL PROPERTIES
Physical state @ 59°F/15°C and 1 atm: Liquid.
Molecular weight: 189.71
Boiling point @ 1 atm: 277°F/136°C/409°K.
Melting/Freezing point: –11°F/–24°C/249°K.
Specific gravity (water = 1): 1.726 @ 68°F/20°C (liquid).
Relative vapor density (air = 1): 6.49
Ratio of specific heats of vapor (gas): (estimate) 1.221
Latent heat of vaporization: 79.7 Btu/lb = 44.3 cal/g = 1.86×10^5 J/kg.
Heat of solution: –482.8 Btu/lb = –268.2 cal/g = -11.22×10^5 J/kg.
Vapor pressure: 9.65 mm.

TOLUENE REC. T:1600

SYNONYMS: ANTISAL LA; BENZENE, METHYL-; EEC No. 601-021-00-3; METHACIDE; METHANE, PHENYL-; METHYL BENZENE; METHYLBENZOL; TOLUENO (Spanish); TOLUOL; TOLUOLO; TOLU-SOL; RCRA No. U220

IDENTIFICATION
CAS Number: 108-88-3
Formula: C_7H_8; $C_6H_5CH_3$
DOT ID Number: UN 1294; DOT Guide Number: 128
Proper Shipping Name: Toluene
Reportable Quantity (RQ): **(CERCLA)** 1000 lb/454 kg

DESCRIPTION: Colorless watery liquid. Pleasant odor, like model airplane glue. Floats on water surface; insoluble. Flammable, irritating vapor is produced.

Highly flammable • Containers may BLEVE when exposed to fire • Vapors may form explosive mixture with air • Vapors are heavier than air and will collect and stay in low areas • Vapors may travel long distances to ignition sources and flashback • Vapors in confined areas (e.g., tanks, sewers, buildings) may explode when exposed to fire • Irritating to the skin, eyes, and respiratory tract • Toxic products of combustion may include carbon monoxide.

Hazard Classification (based on NFPA-704 Rating System)
Health Hazards (Blue): 2; Flammability (Red): 3; Reactivity (Yellow): 0

EMERGENCY RESPONSE: See Appendix A (128)
Evacuation:
Public safety: Isolate spill area for at least 25 to 50 meters (80 to 160 feet) in all directions.
Spill: Large spill–Consider initial downwind evacuation for at least 300 meters (1000 feet).
Fire: Isolate for 800 meters (½ mile) in all directions, especially if tank, rail car, or tank truck is involved in fire.

EXPOSURE
Short-term effects: *SEEK MEDICAL ATTENTION.* **Vapor:** *HARMFUL IF ABSORBED THROUGH THE SKIN.* Irritating to eyes, nose, and throat. *IF INHALED*, will, will cause nausea, vomiting, headache, dizziness, difficult breathing, or loss of consciousness. Move to fresh air. *IF BREATHING HAS STOPPED*, give artificial respiration. IF breathing difficult, administer oxygen. **Liquid:** Irritating to skin and eyes. *IF SWALLOWED*, will cause nausea, vomiting, or loss of consciousness. May cause pneumonia if drops enter lungs. Remove contaminated clothing and shoes. Flush affected areas with plenty of water. *IF IN EYES*, hold eyelids open and flush with plenty of water. *IF SWALLOWED* and victim is *CONSCIOUS AND ABLE TO SWALLOW*, have victim drink 4 to 8 ounces of water. **Do NOT induce vomiting.**

HEALTH HAZARDS
Personal protective equipment (PPE): B-Level PPE. OSHA Table Z-1-A air contaminant. OSHA Table Z-2 air contaminant. Organic vapor respirator. Air–supplied mask; goggles or face shield; gloves. Chemical protective material(s) reported to have good to excellent resistance: PV alcohol, Teflon®, Viton®, Viton®/neoprene, Viton®/chlorophenol rubber, Silvershield®. Also, chlorinated polyethylene offers limited protection.
Recommendations for respirator selection: NIOSH/OSHA
500 ppm: CCRFOV* [any air-purifying, full-facepiece respirator (gas mask) with a chin-style, front- or back-mounted acid gas canister]; or PAPROV* [any powered, air-purifying respirator with organic vapor cartridge(s)]; or GMFOV [any air-purifying, full-facepiece respirator (gas mask) with a chin-style, front- or back-mounted organic vapor canister]; or SA* (any supplied-air respirator); or SCBAF (any self-contained breathing apparatus with a full facepiece). *EMERGENCY OR PLANNED ENTRY INTO UNKNOWN CONCENTRATIONS OR IDLH CONDITIONS:* SCBAF:PD,PP (any self-contained breathing apparatus that has a full facepiece and is operated in a pressure-demand or other positive-pressure mode); or SAF:PD,PP:ASCBA (any supplied-air respirator that has a full facepiece and is operated in a pressure-demand or other positive-pressure mode in combination with an auxiliary self-contained breathing apparatus operated in a pressure-demand or other positive pressure mode). *ESCAPE:* GMFOV [any air-purifying, full-facepiece respirator (gas mask) with a chin-style, front- or back-mounted organic vapor canister]; or SCBAE (any appropriate escape-type, self-contained breathing apparatus).
**Note:* Substance reported to cause eye irritation or damage; may require eye protection.
Exposure limits (TWA unless otherwise noted): ACGIH TLV 50 ppm (188 mg/m³); OSHA PEL 200 ppm; NIOSH REL 100 ppm (375 mg/m³) skin contact contributes significantly in overall exposure.

1058 *m*-Toluene diamine

Short-term exposure limits (15-minute TWA): NIOSH STEL 150 ppm (560 mg/m^3); OSHA ceiling 300 ppm, maximum peak 500 ppm [10 minute].
Toxicity by ingestion: Grade 2; LD$_{50}$ = 0.5 to 5 g/kg.
Long-term health effects: Kidney and liver damage may follow ingestion. IARC rating 3; inadequate human and animal data. Chronic exposure can result in permanent neuropsychiatric effects. Disorders of the muscles, cardiovascular effects, renal tubular damage, and sudden death have occurred in chronic abusers of toluene (e.g., glue sniffers).
Vapor (gas) irritant characteristics: Eye and respiratory tract irritant. Vapors cause smarting of the eyes or respiratory system if present in high concentrations. The effect is temporary.
Liquid or solid irritant characteristics: Eye and skin irritant. If spilled on clothing and allowed to remain, may cause smarting and reddening of the skin.
Odor threshold: 0.17 ppm.
IDLH value: 500 ppm.

FIRE DATA
Flash point: 55°F/13°C (oc); 40°F/4°C (cc).
Flammable limits in air: LEL: 1.1%; UEL: 7.1%.
Fire extinguishing agents not to be used: Water may be ineffective.
Autoignition temperature: 905°F/485°C/758°K.
Electrical hazard: Class I, Group D. Flow or agitation of substance may generate electrostatic charges due to low conductivity.
Burning rate: 5.7 mm/min

CHEMICAL REACTIVITY
Binary reactants: Violent reaction with oxidizers. Rubber will swell, soften and deteriorate.
Reactivity group: 32
Compatibility class: Aromatic hydrocarbons

ENVIRONMENTAL DATA
Food chain concentration potential: Log POw = 2.6. Unlikely to accumulate.
Water pollution: Dangerous to aquatic life in high concentrations. Fouling to shoreline. May be dangerous if it enters nearby water intakes; notify operators. Notify local health and wildlife officials.
Response to discharge: Issue warning–high flammability. Evacuate area.

SHIPPING INFORMATION
Grades of purity: Research, reagent, nitration-all 99.8+%; industrial: contains 94+%, with 5% xylene and small amounts of benzene and nonaromatic hydrocarbons; 90/120: Less pure than industrial; **Storage temperature:** Ambient; **Inert atmosphere:** None; **Venting:** Open (flame arrester) or pressure-vacuum; **Stability during transport:** Stable.

NAS HAZARD CLASSIFICATION FOR BULK WATER TRANSPORTATION
FIRE: 3
HEALTH: Vapor irritant: 1; Liquid or solid irritant: 1; Poisons: 2
WATER POLLUTION: Human toxicity: 1; Aquatic toxicity: 3; Aesthetic effect: 2
REACTIVITY: Other chemicals: 1; Water: 0; Self-reaction: 0

PHYSICAL AND CHEMICAL PROPERTIES
Physical state @ 59°F/15°C and 1 atm: Liquid.
Molecular weight: 92.14
Boiling point @ 1 atm: 231.1°F/110.6°C/383.8°K.
Melting/Freezing point: –139°F/–95.0°C/178.2°K.
Critical temperature: 605.4°F/318.6°C/591.8°K.
Critical pressure: 596.1 psia = 40.55 atm = 4.108 MN/m^2.
Specific gravity (water = 1): 0.867 @ 68°F/20°C (liquid).
Liquid surface tension: 29.0 dynes/cm = 0.0290 N/m @ 68°F/20°C.
Liquid water interfacial tension: 36.1 dynes/cm = 0.0361 N/m @ 77°F/25°C.
Relative vapor density (air = 1): 3.2
Ratio of specific heats of vapor (gas): 1.089
Latent heat of vaporization: 155 Btu/lb = 86.1 cal/g = 3.61 x 10^5 J/kg.
Heat of combustion: –17,430 Btu/lb = –9686 cal/g = –405.5 x 10^5 J/kg.
Heat of fusion: 17.17 cal/g.
Vapor pressure: (Reid) 1.1 psia; 22 mm.

---

m-TOLUENE DIAMINE   REC. T:1650

SYNONYMS: AZOGEN DEVELOPER-H; BENZENEDIAMINE,AR-METHYL-; C.I. 76035; C.I. OXIDATION BASE; 2,4-DIAMINOTOLUENE; 2,4-DIAMINOTOLUENO (Spanish); 2,4-TOLAMINE; 4-*m*-TOLYLENEDIAMINE; 2,4-TOLUENEDIAMINE; 2,4-TOLUENE-2,4-DIAMINE; *m*-TOLUYLENEDIAMINE; MTD; PELAGOL GREY J; TDA; ZOBA GKE; ZOGEN DEVELOPER-H; RCRA No. U221

IDENTIFICATION
CAS Number: 95-80-7
Formula: C$_7$H$_{10}$N$_2$; CH$_3$C$_6$H$_3$(NH$_2$)$_2$
DOT ID Number: UN 1709; DOT Guide Number: 151
Proper Shipping Name: 2,4-Toluylenediamine; 2,4-toluenediamine
Reportable Quantity (RQ): **(CERCLA)** 10 lb/4.54 kg

DESCRIPTION: Colorless crystalline solid. Odorless. Floats on the surface of water; soluble.

Combustible solid • Containers may BLEVE when exposed to fire • Concentrated dust in confined areas (e.g., tanks, sewers, buildings) may explode when exposed to fire • Irritating to the skin, eyes, and respiratory tract • May interfere with the body's ability to use oxygen • Toxic products of combustion may include nitrogen oxides.

Hazard Classification (based on NFPA-704 Rating System)
Health Hazards (Blue): 3; Flammability (Red): 1; Reactivity (Yellow): 0

EMERGENCY RESPONSE: See Appendix A (151)
Evacuation:
Public safety: Isolate the area of spill or leak for at least 25 to 50 meters (80 to 160 feet) in all directions.
Spill: Increase, in the downwind direction, as necessary, the distance shown under "Public Safety."
Fire: If tank, rail car, or tank truck is involved in fire, isolate for at least 800 meters (½ mile) in all directions; also, consider initial evacuation for 800 meters (½ mile) in all directions.

EXPOSURE
Short-term effects: *SEEK MEDICAL ATTENTION.* **Dust:** *TOXIC BY INHALATION, SKIN ABSORPTION, AND INGESTION.* Irritating to eyes and skin. *IF BREATHING HAS STOPPED*, give artificial respiration; *avoid mouth-to-mouth resuscitation; use*

bag/mask apparatus. IF IN EYES, hold eyelids open and flush with water for at least 15 minutes. IF on skin, remove clothing and shower thoroughly with soap and water and put on clean clothing. **Solid:** Irritating to eyes and skin. *TOXIC IF SWALLOWED. IF IN EYES,* hold eyelids open and flush with plenty of water for at least 15 minutes. IF on skin, remove clothing and shower thoroughly with soap and water. *IF SWALLOWED* and victim is *UNCONSCIOUS OR HAVING CONVULSIONS,* do nothing except keep victim warm.

HEALTH HAZARDS
Personal protective equipment (PPE): B-Level PPE. Hat and goggles, respirator with combination dust–acid–gas–organic vapor cartridge, gauntlet vinyl gloves taped to jacket, and long-sleeved underwear, vinyl apron, and rubber footwear.
Recommendations for respirator selection: NIOSH
At any concentrations above the NIOSH REL, or where there is no REL, at any detectable concentration: SCBAF:PD,PP (any self-contained breathing apparatus that has a full facepiece and is operated in a pressure-demand or other positive-pressure mode); or SAF:PD,PP:ASCBA (any supplied-air respirator that has a full facepiece and is operated in a pressure-demand or other positive-pressure mode in combination with an auxiliary self-contained breathing apparatus operated in a pressure-demand or other positive pressure mode). *ESCAPE:* GMFOV [any air-purifying, full-facepiece respirator (gas mask) with a chin-style, front- or back-mounted organic vapor canister]; or, SCBAE (any appropriate escape-type, self-contained breathing apparatus).
Exposure limits (TWA unless otherwise noted): NIOSH REL potential human carcinogen; reduce exposure to lowest feasible levels.
Exposure limits (TWA unless otherwise noted): 0.005 ppm [skin] (AIHA WEEL).
Toxicity by ingestion: GRADE 3; LD_{50} = 100 mg/kg.
Long-term health effects: This compound is extremely dangerous. Toxic to the liver and central nervous system. Confirmed animal carcinogen. It causes jaundice and anemia; cancer in rats and female mice in feeding studies. It possesses mutagenic activities and it causes fatty degeneration of the liver. It was found to induce liver tumors in rats when fed at levels up to one percent in the diet.
Vapor (gas) irritant characteristics: Eye and respiratory tract irritant.
Liquid or solid irritant characteristics: Irritates eyes and skin, and it can cause skin blistering in sensitive individuals.

FIRE DATA
Flash point: 300°F/149°C (cc); 410°F/210°C (oc).
Autoignition temperature: More than 887°F/475°C/748°K.

CHEMICAL REACTIVITY
Binary reactants: Incompatible with oxidizers, strong acids, chloroformates, organic anhydrides, isocyanates, aldehydes, isopropyl chlorocarbonate, nitrosyl perchlorate. Contact with diisopropyl perdicarbonate may cause explosion. Attacks aluminum, brass, bronze, copper, zinc.
Reactivity group: 9
Compatibility class: Aromatic amines

ENVIRONMENTAL DATA
Food chain concentration potential: Log P_{ow} = 0.34. Unlikely to accumulate.
Water pollution: Effect of low concentration on aquatic life is unknown. May be dangerous if enters water intakes. Notify local health and wildlife officials. **Response to discharge:** Issue warning–water contaminant. Should be removed. Chemical and physical treatment.

SHIPPING INFORMATION
Grades of purity: 98-99+%; **Stability during transport:** Stable.

PHYSICAL AND CHEMICAL PROPERTIES
Physical state @ 59°F/15°C and 1 atm: Solid.
Molecular weight: 122.17
Boiling point @ 1 atm: 558°F/292°C/565°K.
Melting/Freezing point: 210°F/99°C/372°K.
Specific gravity (water = 1): 1.0
Vapor pressure: 1 mm @ 107°F

TOLUENE-2,4-DIISOCYANATE REC. T:1700

SYNONYMS: BENZENE,2,4-DIISOCYANATOMETHYL-; BENZENE,2,4-DIISOCYANATO-1-METHYL-; CRESORCINOL DIISOCYANATE; DESMODUR T80; DI-ISOCYANATE de TOLUYLENE (French); DI-ISOCYANATOLUENE; 2,4-DIISOCYANATO-1-METHYLBENZENE; 2,4-DIISOCYANATOTOLUENE; DIISOCYANAT-TOLUOL (German); EEC No. 615-006-00-4; HYLENE T; HYLENE TCPA; HYLENE TLC; HYLENE TM; HYLENE TM-65; HYLENE TRF; ISOCYANIC ACID, METHYLPHENYLENE ESTER; ISOCYANIC ACID, 4-METHYL-M-PHENYLENE ESTER; 4-METHYL-PHENYLENE DIISOCYANATE; 4-METHYL-PHENYLENE ISOCYANATE; MONDUR TDS; NACCONATE IOO; NIAX TDI; NIAX TDI-P; TDI; 2,4-TDI; TDI-80; TOLUEN-2,4-DIISOCIATO (Spanish); TOLUENE DIISOCYANATE; TOLUENE DIISOCYANATE; TOLUENE-2,4-DIISOCYANATE; 2,4-TOLUENEDIISOCYANATE; TOLUYLENE-2,4-DIISOCYANATE; TOLYENE 2,4-DIISOCYANATE; TOLYLENE-2,4-DIISOCYANATE 2,4-TOLYLENEDIISOCYANATE; 2,4-TOLYLENE DIISOCYANATE; TOLUENE DI-ISOCYANATE; TULUYLENE-2,4-DIISOCYANATE; TULUYLEN DIISOCYANAT (German); RCRA No. U223

IDENTIFICATION
CAS Number: 584-84-9 (2,4-isomer); 91-08-7 (2,6-isomer); 26471-62-5 (mixture)
Formula: $C_9H_6N_2O_2$
DOT ID Number: UN 2078; DOT Guide Number: 156
Proper Shipping Name: Toluene diisocyanate
Reportable Quantity (RQ): **(CERCLA)** 100 lb/45.4 kg

DESCRIPTION: Colorless to pale yellow solid or liquid. Sharp, pungent odor. Sinks in water; reacts slowly forming CO_2 and polyureas. Freezing point is 68-72°F/20-22°C. *Warning:* Odor is not a reliable indicator of the presence of toxic amounts of toluene diisocyanate.

Poison! • Corrosive • Breathing the vapor, skin contact, or swallowing the material can kill you; inhalation symptoms may be delayed • Firefighting gear (including SCBA) does not provide adequate protection. If exposure occurs, remove and isolate gear immediately and thoroughly decontaminate personnel • Polymerization hazard • May react with itself blocking relief valves; leading to tank explosions • Containers may BLEVE when exposed to fire • Vapors or Concentrated dust in confined areas (e.g., tanks, sewers, buildings) may explode when exposed to fire • Irritating to the skin, eyes, and respiratory tract • Toxic products of combustion may include nitrogen oxides and cyanide vapors.

Toluene-2,4-diisocyanate

Hazard Classification (based on NFPA-704 Rating System)
Health Hazards (Blue): 3; Flammability (Red): 1; Reactivity (Yellow): 3; Special Notice (White): Water Reactive

EMERGENCY RESPONSE: See Appendix A (156)
Evacuation:
Public safety: Isolate the area of spill or leak for at least 50 to 100 meters (160 to 330 feet) in all directions.
Spill: Increase, in the downwind direction, as necessary, the distance shown under "Public Safety."
Fire: If tank, rail car, or tank truck is involved in fire, isolate for at least 800 meters (½ mile) in all directions; also, consider initial evacuation for 800 meters (½ mile) in all directions.

EXPOSURE
Short-term effects: *SEEK MEDICAL ATTENTION*. If artificial respiration is administered, *avoid mouth-to-mouth resuscitation; use bag/mask apparatus*. **Liquid:** *POISONOUS IF SWALLOWED*. Will burn skin and eyes. Remove contaminated clothing and shoes. Flush affected areas with plenty of water. *IF IN EYES*, hold eyelids open and flush with plenty of water. *IF SWALLOWED* and victim is *CONSCIOUS AND ABLE TO SWALLOW*, have victim drink 4 to 8 ounces of water. **Do NOT induce vomiting.**
Note to physician or authorized medical personnel: Consider the use of amyl nitrite perles if symptoms of cyanide poisoning develop. If symptoms indicate, initial treatment includes the cyanide antidote kit. In all cases, break an amyl nitrite perle in a gauze pad and hold lightly under victim's nose for 15 seconds, repeating 5 times at about 15-second intervals; if necessary (and if sodium nitrite infusions will be delayed), repeat procedure every 3 minutes with fresh pearls until 3 or 4 have been used. Avoid breathing the vapor while administering it to the victim. Administer sodium nitrite IV, ASAP. The usual adult dose is 10 to 20 mL of a 3% solution infused over no less than 5 minutes; the average child dose is 0.15 to 0.20 mL/kg. Monitor blood pressure during administration, and slow the rate of infusion if hypotention develops. Next, infuse sodium thiosulfate IV. The usual adult dose is 50 mL of a 25% solution infused over 10 to 20 minutes; the average child dose is 1.65 mL/kg. Repeat with nitrite and thiosulfate as required.

HEALTH HAZARDS
Personal protective equipment (PPE): A-Level PPE. OSHA Table Z-1-A air contaminant. Chemical protective material(s) reported to have good to excellent resistance: butyl rubber, nitrile, PV alcohol, Viton®, Saranex®.
Recommendations for respirator selection: NIOSH
At any concentrations above the NIOSH REL, or where there is no REL, at any detectable concentration: SCBAF:PD,PP (any self-contained breathing apparatus that has a full facepiece and is operated in a pressure-demand or other positive-pressure mode); or SAF:PD,PP:ASCBA (any supplied-air respirator that has a full facepiece and is operated in a pressure-demand or other positive-pressure mode in combination with an auxiliary self-contained breathing apparatus operated in a pressure-demand or other positive pressure mode). *ESCAPE:* GMFOV [any air-purifying, full-facepiece respirator (gas mask) with a chin-style, front- or back-mounted organic vapor canister]; or SCBAE (any appropriate escape-type, self-contained breathing apparatus).
Exposure limits (TWA unless otherwise noted): ACGIH TLV 0.005 ppm; NIOSH REL potential human carcinogen.
Short-term exposure limits (15-minute TWA): ACGIH STEL 0.02 ppm; OSHA ceiling 0.02 ppm (0.14 mg/m^3).
Toxicity by ingestion: Grade 2; LD_{50} = 0.5 to 5 g/kg.
Long-term health effects: IARC possible human carcinogen, rating 2B; sufficient animal data. Allergen; chronic lung disease and skin problems. Mutagen.
Vapor (gas) irritant characteristics: Eye and respiratory tract irritant. Vapor is irritating such that personnel will not usually tolerate moderate or high vapor concentrations.
Liquid or solid irritant characteristics: Severe skin and eye irritant; may cause pain and second-degree burns after a few minutes of contact. Multiple exposures should be avoided.
Odor threshold: 0.4-2.2 ppm. *Note*: Above the Exposure limits.
IDLH value: 2.5 ppm; potential human carcinogen.

FIRE DATA
Flash point: 260°F/127°C (cc).
Flammable limits in air: LEL: 0.9%; UEL 9.5%.
Fire extinguishing agents not to be used: Do not allow water to come in contact with material. Water or foam may cause frothing.
Behavior in fire: Containers may explode.
Autoignition temperature: 1150°F/621°C/894°K.
Electrical hazard: Class I, Group D.

CHEMICAL REACTIVITY
Reactivity with water: Reacts exothermically with water. Forms CO_2 gas and an organic base. Vigorous reaction with warm water.
Binary reactants: Reacts with strong oxidizers, ammonia, amines, acids and alcohols. Attacks some copper and its alloys; plastics; polyethylene.
Polymerization: May be caused by contact with bases.
Reactivity group: 12
Compatibility class: Isocyanates

ENVIRONMENTAL DATA
Water pollution: Effect of low concentrations on aquatic life is unknown. May be dangerous if it enters nearby water intakes; notify operators. Notify local health and wildlife officials. **Response to discharge:** Issue warning–water contaminant. Restrict access. Should be removed. Chemical and physical treatment.

SHIPPING INFORMATION
Grades of purity: Commercial distilled, 99% total diisocyanate. The following isomer ratios are shipped: (a) 100% 2,4-isomer; (b) 80% 2,4-isomer; 20% 2,6-isomer (most common); (c) 65% 2,4-isomer; 35% 2,6-isomer. All mixtures have about the same hazard characteristics; **Storage temperature:** 75–100°F/24–38°C; **Inert atmosphere:** Inerted; **Venting:** Pressure-vacuum; **Stability during transport:** Stable.

NAS HAZARD CLASSIFICATION FOR BULK WATER TRANSPORTATION
FIRE: 1
HEALTH: Vapor irritant: 3; Liquid or solid irritant: 3; Poisons: 4
WATER POLLUTION: Human toxicity: 2; Aquatic toxicity:–; Aesthetic effect: 4
REACTIVITY: Other chemicals: 3; Water: 3; Self-reaction: 3

PHYSICAL AND CHEMICAL PROPERTIES
Physical state @ 59°F/15°C and 1 atm: Solid.
Molecular weight: 174.16
Boiling point @ 1 atm: 482°F/250°C/523°K.
Melting/Freezing point: 68-72°F/20-22°C/293-295°K.
Specific gravity (water = 1): 1.22 @ 77°F/25°C (liquid).
Liquid surface tension: (estimate) 25 dynes/cm = 0.025 N/m @ 77°F/25°C.
Liquid water interfacial tension: (estimate) 45 dynes/cm = 0.045 N/m @ 77°F/25°C.
Relative vapor density (air = 1): 6.0

p-TOLUENESULFONIC ACID   REC. T:1750

SYNONYMS: ACIDO p-TOLUENSULFONICO (Spanish); EEC No. 016-030-00-2; METHYLBENZENESULFONIC ACID; 4-METHYLBENZENESULFONIC ACID; 4-TOLUENESULFONIC ACID; TOSIC ACID; p-TSA

IDENTIFICATION
CAS Number: 104-15-4
Formula: $C_7H_8O_3S$; $CH_3C_6H_4SO_3H$
DOT ID Number: UN 2583 (solid); UN 2585 (solid); UN 2584 (liquid); UN 2586 (liquid); DOT Guide Number: 153
PROPER SHIPPING NAMES: Alkyl sulfonic acids, solid; Aryl sulfonic acids, solid, with more than 5% free sulfuric acid (UN 2583); Alkyl sulfonic acids, liquid or Aryl sulfonic acids, liquid with more than 5% free sulfuric acid (UN 2584); Alkyl sulfonic acids, solid or Aryl sulfonic acids, solid with not more than 5% free sulfuric acid (UN 2585); Alkyl sulfonic acids, liquid or Aryl sulfonic acids, liquid with not more than 5% free sulfuric acid (UN 2586).

DESCRIPTION: White to brown to black; yellow to amber solid. No odor when pure; technical grade has slight aromatic odor. Mixes with water forming a strong acid.

Combustible • More than 5% sulfuric acid, this material is corrosive to skin., eyes, and respiratory tract • Less than 5% sulfuric acid, irritating to the skin, eyes, and respiratory tract • Extremely irritating to skin, eyes, and respiratory tract; contact may cause severe burns, impaired vision, or blindness; inhalation symptoms may be delayed • Containers may BLEVE when exposed to fire •Vapors may form explosive mixture with air • Vapors in confined areas (e.g., tanks, sewers, buildings) may explode when exposed to fire • Attacks metals forming flammable and explosive hydrogen gas • Toxic products of combustion may include and corrosive sulfur oxides.

Hazard Classification (based on NFPA-704 Rating System)
Health Hazards (Blue):; Flammability (Red): 1; Reactivity (Yellow):

EMERGENCY RESPONSE: See Appendix A (153)
Evacuation:
Public safety: Isolate the area of spill or leak for at least 25 to 50 meters (80 to 160 feet) in all directions.
Spill: Increase, in the downwind direction, as necessary, the distance shown under "Public Safety."
Fire: If tank, rail car, or tank truck is involved in fire, isolate for at least 800 meters (½ mile) in all directions; also, consider initial evacuation for 800 meters (½ mile) in all directions.

EXPOSURE
Short-term effects: *SEEK MEDICAL ATTENTION. IF BREATHING HAS STOPPED*, give artificial respiration; *avoid mouth-to-mouth resuscitation; use bag/mask apparatus.* **Solid:** Irritating to skin and eyes. Harmful if swallowed. Remove contaminated clothing and shoes. Flush affected areas with plenty of water. *IF IN EYES*, hold eyelids open and flush with plenty of water. *IF SWALLOWED* and victim is *CONSCIOUS AND ABLE TO SWALLOW*, have victim drink 4 to 8 ounces of water.

Note to physician or authorized medical personnel: Medical observation is recommended for 24 to 48 hours after breathing overexposure, as pulmonary edema may be delayed. As first aid for pulmonary edema, consider administering a corticosteroid spray. Cigarette smoking may exacerbate pulmonary injury and should be discouraged for at least 72 hours following exposure.

HEALTH HAZARDS
Personal protective equipment (PPE): B-Level PPE. Chemical goggles or face shield; rubber gloves. May contain sulfuric acid. Chemical protective material(s) reported to have good to excellent resistance: Chlorinated polyethylene, neoprene, PVC.
Toxicity by ingestion: Grade 3; oral LD_{50} = 400 mg/kg (rat).
Vapor (gas) irritant characteristics: Eye and respiratory tract irritant.
Liquid or solid irritant characteristics: Severe eye and skin irritant.

FIRE DATA
Flash point: 367°F/186°C/459°K.
Electrical hazard: Flow or agitation of substance may generate electrostatic charges due to low conductivity.

CHEMICAL REACTIVITY
Reactivity with water: Solution is corrosive; strong acid.
Binary reactants: A strong acid which can react with strong bases. In presence of moisture attacks common metals yielding hydrogen gas.
Neutralizing agents for acids and caustics: Flush with water, rinse with dilute sodium bicarbonate or lime solution.

ENVIRONMENTAL DATA
Food chain concentration potential: Negative; unlikely to accumulate.
Water pollution: Effect of low concentrations on aquatic life is unknown. May be dangerous if it enters nearby water intakes; notify operators. Notify local health and wildlife officials. **Response to discharge:** Issue warning–corrosive. Disperse and flush.

SHIPPING INFORMATION
Grades of purity: 93+%; **Storage temperature:** Ambient; **Inert atmosphere:** None; **Venting:** Open; **Stability during transport:** Stable.

PHYSICAL AND CHEMICAL PROPERTIES
Physical state @ 59°F/15°C and 1 atm: Solid.
Molecular weight: 172.2
Boiling point @ 1 atm: Decomposes. 291°F/142°C/415°K.
Melting/Freezing point: 219–221°F/104–105°C/377–378°K.
Specific gravity (water = 1): 1.45 @ 77°F/25°C (solid).
Heat of solution: -50 Btu/lb = -28 cal/g = -1.2×10^5 J/kg.
Vapor pressure: 20 mm.

m-TOLUIDINE   REC. T:1800

SYNONYMS: 3-AMINO-1-METHYLBENZENE; 3-AMINOPHENYLMETHANE; m-AMINOTOLUENE; 3-AMINOTOLUENE; ANILINE, 3-METHYL-; BENZENEAMINE, 3-METHYL-; EEC No. 601-021-00-3; m-METHYLANILINE; 3-METHYLANILINE; m-METHYLBENZENAMINE; m-TOLUIDINA (Spanish); 3-TOLUIDINE; m-TOLYLAMINE

IDENTIFICATION
CAS Number: 108-44-1

o-Toluidine

Formula: C_7H_9N; $3\text{-}CH_3C_6H_4NH_2$
DOT ID Number: UN 1708; DOT Guide Number: 153
Proper Shipping Name: Toluidines, Liquid.

DESCRIPTION: Colorless to reddish-brown liquid. Aniline-like, aromatic odor. May float or sink in water; slightly soluble.

Poison! • May interfere with the body's ability to use oxygen • Irritating to skin, eyes, and respiratory tract • Firefighting gear (including SCBA) does not provide adequate protection. If exposure occurs, remove and isolate gear immediately and thoroughly decontaminate personnel • Containers may BLEVE when exposed to fire • Vapors are heavier than air and will collect and stay in low areas • Vapors in confined areas (e.g., tanks, sewers, buildings) may explode when exposed to fire • Toxic products of combustion may include nitrogen oxides • Do not put yourself in danger by entering a contaminated area to rescue a victim

Hazard Classification (based on NFPA-704 Rating System) (*o*- and *p*-isomers)
Health Hazards (Blue): 3; Flammability (Red): 2; Reactivity (Yellow): 0

EMERGENCY RESPONSE: See Appendix A (153)
Evacuation:
Public safety: Isolate the area of spill or leak for at least 25 to 50 meters (80 to 160 feet) in all directions.
Spill: Increase, in the downwind direction, as necessary, the distance shown under "Public Safety."
Fire: If tank, rail car, or tank truck is involved in fire, isolate for at least 800 meters (½ mile) in all directions; also, consider initial evacuation for 800 meters (½ mile) in all directions.

EXPOSURE
Short-term effects: *SEEK MEDICAL ATTENTION*. **Liquid:** Irritating to skin and eyes. *IF SWALLOWED*, will cause nausea, vomiting, or loss of consciousness. Remove contaminated clothing and shoes. Flush affected areas with plenty of water. *IF BREATHING HAS STOPPED*, give artificial respiration; *avoid mouth-to-mouth resuscitation; use bag/mask apparatus. IF IN EYES*, hold eyelids open and flush with plenty of water. *IF SWALLOWED* and victim is *CONSCIOUS AND ABLE TO SWALLOW*, have victim drink 4 to 8 ounces of water and induce vomiting. *IF SWALLOWED* and victim is *UNCONSCIOUS OR HAVING CONVULSIONS*, do nothing except keep victim warm.

HEALTH HAZARDS
Personal protective equipment (PPE): A-Level PPE. OSHA Table Z-1-A air contaminant. Chemical safety goggles; face shield; approved respirator; butyl rubber gloves.
Exposure limits (TWA unless otherwise noted): ACGIH 2 ppm (8.8 mg/m³); NIOSH PROPOSED 2 ppm; skin contact contributes significantly in overall exposure.
Toxicity by ingestion: Grade 3; LD_{50} = 450 mg/kg (rat).
Long-term health effects: Kidney, bladder and blood damage.
Vapor (gas) irritant characteristics: Eye and respiratory tract irritant. Vapors are irritating such that personnel will not tolerate moderate or high concentrations.
Liquid or solid irritant characteristics: Eye and skin irritant. Causes smarting of the skin and first-degree burns on short exposure; may cause second-degree burns on long exposure.
Odor threshold: 0.46–6.0 ppm.

FIRE DATA
Flash point: 188°F/87°C (cc).
Flammable limits in air: LEL: 1.1; UEL: 6.6%.
Autoignition temperature: 899°F/481°C/754°K.
Electrical hazard: Class I, Group D.
Electrical hazard: Flow or agitation of substance may generate electrostatic charges due to low conductivity.

CHEMICAL REACTIVITY
Binary reactants: React violently with strong oxidizers and acids

ENVIRONMENTAL DATA
Food chain concentration potential: Log P ow = 1.3. Unlikely to accumulate.
Water pollution: Effect of low concentration on aquatic life is unknown. Fouling to shoreline. May be dangerous if it enters nearby water intakes; notify operators. Notify local health and wildlife officials. **Response to discharge:** Issue warning–water contaminant. Restrict access. Mechanical containment. Should be removed. Chemical and physical treatment.

SHIPPING INFORMATION
Grades of purity: Commercial, 99.5+%; **Storage temperature:** Ambient; **Stability during transport:** Stable.

NAS HAZARD CLASSIFICATION FOR BULK WATER TRANSPORTATION
FIRE: 1
HEALTH: Vapor irritant: 2; Liquid or solid irritant: 1; Poisons: 3
WATER POLLUTION: Human toxicity: 3; Aquatic toxicity: 3; Aesthetic effect: 2
REACTIVITY: Other chemicals: 2; Water: 0; Self-reaction: 0

PHYSICAL AND CHEMICAL PROPERTIES
Physical state @ 59°F/15°C and 1 atm: Liquid.
Molecular weight: 107.2
Boiling point @ 1 atm: 397°F/203°C/476°K.
Melting/Freezing point: –22.7°F/–30.4°C/242.8°K.
Specific gravity (water = 1): 0.989 @ 68°F/20°C (liquid).
Liquid surface tension: 36.9 dynes/cm = 0.037 N/m @ 68°F/20°C.
Relative vapor density (air = 1): 3.9
Heat of combustion: –15,883 Btu/lb = –8824 cal/g = –369 x 10^5 J/kg.
Vapor pressure: (Reid) Very low; 1mm @ 108°F/42°C.

o-TOLUIDINE   REC. T:1850

SYNONYMS: 1-AMINO-2-METHYLBENZENE; *o*-AMINOTOLUENE; 2-AMINO-1-METHYLBENZENE; 2-AMINOTOLUENE; BIANISIDINE; 1,1'-BIPHENYL-4,4'-DIAMINE,3,3'-DIMETHYL-; 4,4'-BI-*o*-TOLUIDINE; C.I. AZOIC DIAZO COMPONENT 113; 4,4'-DIAMINO-3,3'-DIMETHYLBIPHENYL; 4,4'-DIAMINO-3,3'-DIMETHYLDIPHENYL; DIAMINODITOLYL; 3,3'-DIMETHYLBENZIDIN; 3,3'-DIMETHYLBENZIDINE; 3,3'-DIMETHYL-4,4'-BIPHENYLDIAMINE; 3,3'-DIMETHYLBIPHENYL-4,4'-DIAMINE; 3,3'-DIMETHYL-4,4'-DIPHENYLDIAMINE; 4,4'-DI-*o*-TOLUIDINE; DMB; EEC No. 612-024-00-4; FAST DARK BLUE BASE R; 2-METHYLANILINE; *o*-METHYLANILINE; 1-METHYL-1,2-AMINO-BENZENE; 2-METHYLBENZENAMINE; *o*-TOLIDIN; 2-TOLIDIN (German); TOLIDINE; *o*-TOLUIDINA (Spanish); *o*,*o*'-TOLIDINE; 2-TOLIDINE; 3,3'-TOLIDINE; RCRA No. U328

o-Toluidine

IDENTIFICATION
CAS Number: 95-53-4
Formula: 1, 2-$CH_3C_6H_4NH_2$
DOT ID Number: UN 1708; DOT Guide Number: 153
Proper Shipping Name: Toluidines, Liquid
Reportable Quantity (RQ): **(CERCLA)** 100 lb/45.4 kg

DESCRIPTION: Clear to light yellow; turns yellow, brown or deep red on exposure to air and light. Chemical odor, like aniline. May float or sink in water.

Poison! • May interfere with the body's ability to use oxygen • Irritating to skin, eyes, and respiratory tract • Firefighting gear (including SCBA) does not provide adequate protection. If exposure occurs, remove and isolate gear immediately and thoroughly decontaminate personnel • Containers may BLEVE when exposed to fire • Vapors are heavier than air and will collect and stay in low areas • Vapors in confined areas (e.g., tanks, sewers, buildings) may explode when exposed to fire • Toxic products of combustion may include nitrogen oxides • Do not put yourself in danger by entering a contaminated area to rescue a victim

Hazard Classification (based on NFPA-704 Rating System)
Health Hazards (Blue): 3; Flammability (Red): 2; Reactivity (Yellow): 0

EMERGENCY RESPONSE: See Appendix A (153)
Evacuation:
Public safety: Isolate the area of spill or leak for at least 25 to 50 meters (80 to 160 feet) in all directions.
Spill: Increase, in the downwind direction, as necessary, the distance shown under "Public Safety."
Fire: If tank, rail car, or tank truck is involved in fire, isolate for at least 800 meters (½ mile) in all directions; also, consider initial evacuation for 800 meters (½ mile) in all directions.

EXPOSURE
Short-term effects: *SEEK MEDICAL ATTENTION.* **Liquid:** *HARMFUL IF ABSORBED THROUGH THE SKIN.* Irritating to skin and eyes. *IF SWALLOWED*, will cause nausea, vomiting, or loss of consciousness. Remove contaminated clothing and shoes. Flush affected areas with plenty of water. *IF BREATHING HAS STOPPED,* give artificial respiration; *avoid mouth-to-mouth resuscitation; use bag/mask apparatus. IF IN EYES*, hold eyelids open and flush with plenty of water. *IF SWALLOWED* and victim is *CONSCIOUS AND ABLE TO SWALLOW*, have victim drink 4 to 8 ounces of water and have victim induce vomiting. *IF SWALLOWED* and victim is *UNCONSCIOUS OR HAVING CONVULSIONS,* do nothing except keep victim warm.

HEALTH HAZARDS
Personal protective equipment (PPE): A-Level PPE. OSHA Table Z-1-A air contaminant. Chemical safety goggles; face shield; Bureau of Mines approved respirator; leather or rubber safety shoes; gloves. Chemical protective material(s) reported to have good to excellent resistance: Teflon®. Also, Saranes® offers limited protection.
Recommendations for respirator selection: NIOSH
At any concentrations above the NIOSH REL, or where there is no REL, at any detectable concentration: SCBAF:PD,PP (any self-contained breathing apparatus that has a full facepiece and is operated in a pressure-demand or other positive-pressure mode); or SAF:PD,PP:ASCBA (any supplied-air respirator that has a full facepiece and is operated in a pressure-demand or other positive-pressure mode in combination with an auxiliary self-contained breathing apparatus operated in a pressure-demand or other positive pressure mode). *ESCAPE:* GMFOV [any air-purifying, full-facepiece respirator (gas mask) with a chin-style, front-or back-mounted organic vapor canister]; or SCBAE (any appropriate escape-type, self-contained breathing apparatus).
Exposure limits (TWA unless otherwise noted): ACGIH TLV 2 ppm (8.8 mg/m^3); OSHA PEL 5 ppm; potential human carcinogen; skin contact contributes significantly in overall exposure.
Toxicity by ingestion: Grade 2; oral LD_{50} = 900 mg/kg (rat).
Long-term health effects: ACGIH potential human carcinogen. Causes tumors in urinary bladder of rats; OSHA specifically regulated carcinogen. IARC possible carcinogen, rating 2B; sufficient animal data. Kidney, and blood damage.
Liquid or solid irritant characteristics: Severe eye and skin irritant.
Odor threshold: 0.024–6.5 ppm.
IDLH value: Potential human carcinogen; 50 ppm.

FIRE DATA
Flash point: 185°F/85°C (cc).
Fire extinguishing agents not to be used: Water may be ineffective.
Behavior in fire: Containers may explode.
Autoignition temperature: 900°F/482°C/755°K.
Electrical hazard: Class I, Group D.
Burning rate: 3.62 mm/min

CHEMICAL REACTIVITY
Binary reactants: Violent reaction with strong oxidizers including red fuming nitric acid. Incompatible with strong acids, acid chlorides, acid anhydrides. Incompatible with organic anhydrides, acrylates, alcohols, aldehydes, alkylene oxides, substituted allyls, cellulose nitrate, cresols, caprolactam solution, epichlorohydrin, ethylene dichloride, isocyanates, ketones, glycols, nitrates, phenols, vinyl acetate. Exothermic decomposition with maleic anhydride. Increases the explosive sensitivity of nitromethane. Reacts with nitroalkanes forming explosive products. Attacks some plastics, rubber, and coatings.

ENVIRONMENTAL DATA
Food chain concentration potential: Log P_{ow} = variably reported at 1.3–2.3. Unlikely to accumulate.
Water pollution: Effect of low concentrations on aquatic life is unknown. Fouling to shoreline. May be dangerous if it enters nearby water intakes; notify operators. Notify local health and wildlife officials. **Response to discharge:** Issue warning–water contaminant. Restrict access. Mechanical containment. Should be removed. Chemical and physical treatment.

SHIPPING INFORMATION
Grades of purity: Commercial, 99.5+%; **Storage temperature:** Ambient; **Inert atmosphere:** None; **Venting:** Open (flame arrester); **Stability during transport:** Stable.

PHYSICAL AND CHEMICAL PROPERTIES
Physical state @ 59°F/15°C and 1 atm: Liquid.
Molecular weight: 107.2
Boiling point @ 1 atm: 392°F/200°C/473°K.
Melting/Freezing point: –3°F/–16°C/257°K.
Critical temperature: 790°F/421°C/694°K.
Critical pressure: 544 psia = 37.0 atm = 3.75 MN/m^2.
Specific gravity (water = 1): 1.01 @ 68°F/20°C (liquid).
Liquid surface tension: 43.55 dynes/cm = 0.04355 N/m @ 68°F/20°C.

Relative vapor density (air = 1): 3.7
Latent heat of vaporization: 179.1 Btu/lb = 99.5 cal/g = 4.16 x 10^5 J/kg.
Heat of combustion: −16,180 Btu/lb = −8990 cal/g = −376 x 10^5 J/kg.
Vapor pressure: 0.3 mm.

p-TOLUIDINE   REC. T:1900

SYNONYMS: 4-AMINO-1-METHYLBENZENE; 4-AMINOTOLUENE; *p*-AMINOTOLUENE; 4-METHYLANILINE; C.I. AZOIC COUPLING COMPONENT-107; *p*-METHYLANILINE; 4-METHYLBENZENEAMINE; NAPHTOL AS-KG; 4-TOLUIDINE; *p*-TOLUIDINA (Spanish); TOLYLAMINE; *p*-TOLYLAMINE; RCRA No. U353

IDENTIFICATION
CAS Number: 106-49-0
Formula: C_7H_9N; 4-$CH_3C_6H_4NH_2$
DOT ID Number: UN 1708; DOT Guide Number: 153
Proper Shipping Name: Toluidines, solid
Reportable Quantity (RQ): **(CERCLA)** 100 lb/45.4 kg

DESCRIPTION: Colorless flammable gas. No odor listed. Sinks and mixes with water.

Poison! • May interfere with the body's ability to use oxygen • Irritating to skin, eyes, and respiratory tract • Firefighting gear (including SCBA) does not provide adequate protection. If exposure occurs, remove and isolate gear immediately and thoroughly decontaminate personnel • Containers may BLEVE when exposed to fire • Vapors are heavier than air and will collect and stay in low areas • Vapors in confined areas (e.g., tanks, sewers, buildings) may explode when exposed to fire • Toxic products of combustion may include nitrogen oxides • Do not put yourself in danger by entering a contaminated area to rescue a victim

Hazard Classification (based on NFPA-704 Rating System)
Health Hazards (Blue): 3; Flammability (Red): 2; Reactivity (Yellow): 0

EMERGENCY RESPONSE: See Appendix A (153)
Evacuation:
Public safety: Isolate the area of spill or leak for at least 25 to 50 meters (80 to 160 feet) in all directions.
Spill: Increase, in the downwind direction, as necessary, the distance shown under "Public Safety."
Fire: If tank, rail car, or tank truck is involved in fire, isolate for at least 800 meters (½ mile) in all directions; also, consider initial evacuation for 800 meters (½ mile) in all directions.

EXPOSURE
Short-term effects: *SEEK MEDICAL ATTENTION.* **Liquid:** Irritating to skin and eyes. *IF SWALLOWED*, will cause nausea, vomiting, or loss of consciousness. Remove contaminated clothing and shoes. Flush affected areas with plenty of water. *IF BREATHING HAS STOPPED*, give artificial respiration; *avoid mouth-to-mouth resuscitation; use bag/mask apparatus. IF IN EYES,* hold eyelids open and flush with plenty of water. *IF SWALLOWED* and victim is *CONSCIOUS AND ABLE TO SWALLOW,* have victim drink 4 to 8 ounces of water and induce vomiting. *IF SWALLOWED* and victim is *UNCONSCIOUS OR HAVING CONVULSIONS,* do nothing except keep victim warm.

HEALTH HAZARDS
Personal protective equipment (PPE): B-Level PPE. OSHA Table Z-1-A air contaminant. Chemical safety goggles; face shield; approved respirator; leather or rubber safety shoes; butyl rubber gloves.
Recommendations for respirator selection: NIOSH
At any concentrations above the NIOSH REL, or where there is no REL, at any detectable concentration: SCBAF:PD,PP (any self-contained breathing apparatus that has a full facepiece and is operated in a pressure-demand or other positive-pressure mode); or SAF:PD,PP:ASCBA (any supplied-air respirator that has a full facepiece and is operated in a pressure-demand or other positive-pressure mode in combination with an auxiliary self-contained breathing apparatus operated in a pressure-demand or other positive pressure mode). *ESCAPE:* GMFOV [any air-purifying, full-facepiece respirator (gas mask) with a chin-style, front-or back-mounted organic vapor canister]; or SCBAE (any appropriate escape-type, self-contained breathing apparatus).
Exposure limits (TWA unless otherwise noted): ACGIH TLV 2 ppm (8.8 mg/m^3); NIOSH REL potential human carcinogen; skin contact contributes significantly in overall exposure.
Toxicity by ingestion: Grade 3: LD_{50} = 330 mg/kg (mouse).
Long-term health effects: ACGIH Suspected human carcinogen; OSHA specifically regulated carcinogen.
Vapor (gas) irritant characteristics: Vapors are irritating such that personnel will not usually tolerate moderate or high concentrations.
Liquid or solid irritant characteristics: Severe eye and skin irritant. Causes smarting of the skin and first-degree burns on short exposure; may cause second-degree burns on long exposure.
Odor threshold: Potent human carcinogen; 0.025–3.3 ppm.

FIRE DATA
Flash point: 188°F/87°C (cc).
Flammable limits in air: LEL: 1.1; UEL: 6.6%.
Fire extinguishing agents not to be used: Water may not be effective.
Behavior in fire: Containers may explode.
Autoignition temperature: 899°F/482°C/755°K.
Electrical hazard: Class I, Group D.

CHEMICAL REACTIVITY
Binary reactants: Violent reaction with strong oxidizers including red fuming nitric acid. Incompatible with strong acids, acid chlorides, acid anhydrides. Incompatible with organic anhydrides, acrylates, alcohols, aldehydes, alkylene oxides, substituted allyls, cellulose nitrate, cresols, caprolactam solution, epichlorohydrin, ethylene dichloride, isocyanates, ketones, glycols, nitrates, phenols, vinyl acetate. Exothermic decomposition with maleic anhydride. Increases the explosive sensitivity of nitromethane. Reacts with nitroalkanes forming explosive products. Attacks some plastics, rubber, and coatings.

ENVIRONMENTAL DATA
Water pollution: Effect of low concentration on aquatic life is unknown. May be dangerous if it enters nearby water intakes; notify operators. Notify local health and wildlife officials. **Response to discharge:** Issue warning–water contaminant. Restrict access. Mechanical containment. Should be removed. Chemical and physical treatment.

SHIPPING INFORMATION
Grades of purity: Commercial, 99.5+%; **Storage temperature:** Ambient; **Stability during transport:** Stable.

NAS HAZARD CLASSIFICATION FOR BULK WATER TRANSPORTATION
FIRE: 1
HEALTH: Vapor irritant: 2; Liquid or solid irritant: 2; Poisons: 3
WATER POLLUTION: Human toxicity: 3; Aquatic toxicity: 3; Aesthetic effect: 1
REACTIVITY: Other chemicals: 2; Water: 0; Self-reaction: 0

PHYSICAL AND CHEMICAL PROPERTIES
Physical state @ 59°F/15°C and 1 atm: Solid.
Molecular weight: 107.2
Boiling point @ 1 atm: 393°F/200.6°C/473.6°K.
Melting/Freezing point: 112.1°F/44.5°C/317.7°K.
Specific gravity (water = 1): 1.05 @ 68°F/20°C.
Liquid surface tension: 34.6 dyne/cm = 0.035 N/m at 50°C.
Relative vapor density (air = 1): 3.9
Heat of combustion: −15883 Btu/lb = −8824 cal/g = −369 x 10^5 J/kg.
Heat of fusion: 71.8 Btu/lb = 39.9 cal/g = 1.7 x 10^5 J/kg.
Vapor pressure: (Reid) Very low; 1 mm @ 108°F/42°C.

TOXAPHENE   REC. T:1950

SYNONYMS: AGRICIDE MAGGOT KILLER (F); ALLTEX; ALLTOX; ATTAC-2; ATTAC 6; ATTAC 6-3; CAMPHECHLOR; CAMPHENE, OCTACHLORO-; CAMPHOCHLOR; CAMPHOCLOR; CAMPHOFENE HUILEUX; CHEM-PHENE; CHLORINATED CAMPHENE; CHLOROCAMPHENE; CLOR CHEM T-590; COMPOUND 3956; CRESTOXO; CRISTOXO 90; ESTONOX; FASCO-TERPENE; GENIPHENE; GY-PHENE; HERCULES 3956; HERCULES TOXAPHENE; KAMFOCHLOR; M 5055; MELIPAX; MOTOX; OCTACHLOROCAMPHENE; PCC; PHENACIDE; PHENATOX; POLYCHLORCAMPHENE; POLYCHLORINATED CAMPHENE; POLYCHLOROCAMPHENE; STROBANE-T; STROBANE-T-90; SYNTHETIC 3956; TOXADUST; TOXAFENO (Spanish); TOXAKIL; TOXASPRAY; TOXON 63; TOXYPHEN; VERTAC 90%; VERTAC TOXAPHENE 90; RCRA No. P123; TECS No. XW5250000

IDENTIFICATION
CAS Number: 8001-35-2
Formula: $C_{10}H_{10}Cl_{18}$
DOT ID Number: UN 2761; **DOT Guide Number:** 151
Proper Shipping Name: Toxaphene
Reportable Quantity (RQ): (CERCLA) 1 lb/0.454 kg

DESCRIPTION: Amber, waxy solid that may be commercially available as dust, granule, wettable powder, or emulsifiable concentrate. Also available dissolved in a flammable liquid (i.e., xylene). Mild, piney, camphor- and chlorine-type odor. Solid sinks in water, solution Floats on water surface; wettable forms may dissolve.

Poison! (organochlorine/chlorinated camphene) • Noncombustible solid but solution is flammable (see xylene) • Breathing the vapor, skin or eye contact, or swallowing the material can kill you; easily absorbed through the skin • Firefighting gear (including SCBA) does not provide adequate protection. If exposure occurs, remove and isolate gear immediately and thoroughly decontaminate personnel • Toxic products of combustion may include hydrogen chloride and carbon monoxide • Do not put yourself in danger by entering a contaminated area to rescue a victim.

Hazard Classification (based on NFPA-704 Rating System)
Health Hazards (Blue): 3; Flammability (Red): 0; Reactivity (Yellow): 0

EMERGENCY RESPONSE: See Appendix A (151)
Evacuation:
Public safety: Isolate the area of spill or leak for at least 25 to 50 meters (80 to 160 feet) in all directions.
Spill: Increase, in the downwind direction, as necessary, the distance shown under "Public Safety."
Fire: If tank, rail car, or tank truck is involved in fire, isolate for at least 800 meters (½ mile) in all directions; also, consider initial evacuation for 800 meters (½ mile) in all directions.

EXPOSURE
Short-term effects: *SEEK MEDICAL ATTENTION.* **Solid or solution:** *POISONOUS IF SWALLOWED.* Irritating to skin and eyes. Flush affected areas with plenty of water. *IF BREATHING HAS STOPPED,* give artificial respiration; *avoid mouth-to-mouth resuscitation; use bag/mask apparatus. IF IN EYES*, hold eyelids open and flush with plenty of water. *IF SWALLOWED* and victim is *CONSCIOUS AND ABLE TO SWALLOW,* have victim drink 4 to 8 ounces of water and have victim induce vomiting, possibly with warm salt or soapy water. *IF SWALLOWED* and victim is *UNCONSCIOUS OR HAVING CONVULSIONS,* do nothing except keep victim quiet and warm.

HEALTH HAZARDS
Personal protective equipment (PPE): A-Level PPE. OSHA Table Z-1-A air contaminant.
Recommendations for respirator selection: NIOSH
At any concentrations above the NIOSH REL, or where there is no REL, at any detectable concentration: SCBAF:PD,PP (any self-contained breathing apparatus that has a full facepiece and is operated in a pressure-demand or other positive-pressure mode); or SAF:PD,PP:ASCBA (any supplied-air respirator that has a full facepiece and is operated in a pressure-demand or other positive-pressure mode in combination with an auxiliary self-contained breathing apparatus operated in a pressure-demand or other positive pressure mode).
ESCAPE: GMFOVHiE [any air-purifying, full-facepiece respirator (gas mask) with a chin-style, front- or back-mounted organic vapor canister having a high-efficiency particulate filter]; or SCBAE (any appropriate escape-type, self-contained breathing apparatus).
Exposure limits (TWA unless otherwise noted): ACGIH TLV 0.5 mg/m³; OSHA PEL 0.5 mg/m³; NIOSH REL as chlorinated camphene; potential human carcinogen; reduce exposure to lowest feasible level; skin contact contributes significantly in overall exposure.
Toxicity by ingestion: Grade 4; LD_{50} below 50 mg/kg (dog).
Long-term health effects: IARC possible carcinogen, rating 2B; sufficient animal data. Skin allergies.
Vapor (gas) irritant characteristics: The solid is nonvolatile. For solutions, see m-xylene.
Liquid or solid irritant characteristics: Eye and skin irritant. If spilled on clothing and allowed to remain, may cause smarting and reddening of the skin.
Odor threshold: 0.140 ppm.
IDLH value: Potential human carcinogen; 200 mg/m³.

FIRE DATA
Flash point: 84°F/29°C (cc) (solution).
Flammable limits in air: LEL: 1.1%; UEL: 6.4% (solvent only).

Fire extinguishing agents not to be used: Water may be ineffective. Water streams applied to adjacent fires will spread contamination of pesticide over wide area.
Behavior in fire: Solution in xylene may produce corrosive products when heated.
Autoignition temperature: 986°F/530°C/803°K (solution).
Burning rate: 5.8 mm/min

CHEMICAL REACTIVITY
Binary reactants: Strong oxidizers. Slightly corrosive to metals under moist conditions.

ENVIRONMENTAL DATA
Food chain concentration potential: Log K_{ow} = 3.4. Values above 3.0 are very likely to accumulate in living tissues; in fats.
Water pollution: DOT Appendix B, §172.101–marine pollutant. Harmful to aquatic life in very low concentrations. Solution is fouling to shoreline. May be dangerous if it enters nearby water intakes; notify operators. Notify local health and wildlife officials.
Response to discharge: Issue warning–poison, water contaminant. Should be removed. Chemical and physical treatment.

SHIPPING INFORMATION
Grades of purity: Technical; 40% dust concentrate; 90% solution in xylene; **Storage temperature:** Ambient; **Inert atmosphere:** None; **Venting:** Sealed containers in well-ventilated area; **Stability during transport:** Stable.

PHYSICAL AND CHEMICAL PROPERTIES
Physical state @ 59°F/15°C and 1 atm: Waxy solid.
Molecular weight: 414 (avg.).
Boiling point @ 1 atm: Decomposes.
Melting/Freezing point: 149–194°F/65–90°C/338–363°K.
Specific gravity (water = 1): 1.6 @ 15°C (solid).
Vapor pressure: 0.4 mm @ 77°F/25°C.

TRIBUTYL PHOSPHATE   **REC. T:2000**

SYNONYMS: BUTYL PHOSPHATE, TRI-; CELLUPHOS 4; EEC No. 015-014-00-2; FOSFITO de TRIBUTILO (Spanish); PHOSPHORIC ACID TRIBUTYL ESTER; TBP; TRIBUTYLE (PHOSPHATE de) (French); TRI-*n*-BUTYL PHOSPHATE

IDENTIFICATION
CAS Number: 126-73-8
Formula: $C_{12}H_{27}O_4P$; $(n-C_4H_9O)_3PO$
DOT ID Number: UN 3287; DOT Guide Number: 151
Proper Shipping Name: Toxic liquid, inorganic, n.o.s.

DESCRIPTION: Colorless to pale yellow liquid. Odorless.

Poison! • Irritates eyes, skin, and respiratory tract • Firefighting gear (including SCBA) does not provide adequate protection. If exposure occurs, remove and isolate gear immediately and thoroughly decontaminate personnel • Containers may BLEVE when exposed to fire • Vapors are heavier than air and will collect and stay in low areas • Vapors in confined areas (e.g., tanks, sewers, buildings) may explode when exposed to fire • Toxic products of combustion may include phosphorus oxides.

Hazard Classification (based on NFPA-704 Rating System)
Health Hazards (Blue): 2; Flammability (Red): 1; Reactivity (Yellow): 0

EMERGENCY RESPONSE: See Appendix A (151)
Evacuation:
Public safety: Isolate the area of spill or leak for at least 25 to 50 meters (80 to 160 feet) in all directions.
Spill: Increase, in the downwind direction, as necessary, the distance shown under "Public Safety."
Fire: If tank, rail car, or tank truck is involved in fire, isolate for at least 800 meters (½ mile) in all directions; also, consider initial evacuation for 800 meters (½ mile) in all directions.

EXPOSURE
Short-term effects: SEEK MEDICAL ATTENTION. **Vapor:** Irritating to the eyes, nose, and throat. May be harmful if inhaled. Remove to fresh air. *IF BREATHING HAS STOPPED*, give artificial respiration; *avoid mouth-to-mouth resuscitation; use bag/mask apparatus*. IF breathing is difficult, administer oxygen. **Liquid:** Irritating to skin and eyes. May be harmful if swallowed or absorbed through skin. *IF IN EYES*, hold eyelids open and flush with plenty of water. Remove contaminated clothing, flush affected areas with plenty of water.

HEALTH HAZARDS
Personal protective equipment (PPE): B-Level PPE. OSHA Table Z-1-A air contaminant. Approved respirator, gloves, safety goggles. Sealed chemical protective materials afford limited protection: Viton/neoprene, Butyl rubber/neoprene.
Recommendations for respirator selection: NIOSH
2 ppm: SA (any supplied-air respirator). *5 ppm:* SA:CF (any supplied-air respirator operated in a continuous-flow mode). *10 ppm:* SCBAF (any self-contained breathing apparatus with a full facepiece); or SAF (any supplied-air respirator with a full facepiece). *50 ppm:* SAF:PD,PP (any supplied-air respirator that has a full facepiece and is operated in a pressure-demand or other positive-pressure mode). *EMERGENCY OR PLANNED ENTRY INTO UNKNOWN CONCENTRATIONS OR IDLH CONDITIONS:* SCBAF:PD,PP (any self-contained breathing apparatus that has a full facepiece and is operated in a pressure-demand or other positive-pressure mode); or SAF:PD,PP:ASCBA (any supplied-air respirator that has a full facepiece and is operated in a pressure-demand or other positive-pressure mode in combination with an auxiliary self-contained breathing apparatus operated in a pressure-demand or other positive pressure mode). *ESCAPE:* GMFOVHiE [any air-purifying, full-facepiece respirator (gas mask) with a chin-style, front- or back-mounted organic vapor canister having a high-efficiency particulate filter]; or SCBAE (any appropriate escape-type, self-contained breathing apparatus).
Exposure limits (TWA unless otherwise noted): ACGIH TLV 0.2 ppm (2.2 mg/m³); OSHA PEL 5 mg/m³; NIOSH REL 0.2 ppm (2.5 mg/m³).
Toxicity by ingestion: Grade 2: LD_{50} = 3 g/kg (rat).
Long-term health effects: Reproductive data
Vapor (gas) irritant characteristics: Eye and respiratory tract irritant. Vapors are moderately irritating such that personnel will not usually tolerate moderate or high concentrations.
Liquid or solid irritant characteristics: Eye and skin irritant. If spilled on clothing and allowed to remain, may cause smarting and reddening of skin.
IDLH value: 30 ppm.

FIRE DATA
Flash point: 295°F/146°C (oc); 380°F/193°C (cc).
Fire extinguishing agents not to be used: Water may be ineffective.
Behavior in fire: Decomposes at 550°F/288°C/561°K.
Autoignition temperature: 770°F/410°C/683°K.

CHEMICAL REACTIVITY
Reactivity with water: Decomposes slowly forming butanol and phosphoric acid.
Binary reactants: Reacts with strong oxidizers and bases.
Neutralizing agents for acids and caustics: Dry lime or soda ash
Reactivity group: 34
Compatibility class: Esters.

ENVIRONMENTAL DATA
Water pollution: Effect of low concentrations on aquatic life is unknown. May be dangerous if it enters nearby water intakes; notify operators. Notify local health and wildlife officials. Notify operators of local water intakes. **Response to discharge:** Mechanical containment. Should be removed. Chemical and physical treatment.

SHIPPING INFORMATION
Grades of purity: 99%; **Stability during transport:** Stable.

NAS HAZARD CLASSIFICATION FOR BULK WATER TRANSPORTATION
FIRE: 1
HEALTH: Vapor irritant: 2; Liquid or solid irritant: 1; Poisons: 2
WATER POLLUTION: Human toxicity: 2; Aquatic toxicity: 3; Aesthetic effect: 2
REACTIVITY: Other chemicals:–; Water: 1; Self-reaction: 0

PHYSICAL AND CHEMICAL PROPERTIES
Physical state @ 59°F/15°C and 1 atm: Liquid.
Molecular weight: 266.32
Boiling point @ 1 atm: 552°F/289°C/562°K (decomposes).
Melting/Freezing point: Less than –112°F/–80°C/193°K.
Specific gravity (water = 1): 0.982 @ 68°F/20°C.
Relative vapor density (air = 1): 9.20
Vapor pressure: 0.004 mm @ 77°F/25°C.

TRICHLORFON   REC. T:2050

SYNONYMS: AEROL 1 PESTICIDE; ANTHON; BAY 15922; BAYER 13/59; BRITON; BRITTEN; CHLOROPHOS; DIMETOX; DIPTEREX; DYREX; DYVON; DIPTEREX; DYLOX; O,O-DIMETHYL-(1-HYDROXY-2,2,2-TRICHLORO)ETHYL PHOSPHATE; FOROTOX; FOSCHLOR; HYPODERMACIDE; LEIVASOM; LOISOL; MAZOTEN; METHYL CHLOROPHOS; NEGUVON; PROXOL; (2,2,2-TRICHLORO-1-HYDROXYETHYL)DIMETHYLPHOSPHONATE; TRICHLOROPHON; TRINEX; TRICLORFON (Spanish); TUGON; VERMICIDE BAYER 2349; VOLFARTOL; VOTEXIT; WEC-50; WOTEXIT

IDENTIFICATION
CAS Number: 52-68-6
Formula: $C_4H_8Cl_3O_4P$
DOT ID Number: NA 2783; DOT Guide Number: 152
Proper Shipping Name: Trichlorfon
Reportable Quantity (RQ): **(CERCLA)** 100 lb/45.4 kg

DESCRIPTION: White to pale yellow, crystalline solid. Sweet ether odor. Sinks and mixes with water.

Poison! (organophosphate) • A tear gas • Breathing the dust or vapor, skin or eye contact, or swallowing the material can cause illness and possible death • Firefighting gear (including SCBA) does not provide adequate protection. If exposure occurs, remove and isolate gear immediately and thoroughly decontaminate personnel • Containers may BLEVE when exposed to fire • Toxic products of combustion may include chlorine, hydrogen chloride, and phosphorus oxides • Do not put yourself in danger by entering a contaminated area to rescue a victim.

Hazard Classification (based on NFPA-704 Rating System)
Health Hazards (Blue): 2; Flammability (Red): 1; Reactivity (Yellow): 0

EMERGENCY RESPONSE: See Appendix A (152)
Evacuation:
Public safety: Isolate the area of spill or leak for at least 25 to 50 meters (80 to 160 feet) in all directions.
Spill: Increase, in the downwind direction, as necessary, the distance shown under "Public Safety."
Fire: If tank, rail car, or tank truck is involved in fire, isolate for at least 800 meters (½ mile) in all directions; also, consider initial evacuation for 800 meters (½ mile) in all directions.

EXPOSURE
Short-term effects: *SEEK MEDICAL ATTENTION.* **Solid or dust:** *POISONOUS IF SWALLOWED OR IF SKIN IS EXPOSED.* Eye pupils are small; blurred vision; runny nose; cough; shortness of breath; pain; diarrhea, nausea and vomiting; increased blood pressure, hypermotility, hallucinations; loss of consciousness; convulsions; breathing stops; death. *IF BREATHING HAS STOPPED,* give artificial respiration; *avoid mouth-to-mouth resuscitation; use bag/mask apparatus.* **Skin:** *Remove and double-bag contaminated clothing and shoes and leave in Hot Zone for later incineration by hazardous materials experts.* Flush affected areas with plenty of water. *Skin can also be decontaminated with diluted hypochlorite solution, U.S. Army M291 kit, and M258(A1) skin decontamination kit.* **Eyes:** hold eyelids open and flush with plenty of water. *IF SWALLOWED* and victim is *CONSCIOUS,* **do NOT induce vomiting but immediately administer slurry of activated charcoal**. **If swallowed** and victim is *CONSCIOUS AND ABLE TO SWALLOW,* have victim drink 4 to 8 ounces of water. **Do NOT induce vomiting but immediately administer slurry of activated charcoal**. *IF SWALLOWED* and victim is *UNCONSCIOUS OR HAVING CONVULSIONS,* do nothing except keep victim warm.

Note to physician or authorized medical personnel: Administer atropine, 2 mg (1/30 gr) intramuscularly or intravenously as soon as any local or systemic signs or symptoms of an intoxication are noted; repeat the administration of atropine every 3 to 8 minutes until signs of atropinization (mydriasis, dry mouth, rapid pulse, hot and dry skin) occur; initiate treatment in children with 0.05 mg/kg of atropine; repeat at 5- to 10-minute intervals. Watch respiration, and remove bronchial secretions if they appear to be obstructing the airway; intubate if necessary. Pralidoxime must be administered within minutes to a few hours following exposure (depending on the specific agent) to be effective. Give 2-PAMCI (Pralidoxime; Protopam), 2.5 g in 100 mL of sterile water or in 5% dextrose and water, intravenously, slowly, in 15 to 30 minutes; if sufficient fluid is not available, give 1 g of 2-PAMCI in 3 mL of distilled water by deep intramuscular injection; repeat this every half hour if respiration weakens or if muscle fasciculation or convulsions recur. Also Diazepam, an anticonvulsant, might be needed.

HEALTH HAZARDS
Personal protective equipment (PPE): B-Level PPE. Safety glasses, all-purpose canister mask, gloves and, long boots. Wear sealed chemical suit of butyl rubber or polycarbonate.

Toxicity by ingestion: Grade 2; LD_{50} = 0.5 to 5 g/kg.
Long-term health effects: Mutagenic, teratogenic, carcinogenic, hepatotoxic and hematotoxic. IARC rating 3; animal data: inadequate.
Liquid or solid irritant characteristics: Eye irritant.

FIRE DATA
Behavior in fire: Water streams applied to adjacent fires will spread contamination of pesticide over wide area.

CHEMICAL REACTIVITY
Binary reactants: Reacts with strong oxidizers.

ENVIRONMENTAL DATA
Food chain concentration potential: Log K_{ow} = 5.8. Values above 3.0 are very likely to accumulate in living tissues; in fats.
Water pollution: DOT Appendix B, §172.101–marine pollutant. Harmful to aquatic life in very low concentrations. May be dangerous if it enters nearby water intakes; notify operators. Notify local health and wildlife officials. For weapons testing, use an M272 Water Detection Kit. **Response to discharge:** Issue warning–poison, water contaminant. Restrict access. Should be removed. Chemical and physical treatment.

SHIPPING INFORMATION
Grades of purity: 1% dry bait; 50% wettable powder; 5% dusts and granules

PHYSICAL AND CHEMICAL PROPERTIES
Physical state @ 59°F/15°C and 1 atm: Solid.
Molecular weight: 257.45
Boiling point @ 1 atm: 212°F/100°C/373.2°K.
Melting/Freezing point: 183.2°F/84°C/357.2°K.
Specific gravity (water = 1): 1.73 @ 68°F/20°C.

TRICHLOROACETALDEHYDE REC. T:2100

SYNONYMS: ACETALDEHYDE, TRICHLORO; CHLORAL; ETHANAL, TRICHLORO-; TRICLORACETALDEHIDO (Spanish); RCRA No. U034

IDENTIFICATION
CAS Number: 75-87-6
Formula: C_2HCl_3O; CCl_3CHO
DOT ID Number: UN 2075; DOT Guide Number: 153
Proper Shipping Name: Chloral, anhydrous, inhibited
Reportable Quantity (RQ): **(CERCLA)** 5000 lb/2270 kg

DESCRIPTION: Colorless, oily liquid. Pungent, irritating odor. Sinks and mixes with water; combines water to yield chloral hydrate.

Poisonous if inhaled or absorbed through the skin • Highly irritating to the eyes and respiratory tract; this material has anesthetic properties • Firefighting gear (including SCBA) does not provide adequate protection. If exposure occurs, remove and isolate gear immediately and thoroughly decontaminate personnel • Containers may BLEVE when exposed to fire • Vapors are heavier than air and will collect and stay in low areas • Toxic products of combustion may include phosgene gas • Do not put yourself in danger by entering a contaminated area to rescue a victim.

Hazard Classification (based on NFPA-704 Rating System)
Health Hazards (Blue): 3; Flammability (Red): -; Reactivity (Yellow): -

EMERGENCY RESPONSE: See Appendix A (153)
Evacuation:
Public safety: Isolate the area of spill or leak for at least 25 to 50 meters (80 to 160 feet) in all directions.
Spill: Increase, in the downwind direction, as necessary, the distance shown under "Public Safety."
Fire: If tank, rail car, or tank truck is involved in fire, isolate for at least 800 meters (½ mile) in all directions; also, consider initial evacuation for 800 meters (½ mile) in all directions.

EXPOSURE
Short-term effects: *SEEK MEDICAL ATTENTION.* **Vapor:** *POISONOUS. MAY BE FATAL IF INHALED.* Irritating to eyes, skin, and respiratory tract. Inhalation causes sore throat, shortness of breath, drowsiness, irritation of respiratory tract, unconsciousness. Move victim to fresh air. *IF BREATHING HAS STOPPED,* give artificial respiration; *avoid mouth-to-mouth resuscitation; use bag/mask apparatus.* IF breathing is difficult, administer oxygen. **Liquid:** *POISONOUS. MAY BE FATAL IF SWALLOWED OR ABSORBED THROUGH SKIN.* May burn skin and eyes. *IF IN EYES OR ON SKIN,* immediately flush contaminated area with running water for at least 15 minutes; hold upper and lower eyelids open occasionally if appropriate. *Speed in removing material from skin is extremely important.* Remove and double bag contaminated clothing at the site. Effects may be delayed. Keep victim under observation. *IF SWALLOWED* and victim is *UNCONSCIOUS OR HAVING CONVULSIONS,* do nothing except keep victim warm.

HEALTH HAZARDS
Personal protective equipment (PPE): B-Level PPE. Wear positive pressure breathing apparatus and special chemical protective clothing. Sealed chemical protective materials afford limited protection: butyl rubber, PV alcohol, Viton®.
Toxicity by ingestion: Grade 4; LD_{50} = 23 mg/kg.
Long-term health effects: Chronic respiratory exposure in animals caused decreases in kidney function, liver function, growth rate and serum transaminase activity along with changes in the central nervous system and in blood factors.
Vapor (gas) irritant characteristics: Vapors cause severe irritation of eyes and throat and can cause eye and lung injury. They cannot be tolerated even at low concentrations.
Liquid or solid irritant characteristics: Severe skin irritant and is very injurious to the eyes. Contact may cause burns to skin and eyes.
Odor threshold: 0.047 ppm.

FIRE DATA
Flash point: 167°F/75°C (procedure not identified).
Behavior in fire: Decomposes in the presence of heat of fire to produce toxic and irritating gases.

CHEMICAL REACTIVITY
Reactivity with water: Reacts with water evolving large quantities of heat and forming chloral hydrate.
Binary reactants: Elevated temperatures, hot surfaces, welding arc generates corrosive gasses. Reacts with oxidizers, with a risk of fire or explosions. Reacts with alkaline materials producing chloroform.
Polymerization: Contact with acids or exposure to light may cause polymerization.

ENVIRONMENTAL DATA
Food chain concentration potential: It is estimated that fish in rivers, ponds, lakes, and reservoirs will bioconcentrate chloral 6.7 times the water concentration.
Water pollution: Effects of low concentration on aquatic life is unknown. May be dangerous if it enters nearby water intakes; notify operators. Notify local health and wildlife officials.
Response to discharge: Issue warning–poison, water contaminant. Restrict access. Should be removed.

SHIPPING INFORMATION
Grades of purity: 40% Aqueous solution; **Storage temperature:** Ambient; **Stability during transport:** Stable (avoid exposure to sunlight).

PHYSICAL AND CHEMICAL PROPERTIES
Physical state @ 59°F/15°C and 1 atm: Liquid.
Molecular weight: 147.38
Boiling point @ 1 atm: 207.9°F/97.7°C/370.9°K.
Melting/Freezing point: –71.5°F/–57.5°C/215.7°K.
Specific gravity (water = 1): 1.510 @ 68°F/20°C.
Liquid surface tension: 25.34 dynes/cm = 0.0253 N/m at 19.4°C.
Relative vapor density (air = 1): 5.1
Latent heat of vaporization: 103.4 Btu/lb = 57.5 cal/g = 2.4×10^5 J/kg.

1,2,3-TRICHLOROBENZENE   REC. T:2150

SYNONYMS: BENZENE, 1,2,3-TRICHLORO-; PYRANOL 1478; 1,2,3-TRICLOROBENCENO (Spanish); *v*-TRICHLOROBENZENE; *vic*-TRICHLOROBENZENE

IDENTIFICATION
CAS Number: 87-61-6; 12002-48-1
Formula: $C_6H_3Cl_3$
DOT ID Number: UN 2321; DOT Guide Number: 153
Proper Shipping Name: Trichlorobenzenes, Liquid.

DESCRIPTION: White, crystalline solid. Sharp, chlorobenzene odor. Sinks in water; insoluble.

Poison! • Irritating to the eyes, skin, and respiratory tract; poisonous on contact with eyes or skin, of if swallowed • Firefighting gear (including SCBA) does not provide adequate protection. If exposure occurs, remove and isolate gear immediately and thoroughly decontaminate personnel • Containers may BLEVE when exposed to fire • Vapors are heavier than air and will collect and stay in low areas • Vapors in confined areas (e.g., tanks, sewers, buildings) may explode when exposed to fire • Toxic products of combustion may include chlorine.

Hazard Classification (based on NFPA-704 Rating System)
Health Hazards (Blue): 2; Flammability (Red): 1; Reactivity (Yellow): 0

EMERGENCY RESPONSE: See Appendix A (153)
Evacuation:
Public safety: Isolate the area of spill or leak for at least 25 to 50 meters (80 to 160 feet) in all directions.
Spill: Increase, in the downwind direction, as necessary, the distance shown under "Public Safety."
Fire: If tank, rail car, or tank truck is involved in fire, isolate for at least 800 meters (½ mile) in all directions; also, consider initial evacuation for 800 meters (½ mile) in all directions.

EXPOSURE
Short-term effects: *SEEK MEDICAL ATTENTION*. **Vapor:** May be irritating to eyes, skin, and respiratory tract. Move to fresh air. *IF BREATHING HAS STOPPED,* give artificial respiration; *avoid mouth-to-mouth resuscitation; use bag/mask apparatus.* IF breathing is difficult, administer oxygen. **Liquid:** May irritate skin and eyes.
POISONOUS IF SWALLOWED. IF IN EYES OR ON SKIN, flush with running water for at least 15 minutes; hold eyelids open if necessary. Remove and double bag contaminated clothing and shoes at the site. Wash skin with soap and water. *IF SWALLOWED* and victim is *CONSCIOUS AND ABLE TO SWALLOW*, have victim drink 4 to 8 ounces of water or milk, and induce vomiting. *IF SWALLOWED* and victim is *UNCONSCIOUS OR HAVING CONVULSIONS,* do nothing except keep victim warm.

HEALTH HAZARDS
Personal protective equipment (PPE): B-Level PPE. Wear self-contained positive pressure breathing apparatus and protective clothing.
Toxicity by ingestion: GRADE 2; LD_{50} = 756–766 mg/kg (rat, mouse).
Long-term health effects: Has caused liver damage in animals; carcinogenic.
Vapor (gas) irritant characteristics: Eye and respiratory tract irritant. Vapors cause moderate irritation. Personnel will find high concentrations unpleasant. The affect is temporary.
Liquid or solid irritant characteristics: Eye and skin irritant. May cause irritation if spilled on clothing and allowed to remain.

FIRE DATA
Flash point: 230°F/110°C (oc); 210°F/99°C (cc).

CHEMICAL REACTIVITY
Binary reactants: Reacts violently with oxidizers, acids, acid fumes, steam. Attacks most rubbers.

ENVIRONMENTAL DATA
Food chain concentration potential: Log K_{ow} = 4.05. Positive. Values above 3.0 are very likely to accumulate in living tissues and especially in fats.
Water pollution: DOT Appendix B, §172.101–marine pollutant. Harmful to aquatic life in very low concentrations. May be dangerous if it enters nearby water intakes; notify operators. Notify local health and wildlife officials. **Response to discharge:** Issue warning–poison; water contaminant. Should be removed. Chemical and physical treatment.

SHIPPING INFORMATION
Grades of purity: 99%; **Storage temperature:** Ambient; **Stability during transport:** Stable.

PHYSICAL AND CHEMICAL PROPERTIES
Physical state @ 59°F/15°C and 1 atm: Solid.
Molecular weight: 181.5
Boiling point @ 1 atm: 425.3°F/218.5°C/491.7°K.
Melting/Freezing point: 126.5°F/52.5°C/325.7°K.
Specific gravity (water = 1): 1.69 @ 77°F/25°C (solid).
Relative vapor density (air = 1): 6.26 (estimate).
Latent heat of vaporization: 113 btu/lb = 63 cal/g = 2.62×10^5 J/kg.

1,2,4-TRICHLOROBENZENE   REC. T:2200

SYNONYMS: BENZENE, 1,2,4-TRICHLORO-; 1,2,4-TRICLOROBENCENO (Spanish); *unsym*-TRICHLOROBENZENE; 1,2,4-TRICHLOROBENZOL

IDENTIFICATION
CAS Number: 120-82-1; 12002-48-1
Formula: $C_6H_3Cl_3$
DOT ID Number: UN 2321; DOT Guide Number: 153 (liquid)
Proper Shipping Name: Trichlorobenzenes, Liquid
Reportable Quantity (RQ): **(CERCLA)** 100 lb/45.4 kg

DESCRIPTION: Liquid or solid. Colorless. Sharp, chlorobenzene odor.

Combustible • Corrosive to skin, eyes, and respiratory tract; contact with skin or eyes can cause burns and impaired vision; inhalation symptoms may be delayed • Heat or flame may cause explosion • Containers may BLEVE when exposed to fire • Vapors or Concentrated dust in confined areas (e.g., tanks, sewers, buildings) may explode when exposed to fire • Toxic products of combustion may include phosgene and hydrogen chloride.

Hazard Classification (based on NFPA-704 Rating System)
Health Hazards (Blue): 2; Flammability (Red): 1; Reactivity (Yellow): 0

EMERGENCY RESPONSE: See Appendix A (153)
Evacuation:
Public safety: Isolate the area of spill or leak for at least 25 to 50 meters (80 to 160 feet) in all directions.
Spill: Increase, in the downwind direction, as necessary, the distance shown under "Public Safety."
Fire: If tank, rail car, or tank truck is involved in fire, isolate for at least 800 meters (½ mile) in all directions; also, consider initial evacuation for 800 meters (½ mile) in all directions.

EXPOSURE
Short-term effects: *SEEK MEDICAL ATTENTION.* **Vapor:** May be irritating to eyes, skin, and respiratory tract. Move to fresh air. *IF BREATHING HAS STOPPED,* give artificial respiration; *avoid mouth-to-mouth resuscitation; use bag/mask apparatus.* IF breathing is difficult, administer oxygen. Lung edema may develop. **Liquid:** May irritate skin and eyes.
POISONOUS IF SWALLOWED. IF IN EYES OR ON SKIN, flush with running water for at least 15 minutes; hold eyelids open if necessary. Remove and double bag contaminated clothing and shoes at the site. *IF SWALLOWED* and victim is *CONSCIOUS AND ABLE TO SWALLOW,* have victim drink 4 to 8 ounces of water and induce vomiting. *IF SWALLOWED* and victim is *UNCONSCIOUS OR HAVING CONVULSIONS,* do nothing except keep victim warm. Drops entering the lungs may cause pneumonia.
Note to physician or authorized medical personnel: Medical observation is recommended for 24 to 48 hours after breathing overexposure, as pulmonary edema may be delayed. As first aid for pulmonary edema, consider administering a corticosteroid spray. Cigarette smoking may exacerbate pulmonary injury and should be discouraged for at least 72 hours following exposure.

HEALTH HAZARDS
Personal protective equipment (PPE): B-Level PPE. OSHA Table Z-1-A air contaminant. Wear self-contained positive pressure breathing apparatus and full protective clothing. Chemical protective material(s) reported to have good to excellent resistance: neoprene, PV alcohol, Teflon®.
Exposure limits (TWA unless otherwise noted): NIOSH REL ceiling 5 ppm (40 mg/m^3).
Short-term exposure limits (15-minute TWA): ACGIH ceiling 5 ppm (37 mg/m^3).
Toxicity by ingestion: Grade 3; LD_{50} = 300 mg/Kg (mouse).
Long-term health effects: May cause lung, liver, and/or kidney damage. Causes teratogenic effects in the rat.
Vapor (gas) irritant characteristics: Eye and respiratory tract irritant. Vapors cause moderate irritation. Personnel may find high concentrations unpleasant. The affect is temporary.
Liquid or solid irritant characteristics: Eye and skin irritant. Causes smarting of the skin and first-degree burns on short exposure; may cause second-degree burns on long exposure.
Odor threshold: 3 ppm.

FIRE DATA
Flash point: 230°F/110°C (oc); 222°F/105°C (cc).
Flammable limits in air: LEL: 1.3%; UEL: 7.1%; LEL: 2.5%; UEL: 6.6% both @ 302°F/150°C.
Autoignition temperature: 1060°F/571°C/844°K.
Electrical hazard: Class I, Group D.

CHEMICAL REACTIVITY
Binary reactants: Reacts with oxidizers. Incompatible with most rubbers.
Reactivity group: 36
Compatibility class: Halogenated hydrocarbons.

ENVIRONMENTAL DATA
Food chain concentration potential: Low potential
Water pollution: DOT Appendix B, §172.101–marine pollutant. Harmful to aquatic life in very low concentrations. May be dangerous if it enters nearby water intakes; notify operators. Notify local health and wildlife officials. **Response to discharge:** Issue warning–poison, water contaminant. Should be removed. Chemical and physical treatment. Clean Water Act.

SHIPPING INFORMATION
Grades of purity: Purified (99%); Technical: 75% 1,2,4-Trichlorobenzene and 25% 1,2,3-Trichlorobenzene; **Storage temperature:** Ambient; **Stability during transport:** Stable.

PHYSICAL AND CHEMICAL PROPERTIES
Physical state @ 59°F/15°C and 1 atm: Solid.
Molecular weight: 181.4
Boiling point @ 1 atm: 415°F/213°C/486°K.
Melting/Freezing point: 61.7°F/16.5°C/289.5°K.
Specific gravity (water = 1): 1.454 @ 68°F/20°C (liquid).
Relative vapor density (air = 1): 6.25
Latent heat of vaporization: 113 Btu/lb = 62.9 cal/g = 2.64 x 10^5 J/kg.
Vapor pressure: 1 mm.

TRICHLOROETHANE   REC. T:2250

SYNONYMS: AEROTHENE; AEROTHENE TT; CHLOROETHANE NU; CHLOROTHENE; EEC No. 602-013-00-2; METHYL CHLOROFORM; METHYLTRICHLOROETHANE; α-TCE; 1,1,1-TCE; TCE; 1,1,1-TRICHLOROETHANE; α-TRICHLOROETHANE; 1,1,1-TRICLOROETANO (Spanish); TRI-ETHANE; TRICHLOROETHAND, STABILIZED; RCRA U226

Trichloroethane

IDENTIFICATION
CAS Number: 71-55-6; 25323-89-1
Formula: $C_2H_3Cl_3$; CH_3CCl_3
DOT ID Number: UN 2831; DOT Guide Number: 160
Proper Shipping Name: 1,1,1-Trichloroethane
Reportable Quantity (RQ): **(CERCLA)** 1000 lb/454 kg

DESCRIPTION: Colorless, watery liquid. Sweet, chloroform-like odor. Sinks in water; insoluble. Large amounts of vapor is produced.

Combustible • Heat or flame may cause explosion • Begins to decompose at 350°F/177°C • Containers may BLEVE when exposed to fire • Vapors are heavier than air and will collect and stay in low areas • Vapors in confined areas (e.g., tanks, sewers, buildings) may explode when exposed to fire • Irritating to the skin, eyes, and respiratory tract • Toxic products of combustion may include hydrogen chloride and phosgene. Dichloroacetylene (DCA) is also a decomposition product.

Hazard Classification (based on NFPA-704 Rating System)
Health Hazards (Blue): 2; Flammability (Red): 1; Reactivity (Yellow): 0

EMERGENCY RESPONSE: See Appendix A (160)
Evacuation:
Public safety: Isolate the area of spill or leak for at least 25 to 50 meters (80- 160 feet) in all directions.
Spill: Consider initial downwind evacuation for at least 100 meters (330 feet); increase, in the downwind direction, as necessary.
Fire: If tank, rail car, or tank truck is involved in fire, isolate for at least 800 meters (½ mile) in all directions; also, consider initial evacuation for 800 meters (½ mile) in all directions.

EXPOSURE
Short-term effects: *SEEK MEDICAL ATTENTION.* **Vapor:** Irritating to eyes, nose, and throat. *IF INHALED*, will, will cause dizziness or difficult breathing. Move to fresh air. *IF BREATHING HAS STOPPED*, give artificial respiration. IF breathing is difficult, administer oxygen. **Liquid:** *HARMFUL IF ABSORBED THROUGH THE SKIN.* Irritating to skin and eyes. *IF SWALLOWED*, may produce nausea. Remove contaminated clothing and shoes. Flush affected areas with plenty of water. *IF IN EYES*, hold eyelids open and flush with plenty of water. *IF SWALLOWED* and victim is *CONSCIOUS AND ABLE TO SWALLOW*, **do not induce vomiting.** If the victim is alert, asymptomatic, and has a gag reflex, administer a slurry of activated charcoal at 1 gm/kg (usual adult dose 60–90 g, child dose 25–50 g). A soda can and straw may be of assistance when offering charcoal to a child. *IF SWALLOWED* and victim is *UNCONSCIOUS OR HAVING CONVULSIONS*, do nothing except keep victim warm.

HEALTH HAZARDS
Personal protective equipment (PPE): OSHA Table Z-1-A air contaminant. Chemical-protective clothing is not generally required when only vapor exposure is expected: 1,1,1-Trichloroethane vapor is only mildly irritating and is not absorbed well through the skin. Chemical-protective clothing is recommended (B-Level PPE) when extensive skin contact with the liquid might occur. Chemical protective material(s) reported to have good to excellent resistance: PV alcohol, Teflon®, Viton®.
Recommendations for respirator selection: NIOSH/OSHA
700 ppm: SA* (any supplied-air respirator); or SCBAF* (any self-contained breathing apparatus with a full facepiece).
EMERGENCY OR PLANNED ENTRY INTO UNKNOWN CONCENTRATIONS OR IDLH CONDITIONS: SCBAF:PD,PP (any self-contained breathing apparatus that has a full facepiece and is operated in a pressure-demand or other positive-pressure mode); or SAF:PD,PP:ASCBA (any supplied-air respirator that has a full facepiece and is operated in a pressure-demand or other positive-pressure mode in combination with an auxiliary self-contained breathing apparatus operated in a pressure-demand or other positive pressure mode). *ESCAPE:* GMFOV [any air-purifying, full-facepiece respirator (gas mask) with a chin-style, front- or back-mounted acid gas canister]; or SCBAE (any appropriate escape-type, self-contained breathing apparatus). *Note:* Substance reported to cause eye irritation or damage; may require eye protection.
Exposure limits (TWA unless otherwise noted): ACGIH TLV 350 ppm (1910 mg/m^3); OSHA PEL 350 ppm (1900 mg/m^3); potential carcinogen.
Short-term exposure limits (15-minute TWA): ACGIH STEL 450 ppm (2460 mg/m^3); NIOSH ceiling 350 ppm (1900 mg/m^3)/15 min
Toxicity by ingestion: Grade 1; LD_{50} = 5 to 15 g/kg (rat, mouse, rabbit, guinea pig).
Long-term health effects: IARC rating 3; inadequate animal data. Mutagen.
Vapor (gas) irritant characteristics: Eye and respiratory tract irritant. Vapors cause smarting of the eyes or respiratory system if present in high concentrations. The effect is temporary.
Liquid or solid irritant characteristics: Eye and skin irritant. If spilled on clothing and allowed to remain, may cause smarting and reddening of the skin.
Odor threshold: 100 ppm.
IDLH value: 700 ppm.

FIRE DATA
Flammable limits in air: LEL: 7%; UEL: 12.5%.
Behavior in fire: Begins to decompose at 350°F/177°C.
Autoignition temperature: 932°F/500°C/773°K.
Electrical hazard: Class I, Group D; Does not burn readily, but can produce a dangerous "flash" when vapors are exposed to high-energy spark sources in a confined space.
Burning rate: (estimate) 2.9 mm/min

CHEMICAL REACTIVITY
Reactivity with water: Reacts slowly, releasing corrosive hydrochloric acid and hydrogen chloride fumes.
Binary reactants: Reacts, possibly violently with strong caustics, strong oxidizers, acetone, chemically active metals, metal powders of aluminum, bronze, copper, magnesium, manganese, sodium, zinc and their alloys; aluminum methyl, aluminum tripropyl, antimony triethyl, antimony, trimethyl, dimethylformamide, trimethyl aluminum. Mixtures with nitrogen tetroxide ar explosive. Contact with hot metal or exposure to ultraviolet radiation, will cause decomposition, forming hydrogen chloride, phosgene, and dichloroacetylene gasses. Attacks some plastics, rubber, and coatings. The uninhibited grade is corrosive to aluminum. Attacks aluminum and its alloys, other metals, especially in the presence of moisture.
Polymerization: Caustics or high heat may cause polymerization unless inhibited.
Reactivity group: 36
Compatibility class: Halogenated hydrocarbons

ENVIRONMENTAL DATA
Food chain concentration potential: Log K_{ow} = 2.47 to 2.49.

Values above 3.0 are very likely to accumulate in living tissues and especially in fats.
Water pollution: Effect of low concentrations on aquatic life is unknown. May be dangerous if it enters nearby water intakes; notify operators. Notify local health and wildlife officials.
Response to discharge: Should be removed. Chemical and physical treatment.

SHIPPING INFORMATION
Grades of purity: Uninhibited; inhibited; industrial inhibited; white room; cold cleaning; **Storage temperature:** Ambient; **Inert atmosphere:** None; **Venting:** Pressure-vacuum; **Stability during transport:** Stable.

NAS HAZARD CLASSIFICATION FOR BULK WATER TRANSPORTATION
FIRE: 1
HEALTH: Vapor irritant: 1; Liquid or solid irritant: 1; Poisons: 2
WATER POLLUTION: Human toxicity: 1; Aquatic toxicity: 3; Aesthetic effect: 2
REACTIVITY: Other chemicals: 1; Water: 0; Self-reaction: 0

PHYSICAL AND CHEMICAL PROPERTIES
Physical state @ 59°F/15°C and 1 atm: Liquid.
Molecular weight: 133.41
Boiling point @ 1 atm: 165°F/74°C/347°K.
Melting/Freezing point: –23°F/–31°C/ 242°K.
Specific gravity (water = 1): 1.34 @ 68°F/20°C (liquid).
Liquid surface tension: 25.4 dynes/cm = 0.0254 N/m @ 68°F/20°C.
Liquid water interfacial tension: (estimate) 45 dynes/cm = 0.045 N/m @ 68°F/20°C.
Relative vapor density (air = 1): 4.6
Ratio of specific heats of vapor (gas): 1.104
Latent heat of vaporization: 100 Btu/lb = 58 cal/g = 2.4×10^5 J/kg.
Heat of combustion: (estimate) 4700 Btu/lb = 2600 cal/g = 110×10^5 J/kg.
Vapor pressure: (Reid) 4.0 psia; 100 mm.

1,1,2-TRICHLOROETHANE   REC. T:2300

SYNONYMS: EEC No. 602-014-00-8; ETHANE TRICHLORIDE; ETHANE,1,1,2-TRICHLORO-; β-T; 1,1,2-TRICHLORETHANE; 1,2,2-TRICHLOROETHANE; β-TRICHLOROETHANE; β-TRICHLOROETHANE; 1,2,2-TRICLOROETANO (Spanish); VINYL TRICHLORIDE; RCRA No. U227

IDENTIFICATION
CAS Number: 79-00-5; 25323-89-1
Formula: $C_2H_3Cl_3$; $CHCl_2CH_2Cl$
DOT ID Number: UN 3082; DOT Guide Number: 171
Proper Shipping Name: Environmentally hazardous substances, liquid, n.o.s.
Reportable Quantity (RQ): **(CERCLA)** 100 lb/45.4 kg

DESCRIPTION: Colorless liquid. Sweet, chloroform-like odor. Sinks in water; very slightly soluble.

Combustible liquid, forms dense soot • Heat, flame, or high energy source may cause closed containers to explode • Containers may BLEVE when exposed to fire • Vapors are heavier than air and will collect and stay in low areas • Vapors in confined areas (e.g., tanks, sewers, buildings) may explode when exposed to fire • Irritating to the skin, eyes, and respiratory tract • Combustion or contact with hot surfaces cause decomposition producing hydrogen chloride, and small amounts of phosgene and chlorine.

Hazard Classification (based on NFPA-704 Rating System)
Health Hazards (Blue): 2; Flammability (Red): 1; Reactivity (Yellow): 0

EMERGENCY RESPONSE: See Appendix A (171)
Evacuation:
Public safety: Isolate the area of spill or leak for at least 10 to 25 meters (30 to 80 feet) in all directions.
Spill: Increase, in the downwind direction, as necessary, the distance shown under "Public Safety."
Fire: If any large container is involved in fire, isolate for at least 800 meters (½ mile) in all directions; also, consider initial evacuation for 800 meters (½ mile) in all directions.

EXPOSURE
Short-term effects: *SEEK MEDICAL ATTENTION*. **Vapor:** Irritating to eyes, nose, throat, lungs and skin; may cause defatting dermatitis.
Highly toxic; death may result from respiratory failure. *IF INHALED*, will, anesthetic or narcotic effect may occur. Move to fresh air. *IF BREATHING HAS STOPPED,* give artificial respiration. IF breathing is difficult, administer oxygen. **Liquid:** Irritating to skin and eyes; severe irritant to gastrointestinal tract. Highly toxic. *IF SWALLOWED*, may cause liver or kidney damage and may increase myocardial irritability. May cause chemical pneumonia if aspirated into lungs. *IF IN EYES OR ON SKIN*, hold eyelids open and flush with water for at least 15 minutes; hold eyelids open if necessary. Remove and double bag contaminated clothing and shoes at the site. *IF SWALLOWED* and victim is *CONSCIOUS AND ABLE TO SWALLOW*, have victim drink water and induce vomiting. *IF SWALLOWED AND VICTIM UNCONSCIOUS OR HAVING CONVULSIONS,* just keep victim warm.

HEALTH HAZARDS
Personal protective equipment (PPE): B-Level PPE. OSHA Table Z-1-A air contaminant. Self-contained positive pressure breathing apparatus and full protective clothing. Chemical protective material(s) reported to have good to excellent resistance: Teflon®, Viton®.
Recommendations for respirator selection: NIOSH
At any concentrations above the NIOSH REL, or where there is no REL, at any detectable concentration: SCBAF:PD,PP (any self-contained breathing apparatus that has a full facepiece and is operated in a pressure-demand or other positive-pressure mode); or SAF:PD,PP:ASCBA (any supplied-air respirator that has a full facepiece and is operated in a pressure-demand or other positive-pressure mode in combination with an auxiliary self-contained breathing apparatus operated in a pressure-demand or other positive pressure mode). *ESCAPE:* GMFOV [any air-purifying, full-facepiece respirator (gas mask) with a chin-style, front-or back-mounted organic vapor canister]; or SCBAE (any appropriate escape-type, self-contained breathing apparatus).
Exposure limits (TWA unless otherwise noted): ACGIH TLV 10 ppm (55 mg/m³); OSHA PEL 10 ppm (45 mg/m³); NIOSH REL potential human carcinogen, reduce exposure to lowest feasible levels; 10 ppm (45 mg/m³); skin contact contributes significantly in overall exposure.
Toxicity by ingestion: Grade 2; LD_{50} = 580 mg/kg (rat).

Long-term health effects: Causes liver and kidney damage; may increase myocardial irritability. It is a central nervous system depressant. Animal carcinogen. IARC rating 3; animal data. May cause chemical pneumonia if aspirated into the lungs.
Vapor (gas) irritant characteristics: Eye and respiratory tract irritant. Vapors cause moderate irritation such that personnel will not tolerate moderate or high concentrations.
Liquid or solid irritant characteristics: Eye and skin irritant. If spilled on skin and allowed to remain, may cause smarting and reddening of the skin.
Odor threshold: 0.5–165 ppm.
IDLH value: 100 ppm; potential human carcinogen.

FIRE DATA
Flash point: None, but material is combustible.
Flammable limits in air: LEL: 6%; UEL: 15.5%.
Behavior in fire: Forms a flammable vapor–air mixture at 109°F/43°C and higher.

CHEMICAL REACTIVITY
Binary reactants: Incompatible with oxidizing materials, strong acids or powders of chemically active metals. Attacks aluminum, some forms of plastics, rubber, and coatings.
Reactivity group: 36
Compatibility class: Halogenated hydrocarbons

ENVIRONMENTAL DATA
Food chain concentration potential: Log K_{ow} = 2.17. Unlikely to accumulate.
Water pollution: Harmful to aquatic life in very low concentrations. May be dangerous if it enters nearby water intakes; notify operators. Notify local health and wildlife officials.
Response to discharge: Should be removed. Chemical and physical treatment. Clean Water Act.

SHIPPING INFORMATION
Grades of purity: Technical grade; stabilized; 95%; **Stability during transport:** Stable (no aluminum or plastic packaging).

PHYSICAL AND CHEMICAL PROPERTIES
Physical state @ 59°F/15°C and 1 atm: Liquid.
Molecular weight: 133.41
Boiling point @ 1 atm: 236.6°F/113.7°C/386.9°K.
Melting/Freezing point: 34°F/–37°C/236°K.
Specific gravity (water = 1): 1.44 @ 68°F/20°C (liquid).
Liquid surface tension: 33.75 dynes/cm = 0.0338 N/m @ 68°F/20°C.
Relative vapor density (air = 1): 4.6
Vapor pressure: 19 mm.

TRICHLOROETHYLENE　　　　　　　　　　　　**REC. T:2350**

SYNONYMS: ACETYLENE TRICHLORIDE; ALGYLEN; ANAMENTH; BENZINOL; BLACOSOLV; CECOLENE; CHLORYLEA; CHORYLEN; CIRCOSOLV; CRAWHASPOL; DENSINFLUAT; DOW-TRI; DUKERON; EEC No. 602-027-00-9; ETHINYL TRICHLORIDE; ETHYLENE TRICHLORIDE; FLECK-FLIP; FLUATE; GERMALGENE; LANADIN; LETHURIN; NARCOGEN; NARKOSOID; NIALK; PERM-A-CHLOR; PETZINOL; TCE; THRETHYLENE; TRI; TRICHLOR; TRIAD; TRIASOL; TRICHLORAETHEN (German); TRICHLORORAN; TRICHLORETHENE (French); TRICHLOROETHYLENE TRI (French); TRICHLOROETHENE; TRI-CLENE; 1,1,2-TRICHLOROETHYLENE; TRICLOROETILENO (Spanish); TRILENE; TRIMAR; TRI-PLUS; VESTROL; VITRAN; WESTROSOL; RCRA No. U228

IDENTIFICATION
CAS Number: 79-01-6
Formula: C_2HCl_3; $CHCl = CCl_2$
DOT ID Number: UN 1710; DOT Guide Number: 160
Proper Shipping Name: Trichloroethylene
Reportable Quantity (RQ): **(CERCLA)** 100 lb/45.4 kg

DESCRIPTION: Colorless, watery liquid. Sweet odor, like ether or chloroform. Sinks in water; insoluble. Irritating vapor is produced.

Combustible • Heat, flame, or high energy source may cause closed containers to explode • Containers may BLEVE when exposed to fire • Vapors in confined areas (e.g., tanks, sewers, buildings) may explode when exposed to fire • Irritating to the skin, eyes, and respiratory tract • Combustion or contact with hot surfaces cause decomposition producing hydrogen chloride, and small amounts of phosgene and hydrogen chloride. Dichloroacetylene (DCA) is also a possible decomposition product.

Hazard Classification (based on NFPA-704 Rating System)
Health Hazards (Blue): 2; Flammability (Red): 1; Reactivity (Yellow): 0

EMERGENCY RESPONSE: See Appendix A (160)
Evacuation:
Public safety: Isolate the area of spill or leak for at least 25 to 50 meters (80- 160 feet) in all directions.
Spill: Consider initial downwind evacuation for at least 100 meters (330 feet); increase, in the downwind direction, as necessary.
Fire: If tank, rail car, or tank truck is involved in fire, isolate for at least 800 meters (½ mile) in all directions; also, consider initial evacuation for 800 meters (½ mile) in all directions.

EXPOSURE
Short-term effects: *SEEK MEDICAL ATTENTION.* **Vapor:** Irritating to eyes, nose, and throat. *IF INHALED*, will, will cause nausea, vomiting, difficult breathing, or loss of consciousness. Move to fresh air. *IF BREATHING HAS STOPPED,* give artificial respiration. IF breathing is difficult, administer oxygen. **Liquid:** *HARMFUL IF ABSORBED THROUGH THE SKIN.* Irritating to skin and eyes. *IF SWALLOWED*, will cause nausea, vomiting, difficult breathing, or loss of consciousness. Remove contaminated clothing and shoes. Flush affected areas with plenty of water. *IF IN EYES*, hold eyelids open and flush with plenty of water. *IF SWALLOWED* and victim is *CONSCIOUS AND ABLE TO SWALLOW*, have victim drink 4 to 8 ounces of water and have victim induce vomiting. **Do NOT induce vomiting.** Administer a slurry of activated charcoal. *IF SWALLOWED* and victim is *UNCONSCIOUS OR HAVING CONVULSIONS,* do nothing except keep victim warm.

HEALTH HAZARDS
Personal protective equipment (PPE): B-Level PPE. OSHA Table Z-1-A air contaminant. Organic vapor-acid gas canister; self-contained breathing apparatus for emergencies; gloves; chemical safety goggles; face-shield; safety shoes; suit or apron for splash protection. Chemical protective material(s) reported to have good to excellent resistance:PV alcohol, Viton®, Silvershield®. Also, chlorobutyl rubber offers limited protection.
Recommendations for respirator selection: NIOSH
At any concentrations above the NIOSH REL, or where there is no

REL, at any detectable concentration: SCBAF:PD,PP (any self-contained breathing apparatus that has a full facepiece and is operated in a pressure-demand or other positive-pressure mode); or SAF:PD,PP:ASCBA (any supplied-air respirator that has a full facepiece and is operated in a pressure-demand or other positive-pressure mode in combination with an auxiliary self-contained breathing apparatus operated in a pressure-demand or other positive pressure mode). *ESCAPE:* GMFOV [any air-purifying, full-facepiece respirator (gas mask) with a chin-style, front-or back-mounted organic vapor canister]; or SCBAE (any appropriate escape-type, self-contained breathing apparatus).
Exposure limits (TWA unless otherwise noted): ACGIH TLV 50 ppm (269 mg/m^3) not suspected as a human carcinogen; OSHA PEL 100 ppm, ceiling 200 ppm,(5 minute maximum peak in any 2 hours) 300 ppm; NIOSH REL potential human carcinogen.
Short-term exposure limits (15-minute TWA): ACGIH STEL 100 ppm (537 mg/m^3).
Toxicity by ingestion: Grade 3; LD$_{50}$ = 50 to 500 mg/kg.
Long-term health effects: Suspected carcinogen. IARC animal carcinogen. NCI animal carcinogen. Liver, kidney and other organ damage. Can be additive.
Vapor (gas) irritant characteristics: Eye and respiratory tract irritant. Vapors cause smarting of the eyes or respiratory system if present in high concentrations. May have a narcotic and anesthetic effect.
Liquid or solid irritant characteristics: Severe eye and skin irritant. If spilled on clothing and allowed to remain, may cause smarting and reddening of the skin.
Odor threshold: 50 ppm.
IDLH value: Potential human carcinogen; 1000 ppm.

FIRE DATA
Flash point: None, but closed containers may explode if subjected to a source of high energy.
Flammable limits in air: LEL: 8.0%; UEL: 10.5% (both @ 77°F/25°C).
Autoignition temperature: 770°F/410°C/683°K.
Electrical hazard: Class I, Group D. Containers may explode if subjected to a high energy source. Due to low electric conductivity, this substance may generate electrostatic charges as a result of agitation and flow.

CHEMICAL REACTIVITY
Reaction with water: Slowly decomposes forming hydrochloric acid.
Binary reactants: Violent reaction with strong caustics (e.g., lye, potassium hydroxide, sodium hydroxide, etc.). Caustics, epichlorohydrin, epoxides produce spontaneously explosive dichloroacetylene. Forms an explosive mixture with nitrogen tetroxide. Violent reaction with finely divided chemically active metals: aluminum, titanium, magnesium; alkaline earth metals may cause ignition upon contact. Explosive reaction with sodium, potassium, lithium. High temperatures, contact with hot metals, open flame, and high intensity ultraviolet light can cause the formation of chlorine gas, hydrogen chloride gas, and phosgene. Reacts, possibly violently, with aluminum methyl, aluminum tripropyl, antimony triethyl, antimony, trimethyl, dimethylformamide, liquid oxygen, ozone, potassium nitrate, trimethyl aluminum. Attacks metals, coatings, and plastics in the presence of moisture.
Polymerization: Contact with aluminum may cause self-accelerating polymerization.
Reactivity group: 36
Compatibility class: Halogenated hydrocarbons

ENVIRONMENTAL DATA
Food chain concentration potential: Log K$_{ow}$ = 2.29. Unlikely to accumulate.
Water pollution: Effect of low concentrations on aquatic life is unknown. May be dangerous if it enters nearby water intakes; notify operators. Notify local health and wildlife officials. **Response to discharge:** Should be removed. Chemical and physical treatment. Clean Water Act.

SHIPPING INFORMATION
Grades of purity: Technical; dry cleaning; degreasing; extraction; **Storage temperature:** Ambient; **Inert atmosphere:** None; **Venting:** Pressure-vacuum; **Stability during transport:** Stable.

NAS HAZARD CLASSIFICATION FOR BULK WATER TRANSPORTATION
FIRE: 1
HEALTH: Vapor irritant: 1; Liquid or solid irritant: 1; Poisons: 2
WATER POLLUTION: Human toxicity: 1; Aquatic toxicity: 2; Aesthetic effect: 2
REACTIVITY: Other chemicals: 1; Water: 0; Self-reaction: 1

PHYSICAL AND CHEMICAL PROPERTIES
Physical state @ 59°F/15°C and 1 atm: Liquid.
Molecular weight: 131.39
Boiling point @ 1 atm: 189°F/87°C/360°K.
Melting/Freezing point: –123.5°F/–86.4°C/186.8°K.
Specific gravity (water = 1): 1.46 @ 68°F/20°C (liquid).
Liquid surface tension: 29.3 dynes/cm = 0.0293 N/m @ 68°F/20°C.
Liquid water interfacial tension: 34.5 dynes/cm = 0.0345 N/m at 24°C.
Relative vapor density (air = 1): 4.5
Ratio of specific heats of vapor (gas): 1.116
Latent heat of vaporization: 103 Btu/lb = 57.2 cal/g = 2.4 x 10^5 J/kg.
Vapor pressure: (Reid) 2.5 psia; 58 mm @ 68°F/20°C.

TRICHLOROFLUOROMETHANE   REC. T:2400

SYNONYMS: ALGOFRENE TYPE 1; ARCTRON 9; CFC-11; ELECTRO-CF 11; ESKIMON 11; F-11; FC-11; FLUOROCARBON 11; FLUOROTRICHLOROMETHANE; FREON 11; FREON MF; GENETRON 11; HALOCARBON 11; ISCEON 131; ISOTRON 11; LEDON 11; MONOFLUROTRICHLOROMETHANE; METHANE, TRICHLOROFLUORO-; PROPELLANT 11; R 11; REFRIGERANT 11; TRICHLOROMONOFLUOROMETHANE; TRICLOROFLUOMETANO (Spanish); UCON REFRIGERANT 11; FRIGEN 11; UCON 11; RCRA No. U121

IDENTIFICATION
CAS Number: 75-69-4
Formula: CCl$_3$F
DOT ID Number: UN 3082; DOT Guide Number: 171
Proper Shipping Name: Environmentally hazardous substances, liquid, n.o.s.
Reportable Quantity (RQ): **(CERCLA)** 5000 lb/2270 kg

DESCRIPTION: Colorless liquid. Odorless; weak chlorinated solvent. Sinks in cold water. Harmful vapor is produced.

Containers may BLEVE when exposed to fire • Vapors are heavier than air and will collect and stay in low areas • Irritating to the skin, eyes, and respiratory tract • Combustion or contact with hot surfaces cause decomposition, producing hydrogen chloride and fluorine vapors.

Hazard Classification (based on NFPA-704 Rating System)
Health Hazards (Blue): 1; Flammability (Red): 0; Reactivity (Yellow): 0

EMERGENCY RESPONSE: See Appendix A (171)
Evacuation:
Public safety: Isolate the area of spill or leak for at least 10 to 25 meters (30 to 80 feet) in all directions.
Spill: Increase, in the downwind direction, as necessary, the distance shown under "Public Safety."
Fire: If any large container is involved in fire, isolate for at least 800 meters (½ mile) in all directions; also, consider initial evacuation for 800 meters (½ mile) in all directions.

EXPOSURE
Short-term effects: *SEEK MEDICAL ATTENTION.* **Vapor:** If inhaled, will cause dizziness or difficult breathing. Move to fresh air. *IF BREATHING HAS STOPPED,* give artificial respiration. IF breathing is difficult, administer oxygen.

HEALTH HAZARDS
Personal protective equipment (PPE): B-Level PPE. OSHA Table Z-1-A air contaminant. Air line respirator; rubber gloves; face mask. Wear thermal protective clothing. Chemical protective material(s) reported to have good to excellent resistance: neoprene. Also, Viton and Viton®/neoprene offers limited protection. Also, nitrile rubber is generally suitable for freons.
Recommendations for respirator selection: NIOSH/OSHA
2000 ppm: SA (any supplied-air respirator); or SCBAF (any self-contained breathing apparatus with a full facepiece). *EMERGENCY OR PLANNED ENTRY INTO UNKNOWN CONCENTRATIONS OR IDLH CONDITIONS:* SCBAF:PD,PP (any self-contained breathing apparatus that has a full facepiece and is operated in a pressure-demand or other positive-pressure mode); or SAF:PD,PP:ASCBA (any supplied-air respirator that has a full facepiece and is operated in a pressure-demand or other positive-pressure mode in combination with an auxiliary self-contained breathing apparatus operated in a pressure-demand or other positive pressure mode). *ESCAPE:* GMFOV [any air-purifying, full-facepiece respirator (gas mask) with a chin-style, front- or back-mounted organic vapor cannister]; or SCBAE (any appropriate escape-type, self-contained breathing apparatus).
Exposure limits (TWA unless otherwise noted): OSHA PEL 1000 ppm (5600 mg/m^3)
Short-term exposure limits (15-minute TWA): ACGIH ceiling 1000 ppm (5620 mg/m^3); NIOSH ceiling 1000 ppm (5600 mg/m^3).
Long-term health effects: May cause heart and other organ damage.
Vapor (gas) irritant characteristics: Nonirritating
Liquid or solid irritant characteristics: May cause frostbite.
Odor threshold: 5–100 ppm.
IDLH value: 2000 ppm.

CHEMICAL REACTIVITY
Binary reactants: Violent reaction with chemically active metals, especially powders of aluminum, barium, lithium. Attacks other metals such as zinc. Decomposes on contact with flame or hot surfaces.

ENVIRONMENTAL DATA
Food chain concentration potential: Log K_{ow} = 2.53 Values above 3.0 are very likely to accumulate in living tissues and especially in fats. Unlikely to accumulate.
Water pollution: Not harmful to aquatic life. May be dangerous if it enters nearby water intakes; notify operators. Notify local health and wildlife officials. **Response to discharge:** Should be removed. Chemical and physical treatment. Clean Water Act.

SHIPPING INFORMATION
Grades of purity: Technical; **Storage temperature:** Ambient; **Inert atmosphere:** None; **Venting:** Safety relief; **Stability during transport:** Stable.

PHYSICAL AND CHEMICAL PROPERTIES
Molecular weight: 137.4
Boiling point @ 1 atm: 75°F/24°C/297°K.
Melting/Freezing point: –166°F/–110°C/163°K.
Specific gravity (water = 1): 1.47 liquid @ 75°F/24°C.
Relative vapor density (air = 1): 4.78
Vapor pressure: 690 mm.

TRICHLOROPHENOL   REC. T:2450

SYNONYMS: DOWICIDE 2; DOWICIDE B; NURELLE; OMAL; PHENACHLOR; PREVENTOL-1; RCRA No. U230; 2,4,5-TRICHLOROPHENOL; TRICLOROFENOL (Spanish)

IDENTIFICATION
CAS Number: 95-95-4; may also apply to 933-75-5 (2,3,6-isomer); 95-95-4 (2,4,5-isomer)
Formula: $C_6H_3Cl_3O$; 1-$HOC_6H_2Cl_3$-2,4,5
DOT ID Number: UN 2020; DOT Guide Number: 153
Proper Shipping Name: Chlorophenols, solid
Reportable Quantity (RQ): **(CERCLA)** 10 lb/4.54 lb

DESCRIPTION: Colorless to gray or yellow crystalline solid or flakes. Strong disinfectant odor. Sinks in water; insoluble.

Poison! • 2,4,5-isomer may explode when heated to decomposition • Breathing the vapor, skin or eye contact, or swallowing the material can kill you • Firefighting gear (including SCBA) does not provide adequate protection. If exposure occurs, remove and isolate gear immediately and thoroughly decontaminate personnel • When heated to decomposition it may explode • Containers may BLEVE when exposed to fire • Toxic products of combustion may include hydrogen chloride • Do not put yourself in danger by entering a contaminated area to rescue a victim:

Hazard Classification (based on NFPA-704 Rating System)
Health Hazards (Blue): 2; Flammability (Red): 0; Reactivity (Yellow): 0

EMERGENCY RESPONSE: See Appendix A (153)
Evacuation:
Public safety: Isolate the area of spill or leak for at least 25 to 50 meters (80 to 160 feet) in all directions.
Spill: Increase, in the downwind direction, as necessary, the distance shown under "Public Safety."
Fire: If tank, rail car, or tank truck is involved in fire, isolate for at least 800 meters (½ mile) in all directions; also, consider initial evacuation for 800 meters (½ mile) in all directions.

EXPOSURE
Short-term effects: *SEEK MEDICAL ATTENTION.* **Dust or solid:** Irritating to eyes, nose, and throat. Move to fresh air. *IF*

1076   2,4,5-trichlorophenoxyacetic acid

BREATHING HAS STOPPED, give artificial respiration; *avoid mouth-to-mouth resuscitation; use bag/mask apparatus. IF IN EYES,* hold eyelids open and flush with plenty of water.

HEALTH HAZARDS
Personal protective equipment (PPE): A-Level PPE. Approved dust respirator for toxic dusts; goggles; protective clothing to prevent contact with skin.
Wear sealed chemical suit of butyl rubber, polycarbonate.
Toxicity by ingestion: 20% solution in fuel oil: Grade 2; LD_{50} = 0.5 to 5 g/kg (rat).
Long-term health effects: IARC rating 3; animal data: inadequate. Repeated or prolonged dermal exposure may cause eczema. Liver, kidney and other organ damage.
Vapor (gas) irritant characteristics: Essentially nonvolatile at ordinary temperatures.
Liquid or solid irritant characteristics: May cause injury to eye. Prolonged contact with skin causes a slight burn. Dust irritates nose and throat.

FIRE DATA
Behavior in fire: Containers may explode.

CHEMICAL REACTIVITY
Binary reactants: Reacts with strong oxidizers.

ENVIRONMENTAL DATA
Food chain concentration potential: Log K_{ow} = 3.4–4.1. Values > 3.0 are likely to bioconcentrate in aquatic organisms and other living tissue, especially in fats.
Water pollution: Harmful to aquatic life in very low concentrations. May be dangerous if it enters nearby water intakes; notify operators. Notify local health and wildlife officials.
Response to discharge: Issue warning–water contaminant. Should be removed. Chemical and physical treatment.

SHIPPING INFORMATION
Grades of purity: Technical: 95%; **Storage temperature:** Ambient; **Inert atmosphere:** None; **Venting:** Open; **Stability during transport:** Stable; **Stability during transport:** Stable.

PHYSICAL AND CHEMICAL PROPERTIES
Physical state @ 59°F/15°C and 1 atm: Solid.
Molecular weight: 197.5
Boiling point @ 1 atm: 485°F/252°C/525°K.
Melting/Freezing point: 135°F/57°C/330°K.
Specific gravity (water = 1): 1.7 @ 77°F/25°C (solid).
Relative vapor density (air = 1): 6.7
Vapor pressure: 0.03 mm; 0.027 mm @ 68°F/20°C.

2,4,5-TRICHLOROPHENOXYACETIC ACID
                                              REC. T:2500

SYNONYMS: ACETIC ACID, (2,4,8-TRICHLOROPHYNOXY)-; ACIDE 2,4,5-TRICHLOROPHENOXYACETIQUE (French); BCF-BUSHKILLER; BRUSH-OFF 445 LOW VOLATILE BRUSH KILLER; BRUSH RHAP; BRUSHTOX; T-5 BRUSH KIL; DACAMINE 4T; DEBROUSSAILLANT CONCENTRE; DEBROUSSAILLANT SUPER CONCENTRE; DECAMINE 4T; DED-WEED BRUSH KILLER; DED-WEED LV-6 BRUSH KIL; DINOXOL; EEC No. 607-041-00-9; ENVERT-T; ESTERCIDE T-2; ESTERCIDE T-245; ESTERON 245; ESTERON BRUSH KILLER; FENCE RIDER; FORRON; FORSTU 46; FORTEX; FRUITONE A; INVERTON 245; LINE RIDER; PHORTOX; REDDON; REDDOX; SPONTOX; SUPER D WEEDONE; 2,4,5-T; 2,4,5-T, ACID; 2,4,5-TRICLOROFENOXIACETICO (Spanish); TIPPON; TORMONA; TRANSAMINE; TRIBUTON; (2,4,5-TRICHLOR-PHENOXY)-ESSIGSAEURE (German); TRINOXOL; TRIOXON; TRIOXONE; U 46; VEON; VEON 245; VERTON 2T; VISKO RHAP LOW VOLATILE ESTER; WEEDAR; WEEDONE; RCRA No. U232

IDENTIFICATION
CAS Number: 93-76-5
Formula: $C_8H_5Cl_3O_3$; 2,4,5-$Cl_3C_6H_2OCH_2COOH$
DOT ID Number: UN 2765; DOT Guide Number: 152
Proper Shipping Name: 2,4,5-Trichlorophenoxyacetic acid; Phenoxy pesticides, solid, toxic
Reportable Quantity (RQ): **(CERCLA)** 1000 lb/454 kg

DESCRIPTION: White solid. Odorless. Sinks in water.

Poison! (phenoxy pesticide) • Breathing the vapor, skin or eye contact, or swallowing the material can kill you • Firefighting gear (including SCBA) does not provide adequate protection. If exposure occurs, remove and isolate gear immediately and thoroughly decontaminate personnel • Combustible • Containers may BLEVE when exposed to fire • Toxic products of combustion may include phosgene and hydrogen chloride • Do not put yourself in danger by entering a contaminated area to rescue a victim:

Hazard Classification (based on NFPA-704 Rating System)
Health Hazards (Blue): 2; Flammability (Red): 1; Reactivity (Yellow): 0

EMERGENCY RESPONSE: See Appendix A (152)
Evacuation:
Public safety: Isolate the area of spill or leak for at least 25 to 50 meters (80 to 160 feet) in all directions.
Spill: Increase, in the downwind direction, as necessary, the distance shown under "Public Safety."
Fire: If tank, rail car, or tank truck is involved in fire, isolate for at least 800 meters (½ mile) in all directions; also, consider initial evacuation for 800 meters (½ mile) in all directions.

EXPOSURE
Short-term effects: *SEEK MEDICAL ATTENTION.* **Solid:** *POISONOUS IF SWALLOWED.* Irritating to skin and eyes. Remove contaminated clothing and shoes. Flush affected areas with plenty of water. *IF BREATHING HAS STOPPED,* give artificial respiration; *avoid mouth-to-mouth resuscitation; use bag/mask apparatus. IF IN EYES,* hold eyelids open and flush with plenty of water. *IF SWALLOWED* and victim is *CONSCIOUS AND ABLE TO SWALLOW,* have victim drink 4 to 8 ounces of water and have victim induce vomiting. *IF SWALLOWED* and victim is *UNCONSCIOUS OR HAVING CONVULSIONS,* do nothing except keep victim warm.

HEALTH HAZARDS
Personal protective equipment (PPE): A-Level PPE. OSHA Table Z-1-A air contaminant.
Recommendations for respirator selection: NIOSH
50 mg/m³: DM (any dust and mist respirator). *100 mg/m³:* DMXSQ (any dust and mist respirator except single-use and quarter mask respirators); or SA (any supplied-air respirator). *250 mg/m³:* SA:CF (any supplied-air respirator operated in a continuous-flow mode); or HiEF (any air-purifying, full-facepiece respirator with a high-efficiency particulate filter); or PAPRDM, *If not present as a fume* (any powered, air-purifying respirator with a dust and mist filter);

or SCBAF (any self-contained breathing apparatus with a full facepiece); or SAF (any supplied-air respirator with a full facepiece). *EMERGENCY OR PLANNED ENTRY INTO UNKNOWN CONCENTRATIONS OR IDLH CONDITIONS:* SCBAF:PD,PP (any self-contained breathing apparatus that has a full facepiece and is operated in a pressure-demand or other positive-pressure mode); or SAF:PD,PP:ASCBA (any supplied-air respirator that has a full facepiece and is operated in a pressure-demand or other positive-pressure mode in combination with an auxiliary self-contained breathing apparatus operated in a pressure-demand or other positive pressure mode). *ESCAPE:* HiEF (any air-purifying, full-facepiece respirator with a high-efficiency particulate filter); or SCBAE (any appropriate escape-type, self-contained breathing apparatus).
Exposure limits (TWA unless otherwise noted): ACGIH TLV 10 mg/m^3; NIOSH/OSHA 10 mg/m^3.
Toxicity by ingestion: Grade 3; oral LD$_{50}$ = 500 mg/kg (rat).
Long-term health effects: Birth defects in rats and mice. Causes an acne-like skin eruption among human workers. suspected carcinogen. Organ damage including liver and kidney.
Vapor (gas) irritant characteristics: Eye and respiratory tract irritation.
Liquid or solid irritant characteristics: Eye and skin irritant.
IDLH value: 250 mg/m^3.

FIRE DATA
Flash point: Combustible solid, but burns with difficulty.
Behavior in fire: Water streams applied to adjacent fires will spread contamination of pesticide over wide area.

CHEMICAL REACTIVITY
Binary reactants: Strong oxidizers. Can be corrosive to common metals

ENVIRONMENTAL DATA
Food chain concentration potential: Log K$_{ow}$ = 4.8. Log values of octanol/water partition coefficients above 3.0 are very likely to accumulate in living tissues and especially in fats.
Water pollution: Harmful to aquatic life in very low concentrations. May be dangerous if it enters nearby water intakes; notify operators. Notify local health and wildlife officials.
Response to discharge: Issue warning–poison, water contaminant. Restrict access. Should be removed. Chemical and physical treatment. Clean Water Act.

SHIPPING INFORMATION
Grades of purity: Commercial; **Storage temperature:** Ambient; **Inert atmosphere:** None; **Venting:** Open (flame arrester); **Stability during transport:** Stable.

PHYSICAL AND CHEMICAL PROPERTIES
Physical state @ 59°F/15°C and 1 atm: Solid.
Molecular weight: 255.5
Boiling point @ 1 atm: Decomposes.
Melting/Freezing point: 316°F/158°C/431°K.
Specific gravity (water = 1): 1.803 @ 68°F/20°C (solid).
Heat of combustion: (estimate) –6500 Btu/lb = –3600 cal/g = –150 x 10^5 J/kg.
Vapor pressure: Less than 1 x 10^{-7} mm.

2,4,5-TRICHLOROPHENOXYACETIC ACID ESTERS
REC. T:2550

SYNONYMS: ARBORICID; BUTYL 2,4,5-T; BUTYL 2,4,5-TRICHLOROPHENOXYACETATE; BUTOXYPROPYL TRICHLOROPHENOXYACETATE; ISOOCTYL TRICHLOROPHENOXYACETATE; KILEX-3; TORMONA; TRIOXONE; U46KW

IDENTIFICATION
CAS Number: 93-79-8
Formula: C$_{12}$H$_{13}$Cl$_3$O$_3$; 2,4,5-Cl$_3$C$_6$H$_2$OCH$_2$COOR where R = C$_4$H$_9$, C$_8$H$_{17}$, etc
DOT ID Number: UN 2765; DOT Guide Number: 152
Proper Shipping Name: Phenoxy pesticides, solid, toxic
Reportable Quantity (RQ): **(CERCLA)** 1000 lb/454 kg

DESCRIPTION: Colorless to yellowish to amber brown liquid. Mild odor; mixtures with kerosene or diesel oil have odor of the solvent. Sinks in water.

Poison! (chlorophenoxy pesticide) • Combustible • Breathing or swallowing the dust can cause illness • Containers may BLEVE when exposed to fire • Concentrated dust in confined areas (e.g., tanks, sewers, buildings) may explode when exposed to fire • Toxic products of combustion may include hydrogen chloride.

Hazard Classification (based on NFPA-704 Rating System)
Health Hazards (Blue): 2; Flammability (Red): 1; Reactivity (Yellow): 0

EMERGENCY RESPONSE: See Appendix A (152)
Evacuation:
Public safety: Isolate the area of spill or leak for at least 25 to 50 meters (80 to 160 feet) in all directions.
Spill: Increase, in the downwind direction, as necessary, the distance shown under "Public Safety."
Fire: If tank, rail car, or tank truck is involved in fire, isolate for at least 800 meters (½ mile) in all directions; also, consider initial evacuation for 800 meters (½ mile) in all directions.

EXPOSURE
Short-term effects: *SEEK MEDICAL ATTENTION.* **Liquid:** Irritating to skin and eyes. *IF SWALLOWED*, will cause nausea and vomiting. Remove contaminated clothing and shoes. Flush affected areas with plenty of water. *IF BREATHING HAS STOPPED*, give artificial respiration; *avoid mouth-to-mouth resuscitation; use bag/mask apparatus*. *IF IN EYES*, hold eyelids open and flush with plenty of water. *IF SWALLOWED* and victim is *CONSCIOUS AND ABLE TO SWALLOW*, have victim drink 4 to 8 ounces of water and have victim induce vomiting. *IF SWALLOWED* and victim is *UNCONSCIOUS OR HAVING CONVULSIONS*, do nothing except keep victim warm.

HEALTH HAZARDS
Personal protective equipment (PPE): A-Level PPE.
Recommendations for respirator selection: NIOSH as 2,4,5-T 50 mg/m^3: DM (any dust and mist respirator). 100 mg/m^3: DMXSQ (any dust and mist respirator except single-use and quarter mask respirators); or SA (any supplied-air respirator). 250 mg/m^3: SA:CF (any supplied-air respirator operated in a continuous-flow mode); or HiEF (any air-purifying, full-facepiece respirator with a high-efficiency particulate filter); or PAPRDM, *If not present as a fume* (any powered, air-purifying respirator with a dust and mist filter); or SCBAF (any self-contained breathing apparatus with a full facepiece); or SAF (any supplied-air respirator with a full facepiece). *EMERGENCY OR PLANNED ENTRY INTO UNKNOWN CONCENTRATIONS OR IDLH CONDITIONS:* SCBAF:PD,PP (any self-contained breathing apparatus that has a

full facepiece and is operated in a pressure-demand or other positive-pressure mode); or SAF:PD,PP:ASCBA (any supplied-air respirator that has a full facepiece and is operated in a pressure-demand or other positive-pressure mode in combination with an auxiliary self-contained breathing apparatus operated in a pressure-demand or other positive pressure mode). *ESCAPE:* HiEF (any air-purifying, full-facepiece respirator with a high-efficiency particulate filter); or SCBAE (any appropriate escape-type, self-contained breathing apparatus).
Toxicity by ingestion: Grade 3; LD_{50} = 50 to 500 mg/kg.
Exposure limits (TWA unless otherwise noted): NIOSH/OSHA 10 mg/m^3 as 2,4,5-T.
Liquid or solid irritant characteristics: Eye and skin irritant.
IDLH value: 250 mg/m^3 as 2,4,5-T.

FIRE DATA
Flash point: 265-420°F/129-216°C (oc).
Fire extinguishing agents not to be used: Water or foam may cause frothing.
Behavior in fire: Water streams applied to adjacent fires will spread contamination of pesticide over wide area.

CHEMICAL REACTIVITY
Reactivity with water: Mild foaming action.
Binary reactants: Strong oxidizers. May attack some forms of plastics

ENVIRONMENTAL DATA
Food chain concentration potential: Log K_{ow} = 4.8. Log values of octanol/water partition coefficients above 3.0 are very likely to accumulate in living tissues and especially in fats.
Water pollution: Effect of low concentrations on aquatic life is unknown. May be dangerous if it enters nearby water intakes; notify operators. Notify local health and wildlife officials.
Response to discharge: Issue warning–water contaminant. Should be removed. Chemical and physical treatment. Clean Water Act.

SHIPPING INFORMATION
Grades of purity: Technical, 96-99%; 55-65% solutions in kerosene or diesel oil, which are combustible; **Storage temperature:** Ambient; **Inert atmosphere:** None; **Venting:** Open; **Stability during transport:** Stable.

PHYSICAL AND CHEMICAL PROPERTIES
Physical state @ 59°F/15°C and 1 atm: Liquid.
Molecular weight: Mixtures, all greater than 300
Boiling point @ 1 atm: butyl: 639°F/337°C/610°K; butoxypropyl: 651°F/344°C/617°K; isooctyl: 770°F/410°C/683°K; 2-ethylhexyl: 770°F/410°C/683°K.
Specific gravity (water = 1): 1.2 @ 68°F/20°C (liquid).
Vapor pressure: Less than 1 x 10^{-7} mm (2,4,5-isomer).

2,4,5-TRICHLOROPHENOXYACETIC ACID, SODIUM SALT   REC. T:2600

SYNONYMS: 2,4,5-T, Sodium salt

IDENTIFICATION
CAS Number: 93-79-8
Formula: $C_{12}H_{13}Cl_3NaO_3$; $C_6H_2Cl_3OCH_2COONa$
DOT ID Number: UN 2765; DOT Guide Number: 152
Proper Shipping Name: Phenoxy pesticides, solid, toxic
Reportable Quantity (RQ): **(CERCLA)** 1000 lb/454 kg

DESCRIPTION: Tan powder formulated as amber to dark-brown liquid. Sinks in water; soluble.

Poison! (phenoxy) • Combustible solid • Breathing or swallowing the dust can cause illness • Containers may BLEVE when exposed to fire • Concentrated dust in confined areas (e.g., tanks, sewers, buildings) may explode when exposed to fire • Toxic products of combustion may include hydrogen chloride.

Hazard Classification (based on NFPA-704 Rating System)
Health Hazards (Blue): 2; Flammability (Red): 1; Reactivity (Yellow): 0

EMERGENCY RESPONSE: See Appendix A (152)
Evacuation:
Public safety: Isolate the area of spill or leak for at least 25 to 50 meters (80 to 160 feet) in all directions.
Spill: Increase, in the downwind direction, as necessary, the distance shown under "Public Safety."
Fire: If tank, rail car, or tank truck is involved in fire, isolate for at least 800 meters (½ mile) in all directions; also, consider initial evacuation for 800 meters (½ mile) in all directions.

EXPOSURE
Short-term effects: *SEEK MEDICAL ATTENTION.* Irritating to skin and eyes. Harmful if swallowed. Remove contaminated clothing and shoes. Flush affected areas with plenty of water. *IF BREATHING HAS STOPPED,* give artificial respiration; *avoid mouth-to-mouth resuscitation; use bag/mask apparatus. IF IN EYES,* hold eyelids open and flush with plenty of water. *IF SWALLOWED* and victim is *CONSCIOUS AND ABLE TO SWALLOW,* have victim drink 4 to 8 ounces of water and have victim induce vomiting.

HEALTH HAZARDS
Personal protective equipment (PPE): B-Level PPE. Wear rubber gloves and boots, safety goggles or face mask, hooded suit and a self-contained breathing apparatus.
Recommendations for respirator selection: NIOSH as 2,4,5-T. *50 mg/m^3:* DM (any dust and mist respirator). *100 mg/m^3:* DMXSQ (any dust and mist respirator except single-use and quarter mask respirators); or SA (any supplied-air respirator). *250 mg/m^3:* SA:CF (any supplied-air respirator operated in a continuous-flow mode); or HiEF (any air-purifying, full-facepiece respirator with a high-efficiency particulate filter); or PAPRDM, *If not present as a fume* (any powered, air-purifying respirator with a dust and mist filter); or SCBAF (any self-contained breathing apparatus with a full facepiece); or SAF (any supplied-air respirator with a full facepiece). *EMERGENCY OR PLANNED ENTRY INTO UNKNOWN CONCENTRATIONS OR IDLH CONDITIONS:* SCBAF:PD,PP (any self-contained breathing apparatus that has a full facepiece and is operated in a pressure-demand or other positive-pressure mode); or SAF:PD,PP:ASCBA (any supplied-air respirator that has a full facepiece and is operated in a pressure-demand or other positive-pressure mode in combination with an auxiliary self-contained breathing apparatus operated in a pressure-demand or other positive pressure mode). *ESCAPE:* HiEF (any air-purifying, full-facepiece respirator with a high-efficiency particulate filter); or SCBAE (any appropriate escape-type, self-contained breathing apparatus).
Exposure limits (TWA unless otherwise noted): NIOSH/OSHA 10 mg/m^3 as 2,4,5-T.
Toxicity by ingestion: (estimate) Grade 2; LD_{50} = 300 to more than 1000 mg/kg Based on acute toxicities.

Toxicity by ingestion: of chlorophenoxy herbicides and their esters and salts.
Long-term health effects: Causes birth defects in laboratory animals. This may be caused by the contaminant dioxin. Suspected human carcinogen.
Liquid or solid irritant characteristics: Causes smarting of the skin and first-degree burns on short exposure; may cause second-degree burns on long exposure.
IDLH value: 250 mg/m^3 as 2,4,5-T.

FIRE DATA
Behavior in fire: Water streams applied to adjacent fires will spread contamination of pesticide over wide area.

CHEMICAL REACTIVITY
Binary reactants: Strong oxidizers, strong acids.
Reactivity group: 34
Compatibility class: Esters.

ENVIRONMENTAL DATA
Water pollution: Harmful to aquatic life at very low concentrations. May be dangerous if it enters nearby water intakes; notify operators. Notify local health and wildlife officials.
Response to discharge: Issue warning–water contaminant. Should be removed. Chemical and physical treatment.

SHIPPING INFORMATION
Storage temperature: Away from heat.

PHYSICAL AND CHEMICAL PROPERTIES
Molecular weight: 254.48.
Boiling point @ 1 atm: Decomposes, based on 2,4,5-T

2-(2,4,5-TRICHLOROPHENYOXY)PROPANOIC ACID
REC. T:2650

SYNONYMS: AQUA-VEX; COLOR-SET; DED-WEED; FENOPROP; FENOMORE; FRUIT-O-NET; KURAN; KURON; KUROSAL; KUROSALG; RCRA No. U233; SILVI-RHAP; STA-FAST; SILVEX; SILVEX (2,4,5-TP); 2,4,5-TP; WEED-B-GONE

IDENTIFICATION
CAS Number: 93-72-1
Formula: $C_9H_7Cl_3O_3$; $(C_6H_2Cl_3)OCH(CH_3)COOH$
DOT ID Number: UN 2765; DOT Guide Number: 152
Proper Shipping Name: Phenoxy pesticides, solid, toxic
Reportable Quantity (RQ): **(CERCLA)** 100 lb/45.4 kg

DESCRIPTION: White crystalline solid or powder. Sinks and mixes slowly with water.

Poison! (phenoxy) • Combustible solid • Breathing or swallowing the dust can cause illness • Containers may BLEVE when exposed to fire • Concentrated dust in confined areas (e.g., tanks, sewers, buildings) may explode when exposed to fire • Toxic products of combustion may include hydrogen chloride.

Hazard Classification (based on NFPA-704 Rating System)
Health Hazards (Blue): 2; Flammability (Red): 1; Reactivity (Yellow): 0

EMERGENCY RESPONSE: See Appendix A (152)
Evacuation:
Public safety: Isolate the area of spill or leak for at least 25 to 50 meters (80 to 160 feet) in all directions.
Spill: Increase, in the downwind direction, as necessary, the distance shown under "Public Safety."
Fire: If tank, rail car, or tank truck is involved in fire, isolate for at least 800 meters (½ mile) in all directions; also, consider initial evacuation for 800 meters (½ mile) in all directions.

EXPOSURE
Short-term effects: *SEEK MEDICAL ATTENTION.* **Solid or dust:** Harmful if swallowed or inhaled. Irritating to skin and eyes. Move to fresh air. Remove contaminated clothing and shoes. Flush affected areas with plenty of water. *IF BREATHING HAS STOPPED*, give artificial respiration; *avoid mouth-to-mouth resuscitation; use bag/mask apparatus. IF IN EYES*, hold eyelids open and flush with plenty of water. *IF SWALLOWED* and victim is *CONSCIOUS AND ABLE TO SWALLOW*, have victim drink 4 to 8 ounces of water, and have victim induce vomiting. *IF SWALLOWED* and victim is *UNCONSCIOUS OR HAVING CONVULSIONS*, do nothing except keep warm.

HEALTH HAZARDS
Personal protective equipment (PPE): A-Level PPE. Self-contained breathing apparatus, rubber gloves, hats, suits and boots, and goggles.
Recommendations for respirator selection: NIOSH as 2,4,5-T. *50 mg/m^3:* DM (any dust and mist respirator). *100 mg/m^3:* DMXSQ (any dust and mist respirator except single-use and quarter mask respirators); or SA (any supplied-air respirator). *250 mg/m^3:* SA:CF (any supplied-air respirator operated in a continuous-flow mode); or HiEF (any air-purifying, full-facepiece respirator with a high-efficiency particulate filter); or PAPRDM, *If not present as a fume* (any powered, air-purifying respirator with a dust and mist filter); or SCBAF (any self-contained breathing apparatus with a full facepiece); or SAF (any supplied-air respirator with a full facepiece). *EMERGENCY OR PLANNED ENTRY INTO UNKNOWN CONCENTRATIONS OR IDLH CONDITIONS:* SCBAF:PD,PP (any self-contained breathing apparatus that has a full facepiece and is operated in a pressure-demand or other positive-pressure mode); or SAF:PD,PP:ASCBA (any supplied-air respirator that has a full facepiece and is operated in a pressure-demand or other positive-pressure mode in combination with an auxiliary self-contained breathing apparatus operated in a pressure-demand or other positive pressure mode). *ESCAPE:* HiEF (any air-purifying, full-facepiece respirator with a high-efficiency particulate filter); or SCBAE (any appropriate escape-type, self-contained breathing apparatus).
Exposure limits (TWA unless otherwise noted): NIOSH/OSHA 10 mg/m^3 as 2,4,5-T
Toxicity by ingestion: Grade 2; LD_{50} = 0.5 to 5 g/kg.
Long-term health effects: A decrease in fetal weight and cleft palates were observed in laboratory mice. Possible liver and kidney damage. A suspected human carcinogen.
Vapor (gas) irritant characteristics: Eye and respiratory tract irritant.
Liquid or solid irritant characteristics: Eye and skin irritant. If powder remains on skin, may cause smarting and reddening of skin.
IDLH value: 250 mg/m^3 as 2,4,5-T

FIRE DATA
Behavior in fire: Water streams applied to adjacent fires will spread contamination of pesticide over wide area.

CHEMICAL REACTIVITY
Binary reactants: Strong oxidizers.

ENVIRONMENTAL DATA
Food chain concentration potential: Based on the literature, Unlikely to be significant.
Water pollution: Harmful to aquatic life in very low concentrations. May be dangerous if it enters nearby water intakes; notify operators. Notify local health and wildlife officials.
Response to discharge: Issue warning–water contaminant. Should be removed. Chemical and physical treatment.

SHIPPING INFORMATION
Grades of purity: 59–65% emulsifiable concentration 10.4% amine salt solution 98%; **Storage temperature:** Ambient; **Stability during transport:** Stable.

PHYSICAL AND CHEMICAL PROPERTIES
Physical state @ 59°F/15°C and 1 atm: Solid.
Molecular weight: 269.51.
Boiling point @ 1 atm: More than 300°F/149°C 422°K.
Melting/Freezing point: 358.9°F/181.6°C/454.75°K.
Specific gravity (water = 1): 1.2085 @ 68°F/20°C.
Relative vapor density (air = 1): 9.29

2-(2,4,5-TRICHLOROPHENOXY)PROPANOIC ACID, ISOOCTYL ESTER   REC. T:2700

SYNONYMS: SILVEX, ISOOCTYL ESTER; 2,4,5-TP ACID ESTERS

IDENTIFICATION
CAS Number: 25168-15-4
Formula: $C_{16}H_{21}Cl_3O_3$
DOT ID Number: UN 2765; DOT Guide Number: 152
Proper Shipping Name: Phenoxy pesticides, solid, toxic
Reportable Quantity (RQ): **(CERCLA)** 100 lb/45.4 kg

DESCRIPTION: Amber to dark brown oily liquid. Sinks in water.

Poison! (chlorophenoxy pesticide) • Combustible • Breathing or swallowing the dust can cause illness • Containers may BLEVE when exposed to fire • Concentrated dust in confined areas (e.g., tanks, sewers, buildings) may explode when exposed to fire • Toxic products of combustion may include hydrogen chloride.

Hazard Classification (based on NFPA-704 Rating System)
Health Hazards (Blue): 2; Flammability (Red): 1; Reactivity (Yellow): 0

EMERGENCY RESPONSE: See Appendix A (152)
Evacuation:
Public safety: Isolate the area of spill or leak for at least 25 to 50 meters (80 to 160 feet) in all directions.
Spill: Increase, in the downwind direction, as necessary, the distance shown under "Public Safety."
Fire: If tank, rail car, or tank truck is involved in fire, isolate for at least 800 meters (½ mile) in all directions; also, consider initial evacuation for 800 meters (½ mile) in all directions.

EXPOSURE
Short-term effects: *SEEK MEDICAL ATTENTION.* **Liquid:** Harmful if swallowed. Irritating to skin and eyes. Remove contaminated clothing and shoes. Flush affected areas with plenty of water. *IF BREATHING HAS STOPPED,* give artificial respiration; *avoid mouth-to-mouth resuscitation; use bag/mask apparatus. IF IN EYES,* hold eyelids open and flush with plenty of water. *IF SWALLOWED* and victim is *CONSCIOUS AND ABLE TO SWALLOW,* have victim drink 4 to 8 ounces of water, and have victim induce vomiting. *IF SWALLOWED* and victim is *UNCONSCIOUS OR HAVING CONVULSIONS,* do nothing except keep warm.

HEALTH HAZARDS
Personal protective equipment (PPE): A-Level PPE. Rubber gloves and boots, safety goggles or face mask, protective clothing and a NIOSH approved respirator.
Recommendations for respirator selection: NIOSH as 2,4,5-T $50 mg/m^3$: DM (any dust and mist respirator). $100 mg/m^3$: DMXSQ (any dust and mist respirator except single-use and quarter mask respirators); or SA (any supplied-air respirator). $250 mg/m^3$: SA:CF (any supplied-air respirator operated in a continuous-flow mode); or HiEF (any air-purifying, full-facepiece respirator with a high-efficiency particulate filter); or PAPRDM, *If not present as a fume* (any powered, air-purifying respirator with a dust and mist filter); or SCBAF (any self-contained breathing apparatus with a full facepiece); or SAF (any supplied-air respirator with a full facepiece). *EMERGENCY OR PLANNED ENTRY INTO UNKNOWN CONCENTRATIONS OR IDLH CONDITIONS*: SCBAF:PD,PP (any self-contained breathing apparatus that has a full facepiece and is operated in a pressure-demand or other positive-pressure mode); or SAF:PD,PP:ASCBA (any supplied-air respirator that has a full facepiece and is operated in a pressure-demand or other positive-pressure mode in combination with an auxiliary self-contained breathing apparatus operated in a pressure-demand or other positive pressure mode). *ESCAPE:* HiEF (any air-purifying, full-facepiece respirator with a high-efficiency particulate filter); or SCBAE (any appropriate escape-type, self-contained breathing apparatus).
Exposure limits (TWA unless otherwise noted): NIOSH/OSHA $10 mg/m^3$.
Toxicity by ingestion: Grade 2; LD_{50} = 0.5 to 5 g/kg.
Long-term health effects: Possible teratogen. Dioxin is, at least partially, the cause of defects. Suspected carcinogen.
Liquid or solid irritant characteristics: Eye and skin irritant. If spilled on clothing and allowed to remain, may cause smarting and reddening of skin.
IDLH value: $250 mg/m^3$.

FIRE DATA
Flash point: 405°F/207°C (oc).
Behavior in fire: May liberate hydrogen chloride.

CHEMICAL REACTIVITY
Binary reactants: Oxidizing materials.

ENVIRONMENTAL DATA
Water pollution: Harmful to aquatic life in very low concentrations. May be dangerous if it enters nearby water intakes; notify operators. Notify local health and wildlife officials. **Response to discharge:** Issue warning–water contaminant. Should be removed. Chemical and physical treatment.

SHIPPING INFORMATION
Grades of purity: 95–97%; **Storage temperature:** Ambient; **Stability during transport:** Stable.

PHYSICAL AND CHEMICAL PROPERTIES
Physical state @ 59°F/15°C and 1 atm: Liquid.
Molecular weight: 381.7259 (calculated).

Boiling point @ 1 atm: 320°F/160°C/433.2°K.
Specific gravity (water = 1): 1.183 @ 68°F/20°C.

1,2,3-TRICHLOROPROPANE   REC. T:2750

SYNONYMS: ALLYL TRICHLORIDE; GLYCEROL TRICHLORHYDRIN; GLYCERYL TRICHLORHYDRIN; PROPANE, 1,2,3-TRICHLORO; TRICHLOROHYDRIN; 1,2,3-TRICLOROPROPANO (Spanish)

IDENTIFICATION
CAS Number: 96-18-4
Formula: $C_3H_5Cl_3$; $CH_2ClCHClCH_2Cl$
DOT ID Number: UN 2810; DOT Guide Number: 153
Proper Shipping Name: Toxic liquids, organic, n.o.s.

DESCRIPTION: Colorless to light yellow liquid. Strong acrid odor, like chloroform. Liquid sinks in water; soluble.

Poison! • Combustible • Breathing the vapor, skin or eye contact, or swallowing the material can cause permanent injury or death • Firefighting gear (including SCBA) does not provide adequate protection. If exposure occurs, remove and isolate gear immediately and thoroughly decontaminate personnel • Containers may BLEVE when exposed to fire • Vapors are heavier than air and will collect and stay in low areas • Vapors in confined areas (e.g., tanks, sewers, buildings) may explode when exposed to fire • Toxic products of combustion may include hydrogen chloride.

Hazard Classification (based on NFPA-704 Rating System)
Health Hazards (Blue): 3; Flammability (Red): 2; Reactivity (Yellow): 0

EMERGENCY RESPONSE: See Appendix A (153)
Evacuation:
Public safety: Isolate the area of spill or leak for at least 25 to 50 meters (80 to 160 feet) in all directions.
Spill: Increase, in the downwind direction, as necessary, the distance shown under "Public Safety."
Fire: If tank, rail car, or tank truck is involved in fire, isolate for at least 800 meters (½ mile) in all directions; also, consider initial evacuation for 800 meters (½ mile) in all directions.

EXPOSURE
Short-term effects: *SEEK MEDICAL ATTENTION.* **Vapor:** Irritating to eyes, nose, throat, and skin. *IF INHALED*, will, will cause anesthesia, dizziness, and nausea. Move to fresh air. *IF BREATHING HAS STOPPED,* give artificial respiration; *avoid mouth-to-mouth resuscitation; use bag/mask apparatus.* IF breathing is difficult, administer oxygen. **Liquid:** *HARMFUL IF ABSORBED THROUGH THE SKIN.* Harmful if swallowed. Irritating to skin and eyes. Flush affected areas with plenty of water. Remove from skin with soap and water. Remove contaminated clothing. *IF IN EYES*, hold eyelids open and flush with water for 15 minutes. *IF SWALLOWED* and victim is *CONSCIOUS AND ABLE TO SWALLOW,* give two glasses of water and have victim induce vomiting. *IF SWALLOWED* and victim is *UNCONSCIOUS OR HAVING CONVULSIONS,* do nothing except keep victim warm.

HEALTH HAZARDS
Personal protective equipment (PPE): A-Level PPE. OSHA Table Z-1-A air contaminant. Full face organic vapor respirator. Chemical protective material(s) reported to have good to excellent resistance: butyl rubber, PV alcohol, Viton®.
Recommendations for respirator selection: NIOSH
At any concentrations above the NIOSH REL, or where there is no REL, at any detectable concentration: SCBAF:PD,PP (any self-contained breathing apparatus that has a full facepiece and is operated in a pressure-demand or other positive-pressure mode); or SAF:PD,PP:ASCBA (any supplied-air respirator that has a full facepiece and is operated in a pressure-demand or other positive-pressure mode in combination with an auxiliary self-contained breathing apparatus operated in a pressure-demand or other positive pressure mode). *ESCAPE:* GMFOV [any air-purifying, full-facepiece respirator (gas mask) with a chin-style, front-or back-mounted organic vapor canister]; or SCBAE (any appropriate escape-type, self-contained breathing apparatus).
Exposure limits (TWA unless otherwise noted): ACGIH TLV 10 ppm (60 mg/m³); OSHA PEL 50 ppm (300 mg/m³); NIOSH REL potential human carcinogen, 10 ppm (60 mg/m³); skin contact contributes significantly in overall exposure.
Toxicity by ingestion: Grade: 3; LD_{50} = 320 to 505 mg/kg (rat).
Long-term health effects: Animal carcinogen. Toxicity is cumulative. Causes damage to heart, liver, and kidneys in humans. May cause death or permanent injury after very short exposure to small quantities.
Vapor (gas) irritant characteristics: Severe irritant to eyes and skin. High irritant via oral and inhalation routes.
Liquid or solid irritant characteristics: Eye and skin irritant. Rated 4 on scale of 1-10 when tested externally on eyes of rabbits. Dermal LD_{50} (rabbit) = 1.77 g/kg.
Odor threshold: 100 ppm.
IDLH value: Potential human carcinogen; 100 ppm.

FIRE DATA
Flash point: 174°F/79°C (oc); 164°F/74°C (cc).
Flammable limits in air: LEL: 3.2% @ 248°F/120°C; UEL: 12.6% @ 302°F/150°C.
Autoignition temperature: 579°F/304°C/577°K.

CHEMICAL REACTIVITY
Reactivity with water: Not reactive.
Binary reactants: Can react vigorously with oxidizing materials. Avoid bases. Decomposition reaction may
be initiated by chemically active metals; aluminum, magnesium and their alloys.
Reactivity group: 36
Compatibility class: Halogenated hydrocarbons

ENVIRONMENTAL DATA
Food chain concentration potential: Unlikely to accumulate.
Water pollution: Effect of low concentrations on aquatic life is unknown. May be dangerous if it enters nearby water intakes; notify operators. Notify local health and wildlife officials. **Response to discharge:** Issue warning-air contaminant. Restrict access. Should be removed. Chemical and physical treatment.

SHIPPING INFORMATION
Grades of purity: 90–99+%; **Storage temperature:** Store out of direct sunlight; **Stability during transport:** Stable.

PHYSICAL AND CHEMICAL PROPERTIES
Physical state @ 59°F/15°C and 1 atm: Liquid.
Molecular weight: 147.43
Boiling point @ 1 atm: 314.33°F/156.85°C/429.85°K.
Melting/Freezing point: 6°F/–14°C/259°K.
Specific gravity (water = 1): 1.3889 @ 68°F/20°C (liquid).

Liquid surface tension: 37.8 dynes/cm = 0.0378 N/m @ 68°F/20°C.
Relative vapor density (air = 1): 5.0
Vapor pressure: 3 mm.

TRICHLOROSILANE   REC. T:2800

SYNONYMS: SILICOCHLOROFORM; TRICHLOROMONOSILANE; TRICLOROSILANO (Spanish)

IDENTIFICATION
CAS Number: 10025-78-2
Formula: $SiHCl_3$
DOT ID Number: UN 1295; DOT Guide Number: 139
Proper Shipping Name: Trichlorosilane

DESCRIPTION: Colorless liquid that fumes on contact with air. Sharp choking odor; smells like hydrochloric acid. Reacts violently with water forming hydrochloric acid and explosive hydrogen gas. Boils at 90°F/32°C.

Extremely flammable • Do NOT use water • Ignites spontaneously in air • Corrosive • Poison! • Breathing the vapors can kill; skin and eye contact causes severe burns and blindness; inhalation symptoms may be delayed • Firefighting gear (including SCBA) does not provide adequate protection. If exposure occurs, remove and isolate gear immediately and thoroughly decontaminate personnel • Vapors are heavier than air and will collect and stay in low areas • Vapors may travel long distances to source of ignition and flashback • Often shipped and stored dissolved in acetone • Exposure of cylinders to elevated temperatures, fire, and flame may cause cylinders to rupture or cause frangible disk to burst, releasing entire contents of cylinder • When combined with surface moisture this material is corrosive to most common metals, forming flammable and explosive hydrogen gas • Toxic products of combustion may include hydrogen chloride • Severely irritating to skin, eyes, and respiratory tract; prolonged contact with the skin causes burns • Do NOT attempt rescue.

Hazard Classification (based on NFPA-704 Rating System)
Health Hazards (Blue): 3; Flammability (Red): 4; Reactivity (Yellow): 2; Special Notice (White): Water reactive

EMERGENCY RESPONSE: See Appendix A (139)
Evacuation:
Public safety: Isolate the area of spill or leak for at least 100 to 150 meters (330 to 490 feet) in all directions.
Spill: *IF SPILLED IN WATER*: Small spill–First: Isolate in all directions 30 meters (100 feet); Then: Protect persons downwind. DAY: 0.2 km (0.1 mile); NIGHT: 0.3 km (0.2 mile). Large spill–First: Isolate in all directions 125 meters (400 feet) Then: Protect persons downwind, DAY: 1.3 km (0.8 mile); NIGHT: 3.2 km (2.0 miles).
Fire: If any large container is involved in fire, isolate for at least 800 meters (½ mile) in all directions; also, consider initial evacuation for 800 meters (½ mile) in all directions.

EXPOSURE
Short-term effects: *SEEK MEDICAL ATTENTION*. **Vapor:** Irritating to eyes, nose, and throat. Harmful if inhaled. Move victim to fresh air. *IF BREATHING HAS STOPPED*, give artificial respiration. IF breathing is difficult, administer oxygen. Lung edema may develop. **Liquid:** Will burn skin and eyes. Harmful if swallowed. Remove contaminated clothing and shoes. Flush affected areas with plenty of water. *IF IN EYES*, hold eyelids open and flush with plenty of water. *IF SWALLOWED* and victim is *CONSCIOUS AND ABLE TO SWALLOW*, have victim drink 4 to 8 ounces of water. **Do NOT induce vomiting.**
Note to physician or authorized medical personnel: Medical observation is recommended for 24 to 48 hours after breathing overexposure, as pulmonary edema may be delayed. As first aid for pulmonary edema, consider administering a corticosteroid spray. Cigarette smoking may exacerbate pulmonary injury and should be discouraged for at least 72 hours following exposure.

HEALTH HAZARDS
Personal protective equipment (PPE): A-Level PPE. Acid-vapor-type respiratory protection; rubber gloves; chemical worker's goggles; other protective equipment as necessary to protect skin and eyes.
Toxicity by ingestion: Grade 2; oral LD_{50} = 1000 mg/kg (rat).
Vapor (gas) irritant characteristics: Vapors cause severe irritation of eyes and throat and can cause eye and lung injury. They cannot be tolerated even at low concentrations.
Liquid or solid irritant characteristics: Severe eye and skin irritant. Causes second- and third-degree burns on short contact and is very injurious to the eyes.

FIRE DATA
Flash point: 7°F/–14°C (oc).
Flammable limits in air: LEL: 1.2%; UEL: 90.5%.
Fire extinguishing agents not to be used: Water, foam
Behavior in fire: Difficult to extinguish; re-ignition may occur. Containers may explode.
Autoignition temperature: 220°F/104°C/377°K.

CHEMICAL REACTIVITY
Reactivity with water: Reacts violently to form hydrogen chloride (hydrochloric acid) fumes.
Binary reactants: Reacts with strong oxidizers and acids; with surface moisture to form hydrochloric acid, which corrodes common metals and forms flammable hydrogen gas.
Neutralizing agents for acids and caustics: Flush with water, rinse with sodium bicarbonate or lime solution.

ENVIRONMENTAL DATA
Food chain concentration potential: Unlikely to accumulate.
Water pollution: Effect of low concentrations on aquatic life is unknown. May be dangerous if it enters nearby water intakes; notify operators. Notify local health and wildlife officials. **Response to discharge:** Issue warning–high flammability, corrosive, air contaminant. Restrict access. Evacuate area. Disperse and flush.

SHIPPING INFORMATION
Grades of purity: 99+%; **Storage temperature:** Ambient; **Inert atmosphere:** None; **Venting:** Pressure-vacuum; **Stability during transport:** Stable.

NAS HAZARD CLASSIFICATION FOR BULK WATER TRANSPORTATION
FIRE: 4
HEALTH: Vapor irritant: 4; Liquid or solid irritant: 4; Poisons: 3
WATER POLLUTION: Human toxicity: 3; Aquatic toxicity: 3; Aesthetic effect: 2
REACTIVITY: Other chemicals: 3; Water: 4; Self-reaction: 1

PHYSICAL AND CHEMICAL PROPERTIES
Physical state @ 59°F/15°C and 1 atm: Liquid.
Molecular weight: 135.5

Boiling point @ 1 atm: 90°F/32°C/305°K.
Melting/Freezing point: −197°F/−127°C/146°K.
Specific gravity (water = 1): 1.344 @ 68°F/20°C (liquid).
Liquid surface tension: (estimate) 18.3 dynes/cm = 0.0183 N/m @ 68°F/20°C.
Relative vapor density (air = 1): 4.9
Latent heat of vaporization: 85 Btu/lb = 47 cal/g = 2.0 x 10^5 J/kg.
Vapor pressure: 500 mm.

TRICHLORO-*s*-TRIAZINETRIONE   REC. T:2850

SYNONYMS: ACL 85; CBD 90; EEC No. 613-031-00-5; FICHLOR 91; FI CLOR 91; ISOCYANURIC CHLORIDE; SYMCLOSEN; SYMCLOSENE; TRICHLORINATED ISOCYANURIC ACID; TRICHLOROCYANURIC ACID; TRICHLOROISOCYANIC ACID; TRICHLOROISOCYANURIC ACID; N,N',N''-TRICHLOROISOCYANURIC ACID; 1,3,5-TRICHLOROISOCYANURIC ACID; 1,3,5-TRIAZINE-2,4,6(1H,3H,5H)-TRIONE, 1,3,5-TRICHLORO-; TRICHLORO-*s*-TRIAZINETRIONE; 1,3,5-TRICHLORO-*s*-TRIAZINE-2,4,6(1H,3H,5H)-TRIONE; 1,3,5-TRICHLORO-2,4,6-TRIOXOHEXAHYDRO-*s*-TRIAZINE; 1,3,5,TRICHLORO-1,2,5-TRIAZINE-2,4,6(1H,3H,5H)-TRIONE; TRICHLORO-*s*-TRIAZINE-2,4,6(1H,3H,5H)-TRIONE; TRICLORO-*s*-TRIAZINATRIONA (Spanish); TRICHLOROIMINOISOCYANURIC ACID

IDENTIFICATION
CAS Number: 87-90-1
Formula: $C_3Cl_3N_3O_3$; $Cl_3(NCO)_3$
DOT ID Number: UN 2468; DOT Guide Number: 140
Proper Shipping Name: Trichloroisocyanuric acid, dry

DESCRIPTION: White solid. Bleach- or chlorine-like odor. Sinks and mixes slowly with water, forming a bleach solution.

Thermally unstable; containers may explode in fire • Strong oxidizer which may react spontaneously with low flash point organics or reducing agents. Heat forms oxygen; will increase the activity of an existing fire • May cause fire or explosion on contact with combustibles (wood, paper, oil, clothing, etc.) • Severely Irritating to the skin, eyes, and respiratory tract • Decomposes at 435°F/224°C. Toxic products of combustion or decomposition include chlorine gas, nitrogen chloride and/or nitrogen trichloride.

Hazard Classification (based on NFPA-704 Rating System)
Health Hazards (Blue): 2; Flammability (Red): 0; Reactivity (Yellow): 2; Special Notice (White): OXY

EMERGENCY RESPONSE: See Appendix A (140)
Evacuation:
Public safety: Isolate the area of spill or leak for at least 10 to 25 meters (30 to 80 feet) in all directions.
Spill: Consider initial downwind evacuation for at least 100 meters (330 feet).
Fire: If any large container is involved in fire, isolate for at least 800 meters (½ mile) in all directions; also, consider initial evacuation for 800 meters (½ mile) in all directions.

EXPOSURE
Short-term effects: *SEEK MEDICAL ATTENTION.* **Dust:** Irritating to eyes, nose, and throat. *IF INHALED*, will, will cause coughing or difficult breathing. Move victim to fresh air. *IF IN EYES*, hold eyelids open and flush with plenty of water. IF breathing is difficult, administer oxygen. **Solid:** Irritating to skin and eyes. Harmful if swallowed. Remove contaminated clothing and shoes. Flush affected areas with plenty of water. *IF IN EYES*, hold eyelids open and flush with plenty of water. *IF SWALLOWED* and victim is *CONSCIOUS AND ABLE TO SWALLOW*, have victim drink 4 to 8 ounces of water and have victim induce vomiting. *IF SWALLOWED* and victim is *UNCONSCIOUS OR HAVING CONVULSIONS*, do nothing except keep victim warm.

HEALTH HAZARDS
Personal protective equipment (PPE): B-Level PPE. Dust mask or chlorine canister mask; goggles; rubber gloves.
Toxicity by ingestion: Grade 2; oral LD_{50} = 750 mg/kg (rat).
Long-term health effects: Organ damage including liver and kidney.
Liquid or solid irritant characteristics: Eye and skin irritant.

FIRE DATA
Behavior in fire: Containers may rupture and explode.

CHEMICAL REACTIVITY
Reactivity with water: Reacts to form a bleach solution. The reaction is not hazardous.
Binary reactants: Reacts with nitrogen compounds; forms explosive material. Contact with most foreign material, organic matter, or easily chlorinated or oxidized materials may result in fire. Avoid oil, grease, sawdust, floor
sweepings, other easily oxidized organic compounds.

ENVIRONMENTAL DATA
Food chain concentration potential: Negative; unlikely to accumulate.
Water pollution: Effect of low concentrations on aquatic life is unknown. May be dangerous if it enters nearby water intakes; notify operators. Notify local health and wildlife officials. **Response to discharge:** Issue warning–oxidizing material, water contaminant. Restrict access. Disperse and flush.

SHIPPING INFORMATION
Grades of purity: 39-90% available chlorine; **Storage temperature:** Ambient. Avoid elevated temperatures; **Inert atmosphere:** None if dry; **Venting:** Pressure-vacuum; **Stability during transport:** Stable.

PHYSICAL AND CHEMICAL PROPERTIES
Physical state @ 59°F/15°C and 1 atm: Solid.
Molecular weight: 232.5
Boiling point @ 1 atm: Decomposes.
Melting/Freezing point: Decomposes. 395°F/202°C/475°K.
Specific gravity (water = 1): (estimate) more than 1 @ 68°F/20°C (solid).

1,1,2-TRICHLORO-1,2,2-TRIFLUOROETHANE   REC. T:2900

SYNONYMS: ARCTON 63; ARKLONE P; CFC-113; DAIFLON S 3; FREON 113; FREON 113 TF; FREON TF; FRIGEN 113A; TTE; GENETRON 113; HALOCARBON 113; ISCEON 113; R 113; REFRIGERANT 113; 1,1,2-TRICLOROFLUOETANO (Spanish); UCON 113; UCON 113 HALOCARBON

IDENTIFICATION
CAS Number: 76-13-1

1084   Tricresyl phosphate

Formula: $C_2Cl_3F_3$; FCl_2CCF_2Cl
DOT ID Number: UN 3082; DOT Guide Number: 171
Proper Shipping Name: Environmentally hazardous substances, liquid, n.o.s.

DESCRIPTION: Colorless gas or volatile liquid. Practically odorless; high concentrations have sweet odor like chloroform. Sinks in water; soluble.

Contact with the liquid may cause frostbite. Decomposition products upon heating may include fluorine and phosgene.

Hazard Classification (based on NFPA-704 Rating System)
Health Hazards (Blue): 0; Flammability (Red): 0; Reactivity (Yellow): 0

EMERGENCY RESPONSE: See Appendix A (171)
Evacuation:
Public safety: Isolate the area of spill or leak for at least 10 to 25 meters (30 to 80 feet) in all directions.
Spill: Increase, in the downwind direction, as necessary, the distance shown under "Public Safety."
Fire: If any large container is involved in fire, isolate for at least 800 meters (½ mile) in all directions; also, consider initial evacuation for 800 meters (½ mile) in all directions.

EXPOSURE
Short-term effects: *SEEK MEDICAL ATTENTION*. **Vapor:** If inhaled, will cause dizziness or difficult breathing. Move to fresh air. *IF BREATHING HAS STOPPED*, give artificial respiration. IF breathing is difficult, administer oxygen. **Liquid:** Not harmful. Flush affected areas with plenty of water. *IF IN EYES*, hold eyelids open and flush with plenty of water. *IF SWALLOWED*, **Do NOT induce vomiting.**

HEALTH HAZARDS
Personal protective equipment (PPE): B-Level PPE. OSHA Table Z-1-A air contaminant. Butyl gloves, splash goggles. If spill is large, use self-contained breathing apparatus. Wear thermal protective clothing. Chemical protective material(s) reported to have good to excellent resistance: nitrile, Teflon®, Viton®, Silvershield®. Also, PV alcohol, neoprene offers limited protection
Recommendations for respirator selection: NIOSH
2000 ppm: SA (any supplied-air respirator); or SCBAF (any self-contained breathing apparatus with a full facepiece). *EMERGENCY OR PLANNED ENTRY INTO UNKNOWN CONCENTRATIONS OR IDLH CONDITIONS*: SCBAF:PD,PP (any self-contained breathing apparatus that has a full facepiece and is operated in a pressure-demand or other positive-pressure mode); or SAF:PD,PP:ASCBA (any supplied-air respirator that has a full facepiece and is operated in a pressure-demand or other positive-pressure mode in combination with an auxiliary self-contained breathing apparatus operated in a pressure-demand or other positive pressure mode). *ESCAPE:* GMFOV [any air-purifying, full-facepiece respirator (gas mask) with a chin-style, front- or back-mounted organic vapor canister]; or SCBAE (any appropriate escape-type, self-contained breathing apparatus).
Exposure limits (TWA unless otherwise noted): ACGIH TLV 1000 ppm (7670 mg/m^3); NIOSH/OSHA 1000 ppm (7600 mg/m^3).
Short-term exposure limits (15-minute TWA): ACGIH STEL 1250 ppm (9590 mg/m^3); NIOSH STEL 1250 ppm (9500 mg/m^3).
Toxicity by ingestion: Grade 1: LD_{50} = 43 g/kg (rat).
Liquid or solid irritant characteristics: May cause frostbite.
IDLH value: 2000 ppm.

FIRE DATA
Behavior in fire: Containers may rupture.
Autoignition temperature: 1256°F/680°C/953°K.

CHEMICAL REACTIVITY
Binary reactants: May react violently with some metals: calcium, powder aluminum, zinc, magnesium, beryllium, contact alloys containing more than 2% magnesium.
Reactivity group: 36
Compatibility class: Halogenated hydrocarbons.

ENVIRONMENTAL DATA
Food chain concentration potential: Log $_{Kow}$ = 1.7. Unlikely to accumulate.
Water pollution: May be dangerous if it enters nearby water intakes; notify operators. Notify local health and wildlife officials.
Response to discharge: Should be removed. Chemical and physical treatment.

SHIPPING INFORMATION
Grades of purity: 100%; **Storage temperature:** Ambient; **Venting:** Open; **Stability during transport:** Stable.

PHYSICAL AND CHEMICAL PROPERTIES
Physical state @ 59°F/15°C and 1 atm: Liquid.
Molecular weight: 187.38
Boiling point @ 1 atm: 118°F/47.7°C/320.85°K.
Melting/Freezing point: –33.52°F/–36.4°C/236.8°K.
Specific gravity (water = 1): 1.57
Liquid surface tension: 19 dyne/cm = 0.019 N/m
Relative vapor density (air = 1): 6.4
Latent heat of vaporization: 63.0 Btu/lb = 35.07 cal/g = 1.5 x 10^5 J/kg.
Vapor pressure: (Reid) 10.747 psia; 285 mm.

TRICRESYL PHOSPHATE   REC. T:2950

SYNONYMS: CELLUFLEX 179C; *o*-CRESYL PHOSPHATE; DISFLAMOLL TKP; DURAD; FOSFATO de TRICRESILO (Spanish); IMOL-140; LINDOL; PHOSPHORIC ACID, TRIS(METHYL PHENYL)ESTER; PHOSPHORIC ACID, TRITOLYL ESTER; TCP; TOCP; TOLYPHOSPHATE; TRI-*o*-CRESYL PHOSPHATE; TRI-*o*-TOLYL PHOSPHATE; TRITOLY PHOSPHATE *o*-TOLY PHOSPHATE

IDENTIFICATION
CAS Number: 1330-78-5
Formula: $C_{21}H_{21}O_4P$; $(p$-$CH_3C_6H_4O)_3PO$
DOT ID Number: UN 2574; DOT Guide Number: 151
Proper Shipping Name: Tricresyl phosphate

DESCRIPTION: Colorless liquid. Odorless. Sinks in water; soluble.

Combustible • Heat or flame may cause explosion • Containers may BLEVE when exposed to fire • Vapors in confined areas (e.g., tanks, sewers, buildings) may explode when exposed to fire • Irritating to the skin, eyes, and respiratory tract • Toxic products of combustion may include phosphorus oxides.

Hazard Classification (based on NFPA-704 Rating System)
Health Hazards (Blue): 2; Flammability (Red): 1; Reactivity (Yellow): 0

EMERGENCY RESPONSE: See Appendix A (151)
Evacuation:
Public safety: Isolate the area of spill or leak for at least 25 to 50 meters (80 to 160 feet) in all directions.
Spill: Increase, in the downwind direction, as necessary, the distance shown under "Public Safety."
Fire: If tank, rail car, or tank truck is involved in fire, isolate for at least 800 meters (½ mile) in all directions; also, consider initial evacuation for 800 meters (½ mile) in all directions.

EXPOSURE
Short-term effects: Liquid: Harmful if swallowed. *IF BREATHING HAS STOPPED,* give artificial respiration; *avoid mouth-to-mouth resuscitation; use bag/mask apparatus. IF SWALLOWED* and victim is *CONSCIOUS AND ABLE TO SWALLOW,* have victim drink 4 to 8 ounces of water.

HEALTH HAZARDS
Personal protective equipment (PPE): B-Level PPE. Goggles or face shield. Chemical protective material(s) reported to have good to excellent resistance: butyl rubber, neoprene, nitrile, nitrile+PVC, polyethylene, PV alcohol, PVC, Viton®. Also, styrene-butadiene, butyl rubber/neoprene, chlorinated polyethylene, Viton®/neoprene offers limited protection
Toxicity by ingestion: Grade 2; LD_{50} = 0.5 to 5 g/kg (chicken LD_{50} more than 2 g/kg).
Vapor (gas) irritant characteristics: Vapors cause a smarting of the eyes or respiratory system if present in high concentrations. The effect is temporary. The compound is nonvolatile for all practical purposes.
Liquid or solid irritant characteristics: Eye and skin irritant.

FIRE DATA
Flash point: 437°F/225°C (cc).
Fire extinguishing agents not to be used: Water or foam may cause frothing.
Autoignition temperature: 725°F/385°C/658°K.
Electrical hazard: Class I, Group D.

CHEMICAL REACTIVITY
Binary reactants: Reacts with oxidizing materials when heated to decomposition.

ENVIRONMENTAL DATA
Food chain concentration potential: Log K_{ow} = 5.11. Values above 3.0 are very likely to accumulate in living tissues and especially in fats.
Water pollution: DOT Appendix B, §172.101–marine pollutant. Low concentrations may be dangerous to aquatic life. May be dangerous if it enters nearby water intakes; notify operators. Notify local health and wildlife officials. **Response to discharge:** Should be removed. Chemical and physical treatment.

SHIPPING INFORMATION
Grades of purity: Consists primarily of the para isomer, but several commercial grades may contain a significant proportion of tri-orthocresyl phosphate. Latter is considerably more toxic than the para-isomer if ingested; **Storage temperature:** Ambient; **Inert atmosphere:** None; **Venting:** Open (flame arrester); **Stability during transport:** Stable.

PHYSICAL AND CHEMICAL PROPERTIES
Physical state @ 59°F/15°C and 1 atm: Liquid.
Molecular weight: 368
Boiling point @ 1 atm: 770°F/410°C/683°K.
Melting/Freezing point: –27°F/–33°C/240°K.
Specific gravity (water = 1): 1.16 @ 68°F/20°C (liquid).
Liquid surface tension: 44 dynes/cm = 0.044 N/m @ 77°F/25°C.
Latent heat of vaporization: (estimate) 80.0 Btu/lb = 44.5 cal/g = 1.86×10^5 J/kg.

TRIDECANE   **REC. T:3000**

SYNONYMS: *n*-TRIDECANE; *n*-TRIDECANO (Spanish)

IDENTIFICATION
CAS Number: 629-50-5
Formula: $C_{13}H_{28}$; $CH_3(CH_2)_{11}CH_3$
DOT ID Number: NA 1993; DOT Guide Number: 128
Proper Shipping Name: Combustible liquid, n.o.s.

DESCRIPTION: Colorless liquid. Floats on water surface; insoluble.

Combustible • Containers may BLEVE when exposed to fire • Vapors may form explosive mixture with air • Vapors are heavier than air and will collect and stay in low areas • Vapors in confined areas (e.g., tanks, sewers, buildings) may explode when exposed to fire • Irritating to the skin, eyes, and respiratory tract • Toxic products of combustion may include carbon monoxide.

Hazard Classification (based on NFPA-704 Rating System)
Health Hazards (Blue): 0; Flammability (Red): 2; Reactivity (Yellow): 0

EMERGENCY RESPONSE: See Appendix A (128)
Evacuation:
Public safety: Isolate spill area for at least 25 to 50 meters (80 to 160 feet) in all directions.
Spill: Large spill–Consider initial downwind evacuation for at least 300 meters (1000 feet).
Fire: Isolate for 800 meters (½ mile) in all directions, especially if tank, rail car, or tank truck is involved in fire.

EXPOSURE
Short-term effects: *SEEK MEDICAL ATTENTION.* **Liquid:** Irritating to skin and eyes. Remove contaminated clothing and shoes. Flush affected areas with plenty of water. *IF IN EYES*, hold eyelids open and flush with plenty of water. May cause central nervous system depression.

HEALTH HAZARDS
Personal protective equipment (PPE): B-Level PPE. Self-contained breathing apparatus, rubber boots, and heavy rubber gloves.
Vapor (gas) irritant characteristics: Vapors cause a slight smarting of the eyes or respiratory system if present in high concentrations. The effect is temporary.
Liquid or solid irritant characteristics: Minimum hazard. If spilled on clothing and allowed to remain, may cause smarting and reddening of skin.

FIRE DATA
Flash point: 175°F/79°C (cc).

CHEMICAL REACTIVITY
Binary reactants: Violent reaction with strong oxidizers. Incompatible with strong acids, nitrates.

Reactivity group: 31
Compatibility class: Paraffins

ENVIRONMENTAL DATA
Water pollution: Effect of low concentrations on aquatic life is unknown. Fouling to shorelines. May be dangerous if it enters nearby water intakes; notify operators. Notify local health and wildlife officials. **Response to discharge:** Restrict access. Should be removed. Mechanical containment. Chemical and physical treatment.

SHIPPING INFORMATION
Grades of purity: 99%; **Storage temperature:** Ambient; **Stability during transport:** Stable.

PHYSICAL AND CHEMICAL PROPERTIES
Physical state @ 59°F/15°C and 1 atm: Liquid.
Molecular weight: 184.37
Boiling point @ 1 atm: 453.2°F/234°C/507.2°K.
Melting/Freezing point: 23–25°F/–5 to –4°C/268–269°K.
Specific gravity (water = 1): 0.756
Relative vapor density (air = 1): 6.4
Vapor pressure: (Reid) Less than 0.01 psia.

TRIDECANOL REC. T:3050

SYNONYMS: ALCOHOL C-13; ISOTRIDECANOL; ISOTRIDECYL ALCOHOL; TRIDECYLALCOHOL; OXOTRIDECYL ALCOHOL; 1-TRIDECANOL; TRIDECYL ALCOHOL

IDENTIFICATION
CAS Number: 112-70-9
Formula: $C_{13}H_{28}O$; $C_{12}H_{25}CH_2OH$

DESCRIPTION: Colorless oily liquid. Mild, pleasant, alcohol odor. Floats on the surface of water.

Combustible • Heat or flame may cause explosion • Containers may BLEVE when exposed to fire • Vapors are heavier than air and will collect and stay in low areas • Vapors in confined areas (e.g., tanks, sewers, buildings) may explode when exposed to fire • Irritating to the skin, eyes, and respiratory tract • Toxic products of combustion may include carbon monoxide.

Hazard Classification (based on NFPA-704 Rating System)
Health Hazards (Blue): 0; Flammability (Red): 1; Reactivity (Yellow): 0

EMERGENCY RESPONSE: See Appendix A (171)
Evacuation:
Public safety: Isolate the area of spill or leak for at least 10 to 25 meters (30 to 80 feet) in all directions.
Spill: Increase, in the downwind direction, as necessary, the distance shown under "Public Safety."
Fire: If any large container is involved in fire, isolate for at least 800 meters (½ mile) in all directions; also, consider initial evacuation for 800 meters (½ mile) in all directions.

HEALTH HAZARDS
Personal protective equipment (PPE): B-Level PPE. Synthetic rubber gloves; chemical goggles. Natural rubber, neoprene, and nitrile rubber may offer some protection from alcohols.
Vapor (gas) irritant characteristics: May cause eye and respiratory tract irritation.
Liquid or solid irritant characteristics: May cause eye and skin irritation.

FIRE DATA
Flash point: 250°F/121°C (oc).
Fire extinguishing agents not to be used: Water or foam may cause frothing.
Electrical hazard: Class I, Group D.

CHEMICAL REACTIVITY
Reactivity with water: Foaming action.
Binary reactants: Reacts with strong oxidizers. Attacks aluminum.
Reactivity group: 20
Compatibility class: Alcohols, glycols

ENVIRONMENTAL DATA
Food chain concentration potential: Negative; unlikely to accumulate.
Water pollution: Effect of low concentrations on aquatic life is unknown. Fouling to shoreline. May be dangerous if it enters nearby water intakes; notify operators. Notify local health and wildlife officials. **Response to discharge:** Mechanical containment. Should be removed. Chemical and physical treatment.

SHIPPING INFORMATION
Grades of purity: Mixed isomers; 99+%; **Storage temperature:** Ambient; **Inert atmosphere:** None; **Venting:** Open (flame arrester); **Stability during transport:** Stable.

NAS HAZARD CLASSIFICATION FOR BULK WATER TRANSPORTATION
FIRE: 1
HEALTH: Vapor irritant: 0; Liquid or solid irritant: 0; Poisons: 0
WATER POLLUTION: Human toxicity: 0; Aquatic toxicity: 0; Aesthetic effect: 3
REACTIVITY: Other chemicals: 2; Water: 0; Self-reaction: 0

PHYSICAL AND CHEMICAL PROPERTIES
Physical state @ 59°F/15°C and 1 atm: Liquid.
Molecular weight: 200.37
Boiling point @ 1 atm: 525°F/274°C/547°K.
Melting/Freezing point: 91°F/33°C/306°K.
Specific gravity (water = 1): 0.846 @ 68°F/20°C (liquid).
Liquid surface tension: (estimate) 30 dynes/cm = 0.03 N/m @ 68°F/20°C.
Liquid water interfacial tension: (estimate) 30 dynes/cm = 0.03 N/m @ 68°F/20°C.
Relative vapor density (air = 1): 6.9
Ratio of specific heats of vapor (gas): 1.027
Latent heat of vaporization: 120 Btu/lb = 64 cal/g = 2.7×10^5 J/kg.
Heat of combustion: (estimate) –12,200 Btu/lb = –6790 cal/g = -284×10^5 J/kg.
Vapor pressure: (Reid) Low; 80 mm.

1-TRIDECENE REC. T:3100

SYNONYMS: OLEFIN C13; UNDECYLETHYLENE

IDENTIFICATION
CAS Number: 112-70-9
Formula: $C_{13}H_{28}O$; $CH_3(CH_2)_{10}CH=CH_2$

DOT ID Number: NA 1993; DOT Guide Number: 128
Proper Shipping Name: Combustible liquid, n.o.s.

DESCRIPTION: Colorless watery liquid. Mild, pleasant odor. Floats on the surface of water.

Combustible • Containers may BLEVE when exposed to fire • Vapors may form explosive mixture with air • Vapors are heavier than air and will collect and stay in low areas • Vapors in confined areas (e.g., tanks, sewers, buildings) may explode when exposed to fire • Irritating to the skin, eyes, and respiratory tract • Toxic products of combustion may include carbon monoxide.

Hazard Classification (based on NFPA-704 Rating System)
Health Hazards (Blue): 0; Flammability (Red): 2; Reactivity (Yellow): 0

EMERGENCY RESPONSE: See Appendix A (128)
Evacuation:
Public safety: Isolate spill area for at least 25 to 50 meters (80 to 160 feet) in all directions.
Spill: Large spill–Consider initial downwind evacuation for at least 300 meters (1000 feet).
Fire: Isolate for 800 meters (½ mile) in all directions, especially if tank, rail car, or tank truck is involved in fire.

EXPOSURE
Short-term effects: Liquid: Irritating to eyes. *IF IN EYES*, hold eyelids open and flush with plenty of water.

HEALTH HAZARDS
Personal protective equipment (PPE): B-Level PPE. Goggles or face shield.
Vapor (gas) irritant characteristics: Nonvolatile
Liquid or solid irritant characteristics: May cause eye irritation.

FIRE DATA
Flash point: 175°F/79°C (approximate).
Fire extinguishing agents not to be used: Water may be ineffective.
Electrical hazard: Class I, Group D.

CHEMICAL REACTIVITY
Reactivity group: 30
Compatibility class: Olefins

ENVIRONMENTAL DATA
Food chain concentration potential: Negative; unlikely to accumulate.
Water pollution: Effect of low concentrations on aquatic life is unknown. Fouling to shoreline. May be dangerous if it enters nearby water intakes; notify operators. Notify local health and wildlife officials. **Response to discharge:** Mechanical containment. Should be removed. Chemical and physical treatment.

SHIPPING INFORMATION
Grades of purity: Technical: 95%; **Storage temperature:** Ambient; **Inert atmosphere:** None; **Venting:** Open (flame arrester); **Stability during transport:** Stable.

PHYSICAL AND CHEMICAL PROPERTIES
Physical state @ 59°F/15°C and 1 atm: Liquid.
Molecular weight: 182.35
Boiling point @ 1 atm: 451°F/233°C/506°K.
Melting/Freezing point: –11°F/–24°C/249°K.
Specific gravity (water = 1): 0.765 @ 68°F/20°C (liquid).
Liquid surface tension: 24.5 dynes/cm = 0.0245 N/m @ 68°F/20°C.
Ratio of specific heats of vapor (gas): 1.029
Latent heat of vaporization: 110 Btu/lb = 59 cal/g = 2.5 x 10^5 J/kg.
Heat of combustion: –19,048 Btu/lb = –10,582 cal/g = –443.05 x 10^5 J/kg.
Vapor pressure: (Reid) Low.

TRIDECYLBENZENE REC. T:3150

SYNONYMS: 1-PHENYLTRIDECANE; TRIDECILBENCENO (Spanish)

IDENTIFICATION
CAS Number: 123-02-4
Formula: $C_6NO_5(CH_2)_{12}CH_3$

DESCRIPTION: Colorless liquid.

Combustible • Containers may BLEVE when exposed to fire • Vapors may form explosive mixture with air • Vapors in confined areas (e.g., tanks, sewers, buildings) may explode when exposed to fire • Irritating to the skin, eyes, and respiratory tract • Toxic products of combustion may include carbon monoxide.

Hazard Classification (based on NFPA-704 Rating System)
Health Hazards (Blue): 1; Flammability (Red): 1; Reactivity (Yellow): 0

EMERGENCY RESPONSE: See Appendix A (171)
Evacuation:
Public safety: Isolate the area of spill or leak for at least 10 to 25 meters (30 to 80 feet) in all directions.
Spill: Increase, in the downwind direction, as necessary, the distance shown under "Public Safety."
Fire: If any large container is involved in fire, isolate for at least 800 meters (½ mile) in all directions; also, consider initial evacuation for 800 meters (½ mile) in all directions.

EXPOSURE
Short-term effects: *SEEK MEDICAL ATTENTION*. Liquid: Irritating to skin and eyes. Harmful if swallowed. Remove contaminated clothing and shoes. Flush affected areas with plenty of water. *IF IN EYES*, hold eyelids open and flush with plenty of water.

HEALTH HAZARDS
Personal protective equipment (PPE): B-Level PPE. Self-contained breathing apparatus, rubber boots, and heavy rubber gloves.
Vapor (gas) irritant characteristics: Vapors cause a slight smarting of the eyes or respiratory system if present in high concentrations. The effect is temporary.
Liquid or solid irritant characteristics: If spilled on clothing and allowed to remain, may cause smarting and reddening of skin.

FIRE DATA
Flash point: More than 230°F/110°C (cc).

CHEMICAL REACTIVITY
Binary reactants: Incompatible with strong oxidizers, nitric acid.

Reactivity group: 32
Compatibility class: Aromatic hydrocarbons

ENVIRONMENTAL DATA
Water pollution: Effect of low concentrations on aquatic life is unknown. Fouling to shoreline. May be dangerous if it enters nearby water intakes; notify operators. Notify local health and wildlife officials. **Response to discharge:** Mechanical containment. Should be removed. Chemical and physical treatment.

SHIPPING INFORMATION
Grades of purity: 99%; **Storage temperature:** Ambient; **Stability during transport:** Stable.

PHYSICAL AND CHEMICAL PROPERTIES
Physical state @ 59°F/15°C and 1 atm: Liquid.
Molecular weight: 260.47
Boiling point @ 1 atm: 654.8°F/346°C/619.2°K.
Melting/Freezing point: 50°F/10°C/283.2°K.
Specific gravity (water = 1): 0.881
Relative vapor density (air = 1): 8.98

TRIETHANOLAMINE   REC. T:3200

SYNONYMS: DALTOGEN; 2,2',2"-NITRILO-TRIETHANOL; TEA; THIOFACO T-35; TRIETHYLOLAMINE; TROLAMINE; T R I E T A N O L A M I N A   (S p a n i s h); T R I (H Y D R O X Y T R I E T H Y L) A M I N E;   T R I (2 - HYDROXYETHYL)AMINE; tris(HYDROXYETHYL)AMINE; TROLAMINE

IDENTIFICATION
CAS Number: 102-71-6
Formula: $C_6H_{15}NO_3$; $(HOCH_2CH_2)_3N$
DOT ID Number: UN 3267; DOT Guide Number: 153
Proper Shipping Name: Corrosive, liquid, basic, organic, n.o.s.

DESCRIPTION: Colorless thick, oily liquid. Turns yellow and then brown on contact with air. Mild ammonia odor. Sinks and mixes with water; forms a medium-strong base.

Combustible • Corrosive • Heat or flame may cause explosion • Containers may BLEVE when exposed to fire • Vapors are heavier than air and will collect and stay in low areas • Vapors in confined areas (e.g., tanks, sewers, buildings) may explode when exposed to fire • Extremely Irritating to the skin, eyes, and respiratory tract; effects of inhalation exposure may be delayed • Begins to decompose at 450°F/232°C. Toxic products of combustion may include nitrogen oxides and cyanide.

Hazard Classification (based on NFPA-704 Rating System)
Health Hazards (Blue): 2; Flammability (Red): 1; Reactivity (Yellow): 1

EMERGENCY RESPONSE: See Appendix A (153)
Evacuation:
Public safety: Isolate the area of spill or leak for at least 25 to 50 meters (80 to 160 feet) in all directions.
Spill: Increase, in the downwind direction, as necessary, the distance shown under "Public Safety."
Fire: If tank, rail car, or tank truck is involved in fire, isolate for at least 800 meters (½ mile) in all directions; also, consider initial evacuation for 800 meters (½ mile) in all directions.

EXPOSURE
Short-term effects: *SEEK MEDICAL ATTENTION*. If artificial respiration is administered, *avoid mouth-to-mouth resuscitation; use bag/mask apparatus*. **Liquid:** Irritating to skin and eyes. Harmful if swallowed. Remove contaminated clothing and shoes. Flush affected areas with plenty of water. *IF IN EYES*, hold eyelids open and flush with plenty of water. *IF SWALLOWED* and victim is *CONSCIOUS AND ABLE TO SWALLOW*, have victim drink 4 to 8 ounces of water.

Note to physician or authorized medical personnel: Consider the use of amyl nitrite perles if symptoms of cyanide poisoning develop. If symptoms indicate, initial treatment includes the cyanide antidote kit. In all cases, break an amyl nitrite perle in a gauze pad and hold lightly under victim's nose for 15 seconds, repeating 5 times at about 15-second intervals; if necessary (and if sodium nitrite infusions will be delayed), repeat procedure every 3 minutes with fresh pearls until 3 or 4 have been used. Avoid breathing the vapor while administering it to the victim. Administer sodium nitrite IV, ASAP. The usual adult dose is 10 to 20 mL of a 3% solution infused over no less than 5 minutes; the average child dose is 0.15 to 0.20 mL/kg. Monitor blood pressure during administration, and slow the rate of infusion if hypotention develops. Next, infuse sodium thiosulfate IV. The usual adult dose is 50 mL of a 25% solution infused over 10 to 20 minutes; the average child dose is 1.65 mL/kg. Repeat with nitrite and thiosulfate as required.

Note to physician or authorized medical personnel: Medical observation is recommended for 24 to 48 hours after breathing overexposure, as pulmonary edema may be delayed. As first aid for pulmonary edema, consider administering a corticosteroid spray. Cigarette smoking may exacerbate pulmonary injury and should be discouraged for at least 72 hours following exposure.

HEALTH HAZARDS
Personal protective equipment (PPE): B-Level PPE. Goggles or face shield; rubber gloves and boots. Chemical protective material(s) reported to have good to excellent resistance: butyl rubber, natural rubber, neoprene, nitrile, PV alcohol, PVC. Also, nitrile+PVC offers limited protection.
Exposure limits (TWA unless otherwise noted): ACGIH TLV 5 mg/m^3.
Toxicity by ingestion: Grade 2; LD_{50} = 0.5 to 5 g/kg (guinea pig).
Long-term health effects: Repeated or prolonged dermal exposure causes eczema.
Vapor (gas) irritant characteristics: Nonvolatile
Liquid or solid irritant characteristics: Eye, skin and respiratory tract irritant. If spilled on clothing and allowed to remain, may cause smarting and reddening of the skin.

FIRE DATA
Flash point: 375°F/191°C (oc); 355°F/180°C (cc).
Fire extinguishing agents not to be used: Water or foam may cause frothing.
Autoignition temperature: 644°F/340°C/613°K.
Electrical hazard: Class I, Group C.

CHEMICAL REACTIVITY
Reactivity with water: Solution is caustic.
Binary reactants: Reacts with strong oxidizers and acids. Exothermic decomposition with maleic anhydride. Corrodes aluminum, copper, copper alloys, tin, zinc.
Neutralizing agents for acids and caustics: Dilute with water.
Reactivity group: 8
Compatibility class: Alkanolamines

ENVIRONMENTAL DATA
Food chain concentration potential: Log K_{ow} = –1.49. Negative. Unlikely to accumulate.
Water pollution: Effect of low concentrations on aquatic life is unknown. May be dangerous if it enters nearby water intakes; notify operators. Notify local health and wildlife officials.
Response to discharge: Disperse and flush.

SHIPPING INFORMATION
Grades of purity: 85–99%; **Storage temperature:** Ambient; **Inert atmosphere:** None; **Venting:** Open; **Stability during transport:** Stable.

NAS HAZARD CLASSIFICATION FOR BULK WATER TRANSPORTATION
FIRE: 1
HEALTH: Vapor irritant: 0; Liquid or solid irritant: 1; Poisons: 1
WATER POLLUTION: Human toxicity: 1; Aquatic toxicity: 1; Aesthetic effect: 2
REACTIVITY: Other chemicals: 3; Water: 0; Self-reaction: 0

PHYSICAL AND CHEMICAL PROPERTIES
Physical state @ 59°F/15°C and 1 atm: Liquid.
Molecular weight: 149.19
Boiling point @ 1 atm: 650°F/343°C/616°K.
Melting/Freezing point: 70.9°F/21.6°C/294.8°K The commercial product may contain up to 25% diethanolamine and 5% ethanolamine. These chemicals are added partly to lower the high freezing point. The resulting mixture may have properties that vary from those shown.
Specific gravity (water = 1): 1.13 @ 68°F/20°C (liquid).
Relative vapor density (air = 1): 5.09
Ratio of specific heats of vapor (gas): 1.036
Latent heat of vaporization: 176 Btu/lb = 97.8 cal/g = 4.10 x 10^5 J/kg.
Heat of combustion: –11,050 Btu/lb = –6140 cal/g = –257 x 10^5 J/kg.
Heat of solution: (estimate) –20 Btu/lb = –12 cal/g = –0.5 x 10^5 J/kg.
Vapor pressure: (Reid) Low; 0.019 mm.

TRIETHYL ALUMINUM   REC. T:3250

SYNONYMS: ALUMINUM, TRIETHYL; ATE; EEC No. 013-004-00-2; TEA; TRIETILALUMINIO (Spanish)

IDENTIFICATION
CAS Number: 97-93-8
Formula: $C_6H_{15}Al$; $(C_2NO_5)_3Al$
DOT ID Number: UN 3051; DOT Guide Number: 135
Proper Shipping Name: Aluminum alkyls

DESCRIPTION: Clear, colorless, liquid. May be available as a solution (20% or less) in hexane, benzene, or heptane. Explodes in cold water; flammable hydrocarbon gas is produced.

Extremely flammable • Pure material is spontaneously combustible when exposed to air; solution in hydrocarbon solvent will explode if solvent evaporates • Corrosive to skin, eyes, and respiratory tract; skin and eye contact causes severe burns and blindness; inhalation symptoms may be delayed • Do not use water, foam, or Halon extinguishing agents • Firefighting gear (including SCBA) does not provide adequate protection. If exposure occurs, remove and isolate gear immediately and thoroughly decontaminate personnel • Containers may BLEVE when exposed to fire • Vapors may form explosive mixture with air • Vapors are heavier than air and will collect and stay in low areas • Vapors may travel long distances to ignition sources and flashback • Vapors in confined areas (e.g., tanks, sewers, buildings) may explode when exposed to fire • Toxic products of combustion may include carbon monoxide. Heavy smoke may cause metal fume fever • Violent reaction with water producing flammable hydrocarbons • May re-ignite after fire is extinguished • Do not put yourself in danger by entering a contaminated area to rescue a victim.

Hazard Classification (based on NFPA-704 Rating System) (Up to 20% by wt. in hydrocarbon solution)
Health Hazards (Blue): 3; Flammability (Red): 4; Reactivity (Yellow): 3; Special Notice (White): Water reactive

EMERGENCY RESPONSE: See Appendix A (135)
Evacuation:
Public safety: Isolate spill area for at least 100 to 150 meters (330 to 490 feet) in all directions.
Spill: Increase downwind, the distance shown above in "Public safety."
Fire: Isolate for 800 meters (½ mile) in all directions, especially if tank, rail car, or tank truck is involved in fire.

EXPOSURE
Short-term effects: *SEEK MEDICAL ATTENTION*. **Vapor/fumes:** Lung edema may develop. **Liquid:** Will burn skin and eyes. Harmful if swallowed. Remove contaminated clothing and shoes. Flush affected areas with plenty of water. *IF IN EYES*, hold eyelids open and flush with plenty of water. *IF SWALLOWED* and victim is *CONSCIOUS AND ABLE TO SWALLOW*, have victim drink 4 to 8 ounces of water. **Do NOT induce vomiting**.
Note to physician or authorized medical personnel: Medical observation is recommended for 24 to 48 hours after breathing overexposure, as pulmonary edema may be delayed. As first aid for pulmonary edema, consider administering a corticosteroid spray. Cigarette smoking may exacerbate pulmonary injury and should be discouraged for at least 72 hours following exposure.

HEALTH HAZARDS
Personal protective equipment (PPE): B-Level PPE. Full protective clothing, preferably of aluminized glass cloth; goggles; face shield; gloves. In case of fire, all-purpose canister); or self-contained breathing apparatus).
Exposure limits (TWA unless otherwise noted): ACGIH TLV 2 mg/m³; NIOSH REL 2 mg/m³ as aluminum
Liquid or solid irritant characteristics: Severe eye, skin and respiratory tract irritant. Causes second-and third-degree burns on short contact and is very injurious to the eyes. Remove contaminated clothing immediately.

FIRE DATA
Flash point: Ignites spontaneously in air at all temperatures.
Flammable limits in air: (solution in hexane) LEL: 1.1%; UEL: 7.5%.
Fire extinguishing agents not to be used: Water, foam, halogenated extinguishing agents. Contact with water applied to adjacent fires causes violent reaction producing toxic and flammable gases. Often reignites after fire has been extinguished.
Autoignition temperature: –51°F/–46°C/227°K (self-ignites at ambient temperature). 460°F/238°C/511°K (solution). Properties may depend on solvent.
Electrical hazard: Due to low electric conductivity, this substance may generate electrostatic charges as a result of agitation and flow.

CHEMICAL REACTIVITY
Reactivity with water: Reacts violently to form flammable ethane gas.
Binary reactants: Reacts violently with air, alcohols, amines, sulfur oxide, carbon tetrachloride, halogenated hydrocarbons, and other materials.

ENVIRONMENTAL DATA
Food chain concentration potential: Negative; unlikely to accumulate.
Water pollution: Effect of low concentrations on aquatic life is unknown. May be dangerous if it enters nearby water intakes; notify operators. Notify local health and wildlife officials.
Response to discharge: Issue warning–high flammability. Restrict access. Evacuate area. Disperse and flush with care; beware of spatter.

SHIPPING INFORMATION
Grades of purity: 92+%; 20% or less by weight in benzene, hexane, or heptane. Solutions are not pyrophoric; **Storage temperature:** Ambient; **Inert atmosphere:** Inerted; dry nitrogen at 5 psi; **Venting:** Safety relief, with rupture disc; **Stability during transport:** Stable.

NAS HAZARD CLASSIFICATION FOR BULK WATER TRANSPORTATION
FIRE: 4
HEALTH: Vapor irritant:; Liquid or solid irritant: 4; Poisons: 3
WATER POLLUTION: Human toxicity: 0; Aquatic toxicity: 2; Aesthetic effect: 3
REACTIVITY: Other chemicals: 4; Water: 4; Self-reaction: 0

PHYSICAL AND CHEMICAL PROPERTIES
Physical state @ 59°F/15°C and 1 atm: Liquid.
Molecular weight: 114.2
Boiling point @ 1 atm: 367.9°F/186.6°C/459.8°K.
Melting/Freezing point: –51°F/–46°C/227°K.
Critical temperature: 761°F/405°C/678°K.
Critical pressure: 1970 psia = 134 atm = 13.6 MN/m^2.
Specific gravity (water = 1): 0.836 @ 68°F/20°C (liquid).
Liquid surface tension: 26.1 dynes/cm = 0.0261 N/m at 28°C.
Relative vapor density (air = 1): 3.9
Latent heat of vaporization: 216 Btu/lb = 120 cal/g = 5.02 x 10^5 J/kg.
Heat of combustion: –18,364 Btu/lb = –10,202 cal/g = –426.85 x 10^5 J/kg.
Heat of solution: –1,995 Btu/lb = –1,109 cal/g = –46.40 x 10^5 J/kg.
Vapor pressure: 0.7 mm.

TRIETHYLAMINE REC. T:3300

SYNONYMS: (DIETHYLAMINO)ETHANE; *N,N*-DIETHYLETHANEAMINE; ETHANAMINE,*N,N*-DIETHYL-; EEC No. 612-004-00-5; TEA; TEN; TRIAETHYLAMIN (German); TRIETILAMINA (Spanish)

IDENTIFICATION
CAS Number: 121-44-8
Formula: $C_6H_{15}N$; $(C_2NO_5)_3N$
DOT ID Number: UN 1296; DOT Guide Number: 132
Proper Shipping Name: Triethylamine
Reportable Quantity (RQ): **(CERCLA)** 5000 lb/2270 kg

DESCRIPTION: Colorless watery liquid. Fishy odor at low concentrations; ammonia-like odor at high concentrations. Floats on water surface; moderately soluble. Produces a large amount of vapor.

Highly flammable • Corrosive to the skin, eyes, and respiratory tract; skin or eye contact causes burns, impaired vision, and blindness; inhalation symptoms may be delayed • Containers may BLEVE when exposed to fire • Vapors may form explosive mixture with air • Vapors are heavier than air and will collect and stay in low areas • Vapors may travel long distances to ignition sources and flashback • Vapors in confined areas (e.g., tanks, sewers, buildings) may explode when exposed to fire • Liquid attacks same forms of plastics, rubber, and coatings • Toxic products of combustion may include nitrogen oxides and carbon monoxide • Do not put yourself in danger by entering a contaminated area to rescue a victim.

Hazard Classification (based on NFPA-704 Rating System)
Health Hazards (Blue): 3; Flammability (Red): 3; Reactivity (Yellow): 0

EMERGENCY RESPONSE: See Appendix A (132)
Evacuation:
Public safety: Isolate spill area for at least 50 to 100 meters (160 to 330 feet) in all directions.
Spill: Increase, as necessary, the isolation distance shown above, in "Public safety."
Fire: Isolate for 800 meters (½ mile) in all directions, especially if tank, rail car, or tank truck is involved in fire.

EXPOSURE
Short-term effects: *SEEK MEDICAL ATTENTION.* **Vapor:** Irritating to eyes, nose, and throat. *IF INHALED*, will, will cause coughing, difficult breathing, or loss of consciousness. Move to fresh air. *IF BREATHING HAS STOPPED*, give artificial respiration. IF breathing is difficult, administer oxygen. Lung edema may develop. **Liquid:** Will burn skin and eyes. *HARMFUL IF ABSORBED THROUGH THE SKIN.* Harmful if swallowed. Remove contaminated clothing and shoes. Flush affected areas with plenty of water. *IF IN EYES*, hold eyelids open and flush with plenty of water. *IF SWALLOWED* and victim is *CONSCIOUS AND ABLE TO SWALLOW*, have victim drink 4 to 8 ounces of water and have victim induce vomiting. *IF SWALLOWED* and victim is *UNCONSCIOUS OR HAVING CONVULSIONS*, do nothing except keep victim warm.

Note to physician or authorized medical personnel: Medical observation is recommended for 24 to 48 hours after breathing overexposure, as pulmonary edema may be delayed. As first aid for pulmonary edema, consider administering a corticosteroid spray. Cigarette smoking may exacerbate pulmonary injury and should be discouraged for at least 72 hours following exposure.

HEALTH HAZARDS
Personal protective equipment (PPE): B-Level PPE. OSHA Table Z-1-A air contaminant. Air–supplied mask; goggles or face shield; rubber gloves.
Protective materials with good to excellent resistance: Chlorinated polyethylene, nitrile, Viton®, Saranex®.
Recommendations for respirator selection: OSHA
200 ppm: SA:CF (any supplied-air respirator operated in a continuous-flow mode); or SCBAF (any self-contained breathing apparatus with a full facepiece); or SAF (any supplied-air respirator with a full facepiece). *EMERGENCY OR PLANNED ENTRY INTO UNKNOWN CONCENTRATIONS OR IDLH CONDITIONS:* SCBAF:PD,PP (any self-contained breathing apparatus that has a

full facepiece and is operated in a pressure-demand or other positive-pressure mode); or SAF:PD,PP:ASCBA (any supplied-air respirator that has a full facepiece and is operated in a pressure-demand or other positive-pressure mode in combination with an auxiliary self-contained breathing apparatus operated in a pressure-demand or other positive pressure mode). *ESCAPE:* GMFS [any air-purifying, full-facepiece respirator (gas mask) with a chin-style, front- or back-mounted canister providing protection against the compound of concern]; or SCBAE (any appropriate escape-type, self-contained breathing apparatus).
Exposure limits (TWA unless otherwise noted): ACGIH TLV proposed 1 ppm (4.1 mg/m^3); OSHA PEL 25 ppm (100 mg/m^3).
Short-term exposure limits (15-minute TWA): ACGIH STEL 3 ppm (20.7 mg/m^3).
Toxicity by ingestion: Grade 3; LD$_{50}$ = 50 to 500 mg/kg (rat-LD$_{50}$ = 460 mg/kg).
Long-term health effects: Organ damage: Heart, liver, kidney. Possible asthma.
Vapor (gas) irritant characteristics: Eye and respiratory tract irritant. Vapors cause irritation, such that personnel will find high concentrations unpleasant. The effect is temporary.
Liquid or solid irritant characteristics: Eye and skin irritant. Causes smarting of the skin and first-degree burns on short exposure; may cause secondary burns on long exposure.
Odor threshold: 0.1–0.7 ppm.
IDLH value: 200 ppm.

FIRE DATA
Flash point: 20°F/–7°C (oc); 16°F/–9°C (cc).
Flammable limits in air: LEL: 1.2%; UEL 8.0%.
Fire extinguishing agents not to be used: Water may be ineffective.
Autoignition temperature: 482°F/250°C/523°K.
Electrical hazard: Class I, Group C.
Burning rate: 6.2 mm/min

CHEMICAL REACTIVITY
Binary reactants: Strong acids, strong oxidizers, chlorine, hypochlorite, halogenated compounds, nitroparaffins. Copper and its alloys are incompatible with this material.
Neutralizing agents for acids and caustics: Dilute with water.
Reactivity group: 7
Compatibility class: Aliphatic amines

ENVIRONMENTAL DATA
Food chain concentration potential: Log K$_{ow}$ = 1.5. Unlikely to accumulate.
Water pollution: Harmful to aquatic life in very low concentrations. Fouling to shoreline. May be dangerous if it enters nearby water intakes; notify operators. Notify local health and wildlife officials. **Response to discharge:** Issue warning–high flammability. Evacuate area. Disperse and flush. Clean Water Act.

SHIPPING INFORMATION
Grades of purity: 98.5+%; **Storage temperature:** Ambient; **Inert atmosphere:** None; **Venting:** Open (flame arrester); **Stability during transport:** Stable.

NAS HAZARD CLASSIFICATION FOR BULK WATER TRANSPORTATION
FIRE: 3
HEALTH: Vapor irritant: 2; Liquid or solid irritant: 2; Poisons: 2
WATER POLLUTION: Human toxicity: 3; Aquatic toxicity: 3; Aesthetic effect: 2
REACTIVITY: Other chemicals: 3; Water: 0; Self-reaction: 0

PHYSICAL AND CHEMICAL PROPERTIES
Physical state @ 59°F/15°C and 1 atm: Liquid.
Molecular weight: 101.19
Boiling point @ 1 atm: 193.1°F/89.5°C/362.7°K.
Melting/Freezing point: –174.5°F/–114.7°C/158.5°K.
Critical temperature: 504°F/262°C/535°K.
Critical pressure: 440 psia = 30 atm = 3.0 MN/m^2.
Specific gravity (water = 1): 0.729 @ 68°F/20°C (liquid).
Liquid surface tension: 20.7 dynes/cm = 0.0207 N/m @ 68°F/20°C.
Relative vapor density (air = 1): 3.5
Ratio of specific heats of vapor (gas): 1.055
Latent heat of vaporization: 140 Btu/lb = 80 cal/g = 3.3 x 10^5 J/kg.
Heat of combustion: –17,040 Btu/lb = –9,466 cal/g = –396.3 x 10^5 J/kg.
Heat of solution: –180 Btu/lb = –99 cal/g = –4.1 x 10^5 J/kg.
Vapor pressure: (Reid) 2.3 psia; 54 mm.

TRIETHYLBENZENE   REC. T:3350

SYNONYMS: 1,3,5-TRIETHYLBENZENE; sym-TRIETHYLBENZENE; TRIETILBENCENO (Spanish)

IDENTIFICATION
CAS Number: 102-25-0; 25340-18-5 (mixed isomers)
Formula: C$_9$H$_{12}$; C$_{12}$H$_{18}$; 1,3,5-C$_6$H$_3$(C$_2$NO$_5$)$_3$
DOT ID Number: UN 3082; DOT Guide Number: 171
Proper Shipping Name: Environmentally hazardous substances, liquid, n.o.s.

DESCRIPTION: Colorless liquid. Odorless. Floats on water surface; insoluble.

Combustible • Containers may BLEVE when exposed to fire • Vapors may form explosive mixture with air • Vapors are heavier than air and will collect and stay in low areas • Vapors in confined areas (e.g., tanks, sewers, buildings) may explode when exposed to fire • Irritating to the skin, eyes, and respiratory tract • Toxic products of combustion may include carbon monoxide.

Hazard Classification (based on NFPA-704 Rating System)
Health Hazards (Blue): 1; Flammability (Red): 2; Reactivity (Yellow): 0

EMERGENCY RESPONSE: See Appendix A (171)
Evacuation:
Public safety: Isolate the area of spill or leak for at least 10 to 25 meters (30 to 80 feet) in all directions.
Spill: Increase, in the downwind direction, as necessary, the distance shown under "Public Safety."
Fire: If any large container is involved in fire, isolate for at least 800 meters (½ mile) in all directions; also, consider initial evacuation for 800 meters (½ mile) in all directions.

EXPOSURE
Short-term effects: *SEEK MEDICAL ATTENTION.* **Liquid:** Irritating to skin and eyes. Remove contaminated clothing and shoes. Flush affected areas with plenty of water. *IF IN EYES*, hold eyelids open and flush with plenty of water.

HEALTH HAZARDS
Personal protective equipment (PPE): B-Level PPE. OSHA Table Z-1-A air contaminant. Goggles or face shield; rubber gloves.

Triethylene glycol

Vapor (gas) irritant characteristics: Eye and respiratory tract irritation. Vapors cause a slight smarting of the eyes or respiratory system if present in high concentrations. The effect is temporary.
Liquid or solid irritant characteristics: Eye and skin irritant. If spilled on clothing and allowed to remain, may cause smarting and reddening of the skin.

FIRE DATA
Flash point: 181°F/83°C (oc) (1,2,4-isomer).
Flammable limits in air: LEL: not listed; UEL: 56% @ 240°F/116°C.
Fire extinguishing agents not to be used: Water may be ineffective.
Electrical hazard: Class I, Group D.

CHEMICAL REACTIVITY
Binary reactants: A strong reducing agent and organic base. Reacts violently with strong oxidizers, nitroparaffins, nitrogen tetroxide, permanganates, peroxides, ammonium persulfate, bromine dioxide, sulfuric acid, nitric acid. Incompatible with organic anhydrides, acrylates, alcohols, aldehydes, alkylene oxides, substituted allyls, cellulose nitrate, cresols, caprolactam solution, epichlorohydrin, ethylene dichloride, glycols, isocyanates, ketones, nitrates, nitrogen tetroxide, phenols, vinyl acetate. Incompatible with maleic anhydride, methyl trichloroacetate. Form explosive materials with triethynyl aluminum. Increases the explosive sensitivity of nitromethane. Attacks aluminum, copper, lead, tin, zinc, and their alloys and some plastics, rubber, and coatings.
Reactivity group: 32
Compatibility class: Aromatic hydrocarbons

ENVIRONMENTAL DATA
Water pollution: DOT Appendix B, §172.101–marine pollutant. Effect of low concentrations on aquatic life is unknown. Fouling to shoreline. May be dangerous if it enters nearby water intakes; notify operators. Notify local health and wildlife officials.
Response to discharge: Mechanical containment. Should be removed. Chemical and physical treatment.

SHIPPING INFORMATION
Storage temperature: Ambient; **Inert atmosphere:** None; **Venting:** Open (flame arrester); **Stability during transport:** Stable.

NAS HAZARD CLASSIFICATION FOR BULK WATER TRANSPORTATION
FIRE: 1
HEALTH: Vapor irritant: 1; Liquid or solid irritant: 1; Poisons: 1
WATER POLLUTION: Human toxicity: 1; Aquatic toxicity: 2; Aesthetic effect: 2
REACTIVITY: Other chemicals: 1; Water: 0; Self-reaction: 0

PHYSICAL AND CHEMICAL PROPERTIES
Physical state @ 59°F/15°C and 1 atm: Liquid.
Molecular weight: 162.27
Boiling point @ 1 atm: 421°F/216°C/489°K.
Melting/Freezing point: –94°F/–70°C/203°K.
Specific gravity (water = 1): 0.861 @ 68°F/20°C (liquid).
Relative vapor density (air = 1): 5.6
Ratio of specific heats of vapor (gas): 1.039
Latent heat of vaporization: (estimate) 120 Btu/lb = 65 cal/g = 2.7×10^5 J/kg.
Vapor pressure: (Reid) 0.03 psia; 0.05 psia @115°F

TRIETHYLENE GLYCOL REC. T:3400

SYNONYMS: DICAPROATE; DI-β-HYDROXYETHOXYETHANE; 3,6-DIOXA-1,8-DIOL; 2,2'-[1,2-ETHANEDIYLBIS(OXY)] BISEHANOL; ETHANOL, 2,2'-(ETHYLENEDIOXY)DI-; 2,2'-ETHYLENEDIOXYDIETHANOL; 2,2'-ETHYLENEDIOXYETHANOL; ETHYLENE GLYCOL-BIS-(2-HYDROXYETHYL ETHER); ETHYLENE GLYCOL DIHYDROXYDIETHYL ETHER; GLYCOL BIS(HYDROXYETHYL) ETHER; 1,2-BIS(2-HYDROXYETHOXY)ETHANE; TEG; TRIETILENGLICOL (Spanish); TRIGEN; TRIGLYCOL

IDENTIFICATION
CAS Number: 112-27-6
Formula: $C_6H_{14}O_4$; $HO(CH_2CH_2O)_3CH$

DESCRIPTION: Colorless liquid. Mild, sweet odor. Sinks and mixes with water.

Combustible • Heat or flame may cause explosion • Containers may BLEVE when exposed to fire • Vapors in confined areas (e.g., tanks, sewers, buildings) may explode when exposed to fire • Irritating to the skin, eyes, and respiratory tract • Toxic products of combustion may include carbon monoxide, carbon monoxide.

Hazard Classification (based on NFPA-704 Rating System)
Health Hazards (Blue): 1; Flammability (Red): 1; Reactivity (Yellow): 0

EMERGENCY RESPONSE: See Appendix A (171)
Evacuation:
Public safety: Isolate the area of spill or leak for at least 10 to 25 meters (30 to 80 feet) in all directions.
Spill: Increase, in the downwind direction, as necessary, the distance shown under "Public Safety."
Fire: If any large container is involved in fire, isolate for at least 800 meters (½ mile) in all directions; also, consider initial evacuation for 800 meters (½ mile) in all directions.

HEALTH HAZARDS
Personal protective equipment (PPE): B-Level PPE. Goggles; gloves. Chemical protective material(s) reported to have good to excellent resistance: butyl rubber.
Toxicity by ingestion: Grade 1; LD_{50} = 5 to 15 g/kg (guinea pig).
Vapor (gas) irritant characteristics: Vapors are nonirritating to the eyes and throat.
Liquid or solid irritant characteristics: No appreciable hazard. Practically harmless to the skin.

FIRE DATA
Flash point: 330°F/166°C (oc); 350°F/177°C (cc).
Flammable limits in air: LEL: 0.9%; UEL: 9.2%.
Fire extinguishing agents not to be used: Water or foam may cause frothing.
Behavior in fire: Moderate explosion hazard.
Autoignition temperature: 700°F/371°C/644°K.
Electrical hazard: Class I, Group C.
Burning rate: 1.7 mm/min

CHEMICAL REACTIVITY
Binary reactants: Incompatible with strong oxidizers, isocyanates, permanganates, peroxides, ammonium persulfate, bromine dioxide, strong acids: sulfuric acid, nitric acid, perchloric acid.

Reactivity group: 40
Compatibility class: Glycol ethers

ENVIRONMENTAL DATA
Food chain concentration potential: Log K_{ow} = –2.0. Unlikely to accumulate.
Water pollution: Effect of low concentrations on aquatic life is unknown. May be dangerous if it enters nearby water intakes; notify operators. Notify local health and wildlife officials.
Response to discharge: Disperse and flush.

SHIPPING INFORMATION
Grades of purity: High purity; air treatment; commercial; **Storage temperature:** Ambient; **Inert atmosphere:** None; **Venting:** Open (flame arrester); **Stability during transport:** Stable.

NAS HAZARD CLASSIFICATION FOR BULK WATER TRANSPORTATION
FIRE: 1
HEALTH: Vapor irritant: 0; Liquid or solid irritant: 0; Poisons: 0
WATER POLLUTION: Human toxicity: 0; Aquatic toxicity: 1; Aesthetic effect: 1
REACTIVITY: Other chemicals: 2; Water: 0; Self-reaction: 0

PHYSICAL AND CHEMICAL PROPERTIES
Physical state @ 59°F/15°C and 1 atm: Liquid.
Molecular weight: 150.17
Boiling point @ 1 atm: 550°F/288°C/561°K.
Melting/Freezing point: 24.3°F/–4.3°C/268.9°K.
Specific gravity (water = 1): 1.125 @ 68°F/20°C (liquid).
Liquid surface tension: 45.2 dynes/cm = 0.0452 N/m @ 68°F/20°C.
Relative vapor density (air = 1): 5.17
Ratio of specific heats of vapor (gas): 1.039
Latent heat of vaporization: 180 Btu/lb = 99 cal/g = 4.1×10^5 J/kg.
Heat of combustion: –10,190 Btu/lb = –5,660 cal/g = -237.0×10^5 J/kg.
Heat of solution: (estimate) –13 Btu/lb = –7 cal/g = $-.3 \times 10^5$ J/kg.
Vapor pressure: (Reid) Very low; 0.01 mm.

TRIETHYLENE GLYCOL DI-(2-ETHYLBUTYRATE)
REC. T:3450

SYNONYMS: DI(2-ETILBUTIRATO) de TRIETILENGLICOL (Spanish); TRIGLYCOL DICAPROATE; TRIGLYCOL DIHEXOATE

IDENTIFICATION
CAS Number: 95-08-9
Formula: $C_{18}H_{34}O_6$; $(C_5H_{11}COOCH_2CH_2OCH_2)_2$

DESCRIPTION: Colorless liquid. Soluble in water.

Combustible • Containers may BLEVE when exposed to fire • Vapors may form explosive mixture with air • Vapors in confined areas (e.g., tanks, sewers, buildings) may explode when exposed to fire • Irritating to the skin, eyes, and respiratory tract • Toxic products of combustion may include carbon monoxide or CO_2

Hazard Classification (based on NFPA-704 Rating System)
Health Hazards (Blue): 0; Flammability (Red): 1; Reactivity (Yellow): 0

EMERGENCY RESPONSE: See Appendix A (171)
Evacuation:
Public safety: Isolate the area of spill or leak for at least 10 to 25 meters (30 to 80 feet) in all directions.
Spill: Increase, in the downwind direction, as necessary, the distance shown under "Public Safety."
Fire: If any large container is involved in fire, isolate for at least 800 meters (½ mile) in all directions; also, consider initial evacuation for 800 meters (½ mile) in all directions.

EXPOSURE
Short-term effects: *SEEK MEDICAL ATTENTION.* **Vapor:** Move victim to fresh air. *IF BREATHING HAS STOPPED,* give artificial respiration. IF breathing is difficult, administer oxygen. **Liquid:** Remove contaminated clothing and shoes. Flush affected areas with water. *IF IN EYES,* hold eyelids open and flush with plenty of water.

HEALTH HAZARDS
Personal protective equipment (PPE): B-Level PPE. Full impervious protective clothing, including boots and gloves. Where splashing is possible wear full face shield or chemical safety goggles. Use approved respirator to protect against vapors. Chemical protective material(s) reported to have good to excellent resistance: butyl rubber.
Toxicity by ingestion: Grade 2; oral rat LD_{50} = 6.0 g/kg.
Vapor (gas) irritant characteristics: Vapors cause a smarting of the eyes or respiratory system if present in high concentrations. The effect is temporary.
Liquid or solid irritant characteristics: Eye irritant. If spilled on clothing and allowed to remain, may cause smarting and reddening of the skin.

FIRE DATA
Flash point: 385°F/196°C (cc).
Fire extinguishing agents not to be used: Water.
Stoichiometric air-to-fuel ratio: 111.9

CHEMICAL REACTIVITY
Binary reactants: Strong acids.
Reactivity group: 34
Compatibility class: Esters.

ENVIRONMENTAL DATA
Water pollution: Effect of low concentrations on aquatic life is unknown. May be dangerous if it enters nearby water intakes; notify operators. Notify local health and wildlife officials. **Response to discharge:** Should be removed.

SHIPPING INFORMATION
Grades of purity: Technical grades; **Storage temperature:** Ambient; **Inert atmosphere:** None; **Venting:** Open; **Stability during transport:** Stable.

PHYSICAL AND CHEMICAL PROPERTIES
Physical state @ 59°F/15°C and 1 atm: Liquid.
Molecular weight: 346.52
Specific gravity (water = 1): 0.9946 @ 20°C.

TRIETHYLENE GLYCOL ETHYL ETHER   REC. T:3500

SYNONYMS: ETHOXYTRIGLYCOL; ETHOXYTRIETHYLENE GLYCOL; DOWANOL®-TE; POLY-SOLV TE; TRIETHYLENE GLYCOL MONOETHYLETHER;

TRIETILENGLICOLMONOETIL ETER (Spanish); TRIGLYCOL MONOETHYL ETHER; 3,6,9-TRIOXAUNDECAN-1-OL; TRIOXYTOL®

IDENTIFICATION
CAS Number: 112-50-5
Formula: $C_8H_{18}O_4$; $C_2H_5O(CH_2)_2O(CH_2)_2OCH_2CH_2OH$

DESCRIPTION: Colorless liquid. Odorless. Sinks in water; highly soluble.

Combustible • Heat or flame may cause explosion • Containers may BLEVE when exposed to fire • Vapors are heavier than air and will collect and stay in low areas • Vapors in confined areas (e.g., tanks, sewers, buildings) may explode when exposed to fire • Irritating to the skin and eyes • Toxic products of combustion may include carbon monoxide.

Hazard Classification (based on NFPA-704 Rating System)
Health Hazards (Blue): 0; Flammability (Red): 1; Reactivity (Yellow): 0

EMERGENCY RESPONSE: See Appendix A (171)
Evacuation:
Public safety: Isolate the area of spill or leak for at least 10 to 25 meters (30 to 80 feet) in all directions.
Spill: Increase, in the downwind direction, as necessary, the distance shown under "Public Safety."
Fire: If any large container is involved in fire, isolate for at least 800 meters (½ mile) in all directions; also, consider initial evacuation for 800 meters (½ mile) in all directions.

HEALTH HAZARDS
Personal protective equipment (PPE): B-Level PPE. Chemical safety goggles and adequate protective clothing. Chemical protective material(s) reported to have good to excellent resistance: butyl rubber.
Toxicity by ingestion: Grade 1; LD_{50} = 10.61 g/kg (rat).
Vapor (gas) irritant characteristics: Vapors are nonirritating to the eyes and throat.
Liquid or solid irritant characteristics: Eye irritant.

FIRE DATA
Flash point: 275°F/135°C (oc).
Fire extinguishing agents not to be used: Water or foam may cause frothing.
Stoichiometric air-to-fuel ratio: 50.0

CHEMICAL REACTIVITY
Reactivity with common materials: Violent reaction with oxidizers, permanganates, peroxides, ammonium persulfate, bromine dioxide, sulfuric acid, nitric acid, perchloric acid, and other strong acids. Incompatible with acyl halides, aliphatic amines, alkalis, boranes, isocyanates.
Reactivity group: 40
Compatibility class: Glycol ethers

ENVIRONMENTAL DATA
Food chain concentration potential: Negative; unlikely to accumulate.
Water pollution: Effect of low concentrations on aquatic life is unknown. May be dangerous if it enters nearby water intakes; notify operators. Notify local health and wildlife officials.
Response to discharge: Disperse and flush.

SHIPPING INFORMATION
Storage temperature: Ambient; **Inert atmosphere:** None; **Stability during transport:** Stable.

NAS HAZARD CLASSIFICATION FOR BULK WATER TRANSPORTATION
FIRE: 1
HEALTH: Vapor irritant: 0; Liquid or solid irritant: 0; Poisons: 0
WATER POLLUTION: Human toxicity: 0; Aquatic toxicity: 1; Aesthetic effect: 1
REACTIVITY: Other chemicals: 2; Water: 0; Self-reaction: 0

PHYSICAL AND CHEMICAL PROPERTIES
Physical state @ 59°F/15°C and 1 atm: Liquid.
Molecular weight: 178.26
Boiling point @ 1 atm: 493°F/256°C/529°K.
Melting/Freezing point: −1.7°F/−18.7°C/−254.5°K.
Specific gravity (water = 1): 1.020 @ 68°F/20°C (liquid).
Relative vapor density (air = 1): 6.0
Ratio of specific heats of vapor (gas): 1.033
Latent heat of vaporization: (estimate) 125 Btu/lb = 69 cal/g = 2.9 x 10^5 J/kg.
Heat of combustion: (estimate) = −11,000 Btu/lb = −6,170 cal/g = −258 x 10^5 J/kg.
Vapor pressure: 0.05 mm.

TRIETHYLENE GLYCOL METHYL ETHER   REC. T:3550

SYNONYMS: TRIETHYLENE GLYCOL MONOMETHYL ETHER; TRIETILENGLICOLMETIL ETER (Spanish); TRIGLYCOL METHYL ETHER

IDENTIFICATION
CAS Number: 112-35-6
Formula: $C_7H_{16}O_4$; $CH_3O(CH_2)_2O(CH_2)_2OCH_2CH_2OH$

DESCRIPTION: Colorless liquid. Odorless. Mixes with water.

Combustible • Heat or flame may cause explosion • Containers may BLEVE when exposed to fire • Vapors are heavier than air and will collect and stay in low areas • Vapors in confined areas (e.g., tanks, sewers, buildings) may explode when exposed to fire • Irritating to the skin, eyes, and respiratory tract • Toxic products of combustion may include carbon monoxide.

Hazard Classification (based on NFPA-704 Rating System)
Health Hazards (Blue): 0; Flammability (Red): 1; Reactivity (Yellow): 0

EMERGENCY RESPONSE: See Appendix A (171)
Evacuation:
Public safety: Isolate the area of spill or leak for at least 10 to 25 meters (30 to 80 feet) in all directions.
Spill: Increase, in the downwind direction, as necessary, the distance shown under "Public Safety."
Fire: If any large container is involved in fire, isolate for at least 800 meters (½ mile) in all directions; also, consider initial evacuation for 800 meters (½ mile) in all directions.

EXPOSURE
Short-term effects: *SEEK MEDICAL ATTENTION.* **Vapor:** Move victim to fresh air. *IF BREATHING HAS STOPPED,* give artificial respiration. If breathing is difficult, administer oxygen.

Liquid: Remove contaminated clothing and shoes. Wash affected areas with soap and water. *IF IN EYES*, hold eyelids open and flush with plenty of water.

HEALTH HAZARDS
Personal protective equipment (PPE): B-Level PPE. Chemical safety goggles and adequate protective clothing. Chemical protective material(s) reported to have good to excellent resistance: butyl rubber.
Toxicity by ingestion: Grade 1; LD_{50} = 11.3 g/kg (rat).
Vapor (gas) irritant characteristics: Vapors are nonirritating to the eyes and throat.
Liquid or solid irritant characteristics: No appreciable hazard. Practically harmless to the skin.

FIRE DATA
Flash point: 275°F/135°C (oc); 230°F/110°C (cc).
Fire extinguishing agents not to be used: Water or foam may cause frothing.
Stoichiometric air-to-fuel ratio: 42.9

ENVIRONMENTAL DATA
Water pollution: Effect of low concentrations on aquatic life is unknown. May be dangerous if it enters nearby water intakes; notify operators. Notify local health and wildlife officials.
Response to discharge: Should be removed.

SHIPPING INFORMATION
Storage temperature: Ambient; **Inert atmosphere:** None; **Stability during transport:** Stable.

PHYSICAL AND CHEMICAL PROPERTIES
Physical state @ 59°F/15°C and 1 atm: Liquid.
Molecular weight: 164.23
Boiling point @ 1 atm: 489°F/254°C/527°K.
Specific gravity (water = 1): 1.026

TRIETHYLENE TETRAMINE   **REC. T:3600**

SYNONYMS: *N,N'*-BIS(2-AMINOETHYL)-1,2-DIAMINOETHANE; *N,N'*-BIS(2-AMINOETHYL)ETHYLENEDIAMINE; 3,6-DIAZA-1,8-DIAMINE; TECZA; TETA; TRIEN; TRIENTINE; TRIETILENTETRAMINA (Spanish)

IDENTIFICATION
CAS Number: 112-24-3
Formula: $C_6H_{18}N_4$; $NH_2(CH_2)_2NH(CH_2)_2NH(CH_2)_2NH_2$
DOT ID Number: UN 2259; DOT Guide Number: 153
Proper Shipping Name: Triethylenetetramine

DESCRIPTION: Light straw to amber oily liquid. A medium strong base. Ammonia odor. Floats on the surface of water; highly soluble forming a medium strong base.

Combustible • Corrosive to the skin, eyes, and respiratory tract; contact with skin or eyes can cause burns and impaired vision; inhalation exposure may be delayed • Containers may BLEVE when exposed to fire • Vapors are heavier than air and will collect and stay in low areas • Vapors in confined areas (e.g., tanks, sewers, buildings) may explode when exposed to fire • Toxic products of combustion may include nitrogen oxides.

Hazard Classification (based on NFPA-704 Rating System)
Health Hazards (Blue): 3; Flammability (Red): 1; Reactivity (Yellow): 0

EMERGENCY RESPONSE: See Appendix A (153)
Evacuation:
Public safety: Isolate the area of spill or leak for at least 25 to 50 meters (80 to 160 feet) in all directions.
Spill: Increase, in the downwind direction, as necessary, the distance shown under "Public Safety."
Fire: If tank, rail car, or tank truck is involved in fire, isolate for at least 800 meters (½ mile) in all directions; also, consider initial evacuation for 800 meters (½ mile) in all directions.

EXPOSURE
Short-term effects: *SEEK MEDICAL ATTENTION*. **Liquid:** *HARMFUL IF ABSORBED THROUGH THE SKIN*. A medium strong base. Will burn skin and eyes. Harmful if swallowed. Remove contaminated clothing and shoes. Flush affected areas with plenty of water. *IF BREATHING HAS STOPPED*, give artificial respiration; *avoid mouth-to-mouth resuscitation; use bag/mask apparatus*. *IF IN EYES*, hold eyelids open and flush with plenty of water. *IF SWALLOWED* and victim is *CONSCIOUS AND ABLE TO SWALLOW*, have victim drink 4 to 8 ounces of water. **Do NOT induce vomiting**.
Note to physician or authorized medical personnel: Medical observation is recommended for 24 to 48 hours after breathing overexposure, as pulmonary edema may be delayed. As first aid for pulmonary edema, consider administering a corticosteroid spray. Cigarette smoking may exacerbate pulmonary injury and should be discouraged for at least 72 hours following exposure.

HEALTH HAZARDS
Personal protective equipment (PPE): B-Level PPE. Amine-type canister; goggles or face shield; rubber gloves. butyl rubber, neoprene, nitrile, Viton®, Silversheild®.
Toxicity by ingestion: Grade 2; LD_{50} = 0.5 to 5 g/kg (rat).
Long-term health effects: May cause dermatitis, asthma and other allergic reactions in humans.
Vapor (gas) irritant characteristics: Eye and respiratory tract irritant. Vapors cause moderate irritation such that personnel will find high concentrations unpleasant. The effect is temporary.
Liquid or solid irritant characteristics: Eye and skin irritant. Causes smarting of the skin and first-degree burns on short exposure; may cause secondary burns on long exposure.

FIRE DATA
Flash point: 290°F/143°C (oc); 275°F/135°C (cc).
Flammable limits in air: LEL: 0.7%; UEL: 7.2%.
Fire extinguishing agents not to be used: Water or foam may cause frothing. Reacts with halon fire extinguishers.
Autoignition temperature: 640°F/338°C/611°K.
Electrical hazard: Class I, Group C.

CHEMICAL REACTIVITY
Reactivity with water: Causes foaming; caustic.
Binary reactants: Reacts with nitrogen-containing compounds; may cause violent decomposition. Reacts violently with strong oxidizers, nitroparaffins, nitrogen tetroxide, permanganates, peroxides, ammonium persulfate, bromine dioxide, sulfuric acid, nitric acid. Incompatible with organic anhydrides, acrylates, alcohols, aldehydes, alkylene oxides, substituted allyls, cellulose nitrate, cresols, caprolactam solution, epichlorohydrin, ethylene dichloride, glycols, halons, halogenated hydrocarbons, isocyanates, ketones, nitrates, nitrogen tetroxide, phenols, urea, vinyl acetate.

Incompatible with maleic anhydride, methyl trichloroacetate. Increases the explosive sensitivity of nitromethane. Attacks aluminum, cobalt, copper, lead, nickel, tin, zinc and alloys, and some plastics, rubber, and coatings.
Neutralizing agents for acids and caustics: After dilution with water, can be neutralized with acetic acid.
Reactivity group: 7
Compatibility class: Aliphatic amines

ENVIRONMENTAL DATA
Food chain concentration potential: Log K_{ow} = –1.69. Unlikely to accumulate.
Water pollution: Effect of low concentrations on aquatic life is unknown. May be dangerous if it enters nearby water intakes; notify operators. Notify local health and wildlife officials.
Response to discharge: Disperse and flush.

SHIPPING INFORMATION
Grades of purity: 99+%; **Storage temperature:** Ambient; **Inert atmosphere:** None; **Venting:** Open; **Stability during transport:** Stable.

NAS HAZARD CLASSIFICATION FOR BULK WATER TRANSPORTATION
FIRE: 1
HEALTH: Vapor irritant: 2; Liquid or solid irritant: 2; Poisons: 1
WATER POLLUTION: Human toxicity: 1; Aquatic toxicity: 1; Aesthetic effect: 3
REACTIVITY: Other chemicals: 3; Water: 0; Self-reaction: 0

PHYSICAL AND CHEMICAL PROPERTIES
Physical state @ 59°F/15°C and 1 atm: Liquid.
Molecular weight: 146.24
Boiling point @ 1 atm: 531.3°F/277.4°C/550.6°K.
Melting/Freezing point: –31°F/–35°C/238°K.
Critical temperature: 860°F/460°C/733°K.
Critical pressure: 470 psia = 32 atm = 3.2 MN/m^2.
Specific gravity (water = 1): 0.982 @ 68°F/20°C (liquid).
Relative vapor density (air = 1): 5.04
Ratio of specific heats of vapor (gas): 1.037
Heat of combustion: (estimate) –13,500 Btu/lb = –7530 cal/g = –315 x 10^5 J/kg.
Heat of solution: (estimate) –13 Btu/lb = –7 cal/g = –0.3 x 10^5 J/kg.
Vapor pressure: Less than 0.07 mm.

TRIETHYL PHOSPHATE   REC. T:3650

SYNONYMS: ETHYL PHOSPHATE; FOSFATO de TRIETILO (Spanish); PHOSPHORIC ACID, TRIETHYL ESTER; TEP

IDENTIFICATION
CAS Number: 78-40-0
Formula: $C_6H_{15}O_4P$; $(C_2H_5O)_3PO$
DOT ID Number: UN 3278; DOT Guide Number: 151
Proper Shipping Name: Organophosphorus compound, toxic, n.o.s.

DESCRIPTION: Colorless liquid. Odorless. Soluble in water.

Combustible • Poison! (organophosphate) • Containers may BLEVE when exposed to fire • Vapors are heavier than air and will collect and stay in low areas • Vapors in confined areas (e.g., tanks, sewers, buildings) may explode when exposed to fire • Irritating to the skin, eyes, and respiratory tract • Toxic products of combustion may include phosphorus oxides.

Hazard Classification (based on NFPA-704 Rating System)
Health Hazards (Blue): 1; Flammability (Red): 1; Reactivity (Yellow): 1

EMERGENCY RESPONSE: See Appendix A (151)
Evacuation:
Public safety: Isolate the area of spill or leak for at least 25 to 50 meters (80 to 160 feet) in all directions.
Spill: Increase, in the downwind direction, as necessary, the distance shown under "Public Safety."
Fire: If tank, rail car, or tank truck is involved in fire, isolate for at least 800 meters (½ mile) in all directions; also, consider initial evacuation for 800 meters (½ mile) in all directions.

EXPOSURE
Short-term effects: *SEEK MEDICAL ATTENTION*. **Vapor or mist:** May cause irritation. Move to fresh air. *IF BREATHING HAS STOPPED,* give artificial respiration; *avoid mouth-to-mouth resuscitation; use bag/mask apparatus*. IF breathing is difficult give oxygen. **Liquid:** May cause irritation. May be harmful if swallowed. Remove contaminated clothing. Flush affected areas with soap and plenty of water for 15 minutes. *IF IN EYES* hold eyelids open and flush with plenty of water for 15 minutes.

HEALTH HAZARDS
Personal protective equipment (PPE): B-Level PPE. Chemical-protective clothing and butyl rubber gloves are recommended when skin contact is possible because liquid is rapidly absorbed through the skin and may cause systemic toxicity.
Toxicity by ingestion: Grade 2: LD_{50} = 1.6 g/kg (mouse).
Long-term health effects: May cause nerve damage, but to a lesser extent than other cholinesterase inhibitors. Mutation data. In male rats, oral administration of TDLo g/kg for 63 days adversely affected reproductive organs and in female rats. Oral administration of TDLo 5.7 g/kg for 92 days before mating and 1 to 22 days during gestation adversely affected live birth index. Mutagenic in D. melanogaster at oral dose of 10 mole/L.
Vapor (gas) irritant characteristics: May cause eye and respiratory tract irritation.
Liquid or solid irritant characteristics: May cause eye and skin irritation.

FIRE DATA
Flash point: 240°F/116°C (oc); 210°F/99°C (cc).
Flammable limits in air: LEL: 1.7%; UEL: 10%.
Autoignition temperature: 845°F/452°C/725°K.

CHEMICAL REACTIVITY
Reactivity with water: Slow reaction; slight decomposition
Binary reactants: Violent reaction with strong oxidizers, antimony(V) pentafluoride. Incompatible with strong acids, lead diacetate, magnesium, nitrates, silver nitrate.
Reactivity group: 34
Compatibility class: Esters.

ENVIRONMENTAL DATA
Water pollution: Effect of low concentrations on aquatic life is unknown. May be dangerous if it enters nearby water intakes; notify operators. Notify local health and wildlife officials. **Response to discharge:** Should be removed. Mechanical and physical treatment.

SHIPPING INFORMATION
Grades of purity: (approximately) 100%; **Storage temperature:** Ambient; **Stability during transport:** Stable.

PHYSICAL AND CHEMICAL PROPERTIES
Physical state @ 59°F/15°C and 1 atm: Liquid.
Molecular weight: 182.16
Boiling point @ 1 atm: 408°F/209°C/482°K.
Specific gravity (water = 1): 1.068
Relative vapor density (air = 1): 6.28
Vapor pressure: (Reid) 0.0165 psia.

TRIETHYL PHOSPHITE   REC. T:3700

SYNONYMS: Phosphorus ACID, TRIETHYL ESTER; FOSFITO de TRIETILO (Spanish)

IDENTIFICATION
CAS Number: 122-52-1
Formula: $(C_2H_5O)_3P$
DOT ID Number: UN 2323; DOT Guide Number: 129
Proper Shipping Name: Triethyl phosphite

DESCRIPTION: Colorless liquid. Odorless.

Flammable • Corrosive to the skin, eyes, and respiratory tract; contact with skin or eyes can cause burns and impaired vision; inhalation exposure may be delayed • Containers may BLEVE when exposed to fire • Vapors may form explosive mixture with air • Vapors are heavier than air and will collect and stay in low areas • Vapors may travel long distances to ignition sources and flashback • Vapors in confined areas (e.g., tanks, sewers, buildings) may explode when exposed to fire • Toxic products of combustion may include carbon monoxide and oxides of phosphorus.

Hazard Classification (based on NFPA-704 Rating System)
Health Hazards (Blue): 1; Flammability (Red): 2; Reactivity (Yellow): 0

EMERGENCY RESPONSE: See Appendix A (129)
Evacuation:
Public safety: Isolate spill area for at least 50 to 100 meters (160 to 330 feet) in all directions.
Spill: Large spill–Consider initial downwind evacuation for at least 300 meters (1000 feet).
Fire: Isolate for 800 meters (½ mile) in all directions, especially if tank, rail car, or tank truck is involved in fire.

EXPOSURE
Short-term effects: *SEEK MEDICAL ATTENTION.* **Vapor or mist:** Harmful if inhaled. Move to fresh air. *IF BREATHING HAS STOPPED,* give artificial respiration. IF breathing is difficult, administer oxygen. **Liquid:** Harmful if absorbed through the skin or swallowed. Remove contaminated clothing. Flush affected areas with soap and plenty of water. *IF IN EYES,* hold eyelids open and flush with water for 15 minutes. *IF SWALLOWED* and victim is *CONSCIOUS AND ABLE TO SWALLOW,* have victim drink 1 to 2 glasses of water or milk and induce vomiting.

HEALTH HAZARDS
Personal protective equipment (PPE): B-Level PPE. Self-contained breathing apparatus, rubber gloves and rubber boots.
Toxicity by ingestion: Grade 2: LD_{50} = 3.2 g/kg (rat).
Long-term health effects: Prolonged exposures may cause chemical pneumonitis.
Vapor (gas) irritant characteristics: Vapors cause a slight smarting of the eyes or respiratory system if present in high concentrations. The effect is temporary.
Liquid or solid irritant characteristics: If spilled on clothing and allowed to remain, may cause smarting and reddening of skin.

FIRE DATA
Flash point: 115°F/46°C (cc); 130°F/55°C (oc).

CHEMICAL REACTIVITY
Binary reactants: Violent reaction with oxidizers. Incompatible with acids, nitrates, oxidizers, magnesium.
Reactivity group: 34
Compatibility class: Esters.

ENVIRONMENTAL DATA
Water pollution: Effect of low concentration on aquatic life is unknown. May be dangerous if it enters nearby water intakes; notify operators. Notify health and wildlife officials. **Response to discharge:** Evacuate area. Should be removed. Mechanical and physical treatment.

SHIPPING INFORMATION
Grades of purity: (approximately) 100%; **Storage temperature:** Ambient; **Stability during transport:** Stable.

PHYSICAL AND CHEMICAL PROPERTIES
Physical state @ 59°F/15°C and 1 atm: Liquid.
Molecular weight: 166.16
Boiling point @ 1 atm: 311°F/155°C/428.2°K.
Specific gravity (water = 1): 0.969
Relative vapor density (air = 1): 5.73

TRIFLUOROCHLOROETHYLENE   REC. T:3750

SYNONYMS: CHLOROTRIFLUOROETHYLENE; CTFE; DAIFLON; FLUOROPLAST-3; GENETRON 1113; KEL F MONOMER; R-1113; 1,1,2-TRIFLUORO-2-CHLOROETHYLENE; TRIFLUOCLOROETILENO (Spanish); TRIFLUOROMONOCHLOROETHYLENE; TRIFLUOROVINYL CHLORIDE; TRITHENE

IDENTIFICATION
CAS Number: 79-38-9
Formula: C_2ClF_3; $F_2C = CFCl$
DOT ID Number: UN 1082; DOT Guide Number: 119P
Proper Shipping Name: Trifluorochloroethylene, inhibited

DESCRIPTION: Colorless gas. Shipped and stored as a liquefied compressed gas. A liquid below –18°F/–28°C. Faint ether-like odor. Sinks and boils in water; flammable visible vapor cloud is produced.

Extremely flammable • Forms explosive mixture with air • Reacts explosively with many materials • Gas from liquefied gas are initially heavier than air and spread along ground • Gas may travel to source of ignition and flashback • Gas in confined areas (e.g., tanks, sewers, buildings) may explode when exposed to fire • Ruptured or venting cylinders may rocket through buildings and/or travel a considerable distance • Polymerization hazard • Heat can induce polymerization with rapid release of energy; sealed containers may rupture explosively • May react with itself blocking relief valves; leading to tank explosions • Irritating to the skin, eyes, and respiratory tract • Contact with liquid may cause frostbite • Toxic products of combustion may include hydrochloric acid and hydrofluoric acid vapors.

Hazard Classification (based on NFPA-704 Rating System)
Health Hazards (Blue): -; Flammability (Red): 4; Reactivity (Yellow): 0

EMERGENCY RESPONSE: See Appendix A (119)
Evacuation:
Public safety: See below.
Spill: Small spill–First: Isolate in all directions 30 meters (100 feet); Then: Protect persons downwind, DAY: 0.2 km (0.1 mile); NIGHT: 0.2 km (0.1 mile). Large spill–First: Isolate in all directions 30 meters (100 feet); Then: Protect persons downwind, DAY: 0.3 km (0.2 mile); NIGHT: 0.8 km (0.5 mile).
Fire: Isolate for 1600 meters (1 mile) in all directions, especially if tank, rail car, or tank truck is involved in fire.

EXPOSURE
Short-term effects: *SEEK MEDICAL ATTENTION*. **Vapor:** IF inhaled, will cause dizziness, nausea, or vomiting. Move victim to fresh air. *IF BREATHING HAS STOPPED*, give artificial respiration; *avoid mouth-to-mouth resuscitation; use bag/mask apparatus*. If breathing is difficult, administer oxygen. **Liquid:** Will cause frostbite. Flush affected areas with plenty of water. *DO NOT RUB AFFECTED AREAS*.

HEALTH HAZARDS
Personal protective equipment (PPE): B-Level PPE. Self-contained breathing apparatus; goggles; rubber gloves. Wear thermal protective clothing.
Exposure limits (TWA unless otherwise noted): 5 ppm (AIHA WEEL).
Long-term health effects: Organ damage including kidney damage.
Liquid or solid irritant characteristics: May cause frostbite.

FIRE DATA
Flash point: Gas.
Flammable limits in air: LEL: 8.4%; UEL: 16%.
Behavior in fire: Containers may explode.
Stoichiometric air-to-fuel ratio: 2.946 (estimate).

CHEMICAL REACTIVITY
Polymerization: Can occur; **Inhibitor of polymerization:** Terpenes; Tributylamine (1%).

ENVIRONMENTAL DATA
Food chain concentration potential: Negative; unlikely to accumulate.
Water pollution: Not harmful to aquatic life. **Response to discharge:** Issue warning–high flammability, air contaminant. Restrict access. Evacuate area

SHIPPING INFORMATION
Grades of purity: Polymerization grade, 99.0+%; **Storage temperature:** Ambient, but less than 150°F; **Inert atmosphere:** Air must be excluded; **Venting:** Safety relief; **Stability during transport:** Stable.

PHYSICAL AND CHEMICAL PROPERTIES
Physical state @ 59°F/15°C and 1 atm: Gas.
Molecular weight: 116.5
Boiling point @ 1 atm: –18°F/–28°C/245°K.
Critical temperature: (estimate) 223.2°F/106.2°C/379.4°K.
Critical pressure: (estimate) 592 psia = 40.2 atm = 4.08 MN/m^2.
Specific gravity (water = 1): 1.307 @ 68°F/20°C (liquid).
Liquid surface tension: (estimate) 12 dynes/cm = 0.012 N/m @ 68°F/20°C.
Relative vapor density (air = 1): 4.02
Latent heat of vaporization: 83 Btu/lb = 46 cal/g = 1.92 x 10^5 J/kg.

TRIFLURALIN   REC. T:3800

SYNONYMS: AGREFLAN; AGRIFLAN 24; BENZENAMINE, 2,6-DINITRO-*N,N*-DIPROPYL-4-(TRIFLUOROMETHYL-); CRISALIN; DIGERMIN; 2,6-DINITRO-*N,N*-DIPROPYL-4-(TRIFLUOROMETHYL)BENZENAMINE; 2,6-DINITRO-*N,N*-DIN-PROPYL-α,α,α-TRIFLUORO-*p*-TOLUIDINE; 4-(DIN-PROPYLAMINO)-3,5-DINITRO-1-TRIFLUOROMETHYLBENZENE; *N,N*-DIN-PROPYL-2,6-DINITRO-4-TRIFLUOROMETHYLANILINE; 2,6-DINITRO-4-TRIFLUORMETHYL-*N,N*-DIPROPYLANILIN (German); *N,N*-DIPROPYL-4-TRIFLUOROMETHYL-2,6-DINITROANILINE; ELANCOLAN; ETHANE, TRIFLUORO-; 2,6-DINITRO-*N,N*-DIPROPYL-4-(TRIFLUOROMETHYL)ANILINE; IPERSAN; L-36352; LILLY 36,352; MARKSMAN 2, TRIGARD; NITRAN; OLITREF; SINFLOWAN; SU SEGURO CARPIDOR; *p*-TOLUIDINE,α,α,α-TRIFLUORO-2,6-DINITRO-*N,N*-DIPROPYL-; TREFANOCIDE; TREFICON; TREFLAN; TREFLANOCIDE ELANCOLAN; TRIFLUORALIN; TRISTAR; α,α,α-TRIFLUORO-2,6-DINITRO-*N,N*-DIPROPYL-*p*-TOLUIDINE; TRIFLURALINA (Spanish); TRIFUREX; TRIKEPIN; TRIM

IDENTIFICATION
CAS Number: 1582-09-8
Formula: $C_{13}H_{16}F_3N_3O_4$
DOT ID Number: UN 2588; DOT Guide Number: 151
Proper Shipping Name: Pesticide, solid, poisonous, n.o.s.
Reportable Quantity (RQ): **(CERCLA)** 1 lb/4.54 kg

DESCRIPTION: Yellow-orange solid. May be dissolved in an organic solvent (acetone, ethanol, xylene). Sinks in water; insoluble.

Poison! (Fluorodinitrotoluidine herbicide) Combustible solid • Containers may BLEVE when exposed to fire •Vapors may form explosive mixture with air • Vapors in confined areas (e.g., tanks, sewers, buildings) may explode when exposed to fire • Irritating to the skin, eyes, and respiratory tract • Toxic products of combustion may include nitrogen oxides and hydrogen fluoride.

Hazard Classification (based on NFPA-704 Rating System)
Health Hazards (Blue): 1; Flammability (Red): 1; Reactivity (Yellow): 0

EMERGENCY RESPONSE: See Appendix A (151)
Evacuation:
Public safety: Isolate the area of spill or leak for at least 25 to 50 meters (80 to 160 feet) in all directions.
Spill: Increase, in the downwind direction, as necessary, the distance shown under "Public Safety."
Fire: If tank, rail car, or tank truck is involved in fire, isolate for at least 800 meters (½ mile) in all directions; also, consider initial evacuation for 800 meters (½ mile) in all directions.

EXPOSURE
Short-term effects: *SEEK MEDICAL ATTENTION.* **Dust:** *POISONOUS IF INHALED.* Move victim to fresh air. *IF IN EYES*, hold eyelids open and flush with plenty of water. IF breathing is difficult, administer oxygen. **Solid:** *POISONOUS IF SWALLOWED.* Irritating to skin and eyes. Remove contaminated clothing and shoes. Flush affected areas with plenty of water. *IF IN EYES*, hold eyelids open and flush with plenty of water. *IF SWALLOWED* and victim is *CONSCIOUS AND ABLE TO SWALLOW*, have victim drink 4 to 8 ounces of water and have victim induce vomiting. *IF SWALLOWED* and victim is *UNCONSCIOUS OR HAVING CONVULSIONS*, do nothing except keep victim warm.
Note to medical personnel: Watch for lung damage from inhalation of solvent.

HEALTH HAZARDS
Personal protective equipment (PPE): B-Level PPE. Protective gloves; goggles; dust mask. Chemical protective material(s) reported to offer minimal to poor protection: nitrile.
Toxicity by ingestion: Grade 3; oral LD_{50} = 500 mg/kg (rat).
Long-term health effects: NCI animal carcinogen.
Liquid or solid irritant characteristics: Dust may cause eye irritation.

FIRE DATA
Flash point: More than 185°F/85°C (oc).

ENVIRONMENTAL DATA
Water pollution: Harmful to aquatic life in very low concentrations. May be dangerous if it enters nearby water intakes; notify operators. Notify local health and wildlife officials.
Response to discharge: Issue warning–poison, water contaminant. Restrict access. Should be removed. Chemical and physical treatment.

SHIPPING INFORMATION
Grades of purity: Technical: 95%. Emulsifiable concentrate in flammable solvents; **Storage temperature:** Ambient; **Inert atmosphere:** None; **Venting:** Pressure-vacuum; **Stability during transport:** Stable.

PHYSICAL AND CHEMICAL PROPERTIES
Physical state @ 59°F/15°C and 1 atm: Solid.
Molecular weight: 335.3
Boiling point @ 1 atm: Decomposes. 282°F/139°C/412°K.
Melting/Freezing point: 108°F/42°C/315°K.
Specific gravity (water = 1): 1.294 @ 77°F/25°C (solid).
Heat of combustion: (estimate) –9,040 Btu/lb = –5,020 cal/g = -210×10^5 J/kg.

TRIISOBUTYLALUMINUM   REC. T:3850

SYNONYMS: ALUMINUM, TRIISOBUTYL-; TRIISOBUTILALUMINIO (Spanish); ALUMINUM, TRIS(2-METHYLPROPYL)-; EEC No. 013-004-00-2; TIBA; TIBAL; TRIISOBUTYLALANE

IDENTIFICATION
CAS Number: 100-99-2
Formula: $(C_4H_9)_3Al$; $(iso\text{-}C_4H_9)_3Al$
DOT ID Number: UN 3051; DOT Guide Number: 135
Proper Shipping Name: Aluminum alkyls

DESCRIPTION: Clear, colorless liquid. Reacts with water forming flammable gas is. Aluminum alkyls may be shipped as a 15-30% solution in a hydrocarbon solvent that may alter its physical properties and make the material less reactive and not spontaneously reactive in air. Freezes at 34°F/–37°C.

Spontaneously combustible in air if hydrocarbon solvent evaporates • Solution is extremely flammable • Do NOT use water • Poison! • Corrosive • Breathing the vapor can kill you; skin or eye contact causes severe burns, impaired vision, or blindness; inhalation exposure may be delayed • Firefighting gear (including SCBA) does not provide adequate protection. If exposure occurs, remove and isolate gear immediately and thoroughly decontaminate personnel • Containers may BLEVE when exposed to fire • Vapors may form explosive mixture with air • Vapors are heavier than air and will collect and stay in low areas • Vapors may travel long distances to ignition sources and flashback • Vapors in confined areas (e.g., tanks, sewers, buildings) may explode when exposed to fire • Toxic products of combustion may include dense smoke and aluminum oxide vapors which can cause metal fume fever with flu-like symptoms • May re-ignite after fire is extinguished • Do not put yourself in danger by entering a contaminated area to rescue a victim.

Hazard Classification (based on NFPA-704 Rating System) (20% or less by weight in hydrocarbon solution) Health Hazards (Blue): 3; Flammability (Red): 4; Reactivity (Yellow): 3; Special Notice (White): Water Reactive

EMERGENCY RESPONSE: See Appendix A (135)
Evacuation:
Public safety: Isolate spill area for at least 100 to 150 meters (330 to 490 feet) in all directions.
Spill: Increase downwind, the distance shown above in "Public safety."
Fire: Isolate for 800 meters (½ mile) in all directions, especially if tank, rail car, or tank truck is involved in fire.

EXPOSURE
Short-term effects: *SEEK MEDICAL ATTENTION.* **Liquid:** Will burn skin and eyes. Harmful if swallowed. Remove contaminated clothing and shoes. Flush affected areas with plenty of water. *IF IN EYES*, hold eyelids open and flush with plenty of water. *IF SWALLOWED* and victim is *CONSCIOUS AND ABLE TO SWALLOW*, have victim drink 4 to 8 ounces of water. **Do NOT induce vomiting**.
Note to physician or authorized medical personnel: Pulmonary edema may be delayed. Medical observation is recommended for 24 to 48 hours after inhalation overexposure. As first aid for pulmonary edema, a physician or authorized medical personnel may consider administering a corticosteroid spray. Cigarette smoking may exacerbate pulmonary injury and should be discouraged for at least 72 hours following exposure.

HEALTH HAZARDS
Personal protective equipment (PPE): A-Level PPE. Full protective clothing, preferably of aluminized glass cloth.
Exposure limits (TWA unless otherwise noted): ACGIH TLV 2 mg/m³ as aluminum
Long-term health effects: May cause lung damage.
Vapor (gas) irritant characteristics: Severe eye and respiratory tract irritant.
Liquid or solid irritant characteristics: Severe eye and skin irritant. Causes second- and third-degree burns on short contact and is very injurious to the eyes.

1100 Triisobutylene

FIRE DATA
Flash point: Ignites spontaneously in air; –14°F/–26°C (15% solution in hexane).
Flammable limits in air: LEL: 1.1%; UEL: 7.5% (15% solution in hexane).
Fire extinguishing agents not to be used: Water, foam, halogenated extinguishing agents, CO_2.
Autoignition temperature: Ignites spontaneously under ambient conditions. Can re-ignite following extinguishing; 464°F/240°C/513°K (15% solution in hexane).

CHEMICAL REACTIVITY
Reactivity with water: Reacts violently to form flammable hydrocarbon gases; risk of fire and explosions.
Binary reactants: Violent reaction with oxidizers. Reacts with CO_2, alcohols, ammonia, halogenated hydrocarbons; other materials; risk of fire and explosions. Not compatible with silicone rubber or urethane rubber.

ENVIRONMENTAL DATA
Food chain concentration potential: Negative; unlikely to accumulate.
Water pollution: Effect of low concentrations on aquatic life is unknown. May be dangerous if it enters nearby water intakes; notify operators. Notify local health and wildlife officials.
Response to discharge: Issue warning–high flammability. Restrict access. Evacuate area. Disperse and flush with care.

SHIPPING INFORMATION
Grades of purity: Technical, 95+%; 20% or less by weight in benzene, hexane, or heptane (hydrocarbon solutions are not pyrophoric); electronic grade; **Storage temperature:** Ambient; **Inert atmosphere:** Inerted; dry nitrogen; **Inert atmosphere:** At 5 psig; **Venting:** Safety relief, with rupture disc; **Stability during transport:** Stable.

PHYSICAL AND CHEMICAL PROPERTIES
Physical state @ 59°F/15°C and 1 atm: Liquid.
Molecular weight: 198.3
Boiling point @ 1 atm: 414°F/212°C/485°K; 156°F/69°C/342°K (15% solution in hexane).
Melting/Freezing point: 33.8°F/1.0°C/274.2°K; –139°F/–95°C/178°K.
Specific gravity (water = 1): 0.788 @ 68°F/20°C (liquid).
Liquid surface tension: (estimate) 24 dynes/cm = 0.024 N/m @ 68°F/20°C.
Relative vapor density (air = 1): 6.3; 3.0 (15% solution in hexane).
Latent heat of vaporization: 101 Btu/lb = 56 cal/g = 2.3×10^5 J/kg.
Heat of combustion: –18,423 Btu/lb = –10,235 cal/g = $–428.23 \times 10^5$ J/kg.
Vapor pressure: 0.7 mm.

TRIISOBUTYLENE REC. T:3900

SYNONYMS: ISOBUTENE TRIMER; 1-PROPENE, 2-METHYL TRIMER; TRIISOBUTENE

IDENTIFICATION
CAS Number: 7756-94-7
Formula: $C_{12}H_{24}$; $(C_4H_8)_3$
DOT ID Number: UN 2324; DOT Guide Number: 128
Proper Shipping Name: Triisobutylene

DESCRIPTION: Pale yellow liquid. Odorless. Floats on water surface; insoluble.

Flammable • Containers may BLEVE when exposed to fire • Vapors may form explosive mixture with air • Vapors are heavier than air and will collect and stay in low areas • Vapors may travel long distances to ignition sources and flashback • Vapors in confined areas (e.g., tanks, sewers, buildings) may explode when exposed to fire • Irritating to the skin, eyes, and respiratory tract • Toxic products of combustion may include carbon monoxide.

Hazard Classification (based on NFPA-704 Rating System) (FEMA) Health Hazards (Blue): 1; Flammability (Red): 2; Reactivity (Yellow): 0

EMERGENCY RESPONSE: See Appendix A (128)
Evacuation:
Public safety: Isolate spill area for at least 25 to 50 meters (80 to 160 feet) in all directions.
Spill: Large spill–Consider initial downwind evacuation for at least 300 meters (1000 feet).
Fire: Isolate for 800 meters (½ mile) in all directions, especially if tank, rail car, or tank truck is involved in fire.

EXPOSURE
Short-term effects: *SEEK MEDICAL ATTENTION*. **Vapor:** If inhaled, will cause dizziness, difficult breathing, or loss of consciousness. Move to fresh air. *IF BREATHING HAS STOPPED*, give artificial respiration. IF breathing is difficult, administer oxygen. **Liquid:** Irritating to skin and eyes. Harmful if swallowed. Remove contaminated clothing and shoes. Flush affected areas with plenty of water. *IF IN EYES*, hold eyelids open and flush with plenty of water. *IF SWALLOWED* and victim is *CONSCIOUS AND ABLE TO SWALLOW*, induce vomiting immediately.

HEALTH HAZARDS
Personal protective equipment (PPE): B-Level PPE. Wear approved respirator, chemical resistant gloves or mask, apron, safety goggles and boots.
Vapor (gas) irritant characteristics: Vapors cause irritation such that personnel will find high concentrations unpleasant.
Liquid or solid irritant characteristics: Eye and skin irritant. Causes smarting of the skin and first-degree burns on short exposure; may cause second-degree burns on long exposure.

FIRE DATA
Flash point: Cannot be found in the literature.
Fire extinguishing agents not to be used: Water may be ineffective.

CHEMICAL REACTIVITY
Reactivity group: 30
Compatibility class: Olefins

ENVIRONMENTAL DATA
Water pollution: Effect of low concentrations on aquatic life is unknown. Fouling to shoreline. May be dangerous if it enters nearby water intakes; notify operators. Notify local health and wildlife officials. **Response to discharge:** Restrict access. Should be removed. Mechanical containment. Chemical and physical treatment.

SHIPPING INFORMATION
Inert atmosphere: Ambient; **Stability during transport:** Stable.

PHYSICAL AND CHEMICAL PROPERTIES
Physical state @ 59°F/15°C and 1 atm: Liquid.
Molecular weight: 168.32
Boiling point @ 1 atm: 350.6°F/177°C/450.2°K.
Melting/Freezing point: −104.8°F/−76°C/197.2°K.
Specific gravity (water = 1): 0.77
Relative vapor density (air = 1): 5.8
Vapor pressure: (Reid) 0.0754 psia.

TRIISOPROPANOLAMINE	REC. T:3950

SYNONYMS: 1,1',1"-NITROLOTRI-2-PROPANOL; 2-PROPANOL 1,1',1"-NITRILOTRI-; TRIISOPROPANOLAMINA (Spanish); TRIS(2-HYDROXYPROPYL) AMINE

IDENTIFICATION
CAS Number: 122-20-3
Formula: $C_9H_{21}NO_3$; $[(CH_3CHOH)CH_2]_3N$

DESCRIPTION: White solid. Slight ammonia odor. Sinks and dissolves in water.

Combustible • Corrosive to the skin, eyes, and respiratory tract; inhalation symptoms may be delayed • Heat or flame may cause explosion • Containers may BLEVE when exposed to fire • Vapors in confined areas (e.g., tanks, sewers, buildings) may explode when exposed to fire • Toxic products of combustion may include nitrogen oxides and cyanide.

Hazard Classification (based on NFPA-704 Rating System)
Health Hazards (Blue): 2; Flammability (Red): 1; Reactivity (Yellow): 0

EMERGENCY RESPONSE: See Appendix A (171)
Evacuation:
Public safety: Isolate the area of spill or leak for at least 10 to 25 meters (30 to 80 feet) in all directions.
Spill: Increase, in the downwind direction, as necessary, the distance shown under "Public Safety."
Fire: If any large container is involved in fire, isolate for at least 800 meters (½ mile) in all directions; also, consider initial evacuation for 800 meters (½ mile) in all directions.

EXPOSURE
Short-term effects: *SEEK MEDICAL ATTENTION*. Solid: Irritating to eyes and skin. May cause slight corneal injury or burn. Low to moderately toxic by oral routes. Remove contaminated clothing and shoes. Flush affected areas with plenty of water. *IF IN EYES*, hold eyelids open and irrigate with plenty of water for 15 minutes. *IF SWALLOWED* and victim is *UNCONSCIOUS OR HAVING CONVULSIONS*, do nothing except keep victim warm. *Note to physician or authorized medical personnel*: Medical observation is recommended for 24 to 48 hours after breathing overexposure, as pulmonary edema may be delayed. As first aid for pulmonary edema, consider administering a corticosteroid spray. Cigarette smoking may exacerbate pulmonary injury and should be discouraged for at least 72 hours following exposure.

HEALTH HAZARDS
Personal protective equipment (PPE): B-Level PPE. Clean, body-covering clothing, rubber gloves, apron, boots, and face shield as dictated by circumstances. Approved, full-face mask or amine vapor mask only if required during a fire.
Toxicity by ingestion: Grade 2; LD_{50} = 1.08 to 3.6 g/kg (guinea pigs).
Vapor (gas) irritant characteristics: Eye and respiratory tract irritant. Vapor causes moderate irritation such that personnel will find high concentration unpleasant.
Liquid or solid irritant characteristics: Eye and skin irritant. If spilled on clothing and allowed to remain, may cause smarting and reddening of skin. Dust causes eye irritation.

FIRE DATA
Flash point: 320°F/160°C (oc).
Flammable limits in air: LEL: 0.8% (calculated); UEL: 5.1% (estimate).
Autoignition temperature: 608°F/320°C/593°K.

CHEMICAL REACTIVITY
Binary reactants: Avoid oxidizing materials and strong acids.
Neutralizing agents for acids and caustics: Flush heavily with water.
Reactivity group: 8
Compatibility class: Alkanolamines

ENVIRONMENTAL DATA
Water pollution: Effect of low concentration on aquatic life is unknown. May be dangerous if it enters nearby water intakes; notify operators. Notify local health and wildlife officials. **Response to discharge:** Disperse and flush.

SHIPPING INFORMATION
Grades of purity: 99%; **Stability during transport:** Stable.

PHYSICAL AND CHEMICAL PROPERTIES
Physical state @ 59°F/15°C and 1 atm: Solid.
Molecular weight: 191.27
Boiling point @ 1 atm: 572–584°F/300–307°C/573–580°K.
Melting/Freezing point: 136°F/58°C/331°K.
Specific gravity (water = 1): 1.0200 @ 68°F/20°C.
Relative vapor density (air = 1): 6.60

TRIMETHYLACETIC ACID	REC. T:4000

SYNONYMS: ACIDO TRIMETILACETICO (Spanish); 2,2-DIMETHYLPROPIONIC ACID; α,α-DIMETHYLPROPIONIC ACID; NEOPENTANOIC ACID; PIVALIC ACID; *tert*-PROPANOIC ACID; PROPANOIC ACID; PROPANOIC ACID, 2,2-DI-METHYL-; TRIMETHYLACETIC ACID

IDENTIFICATION
CAS Number: 75-98-9
Formula: $C_5H_{10}O_2$; $(CH_3)_3CCOOH$
DOT ID Number: UN 1325 (pure material only); DOT Guide Number: 133
Proper Shipping Name: Flammable solids, organic, n.o.s.

DESCRIPTION: Colored crystalline solid. Floats on and slowly mixes with water.

Flammable solid • Containers may BLEVE when exposed to fire • Dust may form explosive mixture with air • Concentrated dust in confined areas (e.g., tanks, sewers, buildings) may explode when exposed to fire • Irritating to the skin, eyes, and respiratory tract • Toxic products of combustion may include carbon monoxide.

Hazard Classification (based on NFPA-704 Rating System)
Health Hazards (Blue): 2; Flammability (Red): 2; Reactivity (Yellow): 0

EMERGENCY RESPONSE: See Appendix A (133)
Evacuation:
Public safety: Isolate spill area for at least 10 to 25 meters (30 to 80 feet) in all directions.
Spill: Consider initial downwind evacuation of at least 100 meters (330 feet).
Fire: Isolate for 800 meters (½ mile) in all directions, especially if tank, rail car, or tank truck is involved in fire.

EXPOSURE
Short-term effects: *SEEK MEDICAL ATTENTION*. **Solid:** Irritating to eyes and skin. Harmful if swallowed. The vapor is irritating at elevated temperatures. Can cause considerable discomfort by oral routes; may cause reversible or irreversible changes to exposed tissue (skin or eyes), not permanent injury or death *IF IN EYES OR ON SKIN*, flush with running water for at least 15 minutes; hold eyelids open if necessary. Wash skin with soap and water. Remove and double bag contaminated clothing and shoes at the site. *IF SWALLOWED* and victim is *UNCONSCIOUS OR HAVING CONVULSIONS*, do nothing except keep victim warm.

HEALTH HAZARDS
Personal protective equipment (PPE): B-Level PPE. Wear self-contained positive breathing apparatus and full protective clothing.
Toxicity by ingestion: Grade 2; LD_{50} = 900 mg/kg (rat).
Long-term health effects: Tumorigenic in animal studies (mouse).
Vapor (gas) irritant characteristics: At elevated temperatures, vapor is irritating to eyes and skin.
Liquid or solid irritant characteristics: Solid is irritating to eyes and skin.

FIRE DATA
Flash point: 147°F/64°C (method unspecified).

CHEMICAL REACTIVITY
Neutralizing agents for acids and caustics: Cover contaminated surfaces with soda ash or sodium bicarbonate; add water to form slurry. Remove slurry and rinse area with soda ash solution.
Reactivity group: 4
Compatibility class: Organic acids

ENVIRONMENTAL DATA
Food chain concentration potential: Log P_{ow} = 1.42 (estimate). Unlikely to accumulate.
Water pollution: Dangerous to aquatic life in high concentrations. May be dangerous if it enter water intakes. Notify local health and wildlife officials. **Response to discharge:** Issue warning–water contaminant. Should be removed. Chemical and physical treatment.

SHIPPING INFORMATION
Grades of purity: 99%; **Stability during transport:** Stable.

PHYSICAL AND CHEMICAL PROPERTIES
Physical state @ 59°F/15°C and 1 atm: Solid.
Molecular weight: 102.13
Boiling point @ 1 atm: 325–327°F/163–164°C/436–437°K.
Melting/Freezing point: 91–95°F/33–35°C/306–308°K.
Specific gravity (water = 1): 0.905 at 50°C.
Relative vapor density (air = 1): 3.5 (estimate).

TRIMETHYLAMINE REC. T:4050

SYNONYMS: EEC No. 612-001-00-9; *N,N*-DIMETHYLMETHANAMINE; TMA; TRIMETILAMINA (Spanish)

IDENTIFICATION
CAS Number: 75-50-3
Formula: $(CH_3)_3N$
DOT ID Number: UN 1083 (anhydrous); UN 1297 (aqueous solution); DOT Guide Number: 118 (UN 1083); 132 (UN 1297)
Proper Shipping Name: Trimethylamine, anhydrous; trimethylamine, aqueous solution
Reportable Quantity (RQ): **(CERCLA)** 100 lb/45.4 kg

DESCRIPTION: Colorless, compressed gas. Fishy odor. Floats and mixes and boils on water. Poisonous, flammable visible vapor cloud is produced. May be shipped as an aqueous solution which is highly flammable and will generate large amounts of flammable gas. A liquid below 37°F/2.7°C.

Extremely flammable • Poisonous if inhaled • Corrosive to skin, eyes, and respiratory tract; skin and eye contact can cause severe burns and blindness; inhalation symptoms may be delayed • Firefighting gear (including SCBA) does not provide adequate protection. If exposure occurs, remove and isolate gear immediately and thoroughly decontaminate personnel • Containers may BLEVE when exposed to fire • Vapors may form explosive mixture with air • Gas is heavier than air and will collect and stay in low areas • Gas may travel long distances to ignition sources and flashback • Gas in confined areas (e.g., tanks, sewers, buildings) may explode when exposed to fire • Contact with the liquid may cause frostbite • Toxic products of combustion may include nitrogen oxides.

Hazard Classification (based on NFPA-704 Rating System)
Health Hazards (Blue): 3; Flammability (Red): 4; Reactivity (Yellow): 0

EMERGENCY RESPONSE, gas: See Appendix A (118)
Evacuation:
Public safety: Isolate spill area for at least 100 to 200 meters (330 to 660 feet) in all directions.
Spill: Consider initial downwind evacuation for at least 800 meters (½ mile).
Fire: Isolate for 1600 meters (1 mile) in all directions, especially if tank, rail car, or tank truck is involved in fire.

EMERGENCY RESPONSE, liquid: See Appendix A (132)
Evacuation:
Public safety: Isolate spill area for at least 50 to 100 meters (160 to 330 feet) in all directions.
Spill: Increase, as necessary, the isolation distance shown above, in "Public safety."
Fire: Isolate for 800 meters (½ mile) in all directions, especially if tank, rail car, or tank truck is involved in fire.

EXPOSURE
Short-term effects: *SEEK MEDICAL ATTENTION*. **Vapor:** *POISONOUS IF INHALED*. Irritating to eyes, nose, and throat. Move to fresh air. *IF IN EYES*, hold eyelids open and flush with plenty of water. *IF BREATHING HAS STOPPED*, give artificial

respiration. IF breathing is difficult, administer oxygen. Lung edema may develop. **Liquid:** Will burn skin and eyes. Harmful if swallowed. Remove contaminated clothing and shoes. Flush affected areas with plenty of water. *IF IN EYES*, hold eyelids open and flush with plenty of water. *IF SWALLOWED* and victim is *CONSCIOUS AND ABLE TO SWALLOW*, have victim drink 4 to 8 ounces of water.

Note to physician or authorized medical personnel: Medical observation is recommended for 24 to 48 hours after breathing overexposure, as pulmonary edema may be delayed. As first aid for pulmonary edema, consider administering a corticosteroid spray. Cigarette smoking may exacerbate pulmonary injury and should be discouraged for at least 72 hours following exposure.

HEALTH HAZARDS
Personal protective equipment (PPE): B-Level PPE. OSHA Table Z-1-A air contaminant. Vapor-proof goggles and face shield; rubber gloves; air–supplied mask. Wear thermal protective clothing.
Exposure limits (TWA unless otherwise noted): ACGIH TLV 5 ppm (12 mg/m^3); NIOSH REL 10 ppm (24 mg/m^3).
Short-term exposure limits (15-minute TWA): ACGIH STEL 15 ppm (36 mg/m^3); NIOSH STEL 15 ppm (36 mg/m^3).
Vapor (gas) irritant characteristics: Eye and respiratory tract irritant. Vapor is irritating such that personnel will not usually tolerate moderate or high concentrations.
Liquid or solid irritant characteristics: Eye and skin irritant. Causes smarting of the skin and first-degree burns on short exposure; may cause secondary burns on long exposure. rapid evaporation of liquid may cause frostbite.
Odor threshold: 0.0001–0.9 ppm.

FIRE DATA
Flash point: Flammable gas; <0°F/<–18°C; 20°F/–7°C (oc); 10°F/–12°C (cc) (aqueous solution).
Flammable limits in air: LEL: 2.0%; UEL: 11.6%.
Autoignition temperature: 374°F/190°C/463°K.
Burning rate: 8 mm/min
Stoichiometric air-to-fuel ratio: 12.19 (estimate)

CHEMICAL REACTIVITY
Binary reactants: A medium strong organic base. Violent reaction with strong oxidizers, bromine, strong acids, ethylene oxide, halogenated compounds, nitrosating compounds, triethynylaluminum. Incompatible with methyl trichloroacetate. Reacts with mercury, forming shock-sensitive explosive material. Attacks chemically active metals: aluminum, copper, tin, zinc, and their alloys.
Neutralizing agents for acids and caustics: Although water solutions may be neutralized with acetic acid, simple evaporation will remove all of the compound.

ENVIRONMENTAL DATA
Food chain concentration potential: Negative; unlikely to accumulate.
Food chain concentration potential: Log P_{ow} = 0.29. Will not accumulate.
Water pollution: Effect of low concentrations on aquatic life is unknown. May be dangerous if it enters nearby water intakes; notify operators. Notify local health and wildlife officials.
Response to discharge: Issue warning–high flammability, air contaminant. Restrict access. Evacuate area. Clean Water Act.

SHIPPING INFORMATION
Grades of purity: Anhydrous, 98.5+%; also shipped as 25–30% solution in water; **Storage temperature:** Ambient; **Inert atmosphere:** None; **Venting:** Safety relief; **Stability during transport:** Stable.

NAS HAZARD CLASSIFICATION FOR BULK WATER TRANSPORTATION
FIRE: 4
HEALTH: Vapor irritant: 3; Liquid or solid irritant: 2; Poisons: 3
WATER POLLUTION: Human toxicity: 2; Aquatic toxicity: 3; Aesthetic effect: 2
REACTIVITY: Other chemicals: 3; Water: 0; Self-reaction: 0

PHYSICAL AND CHEMICAL PROPERTIES (Physical properties apply to anhydrous material.)
Physical state @ 59°F/15°C and 1 atm: Gas; 86°F/30°C/303°K (40% aqueous solution).
Molecular weight: 59.11
Boiling point @ 1 atm: 37.2°F/2.9°C/276.1°K (gas).
Melting/Freezing point: –178.8°F/–117.1°C/156.1°K (gas); 36°F/2°C/275°K (40% aqueous solution).
Critical temperature: 320.2°F/160.1°C/433.3°K.
Critical pressure: 591 psia = 40.2 atm = 4.07 MN/m^2.
Specific gravity (water = 1): 0.633 @ 68°F/20°C; 0.7 (aqueous solution).
Liquid surface tension: 17.4 dynes/cm = 0.0174 N/m at –4°C.
Relative vapor density (air = 1): 2.09
Ratio of specific heats of vapor (gas): 1.139
Latent heat of vaporization: 174 Btu/lb = 96.5 cal/g = 4.04 x 10^5 J/kg.
Heat of combustion: –17,660 Btu/lb = –9810 cal/g = –410.7 x 10^5 J/kg.
Heat of solution: –385 Btu/lb = –214 cal/g = –8.96 x 10^5 J/kg.
Heat of fusion: 26.47 cal/g.
Vapor pressure: 1454 mm @ 70°F/21°C/294°K; 505 mm @ 70°F/21°C/294°K (aqueous solution).

1,2,4-TRIMETHYLBENZENE   REC. T:4100

SYNONYMS: asym-TRIMETHYL BENZENE; ψ-CUMENE; psi-CUMENE; PSEUDOCUMENE; PSEUDOCUMOL; 1,2,4-TRIMETILBENCENO (Spanish)

IDENTIFICATION
CAS Number: 95-63-6; 108-67-8 (1,3,5-isomer)
Formula: C$_9$H$_{12}$; C$_6$H$_3$(CH$_3$)$_3$
DOT ID Number: UN 1993; UN 2325 (1,3,5-isomer); DOT Guide Number: 128; 129 (1,3,5-isomer)
Proper Shipping Name: Flammable liquids, n.o.s; 1,3,5-trimethylbenzene

DESCRIPTION: Colorless liquid. Irritant. Inoluble in water.

Flammable • Containers may BLEVE when exposed to fire • Vapors may form explosive mixture with air • Vapors are heavier than air and will collect and stay in low areas • Vapors may travel long distances to ignition sources and flashback • Vapors in confined areas (e.g., tanks, sewers, buildings) may explode when exposed to fire • Irritating to the skin, eyes, and respiratory tract • Toxic products of combustion may include carbon monoxide.

Hazard Classification (based on NFPA-704 Rating System)
Health Hazards (Blue): 0; Flammability (Red): 2; Reactivity (Yellow): 0

EMERGENCY RESPONSE (1,2,4-isomer): See Appendix A (128)
Evacuation:
Public safety: Isolate spill area for at least 25 to 50 meters (80 to 160 feet) in all directions.
Spill: Large spill–Consider initial downwind evacuation for at least 300 meters (1000 feet).
Fire: Isolate for 800 meters (½ mile) in all directions, especially if tank, rail car, or tank truck is involved in fire.

EMERGENCY RESPONSE (1,3,5-isomer): See Appendix A (129)
Evacuation:
Public safety: Isolate spill area for at least 50 to 100 meters (160 to 330 feet) in all directions.
Spill: Large spill–Consider initial downwind evacuation for at least 300 meters (1000 feet).
Fire: Isolate for 800 meters (½ mile) in all directions, especially if tank, rail car, or tank truck is involved in fire.

EXPOSURE
Short-term effects: *CALL FOR MEDICAL AID* **Liquid:** Irritating to skin and eyes and respiratory tract. Remove contaminated clothing and shoes. Flush affected areas with plenty of water. *IF IN EYES*, hold eyelids open and flush with plenty of water. *IF SWALLOWED*, Do NOT induce vomiting.
Vapors or mist: Irritating to skin, eyes and respiratory tract. Move to fresh air. *IF BREATHING HAS STOPPED*, give artificial respiration. IF breathing is difficult, administer oxygen. *IF IN EYES*, hold eyelids open and flush with plenty of water.

HEALTH HAZARDS
Personal protective equipment (PPE): B-Level PPE. Wear self contained breathing apparatus, rubber boots, and heavy rubber gloves.
Exposure limits (TWA unless otherwise noted): ACGIH TLV 25 (123 mg/m^3); NIOSH REL 25 ppm (125 mg/m^3).
Toxicity by ingestion: Grade 2: LD_{50} = 5 g/kg (rat).
Vapor (gas) irritant characteristics: Eye and respiratory tract irritant. Vapors cause smarting of the eyes or respiratory system; if present in high concentrations the sensation is unpleasant.
Liquid or solid irritant characteristics: Eye and skin irritant. If spilled on clothing and allowed to remain, may cause smarting and reddening of skin.

FIRE DATA
Flash point: 112°F/44°C (cc).
Flammable limits in air: LEL: 0.9%; UEL: 6.4%.
Autoignition temperature: 959°F/515°C/788°K.

CHEMICAL REACTIVITY
Binary reactants: Oxidizers, nitric acid
Reactivity group: 32
Compatibility class: Aromatic hydrocarbons

ENVIRONMENTAL DATA
Water pollution: DOT Appendix B, §172.101–marine pollutant. Effect of low concentration on aquatic life is unknown. Fouling to shoreline. May be dangerous if it enters nearby water intakes; notify operators. Notify local health and wildlife officials.
Response to discharge: Restrict access. Mechanical containment. Should be removed. Chemical and physical treatment.

SHIPPING INFORMATION
Grades of purity: 99%; **Storage temperature:** Ambient

PHYSICAL AND CHEMICAL PROPERTIES
Physical state @ 59°F/15°C and 1 atm: Liquid.
Molecular weight: 120.20
Boiling point @ 1 atm: 334°F/168°C/441°K; 329°F/165°C/438°K (1,3,5-isomer).
Melting/Freezing point: −77°F/−61°C/212°K; −64°F/−53°C/220°K (1,3,5-isomer).
Specific gravity (water = 1): 0.889; 0863 (1,3,5-isomer).
Relative vapor density (air = 1): 4.2
Vapor pressure: (Reid) 0.0948 psia; 1 mm @ 56°F

TRIMETHYLCHLOROSILANE REC. T:4150

SYNONYMS: CHLOROTRIMETHYL SILANE; CHLOROTRIMETHYLSILICANE; SILANE, CHLOROTRIMETHYL-; SILANE TRIMETHYLCHLORO-; SILICANE, CHLOROTRIMETHYL-; TL 1163; CHLOROTRIMETHYLSILANE; TRIMETHYLSILYL CHLORIDE; TRIMETILCLOROSILANO (Spanish)

IDENTIFICATION
CAS Number: 75-77-4
Formula: C_3H_9ClSi; $(CH_3)_3SiCl$
DOT ID Number: UN 1298; DOT Guide Number: 155
Proper Shipping Name: Trimethylchlorosilane
Reportable Quantity (RQ): **(EHS)** 1 lb/0.454 kg

DESCRIPTION: Colorless fuming liquid. Sharp, irritating odor, like hydrochloric acid. Reacts violently with water; forming toxic hydrochloric acid and hydrogen chloride fumes.

Poison! • Do NOT use water • Breathing the vapor can kill you; skin or eye contact causes severe burns, impaired vision, or blindness • Firefighting gear (including SCBA) does not provide adequate protection. If exposure occurs, remove and isolate gear immediately and thoroughly decontaminate personnel • Containers may BLEVE when exposed to fire • Toxic products of combustion may include hydrogen chloride • Corrosive to metals forming explosive hydrogen gas • Do not put yourself in danger by entering a contaminated area to rescue a victim.

Hazard Classification (based on NFPA-704 Rating System)
Health Hazards (Blue): 3; Flammability (Red): 3; Reactivity (Yellow): 2; Special Notice (White): Water Reactive

EMERGENCY RESPONSE: See Appendix A (155)
Evacuation:
Public safety: Isolate the area of spill or leak for at least 50 to 100 meters (160 to 330 feet) in all directions.
Spill: Increase, in the downwind direction, as necessary, the distance shown under "Public Safety."
Fire: If tank, rail car, or tank truck is involved in fire, isolate for at least 800 meters (½ mile) in all directions; also, consider initial evacuation for 800 meters (½ mile) in all directions.

EXPOSURE
Short-term effects: *SEEK MEDICAL ATTENTION.* **Vapor:** Irritating to eyes, nose, and throat. Harmful if inhaled. Move victim to fresh air. *IF BREATHING HAS STOPPED*, give artificial respiration; *avoid mouth-to-mouth resuscitation; use bag/mask apparatus.* IF breathing is difficult, administer oxygen. **Liquid:** Will burn skin and eyes. Harmful if swallowed. Remove contaminated clothing and shoes. Flush affected areas with plenty of water. *IF IN EYES*, hold eyelids open and flush with plenty of

water. *IF SWALLOWED* and victim is *CONSCIOUS AND ABLE TO SWALLOW*, have victim drink 4 to 8 ounces of water. **Do NOT induce vomiting.**

HEALTH HAZARDS
Personal protective equipment (PPE): A-Level PPE. Acid-vapor-type respiratory protection; rubber gloves; chemical worker's goggles; other protective equipment as necessary to protect skin and eyes.
Toxicity by ingestion: Grade 3; LD_{50} = 0.5 to 5 g/kg.
Vapor (gas) irritant characteristics: Vapors cause severe irritation of eyes and throat and can cause eye and lung injury. They cannot be tolerated even at low concentrations.
Liquid or solid irritant characteristics: Severe eye and skin irritant. Causes second- and third-degree burns on short contact and is very injurious to the eyes.

FIRE DATA
Flash point: –18KF = –28°C (cc).
Flammable limits in air: LEL: 1.8%; UEL: 6.0%.
Fire extinguishing agents not to be used: Water, foam
Behavior in fire: Difficult to extinguish; re-ignition may occur.
Autoignition temperature: 743°F/395°C/668°K.
Burning rate: 5.3 mm/min

CHEMICAL REACTIVITY
Reactivity with water: Reacts vigorously, evolving hydrochloric acid and hydrogen chloride gas.
Binary reactants: In the presence of moisture, corrodes common metals and form flammable hydrogen gas. Reaction with air yields hydrochloric acid fumes.
Neutralizing agents for acids and caustics: Flush with water, rinse with sodium bicarbonate or lime solution.

ENVIRONMENTAL DATA
Food chain concentration potential: Negative; unlikely to accumulate.
Water pollution: Effect of low concentrations on aquatic life is unknown. May be dangerous if it enters nearby water intakes; notify operators. Notify local health and wildlife officials.
Response to discharge: Issue warning–high flammability, air contaminant, corrosive. Restrict access. Evacuate area. Disperse and flush with care.

SHIPPING INFORMATION
Grades of purity: 98+%; **Storage temperature:** Ambient; **Inert atmosphere:** None; **Venting:** Pressure-vacuum; **Stability during transport:** Stable.

NAS HAZARD CLASSIFICATION FOR BULK WATER TRANSPORTATION
FIRE: 3
HEALTH: Vapor irritant: 4; Liquid or solid irritant: 4; Poisons: 3
WATER POLLUTION: Human toxicity: 3; Aquatic toxicity: 3; Aesthetic effect: 2
REACTIVITY: Other chemicals: 3; Water: 4; Self-reaction: 1

PHYSICAL AND CHEMICAL PROPERTIES
Physical state @ 59°F/15°C and 1 atm: Liquid.
Molecular weight: 108.7
Boiling point @ 1 atm: 135°F/57°C/330°K.
Specific gravity (water = 1): 0.846 @ 77°F/25°C (liquid).
Liquid surface tension: (estimate) 17.8 dynes/cm = 0.0178 N/m @ 68°F/20°C.
Relative vapor density (air = 1): 3.7
Ratio of specific heats of vapor (gas): (estimate) 1.0683
Latent heat of vaporization: 126 Btu/lb = 70 cal/g = 2.9 x 10^5 J/kg.
Heat of combustion: (estimate) –10,300 Btu/lb = –5700 cal/g = –240 x 10^5 J/kg.
Vapor pressure: 200 mm.

TRIMETHYL HEXAMETHYLENE DIAMINE
REC. T:4200

SYNONYMS: 1,6-HEXANEDIAMINE,2,2,4 TRIMETHYL-; 1,6-HEXANEDIAMINE,2,4,4-TRIMETHYL-; 1,6-DIAMINO-2,2,4 TRIMETHYLHEXANE; 1,6-DIAMINO-2,4,4-TRIMETHYLHEXANE; TRIMETILHEXAMETILENDIAMINO (Spanish)

IDENTIFICATION
CAS Number: 25513-64-8
Formula: $C_9H_{22}N_2$
DOT ID Number: UN 2327; UN 1783; DOT Guide Number: 153
Proper Shipping Name: Trimethylhexamethylenediamines; Hexamethylenediamine solution

DESCRIPTION: Colorless liquid. Faint amine odor. Floats on the surface of water; soluble.

Combustible • Heat or flame may cause explosion • Corrosive to skin, eyes, and respiratory tract; contact with skin or eyes may cause burns or impair vision; inhalation symptoms may be delayed • Containers may BLEVE when exposed to fire • Vapors in confined areas (e.g., tanks, sewers, buildings) may explode when exposed to fire • Toxic products of combustion may include nitrogen oxides.

Hazard Classification (based on NFPA-704 Rating System)
Health Hazards (Blue): 1; Flammability (Red): 1; Reactivity (Yellow): 0

EMERGENCY RESPONSE: See Appendix A (153)
Evacuation:
Public safety: Isolate the area of spill or leak for at least 25 to 50 meters (80 to 160 feet) in all directions.
Spill: Increase, in the downwind direction, as necessary, the distance shown under "Public Safety."
Fire: If tank, rail car, or tank truck is involved in fire, isolate for at least 800 meters (½ mile) in all directions; also, consider initial evacuation for 800 meters (½ mile) in all directions.

EXPOSURE
Short-term effects: *SEEK MEDICAL ATTENTION*. **Vapor:** Irritating to eyes, mucous membranes and skin. Overexposure causes coughing and nausea. Move to fresh air. *IF BREATHING HAS STOPPED*, give artificial respiration; *avoid mouth-to-mouth resuscitation; use bag/mask apparatus*. IF breathing is difficult, administer oxygen. Lung edema may develop. **Liquid:** Will burn eyes and skin. May be harmful if swallowed. Remove and double bag contaminated clothing and shoes. Flush affected areas with running water for at least 15 minutes. *IF IN EYES* hold eyelids open while flushing with water. *IF SWALLOWED* and victim is *UNCONSCIOUS OR HAVING CONVULSIONS*, do nothing except keep victim warm.
Note to physician or authorized medical personnel: Medical observation is recommended for 24 to 48 hours after breathing overexposure, as pulmonary edema may be delayed. As first aid for

pulmonary edema, consider administering a corticosteroid spray. Cigarette smoking may exacerbate pulmonary injury and should be discouraged for at least 72 hours following exposure.

HEALTH HAZARDS
Personal protective equipment (PPE): B-Level PPE. Wear self-contained (positive pressure if available) breathing apparatus (with acid filter like that used for ammonia) and full protective clothing.
Vapor (gas) irritant characteristics: Eye and respiratory tract irritant. Overexposure causes coughing and nausea.
Liquid or solid irritant characteristics: Severe eye and skin irritant. Contact causes burns to skin and eyes. Visible necrosis of intact skin occurs within a period of 1 to 4 hours.

FIRE DATA
Flash point: 261°F/127°C (oc).
Behavior in fire: May generate toxic and irritating gases.

CHEMICAL REACTIVITY
Neutralizing agents for acids and caustics: Sodium bisulfate
Reactivity group: 7
Compatibility class: Aliphatic amines

ENVIRONMENTAL DATA
Water pollution: Effect of low concentration on aquatic life is unknown. May be dangerous if it enters nearby water intakes; notify operators. Notify local health and wildlife officials.
Response to discharge: Issue warning–water contaminant. Restrict access. Should be removed. Chemical and physical treatment.

SHIPPING INFORMATION
Grades of purity: More than 99.7%; **Stability during transport:** Stable.

PHYSICAL AND CHEMICAL PROPERTIES
Molecular weight: 158.29 (calculated).
Boiling point @ 1 atm: 449.6°F/232°C/505.2°K.
Specific gravity (water = 1): 0.867 @ 68°F/20°C.
Relative vapor density (air = 1): 5.47

TRIMETHYLHEXAMETHYLENE DIISOCYANATE
REC. T:4250

SYNONYMS: HEXANE,1,6-DIISOCYANATO-2,2,4-TRIMETHYL-; TRIMETILHEXAMETILENDIISOCIATO (Spanish)

IDENTIFICATION
CAS Number: 16938-22-0 (2,2,4-isomer)
Formula: $C_{11}H_{18}N_2O_2$
DOT ID Number: UN 2328; DOT Guide Number: 156
Proper Shipping Name: Trimethylhexamethylene diisocyanate

DESCRIPTION: Colorless or yellowish liquid. Reacts with water; producing CO_2 and the diamine.

Flammable • Poison! • Breathing the vapor, skin or eye contact, or swallowing the material can kill you • Firefighting gear (including SCBA) does not provide adequate protection. If exposure occurs, remove and isolate gear immediately and thoroughly decontaminate personnel • Containers may BLEVE when exposed to fire • Toxic products of combustion may include nitrogen oxides • Do not put yourself in danger by entering a contaminated area to rescue a victim:

Hazard Classification (based on NFPA-704 Rating System)
Health Hazards (Blue): 2; Flammability (Red): 2; Reactivity (Yellow): 1; Special Notice (White): Water reactive

EMERGENCY RESPONSE: See Appendix A (156)
Evacuation:
Public safety: Isolate the area of spill or leak for at least 50 to 100 meters (160 to 330 feet) in all directions.
Spill: Increase, in the downwind direction, as necessary, the distance shown under "Public Safety."
Fire: If tank, rail car, or tank truck is involved in fire, isolate for at least 800 meters (½ mile) in all directions; also, consider initial evacuation for 800 meters (½ mile) in all directions.

EXPOSURE
Short-term effects: SEEK MEDICAL ATTENTION. **Vapor:** *POISONOUS. MAY BE FATAL IF INHALED OR ABSORBED THOUGH SKIN.* Contact may cause burns to skin and eyes. Move to fresh air. *IF BREATHING HAS STOPPED*, give artificial respiration; *avoid mouth-to-mouth resuscitation; use bag/mask apparatus.* IF breathing is difficult, administer oxygen. **Liquid:** *POISONOUS. MAY BE FATAL IF SWALLOWED OR ABSORBED THROUGH SKIN.* Contact may burn skin and eyes. Immediately flush skin or eyes with running water for at least 15 minutes. Speed in removing from skin is of extreme importance. Remove and double bag contaminated clothing and shoes at the site. Keep victim quiet and maintain normal body temperature.
Effects may be delayed; keep victim under observation. *IF SWALLOWED* and victim is *UNCONSCIOUS OR HAVING CONVULSIONS*, do nothing except keep victim warm.

HEALTH HAZARDS
Personal protective equipment (PPE): A-Level PPE. Wear positive pressure breathing apparatus and special protective clothing.

CHEMICAL REACTIVITY
Reactivity with water: Can react with water to produce CO_2 and the diamine.
Binary reactants: May react violently with alcohols in the presence of a base.
Reactivity group: 12
Compatibility class: Isocyanates

ENVIRONMENTAL DATA
Water pollution: Effect of low concentrations on aquatic life is unknown. May be dangerous if it enters nearby water intakes; notify operators. Notify local health and wildlife officials. **Response to discharge:** Issue warning–poison. Restrict access. Should be removed.

PHYSICAL AND CHEMICAL PROPERTIES
Physical state @ 59°F/15°C and 1 atm: Liquid.
Molecular weight: 210.27
Relative vapor density (air = 1): 7.3

TRIMETHYLHEXAMETHYLENE DIISOCYANATE
REC. T:4300

SYNONYMS: HEXANE,1,6-DIISOCYANATO-2,4,4-TRIMETHYL-; TRIMETILHEXAMETILENDIISOCIATO (Spanish)

IDENTIFICATION
CAS Number: 15646-96-5
Formula: $C_{11}H_{18}N_2O_2$
DOT ID Number: UN 2328; DOT Guide Number: 156
Proper Shipping Name: Trimethylhexamethylene diisocyanate

DESCRIPTION: Colorless or yellowish liquid. Reacts with water; producing CO_2 and the diamine.

Flammable • Poison! • Breathing the vapor, skin or eye contact, or swallowing the material can kill you • Firefighting gear (including SCBA) does not provide adequate protection. If exposure occurs, remove and isolate gear immediately and thoroughly decontaminate personnel • Containers may BLEVE when exposed to fire • Toxic products of combustion may include nitrogen oxides • Do not put yourself in danger by entering a contaminated area to rescue a victim:

Hazard Classification (based on NFPA-704 Rating System)
Health Hazards (Blue): 2; Flammability (Red): 2; Reactivity (Yellow): 1; Special Notice (White): Water reactive

EMERGENCY RESPONSE: See Appendix A (156)
Evacuation:
Public safety: Isolate the area of spill or leak for at least 50 to 100 meters (160 to 330 feet) in all directions.
Spill: Increase, in the downwind direction, as necessary, the distance shown under "Public Safety."
Fire: If tank, rail car, or tank truck is involved in fire, isolate for at least 800 meters (½ mile) in all directions; also, consider initial evacuation for 800 meters (½ mile) in all directions.

EXPOSURE
Short-term effects: *SEEK MEDICAL ATTENTION.* **Vapor:** *POISONOUS. MAY BE FATAL IF INHALED OR ABSORBED THOUGH SKIN.* Contact may cause burns to skin and eyes. Move to fresh air. *IF BREATHING HAS STOPPED,* give artificial respiration; *avoid mouth-to-mouth resuscitation; use bag/mask apparatus.* IF breathing is difficult, administer oxygen. **Liquid:** *POISONOUS. MAY BE FATAL IF SWALLOWED OR ABSORBED THROUGH SKIN.* Contact may burn skin and eyes. Immediately flush skin or eyes with running water for at least 15 minutes. Speed in removing from skin is of extreme importance. Remove and double bag contaminated clothing and shoes at the site. Keep victim quiet and maintain normal body temperature. Effects may be delayed; keep victim under observation. *IF SWALLOWED* and victim is *UNCONSCIOUS OR HAVING CONVULSIONS,* do nothing except keep victim warm.

HEALTH HAZARDS
Personal protective equipment (PPE): A-Level PPE. Wear positive pressure breathing apparatus and special protective clothing.

CHEMICAL REACTIVITY
Reactivity with water: Can react with water to produce CO_2 and the diamine.
Reactivity group: 12
Compatibility class: Isocyanates

ENVIRONMENTAL DATA
Water pollution: Effect of low concentrations on aquatic life is unknown. May be dangerous if it enters nearby water intakes; notify operators. Notify local health and wildlife officials.

Response to discharge: Issue warning–poison. Restrict access. Should be removed.

PHYSICAL AND CHEMICAL PROPERTIES
Physical state @ 59°F/15°C and 1 atm: Liquid.
Molecular weight: 210.27
Relative vapor density (air = 1): 7.3

2,2,4-TRIMETHYL-1,3-PENTANEDIOL-1-ISOBUTYRATE
REC. T:4350

SYNONYMS: MONOISOBUTIRATO de 2,4,4-TRIMETIL-1,3-PENTANODIOL (Spanish); TEXANOL; 2,2,4-TRIMETHYL-1,3-PENTANEDIOLMONOISOBUTYRATE

IDENTIFICATION
CAS Number: 25265-77-4; 6846-50-0
Formula: $C_{12}H_{24}O_3$

DESCRIPTION: Colorless liquid. Mild, characteristic odor. Floats on the surface of water.

Combustible • Heat or flame may cause explosion • Containers may BLEVE when exposed to fire • Vapors in confined areas (e.g., tanks, sewers, buildings) may explode when exposed to fire • Irritating to the skin, eyes, and respiratory tract • Toxic products of combustion may include carbon monoxide.

Hazard Classification (based on NFPA-704 Rating System)
Health Hazards (Blue): 0; Flammability (Red): 1; Reactivity (Yellow): 0

EMERGENCY RESPONSE: See Appendix A (171)
Evacuation:
Public safety: Isolate the area of spill or leak for at least 10 to 25 meters (30 to 80 feet) in all directions.
Spill: Increase, in the downwind direction, as necessary, the distance shown under "Public Safety."
Fire: If any large container is involved in fire, isolate for at least 800 meters (½ mile) in all directions; also, consider initial evacuation for 800 meters (½ mile) in all directions.

EXPOSURE
Short-term effects: *SEEK MEDICAL ATTENTION.* **Vapor:** Move victim to fresh air. *IF BREATHING HAS STOPPED,* give artificial respiration. If breathing is difficult, administer oxygen. **Liquid:** Remove contaminated clothing and shoes. Wash affected areas with soap and water. *IF IN EYES,* hold eyelids open and flush with plenty of water.

HEALTH HAZARDS
Personal protective equipment (PPE): B-Level PPE. Where splashing is possible wear full face shield or chemical safety goggles.
Vapor (gas) irritant characteristics: Vapors cause a slight smarting of the eyes or respiratory system if present in high concentrations. The effect is temporary.
Liquid or solid irritant characteristics: Minimum hazard. If spilled on clothing and allowed to remain, may cause smarting and reddening of the skin.

FIRE DATA
Flash point: 248°F/120°C (oc).

Trimethyl phosphite

Flammable limits in air: LEL: 0.62% @ 300°F/149°C; UEL: 4.24% @ 393°F/201°C.
Fire extinguishing agents not to be used: Since material is lighter than water and insoluble, fire could be spread by using water in an uncontained area.
Autoignition temperature: 743°F/395°C/668°K.
Stoichiometric air-to-fuel ratio: 75.0

CHEMICAL REACTIVITY
Binary reactants: Can react vigorously with oxidizing agents.
Reactivity group: 34
Compatibility class: Esters.

ENVIRONMENTAL DATA
Water pollution: Effect of low concentrations on aquatic life is unknown. May be dangerous if it enters nearby water intakes; notify operators. Notify local health and wildlife officials.
Response to discharge: Should be removed.

SHIPPING INFORMATION
Grades of purity: Technical grades; **Storage temperature:** Ambient; **Inert atmosphere:** None; **Venting:** Open; **Stability during transport:** Stable.

PHYSICAL AND CHEMICAL PROPERTIES
Physical state @ 59°F/15°C and 1 atm: Liquid.
Molecular weight: 216.31
Boiling point @ 1 atm: 471°F/244°C/517°K.
Specific gravity (water = 1): 0.95 @ 20°C.
Relative vapor density (air = 1): 7.45

TRIMETHYL PHOSPHITE   REC. T:4400

SYNONYMS: FOSFITO de TRIMETILO (Spanish); METHYL PHOSPHITE; Phosphorus ACID,TRIMETHYL ESTER; PHTHALIC ACID, BIS(2-METHOXYETHYL)ESTER; TMP; TRIMETHOXYPHOSPHINE; TRIMETHYLPHOSPHITE

IDENTIFICATION
CAS Number: 121-45-9
Formula: $C_3H_9O_3P$; $(CH_3O)_3P$
DOT ID Number: UN 2329; DOT Guide Number: 129
Proper Shipping Name: Trimethyl phosphite

DESCRIPTION: Colorless liquid. Sickly, pyridine-like odor. Insoluble in water.

Flammable • Containers may BLEVE when exposed to fire • Vapors may form explosive mixture with air • Vapors are heavier than air and will collect and stay in low areas • Vapors may travel long distances to ignition sources and flashback • Vapors in confined areas (e.g., tanks, sewers, buildings) may explode when exposed to fire • Irritating to the skin, eyes, and respiratory tract • Toxic products of combustion may include phosphorus oxides.

Hazard Classification (based on NFPA-704 Rating System)
Health Hazards (Blue): 0; Flammability (Red): 2; Reactivity (Yellow): 0

EMERGENCY RESPONSE: See Appendix A (129)
Evacuation:
Public safety: Isolate spill area for at least 50 to 100 meters (160 to 330 feet) in all directions.
Spill: Large spill–Consider initial downwind evacuation for at least 300 meters (1000 feet).
Fire: Isolate for 800 meters (½ mile) in all directions, especially if tank, rail car, or tank truck is involved in fire.

EXPOSURE
Short-term effects: *SEEK MEDICAL ATTENTION*. **Vapor:** *MAY BE POISONOUS IF INHALED*. Vapors may cause dizziness or suffocation. Fire may produce irritating or poisonous gases. Move victim to fresh air. If not breathing, give artificial respiration. IF breathing is difficult, administer oxygen. **Liquid:** Extremely destructive to upper respiratory tract, eyes, and skin. Harmful if absorbed through skin and eyes. Wash skin with soap and water. Remove and double bag contaminated clothing and shoes at the site. *IF IN EYES*: immediately flush eyes with running water for at least 15 minutes. *IF SWALLOWED* and victim is CONSCIOUS: Have victim drink 4 to 8 ounces of water. **Do NOT induce vomiting**. IF SWALLOWED and victim is *UNCONSCIOUS OR HAVING CONVULSIONS*: Do nothing except keep victim warm.

HEALTH HAZARDS
Personal protective equipment (PPE): B-Level PPE. OSHA Table Z-1-A air contaminant. Approved respirator, safety goggles, rubber gloves, full protective clothing.
Exposure limits (TWA unless otherwise noted): ACGIH TLV 2 ppm (10 mg/m^3); NIOSH 2 ppm (10 mg/m^3); skin contact contributes significantly in overall exposure.
Toxicity by ingestion: Grade 2: LD_{50} = 1.6 g/kg (rat).
Long-term health effects: Hydrolyzes to dimethyl hydrogen phosphite, a known animal carcinogen. NCI carcinogenic in animals.
Vapor (gas) irritant characteristics: Vapors cause severe irritation of the eyes and throat and can cause eye and lung injury. They cannot be tolerated even at low concentrations.
Liquid or solid irritant characteristics: Severe eye and skin irritant. Causes second- and third-degree burns on short contact and is very injurious to the eyes.
Odor threshold: 0.001 ppm.

FIRE DATA
Flash point: 130°F/55°C (oc); 82°F/28°C (cc).

CHEMICAL REACTIVITY
Binary reactants: Reacts with strong oxidizers; magnesium salts of perchloric acid (explosive).
Reactivity group: 34
Compatibility class: Esters.

ENVIRONMENTAL DATA
Water pollution: Effects of low concentrations on aquatic life are not known. May be harmful if it enters water intakes. Notify local health and wildlife officials. Notify operators of local water intakes.
Response to discharge: Issue warning–high flammability. Evacuate area. Should be removed. Chemical and physical treatment.

SHIPPING INFORMATION
Grades of purity: 99+%; **Inert atmosphere:** Store under nitrogen; **Stability during transport:** Stable.

NAS HAZARD CLASSIFICATIONS FOR BULK WATER TRANSPORTATION
FIRE: 2
HEALTH: Vapor irritant: 4; Liquid or solid irritant: 4; Poisons: 2

WATER POLLUTION: Human toxicity: 2; Aquatic toxicity: 3; Aesthetic effect: 3
REACTIVITY: Other chemicals: 3; Water: 1; Self-reaction: 0

PHYSICAL AND CHEMICAL PROPERTIES
Physical state @ 59°F/15°C and 1 atm: Liquid.
Molecular weight: 124.08
Boiling point @ 1 atm: 232°F/111°C/384°K.
Melting/Freezing point: −108°F/−78°C/195°K.
Specific gravity (water = 1): 1.046 @ 68°F/20°C.
Relative vapor density (air = 1): 4.3
Vapor pressure: 24 mm @ 77°F/25°C.

TRIORTHOCRESYL PHOSPHATE REC. T:4450

SYNONYMS: EEC No. 015-015-00-8; *o*-CRESYL PHOSPHATE; PHOSPHORIC ACID, TRIS(2-METHYLPHENYL) ESTER; PHOSFLEX 179-C; TOCP; TOFK; TOTP; *o*-TOLYL PHOSPHATE; *o*-TOLYL PHOSPHATE PHOSPHORIC ACID; TRICRESYL PHOSPHATE; TRI-*o*-CRESYL ESTER; TRI-*o*-CRESILFOSFATO (Spanish)

IDENTIFICATION
CAS Number: 78-30-8
Formula: $C_{21}H_{21}O_4P$; $(CH_3C_6H_4O)_3PO$
DOT ID Number: UN 2574; DOT Guide Number: 151
Proper Shipping Name: Tricresyl phosphate

DESCRIPTION: Colorless liquid. Odorless. Sinks in water forming phosphoric acid solution.

Combustible • Heat or flame may cause explosion • Containers may BLEVE when exposed to fire • Vapors in confined areas (e.g., tanks, sewers, buildings) may explode when exposed to fire • Irritating to the skin, eyes, and respiratory tract • Decomposes above 752°F/400°C; toxic products of combustion or decomposition include phosphorus oxides.

Hazard Classification (based on NFPA-704 Rating System)
Health Hazards (Blue): 2; Flammability (Red): 1; Reactivity (Yellow): 0

EMERGENCY RESPONSE: See Appendix A (151)
Evacuation:
Public safety: Isolate the area of spill or leak for at least 25 to 50 meters (80 to 160 feet) in all directions.
Spill: Increase, in the downwind direction, as necessary, the distance shown under "Public Safety."
Fire: If tank, rail car, or tank truck is involved in fire, isolate for at least 800 meters (½ mile) in all directions; also, consider initial evacuation for 800 meters (½ mile) in all directions.

EXPOSURE
Short-term effects: *CALL FOR MEDICAL AID.* **Vapor or liquid:** *POISONOUS. MAY BE FATAL IF INHALED, IF SWALLOWED, OR ABSORBED THROUGH SKIN.* Exposure causes nausea, vomiting, diarrhea and abdominal pain. Delayed effects begin in 1-3 weeks after initial effects. **Vapor:** Move to fresh air. *IF IN EYES*, hold eyelids open and flush with running water for 15 minutes. *IF BREATHING HAS STOPPED*, give artificial respiration; *avoid mouth-to-mouth resuscitation; use bag/mask apparatus*. IF breathing is difficult, administer oxygen. **Liquid:** Remove contaminated clothing and shoes and isolate. *IF IN EYES*, flush with running water for at least 15 minutes; lift upper and lower eyelids occasionally. *IF ON SKIN*, wash with soap and mild detergent and flush with running water for at least 15 minutes. *IF SWALLOWED* and victim is *CONSCIOUS AND ABLE TO SWALLOW*, have victim drink water and induce vomiting by touching back of throat with finger. *IF SWALLOWED* and victim is *UNCONSCIOUS OR HAVING CONVULSIONS*, do nothing except keep victim quiet and maintain body temperature.

HEALTH HAZARDS
Personal protective equipment (PPE): B-Level PPE. OSHA Table Z-1-A air contaminant. Wear positive pressure breathing apparatus and special chemical protective clothing.
Recommendations for respirator selection: NIOSH/OSHA
0.5 mg/m³: DM (any dust and mist respirator). *1 mg/m³:* DMXSQ (any dust and mist respirator except single-use and quarter mask respirators); or SA (any supplied-air respirator). *2.5 mg/m³:* SA:CF (any supplied-air respirator operated in a continuous-flow mode); or PAPRDM (any powered, air-purifying respirator with a dust and mist filter). *5 mg/m³:* HiEF (any air-purifying, full-facepiece respirator with a high-efficiency particulate filter); or SAT:CF (any supplied-air respirator that has a tight-fitting facepiece and is operated in a continuous-flow mode); or PAPRTHiE (any powered, air-purifying respirator with a tight-fitting facepiece and a high-efficiency particulate filter); SCBAF (any self-contained breathing apparatus with a full facepiece); or SAF (any supplied-air respirator with a full facepiece). *40 mg/m³:* SAF:PD,PP (any supplied-air respirator that has a full facepiece and is operated in a pressure-demand or other positive-pressure mode). *EMERGENCY OR PLANNED ENTRY INTO UNKNOWN CONCENTRATIONS OR IDLH CONDITIONS:* SCBAF:PD,PP (any self-contained breathing apparatus that has a full facepiece and is operated in a pressure-demand or other positive-pressure mode); or SAF:PD,PP:ASCBA (any supplied-air respirator that has a full facepiece and is operated in a pressure-demand or other positive-pressure mode in combination with an auxiliary self-contained breathing apparatus operated in a pressure-demand or other positive pressure mode). *ESCAPE:* HiEF (any air-purifying, full-facepiece respirator with a high-efficiency particulate filter); or SCBAE (any appropriate escape-type, self-contained breathing apparatus).
Exposure limits (TWA unless otherwise noted): ACGIH TLV 0.1 mg/m³; NIOSH/OSHA 0.1 mg/m³; skin contact contributes significantly in overall exposure.
Toxicity by ingestion: Grade 2; LD_{50} = 3.0 g/kg (rat).
Long-term health effects: Delayed neurotoxin. After a symptom-free period of 3 to 28 days following initial effects, paralysis, especially of the lower arm and legs, may develop. Recovery may take months or years; permanent residual effects occur in 25 to 30% of the cases.
Vapor (gas) irritant characteristics: Readily absorbed through the skin without local irritant effects. Not known to be an eye irritant.
Liquid or solid irritant characteristics: Readily absorbed through the skin without local irritant effects.
IDLH value: 40 mg/m³.

FIRE DATA
Flash point: 437°F/225°C (cc).
Autoignition temperature: 725°F/385°C/658°K.

CHEMICAL REACTIVITY
Binary reactants: Violent reaction with strong oxidizers. Contact with magnesium may cause explosion. The aqueous solution is incompatible with sulfuric acid, caustics, ammonia, aliphatic amines, alkanolamines, isocyanates, alkylene oxides, epichlorohydrin, nitromethane. Attacks some plastics, rubber and coatings.

Reactivity group: 34
Compatibility class: Esters.

ENVIRONMENTAL DATA
Water pollution: Effect of low concentrations on aquatic life is unknown. May be dangerous if it enters nearby water intakes; notify operators. Notify local health and wildlife officials.
Response to discharge: Should be removed. Chemical and physical treatment.

SHIPPING INFORMATION
Grades of purity: Commercial grades are usually mixtures of isomers (meta, ortho and para); **Venting:** Pressure vacuum; **Stability during transport:** Stable.

PHYSICAL AND CHEMICAL PROPERTIES
Physical state @ 59°F/15°C and 1 atm: Liquid.
Molecular weight: 368.37
Boiling point @ 1 atm: 770°F/410°C/683.2°K (decomposes).
Melting/Freezing point: –18°F/–28°C/245°K.
Specific gravity (water = 1): 1.162 @ 77°F/25°C.
Relative vapor density (air = 1): 12.7
Vapor pressure: 1.5 x 106 mm @ 77°F/25°C.

TRIPROPYLAMINE REC. T:4500

SYNONYMS: *N,N*-DIPROPYL-1-PROPANAMINE; TNPA; TRIPROPILAMINA (Spanish); TRI*N*-PROPYLAMINE; TX1575000

IDENTIFICATION
CAS Number: 102-69-2
Formula: $C_9H_{21}N$; $(CH_3CH_2CH_2)_3N$
DOT ID Number: UN 2260; DOT Guide Number: 132
Proper Shipping Name: Tripropylamine

DESCRIPTION: Water white liquid. Amine, fishy odor. Practically insoluble in water.

Poison! • Corrosive to the skin, eyes, and respiratory tract; skin or eye contact may cause burns or impaired vision; inhalation symptoms may be delayed • Highly flammable • Firefighting gear (including SCBA) does not provide adequate protection. If exposure occurs, remove and isolate gear immediately and thoroughly decontaminate personnel • Containers may BLEVE when exposed to fire • Vapors may form explosive mixture with air • Vapors are heavier than air and will collect and stay in low areas • Vapors may travel long distances to ignition sources and flashback • Vapors in confined areas (e.g., tanks, sewers, buildings) may explode when exposed to fire • Toxic products of combustion may include nitrogen oxides.

Hazard Classification (based on NFPA-704 Rating System)
Health Hazards (Blue): 2; Flammability (Red): 2; Reactivity (Yellow): 0

EMERGENCY RESPONSE: See Appendix A (132)
Evacuation:
Public safety: Isolate spill area for at least 50 to 100 meters (160 to 330 feet) in all directions.
Spill: Increase, as necessary, the isolation distance shown above, in "Public safety."
Fire: Isolate for 800 meters (½ mile) in all directions, especially if tank, rail car, or tank truck is involved in fire.

EXPOSURE
Short-term effects: *CALL FOR MEDICAL AID. TOXIC BY INHALATION OR INGESTION.* **Vapor:** Move victim to fresh air. *IF BREATHING HAS STOPPED,* give artificial respiration. If breathing is difficult, administer oxygen. **Liquid:** Remove contaminated clothing and shoes. Flush affected areas with water. *IF IN EYES,* hold eyelids open and flush with plenty of water.
Note to physician or authorized medical personnel: Medical observation is recommended for 24 to 48 hours after breathing overexposure, as pulmonary edema may be delayed. As first aid for pulmonary edema, consider administering a corticosteroid spray. Cigarette smoking may exacerbate pulmonary injury and should be discouraged for at least 72 hours following exposure.

HEALTH HAZARDS
Personal protective equipment (PPE): B-Level PPE. Full impervious protective clothing, including boots and gloves. Where splashing is possible wear full face shield or chemical safety goggles. Use approved respirator to protect against vapors. Chemical protective material(s) reported to offer minimal to poor protection: neoprene, nitrile, PV alcohol, Viton®
Toxicity by ingestion: Grade 3; oral rat LD_{50} = 72 mg/kg.
Vapor (gas) irritant characteristics: Eye and respiratory tract irritant. Vapors cause smarting of the eyes or respiratory system if present in high concentrations. The effect is temporary.
Liquid or solid irritant characteristics: Severe eye and skin irritant. If spilled on clothing and allowed to remain, may cause smarting and reddening of the skin.

FIRE DATA
Flash point: 105°F/41°C (cc).
Flammable limits in air: LEL: 0.7%; UEL: 5.6%.
Fire extinguishing agents not to be used: Water.
Autoignition temperature: 356°F/180°C/453°K.
Stoichiometric air-to-fuel ratio: 67.9

CHEMICAL REACTIVITY
Binary reactants: A medium-strong organic base. Violent reaction with strong acids, strong oxidizers. Incompatible with organic anhydrides, acrylates, alcohols, aldehydes, alkylene oxides, substituted allyls, cellulose nitrate, cresols, caprolactam solution, epichlorohydrin, ethylene dichloride, isocyanates, ketones, glycols, nitrates, phenols, vinyl acetate. Exothermic decomposition with maleic anhydride. Attacks aluminum, copper, zinc, and their alloys.

ENVIRONMENTAL DATA
Food chain concentration potential: Log P_{ow} = 2.8. May accumulate slightly.
Water pollution: Effect of low concentrations on aquatic life is unknown. May be dangerous if it enters nearby water intakes; notify operators. Notify local health and wildlife officials. **Response to discharge:** Should be removed.

SHIPPING INFORMATION
Grades of purity: 99%; technical; **Storage temperature:** Ambient; **Inert atmosphere:** None; **Stability during transport:** Stable.

PHYSICAL AND CHEMICAL PROPERTIES
Physical state @ 59°F/15°C and 1 atm: Liquid.
Molecular weight: 143.31
Boiling point @ 1 atm: 302–313°F/150–156°C/423–429°K.
Specific gravity (water = 1): 0.754 @ 20°F
Relative vapor density (air = 1): 5.0
Vapor pressure: 3 mm.

TRIPROPYLENE GLYCOL
REC. T:4550

SYNONYMS: 4,8-DIOXY-UNDECANE DIOL,-1,11; TRIPROPILENGLICOL (Spanish)

IDENTIFICATION
CAS Number: 24800-44-0
Formula: $C_9H_{20}O_4$; $HO(C_3H_6O)_2C_3H_6OH$

DESCRIPTION: Colorless liquid. Characteristic odor. Soluble in water.

Combustible • Heat or flame may cause explosion • Containers may BLEVE when exposed to fire • Vapors in confined areas (e.g., tanks, sewers, buildings) may explode when exposed to fire • Affects the central nervous system; irritating to the skin, eyes, and respiratory tract • Toxic products of combustion may include carbon monoxide.

Hazard Classification (based on NFPA-704 Rating System)
Health Hazards (Blue): 0; Flammability (Red): 1; Reactivity (Yellow): 0

EMERGENCY RESPONSE: See Appendix A (171)
Evacuation:
Public safety: Isolate the area of spill or leak for at least 10 to 25 meters (30 to 80 feet) in all directions.
Spill: Increase, in the downwind direction, as necessary, the distance shown under "Public Safety."
Fire: If any large container is involved in fire, isolate for at least 800 meters (½ mile) in all directions; also, consider initial evacuation for 800 meters (½ mile) in all directions.

EXPOSURE
Short-term effects:
Liquid or vapors: Affects the central nervous system; irritating to the skin, eyes, and respiratory tract

HEALTH HAZARDS
Personal protective equipment (PPE): B-Level PPE. Gloves; safety glasses or face shield. Chemical protective material(s) reported to have good to excellent resistance: butyl rubber.
Toxicity by ingestion: Grade 2; oral LD_{50} = 3000 mg/kg (rat).

FIRE DATA
Flash point: 285°F/141°C (oc).
Flammable limits in air: LEL: 0.8%; UEL: 5.0% (estimate).
Fire extinguishing agents not to be used: Water may be ineffective

CHEMICAL REACTIVITY
Binary reactants: Reacts with strong oxidizers. May attack some forms of plastics
Reactivity group: 40
Compatibility class: Glycol ethers

ENVIRONMENTAL DATA
Food chain concentration potential: Negative; unlikely to accumulate.
Food chain concentration potential: Log P_{ow} = –1.59. Negative. Will not accumulate.
Water pollution: Effect of low concentrations on aquatic life is unknown. May be dangerous if it enters nearby water intakes; notify operators. Notify local health and wildlife officials.
Response to discharge: Disperse and flush.

SHIPPING INFORMATION
Grades of purity: Commercial, 99%; **Storage temperature:** Ambient; **Inert atmosphere:** None; **Venting:** Open (flame arrester); **Stability during transport:** Stable.

NAS HAZARD CLASSIFICATION FOR BULK WATER TRANSPORTATION
FIRE: 1
HEALTH: Vapor irritant: 0; Liquid or solid irritant: 0; Poisons: 0
WATER POLLUTION: Human toxicity: 0; Aquatic toxicity: 0; Aesthetic effect: 0
REACTIVITY: Other chemicals: 1; Water: 0; Self-reaction: 0

PHYSICAL AND CHEMICAL PROPERTIES
Physical state @ 59°F/15°C and 1 atm: Liquid.
Molecular weight: 192.26
Boiling point @ 1 atm: 523°F/273°C/546°K.
Melting/Freezing point: (sets to glass) –49°F/–45°C/228°K.
Specific gravity (water = 1): 1.022 @ 68°F/20°C (liquid).
Relative vapor density (air = 1): 6.5
Heat of combustion: (estimate) –13,700 Btu/lb = –7610 cal/g = –318 x 10^5 J/kg.
Vapor pressure: 0.07 mm.

TRIPROPYLENE GLYCOL METHYL ETHER
REC. T:4600

SYNONYMS: DOWANOL TPM; PROPANOL, 3-[3-(3-METHOXY PROPOXY)PROPOXY]-; TRIPROPILENGLICOLMETIL ETER (Spanish)

IDENTIFICATION
CAS Number: 25498-49-1
Formula: $C_{10}H_{22}O_4$; $CH_3[CH_2CH(CH_3)O]_3OH$

DESCRIPTION: Colorless liquid. Mild odor.

Combustible • Heat or flame may cause explosion • Containers may BLEVE when exposed to fire • Vapors in confined areas (e.g., tanks, sewers, buildings) may explode when exposed to fire • Irritating to the skin, eyes, and respiratory tract • Toxic products of combustion may include carbon monoxide.

Hazard Classification (based on NFPA-704 Rating System)
Health Hazards (Blue): 0; Flammability (Red): 1; Reactivity (Yellow): 0

EMERGENCY RESPONSE: See Appendix A (171)
Evacuation:
Public safety: Isolate the area of spill or leak for at least 10 to 25 meters (30 to 80 feet) in all directions.
Spill: Increase, in the downwind direction, as necessary, the distance shown under "Public Safety."
Fire: If any large container is involved in fire, isolate for at least 800 meters (½ mile) in all directions; also, consider initial evacuation for 800 meters (½ mile) in all directions.

EXPOSURE
Short-term effects: *SEEK MEDICAL ATTENTION.* **Liquid:** May be irritating to skin and eyes. Remove contaminated clothing and shoes. Flush affected areas with plenty of water. *IF IN EYES*, hold eyelids open and flush with plenty of water. *IF SWALLOWED* and victim is *CONSCIOUS AND ABLE TO SWALLOW*, have victim drink 2 glasses of water and immediately induce vomiting.

HEALTH HAZARDS
Personal protective equipment (PPE): B-Level PPE. Chemical goggles, rubber boots and gloves, and self contained breathing apparatus). Chemical protective material(s) reported to have good to excellent resistance: butyl rubber.
Eyes: May cause slight transient (temporary) eye irritation. Corneal injury is unlikely. **Skin:** Prolonged or repeated exposure is not likely to cause significant skin irritation. Repeated prolonged exposure may cause sleepiness. Single prolonged exposure is not likely to result in absorption of harmful amount through skin.
Ingestion: Low oral toxicity. Large amount may cause injury. Exposure may have anesthetic or narcotic effects.
Toxicity by ingestion: Grade 2: LD_{50} = 3.3g/kg (rat).
Long-term health effects: Prolonged and repeated exposure to high concentration may cause kidney and neural disfunction.
Liquid or solid irritant characteristics: Eye irritant. Practically harmless to the skin of most persons.

FIRE DATA
Flash point: 250°F/121°C (cc)

CHEMICAL REACTIVITY
Reactivity group: 40
Compatibility class: Glycol ethers

ENVIRONMENTAL DATA
Water pollution: Effect of low concentrations on aquatic life is unknown. May be dangerous if it enters nearby water intakes; notify operators. Notify local health and wildlife officials.
Response to discharge: Chemical and physical treatment.

SHIPPING INFORMATION
Storage temperature: Ambient; **Stability during transport:** Stable.

PHYSICAL AND CHEMICAL PROPERTIES
Physical state @ 59°F/15°C and 1 atm: Liquid.
Molecular weight: 206.3
Boiling point @ 1 atm: 468°F/242.4°C/515.2°K.
Melting/Freezing point: –110°F/–78.89°C/194.3°K.
Specific gravity (water = 1): 0.965
Liquid surface tension: 30 dynes/cm = 0.030 N/m
Relative vapor density (air = 1): 7.15

TRIS(AZIRIDINYL)PHOSPHINE OXIDE   REC. T:4650

SYNONYMS: APHOXIDE; APO; CBC-906288; IMPERON FIXER T; SK-3818; TEF; TEPA; TRIS(1-AZIRIDINYL)PHOSPHINE OXIDE; PHOSPHORIC ACID TRIETHYLENEIMIDE; TRIETHYLENEPHOSPHORAMIDE; TRIETILENFOSFORAMIDA (Spanish)

IDENTIFICATION
CAS Number: 545-55-1
Formula: $C_6H_{12}N_3PO$; $(CH_2CH_2N)_3PO$
DOT ID Number: UN 2501; DOT Guide Number: 152
Proper Shipping Name: Tris-(1-aziridinyl)phosphine oxide, solution

DESCRIPTION: White solid. Mixes with water.

Corrosive • Poison! • Breathing the vapor can kill you; skin or eye contact causes severe burns, impaired vision, or blindness • Firefighting gear (including SCBA) does not provide adequate protection. If exposure occurs, remove and isolate gear immediately and thoroughly decontaminate personnel • Containers may BLEVE when exposed to fire • Toxic products of combustion may include phosphorus and nitrogen oxides • Do not put yourself in danger by entering a contaminated area to rescue a victim.

Hazard Classification (based on NFPA-704 Rating System)
Health Hazards (Blue): 3; Flammability (Red): 0; Reactivity (Yellow): 2

EMERGENCY RESPONSE: See Appendix A (152)
Evacuation:
Public safety: Isolate the area of spill or leak for at least 25 to 50 meters (80 to 160 feet) in all directions.
Spill: Increase, in the downwind direction, as necessary, the distance shown under "Public Safety."
Fire: If tank, rail car, or tank truck is involved in fire, isolate for at least 800 meters (½ mile) in all directions; also, consider initial evacuation for 800 meters (½ mile) in all directions.

EXPOSURE
Short-term effects: *SEEK MEDICAL ATTENTION*. **Solid:** Irritating to skin and eyes. Harmful if swallowed. Remove contaminated clothing and shoes. Flush affected areas with plenty of water. *IF BREATHING HAS STOPPED*, give artificial respiration; *avoid mouth-to-mouth resuscitation; use bag/mask apparatus. IF IN EYES*, hold eyelids open and flush with plenty of water. *IF SWALLOWED* and victim is *CONSCIOUS AND ABLE TO SWALLOW*, have victim drink 4 to 8 ounces of water.
Note to physician or authorized medical personnel: Medical observation is recommended for 24 to 48 hours after breathing overexposure, as pulmonary edema may be delayed. As first aid for pulmonary edema, consider administering a corticosteroid spray. Cigarette smoking may exacerbate pulmonary injury and should be discouraged for at least 72 hours following exposure.

HEALTH HAZARDS
Personal protective equipment (PPE): B-Level PPE. Protective clothing and gloves to prevent contact with skin; goggles.
Toxicity by ingestion: Grade 4; oral rat LD_{50} = 37 mg/kg.
Long-term health effects: IARC rating 3; animal data: inadequate. Amides may cause liver, kidney, and brain damage.
Liquid or solid irritant characteristics: May cause eye and skin irritation.

CHEMICAL REACTIVITY, unless in presence of acids or strong caustics.
Polymerization: Violent polymerization occurs at about 255°F/124°C. Acid fumes also cause polymerization at ordinary temperatures; **Inhibitor of polymerization:** None used

ENVIRONMENTAL DATA
Waterfowl toxicity: 8.5-13 mg/kg LD_{50}
Food chain concentration potential: Negative; unlikely to accumulate.
Water pollution: Effect of low concentrations on aquatic life is unknown. May be dangerous if it enters nearby water intakes; notify operators. Notify local health and wildlife officials. **Response to discharge:** Issue warning–corrosive, water contaminant. Disperse and flush.

SHIPPING INFORMATION
Grades of purity: 85% solution in acetone-methylene chloride; **Storage temperature:** Below 100°F/38°C; **Inert atmosphere:** None; **Venting:** Open; **Stability during transport:** Stable if cool.

PHYSICAL AND CHEMICAL PROPERTIES
Physical state @ 59°F/15°C and 1 atm: Solid.
Molecular weight: 173.16 (decomposes).
Melting/Freezing point: 106°F/41°C/314°K.
Specific gravity (water = 1): (estimate) more than 1 @ 68°F/20°C (solid).

TRIXYLENYL PHOSPHATE   REC. T:4700

SYNONYMS: COALITE NTP; DIMETHYLPHENOL PHOSPHATE; FOSFATO de TRIXILENILO (Spanish); REOFOS 95; TRIXYLYL PHOSPHATE; TRIDIMETHYLPHENYL PHOSPHATE; XYLENOL, PHOSPHATE; XYLYL PHOSPHATE

IDENTIFICATION
CAS Number: 25155-23-1; 121-06-2 (2,6-isomer)
Formula: $C_{24}H_{27}O_4P$

DESCRIPTION: Slightly colored liquid. Slight odor. Sinks in water; insoluble.

Combustible • Heat or flame may cause explosion • Containers may BLEVE when exposed to fire • Vapors in confined areas (e.g., tanks, sewers, buildings) may explode when exposed to fire • Irritating to the skin, eyes, and respiratory tract • Toxic products of combustion may include phosphorus oxides.

Hazard Classification (based on NFPA-704 Rating System)
Health Hazards (Blue): 2; Flammability (Red): 1; Reactivity (Yellow): 0

EMERGENCY RESPONSE: See Appendix A (171)
Evacuation:
Public safety: Isolate the area of spill or leak for at least 10 to 25 meters (30 to 80 feet) in all directions.
Spill: Increase, in the downwind direction, as necessary, the distance shown under "Public Safety."
Fire: If any large container is involved in fire, isolate for at least 800 meters (½ mile) in all directions; also, consider initial evacuation for 800 meters (½ mile) in all directions.

EXPOSURE
Short-term effects: *SEEK MEDICAL ATTENTION*. **Liquid:** Harmful if swallowed. *IF SWALLOWED* and victim is *CONSCIOUS AND ABLE TO SWALLOW*, have victim drink 4 to 8 ounces of water and induce vomiting. Remove clothing and wash skin with soap and water.

HEALTH HAZARDS
Personal protective equipment (PPE): B-Level PPE. Self contained breathing apparatus.
Toxicity by ingestion: Grade 1: LD_{50} = 11.8 g/kg (mouse).
Vapor (gas) irritant characteristics: Vapors/mists cause smarting of the eyes or respiratory system if present in high concentrations. The effect is temporary.
Liquid or solid irritant characteristics: Eye and skin irritant. If spilled on clothing and allowed to remain, may cause smarting and reddening of skin.

FIRE DATA
Flash point: 390°F/199°C (cc).
Autoignition temperature: 650°F/343°C/616°K.

CHEMICAL REACTIVITY
Reactivity group: 34
Compatibility class: Esters.

ENVIRONMENTAL DATA
Food chain concentration potential: Log K_{ow} = 5.62. Values above 3.0 are very likely to accumulate in living tissues and especially in fats.
Water pollution: DOT Appendix B, §172.101–marine pollutant. Harmful to aquatic life. May be dangerous if it enters nearby water intakes; notify operators. Notify local health and wildlife officials.
Response to discharge: Should be removed. Chemical and physical treatment.

SHIPPING INFORMATION
Storage temperature: Ambient; **Inert atmosphere:** Nitrogen Atmosphere; **Venting:** Pressure venting; **Stability during transport:** Stable.

PHYSICAL AND CHEMICAL PROPERTIES
Physical state @ 59°F/15°C and 1 atm: Liquid.
Molecular weight: 410.4
Boiling point @ 1 atm: 480-510°F/248-265°C/521-538°K.
Melting/Freezing point: –4°F/–20°C/253.2°K (pour point).
Specific gravity (water = 1): 1.130-1.155
Relative vapor density (air = 1): 14.2

TURPENTINE   REC. T:4750

SYNONYMS: EEC No. 650-002-00-6; GUM SPIRITS; GUM TURPENTINE; OIL OF TURPENTINE; SPIRITS OF TURPENTINE; SULFATE WOOD TURPENTINE; STEAM DISTILLED TURPENTINE; TEREBENTHINE (French); TERPENTIN OEL (German); TURPENTINE STEAM DISTILLED; TURPS; WOOD TURPENTINE; SULFATE TURPENTINE

IDENTIFICATION
CAS Number: 8006-64-2
Formula: $C_{10}H_{16}$
DOT ID Number: UN 1299; DOT Guide Number: 128
Proper Shipping Name: Turpentine

DESCRIPTION: Colorless watery liquid. Penetrating, characteristic odor. Floats on the surface of water. Irritating vapor is produced.

Highly flammable • Containers may BLEVE when exposed to fire • Vapors may form explosive mixture with air • Vapors are heavier than air and will collect and stay in low areas • Vapors may travel long distances to ignition sources and flashback • Vapors in confined areas (e.g., tanks, sewers, buildings) may explode when exposed to fire • Irritating to the skin, eyes, and respiratory tract • Toxic products of combustion may include carbon monoxide.

Hazard Classification (based on NFPA-704 Rating System)
Health Hazards (Blue): 1; Flammability (Red): 3; Reactivity (Yellow): 0

EMERGENCY RESPONSE: See Appendix A (128)
Evacuation:
Public safety: Isolate spill area for at least 25 to 50 meters (80 to 160 feet) in all directions.

Spill: Large spill–Consider initial downwind evacuation for at least 300 meters (1000 feet).
Fire: Isolate for 800 meters (½ mile) in all directions, especially if tank, rail car, or tank truck is involved in fire.

EXPOSURE
Short-term effects: *SEEK MEDICAL ATTENTION.* The use of alcoholic beverages may enhance toxic effects. **Vapor:** Irritating to eyes, nose, and throat. *IF INHALED*, will, will cause nausea, vomiting, headache, difficult breathing, or loss of consciousness. Move to fresh air. *IF BREATHING HAS STOPPED,* give artificial respiration. IF breathing is difficult, administer oxygen. Lung edema may develop. **Liquid:** *POISONOUS IF SWALLOWED.* Irritating to skin and eyes. Remove contaminated clothing and shoes. Flush affected areas with plenty of water. *IF IN EYES*, hold eyelids open and flush with plenty of water. *IF SWALLOWED* and victim is *CONSCIOUS AND ABLE TO SWALLOW*, have victim drink 4 to 8 ounces of water.
Note to physician or authorized medical personnel: Medical observation is recommended for 24 to 48 hours after breathing overexposure, as pulmonary edema may be delayed. As first aid for pulmonary edema, consider administering a corticosteroid spray. Cigarette smoking may exacerbate pulmonary injury and should be discouraged for at least 72 hours following exposure.

HEALTH HAZARDS
Personal protective equipment (PPE): B-Level PPE. OSHA Table Z-1-A air contaminant. Organic canister); or air–supplied mask; goggles or face shield; rubber gloves. Chemical protective material(s) reported to have good to excellent resistance: polyurethane, PV alcohol, nitrile, Teflon®. Also, nitrile, chlorinated polyethylene, Viton®/neoprene offers limited protection
Recommendations for respirator selection: NIOSH/OSHA
800 ppm: SA:CF* (any supplied-air respirator operated in a continuous-flow mode); or PAPROV* [any powered, air-purifying respirator with organic vapor cartridge(s)]; or CCRFOV [any air-purifying, full-facepiece respirator (gas mask) with a chin-style, front- or back-mounted acid gas canister]; or GMFOV [any air-purifying, full-facepiece respirator (gas mask) with a chin-style, front- or back-mounted acid gas canister]; or SCBAF (any self-contained breathing apparatus with a full facepiece); or SAF (any supplied-air respirator with a full facepiece). *EMERGENCY OR PLANNED ENTRY INTO UNKNOWN CONCENTRATIONS OR IDLH CONDITIONS:* SCBAF:PD,PP (any self-contained breathing apparatus that has a full facepiece and is operated in a pressure-demand or other positive-pressure mode); or SAF:PD,PP:ASCBA (any supplied-air respirator that has a full facepiece and is operated in a pressure-demand or other positive-pressure mode in combination with an auxiliary self-contained breathing apparatus operated in a pressure-demand or other positive pressure mode). *ESCAPE:* GMFOV [any air-purifying, full-facepiece respirator (gas mask) with a chin-style, front- or back-mounted organic vapor canister]; or SCBAE (any appropriate escape-type, self-contained breathing apparatus). *Note:* Substance causes eye irritation or damage; eye protection needed.
Exposure limits (TWA unless otherwise noted): ACGIH TLV 100 ppm (556 mg/m^3); NIOSH/OSHA 100 ppm (560 mg/m^3). A component of castor oil.
Toxicity by ingestion: Grade 2; LD_{50} = 0.5 to 5 g/kg.
Long-term health effects: Organ damage, including bladder and kidney. May cause anemia.
Vapor (gas) irritant characteristics: Vapors cause smarting of the eyes or respiratory system if present in high concentrations. The effect is temporary.

Liquid or solid irritant characteristics: Eye irritant. If spilled on clothing and allowed to remain, may cause smarting and reddening of the skin.
Odor threshold: 50–200 ppm.
IDLH value: 800 ppm.

FIRE DATA
Flash point: 95°F/35°C (cc).
Flammable limits in air: LEL: 0.8%; UEL: 6.0%.
Fire extinguishing agents not to be used: Water may be ineffective.
Behavior in fire: Forms heavy black smoke and soot
Autoignition temperature: 488°F/253°C/526°K.
Electrical hazard: Due to low electric conductivity, this substance may generate electrostatic charges as a result of agitation and flow.
Burning rate: 2.4 mm/min

CHEMICAL REACTIVITY
Binary reactants: Strong oxidizers, chlorine, chromic anhydride, stannic chloride, chromyl chloride, rubber.
Reactivity group: 30
Compatibility class: Olefins

ENVIRONMENTAL DATA
Food chain concentration potential: Negative; unlikely to accumulate.
Water pollution: DOT Appendix B, §172.101–marine pollutant. Dangerous to aquatic life in high concentrations. Fouling to shoreline. May be dangerous if it enters nearby water intakes; notify operators. Notify local health and wildlife officials. **Response to discharge:** Mechanical containment. Should be removed. Chemical and physical treatment.

SHIPPING INFORMATION
Grades of purity: A wide variety of grades and purities are shipped. All have about the same hazardous properties; **Storage temperature:** Ambient; **Inert atmosphere:** None; **Venting:** Open (flame arrester); **Stability during transport:** Stable.

NAS HAZARD CLASSIFICATION FOR BULK WATER TRANSPORTATION
FIRE: 3
HEALTH: Vapor irritant: 1; Liquid or solid irritant: 1; Poisons: 1
WATER POLLUTION: Human toxicity: 2; Aquatic toxicity: 3; Aesthetic effect: 2
REACTIVITY: Other chemicals: 1; Water: 0; Self-reaction: 0

PHYSICAL AND CHEMICAL PROPERTIES
Physical state @ 59°F/15°C and 1 atm: Liquid.
Molecular weight: 136 (approximate).
Boiling point @ 1 atm: 302–320°F/150–160°C/423–433°K.
Melting/Freezing point: –58°F to –76°F
Specific gravity (water = 1): 0.86 @ 15°C (liquid).
Liquid water interfacial tension: 14 dynes/cm = 0.014 N/m at 22.7°C.
Relative vapor density (air = 1): 4.6
Vapor pressure: (Reid) 0.26 psia; 4 mm.

UNDECANOIC ACID   REC. U:0100

SYNONYMS: ACIDO UNDECANOICO (Spanish); 1-DECANECARBOXYLIC ACID; HENDECANOIC ACID; *n*-UNDECANOIC ACID; *n*-UNDECYLIC ACID

IDENTIFICATION
CAS Number: 112-37-8
Formula: $C_{11}H_{22}O_2$; $CH_3(CH_2)_9CO_2H$

DESCRIPTION: White crystalline Solid. Soluble in water.

Combustible • Corrosive to skin, eyes, and respiratory tract; skin or eye contact causes burns or vision impairment; inhalation symptoms may be delayed • Heat or flame may cause explosion • Containers may BLEVE when exposed to fire • Vapors in confined areas (e.g., tanks, sewers, buildings) may explode when exposed to fire • Toxic products of combustion may include carbon monoxide.

Hazard Classification (based on NFPA-704 Rating System)
Health Hazards (Blue): 1; Flammability (Red): 1; Reactivity (Yellow): 0

EMERGENCY RESPONSE: See Appendix A (171)
Evacuation:
Public safety: Isolate the area of spill or leak for at least 10 to 25 meters (30 to 80 feet) in all directions.
Spill: Increase, in the downwind direction, as necessary, the distance shown under "Public Safety."
Fire: If any large container is involved in fire, isolate for at least 800 meters (½ mile) in all directions; also, consider initial evacuation for 800 meters (½ mile) in all directions.

EXPOSURE
Short-term effects: *SEEK MEDICAL ATTENTION.* **Vapor/mist or dust:** Irritating to eyes, nose, and throat. *IF INHALED,* will, will cause coughing or difficult breathing. *IF IN EYES,* hold eyelids open and flush with plenty of water. *IF BREATHING HAS STOPPED,* give artificial respiration. IF breathing is difficult, administer oxygen. **Liquid or solid:** Will burn skin and eyes. *IF SWALLOWED* may cause nausea and vomiting. Remove contaminated clothing and shoes. Flush affected areas with plenty of water. *IF IN EYES,* hold eyelids open and flush with plenty of water.
Note to physician or authorized medical personnel: Medical observation is recommended for 24 to 48 hours after breathing overexposure, as pulmonary edema may be delayed. As first aid for pulmonary edema, consider administering a corticosteroid spray. Cigarette smoking may exacerbate pulmonary injury and should be discouraged for at least 72 hours following exposure.

HEALTH HAZARDS
Personal protective equipment (PPE): B-Level PPE. Wear full protective clothing and respiratory protection. Butyl rubber is generally suitable for carbooxylic acid compounds.
Vapor (gas) irritant characteristics: Vapors, dust, or mist cause severe irritation of the eyes and throat and can cause eye and lung injury. They cannot be tolerated even at low concentrations.
Liquid or solid irritant characteristics: Severe skin irritant. Causes second- and third-degree burns on short contact and is very injurious to the eye.

FIRE DATA
Flash point: More than 230°F/110°C (cc).
Fire extinguishing agents not to be used: Water may not be effective.

CHEMICAL REACTIVITY
Binary reactants: Incompatible with oxidizers, sulfuric acid, caustics, ammonia and amines. May attack some common metals.
Neutralizing agents for acids and caustics: Sodium bicarbonate solution
Reactivity group: 4
Compatibility class: Organic acids.

ENVIRONMENTAL DATA
Water pollution: May be dangerous to aquatic life in high concentrations. May be dangerous if it enters nearby water intakes; notify operators. Notify local health and wildlife officials. **Response to discharge:** Restrict access. Mechanical containment. Should be removed.

SHIPPING INFORMATION
Grades of purity: 99%; **Storage temperature:** Ambient; **Stability during transport:** Stable.

PHYSICAL AND CHEMICAL PROPERTIES
Physical state @ 59°F/15°C and 1 atm: Solid.
Molecular weight: 186.3
Boiling point @ 1 atm: 442.2°F/228°C/501.2°K (at 160 mmHg = 0.211 atm).
Melting/Freezing point: 83.3°F/28.5°C/301.7°K.
Specific gravity (water = 1): 0.891
Relative vapor density (air = 1): 6.42

UNDECANOL   REC. U:0200

SYNONYMS: ALCOHOL C-11 (UNDECYLIC); HENDECANOIC ALCOHOL; 1-HENDECANOL; 1-UNDECANOL; UNDECYL ALCOHOL

IDENTIFICATION
CAS Number: 103-08-2
Formula: $CH_3(CH_2)_9CH_2OH$

DESCRIPTION: Colorless liquid or solid. Mild alcoholic odor. Floats on the surface of water.

Combustible • Heat or flame may cause explosion • Containers may BLEVE when exposed to fire • Vapors in confined areas (e.g., tanks, sewers, buildings) may explode when exposed to fire • Irritating to the skin, eyes, and respiratory tract • Toxic products of combustion may include carbon monoxide.

Hazard Classification (based on NFPA-704 Rating System)
Health Hazards (Blue): 1; Flammability (Red): 1; Reactivity (Yellow): 0

EMERGENCY RESPONSE: See Appendix A (171)
Evacuation:
Public safety: Isolate the area of spill or leak for at least 10 to 25 meters (30 to 80 feet) in all directions.
Spill: Increase, in the downwind direction, as necessary, the distance shown under "Public Safety."
Fire: If any large container is involved in fire, isolate for at least 800 meters (½ mile) in all directions; also, consider initial evacuation for 800 meters (½ mile) in all directions.

EXPOSURE
Short-term effects: *SEEK MEDICAL ATTENTION.* **Liquid:** Irritating to eyes. *IF IN EYES,* hold eyelids open and flush with plenty of water.

HEALTH HAZARDS
Personal protective equipment (PPE): B-Level PPE. Goggles or face shield. Chemical protective material(s) reported to have good to excellent resistance: butyl rubber, neoprene, nitrile.
Toxicity by ingestion: Grade 2; LD_{50} = 0.5 to 5 g/kg.
Liquid or solid irritant characteristics: May cause eye irritation.

FIRE DATA
Flash point: 235°F/113°C (oc).
Fire extinguishing agents not to be used: Water or foam may cause frothing.

CHEMICAL REACTIVITY
Binary reactants: Oxidizers, strong acids, caustics, amines.
Reactivity group: 20
Compatibility class: Alcohols, glycols.

ENVIRONMENTAL DATA
Food chain concentration potential: Negative; unlikely to accumulate.
Water pollution: DOT Appendix B, §172.101–marine pollutant. Effect of low concentrations on aquatic life is unknown. Fouling to shoreline. May be dangerous if it enters nearby water intakes; notify operators. Notify local health and wildlife officials.
Response to discharge: Mechanical containment. Should be removed. Chemical and physical treatment.

SHIPPING INFORMATION
Grades of purity: Technical; **Storage temperature:** Ambient; **Inert atmosphere:** None; **Venting:** Open (flame arrester); **Stability during transport:** Stable.

PHYSICAL AND CHEMICAL PROPERTIES
Physical state @ 59°F/15°C and 1 atm: Liquid.
Molecular weight: 172.30
Boiling point @ 1 atm: 437°F/225°C/490°K.
Melting/Freezing point: 60.6°F/15.9°C/289.1°K.
Critical temperature: 739°F/393°C/666°K.
Critical pressure: 308 psia = 21 atm = 2.1 MN/m^2.
Specific gravity (water = 1): 0.835 @ 68°F/20°C (liquid).
Liquid surface tension: 26.5 dynes/cm = 0.0265 N/m @ 68°F/20°C.
Liquid water interfacial tension: (estimate) 40 dynes/cm = 0.04 N/m @ 68°F/20°C.
Ratio of specific heats of vapor (gas): 1.032
Heat of combustion: (estimate) –18,000 Btu/lb = –10,000 cal/g = –419 x 10^5 J/kg.

1-UNDECENE   REC. U:0300

SYNONYMS: *n*-NONYLETHYLENE

IDENTIFICATION
Formula: $CH_3(CH_2)_8CH=CH_2$

DESCRIPTION: Colorless liquid. Mild, pleasant odor. Floats on the surface of water.

Combustible • Containers may BLEVE when exposed to fire • Vapors may form explosive mixture with air • Vapors are heavier than air and will collect and stay in low areas • Vapors may travel long distances to ignition sources and flashback • Vapors in confined areas (e.g., tanks, sewers, buildings) may explode when exposed to fire • Irritating to the skin, eyes, and respiratory tract • Toxic products of combustion may include smoek and irritating vapors.

EMERGENCY RESPONSE: See Appendix A (171)
Evacuation:
Public safety: Isolate the area of spill or leak for at least 10 to 25 meters (30 to 80 feet) in all directions.
Spill: Increase, in the downwind direction, as necessary, the distance shown under "Public Safety."
Fire: If any large container is involved in fire, isolate for at least 800 meters (½ mile) in all directions; also, consider initial evacuation for 800 meters (½ mile) in all directions.

EXPOSURE
Short-term effects: *SEEK MEDICAL ATTENTION*. **Liquid:** Irritating to skin and eyes. Harmful if swallowed. Remove contaminated clothing and shoes. Flush affected areas with plenty of water. *IF IN EYES*, hold eyelids open and flush with plenty of water. *IF SWALLOWED* and victim is *CONSCIOUS AND ABLE TO SWALLOW*, have victim drink 4 to 8 ounces of water. **Do NOT induce vomiting.**

HEALTH HAZARDS
Personal protective equipment (PPE): B-Level PPE. Goggles or face shield; rubber gloves.
Vapor (gas) irritant characteristics: Smarting of eyes and respiratory system at high concentrations. The effect may be temporary.
Liquid or solid irritant characteristics: Minimum hazard. If spilled on clothing and allowed to remain, may cause smarting and reddening of the skin.

FIRE DATA
Flash point: 160°F/71°C (oc).
Fire extinguishing agents not to be used: Water may be ineffective.
Burning rate: 4.8 mm/min

CHEMICAL REACTIVITY
Binary reactants: Oxidizers, strong acids.
Reactivity group: 30
Compatibility class: Olefins

ENVIRONMENTAL DATA
Food chain concentration potential: Negative; unlikely to accumulate.
Water pollution: Effect of low concentrations on aquatic life is unknown. Fouling to shoreline. May be dangerous if it enters nearby water intakes; notify operators. Notify local health and wildlife officials. **Response to discharge:** Mechanical containment. Should be removed. Chemical and physical treatment.

SHIPPING INFORMATION
Grades of purity: Technical: 99%; **Storage temperature:** Ambient; **Inert atmosphere:** None; **Venting:** Open (flame arrester); **Stability during transport:** Stable.

PHYSICAL AND CHEMICAL PROPERTIES
Physical state @ 59°F/15°C and 1 atm: Liquid.
Molecular weight: 154.2
Boiling point @ 1 atm: 379°F/193°C/466°K.
Melting/Freezing point: –56°F/49°C/224°K.
Specific gravity (water = 1): 0.750 @ 68°F/20°C (liquid).

Liquid surface tension: 23.4 dynes/cm = 0.0234 N/m @ 68°F/20°C.
Liquid water interfacial tension: (estimate) 50 dynes/cm = 0.050 N/m @ 68°F/20°C.
Ratio of specific heats of vapor (gas): 1.035
Latent heat of vaporization: 154 Btu/lb = 85.8 cal/g = 3.59 10^5 J/kg.
Heat of combustion: $-19,084$ Btu/lb = $-10,602$ cal/g = -443.89×10^5 J/kg.

n-UNDECYLBENZENE   REC. U:0400

SYNONYMS: 1-PHENYLUNDECANE

IDENTIFICATION
Formula: $C_6NO_5(CH_2)_{10}CH_3$

DESCRIPTION: Colorless liquid. Mild odor. Floats on the surface of water.

Combustible • Containers may BLEVE when exposed to fire • Vapors in confined areas (e.g., tanks, sewers, buildings) may explode when exposed to fire • Irritating to the skin, eyes, and respiratory tract • Toxic products of combustion may include carbon monoxide.

Hazard Classification (based on NFPA-704 Rating System)
Health Hazards (Blue): 1; Flammability (Red): 1; Reactivity (Yellow): 0

EMERGENCY RESPONSE: See Appendix A (171)
Evacuation:
Public safety: Isolate the area of spill or leak for at least 10 to 25 meters (30 to 80 feet) in all directions.
Spill: Increase, in the downwind direction, as necessary, the distance shown under "Public Safety."
Fire: If any large container is involved in fire, isolate for at least 800 meters (½ mile) in all directions; also, consider initial evacuation for 800 meters (½ mile) in all directions.

EXPOSURE
Short-term effects: *SEEK MEDICAL ATTENTION.* **Liquid:** Irritating to skin and eyes. *IF SWALLOWED*, will cause nausea and vomiting. Remove contaminated clothing and shoes. Flush affected areas with plenty of water. *IF IN EYES*, hold eyelids open and flush with plenty of water. *IF SWALLOWED* and victim is *CONSCIOUS AND ABLE TO SWALLOW*, have victim drink 4 to 8 ounces of water and have victim induce vomiting. *IF SWALLOWED* and victim is *UNCONSCIOUS OR HAVING CONVULSIONS,* do nothing except keep victim warm.

HEALTH HAZARDS
Personal protective equipment (PPE): B-Level PPE. Goggles or face shield and rubber gloves.
FIRE DATA
Flash point: 285°F/141°C (cc).
Fire extinguishing agents not to be used: Water may be ineffective.

CHEMICAL REACTIVITY
Binary reactants: Strong oxidizers. May attack some forms of plastics
Reactivity group: 32
Compatibility class: Aromatic hydrocarbons

ENVIRONMENTAL DATA
Food chain concentration potential: Negative; unlikely to accumulate.
Water pollution: Effect of low concentrations on aquatic life is unknown. Fouling to shoreline. May be dangerous if it enters nearby water intakes; notify operators. Notify local health and wildlife officials. **Response to discharge:** Mechanical containment. Should be removed. Chemical and physical treatment.

SHIPPING INFORMATION
Grades of purity: Mixture with decylbenzene and dodecylbenzene, all of which have same general properties; **Storage temperature:** Ambient; **Inert atmosphere:** None; **Venting:** Open (flame arrester); **Stability during transport:** Stable.

PHYSICAL AND CHEMICAL PROPERTIES
Physical state @ 59°F/15°C and 1 atm: Liquid.
Molecular weight: 232.4
Boiling point @ 1 atm: 601°F/316°C/589°K.
Melting/Freezing point: 23°F/–5°C/268°K.
Critical temperature: 918.1°F/492.3°C/765.5°K.
Critical pressure: 234 psia = 15.9 atm = 1.61 MN/m².
Specific gravity (water = 1): 0.855 @ 68°F/20°C (liquid).
Latent heat of vaporization: 101.27 Btu/lb = 56.26 cal/g = 2.354×10^5 J/kg.
Heat of combustion: $-19,490$ Btu/lb = $-10,830$ cal/g = -453.1×10^5 J/kg.

URANIUM PEROXIDE   REC. U:0500

SYNONYMS: PEROXIDO de URANIO (Spanish); URANIUM OXIDE; URANIUM OXIDE PEROXIDE

IDENTIFICATION
CAS Number: 12036-71-4
Formula: $UO_4 \cdot 2H_2O$
DOT ID Number: UN 2982; DOT Guide Number: 163
Proper Shipping Name: Radioactive material, n.o.s.

DESCRIPTION: Yellow solid crystals. Sinks in water.

Radioactive ["**priorities for rescue: lifesaving, first aid, and control of fire and other hazards are higher than the priority for measuring radiation levels**" (ERG 2000)] • Poison! • Able to form unstable peroxides • Breathing the vapor, skin or eye contact, or swallowing the material can kill you • Firefighting gear (including SCBA) does not provide adequate protection. If exposure occurs, remove and isolate gear immediately and thoroughly decontaminate personnel • Decomposition products upon heating, forming $U2O_7$ then to uranium oxide and oxygen • Do not put yourself in danger by entering a contaminated area to rescue a victim.

Hazard Classification (based on NFPA-704 Rating System)
Health Hazards (Blue): 4; Flammability (Red): 0; Reactivity (Yellow): 0; Special Notice (White): Radioactive

EMERGENCY RESPONSE: See Appendix A (163)
Evacuation:
Public safety: Isolate the area of spill or leak for at least 25 to 50 meters (80 to 160 feet) in all directions. Stay upwind.
Spill: Consider initial downwind evacuation for at least 100 meters (330 feet); increase, in the downwind direction, as necessary.

Fire: When a large quantity of this material is involved in a major fire consider evacuation distance of 300 meters (1000 feet) in all directions.

EXPOSURE
Short-term effects: *SEEK MEDICAL ATTENTION.* Medical problems take priority over radiological concerns. Ensure that medical personnel are aware of the material(s) involved; take precautions to protect themselves. **Dust:** *POISONOUS IF INHALED.* Move to fresh air. Keep victim quiet and warm. **Solid:** Harmful if swallowed. Remove contaminated clothing and shoes. Flush affected areas with plenty of water. *IF IN EYES,* hold eyelids open and flush with plenty of water.
Note to physician or authorized medical personnel: If ingested, consider the administration of large doses of sodium bicarbonate that will convert the uranium salt to the less toxic bicarbonate.

HEALTH HAZARDS
Personal protective equipment (PPE): A-Level PPE. OSHA Table Z-1-A ir contaminant. Approved respirator, leather, gloves and protective goggles.
Recommendations for respirator selection: NIOSH *as insoluble uranium*
At any concentrations above the NIOSH REL, or where there is no REL, at any detectable concentration: SCBAF:PD,PP (any self-contained breathing apparatus that has a full facepiece and is operated in a pressure-demand or other positive-pressure mode); or SAF:PD,PP:ASCBA (any supplied-air respirator that has a full facepiece and is operated in a pressure-demand or other positive-pressure mode in combination with an auxiliary self-contained breathing apparatus operated in a pressure-demand or other positive pressure mode). *ESCAPE:* HiEF (any air-purifying, full-facepiece respirator with a high-efficiency particulate filter); or SCBAE (any appropriate escape-type, self-contained breathing apparatus).
Exposure limits (TWA unless otherwise noted): ACGIH TLV 0.2 mg/m^3 as uranium; OSHA PEL 0.05 mg/m^3 as uranium; suspected human carcinogen, reduce exposure to lowest feasible level.
Short-term exposure limits (15-minute TWA): ACGIH STEL 0.6 mg/m^3 as uranium.
Toxicity by ingestion: Grade 2; LD$_{50}$ = 0.5 to 5 g/kg.
Long-term health effects: Retained in the lungs where it causes radiation injuries, lung cancer may develop. May cause kidney damage.
Liquid or solid irritant characteristics: May cause eye and skin irritation.
IDLH value: 15 mg/m^3 as uranium

FIRE DATA
Behavior in fire: Decomposes to form U_2O_7 then to uranium oxide and oxygen. *Note:* Contact the local, state, or United States Department of Energy radiological response team.

ENVIRONMENTAL DATA
Water pollution: Effect of low concentrations on aquatic life is unknown. May be dangerous if it enters nearby water intakes; notify operators. Notify local health and wildlife officials. See also environmental data. **Response to discharge:** Issue warning–water contaminant. Restrict access. Should be removed.

SHIPPING INFORMATION
Stability during transport: Stable at ambient temperature.

CHEMICAL REACTIVITY
Binary reactants: Nitric acid.

PHYSICAL AND CHEMICAL PROPERTIES
Physical state @ 59°F/15°C and 1 atm: Solid.
Molecular weight: 338.06
Boiling point @ 1 atm: Decomposes.
Melting/Freezing point: Decomposes 239°F/115°C/388°K.
Relative vapor density (air = 1): (calculated) 11.66
Heat of decomposition: (estimate @ > 194°F/90°C) 98.0 Btu/lb = 54.45 cal/g = 2.28 x 10^5 J/kg.

URANYL ACETATE   REC. U:0600

SYNONYMS: ACETATO de URANILO (Spanish); URANYL ACETATE DIHYDRATE; URANIUM ACETATE; URANIUM OXYACETATE; URANIUM OXYACETATE, DIHYDRATE; BIS(ACETATE) DIOXOURANIUM; URANIUM ACETATE, DIHYDRATE

IDENTIFICATION
CAS Number: 541-09-3
Formula: $C_4H_6O_6U \cdot 2H_2O$
DOT ID Number: UN 9180; DOT Guide Number: 162
Proper Shipping Name: Uranyl acetate
Reportable Quantity (RQ): **(CERCLA)** 100 lb/45.5 kg

DESCRIPTION: Yellow solid. Slight vinegar odor. Sinks and mixes slowly with water.

Radioactive [**"priorities for rescue: lifesaving, first aid, and control of fire and other hazards are higher than the priority for measuring radiation levels" (ERG 2000)**] • Toxic products of combustion may include carbon monoxide and uranium fumes.

Hazard Classification (based on NFPA-704 Rating System)
Health Hazards (Blue): 4; Flammability (Red): 0; Reactivity (Yellow): 1; Special Notice (White): Radioactive

EMERGENCY RESPONSE See Appendix A (162)
Evacuation:
Public safety: Isolate the area of spill or leak for at least 25 to 50 meters (80 to 160 feet) in all directions. Stay upwind.
Spill: Consider initial downwind evacuation for at least 100 meters (330 feet); increase, in the downwind direction, as necessary.
Fire: When a large quantity of this material is involved in a major fire consider evacuation distance of 300 meters (1000 feet) in all directions.

EXPOSURE
Short-term effects: *SEEK MEDICAL ATTENTION.* **Dust:** Irritating to eyes, nose, and throat. *IF INHALED,* will, will cause coughing or difficult breathing. *IF IN EYES,* hold eyelids open and flush with plenty of water. *IF BREATHING HAS STOPPED,* give artificial respiration. IF breathing is difficult, administer oxygen. **Solid:** Irritating to skin and eyes. Harmful if swallowed. Remove contaminated clothing and shoes. Flush affected areas with plenty of water. *IF IN EYES,* hold eyelids open and flush with plenty of water. *IF SWALLOWED* and victim is *CONSCIOUS AND ABLE TO SWALLOW,* have victim drink 4 to 8 ounces of water and have victim induce vomiting. *IF SWALLOWED* and victim is *UNCONSCIOUS OR HAVING CONVULSIONS,* do nothing except keep victim warm.
Note to physician or authorized medical personnel: If ingested, consider the administration of large doses of sodium bicarbonate that will convert the uranium salt to the less toxic bicarbonate.

HEALTH HAZARDS
Personal protective equipment (PPE): B-Level PPE. OSHA Table Z-1-A air contaminant. Approved dust respirator; goggles or face shield; protective clothing.
Recommendations for respirator selection: NIOSH
Any detectable concentration: SCBAF:PD,PP (any self-contained breathing apparatus that has a full facepiece and is operated in a pressure-demand or other positive-pressure mode); or SAF:PD,PP:ASCBA (any supplied-air respirator that has a full facepiece and is operated in a pressure-demand or other positive-pressure mode in combination with an auxiliary self-contained breathing apparatus operated in a pressure-demand or other positive pressure mode). *ESCAPE:* HiEF (any air-purifying, full-facepiece respirator with a high-efficiency particulate filter); or SCBAE (any appropriate escape-type, self-contained breathing apparatus).
Exposure limits (TWA unless otherwise noted): ACGIH TLV 0.2 mg/m^3 as uranium; OSHA PEL 0.05 mg/m^3 as uranium; suspected human carcinogen, reduce exposure to lowest feasible level.
Short-term exposure limits (15-minute TWA): ACGIH STEL 0.6 mg/m^3 as uranium.
Toxicity by ingestion: Grade 1; LD_{50} = 5 to 15 g/kg.
Vapor (gas) irritant characteristics: May cause eye and respiratory system irritation.
Liquid or solid irritant characteristics: May cause eye and skin irritation.
IDLH value: 10 mg/m^3 as uranium.

FIRE DATA
Note: Contact the local, state, or United States Department of Energy radiological response team.

CHEMICAL REACTIVITY
Reactivity with water: Dissolves and reacts to give a milky solution. The reaction may not hazardous.

ENVIRONMENTAL DATA
Disposal of wastes containing uranium: Follow guidelines set forth by the Nuclear Regulatory Commission (uranium and compounds). Notify local and state health authorities, local solid waste disposal authorities, supplier, shipper, and United States Department of Energy.
Water pollution: Harmful to aquatic life in very low concentrations. May be dangerous if it enters nearby water intakes; notify operators. Notify local health and wildlife officials. See also environmental data. **Response to discharge:** Issue warning–water contaminant. Restrict access. Disperse and flush.

SHIPPING INFORMATION
Grades of purity: Commercial; Reagent; **Storage temperature:** Ambient; **Inert atmosphere:** None; **Venting:** Open; **Stability during transport:** Stable.

PHYSICAL AND CHEMICAL PROPERTIES
Physical state @ 59°F/15°C and 1 atm: Solid.
Molecular weight: 424.2
Boiling point @ 1 atm: Decomposes.
Specific gravity (water = 1): 2.89 @ 68°F/20°C (solid).

URANYL NITRATE REC. U:0700

SYNONYMS: NITRATO de URANILO (Spanish); BIS(NITRATO-*O,O*)DIOXO URANIUM; URANIUM NITRATE

IDENTIFICATION
CAS Number: 10102-06-4; 36478-76-9; 13520-83-7 (hexahydrate)
Formula: N_2O_8U; $UO_2(NO_3)_2 \cdot 6H_2O$ (hexahydrate)
DOT ID Number: UN 2981; DOT Guide Number: 162
Proper Shipping Name: Uranyl nitrate, solid
Reportable Quantity (RQ): **(CERCLA)** 100 lb/45.4 kg

DESCRIPTION: Light yellow solid. Odorless. Mixes with water. Radioactive [**"priorities for rescue: lifesaving, first aid, control of fire, and other hazards are higher than the priority for measuring radiation levels" (ERG 2000)**] • Corrosive • Poison! • A powerful explosive • Breathing the vapor can kill you; skin or eye contact causes severe burns, impaired vision, or blindness • Firefighting gear (including SCBA) does not provide adequate protection. If exposure occurs, remove and isolate gear immediately and thoroughly decontaminate personnel • Strong oxidizer which may react spontaneously with low flash point organics or reducing agents. Heat forms oxygen; will increase the activity of an existing fire • May cause fire or explosion on contact with combustibles (wood, paper, oil, clothing, etc.) • Toxic products of combustion may include nitrogen oxide and uranium:

Hazard Classification (based on NFPA-704 Rating System)
Health Hazards (Blue): 4; Flammability (Red): 0; Reactivity (Yellow): 0; Special Notice (White): OXY, Radioactive

EMERGENCY RESPONSE: See Appendix A (162)
Evacuation:
Public safety: Isolate the area of spill or leak for at least 25 to 50 meters (80 to 160 feet) in all directions. Stay upwind.
Spill: Consider initial downwind evacuation for at least 100 meters (330 feet); increase, in the downwind direction, as necessary.
Fire: When a large quantity of this material is involved in a major fire consider evacuation distance of 300 meters (1000 feet) in all directions.

EXPOSURE
Short-term effects: *SEEK MEDICAL ATTENTION.* **Dust:** Irritating to eyes, nose, and throat. Harmful if inhaled. Move victim to fresh air. *IF IN EYES*, hold eyelids open and flush with plenty of water. **Solid:** Irritating to skin and eyes. Harmful if swallowed. Remove contaminated clothing and shoes. Flush affected areas with plenty of water. *IF IN EYES*, hold eyelids open and flush with plenty of water. *IF SWALLOWED* and victim is *CONSCIOUS AND ABLE TO SWALLOW*, have victim drink 4 to 8 ounces of water.
Note to physician or authorized medical personnel: **Ingestion:** Consider the administration of large doses of sodium bicarbonate that will convert the uranium salt to the less toxic bicarbonate. Additional treatment is symptomatic; *GET MEDICAL ATTENTION.*
Note to physician or authorized medical personnel: Medical observation is recommended for 24 to 48 hours after breathing overexposure, as pulmonary edema may be delayed. As first aid for pulmonary edema, consider administering a corticosteroid spray. Cigarette smoking may exacerbate pulmonary injury and should be discouraged for at least 72 hours following exposure.

HEALTH HAZARDS
Personal protective equipment (PPE): B-Level PPE. OSHA Table Z-1-A air contaminant. Dust mask, gloves, goggles
Recommendations for respirator selection: NIOSH
Any detectable concentration: SCBAF:PD,PP (any self-contained breathing apparatus that has a full facepiece and is operated in a pressure-demand or other positive-pressure mode); or

1120 Uranyl sulfate

SAF:PD,PP:ASCBA (any supplied-air respirator that has a full facepiece and is operated in a pressure-demand or other positive-pressure mode in combination with an auxiliary self-contained breathing apparatus operated in a pressure-demand or other positive pressure mode). *ESCAPE:* (nonhalides) HiEF (any air-purifying, full-facepiece respirator with a high-efficiency particulate filter); or SCBAE (any appropriate escape-type, self-contained breathing apparatus).
Exposure limits (TWA unless otherwise noted): ACGIH TLV 0.2 mg/m^3 as uranium; OSHA PEL 0.05 mg/m^3 as uranium; suspected human carcinogen, reduce exposure to lowest feasible level.
Short-term exposure limits (15-minute TWA): ACGIH STEL 0.6 mg/m^3 as uranium.
Toxicity by ingestion: Grade 3; LD$_{50}$ = 50 to 500 mg/kg.
Long-term health effects: Delayed inflammation of kidneys. Airborne radioactive particles have apparently been responsible for a significantly increased death rate from lung cancer among long-term uranium miners.
Vapor (gas) irritant characteristics: Mays cause eye and respiratory irritation.
Liquid or solid irritant characteristics: Causes eye irritation. May cause skin irritation.
IDLH value: 10 mg/m^3 as uranium

FIRE DATA, but may cause fire on contact with combustibles.
Behavior in fire: When large quantities are involved, nitrate may fuse or melt; application of water may then cause extensive scattering of molten material.
Note: Contact the local, state, or United States Department of Energy radiological response team.

CHEMICAL REACTIVITY
Reactivity with water: Dissolves, forming weak solution of nitric acid.
Binary reactants: Contact with steam my cause explosion. A powerful oxidizer. Violent reaction with many materials including reducing agents, combustible materials, ethers, fuels, organic substances, powdered metals. Incompatible with cellulose. Attacks many plastics, rubber, and coatings. Attacks most metals in the presence of moisture.
Neutralizing agents for acids and caustics: Wash with water.

ENVIRONMENTAL DATA
Water pollution: Harmful to aquatic life in very low concentrations. May be dangerous if it enters nearby water intakes; notify operators. Notify local health and wildlife officials. See also environmental data. **Response to discharge:** Issue warning-radioactive, oxidizing material, water contaminant. Restrict access. Disperse and flush
DISPOSAL OF WASTES CONTAINING URANIUM: Follow guidelines set forth by the Nuclear Regulatory Commission (uranium and compounds). Notify local and state health authorities, local solid waste disposal authorities, supplier, shipper, and United States Department of Energy.

SHIPPING INFORMATION
Grades of purity: Analytical reagent; **Storage temperature:** Ambient; **Inert atmosphere:** None; **Venting:** Open; **Stability during transport:** Stable.

PHYSICAL AND CHEMICAL PROPERTIES
Physical state @ 59°F/15°C and 1 atm: Solid.
Molecular weight: 502.13
Boiling point: Decomposes

Melting/Freezing point: 140°F/60°C/333°K.
Specific gravity (water = 1): 2.81 at 13°C (solid).

URANYL SULFATE   REC. U:0800

SYNONYMS: SULFATO de URANILO (Spanish); URANYL SULFATE TRIHYDRATE; URANIUM SULFATE; URANIUM SULFATE TRIHYDRATE

IDENTIFICATION
CAS Number: 1314-64-3
Formula: UO$_2$SO$_4$·3H$_2$O
DOT ID Number: UN 2982; DOT Guide Number: 163
Proper Shipping Name: Radioactive material, n.o.s.

DESCRIPTION: Yellow solid. Odorless. Sinks and mixes with water.

Radioactive [**"priorities for rescue: lifesaving, first aid, and control of fire and other hazards are higher than the priority for measuring radiation levels" (ERG 2000)**] • Poison! • Breathing the vapor, skin or eye contact, or swallowing the material can kill you • Firefighting gear (including SCBA) does not provide adequate protection. If exposure occurs, remove and isolate gear immediately and thoroughly decontaminate personnel Toxic products of combustion may include sulfur oxides • Do not put yourself in danger by entering a contaminated area to rescue a victim.

Hazard Classification (based on NFPA-704 Rating System)
Health Hazards (Blue): 4; Flammability (Red): 0; Reactivity (Yellow): 0; Special Notice (White): Radioactive

EMERGENCY RESPONSE. See Appendix A (163)
Evacuation:
Public safety: Isolate the area of spill or leak for at least 25 to 50 meters (80 to 160 feet) in all directions. Stay upwind.
Spill: Consider initial downwind evacuation for at least 100 meters (330 feet); increase, in the downwind direction, as necessary.
Fire: When a large quantity of this material is involved in a major fire consider evacuation distance of 300 meters (1000 feet) in all directions.

EXPOSURE
Short-term effects: *SEEK MEDICAL ATTENTION*. Medical problems take priority over radiological concerns. Ensure that medical personnel are aware of the material(s) involved; take precautions to protect themselves. **Dust:** Irritating to eyes, nose, and throat. *IF INHALED*, will, will cause coughing or difficult breathing. *IF IN EYES*, hold eyelids open and flush with plenty of water. *IF BREATHING HAS STOPPED,* give artificial respiration. If breathing is difficult, administer oxygen. **Solid:** Irritating to skin and eyes. Harmful if swallowed. Remove contaminated clothing and shoes. Flush affected areas with plenty of water. *IF IN EYES*, hold eyelids open and flush with plenty of water. *IF SWALLOWED* and victim is *CONSCIOUS AND ABLE TO SWALLOW*, have victim drink 4 to 8 ounces of water and have victim induce vomiting. *IF SWALLOWED* and victim is *UNCONSCIOUS OR HAVING CONVULSIONS,* do nothing except keep victim warm.
Note to physician or authorized medical personnel: **Ingestion:** Consider the administration of large doses of sodium bicarbonate that will convert the uranium salt to the less toxic bicarbonate. Additional treatment is symptomatic; *GET MEDICAL ATTENTION.*

HEALTH HAZARDS

Personal protective equipment (PPE): B-Level PPE. OSHA Table Z-1-A air contaminant. Approved dust respirator; goggles or face shield; protective clothing

Recommendations for respirator selection: NIOSH

Any detectable concentration: SCBAF:PD,PP (any self-contained breathing apparatus that has a full facepiece and is operated in a pressure-demand or other positive-pressure mode); or SAF:PD,PP:ASCBA (any supplied-air respirator that has a full facepiece and is operated in a pressure-demand or other positive-pressure mode in combination with an auxiliary self-contained breathing apparatus operated in a pressure-demand or other positive pressure mode). *ESCAPE:* HiEF (any air-purifying, full-facepiece respirator with a high-efficiency particulate filter); or SCBAE (any appropriate escape-type, self-contained breathing apparatus).

Exposure limits (TWA unless otherwise noted): ACGIH TLV 0.2 mg/m^3 as uranium; OSHA PEL 0.05 mg/m^3 as uranium; suspected human carcinogen, reduce exposure to lowest feasible level.

Short-term exposure limits (15-minute TWA): ACGIH STEL 0.6 mg/m^3 as uranium

Toxicity by ingestion: Grade 1; LD$_{50}$ = 5 to 15 g/kg.

Vapor (gas) irritant characteristics: May cause eye and respiratory tract irritation.

Liquid or solid irritant characteristics: Causes eyes and skin irritation.

IDLH value: 10 mg/m^3 as uranium.

FIRE DATA

Note: Contact the local, state, or United States Department of Energy radiological response team.

CHEMICAL REACTIVITY

Binary reactants: Violent reaction with aluminum, magnesium.

ENVIRONMENTAL DATA

Water pollution: Effect of low concentrations on aquatic life is unknown. May be dangerous if it enters nearby water intakes; notify operators. Notify local health and wildlife officials. See also environmental data. **Response to discharge:** Issue warning–water contaminant. Restrict access. Disperse and flush.

Disposal of wastes containing uranium: Follow guidelines set forth by the Nuclear Regulatory Commission (uranium and compounds). Notify local and state health authorities, local solid waste disposal authorities, supplier, shipper, and United States Department of Energy.

SHIPPING INFORMATION

Grades of purity: Commercial; Pure; **Storage temperature:** Ambient; **Inert atmosphere:** None; **Venting:** Open; **Stability during transport:** Stable.

PHYSICAL AND CHEMICAL PROPERTIES

Physical state @ 59°F/15°C and 1 atm: Solid.
Molecular weight: 420.2
Boiling point @ 1 atm: Decomposes.
Specific gravity (water = 1): 3.28 @ 68°F/20°C (solid).

UREA REC. U:0900

SYNONYMS: B-I-K; CARBAMIDE; CARBAMIDE ACID; CARBAMIDE RESIN; CARBONYL DIAMINE; CARBAMIMIDIC ACID; CARBONYL DIAMIDE; CARBONYLDIAMIDE; ISOUREA; PRESPERSION, 75 UREA; PSEUDOUREA; SUPERCEL 3000; UREAPHIL; UREOPHIL; UREVERT; VARIOFORM II

IDENTIFICATION

CAS Number: 57-13-6
Formula: CH_4N_2O; NH_2CONH_2

DESCRIPTION: White crystalline solid, pellets or solution. Slight ammonia odor. Sinks and mixes with water.

Combustible solid • Heat or flame may cause explosion • Containers may BLEVE when exposed to fire • Concentrated dust in confined areas (e.g., tanks, sewers, buildings) may explode when exposed to fire • Irritating to the skin, eyes, and respiratory tract • Toxic products of combustion may include nitrogen oxides, ammonia and other products. The decomposition may not be explosive.

Hazard Classification (based on NFPA-704 Rating System)
Health Hazards (Blue): 0; Flammability (Red): 0; Reactivity (Yellow): 0

EMERGENCY RESPONSE: See Appendix A (162)

Evacuation:
Public safety: Isolate the area of spill or leak for at least 10 to 25 meters (30 to 80 feet) in all directions.
Spill: Increase, in the downwind direction, as necessary, the distance shown under "Public Safety."
Fire: If any large container is involved in fire, isolate for at least 800 meters (½ mile) in all directions; also, consider initial evacuation for 800 meters (½ mile) in all directions.

EXPOSURE

Short-term effects: Skin irritant.

HEALTH HAZARDS

Personal protective equipment (PPE): B-Level PPE. Goggles or face shield; dust mask. Chemical protective material(s) reported to offer limited protection: natural rubber, neoprene, nitrile+PVC, nitrile, PVC.

Exposure limits (TWA unless otherwise noted): 10 mg/m^3 (AIHA WEEL).

Long-term health effects: Mutation and reproductive data. Amides may cause liver, kidney, and brain damage.

Vapor (gas) irritant characteristics: Nonvolatile, but dust may cause irritation of the nose and throat, sore throat and shortness of breath.

Liquid or solid irritant characteristics: Irritates eyes, skin, and respiratory tract.

CHEMICAL REACTIVITY

Binary reactants: Reacts with nitrosyl perchlorate. Forms explosive with nitrates and nitrogen trichloride with sodium or calcium hypochlorite. Reacts with oxidizers and chlorinating materials.

ENVIRONMENTAL DATA

Food chain concentration potential: Negative; unlikely to accumulate.

Water pollution: Effect of low concentrations on aquatic life is unknown. May be dangerous if it enters nearby water intakes; notify operators. Notify local health and wildlife officials. **Response to discharge:** Disperse and flush.

SHIPPING INFORMATION
Grades of purity: Various grades and purities, which depend on manufacturing process and intended use. All have essentially the same hazardous properties; **Storage temperature:** Ambient; **Inert atmosphere:** None; **Venting:** Open; **Stability during transport:** Stable.

PHYSICAL AND CHEMICAL PROPERTIES
Physical state @ 59°F/15°C and 1 atm: Solid.
Molecular weight: 60.06
Boiling point @ 1 atm: Decomposes.
Melting/Freezing point: 271°F/133°C/406°K.
Specific gravity (water = 1): 1.34 @ 68°F/20°C (solid).
Heat of combustion: -3913 Btu/lb = -2174 cal/g = -91.02×10^5 J/kg.
Heat of solution: -108 Btu/lb = -60.1 cal/g = -2.52×10^5 J/kg.

UREA, AMMONIUM NITRATE SOLUTION (in aqueous ammonia)   REC. U:1000

SYNONYMS: KRENITE® 10; LIQUAMON 28; SOLAR NITROGEN SOLUTION; UAN-NITROGEN SOLUTION; UAN SOLUTION; URAN, RUSTICA

IDENTIFICATION
CAS Number: 15978-77-5
Formula: $H_2ONH_3HNO_3CO(NH_2)_2$

DESCRIPTION: Clear liquid. Slight ammonia odor. Mixes with water.

Irritating to the skin, eyes, and respiratory tract • Toxic products of combustion may include nitrogen oxides.

Hazard Classification (based on NFPA-704 Rating System)
Health Hazards (Blue): 1; Flammability (Red): 0; Reactivity (Yellow): 0

EMERGENCY RESPONSE: See Appendix A (171)
Evacuation:
Public safety: Isolate the area of spill or leak for at least 10 to 25 meters (30 to 80 feet) in all directions.
Spill: Increase, in the downwind direction, as necessary, the distance shown under "Public Safety."
Fire: If any large container is involved in fire, isolate for at least 800 meters (½ mile) in all directions; also, consider initial evacuation for 800 meters (½ mile) in all directions.

EXPOSURE
Short-term effects: *SEEK MEDICAL ATTENTION*. **Liquid:** Irritating to skin and eyes. Harmful if swallowed. Remove contaminated clothing and shoes. *IF ON SKIN*, flush skin thoroughly and immediately with water. IF irritation persists obtain medical aid. *IF IN EYES*, flush eyes with water for 15 minutes or until irritation subsides. *IF SWALLOWED* and victim is *CONSCIOUS AND ABLE TO SWALLOW*, have victim drink 4 to 8 ounces of water.
Note to physician or authorized medical personnel: Give milk and demulcents, induce emesis or perform gastric lavage: Give fluids: observe for methemoglobinemia. If needed, give methylene blue as a 1% solution intravenously, 1–2 mg/kg; an oral dose of 3–5 mg/kg. If severe, consider exchange transfusion with whole blood.

HEALTH HAZARDS
Personal protective equipment (PPE): B-Level PPE. Rubber gloves, safety glasses, clothes that minimize skin exposure.
Toxicity by ingestion: Grade 2; LD_{50} = 3.0 g/kg (female rat).
Vapor (gas) irritant characteristics: Vapors may irritate eyes and throat.
Liquid or solid irritant characteristics: May cause eye and skin irritation. If spilled on clothing and allowed to remain, may cause smarting and reddening of skin.
Odor threshold: About 45 ppm.

FIRE DATA
Fire extinguishing agents not to be used: Do not use CO_2, dry chemical or foam extinguisher.
Behavior in fire: Organic and oxidizable materials can sensitize DRY ammonium nitrate to readily explodable state; can detonate if heated under confinement with high pressure.
Electrical hazard: Class I, Group D.

CHEMICAL REACTIVITY
Binary reactants: Copper or copper alloys are prohibited materials. Reacts with oxidizers, combustible materials, wood chips, organic materials, sulfur, metal, finely divided lead, zinc, galvanized iron, acids.
Reactivity group: 6
Compatibility class: Ammonia

ENVIRONMENTAL DATA
Water pollution: Effect of low concentrations on aquatic life is unknown. May be dangerous if it enters nearby water intakes; notify operators. Notify local health and wildlife officials. **Response to discharge:** Issue warning. Disperse and flush.

SHIPPING INFORMATION
Grades of purity: Ammonium Nitrate: 44–45% by weight; Urea: 34–35% by weight; Water: 20–22% by weight; **Storage temperature:** Ambient; **Venting:** Pressure vacuum; **Stability during transport:** Stable.

PHYSICAL AND CHEMICAL PROPERTIES
Physical state @ 59°F/15°C and 1 atm: Liquid.

Boiling point @ 1 atm: 225°F/107°C/380.2°K.
Specific gravity (water = 1): 1.326 at 15.56°C.

UREA PEROXIDE   REC. U:1100

SYNONYMS: CARBAMIDE PEROXIDE; CARBONYL DIAMINE PEROXIDE; GLY-OXIDE; HYDROGEN PEROXIDE CARBAMIDE; HYDROGEN PEROXIDE WITH UREA; HYPEROL; ORTIZON; PERCARBAMIDE; PERCARBAMITE; PEROXIDO de UREA (Spanish); UREA DIOXIDE; UREA HYDROGEN PEROXIDE; UREA, HYDROGEN PEROXIDE SALT

IDENTIFICATION
CAS Number: 124-43-6
Formula: $CO(NH_2)_2 \cdot H_2O_2$
DOT ID Number: UN 1511; DOT Guide Number: 140
Proper Shipping Name: Urea hydrogen peroxide

DESCRIPTION: White solid. Odorless. Mixes with water, releasing hydrogen peroxide.

Strong oxidizer which may react spontaneously with low flash point organics or reducing agents. Heat forms oxygen; will increase the activity of an existing fire • May cause fire or explosion on contact with combustibles (wood, paper, oil, clothing, etc.). Irritating to the skin, eyes, and respiratory tract; prolonged contact with skin may cause burns • Toxic products of combustion may include nitrogen oxides.

EMERGENCY RESPONSE: See Appendix A (140)
Evacuation:
Public safety: Isolate the area of spill or leak for at least 10 to 25 meters (30 to 80 feet) in all directions.
Spill: Consider initial downwind evacuation for at least 100 meters (330 feet).
Fire: If any large container is involved in fire, isolate for at least 800 meters (½ mile) in all directions; also, consider initial evacuation for 800 meters (½ mile) in all directions.

EXPOSURE
Short-term effects: *SEEK MEDICAL ATTENTION.* **Solid:** Irritating to skin and eyes. Remove contaminated clothing and shoes. Flush affected areas with plenty of water. *IF IN EYES,* hold eyelids open and flush with plenty of water. *IF SWALLOWED* and victim is *CONSCIOUS AND ABLE TO SWALLOW,* have victim drink 4 to 8 ounces of water.

HEALTH HAZARDS
Personal protective equipment (PPE): B-Level PPE. Rubber gloves and protective goggles. Chemical protective material(s) reported to have good to excellent resistance: butyl rubber.
Vapor (gas) irritant characteristics: May cause eye and nose irritation.
Liquid or solid irritant characteristics: May cause severe irritation of the eyes and skin.
Long-term health effects: Amides may cause liver, kidney, and brain damage.

FIRE DATA
Flash point: Combustible solid.
Behavior in fire: Containers may explode.
Autoignition temperature: More than 680°F/360°C/633°K.

CHEMICAL REACTIVITY
Reactivity with water: Forms solution of hydrogen peroxide (nonhazardous reaction).
Binary reactants: No significant reaction at ordinary temperatures. At 122°F/50°C reacts with dust and organic materials including rubbish.

ENVIRONMENTAL DATA
Food chain concentration potential: Negative; unlikely to accumulate.
Water pollution: Effect of low concentrations on aquatic life is unknown. May be dangerous if it enters nearby water intakes; notify operators. Notify local health and wildlife officials.
Response to discharge: Issue warning–oxidizing material. Restrict access. Disperse and flush.

SHIPPING INFORMATION
Grades of purity: 98–100%; **Storage temperature:** Below 140°F/60°C; **Inert atmosphere:** None; **Venting:** Open (flame arrester); **Stability during transport:** Stable below 140°F/60°C.

PHYSICAL AND CHEMICAL PROPERTIES
Physical state @ 59°F/15°C and 1 atm: Solid.
Molecular weight: 94.1
Boiling point @ 1 atm: Decomposes.
Specific gravity (water = 1): 0.8 @ 68°F/20°C (solid).
Heat of decomposition: –540 Btu/lb = –300 cal/g = –12.5 x 10^5 J/kg.

VALERALDEHYDE   REC. V:0100

SYNONYMS: AMYL ALDEHYDE; BUTYL FORMAL; PENTANAL; *n*-PENTANAL; VALERILADEHIDO (Spanish); VALERIC ALDEHYDE; VALERAL; VALERIANIC ALDEHYDE; VALERIC ALDEHYDE; VALERIC ACID ALDEHYDE; *n*-VALERIC ALDEHYDE; *n*-VALERALDEHYDE

IDENTIFICATION
CAS Number: 110-62-3
Formula: $C_5H_{10}O$; $CH_3(CH_2)_2CHO$
DOT ID Number: UN 2058; DOT Guide Number: 129
Proper Shipping Name: Valeraldehyde

DESCRIPTION: Colorless watery liquid. Pungent, acrid odor and fruity taste. Floats on water surface; dissolves slowly. Flammable, irritating vapor is produced.

Extremely flammable • Containers may BLEVE when exposed to fire •Vapors may form explosive mixture with air • Vapors are heavier than air and will collect and stay in low areas • Vapors may travel long distances to ignition sources and flashback • Vapors in confined areas (e.g., tanks, sewers, buildings) may explode when exposed to fire • Highly irritating to the eyes and respiratory tract; this material has anesthetic properties • Toxic products of combustion may include carbon monoxide.

Hazard Classification (based on NFPA-704 Rating System)
Health Hazards (Blue): 1; Flammability (Red): 3; Reactivity (Yellow): 0

EMERGENCY RESPONSE: See Appendix A (129)
Evacuation:
Public safety: Isolate spill area for at least 50 to 100 meters (160 to 330 feet) in all directions.
Spill: Large spill–Consider initial downwind evacuation for at least 300 meters (1000 feet).
Fire: Isolate for 800 meters (½ mile) in all directions, especially if tank, rail car, or tank truck is involved in fire.

EXPOSURE
Short-term effects: **Vapor:** Irritating to eyes, nose, and throat. Move to fresh air. Slightly narcotic effects. **Liquid:** Irritating to skin and eyes.
Flush affected areas with plenty of water.

HEALTH HAZARDS
Personal protective equipment (PPE): B-Level PPE. OSHA Table Z-1-A Air contaminant. Goggles or face shield; gloves and boots. Chemical protective material(s) reported to have good to excellent resistance: butyl rubber.
Exposure limits (TWA unless otherwise noted): ACGIH TLV 50 ppm (176 m/m3); NIOSH REL 50 ppm (175 mg/m³).
Toxicity by ingestion: Grade 1; LD_{50} = 5 to 15 g/kg (mouse); LD_{50} = 3.2 g/kg (rat).
Vapor (gas) irritant characteristics: Severe irritation of respiratory tract. Vapors cause smarting of the eyes or respiratory system if present in high concentrations.

VANADIUM OXIDE   REC. V:0200

SYNONYMS: TRIOXIDO de VANADIO (Spanish); VANADIC OXIDE; VANADIUM (III) OXIDE; VANADIUM SESQUIOXIDE; VANADIUM TRIOXIDE

IDENTIFICATION
CAS Number: 1314-34-7
Formula: V_2O_5
DOT ID Number: UN 3285; DOT Guide Number: 151
Proper Shipping Name: Vanadium compound, toxic, n.o.s.

DESCRIPTION: Black crystalline solid Gradually turns indigo blue on contact with air. Odorless. Slightly soluble in water.

Poison! • May explode in fire; ignites on heating in air • Breathing the vapor or swallowing the material can cause permanent injury; can be absorbed through the skin; skin contact causes burns • Firefighting gear (including SCBA) does not provide adequate protection. If exposure occurs, remove and isolate gear immediately and thoroughly decontaminate personnel • Toxic products of combustion may include vanadium oxides.

Hazard Classification (based on NFPA-704 Rating System)
Health Hazards (Blue): 1; Flammability (Red): 0; Reactivity (Yellow): 0

EMERGENCY RESPONSE: See Appendix A (151)
Evacuation:
Public safety: Isolate the area of spill or leak for at least 25 to 50 meters (80 to 160 feet) in all directions.
Spill: Increase, in the downwind direction, as necessary, the distance shown under "Public Safety."
Fire: If tank, rail car, or tank truck is involved in fire, isolate for at least 800 meters (½ mile) in all directions; also, consider initial evacuation for 800 meters (½ mile) in all directions.

EXPOSURE
Short-term effects: *SEEK MEDICAL ATTENTION*. **Dust or Solid:** Irritating to skin and eyes. Move victim to fresh air. *IF BREATHING HAS STOPPED*, give artificial respiration; *avoid mouth-to-mouth resuscitation; use bag/mask apparatus*. If breathing is difficult, administer oxygen. Remove contaminated clothing and shoes. Wash affected areas with soap and water. *IF IN EYES*, hold eyelids open and flush with plenty of water. *IF SWALLOWED* and victim is *CONSCIOUS AND ABLE TO SWALLOW*, give two glasses of water and induce vomiting.

HEALTH HAZARDS
Personal protective equipment (PPE): B-Level PPE. OSHA Table Z-1-A air contaminant. Full covering clothing, chemical protective gloves and approved respirator.
Recommendations for respirator selection: NIOSH/OSHA as vanadium
0.5 mg/m³: HiE* (any air-purifying, respirator with a high-efficiency particulate filter); or SA* (any supplied-air respirator). *1.25 mg/m³:* SA:CF (any supplied-air respirator operated in a continuous-flow mode); or PAPRHiE* (any powered, air-purifying respirator with a high-efficiency particulate filter). *2.5 mg/m³:* HiEF (any air-purifying, full-facepiece respirator with a high-efficiency particulate filter); or PAPRHiE (any powered, air-purifying respirator with a high-efficiency particulate filter); or SCBAF (any self-contained breathing apparatus with a full facepiece); or SAF (any supplied-air respirator with a full facepiece). *35 mg/m³* SAF:PD,PP (any supplied-air respirator that has a full facepiece and

Liquid or solid irritant characteristics: Severe eye and skin irritant. If spilled on clothing and allowed to remain, may cause smarting and reddening of the skin.
Odor threshold: 0.00061–8.5 ppm.

FIRE DATA
Flash point: 54°F/12°C (oc).
Fire extinguishing agents not to be used: Water may be ineffective.
Behavior in fire: Forms acrid carbon monoxide.
Autoignition temperature: 432°F/222°C/495°K.
Electrical hazard: Class I, Group C.
Burning rate: 1.9 mm/min
Stoichiometric air-to-fuel ratio: 26.2

CHEMICAL REACTIVITY
Binary reactants: Violent reaction with strong oxidizers, strong acids, bromines, ketones. Incompatible with caustics, ammonia, amines.
Reactivity group: 19
Compatibility class: Aldehydes

ENVIRONMENTAL DATA
Food chain concentration potential: Negative; unlikely to accumulate.
Water pollution: DOT Appendix B, §172.101–marine pollutant. Effect of low concentrations on aquatic life is unknown. Fouling to shoreline. May be dangerous if it enters nearby water intakes; notify operators. Notify local health and wildlife officials.
Response to discharge: Issue warning–high flammability. Evacuate area. Disperse and flush.

SHIPPING INFORMATION
Grades of purity: 98.5+%; **Storage temperature:** Ambient; **Inert atmosphere:** None; **Venting:** Open (flame arrester) or pressure-vacuum valve; **Stability during transport:** Stable.

NAS HAZARD CLASSIFICATION FOR BULK WATER TRANSPORTATION
FIRE: 3
HEALTH: Vapor irritant: 1; Liquid or solid irritant: 1; Poisons: 2
WATER POLLUTION: Human toxicity: 1; Aquatic toxicity: 2; Aesthetic effect: 3
REACTIVITY: Other chemicals: 2; Water: 0; Self-reaction: 1

PHYSICAL AND CHEMICAL PROPERTIES
Physical state @ 59°F/15°C and 1 atm: Liquid.
Molecular weight: 86.13
Boiling point @ 1 atm: 217.4°F/103.0°C/376.2°K.
Melting/Freezing point: –132°F/–91°C/182°K.
Critical temperature: 538°F/281°C/554°K.
Critical pressure: 514 psia = 35 atm = 3.5 MN/m².
Specific gravity (water = 1): 0.811 @ 68°F/20°C (liquid).
Liquid surface tension: (estimate) 30 dynes/cm = 0.03 N/m @ 68°F/20°C.
Liquid water interfacial tension: (estimate) 30 dynes/cm = 0.03 N/m @ 68°F/20°C.
Relative vapor density (air = 1): 3.0
Ratio of specific heats of vapor (gas): 1.072
Latent heat of vaporization: 170 Btu/lb = 93 cal/g = 3.9 x 10⁵ J/kg.
Heat of combustion: –15,500 Btu/lb = –8610 cal/g = –360.5 x 10⁵ J/kg.
Vapor pressure: 26 mm.

is operated in a pressure-demand or other positive-pressure mode). *EMERGENCY OR PLANNED ENTRY INTO UNKNOWN CONCENTRATIONS OR IDLH CONDITIONS*: SCBAF:PD,PP (any self-contained breathing apparatus that has a full facepiece and is operated in a pressure-demand or other positive-pressure mode); or SAF:PD,PP:ASCBA (any supplied-air respirator that has a full facepiece and is operated in a pressure-demand or other positive-pressure mode in combination with an auxiliary self-contained breathing apparatus operated in a pressure-demand or other positive pressure mode). *ESCAPE:* HiEF (any air-purifying, full-facepiece respirator with a high-efficiency particulate filter); or SCBAE (any appropriate escape-type, self-contained breathing apparatus). *Note*: Substance reported to cause eye irritation or damage; may require eye protection.
Exposure limits (TWA unless otherwise noted): ACGIH TLV 0.05 mg/m^3 as vanadium; (dust) OSHA PEL ceiling 0.05 mg/m^3 (resp); (fume) ceiling 0.1 mg/m^3 as vanadium pentoxide
Short-term exposure limits (15-minute TWA): NIOSH ceiling (15 minutes) 0.05 mg/m^3 as vanadium.
Toxicity by ingestion: Grade 3; oral rat LD$_{50}$ more than 70 mg/kg.
Long-term health effects: Repeated or prolonged exposure may cause an allergic skin rash. Poisoning may affect the nervous system, kidneys and lungs.
Vapor (gas) irritant characteristics: May cause irritation and difficult breathing.
Liquid or solid irritant characteristics: Causes eye and skin irritation. If spilled on clothing and allowed to remain, may cause smarting and reddening of the skin.
IDLH value: 35 mg/m^3 as vanadium

CHEMICAL REACTIVITY
Binary reactants: A strong oxidizer. Reacts with strong acids, calcium, chlorine trifluoride, peroxyformic acid, combustible materials, organic substances, sulfur, water. Reacts with lithium at elevated temperatures. Contact with alkalis forms water soluble vanadates. Aqueous solution is acidic; incompatible with sulfuric acid, alkalis, ammonia, aliphatic amines, alkanolamines, alkylene oxides, amides, chlorine trifluoride, epichlorohydrin, nitromethane, organic anhydrides, isocyanates, peroxyformic acid, vinyl acetate.

ENVIRONMENTAL DATA
Water pollution: Effect of low concentrations on aquatic life is unknown. May be dangerous if it enters nearby water intakes; notify operators. Notify local health and wildlife officials.
Response to discharge: Should be removed.

SHIPPING INFORMATION
Grades of purity: Technical; **Storage temperature:** Ambient; **Inert atmosphere:** None; **Stability during transport:** Stable.

PHYSICAL AND CHEMICAL PROPERTIES
Physical state @ 59°F/15°C and 1 atm: Solid.
Molecular weight: 181.88
Boiling point @ 1 atm: Decomposes 3182°F/1750°C/2023°K.
Melting/Freezing point: 1216.4°F/658°C/931°K.
Specific gravity (water = 1): 3.36 @ 68°F/20°C (solid).
Heat of fusion: 85.5 cal/g.
Vapor pressure: 0 mm (approximate).

VANADIUM OXYTRICHLORIDE   REC. V:0300

SYNONYMS: OXITRICLORURO de VANADIO (Spanish); TRICHLOROOXOVANADIUM; VANADIUM TRICHLORIDE OXIDE; VANADYL TRICHLORIDE; VANADYL CHLORIDE

IDENTIFICATION
CAS Number: 7727-18-6
Formula: Cl$_3$OV; VOCl$_3$
DOT ID Number: UN 2443; DOT Guide Number: 137
Proper Shipping Name: Vanadium oxytrichloride

DESCRIPTION: Lemon yellow liquid that becomes thick and blood-red when mixed with water. Emits red fumes on contact with air. Sharp, acrid, unpleasant odor. Sinks and mixes violently with water forming hydrochloric acid and releasing hydrogen chloride vapors.

Poison! • Highly corrosive • Do not use water • Breathing the vapor can kill you; skin or eye contact causes severe burns, impaired vision, or blindness • Firefighting gear (including SCBA) does not provide adequate protection. If exposure occurs, remove and isolate gear immediately and thoroughly decontaminate personnel • Containers may BLEVE when exposed to fire • Toxic products of combustion may include hydrogen chloride and vanadium oxides • Corrosive to most metals in the presence of moisture • Do not put yourself in danger by entering a contaminated area to rescue a victim: *Note*: Persons with a history of asthma or who may have developed asthma symptoms from previous exposure should not be allowed to come in contact with this material.

Hazard Classification (based on NFPA-704 Rating System)
Health Hazards (Blue): 3; Flammability (Red): 0; Reactivity (Yellow): 2; Special Notice (White): Water reactive

EMERGENCY RESPONSE: See Appendix A (137)
Evacuation:
Public safety: Isolate the area of spill or leak for at least 50 to 100 meters (160 to 330 feet) in all directions.
Spill: Increase, in the downwind direction, as necessary, the distance shown under "Public Safety."
Fire: If any large container is involved in fire, isolate for at least 800 meters (½ mile) in all directions; also, consider initial evacuation for 800 meters (½ mile) in all directions.

EXPOSURE
Short-term effects: *SEEK MEDICAL ATTENTION*. **Vapor:** Irritating to eyes, nose, and throat. *IF INHALED*, will, will cause coughing or difficult breathing. *IF IN EYES*, hold eyelids open and flush with plenty of water. *IF BREATHING HAS STOPPED*, give artificial respiration respiration; *avoid mouth-to-mouth resuscitation; use bag/mask apparatus*. IF breathing is difficult, administer oxygen. **Liquid:** *POISONOUS IF SWALLOWED OR IF SKIN IS EXPOSED*. Irritating to skin and eyes. Harmful if swallowed. Remove contaminated clothing and shoes. Flush affected areas with plenty of water. *IF IN EYES*, hold eyelids open and flush with plenty of water. *IF SWALLOWED* and victim is *CONSCIOUS AND ABLE TO SWALLOW*, have victim drink 4 to 8 ounces of water. *IF SWALLOWED* and victim is *UNCONSCIOUS OR HAVING CONVULSIONS*, do nothing except keep victim warm.
Note to medical personnel: Asthma symptoms may be delayed and develop several hours following exposure: Physical exertion. may aggravate the condition. Consider hospitalization and observation. Once exposed, victim should avoid all contact with this material.
Note to physician or authorized medical personnel: Medical observation is recommended for 24 to 48 hours after breathing overexposure, as pulmonary edema may be delayed. As first aid for pulmonary edema, consider administering a corticosteroid spray. Cigarette smoking may exacerbate pulmonary injury and should be discouraged for at least 72 hours following exposure.

HEALTH HAZARDS

Personal protective equipment (PPE): A-Level PPE. OSHA Table Z-1-A air contaminant. Acid vapor mask; rubber gloves; face shield; acid-resistant clothing.

Recommendations for respirator selection: NIOSH/OSHA as vanadium

0.5 mg/m³: HiE (any air-purifying, respirator with a high-efficiency particulate filter); or SA (any supplied-air respirator). *1.25 mg/m³:* SA:CF (any supplied-air respirator operated in a continuous-flow mode); or PAPRHiE (any powered, air-purifying respirator with a high-efficiency particulate filter). *2.5 mg/m³:* HiEF (any air-purifying, full-facepiece respirator with a high-efficiency particulate filter); or PAPRHiE (any powered, air-purifying respirator with a high-efficiency particulate filter); or SCBAF (any self-contained breathing apparatus with a full facepiece); or SAF (any supplied-air respirator with a full facepiece). *EMERGENCY OR PLANNED ENTRY INTO UNKNOWN CONCENTRATIONS OR IDLH CONDITIONS:* SCBAF:PD,PP (any self-contained breathing apparatus that has a full facepiece and is operated in a pressure-demand or other positive-pressure mode); or SAF:PD,PP:ASCBA (any supplied-air respirator that has a full facepiece and is operated in a pressure-demand or other positive-pressure mode in combination with an auxiliary self-contained breathing apparatus operated in a pressure-demand or other positive pressure mode). *ESCAPE:* HiEF (any air-purifying, full-facepiece respirator with a high-efficiency particulate filter); or SCBAE (any appropriate escape-type, self-contained breathing apparatus). *Note*: Substance reported to cause eye irritation or damage; may require eye protection.

Exposure limits (TWA unless otherwise noted): ACGIH TLV 0.05 mg/m³ as vanadium; (dust) OSHA PEL ceiling 0.05 mg/m³ (resp); (fume) ceiling 0.1 mg/m³ as vanadium pentoxide

Short-term exposure limits (15-minute TWA): NIOSH ceiling (15 minute) 0.05 mg/m³ as vanadium.

Toxicity by ingestion: Grade 3; oral rat LD_{50} = 140 mg/kg.

Long-term health effects: Repeated exposures may cause discoloration of tongue, loss of appetite, anemia, kidney disorders, and blindness.

Vapor (gas) irritant characteristics: Causes irritation of the eyes, skin, and respiratory tract.

Liquid or solid irritant characteristics: Causes eye and skin irritation.

IDLH value: 35 mg/m³ as vanadium

FIRE DATA

Fire extinguishing agents not to be used: Water, unless in flooding amounts, should not be used on adjacent fires.

CHEMICAL REACTIVITY

Reactivity with water: Reacts to form a solution of hydrochloric acid.

Binary reactants: Forms hydrogen chloride. In presence of moisture corrodes most metals. Violent reaction with strong bases, sodium or potassium.

Neutralizing agents for acids and caustics: Flush with water and sprinkle with powdered limestone or rinse with dilute solution of sodium bicarbonate or soda ash.

ENVIRONMENTAL DATA

Water pollution: Effect of low concentrations on aquatic life is unknown. May be dangerous if it enters nearby water intakes; notify operators. Notify local health and wildlife officials.

Response to discharge: Issue warning–corrosive, air contaminant. Restrict access. Disperse and flush.

SHIPPING INFORMATION

Grades of purity: Technical, 99%; **Storage temperature:** Ambient; **Inert atmosphere:** None; **Venting:** Pressure-vacuum; **Stability during transport:** Stable.

PHYSICAL AND CHEMICAL PROPERTIES

Physical state @ 59°F/15°C and 1 atm: Liquid.
Molecular weight: 173.3
Boiling point @ 1 atm: 259°F/126°C/399°K.
Melting/Freezing point: –107°F/–77°C/196°K.
Specific gravity (water = 1): 1.83 @ 68°F/20°C (liquid).
Relative vapor density (air = 1): 5.98
Vapor pressure: 6.0 mm (approximate).

VANADIUM PENTOXIDE   REC. V:0400

SYNONYMS: C.I. 77938; EEC No. 023-001-00-8; PENTOXIDO de VANADIO (Spanish); VANADIC ANHYDRIDE; VANADIUM OXIDE; VANADIUM(V) OXIDE FUME; VANADIUM(5+) OXIDE; VANADIUM PENTOXIDE DUST; VANADIUM PENTOXIDE (Fume); RCRA No. P120; RCRA No. YW2450000

IDENTIFICATION

CAS Number: 1314-62-1
Formula: $O_5 V_2$; V_2O_5
DOT ID Number: UN 2862; DOT Guide Number: 151
Proper Shipping Name: Vanadium pentoxide
Reportable Quantity (RQ): **(CERCLA)** 1000 lb/454 kg

DESCRIPTION: Yellow to red crystalline solid; yellowish-orange powder. Odorless. Sinks in water; insoluble.

Poison! • Corrosive • Breathing the dust or vapor can kill you; skin or eye contact causes severe burns, impaired vision, or blindness; symptoms of inhalation exposure may be delayed • Firefighting gear (including SCBA) does not provide adequate protection. If exposure occurs, remove and isolate gear immediately and thoroughly decontaminate personnel • Strong oxidizer which may react spontaneously with low flash point organics or reducing agents. Heat forms oxygen; will increase the activity of an existing fire • May cause fire or explosion contact with combustibles (wood, paper, oil, clothing, etc.) • Toxic products of combustion may include vanadium oxide.

Hazard Classification (based on NFPA-704 Rating System)
Health Hazards (Blue): 1; Flammability (Red): 0; Reactivity (Yellow): 0; Special Notice (White): OXY

EMERGENCY RESPONSE: See Appendix A (151)
Evacuation:
Public safety: Isolate the area of spill or leak for at least 25 to 50 meters (80 to 160 feet) in all directions.
Spill: Increase, in the downwind direction, as necessary, the distance shown under "Public Safety."
Fire: If tank, rail car, or tank truck is involved in fire, isolate for at least 800 meters (½ mile) in all directions; also, consider initial evacuation for 800 meters (½ mile) in all directions.

EXPOSURE

Short-term effects: *SEEK MEDICAL ATTENTION.* **Dust:** Irritating to eyes, nose, and throat. *IF INHALED*, will, will cause coughing or difficult breathing. *IF IN EYES*, hold eyelids open and flush with plenty of water. *IF BREATHING HAS STOPPED*, give artificial respiration; *avoid mouth-to-mouth resuscitation; use* *bag/mask apparatus*. IF breathing is difficult, administer oxygen. Lung edema may develop. **Solid:** Irritating to skin and eyes. *IF SWALLOWED*, will cause nausea. Remove contaminated clothing and shoes. Flush affected areas with plenty of water. *IF IN EYES*, hold eyelids open and flush with plenty of water. *IF SWALLOWED and victim is CONSCIOUS AND ABLE TO SWALLOW*, have victim drink 4 to 8 ounces of water and have victim induce vomiting. *IF SWALLOWED and victim is UNCONSCIOUS OR HAVING CONVULSIONS*, do nothing except keep victim warm. *Note to physician or authorized medical personnel*: Medical observation is recommended for 24 to 48 hours after breathing overexposure, as pulmonary edema may be delayed. As first aid for pulmonary edema, consider administering a corticosteroid spray. Cigarette smoking may exacerbate pulmonary injury and should be discouraged for at least 72 hours following exposure.

HEALTH HAZARDS

Personal protective equipment (PPE): A-Level PPE. OSHA Table Z-1-A air contaminant. Bureau of Mines approved respirator; rubber gloves; goggles for prolonged exposure.
Recommendations for respirator selection: NIOSH/OSHA as vanadium
0.5 mg/m³: HiE (any air-purifying, respirator with a high-efficiency particulate filter); or SA (any supplied-air respirator). *1.25 mg/m³*: SA:CF (any supplied-air respirator operated in a continuous-flow mode); or PAPRHiE (any powered, air-purifying respirator with a high-efficiency particulate filter). *2.5 mg/m³*: HiEF (any air-purifying, full-facepiece respirator with a high-efficiency particulate filter); or PAPRHiE (any powered, air-purifying respirator with a high-efficiency particulate filter); or SCBAF (any self-contained breathing apparatus with a full facepiece); or SAF (any supplied-air respirator with a full facepiece). *EMERGENCY OR PLANNED ENTRY INTO UNKNOWN CONCENTRATIONS OR IDLH CONDITIONS*: SCBAF:PD,PP (any self-contained breathing apparatus that has a full facepiece and is operated in a pressure-demand or other positive-pressure mode); or SAF:PD,PP:ASCBA (any supplied-air respirator that has a full facepiece and is operated in a pressure-demand or other positive-pressure mode in combination with an auxiliary self-contained breathing apparatus operated in a pressure-demand or other positive pressure mode). *ESCAPE:* HiEF (any air-purifying, full-facepiece respirator with a high-efficiency particulate filter); or SCBAE (any appropriate escape-type, self-contained breathing apparatus). *Note*: Substance reported to cause eye irritation or damage; may require eye protection.
Exposure limits (TWA unless otherwise noted): ACGIH TLV 0.05 mg/m³ as vanadium dust or fume.
Short-term exposure limits (15-minute TWA): NIOSH ceiling 0.05 mg/m³/15 min as vanadium; OSHA ceiling (resp) 0.05 mg/m³; ceiling (fume) 0.1 mg/m³ as vanadium pentoxide.
Toxicity by ingestion: Poison. Grade 4; oral LD_{50} = 23 mg/kg (mouse).
Long-term health effects: Repeated exposures may cause discoloration of tongue, loss of appetite, anemia, kidney disorders, lung disorders and blindness. Mutation and reproductive data reported.
Vapor (gas) irritant characteristics: May cause irritation of the eyes and respiratory tract; difficult breathing and lung edema.
Liquid or solid irritant characteristics: May cause eye and skin irritation.
IDLH value: 35 mg/m³ as vanadium

CHEMICAL REACTIVITY

Reactivity with water: Aqueous solution is acetic.
Binary reactants: Lithium, chlorine trifluoride; peroxyformic acid. Aqueous system reacts with bases.

ENVIRONMENTAL DATA

Food chain concentration potential: Negative; unlikely to accumulate.
Water pollution: Harmful to aquatic life in very low concentrations. May be dangerous if it enters nearby water intakes; notify operators. Notify local health and wildlife officials. **Response to discharge:** Should be removed. Chemical and physical treatment.

SHIPPING INFORMATION

Grades of purity: Commercial, 98-99%; **Storage temperature:** Ambient; **Inert atmosphere:** None; **Venting:** Open; **Stability during transport:** Stable.

PHYSICAL AND CHEMICAL PROPERTIES

Physical state @ 59°F/15°C and 1 atm: Solid.
Molecular weight: 181.88
Boiling point @ 1 atm: 3182°F/1750°C/2023°K (decomposes) (decomposes).
Melting/Freezing point: 1265°F/685°C/958°K.
Specific gravity (water = 1): 3.36 @ 68°F/20°C (solid).
Heat of fusion: 85.5 cal/g.
Vapor pressure: 0 mm (approximate).

VANADYL SULFATE REC. V:0500

SYNONYMS: C.I. 77940; OXYSULFATOVANADIUM; SULFATO de VANADILO (Spanish); VANADIUM OXYSULFATE; VANADYL SULFATE DIHYDRATE

IDENTIFICATION

CAS Number: 27774-13-6
Formula: O_5SV; $VOSO_4 \cdot 2H_2O_2$
DOT ID Number: UN 2931; DOT Guide Number: 151
Proper Shipping Name: Vanadyl sulfate
Reportable Quantity (RQ): **(CERCLA)** 1000 lb/454 kg

DESCRIPTION: Pale blue crystalline solid or powder. Odorless. Sinks and mixes with water.

Severely irritating to eyes, nose, and throat • Firefighting gear (including SCBA) does not provide adequate protection. If exposure occurs, remove and isolate gear immediately and thoroughly decontaminate personnel • Containers may BLEVE when exposed to fire • Toxic products of combustion may include sulfur oxides and vanadium oxides.

Hazard Classification (based on NFPA-704 Rating System)
Health Hazards (Blue): 1; Flammability (Red): 0; Reactivity (Yellow): 0

EMERGENCY RESPONSE: See Appendix A (151)
Evacuation:
Public safety: Isolate the area of spill or leak for at least 25 to 50 meters (80 to 160 feet) in all directions.
Spill: Increase, in the downwind direction, as necessary, the distance shown under "Public Safety."
Fire: If tank, rail car, or tank truck is involved in fire, isolate for at least 800 meters (½ mile) in all directions; also, consider initial evacuation for 800 meters (½ mile) in all directions.

EXPOSURE
Short-term effects: *SEEK MEDICAL ATTENTION*. **Dust:** Irritating to eyes, nose, and throat. *IF INHALED*, will, will cause coughing or difficult breathing. *IF IN EYES*, hold eyelids open and flush with plenty of water. *IF BREATHING HAS STOPPED*, give artificial respiration; *avoid mouth-to-mouth resuscitation; use bag/mask apparatus*. IF breathing is difficult, administer oxygen. **Solid:** Irritating to skin and eyes. *IF SWALLOWED*, will cause nausea or coughing. Remove contaminated clothing and shoes. Flush affected areas with plenty of water. *IF IN EYES*, hold eyelids open and flush with plenty of water. *IF SWALLOWED* and victim is *CONSCIOUS AND ABLE TO SWALLOW*, have victim drink 4 to 8 ounces of water and have victim induce vomiting. *IF SWALLOWED* and victim is *UNCONSCIOUS OR HAVING CONVULSIONS*, do nothing except keep victim warm.

HEALTH HAZARDS
Personal protective equipment (PPE): B-Level PPE. OSHA Table Z-1-A air contaminant. Dust mask; goggles or face shield; protective gloves.
Recommendations for respirator selection: NIOSH/OSHA as vanadium
$0.5\ mg/m^3$: HiE (any air-purifying, respirator with a high-efficiency particulate filter); or SA (any supplied-air respirator). $1.25\ mg/m^3$: SA:CF (any supplied-air respirator operated in a continuous-flow mode); or PAPRHiE (any powered, air-purifying respirator with a high-efficiency particulate filter). $2.5\ mg/m^3$: HiEF (any air-purifying, full-facepiece respirator with a high-efficiency particulate filter); or PAPRHiE (any powered, air-purifying respirator with a high-efficiency particulate filter); or SCBAF (any self-contained breathing apparatus with a full facepiece); or SAF (any supplied-air respirator with a full facepiece). *EMERGENCY OR PLANNED ENTRY INTO UNKNOWN CONCENTRATIONS OR IDLH CONDITIONS*: SCBAF:PD,PP (any self-contained breathing apparatus that has a full facepiece and is operated in a pressure-demand or other positive-pressure mode); or SAF:PD,PP:ASCBA (any supplied-air respirator that has a full facepiece and is operated in a pressure-demand or other positive-pressure mode in combination with an auxiliary self-contained breathing apparatus operated in a pressure-demand or other positive pressure mode). *ESCAPE*: HiEF (any air-purifying, full-facepiece respirator with a high-efficiency particulate filter); or SCBAE (any appropriate escape-type, self-contained breathing apparatus). *Note*: Substance reported to cause eye irritation or damage; may require eye protection.
Exposure limits (TWA unless otherwise noted): ACGIH TLV $0.05\ mg/m^3$ as vanadium dust or fume.
Short-term exposure limits (15-minute TWA): NIOSH ceiling $0.05\ mg/m^3$/15 min as vanadium; OSHA ceiling (resp) $0.05\ mg/m^3$; ceiling (fume) $0.1\ mg/m^3$ as vanadium pentoxide.
Toxicity by ingestion: Grade 3; LD_{50} = 50 to 500 mg/kg.
Long-term health effects: Repeated exposures may cause discoloration of tongue, loss of appetite, anemia, kidney disorders, and blindness. Mutation data.
Liquid or solid irritant characteristics: May cause eye, skin and respiratory tract irritation.
IDLH value: $35\ mg/m^3$ as vanadium.

CHEMICAL REACTIVITY
Binary reactants: Violent reaction with aluminum, magnesium.

ENVIRONMENTAL DATA
Water pollution: Harmful to aquatic life in very low concentrations. May be dangerous if it enters nearby water intakes; notify operators. Notify local health and wildlife officials.

Response to discharge: Issue warning–water contaminant. Disperse and flush.

SHIPPING INFORMATION
Grades of purity: Commercial; Pure; **Storage temperature:** Ambient; **Inert atmosphere:** None; **Venting:** Open; **Stability during transport:** Stable.

PHYSICAL AND CHEMICAL PROPERTIES
Physical state @ 59°F/15°C and 1 atm: Solid.
Molecular weight: 199.1
Boiling point @ 1 atm: Decomposes.
Specific gravity (water = 1): (approximate) 2.5 @ 68°F/20°C (solid).

VANILLAN BLACK LIQUOR REC. V:0600

SYNONYMS: UF OXYLIGNIN

IDENTIFICATION
CAS Number: 68514-06-7
Formula: $C_8H_8O_3$
DOT ID Number: UN 1760; DOT Guide Number: 154
Proper Shipping Name: Corrosive liquids, n.o.s.

DESCRIPTION: Liquid. Brown. Sweet odor. Soluble in water.

Corrosive • Skin or eye contact causes severe burns; effects of contact or inhalation may be delayed • Firefighting gear (including SCBA) does not provide adequate protection. If exposure occurs, remove and isolate gear immediately and thoroughly decontaminate personnel • Containers may BLEVE when exposed to fire • Toxic products of combustion may include caustic vapor, carbon monoxide, and CO_2.

Hazard Classification (based on NFPA-704 Rating System)
Health Hazards (Blue): 1; Flammability (Red): 0; Reactivity (Yellow): 0

EMERGENCY RESPONSE: See Appendix A (154)
Evacuation:
Public safety: Isolate the area of spill or leak for at least 25 to 50 meters (80 to 160 feet) in all directions.
Spill: Increase, in the downwind direction, as necessary, the distance shown under "Public Safety."
Fire: If tank, rail car, or tank truck is involved in fire, isolate for at least 800 meters (½ mile) in all directions; also, consider initial evacuation for 800 meters (½ mile) in all directions.

EXPOSURE
Short-term effects: *SEEK MEDICAL ATTENTION*. **Vapor:** Move victim to fresh air. *POISONOUS IF SWALLOWED*. If breathing is difficult, administer oxygen. **Liquid:** Corrosive to skin, eyes and respiratory tract. Remove contaminated clothing and shoes. Wash affected areas with soap and water. *IF IN EYES*, hold eyelids open and flush with plenty of water. *IF SWALLOWED* and victim is *CONSCIOUS AND ABLE TO SWALLOW*, drink lots of water. **Do NOT induce vomiting.**
Note to physician or authorized medical personnel: Medical observation is recommended for 24 to 48 hours after breathing overexposure, as pulmonary edema may be delayed. As first aid for pulmonary edema, consider administering a corticosteroid spray. Cigarette smoking may exacerbate pulmonary injury and should be discouraged for at least 72 hours following exposure.

HEALTH HAZARDS

Personal protective equipment (PPE): B-Level PPE. Full impervious protective clothing, including boots and gloves. Where splashing is possible wear full face shield or chemical safety goggles. Use approved respirator to protect against vapors.
Long-term health effects: Prolonged exposure may cause dermatitis and permanent scarring.
Vapor (gas) irritant characteristics: Vapors cause severe irritation of eyes and throat and can cause eye and lung injury. They cannot be tolerated even at low concentrations. **Liquid or solid irritant characteristics:** Severe skin irritant. Causes second- and third-degree burns on short contact and is very injurious to the eyes.

CHEMICAL REACTIVITY

Reactivity with water: May generate heat.
Binary reactants: Not compatible with aluminum, zinc and tin. Reaction with these can produce hydrogen gas and heat. Contact with acids will produce CO_2 and may create alkaline mists.
Neutralizing agents for acids and caustics: Dilute acid.
Reactivity group: 5
Compatibility class: Caustics (mixture).

ENVIRONMENTAL DATA

Water pollution: Effect of low concentrations on aquatic life is unknown. May be dangerous if it enters nearby water intakes; notify operators. Notify local health and wildlife officials.
Response to discharge: Issue warning–corrosive. Chemical and physical treatment. Should be removed.

SHIPPING INFORMATION

Grades of purity: Technical in varying concentrations of components. Mixture includes sodium carbonate, sodium hydroxide, sodium sulfate, and sodium lignosulfonate; **Storage temperature:** Ambient; **Inert atmosphere:** None; **Venting:** Open; **Stability during transport:** Stable.

PHYSICAL AND CHEMICAL PROPERTIES

Physical state @ 59°F/15°C and 1 atm: Liquid.
Boiling point @ 1 atm: 225°F/107°C/380°K.
Specific gravity (water = 1): 1.3

VINYL ACETATE   REC. V:0700

SYNONYMS: ACETATE de VINYLE (French); ACETO de VINILO (Spanish); ACETIC ACID, ETHENYL ESTER; ACETIC ACID, ETHENYL ESTER; ACETIC ACID, VINYL ESTER; 1-ACETOXYETHYLENE; EEC No. 607-023-00-0; ETHENYL ACETATE; ETHENYL ETHANOATE; ETHENYLETHANOATE; VAC; VAM; VINYLACETAT (German); VINYL ACETATE H.Q; VINYL A MONOMER; VINYL ACETATE MONOMER; VINYLE (ACETATE de) (French); VyAc; ZESET T; RCRA No. AK0875000

IDENTIFICATION
CAS Number: 108-05-4
Formula: $C_4H_6O_2$; $CH_3COOCH=CH_2$
DOT ID Number: UN 1301; DOT Guide Number: 129P
Proper Shipping Name: Vinyl acetate, inhibited
Reportable Quantity (RQ): **(CERCLA)** 5000 lb/2270 kg

DESCRIPTION: Colorless, watery liquid. Pleasant, fruity odor. Floats on water surface; slightly soluble. Large amounts of flammable, irritating vapor is produced.

Highly flammable • Polymerization hazard • May undergo spontaneous exothermic polymerization; sealed containers may rupture explosively • May react with itself blocking relief valves; leading to tank explosions •Containers may BLEVE when exposed to fire •Vapors may form explosive mixture with air • Vapors are heavier than air and will collect and stay in low areas • Vapors may travel long distances to ignition sources and flashback • Vapors in confined areas (e.g, tanks, sewers, buildings) may explode when exposed to fire • Irritating to the skin, eyes, and respiratory tract; prolonged contact with skin caused burns • Toxic products of combustion may include carbon monoxide.

Hazard Classification (based on NFPA-704 Rating System)
Health Hazards (Blue): 2; Flammability (Red): 3; Reactivity (Yellow): 2

EMERGENCY RESPONSE: See Appendix A (129)
Evacuation:
Public safety: Isolate spill area for at least 50 to 100 meters (160 to 330 feet) in all directions.
Spill: Large spill–Consider initial downwind evacuation for at least 300 meters (1000 feet).
Fire: Isolate for 800 meters (½ mile) in all directions, especially if tank, rail car, or tank truck is involved in fire.

EXPOSURE

Short-term effects: *SEEK MEDICAL ATTENTION.* **Vapor:** Irritating to eyes, nose, and throat. *IF INHALED*, will, will cause dizziness or difficult breathing. Move to fresh air. *IF BREATHING HAS STOPPED*, give artificial respiration. IF breathing is difficult, administer oxygen. **Liquid:** Irritating to skin and eyes. Harmful if swallowed or if spilled on skin. Remove contaminated clothing and shoes. Flush affected areas with plenty of water. *IF IN EYES*, hold eyelids open and flush with plenty of water. *IF SWALLOWED* and victim is *CONSCIOUS AND ABLE TO SWALLOW*, have victim drink 4 to 8 ounces of water. The use of alcoholic beverages may enhance the toxic effect.

HEALTH HAZARDS

Personal protective equipment (PPE): B-Level PPE. OSHA Z-1-A air contaminant. Approved canister); or air–supplied mask; goggles or face shield; gloves. Chemical protective material(s) reported to offer limited protecion: Teflon®, chlorinated polyethylene.
Recommendations for respirator selection: NIOSH/OSHA
40 ppm: CCRFOV* [any chemical cartridge respirator with a full facepiece and organic vapor cartridges(s)]; or SA* (any supplied-air respirator). *100 ppm*: SA:CF* (any supplied-air respirator operated in a continuous-flow mode); or PAPROV* [any powered, air-purifying respirator with organic vapor cartridge(s)]. *200 ppm*: CCRFOV [any chemical cartridge respirator with a full facepiece and organic vapor cartridges(s)]; or GMFOV [any air-purifying, full-facepiece respirator (gas mask) with a chin-style, front- or back-mounted acid gas canister]; or PAPRTOV* [any powered, air-purifying respirator with a tight-fitting facepiece and organic vapor cartridges(s)]; or SCBAF (any self-contained breathing apparatus with a full facepiece); or SAF (any supplied-air respirator with a full facepiece). *4000 ppm:* SA:PD,PP (any supplied-air respirator operated in a pressure-demand or other positive-pressure mode). *EMERGENCY OR PLANNED ENTRY INTO UNKNOWN CONCENTRATIONS OR IDLH CONDITIONS:* SCBAF:PD,PP (any self-contained breathing apparatus that has a full facepiece and is operated in a pressure-demand or other positive-pressure mode); or SAF:PD,PP:ASCBA (any supplied-air respirator that has a full facepiece and is operated in a pressure-demand or other positive-

pressure mode in combination with an auxiliary self-contained breathing apparatus operated in a pressure-demand or other positive pressure mode). *ESCAPE:* GMFOV [any air-purifying, full-facepiece respirator (gas mask) with a chin-style, front- or back-mounted organic vapor cannister]; or SCBAE (any appropriate escape-type, self-contained breathing apparatus).
Note: Substance reported to cause eye irritation or damage; may require eye protection.
Exposure limits (TWA unless otherwise noted): ACGIH TLV 10 ppm.
Short-term exposure limits (15-minute TWA): ACGIH STEL 15 ppm; NIOSH ceiling 4 ppm (15 mg/m^3)/15 min.
Toxicity by ingestion: Toxic. Grade 2; LD_{50} = 0.5 to 5 g/kg (rat).
Long-term health effects: Suspected human carcinogen.
Vapor (gas) irritant characteristics: May cause irritation of the eyes, skin, and respiratory tract. Vapors cause smarting of the eyes or respiratory system if present in high concentrations. The effect is temporary.
Liquid or solid irritant characteristics: Can irritate eyes and skin. If spilled on clothing and allowed to remain, may cause smarting and reddening of the skin.
Odor threshold: 0.12 ppm.

FIRE DATA
Flash point: 23°F/–5°C (oc); 18°F/–8°C (cc).
Flammable limits in air: LEL: 2.6%; UEL 13.4%.
Fire extinguishing agents not to be used: Water may be ineffective.
Behavior in fire: May polymerize and rupture container.
Autoignition temperature: 752°F/400°C/673°K.
Electrical hazard: Class I, Group D.
Burning rate: 3.8 mm/min

CHEMICAL REACTIVITY
Reactivity with water:
Binary reactants: Highly reactive with many materials. Reacts with acids, bases, silica gel, alumina, strong oxidizers (may be violent), azo compounds, ozone. Can react with oxygen or light forming explosive peroxides: exothermic polymerization reaction and explosion may result.
Polymerization: Can occur when in contact with peroxides and strong acids; only under extreme conditions; **Inhibitor of polymerization:** 3–5 ppm or 14–17 ppm hydroquinone. Shipments usually also contain 200 ppm of diphenylamine. With hydroquinone storage is limited to 60 days (Grade H). Diphenylamine may be used for indefinite storage (Grade A).
Reactivity group: 13
Compatibility class: Vinyl acetate

ENVIRONMENTAL DATA
Food chain concentration potential: Log K_{ow} = 0.73. Values above 3.0 are very likely to accumulate in living tissues and especially in fats. Unlikely to accumulate.
Water pollution: Harmful to aquatic life in very low concentrations. Fouling to shoreline. May be dangerous if it enters nearby water intakes; notify operators. Notify local health and wildlife officials. **Response to discharge:** Issue warning–high flammability, air contaminant. Evacuate area.

SHIPPING INFORMATION
Grades of purity: Grade A (diphenylamine-inhibited): 99.8%; Grade H (hydroquinone-inhibited): 99.8%; **Storage temperature:** Ambient; **Inert atmosphere:** None; **Venting:** Pressure-vacuum; **Stability during transport:** Stable.

NAS HAZARD CLASSIFICATION FOR BULK WATER TRANSPORTATION
FIRE: 3
HEALTH: Vapor irritant: 1; Liquid or solid irritant: 1; Poisons: 2
WATER POLLUTION: Human toxicity: 2; Aquatic toxicity: 1; Aesthetic effect: 2
REACTIVITY: Other chemicals: 2; Water: 0; Self-reaction: 3

PHYSICAL AND CHEMICAL PROPERTIES
Physical state @ 59°F/15°C and 1 atm: Liquid.
Molecular weight: 86.09
Boiling point @ 1 atm: 163.2°F/72.9°C/346.1°K.
Melting/Freezing point: –135.0°F/–92.8°C/180.4°K.
Critical temperature: 486°F/252°C/525°K.
Critical pressure: 617 atm = 42 psia = 4.25 MN/m^2.
Specific gravity (water = 1): 0.934 @ 68°F/20°C (liquid).
Liquid surface tension: 23.95 dynes/cm = 0.02395 N/m @ 68°F/20°C.
Liquid water interfacial tension: (estimate) 30 dynes/cm = 0.03 N/m @ 68°F/20°C.
Relative vapor density (air = 1): 2.99
Ratio of specific heats of vapor (gas): 1.103
Latent heat of vaporization: 163 Btu/lb = 90.6 cal/g = 3.79 x 10^5 J/kg.
Heat of combustion: –9754 Btu/lb = –5419 cal/g = –226.9 x 10^5 J/kg.
Heat of polymerization: –439 Btu/lb = –244 cal/g = –10.2 x 10^5 J/kg.
Vapor pressure: (Reid) 3.7 psia; 83 mm.

VINYL CHLORIDE   REC. V:0800

SYNONYMS: CHLOROETHYLENE; CHLOROETHENE; CHLOROETHYLENE; CHLORURE de VINYLE (French); CLORURO de VINILO (Spanish); EEC No. 602-023-00-7; ETHYLENE MONOCHLORIDE; MONOCHLOROETHENE; MONOCHLOROETHYLENE; TROVIDUER; VC; VCL; VCM; VINYL CHLORIDE MONOMER; VINYL C MONOMER; RCRA No. U043

IDENTIFICATION
CAS Number: 75-01-4
Formula: C_2H_3Cl; $CH_2=CHCl$
DOT ID Number: UN 1086; DOT Guide Number: 116P
Proper Shipping Name: Vinyl chloride; Vinyl chloride, inhibited; Vinyl chloride, stabilized
Reportable Quantity (RQ): **(CERCLA)** 1 lb/0.454 kg

DESCRIPTION: Colorless compressed gas. Shipped as liquefied compressed gas. Sweet, pleasant odor. Liquid floats and boils on water surface; insoluble. Combustion produces a visible, toxic, irritating vapor cloud. Liquid below 7°F/–14°C.

Highly flammable • Polymerization hazard • May undergo spontaneous exothermic polymerization; sealed containers may rupture explosively • May react with itself blocking relief valves; leading to tank explosions •Vapors may cause dizziness or asphyxiation without warning • Contact with liquid may cause frostbite • Containers may BLEVE when exposed to fire •Vapors may form explosive mixture with air • Vapors are heavier than air and will collect and stay in low areas • Vapors may travel long distances to ignition sources and flashback • Vapors in confined areas (e.g. tanks, sewers, buildings) may explode when exposed to fire • Irritating to the skin, eyes, and respiratory tract; prolonged contact with skin caused burns • Toxic products of combustion may include hydrogen chloride, carbon monoxide, CO_2, and traces of phosgene.

Hazard Classification (based on NFPA-704 Rating System)
Health Hazards (Blue): 2; Flammability (Red): 4; Reactivity (Yellow): 2

EMERGENCY RESPONSE: See Appendix A (116)
Evacuation:
Public safety: Isolate spill area for at least 100 meters (330 feet) in all directions.
Spill: Consider initial downwind evacuation for at least 800 meters (½ mile).
Fire: Isolate for 1600 meters (1 mile) in all directions, especially if tank, rail car, or tank truck is involved in fire.

EXPOSURE
Short-term effects: *SEEK MEDICAL ATTENTION.* **Vapor:** Irritating to eyes, nose, and throat. *IF INHALED*, will, will cause dizziness or difficult breathing. Move to fresh air. *IF BREATHING HAS STOPPED*, give artificial respiration. IF breathing is difficult, administer oxygen. **Liquid:** Will cause frostbite. Flush affected areas with plenty of water. *DO NOT RUB AFFECTED AREAS.*

HEALTH HAZARDS
Personal protective equipment (PPE): B-Level PPE. OSHA Table Z-1-A air contaminant. Rubber gloves and shoes; gas-tight goggles; organic vapor canister); or self-contained breathing apparatus. Wear thermal protective clothing. Chemical protective material(s) reported to offer limited protecion: Chlorinated polyethylene, Viton®, Silvershield®.
Recommendations for respirator selection: NIOSH
At any detectable concentration: SCBAF:PD,PP (any MSHA/NIOSH approved self-contained breathing apparatus that has a full facepiece and is operated in a pressure-demand or other positive-pressure mode); or SAF:PD,PP:ASCBA (any supplied-air respirator that has a full facepiece and is operated in a pressure-demand or other positive-pressure mode in combination with an auxiliary, self-contained breathing apparatus operated in a pressure-demand or other positive pressure mode). *ESCAPE:* GMFS [any air-purifying, full-facepiece respirator (gas mask) with a chin-style, front- or back-mounted canister providing protection against the compound of concern]; or SCBAE (any appropriate escape-type, self-contained breathing apparatus).
Exposure limits (TWA unless otherwise noted): ACGIH TLV 1 ppm; OSHA PEL [1910.1017] 1 ppm; confirmed human carcinogen.
Short-term exposure limits (15-minute TWA): OSHA ceiling 5 ppm/15 min.
Toxicity by ingestion: Toxic.
Long-term health effects: Chronic exposure may cause blood, skin, liver, kidney, heart, and bone damage. Confirmed human carcinogen. OSHA specifically regulated carcinogen; IARC carcinogen; NTP known carcinogen.
Vapor (gas) irritant characteristics: Vapors cause irritation such that personnel will find high concentrations unpleasant. The effect may be temporary.
Liquid or solid irritant characteristics: Minimum hazard. If spilled on clothing and allowed to remain, may cause smarting and reddening of skin. Contact with liquid may cause burns and frostbite due to rapid evaporation.
Odor threshold: 10-20 ppm. Detectable odor is greater than the Exposure limits. Exposure to potentially dangerous vapor concentrations can occur before the vapor is detected by smell.

FIRE DATA
Flash point: –110°F/–79°C (oc); flammable gas.
Flammable limits in air: LEL: 3.6%; UEL: 33.0%.
Behavior in fire: Container may explode in fire.
Autoignition temperature: 882°F/472°C/745°K.
Electrical hazard: Class I, Group D.
Burning rate: 4.3 mm/min
Stoichiometric air-to-fuel ratio: 5.490 (estimate).

CHEMICAL REACTIVITY
Binary reactants: Forms unstable peroxides with atmospheric oxygen, strong oxidizers, and various contaminants; possible violent polymerization. Violent reaction with strong oxidizers or oxides of nitrogen. Contact with copper or other acetylide-forming metals form sensitive explosive compounds. Reacts with aluminum, copper and their alloys. Attacks iron and steel in the presence of moisture. The uninhibited monomer vapor may block vents and confined spaces by forming a solid polymer material.
Polymerization: Polymerizes in presence of air, sunlight, or heat unless stabilized by inhibitors; **Inhibitor of polymerization:** Then 40–100 ppm of phenol when high temperatures are expected.
Reactivity group: 35
Compatibility class: Vinyl halides

ENVIRONMENTAL DATA
Food chain concentration potential: Log K_{ow} = 0.6 (estimate). Values above 3.0 are very likely to accumulate in living tissues and especially in fats. This material is unlikely to accumulate.
Water pollution: Response to discharge: Issue warning–high flammability. Should be removed. Clean Water Act.

SHIPPING INFORMATION
Grades of purity: Commercial or technical 99+%; **Storage temperature:** Under pressure: Ambient; At atmospheric pressure: low; **Inert atmosphere:** None; **Venting:** Under pressure; safety relief at atmospheric pressure; pressure-vacuum; **Stability during transport:** Stable.

NAS HAZARD CLASSIFICATION FOR BULK WATER TRANSPORTATION
FIRE: 4
HEALTH: Vapor irritant: 2; Liquid or solid irritant: 1; Poisons: 2
WATER POLLUTION: Human toxicity: 0; Aquatic toxicity: 0; Aesthetic effect: 0
REACTIVITY: Other chemicals: 2; Water: 0; Self-reaction: 2

PHYSICAL AND CHEMICAL PROPERTIES
Physical state @ 59°F/15°C and 1 atm: Gas.
Molecular weight: 62.50
Boiling point @ 1 atm: 7.2°F/13.8°C/259.4°K.
Melting/Freezing point: –244.8°F/–153.8°C/–119.4°K.
Critical temperature: 317.1°F/158.4°C/431.6°K.
Critical pressure: 775 psia = 52.7 atm = 5.34 MN/m^2.
Specific gravity (water = 1): 0.969 at –13°C (liquid).
Liquid surface tension: 16.0 dynes/cm = 0.0160 N/m @ 77°F/25°C.
Liquid water interfacial tension: (estimate) 30 dynes/cm = 0.03 N/m @ 68°F/20°C.
Relative vapor density (air = 1): 2.21
Ratio of specific heats of vapor (gas): 1.186
Latent heat of vaporization: 160 Btu/lb = 88 cal/g = 3.7 x 10^5 J/kg..
Heat of combustion: –8136 Btu/lb = –4520 cal/g = –189.1 x 10^5 J/kg..

Heat of polymerization: −729 Btu/lb = −405 cal/g = 16.9 x 10^5 J/kg.
Heat of fusion: 18.14 cal/g.
Vapor pressure: (Reid) 75 psia; 3.3 atm; 2585 mm.

VINYL CYCLOHEXENE   **REC. V:0900**

SYNONYMS: BUTADIENE DIMER; CYCLOHEXENYLETHYLENE; VINILCICLOHEXANO (Spanish); VINYL-1-CYCLOHEXENE; 4-VINYL-1-CYCLOHEXENE; 4-VINYL CYCLOHEXENE

IDENTIFICATION
CAS Number: 100-40-3; 106-87-6 (dioxide)
Formula: C_8H_{12}; $C_6H_9CH=CH_2$; $C_8H_{12}O_2$ (dioxide)
DOT ID Number: UN 1993; DOT Guide Number: 128
Proper Shipping Name: Flammable liquid, n.o.s.

DESCRIPTION: Colorless liquid. Floats on the surface of water; slowly hydrolyzes.

Highly flammable • Containers may BLEVE when exposed to fire • Vapors may form explosive mixture with air • Vapors are heavier than air and will collect and stay in low areas • Vapors may travel long distances to ignition sources and flashback • Vapors in confined areas (e.g., tanks, sewers, buildings) may explode when exposed to fire • Irritating to the skin, eyes, and respiratory tract • Toxic products of combustion may include acrid carbon monoxide.

Hazard Classification (based on NFPA-704 Rating System)
Health Hazards (Blue): 2; Flammability (Red): 3; Reactivity (Yellow): 2

EMERGENCY RESPONSE: See Appendix A (128)
Evacuation:
Public safety: Isolate spill area for at least 25 to 50 meters (80 to 160 feet) in all directions.
Spill: Large spill–Consider initial downwind evacuation for at least 300 meters (1000 feet).
Fire: Isolate for 800 meters (½ mile) in all directions, especially if tank, rail car, or tank truck is involved in fire.

EXPOSURE
Short-term effects: *SEEK MEDICAL ATTENTION.* **Vapor:** Move victim to fresh air. *IF BREATHING HAS STOPPED,* give artificial respiration. If breathing is difficult, administer oxygen. **Liquid:** Irritating to skin and eyes. Overexposures may have a narcotic effect. Remove contaminated clothing and shoes. Flush affected areas with water. *IF IN EYES*, hold eyelids open and flush with plenty of water.
HEALTH HAZARDS
Personal protective equipment (PPE): B-Level PPE. Full impervious protective clothing, including boots and gloves. Where splashing is possible wear full face shield or chemical safety goggles. Chemical protective material(s) reported to have good to excellent resistance: nitrile, Viton®. Also, PV alcohol offers limited protection.
Recommendations for respirator selection: NIOSH as vinylcyclohexene dioxide.
At any detectable concentration: SCBAF:PD,PP (any self-contained breathing apparatus that has a full facepiece and is operated in a pressure-demand or other positive-pressure mode); or SAF:PD,PP:ASCBA (any supplied-air respirator that has a full facepiece and is operated in a pressure-demand or other positive-pressure mode in combination with an auxiliary self-contained breathing apparatus operated in a pressure-demand or other positive pressure mode). *ESCAPE:* GMFOV [any air-purifying, full-facepiece respirator (gas mask) with a chin-style, front- or back-mounted organic vapor canister]; or SCBAE (any appropriate escape-type, self-contained breathing apparatus).
Exposure limits (TWA unless otherwise noted): ACGIH TLV 0.1 ppm (0.4 mg/m^3); NIOSH 10 ppm; suspected human carcinogen, as vinyl cyclohexene dioxide.
Toxicity by ingestion: Grade 2; oral rat LD$_{50}$ = 2.563 g/kg.
Long-term health effects: Confirmed carcinogen. NTP, clear evidence: mouse
Vapor (gas) irritant characteristics: Vapors cause smarting of the eyes or respiratory system if present in high concentrations. The effect is temporary.
Liquid or solid irritant characteristics: If spilled on clothing and allowed to remain, may cause smarting and reddening of the skin.

FIRE DATA
Flash point: 60°F/15°C (oc); 230°F/110°C (oc) (dioxide)
Fire extinguishing agents not to be used: Water.
Autoignition temperature: 517°F/269°C/542°K.
Stoichiometric air-to-fuel ratio: 52.4

CHEMICAL REACTIVITY
Reactivity with water: Hydrolyzes in water.
Binary reactants: Reacts with alcohols, amines.

ENVIRONMENTAL DATA
Water pollution: Effect of low concentrations on aquatic life is unknown. May be dangerous if it enters nearby water intakes; notify operators. Notify local health and wildlife officials. **Response to discharge:** Issue warning–high flammability. Should be removed.

SHIPPING INFORMATION
Grades of purity: 99%; Research grade; technical 95%; **Storage temperature:** Ambient; **Inert atmosphere:** None; **Stability during transport:** Stable.

PHYSICAL AND CHEMICAL PROPERTIES
Physical state @ 59°F/15°C and 1 atm: Liquid.
Molecular weight: 108.18; 140.2 (dioxide)
Boiling point @ 1 atm: 262°F/128°C/401°K; 441°F/227°C/500°K (dioxide)
Melting/Freezing point: −164°F/−109°C/164°K.
Specific gravity (water = 1): 0.8303 @ 20°C.
Relative vapor density (air = 1): 3.69
Vapor pressure: 0.1 mmHg (dioxide)

VINYL ETHYL ETHER   **REC. V:1000**

SYNONYMS: ETHER, VINYL ETHYL; ETOXYETHENE; ETHYL VINYL ETHER; EVE; VINAMAR; VINIL ETIL ETER (Spanish)

IDENTIFICATION
CAS Number: 109-92-2
Formula: C_4H_8O; $CH_2CHOC_2NO_5$
DOT ID Number: UN 1302; DOT Guide Number: 127P
Proper Shipping Name: Vinyl ethyl ether, inhibited

DESCRIPTION: Colorless liquid. Ether-like odor. Floats on water surface; insoluble. Boils at 96°F/36°C.

Extremely flammable • Polymerization hazard • Containers may BLEVE when exposed to fire •Vapors may form explosive mixture with air • Vapors are heavier than air and will collect and stay in low areas • Vapors may travel long distances to ignition sources and flashback • Vapors in confined areas (e.g., tanks, sewers, buildings) may explode when exposed to fire • Irritating to the skin, eyes, and respiratory tract • Toxic products of combustion may include carbon monoxide.

Hazard Classification (based on NFPA-704 Rating System)
Health Hazards (Blue): 2; Flammability (Red): 4; Reactivity (Yellow): 2

EMERGENCY RESPONSE: See Appendix A (127)
Evacuation:
Public safety: Isolate spill area for at least 25 to 50 meters (80 to 160 feet) in all directions.
Spill: Large spill–Consider initial downwind evacuation for at least 300 meters (1000 feet).
Fire: Isolate for 800 meters (½ mile) in all directions, especially if tank, rail car, or tank truck is involved in fire.

EXPOSURE
Short-term effects: *SEEK MEDICAL ATTENTION*. **Liquid or vapor:** Irritating to skin and eyes. Harmful if swallowed or inhaled. Move to fresh air. *IF BREATHING HAS STOPPED*, give artificial respiration. IF breathing is difficult, administer oxygen. Remove contaminated clothing and shoes. Flush affected areas with plenty of water. *IF IN EYES*, hold eyelids open and flush with plenty of water. *IF SWALLOWED* and victim is *CONSCIOUS AND ABLE TO SWALLOW*, have victim drink 4 to 8 ounces of water.

HEALTH HAZARDS
Personal protective equipment (PPE): B-Level PPE. Full face mask, self-contained breathing apparatus, eye protection, and butyl rubber gloves. Chemical protective material(s) reported to have good to excellent resistance: butyl rubber.
Toxicity by ingestion: Grade 1; LD_{50} = 5 to 15 g/kg.
Long-term health effects: Prolonged contact with skin may cause dermatitis. Possible liver damage may occur with repeated use.
Vapor (gas) irritant characteristics: Vapors cause smarting of the eyes or respiratory system if present in high concentrations. The effect is temporary.
Liquid or solid irritant characteristics: If spilled on clothing and allowed to remain, may cause reddening of skin.

FIRE DATA
Flash point: Less than –50°F/–46°C (cc).
Flammable limits in air: LEL: 1.7%; UEL: 28%.
Fire extinguishing agents not to be used: Water may be ineffective
Behavior in fire: Explosive hazard
Autoignition temperature: 395°F/202°C/475°K.

CHEMICAL REACTIVITY
Binary reactants: Strong acids. Contact with methane sulfonic acid may cause explosive polymerization.
Polymerization: Peroxides may accumulate after prolonged storage in presence of air. Peroxides may be detonated by heating, impact, or friction

ENVIRONMENTAL DATA
Food chain concentration potential: Negative; unlikely to accumulate.

Water pollution: Effects of low concentrations on aquatic life are unknown. May be dangerous if it enters nearby water intakes; notify operators. Notify local health and wildlife officials. **Response to discharge:** Issue warning–high flammability. Restrict access. Evacuate area.

PHYSICAL AND CHEMICAL PROPERTIES
Physical state @ 59°F/15°C and 1 atm: Liquid.
Molecular weight: 72.104
Boiling point @ 1 atm: 96°F/36°C/309°K.
Melting/Freezing point: –175°F/–115°C/158°K.
Specific gravity (water = 1): 0.7589 @ 68°F/20°C.
Relative vapor density (air = 1): 2.5
Latent heat of vaporization: (estimated) 165 Btu/lb = 91.8 cal/g = 3.84 x 10^5 J/kg.
Heat of combustion: –14.326 Btu/lb = –7959 cal/g = –333 x 10^5 J/kg.

VINYL FLUORIDE   REC. V:1100

SYNONYMS: FLUOROETHENE; FLURETHYLENE; FLUORURO de VINILO (Spanish); MONOFLUROETHYLENE; VINYL FLUORIDE MONOMER

IDENTIFICATION
CAS Number: 75-02-5
Formula: C_2H_3F; CH_2=CHF
DOT ID Number: UN 1860; DOT Guide Number: 116P
Proper Shipping Name: Vinyl fluoride, inhibited

DESCRIPTION: Colorless compressed gas. Sweet, ether-like odor. Floats and boils on water surface; insoluble.

Extremely flammable • Polymerization hazard • May react with itself blocking relief valves; leading to tank explosions • Vapors may cause dizziness or asphyxiation without warning • Contact with liquid may cause frostbite • Forms explosive mixture with air • Gas from liquefied gas are initially heavier than air and spread along ground • Gas may travel to source of ignition and flashback • Gas in confined areas (e.g., tanks, sewers, buildings) may explode when exposed to fire • Ruptured or venting cylinders may rocket through buildings and/or travel a considerable distance •Irritating to the skin, eyes, and respiratory tract • Toxic products of combustion may include hydrogen fluoride.

Hazard Classification (based on NFPA-704 Rating System)
Health Hazards (Blue): 1; Flammability (Red): 4; Reactivity (Yellow): 2

EMERGENCY RESPONSE: See Appendix A (116)
Evacuation:
Public safety: Isolate spill area for at least 100 meters (330 feet) in all directions.
Spill: Consider initial downwind evacuation for at least 800 meters (½ mile).
Fire: Isolate for 1600 meters (1 mile) in all directions, especially if tank, rail car, or tank truck is involved in fire.

EXPOSURE
Short-term effects: *SEEK MEDICAL ATTENTION*. **Vapor:** IF inhaled, will cause headache, or dizziness. Move victim to fresh air. If breathing is difficult, administer oxygen. **Liquid:** Will cause frostbite. Flush affected areas with plenty of water. *DO NOT RUB AFFECTED AREAS.*

HEALTH HAZARDS
Personal protective equipment (PPE): B-Level PPE. Protective gloves; safety glasses; self-contained breathing apparatus. Wear thermal protective clothing.
Recommendations for respirator selection: NIOSH
10 ppm: CCROV [any chemical cartridge respirator with organic vapor cartridge(s)]; or SA (any supplied-air respirator). *25 ppm:* SA:CF (any supplied-air respirator operated in a continuous-flow mode); or PAPROV [any powered, air-purifying respirator with organic vapor cartridge(s)]. *50 ppm:* CCRFOV [any chemical cartridge respirator with a full facepiece and organic vapor cartridges(s)]; or GMFOV [any air-purifying, full-facepiece respirator (gas mask) with a chin-style, front- or back-mounted acid gas canister]; or PAPRTOV [any powered, air-purifying respirator with a tight-fitting facepiece and organic vapor cartridges(s)]; or SCBAF (any self-contained breathing apparatus with a full facepiece); or SAF (any supplied-air respirator with a full facepiece). *200 ppm:* SAF:PD,PP (any supplied-air respirator that has a full facepiece and is operated in a pressure-demand or other positive-pressure mode). *EMERGENCY OR PLANNED ENTRY INTO UNKNOWN CONCENTRATIONS OR IDLH CONDITIONS:* SCBAF:PD,PP (any self-contained breathing apparatus that has a full facepiece and is operated in a pressure-demand or other positive-pressure mode); or SAF:PD,PP:ASCBA (any supplied-air respirator that has a full facepiece and is operated in a pressure-demand or other positive-pressure mode in combination with an auxiliary self-contained breathing apparatus operated in a pressure-demand or other positive pressure mode). *ESCAPE:* GMFOV [any air-purifying, full-facepiece respirator (gas mask) with a chin-style, front- or back-mounted organic vapor cannister]; or SCBAE (any appropriate escape-type, self-contained breathing apparatus).
Exposure limits (TWA unless otherwise noted): NIOSH REL 1 ppm; NIOSH REL 1 ppm.
Toxicity by ingestion: Poisonous.
Short-term exposure limits (15-minute TWA): NIOSH ceiling 5 ppm [use 1910.1017].
Vapor (gas) irritant characteristics: Causes irritation of the eyes respiratory and nervous systems.
Liquid or solid irritant characteristics: Causes irritation of the eyes and skin. Possible frostbite from rapid evaporation of liquid.

FIRE DATA
Flash point: Flammable, compressed liquefied gas.
Flammable limits in air: LEL: 2.6%; UEL: 21.7%.
Behavior in fire: Containers may explode.
Autoignition temperature: 725°F/385°C/658°K.
Electrical hazard: Flow or agitation of substance may generate electrostatic charges due to low conductivity.
Stoichiometric air-to-fuel ratio: 8.189 (estimate).

CHEMICAL REACTIVITY
Binary reactants: Violent reaction with oxidizers. Heat of combustion forms toxic hydrogen fluoride gas. May accumulate static electrical charges, and may cause ignition of its vapors.
Polymerization: Unless inhibited, polymerization can occur;
Inhibitor of polymerization: 0.2% terpene B-0.

ENVIRONMENTAL DATA
Food chain concentration potential: Log P_{ow} = 1.2. Unlikely to accumulate.
Water pollution: Poisonous. May be harmful to aquatic life.
Response to discharge: Issue warning–high flammability, air contaminant. Restrict access. Evacuate area.

SHIPPING INFORMATION
Grades of purity: 99.9+%; **Storage temperature:** Ambient; **Inert atmosphere:** None; **Venting:** Safety relief; **Stability during transport:** Stable.

PHYSICAL AND CHEMICAL PROPERTIES
Physical state @ 59°F/15°C and 1 atm: Gas.
Molecular weight: 46.1
Boiling point @ 1 atm: –98°F/–72°C/201°K.
Melting/Freezing point: –258°F/–161°C/112°K.
Critical temperature: 131°F/55°C/328°K.
Critical pressure: 760 psia = 51.6 atm = 5.24 MN/m^2.
Specific gravity (water = 1): 0.707 at 0°C (liquid).
Liquid surface tension: 5 dynes/cm = 0.005 N/m @ 15°C.
Relative vapor density (air = 1): 1.6
Ratio of specific heats of vapor (gas): 1.2097
Latent heat of vaporization: 156 Btu/lb = 86.5 cal/g = 3.62 x 10^5 J/kg.
Heat of combustion: (estimate) –6500 Btu/lb = –3600 cal/g = –150 x 10^5 J/kg.
Vapor pressure: 25.2 atm; 20 mm (approximate).

VINYLIDENE CHLORIDE   REC. V:1200

SYNONYMS: CLORURO de VINILDENO (Spanish); CHLORURE de VINYLIDENE (French); 1,1-DCE; 1,1-DICHLOROETHENE; 1,1-DICHLOROETHYLENE; ETHENE, 1,1-DICHLORO-; SCONATEX; VDC; VINYLIDENE CHLORIDE(II); VINYLIDENE DICHLORIDE; VINYLIDINE CHLORIDE MONOMER; *unsym*-DICHLOROETHYLENE; RCRA No. U078

IDENTIFICATION
CAS Number: 75-35-4
Formula: $C_2H_2Cl_2$; $CH_2 = CCl_2$
DOT ID Number: UN 1303; DOT Guide Number: 129P
Proper Shipping Name: Vinylidene chloride, inhibited
Reportable Quantity (RQ): **(CERCLA)** 100 lb/45.4 kg

DESCRIPTION: Colorless, clear, watery liquid or gas (above 89°F/32°C). Sweet odor, like chloroform. Sinks in water; slightly soluble. Forms white solid deposits on storage which may explode with shock or heat.

Extremely flammable • Severely irritating to the skin, eyes, and respiratory tract; inhalation symptoms may be delayed • Polymerization hazard • Heat can induce polymerization with rapid release of energy; sealed containers may rupture explosively • May react with itself blocking relief valves; leading to tank explosions • Containers may BLEVE when exposed to fire •Vapors may form explosive mixture with air • Vapors are heavier than air and will collect and stay in low areas • Vapors may travel long distances to ignition sources and flashback • Vapors in confined areas (e.g., tanks, sewers, buildings) may explode when exposed to fire • Toxic products of combustion may include hydrogen chloride and phosgene which may be more toxic than the material itself • Do not put yourself in danger by entering a contaminated area to rescue a victim.

Hazard Classification (based on NFPA-704 Rating System)
Health Hazards (Blue): 4; Flammability (Red): 4; Reactivity (Yellow): 2

Vinyl methyl ether

EMERGENCY RESPONSE: See Appendix A (129)
Evacuation:
Public safety: Isolate spill area for at least 50 to 100 meters (160 to 330 feet) in all directions.
Spill: Large spill–Consider initial downwind evacuation for at least 300 meters (1000 feet).
Fire: Isolate for 800 meters (½ mile) in all directions, especially if tank, rail car, or tank truck is involved in fire.

EXPOSURE
Short-term effects: *SEEK MEDICAL ATTENTION.* **Vapor:** Irritating to eyes, nose, and throat. *IF INHALED,* will, will cause dizziness or difficult breathing. Move to fresh air. *IF BREATHING HAS STOPPED,* give artificial respiration. IF breathing is difficult, administer oxygen. May cause lung edema; physical exertion will aggravate this condition. **Liquid:** Will burn skin and eyes. Harmful if swallowed. Remove contaminated clothing and shoes. Flush affected areas with plenty of water. *IF IN EYES,* hold eyelids open and flush with plenty of water. *IF SWALLOWED* and victim is *CONSCIOUS AND ABLE TO SWALLOW,* have victim drink 4 to 8 ounces of water.
Note to physician or authorized medical personnel: Medical observation is recommended for 24 to 48 hours after breathing overexposure, as pulmonary edema may be delayed. As first aid for pulmonary edema, consider administering a corticosteroid spray. Cigarette smoking may exacerbate pulmonary injury and should be discouraged for at least 72 hours following exposure.

HEALTH HAZARDS
Personal protective equipment (PPE): A-Level PPE. Chemical protective material(s) reported to have good to excellent resistance: PV alcohol, Teflon®.
Recommendations for respirator selection: NIOSH
At any concentrations above the NIOSH REL, or where there is no REL, at any detectable concentration: SCBAF:PD,PP (any self-contained breathing apparatus that has a full facepiece and is operated in a pressure-demand or other positive-pressure mode); or SAF:PD,PP:ASCBA (any supplied-air respirator that has a full facepiece and is operated in a pressure-demand or other positive-pressure mode in combination with an auxiliary self-contained breathing apparatus operated in a pressure-demand or other positive pressure mode). *ESCAPE:* GMFOV [any air-purifying, full-facepiece respirator (gas mask) with a chin-style, front-or back-mounted organic vapor canister]; or SCBAE (any appropriate escape-type, self-contained breathing apparatus).
Exposure limits (TWA unless otherwise noted): ACGIH TLV 5 ppm (20 mg/m^3); NIOSH REL potential human carcinogen; reduce exposure to lowest feasible level.
Short-term exposure limits (15-minute TWA): ACGIH STEL 20 ppm (79 mg/m^3).
Toxicity by ingestion: Grade 3; oral LD_{50} = 24 hr = 84 mg/kg (adrenalectomized rat).
Long-term health effects: Suspected carcinogen. IARC rating 3. Mutation data.
Vapor (gas) irritant characteristics: Vapors cause irritation such that personnel will find high concentrations unpleasant. The effect may be temporary.
Liquid or solid irritant characteristics: Causes smarting of the skin and first-degree burns on short exposure; may cause secondary burns on long exposure.
Odor threshold: 190 ppm.

FIRE DATA
Flash point: –2°F/–19°C (oc); –19°F/–28°C (cc).
Flammable limits in air: LEL: 6.5%; UEL: 15.5%.

Fire extinguishing agents not to be used: Water may be ineffective.
Behavior in fire: Containers may rupture and explode in fire due to polymerization.
Autoignition temperature: 1062°F/572°C/845°K.
Electrical hazard: Class I, Group D.
Burning rate: 2.7 mm/min

CHEMICAL REACTIVITY
Binary reactants: Reacts with sunlight, air, heat. Contact with copper and its annoys or aluminum or its alloys can cause polymerization.
Polymerization: Can occur if exposed to oxidizers, chlorosulfonic acid, oleum, sunlight, air, copper, aluminum, heat; **Inhibitor of polymerization:** 200 ppm monomethyl ether of hydroquinone; 0.6–0.8% phenol.
Reactivity group: 35
Compatibility class: Vinyl halides (halogen compounds)

ENVIRONMENTAL DATA
Food chain concentration potential: Negative; unlikely to accumulate.
Water pollution: DOT Appendix B, §172.101–marine pollutant. Effect of low concentrations on aquatic life is unknown. May be dangerous if it enters nearby water intakes; notify operators. Notify local health and wildlife officials. **Response to discharge:** Issue warning–high flammability. Evacuate area.

SHIPPING INFORMATION
Grades of purity: 99%; **Storage temperature:** Ambient; **Inert atmosphere:** Padded; **Venting:** Pressure-vacuum; **Stability during transport:** Stable.

NAS HAZARD CLASSIFICATION FOR BULK WATER TRANSPORTATION
FIRE: 3
HEALTH: Vapor irritant: 2; Liquid or solid irritant: 2; Poisons: 3
WATER POLLUTION: Human toxicity: 0; Aquatic toxicity: 2; Aesthetic effect: 2
REACTIVITY: Other chemicals: 2; Water: 0; Self-reaction: 3

PHYSICAL AND CHEMICAL PROPERTIES
Physical state @ 59°F/15°C and 1 atm: Liquid.
Molecular weight: 96.95
Boiling point @ 1 atm: 89°F/32°C/305°K.
Melting/Freezing point: –188°F/122°C/152°K.
Specific gravity (water = 1): 1.21 @ 68°F/20°C (liquid).
Liquid surface tension: 24 dynes/cm = 0.024 N/m @ 15°C.
Liquid water interfacial tension: 37 dynes/cm = 0.037 N/m at 22.7°C.
Relative vapor density (air = 1): 3.3
Latent heat of vaporization: 130 Btu/lb = 72 cal/g = 3.0×10^5 J/kg.
Heat of combustion: –4860 Btu/lb = –2700 cal/g = $–113.0 \times 10^5$ J/kg.
Heat of polymerization: –333 Btu/lb = –185 cal/g = $–7.75 \times 10^5$ J/kg.
Vapor pressure: (Reid) 18.3 psia; 500 mm.

VINYL METHYL ETHER   REC. V:1300

SYNONYMS: EEC No. 603-021-00-9; METHYL VINYL ETHER; METHOXYETHYLENE; VINIL METIL ETER (Spanish)

IDENTIFICATION
CAS Number: 107-25-5
Formula: C_3H_6O; $CH_2=CH-O-CH_3$
DOT ID Number: UN 1087; DOT Guide Number: 116P
Proper Shipping Name: Vinyl methyl ether; Vinyl methyl ether, inhibited

DESCRIPTION: Colorless gas. Shipped and stored as a compressed gas. Sweet pleasant odor. Liquid floats on the surface of water; slightly soluble; and may boil on water. A liquid below 42°F/6°C.

Extremely flammable • Polymerization hazard • Vapors may cause dizziness or asphyxiation without warning • Contact with liquid may cause frostbite • Heat can induce polymerization with rapid release of energy; sealed containers may rupture explosively • May react with itself blocking relief valves; leading to tank explosions • Containers may BLEVE when exposed to fire • Vapors may form explosive mixture with air • Vapors are heavier than air and will collect and stay in low areas • Vapors may travel long distances to ignition sources and flashback • Vapors in confined areas (e.g., tanks, sewers, buildings) may explode when exposed to fire • Irritating to the skin, eyes, and respiratory tract • Toxic products of combustion may include carbon monoxide.

Hazard Classification (based on NFPA-704 Rating System)
Health Hazards (Blue): 2; Flammability (Red): 4; Reactivity (Yellow): 2

EMERGENCY RESPONSE: See Appendix A (116)
Evacuation:
Public safety: Isolate spill area for at least 100 meters (330 feet) in all directions.
Spill: Consider initial downwind evacuation for at least 800 meters (½ mile).
Fire: Isolate for 1600 meters (1 mile) in all directions, especially if tank, rail car, or tank truck is involved in fire.

EXPOSURE
Short-term effects: *SEEK MEDICAL ATTENTION.* **Vapor:** Irritating to eyes, nose, and throat. *IF INHALED*, will, will cause headache, dizziness, or loss of consciousness. Move victim to fresh air. If breathing is difficult, administer oxygen. **Liquid:** Irritating to eyes. Will cause frostbite. Harmful if swallowed. *DO NOT RUB AFFECTED AREAS.* Flush affected areas with plenty of water. *IF IN EYES*, hold eyelids open and flush with plenty of water. *IF SWALLOWED* and victim is *CONSCIOUS AND ABLE TO SWALLOW*, have victim drink 4 to 8 ounces of water. **Do NOT induce vomiting.**

HEALTH HAZARDS
Personal protective equipment (PPE): B-Level PPE. Organic-vapor mask; plastic or rubber gloves; safety glasses. Wear thermal protective clothing. Chemical protective material(s) reported to have good to excellent resistance: butyl rubber.
Toxicity by ingestion: Grade 2; LD_{50} = 0.5 to 5 g/kg.
Vapor (gas) irritant characteristics: Causes irritation of the eyes and respiratory system.
Liquid or solid irritant characteristics: Causes irritation of the eyes and skin. Possible frostbite from rapid evaporation of liquid.

FIRE DATA
Flash point: −69°F/−56°C (oc).
Flammable limits in air: LEL: 2.6%; UEL: 39%.
Fire extinguishing agents not to be used: Water may be ineffective.
Behavior in fire: Containers may explode.
Autoignition temperature: 545°F/285°C/558°K.
Stoichiometric air-to-fuel ratio: 9.451 (estimate).

CHEMICAL REACTIVITY
Reactivity with water: Reacts slowly to form acetaldehyde; reaction is not hazardous unless water is hot or acids are present.
Binary reactants: Acids will cause polymerization. Reacts with halogens (e.g., bromine, chlorine); may cause explosion.
Polymerization: Can polymerize in the presence of acids;
Inhibitor of polymerization: Dioctylamine; triethanolamine; solid potassium hydroxide

ENVIRONMENTAL DATA
Food chain concentration potential: Negative; unlikely to accumulate.
Water pollution: Effect of low concentrations on aquatic life is unknown. May be dangerous if it enters nearby water intakes; notify operators. Notify local health and wildlife officials. **Response to discharge:** Issue warning–high flammability, air contaminant. Restrict access. Evacuate area.

SHIPPING INFORMATION
Grades of purity: 99.7+%; **Storage temperature:** Ambient; **Inert atmosphere:** None; **Venting:** Safety relief; **Stability during transport:** Stable if kept free from acids.

PHYSICAL AND CHEMICAL PROPERTIES
Physical state @ 59°F/15°C and 1 atm: Gas.
Molecular weight: 58.1
Boiling point @ 1 atm: 42°F/6°C/279°K.
Melting/Freezing point: −188°F/−122°C/151°K.
Specific gravity (water = 1): 0.777 at 0°C (liquid).
Liquid surface tension: (estimate) 10 dynes/cm = 0.010 N/m at 0°C.
Liquid water interfacial tension: (estimate) 25 dynes/cm = 0.025 N/m at 0°C.
Relative vapor density (air = 1): 2.0
Ratio of specific heats of vapor (gas): (estimate) 1.1473
Latent heat of vaporization: (estimate) 180 Btu/lb = 100 cal/g = 4.2×10^5 J/kg.
Heat of combustion: (estimate) −14,200 Btu/lb = −7900 cal/g = $−330 \times 10^5$ J/kg.
Vapor pressure: 1115 mm.

VINYL NEODECANOATE   REC. V:1400

SYNONYMS: NEODECANOIC ACID, VINYL ESTER; VV 10 VINYL MONOMER

IDENTIFICATION
Formula: $C_{12}H_{22}O_2$

DESCRIPTION: Colorless liquid. Pleasant odor. Floats on the surface of water.

Combustible • Extremely flammable • Polymerization hazard • Heat can induce polymerization with rapid release of energy; sealed containers may rupture explosively • May react with itself blocking relief valves; leading to tank explosions • Containers may BLEVE when exposed to fire • Vapors are heavier than air and will collect and stay in low areas • Vapors in confined areas (e.g., tanks, sewers, buildings) may explode when exposed to fire • Irritating to the skin, eyes, and respiratory tract • Toxic products of combustion may include carbon monoxide.

Hazard Classification (based on NFPA-704 Rating System)
Health Hazards (Blue): 2; Flammability (Red): 2; Reactivity (Yellow): 1

EMERGENCY RESPONSE: See Appendix A (171)
Evacuation:
Public safety: Isolate the area of spill or leak for at least 10 to 25 meters (30 to 80 feet) in all directions.
Spill: Increase, in the downwind direction, as necessary, the distance shown under "Public Safety."
Fire: If any large container is involved in fire, isolate for at least 800 meters (½ mile) in all directions; also, consider initial evacuation for 800 meters (½ mile) in all directions.

EXPOSURE
Short-term effects:
Prolonged or repeated contact of this material with the skin should be avoided.

HEALTH HAZARDS
Personal protective equipment (PPE): B-Level PPE. Chemical protective material(s) reported to offer limited protecion: Teflon®, chlorinated polyethylene.
Toxicity by ingestion: Grade 1; LD_{50} = 23.1 g/kg (rat).
Liquid or solid irritant characteristics: If spilled on clothing and allowed to remain, may cause smarting and reddening of skin.

FIRE DATA
Flash point: More than 175°F/79°C.
Autoignition temperature: 588°F/309°C.

CHEMICAL REACTIVITY
Binary reactants: The monomer is supplied in bulk or resin lined drums, and could be safely stored in tin-lined or stainless steel containers. Storage in plastic or other vessels is not recommended. Copper will inhibit polymerization, and zinc will promote discoloration.
Polymerization: Commercial application is a modifying monomer in polymerization reactions; **Inhibitor of polymerization:** Monomethyl ether of hydroquinone
Reactivity group: 13
Compatibility class: Vinyl acetate

ENVIRONMENTAL DATA
Water pollution: Effect of low concentrations on aquatic life is unknown. May be dangerous if it enter water intakes. Notify local health and wildlife officials. **Response to discharge:** Mechanical containment. Should be removed.

SHIPPING INFORMATION
The product is a mixture of compounds; **Storage temperature:** Ambient; **Stability during transport:** Stable.

PHYSICAL AND CHEMICAL PROPERTIES
Physical state @ 59°F/15°C and 1 atm: Liquid.
Molecular weight: 198.3
Melting/Freezing point: –4°F/–20°C/253°K.
Relative vapor density (air = 1): 11.107 (calculated).
Latent heat of vaporization: 106.2 Btu/lb = 59 cal/g = 2.5×10^5 J/kg..
Heat of polymerization: 208.7 Btu/lb = 116 cal/g = 4.86×10^5 J/kg.

VINYL TOLUENE   REC. V:1500

SYNONYMS: METHYL STYRENE; 3-METHYL STYRENE; 4-METHYL STYRENE; *m-* METHYL STYRENE; *p-*METHYL STYRENE; VINILTOLUENO (Spanish); VINYL TOLUENE, mixed isomers; *m-*VINYL TOLUENE; *p-*VINYL TOLUENE

IDENTIFICATION
CAS Number: 25013-15-4
Formula: C_9H_{10}; $CH_3C_6H_4CH=CH_2$
DOT ID Number: UN 2618; DOT Guide Number: 128
Proper Shipping Name: Vinyl toluene, inhibited mixed isomers

DESCRIPTION: Colorless watery liquid. Strong, unpleasant odor. Floats on water surface; insoluble.

Highly flammable • Polymerization hazard • Heat can induce polymerization with rapid release of energy; sealed containers may rupture explosively • May react with itself blocking relief valves; leading to tank explosions • Containers may BLEVE when exposed to fire • Vapors may form explosive mixture with air • Vapors are heavier than air and will collect and stay in low areas • Vapors may travel long distances to ignition sources and flashback • Vapors in confined areas (e.g., tanks, sewers, buildings) may explode when exposed to fire • Irritating to the skin, eyes, and respiratory tract • Toxic products of combustion may include carbon monoxide.

Hazard Classification (based on NFPA-704 Rating System)
Health Hazards (Blue): 2; Flammability (Red): 2; Reactivity (Yellow): 2

EMERGENCY RESPONSE: See Appendix A (128)
Evacuation:
Public safety: Isolate spill area for at least 25 to 50 meters (80 to 160 feet) in all directions.
Spill: Large spill–Consider initial downwind evacuation for at least 300 meters (1000 feet).
Fire: Isolate for 800 meters (½ mile) in all directions, especially if tank, rail car, or tank truck is involved in fire.

EXPOSURE
Short-term effects: *SEEK MEDICAL ATTENTION.* **Liquid:** Irritating to skin and eyes. Harmful if swallowed. Remove contaminated clothing and shoes. Flush affected areas with plenty of water. *IF IN EYES*, hold eyelids open and flush with plenty of water. *IF SWALLOWED* and victim is *CONSCIOUS AND ABLE TO SWALLOW*, have victim drink 4 to 8 ounces of water. The use of alcoholic beverages may enhance the toxic effect.

HEALTH HAZARDS
Personal protective equipment (PPE): A-Level PPE. OSHA Table Z-1-A air contaminant. Air–supplied mask; goggles or face shield; plastic gloves.
Recommendations for respirator selection: NIOSH/OSHA
400 ppm: CCROV* [any chemical cartridge respirator with organic vapor cartridge(s)]; or PAPROV* [any powered, air-purifying respirator with organic vapor cartridge(s)]; or GMFOV [any air-purifying, full-facepiece respirator (gas mask) with a chin-style, front- or back-mounted organic vapor canister]; or SA* (any supplied-air respirator); or SCBAF (any self-contained breathing apparatus with a full facepiece). *EMERGENCY OR PLANNED*

1138 Vinyl trichlorosilane

ENTRY INTO UNKNOWN CONCENTRATIONS OR IDLH CONDITIONS: SCBAF:PD,PP (any self-contained breathing apparatus that has a full facepiece and is operated in a pressure-demand or other positive-pressure mode); or SAF:PD,PP:ASCBA (any supplied-air respirator that has a full facepiece and is operated in a pressure-demand or other positive-pressure mode in combination with an auxiliary self-contained breathing apparatus operated in a pressure-demand or other positive pressure mode). *ESCAPE:* GMFOV [any air-purifying, full-facepiece respirator (gas mask) with a chin-style, front- or back-mounted organic vapor cannister]; or SCBAE (any appropriate escape-type, self-contained breathing apparatus). *Note:* Substance reported to cause eye irritation or damage; may require eye protection.
Exposure limits (TWA unless otherwise noted): ACGIH TLV 50 ppm (242 mg/m^3); NIOSH/OSHA 100 ppm (480 mg/m^3).
Short-term exposure limits (15-minute TWA): ACGIH STEL 100 ppm (483 mg/m^3).
Toxicity by ingestion: Grade 2; LD$_{50}$ = 0.5 to 5 g/kg (rat).
Vapor (gas) irritant characteristics: Eye and respiratory tract irritant. Vapors cause irritation such that personnel will find high concentrations unpleasant. The effect is temporary.
Liquid or solid irritant characteristics: Eye and skin irritant. If spilled on clothing and allowed to remain, may cause smarting and reddening of the skin.
Odor threshold: 10 ppm.
IDLH value: 400 ppm.

FIRE DATA
Flash point: 137°F/58°C (oc); 127°F/53°C (cc).
Flammable limits in air: LEL: 0.8%; UEL: 11%.
Behavior in fire: Containers may explode or rupture in a fire due to polymerization.
Autoignition temperature: 995°F/535°C/808°K.
Electrical hazard: Class I, Group D.
Burning rate: 6.0 mm/min

CHEMICAL REACTIVITY
Binary reactants: Oxidizers (may be violent), peroxides, strong acids, iron or aluminum salts.
Polymerization: Slow at ordinary temperatures but when hot may rupture container. Also polymerized by metal salts such as those of iron or aluminum; **Inhibitor of polymerization:** 10–50 ppm *tert*-butylcatechol
Reactivity group: 30
Compatibility class: Olefins

ENVIRONMENTAL DATA
Food chain concentration potential: Log P_{ow} = 3.35. Values > 3.0 are likely to bioconcentrate in aquatic organisms and other living tissue, especially in fats.
Water pollution: DOT Appendix B, §172.101–marine pollutant. Effect of low concentrations on aquatic life is unknown. Fouling to shoreline. May be dangerous if it enters nearby water intakes; notify operators. Notify local health and wildlife officials.
Response to discharge: Mechanical containment. Should be removed. Chemical and physical treatment.

SHIPPING INFORMATION
Grades of purity: 99.2+%; **Storage temperature:** Ambient; **Inert atmosphere:** None; **Venting:** Open (flame arrester); **Stability during transport:** Stable.

NAS HAZARD CLASSIFICATION FOR BULK WATER TRANSPORTATION
FIRE: 2
HEALTH: Vapor irritant: 2; Liquid or solid irritant: 1; Poisons: 1
WATER POLLUTION: Human toxicity: 1; Aquatic toxicity: 3; Aesthetic effect: 2
REACTIVITY: Other chemicals: 2; Water: 0; Self-reaction: 3

PHYSICAL AND CHEMICAL PROPERTIES
Physical state @ 59°F/15°C and 1 atm: Liquid.
Molecular weight: 118.18
Boiling point @ 1 atm: 333.9°F/167.7°C/440.9°K.
Melting/Freezing point: –106.6°F/–77.0°C/196.2°K.
Specific gravity (water = 1): 0.897 @ 68°F/20°C (liquid).
Liquid surface tension: 31.53 dynes/cm = 0.03153 N/m @ 68°F/20°C.
Liquid water interfacial tension: (estimate) 45 dynes/cm = 0.045 N/m @ 68°F/20°C.
Relative vapor density (air = 1): 4.9
Ratio of specific heats of vapor (gas): (estimate) 1.060
Latent heat of vaporization: 150 Btu/lb = 83.5 cal/g = 3.50 x 10^5 J/kg.
Heat of combustion: –17.710 Btu/lb = –9840 cal/g = 412.0 x 10^5 J/kg.
Heat of polymerization: –243 Btu/lb = –135 cal/g = –5.65 x 10^5 J/kg.
Vapor pressure: (Reid) 0.07 psia; 1.1 mmHg @ 68°F/20°C.

VINYL TRICHLOROSILANE   REC. V:1600

SYNONYMS: SILANE, VINYL TRICHLORO 1-150; TRICHLORO(VINYL)SILANE; TRICHLOROVINYL SILICANE; UNION CARBIDE A-150; VINILTRICLOROSILANO (Spanish); VINYLSILICON TRICHLORIDE

IDENTIFICATION
CAS Number: 75-94-5
Formula: $C_2H_3Cl_3Si$; $CH_2=CHSiCl_3$
DOT ID Number: UN 1305; DOT Guide Number: 155
Proper Shipping Name: Vinyltrichlosilane

DESCRIPTION: Clear, colorless to light yellow, fuming liquid. Sharp choking odor; like hydrochloric acid. Reacts violently with water; forming hydrochloric acid and hydrogen chloride vapors.

Poison! • Do NOT use water • Breathing the vapor can kill you • Firefighting gear (including SCBA) does not provide adequate protection. If exposure occurs, remove and isolate gear immediately and thoroughly decontaminate personnel • Highly flammable Polymerization hazard • Heat can induce polymerization with rapid release of energy; sealed containers may rupture explosively • Containers may BLEVE when exposed to fire •Vapors may form explosive mixture with air • Vapors are heavier than air and will collect and stay in low areas • Vapors may travel long distances to ignition sources and flashback •Vapors in confined areas (e.g., tanks, sewers, buildings) may explode when exposed to fire • Severely irritating to skin, eyes, and respiratory tract; prolonged contact with the skin causes burns • Toxic products of combustion may include hydrogen chloride and phosgene • Reacts with metals in the presence of moisture forming explosive hydrogen gas • Do not put yourself in danger by entering a contaminated area to rescue a victim.

Hazard Classification (based on NFPA-704 Rating System)
Health Hazards (Blue): 3; Flammability (Red): 3; Reactivity (Yellow): 2; Special Notice (White): Water reactive

EMERGENCY RESPONSE: See Appendix A (155)
Evacuation:
Public safety: Isolate the area of spill or leak for at least 50 to 100 meters (160 to 330 feet) in all directions.
Spill: Increase, in the downwind direction, as necessary, the distance shown under "Public Safety."
Fire: If tank, rail car, or tank truck is involved in fire, isolate for at least 800 meters (½ mile) in all directions; also, consider initial evacuation for 800 meters (½ mile) in all directions.

EXPOSURE
Short-term effects: *SEEK MEDICAL ATTENTION.* **Vapor:** Irritating to eyes, nose, and throat. Harmful if inhaled. Move victim to fresh air. *IF BREATHING HAS STOPPED,* give artificial respiration; *avoid mouth-to-mouth resuscitation; use bag/mask apparatus.* If breathing is difficult, administer oxygen. **Liquid:** Will burn skin and eyes. Harmful if swallowed. Remove contaminated clothing and shoes. Flush affected areas with plenty of water. *IF IN EYES*, hold eyelids open and flush with plenty of water. *IF SWALLOWED* and victim is *CONSCIOUS AND ABLE TO SWALLOW*, have victim drink 4 to 8 ounces of water. **Do NOT induce vomiting.**

HEALTH HAZARDS
Personal protective equipment (PPE): A-Level PPE. Acid-vapor-type respiratory protection; rubber gloves; chemical worker's goggles; other protective equipment as necessary to protect skin and eyes.
Toxicity by ingestion: Grade 2; oral LD_{50} = 1280 mg/kg (rat).
Vapor (gas) irritant characteristics: Vapors cause severe irritation of eyes and throat and can cause eye and lung injury. They cannot be tolerated even at low concentrations.
Liquid or solid irritant characteristics: Severe skin irritant. Causes second- and third-degree burns on short contact and is very injurious to the eyes.

FIRE DATA
Flash point: 70°F/21°C (oc); 52°F/11°C (cc).
Fire extinguishing agents not to be used: Water, foam
Behavior in fire: Difficult to extinguish; re-ignition may occur.
Autoignition temperature: 505°F/263°C 536°K.
Burning rate: 2.9 mm/min

CHEMICAL REACTIVITY
Reactivity with water: Reacts vigorously, evolving hydrogen chloride (hydrochloric acid).
Binary reactants: May self-ignite in air. Violent reaction with water, steam, alcohols, forming hydrogen chloride. Violent reaction with strong oxidizers, ammonia. Incompatible with alkalis, strong acids, aliphatic amines, alkanolamines, isocyanates, alkylene oxides, epichlorohydrin, halogenated compounds, nitrogen oxides. Corrodes common metals in the presence of moisture and produces flammable hydrogen. The uninhibited monomer vapor may block vents and confined spaces by forming a solid polymer material.
Neutralizing agents for acids and caustics: Flush with water, rinse with sodium bicarbonate or lime solution.
Polymerization: Polymerizes easily. May occur in absence of inhibitor; **Inhibitor of polymerization:** Diphenylamine; Hydroquinone.

ENVIRONMENTAL DATA
Food chain concentration potential: Negative; unlikely to accumulate.
Water pollution: Effect of low concentrations on aquatic life is unknown. May be dangerous if it enters nearby water intakes; notify operators. Notify local health and wildlife officials. **Response to discharge:** Issue warning–high flammability, corrosive, air contaminant. Restrict access. Disperse and flush with care.

SHIPPING INFORMATION
Grades of purity: 96+%; 98.5+%; **Storage temperature:** Ambient; **Inert atmosphere:** None; **Venting:** Pressure-vacuum; **Stability during transport:** Stable if protected from moisture.

NAS HAZARD CLASSIFICATION FOR BULK WATER TRANSPORTATION
FIRE: 3
HEALTH: Vapor irritant: 4; Liquid or solid irritant: 4; Poisons: 3
WATER POLLUTION: Human toxicity: 3; Aquatic toxicity: 3; Aesthetic effect: 3
REACTIVITY: Other chemicals: 3; Water: 4; Self-reaction: 3

PHYSICAL AND CHEMICAL PROPERTIES
Physical state @ 59°F/15°C and 1 atm: Liquid.
Molecular weight: 161.5
Boiling point @ 1 atm: 195°F/91°C/364°K.
Melting/Freezing point: –139°F/–95°C/178°K.
Specific gravity (water = 1): 1.26 @ 68°F/20°C (liquid).
Liquid surface tension: (estimate) 28 dynes/cm = 0.028 N/m @ 68°F/20°C.
Relative vapor density (air = 1): 5.61
Latent heat of vaporization: 88 Btu/lb = 49 cal/g = 2.0×10^5 J/kg.
Heat of combustion: (estimate) –4300 Btu/lb = –2400 cal/g = –100 $\times 10^5$ J/kg.

VX REC. V:1650

SYNONYMS: s-(2-DIISOPROPYLAMINOETHYL)-*O*-ETHYL METHYL PHOSPHONOTHIOLATE; ETHYL-*s*-DIISOPROPYLAMINOETHYL METHYLTHIOPHOSPHONATE; *O*-ETHYL-*s*-(2-DIISOPROPYLAMINOETHYL METHYLPHOSPHONOTHIOTE)(DOT); METHYLPHOSPHONOTHIOIC ACID-*s*-[2-(BIS(METHYLETHYL)AMINO)ETHYL]O-ETHYL ESTER; PHOSPHONOTHIOIC ACID, METHYL-, s-2-[BIS(1-METHYLETHYL)AMINOETHYL] *O*-ETHYL ESTER; VX NERVE GAS

IDENTIFICATION:
CAS Registry Number: 50782-69-9
Formula: $C_{11}H_{26}NO_2PS$
DOT ID Number: UN 2810; DOT Guide Number: 153
Proper Shipping Name: VX; *O*-ethyl-*S*-(2-diisopropylaminoethyl methylphosphonothiote)
Reportable Quantity (RQ): **(EHS):** 1 lb (0.454 kg)

DESCRIPTION: Amber colored oily liquid. Odorless and tasteless. *No warning properties.* Sinks in water; slightly soluble. Hydrolysis of VX produces a class B poison. *Note:* If used as a weapon, notify U.S. Department of Defense: Army. Damage and/or death may occur before chemical detection can take place. Use M8 Paper (Detection: Dark green) or M256-A1 Detector Kit (Detection limit: 0.02 mg/m³) if available.

Combustible • Highly persistent poison! (organophosphate) • Combustible • Breathing the vapor, skin or eye contact, or swallowing the material can kill you; symptoms may be delayed for several hours • Firefighting gear (including SCBA) provide NO protection. If exposure occurs, remove and isolate gear immediately and thoroughly decontaminate personnel • Containers may BLEVE when exposed to fire • Vapors are heavier than air and will collect and stay in low areas • Vapors in confined areas (e.g., tanks, sewers, buildings) may explode when exposed to fire • *Combustion products are less deadly than the material itself.* Toxic products of combustion may include carbon monoxide, sulfur oxide, phosphorus oxide, and nitrogen oxides • Do NOT attempt rescue. *Warning:* Odor is not a reliable indicator of the presence of toxic amounts of VX gas.

Hazard Classification (based on NFPA-704 Rating System)

Health Hazards (Blue): 4; Flammability (Red): 1; Reactivity (Yellow): 0

EMERGENCY RESPONSE: See Appendix A (153) with notes.
Evacuation: See below.
Spill: For VX [weaponized]: Small spill–First: Isolate in all directions 30 meters (100 feet); Then: Protect persons downwind, DAY: 0.2 km (0.1 mile); NIGHT: 0.2 km (0.1 mile). Large spill–First: Isolate in all directions 60 meters (200 feet); Then: Protect persons downwind, DAY: 0.6 km (0.4 mile); NIGHT: 1.0 km (0.6 mile).
Fire: If tank, rail car, or tank truck is involved in fire, isolate for at least 800 meters (½ mile) in all directions; also, consider initial evacuation for 800 meters (½ mile) in all directions.

EXPOSURE
Short-term effects: *SEEK MEDICAL ATTENTION.* **Liquid:** *POISONOUS BY SKIN CONTACT OR IF SWALLOWED.* Will remain as a liquid for more than 24 hours. **Vapor or Liquid:** *POISONOUS IF INHALED, IF SWALLOWED, OR IF SKIN IS EXPOSED.* Eye pupils are small; blurred vision; runny nose; cough; shortness of breath; pain; diarrhea, nausea and vomiting; increased blood pressure, hypermotility, hallucinations; loss of consciousness; convulsions; breathing stops; death. *IF BREATHING HAS STOPPED,* give artificial respiration; *avoid mouth-to-mouth resuscitation; use bag/mask apparatus.* **Skin:** *Remove and double-bag contaminated clothing (including shoes) and leave them in the Hot Zone*; wash skin with soap and water and large volumes of water for at least 15 minutes. *Skin can also be decontaminated with diluted hypochlorite solution, U.S. Army M291 kit, and M258(A1) skin decontamination kit.* **Eye:** Rinse with large volumes of water or saline for at least 15 minutes. *IF SWALLOWED,* will cause nausea, vomiting, or loss of consciousness. *Remove and double-bag contaminated clothing and shoes and leave in Hot Zone for later incineration by hazardous materials experts.* Flush affected areas with plenty of water. **If swallowed** and victim is *CONSCIOUS AND ABLE TO SWALLOW,* have victim drink 4 to 8 ounces of water. **Do NOT induce vomiting but immediately administer slurry of activated charcoal.** *IF SWALLOWED* and victim is *UNCONSCIOUS OR HAVING CONVULSIONS,* do nothing except keep victim warm.
Note to physician and medical personnel: If symptoms indicate, initial treatment for an adult includes atropine, 2 mg (1/30 gr) intramuscularly or intravenously as soon as any local or systemic signs or symptoms of an intoxication are noted; repeat the administration of atropine every 3 to 8 minutes until signs of atropinization (mydriasis, dry mouth, rapid pulse, hot and dry skin) occur; initiate treatment in children with 0.05 mg/kg of atropine; repeat at 5- to 10-minute intervals. Watch respiration, and remove bronchial secretions if they appear to be obstructing the airway; intubate if necessary. Pralidoxime must be administered within minutes to a few hours following exposure (depending on the specific agent) to be effective. Give 2-PAMCI (Pralidoxime; Protopam), 2.5 g in 100 mL of sterile water or in 5% dextrose and water, intravenously, slowly, in 15 to 30 minutes; if sufficient fluid is not available, give 1 g of 2-PAMCI in 3 mL of distilled water by deep intramuscular injection; repeat this every half hour if respiration weakens or if muscle fasciculation or convulsions recur. Also Diazepam, an anticonvulsant, might be needed. For adults, the dose is 5 to 10 mg (slow IV), repeated every 12 to 15 minutes up to three (3) doses maximum. For children, the dose is 0.2 to 0.5 mg/kg.
Note to physician or authorized medical personnel: Administer atropine. *IF INHALED*, will, medical observation is recommended for 24 to 48 hours after breathing overexposure, as pulmonary edema may be delayed. As first aid for pulmonary edema, consider administering a corticosteroid spray. Cigarette smoking may exacerbate pulmonary injury and should be discouraged for at least 72 hours following exposure.

HEALTH HAZARDS
Personal protective equipment (PPE): A-Level PPE. Rubber gloves, protective clothing, goggles, respirators. Butyl rubber gloves and Tyvek® "F" decontamination suit provide barrier protection against chemical warfare agents. Airtight, impermeable clothing was developed for personnel who must enter heavily contaminated areas. This clothing is made of butyl rubber or a coated fabric such as Tyvek "F" and provide barrier protection against liquid chemical warfare agents. Although resistant to liquid chemical agents, impermeable protective clothing may be penetrated after a few hours of exposure to heavy concentration of agent. Consequently, liquid contamination on the clothing must be neutralized or removed as soon as possible. If the proper equipment is not available, or if the rescuers have not been trained in its use, call for assistance from the U.S. Soldier and Biological Chemical Command–Edgewood Research Development and Engineering Center (from 0700-1630 EST call 410-671-4411, and from 1630-0700 EST call 410-278-5201; ask for the Staff Duty Officer).
Recommendations for respirator selection: *EMERGENCY OR PLANNED ENTRY INTO UNKNOWN CONCENTRATIONS OR IDLH CONDITIONS:* SCBAF:PD,PP (any self-contained breathing apparatus that has a full facepiece and is operated in a pressure-demand or other positive-pressure mode); or SAF:PD,PP:ASCBA (any supplied-air respirator that has a full facepiece and is operated in a pressure-demand or other positive-pressure mode in combination with an auxiliary self-contained breathing apparatus operated in a pressure-demand or other positive pressure mode). *ESCAPE:* GMFOVHiE [any air-purifying, full-facepiece respirator (gas mask) with a chin-style, front-or back-mounted canister having a high efficiency particulate filter]; or SCBAE (any appropriate escape-type, self-contained breathing apparatus). The U.S. Army standard M17A1 (which is being replaced by the M40 protective mask) mask provides complete respiratory protection against all known military toxic chemical agents, but it cannot be used in an oxygen deficient environment and it does not afford protection against industrial toxics such as ammonia and carbon monoxide. *It is not approved for civilian use.*
Exposure limits (TWA unless otherwise noted): Workplace: 0.00001 mg/m^3; General population limits (as recommended by the Surgeon General's Working Group, U.S. Department of Health): 0.000003 mg/m^3.
Toxicity: The median lethal dosage (vapor/respiratory) LCt_{50} = 10 mg-min/m^3 for humans. The LD$_{50}$ [skin] = 10 mg/70 kg/156.8 lb [man]. (*Medical Aspects of Chemical and Biological Warfare, Part I,* Walter Reed Medical Center, 1997).

FIRE DATA
Flash point: 318.2°F/159°C.

CHEMICAL REACTIVITY
Reaction with water: Hydrolysis of VX produces a Class B poison.

ENVIRONMENTAL DATA
Food chain concentration potential: Broken down in water moist soil quickly. Not expected to accumulate in the food chain.
Water pollution: If used as a weapon, utilize use an M272 Water Detection Kit (Detection limit: 0.02 mg/L). Dangerous to aquatic life in high concentrations. May be dangerous if it enters water intakes. Notify local health and pollution control officials. Notify operators of nearby water intakes. This material will be broken down in water quickly, but small amounts may evaporate. This material will be broken down in moist soil quickly. Small amounts may evaporate into the air or travel below the soil surface and contaminated groundwater. Depending on conditions, soil persistence may be 2-7 days.

PHYSICAL AND CHEMICAL PROPERTIES
Physical state @ 59°F/15°C and 1 atm: Liquid.
Molecular weight: 267.41
Boiling point @ 1 atm: 568.4°F/298°C/571°K.
Melting/Freezing point: −59.8°F/−51°C/222°K
Relative vapor density (air = 1): 9.2
Specific Gravity (water = 1): 1.008
Solubility in water: 3g/100 g (miscible below 48.9°F/9.4°C)
Liquid density: 1.008 g/mL @ 68°F/20°C.
Vapor pressure: 0.0007 mm @ 68°F/20°C.
Volatility: 10.45 mg/m3 @ 77°F/25°C.

WAXES: CARNAUBA   **REC. W:0100**

SYNONYMS: BRAZIL WAX; CARNAUBA WAX

IDENTIFICATION

DESCRIPTION: Yellow to dark brownish green solid. Odorless. Floats on water and solidifies; insoluble.

Combustible • Containers may BLEVE when exposed to fire • Irritating to the skin, eyes, and respiratory tract • Toxic products of combustion may include carbon monoxide.

Hazard Classification (based on NFPA-704 Rating System)
Health Hazards (Blue): 0; Flammability (Red): 1; Reactivity (Yellow): 0

EMERGENCY RESPONSE: See Appendix A (171)
Evacuation:
Public safety: Isolate the area of spill or leak for at least 10 to 25 meters (30 to 80 feet) in all directions.
Spill: Increase, in the downwind direction, as necessary, the distance shown under "Public Safety."
Fire: If any large container is involved in fire, isolate for at least 800 meters (½ mile) in all directions; also, consider initial evacuation for 800 meters (½ mile) in all directions.

EXPOSURE
Short-term effects: Liquid: Will burn skin and eyes. Remove wax. Flush affected areas with plenty of water.

HEALTH HAZARDS
Personal protective equipment (PPE): B-Level PPE. Goggles or face shield; protective gloves and clothing for hot liquid wax.
Skin or eye contact: Remove solidified wax from skin, wash with soap and water; *IF IN EYES*, or if skin is burned, *CALL A DOCTOR*.
Vapor (gas) irritant characteristics: Not volatile
Liquid or solid irritant characteristics: Hot wax can burn skin and eyes.

FIRE DATA
Flash point: 595°F/313°C (oc); 540°F/282°C (cc).
Fire extinguishing agents not to be used: Water or foam may cause frothing.

CHEMICAL REACTIVITY
Binary reactants: Strong oxidizers.

ENVIRONMENTAL DATA
Food chain concentration potential: Negative; unlikely to accumulate.
Water pollution: Effect of low concentrations on aquatic life is unknown. Fouling to shoreline. May be dangerous if it enters nearby water intakes; notify operators. Notify local health and wildlife officials. **Response to discharge:** Mechanical containment. Should be removed. Chemical and physical treatment.

SHIPPING INFORMATION
Grades of purity: Shipped in a variety of grades, depending on source of wax and intended use. All have about the same hazardous properties; **Storage temperature:** Elevated; **Inert atmosphere:** None; **Venting:** Open (flame arrester); **Stability during transport:** Stable.

PHYSICAL AND CHEMICAL PROPERTIES
Physical state @ 59°F/15°C and 1 atm: Solid.
Boiling point @ 1 atm: Very high.
Melting/Freezing point: 176–187°F/80–86°C/353–359°K.
Specific gravity (water = 1): 0.998 @ 77°F/25°C (solid).
Liquid surface tension: 32 dynes/cm = 0.032 N/m at 100°C.
Vapor pressure: (Reid) Low.

WHITE SPIRIT [low (15-20%) aromatic]   **REC. W:0200**

SYNONYMS: SKDN

IDENTIFICATION
CAS Number: 63394-00-3
DOT ID Number: UN 2319; DOT Guide Number: 128
Proper Shipping Name: Terpene hydrocarbons, n.o.s.

DESCRIPTION: Colorless liquid.

Highly flammable • Containers may BLEVE when exposed to fire • Vapors may form explosive mixture with air • Vapors are heavier than air and will collect and stay in low areas • Vapors may travel long distances to ignition sources and flashback • Vapors in confined areas (e.g., tanks, sewers, buildings) may explode when exposed to fire • Irritating to the skin, eyes, and respiratory tract; may have a narcotic effect in high concentrations • Toxic products of combustion may include carbon monoxide.

Hazard Classification (based on NFPA-704 Rating System)
Health Hazards (Blue): 0; Flammability (Red): 2; Reactivity (Yellow): 0

EMERGENCY RESPONSE: See Appendix A (128)
Evacuation:
Public safety: Isolate spill area for at least 25 to 50 meters (80 to 160 feet) in all directions.
Spill: Large spill–Consider initial downwind evacuation for at least 300 meters (1000 feet).
Fire: Isolate for 800 meters (½ mile) in all directions, especially if tank, rail car, or tank truck is involved in fire.

EXPOSURE
Short-term effects: *SEEK MEDICAL ATTENTION.* **Vapor and liquid:** May cause narcotic effects at high concentration, may cause irritating eyes and skin, and may induce coughing. *IF INHALED*, will, move to fresh air and treat symptoms. Flush affected areas with water and soap. *IF IN EYES*, hold eyelids open and flush with plenty of water. May be toxic if swallowed. **If swallowed** and victim is *CONSCIOUS AND ABLE TO SWALLOW*, have victim drink 4 to 8 ounces of water.

HEALTH HAZARDS
Personal protective equipment (PPE): B-Level PPE. Self-contained breathing apparatus). Chemical protective material(s) reported to have good to excellent resistance: neoprene.
Toxicity by ingestion: Grade 2: LD_{50} more than 6.0 g/kg (rat).
Long-term health effects: May be a teratogen.
Vapor (gas) irritant characteristics: Vapors cause irritation such that personnel will find high concentrations unpleasant. The effect is temporary.
Liquid or solid irritant characteristics: If spilled on clothing and allowed to remain, may cause smarting and reddening of skin.

FIRE DATA
Flash point: 104°F/40°C (cc).
Flammable limits in air: LEL: 0.6%: UEL: 6.5%.
Fire extinguishing agents not to be used: Water may not be effective.
Autoignition temperature: 410°F/210°C/483°K.
Electrical hazards: Flow or agitation of substance may generate electrostatic charges due to low conductivity.

CHEMICAL REACTIVITY
Binary reactants: Oxidizers, nitric acid.
Reactivity group: 33
Compatibility class: Miscellaneous hydrocarbon mixtures.

ENVIRONMENTAL DATA
Water pollution: DOT Appendix B, §172.101–marine pollutant. Effect of low concentration on aquatic life is not known. May be dangerous if it enters nearby water intakes; notify operators. Notify local health and wildlife officials. **Response to discharge:** Restrict access. Mechanical containment. Should be removed. Chemical and physical treatment.

SHIPPING INFORMATION
Storage temperature: Ambient; **Stability during transport:** Stable.

PHYSICAL AND CHEMICAL PROPERTIES
Boiling point @ 1 atm: 302–374°F/150–190°C/423–463°K.
Specific gravity (water = 1): 0.780

m-XYLENE   REC. X:0100

SYNONYMS: *m*-DIMETHYLBENZENE; 1,3-DIMETHYLBENZENE; EEC No. 601-039-00-1; *m*-XILENO (Spanish); 1,3-XYLENE; *m*-XYLOL; RCRA No. U239

IDENTIFICATION
CAS Number: 108-38-3; 1330-20-7 (mixed isomers)
Formula: C_8H_{10}; *m*-$C_6H_4(CH_3)_2$
DOT ID Number: UN 1307; DOT Guide Number: 128
Proper Shipping Name: Xylenes
Reportable Quantity (RQ): **(CERCLA)** 1000 lb/454 kg

DESCRIPTION: Clear, colorless, watery liquid. Sweet odor, like benzene. Floats on water surface; insoluble. Flammable, irritating vapor is produced.

Highly flammable • Containers may BLEVE when exposed to fire • Vapors may form explosive mixture with air • Vapors are heavier than air and will collect and stay in low areas • Vapors may travel long distances to ignition sources and flashback • Vapors in confined areas (e.g., tanks, sewers, buildings) may explode when exposed to fire • Irritating to the skin, eyes, and respiratory tract • Toxic products of combustion may include carbon monoxide.

Hazard Classification (based on NFPA-704 Rating System)
Health Hazards (Blue): 2; Flammability (Red): 3; Reactivity (Yellow): 0

EMERGENCY RESPONSE: See Appendix A (128)
Evacuation:
Public safety: Isolate spill area for at least 25 to 50 meters (80 to 160 feet) in all directions.
Spill: Large spill–Consider initial downwind evacuation for at least 300 meters (1000 feet).
Fire: Isolate for 800 meters (½ mile) in all directions, especially if tank, rail car, or tank truck is involved in fire.

EXPOSURE
Short-term effects: *SEEK MEDICAL ATTENTION.* **Vapor:** Irritating to eyes, nose, and throat. *IF INHALED*, will, will cause headache, difficult breathing, or loss of consciousness. Move to fresh air. *IF BREATHING HAS STOPPED,* give artificial respiration. IF breathing is difficult, administer oxygen. **Liquid:** Irritating to skin and eyes. *IF SWALLOWED*, will cause nausea, vomiting, or loss consciousness. Remove contaminated clothing and shoes. Flush affected areas with plenty of water. *IF IN EYES*, hold eyelids open and flush with plenty of water. *IF SWALLOWED* and victim is *CONSCIOUS AND ABLE TO SWALLOW*, have victim drink 4 to 8 ounces of water. **Do NOT induce vomiting**. Administer a slurry of activated charcoal. The use of alcoholic beverages may enhance the toxic effect.

HEALTH HAZARDS
Personal protective equipment (PPE): B-Level PPE. Approved canister); or air–supplied mask; goggles or face shield; plastic gloves and boots. Chemical protective material(s) reported to have good to excellent resistance: PV alcohol, Viton®, nitrile, Teflon®, Viton®, Silvershield®.
Recommendations for respirator selection: NIOSH/OSHA
900 ppm: CCROV [any chemical cartridge respirator with organic vapor cartridge(s)]; or PAPROV [any powered, air-purifying respirator with organic vapor cartridge(s)]; or SA (any supplied-air respirator); or SCBAF (any self-contained breathing apparatus with a full facepiece). *EMERGENCY OR PLANNED ENTRY INTO*

UNKNOWN CONCENTRATIONS OR IDLH CONDITIONS:
SCBAF:PD,PP (any self-contained breathing apparatus that has a full facepiece and is operated in a pressure-demand or other positive-pressure mode); or SAF:PD,PP:ASCBA (any supplied-air respirator that has a full facepiece and is operated in a pressure-demand or other positive-pressure mode in combination with an auxiliary self-contained breathing apparatus operated in a pressure-demand or other positive pressure mode). *ESCAPE:* GMFOV [any air-purifying, full-facepiece respirator (gas mask) with a chin-style, front- or back-mounted organic vapor cannister]; or SCBAE (any appropriate escape-type, self-contained breathing apparatus).
*Note: Substance reported to cause eye irritation or damage; may require eye protection.
Exposure limits (TWA unless otherwise noted): ACGIH TLV 100 ppm (434 mg/m^3); NIOSH/OSHA PEL 100 ppm (435 mg/m^3).
Short-term exposure limits (15-minute TWA): ACGIH STEL 150 ppm (651 mg/m^3); NIOSH STEL 150 ppm (655 mg/m^3).
Toxicity by ingestion: Grade 3; LD$_{50}$ = 50 to 500 g/kg.
Long-term health effects: Kidney and liver damage; skin disorders. IARC (3 rating) unclassifiable substance.
Vapor (gas) irritant characteristics: Vapors cause smarting of the eyes or respiratory system if present in high concentrations. The effect may be temporary.
Liquid or solid irritant characteristics: If spilled on clothing and allowed to remain, may cause smarting and reddening of the skin.
Odor threshold: 0.05 ppm.
IDLH value: 900 ppm.

FIRE DATA
Flash point: 82°F/28°C (cc).
Flammable limits in air: LEL: 1.1%; UEL: 7.1%.
Fire extinguishing agents not to be used: Water may be ineffective.
Behavior in fire: Containers may rupture and explode.
Autoignition temperature: 986°F/530°C/803°K.
Electrical hazard: Class I, Group D. Flow or agitation of substance may generate electrostatic charges due to low conductivity.
Burning rate: 5.8 mm/min

CHEMICAL REACTIVITY
Binary reactants: Strong oxidizers, strong acids
Reactivity group: 32
Compatibility class: Aromatic hydrocarbons

ENVIRONMENTAL DATA
Food chain concentration potential: Log P$_{ow}$ = 3.2 (mixed isomers). Values > 3.0 are likely to bioconcentrate in aquatic organisms and other living tissue, especially in fats.
Water pollution: Harmful to aquatic life in very low concentrations. Fouling to shoreline. May be dangerous if it enters nearby water intakes; notify operators. Notify local health and wildlife officials. **Response to discharge:** Issue warning–high flammability. Evacuate area. Should be removed. Chemical and physical treatment.

SHIPPING INFORMATION
Grades of purity: Research: 99.99%; Pure: 99.9%; Technical: 99.2%; **Storage temperature:** Ambient; **Inert atmosphere:** None; **Venting:** Open (flame arrester) or pressure-vacuum; **Stability during transport:** Stable.

NAS HAZARD CLASSIFICATION FOR BULK WATER TRANSPORTATION
FIRE: 3
HEALTH: Vapor irritant: 1; Liquid or solid irritant: 1; Poisons: 2
WATER POLLUTION: Human toxicity: 1; Aquatic toxicity: 3; Aesthetic effect: 2
REACTIVITY: Other chemicals: 1; Water: 0; Self-reaction: 0

PHYSICAL AND CHEMICAL PROPERTIES
Physical state @ 59°F/15°C and 1 atm: Liquid.
Molecular weight: 106.16
Boiling point @ 1 atm: 284°F/140°F/413°K.
Melting/Freezing point: –54.2°F/–47.9°C/225.3°K.
Critical temperature: 650.8°F/343.8°C/617.0°K.
Critical pressure: 513.8 atm = 34.95 psia = 3.540 MN/m^2.
Specific gravity (water = 1): 0.864 @ 68°F/20°C (liquid).
Liquid surface tension: 28.6 dynes/cm = 0.0286 N/m @ 68°F/20°C.
Liquid water interfacial tension: 36.4 dynes/cm = 0.0364 N/m at 30°C.
Ratio of specific heats of vapor (gas): 1.071
Latent heat of vaporization: 147 Btu/lb = 81.9 cal/g = 3.43 x 10^5 J/kg.
Heat of combustion: –17,554 Btu/lb = –9752.4 cal/g = –408.31 x 10^5 J/kg.
Heat of fusion: 26.01 cal/g.
Vapor pressure: (Reid) 0.34 psia; 9 mm.

o-XYLENE   REC. X:0200

SYNONYMS: *o*-DIMETHYLBENZENE; 1,2-DIMETHYLBENZENE; *o*-METHYLTOLUENE; *o*-XILENO (Spanish); 1,2-XYLENE; ORTHO-XYLENE; *o*-XYLOL; RCRA No. U239

IDENTIFICATION
CAS Number: 95-47-6
Formula: *o*-C$_6$H$_4$(CH$_3$)$_2$
DOT ID Number: UN 1307; DOT Guide Number: 128
Proper Shipping Name: Xylenes
Reportable Quantity (RQ): **(CERCLA)** 1000 lb/454 kg

Highly flammable • Containers may BLEVE when exposed to fire • Vapors may form explosive mixture with air • Vapors are heavier than air and will collect and stay in low areas • Vapors may travel long distances to ignition sources and flashback • Vapors in confined areas (e.g., tanks, sewers, buildings) may explode when exposed to fire • Irritating to the skin, eyes, and respiratory tract • Toxic products of combustion may include carbon monoxide.

DESCRIPTION: Colorless, watery liquid. Sweet aromatic odor, like benzene. Floats on the surface of water. Flammable, irritating vapor is produced.

Hazard Classification (based on NFPA-704 Rating System)
Health Hazards (Blue): 2; Flammability (Red): 3; Reactivity (Yellow): 0

EMERGENCY RESPONSE: See Appendix A (128)
Evacuation:
Public safety: Isolate spill area for at least 25 to 50 meters (80 to 160 feet) in all directions.
Spill: Large spill–Consider initial downwind evacuation for at least 300 meters (1000 feet).
Fire: Isolate for 800 meters (½ mile) in all directions, especially if tank, rail car, or tank truck is involved in fire.

EXPOSURE

Short-term effects: *SEEK MEDICAL ATTENTION.* **Vapor:** Irritating to eyes, nose, and throat. *IF INHALED*, will, will cause headache, difficult breathing, or loss of consciousness. Move to fresh air. *IF BREATHING HAS STOPPED,* give artificial respiration. IF breathing is difficult, administer oxygen. **Liquid:** Irritating to skin and eyes. *IF SWALLOWED*, will cause nausea, vomiting, or loss of consciousness. Remove contaminated clothing and shoes. Flush affected areas with plenty of water. *IF IN EYES*, hold eyelids open and flush with plenty of water. *IF SWALLOWED* and victim is *CONSCIOUS AND ABLE TO SWALLOW*, have victim drink 4 to 8 ounces of water. **Do NOT induce vomiting.** Administer a slurry of activated charcoal.

HEALTH HAZARDS

Personal protective equipment (PPE): B-Level PPE. Approved canister); or air–supplied mask; goggles or face shield; plastic gloves and boots. Chemical protective material(s) reported to have good to excellent resistance: PV alcohol, nitrile, Teflon®, Viton®, Silvershield®.

Recommendations for respirator selection: NIOSH/OSHA *900 ppm:* CCROV* [any chemical cartridge respirator with organic vapor cartridge(s)]; or PAPROV* [any powered, air-purifying respirator with organic vapor cartridge(s)]; or SA* (any supplied-air respirator); or SCBAF* (any self-contained breathing apparatus with a full facepiece). *EMERGENCY OR PLANNED ENTRY INTO UNKNOWN CONCENTRATIONS OR IDLH CONDITIONS:* SCBAF:PD,PP (any self-contained breathing apparatus that has a full facepiece and is operated in a pressure-demand or other positive-pressure mode); or SAF:PD,PP:ASCBA (any supplied-air respirator that has a full facepiece and is operated in a pressure-demand or other positive-pressure mode in combination with an auxiliary self-contained breathing apparatus operated in a pressure-demand or other positive pressure mode). *ESCAPE:* GMFOV [any air-purifying, full-facepiece respirator (gas mask) with a chin-style, front- or back-mounted organic vapor cannister]; or SCBAE (any appropriate escape-type, self-contained breathing apparatus).* *Note:* Substance reported to cause eye irritation or damage; may require eye protection.

Exposure limits (TWA unless otherwise noted): ACGIH TLV 100 ppm (434 mg/m^3); NIOSH/OSHA PEL 100 ppm (435 mg/m^3).

Short-term exposure limits (15-minute TWA): ACGIH STEL 150 ppm (651 mg/m^3); NIOSH STEL 150 ppm (655 mg/m^3).

Toxicity by ingestion: Grade 3; LD_{50} = 50 to 500 mg/kg.

Long-term health effects: Kidney and liver damage; skin disorders. IARC (3 rating) unclassifiable substance.

Vapor (gas) irritant characteristics: Vapors cause smarting of the eyes or respiratory system if present in high concentrations. The effect may be temporary.

Liquid or solid irritant characteristics: If spilled on clothing and allowed to remain, may cause smarting and reddening of the skin.

Odor threshold: 0.05 ppm.

IDLH value: 900 ppm.

FIRE DATA

Flash point: 75°F/24°C (oc); 90°F/32°C (cc).

Flammable limits in air: LEL: 0.9%; UEL: 6.7%.

Fire extinguishing agents not to be used: Water may be ineffective.

Behavior in fire: Containers may rupture and explode.

Autoignition temperature: 869°F/465°F/738°K.

Electrical hazard: Class I, Group D. Flow or agitation of substance may generate electrostatic charges due to low conductivity.

Burning rate: 5.8 mm/min

CHEMICAL REACTIVITY

Binary reactants: Strong oxidizers; strong acids

Reactivity group: 32

Compatibility class: Aromatic hydrocarbons

ENVIRONMENTAL DATA

Food chain concentration potential: Log P_{ow} = 3.2 (mixed isomers). Values > 3.0 are likely to bioconcentrate in aquatic organisms and other living tissue, especially in fats.

Water pollution: Dangerous to aquatic life in high concentrations. Fouling to shoreline. May be dangerous if it enters nearby water intakes; notify operators. Notify local health and wildlife officials.

Response to discharge: Issue warning–high flammability. Evacuate area. Should be removed. Chemical and physical treatment.

SHIPPING INFORMATION

Grades of purity: Research: 99.99%; Pure: 99.7%; Commercial: 95+%; **Storage temperature:** Ambient; **Venting:** Open (flame arrester) or pressure-vacuum; **Stability during transport:** Stable.

NAS HAZARD CLASSIFICATION FOR BULK WATER TRANSPORTATION

FIRE: 3

HEALTH: Vapor irritant: 1; Liquid or solid irritant: 1; Poisons: 2

WATER POLLUTION: Human toxicity: 1; Aquatic toxicity: 3; Aesthetic effect: 2

REACTIVITY: Other chemicals: 1; Water: 0; Self-reaction: 0

PHYSICAL AND CHEMICAL PROPERTIES

Physical state @ 59°F/15°C and 1 atm: Liquid.

Molecular weight: 106.16

Boiling point @ 1 atm: 291.9°F/144.4°C/417.6°K.

Melting/Freezing point: –13.3°F/–25.2°C/248.0°K.

Critical temperature: 674.8°F/357.1°C/630.3°K.

Critical pressure: 541.5 atm = 36.84 psia = 3.732 MN/m^2.

Specific gravity (water = 1): 0.880 @ 68°F/20°C (liquid).

Liquid surface tension: 30.53 dynes/cm = 0.03053 N/m at 15.5°C.

Liquid water interfacial tension: 36.06 dynes/cm = 0.03606 N/m @ 68°F/20°C.

Relative vapor density (air = 1): 3.69

Ratio of specific heats of vapor (gas): 1.068

Latent heat of vaporization: 149 Btu/lb = 82.9 cal/g = 3.47 x 10^5 J/kg.

Heat of combustion: –17,558 Btu/lb = –9754.7 cal/g = –408.41 x 10^5 J/kg.

Heat of fusion: 30.64 cal/g.

Vapor pressure: (Reid) 0.28 psia; 7 mm.

---

p-XYLENE   REC. X:0300

SYNONYMS: CHROMAR; *p*-DIMETHYLBENZENE; 1,4-DIMETHYLBENZENE; EEC No. 601-040-00-7; *p*-METHYLTOLUENE; SCINTILLAR; *p*-XILENO (Spanish); PARA-XYLENE; 1,4-XYLENE; *p*-XYLOL; RCRA No. U239

IDENTIFICATION

CAS Number: 106-42-3

Formula: *p*-C$_6$H$_4$(CH$_3$)$_2$

DOT ID Number: UN 1307; DOT Guide Number: 128

Proper Shipping Name: Xylenes

Reportable Quantity (RQ): **(CERCLA)** 1000 lb/454 kg

DESCRIPTION: Colorless, watery liquid. Sweet aromatic odor, like benzene. Floats on the surface of water. Flammable, irritating vapor is produced.

Highly flammable • Containers may BLEVE when exposed to fire • Vapors may form explosive mixture with air • Vapors are heavier than air and will collect and stay in low areas • Vapors may travel long distances to ignition sources and flashback • Vapors in confined areas (e.g., tanks, sewers, buildings) may explode when exposed to fire • Irritating to the skin, eyes, and respiratory tract • Toxic products of combustion may include carbon monoxide.

Hazard Classification (based on NFPA-704 Rating System)
Health Hazards (Blue): 2; Flammability (Red): 3; Reactivity (Yellow): 0

EMERGENCY RESPONSE: See Appendix A (128)
Evacuation:
Public safety: Isolate spill area for at least 25 to 50 meters (80 to 160 feet) in all directions.
Spill: Large spill–Consider initial downwind evacuation for at least 300 meters (1000 feet).
Fire: Isolate for 800 meters (½ mile) in all directions, especially if tank, rail car, or tank truck is involved in fire.

EXPOSURE
Short-term effects: *SEEK MEDICAL ATTENTION.* **Vapor:** Irritating to eyes, nose, and throat. *IF INHALED*, will, will cause dizziness, difficult breathing, or loss of consciousness. Move to fresh air. *IF BREATHING HAS STOPPED,* give artificial respiration. IF breathing is difficult, administer oxygen. **Liquid:** Irritating to skin and eyes. *IF SWALLOWED*, will cause nausea, vomiting, loss of consciousness. Remove contaminated clothing and shoes. Flush affected areas with plenty of water. *IF IN EYES*, hold eyelids open and flush with plenty of water. *IF SWALLOWED* and victim is *CONSCIOUS AND ABLE TO SWALLOW*, have victim drink 4 to 8 ounces of water. **Do NOT induce vomiting.** Administer a slurry of activated charcoal.

HEALTH HAZARDS
Personal protective equipment (PPE): B-Level PPE. Approved canister); or air–supplied mask; goggles or face shield; plastic gloves and boots. Chemical protective material(s) reported to have good to excellent resistance: PV alcohol, Viton®, nitrile, Teflon®, Viton®, Silvershield®.
Recommendations for respirator selection: NIOSH/OSHA *900 ppm:* CCROV [any chemical cartridge respirator with organic vapor cartridge(s)]; or PAPROV [any powered, air-purifying respirator with organic vapor cartridge(s)]; or SA (any supplied-air respirator); or SCBAF (any self-contained breathing apparatus with a full facepiece). *EMERGENCY OR PLANNED ENTRY INTO UNKNOWN CONCENTRATIONS OR IDLH CONDITIONS:* SCBAF:PD,PP (any self-contained breathing apparatus that has a full facepiece and is operated in a pressure-demand or other positive-pressure mode); or SAF:PD,PP:ASCBA (any supplied-air respirator that has a full facepiece and is operated in a pressure-demand or other positive-pressure mode in combination with an auxiliary self-contained breathing apparatus operated in a pressure-demand or other positive pressure mode). *ESCAPE:* GMFOV [any air-purifying, full-facepiece respirator (gas mask) with a chin-style, front- or back-mounted organic vapor cannister]; or SCBAE (any appropriate escape-type, self-contained breathing apparatus). *Note:* Substance reported to cause eye irritation or damage; may require eye protection.

Exposure limits (TWA unless otherwise noted): ACGIH TLV 100 ppm (434 mg/m^3); NIOSH/OSHA PEL 100 ppm (435 mg/m^3).
Short-term exposure limits (15-minute TWA): ACGIH STEL 150 ppm (651 mg/m^3); NIOSH STEL 150 ppm (655 mg/m^3)/15 minutes
Toxicity by ingestion: Grade 3; LD_{50} = 50 to 500 mg/kg.
Long-term health effects: Kidney and liver damage; skin disorders. IARC unclassifiable substance; 3 rating
Vapor (gas) irritant characteristics: Vapors cause smarting of the eyes or respiratory system if present in high concentrations. The effect may be temporary.
Liquid or solid irritant characteristics: If spilled on clothing and allowed to remain, may cause smarting and reddening of the skin.
Odor threshold: 0.05 ppm.
IDLH value: 900 ppm.

FIRE DATA
Flash point: 81°F/27°C (cc).
Flammable limits in air: LEL: 1.1%; UEL: 6.9%.
Fire extinguishing agents not to be used: Water may be ineffective.
Autoignition temperature: 986°F/530°C/803°K.
Electrical hazard: Class I, Group D. Flow or agitation of substance may generate electrostatic charges due to low conductivity.
Burning rate: 5.8 mm/min

CHEMICAL REACTIVITY
Binary reactants: Strong oxidizers, strong acids
Reactivity group: 32
Compatibility class: Aromatic hydrocarbons

ENVIRONMENTAL DATA
Food chain concentration potential: Log P_{ow} = 3.2 (mixed isomers). Values > 3.0 are likely to bioconcentrate in aquatic organisms and other living tissue, especially in fats.
Water pollution: Harmful to aquatic life in very low concentrations. Fouling to shoreline. May be dangerous if it enters nearby water intakes; notify operators. Notify local health and wildlife officials. **Response to discharge:** Issue warning–high flammability. Evacuate area. Should be removed. Chemical and physical treatment.

SHIPPING INFORMATION
Grades of purity: Research: 99.99%; Pure: 99.8%; Technical: 99.0%; **Storage temperature:** Ambient; **Inert atmosphere:** None; **Venting:** Open (flame arrester) or pressure-vacuum; **Stability during transport:** Stable.

NAS HAZARD CLASSIFICATION FOR BULK WATER TRANSPORTATION
FIRE: 3
HEALTH: Vapor irritant: 1; Liquid or solid irritant: 1; Poisons: 2
WATER POLLUTION: Human toxicity: 1; Aquatic toxicity: 3; Aesthetic effect: 2
REACTIVITY: Other chemicals: 1; Water: 0; Self-reaction: 0

PHYSICAL AND CHEMICAL PROPERTIES
Physical state @ 59°F/15°C and 1 atm: Liquid.
Molecular weight: 106.16
Boiling point @ 1 atm: 280.9°F/138.3°C/411.5°K.
Melting/Freezing point: 55.9°F/13.3°C/286.5°K.
Critical temperature: 649.4°F/343.0°C/616.2°K.
Critical pressure: 509.4 atm = 34.65 psia = 3.510 MN/m^2.
Specific gravity (water = 1): 0.861 @ 68°F/20°C (liquid).

Liquid surface tension: 28.3 dynes/cm = 0.0283 N/m @ 68°F/20°C.
Liquid water interfacial tension: 37.8 dynes/cm = 0.0378 N/m @ 68°F/20°C.
Ratio of specific heats of vapor (gas): 1.071
Latent heat of vaporization: 150 Btu/lb = 81 cal/g = 3.4×10^5 J/kg.
Heat of combustion: −17,559 Btu/lb = −9754.7 cal/g = -408.41×10^5 J/kg.
Heat of fusion: 37.83 cal/g.
Vapor pressure: (Reid) 0.34 psia; 6.2 mm.

XYLENOL   REC. X:0400

SYNONYMS: DIMETHYLPHENOL; PHENOL, DIMETHYL-; XILENOL (Spanish); 2,6-XYLENOL; 2-HYDROXY-*m*-XYLENE

IDENTIFICATION
CAS Number: 1300-71-6; 526-75-0 (2,3-isomer); 105-67-9 (2,4-isomer); 95-87-4 (2,5-isomer); 108-68-9 (3,5-isomer)
Formula: $C_8H_{10}O$; $2,6\text{-}(CH_3)_2C_6H_3OH$
DOT ID Number: UN2261; DOT Guide Number: 153
Proper Shipping Name: Xylenols
Reportable Quantity (RQ): **(CERCLA)** 1000 lb/454 kg

DESCRIPTION: Light yellowish-brown solid or liquid. Sweet tarry odor. May float or sink in water; slightly soluble.

Combustible • Containers may BLEVE when exposed to fire • Vapors may form explosive mixture with air • Dust or vapors in confined areas (e.g., tanks, sewers, buildings) may explode when exposed to fire • Irritating to the skin, eyes, and respiratory tract • Toxic products of combustion may include carbon monoxide and toxic vapors of unburned material.

Hazard Classification (based on NFPA-704 Rating System)
Health Hazards (Blue): 1; Flammability (Red): 1; Reactivity (Yellow): 0

EMERGENCY RESPONSE: See Appendix A (153)
Evacuation:
Public safety: Isolate the area of spill or leak for at least 25 to 50 meters (80 to 160 feet) in all directions.
Spill: Increase, in the downwind direction, as necessary, the distance shown under "Public Safety."
Fire: If tank, rail car, or tank truck is involved in fire, isolate for at least 800 meters (½ mile) in all directions; also, consider initial evacuation for 800 meters (½ mile) in all directions.

EXPOSURE
Short-term effects: *SEEK MEDICAL ATTENTION.* **Dust:** Irritating to eyes, nose, and throat. Harmful if inhaled. Move victim to fresh air. *IF IN EYES*, hold eyelids open and flush with plenty of water. *IF BREATHING HAS STOPPED*, give artificial respiration; *avoid mouth-to-mouth resuscitation; use bag/mask apparatus*. IF breathing is difficult, administer oxygen. **Liquid or solid:** Irritating to skin and eyes. *IF SWALLOWED* or skin is exposed will cause nausea and vomiting. Remove contaminated clothing and shoes. *IF ON SKIN*, wash affected areas with large quantities of water or soapy water until all odor is gone; then wash with alcohol or 20% glycerin solution and more water; keep patient warm, but not hot; cover chemical burns continuously with compresses wet with saturated solution of sodium thiosulfate; apply no salves or ointments for 24 hrs after injury. Flush affected areas with plenty of water. *IF IN EYES*, hold eyelids open and flush with plenty of water. Medical personnel may want to instill with 2–3 drops of 0.5% pontocaine or equivalent after first 15 minutes; do not use oils or oily ointments unless ordered by physician. Do not use oils or oily ointments unless ordered by physician. *IF SWALLOWED* and victim is *CONSCIOUS AND ABLE TO SWALLOW*, have victim drink 4 to 8 ounces of water and have victim induce vomiting. *IF SWALLOWED* and victim is *UNCONSCIOUS OR HAVING CONVULSIONS*, do nothing except keep victim warm.

HEALTH HAZARDS
Personal protective equipment (PPE): B-Level PPE. Organic canister mask; goggles or face shield; rubber gloves; other protective clothing to prevent contact with skin. Chemical protective material(s) reported to have good to excellent resistance: Teflon®.
Toxicity by ingestion: Grade 2; oral LD_{50} = 1070 mg/kg (mouse).
Long-term health effects: Damage to heart muscle, and changes in liver, kidney, and spleen in rats
Vapor (gas) irritant characteristics: Causes irritation of the eyes, and respiratory system. See symptoms, above.
Liquid or solid irritant characteristics: Causes irritation of the eyes, skin, and respiratory system.

FIRE DATA
Flash point: 186°F/86°C (cc).
Flammable limits in air: LEL: 1.4% (mixed isomers).
Autoignition temperature: 1110°F/599°C/872°K.

CHEMICAL REACTIVITY
Binary reactants: A weak organic acid. Keep away from oxidizers, sulfuric acid, caustics, ammonia, aliphatic amines, alkanolamines, isocyanates, alkylene oxides, epichlorohydrin.

ENVIRONMENTAL DATA
Food chain concentration potential: Log P_{ow} = 2.2–2.5. Unlikely to accumulate.
Water pollution: DOT Appendix B, §172.101–marine pollutant. Harmful to aquatic life in very low concentrations. Fouling to shoreline. May be dangerous if it enters nearby water intakes; notify operators. Notify local health and wildlife officials. **Response to discharge:** Issue warning–water contaminant. Restrict access. Mechanical containment. Should be removed. Chemical and physical treatment.

SHIPPING INFORMATION
Grades of purity: 99% 2,6-xylenol. Other commercial Xylenols include 2,3-; 2,4-; 2,5-; 3,4-; 3,5-; and various mixtures of these. Properties are similar to those of the 2,6-isomer; **Storage temperature:** Ambient; **Inert atmosphere:** None; **Venting:** Open (flame arrester); **Stability during transport:** Stable.

PHYSICAL AND CHEMICAL PROPERTIES
Physical state @ 59°F/15°C and 1 atm: Solid or Liquid.
Molecular weight: 122.2
Boiling point @ 1 atm: 413°F/212°C/485°K; 424°F/218°C/491°K (2,3-isomer); 414°F/212°C/485°K (2,4-isomer), (2,5-isomer). 428°F/220°C/493°K (3,5-isomer).
Melting/Freezing point: −40 to +106°F/−40 to +45°C/233 to 318°K; 167°F/75°C/348°K (2,3-isomer), (2,5-isomer); 79°F/26°C/299°K (2,4-isomer); 147°F/64°C/337°K (3,5-isomer).
Specific gravity (water = 1): 1.01 @ 68°F/20°C (liquid).
Liquid surface tension: (estimate) 30 dynes/cm = 0.030 N/m at 30°C.

Liquid water interfacial tension: (estimate) 25 dynes/cm = 0.025 N/m @ 77°F/25°C.
Latent heat of vaporization: 212.74 Btu/lb = 118.19 cal/g = 4.9451 x 10^5 J/kg @ 77°F/25°C.
Heat of combustion: –15,310 Btu/lb = –8500 cal/g = –356 x 10^5 J/kg.
Vapor pressure: 0.014-0.18 torr @ 77°F/25°C.

o-XYLIDINE   REC. X:0450

SYNONYMS: AMINODIMETHYLBENZENE; ANILINE, 2,6-DIMETHYL; 2,6-DIMETHYLANILINE; 2,6-DIMETHYLBENZENAMINE; 2,6-XYLIDINE; 2,6-XYLYLAMINE

IDENTIFICATION
CAS Number: 87-62-7; 1300-73-8 (mixed isomers); health information may apply to 95-78-3 (*p*-isomer); 95-68-1 (*m*-isomer); 121-69-7 (*N,N*-dimethylaniline)
Formula: $C_8H_{11}N$; $(CH_3)_2C_6H_3NH_2$
DOT ID Number: UN 1711; DOT Guide Number: 153
Proper Shipping Name: Xylidines

DESCRIPTION: Pale yellow liquid. Floats; sparingly soluble in water.

Combustible • Poison! • Forms methemoglobin; may interfere with the body's ability to use oxygen. Irritating to the skin, eyes, and respiratory tract; prolonged contact may cause burns • Firefighting gear (including SCBA) does not provide adequate protection. If exposure occurs, remove and isolate gear immediately and thoroughly decontaminate personnel • Containers may BLEVE when exposed to fire • Vapors are heavier than air and will collect and stay in low areas • Vapors in confined areas (e.g., tanks, sewers, buildings) may explode when exposed to fire • Irritating to the skin, eyes, and respiratory tract • Toxic products of combustion may include nitrogen oxides.

Hazard Classification (based on NFPA-704 Rating System)
Health Hazards (Blue): 3; Flammability (Red): 1; Reactivity (Yellow): 0

EMERGENCY RESPONSE: See Appendix A (153)
Evacuation:
Public safety: Isolate the area of spill or leak for at least 25 to 50 meters (80 to 160 feet) in all directions.
Spill: Increase, in the downwind direction, as necessary, the distance shown under "Public Safety."
Fire: If tank, rail car, or tank truck is involved in fire, isolate for at least 800 meters (½ mile) in all directions; also, consider initial evacuation for 800 meters (½ mile) in all directions.

EXPOSURE
Short-term effects: *SEEK MEDICAL ATTENTION*. **Vapor/mist:** *POISONOUS IF INHALED*. Irritating to eyes, nose, and throat. Move the victim to fresh air. *IF IN EYES*, hold the eyes open and flush with plenty of water.*IF BREATHING HAS STOPPED*, give artificial respiration; *avoid mouth-to-mouth resuscitation; use bag/mask apparatus*. If breathing is difficult, administer oxygen.
Liquid: *HARMFUL IF ABSORBED THROUGH THE SKIN*. Irritating to skin and eyes. *IF SWALLOWED*, will cause nausea, dizziness and headaches. Remove contaminated clothing and shoes. Flush affected areas with plenty of water. *IF IN EYES*, hold eyelids open and flush with plenty of water.

HEALTH HAZARDS
Personal protective equipment (PPE): A-Level PPE. Full face organic vapor respirator, rubber boots, heavy rubber gloves, and protective clothing.
Recommendations for respirator selection: NIOSH as xylidines *20 ppm:* CCROV [any chemical cartridge respirator with organic vapor cartridge(s)]; or SA (any supplied-air respirator). *50 ppm:* SA:CF (any supplied-air respirator operated in a continuous-flow mode); or CCRFOV [any chemical cartridge respirator with a full facepiece and organic vapor cartridges(s)]; or, GMFOV [any air-purifying, full-facepiece respirator (gas mask) with a chin-style, front- or back-mounted acid gas canister]; or PAPROV [any powered, air-purifying respirator with organic vapor cartridge(s)]; or SCBAF (any self-contained breathing apparatus with a full facepiece); or SAF (any supplied-air respirator with a full facepiece). *EMERGENCY OR PLANNED ENTRY INTO UNKNOWN CONCENTRATIONS OR IDLH CONDITIONS:* SCBAF:PD,PP (any self-contained breathing apparatus that has a full facepiece and is operated in a pressure-demand or other positive-pressure mode); or SAF:PD,PP:ASCBA (any supplied-air respirator that has a full facepiece and is operated in a pressure-demand or other positive-pressure mode in combination with an auxiliary self-contained breathing apparatus operated in a pressure-demand or other positive pressure mode). *ESCAPE:* GMFOV [any air-purifying, full-facepiece respirator (gas mask) with a chin-style, front- or back-mounted organic vapor canister]; or SCBAE (any appropriate escape-type, self-contained breathing apparatus). *Note:* Substance causes eye irritation or damage; eye protection needed.
Exposure limits (TWA unless otherwise noted): ACGIH TLV 0.5 ppm as xylidine; NIOSH 2 ppm (10 mg/m^3); OSHA 5 ppm (25 mg/m^3); skin contact contributes significantly in overall exposure as xylidines.
Short-term exposure limits (15-minute TWA): ACGIH STEL 10 ppm (50 mg/m^3).
Toxicity by ingestion: Grade 2: LD_{50} = 707 mg/kg (mouse).
Long-term health effects: Possible carcinogen. May cause blood problems.
Vapor (gas) irritant characteristics: Eye and skin irritant and poison. Vapors are moderately irritating such that personnel will not usually tolerate moderate or high concentrations.
Liquid or solid irritant characteristics: Severe eye and skin irritant and poison. May cause pain and second-degree burns after a few minutes of contact.
Odor threshold: 0.22 ppm.
IDLH value: 100 ppm; 50 ppm as xylidines.

FIRE DATA
Flash point: 206°F/97°C (cc); 145°F/63°C (*N,N*-)
Flammable limits in air: LEL: 1.0%; UEL: 7.0%.
Fire extinguishing agents not to be used: Water may be ineffective.

CHEMICAL REACTIVITY
Binary reactants: Oxidizers, strong acids, isocyanates.
Reactivity group: 9
Compatibility class: Aromatic amines

ENVIRONMENTAL DATA
Food chain concentration potential: Log P_{ow} = 1.8 (mixed isomers); 2.5 (*N,N*-dimethylaniline); 2.2 (2,6-isomer). Unlikely to accumulate.
Water pollution: Effects of low concentrations on aquatic life is unknown. Fouling to shoreline. May be dangerous if it enters nearby water intakes; notify operators. Notify local health and wildlife officials. **Response to discharge:** Issue warning. Restrict

access. Evacuate area. Mechanical containment. Should be removed. Chemical and physical treatment.

SHIPPING INFORMATION
Grades of purity: 99%; **Storage temperature:** Ambient; **Stability during transport:** Stable.

PHYSICAL AND CHEMICAL PROPERTIES
Physical state @ 59°F/15°C and 1 atm: Liquid.
Molecular weight: 121.18
Boiling point @ 1 atm: 417.2°F/214°C/487.2°K (at 739 mmHg = 0.972 atm); 379°F/193°C/466°K (*N,N-*)
Melting/Freezing point: 36°F/2°C/275°K.
Specific gravity (water = 1): 0.984
Relative vapor density (air = 1): 4.18; 4.17 (*N,N-*)
Vapor pressure: 0.5 mm; 1 mm (*N,N-*)

ZECTRAN   **REC. Z:0100**

SYNONYMS: DOWCO 139; MEXACARBATE; MEXACARBATO (Spanish); ZECTANE; ZEXTRAN

IDENTIFICATION
CAS Number: 315-18-4
Formula: $C_{12}H_{18}N_2O_2$
DOT ID Number: UN 2757; DOT Guide Number: 151
Proper Shipping Name: Carbamate pesticides, solid, toxic, n.o.s.
Reportable Quantity (RQ): **(CERCLA)** 1000 lb/454 kg

DESCRIPTION: White to tan crystals or solution. Odorless.

Combustible solid • Poison! (carbamate) • Corrosive • Heat or flame may cause explosion of dust cloud • Breathing the vapor, skin or eye contact, or swallowing the material can kill you • Firefighting gear (including SCBA) does not provide adequate protection. If exposure occurs, remove and isolate gear immediately and thoroughly decontaminate personnel • Containers may BLEVE when exposed to fire • Concentrated dust in confined areas (e.g., tanks, sewers, buildings) may explode when exposed to fire • Highly Irritating to the skin, eyes, and respiratory tract • Toxic products of combustion may include nitrogen oxides.

Hazard Classification (based on NFPA-704 Rating System)
Health Hazards (Blue): 3; Flammability (Red): 1; Reactivity (Yellow): 0

EMERGENCY RESPONSE: See Appendix A (151)
Evacuation:
Public safety: Isolate the area of spill or leak for at least 25 to 50 meters (80 to 160 feet) in all directions.
Spill: Increase, in the downwind direction, as necessary, the distance shown under "Public Safety."
Fire: If tank, rail car, or tank truck is involved in fire, isolate for at least 800 meters (½ mile) in all directions; also, consider initial evacuation for 800 meters (½ mile) in all directions.

EXPOSURE
Short-term effects: *SEEK MEDICAL ATTENTION.* An acetylcholinesterase inhibitor. **Solid or solution:** *POISONOUS IF SWALLOWED, INHALED OR IF SKIN IS EXPOSED.* Irritating to eyes. Move to fresh air. *IF BREATHING HAS STOPPED,* give artificial respiration; *avoid mouth-to-mouth resuscitation; use bag/mask apparatus.* IF breathing is difficult, administer oxygen. Remove contaminated clothing and shoes. Flush affected areas with plenty of water. *IF IN EYES,* hold eyelids open and flush. with plenty of water. *IF SWALLOWED* and victim is *CONSCIOUS AND ABLE TO SWALLOW,* have victim drink 4 to 8 ounces of water.
Note to physician or authorized medical personnel: Administer atropine, 2 mg (1/30 gr) intramuscularly or intravenously as soon as any local or systemic signs or symptoms of an intoxication are noted; repeat the administration of atropine every 3 to 8 minutes until signs of atropinization (mydriasis, dry mouth, rapid pulse, hot and dry skin) occur; initiate treatment in children with 0.05 mg/kg of atropine; repeat at 5- to 10-minute intervals. Watch respiration, and remove bronchial secretions if they appear to be obstructing the airway; intubate if necessary. Pralidoxime must be administered within minutes to a few hours following exposure (depending on the specific agent) to be effective. Give 2-PAMCI (Pralidoxime; Protopam), 2.5 g in 100 mL of sterile water or in 5% dextrose and water, intravenously, slowly, in 15 to 30 minutes; if sufficient fluid is not available, give 1 g of 2-PAMCI in 3 mL of distilled water by deep intramuscular injection*; repeat this every half hour if respiration weakens or if muscle fasciculation or convulsions recur. Also Diazepam, an anticonvulsant, might be needed.
*Note: Due to the rapid regeneration of chlolinesterase and the fact that 2-PAMCI may be contraindicated in the case of some carbamate poisonings, 2-PAMCI (Pralidoxime; Protopam) may not be needed.
Note to physician or authorized medical personnel: Medical observation is recommended for 24 to 48 hours after breathing overexposure, as pulmonary edema may be delayed. As first aid for pulmonary edema, consider administering a corticosteroid spray. Cigarette smoking may exacerbate pulmonary injury and should be discouraged for at least 72 hours following exposure.

HEALTH HAZARDS
Personal protective equipment (PPE): A-Level PPE. Natural rubber gloves.
Recommendations for respirator selection: SCBAF:PD,PP (any self-contained breathing apparatus that has a full facepiece and is operated in a pressure-demand or other positive-pressure mode); or SAF:PD,PP:ASCBA (any supplied-air respirator that has a full facepiece and is operated in a pressure-demand or other positive-pressure mode in combination with an auxiliary, self-contained breathing apparatus operated in a pressure-demand or other positive pressure mode). *ESCAPE:* GMFOVHiE [any air-purifying, full-facepiece respirator (gas mask) with a chin-style, front- or back-mounted organic vapor canister having a high-efficiency particulate filter]; or SCBAE (any appropriate escape-type, self-contained breathing apparatus).
Toxicity by ingestion: Grade 4; LD_{50} below 50 mg/kg.
Long-term health effects: May produce carcinogenic material when metabolized.

FIRE DATA
Behavior in fire: May be unstable

CHEMICAL REACTIVITY
Binary reactants: Strong oxidizers.

ENVIRONMENTAL DATA
Food chain concentration potential: Carbamates may bioconcentrate in fish.
Water pollution: DOT Appendix B, §172.101–marine pollutant. Harmful to aquatic life in very low concentrations. May be dangerous if it enters nearby water intakes; notify operators. Notify local health and wildlife officials. **Response to discharge:** Issue warning–water contaminant. Should be removed.

SHIPPING INFORMATION
Grades of purity: 93.3%; 91% (technical); **Stability during transport:** Stable.

PHYSICAL AND CHEMICAL PROPERTIES
Physical state @ 59°F/15°C and 1 atm: Solid.
Molecular weight: 222.29 (decomposes).
Melting/Freezing point: 185°F/85°C/358°K.
Relative vapor density (air = 1): 7.67
Vapor pressure: Less than 0.1 mm @ 139°C.

ZINC ACETATE   REC. Z:0200

SYNONYMS: ACETATO de ZINC (Spanish); ACETIC ACID, ZINC SALT; DICARBOMETHOXY ZINC; ZINC ACETATE DIHYDRATE; ZINC DIACETATE

IDENTIFICATION
CAS Number: 557-34-6
Formula: $C_4H_6O_4 \cdot Zn$; $Zn(C_2H_3O_2)_2$; $Zn(C_2H_3O_2)_2 \cdot 2H_2O_2$
DOT ID Number: UN 9153; DOT Guide Number: 171
Proper Shipping Name: Zinc acetate
Reportable Quantity (RQ): **(CERCLA)** 1000 lb/454 kg

DESCRIPTION: White solid. Faint vinegar or acetic acid odor. Sinks and mixes with water.

Poisonous • Toxic products of combustion may include zinc oxide.

Hazard Classification (based on NFPA-704 Rating System)
Health Hazards (Blue): 2; Flammability (Red): 0; Reactivity (Yellow): 0

EMERGENCY RESPONSE: See Appendix A (171)
Evacuation:
Public safety: Isolate the area of spill or leak for at least 10 to 25 meters (30 to 80 feet) in all directions.
Spill: Increase, in the downwind direction, as necessary, the distance shown under "Public Safety."
Fire: If any large container is involved in fire, isolate for at least 800 meters (½ mile) in all directions; also, consider initial evacuation for 800 meters (½ mile) in all directions.

EXPOSURE
Short-term effects: *SEEK MEDICAL ATTENTION.* **Dust:** Irritating to eyes, nose, and throat. *IF INHALED*, will, will cause coughing or difficult breathing. *IF IN EYES*, hold eyelids open and flush with plenty of water. *IF BREATHING HAS STOPPED,* give artificial respiration. IF breathing is difficult, administer oxygen.
Solid: Irritating to skin and eyes. *IF SWALLOWED*, will cause nausea and vomiting. Remove contaminated clothing and shoes. Flush affected areas with plenty of water. *IF IN EYES*, hold eyelids open and flush. with plenty of water. *IF SWALLOWED* and victim is *CONSCIOUS AND ABLE TO SWALLOW*, have victim drink 4 to 8 ounces of water and have victim induce vomiting. *IF SWALLOWED* and victim is *UNCONSCIOUS OR HAVING CONVULSIONS,* do nothing except keep victim warm.

HEALTH HAZARDS
Personal protective equipment (PPE): B-Level PPE. Bureau of Mines approved respirator; rubber gloves; chemical goggles
Toxicity by ingestion: Grade 2; LD_{50} = 0.5 to 5 g/kg.
Liquid or solid irritant characteristics: Causes irritation of the eyes and skin.

CHEMICAL REACTIVITY
Reaction with water: Moisture may cause hydrolysis/decomposition.
Binary reactants: Incompatible with strong acids, strong bases.

ENVIRONMENTAL DATA
Food chain concentration potential: Zinc is not considered to be bioconcentrative even though it accumulates in some organisms.
Water pollution: Harmful to aquatic life in very low concentrations. May be dangerous if it enters nearby water intakes; notify operators. Notify local health and wildlife officials. **Response to discharge:** Disperse and flush.

SHIPPING INFORMATION
Grades of purity: Reagent, 99%; Commercial, 98.4%; **Storage temperature:** Ambient; **Inert atmosphere:** None; **Venting:** Open; **Stability during transport:** Stable.

PHYSICAL AND CHEMICAL PROPERTIES
Physical state @ 59°F/15°C and 1 atm: Solid.
Molecular weight: 219.49 (decomposes).
Specific gravity (water = 1): 1.74 @ 68°F/20°C (solid).
Heat of solution: (approximate) –0.5 Btu/lb = –0.3 cal/g = –0.01 x 10^5 J/kg.

ZINC AMMONIUM CHLORIDE   REC. Z:0300

SYNONYMS: AMMONIUM PENTACHLOROZINCATE; AMMONIUM ZINC CHLORIDE; CLORURO de ZINC y AMONIO (Spanish)

IDENTIFICATION
CAS Number: 52628-25-8; 14639-97-5; 14639-98-6
Formula: $ZnCl_2 \cdot 3NH_4Cl$
DOT ID Number: UN 9154; DOT Guide Number: 171
Proper Shipping Name: Zinc ammonium chloride
Reportable Quantity (RQ): **(CERCLA)** 1000 lb/454 kg

DESCRIPTION: White solid. Odorless. Sinks and mixes with water.

Corrosive to the skin, eyes, and respiratory tract; contact with skin or eyes can cause burns and vision impairment; inhalation symptoms may be delayed • Toxic products of combustion may include ammonia, hydrogen chloride, and zinc.

Hazard Classification (based on NFPA-704 Rating System)
Health Hazards (Blue): 2; Flammability (Red): 0; Reactivity (Yellow): 0

EMERGENCY RESPONSE: See Appendix A (171)
Evacuation:
Public safety: Isolate the area of spill or leak for at least 10 to 25 meters (30 to 80 feet) in all directions.
Spill: Increase, in the downwind direction, as necessary, the distance shown under "Public Safety."
Fire: If any large container is involved in fire, isolate for at least 800 meters (½ mile) in all directions; also, consider initial evacuation for 800 meters (½ mile) in all directions.

EXPOSURE
Short-term effects: *SEEK MEDICAL ATTENTION.* **Dust:** Irritating to eyes, nose, and throat. *IF INHALED*, will, will cause coughing or difficult breathing. *IF IN EYES*, hold eyelids open and

flush. with plenty of water. *IF BREATHING HAS STOPPED,* give artificial respiration. IF breathing is difficult, administer oxygen.
Solid: Irritating to skin and eyes. *IF SWALLOWED,* will cause nausea and vomiting. Remove contaminated clothing and shoes. Flush affected areas with plenty of water. *IF IN EYES,* hold eyelids open and flush. with plenty of water. *IF SWALLOWED* and victim is *CONSCIOUS AND ABLE TO SWALLOW,* have victim drink 4 to 8 ounces of water and have victim induce vomiting. *IF SWALLOWED* and victim is *UNCONSCIOUS OR HAVING CONVULSIONS,* do nothing except keep victim warm.

HEALTH HAZARDS
Personal protective equipment (PPE): B-Level PPE. Dust mask; goggles or face shield; protective gloves.
Toxicity by ingestion: Grade 2; LD_{50} = 0.5 to 5 g/kg.
Liquid or solid irritant characteristics: Irritates the eyes, skin, and respiratory tract.

CHEMICAL REACTIVITY
Binary reactants: Strong acids.

ENVIRONMENTAL DATA
Food chain concentration potential: Zinc accumulates in some organisms but is not considered to be bioconcentrative.
Water pollution: Effect of low concentrations on aquatic life is unknown. May be dangerous if it enters nearby water intakes; notify operators. Notify local health and wildlife officials.
Response to discharge: Disperse and flush.

SHIPPING INFORMATION
Grades of purity: Commercial; **Storage temperature:** Ambient; **Inert atmosphere:** None; **Venting:** Open; **Stability during transport:** Stable.

PHYSICAL AND CHEMICAL PROPERTIES
Physical state @ 59°F/15°C and 1 atm: Solid.
Molecular weight: 296.8
Boiling point @ 1 atm: (sublimes) 644°F/340°C/613°K.
Specific gravity (water = 1): 1.81 @ 68°F/20°C (solid).

ZINC ARSENATE   **REC. Z:0400**

SYNONYMS: ARSENIATO de ZINC (Spanish); ZINC ACID, ZINC SALT

IDENTIFICATION
CAS Number: 1303-39-5
Formula: $As_4O_{15} \cdot 5Zn$
DOT ID Number: UN 1712; DOT Guide Number: 151
Proper Shipping Name: Zinc arsenate; Zinc arsenate and Zinc arsenite mixtures

DESCRIPTION: White solid. Odorless. Sinks in water.

Poison! • Breathing the dust, skin or eye contact, or swallowing the material can cause permanent injury, and possible death • Firefighting gear (including SCBA) does not provide adequate protection. If exposure occurs, remove and isolate gear immediately and thoroughly decontaminate personnel Toxic products of combustion may include arsenic and zinc oxide.

Hazard Classification (based on NFPA-704 Rating System)
Health Hazards (Blue): 3; Flammability (Red): 0; Reactivity (Yellow): 0

EMERGENCY RESPONSE: See Appendix A (151)
Evacuation:
Public safety: Isolate the area of spill or leak for at least 25 to 50 meters (80 to 160 feet) in all directions.
Spill: Increase, in the downwind direction, as necessary, the distance shown under "Public Safety."
Fire: If tank, rail car, or tank truck is involved in fire, isolate for at least 800 meters (½ mile) in all directions; also, consider initial evacuation for 800 meters (½ mile) in all directions.

EXPOSURE
Short-term effects: *SEEK MEDICAL ATTENTION.* **Dust:** *POISONOUS IF INHALED.* Irritating to eyes, nose, and throat. Move victim to fresh air. *IF IN EYES,* hold eyelids open and flush. with plenty of water. *IF BREATHING HAS STOPPED,* give artificial respiration; *avoid mouth-to-mouth resuscitation; use bag/mask apparatus.* If breathing is difficult, administer oxygen.
Solid: *POISONOUS IF SWALLOWED.* Irritating to skin and eyes. Remove contaminated clothing and shoes. Flush affected areas with plenty of water. *IF IN EYES,* hold eyelids open and flush. with plenty of water. *IF SWALLOWED* and victim is *CONSCIOUS AND ABLE TO SWALLOW,* have victim drink 4 to 8 ounces of water and have victim induce vomiting gastric lavage, and catharsis. *IF SWALLOWED* and victim is *UNCONSCIOUS OR HAVING CONVULSIONS,* do nothing except keep victim warm. *Note to physician or authorized medical personnel*: Consider prompt administration of BAL.

HEALTH HAZARDS
Personal protective equipment (PPE): B-Level PPE. OSHA Table Z-1-A air contaminant. Dust mask; rubber **gloves.**
Recommendations for respirator selection: NIOSH
At any concentrations above the NIOSH REL, or where there is no REL, at any detectable concentration: SCBAF:PD,PP (any self-contained breathing apparatus that has a full faceplate and is operated in a pressure-demand or other positive-pressure mode); or SAF:PD,PP:ASCBA (any supplied-air respirator that has a full facepiece and is operated in a pressure-demand or other positive-pressure mode in combination with an auxiliary self-contained breathing apparatus operated in a pressure-demand or other positive-pressure mode). *ESCAPE:* GMFAGHiE [any air-purifying, full-facepiece respirator (gas mask) with a chin-style, front-or back-mounted acid gas canister having a high-efficiency particulate filter]; or SCBAE (any appropriate escape-type, self-contained breathing apparatus).
Exposure limits (TWA unless otherwise noted): ACGIH TLV 0.01 mg/m³; OSHA PEL [1910.1080] 0.010 mg/m³; NIOSH REL ceiling 0.002 mg/m³/15 min, as arsenic; potential human carcinogen; reduce exposure to lowest feasible level
Toxicity by ingestion: Poisonous.
Long-term health effects: May be carcinogenic. Arsenic poisoning may develop.
Liquid or solid irritant characteristics: May cause eye irritation.
IDLH value: Potential human carcinogen 5 mg/m³ as arsenic.

ENVIRONMENTAL DATA
Water pollution: Effect of low concentrations on aquatic life is unknown. May be dangerous if it enters nearby water intakes; notify operators. Notify local health and wildlife officials. **Response to discharge:** Issue warning–poison, water contaminant. Restrict access. Should be removed. Chemical and physical treatment.

SHIPPING INFORMATION
Grades of purity: Technical; **Storage temperature:** Ambient;

Inert atmosphere: None; **Venting:** Open; **Stability during transport:** Stable.

PHYSICAL AND CHEMICAL PROPERTIES
Physical state @ 59°F/15°C and 1 atm: Solid.
Molecular weight: 866 (approximate).
Boiling point @ 1 atm: (decomposes).
Specific gravity (water = 1): 3.31 @ 15°C (solid).

ZINC BICHROMATE   REC. Z:0500

SYNONYMS: ZINC DICHROMATE; DICROMATO de ZINC (Spanish)

IDENTIFICATION
CAS Number: 14018-95-2
Formula: $Cr_2O_7 \cdot Zn$; $ZnCr_2O_7 \cdot 3H_2O$
DOT ID Number: UN 1479; DOT Guide Number: 140
Proper Shipping Name: Oxidizing solid, n.o.s.

DESCRIPTION: Yellow-orange crystalline powder. Reddish-brown or orange-yellow. Mixes with water.

Poison! • Corrosive • Breathing the dust, or swallowing the material can cause permanent injury; Extremely Irritating to the skin, eyes, and respiratory tract • Toxic products of combustion may include chromium and zinc • Strong oxidizer which may react spontaneously with low flash point organics or reducing agents. Heat forms oxygen; will increase the activity of an existing fire • May cause fire or explosion on contact with combustibles (wood, paper, oil, clothing, etc.).

Hazard Classification (based on NFPA-704 Rating System)
Health Hazards (Blue): 1; Flammability (Red): 0; Reactivity (Yellow): 1; Special Notice (White): OXY

EMERGENCY RESPONSE: See Appendix A (140)
Evacuation:
Public safety: Isolate the area of spill or leak for at least 10 to 25 meters (30 to 80 feet) in all directions.
Spill: Consider initial downwind evacuation for at least 100 meters (330 feet).
Fire: If any large container is involved in fire, isolate for at least 800 meters (½ mile) in all directions; also, consider initial evacuation for 800 meters (½ mile) in all directions.

EXPOSURE
Short-term effects: *SEEK MEDICAL ATTENTION.* **Dust:** Irritating to eyes, nose, and throat. Harmful if inhaled. Move to fresh air. *IF BREATHING HAS STOPPED,* give artificial respiration. IF breathing is difficult, administer oxygen. **Solid:** Will burn skin and eyes. *IF SWALLOWED* can cause dizziness, nausea, convulsions, and coma. Remove contaminated clothing and shoes. Flush affected areas with plenty of water. *IF IN EYES,* hold eyelids open and flush. with plenty of water. *IF SWALLOWED* and victim is *CONSCIOUS AND ABLE TO SWALLOW,* have victim drink 4 to 8 ounces of water.
Note to physician or authorized medical personnel: Medical observation is recommended for 24 to 48 hours after breathing overexposure, as pulmonary edema may be delayed. As first aid for pulmonary edema, consider administering a corticosteroid spray. Cigarette smoking may exacerbate pulmonary injury and should be discouraged for at least 72 hours following exposure.

HEALTH HAZARDS
Personal protective equipment (PPE): B-Level PPE. Rubber gloves, face shield or goggles, approved dust mask.
Recommendations for respirator selection: NIOSH
At any concentrations above the NIOSH REL, or where there is no REL, at any detectable concentration: SCBAF:PD,PP (any self-contained breathing apparatus that has a full facepiece and is operated in a pressure-demand or other positive-pressure mode); or SAF:PD,PP:ASCBA (any supplied-air respirator that has a full facepiece and is operated in a pressure-demand or other positive-pressure mode in combination with an auxiliary self-contained breathing apparatus operated in a pressure-demand or other positive pressure mode). *ESCAPE:* HiEF (any air-purifying, full-facepiece respirator with a high-efficiency particulate filter); or SCBAE (any appropriate escape-type, self-contained breathing apparatus).
Exposure limits (TWA unless otherwise noted): ACGIH TLV 0.01 mg/m^3 as zinc chromates; NIOSH 0.001 mg/m^3 as chromates; potential human carcinogen; reduce exposure to lowest feasible level; confirmed human carcinogen
Short-term exposure limits (15-minute TWA): OSHA ceiling 0.1 mg/m^3 as chromates.
Toxicity by ingestion: Grade 3; LD_{50} = 50 to 500 mg/kg.
Long-term health effects: OSHA specifically regulated carcinogen; IARC sufficient animal evidence
Liquid or solid irritant characteristics: Causes smarting of the skin and first-degree burns on short-exposure; may cause second-degree burns on long exposure.
IDLH value: 15 mg/m^3 as chromium(VI).

CHEMICAL REACTIVITY
Binary reactants: Oxidizer. Violent reaction with reducing agents. Keep away from all combustible materials.

ENVIRONMENTAL DATA
Food chain concentration potential: Rainbow trout can accumulate chromium from water containing as little as 1 μgm/L. Zn can accumulate in some organisms.
Water pollution: Harmful to aquatic life in very low concentrations. May be dangerous if it enters nearby water intakes; notify operators. Notify local health and wildlife officials. **Response to discharge:** Issue warning–water contaminant, oxidizing material. Chemical and physical treatment. Disperse and flush.

SHIPPING INFORMATION
Stability during transport: Stable but hygroscopic.

PHYSICAL AND CHEMICAL PROPERTIES
Physical state @ 59°F/15°C and 1 atm: Solid.
Molecular weight: 335.45
Relative vapor density (air = 1): 11.57 (calculated).

ZINC BORATE   REC. Z:0600

SYNONYMS: BORATO de ZINC (Spanish); BORAX 2335; BORIC ACID, ZINC SALT; FIREBRAKE ZB; ZB 112; ZB 237; ZN 100

IDENTIFICATION
CAS Number: 1332-07-6
Formula: (approximate)
Formula: $2ZnO \cdot 3B_2O_3 \cdot 3.5H_2O$
DOT ID Number: UN 3077; DOT Guide Number: 171

1152   Zinc bromide

Proper Shipping Name: Environmentally hazardous substance, solid, n.o.s.
Reportable Quantity (RQ): **(CERCLA)** 1000 lb/454 kg

DESCRIPTION: White amorphous powder. Odorless. Sinks in water; slightly soluble.

Poison! • Breathing the dust, or swallowing the material can kill you; extremely Irritating to the skin, eyes, and respiratory tract • Toxic products of combustion may include zinc and boron.

Hazard Classification (based on NFPA-704 Rating System)
Health Hazards (Blue): 2; Flammability (Red): 0; Reactivity (Yellow): 0

EMERGENCY RESPONSE: See Appendix A (171)
Evacuation:
Public safety: Isolate the area of spill or leak for at least 10 to 25 meters (30 to 80 feet) in all directions.
Spill: Increase, in the downwind direction, as necessary, the distance shown under "Public Safety."
Fire: If any large container is involved in fire, isolate for at least 800 meters (½ mile) in all directions; also, consider initial evacuation for 800 meters (½ mile) in all directions.

EXPOSURE
Short-term effects: *SEEK MEDICAL ATTENTION.* **Dust:** Irritating to eyes, nose, and throat. *IF INHALED*, will, will cause coughing or difficult breathing. *IF IN EYES*, hold eyelids open and flush. with plenty of water. *IF BREATHING HAS STOPPED*, give artificial respiration. IF breathing is difficult, administer oxygen. **Solid:** Irritating to skin and eyes. *IF SWALLOWED*, will cause nausea and vomiting. Remove contaminated clothing and shoes. Flush affected areas with plenty of water. *IF IN EYES*, hold eyelids open and flush. with plenty of water. *IF SWALLOWED* and victim is *CONSCIOUS AND ABLE TO SWALLOW*, have victim drink 4 to 8 ounces of water and have victim induce vomiting. *IF SWALLOWED* and victim is *UNCONSCIOUS OR HAVING CONVULSIONS*, do nothing except keep victim warm.

HEALTH HAZARDS
Personal protective equipment (PPE): B-Level PPE. Dust mask; goggles or face shield; protective gloves.
Recommendations for respirator selection: NIOSH/OSHA as boric oxide
50 mg/m³: DM (any dust and mist respirator). *100 mg/m³*: DMXSQ, *If not present as a fume* (any dust and mist respirator except single-use and quarter mask respirators); or SA (any supplied-air respirator). *250 mg/m³*: SA:CF (any supplied-air respirator operated in a continuous-flow mode); or PAPRDM, *If not present as a fume* (any powered, air-purifying respirator with a dust and mist filter); or
500 mg/m³: HiEF (any air-purifying, full-facepiece respirator with a high-efficiency particulate filter); or PAPRTHiE (any powered, air-purifying respirator with a tight-fitting facepiece and a high-efficiency particulate filter); or SCBAF (any self-contained breathing apparatus with a full facepiece); or SAF (any supplied-air respirator with a full facepiece). *2000 mg/m³*: SAF:PD,PP (any supplied-air respirator that has a full facepiece and is operated in a pressure-demand or other positive-pressure mode). *EMERGENCY OR PLANNED ENTRY INTO UNKNOWN CONCENTRATIONS OR IDLH CONDITIONS*: SCBAF:PD,PP (any self-contained breathing apparatus that has a full facepiece and is operated in a pressure-demand or other positive-pressure mode); or SAF:PD,PP:ASCBA (any supplied-air respirator that has a full facepiece and is operated in a pressure-demand or other positive-pressure mode in combination with an auxiliary self-contained breathing apparatus operated in a pressure-demand or other positive pressure mode). *ESCAPE:* HiEF (any air-purifying, full-facepiece respirator with a high-efficiency particulate filter); or SCBAE (any appropriate escape-type, self-contained breathing apparatus). *Note*: Substance reported to cause eye irritation or damage; may require eye protection.
Exposure limits (TWA unless otherwise noted): ACGIH TLV 10 mg/m³ as boron oxide; OSHA PEL 15 mg/m³; NIOSH 10 mg/m³ as boric oxide
Toxicity by ingestion: Grade 1; LD_{50} = 5 to15 g/kg.
Liquid or solid irritant characteristics: Causes irritation of the eyes and skin.

CHEMICAL REACTIVITY
Binary reactants: Strong oxidizers, fluorine.

ENVIRONMENTAL DATA
Food chain concentration potential: Zinc is concentrated by some organisms but is not considered to be bioconcentrative in a spill situation.
Water pollution: Effect of low concentrations on aquatic life is unknown. May be dangerous if it enters nearby water intakes; notify operators. Notify local health and wildlife officials. **Response to discharge:** Should be removed. Chemical and physical treatment.

SHIPPING INFORMATION
Grades of purity: Commercial grades contain 45–52% zinc oxide, 29–35% boric anhydride; **Storage temperature:** Ambient; **Inert atmosphere:** None; **Venting:** Open; **Stability during transport:** Stable.

PHYSICAL AND CHEMICAL PROPERTIES
Physical state @ 59°F/15°C and 1 atm: Solid.
Molecular weight: 434.75
Boiling point @ 1 atm: (decomposes).
Specific gravity (water = 1): 2.7 @ 68°F/20°C (solid).

ZINC BROMIDE REC. Z:0700

SYNONYMS: BROMURO de ZINC (Spanish); ZINC DIBROMIDE

IDENTIFICATION
CAS Number: 7699-45-8
Formula: Br_2Zn; $ZnBr_2$
DOT ID Number: UN 9156; DOT Guide Number: 171
Proper Shipping Name: Zinc bromide
Reportable Quantity (RQ): **(CERCLA)** 1000 lb/454 kg

DESCRIPTION: White solid. Odorless. Sinks and mixes with water form an acid solution.

Corrosive • Breathing the dust can kill you; skin or eye contact causes severe burns, impaired vision, or blindness; effects of inhalation may be delayed • Toxic products of combustion may include zinc oxide and bromine.

Hazard Classification (based on NFPA-704 Rating System)
Health Hazards (Blue): 1; Flammability (Red): 0; Reactivity (Yellow): 0

EMERGENCY RESPONSE: See Appendix A (171)
Evacuation:
Public safety: Isolate the area of spill or leak for at least 10 to 25 meters (30 to 80 feet) in all directions.
Spill: Increase, in the downwind direction, as necessary, the distance shown under "Public Safety."
Fire: If any large container is involved in fire, isolate for at least 800 meters (½ mile) in all directions; also, consider initial evacuation for 800 meters (½ mile) in all directions.

EXPOSURE
Short-term effects: *SEEK MEDICAL ATTENTION.* **Dust:** Irritating to eyes, nose, and throat. *IF INHALED*, will, will cause coughing or difficult breathing. *IF IN EYES*, hold eyelids open and flush. with plenty of water. *IF BREATHING HAS STOPPED,* give artificial respiration. IF breathing is difficult, administer oxygen. **Solid:** Irritating to skin and eyes. *IF SWALLOWED*, will cause nausea and vomiting. Remove contaminated clothing and shoes. Flush affected areas with plenty of water. *IF IN EYES*, hold eyelids open and flush with plenty of water. *IF SWALLOWED* and victim is *CONSCIOUS AND ABLE TO SWALLOW*, have victim drink 4 to 8 ounces of water and have victim induce vomiting. *IF SWALLOWED* and victim is *UNCONSCIOUS OR HAVING CONVULSIONS,* do nothing except keep victim warm.
Note to physician or authorized medical personnel: Medical observation is recommended for 24 to 48 hours after breathing overexposure, as pulmonary edema may be delayed. As first aid for pulmonary edema, consider administering a corticosteroid spray. Cigarette smoking may exacerbate pulmonary injury and should be discouraged for at least 72 hours following exposure.

HEALTH HAZARDS
Personal protective equipment (PPE): B-Level PPE. Chemical goggles or face shield; rubber gloves; dust mask
Toxicity by ingestion: Grade 2; LD_{50} = 0.5 to 5 g/kg.
Liquid or solid irritant characteristics: Causes irritation of the eyes and skin.
Long-term health effects: Prolonged contact or inhalation of inorganic bromides causes acne-like bromoderma (bromide rash), especially of the hands and face. Other effects include emaciation, depression; and, in severe cases, psychosis and mental deterioration.

CHEMICAL REACTIVITY
Binary reactants: Violent reaction with alkali metals including metallic sodium, potassium. Store above 32°F/0°C.

ENVIRONMENTAL DATA
Food chain concentration potential: Zinc is concentrated by some organisms but is not considered to be bioconcentrative in a spill situation.
Water pollution: DOT Appendix B, §172.101–marine pollutant. Effect of low concentrations on aquatic life is unknown. May be dangerous if it enters nearby water intakes; notify operators. Notify local health and wildlife officials. **Response to discharge:** Issue warning–water contaminant. Disperse and flush.

SHIPPING INFORMATION
Grades of purity: Technical, 98+%; **Storage temperature:** Ambient, best above 32°F/0°C.; **Inert atmosphere:** None; **Venting:** Open; **Stability during transport:** Stable.

PHYSICAL AND CHEMICAL PROPERTIES
Physical state @ 59°F/15°C and 1 atm: Solid.
Molecular weight: 225.18
Boiling point @ 1 atm: Decomposes.
Specific gravity (water = 1): 4.22 @ 68°F/20°C (solid).

ZINC CARBONATE   REC. Z:0800

SYNONYMS: CALAMINE; CARBONATO de ZINC (Spanish); CARBONIC ACID, ZINC SALT; SMITHSONITE; ZINC CARBONATE

IDENTIFICATION
CAS Number: 3486-35-9
Formula: $ZnCO_3$
DOT ID Number: UN 9157; DOT Guide Number: 171
Proper Shipping Name: Zinc carbonate
Reportable Quantity (RQ): **(CERCLA)** 1000 lb/454 kg

DESCRIPTION: White crystalline powder. Odorless. Sinks in water; practically insoluble.

Corrosive • Skin or eye contact causes severe burns, impaired vision, or blindness; symptoms of inhalation exposure may be delayed • Toxic products of combustion may include carbon monoxide and zinc oxide.

Hazard Classification (based on NFPA-704 Rating System)
Health Hazards (Blue): 1; Flammability (Red): 0; Reactivity (Yellow): 0

EMERGENCY RESPONSE: See Appendix A (171)
Evacuation:
Public safety: Isolate the area of spill or leak for at least 10 to 25 meters (30 to 80 feet) in all directions.
Spill: Increase, in the downwind direction, as necessary, the distance shown under "Public Safety."
Fire: If any large container is involved in fire, isolate for at least 800 meters (½ mile) in all directions; also, consider initial evacuation for 800 meters (½ mile) in all directions.

EXPOSURE
Short-term effects: *SEEK MEDICAL ATTENTION.* **Solid:** If swallowed, may cause nausea and vomiting. *IF SWALLOWED* and victim is *CONSCIOUS AND ABLE TO SWALLOW*, have victim drink 4 to 8 ounces of water or milk and remove by gastric lavage. To relieve irritation administer demulcents.
Note to physician or authorized medical personnel: Medical observation is recommended for 24 to 48 hours after breathing overexposure, as pulmonary edema may be delayed. As first aid for pulmonary edema, consider administering a corticosteroid spray. Cigarette smoking may exacerbate pulmonary injury and should be discouraged for at least 72 hours following exposure.

HEALTH HAZARDS
Personal protective equipment (PPE): B-Level PPE. Chemical goggles or face shield; rubber gloves; dust mask
Liquid or solid irritant characteristics: Inhalation: Dust or fumes may cause dry throat, cough and chest discomfort. Fever and sweating. **Ingestion:** May cause nausea and vomiting. **Skin:** Astringent.

ENVIRONMENTAL DATA
Food chain concentration potential: Zinc is concentrated by some organisms but is not considered to be bioconcentrative in a spill situation.

Water pollution: Harmful to aquatic life in very low concentrations. May be dangerous if it enters nearby water intakes; notify operators. Notify local health and wildlife officials.
Response to discharge: Should be removed. Chemical and physical treatment.

SHIPPING INFORMATION
Grades of purity: 70% (as ZnO); **Storage temperature:** Ambient; **Stability during transport:** Stable.

PHYSICAL AND CHEMICAL PROPERTIES
Physical state @ 59°F/15°C and 1 atm: Solid.
Molecular weight: 125.4
Melting/Freezing point: Loses CO_2 @ 572°F/300°C/573°K.
Specific gravity (water = 1): 4.398 @ room temperature

ZINC CHLORIDE   REC. Z:0900

SYNONYMS: BUTTER OF ZINC; CHLORURE de ZINC (French); CLORURO de ZINC (Spanish); EEC No. 030-003-00-2; TINNING GLUX; ZINC BUTTER; ZINC CHLORIDE FUME; ZINC (CHLORURE de) (French); ZINC MURIATE SOLUTION; ZINC DICHLORIDE; ZINKCHLORID (German)

IDENTIFICATION
CAS Number: 7646-85-7
Formula: $ZnCl_2$
DOT ID Number: UN2331 (anhydrous); UN 1840 (solution); DOT Guide Number: 154
Proper Shipping Name: Zinc chloride (anhydrous); Zinc chloride, solution
Reportable Quantity (RQ): **(CERCLA)** 1000 lb/454 kg

DESCRIPTION: White crystalline solid. Odorless. Solid sinks and mixes with water forming a strong acid.

Corrosive to skin, eyes, and respiratory tract; skin or eye contact causes severe burns, impaired vision, or blindness; effects of inhalation may be delayed • Firefighting gear (including SCBA) does not provide adequate protection. If exposure occurs, remove and isolate gear immediately and thoroughly decontaminate personnel •Toxic products of combustion may include zinc oxide and hydrogen chloride.

Hazard Classification (based on NFPA-704 Rating System)
Health Hazards (Blue): 1; Flammability (Red): 0; Reactivity (Yellow): 0

EMERGENCY RESPONSE: See Appendix A (154)
Evacuation:
Public safety: Isolate the area of spill or leak for at least 25 to 50 meters (80 to 160 feet) in all directions.
Spill: Increase, in the downwind direction, as necessary, the distance shown under "Public Safety."
Fire: If tank, rail car, or tank truck is involved in fire, isolate for at least 800 meters (½ mile) in all directions; also, consider initial evacuation for 800 meters (½ mile) in all directions.

EXPOSURE
Short-term effects: *SEEK MEDICAL ATTENTION.* **Solid or solution:**
Irritating to skin and eyes. *IF SWALLOWED*, will cause nausea or vomiting. Flush affected areas with plenty of water. *IF BREATHING HAS STOPPED,* give artificial respiration; *avoid mouth-to-mouth resuscitation; use bag/mask apparatus. IF IN EYES*, hold eyelids open and flush. with plenty of water. *IF SWALLOWED* and victim is *CONSCIOUS AND ABLE TO SWALLOW*, have victim drink 4 to 8 ounces of water and have victim induce vomiting. *IF SWALLOWED* and victim is *UNCONSCIOUS OR HAVING CONVULSIONS*, do nothing except keep victim warm.
Note to physician or authorized medical personnel: Medical observation is recommended for 24 to 48 hours after breathing overexposure, as pulmonary edema may be delayed. As first aid for pulmonary edema, consider administering a corticosteroid spray. Cigarette smoking may exacerbate pulmonary injury and should be discouraged for at least 72 hours following exposure.

HEALTH HAZARDS
Personal protective equipment (PPE): B-Level PPE. OSHA Table Z-1-A air contaminant. Goggles or face shield.
Recommendations for respirator selection: NIOSH/OSHA fume *10 mg/m³:* DMFu* (any dust, mist and fume respirator); or SA* (any supplied-air respirator). *25 mg/m³:* SA:CF* (any supplied-air respirator operated in a continuous-flow mode); or PAPRDMFu* (any powered, air-purifying respirator with a dust, mist, and fume filter). *50 mg/m³:* HiEF (any air-purifying, full-facepiece respirator with a high-efficiency particulate filter); or PAPRHiE* (any powered, air-purifying respirator with a high-efficiency particulate filter); or SCBAF (any self-contained breathing apparatus with a full facepiece); or SAF (any supplied-air respirator with a full facepiece). *EMERGENCY OR PLANNED ENTRY INTO UNKNOWN CONCENTRATIONS OR IDLH CONDITIONS:* SCBAF:PD,PP (any self-contained breathing apparatus that has a full facepiece and is operated in a pressure-demand or other positive-pressure mode); or SAF:PD,PP:ASCBA (any supplied-air respirator that has a full facepiece and is operated in a pressure-demand or other positive-pressure mode in combination with an auxiliary self-contained breathing apparatus operated in a pressure-demand or other positive pressure mode). *ESCAPE:* HiEF (any air-purifying, full-facepiece respirator with a high-efficiency particulate filter); or SCBAE (any appropriate escape-type, self-contained breathing apparatus). **Note*: Substance reported to cause eye irritation or damage; may require eye protection.
Exposure limits (TWA unless otherwise noted): ACGIH TLV 1 mg/m³ (fume); NIOSH/OSHA PEL 1 mg/m³ (fume).
Short-term exposure limits (15-minute TWA): ACGIH STEL 2 mg/m³; NIOSH STEL 2 mg/m³.
Toxicity by ingestion: Grade 3; LD_{50} = 50 to 500 mg/kg.
Vapor (gas) irritant characteristics: Nonvolatile
Liquid or solid irritant characteristics: Corrosive to eyes; irritates skin.
IDLH value: 50 mg/m³ (fume).

CHEMICAL REACTIVITY
Reactivity with water: Reacts with water forming an acidic solution (pH about 4); zinc oxychloride may be formed with large amounts of water.
Binary reactants: Violent reaction with strong bases, potassium. Attacks metals as fume or in the presence of moisture.

ENVIRONMENTAL DATA
Food chain concentration potential: Zinc is concentrated by some organisms but is not considered to be bioconcentrative in a spill situation.
Water pollution: Harmful to aquatic life in very low concentrations. May be dangerous if it enters nearby water intakes; notify operators. Notify local health and wildlife officials.

Response to discharge: Issue warning–water contaminant. Disperse and flush.

SHIPPING INFORMATION
Grades of purity: Reagent; USP; technical; 50% solution in water; **Storage temperature:** Ambient; **Inert atmosphere:** None

PHYSICAL AND CHEMICAL PROPERTIES
Physical state @ 59°F/15°C and 1 atm: Solid.
Molecular weight: 136.28
Boiling point @ 1 atm: 1346°F/730°C/1003°K.
Melting/Freezing point: 541°F/283°C/556°K.
Specific gravity (water = 1): 2.91 @ 77°F/25°C (solid).
Heat of fusion: 40.6 cal/g.
Vapor pressure: 1 mm.

ZINC CHROMATE REC. Z:1000

SYNONYMS: BASIC ZINC CHROMATE; BASIC ZINC CHROMATE X-2259; CROMATO de ZINC (Spanish); BUTTERCUP YELLOW; CHROMIC ACID, ZINC SALT; CHROMIUM ZINC OXIDE; C.I. 77955; C.I. PIGMENT YELLOW 36; CITRON YELLOW; C.P. ZINC YELLOW X-883; PRIMROSE YELLOW; PURE ZINC CHROME; PURE ZINC YELLOW; ZINC CHROMATE C; ZINC CHROMATE O; ZINC CHROMATE T; ZINC CHROMATE(VI) HYDROXIDE; ZINC CHROMATE Z; ZINC CHROME; ZINC CHROME YELLOW; ZINC CHROMIUM OXIDE; ZINC HYDROXYCHROMATE; ZINC TETRAOXYCHROMATE; ZINC TETRAOXYCHROMATE 76A; ZINC TETRAOXYCHROMATE 780B; ZINC YELLOW; ZINC YELLOW 1; ZINC YELLOW 1425; ZINC YELLOW 40-9015; ZINC YELLOW AZ-16; ZINC YELLOW AZ-18; ZINC YELLOW KSH; ZINC YELLOW 386N; ZINC YELLOW

IDENTIFICATION
CAS Number: 13530-65-9; 12018-19-8
Formula: CrH_2O_4Zn; Cr_2O_4Zn; $Cr_2O_7 \cdot Zn$; $CrH_2O_4 \cdot Zn$
DOT ID Number: UN 3077; DOT Guide Number: 171
Proper Shipping Name: Environmentally hazardous substances, solid, n.o.s.

DESCRIPTION: Yellow crystalline solid or powder. Odorless. Sinks in water; soluble.

Poison! • Breathing the dust can cause permanent injury, and possible death • Firefighting gear (including SCBA) does not provide adequate protection. If exposure occurs, remove and isolate gear immediately and thoroughly decontaminate personnel • Strong oxidizer which may react spontaneously with low flash point organics or reducing agents. Heat forms oxygen; will increase the activity of an existing fire • May cause fire or explosion contact with combustibles (wood, paper, oil, clothing, etc.) • Toxic products of combustion may include zinc oxides, chromic acid, and chromates.

Hazard Classification (based on NFPA-704 Rating System)
Health Hazards (Blue): 2; Flammability (Red): 0; Reactivity (Yellow): 0; Special Notice (White): OXY

EMERGENCY RESPONSE: See Appendix A (171)
Evacuation:
Public safety: Isolate the area of spill or leak for at least 10 to 25 meters (30 to 80 feet) in all directions.
Spill: Increase, in the downwind direction, as necessary, the distance shown under "Public Safety."
Fire: If any large container is involved in fire, isolate for at least 800 meters (½ mile) in all directions; also, consider initial evacuation for 800 meters (½ mile) in all directions.

EXPOSURE
Short-term effects: *SEEK MEDICAL ATTENTION*. **Dust:** Irritating to eyes, nose, and throat. *IF INHALED*, will, will cause coughing or difficult breathing. *IF IN EYES*, hold eyelids open and flush. with plenty of water. *IF BREATHING HAS STOPPED*, give artificial respiration: *avoid mouth-to-mouth resuscitation; use bag/mask apparatus*. IF breathing is difficult, administer oxygen. **Solid:** Irritating to skin and eyes. *IF SWALLOWED*, will cause nausea and vomiting. Remove contaminated clothing and shoes. Flush affected areas with plenty of water. *IF IN EYES*, hold eyelids open and flush with plenty of water. *IF SWALLOWED* and victim is *CONSCIOUS AND ABLE TO SWALLOW*, have victim drink 4 to 8 ounces of water and have victim induce vomiting. *IF SWALLOWED* and victim is *UNCONSCIOUS OR HAVING CONVULSIONS*, do nothing except keep victim warm.

HEALTH HAZARDS
Personal protective equipment (PPE): B-Level PPE. OSHA Table Z-1-A air contaminant. Rubber gloves; chemical goggles or face shield.
Recommendations for respirator selection: NIOSH
At any concentrations above the NIOSH REL, or where there is no REL, at any detectable concentration: SCBAF:PD,PP (any self-contained breathing apparatus that has a full facepiece and is operated in a pressure-demand or other positive-pressure mode); or SAF:PD,PP:ASCBA (any supplied-air respirator that has a full facepiece and is operated in a pressure-demand or other positive-pressure mode in combination with an auxiliary self-contained breathing apparatus operated in a pressure-demand or other positive pressure mode). *ESCAPE:* HiEF (any air-purifying, full-facepiece respirator with a high-efficiency particulate filter); or SCBAE (any appropriate escape-type, self-contained breathing apparatus).
Exposure limits (TWA unless otherwise noted): ACGIH TLV 0.01 mg/m^3; OSHA PEL ceiling 0.1 mg/m^3; NIOSH 0.025 mg/m^3; ceiling 0.05 mg/m^3/15 minutes; confirmed human carcinogen
Toxicity by ingestion: Grade 2; LD_{50} = 0.5 to 5 g/kg.
Long-term health effects: May cause lung tumors and cancer. NTP known carcinogen
Liquid or solid irritant characteristics: Irritates eyes and skin.
IDLH value: 15 mg/m^3 as chromium(VI).

ENVIRONMENTAL DATA
Food chain concentration potential: Both chromium and zinc are concentrated by some organisms but are not considered to be bioconcentrative in a spill situation.
Water pollution: Effect of low concentrations on aquatic life is unknown. May be dangerous if it enters nearby water intakes; notify operators. Notify local health and wildlife officials. **Response to discharge:** Should be removed. Chemical and physical treatment.

SHIPPING INFORMATION
Grades of purity: Commercial, 100%; **Storage temperature:** Ambient; **Inert atmosphere:** None; **Venting:** Open; **Stability during transport:** Stable.

PHYSICAL AND CHEMICAL PROPERTIES
Physical state @ 59°F/15°C and 1 atm: Solid.
Molecular weight: 874 (approximate).

Boiling point @ 1 atm: Decomposes.
Specific gravity (water = 1): 3.43 @ 68°F/20°C (solid).

ZINC CYANIDE   REC. Z:1100

SYNONYMS: CIANURO de ZINC (Spanish); CYANURE de ZINC (French); ZINC DICYANIDE; CYANIDE OF ZINC; RCRA P121

IDENTIFICATION
CAS Number: 557-21-1
Formula: $Zn(CN)_2$
DOT ID Number: UN 1713; DOT Guide Number: 151
Proper Shipping Name: Zinc cyanide
Reportable Quantity (RQ): **(CERCLA)** 10 lb/4.54 kg

DESCRIPTION: Colorless crystalline solid or white powder. Slight bitter almond odor. Sinks in water; insoluble.

Poison! • Breathing the dust can cause permanent injury and possible death • Firefighting gear (including SCBA) does not provide adequate protection. If exposure occurs, remove and isolate gear immediately and thoroughly decontaminate personnel Toxic products of combustion may include cyanide, zinc oxide, and nitrogen oxides • Do not put yourself in danger by entering a contaminated area to rescue a victim.

Hazard Classification (based on NFPA-704 Rating System)
Health Hazards (Blue): 3; Flammability (Red): 0; Reactivity (Yellow): 0

EMERGENCY RESPONSE: See Appendix A (151)
Evacuation:
Public safety: Isolate the area of spill or leak for at least 25 to 50 meters (80 to 160 feet) in all directions.
Spill: Increase, in the downwind direction, as necessary, the distance shown under "Public Safety."
Fire: If tank, rail car, or tank truck is involved in fire, isolate for at least 800 meters (½ mile) in all directions; also, consider initial evacuation for 800 meters (½ mile) in all directions.

EXPOSURE
Short-term effects: *SEEK MEDICAL ATTENTION.* **Dust:** *POISONOUS IF INHALED OR IF SKIN IS EXPOSED.* Move to fresh air. *IF BREATHING HAS STOPPED,* give artificial respiration; *avoid mouth-to-mouth resuscitation; use bag/mask apparatus.* IF breathing is difficult, administer oxygen. **Solid:** *POISONOUS IF SWALLOWED OR IF SKIN IS EXPOSED.* Irritating to eyes. Remove contaminated clothing and shoes. Flush affected areas with plenty of water. *IF IN EYES*, hold eyelids open and flush with plenty of water. *IF SWALLOWED* and victim is *CONSCIOUS AND ABLE TO SWALLOW*, have victim drink 4 to 8 ounces of water and have victim induce vomiting. *IF SWALLOWED* and victim is *UNCONSCIOUS OR HAVING CONVULSIONS,* do nothing except keep victim warm.
Note to physician or authorized medical personnel: Consider the use of amyl nitrite perles if symptoms of cyanide poisoning develop. If symptoms indicate, initial treatment includes the cyanide antidote kit. In all cases, break an amyl nitrite perle in a gauze pad and hold lightly under victim's nose for 15 seconds, repeating 5 times at about 15-second intervals; if necessary (and if sodium nitrite infusions will be delayed), repeat procedure every 3 minutes with fresh pearls until 3 or 4 have been used. Avoid breathing the vapor while administering it to the victim. Administer sodium nitrite IV, ASAP. The usual adult dose is 10 to 20 mL of a 3% solution infused over no less than 5 minutes; the average child dose is 0.15 to 0.20 mL/kg. Monitor blood pressure during administration, and slow the rate of infusion if hypotention develops. Next, infuse sodium thiosulfate IV. The usual adult dose is 50 mL of a 25% solution infused over 10 to 20 minutes; the average child dose is 1.65 mL/kg. Repeat with nitrite and thiosulfate as required.

HEALTH HAZARDS
Personal protective equipment (PPE): B-Level PPE.
Recommendations for respirator selection: NIOSH as cyanides 25 mg/m^3: SA (any supplied-air respirator); or SCBAF (any self-contained breathing apparatus with full facepiece) *EMERGENCY OR PLANNED ENTRY INTO UNKNOWN CONCENTRATIONS OR IDLH CONDITIONS*: SCBAF:PD,PP (any self-contained breathing apparatus that has a full facepiece and is operated in a pressure-demand or other positive-pressure mode); or SAF:PD,PP:ASCBA (any supplied-air respirator that has a full facepiece and is operated in a pressure-demand or other positive-pressure mode in combination with an auxiliary self-contained breathing apparatus operated in a pressure-demand or other positive pressure mode). *ESCAPE:* GMFSHiE [any air-purifying, full-facepiece respirator (gas mask) with a chin-style, front- or back-mounted canister providing protection against the compound of concern and having a high efficiency particulate filter); or SCBAE (any appropriate escape-type, self-contained breathing apparatus).
Exposure limits (TWA unless otherwise noted): ACGIH TLV ceiling 4.7 ppm (5 mg/m^3) as CN; NIOSH ceiling/10 minute 5 mg/m^3 (4.7 ppm); OSHA 5 mg/m^3 (4.7 ppm).
Toxicity by ingestion: Grade 4; LD_{50} = < 50 mg/kg.
Long-term health effects: Chronic exposure may cause headache, lack of appetite, weakness and inflammation of the skin with small pimples or blistery spots.
Liquid or solid irritant characteristics: Causes smarting of the skin and first-degree burns on short exposure; may cause second-degree burns on long exposure.
IDLH value: 50 mg/m^3 as cyanide

CHEMICAL REACTIVITY
Binary reactants: Contact with acids or acid salts will liberate highly toxic and flammable hydrogen cyanide gas. Reacts with reducing agents, alcohols, combustible materials, ethers, hydrazines, organic substances, metal powders.
Neutralizing agents for acids and caustics: Hypochlorite solution to destroy the cyanide.

ENVIRONMENTAL DATA
Food chain concentration potential: Zinc is concentrated by some organisms but is not considered to be bioconcentrative in a spill situation.
Water pollution: DOT Appendix B, §172.101–marine pollutant. Harmful to aquatic life in very low concentrations. May be dangerous if it enters nearby water intakes; notify operators. Notify local health and wildlife officials. **Response to discharge:** Issue warning–water contaminant, poison. Restrict access. Evacuate area. Chemical and physical treatment.

SHIPPING INFORMATION
Grades of purity: 55% Zn; 40% Cn; **Storage temperature:** Ambient; **Stability during transport:** Stable.

PHYSICAL AND CHEMICAL PROPERTIES
Physical state @ 59°F/15°C and 1 atm: Solid.

Molecular weight: 117.42
Melting/Freezing point: Decomposes at 1470°F/800°C/973°K.
Specific gravity (water = 1): 1.85 at room temperature

ZINC DIALKYLDITHIOPHOSPHATE REC. Z:1200

SYNONYMS: DIALQUILDITIOFOSFATO de ZINC (Spanish); ZINC DIHEXYLDITHIOPHOSPHATE; ZINC DIHEXYLPHOSPHORODITHIOATE; ZINC o,o-DI-n-BUTYLPHOSPHORODITHIOATE

IDENTIFICATION
Formula: $[(RO)_2PSS]_{2Zn}$ where $R=C_4H_9$, etc
DOT ID Number: UN 3077; DOT Guide Number: 171
Proper Shipping Name: Environmentally hazardous substances, solid, n.o.s.

DESCRIPTION: Straw yellow to yellow-green solid or liquid. Sweet, alcohol-like odor. Sinks in water.

Poison! • Breathing the dust, skin or eye contact, or swallowing the material can cause permanent injury and possible death • Firefighting gear (including SCBA) does not provide adequate protection. If exposure occurs, remove and isolate gear immediately and thoroughly decontaminate personnel • Containers may BLEVE when exposed to fire • Concentrated dust in confined areas (e.g., tanks, sewers, buildings) may explode when exposed to fire • Toxic products of combustion may include oxides of zinc, phosphorus, and sulfur • Do not put yourself in danger by entering a contaminated area to rescue a victim.

Hazard Classification (based on NFPA-704 Rating System)
Health Hazards (Blue): 3; Flammability (Red): 1; Reactivity (Yellow): 0

EMERGENCY RESPONSE: See Appendix A (171)
Evacuation:
Public safety: Isolate the area of spill or leak for at least 10 to 25 meters (30 to 80 feet) in all directions.
Spill: Increase, in the downwind direction, as necessary, the distance shown under "Public Safety."
Fire: If any large container is involved in fire, isolate for at least 800 meters (½ mile) in all directions; also, consider initial evacuation for 800 meters (½ mile) in all directions.

EXPOSURE
Short-term effects: *SEEK MEDICAL ATTENTION.* **Dust:** Irritating to eyes, nose, and throat. *IF INHALED*, will, will cause coughing or difficult breathing. *IF IN EYES*, hold eyelids open and flush. with plenty of water. *IF BREATHING HAS STOPPED,* give artificial respiration. IF breathing is difficult, administer oxygen. **Liquid or solid:** Irritating to skin and eyes. *IF SWALLOWED*, will cause nausea and vomiting. Remove contaminated clothing and shoes. Flush affected areas with plenty of water. *IF IN EYES*, hold eyelids open and flush. with plenty of water. *IF SWALLOWED* and victim is *CONSCIOUS AND ABLE TO SWALLOW*, have victim drink 4 to 8 ounces of water and have victim induce vomiting. *IF SWALLOWED* and victim is *UNCONSCIOUS OR HAVING CONVULSIONS,* do nothing except keep victim warm.

HEALTH HAZARDS
Personal protective equipment (PPE): B-Level PPE. Rubber gloves; safety glasses or face shield; dust respirator for solid form.
Toxicity by ingestion: Grade 2; LD_{50} = 0.5 to 5 g/kg.

Liquid or solid irritant characteristics: Causes eye and skin irritation.

FIRE DATA
Flash point: 360°F/182°C (cc).

ENVIRONMENTAL DATA
Food chain concentration potential: Zinc is concentrated by some organisms but is not considered to be bioconcentrative in a spill situation.
Water pollution: Effect of low concentrations on aquatic life is unknown. May be dangerous if it enters nearby water intakes; notify operators. Notify local health and wildlife officials. **Response to discharge:** Issue warning–water contaminant. Restrict access. Should be removed. Chemical and physical treatment.

SHIPPING INFORMATION
Grades of purity: Technical; 62% on inert filler; **Storage temperature:** Below 66°C (150°F); **Inert atmosphere:** None; **Venting:** Open; **Stability during transport:** Stable (flame arrester); **Stability during transport:** Stable.

PHYSICAL AND CHEMICAL PROPERTIES
Physical state @ 59°F/15°C and 1 atm: Solid or Liquid.
Molecular weight: 548 (approximate) (decomposes).
Specific gravity (water = 1): 1.12-1.26 @ 68°F/20°C (liquid); 1.6 @ 68°F/20°C (solid).
Vapor pressure: (Reid) Low

ZINC FLUORIDE REC. Z:1300

SYNONYMS: FLUORURO de ZINC (Spanish); ZINC DIFLUORIDE; ZINC FLUORURE (French).

IDENTIFICATION
CAS Number: 7783-49-5
Formula: ZnF_2
DOT ID Number: UN 9158; DOT Guide Number: 151
Proper Shipping Name: Zinc fluoride
Reportable Quantity (RQ): **(CERCLA)** 1000 lb/454 kg

DESCRIPTION: Colorless to white needles or crystalline solid. Sinks and mixes with water.

Poison! • Breathing the dust, skin or eye contact, or swallowing the material can cause permanent damage • Firefighting gear (including SCBA) does not provide adequate protection. If exposure occurs, remove and isolate gear immediately and thoroughly decontaminate personnel Toxic products of combustion may include zinc oxide and fluorine.

Hazard Classification (based on NFPA-704 Rating System)
Health Hazards (Blue): 2; Flammability (Red): 0; Reactivity (Yellow): 0

EMERGENCY RESPONSE: See Appendix A (151)
Evacuation:
Public safety: Isolate the area of spill or leak for at least 25 to 50 meters (80 to 160 feet) in all directions.
Spill: Increase, in the downwind direction, as necessary, the distance shown under "Public Safety."
Fire: If tank, rail car, or tank truck is involved in fire, isolate for at least 800 meters (½ mile) in all directions; also, consider initial evacuation for 800 meters (½ mile) in all directions.

Zinc fluoroborate

EXPOSURE
Short-term effects: *SEEK MEDICAL ATTENTION.* **Solid or dust:** Irritating to eyes and nose. Harmful if swallowed. Move to fresh air. *IF BREATHING HAS STOPPED,* give artificial respiration; *avoid mouth-to-mouth resuscitation; use bag/mask apparatus. IF IN EYES,* hold eyelids open and flush. with plenty of water. *IF SWALLOWED* and victim is *CONSCIOUS AND ABLE TO SWALLOW,* have victim drink 4 to 8 ounces of water.

HEALTH HAZARDS
Personal protective equipment (PPE): B-Level PPE. Approved dust and fume respirator, skin and eye protection.
Recommendations for respirator selection: NIOSH/OSHA as F. *12.5 mg/m³:* DM (any dust and mist respirator). *25 mg/m³:* DMXSQ* (any dust and mist respirator except single-use and quarter-mask respirators); or SA* (any supplied-air respirator). *62.5 mg/m³:* SA:CF* [any supplied-air respirator operated in a continuous-flow mode)]; or PAPRDM*⁺ *if not present as a fume* (any powered, air-purifying respirator with a dust and mist filter). *125 mg/m³:* HiEF + (any air-purifying, full-facepiece respirator with a high-efficiency particulate filter); or SCBAF (any self-contained breathing apparatus with a full facepiece); or SAF (any supplied-air respirator with a full facepiece). *250 mg/m³:* SA:PD,PP (any supplied-air respirator operated in a pressure-demand or other positive-pressure mode). *EMERGENCY OR PLANNED ENTRY INTO UNKNOWN CONCENTRATIONS OR IDLH CONDITIONS:* SCBAF:PD,PP (any self-contained breathing apparatus that has a full faceplate and is operated in a pressure-demand or other positive-pressure mode); or SAF:PD,PP:ASCBA (any supplied-air respirator that has a full facepiece and is operated in a pressure-demand or other positive-pressure mode in combination with an auxiliary, self-contained breathing apparatus operated in a pressure-demand or other positive-pressure mode). *ESCAPE:* HiEF+ (any air-purifying, full-facepiece respirator with a high-efficiency particulate filter); or SCBAE (any appropriate escape-type, self-contained breathing apparatus). *Notes:* *Substance reported to cause eye irritation or damage; may require eye protection. ⁺May need acid gas sorbent.
Exposure limits (TWA unless otherwise noted): ACGIH TLV 2.5 mg/m³ as fluorides; NIOSH/OSHA 2.5 mg/m³ as inorganic fluorides.
Toxicity by ingestion: Grade 3; LD_{50} = 50 to 500 mg/kg.
Long-term health effects: Repeated exposures to excessive concentrations of fluorides may increase radiographic density of bones and eventually may be responsible for anatomical abnormalities.
Liquid or solid irritant characteristics: Causes irritation of eyes and skin.
IDLH value: 250 mg/m³ as F.

ENVIRONMENTAL DATA
Food chain concentration potential: Zinc is concentrated by some organisms but is not considered to be bioconcentrative in a spill situation. Fluorine is accumulated by aquatic animals.
Water pollution: Harmful to aquatic life in very low concentrations. May be dangerous if it enters nearby water intakes; notify operators. Notify local health and wildlife officials.
Response to discharge: Disperse and flush.

PHYSICAL AND CHEMICAL PROPERTIES
Physical state @ 59°F/15°C and 1 atm: Solid.
Molecular weight: 103.38 (anhydrous salt).
Boiling point @ 1 atm: 2732°F/1500°C/1773°K.
Melting/Freezing point: (Anhydrous salt) 1602°F/872°C/1145°K.
Specific gravity (water = 1): 4.84 @ 15°C 4.95 @ 77°F/25°C.

Latent heat of vaporization: Estimated at BP 1500°C. 958.4 Btu/lb = 532.5 cal/g = 22.3 x 10⁵ J/kg.
Heat of solution: Exothermic for anhydrous salt –227.8 Btu/lb = –126.5 cal/g = –5.3 x 10⁵ J/kg.

ZINC FLUOROBORATE   REC. Z:1400

SYNONYMS: FLUOBORATO de ZINC (Spanish); ZINC FLUOBORATE SOLUTION

IDENTIFICATION
CAS Number: 13597-54-1
Formula: $Zn(BF_4)_2 \cdot H_2O$
DOT ID Number: UN 3077; DOT Guide Number: 171
Proper Shipping Name: Environmentally hazardous substances, solid, n.o.s.

DESCRIPTION: Colorless liquid. Odorless. Sinks and mixes with water.

Poison! • Breathing the dust, skin or eye contact, or swallowing the material can cause permanent damage • Firefighting gear (including SCBA) does not provide adequate protection. If exposure occurs, remove and isolate gear immediately and thoroughly decontaminate personnel • Toxic products of combustion may include zinc oxide and boron.

Hazard Classification (based on NFPA-704 Rating System)
Health Hazards (Blue): 2; Flammability (Red): 0; Reactivity (Yellow): 0

EMERGENCY RESPONSE: See Appendix A (171)
Evacuation:
Public safety: Isolate the area of spill or leak for at least 10 to 25 meters (30 to 80 feet) in all directions.
Spill: Increase, in the downwind direction, as necessary, the distance shown under "Public Safety."
Fire: If any large container is involved in fire, isolate for at least 800 meters (½ mile) in all directions; also, consider initial evacuation for 800 meters (½ mile) in all directions.

EXPOSURE
Short-term effects: *SEEK MEDICAL ATTENTION.* **Liquid:** Irritating to skin and eyes. *IF SWALLOWED,* will cause nausea and vomiting. Remove contaminated clothing and shoes. Flush affected areas with plenty of water. *IF IN EYES,* hold eyelids open and flush. with plenty of water. *IF SWALLOWED* and victim is *CONSCIOUS AND ABLE TO SWALLOW,* have victim drink 4 to 8 ounces of water and have victim induce vomiting. *IF SWALLOWED* and victim is *UNCONSCIOUS OR HAVING CONVULSIONS,* do nothing except keep victim warm.

HEALTH HAZARDS
Personal protective equipment (PPE): B-Level PPE. Rubber gloves; safety glasses or face shield.
Toxicity by ingestion: Grade 2; LD_{50} = 0.5 to 5 g/kg.
Liquid or solid irritant characteristics: Causes irritation of eyes and skin.

ENVIRONMENTAL DATA
Food chain concentration potential: Zinc is concentrated by some organisms but is not considered to be bioconcentrative in a spill situation.

Water pollution: Effect of low concentrations on aquatic life is unknown. May be dangerous if it enters nearby water intakes; notify operators. Notify local health and wildlife officials. **Response to discharge:** Issue warning–water contaminant. Disperse and flush.

SHIPPING INFORMATION
Grades of purity: Purified, 41% solution in water; **Storage temperature:** Ambient; **Inert atmosphere:** None; **Venting:** Open; **Stability during transport:** Stable.

PHYSICAL AND CHEMICAL PROPERTIES
Physical state @ 59°F/15°C and 1 atm: Liquid.
Molecular weight: 238.98 (solute only).
Boiling point @ 1 atm: (approximate) 212°F/100°C/373°K.
Specific gravity (water = 1): 1.45 @ 68°F/20°C (liquid).

ZINC FORMATE   REC. Z:1500

SYNONYMS: FORMIATO de ZINC (Spanish); FORMIC ACID, ZINC SALT; ZINC DIFORMATE

IDENTIFICATION
CAS Number: 557-41-5
Formula: $Zn(HCOO)_2 \cdot 2H_2O_2$; $Zn(CHO_2)_2$
DOT ID Number: UN 9159; DOT Guide Number: 171
Proper Shipping Name: Zinc formate
Reportable Quantity (RQ): **(CERCLA)** 1000 lb/454 kg

DESCRIPTION: White crystalline solid. Sinks and mixes with water.

Toxic products of combustion may include zinc oxide.

Hazard Classification (based on NFPA-704 Rating System)
Health Hazards (Blue): 1; Flammability (Red): 0; Reactivity (Yellow): 0

EMERGENCY RESPONSE: See Appendix A (171)
Evacuation:
Public safety: Isolate the area of spill or leak for at least 10 to 25 meters (30 to 80 feet) in all directions.
Spill: Increase, in the downwind direction, as necessary, the distance shown under "Public Safety."
Fire: If any large container is involved in fire, isolate for at least 800 meters (½ mile) in all directions; also, consider initial evacuation for 800 meters (½ mile) in all directions.

EXPOSURE
Short-term effects: *SEEK MEDICAL ATTENTION.* **Solid:** Irritating to skin and eyes. *IF SWALLOWED* may cause nausea and vomiting.
Flush affected areas with plenty of water. *IF IN EYES*, hold eyelids open and flush. with plenty of water. *IF SWALLOWED* and victim is *CONSCIOUS AND ABLE TO SWALLOW*, have victim drink 4 to 8 ounces of water and have victim induce vomiting.

HEALTH HAZARDS
Personal protective equipment (PPE): B-Level PPE. Approved respirator, rubber gloves, and safety goggles.
Toxicity by ingestion: Grade 2; LD_{50} = 0.5 to 5 g/kg.
Liquid or solid irritant characteristics: May cause irritation of eyes and skin.

FIRE DATA
Behavior in fire: Moderate hazard. May emit toxic zinc oxide fumes.

CHEMICAL REACTIVITY
Neutralizing agents for acids and caustics: Neutralize with sodium hydroxide.

ENVIRONMENTAL DATA
Food chain concentration potential: Zinc is concentrated by some organisms but is not considered to be bioconcentrative in a spill situation.
Water pollution: Harmful to aquatic life in very low concentrations. May be dangerous if it enters nearby water intakes; notify operators. Notify local health and wildlife officials. **Response to discharge:** Issue warning–water contaminant. Chemical and physical treatment. Disperse and flush.

SHIPPING INFORMATION
Storage temperature: Ambient; **Stability during transport:** Stable.

PHYSICAL AND CHEMICAL PROPERTIES
Physical state @ 59°F/15°C and 1 atm: Solid.
Molecular weight: 191.45 Dihydrate; 155.41 anhydrous
Boiling point @ 1 atm: Decomposes. Loses $2H_2O$ at 284°F/140°C.
Specific gravity (water = 1): 2.207 @ 68°F/20°C dihydrate; 2.368 anhydrous
Heat of solution: Exothermic –46.3 Btu/lb = –25.7 cal/g = –1.08 x 10^5 J/kg @ 77°F/25°C –52 Btu/lb = –29 cal/g = 1.2 x 10^5 J/kg @ 15°C.

ZINC HYDROSULFITE   REC. Z:1600

SYNONYMS: HIDROSULFITO de ZINC (Spanish); DITTHIONOUS ACID, ZINC SALT; ZINC DITHIONITE

IDENTIFICATION
CAS Number: 7779-86-4
Formula: $O_4S_2 \cdot Zn$; ZnS_2O_4
DOT ID Number: UN 1931; DOT Guide Number: 171
Proper Shipping Name: Zinc dithionite
Reportable Quantity (RQ): **(CERCLA)** 1000 lb/454 kg

DESCRIPTION: Amorphous, white solid. Slight sulfur dioxide odor. Mixes with water.

Dust is Irritating to the skin, eyes, and respiratory tract • Toxic products of combustion may include zinc oxide and sulfur oxide.

Hazard Classification (based on NFPA-704 Rating System)
Health Hazards (Blue): -; Flammability (Red): 0; Reactivity (Yellow): 1

EMERGENCY RESPONSE: See Appendix A (171)
Evacuation:
Public safety: Isolate the area of spill or leak for at least 10 to 25 meters (30 to 80 feet) in all directions.
Spill: Increase, in the downwind direction, as necessary, the distance shown under "Public Safety."
Fire: If any large container is involved in fire, isolate for at least 800 meters (½ mile) in all directions; also, consider initial evacuation for 800 meters (½ mile) in all directions.

EXPOSURE

Short-term effects: *SEEK MEDICAL ATTENTION.* **Dust:** Irritating to eyes, nose, and throat. Move to fresh air. **Solid:** Irritating to skin and eyes. *IF SWALLOWED*, will cause nausea and vomiting.
Flush affected areas with plenty of water. *IF IN EYES*, hold eyelids open and flush. with plenty of water. *IF SWALLOWED* and victim is *CONSCIOUS AND ABLE TO SWALLOW*, have victim drink plenty of water or milk and have victim induce vomiting.

HEALTH HAZARDS

Personal protective equipment (PPE): B-Level PPE. Dust mask, chemical workers goggles, and rubber gloves.
Toxicity by ingestion: Grade 2; LD_{50} = 0.5 to 5 g/kg.
Liquid or solid irritant characteristics: Causes eye and skin irritation.

CHEMICAL REACTIVITY

Reactivity with water: Contact with water liberates irritating oxides of sulfur.
Binary reactants: Oxidizing agents and acids
Neutralizing agents for acids and caustics: Neutralize with soda ash and NaOH.

ENVIRONMENTAL DATA

Food chain concentration potential: Zinc is concentrated by some organisms but is not considered to be bioconcentrative in a spill situation.
Water pollution: Harmful to aquatic life in very low concentrations. May be dangerous if it enters nearby water intakes; notify operators. Notify local health and wildlife officials.
Response to discharge: Issue warning–water contaminant. Restrict access. Chemical and physical treatment. Disperse and flush.

SHIPPING INFORMATION

Grades of purity: 86–88% ZnS_2O_4; **Storage temperature:** Cool; **Stability during transport:** Hygroscopic should be protected from moisture and heat.

PHYSICAL AND CHEMICAL PROPERTIES

Physical state @ 59°F/15°C and 1 atm: Solid.
Molecular weight: 193.45

ZINC NITRATE   REC. Z:1700

SYNONYMS: NITRATO de ZINC (Spanish); NITRATE de ZINC (French); NITRIC ACID, ZINC SALT; ZINC NITRATE HEXAHYDRATE

IDENTIFICATION

CAS Number: 7779-88-6
Formula: $N_2O_6 \cdot Zn$; $Zn(NO_3)_2 \cdot 6H_2O$
DOT ID Number: UN 1514; DOT Guide Number: 140
Proper Shipping Name: Zinc nitrate
Reportable Quantity (RQ): **(CERCLA)** 1000 lb/454 kg

DESCRIPTION: Colorless crystalline solid. Odorless. Sinks and mixes with water.

Corrosive to skin, eyes, and respiratory tract; skin or eye contact can cause burns or impaired vision; inhalation symptoms may be delayed • Strong oxidizer which may react spontaneously with low flash point organics or reducing agents. Heat forms oxygen; will increase the activity of an existing fire • May cause fire or explosion on contact with combustibles (wood, paper, oil, clothing, etc.). Decomposes above 284°F/140°C; toxic products of combustion or decomposition include oxides of zinc and nitrogen.

Hazard Classification (based on NFPA-704 Rating System)
Health Hazards (Blue): 1; Flammability (Red): 0; Reactivity (Yellow): 0; Special Notice (White): OXY

EMERGENCY RESPONSE: See Appendix A (140)
Evacuation:
Public safety: Isolate the area of spill or leak for at least 10 to 25 meters (30 to 80 feet) in all directions.
Spill: Consider initial downwind evacuation for at least 100 meters (330 feet).
Fire: If any large container is involved in fire, isolate for at least 800 meters (½ mile) in all directions; also, consider initial evacuation for 800 meters (½ mile) in all directions.

EXPOSURE

Short-term effects: *SEEK MEDICAL ATTENTION.* **Dust:** Irritating to eyes, nose, and throat. *IF INHALED*, will, will cause coughing or difficult breathing. *IF IN EYES*, hold eyelids open and flush. with plenty of water. *IF BREATHING HAS STOPPED*, give artificial respiration. IF breathing is difficult, administer oxygen. Lung edema may develop. **Solid:** Irritating to skin and eyes. *IF SWALLOWED*, will cause nausea and vomiting. Remove contaminated clothing and shoes. Flush affected areas with plenty of water. *IF IN EYES*, hold eyelids open and flush. with plenty of water. *IF SWALLOWED* and victim is *CONSCIOUS AND ABLE TO SWALLOW*, have victim drink 4 to 8 ounces of water and have victim induce vomiting. *IF SWALLOWED* and victim is *UNCONSCIOUS OR HAVING CONVULSIONS*, do nothing except keep victim warm.
Note to physician or authorized medical personnel: Medical observation is recommended for 24 to 48 hours after breathing overexposure, as pulmonary edema may be delayed. As first aid for pulmonary edema, consider administering a corticosteroid spray. Cigarette smoking may exacerbate pulmonary injury and should be discouraged for at least 72 hours following exposure.

HEALTH HAZARDS

Personal protective equipment (PPE): B-Level PPE. Dust mask; goggles or face shield; protective gloves.
Toxicity by ingestion: Grade 2; oral LD_{50} = 2500 mg/kg (rat).
Long-term health effects: Causes enlarged liver, spleen, and bone marrow in rabbits
Liquid or solid irritant characteristics: Corrosive to the eyes, skin, and respiratory tract.

FIRE DATA

Behavior in fire: Noncombustible, but will enhance the combustibility of other materials.

CHEMICAL REACTIVITY

Binary reactants: Many chemical reactions can cause fire and explosions. A strong oxidizer. Violent reaction with reducing agents, strong oxidizers, combustible materials, organic substances, metallic powders, acetic anhydride, carbon, dimethylformamide, metal cyanides, metal sulfides, phosphorus, sodium acetylide, sulfur, thiocyanates. Incompatible with amines, ammonium hexacyanoferrate(II), boranes, cyanides, citric acid, esters, hydrazinium perchlorate, isopropyl chlorocarbonate, nitrosyl perchlorate, organic azides, organic bases, sodium thiosulfate, sulfamic acid. Attacks metals in the presence of moisture.

ENVIRONMENTAL DATA
Food chain concentration potential: Zinc is concentrated by some organisms but is not considered to be bioconcentrative in a spill situation.
Water pollution: Harmful to aquatic life in very low concentrations. May be dangerous if it enters nearby water intakes; notify operators. Notify local health and wildlife officials.
Response to discharge: Disperse and flush.

SHIPPING INFORMATION
Grades of purity: Reagent; Technical; **Storage temperature:** Ambient; **Inert atmosphere:** None; **Venting:** Open; **Stability during transport:** Stable.

PHYSICAL AND CHEMICAL PROPERTIES
Physical state @ 59°F/15°C and 1 atm: Solid.
Molecular weight: 297.47
Boiling point @ 1 atm: 284°F/140°C/413°K (decomposes).
Melting/Freezing point: 97°F/36°C/309°K.
Specific gravity (water = 1): 2.07 @ 68°F/20°C (solid).

ZINC PHENOLSULFONATE   REC. Z:1800

SYNONYMS: FENOLSULFONATO de ZINC (Spanish); ZINC *p*-PHENOLSULFONATE; ZINC SULFOCARBOLATE; ZINC SULFOPHENATE; ZINC PHENOLSULFONATE OCTAHYDRATE

IDENTIFICATION
CAS Number: 127-82-2
Formula: $C_{12}H_{12}O_8S_2 \cdot Zn$; $(1,4-HOC_6H_4SO_3)_2Zn \cdot 8H_2O$
DOT ID Number: UN 9160; DOT Guide Number: 171
Proper Shipping Name: Zinc phenolsulfonate
Reportable Quantity (RQ): **(CERCLA)** 5000 lb/2270 kg

DESCRIPTION: Clear, colorless crystals or white powder. Odorless. Sinks and mixes with water, forming a basic solution.

Corrosive to skin, eyes, and respiratory tract; skin or eye contact can cause burns or impaired vision; inhalation symptoms may be delayed • Toxic products of combustion may include oxides of zinc and sulfur.

Hazard Classification (based on NFPA-704 Rating System)
Health Hazards (Blue): 1; Flammability (Red): 0; Reactivity (Yellow): 0

EMERGENCY RESPONSE: See Appendix A (171)
Evacuation:
Public safety: Isolate the area of spill or leak for at least 10 to 25 meters (30 to 80 feet) in all directions.
Spill: Increase, in the downwind direction, as necessary, the distance shown under "Public Safety."
Fire: If any large container is involved in fire, isolate for at least 800 meters (½ mile) in all directions; also, consider initial evacuation for 800 meters (½ mile) in all directions.

EXPOSURE
Short-term effects: *SEEK MEDICAL ATTENTION.* **Dust:** Irritating to eyes, nose, and throat. *IF INHALED*, will, will cause coughing or difficult breathing. *IF IN EYES*, hold eyelids open and flush. with plenty of water. *IF BREATHING HAS STOPPED,* give artificial respiration. IF breathing is difficult, administer oxygen. **Solid:** Irritating to skin and eyes. *IF SWALLOWED*, will cause nausea and vomiting. Remove contaminated clothing and shoes. Flush affected areas with plenty of water. *IF IN EYES*, hold eyelids open and flush. with plenty of water. *IF SWALLOWED* and victim is *CONSCIOUS AND ABLE TO SWALLOW*, have victim drink 4 to 8 ounces of water and have victim induce vomiting. *IF SWALLOWED* and victim is *UNCONSCIOUS OR HAVING CONVULSIONS,* do nothing except keep victim warm.
Note to physician or authorized medical personnel: Medical observation is recommended for 24 to 48 hours after breathing overexposure, as pulmonary edema may be delayed. As first aid for pulmonary edema, consider administering a corticosteroid spray. Cigarette smoking may exacerbate pulmonary injury and should be discouraged for at least 72 hours following exposure.

HEALTH HAZARDS
Personal protective equipment (PPE): B-Level PPE. Dust mask; goggles or face shield; protective gloves.
Toxicity by ingestion: Grade 2; LD_{50} = 0.5 to 5 g/kg.
Liquid or solid irritant characteristics: Causes irritation of the eyes and skin.

CHEMICAL REACTIVITY
Binary reactants: Acids or acid fumes form toxic sulfur oxides.

ENVIRONMENTAL DATA
Food chain concentration potential: Zinc is concentrated by some organisms but is not considered to be bioconcentrative in a spill situation.
Water pollution: Effect of low concentrations on aquatic life is unknown. May be dangerous if it enters nearby water intakes; notify operators. Notify local health and wildlife officials. **Response to discharge:** Issue warning–water contaminant. Disperse and flush.

SHIPPING INFORMATION
Grades of purity: Purified; **Storage temperature:** Ambient; **Inert atmosphere:** None; **Venting:** Open; **Stability during transport:** Stable.

PHYSICAL AND CHEMICAL PROPERTIES
Physical state @ 59°F/15°C and 1 atm: Solid.
Molecular weight: 555.8
Boiling point @ 1 atm: (decomposes) 248°F/120°C/393°K.
Specific gravity (water = 1): More than 1 @ 68°F/20°C (solid).

ZINC PHOSPHIDE   REC. Z:1900

SYNONYMS: FOSFURO de ZINC (Spanish); BLUE-OX; KILRAT; PHOSVIN; RUMETAN; ZINC-TOX; Z-P; RCRA No. P122

IDENTIFICATION
CAS Number: 1314-84-7
Formula: Zn_3P_2
DOT ID Number: UN 1714; DOT Guide Number: 139
Proper Shipping Name: Zinc phosphide
Reportable Quantity (RQ): **(CERCLA)** 100 lb/45.4 kg

DESCRIPTION: Gray to black solid. Faint, garlic-like odor. Sinks in water; insoluble; reacts, releasing highly toxic and flammable phosphine gas.

Poison! • Do NOT use water • Breathing the dust can kill you; skin or eye contact causes severe burns, impaired vision, or blindness •

Firefighting gear (including SCBA) does not provide adequate protection. If exposure occurs, remove and isolate gear immediately and thoroughly decontaminate personnel • Containers may BLEVE when exposed to fire • Toxic products of combustion may include phosphorus oxides and zinc fumes that can cause metal fume fever • Do not put yourself in danger by entering a contaminated area to rescue a victim.

Hazard Classification (based on NFPA-704 Rating System)
Health Hazards (Blue): 3; Flammability (Red): 3; Reactivity (Yellow): 1

EMERGENCY RESPONSE: See Appendix A (139)
Evacuation:
Public safety: Isolate the area of spill or leak for at least 100 to 150 meters (330 to 490 feet) in all directions.
Spill: *IF SPILLED IN WATER*: Small spill–First: Isolate in all directions 30 meters (100 feet); Then: Protect persons downwind, DAY: 0.2 km (0.1 mile); NIGHT: 0.3 km (0.2 mile). Large spill–First: Isolate in all directions 125 meters (400 feet); Then: Protect persons downwind, DAY: 1.3 km (0.8 mile); NIGHT: 3.2 km (2.0 miles).
Fire: If any large container is involved in fire, isolate for at least 800 meters (½ mile) in all directions; also, consider initial evacuation for 800 meters (½ mile) in all directions.

EXPOSURE
Short-term effects: *SEEK MEDICAL ATTENTION*. **Dust:** Irritating to eyes, nose, and throat. *IF INHALED*, will, will cause dizziness, difficult breathing, or loss of consciousness. *IF IN EYES*, hold eyelids open and flush. with plenty of water. *IF BREATHING HAS STOPPED*, give artificial respiration. IF breathing is difficult, administer oxygen. **Solid:** *POISONOUS IF SWALLOWED*. Irritating to skin and eyes. *IF SWALLOWED*, will cause dizziness, nausea, vomiting, or loss of consciousness. Remove contaminated clothing and shoes. Flush affected areas with plenty of water. *IF IN EYES*, hold eyelids open and flush. with plenty of water. *IF SWALLOWED* and victim is *CONSCIOUS AND ABLE TO SWALLOW*, have victim drink 4 to 8 ounces of water and have victim induce vomiting. *IF SWALLOWED* and victim is *UNCONSCIOUS, OR HAVING CONVULSIONS*, do nothing except keep victim warm.

HEALTH HAZARDS
Personal protective equipment (PPE): A-Level PPE. Dust mask or self-contained breathing apparatus; goggles or face shield; protective gloves.
Toxicity by ingestion: Grade 4; oral LD_{50} = 40 mg/kg (rat).
Liquid or solid irritant characteristics: Causes irritation of the eyes and skin.

FIRE DATA
Fire extinguishing agents not to be used: Do not use agents with an acid reaction (e.g., CO_2 or halogenated agents); will liberate phosphine, a toxic and spontaneously flammable gas.

CHEMICAL REACTIVITY
Reactivity with water: Reacts slowly with water, more rapidly with dilute acid, to form phosphine gas, which is toxic and spontaneously flammable.
Binary reactants: Reacts with oxidizers and acids.

ENVIRONMENTAL DATA
Food chain concentration potential: Zinc is concentrated by some organisms but is not considered to be bioconcentrative in a spill situation.
Water pollution: Effect of low concentrations on aquatic life is unknown. May be dangerous if it enters nearby water intakes; notify operators. Notify local health and wildlife officials. **Response to discharge:** Issue warning–poison, water contaminant, air contaminant. Restrict access. Should be removed. Chemical and physical treatment.

SHIPPING INFORMATION
Grades of purity: Technical, 94+%; **Storage temperature:** Ambient; **Inert atmosphere:** None; **Venting:** Pressure-vacuum; **Stability during transport:** Stable unless exposed to moisture; toxic phosphine gas may then be released and collect in closed spaces.

PHYSICAL AND CHEMICAL PROPERTIES
Physical state @ 59°F/15°C and 1 atm: Solid.
Molecular weight: 258.10
Boiling point @ 1 atm: 2012°F/1110°C/1373°K.
Melting/Freezing point: (sublimes) 788°F/420°C/693°K.
Specific gravity (water = 1): 4.55 at 13°C (solid).
Heat of combustion: -4100 Btu/lb = -2270 cal/g = -95×10^5 J/kg.

ZINC POTASSIUM CHROMATE   REC. Z:2000

SYNONYMS: CROMATO de ZINC y POTASIO (Spanish); ZINC YELLOW Y-539-D; POTASSIUM ZINC CHROMATE

IDENTIFICATION
CAS Number: 1103-86-9
Formula: $K_2CrO_4 \cdot ZnCrO_4$
DOT ID Number: UN 3077; DOT Guide Number: 171
Proper Shipping Name: Environmentally hazardous substances, solid, n.o.s.

DESCRIPTION: Yellow powder. Odorless. Sinks and mixes with water.

Combustible solid • Containers may BLEVE when exposed to fire • Concentrated dust in confined areas (e.g., tanks, sewers, buildings) may explode when exposed to fire • Toxic products of combustion may include zinc oxides, potassium oxide, and chromium.

Hazard Classification (based on NFPA-704 Rating System)
Health Hazards (Blue): 2; Flammability (Red): 1; Reactivity (Yellow): 0

EMERGENCY RESPONSE: See Appendix A (171)
Evacuation:
Public safety: Isolate the area of spill or leak for at least 10 to 25 meters (30 to 80 feet) in all directions.
Spill: Increase, in the downwind direction, as necessary, the distance shown under "Public Safety."
Fire: If any large container is involved in fire, isolate for at least 800 meters (½ mile) in all directions; also, consider initial evacuation for 800 meters (½ mile) in all directions.

EXPOSURE
Short-term effects: *SEEK MEDICAL ATTENTION*. **Solid and dust:** Irritating to skin, eyes, nose, and throat. Harmful if inhaled or swallowed. Move to fresh air. Remove contaminated clothing and shoes. Flush affected areas with plenty of water. *IF IN EYES*, hold

eyelids open and flush. with plenty of water. *IF SWALLOWED* and victim is *CONSCIOUS AND ABLE TO SWALLOW*, have victim drink 4 to 8 ounces of water and have victim induce vomiting.

HEALTH HAZARDS
Personal protective equipment (PPE): B-Level PPE. Respirator, gloves, glasses and protective clothing.
Recommendations for respirator selection: NIOSH
At any concentrations above the NIOSH REL, or where there is no REL, at any detectable concentration: SCBAF:PD,PP (any self-contained breathing apparatus that has a full facepiece and is operated in a pressure-demand or other positive-pressure mode); or SAF:PD,PP:ASCBA (any supplied-air respirator that has a full facepiece and is operated in a pressure-demand or other positive-pressure mode in combination with an auxiliary self-contained breathing apparatus operated in a pressure-demand or other positive pressure mode). *ESCAPE:* HiEF (any air-purifying, full-facepiece respirator with a high-efficiency particulate filter); or SCBAE (any appropriate escape-type, self-contained breathing apparatus).
Exposure limits (TWA unless otherwise noted): ACGIH TLV 0.05 mg/m^3 as water soluble Cr(VI) compounds; OSHA PEL 0.1 mg/m^3 as CrO$_3$; NIOSH 0.001 mg/m^3 as chromates; potential human carcinogen; reduce exposure to lowest feasible level.
Toxicity by ingestion: Grade 3; LD$_{50}$ = 50 to 500 mg/kg.
Long-term health effects: Chromates may cause lung cancer. IARC rating 3
Liquid or solid irritant characteristics: Causes irritation of the eyes and skin.
IDLH value: Potential human carcinogen; 15 mg/m^3 as chromium(VI).

ENVIRONMENTAL DATA
Food chain concentration potential: Both Cr and Zn are concentrated by some organisms but are not considered to be bioconcentrative in a spill situation.
Water pollution: Harmful to aquatic life in very low concentrations. May be dangerous if it enters nearby water intakes; notify operators. Notify local health and wildlife agencies.
Response to discharge: Disperse and flush.

SHIPPING INFORMATION
Grades of purity: Zn 30.8%; Cr 23.5%; K 9.1%; **Storage temperature:** Ambient; **Stability during transport:** Stable

PHYSICAL AND CHEMICAL PROPERTIES
Physical state @ 59°F/15°C and 1 atm: Solid.
Specific gravity (water = 1): 3.40 to 3.60

ZINC SILICOFLUORIDE   REC. Z:2100

SYNONYMS: SILICOFLUORURO de ZINC (Spanish); ZINC HEXAFLUOROSILICATE; ZINC FLUOROSILICATE; ZINC SILICOFLUORIDE HEXAHYDRATE

IDENTIFICATION
CAS Number: 16871-71-9
Formula: ZnSiF$_6$·6H$_2$O
DOT ID Number: UN 2855; DOT Guide Number: 151
Proper Shipping Name: Zinc fluorosilicate

DESCRIPTION: White, transparent solid. Odorless. Sinks and mixes with water.

Dust is Irritating to the skin, eyes, and respiratory tract • Toxic products of combustion may include hydrogen fluoride and silicon tetrafluoride

Hazard Classification (based on NFPA-704 Rating System)
Health Hazards (Blue): 2; Flammability (Red): 0; Reactivity (Yellow): 0

EMERGENCY RESPONSE: See Appendix A (151)
Evacuation:
Public safety: Isolate the area of spill or leak for at least 25 to 50 meters (80 to 160 feet) in all directions.
Spill: Increase, in the downwind direction, as necessary, the distance shown under "Public Safety."
Fire: If tank, rail car, or tank truck is involved in fire, isolate for at least 800 meters (½ mile) in all directions; also, consider initial evacuation for 800 meters (½ mile) in all directions.

EXPOSURE
Short-term effects: *SEEK MEDICAL ATTENTION.* **Dust:** Irritating to eyes, nose, and throat. *IF INHALED*, will, will cause coughing or difficult breathing. *IF IN EYES*, hold eyelids open and flush. with plenty of water. *IF BREATHING HAS STOPPED*, give artificial respiration; *avoid mouth-to-mouth resuscitation; use bag/mask apparatus*. IF breathing is difficult, administer oxygen.
Solid: *POISONOUS IF SWALLOWED.* Irritating to skin and eyes. *IF SWALLOWED*, will cause nausea, vomiting, or loss of consciousness. Remove contaminated clothing and shoes. Flush affected areas with plenty of water. *IF IN EYES*, hold eyelids open and flush. with plenty of water. *IF SWALLOWED* and victim is *CONSCIOUS AND ABLE TO SWALLOW*, have victim drink 4 to 8 ounces of water and have victim induce vomiting. *IF SWALLOWED* and victim is *UNCONSCIOUS OR HAVING CONVULSIONS,* do nothing except keep victim warm.

HEALTH HAZARDS
Personal protective equipment (PPE): B-Level PPE. Respirator; chemical goggles or face shield; protective **gloves.**
Recommendations for respirator selection: NIOSH/OSHA as F. *12.5 mg/m^3:* DM (any dust and mist respirator). *25 mg/m^3:* DMXSQ* (any dust and mist respirator except single-use and quarter-mask respirators); or SA* (any supplied-air respirator). *62.5 mg/m^3:* SA:CF* [any supplied-air respirator operated in a continuous-flow mode)]; or PAPRDM*$^+$ *if not present as a fume* (any powered, air-purifying respirator with a dust and mist filter). *125 mg/m^3:* HiEF + (any air-purifying, full-facepiece respirator with a high-efficiency particulate filter); or SCBAF (any self-contained breathing apparatus with a full facepiece); or SAF (any supplied-air respirator with a full facepiece). *250 mg/m^3:* SA:PD,PP (any supplied-air respirator operated in a pressure-demand or other positive-pressure mode). *EMERGENCY OR PLANNED ENTRY INTO UNKNOWN CONCENTRATIONS OR IDLH CONDITIONS:* SCBAF:PD,PP (any self-contained breathing apparatus that has a full faceplate and is operated in a pressure-demand or other positive-pressure mode); or SAF:PD,PP:ASCBA (any supplied-air respirator that has a full facepiece and is operated in a pressure-demand or other positive-pressure mode in combination with an auxiliary, self-contained breathing apparatus operated in a pressure-demand or other positive-pressure mode). *ESCAPE:* HiEF+ (any air-purifying, full-facepiece respirator with a high-efficiency particulate filter); or SCBAE (any appropriate escape-type, self-contained breathing apparatus). *Notes:* *Substance reported to cause eye irritation or damage; may require eye protection. $^+$May need acid gas sorbent.

Exposure limits (TWA unless otherwise noted): ACGIH TLV 2.5 mg/m^3 as fluorides; NIOSH/OSHA 2.5 mg/m^3 as inorganic fluorides.
Toxicity by ingestion: Oral LD$_{Lo}$ = 100 mg/kg (rat).
Liquid or solid irritant characteristics: Causes irritation of the eyes and skin.

ENVIRONMENTAL DATA
Food chain concentration potential: Zinc is concentrated by some organisms but is not considered to be bioconcentrative in a spill situation.
Water pollution: Effect of low concentrations on aquatic life is unknown. May be dangerous if it enters nearby water intakes; notify operators. Notify local health and wildlife officials.
Response to discharge: Issue warning–water contaminant. Disperse and flush.

SHIPPING INFORMATION
Grades of purity: Technical, 98–99%; **Storage temperature:** Ambient; **Inert atmosphere:** None; **Venting:** Open; **Stability during transport:** Stable.

PHYSICAL AND CHEMICAL PROPERTIES
Physical state @ 59°F/15°C and 1 atm: Solid.
Molecular weight: 315.5
Boiling point @ 1 atm: (decomposes) 122–158°F/50–70°C/323–343°K.
Specific gravity (water = 1): 2.10 @ 68°F/20°C (solid).

ZINC SULFATE   REC. Z:2200

SYNONYMS: BONAZEN; BUFOPTO ZINC SULFATE; OP-THAL-ZIN; SULFATO de ZINC (Spanish); SULFATE de ZINC (French); SULFURIC ACID, ZINC SALT; VERAZINC; WHITE COPPERAS; WHITE VITRIOL; ZINC SULFATE HEPTAHYDRATE; ZINC VITRIOL; ZINKOSITE

IDENTIFICATION
CAS Number: 7733-02-0
Formula: O$_4$S·Zn; ZnSO$_4$·7H$_2$O
DOT ID Number: UN 9161; DOT Guide Number: 171
Proper Shipping Name: Zinc sulfate
Reportable Quantity (RQ): **(CERCLA)** 1000 lb/454 kg

DESCRIPTION: Colorless crystalline solid or white powder. Odorless. Sinks and mixes with water.

Corrosive to skin, eyes, and respiratory tract; skin or eye contact can cause burns or impaired vision; inhalation symptoms may be delayed • Toxic products of combustion may include oxides of sulfur and zinc.

Hazard Classification (based on NFPA-704 Rating System)
Health Hazards (Blue): 0; Flammability (Red): 0; Reactivity (Yellow): 0

EMERGENCY RESPONSE: See Appendix A (171)
Evacuation:
Public safety: Isolate the area of spill or leak for at least 10 to 25 meters (30 to 80 feet) in all directions.
Spill: Increase, in the downwind direction, as necessary, the distance shown under "Public Safety."
Fire: If any large container is involved in fire, isolate for at least 800 meters (½ mile) in all directions; also, consider initial evacuation for 800 meters (½ mile) in all directions.

EXPOSURE
Short-term effects: *SEEK MEDICAL ATTENTION.* **Dust:** Irritating to eyes, nose, and throat. *IF INHALED*, will, will cause coughing or difficult breathing. *IF IN EYES*, hold eyelids open and flush. with plenty of water. *IF BREATHING HAS STOPPED*, give artificial respiration. IF breathing is difficult, administer oxygen. Lung edema may develop. **Solid:** Irritating to skin and eyes. *IF SWALLOWED*, will cause nausea and vomiting. Remove contaminated clothing and shoes. Flush affected areas with plenty of water. *IF IN EYES*, hold eyelids open and flush. with plenty of water. *IF SWALLOWED* and victim is *CONSCIOUS AND ABLE TO SWALLOW*, have victim drink 4 to 8 ounces of water and have victim induce vomiting. *IF SWALLOWED* and victim is *UNCONSCIOUS OR HAVING CONVULSIONS*, do nothing except keep victim warm.
Note to physician or authorized medical personnel: Medical observation is recommended for 24 to 48 hours after breathing overexposure, as pulmonary edema may be delayed. As first aid for pulmonary edema, consider administering a corticosteroid spray. Cigarette smoking may exacerbate pulmonary injury and should be discouraged for at least 72 hours following exposure.

HEALTH HAZARDS
Personal protective equipment (PPE): B-Level PPE. Dust mask; goggles or face shield; protective gloves.
Toxicity by ingestion: Grade 2; LD$_{50}$ = 0.5 to 5 g/kg.
Liquid or solid irritant characteristics: Causes irritation of the eyes and skin.

CHEMICAL REACTIVITY
Binary reactants: Violent reaction with aluminum, magnesium, strong bases.

ENVIRONMENTAL DATA
Food chain concentration potential: Zinc is concentrated by some organisms but is not considered to be bioconcentrative in a spill situation.
Water pollution: Harmful to aquatic life in very low concentrations. May be dangerous if it enters nearby water intakes; notify operators. Notify local health and wildlife officials. **Response to discharge:** Disperse and flush.

SHIPPING INFORMATION
Grades of purity: Reagent; Technical; **Storage temperature:** Ambient; **Inert atmosphere:** None; **Venting:** Open; **Stability during transport:** Stable.

PHYSICAL AND CHEMICAL PROPERTIES
Physical state @ 59°F/15°C and 1 atm: Solid.
Molecular weight: 287.54
Boiling point @ 1 atm: (decomposes).
Melting/Freezing point: (decomposes) 122–212°F/50–100°C/323–373°K.
Specific gravity (water = 1): 1.96 @ 68°F/20°C (solid).

ZIRCONIUM ACETATE   REC. Z:2300

SYNONYMS: ACETATO de ZIRCONIO (Spanish); ZIRCONIUM ACETATE SOLUTION

IDENTIFICATION
CAS Number: 7585-20-8
Formula: $Zr(C_2H_3O_2)_4 \cdot H_2O$

DESCRIPTION: Colorless liquid. Weak vinegar odor. Sinks and mixes with water.

Dust is Irritating to the skin, eyes, and respiratory tract. Toxic products of combustion may include carbon monoxide, and zinc oxide.

Hazard Classification (based on NFPA-704 Rating System)
Health Hazards (Blue): 1; Flammability (Red): 0; Reactivity (Yellow): 0

EMERGENCY RESPONSE: See Appendix A (171)
Evacuation:
Public safety: Isolate the area of spill or leak for at least 10 to 25 meters (30 to 80 feet) in all directions.
Spill: Increase, in the downwind direction, as necessary, the distance shown under "Public Safety."
Fire: If any large container is involved in fire, isolate for at least 800 meters (½ mile) in all directions; also, consider initial evacuation for 800 meters (½ mile) in all directions.

EXPOSURE
Short-term effects: *SEEK MEDICAL ATTENTION.* **Liquid:** Irritating to skin and eyes. Harmful if swallowed. Remove contaminated clothing and shoes. Flush affected areas with plenty of water. *IF IN EYES*, hold eyelids open and flush. with plenty of water. *IF SWALLOWED* and victim is *CONSCIOUS AND ABLE TO SWALLOW*, have victim drink 4 to 8 ounces of water. *IF SWALLOWED* and victim is *UNCONSCIOUS OR HAVING CONVULSIONS,* do nothing except keep victim warm.

HEALTH HAZARDS
Personal protective equipment (PPE): B-Level PPE. OSHA Table Z-1-A air contaminant. Rubber gloves; chemical goggles or face shield
Recommendations for respirator selection: NIOSH/OSHA as zirconium.
25 mg/m³: DM (any dust and mist respirator). *50 mg/m³:* DMXSQ, *If not present as a fume* (any dust and mist respirator except single-use and quarter mask respirators); or SA (any supplied-air respirator); or PAPRDM, *If not present as a fume* (any powered, air-purifying respirator with a dust and mist filter); or HiEF (any air-purifying, full-facepiece respirator with a high-efficiency particulate filter); or SA (any supplied-air respirator); or SCBAF (any self-contained breathing apparatus with a full facepiece). *EMERGENCY OR PLANNED ENTRY INTO UNKNOWN CONCENTRATIONS OR IDLH CONDITIONS:* SCBAF:PD,PP (any self-contained breathing apparatus that has a full facepiece and is operated in a pressure-demand or other positive-pressure mode); or SAF:PD,PP:ASCBA (any supplied-air respirator that has a full facepiece and is operated in a pressure-demand or other positive-pressure mode in combination with an auxiliary self-contained breathing apparatus operated in a pressure-demand or other positive pressure mode). *ESCAPE:* HiEF (any air-purifying, full-facepiece respirator with a high-efficiency particulate filter); or SCBAE (any appropriate escape-type, self-contained breathing apparatus).
Exposure limits (TWA unless otherwise noted): ACGIH TLV 5 mg/m³; NIOSH/OSHA 5 mg/m³ as zirconium.
Short-term exposure limits (15-minute TWA): NIOSH & ACGIH STEL 10 mg/m³ as zirconium.

Toxicity by ingestion: Grade 2; LD_{50} = 0.5 to 5 g/kg (rat).
Liquid or solid irritant characteristics: May cause irritation of the eyes and skin.
IDLH value: 50 mg/m³ as zirconium.

ENVIRONMENTAL DATA
Water pollution: Effect of low concentrations on aquatic life is unknown. May be dangerous if it enters nearby water intakes; notify operators. Notify local health and wildlife officials. **Response to discharge:** Issue warning–water contaminant. Disperse and flush.

SHIPPING INFORMATION
Grades of purity: 25% solution in water; **Storage temperature:** Ambient; **Inert atmosphere:** None; **Venting:** Open; **Stability during transport:** Stable.

PHYSICAL AND CHEMICAL PROPERTIES
Physical state @ 59°F/15°C and 1 atm: Liquid.
Molecular weight: 327 (solute only).
Specific gravity (water = 1): 1.37 @ 68°F/20°C (liquid).

ZIRCONIUM NITRATE   REC. Z:2400

SYNONYMS: NITRATO de ZIRCONIO (Spanish); ZIRCONIUM NITRATE PENTAHYDRATE

IDENTIFICATION
CAS Number: 13746-89-9
Formula: $Zr(NO_3)_4 \cdot 5H_2O$
DOT ID Number: UN 2728; DOT Guide Number: 140
Proper Shipping Name: Zirconium nitrate
Reportable Quantity (RQ): **(CERCLA)** 5000 lb/2270 kg

DESCRIPTION: White crystalline solid. Odorless. Sinks and mixes with water; forms an acid solution.

Corrosive • Poison! • Dust is corrosive to the skin, eyes, and respiratory tract • Strong oxidizer which may react explosively with low flash point organics or reducing agents. Heat forms oxygen; will increase the activity of an existing fire • May cause fire or explosion on contact with combustibles (wood, paper, oil, clothing, etc.) • Irritating to the skin, eyes, and respiratory tract • Toxic products of combustion may include nitrogen oxides.

Hazard Classification (based on NFPA-704 Rating System)
Health Hazards (Blue): 1; Flammability (Red): 0; Reactivity (Yellow): 0; Special Notice (White): OXY

EMERGENCY RESPONSE: See Appendix A (140)
Evacuation:
Public safety: Isolate the area of spill or leak for at least 10 to 25 meters (30 to 80 feet) in all directions.
Spill: Consider initial downwind evacuation for at least 100 meters (330 feet).
Fire: If any large container is involved in fire, isolate for at least 800 meters (½ mile) in all directions; also, consider initial evacuation for 800 meters (½ mile) in all directions.

EXPOSURE
Short-term effects: *SEEK MEDICAL ATTENTION.* **Dust:** Irritating to eyes, nose, and throat. *IF INHALED*, will, will cause coughing or difficult breathing. *IF IN EYES*, hold eyelids open and flush. with plenty of water. *IF BREATHING HAS STOPPED,* give

1166 Zirconium oxychloride

artificial respiration. IF breathing is difficult, administer oxygen.
Solid: Irritating to skin and eyes. *IF SWALLOWED*, will cause nausea and vomiting. Remove contaminated clothing and shoes. Flush affected areas with plenty of water. *IF IN EYES*, hold eyelids open and flush. with plenty of water. *IF SWALLOWED* and victim is *CONSCIOUS AND ABLE TO SWALLOW*, have victim drink 4 to 8 ounces of water. *IF SWALLOWED* and victim is *UNCONSCIOUS OR HAVING CONVULSIONS*, do nothing except keep victim warm.
Note to physician or authorized medical personnel: Medical observation is recommended for 24 to 48 hours after breathing overexposure, as pulmonary edema may be delayed. As first aid for pulmonary edema, consider administering a corticosteroid spray. Cigarette smoking may exacerbate pulmonary injury and should be discouraged for at least 72 hours following exposure.

HEALTH HAZARDS
Personal protective equipment (PPE): B-Level PPE. OSHA Table Z-1-A air contaminant. Dust mask; goggles or face shield; protective **gloves.**
Recommendations for respirator selection: NIOSH/OSHA as zirconium compounds
25 mg/m³: DM (any dust and mist respirator). *50 mg/m³:* DMXSQ, *If not present as a fume* (any dust and mist respirator except single-use and quarter mask respirators); or SA (any supplied-air respirator); or PAPRDM, *If not present as a fume* (any powered, air-purifying respirator with a dust and mist filter); or HiEF (any air-purifying, full-facepiece respirator with a high-efficiency particulate filter); or SA (any supplied-air respirator); or SCBAF (any self-contained breathing apparatus with a full facepiece). *EMERGENCY OR PLANNED ENTRY INTO UNKNOWN CONCENTRATIONS OR IDLH CONDITIONS:* SCBAF:PD,PP (any self-contained breathing apparatus that has a full facepiece and is operated in a pressure-demand or other positive-pressure mode); or SAF:PD,PP:ASCBA (any supplied-air respirator that has a full facepiece and is operated in a pressure-demand or other positive-pressure mode in combination with an auxiliary self-contained breathing apparatus operated in a pressure-demand or other positive pressure mode). *ESCAPE:* HiEF (any air-purifying, full-facepiece respirator with a high-efficiency particulate filter); or SCBAE (any appropriate escape-type, self-contained breathing apparatus).
Exposure limits (TWA unless otherwise noted): ACGIH TLV 5 mg/m³; NIOSH/OSHA 5 mg/m³ as zirconium.
Short-term exposure limits (15-minute TWA): NIOSH & ACGIH STEL 10 mg/m³ **as zirconium.**
Toxicity by ingestion: Grade 2; oral LD_{50} = 2.5 g/kg (rat).
Liquid or solid irritant characteristics: Causes eye and skin irritation.
IDLH value: 50 mg/m³ as zirconium.

CHEMICAL REACTIVITY
Reactivity with water: Dissolves to give an acid solution of zirconium oxychloride octahydrate.
Binary reactants: Corrodes most metals
Neutralizing agents for acids and caustics: Flush with water.

ENVIRONMENTAL DATA
Water pollution: Effect of low concentrations on aquatic life is unknown. May be dangerous if it enters nearby water intakes; notify operators. Notify local health and wildlife officials.
Response to discharge: Issue warning–water contaminant. Disperse and flush.

SHIPPING INFORMATION
Grades of purity: Commercial; 99+%; **Storage temperature:** Ambient; **Inert atmosphere:** None; **Venting:** Open; **Stability during transport:** Stable.

PHYSICAL AND CHEMICAL PROPERTIES
Physical state @ 59°F/15°C and 1 atm: Solid.
Molecular weight: 429.3
Boiling point @ 1 atm: (decomposes).
Specific gravity (water = 1): More than 1 @ 68°F/20°C (solid).

ZIRCONIUM OXYCHLORIDE   REC. Z:2500

SYNONYMS: BASIC ZIRCONIUM CHLORIDE; OXICLORURO de ZIRCONIO (Spanish); ZIRCONYL CHLORIDE; ZIRCONIUM OXYCHLORIDE HYDRATE; ZIRCONIUM OXIDE CHLORIDE

IDENTIFICATION
CAS Number: 7699-43-6
Formula: Cl_2OZr; $ZrOCl_2 \cdot 8H_2O$

DESCRIPTION: White to yellow solid. Odorless. Sinks and mixes with water forming highly acidic zirconium oxychloride octahydrate.

Corrosive to the skin, eyes, and respiratory tract; contact with skin or eyes can cause burns and vision impairment; inhalation symptoms may be delayed • Toxic products of combustion may include hydrogen chloride.

Hazard Classification (based on NFPA-704 Rating System)
Health Hazards (Blue): 3; Flammability (Red): 0; Reactivity (Yellow): 0

EMERGENCY RESPONSE: See Appendix A (171)
Evacuation:
Public safety: Isolate the area of spill or leak for at least 10 to 25 meters (30 to 80 feet) in all directions.
Spill: Increase, in the downwind direction, as necessary, the distance shown under "Public Safety."
Fire: If any large container is involved in fire, isolate for at least 800 meters (½ mile) in all directions; also, consider initial evacuation for 800 meters (½ mile) in all directions.

EXPOSURE
Short-term effects: *SEEK MEDICAL ATTENTION.* **Dust:** Irritating to eyes, nose, and throat. *IF INHALED*, will, will cause coughing or difficult breathing. *IF IN EYES*, hold eyelids open and flush. with plenty of water. *IF BREATHING HAS STOPPED*, give artificial respiration. IF breathing is difficult, administer oxygen.
Solid: Irritating to skin and eyes. *IF SWALLOWED*, will cause nausea and vomiting. Remove contaminated clothing and shoes. Flush affected areas with plenty of water. *IF IN EYES*, hold eyelids open and flush. with plenty of water. *IF SWALLOWED* and victim is *CONSCIOUS AND ABLE TO SWALLOW*, have victim drink 4 to 8 ounces of water. *IF SWALLOWED* and victim is *UNCONSCIOUS OR HAVING CONVULSIONS*, do nothing except keep victim warm.
Note to physician or authorized medical personnel: Medical observation is recommended for 24 to 48 hours after breathing overexposure, as pulmonary edema may be delayed. As first aid for pulmonary edema, consider administering a corticosteroid spray.

Cigarette smoking may exacerbate pulmonary injury and should be discouraged for at least 72 hours following exposure.

HEALTH HAZARDS
Personal protective equipment (PPE): B-Level PPE. OSHA Table Z-1-A air contaminant. Safety glasses or face shield; protective gloves; dust mask.
Recommendations for respirator selection: NIOSH/OSHA as zirconium.
25 mg/m³: DM (any dust and mist respirator). *50 mg/m³:* DMXSQ, *If not present as a fume* (any dust and mist respirator except single-use and quarter mask respirators); or SA (any supplied-air respirator); or PAPRDM, *If not present as a fume* (any powered, air-purifying respirator with a dust and mist filter); or HiEF (any air-purifying, full-facepiece respirator with a high-efficiency particulate filter); or SA (any supplied-air respirator); or SCBAF (any self-contained breathing apparatus with a full facepiece). *EMERGENCY OR PLANNED ENTRY INTO UNKNOWN CONCENTRATIONS OR IDLH CONDITIONS:* SCBAF:PD,PP (any self-contained breathing apparatus that has a full facepiece and is operated in a pressure-demand or other positive-pressure mode); or SAF:PD,PP:ASCBA (any supplied-air respirator that has a full facepiece and is operated in a pressure-demand or other positive-pressure mode in combination with an auxiliary self-contained breathing apparatus operated in a pressure-demand or other positive pressure mode). *ESCAPE:* HiEF (any air-purifying, full-facepiece respirator with a high-efficiency particulate filter); or SCBAE (any appropriate escape-type, self-contained breathing apparatus).
Exposure limits (TWA unless otherwise noted): ACGIH TLV 5 mg/m³; NIOSH/OSHA 5 mg/m³ as zirconium.
Short-term exposure limits (15-minute TWA): NIOSH & ACGIH STEL 10 mg/m³ as zirconium.
Toxicity by ingestion: Grade 2; oral LD_{50} = 3.5 g/kg (rat).
Liquid or solid irritant characteristics: may cause irritation of eyes and skin.
IDLH value: 50 mg/m³ as zirconium.

ENVIRONMENTAL DATA
Water pollution: Effect of low concentrations on aquatic life is unknown. May be dangerous if it enters nearby water intakes; notify operators. Notify local health and wildlife officials.
Response to discharge: Issue warning–water contaminant. Disperse and flush.

SHIPPING INFORMATION
Grades of purity: Technical; Pure; **Storage temperature:** Ambient; **Inert atmosphere:** None; **Venting:** Open; **Stability during transport:** Stable.

PHYSICAL AND CHEMICAL PROPERTIES
Physical state @ 59°F/15°C and 1 atm: Solid.
Molecular weight: 322.3
Boiling point @ 1 atm: (decomposes).
Specific gravity (water = 1): More than 1 @ 68°F/20°C (solid).

ZIRCONIUM POTASSIUM FLUORIDE   REC. Z:2600

SYNONYMS: FLUORURU de ZIRCONIO y POTASIO (Spanish); POTASSIUM FLUOZIRCONATE; POTASSIUM HEXAFLUOROZIRCONATE; POTASSIUM ZIRCONIUM FLUORIDE

IDENTIFICATION
CAS Number: 16923-95-8
Formula: K_2ZrF_6
DOT ID Number: UN 9162; DOT Guide Number: 171
Proper Shipping Name: Zirconium potassium fluoride
Reportable Quantity (RQ): **(CERCLA)** 1000 lb/454 kg

DESCRIPTION: Colorless crystalline solid or white powder. Sinks and mixes with water.

Dust is Irritating to the skin, eyes, and respiratory tract • Toxic products of combustion may include fluorine and potassium oxide.

Hazard Classification (based on NFPA-704 Rating System)
Health Hazards (Blue): 2; Flammability (Red): 0; Reactivity (Yellow): 0

EMERGENCY RESPONSE: See Appendix A (171)
Evacuation:
Public safety: Isolate the area of spill or leak for at least 10 to 25 meters (30 to 80 feet) in all directions.
Spill: Increase, in the downwind direction, as necessary, the distance shown under "Public Safety."
Fire: If any large container is involved in fire, isolate for at least 800 meters (½ mile) in all directions; also, consider initial evacuation for 800 meters (½ mile) in all directions.

EXPOSURE
Short-term effects: *SEEK MEDICAL ATTENTION.* **Dust:** Harmful if inhaled. Move to fresh air. **Solid:** Harmful if swallowed. Irritating to skin.
Flush affected areas with plenty of water. *IF IN EYES*, hold eyelids open and flush. with plenty of water. *IF SWALLOWED* and victim is *CONSCIOUS AND ABLE TO SWALLOW*, have victim drink 4 to 8 ounces of water and have victim induce vomiting.

HEALTH HAZARDS
Personal protective equipment (PPE): B-Level PPE. OSHA Table Z-1-A air contaminant. Rubber gloves, safety glasses.
Recommendations for respirator selection: NIOSH/OSHA as zirconium.
25 mg/m³: DM (any dust and mist respirator). *50 mg/m³:* DMXSQ, *If not present as a fume* (any dust and mist respirator except single-use and quarter mask respirators); or SA (any supplied-air respirator); or PAPRDM, *If not present as a fume* (any powered, air-purifying respirator with a dust and mist filter); or HiEF (any air-purifying, full-facepiece respirator with a high-efficiency particulate filter); or SA (any supplied-air respirator); or SCBAF (any self-contained breathing apparatus with a full facepiece). *EMERGENCY OR PLANNED ENTRY INTO UNKNOWN CONCENTRATIONS OR IDLH CONDITIONS:* SCBAF:PD,PP (any self-contained breathing apparatus that has a full facepiece and is operated in a pressure-demand or other positive-pressure mode); or SAF:PD,PP:ASCBA (any supplied-air respirator that has a full facepiece and is operated in a pressure-demand or other positive-pressure mode in combination with an auxiliary self-contained breathing apparatus operated in a pressure-demand or other positive pressure mode). *ESCAPE:* HiEF (any air-purifying, full-facepiece respirator with a high-efficiency particulate filter); or SCBAE (any appropriate escape-type, self-contained breathing apparatus).
Exposure limits (TWA unless otherwise noted): ACGIH TLV 5 mg/m³; NIOSH/OSHA 5 mg/m³ as zirconium.
Short-term exposure limits (15-minute TWA): NIOSH & ACGIH STEL 10 mg/m³ as zirconium.

Toxicity by ingestion: Grade 3; LD_{50} = 50 to 500 mg/kg.
Long-term health effects: Suppressed growth and caused thickening interalveolar lung septa in rats. May cause erythematous, papular and granulomatous reactions on skin.
Liquid or solid irritant characteristics: May cause irritation of eyes and skin.
IDLH value: 50 mg/m^3 as zirconium.

ENVIRONMENTAL DATA
Food chain concentration potential: Fluorine is concentrated by aquatic animals.
Water pollution: Harmful to aquatic life in very low concentrations. May be dangerous if it enters nearby water intakes; notify operators. Notify local health and wildlife officials.
Response to discharge: Chemical and physical treatment. Disperse and flush.

PHYSICAL AND CHEMICAL PROPERTIES
Physical state @ 59°F/15°C and 1 atm: Solid.
Molecular weight: 283.41
Melting/Freezing point: 1724°F/840°C/1113.2°K.
Specific gravity (water = 1): 3.48 at room temperature

ZIRCONIUM SULFATE   REC. Z:2700

SYNONYMS: DISULFATOZIRCONIC ACID; SULFATO de ZIRCONIO (Spanish); ZIRCONIUM(IV) SULFATE; ZIRCONIUM SULFATE TETRAHYDRATE; ZIRCONYL SULFATE

IDENTIFICATION
CAS Number: 14644-61-2
Formula: $O_8S_2 \cdot Zr$; $Zr(SO_4)_2 \cdot 4H_2O$
DOT ID Number: UN 9163; DOT Guide Number: 171
Proper Shipping Name: Zirconium sulfate
Reportable Quantity (RQ): **(CERCLA)** 5000 lb/2270 kg

DESCRIPTION: White crystalline solid. Odorless. Sinks and mixes with water.

Dust is Irritating to the skin, eyes, and respiratory tract • Toxic products of combustion may include sulfur oxides.

Hazard Classification (based on NFPA-704 Rating System)
Health Hazards (Blue): 0; Flammability (Red): 0; Reactivity (Yellow): 0

EMERGENCY RESPONSE: See Appendix A (171)
Evacuation:
Public safety: Isolate the area of spill or leak for at least 10 to 25 meters (30 to 80 feet) in all directions.
Spill: Increase, in the downwind direction, as necessary, the distance shown under "Public Safety."
Fire: If any large container is involved in fire, isolate for at least 800 meters (½ mile) in all directions; also, consider initial evacuation for 800 meters (½ mile) in all directions.

EXPOSURE
Short-term effects: *SEEK MEDICAL ATTENTION.* **Dust:** Irritating to eyes, nose, and throat. *IF INHALED*, will, will cause coughing or difficult breathing. *IF IN EYES*, hold eyelids open and flush. with plenty of water. *IF BREATHING HAS STOPPED*, give artificial respiration. IF breathing is difficult, administer oxygen. **Solid:** Irritating to skin and eyes. *IF SWALLOWED*, will cause nausea and vomiting. Remove contaminated clothing and shoes. Flush affected areas with plenty of water. *IF IN EYES*, hold eyelids open and flush. with plenty of water. *IF SWALLOWED* and victim is *CONSCIOUS AND ABLE TO SWALLOW*, have victim drink 4 to 8 ounces of water. *IF SWALLOWED* and victim is *UNCONSCIOUS OR HAVING CONVULSIONS*, do nothing except keep victim warm.

HEALTH HAZARDS
Personal protective equipment (PPE): B-Level PPE. OSHA Table Z-1-A air contaminant. Dust mask; goggles or face shield; protective gloves.
Recommendations for respirator selection: NIOSH/OSHA as zirconium compounds
25 mg/m^3: DM (any dust and mist respirator). *50 mg/m^3:* DMXSQ, *If not present as a fume* (any dust and mist respirator except single-use and quarter mask respirators); or SA (any supplied-air respirator); or PAPRDM, *If not present as a fume* (any powered, air-purifying respirator with a dust and mist filter); or HiEF (any air-purifying, full-facepiece respirator with a high-efficiency particulate filter); or SA (any supplied-air respirator); or SCBAF (any self-contained breathing apparatus with a full facepiece). *EMERGENCY OR PLANNED ENTRY INTO UNKNOWN CONCENTRATIONS OR IDLH CONDITIONS:* SCBAF:PD,PP (any self-contained breathing apparatus that has a full facepiece and is operated in a pressure-demand or other positive-pressure mode); or SAF:PD,PP:ASCBA (any supplied-air respirator that has a full facepiece and is operated in a pressure-demand or other positive-pressure mode in combination with an auxiliary self-contained breathing apparatus operated in a pressure-demand or other positive pressure mode). *ESCAPE:* HiEF (any air-purifying, full-facepiece respirator with a high-efficiency particulate filter); or SCBAE (any appropriate escape-type, self-contained breathing apparatus).
Exposure limits (TWA unless otherwise noted): ACGIH TLV 5 mg/m^3 NIOSH/OSHA 5 mg/m^3 as zirconium.
Short-term exposure limits (15-minute TWA): NIOSH & ACGIH STEL 10 mg/m^3 **as zirconium.**
Toxicity by ingestion: Grade 2; oral LD_{50} = 3.5 g/kg (rat).
Long-term health effects: Reproductive and mutation data.
Liquid or solid irritant characteristics: Causes irritation of eyes and skin.
IDLH value: 50 mg/m^3 as zirconium.

FIRE DATA
Fire extinguishing agents not to be used: Water streams applied to adjacent fires will spread contamination of environmentally hazardous substance over wide area.

CHEMICAL REACTIVITY
Binary reactants: Incompatible with aluminum, magnesium. Dusts of zirconium compounds ignite and explode in a Carbon dioxide atmosphere.

ENVIRONMENTAL DATA
Water pollution: Effect of low concentrations on aquatic life is unknown. May be dangerous to water intakes. Notify local health and wildlife officials. **Response to discharge:** Issue warning–water contaminant. Disperse and flush.

SHIPPING INFORMATION
Grades of purity: Technical; **Storage temperature:** Ambient; **Inert atmosphere:** None; **Venting:** Open; **Stability during transport:** Stable.

PHYSICAL AND CHEMICAL PROPERTIES
Physical state @ 59°F/15°C and 1 atm: Solid.
Molecular weight: 355.4
Specific gravity (water = 1): (approximate) 3.0 @ 68°F/20°C (solid).

ZIRCONIUM TETRACHLORIDE   REC. Z:2800

SYNONYMS: ZIRCONIUM CHLORIDE; ZIRCONIUM(IV) CHLORIDE; TETRACHLORO ZIRCONIUM; TETRACLORURO de ZIRCONIO (Spanish)

IDENTIFICATION
CAS Number: 10026-11-6
Formula: $ZrCl_4$
DOT ID Number: UN 2503; DOT Guide Number: 137
Proper Shipping Name: Zirconium tetrachloride
Reportable Quantity (RQ): **(CERCLA)** 5000 lb/2270 kg

DESCRIPTION: White, lustrous crystalline solid. Sinks and decomposes in water; reacts vigorously, forming heat, hydrochloric acid, and zirconium hydroxide fumes.

Corrosive to the skin, eyes, and respiratory tract; contact with skin or eyes can cause burns and vision impairment; inhalation symptoms may be delayed • Firefighting gear (including SCBA) does not provide adequate protection. If exposure occurs, remove and isolate gear immediately and thoroughly decontaminate personnel • Toxic products of combustion may include hydrogen chloride.

Hazard Classification (based on NFPA-704 Rating System)
Health Hazards (Blue): 3; Flammability (Red): 0; Reactivity (Yellow): 2; Special Notice (White): Water reactive

EMERGENCY RESPONSE: See Appendix A (137)
Evacuation:
Public safety: Isolate the area of spill or leak for at least 50 to 100 meters (160 to 330 feet) in all directions.
Spill: Increase, in the downwind direction, as necessary, the distance shown under "Public Safety."
Fire: If any large container is involved in fire, isolate for at least 800 meters (½ mile) in all directions; also, consider initial evacuation for 800 meters (½ mile) in all directions.

EXPOSURE
Short-term effects: *SEEK MEDICAL ATTENTION.* **Vapor:** Irritating to eyes, nose, and throat. Harmful if inhaled. Move to fresh air. *IF BREATHING HAS STOPPED,* give artificial respiration respiration; *avoid mouth-to-mouth resuscitation; use bag/mask apparatus.* IF breathing is difficult, administer oxygen. **Solid:** Irritation to skin and eyes. Harmful if swallowed. Remove contaminated clothing and shoes. Flush affected areas with plenty of water. *IF IN EYES,* hold eyelids open and flush. with plenty of water. *IF SWALLOWED* and victim is *CONSCIOUS AND ABLE TO SWALLOW,* have victim drink 4 to 8 ounces of water. **Do NOT induce vomiting.**
Note to physician or authorized medical personnel: Medical observation is recommended for 24 to 48 hours after breathing overexposure, as pulmonary edema may be delayed. As first aid for pulmonary edema, consider administering a corticosteroid spray. Cigarette smoking may exacerbate pulmonary injury and should be discouraged for at least 72 hours following exposure.

HEALTH HAZARDS
Personal protective equipment (PPE): A-Level PPE. OSHA Table Z-1-A air contaminant. Helmet, self-contained breathing apparatus, rubber boots, gloves, bands around legs, arms and waist. Facemask as well as covering rest of head.
Recommendations for respirator selection: NIOSH/OSHA as zirconium.
25 mg/m^3: DM (any dust and mist respirator). *50 mg/m^3:* DMXSQ, *If not present as a fume* (any dust and mist respirator except single-use and quarter mask respirators); or SA (any supplied-air respirator); or PAPRDM, *If not present as a fume* (any powered, air-purifying respirator with a dust and mist filter); or HiEF (any air-purifying, full-facepiece respirator with a high-efficiency particulate filter); or SA (any supplied-air respirator); or SCBAF (any self-contained breathing apparatus with a full facepiece). *EMERGENCY OR PLANNED ENTRY INTO UNKNOWN CONCENTRATIONS OR IDLH CONDITIONS:* SCBAF:PD,PP (any self-contained breathing apparatus that has a full facepiece and is operated in a pressure-demand or other positive-pressure mode); or SAF:PD,PP:ASCBA (any supplied-air respirator that has a full facepiece and is operated in a pressure-demand or other positive-pressure mode in combination with an auxiliary self-contained breathing apparatus operated in a pressure-demand or other positive pressure mode). *ESCAPE:* HiEF (any air-purifying, full-facepiece respirator with a high-efficiency particulate filter); or SCBAE (any appropriate escape-type, self-contained breathing apparatus).
Exposure limits (TWA unless otherwise noted): ACGIH TLV 5 mg/m^3; NIOSH/OSHA 5 mg/m^3 as zirconium.
Short-term exposure limits (15-minute TWA): NIOSH & ACGIH STEL 10 mg/m^3 as zirconium.
Toxicity by ingestion: Grade 2; LD_{50} = 0.5 to 5.0 g/kg.
Long-term health effects: May cause granuloma in skin. A mild lung irritant.
Vapor (gas) irritant characteristics: Vapors of hydrogen chloride given off cause severe irritation of eyes and throat and can cause lung injury. They cannot be tolerated even at low concentrations.
Liquid or solid irritant characteristics: Fairly severe skin irritant, may cause pain and second-degree burns after a few minutes of contact.
IDLH value: 50 mg/m^3 as zirconium.

FIRE DATA
Fire extinguishing agents not to be used: Water reactive. Water streams applied to adjacent fires will spread contamination of environmentally hazardous substance over wide area.
Behavior in fire: Will not burn. Sublimes above 626°F/331°C.

CHEMICAL REACTIVITY
Reactivity with water: Decomposes in the presence of moist air forming a dense white cloud and hydrogen chloride fumes; may ignite spontaneously.
Binary reactants: Reacts with water, steam releasing heat, a dense white cloud, and hydrogen chloride. Violent reaction with bromine pentafluoride, lithium, oxygen difluoride, tetrahydrofuran. Contact with alkali metals may cause explosions. Incompatible with ammonium-*N*-nitrosophenylhydroxylamine. Attacks many plastics, rubber, and coatings. If moist will form hydrochloric acid which is corrosive to many metals.
Neutralizing agents for acids and caustics: Sodium bicarbonate and ammonia.

ENVIRONMENTAL DATA
Water pollution: Dangerous to aquatic life in high concentrations. May be a danger if it enters nearby water intakes; notify operators.

Notify local health and wildlife officials. **Response to discharge:** Issue warning–corrosive. Restrict access. Chemical and physical treatment. Disperse and flush.

SHIPPING INFORMATION
Storage temperature: Cool; **Stability during transport:** Stable

PHYSICAL AND CHEMICAL PROPERTIES
Physical state @ 59°F/15°C and 1 atm: Solid.
Molecular weight: 233.05
Boiling point @ 1 atm: Sublimes 628°F/331°C/604°K.
Melting/Freezing point: 572°F/300°C/573°K.
Specific gravity (water = 1): 2.80
Latent heat of vaporization: (sublimation) 195.3 btu/lb = 108.5 cal/g at 311°C/4.54 x 10^5 J/kg.

Appendix A

Emergency Response

DOT GUIDE NUMBER: 113 EMERGENCY RESPONSE: Stay upwind and uphill • Determine the extent of the problem • Isolate the area of release or fire and deny entry • Remove all ignition sources • IF material or contaminated runoff from fire control enters waterways, notify downstream users of potentially contaminated water • Wear positive-pressure, self-contained breathing apparatus (SCBA). Wear chemical protective clothing which is specifically recommended by the manufacturer • **RELEASE NO FIRE** • *DO NOT LET THIS MATERIAL DRY OUT.* Small spill: Flush areas with flooding quantities of water; dike for later disposal. Large spill: Wet down with water and dike for later disposal. *KEEP "WETTED" PRODUCT WET BY SLOWLY ADDING FLOODING QUANTITIES OF WATER.* Ventilate confined area if it can be done without placing personnel at risk • **FIRE** • Cargo: *Do NOT attempt to extinguish when fire reaches cargo! Cargo may explode!* Evacuate (see Fire, below) and let burn. Do not move cargo or vehicle if cargo has been exposed to heat. Use flooding quantities of water; if water is not available, use CO_2, *dry* chemical, *dry* sand, or *dry* earth. If possible, and *WITHOUT RISK*, cool exposed containers with large quantities of water from maximum distance and unattended equipment until well after fire is out. Always stay away from ends of tanks, especially when engulfed in flames. If employees are expected to fight fires, they must be trained and equipped in OSHA 1910.156.

DOT GUIDE NUMBER: 115 EMERGENCY RESPONSE: Stay upwind and uphill • Determine the extent of the problem • Isolate the area of release or fire and deny entry • Remove all ignition sources • IF material or contaminated runoff from fire control enters waterways, notify downstream users of potentially contaminated water • Wear positive-pressure, self-contained breathing apparatus (SCBA). Wear chemical protective clothing which is specifically recommended by the manufacturer • **RELEASE NO FIRE** • Stop the release if it can be done safely from a distance • If possible, turn leaking containers so that gas escapes rather than liquid • Use large amount of water to disperse gas–contain runoff. Ventilate confined area if it can be done without placing personnel at risk • **FIRE** • *DO NOT EXTINGUISH A LEAKING GAS FIRE UNLESS THE FLOW OF THE GAS CAN BE STOPPED* and any remaining gas is out of the line. Specially trained personnel may use fog lines to cool exposures and let the fire burn itself out • Small fire: Use *dry* chemical or CO_2. Large fire: Move containers from fire area if it can be done without risk. Use water spray or fog. Cool exposed containers with large quantities of water from unattended equipment until well after fire is out • Do not direct water at source of leak or safety devices; icing may occur • If cooling streams are ineffective (*venting sound increases in volume and pitch, tank discolors or shows signs of deforming*), withdraw immediately to a secure location. Always stay away from ends of tanks, especially when engulfed in flames. *Caution:* When in contact with refrigerated/cryogenic liquids, many materials become brittle and are likely to break without warning. If employees are expected to fight fires, they must be trained and equipped in OSHA 1910.156.

DOT GUIDE NUMBER: 116 EMERGENCY RESPONSE: Restrict access to the spill and isolate area until gas has dispersed • Stay upwind and uphill • Determine the extent of the problem • Isolate the area of release or fire and deny entry • Remove all ignition sources • IF material or contaminated runoff from fire control enters waterways, notify downstream users of potentially contaminated water • Wear positive-pressure, self-contained breathing apparatus (SCBA). Wear chemical protective clothing which is specifically recommended by the manufacturer • **RELEASE NO FIRE** • Stop the release if it can be done safely from a distance • If possible, turn leaking containers so that gas escapes rather than liquid • Use large amount of water to disperse gas–contain runoff. Ventilate confined area if it can be done without placing personnel at risk • **FIRE** • *DO NOT EXTINGUISH A LEAKING GAS FIRE UNLESS THE FLOW OF THE GAS CAN BE STOPPED* and any remaining gas is out of the line. Specially trained personnel may use fog lines to cool exposures and let the fire burn itself out • Small fire: Use *dry* chemical or CO_2. Large fire: Move containers from fire area if it can be done without risk. Use water spray, fog, or regular foam. Cool exposed containers with large quantities of water from unattended equipment until well after fire is out • Do not direct water at source of leak or safety devices; icing may occur • If cooling streams are ineffective (*venting sound increases in volume and pitch, tank discolors or shows signs of deforming*), withdraw immediately to a secure location. Always stay away from ends of tanks, especially when engulfed in flames. *Caution:* When in contact with refrigerated/cryogenic liquids, many materials become brittle and are likely to break without warning. If employees are expected to fight fires, they must be trained and equipped in OSHA 1910.156.

DOT GUIDE NUMBER: 117 EMERGENCY RESPONSE: Restrict access to the spill and isolate area until gas has dispersed • Stay upwind and uphill • Determine the extent of the problem • Isolate the area of release or fire and deny entry • Remove all ignition sources • IF material or contaminated runoff from fire control enters waterways, notify downstream users of potentially contaminated water • Wear positive-pressure, self-contained breathing apparatus (SCBA). Wear chemical protective clothing which is specifically recommended by the manufacturer • **RELEASE NO FIRE** • Stop the release if it can be done safely from a distance • If possible, turn leaking containers so that gas escapes rather than liquid • Use large amount of water to disperse gas–contain runoff. Ventilate confined area if it can be done without placing personnel at risk • **FIRE** • *DO NOT EXTINGUISH A LEAKING GAS FIRE UNLESS THE FLOW OF THE GAS CAN BE STOPPED* and any remaining gas is out of the line. Specially trained personnel may use fog lines to cool exposures and let the fire burn itself out • Small fire: Use *dry* chemical, CO_2, or regular foam. Large fire: Move containers from fire area if it can be done without risk. *Damaged cylinders should be handled only by specialists.* Use water spray, fog, or regular foam. Cool exposed containers with large quantities of water from unattended equipment until well after fire is out • Do not direct water at source of leak or safety devices; icing may occur • If cooling streams are ineffective (*venting sound increases in volume and pitch, tank discolors or shows signs of deforming*), withdraw immediately to a secure location. *Caution:* When in contact with refrigerated/cryogenic liquids, many materials become brittle and are likely to break without warning. Always stay away from ends of tanks, especially when engulfed in flames. If employees are expected to fight fires, they must be trained and equipped in OSHA 1910.156.

DOT GUIDE NUMBER: 118 EMERGENCY RESPONSE: Stay upwind and uphill • Determine the extent of the problem • Isolate the area of release or fire and deny entry • Remove all ignition sources • Wear positive-pressure, self-contained breathing apparatus (SCBA). Wear fully encapsulating, vapor protective chemical protective clothing which is specifically recommended by the manufacturer • Structural firefighters' protective clothing offers limited protection and not effective in fire situations • **RELEASE, NO FIRE** • Stop the release if it can be done safely from a distance • Use large amount of water to disperse gas–contain runoff. If possible, turn leaking containers so that gas escapes rather than liquid. Ventilate confined area if it can be done without placing personnel at risk • **FIRE** • *DO NOT EXTINGUISH THE FIRE UNLESS THE FLOW OF THE GAS CAN BE STOPPED AND ANY REMAINING GAS IS OUT OF THE LINE* • Stop the release if it can be done safely from a distance • Small fire: Use *dry* chemical or CO_2. Specially trained personnel may use fog lines to cool exposures and let the fire burn itself out • Large fire: Use water spray, fog, or regular foam. Cool exposed containers with large quantities of water from unattended equipment until well after fire is out, or remove intact containers if it can be done safely • Do not direct water at source of leak or safety devices; icing may occur • If cooling streams are ineffective (*venting sound increases in volume and pitch, tank discolors or shows signs of deforming*), withdraw immediately to a secure location. Always stay away from ends of tanks, especially when engulfed in flames. If employees are expected to fight fires, they must be trained and equipped in OSHA 1910.156.

DOT GUIDE NUMBER: 119 *CHLOROSILANES ONLY* (See below for other chemicals) **EMERGENCY RESPONSE:** Stay upwind and uphill • Isolate the area of release or fire and deny entry • Remove all ignition sources. All equipment used to handle this material must be grounded • IF material or contaminated runoff from fire control enters waterways, notify downstream users of potentially contaminated water • Wear positive-pressure, self-contained breathing apparatus (SCBA). Wear chemical protective clothing which is specifically recommended by the manufacturer. Structural firefighter's protective clothing provides only limited protection • **RELEASE, NO FIRE** • Small spill: Cover with *dry* earth, *dry* sand, or other noncombustible material followed with plastic sheet to minimize spreading or contact with water or rain. Use clean nonsparking tools to collect material and place material into loosely covered plastic containers for later disposal • Stop the release if it can be done safely from a distance • Prevent material and runoff from entering sewers and waterways if it can be done safely well ahead of the release • For chlorosilanes, apply alcohol-resistanttant (AFFF) medium expansion foam to large areas of spilled liquid to control vapors • Ventilate confined area if it can be done without placing personnel at risk • **FIRE** • For chlorosilanes, do NOT use water. Small fire: Extinguish with *dry* chemical, CO_2, *dry* sand, or alcohol-resistant foam. *CAUTION:* Most foams will react with the material and release corrosive/toxic gases. May react with foams releasing corrosive/toxic gases • Large fire: Use alcohol-resistant medium expansion foam. *CAUTION:* Most foams will react with the material and release corrosive/toxic gases • Cool closed, exposed containers with large quantities of water from unattended equipment or remove intact containers if it can be done safely. DO NOT GET WATER ON SPILLED SUBSTANCE OR INSIDE CONTAINERS • If cooling streams are ineffective (*venting sound increases in volume and pitch, tank discolors or shows signs of deforming*), withdraw immediately to a secure location. Always stay away from ends of tanks, especially when engulfed in flames. If employees are expected to fight fires, they must be trained and equipped in OSHA 1910.156.

DOT GUIDE NUMBER: 119 EMERGENCY RESPONSE: Stay upwind and uphill • Determine the extent of the problem • Isolate the area of release or fire and deny entry • Remove all ignition sources • IF material or contaminated runoff from fire control enters waterways, notify downstream users of potentially contaminated water • Wear positive-pressure, self-contained breathing apparatus (SCBA). Wear fully encapsulating, vapor protective chemical protective clothing which is specifically recommended by the manufacturer • Structural firefighters' protective clothing offers limited protection and not effective in fire situations • **RELEASE, NO FIRE** • Stop the release if it can be done safely from a distance • Use water spray to disperse gas–contain runoff. If possible, turn leaking containers so that gas escapes rather than liquid. Ventilate confined area if it can be done without placing personnel at risk • **FIRE** • *DO NOT EXTINGUISH THE FIRE UNLESS THE FLOW OF THE GAS CAN BE STOPPED AND ANY REMAINING GAS IS OUT OF THE LINE* • Stop the release if it can be done safely from a distance • Small fire: Use *dry* chemical, CO_2, water spray, or alcohol-resistant foam. Specially trained personnel may use fog lines to cool exposures and let the fire burn itself out • Large fire: Use water spray, fog, or alcohol-resistant foam. Cool exposed containers with large quantities of water from unattended equipment until well after fire is out, or remove intact containers if it can be done safely • Do not direct water at source of leak or safety devices; icing may occur • If cooling streams are ineffective (*venting sound increases in volume and pitch, tank discolors or shows signs of deforming*), withdraw immediately to a secure location. Always stay away from ends of tanks, especially when engulfed in flames. If employees are expected to fight fires, they must be trained and equipped in OSHA 1910.156.

DOT GUIDE NUMBER: 120 EMERGENCY RESPONSE: Stay upwind and uphill • Determine the extent of the problem • Isolate the area of release or fire and deny entry • Remove all ignition sources • Wear positive-pressure, self-contained breathing apparatus (SCBA). Wear fully encapsulating, vapor protective chemical protective clothing which is specifically recommended by the manufacturer • **RELEASE, NO FIRE** • Stop the release if it can be done safely from a distance • Use large amount of water to disperse gas–contain runoff. If possible, turn leaking containers so that gas escapes rather than liquid. Ventilate confined area if it can be done without placing personnel at risk • **FIRE** • Stop the release if it can be done safely from a distance • Use extinguishing agent suitable for type of surrounding fire • Cool exposed containers with large quantities of water from unattended equipment until well after fire is out, or remove intact containers if it can be done safely • Damaged cylinders should be handled only by specialists • Do not direct water at source of leak or safety devices; icing may occur • *Caution:* When in contact with refrigerated/cryogenic liquids, many materials become brittle and are likely to break without warning. If cooling streams are ineffective (*venting sound increases in volume and pitch, tank discolors or shows signs of deforming*), withdraw immediately to a secure location. Always stay away from ends of tanks, especially when engulfed in flames. If employees are expected to fight fires, they must be trained and equipped in OSHA 1910.156.

DOT GUIDE NUMBER: 121 EMERGENCY RESPONSE: Stay upwind and uphill • Determine the extent of the problem • Isolate the area of release or fire and deny entry • Remove all ignition sources • Wear positive-pressure, self-contained breathing apparatus (SCBA). Wear fully encapsulating, vapor protective chemical protective clothing which is specifically recommended by the manufacturer • **RELEASE, NO FIRE** • Stop the release if it can be done safely from a distance • Use large amount of water spray to disperse gas–contain runoff. If possible, turn leaking containers so that gas escapes rather than liquid. Ventilate confined area if it can be done without placing personnel at risk • **FIRE** • Stop the release if it can be done safely from a distance • Use extinguishing agent suitable for type of surrounding fire • Cool exposed containers with large quantities of water from unattended equipment until well after fire is out, or remove intact containers if it can be done safely • Damaged cylinders should be handled only by specialists • If tanks are involved or in case of Large fire: Fight fire from maximum distances or use unmanned hose holders or monitor nozzles; if this is impossible, withdraw from area and let fire burn • Do not direct water at source of leak or safety devices; icing may occur. *Caution:* When in contact with refrigerated/cryogenic liquids, many materials become brittle and are likely to break without warning • If cooling streams are ineffective (*venting sound increases in volume and pitch, tank discolors or shows signs of deforming*), withdraw immediately to a secure location. Always stay away from ends of tanks, especially when engulfed in flames. If employees are expected to fight fires, they must be trained and equipped in OSHA 1910.156.

DOT GUIDE NUMBER: 122 EMERGENCY RESPONSE: Stay upwind and uphill • Determine the extent of the problem • Isolate the area of release or fire and deny entry • Remove all ignition sources • May react explosively with fuels. Runoff may create explosion hazard • Wear positive-pressure, self-contained breathing apparatus (SCBA). Wear fully encapsulating, vapor protective chemical protective clothing which is specifically recommended by the manufacturer • Structural firefighters' protective clothing offers limited protection and not effective in fire situations • **RELEASE, NO FIRE** • Stop the release if it can be done safely from a distance • Use large amount of water spray to disperse gas–contain runoff. If possible, turn leaking containers so that gas escapes rather than liquid. Ventilate confined area if it can be done without placing personnel at risk • **FIRE** • Stop the release if it can be done safely from a distance • Use extinguishing agent suitable for type of surrounding fire • Small fire: Use *dry* chemical or CO_2. Large fire: Use water spray, fog, or regular foam. Cool exposed containers with large quantities of water from unattended equipment until well after fire is out, or remove intact containers if it can be done safely • Damaged cylinders should be handled only by specialists • If tanks are involved or in case of Large fire: Fight fire from maximum distances or use unmanned hose holders or monitor nozzles; if this is impossible, withdraw from area and let fire burn • Do not direct water at source of leak or safety devices; icing may occur • If cooling streams are ineffective (*venting sound increases in volume and pitch, tank discolors or shows signs of deforming*), withdraw immediately to a secure location. *Caution:* When in contact with refrigerated/cryogenic liquids, many materials become brittle and are likely to break without warning •Always stay away from ends of tanks, especially when engulfed in flames • If employees are expected to fight fires, they must be trained and equipped in OSHA 1910.156.

DOT GUIDE NUMBER: 123 EMERGENCY RESPONSE: Stay upwind and uphill • Determine the extent of the problem • Isolate the area of release or fire and deny entry • Remove all ignition sources • IF material or contaminated runoff from fire control enters waterways, notify downstream users of potentially contaminated water • Wear positive-pressure, self-contained breathing apparatus (SCBA). Wear fully encapsulating, vapor protective chemical protective clothing which is specifically recommended by the manufacturer • Structural firefighters' protective clothing offers limited protection and not effective in spill situations • **RELEASE, NO FIRE** • Stop the release if it can be done safely from a distance • Use large amount of water to disperse gas–contain runoff. If possible, turn leaking containers so that gas escapes rather than liquid. Ventilate confined area if it can be done without placing personnel at risk • **FIRE** • Stop the release if it can be done safely from a distance • Small fire: Use *dry* chemical or CO_2. Specially trained personnel may use fog lines to cool exposures and let the fire burn itself out • Large fire: Use water spray, fog, or regular foam. Cool exposed containers with large quantities of water from unattended equipment until well after fire is out, or remove intact containers if it can be done safely • Damaged cylinders should be handled only by specialists • Do not direct water at source of leak or safety devices; icing may occur • If cooling streams are ineffective (*venting sound increases in volume and pitch, tank discolors or shows signs of deforming*), withdraw immediately to a secure location. Always stay away from ends of tanks, especially when engulfed in flames. If employees are expected to fight fires, they must be trained and equipped in OSHA 1910.156.

DOT GUIDE NUMBER: 124 EMERGENCY RESPONSE: Stay upwind and uphill • Determine the extent of the problem • Isolate the area of release or fire and deny entry • Remove all ignition sources • IF material or contaminated runoff from fire control enters waterways, notify downstream users of potentially contaminated water • Wear positive-pressure, self-contained breathing apparatus (SCBA). Wear fully encapsulating, vapor protective chemical protective clothing which is specifically recommended by the manufacturer • Structural firefighters' protective clothing offers limited protection and not effective in fire situations • **RELEASE, NO FIRE** • Stop the release if it can be done safely from a distance • Use large amount of water to disperse gas–contain runoff. If possible, turn leaking containers so that gas escapes rather than liquid. Ventilate confined area if it can be done without placing personnel at risk • **FIRE** • Stop the release if it can be done safely from a distance • Use water ONLY; NO *dry* chemical, CO_2, or Halon. Contain fire and let burn. If fire must be fought, water spray or fog is recommended • If tanks are involved or in case of Large fire: Fight fire from maximum distances or use unmanned hose holders or monitor nozzles; if this is impossible, withdraw from area and let fire burn. Cool exposed containers with large quantities of water from unattended equipment until well after fire is out, or remove intact containers if it can be done safely • Damaged cylinders should be handled only by specialists • Do not direct water at source of leak or safety devices; icing may occur • If cooling streams are ineffective (*venting sound increases in volume and pitch, tank discolors or shows signs of deforming*), withdraw immediately to a secure location. Always stay away from ends of tanks, especially when engulfed in flames. If employees are expected to fight fires, they must be trained and equipped in OSHA 1910.156.

DOT GUIDE NUMBER: 125 EMERGENCY RESPONSE: Stay upwind and uphill • Determine the extent of the problem • Isolate the area of release or fire and deny entry • Remove all ignition sources • IF material or contaminated runoff from fire control enters waterways, notify downstream users of potentially contaminated water • Wear positive-pressure, self-contained breathing apparatus (SCBA). Wear fully encapsulating, vapor protective chemical protective clothing which is specifically recommended by the manufacturer • Structural firefighters' protective clothing offers limited protection and not effective in fire situations • **RELEASE, NO FIRE** • Stop the release if it can be done safely from a distance • Use large amount of water to disperse gas–contain runoff. If possible, turn leaking containers so that gas escapes rather than liquid. Ventilate confined area if it can be done without placing personnel at risk • **FIRE** • Stop the release if it can be done safely from a distance • Small fire: Use *dry* chemical, CO_2. Large fire: Use water spray, fog, or regular foam • If tanks are involved or in case of Large fire: Fight fire from maximum distances or use unmanned hose holders or monitor nozzles; if this is impossible, withdraw from area and let fire burn. Cool exposed containers with large quantities of water from unattended equipment until well after fire is out, or remove intact containers if it can be done safely • Damaged cylinders should be handled only by specialists • Do not direct water at source of leak or safety devices; icing may occur • If cooling streams are ineffective (*venting sound increases in volume and pitch, tank discolors or shows signs of deforming*), withdraw immediately to a secure location. Always stay away from ends of tanks, especially when engulfed in flames. If employees are expected to fight fires, they must be trained and equipped in OSHA 1910.156.

DOT GUIDE NUMBER: 126 EMERGENCY RESPONSE: Stay upwind and uphill • Determine the extent of the problem • Isolate the area of release or fire and deny entry • Remove all ignition sources • Wear positive-pressure, self-contained breathing apparatus (SCBA). Wear fully encapsulating, vapor protective chemical protective clothing which is specifically recommended by the manufacturer • Structural firefighters' protective clothing offers only limited protection • **RELEASE NO FIRE** • Stop the release if it can be done safely from a distance • Use large amount of water to disperse gas–contain runoff. If possible, turn leaking containers so that gas escapes rather than liquid. Ventilate confined area if it can be done without placing personnel at risk • **FIRE** • Stop the release if it can be done safely from a distance • Small fire: Use *dry* chemical or CO_2. Large fire: Use water spray, fog, or regular foam • If tanks are involved or in case of Large fire: Fight fire from maximum distances or use unmanned hose holders or monitor nozzles; if this is impossible, withdraw from area and let fire burn. Cool exposed containers with large quantities of water from unattended equipment until well after fire is out, or remove intact containers if it can be done safely • Damaged cylinders should be handled only by specialists • Do not direct water at source of leak or safety devices; icing may occur • If cooling streams are ineffective (*venting sound increases in volume and pitch, tank discolors or shows signs of deforming*), withdraw immediately to a secure location. Always stay away from ends of tanks, especially when engulfed in flames. If employees are expected to fight fires, they must be trained and equipped in OSHA 1910.156.

DOT GUIDE NUMBER: 127 EMERGENCY RESPONSE: Stay upwind and uphill • Isolate the area of release or fire and deny entry • Remove all ignition sources • Wear positive-pressure, self-contained breathing apparatus (SCBA). Wear chemical protective clothing which is specifically recommended by the manufacturer • Structural firefighters' protective clothing offers only limited protection • **RELEASE, NO FIRE** • Stop the release if it can be done safely from a distance • Prevent material and runoff from entering sewers and waterways if it can be done safely well ahead of the release • Use large amounts of water well away from release to disperse vapors–contain runoff • Consider application of alcohol-resistanttant (AFFF) foam to large areas of spilled liquid to control vapors • Ventilate confined area if it can be done without placing personnel at risk • **FIRE** • Caution, this material has very low flash point: Use of water spray when fighting fire may be inefficient • Small fire: Use *dry* chemical, alcohol-resistant foam, CO_2, or water spray • Large fire: Use water spray, fog, or alcohol-resistant foam. Specially trained personnel operating from a safe distance can fight fires using alcohol-resistanttant (AFFF) foam or *dry* chemical if it is available in sufficient amounts • Under favorable conditions, experienced crews can use coordinated fog streams to sweep the flames off the surface of the burning liquid. Keep exposures cool to protect against re-ignition. Do not direct straight streams into the liquid • Cool exposed containers with large quantities of water from unattended equipment or remove intact containers if it can be done safely. DO NOT ALLOW WATER TO GET INSIDE CONTAINERS • If cooling streams are ineffective (*venting sound increases in volume and pitch, tank discolors or shows signs of deforming*), withdraw immediately to a secure location. Always stay away from ends of tanks, especially when engulfed in flames. If employees are expected to fight fires, they must be trained and equipped in OSHA 1910.156.

DOT GUIDE NUMBER: 128 EMERGENCY RESPONSE: Stay upwind and uphill • Isolate the area of release or fire and deny entry • Remove all ignition sources • Wear positive-pressure, self-contained breathing apparatus (SCBA). Wear chemical protective clothing which is specifically recommended by the manufacturer • Structural firefighters' protective clothing offers only limited protection • **RELEASE, NO FIRE** • Stop the release if it can be done safely from a distance • Prevent material and runoff from entering sewers and waterways if it can be done safely well ahead of the release • Use large amounts of water well away from release to disperse vapors–contain runoff • Water spray may reduce vapor; but may not prevent ignition in closed spaces • Ventilate confined area if it can be done without placing personnel at risk • **FIRE** • Caution, this material has very low flash point: Use of water spray when fighting fire may be inefficient • Small fire: Use *dry* chemical, regular foam, CO_2, or water spray • Large fire: Use water spray, fog (do not direct straight streams into the liquid), or regular foam • Under favorable conditions, experienced crews can use coordinated fog streams to sweep the flames off the surface of the burning liquid. Keep exposures cool to protect against reignition • Cool exposed containers with large quantities of water from unattended equipment or remove intact containers if it can be done safely. Always stay away from tanks engulfed in flames. DO NOT ALLOW WATER TO GET INSIDE CONTAINERS • If cooling streams are ineffective (*venting sound increases in volume and pitch, tank discolors or shows signs of deforming*), withdraw immediately to a secure location. Always stay away from ends of tanks, especially when engulfed in flames. If employees are expected to fight fires, they must be trained and equipped in OSHA 1910.156.

DOT GUIDE NUMBER: 129 EMERGENCY RESPONSE: Stay upwind and uphill • Isolate the area of release or fire and deny entry • Remove all ignition sources • Wear positive-pressure, self-contained breathing apparatus (SCBA). Wear chemical protective clothing which is specifically recommended by the manufacturer • Structural firefighters' protective clothing offers only limited protection • **RELEASE, NO FIRE** • Stop the release if it can be done safely from a distance • Prevent material and runoff from entering sewers and waterways if it can be done safely well ahead of the release • Use large amounts of water well away from release to disperse vapors–contain runoff • Water spray may reduce vapor; but may not prevent ignition in closed spaces • Ventilate confined area if it can be done without placing personnel at risk • **FIRE** • Caution, this material has very low flash point: Use of water spray when fighting fire may be inefficient • Small fire: Use *dry* chemical, CO_2, water spray, or alcohol-resistant foam • Large fire: Use water spray, fog, or alcohol-resistant foam. Do NOT use straight streams • Under favorable conditions, experienced crews can use coordinated fog streams to sweep the flames off the surface of the burning liquid. Keep exposures cool to protect against reignition • Cool exposed containers with large quantities of water from unattended equipment or remove intact containers if it can be done safely. DO NOT ALLOW WATER TO GET INSIDE CONTAINERS • If cooling streams are ineffective (*venting sound increases in volume and pitch, tank discolors or shows signs of deforming*), withdraw immediately to a secure location. Always stay away from ends of tanks, especially when engulfed in flames.

If employees are expected to fight fires, they must be trained and equipped in OSHA 1910.156.

DOT GUIDE NUMBER: 130 EMERGENCY RESPONSE: Stay upwind and uphill • Isolate the area of release or fire and deny entry • Remove all ignition sources • Wear positive-pressure, self-contained breathing apparatus (SCBA). Wear chemical protective clothing which is specifically recommended by the manufacturer • Structural firefighters' protective clothing offers only limited protection • **RELEASE, NO FIRE** • Stop the release if it can be done safely from a distance • Prevent material and runoff from entering sewers and waterways if it can be done safely well ahead of the release • Use large amounts of water well away from release to disperse vapors–contain runoff • Water spray may reduce vapor; but may not prevent ignition in closed spaces • Ventilate confined area if it can be done without placing personnel at risk • **FIRE** • Caution, this material has very low flash point: Use of water spray when fighting fire may be inefficient • Small fire: Use *dry* chemical, CO_2, water spray, or regular foam • Large fire: Use water spray, fog (do not direct straight streams into the liquid), or regular foam • Under favorable conditions, experienced crews can use coordinated fog streams to sweep the flames off the surface of the burning liquid. Keep exposures cool to protect against reignition • Cool exposed containers with large quantities of water from unattended equipment or remove intact containers if it can be done safely. **SHIPPING INFORMATION** engulfed in flames. DO NOT ALLOW WATER TO GET INSIDE CONTAINERS • If cooling streams are ineffective (*venting sound increases in volume and pitch, tank discolors or shows signs of deforming*), withdraw immediately to a secure location. Always stay away from ends of tanks, especially when engulfed in flames. If employees are expected to fight fires, they must be trained and equipped in OSHA 1910.156.

DOT GUIDE NUMBER: 131 EMERGENCY RESPONSE: Stay upwind and uphill • Isolate the area of release or fire and deny entry • Remove all ignition sources • IF material or contaminated runoff from fire control enters waterways, notify downstream users of potentially contaminated water • Wear positive-pressure, self-contained breathing apparatus (SCBA). Wear chemical protective clothing which is specifically recommended by the manufacturer • Structural firefighters' protective clothing offers only limited protection • **RELEASE, NO FIRE** • Wear fully encapsulating, vapor protective clothing. Stop the release if it can be done safely from a distance • Prevent material and runoff from entering sewers and waterways if it can be done safely well ahead of the release • Small spill: absorb with earth, sand, or other noncombustible material and transfer to containers with clean, nonsparking tools for later disposal. Large spill: Dike far ahead of liquid spill for later disposal • Water spray may reduce vapor; but may not prevent ignition in closed spaces • Ventilate confined area if it can be done without placing personnel at risk • **FIRE** • Caution, this material has very low flash point: Use of water spray when fighting fire may be inefficient • Small fire: Use *dry* chemical, CO_2, water spray, or alcohol-resistant foam • Large fire: Use water spray, fog (do not direct straight streams into the liquid), or alcohol-resistant foam • Under favorable conditions, experienced crews can use coordinated fog streams to sweep the flames off the surface of the burning liquid. Keep exposures cool to protect against reignition • Cool exposed containers with large quantities of water from unattended equipment or remove intact containers if it can be done safely. Always stay away from tanks engulfed in flames. DO NOT ALLOW WATER TO GET INSIDE CONTAINERS • If cooling streams are ineffective (*venting sound increases in volume and pitch, tank discolors or shows signs of deforming*), withdraw

DOT GUIDE NUMBER: 132 EMERGENCY RESPONSE: Stay upwind and uphill • Isolate the area of release or fire and deny entry • Remove all ignition sources • IF material or contaminated runoff from fire control enters waterways, notify downstream users of potentially contaminated water • Wear positive-pressure, self-contained breathing apparatus (SCBA). Wear chemical protective clothing which is specifically recommended by the manufacturer • Structural firefighters' protective clothing offers only limited protection • **RELEASE NO FIRE** • Wear fully encapsulating, vapor protective clothing. Stop the release if it can be done safely from a distance • Prevent material and runoff from entering sewers and waterways if it can be done safely well ahead of the release • Small spill: Absorb with earth, sand, or other noncombustible material and transfer to containers with clean, nonsparking tools for later disposal. Large spill: Dike far ahead of liquid spill for later disposal • Water spray may reduce vapor; but may not prevent ignition in closed spaces • Ventilate confined area if it can be done without placing personnel at risk • **FIRE** • Check to make certain this material does not react with water. Small fire: Use *dry* chemical, CO_2, water spray, or alcohol-resistant foam • Large fire: Use water spray, fog (do not direct straight streams into the liquid), or alcohol-resistant foam • Under favorable conditions, experienced crews can use coordinated fog streams to sweep the flames off the surface of the burning liquid. Keep exposures cool to protect against reignition • Cool exposed containers with large quantities of water from unattended equipment or remove intact containers if it can be done safely. *DO NOT ALLOW WATER TO GET INSIDE CONTAINERS* • If cooling streams are ineffective (*venting sound increases in volume and pitch, tank discolors or shows signs of deforming*), withdraw immediately to a secure location. Always stay away from ends of tanks, especially when engulfed in flames. If employees are expected to fight fires, they must be trained and equipped in OSHA 1910.156.

DOT GUIDE NUMBER: 133 EMERGENCY RESPONSE: Stay upwind and uphill • Isolate the area of release or fire and deny entry • Remove all ignition sources • IF material or contaminated runoff from fire control enters waterways, notify downstream users of potentially contaminated water • Wear positive-pressure, self-contained breathing apparatus (SCBA). Wear chemical protective clothing which is specifically recommended by the manufacturer • Structural firefighters' protective clothing offers only limited protection • **RELEASE NO FIRE** • Stop the release if it can be done safely from a distance • Small spill: Absorb with earth, sand, or other noncombustible material and transfer to containers with clean, nonsparking tools for later disposal. Large spill: Dike far ahead of liquid spill for later disposal • Water spray may reduce vapor; but may not prevent ignition in closed spaces • Ventilate confined area if it can be done without placing personnel at risk • **FIRE** • Check to make certain this material does not react with water. Small fire: Use *dry* chemical, CO_2, water spray, or regular foam • Large fire: Use water spray, fog (do not direct straight streams into the liquid), or regular foam • Keep exposures cool to protect against reignition • Cool exposed containers with large quantities of water from unattended equipment or remove intact containers if it can be done safely. *DO NOT ALLOW WATER TO GET INSIDE CONTAINERS* • If cooling streams are ineffective (*venting sound increases in volume and pitch, tank discolors or shows signs of deforming*), withdraw immediately to a secure location. Always stay away from ends of tanks, especially when engulfed in flames. If employees are expected to fight fires, they must be trained and equipped in OSHA 1910.156.

DOT GUIDE NUMBER: 134 EMERGENCY RESPONSE: Stay upwind and uphill • Isolate the area of release or fire and deny entry • Remove all ignition sources • IF material or contaminated runoff from fire control enters waterways, notify downstream users of potentially contaminated water • Wear positive-pressure, self-contained breathing apparatus (SCBA). Wear chemical protective clothing which is specifically recommended by the manufacturer • Structural firefighters' protective clothing offers only limited protection • **RELEASE NO FIRE** • Fully encapsulating, vapor protective clothing should be worn for spills and leaks with no fire. Prevent material and runoff from entering sewers and waterways if it can be done safely well ahead of the release • With clean shovel place material in clean, *dry* container and cover loosely; move containers from spill area • **FIRE** • Small fire: Use *dry* chemical, CO_2, water spray, or alcohol-resistant foam • Large fire: Use water spray, fog, or alcohol-resistant foam • Cool exposed containers with large quantities of water from unattended equipment or remove intact containers if it can be done safely. *DO NOT ALLOW WATER TO GET INSIDE CONTAINERS* • If cooling streams are ineffective (*venting sound increases in volume and pitch, tank discolors or shows signs of deforming*), withdraw immediately to a secure location. Always stay away from ends of tanks, especially when engulfed in flames. If employees are expected to fight fires, they must be trained and equipped in OSHA 1910.156.

DOT GUIDE NUMBER: 135 EMERGENCY RESPONSE: see note for triisobutylaluminum (UN 3051). Stay upwind and uphill • Remove all ignition sources • Wear positive-pressure, self-contained breathing apparatus (SCBA). Wear fully encapsulating vapor protective clothing which is specifically recommended by the manufacturer • Structural firefighter's protective clothing provides only limited protection • **RELEASE, NO FIRE** • Fully encapsulating, vapor protective clothing should be worn for spills and leaks with no fire. Stop the release if it can be done safely from a distance • Small spill: Cover with *dry* earth or sand, or other *dry*, noncombustible material followed with plastic sheet to minimize spreading or contact with rain • Ventilate confined area if it can be done without placing personnel at risk • **FIRE** • Approach fire with extreme caution: If possible let fire burn; there is a good chance of explosive reignition. Do NOT use water, foam, or CO_2 on material itself. Do NOT use water on adjacent fires. Small fire: Use *dry* chemical, *dry* graphite, soda ash, lime, or *dry* sand • Large fire: Specially trained personnel operating from a safe distance can fight fire using *dry* sand, *dry* chemical, soda ash, lime, or other inert material if it is available in sufficient amounts; or, withdraw from area and let fire burn. • IF material is not leaking, cool exposed containers with large quantities of water from unattended equipment or remove intact containers if it can be done safely. DO NOT ALLOW WATER TO GET INSIDE CONTAINERS • If cooling streams are ineffective (*venting sound increases in volume and pitch, tank discolors or shows signs of deforming*), withdraw immediately to a secure location. Always stay away from ends of tanks, especially when engulfed in flames. If employees are expected to fight fires, they must be trained and equipped in OSHA 1910.156.

Triisobutylaluminum: IF material is on fire and if conditions permit, DO NOT EXTINGUISH. (FEMA). Specially trained personnel operating from a safe distance can fight fire using *dry* chemical if available in sufficient amounts. Under favorable conditions, experienced crews can use coordinated fog streams to sweep the flames off the surface of the burning liquid. Keep exposures cool to protect against reignition. Do not direct straight streams into the liquid. IF material is not leaking, cool exposed containers with large quantities of water from unattended equipment or remove intact containers if it can be done safely. DO NOT ALLOW WATER TO GET INSIDE CONTAINERS • If cooling streams are ineffective (*venting sound increases in volume and pitch, tank discolors or shows signs of deforming*), withdraw immediately to a secure location. Always stay away from ends of tanks, especially when engulfed in flames. If employees are expected to fight fires, they must be trained and equipped in OSHA 1910.156.

DOT GUIDE NUMBER: 136 EMERGENCY RESPONSE: Stay upwind and uphill • Remove all ignition sources • IF material or contaminated runoff from fire control enters waterways, notify downstream users of potentially contaminated water • Wear positive-pressure, self-contained breathing apparatus (SCBA). Wear fully encapsulating vapor protective clothing which is specifically recommended by the manufacturer • Structural firefighter's protective clothing provides only limited protection • **RELEASE, NO FIRE** • Fully encapsulating, vapor protective clothing should be worn for spills and leaks with no fire. Stop the release if it can be done safely from a distance • Small spill: Cover with water, sand, or earth. Shovel into metal containers and keep material under water. Small spills can be rendered harmless with copper sulfate solution; or cover with water, sand, or earth; keep damp with water; transfer to drum and cover with water; do not dump in sewer • Large spill: Dike for later disposal and cover with wet sand or earth. Prevent material and runoff from entering sewers and waterways if it can be done safely well ahead of the release • Ventilate confined area if it can be done without placing personnel at risk • **FIRE** • Approach fire with extreme caution: If possible let fire burn; there is a good chance of explosive reignition. Small fire: Use water spray, wet sand, or wet earth • Large fire: Specially trained personnel operating from a safe distance can fight fire using water spray or fog. Do not scatter spilled material with high pressure water streams • IF material is not leaking, cool exposed containers with large quantities of water from unattended equipment or remove intact containers if it can be done safely • If cooling streams are ineffective (*venting sound increases in volume and pitch, tank discolors or shows signs of deforming*), withdraw immediately to a secure location. Always stay away from ends of tanks, especially when engulfed in flames. If employees are expected to fight fires, they must be trained and equipped in OSHA 1910.156.

DOT GUIDE NUMBER: 137 EMERGENCY RESPONSE: Stay upwind and uphill • Isolate the area of release or fire and deny entry • Remove all ignition sources • IF material or contaminated runoff from fire control enters waterways, notify downstream users of potentially contaminated water • Wear positive-pressure, self-contained breathing apparatus (SCBA). Wear chemical protective clothing which is specifically recommended by the manufacturer. Structural firefighter's protective clothing provides only limited protection • **RELEASE, NO FIRE** • Fully encapsulating, vapor protective clothing should be worn for spills or leaks with no fire • *When material is not involved in fire, do not use water on material itself* • Small spill: Cover with *dry* earth, *dry* sand, or other noncombustible material followed with plastic sheet to minimize spreading or contact with water or rain. Use clean nonsparking tools to collect material and place material into loosely covered plastic containers for later disposal. Prevent entry of material into waterways, sewers, basements or confined areas • Large Spill: Stop the release if it can be done safely from a distance • Use water spray to reduce vapors; do not put water directly on leak, spill, or inside containers. Prevent material and runoff from entering sewers and waterways if it can be done safely well ahead of the release • Ventilate confined area if it can be done without placing personnel at risk • **FIRE** • Small fire: Extinguish with *dry* chemical or CO_2. Large fire: If water must be used, use in flooding quantities. Flood area with large amounts of water; while knocking down vapors with water fog. If insufficient water supply; knock down vapors only • IF material is not leaking, cool exposed containers with large quantities of water from unattended equipment or remove intact containers if it can be done safely. DO NOT ALLOW WATER TO GET INSIDE CONTAINERS • If cooling streams are ineffective (*venting sound increases in volume and pitch, tank discolors or shows signs of deforming*), withdraw immediately to a secure location. Always stay away from ends of tanks, especially when engulfed in flames. If employees are expected to fight fires, they must be trained and equipped in OSHA 1910.156.

DOT GUIDE NUMBER: 138 EMERGENCY RESPONSE:
Calcium, calcium carbide: Stay upwind and uphill • Isolate the area of release or fire and deny entry • Remove all ignition sources • IF material or contaminated runoff from fire control enters waterways, notify downstream users of potentially contaminated water • Wear positive-pressure, self-contained breathing apparatus (SCBA). Wear chemical protective clothing which is specifically recommended by the manufacturer. Structural firefighter's protective clothing provides only limited protection • **RELEASE, NO FIRE** • Fully encapsulating, vapor protective clothing should be worn for spills or leaks with no fire • *When material is not involved in fire, do not use water on material itself* • Small spill: Cover with *dry* earth, *dry* sand, or other noncombustible material. Powder spill: Keep dry. Cover with plastic sheet to minimize spreading or contact with wind, water spray, or rain • Large Spill: Stop the release if it can be done safely from a distance • *DO NOT CLEAN UP OR DISPOSE OF, EXCEPT UNDER THE SUPERVISION OF A SPECIALIST* • Prevent material and any runoff from entering sewers and waterways if it can be done safely well ahead of the release • Ventilate confined area if it can be done without placing personnel at risk • **FIRE** • *DO NOT USE WATER, FOAM, CO$_2$ OR VAPORIZING LIQUIDS ON FIRE*. Small fire: Use *dry* chemical, soda ash, lime, or sand • Large fire: Use *dry* sand, *dry* chemical, soda ash, or lime, or withdraw from area and let fire burn • IF material is not leaking, cool exposed containers with large quantities of water from unattended equipment or remove intact containers if it can be done safely. *DO NOT ALLOW WATER TO GET INSIDE CONTAINERS* • If cooling streams are ineffective (*venting sound increases in volume and pitch, tank discolors or shows signs of deforming*), withdraw immediately to a secure location. Always stay away from tanks engulfed in flames. If employees are expected to fight fires, they must be trained and equipped in OSHA 1910.156.
Lithium fire: (Lithium, lithium aluminum hydride, lithium hydride): Stay upwind and uphill • Isolate the area of release or fire and deny entry • Remove all ignition sources • IF material or contaminated runoff from fire control enters waterways, notify downstream users of potentially contaminated water • Wear positive-pressure, self-contained breathing apparatus (SCBA). Wear chemical protective clothing which is specifically recommended by the manufacturer. Structural firefighter's protective clothing provides only limited protection • **RELEASE, NO FIRE** • Fully encapsulating, vapor protective clothing should be worn for spills or leaks with no fire • *When material is not involved in fire, do not use water on material itself* • Small spill: Cover with *dry* earth, *dry* sand, or other noncombustible material. Powder spill: Keep dry. Cover with plastic sheet to minimize spreading or contact with wind, water spray, or rain • Large Spill: Stop the release if it can be done safely from a distance • *DO NOT CLEAN UP OR DISPOSE OF, EXCEPT UNDER THE SUPERVISION OF A SPECIALIST* • Prevent material and any runoff from entering sewers and waterways if it can be done safely well ahead of the release • Ventilate confined area if it can be done without placing personnel at risk • **FIRE** • *DO NOT USE WATER, FOAM, CO$_2$ OR VAPORIZING LIQUIDS ON FIRE*. Small fire: Use *dry* chemical, soda ash, lime, or sand • Large fires (Lithium): Use *dry* sand, sodium chloride powder; graphite powder, copper powder or Lith-X® powder, or withdraw from area and let fire burn • IF material is not leaking, cool exposed containers with large quantities of water from unattended equipment until well after fire is out. *DO NOT ALLOW WATER TO GET INSIDE CONTAINERS* • If cooling streams are ineffective (*venting sound increases in volume and pitch, tank discolors or shows signs of deforming*), withdraw immediately to a secure location. Always stay away from tanks engulfed in flames. If employees are expected to fight fires, they must be trained and equipped in OSHA 1910.156.
Magnesium: Stay upwind and uphill • Isolate the area of release or fire and deny entry • Remove all ignition sources • Wear chemical protective clothing which is specifically recommended by the manufacturer • **RELEASE, NO FIRE** • *When material is not involved in fire, do not use water on material itself* • Small spill: Cover with *dry* earth, *dry* sand, or other noncombustible material. Powder spill: Keep dry. Cover with plastic sheet to minimize spreading or contact with wind, water spray, or rain • Large Spill: Stop the release if it can be done safely from a distance • *DO NOT CLEAN UP OR DISPOSE OF, EXCEPT UNDER THE SUPERVISION OF A SPECIALIST* • Prevent material and any runoff from entering sewers and waterways if it can be done safely well ahead of the release • **FIRE** • *DO NOT USE WATER, FOAM, CO$_2$, HALOGENS, or OR VAPORIZING LIQUIDS ON FIRE*. Small fire: Use *dry* chemical, soda ash, lime, or sand • Large fire: Use *dry* sand, sodium chloride powder; graphite powder, Met-L-X® powder, or withdraw from area and let fire burn • IF material is not leaking, cool exposed containers with large quantities of water from unattended equipment until long after fire is out. *DO NOT ALLOW WATER TO GET INSIDE CONTAINERS* • If cooling streams are ineffective (*venting sound increases in volume and pitch, tank discolors or shows signs of deforming*), withdraw immediately to a secure location. Always stay away from ends of tanks, especially when engulfed in flames. If employees are expected to fight fires, they must be trained and equipped in OSHA 1910.156.
Potassium: Stay upwind and uphill • Isolate the area of release or fire and deny entry • Remove all ignition sources • IF material or contaminated runoff from fire control enters waterways, notify downstream users of potentially contaminated water • Wear positive-pressure, self-contained breathing apparatus (SCBA). Wear chemical protective clothing which is specifically recommended by the manufacturer. Structural firefighter's protective clothing provides only limited protection • **RELEASE, NO FIRE** • Fully encapsulating, vapor protective clothing should be worn for spills or leaks with no fire • *When material is not involved in fire, do not use water on material itself* • Small spill: Cover with *dry* earth, *dry* sand, or other noncombustible material. Powder spill: Keep dry. Cover with plastic sheet to minimize spreading or contact with wind, water spray, or rain • Large Spill: Stop the release if it can be done safely from a distance • *DO NOT CLEAN UP OR DISPOSE OF, EXCEPT UNDER THE SUPERVISION OF A SPECIALIST* • Prevent material and any runoff from entering sewers and waterways if it can be done safely well ahead of the release • Ventilate confined area if it can be done without placing personnel at risk • **FIRE** • *DO NOT USE WATER, FOAM, CO$_2$ OR VAPORIZING LIQUIDS ON FIRE*. Small fire: Use *dry* chemical, soda ash, lime, or sand • Large fire: Use *dry* sand, *dry* chemical, soda ash, or lime, or withdraw from area and let fire burn • IF material is not leaking, cool exposed containers with large quantities of water from unattended equipment or

remove intact containers if it can be done safely. *DO NOT ALLOW WATER TO GET INSIDE CONTAINERS* • If cooling streams are ineffective (*venting sound increases in volume and pitch, tank discolors or shows signs of deforming*), withdraw immediately to a secure location. Always stay away from tanks engulfed in flames • If employees are expected to fight fires, they must be trained and equipped in OSHA 1910.156.

Sodium, sodium borohydride: Stay upwind and uphill • Isolate the area of release or fire and deny entry • Remove all ignition sources • IF material or contaminated runoff from fire control enters waterways, notify downstream users of potentially contaminated water • Wear positive-pressure, self-contained breathing apparatus (SCBA). Wear chemical protective clothing which is specifically recommended by the manufacturer. Structural firefighter's protective clothing provides only limited protection • **RELEASE, NO FIRE** • Fully encapsulating, vapor protective clothing should be worn for spills or leaks with no fire • *When material is not involved in fire, do not use water on material itself* • Small spill: Cover with *dry* earth, *dry* sand, or other noncombustible material. Powder spill: Keep dry. Cover with plastic sheet to minimize spreading or contact with wind, water spray, or rain • Large Spill: Stop the release if it can be done safely from a distance • *DO NOT CLEAN UP OR DISPOSE OF, EXCEPT UNDER THE SUPERVISION OF A SPECIALIST* • Prevent material and any runoff from entering sewers and waterways if it can be done safely well ahead of the release • Ventilate confined area if it can be done without placing personnel at risk • **FIRE** • *DO NOT USE WATER, FOAM, CO$_2$ OR VAPORIZING LIQUIDS ON FIRE*. Small fire: Use *dry* chemical, soda ash, lime, or sand • Large fire: Use *dry* sand, *dry* chemical, soda ash, or lime, or withdraw from area and let fire burn • IF material is not leaking, cool exposed containers with large quantities of water from unattended equipment until well after the fire is out; or, remove intact containers if it can be done safely. *DO NOT ALLOW WATER TO GET INSIDE CONTAINERS* • If cooling streams are ineffective (*venting sound increases in volume and pitch, tank discolors or shows signs of deforming*), withdraw immediately to a secure location. Always stay away from tanks engulfed in flames • If employees are expected to fight fires, they must be trained and equipped in OSHA 1910.156.

DOT GUIDE NUMBER: 139 EMERGENCY RESPONSE: *CHLOROSILANES* (For other chemicals, see below) Restrict access to the spill area • Stay upwind and uphill • Isolate the area of release or fire and deny entry • Evacuate the immediate area and downwind for a large release • Remove all ignition sources. All equipment used to handle this material must be grounded • Notify local health and fire officials and pollution control agencies • IF material or contaminated runoff from fire control enters waterways, notify downstream users of potentially contaminated water • Wear positive-pressure, self-contained breathing apparatus (SCBA). Wear chemical protective clothing which is specifically recommended by the manufacturer. Structural firefighter's protective clothing provides only limited protection • **RELEASE, NO FIRE** • Fully encapsulating, vapor protective clothing should be worn for spills or leaks with no fire • Stop the release if it can be done safely from a distance • Prevent material and runoff from entering sewers and waterways if it can be done safely well ahead of the release • For *CHLOROSILANES* use AFFF medium-expansion foam to disperse vapors • Ventilate confined area if it can be done without placing personnel at risk • Small spill: Cover with *dry* earth, *dry* sand, or other *dry* noncombustible material followed with plastic sheet to minimize spreading or contact with water or rain. Use clean nonsparking tools to collect material and place material into loosely covered plastic containers for later disposal. Powder Spill: Cover spill with plastic sheet or tarp to minimize spreading and keep powder dry. *DO NOT CLEAN UP OR DISPOSE OF THIS MATERIAL EXCEPT UNDER SUPERVISION OF A SPECIALIST* • **FIRE** • Small fire: *DO NOT USE WATER*. Use *dry* chemical, *dry* sand, lime, or soda ash. Large fire: *DO NOT USE WATER* . Chlorosilanes react with water but can be extinguished with medium-expansion alcohol-resistant (AFFF) foam if available in sufficient amounts. DO NOT USE *dry* chemicals, soda ash, or lime on fires as they release large amounts of hydrogen gas which may explode. IF material is on fire and conditions permit, *DO NOT EXTINGUISH*. IF material is not leaking, cool exposed containers with flooding quantities of water from unattended equipment. *DO NOT ALLOW WATER TO GET INSIDE CONTAINERS* • If cooling streams are ineffective (*venting sound increases in volume and pitch, tank discolors or shows signs of deforming*), withdraw immediately to a secure location. Always stay away from ends of tanks, especially when engulfed in flames. If employees are expected to fight fires, they must be trained and equipped in OSHA 1910.156.

DOT GUIDE NUMBER: 139 EMERGENCY RESPONSE (For CHLOROSILANES see above; for zinc phosphide see below): Stay upwind and uphill • Isolate the area of release or fire and deny entry • Remove all ignition sources. All equipment used to handle this material must be grounded • IF material or contaminated runoff from fire control enters waterways, notify downstream users of potentially contaminated water • Wear positive-pressure, self-contained breathing apparatus (SCBA). Wear chemical protective clothing which is specifically recommended by the manufacturer. Structural firefighter's protective clothing provides only limited protection • **RELEASE, NO FIRE** • Fully encapsulating, vapor protective clothing should be worn for spills or leaks with no fire • Stop the release if it can be done safely from a distance • Prevent material and runoff from entering sewers and waterways if it can be done safely well ahead of the release • Ventilate confined area if it can be done without placing personnel at risk • Small spill: Cover with *dry* earth, *dry* sand, or other *dry* noncombustible material followed with plastic sheet to minimize spreading or contact with water or rain. Use clean nonsparking tools to collect material and place material into loosely covered plastic containers for later disposal. Powder Spill: Cover spill with plastic sheet or tarp to minimize spreading and keep powder dry. *DO NOT CLEAN UP OR DISPOSE OF THIS MATERIAL EXCEPT UNDER SUPERVISION OF A SPECIALIST* • **FIRE** • *DO NOT USE WATER OR FOAM*. Small fire: Use *dry* chemical, *dry* sand, lime, or soda ash. For Large fire: Use *dry* sand, *dry* chemicals, soda ash, or lime or withdraw from area and let fire burn. IF material is on fire and conditions permit, *DO NOT EXTINGUISH*. IF material is *NOT LEAKING*, cool exposed containers with flooding quantities of water from unattended equipment until well after the fire is out. *DO NOT ALLOW WATER TO GET INSIDE CONTAINERS* • If cooling streams are ineffective (*venting sound increases in volume and pitch, tank discolors or shows signs of deforming*), withdraw immediately to a secure location. Always stay away from ends of tanks, especially when engulfed in flames. If employees are expected to fight fires, they must be trained and equipped in OSHA 1910.156.

Zinc Phosphide: IF material is on fire and conditions permit, DO NOT EXTINGUISH. Do not allow water to come in contact with material. Use Class D extinguisher.

DOT GUIDE NUMBER: 140 EMERGENCY RESPONSE: Stay upwind and uphill • Isolate the area of release or fire and deny entry • Keep material away from combustible materials, fuels, etc. Remove all ignition sources • IF material or contaminated runoff from fire control enters waterways, notify downstream users of potentially contaminated water • Wear positive-pressure, self-contained breathing apparatus (SCBA). Wear chemical protective clothing which is specifically recommended by the manufacturer. Structural firefighter's protective clothing provides only limited protection • **RELEASE, NO FIRE** • Keep combustibles (wood, paper, oil, etc.) away from spilled material • Small dry spill: Use clean shovel to collect material and place material into loosely covered *dry* containers for later disposal. Small Liquid Spill: Use noncombustible material like vermiculite, sand, or earth to soak up material and place in a container for later disposal. Large Spill: Dike far ahead of spill for later disposal. Following product recovery, flush with water. Prevent material and runoff from entering sewers and waterways if it can be done safely well ahead of the release • Ventilate confined area if it can be done without placing personnel at risk • **FIRE** • Small fire: Use water only. Do NOT use *dry* chemicals, CO_2, Halon®, or foams • Large fire: Flood area with water from a distance • Move containers from fire area it it can be done without risk. Do not move cargo or vehicle if cargo has been exposed to heat. IF material is not leaking, cool exposed containers with large quantities of water from unattended equipment or remove intact containers if it can be done safely. DO NOT ALLOW WATER TO GET INSIDE CONTAINERS • If cooling streams are ineffective (*venting sound increases in volume and pitch, tank discolors or shows signs of deforming*), withdraw immediately to a secure location. Always stay away from ends of tanks, especially when engulfed in flames. If employees are expected to fight fires, they must be trained and equipped in OSHA 1910.156.

DOT GUIDE NUMBER: 141 EMERGENCY RESPONSE: Stay upwind and uphill • Isolate the area of release or fire and deny entry • Keep material away from combustible materials, fuels, etc. Remove all ignition sources • IF material or contaminated runoff from fire control enters waterways, notify downstream users of potentially contaminated water • Wear positive-pressure, self-contained breathing apparatus (SCBA). Wear chemical protective clothing which is specifically recommended by the manufacturer. Structural firefighter's protective clothing provides only limited protection • **RELEASE, NO FIRE** • Keep combustibles (wood, paper, oil, etc.) away from spilled material • Small dry spill: Use clean shovel to collect material and place material into loosely covered *dry* containers for later disposal. Small liquid spill: Use noncombustible material like vermiculite, sand, or earth to soak up material and place in a container for later disposal. Large Spill: Dike far ahead of spill for later disposal • Ventilate confined area if it can be done without placing personnel at risk • **FIRE** • Small fire: Use water only. Do NOT use *dry* chemicals or foam. CO_2 or Halon® may provide limited control • Large fire: Flood area with water from a distance • Do not move cargo or vehicle if cargo has been exposed to heat. IF material is not leaking, cool exposed containers with large quantities of water from unattended equipment until long after fire is out; or, remove intact containers if it can be done safely. *DO NOT ALLOW WATER TO GET INSIDE CONTAINERS* • If cooling streams are ineffective (*venting sound increases in volume and pitch, tank discolors or shows signs of deforming*), withdraw immediately to a secure location. Always stay away from ends of tanks, especially when engulfed in flames. If employees are expected to fight fires, they must be trained and equipped in OSHA 1910.156.

DOT GUIDE NUMBER: 143 EMERGENCY RESPONSE: Stay upwind and uphill • Isolate the area of release or fire and deny entry • Keep material away from combustible materials, fuels, etc. Remove all ignition sources • IF material or contaminated runoff from fire control enters waterways, notify downstream users of potentially contaminated water • Wear positive-pressure, self-contained breathing apparatus (SCBA). Wear chemical protective clothing which is specifically recommended by the manufacturer. Structural firefighter's protective clothing provides only limited protection • **RELEASE, NO FIRE** • Keep combustibles (wood, paper, oil, etc.) away from spilled material • Small dry spill: Use clean shovel to collect material and place material into loosely covered *dry* containers for later disposal. Small liquid spill: Use noncombustible material like vermiculite, sand, or earth to soak up material and place in a container for later disposal. Large Spill: Dike far ahead of spill for later disposal • Ventilate confined area if it can be done without placing personnel at risk • **FIRE** • Small fire: Use water only. Do NOT use *dry* chemicals or foams. CO_2 or Halon® may provide limited control • Large fire: Flood area with water from a distance • Large fire: Flood area with water from a distance • Do not move cargo or vehicle if cargo has been exposed to heat. IF material is not leaking, cool exposed containers with large quantities of water from unattended equipment or remove intact containers if it can be done safely. *DO NOT ALLOW WATER TO GET INSIDE CONTAINERS* • If cooling streams are ineffective (*venting sound increases in volume and pitch, tank discolors or shows signs of deforming*), withdraw immediately to a secure location. Always stay away from ends of tanks, especially when engulfed in flames. If employees are expected to fight fires, they must be trained and equipped in OSHA 1910.156.

DOT GUIDE NUMBER: 144 EMERGENCY RESPONSE: Stay upwind and uphill • Isolate the area of release or fire and deny entry • Keep material away from combustible materials, fuels, etc. Remove all ignition sources • IF material or contaminated runoff from fire control enters waterways, notify downstream users of potentially contaminated water • Wear positive-pressure, self-contained breathing apparatus (SCBA). Wear chemical protective clothing which is specifically recommended by the manufacturer. Structural firefighter's protective clothing provides only limited protection • **RELEASE, NO FIRE** • Keep combustibles (wood, paper, oil, etc.) away from spilled material • Small spill: Cover with *dry* sand or sand, or other *dry*, noncombustible material followed with plastic sheet to minimize spreading or contact with rain • Large Spill: *DO NOT CLEAN UP OR DISPOSE OF MATERIAL, EXCEPT UNDER SUPERVISION OF A SPECIALIST* • **FIRE** • *DO NOT USE WATER OR FOAM*. Small fire: Use *dry* chemicals, soda ash or lime. Remove intact containers if it can be done safely • Large fire: Use *dry* sand, *dry* chemical, soda ash, or lime or withdraw from area and let burn. IF material is not leaking, cool exposed containers with large quantities of water from unattended equipment until well after fire is out. *DO NOT ALLOW WATER TO GET INSIDE CONTAINERS* • If cooling streams are ineffective (*venting sound increases in volume and pitch, tank discolors or shows signs of deforming*), withdraw immediately to a secure location. Always stay away from ends of tanks, especially when engulfed in flames. If employees are expected to fight fires, they must be trained and equipped in OSHA 1910.156.

DOT GUIDE NUMBER: 145 EMERGENCY RESPONSE: Stay upwind and uphill • Remove all ignition sources • IF material or contaminated runoff from fire control enters waterways, notify downstream users of potentially contaminated water • Wear positive-pressure, self-contained breathing apparatus (SCBA). Wear chemical protective clothing which is specifically recommended by the manufacturer. Structural firefighter's protective clothing provides only limited protection • **RELEASE, NO FIRE** • Small spill: Cover with inert, damp noncombustible material using nonsparking tools and place into loosely covered plastic containers for later disposal. Large Spill: Wet down with water and dike for later disposal. *DO NOT CLEAN UP OR DISPOSE OF, EXCEPT UNDER SUPERVISION OF A SPECIALIST* • Prevent material and runoff from entering sewers and waterways if it can be done safely well ahead of the release • Ventilate confined area if it can be done without placing personnel at risk • **FIRE** • Small fire: Water spray or fog is preferred. If water is unavailable use *dry* chemical, CO_2, or regular foam. Large fire: Flood fire area with water from a distance. Do not use straight streams • IF material is not leaking, cool exposed containers with large quantities of water from unattended equipment or remove intact containers if it can be done safely • If cylinders are exposed to excessive heat from fire or flame contact, withdraw immediately to secure location. Always stay away from ends of tanks, especially when engulfed in flames. If employees are expected to fight fires, they must be trained and equipped in OSHA 1910.156.

DOT GUIDE NUMBER: 146 EMERGENCY RESPONSE: Stay upwind and uphill • Remove all ignition sources • IF material or contaminated runoff from fire control enters waterways, notify downstream users of potentially contaminated water • Wear positive-pressure, self-contained breathing apparatus (SCBA). Wear chemical protective clothing which is specifically recommended by the manufacturer. Structural firefighter's protective clothing provides only limited protection • **RELEASE, NO FIRE** • Small spill: Cover with inert, damp, noncombustible material using nonsparking tools and place into loosely covered plastic containers for later disposal. Large Spill: Wet down with water and dike for later disposal. *DO NOT CLEAN UP OR DISPOSE OF, EXCEPT UNDER SUPERVISION OF A SPECIALIST* • Use large amounts of water well away from the material to disperse vapors–contain runoff. Prevent material and runoff from entering sewers and waterways if it can be done safely well ahead of the release • Ventilate confined area if it can be done without placing personnel at risk • **FIRE** • Small fire: Water spray if preferred; if water is not available use *dry* chemical, CO_2, or regular foam. Large fire: Flood fire with water from a distance; do not use straight streams • Do not move cargo or vehicles if cargo has been exposed to heat • IF material is not leaking, cool exposed containers with large quantities of water from unattended equipment or remove intact containers if it can be done safely. *DO NOT ALLOW WATER TO GET INSIDE CONTAINERS* • If cylinders are exposed to excessive heat from fire or flame contact, withdraw immediately to secure location. Always stay away from ends of tanks, especially when engulfed in flames. If employees are expected to fight fires, they must be trained and equipped in OSHA 1910.156.

DOT GUIDE NUMBER: 147 EMERGENCY RESPONSE: Stay upwind and uphill • Remove all ignition sources • IF material or contaminated runoff from fire control enters waterways, notify downstream users of potentially contaminated water • Wear positive-pressure, self-contained breathing apparatus (SCBA). Wear chemical protective clothing which is specifically recommended by the manufacturer. Structural firefighter's protective clothing provides only limited protection • **RELEASE, NO FIRE** • Small spill: Cover with inert, damp, noncombustible material using nonsparking tools and place into loosely covered plastic containers for later disposal. Large Spill: Wet down with water and dike for later disposal. *DO NOT CLEAN UP OR DISPOSE OF, EXCEPT UNDER SUPERVISION OF A SPECIALIST* • Use large amounts of water well away from the material to disperse vapors–contain runoff. Prevent material and runoff from entering sewers and waterways if it can be done safely well ahead of the release • Ventilate confined area if it can be done without placing personnel at risk • **FIRE** • Small fire: Water spray if preferred; if water is not available use *dry* chemical, CO_2, or regular foam. Large fire: Flood fire with water from a distance; do not use straight streams • Do not move cargo or vehicles if cargo has been exposed to heat • IF material is not leaking, cool exposed containers with large quantities of water from unattended equipment or remove intact containers if it can be done safely. *DO NOT ALLOW WATER TO GET INSIDE CONTAINERS* • If cylinders are exposed to excessive heat from fire or flame contact, withdraw immediately to secure location. Always stay away from ends of tanks, especially when engulfed in flames. If employees are expected to fight fires, they must be trained and equipped in OSHA 1910.156.

DOT GUIDE NUMBER: 148 EMERGENCY RESPONSE: *DO NOT ALLOW THIS MATERIAL TO WARM UP. IF LIQUID NITROGEN OR DRY ICE CANNOT BE OBTAINED FOR COOLING, EVACUATE AREA IMMEDIATELY.* Stay upwind and uphill • Remove all ignition sources • IF material or contaminated runoff from fire control enters waterways, notify downstream users of potentially contaminated water • Wear positive-pressure, self-contained breathing apparatus (SCBA). Wear chemical protective clothing which is specifically recommended by the manufacturer. Structural firefighter's protective clothing provides only limited protection • **RELEASE, NO FIRE** • Small spill: Cover with inert, damp, noncombustible material using clean nonsparking tools and place into loosely covered plastic containers for later disposal. Large Spill: Dike far ahead of liquid for later. *DO NOT CLEAN UP OR DISPOSE OF, EXCEPT UNDER SUPERVISION OF A SPECIALIST* • Prevent material and runoff from entering sewers, basements or other confined areas, and waterways if it can be done safely well ahead of the release • Ventilate confined area if it can be done without placing personnel at risk • **FIRE** • Small fire: Water spray if preferred; if water is not available use *dry* chemical, CO_2, or regular foam. Large fire: Flood fire with water from a distance; do not use straight streams • Do not move cargo or vehicles if cargo has been exposed to heat. The temperature of the substance must be maintained at or below the "control temperature" at all times • IF material is not leaking, cool exposed containers with large quantities of water from unattended equipment until long after the fire is out, or remove intact containers if it can be done safely. *BEWARE OF CONTAINER EXPLOSIONS. DO NOT ALLOW WATER TO GET INSIDE CONTAINERS* • Always stay away from the ends of tanks. If cylinders are exposed to excessive heat from fire or flame contact, withdraw immediately to secure location and let fire burn. Always stay away from ends of tanks, especially when engulfed in flames. If employees are expected to fight fires, they must be trained and equipped in OSHA 1910.156.

DOT GUIDE NUMBER: 151 EMERGENCY RESPONSE: Stay upwind and uphill • IF material or contaminated runoff from fire control enters waterways, notify downstream users of potentially contaminated water • Wear positive-pressure, self-contained breathing apparatus (SCBA). Wear chemical protective clothing which is specifically recommended by the manufacturer • **RELEASE, NO FIRE** • Do not touch damaged containers or spilled material unless wearing appropriate protective material • Cover material to protect from wind, rain, or spray • Prevent material and runoff from entering sewers and waterways if it can be done safely well ahead of the release • Use large amounts of water well away from release to disperse vapors–contain runoff • Ventilate confined area if it can be done without placing personnel at risk. *DO NOT ALLOW WATER TO GET INSIDE CONTAINERS* • **FIRE** • Small fire: Extinguish with *dry* chemical, CO_2 or water spray. Large fire: Use water spray, fog, or regular foam. Do not use straight streams • Cool exposed containers with large quantities of water from unattended equipment or remove intact containers if it can be done safely. DO NOT ALLOW WATER TO GET INSIDE CONTAINERS • If cooling streams are ineffective (*venting sound increases in volume and pitch, tank discolors or shows signs of deforming*), withdraw immediately to a secure location. Always stay away from ends of tanks, especially when engulfed in flames. If employees are expected to fight fires, they must be trained and equipped in OSHA 1910.156.

DOT GUIDE NUMBER: 152 EMERGENCY RESPONSE: Stay upwind and uphill • Remove all ignition sources • IF material or contaminated runoff from fire control enters waterways, notify downstream users of potentially contaminated water • Wear positive-pressure, self-contained breathing apparatus (SCBA). Wear chemical protective clothing which is specifically recommended by the manufacturer. Structural firefighter's protective clothing provides only limited protection • **RELEASE, NO FIRE** • Do not touch damaged containers or spilled material unless wearing appropriate protective clothing. Cover with plastic sheet to prevent spreading and to protect it from wind, rain or spray. Absorb or cover with *dry* sand, sand or other noncombustible material and transfer to containers • Stop the release if it can be done safely from a distance. Prevent material and runoff from entering sewers and waterways if it can be done safely well ahead of the release. Use large amounts of water well away from release to disperse vapors–contain runoff. Water spray may reduce vapor, but may not prevent ignition in enclosed spaces. Ventilate confined area if it can be done without placing personnel at risk . If in a building, evacuate the building and confine vapors by closing doors and shutting down HVAC system • **FIRE** • Small fire: Use *dry* chemical, CO_2, or water spray • Large fire: Use water spray (do not use straight streams), fog, or regular foam. Keep exposures cool to protect against
reignition. Do not direct straight streams into the liquid • Cool exposed containers with large quantities of water from unattended equipment or remove intact containers if it can be done safely. *DO NOT ALLOW WATER TO GET INSIDE CONTAINERS* • If cooling streams are ineffective (*venting sound increases in volume and pitch, tank discolors or shows signs of deforming*), withdraw immediately to a secure location. Always stay away from ends of tanks, especially when engulfed in flames. If employees are expected to fight fires, they must be trained and equipped in OSHA 1910.156.

DOT GUIDE NUMBER: 153 EMERGENCY RESPONSE (see notes below for Sarin, VX, Tabun): Stay upwind and uphill • Remove all ignition sources • Ensure medical personnel are aware of the material(s) involved, and take precautions to protect themselves • IF material or contaminated runoff from fire control enters waterways, notify downstream users of potentially contaminated water • Wear positive-pressure, self-contained breathing apparatus (SCBA). Wear chemical protective clothing which is specifically recommended by the manufacturer. Structural firefighter's protective clothing provides NO protection • **RELEASE, NO FIRE** • (see below for VX) Do not touch damaged containers or spilled material unless wearing appropriate protective clothing. Cover with plastic sheet to prevent spreading and to protect it from wind, rain or spray. Absorb or cover with *dry* sand, sand or other noncombustible material and transfer to containers • Stop the release if it can be done safely from a distance. Prevent material and runoff from entering sewers and waterways if it can be done safely well ahead of the release. Use large amounts of water well away from release to disperse vapors–contain runoff. Water spray may reduce vapor, but may not

prevent ignition in enclosed spaces. Ventilate confined area if it can be done without placing personnel at risk . If in a building, evacuate the building and confine vapors by closing doors and shutting down HVAC system • **FIRE** • Small fire: Use *dry* chemical, CO_2, or water spray • Large fire: Use *dry* chemical, CO_2, alcohol-resistant foam, or water spray (do not use straight streams). Keep exposures cool to protect against
reignition. Do not direct straight streams into the liquid • Cool exposed containers with large quantities of water from unattended equipment or remove intact containers if it can be done safely. *DO NOT ALLOW WATER TO GET INSIDE CONTAINERS* • If cooling streams are ineffective (*venting sound increases in volume and pitch, tank discolors or shows signs of deforming*), withdraw immediately to a secure location. Always stay away from ends of tanks, especially when engulfed in flames. If employees are expected to fight fires, they must be trained and equipped in OSHA 1910.156.

Notes for VX • **RELEASE, NO FIRE** • *BACK OFF! ISOLATE A WIDE AREA AROUND THE RELEASE AND CALL FOR EXPERT HELP.* If in a building, evacuate and confine vapors by closing doors and shutting down HVAC system • **FIRE** • IF material is on fire and conditions permit, DO NOT EXTINGUISH. *Combustion products are less deadly than the material itself.* IF material is involved in a fire which must be extinguished, use an agent appropriate for the burning material, using unattended equipment. *Notes for Sarin, Tabun* • **RELEASE, NO FIRE** • BACK OFF! ISOLATE A WIDE AREA AROUND THE RELEASE AND CALL FOR EXPERT HELP. If in a building, evacuate and confine vapors by closing doors and shutting down HVAC system • **FIRE** • IF material is on fire and conditions permit, DO NOT EXTINGUISH. *Combustion products are less deadly than the material itself.* IF material is involved in a fire which must be extinguished, use an agent appropriate for the burning material, using unattended equipment. Be aware that hydrogen cyanide is produced when the material comes in contact with water.

DOT GUIDE NUMBER: 154 EMERGENCY RESPONSE: Stay upwind and uphill • Isolate the area of release or fire and deny entry • Remove all ignition sources • IF material or contaminated runoff from fire control enters waterways, notify downstream users of potentially contaminated water • Wear positive-pressure, self-contained breathing apparatus (SCBA). Wear chemical protective clothing which is specifically recommended by the manufacturer. Structural firefighter's protective clothing provides only limited protection • **RELEASE, NO FIRE** • Do not touch damaged containers or spilled material unless wearing appropriate protective clothing. Cover with plastic sheet to prevent spreading and to protect it from wind, rain or spray. Absorb or cover with *dry* sand, sand or other noncombustible material and transfer to containers • Stop the release if it can be done safely from a distance. Prevent material and runoff from entering sewers and waterways if it can be done safely well ahead of the release. Use large amounts of water well away from release to disperse vapors–contain runoff. Water spray may reduce vapor, but may not prevent ignition in enclosed spaces. Ventilate confined area if it can be done without placing personnel at risk . If in a building, evacuate the building and confine vapors by closing doors and shutting down HVAC system • **FIRE** • Small fire: Use *dry* chemical, CO_2, or water spray • Large fire: Use *dry* chemical, CO_2, alcohol-resistant foam, or water spray . Keep exposures cool to protect against
reignition. Do not direct straight streams into the liquid • Cool exposed containers with large quantities of water from unattended equipment or remove intact containers if it can be done safely. *DO NOT ALLOW WATER TO GET INSIDE CONTAINERS* • If cooling streams are ineffective (*venting sound increases in volume and*

pitch, tank discolors or shows signs of deforming), withdraw immediately to a secure location. Always stay away from ends of tanks, especially when engulfed in flames. If employees are expected to fight fires, they must be trained and equipped in OSHA 1910.156.

DOT GUIDE NUMBER: 155 EMERGENCY RESPONSE (For trimethylchlorosilane, vinyl trichlorosilane see notes below): Stay upwind and uphill • Remove all ignition sources. All equipment used to handle this material must be grounded • IF material or contaminated runoff from fire control enters waterways, notify downstream users of potentially contaminated water • Wear positive-pressure, self-contained breathing apparatus (SCBA). Wear chemical protective clothing which is specifically recommended by the manufacturer. Structural firefighter's protective clothing provides only limited protection • **RELEASE, NO FIRE** • Small spill: Cover with *dry* earth, *dry* sand, or other noncombustible material followed with plastic sheet to minimize spreading or contact with water or rain. Use clean nonsparking tools to collect material and place material into loosely covered plastic containers for later disposal • Stop the release if it can be done safely from a distance • Prevent material and runoff from entering sewers and waterways if it can be done safely well ahead of the release • Use vapor suppressing foam to disperse vapors–contain runoff. Water spray may reduce vapor, but may not prevent ignition in enclosed spaces • Consider application of alcohol-resistanttant (AFFF) foam to large areas of spilled liquid to control vapors • Ventilate confined area if it can be done without placing personnel at risk • **FIRE** • *FOR CHLOROSILANES DO NOT USE WATER. USE AFFF ALCOHOL-RESISTANT, MEDIUM EXPANSION FOAM*. Small fire: Extinguish with *dry* chemical, CO_2, *dry* sand, or alcohol-resistant foam. May react with foams releasing corrosive/toxic gases • Large fire: Use water spray or alcohol-resistant foam if it is available in sufficient amounts • Under favorable conditions, experienced crews can use coordinated fog streams to sweep the flames off the surface of the burning liquid. Keep exposures cool to protect against reignition. Do not direct straight streams into the liquid • Cool exposed containers with large quantities of water from unattended equipment or remove intact containers if it can be done safely. DO NOT GET WATER ON SPILLED SUBSTANCE OR INSIDE CONTAINERS • If cooling streams are ineffective (*venting sound increases in volume and pitch, tank discolors or shows signs of deforming*), withdraw immediately to a secure location. Always stay away from ends of tanks, especially when engulfed in flames. Always stay away from ends of tanks, especially when engulfed in flames. If employees are expected to fight fires, they must be trained and equipped in OSHA 1910.156.

Trimethylchlorosilane, vinyl trichlorosilane: IF material is on fire and if conditions permit, *DO NOT EXTINGUISH*. Although material reacts with water it can be extinguished with low or medium expansion (AFFF) alcohol-resistant foam.

DOT GUIDE NUMBER: 156 EMERGENCY RESPONSE: Stay upwind and uphill • Isolate the area of release or fire and deny entry • Remove all ignition sources. All equipment used to handle this material must be grounded • IF material or contaminated runoff from fire control enters waterways, notify downstream users of potentially contaminated water • Wear positive-pressure, self-contained breathing apparatus (SCBA). Wear chemical protective clothing which is specifically recommended by the manufacturer. Structural firefighter's protective clothing provides only limited protection • **RELEASE, NO FIRE** • Small spill: Cover with *dry* earth, *dry* sand, or other noncombustible material followed with plastic sheet to minimize spreading or contact with water or rain. Use clean nonsparking tools to collect material and place material into loosely covered plastic containers for later disposal • Stop the release if it can be done safely from a distance • Prevent material and runoff from entering sewers and waterways if it can be done safely well ahead of the release • Use vapor suppressing foam to disperse vapors–contain runoff. Water spray may reduce vapor, but may not prevent ignition in enclosed spaces • Ventilate confined area if it can be done without placing personnel at risk • **FIRE** • *FOR CHLOROSILANES DO NOT USE WATER. USE AFFF ALCOHOL-RESISTANT, MEDIUM EXPANSION FOAM*. Small fire: Use *dry* chemical, CO_2, *dry* sand, or alcohol-resistant foam. May react with foams releasing corrosive/toxic gases • Large fire: Use water spray or alcohol-resistant foam if it is available in sufficient amounts • Under favorable conditions, experienced crews can use coordinated fog streams to sweep the flames off the surface of the burning liquid. Keep exposures cool long after fire is out to protect against reignition. Do not direct straight streams into the liquid • Cool exposed containers with large quantities of water from unattended equipment or remove intact containers if it can be done safely. *DO NOT GET WATER ON SPILLED SUBSTANCE OR INSIDE CONTAINERS* • If cooling streams are ineffective (*venting sound increases in volume and pitch, tank discolors or shows signs of deforming*), withdraw immediately to a secure location. Always stay away from ends of tanks, especially when engulfed in flames. If employees are expected to fight fires, they must be trained and equipped in OSHA 1910.156.

DOT GUIDE NUMBER: 157 EMERGENCY RESPONSE: Stay upwind and uphill • Isolate the area of release or fire and deny entry • Remove all ignition sources. All equipment used to handle this material must be grounded • IF material or contaminated runoff from fire control enters waterways, notify downstream users of potentially contaminated water • Wear positive-pressure, self-contained breathing apparatus (SCBA). Wear chemical protective clothing which is specifically recommended by the manufacturer. Structural firefighter's protective clothing provides only limited protection • **RELEASE, NO FIRE** • Small spill: Cover with *dry* earth, *dry* sand, or other noncombustible material followed with plastic sheet to minimize spreading or contact with water or rain. Use clean nonsparking tools to collect material and place material into loosely covered plastic containers for later disposal • Stop the release if it can be done safely from a distance • Prevent material and runoff from entering sewers and waterways if it can be done safely well ahead of the release • For chlorosilanes alcohol-resistanttant (AFFF) medium expansion foam to large areas of spilled liquid to control vapors • Ventilate confined area if it can be done without placing personnel at risk • **FIRE** • *CAUTION*: MOST FOAMS WILL REACT WITH THE MATERIAL AND RELEASE CORROSIVE/TOXIC GASES. Small fire: Extinguish with *dry* chemical, CO_2, *dry* sand, or alcohol-resistant foam. May react with foams releasing corrosive/toxic gases • Large fire: Use water spray or alcohol-resistant foam if it is available in sufficient amounts • Under favorable conditions, experienced crews can use coordinated fog streams to sweep the flames off the surface of the burning liquid. Keep exposures cool to protect against reignition. Do not direct straight streams into the liquid • Cool exposed containers with large quantities of water from unattended equipment or remove intact containers if it can be done safely. DO NOT GET WATER ON SPILLED SUBSTANCE OR INSIDE CONTAINERS • If cooling streams are ineffective (*venting sound increases in volume and pitch, tank discolors or shows signs of deforming*), withdraw immediately to a secure location. Always stay away from ends of tanks, especially when engulfed in flames. If employees are expected to fight fires, they must be trained and equipped in OSHA 1910.156.

DOT GUIDE NUMBER: 159 EMERGENCY RESPONSE: Stay upwind and uphill • Isolate the area of release or fire and deny entry • Remove all ignition sources. All equipment used to handle this material must be grounded • IF material or contaminated runoff from fire control enters waterways, notify downstream users of potentially contaminated water • Wear positive-pressure, self-contained breathing apparatus (SCBA). Wear chemical protective clothing which is specifically recommended by the manufacturer. Structural firefighter's protective clothing provides only limited protection • **RELEASE, NO FIRE** • Small spill: Cover with earth, sand, or other noncombustible material and place material into containers for later disposal • Stop the release if it can be done safely from a distance • Prevent material and runoff from entering sewers and waterways if it can be done safely well ahead of the release • Use vapor suppressing foam to disperse vapors–contain runoff. Water spray may reduce vapor, but may not prevent ignition in enclosed spaces • Ventilate confined area if it can be done without placing personnel at risk • **FIRE** • Small fire: Extinguish with *dry* chemical, CO_2, water spray or regular foam. Large fire: Use water spray or regular foam if it is available in sufficient amounts • Under favorable conditions, experienced crews can use coordinated fog streams to sweep the flames off the surface of the burning liquid • Cool exposed containers with large quantities of water from unattended equipment. If this impossible, withdraw from area and let fire burn • If cooling streams are ineffective (*venting sound increases in volume and pitch, tank discolors or shows signs of deforming*), withdraw immediately to a secure location. Always stay away from ends of tanks, especially when engulfed in flames. If employees are expected to fight fires, they must be trained and equipped in OSHA 1910.156.

DOT GUIDE NUMBER: 160 EMERGENCY RESPONSE: Stay upwind and uphill • Isolate the area of release or fire and deny entry • Remove all ignition sources. All equipment used to handle this material must be grounded • IF material or contaminated runoff from fire control enters waterways, notify downstream users of potentially contaminated water • Wear positive-pressure, self-contained breathing apparatus (SCBA). Wear chemical protective clothing which is specifically recommended by the manufacturer. Structural firefighter's protective clothing provides only limited protection • **RELEASE, NO FIRE** • Small spill: Cover with earth, sand, or other noncombustible material • Stop the release if it can be done safely from a distance • Prevent material and runoff from entering sewers and waterways if it can be done safely well ahead of the release • Use vapor suppressing foam to disperse vapors–contain runoff. Water spray may reduce vapor, but may not prevent ignition in enclosed spaces • Ventilate confined area if it can be done without placing personnel at risk • **FIRE** • Small fire: Extinguish with *dry* chemical, CO_2, or water spray. Large fire: Use cry chemical, CO_2, water spray, or alcohol-resistant foam if it is available in sufficient amounts • Under favorable conditions, experienced crews can use coordinated fog streams to sweep the flames off the surface of the burning liquid • Cool exposed containers with large quantities of water from unattended equipment. If this impossible, withdraw from area and let fire burn • If cooling streams are ineffective (*venting sound increases in volume and pitch, tank discolors or shows signs of deforming*), withdraw immediately to a secure location. Always stay away from ends of tanks, especially when engulfed in flames. If employees are expected to fight fires, they must be trained and equipped in OSHA 1910.156.

DOT GUIDE NUMBER: 162 Priorities for Presence of radioactive material will not change effectiveness of fire control techniques • Stay upwind and uphill • Isolate the area of release or fire and deny entry • Remove all ignition sources. All equipment used to handle this material must be grounded • IF material or contaminated runoff from fire control enters waterways, notify downstream users of potentially contaminated water • Wear positive-pressure, self-contained breathing apparatus (SCBA). Wear chemical protective clothing which is specifically recommended by the manufacturer. Structural firefighter's protective clothing provides only limited protection • **RELEASE, NO FIRE** • Small spill: Do not touch damaged packages or spilled material. Cover liquid spill with earth, sand, or other noncombustible material • Large liquid spill: Dike and collect. Stop the release if it can be done safely from a distance • Prevent material and runoff from entering sewers and waterways if it can be done safely well ahead of the release • **FIRE** • Do not move damaged packages; move undamaged packages out of fire zone if it can be done without risk • Small fire: Extinguish with *dry* chemical, CO_2, or regular foam. Large fire: Use water spray, fog (flooding amounts). Dike fire control water for later disposal • Cool exposed containers with large quantities of water from unattended equipment • If cooling streams are ineffective (*venting sound increases in volume and pitch, tank discolors or shows signs of deforming*), withdraw immediately to a secure location. Always stay away from ends of tanks, especially when engulfed in flames. If employees are expected to fight fires, they must be trained and equipped in OSHA 1910.156.

DOT GUIDE NUMBER: 163 EMERGENCY RESPONSE: Priorities for Presence of radioactive material will not change effectiveness of fire control techniques • Stay upwind and uphill • Isolate the area of release or fire and deny entry • Remove all ignition sources. All equipment used to handle this material must be grounded • IF material or contaminated runoff from fire control enters waterways, notify downstream users of potentially contaminated water • Wear positive-pressure, self-contained breathing apparatus (SCBA). Wear chemical protective clothing which is specifically recommended by the manufacturer. Structural firefighter's protective clothing provides only limited protection • **RELEASE, NO FIRE** • Small spill: Do not touch damaged packages or spilled material. Cover liquid spill with earth, sand, or other noncombustible material • Stop the release if it can be done safely from a distance • Prevent material and runoff from entering sewers and waterways if it can be done safely well ahead of the release • **FIRE** • Do not move damaged packages; move undamaged packages out of fire zone if it can be done without risk • Small fire: Extinguish with *dry* chemical, CO_2, or regular foam. Large fire: Use water spray, fog (flooding amounts) • Cool exposed containers with large quantities of water from unattended equipment • If cooling streams are ineffective (*venting sound increases in volume and pitch, tank discolors or shows signs of deforming*), withdraw immediately to a secure location. Always stay away from ends of tanks, especially when engulfed in flames. Always stay away from ends of tanks, especially when engulfed in flames. If employees are expected to fight fires, they must be trained and equipped in OSHA 1910.156.

DOT GUIDE NUMBER: 167 EMERGENCY RESPONSE: Stay upwind and uphill • Isolate the area of release or fire and deny entry • Evacuate the immediate area and downwind for a large release • Remove all ignition sources • Notify local health and fire officials and pollution control agencies • IF material or contaminated runoff from fire control enters waterways, notify downstream users of potentially contaminated water • Wear positive-pressure, self-contained breathing apparatus (SCBA). Wear chemical protective clothing which is specifically recommended by the manufacturer; but they may provide little or no thermal protection. Structural firefighter's protective clothing provides only limited protection • **RELEASE, NO FIRE** • Do NOT touch or walk through spilled material. If you are NOT wearing special protective clothing approved for this material, do NOT expose yourself to any risk of this material touching you • Stop the release if it can be done safely from a distance • Prevent material and runoff from entering sewers and waterways if it can be done safely well ahead of the release • Ventilate confined area if it can be done without placing personnel at risk • Small spill: Cover with *dry* earth, *dry* sand, or other noncombustible material followed with plastic sheet to minimize spreading or contact with water or rain. Use clean nonsparking tools to collect material and place material into loosely covered plastic containers for later disposal • *Do not direct water at spill or source of leak.* A fine water spray remotely directed to the edge of the spill pool can be used to direct and maintain a hot flare fire which will burn the spilled material in a controlled manner • **FIRE** • Small fire: Use *dry* chemical, soda ash, lime, or sand. Large fire: Use water spray, flooding amounts of fog • *DO NOT ALLOW WATER TO GET INSIDE CONTAINERS* • Fight fire from maximum distance or use unmanned hose holders or monitor nozzles • Cool containers with flooding quantities of water until well after fire is out • Do NOT direct water at sources of leak or safety devices; icing may occur • Move containers from fire area if it can be done without risk • If cooling streams are ineffective (*venting sound increases in volume and pitch, tank discolors or shows signs of deforming*), withdraw immediately to a secure location. Always stay away from ends of tanks, especially when engulfed in flames. If employees are expected to fight fires, they must be trained and equipped in OSHA 1910.156.

DOT GUIDE NUMBER: 168 EMERGENCY RESPONSE: Stay upwind and uphill • Isolate the area of release or fire and deny entry • Remove all ignition sources • IF material or contaminated runoff from fire control enters waterways, notify downstream users of potentially contaminated water • Wear positive-pressure, self-contained breathing apparatus (SCBA). Wear chemical protective clothing which is specifically recommended by the manufacturer; but they may provide little or no thermal protection. Structural firefighter's protective clothing provides only limited protection • **RELEASE, NO FIRE** • Wear fully encapsulating protective clothing approved for this material, do NOT expose yourself to any risk of this material touching you • Stop the leak if it can be done safely, without risk • Prevent material and runoff from entering sewers and waterways if it can be done safely well ahead of the release • Ventilate confined area if it can be done without placing personnel at risk • Isolate area until gas has dispersed • **Fire** • *DO NOT EXTINGUISH A LEAKING GAS FIRE UNLESS LEAK CAN BE STOPPED.* Small fire: Use *dry* chemical, CO_2, or water spray. Large fire: Use water spray, fog, or regular foam • Fight fire from maximum distance or use unmanned hose holders or monitor nozzles • Cool containers with flooding quantities of water until well after fire is out • Do NOT direct water at sources of leak or safety devices; icing may occur • Move containers from fire area if it can be done without risk • If cooling streams are ineffective (*venting sound increases in volume and pitch, tank discolors or shows signs of deforming*), withdraw immediately to a secure location. Always stay away from ends of tanks, especially when engulfed in flames. If employees are expected to fight fires, they must be trained and equipped in OSHA 1910.156.

DOT GUIDE NUMBER: 171 EMERGENCY RESPONSE: Stay upwind and uphill • Isolate the area of release or fire and deny entry • Remove all ignition sources. All equipment used to handle this material must be grounded • IF material or contaminated runoff from fire control enters waterways, notify downstream users of potentially contaminated water • Wear positive-pressure, self-contained breathing apparatus (SCBA). Wear chemical protective clothing which is specifically recommended by the manufacturer. Structural firefighter's protective clothing provides only limited protection • **RELEASE, NO FIRE** • Prevent dust cloud • Small spill: Cover with sand or other noncombustible material and place in a container for later disposal • Cover powder spills with plastic sheet or tarp to minimize spreading and protect from wind, rain, or spray. Large spill: Stop the release if it can be done safely from a distance • Prevent material and runoff from entering sewers and waterways if it can be done safely well ahead of the release • Use large amounts of water well away from release to disperse vapors–contain runoff • Ventilate confined area if it can be done without placing personnel at risk • **FIRE** • Small fire: Use *dry* chemical, CO_2, water spray, or regular foam. Large fire: Use water spray, fog, or regular foam • Do not scatter spilled material with high pressure water streams • Dike fire-control water for later disposal • IF material is not leaking, cool exposed containers with large quantities of water from unattended equipment until fire is well out, or remove intact containers if it can be done safely • If cooling streams are ineffective (*venting sound increases in volume and pitch, tank discolors or shows signs of deforming*), withdraw immediately to a secure location. Always stay away from ends of tanks, especially when engulfed in flames. If employees are expected to fight fires, they must be trained and equipped in OSHA 1910.156.

DOT GUIDE NUMBER: 172 (Mercury) **EMERGENCY RESPONSE:** Stay upwind and uphill • Isolate the area of release or fire and deny entry • Remove all ignition sources. All equipment used to handle this material must be grounded • IF material or contaminated runoff from fire control enters waterways, notify downstream users of potentially contaminated water • Wear positive-pressure, self-contained breathing apparatus (SCBA). Wear chemical protective clothing which is specifically recommended by the manufacturer. Structural firefighter's protective clothing provides only limited protection • **RELEASE, NO FIRE** • Do not touch or walk through spilled material. Do not touch damaged containers or spilled material unless wearing appropriate protective clothing. Stop release if it can be done without risk • Prevent material and runoff from entering sewers and waterways if it can be done safely well ahead of the release. Do not use steel or aluminum tools or equipment. Cover with earth, sand, or other noncombustible materials followed with a plastic sheet to minimize spreading or contact with rain. Use mercury spill kit. Mercury spill areas may be subsequently treated with calcium sulfide or with sodium thiosulfate to neutralize any residual mercury • **FIRE** • Use extinguishing agent suitable for type of surrounding fire; do not direct water at the heated metal • IF material is not leaking, cool exposed containers with large quantities of water from unattended equipment until fire is well out, or remove intact containers if it can be done safely • If cooling streams are ineffective (*venting sound increases in volume and pitch, tank discolors or shows signs of deforming*), withdraw immediately to a secure location. Always stay away from ends of tanks, especially when engulfed in flames. If employees are expected to fight fires, they must be trained and equipped in OSHA 1910.156.

Appendix B

List of Marine Pollutants (§172.101 - Appendix B)

If the letters "PP" appear in this column for a material, the material is a severe marine pollutant, otherwise it is not. If a material not listed in this appendix meets the criteria for a marine pollutant, as provided in the *General Introduction of the IMDG Code, Guidelines for the Identification of Harmful Substances in Packaged Form*, the material may be transported as a marine pollutant in accordance with the applicable requirements of this subchapter.

Acetal
Acetaldehyde
Acetone cyanohydrin, stabilized
Acetylene tetrabromide
Acetylene tetrachloride
Acraldehyde, inhibited
Acrolein, inhibited
Acrylic aldehyde, inhibited
Alcohol C-12–C-16 poly(1-6)ethoxylate
Alcohol C-13–C-15 poly(1-6)ethoxylate
Alcohol C-6–C-17 (secondary)poly(3-6) ethoxylate
Aldicarb
Aldrin (PP)
Alkyl (C-12–C-14) dimethylamine
Alkyl (C-7–C-9) nitrates
Alkylbenzenesulphonates, branched and straight chain
Alkylphenols, liquid, n.o.s. (including C2–C12 homologues)
Allyl bromide
o-Aminoanisole
Aminocarb
Ammonium dinitro-*o*-cresolate
n-Amylbenzene
Amyl mercaptans
Anisole
Azinphos-ethyl (PP)
Azinphos-methyl (PP)
Barium cyanide
Bendiocarb
Benomyl
Benquinox
Benzaldehyde
Benzyl chlorocarbonate
Benzyl chloroformate
Binapacryl (PP)
N,*N*-Bis(2-hydroxyethyl)oleamide (LOA)
Brodifacoum (PP)
Bromine cyanide
Bromoacetone
Bromoallylene
Bromobenzene
o-Bromobenzyl cyanide
Bromocyane
Bromoform
Bromophos-ethyl (PP)
3-Bromopropene
Bromoxynil
Butanedione
2-Butenal, stabilized
Butyl benzenes
Butyl benzyl phthalate
n-Butyl butyrate
Butyl mercaptans
N-tert-butyl-*N*-cyclopropyl-6-methylthio-1,3,5-triazine-2,4-diamine
Butylphenols, liquid
Butylphenols, solid
p-tert-butyltoluene
Butyraldehyde
Cadmium compounds (PP)
Cadmium sulphide
Calcium arsenate
Calcium arsenate and calcium arsenite, mixtures, solid
Calcium cyanide
Calcium naphthenate
Camphechlor (PP)
Camphor oil
Carbaryl
Carbendazim
Carbofuran
Carbon tetrabromide
Carbon tetrachloride
Carbophenothion (PP)
Cartap hydrochloride
Chlordane (PP)
Chlorfenvinphos
Chlorinated paraffins (C-10– C-13) (PP)
Chlorinated paraffins (C-14– C-17), with > 1% shorter chain length (PP)
Chlorine
Chlorine cyanide, inhibited
Chlormephos
Chloroacetone, stabilized
1-Chloro-2,3-Epoxypropane
2-Chloro-6-nitrotoluene
4-Chloro-2-nitrotoluene
Chloro-*o*-nitrotoluene
2-Chloro-5-trifluoromethylnitrobenzene
p-Chlorobenzyl chloride, liquid or solid
Chlorodinitrobenzenes, liquid or solid
1-Chloroheptane
1-Chlorohexane
Chloronitroanilines
Chloronitrotoluenes liquid
Chloronitrotoluenes, solid
1-Chlorooctane
Chlorophenolates, liquid (PP)
Chlorophenolates, solid (PP)
Chlorophenols, liquid
Chlorophenols, solid
Chlorophenyltrichlorosilane
Chlorotoluenes (*o*-, *m*-, *p*-)
Chlorpyriphos (PP)
Chlorthiophos (PP)
Coal tar
Coal tar naphtha
Cocculus
Coconitrile
Copper acetoarsenite
Copper arsenite
Copper chloride
Copper chloride solution (PP)
Copper cyanide (PP)
Copper metal powder (PP)
Copper sulphate, anhydrous, hydrates (PP)
Coumachlor
Coumaphos (PP)
Creosote (coal tar)
Creosote (wood tar)
Cresols (*o*-, *m*-, *p*-)
Cresyl diphenyl phosphate (PP)

Cresylic acid
Cresylic acid sodium salt
Crotonaldehyde, stabilized
Crotonic aldehyde, stabilized
Crotoxyphos
Cumene
Cupric arsenite
Cupric chloride (PP)
Cupric cyanide (PP)
Cupric sulfate (PP)
Cupriethylenediamine solution
Cuprous chloride (PP)
Cyanide mixtures
Cyanide solutions
Cyanides, inorganic, n.o.s.
Cyanogen bromide
Cyanogen chloride, inhibited
Cyanophos
1,5,9-Cyclododecatriene (PP)
Cyhexatin (PP)
Cymenes (o-, m-, p-) (PP)
Cypermethrin (PP)
2,4-D
DDT (PP)
n-Decaldehyde
n-Decanol
Decyl acrylate
Decycloxytetrahydrothiophene dioxide
DEF
Di-allate
Di-n-Butyl phthalate
Dialifos (PP)
4,4'-Diaminodiphenylmethane
Diazinon (PP)
1,3-Dibromobenzene
Dichlofenthion (PP)
Dichloroanilines
1,3-Dichlorobenzene
1,2-Dichlorobenzene
1,4-Dichlorobenzene
Dichlorobenzene (m-, o-, p-)
2,2-Dichlorodiethyl ether
Dichlorodimethyl ether, symmetrical
Di-(2-chloroethyl) ether
1,1-Dichloroethylene, inhibited
1,6-Dichlorohexane
Dichlorophenols, liquid
Dichlorophenols, solid
2,4-Dichlorophenoxyacetic acid (see also 2,4-D)
2,4-Dichlorophenoxyacetic acid diethanolamine salt
2,4-Dichlorophenoxyacetic acid dimethylamine salt
2,4-Dichlorophenoxyacetic acid triisopropylamine salt
Dichlorophenyltrichlorosilane
Dichlorvos (PP)
Dicrotophos
Dieldrin (PP)
Diethybenzenes (mixed isomers)
Diisopropylbenzenes
Diisopropylnaphthalene
Dimethoate (PP)
Dimethyl disulphide
Dimethyl glyoxal (butanedione)
Dimethylhydrazine, symmetrical
Dimethylhydrazine, unsymmetrical
Dimethyl sulphide
Dimethylphenols, liquid or solid

Dinitro-o-cresol, solid
Dinitro-o-cresol, solution
Dinitrochlorobenzenes, liquid or solid
Dinitrophenol, dry or wetted with < 15 % water, by mass
Dinitrophenol solutions
Dinitrophenol, wetted with not < 15 % water, by mass
Dinitrophenolates alkali metals, dry or wetted with less than 15 % water, by mass
Dinitrophenolates, wetted with not < 15 % water, by mass
Dinobuton
Dinoseb
Dinoseb acetate
Dioxacarb
Dioxathion
Dipentene
Diphacinone
2,4-Di-tert-butylphenol
2,6-Di-tert-butylphenol
Diphenyl
Diphenyl ether
Diphenyl ether/biphenyl phenyl ether mixtures
Diphenyl/diphenyl ether (mixtures)
Diphenyl oxide and biphenyl phenyl ether mixtures
Diphenylamine chloroarsine (PP)
Diphenylchloroarsine, solid or liquid (PP)
Disulfoton
1,4-Di-tert-butylbenzene
DNOC
DNOC (pesticide)
Dodecyl diphenyl oxide disulphonate
Dodecyl hydroxypropyl sulfide
1-Dodecylamine
Dodecylphenol (PP)
Drazoxolon
Edifenphos
Endosulfan (PP)
Endrin (PP)
Epibromohydrin
Epichlorohydrin
EPN (PP)
EPTC (ISO)
Esfenvalerate (PP)
Ethion (PP)
Ethoprophos
Ethyl acrylate, inhibited
Ethyl chlorothioformate
Ethyl fluid
2-Ethylhexaldehyde
2-Ethylbutyraldehyde
Ethyl mercaptan
1-Ethyl-2-methylbenzene
2-Ethylhexyl nitrate
5-Ethyl-2-picoline
Ethyl propenoate, inhibited
2-Ethyl-3-propylacrolein
Ethyl tetraphosphate
Ethyldichloroarsine
Ethylene dibromide and methyl bromide mixtures, liquid
2-Ethylhexenal
Fenaminphos
Fenbutatin oxide (PP)
Fenitrothion (PP)
Fenpropathrin (PP)
Fensulfothion
Fenthion (PP)
Fentin acetate (PP)

Fentin hydroxide (PP)
Ferric arsenate
Ferric arsenite
Ferrous arsenate
Fonofos (PP)
Formetanate
Furathiocarb (ISO) (PP)
gamma-BHC (γ-BHC) (PP)
Gasoline, leaded
Heptachlor (PP)
Heptenophos
n-Heptyl aldehyde (PP)
n-Heptyl chloride
n-Heptylbenzene
Hexachlorobutadiene (PP)
1,3-Hexachlorobutadiene (PP)
2,4-Hexadiene aldehyde
Hexaethyl tetraphosphate, liquid
Hexaethyl tetraphosphate, solid
n-Hexyl chloride
n-Hexylbenzene
Hydrocyanic acid, anhydrous, stabilized, containing less than 3% water
Hydrocyanic acid, anhydrous, stabilized, containing less than 3% water and absorbed in a porous inert material
Hydrocyanic acid, aqueous solutions not > 20% hydrocyanic acid
Hydrogen cyanide solution in alcohol, with not more than 45% hydrogen cyanide
Hydrogen cyanide, stabilized with < 3% water
Hydrogen cyanide, stabilized with < 3% water and absorbed in a porous inert material
Hydroxydimethylbenzenes, liquid or solid
Ioxynil
Iron oxide, spent
Iron sponge, spent
Isoamyl mercaptan
Isobenzan
Isobutyl aldehyde
Isobutyl butyrate
Isobutyl isobutyrate
Isobutyl propionate
Isobutylbenzene
Isobutyraldehyde
Isodecaldehyde
Isodecanol
Isodecyl acrylate
Isodecyl diphenyl phosphate
Isofenphos
Isononanol
Isooctanol
Isooctyl nitrate
Isoprocarb
Isopropenylbenzene
Isopropylbenzene (Cumene)
Isotetramethylbenzene
Isovaleraldehyde
Isoxathion (PP)
Lead acetate
Lead arsenates
Lead arsenites
Lead compounds, soluble, n.o.s.
Lead cyanide
Lead nitrate
Lead perchlorate, solid or solution
Lead tetraethyl
Lead tetramethyl

Lindane (PP)
London Purple
Magnesium arsenate
Malathion
Mancozeb (ISO)
Maneb (12427-38-2)
Maneb preparations with not < 60 % maneb
Maneb preparation, stabilized against self-heating
Maneb stabilized or Maneb preparations, stabilized against self-heating
Manganese ethylene-1,2-bis dithiocarbamate
Manganese ethylene-1,2-bis-dithiocarbamate, stabilized against self-heating
Mephosfolan
Mercaptodimethur
Mecarbam
Mercuric acetate (PP)
Mercuric ammonium chloride (PP)
Mercuric arsenate (PP)
Mercuric benzoate (PP)
Mercuric bisulphate (PP)
Mercuric bromide (PP)
Mercuric chloride (PP)
Mercuric cyanide (PP)
Mercuric gluconate (PP)
Mercuric iodide
Mercuric nitrate (PP)
Mercuric oleate (PP)
Mercuric oxide (PP)
Mercuric oxycyanide, desensitized (PP)
Mercuric potassium cyanide (PP)
Mercuric Sulphate (PP)
Mercuric thiocyanate (PP)
Mercurol (PP)
Mercurous acetate (PP)
Mercurous bisulphate (PP)
Mercurous bromide (PP)
Mercurous chloride (PP)
Mercurous nitrate (PP)
Mercurous salicylate (PP)
Mercurous sulphate (PP)
Mercury acetates (PP)
Mercury ammonium chloride (PP)
Mercury based pesticide, liquid, flammable, toxic (PP)
Mercury based pesticide, liquid, toxic, flammable (PP)
Mercury based pesticide, liquid, toxic (PP)
Mercury based pesticide, solid, toxic (PP)
Mercury benzoate (PP)
Mercury bichloride (PP)
Mercury bisulphates (PP)
Mercury bromides (PP)
Mercury compounds, liquid, n.o.s. (PP)
Mercury compounds, solid, n.o.s. (PP)
Mercury cyanide (PP)
Mercury gluconate (PP)
Mercury (I) (mercurous) compounds (pesticides) (PP)
Mercury (II) (mercuric) compounds (pesticides) (PP)
Mercury iodide
Mercury nucleate (PP)
Mercury oleate (PP)
Mercury oxide (PP)
Mercury oxycyanide, desensitized (PP)
Mercury potassium cyanide (PP)
Mercury potassium iodide (PP)
Mercury salicylate (PP)
Mercury sulfates (PP)

Mercury thiocyanate (PP)
Metam-sodium
Methamidophos
Methanethiol
Methidathion
Methomyl
o- Methoxyaniline
Methyl bromide and ethylene dibromide mixtures, liquid
2-Methylbutyraldehyde
1-Methyl-2-ethylbenzene
1-Methyl-4-ethylbenzene
2-Methyl-5-ethylpyridine
Methyl mercaptan
2-Methyl-2-phenylpropane
Methyl salicylate
3-Methylacrolein, stabilized
Methylchlorobenzenes
Methylnaphthalenes, liquid
Methylnaphthalenes, solid
Methylnitrophenols
α-Methylstyrene
Methyl styrenes, inhibited
Methyltrithion
Methylvinylbenzenes, inhibited
Mevinphos (PP)
Mexacarbate
Mirex
Monocrotophos
Motor fuel anti-knock mixtures
Motor fuel anti-knock mixtures or compounds
Nabam
Naled
Naphthalene, crude or refined
Naphthalene, molten
Naphthenic acids, liquid
Naphthenic acids, solid
N,N-Dimethyldodecylamine (PP)
Nickel carbonyl (PP)
Nickel cyanide (PP)
Nickel tetracarbonyl (PP)
3-Nitro-4-chlorobenzotrifluoride
Nitrobenzene
Nitrobenzotrifluorides, liquid or solid
Nitrocresols
Nitrotoluenes (o-;m-;p-), liquid
Nitrotoluenes (o-; m-; p-), solid
Nitroxylenes, liquid or solid
1-Nonanal
1-Nonanol
Nonylphenol
n-heptaldehyde
n-Octaldehyde
1-Octanol
Oleylamine
Organotin compounds, liquid, n.o.s. (PP)
Organotin compounds (pesticides) (PP)
Organotin compounds, solid, n.o.s. (PP)
Organotin pesticides, liquid, flammable, toxic, n.o.s., flash point < 23°C (PP)
Organotin pesticides, liquid, toxic, flammable, n.o.s. (PP)
Organotin pesticides, liquid, toxic, n.o.s. (PP)
Organotin pesticides, solid, toxic, n.o.s. (PP)
Orthoarsenic acid
Osmium tetroxide (PP)
Oxamyl
Oxydisulfoton

Paraoxon
Parathion (PP)
Parathion-methyl (PP)
PCBs (PP)
Pentachloroethane
Pentachlorophenol (PP)
Pentalin
Pentanethiols
n-Pentylbenzene
Perchloroethylene
Perchloromethylmercaptan
Petrol, leaded
Phenarsazine chloride (PP)
δ-Phenothrin
Phenthoate (PP)
1-Phenylbutane
2-Phenylbutane
Phenylcyclohexane
Phenylethylene, inhibited
Phenylmercuric acetate (PP)
Phenylmercuric compounds, n.o.s. (PP)
Phenylmercuric hydroxide (PP)
Phenylmercuric nitrate (PP)
2-Phenylpropene
Phorate (PP)
Phosalone (PP)
Phosmet
Phosphamidon (PP)
Phosphorus, white, molten (PP)
Phosphorus, white or yellow dry or under water or in solution (PP)
Phosphorus white, or yellow, molten (PP)
Pindone (and salts of)
Phosphorus, yellow, molten (PP)
α-Pinene
Pirimicarb
Pirimiphos-ethyl (PP)
Polychlorinated biphenyls (PP)
Polyhalogenated biphenyls, liquid or Terphenyls liquid (PP)
Polyhalogenated biphenyls, solid or Terphenyls, solid (PP)
Potassium cuprocyanide (PP)
Potassium cyanide, solid
Potassium cyanide, solution
Potassium cyanocuprate(I) (PP)
Potassium cyanomercurate (PP)
Potassium mercuric iodide (PP)
Promecarb
Propachlor
Propanethiols
Propaphos
Propenal, inhibited
Propionaldehyde
Propoxur
n-Propylbenzene
Propylidene dichloride
Prothoate
Prussic acid, anhydrous, stabilized
Prussic acid, anhydrous, stabilized, absorbed in a porous inert material
Pyrazophos (PP)
Quinalphos
Quizalofop (PP)
Quizalofop-p-ethyl (PP)
Rotenone
Salithion
Silver arsenite
Silver cyanide

Silver orthoarsenite
Sodium copper cyanide, solid (PP)
Sodium copper cyanide solution (PP)
Sodium cuprocyanide, solid (PP)
Sodium cuprocyanide, solution (PP)
Sodium cyanide, solid
Sodium cyanide, solution
Sodium dinitro-o-cresolate, dry or wetted with less than 15 % water, by mass
Sodium dinitro-o-cresolate, wetted with not less than 15 % water, by mass
Sodium pentachlorophenate (PP)
Strychnine or Strychnine salts
Styrene monomer, inhibited
Sulfotep
Sulprophos (PP)
Tallow nitrile
Temephos
TEPP
Terbufos (PP)
Tetrabromoethane
Tetrabromomethane
1,1,2,2-Tetrachloroethane
Tetrachloroethylene
Tetrachloromethane
Tetrachlorophenol
Tetrachlorovinfos (PP)
Tetrachlorvinphos
Tetraethyl dithiopyrophosphate
Tetraethyl lead, liquid (PP)
Tetramethrin
n-Tetramethylbenzenes
Tetramethyllead
Thallium chlorate
Thallium compounds, n.o.s.
Thallium compounds (pesticides)
Thallium nitrate
Thallium sulfate
Thallous chlorate
4-Thiapentanal
Thiocarbonyl tetrachloride
Triaryl phosphates, isopropylated
Triaryl phosphates, n.o.s. (PP)
Triazophos
Tribromomethane
Tributyltin compounds (PP)
Trichlorfon
Trichlorobenzenes, liquid
Trichlorobutene
Trichlorobutylene
Trichloromethane sulphuryl chloride
Trichloromethyl sulphochloride
Trichloronat
Tricresyl phosphate (< 1% o-isomer)
Tricresyl phosphate, not < 1% o-isomer but not > 3% o-isomer (PP)
Tricresyl phosphate with > 3 % o-isomer (PP)
Triethylbenzene
Triisopropylated phenyl phosphates
1,2,3-Trimethylbenzene
1,2,4-Trimethylbenzene
1,3,5-Trimethylbenzene
Trimethylene dichloride
Triphenylphosphate (PP)
Triphenyl phosphate/*tert*-butylated triphenyl phosphates mixtures containing 5% to 10% triphenyl phosphates
Triphenyl phosphate/*tert*-butylated triphenyl phosphates mixtures containing 10% to 48% triphenyl phosphates   (PP)
Triphenyltin compounds (PP)
Tritolyl phosphate (< 1% o-isomer)
Tritolyl phosphate (not < 1% o-isomer) (PP)
Trixylenyl phosphate
Turpentine
1-Undecanol
n-Valeraldehyde
Vinylbenzene, inhibited
Vinylidene chloride, inhibited
Vinyltoluenes, inhibited mixed isomers
Warfarin (and salts of Warfarin)
White phosphorus, dry (PP)
White phosphorus, wet (PP)
White spirit, low (15-20%) aromatic
Xylenols
Yellow phosphorus, dry (PP)
Yellow phosphorus, wet (PP)
Zinc bromide
Zinc cyanide

Appendix C

Synonym and Trade Name Index

1080 S:3500
A 361 A:5750
A-10846 D:6200
AA-9 D:8300
AAT P:0200
AATACK T:1450
AATP P:0200
AATREX A:5750
AATREX 80W A:5750
2-AB B:3550
ABESON NAM D:8300
ABSOLUTE ETHANOL E:1900
ABSORBENT OIL O:2900
ABSORPTION OIL O:2900
AC [weaponized] H:1500
AC 3422 P:0200
AC 3422 E:0600
ACARIN D:2600
ACC 3422 P:0200
ACCELERATOR HX E:3325
ACEDE CRESYLIQUE (French) C:5100
ACEITE ABSORBENTE (Spanish) O:2900
ACEITE de ALQUITRAN de HULLA (Spanish) O:3200
ACEITE de CACAHUETE (Spanish) O:1900
ACEITE de CASCARA de NUEZ de ANACARDO (Spanish) O:3100
ACEDE CRESYLIQUE (French) C:5100
ACEITE de CROTON (Spanish) O:3300
ACEITE de CURTIDO (Spanish) O:4800
ACEITE de LINAZA (Spanish) O:3400
ACEITE de LUBRICANTE (Spanish) O:3500
ACEITE de MANTECA de CERDO (Spanish) O:16500
ACEITE de MINERAL (Spanish) O:3600
ACEITE de MOTOR (Spanish) O:3800
ACEITE de NUEZ de COCO (Spanish) O:1300
ACEITE de OLIVA (Spanish) O:1700
ACEITE de PALMA (Spanish) O:1800
ACEITE PARA HUSOS (Spanish) O:4500
ACEITE de PESCADO (Spanish) O:1500
ACEITE de PIE de BUEY (Spanish) O:3900
ACEITE de PINO (Spanish) O:4100
ACEITE de RESINA (Spanish) O:4200
ACEITE de RICINO (Spanish) O:1200
ACEITE de SEMILLA de ALGODON (Spanish) O:1400
ACEITE de SOJA (Spanish) O:2100
ACEITE de TRANSFORMADOR (Spanish) O:4900
ACEITE VEGETALES (Spanish) O:2300
ACETAL A:0100
ACETALDEHIDO (Spanish) A:0150
ACETAL DIETHYLIQUE (French) A:0100
ACETALDEHYDE A:0150
p-ACETALDEHYDE P:0175
ACETALDEHYDE, CHLORO- C:2200
ACETALDEHYDE DIETHYLACETAL A:0100
ACETALDEHYDE, TRICHLORO T:2100
ACETALDEHYDE TRIMER P:0175
ACETAMIDE N,N-DIMETHYL D:4800
ACETATE C-7 H:0350
ACETATE C-9 N:2500
ACETATE d'AMYLE (French) A:4200
ACETATE de BUTYLE (French) B:3100
ACETATE de BUTYLE SECONDAIRE (French) B:3150
ACETATE de CUIVRE (French) C:4000
ACETATE de METHYLE (French) M:1550
ACETATE PHENYLMERCURIQUE (French) P:1400
ACETATE de PROPYLE NORMAL (French) P:3900
ACETATE de VINYLE (French) V:0700
ACETATO de n-AMILO (Spanish) A:4200
ACETATO de sec-AMILO (Spanish) A:4250
ACETATO de terc-AMILO (Spanish) A:4250
ACETATO de AMILO TERCIARIO (Spanish) A:4300
ACETATO AMONICO (Spanish) A:2050
ACETATO de BENCILO (Spanish) B:1150
ACETATO de BUTILO (Spanish) B:3100
ACETATO de sec-BUTILO (Spanish) B:3150
ACETATO de terc-BUTILO (Spanish) B:3200
ACETATO de CICLOHEXANILO (Spanish) C:6150
ACETATO de COBALTO (Spanish) C:3550
ACETATO de COBRE (Spanish) C:4000
ACETATO CROMICO (Spanish) C:3250
ACETATO de DIETILENGLICOL MONOETIL ETER (Spanish) D:3150
ACETATO de DIETILENGLICOL MONOMETIL ETER (Spanish) D:3250
ACETATO de 2-ETILBUTILO (Spanish) E:1600
ACETATO de ETILENGLICOL (Spanish) E:4300
ACETATO de ETILENGLICOL METIL ETER (Spanish) E:4800
ACETATO de ETILENGLICOL MONOBUTIL ETER (Spanish) E:5000
ACETATO de 2-ETILHEXILO (Spanish) E:6200
ACETATO de ETILO (Spanish) E:1700
ACETATO de 2-ETOXIETILO (Spanish) E:0900
ACETATO FENILMERCURICO (Spanish) P:1400
ACETATO de 1-HEPTILO (Spanish) H:0350
ACETATO de sec-HEXILO (Spanish) H:1150
ACETATO de ISOAMILICO (Spanish) I:0100
ACETATO de ISOBUTILO (Spanish) I:0300
ACETATO de NIQUEL (Spanish) N:0650
ACETATO de n-NONILO (Spanish) N:2500
(ACETATO)PHENYL MERCURY P:1400
ACETATOPHENYLMERCURY P:1400
ACETATO de PLATA (Spanish) S:0600
ACETATO de PLOMO (Spanish) L:1700
ACETATO de PROPILENGLICOL MONOMETIL ETER (Spanish) P:4350
ACETATO de TALIO (Spanish) T:1150
ACETATO de ZINC (Spanish) Z:0200
ACETATO de ZIRCONIO (Spanish) Z:2300
ACETDIMETHYLAMIDE D:4800
ACETEHYD (German) A:0150
ACETEHYDE A:0150
ACETENE E:3500
ACETIC ACID A:0200
ACETIC ACID, AMMONIUM SALT A:2050
ACETIC ACID, n-AMYL ESTER A:4200
ACETIC ACID, ANHYDRIDE A:0250
ACETIC ACID, BENZYL ESTER B:1150
ACETIC ACID BROMIDE A:0550
ACETIC ACID, 2-BUTOXYETHYL ESTER E:5000
ACETIC ACID, 2-BUTOXY ESTER B:3150
ACETIC ACID, BUTYL ESTER B:3100
ACETIC ACID, n-BUTYL ESTER B:3100
ACETIC ACID, sec-BUTYL ESTER B:3150
ACETIC ACID, tert-BUTYL ESTER B:3200
ACETIC ACID, CADMIUM SALT C:0150
ACETIC ACID CHLORIDE A:0600
ACETIC ACID, CHROMIUM SALT C:3250
ACETIC ACID, CHROMIUM(3+) SALT C:3250
ACETIC ACID, COBALT(2+) SALT C:3550
ACETIC ACID, COPPER (2+) C:4000
ACETIC ACID, CUPRIC SALT C:4000
ACETIC ACID, CYCLOHEXYL ESTER C:6150
ACETIC ACID, DIMETHYLAMIDE D:4800
ACETIC ACID, 1,1-DIMETHYLETHYL ESTER B:3200
ACETIC ACID-1,3-DIMETHYLBUTYL ESTER M:1950
ACETIC ACID, ETHENYL ESTER V:0700
ACETIC ACID, 2-ETHOXYETHYL ESTER E:0900
ACETIC ACID (ETHYLENEDINITRILO)TETRA- E:3900
ACETIC ACID, ETHYL ESTER E:1600
ACETIC ACID, FLUORO-, SODIUM SALT S:3500
ACETIC ACID, GLACIAL A:0200
ACETIC ACID, HEPTYL ESTER H:0350
ACETIC ACID, HEXYL ESTER H:1150
ACETIC ACID, ISOBUTYL ESTER I:0300
ACETIC ACID, ISOPENTYL ESTER I:0100
ACETIC ACID, ISOPROPYL ESTER I:1200
ACETIC ACID, MERCURY(2+) SALT M:0500
ACETIC ACID, METHYL ESTER M:1550
ACETIC ACID, 1-METHYLETHYL ESTER I:1200
ACETIC ACID, 3-METHOXYBUTYL ESTER M:1450
ACETIC ACID, 2-(2-METHOXY-ETHOXY)ETHYL ESTER D:3250
ACETIC ACID, 1-METHYLPROPYL ESTER B:3150
ACETIC ACID, 2-METHYLPROPYL ESTER I:0300
ACETIC ACID,NICKEL(2+) SALT N:0650
ACETIC ACID,NICKEL(II) SALT N:0650
ACETIC ACID, n-NONYL ESTER N:2500
ACETIC ACID, 2-PENTYL ESTER A:4250
ACETIC ACID, PHENYLMERCURY DERIVATIVE P:1400
ACETIC ACID, PHENYLMETHYL ESTER B:1150
ACETIC ACID, n-PROPYL ESTER P:3900
ACETIC ACID, PROPYL ESTER P:3900
ACETIC ACID, THALLIUM SALT T:1150
ACETIC ACID, THALLOUS SALT T:1150
ACETIC ACID, (2,4,8-TRICHLOROPHENOXY)- T:2500
ACETIC ACID, VINYL ESTER V:0700
ACETIC ACID, ZINC SALT Z:0200
ACETIC ALDEHYDE A:0150
ACETIC ANHYDRIDE A:0250
ACETIC BROMIDE A:0550
ACETIC sec-BUTYL ESTER B:3150
ACETIC CHLORIDE A:0600
ACETIC EHYDE A:0150
ACETIC ESTER E:1600
ACETIC ETHER E:1600
ACETIC METHYL ETHER M:1600
ACETIC OXIDE A:0250
ACETIC PEROXIDE P:0900
ACETIDIN E:1600
ACETILACETONA (Spanish) A:0500
ACETILENO (Spanish) A:0650
ACETILO de BROMURA (Spanish) A:0550
ACETOACETIC ACID, ETHYL ESTER E:1700

ACETOACETIC ACID, METHYL ESTER M:1600
ACETOACETIC ESTER E:1700
ACETOACETONE A:0500
ACETOARSENITE de CUIVRE (French) C:4050
ACETOARSENITO de COBRE (Spanish) C:4050
ACETO CADMIO (Spanish) C:0150
ACETOFENONA (Spanish) A:0450
ACETOMETHYLBENZENE B:1150
ACETONA (Spanish) A:0300
ACETONA, CIANHIDRINA de (Spanish) A:0350
ACETONCYANHYDRIN (German) A:0350
ACETONE A:0300
ACETONE CYANOHYDRIN A:0350
ACETONECYANHYDRINE (French) A:0350
ACETONE, METHYL- M:4100
ACETONIC ACID L:0100
ACETONITRILE A:0400
ACETONITRILO (Spanish) A:0400
ACETONYL BROMIDE B:2400
ACETONYLDIMETHYLCARBINOL D:0900
ACETOPHENONE A:0450
ACETO TETD T:1450
ACETO de VINILO (Spanish) C:0150
ACETOXYETHANE E:1600
1-ACETOXYETHYLENE V:0700
2-ACETOXYPENTANE A:4250
1-ACETOXYPROPANE P:3900
2-ACETOXYPROPANE I:1200
α-ACETOXYTOLUENE B:1150
ACETYL ACETONE A:0500
ACETYL ANHYDRIDE A:0250
ACETYL BENZENE A:0450
ACETYL BROMIDE A:0550
ACETYL CHLORIDE A:0600
ACETYL CHLORIDE, CHLORO- C:2300
ACETYLDIMETHYLAMINE D:4800
ACETYL ETHER A:0250
ACETYL ETHYLENE M:5500
ACETYL HYDROPEROXIDE P:0900
ACETYLMETHYL BROMIDE B:2400
ACETYL OXIDE A:0250
ACETYL PEROXIDE A:0700
ACETYL PEROXIDE SOLUTION A:0700
2-ACETYL PROPANE M:3150
ACETYLEN A:0650
ACETYLENE A:0650
ACETYLENE DICHLORIDE D:1750
cis-ACETYLENE DICHLORIDE D:1750
trans-ACETYLENE DICHLORIDE D:1750
ACETYLENE TRICHLORIDE T:2350
ACETYLENOGEN C:0650
ACID AMMONIUM CARBONATE A:2150
ACID AMMONIUM CARBONATE, MONOAMMONIUM SALT A:2150
ACID AMMONIUM FLUORIDE A:2200
ACID GAS C:1650
ACID PHOSPHORIQUE (French) P:2350
ACIDE ACETIQUE (French) A:0200
ACIDE ARSENIEUX (French) A:5500
ACIDE BROMHYDRIQUE (French) H:1400
ACIDE CARBOLIQUE (French) P:1200
ACIDE CHLORHYDRIQUE (French) H:1450
ACIDE CYANHYDRIQUE (French) H:1500
ACIDE 2,4-DICHLORO PHENOXYACETIQUE (French) D:0100
ACIDE ETHYLENEDIAMINETETRACETIQUE (French) E:3900
ACIDE FORMIQUE (French) F:1350
ACIDE NITRIQUE (French) N:1350

ACIDE OXALIQUE (French) O:5500
ACIDE PERACETIQUE (French) P:0900
ACIDE PHOSPHORIQUE (French) P:1600
ACIDE PROPIONIQUE (French) P:3700
ACIDE SULFHYDRIQUE (French) H:1650
ACIDE 2,4,5-TRICHLOROPHENOXYACETIQUE (French) T:2500
ACIDO ARSENICO (Spanish) A:5300
ACIDO BENZOICO (Spanish) B:0900
ACIDO BORICO (Spanish) B:2100
ACIDO BUTRICO (Spanish) B:4900
ACIDO CACODILICO (Spanish) C:0100
ACIDO CIANHIDRICO (Spanish) H:1500
ACIDO CIANOACETICO (Spanish) C:5600
ACIDO CITRICO (Spanish) C:3500
ACIDO CLORHIDRICO (Spanish) H:1450
ACIDO 3-CLOROPROPIONICO (Spanish) C:2950
ACIDO CLOROSULFONICO (Spanish) C:3000
ACIDO DECANOICO (Spanish) D:0550
ACIDO 2,4-DICLOROFENOXIACETICO (Spanish) D:0100
ACIDO 2,2-DICLOROPROPIONICO (Spanish) D:0200
ACETATO de DIETILENGLICOL MONOBUTIL ETER (Spanish) D:3350
ACIDO DI(2-ETILHEXIL)FOSFORICO (Spanish) D:3350
ACIDO DIFLUOROFOSFORICO (Spanish) D:3950
ACIDO DIMETILOCTANOICO (Spanish) D:5850
ACIDO ESTEARICO (Spanish) S:5900
ACIDO ETILENDIAMINOTETRAACETICO (Spanish) E:3900
ACIDO 2-ETILHEXANOICO (Spanish) E.6000
ACIDO FLUORHIDRICO (Spanish) H:1550
ACIDO FLUOSILICICO (Spanish) H:1300
ACIDO FLUOSULFONICO (Spanish) F:1200
ACIDO FUMARICO (Spanish) F:1400
ACIDO GALLICO (Spanish) T:0250
ACIDO n-HEPTANOICO (Spanish) H:0200
ACIDO HEXANOICO (Spanish) H:0100
ACIDO ISOBUTIRICO (Spanish) I:0500
ACIDO ISOFTALICO (Spanish) I:1100
ACIDO LACTICO (Spanish) L:0100
ACIDO LAURICO (Spanish) L:0300
ACIDO α-METACRILICO (Spanish) M:1200
ACIDO MONOCLORACETICO (Spanish) M:5750
ACIDO NAFTENICO (Spanish) N:0450
ACIDO NEODECANOICO (Spanish) N:0550
ACIDO NITRILOTRIACETICO (Spanish) N:1750
ACIDO OCTANOICO (Spanish) O:0200
ACIDO OLEICO (Spanish) O:5100
ACIDO OXALICO (Spanish) O:5500
ACIDO PENTNOICO (Spanish) P:0700
ACIDO PERACETICO (Spanish) P:0900
ACIDO PERCLORICO (Spanish) P:0950
ACIDO POLIFOSFORICO (Spanish) P:2350
ACIDO PROPRIONICO (Spanish) P:3700
ACIDO SALICILICO (Spanish) S:0200
ACIDO SLFURICO (Spanish) S:6800
ACIDO TANICO (Spanish) T:0250
ACIDO p-TOLUENSULFONICO (Spanish) T:1750
ACIDO TRIMETILACETICO (Spanish) T:4000
ACIDO UNDECANOICO (Spanish) U:0100
ACETATO de n-AMILO (Spanish) A:4200
ACETATO de sec-AMILO (Spanish) A:4250

ACETATO de AMILO TERCIARIO (Spanish) A:4300
ACETATO AMONICO (Spanish) A:2050
ACETATO de BENCILO (Spanish) B:1150
ACETATO de BUTILO (Spanish) B:3100
ACETATO de sec-BUTILO (Spanish) B:3150
ACETATO de terc-BUTILO (Spanish) B:3200
ACETATO de CADMIO (Spanish) C:0150
ACETATO de CICLOHEXANILO (Spanish) C:6150
ACETATO de COBALTO (Spanish) C:3550
ACETATO de COBRE (Spanish) C:4000
ACETATO CROMICO (Spanish) C:3250
ACETATO de DIETILENGLICOL MONOBUTIL ETER (Spanish) D:3350
ACETATO de DIETILENGLICOL MONOETIL ETER (Spanish) D:3150
ACETATO de DIETILENGLICOL MONOMETIL ETER (Spanish) E:4800
ACETATO de 2-ETILBUTILO (Spanish) E:1600
ACETATO de ETILENGLICOL (Spanish) E:4300
ACETATO de ETILENGLICOL METIL ETER (Spanish) E:4800
ACETATO de ETILO (Spanish) E:1700
ACETATO de ETILENGLICOL MONOBUTIL ETER (Spanish) E:5000
ACETATO de 2-ETILHEXILO (Spanish) E:6200
ACETATO de 2-ETOXIETILO (Spanish) E:0900
ACETATO FENILMERCURICO (Spanish) P:1400
ACETATO de 1-HEPTILO (Spanish) H:0350
ACETATO de sec-HEXILO (Spanish) M:1950
ACETATO de ISOAMILICO (Spanish) I:0100
ACETATO de ISOBUTILO (Spanish) I:0300
ACETATO de ISOPROPILO (Spanish) I:1200
ACETATO de NIQUEL (Spanish) N:0650
ACETATO de n-NONILO (Spanish) N:2500
ACETATO de PLATA (Spanish) S:0600
ACETATO de PLOMO (Spanish) L:0600
ACETATO de PROPILENGLICOL MONOMETIL ETER (Spanish) P:4350
ACETATO de TALIO (Spanish) T:1150
ACETATO de URANILO (Spanish) U:0600
ACETATO de VINILO (Spanish) V:0700
ACETATO de ZINC (Spanish) Z:0200
ACETATO de ZIRCONIO (Spanish) Z:2300
ACETOARSENITE de CUIVRE (French) C:4050
ACETOARSENITO de COBRE (Spanish) C:4050
ACETONA (Spanish) A:0300
ACETONA, CIANHIDRINA de (Spanish) A:0350
ACIFLOCTIN A:1000
ACIGENA H:0600
ACINETTEN A:1000
ACINTENE A P:2050
ACINTENE DP D:7150
ACINTENE DP DIPENTENE D:7150
ACL 85 T:2850
ACQUINITE C:2800
ACREHYDE A:0800
ACRIDINE A:0750
ACRIDINA (Spanish) A:0750
ACRILAMIDA (Spanish) A:0850
ACRILONITRILO (Spanish) A:0950
ACROLEIC ACID A:0900
ACROLEIN A:0800
ACROLEINA (Spanish) A:0800

ACROLEINE (French) A:0800
ACRYLALDEHYDE A:0800
ACRYLAMIDE A:0850
ACRYLAMIDE MONOMER A:0850
ACRYLATE d'ETHYLE (French) E:1800
ACRYLATE de METHYLE (French) M:1700
ACRILATO de n-BUTILO (Spanish) B:3250
ACRILATO de iso-BUTILO (Spanish) I:0325
ACRILATO de ETILO (Spanish) E:1800
ACRILATO de 2-ETILHEXILO (Spanish) E:6200
ACRILATO de 2-HIDROXIPROPILO (Spanish) H:1950
ACRILATO de ISODECILO (Spanish) I:0650
ACRILATO de METILO (Spanish) M:1700
ACRYLEHYDE A:0800
ACRYLIC ACID A:0900
ACRYLIC ACID AMIDE (50%) A:0850
ACRYLIC ACID, BUTYL ESTER B:3250
ACRYLIC ACID n-BUTYL ESTER B:3250
ACRYLIC ACID, DECYL ESTER D:0650
ACRYLIC ACID, n-DECYL ESTER D:0650
ACRYLIC ACID, ETHYL ESTER E:1800
ACRYLIC ACID, 2-ETHYLHEXYLESTER E:6300
ACRYLIC ACID, GLACIAL A:0900
ACRYLIC ACID, 2-HYDROXYETHYL ESTER H:1750
ACRYLIC ACID-2-HYDROXYPROPYL ESTER H:1950
ACRYLIC ACID, ISOBUTYL ESTER I:0325
ACRYLIC ACID, ISODECYL ESTER I:0650
ACRYLIC ACID, 2-METHYL- M:1200
ACRYLIC ACID, METHYL ESTER M:1700
ACRYLIC ALDEHYDE A:0800
ACRYLIC AMIDE A:0850
ACRYLIC AMIDE 50% A:0850
ACRYLIC EHYDE A:0800
ACRYLNITRIL (German) A:0950
ACRYLON A:0950
ACRYLONITRILE A:0950
ACRYLONITRILE MONOMER A:0950
ACRYLSAEUREAETHYLESTER (German) E:1800
ACRYLSAEUREMETHYLESTER (German) M:1700
ACTIVATED CARBON C:2000
ACTIVATED CHARCOAL C:2000
ACTIVE ACETYL ACETATE E:1700
ACTYLENOGEN, CALCIUM ACETYLIDE C:0650
ACYTOL E:6700
ADACENE-12 D:8050
ADILAC-TETTEN A:1000
ADIPATO de BUTILO D:3550
ADIPATO de DI(2-ETILHEXILO) (Spanish) D:7000
ADIPATO de DI-n-HEXILO (Spanish) D:4050
ADIPATO de DIISONONILO (Spanish) D:4400
ADIPATO de DIOCTILO (Spanish) D:7000
ADIPATO de METILO (Spanish) D:4900
ADIPIC ACID A:1000
ADIPIC ACID, BIS(2-ETHYLHEXYL) ESTER D:7000
ADIPIC ACID, DIBUTYL ESTER D:3550
ADIPIC ACID, DIHEXYL ESTER D:4050
ADIPIC ACID, DIMETHYL ESTER D:4900
ADIPIC ACID DINITRILE A:1050
ADIPIC ACID NITRILE A:1050
ADIPINIC ACID A:1000
ADIPODINITRILE A:1050
ADIPOL 2EH D:7000
ADIPONITRILE A:1050

ADIPONITRILO (Spanish) A:1050
ADRONAL C:5900
AERO LIQUID HCN H:1500
AEROL 1 PESTICIDE T:2050
AEROSOL GPG D:7050
AEROSOL SURFACTANT D:7050
AEROTEX GLYOXAL 40 G:0700
AEROTHENE T:2250
AEROTHENE TT T:2250
AETHALDIAMIN (German) E:3800
AETHANOL (German) E:1900
AETHER E:5600
AETHYLACETAT (German) E:1600
AETHYLACRYLAT (German) E:1800
AETHYLAMINE (German) E:2200
2-AETHYLAMINO-4-CHLOR-6-ISOPROPYLAMINO-1,3,5-TRIAZIN (German) A:5750
AETHYLBENZOL (German) E:2400
AETHYLCHLORID (German) E:2900
AETHYLDLKOHOL (German) E:1900
AETHYLENBROMID (German) E:4000
AETHYLENEDIAMIN (German) E:3800
AETHYLENGYKOL-MONOMETHYLAETHER (German) E:5100
AETHYLENOXID (German) E:5500
AETHYLIDENCHLORID (German) D:1700
AETHYLIS E:2900
AETHYLIS CHLORIDUM E:2900
AETHYLMETHYLKETON (German) M:4100
AF 101 D:7950
AFICIDE B:0600
AGENAP N:0450
AGENT 504 D:0700
AGENT SA A:5575
AGISTAT D:1400
AGREFLAN T:3800
AGRICIDE MAGGOT KILLER (F) T:1950
AGRIDIP C:5000
AGRIFLAN 24 T:3800
AGRISOL G-20 B:0600
AGRITAN D:0300
A GRO M:4900
AGROCERES H:0100
AGROCIDE or AGROCIDE 2 or 6G or 7 or III or WP B:0600
AGRONEXIT B:0600
AGROSAN P:1400
AGROTECT D:0100
AIM S:5800
AIP A:1700
AK-33X M:3800
AKTIKON A:5750
AKTIKON PK A:5750
AKTINIT A A:5750
AKTINIT PK A:5750
AL-PHOS A:1700
ALANINOL P:3400
β-ALANINOL P:3350
ALATEX D:0200
ALBOLINE O:3600
ALBONE H:1600
ALBUS M:0550
ALCANFOR (Spanish) C:1400
ALCOA SODIUM FLUORIDE S:3400
ALCOHOL E:1900
ALCOHOL AMILICO (Spanish) A:4350
ALCOHOL terc-AMILICO (Spanish) M:4400
ALCOHOL, ANHYDROUS E:1900
ALCOHOL BENCILICO (Spanish) B:1200
ALCOHOL n-BUTILICO (Spanish) B:3350
ALCOHOL sec-BUTILICO (Spanish) B:3400
ALCOHOL terc-BUTILICO (Spanish) B:3450

ALCOHOL C-1 M:1750
ALCOHOL C-2 E:1900
ALCOHOL C-3 P:3950
ALCOHOL C-4 I:0350
ALCOHOL C-4 B:3350
ALCOHOL C-4 B:3400
ALCOHOL C-4 B:3450
ALCOHOL C-4 B:3400
ALCOHOL C-5 A:4350
ALCOHOL C-6 H:1050
ALCOHOL C-8 O:0300
ALCOHOL C-9 N:2400
ALCOHOL C-10 D:0700
ALCOHOL C-11 (undecylic) U:0200
ALCOHOL C-12 D:8000
ALCOHOL C-12 L:2000
ALCOHOL C-13 T:3050
ALCOHOL n-DECILICO (Spanish) D:0700
ALCOHOL, DEHYDRATED E:1900
ALCOHOL ETILICO (Spanish) E:1900
ALCOHOL FURFURILICO (Spanish) F:1550
ALCOHOL ISOAMILICO PRIMARIO (Spanish) I:0150
ALCOHOL ISOAMILICO SECUNDARIO (Spanish) I:0200
ALCOHOL ISOBUTILICO (Spanish) I:0350
ALCOHOL de ISODECILO (Spanish) I:0700
ALCOHOL ISOOCTILICO (Spanish) I:0850
ALCOHOL ISOPENTANO (Spanish) I:0150
ALCOHOL de ISOPROPILO (Spanish) I:1250
ALCOHOL LINEAL (Spanish) L:2000
ALCOHOL MIRISTILICO (Spanish) T:0450
ALCOHOL n-NONILICO (Spanish) N:2400
ALCOHOL PROPARGILICO (Spanish) P:3550
ALCOHOL PROPILICO (Spanish) P:3950
ALCOHOL sec-PROPILICO (Spanish) I:1250
ALCOHOLS, C_{11-15}-SECONDARY, ETHOXYLATED E:1200
ALCOOL ALLYLIQUE (French) A:1200
ALCOOL AMYLIQUE (French) A:4350
ALCOOL BUTYLIQUE (French) B:3350
ALCOOL BUTYLIQUE SECONDAIRE (French) B:3400
ALCOOL BUTYLIQUE TERTIAIRE (French) B:3450
ALCOOL ETHYLIQUE (French) E:1900
ALCOOL METHYL AMYLIQUE (French) M:2000
ALCOOL METHYLIQUE (French) M:1750
ALCOOL PROPYLIQUE (French) P:3950
ALCOPOL O D:7050
ALDEHIDO ISOVALERIANICO (Spanish) I:1700
ALDEHYDE C-6 H:0750
ALDEHYDE C-10 D:0450
ALDEHYDE BUTYRIQUE (French) B:4800
ALDEHYDE-COLLIDINE M:4150
ALDEHYDE FORMIQUE (French) F:1250
ALDEHYDE PROPIONIQUE (French) P:3650
ALDEHYDINE M:4150
ALDIFEN D:6650
ALDREX A:1100
ALDREX-30 A:1100
ALDRIN A:1100
ALDRINA (Spanish) A:1100
ALDRINE (French) A:1100
ALDRITE A:1100
ALDROSOL A:1100
ALFA-TOX D:1000
ALFOL 8 O:0300
ALFOL-12 D:8000
ALGIMYCIN P:1400
ALGOFRENE TYPE 1 T:2400

ALGOFRENE TYPE 2 D:1650
ALGOFRENE TYPE 5 D:2100
ALGOFRENE TYPE 6 M:5800
ALGOFRENE TYPE 67 D:3900
ALGRAIN E:1900
ALGYLEN T:2350
ALILICO ALCOHOL (Spanish) A:1200
ALILTRICLOROSILANO (Spanish) A:1400
ALIMET H:1900
ALIPHATIC PETROLEUM NAPHTHA D:7750
ALKAPOL PPG-1200 P:2450
ALKARSODL S:2700
ALKAWAY LIQUID ALKALINE DERUSTER B:2050
ALKOHOL (German) E:1900
ALKRON P:0200
ALKYLATES, GASOLINE BLENDING STOCKS G:0300
ALKYLBENZENESULFONIC ACIDS A:1150
ALLKYLBENZENESULFONIC ACID, SODIUM SALT S:1400
p-n-ALKYLBENZENESULFONIC ACID DERIVITIVE, SODIUM SALT S:1400
ALLENE–METHYLACETYLENE MIXTURE M:1650
ALLERON P:0200
ALLOMALEIC ACID F:1400
ALLTEX T:1950
ALLTOX T:1950
ALLYLALDEHYDE A:0800
ALLYL ALCOHOL A:1200
ALLYLALKOHOL (German) A:1200
ALLYL BROMIDE A:1250
ALLYLCHLORID (German) A:1300
ALLYL CHLORIDE A:1300
ALLYL CHLOROCARBONATE A:1350
ALLYL CHLOROFORMATE A:1350
ALLYLE (CHLORURE D') (French) A:1300
ALLYL EHYDE A:0800
ALLYLEHYDE A:0800
ALLYLETHYLENE P:0500
ALLYLIC ALCOHOL A:1200
ALLYLSILICONE TRICHLORIDE A:1400
ALLYL TRICHLORIDE T:2750
ALLYLTRICHLOROSILANE A:1400
ALLYL TRICHLOROSILANE A:1400
ALLYL TRICHOROSILANE, STABILIZED A:1400
ALMEDERM H:0600
ALMOND ARTIFICIAL ESSENTIAL OIL B:0500
ALPEROX C L:0400
ALPHASOL OT D:7050
ALQUILADO GASOLINA (Spanish) G:0300
ALROWET D65 D:7050
ALTOX A:1100
ALUM A:1750
ALUMINATO SODICO (Spanish) S:1600
ALUMINUM ALUM A:1750
ALUMINUMCHLORID (German) A:1450
ALUMINUM CHLORIDE A:1450
ALUMINUM CHLORIDE A:1450
ALUMINUM ETHYL DICHLORIDE E:2000
ALUMINUM FLUORIDE A:1550
ALUMINUM LITHIUM HYDRIDE L:2500
ALUMINUM MONOPHOSPHIDE A:1700
ALUMINUM NITRATE A:1650
ALUMINUM(III) NITRATE A:1650
ALUMINUM PHOSPHIDE A:1700
ALUMINUM PHOSPHITE A:1700
ALUMINUM SALT OF NITRIC ACID A:1650
ALUMINUM SULFATE A:1750
ALUMINUM TRICHLORIDE A:1450

ALUMINUM TRICHLORIDE SOLUTION A:1500
ALUMINUM TRIETHYL T:3250
ALUMINUM TRIISOBUTYL- T:3850
ALUMINUM TRINITRATE A:1650
ALUMINUM TRIS(2-METHYLPROPYL)- T:3850
ALUMINUM TRISULFATE A:1750
ALUMINUM TRISULFATE SOLUTION A:1800
ALUNOGENITE A:1750
ALVIT D:2700
AMATIN H:0400
AMCHLOR A:2450
AMCHLORIDE A:2450
AMCIDE A:3800
AMEISENATOD B:0600
AMEISENMITTEL MERCK B:0600
AMEISENSAEURE (German) F:1350
AMERCIDE C:1475
AMERICAN CYANAMID 4,049 M:0250
AMERICAN PALM KERNEL OIL O:2200
AMFO A:3150
AM-FOL A:2000
AMICIDE A:3800
AMIDA SODICO (Spanish) S:1700
AMIDOX D:0100
AMILMERCAPTANO (Spanish) A:4450
n-AMILMETILCETONA (Spanish) A:4500
AMILTRICLOROSILANO (Spanish) A:4700
AMINIC ACID F:1350
AMINOBENZENE A:4750
1-AMINOBUTANE B:3500
2-AMINOBUTANE B:3550
AMINOCAPROIC LACTAM C:1450
1-AMINO-4-CHLOROBENZENE C:2350
1-AMINO-3-CHLORO-6-METHYLBENZENE C:3200
2-AMINO-4-CHLOROTOLUENE C:3200
2-AMINO-5-CHLOROTOLUENE C:3200
AMINOCYCLOHEXANE C:6100
AMINODIMETHYLBENZENE X:0450
2-AMINODIMETHYLETHANOL β-AMINOISOBUTANOL A:1925
AMINOETHANDIAMINE D:3450
AMINOETHANE E:2200
1-AMINOETHANE E:2200
2-AMINOETHANOL M:5950
2-(2-AMINOETHOXY) ETHANOL A:1850
N-(2-AMINOETHYL)ETHYLENEDIAMINE D:3450
AMINOETHYL ALCOHOL M:5950
β-AMINOETHYL ALCOHOL M:5950
β-AMINOETHYLAMINE E:3800
2-[(2-AMINOETHYL) AMINO] ETHANOL A:1850
N,N'-BIS(2-AMINOETHYL)-1,2-DIAMINOETHANE T:3600
AMINOETHYLENE E:5400
AMINOETHYLETHANDIAMINE D:3450
AMINO ETHYL-1,2-ETHANEDIAMINE, 1,4,7,10,13,-PENTAAZATRIDECANE 3-AMINOPHENYLMETHANE T:1800
AMINOETHYLETHANOL AMINE A:1850
N-AMINOETHYLETHANOL AMINE A:1850
N-(2-AMINOETHYL)ETHANOLAMINE A:1850
N,N'-BIS(2-AMINOETHYL)ETHYLENE-DIAMINE T:3600
1-AMINO-2-ETHYLHEXANE E:6500
1-(2-AMINOETHYL)PIPERAZINE A:1900
N-AMINOETHYLPIPERAZINE A:1900

N-(2-AMINOETHYL) PIPERAZINE A:1900
N-(2-AMINOETIL)PIPERAZINA (Spanish) A:1900
1-AMINO-2-FLUOROBENZENE F:0850
1-AMINO-4-FLUOROBENZENE F:0900
AMINOFORM H:0900
AMINOHEXAHYDROBENZENE C:6100
1-AMINO-2-HYDROXYPROPANE I:1300
2-AMINOISOBUTANE B:3600
1-AMINO-2-METHYLBENZENE T:1850
2-AMINO-1-METHYLBENZENE T:1850
3-AMINO-1-METHYLBENZENE T:1800
4-AMINO-1-METHYLBENZENE T:1900
1-AMINO-2-METHYLPROPANE I:0400
2-AMINO-2-METHYLPROPANE B:3600
AMINOMERCURIC CHLORIDE M:0550
AMINOMETHANE M:1850
AMINOMETHANE (cylinder) M:1900
3-AMINOMETHYL-3,5,5-TRIMETHYLCYCLO-HEXYLAMINE I:1000
1-AMINONAPHTHALENE N:0500
1-AMINO-2-NITROBENZENE N:1450
1-AMINO-4-NITROBENZENE N:1450
p-AMINONITROBENZENE N:1450
2-AMINOPENTANE D:2800
AMINOPHEN A:4750
4-AMINOPIRIDINA (Spanish) A:1950
1-AMINOPROPANE P:4000
2-AMINOPROPANE I:1300
1-AMINO-2-PROPANOL M:6000
1-AMINO-PROPANOL-2 I:1300
2-AMINO-1-PROPANOL P:3400
3-AMINO-1-PROPANOL P:3350
γ-AMINOPROPANOL P:3350
4-AMINOPYRIDINE A:1950
α-AMINOPYRIDINE A:1950
p-AMINOPYRIDINE A:1950
2-AMINOTOLUENE T:1850
3-AMINOTOLUENE T:1800
4-AMINOTOLUENE T:1900
α-AMINOTOLUENE B:1250
m-AMINOTOLUENE T:1800
o-AMINOTOLUENE T:1850
p-AMINOTOLUENE T:1900
AMINOTRIACETIC ACID N:1750
AMMAT or AMMATE HERBICIDE A:3800
AMMONERIC A:2450
AMMONIA A:2000
AMMONIA, ANHYDROUS A:2000
AMMONIA SOAP A:3350
AMMONIA WATER A:2850
AMMONIAC (French) A:2000
AMMONIALE (German) A:2000
AMMONIATED MERCURY M:0550
AMMONIO-CUPRIC SULFATE C:4850
AMMONIOFORMALDEHYDE H:0900
AMMONIUM ACETATE A:2050
AMMONIUM ACID FLUORIDE A:2200
AMMONIUM AMIDOSULFONATE A:3800
AMMONIUM AMIDOSULPHATE A:3800
AMMONIUM AMINOFORMATE A:2350
AMMONIUM AMINOSULFONATE A:3800
AMMONIUM BENZOATE A:2100
AMMONIUM BICARBONATE A:2150
AMMONIUM BICHROMATE A:2600
AMMONIUM BIFLUORIDE A:2200
AMMONIUM BIPHOSPHATE A:3600
AMMONIUM BISULFIDE A:3900
AMMONIUM BISULFITE A:2250
AMMONIUM BOROFLUORIDE A:2650
AMMONIUM BROMIDE A:2300
AMMONIUM CARBAMATE A:2350
AMMONIUM CARBONAT (German) A:2400

AMMONIUM CARBAZOATE A:3650
AMMONIUM CARBONATE A:2400
AMMONIUM CHLORIDE A:2450
AMMONIUM CHROMATE A:2500
AMMONIUM CITRATE A:2550
AMMONIUM CITRATE, DIBASIC A:2550
AMMONIUM CUPRIC SULFATE C:4850
AMMONIUM DECABORATE OCTAHYDRATE A:3450
AMMONIUM DICHROMATE A:2600
AMMONIUM DICHROMATE(VI) A:2600
AMMONIUM DIHYDROGEN PHOSPHATE A:3600
AMMONIUM DISULFATONICKELATE (II) N:0700
AMMONIUM FERRIC CITRATE F:0100
AMMONIUM FERRIC OXALATE TRIHYDRATE F:0150
AMMONIUM FERRIOXALATE F:0150
AMMONIUM FERROUS SULFATE F:0550
AMMONIUM FLUOBORATE A:2650
AMMONIUM FLUORIDE A:2700
AMMONIUM FLUOROSILICATE A:3700
AMMONIUM FORMATE A:2750
AMMONIUM GLUCONATE A:2800
AMMONIUM HYDROGEN CARBONATE A:2150
AMMONIUM HYDROGEN DIFLUORIDE A:2200
AMMONIUM HYDROGEN FLUORIDE A:2200
AMMONIUM HYDROGEN SULFIDE A:3900
AMMONIUM HYDROGEN SULFIDE SOLUTION A:3900
AMMONIUM HYDROGEN SULFATE A:3850
AMMONIUM HYDROGEN SULFITE A:2250
AMMONIUM HYDROSULFITE A:2250
AMMONIUM HYDROXIDE A:2000
AMMONIUM HYDROXIDE (less than 28% AQUEOUS AMMONIA) A:2850
AMMONIUM HYPO SOLUTION A:4150
AMMONIUM HYPOPHOSPHITE A:2900
AMMONIUM HYPOSULFITE SOLUTION A:4150
AMMONIUM HYPOSULFITE A:4100
AMMONIUM IODIDE A:2950
AMMONIUM IRON SULFATE F:0550
AMMONIUM LACTATE A:3000
AMMONIUM LACTATE SYRUP A:3000
AMMONIUM MOLYBDATE A:3100
AMMONIUM MONOSULFIDE A:3900
AMMONIUM MONOSULFITE A:2250
AMMONIUM MURIATE A:2450
AMMONIUM NICKEL (II) SALT (2:2:1) N:0700
AMMONIUM NICKEL SULFATE N:0700
AMMONIUM NITRATE A:3150
AMMONIUM NITRATE-PHOSPHATE MIXTURE A:3200
AMMONIUM NITRATE-SULFATE MIXTURE A:3200
AMMONIUM NITRATE-UREA SOLUTION A:3300
AMMONIUM OLEATATE A:3350

AMMONIUM OXALATE A:3400
AMMONIUM OXALATE HYDRATE A:3400
AMMONIUM OXALATE MONOHYDRATE A:3400
AMMONIUM OXALATE MONOHYDRATE A:3400
AMMONIUM PARAMOLYBDATE A:3100
AMMONIUM PENTABORATE TETRAHYDRATE A:3450
AMMONIUM PENTABORATE A:3450
AMMONIUM PENTACHLOROZINCATE Z:0300
AMMONIUM PERCHLORATE A:3500
AMMONIUM PEROXYDISULFATE A:3550
AMMONIUM PERSULFATE A:3550
AMMONIUM PHOSPHATE A:3600
sec-AMMONIUM PHOSPHATE A:3600
AMMONIUM PHOSPHATE, DIBASIC A:3600
AMMONIUM PICRATE, DRY A:3650
AMMONIUM PICRATE, wetted with more than 10% water A:3650
AMMONIUM PICRATE, WET A:3650
AMMONIUM PICRATE (YELLOW) A:3650
AMMONIUM PICRONITRATE A:3650
AMMONIUM RHODANATE A:4050
AMMONIUM RHODANIDE A:4050
AMMONIUM SALZ DER AMIDOSULFONSAURE (German) A:3800
AMMONIUM SILICOFLUORIDE A:3700
AMMONIUM STEARATE A:3750
AMMONIUM SULFAMATE A:3800
AMMONIUM SULFATE A:3850
AMMONIUM SULFHYDRATE SOLUTION A:3900
AMMONIUM SULFIDE A:3900
AMMONIUM SULFITE A:3950
AMMONIUM SULFITE, HYDROGEN A:2250
AMMONIUM SULFOCYANATE A:4050
AMMONIUM SULFOCYANIDE A:4050
AMMONIUM SULPHAMATE A:3800
AMMONIUM SULPHATE A:3850
AMMONIUM TARTRATE A:4000
AMMONIUM TETRAFLUOBORATE A:2650
AMMONIUM THIOCYANATE A:4050
AMMONIUM THIOSULFATE A:4100
AMMONIUM THIOSULFATE SOLUTION (60% OR LESS) A:4150
AMMONIUM TRIOXALATOFERRATE(III) F:0150
AMMONIUM TRIOXALATOFERRATE(III) TRIHYDRATE F:0150
AMMONIUM ZINC CHLORIDE Z:0300
AMOIL D:0975
AMORPHOUS PHOSPHORUS, BLACK P:1650
AMORPHOUS PHOSPHORUS, RED P:1700
AMOXONE D:0100
AMP A:1925
AMP-95 A:1925
AMS A:3800
AMS M:5400
AMSCO TETRAMER P:4500
AMSCO TETRAMER D:8050
AMYAZETAT (German) A:4200
AMYL ACETATE A:4200
n-AMYL ACETATE A:4200
sec-AMYL ACETATE A:4250
tert-AMYL ACETATE A:4300
AMYL ACETATE, iso- I:0100
AMYL ACETATE, mixed isomers A:4200
AMYLACETIC ESTER A:4250
AMYLACETIC ESTER I:0100
AMYL ACETIC ETHER A:4200
AMYL ALCOHOL A:4350
1-AMYL ALCOHOL A:4350
n-AMYL ALCOHOL A:4350
prim-n-AMYL ALCOHOL A:4350
tert-AMYL ALCOHOL M:4400
AMYL ALCOHOL, normal A:4350
AMYL ALDEHYDE V:0100
AMYL BROMIDE B:2700

AMYL CARBINOL H:1050
n-AMYL CARBINOL H:1050
AMYL CHLORIDE A:4400
n-AMYL CHLORIDE A:4400
AMYLENE P:0800
α-n-AMYLENE P:0800
AMYLENE HYDRATE M:4400
AMYL ETHYL KETONE E:2300
AMYL HYDRIDE P:0650
AMYL HYDROSULFIDE A:4450
AMYL MERCAPTAN A:4450
n-AMYL MERCAPTAN A:4450
AMYL METHYL ALCOHOL M:4500
AMYL-METHYL-CETONE (French) A:4500
AMYL METHYL KETONE A:4500
n-AMYL METHYL KETONE A:4500
n-AMYL NITRATE A:4550
AMYL NITRITE A:4600
iso-AMYL NITRITE A:4600
AMYLOL A:4350
AMYL PHTHALATE D:0975
n-AMYL PROPIONATE P:0850
AMYL SULFHYDRATE A:4450
AMYL THIOALCOHOL A:4450
AMYL TRICHLOROSILANE A:4700
n-AMYL TRICHLOROSILANE A:4700
AN A:0950
ANACARDIC ACID O:3100
ANAESTHETIC ETHER E:5600
ANAMENTH T:2350
ANESTHENYL M:4200
ANESTHESIA ETHER E:5600
ANESTHETIC ETHER E:5600
AN/FO A:3150
ANGLISITE L:1500
ANHIDRIDO ACETICO (Spanish) A:0250
ANHIDRIDO CROMICO (Spanish) C:3300
ANHIDRIDO FTALICO (Spanish) P:2000
ANHIDRIDO PROPIONICO (Spanish) P:3750
ANHYDRIDE ACETIQUE (French) A:0250
ANHYDRIDE of AMMONIUM CARBONATE A:2350
ANHYDRIDE ARSENIEUX (French) A:5500
ANHYDRIDE ARSENIQUE (French) A:5400
ANHYDRIDE CARBONIQUE (French) C:1650
ANHYDRIDE CHROMIQUE (French) C:3300
ANHYDRIDE PHTALIQUE (French) P:2000
ANHYDROFLUORIC ACID H:1550
ANHYDROL E:1900
ANHYDRONE M:0200
ANHYDROUS AMMONIA A:2000
ANHYDROUS CHLORAL T:2100
ANHYDROUS HYDROBROMIC ACID H:1400
ANHYDROUS HYDROCHLORIC ACID H:1450
ANHYDROUS HYDROGEN FLUORIDE H:1550
ANILINA (Spanish) A:4750
ANILINE A:4750
ANILINE, 2,6-DIETHYL D:2850
ANILINE, 2,6-DIMETHYL X:0450
ANILINE, 3-METHYL- T:1800
ANILINE, HEXAHYDRO- C:6100
ANILINE, N-PHENYL D:7250
ANILINE OIL A:4750
ANILINOBENZENE D:7250
ANILINOMETHANE M:2050
ANIMAL CARBON C:2000
o-ANISIC ACID M:5350
ANISOYL CHLORIDE A:4800
p-ANISOYL CHLORIDE A:4800
ANNAMENE S:6300

ANODYNON E:2900
ANOFEX D:0300
ANOL C:5950
ANOL C:5900
ANONE C:5950
ANOZOL D:3750
ANPROLENE E:5500
ANSAR C:0100
ANSAR 160 S:2700
ANSUL ETHER 121 E:4700
ANTAK D:0700
ANTHON T:2050
ANTHRACEN (German) A:4850
ANTHRACENE A:4850
ANTHRACENO (Spanish) A:4850
ANTHRACIN A:4850
ANTI-KNOCK MIXTURE M:6150
ANTI-RUST S:4400
ANTIBULIT S:3400
ANTICARIE H:0400
ANTIFORMIN S:3900
ANTIKNOCK-33 M:3800
ANTIMOINE (TRICHLORURE d') (French) A:5100
ANTIMOINE FLUORURE (French) A:5150
ANTIMONIC CHLORIDE A:4900
ANTIMONIOUS OXIDE A:5200
ANTIMONIUS CHLORIDE A:5100
ANTIMONOUS BROMIDE A:5050
ANTIMONOUS FLUORIDE A:5150
ANTIMONPENTACHLORID (German) A:4900
ANTIMONY BUTTER A:5100
ANTIMONY(III) CHLORIDE A:5100
ANTIMONY(V) CHLORIDE, ANTIMONY PERCHLORIDE A:4900
ANTIMONY FLUORIDE A:4950
ANTIMONY(III) FLUORIDE A:5150
ANTIMONY(5+) FLUORIDE A:4950
ANTIMONY(V) FLUORIDE A:4950
ANTIMONY PENTACHLORIDE A:4900
ANTIMONY PENTAFLUORIDE A:4950
ANTIMONY(5+) PENTAFLUORIDE A:4950
ANTIMONY(V) PENTAFLUORIDE A:4950
ANTIMONY PEROXIDE A:5200
ANTIMONY POTASSIUM TARTRATE A:5000
ANTIMONY SESQUIOXIDE A:5200
ANTIMONY TRIBROMIDE A:5050
ANTIMONY TRICHLORIDE A:5100
ANTIMONY TRIFLUORIDE A:5150
ANTIMONY TRIOXIDE A:5200
ANTIMONY, WHITE A:5200
ANTIMONYL POTASSIUM TARTRATE A:5000
ANTIMUCIN WDR P:1400
ANTIREN P:2100
ANTISAL LA T:1600
AOUARA OIL O:2200
APARASIN B:0600
APAVAP D:2650
APAVINPHOS P:1500
APHAMITE P:0200
APHOXIDE T:4650
APHTIRIA B:0600
APLIDAL B:0600
APO T:4650
AQUA-KLEEN D:0100
AQUA-VEX T:2650
AQUA AMMONIA A:2000
AQUA FORTIS N:1350
AQUACIDE D:7650
AQUALIN A:0800
AQUALINE A:0800

AQUAREX METHYL D:8700
AQUEOUS AMMONIA A:2850
AQUEOUS HYDROGEN CHLORIDE H:1450
AQUEOUS HYDROGEN FLUORIDE H:1550
AR-TOLUENOL C:5100
ARACHIS OIL O:1900
ARALO P:0200
ARASAN T:1450
ARBONIC ACID, THALLIUM SALT T:1200
ARBORICID T:2550
ARBITEX B:0600
ARCOSOLV D:7600
ARCTON-3 M:5900
ARCTON-4 M:5800
ARCTON 6 D:1650
ARCTON 63 T:2900
ARCTON 7 D:2100
ARCTRON 9 T:2400
ARCTUVIN H:1700
ARECA NUT OIL O:2200
ARGENTIC FLUORIDE S:0800
ARGENTOUS FLUORIDE S:0800
ARGENTOUS OXIDE S:1100
ARGEZIN A:5750
ARIZOLE O:4100
ARKLONE P T:2900
ARKOTINE D:0300
AROCLOR P:2200
AROCLOR 1221 P:2200
AROCLOR 1232 P:2200
AROCLOR 1242 P:2200
AROCLOR 1248 P:2200
AROCLOR 1254 P:2200
AROCLOR 1260 P:2200
AROCLOR 1262 P:2200
AROCLOR 1268 P:2200
AROCLOR 2565 P:2200
AROCLOR 4465 P:2200
AROMATIC SOLVENT N:0200
AROSOL E:5200
ARSEN (German) A:5250
ARSENATE A:5300
ARSENATE DE CALCIUM (French) C:0550
ARSENATE OF LEAD L:0700
ARSENIATO CALCICO (Spanish) C:0550
ARSENIATO POTASICO (Spanish) P:2550
ARSENIATO de PLOMO (Spanish) L:0700
ARSENIATO SODICO (Spanish) S:1800
ARSENIATO de ZINC (Spanish) Z:0400
ARSENIC A:5250
ARSENIC 75 A:5250
ARSENIC ACID A:5300
o-ARSENIC ACID A:5300
ARSENIC ACID ANHYDRIDE A:5400
ARSENIC ACID, CALCIUM SALT C:0550
ARSENIC ACID, MONOPOTASSIUM SALT P:2550
ARSENIC ACID, SODIUM SALT S:1800
ARSENICALS A:5250
ARSENIC ANHYDRIDE A:5400
ARSENIC BLACK A:5250
ARSENIC BLANC (French) A:5500
ARSENIC CHLORIDE A:5450
ARSENICO (Spanish) A:5250
ARSENICUM ALBUM A:5500
ARSENIC DISULFIDE A:5350
ARSENIC HYDRIDE A:5575
ARSENIC, METALLIC A:5250
ARSENIC OXIDE A:5400
ARSENIC(III) OXIDE A:5500
ARSENIC(V) OXIDE A:5400
ARSENIC PENTAOXIDE A:5400
ARSENIC PENTOXIDE A:5400

ARSENIC PENTOXIDE A:5300
ARSENIC SESQUIOXIDE A:5500
ARSENIC SESQUISULFIDE A:5550
ARSENIC SULFIDE A:5550
ARSENIC TRICHLORIDE A:5450
ARSENIC TRIHYDRIDE A:5575
ARSENIC TRIOXIDE A:5500
ARSENIC TRISULFIDE A:5550
ARSENIC YELLOW A:5550
ARSENIGEN SAURE (German) A:5500
ARSENIOUS ACID A:5500
ARSENIOUS ACID, POTASSIUM SALT P:2600
ARSENIOUS OXIDE A:5500
ARSENIOUS TRIOXIDE A:5500
ARSENITE A:5500
ARSENIOUS HYDRIDE A:5575
ARSENITO CALCICO (Spanish) C:0600
ARSENITO de COBRE (Spanish) C:4100
ARSENITO POTASICO (Spanish) P:2600
ARSENITO SODICO (Spanish) S:1900
ARSENIURETTED HYDROGEN A:5575
ARSENOLITE A:5500
ARSENOUS ACID A:5500
ARSENOUS ACID ANHYDRIDE A:5500
ARSENOUS ACID, CALCIUM SALT C:0600
ARSENOUS ACID, SODIUM SALT S:1900
ARSENOUS ANHYDRIDE A:5500
ARSENOUS CHLORIDE A:5450
ARSENOUS DICHLORIDE(2-CHLORO-ETHENYL)- B:2025
ARSENOUS OXIDE A:5500
ARSENOUS OXIDE ANHYDRIDE A:5500
ARSENOUS TRICHLORIDE A:5450
ARSENTRIOXIDE A:5500
ARSICODILE S:2700
ARSINE A:5575
ARSINE(2-CHLOROVINYL)DICHLORO- B:2025
ARSODENT A:5500
ARSONIC ACID, COPPER(2+) SALT(1:1) C:4100
ARSONIC ACID, POTASSIUM SALT P:2600
ARSYCODILE S:2700
ARTHODIBROM N:0150
ARTIC M:3450
ARTIFICIAL ALMOND OIL B:0500
ARTIFICIAL ANT OIL F:1500
ARTIFICIAL CINNABAR M:0900
ASEX S:2800
ASFALTO (Spanish) A:5600
ASP 47 T:0600
ASPHALT A:5650
ASPHALT A:5600
ASPHALT BITUMEN A:5600
ASPHALT BLENDING STOCKS: ROOFERS FLUX A:5650
ASPHALT BLENDING STOCKS: STRAIGHT RUN RESIDUE A:5700
ASPHALT CEMENTS A:5600
ASPHALT (CUT BACK) A:5650
ASPHALT, PETROLEUM O:4300
ASPHALTIC BITUMEN A:5600
ASPHALTUM A:5600
ASPHALTUM A:5650
ASPHALTUM OIL A:5650
ASPON-CHLORDANE C:2050
ASTRAL OIL O:0100
ASTROBOT D:2650
ASUNTHOL C:5000
ASUNTOL C:5000
AT-7 H:0600
AT-17 H:0600

ATAZINAX A:5750
ATE T:3250
ATGARD D:2650
ATHYLEN (German) E:3500
ATHYLENGLYKOL-MONOATHYLATHER (German) E:0800
ATLACIDE S:2800
ATLAS-A S:1900
ATRANEX A:5750
ATRASINE A:5750
ATRATOL A A:5750
ATRATOL B-HERBATOX S:2800
ATRAZINA (Spanish) A:5750
ATRAZINE A:5750
ATRED A:5750
ATTAC-2 T:1950
ATTAC 6 T:1950
ATTAC 6-3 T:1950
AULES T:1450
AUSTRALENE P:2050
AVIROL 118 CONC D:8700
AVITROL A:1950
AVLOTHANE H:0550
AVOLIN D:5900
AWPA 1 C:5050
9-AZAANTHRACENE A:0750
10-AZAANTHRACENE A:0750
AZABENZENE P:4800
AZACYCLOHEPTANE H:0850
1-AZACYCLOHEPTANE H:0850
AZACYCLOPROPANE E:5400
1-AZANAPHTHALENE Q:0100
3-AZAPENTANE-1,5-DIAMINE D:3450
AZDEL P:2400
AZIDE S:2000
AZIDA SODICO (Spanish) S:2000
AZINE P:4800
AZINPHOS-METHYL A:5800
AZIRANE E:5400
AZIRIDINA (Spanish) E:5400
AZIRIDINA, 2-METIL (Spanish) P:4400
AZIRIDINE E:5400
AZIRIDINE, 2-METHYL- P:4400
AZIRINE E:5400
1H-AZIRINE, DIHYDRO- E:5400
AZIUM S:2000
AZOENE FAST ORANGE GR SALT N:1450
AZOFOS M:4900
AZOGEN DEVELOPER-H T:1650
AZOIC DIAZO COMPONENT 6 N:1450
AZOIC DIAZO COMPONENT 37 N:1450
AZOPHOS M:4900
AZOTIC ACID N:1350
ZOTOX D:0300
AZOTURE de SODIUM (French) S:2000
AZUCAR (Spanish) S:6400
AZUFRE (Spanish) S:6600
AZUNTHOL C:5000
B D:4650
B-I-K U:0900
B-K LIQUID S:3900
B-K POWDER C:1000
B-SELEKTONON D:0100
B-W S:5100
B 404 P:0200
B32 H:0600
BABULUM OIL O:3900
BACILLOL C:5100
BAKER'S P AND S LIQUID P:1200
BAKER'S P AND S OINTMENT P:1200
BAN-MITE M:0250
BANANA OIL I:0100
BANANA OIL A:4250

BANEX D:1300
BANLEN D:1300
BANVEL D:1300
BANVEL D D:1300
BANVEL HERBICIDE D:1300
BARIUM BINOXIDE, BARIUM DIOXIDE B:0400
BARIUM CARBONATE B:0100
BARIUM CARBONATE B:0100
BARIUM CHLORATE B:0150
BARIUM CHLORATE MONOHYDRATE B:0150
BARIUM CYANIDE B:0200
BARIUM DICYANIDE B:0200
BARIUM DINITRATE B:0250
BARIUM MANGANATE (VIII) B:0350
BARIUM NITRATE B:0250
BARIUM PERCHLORATE B:0300
BARIUM PERCHLORATE TRIHYDRATE B:0300
BARIUM PERMANGANATE B:0350
BARIUM PEROXIDE B:0400
BARIUM SUPEROXIDE B:0400
BARIUMPEROXID (German) B:0400
BASFAPON D:0200
BASFAPON/BASFAPON N D:0200
BASFAPON B D:0200
BASIC BISMUTH CHLORIDE B:1900
BASIC COPPER ACETATE C:4750
BASIC ZINC CHROMATE Z:1000
BASIC ZINC CHROMATE X-2259 Z:1000
BASIC ZIRCONIUM CHLORIDE Z:2500
BASINEX D:0200
BASUDIN® D:1000
BASUDIN® 10 G D:1000
BATTERY ACID S:6800
BAY 1145 M:4900
BAY 9026 M:0450
BAY 9027 A:5800
BAY 10756 D:0800
BAY 15922 T:2050
BAY 19639 D:7700
BAY 37344 M:0450
BAY 70143 C:1500
BAY E-601 M:4900
BAY E-605 P:0200
BAYER-E-393 T:0600
BAYER E-605 P:0200
BAYER 13/59 T:2050
BAYER 21/199 C:5000
BAYER 8169 D:0800
BAYER 17147 A:5800
BAYER 19639 D:7700
BAYER 37344 M:0450
BAYMIX C:5000
BAYMIX 50 C:5000
BAYOL F O:3600
BAZUDEN D:1000
BBH B:0600
BBP B:3650
BCF-BUSHKILLER T:2500
BCS COPPER FUNGICIDE C:4800
BEARING OIL O:4500
BEET SUGAR S:6400
BEHA D:7000
BEHP D:3650
BELL MINE C:0900
BELT C:2050
BENCENO (Spanish) B:0550
BENCIDINA (Spanish) B:0850
BENCILAMINA (Spanish) B:1250
N-BENCILDIMETILAMINA (Spanish) B:1450
BENFOS D:2650

BEN-HEX B:0600
BENSULFOID S:6600
BENTOX 10 B:0600
BENZAL CHLORIDE B:0450
BENZALDEHIDO (Spanish) B:0500
BENZALDEHYDE B:0500
BENZALDEHYDE, α-CHLORO- B:1100
1-BENZAZINE Q:0100
BENZELENE B:0550
BENZENAMINE, 5-CHLORO-2-METHYL- C:3200
BENZENAMINE, 2,6-DINITRO-N,N-DIPROPYL-4-(TRIFLUOROMETHYL-) T:3800
BENZENAMINE, 4-NITRO- N:1450
BENZENE B:0550
BENZENEAMINE A:4750
BENZENEAMINE, 2,6-DIETHYL- D:2850
BENZENEAMINE, 4-FLUORO- F:0900
BENZENEAMINE, N-METHYL- M:2050
BENZENE, ANILINO- D:7250
BENZENE, (BROMOMETHYL)- B:1300
BENZENECARBINOL B:1200
BENZENECARBONAL B:0500
BENZENECARBONYL CHLORIDE B:1100
BENZENECARBOXYLIC ACID B:0900
BENZENE,CHLOROMETHYL- B:1350
BENZENE CARBALDEHYDE B:0500
BENZENE CARCABOXALDEHYDE B:0500
BENZENE CHLORIDE C:2400
BENZENE, CHLORO- C:2400
BENZENE, CHLOROMETHYL- B:1350
BENZENE, 1-CHLORO-2-METHYL-BENZENEAMINE, 3-METHYL- T:1800
BENZENE, 1-CHLORO-2-METHYL C:3100
BENZENE, CYANO- B:0950
BENZENEDIAMINE,AR-METHYL- T:1650
BENZENE-1,3-DICARBOXYLIC ACID I:1100
1,2-BENZENEDICARBOXYLIC ACID D:4500
BENZENE-o-DICARBOXYLIC ACID DI-N-BUTYL ESTER D:1250
1,2-BENZENEDICARBOXYLIC ACID ANHYDRIDE P:2000
1,2-BENZENEDICARBOXYLIC ACID, DIBUTYL ESTER D:1250
1,2-BENZENEDICARBOXYLIC ACID, DIETHYL ESTER D:3750
1,2-BENZENEDICARBOXYLIC ACID, DI-ISONONYL ESTER D:4450
1,2-BENZENEDICARBOXYLIC ACID, DIISOOCTYL ESTER D:4500
1,2-BENZENEDICARBOXYLIC ACID, DI-(2-METHYLPROPYL)ESTER D:4300
1,2-BENZENEDICARBOXYLIC ACID, DI-n-OCTYL ESTER D:5250
1,2-BENZENEDICARBOXYLIC ACID, DI-UNDECYL ESTER D:7900
o-BENZENEDICARBOXYLIC ACID, DIBUTYL ESTER D:1250
1,2-BENZENEDICARBOXYLIC ACID, DIMETHYL ESTER D:5900
1,4-BENZENE DICARBOXYLIC ACID METHYL ESTER D:6250
1,2-BENZENEDICARBOXYLIC ACID, DIPENTYL ESTER D:0975
1,2-BENZENEDICARBOXYLIC ANHYDRIDE P:2000
BENZENE, 1,2-DICHLORO- D:1450
BENZENE, 1,4-DICHLORO- D:1550
BENZENE, DICHLORO METHYL- B:0450
BENZENE, o-DIHYDROXY- C:1950
BENZENE, m-DIHYDROXY- R:0100
BENZENE, DIISOPROPYL D:4650

BENZENE,2,4-DIISOCYANATO-1-METHYL- T:1700
BENZENE,2,4-DIISOCYANATO-METHYL- T:1700
1,2-BENZENEDIOL C:1950
1,3-BENZENEDIOL R:0100
1,4-BENZENEDIOL H:1700
m-BENZENEDIOL R:0100
o-BENZENEDIOL C:1950
p-BENZENEDIOL H:1700
BENZENE FLUORIDE F:0950
BENZENE, FLUORO F:0950
BENZENE HEXACHLORIDE H:0400
BENZENE HEXACHLORIDE, gamma isomer B:0600
γ-BENZENE HEXACHLORIDE B:0600
BENZENE, HEXACHLORO- H:0400
BENZENE HEXAHYDRIDE C:5850
BENZENE, HEXAHYDRO- C:5850
BENZENE,HYDROXY- P:1200
BENZENEMETHTAL B:0500
BENZENE, METHYL- T:1600
BENZENE, 1-METHYL-2,4-DINITRO- D:6800
BENZENE, (1-METHYLETHYL-)- C:5450
BENZENENITRILE B:0950
BENZENE, NITRO- N:1550
BENZENE, 1,1'-OXYBIS- D:7350
BENZENE PHOSPHORUS DICHLORIDE B:0650
BENZENE PHOSPHORUS THIODICHLORIDE B:0700
BENZENE, PROPYL P:4050
BENZENE SULFOCHLORIDE B:0750
BENZENE SULFONE-CHLORIDE B:0750
BENZENE SULFONECHLORIDE B:0750
BENZENESULFONIC ACID CHLORIDE B:0750
BENZENESULFONYL CHLORIDE B:0750
BENZENE SULFONIC ACID, DODECYL-, compd. with 1-AMINO-2-PROPANOL D:8250
BENZENE SULFONIC ACID, DODECYL ESTER D:8150
BENZENE SULFONIC ACID, DODECYL ESTER D:8200
BENZENESULFONIC ACID, DODECYL- D:8200
BENZENESULFONIC ACID, DODECYL- D:8150
BENZENETHIOL B:0800
BENZENETHIOPHOSPHONYL CHLORIDE B:0700
BENZENE, 1,2,3-TRICHLORO- T:2150
BENZENE, 1,2,4-TRICHLORO- T:2200
BENZENE, 1,1'-(2,2,2-TRICHLOROETHYLIDENE)BIS(4-CHLORO- D:0300
BENZENE 1,2,3,5-TETRAMETHYL T:1000
1,2,3-BENZENETRIOL P:4850
BENZENE, VINYL- S:6300
BENZENOSULPHOCHLORIDE B:0750
BENZIDINE B:0850
BENZIN N:0200
BENZIN (German) G:0200
1-BENZINE Q:0100
BENZINOFORM C:1850
BENZINOL T:2350
BENZOATO AMONICO (Spanish) A:2100
BENZO-CHINON (German) B:1050
BENZOFENONA (Spanish) B:1000
BENZONITRILO (Spanish) B:0950
BENZO[b]PYRIDINE Q:0100
BENZO[b]QUINOLINE A:0750
BENZOEPIN E:0100

BENZOFLEX 9-88 D:7550
BENZOFLEX 9-88 SG D:7550
BENZOFLEX 9-98 D:7550
BENZOHYDROQUINONE H:1700
BENZOIC ACID B:0900
BENZOIC ACID, AMMONIUM SALT A:2100
BENZOIC ACID, CHLORIDE B:1100
BENZOIC ACID, 2-METHOXY- M:5350
BENZOIC ACID, METHYL ESTER M:3000
BENZOIC ACID NITRILE B:0950
BENZOIC ALDEHYDE B:0500
BENZOL B:0550
BENZOLE B:0550
BENZOLINE N:0200
BENZONITRILE B:0950
BENZOPHENONE B:1000
BENZOPEROXIDE D:1050
BENZOQUINOL H:1700
2,3-BENZOQUINOLINE A:0750
p-BENZOQUINONA (Spanish) B:1050
BENZOQUINONE B:1050
1,4-BENZOQUINONE B:1050
p-BENZOQUINONE B:1050
2-BENZOTHIAZOLETHIOL, SODIUM SALT S:4000
2-(3h)-BENZOTHIAZOLETHIONE, SODIUM SALT S:4000
BENZOYL BENZENE B:1000
BENZOYL CHLORIDE B:1100
BENZOYL METHIDE HYPNONE A:0450
BENZOYL PEROXIDE D:1050
BENZOYL SUPEROXIDE D:1050
BENZYL ACETATE B:1150
BENZYL ALCOHOL B:1200
BENZYL AMINE B:1250
BENZYL BROMIDE B:1300
BENZYL n-BUTYL PHTHALATE B:3650
BENZYLCARBONYL CHLORIDE B:1400
BENZYLCHLORID (German) B:1350
BENZYL CHLORIDE B:1350
BENZYL CHLOROCARBONATE B:1400
BENZYL CHLOROFORMATE B:1400
BENZYL DICHLORIDE B:0450
BENZYL DIMETHYLAMINE B:1450
N-BENZYLDIMETHYLAMINE B:1450
BENZYL DIMETHYLOCTADECYL AMMONIUM CHLORIDE B:1500
BENZYLDIMETHYLSTEARYL-AMMONIUM CHLORIDE B:1500
BENZYLE (CHLORURE de) (French) B:1350
BENZYLENE CHLORIDE B:0450
BENZYL ETHANOATE B:1150
BENZYL ETHER D:1100
BENZYLIDENE CHLORIDE B:0450
BENZYL OXIDE D:1100
BENZYL TRIMETHYL AMMONIUM CHLORIDE B:1550
BEOSIT E:0100
BERILIO (Spanish) B:1600
BEROL 478 D:7050
BERTHOLITE C:2100
BERTHOLLET'S SALT P:2700
BERYLLIA B:1800
BERYLLIUM B:1600
BERYLLIUM CHLORIDE B:1650
BERYLLIUM DICHLORIDE B:1650
BERYLLIUM DIFLUORIDE B:1700
BERYLLIUM DINITRATE B:1750
BERYLLIUM DUST B:1600
BERYLLIUM FLUORIDE B:1700
BERYLLIUM, METAL POWDER B:1600
BERYLLIUM MONOXIDE B:1800
BERYLLIUM NITRATE B:1750

BERYLLIUM NITRATE TRIHYDRATE B:1750
BERYLLIUM OXIDE B:1800
BERYLLIUM POWDER B:1600
BERYLLIUM SULFATE B:1850
BERYLLIUM SULFATE TETRAHYDRATE B:1850
BETEL NUT OIL O:2200
BETRAPRONE P:3600
BETULA OIL M:5350
BEXOL B:0600
BFV F:1250
BGE B:4150
n-BGE B:4150
BH 2,4-D D:0100
BHC B:0600
γ-BHC B:0600
BH DALAPON D:0200
BH DOCK KILLER M:0400
BIANISIDINE T:1850
4,4'-BIANILINE B:0850
p,p-BIANILINE B:0850
BIBENZENE D:7200
BIBESOL D:2650
BICARBONATO AMONICO (Spanish) A:2150
BICARBURRETTED HYDROGEN E:3500
BICHLORENDO M:5650
BICHLORIDE of MERCURY M:0600
BICHLORURE de MERCURE (French) M:0600
BICHLORURE de PROPYLENE (French) D:2300
BICROMATO de LITIO (Spanish) L:2600
BICHROMATE de SODIUM (French) S:3300
BICHROMATE OF POTASH P:2900
BICHROMATE OF SODA S:3300
BICHROME P:2900
BICYCLO [4.4.0] DECANE D:0400
BICYCLO 2.2.1HEPTAN-2-ONE,1,7,7-TRIMETHYL- C:1400
BICYCLO 221 HEPT-2-ENE, 5-ETHYLIDENE- E:6600
BICYCLOPENTADIENE D:2675
BIEBERITE C:3900
BIETHYLENE B:2850
BIFLUORO SODICO (Spanish) S:2100
BIFLUORURO AMONICO (Spanish) REC. A:2200
BIFORMAL G:0700
BIFORMYL G:0700
BIG DIPPER D:7250
BILEVON H:0600
BILORIN F:1350
BIMETHYL E:0500
BINITROBENZENE D:6400
BIO-SOFT D-40 D:8300
BIO 5,462 E:0100
BIOCIDE A:0800
BIOFLEX 91 D:6950
BIOGAS M:1400
BIOXALATO POTASICO (Spanish) P:2650
BIPHENYL D:7200
BIPHENYL, 4,4'-DIAMINO- B:0850
1,1'-BIPHENYL D:7200
(1,1'-BIPHENYL)-4,4'-DIAMINE B:0850
1,1'-BIPHENYL-4,4'-DIAMINE,3,3'-DIMETHYL- T:1850
4,4'-BIPHENYLDIAMINE B:0850
4,4'-BIPHENYLENEDIAMINE B:0850
BIPHENYL ETHER D:7350
BIPHENYL, MIXED with BIPHENYL OXIDE D:8850
BIPHENYL OXIDE D:7350

BIPHENYL, POLYCHLORO- P:2200
BIPOTASSIUM CHROMATE P:2750
BIRNENOEL A:4200
BIS(ACETATE)DIOXOURANIUM U:0600
BIS(ACETO)COBALT C:3550
BIS(ACETOXY)CADMIUM C:0150
BIS(ACETYLOXY)MERCURY M:0500
BIS(2-AMINOETHYL)AMINE D:3450
BIS(β-AMINOETHYL)AMINE D:3450
BIS(2-BUTOXYETHYL) ETHER D:3050
BIS-(p-CHLOROBENZOYL) PEROXIDE D:1600
BIS(2-CHLOROETHYL) ETHER D:1800
BIS(2-CHLORO-1-METHYLETHYL ETHER) D:1950
BIS(2-CHLOROETHYL)ETHYLAMINE B:2025
BIS(2-CHLOROETHYL)METHYLAMINE B:2025
N,N-BIS(2-CHLOROETHYL)METHYLAMINE B:2025
BIS(β-CHLOROETHYL)METHYLAMINE B:2025
BIS(2-CHLOROETHYL)SULFIDE B:2025
BIS(β-CHLOROETHYL)SULFIDE B:2025
1,1-BIS-(p-CHLOROPHENYL)-2,2,2-TRICHLOROETHANE D:0300
2,2-BIS(p-CHLOROPHENYL)-1,1-TRICHLOROETHANE D:0300
2,2-BIS(p-CHLOROPHENYL)-1,1,1-TRICHLOROETHANE D:0300
1,1-BIS(p-CHLOROPHENYL)-2,2-DICHLOROETHANE D:0250
α,α-BIS(p-CHLOROPHENYL)-β,β,β-TRICHLORETHANE D:0300
1,1-BIS(p-CHLOROPHENYL)-2,2,2-TRICHLOROETHANOL D:2600
BIS(2-CHLOROISOPROPYL) ETHER D:1950
BISCYCLOPENTADIENE D:2675
BIS(o,o-DIAETHYLPHOSPHORSAEURE-ANHYDRID (German) T:0800
BIS[s-(DIETHOXYPHOSPHINOTHIOYL]MERCAPTO)METHANE E:0600
BIS-o,o-DIETHYLPHOSPHORIC ANHYDRIDE T:0800
BIS-o,o-DIETHYLPHOSPHOROTHIONIC ANHYDRIDE T:0600
BIS(DIMETHYLAMINO)CARBONOTHIOYL) DISULPHIDE T:1450
BIS(DIMETHYL-THIOCARBAMOYL) DISULFID (German) T:1450
BIS(DITHIOPHOSPHATEDE o,o-DIETHYLE) de s,s'-METHYLENE (French) E:0600
2,2-BIS(4-(2,3-EPOXYPROPYLOXY)PHENYL)PROPANE B:2000
1,2-BIS(ETHOXYCARBONYL)ETHYL M:0250
BIS(2-ETHYLHEXYL)ADIPATE D:7000
BIS(2-ETHYLHEXYL)HYDROGEN PHOSPHATE D:3600
BIS(2-ETHYLHEXYL)ORTHOPHOSPHORIC ACID D:3600
BIS(2-ETHYLHEXYL)PHTHALATE D:3650
BIS(2-ETHYLHEXYL)PHOSPHORIC ACID D:3600
BIS(2-ETHYLHEXYL)SODIUM SULFOSUCCINATE D:7050
BIS(2-CHLOROETHYL)SULFIDE and BIS[2-(2-CHLOROETHYLTHIO)-ETHYL]ETHER Mixture (SULFUR MUSTARD HT) B:2025
2,2-BIS[4-(2,3-EPOXYPROPYLOXY)PHENYL]PROPANE B:2000
s-[1,2-BIS(ETHOXYCARBONYL)ETHYL]o,o-DIMETHYL DITHIOPHOSPHATE OF DIETHYL MERCAPTOSUCCINATE M:0250
BISFENOL A (Spanish) B:1950
BISFENOL A DIGLICIDAL ETER (Spanish) B:2000
BIS(4-GLYCIDYLOXYPHENYL)DIMETHYAMETHANE B:2000
2,2-BIS(p-GLYCIDYLOXYPHENYL)PROPANE B:2000
BIS(GLYCINATO) COPPER C:4450
1,2-BIS(2-HYDROXYETHOXY) ETHANE T:3400
BIS-[2-(2-HYDROXYETHOXY) ETHYL] ETHER T:0650
BIS(2-HYDROXYETHYL) ETHER D:3000
BIS(2-HYDROXYETHYL)AMINE D:2750
2,2-BIS(4-HYDROXYFENYL) PROPANE B:1950
BIS(4-HYDROXYPHENYL)DIMETHYLMETHANE DIGLYCIDYL ETHER B:2000
2,2-BIS(4-HYDROXYPHENYL)PROPANE,DIGLYCIDYL ETHER B:2000
BIS(2-HYDROXYPROPYL) ETHER D:7500
BIS(4-HYDROXYPHENYL)DIMETHYLMETHANE DIGLYCIDYL ETHER B:2000
2,2-BIS(4-HYDROXYPHENYL) PROPANE B:1950
2,2-BIS(4-HYDROXYPHENYL)PROPANE, DIGLYCIDYL ETHER B:2000
BIS(HYDROXYLAMINE) SULFATE H:1850
2,2-BIS(HYDROXYMETHYL)-1,3-PROPANEDIOL P:0550
BIS(ISOPROPYL)AMINE D:4600
BIS(ISOPROPYL)NAPHTHALENE D:4750
BIS(METHYLCYCLOPENTADIENE) M:3750
BIS(2-METHOXYETHYL)-ETHER D:3100
2,2-bis(p-METHOXYPHENYL)-1,1,1-TRICHLOROETHANE M:1500
BIS(1-METHYLETHYL)-BENZENE D:4650
BIS(6-METHYLHEPTYL)ESTER OF PHTHALIC ACID D:4500
BIS(6-METHYLHEPTYL)PHTHALATE D:4500
N,N-BIS(2-METHYLPROPYL)AMINE D:4100
BIS(2-HYDROXY-3,5,6-TRICHLOROPHENYL)METHANE H:0600
BISMUTH CHLORIDE OXIDE B:1900
BISMUTH OXYCHLORIDE B:1900
BISMUTH SUBCHLORIDE B:1900
BISMUTHYL CHLORIDE B:1900
BIS(NITRATO-o,o')DIOXO URANIUM U:0700
BISOFLEX-81 D:3650
BISOFLEX DOA D:7000
BISOFLEX DOP D:3650
BISPHENOL A B:1950
BISPHENOL A DIGLYCIDYL ETHER B:2000
BISPHENOL A EPICHLOROHYDRIN CONDENSATE B:2000
BIS(THIOCYANATO)-MERCURY M:0950
BISULFITE S:6700
BISULFITE de SODIUM (French) S:2200
BISULFITO AMONICO (Spanish) A:2250
BISULFITO SODICO (Spanish) S:2200
4,4'-BI-o-TOLUIDINE T:1850
n-BUTILAMINA (Italian, Spanish) B:3500
sec-BUTILAMINA (Italian, Spanish) B:3550
BITUMEN (European term) A:5600, A:5650, A:5700
BIVINYL B:2850
BLACK AND WHITE BLEACHING CREAM H:1700
BLACK LEAF N:1200
BLACK LEAF 40 (40% water solution) N:1250
BLACK PHOSPHORUS P:1650
BLACOSOLV T:2350
BLADAFUME T:0600
BLADAFUN T:0600
BLADAN P:0200
BLADAN T:0800
BLADAN E:0600
BLADAN-F P:0200
BLADAN-M M:4900
BLADON T:0800
BLANDLUBE O:3600
BLAUSAEURE (German) H:1500
BLEACH S:3900
BLEACHING POWDER C:1000
BLEISTEARAT (German) L:1400
BLEISULFAT (German) L:1500
BLUE COPPER C:4800
BLUE OIL A:4750
BLUE-OX Z:1900
BLUE STONE C:4800
BLUE VERDIGRIS C:4750
BLUE VITRIOL C:4800
BOILED LINSEED OIL O:3400
BOILER COMPOUND B:2050
BOLETIC ACID F:1400
BONAZEN Z:2200
BONOFORM T:0350
BORACIC ACID B:2100
BORANE, TRIBROMO- B:2150
BORATO SODICO (Spanish) S:2400
BORATO de ZINC (Spanish) Z:0600
BORAX S:2400
BORAX 2335 Z:0600
BORAX, ANHYDROUS S:2400
BORIC ACID B:2100
BORIC ACID, ZINC SALT Z:0600
BORNANE, 2-oxo- C:1400
2-BORNANONE C:1400
BOROFAX B:2100
BOROHIDRURO SODICO (Spanish) S:2500
BOROHYDRIDE SOLUTION S:2600
BOROHYDRURE de SODIUM (French) S:2500
BORON BROMIDE B:2150
BORON CHLORIDE B:2200
BORON FLUORIDE B:2350
BORON HYDRIDE D:0350
BORON TRIBROMIDE B:2150
BORON TRIBROMIDE 6 B:2150
BORON TRICHLORIDE B:2200
BORSAURE (German) B:2100
BOS MH M:0400
BOSAN SUPRA D:0300
BOTTLED GAS P:3300
BOTTLED GAS L:2200
BOV S:6800
BOVIDERMOL D:0300
BOX TOE GUM C:3950
BP D:1050
BPL P:3600
BPO D:1050
BRAZIL WAX W:0100
BRECOLANE NDG D:3000
BREVINYL D:2650
BRICK OIL C:5050

BRIMSTONE S:6600
BRITON T:2050
BRITTEN T:2050
BROCIDE E:4100
BRODAN D:8900
BROM (German) B:2250
BROME (French) B:2250
BROMELITE B:1800
BROMELLITE B:1800
BROM-O-GAS M:3100
BPOMOPROPANE B:2400
BROMALLYLENE A:1250
BROMCHLOPHOS N:0150
BROMEX N:0150
BROMINE B:2250
BROMINE CYANIDE C:5700
BROMINE FLUORIDE B:2350
BROMINE FLUORIDE B:2300
BROMINE PENTAFLUORIDE B:2300
BROMINE TRIFLUORIDE B:2350
BROMO (Spanish) B:2250
1-BROMO BUTANE B:2550
BROMOACETONA (Spanish) B:2400
BROMOACETONE B:2400
BROMOACETYL BROMIDE B:2450
BROMOBENCENO (Spanish) B:2500
BROMOBENZENE B:2500
BROMOBENZOL B:2500
n-BROMOBUTANE B:2550
2-BROMOBUTANE B:2600
1-BROMOBUTANO (Spanish) B:2600
2-BROMOBUTANO (Spanish) B:2600
BROMOCYAN C:5700
BROMOCYANOGEN C:5700
BROMOETHANOYL BROMIDE B:2450
BROMOFORM B:2650
BROMOFORME (French) B:2650
BROMOFUME E:4000
BROMOMETHANE M:3100
(BROMOMETHYL)BENZENE B:1300
p-(BROMOMETHYL)NITROBENZENE B:1300
BROMOMETHYL METHYL KETONE B:2400
2-BROMOPENTANE B:2700
2-BROMOPENTANO (Spanish) B:2700
BROMOPHENYLMETHANE B:1300
1-BROMOPROPANE B:2750
1-BROMOPROPANO (Spanish) B:2750
BROMO-2-PROPANONE B:2400
1-BROMO-2-PROPANONE B:2400
3-BROMO-1-PROPENE A:1250
3-BROMOPROPENE A:1250
3-BROMOPROPYLENE A:1250
α-BROMOTOLUENE B:1300
ω-BROMOTOLUENE B:1300
BROMURE de CYANOGEN (French) C:5700
BROMURO de ALILO (Spanish) A:1250
BROMURO AMONICO (Spanish) A:2300
BROMURO de BENCILO (Spanish) B:1300
BROMURO de CADMIO (Spanish) C:0200
BROMURO de CIANOGENO (Spanish) C:5700
BROMURO de COBALTO (Spanish) C:3600
BROMURO de COBRE (Spanish) C:4150
BROMURO de HIDROGENO (Spanish) H:1400
BROMURO de METILENO (Spanish) D:1150
BROMURO de ZINC (Spanish) Z:0700
BROMWASSERSTOFF (German) H:1400
BRUCINA (Spanish) B:2800
BRUCINE B:2800
(-)BRUCINE B:2800

(-)BRUCINE DIHYDRATE B:2800
BRUCINE HYDRATE B:2800
BRUCINE QUARTERNARY HYDRATE B:2800
BRUSH-OFF 445 LOW VOLATILE BRUSH KILLER T:2500
BRUSH-RHAP D:0100
BRUSH BUSTER D:1300
BRUSH RHAP T:2500
BRUSHTOX T:2500
BSC-REFINED D B:0750
BTMAC B:1550
BUCS E:4900
BUFEN P:1400
BUFOPTO ZINC SULFATE Z:2200
BUHACH P:4750
BUNKER C OIL O:2800
BUNT-CURE H:0400
BUNT-NO-MORE H:0400
BURNED LIME C:1100
BURNT LIME C:1100
BUSH KILLER D:0100
BUTADIENE B:2850
1,2-BUTADIENE B:2850
1,3-BUTADIENE B:2850
α-γ-BUTADIENE B:2850
BUTA-1,3-DIENE B:2850
BUTADIENE DIMER V:0900
1,3-BUTADIENE,1,1,2,3,4,4-HEXACHLORO- H:0450
BUTADIENE DIOXIDE D:2750
1,3-BUTADIENO (Spanish) B:2850
BUTAL B:4800
BUTALDEHYDE B:4800
n-BUTALDEHYDE B:4800
BUTALYDE B:4800
BUTAN-1-OL B:3350
BUTAN-2-OL B:3400
BUTANAL B:4800
BUTANALDEHYDE B:4800
1-BUTANAMINE B:3500
2-BUTANAMINE B:3550
1-BUTANAMINE, n-BUTYL D:1160
n-BUTANE B:2900
1-BUTANECARBOXYLIC ACID P:0700
BUTANE, 1-CHLORO- B:3750
1,4-BUTANEDICARBOXYLIC ACID A:1000
BUTANEDIOIC ACID, DIMETHYL ESTER D:6050
1,3-BUTANEDIOL B:3900
1,4-BUTANEDIOL B:3950
2,3-BUTANEDIOL B:4000
BUTANE-1,3-DIOL B:3900
BUTANE-1,4-DIOL B:3950
BUTANE, 1,2,3,4-DIEPOXY D:2750
BUTANE, 1,4-EPOXY- T:0900
BUTANENITRILE, 4-CHLORO- C:2450
BUTANE-THIOL B:4400
BUTANETHIOL B:4400
n-BUTANETHIOL B:4400
BUTANIC ACID B:4900
BUTANOIC ACID B:4900
BUTANOIC ACID, BUTYL ESTER B:3700
BUTANOIC ACID, 3-oxo-METHYL ESTER M:1600
BUTANOIC ACID, METHYL ESTER M:3400
BUTANO (Spanish) B:2900
BUTANOL B:3350
1-BUTANOL B:3350
2-BUTANOL B:3400
dl-2-BUTANOL B:3400
n-BUTANOL B:3350
sec-BUTANOL B:3400

tert-BUTANOL B:3450
BUTANOL-2 B:3400
1-BUTANOL, 3-METHOXY-, ACETATE M:1450
1-BUTANOL, 3-METHOXYACETATE M:1450
BUTANOL TERTIAIRE (French) B:3450
BUTANONE 2 (French) M:4100
BUTANONE M:4100
2-BUTANONE M:4100
2-BUTANONE PEROXIDE B:3050
BUTANOX M50 B:3050
BUTANOYL CHLORIDE B:5000
3-BUTEN-2-ONE M:5500
1-BUTEN-3-OL, 3-METHYL M:3200
2-BUTENAL C:5400
trans-2-BUTENAL C:5400
BUTENE B:3850
1-BUTENE B:3850
n-BUTENE B:3850
2-BUTENE-1,4-DIOL B:3000
cis-2-BUTENE-1,4-DIOL B:3000
1-BUTENE OXIDE B:4050
BUTENE RESINS P:2150
BUTENEDIOIC ACID, (E)- F:1400
BUTENEDIOIC ACID, (Z)- M:0300
2-BUTENEDIOIC ACID (E) F:1400
(E)-BUTENEDIOIC ACID F:1400
(Z)BUTENEDIOIC ACID M:0300
cis-BUTENEDIOIC ACID M:0300
trans-BUTENEDIOIC ACID F:1400
cis-BUTENEDIOIC ANHYDRIDE M:0350
1,4-BUTENEDIOL B:3000
BUTENONE M:5500
2-BUTENONE M:5500
terc-BUTILAMINA (Italian, Spanish) B:2900
1,3-BUTILENGLICOL (Spanish) B:3900
BUTILENO (Spanish) B:3850
p-terc-BUTILFENOL (Spanish) B:4600
n-BUTILMERCAPTANO (Spanish) B:4400
BUTILTRICLOROSILANO (Spanish) B:4700
1,4-BUTINODIOL (Spanish) B:4750
n-BUTIRALDEHIDO (Spanish) B:4800
BUTIRATO de n-BUTILO (Spanish) B:3700
BUTIRATO de EDTILO (Spanish) E:2800
BUTIRONITRILO (Spanish) B:4950
BUTONIC ACID ETHYL ESTER E:2800
1-BUTOXY BUTANE D:1170
1-BUTOXY-2,3-EPOXYPROPANE B:4150
BUTOXYDIETHYLENE GLYCOL D:3300
BUTOXYDIGLYCOL D:3300
2-BUTOXYETHANOL E:4900
2-BUTOXYETHANOL ACETATE E:5000
2-(2-BUTOXYETHOXY)ETHANOL D:3300
2-(2-BUTOXYETHOXY)ETHANOL ACETATE D:3350
2-(2-BUTOXYETHOXY)ETHYL ACETATE D:3350
2-BUTOXYETHYL ACETATE E:5000
BUTOXYETHYL CELLOSOLVE ACETATE E:5000
BUTOXYL M:1450
BUTOXYPROPYL TRICHLOROPHENOXYACETATE T:2550
BUTRYIC ACID, 2-HYDROXY-4-METHYLTHIO- H:1900
BUTTER OF ANTIMONY A:4900
BUTTER OF ANTIMONY A:5100
BUTTER OF ARSENIC A:5250
BUTTER OF ARSENIC A:5450
BUTTER OF ZINC Z:0900
BUTTERCUP YELLOW Z:1000
BUTTERSAEURE (German) B:4900

BUTYL ACETATE B:3100
1-BUTYL ACETATE B:3100
2-BUTYL ACETATE B:3150
n-BUTYL ACETATE B:3100
sec-BUTYL ACETATE B:3150
tert-BUTYL ACETATE B:3200
BUTYL ACETATE, ISO- I:0300
BUTYLACETIC ACID H:1000
BUTYL ACRYLATE B:3250
n-BUTYL ACRYLATE B:3250
BUTYL ADIPATE D:3550
BUTYL ALCOHOL B:3350
2-BUTYL ALCOHOL B:3400
n-BUTYL ALCOHOL B:3350
sec-BUTYL ALCOHOL B:3400
tert-BUTYL ALCOHOL B:3450
sec-BUTYL ALCOHOL ACETATE B:3150
BUTYL ALDEHYDE B:4800
n-BUTYL ALDEHYDE B:4800
n-BUTYL α-METHYLACRYLATE B:4450
n-BUTYLAMIN (German) B:3500
BUTYLAMINE B:3500
MONO-n-BUTYLAMINE B:3500
n-BUTYLAMINE B:3500
sec-BUTYLAMINE B:3550
tert-BUTYLAMINE B:3600
BUTYLAMINE, TERTIARY B:3600
BUTYL BENZYL PHTHALATE B:3650
BUTYL BROMIDE B:2550
n-BUTYL BROMIDE B:2550
sec-BUTYL BROMIDE B:2600
n-BUTYL-1-BUTANAMINE D:1160
BUTYL BUTANOATE B:3700
n-BUTYL n-BUTANOATE B:3700
2-BUTYLBUTANOIC ACID E:6000
n-BUTYL n-BUTYRATE B:3700
BUTYL BUTYRATE B:3700
n-BUTYL CARBINOL A:4350
n-BUTYLCARBINYL CHLORIDE A:4400
BUTYL CARBITOL D:3300
BUTYL CARBITOL ACETATE D:3350
BUTYL CELLOSOLVE E:4900
BUTYL CELLOSOLVE ACETATE E:5000
BUTYL CHLORIDE B:3750
n-BUTYL CHLORIDE B:3750
n-BUTYL CHLOROCARBONATE B:3800
n-BUTYL CHLOROFORMATE B:3800
BUTYL, DECYL, CETYL, EICOSYL METHACRYLATE MIXTURE M:1250
BUTYL, DECYL, CETYL, EICOSYL 2-METHYL-2-PROPENOATE M:1250
BUTYL DIGLYME D:3050
BUTYLE (ACETATE de) (French) B:3100
BUTYLENE B:3850
1-BUTYLENE B:3850
α-BUTYLENE B:3850
γ-BUTYLENE I:0450
2-BUTYLENE DICHLORIDE D:2000
1,3-BUTYLENE GLYCOL B:3900
1,4-BUTYLENE GLYCOL B:3950
2,3-BUTYLENE GLYCOL B:4000
β-BUTYLENE GLYCOL B:3900
β-BUTYLENE GLYCOL B:3950
BUTYLENE GLYCOL (pseudo) B:4000
BUTYLENE HYDRATE B:3400
1-BUTYLENE OXIDE B:4050
1,2-BUTYLENE OXIDE B:4050
α-BUTYLENE OXIDE B:4050
1,2-BUTYLENE OXIDE, stabilized B:4050
1,4-BUTYLENEGLYCOL B:2950
BUTYLETHYLACETIC ACID E:6000
BUTYLETHYLAMINE E:2600
n-BUTYL ESTER OF ACETIC ACID B:3100

tert-BUTYL ESTER OF ACETIC ACID B:3200
BUTYL ETHANOATE B:3100
BUTYL ETHER D:1170
n-BUTYL ETHER D:1170
BUTYL ETHYL ACETALDEHYDE E:5900
BUTYL ETHYL KETONE E:2700
n-BUTYL ETHYL KETONE E:2700
BUTYL ETHYLENE H:1100
BUTYL FORMAL V:0100
BUTYL FORMATE B:4100
n-BUTYL FORMATE b:4100
n-BUTYL GLYCIDYL ETHER B:4150
BUTYL GLYCOL ACETATE E:5000
BUTYL HYDRIDE B:2900
tert-BUTYL HYDROPEROXIDE B:4200
BUTYL HYDROXIDE B:3350
tert-BUTYL HYDROXIDE B:3450
BUTYL-α-HYDROXYPROPIONATE B:4350
BUTYL 2,4-DISOPROPYL ESTER D:0150
BUTYL KETONE D:1180
BUTYL LACTATE B:4350
n-BUTYL LACTATE B:4350
n-BUTYL MERCAPTAN B:4400
BUTYL METHACRYLATE B:4450
BUTYL 2-METHACRYLATE B:4450
n-BUTYL METHACRYLATE B:4450
BUTYL METHANOATE B:4100
tert-BUTYL METHYL ETHER M:3300
BUTYL METHYL KETONE M:3350
n-BUTYL METHYL KETONE M:3350
BUTYL-2-METHYL-2-PROPENOATE B:4450
BUTYL OXITOL E:4900
BUTYLPHEN B:4600
4-tert-BUTYLPHENOL B:4600
p-tert-BUTYLPHENOL B:4600
BUTYL PHOSPHATE, TRI- T:2000
BUTYL PHTHALATE D:1250
n-BUTYL PHTHALATE D:1250
BUTYL PROPANOATE B:4550
n-BUTYL PROPIONATE B:4550
BUTYL 2-PROPENOATE B:3250
N-BUTYL 2-PROPENOATE B:3250
BUTYL 2,4,5-T T:2550
n-BUTYL THIOALCOHOL B:4400
BUTYL TITANATE T:0300
BUTYL TITANATE MONOMER T:0300
BUTYL TOLUENE B:4650
4-tert-BUTYLTOLUENE B:4650
p-tert-BUTYLTOLUENE B:4650
BUTYL 2,4,5-TRICHLOROPHENOXYACETATE T:2550
BUTYLTRICHLOROSILANE B:4700
n-BUTYLTRICHLOROSILANE B:4700
2-BUTYNE-1,4-DIOL B:4750
1,4-BUTYNEDIOL B:4750
BUTYRAL B:4800
BUTYRAL BUTYRIC ALDEHYDE B:4800
BUTYRALDEHYD (German) B:4800
BUTYRALDEHYDE B:4800
iso-BUTYRALDEHYDE B:4850
n-BUTYRALDEHYDE B:4800
BUTYRIC ACID B:4900
BUTYRIC ACID B:4800
n-BUTYRIC ACID B:4900
BUTYRIC ACID, BUTYL ESTER B:3700
BUTYRIC ACID, ETHYL ESTER E:2800
BUTYRIC ACID, METHYL ESTER M:3400
BUTYRIC ACID NITRILE B:4950
BUTYRIC ALDEHYDE B:4800
BUTYRIC ETHER E:2800
BUTYRONITRILE B:4950
BUTYRONITRILE, 4-CHLORO- C:2450
BUTYROYL CHLORIDE B:5000

BUTYRYL CHLORIDE B:5000
n-BUTYRYL CHLORIDE B:5000
BZCF B:1400
C-1297 L:0300
C-1 ALCOHOL M:1750
C-2 ALCOHOL E:1900
C-3 ALCOHOL P:3950
C-4 ALCOHOL B:3400
C-4 ALCOHOL B:3450
C-4 ALCOHOL I:0350
C-5 ALCOHOL A:4350
C-6 ALCOHOL H:1050
C-8 ACID O:0200
C-8 ALCOHOL O:0300
C-9 ALCOHOL N:2400
C-9 OLEFIN MIXTURE N:2450
C-10 ALCOHOL D:0700
C-11 ALCOHOL (UNDECYLIC) U:0200
C-12 ALCOHOL D:8000
C-13 ALCOHOL T:3050
C-56 H:0500
CAA C:5600
CABLE OIL O:3600
CACALOT L-50 L:2000
CACODILATO SODICO (Spanish) S:2700
CACODYLIC ACID C:0100
CADDY C:0250
CADMIUM ACETATE C:0150
CADMIUM (II) ACETATE C:0150
CADMIUM ACETATE DIHYDRATE C:0150
CADMIUM BROMIDE C:0200
CADMIUM BROMIDE TETRAHYDRATE C:0200
CADMIUM CHLORIDE C:0250
CADMIUM DIACETATE C:0150
CADMIUM DICHLORIDE C:0250
CADMIUM FLUOBORATE C:0300
CADMIUM FLUOROBORATE C:0300
CADMIUM FUME C:0400
CADMIUM MONOXIDE C:0400
CADMIUM(II) NITRATE, TETRAHYDRATE (1:2:4) C:0350
CADMIUM OXIDE C:0400
CADMIUM OXIDE FUME C:0400
CADMIUM SULFATE C:0450
CADMIUM SULPHATE C:0450
CADOX HDP C:6000
CADOX PS D:1600
CADOX TBH B:4200
CAIROX P:3200
CAJEPUTENE D:7150
CAKE ALUM A:1750
CAKE ALUMINUM A:1750
CAL HYPO C:1000
CAL PLUS C:0750
CALAMINE Z:0800
CALCIA C:1100
CALCICAT C:0500
CALCIO (Spanish) C:0500
CALCIUM C:0500
CALCIUM ABIETATE C:1300
CALCIUM ALKYLAROMATIC SULFONATE D:8200
CALCIUM ALKYLBENZENESULFONATE D:8200
CALCIUM ARSENATE C:0550
CALCIUM ARSENITE C:0600
CALCIUM BIPHOSPHATE C:1200
CALCIUM CARBIDE C:0650
CALCIUM CHLORATE C:0700
CALCIUM CHLORIDE C:0750
CALCIUM CHLORIDE HYDRATES C:0750

CALCIUM CHLOROHYDROCHLORITE C:1000
CALCIUM CHROMATE C:0800
CALCIUM CHROMATE(VI) C:0800
CALCIUM CHROMATE DIHYDRATE C:0800
CALCIUM CHROME YELLOW C:0800
CALCIUM CHROMIUM OXIDE C:0800
CALCIUM CYANIDE C:0850
CALCIUM CYANIDE MIXTURE C:0850
CALCIUM DICARBIDE C:0650
CALCIUM DIFLUORIDE C:0900
CALCIUM DIOXIDE C:1150
CALCIUM FLUORIDE C:0900
CALCIUM HYDRATE C:0900
CALCIUM HYDROXIDE C:0900
CALCIUM HYPOCHLORIDE C:1000
CALCIUM HYPOCHLORITE C:1000
CALCIUM LIMED WOOD ROSIN C:1300
CALCIUM METAL, CRYSTALINE C:0500
CALCIUM MONOCHROMATE C:0800
CALCIUM NITRATE C:1050
CALCIUM(II) NITRATE C:1050
CALCIUM NITRATE TETRAHYDRATE C:1050
CALCIUM ORTHOARSENATE C:0550
CALCIUM OXIDE C:1100
CALCIUM OXYCHLORIDE C:1000
CALCIUM PEROXIDE C:1150
CALCIUM PHOSPHATE C:1200
CALCIUM PHOSPHIDE C:1250
CALCIUM PYROPHOSPHATE C:1200
CALCIUM RESINATE C:1300
CALCIUM ROSIN C:1300
CALCIUM SALT of ARSENIC ACID C:0550
CALCIUM SUPEROXIDE C:1150
CALCIUM SUPERPHOSPHATE C:1200
CALMATHION M:0250
CALOCHLOR M:0600
CALOMEL M:1000
CALSOFT F-90 D:8300
CALTAC C:0750
CALX C:1100
2-CAMPHANONE C:1400
CAMPHECHLOR T:1950
CAMPHENE C:1350
CAMPHENE,OCTACHLORO- T:1950
CAMPHOCHLOR T:1950
CAMPHOCLOR T:1950
CAMPHOFENE HUILEUX T:1950
CAMPHOR, NATURAL C:1400
CAMPHOR OIL C:1400
CAMPHOR TAR N:0400
2-CAMPHORONE C:1400
CAMPILIT C:5700
CANFENO (Spanish) C:1350
CANDEX A:5750
CANE SUGAR S:6400
CANOGARD D:2650
CAPORIT C:1000
CAPRALDEHYDE D:0450
CAPRIC ACID D:0550
n-CAPRIC ACID D:0550
CAPRIC ALCOHOL D:0700
CAPRIC ALDEHYDE D:0450
CAPRINIC ACID D:0550
CAPRINIC ALCOHOL D:0700
CAPROALDEHYDE H:0750
n-CAPROIC ACID H:1000
CAPROLACTAM C:1450
ε-CAPROLACTAM C:1450
CAPROLACTAMA (Spanish) C:1450
CAPROLIN C:1550

CAPRONALDEHYDE H:0750
CAPRONIC ACID H:1000
CAPRONIC ALDEHYDE H:0750
CAPROYL ALCOHOL H:1050
n-CAPROYLALDEHYDE H:0750
CAPRYL ALCOHOL O:0300
CAPRYLALDEHYDE O:0500
CAPRYLENE O:0400
n-CAPRYLIC ACID O:0200
CAPRYLIC ALCOHOL O:0300
CAPRYLIC ALDEHYDE O:0500
CAPRYNIC ACID D:0550
CAPTAF C:1475
CAPTAF 85W C:1475
CAPTAN C:1475
CAPTANCAPTENEET 26,538 C:1475
CAPTANE C:1475
CAPTEX C:1475
CARADATE 30 D:7400
CARBACRYL A:0950
CARBAMALDEHYDE F:1300
CARBAMATO AMONICO (Spanish) A:2350
CARBAMIC ACID, AMMONIUM SALT A:2350
CARBAMIC ACID, METHYL-, 2,2-DIMETHYL-2,3-DIHYDROBENZOFURAN-7-YL ESTER C:1500
CARBAMIC ACID, MONOAMMONIUM SALT A:2350
CARBAMIC CHLORIDE, DIMETHYL- D:5050
CARBAMIDE U:0900
CARBAMIDE ACID U:0900
CARBAMIDE PEROXIDE U:1100
CARBAMIDE RESIN U:0900
CARBAMIMIDIC ACID U:0900
CARBAMINE C:1550
CARBAMOYL CHLORIDE, DIMETHYL- D:5050
CARBARILO (Spanish) C:1550
CARBARYL C:1550
CARBATOX C:1550
CARBATOX-60 C:1550
CARBATOX 75 C:1550
CARBAX D:2600
CARBETHOXY MALATHION M:0250
CARBETOVUR M:0250
CARBETOX M:0250
CARBIDE C:0650
CARBINAMINE M:1850
CARBINAMINE SOLUTION M:1900
CARBINOL M:1750
CARBITOL D:3400
CARBITOL D:3000
CARBITOL ACETATE D:3150
CARBITOL CELLOSOLVE D:3400
CARBITOL SOLVENT D:3400
CARBOBENZOXY CHLORIDE B:1400
CARBOFOS M:0250
CARBOFURAN C:1500
CARBOFURANO (Spanish) C:1500
CARBOLIC ACID P:1200
CARBOLSAURE (German) P:1200
2-CARBOMETHOXY-1-METHYLVINYLDIMETHYL PHOSPHATE P:1500
α-2-CARBOMETHOXY-1-METHYLVINYL DIMETHYL PHOSPHATE P:1500
1-CARBOMETHOXY-1-PROPEN-2-Y PHOSPHATE P:1500
CARBON, ACTIVATED C:2000
CARBON ACTIVO (Spanish) C:2000
CARBONATO AMONICO (Spanish) A:2400

CARBONATO BARICO (Spanish) B:0100
CARBONATO de DIETILO (Spanish) D:2950
CARBONATO de PLATA (Spanish) S:0700 S:0700
CARBONATO de TALIO (Spanish) T:1200
CARBONATO de ZINC (Spanish) Z:0800
CARBON BICHLORIDE T:0400
CARBON BISULFIDE C:1700
CARBON BISULPHIDE C:1700
CARBON CHLORIDE C:1850
CARBON DIFLUORIDE OXIDE C:1800
CARBON DIOXIDE C:1650
CARBON DISULFIDE C:1700
CARBON DISULPHIDE C:1700
CARBON FLUORIDE OXIDE C:1800
CARBON HEXACHLORIDE H:0550
CARBON MONOXIDE C:1750
CARBON NAPHTHA B:0550
CARBON NITRIDE C:5650
CARBON OIL B:0550
CARBON OXIDE C:1750
CARBON OXYCHLORIDE P:1550
CARBON OXYFLUORIDE C:1800
CARBON SULFIDE C:1700
CARBON TET C:1850
CARBON TETRACHLORIDE C:1850
CARBONA C:1850
CARBONE(OXYCHLORURE de) (French) P:1550
CARBONE(OXYDE de) (French) C:1750
CARBONE(SUFURE DE) (French) C:1700
CARBONIC ACID, AMMONIUM SALT A:2400
CARBONIC ACID, BARIUM SALT B:0100
CARBONIC ACID, DIAMMONIUM SALT A:2400
CARBONIC ACID, DIETHYL ESTER D:2950
CARBONIC ACID GAS C:1650
CARBONIC ACID, MONOAMMONIUM SALT A:2150
CARBONIC ACID, ZINC SALT Z:0800
CARBONIC ANHYDRIDE C:1650
CARBONIC DICHLORIDE P:1550
CARBONIC DIFLUORIDE C:1800
CARBONIC OXIDE C:1750
CARBONOCHLORIDIC ACID, BUTYL ESTER B:3800
CARBONOCHLORIDIC ACID, 2-PROPENYL ESTER A:1350
CARBONYETHANE P:3700
CARBONYL CHLORIDE P:1550
CARBONYL DIAMIDE U:0900
CARBONYL DIAMINE U:0900
CARBONYL DIAMINE PEROXIDE U:1100
CARBONYL DIFLUORIDE C:1800
CARBONYL FLUORIDE C:1800
CARBONYLCHLORID (German) P:1550
CARBONYLDIAMIDE U:0900
CARBOPHOS M:0250
CARBORAFFIN C:2000
CARBORAFINE C:2000
CARBOSIP 5G C:1500
CARBOXIDE C:0900
CARBOXYETHANE P:3700
CARBOXYLBENZENE B:0900
CARBURO CALCICO (Spanish) C:0650
CARENE C:1900
3-CARENE C:1900
δ-3-CARENO (Spanish) C:1900
CARFENE A:5800
CARNAUBA WAX W:0100
CAROLID AL D:7200
CARPETING MEDIUM A:5700

CARPOLIN C:1550
CARREL-DAKINS SOLUTION S:3900
CARTHAMUS TINCTORIUS OIL O:2000
CARWINATE 125 M D:7400
CARYOLYSIN B:2025
CASHEW NUTSHELL LIQUID O:3100
CASHEW NUTSHELL OIL O:3100
CASINGHEAD, GASOLINE G:0400
CASORON D:1350
CASTER OIL O:1200
CASTOR OIL O:1200
CASTOR OIL, HYDROGENATED O:1200
CATACOL (Spanish) C:1950
CATALYST 9915 B:1450
CATECHIN C:1950
CATECHOL C:1950
CAUSTIC ARSENIC CHLORIDE A:5450
CAUSTIC POTASH P:2950
CAUSTIC SODA S:3800
CAVI-TROL S:3400
CBC-906288 T:4650
CBD 90 T:2850
CCH C:1000
CCS 203 B:3350
CCS 301 B:3400
CD 68 C:2050
CECOLENE T:2350
CEKIURON D:7950
CEKUMETHION M:4900
CEKUSAN D:2650
CEKUSIL P:1400
CEKUZINA-T A:5750
CELANEX B:0600
CELANOL DOS 75 D:7050
CELLOIDIN C:3950
CELLON T:0350
CELLOSOLVE E:0800
CELLOSOLVE ACETATE E:0900
CELLOSOLVE SOLVENT E:0800
CELLUFLEX 179C T:2950
CELLULEX DOP D:5250
CELLULOSE NITRATE SOLUTION C:3950
CELLUPHOS 4 T:2000
CELMER P:1400
CELMIDE E:4000
CELON A E:3900
CELON ATH E:3900
CENTURY-1240 S:5900
CERESAN P:1400
CETYL SODIUM SULFATE H:0650
CETYLENE TETRACHLORIDE T:0350
CETYLTRIMETHYL AMMONIUM CHLORIDE H:0700
CFC-11 T:2400
CFC-12 D:1650
CFC-113 T:2900
CG P:1550
CG-1283 M:5650
CGA24705 M:5550
CHA C:6100
CHALOXYD B:3050
CHAMBER ACID S:6800
CHAMELEON MINERAL P:3200
CHARCOAL C:2000
CHARCOAL, ACTIVATED C:2000
CHARCOAL, SHELL C:2000
CHEELOX E:3900
CHELEN E:2900
CHEM-PHENE T:1950
CHEM-TOL P:0350
CHEM BAM N:0100
CHEMAID S:2700
CHEMATHION M:0250

CHEMCOLOX 340 E:3900
CHEMFORM M:1500
CHEMIFLUOR S:3400
CHEMOX PE D:6650
CHILE SALTPETER S:4300
CHINESE BEAN OIL O:2100
CHINESE RED M:0900
CHINESE TANNIN T:0250
CHINOLEINE Q:0100
CHINOLINE Q:0100
CHINON (German) B:1050
p-CHINON (German) B:1050
CHINONE B:1050
CHIPCO THIRAM 75 T:1450
CHIPCO TURF HERBICIDE "D" D:0100
CHLOPHEN P:2200
CHLOR-METHAN (German) M:3450
CHLOR-O-PIC C:2800
CHLOR (German) C:2100
CHLOR KIL C:2050
CHLORACETIC ACID M:5750
α-CHLORACETIC ACID M:5750
CHLORACETYL CHLORIDE C:2300
CHLORAL T:2100
CHLORALLYLENE A:1300
CHLORATE de CALCIUM (French) C:0700
CHLORATE de POTASSIUM (French) P:2700
CHLORATE OF POTASH P:2700
CHLORATE OF SODA S:2900
CHLORATE OF SODA S:2800
CHLORATE SALT OF SODIUM S:2800
CHLORAX S:2800
CHLORBENZEN C:2400
CHLORCYAN C:5750
CHLORDAN C:2050
γ-CHLORDAN C:2050
CHLORDANE C:2050
CHLORDECONE K:0100
CHLORE (French) C:2100
CHLORESENE B:0600
2-CHLORETHANOL E:3600
δ-CHLORETHANOL E:3600
CHLORETHYL E:2900
β-CHLORETHYL ALCOHOL E:3600
CHLOREX D:1800
CHLOREXTOL P:2200
CHLORIC ACID, CALCIUM SALT C:0700
CHLORIC ACID, POTASSIUM SALT P:2700
CHLORIC ACID, SODIUM SALT S:2800
CHLORID ANTIMONITY A:5100
CHLORIDE OF AMYL A:4400
CHLORIDE OF LIME C:1000
CHLORIDE OF PHOSPHORUS P:1950
CHLORIDUM E:2900
CHLORINATED BIPHENYLS P:2200
CHLORINATED CAMPHENE T:1950
CHLORINATED DIPHENYL P:2200
CHLORINATED DIPHENYLENE P:2200
CHLORINATED HYDROCHLORIC ETHER D:1700
CHLORINATED LIME C:1000
CHLORINDAN C:2050
CHLORINE C:2100
CHLORINE CYANIDE C:5750
CHLORINE FLUORIDE C:2150
CHLORINE MOLECULAR C:2100
CHLORINE TRIFLUORIDE C:2150
CHLORMETHINE B:2025
CHLOROACETALDEHYDE C:2200
2-CHLOROACETALDEHYDE C:2200
CHLOROACETALDEHYDE (40% solution) C:2200

CHLOROACETALDEHYDE MONOMER C:2200
CHLOROACETIC ACID M:5750
CHLOROACETIC ACID CHLORIDE C:2300
CHLOROACETIC ACID, ETHYL ESTER E:3000
CHLOROACETIC ACID, METHYL ESTER M:3500
CHLOROACETIC CHLORIDE C:2300
CHLOROACETO PHENONE C:2250
2-CHLOROACETOPHENONE C:2250
α-CHLOROACETOPHENONE C:2250
ω-CHLOROACETOPHENONE C:2250
CHLOROACETYL CHLORIDE C:2300
CHLOROAETHAN (German) E:2900
2-CHLOROALLYL CHLORIDE D:2450
γ-CHLORO ALLYL CHLORIDE D:2400
α-CHLOROALLYL CHLORIDE D:2400
4-CHLORO-1-AMINOBENZENE C:2350
p-CHLOROAMINOBENZENE C:2350
4-CHLORO-2-AMINOTOLUENE C:3200
5-CHLORO-2-AMINOTOLUENE C:3200
4-CHLOROANILINE C:2350
p-CHLOROANILINE C:2350
CHLOROBEN D:1450
CHLOROBENZAL B:0450
α-CHLOROBENZALDEHYDE B:1100
CHLOROBENZENE C:2400
4-CHLOROBENZENEAMINE C:2350
CHLOROBENZOL C:2400
p-CHLOROBENZOYL PEROXIDE D:1600
p,p'-CHLOROBENZOYL PEROXIDE D:1600
CHLORO 1,1-BIPHENYL P:2200
CHLORO BIPHENYLS P:2200
2-CHLORO-N,N-BIS(2-CHLOROETHYL)ETHANAMINE B:2025
2-CHLORO-1,3-BUTADIENE C:2850
CHLOROBUTADIENE C:2850
2-CHLOROBUTADIENE C:2850
1-CHLOROBUTANE B:3750
4-CHLOROBUTYRONITRILE C:2450
γ-CHLOROBUTYRONITRILE C:2450
CHLOROCAMPHENE T:1950
CHLOROCARBONIC ACID, METHYL ESTER M:3550
CHLOROCARBONIC ACID, n-BUTYL ESTER B:3800
3-CHLOROCHLORDENE H:0100
2-CHLORO-N-(2-CHLOROETHYL)-N-ETHYLETHANAMINE B:2025
CHLOROCHROMIC ANHYDRIDE C:3450
CHLOROCYAN C:5750
CHLOROCYANIDE C:5750
CHLOROCYANOGEN C:5750
CHLORODANE C:2050
CHLORODEN D:1450
2-CHLORO-1,3-DIENE C:2850
CHLORODIFLUOROMETHANE M:5800
CHLORODIPHENYL (54% CHLORINE) P:2200
3-CHLORO-1,2-EPOXYPROPANE C:2550
1-CHLORO-2,3-EPOXYPROPANE C:2550
1-CHLORO-2,3-EPOXYPROPANE E:0300
2-CHLORO-1-ETHANAL C:2200
CHLOROETHANAL C:2200
2-CHLOROETHANAL C:2200
CHLOROETHANE E:2900
CHLOROETHANE NU T:2250
CHLOROETHANOL E:3600
2-CHLOROETHANOL E:3600
CHLOROETHENE V:0800
(2-CHLOROETHENYL)ARSONOUS DICHLORIDE B:2025

2-CHLOROETHYL ALCOHOL E:3600
1-CHLORO-3-ETHYLAMINO-5-ISOPROPYLAMINO-2,4,6-TRIAZINE A:5750
1-CHLORO-3-ETHYLAMINO-5-ISOPROPYLAMINO-s-TRIAZINE A:5750
2-CHLORO-4-ETHYLAMINEISOPROPYLAMINE-s-TRIAZINE A:5750
2-CHLORO-4-ETHYLAMINO-6-ISOPROPYLAMINO-1,3,5-TRIAZINE A:5750
2-CHLORO-4-ETHYLAMINO-6-ISOPROPYLAMINO-s-TRIAZINE A:5750
CHLOROETHYL ETHER D:1800
CHLOROETHYLENE V:0800
6-CHLORO-N-ETHYL-N'-(1-METHYLETHYL)-1,3,5-TRIAZINE-2,4-DIAMINE A:5750
1-CHLORO-2-(-CHLOROETHYLTHIO)ETHANE B:2025
CHLOROFORM C:2500
CHLOROFORME (French) C:2500
CHLOROFORMIC ACID, BENZYL ESTER B:1400
CHLOROFORMIC ACID DIMETHYLAMIDE D:5050
CHLOROFORMIC ACID, ETHYL ESTER E:3100
CHLOROFORMIC ACID, METHYL ESTER M:3550
CHLOROFORMIC ACID, n-BUTYL ESTER B:3800
CHLOROFORMYL CHLORIDE P:1550
CHLOROHYDRIC ACID H:1450
CHLOROHYDRINS C:2550
1-CHLORO-2-HYDROXYBENZENE C:27003
2-CHLORO-1-HYDROXYBENZENE C:2700
4-CHLORO-1-HYDROXYBENZENE C:2750
3-CHLORO-7-HYDROXY-4-METHYL-COUMARIN-o,o-DIETHYLPHOSPHOROTHIONATE C:5000
3-CHLORO-7-HYDROXY-4-METHYL-COUMARIN o-ESTER with o,o-DIETHYL BICYCLO-(2.2.1)HEPTANE C:1350
3-CHLORO-7-HYDROXY-4-METHYL-COUMARIN 3-CHLORO-ALLYL CHLORIDE D:2400
γ-CHLOROISOBUTYLENE M:1350
CHLOROMETHANE M:3450
1-CHLORO-2-METHYL- C:3100
4-CHLORO-2-METHYLANILINE C:3200
1-CHLORO-3-METHYLBENZENE C:3050
2-CHLORO-1-METHYLBENZENE C:3100
3-CHLORO-1-METHYLBENZENE C:3050
4-CHLORO-1-METHYLBENZENE C:3150
CHLOROMETHYLBENZENE B:1350
CHLOROMETHYL METHYL ETHER C:2600
CHLOROMETHYL OXIRANE E:0300
3-CHLORO-4-METHYL-7-HYDROXYCOUMARIN DIETHYL THIOPHOSPHORIC ACID ESTER C:5000
3-CHLORO-4-METHYL-7-COUMARINYLDIETHYLPHOSPHOROTHIOATE C:5000
O-3-CHLORO-4-METHYL-7-COUMARINYL o,o-DIETHYLPHOSPHOROTHIOATE C:5000
O-(3-CHLORO-4-METHYL-2-oxo-(2H)-1-BENZOPYRAN-7-YL)PHOSPHOROTHIOATE C:5000
CHLOROMETHYL PHENYL KETONE C:2250
3-CHLORO-6-METHYLANILINE C:3200
1-CHLORO-4-METHYLBENZENE C:3150
3-CHLORO-2-METHYLPROPENE M:1350
3-CHLORO-4-METHYLUMBELLIFERONEO-ESTER with o,o-DIETHYL PHOSPHOROTHIOATE C:5000
CHLOROMETHYLOXIRANE C:2550
CHLOROMETHYLOXIRANE E:0300
1-CHLORO-2-NITROBENZENE C:2650
2-CHLORO-1-NITROBENZENE C:2650
2-CHLORONITROBENZENE C:2650
1-CHLORO-2(3)-NITROBENZENE C:2650
o-CHLORONITROBENZENE C:2650
CHLORO-o-NITROBENZENE C:2650
CHLORONITROBENZENES C:2650
CHLOROPHEN P:0350
CHLOROPHENATE C:2750
2-CHLOROPHENOL C:2700
4-CHLOROPHENOL C:2750
o-CHLOROPHENOL C:2700
p-CHLOROPHENOL C:2750
CHLOROPHENOTHAN D:0300
CHLOROPHENOTHANE D:0300
CHLOROPHENOTOXUM D:0300
p-CHLOROPHENYL CHLORIDE D:1550
DI-(p-CHLOROPHENYL)TRICHLOROMETHYLCARBINOL D:2600
4-CHLOROPHENYLAMINE C:2350
CHLOROPHENYLMETHANE B:1350
CHLOROPHOS T:2050
CHLOROPICRIN C:2800
CHLOROPICRINE (French) C:2800
CHLOROPRENE C:2850
3-CHLOROPRENE A:1300
β-CHLOROPRENE C:2850
β-CHLOROPRENE C:2850
1-CHLOROPROPANE P:4100
3-CHLOROPROPANOIC ACID C:2950
3-CHLOROPROPENE A:1300
3-CHLORO-1-PROPENE A:1300
3-CHLOROPROPENE-1 A:1300
1-CHLORO PROPENE-2 A:1300
1-CHLORO-2-PROPENE A:1300
3-CHLOROPROPENYL CHLORIDE D:2400
2-CHLOROPROPIONIC ACID C:2900
3-CHLOROPROPIONIC ACID C:2950
α-CHLOROPROPIONIC ACID C:2900
β-CHLOROPROPIONIC ACID C:2950
2-CHLORO-4-(2-PROPYLAMINO)-6-ETHYLAMINO-s-TRIAZINE A:5750
3-CHLOROPROPYLENE A:1300
γ-CHLOROPROPYLENE OXIDE E:0300
3-CHLORO-1,2-PROPYLENE OXIDE E:0300
2-CHLOROPROPYLENE OXIDE E:0300
2-CHLOROPROPYLENE OXIDE C:2550
3-CHLOROPROPYLENE OXIDE E:0300
γ-CHLOROPROPYLENE OXIDE E:0300
CHLOROS S:3900
CHLOROSULFONIC ACID C:3000
CHLOROSULFURIC ACID C:3000
2-CHLORO-1,1,2,2-TETRAFLUOROETHANE M:5850
CHLOROTETRAFLUOROETHANE M:5850
CHLOROTHENE T:2250
2-CHLOROTOLUENE C:3100
3-CHLOROTOLUENE C:3050
4-CHLOROTOLUENE C:3150
α-CHLOROTOLUENE B:1350
m-CHLOROTOLUENE C:3050
o-CHLOROTOLUENE C:3100
ω-CHLOROTOLUENE B:1350
p-CHLOROTOLUENE C:3150
CHLOROTOLUIDINE C:3200
4-CHLORO-o-TOLUIDINE C:3200
5-CHLORO-o-TOLUIDINE C:3200
CHLOROTRIFLUORIDE C:2150
CHLOROTRIFLUOROETHYLENE T:3750
CHLOROTRIFLUOROMETHANE M:5900
CHLOROTRIMETHYL SILANE T:4150
CHLOROTRIMETHYLSILANE T:4150
CHLOROTRIMETHYLSILICANE T:4150
CHLOROVINYLARSINE DICHLORIDE B:2025
β-CHLOROVINYLBICHLOROARSINE B:2025
(2-CHLOROVINYL)DICHLOROARSINE B:2025
2-CHLOROVINYLDICHLOROARSINE B:2025
CHLOROXONE D:0100
1-CHLORPENTANE A:4400
o-CHLORPHENOL (German) C:2700
CHLORPICRINA (Spanish) C:2800
CHLORPIKRIN (German) C:2800
3-CHLORPROPEN (German) A:1300
CHLORPYRIFOS D:8900
CHLORSAURE (GERMAN) S:2800
CHLRTHEPIN E:0100
α-CHLORTOLUOL (German) B:1350
CHLORURE d'ALUMINUM (French) A:1450
CHLORURE de BENZYLE (French) B:1350
CHLORURE de BENZYLIDENE (French) B:0450
CHLORURE de BUTYLE (French) B:3750
CHLORURE de CHLORACETYLE (French) C:2300
CHLORURE de CYANOGENE (French) C:5750
CHLORURE d'ETHYLE (French) E:2900
CHLORURE d'ETHYLIDENE (French) D:1700
CHLORURE MERCURIQUE (French) M:0600
CHLORURE de METHYLE (French) M:3450
CHLORURE de VINYLE (French) V:0800
CHLORURE de VINYLIDENE (French) V:1200
CHLORURE de ZINC (French) Z:0900
CHLORWASSERSTOFF (German) H:1450
CHLORYL E:2900
CHLORYL ANESTHETIC E:2900
CHLORYLEA T:2350
CHOLAXINE S:5700
CHORYLEN T:2350
CHP C:5500
CHROMAR X:0300
CHROMATE OF POTASSIUM P:2750
CHROMATE OF SODA S:3000
CHROMIC ACETATE C:3250
CHROMIC(III) ACETATE C:3250
CHROMIC ACETATE(III) C:3250
CHROMIC ACID C:3300
CHROMIC(VI) ACID C:3300
CHROMIC ACID, CALCIUM SALT C:0800
CHROMIC ACID, DILITHIUM SALT L:2700
CHROMIC ACID, DIPOTASSIUM SALT P:2900
CHROMIC ACID, DIPOTASSIUM SALT P:2750
CHROMIC ACID, STRONTIUM SALT S:6000
CHROMIC ACID, ZINC SALT Z:1000
CHROMIC ACID, DISODIUM SALT S:3300
CHROMIC ANHYDRIDE C:3300
CHROMIC OXIDE C:3300
CHROMIC OXYCHLORIDE C:3450
CHROMIC SULFATE C:3350
CHROMIC SULPHATE C:3350
CHROMIC TRIOXIDE C:3300
CHROMIUM ACETATE C:3250

CHROMIUM(III) ACETATE C:3250
CHROMIUM CHLORIDE OXIDE C:3450
CHROMIUM DICHLORIDE C:3400
CHROMIUM DICHLORIDE DIOXIDE C:3450
CHROMIUM DIOXIDE DICHLORIDE C:3450
CHROMIUM DIOXYCHLORIDE C:3450
CHROMIUM DISODIUM OXIDE S:3000
CHROMIUM LITHIUM OXIDE L:2700
CHROMIUM OXIDE C:3300
CHROMIUM(VI) OXIDE C:3300
CHROMIUM(VI) OXIDE C:3300
CHROMIUM OXYCHLORIDE C:3450
CHROMIUM SODIUM OXIDE S:3300
CHROMIUM SODIUM OXIDE S:3000
CHROMIUM SULFATE C:3350
CHROMIUM(III) SULFATE C:3350
CHROMIUM SULPHATE C:3350
CHROMIUM TRIACETATE C:3250
CHROMIUM TRIOXIDE C:3300
CHROMIUM TRIOXIDE, ANHYDROUS C:3300
CHROMIUM ZINC OXIDE Z:1000
CHROMOCHROMIC ANHYDRIDE C:3450
CHROMOUS CHLORIDE C:3400
CHROMSAUREANHYDRID (German) C:3300
CHROMYL CHLORIDE C:3450
C.I. AZOIC COUPLING COMPONENT-107 T:1900
C.I. AZOIC DIAZO B:0850
C.I. AZOIC DIAZO COMPONENT 112 B:0850
C.I. AZOIC DIAZO COMPONENT 113 T:1850
C.I. AZOIC DIAZO COMPONENT 114 N:0500
C.I. DEVELOPER 4 R:0100
C.I. OXIDATION BASE T:1650
C.I. OXIDATION BASE 26 C:1950
C.I. OXIDATION BASE 31 R:0100
C.I. OXIDATION BASE 32 P:4850
C.I. PIGMENT WHITE 3 L:1500
C.I. PIGMENT WHITE 10 B:0100
C.I. PIGMENT YELLOW 36 Z:1000
C.I. PIGMENT YELLOW 33 C:0800
C.I. PIGMENT YELLOW 32 S:6000
C.I. PIGMENT YELLOW 46 L:2300
C.I. ZINC YELLOW X-883 Z:1000
CIANHIDRINA ETILENICA (Spanish) E:3700
CIANOGENO (Spanish) C:5650
CIANURINA M:0650
CIANURO BARICO (Spanish) B:0200
CIANURO de COBRE (Spanish) C:4300
CIANURO CALCICO (Spanish) C:0850
CIANURO de ETILO (Spanish) P:3800
CIANURO de NIQUEL (Spanish) N:0900
CIANURO SODICO (Spanish) S:3100
CIANURO VINILICO (Spanish) A:0950
CIANURO de ZINC (Spanish) Z:1100
CICHLORODIOXO CHROMIUM C:3450
CICLOHEPTANO (Spanish) C:5800
CICLOHEXANO (Spanish) C:5850
CICLOHEXANOL (Spanish) C:5900
CICLOHEXANONA (Spanish) C:5950
CICLOHEXENILTRICLOROSILANO (Spanish) C:6050
n-CICLOHEXILETILAMINA (Spanish) E:3325
CICLOPENTANO (Spanish) C:6200
CICLOPROPANO (Spanish) C:6300
CICN C:5750
CIDEX G:0600
CIMENO (Spanish) C:6350
CIMEXAN M:0250
CINENE D:7150
CINERIN I P:4750
CINERIN II P:4750
CINNAMENE S:6300

CINNAMENOL S:6300
CINNAMOL S:6300
CIRCOSOLV T:2350
cis-BUTENEDIOIC ANHYDRIDE M:0400
cis-1,4-DICHLORO-2-BUTENE D:2000
cis-PENTADIENE-1,3 P:0450
cis-PHOSDRIN P:1500
CITOX D:0300
CITRATO AMONICO DIBASICO (Spanish) A:2550
CITRATO FERRICO AMONICO (Spanish) F:0100
CLORATO BARICO (Spanish) B:0150
CLORATO CALCICO (Spanish) C:0700
CLORATO POTASICO (Spanish) P:2700
CLORATO SODICO (Spanish) S:2800
CITRIC ACID C:3500
CITRIC ACID, AMMONIUM SALT A:2550
CITRIC ACID, DIAMMONIUM SALT A:2550
CITRON YELLOW Z:1000
CK C:5750
CLAIRSIT P:1050
CLARIFIED OILS (PETROLEUM), CATALYTIC CRACKED O:0700
CLAUDELITE A:5500
CLAUDETITE A:5500
CLEANING SOLVENT N:0300
CLESTOL D:7050
CLOFENOTANE D:0300
CLOPHEN P:2200
CLORAMIN B:2025
CLOR CHEM T-590 T:1950
CLORDAN (Spanish) C:2050
CLORHIDRINA ETILENICA (Spanish) E:3600
CLOROACETALDEHIDO (Spanish) C:2200
CLOROACETATO de ETILO (Spanish) E:3000
p-CLOROANILINA (Spanish) C:2350
CLOROBEN D:1450
CLOROBENCENO (Spanish) C:2400
p-CLOROFENOL (Spanish) C:2750
CLOROFORMATO de ALILO (Spanish) A:1350
CLOROFORMIATO de BENCILO (Spanish) B:1400
CLOROFORMIATO de ETILO (Spanish) E:3100
CLOROFORMO (Spanish) C:2500
CLOROHYDRINA (Spanish) C:2550
CLOROMETIL METIL ETER (Spanish) C:2600
o-CLORONITROBENCENO (Spanish) C:2650
β-CLOROPRENO (Spanish) C:2850
CLOROTIOFORMATO de ETILO (Spanish) E:3200
m-CLOROTOLUENO (Spanish) C:3050
o-CLOROTOLUENO (Spanish) C:3100
p-CLOROTOLUENO (Spanish) C:3150
CLORO-o-TOLUIDINA (Spanish) C:3200
CLOROX S:3900
CLORURO de ACETILO (Spanish) A:0600
CLORURO de ALILO (Spanish) A:1300
CLORURO ALUMINICO ANHIDRO (Spanish) A:1450
CLORURO de n-AMILO (Spanish) REC. A:4400
CLORURO de ANISOILO (Spanish) A:4800
CLORURO de AZUFRE (Spanish) S:7000
CLORURO BENCENOFOSFOROSO (Spanish) B:0650
CLORURO de BENCILO (Spanish) B:1350
CLORURO de BENCILTRIMETILAMONIO (Spanish) B:1550
CLORURO de BENZAL (Spanish) B:0450

CLORURO de BENZOILO (Spanish) B:1100
CLORURO de BERILO (Spanish) B:1650
CLORURO de BISMUTO (Spanish) B:1900
CLORURO de sec-BUTILO (Spanish) B:3750
CLORURO de BUTIRILO (Spanish) B:5000
CLORURO de CADMIO (Spanish) C:0250
CLORURO de CIANOGENO (Spanish) C:5750
CLORURO de CLOROACETILO (Spanish) C:2300
CLORURO COBALTOSO (Spanish) C:3650
CLORURO de COBRE (Spanish) C:4250
CLORURO de CROMILO (Spanish) C:3450
CLORURO de DIMETILCARBAMOILO (Spanish) D:5050
CLORURO de ETILO (Spanish) E:2900
CLORURO FERRICO ANHIDRO (Spanish) F:0200
CLORURO FERROSO (Spanish) F:0600
CLORURO de FOSFOILO (Spanish) P:1950
CLORURO de HEXADECILTRIMETILAMONIO (Spanish) H:0700
CLORURO de NIQUEL (Spanish) N:0850
CLORURO de NITROSILO (Spanish) N:2100
CLORURO de PLOMO (Spanish) L:0800
CLORURO de PROPILO (Spanish) P:4100
CLORURO de VINILDENO (Spanish) V:1200
CLORURO de VINILO (Spanish) V:0800
CLORURO de ZINC (Spanish) Z:0900
CLYCLOHEXADEINEDIONE B:1050
CLYSAR P:2400
CMDP P:1500
CMME C:2600
CN (weaponized) C:2250
CNCI C:5750
CO C:1750
CO-12 D:8000
CO-12 L:2000
CO-1214 L:2000
CO-OP HEXA H:0400
CO-RAL C:5000
COAL NAPHTHA B:0550
COAL NAPHTHA, PHENYL HYDRIDE B:0550
COAL OIL O:2400
COAL OIL O:0800
COAL OIL K:0150
COAL TAR CREOSOTE C:5050
COAL TAR LIGHT OIL N:0350
COAL TAR NAPHTHA B:0550, N:0350
COAL TAR OIL O:3200; N:0350
COAL TAR OIL C:5050
COALITE NTP T:4700
COBALT ACETATE C:3550
COBALT(2+) ACETATE C:3550
COBALT(II) ACETATE C:3550
COBALT(II) CHLORIDE C:3650
COBALT ACETATE TETRAHYDRATE C:3550
COBALT AMINO SULFONATE C:3850
COBALT BROMIDE C:3600
COBALT(II) BROMIDE C:3600
COBALT CHLORIDE C:3650
COBALT(2+) CHLORIDE C:3650
COBALT DIACETATE TETRAHYDRATE C:3550
COBALT DIBROMIDE C:3600
COBALT DIFLUORIDE C:3700
COBALT DIFORMATE C:3750
COBALT FLUORIDE C:3700
COBALT(II) FLUORIDE C:3700
COBALT FORMATE C:3750
COBALT MURIATE C:3650

COBALT NITRATE C:3800
COBALT(2+) NITRATE C:3800
COBALT(II) NITRATE C:3800
COBALT SULFAMATE C:3850
COBALT SULFATE C:3900
COBALT(2+) SULFATE C:3900
COBALT(II) SULFATE C:3900
COBALTOUS ACETATE TETRAHYDRATE C:3550
COBALTOUS BROMIDE C:3600
COBALTOUS CHLORIDE C:3650
COBALTOUS CHLORIDE DIHYDRATE C:3650
COBALTOUS CHLORIDE HEXAHYDRATE C:3650
COBALTOUS FLUORIDE C:3700
COBALTOUS FORMATE C:3750
COBALTOUS NITRATE C:3800
COBALTOUS NITRATE, HEXAHYDRATE C:3800
COBALTOUS SULFAMATE C:3850
COBALTOUS SULFATE HEPTAHYDRATE C:3900
COCOCO C-50 D:8300
COCONUT BUTTER O:1300
COCONUT OIL O:1300
COCONUT OIL, CRUDE O:1300
COCONUT OIL, REFINED O:1300
COCURE 26 P:1400
CODAL M:5550
CODECHINE B:0600
CODOIL O:4200
COLACE D:7050
COLLIDINE ALDEHYDECOLLIDINE M:4150
COLLODION C:3950
COLLODION COTTON C:3950
COLLOIDAL-S S:6600
COLLOIDAL ARSENIC A:5250
COLLOIDAL MERCURY M:1100
COLLOIDAL SULFUR S:6600
COLLOKIT S:6600
COLODION (Spanish) C:3950
COLOGNE SPIRIT E:1900
COLOGNE SPIRITS E:1900
COLONIAL SPIRIT M:1750
COLOR-SET T:2650
COLSUL S:6600
COLUMBIAN SPIRITS M:1750
COMBUSTIBLE de REACTOR, JP-4 (Spanish) Z:0900
COMBUSTION IMPROVER C-12 M:3800
COMBUSTION IMPROVER-2 M:3800
COMITE P:3500
COMMERCIAL 40% T:0800
COMMON VERDIGRIS C:4750
COMPLEMIX D:7050
COMPLEXON II E:3900
COMPONENT 112 B:0850
COMPOUND 118 A:1100
COMPOUND 269 E:0200
COMPOUND 497 D:2700
COMPOUND 604 D:1400
COMPOUND 1080 S:3500
COMPOUND 2046 P:1500
COMPOUND 3422 P:0200
COMPOUND 3956 T:1950
COMPOUND 4049 M:0250
COMPOUND 7744 C:1550
COMPOUND B DICAMBRA D:1300
COMPOUND G-11 H:0600
COMPRESSED PETROLEUM GAS L:2200
CONCO SULFATE WA D:8700
CONDENSED PHOSPHORIC ACID P:2350

CONDY'S CRYSTALS P:3200
CONFECTIONER'S SUGAR S:6400
CONOCO SA 597 D:8150
CONSTONATE D:7050
CONTRA CREME CYANACIDE P:1400
COPPER ACETATE C:4000
COPPER(2+) ACETATE C:4000
COPPER(II) ACETATE C:4000
COPPER ACETOARSENITE C:4050
COPPER AMINOSULFATE C:4850
COPPER AMMONIUM SULFATE C:4850
COPPER ARSENITE C:4100
COPPERAS F:0750
COPPER BOROFLUORIDE SOLUTION C:4350
COPPER BROMIDE C:4150
COPPER BROMIDE (OUS) C:4200
COPPER CHLORIDE C:4250
COPPER(2+) CHLORIDE C:4250
COPPER(II) CHLORIDE C:4250
COPPER CYANIDE C:4300
COPPER DIACETATE C:4000
COPPER(2+) DIACETATE C:4000
COPPER DINITRATE C:4650
COPPER FLUOROBORATE C:4350
COPPER(II) FLUOBORATE SOLUTION C:4350
COPPER FORMATE C:4400
COPPER GLYCINATE C:4450
COPPER IODIDE C:4500
COPPER LACTATE C:4550
COPPER MONOBROMIDE C:4200
COPPER MONOSULFATE C:4800
COPPER NAPHTHENATE C:4600
COPPER NITRATE C:4650
COPPER (2+) NITRATE C:4650
COPPER(II) NITRATE C:4650
COPPER ORTHOARSENITE C:4100
COPPER OXALATE C:4700
COPPER SUBACETATE C:4750
COPPER SULFATE C:4800
COPPER(2+) SULFATE C:4800
COPPER(II) SULFATE C:4800
COPPER SULFATE, AMMONIATED C:4850
COPPER SULFATE PENTAHYDRATE C:4800
COPPER SULPHATE C:4800
COPPER TARTRATE C:4900
COPRA OIL O:1300
COPROL D:7050
CORFLEX 880 D:4500
CORODANE C:2050
COROSUL D S:6600
COROSUL S S:6600
COROTHION P:0200
CORROSIVE MERCURY CHLORIDE M:0600
CORTHION P:0200
CORTHIONE P:0200
CORTILAN-NEU C:2050
COSAN S:6600
COSAN 100 P:1400
COSAN 80 S:6600
COTNION METHYL A:5800
COTOFILM H:0600
COTORAN MULTI M:5550
COTTONSEED OIL O:1400
COTTONSEED OIL, REFINED O:1400
COTTONSEED OIL, UNHYDROGENATED O:1400
COUMAFOS C:5000
COUMAPHOS C:5000
CP BASIC SULFATE C:4800
CRAG SEVIN C:1550

CRANKCASE OIL O:3500
CRANKCASE OIL O:3800
CRAWHASPOL T:2350
CREDO S:3400
CREOSOTA de ALQUITRAN de HULLA (Spanish) C:5050
CREOSOTE, COAL TAR C:5050
CREOSOTE from COAL TAR C:5050
CREOSOTE OIL C:5050
CREOSOTE P1 C:5050
CREOSOTUM C:5050
2-CRESOL C:5100
3-CRESOL C:5100
m-CRESOL C:5100
o-CRESOL C:5100
p-CRESOL C:5100
CRESOL, EPOXYPROPYL ETHER C:5350
CRESOLS C:5100
CRESORCINOL DIISOCYANATE T:1700
CRESTOXO T:1950
TRI-o-CRESYL ESTER T:4450
CRESYL GLICIDE ETHER C:5350
CRESYL GLYCIDYL ETHER C:5350
o-CRESYL PHOSPHATE T:2950
o-CRESYL PHOSPHATE T:4450
TRI-o-CRESYL PHOSPHATE T:2950
CRESYLATE SPENT CAUSTIC SOLUTION C:5300
CRESYLIC ACID C:5100
m-CRESYLIC ACID C:5100
CRESYLIC ACIDS C:5100
CRESYLIC CREOSOTE C:5050
CRISALIN T:3800
CRISAPON D:0200
CRISAZINE A:5750
CRISFURAN C:1500
CRISTOXO 90 T:1950
CRISURON D:7950
CROMATO AMONICO (Spanish) A:2500
CROMATO CALCICO (Spanish) C:0800
CROMATO de LITIO (Spanish) L:2700
CROMATO SODICO (Spanish) S:3000
CROMATO de ZINC y POTASIO (Spanish) Z:2000
CROP RIDER D:0100
CROTENALDEHYDE C:5400
CROTILIN D:0100
CROTONALDEHIDO (Spanish) C:5400
CROTON OIL O:3300
CROTON TIGLIUM OIL O:3300
CROTONALDEHYDE C:5400
CROTONIC ALDEHYDE C:5400
CROTONOEL O:3300
CRUDE ARSENIC A:5500
CRUDE COAL TAR O:3200
CRUDE EPICHLOROHYDRIN C:2550
CRUDE OIL O:0800
CRUDE SOLVENT COAL TAR NAPHTHA N:0350
CRUISULFAN E:0100
CRYPTOGIL OL P:0350
CRYPTOHALITE A:3700
CRYSTAL AMMONIA A:2400
CRYSTALLIZED VERDIGRIS C:4000
CRYSTALS OF VENUS C:4000
CRYSTAMET S:5100
CRYSTEX S:6600
CRYSTHYON A:5800
CRYSTHION 2L A:5800
CRYSTOSOL O:3600
CTF C:2150
CTFE T:3750
CUBIC NITER S:4300

CUCUMBER DUST C:0550
CUDEX G:0600
CUMAFOS (Spanish) C:5000
CUMENE C:5450
ψ-CUMENE T:4100
CUMENE BOTTOMS D:4650
CUMENE HYDROPEROXIDE C:5500
CUMENO (Spanish) C:5450
α-CUMENYL HYDROPEROXIDE C:5500
CUMOL C:5450
CUPRAMMONIUM SULFATE C:4850
CUPRIC ACETATE C:4000
CUPRIC ACETATE, BASIC C:4750
CUPRIC ACETATE MONOHYDRATE C:4000
CUPRIC ACETOARSENITE C:4050
CUPRIC AMINE SULFATE C:4850
CUPRIC AMINO ACETATE C:4450
CUPRIC ARSENITE C:4100
CUPRIC BROMIDE, ANHYDROUS C:4150
CUPRIC CHLORIDE C:4250
CUPRIC CHLORIDE DIHYDRATE C:4250
CUPRIC DIACETATE C:4000
CUPRIC DIFORMATE C:4400
CUPRIC DINITRATE C:4650
CUPRIC FLUOBORATE SOLUTION C:4350
CUPRIC GREEN C:4100
CUPRIC LACTATE C:4550
CUPRIC NITRATE C:4650
CUPRIC NITRATE TRIHYDRATE C:4650
CUPRIC OXALATE C:4700
CUPRIC OXALATE HEMIHYDRATE C:4700
CUPRIC SULFATE ANHYDROUS C:4800
CUPRIC SULPHATE C:4800
CUPRIC TARTRATE C:4900
CUPRICIN C:4300
CUPRIETHYLENE DIAMINE HYDROXIDE SOLUTION C:5550
CUPRIETHYLENE DIAMINE SOLUTION C:5550
CUPRIETILENDIAMINA (Spanish) C:5550
CUPRINOL C:4600
CUPROUS CYANIDE C:4300
CUPROUS IODIDE C:4500
CURATERR C:1500
CURITHANE 103 M:1700
CUTTING OIL O:3600
CYANACETIC ACID C:5600
CYANHYDRINE d'ACETONE (French) A:0350
CYANIDE P:2800
CYANIDE of CALCIUM C:0850
CYANIDE of POTASSIUM P:2800
CYANIDE of SODIUM S:3100
CYANIDE of ZINC Z:1100
2-CYANO-1-PROPENE M:1300
2-CYANO-2-PROPONAL A:0350
CYANOACETIC ACID C:5600
CYANOACETONITRILE P:3450
CYANOBENZENE B:0950
CYANOBRIK S:3100
CYANOBROMIDE C:5700
CYANOETHANE P:3800
2-CYANOETHANOL E:3700
2-CYANOETHYL ALCOHOL E:3700
CYANOETHYLENE A:0950
CYANOGAS A-DUST C:0850
CYANOGAS G-FUMIGANT C:0850
CYANOGEN C:5650
CYANOGEN BROMIDE C:5700
CYANOGEN CHLORIDE C:5750
CYANOGEN MONOBROMIDE C:5700
CYANOGENE (French) C:5650

CYANOGRAN S:3100
CYANOMETHANE A:0400
CYANOPROPANE B:4950
2-CYANOPROPANE I:0550
2-CYANOPROPENE-1 M:1300
CYANURE de MERCURE (French) M:0650
CYANURE de METHYL (French) A:0400
CYANURE de POTASSIUM (French) P:2800
CYANURE de SODIUM (French) S:3100
CYANURE de VINYLE (French) A:0950
CYANURE de ZINC (French) Z:1100
CYANWASSERSTOFF (German) H:1500
CYAZIN A:5750
N-CYCLO-HEXYLDIMETHYLAMINE D:5100
CYCLODAN E:0100
CYCLOHEPTANE C:5800
2,5-CYCLOHEXADIENE-1,4-DIONE B:1050
1,4-CYCLOHEXADIENEDIONE B:1050
CYCLOHEXAMETHYENEIMINE H:0850
CYCLOHEXAN (German) C:5850
CYCLOHEXANAMINE C:6100
CYCLOHEXANE C:5850
CYCLOHEXANE, METHYL- M:3600
CYCLOHEXANOL C:5900
1-CYCLOHEXANOL C:5900
CYCLOHEXANONE C:5950
CYCLOHEXANONE PEROXIDE C:6000
CYCLOHEXANYL ACETATE C:6150
CYCLOHEXATRIENE B:0550
CYCLOHEXENE C:5900
CYCLOHEXENE, 1-METHYL-4-(1-METHYLETHENYL)- D:7150
CYCLOHEXENYLETHYLENE V:0900
CYCLOHEXENYL TRICHLOROSILANE C:6050
CYCLOHEXENYLTRICHLOROSILANE C:6050
CYCLOHEXYL ACETATE C:6150
CYCLOHEXYL ALCOHOL C:5900
CYCLOHEXYL ALCOHOL C:5950
CYCLOHEXYLAMINE C:6100
CYCLOHEXYLAMINE, N-ETHYL E:3325
CYCLOHEXYLAMINE, N,N-DIMETHYL D:5100
2-CYCLOHEXYL-4,6-DINITROPHENOL D:6600
CYCLOHEXYLDIMETHYLAMINE D:5100
CYCLOHEXYL ETHANE E:3300
N-CYCLOHEXYLETHYLAMINE E:3325
CYCLOHEXYL KETONE C:5950
CYCLOHEXYLMETHANE M:3600
CYCLON H:1500
CYCLONE B H:1500
1,3-CYCLOPENTADIENE, DIMER D:2675
CYCLOPENTANE C:6200
CYCLOPENTANE, METHYL- M:3850
CYCLOPENTENE C:6250
CYCLOPROPANE C:6300
CYCLOPROPANE, LIQUEFIED C:6300
CYCLOTETRAMETHYLENE OXIDE T:0900
CYMAG S:3100
CYMENE C:6350
p-CYMENE C:6350
CYMOL C:6350
CYPONA D:2650
CYTHION M:0250
CYURAM DS T:1450
D 1221 C:1500
D 50 D:0100
2,4-D D:0100
2,4-D (ESTERS) D:0150
2,4-D ACID D:0100

2,4-D, SALTS AND ESTERS D:0100
DACAMINE D:0100
DACAMINE 4T T:2500
DAF 68 D:3650
DAIFLON T:3750
DAIFLON S 3 T:2900
DAILON D:7950
DAIMMONIUM OXALATE A:3400
DAKINS SOLUTION S:3900
DALAPON D:0200
DALAPON 85 D:0200
DALF M:4900
DALMATION-INSECT POWDER P:4750
DALTOGEN T:3200
DANTHION P:0200
DAR-CHEM S:5900
DAWSON 100 M:3100
DAXAD-32S A:2000
DAZZEL D:1000
DBA D:1160
DBD A:5800
DBE E:4000
DBH B-0600
DBP D:1250
DCB D:2000
1,4-DCB D:2000
p-DCB D:1550
1,1-DCE V:1200
1,2-DCE D:1750
DCEE D:1800
2,2'-DCEE D:1800
DCMU D:7950
DCP C:1200
DCP D:2200
2,4-DCP D:2200
DCPD D:2675
DD MIXTURE D:2500
DDBSA D:8150
DDC D:5050
DDD D:0250
p,p'-DDD D:0250
D-D SOIL FUMIGANT D:2500
DDT D:0300
4,4' DDT D:0300
p,p'-DDT D:0300
DDVP D:2650
DE-FOL-ATE S:2800
D-GLUCITOL S:5700
DE KALIN D:0400
DEA D:2750
DEAD OIL C:5050
DEAE D:3500
DEANOL D:5175
DEBROUSSAILLANT 600 D:0100
DEBROUSSAILLANT CONCENTRE T:2500
DEBROUSSAILLANT SUPER CONCENTRE T:2500
DEC D:0400
DECABORANE D:0350
DECABORANE(14) D:0350
DECABORANO (Spanish) D:0350
DECACHLOROKETONE K:0100
DECACHLOROOCTAHYDRO-1,3,4-METHENO-2H-CYCLOBUTA(CD)-PENTALEN-2-ONE K:0100
DECACHLOROOCTAHYDRO-4,7-METHANOINDENEONE K:0100
DECACHLOROOCTAHYDRO-KEPONE-2-ONE K:0100
DECAHIDRONAFTALENO (Spanish) D:0400
DECAHYDRONAPHTHALENE (cis-/trans-) D:0400

DECALDEHIDO (Spanish) D:0450
DECALDEHYDE D:0450
DECALIN (cis-/trans-) D:0400
DECAMINE D:0100
DECAMINE 4T T:2500
DECANAL D:0450
1-DECANAL D:0450
DECANAL (mixed Isomers) D:0450
DECANAL DIMETHYL ACETAL D:0700
DECANE D:0500
n-DECANE D:0500
1-DECANECARBOXYLIC ACID U:100
n-DECANO (Spanish) D:0500
DECANOIC ACID D:0550
n-DECANOIC ACID D:0550
DECANOL D:0700
L-DECANOL D:0700
DECARBORON TETRADECAHYDRIDE D:0350
n-DECATYL ALCOHOL D:0700
DECENE D:0600
1-DECENE D:0600
α-DECENE D:0600
α-DECENO (Spanish) D:0600
DECHLORANE M:5650
DECHLORANE-4070 M:5650
DECIL ACRILATO (Spanish) D:0650
DECILBENCENO (Spanish) D:0750
n-DECOIC ACID D:0550
n-DECYL ACRYLATE D:0650
n-DECYL ALCOHOL D:0700
1-DECYL ALDEHYDE D:0450
n-DECYL ALDEHYDE D:0450
DECYLBENZENE D:0750
n-DECYLBENZENE D:0750
DECYLBENZENESULFONIC ACID A:1150
n-DECYLIC ACID D:0550
DECYLIC ALCOHOL D:0700
DED-WEED D:0100
DED-WEED D:0200
DED-WEED T:2650
DED-WEED BRUSH KILLER T:2500
DED-WEED LV-6 BRUSH KILL T:2500
DED-WEED LV-69 D:0100
DEDELO D:0300
DEEP LEMON YELLOW S:6000
DEFILIN D:7050
DEG D:3000
300-DEGREE OIL O:3700
D.E.H. 26 T:0700
DEHA D:7000
DEHP D:3650
DEHPA D:3600
DEHPA EXTRACT D:3600
DEHYDRITE M:0200
DEIQUAT D:7650
DEK D:3700
DELICIA A:1700
DELICIA GASTOXIN A:1700
DELTAN D:6200
DEMASORB D:6200
DEMAVET D:6200
DEMESO D:6200
4-(1,1-DIMETHYLETHYL) PHENOL B:4600
DEMETON D:0800
DEMETONA (Spanish) D:0800
DEMETON O+DEMETON S D:0800
DEMOX D:0800
DEMSODROX D:6200
DEN D:2800
DENAPON C:1550
DENOX D:0800
DENSINFLUAT T:2350

DEOBASE K:0150
DEODORIZED, WINTERIZED COTTONSEED OIL O:1400
DEOVAL D:0300
DEP D:3750
D.E.R. 332 B:2000
DERABAN D:2650
DERMADEX H:0600
DERMASORB D:6200
DERRIBANTE D:2650
DESMODUR-44 D:7400
DESMODUR T80 T:1700
DESOLET S:2800
DESORMONE D:0100
DETA D:3450
DETERGENT 66 D:8700
DETERGENT ALKYLATE No. 2 D:8100
DETERGENT ALKYLATES D:8100
DETERGENT HD-90 D:8300
DETIA A:1700
DETIA-EX-B A:1700
DETIA GAS EX-B A:1700
DETMOL-EXTRAKT B:0600
DETMOL MA M:0250
DETMOL MA 96% M:0250
DETOX D:0300
DETOX 25 B:0600
DETOXAN D:0300
DEVELOPER O R:0100
DEVELOPER P N:1450
DEVELOPER R R:0100
DEVELOPER RS R:0100
DEVIKOL D:2650
DEVIPON D:0200
DEVITHION M:4900
DEVORAN B:0600
DEVOTON M:1550
DEXOL STUMP REMOVER P:3050
DEXTRONE D:7650
DFA D:7250
DFP D:0850
DIACETATO de ETILENGLICOL (Spanish) E:4400
DIACETIC ETHER E:1700
DIACETON-ALCOHOL (Spanish) D:0900
DIACETONALKOHOL (German) D:0900
DIACETONE-ALCOOL (French) D:0900
DIACETONE ALCOHOL D:0900
DIACETONE,4-HYDROXY-4-METHYL-2-PENTANONE,2-METHYL-2-PENTANOL-4-ONE D:0900
DIACETOXYMERCURY M:0500
DIACETYL METHANE A:0500
DIACETYL PEROXIDE SOLUTION A:0700
DIAETHANOLAMIN (German) D:2750
1,1-DIAETHOXY-AETHAN (German) A:0100
o,o-DIAETHYL-o-(2-ISOPROPYL-4-METHYL-PYRIMIDIN-6-YL)MONOTHIOPHOSPHAT (German) D:1000
o,o-DIAETHYL-o-(3-CHLOR-4-METHYL-CUMARIN-7-YL)-MONOTHIOPHOSPHAT (German) C:5000
o,o-DIAETHYL-s-(2-AETHYLTHIO-AETHYL)-DITHIOPHOSPHAT (German) D:7700
o,o-DIAETHYL-s-(3-THIA-PENTYL)-DITHIOPHOSPHAT (German) D:7700
DIAETHYLACETAL (German) A:0100
DIAETHYLAETHER (German) E:5600
DIAETHYLAMINOAETHANOL (German) D:3500
DIAETHYLSULFAT (German) D:3800
DIAK-S H:1700

DIAKARMON S:5700
DIAKON M:4800
DIALQUILDITIOFOSFATO de ZINC (Spanish) Z:1200
DIALUMINUM SULFATE A:1750
DIALUMINUM SULFATE SOLUTION A:1800
DIALUMINUM TRISULFATE A:1750
DIAMIDE H:1250
DIAMINE H:1250
DIAMINE, HYDRAZINE BASE H:1250
1,2-DIAMINOAETHAN (German) E:3800
4,4'-DIAMINOBIPHENYL B:0850
p,p'-DIAMINOBIPHENYL B:0850
4,4'-DIAMINO-1,1'-BIPHENYL B:0850
2,2'-DIAMINODIETHYLAMINE D:3450
4,4'-DIAMINO-3,3'-DIMETHYLBIPHENYL T:1850
4,4'-DIAMINODIPHENYL B:0850
p-DIAMINODIPHENYL B:0850
DIAMINODITOLYL T:1850
1,2-DIAMINOETHANE, ANHYDROUS E:3800
1,2-DIAMINOETHANE COPPER COMPLEX C:5550
1,6-DIAMINOHEXANE H:0800
2,4-DIAMINOTOLUENE T:1650
2,4-DIAMINOTOLUENO (Spanish) T:1650
DIAMINO-2,2,4 TRIMETHYLHEXANE T:4200
1,11-DIAMINO-3,6,9 TRIAZAUNDECANE T:0700
DIAMMONIUM CARBONATE A:2400
DIAMMONIUM CHROMATE A:2500
DIAMMONIUM CITRATE A:2550
DIAMMONIUM HYDROGEN PHOSPHATE A:3600
DIAMMONIUM MOLYBDATE A:3100
DIAMMONIUM ORTHOPHOSPHATE A:3600
DIAMMONIUM PHOSPHATE A:3600
DIAMMONIUM SALT OF ZINC EDTA D:0950
DIAMMONIUM SULFATE A:3850
DIAMMONIUM SULFIDE A:3900
DIAMMONIUM THIOSULFATE A:4150
DIAMYL PHTHALATE D:0975
DI-n-AMYL PHTHALATE D:0975
DIANATE D:1300
p,p'-DIANILINE B:0850
DIANON D:1000
DIANTIMONY TRIOXIDE A:5200
DIAREX HF 77 S:6300
DIARSENIC PENTOXIDE A:5400
DIARSENIC TRIOXIDE A:5500
DIARSENIC TRISULFIDE A:5550
DIATER D:7950
DIATERR-FOS D:1000
1,4-DIAZACYCLOHEXANE P:2100
DIAZAJET D:1000
3,6-DIAZAOCTANE-1,8-DIAMINE T:3600
3,6-DIAZAOCTANEDIOIC ACID,3,6-BIS(CARBOXYMETHYL)- E:3900
DIAZATOL D:1000
DIAZIDE D:1000
DIAZINON D:1000
DIAZINONE D:1000
DIAZITOL D:1000
DIAZOL D:1000
DIBASIC LEAD ARSENATE L:0700
DIBENSOATO de DIPROPILENGLICOL (Spanish) D:7550
DIBENZO[b,e]PYRIDINE A:0750
DIBENZOL DIPROPYLENE GLYCOL ESTER D:7550

DIBENZOYL PEROXIDE D:1050
DIBENZYL ETHER D:1100
DIBK D:4250
DIBOVAN D:0300
DIBP D:4300
DIBROM N:0150
o-(1,2-DIBROM-2,2-DICHLOR-AETHYL)-o,o-DIMETHYL-PHOSPHAT (German)
1,2-DIBROMAETHAN (German) E:4000
DIBROMOETHANE E:4000
1,2-DIBROMO-2,2-DICHLOROETHYL DIMETHYL PHOSPHATE N:0150
DIBROMOMETANO (Spanish) D:1150
1,2-DIBROMOETANO (Spanish) E:4000
1,2-DIBROMOETHANE E:4000
α,β-DIBROMOETHANE E:4000
sym-DIBROMOETHANE E:4000
DIBROMOMETHANE D:1150
DIBROMUO de ETILENO (Spanish) E:4000
DIBROMURE d'ETHYLENE (French) E:4000
DI-n-BUTILAMINA (Spanish) D:1160
DIBUTILCETONA (Spanish) D:1180
DIBUTILFENOL (Spanish) D:1200
1,2-DIBUTOXYETHANE E:4500
2,2'-DIBUTOXYETHYL ETHER D:3050
DIBUTYL-1,2-BENZENEDICARBOXYLATE D:1250
DIBUTYL ADIPINATE D:3550
DIBUTYLAMINE D:1160
DI-n-BUTYLAMINE D:1160
n-DIBUTYLAMINE D:1160
DIBUTYL CARBITOL D:3050
DIBUTYL CELLOSOLVE E:4500
DIBUTYL ETHER D:1170
DI-n-BUTYL ETHER D:1170
n-DIBUTYL ETHER D:1170
DIBUTYL HEXANEDIOATE D:3550
DIBUTYL KETONE D:1180
DI-n-BUTYL KETONE D:1180
DIBUTYL OXIDE D:1170
DIBUTYL PHTHALATE D:1250
DI-n-BUTYL PHTHALATE D:1250
DIBUTYLPHENOL D:1200
DICALCIUM PHOSPHATE C:1200
DICAMBA D:1300
DICAPROATE T:3400
DICARBAM C:1550
DICARBOMETHOXY ZINC Z:0200
o-DICARBOXYBENZENE D:4450
DICARBURRETTED HYDROGEN E:3500
2,4-DI-tert-BUTYLPHENOL D:1200
2,6-DI-tert-BUTYLPHENOL D:1200
DICHLOBENIL D:1350
DICHLONE D:1400
1,1-DICHLORAETHAN (German) D:1700
1,2-DICHLOR-AETHEN (German) D:1750
DICHLORAMINE B:2025
o-DICHLORBENZENE D:1450
1,4-DICHLOR-BENZOL (German) D:1550
o-DICHLOR BENZOL D:1450
p-DICHLORBENZOL (German) D:1550
DICHLOREN (German) B:2025
DICHLORFENIDIM D:7950
DICHLORICIDE D:1450
DI-CHLORICIDE D:1550
DICHLORINE C:2100
DI-(4-CHLOROBENZOYL) PEROXIDE D:1600
DI-(p-CHLOROBENZOYL) PEROXIDE D:1600
DICHLORO(2-CHLOROVINYL)ARSINE B:2025
DICHLORODIETHYL SULFIDE B:2025

DI-(2-CHLOROETHYL) ETHER D:1800
(2,4-DICHLOR-PHENOXY)-ESSIG SAEURE (German) D:0100
3-(3,4-DICHLOR-PHENYL)-1,1-DIMETHYL-HARNSTOFF (German) D:7950
DICHLOR-s-TRIAZIN-2,4,6(1H,3H,5H)TRIONE POTASSIUM P:2850
1,4-DICHLORO-2-BUTENE D:2000
trans-1,4-DICHLORO-2-BUTENE D:2000
1,4-DICHLORO-2-BUTYLENE D:2000
1,1-DICHLORO-2,2-DICHLOROETHANE T:0350
β,β'-DICHLORODIETHYL-N-METHYLAMINE B:2025
2,2'-DICHLORODIETHYL SULFIDE B:2025
1,1-DICHLOROETHANE D:1700
DICHLORO-1,2-ETHYLENE (French) D:1750
1,1-DICHLORO-1-NITROETHANE D:1900
2,3-DICHLORO-1-PROPANE D:2450
DICHLORO-1,2-PROPANE D:2300
1,3-DICHLORO-1-PROPENE D:2400
2,3-DICHLORO-1-PROPENE D:2450
4,4'-DICHLORO-α-TRICHLOROMETHYLBENZHYDROL D:2600
DICHLORO-DIOXOCHROMIUM C:3450
3,6-DICHLORO-o-ANISIC ACID D:1300
DICHLORO-s-TRIAZINE-2,4,6(1H,3H,5H)-TRIONE POTASSIUM DERIV P:2850
1,3-DICHLORO-s-TRIAZINE-2,4,6(1H,3H,5H)TRIONE POTASSIUM SALT P:2850
DICHLOROACETIC ACID, METHYL ESTER M:3900
1,2-DICHLOROBENZENE D:1450
1,3-DICHLOROBENZENE D:1500
1,4-DICHLOROBENZENE D:1550
m-DICHLOROBENZENE D:1500
meta-DICHLOROBENZENE D:1500
o-DICHLOROBENZENE D:1450
p-DICHLOROBENZENE D:1550
DICHLOROBENZENE, ORTHO, LIQUID D:1450
DICHLOROBENZENE, PARA, SOLID D:1550
o-DICHLOROBENZOL D:1450
p-DICHLOROBENZOL D:1550
2,6-DICHLOROBENZONITRILE D:1350
p,p'-DICHLOROBENZOYL PEROXIDE D:1600
DICHLOROBUTENE D:2000
DICHLOROCHLORDENE C:2050
1,1-DICHLORO-2,2-bis(p-CHLORO-PHENYL) ETHANE D:0250
DICHLOROCIDE D:1550
DICHLORODIETHYL ETHER D:1800
2,2'-DICHLORODIETHYL ETHER D:1800
DICHLORODIETHYL OXIDE D:1800
DICHLORODIFLUOROMETHANE D:1650
DICHLORODIISOPROPYL ETHER D:1950
2,2'-DICHLORO-N-METHYLDIETHYLAMINE B:2025
DICHLORODIMETHYL SILANE D:5150
DICHLORODIOXOCHROMIUM C:3450
DICHLORODIPHENYLDICHLORO ETHANE D:0250
DICHLORODIPHENYLSILANE D:7300
DICHLORODIPHENYLSILICANE D:7300
DICHLORODIPHENYL TRICHLOROETHANE D:0300
DICHLORODIPHENYLTRICHLOROETHANE D:0300
4,4'-DICHLORO-DIPHENYLTRICHLOROETHANE D:0300

p,p'-DICHLORO-DIPHENYLTRICHLOROETHANE D:0300
1,1-DICHLOROETHANE D:1700
1,2-DICHLOROETHANE E:4100
asymmetrical-DICHLOROETHANE D:1700
asym-DICHLOROETHANE D:1700
1,1-DICHLOROETHENE V:1200
DICHLOROETHER D:1800
2,2'-DICHLOROETHYL ETHER D:1800
DICHLOROETHYLENE D:1750
unsym-DICHLOROETHYLENE V:1200
1,1-DICHLOROETHYLENE V:1200
1,2-DICHLOROETHYLENE D:1750
cis-1,2-DICHLOROETHYLENE D:1750
trans-1,2-DICHLOROETHYLENE D:1750
DI(2-CHLOROETHYL)METHYLAMINE B:2025
DICHLOROETHYLPHENYLSILANE E:7300
DICHLOROETHYLSILANE E:3400
DI-2-CHLOROETHYL SULFIDE B:2025
2,2'-DICHLOROETHYL SULFIDE B:2025
DICHLOROFLUOROMETHANE D:2100
1,6-DICHLOROHEXANE D:1850
DICHLOROISOCYANURIC ACID POTASSIUM SALT P:2850
DICHLOROISOPROPYL ETHER D:1950
2,2'-DICHLOROISOPROPYL ETHER D:1950
DICHLOROMETHANE D:2050
DICHLOROMETHYLETHANE D:1700
DICHLOROMETHYLSILANE M:3950
DICHROMIC ACID, DIPOTASSIUM SALT P:2900
DICHROMIUM SULFATE C:3350
DICHROMIUM SULPHATE C:3350
DICHROMIUM TRISULFATE C:3350
DICHROMIUM TRISULPHATE C:3350
DICHLOROMONOFLUOROMETHANE D:2100
DICHLORONITROETHANE D:1900
2,4-DICHLOROPHENOL D:2200
3-(3,4-DICHLOROPHENOL)-1,1-DIMETHYL UREA D:7950
(2,4-DICHLOROPHENOXY)ACETIC ACID, ISOPROPYL ESTER D:0150
DICHLOROPHENOXYACETIC ACID D:0100
2,4-DICHLOROPHENOXYACETATE D:0150
2,4-DICHLOROPHENOXYACETIC ACID, BUTOXYETHYL ESTER D:0150
2,4-DICHLOROPHENOXYACETIC ACID, SALTS AND ESTERS D:0100
3-(3,4-DICHLOROPHENYL)-1,1-DIMETHYLUREA D:7950
1-(3,4-DICHLOROPHENYL)-3,3-DIMETHYLUREE (French) D:7950
N'-(3,4-DICHLOROPHENYL)-N,N-DIMETHYLUREA D:7950
DICHLOROPHENYLARSINE P:1250
DICHLOROPHENYLPHOSPHINE B:0650
DICHLOROPHOS D:2650
DICHLOROPHOSPHORIC ACID, ETHYL ESTER E:7500
DICHLOROPROPANE D:2300
1,1-DICHLOROPROPANE D:2250
1,2-DICHLOROPROPANE D:2300
1,3-DICHLOROPROPANE D:2350
α,β-DICHLOROPROPANE D:2300
2,2-DICHLOROPROPANOIC ACID D:0200
DICHLOROPROPENE D:2400
1,3-DICHLOROPROPENE D:2400
2,3-DICHLOROPROPENE D:2450
1,3-DICHLOROPROPENE-1 D:2400
1,3-DICHLOROPROPENE AND 1,2-DICHLOROPROPANE MIXTURE D:2500

DICHLOROPROPENE-DICHLOROPROPANE MIXTURE D:2500
2,2-DICHLOROPROPIONIC ACID D:0200
α-DICHLOROPROPIONIC ACID D:0200
α,α-DICHLOROPROPIONIC ACID D:0200
1,3-DICHLOROPROPYLENE D:2400
2,3-DICHLOROPROPYLENE D:2450
α,γ-DICHLOROPROPYLENE D:2400
DICHLOROTETRAFLUOROETHANE D:2550
1,2-DICHLOROTETRAFLUOROETHANE D:2550
DICHLOROTHIOCARBONYL T:1400
α,α-DICHLOROTOLUENE B:0450
2,2'-DICHLOROTRIETHYLAMINE B:2025
2,2-DICHLOROVINYL DIMETHYL PHOSPHATE D:2650
2,2-DICHLOROVINYL o,o-DIMETHYL PHOSPHATE D:2650
2,4-DICHLORPHENOXYACETIC ACID D:0100
DICHLORPROPAN-DICHLORPROPEN GEMISCH (German) D:2500
DICHLORVOS D:2650
m-DICLOROBENCENO (Spanish) D:1450
o-DICLOROBENCENO (Spanish) D:1500
p-DICLOROBENCENO (Spanish) D:1550
1,1-DICLORO-2,2-BIS(p-ETILFENIL)ETANO (Spanish) D:0250
DICLOROBUTENO (Spanish) D:2000
DICLORODIFENILTRICLOROETANO (Spanish) D:0300
1,2-DICLOROETENO (Spanish) D:1750
1,1-DICLOROETANO (Spanish) D:1700
1,2-DICLOROETANO (Spanish) E:4100
DICLOROFENOL (Spanish) D:2200
DICLORODIFLUOMETANO (Spanish) D:1650
1,6-DICLOROHEXANO (Spanish) D:1850
DICLOROMETANO (Spanish) D:2050
DICLOROMONOFLUOMETANO (Spanish) D:2100
DICLONA (Spanish) D:1400
DICLORONITROETANO (Spanish) D:1900
DICICLOPENTADIENO (Spanish) D:2675
1,1-DICLOROPROPANO (Spanish) D:2250
1,2-DICLOROPROPANO (Spanish) D:2300
1,3-DICLOROPROPANO (Spanish) D:2350
1,3-DICLOROPROPENO (Spanish) D:2400
2,3-DICLOROPROPENO (Spanish) D:2450
DICLOROTETRAFLUOETANO (Spanish) D:2550
DICLORURO BENCENOFOSFOROSO (Spanish) B:0650
DICLORURO CROMOSO (Spanish) C:3400
DICLORURO de ETILALUMINIO (Spanish) E:2000
DICLORURO de ETILENO (Spanish) E:4100
DICLORURO de METILFOSFONICO (Spanish) M:5050
DICLORVOS (Spanish) D:2650
DICOFOL D:2600
DICOL D:3000
DICOPHANE D:0300
DICOPUR D:0100
DICOTOX D:0100
DICROMO AMONICO (Spanish) A:2600
DICROMATO SODICO (Spanish) S:3000
DICROMATO de ZINC (Spanish) Z:0500
DICYAN C:5650
DICYAN C:5650
1,4-DICYANOBUTANE A:1050
DICYANOGEN C:5650
DICYCLOHEXANONE DIPEROXIDE C:6000
DICYCLOPENTADIENE D:2675

1,3-DICYCLOPENTADIENE DIMER D:2675
DIDIGAM D:0300
DIDIMAC D:0300
DIELDREX D:2700
DIELDRIN D:2700
DIELDRINA (Spanish) D:2700
DIELDRINE (French) D:2700
DIELDRITE D:2700
DIESEL FUEL O:0900
DIESEL FUEL 1-D O:1000
DIESEL IGNITION IMPROVER A:4550
DIESEL OIL O:0900
DIESEL OIL (LIGHT) O:1000
DIESEL OIL, MEDIUM O:1100
DIETANOLAMINA (Spanish) D:2750
DIETHAMINE D:2800
DIETHANOLAMINE D:2750
N,N-DIETHANOLAMINE D:2750
DIETHANOLAMINE LAURYL SULFATE SOLUTION D:8600
DIETHION E:0600
1,1-DIETHOXYETHANE A:0100
1,2-DIETHOXYETHANE E:4600
DIETHYL B:2900
DIETHYL ACETAL A:0100
DIETHYL ACETAL A:0100
DIETHYLAMINE D:2800
N,N-DIETHYLAMINE D:2800
DIETHYLAMINE,2,2'-DIHYDROXY- D:2750
2-DIETHYLAMINO- D:3500
2-(DIETHYLAMINO)ETHYL ALCOHOL D:3500
(DIETHYLAMINO)ETHANE T:3300
2-(DIETHYLAMINO)ETHANOL D:3500
2-N-DIETHYLAMINOETHANOL D:3500
β-DIETHYLAMINOETHANOL D:3500
N-DIETHYLAMINOETHANOL D:3500
β-DIETHYLAMINOETHYL ALCOHOL D:3500
2,6-DIETHYL ANILINE D:2850
2,6-DIETHYLBENZENAMINE D:2850
DIETHYLBENZENE D:2900
DIETHYLBENZOL D:2900
DIETHYL CARBONATE D:2950
DIETHYLCARBONAT (German) D:2950
DIETHYL CELLOSOLVE E:4600
DIETHYL-3-CHLORO-4-METHYLUMBELLIFERYL THIONOPHOSPHATE C:5000
o,o-DIETHYL3-CHLORO-4-METHYL-7-UMBELLIFERONE THIOPHOSPHATE C:5000
o,o-DIETHYLO-(3-CHLORO-4-METHYLCOUMARINYL-7) THIOPHOSPHATE C:5000
o,o-DIETHYLO-(3-CHLORO-4-METHYLUMBELLIFERYL)PHOSPHOROTHIOATE C:5000
o,o-DIETHYLO-(3-CHLORO-4-METHYL-2-oxo-2H-BENZOPYRAN-7-YL)PHOSPHOROTHIOATE C:5000
o,o-DIETHYLO-(3-CHLORO-4-METHYL-7-COUMARINYL)PHOSPHOROTHIOATE C:5000
DIETHYLCETONE (French) D:3700
DIETHYLENE-1,4-DIOXIDE D:7100
DIETHYLENE DIOXIDE D:7100
DIETHYLENE ETHER D:7100
DIETHYLENE GLYCOL D:3000
DIETHYLENE GLYCOL BUTYL ETHER D:3300
DIETHYLENE GLYCOL n-BUTYL ETHER D:3300

DIETHYLENE GLYCOL BUTYL ETHER ACETATE D:3350
DIETHYLENE GLYCOL DI-n-BUTYL ETHER D:3050
DIETHYLENE GLYCOL DIBUTYL ETHER D:3050
DIETHYLENE GLYCOL DIMETHYL ETHER D:3100
DIETHYLENE GLYCOL ETHYL ETHER D:3000
DIETHYLENE GLYCOL ETHYL ETHER D:3400
DIETHYLENE GLYCOL ETHYL ETHER ACETATE D:3150
DIETHYLENE GLYCOL n-HEXYL ETHER D:3200
DIETHYLENE GLYCOL METHYL ETHER D:3415
DIETHYLENE GLYCOL METHYL ETHER ACETATE D:3250
DIETHYLENE GLYCOL MONOBUTYL ETHER ACETATE D:3350
DIETHYLENE GLYCOL MONOBUTYL ETHER D:3300
DIETHYLENE GLYCOL MONOETHYL ETHER D:3400
DIETHYLENE GLYCOL MONOETHYL ETHER D:3415
DIETHYLENE GLYCOL MONOMETHYL ETHER D:3000
DIETHYLENE IMIDOXIDE M:6100
DIETHYLENE OXIDE T:0900
DI-(ETHYLENE OXIDE) D:7100
DIETHYLENE OXIMIDE M:6100
DIETHYLENE TRIAMINE D:3450
1,4-DIETHYLENEDIAMINE P:2100
1,4-DIETHYLENEDIOXIDE D:7100
DIETHYLENEIMIDE OXIDE M:6100
DIETHYLENEIMINE P:2100
DIETHYLENIMIDE OXIDE M:6100
N,N-DIETHYLETHANEAMINE T:3300
DIETHYLETHANOLAMINE D:3500
N,N-DIETHYLETHANOLAMINE D:3500
DIETHYL ESTER OF PHTHALIC ACID D:3750
DIETHYL ESTER SULFURIC ACID D:3800
DIETHYL ETHER E:5600
o,o-DIETHYL-s-2-(ETHYLTHIO)ETHYL PHOSPHODITHIOATE D:7700
o,o-DIETHYL-2-ETHYLTHIOETHYLPHOSPHORODITHIOATE D:7700
o,o[DIETHYL-o (and s)-]2-(ETHYLTHIO)ETHYLPHOSPHOROTHIOATES D:0800
o,o-DIETHYL s-(2-ETHTHIOETHYL)PHOSPHORODITHIOATE D:7700
o,o-DIETHYL s-2-(ETHYLTHIO)ETHYLPHOSPHORODITHIOATE D:7700
o,o-DIETHYL s-(2-ETHTHIOETHYL)THIOTHIONOPHOSPHATE D:7700
o,o-DIETHYL-2-ETHYLMERCAPTOETHYL THIOPHOSPHATE,DIETHOXYTHIOPHOSPHORIC ACID D:0800
o,o-DIETHYL s-(2-ETHYLMERCAPTOETHYL)DITHIOPHOSPHATE D:7700
DIETHYL GLYCOL DIMETHYL ETHER D:3100
DI-(2-ETHYLHEXYL) ADIPATE D:7000

N,N-DIETHYL-2-HYDROXYETHYLAMINE D:3500
N,11-DIETHYL-N-(8-HYDROXY-ETHYL)AMINE D:3500
N,N-DIETHYL-N-(β-HYDROXYETHYL)AMINE D:3500
o,o-DIETHYLO-(2-ISOPROPYL-6-METHYL-4-PYRIMIDINYL)PHOSPHOROTHIOATE D:1000
DIETHYL KETONE D:3700
DIETHYL MONOSULFATE D:3800
o,o-DIETHYL-o,p-NITROPHENYL PHOSPHOROTHIOATE P:0200
DIETHYL 4-NITROPHENYL PHOSPHOROTHIONATE P:0200
DIETHYL p-NITROPHENYL THIONOPHOSPHATE P:0200
DIETHYL OXIDE E:5600
DIETHYLPARATHION P:0200
DIETHYL PHTHALATE D:3750
DI-(2-ETHYLHEXYL) PHOSPHATE D:3600
DI-(2-ETHYLHEXYL) PHOSPHORIC ACID D:3600
DI-(2-ETHYLHEXYL) PHTHALATE D:3650
DI-(2-ETHYLHEXYL)PHOSPHORIC ACID D:3600
DI-(2-ETHYLHEXYL)SULFOSUCCINATE, SODIUM SALT D:7050
DIETHYL SULFATE D:3800
DIETHYL SULPHATE D:3800
DIETHYLTHIOPHOSPHORIC ACID ESTER OF 3-CHLORO-4-METHYL-7-HYDROXYCOUMARIN C:5000
DIETHYL ZINC D:3850
DIETILAMINA (Spanish) D:2800
2,6-DIETILANILINA (Spanish) D:2850
N,N-DIETILANILINA (Spanish) D:2850
DIETILBENCENO (Spanish) D:2900
DI(2-ETILBUTIRATO) de TRIETILENGLICOL (Spanish) T:3450
DIETILCETONA (Spanish) D:3700
N,N-DIETILETANOLAMINA (Spanish) D:3500
DIETILENGLICOL (Spanish) D:3000
DIETILENGLICOLDIBUTIL ETER (Spanish) D:3050
DIETILENGLICOLDIMETIL ETER (Spanish) D:3100
DIETILENGLICOL n-HEXIL ETER (Spanish) D:3200
DIETILENGLICOL MONOBUTIL ETER (Spanish) D:3300
DIETILENGLICOL MONOETIL ETER (Spanish) D:3415
DIETILENGLICOL MONOMETIL ETER (Spanish) D:3000
DIETILENTRIAMINA (Spanish) D:3450
DIETILZINC (Spanish) D:3850
DIEYANOMETHANE P:3450
DIFENILAMINA (Spanish) D:7250
DIFENOL (Spanish) D:7200
DIFENOL DICLOROSILANO (Spanish) D:7300
DIFENYLOL PROPANE B:1950
DIFLUPYL D:0850
DIFLUROCHLOROMETHANE M:5800
DIFLUORODICHLOROMETHANE D:1650
1,1-DIFLUORETANO (Spanish) D:3900
DIFLUROETHANE D:3900
1,1-DIFLUOROETHANE D:3900
DIFLUROPHATE D:0850
DIFLUOROPHOSPHORIC ACID D:3950

DIFLUOROPHOSPHORUS ACID, anhydrous D:3950
1,3-DIFORMAL PROPANE G:0600
DIFORMYL G:0700
DIGALLIC ACID T:0250
DIGERMIN T:3800
DIGGE D:3000
DIGLYCIDYL ETHER of BISPHENOL A B:2000
DIGLYCOL D:3000
DIGLYCOL MONOBUTYL ETHER D:3300
DIGLYCOL MONOBUTYL ETHER ACETATE D:3350
DIGLYCOL MONOETHYL ETHER D:3415
DIGLYCOL MONOETHYL ETHER ACETATE D:3150
DIGLYME D:3100
DIHEPTYL PHTHALATE D:4000
DI-N-HEPTL PHTHALATE D:0975
DIHEXYL ADIPATE (plasticizer) D:4050
DI-n-HEXYL ADIPATE D:4050
DIHEXYL HEXANEDIOATE D:4050
2,5-DIHIDROPEROXIDO de 2,5-DIMETILHEXANO (Spanish) D:5650
DIHYDROAZIRINE E:5400
DIHYDRO-1-AZIRINE E:5400
9,10-DIHYDRO-8a,10,-DIAZONIAPHENANTHRENE DIBROMIDE D:7650
9,10-DIHYDRO-8a,10a-DIAZONIAPHENANTHRENE(1,1'-ETHYLENE-2,2'-BIPYRIDYLIUM)DIBROMIDE D:7650
DIHYDRO-2,5-DIOXOFURAN M:0350
5,6-DIHYDRO-DIPYRIDO(1,2a,2,1c)PYRAZINIUM DIBROMIDE D:7650
DIHYDROOXIRENE E:5500
DIHYDROPENTABORANE(9) P:0250
2,5-DIHYDROPEROXY-2,5-DIMETHYLHEXANE D:5650
1,2-DIHYDRO-3,6-PYRIDAZINEDIONE M:0400
1,2-DIHYDROPYRIDAZINE-3,6-DIONE M:0400
6,7-DIHYDROPYRIDO(1,2a,2':1'-c) PYRAZINEDIUM DIBROMIDE D:7650
6,7-DIHYDROPYRIDO[1,2a:2',1'-c] PYRAZINEDINIUM ION D:7650
DIHYDROXYBENZENE H:1700
1,2-DIHYDROXYBENZENE C:1950
1,3-DIHYDROXYBENZENE R:0100
1,4-DIHYDROXYBENZENE H:1700
m-DIHYDROXYBENZENE R:0100
o-DIHYDROXYBENZENE C:1950
p-DIHYDROXYBENZENE H:1700
1,4-DIHYDROXY-BENZOL (German) H:1700
m-DIHYDROXYBENZOL R:0100
1,3-DIHYDROXYBUTANE B:3900
1,4-DIHYDROXYBUTANE B:2950
2,3-DIHYDROXY BUTANE B:4000
1,4-DIHYDROXY-2-BUTENE B:3000
1,4-DIHYDROXY-2-BUTYNE B:4750
2,2'-DIHYDROXYDIETHYLAMINE D:2750
β,β-DIHYDROXYDIETHYL ETHER D:3000
p,p'-DIHYDROXYDIPHENYL-DIMETHYLMETHANE B:1950
2,2'-DIHYDROXYDIPROPYL ETHER D:7500
2,2'-DIHYDROXYDIPROPYLAMINE D:4550
1,2-DIHYDROXYETHANE E:4200
DI-β-HYDROXYETHOXYETHANE T:3400
2,2'-DIHYDROXYETHYLAMINE D:2750
DI-(2-HYDROXYETHYL)AMINE D:2750

2,2'-DIHYDROXY-3,3',5,5',6,6'-HEXACHLORODIPHENYLMETHANE H:0600
2,2'-DIHYDROXY-3,5,6,3',5',6'-HEXACHLORODIPHENYLMEXOFENE H:0600
2,4-DIHYDROXY-2-METHYLPENTANE H:1200
1,2-DIHYDROXYPROPANE P:4200
DIIRONTRISULFATE F:0400
DIISIPROPYL-ACETONE D:4250
DIISOBUTILAMINA (Spanish) D:4100
DIISOBUTILCARBINOL (Spanish) D:4150
DIISOBUTILCETONA (Spanish) D:4250
DIISOBUTILENO (Spanish) D:4200
DIISOBUTYL CARBINOL D:4150
DI-ISOBUTYLCETONE (French) D:4250
DIISOBUTYLAMINE D:4100
DIISOBUTYLENE D:4200
DIISOBUTYLKETON (German) D:4250
DIISOBUTYL KETONE D:4250
DIISOBUTYL PHTHALATE D:4300
4,4'-DIISOCIANAT de DIFENILMETANO (Spanish) D:7400
DI-ISOCYANATE de TOLUYLENE (French) T:1700
DIISOCIANATO de ISOFORONA (Spanish) I:1050
N,N-DI-iso-CYANATOLUENE T:1700
DIISOCYANAT-TOLUOL (German) T:1700
2,4-DIISOCYANATO-1-METHYLBENZENE T:1700
2,4-DIISOCYANATOTOLUENE T:1700
DIISODECYL PHTHALATE D:4350
DIISONONYL ADIPATE D:4400
DIISONONYL PHTHALATE D:4450
DIISOOCTYL ACID PHOSPHATE D:3600
DIISOOCTYL PHTHALATE D:4500
DIISOPROPANOLAMINA (Spanish) D:4550
DIISOPROPANOLAMINE D:4550
DIISOPROPILAMINA (Spanish) D:4550
DIISOPROPILBENCENO (Spanish) D:4650
2,6-DIISOPROPIL NAFTALENO (Spanish) D:4750
DIISOPROPOXYPHOSPHORYL FLUORIDE D:0850
DIISOPROPYL ACETONE D:4250
5-DIISOPROPYL ACETONE D:4250
DIISOPROPYL AMINE D:4600
2-DIISOPROPYL AMINOETHYL)-O-ETHYL METHYL PHOSPHONOTHIOLATE V:1650
DIISOPROPYL BENZENE HYDROPEROXIDE D:4700
DIISOPROPYL BENZENE (all isomers) D:4650
DIISOPROPYL ETHER I:1400
o,o-DIISOPROPYL FLUOROPHOSPHATE D:0850
DIISOPROPYL FLUOROPHOSPHONATE D:0850
DIISOPROPYL FLUOROPHOSPHORIC ACID ESTER D:0850
DIISOPROPYL FLUORPHOSPHORSAEUREESTER (German) D:0850
DIISOPROPYL NAPHTHALENE D:4750
2,6-DIISOPROPYL NAPHTHALENE D:4750
DIISOPROPYL OXIDE I:1400
DIISOPROPYL PERCARBONATE I:1600
DIISOPROPYL PEROXYDICARBONATE I:1600
DIISOPROPYL PHENYLHYDROPEROXIDE D:4700

DIISOPROPYL PHOSPHOFLUORIDATE D:0850
o,o'-DIISOPROPYL PHOSPHORYL FLUORIDE D:0850
DIKETONE ALCOHOL D:0900
DILANTIN DB D:1450
DILATIN DB D:1450
DILAUROYL PEROXIDE L:0400
DILENE D:0250
DILITHIUM CHROMATE L:2700
DILUTE SULFURIC ACID S:6900
DIMAZ D:7700
DIMAZINE D:5700
DI-p-MENTHA-1,8-DIENE D:7150
DIMER OF NITROGEN DIOXIDE N:1700
2,7:3,6-DIMETHANONAPTH 2,3-B OXIRENE, 3,4,5,6,9,9-HEXACHLORO-1A,2,2A,3,6,6A,7,7A-OCTAHYDRO-(1A α,2 β,2A α,3 β,6 β,6Aα,7 β,7Aα)- D:2700
DIMETHICONE 350 L:0200
p,p'-DIMETHOXYDIPHENYLTRICHLORO-ETHANE M:1500
1,2-DIMETHOXYETHANE E:4700
DIMETHOXYMETHANE M:4200
2,3-DIMETHOXYSTRICHNIDIN-10-ONE B:2800
DIMETHOXY STRYCHNINE B:2800
10,11-DIMETHOXYSTRYCHNINE B:2800
2,3-DIMETHOXYSTRYCHNINE B:2800
DIMETHYLACETAL FORMALDEHYDE M:4200
DIMETHYLACETAMIDE D:4800
DIMETHYL ACETAMIDE D:4800
N,N-DIMETHYL ACETAMIDE D:4800
DIMETHYLACETIC ACID I:0500
DIMETHYLACETONE D:3700
DIMETHYLACETONE AMIDE D:4800
DIMETHYLACETONYLCARBINOL D:0900
DIMETHYLACETYLENECARBINOL M:4400
DIMETHYLADIPATE D:4900
DIMETHYLAMIDE ACETATE D:4800
DIMETHYLAMINE D:4950
DIMETHYLAMIDOETHOXYPHOSPHORYL CYANIDE T:0050
N,N-DIMETHYLAMINE D:4950
DIMETHYLAMINE, ANHYDROUS D:4950
DIMETHYLAMINE, AQUEOUS SOLUTION D:4950
DIMETHYLAMINE SOLUTION D:4950
2-(DIMETHYLAMINO)ETHANOL D:5175
α-(DIMETHYLAMINO)TOLUENE B:1450
DIMETHYLAMINO CARBONYL CHLORIDE D:5050
DIMETHYLAMINOCYANPHOSPHORSAEU REAETHYLESTER (German) T:0050
N,N-DIMETHYLAMINOCYCLOHEXANE D:5100
β-DIMETHYLAMINOETHYL ALCOHOL D:5175
2,6-DIMETHYLANILINE X:0450
DIMETHYLARSINIC ACID C:0100
2,6-DIMETHYLBENZENAMINE X:0450
1,2-DIMETHYLBENZENE X:0200
1,3-DIMETHYLBENZENE X:0100
1,4-DIMETHYLBENZENE X:0300
m-DIMETHYLBENZENE X:0100
o-DIMETHYLBENZENE X:0200
p-DIMETHYLBENZENE X:0300
DIMETHYL 1,2-BENZENEDI-CARBOXYLATE D:5900
alpha,α-DIMETHYLBENZENE HYDROPEROXIDE C:5500

N,N-DIMETHYL BENZENE METHANAMINE B:1450
DIMETHYL BENZENEORTHODICARBOXYLATE D:5900
3,3'-DIMETHYLBENZIDIN T:1850
3,3'-DIMETHYLBENZIDINE T:1850
DIMETHYLBENZYL HYDROPEROXIDE C:5500
alpha,α-DIMETHYLBENZYL HYDROPEROXIDE C:5500
3,3'-DIMETHYLBIPHENYL-4,4'-DIAMINE T:1850
3,3'-DIMETHYL-4,4'-BIPHENYLDIAMINE T:1850
DIMETHYL BUTANEDIOATE D:6050
1,3-DIMETHYL BUTANOL M:4500
2,2-DIMETHYLBUTANE N:0600
DIMETHYL BUTANEDIOATE D:6050
1,3-DIMETHYLBUTANOL M:2000
3,3-DIMETHYL-2-BUTANOL METHYLPHOSPHONOFLUORIDATE S:5650
1,3-DIMETHYLBUTYL ACETATE M:1950
3,3-DIMETHYL-2-BUTYL METHYLPHOSPHONOFLUORIDATE S:5650
3,3-DIMETHYL-n-BUT-2-YL METHYLPHOSPHONOFLUORIDATE S:5650
2,2-DIMETHYLCAPRYLIC ACID D:5850
DIMETHYLCARBAMIC ACID CHLORIDE D:5050
DIMETHYLCARBAMIC CHLORIDE D:5050
DIMETHYLCARBAMIDOYL CHLORIDE D:5050
N,N-DIMETHYLCARBAMOYL CHLORIDE D:5050
DIMETHYL CARBAMOYL CHLORIDE D:5050
DIMETHYLCARBAMYL CHLORIDE D:5050
N,N-DIMETHYLCARBAMYL CHLORIDE D:5050
DIMETHYLCARBINOL I:1250
DIMETHYLCHLOROETHER C:2600
N,N-DIMETHYLCHLOROFORMAMIDE D:5050
N-DIMETHYLCYCLOHEXANAMINE D:5100
N,N-DIMETHYLCYCLOHEXYLAMINE D:5100
DIMETHYL CELLOSOLVE E:4700
DIMETHYL 1,2-DIBROMO-2,2-DICHLOROETHYLPHOSPHATE N:0150
o,o-DIMETHYL-O-(1,2-DIBROMO-2,2-DICHLOROETHYL)PHOSPHATE N:0150
o,o-DIMETHYLS-(1,2-DICARBETHOXY-ETHYL)PHOSPHORO-DITHIOCITE M:0250
DIMETHYLDICHLOROSILAN D:5150
DIMETHYLDICHLOROSILANE D:5150
1,1'-DIMETHYLDIETHYLENE GLYCOL D:7500
o,o-DIMETHYL o-2,2-DICHLORO-1,2-DIBROMOETHYL PHOSPHATE N:0150
1,1-DIMETHYL-3-(3,4-DICHLORO-PHENYL)UREA D:7950
2,2-DIMETHYL-2,2-DIHYDRO-BENZOFURANYL-7 N-METHYLCARBAMATE C:1500
3,3'-DIMETHYL-4,4'-DIPHENYLDIAMINE T:1850
o,o-DIMETHYL DITHIOPHOSPHATE of DIETHYL MERCAPTOSUCCINATE M:0250
DIMETHYLENEDIAMINE E:3800
DIMETHYLENE GLYCOL B:4000
DIMETHYLENEIMINE E:5400
3,3-DIMETHYLENENORCAMPHENE C:1350

DIMETHYLENE OXIDE E:5500
DIMETHYLENIMINE E:5400
DIMETHYL ESTER OF SULFURIC ACID D:6100
1,1-DIMETHYLETHANE I:0250
1,1-DIMETHYLETHANOL B:3450
DIMETHYLETHANOLAMINE D:5175
N,N-DIMETHYLETHANOLAMINE D:5175
DIMETHYL ETHER D:5200
1,1-DIMETHYLETHYLAMINE B:3600
DIMETHYLETHYLCARBINOL M:4400
1,1-DIMETHYLETHYL HYDROPEROXIDE B:4200
DIMETHYL FORMAL M:4200
DIMETHYLFORMALDEHYDE A:0300
DIMETHYLFORMAMID (German) D:5550
DIMETHYL FORMAMIDE D:5550
N,N-DIMETHYLFORMAMIDE D:5550
N,N-DIMETHYL FORMAMIDE D:5550
DIMETHYLFORMEHYDE A:0300
DIMETHYL GLUTARATE D:5600
2,6-DIMETHYLHEPTAN-4-ONE D:4250
2,6-DIMETHYL-HEPTAN-4-ON (German) D:4250
2,6-DIMETHYL-4-HEPTANE D:4250
2,6-DIMETHYL-4-HEPTANOL D:4150
2,6-DIMETHYLHEPTANONE D:4250
2,6-DIMETHYL-4-HEPTANONE D:4250
DI(6-METHYLHEPTYL)PHTHALATE D:4500
DIMETHYL HEXANEDIOATE D:4900
DIMETHYL-1-HEXANOLS I:0850
DIMETHYLHEXANALS I:0800
2,5-DIMETHYLHEXANE-2,5-DIHY-DROPEROXIDE D:5650
DIMETHYLHEXANE DIHYDROPEROXIDE D:5650
DIMETHYLHEXANE DIHYDROPEROXIDE (with 18% or more water) D:5650
2,5-DIMETHYL-2,5-DIHYDROPEROXY HEXANE, not more than 82% with water D:5650
1,1-DIMETHYLHYDRAZIN (German) D:5700
1,2-DIMETHYLHYDRAZIN (German) D:5750
DIMETHYLHYDRAZINE D:5700
1,1-DIMETHYLHYDRAZINE D:5700
1,2-DIMETHYLHYDRAZINE D:5750
asym-DIMETHYLHYDRAZINE D:5700
N,N-DIMETHYLHYDRAZINE D:5750
N,N-DIMETHYLHYDRAZINE D:5700
sym-DIMETHYLHYDRAZINE D:5750
unsym-DIMETHYLHYDRAZINE D:5700
DIMETHYLHYDRAZINE, SYMMETRICAL D:5750
DIMETHYLHYDRAZINE, UNSYMMETRICAL D:5700
DIMETHYL HYDROGEN PHOSPHITE D:5800
N,N-DIMETHYL-N-(2-HYDROXYETHYL) AMINE D:5175
o,o-DIMETHYL-(1-HYDROXY-2,2,2-TRICHLORO)ETHYL PHOSPHATE T:2050
DIMETHYLKETAL A:0300
N,N-DIMETHYLMETHANAMINE T:4050
DIMETHYLMETHANE P:3300
N,N-DIMETHYLMETHANIDE D:5550
3,3-DIMETHYL-2-METHYLENE NORCAMPHANE C:1350
2,2-DIMETHYL-3-METHYLENE- C:1350
2-2-DIMETHYL-3-METHYLENE NORBORANE C:1350
3,5-DIMETHYL-4-(METHYLTHIO)-,METHYLCARBAMATE M:0450

3,5-DIMETHYL-4-(METHYLTHIO)PHENOLMETHYLCARBAMATE M:0450
3,5-DIMETHYL-4-METHYLTHIOPHENYL N-METHYLCARBAMATE M:0450
DIMETHYL MONOSULFATE D:6100
DIMETHYLNITROMETHANE N:2050
o,o-DIMETHYL O-4-NITROPHENYL PHOSPHOROTHIOATE M:4900
o,o-DIMETHYL o-p-NITROPHENYLPHOSPHOROTHIOATE M:4900
2,2-DIMETHYLOCTANOIC ACID D:5850
2,2-DIMETHYL OCTANOIC ACID N:0550
N,N-DIMETHYL-N-OCTYLBENZENEMETHANAMINIUMCHLORIDE B:1500
DI(7-METHYLOCTYL)PHTHALATE D:4450
DIMETHYLPHOSPHORAMIDOCYANIDIC ACID, ETHYL ESTER T:0050
DIMETHYLOL PROPANE D:6000
o,o-DIMETHYL-s-(4-oxo-1,2,3-BEZOTRIAZIN-3(4H)-YL METHYL)PHOSPHORODITHIOATE A:5800
DIMETHYLPHENOL X:0400
DIMETHYLPHENOL PHOSPHATE T:4700
3-[(DIMETHYLPHOSPHINYL)OXY]-2-BUTENOIC ACID METHYL ESTHER P:1500
DIMETHYLPOLYSILOXANE D:5950
2,2-DIMETHYLPROPANE-1,3-DIOL D:6000
2,2-DIMETHYLPROPIONIC ACID T:4000
α,α-DIMETHYLPROPIONIC ACID T:4000
DIMETHYL PHOSPHITE D:5800
DIMETHYL PHOSPHONATE D:5800
o,o-DIMETHYL-PHOSPHORODITHIOATE M:0250
o,o-DIMETHYL PHOSTENE P:1500
DIMETHYL PHTHALATE D:5900
2,2-DIMETHYL-1,3-PROPANEDIOL D:6000
DIMETHYL SILICONE D:5950
DIMETHYL SILICONE FLUIDS D:5950
DIMETHYL SILICONE OIL D:5950
10,11-DIMETHYLSTRYCHNINE B:2800
DIMETHYL SUCCINATE D:6050
DIMETHYL SULFATE D:6100
DIMETHYL SULFIDE D:6150
DIMETHYL SULFOXIDE D:6200
DIMETHYL SULPHATE D:6100
DIMETHYL SULPHIDE D:6150
DIMETHYL SULPHOXIDE D:6200
DIMETHYL TEREPHTHALATE D:6250
N,N-(DIMETHYL)-α-TOLUENEAMINE B:1450
DIMETHYLTRIMETHYLENE GLYCOL D:6000
DIMETHYLZINC D:6300
N,N-DIMETILACETAMIDA (Spanish) D:4800
DIMETILAMINA (Spanish) D:4950
N,N-DIMETILCICLOHEXILAMINA (Spanish) D:5100
DIMETILDICLOROSILANO (Spanish) D:5150
DIMETILETANOAMINA (Spanish) D:5175
N,N-DIMETILFORMAMIDA (Spanish) D:5550
1,1-DIMETILHIDRAZINA (Spanish) D:5700
1,2-DIMETILHIDRAZINA (Spanish) D:5750
DIMETILPOLISILOXANO (Spanish) D:5950
2,2-DIMETIL-1,3-PROPANDIOL (Spanish) D:6000
DIMETILZINC (Spanish) D:6300
DIMETOX T:2050
1-(1,1-DIMETYLETHYL)-4-METHYLBENZENE B:4650
DIMEXIDE D:6200

DIMPYLATE D:1000
DINIL D:8850
2,4-DINITRANILINE D:6350
DINITROANILINE D:6350
2,4-DINITROANILINE D:6350
m-DINITROBENCENO (Spanish) D:6400
o-DINITROBENCENO (Spanish) D:6450
p-DINITROBENCENO (Spanish) D:6500
2,4-DINITROBENZENAMIME D:6350
1,2-DINITROBENZENE D:6450
1,3-DINITROBENZENE D:6400
1,4-DINITROBENZENE D:6500
2,4-DINITROBENZENE D:6400
m-DINITROBENZENE D:6400
o-DINITROBENZENE D:6450
p-DINITROBENZENE D:6500
p-DINITROBENZOL D:6400
1,3-DINITROBENZOL D:6400
o-DINITROBENZOL D:6450
2,4-DINITROCICLOHEXILFENOL (Spanish) D:6600
DINITROCRESOL D:6550
DINITRO-o-CRESOL D:6550
2,6-DINITRO-o-CRESOL D:6550
3,5-DINITRO-o-CRESOL D:6550
4,6-DINITRO-o-CRESOL D:6550
DINITROCYCLOHEXYLPHENOL D:6600
DINITRO-o-CYCLOHEXYLPHENOL D:6600
2,4-DINITRO-6-CYCLOHEXYLPHENOL D:6600
4,6-DINITRO-o-CYCLOHEXYL PHENOL D:6600
2,6-DINITRO-N,N-DI-N-PROPYL-α,α,α-TRIFLUORO-p-TOLUIDINE T:0700
2,6-DINITRO-N,N-DIPROPYL-4-(TRIFLUOROMETHYL)BENZENAMINE T:3800
2,6-DINITRO-N,N-DIPROPYL-4-(TRIFLUOROMETHYL)ANILINE T:3800
2,4-DINITROFENOL (Spanish) D:6650
2,5-DINITROFENOL (Spanish) D:6700
2,6-DINITROFENOL (Spanish) D:6750
DINITROGEN MONOXIDE N:2300
DINITROGEN TETROXIDE N:1700
3,5-DINITRO-2-HYDROXYTOLUENE D:6550
4,6-DINITRO-2-METHYL PHENOL D:6550
2,4-DINITROANILIN (German) D:6350
2,4-DINITROANILINA (Spanish) D:6350
2,4-DINITROPHENOL D:6650
2,5-DINITROPHENOL D:6700
2,6-DINITROPHENOL D:6750
α-DINITROPHENOL D:6650
β-DINITROPHENOL D:6750
γ-DINITROPHENOL D:6700
O-O-DINITROPHENOL D:6750
2,4-DINITROTOLUENE D:6800
2,6-DINITROTOLUENE D:6850
3,4-DINITROTOLUENE D:6900
2,4-DINITROTOLUENO (Spanish) D:6800
2,6-DINITROTOLUENO (Spanish) D:6850
3,4-DINITROTOLUENO (Spanish) D:6900
DINITROTOLUOL D:6800
2,4-DINITROTOLUOL D:6800
2,6-DINITRO-4-TRIFLUORMETHYL-N,N-DIPROPYLANILIN T:3800
DINOFAN D:6650
DINONYL 1,2-BENZENEDICARBOXYLATE D:6950
DINONYL PHTHALATE D:6950
DINONYL-n-PHTHALATE D:6950
DI-n-NONYL PHTHALATE D:6950
DINOPOL NOP D:5250

DINOXOL T:2500
DINOXOL D:0100
DINYL D:8850
DIOCTYAL D:7050
DIOCTYL ADIPATE D:7000
DIOCTYL PHTHALATE D:5250
DIOCTYL PHTHALATE, secondary D:3650
DI-sec-OCTYL PHTHALATE D:3650
DIOCTYL SODIUM SULFOSUCCINATE D:7050
DIOCTYN D:7050
DIOFORM D:1750
DIOKAN D:7100
DIOLAMINE D:2750
DIOLANE H:1200
DIOMEDICONE D:7050
DI-ON D:7950
DIOP D:4500
DIOSUCCIN D:7050
DIOTILAN D:7050
DIOVAC D:7050
1,4-DIOXACYCLOHEXANE D:7100
DIOXAN D:7100
DIOXAN-1,4 (German) D:7100
p-DIOXAN, TETRAHYDRO- D:7100
DIOXANE D:7100
DIOXANE-1,4 D:7100
1,4-DIOXANE D:7100
p-DIOXANE D:7100
DIOXANNE (French) D:7100
3,6-DIOXAOCTANE-1,8-DIOL T:3400
DIOXIDO de AZUFRE (Spanish) S:6700
DIOXIDO de CARBONO (Spanish) C:1650
DIOXIDO de SELENIO (Spanish) S:0300
DIOXITOL T:3400
p-DIOXOBENZENE H:1700
DIOXODICHLOROCHROMIUM C:3450
DIOXONIUM PERCHLORATE SOLUTION P:0950
1,3-DIOXOPHTHALAN P:2000
4,8-DIOXY-UNDECANE DIOL,-1,11 T:4550
1,4-DIOXYBENZENE B:1050
m-DIOXYBENZENE R:0100
o-DIOXYBENZENE C:1950
DIOXYDE de BARYUM (French) B:0400
DIOXYETHYLENE ETHER D:7100
DIPA D:4600
DIPA D:4550
DIPANOL D:7150
DIPENTENE D:7150
DIPENTENO (Spanish) D:7150
DIPENTYL PHTHALATE D:0975
DI-n-PENTYLPHTHALATE D:0975
o-DIPHENOL C:1950
DIPHENYL D:7200
DIPHENYLAMINE D:7250
N,N-DIPHENYLAMINE D:7250
DIPHENYLDICHLOROSILANE D:7300
4,4'-DIPHENYLENEDIAMINE B:0850
DIPHENYL-DIPHENYL ETHER MIXTURE D:8850
DIPHENYL ETHER D:7350
DIPHENYL KETONE B:1000
DIPHENYL METHANONE B:1000
DIPHENYL MIXED with DIPHENYL ETHER D:8850
DIPHENYL OXIDE D:7350
DIPHENYLMETHANE-4,4'-DIISOCYANATE D:7400
DIPHENYLMETHANE DIISOCYANATE D:7400
4,4'-DIPHENYLMETHANE DIISOCYANATE D:7400

DIPHENYLSILICON DICHLORIDE D:7300
DIPHENYLTRICHLOROETHANE D:0300
DIPHOSGENE P:1550
DIPHOSPHORIC ACID, TETRAETHYL ESTER T:0800
DIPHYL D:8850
DIPIRATRIL-TROPICO D:6200
DIPOFENE D:1000
DIPOTASSIUM CHROMATE P:2750
DIPOTASSIUM DICHROMATE P:2900
DIPOTASSIUM MONOCHROMATE P:2750
DIPPING ACID S:6800
DIPROPAL METHANE H:0150
DIPROPANEDIOL DIBENZOATE D:7550
DI-n-DIPROPILAMINA (Spanish) D:7450
DIPROPILENGLICOL (Spanish) D:7500
DIPROPILENGLICOL MONOMETIL ETER (Spanish) D:7600
DIPROPYLAMINE D:7450
DI-n-PROPYLAMINE D:7450
4-(DI-N-PROPYLAMINO)-3,5-DINITRO-1-TRIFLUOROMETHYLBENZENE T:3800
N,N-DIPROPYLANILINE N:1300
DI-N-PROPYL-2,6-DINITRO-4-TRIFLUOROMETHYLANILINE T:3800
DIPROPYL ETHER P:4600
DIPROPYLENE GLYCOL D:7500
DIPROPYLENE GLYCOL DIBENZOATE D:7550
DIPROPYLENE GLYCOL METHYL ETHER D:7600
DIPROPYLENE GLYCOL MONOMETHYL ETHER D:7600
DIPROPYL METHANE H:0150
DIPROPYL OXIDE P:4600
N,N-DIPROPYL-1-PROPANAMINE T:4500
N,N-DIPROPYL-4-TRIFLUOROMETHYL-2,6-DINITROANILINE T:3800
DIPTEREX T:2050
DIQUAT D:7650
DIQUAT DIBROMIDE D:7650
DISB D:4300
DISFLAMOLL TKP T:2950
DISILVER OXIDE S:1100
DISODIUM ARSENATE HEPTAHYDRATE S:1800
DISODIUM CHROMATE S:3000
DISODIUM DICHROMATE S:3300
DISODIUM DIFLUORIDE S:3400
DISODIUM DIHYDROGEN PYROPHOSPHATE S:4800
DISODIUM ETHYLENEBIS-[DITHIOCARBAMATE] N:0100
DISODIUM METASILICATE S:5100
DISODIUM MONOSILICATE S:5100
DISODIUM MONOSULFIDE S:5300
DISODIUM NITRILOTRIACETATE N:1750
DISODIUM SELENITE S:5000
DISODIUM SULFIDE S:5300
DISODIUM SULFITE S:5400
DISPERMINE P:2100
DISTILLATES: FLASHED FEED STOCKS D:7750
DISTILLATES: STRAIGHT RUN D:7800
DISTOKAL H:0550
DISTOPAN H:0550
DISULFATON D:7700
DISULFATOZIRCONIC ACID Z:2700
DISULFOTON D:7700
DISULFUR DICHLORIDE S:7000
DISULFURE de TETRAMETHYL-THIOURAME (French) T:1450

DISULFURIC ACID O:5400
DISULFURO de ARSENICO (Spanish) A:5350
DISULFURO de CARBONO (Spanish) C:1700
DI-SYSTON D:7700
DISYSTOX D:7700
DITHALLIUM CARBONATE T:1200
DITHANE N:0100
DITHANE A-40 N:0100
DITHANE D-14 N:0100
DITHIO T:0600
α,α'-DITHIOBIS(DIMETHYLTHIO)-FORMAMIDE T:1450
DITHIOCARBONIC ANHYDRIDE C:1700
DITHIODEMETON D:7700
DITHIODIPHOSPHORIC ACID, TETRAETHYL ESTER T:0600
DITHIOFOS T:0600
DI(THIOPHOSPHORIC) ACID, TETRAETHYL ESTER T:0600
DITHIOPIROFOSFATO de TETRAETILO (Spanish) T:0600
DITHION T:0600
DITHIONE T:0600
N,N'-(DITHIODICARBOROTHIOYL)-BIS(N-METHYL-METHANAMINE) T:1450
DITHIONIC ACID O:5400
DITHIOPHOS T:0600
DITHIOPHOSPHATE de o,o-DIETHYLE ET DE S-(2-ETHYLTHIO-ETHYLE) (French) D:7700
DITHIOPYROPHOSPHATE de TETRAETHYLE (French) T:0600
DITHIOPYROPHOSPHORIC ACID,o,o,o,o-TETRAETHYL ESTER T:0600
DITHIOSYSTOX D:7700
DITHIOTEP T:0600
4,4'-DI-o-TOLUIDINE T:1850
DITRIDECYL PHTHALATE D:7850
DITTHIONOUS ACID, ZINC SALT Z:1600
DIUNDECYL PHTHALATE D:7900
DIUREX D:7950
DIUROL D:7950
DIURON D:7950
DIURON 4L D:7950
DIVINYL B:2850
DIVINYLENE OXIDE F:1450
DIVINYMETHANE P:0500
DIVIPAN D:2650
DIZENE D:1450
DIZINON D:1000
dl-LACTIC ACID, AMMONIUM SALT A:3000
dl-LIMONENE D:7150
DMA D:4950
DMA D:4800
DMA-4 D:0100
DMAC D:4800
DMAE D:5175
DMB T:1850
DMCC D:5050
DMDT M:1500
DME D:5200
DMF D:5550
DMFA D:5550
DMH D:5750
DMH D:5700
DMP D:5900
DMS (METHYL SULFATE) D:6100
DMS D:6150
DMS D:6100
DMS-70 D:6200
DMS-90 D:6200
DMSO D:6200
DMT D:6250

DMU D:7950
DN D:6550
DNA D:6350
DNBA D:1160
DNC D:6550
DNOC D:6550
DNOP D:5250
DNP D:6750
2,4-DNP D:6650
2,5-DNP D:6700
2,4-DNT D:6800
2,6-DNT D:6850
3,4-DNT D:6900
DNTP P:0200
DO A:1150
DO 14 P:3500
DOCUSATE SODIUM D:7050
DODAT D:0300
1-DODECANETHIOL L:0500
DODECANOIC ACID L:0300
n-DODECANOIC ACID L:0300
DODECANOL D:8000, L:2000
1-DODECANOL D:8000, L:2000
n-DODECANOL D:8000, L:2000
DODECANOYL PEROXIDE L:0400
DODECENE D:8050
DODECENE (non-linear) P:4500
DODECENE (non-linear) D:8050
1-DODECENE D:8050
DODECILBENCENO (Spanish) D:8100
DODECILFENOL (Spanish) D:8550
DODECONIC ACID L:0300
DODECYL ALCOHOL D:8000
DODECYL ALCOHOL L:2000
N-DODECYL ALCOHOL D:8000
DODECYLBENZENE D:8100
n-DODECYLBENZENE D:8100
DODECYLBENZENESULFONATE SODIUM SALT D:8300
DODECYL BENZENESULFONATE D:8200
DODECYL BENZENESULFONATE D:8150
DODECYLBENZENESULFONIC ACID A:1150
DODECYLBENZENESULFONIC ACID D:8150
DODECYLBENZENESULFONIC ACID, CALCIUM SALT D:8200
DODECYL BENZENE SULFONIC ACID, SODIUM SALT D:8300
DODECYLBENZENESULFONIC ACID, ISOPROPYLAMINE SALT D:8250
DODECYLBENZENESULFONIC ACID, TRIETHANOLAMINE SALT D:8350
DODECYL DIPHENYL ETHER DISULFONATE SOLUTION D:8400
DODECYL DIPHENYL ETHER SULFONATE, DISODIUM SALT, AQUEOUS SOLUTION D:8400
DODECYL MERCAPTAN L:0500
DODECYL-2-METHYL-2-PROPENOATE D:8450
DODECYL/PENTADECYL METHACRYLATE D:8500
DODECYL PHENOL D:8550
DODECYL SODIUM SULFATE D:8700
DODECYL SULFATE, AMMONIUM SALT A:3050
DODECYL SULFATE, DIETHANOLAMINE SALT D:8600
DODECYL SULFATE, MAGNESIUM SALT D:8650
DODECYL SULFATE, SODIUM SALT D:8700

DODECYL SULFATE, TRIETHANOLAMINE SALT D:8750
DODECYL TRICHLOROSILANE D:8800
DODECYLENE D:8050
DODECYLENE-α D:8050
α-DODECYLENE D:8050
DODECYLETHYLENE T:0500
DODECYLMETHACRYLATE D:8450
DODECYLTRICHLOROSILANE D:8800
DOL GRANULE B:0600
DOLCO MOUSE CEREAL S:6200
DOLCYMENE C:6350
DOLEN-PUR H:0450
DOLICUR D:6200
DOMOSO D:6200
DOP D:5250
DOP D:3650
DORMANT® FUNGICIDE (DOW) S:4700
DORMANT OIL O:4600
DORMONE D:0100
DOW-PER T:0400
DOW-TRI T:2350
DOW PENTACHLOROPHENOL DP-2 ANTIMICROBIAL P:0350
DOWANOL® D:3400
DOWANOL® 50B D:7600
DOWANOL® DB D:3300
DOWANOL® DE D:3400
DOWANOL® DM D:3415
DOWANOL® DPM D:7600
DOWANOL® E E:0800
DOWANOL® EB E:4900
DOWANOL® EE E:0800
DOWANOL® EIPAT I:1175
DOWANOL® EM E:5100
DOWANOL® EP E:5200
DOWANOL® EPH E:5200
DOWANOL® PM P:4300
DOWANOL® TE E:1500
DOWANOL® TE T:3500
DOWANOL® TPM T:4600
DOWCHLOR® C:2050
DOWCIDE® 7 P:0350
DOWCIDE® G-ST S:4700
DOWCO® 179 D:8900
DOWCO® 139 Z:0100
DOW CORNING® 346 L:0200
DOWFAX® 2A1 D:8400
DOWFLAKE® C:0750
DOWFROST® P:4200
DOWFUME® M:3100
DOWFUME® 40 E:4000
DOWFUME® EDB E:4000
DOWFUME® N D:2500
DOWFUME® W-8 E:4000
DOWFUME® W-10 E:4000
DOWFUME® W-15 E:4000
DOWFUME® W-40 E:4000
DOWFUME® W-85 E:4000
DOWICIDE® 2 T:2450
DOWICIDE® 7 P:0350
DOWICIDE® B T:2450
DOWICIDE® EC-7 P:0350
DOWICIDE® G P:0350
DOWPON® D:0200
DOWPON® M D:0200
DOWTHERM® D:8850
DOWTHERM® D:7350
DOWTHERM® 209 P:4300
DOWTHERM® A D:8850
DOWTHERM® E® D:1450
DOWTHERM® SR-1 E:4200
DOXINATE D:7050

DOXOL D:7050
DP [weaponized] P:1550
DPA D:7250
DPP P:0200
DPP D:0975
DPY-97 F L:0400
DRACYCLIC ACID B:0900
DRAKEOL O:3600
DRAZA M:0450
DREFT D:8700
DREXEL D:7950
DREXEL DEFOL S:2800
DREXEL PARATHION 8E P:0200
DRI-TRI S:4900
DRILL TOX-SPEZIAL B:0600
DRINOX A:1100
DRINOX H:0100
DRINOX H-34 H:0100
DROP LEAF S:2800
DRY ICE (SOLID) C:1650
DRYCLEANER NAPHTHA N:0300
DRYCLEANING SAFETY SOLVENT N:0300
DRYING OIL EPOXIDES E:0400
DS D:3800
DSE N:0100
DSS D:7050
DTDP D:7850
DTMC D:2600
DU-SPREX D:1350
DUAL M:5550
DUKERON T:2350
DULSIVAC D:7050
DUODECYLIC ACID L:0300
DUO-KILL D:2650
DUODEX S:4000
DUOSOL D:7050
DUPONOL D:8700
DURAD T:2950
DURAFUR DEVELOPER C C:1950
DURAFUR DEVELOPER G R:0100
DURALYN T:1450
DURAMITEX M:0250
DURAN D:7950
DURAPHOS P:1500
DURASORB D:6200
DURAVOS D:2650
DURETTER F:0750
DUROFERON F:0750
DUROTOX P:0350
DURSBAN D:8900
DURSBAN F D:8900
DUST-LAYING OIL A:5650
DUTCH LIQUID E:4100
DUTCH TREAT S:2700
DYFLOS D:0850
DYKANOL P:2200
DYKOL D:0300
DYLOX T:2050
DYNEX D:7950
DYREX T:2050
DYTOL J-68 L:2000
DYTOL M-83 O:0300
DYTOL S-91 D:0700
DYVON T:2050
DYZOL D:1000
E-1 E:5400
E393 T:0600
E 605 P:0200
E-1059 D:0800
E 3314 H:0100
EA E:2200
EA1034 B:2025
EAA E:1700

EADC E:2000
EAK E:2300
EARTHNUT OIL O:1900
EASC E:2100
EB E:2400
EBDC, SODIUM SALT N:0100
EBRAGOM TB T:1450
EC 300 M:0400
ECATOX P:0200
ECF E:3100
ECH E:0300
ECRINITRIT S:4400
EDATHAMIL E:3900
EDB E:4000
EDB-85 E:4000
E-D-BEE E:4000
EDC E:4100
EDCO M:3100
EDETIC E:3900
EDETIC ACID E:3900
EDIBLE TALLOW T:0150
EDTA E:3900
EDTA ACID E:3900
EDTA-ZINC D:0950
EDTA-ZINC COMPLEX D:0950
EDTA ZINC SALT D:0950
2EE E:0800
EEC NO. 001-001-00-9 H:1350
EEC NO. 001-002-00-4 L:2500
EEC NO. 003-001-00-4 L:2400
EEC NO. 004-001-00-7 B:1600
EEC NO. 004-002-00-2 B:1800
EEC NO. 005-002-00-5 B:2200
EEC NO. 006-001-00-2 C:1750
EEC NO. 006-002-00-8 P:1550
EEC NO. 006-003-00-3 C:1700
EEC NO. 006-004-00-9 C:0650
EEC NO. 006-006-00-X H:1500
EEC NO. 006-007-00-5 S:3100
EEC NO. 006-007-00-5 P:2800
EEC NO. 006-011-00-7 C:1550
EEC NO. 007-001-00-5 A:2000
EEC NO. 007-002-00-0 N:1700
EEC NO. 007-004-00-1 N:1350
EEC NO. 007-004-01-9 N:1350
EEC NO. 007-010-00-4 S:4400
EEC NO. 007-012-00-5 D:5700
EEC NO. 008-003-00-9 H:1600
EEC NO. 009-001-00-0 F:0800
EEC NO. 009-002-00-6 H:1550
EEC NO. 009-004-00-7 S:3400
EEC NO. 009-006-00-8 A:2700
EEC NO. 011-001-00-0 S:1300
EEC NO. 011-002-00-6 S:3800
EEC NO. 011-002-01-3 S:3800
EEC NO. 011-004-00-7 S:2000
EEC NO. 012-002-00-9 M:0100
EEC NO. 013-003-00-7 A:1450
EEC NO. 013-004-00-2 T:3250
EEC NO. 013-004-00-2 T:3850
EEC NO. 014-002-00-4 S:0500
EEC NO. 014-003-00-X D:5150
EEC NO. 014-004-00-5 M:5450
EEC NO. 014-005-0 E:7800
EEC NO. 015-001-00-1 P:1750
EEC NO. 015-002-00-7 P:1700
EEC NO. 015-003-00-2 C:1250
EEC NO. 015-004-00-8 A:1700
EEC NO. 015-007-00-4 P:1950
EEC NO. 015-009-00-5 P:1800
EEC NO. 015-011-00-6 P:1600
EEC NO. 015-014-00-X M:0250
EEC NO. 015-014-00-2 T:2000

EEC NO. 015-015-00-8 T:4450
EEC NO. 015-034-00-1 P:0200
EEC NO. 015-103-00-6 P:1900
EEC NO. 016-001-00-4 H:1650
EEC NO. 016-009-00-8 S:5300
EEC NO. 016-011-00-9 S:6700
EEC NO. 016-011-00-9 S:5000
EEC NO. 016-012-00-4 S:7000
EEC NO. 016-016-00-6 S:7100
EEC NO. 016-017-00-1 C:3000
EEC NO. 016-019-00-2 O:5400
EEC NO. 016-020-00-8 S:6900
EEC NO. 016-020-00-8 S:6800
EEC NO. 016-021-00-3 M:4750
EEC NO. 016-022-00-9 E:6800
EEC NO. 016-023-00-4 D:6100
EEC NO. 016-027-00-6 D:3800
EEC NO. 016-030-00-2 T:1750
EEC NO. 017-001-00-7 C:2100
EEC NO. 017-002-00-2 H:1450
EEC NO. 017-003-00-8 B:0150
EEC NO. 017-004-00-3 P:2700
EEC NO. 017-005-00-9 S:2800
EEC NO. 017-005-00-9 S:2900
EEC NO. 017-006-00-4 P:0950
EEC NO. 017-007-00-X B:0300
EEC NO. 017-011-00-1 S:3900
EEC NO. 017-011-01-9 S:3900
EEC NO. 017-012-00-7 C:1000
EEC NO. 017-013-00-2 C:0750
EEC NO. 017-014-00-8 A:2450
EEC NO. 019-001-00-2 P:2500
EEC NO. 019-002-00-8 P:2950
EEC NO. 020-001-00-X C:0500
EEC NO. 022-001-00-5 T:1550
EEC NO. 023-001-00-8 V:0400
EEC NO. 024-001-00-0 C:3300
EEC NO. 024-002-00-6 P:2900
EEC NO. 024-003-00-1 A:2600
EEC NO. 024-004-00-7 S:3300
EEC NO. 024-006-00-8 P:2750
EEC NO. 025-002-00-9 P:3200
EEC NO. 028-001-00-1 N:0800
EEC NO. 030-003-00-2 Z:0900
EEC NO. 033-002-00-5 A:5450
EEC NO. 033-003-00-0 A:5500
EEC NO. 034-002-00-8 S:0400
EEC NO. 034-002-00-8 S:0300
EEC NO. 035-001-00-5 B:2250
EEC NO. 035-002-00-0 H:1400
EEC NO. 047-001-00-2 S:1000
EEC NO. 048-001-00-5 C:0150
EEC NO. 048-002-00-0 C:0400
EEC NO. 051-003-00-9 A:5050
EEC NO. 051-022-00-3 A:4900
EEC NO. 056-001-00-1 B:0400
EEC NO. 056-002-00-7 B:0250
EEC NO. 056-002-00-7 B:0100
EEC NO. 080-001-00-0 M:1100
EEC NO. 080-002-00-6 M:0500
EEC NO. 080-002-00-6 M:0950
EEC NO. 080-002-00-6 M:0750
EEC NO. 080-002-00-6 M:0600
EEC NO. 080-002-00-6 M:0800
EEC NO. 080-002-00-6 M:0850
EEC NO. 082-001-00-6 L:0600
EEC NO. 082-001-00-6 L:1300
EEC NO. 082-002-00-1 T:1050
EEC NO. 082-002-00-3 T:0750
EEC NO. 601-001-00-4 M:1400
EEC NO. 601-002-00-X E:0500
EEC NO. 601-003-00-5 P:3300
EEC NO. 601-004-00-0 I:0250
EEC NO. 601-004-00-0 B:2900
EEC NO. 601-006-00-1 I:0900
EEC NO. 601-006-00-1 P:0650
EEC NO. 601-008-00-2 H:0150
EEC NO. 601-009-00-8 O:0100
EEC NO. 601-010-00-3 E:3500
EEC NO. 601-011-00-9 P:4150
EEC NO. 601-012-00-4 I:0450
EEC NO. 601-013-00-X B:2850
EEC NO. 601-014-00-5 I:1150
EEC NO. 601-015-00-0 A:0650
EEC NO. 601-016-00-6 C:6300
EEC NO. 601-017-00-1 C:5850
EEC NO. 601-018-00-7 M:3600
EEC NO. 601-020-00-8 B:0550
EEC NO. 601-021-00-3 T:1800
EEC NO. 601-021-00-3 T:1600
EEC NO. 601-023-00-4 E:2400
EEC NO. 601-024-00-X C:5450
EEC NO. 601-026-00-0 S:6300
EEC NO. 601-027-00-6 M:5400
EEC NO. 601-029-00-7 D:7150
EEC NO. 601-030-00-2 C:6200
EEC NO. 601-031-00-8 D:4200
EEC NO. 601-037-00-0 H:0950
EEC NO. 601-039-00-1 X:0100
EEC NO. 601-040-00-7 X:0300
EEC NO. 602-001-00-7 M:3450
EEC NO. 602-002-00-3 M:3100
EEC NO. 602-004-00-3 D:2050
EEC NO. 602-005-00-9 M:4450
EEC NO. 602-006-00-4 C:2500
EEC NO. 602-008-00-5 C:1850
EEC NO. 602-009-00-0 E:2900
EEC NO. 602-010-00-6 E:4000
EEC NO. 602-011-00-1 D:1700
EEC NO. 602-012-00-7 E:4100
EEC NO. 602-013-00-2 T:2250
EEC NO. 602-014-00-8 T:2300
EEC NO. 602-015-00-3 T:0350
EEC NO. 602-020-00-0 D:2300
EEC NO. 602-023-00-7 V:0800
EEC NO. 602-027-00-9 T:2350
EEC NO. 602-028-00-4 T:0400
EEC NO. 602-029-00-x A:1300
EEC NO. 602-030-00-5 D:2400
EEC NO. 602-033-00-1 C:2400
EEC NO. 602-036-00-8 C:2850
EEC NO. 602-036-00-2 D:1550
EEC NO. 602-037-00-3 B:1350
EEC NO. 602-043-00-6 B:0600
EEC NO. 602-057-00-2 B:1300
EEC NO. 602-058-00-8 B:0450
EEC NO. 602-060-00-9 B:2500
EEC NO. 603-001-00-X M:1750
EEC NO. 603-002-00-5 E:1900
EEC NO. 603-003-00-0 I:1250
EEC NO. 603-003-00-0 P:3950
EEC NO. 603-004-00-6 I:0350
EEC NO. 603-004-00-6 B:3350
EEC NO. 603-005-00-1 B:3450
EEC NO. 603-006-00-7 A:4350
EEC NO. 603-006-00-7 I:0150
EEC NO. 603-007-00-2 M:4400
EEC NO. 606-004-00-4 M:4550
EEC NO. 603-008-00-8 M:2000
EEC NO. 603-008-00-8 M:4500
EEC NO. 603-009-00-3 C:5900
EEC NO. 603-011-00-4 E:5100
EEC NO. 603-012-00-X E:0800
EEC NO. 603-014-00-0 E:4900
EEC NO. 603-015-00-6 A:1200
EEC NO. 603-016-00-1 D:0900
EEC NO. 603-018-00-2 F:1550
EEC NO. 603-019-00-8 D:5200
EEC NO. 603-021-00-9 V:1300
EEC NO. 603-022-00-4 E:5600
EEC NO. 603-023-00-X E:5500
EEC NO. 603-024-00-5 D:7100
EEC NO. 603-025-00-0 T:0900
EEC NO. 603-026-00-6 C:2550
EEC NO. 603-026-00-6 E:0300
EEC NO. 603-027-00-1 E:4200
EEC NO. 603-028-00-7 E:3600
EEC NO. 603-029-00-2 D:1800
EEC NO. 603-030-00-8 M:5950
EEC NO. 603-031-00-3 E:4700
EEC NO. 603-039-00-7 B:4150
EEC NO. 603-040-00-2 S:4200
EEC NO. 603-045-00-X I:1400
EEC NO. 603-047-00-0 D:5175
EEC NO. 603-048-00-6 D:3500
EEC NO. 603-053-00-3 H:1200
EEC NO. 603-054-00-9 D:1170
EEC NO. 603-055-00-4 P:4450
EEC NO. 603-057-00-5 B:1200
EEC NO. 603-059-00-6 H:1050
EEC NO. 603-064-00-3 P:4300
EEC NO. 603-075-00-3 C:2600
EEC NO. 603-080-00-0 M:6050
EEC NO. 604-001-00-2 P:1200
EEC NO. 604-002-00-8 P:0350
EEC NO. 604-004-00-9 C:5100
EEC NO. 604-005-00-4 H:1700
EEC NO. 604-008-00-0 C:2750
EEC NO. 604-008-00-0 C:2700
EEC NO. 604-009-00-6 P:4850
EEC NO. 604-010-00-1 R:0100
EEC NO. 604-010-00-1 R:0100
EEC NO. 604-015-00-9 H:0600
EEC NO. 605-001-00-5 F:1250
EEC NO. 605-003-00-6 A:0150
EEC NO. 605-004-00-1 P:0175
EEC NO. 605-006-00-2 B:4800
EEC NO. 605-009-00-9 C:5400
EEC NO. 605-010-00-4 F:1500
EEC NO. 605-012-00-5 B:0500
EEC NO. 605-015-00-1 A:0100
EEC NO. 605-016-00-7 G:0700
EEC NO. 605-018-00-8 P:3650
EEC NO. 606-001-00-8 A:0300
EEC NO. 606-002-00-3 M:4100
EEC NO. 606-005-00-X D:4250
EEC NO. 606-006-00-5 D:3700
EEC NO. 606-009-00-1 M:1150
EEC NO. 606-010-00-7 C:5950
EEC NO. 606-012-00-8 I:0950
EEC NO. 606-013-00-3 B:1050
EEC NO. 606-020-00-1 E:2300
EEC NO. 606-021-00-7 M:5300
EEC NO. 606-029-00-0 A:0500
EEC NO. 606-030-00-6 M:3350
EEC NO. 606-031-00-1 P:3600
EEC NO. 607-001-00-1 F:1350
EEC NO. 607-002-00-6 A:0200
EEC NO. 607-002-01-3 A:0200
EEC NO. 607-003-00-1 M:5750
EEC NO. 607-006-00-8 O:5500
EEC NO. 607-007-00-3 A:3400
EEC NO. 607-008-00-9 A:0250
EEC NO. 607-009-00-4 P:2000
EEC NO. 607-010-00-X P:3750
EEC NO. 607-011-00-5 A:0600
EEC NO. 607-012-00-0 B:1100
EEC NO. 607-014-00-1 M:4250
EEC NO. 607-015-00-7 E:5800

EEC NO. 607-017-00-8 B:4100
EEC NO. 607-019-00-9 M:3550
EEC NO. 607-020-00-4 E:3100
EEC NO. 607-021-00-X M:1550
EEC NO. 607-022-00-5 E:1600
EEC NO. 607-023-00-0 V:0700
EEC NO. 607-024-00-6 P:3900
EEC NO. 607-024-00-6 I:1200
EEC NO. 607-025-00-1 B:3100
EEC NO. 607-026-00-7 B:3150
EEC NO. 607-026-00-7 I:0300
EEC NO. 607-028-00-8 E:7600
EEC NO. 607-029-00-3 B:4550
EEC NO. 607-031-00-4 B:3700
EEC NO. 607-032-00-X E:1800
EEC NO. 607-033-00-5 B:4450
EEC NO. 607-034-00-0 M:1700
EEC NO. 607-035-00-6 M:4800
EEC NO. 607-037-00-7 E:0900
EEC NO. 607-039-00-8 D:0100
EEC NO. 607-041-00-9 T:2500
EEC NO. 607-061-00-8 A:0900
EEC NO. 607-061-00-8 A:0800
EEC NO. 607-062-00-3 B:3250
EEC NO. 607-070-00-7 E:3000
EEC NO. 607-080-00-1 C:2300
EEC NO. 607-088-00-5 M:1200
EEC NO. 607-089-00-0 P:3700
EEC NO. 607-094-00-8 P:0900
EEC NO. 607-095-00-3 M:0300
EEC NO. 607-096-00-9 M:0350
EEC NO. 607-107-00-7 E:6300
EEC NO. 607-130-00-2 A:4200
EEC NO. 607-130-00-2 A:4300
EEC NO. 607-130-00-2 A:4250
EEC NO. 607-135-00-X B:4900
EEC NO. 607-136-00-5 B:5000
EEC NO. 607-139-00-1 C:2950
EEC NO. 607-139-00-1 C:2900
EEC NO. 607-144-00-9 A:1000
EEC NO. 607-146-00-X F:1400
EEC NO. 608-001-00-3 A:0400
EEC NO. 608-003-00-4 A:0950
EEC NO. 608-004-00-X A:0350
EEC NO. 608-005-00-5 B:4950
EEC NO. 608-010-00-2 M:1300
EEC NO. 608-012-00-3 B:0950
EEC NO. 609-001-00-6 N:2000
EEC NO. 609-002-00-1 N:2050
EEC NO. 609-003-00-7 N:1550
EEC NO. 609-006-00-3 N:2200
EEC NO. 609-006-00-3 N:2250
EEC NO. 609-006-00-3 N:2150
EEC NO. 609-015-00-2 N:1900
EEC NO. 609-015-00-0 N:1950
EEC NO. 609-016-00-8 D:6650
EEC NO. 609-035-00-1 N:1600
EEC NO. 609-036-00-7 N:1800
EEC NO. 610-001-00-3 C:2800
EEC NO. 612-001-00-9 M:1850
EEC NO. 612-001-00-9 D:4950
EEC NO. 612-001-00-9 M:1900
EEC NO. 612-001-00-9 T:4050
EEC NO. 612-002-00-4 E:2200
EEC NO. 612-003-00-X D:2800
EEC NO. 612-004-00-5 T:3300
EEC NO. 612-005-00-0 B:3500
EEC NO. 612-006-00-6 E:3800
EEC NO. 612-007-00-1 I:1300
EEC NO. 612-008-00-7 A:4750
EEC NO. 612-010-00-8 C:2350
EEC NO. 612-012-00-9 N:1450
EEC NO. 612-015-00-5 M:2050
EEC NO. 612-023-00-9 P:1300
EEC NO. 612-024-00-4 T:1850
EEC NO. 612-040-00-1 D:6350
EEC NO. 612-042-00-2 B:0850
EEC NO. 612-049-00-0 D:1160
EEC NO. 612-050-00-6 C:6100
EEC NO. 612-057-00-4 P:2100
EEC NO. 612-058-00-X D:3450
EEC NO. 612-060-00-0 T:0700
EEC NO. 612-065-00-8 A:1900
EEC NO. 613-001-00-1 E:5400
EEC NO. 613-002-00-7 P:4800
EEC NO. 613-028-00-9 M:6100
EEC NO. 613-031-00-5 T:2850
EEC NO. 613-036-00-2 M:5150
EEC NO. 613-037-00-8 M:5200
EEC NO. 613-037-00-8 M:5250
EEC NO. 614-001-00-4 N:1200
EEC NO. 614-006-00-1 B:2800
EEC NO. 615-001-00-7 M:4600
EEC NO. 615-005-01-6 D:7400
EEC NO. 615-006-00-4 T:1700
EEC NO. 616-001-00-X D:5550
EEC NO. 616-003-00-0 A:0850
EEC NO. 616-011-00-4 D:4800
EEC NO. 617-002-00-8 C:5500
EEC NO. 650-001-01-8 G:0200
EEC NO. 650-001-02-5 K:0150
EEC NO. 650-001-02-5 N:0300
EEC NO. 650-001-02-5 J:0100
EEC NO. 650-002-00-6 T:4750
EFFEMOLL DOA D:7000
EG E:4200
EGBE E:4900
EGBEA E:5000
EGDME E:4700
EGEE E:0800
EGEEA E:0900
EGITOL H:0550
EGM E:5100
EGME E:5100
EGMEA E:4800
EHYDE ACETIQUE (French) A:0150
EHYDE ACRYLIQUE (French) A:0800
EINECS No. 231-635-3 A:2000
EINECS No. 232-188-7 C:0900
EKAGOM TB T:1450
EKATIN WF & WF ULV P:0200
EKATOX P:0200
EKTASOLVE DB ACETATE D:3350
EKTASOLVE EB ACETATE E:5000
EKTASOLVE EB SOLVENT E:4900
EKTASOLVE EE E:0800
EKTASOLVE EP E:5300
EL 4049 M:0250
ELAIC ACID O:5100
ELANCOLAN T:3800
ELALDEHYDE P:0175
ELAOL D:1250
ELAYL E:3500
ELDOPOQUE H:1700
ELDOQUIN H:1700
ELECTRICAL INSULATING OIL O:4900
ELECTRO-CF 11 T:2400
ELECTRO-CF 12 D:1650
ELEMENTAL WHITE PHOSPHORUS P:1750
ELPON P:2400
EMBAFUME M:3100
EMBATHION E:0600
EMBICHIN B:2025
EMERALD GREEN C:4050
EMERSAL 6400 D:8700
EMERSOL-120 S:5900
EMERSOL 210 O:5100
EMERSSENCE 1160 E:5200
EMERY 6705 E:5200
EMETIQUE (French) A:5000
EMMATOS M:0250
EMO-NIK N:1200
EMOTOS EXTRA M:0250
EMULSAMINE BK D:0100
EMULSAMINE E-3 D:0100
ENANTHIC ACID H:0200
ENANTHIC ALCOHOL H:0250
ENB E:6600
ENDOCEL E:0100
1,4-endo- 5,8-DIMETHANONAPHTHALENE A:1100
ENDOSULFAN E:0100
ENDOSULPHAN E:0100
ENDRATE E:3900
ENDREX E:0200
ENDRIN E:0200
ENDRINE (French) E:0200
ENGRAVERS ACID N:1350
ENSURE E:0100
ENTOMOXAN B:0600
ENVERT-T T:2500
ENVERT 171 D:0100
ENVERT DT D:0100
E.O. E:5500
EP-161E M:4700
EP 30 P:0350
EPAL 6 H:1050
EPAL 10 D:0700
EPAL 12 D:8000
EPICLORHIDRINA (Spanish) E:0300
EPICHLOROHYDRIN E:0300
α-EPICHLOROHYDRIN E:0300
EPI-REZ 508 B:2000
EPI-REZ 510 B:2000
EPON 828 B:2000
EPOXIDE A B:2000
EPOXIDIZED DRYING OILS E:0400
EPOXIDIZED OILS E:0400
EPOXIDIZED TALL OIL, OCTYL ESTER O:0600
EPOXIDIZED VEGETABLE OILS E:0400
1,2-EPOXYAETHAN (German) E:5500
1,2-EPOXYBUTANE B:4050
1,4-EPOXYBUTANE T:0900
1,2-EPOXY-3-BUTOXY PROPANE B:4150
1,2-EPOXY-3-CHLOROPROPANE E:0300
1,2-EPOXY ETHANE E:5500
1,2-EPOXY-3-ISOPROPOXYPROPANE I:1450
EPOXYPROPANE P:4450
1,2-EPOXYPROPANE P:4450
2,3-EPOXYPROPANE P:4450
2,3-EPOXYPROPYL BUTYL ETHER B:4150
EPTACLORO (Spanish) H:0100
EPAL 12 L:2000
EQUIGEL D:2650
ERADEX D:8900
ERGOPLAST AdDO D:7000
ERL-27774 B:2000
ERYTHRENE B:2850
ESEN P:2000
ESKIMON 11 T:2400
ESKIMON 12 D:1650
ESKIMON 22 M:5800
ESSENCE (French) G:0200
ESSENCE OF MIRBANE N:1550
ESSENCE OF MYRBANE N:1550
ESSENCE OF NIOBE M:3000
ESSIGESTER (German) E:1600

ESSIGSAEURE (German) A:0200
ESSIGSAEUREANHYDRID (German) A:0250
ESSOFUNGICIDE 406 C:1475
ETANO (Spanish) E:0500
ETANOLAMINA (Spanish) M:5950
ESTAR O:3200
ESTEARATO AMONICO (Spanish) A:3750
ESTEARATO de PLOMO (Spanish) L:1400
ESTERCIDE T-2 T:2500
ESTERCIDE T-245 T:2500
ESTERON 245 T:2500
ESTERON 44 D:0150
ESTERON 76 BE D:0100
ESTERON 99 D:0100
ESTERON BRUSH KILLER T:2500
ESTERONE FOUR D:0100
ESTIRENO (Spanish) S:6300
ESTOL-1550 D:3750
ESTONATE D:0300
ESTONE D:0100
ESTONOX T:1950
ESTRICNINA (Spanish) S:6200
ESTROSEL D:2650
ESTROSOL D:2650
ETENO (Spanish) E:3500
ETER n-DIBUTILICO (Spanish) D:1170
ETER sim-DICLOROETILO (Spanish) D:1800
ETER DICLOROISOPROPILICO (Spanish) D:1950
ETER DIFENILICO (Spanish) D:7350
ETER ETILICO (Spanish) E:5600
ETHANAL A:0150
ETHANAL, TRICHLORO- T:2100
ETHANAMINE E:2200
ETHANAMINE,N,N-DIETHYL- T:3300
ETHANDIAL G:0700
ETHANE E:0500
ETHANE CARBOXYLIC ACID P:3700
ETHANEDIAL G:0700
1,2-ETHANEDIAMINE E:3800
1,2-ETHANEDIAMINE, N-(2-AMINOETHYL)- D:3450
1,2-ETHANEDIAMINE,N-(2-AMINOETHYL)-N'-[2-(2-AMINOETHYL)ETHYL]- T:0700
1,2-ETHANEDIAMINE,N-(2-AMINOETHYL)-N'-(2-AMINOETHYL)AMINOETHYL- T:0700
ETHANE, 1,2-DIBUTOXY E:4500
ETHANE DICHLORIDE E:4100
ETHANE,1,1-DICHLORO- D:1700
ETHANE DINITRILE C:5650
ETHANEDIOIC ACID O:5500
ETHANEDIOIC ACID, DISODIUM SALT S:4600
ETHANEDIOIC ACID O:5500
ETHANEDIOIC ACID, COPPER(II) SALT (1:1) C:4700
1,2-ETHANEDIOL E:4200
1,2-ETHANEDIOL DIACETATE E:4400
1,2-ETHANEDIOL, MONOACETATE E:4300
1,2-ETHANEDIONE G:0700
2,2'-[1,2-ETHANEDIYLBIS(OXY)]BISEHANOL T:3400
ETHANE HEXACHLORIDE H:0550
ETHANENITRILE A:0400
ETHANE,1,1'-OXYBIS- E:5600
ETHANE PENTACHLORIDE P:0300
ETHANE, PENTACHLORO- P:0300
ETHANEPEROXIC ACID P:0900
ETHANE TRICHLORIDE T:2300
ETHANE,1,1,2-TRICHLORO- T:2300
ETHANE,1,1,2,2-TETRACHLORO- T:0350
ETHANE, 1,1'-THIOBIS(2-CHLORO- B:2025

ETHANETHIOL E:6800
ETHANE, TRIFLUORO- T:3800
ETHANOIC ACID A:0200
ETHANOIC ANHYDRATE A:0250
ETHANOL E:1900
ETHANOLAMINE M:5950
ETHANOL,1,2-DIBROMO-2,2-DICHLORO-, DIMETHYL PHOSPHATE N:0150
ETHANOL,2-(DIETHYLAMINO)- D:3500
ETHANOL,2-ETHOXY- E:0800
ETHANOL, 2,2'-(ETHYLENEDIOXY)DI- T:3400
ETHANOL, 2,2'-IMINOBIS- D:2750
ETHANOL, 2-ISOPROPOXY I:1175
ETHANOL,1-PHENYL- M:3050
ETHANOL, 200 PROOF E:1900
ETHANOYL BROMIDE A:0550
ETHANOYL CHLORIDE A:0600
ETHENE E:3500
ETHENE, 1,2-DICHLORO- D:1750
ETHENE,1,1-DICHLORO- V:1200
1,2-ETHENEDICARBOXYLIC ACID, trans- F:1400
ETHENYL ACETATE V:0700
ETHENYL ETHANOATE V:0700
ETHENYLETHANOATE V:0700
ETHER E:5600
ETHER,BIS(2-CHLORO-1-METHYLETHYL) D:1950
ETHER BUTYLIQUE (French) D:1170
ETHER CHLORATUS E:2900
ETHER CYANATUS P:3800
ETHER CYANIDE P:3800
ETHER, DIMETHYLCHLORO C:2600
ETHER, ETHYL E:5600
ETHER ETHYLENE GLYCOL DIBUTYL E:4500
ETHER ETHYLIQUE (French) E:5600
ETHER HYDROCHLORIC E:2900
ETHER METHYLIQUE MONOCHLORE (French) C:2600
ETHER MONOETHYLIQUE de L'ETHYLENE-GLYCOL (French) E:0800
ETHER MONOMETHYLIQUE de L'ETHYLENE-GLYCIL (French) E:5100
ETHER MURIATIC E:2900
ETHER, VINYL ETHYL V:1000
ETHIDE D:1900
ETHINE A:0650
ETHINYL TRICHLORIDE T:2350
ETHIOL E:0600
ETHIOLACAR M:0250
ETHION E:0600
ETHIOPS MINERAL M:0900
ETHLON P:0200
ETHODAN E:0600
ETHOXYCARBONYLETHYLENE E:1800
ETHOXY DIGLYCOL D:3400
2-ETHOXY-2,3-DIHYDRO-γ-PYRAN E:0700
2-ETHOXY-3,4-DIHYDRO-γ-PYRAN E:0700
2-ETHOXY-3,4-DIHYDRO-1,2-PYRAN E:0700
2-ETHOXY-3,4-DIHYDRO-2H-PYRAN E:0700
ETHOXY DIHYDROPYRAN E:0700
2-ETHOXY DIHYDROPYRAN E:0700
ETHOXYETHANE E:5600
2-ETHOXYETHANOL E:0800
2-(2-ETHOXYETHOXY)ETHANOL D:3400
2-(2-ETHOXYETHOXY)ETHANOL D:3000
2-(2-ETHOXYETHOXY)ETHANOL ACETATE D:3150
ETHOXYETHYL ACETATE E:0900

2-ETHOXYETHYL ACETATE E:0900
ETHOXYFORMIC ANHYDRIDE D:2950
ETHOXYLATED DODECANOL E:1000
ETHOXYLATED DODECYL ALCOHOL E:1000
ETHOXYLATED LAURYL ALCOHOL E:1000
ETHOXYLATED MYRISTYL ALCOHOL E:1400
ETHOXYLATED NONYLPHENOL E:1100
ETHOXYLATED PENTADECANOL E:1200
ETHOXYLATED PENTADE-CYLALCOHOL E:1200
ETHOXYLATED TETRADECANOL E:1400
ETHOXYLATED TETRADECYL ALCOHOL E:1400
ETHOXYLATED TRIDECANOL E:1300
ETHOXYLATED TRIDECYL ALCOHOL E:1300
1-ETHOXY-2-PROPANOL P:4250
ETHOXY PROPIONIC ACID, ETHYL ESTER E:5700
ETHOXYTRIETHYLENE GLYCOL E:1500
ETHOXYTRIETHYLENE GLYCOL T:3500
ETHOXYTRIGLYCOL T:3500
ETHOXYTRIGLYCOL E:1500
ETHYL ACETATE E:1600
ETHYLACETIC ACID B:4900
ETHYL ACETIC ESTER E:1600
ETHYL ACETOACETATE E:1700
ETHYL ACETONE M:5100
ETHYL ACETYL ACETATE E:1700
ETHYL ACETYLACETATE E:1700
ETHYL ACRYLATE E:1800
ETHYL ALCOHOL E:1900
ETHYL ALCOHOL ANHYDRO-S E:1900
ETHYL ALDEHYDE A:0150
ETHYL ALUMINUM DICHLORIDE E:2000
ETHYLALUMINUM SESQUICHLORIDE E:2100
ETHYLAMINE E:2200
ETHYL AMYL KETONE E:2300
ETHYLBENZENE E:2400
ETHYLBENZOL E:2400
ETHYLBIS(2-CHLOROETHYL)AMINE B:2025
ETHYL BUTANOATE E:2800
ETHYLBUTANOL E:2500
2-ETHYL BUTANOL E:2500
2-ETHYL BUTANOL-1 E:2500
2-ETHYL-1-BUTANOL E:2500
ETHYL BUTYLACETALDEHYDE E:5900
2-ETHYLBUTYL ALCOHOL E:2500
ETHYLBUTYLAMINE E:2600
n-ETHYL-n-BUTYLAMINE E:2600
ETHYL BUTYL KETONE E:2700
ETHYL BUTYRATE E:2800
ETHYL-n-BUTYRATE E:2800
2-ETHYLCAPROALDEHYDE E:5900
α-ETHYLCAPROALDEHYDE E:5900
α-ETHYLCAPROIC ACID E:6000
ETHYL CARBONATE D:2950
ETHYL CARBINOL P:3950
ETHYL CARBITOL D:3400
ETHYL CELLOSOLVE E:0800
ETHYL CHLORIDE E:2900
ETHYL CHLOROACETATE E:3000
ETHYL-α-CHLOROACETATE E:3000
ETHYL CHLOROCARBONATE E:3100
ETHYL CHLOROETHANOATE E:3000
ETHYL-2-CHLOROETHANOATE E:3000
ETHYL CHLOROFORMATE E:3100
ETHYL DICHLOROPHOSPHATE E:7500

ETHYLDICHLOROSILANE E:3400
ETHYL DICHLOROSILANE E:3400
ETHYL CHLOROTHIOFORMATE E:3200
ETHYL CHLOROTHIOLFORMATE E:3200
ETHYL CYANIDE P:3800
N-ETHYLCYCLOHEXANAMINE E:3325
ETHYL CYCLOHEXANE E:3300
N-ETHYL(CYCLO)HEXYLAMINE E:3325
ETHYL DIETHYLENE GLYCOL D:3400
ETHYL-s-DIISOPROPYLAMINOETHYL METHYLTHIOPHOSPHONATE V:1650
O-ETHYL-s-(2-DIISOPROPYLAMINOETHYL METHYLPHOSPHONOTHIOTE)(DOT) V:1650
ETHYL DIMETHYLAMIDO-CYANOPHOSPHATE T:0050
ETHYL N,N-DIMETHYLAMINO CYANOPHOSPHATE T:0050
ETHYL DIMETHYL METHANE I:0900
ETHYL DIMETHYLPHOSPHOR-AMIDOCYANIDATE T:0050
ETHYL-N,N-DIMETHYL-PHOSPHORAMIDOCYANIDATE T:0050
ETHYL α-HYDROXYPROPIONATE E:6700
ETHYL DL-LACTATE E:6700
ETHYLE (ACETATE d') (French) E:1600
ETHYLENE E:3500
ETHYLENE ACETATE E:4400
ETHYLENE ALCOHOL E:4200
ETHYLENE ALDEHYDE A:0800
1,1'-ETHYLENE-2,2'-BIPYRIDYLIUM-DIBROMIDE D:7650
ETHYLENEBIS(DITHIOCARBAMIC ACID), DISODIUM SALT N:0100
ETHYLENEBIS(IMINODIACETIC ACID) E:3900
ETHYLENE BROMIDE E:4000
ETHYLENECARBOXAMIDE A:0850
ETHYLENE CARBOXYLIC ACID A:0900
ETHYLENE CHLORHYDRIN E:3600
ETHYLENE CHLORIDE E:4100
ETHYLENE CHLOROHYDRIN E:3600
ETHYLENE CYANOHYDRIN E:3700
ETHYLENE DIACETATE E:4400
ETHYLENE DIAMINE E:3800
ETHYLENEDIAMINE D:3450
1,2-ETHYLENEDIAMINE E:3800
ETHYLENEDIAMINE-N,N,N',N'-TETRAACETIC ACID E:3900
ETHYLENEDIAMINETETRACETIC ACID (EDTA) E:3900
ETHYLENEDIAMINETETRAACETIC ACID E:3900
ETHYLENEDIAMINETETRAACETATE E:3900
ETHYLENE DIBROMIDE E:4000
1,2-ETHYLENE DIBROMIDE E:4000
1,2-ETHYLENEDICARBOXYLIC ACID, (Z) M:0300
1,2-ETHYLENEDICARBOXYLIC ACID, (E) F:1400
cis-1,2-ETHYLENEDICARBOXYLIC ACID M:0300
trans-1,2-ETHYLENEDICARBOXYLIC ACID F:1400
ETHYLENE DICHLORIDE E:4100
1,2-ETHYLENE DICHLORIDE E:4100
ETHYLENE DIGLYCOL D:3000
ETHYLENE DIGLYCOL MONOMETHYL ETHER D:3415
ETHYLENE DIHYDRATE E:4200
ETHYLENE DIMETHYL ETHER E:4700

(ETHYLENEDINITRILO)TETRAACETIC ACID E:3900
ETHYLENEDINITRILOTETRAACETIC ACID E:3900
2,2'-ETHYLENEDIOXYDIETHANOL T:3400
2,2'-ETHYLENEDIOXYETHANOL T:3400
ETHYLENE DIPYRIDYLIUM DIBROMIDE D:7650
1,1'-ETHYLENE-2,2'-DIPYRIDYLIUM-DIBROMIDE D:7650
1,1-ETHYLENE 2,2-DIPYRIDYLIUM DIBROMIDE D:7650
ETHYLENE EHYDE A:0800
ETHYLENE FLUORIDE D:3900
ETHYLENE GLYCOL E:4200
ETHYLENE GLYCOL ACETATE E:4400
ETHYLENE GLYCOL ACETATE E:4300
ETHYLENE GLYCOL-BIS-(2-HYDROXYETHYL ETHER) T:3400
ETHYLENE GLYCOL DIACETATE E:4400
ETHYLENE GLYCOL DIBUTYL ETHER E:4500
ETHYLENE GLYCOL DIETHYL ETHER E:4600
ETHYLENE GLYCOL DIHYDROXYDIETHYL ETHER T:3400
ETHYLENE GLYCOL DIMETHYL ETHER E:4700
ETHYLENE GLYCOL ETHYL ETHER E:0800
ETHYLENE GLYCOL ETHYL ETHER ACETATE E:0900
ETHYLENE GLYCOL ISOPROPYL ETHER I:1175
ETHYLENE GLYCOL METHYL ETHER E:5100
ETHYLENE GLYCOL METHYL ETHER ACETATE E:4800
ETHYLENE GLYCOL MONOACETATE E:4300
ETHYLENE GLYCOL MONOBUTYL ETHER E:4900
ETHYLENE GLYCOL MONOBUTYL ETHER ACETATE E:5000
ETHYLENE GLYCOL MONOETHYL ETHER E:0800
ETHYLENE GLYCOL MONOETHYL ETHER ACETATE E:0900
ETHYLENE GLYCOL MONOETHYL ETHER MONOACETATE E:0900
ETHYLENE GLYCOL MONOISOPROPYL ETHER I:1175
ETHYLENE GLYCOL MONOMETHYL ETHER E:5100
ETHYLENE GLYCOL MONOMETHYL ETHER ACETATE E:4800
ETHYLENE GLYCOL MONOPHENYL ETHER E:5200
ETHYLENE GLYCOL MONOPROPYL ETHER E:5300
ETHYLENE GLYCOL PHENYL ETHER E:5200
ETHYLENE GLYCOL PROPYL ETHER E:5300
ETHYLENE HEXACHLORIDE H:0550
ETHYLENEIMINE E:5400
ETHYLENE MONOCHLORIDE V:0800
ETHYLENE MONOCLINIC TABLETS CARBOXAMIDE A:0850
ETHYLENE OXIDE E:5500
ETHYLENE OXIDE, METHYL- P:4450
ETHYLENE (OXYDE d') (French) E:5500
ETHYLENE, PHENYL- S:6300
ETHYLENE TETRACHLORIDE T:0400

ETHYLENE, TETRAFLUORO- T:0850
ETHYLENE TRICHLORIDE T:2350
ETHYLENIMINE E:5400
N-ETHYLETHANAMINE D:2800
ETHYL ETHANOATE E:1600
ETHYL ETHER E:5600
ETHYL-3-ETHOXYPROPIONATE E:5700
ETHYL β-ETHOXYPROPIONATE E:5700
ETHYLETHYLENE B:3850
ETHYL EHYDE A:0150
ETHYL ESTER OF FORMIC ACID E:5800
ETHYL FORMATE E:5800
ETHYLFORMIC ACID P:3700
ETHYL FORMIC ESTER E:5800
ETHYL GLYME E:4600
ETHYLHEXALDEHYDE E:5900
2-ETHYLHEXALDEHYDE E:5900
2-ETHYLHEXANAL E:5900
2-ETHYL-1-HEXANAL E:5900
2-ETHYLHEXANOIC ACID E:6000
2-ETHYL HEXANOL E:6100
2-ETHYL-1-HEXANOL E:6100
2-ETHYL-1-HEXANOL HYDROGEN PHOSPHATE D:3600
2-ETHYL-2-HEXENAL E:7700
2-ETHYLHEXOIC ACID E:6000
2-ETHYLHEXYL, 2-PROPENOATE E:6300
2-ETHYLHEXYL ACETATE E:6200
2-ETHYLHEXYL ACRYLATE E:6300
2-ETHYLHEXYL ALCOHOL E:6100
β-ETHYLHEXYLAMINE E:6500
2-ETHYL HEXYLAMINE E:6500
2-ETHYLHEXYLAMINE-1 E:6500
2-ETHYL-1-HEXYLAMINE E:6500
ETHYL HYDRATE E:1900
ETHYL HYDRIDE E:0500
ETHYL HYDROSULFIDE E:6800
ETHYL HYDROXIDE E:1900
ETHYL-2-HYDROXYPROPANOATE E:6700
ETHYL-2-HYDROXYPROPIONATE E:6700
ETHYLIC ACID A:0200
5-ETHYLIDENEBICYCLO(2.2.1)HEPT-2-ENE E:6600
ETHYLIDENE CHLORIDE D:1700
ETHYLIDENE DIETHYL ETHER A:0100
ETHYLIDENE DICHLORIDE D:1700
1,1-ETHYLIDENE DICHLORIDE D:1700
ETHYLIDENE DIFLUORIDE D:3900
ETHYLIDENE FLUORIDE D:3900
ETHYLIDENE LACTIC ACID L:0100
ETHYLIDENE NORBORNENE E:6600
5-ETHYLIDENE-2-NORBORNENE E:6600
ETHYLIDENENORCAMPHENE E:6600
ETHYLIDENENORBORNYLENE E:6600
ETHYLIMINE E:5400
ETHYL KETONE D:3700
ETHYL LACTATE E:6700
ETHYL MERCAPTAN E:6800
ETHYL METHACRYLATE E:6900
ETHYL-1-2-METHACRYLATE E:6900
ETHYL-2-METHACRYLATE E:6900
ETHYL METHANOATE E:5800
ETHYLMETHANOL P:3550
ETHYL METHYL ACRYLATE E:6900
ETHYL-α-METHYLACRYLATE E:6900
6-ETHYL-2-METHYLANILINE M:4050
o-ETHYLMETHYLBENZENE E:7900
1-ETHYL-2-METHYLBENZENE E:7900
ETHYL METHYL CARBINOL B:3400
ETHYL METHYL CETONE (French) M:4100
ETHYL METHYLENE PHOSPHORO-DITHIOATE E:0600
ETHYL METHYL KETONE M:4100

ETHYL METHYL KETONE PEROXIDE B:3050
2ETHYL-2-METHYL-2-PROPENOATE E:6900
5-ETHYL-2-METHYLPYRIDINE M:4150
ETHYL MONOCHLOROACETATE E:3000
N-ETHYLMORPHOLINE E:7000
4-ETHYLMORPHOLINE E:7000
ETHYL NITRIL A:0400
ETHYL NITRILE A:0400
ETHYL NITRITE E:7100
ETHYL NITRITE SOLUTION E:7100
ETHYL ORTHOSILICATE E:7800
ETHYL-3-OXOBUTANOATE E:1700
ETHYLOXITOL ACETATE E:0900
ETHYL PARATHION P:0200
ETHYLPHENOL E:7200
2-ETHYLPHENOL E:7200
o-ETHYLPHENOL E:7200
ETHYLPHENYLDICHLOROSILANE E:7300
ETHYL PHOSPHATE T:3650
ETHYL PHOSPHONOTHIOIC DICHLORIDE E:7400
ETHYL PHOSPHORODICHLORIDATE E:7500
ETHYL PHOSPHORODICHLO-RIDOTHIONATE E:7400
ETHYL PHTHALATE D:3750
5-ETHYL-2-PICOLINE M:4150
ETHYL PROPANOATE E:7600
ETHYL PROPENOATE E:1800
ETHYL-2-PROPENOATE E:1800
ETHYL PROPIONATE E:7600
ETHYL PROPIONYL D:3700
2-ETHYL-3-PROPYLACROLEIN E:7700
2-ETHYL-3-PROPYL ACRYLALDEHYDE E:7700
ETHYL PYROPHOSPHATE, TETRA- T:0800
ETHYL-S B:2025
ETHYL SILICATE E:7800
ETHYL SILICATE 40 E:7800
ETHYL SILICON TRICHLORIDE E:8000
ETHYL SULFATE D:3800
ETHYL SULFHYDRATE E:6800
ETHYL THIOALCOHOL E:6800
s-2-(ETHYLTHIO)ETHYL o,o-DIETHYL ESTER OF PHOSPHORODITHIOIC ACID D:7700
o,o-ETHYL-s-2(ETHYLTHIO)ETHYL PHOSPHORODITHIOATE D:7700
ETHYL THIONOPHOSPHORYL DICHLORIDE E:7400
ETHYL THIOPYROPHOSPHATE T:0600
2-ETHYL TOLUENE E:7900
ETHYL TRICHLOROSILANE E:8000
6-ETHYL-o-TOLUIDINE M:4050
o-ETHYLTOLUENE E:7900
ETHYLTRICHLORO SILANE E:8000
ETHYL VINYL ETHER V:1000
ETHYLZINC D:3850
ETHYNE A:0650
ETHYNYL CARBINOL P:3550
ETHYNYLCARBINOL P:3550
ETHYNYLDIMETHYLCARBINOL M:3250
ETILAMINA (Spanish) E:2200
ETIL AMIL CETONA (Spanish) E:2300
ETILBENCENO (Spanish) E:2400
n-ETIL BUTILAMINA (Spanish) E:2600
ETIL BUTIL CETONA (Spanish) E:2700
2-ETILBUTINOL (Spanish) E:2500
ETIL CICLOHEXANO (Spanish) E:3300
ETILDICLOROSILANO (Spanish) E:3400
ETILENDIAMINA (Spanish) E:3800

ETILENGLICOL (Spanish) E:4200
ETILENGLICOL DIBUTIL ETER (Spanish) E:4500
ETILENGLICOL DIETIL ETER (Spanish) E:4600
ETILENGLICOL DIMETIL ETER (Spanish) E:4700
ETILENGLICOL FENIL ETER (Spanish) E:5200
ETILENGLICOL MONOBUTIL ETER (Spanish) E:4900
ETILENGLICOL MONOMETIL ETER (Spanish) E:5100
ETILENGLICOL PROPIL ETER (Spanish) E:5300
ETILENIMIDA (Spanish) E:5400
ETILENO (Spanish) E:3500
ETILFENILDICLOROSILANO (Spanish) E:7300
ETILFENOL (Spanish) E:7200
2-ETILHEXALDEHIDO (Spanish) E:5900
2-ETILHEXANOL (Spanish) E:6100
2-ETILHEXILAMINA (Spanish) E:6500
5-ETILIDENO-2-NORBORNENO (Spanish) E:6600
ETILMERCAPTANO (Italian, Spanish) E:6800
4-ETILMORFOLINA (Spanish) E:7000
2-ETIL-3-PROPILACROLEINA (Spanish) E:7700
2-ETILTOLUENO (Spanish) E:7900
ETILTRICLOROSILANO (Spanish) E:8000
ETIOL M:0250
ETION (Spanish) E:0600
ETO E:5500
2-ETOXI-3,4-DIHIDRO-2h-PIRANO (Spanish) E:0700
2-ETOXIETHANOL (Spanish) E:0800
ETOXYETHENE V:1000
2-ETYYLCAPROIC ACID E:6000
EUFIN D:2950
EUNATROL O:5300
EVAU-SUPERFALL S:2800
EVE V:1000
EVOLA D:1550
EXAGAMA B:0600
EXITELITE A:5200
EXOPHENE H:0600
EXPERIMENTAL INSECTICIDE 7744 C:1550
EXPLOSIVE D A:3650
EXSICCATED FERROUS SULFATE F:0750
EXSICCATED FERROUS SULPHATE F:0750
EXTERMATHION M:0250
EXTRACT-S D:6150
EXTREMA E:7800
EXTREMA S:0500
F 11 T:2400
F 12 D:1650
F 13 M:5900
F 21 D:2100
F 22 M:5800
F 114 D:2550
F 124 M:5850
FA F:1250
FACTITIOUS AIR N:2300
FALITIRAM T:1450
FALKITOL, FASCIOLIN H:0550
FANNOFORM F:1250
FASCIOLIN C:1850
FASCO-TERPENE T:1950
FAST CORINTH BASE B B:0850
FAST DARK BLUE BASE R T:1850
FAST GARNET B BASE N:0500

FAST GARNET BASE B N:0500
FAST RED 2G BASE N:1450
FAST RED BASE N:1450
FAST RED GG BASE N:1450
FAST RED IG BASE N:1450
FAST RED TR BASE C:3200
FAST WHITE L:1500
FATSCO ANT POISON S:1800
FB/2 D:7650
FC-11 T:2400
FC 12 D:1650
FDA P:1250
FECAMA D:2650
FEGLOX D:7650
FEKABIT P:2700
FEMMA P:1400
FENAMIN A:5750
FENAMINE A:5750
FENATROL A:5750
FENCE RIDER T:2500
FENCLOR P:2200
FENILDICLOROARSINA (Spanish) P:1250
FENILHIDRIZINA (Spanish) P:1300
FENILMERCAPTANO (Spanish) B:0800
1-FENIL-1-XILILO ETANO (Spanish) P:1450
FENOLSULFONATO de ZINC (Spanish) Z:1800
FENOMORE T:2650
FENOPROP T:2650
FENOXYL CARBON N D:6650
FEOSOL F:0750
FEOSPAN F:0750
FER-IN-SOL F:0750
FERMENICIDE S:6700
FERMENTATION ALCOHOL E:1900
FERMENTATION AMYL ALCOHOL I:0150
FERMENTATION BUTYL ALCOHOL I:0350
FERMINE D:5900
FERNACOL T:1450
FERNASAN T:1450
FERNESTA D:0100
FERNIDE T:1450
FERNIMINE D:0100
FERNOXONE D:0100
FERRALYN F:0750
FERRIAMICIDE M:5650
FERRIC AMMONIUM CITRATE F:0100
FERRIC AMMONIUM CITRATE, BROWN F:0100
FERRIC AMMONIUM CITRATE, GREEN F:0100
FERRIC AMMONIUM OXALATE F:0150
FERRIC CHLORIDE F:0200
FERRIC CHLORIDE, HEXAHYDRATE F:0200
FERRIC FLUORIDE F:0250
FERRIC GLYCEROPHOSPHATE F:0300
FERRIC NITRATE F:0350
FERRIC NITRATE, NONHYDRATE F:0350
FERRIC SULFATE F:0400
FERROFOSFORO (Spanish) F:0450
FERRO-GRADUMET F:0750
FERRO-THERON F:0750
FERROPHOSPHORUS F:0450
FERROSILICIO (Spanish) F:0500
FERROSILICON F:0500
FERROSILICON, containing more than 30% but less than 90% silicon F:0500
FERROSULFAT (German) F:0750
FERROSULFATE F:0750
FERROUS AMMONIUM SULFATE F:0550
FERROUS AMMONIUM SULFATE HEXAHYDRATE F:0550

FERROUS BOROFLUORIDE F:0650
FERROUS CHLORIDE F:0600
FERROUS(III) CHLORIDE F:0200
FERROUS CHLORIDE TETRAHYDRATE F:0600
FERROUS FLUOROBORATE F:0650
FERROUS OXALATE F:0700
FERROUS OXALATE DIHYDRATE F:0700
FERROUS SULFATE F:0750
FERROUS SULFATE F:0750
FERROUS SULPHATE F:0750
FERROX F:0700
FERSOLATE F:0750
FERTILIZER ACID S:6800
FERXONE D:0100
FI CLOR 91 T:2850
FICHLOR 91 T:2850
FILMERINE S:4400
FILTER ALUM A:1750
FIREBRAKE ZB Z:0600
FIRE DAMP M:1400
FIRMOTOX P:4750
FISH OIL O:1500
FLAXSEED OIL O:3400
FLECK-FLIP T:2350
FLEXOL A-26 D:7000
FLEXOL PLASTICIZER P D:4500
FLIT 406 C:1475
FLO-MORE P:0150
FLO PRO T SEED PROTECTANT T:1450
FLORES MARTIS F:0200
FLORIDINE S:3400
FLOROCID S:3400
FLOROPRYL D:0850
FLOUR SULPHUR S:6600
FLUORISTAN S:5800
FLOWERS OF ANTIMONY A:5200
FLOWERS OF SULPHUR S:6600
FLOZENGES S:3400
FLUATE T:2350
FLUE GAS C:1750
FLUKOIDS C:1850
FLUOACETATO SODICO (Spanish) S:3500
FLUOBORATO de CADMIO (Spanish) C:0300
FLUOBORATO de COBRE (Spanish) C:4350
FLUOBORATO de PLOMO (Spanish) L:1000
FLUOBORATO de NIQUEL (Spanish) N:0950
FLUOBORATO de ZINC (Spanish) Z:1400
FLUOPHOSGENE C:1800
FLUOPHOSPHORIC ACID, DIISOPROPYL ESTER D:0850
FLUOR (French, German, Spanish) F:0800
FLUORAL S:3400
FLUORANILINA (Spanish) F:0850
p-FLUORANILINA (Spanish) F:0900
FLUORANE 114 D:2550
FLUORBENCENO (Spanish) F:0950
FLUORETHYLENE V:1100
FLUORINE F:0800
FLUORINE-19 F:0800
FLUOROACETIC ACID, SODIUM SALT S:3500
FLUOURO AMONICO (Spanish) A:2700
2-FLUOROANILINE F:0850
4-FLUOROANILINE F:0900
o-FLUOROANILINE F:0850
p-FLUOROANILINE F:0900
2-FLUOROBENZENAMINE F:0850
4-FLUOROBENZENAMINE F:0900
FLUOROBENZENE F:0950
FLUOROCARBON 11 T:2400
FLUOROCARBON 12 D:1650
FLUORODICHLOROMETHANE D:2100

FLUORODIISOPROPYL PHOSPHATE D:0850
FLUOROETHENE V:1100
FLUOROFORMYL FLUORIDE C:1800
FLUOROISOPROPOXYMETHYLPHOSPHINE OXIDE S:0250
1-FLUORO-2-METHYLBENZENE F:1050
1-FLUORO-3-METHYLBENZENE F:1100
1-FLUORO-4-METHYLBENZENE F:1150
2-FLUORO-1-METHYLBENZENE F:1050
4-FLUORO-1-METHYLBENZENE F:1150
FLUOROMETHYL(1,2,2-TRIMETHYL-PROPOXY)PHOSPHINE OXIDE S:5650
2-FLUOROPHENYLAMINE F:0850
4-FLUOROPHENYLAMINE F:0900
FLUOROPHOSGENE C:1800
FLUOROPLAST-3 T:3750
FLUOROPRYL D:0850
FLUOROSILICIC ACID H:1300
FLUOROSILICIC ACID F:1000
FLUOROSULFONIC ACID F:1200
FLUOROSULFURIC ACID F:1200
2-FLUOROTOLUENE F:1050
3-FLUOROTOLUENE F:1100
4-FLUOROTOLUENE F:1150
m-FLUOROTOLUENE F:1100
o-FLUOROTOLUENE F:1050
p-FLUOROTOLUENE F:1150
FLUOROTRICHLOROMETHANE T:2400
FLUORSPAR F:0900
m-FLUORTOLUENO (Spanish) F:1050
o-FLUORTOLUENO (Spanish) F:1100
p-FLUORTOLUENO (Spanish) F:1150
FLUORURES ACIDE (French) F:0800
FLUORURO ALUMINICO ANHIDRO (Spanish) A:1550
FLUORURO de BERILIO (Spanish) B:1700
FLUORURO de BROMURO B:2300
FLUORURO CALCICO (Spanish) C:0900
FLUORURO de CARBONILO (Spanish) C:1800
FLUORURO COBALTICO (Spanish) C:3700
FLUORURO ESTANNOSO (Spanish) S:5800
FLUORURO FERRICO (Spanish) F:0250
FLUORURO de PLATA (Spanish) S:0800
FLUORURO de PLOMO (Spanish) L:0900
FLUORURO de VINILO (Spanish) V:1100
FLUORURO SODICO (Spanish) S:3400
FLUORURO de ZINC (Spanish) Z:1300
FLUORURU de ZIRCONIO y POTASIO (Spanish) Z:2600
FLUOSILIC ACID F:1000
FLUOSILICATE de AMMONIUM (French) A:3700
FLUOSILICIC ACID H:1300
FLUOSPAR C:0900
FLUOSTIGMINE D:0850
FLUOSULFONIC ACID F:1200
FLUROPLAST 4 T:0850
FLUORTOLUENO (Spanish) F:1050
FLUX MAAY N:1200
FLUXING OIL A:5650
FLY-DIE D:2650
FLY FIGHTER D:2650
FMA P:1400
FMC-1240 E:0600
FMC-5462 E:0100
FMC-10242 C:1500
FOLIAGE OIL O:4600
FOLIDOL P:0200
FOLIDOL E&E 605 P:0200
FOLIDOL E605 P:0200
FOLIDOL M M:4900
FOMAC H:0600

FONOLINE O:3600
FOREDEX 75 D:0100
FORLIN B:0600
FORMAGENE P:0150
FORMAL M:4200
FORMAL M:0250
FORMALDEHIDO (Spanish) F:1250
FORMALDEHYDE DIMETHYLACETAL M:4200
FORMALDEHYDE POLYMER P:0150
FORMALDEHYDE SOLUTION F:1250
FORMALIN F:1250
FORMALIN 40 F:1250
FORMALINE (German) F:1250
FORMALIN LOESUNGEN (German) F:1250
FORMALITH F:1250
FORMAMIDA (Spanish) F:1300
FORMAMIDE F:1300
FORMAMIDE, N,N-DIMETHYL- D:5550
FORMIATE de METHYLE (French) M:4250
FORMIATO AMONICO (Spanish) A:2750
FORMIATO de BUTILO (Spanish) B:4100
FORMIATO COBALTOSO (Spanish) C:3750
FORMIATO de COBRE (Spanish) C:4400
FORMIATO de ETILO (Spanish) E:5800
FORMIATO de METILO (Spanish) M:4250
FORMIATO de NIQUEL (Spanish) N:1000
FORMIATO de ZINC (Spanish) Z:1500
FORMIC ACID F:1350
FORMIC ACID, AMIDE F:1300
FORMIC ACID, AMMONIUM SALT A:2750
FORMIC ACID, BUTYL ESTER B:4100
FORMIC ACID, ETHYL ESTER E:5800
FORMIC ACID, CHLORO-, BENZYL ESTER B:1400
FORMIC ACID, METHYL ESTER M:4250
FORMIC ACID, ZINC SALT Z:1500
FORMIC ALDEHYDE F:1250
FORMIC ETHER E:5800
FORMIMIDIC ACID F:1300
FORMOL F:1250
FORMONITRILE H:1500
FORMOSA CAMPHOR C:1400
FORMULA 40 D:0100
N-FORMYLDIMETHYLAMINE D:5550
FORMYLFORMIC ACID G:0750
FORMYLIC ACID F:1350
2-FORMYLPHENOL S:0100
α-FORMYLPHENOL S:0100
FORMYL TRIBROMIDE B:2650
FORMYL TRICHLORIDE C:2500
FOROTOX T:2050
FORRON T:2500
FORSTU 46 T:2500
FORTEX T:2500
FORTHION M:0250
FOSCHLOR T:2050
FOSDRIN P:1500
FOSFATO AMONICO (Spanish) A:3600
FOSFATO CALCICO (Spanish) C:1200
FOSFATO SODICO (Spanish) S:4800
FOSFATO TRIBASICO SODICO (Spanish) S:4900
FOSFATO de TRICRESILO (Spanish) T:2950
FOSFATO de TRIETILO (Spanish) T:3650
FOSFERMO P:0200
FOSFEX P:0200
FOSFITO de DIMETILO (Spanish) D:5800
FOSFITO de TRIETILO (Spanish) T:3700
FOSFITO de TRIMETILO (Spanish) T:4400
FOSFIVE
FOSFONO 50 E:0600
FOSFORO BLANCO (Spanish) P:1750

FOSFURO CALCICO (Spanish) C:1250
FOSFORO NEGRO (Spanish) P:1650
FOSFORO ROJO (Spanish) P:1700
FOSFOTHION M:0250
FOSFURO ALUMINICO (Spanish) A:1700
FOSFURO de ZINC (Spanish) Z:1900
FOSGENO (Spanish) P:1550
FOSOVA P:0200
FOSTERN P:0200
FOSTOX P:0200
FOSTRIL H:0600
FOSVEX T:0800
FOTOX A:5400
FOURAMINE PCH C:1950
FOURAMINE RS R:0100
FOURRINE 68 C:1950
FOURRINE 79 R:0100
FOURRINE EW R:0100
FOURRINE PG P:4850
FOWLER'S SOLUTION P:2600
FRATOL S:3500
FREEMANS WHITE LEAD L:1500
FRENCH VERDIGRIS C:4750
FREON 10 C:1850
FREON 11 T:2400
FREON 12 D:1650
FREON 13 M:5900
FREON 20 C:2500
FREON 21 D:2100
FREON 22 M:5800
FREON 113 T:2900
FREON 113TF T:2900
FREON 114 D:2550
FREON 152 D:3900
FREON F-12 D:1650
FREON MF T:2400
FREON TF T:2900
FRIDEX E:4200
FRIGEN 11 T:2400
FRIGEN 12 D:1650
FRIGEN 113A T:2900
FRUIT-O-NET T:2650
FRUITONE A T:2500
FRUMIN AL D:7700
FRUMIN G D:7700
FTALATO de AMILO D:0975
FTALATO de BUTILBENCILO (Spanish) B:3650
FTALATO de DI(2-ETILHEXILO) (Spanish) D:5250
FTALATO de DIETILO (Spanish) D:3750
FTALATO de DIISOBUTILO (Spanish) D:4300
FTALATO de DIISOCTILO (Spanish) D:4500
FTALATO de DIISODECILO (Spanish) D:4350
FTALATO de DIISONONILO (Spanish) D:4450
FTALATO de DINONILO (Spanish) D:6950
FTALATO de DITETRADECILO (Spanish) D:7850
FTALATO de DITRIDECILO (Spanish) D:7900
FTALATO de ETILHEXILO (Spanish) E:6550
FTALATO de METILO (Spanish) D:5900
FTALATO de n-OCTIL-n-DECILO (Spanish) O:0550
FUEL OIL 1-D O:0900
FUEL OIL 2-D O:0900
FUEL OIL No. 1 O:2400
FUEL OIL No. 6 O:2800
FUMARIC ACID F:1400
FUMIGRAIN® A:0950
FUMING LIQUID ARSENIC A:5450
FUMING SULFURIC ACID O:5400
FUMO-GAS E:4000

FUNGCHEX® M:0600
FUNGIFEN P:0350
FUNGITOX OR P:1400
FUNGUS BAN TYPE II C:1475
FURADAN C:1500
FURAL F:1500
FURAL/PYROMUCIC ALDEHYDE F:1500
2-FURALDEHYDE F:1500
FURALE F:1500
FURAN F:1450
FURANO (Spanish) F:1450 F:1450
FURAN, TETRAHYDRO- T:0900
2-FURANALDEHYDE F:1500
2-FURANCARBONAL F:1500
2-FURANCARBINOL F:1550
2-FURANCARBOXALDEHYDE F:1500
2,5-FURANDIONE M:0350
2,5-FURANEDIONE M:0350
FURANIDINE P:0900
2-FURANMETHANOL F:1550
FURATOL S:3500
FURFURAL F:1500
FURFURAL ALCOHOL F:1550
FURFURALCOHOL F:1550
FURFURALDEHYDE F:1500
FURFURAN F:1450
FURFUROL F:1500
FURFUROLE F:1500
FURFURYL ALCOHOL F:1550
FURODAN C:1500
FUROLE F:1500
α-FUROLE F:1500
2-FURYL-METHANAL F:1500
FURYLALCOHOL F:1550
2-FURYLCARBINOL F:1550
FUSEL OIL I:0150
FYDE F:1250
G-11 H:0600
G-301 D:1000
G-24480 D:1000
G-30027 A:5750
GA (nerve gas) T:0050
GALENA L:1600
GALLIC ACID G:0100
GALLIC ACID MONOHYDRATE G:0100
GALLOGAMA B:0600
GALLOTANNIC ACID T:0250
GALLOTANNIN T:0250
GALLOTOX P:1400
GAMACID B:0600
GAMAPHEX B:0600
GAMASOL-90 D:6200
GAMENE B:0600
GAMMAHEXA B:0600
GAMMAHEXANE B:0600
GAMMALIN B:0600
GAMMALIN 20 B:0600
GAMMATERR B:0600
GAMMEX B:0600
GAMMEXANE B:0600
GAMMOPAZ B:0600
GAMONIL C:1550
GAMOPHEN H:0600
GAMOPHENE H:0600
GARDENTOX D:1000
GAS MOSTAZA (Spanish) B:2025
GAS, NATURAL L:2100
GAS NATURAL LICUADO (Spanish) L:2100
GAS OIL G:0150
GAS OIL, CRACKED G:0150
GAS OILS (PETROLEUM), LIGHT VACUUM G:0150

GAS OILS (PETROLEUM), STRAIGHT RUN D:7800
GASOLINA (Spanish) G:0200
GASOLINA RECTIFICADA (Spanish) G:0400
GASOLINA de AVIACION (Spanish) G:0250
GASOLINE, AVIATION (less than 4.86g lead/gal) G:0250
GASOLINE, AVIATION GRADE (100-130 OCTANE) G:0250
GASOLINE, AVIATION GRADE (115-145 OCTANE) G:0250
GASOLINE BLENDING STOCKS: ALKYLATES G:0300
GASOLINE BLENDING STOCKS: REFORMATES G:0350
GASOLINE, NATURAL G:0400
GASOLINE, STRAIGHT RUN, TOPPING-PLANT G:0500
GASOLINES: AUTOMOTIVE (less than 4.23g lead/gal) G:0200
GASOLINES: AVIATION G:0250
GASOLINES: CASINGHEAD G:0400
GASOLINES: POLYMER G:0450
GASOLINES: STRAIGHT RUN G:0500
GAS de PETROLEO LICUADO (Spanish) L:2100
GAULTHERIA OIL M:5350
GB (nerve gas) S:0250
GC-1189 K:0100
GD S:5650
GEARPHOS M:4900
GEARPHOS P:0200
GEIGY 24,480 D:1000
GEIGY 30,027 A:5750
GELAN I T:0050
GELBIN YELLOW ULTRAMARINE C:0800
G-ELEVEN COMPOUND H:0600
GENERAL CHEMICALS 1189 K:0100
GENETRON 11 T:2400
GENETRON 12 D:1650
GENETRON 13 M:5900
GENETRON 21 D:2100
GENETRON 22 M:5800
GENETRON 100 D:3900
GENETRON 113 T:2900
GENETRON 114 D:2550
GENETRON 1113 T:3750
GENIPHENE T:1950
GENITHION P:0200
GENITOX D:0300
GEON L:0200
GERANIUM CRYSTALS D:7350
GERFIL P:2400
GERMA-MEDICA H:0600
GERMAIN'S C:1550
GERMALGENE T:2350
GESAFID D:0300
GESAPON D:0300
GESAPRIM A:5750
GESAREX D:0300
GESAROL D:0300
GESFID P:1500
GESOPRIM A:5750
GESTID P:1500
GEXANE B:0600
GLACIAL ACETIC ACID A:0200
GLACIAL ACRYLIC ACID A:0900
GLAZED PENTA P:0350
GLICERINA (Spanish) G:0650
GLICEROFOSFATO FERRICO (Spanish) F:0300
GLICINATO de COBRE (Spanish) C:4450
GLIOXAL (Spanish) G:0700

GLUCINIUM B:1600
GLUCITOL S:5700
GLUCONATO AMONICO (Spanish) A:2800
GLUTAMIC DIALDEHYDE G:0600
GLUTARAL G:0600
GLUTARALDEHIDO (Spanish) G:0600
GLUTARALDEHYDE SOLUTION G:0600
GLUTARD DIALDEHYDE G:0600
GLUTARIC ACID DIALDEHYSE G:0600
GLUTARIC DIALDEHYDE G:0600
GLY-OXIDE U:1100
GLYCERINE G:0650
GLYCERIN (90 TECHNICAL) G:0650
GLYCERIN, ANHYDROUS G:0650
GLYCERIN, SYNTHETIC G:0650
GLYCERITE T:0250
GLYCERITOL G:0650
GLYCEROL G:0650
GLYCEROL EPICHLOROHYDRIN E:0300
GLYCERYL TRICHLORHYDRIN T:2750
GLYCIDYL-CHLORIDE E:0300
GLYCIDYL ISOPROPYL ETHER I:1450
GLYCIDYL METHACRYLATE G:0675
GLYCIDYL α-METHYL ACRYLATE G:0675
GLYCIDYL METHYLPHENYL ETHER C:5350
GLYCINE COPPER COMPLEX C:4450
GLYCINE, N,N'-1,2-ETHANEDIYLBIS(N-(CARBOXYMETHYL)- E:3900
GLYCINOL M:5950
GLYCOCOLL-COPPER C:4450
GLYCOL E:4200
GLYCOL ALCOHOL E:4200
GLYCOLBIS(HYDROXYETHYL) ETHER T:3400
GLYCOL BROMIDE E:4000
GLYCOL BUTYL ETHER E:4900
GLYCOL CHLOROHYDRIN E:3600
GLYCOL CYANOHYDRIN E:3700
GLYCOL DIACETATE E:4400
GLYCOL DIBROMIDE E:4000
GLYCOL DICHLORIDE E:4100
GLYCOL DIMETHYL ETHER E:4700
GLYCOL ETHER D:3000
GLYCOL ETHER DB ACETATE D:3350
GLYCOL ETHER DE D:3000
GLYCOL ETHER E:0800
GLYCOL ETHER EM E:5100
GLYCOL ETHYL ETHER E:0800
GLYCOL ETHYL ETHER D:3000
GLYCOL ETHYLENE ETHER D:7100
GLYCOL METHYL ETHER E:5100
GLYCOL MONOACETATE E:4300
GLYCOL-MONOACETIN E:4300
GLYCOL MONOBUTYL ETHER ACETATE E:5000
GLYCOL MONOETHYL ETHER E:0800
GLYCOL MONOETHYL ETHER ACETATE E:0900
GLYCOL MONOMETHYL ETHER E:5100
GLYCOL MONOMETHYL ETHER ACETATE E:4800
GLYCON DP S:5900
GLYCON RO O:5100
GLYCON TP S:5900
GLYCON WO O:5100
GLYCYL ALCOHOL G:0650
GLYME E:4700
GLYMOL O:3600
GLYODEX 3722 C:1475
GLYOXAL G:0700
GLYOXYALDEHYDE G:0700
GLYOXYLIC ACID (50% or less) G:0750

GOOD-RITE L:0200
GP-40-66:120 H:0450
GPKH H:0100
GRAIN ALCOHOL E:1900
GRAIN SORGHUM HARVEST AID S:2800
GRAMEVIN D:0200
GRANEX OK S:2800
GRANOX MN H:0400
GRANULATED SUGAR S:6400
GREEN NICKEL OXIDE N:1050
GREEN OIL A:4850
GREEN VERDIGRIS C:4750
GREEN VITRIOL IRON MONOSULFATE F:0750
GREY ARSENIC A:5250
GRISOL T:0800
GROCO O:3400
GROCO-54 S:5900
GROCOLENE G:0650
GROUND VOCLE SULPHUR S:6600
GROUNDNUT OIL O:1900
GRUNDIER ARBEZOL P:0350
GUESAROL D:0300
GUM L:0200
GUM CAMPHOR C:1400
GUM SPIRITS T:4750
GUM TURPENTINE T:4750
GUN COTTON C:3950
GUSATHION A:5800
GUSATHION INSECTICIDE A:5800
GUSATHION M A:5800
GUTHION A:5800
GUTHION INSECTICIDE A:5800
GY-PHENE T:1950
GYLCIDY BUTYL ETHER B:4150
GYRON D:0300
H (sulfur mustard agent) B:2025
H 34 H:0100
H 321 M:0450
HAIMASED S:5500
HALOCARBON 11 T:2400
HALOCARBON 12 D:1650
HALOCARBON 13 M:5900
HALOCARBON 21 D:2100
HALOCARBON 22 M:5800
HALOCARBON 113 T:2900
HALOCARBON 152a D:3900
HALOGENATED WAXES P:2200
HALON D:1650
HALON 104 C:1850
HALON 112 D:2100
HALON 122 D:1650
HALON 241 M:5850
HALON 242 D:2550
HALON 1001 M:3100
HALON 10001 M:4450
HARD WAX P:0100
HARTSHORN A:2400
HARVEST-AID S:2800
HATCOL XP P:1450
HAVERO-EXTRA D:0300
HAVIDOTE E:3900
HCB H:0400
HCBD H:0450
HCCH B:0600
HCCPD H:0500
HCDB T:1450
HCH B:0600
γ-HCH B:0600
HCL H:1450
HCN H:1500
HCP H:0600
HCS 3260 C:2050

HD (sulfur mustard agent) B:2025
HDEHP D:3600
HEA H:1750
HEAT TREATING OIL O:3600
HEAVY INDUSTRIAL FUEL OIL O:2800
HEAVY OIL C:5050
HECLOTOX B:0600
HEDONAL D:0100
HENDECANOIC ACID U:100
HENDECANOIC ALCOHOL U:0200
1-HENDECANOL U:0200
HEOD D:2700
HEOD-endo,exo-1,2,3,4,10,10- D:2700
HEPATIC GAS H:1650
HEPT T:0800
HEPTACHLOR H:0100
HEPTACHLORE (French) H:0100
1,4,5,6,7,8,8-HEPTACHLOR-3A,4,7,7,A-TETRAHYDRO-4,7-endo- METHANO-INDEN (German) H:0100
3,4,5,6,7,8,8-HEPTACHLORO-DICYCLOPENTADIENE H:0100
1,4,5,6,7,8,8-HEPTACHLORO-3a,4,7,7a-TETRAHYDRO-4,7-ENDOMETHANOINDENE H:0100
1,4,5,6,7,10,10-HEPTACHLORO-4,7,8,9-TETRAHYDRO-4,7-ENDOMETHYLENEINDENE H:0100
1(3a),4,5,6,7,8,8-HEPTACHLORO-3a(1),4,7,7a-TETRAHYDRO-4,7-METHANOINDENE H:0100
1,4,5,6,7,8,8A-HEPTACHLORO-3a,4,7,7a-TETRAHYDRO-4,7-METHANOINDANE H:0100
1,4,5,6,7,8,8-HEPTACHLORO-3a,4,7,7a-TETRAHYDRO-4,7-METHYLENE INDENE H:0100
1,4,5,6,7,8,8-HEPTACHLORO-3a,4,7,7a-TETRAHYDRO-4,7-METHANOINDENE H:0100
1,4,5,6,7,8,8-HEPTACHLORO-3a,4,7,7a-TETRAHYDRO-4,7-METHANOL-1H-INDENE H:0100
1,4,5,6,7,10,10-HEPTACHLORO-4,7,8,9-TETRAHYDRO-4,7-METHYLENEINDENE H:0100
HEPTACLORO (Spanish) H:0100
1-HEPTADECANECARBOXYLIC ACID S:5900
cis-8-HEPTADECYLENECARBOXYLIC ACID O:5100
HEPTAGRAN H:0100
HEPTAMETHYLENE C:5800
HEPTAMUL H:0100
HEPTANAPHTHENE M:3600
HEPTANE H:0150
n-HEPTANE H:0150
1-HEPTANECARBOXYLIC ACID O:0200
3-HEPTANECARBOXYLIC ACID E:6000
n-HEPTANO (Spanish) H:0150
HEPTANOIC ACID H:0200
HEPTANOL H:0250
1-HEPTANOL H:0250
HEPTANOL-1 H:0250
n-HEPTANOL H:0250
2-HEPTANONE A:4500
3-HEPTANONE E:2700
HEPTAN-3-ONE E:2700
4-HEPTANONE,2,6-DIMETHYL- D:4250
HEPTANYL ACETATE H:0350
HEPTENE H:0300
1-HEPTENE H:0300
n-HEPTENE H:0300

n-HEPTENO (Spanish) H:0300
HEPTHLIC ACID H:0200
n-HEPTOIC ACID H:0200
HEPTYL ACETATE H:0350
1-HEPTYL ACETATE H:0350
n-HEPTYL ACETATE H:0350
HEPTYL ALCOHOL H:0250
HEPTYL CARBINOL O:0300
HEPTYL HYDRIDE H:0150
HEPTYL PHTHALATE D:4000
HEPTYL PHTHALATE D:0975
HEPTYLENE H:0300
1-HEPTYLENE H:0300
n-HEPTYLETHYLENE N:2450
n-HEPTYLIC ACID H:0200
HERBATOX D:7950
HERBIDAL D:0100
HERCULES 3956 T:1950
HERCULES TOXAPHENE T:1950
HERKAL D:2650
HERMAL T:1450
HERMAT TMT T:1450
HERYL T:1450
n-HETANOL-1 H:0250
HEXA B:0600
HEXA H:0900
HEXA C. B. H:0400
HEXABALM H:0600
HEXABENZENESULFONIC ACID A:1150
1,2,3,4,5,6-HEXACHLOR-CYCLOHEXANE B:0600
HEXACHLORAN B:0600
γ-HEXACHLORAN B:0600
HEXACHLORANE B:0600
γ-HEXACHLORANE B:0600
HEXACHLOROBENZENE H:0400
γ-HEXACHLOROBENZENE B:0600
HEXACHLOROBENZOL (German) H:0400
1,2,3,4,7,7-HEXACHLORO-BICYCLO(2,2,1)HEPTEN-5,6-BIOXYMETHYLENESULFITE E:0100
HEXACHLOROBUTADIENE H:0450
HEXACHLORO-1,3-BUTADIENE H:0450
1-α,2-α,3-β,4-α,5-α,6-β-HEXACHLORO-CYCLOHEXANE B:0600
γ-1,2,3,4,5,6-HEXACHLOROCYCLOHEXANE B:0600
γ-HEXACHLOROCYCLOHEXANE B:0600
HEXACHLOROCYCLOHEXANE, gamma isomer B:0600
1,2,3,4,5,6-HEXACHLOROCYCLOHEXANE, γ-isomer B:0600
HEXACHLOROCYCLOPENTADIENE H:0500
HEXACHLORO-1,3-CYCLOPENTADIENE H:0500
HEXACHLOROCYCLOPENTADIENE DIMER M:5650
1,2,3,4,5,5-HEXACHLORO-1,3-CYCLOPENTADIENE H:0500
2,2',3,3',5,5'-HEXACHLORO-6,6'-DIHYDROXYDIPHENYLMETHANE H:0600
HEXACHLOROEPOXYOCTAHYDRO-endo,endo-DIMETHANONAPTHALENE E:0200
HEXACHLOROEPOXYOCTAHYDRO-endo,exo-DIMETHANONAPHTHALENE D:2700
HEXACHLORO-6,7-EXPOXY-1,4,4a,5,6,7,8,8a-OCTAHYDRO-1,4:5,8-DIMETHANONAPHTHALENE D:2700
1,2,3,4,10,10-HEXACHLORO-6,7-EPOXY-1,4,4a,5,6,7,8,8a-OCTAHYDRO-1,4-endo-

endo-1,4,5,8-DIMETHANONAPHTHALENE E:0200
1,2,3,4,10,10-HEXACHLORO-6,7-EPOXY-1,4,4a,5,6,7,8,8a-OCTAHYDRO-1,4-endo-exo-5,8-DI-METHANONAPHTHALENE D:2700
HEXACHLOROETHANE H:0550
1,1,1,2,2,2-HEXACHLOROETHANE H:0550
HEXACHLOROHEXAHYDRO-endo- exo-DIMETH ANONAPHTHALENE A:1100
1,2,3,4,10,10-HEXACHLORO-1,4,4a,5,8,8a-HEXAHYDRO-1,4,5,8-DIMETHANONAPHTHALENE A:1100
1,2,3,4,10-10-HEXACHLORO-1,4,4a,5,8,8a-HEXAHYDRO-1,4,5,8-endo,exo-DIMETHANONAPHTHALENE A:1100
1,2,3,4,10,10-HEXACHLORO-1,4,4a,5,8,8a-HEXAHYDRO-1,4-endo exo-5,8-DIMETHANONAPHTHALENE A:1100
6,7,8,9,10-HEXACHLORO-1,5,5a,6,9,9a,-HEXAHYDRO-6,9-METHANO-2,4,3-BENZODIOXATHIEPIN-3-OXIDE E:0100
3,4,5,6,9,9-HEXACHLORO-1A,2,2A,3,6,6A,7,7A-OCTAHYDRO-2,7:3,6-DIMETHANO D:2700
HEXACHLOROPHANE H:0600
HEXACHLOROPHENE H:0600
HEXACID 698 H:1000
HEXACID 898 O:0200
HEXACID 1095 D:0550
HEXACLOROBENCENO (Spanish) H:0400
HEXACLOROBUTADIENO (Spanish) H:0450
HEXACLOROCICLOPENTADIENO (Spanish) H:0500
HEXACLOROETANO (Spanish) H:0550
HEXACLOROFENO (Spanish) H:0600
HEXADECYL SULFATE, SODIUM SALT H:0650
HEXADECYLTRIMETHYLAMMONIUM CHLORIDE H:0700
HEXADRIN E:0200
HEXAFEN H:0600
HEXAFLUOSILICIC ACID F:1000
HEXAFLUOSILICIC ACID H:1300
HEXAHYDRATE M:0150
HEXAHYDRIC ALCOHOL S:5700
HEXAHYDROANILINE C:6100
HEXAHYDROAZEPINE H:0850
HEXAHYDRO-2h-AZEPINE-2-ONE C:1450
HEXAHYDROBENZENAMINE C:6100
HEXAHYDROBENZENE C:5850
HEXAHYDROBENZENE, HEXAMETHYLENE C:5850
HEXAHYDROCRESOL M:3650
HEXAHYDROCUMENE I:1350
HEXAHYDRO-1,4-DIAZINE P:2100
HEXAHYDROMETHYLPHENOL M:3650
HEXAHYDROPHENOL C:5900
HEXAHYDROPYRAZINE P:2100
HEXAHYDROTOLUENE M:3600
n-HEXALDEHIDO (Spanish) H:0750
HEXALDEHYDE H:0750
n-HEXALDEHYDE H:0750
HEXALENGLICOL (Spanish) H:1200
HEXALIN C:5900
HEXAMETHYLENE C:5850
HEXAMETHYLENEDIAMINE H:0800
HEXAMETHYLENEIMINE H:0850
HEXAMETHYLENETETRAMINE H:0900
HEXAMETILENDIAMINA (Spanish) H:0800
HEXAMETILENIMINA (Spanish) H:0850
HEXAMETILENTETRAMINA (Spanish) H:0900
HEXAMINE H:0900

HEXAMITE T:0800
HEXAMOL SLS D:8700
HEXANAL H:0750
1-HEXANAL H:0750
HEXANAL, 2-ETHYL E:5900
HEXANAPHTHENE C:5850
HEXANE H:0950
n-HEXANE H:0950
HEXANE CARBOXYLIC ACID H:0200
1-HEXANECARCOXYLIC ACID H:0200
1,6-HEXANEDIAMINE H:0800
1,6-HEXANEDIAMINE,2,2,4 TRIMETHYL- T:4200
HEXANE,1,6-DIISOCYANATO-2,2,4-TRIMETHYL- T:4250
HEXANEDINITRILE A:1050
HEXANEDIOIC ACID A:1000
1,6-HEXANEDIOIC ACID A:1000
HEXANEDIOIC ACID, DIBUTYL ESTER D:3550
HEXANEDIOIC ACID, DIHEXYL ESTER D:4050
HEXANEDIOIC ACID, DIMETHYL ESTER D:4900
HEXANEDIOIC ACID, DINITRILE A:1050
1,2-HEXANEDIOL H:1200
1,2,3,4,5,6,-HEXANNEHEXOL S:5700
HEXANO (Spanish) H:0950
HEXANOIC ACID H:1000
n-HEXANOIC ACID H:1000
HEXANOIC ACID, 2-ETHYL- E:6000
HEXANOL H:1050
1-HEXANOL H:1050
n-HEXANOL H:1050
sec-HEXANOL E:2500
HEXANON C:5950
2-HEXANONE M:3350
HEXANONE-2 M:3350
HEXAPLAS DIOP D:4500
HEXAPLAS M-1B D:4300
HEXAPLAS M/B D:4300
HEXAPLAS M/B D:1250
HEXAPLAS M/O D:4500
HEXASUL S:6600
HEXATHIR T:1450
HEXATOX B:0600
HEXAVERM B:0600
HEXAVIN C:1550
HEXENE H:1100
1-HEXENE H:1100
α-HEXENE H:1100
iso-HEXENE M:4950
1-HEXENEO (Spanish) H:1100
HEXICIDE B:0600
HEXIDE H:0600
n-HEXOIC ACID H:1000
HEXONE M:4550
HEXOPHENE H:0600
HEXOSAN H:0600
HEXYCLAN B:0600
HEXYL ACETATE H:1150
1-HEXYL ACETATE H:1150
n-HEXYL ACETATE H:1150
sec-HEXYL ACETATE M:1950
HEXYL ALCOHOL H:1050
n-HEXYL ALCOHOL H:1050
sec-HEXYL ALCOHOL E:2500
HEXYL ALCOHOL, ACETATE H:1150
HEXYL CARBITOL D:3200
HEXYL ETHANOATE H:1150
HEXYL HYDRIDE H:0950
HEXYLENE H:1100
HEXYLENE GLYCOL H:1200

HF H:1550
HF-A H:1550
HGI B:0600
HHDN A:1100
HIDRAZINA (Spanish) H:1250
HIDRIXIETILETILENIMIA (Spanish) A:1850
HIDROGENO (Spanish) H:1350
HIDROPEROXIDO de terc-BUTILO (Spanish) B:4200
HIDROPEROXIDO de CUMENO (Spanish) C:5500
HIDROPEROXIDO de DIISOPROPILBENCENO (Spanish) D:4700
HIDROQUINONA (Spanish) H:1700
HIDROSULFITO de ZINC (Spanish) Z:1600
HIDROSULFURO SODICO (Spanish) S:3700
2-HIDROXIETILACRILATO (Spanish) H:1750
HIDROXIDO AMONICO (Spanish) A:2850
HIDROXIDO CALCICO (Spanish) C:0950
HIDROXIDO NIQUEL (Spanish) N:1050
HIDROXIDO POTASICO (Spanish) P:2950
HIDROXIDO SODICO (Spanish) S:3800
HIDROXILAMINA (Spanish) H:1850
HYDRURO de LITIO (Spanish) L:2800
HIDRURO de LITIO y ALUMINIO (Spanish) L:2500
HIDRURO SODICO (Spanish) S:3600
HI-DRY T:0650
HIGH SOLVENT NAPHTHA N:0350
HIGH SPEED BEARING OIL O:4500
HIGH STRENGTH HYDROGEN PEROXIDE H:1600
HIGHER FATTY ALCOHOL T:0200
HILDAN E:0100
HILDIT D:0300
HILTHION M:0250
HILTHION 25WDP M:0250
HIPOCLORITO CALCICO (Spanish) C:1000
HIPOCLORITO SODICO (Spanish) S:3900
HIPOFOSFITO AMONICO (Spanish) A:2900
HI-POINT 90 B:3050
HITTORF'S PHOSPHORUS P:1700
HL (mustard lewisite) B:2025
HL-331 P:1400
HMDA H:0800
HN-1 (nitrogen mustard agent) B:2025
HN-2 (nitrogen mustard agent) B:2025
HN-3 (nitrogen mustard agent) B:2025
HOE-2671 E:0100
HOME-HEATING OIL No. 2 O:2500
HOMOPIPERIDINE H:0850
HONG KIEN P:1400
HOOF OIL O:3900
HORTEX B:0600
HOSTAQUICK P:1400
HOUSEHOLD AMMONIA A:2850
HPA H:1950
β-HPN E:3700
HRS 1276 M:5650
HS H:1850
HSDB 5700 H:1900
HT (sulfur mustard agent) B:2025
HTH C:1000
HTH DRY CHLORINE C:1000
HUILE de CAMPHRE (French) C:1400
HUMIFEN WT-27G D:7050
HUNGAZIN A:5750
HUNGAZIN PK A:5750
HW 920 D:7950
HYADUR D:6200
HYCAR L:0200
HY-CHLOR C:1000
HYCLORITE S:3900

HYDRACRYLIC ACID, β-LACTONE P:3600
HYDRACRYLONITRILE E:3700
HYDRALIN C:5950
HYDRALIN C:5900
HYDRARGYRUM BIJODATUM (German) M:0700
HYDRATED KEMIKAL C:0900
HYDRAULIC OIL O:3600
HYDRAZINE H:1250
HYDRAZINE-BENZENE P:1300
HYDRAZINE, 1,1-DIMETHYL- D:5700
HYDRAZINE, 1,2-DIMETHYL- D:5750
HYDRAZINE, ANHYDROUS H:1250
HYDRAZINE BASE H:1250
HYDRAZINE, METHYL- M:4350
HYDRAZINOBENZENE P:1300
HYDRAZOIC ACID, SODIUM SALT S:2000
HYDRAZOMETHANE M:4350
HYDROBROMIC ACID H:1400
HYDROBROMIC ACID, ANHYDROUS H:1400
HYDROBROMIC ACID MONO-AMMONIATE A:2300
HYDROCHLORIC ACID H:1450
HYDROCHLORIC ACID, ANHYDROUS H:1450
HYDROCHLORIC ETHER E:2900
HYDROCHLORIDE H:1450
HYDROCYANIC ACID H:1500
HYDROCYANIC ACID, POTASSIUM SALT P:2800
HYDROCYANIC ACID, SODIUM SALT S:3100
HYDROCYANIC ETHER P:3800
HYDROFOL ACID 1255 L:0300
HYDROFLUORIC ACID H:1550
HYDROFLUORIDE H:1550
HYDROFLUOROSILICIC ACID (25% OR LESS) H:1300
HYDROFLUOSILIC ACID F:1000
HYDROFURAN T:0900
HYDROGEN H:1350
HYDROGEN, compressed H:1350
HYDROGEN, refrigerated liquid H:1350
HYDROGEN ARSENIC A:5575
HYDROGEN ARSENIDE A:5575
HYDROGEN BROMIDE H:1400
HYDROGEN BROMIDE, ANHYDROUS H:1400
HYDROGEN CARBOXYLIC ACID F:1350
HYDROGEN CHLORIDE H:1450
HYDROGEN CYANIDE H:1500
HYDROGEN DIOXIDE H:1600
HYDROGENE SULFURE (French) H:1650
HYDROGEN FLUORIDE H:1550
HYDROGEN HEXAFLUOROSILICATE F:1000
HYDROGEN NITRATE N:1350
HYDROGEN PEROXIDE H:1600
HYDROGEN PEROXIDE CARBAMIDE U:1100
HYDROGEN PEROXIDE with UREA U:1100
HYDROGEN SULFIDE H:1650
HYDROGEN SULFITE SODIUM S:2200
HYDROOXYCYCLOHEXANE C:5950
1-HYDROXYCYCLOHEXYL PEROXIDE C:6000
HYDROPEROXIDE H:1600
HYDROPEROXIDE, ACETYL P:0900
HYDROPEROXIDE,1-METHYL-1-PHENYLETHYL- C:5500
1-HYDROPEROXYCYCLOHEXYL C:6000

HYDROPEROXY-2-METHYL PROPANE B:4200
HYDROPHENOL C:5900
HYDROQUINOL H:1700
HYDROQUINOLE H:1700
HYDROQUINONE H:1700
α-HYDROQUINONE H:1700
m-HYDROQUINONE R:0100
o-HYDROQUINONE C:1950
p-HYDROQUINONE H:1700
HYDROSULFURIC ACID H:1650
N-HYDROXETHYL-1,2-ETHANEDIAMINE A:1850
HYDROXIDE de POTASSIUM (French) P:2950
2-HYDROXYBENZALDEHYDE S:0100
o-HYDROXYBENZALDEHYDE S:0100
HYDROXYBENZENE P:1200
2-HYDROXYBENZOIC ACID S:0200
o-HYDROXYBENZOIC ACID S:0200
1-HYDROXYBUTANE B:3350
2-HYDROXYBUTANE B:3400
1-HYDROXY-4-tert-BUTYLBENZENE B:4600
2-HYDROXYCHLOROBENZENE C:2700
4-HYDROXYCHLOROBENZENE C:2750
3-HYDROXYCROTONIC ACID METHYL ESTER DIMETHYL PHOSPHATE P:1500
1-HYDROXY-2-CYANOETHANE E:3700
3-HYDROXYCYCLOHEXADIEN-1-ONE R:0100
HYDROXYCYCLOHEXANE C:5900
HYDROXYDE de SODIUM (French) S:3800
1-HYDROXY-2,4-DINITRO-BENZENE D:6650
HYDROXYDIMETHYLARSINE OXIDE C:0100
HYDROXYDIMETHYLARSINE OXIDE, SODIUM SALT S:2700
2-HYDROXYETHANOL E:4200
HYDROXY ETHER E:0900
HYDROXY ETHER E:0800
2-HYDROXYETHYL ACETATE E:4300
2-HYDROXYETHYL ACRYLATE H:1750
β-HYDROXYETHYL ACRYLATE H:1750
2-HYDROXYETHYLAMINE M:5950
1-(2-HYDROXYETHYLAMINO)-2-AMINOETHANE A:1850
2-(2 HYDROXYETHYLAMINO)ETHYLAMINE A:1850
(1-HYDROXYETHYL)BENZENE M:3050
β-HYDROXYETHYLDIMETHYLAMINE D:5175
n-β-HYDROXYETHYLETHYLENEDIAMINE A:1850
HYDROXYETHYLETHYLENEDIAMINE, n-β- A:1850
β-HYDROXYETHYL ISOPROPYL ETHER I:1175
2-HYDROXYETHYL 2-PROPENOATE H:1750
2-(HYDROXYETHYL) METHYLAMINE M:6050
1-HYDROXYHEPTANE H:0250
1-HYDROXHEXANE H:1050
HYDROXY ISOBUTYRONITRITE A:0350
2-HYDROXYISOBUTYRONITRILE A:0350
α-HYDROXY ISOBUTYRONITRILE A:0350
4-HYDROXY-2-keto-4-METHYLPENTANE D:0900
4-HYDROXYL-2-keto-4-METHYLPENTANE D:0900

HYDROXYLAMINE H:1800
HYDROXYLAMINE SULFATE H:1850
HYDROXYLAMMONIUM SULFATE H:1850
2-HYDROXY-2-METHYL-3-BUTYNE M:3250
2-HYDROXYMETHYLFURAN F:1550
4-HYDROXY-4-METHYL-2-PENTANONE D:0900
4-HYDROXY-4-METHYLPENTAN-2-ONE D:0900
4-HYDROXY-4-METHYL-PENTAN-2-ON (German) D:0900
2-HYDROXY-2-METHYLPROPIONITRILE A:0350
1-HYDROXYMETHYLPROPANE I:0350
2-HYDROXY-4-(METHYLTHIO)-BUTANOIC ACID H:1900
2-HYDROXYNITROBENZENE N:1850
3-HYDROXYNITROBENZENE N:1900
4-HYDROXYNITROBENZENE N:1950
m-HYDROXYNITROBENZENE N:1900
1-HYDROXYOCTANE O:0300
2-HYDROXYPHENOL C:1950
3-HYDROXYPHENOL R:0100
m-HYDROXYPHENOL R:0100
o-HYDROXYPHENOL C:1950
p-HYDROXYPHENOL H:1700
1-HYDROXYPROPANE P:3950
3-HYDROXYPROPANENITRILE E:3700
2-HYDROXY-1,2,3-PROPANE-TRICARBOXYLIC ACID C:3500
2-HYDROXYPROPANOIC ACID L:0100
α-HYDROXYPROPIONIC ACID L:0100
2-HYDROXYPROPANOL P:4200
3-HYDROXYPROPENE A:1200
3-HYDROXY-β-PROPIOLACTONE P:3600
HYDROXYPROPYL ACRYLATE H:1950
β-HYDROXYPROPYL ACRYLATE H:1950
HYDROXYPROPYL METHACRYLATE H:2000
2-HYDROXYPROPYLAMINE I:1300
2-HYDROXYPROPYLAMINE M:6000
3-HYDROXYPROPYLAMINE P:3350
6-HYDROXY-3(2h)-PYRIDAZINONE M:0400
3-HYDROXYTOLUENE C:5100
4-HYDROXYTOLUENE C:5100
α-HYDROXYTOLUENE B:1200
o-HYDROXYTOLUENE C:5100
HYDROXYTOLUENES C:5100
HYDROXYTOLUOLE (German) C:5100
β-HYDROXYTRICARBALLYLIC ACID C:3500
β-HYDROXYTRICARBOXYLIC ACID C:3500
tri(HYDROXYTRIETHYL)AMINE T:3200
2-HYDROXYTRIETHYLAMINE D:3500
2-HYDROXY-m-XYLENE X:0400
HYDRURE de LITHIUM (French) L:2800
HYLEMOX E:0600
HYLENE M50 D:7400
HYLENE T T:1700
HYLENE TCPA T:1700
HYLENE TLC T:1700
HYLENE TM T:1700
HYLENE TM-65 T:1700
HYLENE TRF T:1700
HYOXYL H:1600
HYPEROL U:1100
HY-PHI 2066 O:5100
HY-PHY 1055 O:5100
HYPNONE A:0450
HYPOCHLOROUS ACID, CALCIUM C:1000
HYPOCHLOROUS ACID, SODIUM SALT S:3900
HYPODERMACIDE T:2050
HYPONITROUS ACID ANHYDRIDE N:2300
HYSTRENE 9512 L:0300
HYTROL O C:5950
IBA I:0350
IBN I:0550
ICEON 22 M:5800
IDI I:1050
IGE I:1450
IKURIN A:3800
ILLOXOL D:2700
ILLUMINATING OIL K:0150
2,2'-IMINOBISETHANOL D:2750
2,2'-IMINOBISETHYLAMINE D:3450
1,1'-IMINODI-2-PROPANOL D:4550
2,2'-IMINODIETHANOL D:2750
IMOL-140 T:2950
IMPERIAL GREEN C:4050
IMPERON FIXER T T:4650
IMPF S:0250
INACTIVE LIMONENE D:7150
INAKOR A:5750
INDOPOL L 50 P:2150
INEDIBLE TALLOW T:0150
INERTEEN P:2200
INEXIT B:0600
INFILTRINA D:6200
INHIBINE H:1600
INSULATING OIL O:4900
INVERTON 245 T:2500
IODOMETHANE M:4450
IODURE DE METHYLE (French) M:4450
IOPEZITE P:2900
IPA I:1250
IPANER D:0100
IPDI I:1050
IPE I:1175
IPERSAN T:3800
IPRIT B:2025
IRADICAR S:5800
IRADICAV S:5800
IRIUM D:8700
IRON AMMONIUM SULFATE F:0550
IRON ALLOY, BASE F:0450
IRON(II) CHLORIDE F:0600
IRON (III) CHLORIDE F:0200
IRON DICHLORIDE F:0600
IRON FLUORIDE F:0250
IRON NITRATE F:0350
IRON(III) NITRATE, ANHYDROUS F:0350
IRON PERCHLORIDE F:0200
IRON PERSULFATE F:0400
IRON PROTOCHLORIDE F:0600
IRON PROTOSULFATE F:0750
IRON PROTOXALATE F:0700
IRON SESQUISULFATE F:0400
IRON-SILICON ALLOY F:0500
IRON(2+) SULFATE F:0750
IRON(2+) SULFATE F:0750
IRON(3+) SULFATE F:0400
IRON(II) SULFATE F:0750
IRON(III) SULFATE F:0400
IRON(OUS) SULFATE F:0750
IRON SULFATE F:0750
IRON SULFATE F:0400
IRON TERSULFATE F:0400
IRON TRICHLORIDE F:0200
IRON TRINITRATE F:0350
IRON VITRIOL F:0750
IROSPAN F:0750
ISCEON 113 T:2900
ISCEON 122 D:1650
ISCEON 131 T:2400
ISCOBROME M:3100
ISCOBROME D E:4000
ISOACETOPHORONE I:0950
ISOAMYLACETATE I:0100
ISOAMYL ALCOHOL, PRIMARY I:0150
ISOAMYL ALCOHOL, SECONDARY I:0200
sec-ISOAMYL ALCOHOL I:0200
ISOAMYL ETHANOATE A:4250
ISOAMYL ETHANOATE I:0100
ISOAMYL HYDRIDE I:0900
ISOAMYOL I:0150
ISOBAC 20 H:0600
ISOBENZOFURAN,1,3-DIHYDRO-1,3-DIOXO- P:2000
1,3-ISOBENZOFURANDIONE P:2000
ISOBUTANAL B:4850
ISOBUTANE I:0250
ISOBUTANO (Spanish) I:0250
ISOBUTANOL I:0350
ISOBUTANOL AMINE A:1925
ISOBUTANOL-2-AMINE A:1925
ISOBUTENE I:0450
ISOBUTENE TRIMER T:3900
ISOBUTENYL METHYL KETONE M:1150
ISOBUTILAMINA (Spanish) I:0400
ISOBUTILENO (Spanish) I:0450
ISOBUTIRONITRILO (Spanish) I:0550
ISOBUTYL ACETATE I:0300
ISOBUTYL ACRYLATE I:0325
ISOBUTYL ALCOHOL I:0350
ISOBUTYL ALDEHYDE B:4850
ISOBUTYLAMINE I:0400
ISOBUTYL CARBINOL I:0350
ISOBUTYL CARBINOL I:0150 xxx
ISOBUTYLENE I:0450
ISOBUTYLETHENE M:5000
ISOBUTYL ISOBUTYRATE I:0475
ISOBUTYL KETONE D:4250
ISOBUTYL METHACRYLATE I:0485
ISOBUTYL α-METHACRYLATE I:0485
ISOBUTYLMETHYLCARBINOL M:4500
ISOBUTYLMETHYL CARBINOL M:2000
ISOBUTYL METHYL KETONE M:4550
ISOBUTYLMETHYLMETHANOL M:2000
ISOBUTYL 2-METHYL-2-PROPENOATE I:0485
ISOBUTYL PHTHALATE D:4300
ISOBUTYL PROPENOATE I:0325
ISOBUTYL 2-PROPENOATE I:0325
ISOBUTYRALDEHYDE B:4850
ISOBUTYRIC ACID I:0500
ISOBUTYRIC ACID, ISOBUTYL ESTER I:0475
ISOBUTYRIC ALDEHYDE B:4850
ISOBUTYRONITRILE I:0550
ISOCIANATO de METILO (Spanish) M:4600
ISOCTYL TRICHLOROPHENOXY-ACETATE T:2550
ISOCUMENE P:4050
ISOCYANATE de METHYLE (French) M:4600
ISOCYANATE METHANE M:4600
ISO-CYANATOMETHANE M:4600
ISOCYANIC ACID, METHYL ESTER M:4600
3-ISOCYANATOMETHYL-3,5,5-TRIMETHYLCYCLOHEXYL-ISOCYATE I:1050
ISOCYANIC ACID, METHYLPHENYLENE ESTER T:1700
ISOCYANIC ACID, 4-METHYL-m-PHENYLENE ESTER T:1700

ISOCYANURIC ACID, DICHLORO-, POTASSIUM SALT P:2850
ISOCYANURIC CHLORIDE T:2850
ISODECALDEHIDO (Spanish) I:0600
ISODECALDEHYDE I:0600
ISODECALDEHYDE, MIXED Isomers I:0600
ISODECANOIC ACID D:5850
ISODECANOL I:0700
ISODECYL ACRYLATE I:0650
ISODECYL ALCOHOL I:0700
ISODECYL PROPENOATE I:0650
ISODIPRENE C:1900
ISODURENE T:1000
ISOFLUROPHATE D:0850
ISOFORONA (Spanish) I:0950
ISOFORONA DIAMINA (Spanish) I:1000
ISOHEXANE I:0750
ISOHEXANO (Spanish) I:0750
ISOHEXENE M:5000
ISOHEXYL ALCOHOL M:4500
ISOL H:1200
ISONATE D:7400
ISONITROPROPANE N:2050
ISOOCTALDEHYDE I:0800
ISOOCTANOL I:0850
ISOOCTENE D:4200
ISOOCTYL ALCOHOL I:0850
ISOOCTYLALDEHYDE I:0800
ISOOCTYL PHTHALATE D:4500
ISOPENTALDEHYDE I:1700
ISOPENTANE I:0900
ISOPENTYL ACETATE A:4250
ISOPENTYL ACETATE I:0100
ISOPENTYL ALCOHOL I:0150
ISOPENTYL NITRITE A:4600
ISOPHORONE I:0950
ISOPHORONE DIAMINE I:1000
ISOPHORONE DIAMINE DIISOCYANATE I:1050
ISOPHORONE DIISOCYANATE I:1050
ISOPHTHALIC ACID I:1100
ISOPRENE I:1150
ISOPRENO (Spanish) I:1150
ISOPROPANOL I:1250
ISOPROPANOLAMINA (Spanish) M:6000
ISOPROPANOLAMINE M:6000
ISOPROPENE CYANIDE M:1300
ISOPROPENYL BENZENE M:5400
ISOPROPENYL CARBINOL M:1800
ISOPROPENYL METHYL KETONE M:4650
ISOPROPENYLNITRILE M:1300
ISOPROPILAMINA (Spanish) I:1300
o-ISOPROPILFENOL (Spanish) I:1650
ISOPROPIL GLICIDIL ETER (Spanish) I:1450
ISOPROPILICO (Spanish) I:1400
ISOPROPILMERCAPTANO (Spanish) I:1500
ISOPROPOXIETANOL (Spanish) I:1175
2-ISOPROPOXYETHANOL I:1175
ISOPROPOXYMETHYLPHORYL, FLUORIDE S:0250
2-ISOPROPOXY PROPANE I:1400
ISOPROPYL ACETATE I:1200
ISOPROPYLACETONE M:4550
ISOPROPYL ALCOHOL I:1250
ISOPROPYLAMINE I:1300
ISOPROPYLAMINEDODECYLBENZENES ULFONATE D:8250
ISOPROPYL BENZENE C:5450
ISOPROPYLBENZENE HYDROPEROXIDE C:5500
ISOPROPYLBENZOL C:5450
ISOPROPYLCARBINOL I:0350
ISOPROPYL CELLOSOLVE I:1175
ISOPROPYLCUMYLHYDROPEROXIDE D:4700
ISOPROPYL CYANIDE I:0550
ISOPROPYL CYCLOHEXANE I:1350
ISOPROPYL 2,4-DICHLOROPHENOXY ACETATE D:0150
ISOPROPYL EPOXYPROPYL ETHER I:1450
ISOPROPYL ESTER OF ACETIC ACID I:1200
ISOPROPYL ETHER I:1400
ISOPROPYL FLUOPHOSPHATE D:0850
ISOPROPYLFORMIC ACID I:0500
ISOPROPYL GLYCIDYL ETHER I:1450
ISOPROPYL GLYCOL I:1175
4,4'-ISOPROPYLIDENDIPHENOL B:1950
ISOPROPYLIDENEACETONE M:1150
4,4'-ISOPROPYLIDENEDIPHENO EPICHLOROHYDRIN RESIN B:2000
4,4'-ISOPROPYLIDENEDIPHENOL DIGLYCIDYL ETHER B:2000
ISOPROPYL MERCAPTAN I:1500
ISOPROPYL METHANEFLUOROPHOSPHONATE ISOPROPHYL S:0250
4-ISOPROPYL-1-METHYL BENZENE C:6350
ISOPROPYL METHYLFLUOROPHOSPHATE S:0250
ISOPROPYL METHYL KETONE M:3150
ISOPROPYL PERCARBONATE I:1600
ISOPROPYL PEROXYDICARBONATE I:1600
2-ISOPROPYL PHENOL I:1650
o-ISOPROPYL PHENOL I:1650
ISOPROPYL METHYLPHOSPHONO-FLUORIDATE S:0250
o-ISOPROPYL METHYLPHOSPHONO-FLUORIDATE S:0250
ISOPROPYL-METHYL-PHOSPHORYL FLUORIDE S:0250
ISOPROPYLOXITOL I:1175
ISOPROPYLTHIOL I:1500
4-ISOPROPYL TOLUENE C:6350
p-ISOPROPYLTOLUENE C:6350
ISOPROPYLTOLUOL C:6350
(ISOPROXYMETHYL)OXIRANE I:1450
ISOTHIOCYANATE de METHYLE (French) M:4700
ISOTHIOCYANIC ACID, METHYL ESTER M:4700
ISOTHIOCYANOMETHANE M:4700
ISOTHIOUREA T:1350
ISOTOX B:0600
ISOTRIDECANOL T:3050
ISOTRIDECYL ALCOHOL T:3050
ISOTRON 11 T:2400
ISOTRON 12 D:1650
ISOTRON-22 M:5800
ISOUREA U:0900
ISOVALERAL I:1700
ISOVALERALDEHYDE I:1700
ISOVALERIC ALDEHYDE I:1700
ISOVALERONE D:4250
p-ISOXAZINE,TETRAHYDRO- M:6100
ITOPAZ E:0600
IVALON F:1250
IVORAN D:0300
IXODEX D:0300
JACUTIN B:0600
JAPAN CAMPHOR C:1400
JASMOLIN I P:4750
JASMOLIN II P:4750
JAVELLE WATER S:3900
JAYFLEX DTDP D:7850
JAYSOL S E:1900
JDB-50-T C:6000
JEFFERSOL EB E:4900
JEFFERSOL EE E:0800
JEFFERSOL EM E:5100
JEFFOX P:2450
JET A J:0100
JET A-1 J:0100
JET FUEL O:2400
JET FUEL: JP-1 K:0150
JET FUELS: JP-1 J:0100
JET FUELS: JP-4 J:0200
JET FUELS: JP-5 J:0250
JP-1 O:2400
JUDEAN PITCH A:5650
JUDEAN PITCH A:5600
JULIN'S CARBON CHLORIDE H:0400
K-52 O:5100
K 113 D:4750
K-FLEX DP D:7550
KADMIUM CHLORID (German) C:0250
KALIUM P:2500
KALIUM-CYANID (German) P:2800
KALIUMCHLORAT (German) P:2700
KALIUMDICHROMAT (German) P:2900
KALIUMHYDROXID (German) P:2950
KALIUMPERMANGANAT (German) P:3200
KALPHOS P:0200
KALZIUMARSENIAT (German) C:0550
KAM 1000 S:5900
KAM 2000 S:5900
KAMFOCHLOR T:1950
KAMPFER (German) C:1400
KAMPSTOFF LOST B:2025
KANECHLOR P:2200
KANECHLOR 300 P:2200
KANECHLOR 400 P:2200
KANECHLOR 500 P:2200
KARBASPRAY C:1550
KARBATOX C:1550
KARBOFOS M:0250
KARBOSEP C:1550
KARMEX D:7950
KARMEX DIURON HERBICIDE D:7950
KARMEX DW D:7950
KARSAN F:1250
KATCHUNG OIL O:1900
KAUTSCHIN D:7150
KAYAFUME M:3100
KAYAZINON D:1000
KAYAZOL D:1000
KAZOE S:2000
KCN P:2800
KELENE E:2900
KEL F MONOMER T:3750
KELTHANE D:2600
p,p-KELTHANE D:2600
KELTHANETHANOL D:2600
KENAPON D:0200
KEPONE K:0100
KERALYT S:0200
KEROSENE O:2400
KEROSENE K:0150
KEROSENE, HEAVY O:4600
KEROSENE, HEAVY J:0250
KEROSENO (Spanish) K:0150
KEROSINE O:2400
KEROSINE K:0150
2-KETO-1,7,7-TRIMETHYLNORCAMPHANE C:1400
2-KETOHEPTANE A:4500
KETOHEXAMETHYLENE C:5950
2-KETOHEXAMETHYLENIMINE C:1450
KETONE A:0300
KETONE, BUTYL METHYL M:3350

KETONE, DIMETHYL A:0300
KETONE, ETHYL METHYL M:4100
KETONE, HEPTYL METHYL M:4300
KETONE, METHYL PHENYL A:0450
KETONE PROPANE A:0300
KETONOX B:3050
β-KETOPROPANE A:0300
KETTLE RENDERED LARD O:1600
KILEX-3 T:2550
KILL-ALL S:1900
KILLAX T:0800
KILLGERM DETHLAC INSECTICIDAL LAQUER D:2700
KILLMASTER D:8900
KILMITE 40 T:0800
KILRAT Z:1900
KING'S GOLD A:5550
KING'S GREEN C:4050
KING'S YELLOW A:5550
KLOREX S:2800
KMC-R 113 D:4750
KOBALT CHLORID (German) C:3650
KODAFLEX DOA D:7000
KOH P:2950
KOHLENDIOXYD (German) C:1650
KOHLENDISULFID (SCHWEFELKOHLENSTOFF) (German) C:1700
KOHLENMONOXID (German) C:1750
KOHLENSAURE (German) C:1650
KOKOTINE B:0600
KOLOFOG S:6600
KOLOSPRAY S:6600
KONLAX D:7050
KOP-THIODAN E:0100
KOP-THION M:0250
KOPFUME E:4000
KOPSOL D:0300
KOSATE D:7050
KRECALVIN D:2650
KREGASAN T:1450
KRESOLE (German) C:5100
KROTILINE D:0100
KUMULUS S:6600
KUPPERSULFAT (German) C:4800
KURAN T:2650
KURON T:2650
KUROSAL T:2650
KUROSALG T:2650
KUSA-TOHRUKUSATOL S:2800
KWELL B:0600
KWIK-KIL S:6200
KWIKSAN P:1400
KWIT E:0600
KYPCHLOR C:2050
KYPFOS M:0250
KYPTHION P:0200
L-310 O:3400
L-36352 T:3800
LACTATE d'ETHYLE (French) E:6700
LACTATO de n-BUTILO (Spanish) B:4350
LACTATO de COBRE (Spanish) C:4550
LACTATO de ETILO (Spanish) E:6700
LACTIC ACID L:0100
LD-LACTIC ACID L:0100
LACTIC ACID, BUTYL ESTER B:4350
LACTIC ACID, ETHYL ESTER E:6700
LAMBETH P:2400
LAH L:2500
LANADIN T:2350
LANARKITE L:1500
LANETTE WAX-S D:8700
LAPIS INFERNALIS S:1000

LARD O:1600
LARD OIL, ANIMAL O:1600
LARD OIL, COMMERCIAL O:1600
LARD OIL, MINERAL O:1600
LARD OIL NO. 1 O:1600
LARD OIL NO. 2 O:1600
LARD OIL, PURE O:1600
LARVACIDE C:2800
LATEX L:0200
LATEX, LIQUID SYNTHETIC L:0200
LAUGHING GAS N:2300
LAUREL CAMPHOR C:1400
LAURIC ACID L:0300
LAURIC ALCOHOL D:8000
LAURIC ALCOHOL L:2000
LAURILMERCAPTANO (Spanish) L:0500
LAUROSTEARIC ACID L:0300
LAUROX L:0400
LAUROYL PEROXIDE L:0400
LAURYDOL L:0400
LAURYL ALCOHOL D:8000
n-LAURYL ALCOHOL D:8000
LAURYL AMMONIUM SULFATE A:3050
LAURYL MAGNESIUM SULFATE D:8650
n-LAURYL MERCAPTAN L:0500
LAURYL METHACRYLATE D:8450
LAURYL SODIUM SULFATE D:8700
LAURYL SULFATE, DIETHANOLAMINE SALT SOLUTION D:8600
LAURYL SULFATE, MAGNESIUM SALT D:8650
LAURYL SULFATE, SODIUM SALT D:8700
LAURYL SULFATE, TRIETHANOLAMINE SALT D:8750
LAURYLBENZENE D:8100
LAURYLBENZENESULFONIC ACID D:8150
LAUXTOL P:0350
LAUXTOL A P:0350
LAVATAR O:3200
LAWN-KEEP D:0100
LAWRENCITE F:0600
LAXINATE D:7050
Le-100 T:0050
LEAD ACETATE L:0600
LEAD(4+) ACETATE L:1700
LEAD(2+) ACETATE L:0600
LEAD(II) ACETATE L:0600
LEAD(IV) ACETATE L:1700
LEAD ACETATE TRIHYDRATE L:0600
LEAD ACETATE(II), TRIHYDRATE L:0600
LEAD ACID ARSENATE L:0700
LEAD ALKYL MIXTURE M:6150
LEAD ARSENATE L:0700
LEAD BOTTOMS L:1500
LEAD CHLORIDE L:0800
LEAD(2+) CHLORIDE L:0800
LEAD(II) CHLORIDE L:0800
LEAD DIACETATE L:0600
LEAD DICHLORIDE L:0800
LEAD DIFLUORIDE L:0900
LEAD DINITRATE L:1300
LEAD FLUORIDE L:0900
LEAD(2+) FLUORIDE L:0900
LEAD(II) FLUORIDE L:0900
LEAD FLUOROBORATE L:1000
LEAD FLUOROBORATE SOLUTION L:1000
LEAD HYPOSULFITE L:1200
LEAD IODIDE L:1100
LEAD IOSULFATE L:1200
LEAD MONOXIDE L:2300
LEAD NITRATE L:1300
LEAD(2+) NITRATE L:1300
LEAD OXIDE L:2300

LEAD(2+) OXIDE L:2300
LEAD OXIDE YELLOW L:2300
LEAD(II) OXIDE L:2300
LEAD PROTOXIDE L:2300
LEAD STEARATE L:1400
LEAD SULFATE L:1500
LEAD(II) SULFATE L:1500
LEAD SULFIDE L:1600
LEAD SULFOCYANATE L:1800
LEAD TETRAACETATE L:1700
LEAD TETRAETHYL T:0750
LEAD TETRAMETHYL T:1050
LEAD THIOCYANATE L:1800
LEAD TUNGSTATE L:1900
LEAD WOLFRAMATE L:1900
LEAF LARD O:1600
LE CAPTANE (French) C:1475
LEDON 11 T:2400
LEDON 12 D:1650
LEIVASOM T:2050
LEMONENE D:7200
LENDINE B:0600
LENTOX B:0600
LETHALAIRE G-52 T:0800
LETHALAIRE G-54 P:0200
LETHALAIRE G-57 T:0600
LETHURIN T:2350
LEUCOL Q:0100
LEUCOLINE Q:0100
LEUKOL Q:0100
LEVEPOX HARDENER T3 P:0600
LEWISITE (ARSENIC COMPOUND) B:2025
LEWISITE (L) B:2025
LEWIS-RED DEVIL LYE S:3800
LEYTOSAN P:1400
LICHENIC ACID F:1400
LIDENAL B:0600
LIGHT NAPHTHA N:0250
LIGHT OIL O:3200
LIGROIN (petroleum benzene) N:0200
LIGROIN M:5600
LIGROINA (Spanish) M:5600
LILLY 36,352 T:3800
LIME C:1100
LIME, BURNED C:1100
LIME, BURNT C:1100
LIME CHLORIDE C:1000
LIME SALTPETER C:1050
LIME, UNSLAKED C:1100
LIME WATER C:0900
LIMED ROSIN C:1300
LIMONENE D:7150
LINDAFOR B:0600
LINDAGAM B:0600
LINDAGRAIN B:0600
LINDAGRANOX B:0600
LINDANE B:0600
γ-LINDANE B:0600
LINDAPOUDRE B:0600
LINDATOX B:0600
LINDOL T:2950
LINDOSEP B:0600
LINEAR ALCOHOLS L:2000
LINE RIDER T:2500
LINSEED OIL O:3400
LINTOX B:0600
LNG L:2100
LIQUAMON 28 U:1000
LIQUID AMMONIA A:2000
LIQUID ASPHALT O:4300
LIQUID ASPHALTUM A:5650
LIQUID BLEACH S:3900
LIQUID CAMPHOR C:1400

LIQUID GUM CAMPHOR C:1400
LIQUID HYDROGEN H:1350
LIQUID IMPURE CAMPHOR C:1400
LIQUID NITROGEN N:1650
LIQUIDOW C:0750
LIQUID OXYGEN O:5600
LIQUID PETROLATUM O:3600
LIQUID PITCH OIL C:5050
LIQUID ROSIN T:0100
LIQUID ROSIN O:4700
LIQUEFIED HYDROCARBON GAS L:2200
LIQUEFIED NATURAL GAS L:2100
LIQUEFIED PETROLEUM GAS B:2900
LIQUEFIED PETROLEUM GAS L:2200
LIQUEFIED PETROLEUM GAS P:3300
LIQUIPHENE P:1400
LIROHEX T:0800
LIROPON D:0200
LIROPREM P:0350
LIROTHION P:0200
LITARGIRIO (Spanish) L:2300
LITHARGE L:2300
LITHARGE YELLOW L-28 L:2300
LITHIUM L:2400
LITHIUM ALUMINUM HYDRIDE L:2500
LITHIUM BICHROMATE DIHYDRATE L:2600
LITHIUM BICHROMATE L:2600
LITHIUM CHROMATE L:2700
LITHIUM DICHROMATE L:2600
LITHIUM HYDRIDE L:2800
LITHIUM METAL L:2400
LITHIUM MONOHYDRIDE L:2400
LITHIUM MONOHYDRIDE L:2800
LITHIUM TETRAHYDROALUMINATE L:2500
LITIO (Spanish) L:2400
LN N:1650
LO-BAX C:1000
LOISOL T:2050
LONG-TIME BURNING OIL O:3700
LOREX S:2800
LOREXANE B:0600
LOROL D:8000
LOROL L:2000
LOROL-20 O:0300
LOROL-22 D:0700
LORSBAN D:8900
LOSANTIN C:1000
LOST B:2025
N-LOST (German) B:2025
S-LOST B:2025
LOSUNGSMITTEL APV D:3400
LOX O:5600
LPG L:2200
LPG ETHYL MERCAPTAN 1010 E:6800
LPT B:3050
LUBRICATING OIL O:3500
LUBRICATING OIL O:3600
LUBRICATING OIL O:3800
LUBRICATING OIL, TURBINE O:5000
LUCIDOL D:1050
LUMBRICAL P:2100
LUMBUCAL P:2100
LUNAR CAUSTIC S:1000
LUPERCO C:6000
LUPERSOL B:3050
LUTROL-9 E:4200
LYE S:3800
LYE P:2950
LYE SOLUTION S:3800
LYP L:0400
LYP 97 L:0400

LYSOFORM F:1250
M-74 D:7700
M105 B:3050
M 140 C:2050
M-176 D:6200
M 410 C:2050
M 5055 T:1950
MA M:2050
MAA M:0350
MA-1214 L:2000
MA 1214 D:8000
MAA M:4500
MAAC M:1950
MACE C:2250
MACH-NIC N:1200
MACQUER'S SALT P:2550
MACROGOL 400 BPC E:4200
MACRONDRAY D:0100
MADERA CARBON (Spanish) C:2000
MAFU D:2650
MAGNESIO (Spanish) M:0100
MAGNESIUM M:0100
MAGNESIUM DODECYL SULFATE D:8650
MAGNESIUM LAURYL SULFATE D:8650
MAGNESIUM METAL M:0100
MAGNESIUM NITRATE M:0150
MAGNESIUM(2+) NITRATE M:0150
MAGNESIUM(II) NITRATE M:0150
MAGNESIUM PERCHLORATE HEXAHYDRATE M:0200
MAGNESIUM PERCHLORATE M:0200
MAGNESIUM PERCHLORATE, ANHYDROUS M:0200
MAGNESIUM RIBBONS M:0100
MAGNETIC 70 S:6600
MAGNETIC 90 S:6600
MAGNETIC 95 S:6600
MALACIDE M:0250
MALAFOR M:0250
MALAGRAN M:0250
MALAKILL M:0250
MALAMAR M:0250
MALAMAR 50 M:0250
MALAPHELE M:0250
MALAPHOS M:0250
MALASOL M:0250
MALASPRAY M:0250
MALATHION M:0250
MALATHIOZOO M:0250
MALATHON M:0250
MALATOL M:0250
MALATOX M:0250
MALAZIDE M:0400
MALDISON M:0250
MALEIC ACID M:0300
MALEIC ACID ANHYDRIDE M:0350
MALEIC ACID HYDRAZIDE M:0400
MALEIC ANHYDRIDE M:0350
MALEIC HYDRAZIDE M:0400
MALEINIC ACID M:0300
MALENIC ACID M:0300
N,N-MALEOYLHYDRAZINE M:0400
MALIX E:0100
MALMED M:0250
MALONIC DINITRILE P:3450
MALONIC MONONITRILE C:5600
MALONONITRILE P:3450
MALPHOS M:0250
MALTOX M:0250
MANGANESE CYCLOPENTADIENYL TRICARBONYL M:3800
MANGANESE TRICARBONYL METHYLCYCLOPENTADIENYL M:3800

MANOXAL OT D:7050
MAOH M:4500
MAOH M:2000
MAPP GAS M:1650
MAPROFIX 563 D:8700
MARKSMAN 2, TRIGARD T:3800
MARLATE M:1500
MARLATE 50 M:1500
MARMER D:7950
MAROXOL-50 D:6650
MARSH GAS M:1400
MARSHITE C:4500
MARVEX D:2650
MASSICOT L:2300
MATRICARIA CAMPHOR C:1400
MATTING ACID S:6800
MAZIDE M:0400
MAZOTEN T:2050
M-B-C FUMIGANT M:3100
MBA B:2025
MBDA D:1300
MBI D:7400
MBK M:3350
MC M:0600
MCA M:5750
MCB C:2400
MCE T:0050
MCF M:3550
MCP C:1200
MDB D:1500
MDCB D:1500
ME-1700 D:0250
MEADOW GREEN C:4050
MECB D:3415
MECHLORETHAMINE B:2025
MECS E:5100
MEDIBEN D:1300
MEETCO M:4100
MEG E:4200
MEK M:4100
MEK PEROXIDE B:3050
MEKP B:3050
MEKP-HA 1 B:3050
MEKP-LA 1 B:3050
MELDANE C:5000
MELDONE C:5000
MELIPAX T:1950
MENDRIN E:0200
MENIPHOS P:1500
MENITE P:1500
p-MENTHA-1, 8-DIENE D:7150
p-MENTHA-1,8-DIENE, DL- D:7150
1,8(9)-p-MENTHADIENE D:7150
MEP M:4150
ME-PARATHION M:4900
MEPATON M:4900
MEPOX M:4900
MER M:4800
MERCAPTAN METHYLIQUE (French) M:4750
MERCAPTOBENZENE B:0800
2-MERCAPTOBENZOTHIAZOL, SODIUM SALT S:4000
2-MERCAPTOBENZOTHIAZOL SOLUTION S:4000
2-MERCAPTOBENZOTIAZOL SODICO (Spanish) S:4000
1-MERCAPTOBUTANE B:4400
MERCAPTODIMETHUR M:0450
1-MERCAPTODODECANE L:0500
MERCAPTOETHANE E:6800
MERCAPTOMETHANE M:4750
MERCAPTOPHOS D:0800

2-MERCAPTOPROPANE I:1500
3-MERCAPTOPROPANE P:4650
3-MERCAPTOPROPANOL P:4650
MERCAPTOTHION M:0250
MERCOL 25 D:8300
MERCURAM T:1450
MERCURE (French) M:1100
MERCURIACETATE M:0500
MERCURIALIN M:1850
MERCURIALIN SOLUTION M:1900
MERCURIC ACETATE M:0500
MERCURIC AMMONIUM CHLORIDE M:0550
MERCURIC BICHLORIDE M:0600
MERCURIC CHLORIDE M:0600
MERCURIC CHLORIDE, AMMONIATED M:0550
MERCURIC CYANIDE M:0650
MERCURIC DIACETATE M:0500
MERCURIC IODIDE M:0700
MERCURIC IODIDE, RED M:0700
MERCURIC NITRATE M:0750
MERCURIC OXIDE M:0800
MERCURIC OXIDE, RED M:0800
MERCURIC OXIDE, YELLOW M:0800
MERCURIC RHODANIDE M:0950
MERCURIC SULFATE M:0850
MERCURIC SULFIDE M:0900
MERCURIC SULFIDE, BLACK M:0900
MERCURIC SULFIDE, RED M:0900
MERCURIC SULFOCYANATE M:0950
MERCURIC SULFOCYANIDE M:0950
MERCURIC THIOCYANATE M:0950
MERCURIO (Spanish) M:1100 M:1100
MERCUROUS CHLORIDE M:1000
MERCUROUS NITRATE M:1050
MERCUROUS NITRATE MONOHYDRATE M:1050
MERCURY M:1100
MERCURY ACETATE M:0500
MERCURY(2+) ACETATE M:0500
MERCURY(II) ACETATE M:0500
MERCURY AMIDE CHLORIDE M:0550
MERCURY AMMONIUM CHLORIDE M:0550
MERCURY BICHLORIDE M:0600
MERCURY BINIODIDE M:0700
MERCURY BISULFATE M:0850
MERCURY(2+) CHLORIDE M:0600
MERCURY(II) CHLORIDE M:0600
MERCURY CYANIDE M:0650
MERCURY(2+) CYANIDE M:0650
MERCURY(II) CYANIDE M:0650
MERCURY DIACETATE M:0500
MERCURY, ELEMENTAL M:1100
MERCURY(2+) IODIDE M:0700
MERCURY(II) IODIDE M:0700
MERCURY, METALLIC M:1100
MERCURY MONOCHLORIDE M:1000
MERCURY NITRATE M:1050
MERCURY(II) NITRATE M:0750
MERCURY NITRATE MONOHYDRATE M:0750
MERCURY OXIDE M:0800
MERCURY PERCHLORIDE M:0600
MERCURY PERNITRATE M:0750
MERCURY PERSULFATE M:0850
MERCURY PROTOCHLORIDE M:1000
MERCURY PROTONITRATE M:1050
MERCURY RHODANIDE M:0950
MERCURY SUBCHLORIDE M:1000
MERCURY(II) SULFATE M:0850
MERCURY(2+) SULFATE M:0850

MERCURY(II) THIOCYANATE M:0950
MERCURYL ACETATE M:0500
MEREX K:0100
MERPAN C:1475
MERVAMINE D:7050
MESITYL OXIDE M:1150
MESUROL M:0450
MET-SPAR C:0900
METACETONE D:3700
METACETONIC ACID P:3700
METACIDE M:4900
METACRILATO de n-BUTILO (Spanish) B:4450
METACRILATO de iso-BUTILO (Spanish) I:0485
METACRILATO de ETILO (Spanish) E:6900
METACRILATO de GLICIDILO (Spanish) G:0675
METACRILATO de HIDROXIPROPILO (Spanish) H:2000
METACRILATO de METILO (Spanish) M:4800 G:0675
METACRILONITRILO (Spanish) M:1300
METAFOS M:4900
METALLIC ARSENIC A:5250
METALLIC MERCURY M:1100
METALLIC RESINATE C:1300
METANO (Spanish) M:1400
METANOL (Spanish) M:1750
METAPHOR M:4900
METAPHOS M:4900
METAPHOSPHORIC ACID P:2350
METAQUEST A E:3900
METAUPON O:5100
METELILACHLOR M:5550
METHACETONE D:3700
METHACIDE T:1600
METHACRYLATE de BUTYLE (French) B:4450
METHACRYLATE DE METHYLE (French) M:4800
METHACRYLIC ACID M:1200
2-METHACRYLIC ACID M:1200
α-METHACRYLIC ACID M:1200
METHACRYLIC ACID, 2,3-EPOXY PROPYL ESTER G:0675
METHACRYLIC ACID, BUTYL ESTER B:4450
METHACRYLIC ACID, BUTYL, DECYL, CETYL AND EICOSYL ESTER MIX M:1250
METHACRYLIC ACID, DODECYL ESTER D:8450
METHACRYLIC ACID, DODECYL AND PENTADECYL ESTER MIX D:8500
1-2-METHACRYLIC ACID, ETHYL ESTER E:6900
METHACRYLIC ACID, GLACIAL M:1200
METHACRYLIC ACID, ISOBUTY ESTER I:0485
METHACRYLIC ACID, LAURYL AND PENTADECYL ESTER MIX D:8500
METHACRYLIC ACID, METHYL ESTER M:4800
METHACRYLONITRILE M:1300
METHACRYLSAEURE BUTYL ESTER (German) B:4450
METHACRYLSAEUREMETHYL ESTER (German) M:4800
METHALLYL ALCOHOL M:1800
METHALLYL CHLORIDE M:1350
2-METHALLYL CHLORIDE M:1350
α-METHALLYL CHLORIDE M:1350
β-METHALLYL CHLORIDE M:1350

METHANAL F:1250
METHANAL SOLUTION F:1250
METHANAMIDE F:1300
METHANAMINE M:1850
METHANAMINE M:1850
METHANAMINE, N-METHYL- D:4950
METHANAMINE SOLUTION M:1900
METHANE M:1400
METHANEARSONIC ACID, SODIUM SALT M:1425
METHANE, BIS(2,3,5-TRICHLORO-6-HYDROXYPHENYL) H:0600
METHANECARBONITRIL A:0400
METHANECARBONITRILE A:0400
METHANE CARBOXYLIC ACID A:0200
METHANE, CHLORO- M:3450
METHANE, CHLOROMETHOXY- C:2600
METHANE, CYANO- A:0400
METHANE, DIBROMO- D:1150
METHANE, DICHLORODIFLUORO- D:1650
METHANE, DIMETHOXY- M:4200
METHANEDITHIOL, s,s-DIESTER with o,o-DIETHYL PHOSPHORODITHIOATE ACID E:0600
METHANEETHIOL M:4750
METHANETHIOMETHANE D:6150
METHANE GAS M:1400
METHANE, IODO- M:4450
METHANE, ISOCYANATO- M:4600
METHANE, ISOTHIOCYANATO- M:4700
METHANE, NITRO- N:1800
METHANE, PHENYL- T:1600
METHANE, REFRIGERATED LIQUID L:2100
METHANE TETRACHLORIDE C:1850
METHANE, TETRACHLORO- C:1850
METHANE, TETRANITRO- T:1100
METHANETHIOL M:4750
METHANE, TRIBROMO- B:2650
METHANE TRICHLORIDE C:2500
METHANE, TRICHLORO- C:2500
METHANE, TRICHLOROFLUORO- T:2400
METHANE, TRICHLORONITRO- C:2800
4,7-METHANO-1H-INDENE D:2675
4,7-METHANO-1H-INDENE,1,2,4,5,6,7,8,8-OCTACHLORO-2,3,3A,4,7,7A-HEXAHYDRO- C:2050
METHANOIC ACID F:1350
METHANOIC ACID, AMIDE F:1300
4,7-METHANOINDENE, 3A,4,7,7A-TETRAHYDRODIMETHYL M:3750
METHANOL M:1750
METHANOLACETONITRILE E:3700
METHANOL, ETHYNYL P:3550
METHANOL, SODIUM SALT S:4200
METHANOL, TRIMETHYL- B:3450
METHANTHIOL (German) M:4750
METHENEAMINE H:0900
METHENYL TRIBROMIDE B:2650
METHENYL TRICHLORIDE C:2500
METHIONINE HYDROXY ANALOG H:1900
METHMERCAPTURON M:0450
METHOGAS M:3100
o-METHOXYBENZOIC ACID M:5350
METHOXYBENZOYL CHLORIDE A:4800
1-METHOXYBUTANE-1,3-DIONE M:1600
3-METHOXY-1-BUTANOL ACETATE M:1450
3-METHOXYBUTYL ACETATE M:1450
1-METHOXYCARBONYL-1-PROPEN-2-YLDIMETHYL PHOSPHATE P:1500
METHOXYCARBONYL CHLORIDE M:3550
METHOXYCARBONYLETHYLENE M:1700
METHOXYCHLOR M:1500

METHOXYDIGLYCOL D:3415
METHOXY DDT M:1500
2-METHOXYETHANOL E:5100
METHOXY ETHER OF PROPYLENE GLYCOL P:4300
2-(2-METHOXYETHOXY)-ETHANOL D:3415
2-(2-METHOXYETHOXY)ETHANOL ACETATE D:3250
2-METHOXYETHYL ACETATE E:4800
2-METHOXYETHYL ACRYLATE E:4800
METHOXYETHYLENE V:1300
1-METHOXY-2-HYDROXYPROPANE P:4300
METHOXYMETHYL CHLORIDE C:2600
2-METHOXY-1-METHYLETHANOL P:4300
METHOXYMETHYL ETHER M:4200
2-METHOXY-2-METHYL PROPANE M:3300
1-METHOXY-2-PROPANOL P:4300
1-METHOXY-2-PROPANOL ACETATE P:4350
4-METHY-2-PENTYL ALCOHOL M:2000
N-(1-METHYETHYL)-2-PROPANAMINE D:4600
METHYLACETALDEHYDE P:3650
METHYLACETAT (German) M:1550
METHYL ACETATE M:1550
METHYLACETIC ACID P:3700
METHYLACETIC ACID ANHYDRIDE P:3750
METHYL ACETIC ESTER M:1550
METHYL ACETOACETATE M:1600
METHYL ACETONE M:4100
METHYL ACETYLACETATE M:1600
METHYL ACETYLACETONATE M:1600
METHYL ACETYLENE-ALLENE MIXTURE M:1650
METHYLACETYLENE–PROPADIENE MIXTURE M:1650
β-METHYL ACROLEIN C:5400
METHYL ACRYLAT (German) M:1700
METHYL ACRYLATE M:1700
2-METHYLACRYLONITRILE M:1300
α-METHYLACRYLONITRILE M:1300
2-METHYLACTONITRILE A:0350
METHYL ADIPATE D:4900
METHYLAL M:4200
METHYL ALCOHOL M:1750
METHYL ALCOHOL, SODIUM SALT S:4200
METHYL ALDEHYDE F:1250
METHYLALKOHOL (German) M:1750
METHYL ALLYL ALCOHOL M:1800
β-METHYLALLYL CHLORIDE M:1350
METHYLAMINE M:1850
METHYLAMINE M:1900
(METHYLAMINO)BENZENE M:2050
2-(METHYLAMINO)ETHANOL M:6050
N-METHYLAMINOBENZENE M:2050
1-METHYL-1,2-AMINO-BENZENE T:1850
n-METHYLAMINOETHANOL M:6050
METHYLAMYL ACETATE H:1150
METHYL AMYL ACETATE M:1950
METHYL AMYL ALCOHOL M:2000
METHYL AMYL-CETONE (French) A:4500
METHYL AMYL KETONE A:4500
METHYL n-AMYL KETONE A:4500
METHYL ANILINE M:2050
N-METHYLANILINE M:2050
2-METHYLANILINE T:1850
3-METHYLANILINE T:1800
4-METHYLANILINE T:1900
m-METHYLANILINE T:1800
o-METHYLANILINE T:1850
p-METHYLANILINE T:1900

METHYLANILINE (MONO) M:2050
2-METHYLAZACLYCLOPROPANE P:4400
METHYL AZINPHOS A:5800
2-METHYLAZIRIDINE P:4400
2-METHYLBENZENAMINE T:1850
3-METHYLBENZENAMINE T:1800
4-METHYLBENZENEAMINE T:1900
N-METHYLBENZENAMINE M:2050
m-METHYLBENZENAMINE T:1800
METHYL BENZENE T:1600
α-METHYL BENZENE METHANOL M:3050
METHYL BENZENECARBOXYLATE M:3000
METHYLBENZENESULFONIC ACID T:1750
4-METHYLBENZENESULFONIC ACID T:1750
METHYL BENZOATE M:3000
METHYLBENZOL T:1600
α-METHYLBENZYL ALCOHOL M:3050
METHYLBIS(2-CHLOROETHYL)AMINE B:2025
N-METHYL-BIS(2-CHLOROETHYL)AMINE (MAK) B:2025
N-METHYL-BIS-CHLORAETHYLAMIN (German) B:2025
α-METHYLBIVINYL P:0450
β-METHYLBIVINYL I:1150
METHYL BROMIDE M:3100
1-METHYLBUTADIENE P:0450
2-METHYL-1,3-BUTADIENE I:1150
3-METHYL-1,3-BUTADIENE I:1150
3-METHYL BUTAN-2-ONE M:3150
3-METHYLBUTANAL I:1700
METHYL-n-BUTANOATE M:3400
2-METHYL-2-BUTANOL M:4400
3-METHYL-2-BUTANOL I:0200
3-METHYL-1-BUTANOL (primary) I:0150
3-METHYL-1-BUTANOL ACETATE A:4250
3-METHYL-1-BUTANOL ACETATE I:0100
3-METHYL-2-BUTANONE M:3150
2-METHYLBUTANE I:0900
3-METHYL-BUTEN-(1)-OL-(3) M:3200
3-METHYL-1-BUTEN-3-OL M:3200
2-METHYL-3-BUTEN-2-OL M:3200
3-METHYL-3-BUTEN-2-ON (German) M:4650
2-METHYL-1-BUTENE-3-ONE M:4650
METHYL BUTENOL M:3200
3-METHYL-1-BUTYL ACETATE I:0100
1-METHYLBUTYL ACETATE A:4250
2-METHYL-BUTYLACRYLATE B:4450
1-METHYL-4-tert-BUTYLBENZENE B:4650
p-METHYL-tert-BUTYLBENZENE B:4650
METHYL-1,3-BUTYLENE GLYCOL ACETATE M:1450
3-METHYLBUTYL ESTER OF ACETIC ACID I:0100
1-METHYLBUTYL ETHANOATE A:4250
3-METHYLBUTYL ETHANOATE I:0100
METHYL BUTYL KETONE M:3350
3-METHYLBUTYL NITRITE A:4600
2-METHYL-3-BUTYN-2-OL M:3250
METHYL BUTYNOL M:3250
METHYL BUTYNOL M:4400
2-BUTYNOL M:4400
3-METHYLBUTYRALDEHYDE I:1700
METHYL BUTYRATE M:3400
METHYL n-BUTYL KETONE M:3350
METHYL n-BUTYRATE M:3400
METHYL tert-BUTYL ETHER M:3300
METHYLCARBAMATE C:1500
METHYLCARBAMATE 1-NAPHTHALENOL, METHYCARBAMATE C:1550

METHYLCARBAMATE 1-NAPHTHALENOL C:1550
N-METHYLCARBAMATE de 1-NAPHTYLE (French) C:1550
METHYLCARBAMIC ACID, 1-NAPHTHYL ESTER C:1550
METHYL CARBAMIC ACID 4-(METHYLTHIO)-3,5-XYLYL ESTER 4-METHYLMERCAPTO-3,5-DIMETHYL-PHENYL N-METHYLCARBAMATE M:0450
METHYLCARBAMYL AMINE M:4600
METHYL CARBINOL E:1900
METHYL CARBITOL D:3415
METHYL CARBITOL ACETATE D:3250
METHYL-4-CARBOMETHOXY BENZOATE D:6250
METHYL CARBONIMIDE M:4600
METHYL CELLOSOLVE E:5100
METHYL CELLOSOLVE ACETATE E:4800
METHYLCHLORID (German) M:3450
METHYL CHLORIDE M:3450
METHYLCHLOROMETHYL ETHER C:2600
METHYLCHLOROMETHYL ETHER, ANHYDROUS C:2600
METHYL CHLOROACETATE M:3500
2-METHYLCHLOROBENZENE C:3100
1-METHYL-2-CHLOROBENZENE C:3100
METHYL CHLOROCARBONATE M:3550
METHYL CHLOROFORM T:2250
METHYL CHLOROFORMATE M:3550
METHYL CHLOROPHOS T:2050
METHYL CHLOROSILANE M:5450
METHYL CYANIDE A:0400
1-METHYLCYCLOHEXAN-2-ONE M:3700
METHYL CYCLOHEXANE M:3600
METHYLCYCLOHEXANE M:3650
METHYLCYCLOHEXANOL M:3650
2-METHYLCYCLOHEXANOL M:3650
2-METHYLCYCLOHEXANONE M:3700
o-METHYLCYCLOHEXANONE M:3700
METHYLCYCLOPENTADIENE DIMER M:3750
METHYLCYCLOPENTADIENYL MANGANESE TRICARBONYL M:3800
METHYLCYCLOPENTANE M:3850
METHYL CYCLOPENTANE M:3850
METHYL DICHLOROACETATE M:3900
METHYL DICHLOROETHANOATE M:3900
N-METHYL-2,2'-DICHLORODIETHYLAMINE B:2025
METHYL DIETHANOLAMINE M:4000
N-METHYLDIETHANOLAMINE M:4000
METHYLDICHLOROSILANE M:3950
METHYL 1,1-DIMETHYLETHYL ETHER M:3300
METHYL 3-(DIMETHOXYPHOSPHINYL-OXYCROTONATE P:1500
2-METHYL-1,3-DINIROBENZENE D:6850
1-METHYL-2,4-DINITROBENZENE D:6800
4-METHYL-1,2-DINITROBENZENE D:6900
METHYL-E 605 M:4900
METHYLE (ACETATE de) (French) M:1550
METHYLE (SULFATE de) (French) D:6100
METHYLE FORMIATE de (French) M:4250
METHYLENE ACETONE M:5500
METHYLENE-s,s'-BIS(o,o-DIAETHYL-DITHIOPHOSPHAT) (German) E:0600
METHYLENE BISPHENYL ISOCYANATE D:7400
METHYLENE BIS(PHENYLISOCYANATE) D:7400
METHYLENEBIS(4-PHENYLISOCYANATE) D:7400

2,2'-METHYLENEBIS(3,4,6-TRICHLOROPHENOL) H:0600
METHYLENE BROMIDE D:1150
METHYLENE CHLORIDE D:2050
METHYLENE CYANIDE P:3450
METHYLENE DI-p-PHENYLENE ESTER OF ISOCYANIC ACID D:7400
METHYLENE DIMETHYL ETHER M:4200
METHYLENE DIBROMIDE D:1150
METHYLENE DICHLORIDE D:2050
4,4-METHYLENEDIPHENYL DIISOCYANATE D:7400
METHYLENEDIPHENYL DIISOCYANATE URETHANE POLYMER P:2300
METHYLENE GLYCOL F:1250
3-METHYLENE-7-METHYL 1,6-OCTADIENE M:6200
s,s'-METHYLENE O,O,O',O'-TETRAETHYL PHOSPHORODITHIOATE E:0600
s,s'-METHYLENE O,O,O',O'-TETRAETHYL ESTER PHOSPHORODITHIOIC ACID E:0600
METHYLENE OXIDE F:1250
METHYLENE TRIBROMIDE B:2650
METHYLETHENE P:4150
METHYL ESTER OF ACETIC ACID M:1550
METHYL ESTER OF ACRYLIC ACID M:1700
METHYL ESTER OF FORMIC ACID M:4250
METHYL ESTER OF ISOCYANIC ACID M:4600
METHYL ESTER OF SULFURIC ACID D:6100
METHYL ETHANOATE M:1550
n-METHYL ETHANOLAMINE M:6050
METHYL ETHER D:5200
METHYL ETHOXOL E:5100
1-METHYLETHYL ACETATE I:1200
2-METHYL-6-ETHYL ANILINE M:4050
(1-METHYLETHYL) BENZENE C:5450
1-(METHYLETHYL) BENZENE M:5400
o-METHYLETHYLBENZENE E:7900
2-METHYL-6-ETHYL BENZENEAMINE M:4050
METHYL ETHYL BROMO-METHANE B:2600
METHYL ETHYL CARBINOL B:3400
1-METHYLETHYLCYCLOHEXANE I:1350
METHYLETHYLENE P:4150
METHYLETHYLENE GLYCOL P:4200
METHYL ETHYLENE OXIDE P:4450
2-METHYLETHYLENIMINE P:4400
DI-(1-METHYLETHYL)ETHER I:1400
METHYL ETHYL KETONE PEROXIDE B:3050
METHYL ETHYL KETONE HYDROPEROXIDE B:3050
METHYL ETHYL KETONE M:4100
METHYL ETHYL METHANE B:2900
METHYLETHYLPYRIDINE M:4150
2-METHYL-5-ETHYLPYRIDINE M:4150
1-METHYL-2-FLUOROBENZENE F:1050
1-METHYL-3-FLUOROBENZENE F:1100
METHYL FORMAL M:4200
METHYL FORMATE M:4250
METHYLFORMIAT (German) M:4250
METHYL FOSFERNO M:4900
METHYL GLYCOL E:5100
METHYL GLYCOL P:4200
METHYL GLYCOL ACETATE E:4800
METHYL GUTHION A:5800
6-METHYL-1-HEPTANAL I:0800
6-METHYL-1-HEPTANOL I:0850
5-METHYL-3-HEPTANONE E:2300

METHYL HEPTYL KETONE M:4300
METHYL HYDRATE M:1750
1-METHYLHYDRAZINE M:4350
METHYLHYDRAZINE M:4350
p-METHYLHYDROXYBENZENE C:5100
METHYL HYDRIDE M:1400
METHYL HYDROXIDE M:1750
4-METHYL-4-HYDROXY-2-PENTANONE D:0900
2-METHYL-2-HYDROXY-3-BUTYNE M:4400
1-METHYL-2-HYDROXY-ETHYLAMINE P:3400
METHYL IODIDE M:4450
METHYL ISOAMYL ACETATE M:1950
METHYL ISOBUTENYL KETONE M:1150
METHYL ISOBUTYL CARBINOL M:2000
METHYL ISOBUTYL CARBINOL M:4500
METHYL-ISOBUTYL-CETONE (French) M:4550
METHYL ISOBUTYL KETONE M:4550
METHYL ISOCYANAT (German) M:4600
METHYL ISOCYANATE M:4600
1-METHYL-4-ISOPROPYLBENZENE C:6350
1-METHYL-4-ISOPROPENYL-1-CYCLOHEXENE D:7150
METHYL ISOPROPENYL KETONE M:4650
METHYL ISOPROPYL KETONE M:3150
METHYL ISOTHIOCYANATE M:4700
2,2'-METHYLIMINODIETHANOL M:4000
METHYLISOBUTYLCARBINOL ACETATE M:1950
METHYLISOBUTYLCARBINYL ACETATE M:1950
METHYLJODID (German) M:4450
METHYL KETONE A:0300
N-METHYL-LOST B:2025
METHYL MERCAPTAN M:4750
METHYL MERCAPTANE M:4750
4-METHYLMERCAPTO-3,5-XYLYLMETHYLCARBAMATE M:0450
METHYL-α-METHYLACRYLATE M:4800
N-METHYLMETHANAMINE D:4950
METHYLMETHANE E:0500
METHYL METHACRYLATE M:4800
METHYL METHANOATE M:4250
2-METHYL-2-METHOXY PROPANE M:3300
2-METHYL-6-METHYLENE-2,7-OCTADIENE M:6200
METHYL MONOCHLOROACETATE M:3500
METHYL MUSTARD OIL M:4700
1-METHYLNAPHTHALENE M:4850
α-METHYL NAPHTHALENE M:4850
N-METHYL-1-NAPHTHYL-CARBAMAT (German) C:1550
N-METHYL-1-NAPHTHYL CARBAMATE C:1550
N-METHYL-α-NAPHTHYLCARBAMATE C:1550
N-METHYL-α-NAPHTHYLURETHAN C:1550
METHYL NIRAN M:4900
METHYL NITROBENZENE N:2200
2-METHYLNITROBENZENE N:2200
3-METHYLNITROBENZENE N:2150
3-METHYLNITROBENZENE N:1900
4-METHYLNITROBENZENE N:2250
m-METHYLNITROBENZENE N:2150
o-METHYLNITROBENZENE N:2200
p-METHYL NITROBENZENE N:2250
METHYLOL M:1750
2-METHYLOLPENTANE E:2500
METHYLOLPROPANE B:3350

METHYL OXIRANE P:4450
METHYL OXITOL E:5100
METHYL 3-OXOBUTYRATE M:1600
METHYL PARATHION M:4900
2-METHYL PENTANE I:0750
2-METHYLPENTANE-2,4-DIOL H:1200
2-METHYL-2,4-PENTANEDIOL H:1200
4-METHYL-2,4-PENTANEDIOL H:1200
2-METHYL-4-PENTANOL M:2000
4-METHYL-2-PENTANOL M:2000
4-METHYL-2-PENTANOL M:4500
4-METHYLPENTANOL-2 M:2000
4-METHYL-2-PENTANOL ACETATE M:1950
2-METHYL-2-PENTANOL-4-ONE D:0900
2-METHYL-4-PENTANONE M:4550
4-METHYL-2-PENTANONE M:4550
2-METHYL-1-PENTENE M:4950
4-METHYL-PENTENE-1 M:5000
4-METHYL-4-PENTENE M:4950
4-METHYL-1-PENTENE M:5000
2-METHYL PENTENE-1 M:4950
4-METHYL-3-PENTENE-2-ONE M:1150
4-METHYL-2-PENTYL ACETATE M:1950
METHYL PENTYL KETONE A:4500
METHYLPHENOL C:5100
2-METHYL PHENOL C:5100
3-METHYL PHENOL C:5100
4-METHYL PHENOL C:5100
m-METHYLPHENOL C:5100
p-METHYLPHENOL C:5100
METHYLPHENOLS C:5100
1-METHYL-1-PHENYL-ETHENE M:5400
1-METHYL-1-PHENYL-ETHYLENE M:5400
METHYLPHENYL AMINE M:2050
METHYL PHENYL KETONE A:0450
METHYLPHENYL METHANOL M:3050
N-METHYLPHENYLAMINE M:2050
4-METHYL-PHENYLENE DIISOCYANATE T:1700
4-METHYL-PHENYLENE ISOCYANATE T:1700
METHYL PHOSPHITE T:4400
METHYLPHOSPHONOFLUORIDATE S:0250
METHYLPHOSPHONOTHIOIC ACID-s-[2-(BIS(METHYLETHYL)AMINO)ETHYL]O-ETHYL ESTER V:1650
METHYLPHOSPHONOFLUORIDIC ACID, 3,3-DIMETHYL-2-BUTYL ESTER S:5650
METHYLPHOSPHONOFLUORIDIC ACID ISOPROPYL ESTER S:0250
METHYLPHOSPHONOFLUORIDIC ACID-1-METHYLETHYL ESTER S:0250
METHYLPHOSPHONOFLUORIDIC ACID 1,2,2-TRIMETHYLPROPYL ESTER S:5650
METHYL PHOSPHONOTHIOIC DICHLORIDE M:5050
METHYL PHOSPHONOUS DICHLORIDE M:5050
METHYLFLUOROPHOSPHORIC ACID, ISOPROPYL ESTER S:0250
METHYLFLUORPHOSPHORSAEUREISOPROPYLESTER (German) S:0250
METHYLFLUORPHOSPHORSAEUREPINAKOLYLESTER (German) S:5650
METHYL PHTHALATE D:5900
METHYL PINACOLYLOXY PHOSPHORYLFLUORIDE S:5650
2-METHYLPROPANAL B:4850
2-METHYL-2-PROPANAMINE B:3600
2-METHYL-n-(2-METHYLPROPYL)-1-PROPANAMINE D:4100
2-METHYLPROPANE I:0250
2-METHYLPROPANENITRILE I:0550

METHYL 2-METHYLPROPENOATE M:4800
2-METHYLPROPANOIC ACID I:0500
2-METHYL-2-PROPANOIC ACID, ETHYL ESTER E:6900
2-METHYL-2-PROPANOL B:3450
2-METHYL-1-PROPANOL I:0350
2-METHYL-2-PROPEN-1-OL M:1800
METHYL PROPENATE M:1700
2-METHYLPROPENE I:0450
2-METHYL-2-PROPENENITRILE M:1300
2-METHYLPROPENENITRILE M:1300
2-METHYL PROPENIC ACID M:1200
2-METHYL PROPENIC ACID, METHYL ESTER M:4800
METHYL-2-PROPENOATE M:1700
METHYL 2-METHYL-2-PROPENOATE M:4800
α-METHYLPROPIONIC ACID I:0500
2-METHYLPROPIONIC ACID M:1200
2-METHYLPROPIONITRILE I:0550
1-METHYL PROPYL ACETATE B:3150
2-METHYLPROPYL ACETATE I:0300
2-METHYL-1-PROPYL ACETATE I:0300
2-METHYL-1-PROPYL ACRYLATE I:0325
METHYL PROPYL BENZENE C:6350
2-METHYLPROPYL ESTER I:0475
β-METHYLPROPYL ETHANOATE I:0300
METHYL PROPYL KETONE M:5100
METHYL-n-PROPYL KETONE M:5100
1-METHYLPROPYLAMINE B:3550
2-METHYLPROPYLAMINE I:0400
2-METHYL-2-PROPYLETHANOL M:4500
1-METHYL-1-PROPYLETHYLENE M:4950
2-METHYLPROPYLISOBUTYRATE I:0475
2-METHYLPYRIDINE M:5150
3-METHYLPYRIDINE M:5200
4-METHYLPYRIDINE M:5250
α-METHYLPYRIDINE M:5150
1-METHYL-2-(3-PYRIDYL)PYRROLIDINE N:1200
1-METHYLPYRROLIDINONE M:5300
1-METHYL-2-PYRROLIDINONE M:5300
N-METHYLPYRROLIDINONE M:5300
1-METHYLPYRROLIDONE M:5300
N-METHYL-α-PYRROLIDONE M:5300
3-(1-METHYL-2-PYRROLIDYL)PYRIDINE N:1200
(S)3-(1-METHYL-2-PYRROLIDYL) PYRIDINE N:1250
METHYL SALICYLATE M:5350
METHYLSENFOEL (German) M:4700
METHYL SILICONE L:0200
METHYL STYRENE V:1500
3-METHYL STYRENE V:1500
4-METHYL STYRENE V:1500
α-METHYL STYRENE M:5400
m-METHYL STYRENE V:1500
p-METHYL STYRENE V:1500
METHYL SULFATE D:6100
METHYL SULFHYDRATE M:4750
METHYL SULFIDE D:6150
4-(METHYLSULFONYL)-2,6-DINITRO N:1300
METHYL SULFOXIDE D:6200
METHYL SULPHIDE D:6150
METHYL THIRAM T:1450
METHYL THIURAMDISULFIDE T:1450
METHYL TRIBROMIDE B:2650
METHYL TRICHLORIDE C:2500
METHYL TRICHLOROSILANE M:5450
METHYL TUADS T:1450
4-METHYLTHIO-3,5-DIMETHYL-PHENYLMETHYLCARBAMATE M:0450

4-(METHYLTHIO)-3,5-XYLYLMETHYL-CARBAMATE M:0450
METHYLTHIOALCOHOL M:4750
METHYLTHIOPHOS M:4900
o-METHYLTOLUENE X:0200
p-METHYLTOLUENE X:0300
METHYLTRICHLOROETHANE T:2250
METHYLTRIMETHYLENE GLYCOL B:3900
METHYL VINYL ETHER V:1300
METHYL VINYL KETONE M:5500
METHYLZINC D:6300
1-METHYPROPYL ALCOHOL B:3400
METIL n-AMIL CETONA (Spanish) A:4500
METILATO SODICO (Spanish) S:4200
METIL AZINFOS (Spanish) A:5800
METILENBIS(FENILISOCIANATO) (Spanish) D:7400
α-METILESTIRENO (Spanish) V:1500
METIL ETIL CETONA (Spanish) M:4100
METILHIDRAZINA (Spanish) M:4350
METIL ISOBUTIL CETONA (Spanish) M:4550
METILMERCAPTANO (Spanish) M:4750
METILPARATIONA (Spanish) M:4900
METILPIRIDINA (Spanish) M:5150
3-METILPIRIDINA (Spanish) M:5200
4-METILPIRIDINA (Spanish) M:5250
n-METIL-2-PIRROLIDONA (Spanish) M:5300
METILTRIAZOTION A:5800
METILTRICLOROSILANO (Spanish) M:5450
METIL VINIL CETONA (Spanish) M:5500
METIOCARB (Spanish) M:0450
METMERCAPTURON M:0450
METOLACHLOR M:5550
METOX M:1500
METOXICLORO (Spanish) M:1500
METRON M:4900
METSO 20 S:5100
METSO PENTABEAD 20 S:5100
MEVINFOS (Spanish) P:1500
MEVINOX P:1500
MEVINPHOS P:1500
MEXACARBATE Z:0100
MEXACARBATO (Spanish) Z:0100
MFB F:0950
MFI S:0250
MH M:0400
MHA ACID H:1900
MHA-FA H:1900
MIBC M:2000
MIBC M:4500
MIBK M:4550
MIC M:2000
MIC M:4500
MIC M:4700
MIC M:4600
3-MIC M:2000
MICRO DDT 75 D:0300
MICROFLOTOX S:6600
MIGHTY 150 N:0400
MIK M:4550
MILBOL 49 B:0600
MILD MERCURY CHLORIDE M:1000
MULHOUSE WHITE L:1500
MILK ACID L:0100
MILK WHITE L:1500
MILLER'S FUMIGRAIN A:0950
MILOCEP M:5550
MILTON S:3900
MINERAL CARBON C:2000
MINERAL COLZA OIL O:3700
MINERAL NAPHTHA B:0550
MINERAL OIL O:3600
MINERAL PITCH A:5650

MINERAL SEAL OIL O:3700
MINERAL SOLVENT M:5600
MINERAL SPIRITS N:0300
MINERAL TERPENTINE M:5600
MINERAL THINNER M:5600
MIPAX D:5900
MIPK M:3150
MIRACLE D:0100
MIRBANE OIL N:1550
MIREX M:5650
MIT M:4700
MITC M:4700
MITIS GREEN C:4050
MIXED PRIMARY AMYL NITRATES A:4550
MIXTURE OF BENZENE, TOLUENE, XYLENES N:0350
MIXTURE OF PENTANE, HEXANE, HEPTANE N:0350
MLA (mixed lead alkyls) M:6150
MLT M:0250
MME M:4800
MMH M:4350
2-MMT M:3800
MNBK M:3350
MNT N:2150
MODANE SOFT D:7050
MOHR'S SALT F:0550
MOLASSES ALCOHOL E:1900
MOLATOC D:7050
MOLCER D:7050
MOLE-DEATH S:6200
MOLECULAR BROMINE B:2250
MOLECULAR CHLORINE C:2100
MOLIBDATO AMONICO (Spanish) A:3100
MOLOFAC D:7050
MOLOL O:3600
MOLYBDENUM(VI) OXIDE M:5700
MOLYBDENUM TRIOXIDE M:5700
MOLYBDIC ACID (58%) A:3100
MOLYBDIC ACID DIAMMONIUM SALT A:3100
MOLYBDIC ANHYDRIDE M:5700
MOLYBDIC TRIOXIDE M:5700
MONDUR TDS T:1700
MONO PE P:0550
MONOAMMONIUM ORTHOPHOSPHATE A:3600
MONOAMMONIUM PHOSPHATE A:3600
MONOAMMONIUM SALT OF SULFAMIC ACID A:3800
MONOAMMONIUM SULFAMATE A:3800
MONOAMMONIUM SULFITE A:2250
MONOBROMOACETONE B:2400
MONOBROMOBENZENE B:2500
MONOBROMOMETHANE M:3100
MONOBUTYLAMINE B:3500
MONOCALCIUM PHOSPHATE MONOHYDRATE C:1200
MONOCHLORBENZENE C:2400
MONOCHLORBENZOL (German) C:2400
MONOCHLORETHANE E:2900
MONOCHLORETHANOIC ACID, ETHYL ESTER E:3000
MONOCHLOROACETALDEHYDE C:2200
MONOCHLOROACETIC ACID M:5750
MONOCHLOROACETIC ACID, METHYL ESTER M:3500
MONOCHLOROACETYL CHLORIDE C:2300
MONOCHLOROBENZENE C:2400
MONOCHLORODIFLUOROMETHANE M:5800
MONOCHLORODIMETHYL ETHER C:2600
MONOCHLOROETHANOIC ACID M:5750

2-MONOCHLOROETHANOL E:3600
MONOCHLOROETHENE V:0800
MONOCHLOROETHYLENE V:0800
MONOCHLOROMETHANE M:3450
MONOCHLOROMETHYL ETHER C:2600
o-MONOCHLOROPHENOL C:2700
β-MONOCHLOROPROPIONIC ACID C:2950
MONOCHLOROSULFURIC ACID C:3000
MONOCHLOROTETRAFLUOROETHANE M:5850
MONOCHLOROTRIFLUOROMETHANE M:5900
MONOCHROMIUM OXIDE C:3300
MONOCHROMIUM TRIOXIDE C:3300
"MONOCITE" METHACRYLATE MONOMER M:4800
MONOCLORODIFLUOMETANO (Spanish) M:5800
MONOCLOROTETRAFLUORETANO (Spanish) M:5850
MONOCYANOGEN C:5650
MONOETANOLAMINA (Spanish) M:5950
MONOETHANOLAMINE M:5950
MONOETHYLAMINE E:2200
MONOETHYLENE GLYCOL E:4200
MONOETHYLENE GLYCOL ETHER E:4700
MONOFLUOROBENZENE F:0950
MONOFLUROETHYLENE V:1100
MONOFLUROTRICHLOROMETHANE T:2400
MONOGLYME E:4700
MONOHYDROXY BENZENE P:1200
MONOHYDROXYMETHANE M:1750
MONOIODOMETHANE M:4450
MONOISOBUTIRATO de 2,4,4-TRIMETIL-1,3-PENTANODIOL (Spanish) T:4350
MONOISOBUTYLAMINE I:0400
MONOISOPROPANOLAMINE M:6000
MONOISOPROPYLAMINE I:1300
MONOMETHYL ANILINE M:2050
MONOMETHYL ETHANOLAMINE M:6050
MONOMETHYL HYDRAZINE M:4350
MONOMETHYLAMINE M:1850
MONOMETHYLAMINE SOLUTION M:1900
MONOMETHYLAMINOETHANOL M:6050
n-MONOMETHYLANILINE M:2050
MONONITROGEN MONOXIDE N:1400
MONOPHENYLHYDRAZINE P:1300
MONOPLEX DOA D:7000
MONOPOTASSIUM ARSENATE P:2550
MONOPOTASSIUM DIHYDROGEN ARSENATE P:2550
MONOPROPYLAMINE P:4000
MONO-n-PROPYLAMINE P:4000
MONOPROPYLENE GLYCOL P:4200
MONOSAN D:0100
MONOSODIUM DIHYDROGEN PHOSPHATE S:4800
MONOSODIUM PHOSPHATE S:4800
MONOSODIUM SALT OF SULFUROUS ACID S:2200
MONOXIDE C:1750
MONOXIDO de CARBONO (Spanish) C:1750
MONOXIDO de NITROGENO (Spanish) N:1400
MONTAR P:2200
MOON G:0650
MOPARI D:2650
MORBICID F:1250
MORPHOLINE M:6100
MORTON WP-161-E M:4700
MORTOPAL T:0800
MOSCARDA M:0250
MOSS GREEN C:4050
MOTH BALLS N:0400
MOTH FLAKES N:0400
MOTOPAL T:0800
MOTOR BENZOL B:0550
MOTOR FUEL G:0200
MOTOR FUEL ANTI-KNOCK COMPOUND T:0750
MOTOR FUEL ANTI-KNOCK COMPOUND T:1050
MOTOR FUEL ANTI-KNOCK COMPOUNDS CONTAINING LEAD ALKYLS M:6150
MOTOR SPIRIT G:0200
MOTOX T:1950
MOUSE-TOX S:6200
MOXIE D:0100
MOXONE D:0100
MPK M:5100
MPTD M:5050
MSP S:4800
MSZYCOL B:0600
MTD T:1650
MULTRATHANE M D:7400
MURFOS P:0200
MURIATIC ACID H:1450
MURIATIC ETHER E:2900
MUSCATOX C:5500
MUSTARD HD B:2025
MUSTARD GASSES B:2025
MUSTARD-LEWISITE (HL) B:2025
MUSTARGEN B:2025
MUSTINE B:2025
MUTAGEN B:2025
MUTOXIN D:0300
MYCROLYSIN C:2800
MYRCENE M:6200
MYRISTIC ALCOHOL T:0450
MYRISTYL ALCOHOL T:0450
MYRISTYL ALCOHOL L:2000
MYRISTYL ALCOHOL, mixed isomers L:2000
NABAC H:0600
NABAM N:0100
NABAME N:0100
NAC C:1550
NACAP S:4000
NACCANOL NR D:8300
NACCANOL SW D:8300
NACCONATE 300 D:7400
NACCONATE 1OO T:1700
NACCONOL 988 A D:8150
NADONE C:5950
NAFTALENO (Spanish) N:0400
NAFTA de PEROLEO (Spanish) P:1150
NAFTENATO de COBRE (Spanish) C:4600
1-NAFTILAMINA (Spanish) N:0500
NAH 80 S:3600
NAKO TGG R:0100
NALED N:0150
NAPCHLOR-G S:4700
NAPHID N:0450
NAPHTHA N:0350
NAPHTHA (<3% aromatics, 120–200°C) N:0200
NAPHTHA (15-20% aromatics, 150–200°C) K:0150
NAPHTHA: COAL TAR N:0350
NAPHTHA: SOLVENT N:0250
NAPHTHA: STODDARD SOLVENT N:0300
NAPHTHA: VM&P N:0200
NAPHTHA PETROLEUM N:0200
NAPHTHA SAFETY SOLVENT N:0300
NAPHTHA SOLVENT N:0200
1-NAPHTHALENAMINE, TECHNICAL GRADE N:0500
NAPHTHALENE N:0400
NAPHTHALENE,BIS(1-METHYLETHYL)- D:4750
NAPHTHALENE, 1-METHYL- M:4850
NAPHTHALENE OIL C:5050
NAPHTHALIDAM N:0500
NAPHTHALIDINE N:0500
NAPHTHANE D:0400
NAPHTHENIC ACID, COPPER SALT C:4600
NAPHTHENIC ACIDS N:0450
2,3-DICHLORO-1,4-NAPHTHO-QUINONE D:1400
1-NAPHTHOL C:1550
1-NAPHTHOL N-METHYLCARBAMATE C:1550
α-NAPHTHYL N-METHYL-CARBAMATE C:1550
1-NAPHTHYLAMINE N:0500
α-NAPHTHYLAMINE N:0500
1-NAPHTHYLMETHYLCARBAMATE C:1550
NAPHTOL AS-KG T:1900
1-NAPHTYL N-METHYL-CARBAMATE C:1550
NAPTHALANE (do not confuse with naphthalene) D:0400
NAPTHALIN N:0400
NAPTHALINE N:0400
NARCOGEN T:2350
NARCOTILE E:2900
NARCYLEN A:0650
NARKOSOID T:2350
NATRIUM S:1300
NATRIUMAZID (German) S:2000
NATRIUMCHLORAT (GERMAN) S:2800
NATRIUMDICHROMAT (German) S:3300
NATRIUMHYDROXID (German) S:3800
NATRIUMRHODANID (German) S:5500
NATURAL GAS M:1400
NATURAL GAS L:2100
NATURAL GAS, REFRIGERATED LIQUID L:2100
NATURAL GASOLINE G:0400
NATURAL LEAD SULFIDE L:1600
NAUGATUCK DO-014 P:3500
NAUGATUCK DO-14 P:3500
NAXOL C:5900
NBA B:3350
NEANTINE D:3750
NEATSFOOT OIL O:3900
NECATORINA C:1850
NECATORINE C:1850
NECIDOL D:1000
NEGUVON T:2050
NEKTAL D:7050
NEMAFENE D:2500
NEMEX T:2400
NENDRIN E:0200
NEOCID D:0300
NEOCIDOL D:1000
NEODECANOIC ACID N:0550
NEODECANOIC ACID, VINYL ESTER V:1400
NEO-FAT 8 O:0200
NEO-FAT 10 D:0550
NEOFAT 12 L:0300
NEO-FAT 12-43 L:0300
NEO-FAT 18-61 S:5900
NEO-FAT 90-04 O:5100
NEOGLAUCIT D:0850
NEOHEXANE N:0600
NEOHEXANO (Spanish) N:0600

NEOL D:6000
NEOLID O:1200
NEOPENTANOIC ACID T:4000
NEOPENTYL GLYCOL D:6000
NEOPENTYLENE GLYCOL D:6000
NEOPRENE C:2850
NEO-SCABICIDOL B:0600
NEOSEPT H:0600
NEPHIS E:4000
NERKOL D:2650
NERVANAID B ACID E:3900
NESOL D:7150
NETAGRONE D:0100
NETAGRONE 600 D:0100
NEUTRAL AMMONIUM FLUORIDE A:2700
NEUTRAL ANHYDROUS CALCIUM HYPOCHLORITE C:1000
NEUTRAL LEAD ACETATE L:0600
NEUTRAL LEAD STEARATE L:1400
NEUTRAL POTASSIUM CHROMATE P:2750
NEUTRAL SODIUM CHROMATE S:3000
NEUTRAL VERDIGRIS C:4000
NEUTRAZYME D:8700
NEVAX D:7050
NEX C:1500
NEXEN FB B:0600
NEXIT B:0600
NEXIT-STARK B:0600
NEXOL-E B:0600
NIA 5462 E:0100
NIA 5996 D:1350
NIA 10242 C:1500
NIAGARA 1240 E:0600
NIAGARA 5462 E:0100
NIAGARA 10242 C:1500
NIALATE E:0600
NIALK T:2350
NIAX TDI T:1700
NIAX TDI-P T:1700
NICKEL(II) ACETATE N:0650
NICKEL ACETATE N:0650
NICKEL ACETATE TETRAHYDRATE N:0650
NICKEL AMMONIUM SULFATE N:0700
NICKEL AMMONIUM SULFATE HEXAHYDRATE N:0700
NICKEL BOROFLUORIDE N:0950
NICKEL BROMIDE N:0750
NICKEL BROMIDE TRIHYDRATE N:0750
NICKEL CARBONYL N:0800
NICKEL CARBONYLE (French) N:0800
NICKEL CHLORIDE N:0850
NICKEL(II) CHLORIDE N:0850
NICKEL CHLORIDE HEXAHYDRATE N:0850
NICKEL CYANIDE N:0900
NICKEL(II) CYANIDE N:0900
NICKEL DIHYDROXIDE N:1050
NICKEL FLUOROBORATE N:0950
NICKEL FLUOROBORATE SOLUTION N:0950
NICKEL(II) FLUOBORATE N:0950
NICKEL FORMATE N:1000
NICKEL FORMATE DIHYDRATE N:1000
NICKEL HYDROXIDE N:1050
NICKEL(II) HYDROXIDE N:1050
NICKEL NITRATE HEXAHYDRATE N:1100
NICKEL(2+) NITRATE, HEXAHYDRATE N:1100
NICKEL(II) NITRATE, HEXAHYDRATE N:1100
NICKEL SULFATE N:1150
NICKEL(2+) SULFATE(1:1) N:1150
NICKEL(II) SULFATE N:1150
NICKEL TETRACARBONYLE (French) N:0800
NICKEL TETRACARBONYL N:0800
NICKELOUS ACETATE N:0650
NICKELOUS CHLORIDE N:0850
NICKELOUS HYDROXIDE N:1050
NICKELOUS SULFATE N:1150
NICKELOUS TETRAFLUOROBORATE N:0950
NICOCHLORAN B:0600
NICOCIDE N:1200
NICODUST N:1200
NICOFUME N:1200
NICOTINA (Spanish) N:1200
NICOTINE N:1200
NICOTINE SULFATE N:1250
NIFOS T:0800
NIFOS T T:0800
NIFROST T:0800
NIKKOL OTP 70 D:7050
NINOL AA-62 EXTRA L:0300
NIOBE OIL M:3000
NIPAR S-20 N:2050
NIPAR S-30 N:2050
NIPHEN N:1950
NIPSAN D:1000
NIQUEL CARBONILO (Spanish) N:0800
NIRAN C:2050
NIRAN P:0200
NIRAN E-4 P:0200
NITAL N:1350
NITER P:3050
NITRALIN N:1300
NITRAM A:3150
NITRAM A:1650
NITRAN T:3800
NITRAN M:4900
4-NITRANBINE N:1450
2-NITRANILINE N:1450
NITRATE d'AMYLE (French) A:4550
NITRATE d'ARGENT (French) S:1000
NITRATE de BARYUM (French) B:0250
NITRATE de SODIUM (French) S:4300
NITRATE de ZINC (French) Z:1700
NITRATE MERCUREUX (French) M:1050
NITRATE MERCURIQUE (French) M:0750
NITRATINE S:4300
NITRATO ALUMINICO (Spanish) A:1650
NITRATO de AMILO (Spanish) A:4550
NITRATO BARICO (Spanish) B:0250
NITRATO de BERILO (Spanish) B:1750
NITRATO de CADMIO(II) (Spanish) C:0350
NITRATO CALCICO (Spanish) C:1050
NITRATO COBALTOSO (Spanish) C:3800
NITRATO de COBRE (Spanish) C:4650
NITRATO de ESTRONCIO (Spanish) S:6100
NITRATO FERRICO (Spanish) F:0350
NITRATO de NIQUEL (Spanish) N:1100
NITRATO de PLATA (Spanish) S:1000
NITRATO de PLOMO (Spanish) L:1300
NITRATO POTASICO (Spanish) P:3050
NITRATO de n-PROPILO (Spanish) P:4700
NITRATO de TALIO (Spanish) T:1250
NITRATO de TORIO (Spanish) T:1500
NITRATO de URANILO (Spanish) U:0700
NITRATO de ZINC (Spanish) Z:1700
NITRATO de ZIRCONIO (Spanish) Z:2400
NITREX NITROGEN SOLUTIONS (NON-PRESSURE) A:3300
NITRIC ACID N:1350
NITRIC ACID, ALUMINUM(3+) A:1650
NITRIC ACID, ALUMINUM SALT A:1650
NITRIC ACID, AMMONIUM SALT A:3150
NITRIC ACID, BARIUM SALT B:0250
NITRIC ACID, BERYLLIUM SALT B:1750
NITRIC ACID, CADMIUM SALT, TETRAHYDRATE C:0350
NITRIC ACID, CALCIUM SALT C:1050
NITRIC ACID, COBALT(2+) SALT C:3800
NITRIC ACID, COPPER (2+) SALT C:4650
NITRIC ACID, COPPER (II) SALT C:4650
NITRIC ACID, IRON(3+) SALT F:0350
NITRIC ACID, IRON(III) SALT F:0350
NITRIC ACID, LEAD(2+) SALT L:1300
NITRIC ACID, LEAD(II) SALT L:1300
NITRIC ACID, MAGNESIUM SALT M:0150
NITRIC ACID, MERCURY SALT M:1050
NITRIC ACID, MERCURY(II) SALT M:0750
NITRIC ACID, NICKEL(II) SALT, HEXAHYDRATE N:1100
NITRIC ACID, NICKEL(2+) SALT, HEXAHYDRATE N:1100
NITRIC ACID, PENTYL ESTER A:4550
NITRIC ACID, POTASSIUM SALT P:3050
NITRIC ACID, PROPYL ESTER P:4700
NITRIC ACID, RED FUMING N:1350
NITRIC ACID, SODIUM SALT S:4300
NITRIC ACID, STRONTIUM SALT S:6100
NITRIC ACID, THALLOUS SALT T:1250
NITRIC ACID, THORIUM(IV) SALT T:1500
NITRIC ACID, THORIUM(4+) SALT T:1500
NITRIC ACID, WHITE FUMING N:1350
NITRIC ACID, ZINC SALT Z:1700
NITRIC OXIDE N:1400
NITRILE ACRYLIQUE (French) A:0950
2,2',2''-NITRILO-TRIETHANOL T:3200
NITRILOACETONITRILE C:5650
NITRILOTRIACETIC ACID, AND SALTS N:1750
NITRITO de AMILO (Spanish) A:4600
NITRITO de ETILO (Spanish) E:7100
NITRITO SODICO (Spanish) S:4400
o-NITROANILINA (Spanish) N:1450
p-NITROANILINA (Spanish) N:1450
4-NITROANILINE N:1450
o-NITROANILINE N:1450
p-NITROANILINE N:1450
NITROBARITE B:0250
NITROBENCENO (Spanish) N:1550
4-NITROBENZENAMINE N:1450
NITROBENZENE N:1550
NITROBENZOL N:1550
NITROBENZOL,L N:1550
NITROCALCITE C:1050
NITROCARBOL N:1800
NITROCELLULOSE C:3950
NITROCELLULOSE GUM C:3950
NITROCELLULOSE SOLUTION C:3950
o-NITROCHLOROBENZENE C:2650
NITROCHLOROFORM C:2800
NITROETANO (Spanish) N:1600
NITROETHANE N:1600
NITROFAN D:6550
m-NITROFENOL (Spanish) N:1900
o-NITROFENOL (Spanish) N:1850
p-NITROFENOL (Spanish) N:1950
NITROGEN N:1650
NITROGEN CHLORIDE OXIDE N:2100
NITROGEN, COMPRESSED N:1650
NITROGEN GAS N:1650
NITROGEN MONOXIDE N:1400
NITROGEN MUSTARD B:2025
NITROGEN MUSTARD HN-1 B:2025
NITROGEN MUSTARD HN-2 B:2025
NITROGEN MUSTARD HN-3 B:2025

NITROGEN OXIDE N:2300
NITROGEN OXYCHLORIDE N:2100
NITROGEN TETROXIDE N:1700
NITROISOPROPANE N:2050
NITRO KLEENUP D:6650
NITRO, LIQUID N:1550
1,1',1"-NITROLOTRI-2-PROPANOL T:3950
NITROMETANO (Spanish) N:1800
NITROMAGNESITE M:0150
NITROMETHANE N:1800
1-NITROPENTANE A:4600
2-NITROPHENOL N:1850
3-NITROPHENOL N:1900
4-NITROPHENOL N:1950
m-NITROPHENOL N:1900
o-NITROPHENOL N:1850
p-NITROPHENOL N:1950
p-NITROPHENYLAMINE N:1450
NITROPROPANE N:2000
1-NITROPROPANE N:2000
2-NITROPROPANE N:2050
β-NITROPROPANE N:2050
sec-NITROPROPANE N:2050
1-NITROPROPANO (Spanish) N:2000
NITRO-SIL A:2000
NITROSTIGMIN (German) P:0200
NITROSTIGMINE P:0200
NITROSYL CHLORIDE N:2100
NITROSYL ETHOXIDE E:7100
2-NITROTOLUENE N:2200
3-NITROTOLUENE N:1900
3-NITROTOLUENE N:2150
4-NITROTOLUENE N:2250
m-NITROTOLUENE N:2150
o-NITROTOLUENE N:2200
p-NITROTOLUENE N:2250
m-NITROTOLUENO (Spanish) N:2150
o-NITROTOLUENO (Spanish) N:2200
p-NITROTOLUENO (Spanish) N:2250
2-NITROTOLUOL N:2200
3-NITROTOLUOL N:2150
4-NITROTOLUOL N:2250
NITROTRICHLOROMETHANE C:2800
NITROUS ACID ETHYL ESTER E:7100
NITROUS ACID, PENTYL ESTER A:4600
NITROUS ACID, SODIUM SALT S:4400
NITROUS ETHER E:7100
NITROUS ETHYL ETHER E:7100
NITROUS FUMES N:1350
NITROUS OXIDE N:2300
NITROX M:4900
NITRYL HYDROXIDE N:1350
NIUIF-100 P:0200
NMP M:5300
NO N:1400
NO-PEST STRIP D:2650
NO BUNT LIQUID H:0400
NO SCALD D:7250
NOBECUTAN T:1450
NOBLEN P:2400
NOFLAMOL P:2200
NOGOS D:2650
NOMERSAN T:1450
NONAN-1-OL N:2400
NONAN-2-ONE M:4300
NONANE N:2350
n-NONANE N:2350
NONANE and TRIMETHYLBENZENE MIXTURE (85:15) N:0300
1-NONANECARBOXYLIC ACID D:0550
n-NONANO (Spanish) N:2350
NONANOL N:2400
1-NONANOL N:2400

NONANOL ACETATE N:2500
2-NONANONE M:4300
5-NONANONE D:1180
NONENE P:4550
1-NONENE N:2450
NONENE (NONLINEAR) P:4550
NONENO (Spanish) N:2450
NONILFENOL N:2550
NONYL ACETATE N:2500
n-NONYL ACETATE N:2500
NONYL ALCOHOL N:2400
n-NONYL ALCOHOL N:2400
sec-NONYL ALCOHOL D:4150
NONYL PHENOL N:2550
NONYLCARBINOL D:0700
NONYLENE N:2450
1-NONYLENE N:2450
n-NONYLETHYLENE U:0300
NONYLHYDRIDE N:2350
5-NORANONE D:1180
2-NORBORNENE,5-ETHYLIDENE- E:6600
NORCAMPHOR, SYNTHETIC CAMPHOR C:1400
NORFORMS P:1400
NORKOOL E:4200
NORMAL LEAD ACETATE L:0600
NORMENTHANE I:1350
NORMERSAN T:1450
NORVAL D:7050
NORVALAMINE B:3500
NORWAY SALTPETER A:3150
NORWAY SALTPETER A:1650
NORWEGIAN SALTPETER C:1050
NOURITHION P:0200
NOVIGAM B:0600
NOWERGAN T:1450
1-NP N:2000
2-NP N:2050
NTA N:1750
NTM D:5900
NTO N:1700
NUCHAR 722 C:2000
NUCIDOL D:1000
NUJOL O:3600
NULLAPON B ACID E:3900
NULLAPON BF ACID E:3900
NUOPLAZ D:7850
NURELLE T:2450
NUTRIFOS STP S:4900
NUVA D:2650
NUX-VOMICA S:6200
NXX N:1650
NYMERATE P:1400
O-PENTADECADIENYL SALICYLIC ACID O:3100
OBELINE PICRATE A:3650
OBSTON D:7050
OCTACHLOR C:2050
OCTACHLOROCAMPHENE T:1950
OCTACHLORODIHYDRODICYCLOPENTADIENE C:2050
1,2,4,5,6,7,8,8-OCTACHLORO-2,3,3a,4,7,7a-HEXAHYDRO-4,7-METHANOINDENE C:2050
1,2,4,5,6,7,8,8-OCTACHLORO-3a,4,7,7a-HEXAHYDRO-4,7-METHYLENE INDANE C:2050
1,2,4,5,6,7,8,8-OCTACHLORO-2,3,3A,4,7,7a-HEXAHYDRO-4,7-METHANO-1H-INDENE C:2050
OCTACHLORO-4,7-METHANO-HYDROINDANE C:2050

1,2,4,5,6,7,10,10-OCTACHLORO-4,7-METHANOTETRAHYDROINDANE C:2050
1,2,4,5,6,7,8,8-OCTACHLORO-4,7-METHANO-3A,4,7,7a-TETRAHYDROINDANE C:2050
1,2,4,5,6,7,8,8-OCTACHLOR-3a,4,7,7a-TETRAHYDRO-4,7-endo-METHANO-INDAN (German) C:2050
1,2,4,5,6,7,8,8-OCTACHLORO-3a,4,7,7a-TETRAHYDRO-4,7-METHANOINDANE C:2050
1,2,4,5,6,7,8,8-OCTACHLORO-3a,4,7,7a-TETRAHYDRO-4,7-METHANOINDAN C:2050
OCTACHLORO-4,7,8,9-TETRAHYDRO-4,7-METHYLENEINDANE C:2050
OCTADECANOIC ACID S:5900
OCTADECANOIC ACID, LEAD SALT L:1400
1-OCTADECANOL, CRUDE T:0200
9,10-OCTADECENOIC ACID O:5100
cis-9-OCTADECENOIC ACID O:5100
n-OCTADECYLIC ACID S:5900
1,6-OCTADIENE,7-METHYL-3-METHYLENE M:6200
OCTA-KLOR C:2050
OCTALENE A:1100
OCTALOX D:2700
1-OCTANAL O:0500
OCTANALDEHYDE O:0500
OCTANE O:0100
n-OCTANE O:0100
n-OCTANO (Spanish) O:0100
OCTANOIC ACID O:0200
OCTANOL O:0300
1-OCTANOL O:0300
n-OCTANOL O:0300
3-OCTANONE E:2300
1-OCTENE O:0400
1-OCTENO (Spanish) O:0400
OCTILIN O:0300
n-OCTOIC ACID O:0200
OCTOIL D:5250
OCTYL ACETATE E:6200
OCTYL ACRYLATE E:6300
OCTYL ADIPATE D:7000
OCTYL ALCOHOL E:6100
n-OCTYL ALCOHOL O:0300
OCTYL ALDEHYDE E:5900
OCTYL ALDEHYDES O:0500
n-OCTYL ALDEHYDE O:0500
OCTYL DECYL PHTHALATE O:0550
n-OCTYL-n-DECYL PHTHALATE O:0550
OCTYL CARBINOL N:2400
OCTYL EPOXY TALLATE O:0600
OCTYL PHTHALATE D:3650
α-OCTYLENE O:0400
ODB D:1450
ODCB D:1450
OIL, FUEL: 2-D O:1100
OIL MIST, MINERAL O:3600
OIL OF BITTER ALMOND B:0500
OIL OF CASHEW NUTSHELL O:3100
OIL OF MIRBANE N:1550
OIL OF MYRBANE N:1550
OIL OF NIOBE M:3000
OIL OF PALMA CHRISTI O:1200
OIL OF PINE O:4100
OIL OF TURPENTINE T:4750
OIL OF VITRIOL S:6800
OIL, PALM O:1800
OILS, CLARIFIED O:0700
OILS, CRUDE O:0800
OILS, DIESEL O:0900

OILS, DIESEL FUEL: 1-D O:1000
OILS, DIESEL FUEL: 2-D O:1100
OILS, EDIBLE: CASTOR O:1200
OILS, EDIBLE: COCONUT O:1300
OILS, EDIBLE: COTTONSEED O:1400
OILS, EDIBLE: FISH O:1500
OILS, EDIBLE: LARD O:1600
OILS, EDIBLE: OLIVE O:1700
OILS, EDIBLE: PALM O:1800
OILS, EDIBLE: PEANUT O:1900
OILS, EDIBLE: SAFFLOWER O:2000
OILS, EDIBLE: SOYA BEAN O:2100
OILS, EDIBLE: TUCUM O:2200
OILS, EDIBLE: VEGETABLE O:2300
OILS, FUEL: 1-D O:1000
OILS, FUEL: NO. 1 O:2400
OILS, FUEL: NO. 2 O:2500
OILS, FUEL: NO. 4 O:2600
OILS, FUEL: NO. 5 O:2700
OILS, FUEL: NO. 6 O:2800
OILS, MISCELLANEOUS: ABSORPTION O:2900
OILS, MISCELLANEOUS: COAL TAR O:3200
OILS, MISCELLANEOUS: CROTON O:3300
OILS, MISCELLANEOUS: CASHEW NUT SHELL O:3100
OILS, MISCELLANEOUS: LINSEED O:3400
OILS, MISCELLANEOUS: LUBRICATING O:3500
OILS, MISCELLANEOUS: MOTOR O:3800
OILS, MISCELLANEOUS: MINERAL O:3600
OILS, MISCELLANEOUS: MINERAL SEAL O:3700
OILS, MISCELLANEOUS: NEATSFOOT O:3900
OILS, MISCELLANEOUS: PENETRATING O:4000
OILS, MISCELLANEOUS: PINE O:4100
OILS, MISCELLANEOUS: RESIN O:4200
OILS, MISCELLANEOUS: ROAD O:4300
OILS, MISCELLANEOUS: SPINDLE O:4500
OILS, MISCELLANEOUS: SPERM O:4400
OILS, MISCELLANEOUS: SPRAY O:4600
OILS, MISCELLANEOUS: TALL O:4700
OILS, MISCELLANEOUS: TANNER'S O:4800
OILS, MISCELLANEOUS: TRANSFORMER O:4900
OILS, MISCELLANEOUS: TURBINE O:5000
OKO D:2650
OKTATERR C:2050
OLAMINE M:5950
OLEATO AMONICO (Spanish) A:3350
OLEFIANT GAS E:3500
OLEFIN C-13 T:3100
OLEIC ACID O:5100
OLEIC ACID, AMMONIUM SALT A:3350
OLEIC ACID, DISTILLED O:5100
OLEIC ACID, POTASSIUM SALT P:3100
OLEIC ACID, POTASSIUM SALT O:5200
OLEIC ACID, SODIUM SALT O:5300
OLEINIC ACID O:5100
OLEOFOS-20 P:0200
OLEOGESAPRIM A:5750
OLEOPARATHION P:0200
OLEOVOFOTOX M:4900
OLEUM O:5400
OLEUM ABIETIS O:4100
OLITREF T:3800
OLIVE OIL O:1700
OMAIT P:3500
OMAL T:2450

OMITE P:3500
OMNITOX B:0600
OMS-14 D:2650
OMS-29 C:1550
OMS-93 M:0450
OMS-570 E:0100
OMS-971 D:8900
ONA N:1450
ONCB C:2650
ONP N:1850
ONT N:2200
ONTRACK 8E M:5550
o,o-DIETHYL C:5000
OP-THAL-ZIN Z:2200
OPTAL P:3950
ORANGE BASE CIBA 2 N:1450
ORDINARY LACTIC ACID L:0100
ORPIMENT A:5550
ORTHO-KLOR C:2050
ORTHO 4355 N:0150
ORTHO C-1 DEFOLIANT & WEED KILLER S:2800
ORTHOARSENIC ACID A:5300
ORTHOBORIC ACID B:2100
ORTHOCIDE 7.5 C:1475
ORTHOCIDE 50 C:1475
ORTHOCIDE 406 C:1475
ORTHOCIDE C:1475
ORTHODIBROM N:0150
ORTHODIBROMO N:0150
ORTHODICHLOROBENZENE D:1450
ORTHODICHLOROBENZOL D:1450
ORTHOPHOS P:0200
ORTHOPHOSPHORIC ACID P:1600
ORTHOPHOSPHORIC ACID P:2350
ORTHOSIL S:5100
ORTHOTITANIC ACID, TETRABUTYL ESTER T:0300
ORTIZON U:1100
ORVINYLCARBINOL A:1200
OS 2046 P:1500
OSMOSOL EXTRA P:3950
OWANOL 33B P:4300
3-OXA-1, 5-PENTANEDIOL D:3000
OXACETIC ACID G:0750
OXACYCLOPENTADIENE F:1450
OXACYCLOPENTANE T:0900
OXACYCLOPROPANE E:5500
OXAL G:0700
OXALATO AMONICO (Spanish) A:3400
OXALATO de COBRE (Spanish) C:4700
OXALATO FERRICO AMONICO (Spanish) F:0150
OXALATO POTASICO (Spanish) P:3150
OXALDEHYDE G:0700
OXALIC ACID O:5500
OXALIC ACID, COPPER(II) SALT (1:1) C:4700
OXALIC ACID, DIAMMONIUM SALT A:3400
OXALIC ACID DIHYDRATE O:5500
OXALIC ACID DINITRILE C:5650
OXALIC ACID, FERROUS SALT F:0700
OXALIC ALDEHYDE G:0700
OXALONITRILE C:5650
OXALSAEURE (German) O:5500
OXALYL CYANIDE C:5650
OXAMMONIUM H:1800
OXAMMONIUM SULFATE H:1850
OXANE E:5500
1-OXA-4-AZACYCLOHEXANE M:6100
2H-1,4-OXAZINE, TETRAHYDRO- M:6100
2-OXETANONE P:3600

OXICLUORO de FOSFORO (Spanish) P:1800
OXICLORURO de ZIRCONIO (Spanish) Z:2500
OXIDATE LE M:3000
OXIDE C:2550
OXIDES OF NITROGEN N:1700
OXIDO de BERILIO (Spanish) B:1800
OXIDO de 1,2-BUTILENO (Spanish) B:4050
OXIDO CALCICO (Spanish) C:1100
OXIDOETHANE E:5500
α,β-OXIDOETHANE E:5500
OXIDO de ETILENO (Spanish) E:5500
OXIDO NITRICO (Spanish) S:1000
OXIDO de PLATA (Spanish) S:1100
OXIDO de PROPILENO (Spanish) P:4450
OXIGENO (Spanish) O:5600
OXIRAN E:5500
OXIRANE E:5500
OXIRANE(CHLOROMETHYL)- C:2550
OXIRANE, 2-(CHLOROMETHYL) E:0300
OXIRANE, 2-(CHLOROMETHYL)- E:0300
OXIRANE, METHYL- P:4450
OXIRENE, DIHYDRO- E:5500
OXITOL E:0800
OXITRICLORURO de VANADIO (Spanish) V:0300
3-OXOBUTANOIC ACID ETHYL ESTER E:1700
3-OXOBUTANOIC ACID METHYL ESTER M:1600
α-OXODIPHENYLMETHANE B:1000
α-OXODITANE B:1000
OXOETHANOIC ACID G:0750
2-OXOHEXAMETHYLENIMINE C:1450
OXOLANE T:0900
OXOLE F:1450
OXOMETHANE F:1250
OXOOCTALDEHYDE I:0800
OXOOCTYL ALCOHOL I:0850
OXOTRIDECYL ALCOHOL T:3050
OXRALOX D:2700
OXYBENZENE P:1200
1,1-OXYBIS(2-CHLOROETHANE) D:1800
1,1'-OXYBIS(BUTANE) D:1170
2,2'-[OXYBIS(ETHYLENEOXY)]DIETHANOL T:0650
1,1'-[OXYBIS(METHYLENE)]BISBENZENE D:1100
OXYBISMETHANE D:5200
1,1-OXYBISBENZENE D:7350
2, 2'OXYBISETHANOL D:3000
2,2'-OXYBISPROPANE I:1400
OXYCIL S:2800
OXYDE D'ETHYLE (French) E:5600
OXYDE de CALCIUM (French) C:1100
OXYDE de CARBONE (French) C:1750
OXYDE de MESITYLE (French) M:1150
OXYDE de PROPYLENE (French) P:4450
1,1'-OXYDI-2-PROPANOL D:7500
2,2'-OXYDIETHANOL D:3000
OXYDOL H:1600
OXYGEN O:5600
OXYGEN, LIQUID O:5600
OXYLITE D:1050
OXYMETHYLENE F:1250
OXYMURIATE OF POTASH P:2700
3-OXYPENTANE-1,5-DIOL D:3000
OXYPHENIC ACID C:1950
OXYSULFATOVANADIUM V:0500
OXYTOLUENES C:5100
PAA P:0900
PAC P:0200
PAINT DRIER C:4600

PAINTERS NAPHTHA N:0200
PALATINOL A D:3750
PALATINOL C D:1250
PALATINOL IC D:4300
PALATINOL M D:5900
PALM BUTTER O:1800
PALM FRUIT OIL O:1800
PALM KERNEL OIL O:2200
PALM NUT OIL O:2200
PALM OIL O:1800
PALM SEED OIL O:2200
PAMISAN P:1400
PAMOLYN O:5100
PAN P:2000
PANAM C:1550
PANORAM D:2700
PANORAM 75 T:1450
PANORAM D-31 D:2700
PANTHION P:0200
PAPER MAKER'S ALUM A:1800
PAPER MAKER'S ALUM A:1750
PAPI P:2300
PARAACETALDEHYDE P:0175
PARA CRYSTALS D:1550
PARACYMENE C:6350
PARACYMOL C:6350
PARACHLOROCIDUM D:0300
PARACHLOROPHENOL C:2750
PARACIDE D:1550
PARADI D:1550
PARADICHLOROBENZENE D:1550
PARADICHLOROBENZOL D:1550
PARADOW D:1550
PARADUST P:0200
PARAFFIN P:0100
PARAFFIN JELLY P:1100
PARAFFIN OIL O:3500
PARAFFIN WAX P:0100
PARAFINA (Spanish) P:0100
PARAFORM F:1250
PARAFORM P:0150
PARAFORM 3 P:0150
PARAFORMALDEHIDO (Spanish) P:0150
PARAFORMALDEHYDE P:0150
PARA HYDROGEN H:1350
PARAL P:0175
PARALDEHIDO (Spanish) P:0175
PARALDEHYD (German) P:0175
PARALDEHYDE P:0175
PARAMAR P:0200
PARAMAR 50 P:0200
PARAMOTH D:1550
PARANAPHTHALENE A:4850
PARANITROPHENOL (French, German) N:1950
PARANUGGETS D:1550
PARAPEST M-50 M:4900
PARAPHOS P:0200
PARATAF M:4900
PARATHENE P:0200
PARATHION P:0200
m-PARATHION M:4900
PARATHION ETHYL P:0200
PARATHION METHYL M:4900
PARATIONA (Spanish) P:0200
PARATOX M:4900
PARAWET P:0200
PARAZENE D:1550
PARIDOL M:4900
PARIS GREEN C:4050
PAROL O:3600
PARROT GREEN C:4050
PARZATE N:0100

PATENT ALUM A:1750
PATENT ALUMINUM A:1750
PATRON M M:4900
PBI CROP SAVER M:0250
PCB P:2200
PCC T:1950
PCHO P:0175
PCL H:0500
PCM P:1050
PCP P:0350
PDB D:1550
PDCB D:1550
PE P:0550
PEANUT OIL O:1900
PEAR OIL A:4200
PEAR OIL I:0100
PEAR OIL A:4250
PEARL ASH P:2700
PEARL STEARIC S:5900
PEARL WHITE B:1900
PEARSALL A:1450
PEB1 D:0300
PEBBLE LIME C:1100
PELADOW C:0750
PELAGOL GREY C C:1950
PELAGOL GREY J T:1650
PELAGOL GREY RS R:0100
PELAGOL RS R:0100
PELARGONIC ALCOHOL N:2400
PENATROL A:5750
PENCHLOROL P:0350
PENETRATING OIL O:4000
PENITE S:1900
PENNAMINE D:0100
PENNAMINE D D:0100
PENNFLOAT M L:0500
PENNFLOAT S L:0500
PENNCAP-M M:4900
PENT ACETATE A:4200
PENTA P:0350
PENTA A:1150
1,4,7,10,13-PENTAAZATRIDECANE T:0700
PENTABORANE P:0250
PENTABORANE(9) P:0250
PENTABORANE UNDECAHYDRIDE P:0250
PENTABORANO (Spanish) P:0250
PENTABORATO AMONICO (Spanish) A:3450
(9)-PENTABORON NONAHYDRIDE P:0250
PENTABORON NONAHYDRIDE P:0250
PENTACHLORAETHAN (German) P:0300
PENTACHLORETHANE (French) P:0300
PENTACHLORIN D:0300
PENTACHLOROANTIMONY A:4900
PENTACHLOROETHANE P:0300
PENTACHLOROFENOL P:0350
PENTACHLOROPHENATE P:0350
PENTACHLOROPHENOL P:0350
2,3,4,5,6-PENTACHLOROPHENOL P:0350
PENTACHLOROPHENYL CHLORIDE H:0400
PENTACHLOROPHENOL, DOWICIDE EC-7 P:0350
PENTACHLOROPHENOL, DP-2 P:0350
PENTACLOROETANO (Spanish) P:0300
PENTACLOROFENATO SODICO (Spanish) S:4700
PENTACLOROFENOL (Spanish) P:0350
PENTACLORURO de ANTIMONIO (Spanish) A:4900
PENTACON P:0350
PENTADECANOL L:2000
PENTADECANOL P:0400

1-PENTADECANOL P:0400
PENTADECYL ALCOHOL P:0400
1,3-PENTADIENE P:0450
1,4-PENTADIENE P:0500
trans-PENTADIENE-1,3 P:0450
PENTA-1,4-DIENE P:0500
1,3-PENTADIENO (Spanish) P:0450
1,4-PENTADIENO (Spanish) P:0500
PENTAERITRITA (Spanish) P:0550
PENTAERYTHRITE P:0550
PENTAERYTHRITOL P:0550
PENTAETHYLENE HEXAMINE P:0600
PENTAETHYLENEHEXAMINE P:0600
PENTAFLUOROANTIMONY A:4950
PENTAFLUORURO de ANTIMONIO (Spanish) A:4950
PENTAFLUORURO de BROMO (Spanish) B:2300
PENTA-KIL P:0350
PENTALIN P:0300
PENTAMETHYLENE C:6200
PENTAN P:0650
PENTAN-1-OL A:4350
PENTAN-2,4-DIONE A:0500
PENTANAL V:0100
n-PENTANAL V:0100
PENTANE P:0650
n-PENTANE P:0650
PENTANEDIAL G:0600
1,5-PENTANEDIAL G:0600
2,4-PENTANEDIOL, 2-METHYL- H:1200
PENTANEDIONE A:0500
1,5-PENTANEDIONE G:0600
2,4-PENTANEDIONE A:0500
1-PENTANETHIOL A:4450
n-PENTANO (Spanish) P:0650
PENTANOIC ACID P:0700
n-PENTANOIC ACID P:0700
PENTANOL A:4350
1-PENTANOL A:4350
n-PENTANOL A:4350
tert-PENTANOL M:4400
PENTANOL-1 A:4350
2-PENTANOL, 4-METHYL- M:2000
1-PENTANOL ACETATE A:4200
2-PENTANOL ACETATE A:4250
2-PENTANONA (Spanish) M:5100
3-PENTANONA (Spanish) D:3700
2-PENTANONE M:5100
3-PENTANONE D:3700
PENTANONE-3 D:3700
2-PENTANONE, 4-HYDROXY-4-METHYL- D:0900
2-PENTANONE, 4-METHYL- M:4550
3-PENTANONE DIMETHYL ACETONE D:3700
PENTASOL (amyl alcohol) A:4350
PENTASOL (pentachlorophenol) P:0350
PENTASULFURE de PHOSPHORE (French) P:1850
PENTASULFURO de FOSFORO (Spanish) P:1850
PENTECH D:0300
PENTEK P:0550
1-PENTENE P:0800
PENTENO (Spanish) P:0800
PENTIFORMIC ACID H:1000
PENTOXIDO de ARSENICO (Spanish) A:5300
PENTOXIDO de VANADIO (Spanish) V:0400
2-PENTYL ACETATE A:4250
tert-PENTYL ACETATE A:4300
PENTYL ACETATES A:4200
PENTYL ALCOHOL A:4350

PENTYL CARBINOL H:1050
1-PENTYL CHLORIDE A:4400
PENTYLENE P:0800
PENTYL ESTER OF ACETIC ACID A:4200
PENTYL MERCAPTAN A:4450
PENTYL METHYL KETONE A:4500
PENTYL NITRITE A:4600
PENTYL PROPIONATE P:0850
n-PENTYL PROPIONATE P:0850
2-PENTYLACETATE A:4250
2-PENTYLBROMIDE B:2700
3-PENTYLCARBINOL E:2500
sec-PENTYLCARBINOL E:2500
PENTYLFORMIC ACID H:1000
PENTYLSILICON TRICHLORIDE A:4700
PENTYLTRICHLOROSILANE A:4700
PENWAR P:0350
PER-CHLORATE SOLUTION P:0950
PERACETIC ACID P:0900
PERATOX P:0350
PERCARBAMIDE U:1100
PERCARBAMITE U:1100
PERCARBONATO de ISOPROPILO (Spanish) I:1600
PERCHLOR T:0400
PERCHLORATE de MAGNESIUM (French) M:0200
PERCHLORIC ACID P:0950
PERCHLORIC ACID, AMMONIUM SALT A:3500
PERCHLORIC ACID, BARIUM SALT-3H MO B:0300
PERCHLORIC ACID, MAGNESIUM SALT M:0200
PERCHLORIC ACID SOLUTION P:0950
PERCHLORIDE OF MERCURY M:0600
PERCHLOROBENZENE H:0400
PERCHLOROBUTADIENE H:0450
PERCHLOROCYCLOPENTADIENE H:0500
PERCHLORODIHOMOCUBANE M:5650
PERCHLOROETHANE H:0550
PERCHLOROETHYLENE T:0400
PERCHLOROMETHANE C:1850
PERCHLOROMETHYLMERCAPTAN P:1050
PERCHLORON C:1000
PERCHLOROPENTACLYCLODECANE M:5650
PERCHLORURE D'ANTIMOINE (French) A:4900
PERCHLORURE DE FER (French) F:0200
PERCLENE T:0400
PERCLORATO AMONICO (Spanish) A:3500
PERCLORATO BARICO (Spanish) B:0300
PERCLOROETILENO (Spanish) T:0400
PERCLOROMETILMERCAPTANO (Spanish) P:1050
PERFLUROETHYLENE T:0850
PERHYDROAZEPINE H:0850
PERHYDROL H:1600
PERHYDRONAPTHALENE D:0400
PERK T:0400
PERKLONE T:0400
PERL ALUM A:1750
PERM-A-CHLOR T:2350
PERMA KLEER 50 ACID E:3900
PERMACIDE P:0350
PERMAGARD P:0350
PERMANGANATE OF POTASH P:3200
PERMANGANATE de POTASSIUM (French) P:3200
PERMANGANATO BARICO (Spanish) B:0350

PERMANGANATO POTASICO (Spanish) P:3200
PERMANGANIC ACID, BARIUM SALT B:0350
PERMASAN P:0350
PERMATOX DP-2 P:0350
PERMATOX PENTA P:0350
PERMITE P:0350
PERONE H:1600
PEROXAN H:1600
PEROXIDE H:1600
PEROXIDO de ACETILO (Spanish) A:0700
PEROXIDO BARICO (Spanish) B:0400
PEROXIDO CALCICO (Spanish) C:1150
PEROXIDO de CICLOHEXANONA (Spanish) C:6000
PEROXIDO de DIBENZOILO (Spanish) D:1050
PEROXIDO de HIDROGENO (Spanish) H:1600
PEROXIDO de LAUROILO (Spanish) L:0400
PEROXIDO de METIL ETIL CETONA (Spanish) B:3050
PEROXIDO POTASICO (Spanish) P:3250
PEROXIDO de URANIO (Spanish) U:0500
PEROXIDO de UREA (Spanish) U:1100
PEROXYACETIC ACID P:0900
PEROXYDE de BARYUM (French) B:0400
PEROXYDICARBONIC ACID, BIS(1-METHYLETHYL) ESTER I:1600
PEROXYDISULFANIC ACID, DIAMMONIUM SALT A:3550
PERSEC T:0400
PERSIA-PERAZOL D:1550
PERSIAN-INSECT POWDER P:4750
PERSULFATE d'AMMONIUM (French) A:3550
PESTMASTER EDB-85 E:4000
PESTOX PLUS P:0200
PETHION P:0200
PETROHOL I:1250
PETROL (Britain) G:0200
PETROLATUM P:1100
PETROLATUM JELLY P:1100
PETROLEUM O:0800
PETROLEUM ASPHALT O:4300
PETROLEUM ASPHALT A:5600
PETROLEUM BENZIN N:0200
PETROLEUM CRUDE OIL O:0800
PETROLEUM DISTILLATE D:7750
PETROLEUM DISTILLATE D:7800
PETROLEUM DISTILLATES, HYDRO-TREATED MIDDLE O:3700
PETROLEUM ETHER N:0200
PETROLEUM GAS, LIQUEFIED L:2200
PETROLEUM INSULATING OIL O:4900
PETROLEUM JELLY P:1100
PETROLEUM NAPHTHA P:1150
PETROLEUM NAPHTHA D:7750
PETROLEUM NAPHTHA N:3050
PETROLEUM OIL O:0900
PETROLEUM PITCH A:5700
PETROLEUM PITCH A:5600
PETROLEUM PITCH A:5650
PETROLEUM RESIDUE A:5700
PETROLEUM ROOFING TAR O:4300
PETROLEUM SOLVENT N:0250
PETROLEUM SOLVENT N:0300
PETROLEUM SOLVENT P:1150
PETROLEUM SOLVENT N:0200
PETROLEUM SPIRITS N:0200
PETROLEUM SPIRITS M:5600
PETROLEUM SPIRITS N:0300

PETROLEUM TAILINGS A:5650
PETROLEUM THINNER N:0300
PETROLEUM WAX P:0100
PETZINOL T:2350
PEZIFILM T:1450
PF-3 D:0850
PFLANZOL B:0600
PG12 P:4200
PGME P:4300
PHELLANDRENE D:7150
PHENACHLOR T:2450
PHENACIDE T:1950
PHENACYL CHLORIDE C:2250
PHENADOR-X D:7200
PHENATOX T:1950
PHENE B:0550
α-PHENETHYL ALCOHOL M:3050
α-PHENETHYL ALCOHOL M:3050
PHENIC ACID P:1200
PHENMAD P:1400
PHENOCHLOR P:2200
PHENOCLOR P:2200
PHENOL P:1200
PHENOL, 2-CHLORO- C:2700
PHENOL, o-CHLORO C:2700
PHENOL, o-ETHYL E:7200
PHENOL, DIMETHYL- X:0400
PHENOL, HEXAHYDRO- C:5900
PHENOL, m-HYDROXY- R:0100
PHENOL, m-HYDROXY- R:0100
PHENOL, METHYL- C:5100
PHENOL, 2-NITRO- N:1850
PHENOL, 4-NITRO N:1950
PHENOL, o-NITRO N:1850
PHENOL, p-NITRO- N:1950
PHENOL, PENTACHLORO- P:0350
PHENOL, PENTACHLORO-,SODIUM SALT S:4700
PHENOL,2,4,6-TRINITRO-, AMMONIUM SALT A:3650
PHENOLE (German) P:1200
PHENOMERCURIC ACETATE P:1400
PHENOMERCURY ACETATE P:1400
PHENOOXY ALCOHOL E:5200
PHENOX D:0100
PHENOXY BENZENE D:7350
1-HYDROXY-2-PHENOXYETHANE E:5200
2-PHENOXYETHANOL E:5200
PHENOXYTOL E:5200
PHENYLAMINE A:4750
PHENYLANILINE D:7250
N-PHENYLANILINE D:7250
PHENYLARSENIC DICHLORIDE P:1250
PHENYLARSENOUS DICHLORIDE P:1250
N-PHENYLBENZENAMINE D:7250
PHENYL BENZENE D:7200
PHENYL BROMIDE B:2500
PHENYLCARBINOL B:1200
PHENYL CELLOSOLVE E:5200
PHENYL CHLORIDE C:2400
PHENYL CHLOROMETHYL KETONE C:2250
PHENYL CYANIDE B:0950
L-PHENYLDECANE D:0750
PHENYLDICHLOROARSINE P:1250
PHENYLDODECAN D:8100
1-PHENYLDODECANE D:8100
o-PHENYLENEDIOL C:1950
PHENYLETHANE E:2400
1-PHENYLETHANOL M:3050
1-PHENYLETHANONE A:0450
PHENYLETHENE S:6300
PHENYL ETHER D:7350

PHENYL ETHER-BIPHENYL MIXTURE D:8850
1-PHENYL ETHYL ALCOHOL M:3050
α-PHENYL ETHYL ALCOHOL M:3050
PHENYLETHYLDICHLOROSILANE E:7300
PHENYLETHYLENE S:6300
PHENYL FLUORIDE F:0950
PHENYL FLUORIDE F:0950
PHENYL HYDRATE P:1200
PHENYLHYDRAZINE P:1300
PHENYLHYDRAZINE HYDROCHLORIDE P:1350
PHENYLHYDRAZINE MONOHYDROCHLORIDE P:1350
PHENYLHYDRAZINIUM CHLORIDE P:1350
PHENYL HYDRIDE B:0550
PHENYL HYDROXIDE P:1200
PHENYLIC ACID P:1200
PHENYL KETONE B:1000
PHENYL MERCAPTAN B:0800
PHENYLMERCURIC ACETATE P:1400
PHENYLMETHANAL B:0500
PHENYL METHANE T:1600
PHENYLMETHANOL B:1200
PHENYLMETHYL ACETATE B:1150
PHENYLMETHYL ALCOHOL B:1200
PHENYLMETHYL AMINE B:1250
PHENYLMETHYL CARBINOL M:3050
PHENYL METHYL KETONE A:0450
N-PHENYLMETHYLAMINE M:2050
PHENYL MONOGLYCOL ETHER E:5200
PHENYL OXIDE D:7350
PHENYL PERCHLORYL H:0400
PHENYLPHOSPHINE DICHLORIDE B:0650
PHENYLPHOSPHINE THIODICHLORIDE B:0700
PHENYLPHOSPHONOTHIOIC DICHLORIDE B:0700
PHENYLPHOSPHONOUS DICHLORIDE B:0650
PHENYLPHOSPHORUS DICHLORIDE B:0650
2-PHENYLPROPANE C:5450
1-PHENYLPROPANE P:4050
PHENYLPROPYLENE M:5400
2-PHENYLPROPYLENE M:5400
β-PHENYLPROPYLENE M:5400
1-PHENYLTETRADECANE T:0550
PHENYLTHIOL B:0800
1-PHENYLTRIDECANE T:3150
1-PHENYLUNDECANE U:0400
PHENYL XYLYL ETHANE P:1450
1-PHENYL-1-XYLYL ETHANE P:1450
PHISODAN H:0600
PHISOHEX H:0600
PHLOROL E:7200
PHORSAEURELOESUNGEN (German) P:2350
PHORTOX T:2500
PHOSDRIN P:1500
PHOSDRIN 24 P:1500
PHOSFENE P:1500
PHOSFLEX 179-C T:4450
PHOSGEN (German) P:1550
PHOSGENE P:1550
PHOSKIL P:0200
PHOSPHATE de o,o-DIMETHLE et de o-(1,2-DIBROMO-2,2-DICHLORETHYLE) (French) N:0150
PHOSPHENE (French) P:1500
PHOSPHENYL CHLORIDE B:0650
PHOSPHINIC ACID, AMMONIUM SALT A:2900
PHOSPHONIC ACID, DIMETHYL ESTER D:5800
PHOSPHONOTHIOIC ACID, METHYL-, S-2-[BIS(1-METHYLETHYL)AMINOETHYL] O-ETHYL ESTER V:1650
PHOSPHORE (TRICHLORURE de) (French) P:1950
PHOSPHORIC ACID P:1600
PHOSPHORIC ACID P:2350
meta-PHOSPHORIC ACID P:1600
PHOSPHORIC ACID, o,o-DIETHYL o-6-METHYL-2-(1-METHYLETHYL)-4-PYRIMIDINYL ESTER D:1000
PHOSPHORIC ACID, TETRAETHYL ESTER T:0800
PHOSPHORIC ACID, TRIETHYL ESTER T:3650
PHOSPHORIC ACID TRIETHYLENEIMIDE T:4650
PHOSPHORIC ACID TRIBUTYL ESTER T:2000
PHOSPHORIC ACID, TRIS(METHYL PHENYL)ESTER T:2950
PHOSPHORIC ACID, TRISODIUM SALT S:4900
PHOSPHORIC ACID, TRIS(2-METHYLPHENYL) ESTER T:4450
PHOSPHORIC ACID, TRITOLYL ESTER T:2950
PHOSPHORIC CHLORIDE P:1750
PHOSPHORIC SULFIDE P:1850
PHOSPHORODICHLORIDIC ACID,ETHYL ESTER E:7500
PHOSPHORODIFLUORIDIC ACID D:3950
PHOSPHORODITHIOIC ACID, o,o-DIETHYL ESTER, s,s-DIESTER with METHANEDITHIOL E:0600
PHOSPHORODITHIONIC ACID,S-2-(ETHYLTHIO)ETHYL-o,o-DIETHYL ESTER D:7700
PHOSPHOROFLUORIDIC ACID, DIISOPROPYL ESTER D:0850
PHOSPHOROTHIOATE C:5000
PHOSPHOROTHIOIC ACID, o,o-DIETHYL ESTER, O-ESTER with 3-CHLORO-7-HYDROXY-4-METHYLCOUMARIN C:5000
PHOSPHOROTHIOIC ACID, o,o-DIETHYL O-2-(ETHYLTHIO)ETHYL ESTER, MIXED with o,o-DIETHYL s-2-(ETHYLTHIO)ETHYL PHOSPHOROTHIOATE D:0800
PHOSPHOROTHIOIC ACID, o,o-DIETHYL o-p-NITROPHENYL ESTER P:0200
PHOSPHOROTHIOIC ACID, o,o,-DIETHYLO-(p-NITROPHENYL)ESTER P:0200
PHOSPHOROTHIOIC ACID o,o-DIMETHYL O-(4-NITROPHENYL)ESTER M:4900
PHOSPHOROUS ACID, TRIETHYL ESTER T:3700
PHOSPHOROUS ACID, TRIMETHYL ESTER T:4400
PHOSPHORUS, AMORPHOUS, BLACK P:1650
PHOSPHORUS, AMORPHOUS, RED P:1700
PHOSPHORUS, BLACK P:1650
PHOSPHORUS BROMIDE P:1900
PHOSPHORTRICHLORID (German) P:1950
PHOSPHOROUS CHLORIDE P:1950
PHOSPHORUS CHLORIDE P:1800
PHOSPHORUS CHLORIDE OXIDE P:1750
PHOSPHORUS CHLORIDE P:1950
PHOSPHORUS OXYCHLORIDE P:1750
PHOSPHORUS OXYCHLORIDE P:1800
PHOSPHORUS OXYTRICHLORIDE P:1800
PHOSPHORUS OXYTRICHLORIDE P:1750
PHOSPHORUS PENTASULFIDE P:1850
PHOSPHORUS PERSULFIDE P:1850
PHOSPHORUS, RED P:1700
PHOSPHORUS SULFIDE P:1850
PHOSPHORUS TRIBROMIDE P:1900
PHOSPHORUS TRICHLORIDE P:1950
PHOSPHORUS, YELLOW or WHITE P:1750
PHOSPHORYL CHLORIDE P:1800
PHOSPHORYL CHLORIDE P:1750
PHOSPHOSTIGMINE P:0200
PHOSPHOTOX E E:0600
PHOSPHURES D'ALUMIUM (French) PHOSTOXIN A:1700
PHOSVIN Z:1900
PHOSVIT D:2650
PHOTOPHOR C:1250
PHPH D:7200
PHTHALANDIONE P:2000
1,3-PHTHALANDIONE P:2000
m-PHTHALIC ACID I:1100
PHTHALIC ACID ANHYDRIDE P:2000
PHTHALIC ACID, BENZYL BUTYL ETHER B:3650
PHTHALIC ACID, BIS-(7-METHYLOCTYL) ESTER D:4450
PHTHALIC ACID, BIS(2-METHOXYETHYL)ESTER T:4400
PHTHALIC ACID, BIS(2-ETHYLHEXYL ESTER) D:3650
PHTHALIC ACID, BIS(2-ETHYLHEXYL ESTER) D:5250
PHTHALIC ACID, BIS(8-METHYL-NONYL) ESTER D:4350
PHTHALIC ACID, DIAMYL ESTER D:0975
PHTHALIC ACID, DIBUTYL ESTERCELLUFLEX DBP D:1250
PHTHALIC ACID, DIBUTYL ESTER D:1250
PHTHALIC ACID, DIETHYL ESTER D:3750
PHTHALIC ACID, DIHEPTYL ESTER D:4000
PHTHALIC ACID, DIISOBUTYL ESTER D:4300
PHTHALIC ACID, DIISODECYL ESTER D:4350
PHTHALIC ACID, DIMETHYL ESTER D:5900
PHTHALIC ACID, DINONYL ESTER D:6950
PHTHALIC ACID, DIOCTYL ESTER D:5250
PHTHALIC ACID, DIPENTYL ESTER D:0975
PHTHALIC ACID, DITRIDECYL ESTER D:7850
PHTHALIC ACID, DIUNDECYL ESTER D:7900
PHTHALIC ANHYDRIDE P:2000
PHTHALOL D:3750
PHTHALSAEUREANHYDRID (German) P:2000
PHTHALSAEUREDIMETHYLESTER (German) D:5900
PHYGON® D:1400
PHYGON®-XL D:1400
PHYTAR S:2700
PHYTAR-560 S:2700
PIC-CHLOR C:2800
PICFUME C:2800
PICKEL ALUM A:1750
2-PICOLINE M:5150
3-PICOLINE M:5200
4-PICOLINE M:5250
α-PICOLINE M:5150
β-PICOLINE M:5200
γ-PICOLINE M:5250
m-PICOLINE M:5200

PICRATO AMONICO (Spanish) A:3650
PICRIC ACID, AMMONIUM SALT A:3650
PICRIDE C:2800
PICTAROL A:3650
PIELIK D:0100
PILITRE (Spanish) P:4750
PILLARFURAN C:1500
PILOT HD-90 D:8300
PILOT SF-40 D:8300
PIMELIC KETONE C:5950
PINACOLOXYMETHYLPHOSPHORYL FLUORIDE S:5650
PINACOLYL METHYLFLUORO-PHOSPHONATE S:5650
PINACOLYL METHYLPHOSPHONO-FLUORIDE S:5650
PINACOLYL METHYLPHOSPHONO-FLUORIDATE S:5650
PINACOLYLOXY METHYLPHOSPHORYL FLUORIDE S:5650
PINAKON H:1200
PINE OIL O:4100
PINE OIL, STEAM DISTILLED O:4100
PINENE P:2050
2-PINENE P:2050
α-PINENE P:2050
α-PINEO (Spanish) P:2050
PINON WOOD CHARCOAL C:2000
PIPERAZIDINE P:2100
PIPERAZIN (German) P:2100
PIPERAZINE P:2100
PIPERAZINE, ANHYDROUS P:2100
1-PIPERAZINE ETHANAMINE A:1900
PIPERYLENE P:0450
PIRETINA (Spanish) P:4750
PIROFOS T:0600
PIROFOSFATO de TETRAETILO (Spanish) T:0800
PITTCHLOR C:1000
PITTCIDE C:1000
PIVALIC ACID T:4000
PLACIDOL E D:3750
PLANAVIN N:1300
PLANOTOX D:0100
PLANT DITHIO AEROSOL T:0600
PLANT SPRAY OIL O:4600
PLANTFUME 103 SMOKE GENERATOR T:0600
PLANTGARD D:0100
PLASTICIZED DDP D:4350
PLASTIC LATEX L:0200
PLASTOMOLL DOA D:7000
PLEOPARAPHENE P:0200
PLOMB FLUORURE (French) L:0900
PLUMBANE, TETRAETHYL- T:0750
PLUMBANE, TETRAMETHYL- T:1050
PLUMBOUS ARSENATE L:0700
PLUMBOUS CHLORIDE L:0800
PLUMBOUS FLUORIDE L:0900
PLUMBOUS OXIDE L:2300
PLUMBOUS SULFIDE L:1600
PLURACOL POLYOL P:2450
PMA P:1400
PMAC P:1400
PMACETATE P:1400
PMAL P:1400
PMAS P:1400
PMM P:1050
PMFP S:5650
PNA N:1450
PNP N:1950
PNT N:2250
PO12 A:5500

POLIPROPILENGLICOL (Spanish) P:4200
POLIPROPILENO (Spanish) P:2400
POLYBUTENE P:2150
POLYCAT-8 D:5100
POLYCHLORCAMPHENE T:1950
POLYCHLORINATED BIPHENYLS P:2200
POLYCHLORINATED CAMPHENE T:1950
POLYCHLOROBIPHENYL P:2200
POLYCHLOROCAMPHENE T:1950
POLYCIZER 962-BPA D:7850
POLYCIZER DBP D:1250
POLY(DIMETHYLSILOXANE) D:5950
POLYDIMETHYL SILYLENE L:0200
POLY(ETHYLENEIMINE) P:2250
POLYETHYLENE POLYAMINES P:2250
POLYETHYLENEIMINE P:2250
POLYFORMALDEHYDE P:0150
POLYISOBUTYLENE PLASTICS, RESINS, and WAXES P:2150
POLYMER GASOLINES G:0450
POLYMETHYLENE POLYPHENYL ISOCYANATE P:2300
POLY(OXYETHYL)DODECYL ETHER E:1000
POLY(OXYETHYL)LAURYL ETHER E:1000
POLY(OXYETHYL)MYRISTYL ETHER E:1400
POLY(OXYETHYL)PENTADECYL ETHER E:1200
POLY(OXYETHYL)TETRADECYL ETHER E:1400
POLY(OXYETHYL)TRIDECYL ETHER E:1300
POLY PROPENE P:2400
POLYOXPROPYLENE GLYCOL P:2450
POLY(PROPYLENE GLYCOL)METHYL ETHER P:4300
POLYOXYMETHYLENE P:0150
POLYOXYMETHYLENE GLYCOL P:0150
POLYOXYMETHYLENE GLYCOLS F:1250
POLYOXYPROPYLENE GLYCOL METHYL ETHER P:4300
POLYOXYPROPYLENE GLYCOL P:2450
POLYPHOSPHORIC ACID P:2350
POLYPROPYLENE P:2400
POLYPROPYLENE GLYCOL P:2450
POLYPROPYLENE GLYCOL 750 P:2450
POLYPROPYLENE GLYCOL METHYL ETHER P:4300
POLYRAM ULTRA T:1450
POLY-SOLV D:3100
POLY-SOLV D:3400
POLY-SOLV DB D:3300
POLY-SOLV DM D:3415
POLY-SOLV E E:0800
POLY-SOLV EB E:4900
POLY-SOLV EE E:0800
POLY-SOLV EE ACETATE E:0900
POLY-SOLV EM E:5100
POLY-SOLV TE E:1500
POLY-SOLV TE T:3500
POLY-SOLVE MPM P:4300
POMARSOL T:1450
POMASOL T:1450
PORAL P:0175
POTASH CHLORATE P:2700
POTASH NITRATE P:3050
POTASH SOAP P:3100
POTASSA P:2950
POTASSE CAUSTIQUE (French) P:2950
POTASSIO (Spanish) P:2500
POTASSIUM P:2500
POTASSIUM ACID ARSENATE P:2550

POTASSIUM ACID OXALATE P:2650
POTASSIUM ANTIMONYL TARTRATE A:5000
POTASSIUM ANTIMONYL-d-TARTRATE A:5000
POTASSIUM ARSENATE P:2550
POTASSIUM ARSENATE, MONOBASIC P:2550
POTASSIUM ARSENITE P:2600
POTASSIUM BICHROMATE P:2900
POTASSIUM BINOXALATE P:2650
POTASSIUM CHLORATE P:2700
POTASSIUM (CHLORATE de) (French) P:2700
POTASSIUM CHROMATE P:2750
POTASSIUM CHROMATE(VI) P:2750
POTASSIUM cis-9-OCTADECENOIC ACID O:5200
POTASSIUM CYANIDE P:2800
POTASSIUM CYANIDE SOLUTION P:2800
POTASSIUM DICHLORO-s-TRIAZINETRIONE P:2850
POTASSIUM DICHLORO ISOCYANURATE P:2850
POTASSIUM DICHLOROISOCYANURATE P:2850
POTASSIUM DICHROMATE(VI) P:2900
POTASSIUM DICHROMATE P:2900
POTASSIUM DIHYDROGEN ARSENATE P:2550
POTASSIUM, ELEMENTAL P:2500
POTASSIUM FLUOZIRCONATE Z:2600
POTASSIUM HEXAFLUOROZIRCONATE Z:2600
POTASSIUM HYDRATE P:2950
POTASSIUM (HYDRIXYDE de) (French) P:2950
POTASSIUM HYDROGEN ARSENATE P:2550
POTASSIUM HYDROXIDE P:2950
POTASSIUM IODIDE P:3000
POTASSIUM METAARSENITE P:2600
POTASSIUM METAL P:2500
POTASSIUM NITRATE P:3050
POTASSIUM OLEATE P:3100
POTASSIUM OLEATE O:5200 xxx
POTASSIUM OXALATE P:3150
POTASSIUM OXALATE MONOHYDRATE P:3150
POTASSIUM OXYMURIATE P:2700
POTASSIUM PERMANGANATE P:3200
POTASSIUM (PERMANGANATE de) (French) P:3200
POTASSIUM PEROXIDE P:3250
POTASSIUM SALT OF HYDROCYANIC ACID P:2800
POTASSIUM SUPEROXIDE P:3250
POTASSIUM TROCLOSENE P:2850
POTASSIUM ZINC CHROMATE Z:2000
POTASSIUM ZIRCONIUM FLUORIDE Z:2600
POTATO ALCOHOL E:1900
POTATO SPIRIT OIL I:0150
POTCRATE P:2700
POTENTIATED ACID GLUTARALDEHYDE G:0600
PRECIPITATED SULFUR S:6600
PREEGLONE D:7650
PRESERV-O-SOTE C:5050
PRESERVATIVE OIL O:4000
PRESPERSION, 75 UREA U:0900
PREVENTOL-1 T:2450
PRILTOX P:0350

PRIMAGRAM M:5550
PRIMARY AMYL ACETATE A:4200
PRIMARY AMYL ALCOHOL A:4350
PRIMARY DECYL ALCOHOL D:0700
PRIMARY OCTYL ALCOHOL O:0300
PRIMATOL A:5750
PRIMATOL A A:5750
PRIMAZE A:5750
PRIME STEAM LARD O:1600
PRIMEXTRA M:5550
PRIMROSE YELLOW Z:1000
PRINDINA (Spanish) P:4800
PRIODERM M:0250
PRIST E:5100
PRODOX 131 I:1650
PRODUCT 161 D:8700
PROFAX P:2400
PROFUME A C:2800
PROP-2-EN-1-AL A:0800
PROPADIENE–ALLENE MIXTURE M:1650
PROPADIENE–METHYLACETYLENE MIXTURE M:1650
PROPALDEHYDE P:3650
PROPANAL P:3650
PROPANAMINE P:4000
1-PROPANAMINE P:4000
2-PROPANAMINE I:1300
1-PROPANAMINE, 2-METHYL-N-(2-METHYL PROPYL)- D:4100
2-PROPANAMINE, N-(1-METHYLETHYL)- D:4600
PROPANDINITRILO (Spanish) P:3450
PROPANE P:3300
n-PROPANE P:3300
PROPANE-BUTANE-(PROPYLENE) L:2200
PROPANE-2-CARBOXYLIC ACID I:0500
PROPANE-1,2-DIOL P:4200
PROPANE, 1-CHLORO-2,3-EPOXY C:2550
PROPANE, 1-NITRO- N:2000
PROPANE, 1,2-DICHLORO- D:2300
PROPANE, 1,2,3-TRICHLORO T:2750
PROPANE, 2-METHYL I:0250
PROPANE, 2-NITRO N:2050
PROPANE, CHLORO- P:4100
PROPANE CYANOHYDRIN A:0350
PROPANE, EPOXY- P:4450
1,2,3-PROPANE TRICARBOXYLIC ACID, 2-HYDROXY-, AMMONIUM SALT A:2550
PROPANE,1,1-DICHLORO- D:2250
PROPANE,2-METHOXY-2-METHYL M:3300
PROPANEACID A:0900
1-PROPANECARBOXYIC ACID B:4900
PROPANEDINITRILE P:3450
1,2-PROPANEDIOL P:4200
1,2-PROPANEDIOL-1-ACRYLATE H:1950
1,2-PROPANEDIOL 1-METHACRYLATE H:2000
1,3-PROPANEDIOL, 2,2-DIMETHYL D:6000
PROPANENITRILE P:3800
PROPANENITRILE,2-HYDROXY-2-METHYL- A:0350
PROPANETHIOL P:4650
1-PROPANETHIOL P:4650
2-PROPANETHIOL I:1500
PROPANE-1-THIOL P:4650
PROPANE-2-THIOL I:1500
1,2,3-PROPANETRIOL G:0650
PROPANEACID A:0900
PROPANE-BUTANE-(PROPYLENE) L:2200
1-PROPANECARBOXYIC ACID B:4900
PROPANE-2-CARBOXYLIC ACID I:0500
PROPANE, CHLORO- P:4100
PROPANE, 1-CHLORO-2,3-EPOXY C:2550

PROPANE CYANOHYDRIN A:0350
1,2-PROPANEDIOL P:4200
PROPANE-1,2-DIOL P:4200
1,2-PROPANEDIOL-1-ACRYLATE H:1950
1,2-PROPANEDIOL 1-METHACRYLATE H:2000
1,3-PROPANEDIOL, 2,2-DIMETHYL D:6000
PROPANE,1,1-DICHLORO- D:2250
PROPANE, 1,2-DICHLORO- D:2300
PROPANEDINITRILE P:3450
PROPANE, EPOXY- P:4450
PROPANE,2-METHOXY-2-METHYL M:3300
PROPANE, 2-METHYL I:0250
PROPANENITRILE P:3800
PROPANENITRILE,2-HYDROXY-2-METHYL- A:0350
PROPANE, 2-NITRO N:2050
PROPANE, 1-NITRO- N:2000
PROPANETHIOL P:4650
1-PROPANETHIOL P:4650
2-PROPANETHIOL I:1500
PROPANE-1-THIOL P:4650
PROPANE-2-THIOL I:1500
1,2,3-PROPANE TRICARBOXYLIC ACID, 2-HYDROXY-, AMMONIUM SALT 2- A:2550
PROPANE, 1,2,3-TRICHLORO T:2750
1,2,3-PROPANETRIOL G:0650
PROPANO (Spanish) P:3300
PROPANOIC ACID P:3700
PROPANOIC ACID T:4000
tert-PROPANOIC ACID T:4000
PROPANOIC ACID BUTYL ESTER B:4550
PROPANOIC ACID BUTYL ESTER B:4550
PROPANOIC ACID, 2-CHLORO- C:2900
PROPANOIC ACID, 2,2-DICHLORO- D:0200
PROPANOIC ACID, 2,2-DI-METHYL- T:4000
PROPANOIC ACID, ETHYL ESTER E:7600
2-PROPANOIC ACID, 1-METHYL-, ETHYL ESTER E:6900
1-PROPANOL P:3950
2-PROPANOL I:1250
PROPANOL-1 P:3950
n-PROPANOL P:3950
2-PROPANOL-1-ETHOXY P:4250
2-AMINO-2-METHYL-1-PROPANOL (90% or less) A:1925
2-PROPANOLAMINA (Spanish) P:3400
3-PROPANOLAMINA (Spanish) P:3350
1,3-PROPANOLAMINE P:3350
2-PROPANOLAMINE P:3400
3-PROPANOLAMINE P:3350
n-PROPANOLAMINE P:3350
1-PROPANOL, 2-AMINO- P:3400
1-PROPANOL, 3-AMINO P:3350
1-PROPANOL, 2-AMINO-2-METHYL- A:1925
2-PROPANOL, 2-METHYL- B:3450
PROPANOL, 3-(3-(3-METHOXY PROPOXY)PROPOXY)- T:4600
2-PROPANOL, 1,1',1''-NITRILOTRI- T:3950
PROPANOL, OXYBIS-, METHYL ETHER D:7600
PROPANOLE (German) P:3950
PROPANOLIDE P:3600
PROPANONE A:0300
2-PROPANONE A:0300
PROPARGIL P:3500
PROPARGITA (Spanish) P:3500
PROPARGITE P:3500
PROPARGYL ALCOHOL P:3550
PROPASOL SOLVENT M P:4300
PROPELLANT 11 T:2400
PROPELLANT 12 D:1650

PROPELLANT 14 D:2550
PROPELLENT-22 M:5800
PROPEN-1-OL-3 A:1200
2-PROPEN-1-OL A:1200
2-PROPEN-1-ONE A:0800
1-PROPEN-3-OL A:1200
PROPENAL A:0800
2-PROPENAL A:0800
PROPENAMIDE A:0850
2-PROPENAMIDE A:0850
1-PROPENE P:4150
1-PROPENE, 2-METHYL TRIMER T:3900
PROPENE OXIDE P:4450
PROPENE POLYMER P:2400
PROPENE TETRAMER D:8050
PROPENE TETRAMER P:4500
PROPENE, TRIMER P:4550
PROPENENITRILE A:0950
2-PROPENENITRILE A:0950
2-PROPENENITRILE, 2-METHYL- M:1300
PROPENEOXIDE P:4450
PROPENE TRIMER N:2450
2-PROPENIC ACID, 2-METHYL-, BUTYL ESTER B:4450
2-PROPENOIC ACID A:0900
2-PROPENOIC ACID, BUTYL ESTER B:3250
2-PROPENOIC ACID, DECYL ESTER D:0650
2-PROPENOIC ACID, ETHYL ESTER E:1800
PROPENOIC ACID, METHYL ESTER M:1700
2-PROPENOIC ACID, METHYL ESTER M:1700
2-PROPENOIC ACID, 2-METHYL-, METHYL ESTER M:4800
PROPENOL A:1200
2-PROPENOL A:1200
PROPENONITRILO (Spanish) A:0950
PROPENYL ALCOHOL A:1200
2-PROPENYL ALCOHOL A:1200
2-PROPENYL CHLORFORMATE A:1350
n-PROPILAMINA (Spanish) P:4000
n-PROPILBENCENO (Spanish) P:4050
PROPILENGLICOL (Spanish) P:4200
PROPILENGLICOL MONOETIL ETER (Spanish) P:4250
PROPILENGLICOL MONOMETIL ETER (Spanish) P:4300
PROPIL ETER (Spanish) P:4600
n-PROPILMERCAPTANO (Spanish) P:4650
β-PROPIOLACTONE P:3600
β-PROPIOLACTONA (Spanish) P:3600
PROPIOLIC ALCOHOL P:3550
PROPIONALDEHIDO (Spanish) P:3650
PROPIONALDEHYDE P:3650
PROPIONATO de n-BUTILO (Spanish) B:4550
PROPIONATO de ETILO (Spanish) E:7600
PROPIONE D:3700
PROPIONIC ACID P:3700
PROPIONIC ACID, 2-METHYLENE- M:1200
PROPIONIC ACID, 3-CHLORO- C:2950
PROPIONIC ACID, 3-ETHOXYETHYL ESTER E:5700
PROPIONIC ACID ANHYDRIDE P:3750
PROPIONIC ACID BUTYL ESTER B:4550
PROPIONIC ACID GRAIN PRESERVER P:3700
PROPIONIC ALDEHYDE P:3650
PROPIONIC ANHYDRIDE P:3750
PROPIONIC ETHER E:7600
PROPIONIC NITRILE P:3800
PROPIONITRILE P:3800
3-PROPIONOLACTONE P:3600
β-PROPIONOLACTONE P:3600
PROPIONYL OXIDE P:3750

n-PROPOXIPROPANOL (Spanish) P:3850
2-PROPOXYETHANOL E:5300
n-PROPOXYPROPANOL P:3850
PROPROP D:0200
n-PROPYL-1-PROPANAMINE D:7450
PROPYL ACETATE P:3900
1-PROPYL ACETATE P:3900
2-PROPYL ACETATE I:1200
n-PROPYL ACETATE P:3900
sec-PROPYL ACETATE I:1200
PROPYL ALCOHOL P:3950
1-PROPYL ALCOHOL P:3950
n-PROPYL ALCOHOL P:3950
sec-PROPYL ALCOHOL I:1250
PROPYL ALDEHYDE P:3650
n-PROPYL ALKOHOL (German) P:3950
PROPYLACETIC ACID P:0700
PROPYLACETONE M:3350
PROPYLAMINE P:4000
2-PROPYLAMINE P:4000
n-PROPYLAMINE P:4000
sec-PROPYLAMINE I:1300
PROPYL BENZENE P:4050
n-PROPYLBENZENE P:4050
PROPYLBROMIDE B:2750
n-PROPYLBROMIDE B:2750
PROPYL CARBINOL B:3350
n-PROPYL CARBINOL B:3350
n-PROPYLCARBINYL CHLORIDE B:3750
PROPYL CELLOSOLVE E:5300
n-PROPYL CHLORIDE P:4100
PROPYL CYANIDE B:4950
1,2-PROPYLEN IMINE P:4400
PROPYLENE P:4150
1-PROPYLENE P:4150
PROPYLENE ALDEHYDE C:5400
PROPYLENE CHLORIDE D:2300
PROPYLENE DICHLORIDE D:2300
α,β-PROPYLENE DICHLORIDE D:2300
PROPYLENE GLYCOL P:4200
1,2-PROPYLENE GLYCOL P:4200
PROPYLENE GLYCOL ETHYL ETHER P:4250
PROPYLENE GLYCOL METHYL ETHER ACETATE P:4350
PROPYLENE GLYCOL MONOACRYLATE H:1950
PROPYLENE GLYCOL MONOETHYL ETHER P:4250
PROPYLENE GLYCOL MONOMETHACRYLATE H:2000
PROPYLENE GLYCOL MONOMETHYL ETHER P:4300
α-PROPYLENE GLYCOL MONOMETHYL ETHER P:4300
PROPYLENE IMINE P:4400
PROPYLENE OXIDE P:4450
1,2-PROPYLENE OXIDE P:4450
PROPYLENE TETRAMER P:4500
PROPYLENE TETRAMER D:8050
PROPYLENE TRIMER P:4550
PROPYLENGLYKOL-MONOMETHYLAETHER (German) P:4300
1,2-PROPYLENIMINE P:4400
n-PROPYL ESTER of ACETIC ACID P:3900
PROPYL ESTER OF NITRIC ACID P:4700
PROPYL ETHER P:4600
DI-n-PROPYL ETHER P:4600
n-PROPYL ETHER P:4600
PROPYLETHYLENE P:0800
PROPYLFORMIC ACID B:4900
PROPYL HYDRATE P:3300
PROPYL HYDRIDE P:3300

PROPYLIC ALCOHOL P:3950
PROPYLIC ALDEHYDE P:3650
PROPYLIDENE CHLORIDE D:2250
PROPYLIDENE DICHLORIDE D:2250
PROPYLMETHANOL B:3350
PROPYL MERCAPTAN P:4650
n-PROPYL MERCAPTAN P:4650
PROPYL NITRATE P:4700
n-PROPYL NITRATE P:4700
PROPYLNITRILE P:3800
2-PROPYN-1-OL P:3550
1-PROPYNE-3-OL P:3550
PROPYNE–ALLENE mixture M:1650
PROPYNE–PROPADIENE mixture M:1650
PROPYNE MIXED with PROPADIENE M:1650
3-PROPYNOL P:3550
2-PROPYNYL ALCOHOL P:3550
PROTIUM H:1350
PROXOL T:2050
PROZOIN P:3700
PRUSSIC ACID H:1500
PRUSSITE C:5650
PS C:2800
PSEUDOACETIC ACID P:3700
PSEUDOCUMENE T:4100
PSEUDOCUMOL T:4100
PSEUDOHEXYL ALCOHOL E:2500
PSEUDOTHIOUREA T:1350
PSEUDOUREA U:0900
PT 3 N:2450
PURALIN T:1450
PURASAN-SC-10 P:1400
PURATRONIC CHROMIUM TRIOXIDE C:3300
PURATURF 10 P:1400
PURE GRAIN ALCOHOL E:1900
PURE ZINC CHROME Z:1000
PURE ZINC YELLOW Z:1000
PURPLE SALT P:3200
PX-238 D:7000
PX 104 D:1250
PYNACOLYL METHYLFLUORO-PHOSPHONATE S:5650
PYRALENE P:2200
PYRANOL P:2200
PYRANOL 1478 T:2150
PYRAZINE HEXAHYDRIDE P:2100
PYRAZINE, HEXAHYDRO- P:2100
PYRETHRINS P:4750
PYRETHRUM P:4750
PYRETHRUM FLOWERS P:4750
PYRETHRUM I P:4750
PYRETHRUM II P:4750
PYRIDIN (German) P:4800
4-PYRIDINAMINE A:1950
PYRIDINE P:4800
PYRIDINE, 3-(1-METHYL-2-PYRROLIDINYL)- N:1200
PYRIDINE, (S)-3-(1-METHYL-2-PYRROLIDINYL)-AND SALTS N:1200
PYRIDINE, 3-METHYL M:5200
4-PYRIDYLAMINE A:1950
PYRINEX D:8900
PYROACETIC ACID A:0300
PYROACETIC ETHER A:0300
PYROBENZOL B:0550
PYROCATECHIN C:1950
PYROCATECHINE C:1950
PYROCATECHINIC ACID C:1950
PYROCATECHOL C:1950
PYROCATECHUIC ACID C:1950
PYROFAX L:2200

PYROGALLIC ACID P:4850
PYROGALLOL P:4850
PYROGENTISIC ACID H:1700
M-PYROL M:5300
PYROLIGNEOUS SPIRIT M:1750
PYROMUCIC ALDEHYDE F:1500
PYROPHOSPHATE de TETRAETHYLE (French) T:0800
PYROPHOSPHORODITHIOIC ACID, TETRAETHYL ESTER T:0600
PYROPHOSPHORODITHIOIC ACID,o,o,o,o-TETRAETHYL ESTER T:0600
PYROSULPHURIC ACID O:5400
PYROXYLIC SPIRIT M:1750
PYROXYLIN SOLUTION C:3950
PYRROLYLENE B:2850
QUAKERAL F:1500
QUECKSILBER (German) M:1100
QUECKSILBER CHLORID (German) M:0600
QUELLADA B:0600
QUICK SILVER M:1100
QUICKLIME C:1100
QUICKPHOS A:1700
QUICKSAN P:1400
QUICKSET EXTRA B:3050
QUINOL H:1700
β-QUINOL H:1700
QUINOLEINA (Spanish) Q:0100
QUINOLINE Q:0100
QUINONE B:1050
p-QUINONE B:1050
QUINTAR D:1400
QUINTAR 504F D:1400
QUINTOX D:2700
QUOLAC EX-UB D:8700
R-10 C:1850
R-11 T:2400
R-12 D:1650
R-13 M:5900
R-14 D:2550
R-20 (REFRIGERANT) C:2500
R-21 D:2100
R-22 M:5800
R-113 T:2900
R-124 M:5850
R-717 A:2000
R-152A D:3900
R-1113 T:3750
R-1582 A:5800
R-40B1 M:3100
RACEMIC-LACTIC ACID L:0100
RAD-E-CATE 16 S:2700
RADAPON D:0200
RADAZIN A:5750
RADIZINE A:5750
RAFEX D:6550
RANGE OIL O:2400
RANGE OIL K:0150
RAPISOL D:7050
RASIKAL S:2800
RASPITE L:1900
RATBANE-1080 S:3500
RATOX T:1300
RAVYON C:1550
RAW LINSEED OIL O:3400
RB P:0200
RC PLASTICIZER DBP D:1250
RCRA No. D026 C:5300
RCRA No. P003 A:0800
RCRA No. P004 A:1100
RCRA No. P005 A:1200
RCRA No. P006 A:1700
RCRA No. P008 A:1950

RCRA No. P009 A:3650	RCRA No. U021 R:0100	RCRA No. U140 I:0350
RCRA No. P009 A:3650	RCRA No. U024 S:0300	RCRA No. U142 K:0100
RCRA No. P010 A:5300	RCRA No. U025 D:1800	RCRA No. U144 L:0600
RCRA No. P011 A:5400	RCRA No. U027 D:1950	RCRA No. U147 M:0350
RCRA No. P013 B:0200	RCRA No. U028 D:3650	RCRA No. U148 M:0400
RCRA No. P014 B:0800	RCRA No. U029 M:3100	RCRA No. U149 P:3450
RCRA No. P015 B:1600	RCRA No. U031 B:3350	RCRA No. U151 M:1100
RCRA No. P017 B:2400	RCRA No. U032 C:0800	RCRA No. U152 M:1300
RCRA No. P018 B:2800	RCRA No. U033 C:1800	RCRA No. U153 M:4750
RCRA No. P021 C:0850	RCRA No. U034 T:2100	RCRA No. U154 M:1750
RCRA No. P022 C:1700	RCRA No. U036 C:2050	RCRA No. U156 M:3550
RCRA No. P023 C:2200	RCRA No. U037 D:2700	RCRA No. U159 M:4100
RCRA No. P028 B:1350	RCRA No. U037 C:2400	RCRA No. U160 B:3050
RCRA No. P029 C:4300	RCRA No. U041 E:0300	RCRA No. U161 M:4550
RCRA No. P030 C:5600	RCRA No. U041 C:2550	RCRA No. U162 M:4800
RCRA No. P031 C:5650	RCRA No. U043 V:0800	RCRA No. U165 N:0400
RCRA No. P033 C:5750	RCRA No. U044 C:2500	RCRA No. U167 N:0500
RCRA No. P034 D:6600	RCRA No. U045 M:3450	RCRA No. U169 N:1550
RCRA No. P036 P:1250	RCRA No. U046 C:2600	RCRA No. U170 N:1950
RCRA No. P039 D:7700	RCRA No. U048 C:2700	RCRA No. U171 N:2050
RCRA No. P043 D:0850	RCRA No. U051 C:5050	RCRA No. U182 P:0175
RCRA No. P048 D:6650	RCRA No. U052 C:5100	RCRA No. U188 P:1200
RCRA No. P050 E:0100	RCRA No. U053 C:5400	RCRA No. U190 P:2000
RCRA No. P051 E:0200	RCRA No. U055 C:5450	RCRA No. U191 M:5150
RCRA No. P054 E:5400	RCRA No. U056 C:5850	RCRA No. U194 P:4000
RCRA No. P056 F:0800	RCRA No. U057 C:5950	RCRA No. U196 P:4800
RCRA No. P058 S:3500	RCRA No. U061 D:0300	RCRA No. U197 B:1050
RCRA No. P059 H:0100	RCRA No. U067 E:4000	RCRA No. U209 T:0350
RCRA No. P063 H:1500	RCRA No. U068 D:1150	RCRA No. U210 T:0400
RCRA No. P064 M:4600	RCRA No. U069 D:1250	RCRA No. U213 T:0900
RCRA No. P067 P:4400	RCRA No. UO7O D:1450	RCRA No. U214 T:1150
RCRA No. P068 M:4350	RCRA No. U071 D:1500	RCRA No. U215 T:1200
RCRA No. P069 A:0350	RCRA No. U072 D:1550	RCRA No. U217 T:1250
RCRA No. P071 M:4900	RCRA No. U074 D:2000	RCRA No. U219 T:1350
RCRA No. P074 N:0900	RCRA No. U075 D:1650	RCRA No. U220 T:1600
RCRA No. P075 N:1200	RCRA No. U076 D:1700	RCRA No. U221 T:1650
RCRA No. P076 N:1400	RCRA No. U077 E:4100	RCRA No. U223 T:1700
RCRA No. P077 N:1450	RCRA No. U078 V:1200	RCRA No. U225 B:2650
RCRA No. P089 P:0200	RCRA No. U080 P:2250	RCRA No. U226 T:2250
RCRA No. P092 P:1400	RCRA No. U081 D:2200	RCRA No. U227 T:2300
RCRA No. P095 P:1550	RCRA No. U083 D:2300	RCRA No. U228 T:2350
RCRA No. P098 P:2800	RCRA No. U084 D:2400	RCRA No. U230 T:2450
RCRA No. P101 P:3800	RCRA No. U088 D:3750	RCRA No. U232 T:2500
RCRA No. P102 P:3550	RCRA No. U092 D:4950	RCRA No. U233 T:2650
RCRA No. P105 S:2000	RCRA No. U097 D:5050	RCRA No. U239 X:0300
RCRA No. P106 S:3100	RCRA No. U098 D:5700	RCRA No. U239 X:0200
RCRA No. P108 S:6200	RCRA No. U099 D:5750	RCRA No. U239 X:0100
RCRA No. P109 T:0600	RCRA No. U102 D:5900	RCRA No. U240 D:0100
RCRA No. P110 T:0750	RCRA No. U105 T:6800	RCRA No. U242 P:0350
RCRA No. P111 T:0800	RCRA No. U106 D:6850	RCRA No. U244 T:1450
RCRA No. P112 T:1100	RCRA No. U106 D:6850	RCRA No. U246 C:5700
RCRA No. P118 P:1050	RCRA No. U108 D:7100	RCRA No. U274 M:1500
RCRA No. P120 V:0400	RCRA No. U109 B:0550	RCRA No. U328 T:1850
RCRA No. P121 Z:1100	RCRA No. U110 D:7450	RCRA No. U353 T:1900
RCRA No. P122 Z:1900	RCRA No. U112 E:1600	RCRA No. U359 E:0800
RCRA No. P123 T:1950	RCRA No. U113 E:1800	RE-4355 N:0150
RCRA No. PO47 D:6550	RCRA No. U115 E:5500	REALGAR A:5350
RCRA No. PO48 D:6650	RCRA No. U117 E:5600	REALGAR A:5250
RCRA No. U001 A:0150	RCRA No. U118 E:6900	RED ARSENIC GLASS A:5350
RCRA No. U002 A:0300	RCRA No. U121 T:2400	RED ARSENIC SULFIDE A:5350
RCRA No. U003 A:0400	RCRA No. U122 F:1250	RED CHROMATE OF POTASH P:2900
RCRA No. U004 A:0450	RCRA No. U123 F:1350	RED FUMING NITRIC ACID N:1350
RCRA No. U006 A:0600	RCRA No. U124 F:1450	RED MERCURIC IODIDE M:0700
RCRA No. U007 A:0850	RCRA No. U125 F:1500	RED OIL O:5100
RCRA No. U008 A:0900	RCRA No. U127 H:0400	RED ORPIMENT A:5350
RCRA No. U009 A:0950	RCRA No. U128 H:0450	RED OXIDE OF MERCURY M:0800
RCRA No. U012 A:4750	RCRA No. U130 H:0500	RED OXIDE OF NITROGEN N:1700
RCRA No. U013 D:6100	RCRA No. U131 H:0550	RED PHOSPHORUS P:1700
RCRA No. U017 D:5250	RCRA No. U132 H:0600	RED TR BASE C:3200
RCRA No. U017 B:0450	RCRA No. U133 H:1250	REDDON T:2500
RCRA No. U020 B:0750	RCRA No. U134 H:1550	REFINED SOLVENT NAPHTHA N:0200
RCRA No. U021 B:0850	RCRA No. U135 H:1650	REFORMATES, GASOLINE BLENDING STOCKS G:0350
RCRA No. U021 R:0100	RCRA No. U138 M:4450	

REFRIGERANT 11 T:2400
REFRIGERANT 12 D:1650
REFRIGERANT 21 D:2100
REFRIGERANT 113 T:2900
REFRIGERANT 114 D:2550
REFRIGERANT 152a D:3900
REGALON D:7650
REGLON D:7650
REGLONE D:7650
REGLOX D:7650
REGULOX M:0400
REGUTOL D:7050
REOFOS 95 T:4700
REOMOL DOA D:7000
REQUTOL D:7050
RESIDUAL ASPHALT A:5700
RESIDUAL FUEL OIL NO. 4 O:2600
RESIDUAL FUEL OIL NO. 5 O:2700
RESIDUAL FUEL OIL NO. 6 O:2800
RESIDUAL OIL A:5650
RESINATO CALCICO (Spanish) C:1300
RESIN OIL O:4200
RESITOX C:5000
RESORCIN R:0100
RESORCINA (Spanish) R:0100
RESORCINE R:0100
RESORCINOL R:0100
RETARDER AK P:2000
RETARDER ESEN P:2000
RETARDER PD P:2000
RETARDER W S:0200
RETINOL O:4200
REVAC D:7050
REVENGE D:0200
REXENE P:2400
REZIFILM T:1450
RFNA N:1350
RHODANATE S:5600
RHODANATE S:5500
RHODIA D:0100
RHODIACHLOR H:0100
RHODIASOL P:0200
RHODIATOX P:0200
RHODIATROX P:0200
RHOPLEX AC-33 E:6900
RHOTHANE D:0250
RICHONATE 1850 D:8300
RICHONOL C D:8700
RICINUS OIL O:1200
RIMSO-50 D:6200
RITOSEPT H:0600
ROACH SALT S:3400
ROAD ASPHALT A:5650
ROAD ASPHALT O:4300
ROAD ASPHALT A:5600
ROAD BINDER A:5700
ROAD OIL A:5650
ROAD TAR A:5650
ROAD TAR A:5600
ROCK CANDY S:6400
ROCK OIL O:0800
RODOCID E:0600
ROMAN VITRIOL C:4800
ROSE ETHER E:5200
ROSIN OIL O:4200
ROSINOL O:4200
ROTOX M:3100
ROYAL MH 30 M:0400
ROYALTAC D:0700
ROYAL TMTD T:1450
RP 8167 E:0600
RUBBING ALCOHOL I:1250
RUBINATE D:7400
RUBY ARSENIC A:5350
RUBY ARSENIC A:5250
RUCOFLEX PLASTICIZER DOA D:7000
RUKSEAM D:0300
RUMETAN Z:1900
S 276 D:7700
S95 D:0200
SA (warfare agent) A:5575
SA (salicylic acid) S:0200
SACCHAROSE S:6400
SACCHARUM S:6400
SACAROSA (Spanish) S:6400
SADOFOS M:0250
SADOPHOS M:0250
SADOPLON T:1450
SAEURE FLUORIDE (German) F:0800
SAFFLOWER OIL O:2000
SAFFLOWER SEED OIL O:2000
SAH S:0100
SAL ACETOSELLA P:2650
SAL AMMONIAC A:2450
SALICILALDEHIDO (Spanish) S:0100
SALICILATO de METILO (Spanish) M:5350
SALICYLAL S:0100
SALICYLALDEHYDE S:0100
SALICYLIC ACID S:0200
SALICYLIC ALDEHYDE S:0100
SALMIAC A:2450
SALPETERSAURE (German) N:1350
SALT OF SATURN L:0600
SALT OF SORREL P:2650
SALT OF TARTER P:2700
SALTPETER P:3050
SALUFER S:5200
SALVO D:0100
SAL VOLATILE A:2400
SAND ACID F:1050
SAND ACID H:1300
SAND ACID F:1000
SANG GAMMA B:0600
SANITIZED SPG P:1400
SANMORIN OT 70 D:7050
SANOCIDE H:0400
SANQUINON D:1400
SANTAR M:0800
SANTICIZER 711 D:7900
SANTOBANE D:0300
SANTOBRITE P:0350
SANTOCHLOR D:1550
SANTOMERSE-3 D:8300
SANTOPHEN P:0350
SANTOPHEN 20 P:0350
SANTOTHERM P:2200
SANTOTHERM FR P:2200
SARIN S:0250
SARIN II S:0250
SAROLEX D:1000
SATICIZER 160 B:3650
SAX S:0200
SBA B:3400
S.B.A. B:3400
SBO D:7050
SC-110 P:1400
SCALDIP D:7250
SCHEELE'S GREEN C:4100
SCHEELE'S MINERAL C:4100
SCHWEFELDDIOXYD (German) S:6700
SCHWEFELKOHLENSTOFF (German) C:1700
SCHWEFEL LOST B:2025
SCHWEFELWASSERSTOFF (German) H:1650
SCHWEINFURTH GREEN C:4050
SCINTILLAR X:0300
SCONATEX V:1200
SCYAN S:5600
SCYAN S:5500
SD-1750 D:2650
SD 5532 C:2050
SD ALCOHOL 23-HYDROGEN E:1900
SD8927000 P:0900
SDMH D:5750
SEAL-COATING MATERIAL A:5700
SEBO (Spanish) T:0150
SEDRIN A:1100
SEEDRIN A:1100
SEEDTOX P:1400
SELENIC ANHYDRIDE S:0400
SELENIOUS ACID, DISODIUM SALT S:5000
SELENIOUS ANHYDRIDE S:0300
SELENIUM(IV) DIOXIDE S:0300
SELENIUM(VI) OXIDE S:0400
SELENIUM DIOXIDE S:0300
SELENIUM OXIDE S:0300
SELENIUM TRIOXIDE S:0400
SELEPHOS P:0200
SENFGAS B:2025
SILICATO de ETILO (Spanish) E:7800 E:7800
SILICATO SODICO (Spanish) S:5100
SENARMONTITE A:5200
SENECA OIL O:0800
SENTRY C:1000
SENTRY GRAIN PRESERVER P:3700
SEPTENE C:1550
SEPTISOL H:0600
SEPTOFEN H:0600
SEQ-100 E:3900
SEQUESTRENE AA E:3900
SEQUESTRIC ACID E:3900
SEQUESTROL E:3900
SESQUICLORURO de ETILALUMINIO (Spanish) E:2100
SEVIMOL C:1550
SEVIN C:1550
SEVIN 50W C:1550
SEWER GAS H:1650
SEXTONE C:5950
SEXTONE B M:3600
SF 60 M:0250
SFA S:3500
SHED-A-LEAF S:2800
SHED-A-LEAF "L" S:2800
SCHEELITE L:1900
SHELL 5520 P:2400
SHELL CHARCOAL C:2000
SHELL MIBK M:4550
SHELL SD-5532 C:2050
SHELLSOL 140 N:2350
SICOL 160 B:3650
SICOL 250 D:7000
SIGNAL OIL O:3700
SILANE, CHLOROTRIMETHYL- T:4150
SILANE, DICHLORODIMETHYL- D:5150
SILANE, DICHLORODIPHENYL- D:7300
SILANE, DICHLOROETHYL- E:3400
SILANE, DICHLOROETHYLPHENYL- E:7300
SILANE, DICHLOROMETHYL- M:3950
SILANE, DODECYLTRICHLORO- D:8800
SILANE, METHYLTRICHLORO- M:5450
SILANE, PENTYLTRICHLORO- A:4700
SILANE, TRICHLORO-2-PROPENYL- A:1400
SILANE, TRICHLOROALLYL- A:1400
SILANE, TRICHLORODODECYL- D:8800
SILANE, TRICHLOROETHYL- E:8000
SILANE, TRICHLOROMETHYL- M:5450
SILANE, TRICHLOROPENTYL- A:4700
SILANE, TRIMETHYLCHLORO- T:4150

SILANE, VINYL TRICHLORO 1-150 V:1600
SILBERNITRAT (German) S:1000
SILIBOND E:7800
SILICANE, CHLOROTRIMETHYL- T:4150
SILICANE, TRICHLOROETHYL- E:8000
SILICATE d'ETHYLE (French) E:7800
SILICIC ACID TETRAETHYL ESTER E:7800
SILICOCHLOROFORM T:2800
SILICOFLUORIC ACID F:1000
SILICOFLUORIC ACID H:1300
SILICOFLURURO AMONICO (Spanish) A:3700
SILICOFLUORURO SODICO (Spanish) S:5200
SILICOFLUORURO de ZINC (Spanish) Z:2100
SILICON CHLORIDE S:0500
SILICON TETRACHLORIDE S:0500
SILICONE FLUIDS D:5950
SILIWAX D:7050
SILVANOL B:0600
SILVER ACETATE S:0600
SILVER CARBONATE S:0700
SILVER DIFLUORIDE S:0800
SILVER FLUORIDE S:0800
SILVER(II) FLUORIDE S:0800
SILVER(2+) FLUORIDE S:0800
SILVER IODATE S:0900
SILVER NITRATE S:1000
SILVER OXIDE S:1100
SILVER SULFATE S:1200
SILVEX T:2650
SILVEX (2,4,5-TP) T:2650
SILVEX HERBICIDE T:2650
SILVEX, ISOOCTYL ESTER T:2700
SILVIRHAP T:2650
SILVISAR S:2700
SILVISAR 510 C:0100
SINAFID M-48 M:4900
SINFLOWAN T:3800
SINITUHO P:0350
SINOX D:6550
SIONIT S:5700
SIONON S:5700
SIPEX OP D:8700
SIPOL L8 O:0300
SIPOL L10 D:0700
SIPOL L12 D:8000
SIPON WD D:8700
SIPTOX 1 M:0250
SIRLENE P:4200
SK-3818 T:4650
SKDN W:0200
SKELLY-SOLVE C H:0150
SKELLYSOLVE-B H:0950
SLAKED LIME C:0900
SLIMICIDE A:0800
SLO-GRO M:0400
SLOW-FE F:0750
SLOW CURING ASPHALT O:4300
SLS D:8700
SMITHSONITE Z:0800
SMUT-GO H:0400
SNIECIOTOX H:0400
SNOMELT C:0750
SNP P:0200
SOBITOL D:7050
SOBUTANOIC ACID I:0500
SODA CHLORATE S:2800
SODA CHLORIC ACID, SODIUM SALT S:2900
SODA LYE S:3800
SODA NITER S:4300
SODAMIDE S:1700

SODANIT S:1900
SODIUM S:1300
SODIUM(DICHROMATE de) (French) S:3300
SODIUM(HYDROXYDE de) (French) S:3800
SODIUM (CHLORATE de) (French) S:2800
SODIUM DIFLUORIDE S:2100
SODIUM 2-BENZOTHIAZOLETHIOATE S:4000
SODIUM 2-MERCAPTOBENZOTHIAZOL S:4000
SODIUM ACID BISULFITE S:2200
SODIUM ACID BISULFITE SOLUTION S:2300
SODIUM ACID FLUORIDE S:2100
SODIUM ACID PYROPHOSPHATE S:4800
SODIUM ACID SULFITE S:2200
SODIUM ALKYL SULFATES S:1500
SODIUM ALKYLBENZENESULFONATES S:1400
SODIUM ALUMINATE SOLUTION (45% or less) S:1600
SODIUM AMIDE S:1700
SODIUM ARSENATE S:1800
SODIUM ARSENATE, DIBASIC S:1800
SODIUM ARSENITE S:1900
SODIUM AZIDE S:2000
SODIUM, AZOTURE de- (French) S:2000
SODIUM BIBORATE S:2400
SODIUM BICHROMATE S:3300
SODIUM BIFLUORIDE S:2100
SODIUM BIFLUORIDE (VAN) S:2100
SODIUM BISULFIDE S:3700
SODIUM BISULFITE, SOLID S:2200
SODIUM BISULFITE, SOLUTION S:2300
SODIUM BORATE S:2400
SODIUM BOROHYDRIDE S:2500
SODIUM BOROHYDRIDE (15% or less) and SODIUM HYDROXIDE SOLUTION S:2600
SODIUM CACODYLATE S:2700
SODIUM CETYL SULFATE H:0650
SODIUM CHLORATE S:2800
SODIUM CHLORATE SOLUTION S:2900
SODIUM CHROMATE(DI-) S:3300
SODIUM CHROMATE S:3000
SODIUM CHROMATE(VI) S:3000
SODIUM CHROMATE, ANHYDROUS S:3000
SODIUM CYANIDE S:3100
SODIUM DICHLORO-s-TRIAZINETRIONE S:3200
SODIUM DICHLORO-s-TRIAZINETRIONE DIHYDRATE S:3200
SODIUM DICHLOROISOCYANURATE S:3200
SODIUM DICHROMATE S:3300
SODIUM DICHROMATE(VI) S:3300
SODIUM DIHYDROGEN PHOSPHATE (1:2:1) S:4800
SODIUM DIMETHYLARSENATE S:2700
SODIUM DODECYL SULFATE D:8700
SODIUM DODECYLBENZENE-SULFONATE D:8300
SODIUM FLUOACETIC ACID S:3500
SODIUM FLUORIDE S:3400
SODIUM FLUOROACETATE S:3500
SODIUM FLUOSILICATE S:5200
SODIUM HEXAFLUOROSILICATE S:5200
SODIUM HYDRATE S:3800
SODIUM HYDRIDE S:3600
SODIUM HYDROGEN ALKYL SULFATE S:1500
SODIUM HYDROGEN DIFLUORIDE S:2100
SODIUM HYDROGEN FLUORIDE S:2100

SODIUM HYDROGEN SULFIDE S:3700
SODIUM HYDROGEN SULFITE S:2200
SODIUM HYDROSULFIDE S:3700
SODIUM HYDROXIDE S:3800
SODIUM HYPOCHLORITE S:3900
SODIUM ISOTHIOCYANATE S:5500
SODIUM LAURYL SULFATE D:8700
SODIUM MERCAPTAN S:3700
SODIUM META ARSENITE S:1900
SODIUM METAARSENATE S:1800
SODIUM METABISULFITE S:2200
SODIUM METAL S:1300
SODIUM, METAL LIQUID ALLOY S:1300
SODIUM METASILICATE, ANHYDROUS S:5100
SODIUM METHOXIDE S:4200
SODIUM METHYLATE S:4200
SODIUM MONOFLUOROACETATE S:3500
SODIUM MONOHYDRIDE S:3600
SODIUM MONOSULFIDE S:5300
SODIUM NITRATE S:4300
SODIUM NITRATE S:4300
SODIUM NITRITE S:4400
SODIUM NITRITE LIQUOR S:4500
SODIUM OF CACODYLIC ACID S:2700
SODIUM OLEATE O:5300
SODIUM ORTHO ARSENITE S:1900
SODIUM ORTHOARSENATE S:1800
SODIUM OXALATE S:4600
SODIUM OXYCHLORIDE S:3900
SODIUM PCP S:4700
SODIUM PENTACHLOROPHENATE S:4700
SODIUM PENTALCHLOROPHENOL S:4700
SODIUM PHOSPHATE S:4800
SODIUM PHOSPHATE DIBASIC S:4800
SODIUM PHOSPHATE, MONOBASIC S:4800
SODIUM PHOSPHATE, TRIBASIC S:4800
SODIUM PHOSPHATE, TRIBASIC S:4900
SODIUM PYROBORATE S:2400
SODIUM PYROSULFITE S:2200
SODIUM RHODANIDE S:5600
SODIUM RHODANIDE S:5500
SODIUM SALT OF HYDRAZOIC ACID S:2000
SODIUM SELENITE S:5000
SODIUM SILICATE S:5100
SODIUM SILICOFLUORIDE S:5200
SODIUM SULFHYDRATE S:3700
SODIUM SULFHYDRATE S:2200
SODIUM SULFIDE S:5300
SODIUM SULFIDE, ANHYDROUS S:5300
SODIUM SULFITE S:5400
SODIUM SULFOCYANATE S:5500
SODIUM SULFOCYANATE S:5600
SODIUM SULFURET S:5300
SODIUM SULPHIDE S:5300
SODIUM TETRABORATE S:2400
SODIUM TETRABORATE DECAHYDRATE S:2400
SODIUM TETRAHYDROBORATE(1-) S:2500
SODIUM THIOCYANATE S:5500
SODIUM THIOCYANATE SOLUTION (56% or less) S:5600
SODIUM THIOCYANIDE S:5500
SOFRIL S:6600
SOFTIL D:7050
SOILBROM-40, or -85, or -90EC E:4000
SOILFUME E:4000
SOK C:1550
SOLACTOL E:6700
SOLAR-40 D:8300
SOLAR NITROGEN SOLUTION U:1000
SOLAR NITROGEN SOLUTIONS A:3300

Appendix C, Synonym and Trade Name Index 1247

SOLAR WINTER BAN P:4200
SOLFO BLACK B D:6650
SOLFO BLACK 2B SUPRA D:6650
SOLFO BLACK BB D:6650
SOLFO BLACK G D:6650
SOLFO BLACK SB D:6650
SOLSOL NEEDLES D:8700
SOLUBLE GLASS S:5100
SOLUSOL-100% D:7050
SOLUSOL-75% D:7050
SOLVANOL D:3750
SOLVANOM D:5900
SOLVARONE D:5900
SOLVENT ETHER E:5600
SOLVENT NAPHTHA N:0200
SOLVIREX D:7700
SOLVOSOL D:3400
SOMAN (nerve gas) S:5650
SOMI-PRONT D:6200
SONACIDE G:0600
SOPRABEL L:0700
SOPRATHION E:0600
SOPRATHION P:0200
SORBITE S:5700
SORBITOL S:5700
SORBO S:5700
SORBOL S:5700
SOVOL P:2200
SOY BEAN OIL O:2100
SOY OIL O:2100
SOYABEAN OIL O:2100
SPECTRACIDE D:1000
SPERLOX-S S:6600
SPERM OIL O:4400
SPERM OIL NO. 1 O:4400
SPERM OIL NO. 2 O:4400
SPERSUL S:6600
SPERSUL THIOVIT S:6600
SPINDLE OIL O:4500
SPIRIT E:1900
SPIRIT OF ETHER NITRITE E:7100
SPIRIT OF HARTSHORN A:2000
SPIRITS OF SALT H:1450
SPIRITS OF TURPENTINE T:4750
SPIRITS OF WINE E:1900
SPONTOX T:2500
SPOR-KIL P:1400
SPOTRETE T:1450
SPOTTING NAPHTHA N:0300
SPRAY OIL O:4600
SPRAYSET MEKP B:3050
SPRING-BAK N:0100
SPRITZ-HORMIN/2,4-D D:0100
SPRITZ-HORMIT/2,4-D D:0100
SPRITZ-RAPIDIN B:0600
SPRUEHPFLANZOL B:0600
SQ 1489 T:1450
SQ 9453 D:6200
SR406 C:1475
STA-FAST T:2650
STABLE PENTABORANE P:0250
STAFLEX DBP D:1250
STAFLEX DTDP D:7850
STANCARE S:5800
STANDAPOL 112 COND D:8700
STANIDE S:5800
STANNOUS FLUORIDE S:5800
STATHION P:0200
STAUFFER CAPTAN C:1475
STCC C:1850
STEAM DISTILLED TURPENTINE T:4750
STEAM TURBINE LUBE OIL O:5000
STEAM TURBINE OIL O:5000

STEAREX S:5900
STEARIC ACID S:5900
STEARIC ACID, AMMONIUM SALT A:3750
STEARIC ACID, LEAD SALT L:1400
STEAROPHANIC ACID S:5900
STEARYL ALCOHOL, CRUDE T:0200
STEARYLDIMETHYLBENZYL-AMMONIUM CHLORIDE B:1500
STEINBUHL YELLOW C:0800
STEPANOL WAQ D:8700
STERAL H:0600
STERASKIN H:0600
STERLING WAQ-COSMETIC D:8700
STIBINE, TRIBROMO- A:5050
STIBINE, TRICHLORO- A:5100
STICKDIOXYD (German) N:2300
STINK DAMP H:1650
STOLZITE L:1900
STOP-GRO M:0400
STRAIGHT RUN GASOLINE D:7800
STRAIGHT RUN GASOLINES G:0500
STRAIGHT RUN KEROSENE K:0150
STRATHION P:0200
STRAZINE A:5750
STREUNEX B:0600
STROBANE-T T:1950
STROBANE-T-90 T:1950
STRONTIUM(II) NITRATE S:6100
STRONTIUM CHROMATE S:6000
STRONTIUM CHROMATE A S:6000
STRONTIUM CHROMATE(VI) S:6000
STRONTIUM CHROMATE 12170 S:6000
STRONTIUM CHROMATE X-2396 S:6000
STRONTIUM NITRATE S:6100
STRONTIUM YELLOW S:6000
STRYCHNINE S:6200
STRYCHNOS S:6200
STUNTMAN M:0400
STYRALLYL ALCOHOL M:3050
STYRALYL ALCOHOL M:3050
STYRENE S:6300
STYRENE MONOMER S:6300
STYROL (German) S:6300
STYROLE S:6300
STYROLENE S:6300
STYRON S:6300
STYROPOL S:6300
STYROPOR S:6300
SUBACETATO de COBRE (Spanish) C:4750
SUBERANE C:5800
SUBLIMED SULFUR S:6600
SUCCINIC ACID, DIMETHYL ESTER D:6050
SUCROSE S:6400
SUGAR S:6400
SUGAR OF LEAD L:0600
SULFAMIC ACID, COBALT SALT C:3850
SULFAMIC ACID, MONOAMMONIUM SALT A:3800
SULFAMINSAURE (German) A:3800
SULFAPOL D:8300
SULFATE H:1850
SULFATE de CUIVRE (French) C:4800
SULFATE DIMETHYLIQUE (French) D:6100
SULFATE MERCURIQUE (French) M:0850
SULFATE de METHYLE (French) D:6100
SULFATE de PLOMB (French) L:1500
SULFATE de ZINC (French) Z:2200
SULFATE OF COPPER C:4800
SULFATE TURPENTINE T:4750
SULFATE WOOD TURPENTINE T:4750
SULFATED NEATSFOOT OIL O:4800
SULFATEP T:0600
SULFATO ALUMINICO (Spanish) A:1750

SULFATO AMONICO (Spanish) A:1750
SULFATO de BERILIO (Spanish) B:1850
SULFATO de CADMIO (Spanish) C:0450
SULFATO CHROMICO (Spanish) C:3350
SULFATO de COBALTO (Spanish) C:3900
SULFATO de COBRE (Spanish) C:4800
SULFATO de COBRE AMONIACAL (Spanish) C:4850
SULFATO de DIETILO (Spanish) C:4850
SULFATO de DIMETILO (Spanish) D:6100
SULFATO FERRICO (Spanish) F:0400
SULFATO FERROSO AMONICO (Spanish) F:0550
SULFATO de HIDROXILAMINA (Spanish) H:1850
SULFATO MERCURICO (Spanish) M:0850
SULFATO de NICOTINA (Spanish) N:1250
SULFATO de NIQUEL (Spanish) N:1150
SULFATO de NIQUEL y AMONIO (Spanish) N:0700
SULFATO de PLATA (Spanish) S:1200
SULFATO de PLOMO (Spanish) L:1500
SULFATO de TALIO (Spanish) T:1300
SULFATO de URANILO (Spanish) U:0800
SULFATO de VANADILO (Spanish) V:0500
SULFATO de ZINC (Spanish) Z:2200
SULFATO de ZIRCONIO (Spanish) Z:2700
SULFERROUS F:0750
SULFICYL BIS(METHANE) D:6200
SULFIDAL S:6600
SULFIDE, BIS(2-CHLOROETHYL) B:2025
SULFIMEL DOS D:7050
SULFITO AMONICO (Spanish) A:2250
SULFITO SODICO (Spanish) S:5400
SULFOLANE S:6500
SULFOLANE-W S:6500
SULFOLANO (Spanish) S:6500
SULFONATED ALKYLBENZENE, SODIUM SALT S:1400
SULFONIC ACID, MONOCHLORIDE C:3000
SULFONYL CHLORIDE S:7100
SULFOPON WA-1 D:8700
SULFORON S:6600
SULFOSUCCINATO de DIOCTILO y SODIO (Spanish) D:7050
SULFOTEP T:0600
SULFOTEPP T:0600
SULFOXIDO de DIMETILO (Spanish) D:6200
SULFRAMIN 40 D:8300
SULFRAMIN 85 D:8300
SULFTECH S:5400
SULFUR S:6600
SULFUR CHLORIDE S:7000
SULFUR DIOXIDE S:6700
SULFUR DIOXIDE, LIQUEFIED S:6700
SULFUR HYDRIDE H:1650
SULFUR HYDROXIDE H:1650
SULFURE de METHYLE (French) D:6150
SULFUR, MOLTEN S:6600
SULFUR MONOCHLORIDE S:7000
SULFUR MUSTARD B:2025
SULFUR MUSTARD GAS B:2025
SULFUR MUSTARD/LEWISITE B:2025
SULFUR OXIDE S:6700
SULFUR PHOSPHIDE P:1850
SULFUR SUBCHLORIDE S:7000
SULFURETTED HYDROGEN H:1650
SULFURIC ACID S:6800
SULFURIC ACID, ALUMINUM SALT A:1750
SULFURIC ACID, AMMONIUM IRON(2+) SALT F:0550
SULFURIC ACID, AMMONIUM NICKEL (2+) SALT (2:2:1) N:0700

SULFURIC ACID, CHROMIUM(3+) SALT C:3350
SULFURIC ACID, COBALT(2+) SALT C:3900
SULFURIC ACID, COBALT(II) SALT C:3900
SULFURIC ACID, COPPER(2+) SALT C:4800
SULFURIC ACID, DIAMMONIUM SALT A:3850
SULFURIC ACID, DIETHYL ESTER D:3800
SULFURIC ACID, DIMETHYL ESTER D:6100
SULFURIC ACID, FUMING O:5400
SULFURIC ACID, IRON(2+) SAL F:0750
SULFURIC ACID, IRON(3+) SALT F:0400
SULFURIC ACID IRON SALT F:0750
SULFURIC ACID, LEAD (2+) SALT L:1500
SULFURIC ACID, LAURYL ESTER, AMMONIUM SALT A:3050
SULFURIC ACID, MERCURY(2+) SALT M:0850
SULFURIC ACID MIXED with SULFUR TRIOXIDE O:5400
SULFURIC ACID, MONODODECYL ESTER, AMMONIUM SALT A:3050
SULFURIC ACID, MONODODECYL ESTER, SODIUM SALT D:8700
SULFURIC ACID, NICKEL(2+) SALT N:1150
SULFURIC ACID, NICKEL(II) SALT N:1150
SULFURIC ACID, SPENT S:6900
SULFURIC ACID, THALLIUM SALT T:1300
SULFURIC ACID, ZINC SALT Z:2200
SULFURIC CHLOROHYDRIN C:3000
SULFURIC ETHER E:5600
SULFURIC OXYCHLORIDE S:7100
SULFURO de ARSENICO (Spanish) A:5550
SULFURO de HIDROGENO (Spanish) H:1650
SULFURO de DIMETILO (Spanish) D:6150
SULFURO de PLOMO (Spanish) L:1600
SULFURO SODICO (Spanish) S:5300
SULFUROUS ACID ANHYDRIDE S:6700
SULFUROUS ACID, 2-[4-(1,1-DIMETHYLETHYL)PHENOXY]CYCLOHEXYL 2-PROPYNYL ESTER P:3500
SULFUROUS ACID, MONOAMMONIUM SALT A:2250
SULFUROUS ACID, MONOSODIUM SALT S:2200
SULFUROUS ACID, SODIUM SALT S:5400
SULFUROUS ANHYDRIDE S:6700
SULFUROUS OXIDE S:6700
SULFURYL CHLORIDE S:7100
SULKOL S:6600
SULOUREA T:1350
SULPHOCARBONIC ANHYDRIDE C:1700
SULPHOS P:0200
SULPHUR S:6600
SULPHURETTED HYDROGEN H:1650
SULPHURIC ACID, CADMIUM SALT C:0450
SULSOL S:6600
SUMITOX M:0250
SUNATIPIC ACID-C N:0450
SUNTOL C:5000
SUPARI (India) O:2200
SUP'R FLO D:7950
SUPER-DE-SPROUT M:0400
SUPER COSAN S:6600
SUPER D WEEDONE D:0100
SUPER D WEEDONE T:2500
SUPER RODIATOX P:0200
SUPERCEL 3000 U:0900
SUPERFLAKE, ANHYDROUS C:0750
SUPERLYSOFORM F:1250
SUPERORMONE CONCENTRE D:0100
SUPEROXOL H:1600

SURCHLOR S:3900
SURGI-CEN H:0600
SURGI-CIN H:0600
SUROFENE H:0600
SU SEGURO CARPIDOR T:3800
SWEDISH GREEN C:4100
SWEENEY'S ANT-GO S:1800
SWEET BIRCH OIL M:5350
SWEET OIL O:1700
SWEET SPIRITS OF NITRE E:7100
SYMCLOSEN T:2850
SYMCLOSENE T:2850
SYNKLOR C:2050
SYNTEXAN D:6200
SYNTHETIC 3956 T:1950
SYNTHETIC GLYCERIN G:0650
SYNTHETIC RUBBER LATEX L:0200
SYSTEMOX D:0800
SYSTOX D:0800
SYSTOX AND ISOSYSTOX MIXTURE D:0800
2,4,5-T T:2500
T-47 P:0200
T-1703 D:0850
T-144 S:0250
T-2104 T:0050
T-2106 S:0250
β-T T:2300
T-STUFF H:1600
2,4,5-T, ACID T:2500
2,4,5-T, SODIUM SALT T:2600
T5 BRUSH KIL T:2500
TABLE SUGAR S:6400
TABOON A T:0050
TABUN T:0050
TAG fungicide P:1400
TALLEOL T:0100
TALLEOL O:4700
TALL OIL O:4700
TALL OIL, FATTY ACID T:0100
TALLOL T:0100
TALLOL O:4700
TALLOW T:0150
TALLOW BENZYL DIMETHYLAMMONIUM CHLORIDE B:1500
TALLOW FATTY ALCOHOL T:0200
TALLOW OIL T:0150
TANGANTANGAN OIL O:1200
TANNERS OIL O:4800
TANNIC ACID T:0250
TANNIN T:0250
TANOL SECONDAIRE (French) B:3400
TAP 9VP D:2650
TAP 85 B:0600
TAR O:3200
TAR ACIDS C:5100
TAR CAMPHOR N:0400
TAR, COAL O:3200
TAR, LIQUID O:3200
TAR OIL C:5050
TARAPACAITE P:2750
TARAPON K 12 D:8700
TARS, LIQUID A:5600
TARTAR EMETIC A:5000
1-TARTARIC ACID, AMMONIUM SALT A:4000
TARTARIC ACID, ANTIMONY POTASSIUM SALT A:5000
TARTARIC ACID, COPPER SALT C:4900
TARTARIZED ANTIMONY A:5000
TARTRATED ANTIMONY A:5000
TARTRATO AMONICO (Spanish) A:4000

TARTRATO de ANTIMONIO y POTASIO (Spanish) A:5000
TASK D:2650
TASK TABS D:2650
TASTOX A:5000
TAT CHLOR 4 C:2050
TBA B:3600
TBA B:3450
TBHP-70 B:4200
TBP T:2000
TBT B:4650
TCE T:2350
TCE T:2250
TCE T:0350
1,1,1-TCE T:2250
α-TCE T:2250
TCM C:2500
TCP T:2950
TDA T:1650
TDE D:0250
p,p'-TDE D:0250
TDI T:1700
2,4-TDI T:1700
TDI-80 T:1700
TEA P:4000
TEA T:3200
TEA T:3250
TEA T:3300
TEABERRY Oil M:5350
TEAR GAS C:2250
TECHNETIUM TC 99M SULFUR COLLOID S:6600
90 TECHNICAL GLYCERIN G:0650
TECQUINOL H:1700
TECZA T:3600
TEDP T:0600
TEDTP T:0600
TEF T:4650
TEFLON MONOMER T:0850
TEG T:3400
TEG T:0650
TEKRESOL C:5100
TEKWAISA M:4900
TEL T:0750
TELONE C D:2400
TELONE D:2400
TELONE D:2500
TELONE II SOIL FUMIGANT D:2400
TELVAR DIURON WEED KILLER D:7950
TEN T:3300
TENAC D:2650
TENDUST N:1200
TENITE 423 P:2400
TENOX HQ H:1700
TENOX P GRAIN PRESERVATIVE P:3700
TENTACHLORURE D'ANTIMOINE (French) A:4900
TEOS E:7800
TEP T:3650
TEP T:0800
TEPA T:4650
TEPP T:0800
TERABOL M:3100
TERCYL C:1550
TEREBENTHINE (French) T:4750
TEREFTALATO de DIMETILO (Spanish) D:6250
TEREPHTHALIC ACID, DIMETHYL ESTER D:6250
TEREPHTHALIC ACID, METHYL ESTER D:6250
TERETON M:1550
TERM-I-TROL P:0350

TERMITKIL D:1450
TERPENTIN OEL (German) T:4750
TERPENTIN OEL (German) O:4100
TERPENTINE SUBSTITUTE M:5600
TERPINENE D:7150
δ-1,8-TERPODIENE D:7150
TERR-O-GAS 100 M:3100
TERSAN T:1450
TERSASEPTIC H:0600
TERTROSULPHUR BLACK PB D:6650
TERTROSULSULPHUR PBR D:6650
TESCOL E:4200
TESCOL E:1900
TESULOID S:6600
TETA T:3600
TETAN T:1100
TETARCARBONYL NICKEL N:0800
TETRA A:1150
TETRAACETATO de PLOMO (Spanish) L:1700
o,o,o',o'-TETRAAETHYL-BIS(DITHIO-PHOSPHAT) (German) E:0600
o,o,o,o-TETRAAETHYL-DIPHOSPHAT, TETRACAP T:0400
TETRAAMINE COPPER SULFATE C:4850
3,6,9,12-TETRAAZATETRADECANE-1,14-DIAMINE P:0600
TETRACHLOORMETAN C:1850
1,1,2,2-TETRACHLORAETHAN (German) T:0350
TETRACHLORETHANE T:0350
TETRACHLORKOHLENSTOFF, TETRA (German) C:1850
TETRACHLORMETHAN (German) C:1850
1,1,2,2-TETRACHLOROETHANE T:0350
TETRACHLORO ZIRCONIUM Z:2800
TETRACHLOROCARBON C:1850
TETRACHLORODIPHENYLETHANE D:0250
TETRACHLOROETHANE T:0350
sym-TETRACHLOROETHANE T:0350
TETRACHLOROETHYLENE T:0400
TETRACHLOROMETHANE C:1850
TETRACHLOROSILANE S:0500
TETRACHLORURE D'ACETYLENE (French) T:0350
TETRACHLORURE de CARBONE (French) C:1850
1,1,2,2-TETRACLOROETANO (Spanish) T:0350
TETRACLOROETANO (Spanish) T:0350
TETRACLOROETILENO (Spanish) T:0400
TETRACLORURO de CARBONO (Spanish) C:1850
TETRACLORURO de SILICO (Spanish) S:0500
TETRACLORURO de TITANIO (Spanish) T:1550
TETRACLORURO de ZIRCONIO (Spanish) Z:2800
TETRADECANOL L:2000
TETRADECANOL T:0450
TETRADECANOL, mixed isomers L:2000
1-TETRADECANOL T:0450
1-TETRADECENE T:0500
TETRADECENO (Spanish) T:0500
TETRADECILBENCENO (Spanish) T:0550
TETRADECYL ALCOHOL L:2000
n-TETRADECYL ALCOHOL T:0450
TETRADECYLBENZENE T:0550
TETRAETHOXYSILANE E:7800
TETRAETHYL DITHIOPYROPHOSPHATE T:0600
TETRAETHYL LEAD T:0750

TETRAETHYL ORTHOSILICATE E:7800
TETRAETHYL PYROPHOSPHATE T:0800
TETRAETHYL SILICATE E:7800
TETRAETHYL s,s'-METHYLENE BIS(PHOSPHOROTHIOLOTHIONATE) E:0600
o,o,o',o'-TETRAETHYL s,s'-METHYLENE DI(PHOSPHORODITHIOATE) E:0600
o,o,o,o-TETRAETHYL s,s'-METHYLENE-BIS(DITHIOPHOSPHATE) E:0600
o,o,o',o'-TETRAETHYL s,s'-METHYLENEBISPHOSPHORDITHIOATE E:0600
TETRAETHYLDITHIO PYROPHOSPHATE T:0600
o,o,o,o-TETRAETHYLDITHIO-PYROPHOSPHATE T:0600
TETRAETHYLE PLUMB T:0750
TETRAETHYLENE GLYCOL T:0650
TETRAETHYLENE PENTAMINE T:0700
TETRAETHYLPLUMBANE T:0750
TETRAETHYLPYROPHOSPHATE T:0800
TETRAETILENGLICOL (Spanish) T:0650
TETRAETILENPENTAMINA (Spanish) T:0700
TETRAFINOL C:1850
TETRAFLUOETILENO (Spanish) T:0850
TETRAFLUROBORATE(1-) LEAD(2+) L:1000
TETRAFLUOROETHYLENE T:0850
TETRAFLUROETHENE T:0850
TETRAFORM C:1850
TETRAHIDROFURANO (Spanish) T:0900
TETRAHIDRONAFTALENO (Spanish) T:0950
TETRAHYDRO-1,4-DIOXIN T:7100
TETRAHYDRO-1,4-ISOXAZINE M:6100
TETRAHYDRO-1,4-OXAZINE M:6100
TETRAHYDRO-2H-1,4-OXAZINE M:6100
3a,4,7,7a-TETRAHYDRO-4,7-METHANOINDENE D:2675
TETRAHYDRO-p-DIOXIN D:7100
TETRAHYDRO-p-OXAZINE M:6100
3a,4,7,7a-TETRAHYDRODIMETHYL-4,7-METHANOINDENE M:3750
TETRAHYDROFURAN T:0900
TETRAHYDROFURANNE (French) T:0900
TETRAHYDRONAPHTHALENE T:0950
1,2,3,4-TETRAHYDRONAPHTHALENE T:0950
TETRAHYDROTHIOPHENE-1,1-DIOXIDE S:6500
TETRAHYDROXYMETHYLMETHANE P:0550
TETRAMETHYLOMETHANE P:0550
TETRALENO T:0400
TETRALEX T:0400
TETRALIN T:0950
TETRALINE T:0950
TETRAMERO de PROPILENO (Spanish) P:4500
TETRAMETHYL-THIRAM DISULFID (German) T:1450
TETRAMETHYL LEAD T:1050
TETRAMETHYL PLUMBANE T:1050
TETRAMETHYL THIURAM DISULFIDE T:1450
N,N,N',N'-TETRAMETHYL THIURAM DISULFIDE T:1450
TETRAMETHYL THIURANE DISULFIDE T:1450
1,2,3,5-TETRAMETHYLBENZENE T:1000
TETRAMETHYLDIURANE SULPHITE T:1450

TETRAMETHYLENE CYANIDE A:1050
TETRAMETHYLENE DICHLORIDE, DCB D:2000
TETRAMETHYLENE GLYCOL B:2950
1,4-TETRAMETHYLENE GLYCOL B:3950
1,4-TETRAMETHYLENE GLYCOL B:2950
TETRAMETHYLENE OXIDE T:0900
TETRAMETHYLENE SULFONE S:6500
TETRAMETHYLENETHIURAM DISULPHIDE T:1450
TETRAMETHYLTHIOCARBAMOYLDISULPHIDE T:1450
TETRAMETHYLTHIOPEROXYDICARBONIC DIAMIDE T:1450
TETRAMETHYLTHIURAM BISULFIDE T:1450
TETRAMETHYLTHIURAM DISULFIDE T:1450
1,2,3,5-TETRAMETIL BENCENO (Spanish) T:1000
TETRAMETILPLOMO (Spanish) T:1050
TETRAMP T:0950
TETRANAP T:0950
TETRANITROMETANO (Spanish) T:1100
TETRANITROMETHANE T:1100
TETRA OLIVE N2G A:4850
TETRAOXYMETHYLENE F:1250
TETRAPROPYLENE P:4500
TETRAPROPYLENE D:8050
TETRASOL C:1850
TETRASTIGMINE T:0800
TETRAVOS D:2650
TETRINE ACID E:3900
TETROLE F:1450
TETRON T:0800
TETRON-100 T:0800
TETROPIL T:0400
TETROSIN LY D:7200
TEX-WET D:7050
TEXACO® LEAD APPRECIATOR B:3200
TEXANOL T:4350
TEXAPON ZHC D:8700
TEXWET 1001 D:7050
TFE T:0850
THALLIUM ACETATE T:1150
THALLIUM CARBONATE T:1200
THALLIUM MONOACETATE T:1150
THALLIUM MONONITRATE T:1250
THALLIUM NITRATE T:1250
THALLIUM SULFATE T:1300
THALLOUS ACETATE T:1150
THALLOUS CARBONATE T:1200
THALLOUS NITRATE T:1250
THALLOUS SULFATE T:1300
THANOL PPG P:2450
THERMACURE B:3050
THERMALOX B:1800
THERMINOL FR-1 P:2200
THF T:0900
THIFOR E:0100
THILLATE T:1450
THIMER T:1450
THIMUL E:0100
1,1'-THIOBIS(2-CHLOROETHANE) B:2025
THIOBUTYL ALCOHOL B:4400
THIOCARBAMATE T:1350
THIOCARBAMIDE T:1350
THIOCARBONYL CHLORIDE T:1400
THIOCARBONYL DICHLORIDE T:1400
THIOCARBONYL TETRACHLORIDE P:1050
THIOCYANATE SODIUM S:5600
THIOCYANIC ACID, AMMONIUM SALT A:4050

THIODAN E:0100
THIODEMETON D:7700
THIODEMETRON D:7700
THIOETHANOL E:6800
THIOETHYL ALCOHOL E:6800
THIOFACO T-35 T:3200
THIOFOSGEN T:1400
THIOLUX S:6600
THIOMETHANOL M:4750
THIOMETHYL ALCOHOL M:4750
THIONEX E:0100
THIOPHENIT M:4900
THIOPHENOL B:0800
THIOPHOS P:0200
THIOPHOS 3422 P:0200
THIOPHOSGENE T:1400
THIOPHOSPHATE de o,o-DIETHYLE et de o-(3-CHLORO-4-METHYL-7-COUMARINYLE) (French) C:5000
THIOPHOSPHORIC ANHYDRIDE P:1850
2-THIOPROPANE D:6150
THIOSAN T:1450
THIOSULFURIC ACID, DIAMMONIUM SALT A:4150
THIOSULFURIC ACID, LEAD SALT L:1200
THIOSULFUROUS DICHLORIDE S:7000
THIOTEPP T:0600
THIOTOX T:1450
THIOUREA T:1350
2-THIOUREA T:1350
THIOVIT S:6600
THIRAM T:1450
THIRAMAD T:1450
THIRAME (French) T:1450
THISULFAN E:0100
THIURAM T:1450
THIURAMYL T:1450
THOMPSONS® WOOD FIX P:0350
THORIUM NITRATE T:1500
THORIUM(4+) NITRATE T:1500
THORIUM(IV) NITRATE T:1500
THORIUM NITRATE TETRAHYDRATE T:1500
THORIUM TETRANITRATE T:1500
THREE ELEPHANT B:2100
THRETHYLENE T:2350
THU T:1350
THYLATE T:1450
THYLFAR M-50 M:4900
TIBA T:3850
TIBAL T:3850
TIN BIFLUORIDE S:5800
TIN DIFLUORIDE S:5800
TIN FLUORIDE S:5800
TINNING GLUX Z:0900
TIOCARBAMIDA (Spanish) T:1350
TIOCIANATO AMONICO (Spanish) A:4050
TIOCIANATO de PLOMO (Spanish) L:1800
TIOCIANATO de MERCURIO(II) (Spanish) M:0950
TIOCIANATO SODICO (Spanish) S:5500
TIODICLORURO BENCENOFOSFOROSO (Spanish) B:0700
TIOFOS P:0200
TIOFOSGENO (Spanish) T:1400
TIOSULFATO AMONICO (Spanish) A:4100
TIOVEL E:0100
TIPPON T:2500
TIRAM (Spanish) T:1450
TITANATO de TETRABUTILO (Spanish) T:0300
TITANIUM(4+) CHLORIDE T:1550
TITANIUM(IV) CHLORIDE T:1550

TITANIUM BUTOXIDE T:0300
TITANIUM CHLORIDE T:1550
TITANIUM TETRABUTOXIDE T:0300
TITANIUM TETRACHLORIDE T:1550
TITRIPLEX E:3900
TIURAMYL T:1450
TK M:0250
TL 4N D:3000
TL 69 P:1250
TL 146 B:2025
TL 314 A:0950
TL 337 E:5400
TL 389 D:5050
TL 423 E:3100
TL 466 D:0850
TL 822 C:5700
TL 869 S:3500
TL 898 M:0600
TL 1091 N:0950
TL 1163 T:4150
TL 1450 M:4600
TL 1618 S:0250
TLA B:3200
TM-4049 M:0250
TMA T:4050
TML T:1050
TMP T:4400
TMTD T:1450
TNCS 53 C:4800
TNM T:1100
TNPA T:4500
TOCP T:4450
TOCP T:2950
TOFK T:4450
2,4-TOLAMINE T:1650
o-TOLIDIN T:1850
2-TOLIDIN (German) T:1850
TOLIDINE T:1850
2-TOLIDINE T:1850
3,3'-TOLIDINE T:1850
o,o'-TOLIDINE T:1850
TOLL M:4900
TOLUEN-2,4-DIISOCIATO (Spanish) T:1700
TOLUENE T:1600
m-TOLUENE DIAMINE T:1650
2,4-TOLUENE-2,4-DIAMINE T:1650
TOLUENE-2,4-DIISOCYANATE T:1700
TOLUENE,2,4-DINITRO- D:6800
TOLUENE,2,6-DINITRO- D:6850
TOLUENE-α,α-DICHLORO- B:0450
TOLUENE DI-ISOCYANATE T:1700
TOLUENE DIISOCYANATE T:1700
TOLUENE HEXAHYDRIDE M:3600
TOLUENE,o-CHLORO- C:3100
TOLUENE,o-NITRO- N:2200
TOLUENE,p-NITRO- N:2250
TOLUENE,p-tert-BUTYL B:4650
3,4-TOLUENE,3,4-DINITRO- D:6900
2,4-TOLUENEDIAMINE T:1650
2,4-TOLUENEDIISOCYANATE T:1700
4-TOLUENESULFONIC ACID T:1750
p-TOLUENESULFONIC ACID T:1750
TOLUENO (Spanish) T:1600
m-TOLUIDINA (Spanish) T:1800
o-TOLUIDINA (Spanish) T:1850
p-TOLUIDINA (Spanish) T:1900
3-TOLUIDINE T:1800
2-TOLUIDINE T:1850
4-TOLUIDINE T:1900
m-TOLUIDINE T:1800
o-TOLUIDINE T:1850
p-TOLUIDINE T:1900

p-TOLUIDINE,α,α,α-TRIFLUORO-2,6-DINITRO-N,N-DIPROPYL- T:3800
TOLUOL T:1600
o-TOLUOL C:5100
p-TOLUOL C:5100
TOLUOLO T:1600
TOLUYLENE-2,4-DIISOCYANATE T:1700
m-TOLUYLENEDIAMINE T:1650
TOLU-SOL T:1600
TOLYENE 2,4-DIISOCYANATE T:1700
TOLYL CHLORIDE B:1350
m-TOLYL CHLORIDE C:3050
o-TOLYL CHLORIDE C:3100
p-TOLYL CHLORIDE C:3150
TOLYL EPOXYPROPYL ETHER C:5350
m-TOLYL FLUORIDE F:1100
o-TOLYL FLUORIDE F:1050
p-TOLYL FLUORIDE F:1150
TOLYL GLYCIDYL ETHER C:5350
o-TOLYL PHOSPHATE T:4450
o-TOLYL PHOSPHATE PHOSPHORIC ACID T:4450
TOLYLAMINE T:1900
m-TOLYLAMINE T:1800
o-TOLYLAMINE T:1850
p-TOLYLAMINE T:1900
o-TOLYLCHLORIDE C:3100
TOLYLENE-2,4-DIISOCYANATE 2,4-TOLYLENEDIISOCYANAT E T:1700
2,4-TOLYLENE DIISOCYANATE T:1700
4-m-TOLYLENEDIAMINE T:1650
TOLYPHOSPHATE T:2950
TOPICHLOR 20 C:2050
TOPICLOR C:2050
TOPICLOR 20 C:2050
TOPSYM D:6200
TORMONA T:2500
TOSIC ACID T:1750
TOTP T:4450
TOX 47 P:0200
TOXADUST T:1950
TOXAFENO (Spanish) T:1950
TOXAKIL T:1950
TOXAPHENE T:1950
TOXASPRAY T:1950
TOXICHLOR C:2050
TOXILIC ACID M:0300
TOXILIC ANHYDRIDE M:0350
TOXILIC ANHYDRIDE M:0400
TOXOL (3) P:0200
TOXON 63 T:1950
TOXYPHEN T:1950
2,4,5-TP T:2650
2,4,5-TP ACID ESTERS T:2700
TRABUTYL TITANATE T:0300
TRANSAMINE T:2500
TRANSAMINE D:0100
TRANSFORMER OIL O:4900
TRANSMISSION OIL O:3500
TRAPEX M:4700
TRAPEXIDE M:4700
TRAVEX S:2800
TREFANOCIDE T:3800
TREFICON T:3800
TREFLAN T:3800
TREFLANOCIDE ELANCOLAN T:3800
TREPENAOL WA D:8700
TRI T:2350
TRI A:1150
TRI-6 B:0600
TRIACETALDEHYDE (French) P:0175
TRIAD T:2350
TRIAETHYLAMIN (German) T:3300

TRIANGLE C:4800
TRIASOL T:2350
1,4,7-TRIAZAHEPTANE D:3450
TRIAZINE A1294 A:5750
1,3,5TRIAZINE-2,4-DIAMINE,6-CHLORO-N-ETHYL-N'-(1-METHYLETHYL)- A:5750
1,3,5-TRIAZINE-2,4,6(1H,3H,5H)-TRIONE, 1,3-DICHLORO-, POTASSIUM SALT P:2850
1,3,5-TRIAZINE-2,4,6(1H,3H,5H)-TRIONE, 1,3,5-TRICHLORO- T:2850
s-TRIAZINE-2,4,6(1H,3H,5H)-TRIONE, DICHLORO-, POTASSIUM DERIVATIVE P:2850
s-TRIAZINE, ZEAZIN A:5750
TRIBROMMETHAN (German) B:2650
TRIBROMO-PHOSPHINE P:1900
TRIBROMOMETHANE B:2650
TRIBROMURO de ANTIMONIO (Spanish) A:5050
TRIBROMURO de BORO (Spanish) B:2150
TRIBROMURO de FOSFORO (Spanish) P:1900
TRIBUTON T:2500
TRIBUTON D:0100
TRIBUTYL PHOSPHATE T:2000
TRI-n-BUTYL PHOSPHATE T:2000
TRIBUTYLE (PHOSPHATE de) (French) T:2000
TRICALCIUM ARSENATE C:0550
TRICALCIUM DIPHOSPHIDE C:1250
TRICALCIUM o-ARSENATE C:0550
TRICALCIUM ORTHOARSENATE C:0550
TRICALCIUMARSENAT (German) C:0550
TRICARNAM C:1550
TRICHLOR C:2800
TRICHLOR T:2350
TRICHLORAETHEN (German) T:2350
1,1,1-TRICHLOR-2,2-BIS(4-CHLOR-PHENYL)-AETHAN (German) D:0300
1,1,2-TRICHLORETHANE T:2300
TRICHLORETHENE (French) T:2350
TRICHLORFON T:2050
TRICHLORINATED ISOCYANURIC ACID T:2850
TRICHLORMETHYL SULFUR CHLORIDE P:1050
TRICHLOROACETALDEHYDE T:2100
TRICHLOROALLYLSILANE A:1400
TRICHLOROALUMINUM A:1450
TRICHLOROALUMINUM SOLUTION A:1500
TRICHLOROAMYLSILANE A:4700
TRICHLOROARSINE A:5450
1,2,3-TRICHLOROBENZENE T:2150
1,2,4-TRICHLOROBENZENE T:2200
$unsym$-TRICHLOROBENZENE T:2200
v-TRICHLOROBENZENE T:2150
vic-TRICHLOROBENZENE T:2150
1,2,4-TRICHLOROBENZOL T:2200
TRICHLOROBIS(4-CHLOROPHENYL)ETHANE D:0300
1,1,1-TRICHLORO-2,2-BIS(p-METHOXYPHENYL)ETHANE M:1500
1,1,1-TRICHLORO-2,2-BIS(p-CHLOROPHENYL)ETHANE D:0300
TRICHLOROBORANE B:2200
TRICHLOROBORON B:2200
TRICHLOROCYANURIC ACID T:2850
1,1,1-TRICHLORO-2,2-DI(4-CHLOROPHENYL)-ETHANE D:0300
TRICHLORODODECYLSILANE D:8800
TRICHLOROETHANE T:2250
1,1,1-TRICHLOROETHANE T:2250

1,1,2-TRICHLOROETHANE T:2300
1,2,2-TRICHLOROETHANE T:2300
α-TRICHLOROETHANE T:2250
β-TRICHLOROETHANE T:2300
TRICHLOROETHANE, STABILIZED T:2250
TRICHLOROETHENE T:2350
TRICHLOROETHYLENE T:2350
1,1,2-TRICHLOROETHYLENE T:2350
TRICHLOROETHYLENE TRI (French) T:2350
TRICHLOROETHYLSILANE E:8000
TRICHLOROETHYLSILICANE E:8000
TRICHLOROFLUOROMETHANE T:2400
TRICHLOROFORM C:2500
TRICHLOROHYDRIN T:2750
(2,2,2-TRICHLORO-1-HYDROXY-ETHYL)DIMETHYLPHOSPHONATE T:2050
BIS(3,5,6-TRICHLORO-2-HYDROXYPHENYL)METHANE H:0600
TRICHLOROIMINOISOCYANURIC ACID T:2850
TRICHLOROISOCYANIC ACID T:2850
TRICHLOROISOCYANURIC ACID T:2850
N,N',N''-TRICHLOROISOCYANURIC ACID T:2850
1,3,5-TRICHLOROISOCYANURIC ACID T:2850
TRICHLOROMETHANE C:2500
TRICHLOROMETHANE SULFURYL CHLORIDE P:1050
TRICHLOROMETHANESULFENYL CHLORIDE P:1050
N-[(TRICHLOROMETHYL)THIO]TETRAHYDROPHTHALIMIDE C:1475
N-[(TRICHLOROMETHYL)THIO]-4-CYCLOHEXENE-1,2-DICARBOXIMIDE C:1475
TRICHLOROMETHYL SULFUR CHLORIDE P:1050
TRICHLOROMETHYL SULFOCHLORIDE P:1050
N-TRICHLOROMETHYLMERCAPTO-4-CYCLOHEXENE-1,2-DICARBOXIMIDE C:1475
N-(TRICHLOROMETHYLMERCAPTO)-δ(sup4)-TETRAHYDROPHTHALIMIDE C:1475
TRICHLOROMETHYLSILANE M:5450
N-TRICHLOROMETHYLTHIO-3a,4,7,7a-TETRAHYDROPHTHALIMIDE C:1475
N-TRICHLOROMETHYLTHIOCYCLOHEX-4-ENE-1,2-DICARBOXIMIDE C:1475
N-TRICHLOROMETHYLTHIO-cis-δ(sup 4)-CYCLOHEXENE-1,2-DICARBOXIMIDE C:1475
TRICHLOROMONOFLUOROMETHANE T:2400
TRICHLOROMONOSILANE T:2800
TRICHLORONITROMETHANE C:2800
TRICHLOROOXOVANADIUM V:0300
TRICHLOROPENTYLSILANE A:4700
TRICHLOROPHENE H:0600
TRICHLOROPHENOL T:2450
2,4,5-TRICHLOROPHENOL T:2450
2,4,5-TRICHLOROPHENOXYACETIC ACID T:2500
2,4,5-TRICHLOROPHENOXYACETIC ACID, ESTERS T:2550
2-(2,4,5-TRICHLOROPHENOXY) PROPANOIC ACID, ISOOCTYL ESTER T:2700
2,4,5-TRICHLOROPHENOXYACETIC ACID, SODIUM SALT T:2600

2-(2,4,5-TRICHLOROPHENYOXY)-PROPANOIC ACID T:2650
TRICHLOROPHON T:2050
TRICHLOROPHOSPHINE P:1950
1,2,3-TRICHLOROPROPANE T:2750
3,5,6-TRICHLORO-2-PYRIDOL-o-ESTER with o,o-DIETYLPHOSPHOROTHIOATE D:8900
TRICHLORORAN T:2350
TRICHLOROSILANE T:2800
TRICHLOROSTIBINE A:5100
1,1,2-TRICHLORO-1,2,2-TRIFLUORO-ETHANE T:2900
TRICHLORO-s-TRIAZINETRIONE T:2850
TRICHLORO-s-TRIAZINE-2,4,6(1H,3H,5H)-TRIONE T:2850
1,3,5-TRICHLORO-s-TRIAZINE-2,4,6(1H,3H,5H)-TRIONE T:2850
1,3,5-TRICHLORO-1,2,5-TRIAZINE-2,4,6(1H,3H,5H)-TRIONE T:2850
2,2',2''-TRICHLOROTRIETHYLAMINE B:202
(2,4,5-TRICHLOR-PHENOXY)-ESSIGSAEURE (German) T:2500
TRICHLORURE d' ANTIMOINE (French) A:5100
1,3,5-TRICHLORO-2,4,6-TRIOXOHEXAHYDRO-s-TRIAZINE T:2850
TRICHLORO(VINYL)SILANE V:1600
TRICHLOROVINYL SILICANE V:1600
TRI-CLENE T:2350
TRICLORACETALDEHIDO (Spanish) T:2100
1,2,3-TRICLOROBENCENO (Spanish) T:2150
1,2,4-TRICLOROBENCENO (Spanish) T:2200
1,1,1-TRICLOROETANO (Spanish) T:2250
1,1,2-TRICLOROETANO (Spanish) T:2300
1,2,2-TRICLOROETANO (Spanish) T:2300
2,4,5-TRICLOROFENOXIACETICO (Spanish) T:2500
TRICLOROFLUOMETANO (Spanish) T:2400
TRICLORFON (Spanish) T:2050
TRICLOROETILENO (Spanish) T:2350
TRICLOROFENOL (Spanish) T:2450
1,2,3-TRICLOROPROPANO (Spanish) T:2750
TRICLOROSILANO (Spanish) T:2800
TRICLORO-s-TRIAZINATRIONA (Spanish) T:2850
TRICLORURO de ANTIMONIO (Spanish) A:5100
TRICLORURO de ARSENICO (Spanish) A:5450
TRICLORURO de BORO (Spanish) B:2200
TRICON BW E:3900
TRICRESOL C:5100
TRICRESYL PHOSPHATE T:2950
TRIDECANE T:3000
n-TRIDECANE T:3000
n-TRIDECANO (Spanish) T:3000
TRIDECANOL T:3050
TRIDECANOL L:2000
1-TRIDECANOL T:3050
1-TRIDECANOL, PHTHALATE D:7850
1-TRIDECENE T:3100
TRIDECILBENCENO (Spanish) T:3150
TRIDECYL ALCOHOL T:3050
TRIDECYLBENZENE T:3150
TRIDIMETHYLPHENYL PHOSPHATE T:4700
TRIEN T:3600
TRIENTINE T:3600
TRIESTE FLOWERS P:4750
TRI-ETHANE T:2250
TRIETANOLAMINA (Spanish) T:3200
TRIETHANOLAMINE T:3200

TRIETHANOLAMINE DODECYL-BENZENESULFONATE D:8350
TRIETHANOLAMINE LAURYL SULFATE D:8750
TRIETHYL ALUMINUM T:3250
sym-TRIETHYLBENZENE T:3350
TRIETHYL PHOSPHATE T:3650
TRIETHYL PHOSPHITE T:3700
TRIETHYLAMINE T:3300
TRIETHYLBENZENE T:3350
1,3,5-TRIETHYLBENZENE T:3350
TRIETHYLENE GLYCOL T:3400
TRIETHYLENE GLYCOL DI-(2-ETHYLBUTYRATE) T:3450
TRIETHYLENE GLYCOL ETHYL ETHER T:3500
TRIETHYLENE GLYCOL METHYL ETHER T:3550
TRIETHYLENE GLYCOL MONOETHYLETHER T:3500
TRIETHYLENE GLYCOL MONOETHYLETHER E:1500
TRIETHYLENE GLYCOL MONOMETHYL ETHER T:3550
TRIETHYLENEPHOSPHORAMIDE T:4650
TRIETHYLENE TETRAMINE T:3600
TRIETHYLOLAMINE T:3200
TRIETILALUMINIO (Spanish) T:3250
TRIETILAMINA (Spanish) T:3300
TRIETILBENCENO (Spanish) T:3350
TRIETILENFOSFORAMIDA (Spanish) T:4650
TRIETILENGLICOL (Spanish) T:3400
TRIETILENGLICOLMETIL ETER (Spanish) T:3550
TRIETILENGLICOL MONOETIL ETER (Spanish) T:3500
TRIETILENTETRAMINA (Spanish) T:3600
TRIFLUOCLOROETILENO (Spanish) T:3750
TRIFLUORALIN T:3800
1,1,2-TRIFLUORO-2-CHLOROETHYLENE T:3750
α,α,α-TRIFLUORO-2,6-DINITRO-N,N-DIPROPYL-p-TOLUIDINE T:3800
TRIFLUOROANTIMONY, STIBINE, TRIFLUORO- A:5150
TRIFLUOROBORANE B:2350
TRIFLUOROCHLORINE C:2150
TRIFLUOROCHLOROETHYLENE T:3750
TRIFLUOROCHLOROMETHANE M:5900
TRIFLUOROMETHYL CHLORIDE M:5900
TRIFLUOROMONOCHLOROETHYLENE T:3750
TRIFLUOROVINYL CHLORIDE T:3750
TRIFLUORURE de CHLORE (French) C:2150
TRIFLUORURO de ANTIMONIO (Spanish) A:5150
TRIFLUORURO de BROMO (Spanish) B:2350
TRIFLUORURO de CLORO (Spanish) C:2150
TRIFLURALIN T:3800
TRIFLURALINA (Spanish) T:3800
TRIFORMOL P:0150
TRIFRINA D:6550
TRIFUREX T:3800
TRIGEN T:3400
TRIGLYCINE N:1750
TRIGLYCOL T:3400
TRIGLYCOL DICAPROATE T:3450
TRIGLYCOL DIHEXOATE T:3450
TRIGLYCOL METHYL ETHER T:3550
TRIGLYCOLMONOETHYL ETHER T:3500
TRIGLYCOL MONOETHYL ETHER E:1500
TRIGLYCOLLAMIC ACID N:1750
1,2,3-TRIHYDROXYBENZENE P:4850

3,4,5-TRIHYDROXYBENZOIC ACID G:0100
TRI(2-HYDROXYETHYL)AMINE T:3200
TRIHYDROXYPROPANE G:0650
1,2,3-TRIHYDROXYPROPANE G:0650
TRIISOBUTENE T:3900
TRIISOBUTILALUMINIO (Spanish) T:3850
TRIISOBUTYLALANE T:3850
TRIISOBUTYLALUMINUM T:3850
TRIISOBUTYLENE T:3900
TRIISOCYANATOISOCYANURATE I:1050
TRIISOPROPANOLAMINA (Spanish) T:3950
TRIISOPROPANOLAMINE T:3950
TRIKEPIN T:3800
TRILENE T:2350
TRILON 83 T:0050
TRILON B E:3900
TRILON BW E:3900
TRILONE 46 S:0250
TRIM T:3800
TRIMAR T:2350
TRIMETHOXYPHOSPHINE T:4400
1,7,7-TRIMETHYL- C:1400
TRIMETHYLACETIC ACID T:4000
TRIMETHYLAMINE T:4050
TRIMETHYLAMINOMETHANE B:3600
1,2,4-TRIMETHYLBENZENE T:4100
asym- TRIMETHYL BENZENE T:4100
TRIMETHYLBENZYLAMMONIUM CHLORIDE B:1550
1,7,7-TRIMETHYLBICYCLO(2.2.1)-2-HEPTANONE C:1400
3,7,7-TRIMETHYLBICYCLO[0,1,4]HEPT-3-ENE C:1900
2,6,6-TRIMETHYLBICYCLO[3.1.1]HEPT-2-ENE P:2050
2,6,6-TRIMETHYLBICYCLO[3.1.1]-2-HEPT-2-ENE P:2050
1,7,7-TRIMETHYLBICYCLO(2,2,1)HEPTANONE-2 C:1400
TRIMETHYL CARBINOL B:3450
TRIMETHYLCARBINYLAMINE B:3600
TRIMETHYLCHLOROSILANE T:4150
3,5,5-TRIMETHYL-2-CYCLOHEXANE-1-ONE I:0950
1,1,3-TRIMETHYL-3-CYCLOHEXEN-5-ONE I:0950
3,5,5-TRIMETHYL-2-CYCLOPHENONE I:0950
TRIMETHYLENE C:6300
TRIMETHYLENE CHLORIDE D:2350
TRIMETHYLENE DICHLORIDE D:2350
TRIMETHYL GLYCOL P:4200
α,α,,α'-TRIMETHYLENE GLYCOL H:1200
TRIMETHYLHEPTANALS I:0600
TRIMETHYL HEXAMETHYLENE DIAMINE T:4200
TRIMETHYLHEXAMETHYLENE DIISOCYANATE T:4250
TRIMETHYLMETHANE I:0250
1,7,7-TRIMETHYLNORCAMPHOR C:1400
4,7,7-TRIMETHYL-3-NORCARENE C:1900
2,2,4-TRIMETHYLPENTANE O:0100
2,2,4-TRIMETHYL-1,3-PENTANEDIOL-1-ISOBUTYRATE T:4350
2,2,4-TRIMETHYL-1,3-PENTANE-DIOLMONOISOBUTYRATE T:4350
2,4,4-TRIMETHYL-1-PENTENE D:4200
2,4,4-TRIMETHYLPENTENE-1 D:4200
2,4,4-TRIMETHYLPENTENE-2 D:4200
TRIMETHYLPHOSPHITE T:4400
1,2,2-TRIMETHYLPROPYL METHYLPHOSPHONOFLUORIDATE D:0850

TRIMETHYLSILYL CHLORIDE T:4150
2,4,6-TRIMETHYL-1,3,5-TRIOXANE P:0175
2,4,6-TRIMETHYLTRIOXYMETHYLENE P:0175
s-TRIMETHYLTRIOXYMETHYLENE P:0175
TRIMETILAMINA (Spanish) T:4050
TRIMETILBENCENO (Spanish) T:4100
1,2,4-TRIMETILBENCENO (Spanish) T:4100
1,3,5-TRIMETILBENCENO (Spanish) T:4100
TRIMETILCLOROSILANO (Spanish) T:4150
TRIMETILHEXAMETILENDIISOCIATO (Spanish) T:4300
TRIMETILHEXAMETILENDIAMINO (Spanish) T:4200
TRINEX T:2050
TRINOXOL T:2500
TRIORTHOCRESYL PHOSPHATE T:4450
TRIOXANE F:1250
1,3,5-TRIOXANE,2,4,6-TRIMETHYL- T:0175
5,8,11-TRIOXAPENTADECANE D:3050
3,6,9-TRIOXAUNDECAN-1-OL T:3500
3,6,9-TRIOXAUNDECAN-1,11-DIOL T:0650
3,6,9-TRIOXAUNDECANOL,11-DIOL T:0650
TRIOXIDO de ANTIMONIO (Spanish) A:5200
TRIOXIDO de ARSENICO (Spanish) A:5500
TRIOXIDO de MOLIBDICO (Spanish) M:5700
TRIOXIDO de SELENIO (Spanish) S:0400
TRIOXIDO de VANADIO (Spanish) V:0200
TRIOXON T:2500
TRIOXONE T:2550
TRIOXONE T:2500
TRIOXYMETHYLENE P:0150
TRIOXYTOL T:3500
TRIPHOSPHORIC ACID P:2350
TRI-PLUS T:2350
TRIPROPILAMINA (Spanish) T:4500
TRI-N-PROPYLAMINE T:4500
TRIPROPILENGLICOL (Spanish) T:4550
TRIPROPILENGLICOLMETIL ETER (Spanish) T:4600
TRIPROPILENO (Spanish) P:4550
TRIPROPYLAMINE T:4500
TRIPROPYLENE P:4550
TRIPROPYLENE GLYCOL T:4550
TRIPROPYLENE GLYCOL METHYL ETHER T:4600
TRIS(1-AZIRIDINYL)PHOSPHINE OXIDE T:4650
TRIS(2-HYDROXYPROPYL) AMINE T:3950
TRIS(AZIRIDINYL)PHOSPHINE OXIDE T:4650
TRIS(2-CHLOROETHYL)AMINE B:2025
TRIS(HYDROXYETHYL)AMINE T:3200
TRISODIUM NITRILOTRIACETATE N:1750
TRISODIUM ORTHOPHOSPHATE S:4900
TRISODIUM PHOSPHATE S:4900
TRISTAR T:3800
TRISULFURO de ARSENICO (Spanish) A:5550
TRITHENE T:3750
TRI-o-TOLYL PHOSPHATE T:2950
TRITOLY PHOSPHATE o-TOLY PHOSPHATE T:2950
TRITON GR-5 D:7050
TRIXYLENYL PHOSPHATE T:4700
TRIXYLYL PHOSPHATE T:4700
TROCLOSENE POTASSIUM P:2850
TROLAMINE T:3200
TROMETE S:4900
TRONA B:2150
TROVIDUER V:0800
TRUFLEX DOA D:7000
p-TSA T:1750

TSP S:4900
TTD T:1450
TTE T:2900
TUADS T:1450
TUBERCUPROSE C:4400
TUCUM OIL O:2200
TUEX T:1450
TUFFLITE P:2400
TUGON T:2050
TULISAN T:1450
TULUYLENDIISOCYANAT (German) T:1700
TULUYLENE-2,4-DIISOCYANATE T:1700
TUMBLEAF S:2800
TUNGSTATO de PLOMO (Spanish) L:1900
TURBINE OIL O:5000
TURGEX H:0600
TURKEY-RED OIL (SULFATED CASTOR OIL) O:1200
TURPENTINE T:4750
TURPENTINE STEAM DISTILLED T:4750
TURPS T:4750
TUTANE B:3550
TX1575000 T:4500
TYRANTON D:0900
U-46 D:0100
U-46 T:2500
U-46DP D:0100
U-46KW T:2550
U-1149 F:1400
U-3886 S:2000
U-4224 D:5550
U-5043 D:0100
U-5954 D:4800
U.S. RUBBER 604 D:1400
U.S. RUBBER D-014 P:3500
UAN-NITROGEN SOLUTION U:1000
UAN SOLUTION U:1000
UC 7744 C:1550
UCANE ALKYLATE 12 D:8100
UCAR-17 E:4200
UCAR BISPHENOL HP B:1950
UCAR BUTYLPHENOL 4-T B:4600
UCAR SOLVENT 2LM D:7600
UCON 11 T:2400
UCON 113 T:2900
UCON 113 HALOCARBON T:2900
UCON 12 D:1650
UCONN 13 M:5900
UCONN-22 M:5800
UCONN22 M:5800
UCON REFRIGERANT 11 T:2400
UDMH D:5700
UF OXYLIGNIN V:0600
ULSTRON P:2400
ULTRA SULFATE SL-1 D:8700
ULTRA WET K D:8300
ULV D:0800
UMBETHION C:5000
UN A:1150
UNDECANOIC ACID U:0100
n-UNDECANOIC ACID U:0100
UNDECANOL U:0200
1-UNDECANOL U:0200
1-UNDECENE U:0300
n-UNDECOIC ACID U:0100
UNDECYL ALCOHOL U:0200
n-UNDECYLBENZENE U:0400
UNDECYLETHYLENE T:3100
n-UNDECYLIC ACID U:0100
UNIDRON D:7950
UNIFUME E:4000
UNION CARBIDE 7,744 C:1550
UNION CARBIDE A-150 V:1600

UNIPINE O:4100
UNIPON D:0200
UNIROYAL 604 D:1400
UNIROYAL D-014 P:3500
UNITED CHEMICAL DEFOLIANT NO. 1 S:2800
UNIVERM C:1850
UNSLAKED LIME C:1100
URAN, RUSTICA U:1000
URANIUM ACETATE U:0600
URANIUM ACETATE, DIHYDRATE U:0600
URANIUM NITRATE U:0700
URANIUM OXIDE U:0500
URANIUM OXIDE PEROXIDE U:0500
URANIUM OXYACETATE U:0600
URANIUM OXYACETATE, DIHYDRATE U:0600
URANIUM PEROXIDE U:0500
URANIUM SULFATE U:0800
URANIUM SULFATE TRIHYDRATE U:0800
URANYL ACETATE U:0600
URANYL ACETATE DIHYDRATE U:0600
URANYL NITRATE U:0700
URANYL SULFATE U:0800
URANYL SULFATE TRIHYDRATE U:0800
UREA U:0900
UREA, AMMONIUM NITRATE SOLUTION (W/AQUA AMMONIA) U:1000
UREA DIOXIDE U:1100
UREA, HYDROGEN PEROXIDE SALT U:1100
UREA HYDROGEN PEROXIDE U:1100
UREA PEROXIDE U:1100
UREAPHIL U:0900
UREA,THIO- T:1350
UREOPHIL U:0900
UREVERT U:0900
UROTROPIN H:0900
UROX D D:7950
USR-604 D:1400
VAC V:0700
VAL-DROP S:2800
VALENTINITE A:5200
VALERAL V:0100
VALERALDEHYDE V:0100
n-VALERALDEHYDE V:0100
VALERIANIC ALDEHYDE V:0100
VALERIC ACID P:0700
VALERIC ACID ALDEHYDE V:0100
VALERIC ALDEHYDE V:0100
n-VALERIC ALDEHYDE V:0100
VALERILADEHIDO (Spanish) V:0100
VALERONE D:4250
VALINE ALDEHYDE B:4850
VAM V:0700
VANADIC ANHYDRIDE V:0400
VANADIC OXIDE V:0200
VANADIUM(5+) OXIDE V:0400
VANADIUM(V) OXIDE FUME V:0400
VANADIUM OXIDE V:0400
VANADIUM OXIDE V:0200
VANADIUM OXYSULFATE V:0500
VANADIUM OXYTRICHLORIDE V:0300
VANADIUM PENTOXIDE V:0400
VANADIUM PENTOXIDE (FUME) V:0400
VANADIUM PENTOXIDE DUST V:0400
VANADIUM SESQUIOXIDE V:0200
VANADIUM TRICHLORIDE OXIDE V:0300
VANADIUM TRIOXIDE V:0200
VANADYL CHLORIDE V:0300
VANADYL SULFATE V:0500
VANADYL SULFATE DIHYDRATE V:0500
VANADYL TRICHLORIDE V:0300

VANCIDA TM-95 T:1450
VANCIDE C:1475
VANCIDE 89 C:1475
VANCIDE TM T:1450
VANGARD K C:1475
VANILLAN BLACK LIQUOR V:0600
VAPONA D:2650
VAPONITE D:2650
VAPOPHOS P:0200
VAPOTONE T:0800
VARIOFORM I A:1650
VARIOFORM II U:0900
VARNISH MAKERS AND PAINTERS NAPHTHA N:0200
VARNOLINE N:0300
VARSOL N:0200
VASELINA (Spanish) P:1100
VASELINE P:1100
VATSOL OT D:7050
VC V:0800
VCL V:0800
VCM V:0800
VCN A:0950
VDC V:1200
VECTAL A:5750
VECTAL SC A:5750
VEGETABLE CARBON C:2000
VEGETABLE MIST O:2300
VEGETABLE OIL MIST O:2300
VEGFRU MALATOX M:0250
VEGFRUFOSMITE E:0600
VEGFRUFOSMITE E:0600
VELMOL D:7050
VELSACOL COMPOUND "D" D:1300
VELSICOL 58-CS-11 D:1300
VELSICOL 104 H:0100
VELSICOL 1068 C:2050
VENTOX A:0950
VEON T:2500
VEON 245 T:2500
VERAZINC Z:2200
VERDICAN D:2650
VERDIPOR D:2650
VERGEMASTER D:0100
VERMICIDE BAYER 2349 T:2050
VERMILION M:0900
VERMOESTRICID C:1850
VERSENE E:3900
VERSENE ACID E:3900
VERSENE NTA ACID N:1750
VERTAC 90% T:1950
VERTAC TOXAPHENE 90 T:1950
VERTON D:0100
VERTON 2D D:0100
VERTON 2T T:2500
VERTON D D:0100
VERTRON 2D D:0100
VESTINOL OA D:7000
VESTROL T:2350
VETIOL M:0250
VI-CAD C:0250
VICKNITE P:3050
VIDDEN D D:2500
VIDDEN D D:2400
VIDON 638 D:0100
VIENNA GREEN C:4050
VILLIAUMITE S:3400
VILRATHANE 4300 D:7400
VINAMAR U:1000
VINEGAR ACID A:0200
VINEGAR NAPHTHA E:1600
VINICIZER D:5250
VINILCICLOHEXANO (Spanish) V:0900

VINIL ETIL ETER (Spanish) V:1000
VINILTOLUENO (Spanish) V:1500
VINILTRICLOROSILANO (Spanish) V:1600
VINYL A MONOMER V:0700
VINYLACETAT (German) V:0700
VINYL ACETATE V:0700
VINYL ACETATE H.Q. V:0700
VINYL ACETATE MONOMER V:0700
VINYL AMIDE A:0850
VINYL BENZENE S:6300
VINYLBENZOL S:6300
VINYL C MONOMER V:0800
VINYL CARBINOL A:1200
VINYL CARBINOL,2-PROPENOL A:1200
VINYL CHLORIDE V:0800
VINYL CHLORIDE MONOMER V:0800
VINYL CYANIDE A:0950
VINYL CYANIDE, PROPENENITRILE A:0950
VINYL-1-CYCLOHEXENE V:0900
4-VINYL-1-CYCLOHEXENE V:0900
VINYL CYCLOHEXENE V:0900
4-VINYL CYCLOHEXENE V:0900
VINYLE (ACETATE de) (French) V:0700
VINYL ETHYL ETHER V:1000
VINYLETHYLENE B:2850
VINYL FLUORIDE V:1100
VINYL FLUORIDE MONOMER V:1100
VINYL FORMIC ACID A:0900
VINYLFOS D:2650
VINYLIDENE CHLORIDE V:1200
VINYLIDENE CHLORIDE(II) V:1200
VINYLIDENE DICHLORIDE V:1200
VINYLIDINE CHLORIDE MONOMER V:1200
VINYL METHYL ETHER V:1300
VINYL METHYL KETONE M:5500
VINYL NEODECANOATE V:1400
VINYLOPHOS D:2650
VINYLSILICON TRICHLORIDE V:1600
VINYL TOLUENE V:1500
m-VINYL TOLUENE V:1500
p-VINYL TOLUENE V:1500
VINYL TOLUENE, MIXED Isomers V:1500
VINYL TRICHLORIDE T:2300
VINYL TRICHLOROSILANE V:1600
VISCOL 350P P:2400
VISKO D:0100
VISKO-RHAP D:0100
VISKO-RHAP LOW DRIFT HERBICIDES D:0100
VISKO-RHAP LOW VOLATILE 4L D:0100
VISKO RHAP LOW VOLATILE ESTER T:2500
VITON B:0600
VITRAN T:2350
VITREX P:0200
VITRIOL BROWN ACID S:6800
VM&P NAPHTHA N:0200
VOFATOX M:4900
VOLFARTOL T:2050
VONDALDHYDE M:0400
VONDCAPTAN C:1475
VONDRAX M:0400
VONDURON D:7950
VORLEX M:4700
VORLEX-201 D:2400
VORTEX M:4700
VOTEXIT T:2050
VUAGT-1-4 T:1450
VULCAFOR TMTD T:1450
VULKACIT HX E:3325
VULKACIT MTIC T:1450

VULKACIT THIURAD T:1450
VULKACIT THIURAM/C T:1450
VULKAUT THIRAM T:1450
VULNOC AB A: 2100
VV 10 VINYL MONOMER V:1400
Vyac V:0700
W 101 P:2400
WARKEELATE ACID E:3900
WASH OIL C:5050
WASSERSTOSSPEROXIDE (German) H:1600
WATER DISPLACING OIL O:4000
WATER GLASS S:5100
WAXES: CARNAUBA W:0100
WAXOL D:7050
WEC-50 T:2050
WECOLINE OO O:5100
WECOLINE 1295 L:0300
WEED-AG-BAR D:0100
WEED-B-GON D:0100
WEED-B-GONE T:2650
WEED-RHAP D:0100
WEED DRENCH A:1200
WEED TOX D:0100
WEEDAR T:2500
WEEDAR D:0100
WEEDAR-64 D:0100
WEEDBEADS S:4700
WEEDEX A A:5750
WEEDEZ WONDER BAR D:0100
WEEDONE T:2500
WEEDONE P:0350
WEEDONE D:0100
WEEDONE 128 D:0150
WEEDONE LV4 D:0100
WEEDTRINE-D D:7650
WEEDTROL D:0100
WEEVILTOX C:1700
WEISSPIESSGLANZ A:5200
WELDING GAS A:0650
WESTRON T:0350
WESTROSOL T:2350
WETAID SR D:7050
WEX 1242 P:2400
WFNA N:1350
WHITE ANTIMONY A:5200
WHITE ARSENIC A:5500
WHITE CAMPHOR OIL C:1400
WHITE CAUSTIC S:3800
WHITE COPPERAS Z:2200
WHITE FUMING NITRIC ACID N:1350
WHITE LEAD L:1500
WHITE MERCURY PRECIPITATE M:0550
WHITE MINERAL OIL O:3600
WHITE OIL O:3600
WHITE OIL OF CAMPHOR C:1400
WHITE PHOSPHORIC ACID P:2350
WHITE PHOSPHORIC ACID P:1600
WHITE PHOSPHORUS P:1750
WHITE SPIRIT N:0300
WHITE SPIRIT (Low [15-20%] aromatic) W:0200
WHITE SPIRITS N:0200
WHITE TAR N:0400
WHITE VITRIOL Z:2200
WICKENOL 158 D:7000
WINTERGREEN OIL M:5350
WITCIZER 300 D:1250
WITCIZER 312 D:3650
WITOMOL-320 D:7000
WITTOX-C C:4600
WN 12 M:4700
WOCHEM-320 O:5100
WOFATOS M:4900

WOFATOX M:4900
WONUK A:5750
WOOD ALCOHOL M:1750
WOOD CHARCOAL C:2000
WOOD ETHER D:5200
WOOD NAPHTHA M:1750
WOOD SPIRIT M:1750
WOOD TURPENTINE T:4750
WOTEXIT T:2050
WP P:1750
WT-27 D:7050
XENENE D:7200
m-XILENO (Spanish) X:0100
o-XILENO (Spanish) X:0200
p-XILENO (Spanish) X:0300
2,6-XILIDINA (Spanish) X:0450
XL ALL INSECTICIDE N:1200
1,2-XYLENE X:0200
1,3-XYLENE X:0100
1,4-XYLENE X:0300
m-XYLENE X:0100
o-XYLENE X:0200
ORTHO-XYLENE X:0200
p-XYLENE X:0300
para-XYLENE X:0300
2,3-XYLENOL X:0400
2,4-XYLENOL X:0400
2,5-XYLENOL X:0400
2,6-XYLENOL X:0400
3,5-XYLENOL X:0400
XYLENOL, PHOSPHATE T:4700
3,5-XYLENOL,4-(METHYLTHIO)-, METHYLCARBAMATE M:0450
2,6-XYLIDINE X:0450
m-XYLIDINE X:0450
o-XYLIDINE X:0450
p-XYLIDINE X:0450
XYLODIDIN C:3950
m-XYLOL X:0100
o-XYLOL X:0200
p-XYLOL X:0300
XYLYL PHOSPHATE T:4700
2,6-XYLYLAMINE X:0450
YALTOX C:1500
YARMOR O:4100
YARMOR PINE OIL O:4100
YASO-KNOCK S:3500
YELLOW ARSENIC SULFIDE A:5550
YELLOW CROSS GAS B:2025
YELLOW CROSS LIQUID B:2025
YELLOW LEAD OCHER L:2300
YELLOW OXIDE OF MERCURY M:0800
YELLOW PETROLATUM P:1100
YELLOW PHOSPHORUS P:1750
YPERITE B:2025
s-YPERITE B:2025
YODATO de PLATA (Spanish) S:0900
YODURO AMONICO (Spanish) A:2950
YODURO de COBRE (Spanish) C:4500
YODURO MERCURICO (Spanish) M: 0700
YODURO de PLOMO (Spanish) L:1100
YODURO POTASICO (Spanish) P:3000
ZACLON DISCOIDS H:1500
ZB 112 Z:0600
ZB 237 Z:0600
ZEAZINE A:5750
ZECTANE Z:0100
ZEIDANE D:0300
ZELIO T:1300
ZERDANE D:0300
ZESET T V:0700
ZETAR O:3200
ZEXTRAN Z:0100

ZIARNIK P:1400
ZINC ACETATE Z:0200
ZINC ACETATE DIHYDRATE Z:0200
ZINC ACID, ZINC SALT Z:0400
ZINC AMMONIUM CHLORIDE Z:0300
ZINC ARSENATE Z:0400
ZINC ARSENATE, BASIC Z:0400
ZINC BICHROMATE Z:0500
ZINC BORATE Z:0600
ZINC BROMIDE Z:0700
ZINC BUTTER Z:0900
ZINC CARBONATE Z:0800
ZINC CARBONATE Z:0800
ZINC CHLORIDE Z:0900
ZINC CHLORIDE FUME Z:0900
ZINC (CHLORURE de) (French) Z:0900
ZINC CHROMATE Z:1000
ZINC CHROMATE(VI) HYDROXIDE Z:1000
ZINC CHROMATE C Z:1000
ZINC CHROMATE O Z:1000
ZINC CHROMATE T Z:1000
ZINC CHROMATE Z Z:1000
ZINC CHROME Z:1000
ZINC CHROME YELLOW Z:1000
ZINC CHROMIUM OXIDE Z:1000
ZINC CYANIDE Z:1100
ZINC DIACETATE Z:0200
ZINC DIALKYLDITHIOPHOSPHATE Z:1200
ZINC o,o-DI-N-BUTYLPHOSPHORO-DITHIOATE Z:1200
ZINC DIBROMIDE Z:0700
ZINC DICHLORIDE Z:0900
ZINC DICHROMATE Z:0500
ZINC DICYANIDE Z:1100
ZINC DIETHYL- D:3850
ZINC DIFLUORIDE Z:1300
ZINC DIFORMATE Z:1500
ZINC DIHEXYLDITHIOPHOSPHATE Z:1200
ZINC DIHEXYLPHOSPHORODITHIOATE Z:1200
ZINC DIMETHYL D:6300
ZINC DITHIONITE Z:1600
ZINC ETHIDE D:3850
ZINC ETHYL D:3850
ZINC ETHYLENEDIAMINETETRAACETATE D:0950
ZINC FLUOBORATE SOLUTION Z:1400
ZINC FLUORIDE Z:1300
ZINC FLUOROBORATE Z:1400
ZINC FLUOROSILICATE Z:2100
ZINC FLUORURE (French) Z:1300
ZINC FORMATE Z:1500
ZINC HEXAFLUOROSILICATE Z:2100
ZINC HYDROSULFITE Z:1600
ZINC HYDROXYCHROMATE Z:1000
ZINC METHYL D:6300
ZINC MURIATE SOLUTION Z:0900
ZINC NITRATE Z:1700
ZINC NITRATE HEXAHYDRATE Z:1700
ZINC PHENOLSULFONATE Z:1800
ZINC p-PHENOLSULFONATE Z:1800
ZINC PHENOLSULFONATE OCTAHYDRATE Z:1800
ZINC PHOSPHIDE Z:1900
ZINC POTASSIUM CHROMATE Z:2000
ZINC SILICOFLUORIDE HEXAHYDRATE Z:2100
ZINC SILICOFLUORIDE Z:2100
ZINC SULFATE Z:2200
ZINC SULFATE HEPTAHYDRATE Z:2200
ZINC SULFOCARBOLATE Z:1800
ZINC SULFOPHENATE Z:1800
ZINC TETRAOXYCHROMATE Z:1000
ZINC TETRAOXYCHROMATE 76A Z:1000
ZINC TETRAOXYCHROMATE 780B Z:1000
ZINC TOX Z:1900
ZINC VITRIOL Z:2200
ZINC YELLOW Z:1000
ZINC YELLOW 1 Z:1000
ZINC YELLOW 386N Z:1000
ZINC YELLOW 40-9015 Z:1000
ZINC YELLOW 1425 Z:1000
ZINC YELLOW AZ-16 Z:1000
ZINC YELLOW AZ-18 Z:1000
ZINC YELLOW KSH Z:1000
ZINC YELLOW Y-539-D Z:2000
ZINKCHLORID (German) Z:0900
ZINKOSITE Z:2200
ZIRCONIUM(IV) CHLORIDE Z:2800
ZIRCONIUM(IV) SULFATE Z:2700
ZIRCONIUM ACETATE Z:2300
ZIRCONIUM CHLORIDE Z:2800
ZIRCONIUM NITRATE Z:2400
ZIRCONIUM NITRATE PENTAHYDRATE Z:2400
ZIRCONIUM OXIDE CHLORIDE Z:2500
ZIRCONIUM OXYCHLORIDE HYDRATE Z:2500
ZIRCONIUM OXYCHLORIDE Z:2500
ZIRCONIUM POTASSIUM FLUORIDE Z:2600
ZIRCONIUM SULFATE Z:2700
ZIRCONIUM SULFATE TETRAHYDRATE Z:2700
ZIRCONIUM TETRACHLORIDE Z:2800
ZIRCONYL CHLORIDE Z:2500
ZIRCONYL SULFATE Z:2700
ZITHIOL M:0250
ZN 100 Z:0600
ZOBA GKE T:1650
ZOGEN DEVELOPER-H T:1650
ZOTOX A:5300
Z-P Z:1900
ZYTOX M:3100
VX (nerve gas) V:1650

Appendix D

CAS Number Index

50-00-0 F:1250
50-21-5 L:0100
50-29-3 D:0300
50-70-4 S:5700
51-28-5 D:6650
51-75-2 B:2025
52-68-6 T:2050
54-11-5 N:1200
56-23-5 C:1850
56-38-2 P:0200
56-72-4 C:5000
56-81-5 G:0650
56-93-9 B:1550
57-11-4 S:5900
57-13-6 U:0900
57-14-7 D:5700
57-24-9 S:6200
57-50-1 S:6400
57-55-6 P:4200
57-57-8 P:3600
57-74-9 C:2050
58-89-9 B:0600
59-88-1 P:1350
60-00-4 E:3900
60-29-7 E:5600
60-34-4 M:4350
60-57-1 D:2700
62-38-4 P:1400
62-53-3 A:4750
62-56-6 T:1350
62-73-7 D:2650
62-74-8 S:3500
62-76-0 S:4600
63-25-2 C:1550
64-17-5 E:1900
64-18-6 F:1350
64-19-7 A:0200
64-67-5 D:3800
65-30-5 N:1250
65-85-0 B:0900
66-25-1 H:0750
67-56-1 M:1750
67-63-0 I:1250
67-64-1 A:0300
67-66-3 C:2500
67-68-5 D:6200
67-72-1 H:0550
68-12-2 D:5550
69-72-7 S:0200
70-30-4 H:0600
71-23-8 P:3950
71-36-3 B:3350
71-41-0 A:4350
71-43-2 B:0550
71-48-7 C:3550
71-55-6 T:2250
72-20-8 E:0200
72-43-5 M:1500
72-54-8 D:0250
74-82-8 M:1400
74-82-8 L:2100
74-83-9 M:3100
74-84-0 E:0500
74-85-1 E:3500
74-86-2 A:0650
74-87-3 M:3450
78-88-4 M:4450
74-89-5 M:1850
74-89-5 M:1900
74-90-8 H:1500
74-93-1 M:4750
74-95-3 D:1150
74-98-6 P:3300
75-00-3 E:2900
75-01-4 V:0800
75-02-5 V:1100
75-04-7 E:2200
75-05-8 A:0400
75-07-0 A:0150
75-08-1 E:6800
75-09-2 D:2050
75-12-7 F:1300
75-15-0 C:1700
75-18-3 D:6150
75-19-4 C:6300
75-20-7 C:0650
75-21-8 E:5500
75-25-2 B:2650
75-28-5 I:0250
75-31-0 I:1300
75-33-2 I:1500
75-34-3 D:1700
75-35-4 V:1200
75-36-5 A:0600
75-37-6 D:3900
75-43-4 D:2100
75-44-5 P:1550
75-45-6 M:5800
75-50-3 T:4050
75-52-5 N:1800
75-54-7 M:3950
75-55-8 P:4400
75-56-9 P:4450
75-60-5 C:0100
75-64-9 B:3600
75-65-0 B:3450
75-69-4 T:2400
75-71-8 D:1650
75-72-9 M:5900
75-74-1 T:1050
75-77-4 T:4150
75-78-5 D:5150
75-79-6 M:5450
75-83-2 N:0600
75-85-4 M:4400
75-86-5 A:0350
75-87-6 T:2100
75-91-2 B:4200
75-94-5 V:1600
75-98-9 T:4000
75-99-0 D:0200
76-01-7 P:0300
76-06-2 C:2800
76-13-1 T:2900
76-14-2 D:2550
76-22-2 C:1400
76-44-8 H:0100
77-47-4 H:0500
77-73-6 D:2675
77-78-1 D:6100
77-92-9 C:3500
78-00-2 T:0750
78-10-4 E:7800
78-18-2 C:6000
78-30-8 T:4450
78-40-0 T:3650
78-59-1 I:0950
78-76-2 B:2600
78-78-4 I:0900
78-79-5 I:1150
78-81-9 I:0400
78-82-0 I:0550
78-83-1 I:0350
78-84-2 B:4850
78-87-5 D:2300
78-88-6 D:2450
78-91-1 P:3400
78-92-2 B:3400
78-93-3 M:4100
78-94-4 M:5500
78-96-6 M:6000
78-99-9 D:2250
79-00-5 T:2300
79-01-6 T:2350
79-04-9 C:2300
79-06-1 A:0850
79-09-4 P:3700
79-10-7 A:0900
79-11-8 M:5750
79-20-9 M:1550
79-21-0 P:0900
79-22-1 M:3550
79-24-3 N:1600
79-31-2 I:0500
79-29-8 N:0600
79-34-5 T:0350
79-38-9 T:3750
79-41-4 M:1200
79-44-7 D:5050
79-46-9 N:2000
79-46-9 N:2050
79-92-5 C:1350
80-05-7 B:1950
80-10-4 D:7300
80-15-9 C:5500
80-56-8 P:2050
80-62-6 M:4800
84-66-2 D:3750
84-69-5 D:4300
84-74-2 D:1250
84-76-4 D:6950
85-00-7 D:7650
85-44-9 P:2000
85-68-7 B:3650
86-50-0 A:5800
87-61-6 T:2150
87-62-7 X:0450
87-65-0 D:2200
87-66-1 P:4850
87-68-3 H:0450
87-86-5 P:0350
87-90-1 T:2850
88-69-7 I:1650
88-72-2 N:2200
88-73-3 C:2650
88-74-4 N:1450
88-75-5 N:1850
88-99-3 D:4450
90-00-6 E:7200
90-02-8 S:0100
91-08-7 T:1700
90-12-0 M:4850
91-17-8 D:0400
91-20-3 N:0400
91-22-5 Q:0100
91-66-7 D:2850
92-52-4 D:7200
92-87-5 B:0850
93-58-3 M:3000
93-72-1 T:2650
93-76-5 T:2500
93-79-8 T:2550
93-79-8 T:2600
94-11-1 D:0150
94-17-7 D:1600
94-36-0 D:1050
94-51-9 D:7550
94-75-7 D:0100
95-08-9 T:3450
95-47-6 X:0200
95-48-7 C:5100
95-49-8 C:3100
95-50-1 D:1450
95-51-2 C:2350
95-52-3 F:1050
95-53-4 T:1850
95-57-8 C:2700
95-63-6 T:4100
95-68-1 X:0450
95-78-3 X:0450
95-79-4 C:3200
95-80-7 T:1650
95-87-4 X:0400
95-95-4 T:2450
96-18-4 T:2750
96-14-0 I:0750
96-22-0 D:3700
96-33-3 M:1700
96-34-4 M:3500
96-37-7 M:3850
97-02-9 D:6350
97-11-0 P:4750
97-63-2 E:6900
97-64-3 E:6700
96-76-4 D:1200
97-85-8 I:0475
97-86-9 I:0485
97-88-1 B:4450
97-93-8 T:3250
97-95-0 E:2500
98-00-0 F:1550
98-01-1 F:1500
98-09-9 B:0750
98-51-1 B:4650
98-54-4 B:4600
98-82-8 C:5450
98-83-9 M:5400
98-85-1 M:3050
98-86-2 A:0450
98-87-3 B:0450

1258 Appendix D, CAS Number Index

98-88-4 B:1100	106-47-8 C:2350	108-84-9 M:1950	111-44-4 D:1800
98-94-2 D:5100	106-48-9 C:2750	108-86-1 B:2500	111-46-6 D:3000
98-95-3 N:1550	106-49-0 T:1900	108-87-2 M:3600	111-49-9 H:0850
99-08-1 N:2150	106-51-4 B:1050	108-88-3 T:1600	111-55-7 E:4400
99-09-2 N:1450	106-63-8 I:0325	108-89-4 M:5250	111-65-9 O:0100
99-65-0 D:6400	106-65-0 D:6050	108-90-7 C:2400	111-66-0 O:0400
99-87-6 C:6350	106-68-3 E:2300	108-91-8 C:6100	111-69-3 A:1050
99-99-0 N:2250	106-87-6 V:0900	108-93-0 C:5900	111-70-6 H:0250
100-00-5 C:2650	106-88-7 B:4050	108-94-1 C:5950	111-76-2 E:4900
100-01-6 N:1500	106-89-8 E:0300	108-95-2 P:1200	111-84-2 N:2350
100-02-7 N:1950	106-89-8 C:2550	108-98-5 B:0800	111-87-5 O:0300
100-07-2 A:4800	106-91-2 G:0675	108-99-6 M:5200	111-90-0 D:3400
100-25-4 D:6500	106-93-4 E:4000	109-06-8 M:5150	111-92-2 D:1160
100-37-8 D:3500	106-94-5 B:2750	109-21-7 B:3700	111-96-6 D:3100
100-39-0 B:1300	106-95-6 A:1250	109-52-4 P:0700	112-06-1 H:0350
100-40-3 V:0900	106-97-8 B:2900	109-59-1 I:1175	112-07-2 E:5000
100-41-4 E:2400	106-99-0 B:2850	109-60-4 P:3900	112-15-2 D:3150
100-42-5 S:6300	107-02-8 A:0800	109-65-9 B:2550	112-24-3 T:3600
100-44-7 B:1350	107-03-9 P:4650	109-66-0 P:0650	112-27-6 T:3400
100-46-9 B:1250	107-05-1 A:1300	109-67-1 P:0800	112-30-1 D:0700
100-47-0 B:0950	107-06-2 E:4100	109-69-3 B:3750	112-31-2 D:0450
100-51-6 B:1200	107-07-3 E:3600	109-73-9 B:3500	112-34-5 D:3300
100-52-7 B:0500	107-10-8 P:4000	109-74-0 B:4950	112-35-6 T:3550
100-61-8 M:2050	107-12-0 P:3800	109-77-3 P:3450	112-37-8 U:0100
100-63-0 P:1300	107-13-1 A:0950	109-78-4 E:3700	112-48-1 E:4500
100-74-3 E:7000	107-15-3 E:3800	109-79-5 B:4400	112-50-5 E:1500
100-97-0 H:0900	107-18-6 A:1200	109-83-1 M:6050	112-50-5 T:3500
100-99-2 T:3850	107-19-7 P:3550	109-86-4 E:5100	112-53-8 L:2000
101-68-8 D:7400	107-20-0 C:2200	109-87-5 M:4200	112-53-8 D:8000
101-84-8 D:7350	107-21-1 E:4200	109-89-7 D:2800	112-55-0 L:0500
101-84-8 D:8850	107-22-2 G:0700	109-92-2 V:1000	112-57-2 T:0700
102-25-0 T:3350	107-25-5 V:1300	109-94-4 E:5800	112-59-4 D:3200
102-69-2 T:4500	107-30-2 C:2600	109-95-5 E:7100	112-60-7 T:0650
102-71-6 T:3200	107-31-3 M:4250	109-99-9 T:0900	112-70-9 T:3050
103-08-2 U:0200	107-37-9 A:1400	110-00-9 F:1450	112-72-1 T:0450
103-09-3 E:6200	107-41-5 H:1200	110-00-9 F:1450	112-72-1 L:2000
103-11-7 E:6300	107-49-3 T:0800	110-16-7 M:0300	112-73-2 D:3050
103-23-1 D:3550	107-72-2 A:4700	110-17-8 F:1400	112-80-1 O:5100
103-23-1 D:7000	107-81-3 B:2700	110-19-0 I:0300	112-92-5 T:0200
103-50-4 D:1100	107-83-5 I:0750	110-22-5 A:0700	115-07-1 P:4150
103-65-1 P:4050	107-87-9 M:5100	110-43-0 A:4500	115-10-6 D:5200
103-75-3 E:0700	107-88-0 B:3900	110-46-3 A:4600	115-11-7 I:0450
103-83-3 B:1450	107-92-6 B:4900	110-49-6 E:4800	115-18-4 M:3200
104-15-4 T:1750	107-94-8 C:2950	110-54-3 H:0950	115-19-5 M:3250
104-72-3 D:0750	107-98-2 P:4300	110-62-3 V:0100	115-21-9 E:8000
104-75-6 E:6500	108-01-0 D:5175	110-63-4 B:2950	115-29-7 E:0100
104-76-7 E:6100	108-03-2 N:2000	110-64-5 B:3000	115-32-2 D:2600
104-90-5 M:4150	108-05-4 V:0700	110-65-6 B:4750	115-77-5 P:0550
105-30-6 M:2000	108-10-1 M:4550	110-66-7 A:4450	116-14-3 T:0850
105-37-3 E:7600	108-11-2 M:4500	110-71-4 E:4700	116-54-1 M:3900
105-39-5 E:3000	108-18-9 D:4600	110-80-5 E:0800	117-80-6 D:1400
105-45-3 M:1600	108-20-3 I:1400	110-82-7 C:5850	117-81-7 D:3650
105-46-4 B:3150	108-21-4 I:1200	110-85-0 P:2100	117-81-7 E:6550
105-54-4 E:2800	108-24-7 A:0250	110-86-1 P:4800	117-84-0 D:5250
105-57-7 A:0100	108-31-6 M:0350	110-91-8 M:6100	118-74-1 H:0400
105-58-8 D:2950	108-39-4 C:5100	110-96-3 D:4100	119-06-2 D:7850
105-59-9 M:4000	108-41-8 C:3050	110-97-4 D:4550	119-07-3 O:0550
105-60-2 C:1450	108-42-9 C:2350	110-98-5 D:7500	119-36-8 M:5350
105-64-6 I:1600	108-44-1 T:1800	111-14-8 H:0200	119-61-9 B:1000
105-67-9 X:0400	108-46-3 R:0100	111-15-9 E:0900	119-64-2 T:0950
105-74-8 L:0400	108-60-1 D:1950	111-27-3 H:1050	120-12-7 A:4850
106-35-4 E:2700	108-65-6 P:4350	111-30-8 G:0600	120-61-6 D:6250
106-42-3 X:0300	108-67-8 T:4100	111-40-0 D:3450	120-80-9 C:1950
106-43-4 C:3150	108-68-9 X:0400	111-41-1 A:1850	120-82-1 T:2200
106-44-5 C:5100	108-82-7 D:4150	111-42-2 D:2750	120-83-2 D:2200
106-46-7 D:1550	108-83-8 D:4250	111-43-3 P:4600	121-06-2 T:4700

121-14-2 D:6800	140-31-8 A:1900	463-82-1 I:0250	563-80-4 M:3150
121-21-1 P:4750	140-88-5 E:1800	472-61-4 N:1300	573-56-8 D:6750
121-29-9 P:4750	141-32-2 B:3250	498-15-7 C:1900	576-24-9 D:2200
121-44-8 T:3300	141-43-5 M:5950	501-53-1 B:1400	577-11-7 D:7050
121-45-9 T:4400	141-75-3 B:5000	502-56-7 D:1180	579-66-8 D:2850
121-69-7 X:0450	141-78-6 E:1600	504-24-5 A:1950	583-52-8 P:3150
121-73-3 C:2650	141-79-7 M:1150	504-29-0 A:1950	583-60-8 M:3700
121-75-5 M:0250	141-82-2 M:0300	504-60-9 P:0450	583-78-8 D:2200
121-91-5 I:1100	141-93-5 D:2900	505-60-2 B:2025	583-91-5 H:1900
122-20-3 T:3950	141-97-9 E:1700	506-68-3 C:5700	584-03-2 B:2950
122-39-4 D:7250	142-28-9 D:2350	506-77-4 C:5750	584-79-2 P:4750
122-52-1 T:3700	142-29-0 C:6250	506-87-6 A:2400	584-84-9 T:1700
122-99-6 E:5200	142-59-6 N:0100	506-96-7 A:0550	589-92-4 M:3700
123-01-3 D:8100	142-62-1 H:1000	509-14-8 T:1100	589-82-2 H:0250
123-02-4 T:3150	142-63-2 P:2100	513-42-8 M:1800	590-01-2 B:4550
123-05-7 E:5900	142-71-2 C:4000	513-49-5 B:3550	590-19-2 B: 2850
123-31-9 H:1700	142-82-5 H:0150	513-77-9 B:0100	590-86-3 I:1700
123-33-1 M:0400	142-84-7 D:7450	513-85-9 B:4000	591-24-2 M:3700
123-35-3 M:6200	142-90-5 D:8450	526-75-0 X:0400	591-78-6 M:3350
123-38-6 P:3650	142-92-7 H:1150	527-53-7 T:1000	591-93-5 P:0500
123-42-2 D:0900	142-96-1 D:1170	527-84-4 C:6350	592-01-8 C:0850
123-51-3 I:0150	143-07-7 L:0300	528-29-0 D:6450	592-04-1 M:0650
123-54-6 A:0500	143-08-8 N:2400	528-75-4 I:0200	592-34-7 B:3800
123-62-6 P:3750	143-13-5 N:2500	532-27-4 C:2250	592-41-6 H:1100
123-72-8 B:4800	143-18-0 P:3100	534-16-7 S:0700	592-76-7 H:0300
123-73-9 C:5400	143-18-0 O:5200	534-52-1 D:6550	592-84-7 B:4100
123-86-4 B:3100	143-19-1 O:5300	535-77-3 C:6350	592-85-8 M:0950
123-91-1 D:7100	143-33-9 S:3100	538-07-8 B:2025	592-87-0 L:1800
123-92-2 I:0100	143-50-0 K:0100	540-54-5 P:4100	594-42-3 P:1050
123-96-6 O:0300	144-62-7 O:5500	540-59-0 D:1750	594-72-9 D:1900
124-04-9 A:1000	149-57-5 E:6000	540-69-2 A:2750	594-84-3 N:0600
124-07-2 O:0200	149-91-7 G:0100	540-72-7 S:5500	598-21-0 B:2450
124-09-4 H:0800	151-21-3 D:8700	540-72-7 S:5600	598-31-2 B:2400
124-11-8 N:2450	151-50-8 P:2800	540-73-8 D:5750	598-78-7 C:2900
124-13-0 O:0500	151-56-4 E:5400	540-88-5 B:3200	598-82-3 L:0100
124-17-4 D:3350	156-87-6 P:3350	541-09-3 U:0600	598-98-0 O:0300
124-18-5 D:0500	260-94-6 A:0750	541-25-3 B:2025	606-20-2 D:6850
124-38-9 C:1650	287-92-3 C:6200	541-41-3 E:3100	608-73-1 B:0600
124-40-3 D:4950	291-64-5 C:5800	541-73-1 D:1500	610-39-9 D:6900
124-41-4 S:4200	298-00-0 M:4900	541-85-5 E:2300	611-14-3 E:7900
124-43-6 U:1100	298-04-4 D:7700	542-59-6 E:4300	617-79-8 E:2600
124-65-2 S:2700	298-07-7 D:3600	542-62-1 B:0200	622-45-7 C:6150
124-68-5 A:1925	298-12-4 G:0750	542-75-6 D:2400	623-42-7 M:3400
126-30-7 D:6000	300-76-5 N:0150	543-49-7 H:0250	624-54-4 P:0850
126-33-0 S:6500	301-04-2 L:0600	543-59-9 A:4400	624-83-9 M:4600
126-73-8 T:2000	302-01-2 H:1250	543-90-8 C:0150	625-16-1 A:4300
126-98-7 M:1300	309-00-2 A:1100	544-18-3 C:3750	626-38-0 A:4250
126-99-8 C:2850	315-18-4 Z:0100	544-19-4 C:4400	627-13-4 P:4700
127-18-4 T:0400	319-84-6 B:0600	544-60-5 A:3350	627-93-0 D:4900
127-19-5 D:4800	329-71-5 D:6700	544-92-3 C:4300	628-20-6 C:2450
127-82-2 Z:1800	330-54-1 D:7950	544-97-8 D:6300	628-63-7 A:4200
127-95-7 P:2650	333-41-5 D:1000	545-55-1 T:4650	629-14-1 E:4600
128-39-2 D:1200	334-48-5 D:0550	546-67-8 L:1700	629-38-9 D:3250
131-11-3 D:5900	348-54-9 F:0850	554-84-7 N:1900	629-50-5 T:3000
131-18-0 D:0975	352-32-9 F:1150	556-61-6 M:4700	629-76-5 P:0400
131-52-2 S:4700	352-70-5 F:1100	555-77-1 B:2025	630-08-0 C:1750
131-74-8 A:3650	353-50-4 C:1800	557-19-7 N:0900	631-61-8 A:2050
131-89-5 D:6600	357-57-3 B:2800	557-20-0 D:3850	644-97-3 B:0650
133-06-2 C:1475	371-40-4 F:0900	557-21-1 Z:1100	645-62-5 E:7700
134-32-7 N:0500	372-09-8 C:5600	557-34-6 Z:0200	676-97-1 M:5050
135-01-3 D:2900	373-02-4 N:0650	557-41-5 Z:1500	646-04-8 P:0800
137-26-8 T:1450	460-19-5 C:5650	563-12-2 E:0600	676-98-2 M:5050
138-22-7 B:4350	462-06-6 F:0950	563-43-9 E:2000	691-37-2 M:5000
138-86-3 D:7150	504-08-8 A:1950	563-47-3 M:1350	696-28-6 P:1250
139-13-9 N:1750	463-04-7 A:4600	563-63-3 S:0600	696-29-7 I:1350
140-11-4 B:1150	463-71-8 T:1400	563-68-8 T:1150	763-29-1 M:4950

763-69-9 E:5700	1330-20-7 X:0100	4067-16-7 P:0600	7646-85-7 Z:0900
764-41-0 D:2000	1330-61-6 I:0650	4098-71-9 I:1050	7647-01-0 H:1450
778-75-5 B:1750	1330-78-5 T:2950	4170-30-3 C:5400	7647-18-9 A:4900
814-78-8 M:4650	1331-22-2 M:3700	4435-53-4 M:1450	7664-38-2 P:1600
814-91-4 C:4700	1332-07-6 Z:0600	4484-72-4 D:8800	7664-38-2 P:2350
815-82-7 C:4900	1333-74-0 H:1350	5103-71-9 C:2050	7664-39-3 H:1550
818-61-1 H:1750	1333-82-0 C:3300	5103-74-2 C:2050	7664-41-7 A:2000
821-55-6 M:4300	1333-83-1 S:2100	5459-93-8 E:3325	7664-93-9 S:6800
868-85-9 D:5800	1336-21-6 A:2000	5566-34-7 C:2050	7664-93-9 S:6900
872-05-9 D:0600	1336-21-6 A:2850	5593-70-4 T:0300	7681-11-0 P:3000
872-50-4 M:5300	1336-36-3 P:2200	5893-66-3 C:4700	7681-49-4 S:3400
917-69-1 C:3550	1338-02-9 C:4600	5972-73-6 A:3400	7681-52-9 S:3900
933-75-5 T:2450	1338-23-4 B:3050	5989-27-5 D:7150	7681-65-4 C:4500
959-55-7 B:1500	1338-24-5 N:0450	5989-54-8 D:7150	7697-37-2 N:1350
993-43-1 E:7400	1341-49-7 A:2200	6009-70-7 A:3400	7699-43-6 Z:2500
999-61-1 H:1950	1344-09-8 S:5100	6032-29-7 A:4350	7699-45-8 Z:0700
1002-16-0 A:4550	1344-48-5 M:0900	6046-93-1 C:4000	7704-34-9 S:6600
1002-89-7 A:3750	1344-67-8 C:4250	6080-56-4 L:0600	7705-08-0 F:0200
1066-30-4 C:3250	1401-55-4 T:0250	6147-53-1 C:3550	7718-54-9 N:0850
1066-33-7 A:2150	1498-51-7 E:7500	6196-95-8 P:1450	7719-12-2 P:1950
1103-86-9 Z:2000	1563-66-2 C:1500	6392-89-8 B:2025	7720-78-7 F:0750
1111-78-0 A:2350	1569-02-4 P:4250	6474-81-0 K0150	7722-64-7 P:3200
1119-40-0 D:5600	1573-98-7 L:1500	6484-52-2 A:3150	7722-84-1 H:1600
1125-27-5 E:7300	1582-09-8 T:3800	6484-52-2 A:3250	7723-14-0 P:1650
1185-57-5 F:0100	1600-27-7 M:0500	6484-52-2 A:3300	7723-14-0 P:1700
1194-65-6 D:1350	1634-04-4 M:3300	6487-48-5 P:3150	7723-14-0 P:1750
1299-88-3 C:4050	1675-54-3 B:2000	6533-73-9 T:1200	7726-95-6 B:2250
1300-71-6 X:0400	1678-91-7 E:3300	6834-92-0 S:5100	7727-18-6 V:0300
1300-73-8 X:0450	1758-33-4 B:4050	6842-15-5 P:4500	7727-37-9 N:1650
1300-82-9 D:2900	1762-95-4 A:4050	6842-15-5 D:8050	7727-54-0 A:3550
1302-30-3 L:2500	1789-58-8 E:3400	6846-50-0 T:4350	7733-02-0 Z:2200
1302-42-7 S:1600	1863-63-4 A:2100	7428-48-0 L:1400	7756-94-7 T:3900
1302-97-2 C:4100	1912-24-9 A:5750	7439-92-1 L:1200	7757-79-1 P:3050
1303-28-2 A:5400	1918-00-9 D:1300	7439-93-2 L:2400	7757-93-9 C:1200
1303-32-8 A:5350	2004-70-8 P:0450	7439-95-4 M:0100	7757-83-7 S:5400
1303-33-9 A:5550	2032-65-7 M:0450	7439-97-6 M:1100	7758-23-8 C:1200
1303-39-5 Z:0400	2156-96-9 D:0650	7440-09-7 P:2500	7758-80-7 S:4800
1303-96-4 S:2400	2163-00-0 D:1850	7440-23-5 S:1300	7758-94-3 F:0600
1304-29-6 B:0400	2163-80-6 M:1425	7440-28-0 T:1150	7758-95-4 L:0800
1304-56-9 B:1800	2244-21-5 P:2850	7440-38-2 A:5250	7758-98-7 C:4800
1305-62-0 C:0950	2312-35-8 P:3500	7440-41-7 B:1600	7758-99-8 C:4800
1305-78-8 C:1100	2385-85-5 M:5650	7440-70-2 C:0500	7759-01-5 L:1900
1305-79-9 C:1150	2426-08-6 B:4150	7446-08-4 S:0300	7761-88-8 S:1000
1305-99-3 C:1250	2492-26-4 S:4000	7446-09-5 S:6700	7772-76-1 A:3600
1306-19-0 C:0400	2642-71-9 A:5800	7446-14-2 L:1500	7773-06-0 A:3800
1309-32-6 A: 3700	2696-92-6 N:2100	7446-18-6 T:1300	7774-29-0 M:0700
1309-64-4 A:5200	2768-72-9 D:7650	7446-70-0 A:1450	7775-09-9 S:2800
1310-02-7 S:5200	2807-30-9 E:5300	7446-70-0 A:1500	7775-09-9 S:2900
1310-58-3 P:2950	2855-13-2 I:1000	7447-39-4 C:4250	7775-11-3 S:3000
1310-73-2 S:3800	2893-78-9 S:3200	7487-94-7 M:0600	7775-41-9 S:0800
1313-27-5 M:5700	2921-88-2 D:8900	7521-80-4 B:4700	7778-39-4 A:5300
1313-82-2 S:5300	2937-50-0 A:1350	7546-30-7 M:1000	7778-44-1 C:0550
1314-34-7 V:0200	2812-73-9 E:3200	7550-45-0 T:1550	7778-50-9 P:2900
1314-62-1 V:0400	2944-67-4 F:0150	7558-79-4 S:4800	7778-54-3 C:1000
1314-64-3 U:0800	3012-65-5 A:2550	7580-67-8 L:2800	7779-86-4 Z:1600
1314-80-3 P:1850	3025-88-5 D:5650	7585-20-8 Z:2300	7779-88-6 Z:1700
1314-84-7 Z:1900	3164-29-2 A:4000	7601-54-9 S:4900	7782-41-4 F:0800
1314-87-0 L:1600	3251-23-8 C:4650	7601-90-3 P:0950	7782-44-7 O:5600
1317-36-8 L:2300	3266-23-7 B:4050	7631-89-2 S:1800	7782-50-5 C:2100
1319-77-3 C:5100	3486-35-9 Z:0800	7631-90-5 S:2200	7782-63-0 F:0750
1319-77-3 C:5100	3648-20-2 D:7900	7631-90-5 S:2300	7782-92-5 S:1700
1319-77-3 C:5100	3648-21-3 D:4000	7631-99-4 S:4300	7783-06-4 H:1650
1319-77-3 C:5100	3687-31-8 L:0700	7632-00-0 S:4400	7783-18-8 A:4100
1327-33-9 A:5200	3689-24-5 T:0600	7632-00-0 S:4500	7783-18-8 A:4150
1327-52-2 A:5300	3811-04-9 P:2700	7646-69-7 S:3600	7783-20-2 A:3850
1327-53-3 A:5500	4016-14-2 I:1450	7646-79-9 C:3650	7783-28-0 A:3600

7783-35-9 M:0850	8006-61-9 G:0400	10103-46-5 C:1200	13765-19-0 C:0800
7783-46-2 L:0900	8006-64-2 T:4750	10103-62-5 C:0550	13768-86-0 S:0400
7783-47-3 S:5800	8007-45-2 O:3200	10108-64-2 C:0250	13779-41-4 D:3950
7783-49-5 Z:1300	8007-45-2 C:5050	10119-30-9 S:3200	13814-96-5 L:1000
7783-50-8 F:0250	8008-20-6 J:0100	10124-36-4 C:0450	13823-29-5 T:1500
7783-56-4 A:5150	8008-20-6 K:0150	10124-37-5 C:1050	13826-83-0 A:2650
7783-70-2 A:4950	8008-20-6 O:2400	10124-43-3 C:3900	13843-81-7 L:2600
7783-85-9 F:0550	8008-51-3 C:1400	10124-48-8 M:0550	13952-84-6 B:3550
7783-95-1 S:0800	8012-95-1 P:1100	10124-50-2 P:2600	13987-01-4 P:4550
7783-97-3 S:0900	8012-95-1 O:3600	10137-69-6 C:6050	14018-95-2 Z:0500
7784-18-1 A:1550	8014-95-7 O:5400	10137-74-3 C:0700	14216-75-2 N:1100
7784-27-2 A:1650	8017-16-1 P:2350	10141-05-6 C:3800	14221-47-7 F:0150
7784-34-1 A:5450	8021-27-0 O:4100	10192-30-0 A:2250	14258-49-2 A:3400
7784-40-9 L:0700	8021-29-2 O:4100	10196-04-0 A:3950	14307-35-8 L:2700
7784-41-0 P:2550	8030-30-6 N:0350	10213-15-7 M:0150	14307-43-8 A:4000
7784-42-1 A:5575	8030-30-6 P:1150	10290-12-7 C:4100	14486-19-2 C:0300
7784-46-5 S:1900	8030-31-7 N:0350	10294-33-4 B:2150	14639-97-5 Z:0300
7786-34-7 P:1500	8032-32-4 N:0200	10294-34-5 B:2200	14639-97-5 Z:0300
7786-81-4 N:1150	8032-32-4 M:5550	10294-38-9 B:0150	14639-98-6 Z:0300
7787-36-2 B:0350	8049-17-0 F:0500	10377-60-3 M:0150	14644-61-2 Z:2700
7787-47-5 B:1650	8049-19-2 F:0450	10380-29-7 C:4850	14684-25-4 B:0700
7787-49-7 B:1700	8052-41-3 N:0300	10415-75-5 M:1050	14708-14-6 N:0950
7787-59-9 B:1900	8052-42-4 O:4300	10421-48-4 F:0350	14763-77-0 C:4300
7787-70-4 C:4200	8052-42-4 A:5600	10544-72-6 N:1700	14735-84-3 C:4350
7787-71-5 B:2350	8052-42-4 A:5650	10544-73-7 N:1700	14977-61-8 C:3450
7788-98-9 A:2500	8052-42-4 A:5700	10588-01-9 S:3300	15194-98-6 C:0600
7789-00-6 P:2750	8065-48-3 D:0800	11071-47-9 D:4200	15283-51-9 F:0650
7789-06-2 S:6000	9002-92-0 E:1300	11096-82-5 P:2200	15467-20-6 N:1750
7789-09-5 A:2600	9003-07-0 P:2400	11097-69-1 P:2200	15646-96-5 T:4300
7789-21-1 F:1200	9003-29-6 P:2150	11097-69-1 P:2200	15694-70-9 N:1000
7789-30-2 B:2300	9004-70-0 C:3950	11100-14-4 P:2200	15699-18-0 N:0700
7789-42-6 C:0200	9007-13-0 C:1300	11104-28-2 P:2200	15978-77-5 U:1000
7789-43-7 C:3600	9008-57-5 E:1000	11120-29-9 P:2200	16107-41-3 C:3850
7789-45-9 C:4150	9016-00-6 L:0200	11138-49-1 S:1600	16219-75-3 E:6600
7789-60-8 P:1900	9016-87-9 P:2300	11141-16-5 P:2200	16291-96-6 C:2000
7789-61-9 A:5050	10022-31-8 B:0250	12002-03-8 C:4050	16721-80-5 S:3700
7789-75-5 C:0900	10022-68-1 C:0350	12002-48-1 T:2150	16853-85-3 L:2500
7790-91-2 C:2150	10024-97-2 N:2300	12002-48-1 T:2200	16871-71-9 Z:2100
7790-94-5 C:3000	10025-67-9 S:7000	12018-19-8 Z:1000	16893-85-9 S:4800
7790-98-9 A:3500	10025-78-2 T:2800	12027-06-4 A:2950	16919-19-0 A:3700
7791-13-1 C:3650	10025-87-3 P:1800	12036-71-4 U:0500	16923-95-8 Z:2600
7791-14-2 C:3450	10025-91-9 A:5100	12054-48-7 N:1050	16938-22-0 T:4250
7791-25-5 S:7100	10026-04-7 S:0500	12075-68-2 E:2100	16940-66-2 S:2500
7803-49-8 H:1800	10026-11-6 Z:2800	12079-65-1 M:3800	16940-66-2 S:2600
7803-57-8 H:1250	10026-17-2 C:3700	12108-13-3 M:3800	16961-83-4 F:1000
8000-26-8 O:4100	10026-22-9 C:3800	12124-97-0 A:2300	16961-83-4 H:1300
8002-64-0 O:4800	10028-22-5 F:0400	12124-99-1 A:3900	17014-71-0 P:3250
8001-23-8 O:2000	10031-59-1 T:1300	12125-01-8 A:2700	17702-41-9 D:0350
8001-24-7 O:3100	10034-81-8 M:0200	12125-02-9 A:2450	18662-53-8 N:1750
8001-25-0 O:1700	10035-10-6 H:1400	12125-56-3 N:1050	18994-66-6 N:1750
8001-26-1 O:3400	10039-54-0 H:1850	12135-76-1 A:3900	19287-89-9 S:1200
8001-28-3 O:3300	10042-76-9 S:6100	12519-36-7 D:0950	19469-07-9 F:0700
8001-29-4 O:1400	10042-84-9 N:1750	12672-29-6 P:2200	19624-22-7 P:0250
8001-31-8 O:1300	10043-01-3 A:1750	12737-87-0 P:2200	20667-12-3 S:1100
8001-35-2 T:1950	10043-01-3 A:1800	13106-76-8 A:3100	20859-73-8 A:1700
8001-58-9 C:5050	10043-35-3 B:2100	13138-45-9 N:1100	21908-53-2 M:0800
8001-79-4 O:1200	10043-52-4 C:0750	13360-63-9 E:2600	23255-03-0 N:1750
8002-03-7 O:1900	10045-89-3 F:0550	13426-91-0 C:5550	24157-81-1 D:4750
8002-05-9 N:0250	10045-94-0 M:0750	13462-88-9 N:0750	24549-06-2 M:4050
8002-05-9 D:7750	10049-05-5 C:3400	13463-39-3 N:0800	24800-44-0 T:4550
8002-09-3 O:4100	10061-01-5 D:2400	13464-35-2 P:2600	25013-15-4 V:1500
8002-26-4 T:0100	10061-02-6 D:2400	13465-95-7 B:0300	25154-52-3 N:2550
8002-26-4 O:4700	10099-74-8 L:1300	13466-78-9 C:1900	25154-54-5 D:6500
8002-74-2 P:0100	10101-53-8 C:3350	13473-90-0 A:1650	25154-54-5 D:6450
8002-75-3 O:1800	10101-63-0 L:1100	13477-00-4 B:0150	25154-54-5 D:6400
8003-19-8 D:2500	10101-98-1 N:1150	13478-00-7 N:1100	25155-23-1 T:4700
8003-34-7 P:4750	10102-03-1 N:1700	13478-50-7 L:1200	25155-30-0 D:8300
8003-73-7 P:4750	10102-06-4 U:0700	13510-49-1 B:1850	25167-32-2 D:8400
8004-13-5 D:8850	10102-18-8 S:5000	13520-83-7 U:0700	25167-67-3 B:3850
8006-61-9 G:0200	10102-43-9 N:1400	13530-65-9 Z:1000	25167-70-8 D:2550
8006-61-9 G:0250	10102-44-0 N:1400	13597-54-1 Z:1400	25265-77-4 T:4350
	10102-45-1 T:1250	13597-99-4 B:1750	25321-09-9 D:4650
	10102-48-4 L:0700	13746-89-9 Z:2400	25321-14-6 D:6800

25321-14-6 D:6850	64741-87-3 P:1150
25321-14-6 D:6900	64741-58-8 G:0150
25322-69-4 P:2450	64741-62-4 O:0700
25323-89-1 T:2250	64742-46-7 O:3700
25323-89-1 T:2300	64742-95-6 P:1150
25339-17-7 I:0700	65996-93-2 C:5050
25340-17-4 D:2900	68131-40-8 E:1200
25340-18-5 T:3350	68308-34-9 O:0800
25498-49-1 T:4600	68334-30-5 O:0900
25513-64-8 T:4200	68334-30-5 O:1000
25550-58-7 D:6650	68411-30-3 A:1150
25639-42-3 M:3650	68411-30-3 S:1400
26447-14-3 C:5350	68476-30-2 O:2500
26471-62-5 T:1700	68476-31-3 O:0900
26472-00-4 M:3750	68476-33-5 O:2600
26628-22-8 S:2000	68476-33-5 O:2700
26746-38-3 D:1200	68476-33-5 O:2800
26760-64-5 P:0800	68476-34-6 O:1100
26761-40-0 D:4350	68476-85-7 L:2200
26762-93-6 D:4700	68512-90-3 O:0900
26896-20-8 N:0550	68514-06-7 V:0600
26913-06-4 P:2250	68606-11-1 G:0500
26952-21-6 I:0850	68606-32-2 K:0150
27176-87-0 D:8150	68955-35-1 P:1150
27176-87-0 D:8200	68956-68-3 O:2300
27193-86-8 D:8550	77650-28-3 O:1100
27196-00-5 T:0450	108171-26-2 P:0100
27196-00-5 L:2000	108171-27-3 P:0100
27214-95-8 N:2450	
27323-41-7 D:8350	
27554-26-3 D:4500	
27774-13-6 V:0500	
27813-02-1 H:2000	
28300-74-5 A:5000	
30525-89-4 P:0150	
32612-48-9 A:3050	
34590-94-8 D:7600	
36478-76-9 U:0700	
37286-64-9 P:4300	
37317-41-2 P:2200	
37317-41-2 P:2200	
37324-23-5 P:2200	
37324-24-6 P:2200	
37353-63-2 P:2200	
38640-62-9 D:4750	
42504-46-1 D:8250	
42615-29-2 A:1150	
51218-45-2 M:5550	
52628-25-8 Z:0300	
52652-59-2 L:1400	
52740-16-6 C:0600	
53421-36-6 C:4700	
53469-21-9 P:2200	
56189-09-4 L:1400	
57608-40-9 A:3200	
59355-75-8 M:1650	
61790-24-7 P:3100	
63394-00-3 W:0200	
63449-39-8 P:0100	
63938-10-3 M:5850	
64365-11-3 C:2000	
64475-85-0 N:0200	
64475-85-0 M:5600	
64741-42-0 D:7800	
64741-43-1 D:7800	
64741-44-2 G:0150	
64741-44-2 O:0900	
64741-46-4 P:1150	
64741-54-4 P:1150	
64741-55-5 P:1150	
64741-63-5 P:1150	
64741-68-0 P:1150	
64741-83-9 P:1150	

Appendix E

Chemicals Likely Involved in Terrorist Incidents					
Agent type	**Example of War Agent and Record Number**	**Example of Industrial Chemicals and Record Number**	**Physical Properties**	**Early Symptoms of Exposure**	**Field Detection Using M8 & M256-A1**
Nerve agents	Sarin (GB) **S:0250** Soman (GD) **S:5650** Tabun (GA) **T:0050** VX **V:1650** DFP **D:0850**	Organophosphate Insecticides Malathion **M:0250** Parathion **P:0200** Sevin (carbaryl) **C:1550** Dichlorvos **D:2650** Ethion **E:0600** Methyl parathion **M:4900** Naled **N:0150** Triethyl phosphate **T:3650**	Vapors or liquids Odorless to fruity odor. VX and Tabun are very persistent ¥ (days) Sarin is less persistent ¥ (< 24 hr)	Tearing eyes Blurred vision Tiny pupils Pain in eyes and head Sweating Breathing problems Muscle weakness Abdominal pain Unconsciousness Convulsions Death	G-Agents: Yellow *; 0.005 mg/m^{3†} VX: Dark Green* 0.02 mg/m^{3†}
Blister Agents or Vesicants	Mustard gases **B:2025** (nitrogen mustard, sulfur mustard, lewisites)	Dimethyl sulfate **D:6100**	Oily liquids. Less persistent ¥ than VX and Tabun.	Skin and eye burns Breathing problems	Mustard Agents: Red*; 3.0 mg/m^{3†} lewisite: 14 mg/m^{3†}
"Blood agents" poisons blood cells	CK **C:5750** AC **H:1500**	Cyanogen chloride **C:5750** Hydrogen cyanide **H:1500**	Liquids or gases stored under pressure. Highly volatile and much less persistent ¥ than Sarin	Headache Breathing problems Convulsions Sudden death	Cyanogen chloride: 10 mg/m^{3†} Hydrogen cyanide: 11 mg/m^{3†}
Choking agents or pulmonary agents		Ammonia **A:2000** Chlorine **C:2100** Phosgene **P:1550**	Irritating odor. Gases (may be stored as liquids under pressure). Non-persistent ¥.	Cough Breathing problems	
Irritant (crowd control) agents	Mace **C:2250** CN **C:2250** Pepper spray Tear gas **C:2250**	Chloroacetophenone **C:2250** Chloropicrin **C:2800**	Irritating odor. Dusts or liquids.	Tearing eyes Cough Breathing problems	

Notes: PPE including gloves and mask must be worn when using detection devices.
*M8 Paper Chemical Agent Detector is similar to litmus paper and turns color specified in about 30 seconds following contact. Some petroleum products including antifreeze can cause false readings.
†M256-A1 Chemical Agent Detector Kit uses disposable "cards" that take *approximately 15 minutes* to respond following contact. Some petroleum products can cause false readings. The reaction produces toxic emissions.
¥Persistency depends on many factors including temperature, wind speed, humidity, and landing surface of an agent. Generally speaking a chemical is "persistent" if the liquid takes more than 24 hours to evaporate. The most dangerously persistent chemicals are VX, tabun, sulfur mustard, nitrogen mustard, lewisite, and mustard-lewisite mixture.